P9-EBZ-587

CRC Concise Encyclopedia of MATHEMATICS

CRC Concise Encyclopedia of MATHEMATICS

Eric W. Weisstein

CRC Press
Boca Raton London New York Washington, D.C.

Library of Congress Cataloging-in-Publication Data

Weisstein, Eric W.
The CRC concise encyclopedia of mathematics / Eric W. Weisstein.
p. cm.
Includes bibliographical references and index.
ISBN 0-8493-9640-9 (alk. paper)
1. Mathematics--Encyclopedias. I. Title.
QA5.W45 1998
510'.3—DC21 98-22385
CIP

International Standard Book Number 0-8493-9640-9
Library of Congress Card Number 98-22385
Printed in the United States of America 1 2 3 4 5 6 7 8 9 0
Printed on acid-free paper

Introduction

The CRC Concise Encyclopedia of Mathematics is a compendium of mathematical definitions, formulas, figures, tabulations, and references. It is written in an informal style intended to make it accessible to a broad spectrum of readers with a wide range of mathematical backgrounds and interests. Although mathematics is a fascinating subject, it all too frequently is clothed in specialized jargon and dry formal exposition that make many interesting and useful mathematical results inaccessible to laypeople. This problem is often further compounded by the difficulty in locating concrete and easily understood examples. To give perspective to a subject, I find it helpful to learn why it is useful, how it is connected to other areas of mathematics and science, and how it is actually implemented. While a picture may be worth a thousand words, explicit examples are worth at least a few hundred! This work attempts to provide enough details to give the reader a flavor for a subject without getting lost in minutiae. While absolute rigor may suffer somewhat, I hope the improvement in usefulness and readability will more than make up for the deficiencies of this approach.

The format of this work is somewhere between a handbook, a dictionary, and an encyclopedia. It differs from existing dictionaries of mathematics in a number of important ways. First, the entire text and all the equations and figures are available in searchable electronic form on CD-ROM. Second, the entries are extensively cross-linked and cross-referenced, not only to related entries but also to many external sites on the Internet. This makes locating information very convenient. It also provides a highly efficient way to "navigate" from one related concept to another, a feature that is especially powerful in the electronic version. Standard mathematical references, combined with a few popular ones, are also given at the end of most entries to facilitate additional reading and exploration. In the interests of offering abundant examples, this work also contains a large number of explicit formulas and derivations, providing a ready place to locate a particular formula, as well as including the framework for understanding where it comes from.

The selection of topics in this work is more extensive than in most mathematical dictionaries (e.g., Borowski and Borwein's *HarperCollins Dictionary of Mathematics* and Jeans and Jeans' *Mathematics Dictionary*). At the same time, the descriptions are more accessible than in "technical" mathematical encyclopedias (e.g., Hazewinkel's *Encyclopaedia of Mathematics* and Iyanaga's *Encyclopedic Dictionary of Mathematics*). While the latter remain models of accuracy and rigor, they are not terribly useful to the undergraduate, research scientist, or recreational mathematician. In this work, the most useful, interesting, and entertaining (at least to my mind) aspects of topics are discussed in addition to their technical definitions. For example, in my entry for pi (π), the definition in terms of the diameter and circumference of a circle is supplemented by a great many formulas and series for pi, including some of the amazing discoveries of Ramanujan. These formulas are comprehensible to readers with only minimal mathematical background, and are interesting to both those with and without formal mathematics training. However, they have not previously been collected in a single convenient location. For this reason, I hope that, in addition to serving as a reference source, this work has some of the same flavor and appeal of Martin Gardner's delightful *Scientific American* columns.

Everything in this work has been compiled by me alone. I am an astronomer by training, but have picked up a fair bit of mathematics along the way. It never ceases to amaze me how mathematical connections weave their way through the physical sciences. It frequently transpires that some piece of recently acquired knowledge turns out to be just what I need to solve some apparently unrelated problem. I have therefore developed the habit of picking up and storing away odd bits of information for future use. This work has provided a mechanism for organizing what has turned out to be a fairly large collection of mathematics. I have also found it very difficult to find clear yet accessible explanations of technical mathematics unless I already have some familiarity with the subject. I hope this encyclopedia will provide jumping-off points for people who are interested in the subjects listed here but who, like me, are not necessarily experts.

The encyclopedia has been compiled over the last 11 years or so, beginning in my college years and continuing during graduate school. The initial document was written in *Microsoft Word*® on a Mac Plus® computer, and had reached about 200 pages by the time I started graduate school in 1990. When Andrew Treverrow made his OzTeX program available for the Mac, I began the task of converting all my documents to TeX, resulting in a vast improvement in readability. While undertaking the *Word* to TeX conversion, I also began cross-referencing entries, anticipating that eventually I would be able to convert the entire document to hypertext. This hope was realized beginning in 1995, when the Internet explosion was in full swing and I learned of Nikos Drakos's excellent TEX to HTML converter, LATEX2HTML. After some additional effort, I was able to post an HTML version of my encyclopedia to the World Wide Web, currently located at `www.astro.virginia.edu/~eww6n/math/`.

The selection of topics included in this compendium is not based on any fixed set of criteria, but rather reflects my own random walk through mathematics. In truth, there is no good way of selecting topics in such a work. The mathematician James Sylvester may have summed up the situation most aptly. According to Sylvester (as quoted in the introduction to Ian Stewart's book *From Here to Infinity*), "Mathematics is not a book confined within a cover and bound between brazen clasps, whose contents it needs only patience to ransack; it is not a mine, whose treasures may take long to reduce into possession, but which fill only a limited number of veins and lodes; it is not a soil, whose fertility can be exhausted by the yield of successive harvests; it is not a continent or an ocean, whose area can be mapped out and its contour defined; it is as limitless as that space which it finds too narrow for its aspiration; its possibilities are as infinite as the worlds which are forever crowding in and multiplying upon the astronomer's gaze; it is as incapable of being restricted within assigned boundaries or being reduced to definitions of permanent validity, as the consciousness of life."

Several of Sylvester's points apply particularly to this undertaking. As he points out, mathematics itself cannot be confined to the pages of a book. The results of mathematics, however, are shared and passed on primarily through the printed (and now electronic) medium. While there is no danger of mathematical results being lost through lack of dissemination, many people miss out on fascinating and useful mathematical results simply because they are not aware of them. Not only does collecting many results in one place provide a single starting point for mathematical exploration, but it should also lessen the aggravation of encountering explanations for new concepts which themselves use unfamiliar terminology. In this work, the reader is only a cross-reference (or a mouse click) away from the necessary background material. As to Sylvester's second point, the very fact that the quantity of mathematics is so great means that any attempt to catalog it with any degree of completeness is doomed to failure. This certainly does not mean that it's not worth trying. Strangely, except for relatively small works usually on particular subjects, there do not appear to have been any substantial attempts to collect and display in a place of prominence the treasure trove of mathematical results that have been discovered (invented?) over the years (one notable exception being Sloane and Plouffe's *Encyclopedia of Integer Sequences*). This work, the product of the "gazing" of a single astronomer, attempts to fill that omission.

Finally, a few words about logistics. Because of the alphabetical listing of entries in the encyclopedia, neither table of contents nor index are included. In many cases, a particular entry of interest can be located from a cross-reference (indicated in SMALL CAPS TYPEFACE in the text) in a related article. In addition, most articles are followed by a "see also" list of related entries for quick navigation. This can be particularly useful if you are looking for a specific entry (say, "Zeno's Paradoxes"), but have forgotten the exact name. By examining the "see also" list at bottom of the entry for "Paradox," you will likely recognize Zeno's name and thus quickly locate the desired entry.

The alphabetization of entries contains a few peculiarities which need mentioning. All entries beginning with a numeral are ordered by increasing value and appear before the first entry for "A." In multiple-word entries containing a space or dash, the space or dash is treated as a character which precedes "a," so entries appear in the following order: "Sum," "Sum P...," "Sum-P...," and "Summary." One exception is that in a series of entries where a trailing "s" appears in some and not others, the trailing "s" is ignored in the alphabetization. Therefore, entries involving Euclid would be alphabetized as follows: "Euclid's Axioms," "Euclid Number," "Euclidean Algorithm." Because of the non-standard nomenclature that ensues from naming mathematical results after their discoverers, an important result such as the "Pythagorean Theorem" is written variously as "Pythagoras's Theorem," the "Pythagoras Theorem," etc. In this encyclopedia, I have endeavored to use the most widely accepted form. I have also tried to consistently give entry titles in the singular (e.g., "Knot" instead of "Knots").

In cases where the same word is applied in different contexts, the context is indicated in parentheses or appended to the end. Examples of the first type are "Crossing Number (Graph)" and "Crossing Number (Link)." Examples of the second type are "Convergent Sequence" and "Convergent Series." In the case of an entry like "Euler Theorem," which may describe one of three or four different formulas, I have taken the liberty of adding descriptive words ("Euler's *Something* Theorem") to all variations, or kept the standard name for the most commonly used variant and added descriptive words for the others. In cases where specific examples are derived from a general concept, em dashes (—) are used (for example, "Fourier Series," "Fourier Series—Power Series," "Fourier Series—Square Wave," "Fourier Series—Triangle"). The decision to put a possessive 's at the end of a name or to use a lone trailing apostrophe is based on whether the final "s" is pronounced. "Gauss's Theorem" is therefore written out, whereas "Archimedes' Recurrence Formula" is not. Finally, given the absence of a definitive stylistic convention, plurals of numerals are written without an apostrophe (e.g., 1990s instead of 1990's).

In an endeavor of this magnitude, errors and typographical mistakes are inevitable. The blame for these lies with me alone. Although the current length makes extensive additions in a printed version problematic, I plan to continue updating, correcting, and improving the work.

Eric Weisstein

Charlottesville, Virginia
August 8, 1998

Acknowledgments

Although I alone have compiled and typeset this work, many people have contributed indirectly and directly to its creation. I have not yet had the good fortune to meet Donald Knuth of Stanford University, but he is unquestionably the person most directly responsible for making this work possible. Before his mathematical typesetting program TeX, it would have been impossible for a single individual to compile such a work as this. Had Prof. Bateman owned a personal computer equipped with TeX, perhaps his shoe box of notes would not have had to await the labors of Erdelyi, Magnus, and Oberhettinger to become a three-volume work on mathematical functions. Andrew Trevorrow's shareware implementation of TeX for the Macintosh, OzTeX (`www.kagi.com/authors/akt/oztex.html`), was also of fundamental importance. Nikos Drakos and Ross Moore have provided another building block for this work by developing the LaTeX2HTML program (`www-dsed.llnl.gov/files/programs/unix/latex2html/manual/manual.html`), which has allowed me to easily maintain and update an on-line version of the encyclopedia long before it existed in book form.

I would like to thank Steven Finch of MathSoft, Inc., for his interesting on-line essays about mathematical constants (`www.mathsoft.com/asolve/constant/constant.html`), and also for his kind permission to reproduce excerpts from some of these essays. I hope that Steven will someday publish his detailed essays in book form. Thanks also to Neil Sloane and Simon Plouffe for compiling and making available the printed and on-line (`www.research.att.com/~njas/sequences/`) versions of the *Encyclopedia of Integer Sequences*, an immensely valuable compilation of useful information which represents a truly mind-boggling investment of labor.

Thanks to Robert Dickau, Simon Plouffe, and Richard Schroeppel for reading portions of the manuscript and providing a number of helpful suggestions and additions. Thanks also to algebraic topologist Ryan Budney for sharing some of his expertise, to Charles Walkden for his helpful comments about dynamical systems theory, and to Lambros Lambrou for his contributions. Thanks to David W. Wilson for a number of helpful comments and corrections. Thanks to Dale Rolfsen, compiler James Bailey, and artist Ali Roth for permission to reproduce their beautiful knot and link diagrams. Thanks to Gavin Theobald for providing diagrams of his masterful polygonal dissections. Thanks to Wolfram Research, not only for creating an indispensable mathematical tool in *Mathematica*®, but also for permission to include figures from the *Mathematica*® book and *MathSource* repository for the braid, conical spiral, double helix, Enneper's surfaces, Hadamard matrix, helicoid, helix, Henneberg's minimal surface, hyperbolic polyhedra, Klein bottle, Maeder's "owl" minimal surface, Penrose tiles, polyhedron, and Scherk's minimal surfaces entries.

Sincere thanks to Judy Schroeder for her skill and diligence in the monumental task of proofreading the entire document for syntax. Thanks also to Bob Stern, my executive editor from *CRC Press*, for his encouragement, and to Mimi Williams of CRC Press for her careful reading of the manuscript for typographical and formatting errors. As this encyclopedia's entry on PROOFREADING MISTAKES shows, the number of mistakes that are expected to remain after three independent proofreadings is much lower than the original number, but unfortunately still nonzero. Many thanks to the library staff at the University of Virginia, who have provided invaluable assistance in tracking down many an obscure citation. Finally, I would like to thank the hundreds of people who took the time to e-mail me comments and suggestions while this work was in its formative stages. Your continued comments and feedback are very welcome.

Numerals

0

see ZERO

1

The number one (1) is the first POSITIVE INTEGER. It is an ODD NUMBER. Although the number 1 used to be considered a PRIME NUMBER, it requires special treatment in so many definitions and applications involving primes greater than or equal to 2 that it is usually placed into a class of its own. The number 1 is sometimes also called "unity," so the nth roots of 1 are often called the nth ROOTS OF UNITY. FRACTIONS having 1 as a NUMERATOR are called UNIT FRACTIONS. If only one root, solution, etc., exists to a given problem, the solution is called UNIQUE.

The GENERATING FUNCTION have all COEFFICIENTS 1 is given by

$$\frac{1}{1-x} = 1 + x + x^2 + x^3 + x^4 + \ldots.$$

see also 2, 3, EXACTLY ONE, ROOT OF UNITY, UNIQUE, UNIT FRACTION, ZERO

2

The number two (2) is the second POSITIVE INTEGER and the first PRIME NUMBER. It is EVEN, and is the only EVEN PRIME (the PRIMES other than 2 are called the ODD PRIMES). The number 2 is also equal to its FACTORIAL since $2! = 2$. A quantity taken to the POWER 2 is said to be SQUARED. The number of times k a given BINARY number $b_n \cdots b_2 b_1 b_0$ is divisible by 2 is given by the position of the first $b_k = 1$, counting from the right. For example, $12 = 1100$ is divisible by 2 twice, and $13 = 1101$ is divisible by 2 0 times.

see also 1, BINARY, 3, SQUARED, ZERO

$2x$ mod 1 Map

Let x_0 be a REAL NUMBER in the CLOSED INTERVAL $[0, 1]$, and generate a SEQUENCE using the MAP

$$x_{n+1} \equiv 2x_n \pmod{1}. \tag{1}$$

Then the number of periodic ORBITS of period p (for p PRIME) is given by

$$N_p = \frac{2^p - 2}{p}. \tag{2}$$

Since a typical ORBIT visits each point with equal probability, the NATURAL INVARIANT is given by

$$\rho(x) = 1. \tag{3}$$

see also TENT MAP

References

Ott, E. *Chaos in Dynamical Systems.* Cambridge: Cambridge University Press, pp. 26–31, 1993.

3

3 is the only INTEGER which is the sum of the preceding POSITIVE INTEGERS ($1 + 2 = 3$) and the only number which is the sum of the FACTORIALS of the preceding POSITIVE INTEGERS ($1! + 2! = 3$). It is also the first ODD PRIME. A quantity taken to the POWER 3 is said to be CUBED.

see also 1, 2, $3x + 1$ MAPPING, CUBED, PERIOD THREE THEOREM, SUPER-3 NUMBER, TERNARY, THREE-COLORABLE, ZERO

$3x + 1$ Mapping

see COLLATZ PROBLEM

10

The number 10 (ten) is the basis for the DECIMAL system of notation. In this system, each "decimal place" consists of a DIGIT 0–9 arranged such that each DIGIT is multiplied by a POWER of 10, decreasing from left to right, and with a decimal place indicating the $10^0 = 1$s place. For example, the number 1234.56 specifies

$$1 \times 10^3 + 2 \times 10^2 + 3 \times 10^1 + 4 \times 10^0 + 5 \times 10^{-1} + 6 \times 10^{-2}.$$

The decimal places to the left of the decimal point are 1, 10, 100, 1000, 10000, 10000, 100000, 10000000, 100000000, ... (Sloane's A011557), called one, ten, HUNDRED, THOUSAND, ten thousand, hundred thousand, MILLION, 10 million, 100 million, and so on. The names of subsequent decimal places for LARGE NUMBERS differ depending on country.

Any POWER of 10 which can be written as the PRODUCT of two numbers not containing 0s must be of the form $2^n \cdot 5^n = 10^n$ for n an INTEGER such that neither 2^n nor 5^n contains any ZEROS. The largest known such number is

$$\begin{aligned} 10^{33} &= 2^{33} \cdot 5^{33} \\ &= 8,589,934,592 \cdot 116,415,321,826,934,814,453,125. \end{aligned}$$

A complete list of known such numbers is

$$\begin{aligned} 10^1 &= 2^1 \cdot 5^1 \\ 10^2 &= 2^2 \cdot 5^2 \\ 10^3 &= 2^3 \cdot 5^3 \\ 10^4 &= 2^4 \cdot 5^4 \\ 10^5 &= 2^5 \cdot 5^5 \\ 10^6 &= 2^6 \cdot 5^6 \\ 10^7 &= 2^7 \cdot 5^7 \\ 10^9 &= 2^9 \cdot 5^9 \\ 10^{18} &= 2^{18} \cdot 5^{18} \\ 10^{33} &= 2^{33} \cdot 5^{33} \end{aligned}$$

(Madachy 1979). Since all POWERS of 2 with exponents $n \leq 4.6 \times 10^7$ contain at least one ZERO (M. Cook), no other POWER of ten less than 46 million can be written as the PRODUCT of two numbers not containing 0s.

see also BILLION, DECIMAL, HUNDRED, LARGE NUMBER, MILLIARD, MILLION, THOUSAND, TRILLION, ZERO

References

Madachy, J. S. *Madachy's Mathematical Recreations.* New York: Dover, pp. 127–128, 1979.

Pickover, C. A. *Keys to Infinity.* New York: W. H. Freeman, p. 135, 1995.

Sloane, N. J. A. Sequence A011557 in "An On-Line Version of the Encyclopedia of Integer Sequences."

12

One DOZEN, or a twelfth of a GROSS.

see also DOZEN, GROSS

13

A NUMBER traditionally associated with bad luck. A so-called BAKER'S DOZEN is equal to 13. Fear of the number 13 is called TRISKAIDEKAPHOBIA.

see also BAKER'S DOZEN, FRIDAY THE THIRTEENTH, TRISKAIDEKAPHOBIA

15

see 15 PUZZLE, FIFTEEN THEOREM

15 Puzzle

1	2	3	4
5	6	7	8
9	10	11	12
13	14	15	

A puzzle introduced by Sam Loyd in 1878. It consists of 15 squares numbered from 1 to 15 which are placed in a 4×4 box leaving one position out of the 16 empty. The goal is to rearrange the squares from a given arbitrary starting arrangement by sliding them one at a time into the configuration shown above. For some initial arrangements, this rearrangement is possible, but for others, it is not.

To address the solubility of a given initial arrangement, proceed as follows. If the SQUARE containing the number i appears "before" (reading the squares in the box from left to right and top to bottom) n numbers which are less than i, then call it an inversion of order n, and denote it n_i. Then define

$$N \equiv \sum_{i=1}^{15} n_i = \sum_{i=2}^{15} n_i,$$

where the sum need run only from 2 to 15 rather than 1 to 15 since there are no numbers less than 1 (so n_1 must equal 0). If N is EVEN, the position is possible, otherwise it is not. This can be formally proved using ALTERNATING GROUPS. For example, in the following arrangement

2	1	3	4
5	6	7	8
9	10	11	12
13	14	15	

$n_2 = 1$ (2 precedes 1) and all other $n_i = 0$, so $N = 1$ and the puzzle cannot be solved.

References

Ball, W. W. R. and Coxeter, H. S. M. *Mathematical Recreations and Essays, 13th ed.* New York: Dover, pp. 312–316, 1987.

Bogomolny, A. "Sam Loyd's Fifteen." http://www.cut-the-knot.com/pythagoras/fifteen.html.

Bogomolny, A. "Sam Loyd's Fifteen [History]." http://www.cut-the-knot.com/pythagoras/history15.html.

Johnson, W. W. "Notes on the '15 Puzzle. I.'" *Amer. J. Math.* **2**, 397–399, 1879.

Kasner, E. and Newman, J. R. *Mathematics and the Imagination.* Redmond, WA: Tempus Books, pp. 177–180, 1989.

Kraitchik, M. "The 15 Puzzle." §12.2.1 in *Mathematical Recreations.* New York: W. W. Norton, pp. 302–308, 1942.

Story, W. E. "Notes on the '15 Puzzle. II.'" *Amer. J. Math.* **2**, 399–404, 1879.

16-Cell

A finite regular 4-D POLYTOPE with SCHLÄFLI SYMBOL $\{3, 3, 4\}$ and VERTICES which are the PERMUTATIONS of $(\pm 1, 0, 0, 0)$.

see also 24-CELL, 120-CELL, 600-CELL, CELL, POLYTOPE

17

17 is a FERMAT PRIME which means that the 17-sided REGULAR POLYGON (the HEPTADECAGON) is CONSTRUCTIBLE using COMPASS and STRAIGHTEDGE (as proved by Gauss).

see also CONSTRUCTIBLE POLYGON , FERMAT PRIME, HEPTADECAGON

References

Carr, M. "Snow White and the Seven(teen) Dwarfs." http:// www . math . harvard . edu / ~ hmb / issue2.1 / SEVENTEEN/seventeen.html.

Fischer, R. "Facts About the Number 17." http://tempo. harvard . edu / ~ rfischer / hcssim / 17_facts / kelly / kelly.html.

Lefevre, V. "Properties of 17." http://www.ens-lyon.fr/~vlefevre/d17_eng.html.

Shell Centre for Mathematical Education. "Number 17." http://acorn.educ.nottingham.ac.uk/ShellCent/Number/Num17.html.

18-Point Problem

Place a point somewhere on a LINE SEGMENT. Now place a second point and number it 2 so that each of the points is in a different half of the LINE SEGMENT. Continue, placing every Nth point so that all N points are on different $(1/N)$th of the LINE SEGMENT. Formally, for a given N, does there exist a sequence of real numbers $x_1, x_2, \ldots, x_N$ such that for every $n \in \{1, \ldots, N\}$ and every $k \in \{1, \ldots, n\}$, the inequality

$$\frac{k-1}{n} \leq x_i < \frac{k}{n}$$

holds for some $i \in \{1, \ldots, n\}$? Surprisingly, it is only possible to place 17 points in this manner (Berlekamp and Graham 1970, Warmus 1976).

Steinhaus (1979) gives a 14-point solution (0.06, 0.55, 0.77, 0.39, 0.96, 0.28, 0.64, 0.13, 0.88, 0.48, 0.19, 0.71, 0.35, 0.82), and Warmus (1976) gives the 17-point solution

$$\tfrac{4}{7} \le x_1 < \tfrac{7}{12}, \tfrac{2}{7} \le x_2 < \tfrac{5}{17}, \tfrac{16}{17} \le x_3 < 1, \tfrac{1}{14} \le x_4 < \tfrac{1}{13},$$
$$\tfrac{8}{11} \le x_5 < \tfrac{11}{15}, \tfrac{5}{11} \le x_6 < \tfrac{6}{13}, \tfrac{1}{7} \le x_7 < \tfrac{2}{13}, \tfrac{14}{17} \le x_8 < \tfrac{5}{6},$$
$$\tfrac{3}{8} \le x_9 < \tfrac{5}{13}, \tfrac{11}{17} \le x_{10} < \tfrac{2}{3}, \tfrac{3}{14} \le x_{11} < \tfrac{3}{13},$$
$$\tfrac{15}{17} \le x_{12} < \tfrac{11}{12}, \tfrac{1}{2} \le x_{12} < \tfrac{9}{17}, 0 \le x_{14} < \tfrac{1}{17},$$
$$\tfrac{13}{17} \le x_{15} < \tfrac{4}{5}, \tfrac{5}{16} \le x_{16} < \tfrac{6}{17}, \tfrac{10}{17} \le x_{17} < \tfrac{11}{17}.$$

Warmus (1976) states that there are 768 patterns of 17-point solutions (counting reversals as equivalent).

see also DISCREPANCY THEOREM, POINT PICKING

References

Berlekamp, E. R. and Graham, R. L. "Irregularities in the Distributions of Finite Sequences." *J. Number Th.* **2**, 152–161, 1970.

Gardner, M. *The Last Recreations: Hydras, Eggs, and Other Mathematical Mystifications.* New York: Springer-Verlag, pp. 34–36, 1997.

Steinhaus, H. "Distribution on Numbers" and "Generalization." Problems 6 and 7 in *One Hundred Problems in Elementary Mathematics.* New York: Dover, pp. 12–13, 1979.

Warmus, M. "A Supplementary Note on the Irregularities of Distributions." *J. Number Th.* **8**, 260–263, 1976.

24-Cell

A finite regular 4-D POLYTOPE with SCHLÄFLI SYMBOL $\{3,4,3\}$. Coxeter (1969) gives a list of the VERTEX positions. The EVEN coefficients of the D_4 lattice are 1, 24, 24, 96, ... (Sloane's A004011), and the 24 shortest vectors in this lattice form the 24-cell (Coxeter 1973, Conway and Sloane 1993, Sloane and Plouffe 1995).

see also 16-CELL, 120-CELL, 600-CELL, CELL, POLYTOPE

References

Conway, J. H. and Sloane, N. J. A. *Sphere-Packings, Lattices and Groups, 2nd ed.* New York: Springer-Verlag, 1993.

Coxeter, H. S. M. *Introduction to Geometry, 2nd ed.* New York: Wiley, p. 404, 1969.

Coxeter, H. S. M. *Regular Polytopes, 3rd ed.* New York: Dover, 1973.

Sloane, N. J. A. Sequences A004011/M5140 in "An On-Line Version of the Encyclopedia of Integer Sequences."

Sloane, N. J. A. and Plouffe, S. Extended entry in *The Encyclopedia of Integer Sequences.* San Diego: Academic Press, 1995.

42

According to Adams, 42 is the ultimate answer to life, the universe, and everything, although it is left as an exercise to the reader to determine the actual question leading to this result.

References

Adams, D. *The Hitchhiker's Guide to the Galaxy.* New York: Ballantine Books, 1997.

72 Rule

see RULE OF 72

120-Cell

A finite regular 4-D POLYTOPE with SCHLÄFLI SYMBOL $\{5,3,3\}$ (Coxeter 1969).

see also 16-CELL, 24-CELL, 600-CELL, CELL, POLYTOPE

References

Coxeter, H. S. M. *Introduction to Geometry, 2nd ed.* New York: Wiley, p. 404, 1969.

144

A DOZEN DOZEN, also called a GROSS. 144 is a SQUARE NUMBER and a SUM-PRODUCT NUMBER.

see also DOZEN

196-Algorithm

Take any POSITIVE INTEGER of two DIGITS or more, reverse the DIGITS, and add to the original number. Now repeat the procedure with the SUM so obtained. This procedure quickly produces PALINDROMIC NUMBERS for most INTEGERS. For example, starting with the number 5280 produces (5280, 6105, 11121, 23232). The end results of applying the algorithm to 1, 2, 3, ... are 1, 2, 3, 4, 5, 6, 7, 8, 9, 11, 11, 33, 44, 55, 66, 77, 88, 99, 121, ... (Sloane's A033865). The value for 89 is especially large, being 8813200023188.

The first few numbers not known to produce PALINDROMES are 196, 887, 1675, 7436, 13783, ... (Sloane's A006960), which are simply the numbers obtained by iteratively applying the algorithm to the number 196. This number therefore lends itself to the name of the ALGORITHM.

The number of terms $a(n)$ in the iteration sequence required to produce a PALINDROMIC NUMBER from n (i.e., $a(n) = 1$ for a PALINDROMIC NUMBER, $a(n) = 2$ if a PALINDROMIC NUMBER is produced after a single iteration of the 196-algorithm, etc.) for $n = 1, 2, \ldots$ are 1, 1, 1, 1, 1, 1, 1, 1, 1, 2, 1, 2, 2, 2, 2, 2, 2, 2, 3, 2, 2, 1, ... (Sloane's A030547). The smallest numbers which require $n = 0, 1, 2, \ldots$ iterations to reach a palindrome are 0, 10, 19, 59, 69, 166, 79, 188, ... (Sloane's A023109).

see also ADDITIVE PERSISTENCE, DIGITADITION, MULTIPLICATIVE PERSISTENCE, PALINDROMIC NUMBER, PALINDROMIC NUMBER CONJECTURE, RATS SEQUENCE, RECURRING DIGITAL INVARIANT

References

Gardner, M. *Mathematical Circus: More Puzzles, Games, Paradoxes and Other Mathematical Entertainments from Scientific American.* New York: Knopf, pp. 242–245, 1979.

Gruenberger, F. "How to Handle Numbers with Thousands of Digits, and Why One Might Want to." *Sci. Amer.* **250**, 19–26, Apr. 1984.

Sloane, N. J. A. Sequences A023109, A030547, A033865, and A006960/M5410 in "An On-Line Version of the Encyclopedia of Integer Sequences."

239

Some interesting properties (as well as a few arcane ones not reiterated here) of the number 239 are discussed in Beeler *et al.* (1972, Item 63). 239 appears in MACHIN'S FORMULA

$$\tfrac{1}{4}\pi = 4\tan(\tfrac{1}{5}) - \tan^{-1}(\tfrac{1}{239}),$$

which is related to the fact that

$$2 \cdot 13^4 - 1 = 239^2,$$

which is why 239/169 is the 7th CONVERGENT of $\sqrt{2}$. Another pair of INVERSE TANGENT FORMULAS involving 239 is

$$\begin{aligned}\tan^{-1}(\tfrac{1}{239}) &= \tan^{-1}(\tfrac{1}{70}) - \tan^{-1}(\tfrac{1}{99}) \\ &= \tan^{-1}(\tfrac{1}{408}) + \tan^{-1}(\tfrac{1}{577}).\end{aligned}$$

239 needs 4 SQUARES (the maximum) to express it, 9 CUBES (the maximum, shared only with 23) to express it, and 19 fourth POWERS (the maximum) to express it (see WARING'S PROBLEM). However, 239 doesn't need the maximum number of fifth POWERS (Beeler *et al.* 1972, Item 63).

References
Beeler, M.; Gosper, R. W.; and Schroeppel, R. *HAKMEM.* Cambridge, MA: MIT Artificial Intelligence Laboratory, Memo AIM-239, Feb. 1972.

257-gon

257 is a FERMAT PRIME, and the 257-gon is therefore a CONSTRUCTIBLE POLYGON using COMPASS and STRAIGHTEDGE, as proved by Gauss. An illustration of the 257-gon is not included here, since its 257 segments so closely resemble a CIRCLE. Richelot and Schwendenwein found constructions for the 257-gon in 1832 (Coxeter 1969). De Temple (1991) gives a construction using 150 CIRCLES (24 of which are CARLYLE CIRCLES) which has GEOMETROGRAPHY symbol $94S_1 + 47S_2 + 275C_1 + 0C_2 + 150C_3$ and SIMPLICITY 566.

see also 65537-GON, CONSTRUCTIBLE POLYGON, FERMAT PRIME, HEPTADECAGON, PENTAGON

References
Coxeter, H. S. M. *Introduction to Geometry, 2nd ed.* New York: Wiley, 1969.
De Temple, D. W. "Carlyle Circles and the Lemoine Simplicity of Polygonal Constructions." *Amer. Math. Monthly* **98**, 97–108, 1991.
Dixon, R. *Mathographics.* New York: Dover, p. 53, 1991.
Rademacher, H. *Lectures on Elementary Number Theory.* New York: Blaisdell, 1964.

600-Cell

A finite regular 4-D POLYTOPE with SCHLÄFLI SYMBOL $\{3,3,5\}$. For VERTICES, see Coxeter (1969).

see also 16-CELL, 24-CELL, 120-CELL, CELL, POLYTOPE

References
Coxeter, H. S. M. *Introduction to Geometry, 2nd ed.* New York: Wiley, p. 404, 1969.

666

A number known as the BEAST NUMBER appearing in the *Bible* and ascribed various numerological properties.

see also APOCALYPTIC NUMBER, BEAST NUMBER, LEVIATHAN NUMBER

References
Hardy, G. H. *A Mathematician's Apology, reprinted with a foreword by C. P. Snow.* New York: Cambridge University Press, p. 96, 1993.

2187

The digits in the number 2187 form the two VAMPIRE NUMBERS: $21 \times 87 = 1827$ and $2187 = 27 \times 81$.

References
Gardner, M. "Lucky Numbers and 2187." *Math. Intell.* **19**, 26–29, Spring 1997.

65537-gon

65537 is the largest known FERMAT PRIME, and the 65537-gon is therefore a CONSTRUCTIBLE POLYGON using COMPASS and STRAIGHTEDGE, as proved by Gauss. The 65537-gon has so many sides that it is, for all intents and purposes, indistinguishable from a CIRCLE using any reasonable printing or display methods. Hermes spent 10 years on the construction of the 65537-gon at Göttingen around 1900 (Coxeter 1969). De Temple (1991) notes that a GEOMETRIC CONSTRUCTION can be done using 1332 or fewer CARLYLE CIRCLES.

see also 257-GON, CONSTRUCTIBLE POLYGON, HEPTADECAGON, PENTAGON

References
Coxeter, H. S. M. *Introduction to Geometry, 2nd ed.* New York: Wiley, 1969.
De Temple, D. W. "Carlyle Circles and the Lemoine Simplicity of Polygonal Constructions." *Amer. Math. Monthly* **98**, 97–108, 1991.
Dixon, R. *Mathographics.* New York: Dover, p. 53, 1991.

A

A-Integrable

A generalization of the LEBESGUE INTEGRAL. A MEASURABLE FUNCTION $f(x)$ is called A-integrable over the CLOSED INTERVAL $[a,b]$ if

$$m\{x : |f(x)| > n\} = \mathcal{O}(n^{-1}), \quad (1)$$

where m is the LEBESGUE MEASURE, and

$$I = \lim_{n\to\infty} \int_a^b [f(x)]_n \, dx \quad (2)$$

exists, where

$$[f(x)]_n = \begin{cases} f(x) & \text{if } |f(x)| \leq n \\ 0 & \text{if } |f(x)| > n. \end{cases} \quad (3)$$

References

Titmarsch, E. G. "On Conjugate Functions." *Proc. London Math. Soc.* **29**, 49–80, 1928.

A-Sequence

N.B. A detailed on-line essay by S. Finch was the starting point for this entry.

An INFINITE SEQUENCE of POSITIVE INTEGERS a_i satisfying

$$1 \leq a_1 < a_2 < a_3 < \ldots \quad (1)$$

is an A-sequence if no a_k is the SUM of two or more distinct earlier terms (Guy 1994). Erdős (1962) proved

$$S(A) \equiv \sup_{\text{all } A \text{ sequences}} \sum_{k=1}^{\infty} \frac{1}{a_k} < 103. \quad (2)$$

Any A-sequence satisfies the CHI INEQUALITY (Levine and O'Sullivan 1977), which gives $S(A) < 3.9998$. Abbott (1987) and Zhang (1992) have given a bound from below, so the best result to date is

$$2.0649 < S(A) < 3.9998. \quad (3)$$

Levine and O'Sullivan (1977) conjectured that the sum of RECIPROCALS of an A-sequence satisfies

$$S(A) \leq \sum_{k=1}^{\infty} \frac{1}{\chi_k} = 3.01\ldots, \quad (4)$$

where χ_i are given by the LEVINE-O'SULLIVAN GREEDY ALGORITHM.

see also B_2-SEQUENCE, MIAN-CHOWLA SEQUENCE

References

Abbott, H. L. "On Sum-Free Sequences." *Acta Arith.* **48**, 93–96, 1987.

Erdős, P. "Remarks on Number Theory III. Some Problems in Additive Number Theory." *Mat. Lapok* **13**, 28–38, 1962.

Finch, S. "Favorite Mathematical Constants." `http://www.mathsoft.com/asolve/constant/erdos/erdos.html`.

Guy, R. K. "B_2-Sequences." §E28 in *Unsolved Problems in Number Theory, 2nd ed.* New York: Springer-Verlag, pp. 228–229, 1994.

Levine, E. and O'Sullivan, J. "An Upper Estimate for the Reciprocal Sum of a Sum-Free Sequence." *Acta Arith.* **34**, 9–24, 1977.

Zhang, Z. X. "A Sum-Free Sequence with Larger Reciprocal Sum." Unpublished manuscript, 1992.

AAA Theorem

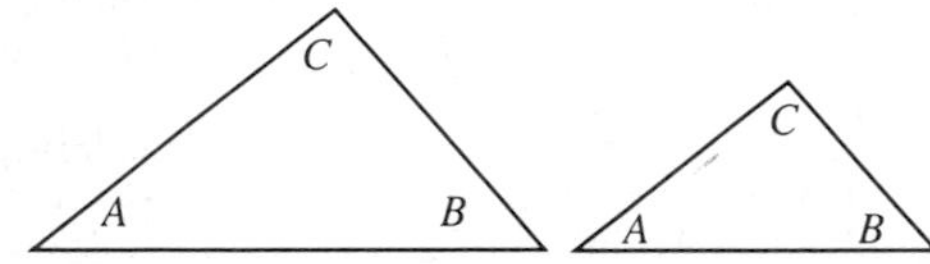

Specifying three ANGLES A, B, and C does not uniquely define a TRIANGLE, but any two TRIANGLES with the same ANGLES are SIMILAR. Specifying two ANGLES of a TRIANGLE automatically gives the third since the sum of ANGLES in a TRIANGLE sums to 180° (π RADIANS), i.e.,

$$C = \pi - A - B.$$

see also AAS THEOREM, ASA THEOREM, ASS THEOREM, SAS THEOREM, SSS THEOREM, TRIANGLE

AAS Theorem

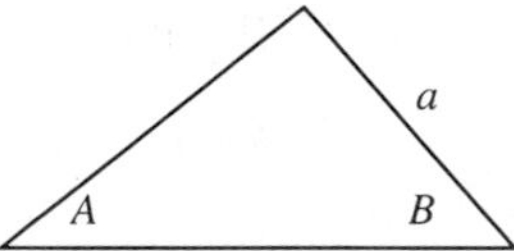

Specifying two angles A and B and a side a uniquely determines a TRIANGLE with AREA

$$K = \frac{a^2 \sin B \sin C}{2 \sin A} = \frac{a^2 \sin B \sin(\pi - A - B)}{2 \sin A}. \quad (1)$$

The third angle is given by

$$C = \pi - A - B, \quad (2)$$

since the sum of angles of a TRIANGLE is 180° (π RADIANS). Solving the LAW OF SINES

$$\frac{a}{\sin A} = \frac{b}{\sin B} \quad (3)$$

for b gives

$$b = a \frac{\sin B}{\sin A}. \quad (4)$$

Finally,

$$c = b \cos A + a \cos B = a(\sin B \cot A + \cos B) \quad (5)$$
$$= a \sin B(\cot A + \cot B). \quad (6)$$

see also AAA THEOREM, ASA THEOREM, ASS THEOREM, SAS THEOREM, SSS THEOREM, TRIANGLE

Abacus

A mechanical counting device consisting of a frame holding a series of parallel rods on each of which beads are strung. Each bead represents a counting unit, and each rod a place value. The primary purpose of the abacus is not to perform actual computations, but to provide a quick means of storing numbers during a calculation. Abaci were used by the Japanese and Chinese, as well as the Romans.

see also ROMAN NUMERAL, SLIDE RULE

References

Boyer, C. B. and Merzbach, U. C. "The Abacus and Decimal Fractions." *A History of Mathematics, 2nd ed.* New York: Wiley, pp. 199–201, 1991.
Fernandes, L. "The Abacus: The Art of Calculating with Beads." `http://www.ee.ryerson.ca:8080/~elf/abacus`.
Gardner, M. "The Abacus." Ch. 18 in *Mathematical Circus: More Puzzles, Games, Paradoxes and Other Mathematical Entertainments from Scientific American.* New York: Knopf, pp. 232–241, 1979.
Pappas, T. "The Abacus." *The Joy of Mathematics.* San Carlos, CA: Wide World Publ./Tetra, p. 209, 1989.
Smith, D. E. "Mechanical Aids to Calculation: The Abacus." Ch. 3 §1 in *History of Mathematics, Vol. 2.* New York: Dover, pp. 156–196, 1958.

abc Conjecture

A CONJECTURE due to J. Oesterlé and D. W. Masser. It states that, for any INFINITESIMAL $\epsilon > 0$, there exists a CONSTANT C_ϵ such that for any three RELATIVELY PRIME INTEGERS a, b, c satisfying

$$a + b = c,$$

the INEQUALITY

$$\max\{|a|,|b|,|c|\} \leq C_\epsilon \prod_{p|abc} p^{1+\epsilon}$$

holds, where $p|abc$ indicates that the PRODUCT is over PRIMES p which DIVIDE the PRODUCT abc. If this CONJECTURE were true, it would imply FERMAT'S LAST THEOREM for sufficiently large POWERS (Goldfeld 1996). This is related to the fact that the abc conjecture implies that there are at least $C \ln x$ WIEFERICH PRIMES $\leq x$ for some constant C (Silverman 1988, Vardi 1991).

see also FERMAT'S LAST THEOREM, MASON'S THEOREM, WIEFERICH PRIME

References

Cox, D. A. "Introduction to Fermat's Last Theorem." *Amer. Math. Monthly* **101**, 3–14, 1994.
Goldfeld, D. "Beyond the Last Theorem." *The Sciences*, 34–40, March/April 1996.
Guy, R. K. *Unsolved Problems in Number Theory, 2nd ed.* New York: Springer-Verlag, pp. 75–76, 1994.
Silverman, J. "Wieferich's Criterion and the abc Conjecture." *J. Number Th.* **30**, 226–237, 1988.
Vardi, I. *Computational Recreations in Mathematica.* Reading, MA: Addison-Wesley, p. 66, 1991.

Abelian

see ABELIAN CATEGORY, ABELIAN DIFFERENTIAL, ABELIAN FUNCTION, ABELIAN GROUP, ABELIAN INTEGRAL, ABELIAN VARIETY, COMMUTATIVE

Abelian Category

An Abelian category is an abstract mathematical CATEGORY which displays some of the characteristic properties of the CATEGORY of all ABELIAN GROUPS.

see also ABELIAN GROUP, CATEGORY

Abel's Curve Theorem

The sum of the values of an INTEGRAL of the "first" or "second" sort

$$\int_{x_0,y_0}^{x_1,y_1} \frac{P\,dx}{Q} + \ldots + \int_{x_0,y_0}^{x_N,y_N} \frac{P\,dx}{Q} = F(z)$$

and

$$\frac{P(x_1,y_1)}{Q(x_1,y_1)}\frac{dx_1}{dz} + \ldots + \frac{P(x_N,y_N)}{Q(x_N,y_N)}\frac{dx_N}{dz} = \frac{dF}{dz},$$

from a FIXED POINT to the points of intersection with a curve depending rationally upon any number of parameters is a RATIONAL FUNCTION of those parameters.

References

Coolidge, J. L. *A Treatise on Algebraic Plane Curves.* New York: Dover, p. 277, 1959.

Abelian Differential

An Abelian differential is an ANALYTIC or MEROMORPHIC DIFFERENTIAL on a COMPACT or closed RIEMANN SURFACE.

Abelian Function

An INVERSE FUNCTION of an ABELIAN INTEGRAL. Abelian functions have two variables and four periods. They are a generalization of ELLIPTIC FUNCTIONS, and are also called HYPERELLIPTIC FUNCTIONS.

see also ABELIAN INTEGRAL, ELLIPTIC FUNCTION

References

Baker, H. F. *Abelian Functions: Abel's Theorem and the Allied Theory, Including the Theory of the Theta Functions.* New York: Cambridge University Press, 1995.
Baker, H. F. *An Introduction to the Theory of Multiply Periodic Functions.* London: Cambridge University Press, 1907.

Abel's Functional Equation

Let $\mathrm{Li}_2(x)$ denote the DILOGARITHM, defined by

$$\mathrm{Li}_2(x) = \sum_{n=1}^{\infty} \frac{x^n}{n^2},$$

then

$$\mathrm{Li}_2(x)+\mathrm{Li}_2(y)+\mathrm{Li}_2(xy)+\mathrm{Li}_2\left(\frac{x(1-y)}{1-xy}\right)+\mathrm{Li}_2\left(\frac{y(1-x)}{1-xy}\right)=3\,\mathrm{Li}_2(1).$$

see also DILOGARITHM, POLYLOGARITHM, RIEMANN ZETA FUNCTION

Abelian Group

N.B. A detailed on-line essay by S. Finch was the starting point for this entry.

A GROUP for which the elements COMMUTE (i.e., $AB = BA$ for all elements A and B) is called an Abelian group. All CYCLIC GROUPS are Abelian, but an Abelian group is not necessarily CYCLIC. All SUBGROUPS of an Abelian group are NORMAL. In an Abelian group, each element is in a CONJUGACY CLASS by itself, and the CHARACTER TABLE involves POWERS of a single element known as a GENERATOR.

No general formula is known for giving the number of nonisomorphic FINITE GROUPS of a given ORDER. However, the number of nonisomorphic Abelian FINITE GROUPS $a(n)$ of any given ORDER n is given by writing n as

$$n=\prod_i {p_i}^{\alpha_i}, \tag{1}$$

where the p_i are distinct PRIME FACTORS, then

$$a(n)=\prod_i P(\alpha_i), \tag{2}$$

where P is the PARTITION FUNCTION. This gives 1, 1, 1, 2, 1, 1, 1, 3, 2, ... (Sloane's A000688). The smallest orders for which $n = 1, 2, 3, \ldots$ nonisomorphic Abelian groups exist are 1, 4, 8, 36, 16, 72, 32, 900, 216, 144, 64, 1800, 0, 288, 128, ... (Sloane's A046056), where 0 denotes an impossible number (i.e., not a product of partition numbers) of nonisomorphic Abelian, groups. The "missing" values are 13, 17, 19, 23, 26, 29, 31, 34, 37, 38, 39, 41, 43, 46, ... (Sloane's A046064). The incrementally largest numbers of Abelian groups as a function of order are 1, 2, 3, 5, 7, 11, 15, 22, 30, 42, 56, 77, 101, ... (Sloane's A046054), which occur for orders 1, 4, 8, 16, 32, 64, 128, 256, 512, 1024, 2048, 4096, 8192, ... (Sloane's A046055).

The KRONECKER DECOMPOSITION THEOREM states that every FINITE Abelian group can be written as a DIRECT PRODUCT of CYCLIC GROUPS of PRIME POWER ORDERS. If the ORDERS of a FINITE GROUP is a PRIME p, then there exists a single Abelian group of order p (denoted Z_p) and no non-Abelian groups. If the ORDERS is a prime squared p^2, then there are two Abelian groups (denoted Z_{p^2} and $Z_p \otimes Z_p$. If the ORDERS is a prime cubed p^3, then there are three Abelian groups (denoted $Z_p \otimes Z_p \otimes Z_p$, $Z_p \otimes Z_{p^2}$, and Z_{p^3}), and five groups total. If the order is a PRODUCT of two primes p and q, then there exists exactly one Abelian group of order pq (denoted $Z_p \otimes Z_q$).

Another interesting result is that if $a(n)$ denotes the number of nonisomorphic Abelian groups of ORDER n, then

$$\sum_{n=1}^{\infty} a(n)n^{-s}=\zeta(s)\zeta(2s)\zeta(3s)\cdots, \tag{3}$$

where $\zeta(s)$ is the RIEMANN ZETA FUNCTION. Srinivasan (1973) has also shown that

$$\sum_{n=1}^{N} a(n)=A_1N+A_2N^{1/2}+A_3N^{1/3}+\mathcal{O}[x^{105/407}(\ln x)^2], \tag{4}$$

where

$$A_k\equiv\prod_{\substack{j=1\\ j\neq k}}\zeta\left(\frac{j}{k}\right)=\begin{cases}2.294856591\ldots & \text{for } k=1\\ -14.6475663\ldots & \text{for } k=2\\ 118.6924619\ldots & \text{for } k=3,\end{cases} \tag{5}$$

and ζ is again the RIEMANN ZETA FUNCTION. [Richert (1952) incorrectly gave $A_3 = 114$.] DeKoninck and Ivic (1980) showed that

$$\sum_{n=1}^{N}\frac{1}{a(n)}=BN+\mathcal{O}[\sqrt{N}(\ln N)^{-1/2}], \tag{6}$$

where

$$B\equiv\prod\left\{1-\sum_{k=2}^{\infty}\left[\frac{1}{P(k-2)}-\frac{1}{P(k)}\right]\frac{1}{p^k}\right\}=0.752\ldots \tag{7}$$

is a product over PRIMES. Bounds for the number of nonisomorphic non-Abelian groups are given by Neumann (1969) and Pyber (1993).

see also FINITE GROUP, GROUP THEORY, KRONECKER DECOMPOSITION THEOREM, PARTITION FUNCTION P, RING

References

DeKoninck, J.-M. and Ivic, A. *Topics in Arithmetical Functions: Asymptotic Formulae for Sums of Reciprocals of Arithmetical Functions and Related Fields.* Amsterdam, Netherlands: North-Holland, 1980.

Erdős, P. and Szekeres, G. "Über die Anzahl abelscher Gruppen gegebener Ordnung und über ein verwandtes zahlentheoretisches Problem." *Acta Sci. Math. (Szeged)* **7**, 95–102, 1935.

Finch, S. "Favorite Mathematical Constants." `http://www.mathsoft.com/asolve/constant/abel/abel.html`.

Kendall, D. G. and Rankin, R. A. "On the Number of Abelian Groups of a Given Order." *Quart. J. Oxford* **18**, 197–208, 1947.

Kolesnik, G. "On the Number of Abelian Groups of a Given Order." *J. Reine Angew. Math.* **329**, 164–175, 1981.

Neumann, P. M. "An Enumeration Theorem for Finite Groups." *Quart. J. Math. Ser. 2* **20**, 395–401, 1969.
Pyber, L. "Enumerating Finite Groups of Given Order." *Ann. Math.* **137**, 203–220, 1993.
Richert, H.-E. "Über die Anzahl abelscher Gruppen gegebener Ordnung I." *Math. Zeitschr.* **56**, 21–32, 1952.
Sloane, N. J. A. Sequence A000688/M0064 in "An On-Line Version of the Encyclopedia of Integer Sequences."
Srinivasan, B. R. "On the Number of Abelian Groups of a Given Order." *Acta Arith.* **23**, 195–205, 1973.

Abel's Identity

Given a homogeneous linear SECOND-ORDER ORDINARY DIFFERENTIAL EQUATION,

$$y'' + P(x)y' + Q(x)y = 0, \tag{1}$$

call the two linearly independent solutions $y_1(x)$ and $y_2(x)$. Then

$$y_1''(x) + P(x)y_1'(x) + Q(x)y_1 = 0 \tag{2}$$

$$y_2''(x) + P(x)y_2'(x) + Q(x)y_2 = 0. \tag{3}$$

Now, take $y_1 \times (3) - y_2 \times (2)$,

$$y_1[y_2'' + P(x)y_2' + Q(x)y_2] - y_2[y_1'' + P(x)y_1' + Q(x)y_1] = 0 \tag{4}$$

$$(y_1y_2'' - y_2y_1'') + P(y_1y_2' - y_1'y_2) + Q(y_1y_2 - y_1y_2) = 0 \tag{5}$$

$$(y_1y_2'' - y_2y_1'') + P(y_1y_2' - y_1'y_2) = 0. \tag{6}$$

Now, use the definition of the WRONSKIAN and take its DERIVATIVE,

$$W \equiv y_1y_2' - y_1'y_2 \tag{7}$$

$$\begin{aligned} W' &= (y_1'y_2' + y_1y_2'') - (y_1'y_2' + y_1''y_2) \\ &= y_1y_2'' - y_1''y_2. \end{aligned} \tag{8}$$

Plugging W and W' into (6) gives

$$W' + PW = 0. \tag{9}$$

This can be rearranged to yield

$$\frac{dW}{W} = -P(x)\,dx \tag{10}$$

which can then be directly integrated to

$$\ln W = -C_1 \int P(x)\,dx, \tag{11}$$

where $\ln x$ is the NATURAL LOGARITHM. A second integration then yields Abel's identity

$$W(x) = C_2 e^{-\int P(x)\,dx}, \tag{12}$$

where C_1 is a constant of integration and $C_2 \equiv e^{C_1}$.

see also ORDINARY DIFFERENTIAL EQUATION—SECOND-ORDER

References

Boyce, W. E. and DiPrima, R. C. *Elementary Differential Equations and Boundary Value Problems, 4th ed.* New York: Wiley, pp. 118, 262, 277, and 355, 1986.

Abel's Impossibility Theorem

In general, POLYNOMIAL equations higher than fourth degree are incapable of algebraic solution in terms of a finite number of ADDITIONS, MULTIPLICATIONS, and ROOT extractions.

see also CUBIC EQUATION, GALOIS'S THEOREM, POLYNOMIAL, QUADRATIC EQUATION, QUARTIC EQUATION, QUINTIC EQUATION

References

Abel, N. H. "Démonstration de l'impossibilité de la résolution algébraique des équations générales qui dépassent le quatrième degré." *Crelle's J.* **1**, 1826.

Abel's Inequality

Let $\{f_n\}$ and $\{a_n\}$ be SEQUENCES with $f_n \geq f_{n+1} > 0$ for $n = 1, 2, \ldots$, then

$$\left|\sum_{n=1}^{m} a_n f_n\right| \leq A f_1,$$

where

$$A = \max\{|a_1|, |a_1 + a_2|, \ldots, |a_1 + a_2 + \ldots + a_m|\}.$$

Abelian Integral

An INTEGRAL of the form

$$\int_0^x \frac{dt}{\sqrt{R(t)}},$$

where $R(t)$ is a POLYNOMIAL of degree > 4. They are also called HYPERELLIPTIC INTEGRALS.

see also ABELIAN FUNCTION, ELLIPTIC INTEGRAL

Abel's Irreducibility Theorem

If one ROOT of the equation $f(x) = 0$, which is irreducible over a FIELD K, is also a ROOT of the equation $F(x) = 0$ in K, then all the ROOTS of the irreducible equation $f(x) = 0$ are ROOTS of $F(x) = 0$. Equivalently, $F(x)$ can be divided by $f(x)$ without a REMAINDER,

$$F(x) = f(x)F_1(x),$$

where $F_1(x)$ is also a POLYNOMIAL over K.

see also ABEL'S LEMMA, KRONECKER'S POLYNOMIAL THEOREM, SCHOENEMANN'S THEOREM

References

Abel, N. H. "Mémoir sur une classe particulière d'équations résolubles algébraiquement." *Crelle's J.* **4**, 1829.
Dörrie, H. *100 Great Problems of Elementary Mathematics: Their History and Solutions.* New York: Dover, p. 120, 1965.

Abel's Lemma

The pure equation

$$x^p = C$$

of PRIME degree p is irreducible over a FIELD when C is a number of the FIELD but not the pth POWER of an element of the FIELD.

see also ABEL'S IRREDUCIBILITY THEOREM, GAUSS'S POLYNOMIAL THEOREM, KRONECKER'S POLYNOMIAL THEOREM, SCHOENEMANN'S THEOREM

References
Dörrie, H. *100 Great Problems of Elementary Mathematics: Their History and Solutions.* New York: Dover, p. 118, 1965.

Abel's Test

see ABEL'S UNIFORM CONVERGENCE TEST

Abel's Theorem

Given a TAYLOR SERIES

$$F(z) = \sum_{n=0}^{\infty} C_n z^n = \sum_{n=0}^{\infty} C_n r^n e^{in\theta}, \quad (1)$$

where the COMPLEX NUMBER z has been written in the polar form $z = re^{i\theta}$, examine the REAL and IMAGINARY PARTS

$$u(r,\theta) = \sum_{n=0}^{\infty} C_n r^n \cos(n\theta) \quad (2)$$

$$v(r,\theta) = \sum_{n=0}^{\infty} C_n r^n \sin(n\theta). \quad (3)$$

Abel's theorem states that, if $u(1,\theta)$ and $v(1,\theta)$ are CONVERGENT, then

$$u(1,\theta) + iv(1,\theta) = \lim_{r\to 1} f(re^{i\theta}). \quad (4)$$

Stated in words, Abel's theorem guarantees that, if a REAL POWER SERIES CONVERGES for some POSITIVE value of the argument, the DOMAIN of UNIFORM CONVERGENCE extends at least up to and including this point. Furthermore, the continuity of the sum function extends at least up to and including this point.

References
Arfken, G. *Mathematical Methods for Physicists, 3rd ed.* Orlando, FL: Academic Press, p. 773, 1985.

Abel Transform

The following INTEGRAL TRANSFORM relationship, known as the Abel transform, exists between two functions $f(x)$ and $g(t)$ for $0 < \alpha < 1$,

$$f(x) = \int_0^x \frac{g(t)\,dt}{(x-t)^\alpha} \quad (1)$$

$$g(t) = -\frac{\sin(\pi\alpha)}{\pi}\frac{d}{dt}\int_0^t \frac{f(x)\,dx}{(x-t)^{1-\alpha}} \quad (2)$$

$$= -\frac{\sin(\pi\alpha)}{\pi}\left[\int_0^t \frac{df}{dx}\frac{dx}{(t-x)^{1-\alpha}} + \frac{f(0)}{t^{1-\alpha}}\right]. \quad (3)$$

The Abel transform is used in calculating the radial mass distribution of galaxies and inverting planetary radio occultation data to obtain atmospheric information.

References
Arfken, G. *Mathematical Methods for Physicists, 3rd ed.* Orlando, FL: Academic Press, pp. 875–876, 1985.
Binney, J. and Tremaine, S. *Galactic Dynamics.* Princeton, NJ: Princeton University Press, p. 651, 1987.
Bracewell, R. *The Fourier Transform and Its Applications.* New York: McGraw-Hill, pp. 262–266, 1965.

Abel's Uniform Convergence Test

Let $\{u_n(x)\}$ be a SEQUENCE of functions. If

1. $u_n(x)$ can be written $u_n(x) = a_n f_n(x)$,
2. $\sum a_n$ is CONVERGENT,
3. $f_n(x)$ is a MONOTONIC DECREASING SEQUENCE (i.e., $f_{n+1}(x) \leq f_n(x)$) for all n, and
4. $f_n(x)$ is BOUNDED in some region (i.e., $0 \leq f_n(x) \leq M$ for all $x \in [a,b]$)

then, for all $x \in [a,b]$, the SERIES $\sum u_n(x)$ CONVERGES UNIFORMLY.

see also CONVERGENCE TESTS

References
Bromwich, T. J. I'a and MacRobert, T. M. *An Introduction to the Theory of Infinite Series, 3rd ed.* New York: Chelsea, p. 59, 1991.
Whittaker, E. T. and Watson, G. N. *A Course in Modern Analysis, 4th ed.* Cambridge, England: Cambridge University Press, p. 17, 1990.

Abelian Variety

An Abelian variety is an algebraic GROUP which is a complete ALGEBRAIC VARIETY. An Abelian variety of DIMENSION 1 is an ELLIPTIC CURVE.

see also ALBANESE VARIETY

References
Murty, V. K. *Introduction to Abelian Varieties.* Providence, RI: Amer. Math. Soc., 1993.

Abhyankar's Conjecture

For a FINITE GROUP G, let $p(G)$ be the SUBGROUP generated by all the SYLOW p-SUBGROUPS of G. If X is a projective curve in characteristic $p > 0$, and if $x_0, \ldots, x_t$ are points of X (for $t > 0$), then a NECESSARY and SUFFICIENT condition that G occur as the GALOIS GROUP of a finite covering Y of X, branched only at the points $x_0, \ldots, x_t$, is that the QUOTIENT GROUP $G/p(G)$ has $2g + t$ generators.

Raynaud (1994) solved the Abhyankar problem in the crucial case of the affine line (i.e., the projective line with a point deleted), and Harbater (1994) proved the full Abhyankar conjecture by building upon this special solution.

see also FINITE GROUP, GALOIS GROUP, QUOTIENT GROUP, SYLOW p-SUBGROUP

References

Abhyankar, S. "Coverings of Algebraic Curves." *Amer. J. Math.* **79**, 825–856, 1957.

American Mathematical Society. "Notices of the AMS, April 1995, 1995 Frank Nelson Cole Prize in Algebra." `http://www.ams.org/notices/199504/prize-cole.html`.

Harbater, D. "Abhyankar's Conjecture on Galois Groups Over Curves." *Invent. Math.* **117**, 1–25, 1994.

Raynaud, M. "Revêtements de la droite affine en caractéristique $p > 0$ et conjecture d'Abhyankar." *Invent. Math.* **116**, 425–462, 1994.

Ablowitz-Ramani-Segur Conjecture

The Ablowitz-Ramani-Segur conjecture states that a nonlinear PARTIAL DIFFERENTIAL EQUATION is solvable by the INVERSE SCATTERING METHOD only if every nonlinear ORDINARY DIFFERENTIAL EQUATION obtained by exact reduction has the PAINLEVÉ PROPERTY.

see also INVERSE SCATTERING METHOD

References

Tabor, M. *Chaos and Integrability in Nonlinear Dynamics: An Introduction.* New York: Wiley, p. 351, 1989.

Abscissa

The x- (horizontal) axis of a GRAPH.

see also AXIS, ORDINATE, REAL LINE, x-AXIS, y-AXIS, z-AXIS

Absolute Convergence

A SERIES $\sum_n u_n$ is said to CONVERGE absolutely if the SERIES $\sum_n |u_n|$ CONVERGES, where $|u_n|$ denotes the ABSOLUTE VALUE. If a SERIES is absolutely convergent, then the sum is independent of the order in which terms are summed. Furthermore, if the SERIES is multiplied by another absolutely convergent series, the product series will also converge absolutely.

see also CONDITIONAL CONVERGENCE, CONVERGENT SERIES, RIEMANN SERIES THEOREM

References

Bromwich, T. J. I'a and MacRobert, T. M. "Absolute Convergence." Ch. 4 in *An Introduction to the Theory of Infinite Series, 3rd ed.* New York: Chelsea, pp. 69–77, 1991.

Absolute Deviation

Let $\bar{u}$ denote the MEAN of a SET of quantities u_i, then the absolute deviation is defined by

$$\Delta u_i \equiv |u_i - \bar{u}|.$$

see also DEVIATION, MEAN DEVIATION, SIGNED DEVIATION, STANDARD DEVIATION

Absolute Error

The DIFFERENCE between the measured or inferred value of a quantity x_0 and its actual value x, given by

$$\Delta x \equiv x_0 - x$$

(sometimes with the ABSOLUTE VALUE taken) is called the absolute error. The absolute error of the SUM or DIFFERENCE of a number of quantities is less than or equal to the SUM of their absolute errors.

see also ERROR PROPAGATION, PERCENTAGE ERROR, RELATIVE ERROR

References

Abramowitz, M. and Stegun, C. A. (Eds.). *Handbook of Mathematical Functions with Formulas, Graphs, and Mathematical Tables, 9th printing.* New York: Dover, p. 14, 1972.

Absolute Geometry

GEOMETRY which depends only on the first four of EUCLID'S POSTULATES and not on the PARALLEL POSTULATE. Euclid himself used only the first four postulates for the first 28 propositions of the *Elements*, but was forced to invoke the PARALLEL POSTULATE on the 29th.

see also AFFINE GEOMETRY, *Elements*, EUCLID'S POSTULATES, GEOMETRY, ORDERED GEOMETRY, PARALLEL POSTULATE

References

Hofstadter, D. R. *Gödel, Escher, Bach: An Eternal Golden Braid.* New York: Vintage Books, pp. 90–91, 1989.

Absolute Pseudoprime

see CARMICHAEL NUMBER

Absolute Square

Also known as the squared NORM. The absolute square of a COMPLEX NUMBER z is written $|z|^2$ and is defined as

$$|z|^2 \equiv zz^*, \tag{1}$$

where z^* denotes the COMPLEX CONJUGATE of z. For a REAL NUMBER, (1) simplifies to

$$|z|^2 = z^2. \tag{2}$$

If the COMPLEX NUMBER is written $z = x + iy$, then the absolute square can be written

$$|x + iy|^2 = x^2 + y^2. \tag{3}$$

An important identity involving the absolute square is given by

$$\begin{aligned} |a \pm be^{-i\delta}|^2 &= (a \pm be^{-i\delta})(a \pm be^{i\delta}) \\ &= a^2 + b^2 \pm ab(e^{i\delta} + e^{-i\delta}) \\ &= a^2 + b^2 \pm 2ab\cos\delta. \end{aligned} \tag{4}$$

If $a = 1$, then (4) becomes

$$\begin{aligned} |1 \pm be^{-i\delta}|^2 &= 1 + b^2 \pm 2b\cos\delta \\ &= 1 + b^2 \pm 2b[1 - 2\sin^2(\tfrac{1}{2}\delta)] \\ &= 1 \pm 2b + b^2 \mp 4b\sin^2(\tfrac{1}{2}\delta) \\ &= (1 \pm b)^2 \mp 4b\sin^2(\tfrac{1}{2}\delta). \end{aligned} \quad (5)$$

If $a = 1$, and $b = 1$, then

$$|1 - e^{-i\delta}|^2 = (1-1)^2 + 4 \cdot 1 \sin^2(\tfrac{1}{2}\delta) = 4\sin^2(\tfrac{1}{2}\delta). \quad (6)$$

Finally,

$$\begin{aligned} |e^{i\phi_1} + e^{i\phi_2}|^2 &= (e^{i\phi_1} + e^{i\phi_2})(e^{-i\phi_1} + e^{-i\phi_2}) \\ &= 2 + e^{i(\phi_2 - \phi_1)} + e^{-i(\phi_2 - \phi_1)} \\ &= 2 + 2\cos(\phi_2 - \phi_1) = 2[1 + \cos(\phi_2 - \phi_1)] \\ &= 4\cos^2(\phi_2 - \phi_1). \end{aligned} \quad (7)$$

Absolute Value

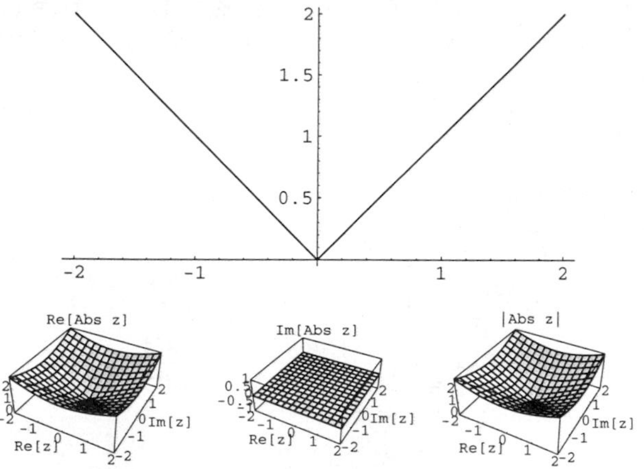

The absolute value of a REAL NUMBER x is denoted $|x|$ and given by

$$|x| = x\,\mathrm{sgn}(x) = \begin{cases} -x & \text{for } x \leq 0 \\ x & \text{for } x \geq 0, \end{cases}$$

where SGN is the sign function.

The same notation is used to denote the MODULUS of a COMPLEX NUMBER $z = x + iy$, $|z| \equiv \sqrt{x^2 + y^2}$, a p-ADIC absolute value, or a general VALUATION. The NORM of a VECTOR $\mathbf{x}$ is also denoted $|\mathbf{x}|$, although $||\mathbf{x}||$ is more commonly used.

Other NOTATIONS similar to the absolute value are the FLOOR FUNCTION $\lfloor x \rfloor$, NINT function $[x]$, and CEILING FUNCTION $\lceil x \rceil$.

see also ABSOLUTE SQUARE, CEILING FUNCTION, FLOOR FUNCTION, MODULUS (COMPLEX NUMBER), NINT, SGN, TRIANGLE FUNCTION, VALUATION

Absolutely Continuous

Let μ be a POSITIVE MEASURE on a SIGMA ALGEBRA M and let λ be an arbitrary (real or complex) MEASURE on M. Then λ is absolutely continuous with respect to μ, written $\lambda \ll \mu$, if $\lambda(E) = 0$ for every $E \in M$ for which $\mu(E) = 0$.

see also CONCENTRATED, MUTUALLY SINGULAR

References

Rudin, W. *Functional Analysis.* New York: McGraw-Hill, pp. 121–125, 1991.

Absorption Law

The law appearing in the definition of a BOOLEAN ALGEBRA which states

$$a \wedge (a \vee b) = a \vee (a \wedge b) = a$$

for binary operators $\vee$ and $\wedge$ (which most commonly are logical OR and logical AND).

see also BOOLEAN ALGEBRA, LATTICE

References

Birkhoff, G. and Mac Lane, S. *A Survey of Modern Algebra, 3rd ed.* New York: Macmillian, p. 317, 1965.

Abstraction Operator

see LAMBDA CALCULUS

Abundance

The abundance of a number n is the quantity

$$A(n) \equiv \sigma(n) - 2n,$$

where $\sigma(n)$ is the DIVISOR FUNCTION. Kravitz has conjectured that no numbers exist whose abundance is an ODD SQUARE (Guy 1994).

The following table lists special classifications given to a number n based on the value of $A(n)$.

$A(n)$	Number
< 0	deficient number
-1	almost perfect number
0	perfect number
1	quasiperfect number
> 0	abundant number

see also DEFICIENCY

References

Guy, R. K. *Unsolved Problems in Number Theory, 2nd ed.* New York: Springer-Verlag, pp. 45–46, 1994.

Abundant Number

An abundant number is an INTEGER n which is not a PERFECT NUMBER and for which

$$s(n) \equiv \sigma(n) - n > n, \tag{1}$$

where $\sigma(n)$ is the DIVISOR FUNCTION. The quantity $\sigma(n) - 2n$ is sometimes called the ABUNDANCE. The first few abundant numbers are 12, 18, 20, 24, 30, 36, ... (Sloane's A005101). Abundant numbers are sometimes called EXCESSIVE NUMBERS.

There are only 21 abundant numbers less than 100, and they are all EVEN. The first ODD abundant number is

$$945 = 3^3 \cdot 7 \cdot 5. \tag{2}$$

That 945 is abundant can be seen by computing

$$s(945) = 975 > 945. \tag{3}$$

Any multiple of a PERFECT NUMBER or an abundant number is also abundant. Every number greater than 20161 can be expressed as a sum of two abundant numbers.

Define the density function

$$A(x) \equiv \lim_{n\to\infty} \frac{|\{n : \sigma(n) \geq xn\}|}{n} \tag{4}$$

for a POSITIVE REAL NUMBER x, then Davenport (1933) proved that $A(x)$ exists and is continuous for all x, and Erdős (1934) gave a simplified proof (Finch). Wall (1971) and Wall *et al.* (1977) showed that

$$0.2441 < A(2) < 0.2909, \tag{5}$$

and Deléglise showed that

$$0.2474 < A(2) < 0.2480. \tag{6}$$

A number which is abundant but for which all its PROPER DIVISORS are DEFICIENT is called a PRIMITIVE ABUNDANT NUMBER (Guy 1994, p. 46).

see also ALIQUOT SEQUENCE, DEFICIENT NUMBER, HIGHLY ABUNDANT NUMBER, MULTIAMICABLE NUMBERS, PERFECT NUMBER, PRACTICAL NUMBER, PRIMITIVE ABUNDANT NUMBER, WEIRD NUMBER

References

Deléglise, M. "Encadrement de la densité des nombres abondants." Submitted.

Dickson, L. E. *History of the Theory of Numbers, Vol. 1: Divisibility and Primality.* New York: Chelsea, pp. 3–33, 1952.

Erdős, P. "On the Density of the Abundant Numbers." *J. London Math. Soc.* **9**, 278–282, 1934.

Finch, S. "Favorite Mathematical Constants." `http://www.mathsoft.com/asolve/constant/abund/abund.html`.

Guy, R. K. *Unsolved Problems in Number Theory, 2nd ed.* New York: Springer-Verlag, pp. 45–46, 1994.

Singh, S. *Fermat's Enigma: The Epic Quest to Solve the World's Greatest Mathematical Problem.* New York: Walker, pp. 11 and 13, 1997.

Sloane, N. J. A. Sequence A005101/M4825 in "An On-Line Version of the Encyclopedia of Integer Sequences."

Wall, C. R. "Density Bounds for the Sum of Divisors Function." In *The Theory of Arithmetic Functions* (Ed. A. A. Gioia and D. L. Goldsmith). New York: Springer-Verlag, pp. 283–287, 1971.

Wall, C. R.; Crews, P. L.; and Johnson, D. B. "Density Bounds for the Sum of Divisors Function." *Math. Comput.* **26**, 773–777, 1972.

Wall, C. R.; Crews, P. L.; and Johnson, D. B. "Density Bounds for the Sum of Divisors Function." *Math. Comput.* **31**, 616, 1977.

Acceleration

Let a particle travel a distance $s(t)$ as a function of time t (here, s can be thought of as the ARC LENGTH of the curve traced out by the particle). The SPEED (the SCALAR NORM of the VECTOR VELOCITY) is then given by

$$\frac{ds}{dt} = \sqrt{\left(\frac{dx}{dt}\right)^2 + \left(\frac{dy}{dt}\right)^2 + \left(\frac{dz}{dt}\right)^2}. \tag{1}$$

The acceleration is defined as the time DERIVATIVE of the VELOCITY, so the SCALAR acceleration is given by

$$\begin{aligned} a &\equiv \frac{dv}{dt} && (2)\\ &= \frac{d^2s}{dt^2} && (3)\\ &= \frac{\frac{dx}{dt}\frac{d^2x}{dt^2} + \frac{dy}{dt}\frac{d^2y}{dt^2} + \frac{dz}{dt}\frac{d^2z}{dt^2}}{\sqrt{\left(\frac{dx}{dt}\right)^2 + \left(\frac{dy}{dt}\right)^2 + \left(\frac{dz}{dt}\right)^2}} && (4)\\ &= \frac{dx}{ds}\frac{d^2x}{dt^2} + \frac{dy}{ds}\frac{d^2y}{dt^2} + \frac{dz}{ds}\frac{d^2z}{dt^2} && (5)\\ &= \frac{d\mathbf{r}}{ds} \cdot \frac{d^2\mathbf{r}}{dt^2}. && (6) \end{aligned}$$

The VECTOR acceleration is given by

$$\mathbf{a} \equiv \frac{d\mathbf{v}}{dt} = \frac{d^2\mathbf{r}}{dt^2} = \frac{d^2s}{dt^2}\hat{\mathbf{T}} + \kappa\left(\frac{ds}{dt}\right)^2 \hat{\mathbf{N}}, \tag{7}$$

where $\hat{\mathbf{T}}$ is the UNIT TANGENT VECTOR, κ the CURVATURE, s the ARC LENGTH, and $\hat{\mathbf{N}}$ the UNIT NORMAL VECTOR.

Let a particle move along a straight LINE so that the positions at times t_1, t_2, and t_3 are s_1, s_2, and s_3, respectively. Then the particle is uniformly accelerated with acceleration a IFF

$$a \equiv 2\left[\frac{(s_2 - s_3)t_1 + (s_3 - s_1)t_2 + (s_1 - s_2)t_3}{(t_1 - t_2)(t_2 - t_3)(t_3 - t_1)}\right] \tag{8}$$

is a constant (Klamkin 1995, 1996).

Consider the measurement of acceleration in a rotating reference frame. Apply the ROTATION OPERATOR

$$\tilde{R} \equiv \left(\frac{d}{dt}\right)_{\text{body}} + \boldsymbol{\omega}\times \tag{9}$$

twice to the RADIUS VECTOR $\mathbf{r}$ and suppress the *body* notation,

$$\begin{aligned}\mathbf{a}_{\text{space}} = \tilde{R}^2\mathbf{r} &= \left(\frac{d}{dt} + \boldsymbol{\omega}\times\right)^2 \mathbf{r}\\ &= \left(\frac{d}{dt} + \boldsymbol{\omega}\times\right)\left(\frac{d\mathbf{r}}{dt} + \boldsymbol{\omega}\times\mathbf{r}\right)\\ &= \frac{d^2\mathbf{r}}{dt^2} + \frac{d}{dt}(\boldsymbol{\omega}\times\mathbf{r}) + \boldsymbol{\omega}\times\frac{d\mathbf{r}}{dt} + \boldsymbol{\omega}\times(\boldsymbol{\omega}\times\mathbf{r})\\ &= \frac{d^2\mathbf{r}}{dt^2} + \boldsymbol{\omega}\times\frac{d\mathbf{r}}{dt} + \mathbf{r}\times\frac{d\boldsymbol{\omega}}{dt} + \boldsymbol{\omega}\times\frac{d\mathbf{r}}{dt}\\ &\quad + \boldsymbol{\omega}\times(\boldsymbol{\omega}\times\mathbf{r}).\end{aligned} \tag{10}$$

Grouping terms and using the definitions of the VELOCITY $\mathbf{v} \equiv d\mathbf{r}/dt$ and ANGULAR VELOCITY $\boldsymbol{\alpha} \equiv d\boldsymbol{\omega}/dt$ give the expression

$$\mathbf{a}_{\text{space}} = \frac{d^2\mathbf{r}}{dt^2} + 2\boldsymbol{\omega}\times\mathbf{v} + \boldsymbol{\omega}\times(\boldsymbol{\omega}\times\mathbf{r}) + \mathbf{r}\times\boldsymbol{\alpha}. \tag{11}$$

Now, we can identify the expression as consisting of three terms

$$\mathbf{a}_{\text{body}} \equiv \frac{d^2\mathbf{r}}{dt^2}, \tag{12}$$

$$\mathbf{a}_{\text{Coriolis}} \equiv 2\boldsymbol{\omega}\times\mathbf{v}, \tag{13}$$

$$\mathbf{a}_{\text{centrifugal}} \equiv \boldsymbol{\omega}\times(\boldsymbol{\omega}\times\mathbf{r}), \tag{14}$$

a "body" acceleration, centrifugal acceleration, and Coriolis acceleration. Using these definitions finally gives

$$\mathbf{a}_{\text{space}} = \mathbf{a}_{\text{body}} + \mathbf{a}_{\text{Coriolis}} + \mathbf{a}_{\text{centrifugal}} + \mathbf{r}\times\boldsymbol{\alpha}, \tag{15}$$

where the fourth term will vanish in a uniformly rotating frame of reference (i.e., $\boldsymbol{\alpha} = \mathbf{0}$). The centrifugal acceleration is familiar to riders of merry-go-rounds, and the Coriolis acceleration is responsible for the motions of hurricanes on Earth and necessitates large trajectory corrections for intercontinental ballistic missiles.

see also ANGULAR ACCELERATION, ARC LENGTH, JERK, VELOCITY

References

Klamkin, M. S. "Problem 1481." *Math. Mag.* **68**, 307, 1995.

Klamkin, M. S. "A Characteristic of Constant Acceleration." Solution to Problem 1481. *Math. Mag.* **69**, 308, 1996.

Accidental Cancellation

see ANOMALOUS CANCELLATION

Accumulation Point

An accumulation point is a POINT which is the limit of a SEQUENCE, also called a LIMIT POINT. For some MAPS, periodic orbits give way to CHAOTIC ones beyond a point known as the accumulation point.

see also CHAOS, LOGISTIC MAP, MODE LOCKING, PERIOD DOUBLING

Achilles and the Tortoise Paradox

see ZENO'S PARADOXES

Ackermann Function

The Ackermann function is the simplest example of a well-defined TOTAL FUNCTION which is COMPUTABLE but not PRIMITIVE RECURSIVE, providing a counterexample to the belief in the early 1900s that every COMPUTABLE FUNCTION was also PRIMITIVE RECURSIVE (Dötzel 1991). It grows faster than an exponential function, or even a multiple exponential function. The Ackermann function $A(x,y)$ is defined by

$$A(x,y) \equiv \begin{cases} y+1 & \text{if } x=0\\ A(x-1,1) & \text{if } y=0\\ A(x-1,A(x,y-1)) & \text{otherwise.}\end{cases} \tag{1}$$

Special values for INTEGER x include

$$A(0,y) = y+1 \tag{2}$$

$$A(1,y) = y+2 \tag{3}$$

$$A(2,y) = 2y+3 \tag{4}$$

$$A(3,y) = 2^{y+3}-3 \tag{5}$$

$$A(4,y) = \underbrace{2^{2^{\cdot^{\cdot^{2}}}}}_{y+3} - 3. \tag{6}$$

Expressions of the latter form are sometimes called POWER TOWERS. $A(0,y)$ follows trivially from the definition. $A(1,y)$ can be derived as follows,

$$\begin{aligned}A(1,y) &= A(0,A(1,y-1)) = A(1,y-1)+1\\ &= A(0,A(1,y-2))+1 = A(1,y-2)+2\\ &= \ldots = A(1,0)+y = A(0,1)+y = y+2.\end{aligned} \tag{7}$$

$A(2,y)$ has a similar derivation,

$$\begin{aligned}A(2,y) &= A(1,A(2,y-1)) = A(2,y-1)+2\\ &= A(1,A(2,y-2))+2 = A(2,y-2)+4 = \ldots\\ &= A(2,0)+2y = A(1,1)+2y = 2y+3.\end{aligned} \tag{8}$$

Buck (1963) defines a related function using the same fundamental RECURRENCE RELATION (with arguments flipped from Buck's convention)

$$F(x,y) = F(x-1,F(x,y-1)), \tag{9}$$

but with the slightly different boundary values

$$F(0,y) = y+1 \quad (10)$$
$$F(1,0) = 2 \quad (11)$$
$$F(2,0) = 0 \quad (12)$$
$$F(x,0) = 1 \quad \text{for } x = 3, 4, \ldots. \quad (13)$$

Buck's recurrence gives

$$F(1,y) = 2+y \quad (14)$$
$$F(2,y) = 2y \quad (15)$$
$$F(3,y) = 2^y \quad (16)$$
$$F(4,y) = \underbrace{2^{2^{\cdot^{\cdot^{\cdot^{2}}}}}}_{y}. \quad (17)$$

Taking $F(4,n)$ gives the sequence 1, 2, 4, 16, 65536, 2^{65536}, ... Defining $\psi(x) = F(x,x)$ for $x = 0, 1, \ldots$ then gives 1, 3, 4, 8, 65536, $\underbrace{2^{2^{\cdot^{\cdot^{\cdot^{2}}}}}}_{m}$, ... (Sloane's A001695), where $m = \underbrace{2^{2^{\cdot^{\cdot^{\cdot^{2}}}}}}_{65536}$, a truly huge number!

see also ACKERMANN NUMBER, COMPUTABLE FUNCTION, GOODSTEIN SEQUENCE, POWER TOWER, PRIMITIVE RECURSIVE FUNCTION, TAK FUNCTION, TOTAL FUNCTION

References

Buck, R. C. "Mathematical Induction and Recursive Definitions." *Amer. Math. Monthly* **70**, 128–135, 1963.

Dötzel, G. "A Function to End All Functions." *Algorithm: Recreational Programming* **2.4**, 16–17, 1991.

Kleene, S. C. *Introduction to Metamathematics.* New York: Elsevier, 1971.

Péter, R. *Rekursive Funktionen.* Budapest: Akad. Kiado, 1951.

Reingold, E. H. and Shen, X. "More Nearly Optimal Algorithms for Unbounded Searching, Part I: The Finite Case." *SIAM J. Comput.* **20**, 156–183, 1991.

Rose, H. E. *Subrecursion, Functions, and Hierarchies.* New York: Clarendon Press, 1988.

Sloane, N. J. A. Sequence A001695/M2352 in "An On-Line Version of the Encyclopedia of Integer Sequences."

Smith, H. J. "Ackermann's Function." `http://www.netcom.com/~hjsmith/Ackerman.html`.

Spencer, J. "Large Numbers and Unprovable Theorems." *Amer. Math. Monthly* **90**, 669–675, 1983.

Tarjan, R. E. *Data Structures and Network Algorithms.* Philadelphia PA: SIAM, 1983.

Vardi, I. *Computational Recreations in Mathematica.* Redwood City, CA: Addison-Wesley, pp. 11, 227, and 232, 1991.

Ackermann Number

A number of the form $\underbrace{n\uparrow\cdots\uparrow n}_{n}$, where ARROW NOTATION has been used. The first few Ackermann numbers are $1\uparrow 1 = 1$, $2\uparrow\uparrow 2 = 4$, and $3\uparrow\uparrow\uparrow 3 = \underbrace{3^{3^{\cdot^{\cdot^{\cdot^{3}}}}}}_{7,625,597,484,987}$.

see also ACKERMANN FUNCTION, ARROW NOTATION, POWER TOWER

References

Ackermann, W. "Zum hilbertschen Aufbau der reellen Zahlen." *Math. Ann.* **99**, 118–133, 1928.

Conway, J. H. and Guy, R. K. *The Book of Numbers.* New York: Springer-Verlag, pp. 60–61, 1996.

Crandall, R. E. "The Challenge of Large Numbers." *Sci. Amer.* **276**, 74–79, Feb. 1997.

Vardi, I. *Computational Recreations in Mathematica.* Redwood City, CA: Addison-Wesley, pp. 11, 227, and 232, 1991.

Acnode

Another name for an ISOLATED POINT.

see also CRUNODE, SPINODE, TACNODE

Acoptic Polyhedron

A term invented by B. Grünbaum in an attempt to promote concrete and precise POLYHEDRON terminology. The word "coptic" derives from the Greek for "to cut," and acoptic polyhedra are defined as POLYHEDRA for which the FACES do not intersect (cut) themselves, making them 2-MANIFOLDS.

see also HONEYCOMB, NOLID, POLYHEDRON, SPONGE

Action

Let $M(X)$ denote the GROUP of all invertible MAPS $X \to X$ and let G be any GROUP. A HOMOMORPHISM $\theta : G \to M(X)$ is called an action of G on X. Therefore, θ satisfies

1. For each $g \in G$, $\theta(g)$ is a MAP $X \to X : x \mapsto \theta(g)x$,
2. $\theta(gh)x = \theta(g)(\theta(h)x)$,
3. $\theta(e)x = x$, where e is the group identity in G,
4. $\theta(g^{-1})x = \theta(g)^{-1}x$.

see also CASCADE, FLOW, SEMIFLOW

Acute Angle

An ANGLE of less than $\pi/2$ RADIANS (90°) is called an acute angle.

see also ANGLE, OBTUSE ANGLE, RIGHT ANGLE, STRAIGHT ANGLE

Acute Triangle

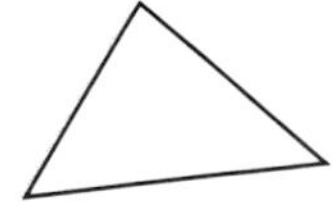

A TRIANGLE in which all three ANGLES are ACUTE ANGLES. A TRIANGLE which is neither acute nor a RIGHT TRIANGLE (i.e., it has an OBTUSE ANGLE) is called an OBTUSE TRIANGLE. A SQUARE can be dissected into as few as 8 acute triangles.

see also OBTUSE TRIANGLE, RIGHT TRIANGLE

Adams-Bashforth-Moulton Method

see ADAMS' METHOD

Adams' Method

Adams' method is a numerical METHOD for solving linear FIRST-ORDER ORDINARY DIFFERENTIAL EQUATIONS of the form

$$\frac{dy}{dx} = f(x, y). \tag{1}$$

Let

$$h = x_{n+1} - x_n \tag{2}$$

be the step interval, and consider the MACLAURIN SERIES of y about x_n,

$$y_{n+1} = y_n + \left(\frac{dy}{dx}\right)_n (x - x_n) + \frac{1}{2}\left(\frac{d^2y}{dx^2}\right)_n (x - x_n)^2 + \ldots \tag{3}$$

$$\left(\frac{dy}{dx}\right)_{n+1} = \left(\frac{dy}{dx}\right)_n + \left(\frac{d^2y}{dx^2}\right)_n (x - x_n)^2 + \ldots. \tag{4}$$

Here, the DERIVATIVES of y are given by the BACKWARD DIFFERENCES

$$q_n \equiv \left(\frac{dy}{dx}\right)_n = \frac{\Delta y_n}{x_{n+1} - x_n} = \frac{y_{n+1} - y_n}{h} \tag{5}$$

$$\nabla q_n \equiv \left(\frac{d^2y}{dx^2}\right)_n = q_n - q_{n-1} \tag{6}$$

$$\nabla^2 q_n \equiv \left(\frac{d^3y}{dx^3}\right)_n = \nabla q_n - \nabla q_{n-1}, \tag{7}$$

etc. Note that by (1), q_n is just the value of $f(x_n, y_n)$.

For first-order interpolation, the method proceeds by iterating the expression

$$y_{n+1} = y_n + q_n h \tag{8}$$

where $q_n \equiv f(x_n, y_n)$. The method can then be extended to arbitrary order using the finite difference integration formula from Beyer (1987)

$$\int_0^1 f_p\,dp = (1 + \tfrac{1}{2}\nabla + \tfrac{5}{12}\nabla^2 + \tfrac{3}{8}\nabla^3 + \tfrac{251}{720}\nabla^4 + \tfrac{95}{288}\nabla^5 + \tfrac{19087}{60480}\nabla^6 + \ldots)f_p \tag{9}$$

to obtain

$$y_{n+1} - y_n = h(q_n + \tfrac{1}{2}\nabla q_{n-1} + \tfrac{5}{12}\nabla^2 q_{n-2} + \tfrac{3}{8}\nabla^3 q_{n-3} + \tfrac{251}{720}\nabla^4 q_{n-4} + \tfrac{95}{288}\nabla^5 q_{n-5} + \ldots). \tag{10}$$

Note that von Kármán and Biot (1940) confusingly use the symbol normally used for FORWARD DIFFERENCES Δ to denote BACKWARD DIFFERENCES ∇.

see also GILL'S METHOD, MILNE'S METHOD, PREDICTOR-CORRECTOR METHODS, RUNGE-KUTTA METHOD

References

Abramowitz, M. and Stegun, C. A. (Eds.). *Handbook of Mathematical Functions with Formulas, Graphs, and Mathematical Tables, 9th printing.* New York: Dover, p. 896, 1972.

Beyer, W. H. *CRC Standard Mathematical Tables, 28th ed.* Boca Raton, FL: CRC Press, p. 455, 1987.

Kármán, T. von and Biot, M. A. *Mathematical Methods in Engineering: An Introduction to the Mathematical Treatment of Engineering Problems.* New York: McGraw-Hill, pp. 14–20, 1940.

Press, W. H.; Flannery, B. P.; Teukolsky, S. A.; and Vetterling, W. T. *Numerical Recipes in FORTRAN: The Art of Scientific Computing, 2nd ed.* Cambridge, England: Cambridge University Press, p. 741, 1992.

Addend

A quantity to be ADDED to another, also called a SUMMAND. For example, in the expression $a+b+c$, a, b, and c are all addends. The first of several addends, or "the one to which the others are added" (a in the previous example), is sometimes called the AUGEND.

see also ADDITION, AUGEND, PLUS, RADICAND

Addition

```
 1 1    ◄ carries
  1 5 8 ◄ addend 1
+ 2 4 9 ◄ addend 2
  4 0 7 ◄ sum
```

The combining of two or more quantities using the PLUS operator. The individual numbers being combined are called ADDENDS, and the total is called the SUM. The first of several ADDENDS, or "the one to which the others are added," is sometimes called the AUGEND. The opposite of addition is SUBTRACTION.

While the usual form of adding two n-digit INTEGERS (which consists of summing over the columns right to left and "CARRYING" a 1 to the next column if the sum exceeds 9) requires n operations (plus carries), two n-digit INTEGERS can be added in about $2\lg n$ steps by n processors using carry-lookahead addition (McGeoch 1993). Here, $\lg x$ is the LG function, the LOGARITHM to the base 2.

see also ADDEND, AMENABLE NUMBER, AUGEND, CARRY, DIFFERENCE, DIVISION, MULTIPLICATION, PLUS, SUBTRACTION, SUM

References

McGeoch, C. C. "Parallel Addition." *Amer. Math. Monthly* **100**, 867–871, 1993.

Addition Chain

An addition chain for a number n is a SEQUENCE $1 = a_0 < a_1 < \ldots < a_r = n$, such that each member after a_0 is the SUM of the two earlier (not necessarily distinct) ones. The number r is called the length of the addition chain. For example,

$$1, 1+1=2, 2+2=4, 4+2=6, 6+2=8, 8+6=14$$

is an addition chain for 14 of length $r = 5$ (Guy 1994).

see also BRAUER CHAIN, HANSEN CHAIN, SCHOLZ CONJECTURE

References

Guy, R. K. "Addition Chains. Brauer Chains. Hansen Chains." §C6 in *Unsolved Problems in Number Theory, 2nd ed.* New York: Springer-Verlag, pp. 111–113, 1994.

Addition-Multiplication Magic Square

46	81	117	102	15	76	200	203
19	60	232	175	54	69	153	78
216	161	17	52	171	90	58	75
135	114	50	87	184	189	13	68
150	261	45	38	91	136	92	27
119	104	108	23	174	225	57	30
116	25	133	120	51	26	162	207
39	34	138	243	100	29	105	152

200	87	95	42	99	1	46	108	170
14	44	10	184	81	85	150	261	19
138	243	17	50	116	190	56	33	5
57	125	232	9	7	66	68	230	54
4	70	22	51	115	216	171	25	174
153	23	162	76	250	58	3	35	88
145	152	75	11	6	63	270	34	92
110	2	28	135	136	69	29	114	225
27	102	207	290	38	100	55	8	21

200	87	95	42	99	1	46	108	170
14	44	10	184	81	85	150	261	19
138	243	17	50	116	190	56	33	5
57	125	232	9	7	66	68	230	54
4	70	22	51	115	216	171	25	174
153	23	162	76	250	58	3	35	88
145	152	75	11	6	63	270	34	92
110	2	28	135	136	69	29	114	225
27	102	207	290	38	100	55	8	21

A square which is simultaneously a MAGIC SQUARE and MULTIPLICATION MAGIC SQUARE. The three squares shown above (the top square has order eight and the bottom two have order nine) have addition MAGIC CONSTANTS (840, 848, 1200) and multiplicative magic constants (2,058,068,231,856,000; 5,804,807,833,440,000; 1,619,541,385,529,760,000), respectively (Hunter and Madachy 1975, Madachy 1979).

see also MAGIC SQUARE

References

Hunter, J. A. H. and Madachy, J. S. "Mystic Arrays." Ch. 3 in *Mathematical Diversions.* New York: Dover, pp. 30–31, 1975.

Madachy, J. S. *Madachy's Mathematical Recreations.* New York: Dover, pp. 89–91, 1979.

Additive Persistence

Consider the process of taking a number, adding its DIGITS, then adding the DIGITS of number derived from it, etc., until the remaining number has only one DIGIT. The number of additions required to obtain a single DIGIT from a number n is called the additive persistence of n, and the DIGIT obtained is called the DIGITAL ROOT of n.

For example, the sequence obtained from the starting number 9876 is (9876, 30, 3), so 9876 has an additive persistence of 2 and a DIGITAL ROOT of 3. The additive persistences of the first few positive integers are 0, 0, 0, 0, 0, 0, 0, 0, 0, 1, 1, 1, 1, 1, 1, 1, 1, 1, 2, 1, ... (Sloane's A031286). The smallest numbers of additive persistence n for $n = 0, 1, \ldots$ are 0, 10, 19, 199, 19999999999999999999999, ... (Sloane's A006050). There is no number $< 10^{50}$ with additive persistence greater than 11.

It is conjectured that the maximum number lacking the DIGIT 1 with persistence 11 is

$$77777733332222222222222222222$$

There is a stronger conjecture that there is a maximum number lacking the DIGIT 1 for each persistence ≥ 2.

The maximum additive persistence in base 2 is 1. It is conjectured that all powers of $2 > 2^{15}$ contain a 0 in base 3, which would imply that the maximum persistence in base 3 is 3 (Guy, 1994).

see also DIGITADITION, DIGITAL ROOT, MULTIPLICATIVE PERSISTENCE, NARCISSISTIC NUMBER, RECURRING DIGITAL INVARIANT

References

Guy, R. K. "The Persistence of a Number." §F25 in *Unsolved Problems in Number Theory, 2nd ed.* New York: Springer-Verlag, pp. 262–263, 1994.

Hinden, H. J. "The Additive Persistence of a Number." *J. Recr. Math.* **7**, 134–135, 1974.

Sloane, N. J. A. Sequences A031286 and A006050/M4683 in "An On-Line Version of the Encyclopedia of Integer Sequences."

Sloane, N. J. A. "The Persistence of a Number." *J. Recr. Math.* **6**, 97–98, 1973.

Adéle

An element of an ADÉLE GROUP, sometimes called a REPARTITION in older literature. Adéles arise in both NUMBER FIELDS and FUNCTION FIELDS. The adéles of a NUMBER FIELD are the additive SUBGROUPS of all elements in $\prod k_\nu$, where ν is the PLACE, whose ABSOLUTE VALUE is < 1 at all but finitely many νs.

Let F be a FUNCTION FIELD of algebraic functions of one variable. Then a MAP r which assigns to every PLACE P of F an element $r(P)$ of F such that there are only a finite number of PLACES P for which $\nu_P(r(P)) < 0$.

see also IDELE

References

Chevalley, C. C. *Introduction to the Theory of Algebraic Functions of One Variable.* Providence, RI: Amer. Math. Soc., p. 25, 1951.

Knapp, A. W. "Group Representations and Harmonic Analysis, Part II." *Not. Amer. Math. Soc.* **43**, 537–549, 1996.

Adéle Group

The restricted topological DIRECT PRODUCT of the GROUP G_{k_ν} with distinct invariant open subgroups G_{0_ν}.

References

Weil, A. *Adéles and Algebraic Groups.* Princeton, NJ: Princeton University Press, 1961.

Adem Relations

Relations in the definition of a STEENROD ALGEBRA which state that, for $i < 2j$,

$$Sq^i \circ Sq^j(x) = \sum_{k=0}^{\lfloor i \rfloor} \binom{j-k-1}{i-2k} Sq^{i+j-k} \circ Sq^k(x),$$

where $f \circ g$ denotes function COMPOSITION and $\lfloor i \rfloor$ is the FLOOR FUNCTION.

see also STEENROD ALGEBRA

Adequate Knot

A class of KNOTS containing the class of ALTERNATING KNOTS. Let $c(K)$ be the CROSSING NUMBER. Then for KNOT SUM $K_1 \# K_2$ which is an adequate knot,

$$c(K_1 \# K_2) = c(K_1) + c(K_2).$$

This relationship is postulated to hold true for all KNOTS.

see also ALTERNATING KNOT, CROSSING NUMBER (LINK)

Adiabatic Invariant

A property of motion which is conserved to exponential accuracy in the small parameter representing the typical rate of change of the gross properties of the body.

see also ALGEBRAIC INVARIANT, LYAPUNOV CHARACTERISTIC NUMBER

Adjacency Matrix

The adjacency matrix of a simple GRAPH is a MATRIX with rows and columns labelled by VERTICES, with a 1 or 0 in position (v_i, v_j) according to whether v_i and v_j are ADJACENT or not.

see also INCIDENCE MATRIX

References

Chartrand, G. *Introductory Graph Theory.* New York: Dover, p. 218, 1985.

Adjacency Relation

The SET E of EDGES of a GRAPH (V, E), being a set of unordered pairs of elements of V, constitutes a RELATION on V. Formally, an adjacency relation is any RELATION which is IRREFLEXIVE and SYMMETRIC.

see also IRREFLEXIVE, RELATION, SYMMETRIC

Adjacent Fraction

Two FRACTIONS are said to be adjacent if their difference has a unit NUMERATOR. For example, 1/3 and 1/4 are adjacent since $1/3 - 1/4 = 1/12$, but 1/2 and 1/5 are not since $1/2 - 1/5 = 3/10$. Adjacent fractions can be adjacent in a FAREY SEQUENCE.

see also FAREY SEQUENCE, FORD CIRCLE, FRACTION, NUMERATOR

References

Pickover, C. A. *Keys to Infinity.* New York: W. H. Freeman, p. 119, 1995.

Adjacent Value

The value nearest to but still inside an inner FENCE.

References

Tukey, J. W. *Explanatory Data Analysis.* Reading, MA: Addison-Wesley, p. 667, 1977.

Adjacent Vertices

In a GRAPH G, two VERTICES are adjacent if they are joined by an EDGE.

Adjoint Curve

A curve which has at least multiplicity $r_i - 1$ at each point where a given curve (having only ordinary singular points and cusps) has a multiplicity r_i is called the adjoint to the given curve. When the adjoint curve is of order $n - 3$, it is called a special adjoint curve.

References

Coolidge, J. L. *A Treatise on Algebraic Plane Curves.* New York: Dover, p. 30, 1959.

Adjoint Matrix

The adjoint matrix, sometimes also called the ADJUGATE MATRIX, is defined by

$$\mathsf{A}^\dagger \equiv (\mathsf{A}^{\mathrm{T}})^*, \tag{1}$$

where the ADJOINT OPERATOR is denoted † and ^{T} denotes the TRANSPOSE. If a MATRIX is SELF-ADJOINT, it is said to be HERMITIAN. The adjoint matrix of a MATRIX product is given by

$$(ab)^\dagger{}_{ij} \equiv [(ab)^{\mathrm{T}}]^*_{ij}. \tag{2}$$

Using the property of transpose products that

$$\begin{aligned}[(ab)^{\mathrm{T}}]^*_{ij} &= (b^{\mathrm{T}} a^{\mathrm{T}})^*_{ij} = (b^{\mathrm{T}}_{ik} a^{\mathrm{T}}_{kj})^* = (b^{\mathrm{T}})^*_{ik} (a^{\mathrm{T}})^*_{kj} \\ &= b^\dagger_{ik} a^\dagger_{kj} = (b^\dagger a^\dagger)_{ij}, \end{aligned} \tag{3}$$

it follows that

$$(\mathsf{AB})^\dagger = \mathsf{B}^\dagger \mathsf{A}^\dagger. \tag{4}$$

Adjoint Operator

Given a SECOND-ORDER ORDINARY DIFFERENTIAL EQUATION

$$\tilde{\mathcal{L}} u(x) \equiv p_0 \frac{du^2}{dx^2} + p_1 \frac{du}{dx} + p_2 u, \tag{1}$$

where $p_i \equiv p_i(x)$ and $u \equiv u(x)$, the adjoint operator $\tilde{\mathcal{L}}^\dagger$ is defined by

$$\begin{aligned}\tilde{\mathcal{L}}^\dagger u &\equiv \frac{d}{dx^2}(p_0 u) - \frac{d}{dx}(p_1 u) + p_2 u \\ &= p_0 \frac{d^2 u}{dx^2} + (2p_0' - p_1)\frac{du}{dx} + (p_0'' - p_1' + p_2)u. \end{aligned} \tag{2}$$

Write the two LINEARLY INDEPENDENT solutions as $y_1(x)$ and $y_2(x)$. Then the adjoint operator can also be written

$$\tilde{\mathcal{L}}^\dagger u = \int (y_2\tilde{\mathcal{L}}y_1 - y_1\tilde{\mathcal{L}}y_2)\,dx = \left[\frac{p_1}{p_0}(y_1{}'y_2 - y_1y_2{}')\right]. \tag{3}$$

see also SELF-ADJOINT OPERATOR, STURM-LIOUVILLE THEORY

Adjugate Matrix

see ADJOINT MATRIX

Adjunction

If a is an element of a FIELD F over the PRIME FIELD P, then the set of all RATIONAL FUNCTIONS of a with COEFFICIENTS in P is a FIELD derived from P by adjunction of a.

Adleman-Pomerance-Rumely Primality Test

A modified MILLER'S PRIMALITY TEST which gives a guarantee of PRIMALITY or COMPOSITENESS. The ALGORITHM's running time for a number N has been proved to be as $\mathcal{O}((\ln N)^{c\ln\ln\ln N})$ for some $c > 0$. It was simplified by Cohen and Lenstra (1984), implemented by Cohen and Lenstra (1987), and subsequently optimized by Bosma and van der Hulst (1990).

References
Adleman, L. M.; Pomerance, C.; and Rumely, R. S. "On Distinguishing Prime Numbers from Composite Number." *Ann. Math.* **117**, 173–206, 1983.
Bosma, W. and van der Hulst, M.-P. "Faster Primality Testing." In *Advances in Cryptology, Proc. Eurocrypt '89, Houthalen, April 10–13, 1989* (Ed. J.-J. Quisquater). New York: Springer-Verlag, 652–656, 1990.
Brillhart, J.; Lehmer, D. H.; Selfridge, J.; Wagstaff, S. S. Jr.; and Tuckerman, B. *Factorizations of $b^n \pm 1$, $b = 2, 3, 5, 6, 7, 10, 11, 12$ Up to High Powers, rev. ed.* Providence, RI: Amer. Math. Soc., pp. lxxxiv–lxxxv, 1988.
Cohen, H. and Lenstra, A. K. "Primality Testing and Jacobi Sums." *Math. Comput.* **42**, 297–330, 1984.
Cohen, H. and Lenstra, A. K. "Implementation of a New Primality Test." *Math. Comput.* **48**, 103–121, 1987.
Mihailescu, P. "A Primality Test Using Cyclotomic Extensions." In *Applied Algebra, Algebraic Algorithms and Error-Correcting Codes* (Proc. AAECC-6, Rome, July 1988). New York: Springer-Verlag, pp. 310–323, 1989.

Adleman-Rumely Primality Test

see ADLEMAN-POMERANCE-RUMELY PRIMALITY TEST

Admissible

A string or word is said to be admissible if that word appears in a given SEQUENCE. For example, in the SEQUENCE *aabaabaabaabaab*..., *a*, *aa*, *baab* are all admissible, but *bb* is inadmissible.

see also BLOCK GROWTH

Affine Complex Plane

The set $\mathbb{A}^2$ of all ordered pairs of COMPLEX NUMBERS.

see also AFFINE CONNECTION, AFFINE EQUATION, AFFINE GEOMETRY, AFFINE GROUP, AFFINE HULL, AFFINE PLANE, AFFINE SPACE, AFFINE TRANSFORMATION, AFFINITY, COMPLEX PLANE, COMPLEX PROJECTIVE PLANE

Affine Connection

see CONNECTION COEFFICIENT

Affine Equation

A nonhomogeneous LINEAR EQUATION or system of nonhomogeneous LINEAR EQUATIONS is said to be affine.

see also AFFINE COMPLEX PLANE, AFFINE CONNECTION, AFFINE GEOMETRY, AFFINE GROUP, AFFINE HULL, AFFINE PLANE, AFFINE SPACE, AFFINE TRANSFORMATION, AFFINITY

Affine Geometry

A GEOMETRY in which properties are preserved by PARALLEL PROJECTION from one PLANE to another. In an affine geometry, the third and fourth of EUCLID'S POSTULATES become meaningless. This type of GEOMETRY was first studied by Euler.

see also ABSOLUTE GEOMETRY, AFFINE COMPLEX PLANE, AFFINE CONNECTION, AFFINE EQUATION, AFFINE GROUP, AFFINE HULL, AFFINE PLANE, AFFINE SPACE, AFFINE TRANSFORMATION, AFFINITY, ORDERED GEOMETRY

References
Birkhoff, G. and Mac Lane, S. "Affine Geometry." §9.13 in *A Survey of Modern Algebra, 3rd ed.* New York: Macmillan, pp. 268–275, 1965.

Affine Group

The set of all nonsingular AFFINE TRANSFORMATIONS of a TRANSLATION in SPACE constitutes a GROUP known as the affine group. The affine group contains the full linear group and the group of TRANSLATIONS as SUBGROUPS.

see also AFFINE COMPLEX PLANE, AFFINE CONNECTION, AFFINE EQUATION, AFFINE GEOMETRY, AFFINE HULL, AFFINE PLANE, AFFINE SPACE, AFFINE TRANSFORMATION, AFFINITY

References
Birkhoff, G. and Mac Lane, S. *A Survey of Modern Algebra, 3rd ed.* New York: Macmillan, p. 237, 1965.

Affine Hull

The IDEAL generated by a SET in a VECTOR SPACE.

see also AFFINE COMPLEX PLANE, AFFINE CONNECTION, AFFINE EQUATION, AFFINE GEOMETRY, AFFINE GROUP, AFFINE PLANE, AFFINE SPACE, AFFINE TRANSFORMATION, AFFINITY, CONVEX HULL, HULL

Affine Plane

A 2-D AFFINE GEOMETRY constructed over a FINITE FIELD. For a FIELD F of size n, the affine plane consists of the set of points which are ordered pairs of elements in F and a set of lines which are themselves a set of points. Adding a POINT AT INFINITY and LINE AT INFINITY allows a PROJECTIVE PLANE to be constructed from an affine plane. An affine plane of order n is a BLOCK DESIGN of the form $(n^2, n, 1)$. An affine plane of order n exists IFF a PROJECTIVE PLANE of order n exists.

see also AFFINE COMPLEX PLANE, AFFINE CONNECTION, AFFINE EQUATION, AFFINE GEOMETRY, AFFINE GROUP, AFFINE HULL, AFFINE SPACE, AFFINE TRANSFORMATION, AFFINITY, PROJECTIVE PLANE

References

Lindner, C. C. and Rodger, C. A. *Design Theory.* Boca Raton, FL: CRC Press, 1997.

Affine Scheme

A technical mathematical object defined as the SPECTRUM $\sigma(A)$ of a set of PRIME IDEALS of a commutative RING A regarded as a local ringed space with a structure sheaf.

see also SCHEME

References

Iyanaga, S. and Kawada, Y. (Eds.). "Schemes." §18E in *Encyclopedic Dictionary of Mathematics.* Cambridge, MA: MIT Press, p. 69, 1980.

Affine Space

Let V be a VECTOR SPACE over a FIELD K, and let A be a nonempty SET. Now define addition $p + \mathbf{a} \in A$ for any VECTOR $\mathbf{a} \in V$ and element $p \in A$ subject to the conditions

1. $p + \mathbf{0} = p$,
2. $(p + \mathbf{a}) + \mathbf{b} = p + (\mathbf{a} + \mathbf{b})$,
3. For any $q \in A$, there EXISTS a unique VECTOR $\mathbf{a} \in V$ such that $q = p + \mathbf{a}$.

Here, $\mathbf{a}, \mathbf{b} \in V$. Note that (1) is implied by (2) and (3). Then A is an affine space and K is called the COEFFICIENT FIELD.

In an affine space, it is possible to fix a point and coordinate axis such that every point in the SPACE can be represented as an n-tuple of its coordinates. Every ordered pair of points A and B in an affine space is then associated with a VECTOR AB.

see also AFFINE COMPLEX PLANE, AFFINE CONNECTION, AFFINE EQUATION, AFFINE GEOMETRY, AFFINE GROUP, AFFINE HULL, AFFINE PLANE, AFFINE SPACE, AFFINE TRANSFORMATION, AFFINITY

Affine Transformation

Any TRANSFORMATION preserving COLLINEARITY (i.e., all points lying on a LINE initially still lie on a LINE after TRANSFORMATION). An affine transformation is also called an AFFINITY. An affine transformation of $\mathbb{R}^n$ is a MAP $F : \mathbb{R}^n \to \mathbb{R}^n$ of the form

$$F(\mathbf{p}) = A\mathbf{p} + \mathbf{q} \tag{1}$$

for all $p \in \mathbb{R}^n$, where A is a linear transformation of $\mathbb{R}^n$. If $\det(A) = 1$, the transformation is ORIENTATION-PRESERVING; if $\det(A) = -1$, it is ORIENTATION-REVERSING.

DILATION (CONTRACTION, HOMOTHECY), EXPANSION, REFLECTION, ROTATION, and TRANSLATION are all affine transformations, as are their combinations. A particular example combining ROTATION and EXPANSION is the rotation-enlargement transformation

$$\begin{aligned}\begin{bmatrix} x' \\ y' \end{bmatrix} &= s \begin{bmatrix} \cos\alpha & \sin\alpha \\ -\sin\alpha & \cos\alpha \end{bmatrix} \begin{bmatrix} x - x_0 \\ y - y_0 \end{bmatrix} \\ &= s \begin{bmatrix} \cos\alpha(x - x_0) + \sin\alpha(y - y_0) \\ -\sin\alpha(x - x_0) + \cos\alpha(y - y_0) \end{bmatrix}. \end{aligned} \tag{2}$$

Separating the equations,

$$x' = (s\cos\alpha)x + (s\sin\alpha)y - s(x_0\cos\alpha + y_0\sin\alpha) \tag{3}$$

$$y' = (-s\sin\alpha)x + (s\cos\alpha)y + s(x_0\sin\alpha - y_0\cos\alpha). \tag{4}$$

This can be also written as

$$x' = ax + by + c \tag{5}$$

$$y' = bx + ay + d, \tag{6}$$

where

$$a = s\cos\alpha \tag{7}$$

$$b = -s\sin\alpha. \tag{8}$$

The scale factor s is then defined by

$$s \equiv \sqrt{a^2 + b^2}, \tag{9}$$

and the rotation ANGLE by

$$\alpha = \tan^{-1}\left(-\frac{b}{a}\right). \tag{10}$$

see also AFFINE COMPLEX PLANE, AFFINE CONNECTION, AFFINE EQUATION, AFFINE GEOMETRY, AFFINE GROUP, AFFINE HULL, AFFINE PLANE, AFFINE SPACE, AFFINE TRANSFORMATION, AFFINITY, EQUIAFFINITY, EUCLIDEAN MOTION

References

Gray, A. *Modern Differential Geometry of Curves and Surfaces.* Boca Raton, FL: CRC Press, p. 105, 1993.

Affinity

see AFFINE TRANSFORMATION

Affix

In the archaic terminology of Whittaker and Watson (1990), the COMPLEX NUMBER z representing $x + iy$.

References

Whittaker, E. T. and Watson, G. N. *A Course in Modern Analysis, 4th ed.* Cambridge, England: Cambridge University Press, 1990.

Aggregate

An archaic word for infinite SETS such as those considered by Georg Cantor.

see also CLASS (SET), SET

AGM

see ARITHMETIC-GEOMETRIC MEAN

Agnesi's Witch

see WITCH OF AGNESI

Agnésienne

see WITCH OF AGNESI

Agonic Lines

see SKEW LINES

Ahlfors-Bers Theorem

The RIEMANN'S MODULI SPACE gives the solution to RIEMANN'S MODULI PROBLEM, which requires an ANALYTIC parameterization of the compact RIEMANN SURFACES in a fixed HOMEOMORPHISM.

Airy Differential Equation

Some authors define a general Airy differential equation as

$$y'' \pm k^2 xy = 0. \tag{1}$$

This equation can be solved by series solution using the expansions

$$y = \sum_{n=0}^{\infty} a_n x^n \tag{2}$$

$$y' = \sum_{n=0}^{\infty} n a_n x^{n-1} = \sum_{n=1}^{\infty} n a_n x^{n-1}$$

$$= \sum_{n=0}^{\infty} (n+1) a_{n+1} x^n \tag{3}$$

$$y'' = \sum_{n=0}^{\infty} (n+1) n a_{n+1} x^{n-1} = \sum_{n=1}^{\infty} (n+1) n a_{n+1} x^{n-1}$$

$$= \sum_{n=0}^{\infty} (n+2)(n+1) a_{n+2} x^n. \tag{4}$$

Specializing to the "conventional" Airy differential equation occurs by taking the MINUS SIGN and setting $k^2 = 1$. Then plug (4) into

$$y'' - xy = 0 \tag{5}$$

to obtain

$$\sum_{n=0}^{\infty} (n+2)(n+1) a_{n+2} x^n - x \sum_{n=0}^{\infty} a_n x^n = 0 \tag{6}$$

$$\sum_{n=0}^{\infty} (n+2)(n+1) a_{n+2} x^n - \sum_{n=0}^{\infty} a_n x^{n+1} = 0 \tag{7}$$

$$2a_2 + \sum_{n=1}^{\infty} (n+2)(n+1) a_{n+2} x^n - \sum_{n=1}^{\infty} a_{n-1} x^n = 0 \tag{8}$$

$$2a_2 + \sum_{n=1}^{\infty} [(n+2)(n+1) a_{n+2} - a_{n-1}] x^n = 0. \tag{9}$$

In order for this equality to hold for all x, each term must separately be 0. Therefore,

$$a_2 = 0 \tag{10}$$

$$(n+2)(n+1) a_{n+2} = a_{n-1}. \tag{11}$$

Starting with the $n = 3$ term and using the above RECURRENCE RELATION, we obtain

$$5 \cdot 4 a_5 = 20 a_5 = a_2 = 0. \tag{12}$$

Continuing, it follows by INDUCTION that

$$a_2 = a_5 = a_8 = a_{11} = \ldots a_{3n-1} = 0 \tag{13}$$

for $n = 1, 2, \ldots$. Now examine terms of the form a_{3n}.

$$a_3 = \frac{a_0}{3 \cdot 2} \tag{14}$$

$$a_6 = \frac{a_3}{6 \cdot 5} = \frac{a_0}{(6 \cdot 5)(3 \cdot 2)} \tag{15}$$

$$a_9 = \frac{a_6}{9 \cdot 8} = \frac{a_0}{(9 \cdot 8)(6 \cdot 5)(3 \cdot 2)}. \tag{16}$$

Again by INDUCTION,

$$a_{3n} = \frac{a_0}{[(3n)(3n-1)][(3n-3)(3n-4)] \cdots [6 \cdot 5][3 \cdot 2]} \tag{17}$$

for $n = 1, 2, \ldots$. Finally, look at terms of the form a_{3n+1},

$$a_4 = \frac{a_1}{4 \cdot 3} \tag{18}$$

$$a_7 = \frac{a_4}{7 \cdot 6} = \frac{a_1}{(7 \cdot 6)(4 \cdot 3)} \tag{19}$$

$$a_{10} = \frac{a_7}{10 \cdot 9} = \frac{a_1}{(10 \cdot 9)(7 \cdot 6)(4 \cdot 3)}. \tag{20}$$

By INDUCTION,

$$a_{3n+1} = \frac{a_1}{[(3n+1)(3n)][(3n-2)(3n-3)]\cdots[7\cdot 6][4\cdot 3]} \tag{21}$$

for $n = 1, 2, \ldots$. The general solution is therefore

$$y = a_0\left[1+\sum_{n=1}^{\infty}\frac{x^{3n}}{(3n)(3n-1)(3n-3)(3n-4)\cdots 3\cdot 2}\right] + a_1\left[x+\sum_{n=1}^{\infty}\frac{x^{3n+1}}{(3n+1)(3n)(3n-2)(3n-3)\cdots 4\cdot 3}\right]. \tag{22}$$

For a general k^2 with a MINUS SIGN, equation (1) is

$$y'' - k^2xy = 0, \tag{23}$$

and the solution is

$$y(x) = \tfrac{1}{3}\sqrt{x}\left[AI_{-1/3}\left(\tfrac{2}{3}kx^{3/2}\right) - BI_{1/3}\left(\tfrac{2}{3}kx^{3/2}\right)\right], \tag{24}$$

where I is a MODIFIED BESSEL FUNCTION OF THE FIRST KIND. This is usually expressed in terms of the AIRY FUNCTIONS $\mathrm{Ai}(x)$ and $\mathrm{Bi}(x)$

$$y(x) = A'\,\mathrm{Ai}(k^{2/3}x) + B'\,\mathrm{Bi}(k^{2/3}x). \tag{25}$$

If the PLUS SIGN is present instead, then

$$y'' + k^2xy = 0 \tag{26}$$

and the solutions are

$$y(x) = \tfrac{1}{3}\sqrt{x}\left[AJ_{-1/3}\left(\tfrac{2}{3}kx^{3/2}\right) + BJ_{1/3}\left(\tfrac{2}{3}kx^{3/2}\right)\right], \tag{27}$$

where $J(z)$ is a BESSEL FUNCTION OF THE FIRST KIND.

see also AIRY-FOCK FUNCTIONS, AIRY FUNCTIONS, BESSEL FUNCTION OF THE FIRST KIND, MODIFIED BESSEL FUNCTION OF THE FIRST KIND

Airy-Fock Functions

The three Airy-Fock functions are

$$\nu(z) = \tfrac{1}{2}\sqrt{\pi}\,\mathrm{Ai}(z) \tag{1}$$

$$w_1(z) = 2e^{i\pi/6}\nu(\omega z) \tag{2}$$

$$w_2(z) = 2e^{-i\pi/6}\nu(\omega^{-1}z), \tag{3}$$

where $\mathrm{Ai}(z)$ is an AIRY FUNCTION. These functions satisfy

$$\nu(z) = \frac{w_1(z) - w_2(z)}{2i} \tag{4}$$

$$[w_1(z)]^* = w_2(z^*), \tag{5}$$

where z^* is the COMPLEX CONJUGATE of z.

see also AIRY FUNCTIONS

References

Hazewinkel, M. (Managing Ed.). *Encyclopaedia of Mathematics: An Updated and Annotated Translation of the Soviet "Mathematical Encyclopaedia."* Dordrecht, Netherlands: Reidel, p. 65, 1988.

Airy Functions

Watson's (1966, pp. 188–190) definition of an Airy function is the solution to the AIRY DIFFERENTIAL EQUATION

$$\Phi'' \pm k^2\Phi x = 0 \tag{1}$$

which is FINITE at the ORIGIN, where Φ' denotes the DERIVATIVE $d\Phi/dx$, $k^2 = 1/3$, and either SIGN is permitted. Call these solutions $(1/\pi)\Phi(\pm k^2, x)$, then

$$\frac{1}{\pi}\Phi(\pm\tfrac{1}{3};x) \equiv \int_0^\infty \cos(t^3 \pm xt)\,dt \tag{2}$$

$$\Phi(\tfrac{1}{3};x) = \tfrac{1}{3}\pi\sqrt{\tfrac{x}{3}}\left[J_{-1/3}\left(\frac{2x^{3/2}}{3^{3/2}}\right) + J_{1/3}\left(\frac{2x^{3/2}}{3^{3/2}}\right)\right] \tag{3}$$

$$\Phi(-\tfrac{1}{3};x) = \tfrac{1}{3}\pi\sqrt{\tfrac{x}{3}}\left[I_{-1/3}\left(\frac{2x^{3/2}}{3^{3/2}}\right) - I_{1/3}\left(\frac{2x^{3/2}}{3^{3/2}}\right)\right], \tag{4}$$

where $J(z)$ is a BESSEL FUNCTION OF THE FIRST KIND and $I(z)$ is a MODIFIED BESSEL FUNCTION OF THE FIRST KIND. Using the identity

$$K_n(x) = \frac{\pi}{2}\frac{I_{-n}(x) - I_n(x)}{\sin(n\pi)}, \tag{5}$$

where $K(z)$ is a MODIFIED BESSEL FUNCTION OF THE SECOND KIND, the second case can be re-expressed

$$\Phi(-\tfrac{1}{3};x) = \tfrac{1}{3}\pi\sqrt{\tfrac{x}{3}}\left(\frac{2}{\pi}\right)\sin\left(\tfrac{1}{3}\pi\right)K_{1/3}\left(\frac{2x^{3/2}}{3^{3/2}}\right) \tag{6}$$

$$= \frac{\pi}{3}\sqrt{\tfrac{1}{3}x}\,\frac{2}{\pi}\frac{\sqrt{3}}{2}K_{1/3}\left(\frac{2x^{3/2}}{3^{3/2}}\right) \tag{7}$$

$$= \tfrac{1}{3}\sqrt{x}K_{1/3}\left(\frac{2x^{3/2}}{3^{3/2}}\right). \tag{8}$$

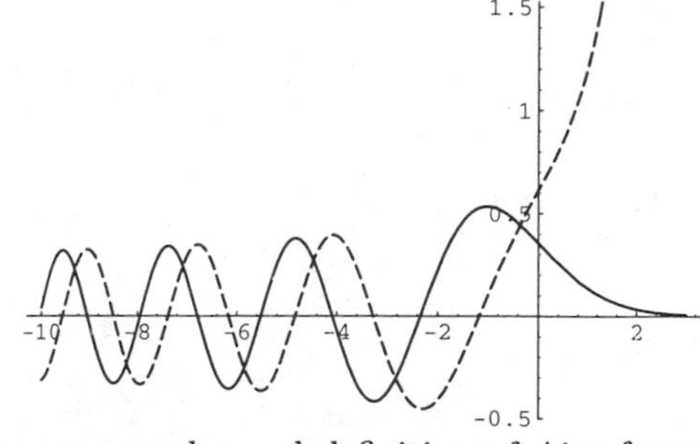

A more commonly used definition of Airy functions is given by Abramowitz and Stegun (1972, pp. 446–447) and illustrated above. This definition identifies the $\mathrm{Ai}(x)$ and $\mathrm{Bi}(x)$ functions as the two LINEARLY INDEPENDENT solutions to (1) with $k^2 = 1$ and a MINUS SIGN,

$$y'' - yz = 0. \tag{9}$$

The solutions are then written

$$y(z) = A\,\mathrm{Ai}(z) + B\,\mathrm{Bi}(z), \tag{10}$$

where

$$\begin{aligned}\mathrm{Ai}(z) &\equiv \frac{1}{\pi}\Phi(-1,z)\\ &= \tfrac{1}{3}\sqrt{x}\,[I_{-1/3}(\tfrac{2}{3}z^{3/2}) - I_{1/3}(\tfrac{2}{3}z^{3/2})]\\ &= \sqrt{\frac{z}{3\pi}}\,K_{1/3}(\tfrac{2}{3}z^{3/2}) \qquad (11)\\ \mathrm{Bi}(z) &\equiv \sqrt{\frac{z}{3}}\,[I_{-1/3}(\tfrac{2}{3}z^{3/2}) + I_{1/3}(\tfrac{2}{3}z^{3/2})]. \qquad (12)\end{aligned}$$

In the above plot, $\mathrm{Ai}(z)$ is the solid curve and $\mathrm{Bi}(z)$ is dashed. For zero argument,

$$\mathrm{Ai}(0) = -\frac{3^{-2/3}}{\Gamma(\frac{2}{3})}, \tag{13}$$

where $\Gamma(z)$ is the GAMMA FUNCTION. This means that Watson's expression becomes

$$(3a)^{-1/3}\pi\,\mathrm{Ai}(\pm(3a)^{-1/3}x) = \int_0^\infty \cos(at^3 \pm xt)\,dt. \tag{14}$$

A generalization has been constructed by Hardy.

The ASYMPTOTIC SERIES of $\mathrm{Ai}(z)$ has a different form in different QUADRANTS of the COMPLEX PLANE, a fact known as the STOKES PHENOMENON. Functions related to the Airy functions have been defined as

$$\mathrm{Gi}(z) \equiv \frac{1}{\pi}\int_0^\infty \sin(\tfrac{1}{3}t^3 + zt)\,dt \tag{15}$$

$$\mathrm{Hi}(z) \equiv \frac{1}{\pi}\int_0^\infty \exp(-\tfrac{1}{3}t^3 + zt)\,dt. \tag{16}$$

see also AIRY-FOCK FUNCTIONS

References

Abramowitz, M. and Stegun, C. A. (Eds.). "Airy Functions." §10.4 in *Handbook of Mathematical Functions with Formulas, Graphs, and Mathematical Tables, 9th printing.* New York: Dover, pp. 446–452, 1972.

Press, W. H.; Flannery, B. P.; Teukolsky, S. A.; and Vetterling, W. T. "Bessel Functions of Fractional Order, Airy Functions, Spherical Bessel Functions." §6.7 in *Numerical Recipes in FORTRAN: The Art of Scientific Computing, 2nd ed.* Cambridge, England: Cambridge University Press, pp. 234–245, 1992.

Spanier, J. and Oldham, K. B. "The Airy Functions Ai(x) and Bi(x)." Ch. 56 in *An Atlas of Functions.* Washington, DC: Hemisphere, pp. 555–562, 1987.

Watson, G. N. *A Treatise on the Theory of Bessel Functions, 2nd ed.* Cambridge, England: Cambridge University Press, 1966.

Airy Projection

A MAP PROJECTION. The inverse equations for ϕ are computed by iteration. Let the ANGLE of the projection plane be θ_b. Define

$$a = \begin{cases} 0 & \text{for } \theta_b = \frac{1}{2}\pi \\ \frac{\ln[\frac{1}{2}\cos(\frac{1}{2}\pi - \theta_b)]}{\tan[\frac{1}{2}(\frac{1}{2}\pi - \theta_b)]} & \text{otherwise.} \end{cases} \tag{1}$$

For proper convergence, let $x_i = \pi/6$ and compute the initial point by checking

$$x_i = \left|\exp[-(\sqrt{x^2+y^2} + a\tan x_i)\tan x_i]\right|. \tag{2}$$

As long as $x_i > 1$, take $x_{i+1} = x_i/2$ and iterate again. The first value for which $x_i < 1$ is then the starting point. Then compute

$$x_i = \cos^{-1}\{\exp[-(\sqrt{x^2+y^2} + a\tan x_i)\tan x_i]\} \tag{3}$$

until the change in x_i between evaluations is smaller than the acceptable tolerance. The (inverse) equations are then given by

$$\phi = \tfrac{1}{2}\pi - 2x_i \tag{4}$$

$$\lambda = \tan^{-1}\left(-\frac{x}{y}\right). \tag{5}$$

Aitken's δ^2 Process

An ALGORITHM which extrapolates the partial sums s_n of a SERIES $\sum_n a_n$ whose CONVERGENCE is approximately geometric and accelerates its rate of CONVERGENCE. The extrapolated partial sum is given by

$$s_n{}' \equiv s_{n+1} - \frac{(s_{n+1} - s_n)^2}{s_{n+1} - 2s_n + s_{n-1}}.$$

see also EULER'S SERIES TRANSFORMATION

References

Abramowitz, M. and Stegun, C. A. (Eds.). *Handbook of Mathematical Functions with Formulas, Graphs, and Mathematical Tables, 9th printing.* New York: Dover, p. 18, 1972.

Press, W. H.; Flannery, B. P.; Teukolsky, S. A.; and Vetterling, W. T. *Numerical Recipes in FORTRAN: The Art of Scientific Computing, 2nd ed.* Cambridge, England: Cambridge University Press, p. 160, 1992.

Aitken Interpolation

An algorithm similar to NEVILLE'S ALGORITHM for constructing the LAGRANGE INTERPOLATING POLYNOMIAL. Let $f(x|x_0, x_1, \ldots, x_k)$ be the unique POLYNOMIAL of kth ORDER coinciding with $f(x)$ at $x_0, \ldots, x_k$. Then

$$f(x|x_0,x_1) = \frac{1}{x_1 - x_0}\begin{vmatrix} f_0 & x_0 - x \\ f_1 & x_1 - x \end{vmatrix}$$

$$f(x|x_0,x_2) = \frac{1}{x_2 - x_0}\begin{vmatrix} f_0 & x_0 - x \\ f_2 & x_2 - x \end{vmatrix}$$

$$f(x|x_0,x_1,x_2) = \frac{1}{x_2 - x_1}\begin{vmatrix} f(x|x_0,x_1) & x_1 - x \\ f(x|x_0,x_2) & x_2 - x \end{vmatrix}$$

$$f(x|x_0,x_1,x_2,x_3) = \frac{1}{x_3 - x_2}\begin{vmatrix} f(x|x_0,x_1,x_2) & x_2 - x \\ f(x|x_0,x_1,x_3) & x_3 - x \end{vmatrix}.$$

see also LAGRANGE INTERPOLATING POLYNOMIAL

References

Abramowitz, M. and Stegun, C. A. (Eds.). *Handbook of Mathematical Functions with Formulas, Graphs, and Mathematical Tables, 9th printing.* New York: Dover, p. 879, 1972.

Acton, F. S. *Numerical Methods That Work, 2nd printing.* Washington, DC: Math. Assoc. Amer., pp. 93–94, 1990.

Press, W. H.; Flannery, B. P.; Teukolsky, S. A.; and Vetterling, W. T. *Numerical Recipes in FORTRAN: The Art of Scientific Computing, 2nd ed.* Cambridge, England: Cambridge University Press, p. 102, 1992.

Ajima-Malfatti Points

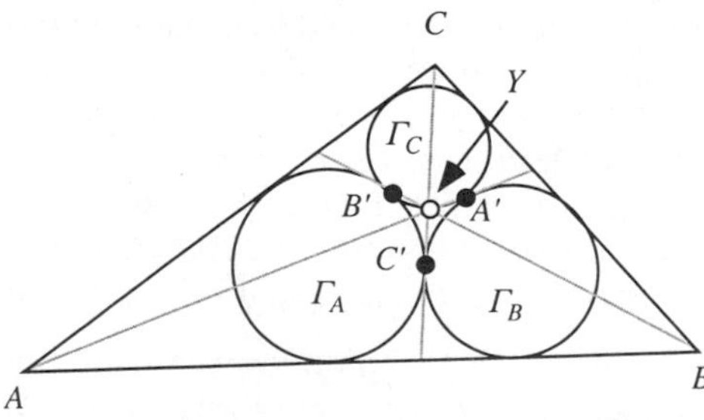

The lines connecting the vertices and corresponding circle-circle intersections in MALFATTI'S TANGENT TRIANGLE PROBLEM coincide in a point Y called the first Ajima-Malfatti point (Kimberling and MacDonald 1990, Kimberling 1994). Similarly, letting A'', B'', and C'' be the excenters of ABC, then the lines $A'A''$, $B'B''$, and $C'C''$ are coincident in another point called the second Ajima-Malfatti point. The points are sometimes simply called the MALFATTI POINTS (Kimberling 1994).

References

Kimberling, C. "Central Points and Central Lines in the Plane of a Triangle." *Math. Mag.* **67**, 163–187, 1994.

Kimberling, C. "1st and 2nd Ajima-Malfatti Points." `http://www.evansville.edu/~ck6/tcenters/recent/ajmalf.html`.

Kimberling, C. and MacDonald, I. G. "Problem E 3251 and Solution. " *Amer. Math. Monthly* **97**, 612–613, 1990.

Albanese Variety

An ABELIAN VARIETY which is canonically attached to an ALGEBRAIC VARIETY which is the solution to a certain universal problem. The Albanese variety is dual to the PICARD VARIETY.

References

Hazewinkel, M. (Managing Ed.). *Encyclopaedia of Mathematics: An Updated and Annotated Translation of the Soviet "Mathematical Encyclopaedia."* Dordrecht, Netherlands: Reidel, pp. 67–68, 1988.

Albers Conic Projection

see ALBERS EQUAL-AREA CONIC PROJECTION

Albers Equal-Area Conic Projection

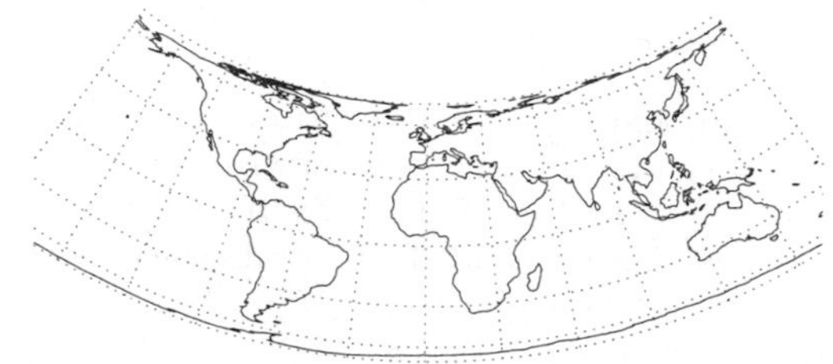

Let ϕ_0 be the LATITUDE for the origin of the CARTESIAN COORDINATES and λ_0 its LONGITUDE. Let ϕ_1 and ϕ_2 be the standard parallels. Then

$$x = \rho \sin\theta \tag{1}$$

$$y = \rho_0 - \rho\cos\theta, \tag{2}$$

where

$$\rho = \frac{\sqrt{C - 2n\sin\phi}}{n} \tag{3}$$

$$\theta = n(\lambda - \lambda_0) \tag{4}$$

$$\rho_0 = \frac{\sqrt{C - 2n\sin\phi_0}}{n} \tag{5}$$

$$C = \cos^2\phi_1 + 2n\sin\phi_1 \tag{6}$$

$$n = \tfrac{1}{2}(\sin\phi_1 + \sin\phi_2). \tag{7}$$

The inverse FORMULAS are

$$\phi = \sin^{-1}\left(\frac{C - \rho^2 n^2}{2n}\right) \tag{8}$$

$$\lambda = \lambda_0 + \frac{\theta}{n}, \tag{9}$$

where

$$\rho = \sqrt{x^2 + (\rho_0 - y)^2} \tag{10}$$

$$\theta = \tan^{-1}\left(\frac{x}{\rho_0 - y}\right). \tag{11}$$

References

Snyder, J. P. *Map Projections—A Working Manual.* U. S. Geological Survey Professional Paper 1395. Washington, DC: U. S. Government Printing Office, pp. 98–103, 1987.

Alcuin's Sequence

The INTEGER SEQUENCE 1, 0, 1, 1, 2, 1, 3, 2, 4, 3, 5, 4, 7, 5, 8, 7, 10, 8, 12, 10, 14, 12, 16, 14, 19, 16, 21, 19, ... (Sloane's A005044) given by the COEFFICIENTS of the MACLAURIN SERIES for $1/(1-x^2)(1-x^3)(1-x^4)$. The number of different TRIANGLES which have INTEGRAL sides and PERIMETER n is given by

$$T(n) = P_3(n) - \sum_{1\leq j\leq \lfloor n/2 \rfloor} P_2(j) \tag{1}$$

$$= \left[\frac{n^2}{12}\right] - \left\lfloor \frac{n}{4} \right\rfloor \left\lfloor \frac{n+2}{4} \right\rfloor \tag{2}$$

$$= \begin{cases} \left[\frac{n^2}{48}\right] & \text{for } n \text{ even} \\ \left[\frac{(n+3)^2}{48}\right] & \text{for } n \text{ odd,} \end{cases} \tag{3}$$

where $P_2(n)$ and $P_3(n)$ are PARTITION FUNCTIONS, with $P_k(n)$ giving the number of ways of writing n as a sum of k terms, $[x]$ is the NINT function, and $\lfloor x \rfloor$ is the FLOOR FUNCTION (Jordan *et al.* 1979, Andrews 1979, Honsberger 1985). Strangely enough, $T(n)$ for n = 3, 4, ... is precisely Alcuin's sequence.

see also PARTITION FUNCTION P, TRIANGLE

References

Andrews, G. "A Note on Partitions and Triangles with Integer Sides." *Amer. Math. Monthly* **86**, 477, 1979.

Honsberger, R. *Mathematical Gems III.* Washington, DC: Math. Assoc. Amer., pp. 39–47, 1985.

Jordan, J. H.; Walch, R.; and Wisner, R. J. "Triangles with Integer Sides." *Amer. Math. Monthly* **86**, 686–689, 1979.

Sloane, N. J. A. Sequence A005044/M0146 in "An On-Line Version of the Encyclopedia of Integer Sequences."

Aleksandrov-Čech Cohomology

A theory which satisfies all the EILENBERG-STEENROD AXIOMS with the possible exception of the LONG EXACT SEQUENCE OF A PAIR AXIOM, as well as a certain additional continuity CONDITION.

References

Hazewinkel, M. (Managing Ed.). *Encyclopaedia of Mathematics: An Updated and Annotated Translation of the Soviet "Mathematical Encyclopaedia."* Dordrecht, Netherlands: Reidel, p. 68, 1988.

Aleksandrov's Uniqueness Theorem

A convex body in EUCLIDEAN n-space that is centrally symmetric with center at the ORIGIN is determined among all such bodies by its brightness function (the VOLUME of each projection).

see also TOMOGRAPHY

References

Gardner, R. J. "Geometric Tomography." *Not. Amer. Math. Soc.* **42**, 422–429, 1995.

Aleph

The SET THEORY symbol ($\aleph$) for the CARDINALITY of an INFINITE SET.

see also ALEPH-0 ($\aleph_0$), ALEPH-1 ($\aleph_1$), COUNTABLE SET, COUNTABLY INFINITE SET, FINITE, INFINITE, TRANSFINITE NUMBER, UNCOUNTABLY INFINITE SET

Aleph-0 ($\aleph_0$)

The SET THEORY symbol for a SET having the same CARDINAL NUMBER as the "small" INFINITE SET of INTEGERS. The ALGEBRAIC NUMBERS also belong to $\aleph_0$. Rather surprising properties satisfied by $\aleph_0$ include

$$\aleph_0{}^r = \aleph_0 \tag{1}$$

$$r\aleph_0 = \aleph_0 \tag{2}$$

$$\aleph_0 + f = \aleph_0, \tag{3}$$

where f is any FINITE SET. However,

$$\aleph_0{}^{\aleph_0} = C, \tag{4}$$

where C is the CONTINUUM.

see also ALEPH-1, CARDINAL NUMBER, CONTINUUM, CONTINUUM HYPOTHESIS, COUNTABLY INFINITE SET, FINITE, INFINITE, TRANSFINITE NUMBER, UNCOUNTABLY INFINITE SET

Aleph-1 ($\aleph_1$)

The SET THEORY symbol for the smallest INFINITE SET larger than ALPHA-0 ($\aleph_0$). The CONTINUUM HYPOTHESIS asserts that $\aleph_1 = c$, where c is the CARDINALITY of the "large" INFINITE SET of REAL NUMBERS (called the CONTINUUM in SET THEORY). However, the truth of the CONTINUUM HYPOTHESIS depends on the version of SET THEORY you are using and so is UNDECIDABLE.

Curiously enough, n-D SPACE has the same number of points (c) as 1-D SPACE, or any FINITE INTERVAL of 1-D SPACE (a LINE SEGMENT), as was first recognized by Georg Cantor.

see also ALEPH-0 ($\aleph_0$), CONTINUUM, CONTINUUM HYPOTHESIS, COUNTABLY INFINITE SET, FINITE, INFINITE, TRANSFINITE NUMBER, UNCOUNTABLY INFINITE SET

Alethic

A term in LOGIC meaning pertaining to TRUTH and FALSEHOOD.

see also FALSE, PREDICATE, TRUE

Alexander-Conway Polynomial

see CONWAY POLYNOMIAL

Alexander's Horned Sphere

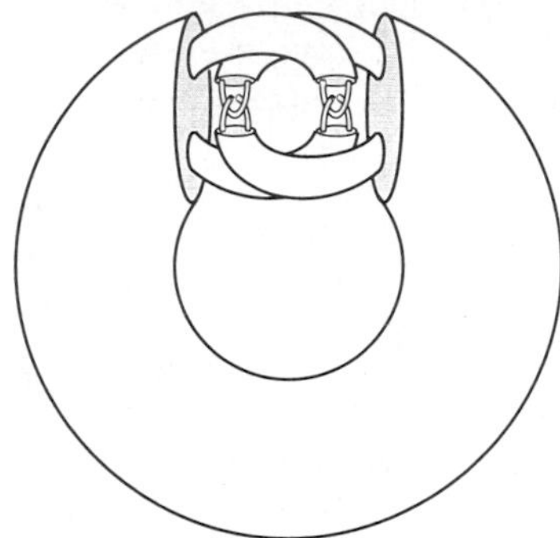

The above solid, composed of a countable UNION of COMPACT SETS, is called Alexander's horned sphere. It is HOMEOMORPHIC with the BALL $\mathbb{B}^3$, and its boundary is therefore a SPHERE. It is therefore an example of a wild embedding in $\mathbb{E}^3$. The outer complement of the solid is not SIMPLY CONNECTED, and its fundamental GROUP is not finitely generated. Furthermore, the set of nonlocally flat ("bad") points of Alexander's horned sphere is a CANTOR SET.

The complement in $\mathbb{R}^3$ of the bad points for Alexander's horned sphere is SIMPLY CONNECTED, making it inequivalent to ANTOINE'S HORNED SPHERE. Alexander's horned sphere has an uncountable infinity of WILD POINTS, which are the limits of the sequences of the horned sphere's branch points (roughly, the "ends" of the horns), since any NEIGHBORHOOD of a limit contains a horned complex.

A humorous drawing by Simon Frazer (Guy 1983, Schroeder 1991, Albers 1994) depicts mathematician John H. Conway with Alexander's horned sphere growing from his head.

see also ANTOINE'S HORNED SPHERE

References

Albers, D. J. Illustration accompanying "The Game of 'Life'." *Math Horizons,* p. 9, Spring 1994.

Guy, R. "Conway's Prime Producing Machine." *Math. Mag.* **56**, 26–33, 1983.

Hocking, J. G. and Young, G. S. *Topology.* New York: Dover, 1988.

Rolfsen, D. *Knots and Links.* Wilmington, DE: Publish or Perish Press, pp. 80–81, 1976.

Schroeder, M. *Fractals, Chaos, Power Law: Minutes from an Infinite Paradise.* New York: W. H. Freeman, p. 58, 1991.

Alexander Ideal

The order IDEAL in Λ, the RING of integral LAURENT POLYNOMIALS, associated with an ALEXANDER MATRIX for a KNOT K. Any generator of a principal Alexander ideal is called an ALEXANDER POLYNOMIAL. Because the ALEXANDER INVARIANT of a TAME KNOT in $\mathbb{S}^3$ has a SQUARE presentation MATRIX, its Alexander ideal is PRINCIPAL and it has an ALEXANDER POLYNOMIAL $\Delta(t)$.

see also ALEXANDER INVARIANT, ALEXANDER MATRIX, ALEXANDER POLYNOMIAL

References

Rolfsen, D. *Knots and Links.* Wilmington, DE: Publish or Perish Press, pp. 206–207, 1976.

Alexander Invariant

The Alexander invariant $H_*(\tilde{X})$ of a KNOT K is the HOMOLOGY of the INFINITE cyclic cover of the complement of K, considered as a MODULE over Λ, the RING of integral LAURENT POLYNOMIALS. The Alexander invariant for a classical TAME KNOT is finitely presentable, and only H_1 is significant.

For any KNOT K^n in $\mathbb{S}^{n+2}$ whose complement has the homotopy type of a FINITE COMPLEX, the Alexander invariant is finitely generated and therefore finitely presentable. Because the Alexander invariant of a TAME KNOT in $\mathbb{S}^3$ has a SQUARE presentation MATRIX, its ALEXANDER IDEAL is PRINCIPAL and it has an ALEXANDER POLYNOMIAL denoted $\Delta(t)$.

see also ALEXANDER IDEAL, ALEXANDER MATRIX, ALEXANDER POLYNOMIAL

References

Rolfsen, D. *Knots and Links.* Wilmington, DE: Publish or Perish Press, pp. 206–207, 1976.

Alexander Matrix

A presentation matrix for the ALEXANDER INVARIANT $H_1(\tilde{X})$ of a KNOT K. If V is a SEIFERT MATRIX for a TAME KNOT K in $\mathbb{S}^3$, then $V^{\mathrm{T}} - tV$ and $V - tV^{\mathrm{T}}$ are Alexander matrices for K, where V^{T} denotes the MATRIX TRANSPOSE.

see also ALEXANDER IDEAL, ALEXANDER INVARIANT, ALEXANDER POLYNOMIAL, SEIFERT MATRIX

References

Rolfsen, D. *Knots and Links.* Wilmington, DE: Publish or Perish Press, pp. 206–207, 1976.

Alexander Polynomial

A POLYNOMIAL invariant of a KNOT discovered in 1923 by J. W. Alexander (Alexander 1928). In technical language, the Alexander polynomial arises from the HOMOLOGY of the infinitely cyclic cover of a KNOT's complement. Any generator of a PRINCIPAL ALEXANDER IDEAL is called an Alexander polynomial (Rolfsen 1976). Because the ALEXANDER INVARIANT of a TAME KNOT in $\mathbb{S}^3$ has a SQUARE presentation MATRIX, its ALEXANDER IDEAL is PRINCIPAL and it has an Alexander polynomial denoted $\Delta(t)$.

Let Ψ be the MATRIX PRODUCT of BRAID WORDS of a KNOT, then

$$\frac{\det(\mathsf{I}-\Psi)}{1+t+\ldots+t^{n-1}}=\Delta_L, \tag{1}$$

where Δ_L is the Alexander polynomial and det is the DETERMINANT. The Alexander polynomial of a TAME KNOT in $\mathbb{S}^3$ satisfies

$$\Delta(t)=\det(V^{\mathrm{T}}-tV), \tag{2}$$

where V is a SEIFERT MATRIX, det is the DETERMINANT, and V^{T} denotes the MATRIX TRANSPOSE. The Alexander polynomial also satisfies

$$\Delta(1)=\pm 1. \tag{3}$$

The Alexander polynomial of a splittable link is always 0. Surprisingly, there are known examples of nontrivial KNOTS with Alexander polynomial 1. An example is the $(-3,5,7)$ PRETZEL KNOT.

The Alexander polynomial remained the *only* known KNOT POLYNOMIAL until the JONES POLYNOMIAL was discovered in 1984. Unlike the Alexander polynomial, the more powerful JONES POLYNOMIAL *does,* in most cases, distinguish HANDEDNESS. A normalized form of the Alexander polynomial symmetric in t and t^{-1} and satisfying

$$\Delta(\text{unknot})=1 \tag{4}$$

was formulated by J. H. Conway and is sometimes denoted ∇_L. The NOTATION $[a+b+c+\ldots$ is an abbreviation for the Conway-normalized Alexander polynomial of a KNOT

$$a+b(x+x^{-1})+c(x^2+x^{-2})+\ldots. \tag{5}$$

For a description of the NOTATION for LINKS, see Rolfsen (1976, p. 389). Examples of the Conway-Alexander polynomials for common KNOTS include

$$\nabla_{\mathrm{TK}}=[1-1=-x^{-1}+1-x \tag{6}$$

$$\nabla_{\mathrm{FEK}}=[3-1=-x^{-1}+3-x \tag{7}$$

$$\nabla_{\mathrm{SSK}}=[1-1+1=x^{-2}-x^{-1}+1-x+x^2 \tag{8}$$

for the TREFOIL KNOT, FIGURE-OF-EIGHT KNOT, and SOLOMON'S SEAL KNOT, respectively. Multiplying through to clear the NEGATIVE POWERS gives the usual Alexander polynomial, where the final SIGN is determined by convention.

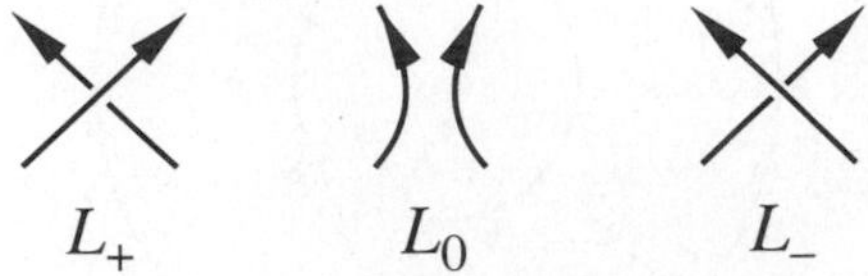

Let an Alexander polynomial be denoted Δ, then there exists a SKEIN RELATIONSHIP (discovered by J. H. Conway)

$$\Delta_{L_+}(t)-\Delta_{L_-}(t)+(t^{-1/2}-t^{1/2})\Delta_{L_0}(t)=0 \tag{9}$$

corresponding to the above LINK DIAGRAMS (Adams 1994). A slightly different SKEIN RELATIONSHIP convention used by Doll and Hoste (1991) is

$$\nabla_{L_+}-\nabla_{L_-}=z\nabla_{L_0}. \tag{10}$$

These relations allow Alexander polynomials to be constructed for arbitrary knots by building them up as a sequence of over- and undercrossings.

For a KNOT,

$$\Delta_K(-1)\equiv\begin{cases}1 \pmod 8 & \text{if } \mathrm{Arf}(K)=0.\\ 5 \pmod 8 & \text{if } \mathrm{Arf}(K)=1,\end{cases} \tag{11}$$

where Arf is the ARF INVARIANT (Jones 1985). If K is a KNOT and

$$|\Delta_K(i)|>3, \tag{12}$$

then K cannot be represented as a closed 3-BRAID. Also, if

$$\Delta_K(e^{2\pi i/5})>\tfrac{13}{2}, \tag{13}$$

then K cannot be represented as a closed 4-braid (Jones 1985).

The HOMFLY POLYNOMIAL $P(a,z)$ generalizes the Alexander polynomial (as well at the JONES POLYNOMIAL) with

$$\nabla(z)=P(1,z) \tag{14}$$

(Doll and Hoste 1991).

Rolfsen (1976) gives a tabulation of Alexander polynomials for KNOTS up to 10 CROSSINGS and LINKS up to 9 CROSSINGS.

see also BRAID GROUP, JONES POLYNOMIAL, KNOT, KNOT DETERMINANT, LINK, SKEIN RELATIONSHIP

References

Adams, C. C. *The Knot Book: An Elementary Introduction to the Mathematical Theory of Knots.* New York: W. H. Freeman, pp. 165–169, 1994.

Alexander, J. W. "Topological Invariants of Knots and Links." *Trans. Amer. Math. Soc.* **30**, 275–306, 1928.

Alexander, J. W. "A Lemma on a System of Knotted Curves." *Proc. Nat. Acad. Sci. USA* **9**, 93–95, 1923.
Doll, H. and Hoste, J. "A Tabulation of Oriented Links." *Math. Comput.* **57**, 747–761, 1991.
Jones, V. "A Polynomial Invariant for Knots via von Neumann Algebras." *Bull. Amer. Math. Soc.* **12**, 103–111, 1985.
Rolfsen, D. "Table of Knots and Links." Appendix C in *Knots and Links.* Wilmington, DE: Publish or Perish Press, pp. 280–287, 1976.
Stoimenow, A. "Alexander Polynomials." `http://www.informatik.hu-berlin.de/~stoimeno/ptab/a10.html`.
Stoimenow, A. "Conway Polynomials." `http://www.informatik.hu-berlin.de/~stoimeno/ptab/c10.html`.

Alexander-Spanier Cohomology

A fundamental result of DE RHAM COHOMOLOGY is that the kth DE RHAM COHOMOLOGY VECTOR SPACE of a MANIFOLD M is canonically isomorphic to the Alexander-Spanier cohomology VECTOR SPACE $H^k(M;\mathbb{R})$ (also called cohomology with compact support). In the case that M is COMPACT, Alexander-Spanier cohomology is exactly "singular" COHOMOLOGY.

Alexander's Theorem

Any LINK can be represented by a closed BRAID.

Algebra

The branch of mathematics dealing with GROUP THEORY and CODING THEORY which studies number systems and operations within them. The word "algebra" is a distortion of the Arabic title of a treatise by Al-Khwarizmi about algebraic methods. Note that mathematicians refer to the "school algebra" generally taught in middle and high school as "ARITHMETIC," reserving the word "algebra" for the more advanced aspects of the subject.

Formally, an algebra is a VECTOR SPACE V, over a FIELD F with a MULTIPLICATION which turns it into a RING defined such that, if $f \in F$ and $x, y \in V$, then

$$f(xy) = (fx)y = x(fy).$$

In addition to the usual algebra of REAL NUMBERS, there are ≈ 1151 additional CONSISTENT algebras which can be formulated by weakening the FIELD AXIOMS, at least 200 of which have been *rigorously* proven to be self-CONSISTENT (Bell 1945).

Algebras which have been investigated and found to be of interest are usually named after one or more of their investigators. This practice leads to exotic-sounding (but unenlightening) names which algebraists frequently use with minimal or nonexistent explanation.

see also ALTERNATE ALGEBRA, ALTERNATING ALGEBRA, B^*-ALGEBRA, BANACH ALGEBRA, BOOLEAN ALGEBRA, BOREL SIGMA ALGEBRA, C^*-ALGEBRA, CAYLEY ALGEBRA, CLIFFORD ALGEBRA, COMMUTATIVE ALGEBRA, EXTERIOR ALGEBRA, FUNDAMENTAL THEOREM OF ALGEBRA, GRADED ALGEBRA, GRASSMANN ALGEBRA, HECKE ALGEBRA, HEYTING ALGEBRA, HOMOLOGICAL ALGEBRA, HOPF ALGEBRA, JORDAN ALGEBRA, LIE ALGEBRA, LINEAR ALGEBRA, MEASURE ALGEBRA, NONASSOCIATIVE ALGEBRA, QUATERNION, ROBBINS ALGEBRA, SCHUR ALGEBRA, SEMISIMPLE ALGEBRA, SIGMA ALGEBRA, SIMPLE ALGEBRA, STEENROD ALGEBRA, VON NEUMANN ALGEBRA

References

Artin, M. *Algebra.* Englewood Cliffs, NJ: Prentice-Hall, 1991.
Bell, E. T. *The Development of Mathematics, 2nd ed.* New York: McGraw-Hill, pp. 35–36, 1945.
Bhattacharya, P. B.; Jain, S. K.; and Nagpu, S. R. (Eds.). *Basic Algebra, 2nd ed.* New York: Cambridge University Press, 1994.
Birkhoff, G. and Mac Lane, S. *A Survey of Modern Algebra, 5th ed.* New York: Macmillan, 1996.
Brown, K. S. "Algebra." `http://www.seanet.com/~ksbrown/ialgebra.htm`.
Cardano, G. *Ars Magna or The Rules of Algebra.* New York: Dover, 1993.
Chevalley, C. C. *Introduction to the Theory of Algebraic Functions of One Variable.* Providence, RI: Amer. Math. Soc., 1951.
Chrystal, G. *Textbook of Algebra, 2 vols.* New York: Dover, 1961.
Dickson, L. E. *Algebras and Their Arithmetics.* Chicago, IL: University of Chicago Press, 1923.
Dickson, L. E. *Modern Algebraic Theories.* Chicago, IL: H. Sanborn, 1926.
Edwards, H. M. *Galois Theory, corrected 2nd printing.* New York: Springer-Verlag, 1993.
Euler, L. *Elements of Algebra.* New York: Springer-Verlag, 1984.
Gallian, J. A. *Contemporary Abstract Algebra, 3rd ed.* Lexington, MA: D. C. Heath, 1994.
Grove, L. *Algebra.* New York: Academic Press, 1983.
Hall, H. S. and Knight, S. R. *Higher Algebra, A Sequel to Elementary Algebra for Schools.* London: Macmillan, 1960.
Harrison, M. A. "The Number of Isomorphism Types of Finite Algebras." *Proc. Amer. Math. Soc.* **17**, 735–737, 1966.
Herstein, I. N. *Noncommutative Rings.* Washington, DC: Math. Assoc. Amer., 1996.
Herstein, I. N. *Topics in Algebra, 2nd ed.* New York: Wiley, 1975.
Jacobson, N. *Basic Algebra II, 2nd ed.* New York: W. H. Freeman, 1989.
Kaplansky, I. *Fields and Rings, 2nd ed.* Chicago, IL: University of Chicago Press, 1995.
Lang, S. *Undergraduate Algebra, 2nd ed.* New York: Springer-Verlag, 1990.
Pedersen, J. "Catalogue of Algebraic Systems." `http://tarski.math.usf.edu/algctlg/`.
Uspensky, J. V. *Theory of Equations.* New York: McGraw-Hill, 1948.
van der Waerden, B. L. *Algebra, Vol. 2.* New York: Springer-Verlag, 1991.
van der Waerden, B. L. *Geometry and Algebra in Ancient Civilizations.* New York: Springer-Verlag, 1983.
van der Waerden, B. L. *A History of Algebra: From Al-Khwarizmi to Emmy Noether.* New York: Springer-Verlag, 1985.
Varadarajan, V. S. *Algebra in Ancient and Modern Times.* Providence, RI: Amer. Math. Soc., 1998.

Algebraic Closure

The algebraic closure of a FIELD K is the "smallest" FIELD containing K which is algebraically closed. For example, the FIELD of COMPLEX NUMBERS $\mathbb{C}$ is the algebraic closure of the FIELD of REALS $\mathbb{R}$.

Algebraic Coding Theory

see CODING THEORY

Algebraic Curve

An algebraic curve over a FIELD K is an equation $f(X,Y)=0$, where $f(X,Y)$ is a POLYNOMIAL in X and Y with COEFFICIENTS in K. A nonsingular algebraic curve is an algebraic curve over K which has no SINGULAR POINTS over K. A point on an algebraic curve is simply a solution of the equation of the curve. A K-RATIONAL POINT is a point (X,Y) on the curve, where X and Y are in the FIELD K.

see also ALGEBRAIC GEOMETRY, ALGEBRAIC VARIETY, CURVE

References

Griffiths, P. A. *Introduction to Algebraic Curves.* Providence, RI: Amer. Math. Soc., 1989.

Algebraic Function

A function which can be constructed using only a finite number of ELEMENTARY FUNCTIONS together with the INVERSES of functions capable of being so constructed.

see also ELEMENTARY FUNCTION, TRANSCENDENTAL FUNCTION

Algebraic Function Field

A finite extension $K=\mathbb{Z}(z)(w)$ of the FIELD $\mathbb{C}(z)$ of RATIONAL FUNCTIONS in the indeterminate z, i.e., w is a ROOT of a POLYNOMIAL $a_0+a_1\alpha+a_2\alpha^2+\ldots+a_n\alpha^n$, where $a_i\in\mathbb{C}(z)$.

see also ALGEBRAIC NUMBER FIELD, RIEMANN SURFACE

Algebraic Geometry

The study of ALGEBRAIC CURVES, ALGEBRAIC VARIETIES, and their generalization to n-D.

see also ALGEBRAIC CURVE, ALGEBRAIC VARIETY, COMMUTATIVE ALGEBRA, DIFFERENTIAL GEOMETRY, GEOMETRY, PLANE CURVE, SPACE CURVE

References

Abhyankar, S. S. *Algebraic Geometry for Scientists and Engineers.* Providence, RI: Amer. Math. Soc., 1990.

Cox, D.; Little, J.; and O'Shea, D. *Ideals, Varieties, and Algorithms: An Introduction to Algebraic Geometry and Commutative Algebra, 2nd ed.* New York: Springer-Verlag, 1996.

Eisenbud, D. *Commutative Algebra with a View Toward Algebraic Geometry.* New York: Springer-Verlag, 1995.

Griffiths, P. and Harris, J. *Principles of Algebraic Geometry.* New York: Wiley, 1978.

Hartshorne, R. *Algebraic Geometry, rev. ed.* New York: Springer-Verlag, 1997.

Lang, S. *Introduction to Algebraic Geometry.* New York: Interscience, 1958.

Pedoe, D. and Hodge, W. V. *Methods of Algebraic Geometry, Vol. 1.* Cambridge, England: Cambridge University Press, 1994.

Pedoe, D. and Hodge, W. V. *Methods of Algebraic Geometry, Vol. 2.* Cambridge, England: Cambridge University Press, 1994.

Pedoe, D. and Hodge, W. V. *Methods of Algebraic Geometry, Vol. 3.* Cambridge, England: Cambridge University Press, 1994.

Seidenberg, A. (Ed.). *Studies in Algebraic Geometry.* Washington, DC: Math. Assoc. Amer., 1980.

Weil, A. *Foundations of Algebraic Geometry, enl. ed.* Providence, RI: Amer. Math. Soc., 1962.

Algebraic Integer

If r is a ROOT of the POLYNOMIAL equation

$$x^n+a_{n-1}x^{n-1}+\ldots+a_1x+a_0=0,$$

where the a_is are INTEGERS and r satisfies no similar equation of degree $<n$, then r is an algebraic INTEGER of degree n. An algebraic INTEGER is a special case of an ALGEBRAIC NUMBER, for which the leading COEFFICIENT a_n need not equal 1. RADICAL INTEGERS are a subring of the ALGEBRAIC INTEGERS.

A SUM or PRODUCT of algebraic integers is again an algebraic integer. However, ABEL'S IMPOSSIBILITY THEOREM shows that there are algebraic integers of degree ≥ 5 which are not expressible in terms of ADDITION, SUBTRACTION, MULTIPLICATION, DIVISION, and the extraction of ROOTS on REAL NUMBERS.

The GAUSSIAN INTEGER are are algebraic integers of $\mathbb{Q}(\sqrt{-1})$, since $a+bi$ are roots of

$$z^2-2az+a^2+b^2=0.$$

see also ALGEBRAIC NUMBER, EUCLIDEAN NUMBER, RADICAL INTEGER

References

Hancock, H. *Foundations of the Theory of Algebraic Numbers, Vol. 1: Introduction to the General Theory.* New York: Macmillan, 1931.

Hancock, H. *Foundations of the Theory of Algebraic Numbers, Vol. 2: The General Theory.* New York: Macmillan, 1932.

Pohst, M. and Zassenhaus, H. *Algorithmic Algebraic Number Theory.* Cambridge, England: Cambridge University Press, 1989.

Wagon, S. "Algebraic Numbers." §10.5 in *Mathematica in Action.* New York: W. H. Freeman, pp. 347–353, 1991.

Algebraic Invariant

A quantity such as a DISCRIMINANT which remains unchanged under a given class of algebraic transformations. Such invariants were originally called HYPERDETERMINANTS by Cayley.

see also DISCRIMINANT (POLYNOMIAL), INVARIANT, QUADRATIC INVARIANT

References
Grace, J. H. and Young, A. *The Algebra of Invariants.* New York: Chelsea, 1965.
Gurevich, G. B. *Foundations of the Theory of Algebraic Invariants.* Groningen, Netherlands: P. Noordhoff, 1964.
Hermann, R. and Ackerman, M. *Hilbert's Invariant Theory Papers.*rookline, MA: Math Sci Press, 1978.
Hilbert, D. *Theory of Algebraic Invariants.* Cambridge, England: Cambridge University Press, 1993.
Mumford, D.; Fogarty, J.; and Kirwan, F. *Geometric Invariant Theory, 3rd enl. ed.* New York: Springer-Verlag, 1994.

Algebraic Knot

A single component ALGEBRAIC LINK.

see also ALGEBRAIC LINK, KNOT, LINK

Algebraic Link

A class of fibered knots and links which arises in ALGEBRAIC GEOMETRY. An algebraic link is formed by connecting the NW and NE strings and the SW and SE strings of an ALGEBRAIC TANGLE (Adams 1994).

see also ALGEBRAIC TANGLE, FIBRATION, TANGLE

References
Adams, C. C. *The Knot Book: An Elementary Introduction to the Mathematical Theory of Knots.* New York: W. H. Freeman, pp. 48–49, 1994.
Rolfsen, D. *Knots and Links.* Wilmington, DE: Publish or Perish Press, p. 335, 1976.

Algebraic Number

If r is a ROOT of the POLYNOMIAL equation

$$a_0x^n + a_1x^{n-1} + \ldots + a_{n-1}x + a_n = 0, \quad (1)$$

where the a_is are INTEGERS and r satisfies no similar equation of degree $< n$, then r is an algebraic number of degree n. If r is an algebraic number and $a_0 = 1$, then it is called an ALGEBRAIC INTEGER. It is also true that if the c_is in

$$c_0x^n + c_1x^{n-1} + \ldots + c_{n-1}x + c_n = 0 \quad (2)$$

are algebraic numbers, then any ROOT of this equation is also an algebraic number.

If α is an algebraic number of degree n satisfying the POLYNOMIAL

$$a(x-\alpha)(x-\beta)(x-\gamma)\cdots, \quad (3)$$

then there are $n-1$ other algebraic numbers β, γ, ... called the conjugates of α. Furthermore, if α satisfies any other algebraic equation, then its conjugates also satisfy the same equation (Conway and Guy 1996).

Any number which is not algebraic is said to be TRANSCENDENTAL.

see also ALGEBRAIC INTEGER, EUCLIDEAN NUMBER, HERMITE-LINDEMANN THEOREM, RADICAL INTEGER, SEMIALGEBRAIC NUMBER, TRANSCENDENTAL NUMBER

References
Conway, J. H. and Guy, R. K. "Algebraic Numbers." In *The Book of Numbers.* New York: Springer-Verlag, pp. 189–190, 1996.
Courant, R. and Robbins, H. "Algebraic and Transcendental Numbers." §2.6 in *What is Mathematics?: An Elementary Approach to Ideas and Methods, 2nd ed.* Oxford, England: Oxford University Press, pp. 103–107, 1996.
Hancock, H. *Foundations of the Theory of Algebraic Numbers. Vol. 1: Introduction to the General Theory.* New York: Macmillan, 1931.
Hancock, H. *Foundations of the Theory of Algebraic Numbers. Vol. 2: The General Theory.* New York: Macmillan, 1932.
Wagon, S. "Algebraic Numbers." §10.5 in *Mathematica in Action.* New York: W. H. Freeman, pp. 347–353, 1991.

Algebraic Number Field

see NUMBER FIELD

Algebraic Surface

The set of ROOTS of a POLYNOMIAL $f(x,y,z) = 0$. An algebraic surface is said to be of degree $n = \max(i+j+k)$, where n is the maximum sum of powers of all terms $a_m x^{i_m} y^{j_m} z^{k_m}$. The following table lists the names of algebraic surfaces of a given degree.

Order	Surface
3	cubic surface
4	quartic surface
5	quintic surface
6	sextic surface
7	heptic surface
8	octic surface
9	nonic surface
10	decic surface

see also BARTH DECIC, BARTH SEXTIC, BOY SURFACE, CAYLEY CUBIC, CHAIR, CLEBSCH DIAGONAL CUBIC, CUSHION, DERVISH, ENDRASS OCTIC, HEART SURFACE, KUMMER SURFACE, ORDER (ALGEBRAIC SURFACE), ROMAN SURFACE, SURFACE, TOGLIATTI SURFACE

References
Fischer, G. (Ed.). *Mathematical Models from the Collections of Universities and Museums.* Braunschweig, Germany: Vieweg, p. 7, 1986.

Algebraic Tangle

Any TANGLE obtained by ADDITIONS and MULTIPLICATIONS of rational TANGLES (Adams 1994).

see also ALGEBRAIC LINK

References
Adams, C. C. *The Knot Book: An Elementary Introduction to the Mathematical Theory of Knots.* New York: W. H. Freeman, pp. 41–51, 1994.

Algebraic Topology

The study of intrinsic qualitative aspects of spatial objects (e.g., SURFACES, SPHERES, TORI, CIRCLES, KNOTS, LINKS, configuration spaces, etc.) that remain invariant under both-directions continuous ONE-TO-ONE (HOMEOMORPHIC) transformations. The discipline of algebraic topology is popularly known as "RUBBER-SHEET GEOMETRY" and can also be viewed as the study of DISCONNECTIVITIES. Algebraic topology has a great deal of mathematical machinery for studying different kinds of HOLE structures, and it gets the prefix "algebraic" since many HOLE structures are represented best by algebraic objects like GROUPS and RINGS.

A technical way of saying this is that algebraic topology is concerned with FUNCTORS from the topological CATEGORY of GROUPS and HOMOMORPHISMS. Here, the FUNCTORS are a kind of filter, and given an "input" SPACE, they spit out something else in return. The returned object (usually a GROUP or RING) is then a representation of the HOLE structure of the SPACE, in the sense that this algebraic object is a vestige of what the original SPACE was like (i.e., much information is lost, but some sort of "shadow" of the SPACE is retained—just enough of a shadow to understand some aspect of its HOLE-structure, but no more). The idea is that FUNCTORS give much simpler objects to deal with. Because SPACES by themselves are very complicated, they are unmanageable without looking at particular aspects.

COMBINATORIAL TOPOLOGY is a special type of algebraic topology that uses COMBINATORIAL methods.

see also CATEGORY, COMBINATORIAL TOPOLOGY, DIFFERENTIAL TOPOLOGY, FUNCTOR, HOMOTOPY THEORY

References

Dieudonné, J. *A History of Algebraic and Differential Topology: 1900–1960.* Boston, MA: Birkhäuser, 1989.

Algebraic Variety

A generalization to n-D of ALGEBRAIC CURVES. More technically, an algebraic variety is a reduced SCHEME of FINITE type over a FIELD K. An algebraic variety V is defined as the SET of points in the REALS $\mathbb{R}^n$ (or the COMPLEX NUMBERS $\mathbb{C}^n$) satisfying a system of POLYNOMIAL equations $f_i(x_1, \ldots, x_n) = 0$ for $i = 1, 2, \ldots$. According to the HILBERT BASIS THEOREM, a FINITE number of equations suffices.

see also ABELIAN VARIETY, ALBANESE VARIETY, BRAUER-SEVERI VARIETY, CHOW VARIETY, PICARD VARIETY

References

Ciliberto, C.; Laura, E.; and Somese, A. J. (Eds.). *Classification of Algebraic Varieties.* Providence, RI: Amer. Math. Soc., 1994.

Algebroidal Function

An ANALYTIC FUNCTION $f(z)$ satisfying the irreducible algebraic equation

$$A_0(z)f^k + A_1(z)f^{k-1} + \ldots + A_k(z) = 0$$

with single-valued MEROMORPHIC functions $A_j(z)$ in a COMPLEX DOMAIN G is called a k-algebroidal function in G.

References

Iyanaga, S. and Kawada, Y. (Eds.). "Algebroidal Functions." §19 in *Encyclopedic Dictionary of Mathematics.* Cambridge, MA: MIT Press, pp. 86–88, 1980.

Algorithm

A specific set of instructions for carrying out a procedure or solving a problem, usually with the requirement that the procedure terminate at some point. Specific algorithms sometimes also go by the name METHOD, PROCEDURE, or TECHNIQUE. The word "algorithm" is a distortion of Al-Khwarizmi, an Arab mathematician who wrote an influential treatise about algebraic methods.

see also 196-ALGORITHM, ALGORITHMIC COMPLEXITY, ARCHIMEDES ALGORITHM, BHASKARA-BROUCKNER ALGORITHM, BORCHARDT-PFAFF ALGORITHM, BRELAZ'S HEURISTIC ALGORITHM, BUCHBERGER'S ALGORITHM, BULIRSCH-STOER ALGORITHM, BUMPING ALGORITHM, CLEAN ALGORITHM, COMPUTABLE FUNCTION, CONTINUED FRACTION FACTORIZATION ALGORITHM, DECISION PROBLEM, DIJKSTRA'S ALGORITHM, EUCLIDEAN ALGORITHM, FERGUSON-FORCADE ALGORITHM, FERMAT'S ALGORITHM, FLOYD'S ALGORITHM, GAUSSIAN APPROXIMATION ALGORITHM, GENETIC ALGORITHM, GOSPER'S ALGORITHM, GREEDY ALGORITHM, HASSE'S ALGORITHM, HJLS ALGORITHM, JACOBI ALGORITHM, KRUSKAL'S ALGORITHM, LEVINE-O'SULLIVAN GREEDY ALGORITHM, LLL ALGORITHM, MARKOV ALGORITHM, MILLER'S ALGORITHM, NEVILLE'S ALGORITHM, NEWTON'S METHOD, PRIME FACTORIZATION ALGORITHMS, PRIMITIVE RECURSIVE FUNCTION, PROGRAM, PSLQ ALGORITHM, PSOS ALGORITHM, QUOTIENT-DIFFERENCE ALGORITHM, RISCH ALGORITHM, SCHRAGE'S ALGORITHM, SHANKS' ALGORITHM, SPIGOT ALGORITHM, SYRACUSE ALGORITHM, TOTAL FUNCTION, TURING MACHINE, ZASSENHAUS-BERLEKAMP ALGORITHM, ZEILBERGER'S ALGORITHM

References

Aho, A. V.; Hopcroft, J. E.; and Ullman, J.D. *The Design and Analysis of Computer Algorithms.* Reading, MA: Addison-Wesley, 1974.

Baase, S. *Computer Algorithms.* Reading, MA: Addison-Wesley, 1988.

Brassard, G. and Bratley, P. *Fundamentals of Algorithmics.* Englewood Cliffs, NJ: Prentice-Hall, 1995.

Cormen, T. H.; Leiserson, C. E.; and Rivest, R. L. *Introduction to Algorithms.* Cambridge, MA: MIT Press, 1990.

Greene, D. H. and Knuth, D. E. *Mathematics for the Analysis of Algorithms, 3rd ed.* Boston: Birkhäuser, 1990.
Harel, D. *Algorithmics: The Spirit of Computing, 2nd ed.* Reading, MA: Addison-Wesley, 1992.
Knuth, D. E. *The Art of Computer Programming, Vol. 1: Fundamental Algorithms, 2nd ed.* Reading, MA: Addison-Wesley, 1973.
Knuth, D. E. *The Art of Computer Programming, Vol. 2: Seminumerical Algorithms, 2nd ed.* Reading, MA: Addison-Wesley, 1981.
Knuth, D. E. *The Art of Computer Programming, Vol. 3: Sorting and Searching, 2nd ed.* Reading, MA: Addison-Wesley, 1973.
Kozen, D. C. *Design and Analysis and Algorithms.* New York: Springer-Verlag, 1991.
Shen, A. *Algorithms and Programming.* Boston: Birkhäuser, 1996.
Skiena, S. S. *The Algorithm Design Manual.* New York: Springer-Verlag, 1997.
Wilf, H. *Algorithms and Complexity.* Englewood Cliffs, NJ: Prentice Hall, 1986. `http://www.cis.upenn.edu/~wilf/`.

Algorithmic Complexity

see BIT COMPLEXITY, KOLMOGOROV COMPLEXITY

Alhazen's Billiard Problem

In a given CIRCLE, find an ISOSCELES TRIANGLE whose LEGS pass through two given POINTS inside the CIRCLE. This can be restated as: from two POINTS in the PLANE of a CIRCLE, draw LINES meeting at the POINT of the CIRCUMFERENCE and making equal ANGLES with the NORMAL at that POINT.

The problem is called the billiard problem because it corresponds to finding the POINT on the edge of a circular "BILLIARD" table at which a cue ball at a given POINT must be aimed in order to carom once off the edge of the table and strike another ball at a second given POINT. The solution leads to a BIQUADRATIC EQUATION of the form

$$H(x^2 - y^2) - 2Kxy + (x^2 + y^2)(hy - kx) = 0.$$

The problem is equivalent to the determination of the point on a spherical mirror where a ray of light will reflect in order to pass from a given source to an observer. It is also equivalent to the problem of finding, given two points and a CIRCLE such that the points are both inside or outside the CIRCLE, the ELLIPSE whose FOCI are the two points and which is tangent to the given CIRCLE.

The problem was first formulated by Ptolemy in 150 AD, and was named after the Arab scholar Alhazen, who discussed it in his work on optics. It was not until 1997 that Neumann proved the problem to be insoluble using a COMPASS and RULER construction because the solution requires extraction of a CUBE ROOT. This is the same reason that the CUBE DUPLICATION problem is insoluble.

see also BILLIARDS, BILLIARD TABLE PROBLEM, CUBE DUPLICATION

References

Dörrie, H. "Alhazen's Billiard Problem." §41 in *100 Great Problems of Elementary Mathematics: Their History and Solutions.* New York: Dover, pp. 197–200, 1965.
Hogendijk, J. P. "Al-Mutaman's Simplified Lemmas for Solving 'Alhazen's Problem'." *From Baghdad to Barcelona/De Bagdad a Barcelona, Vol. I, II (Zaragoza, 1993),* pp. 59–101, Anu. Filol. Univ. Barc., XIX B-2, Univ. Barcelona, Barcelona, 1996.
Lohne, J. A. "Alhazens Spiegelproblem." *Nordisk Mat. Tidskr.* **18**, 5–35, 1970.
Neumann, P. Submitted to *Amer. Math. Monthly.*
Riede, H. "Reflexion am Kugelspiegel. Oder: das Problem des Alhazen." *Praxis Math.* **31**, 65–70, 1989.
Sabra, A. I. "ibn al-Haytham's Lemmas for Solving 'Alhazen's Problem'." *Arch. Hist. Exact Sci.* **26**, 299-324, 1982.

Alhazen's Problem

see ALHAZEN'S BILLIARD PROBLEM

Alias' Paradox

Choose between the following two alternatives:

1. 90% chance of an unknown amount x and a 10% chance of \$1 million, or
2. 89% chance of the same unknown amount x, 10% chance of \$2.5 million, and 1% chance of nothing.

The PARADOX is to determine which choice has the larger expectation value, $0.9x + \$100,000$ or $0.89x + \$250,000$. However, the best choice depends on the unknown amount, even though it is the same in both cases! This appears to violate the INDEPENDENCE AXIOM.

see also INDEPENDENCE AXIOM, MONTY HALL PROBLEM, NEWCOMB'S PARADOX

Aliasing

Given a power spectrum (a plot of power vs. frequency), aliasing is a false translation of power falling in some frequency range $(-f_c, f_c)$ outside the range. Aliasing can be caused by discrete sampling below the NYQUIST FREQUENCY. The sidelobes of any INSTRUMENT FUNCTION (including the simple SINC SQUARED function obtained simply from FINITE sampling) are also a form of aliasing. Although sidelobe contribution at large offsets can be minimized with the use of an APODIZATION FUNCTION, the tradeoff is a widening of the response (i.e., a lowering of the resolution).

see also APODIZATION FUNCTION, NYQUIST FREQUENCY

Aliquant Divisor

A number which does not DIVIDE another exactly. For instance, 4 and 5 are aliquant divisors of 6. A number which is not an aliquant divisor (i.e., one that *does* DIVIDE another exactly) is said to be an ALIQUOT DIVISOR.

see also ALIQUOT DIVISOR, DIVISOR, PROPER DIVISOR

Aliquot Cycle

see SOCIABLE NUMBERS

Aliquot Divisor

A number which DIVIDES another exactly. For instance, 1, 2, 3, and 6 are aliquot divisors of 6. A number which is not an aliquot divisor is said to be an ALIQUANT DIVISOR. The term "aliquot" is frequently used to specifically mean a PROPER DIVISOR, i.e., a DIVISOR of a number other than the number itself.

see also ALIQUANT DIVISOR, DIVISOR, PROPER DIVISOR

Aliquot Sequence

Let

$$s(n) \equiv \sigma(n) - n,$$

where $\sigma(n)$ is the DIVISOR FUNCTION and $s(n)$ is the RESTRICTED DIVISOR FUNCTION. Then the SEQUENCE of numbers

$$s^0(n) \equiv n, s^1(n) = s(n), s^2(n) = s(s(n)), \ldots$$

is called an aliquot sequence. If the SEQUENCE for a given n is bounded, it either ends at $s(1) = 0$ or becomes periodic.

1. If the SEQUENCE reaches a constant, the constant is known as a PERFECT NUMBER.
2. If the SEQUENCE reaches an alternating pair, it is called an AMICABLE PAIR.
3. If, after k iterations, the SEQUENCE yields a cycle of minimum length t of the form $s^{k+1}(n)$, $s^{k+2}(n)$, ..., $s^{k+t}(n)$, then these numbers form a group of SOCIABLE NUMBERS of order t.

It has not been proven that all aliquot sequences eventually terminate and become period. The smallest number whose fate is not known is 276, which has been computed up to $s^{487}(276)$ (Guy 1994).

see also 196-ALGORITHM, ADDITIVE PERSISTENCE, AMICABLE NUMBERS, MULTIAMICABLE NUMBERS, MULTIPERFECT NUMBER, MULTIPLICATIVE PERSISTENCE, PERFECT NUMBER, SOCIABLE NUMBERS, UNITARY ALIQUOT SEQUENCE

References

Guy, R. K. "Aliquot Sequences." §B6 in *Unsolved Problems in Number Theory, 2nd ed.* New York: Springer-Verlag, pp. 60–62, 1994.

Guy, R. K. and Selfridge, J. L. "What Drives Aliquot Sequences." *Math. Comput.* **29**, 101–107, 1975.

Sloane, N. J. A. Sequences A003023/M0062 in "An On-Line Version of the Encyclopedia of Integer Sequences."

Sloane, N. J. A. and Plouffe, S. Extended entry in *The Encyclopedia of Integer Sequences.* San Diego: Academic Press, 1995.

All-Poles Model

see MAXIMUM ENTROPY METHOD

Alladi-Grinstead Constant

N.B. A detailed on-line essay by S. Finch was the starting point for this entry.

Let $N(n)$ be the number of ways in which the FACTORIAL $n!$ can be decomposed into n FACTORS of the form ${p_k}^{b_k}$ arranged in nondecreasing order. Also define

$$m(n) \equiv \max({p_1}^{b_1}), \tag{1}$$

i.e., $m(n)$ is the LEAST PRIME FACTOR raised to its appropriate POWER in the factorization. Then define

$$\alpha(n) \equiv \frac{\ln m(n)}{\ln n} \tag{2}$$

where $\ln(x)$ is the NATURAL LOGARITHM. For instance,

$$\begin{aligned} 9! &= 2\cdot 2\cdot 2\cdot 2\cdot 2\cdot 2^2\cdot 5\cdot 7\cdot 3^4 \\ &= 2\cdot 2\cdot 2\cdot 2\cdot 3\cdot 5\cdot 7\cdot 2^3\cdot 3^3 \\ &= 2\cdot 2\cdot 2\cdot 2\cdot 5\cdot 7\cdot 2^3\cdot 3^2\cdot 3^2 \\ &= 2\cdot 2\cdot 2\cdot 3\cdot 2^2\cdot 2^2\cdot 5\cdot 7\cdot 3^3 \\ &= 2\cdot 2\cdot 2\cdot 2^2\cdot 2^2\cdot 5\cdot 7\cdot 3^2\cdot 3^2 \\ &= 2\cdot 2\cdot 2\cdot 3\cdot 3\cdot 5\cdot 7\cdot 3^2\cdot 2^4 \\ &= 2\cdot 2\cdot 3\cdot 3\cdot 2^2\cdot 5\cdot 7\cdot 2^3\cdot 3^2 \\ &= 2\cdot 2\cdot 3\cdot 3\cdot 3\cdot 3\cdot 5\cdot 7\cdot 2^5 \\ &= 2\cdot 3\cdot 3\cdot 2^2\cdot 2^2\cdot 2^2\cdot 5\cdot 7\cdot 3^2 \\ &= 2\cdot 3\cdot 3\cdot 3\cdot 3\cdot 2^2\cdot 5\cdot 7\cdot 2^4 \\ &= 2\cdot 3\cdot 3\cdot 3\cdot 3\cdot 5\cdot 7\cdot 2^3\cdot 2^3 \\ &= 3\cdot 3\cdot 3\cdot 3\cdot 2^2\cdot 2^2\cdot 5\cdot 7\cdot 2^3, \end{aligned} \tag{3}$$

so

$$\alpha(9) = \frac{\ln 3}{\ln 9} = \frac{\ln 3}{2\ln 3} = \frac{1}{2}. \tag{4}$$

For large n,

$$\lim_{n\to\infty} \alpha(n) = e^{c-1} = 0.809394020534\ldots, \tag{5}$$

where

$$c \equiv \sum_{k=2}^{\infty} \frac{1}{k} \ln\left(\frac{k}{k-1}\right). \tag{6}$$

References

Alladi, K. and Grinstead, C. "On the Decomposition of $n!$ into Prime Powers." *J. Number Th.* **9**, 452–458, 1977.

Finch, S. "Favorite Mathematical Constants." `http://www.mathsoft.com/asolve/constant/aldgrns/aldgrns.html`.

Guy, R. K. "Factorial n as the Product of n Large Factors." §B22 in *Unsolved Problems in Number Theory, 2nd ed.* New York: Springer-Verlag, p. 79, 1994.

Allegory

A technical mathematical object which bears the same resemblance to binary relations as CATEGORIES do to FUNCTIONS and SETS.

see also CATEGORY

References

Freyd, P. J. and Scedrov, A. *Categories, Allegories.* Amsterdam, Netherlands: North-Holland, 1990.

Allometric

Mathematical growth in which one population grows at a rate PROPORTIONAL to the POWER of another population.

References

Cofrey, W. J. *Geography Towards a General Spatial Systems Approach.* London: Routledge, Chapman & Hall, 1981.

Almost All

Given a property P, if $P(x) \sim x$ as $x \to \infty$ (so the number of numbers less than x not satisfying the property P is $o(x)$), then P is said to hold true for almost all numbers. For example, almost all positive integers are COMPOSITE NUMBERS (which is not in conflict with the second of EUCLID'S THEOREMS that there are an infinite number of PRIMES).

see also FOR ALL, NORMAL ORDER

References

Hardy, G. H. and Wright, E. M. *An Introduction to the Theory of Numbers, 5th ed.* Oxford, England: Clarendon Press, p. 8, 1979.

Almost Alternating Knot

An ALMOST ALTERNATING LINK with a single component.

Almost Alternating Link

Call a projection of a LINK an almost alternating projection if one crossing change in the projection makes it an alternating projection. Then an almost alternating link is a LINK with an almost alternating projection, but no alternating projection. Every ALTERNATING KNOT has an almost alternating projection. A PRIME KNOT which is almost alternating is either a TORUS KNOT or a HYPERBOLIC KNOT. Therefore, no SATELLITE KNOT is an almost alternating knot.

All nonalternating 9-crossing PRIME KNOTS are almost alternating. Of the 393 nonalternating with 11 or fewer crossings, all but five are known to be nonalternating (3 of these have 11 crossings). The fate of the remaining five is not known. The $(2,q)$, $(3,4)$, and $(3,5)$-TORUS KNOTS are almost alternating.

see also ALTERNATING KNOT, LINK

References

Adams, C. C. *The Knot Book: An Elementary Introduction to the Mathematical Theory of Knots.* New York: W. H. Freeman, pp. 139–146, 1994.

Almost Everywhere

A property of X is said to hold almost everywhere if the SET of points in X where this property fails has MEASURE 0.

see also MEASURE

References

Sansone, G. *Orthogonal Functions, rev. English ed.* New York: Dover, p. 1, 1991.

Almost Integer

A number which is very close to an INTEGER. One surprising example involving both e and PI is

$$e^\pi - \pi = 19.999099979\ldots, \tag{1}$$

which can also be written as

$$(\pi+20)^i = -0.9999999992 - 0.0000388927i \approx -1 \tag{2}$$

$$\cos(\ln(\pi+20)) \approx -0.9999999992. \tag{3}$$

Applying COSINE a few more times gives

$$\cos(\pi\cos(\pi\cos(\ln(\pi+20)))) \approx -1 + 3.9321609261 \times 10^{-35}. \tag{4}$$

This curious near-identity was apparently noticed almost simultaneously around 1988 by N. J. A. Sloane, J. H. Conway, and S. Plouffe, but no satisfying explanation as to "why" it has been true has yet been discovered.

An interesting near-identity is given by

$$\frac{1}{4}\left[\cos(\tfrac{1}{10}) + \cosh(\tfrac{1}{10}) + 2\cos(\tfrac{1}{20}\sqrt{2})\cosh(\tfrac{1}{20}\sqrt{2})\right] = 1 + 2.480\ldots \times 10^{-13} \tag{5}$$

(W. Dubuque). Other remarkable near-identities are given by

$$\frac{5(1+\sqrt{5})[\Gamma(\frac{3}{4})]^2}{e^{5\pi/6}\sqrt{\pi}} = 1 + 4.5422\ldots \times 10^{-14}, \tag{6}$$

where $\Gamma(z)$ is the GAMMA FUNCTION (S. Plouffe), and

$$e^6 - \pi^4 - \pi^5 = 0.000017673\ldots \tag{7}$$

(D. Wilson).

A whole class of IRRATIONAL "almost integers" can be found using the theory of MODULAR FUNCTIONS, and a few rather spectacular examples are given by Ramanujan (1913–14). Such approximations were also studied by Hermite (1859), Kronecker (1863), and Smith (1965). They can be generated using some amazing (and very deep) properties of the j-FUNCTION. Some of the numbers which are closest approximations to INTEGERS are $e^{\pi\sqrt{163}}$ (sometimes known as the RAMANUJAN CONSTANT and which corresponds to the field $\mathbb{Q}(\sqrt{-163})$ which has CLASS NUMBER 1 and is the IMAGINARY quadratic field of maximal discriminant), $e^{\pi\sqrt{22}}$, $e^{\pi\sqrt{37}}$, and $e^{\pi\sqrt{58}}$, the latter three of which have CLASS NUMBER 2 and are due to Ramanujan (Berndt 1994, Waldschmidt 1988).

The properties of the j-FUNCTION also give rise to the spectacular identity

$$\left[\frac{\ln(640320^3+744)}{\pi}\right]^2 = 163 + 2.32167\ldots \times 10^{-29} \quad (8)$$

(Le Lionnais 1983, p. 152).

The list below gives numbers of the form $x \equiv e^{\pi\sqrt{n}}$ for $n \leq 1000$ for which $\lceil x \rceil - x \leq 0.01$.

$$e^{\pi\sqrt{6}} = 2,197.990\,869\,543\ldots$$
$$e^{\pi\sqrt{17}} = 422,150.997\,675\,680\ldots$$
$$e^{\pi\sqrt{18}} = 614,551.992\,885\,619\ldots$$
$$e^{\pi\sqrt{22}} = 2,508,951.998\,257\,553\ldots$$
$$e^{\pi\sqrt{25}} = 6,635,623.999\,341\,134\ldots$$
$$e^{\pi\sqrt{37}} = 199,148,647.999\,978\,046\,551\ldots$$
$$e^{\pi\sqrt{43}} = 884,736,743.999\,777\,466\ldots$$
$$e^{\pi\sqrt{58}} = 24,591,257,751.999\,999\,822\,213\ldots$$
$$e^{\pi\sqrt{59}} = 30,197,683,486.993\,182\,260\ldots$$
$$e^{\pi\sqrt{67}} = 147,197,952,743.999\,998\,662\,454\ldots$$
$$e^{\pi\sqrt{74}} = 54,551,812,208.999\,917\,467\,885\ldots$$
$$e^{\pi\sqrt{149}} = 45,116,546,012,289,599.991\,830\,287\ldots$$
$$e^{\pi\sqrt{163}} = 262,537,412,640,768,743.999\,999\,999\,999\,250\,072\ldots$$
$$e^{\pi\sqrt{177}} = 1,418,556,986,635,586,485.996\,179\,355\ldots$$
$$e^{\pi\sqrt{232}} = 604,729,957,825,300,084,759.999\,992\,171\,526\ldots$$
$$e^{\pi\sqrt{267}} = 19,683,091,854,079,461,001,445.992\,737\,040\ldots$$

$$e^{\pi\sqrt{326}} = 4,309,793,301,730,386,363,005,719.996\,011\,651\ldots$$
$$e^{\pi\sqrt{386}} = 639,355,180,631,208,421,\cdots \cdots 212,174,016.997\,669\,832\ldots$$
$$e^{\pi\sqrt{522}} = 14,871,070,263,238,043,663,567,\cdots \cdots 627,879,007.999\,848\,726\ldots$$
$$e^{\pi\sqrt{566}} = 288,099,755,064,053,264,917,867,\cdots \cdots 975,825,573.993\,898\,311\ldots$$
$$e^{\pi\sqrt{638}} = 28,994,858,898,043,231,996,779,\cdots \cdots 771,804,797,161.992\,372\,939\ldots$$
$$e^{\pi\sqrt{719}} = 3,842,614,373,539,548,891,490,\cdots \cdots 294,277,805,829,192.999\,987\,249\ldots$$
$$e^{\pi\sqrt{790}} = 223,070,667,213,077,889,794,379,\cdots \cdots 623,183,838,336;437.992\,055\,118\ldots$$
$$e^{\pi\sqrt{792}} = 249,433,117,287,892,229,255,125,\cdots \cdots 388,685,911,710,805.996\,097\,323\ldots$$
$$e^{\pi\sqrt{928}} = 365,698,321,891,389,219,219,142,\cdots \cdots 531,076,638,716,362,775.998\,259\,747\ldots$$
$$e^{\pi\sqrt{986}} = 6,954,830,200,814,801,770,418,837,\cdots\ 940,281,460,320,666,108.994\,649\,611\ldots.$$

Gosper noted that the expression

$$1 - 262537412640768744e^{-\pi\sqrt{163}} - 196884e^{-2\pi\sqrt{163}} + 103378831900730205293632e^{-3\pi\sqrt{163}}. \quad (9)$$

differs from an INTEGER by a mere 10^{-59}.

see also CLASS NUMBER, j-FUNCTION, PI

References

Berndt, B. C. *Ramanujan's Notebooks, Part IV.* New York: Springer-Verlag, pp. 90–91, 1994.

Hermite, C. "Sur la théorie des équations modulaires." *C. R. Acad. Sci. (Paris)* **48**, 1079–1084 and 1095–1102, 1859.

Hermite, C. "Sur la théorie des équations modulaires." *C. R. Acad. Sci. (Paris)* **49**, 16–24, 110–118, and 141–144, 1859.

Kronecker, L. "Über die Klassenzahl der aus Werzeln der Einheit gebildeten komplexen Zahlen." *Monatsber. K. Preuss. Akad. Wiss. Berlin*, 340–345. 1863.

Le Lionnais, F. *Les nombres remarquables.* Paris: Hermann, 1983.

Ramanujan, S. "Modular Equations and Approximations to π." *Quart. J. Pure Appl. Math.* **45**, 350–372, 1913–1914.

Smith, H. J. S. *Report on the Theory of Numbers.* New York: Chelsea, 1965.

Waldschmidt, M. "Some Transcendental Aspects of Ramanujan's Work." In *Ramanujan Revisited: Proceedings of the Centenary Conference* (Ed. G. E. Andrews, B. C. Berndt, and R. A. Rankin). New York: Academic Press, pp. 57–76, 1988.

Almost Perfect Number

A number n for which the DIVISOR FUNCTION satisfies $\sigma(n) = 2n - 1$ is called almost perfect. The only known almost perfect numbers are the POWERS of 2, namely 1, 2, 4, 8, 16, 32, ... (Sloane's A000079). Singh (1997) calls almost perfect numbers SLIGHTLY DEFECTIVE.

see also QUASIPERFECT NUMBER

References

Guy, R. K. "Almost Perfect, Quasi-Perfect, Pseudoperfect, Harmonic, Weird, Multiperfect and Hyperperfect Numbers." §B2 in *Unsolved Problems in Number Theory, 2nd ed.* New York: Springer-Verlag, pp. 16 and 45–53, 1994.

Singh, S. *Fermat's Enigma: The Epic Quest to Solve the World's Greatest Mathematical Problem.* New York: Walker, p. 13, 1997.

Sloane, N. J. A. Sequence A000079/M1129 in "An On-Line Version of the Encyclopedia of Integer Sequences."

Almost Prime

A number n with prime factorization

$$n = \prod_{i=1}^{r} {p_i}^{a_i}$$

is called k-almost prime when the sum of the POWERS $\sum_{i=1}^{r} a_i = k$. The set of k-almost primes is denoted P_k.

The PRIMES correspond to the "1-almost prime" numbers 2, 3, 5, 7, 11, ... (Sloane's A000040). The 2-almost prime numbers correspond to SEMIPRIMES 4, 6, 9, 10, 14, 15, 21, 22, ... (Sloane's A001358). The first few 3-almost primes are 8, 12, 18, 20, 27, 28, 30, 42, 44, 45, 50, 52, 63, 66, 68, 70, 75, 76, 78, 92, 98, 99, ... (Sloane's A014612). The first few 4-almost primes are 16, 24, 36, 40, 54, 56, 60, 81, 84, 88, 90, 100, ... (Sloane's A014613). The first few 5-almost primes are 32, 48, 72, 80, ... (Sloane's A014614).

see also CHEN'S THEOREM, PRIME NUMBER, SEMI-PRIME

References

Sloane, N. J. A. Sequences A014612, A014613, A014614, A000040/M0652, and A001358/M3274 in "An On-Line Version of the Encyclopedia of Integer Sequences."

Alpha

A financial measure giving the difference between a fund's actual return and its expected level of performance, given its level of risk (as measured by BETA). A POSITIVE alpha indicates that a fund has performed better than expected based on its BETA, whereas a NEGATIVE alpha indicates poorer performance

see also BETA, SHARPE RATIO

Alpha Function

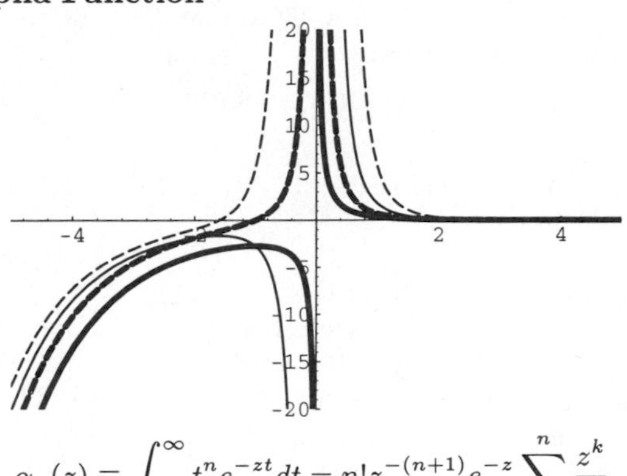

$$\alpha_n(z) \equiv \int_1^\infty t^n e^{-zt}\,dt = n!z^{-(n+1)}e^{-z}\sum_{k=0}^n \frac{z^k}{k!}.$$

The alpha function satisfies the RECURRENCE RELATION

$$z\alpha_n(z) = e^{-z} + n\alpha_{n-1}(z).$$

see also BETA FUNCTION (EXPONENTIAL)

Alpha Value

An alpha value is a number $0 \leq \alpha \leq 1$ such that $P(z \geq z_{\text{observed}}) \leq \alpha$ is considered "SIGNIFICANT," where P is a P-VALUE.

see also CONFIDENCE INTERVAL, P-VALUE, SIGNIFICANCE

Alphabet

A SET (usually of letters) from which a SUBSET is drawn. A sequence of letters is called a WORD, and a set of WORDS is called a CODE.

see also CODE, WORD

Alphamagic Square

A MAGIC SQUARE for which the number of letters in the word for each number generates another MAGIC SQUARE. This definition depends, of course, on the language being used. In English, for example,

$$\begin{array}{ccc} 5 & 22 & 18 \\ 28 & 15 & 2 \\ 12 & 8 & 25 \end{array} \qquad \begin{array}{ccc} 4 & 9 & 8 \\ 11 & 7 & 3 \\ 6 & 5 & 10 \end{array},$$

where the MAGIC SQUARE on the right corresponds to the number of letters in

five	twenty-two	eighteen
twenty-eight	fifteen	two
twelve	eight	twenty-five

References

Sallows, L. C. F. "Alphamagic Squares." *Abacus* **4**, 28–45, 1986.

Sallows, L. C. F. "Alphamagic Squares. 2." *Abacus* **4**, 20–29 and 43, 1987.

Sallows, L. C. F. "Alpha Magic Squares." In *The Lighter Side of Mathematics* (Ed. R. K. Guy and R. E. Woodrow). Washington, DC: Math. Assoc. Amer., 1994.

Alphametic

A CRYPTARITHM in which the letters used to represent distinct DIGITS are derived from related words or meaningful phrases. The term was coined by Hunter in 1955 (Madachy 1979, p. 178).

References

Brooke, M. *One Hundred & Fifty Puzzles in Crypt-Arithmetic.* New York: Dover, 1963.

Hunter, J. A. H. and Madachy, J. S. "Alphametics and the Like." Ch. 9 in *Mathematical Diversions.* New York: Dover, pp. 90–95, 1975.

Madachy, J. S. "Alphametics." Ch. 7 in *Madachy's Mathematical Recreations.* New York: Dover, pp. 178–200 1979.

Alternate Algebra

Let A denote an $\mathbb{R}$-ALGEBRA, so that A is a VECTOR SPACE over R and

$$A \times A \to A \qquad (1)$$

$$(x, y) \mapsto x \cdot y. \qquad (2)$$

Then A is said to be alternate if, for all $x, y \in A$,

$$(x \cdot y) \cdot y = x \cdot (y \cdot y) \qquad (3)$$

$$(x \cdot x) \cdot y = x \cdot (x \cdot y). \qquad (4)$$

Here, VECTOR MULTIPLICATION $x \cdot y$ is assumed to be BILINEAR.

References

Finch, S. "Zero Structures in Real Algebras." `http://www.mathsoft.com/asolve/zerodiv/zerodiv.html`.

Schafer, R. D. *An Introduction to Non-Associative Algebras.* New York: Dover, 1995.

Alternating Algebra

see EXTERIOR ALGEBRA

Alternating Group

EVEN PERMUTATION GROUPS A_n which are NORMAL SUBGROUPS of the PERMUTATION GROUP of ORDER $n!/2$. They are FINITE analogs of the families of simple LIE GROUPS. The lowest order alternating group is 60. Alternating groups with $n \geq 5$ are non-ABELIAN SIMPLE GROUPS. The number of conjugacy classes in the alternating groups A_n for $n = 2, 3, \ldots$ are 1, 3, 4, 5, 7, 9, ... (Sloane's A000702).

see also 15 PUZZLE, FINITE GROUP, GROUP, LIE GROUP, SIMPLE GROUP, SYMMETRIC GROUP

References

Sloane, N. J. A. Sequence A000702/M2307 in "An On-Line Version of the Encyclopedia of Integer Sequences."

Wilson, R. A. "ATLAS of Finite Group Representation." http://for.mat.bham.ac.uk/atlas#alt.

Alternating Knot

An alternating knot is a KNOT which possesses a knot diagram in which crossings alternate between under- and overpasses. Not all knot diagrams of alternating knots need be alternating diagrams.

The TREFOIL KNOT and FIGURE-OF-EIGHT KNOT are alternating knots. One of TAIT'S KNOT CONJECTURES states that the number of crossings is the same for any diagram of a reduced alternating knot. Furthermore, a reduced alternating projection of a knot has the least number of crossings for any projection of that knot. Both of these facts were proved true by Kauffman (1988), Thistlethwaite (1987), and Murasugi (1987).

If K has a reduced alternating projection of n crossings, then the SPAN of K is $4n$. Let $c(K)$ be the CROSSING NUMBER. Then an alternating knot $K_1\#K_2$ (a KNOT SUM) satisfies

$$c(K_1\#K_2) = c(K_1) + c(K_2).$$

In fact, this is true as well for the larger class of ADEQUATE KNOTS and postulated for all KNOTS. The number of PRIME alternating knots of n crossing for $n = 1, 2, \ldots$ are 0, 0, 1, 1, 2, 3, 7, 18, 41, 123, 367, ... (Sloane's A002864).

see also ADEQUATE KNOT, ALMOST ALTERNATING LINK, ALTERNATING LINK, FLYPING CONJECTURE

References

Adams, C. C. *The Knot Book: An Elementary Introduction to the Mathematical Theory of Knots.* New York: W. H. Freeman, pp. 159–164, 1994.

Arnold, B.; Au, M.; Candy, C.; Erdener, K.; Fan, J.; Flynn, R.; Muir, J.; Wu, D.; and Hoste, J. "Tabulating Alternating Knots through 14 Crossings." ftp://chs.cusd.claremont.edu/pub/knot/paper.TeX.txt and ftp://chs.cusd.claremont.edu/pub/knot/AltKnots/.

Erdener, K. and Flynn, R. "Rolfsen's Table of all Alternating Diagrams through 9 Crossings." ftp://chs.cusd.claremont.edu/pub/knot/Rolfsen_table.final.

Kauffman, L. "New Invariants in the Theory of Knots." *Amer. Math. Monthly* **95**, 195–242, 1988.

Murasugi, K. "Jones Polynomials and Classical Conjectures in Knot Theory." *Topology* **26**, 297–307, 1987.

Sloane, N. J. A. Sequence A002864/M0847 in "An On-Line Version of the Encyclopedia of Integer Sequences."

Thistlethwaite, M. "A Spanning Tree Expansion for the Jones Polynomial." *Topology* **26**, 297–309, 1987.

Alternating Knot Diagram

A KNOT DIAGRAM which has alternating under- and overcrossings as the KNOT projection is traversed. The first KNOT which does not have an alternating diagram has 8 crossings.

Alternating Link

A LINK which has a LINK DIAGRAM with alternating underpasses and overpasses.

see also ALMOST ALTERNATING LINK

References

Menasco, W. and Thistlethwaite, M. "The Classification of Alternating Links." *Ann. Math.* **138**, 113–171, 1993.

Alternating Permutation

An arrangement of the elements $c_1, \ldots, c_n$ such that no element c_i has a magnitude between c_{i-1} and c_{i+1} is called an alternating (or ZIGZAG) permutation. The determination of the number of alternating permutations for the set of the first n INTEGERS $\{1, 2, \ldots, n\}$ is known as ANDRÉ'S PROBLEM. An example of an alternating permutation is (1, 3, 2, 5, 4).

As many alternating permutations among n elements begin by rising as by falling. The magnitude of the c_ns does not matter; only the number of them. Let the number of alternating permutations be given by $Z_n = 2A_n$. This quantity can then be computed from

$$2na_n = \sum a_r a_s, \tag{1}$$

where r and s pass through all INTEGRAL numbers such that

$$r + s = n - 1, \tag{2}$$

$a_0 = a_1 = 1$, and

$$A_n = n!a_n. \tag{3}$$

The numbers A_n are sometimes called the EULER ZIGZAG NUMBERS, and the first few are given by 1, 1, 1, 2, 5, 16, 61, 272, ... (Sloane's A000111). The ODD-numbered A_ns are called EULER NUMBERS, SECANT NUMBERS, or ZIG NUMBERS, and the EVEN-numbered ones are sometimes called TANGENT NUMBERS or ZAG NUMBERS.

Curiously enough, the SECANT and TANGENT MACLAURIN SERIES can be written in terms of the A_ns as

$$\sec x = A_0 + A_2 \frac{x^2}{2!} + A_4 \frac{x^4}{4!} + \ldots \quad (4)$$

$$\tan x = A_1 x + A_3 \frac{x^3}{3!} + A_5 \frac{x^5}{5!} + \ldots, \quad (5)$$

or combining them,

$$\sec x + \tan x$$

$$= A_0 + A_1 x + A_2 \frac{x^2}{2!} + A_3 \frac{x^3}{3!} + A_4 \frac{x^4}{4!} + A_5 \frac{x^5}{5!} + \ldots. \quad (6)$$

see also ENTRINGER NUMBER, EULER NUMBER, EULER ZIGZAG NUMBER, SECANT NUMBER, SEIDEL-ENTRINGER-ARNOLD TRIANGLE, TANGENT NUMBER

References

André, D. "Developments de $\sec x$ et $\tan x$." *C. R. Acad. Sci. Paris* **88**, 965–967, 1879.

André, D. "Memoire sur le permutations alternées." *J. Math.* **7**, 167–184, 1881.

Arnold, V. I. "Bernoulli-Euler Updown Numbers Associated with Function Singularities, Their Combinatorics and Arithmetics." *Duke Math. J.* **63**, 537–555, 1991.

Arnold, V. I. "Snake Calculus and Combinatorics of Bernoulli, Euler, and Springer Numbers for Coxeter Groups." *Russian Math. Surveys* **47**, 3–45, 1992.

Bauslaugh, B. and Ruskey, F. "Generating Alternating Permutations Lexicographically." *BIT* **30**, 17–26, 1990.

Conway, J. H. and Guy, R. K. In *The Book of Numbers.* New York: Springer-Verlag, pp. 110–111, 1996.

Dörrie, H. "André's Deviation of the Secant and Tangent Series." §16 in *100 Great Problems of Elementary Mathematics: Their History and Solutions.* New York: Dover, pp. 64–69, 1965.

Honsberger, R. *Mathematical Gems III.* Washington, DC: Math. Assoc. Amer., pp. 69–75, 1985.

Knuth, D. E. and Buckholtz, T. J. "Computation of Tangent, Euler, and Bernoulli Numbers." *Math. Comput.* **21**, 663–688, 1967.

Millar, J.; Sloane, N. J. A.; and Young, N. E. "A New Operation on Sequences: The Boustrophedon Transform." *J. Combin. Th. Ser. A* **76**, 44–54, 1996.

Ruskey, F. "Information of Alternating Permutations." `http://sue.csc.uvic.ca/~cos/inf/perm/Alternating.html`.

Sloane, N. J. A. Sequence A000111/M1492 in "An On-Line Version of the Encyclopedia of Integer Sequences."

Alternating Series

A SERIES of the form

$$\sum_{k=1}^{\infty} (-1)^{k+1} a_k$$

or

$$\sum_{k=1}^{\infty} (-1)^k a_k.$$

see also SERIES

References

Arfken, G. "Alternating Series." §5.3 in *Mathematical Methods for Physicists, 3rd ed.* Orlando, FL: Academic Press, pp. 293–294, 1985.

Bromwich, T. J. I'a and MacRobert, T. M. "Alternating Series." §19 in *An Introduction to the Theory of Infinite Series, 3rd ed.* New York: Chelsea, pp. 55–57, 1991.

Pinsky, M. A. "Averaging an Alternating Series." *Math. Mag.* **51**, 235–237, 1978.

Alternating Series Test

Also known as the LEIBNIZ CRITERION. An ALTERNATING SERIES CONVERGES if $a_1 \geq a_2 \geq \ldots$ and

$$\lim_{k\to\infty} a_k = 0.$$

see also CONVERGENCE TESTS

Alternative Link

A category of LINK encompassing both ALTERNATING KNOTS and TORUS KNOTS.

see also ALTERNATING KNOT, LINK, TORUS KNOT

References

Kauffman, L. "Combinatorics and Knot Theory." *Contemp. Math.* **20**, 181–200, 1983.

Altitude

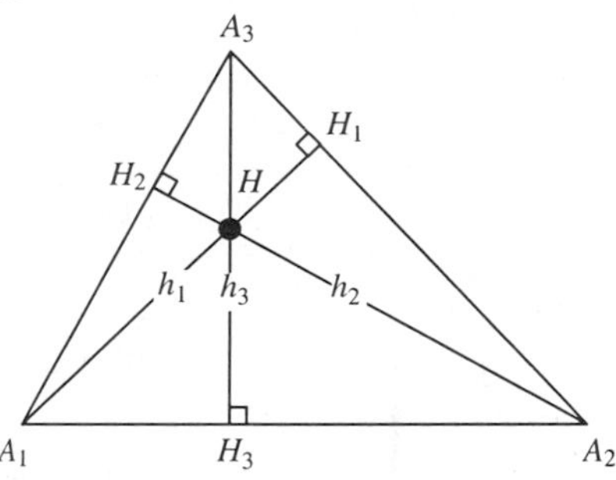

The altitudes of a TRIANGLE are the CEVIANS $A_i H_i$ which are PERPENDICULAR to the LEGS $A_j A_k$ opposite A_i. They have lengths $h_i \equiv \overline{A_i H_i}$ given by

$$h_i = a_{i+1} \sin \alpha_{i+2} = a_{i+2} \sin \alpha_{i+1} \quad (1)$$

$$h_1 = \frac{2\sqrt{s(s-a_1)(s-a_2)(s-a_3)}}{a_1}, \quad (2)$$

where s is the SEMIPERIMETER and $a_i \equiv \overline{A_j A_k}$. Another interesting FORMULA is

$$h_1 h_2 h_3 = 2s\Delta \quad (3)$$

(Johnson 1929, p. 191), where Δ is the AREA of the TRIANGLE. The three altitudes of any TRIANGLE are CONCURRENT at the ORTHOCENTER H. This fundamental fact did not appear anywhere in Euclid's *Elements*.

Other formulas satisfied by the altitude include

$$\frac{1}{h_1} + \frac{1}{h_2} + \frac{1}{h_3} = \frac{1}{r} \quad (4)$$

$$\frac{1}{r_1} = \frac{1}{h_2} + \frac{1}{h_3} - \frac{1}{h_1} \quad (5)$$

$$\frac{1}{r_2} + \frac{1}{r_3} = \frac{1}{r} - \frac{1}{r_1} = \frac{2}{h_1}, \quad (6)$$

where r is the INRADIUS and r_i are the EXRADII (Johnson 1929, p. 189). In addition,

$$HA_1 \cdot HH_1 = HA_2 \cdot HH_2 = HA_3 \cdot HH_3 \quad (7)$$

$$HA_1 \cdot HH_1 = \tfrac{1}{2}({a_1}^2 + {a_2}^2 + {a_3}^2) - 4R^2, \quad (8)$$

where R is the CIRCUMRADIUS.

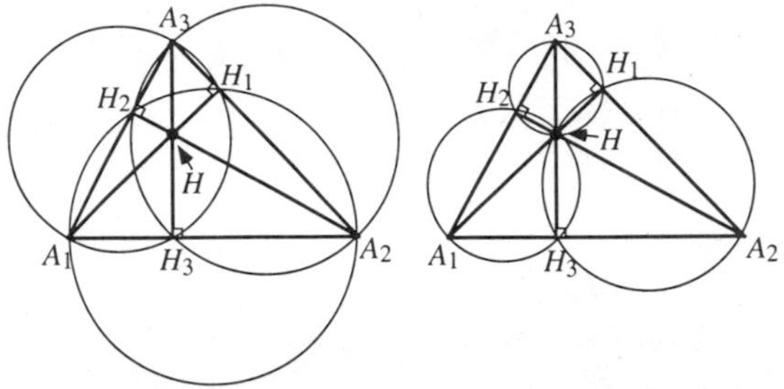

The points A_1, A_3, H_1, and H_3 (and their permutations with respect to indices) all lie on a CIRCLE, as do the points A_3, H_3, H, and H_1 (and their permutations with respect to indices). TRIANGLES $\Delta A_1A_2A_3$ and $\Delta A_1H_2H_3$ are inversely similar.

The triangle $H_1H_2H_3$ has the minimum PERIMETER of any TRIANGLE inscribed in a given ACUTE TRIANGLE (Johnson 1929, pp. 161–165). The PERIMETER of $\Delta H_1H_2H_3$ is $2\Delta/R$ (Johnson 1929, p. 191). Additional properties involving the FEET of the altitudes are given by Johnson (1929, pp. 261–262).

see also CEVIAN, FOOT, ORTHOCENTER, PERPENDICULAR, PERPENDICULAR FOOT

References
Coxeter, H. S. M. and Greitzer, S. L. *Geometry Revisited.* Washington, DC: Math. Assoc. Amer., pp. 9 and 36–40, 1967.
Johnson, R. A. *Modern Geometry: An Elementary Treatise on the Geometry of the Triangle and the Circle.* Boston, MA: Houghton Mifflin, 1929.

Alysoid

see CATENARY

Ambient Isotopy

An ambient isotopy from an embedding of a MANIFOLD M in N to another is a HOMOTOPY of self DIFFEOMORPHISMS (or ISOMORPHISMS, or piecewise-linear transformations, etc.) of N, starting at the IDENTITY MAP, such that the "last" DIFFEOMORPHISM compounded with the first embedding of M is the second embedding of M. In other words, an ambient isotopy is like an ISOTOPY except that instead of distorting the embedding, the whole ambient SPACE is being stretched and distorted and the embedding is just "coming along for the ride."

For SMOOTH MANIFOLDS, a MAP is ISOTOPIC IFF it is ambiently isotopic.

For KNOTS, the equivalence of MANIFOLDS under continuous deformation is independent of the embedding SPACE. KNOTS of opposite CHIRALITY have ambient isotopy, but not REGULAR ISOTOPY.

see also ISOTOPY, REGULAR ISOTOPY

References
Hirsch, M. W. *Differential Topology.* New York: Springer-Verlag, 1988.

Ambiguous

An expression is said to be ambiguous (or poorly defined) if its definition does not assign it a unique interpretation or value. An expression which is *not* ambiguous is said to be WELL-DEFINED.

see also WELL-DEFINED

Ambrose-Kakutani Theorem

For every ergodic FLOW on a nonatomic PROBABILITY SPACE, there is a MEASURABLE SET intersecting almost every orbit in a discrete set.

Amenable Number

A number n which can be built up from INTEGERS a_1, $a_2, \ldots, a_k$ by either ADDITION or MULTIPLICATION such that

$$\sum_{i=1}^{k} a_i = \prod_{i=1}^{k} a_i = n.$$

The numbers $\{a_1, \ldots, a_n\}$ in the SUM are simply a PARTITION of n. The first few amenable numbers are

$$\begin{aligned} 2+2 &= 2\times 2 = 4 \\ 1+2+3 &= 1\times 2\times 3 = 6 \\ 1+1+2+4 &= 1\times 1\times 2\times 4 = 8 \\ 1+1+2+2+2 &= 1\times 1\times 2\times 2\times 2 = 8. \end{aligned}$$

In fact, all COMPOSITE NUMBERS are amenable.

see also COMPOSITE NUMBER, PARTITION, SUM

References
Tamvakis, H. "Problem 10454." *Amer. Math. Monthly* **102**, 463, 1995.

Amicable Numbers

see AMICABLE PAIR, AMICABLE QUADRUPLE, AMICABLE TRIPLE, MULTIAMICABLE NUMBERS

Amicable Pair

An amicable pair consists of two INTEGERS m, n for which the sum of PROPER DIVISORS (the DIVISORS excluding the number itself) of one number equals the other. Amicable pairs are occasionally called FRIENDLY PAIRS, although this nomenclature is to be discouraged since FRIENDLY PAIRS are defined by a different, if related, criterion. Symbolically, amicable pairs satisfy

$$s(m) = n \tag{1}$$

$$s(n) = m, \tag{2}$$

where $s(n)$ is the RESTRICTED DIVISOR FUNCTION or, equivalently,

$$\sigma(m) = \sigma(n) = s(m) + s(n) = m + n, \tag{3}$$

where $\sigma(n)$ is the DIVISOR FUNCTION. The smallest amicable pair is (220, 284) which has factorizations

$$220 = 11 \cdot 5 \cdot 2^2 \tag{4}$$

$$284 = 71 \cdot 2^2 \tag{5}$$

giving RESTRICTED DIVISOR FUNCTIONS

$$s(220) = \sum\{1, 2, 4, 5, 10, 11, 20, 22, 44, 55, 110\} = 284 \tag{6}$$

$$s(284) = \sum\{1, 2, 4, 71, 142\} = 220. \tag{7}$$

The quantity

$$\sigma(m) = \sigma(n) = s(m) + s(n), \tag{8}$$

in this case, $220 + 284 = 504$, is called the PAIR SUM.

In 1636, Fermat found the pair (17296, 18416) and in 1638, Descartes found (9363584, 9437056). By 1747, Euler had found 30 pairs, a number which he later extended to 60. There were 390 known as of 1946 (Scott 1946). There are a total of 236 amicable pairs below 10^8 (Cohen 1970), 1427 below 10^{10} (te Riele 1986), 3340 less than 10^{11} (Moews and Moews 1993), [illegible] less than 2.01×10^{11} (Moews and Moews), and 5001 less than $\approx 3.06 \times 10^{11}$ (Moews and Moews).

The first few amicable pairs are (220, 284), (1184, 1210), (2620, 2924) (5020, 5564), (6232, 6368), (10744, 10856), (12285, 14595), (17296, 18416), (63020, 76084), ... (Sloane's A002025 and A002046). An exhaustive tabulation is maintained by D. Moews.

Let an amicable pair be denoted (m, n) with $m < n$. (m, n) is called a regular amicable pair of type (i, j) if

$$(m, n) = (gM, gN), \tag{9}$$

where $g \equiv \mathrm{GCD}(m, n)$ is the GREATEST COMMON DIVISOR,

$$\mathrm{GCD}(g, M) = \mathrm{GCD}(g, N) = 1, \tag{10}$$

M and N are SQUAREFREE, then the number of PRIME factors of M and N are i and j. Pairs which are not regular are called irregular or exotic (te Riele 1986). There are no regular pairs of type $(1, j)$ for $j \geq 1$. If $m \equiv 0 \pmod 6$ and

$$n = \sigma(m) - m \tag{11}$$

is EVEN, then (m, n) cannot be an amicable pair (Lee 1969). The minimal and maximal values of m/n found by te Riele (1986) were

$$938304290/1344480478 = 0.697893577\ldots \tag{12}$$

and

$$4000783984/4001351168 = 0.9998582519\ldots. \tag{13}$$

te Riele (1986) also found 37 pairs of amicable pairs having the same PAIR SUM. The first such pair is (609928, 686072) and (643336, 652664), which has the PAIR SUM

$$\sigma(m) = \sigma(n) = m + n = 1{,}296{,}000. \tag{14}$$

te Riele (1986) found no amicable n-tuples having the same PAIR SUM for $n > 2$. However, Moews and Moews found a triple in 1993, and te Riele found a quadruple in 1995. In November 1997, a quintuple and sextuple were discovered. The sextuple is (1953433861918, 2216492794082), (1968039941816, 2201886714184), (1981957651366, 2187969004634), (1993501042130, 2176425613870), (2046897812505, 2123028843495), (2068113162038, 2101813493962), all having PAIR SUM 4169926656000. Amazingly, the sextuple is smaller than any known quadruple or quintuple, and is likely smaller than any quintuple.

On October 4, 1997, Mariano Garcia found the largest known amicable pair, each of whose members has 4829 DIGITS. The new pair is

$$N_1 = CM[(P + Q)P^{89} - 1] \tag{15}$$

$$N_2 = CQ[(P - M)P^{89} - 1], \tag{16}$$

where

$$C = 2^{11}P^{89} \tag{17}$$

$$M = 287155430510003638403359267 \tag{18}$$

$$P = 574451143340278962374313859 \tag{19}$$

$$Q = 136272576607912041393307632916794623. \tag{20}$$

P, Q, $(P + Q)P^{89} - 1$, and $(P - M)P^{89} - 1$ are PRIME.

Pomerance (1981) has proved that

$$[\text{amicable numbers } \leq n] < ne^{-[\ln(n)]^{1/3}} \tag{21}$$

for large enough n (Guy 1994). No nonfinite lower bound has been proven.

see also AMICABLE QUADRUPLE, AMICABLE TRIPLE, AUGMENTED AMICABLE PAIR, BREEDER, CROWD, EULER'S RULE, FRIENDLY PAIR, MULTIAMICABLE NUMBERS, PAIR SUM, QUASIAMICABLE PAIR, SOCIABLE NUMBERS, UNITARY AMICABLE PAIR

References

Alanen, J.; Ore, Ø.; and Stemple, J. "Systematic Computations on Amicable Numbers." *Math. Comput.* **21**, 242–245, 1967.

Battiato, S. and Borho, W. "Are there Odd Amicable Numbers not Divisible by Three?" *Math. Comput.* **50**, 633–637, 1988.

Beeler, M.; Gosper, R. W.; and Schroeppel, R. Item 62 in *HAKMEM.* Cambridge, MA: MIT Artificial Intelligence Laboratory, Memo AIM-239, Feb. 1972.

Borho, W. and Hoffmann, H. "Breeding Amicable Numbers in Abundance." *Math. Comput.* **46**, 281–293, 1986.

Bratley, P.; Lunnon, F.; and McKay, J. "Amicable Numbers and Their Distribution." *Math. Comput.* **24**, 431–432, 1970.

Cohen, H. "On Amicable and Sociable Numbers." *Math. Comput.* **24**, 423–429, 1970.

Costello, P. "Amicable Pairs of Euler's First Form." *J. Rec. Math.* **10**, 183–189, 1977–1978.

Costello, P. "Amicable Pairs of the Form $(i,1)$." *Math. Comput.* **56**, 859–865, 1991.

Dickson, L. E. *History of the Theory of Numbers, Vol. 1: Divisibility and Primality.* New York: Chelsea, pp. 38–50, 1952.

Erdős, P. "On Amicable Numbers." *Publ. Math. Debrecen* **4**, 108–111, 1955–1956.

Erdős, P. "On Asymptotic Properties of Aliquot Sequences." *Math. Comput.* **30**, 641–645, 1976.

Gardner, M. "Perfect, Amicable, Sociable." Ch. 12 in *Mathematical Magic Show: More Puzzles, Games, Diversions, Illusions and Other Mathematical Sleight-of-Mind from Scientific American.* New York: Vintage, pp. 160–171, 1978.

Guy, R. K. "Amicable Numbers." §B4 in *Unsolved Problems in Number Theory, 2nd ed.* New York: Springer-Verlag, pp. 55–59, 1994.

Lee, E. J. "Amicable Numbers and the Bilinear Diophantine Equation." *Math. Comput.* **22**, 181–197, 1968.

Lee, E. J. "On Divisibility of the Sums of Even Amicable Pairs." *Math. Comput.* **23**, 545–548, 1969.

Lee, E. J. and Madachy, J. S. "The History and Discovery of Amicable Numbers, I." *J. Rec. Math.* **5**, 77–93, 1972.

Lee, E. J. and Madachy, J. S. "The History and Discovery of Amicable Numbers, II." *J. Rec. Math.* **5**, 153–173, 1972.

Lee, E. J. and Madachy, J. S. "The History and Discovery of Amicable Numbers, III." *J. Rec. Math.* **5**, 231–249, 1972.

Madachy, J. S. *Madachy's Mathematical Recreations.* New York: Dover, pp. 145 and 155–156, 1979.

Moews, D. and Moews, P. C. "A Search for Aliquot Cycles and Amicable Pairs." *Math. Comput.* **61**, 935–938, 1993.

Moews, D. and Moews, P. C. "A List of Amicable Pairs Below 2.01×10^{11}." Rev. Jan. 8, 1993. `http://xraysgi.ims.uconn.edu:8080/amicable.txt`.

Moews, D. and Moews, P. C. "A List of the First 5001 Amicable Pairs." Rev. Jan. 7, 1996. `http://xraysgi.ims.uconn.edu:8080/amicable2.txt`.

Ore, Ø. *Number Theory and Its History.* New York: Dover, pp. 96–100, 1988.

Pedersen, J. M. "Known Amicable Pairs." `http://www.vejlehs.dk/staff/jmp/aliquot/knwnap.htm`.

Pomerance, C. "On the Distribution of Amicable Numbers." *J. reine angew. Math.* **293/294**, 217–222, 1977.

Pomerance, C. "On the Distribution of Amicable Numbers, II." *J. reine angew. Math.* **325**, 182–188, 1981.

Scott, E. B. E. "Amicable Numbers." *Scripta Math.* **12**, 61–72, 1946.

Sloane, N. J. A. Sequences A002025/M5414 and A002046/M5435 in "An On-Line Version of the Encyclopedia of Integer Sequences."

te Riele, H. J. J. "On Generating New Amicable Pairs from Given Amicable Pairs." *Math. Comput.* **42**, 219–223, 1984.

te Riele, H. J. J. "Computation of All the Amicable Pairs Below 10^{10}." *Math. Comput.* **47**, 361–368 and S9–S35, 1986.

te Riele, H. J. J.; Borho, W.; Battiato, S.; Hoffmann, H.; and Lee, E. J. "Table of Amicable Pairs Between 10^10 and 10^{52}." Centrum voor Wiskunde en Informatica, Note NM-N8603. Amsterdam: Stichting Math. Centrum, 1986.

te Riele, H. J. J. "A New Method for Finding Amicable Pairs." In *Mathematics of Computation 1943–1993: A Half-Century of Computational Mathematics (Vancouver, BC, August 9–13, 1993)* (Ed. W. Gautschi). Providence, RI: Amer. Math. Soc., pp. 577–581, 1994.

Weisstein, E. W. "Sociable and Amicable Numbers." `http://www.astro.virginia.edu/~eww6n/math/notebooks/Sociable.m`.

Amicable Quadruple

An amicable quadruple as a QUADRUPLE (a, b, c, d) such that

$$\sigma(a) = \sigma(b) = \sigma(c) = \sigma(d) = a + b + c + d,$$

where $\sigma(n)$ is the DIVISOR FUNCTION.

References

Guy, R. K. *Unsolved Problems in Number Theory, 2nd ed.* New York: Springer-Verlag, p. 59, 1994.

Amicable Triple

Dickson (1913, 1952) defined an amicable triple to be a TRIPLE of three numbers (l, m, n) such that

$$\begin{aligned} s(l) &= m + n \\ s(m) &= l + n \\ s(n) &= l + m, \end{aligned}$$

where $s(n)$ is the RESTRICTED DIVISOR FUNCTION (Madachy 1979). Dickson (1913, 1952) found eight sets of amicable triples with two equal numbers, and two sets with distinct numbers. The latter are (123228768, 103340640, 124015008), for which

$$\begin{aligned} s(12322876) &= 103340640 + 124015008 = 227355648 \\ s(103340640) &= 123228768 + 124015008 = 24724377 \\ s(124015008) &= 123228768 + 10334064 = 226569408, \end{aligned}$$

and (1945330728960, 2324196638720, 2615631953920), for which

$$\begin{aligned} s(1945330728960) &= 2324196638720 + 2615631953920 \\ &= 4939828592640 \\ s(2324196638720) &= 1945330728960 + 2615631953920 \\ &= 4560962682880 \\ s(2615631953920) &= 1945330728960 + 2324196638720 \\ &= 4269527367680. \end{aligned}$$

A second definition (Guy 1994) defines an amicable triple as a TRIPLE (a, b, c) such that

$$\sigma(a) = \sigma(b) = \sigma(c) = a + b + c,$$

where $\sigma(n)$ is the DIVISOR FUNCTION. An example is $(2^2 3^2 5 \cdot 11,\ 2^5 3^2 7,\ 2^2 3^2 71)$.

see also AMICABLE PAIR, AMICABLE QUADRUPLE

References

Dickson, L. E. "Amicable Number Triples." *Amer. Math. Monthly* **20**, 84–92, 1913.

Dickson, L. E. *History of the Theory of Numbers, Vol. 1: Divisibility and Primality.* New York: Chelsea, p. 50, 1952.

Guy, R. K. *Unsolved Problems in Number Theory, 2nd ed.* New York: Springer-Verlag, p. 59, 1994.

Madachy, J. S. *Madachy's Mathematical Recreations.* New York: Dover, p. 156, 1979.

Mason, T. E. "On Amicable Numbers and Their Generalizations." *Amer. Math. Monthly* **28**, 195–200, 1921.

Weisstein, E. W. "Sociable and Amicable Numbers." http://www.astro.virginia.edu/~eww6n/math/notebooks/Sociable.m.

Amortization

The payment of a debt plus accrued INTEREST by regular payments.

Ampersand Curve

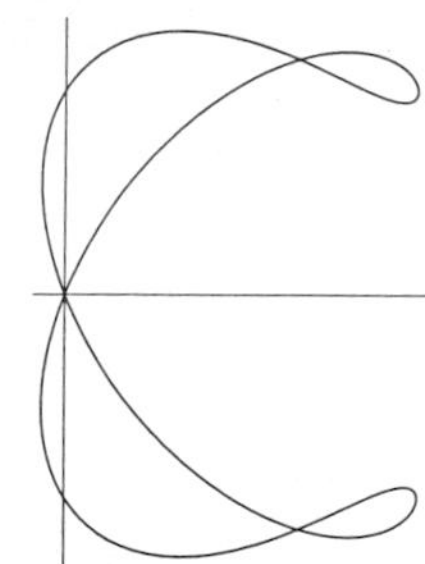

The PLANE CURVE with Cartesian equation

$$(y^2 - x^2)(x - 1)(2x - 3) = 4(x^2 + y^2 - 2x)^2.$$

References

Cundy, H. and Rollett, A. *Mathematical Models, 3rd ed.* Stradbroke, England: Tarquin Pub., p. 72, 1989.

Amphichiral

An object is amphichiral (also called REFLEXIBLE) if it is superposable with its MIRROR IMAGE (i.e., its image in a plane mirror).

see also AMPHICHIRAL KNOT, CHIRAL, DISYMMETRIC, HANDEDNESS, MIRROR IMAGE

Amphichiral Knot

An amphichiral knot is a KNOT which is capable of being continuously deformed into its own MIRROR IMAGE. The amphichiral knots having ten or fewer crossings are 04_{001} (FIGURE-OF-EIGHT KNOT), 06_{003}, 08_{003}, 08_{009}, 08_{012}, 08_{017}, 08_{018}, 10_{017}, 10_{033}, 10_{037}, 10_{043}, 10_{045}, 10_{079}, 10_{081}, 10_{088}, 10_{099}, 10_{109}, 10_{115}, 10_{118}, and 10_{123} (Jones 1985). The HOMFLY POLYNOMIAL is good at identifying amphichiral knots, but sometimes fails to identify knots which are not. No complete invariant (an invariant which always definitively determines if a KNOT is AMPHICHIRAL) is known.

Let b_+ be the SUM of POSITIVE exponents, and b_- the SUM of NEGATIVE exponents in the BRAID GROUP B_n. If

$$b_+ - 3b_- - n + 1 > 0,$$

then the KNOT corresponding to the closed BRAID b is not amphichiral (Jones 1985).

see also AMPHICHIRAL, BRAID GROUP, INVERTIBLE KNOT, MIRROR IMAGE

References

Burde, G. and Zieschang, H. *Knots.* Berlin: de Gruyter, pp. 311–319, 1985.

Jones, V. "A Polynomial Invariant for Knots via von Neumann Algebras." *Bull. Amer. Math. Soc.* **12**, 103–111, 1985.

Jones, V. "Hecke Algebra Representations of Braid Groups and Link Polynomials." *Ann. Math.* **126**, 335–388, 1987.

Amplitude

The variable ϕ used in ELLIPTIC FUNCTIONS and ELLIPTIC INTEGRALS, which can be defined by

$$\phi = \operatorname{am} u \equiv \int \operatorname{dn} u \, du,$$

where $\operatorname{dn}(u)$ is a JACOBI ELLIPTIC FUNCTION. The term "amplitude" is also used to refer to the maximum offset of a function from its baseline level.

see also ARGUMENT (ELLIPTIC INTEGRAL), CHARACTERISTIC (ELLIPTIC INTEGRAL), DELTA AMPLITUDE, ELLIPTIC FUNCTION, ELLIPTIC INTEGRAL, JACOBI ELLIPTIC FUNCTIONS, MODULAR ANGLE, MODULUS (ELLIPTIC INTEGRAL), NOME, PARAMETER

References

Abramowitz, M. and Stegun, C. A. (Eds.). *Handbook of Mathematical Functions with Formulas, Graphs, and Mathematical Tables, 9th printing.* New York: Dover, p. 590, 1972.

Fischer, G. (Ed.). Plate 132 in *Mathematische Modelle/Mathematical Models, Bildband/Photograph Volume.* Braunschweig, Germany: Vieweg, p. 129, 1986.

Anallagmatic Curve

A curve which is invariant under INVERSION. Examples include the CARDIOID, CARTESIAN OVALS, CASSINI OVALS, LIMAÇON, STROPHOID, and MACLAURIN TRISECTRIX.

Anallagmatic Pavement

see HADAMARD MATRIX

Analogy

Inference of the TRUTH of an unknown result obtained by noting its similarity to a result already known to be TRUE. In the hands of a skilled mathematician, analogy can be a very powerful tool for suggesting new and extending old results. However, subtleties can render results obtained by analogy incorrect, so rigorous PROOF is still needed.

see also INDUCTION

Analysis

The study of how continuous mathematical structures (FUNCTIONS) vary around the NEIGHBORHOOD of a point on a SURFACE. Analysis includes CALCULUS, DIFFERENTIAL EQUATIONS, etc.

see also ANALYSIS SITUS, CALCULUS, COMPLEX ANALYSIS, FUNCTIONAL ANALYSIS, NONSTANDARD ANALYSIS, REAL ANALYSIS

References
Bottazzini, U. *The "Higher Calculus": A History of Real and Complex Analysis from Euler to Weierstraß.* New York: Springer-Verlag, 1986.
Bressoud, D. M. *A Radical Approach to Real Analysis.* Washington, DC: Math. Assoc. Amer., 1994.
Ehrlich, P. *Real Numbers, Generalization of the Reals, & Theories of Continua.* Norwell, MA: Kluwer, 1994.
Hairer, E. and Wanner, G. *Analysis by Its History.* New York: Springer-Verlag, 1996.
Royden, H. L. *Real Analysis, 3rd ed.* New York: Macmillan, 1988.
Wheeden, R. L. and Zygmund, A. *Measure and Integral: An Introduction to Real Analysis.* New York: Dekker, 1977.
Whittaker, E. T. and Watson, G. N. *A Course in Modern Analysis, 4th ed.* Cambridge, England: Cambridge University Press, 1990.

Analysis Situs

An archaic name for TOPOLOGY.

Analytic Continuation

A process of extending the region in which a COMPLEX FUNCTION is defined.

see also MONODROMY THEOREM, PERMANENCE OF ALGEBRAIC FORM, PERMANENCE OF MATHEMATICAL RELATIONS PRINCIPLE

References
Arfken, G. *Mathematical Methods for Physicists, 3rd ed.* Orlando, FL: Academic Press, pp. 378–380, 1985.
Morse, P. M. and Feshbach, H. *Methods of Theoretical Physics, Part I.* New York: McGraw-Hill, pp. 389–390 and 392–398, 1953.

Analytic Function

A FUNCTION in the COMPLEX NUMBERS $\mathbb{C}$ is analytic on a region R if it is COMPLEX DIFFERENTIABLE at every point in R. The terms HOLOMORPHIC FUNCTION and REGULAR FUNCTION are sometimes used interchangeably with "analytic function." If a FUNCTION is analytic, it is infinitely DIFFERENTIABLE.

see also BERGMAN SPACE, COMPLEX DIFFERENTIABLE, DIFFERENTIABLE, PSEUDOANALYTIC FUNCTION, SEMIANALYTIC, SUBANALYTIC

References
Morse, P. M. and Feshbach, H. "Analytic Functions." §4.2 in *Methods of Theoretical Physics, Part I.* New York: McGraw-Hill, pp. 356–374, 1953.

Analytic Geometry

The study of the GEOMETRY of figures by algebraic representation and manipulation of equations describing their positions, configurations, and separations. Analytic geometry is also called COORDINATE GEOMETRY since the objects are described as n-tuples of points (where $n = 2$ in the PLANE and 3 in SPACE) in some COORDINATE SYSTEM.

see also ARGAND DIAGRAM, CARTESIAN COORDINATES, COMPLEX PLANE, GEOMETRY, PLANE, QUADRANT, SPACE, x-AXIS, y-AXIS, z-AXIS

References
Courant, R. and Robbins, H. "Remarks on Analytic Geometry." §2.3 in *What is Mathematics?: An Elementary Approach to Ideas and Methods, 2nd ed.* Oxford, England: Oxford University Press, pp. 72–77, 1996.

Analytic Set

A DEFINABLE SET, also called a SOUSLIN SET.

see also COANALYTIC SET, SOUSLIN SET

Anarboricity

Given a GRAPH G, the anarboricity is the maximum number of line-disjoint nonacyclic SUBGRAPHS whose UNION is G.

see also ARBORICITY

Anchor

An anchor is the BUNDLE MAP ρ from a VECTOR BUNDLE A to the TANGENT BUNDLE TB satisfying

1. $[\rho(X), \rho(Y)] = \rho([X, Y])$ and
2. $[X, \phi Y] = \phi[X, Y] + (\rho(X) \cdot \phi)Y$,

where X and Y are smooth sections of A, ϕ is a smooth function of B, and the bracket is the "Jacobi-Lie bracket" of a VECTOR FIELD.

see also LIE ALGEBROID

References
Weinstein, A. "Groupoids: Unifying Internal and External Symmetry." *Not. Amer. Math. Soc.* **43**, 744–752, 1996.

Anchor Ring

An archaic name for the TORUS.

References

Eisenhart, L. P. *A Treatise on the Differential Geometry of Curves and Surfaces.* New York: Dover, p. 314, 1960.

Stacey, F. D. *Physics of the Earth, 2nd ed.* New York: Wiley, p. 239, 1977.

Whittaker, E. T. *A Treatise on the Analytical Dynamics of Particles & Rigid Bodies, 4th ed.* Cambridge, England: Cambridge University Press, p. 21, 1959.

And

A term (PREDICATE) in LOGIC which yields TRUE if one or more conditions are TRUE, and FALSE if any condition is FALSE. A AND B is denoted $A\&B$, $A \wedge B$, or simply AB. The BINARY AND operator has the following TRUTH TABLE:

A	B	$A \wedge B$
F	F	F
F	T	F
T	F	F
T	T	T

A PRODUCT of ANDs (the AND of n conditions) is called a CONJUNCTION, and is denoted

$$\bigwedge_{k=1}^{n} A_k.$$

Two binary numbers can have the operation AND performed bitwise with 1 representing TRUE and 0 FALSE. Some computer languages denote this operation on A, B, and C as `A&&B&&C` or `logand(A,B,C)`.

see also BINARY OPERATOR, INTERSECTION, NOT, OR, PREDICATE, TRUTH TABLE, XOR

Anderson-Darling Statistic

A statistic defined to improve the KOLMOGOROV-SMIRNOV TEST in the TAIL of a distribution.

see also KOLMOGOROV-SMIRNOV TEST, KUIPER STATISTIC

References

Press, W. H.; Flannery, B. P.; Teukolsky, S. A.; and Vetterling, W. T. *Numerical Recipes in FORTRAN: The Art of Scientific Computing, 2nd ed.* Cambridge, England: Cambridge University Press, p. 621, 1992.

André's Problem

The determination of the number of ALTERNATING PERMUTATIONS having elements $\{1, 2, \ldots, n\}$

see also ALTERNATING PERMUTATION

André's Reflection Method

A technique used by André (1887) to provide an elegant solution to the BALLOT PROBLEM (Hilton and Pederson 1991).

References

André, D. "Solution directe du problème résolu par M. Bertrand." *Comptes Rendus Acad. Sci. Paris* **105**, 436–437, 1887.

Comtet, L. *Advanced Combinatorics.* Dordrecht, Netherlands: Reidel, p. 22, 1974.

Hilton, P. and Pederson, J. "Catalan Numbers, Their Generalization, and Their Uses." *Math. Intel.* **13**, 64–75, 1991.

Vardi, I. *Computational Recreations in Mathematica.* Reading, MA: Addison-Wesley, p. 185, 1991.

Andrew's Sine

The function

$$\psi(z) = \begin{cases} \sin\left(\frac{z}{c}\right) & |z| < c\pi \\ 0, & |z| > c\pi \end{cases}$$

which occurs in estimation theory.

see also SINE

References

Press, W. H.; Flannery, B. P.; Teukolsky, S. A.; and Vetterling, W. T. *Numerical Recipes in FORTRAN: The Art of Scientific Computing, 2nd ed.* Cambridge, England: Cambridge University Press, p. 697, 1992.

Andrews Cube

see SEMIPERFECT MAGIC CUBE

Andrews-Curtis Link

The LINK of 2-spheres in $\mathbb{R}^4$ obtained by SPINNING intertwined arcs. The link consists of a knotted 2-sphere and a SPUN TREFOIL KNOT.

see also SPUN KNOT, TREFOIL KNOT

References

Rolfsen, D. *Knots and Links.* Wilmington, DE: Publish or Perish Press, p. 94, 1976.

Andrews-Schur Identity

$$\sum_{k=0}^{n} q^{k^2+ak} \begin{bmatrix} 2n-k+a \\ k \end{bmatrix} = \sum_{k=-\infty}^{\infty} q^{10k^2+(4a-1)k} \begin{bmatrix} 2n+2a+2 \\ n-5k \end{bmatrix} \frac{[10k+2a+2]}{[2n+2a+2]}, \tag{1}$$

where $[x]$ is a GAUSSIAN POLYNOMIAL. It is a POLYNOMIAL identity for $a = 0, 1$ which implies the ROGERS-RAMANUJAN IDENTITIES by taking $n \to \infty$ and applying the JACOBI TRIPLE PRODUCT identity. A variant of this equation is

$$\sum_{k=-\lfloor a/2 \rfloor}^{n} q^{k^2+2ak} \begin{bmatrix} n+k+a \\ n-k \end{bmatrix} = \sum_{-\lfloor (n+2a+2)/5 \rfloor}^{\lfloor n/5 \rfloor} q^{15k^2+(6a+1)k} \begin{bmatrix} 2n+2a+2 \\ 5-5k \end{bmatrix} \times \frac{[10k+2a+2]}{[2n+2a+2]}, \quad (2)$$

where the symbol $\lfloor x \rfloor$ in the SUM limits is the FLOOR FUNCTION (Paule 1994). The RECIPROCAL of the identity is

$$\sum_{k=0}^{\infty} \frac{q^{k^2+2ak}}{(q;q)_{2k+a}} = \prod_{j=0}^{\infty} \frac{1}{(1-q^{2j+1})(1-q^{20j+4a+4})(1-q^{20j-4a+16})} \quad (3)$$

for $a = 0, 1$ (Paule 1994). For $q = 1$, (1) and (2) become

$$\sum_{-\lfloor a/2 \rfloor}^{n} \binom{n+k+a}{n-k} = \sum_{-\lfloor (n+2a+2)/5 \rfloor}^{\lfloor n/5 \rfloor} \binom{2n+2a+2}{n-5k} \frac{5k+a+1}{n+a+1}. \quad (4)$$

References

Andrews, G. E. "A Polynomial Identity which Implies the Rogers-Ramanujan Identities." *Scripta Math.* **28**, 297–305, 1970.

Paule, P. "Short and Easy Computer Proofs of the Rogers-Ramanujan Identities and of Identities of Similar Type." *Electronic J. Combinatorics* **1**, R10, 1–9, 1994. http://www.combinatorics.org/Volume_1/volume1.html#R10.

Andrica's Conjecture

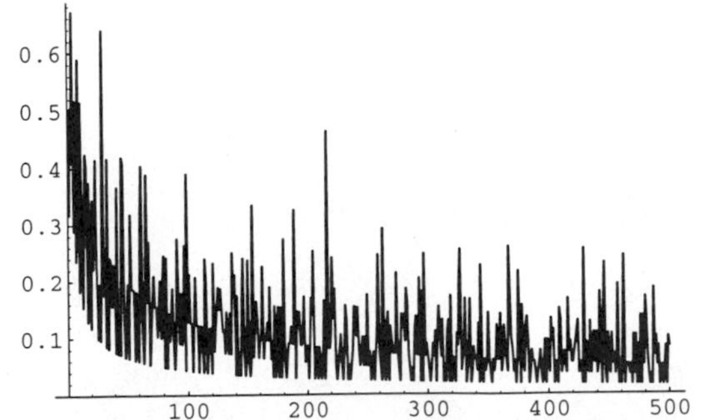

Andrica's conjecture states that, for p_n the nth PRIME NUMBER, the INEQUALITY

$$A_n \equiv \sqrt{p_{n+1}} - \sqrt{p_n} < 1$$

holds, where the discrete function A_n is plotted above. The largest value among the first 1000 PRIMES is for $n = 4$, giving $\sqrt{11} - \sqrt{7} \approx 0.670873$. Since the Andrica function falls asymptotically as n increases so a PRIME GAP of increasing size is needed at large n, it seems likely the CONJECTURE is true. However, it has not yet been proven.

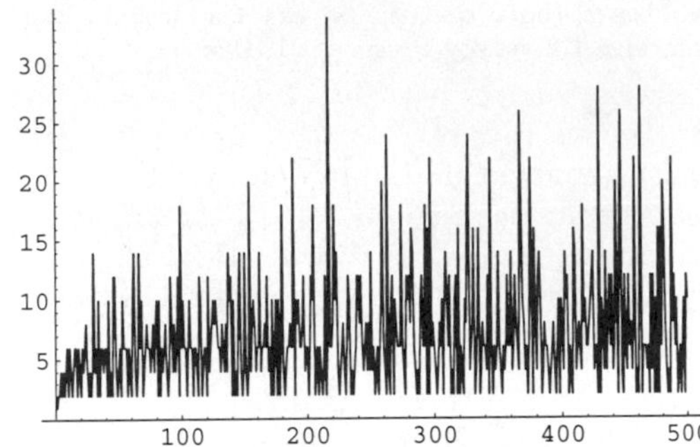

A_n bears a strong resemblance to the PRIME DIFFERENCE FUNCTION, plotted above, the first few values of which are 1, 2, 2, 4, 2, 4, 2, 4, 6, 2, 6, ... (Sloane's A001223).

see also BROCARD'S CONJECTURE, GOOD PRIME, FORTUNATE PRIME, PÓLYA CONJECTURE, PRIME DIFFERENCE FUNCTION, TWIN PEAKS

References

Golomb, S. W. "Problem E2506: Limits of Differences of Square Roots." *Amer. Math. Monthly* **83**, 60–61, 1976.

Guy, R. K. *Unsolved Problems in Number Theory, 2nd ed.* New York: Springer-Verlag, p. 21, 1994.

Rivera, C. "Problems & Puzzles (Conjectures): Andrica's Conjecture." http://www.sci.net.mx/~crivera/ppp/conj_008.htm.

Sloane, N. J. A. Sequence A001223/M0296 in "An On-Line Version of the Encyclopedia of Integer Sequences."

Anger Function

A generalization of the BESSEL FUNCTION OF THE FIRST KIND defined by

$$\mathcal{J}_\nu(z) \equiv \frac{1}{\pi} \int_0^\pi \cos(\nu\theta - z\sin\theta)\, d\theta.$$

If ν is an INTEGER n, then $\mathcal{J}_n(z) = J_n(z)$, where $J_n(z)$ is a BESSEL FUNCTION OF THE FIRST KIND. Anger's original function had an upper limit of 2π, but the current NOTATION was standardized by Watson (1966).

see also BESSEL FUNCTION, MODIFIED STRUVE FUNCTION, PARABOLIC CYLINDER FUNCTION, STRUVE FUNCTION, WEBER FUNCTIONS

References

Abramowitz, M. and Stegun, C. A. (Eds.). "Anger and Weber Functions." §12.3 in *Handbook of Mathematical Functions with Formulas, Graphs, and Mathematical Tables, 9th printing.* New York: Dover, pp. 498–499, 1972.

Watson, G. N. *A Treatise on the Theory of Bessel Functions, 2nd ed.* Cambridge, England: Cambridge University Press, 1966.

Angle

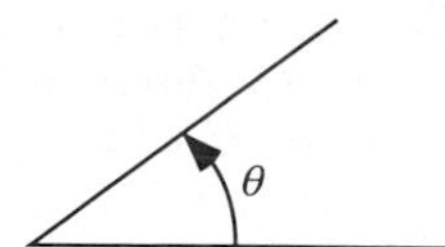

Given two intersecting LINES or LINE SEGMENTS, the amount of ROTATION about the point of intersection (the VERTEX) required to bring one into correspondence with the other is called the angle θ between them. Angles are usually measured in DEGREES (denoted °), RADIANS (denoted rad, or without a unit), or sometimes GRADIANS (denoted grad).

One full rotation in these three measures corresponds to 360°, 2π rad, or 400 grad. Half a full ROTATION is called a STRAIGHT ANGLE, and a QUARTER of a full rotation is called a RIGHT ANGLE. An angle less than a RIGHT ANGLE is called an ACUTE ANGLE, and an angle greater than a RIGHT ANGLE is called an OBTUSE ANGLE.

The use of DEGREES to measure angles harks back to the Babylonians, whose SEXAGESIMAL number system was based on the number 60. 360° likely arises from the Babylonian year, which was composed of 360 days (12 months of 30 days each). The DEGREE is further divided into 60 ARC MINUTES, and an ARC MINUTE into 60 ARC SECONDS. A more natural measure of an angle is the RADIAN. It has the property that the ARC LENGTH around a CIRCLE is simply given by the radian angle measure times the CIRCLE RADIUS. The RADIAN is also the most useful angle measure in CALCULUS because the DERIVATIVE of TRIGONOMETRIC functions such as

$$\frac{d}{dx}\sin x = \cos x$$

does not require the insertion of multiplicative constants like $\pi/180$. GRADIANS are sometimes used in surveying (they have the nice property that a RIGHT ANGLE is exactly 100 GRADIANS), but are encountered infrequently, if at all, in mathematics.

The concept of an angle can be generalized from the CIRCLE to the SPHERE. The fraction of a SPHERE subtended by an object is measured in STERADIANS, with the entire SPHERE corresponding to 4π STERADIANS.

A ruled SEMICIRCLE used for measuring and drawing angles is called a PROTRACTOR. A COMPASS can also be used to draw circular ARCS of some angular extent.

see also ACUTE ANGLE, ARC MINUTE, ARC SECOND, CENTRAL ANGLE, COMPLEMENTARY ANGLE, DEGREE, DIHEDRAL ANGLE, DIRECTED ANGLE, EULER ANGLES, GRADIAN, HORN ANGLE, INSCRIBED ANGLE, OBLIQUE ANGLE, OBTUSE ANGLE, PERIGON, PROTRACTOR, RADIAN, RIGHT ANGLE, SOLID ANGLE, STERADIAN, STRAIGHT ANGLE, SUBTEND, SUPPLEMENTARY ANGLE, VERTEX ANGLE

References

Dixon, R. *Mathographics.* New York: Dover, pp. 99–100, 1991.

Angle Bisector

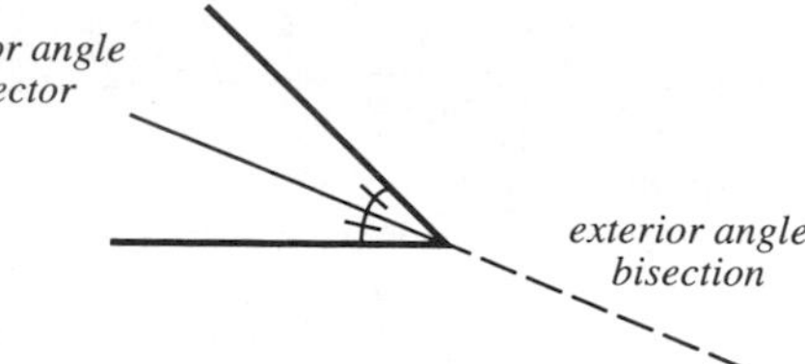

The (interior) bisector of an ANGLE is the LINE or LINE SEGMENT which cuts it into two equal ANGLES on the same "side" as the ANGLE.

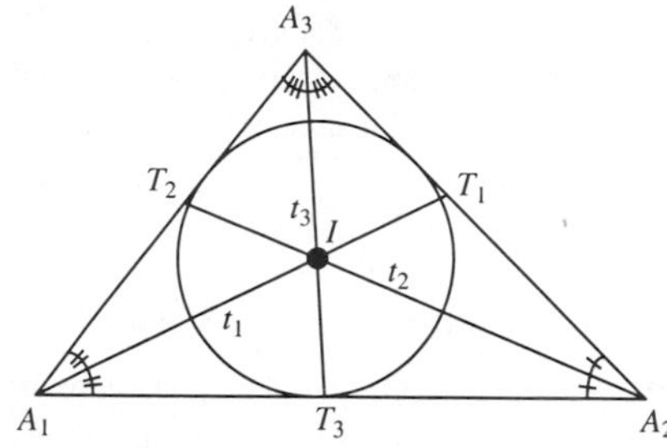

The length of the bisector of ANGLE A_1 in the above TRIANGLE $\Delta A_1A_2A_3$ is given by

$$t_1{}^2 = a_2a_3\left[1 - \frac{a_1{}^2}{(a_2+a_3)^2}\right],$$

where $t_i \equiv \overline{A_iT_i}$ and $a_i \equiv \overline{A_jA_k}$. The angle bisectors meet at the INCENTER I, which has TRILINEAR COORDINATES 1:1:1.

see also ANGLE BISECTOR THEOREM, CYCLIC QUADRANGLE, EXTERIOR ANGLE BISECTOR, ISODYNAMIC POINTS, ORTHOCENTRIC SYSTEM, STEINER-LEHMUS THEOREM, TRISECTION

References

Coxeter, H. S. M. and Greitzer, S. L. *Geometry Revisited.* Washington, DC: Math. Assoc. Amer., pp. 9–10, 1967.

Dixon, R. *Mathographics.* New York: Dover, p. 19, 1991.

Mackay, J. S. "Properties Concerned with the Angular Bisectors of a Triangle." *Proc. Edinburgh Math. Soc.* **13**, 37–102, 1895.

Angle Bisector Theorem

The ANGLE BISECTOR of an ANGLE in a TRIANGLE divides the opposite side in the same RATIO as the sides adjacent to the ANGLE.

Angle Bracket

The combination of a BRA and KET (bra+ket = bracket) which represents the INNER PRODUCT of two functions or vectors,

$$\langle f|g\rangle = \int f(x)g(x)\,dx$$

$$\langle \mathbf{v}|\mathbf{w}\rangle = \mathbf{v}\cdot\mathbf{w}.$$

By itself, the BRA is a COVARIANT 1-VECTOR, and the KET is a COVARIANT ONE-FORM. These terms are commonly used in quantum mechanics.

see also BRA, DIFFERENTIAL k-FORM, KET, ONE-FORM

Angle of Parallelism

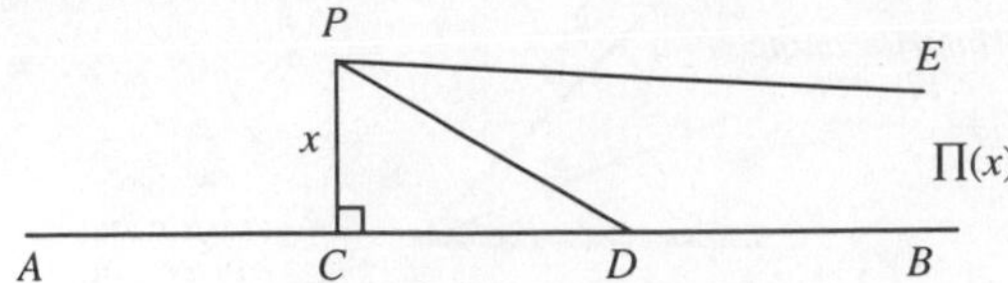

Given a point P and a LINE AB, draw the PERPENDICULAR through P and call it PC. Let PD be any other line from P which meets CB in D. In a HYPERBOLIC GEOMETRY, as D moves off to infinity along CB, then the line PD approaches the limiting line PE, which is said to be parallel to CB at P. The angle $\angle CPE$ which PE makes with PC is then called the angle of parallelism for perpendicular distance x, and is given by

$$\Pi(x) = 2\tan^{-1}(e^{-x}).$$

This is known as LOBACHEVSKY'S FORMULA.

see also HYPERBOLIC GEOMETRY, LOBACHEVSKY'S FORMULA

References

Manning, H. P. *Introductory Non-Euclidean Geometry.* New York: Dover, pp. 31–32 and 58, 1963.

Angle Trisection

see TRISECTION

Angular Acceleration

The angular acceleration $\boldsymbol{\alpha}$ is defined as the time DERIVATIVE of the ANGULAR VELOCITY $\boldsymbol{\omega}$,

$$\boldsymbol{\alpha} \equiv \frac{d\boldsymbol{\omega}}{dt} = \frac{d^2\theta}{dt^2}\hat{\mathbf{z}} = \frac{\mathbf{a}}{r}.$$

see also ACCELERATION, ANGULAR DISTANCE, ANGULAR VELOCITY

Angular Defect

The DIFFERENCE between the SUM of face ANGLES A_i at a VERTEX of a POLYHEDRON and 2π,

$$\delta = 2\pi - \sum_i A_i.$$

see also DESCARTES TOTAL ANGULAR DEFECT, JUMP ANGLE

Angular Distance

The angular distance traveled around a CIRCLE is the number of RADIANS the path subtends,

$$\theta \equiv \frac{\ell}{2\pi r}2\pi = \frac{\ell}{r}.$$

see also ANGULAR ACCELERATION, ANGULAR VELOCITY

Angular Velocity

The angular velocity $\boldsymbol{\omega}$ is the time DERIVATIVE of the ANGULAR DISTANCE θ with direction $\hat{\mathbf{z}}$ PERPENDICULAR to the plane of angular motion,

$$\boldsymbol{\omega} \equiv \frac{d\theta}{dt}\hat{\mathbf{z}} = \frac{\mathbf{v}}{r}.$$

see also ANGULAR ACCELERATION, ANGULAR DISTANCE

Anharmonic Ratio

see CROSS-RATIO

Anisohedral Tiling

A k-anisohedral tiling is a tiling which permits no n-ISOHEDRAL TILING with $n < k$.

References

Berglund, J. "Is There a k-Anisohedral Tile for $k \geq 5$?" *Amer. Math. Monthly* **100**, 585-588, 1993.

Klee, V. and Wagon, S. *Old and New Unsolved Problems in Plane Geometry and Number Theory.* Washington, DC: Math. Assoc. Amer., 1991.

Annihilator

The term annihilator is used in several different ways in various aspects of mathematics. It is most commonly used to mean the SET of all functions satisfying a given set of conditions which is zero on every member of a given SET.

Annulus

The region in common to two concentric CIRCLES of RADII a and b. The AREA of an annulus is

$$A_{\text{annulus}} = \pi(b^2 - a^2).$$

An interesting identity is as follows. In the figure,

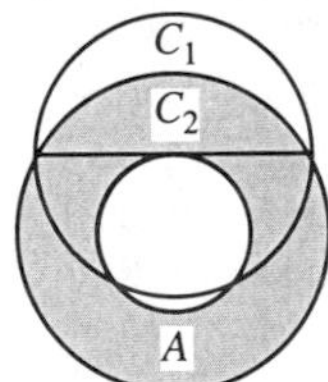

the AREA of the shaded region A is given by

$$A = C_1 + C_2.$$

see also CHORD, CIRCLE, CONCENTRIC CIRCLES, LUNE (PLANE), SPHERICAL SHELL

References

Pappas, T. "The Amazing Trick." *The Joy of Mathematics.* San Carlos, CA: Wide World Publ./Tetra, p. 69, 1989.

Annulus Conjecture

see ANNULUS THEOREM

Annulus Theorem

Let K_1^n and K_2^n be disjoint bicollared knots in $\mathbb{R}^{n+1}$ or $\mathbb{S}^{n+1}$ and let U denote the open region between them. Then the closure of U is a closed annulus $\mathbb{S}^n \times [0,1]$. Except for the case $n = 3$, the theorem was proved by Kirby (1969).

References

Kirby, R. C. "Stable Homeomorphisms and the Annulus Conjecture." *Ann. Math.* **89**, 575–582, 1969.

Rolfsen, D. *Knots and Links.* Wilmington, DE: Publish or Perish Press, p. 38, 1976.

Anomalous Cancellation

The simplification of a FRACTION a/b which gives a correct answer by "canceling" DIGITS of a and b. There are only four such cases for NUMERATOR and DENOMINATORS of two DIGITS in base 10: $64/16 = 4/1 = 4$, $98/49 = 8/4 = 2$, $95/19 = 5/1 = 5$, and $65/26 = 5/2$ (Boas 1979).

The concept of anomalous cancellation can be extended to arbitrary bases. PRIME bases have no solutions, but there is a solution corresponding to each PROPER DIVISOR of a COMPOSITE b. When $b-1$ is PRIME, this type of solution is the only one. For base 4, for example, the only solution is $32_4/13_4 = 2_4$. Boas gives a table of solutions for $b \leq 39$. The number of solutions is EVEN unless b is an EVEN SQUARE.

b	N	b	N
4	1	26	4
6	2	27	6
8	2	28	10
9	2	30	6
10	4	32	4
12	4	34	6
14	2	35	6
15	6	36	21
16	7	38	2
18	4	39	6
20	4		
21	10		
22	6		
24	6		

see also FRACTION, PRINTER'S ERRORS, REDUCED FRACTION

References

Boas, R. P. "Anomalous Cancellation." Ch. 6 in *Mathematical Plums* (Ed. R. Honsberger). Washington, DC: Math. Assoc. Amer., pp. 113–129, 1979.

Ogilvy, C. S. and Anderson, J. T. *Excursions in Number Theory.* New York: Dover, pp. 86–87, 1988.

Anomalous Number

see BENFORD'S LAW

Anonymous

A term in SOCIAL CHOICE THEORY meaning invariance of a result under permutation of voters.

see also DUAL VOTING, MONOTONIC VOTING

Anosov Automorphism

A HYPERBOLIC linear map $\mathbb{R}^n \to \mathbb{R}^n$ with INTEGER entries in the transformation MATRIX and DETERMINANT ± 1 is an ANOSOV DIFFEOMORPHISM of the n-TORUS, called an Anosov automorphism (or HYPERBOLIC AUTOMORPHISM). Here, the term automorphism is used in the GROUP THEORY sense.

Anosov Diffeomorphism

An Anosov diffeomorphism is a C^1 DIFFEOMORPHISM ϕ such that the MANIFOLD M is HYPERBOLIC with respect to ϕ. Very few classes of Anosov diffeomorphisms are known. The best known is ARNOLD'S CAT MAP.

A HYPERBOLIC linear map $\mathbb{R}^n \to \mathbb{R}^n$ with INTEGER entries in the transformation MATRIX and DETERMINANT ± 1 is an Anosov diffeomorphism of the n-TORUS. Not every MANIFOLD admits an Anosov diffeomorphism. Anosov diffeomorphisms are EXPANSIVE, and there are no Anosov diffeomorphisms on the CIRCLE.

It is conjectured that if $\phi : M \to M$ is an Anosov diffeomorphism on a COMPACT RIEMANNIAN MANIFOLD and the NONWANDERING SET $\Omega(\phi)$ of ϕ is M, then ϕ is TOPOLOGICALLY CONJUGATE to a FINITE-TO-ONE FACTOR of an ANOSOV AUTOMORPHISM of a NILMANIFOLD. It has been proved that any Anosov diffeomorphism on the n-TORUS is TOPOLOGICALLY CONJUGATE to an ANOSOV AUTOMORPHISM, and also that Anosov diffeomorphisms are C^1 STRUCTURALLY STABLE.

see also ANOSOV AUTOMORPHISM, AXIOM A DIFFEOMORPHISM, DYNAMICAL SYSTEM

References

Anosov, D. V. "Geodesic Flow on Closed Riemannian Manifolds with Negative Curvature." *Proc. Steklov Inst., A. M. S.* 1969.

Smale, S. "Differentiable Dynamical Systems." *Bull. Amer. Math. Soc.* **73**, 747–817, 1967.

Anosov Flow

A FLOW defined analogously to the ANOSOV DIFFEOMORPHISM, except that instead of splitting the TANGENT BUNDLE into two invariant sub-BUNDLES, they are split into three (one exponentially contracting, one expanding, and one which is 1-dimensional and tangential to the flow direction).

see also DYNAMICAL SYSTEM

Anosov Map
An important example of a ANOSOV DIFFEOMORPHISM.

$$\begin{bmatrix} x_{n+1} \\ y_{n+1} \end{bmatrix} = \begin{bmatrix} 2 & 1 \\ 1 & 1 \end{bmatrix} \begin{bmatrix} x_n \\ y_n \end{bmatrix},$$

where x_{n+1}, y_{n+1} are computed mod 1.

see also ARNOLD'S CAT MAP

ANOVA
"Analysis of Variance." A STATISTICAL TEST for heterogeneity of MEANS by analysis of group VARIANCES. To apply the test, assume random sampling of a variate y with equal VARIANCES, independent errors, and a NORMAL DISTRIBUTION. Let n be the number of REPLICATES (sets of identical observations) within each of K FACTOR LEVELS (treatment groups), and y_{ij} be the jth observation within FACTOR LEVEL i. Also assume that the ANOVA is "balanced" by restricting n to be the same for each FACTOR LEVEL.

Now define the sum of square terms

$$\text{SST} \equiv \sum_{i=1}^{k}\sum_{j=1}^{n}(y_{ij} - \bar{\bar{y}})^2 \tag{1}$$

$$= \sum_{i=1}^{k}\sum_{j=1}^{n} {y_{ij}}^2 - \frac{\left(\sum_{i=1}^{k}\sum_{j=1}^{n} y_{ij}\right)^2}{Kn} \tag{2}$$

$$\text{SSA} \equiv \frac{1}{n}\sum_{i=1}^{k}\left(\sum_{j=1}^{n} y_{ij}\right)^2 - \frac{1}{Kn}\left(\sum_{i=1}^{k}\sum_{j=1}^{n} y_{ij}\right)^2 \tag{3}$$

$$\text{SSE} \equiv \sum_{i=1}^{k}\sum_{j=1}^{n}(y_{ij} - \bar{y}_i)^2 \tag{4}$$

$$= \text{SST} - \text{SSA}, \tag{5}$$

which are the total, treatment, and error sums of squares. Here, $\bar{y}_i$ is the mean of observations within FACTOR LEVEL i, and $\bar{\bar{y}}$ is the "group" mean (i.e., mean of means). Compute the entries in the following table, obtaining the P-VALUE corresponding to the calculated F-RATIO of the mean squared values

$$F = \frac{\text{MSA}}{\text{MSE}}. \tag{6}$$

Category	SS	°Freedom	Mean Squared	F-Ratio
treatment	SSA	$K-1$	$\text{MSA} \equiv \frac{\text{SSA}}{K-1}$	$\frac{\text{MSA}}{\text{MSE}}$
error	SSE	$K(n-1)$	$\text{MSE} \equiv \frac{\text{SSE}}{K(n-1)}$	
total	SST	$Kn-1$	$\text{MST} \equiv \frac{\text{SST}}{Kn-1}$	

If the P-VALUE is small, reject the NULL HYPOTHESIS that all MEANS are the same for the different groups.

see also FACTOR LEVEL, REPLICATE, VARIANCE

Anthropomorphic Polygon
A SIMPLE POLYGON with precisely two EARS and one MOUTH.

References

Toussaint, G. "Anthropomorphic Polygons." *Amer. Math. Monthly* **122**, 31–35, 1991.

Anthyphairetic Ratio
An archaic word for a CONTINUED FRACTION.

References

Fowler, D. H. *The Mathematics of Plato's Academy: A New Reconstruction.* New York: Oxford University Press, 1987.

Antiautomorphism
If a MAP $f : G \to G'$ from a GROUP G to a GROUP G' satisfies $f(ab) = f(a)f(b)$ for all $a, b \in G$, then f is said to be an antiautomorphism.

see also AUTOMORPHISM

Anticevian Triangle
Given a center $\alpha : \beta : \gamma$, the anticevian triangle is defined as the TRIANGLE with VERTICES $-\alpha : \beta : \gamma$, $\alpha : -\beta : \gamma$, and $\alpha : \beta : -\gamma$. If $A'B'C'$ is the CEVIAN TRIANGLE of X and $A''B''C''$ is an anticevian triangle, then X and A'' are HARMONIC CONJUGATE POINTS with respect to A and A'.

see also CEVIAN TRIANGLE

References

Kimberling, C. "Central Points and Central Lines in the Plane of a Triangle." *Math. Mag.* **67**, 163–187, 1994.

Antichain
Let P be a finite PARTIALLY ORDERED SET. An antichain in P is a set of pairwise incomparable elements (a family of SUBSETS such that, for any two members, one is not the SUBSET of another). The WIDTH of P is the maximum CARDINALITY of an ANTICHAIN in P. For a PARTIAL ORDER, the size of the longest ANTICHAIN is called the WIDTH.

see also CHAIN, DILWORTH'S LEMMA, PARTIALLY ORDERED SET, WIDTH (PARTIAL ORDER)

References

Sloane, N. J. A. Sequence A006826/M2469 in "An On-Line Version of the Encyclopedia of Integer Sequences."

Anticlastic
When the GAUSSIAN CURVATURE K is everywhere NEGATIVE, a SURFACE is called anticlastic and is saddle-shaped. A SURFACE on which K is everywhere POSITIVE is called SYNCLASTIC. A point at which the GAUSSIAN CURVATURE is NEGATIVE is called a HYPERBOLIC POINT.

see also ELLIPTIC POINT, GAUSSIAN QUADRATURE, HYPERBOLIC POINT, PARABOLIC POINT, PLANAR POINT, SYNCLASTIC

Anticommutative

An OPERATOR $*$ for which $a * b = -b * a$ is said to be anticommutative.

see also COMMUTATIVE

Anticommutator

For OPERATORS $\tilde{A}$ and $\tilde{B}$, the anticommutator is defined by

$$\{\tilde{A}, \tilde{B}\} \equiv \tilde{A}\tilde{B} + \tilde{B}\tilde{A}.$$

see also COMMUTATOR, JORDAN ALGEBRA

Anticomplementary Triangle

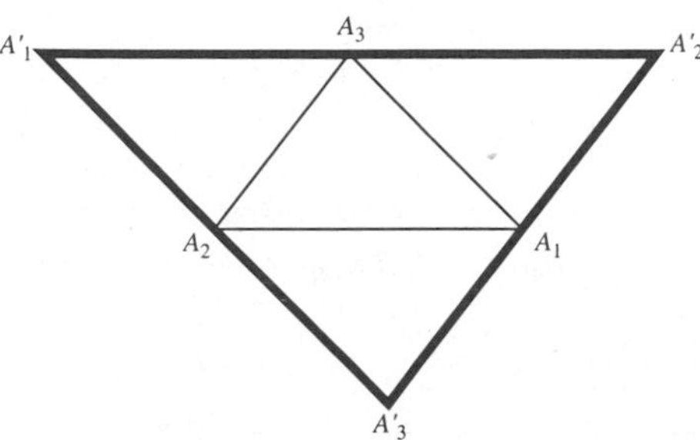

A TRIANGLE $\Delta A'B'C'$ which has a given TRIANGLE ΔABC as its MEDIAL TRIANGLE. The TRILINEAR COORDINATES of the anticomplementary triangle are

$$\begin{aligned} A' &= -a^{-1} : b^{-1} : c^{-1} \\ B' &= a^{-1} : -b^{-1} : c^{-1} \\ C' &= a^{-1} : b^{-1} : -c^{-1}. \end{aligned}$$

see also MEDIAL TRIANGLE

Antiderivative

see INTEGRAL

Antidifferentiation

see INTEGRATION

Antigonal Points

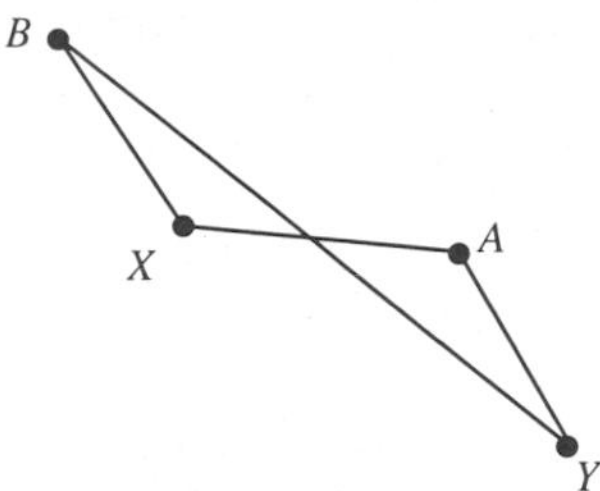

Given $\angle AXB + \angle AYB = \pi$ RADIANS in the above figure, then X and Y are said to be antigonal points with respect to A and B.

Antihomography

A CIRCLE-preserving TRANSFORMATION composed of an ODD number of INVERSIONS.

see also HOMOGRAPHY

Antihomologous Points

Two points which are COLLINEAR with respect to a SIMILITUDE CENTER but are not HOMOLOGOUS POINTS. Four interesting theorems from Johnson (1929) follow.

1. Two pairs of antihomologous points form inversely similar triangles with the HOMOTHETIC CENTER.
2. The PRODUCT of distances from a HOMOTHETIC CENTER to two antihomologous points is a constant.
3. Any two pairs of points which are antihomologous with respect to a SIMILITUDE CENTER lie on a CIRCLE.
4. The tangents to two CIRCLES at antihomologous points make equal ANGLES with the LINE through the points.

see also HOMOLOGOUS POINTS, HOMOTHETIC CENTER, SIMILITUDE CENTER

References

Johnson, R. A. *Modern Geometry: An Elementary Treatise on the Geometry of the Triangle and the Circle.* Boston, MA: Houghton Mifflin, pp. 19–21, 1929.

Antilaplacian

The antilaplacian of u with respect to x is a function whose LAPLACIAN with respect to x equals u. The antilaplacian is never unique.

see also LAPLACIAN

Antilinear Operator

An antilinear OPERATOR satisfies the following two properties:

$$\begin{aligned} \tilde{A}[f_1(x) + f_2(x)] &= \tilde{A}f_1(x) + \tilde{A}f_2(x) \\ \tilde{A}cf(x) &= c^*\tilde{A}f(x), \end{aligned}$$

where c^* is the COMPLEX CONJUGATE of c.

see also LINEAR OPERATOR

Antilogarithm

The INVERSE FUNCTION of the LOGARITHM, defined such that

$$\log_b(\text{antilog}_b\, z) = z = \text{antilog}_b(\log_b z).$$

The antilogarithm in base b of z is therefore b^z.

see also COLOGARITHM, LOGARITHM, POWER

Antimagic Graph

A GRAPH with e EDGES labeled with distinct elements $\{1, 2, \ldots, e\}$ so that the SUM of the EDGE labels at each VERTEX differ.

see also MAGIC GRAPH

References

Hartsfield, N. and Ringel, G. *Pearls in Graph Theory: A Comprehensive Introduction.* San Diego, CA: Academic Press, 1990.

Antimagic Square

15	2	12	4
1	14	10	5
8	9	3	16
11	13	6	7

21	18	6	17	4
7	3	13	16	24
5	20	23	11	1
15	8	19	2	25
14	12	9	22	10

10	25	32	13	16	9
22	7	3	24	21	30
20	27	18	26	11	6
1	31	23	33	17	8
19	5	36	12	15	29
34	14	2	4	35	28

14	3	34	21	47	29	22
43	16	13	25	6	26	44
30	48	24	8	12	9	45
10	5	11	38	49	46	19
4	41	37	36	33	27	1
39	17	40	20	7	35	23
31	42	18	32	28	2	15

49	16	50	10	19	28	24	56
42	43	11	15	44	38	55	5
25	21	48	46	9	37	6	63
29	47	8	40	51	30	52	1
45	22	54	23	20	34	2	62
14	59	18	33	41	26	61	13
36	12	58	32	27	64	3	35
17	39	7	57	53	4	60	31

52	19	81	22	29	15	42	31	76
61	10	67	23	54	79	25	33	16
57	9	71	24	38	1	51	47	75
26	78	7	69	66	77	13	27	12
39	21	74	20	37	17	49	55	64
8	65	4	62	50	34	73	41	40
56	68	2	63	14	72	35	44	6
53	30	60	32	36	3	46	43	58
11	70	5	59	48	80	28	45	18

An antimagic square is an $n \times n$ ARRAY of integers from 1 to n^2 such that each row, column, and main diagonal produces a different sum such that these sums form a SEQUENCE of consecutive integers. It is therefore a special case of a HETEROSQUARE.

Antimagic squares of orders one and two are impossible, and it is believed that there are also no antimagic squares of order three. There are 18 families of antimagic squares of order four. Antimagic squares of orders 4–9 are illustrated above (Madachy 1979).

see also HETEROSQUARE, MAGIC SQUARE, TALISMAN SQUARE

References

Abe, G. "Unsolved Problems on Magic Squares." *Disc. Math.* **127**, 3–13, 1994.

Madachy, J. S. "Magic and Antimagic Squares." Ch. 4 in *Madachy's Mathematical Recreations.* New York: Dover, pp. 103–113, 1979.

Weisstein, E. W. "Magic Squares." http://www.astro.virginia.edu/~eww6n/math/notebooks/MagicSquares.m.

Antimorph

A number which can be represented both in the form ${x_0}^2 - D{y_0}^2$ and in the form $D{x_1}^2 - {y_1}^2$. This is only possible when the PELL EQUATION

$$x^2 - Dy^2 = -1$$

is solvable. Then

$$\begin{aligned} x^2 - Dy^2 &= -({x_0} - D{y_0}^2)({x_n}^2 - D{y_n}^2) \\ &= D(x_0 y_n - y_0 x_n)^2 - (x_0 x_n - D y_0 y_n)^2. \end{aligned}$$

see also IDONEAL NUMBER, POLYMORPH

References

Beiler, A. H. *Recreations in the Theory of Numbers: The Queen of Mathematical Entertains.* New York: Dover, 1964.

Antimorphic Number

see ANTIMORPH

Antinomy

A PARADOX or contradiction.

Antiparallel

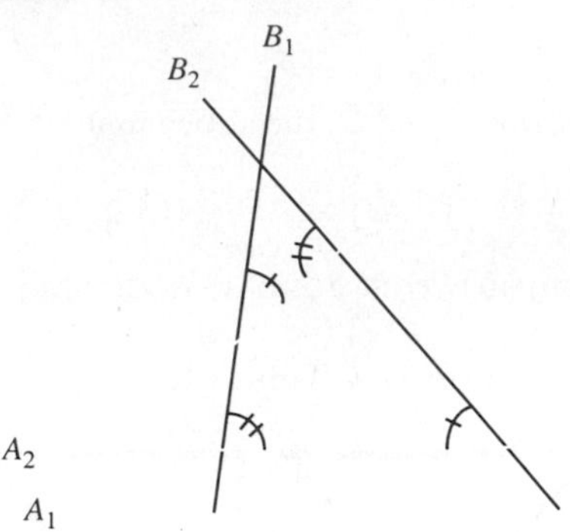

A pair of LINES B_1, B_2 which make the same ANGLES but in opposite order with two other given LINES A_1 and A_2, as in the above diagram, are said to be antiparallel to A_1 and A_2.

see also HYPERPARALLEL, PARALLEL

References

Phillips, A. W. and Fisher, I. *Elements of Geometry.* New York: American Book Co., 1896.

Antipedal Triangle

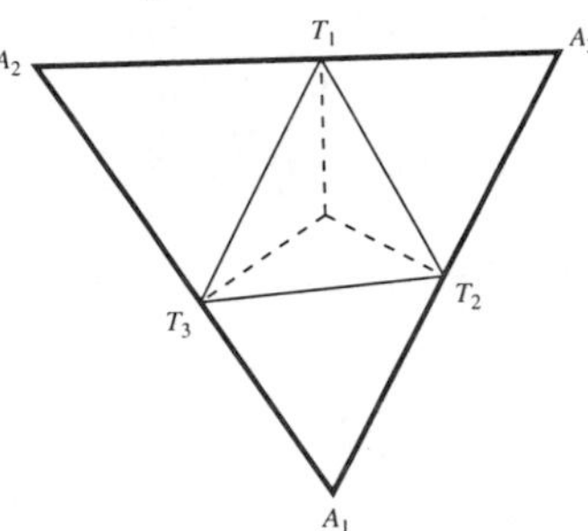

The antipedal triangle A of a given TRIANGLE T is the TRIANGLE of which T is the PEDAL TRIANGLE. For a TRIANGLE with TRILINEAR COORDINATES $\alpha : \beta : \gamma$ and ANGLES A, B, and C, the antipedal triangle has VERTICES with TRILINEAR COORDINATES

$$\begin{aligned} &-(\beta+\alpha\cos C)(\gamma+\alpha\cos B) : (\gamma+\alpha\cos B)(\alpha+\beta\cos C) : \\ &\qquad (\beta+\alpha\cos C)(\alpha+\gamma\cos B) \\ &(\gamma+\beta\cos A)(\beta+\alpha\cos C) : -(\gamma+\beta\cos A)(\alpha+\beta\cos C) : \\ &\qquad (\alpha+\beta\cos C)(\beta+\gamma\cos A) \\ &(\beta+\gamma\cos A)(\gamma+\alpha\cos B) : (\alpha+\gamma\cos B)(\gamma+\beta\cos A) : \\ &\qquad -(\alpha+\gamma\cos B)(\beta+\gamma\cos A). \end{aligned}$$

The ISOGONAL CONJUGATE of the ANTIPEDAL TRIANGLE of a given TRIANGLE is HOMOTHETIC with the original TRIANGLE. Furthermore, the PRODUCT of their AREAS equals the SQUARE of the AREA of the original TRIANGLE (Gallatly 1913).

see also PEDAL TRIANGLE

References
Gallatly, W. *The Modern Geometry of the Triangle, 2nd ed.* London: Hodgson, pp. 56–58, 1913.

Antipersistent Process

A FRACTAL PROCESS for which $H < 1/2$, so $r < 0$.

see also PERSISTENT PROCESS

Antipodal Map

The MAP which takes points on the surface of a SPHERE $\mathbb{S}^2$ to their ANTIPODAL POINTS.

Antipodal Points

Two points are antipodal (i.e., each is the ANTIPODE of the other) if they are diametrically opposite. Examples include endpoints of a LINE SEGMENT, or poles of a SPHERE. Given a point on a SPHERE with LATITUDE δ and LONGITUDE λ, the antipodal point has LATITUDE $-\delta$ and LONGITUDE $\lambda \pm 180°$ (where the sign is taken so that the result is between $-180°$ and $+180°$).

see also ANTIPODE, DIAMETER, GREAT CIRCLE, SPHERE

Antipode

Given a point A, the point B which is the ANTIPODAL POINT of A is said to be the antipode of A.

see also ANTIPODAL POINTS

Antiprism

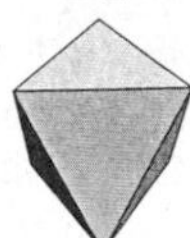 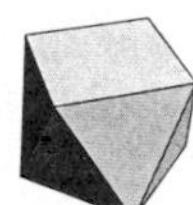 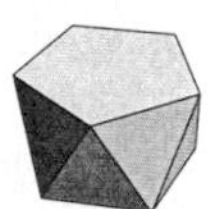 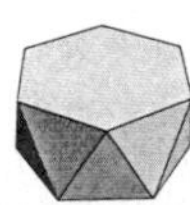

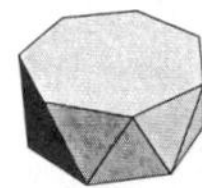 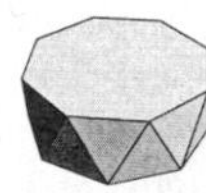 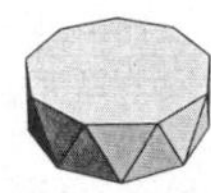 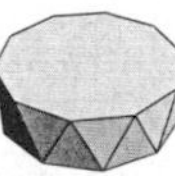

A SEMIREGULAR POLYHEDRON constructed with 2 n-gons and $2n$ TRIANGLES. The 3-antiprism is simply the OCTAHEDRON. The DUALS are the TRAPEZOHEDRA. The SURFACE AREA of a n-gonal antiprism is

$$\begin{aligned} S &= 2A_{n\text{-gon}} + 2nA_\Delta \\ &= 2\left[\tfrac{1}{4}na^2 \cot\left(\frac{\pi}{n}\right)\right] + 2n(\tfrac{1}{4}\sqrt{3}\,a^2) \\ &= \tfrac{1}{2}na^2\left[\cot\left(\frac{\pi}{n}\right) + \sqrt{3}\right]. \end{aligned}$$

see also OCTAHEDRON, PRISM, PRISMOID, TRAPEZOHEDRON

References
Ball, W. W. R. and Coxeter, H. S. M. "Polyhedra." Ch. 5 in *Mathematical Recreations and Essays, 13th ed.* New York: Dover, p. 130, 1987.
Cromwell, P. R. *Polyhedra.* New York: Cambridge University Press, pp. 85–86, 1997.
Weisstein, E. W. "Prisms and Antiprisms." http://www.astro.virginia.edu/~eww6n/math/notebooks/Prism.m.

Antiquity

see GEOMETRIC PROBLEMS OF ANTIQUITY

Antisnowflake

see KOCH ANTISNOWFLAKE

Antisquare Number

A number of the form $p^a \cdot A$ is said to be an antisquare if it fails to be a SQUARE NUMBER for the two reasons that a is ODD and A is a nonsquare modulo p.

see also SQUARE NUMBER

Antisymmetric

A quantity which changes SIGN when indices are reversed. For example, $A_{ij} \equiv a_i - a_j$ is antisymmetric since $A_{ij} = -A_{ji}$.

see also ANTISYMMETRIC MATRIX, ANTISYMMETRIC TENSOR, SYMMETRIC

Antisymmetric Matrix

An antisymmetric matrix is a MATRIX which satisfies the identity

$$\mathsf{A} = -\mathsf{A}^{\mathrm{T}} \tag{1}$$

where A^{T} is the MATRIX TRANSPOSE. In component notation, this becomes

$$a_{ij} = -a_{ji}. \tag{2}$$

Letting $k = i = j$, the requirement becomes

$$a_{kk} = -a_{kk}, \tag{3}$$

so an antisymmetric matrix must have zeros on its diagonal. The general 3×3 antisymmetric matrix is of the form

$$\begin{bmatrix} 0 & a_{12} & a_{13} \\ -a_{12} & 0 & a_{23} \\ -a_{13} & -a_{23} & 0 \end{bmatrix}. \tag{4}$$

Applying A^{-1} to both sides of the antisymmetry condition gives

$$-\mathsf{A}^{-1}\mathsf{A}^{\mathrm{T}} = \mathsf{I}. \tag{5}$$

Any SQUARE MATRIX can be expressed as the sum of symmetric and antisymmetric parts. Write

$$\mathsf{A} = \tfrac{1}{2}(\mathsf{A} + \mathsf{A}^{\mathrm{T}}) + \tfrac{1}{2}(\mathsf{A} - \mathsf{A}^{\mathrm{T}}). \tag{6}$$

$$\mathsf{A} = \begin{bmatrix} a_{11} & a_{12} & \cdots & a_{1n} \\ a_{21} & a_{22} & \cdots & a_{2n} \\ \vdots & \vdots & \ddots & \vdots \\ a_{n1} & a_{n2} & \cdots & a_{nn} \end{bmatrix} \tag{7}$$

$$\mathsf{A}^{\mathrm{T}} = \begin{bmatrix} a_{11} & a_{21} & \cdots & a_{n1} \\ a_{12} & a_{22} & \cdots & a_{n2} \\ \vdots & \vdots & \ddots & \vdots \\ a_{1n} & a_{2n} & \cdots & a_{nn} \end{bmatrix}, \qquad (8)$$

so

$$\mathsf{A} + \mathsf{A}^{\mathrm{T}} = \begin{bmatrix} 2a_{11} & a_{12}+a_{21} & \cdots & a_{1n}+a_{n1} \\ a_{12}+a_{21} & 2a_{22} & \cdots & a_{2n}+a_{n2} \\ \vdots & \vdots & \ddots & \vdots \\ a_{1n}+a_{n1} & a_{2n}+a_{n2} & \cdots & 2a_{nn} \end{bmatrix}, \qquad (9)$$

which is symmetric, and

$$\mathsf{A} - \mathsf{A}^{\mathrm{T}} = \begin{bmatrix} 0 & a_{12}-a_{21} & \cdots & a_{1n}-a_{n1} \\ -(a_{12}-a_{21}) & 0 & \cdots & a_{2n}-a_{n2} \\ \vdots & \vdots & \ddots & \vdots \\ -(a_{1n}-a_{n1}) & -(a_{2n}-a_{n2}) & \cdots & 0 \end{bmatrix}, \qquad (10)$$

which is antisymmetric.

see also SKEW SYMMETRIC MATRIX, SYMMETRIC MATRIX

Antisymmetric Relation

A RELATION R on a SET S is antisymmetric provided that distinct elements are never both related to one another. In other words xRy and yRx together imply that $x = y$.

Antisymmetric Tensor

An antisymmetric tensor is defined as a TENSOR for which

$$A^{mn} = -A^{nm}. \qquad (1)$$

Any TENSOR can be written as a sum of SYMMETRIC and antisymmetric parts as

$$A^{mn} = \tfrac{1}{2}(A^{mn} + A^{nm}) + \tfrac{1}{2}(A^{mn} - A^{nm}). \qquad (2)$$

The antisymmetric part is sometimes denoted using the special notation

$$A^{[ab]} = \tfrac{1}{2}(A^{ab} - A^{ba}). \qquad (3)$$

For a general TENSOR,

$$A^{[a_1\cdots a_n]} \equiv \frac{1}{n!}\epsilon_{a_1\cdots a_n} \sum_{\text{permutations}} A^{a_1\cdots a_n}, \qquad (4)$$

where $\epsilon_{a_1\cdots a_n}$ is the LEVI-CIVITA SYMBOL, a.k.a. the PERMUTATION SYMBOL.

see also SYMMETRIC TENSOR

Antoine's Horned Sphere

A topological 2-sphere in 3-space whose exterior is not SIMPLY CONNECTED. The outer complement of Antoine's horned sphere is not SIMPLY CONNECTED. Furthermore, the group of the outer complement is not even finitely generated. Antoine's horned sphere is inequivalent to ALEXANDER'S HORNED SPHERE since the complement in $\mathbb{R}^3$ of the bad points for ALEXANDER'S HORNED SPHERE is SIMPLY CONNECTED.

see also ALEXANDER'S HORNED SPHERE

References

Alexander, J. W. "An Example of a Simply-Connected Surface Bounding a Region which is not Simply-Connected." *Proc. Nat. Acad. Sci.* **10**, 8–10, 1924.

Rolfsen, D. *Knots and Links.* Wilmington, DE: Publish or Perish Press, pp. 76–79, 1976.

Antoine's Necklace

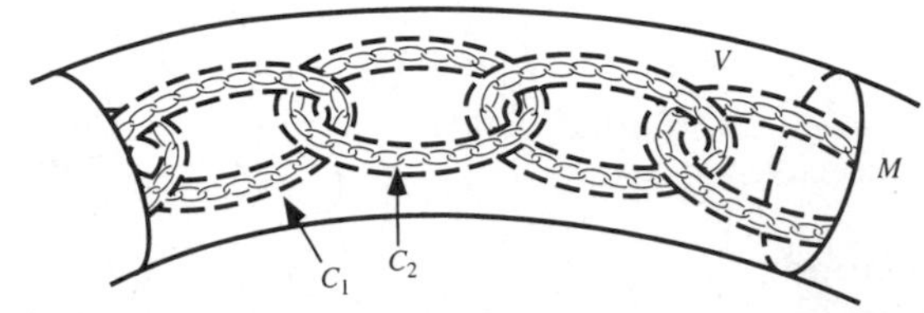

Construct a chain C of $2n$ components in a solid TORUS V. Now form a chain C_1 of $2n$ solid tori in V, where

$$\pi_1(V - C_1) \cong \pi_1(V - C)$$

via inclusion. In each component of C_1, construct a smaller chain of solid tori embedded in that component. Denote the union of these smaller solid tori C_2. Continue this process a countable number of times, then the intersection

$$A = \bigcap_{i=1}^{\infty} C_i$$

which is a nonempty compact SUBSET of $\mathbb{R}^3$ is called Antoine's necklace. Antoine's necklace is HOMEOMORPHIC with the CANTOR SET.

see also ALEXANDER'S HORNED SPHERE, NECKLACE

References

Rolfsen, D. *Knots and Links.* Wilmington, DE: Publish or Perish Press, pp. 73–74, 1976.

Apeirogon

The REGULAR POLYGON essentially equivalent to the CIRCLE having an infinite number of sides and denoted with SCHLÄFLI SYMBOL $\{\infty\}$.

see also CIRCLE, REGULAR POLYGON

References

Coxeter, H. S. M. *Regular Polytopes, 3rd ed.* New York: Dover, 1973.

Schwartzman, S. *The Words of Mathematics: An Etymological Dictionary of Mathematical Terms Used in English.* Washington, DC: Math. Assoc. Amer., 1994.

Apéry's Constant

N.B. A detailed on-line essay by S. Finch was the starting point for this entry.

Apéry's constant is defined by

$$\zeta(3) = 1.2020569\ldots, \tag{1}$$

(Sloane's A002117) where $\zeta(z)$ is the RIEMANN ZETA FUNCTION. Apéry (1979) proved that $\zeta(3)$ is IRRATIONAL, although it is not known if it is TRANSCENDENTAL. The CONTINUED FRACTION for $\zeta(3)$ is [1, 4, 1, 18, 1, 1, 1, 4, 1, ...] (Sloane's A013631). The positions at which the numbers 1, 2, ... occur in the continued fraction are 1, 12, 25, 2, 64, 27, 17, 140, 10, ...

Sums related to $\zeta(3)$ are

$$\zeta(3) = \frac{5}{2}\sum_{n=1}^{\infty}\frac{(-1)^{n-1}}{n^3\binom{2n}{n}} = \frac{5}{2}\sum_{k=1}^{\infty}\frac{(-1)^{k+1}(k!)^2}{(2k)!k^3} \tag{2}$$

(used by Apéry), and

$$\lambda(3) = \sum_{k=0}^{\infty}\frac{1}{(2k+1)^3} = \tfrac{7}{8}\zeta(3) \tag{3}$$

$$\sum_{k=0}^{\infty}\frac{1}{(3k+1)^3} = \frac{2\pi^3}{81\sqrt{3}} + \tfrac{13}{27}\zeta(3) \tag{4}$$

$$\sum_{k=0}^{\infty}\frac{1}{(4k+1)^3} = \frac{\pi^3}{64} + \tfrac{7}{16}\zeta(3) \tag{5}$$

$$\sum_{k=0}^{\infty}\frac{1}{(6k+1)^3} = \frac{\pi^3}{36\sqrt{3}} + \tfrac{91}{216}\zeta(3), \tag{6}$$

where $\lambda(z)$ is the DIRICHLET LAMBDA FUNCTION. The above equations are special cases of a general result due to Ramanujan (Berndt 1985). Apéry's proof relied on showing that the sum

$$a(n) \equiv \sum_{k=0}^{n}\binom{n}{k}^2\binom{n+k}{k}^2, \tag{7}$$

where $\binom{n}{k}$ is a BINOMIAL COEFFICIENT, satisfies the RECURRENCE RELATION

$$(n+1)^3a(n+1) - (34n^3+51n^2+27n+5)a(n) + n^3a(n-1) = 0 \tag{8}$$

(van der Poorten 1979, Zeilberger 1991).

Apéry's constant is also given by

$$\zeta(3) = \sum_{n=1}^{\infty}\frac{S_{n,2}}{n!n}, \tag{9}$$

where $S_{n,m}$ is a STIRLING NUMBER OF THE FIRST KIND. This can be rewritten as

$$\sum_{n=1}^{\infty}\frac{H_n}{n^2} = 2\zeta(3), \tag{10}$$

where H_n is the nth HARMONIC NUMBER. Yet another expression for $\zeta(3)$ is

$$\zeta(3) = \frac{1}{2}\sum_{n=1}^{\infty}\frac{1}{n^2}\left(1+\frac{1}{2}+\ldots+\frac{1}{n}\right) \tag{11}$$

(Castellanos 1988).

INTEGRALS for $\zeta(3)$ include

$$\zeta(3) = \frac{1}{2}\int_0^{\infty}\frac{t^2}{e^t-1}\,dt \tag{12}$$

$$= \frac{8}{7}\left[\tfrac{1}{4}\pi^2\ln 2 + 2\int_0^{\pi/4}x\ln(\sin x)\,dx\right]. \tag{13}$$

Gosper (1990) gave

$$\zeta(3) = \frac{1}{4}\sum_{k=1}^{\infty}\frac{30k-11}{(2k-1)k^3\binom{2k}{k}^2}. \tag{14}$$

A CONTINUED FRACTION involving Apéry's constant is

$$\frac{6}{\zeta(3)} = 5 - \frac{1^6}{117-}\,\frac{2^6}{535-}\cdots\frac{n^6}{34n^3+51n^2+27n+5-}\cdots \tag{15}$$

(Apéry 1979, Le Lionnais 1983). Amdeberhan (1996) used WILF-ZEILBERGER PAIRS (F, G) with

$$F(n,k) = \frac{(-1)^k k!^2(sn-k-1)!}{(sn+k+1)!(k+1)}, \tag{16}$$

$s = 1$ to obtain

$$\zeta(3) = \frac{5}{2}\sum_{n=1}^{\infty}(-1)^{n-1}\frac{1}{\binom{2n}{n}n^3}. \tag{17}$$

For $s = 2$,

$$\zeta(3) = \frac{1}{4}\sum_{n=1}^{\infty}(-1)^{n-1}\frac{56n^2-32+5}{(2n-1)^2}\frac{1}{\binom{3n}{n}\binom{2n}{n}n^3} \tag{18}$$

and for $s = 3$,

$$\zeta(3) = \sum_{n=0}^{\infty}\frac{(-1)^n}{72\binom{4n}{n}\binom{3n}{n}} \times\frac{6120n+5265n^4+13761n^2+13878n^3+1040}{(4n+1)(4n+3)(n+1)(3n+1)^2(3n+2)^2} \tag{19}$$

(Amdeberhan 1996). The corresponding $G(n,k)$ for $s = 1$ and 2 are

$$G(n,k) = \frac{2(-1)^k k!^2 (n-k)!}{(n+k+1)!(n+1)^2} \tag{20}$$

and

$$G(n,k) = \frac{(-1)^k k!^2 (2n-k)!(3+4n)(4n^2+6n+k+3)}{2(2n+k+2)!(n+1)^2(2n+1)^2}. \tag{21}$$

Gosper (1996) expressed $\zeta(3)$ as the MATRIX PRODUCT

$$\lim_{N\to\infty} \prod_{n=1}^{N} M_n = \begin{bmatrix} 0 & \zeta(3) \\ 0 & 1 \end{bmatrix}, \tag{22}$$

where

$$M_n \equiv \begin{bmatrix} \frac{(n+1)^4}{4096(n+\frac{5}{4})^2(n+\frac{7}{4})^2} & \frac{24570n^4+64161n^3+62152n^2+26427n+4154}{31104(n+\frac{1}{3})(n+\frac{1}{2})(n+\frac{2}{3})} \\ 0 & 1 \end{bmatrix} \tag{23}$$

which gives 12 bits per term. The first few terms are

$$M_1 = \begin{bmatrix} \frac{1}{19600} & \frac{2077}{1728} \\ 0 & 1 \end{bmatrix} \tag{24}$$

$$M_2 = \begin{bmatrix} \frac{1}{9801} & \frac{7561}{4320} \\ 0 & 1 \end{bmatrix} \tag{25}$$

$$M_3 = \begin{bmatrix} \frac{9}{67600} & \frac{50501}{20160} \\ 0 & 1 \end{bmatrix}, \tag{26}$$

which gives

$$\zeta(3) \approx \tfrac{423203577229}{352066176000} = 1.20205690315732\ldots. \tag{27}$$

Given three INTEGERS chosen at random, the probability that no common factor will divide them all is

$$[\zeta(3)]^{-1} \approx 1.202^{-1} = 0.832\ldots. \tag{28}$$

B. Haible and T. Papanikolaou computed $\zeta(3)$ to 1,000,000 DIGITS using a WILF-ZEILBERGER PAIR identity with

$$F(n,k) = (-1)^k \frac{n!^6(2n-k-1)!k!^3}{2(n+k+1)!^2(2n)!^3}, \tag{29}$$

$s = 1$, and $t = 1$, giving the rapidly converging

$$\zeta(3) = \sum_{n=0}^{\infty} (-1)^n \frac{n!^{10}(205n^2+250n+77)}{64(2n+1)!^5} \tag{30}$$

(Amdeberhan and Zeilberger 1997). The record as of Aug. 1998 was 64 million digits (Plouffe).

see also RIEMANN ZETA FUNCTION, WILF-ZEILBERGER PAIR

References

Amdeberhan, T. "Faster and Faster Convergent Series for $\zeta(3)$." *Electronic J. Combinatorics* **3**, R13, 1–2, 1996. `http://www.combinatorics.org/Volume_3/volume3.html#R13`.

Amdeberhan, T. and Zeilberger, D. "Hypergeometric Series Acceleration via the WZ Method." *Electronic J. Combinatorics* **4**, No. 2, R3, 1–3, 1997. `http://www.combinatorics.org/Volume_4/wilftoc.html#R03`. Also available at `http://www.math.temple.edu/~zeilberg/mamarim/mamarimhtml/accel.html`.

Apéry, R. "Irrationalité de $\zeta(2)$ et $\zeta(3)$." *Astérisque* **61**, 11–13, 1979.

Berndt, B. C. *Ramanujan's Notebooks: Part I.* New York: Springer-Verlag, 1985.

Beukers, F. "A Note on the Irrationality of $\zeta(3)$." *Bull. London Math. Soc.* **11**, 268–272, 1979.

Borwein, J. M. and Borwein, P. B. *Pi & the AGM: A Study in Analytic Number Theory and Computational Complexity.* New York: Wiley, 1987.

Castellanos, D. "The Ubiquitous Pi. Part I." *Math. Mag.* **61**, 67–98, 1988.

Conway, J. H. and Guy, R. K. "The Great Enigma." In *The Book of Numbers.* New York: Springer-Verlag, pp. 261–262, 1996.

Ewell, J. A. "A New Series Representation for $\zeta(3)$." *Amer. Math. Monthly* **97**, 219–220, 1990.

Finch, S. "Favorite Mathematical Constants." `http://www.mathsoft.com/asolve/constant/apery/apery.html`.

Gosper, R. W. "Strip Mining in the Abandoned Orefields of Nineteenth Century Mathematics." In *Computers in Mathematics* (Ed. D. V. Chudnovsky and R. D. Jenks). New York: Marcel Dekker, 1990.

Haible, B. and Papanikolaou, T. "Fast Multiprecision Evaluation of Series of Rational Numbers." Technical Report TI-97-7. Darmstadt, Germany: Darmstadt University of Technology, Apr. 1997.

Le Lionnais, F. *Les nombres remarquables.* Paris: Hermann, p. 36, 1983.

Plouffe, S. "Plouffe's Inverter: Table of Current Records for the Computation of Constants." `http://lacim.uqam.ca/pi/records.html`.

Plouffe, S. "32,000,279 Digits of Zeta(3)." `http://lacim.uqam.ca/piDATA/Zeta3.txt`.

Sloane, N. J. A. Sequences A013631 and A002117/M0020 in "An On-Line Version of the Encyclopedia of Integer Sequences."

van der Poorten, A. "A Proof that Euler Missed... Apéry's Proof of the Irrationality of $\zeta(3)$." *Math. Intel.* **1**, 196–203, 1979.

Zeilberger, D. "The Method of Creative Telescoping." *J. Symb. Comput.* **11**, 195–204, 1991.

Apoapsis

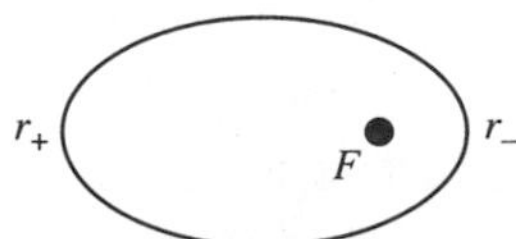

The greatest radial distance of an ELLIPSE as measured from a FOCUS. Taking $v = \pi$ in the equation of an ELLIPSE

$$r = \frac{a(1-e^2)}{1+e\cos v}$$

gives the apoapsis distance

$$r_+ = a(1+e).$$

Apoapsis for an orbit around the Earth is called apogee, and apoapsis for an orbit around the Sun is called aphelion.

see also ECCENTRICITY, ELLIPSE, FOCUS, PERIAPSIS

Apocalypse Number

A number having 666 DIGITS (where 666 is the BEAST NUMBER) is called an apocalypse number. The FIBONACCI NUMBER F_{3184} is an apocalypse number.

see also BEAST NUMBER, LEVIATHAN NUMBER

References

Pickover, C. A. *Keys to Infinity.* New York: Wiley, pp. 97–102, 1995.

Apocalyptic Number

A number of the form 2^n which contains the digits 666 (the BEAST NUMBER) is called an APOCALYPTIC NUMBER. 2^{157} is an apocalyptic number. The first few such powers are 157, 192, 218, 220, ... (Sloane's A007356).

see also APOCALYPSE NUMBER, LEVIATHAN NUMBER

References

Pickover, C. A. *Keys to Infinity.* New York: Wiley, pp. 97–102, 1995.
Sloane, N. J. A. Sequences A007356/M5405 in "An On-Line Version of the Encyclopedia of Integer Sequences."
Sloane, N. J. A. and Plouffe, S. Extended entry in *The Encyclopedia of Integer Sequences.* San Diego: Academic Press, 1995.

Apodization

The application of an APODIZATION FUNCTION.

Apodization Function

A function (also called a TAPERING FUNCTION) used to bring an interferogram smoothly down to zero at the edges of the sampled region. This suppresses sidelobes which would otherwise be produced, but at the expense of widening the lines and therefore decreasing the resolution.

The following are apodization functions for symmetrical (2-sided) interferograms, together with the INSTRUMENT FUNCTIONS (or APPARATUS FUNCTIONS) they produce and a blowup of the INSTRUMENT FUNCTION sidelobes. The INSTRUMENT FUNCTION $I(k)$ corresponding to a given apodization function $A(x)$ can be computed by taking the finite FOURIER COSINE TRANSFORM,

$$I(k) = \int_{-a}^{a} \cos(2\pi kx) A(x)\, dx. \tag{1}$$

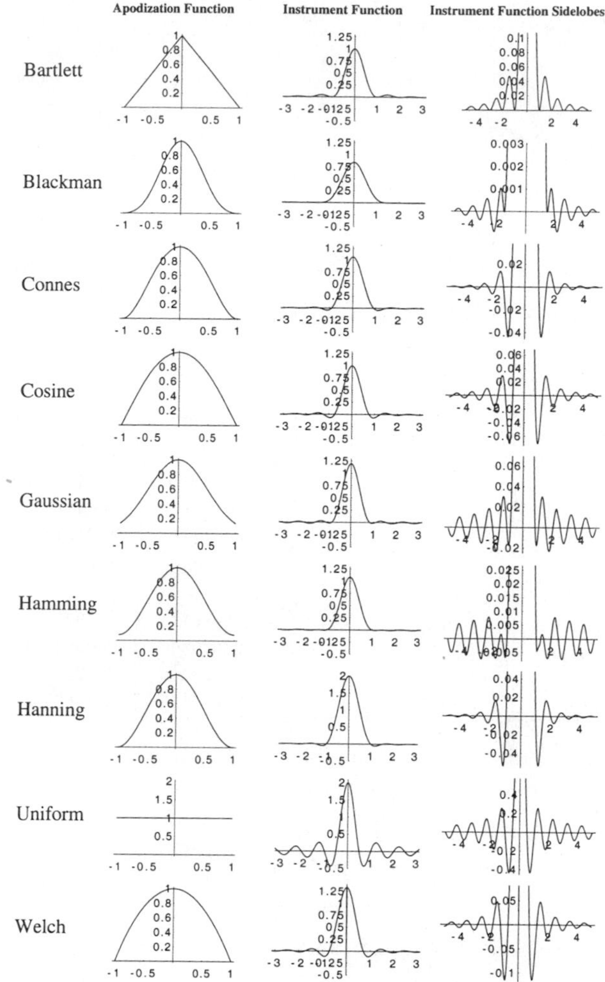

Type	Apodization Function	Instrument Function
Bartlett	$1-\frac{\lvert x\rvert}{a}$	$a\,\text{sinc}^2(\pi ka)$
Blackman	$B_A(x)$	$B_I(k)$
Connes	$\left(1-\frac{x^2}{a^2}\right)^2$	$8a\sqrt{2\pi}\,\frac{J_{5/2}(2\pi ka)}{(2\pi ka)^{5/2}}$
Cosine	$\cos\left(\frac{\pi x}{2a}\right)$	$\frac{4a\cos(2\pi ak)}{\pi(1-16a^2k^2)}$
Gaussian	$e^{-x^2/(2\sigma^2)}$	$2\int_0^a \cos(2\pi kx)e^{-x^2/(2\sigma^2)}\,dx$
Hamming	$Hm_A(x)$	$Hm_I(k)$
Hanning	$Hn_A(x)$	$Hn_I(k)$
Uniform	1	$2a\,\text{sinc}\,(2\pi ka)$
Welch	$1-\frac{x^2}{a^2}$	$W_I(k)$

where

$$B_A(x) = 0.42 + 0.5\cos\left(\frac{\pi x}{a}\right) + 0.08\cos\left(\frac{2\pi x}{a}\right) \tag{2}$$

$$B_I(k) = \frac{a(0.84 - 0.36a^2k^2 - 2.17\times 10^{-19}a^4k^4)\,\text{sinc}(2\pi ak)}{(1-a^2k^2)(1-4a^2k^2)} \tag{3}$$

$$Hm_A(x) = 0.54 + 0.46\cos\left(\frac{\pi x}{a}\right) \tag{4}$$

$$Hm_I(k) = \frac{a(1.08 - 0.64a^2k^2)\,\text{sinc}(2\pi ak)}{1-4a^2k^2} \tag{5}$$

$$Hn_A(x) = \cos^2\left(\frac{\pi x}{2a}\right) \tag{6}$$

$$= \frac{1}{2}\left[1 + \cos\left(\frac{\pi x}{a}\right)\right] \tag{7}$$

$$Hn_I(k) = \frac{a \operatorname{sinc}(2\pi ak)}{1 - 4a^2k^2} \tag{8}$$

$$= a[\operatorname{sinc}(2\pi ka) + \tfrac{1}{2}\operatorname{sinc}(2\pi ka - \pi) + \tfrac{1}{2}\operatorname{sinc}(2\pi ka + \pi)] \tag{9}$$

$$W_I(k) = a2\sqrt{2\pi}\,\frac{J_{3/2}(2\pi ka)}{(2\pi ka)^{3/2}} \tag{10}$$

$$= a\frac{\sin(2\pi ka) - 2\pi ak\cos(2\pi ak)}{2a^3k^3\pi^3}. \tag{11}$$

Type	IF FWHM	IF Peak	Peak (−) S.L. / Peak	Peak (+) S.L. / Peak
Bartlett	1.77179	1	0.00000000	0.0471904
Blackman	2.29880	0.84	−0.00106724	0.00124325
Connes	1.90416	$\frac{16}{15}$	−0.0411049	0.0128926
Cosine	1.63941	$\frac{4}{\pi}$	−0.0708048	0.0292720
Gaussian	—	1	—	—
Hamming	1.81522	1.08	−0.00689132	0.00734934
Hanning	2.00000	1	−0.0267076	0.00843441
Uniform	1.20671	2	−0.217234	0.128375
Welch	1.59044	$\frac{4}{3}$	−0.0861713	0.356044

A general symmetric apodization function $A(x)$ can be written as a FOURIER SERIES

$$A(x) = a_0 + 2\sum_{n=1}^{\infty} a_n \cos\left(\frac{n\pi x}{b}\right), \tag{12}$$

where the COEFFICIENTS satisfy

$$a_0 + 2\sum_{n=1}^{\infty} a_n = 1. \tag{13}$$

The corresponding apparatus function is

$$I(t) \equiv \int_{-b}^{b} A(x)e^{-2\pi ikx}\,dx = 2b\Big\{a_0 \operatorname{sinc}(2\pi kb) + \sum_{n=1}^{\infty}[\operatorname{sinc}(2\pi kb + n\pi) + \operatorname{sinc}(2\pi kb - n\pi)]\Big\}. \tag{14}$$

To obtain an APODIZATION FUNCTION with zero at $ka = 3/4$, use

$$a_0 \operatorname{sinc}(\tfrac{3}{2}\pi) + a_1[\operatorname{sinc}(\tfrac{5}{2}\pi) + \operatorname{sinc}(\tfrac{1}{2}\pi) = 0. \tag{15}$$

Plugging in (13),

$$-(1-2a_1)\frac{2}{3\pi} + a_1\left(\frac{2}{5\pi} + \frac{2}{\pi}\right) = -\tfrac{1}{3}(1-2a_1) + a_1(\tfrac{1}{5}+1) = 0 \tag{16}$$

$$a_1(\tfrac{6}{5} + \tfrac{2}{3}) = \tfrac{1}{3} \tag{17}$$

$$a_1 = \frac{\frac{1}{3}}{\frac{6}{5}+\frac{2}{3}} = \frac{5}{6\cdot 3 + 2\cdot 5} = \tfrac{5}{28} \tag{18}$$

$$a_0 = 1 - 2a_1 = \frac{28 - 2\cdot 5}{28} = \tfrac{18}{28} = \tfrac{9}{14}. \tag{19}$$

The HAMMING FUNCTION is close to the requirement that the APPARATUS FUNCTION goes to 0 at $ka = 5/4$, giving

$$a_0 = \tfrac{25}{46} \approx 0.5435 \tag{20}$$

$$a_1 = \tfrac{21}{92} \approx 0.2283. \tag{21}$$

The BLACKMAN FUNCTION is chosen so that the APPARATUS FUNCTION goes to 0 at $ka = 5/4$ and $9/4$, giving

$$a_0 = \tfrac{3969}{9304} \approx 0.4266 \tag{22}$$

$$a_1 = \tfrac{1155}{4652} \approx 0.2483 \tag{23}$$

$$a_2 = \tfrac{715}{18608} \approx 0.0384. \tag{24}$$

see also BARTLETT FUNCTION, BLACKMAN FUNCTION, CONNES FUNCTION, COSINE APODIZATION FUNCTION, FULL WIDTH AT HALF MAXIMUM, GAUSSIAN FUNCTION, HAMMING FUNCTION, HANN FUNCTION, HANNING FUNCTION, MERTZ APODIZATION FUNCTION, PARZEN APODIZATION FUNCTION, UNIFORM APODIZATION FUNCTION, WELCH APODIZATION FUNCTION

References

Ball, J. A. "The Spectral Resolution in a Correlator System" §4.3.5 in *Methods of Experimental Physics* **12C** (Ed. M. L. Meeks). New York: Academic Press, pp. 55–57, 1976.

Blackman, R. B. and Tukey, J. W. "Particular Pairs of Windows." In *The Measurement of Power Spectra, From the Point of View of Communications Engineering.* New York: Dover, pp. 95–101, 1959.

Brault, J. W. "Fourier Transform Spectrometry." In *High Resolution in Astronomy: 15th Advanced Course of the Swiss Society of Astronomy and Astrophysics* (Ed. A. Benz, M. Huber, and M. Mayor). Geneva Observatory, Sauverny, Switzerland, pp. 31–32, 1985.

Harris, F. J. "On the Use of Windows for Harmonic Analysis with the Discrete Fourier Transform." *Proc. IEEE* **66**, 51–83, 1978.

Norton, R. H. and Beer, R. "New Apodizing Functions for Fourier Spectroscopy." *J. Opt. Soc. Amer.* **66**, 259–264, 1976.

Press, W. H.; Flannery, B. P.; Teukolsky, S. A.; and Vetterling, W. T. *Numerical Recipes in FORTRAN: The Art of Scientific Computing, 2nd ed.* Cambridge, England: Cambridge University Press, pp. 547–548, 1992.

Schnopper, H. W. and Thompson, R. I. "Fourier Spectrometers." In *Methods of Experimental Physics* **12A** (Ed. M. L. Meeks). New York: Academic Press, pp. 491–529, 1974.

Apollonius Circles

There are two completely different definitions of the so-called Apollonius circles:

1. The set of all points whose distances from two fixed points are in a constant ratio $1 : \mu$ (Ogilvy 1990).

2. The eight CIRCLES (two of which are nondegenerate) which solve APOLLONIUS' PROBLEM for three CIRCLES.

Given one side of a TRIANGLE and the ratio of the lengths of the other two sides, the LOCUS of the third VERTEX is the Apollonius circle (of the first type) whose CENTER is on the extension of the given side. For a given TRIANGLE, there are three circles of Apollonius.

Denote the three Apollonius circles (of the first type) of a TRIANGLE by k_1, k_2, and k_3, and their centers L_1, L_2, and L_3. The center L_1 is the intersection of the side A_2A_3 with the tangent to the CIRCUMCIRCLE at A_1. L_1 is also the pole of the SYMMEDIAN POINT K with respect to CIRCUMCIRCLE. The centers L_1, L_2, and L_3 are COLLINEAR on the POLAR of K with regard to its CIRCUMCIRCLE, called the LEMOINE LINE. The circle of Apollonius k_1 is also the locus of a point whose PEDAL TRIANGLE is ISOSCELES such that $\overline{P_1P_2} = \overline{P_1P_3}$.

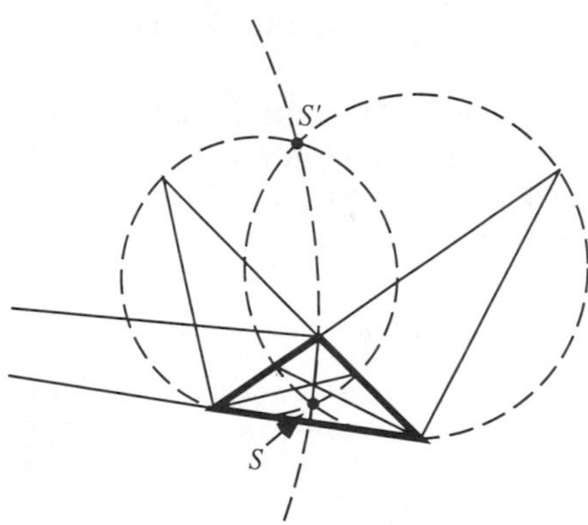

Let U and V be points on the side line BC of a TRIANGLE ΔABC met by the interior and exterior ANGLE BISECTORS of ANGLES A. The CIRCLE with DIAMETER UV is called the A-Apollonian circle. Similarly, construct the B- and C-Apollonian circles. The Apollonian circles pass through the VERTICES A, B, and C, and through the two ISODYNAMIC POINTS S and S'. The VERTICES of the D-TRIANGLE lie on the respective Apollonius circles.

see also APOLLONIUS' PROBLEM, APOLLONIUS PURSUIT PROBLEM, CASEY'S THEOREM, HART'S THEOREM, ISODYNAMIC POINTS, SODDY CIRCLES

References

Johnson, R. A. *Modern Geometry: An Elementary Treatise on the Geometry of the Triangle and the Circle.* Boston, MA: Houghton Mifflin, pp. 40 and 294–299, 1929.

Ogilvy, C. S. *Excursions in Geometry.* New York: Dover, pp. 14–23, 1990.

Apollonius Point

Consider the EXCIRCLES Γ_A, Γ_B, and Γ_C of a TRIANGLE, and the CIRCLE Γ internally TANGENT to all three. Denote the contact point of Γ and Γ_A by A', etc. Then the LINES AA', BB', and CC' CONCUR in this point. It has TRIANGLE CENTER FUNCTION

$$\alpha = \sin^2 A \cos^2[\tfrac{1}{2}(B - C)].$$

References

Kimberling, C. "Apollonius Point." http://www.evansville.edu/~ck6/tcenters/recent/apollon.html.

Kimberling, C. "Central Points and Central Lines in the Plane of a Triangle." *Math. Mag.* **67**, 163–187, 1994.

Kimberling, C.; Iwata, S.; and Hidetosi, F. "Problem 1091 and Solution." *Crux Math.* **13**, 128–129 and 217–218, 1987.

Apollonius' Problem

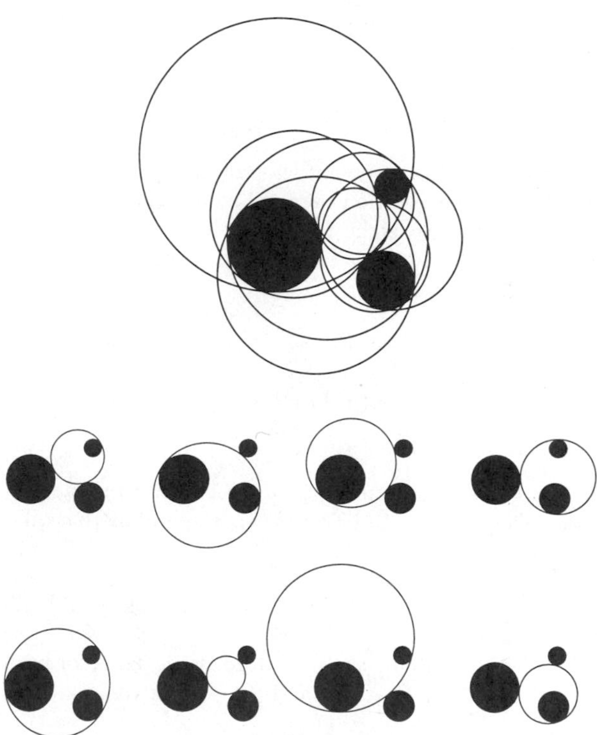

Given three objects, each of which may be a POINT, LINE, or CIRCLE, draw a CIRCLE that is TANGENT to each. There are a total of ten cases. The two easiest involve three points or three LINES, and the hardest involves three CIRCLES. Euclid solved the two easiest cases in his *Elements*, and the others (with the exception of the three CIRCLE problem), appeared in the *Tangencies* of Apollonius which was, however, lost. The general problem is, in principle, solvable by STRAIGHTEDGE and COMPASS alone.

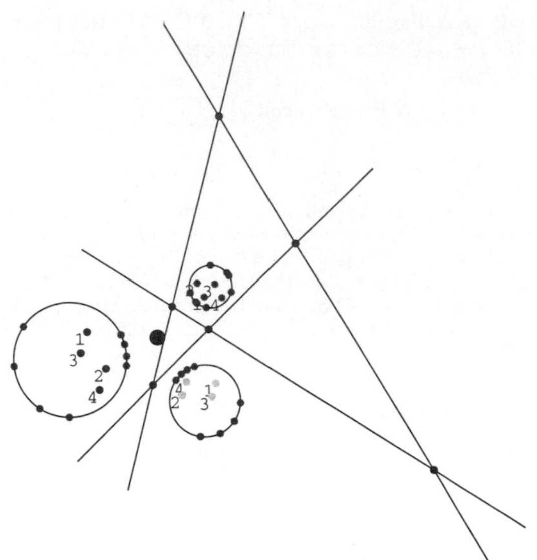

The three-CIRCLE problem was solved by Viète (Boyer 1968), and the solutions are called APOLLONIUS CIRCLES. There are eight total solutions. The simplest solution is obtained by solving the three simultaneous quadratic equations

$$(x - x_1)^2 + (y - y_1)^2 - (r \pm r_1)^2 = 0 \quad (1)$$

$$(x - x_2)^2 + (y - y_2)^2 - (r \pm r_2)^2 = 0 \quad (2)$$

$$(x - x_3)^2 + (y - y_3)^2 - (r \pm r_3)^2 = 0 \quad (3)$$

in the three unknowns x, y, r for the eight triplets of signs (Courant and Robbins 1996). Expanding the equations gives

$$(x^2 + y^2 - r^2) - 2xx_i - 2yy_i \pm 2rr_i + ({x_i}^2 + {y_i}^2 - {r_i}^2) = 0 \quad (4)$$

for $i = 1, 2, 3$. Since the first term is the same for each equation, taking (2) − (1) and (3) − (1) gives

$$ax + by + cr = d \quad (5)$$

$$a'x + b'y + c'r = d', \quad (6)$$

where

$$a = 2(x_1 - x_2) \quad (7)$$

$$b = 2(y_1 - y_2) \quad (8)$$

$$c = \mp 2(r_1 - r_2) \quad (9)$$

$$d = ({x_2}^2 + {y_2}^2 - {r_2}^2) - ({x_1}^2 + {y_1}^2 - {r_1}^2) \quad (10)$$

and similarly for a', b', c' and d' (where the 2 subscripts are replaced by 3s). Solving these two simultaneous linear equations gives

$$x = \frac{b'd - bd' - b'cr + bc'r}{ab' - ba'} \quad (11)$$

$$y = \frac{-a'd + ad' + a'cr - ac'r}{ab' - a'b}, \quad (12)$$

which can then be plugged back into the QUADRATIC EQUATION (1) and solved using the QUADRATIC FORMULA.

Perhaps the most elegant solution is due to Gergonne. It proceeds by locating the six HOMOTHETIC CENTERS (three internal and three external) of the three given CIRCLES. These lie three by three on four lines (illustrated above). Determine the POLES of one of these with respect to each of the three CIRCLES and connect the POLES with the RADICAL CENTER of the CIRCLES. If the connectors meet, then the three pairs of intersections are the points of tangency of two of the eight circles (Johnson 1929, Dörrie 1965). To determine *which* two of the eight Apollonius circles are produced by the three pairs, simply take the two which intersect the original three CIRCLES only in a single point of tangency. The procedure, when repeated, gives the other three pairs of CIRCLES.

If the three CIRCLES are mutually tangent, then the eight solutions collapse to two, known as the SODDY CIRCLES.

see also APOLLONIUS PURSUIT PROBLEM, BEND (CURVATURE), CASEY'S THEOREM, DESCARTES CIRCLE THEOREM, FOUR COINS PROBLEM, HART'S THEOREM, SODDY CIRCLES

References

Boyer, C. B. *A History of Mathematics.* New York: Wiley, p. 159, 1968.

Courant, R. and Robbins, H. "Apollonius' Problem." §3.3 in *What is Mathematics?: An Elementary Approach to Ideas and Methods, 2nd ed.* Oxford, England: Oxford University Press, pp. 117 and 125–127, 1996.

Dörrie, H. "The Tangency Problem of Apollonius." §32 in *100 Great Problems of Elementary Mathematics: Their History and Solutions.* New York: Dover, pp. 154–160, 1965.

Gauss, C. F. *Werke, Vol. 4.* New York: George Olms, p. 399, 1981.

Johnson, R. A. *Modern Geometry: An Elementary Treatise on the Geometry of the Triangle and the Circle.* Boston, MA: Houghton Mifflin, pp. 118–121, 1929.

Ogilvy, C. S. *Excursions in Geometry.* New York: Dover, pp. 48–51, 1990.

Pappas, T. *The Joy of Mathematics.* San Carlos, CA: Wide World Publ./Tetra, p. 151, 1989.

Simon, M. *Über die Entwicklung der Elementargeometrie im XIX Jahrhundert.* Berlin, pp. 97–105, 1906.

Weisstein, E. W. "Plane Geometry." http://www.astro.virginia.edu/~eww6n/math/notebooks/PlaneGeometry.m.

Apollonius Pursuit Problem

Given a ship with a known constant direction and speed v, what course should be taken by a chase ship in pursuit (traveling at speed V) in order to intersect the other ship in as short a time as possible? The problem can be solved by finding all points which can be simultaneously reached by both ships, which is an APOLLONIUS CIRCLE with $\mu = v/V$. If the CIRCLE cuts the path of the pursued ship, the intersection is the point towards which the pursuit ship should steer. If the CIRCLE does not cut the path, then it cannot be caught.

see also APOLLONIUS CIRCLES, APOLLONIUS' PROBLEM, PURSUIT CURVE

References

Ogilvy, C. S. Solved by M. S. Klamkin. "A Slow Ship Intercepting a Fast Ship." Problem E991. *Amer. Math. Monthly* **59**, 408, 1952.

Ogilvy, C. S. *Excursions in Geometry.* New York: Dover, p. 17, 1990.

Steinhaus, H. *Mathematical Snapshots, 3rd American ed.* New York: Oxford University Press, pp. 126–138, 1983.

Apollonius Theorem

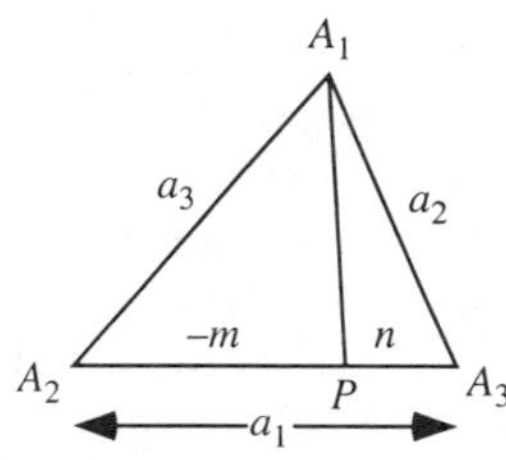

$$m{a_2}^2 + n{a_3}^2 = (m+n)\overline{A_1P}^2 + m\overline{PA_3}^2 + n\overline{PA_2}^2.$$

Apothem

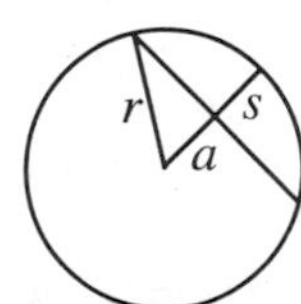

Given a CIRCLE, the PERPENDICULAR distance a from the MIDPOINT of a CHORD to the CIRCLE's center is called the apothem. It is also equal to the RADIUS r minus the SAGITTA s,

$$a = r - s.$$

see also CHORD, RADIUS, SAGITTA, SECTOR, SEGMENT

Apparatus Function

see INSTRUMENT FUNCTION

Appell Hypergeometric Function

A formal extension of the HYPERGEOMETRIC FUNCTION to two variables, resulting in four kinds of functions (Appell 1925),

$$F_1(\alpha;\beta,\beta';\gamma;x,y) = \sum_{m=0}^\infty \sum_{n=0}^\infty \frac{(\alpha)_{m+n}(\beta)_m(\beta')_n}{m!n!(\gamma)_{m+n}} x^m y^n$$

$$F_2(\alpha;\beta,\beta';\gamma,\gamma';x,y) = \sum_{m=0}^\infty \sum_{n=0}^\infty \frac{(\alpha)_{m+n}(\beta)_m(\beta')_n}{m!n!(\gamma)_m(\gamma')_n} x^m y^n$$

$$F_3(\alpha,\alpha';\beta,\beta';\gamma;x,y) = \sum_{m=0}^\infty \sum_{n=0}^\infty \frac{(\alpha)_m(\alpha')_n(\beta)_m(\beta')_n}{m!n!(\gamma)_{m+n}} x^m y^n$$

$$F_4(\alpha;\beta;\gamma,\gamma';x,y) = \sum_{m=0}^\infty \sum_{n=0}^\infty \frac{(\alpha)_{m+n}(\beta)_{m+n}}{m!n!(\gamma)_m(\gamma')_n} x^m y^n.$$

Appell defined the functions in 1880, and Picard showed in 1881 that they may all be expressed by INTEGRALS of the form

$$\int_0^1 u^\alpha(1-u)^\beta(1-xu)^\gamma(1-yu)^\delta\,du.$$

References

Appell, P. "Sur les fonctions hypergéométriques de plusieurs variables." In *Mémoir. Sci. Math.* Paris: Gauthier-Villars, 1925.

Bailey, W. N. *Generalised Hypergeometric Series.* Cambridge, England: Cambridge University Press, p. 73, 1935.

Iyanaga, S. and Kawada, Y. (Eds.). *Encyclopedic Dictionary of Mathematics.* Cambridge, MA: MIT Press, p. 1461, 1980.

Appell Polynomial

A type of POLYNOMIAL which includes the BERNOULLI POLYNOMIAL, HERMITE POLYNOMIAL, and LAGUERRE POLYNOMIAL as special cases. The series of POLYNOMIALS $\{A_n(z)\}_{n=0}^\infty$ is defined by

$$A(t)e^{zt} = \sum_{n=0}^\infty A_n(z)t^n,$$

where

$$A(t) = \sum_{k=0}^\infty a_k t^k$$

is a formal POWER series with $k = 0, 1, \ldots$ and $a_0 \neq 0$.

References

Hazewinkel, M. (Managing Ed.). *Encyclopaedia of Mathematics: An Updated and Annotated Translation of the Soviet "Mathematical Encyclopaedia."* Dordrecht, Netherlands: Reidel, pp. 209–210, 1988.

Appell Transformation

A HOMOGRAPHIC transformation

$$x_1 = \frac{ax+by+c}{a''x+b''y+c}$$

$$y_1 = \frac{a'x+b'y+c'}{a''x+b''y+c''}$$

with t_1 substituted for t according to

$$k\,dt_1 = \frac{dt}{(a''x+b''y+c'')^2}.$$

References

Hazewinkel, M. (Managing Ed.). *Encyclopaedia of Mathematics: An Updated and Annotated Translation of the Soviet "Mathematical Encyclopaedia."* Dordrecht, Netherlands: Reidel, pp. 210–211, 1988.

Apple

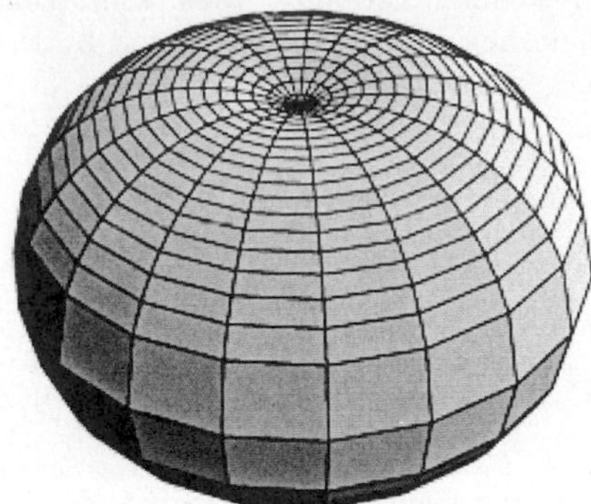

A SURFACE OF REVOLUTION defined by Kepler. It consists of more than half of a circular ARC rotated about an axis passing through the endpoints of the ARC. The equations of the upper and lower boundaries in the x-z PLANE are

$$z_{\pm} = \pm\sqrt{R^2 - (x-r)^2}$$

for $R > r$ and $x \in [-(r+R), r+R]$. It is the outside surface of a SPINDLE TORUS.

see also BUBBLE, LEMON, SPHERE-SPHERE INTERSECTION, SPINDLE TORUS

Approximately Equal

If two quantities A and B are approximately equal, this is written $A \approx B$.

see also DEFINED, EQUAL

Approximation Theory

The mathematical study of how given quantities can be approximated by other (usually simpler) ones under appropriate conditions. Approximation theory also studies the size and properties of the ERROR introduced by approximation. Approximations are often obtained by POWER SERIES expansions in which the higher order terms are dropped.

see also LAGRANGE REMAINDER

References

Achieser, N. I. and Hyman, C. J. *Theory of Approximation.* New York: Dover, 1993.

Akheizer, N. I. *Theory of Approximation.* New York: Dover, 1992.

Cheney, E. W. *Introduction to Approximation Theory.* New York: McGraw-Hill, 1966.

Golomb, M. *Lectures on Theory of Approximation.* Argonne, IL: Argonne National Laboratory, 1962.

Jackson, D. *The Theory of Approximation.* New York: Amer. Math. Soc., 1930.

Natanson, I. P. *Constructive Function Theory, Vol. 1: Uniform Approximation.* New York: Ungar, 1964.

Petrushev, P. P. and Popov, V. A. *Rational Approximation of Real Functions.* New York: Cambridge University Press, 1987.

Rivlin, T. J. *An Introduction to the Approximation of Functions.* New York: Dover, 1981.

Timan, A. F. *Theory of Approximation of Functions of a Real Variable.* New York: Dover, 1994.

Arakelov Theory

A formal mathematical theory which introduces "components at infinity" by defining a new type of divisor class group of INTEGERS of a NUMBER FIELD. The divisor class group is called an "arithmetic surface."

see also ARITHMETIC GEOMETRY

Arbelos

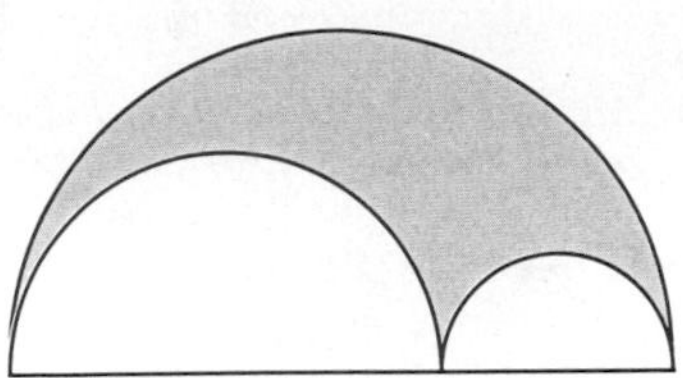

The term "arbelos" means SHOEMAKER'S KNIFE in Greek, and this term is applied to the shaded AREA in the above figure which resembles the blade of a knife used by ancient cobblers (Gardner 1979). Archimedes himself is believed to have been the first mathematician to study the mathematical properties of this figure. The position of the central notch is arbitrary and can be located anywhere along the DIAMETER.

The arbelos satisfies a number of unexpected identities (Gardner 1979).

1. Call the radii of the left and right SEMICIRCLES a and b, respectively, with $a + b \equiv R$. Then the arc length along the bottom of the arbelos is

$$L = 2\pi a + 2\pi b = 2\pi(a+b) = 2\pi R,$$

so the arc lengths along the top and bottom of the arbelos are the same.

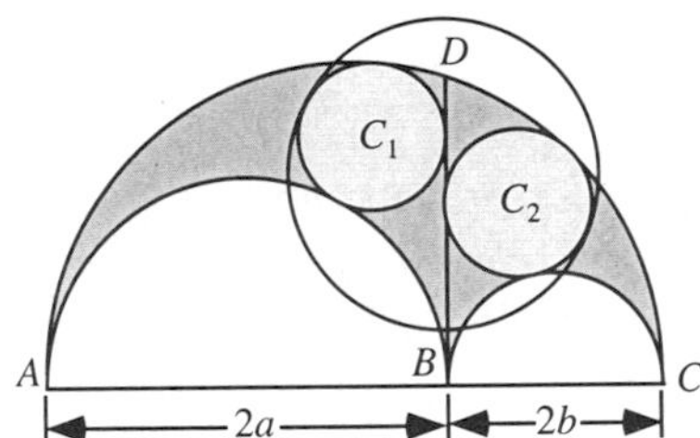

2. Draw the PERPENDICULAR BD from the tangent of the two SEMICIRCLES to the edge of the large CIRCLE. Then the AREA of the arbelos is the same as the AREA of the CIRCLE with DIAMETER BD.
3. The CIRCLES C_1 and C_2 inscribed on each half of BD on the arbelos (called ARCHIMEDES' CIRCLES) each have DIAMETER $(AB)(BC)/(AC)$. Furthermore, the smallest CIRCUMCIRCLE of these two circles has an area equal to that of the arbelos.
4. The line tangent to the semicircles AB and BC contains the point E and F which lie on the lines AD and CD, respectively. Furthermore, BD and EF bisect each other, and the points B, D, E, and F are CONCYCLIC.

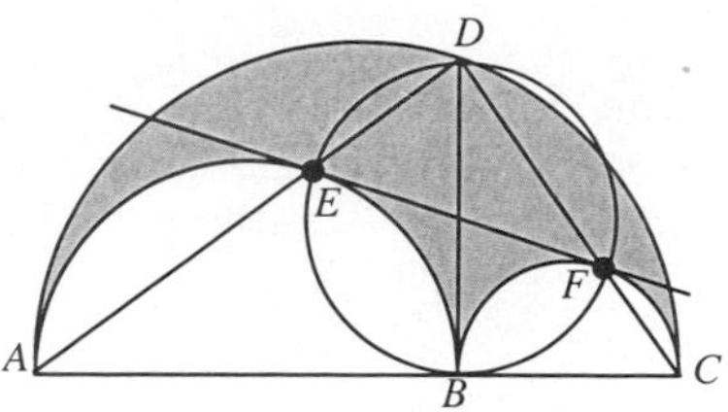

5. In addition to the ARCHIMEDES' CIRCLES C_1 and C_2 in the arbelos figure, there is a third circle C_3 called the BANKOFF CIRCLE which is congruent to these two.

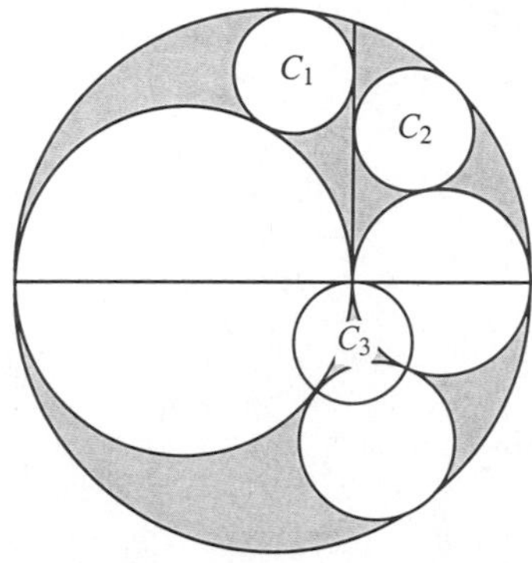

6. Construct a chain of TANGENT CIRCLES starting with the CIRCLE TANGENT to the two small ones and large one. The centers of the CIRCLES lie on an ELLIPSE, and the DIAMETER of the nth CIRCLE C_n is $(1/n)$th PERPENDICULAR distance to the base of the SEMICIRCLE. This result is most easily proven using INVERSION, but was known to Pappus, who referred to it as an ancient theorem (Hood 1961, Cadwell 1966, Gardner 1979, Bankoff 1981). If $r \equiv AB/AC$, then the radius of the nth circle in the PAPPUS CHAIN is

$$r_n = \frac{(1-r)r}{2[n^2(1-r)^2+r]}.$$

This general result simplifies to $r_n = 1/(6+n^2)$ for $r = 2/3$ (Gardner 1979). Further special cases when $AC = 1 + AB$ are considered by Gaba (1940).

7. If B divides AC in the GOLDEN RATIO ϕ, then the circles in the chain satisfy a number of other special properties (Bankoff 1955).

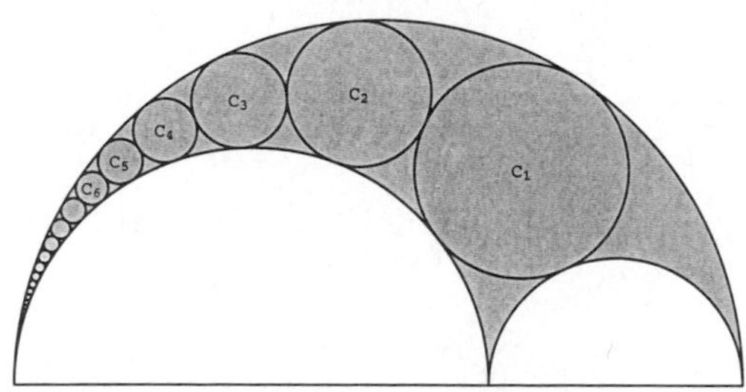

see also ARCHIMEDES' CIRCLES, BANKOFF CIRCLE, COXETER'S LOXODROMIC SEQUENCE OF TANGENT CIRCLES, GOLDEN RATIO, INVERSION, PAPPUS CHAIN, STEINER CHAIN

References

Bankoff, L. "The Fibonacci Arbelos." *Scripta Math.* **20**, 218, 1954.

Bankoff, L. "The Golden Arbelos." *Scripta Math.* **21**, 70–76, 1955.

Bankoff, L. "Are the Twin Circles of Archimedes Really Twins?" *Math. Mag.* **47**, 214–218, 1974.

Bankoff, L. "How Did Pappus Do It?" In *The Mathematical Gardner* (Ed. D. Klarner). Boston, MA: Prindle, Weber, and Schmidt, pp. 112–118, 1981.

Bankoff, L. "The Marvelous Arbelos." In *The Lighter Side of Mathematics* (Ed. R. K. Guy and R. E. Woodrow). Washington, DC: Math. Assoc. Amer., 1994.

Cadwell, J. H. *Topics in Recreational Mathematics.* Cambridge, England: Cambridge University Press, 1966.

Gaba, M. G. "On a Generalization of the Arbelos." *Amer. Math. Monthly* **47**, 19–24, 1940.

Gardner, M. "Mathematical Games: The Diverse Pleasures of Circles that Are Tangent to One Another." *Sci. Amer.* **240**, 18–28, Jan. 1979.

Heath, T. L. *The Works of Archimedes with the Method of Archimedes.* New York: Dover, 1953.

Hood, R. T. "A Chain of Circles." *Math. Teacher* **54**, 134–137, 1961.

Johnson, R. A. *Modern Geometry: An Elementary Treatise on the Geometry of the Triangle and the Circle.* Boston, MA: Houghton Mifflin, pp. 116–117, 1929.

Ogilvy, C. S. *Excursions in Geometry.* New York: Dover, pp. 54–55, 1990.

Arborescence

A DIGRAPH is called an arborescence if, from a given node x known as the ROOT, there is exactly one elementary path from x to every other node y.

see also ARBORICITY

Arboricity

Given a GRAPH G, the arboricity is the MINIMUM number of line-disjoint acyclic SUBGRAPHS whose UNION is G.

see also ANARBORICITY

Arc

In general, any smooth curve joining two points. In particular, any portion (other than the entire curve) of a CIRCLE or ELLIPSE.

see also APPLE, CIRCLE-CIRCLE INTERSECTION, FIVE DISKS PROBLEM, FLOWER OF LIFE, LEMON, LENS, PIECEWISE CIRCULAR CURVE, REULEAUX POLYGON, REULEAUX TRIANGLE, SALINON, SEED OF LIFE, TRIANGLE ARCS, VENN DIAGRAM, YIN-YANG

Arc Length

Arc length is defined as the length along a curve,

$$s \equiv \int_a^b |d\boldsymbol{\ell}|. \tag{1}$$

Defining the line element $ds^2 \equiv |d\boldsymbol{\ell}|^2$, parameterizing the curve in terms of a parameter t, and noting that

ds/dt is simply the magnitude of the VELOCITY with which the end of the RADIUS VECTOR $\mathbf{r}$ moves gives

$$s = \int_a^b ds = \int_a^b \frac{ds}{dt}\,dt = \int_a^b |\mathbf{r}'(t)|\,dt. \tag{2}$$

In POLAR COORDINATES,

$$d\boldsymbol{\ell} = \hat{\mathbf{r}}\,dr + r\hat{\theta}\,d\theta = \left(\frac{dr}{d\theta}\hat{\mathbf{r}} + r\hat{\theta}\right)d\theta, \tag{3}$$

so

$$ds = |d\boldsymbol{\ell}| = \sqrt{r^2 + \left(\frac{dr}{d\theta}\right)^2}\,d\theta \tag{4}$$

$$s = \int |d\boldsymbol{\ell}| = \int_{\theta_1}^{\theta_2} \sqrt{r^2 + \left(\frac{dr}{d\theta}\right)^2}\,d\theta. \tag{5}$$

In CARTESIAN COORDINATES,

$$d\boldsymbol{\ell} = x\hat{\mathbf{x}} + y\hat{\mathbf{y}} \tag{6}$$

$$ds = \sqrt{dx^2 + dy^2} = \sqrt{\left(\frac{dy}{dx}\right)^2 + 1}\,dx. \tag{7}$$

Therefore, if the curve is written

$$\mathbf{r}(x) = x\hat{\mathbf{x}} + f(x)\hat{\mathbf{y}}, \tag{8}$$

then

$$s = \int_a^b \sqrt{1 + f'^2(x)}\,dx. \tag{9}$$

If the curve is instead written

$$\mathbf{r}(t) = x(t)\hat{\mathbf{x}} + y(t)\hat{\mathbf{y}}, \tag{10}$$

then

$$s = \int_a^b \sqrt{x'^2(t) + y'^2(t)}\,dt. \tag{11}$$

Or, in three dimensions,

$$\mathbf{r}(t) = x(t)\hat{\mathbf{x}} + y(t)\hat{\mathbf{y}} + z(t)\hat{\mathbf{z}}, \tag{12}$$

so

$$s = \int_a^b \sqrt{x'^2(t) + y'^2(t) + z'^2(t)}\,dt. \tag{13}$$

see also CURVATURE, GEODESIC, NORMAL VECTOR, RADIUS OF CURVATURE, RADIUS OF TORSION, SPEED, SURFACE AREA, TANGENTIAL ANGLE, TANGENT VECTOR, TORSION (DIFFERENTIAL GEOMETRY), VELOCITY

Arc Minute

A unit of ANGULAR measure equal to 60 ARC SECONDS, or 1/60 of a DEGREE. The arc minute is denoted $'$ (not to be confused with the symbol for feet).

Arc Second

A unit of ANGULAR measure equal to 1/60 of an ARC MINUTE, or 1/3600 of a DEGREE. The arc second is denoted $''$ (not to be confused with the symbol for inches).

Arccosecant

see INVERSE COSECANT

Arccosine

see INVERSE COSINE

Arccotangent

see INVERSE COTANGENT

Arch

A 4-POLYHEX.

References

Gardner, M. *Mathematical Magic Show: More Puzzles, Games, Diversions, Illusions and Other Mathematical Sleight-of-Mind from Scientific American.* New York: Vintage, p. 147, 1978.

Archimedes Algorithm

Successive application of ARCHIMEDES' RECURRENCE FORMULA gives the Archimedes algorithm, which can be used to provide successive approximations to π (PI). The algorithm is also called the BORCHARDT-PFAFF ALGORITHM. Archimedes obtained the first rigorous approximation of π by CIRCUMSCRIBING and INSCRIBING $n = 6 \cdot 2^k$-gons on a CIRCLE. From ARCHIMEDES' RECURRENCE FORMULA, the CIRCUMFERENCES a and b of the circumscribed and inscribed POLYGONS are

$$a(n) = 2n\tan\left(\frac{\pi}{n}\right) \tag{1}$$

$$b(n) = 2n\sin\left(\frac{\pi}{n}\right), \tag{2}$$

where

$$b(n) < C = 2\pi r = 2\pi \cdot 1 = 2\pi < a(n). \tag{3}$$

For a HEXAGON, $n = 6$ and

$$a_0 \equiv a(6) = 4\sqrt{3} \tag{4}$$

$$b_0 \equiv b(6) = 6, \tag{5}$$

where $a_k \equiv a(6 \cdot 2^k)$. The first iteration of ARCHIMEDES' RECURRENCE FORMULA then gives

$$a_1 = \frac{2 \cdot 6 \cdot 4\sqrt{3}}{6 + 4\sqrt{3}} = \frac{24\sqrt{3}}{3 + 2\sqrt{3}} = 24(2 - \sqrt{3}) \tag{6}$$

$$b_1 = \sqrt{24(2 - \sqrt{3}) \cdot 6} = 12\sqrt{2 - \sqrt{3}} = 6(\sqrt{6} - \sqrt{2}). \tag{7}$$

Additional iterations do not have simple closed forms, but the numerical approximations for $k = 0, 1, 2, 3, 4$ (corresponding to 6-, 12-, 24-, 48-, and 96-gons) are

$$3.00000 < \pi < 3.46410 \quad (8)$$

$$3.10583 < \pi < 3.21539 \quad (9)$$

$$3.13263 < \pi < 3.15966 \quad (10)$$

$$3.13935 < \pi < 3.14609 \quad (11)$$

$$3.14103 < \pi < 3.14271. \quad (12)$$

By taking $k = 4$ (a 96-gon) and using strict inequalities to convert irrational bounds to rational bounds at each step, Archimedes obtained the slightly looser result

$$\frac{223}{71} = 3.14084\ldots < \pi < \frac{22}{7} = 3.14285\ldots. \quad (13)$$

References

Miel, G. "Of Calculations Past and Present: The Archimedean Algorithm." *Amer. Math. Monthly* **90**, 17–35, 1983.

Phillips, G. M. "Archimedes in the Complex Plane." *Amer. Math. Monthly* **91**, 108–114, 1984.

Archimedes' Axiom

An AXIOM actually attributed to Eudoxus (Boyer 1968) which states that

$$a/b = c/d$$

IFF the appropriate one of following conditions is satisfied for INTEGERS m and n:

1. If $ma < nb$, then $mc < md$.
2. If $ma = nd$, then $mc = nd$.
3. If $ma > nd$, then $mc > nd$.

ARCHIMEDES' LEMMA is sometimes also known as Archimedes' axiom.

References

Boyer, C. B. *A History of Mathematics.* New York: Wiley, p. 99, 1968.

Archimedes' Cattle Problem

Also called the BOVINUM PROBLEMA. It is stated as follows: "The sun god had a herd of cattle consisting of bulls and cows, one part of which was white, a second black, a third spotted, and a fourth brown. Among the bulls, the number of white ones was one half plus one third the number of the black greater than the brown; the number of the black, one quarter plus one fifth the number of the spotted greater than the brown; the number of the spotted, one sixth and one seventh the number of the white greater than the brown. Among the cows, the number of white ones was one third plus one quarter of the total black cattle; the number of the black, one quarter plus one fifth the total of the spotted cattle; the number of spotted, one fifth plus one sixth the total of the brown cattle; the number of the brown, one sixth plus one seventh the total of the white cattle. What was the composition of the herd?"

Solution consists of solving the simultaneous DIOPHANTINE EQUATIONS in INTEGERS W, X, Y, Z (the number of white, black, spotted, and brown bulls) and w, x, y, z (the number of white, black, spotted, and brown cows),

$$W = \tfrac{5}{6}X + Z \quad (1)$$

$$X = \tfrac{9}{20}Y + Z \quad (2)$$

$$Y = \tfrac{13}{42}W + Z \quad (3)$$

$$w = \tfrac{7}{12}(X + x) \quad (4)$$

$$x = \tfrac{9}{20}(Y + y) \quad (5)$$

$$y = \tfrac{11}{30}(Z + z) \quad (6)$$

$$z = \tfrac{13}{42}(W + w). \quad (7)$$

The smallest solution in INTEGERS is

$$W = 10{,}366{,}482 \quad (8)$$

$$X = 7{,}460{,}514 \quad (9)$$

$$Y = 7{,}358{,}060 \quad (10)$$

$$Z = 4{,}149{,}387 \quad (11)$$

$$w = 7{,}206{,}360 \quad (12)$$

$$x = 4{,}893{,}246 \quad (13)$$

$$y = 3{,}515{,}820 \quad (14)$$

$$z = 5{,}439{,}213. \quad (15)$$

A more complicated version of the problem requires that $W+X$ be a SQUARE NUMBER and $Y+Z$ a TRIANGULAR NUMBER. The solution to this PROBLEM are numbers with 206544 or 206545 digits.

References

Amthor, A. and Krumbiegel B. "Das Problema bovinum des Archimedes." *Z. Math. Phys.* **25**, 121–171, 1880.

Archibald, R. C. "Cattle Problem of Archimedes." *Amer. Math. Monthly* **25**, 411–414, 1918.

Beiler, A. H. *Recreations in the Theory of Numbers: The Queen of Mathematics Entertains.* New York: Dover, pp. 249–252, 1966.

Bell, A. H. "Solution to the Celebrated Indeterminate Equation $x^2 - ny^2 = 1$." *Amer. Math. Monthly* **1**, 240, 1894.

Bell, A. H. "'Cattle Problem.' By Archimedes 251 BC." *Amer. Math. Monthly* **2**, 140, 1895.

Bell, A. H. "Cattle Problem of Archimedes." *Math. Mag.* **1**, 163, 1882–1884.

Calkins, K. G. "Archimedes' *Problema Bovinum.*" `http://www.andrews.edu/~calkins/cattle.html`.

Dörrie, H. "Archimedes' *Problema Bovinum.*" §1 in *100 Great Problems of Elementary Mathematics: Their History and Solutions.* New York: Dover, pp. 3–7, 1965.

Grosjean, C. C. and de Meyer, H. E. "A New Contribution to the Mathematical Study of the Cattle-Problem of Archimedes." In *Constantin Carathéodory: An International Tribute, Vols. 1 and 2* (Ed. T. M. Rassias). Teaneck, NJ: World Scientific, pp. 404–453, 1991.

Merriman, M. "Cattle Problem of Archimedes." *Pop. Sci. Monthly* **67**, 660, 1905.

Rorres, C. "The Cattle Problem." `http://www.mcs.drexel.edu/~crorres/Archimedes/Cattle/Statement.html`.

Vardi, I. "Archimedes' Cattle Problem." *Amer. Math. Monthly* **105**, 305–319, 1998.

Archimedes' Circles

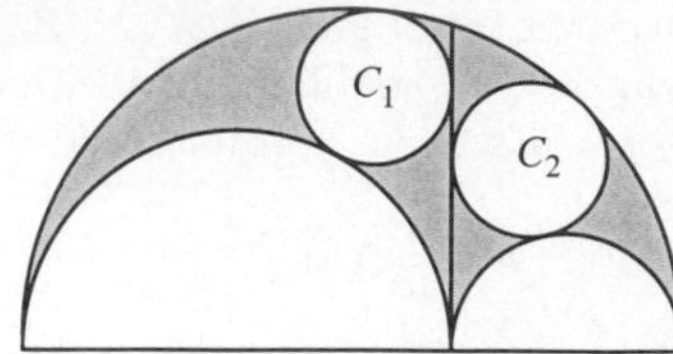

Draw the PERPENDICULAR LINE from the intersection of the two small SEMICIRCLES in the ARBELOS. The two CIRCLES C_1 and C_2 TANGENT to this line, the large SEMICIRCLE, and each of the two SEMICIRCLES are then congruent and known as Archimedes' circles.

see also ARBELOS, BANKOFF CIRCLE, SEMICIRCLE

Archimedes' Constant

see PI

Archimedes' Hat-Box Theorem

Enclose a SPHERE in a CYLINDER and slice PERPENDICULARLY to the CYLINDER's axis. Then the SURFACE AREA of the of SPHERE slice is equal to the SURFACE AREA of the CYLINDER slice.

Archimedes' Lemma

Also known as the continuity axiom, this LEMMA survives in the writings of Eudoxus (Boyer 1968). It states that, given two magnitudes having a ratio, one can find a multiple of either which will exceed the other. This principle was the basis for the EXHAUSTION METHOD which Archimedes invented to solve problems of AREA and VOLUME.

see also CONTINUITY AXIOMS

References

Boyer, C. B. *A History of Mathematics.* New York: Wiley, p. 100, 1968.

Archimedes' Midpoint Theorem

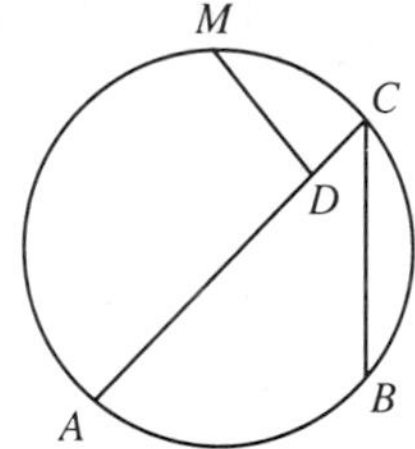

Let M be the MIDPOINT of the ARC AMB. Pick C at random and pick D such that $MD \perp AC$ (where $\perp$ denotes PERPENDICULAR). Then

$$AD = DC + BC.$$

see also MIDPOINT

References

Honsberger, R. *More Mathematical Morsels.* Washington, DC: Math. Assoc. Amer., pp. 31–32, 1991.

Archimedes' Postulate

see ARCHIMEDES' LEMMA

Archimedes' Problem

Cut a SPHERE by a PLANE in such a way that the VOLUMES of the SPHERICAL SEGMENTS have a given RATIO.

see also SPHERICAL SEGMENT

Archimedes' Recurrence Formula

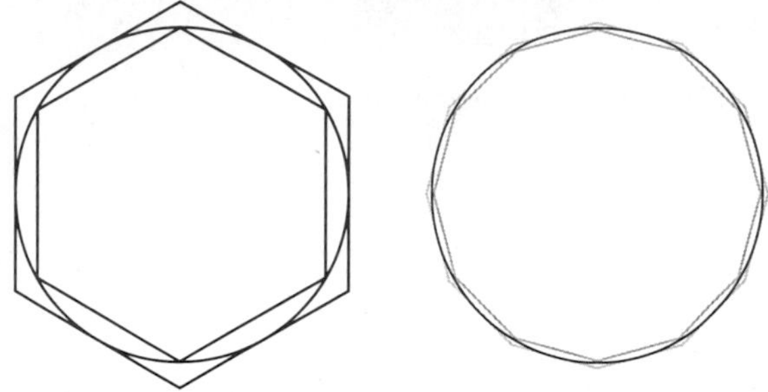

Let a_n and b_n be the PERIMETERS of the CIRCUMSCRIBED and INSCRIBED n-gon and a_{2n} and b_{2n} the PERIMETERS of the CIRCUMSCRIBED and INSCRIBED $2n$-gon. Then

$$a_{2n} = \frac{2a_nb_n}{a_n+b_n} \tag{1}$$

$$b_{2n} = \sqrt{a_{2n}b_n}\,. \tag{2}$$

The first follows from the fact that side lengths of the POLYGONS on a CIRCLE of RADIUS $r=1$ are

$$s_R = 2\tan\left(\frac{\pi}{n}\right) \tag{3}$$

$$s_r = 2\sin\left(\frac{\pi}{n}\right), \tag{4}$$

so

$$a_n = 2n\tan\left(\frac{\pi}{n}\right) \tag{5}$$

$$b_n = 2n\sin\left(\frac{\pi}{n}\right). \tag{6}$$

But

$$\frac{2a_nb_n}{a_n+b_n} = \frac{2\cdot 2n\tan\left(\frac{\pi}{n}\right)\cdot 2n\sin\left(\frac{\pi}{n}\right)}{2n\tan\left(\frac{\pi}{n}\right)+2n\sin\left(\frac{\pi}{n}\right)} = 4n\frac{\tan\left(\frac{\pi}{n}\right)\sin\left(\frac{\pi}{n}\right)}{\tan\left(\frac{\pi}{n}\right)+\sin\left(\frac{\pi}{n}\right)}. \tag{7}$$

Using the identity

$$\tan(\tfrac{1}{2}x) = \frac{\tan x\sin x}{\tan x+\sin x} \tag{8}$$

then gives

$$\frac{2a_nb_n}{a_n+b_n} = 4n\tan\left(\frac{\pi}{2n}\right) = a_{2n}. \tag{9}$$

The second follows from

$$\sqrt{a_{2n}b_n} = \sqrt{4n\tan\left(\frac{\pi}{2n}\right)\cdot 2n\sin\left(\frac{\pi}{n}\right)} \qquad (10)$$

Using the identity

$$\sin x = 2\sin(\tfrac{1}{2}x)\cos(\tfrac{1}{2}x) \qquad (11)$$

gives

$$\sqrt{a_{2n}b_n} = 2n\sqrt{2\tan\left(\frac{\pi}{2n}\right)\cdot 2\sin\left(\frac{\pi}{2n}\right)\cos\left(\frac{\pi}{2n}\right)}$$
$$= 4n\sqrt{\sin^2\left(\frac{\pi}{2n}\right)} = 4n\sin\left(\frac{\pi}{2n}\right) = b_{2n}. \qquad (12)$$

Successive application gives the ARCHIMEDES ALGORITHM, which can be used to provide successive approximations to PI (π).

see also ARCHIMEDES ALGORITHM, PI

References

Dörrie, H. *100 Great Problems of Elementary Mathematics: Their History and Solutions.* New York: Dover, p. 186, 1965.

Archimedean Solid

The Archimedean solids are convex POLYHEDRA which have a similar arrangement of nonintersecting regular plane CONVEX POLYGONS of two or more different types about each VERTEX with all sides the same length. The Archimedean solids are distinguished from the PRISMS, ANTIPRISMS, and ELONGATED SQUARE GYROBICUPOLA by their symmetry group: the Archimedean solids have a spherical symmetry, while the others have "dihedral" symmetry. The Archimedean solids are sometimes also referred to as the SEMIREGULAR POLYHEDRA.

Pugh (1976, p. 25) points out the Archimedean solids are all capable of being circumscribed by a regular TETRAHEDRON so that four of their faces lie on the faces of that TETRAHEDRON. A method of constructing the Archimedean solids using a method known as "expansion" has been enumerated by Stott (Stott 1910; Ball and Coxeter 1987, pp. 139–140).

Let the cyclic sequence $S = (p_1, p_2, \ldots, p_q)$ represent the degrees of the faces surrounding a vertex (i.e., S is a list of the number of sides of all polygons surrounding any vertex). Then the definition of an Archimedean solid requires that the sequence must be the same for each vertex to within ROTATION and REFLECTION. Walsh (1972) demonstrates that S represents the degrees of the faces surrounding each vertex of a semiregular convex polyhedron or TESSELLATION of the plane IFF

1. $q \geq 3$ and every member of S is at least 3,
2. $\sum_{i=1}^{q} \frac{1}{p_i} \geq \frac{1}{2}q - 1$, with equality in the case of a plane TESSELLATION, and
3. for every ODD NUMBER $p \in S$, S contains a subsequence (b, p, b).

Condition (1) simply says that the figure consists of two or more polygons, each having at least three sides. Condition (2) requires that the sum of interior angles at a vertex must be equal to a full rotation for the figure to lie in the plane, and less than a full rotation for a solid figure to be convex.

The usual way of enumerating the semiregular polyhedra is to eliminate solutions of conditions (1) and (2) using several classes of arguments and then prove that the solutions left are, in fact, semiregular (Kepler 1864, pp. 116–126; Catalan 1865, pp. 25–32; Coxeter 1940, p. 394; Coxeter *et al.* 1954; Lines 1965, pp. 202–203; Walsh 1972). The following table gives all possible regular and semiregular polyhedra and tessellations. In the table, 'P' denotes PLATONIC SOLID, 'M' denotes a PRISM or ANTIPRISM, 'A' denotes an Archimedean solid, and 'T' a plane tessellation.

S	Fg.	Solid	Schläfli
(3, 3, 3)	P	tetrahedron	$\{3,3\}$
(3, 4, 4)	M	triangular prism	$t\{2,3\}$
(3, 6, 6)	A	truncated tetrahedron	$t\{3,3\}$
(3, 8, 8)	A	truncated cube	$t\{4,3\}$
(3, 10, 10)	A	truncated dodecahedron	$t\{5,3\}$
(3, 12, 12)	T	(plane tessellation)	$t\{6,3\}$
(4, 4, n)	M	n-gonal Prism	$t\{2,n\}$
(4, 4, 4)	P	cube	$\{4,3\}$
(4, 6, 6)	A	truncated octahedron	$t\{3,4\}$
(4, 6, 8)	A	great rhombicuboct.	$t\begin{Bmatrix}3\\4\end{Bmatrix}$
(4, 6, 10)	A	great rhombicosidodec.	$t\begin{Bmatrix}3\\5\end{Bmatrix}$
(4, 6, 12)	T	(plane tessellation)	$t\begin{Bmatrix}3\\6\end{Bmatrix}$
(4, 8, 8)	T	(plane tessellation)	$t\{4,4\}$
(5, 5, 5)	P	dodecahedron	$\{5,3\}$
(5, 6, 6)	A	truncated icosahedron	$t\{3,5\}$
(6, 6, 6)	T	(plane tessellation)	$\{6,3\}$
(3, 3, 3, n)	M	n-gonal antiprism	$s\begin{Bmatrix}2\\n\end{Bmatrix}$
(3, 3, 3, 3)	P	octahedron	$\{3,4\}$
(3, 4, 3, 4)	A	cuboctahedron	$\begin{Bmatrix}3\\4\end{Bmatrix}$
(3, 5, 3, 5)	A	icosidodecahedron	$\begin{Bmatrix}3\\5\end{Bmatrix}$
(3, 6, 3, 6)	T	(plane tessellation)	$\begin{Bmatrix}3\\6\end{Bmatrix}$
(3, 4, 4, 4)	A	small rhombicuboct.	$r\begin{Bmatrix}3\\4\end{Bmatrix}$
(3, 4, 5, 4)	A	small rhombicosidodec.	$r\begin{Bmatrix}3\\5\end{Bmatrix}$
(3, 4, 6, 4)	T	(plane tessellation)	$r\begin{Bmatrix}3\\6\end{Bmatrix}$
(4, 4, 4, 4)	T	(plane tessellation)	$\{4,4\}$
(3, 3, 3, 3, 3)	P	icosahedron	$\{3,5\}$
(3, 3, 3, 3, 4)	A	snub cube	$s\begin{Bmatrix}3\\4\end{Bmatrix}$
(3, 3, 3, 3, 5)	A	snub dodecahedron	$s\begin{Bmatrix}3\\5\end{Bmatrix}$
(3, 3, 3, 3, 6)	T	(plane tessellation)	$s\begin{Bmatrix}3\\6\end{Bmatrix}$
(3, 3, 3, 4, 4)	T	(plane tessellation)	—
(3, 3, 4, 3, 4)	T	(plane tessellation)	$s\begin{Bmatrix}4\\4\end{Bmatrix}$
(3, 3, 3, 3, 3)	T	(plane tessellation)	$\{3,6\}$

As shown in the above table, there are exactly 13 Archimedean solids (Walsh 1972, Ball and Coxeter 1987).

They are called the CUBOCTAHEDRON, GREAT RHOMBICOSIDODECAHEDRON, GREAT RHOMBICUBOCTAHEDRON, ICOSIDODECAHEDRON, SMALL RHOMBICOSIDODECAHEDRON, SMALL RHOMBICUBOCTAHEDRON, SNUB CUBE, SNUB DODECAHEDRON, TRUNCATED CUBE, TRUNCATED DODECAHEDRON, TRUNCATED ICOSAHEDRON (soccer ball), TRUNCATED OCTAHEDRON, and TRUNCATED TETRAHEDRON. The Archimedean solids satisfy

$$(2\pi - \sigma)V = 4\pi,$$

where σ is the sum of face-angles at a vertex and V is the number of vertices (Steinitz and Rademacher 1934, Ball and Coxeter 1987).

Here are the Archimedean solids shown in alphabetical order (left to right, then continuing to the next row).

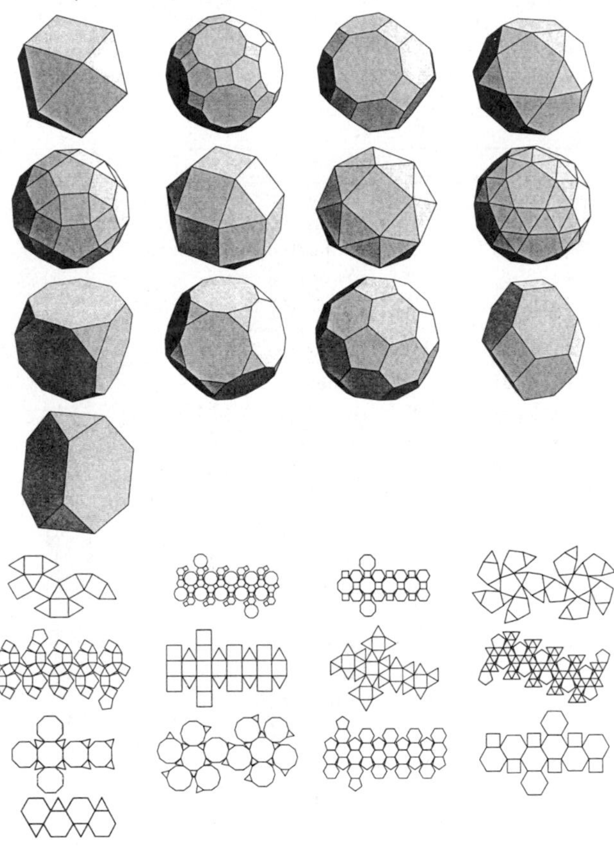

The following table lists the symbol and number of faces of each type for the Archimedean solids (Wenninger 1989, p. 9).

Solid	Schläfli	Wythoff	C&R
cuboctahedron	$\begin{Bmatrix}3\\4\end{Bmatrix}$	2 | 3 4	$(3.4)^2$
great rhombicosidodecahedron	$t\begin{Bmatrix}3\\5\end{Bmatrix}$	2 3 5 |	
great rhombicuboctahedron	$t\begin{Bmatrix}3\\4\end{Bmatrix}$	2 3 4 |	
icosidodecahedron	$\begin{Bmatrix}3\\5\end{Bmatrix}$	2 | 3 5	$(3.5)^2$
small rhombicosidodecahedron	$t\begin{Bmatrix}3\\5\end{Bmatrix}$	3 5 | 2	3.4.5.4
small rhombicuboctahedron	$r\begin{Bmatrix}3\\4\end{Bmatrix}$	3 4 | 2	3.4^3
snub cube	$s\begin{Bmatrix}3\\4\end{Bmatrix}$	| 2 3 4	$3^4.4$
snub dodecahedron	$s\begin{Bmatrix}3\\5\end{Bmatrix}$	| 2 3 5	$3^4.5$
truncated cube	t{4, 3}	2 3 | 4	3.8^2
truncated dodecahedron	t{5, 3}	2 3 | 5	3.10^2
truncated icosahedron	t{3, 5}	2 5 | 3	5.6^2
truncated octahedron	t{3, 4}	2 4 | 3	4.6^2
truncated tetrahedron	t{3, 3}	2 3 | 3	3.6^2

Solid	v	e	f_3	f_4	f_5	f_6	f_8	f_{10}
cuboctahedron	12	24	8	6				
great rhombicos.	120	180		30		20		12
great rhombicub.	48	72		12		8	6	
icosidodecahedron	30	60	20		12			
small rhombicos.	60	120	20	30	12			
small rhombicub.	24	48	8	18				
snub cube	24	60	32	6				
snub dodecahedron	60	150	80		12			
trunc. cube	24	36	8				6	
trunc. dodec.	60	90	20					12
trunc. icosahedron	60	90			12	20		
trunc. octahedron	24	36		6		8		
trunc. tetrahedron	12	18	4			4		

Let r be the INRADIUS, ρ the MIDRADIUS, and R the CIRCUMRADIUS. The following tables give the analytic and numerical values of r, ρ, and R for the Archimedean solids with EDGES of unit length.

Solid	r
cuboctahedron	$\frac{3}{4}$
great rhombicosidodecahedron	$\frac{1}{241}(105 + 6\sqrt{5})\sqrt{31 + 12\sqrt{5}}$
great rhombicuboctahedron	$\frac{3}{97}(14 + \sqrt{2})\sqrt{13 + 6\sqrt{2}}$
icosidodecahedron	$\frac{1}{8}(5 + 3\sqrt{5})$
small rhombicosidodecahedron	$\frac{1}{41}(15 + 2\sqrt{5})\sqrt{11 + 4\sqrt{5}}$
small rhombicuboctahedron	$\frac{1}{17}(6 + \sqrt{2})\sqrt{5 + 2\sqrt{2}}$
snub cube	*
snub dodecahedron	*
truncated cube	$\frac{1}{17}(5 + 2\sqrt{2})\sqrt{7 + 4\sqrt{2}}$
truncated dodecahedron	$\frac{5}{488}(17\sqrt{2} + 3\sqrt{10})\sqrt{37 + 15\sqrt{5}}$
truncated icosahedron	$\frac{9}{872}(21 + \sqrt{5})\sqrt{58 + 18\sqrt{5}}$
truncated octahedron	$\frac{9}{20}\sqrt{10}$
truncated tetrahedron	$\frac{9}{44}\sqrt{22}$

Solid	ρ	R
cuboctahedron	$\frac{1}{2}\sqrt{3}$	1
great rhombicosidodecahedron	$\frac{1}{2}\sqrt{30+12\sqrt{5}}$	$\frac{1}{2}\sqrt{31+12\sqrt{5}}$
great rhombicuboctahedron	$\frac{1}{2}\sqrt{12+6\sqrt{2}}$	$\frac{1}{2}\sqrt{13+6\sqrt{2}}$
icosidodecahedron	$\frac{1}{2}\sqrt{5+2\sqrt{5}}$	$\frac{1}{2}(1+\sqrt{5})$
small rhombicosidodecahedron	$\frac{1}{2}\sqrt{10+4\sqrt{5}}$	$\frac{1}{2}\sqrt{11+4\sqrt{5}}$
small rhombicuboctahedron	$\frac{1}{2}\sqrt{4+2\sqrt{2}}$	$\frac{1}{2}\sqrt{5+2\sqrt{2}}$
snub cube	*	*
snub dodecahedron	*	*
truncated cube	$\frac{1}{2}(2+\sqrt{2})$	$\frac{1}{2}\sqrt{7+4\sqrt{2}}$
truncated dodecahedron	$\frac{1}{4}(5+3\sqrt{5})$	$\frac{1}{4}\sqrt{74+30\sqrt{5}}$
truncated icosahedron	$\frac{3}{4}(1+\sqrt{5})$	$\frac{1}{4}\sqrt{58+18\sqrt{5}}$
truncated octahedron	$\frac{3}{2}$	$\frac{1}{2}\sqrt{10}$
truncated tetrahedron	$\frac{3}{4}\sqrt{2}$	$\frac{1}{4}\sqrt{22}$

*The complicated analytic expressions for the CIRCUMRADII of these solids are given in the entries for the SNUB CUBE and SNUB DODECAHEDRON.

Solid	r	ρ	R
cuboctahedron	0.75	0.86603	1
great rhombicosidodecahedron	3.73665	3.76938	3.80239
great rhombicuboctahedron	2.20974	2.26303	2.31761
icosidodecahedron	1.46353	1.53884	1.61803
small rhombicosidodecahedron	2.12099	2.17625	2.23295
small rhombicuboctahedron	1.22026	1.30656	1.39897
snub cube	1.15766	1.24722	1.34371
snub dodecahedron	2.03987	2.09705	2.15583
truncated cube	1.63828	1.70711	1.77882
truncated dodecahedron	2.88526	2.92705	2.96945
truncated icosahedron	2.37713	2.42705	2.47802
truncated octahedron	1.42302	1.5	1.58114
truncated tetrahedron	0.95940	1.06066	1.17260

The DUALS of the Archimedean solids, sometimes called the CATALAN SOLIDS, are given in the following table.

Archimedean Solid	Dual
rhombicosidodecahedron	deltoidal hexecontahedron
small rhombicuboctahedron	deltoidal icositetrahedron
great rhombicuboctahedron	disdyakis dodecahedron
great rhombicosidodecahedron	disdyakis triacontahedron
truncated icosahedron	pentakis dodecahedron
snub dodecahedron (laevo)	pentagonal hexecontahedron (dextro)
snub cube (laevo)	pentagonal icositetrahedron (dextro)
cuboctahedron	rhombic dodecahedron
icosidodecahedron	rhombic triacontahedron
truncated octahedron	tetrakis hexahedron
truncated dodecahedron	triakis icosahedron
truncated cube	triakis octahedron
truncated tetrahedron	triakis tetrahedron

Here are the Archimedean DUALS (Holden 1971, Pearce 1978) displayed in alphabetical order (left to right, then continuing to the next row).

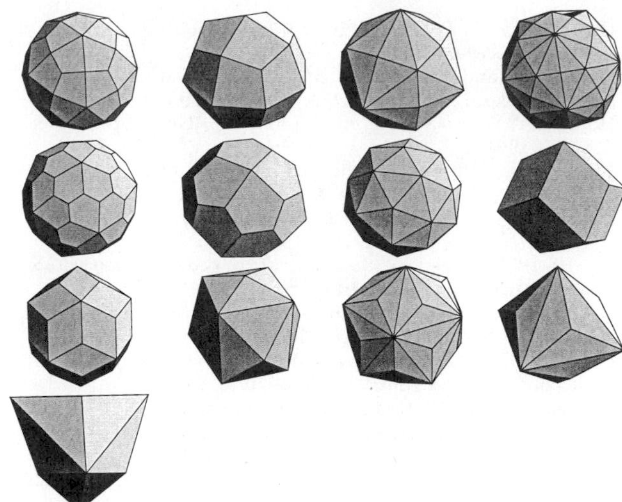

Here are the Archimedean solids paired with their DUALS.

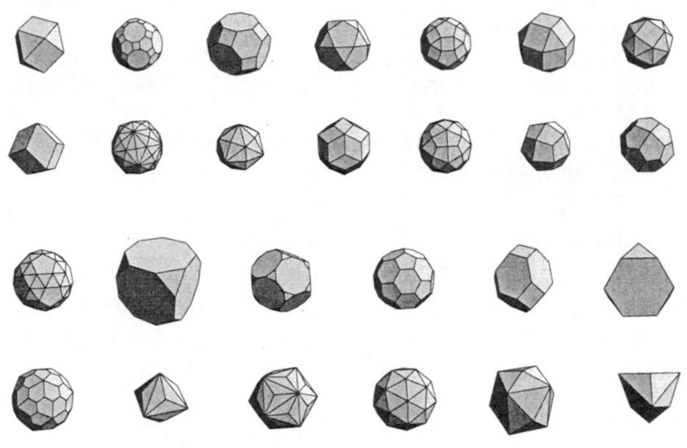

The Archimedean solids and their DUALS are all CANONICAL POLYHEDRA.

see also ARCHIMEDEAN SOLID STELLATION, CATALAN SOLID, DELTAHEDRON, JOHNSON SOLID, KEPLER-POINSOT SOLID, PLATONIC SOLID, SEMIREGULAR POLYHEDRON, UNIFORM POLYHEDRON

References

Ball, W. W. R. and Coxeter, H. S. M. *Mathematical Recreations and Essays, 13th ed.* New York: Dover, p. 136, 1987.

Behnke, H.; Bachman, F.; Fladt, K.; and Kunle, H. (Eds.). *Fundamentals of Mathematics, Vol. 2.* Cambridge, MA: MIT Press, pp. 269–286, 1974.

Catalan, E. "Mémoire sur la Théorie des Polyèdres." *J. l'École Polytechnique (Paris)* **41**, 1–71, 1865.

Coxeter, H. S. M. "The Pure Archimedean Polytopes in Six and Seven Dimensions." *Proc. Cambridge Phil. Soc.* **24**, 1–9, 1928.

Coxeter, H. S. M. "Regular and Semi-Regular Polytopes I." *Math. Z.* **46**, 380–407, 1940.

Coxeter, H. S. M. *Regular Polytopes, 3rd ed.* New York: Dover, 1973.

Coxeter, H. S. M.; Longuet-Higgins, M. S.; and Miller, J. C. P. "Uniform Polyhedra." *Phil. Trans. Roy. Soc. London Ser. A* **246**, 401–450, 1954.

Critchlow, K. *Order in Space: A Design Source Book.* New York: Viking Press, 1970.

Cromwell, P. R. *Polyhedra.* New York: Cambridge University Press, pp. 79–86, 1997.
Cundy, H. and Rollett, A. *Mathematical Models, 3rd ed.* Stradbroke, England: Tarquin Pub., 1989.
Holden, A. *Shapes, Space, and Symmetry.* New York: Dover, p. 54, 1991.
Kepler, J. "Harmonice Mundi." *Opera Omnia, Vol. 5.* Frankfurt, pp. 75–334, 1864.
Kraitchik, M. *Mathematical Recreations.* New York: W. W. Norton, pp. 199–207, 1942.
Le, Ha. "Archimedean Solids." `http://daisy.uwaterloo.ca/~hqle/archimedean.html`.
Pearce, P. *Structure in Nature is a Strategy for Design.* Cambridge, MA: MIT Press, pp. 34–35, 1978.
Pugh, A. *Polyhedra: A Visual Approach.* Berkeley: University of California Press, p. 25, 1976.
Rawles, B. A. "Platonic and Archimedean Solids—Faces, Edges, Areas, Vertices, Angles, Volumes, Sphere Ratios." `http://www.intent.com/sg/polyhedra.html`.
Rorres, C. "Archimedean Solids: Pappus." `http://www.mcs.drexel.edu/~crorres/Archimedes/Solids/Pappus.html`.
Steinitz, E. and Rademacher, H. *Vorlesungen über die Theorie der Polyheder.* Berlin, p. 11, 1934.
Stott, A. B. *Verhandelingen der Konniklijke Akad. Wetenschappen, Amsterdam* **11**, 1910.
Walsh, T. R. S. "Characterizing the Vertex Neighbourhoods of Semi-Regular Polyhedra." *Geometriae Dedicata* **1**, 117–123, 1972.
Wenninger, M. J. *Polyhedron Models.* New York: Cambridge University Press, 1989.

Archimedean Solid Stellation

A large class of POLYHEDRA which includes the DODECADODECAHEDRON and GREAT ICOSIDODECAHEDRON. No complete enumeration (even with restrictive uniqueness conditions) has been worked out.

References

Coxeter, H. S. M.; Longuet-Higgins, M. S.; and Miller, J. C. P. "Uniform Polyhedra." *Phil. Trans. Roy. Soc. London Ser. A* **246**, 401–450, 1954.
Wenninger, M. J. *Polyhedron Models.* New York: Cambridge University Press, pp. 66–72, 1989.

Archimedean Spiral

A SPIRAL with POLAR equation

$$r = a\theta^{1/m},$$

where r is the radial distance, θ is the polar angle, and m is a constant which determines how tightly the spiral is "wrapped." The CURVATURE of an Archimedean spiral is given by

$$\kappa = \frac{n\theta^{1-1/n}(1+n+n^2\theta^2)}{a(1+n^2\theta^2)^{3/2}}.$$

Various special cases are given in the following table.

Name	m
lituus	-2
hyperbolic spiral	-1
Archimedes' spiral	1
Fermat's spiral	2

see also ARCHIMEDES' SPIRAL, DAISY, FERMAT'S SPIRAL, HYPERBOLIC SPIRAL, LITUUS, SPIRAL

References

Gray, A. *Modern Differential Geometry of Curves and Surfaces.* Boca Raton, FL: CRC Press, pp. 69–70, 1993.
Lauweirer, H. *Fractals: Endlessly Repeated Geometric Figures.* Princeton, NJ: Princeton University Press, pp. 59–60, 1991.
Lawrence, J. D. *A Catalog of Special Plane Curves.* New York: Dover, pp. 186 and 189, 1972.
Lee, X. "Archimedean Spiral." `http://www.best.com/~xah/Special Plane Curves _ dir / Archimedean Spiral _ dir / archimedeanSpiral.html`.
Lockwood, E. H. *A Book of Curves.* Cambridge, England: Cambridge University Press, p. 175, 1967.
MacTutor History of Mathematics Archive. "Spiral of Archimedes." `http: // www - groups . dcs . st - and . ac . uk / ~history/Curves/Spiral.html`.
Pappas, T. "The Spiral of Archimedes." *The Joy of Mathematics.* San Carlos, CA: Wide World Publ./Tetra, p. 149, 1989.

Archimedean Spiral Inverse Curve

The INVERSE CURVE of the ARCHIMEDEAN SPIRAL

$$r = a\theta^{1/m}$$

with INVERSION CENTER at the origin and inversion RADIUS k is the ARCHIMEDEAN SPIRAL

$$r = ka\theta^{1/m}.$$

Archimedes' Spiral

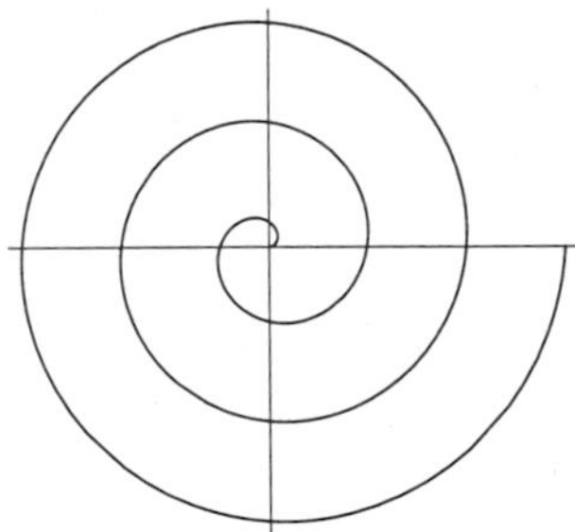

An ARCHIMEDEAN SPIRAL with POLAR equation

$$r = a\theta.$$

This spiral was studied by Conon, and later by Archimedes in *On Spirals* about 225 BC. Archimedes was able to work out the lengths of various tangents to the spiral.

Archimedes' spiral can be used for COMPASS and STRAIGHTEDGE division of an ANGLE into n parts (including ANGLE TRISECTION) and can also be used for CIRCLE SQUARING. In addition, the curve can be used as a cam to convert uniform circular motion into uniform linear motion. The cam consists of one arch of the spiral above the x-AXIS together with its reflection in the x-AXIS. Rotating this with uniform angular velocity about its center will result in uniform linear motion of the point where it crosses the y-AXIS.

see also ARCHIMEDEAN SPIRAL

References

Gardner, M. *The Unexpected Hanging and Other Mathematical Diversions.* Chicago, IL: Chicago University Press, pp. 106–107, 1991.

Gray, A. *Modern Differential Geometry of Curves and Surfaces.* Boca Raton, FL: CRC Press, pp. 69–70, 1993.

Lawrence, J. D. *A Catalog of Special Plane Curves.* New York: Dover, pp. 186–187, 1972.

Lockwood, E. H. *A Book of Curves.* Cambridge, England: Cambridge University Press, pp. 173–164, 1967.

Archimedes' Spiral Inverse

Taking the ORIGIN as the INVERSION CENTER, ARCHIMEDES' SPIRAL $r = a\theta$ inverts to the HYPERBOLIC SPIRAL $r = a/\theta$.

Archimedean Valuation

A VALUATION for which $|x| \leq 1$ IMPLIES $|1+x| \leq C$ for the constant $C = 1$ (independent of x). Such a VALUATION *does not* satisfy the strong TRIANGLE INEQUALITY

$$|x + y| \leq \max(|x|, |y|).$$

Arcsecant

see INVERSE SECANT

Arcsine

see INVERSE SINE

Arctangent

see INVERSE TANGENT

Area

The AREA of a SURFACE is the amount of material needed to "cover" it completely. The AREA of a TRIANGLE is given by

$$A_\Delta = \tfrac{1}{2} lh, \tag{1}$$

where l is the base length and h is the height, or by HERON'S FORMULA

$$A_\Delta = \sqrt{s(s-a)(s-b)(s-c)}, \tag{2}$$

where the side lengths are a, b, and c and s the SEMIPERIMETER. The AREA of a RECTANGLE is given by

$$A_{\text{rectangle}} = ab, \tag{3}$$

where the sides are length a and b. This gives the special case of

$$A_{\text{square}} = a^2 \tag{4}$$

for the SQUARE. The AREA of a regular POLYGON with n sides and side length s is given by

$$A_{n\text{-gon}} = \tfrac{1}{4} ns^2 \cot\left(\frac{\pi}{n}\right). \tag{5}$$

CALCULUS and, in particular, the INTEGRAL, are powerful tools for computing the AREA between a curve $f(x)$ and the x-AXIS over an INTERVAL $[a, b]$, giving

$$A = \int_a^b f(x)\,dx. \tag{6}$$

The AREA of a POLAR curve with equation $r = r(\theta)$ is

$$A = \tfrac{1}{2} \int r^2\,d\theta. \tag{7}$$

Written in CARTESIAN COORDINATES, this becomes

$$A = \frac{1}{2} \int \left(x\frac{dy}{dt} - y\frac{dx}{dt} \right) dt \tag{8}$$

$$= \frac{1}{2} \int (x\,dy - y\,dx). \tag{9}$$

For the AREA of special surfaces or regions, see the entry for that region. The generalization of AREA to 3-D is called VOLUME, and to higher DIMENSIONS is called CONTENT.

see also ARC LENGTH, AREA ELEMENT, CONTENT, SURFACE AREA, VOLUME

References

Gray, A. "The Intuitive Idea of Area on a Surface." §13.2 in *Modern Differential Geometry of Curves and Surfaces.* Boca Raton, FL: CRC Press, pp. 259–260, 1993.

Area Element

The area element for a SURFACE with RIEMANNIAN METRIC

$$ds^2 = E\,du^2 + 2F\,du\,dv + G\,dv^2$$

is

$$dA = \sqrt{EG - F^2}\,du \wedge dv,$$

where $du \wedge dv$ is the WEDGE PRODUCT.

see also AREA, LINE ELEMENT, RIEMANNIAN METRIC, VOLUME ELEMENT

References

Gray, A. "The Intuitive Idea of Area on a Surface." §13.2 in *Modern Differential Geometry of Curves and Surfaces.* Boca Raton, FL: CRC Press, pp. 259–260, 1993.

Area-Preserving Map

A MAP F from $\mathbb{R}^n$ to $\mathbb{R}^n$ is AREA-preserving if

$$m(F(A)) = m(A)$$

for every subregion A of $\mathbb{R}^n$, where $m(A)$ is the n-D MEASURE of A. A linear transformation is AREA-preserving if its corresponding DETERMINANT is equal to 1.

see also CONFORMAL MAP, SYMPLECTIC MAP

Area Principle

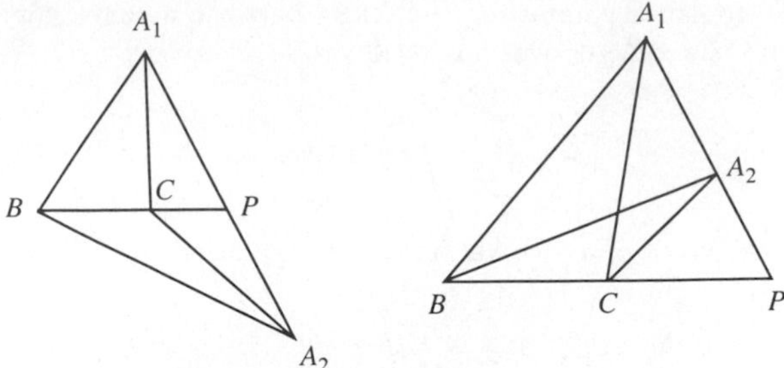

The "AREA principle" states that

$$\frac{|A_1P|}{|A_2P|} = \frac{|A_1BC|}{|A_2BC|}. \quad (1)$$

This can also be written in the form

$$\left[\frac{A_1P}{A_2P}\right] = \left[\frac{A_1BC}{A_2BC}\right], \quad (2)$$

where

$$\left[\frac{AB}{CD}\right] \quad (3)$$

is the ratio of the lengths $[A, B]$ and $[C, D]$ for $AB||CD$ with a PLUS or MINUS SIGN depending on if these segments have the same or opposite directions, and

$$\left[\frac{ABC}{DEFG}\right] \quad (4)$$

is the RATIO of signed AREAS of the TRIANGLES. Grünbaum and Shepard show that CEVA'S THEOREM, HOEHN'S THEOREM, and MENELAUS' THEOREM are the consequences of this result.

see also CEVA'S THEOREM, HOEHN'S THEOREM, MENELAUS' THEOREM, SELF-TRANSVERSALITY THEOREM

References
Grünbaum, B. and Shepard, G. C. "Ceva, Menelaus, and the Area Principle." *Math. Mag.* **68**, 254–268, 1995.

Areal Coordinates

TRILINEAR COORDINATES normalized so that

$$t_1 + t_2 + t_3 = 1.$$

When so normalized, they become the AREAS of the TRIANGLES PA_1A_2, PA_1A_3, and PA_2A_3, where P is the point whose coordinates have been specified.

Arf Invariant

A LINK invariant which always has the value 0 or 1. A KNOT has ARF INVARIANT 0 if the KNOT is "pass equivalent" to the UNKNOT and 1 if it is pass equivalent to the TREFOIL KNOT. If K_+, K_-, and L are projections which are identical outside the region of the crossing diagram, and K_+ and K_- are KNOTS while L is a 2-component LINK with a nonintersecting crossing diagram where the two left and right strands belong to the different LINKS, then

$$a(K_+) = a(K_-) + l(L_1, L_2), \quad (1)$$

where l is the LINKING NUMBER of L_1 and L_2. The Arf invariant can be determined from the ALEXANDER POLYNOMIAL or JONES POLYNOMIAL for a KNOT. For Δ_K the ALEXANDER POLYNOMIAL of K, the Arf invariant is given by

$$\Delta_K(-1) \equiv \begin{cases} 1 \pmod 8 & \text{if } \mathrm{Arf}(K) = 0 \\ 5 \pmod 8 & \text{if } \mathrm{Arf}(K) = 1 \end{cases} \quad (2)$$

(Jones 1985). For the JONES POLYNOMIAL W_K of a KNOT K,

$$\mathrm{Arf}(K) = W_K(i) \quad (3)$$

(Jones 1985), where i is the IMAGINARY NUMBER.

References
Adams, C. C. *The Knot Book: An Elementary Introduction to the Mathematical Theory of Knots.* New York: W. H. Freeman, pp. 223–231, 1994.
Jones, V. "A Polynomial Invariant for Knots via von Neumann Algebras." *Bull. Amer. Math. Soc.* **12**, 103–111, 1985.
Weisstein, E. W. "Knots." http://www.astro.virginia.edu/~eww6n/math/notebooks/Knots.m.

Argand Diagram

A plot of COMPLEX NUMBERS as points

$$z = x + iy$$

using the x-AXIS as the REAL axis and y-AXIS as the IMAGINARY axis. This is also called the COMPLEX PLANE or ARGAND PLANE.

Argand Plane

see ARGAND DIAGRAM

Argoh's Conjecture

Let B_k be the kth BERNOULLI NUMBER. Then does

$$nB_{n-1} \equiv -1 \pmod n$$

IFF n is PRIME? For example, for $n = 1, 2, \ldots, nB_{n-1} \pmod n$ is 0, −1, −1, 0, −1, 0, −1, 0, −3, 0, −1, ... There are no counterexamples less than $n = 5{,}600$. Any counterexample to Argoh's conjecture would be a contradiction to GIUGA'S CONJECTURE, and vice versa.

see also BERNOULLI NUMBER, GIUGA'S CONJECTURE

References
Borwein, D.; Borwein, J. M.; Borwein, P. B.; and Girgensohn, R. "Giuga's Conjecture on Primality." *Amer. Math. Monthly* **103**, 40–50, 1996.

Argument Addition Relation

A mathematical relationship relating $f(x+y)$ to $f(x)$ and $f(y)$.

see also ARGUMENT MULTIPLICATION RELATION, RECURRENCE RELATION, REFLECTION RELATION, TRANSLATION RELATION

Argument (Complex Number)

A COMPLEX NUMBER z may be represented as

$$z \equiv x + iy = |z|e^{i\theta}, \tag{1}$$

where $|z|$ is called the MODULUS of z, and θ is called the argument

$$\arg(x+iy) \equiv \tan^{-1}\left(\frac{y}{x}\right). \tag{2}$$

Therefore,

$$\begin{aligned}\arg(zw) &= \arg(|z|e^{i\theta_z}|w|e^{i\theta_w}) = \arg(e^{i\theta_z}e^{i\theta_w}) \\ &= \arg[e^{i(\theta_z+\theta_w)}] = \arg(z) + \arg(w).\end{aligned} \tag{3}$$

Extending this procedure gives

$$\arg(z^n) = n\arg(z). \tag{4}$$

The argument of a COMPLEX NUMBER is sometimes called the PHASE.

see also AFFIX, COMPLEX NUMBER, DE MOIVRE'S IDENTITY, EULER FORMULA, MODULUS (COMPLEX NUMBER), PHASE, PHASOR

References

Abramowitz, M. and Stegun, C. A. (Eds.). *Handbook of Mathematical Functions with Formulas, Graphs, and Mathematical Tables, 9th printing.* New York: Dover, p. 16, 1972.

Argument (Elliptic Integral)

Given an AMPLITUDE ϕ in an ELLIPTIC INTEGRAL, the argument u is defined by the relation

$$\phi \equiv \operatorname{am} u.$$

see also AMPLITUDE, ELLIPTIC INTEGRAL

Argument (Function)

An argument of a FUNCTION $f(x_1,\ldots,x_n)$ is one of the n parameters on which the function's value depends. For example, the SINE $\sin x$ is a one-argument function, the BINOMIAL COEFFICIENT $\binom{n}{m}$ is a two-argument function, and the HYPERGEOMETRIC FUNCTION ${}_2F_1(a,b;c;z)$ is a four-argument function.

Argument Multiplication Relation

A mathematical relationship relating $f(nx)$ to $f(x)$ for INTEGER n.

see also ARGUMENT ADDITION RELATION, RECURRENCE RELATION, REFLECTION RELATION, TRANSLATION RELATION

Argument Principle

If $f(z)$ is MEROMORPHIC in a region R enclosed by a curve γ, let N be the number of COMPLEX ROOTS of $f(z)$ in γ, and P be the number of POLES in γ, then

$$N - P = \frac{1}{2\pi i}\int_\gamma \frac{f'(z)\,dz}{f(z)}.$$

Defining $w \equiv f(z)$ and $\sigma \equiv f(\gamma)$ gives

$$N - P = \frac{1}{2\pi i}\int_\sigma \frac{dw}{w}.$$

see also VARIATION OF ARGUMENT

References

Duren, P.; Hengartner, W.; and Laugessen, R. S. "The Argument Principle for Harmonic Functions." *Math. Mag.* **103**, 411–415, 1996.

Argument Variation

see VARIATION OF ARGUMENT

Aristotle's Wheel Paradox

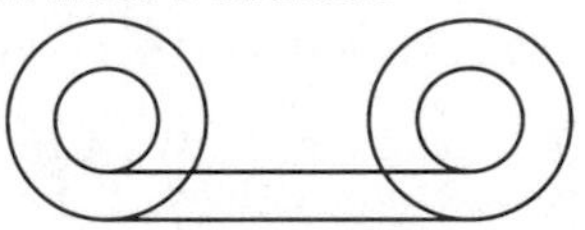

A PARADOX mentioned in the Greek work *Mechanica,* dubiously attributed to Aristotle. Consider the above diagram depicting a wheel consisting of two concentric CIRCLES of different DIAMETERS (a wheel within a wheel). There is a 1:1 correspondence of points on the large CIRCLE with points on the small CIRCLE, so the wheel should travel the same distance regardless of whether it is rolled from left to right on the top straight line or on the bottom one. This seems to imply that the two CIRCUMFERENCES of different sized CIRCLES are equal, which is impossible.

The fallacy lies in the assumption that a 1:1 correspondence of points means that two curves must have the same length. In fact, the CARDINALITIES of points in a LINE SEGMENT of any length (or even an INFINITE LINE, a PLANE, a 3-D SPACE, or an infinite dimensional EUCLIDEAN SPACE) are all the same: $\aleph_1$ (ALEPH-1), so the points of any of these can be put in a ONE-TO-ONE correspondence with those of any other.

see also ZENO'S PARADOXES

References

Ballew, D. "The Wheel of Aristotle." *Math. Teacher* **65**, 507–509, 1972.

Costabel, P. "The Wheel of Aristotle and French Consideration of Galileo's Arguments." *Math. Teacher* **61**, 527–534, 1968.

Drabkin, I. "Aristotle's Wheel: Notes on the History of the Paradox." *Osiris* **9**, 162–198, 1950.

Gardner, M. *Wheels, Life, and other Mathematical Amusements.* New York: W. H. Freeman, pp. 2–4, 1983.

Pappas, T. "The Wheel of Paradox Aristotle." *The Joy of Mathematics.* San Carlos, CA: Wide World Publ./Tetra, p. 202, 1989.

vos Savant, M. *The World's Most Famous Math Problem.* New York: St. Martin's Press, pp. 48–50, 1993.

Arithmetic

The branch of mathematics dealing with INTEGERS or, more generally, numerical computation. Arithmetical operations include ADDITION, CONGRUENCE calculation, DIVISION, FACTORIZATION, MULTIPLICATION, POWER computation, ROOT extraction, and SUBTRACTION.

The FUNDAMENTAL THEOREM OF ARITHMETIC, also called the UNIQUE FACTORIZATION THEOREM, states that any POSITIVE INTEGER can be represented in exactly one way as a PRODUCT of PRIMES.

The LÖWENHEIMER-SKOLEM THEOREM, which is a fundamental result in MODEL THEORY, establishes the existence of "nonstandard" models of arithmetic.

see also ALGEBRA, CALCULUS, FUNDAMENTAL THEOREM OF ARITHMETIC, GROUP THEORY, HIGHER ARITHMETIC, LINEAR ALGEBRA, LÖWENHEIMER-SKOLEM THEOREM, MODEL THEORY, NUMBER THEORY, TRIGONOMETRY

References

Karpinski, L. C. *The History of Arithmetic.* Chicago, IL: Rand, McNally, & Co., 1925.

Maxfield, J. E. and Maxfield, M. W. *Abstract Algebra and Solution by Radicals.* Philadelphia, PA: Saunders, 1992.

Thompson, J. E. *Arithmetic for the Practical Man.* New York: Van Nostrand Reinhold, 1973.

Arithmetic-Geometric Mean

The arithmetic-geometric mean (AGM) $M(a,b)$ of two numbers a and b is defined by starting with $a_0 \equiv a$ and $b_0 \equiv b$, then iterating

$$a_{n+1} = \tfrac{1}{2}(a_n + b_n) \tag{1}$$

$$b_{n+1} = \sqrt{a_n b_n} \tag{2}$$

until $a_n = b_n$. a_n and b_n converge towards each other since

$$\begin{aligned} a_{n+1} - b_{n+1} &= \tfrac{1}{2}(a_n + b_n) - \sqrt{a_n b_n} \\ &= \frac{a_n - 2\sqrt{a_n b_n} + b_n}{2}. \end{aligned} \tag{3}$$

But $\sqrt{b_n} < \sqrt{a_n}$, so

$$2b_n < 2\sqrt{a_n b_n}. \tag{4}$$

Now, add $a_n - b_n - 2\sqrt{a_n b_n}$ to each side

$$a_n + b_n - 2\sqrt{a_n b_n} < a_n - b_n, \tag{5}$$

so

$$a_{n+1} - b_{n+1} < \tfrac{1}{2}(a_n - b_n). \tag{6}$$

The AGM is very useful in computing the values of complete ELLIPTIC INTEGRALS and can also be used for finding the INVERSE TANGENT. The special value $1/M(1, \sqrt{2})$ is called GAUSS'S CONSTANT.

The AGM has the properties

$$\lambda M(a,b) = M(\lambda a, \lambda b) \tag{7}$$

$$M(a,b) = M\left(\tfrac{1}{2}(a+b), \sqrt{ab}\right) \tag{8}$$

$$M(1, \sqrt{1-x^2}) = M(1+x, 1-x) \tag{9}$$

$$M(1,b) = \frac{1+b}{2} M\left(1, \frac{2\sqrt{b}}{1+b}\right). \tag{10}$$

The Legendre form is given by

$$M(1,x) = \prod_{n=0}^{\infty} \tfrac{1}{2}(1+k_n), \tag{11}$$

where $k_0 \equiv x$ and

$$k_{n+1} \equiv \frac{2\sqrt{k_n}}{1+k_n}. \tag{12}$$

Solutions to the differential equation

$$(x^3 - x)\frac{d^2y}{dx^2} + (3x^2 - 1)\frac{dy}{dx} + xy = 0 \tag{13}$$

are given by $[M(1+x, 1-x)]^{-1}$ and $[M(1,x)]^{-1}$.

A generalization of the ARITHMETIC-GEOMETRIC MEAN is

$$I_p(a,b) = \int_0^\infty \frac{x^{p-2}\,dx}{(x^p + a^p)^{1/p}(x^p + b^p)^{(p-1)/p}}, \tag{14}$$

which is related to solutions of the differential equation

$$x(1-x^p)Y'' + [1-(p+1)x^p]Y' - (p-1)x^{p-1}Y = 0. \tag{15}$$

When $p = 2$ or $p = 3$, there is a modular transformation for the solutions of (15) that are bounded as $x \to 0$. Letting $J_p(x)$ be one of these solutions, the transformation takes the form

$$J_p(\lambda) = \mu J_p(x), \tag{16}$$

where

$$\lambda = \frac{1-u}{1+(p-1)u} \tag{17}$$

$$\mu = \frac{1+(p-1)u}{p} \tag{18}$$

and

$$x^p + u^p = 1. \tag{19}$$

The case $p = 2$ gives the ARITHMETIC-GEOMETRIC MEAN, and $p = 3$ gives a cubic relative discussed by Borwein and Borwein (1990, 1991) and Borwein (1996) in which, for $a, b > 0$ and $I(a,b)$ defined by

$$I(a,b) = \int_0^\infty \frac{t\,dt}{[(a^3+t^3)(b^3+t^3)^2]^{1/3}}, \tag{20}$$

$$I(a,b) = I\left(\frac{a+2b}{3}, \left[\frac{b}{3}(a^2+ab+b^2)\right]\right). \tag{21}$$

For iteration with $a_0 = a$ and $b_0 = b$ and

$$a_{n+1} = \frac{a_n + 2b_n}{3} \tag{22}$$

$$b_{n+1} = \frac{b_n}{3}({a_n}^2 + a_n b_n + {b_n}^2), \tag{23}$$

$$\lim_{n\to\infty} a_n = \lim_{n\to\infty} b_n = \frac{I(1,1)}{I(a,b)}. \tag{24}$$

Modular transformations are known when $p = 4$ and $p = 6$, but they do not give identities for $p = 6$ (Borwein 1996).

see also ARITHMETIC-HARMONIC MEAN

References

Abramowitz, M. and Stegun, C. A. (Eds.). "The Process of the Arithmetic-Geometric Mean." §17.6 in *Handbook of Mathematical Functions with Formulas, Graphs, and Mathematical Tables, 9th printing.* New York: Dover, pp. 571 ad 598–599, 1972.

Borwein, J. M. Problem 10281. "A Cubic Relative of the AGM." *Amer. Math. Monthly* **103**, 181–183, 1996.

Borwein, J. M. and Borwein, P. B. "A Remarkable Cubic Iteration." In *Computational Method & Function Theory: Proc. Conference Held in Valparaiso, Chile, March 13–18, 1989*0387527680 (Ed. A. Dold, B. Eckmann, F. Takens, E. B Saff, S. Ruscheweyh, L. C. Salinas, L. C., and R. S. Varga). New York: Springer-Verlag, 1990.

Borwein, J. M. and Borwein, P. B. "A Cubic Counterpart of Jacobi's Identity and the AGM." *Trans. Amer. Math. Soc.* **323**, 691–701, 1991.

Press, W. H.; Flannery, B. P.; Teukolsky, S. A.; and Vetterling, W. T. *Numerical Recipes in FORTRAN: The Art of Scientific Computing, 2nd ed.* Cambridge, England: Cambridge University Press, pp. 906–907, 1992.

Arithmetic Geometry

A vaguely defined branch of mathematics dealing with VARIETIES, the MORDELL CONJECTURE, ARAKELOV THEORY, and ELLIPTIC CURVES.

References

Cornell, G. and Silverman, J. H. (Eds.). *Arithmetic Geometry.* New York: Springer-Verlag, 1986.

Lorenzini, D. *An Invitation to Arithmetic Geometry.* Providence, RI: Amer. Math. Soc., 1996.

Arithmetic-Harmonic Mean

Let

$$a_{n+1} = \tfrac{1}{2}(a_n + b_n) \tag{1}$$

$$b_{n+1} = \frac{2a_n b_2}{a_n + b_n}. \tag{2}$$

Then

$$A(a_0, b_0) = \lim_{n\to\infty} a_n = \lim_{n\to\infty} b_n = \sqrt{a_0 b_0}, \tag{3}$$

which is just the GEOMETRIC MEAN.

Arithmetic-Logarithmic-Geometric Mean Inequality

$$\frac{a+b}{2} > \frac{b-a}{\ln b - \ln a} > \sqrt{ab}.$$

see also NAPIER'S INEQUALITY

References

Nelson, R. B. "Proof without Words: The Arithmetic-Logarithmic-Geometric Mean Inequality." *Math. Mag.* **68**, 305, 1995.

Arithmetic Mean

For a CONTINUOUS DISTRIBUTION function, the arithmetic mean of the population, denoted μ, $\bar{x}$, $\langle x \rangle$, or $A(x)$, is given by

$$\mu = \langle f(x) \rangle \equiv \int_{-\infty}^{\infty} P(x) f(x)\, dx, \tag{1}$$

where $\langle x \rangle$ is the EXPECTATION VALUE. For a DISCRETE DISTRIBUTION,

$$\mu = \langle f(x) \rangle \equiv \frac{\sum_{n=0}^{N} P(x_n) f(x_n)}{\sum_{n=0}^{N} P(x_n)} = \sum_{n=0}^{N} P(x_n) f(x_n). \tag{2}$$

The population mean satisfies

$$\langle f(x) + g(x) \rangle = \langle f(x) \rangle + \langle g(x) \rangle \tag{3}$$

$$\langle c f(x) \rangle = c \langle f(x) \rangle, \tag{4}$$

and

$$\langle f(x) g(y) \rangle = \langle f(x) \rangle \langle g(y) \rangle \tag{5}$$

if x and y are INDEPENDENT STATISTICS. The "sample mean," which is the mean estimated from a statistical sample, is an UNBIASED ESTIMATOR for the population mean.

For small samples, the mean is more efficient than the MEDIAN and approximately $\pi/2$ less (Kenney and Keeping 1962, p. 211). A general expression which often holds approximately is

$$\text{mean} - \text{mode} \approx 3(\text{mean} - \text{median}). \tag{6}$$

Given a set of samples $\{x_i\}$, the arithmetic mean is

$$A(x) \equiv \bar{x} \equiv \mu \equiv \langle x \rangle = \frac{1}{N} \sum_{i=1}^{N} x_i. \tag{7}$$

Hoehn and Niven (1985) show that

$$A(a_1 + c, a_2 + c, \ldots, a_n + c) = c + A(a_1, a_2, \ldots, a_n) \tag{8}$$

for any POSITIVE constant c. The arithmetic mean satisfies

$$A \geq G \geq H, \tag{9}$$

where G is the GEOMETRIC MEAN and H is the HARMONIC MEAN (Hardy *et al.* 1952; Mitrinović 1970; Beckenbach and Bellman 1983; Bullen *et al.* 1988; Mitrinović *et al.* 1993; Alzer 1996). This can be shown as follows. For $a, b > 0$,

$$\left(\frac{1}{\sqrt{a}} - \frac{1}{\sqrt{b}}\right)^2 \geq 0 \tag{10}$$

$$\frac{1}{a} - \frac{2}{\sqrt{ab}} + \frac{1}{b} \geq 0 \tag{11}$$

$$\frac{1}{a} + \frac{1}{b} \geq \frac{2}{\sqrt{ab}} \tag{12}$$

$$\frac{2}{\frac{1}{a} + \frac{1}{b}} \geq \sqrt{ab} \tag{13}$$

$$H \geq G, \tag{14}$$

with equality IFF $b = a$. To show the second part of the inequality,

$$(\sqrt{a} - \sqrt{b})^2 = a - 2\sqrt{ab} + b \geq 0 \tag{15}$$

$$\frac{a+b}{2} \geq \sqrt{ab} \tag{16}$$

$$A \geq H, \tag{17}$$

with equality IFF $a = b$. Combining (14) and (17) then gives (9).

Given n independent random GAUSSIAN DISTRIBUTED variates x_i, each with population mean $\mu_i = \mu$ and VARIANCE ${\sigma_i}^2 = \sigma^2$,

$$\bar{x} \equiv \frac{1}{N} \sum_{i=1}^{N} x_i \tag{18}$$

$$\begin{aligned}\langle \bar{x} \rangle &= \frac{1}{N} \left\langle \sum_{i=1}^{N} x_i \right\rangle = \frac{1}{N} \sum_{i=1}^{N} \langle x_i \rangle \\ &= \frac{1}{N} \sum_{i=1}^{N} \mu = \frac{1}{N}(N\mu) = \mu,\end{aligned} \tag{19}$$

so the sample mean is an UNBIASED ESTIMATOR of population mean. However, the distribution of $\bar{x}$ depends on the sample size. For large samples, $\bar{x}$ is approximately NORMAL. For small samples, STUDENT'S t-DISTRIBUTION should be used.

The VARIANCE of the population mean is independent of the distribution.

$$\begin{aligned}\operatorname{var}(\bar{x}) &= \operatorname{var}\left(\frac{1}{n} \sum_{i=1}^{N} x_i\right) = \frac{1}{N^2} \operatorname{var}\left(\sum_{i=1}^{N} x_i\right) \\ &= \frac{1}{N^2} \sum_{i=1}^{n} \operatorname{var}(x_i) = \left(\frac{1}{N^2}\right) \sum_{i=1}^{N} \sigma^2 = \frac{\sigma^2}{N}.\end{aligned} \tag{20}$$

From k-STATISTICS for a GAUSSIAN DISTRIBUTION, the UNBIASED ESTIMATOR for the VARIANCE is given by

$$\sigma^2 = \frac{N}{N-1} s^2, \tag{21}$$

where

$$s \equiv \frac{1}{N} \sum_{i=1}^{N} (x_i - \bar{x})^2, \tag{22}$$

so

$$\operatorname{var}(\bar{x}) = \frac{s^2}{N-1}. \tag{23}$$

The SQUARE ROOT of this,

$$\sigma_x = \frac{s}{\sqrt{N-1}}, \tag{24}$$

is called the STANDARD ERROR.

$$\operatorname{var}(\bar{x}) \equiv \left\langle \bar{x}^2 \right\rangle - \langle \bar{x} \rangle^2, \tag{25}$$

so

$$\left\langle \bar{x}^2 \right\rangle = \operatorname{var}(\bar{x}) + \langle \bar{x} \rangle^2 = \frac{\sigma^2}{N} + \mu^2. \tag{26}$$

see also ARITHMETIC-GEOMETRIC MEAN, ARITHMETIC-HARMONIC MEAN, CARLEMAN'S INEQUALITY, CUMULANT, GENERALIZED MEAN, GEOMETRIC MEAN, HARMONIC MEAN, HARMONIC-GEOMETRIC MEAN, KURTOSIS, MEAN, MEAN DEVIATION, MEDIAN (STATISTICS), MODE, MOMENT, QUADRATIC MEAN, ROOT-MEAN-SQUARE, SAMPLE VARIANCE, SKEWNESS, STANDARD DEVIATION, TRIMEAN, VARIANCE

References

Abramowitz, M. and Stegun, C. A. (Eds.). *Handbook of Mathematical Functions with Formulas, Graphs, and Mathematical Tables, 9th printing.* New York: Dover, p. 10, 1972.

Alzer, H. "A Proof of the Arithmetic Mean–Geometric Mean Inequality." *Amer. Math. Monthly* **103**, 585, 1996.

Beckenbach, E. F. and Bellman, R. *Inequalities.* New York: Springer-Verlag, 1983.

Beyer, W. H. *CRC Standard Mathematical Tables, 28th ed.* Boca Raton, FL: CRC Press, p. 471, 1987.

Bullen, P. S.; Mitrinović, D. S.; and Vasić, P. M. *Means & Their Inequalities.* Dordrecht, Netherlands: Reidel, 1988.

Hardy, G. H.; Littlewood, J. E.; and Pólya, G. *Inequalities.* Cambridge, England: Cambridge University Press, 1952.

Hoehn, L. and Niven, I. "Averages on the Move." *Math. Mag.* **58**, 151–156, 1985.

Kenney, J. F. and Keeping, E. S. *Mathematics of Statistics, Pt. 1, 3rd ed.* Princeton, NJ: Van Nostrand, 1962.

Mitrinović, D. S.; Pečarić, J. E.; and Fink, A. M. *Classical and New Inequalities in Analysis.* Dordrecht, Netherlands: Kluwer, 1993.

Vasic, P. M. and Mitrinović, D. S. *Analytic Inequalities.* New York: Springer-Verlag, 1970.

Arithmetic Progression

see ARITHMETIC SERIES

Arithmetic Sequence

A SEQUENCE of n numbers $\{d_0 + kd\}_{k=0}^{n-1}$ such that the differences between successive terms is a constant d.

see also ARITHMETIC SERIES, SEQUENCE

Arithmetic Series

An arithmetic series is the SUM of a SEQUENCE $\{a_k\}$, $k = 1, 2, \ldots$, in which each term is computed from the previous one by adding (or subtracting) a constant. Therefore, for $k > 1$,

$$a_k = a_{k-1} + d = a_{k-2} + 2d = \ldots = a_1 + d(k-1). \quad (1)$$

The sum of the sequence of the first n terms is then given by

$$\begin{aligned} S_n &\equiv \sum_{k=1}^{n} a_k = \sum_{k=1}^{n} [a_1 + (k-1)d] \\ &= na_1 + d\sum_{k=1}^{n}(k-1) = na_1 + d\sum_{k=2}^{n}(k-1) \\ &= na_1 + d\sum_{k=1}^{n-1} k \end{aligned} \quad (2)$$

Using the SUM identity

$$\sum_{k=1}^{n} = \tfrac{1}{2}n(n+1) \quad (3)$$

then gives

$$S_n = na_1 + \tfrac{1}{2}d(n-1) = \tfrac{1}{2}n[2a_1 + d(n-1)]. \quad (4)$$

Note, however, that

$$a_1 + a_n = a_1 + [a_1 + d(n-1)] = 2a_1 + d(n-1), \quad (5)$$

so

$$S_n = \tfrac{1}{2}n(a_1 + a_n), \quad (6)$$

or n times the AVERAGE of the first and last terms! This is the trick Gauss used as a schoolboy to solve the problem of summing the INTEGERS from 1 to 100 given as busy-work by his teacher. While his classmates toiled away doing the ADDITION longhand, Gauss wrote a single number, the correct answer

$$\tfrac{1}{2}(100)(1 + 100) = 50 \cdot 101 = 5050 \quad (7)$$

on his slate. When the answers were examined, Gauss's proved to be the only correct one.

see also ARITHMETIC SEQUENCE, GEOMETRIC SERIES, HARMONIC SERIES, PRIME ARITHMETIC PROGRESSION

References

Abramowitz, M. and Stegun, C. A. (Eds.). *Handbook of Mathematical Functions with Formulas, Graphs, and Mathematical Tables, 9th printing.* New York: Dover, p. 10, 1972.

Beyer, W. H. (Ed.). *CRC Standard Mathematical Tables, 28th ed.* Boca Raton, FL: CRC Press, p. 8, 1987.

Courant, R. and Robbins, H. "The Arithmetical Progression." §1.2.2 in *What is Mathematics?: An Elementary Approach to Ideas and Methods, 2nd ed.* Oxford, England: Oxford University Press, pp. 12–13, 1996.

Pappas, T. *The Joy of Mathematics.* San Carlos, CA: Wide World Publ./Tetra, p. 164, 1989.

Armstrong Number

The n-digit numbers equal to sum of nth powers of their digits (a finite sequence), also called PLUS PERFECT NUMBERS. They first few are given by 1, 2, 3, 4, 5, 6, 7, 8, 9, 153, 370, 371, 407, 1634, 8208, 9474, 54748, ... (Sloane's A005188).

see also NARCISSISTIC NUMBER

References

Sloane, N. J. A. Sequence A005188/M0488 in "An On-Line Version of the Encyclopedia of Integer Sequences."

Arnold's Cat Map

The best known example of an ANOSOV DIFFEOMORPHISM. It is given by the TRANSFORMATION

$$\begin{bmatrix} x_{n+1} \\ y_{n+1} \end{bmatrix} = \begin{bmatrix} 1 & 1 \\ 1 & 2 \end{bmatrix} \begin{bmatrix} x_n \\ y_n \end{bmatrix}, \quad (1)$$

where x_{n+1} and y_{n+1} are computed mod 1. The Arnold cat mapping is non-Hamiltonian, nonanalytic, and mixing. However, it is AREA-PRESERVING since the DETERMINANT is 1. The LYAPUNOV CHARACTERISTIC EXPONENTS are given by

$$\begin{vmatrix} 1-\sigma & 1 \\ 1 & 2-\sigma \end{vmatrix} = \sigma^2 - 3\sigma + 1 = 0, \quad (2)$$

so

$$\sigma_\pm = \tfrac{1}{2}(3 \pm \sqrt{5}). \quad (3)$$

The EIGENVECTORS are found by plugging $\sigma_\pm$ into the MATRIX EQUATION

$$\begin{bmatrix} 1-\sigma_\pm & 1 \\ 1 & 2-\sigma_\pm \end{bmatrix} \begin{bmatrix} x \\ y \end{bmatrix} = \begin{bmatrix} 0 \\ 0 \end{bmatrix}. \quad (4)$$

For σ_+, the solution is

$$y = \tfrac{1}{2}(1 + \sqrt{5})x \equiv \phi x, \quad (5)$$

where ϕ is the GOLDEN RATIO, so the unstable (normalized) EIGENVECTOR is

$$\boldsymbol{\xi}_+ = \tfrac{1}{10}\sqrt{50 - 10\sqrt{5}} \begin{bmatrix} 1 \\ \frac{1}{2}(1+\sqrt{5}) \end{bmatrix}. \quad (6)$$

Similarly, for σ_-, the solution is

$$y = -\tfrac{1}{2}(\sqrt{5} - 1)x \equiv \phi^{-1}x, \quad (7)$$

so the stable (normalized) EIGENVECTOR is

$$\boldsymbol{\xi}_- = \tfrac{1}{10}\sqrt{50 + 10\sqrt{5}} \begin{bmatrix} 1 \\ \frac{1}{2}(1-\sqrt{5}) \end{bmatrix}. \quad (8)$$

see also ANOSOV MAP

Arnold Diffusion

The nonconservation of ADIABATIC INVARIANTS which arises in systems with three or more DEGREES OF FREEDOM.

Arnold Tongue

Consider the CIRCLE MAP. If K is NONZERO, then the motion is periodic in some FINITE region surrounding each rational Ω. This execution of periodic motion in response to an irrational forcing is known as MODE LOCKING. If a plot is made of K versus Ω with the regions of periodic MODE-LOCKED parameter space plotted around rational Ω values (the WINDING NUMBERS), then the regions are seen to widen upward from 0 at $K = 0$ to some FINITE width at $K = 1$. The region surrounding each RATIONAL NUMBER is known as an ARNOLD TONGUE.

At $K = 0$, the Arnold tongues are an isolated set of MEASURE zero. At $K = 1$, they form a general CANTOR SET of dimension $d \approx 0.8700$. In general, an Arnold tongue is defined as a resonance zone emanating out from RATIONAL NUMBERS in a two-dimensional parameter space of variables.

see also CIRCLE MAP

Aronhold Process

The process used to generate an expression for a covariant in the first degree of any one of the equivalent sets of COEFFICIENTS for a curve.

see also CLEBSCH-ARONHOLD NOTATION, JOACHIMSTHAL'S EQUATION

References

Coolidge, J. L. *A Treatise on Algebraic Plane Curves.* New York: Dover, p. 74, 1959.

Aronson's Sequence

The sequence whose definition is: "t is the first, fourth, eleventh, ... letter of this sentence." The first few values are 1, 4, 11, 16, 24, 29, 33, 35, 39, ... (Sloane's A005224).

References

Hofstadter, D. R. *Metamagical Themas: Questing of Mind and Pattern.* New York: BasicBooks, p. 44, 1985.

Sloane, N. J. A. Sequence A005224/M3406 in "An On-Line Version of the Encyclopedia of Integer Sequences."

Arrangement

In general, an arrangement of objects is simply a grouping of them. The number of "arrangements" of n items is given either by a COMBINATION (order is ignored) or PERMUTATION (order is significant).

The division of SPACE into cells by a collection of HYPERPLANES is also called an arrangement.

see also COMBINATION, CUTTING, HYPERPLANE, ORDERING, PERMUTATION

Arrangement Number

see PERMUTATION

Array

An array is a "list of lists" with the length of each level of list the same. The size (sometimes called the "shape") of a d-dimensional array is then indicated as $\underbrace{m \times n \times \cdots \times p}_{d}$. The most common type of array encountered is the 2-D $m \times n$ rectangular array having m columns and n rows. If $m = n$, a square array results. Sometimes, the order of the elements in an array is significant (as in a MATRIX), whereas at other times, arrays which are equivalent modulo reflections (and rotations, in the case of a square array) are considered identical (as in a MAGIC SQUARE or PRIME ARRAY).

In order to exhaustively list the number of distinct arrays of a given shape with each element being one of k possible choices, the naive algorithm of running through each case and checking to see whether it's equivalent to an earlier one is already just about as efficient as can be. The running time must be at least the number of answers, and this is so close to $k^{mn\cdots p}$ that the difference isn't significant.

However, finding the *number* of possible arrays of a given shape is much easier, and an exact formula can be obtained using the POLYA ENUMERATION THEOREM. For the simple case of an $m \times n$ array, even this proves unnecessary since there are only a few possible symmetry types, allowing the possibilities to be counted explicitly. For example, consider the case of m and n EVEN and distinct, so only reflections need be included. To take a specific case, let $m = 6$ and $n = 4$ so the array looks like

$$\begin{array}{cccc|ccc} a & b & c & & d & e & f \\ g & h & i & & j & k & l \\ \hline m & n & o & & p & q & r \\ s & t & u & & v & w & x, \end{array}$$

where each $a, b, \ldots, x$ can take a value from 1 to k. The total number of possible arrangements is k^{24} (k^{mn} in general). The number of arrangements which are equivalent to their left-right mirror images is k^{12} (in general, $k^{mn/2}$), as is the number equal to their up-down mirror images, or their rotations through 180°. There are also k^6 arrangements (in general, $k^{mn/4}$) with full symmetry.

In general, it is therefore true that

$$\begin{cases} k^{mn/4} & \text{with full symmetry} \\ k^{mn/2} - k^{mn/4} & \text{with } only \text{ left-right reflection} \\ k^{mn/2} - k^{mn/4} & \text{with } only \text{ up-down reflection} \\ k^{mn/2} - k^{mn/4} & \text{with } only \text{ } 180^\circ \text{ rotation,} \end{cases}$$

so there are

$$k^{mn} - 3k^{mn/2} + 2k^{mn/4}$$

arrangements with no symmetry. Now dividing by the number of images of each type, the result, for $m \neq n$ with m, n EVEN, is

$$\begin{aligned} N(m,n,k) &= \tfrac{1}{4}k^{mn} + (\tfrac{1}{2})(3)(k^{mn/2} - k^{mn/4}) \\ &\quad + \tfrac{1}{4}(k^{mn} - 3k^{mn/2} + 2k^{mn/4}) \\ &= \tfrac{1}{4}k^{mn} + \tfrac{3}{4}k^{mn/2} + \tfrac{1}{2}k^{mn/4}. \end{aligned}$$

The number is therefore of order $\mathcal{O}(k^{mn}/4)$, with "correction" terms of much smaller order.

see also ANTIMAGIC SQUARE, EULER SQUARE, KIRKMAN'S SCHOOLGIRL PROBLEM, LATIN RECTANGLE, LATIN SQUARE, MAGIC SQUARE, MATRIX, MRS. PERKINS' QUILT, MULTIPLICATION TABLE, ORTHOGONAL ARRAY, PERFECT SQUARE, PRIME ARRAY, QUOTIENT-DIFFERENCE TABLE, ROOM SQUARE, STOLARSKY ARRAY, TRUTH TABLE, WYTHOFF ARRAY

Arrow Notation

A NOTATION invented by Knuth (1976) to represent LARGE NUMBERS in which evaluation proceeds from the right (Conway and Guy 1996, p. 60).

$$\begin{array}{ll} m \uparrow n & \underbrace{m \cdot m \cdots m}_{n} \\ m \uparrow\uparrow n & \underbrace{m \uparrow m \uparrow \cdots \uparrow m}_{n} \\ m \uparrow\uparrow\uparrow n & \underbrace{m \uparrow\uparrow m \uparrow\uparrow \cdots \uparrow\uparrow m}_{n} \end{array}$$

For example,

$$m \uparrow n = m^n \tag{1}$$

$$m \uparrow\uparrow n = \underbrace{m \uparrow \cdots \uparrow m}_{n} = \underbrace{m^{m^{\cdot^{\cdot^{\cdot^{m}}}}}}_{n}$$

$$m \uparrow\uparrow 2 = \underbrace{m \uparrow m}_{2} = m \uparrow m = m^m \tag{2}$$

$$\begin{aligned} m \uparrow\uparrow 3 &= \underbrace{m \uparrow m \uparrow m}_{3} = m \uparrow (m \uparrow m) \\ &= m \uparrow m^m = m^{m^m} \end{aligned} \tag{3}$$

$$m \uparrow\uparrow\uparrow 2 = \underbrace{m \uparrow\uparrow m}_{2} = m \uparrow\uparrow m = \underbrace{m^{m^{\cdot^{\cdot^{\cdot^{m}}}}}}_{m} \tag{4}$$

$$\begin{aligned} m \uparrow\uparrow\uparrow 3 &= \underbrace{m \uparrow\uparrow m \uparrow\uparrow m}_{3} = m \uparrow\uparrow \underbrace{m^{m^{\cdot^{\cdot^{\cdot^{m}}}}}}_{m} \\ &= \underbrace{m \uparrow \cdots \uparrow m}_{\underbrace{m^{m^{\cdot^{\cdot^{\cdot^{m}}}}}}_{m}} = \underbrace{m^{m^{\cdot^{\cdot^{\cdot^{m}}}}}}_{\underbrace{m^{m^{\cdot^{\cdot^{\cdot^{m}}}}}}_{m}}. \end{aligned} \tag{5}$$

$m \uparrow\uparrow n$ is sometimes called a POWER TOWER. The values $\underbrace{n \uparrow \cdots \uparrow n}_{n}$ are called ACKERMANN NUMBERS.

see also ACKERMANN NUMBER, CHAINED ARROW NOTATION, DOWN ARROW NOTATION, LARGE NUMBER, POWER TOWER, STEINHAUS-MOSER NOTATION

References

Conway, J. H. and Guy, R. K. *The Book of Numbers.* New York: Springer-Verlag, pp. 59–62, 1996.

Guy, R. K. and Selfridge, J. L. "The Nesting and Roosting Habits of the Laddered Parenthesis." *Amer. Math. Monthly* **80**, 868–876, 1973.

Knuth, D. E. "Mathematics and Computer Science: Coping with Finiteness. Advances in Our Ability to Compute are Bringing Us Substantially Closer to Ultimate Limitations." *Science* **194**, 1235–1242, 1976.

Vardi, I. *Computational Recreations in Mathematica.* Redwood City, CA: Addison-Wesley, pp. 11 and 226–229, 1991.

Arrow's Paradox

Perfect democratic voting is, not just in practice but *in principle,* impossible.

References

Gardner, M. *Time Travel and Other Mathematical Bewilderments.* New York: W. H. Freeman, p. 56, 1988.

Arrowhead Curve

see SIERPIŃSKI ARROWHEAD CURVE

Art Gallery Theorem

Also called CHVÁTAL'S ART GALLERY THEOREM. If the walls of an art gallery are made up of n straight LINES SEGMENTS, then the entire gallery can always be supervised by $\lfloor n/3 \rfloor$ watchmen placed in corners, where $\lfloor x \rfloor$ is the FLOOR FUNCTION. This theorem was proved by V. Chvátal in 1973. It is conjectured that an art gallery with n walls and h HOLES requires $\lfloor (n+h)/3 \rfloor$ watchmen.

see also ILLUMINATION PROBLEM

References

Honsberger, R. "Chvátal's Art Gallery Theorem." Ch. 11 in *Mathematical Gems II.* Washington, DC: Math. Assoc. Amer., pp. 104–110, 1976.

O'Rourke, J. *Art Gallery Theorems and Algorithms.* New York: Oxford University Press, 1987.

Stewart, I. "How Many Guards in the Gallery?" *Sci. Amer.* **270**, 118–120, May 1994.

Tucker, A. "The Art Gallery Problem." *Math Horizons,* pp. 24-26, Spring 1994.

Wagon, S. "The Art Gallery Theorem." §10.3 in *Mathematica in Action.* New York: W. H. Freeman, pp. 333–345, 1991.

Articulation Vertex

A VERTEX whose removal will disconnect a GRAPH, also called a CUT-VERTEX.

see also BRIDGE (GRAPH)

References

Chartrand, G. "Cut-Vertices and Bridges." §2.4 in *Introductory Graph Theory.* New York: Dover, pp. 45–49, 1985.

Artin Braid Group

see BRAID GROUP

Artin's Conjecture

There are at least two statements which go by the name of Artin's conjecture. The first is the RIEMANN HYPOTHESIS. The second states that every INTEGER not equal to -1 or a SQUARE NUMBER is a primitive root modulo p for infinitely many p and proposes a density for the set of such p which are always rational multiples of a constant known as ARTIN'S CONSTANT. There is an analogous theorem for functions instead of numbers which has been proved by Billharz (Shanks 1993, p. 147).

see also ARTIN'S CONSTANT, RIEMANN HYPOTHESIS

References
Shanks, D. *Solved and Unsolved Problems in Number Theory, 4th ed.* New York: Chelsea, pp. 31, 80–83, and 147, 1993.

Artin's Constant

If $n \neq -1$ and n is not a PERFECT SQUARE, then Artin conjectured that the SET $S(n)$ of all PRIMES for which n is a PRIMITIVE ROOT is infinite. Under the assumption of the EXTENDED RIEMANN HYPOTHESIS, Artin's conjecture was solved in 1967 by C. Hooley. If, in addition, n is not an rth POWER for any $r > 1$, then Artin conjectured that the density of $S(n)$ relative to the PRIMES is C_{Artin} (independent of the choice of n), where

$$C_{\text{Artin}} = \prod_{q \text{ prime}} \left[1 - \frac{1}{q(q-1)}\right] = 0.3739558136\ldots,$$

and the PRODUCT is over PRIMES. The significance of this constant is more easily seen by describing it as the fraction of PRIMES p for which $1/p$ has a maximal DECIMAL EXPANSION (Conway and Guy 1996).

References
Conway, J. H. and Guy, R. K. *The Book of Numbers.* New York: Springer-Verlag, p. 169, 1996.
Finch, S. "Favorite Mathematical Constants." `http://www.mathsoft.com/asolve/constant/artin/artin.html`.
Hooley, C. "On Artin's Conjecture." *J. reine angew. Math.* **225**, 209–220, 1967.
Ireland, K. and Rosen, M. *A Classical Introduction to Modern Number Theory, 2nd ed.* New York: Springer-Verlag, 1990.
Ribenboim, P. *The Book of Prime Number Records.* New York: Springer-Verlag, 1989.
Shanks, D. *Solved and Unsolved Problems in Number Theory, 4th ed.* New York: Chelsea, pp. 80–83, 1993.
Wrench, J. W. "Evaluation of Artin's Constant and the Twin Prime Constant." *Math. Comput.* **15**, 396–398, 1961.

Artin L-Function

An Artin L-function over the RATIONALS $\mathbb{Q}$ encodes in a GENERATING FUNCTION information about how an irreducible monic POLYNOMIAL over $\mathbb{Z}$ factors when reduced modulo each PRIME. For the POLYNOMIAL x^2+1, the Artin L-function is

$$L(s, \mathbb{Q}(i)/\mathbb{Q}, \text{sgn}) = \prod_{p \text{ odd prime}} \frac{1}{1 - \left(\frac{-1}{p}\right) p^{-s}},$$

where $(-1/p)$ is a LEGENDRE SYMBOL, which is equivalent to the EULER L-FUNCTION. The definition over arbitrary POLYNOMIALS generalizes the above expression.

see also LANGLANDS RECIPROCITY

References
Knapp, A. W. "Group Representations and Harmonic Analysis, Part II." *Not. Amer. Math. Soc.* **43**, 537–549, 1996.

Artin Reciprocity

see ARTIN'S RECIPROCITY THEOREM

Artin's Reciprocity Theorem

A general RECIPROCITY THEOREM for all orders. If R is a NUMBER FIELD and R' a finite integral extension, then there is a SURJECTION from the group of fractional IDEALS prime to the discriminant, given by the Artin symbol. For some cycle c, the kernel of this SURJECTION contains each PRINCIPAL fractional IDEAL generated by an element congruent to 1 mod c.

see also LANGLANDS PROGRAM

Artinian Group

A GROUP in which any decreasing CHAIN of distinct SUBGROUPS terminates after a FINITE number.

Artinian Ring

A noncommutative SEMISIMPLE RING satisfying the "descending chain condition."

see also GORENSTEIN RING, SEMISIMPLE RING

References
Artin, E. "Zur Theorie der hyperkomplexer Zahlen." *Hamb. Abh.* **5**, 251–260, 1928.
Artin, E. "Zur Arithmetik hyperkomplexer Zahlen." *Hamb. Abh.* **5**, 261–289, 1928.

Artistic Series

A SERIES is called artistic if every three consecutive terms have a common three-way ratio

$$P[a_i, a_{i+1}, a_{i+2}] = \frac{(a_i + a_{i+1} + a_{i+2})a_{i+1}}{a_i a_{i+2}}.$$

A SERIES is also artistic IFF its BIAS is a constant. A GEOMETRIC SERIES with RATIO $r > 0$ is an artistic series with

$$P = \frac{1}{r} + 1 + r \geq 3.$$

see also BIAS (SERIES), GEOMETRIC SERIES, MELODIC SERIES

References
Duffin, R. J. "On Seeing Progressions of Constant Cross Ratio." *Amer. Math. Monthly* **100**, 38–47, 1993.

ASA Theorem

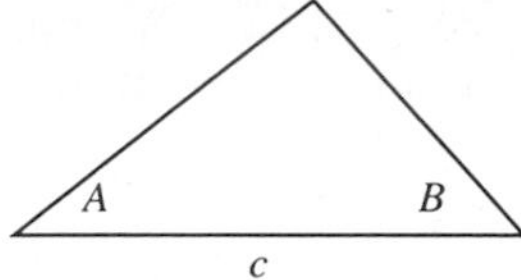

Specifying two adjacent ANGLES A and B and the side between them c uniquely determines a TRIANGLE with AREA

$$K = \frac{c^2}{2(\cot A + \cot B)}. \tag{1}$$

The angle C is given in terms of A and B by

$$C = \pi - A - B, \tag{2}$$

and the sides a and b can be determined by using the LAW OF SINES

$$\frac{a}{\sin A} = \frac{b}{\sin B} = \frac{c}{\sin C} \tag{3}$$

to obtain

$$a = \frac{\sin A}{\sin(\pi - A - B)} c \tag{4}$$

$$b = \frac{\sin B}{\sin(\pi - A - B)} c. \tag{5}$$

see also AAA THEOREM, AAS THEOREM, ASS THEOREM, SAS THEOREM, SSS THEOREM, TRIANGLE

Aschbacher's Component Theorem

Suppose that $E(G)$ (the commuting product of all components of G) is SIMPLE and G contains a SEMISIMPLE INVOLUTION. Then there is some SEMISIMPLE INVOLUTION x such that $C_G(x)$ has a NORMAL SUBGROUP K which is either QUASISIMPLE or ISOMORPHIC to $O^+(4,q)'$ and such that $Q = C_G(K)$ is TIGHTLY EMBEDDED.

see also INVOLUTION (GROUP), ISOMORPHIC GROUPS, NORMAL SUBGROUP, QUASISIMPLE GROUP, SIMPLE GROUP, TIGHTLY EMBEDDED

ASS Theorem

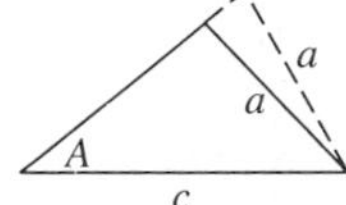

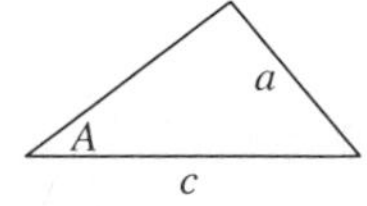

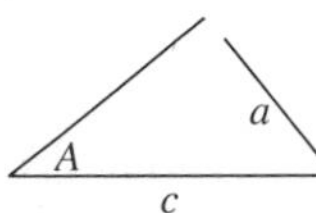

Specifying two adjacent side lengths a and b of a TRIANGLE (taking $a > b$) and one ACUTE ANGLE A opposite a does not, in general, uniquely determine a triangle. If $\sin A < a/c$, there are two possible TRIANGLES satisfying the given conditions. If $\sin A = a/c$, there is one possible TRIANGLE. If $\sin A > a/c$, there are no possible TRIANGLES. Remember: don't try to prove congruence with the ASS theorem or you will make make an ASS out of yourself.

see also AAA THEOREM, AAS THEOREM, SAS THEOREM, SSS THEOREM, TRIANGLE

Associative

In simple terms, let x, y, and z be members of an ALGEBRA. Then the ALGEBRA is said to be associative if

$$x \cdot (y \cdot z) = (x \cdot y) \cdot z, \tag{1}$$

where $\cdot$ denotes MULTIPLICATION. More formally, let A denote an $\mathbb{R}$-algebra, so that A is a VECTOR SPACE over $\mathbb{R}$ and

$$A \times A \to A \tag{2}$$

$$(x, y) \mapsto x \cdot y. \tag{3}$$

Then A is said to be m-associative if there exists an m-D SUBSPACE S of A such that

$$(y \cdot x) \cdot z = y \cdot (x \cdot z) \tag{4}$$

for all $y, z \in A$ and $x \in S$. Here, VECTOR MULTIPLICATION $x \cdot y$ is assumed to be BILINEAR. An n-D n-associative ALGEBRA is simply said to be "associative."

see also COMMUTATIVE, DISTRIBUTIVE

References

Finch, S. "Zero Structures in Real Algebras." `http://www.mathsoft.com/asolve/zerodiv/zerodiv.html`.

Associative Magic Square

1	15	24	8	17
23	7	16	5	14
20	4	13	22	6
12	21	10	19	3
9	18	2	11	25

An $n \times n$ MAGIC SQUARE for which every pair of numbers symmetrically opposite the center sum to $n^2 + 1$. The LO SHU is associative but not PANMAGIC. Order four squares can be PANMAGIC or associative, but not both. Order five squares are the smallest which can be both associative and PANMAGIC, and 16 distinct associative PANMAGIC SQUARES exist, one of which is illustrated above (Gardner 1988).

see also MAGIC SQUARE, PANMAGIC SQUARE

References

Gardner, M. "Magic Squares and Cubes." Ch. 17 in *Time Travel and Other Mathematical Bewilderments.* New York: W. H. Freeman, 1988.

Astroid

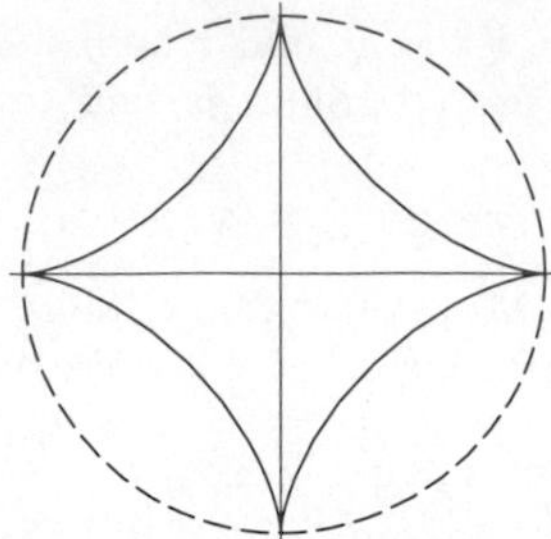

A 4-cusped HYPOCYCLOID which is sometimes also called a TETRACUSPID, CUBOCYCLOID, or PARACYCLE. The parametric equations of the astroid can be obtained by plugging in $n \equiv a/b = 4$ or $4/3$ into the equations for a general HYPOCYCLOID, giving

$$x = 3b\cos\phi + b\cos(3\phi) = 4b\cos^3\phi = a\cos^3\phi \quad (1)$$
$$y = 3b\sin\phi - b\sin(3\phi) = 4b\sin^3\phi = a\sin^3\phi. \quad (2)$$

In CARTESIAN COORDINATES,

$$x^{2/3} + y^{2/3} = a^{2/3}. \quad (3)$$

In PEDAL COORDINATES with the PEDAL POINT at the center, the equation is

$$r^2 + 3p^2 = a^2. \quad (4)$$

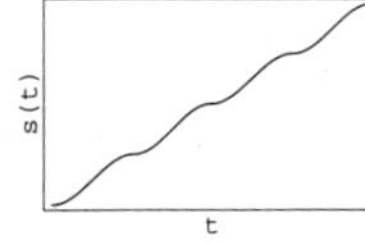

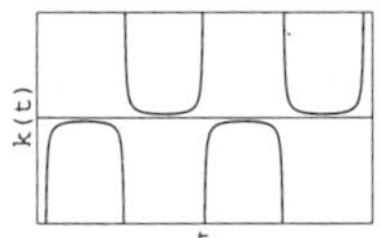

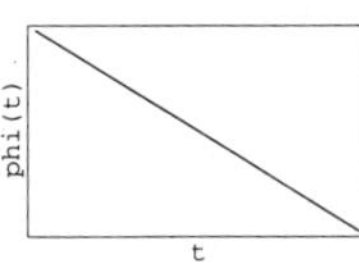

The ARC LENGTH, CURVATURE, and TANGENTIAL ANGLE are

$$s(t) = \tfrac{3}{2}\int_0^t |\sin(2t')|\,dt' = \tfrac{3}{2}\sin^2 t \quad (5)$$
$$\kappa(t) = -\tfrac{2}{3}\csc(2t) \quad (6)$$
$$\phi(t) = -t. \quad (7)$$

As usual, care must be taken in the evaluation of $s(t)$ for $t > \pi/2$. Since (5) comes from an integral involving the ABSOLUTE VALUE of a function, it must be monotonic increasing. Each QUADRANT can be treated correctly by defining

$$n = \left\lfloor \frac{2t}{\pi} \right\rfloor + 1, \quad (8)$$

where $\lfloor x \rfloor$ is the FLOOR FUNCTION, giving the formula

$$s(t) = (-1)^{1+[n \pmod 2]}\tfrac{3}{2}\sin^2 t + 3\left\lfloor \tfrac{1}{2}n \right\rfloor. \quad (9)$$

The overall ARC LENGTH of the astroid can be computed from the general HYPOCYCLOID formula

$$s_n = \frac{8a(n-1)}{n} \quad (10)$$

with $n = 4$,

$$s_4 = 6a. \quad (11)$$

The AREA is given by

$$A_n = \frac{(n-1)(n-2)}{n^2}\pi a^2 \quad (12)$$

with $n = 4$,

$$A_4 = \tfrac{3}{8}\pi a^2. \quad (13)$$

The EVOLUTE of an ELLIPSE is a stretched HYPOCYCLOID. The gradient of the TANGENT T from the point with parameter p is $-\tan p$. The equation of this TANGENT T is

$$x\sin p + y\cos p = \tfrac{1}{2}a\sin(2p) \quad (14)$$

(MacTutor Archive). Let T cut the x-AXIS and the y-AXIS at X and Y, respectively. Then the length XY is a constant and is equal to a.

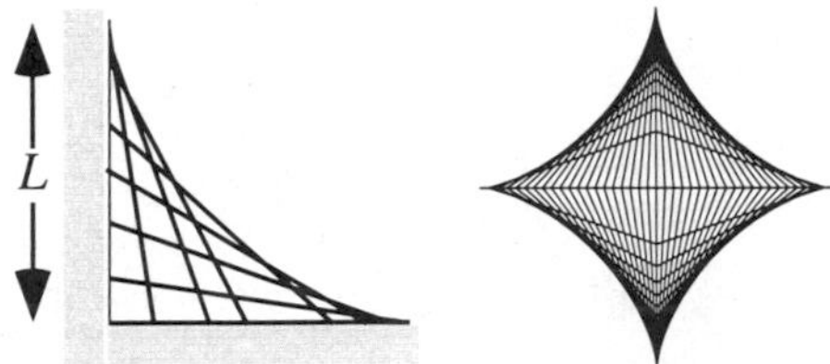

The astroid can also be formed as the ENVELOPE produced when a LINE SEGMENT is moved with each end on one of a pair of PERPENDICULAR axes (e.g., it is the curve enveloped by a ladder sliding against a wall or a garage door with the top corner moving along a vertical track; left figure above). The astroid is therefore a GLISSETTE. To see this, note that for a ladder of length L, the points of contact with the wall and floor are $(x_0, 0)$ and $(0, \sqrt{L^2 - x_0{}^2})$, respectively. The equation of the LINE made by the ladder with its foot at $(x_0, 0)$ is therefore

$$y - 0 = \frac{\sqrt{L^2 - x_0{}^2}}{-x_0}(x - x_0) \quad (15)$$

which can be written

$$U(x, y, x_0) = y + \frac{\sqrt{L^2 - x_0{}^2}}{x_0}(x - x_0). \quad (16)$$

The equation of the ENVELOPE is given by the simultaneous solution of

$$\begin{cases} U(x, y, x_0) = y + \frac{\sqrt{L^2 - x_0{}^2}}{x_0}(x - x_0) = 0 \\ \frac{\partial U}{\partial x_0} = \frac{x_0{}^2 - Lx^2}{x_0{}^2\sqrt{L^2 - x_0{}^2}} = 0, \end{cases} \quad (17)$$

which is

$$x = \frac{x_0{}^3}{L^2} \quad (18)$$
$$y = \frac{(L^2 - x_0{}^2)^{3/2}}{L^2}. \quad (19)$$

Noting that

$$x^{2/3} = \frac{{x_0}^2}{L^{4/3}} \tag{20}$$

$$y^{2/3} = \frac{L^2 - {x_0}^2}{L^{4/3}} \tag{21}$$

allows this to be written implicitly as

$$x^{2/3} + y^{2/3} = L^{2/3}, \tag{22}$$

the equation of the astroid, as promised.

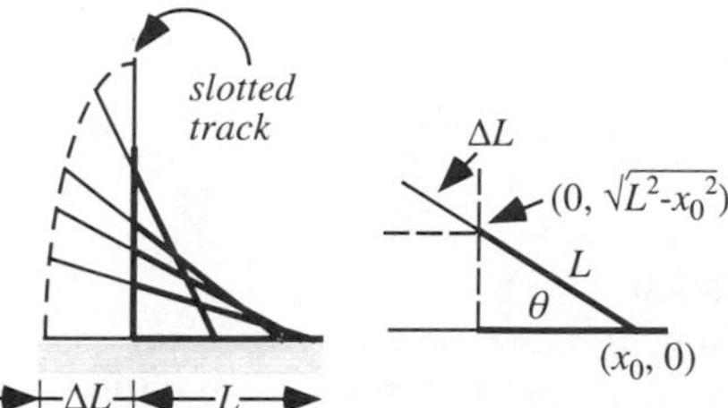

The related problem obtained by having the "garage door" of length L with an "extension" of length ΔL move up and down a slotted track also gives a surprising answer. In this case, the position of the "extended" end for the foot of the door at horizontal position x_0 and ANGLE θ is given by

$$x = -\Delta L \cos\theta \tag{23}$$

$$y = \sqrt{L^2 - {x_0}^2} + \Delta L \sin\theta. \tag{24}$$

Using

$$x_0 = L\cos\theta \tag{25}$$

then gives

$$x = -\frac{\Delta L}{L} x_0 \tag{26}$$

$$y = \sqrt{L^2 - {x_0}^2}\left(1 + \frac{\Delta L}{L}\right). \tag{27}$$

Solving (26) for x_0, plugging into (27) and squaring then gives

$$y^2 = L^2 - \frac{L^2 x^2}{(\Delta L)^2}\left(1 + \frac{\Delta L}{L}\right)^2. \tag{28}$$

Rearranging produces the equation

$$\frac{x^2}{(\Delta L)^2} + \frac{y^2}{(L + \Delta L)^2} = 1, \tag{29}$$

the equation of a (QUADRANT of an) ELLIPSE with SEMIMAJOR and SEMIMINOR AXES of lengths ΔL and $L + \Delta L$.

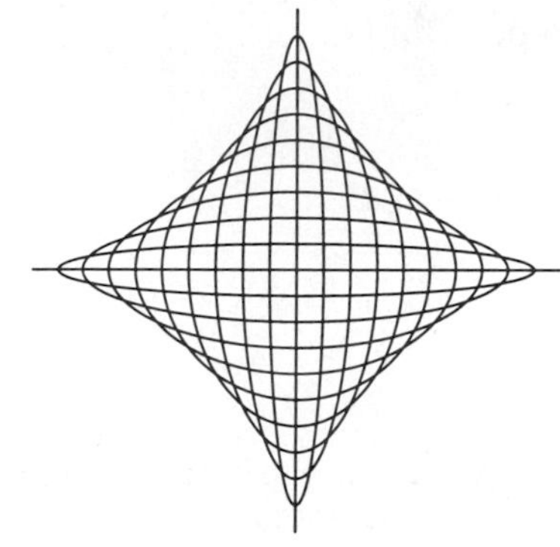

The astroid is also the ENVELOPE of the family of ELLIPSES

$$\frac{x^2}{c^2} + \frac{y^2}{(1-c)^2} - 1 = 0, \tag{30}$$

illustrated above.

see also DELTOID, ELLIPSE ENVELOPE, LAMÉ CURVE, NEPHROID, RANUNCULOID

References

Lawrence, J. D. *A Catalog of Special Plane Curves.* New York: Dover, pp. 172–175, 1972.

Lee, X. "Astroid." `http://www.best.com/~xah/SpecialPlaneCurves_dir/Astroid_dir/astroid.html`.

Lockwood, E. H. "The Astroid." Ch. 6 in *A Book of Curves.* Cambridge, England: Cambridge University Press, pp. 52–61, 1967.

MacTutor History of Mathematics Archive. "Astroid." `http://www-groups.dcs.st-and.ac.uk/~history/Curves/Astroid.html`.

Yates, R. C. "Astroid." *A Handbook on Curves and Their Properties.* Ann Arbor, MI: J. W. Edwards, pp. 1–3, 1952.

Astroid Evolute

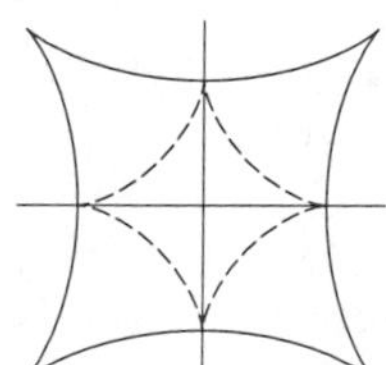

A HYPOCYCLOID EVOLUTE for $n = 4$ is another ASTROID scaled by a factor $n/(n-2) = 4/2 = 2$ and rotated $1/(2 \cdot 4) = 1/8$ of a turn.

Astroid Involute

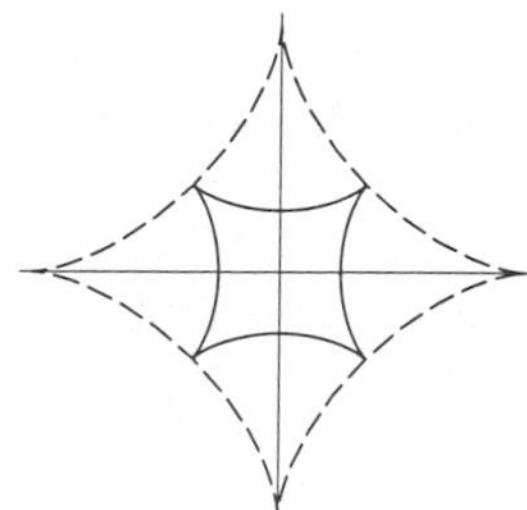

A HYPOCYCLOID INVOLUTE for $n = 4$ is another ASTROID scaled by a factor $(n-2)/2 = 2/4 = 1/2$ and rotated $1/(2 \cdot 4) = 1/8$ of a turn.

Astroid Pedal Curve

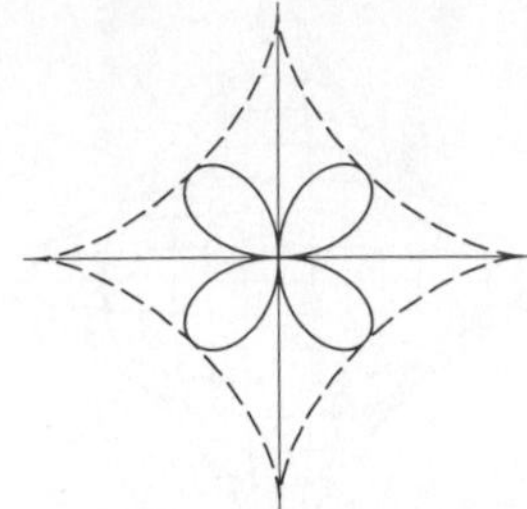

The PEDAL CURVE of an ASTROID with PEDAL POINT at the center is a QUADRIFOLIUM.

Astroid Radial Curve

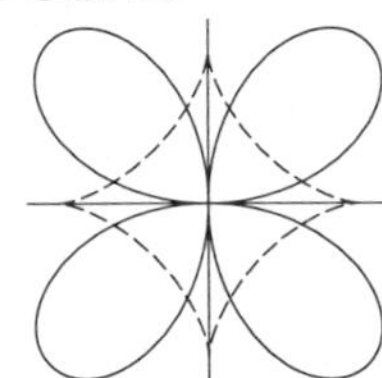

The QUADRIFOLIUM

$$x = x_0 + 3a\cos t - 3a\cos(3t)$$
$$y = y_0 + 3a\sin t + 3a\sin(3t).$$

Astroidal Ellipsoid

The surface which is the inverse of the ELLIPSOID in the sense that it "goes in" where the ELLIPSOID "goes out." It is given by the parametric equations

$$x = (a\cos u\cos v)^3$$
$$y = (b\sin u\cos v)^3$$
$$z = (c\sin v)^3$$

for $u \in [-\pi/2, \pi/2]$ and $v \in [-\pi, \pi]$. The special case $a = b = c = 1$ corresponds to the HYPERBOLIC OCTAHEDRON.

see also ELLIPSOID, HYPERBOLIC OCTAHEDRON

References

Nordstrand, T. "Astroidal Ellipsoid." http://www.uib.no/people/nfytn/asttxt.htm.

Asymptosy

ASYMPTOTIC behavior. A useful yet endangered word, found rarely outside the captivity of the *Oxford English Dictionary*.

see also ASYMPTOTE, ASYMPTOTIC

Asymptote

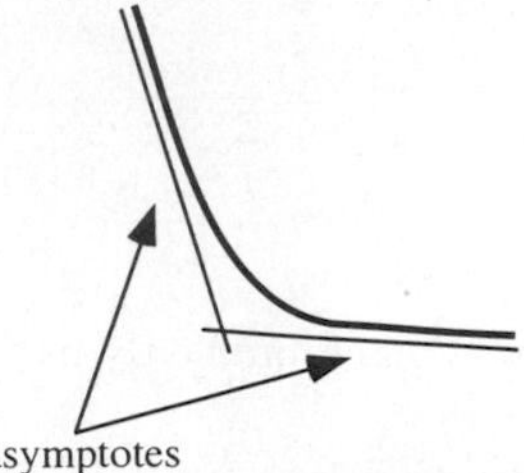

A curve approaching a given curve arbitrarily closely, as illustrated in the above diagram.

see also ASYMPTOSY, ASYMPTOTIC, ASYMPTOTIC CURVE

References

Giblin, P. J. "What is an Asymptote?" *Math. Gaz.* **56**, 274–284, 1972.

Asymptotic

Approaching a value or curve arbitrarily closely (i.e., as some sort of LIMIT is taken). A CURVE A which is asymptotic to given CURVE C is called the ASYMPTOTE of C.

see also ASYMPTOSY, ASYMPTOTE, ASYMPTOTIC CURVE, ASYMPTOTIC DIRECTION, ASYMPTOTIC SERIES, LIMIT

Asymptotic Curve

Given a REGULAR SURFACE M, an asymptotic curve is formally defined as a curve $\mathbf{x}(t)$ on M such that the NORMAL CURVATURE is 0 in the direction $\mathbf{x}'(t)$ for all t in the domain of $\mathbf{x}$. The differential equation for the parametric representation of an asymptotic curve is

$$eu'^2 + 2fu'v' + gv'^2 = 0, \tag{1}$$

where e, f, and g are second FUNDAMENTAL FORMS. The differential equation for asymptotic curves on a MONGE PATCH $(u, v, h(u, v))$ is

$$h_{uu}u'^2 + 2h_{uu}u'v' + h_{vv}v'^2 = 0, \tag{2}$$

and on a polar patch $(r\cos\theta, r\sin\theta, h(r))$ is

$$h''(r)r'^2 + h'(r)r\theta'^2 = 0. \tag{3}$$

The images below show asymptotic curves for the ELLIPTIC HELICOID, FUNNEL, HYPERBOLIC PARABOLOID, and MONKEY SADDLE.

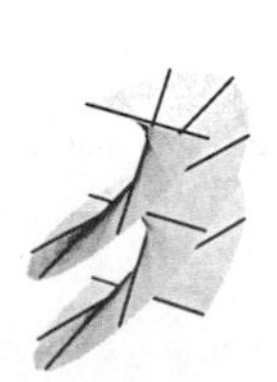

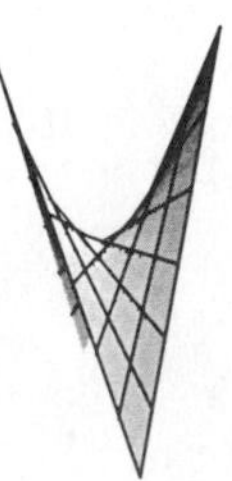

see also RULED SURFACE

References

Gray, A. "Asymptotic Curves," "Examples of Asymptotic Curves," "Using Mathematica to Find Asymptotic Curves." §16.1, 16.2, and 16.3 in *Modern Differential Geometry of Curves and Surfaces.* Boca Raton, FL: CRC Press, pp. 320–331, 1993.

Asymptotic Direction

An asymptotic direction at a point **p** of a REGULAR SURFACE $M \in \mathbb{R}^3$ is a direction in which the NORMAL CURVATURE of M vanishes.

1. There are no asymptotic directions at an ELLIPTIC POINT.
2. There are exactly two asymptotic directions at a HYPERBOLIC POINT.
3. There is exactly one asymptotic direction at a PARABOLIC POINT.
4. Every direction is asymptotic at a PLANAR POINT.

see also ASYMPTOTIC CURVE

References

Gray, A. *Modern Differential Geometry of Curves and Surfaces.*Boca Raton, FL: CRC Press, pp. 270 and 320, 1993.

Asymptotic Notation

Let n be a integer variable which tends to infinity and let x be a continuous variable tending to some limit. Also, let $\phi(n)$ or $\phi(x)$ be a positive function and $f(n)$ or $f(x)$ any function. Then Hardy and Wright (1979) define

1. $f = \mathcal{O}(\phi)$ to mean that $|f| < A\phi$ for some constant A and all values of n and x,
2. $f = o(\phi)$ to mean that $f/\phi \to 0$,
3. $f \sim \phi$ to mean that $f/\phi \to 1$,
4. $f \prec \phi$ to mean the same as $f = o(\phi)$,
5. $f \succ \phi$ to mean $f/\phi \to \infty$, and
6. $f \asymp \phi$ to mean $A_1\phi < f < A_2$ for some positive constants A_1 and A_2.

$f = o(\phi)$ implies and is stronger than $f = \mathcal{O}(\phi)$.

References

Hardy, G. H. and Wright, E. M. "Some Notation." §1.6 in *An Introduction to the Theory of Numbers, 5th ed.* Oxford, England: Clarendon Press, pp. 7–8, 1979.

Asymptotic Series

An asymptotic series is a SERIES EXPANSION of a FUNCTION in a variable x which may converge or diverge (Erdelyi 1987, p. 1), but whose partial sums can be made an arbitrarily good approximation to a given function for large enough x. To form an asymptotic series $R(x)$ of $f(x)$, written

$$f(x) \sim R(x), \tag{1}$$

take

$$x^n R_n(x) = x^n[f(x) - S_n(x)], \tag{2}$$

where

$$S_n(x) \equiv a_0 + \frac{a_1}{x} + \frac{a_2}{x^2} + \ldots + \frac{a_n}{x^n}. \tag{3}$$

The asymptotic series is defined to have the properties

$$\lim_{x\to\infty} x^n R_n(x) = 0 \qquad \text{for fixed } n \tag{4}$$

$$\lim_{n\to\infty} x^n R_n(x) = \infty \qquad \text{for fixed } x. \tag{5}$$

Therefore,

$$f(x) \approx \sum_{n=0}^{\infty} a_n x^{-n} \tag{6}$$

in the limit $x \to \infty$. If a function has an asymptotic expansion, the expansion is unique. The symbol $\sim$ is also used to mean directly SIMILAR.

References

Abramowitz, M. and Stegun, C. A. (Eds.). *Handbook of Mathematical Functions with Formulas, Graphs, and Mathematical Tables, 9th printing.* New York: Dover, p. 15, 1972.

Arfken, G. "Asymptotic of Semiconvergent Series." §5.10 in *Mathematical Methods for Physicists, 3rd ed.* Orlando, FL: Academic Press, pp. 339–346, 1985.

Bleistein, N. and Handelsman, R. A. *Asymptotic Expansions of Integrals.* New York: Dover, 1986.

Copson, E. T. *Asymptotic Expansions.* Cambridge, England: Cambridge University Press, 1965.

de Bruijn, N. G. *Asymptotic Methods in Analysis, 2nd ed.* New York: Dover, 1982.

Dingle, R. B. *Asymptotic Expansions: Their Derivation and Interpretation.* London: Academic Press, 1973.

Erdelyi, A. *Asymptotic Expansions.* New York: Dover, 1987.

Morse, P. M. and Feshbach, H. "Asymptotic Series; Method of Steepest Descent." §4.6 in *Methods of Theoretical Physics, Part I.* New York: McGraw-Hill, pp. 434–443, 1953.

Olver, F. W. J. *Asymptotics and Special Functions.* New York: Academic Press, 1974.

Wasow, W. R. *Asymptotic Expansions for Ordinary Differential Equations.* New York: Dover, 1987.

Atiyah-Singer Index Theorem

A theorem which states that the analytic and topological "indices" are equal for any elliptic differential operator on an n-D COMPACT DIFFERENTIABLE $\mathbb{C}^\infty$ boundaryless MANIFOLD.

see also COMPACT MANIFOLD, DIFFERENTIABLE MANIFOLD

References

Atiyah, M. F. and Singer, I. M. "The Index of Elliptic Operators on Compact Manifolds." *Bull. Amer. Math. Soc.* **69**, 322–433, 1963.

Atiyah, M. F. and Singer, I. M. "The Index of Elliptic Operators I, II, III." *Ann. Math.* **87**, 484–604, 1968.

Petkovšek, M.; Wilf, H. S.; and Zeilberger, D. *A=B.* Wellesley, MA: A. K. Peters, p. 4, 1996.

Atkin-Goldwasser Kilian-Morain Certificate

A recursive PRIMALITY CERTIFICATE for a PRIME p. The certificate consists of a list of

1. A point on an ELLIPTIC CURVE C

$$y^2 = x^3 + g_2 x + g_3 \pmod{p}$$

for some numbers g_2 and g_3.

2. A PRIME q with $q > (p^{1/4} + 1)^2$, such that for some other number k and $m = kq$ with $k \neq 1$, $mC(x, y, g_2, g_3, p)$ is the identity on the curve, but $kC(x, y, g_2, g_3, p)$ is not the identity. This guarantees PRIMALITY of p by a theorem of Goldwasser and Kilian (1986).
3. Each q has its recursive certificate following it. So if the smallest q is known to be PRIME, all the numbers are certified PRIME up the chain.

A PRATT CERTIFICATE is quicker to generate for small numbers. The *Mathematica®* (Wolfram Research, Champaign, IL) task `ProvablePrime[n]` therefore generates an Atkin-Goldwasser-Kilian-Morain certificate only for numbers above a certain limit (10^{10} by default), and a PRATT CERTIFICATE for smaller numbers.

see also ELLIPTIC CURVE PRIMALITY PROVING, ELLIPTIC PSEUDOPRIME, PRATT CERTIFICATE, PRIMALITY CERTIFICATE, WITNESS

References

Atkin, A. O. L. and Morain, F. "Elliptic Curves and Primality Proving." *Math. Comput.* **61**, 29–68, 1993.

Bressoud, D. M. *Factorization and Prime Testing.* New York: Springer-Verlag, 1989.

Goldwasser, S. and Kilian, J. "Almost All Primes Can Be Quickly Certified." *Proc. 18th STOC.* pp. 316-329, 1986.

Morain, F. "Implementation of the Atkin-Goldwasser-Kilian Primality Testing Algorithm." Rapport de Recherche 911, INRIA, Octobre 1988.

Schoof, R. "Elliptic Curves over Finite Fields and the Computation of Square Roots mod p." *Math. Comput.* **44**, 483–494, 1985.

Wunderlich, M. C. "A Performance Analysis of a Simple Prime-Testing Algorithm." *Math. Comput.* **40**, 709–714, 1983.

Atomic Statement

In LOGIC, a statement which cannot be broken down into smaller statements.

Attraction Basin

see BASIN OF ATTRACTION

Attractor

An attractor is a SET of states (points in the PHASE SPACE), invariant under the dynamics, towards which neighboring states in a given BASIN OF ATTRACTION asymptotically approach in the course of dynamic evolution. An attractor is defined as the smallest unit which cannot be itself decomposed into two or more attractors with distinct BASINS OF ATTRACTION. This restriction is necessary since a DYNAMICAL SYSTEM may have multiple attractors, each with its own BASIN OF ATTRACTION.

Conservative systems do not have attractors, since the motion is periodic. For dissipative DYNAMICAL SYSTEMS, however, volumes shrink exponentially so attractors have 0 volume in n-D phase space.

A stable FIXED POINT surrounded by a dissipative region is an attractor known as a SINK. Regular attractors (corresponding to 0 LYAPUNOV CHARACTERISTIC EXPONENTS) act as LIMIT CYCLES, in which trajectories circle around a limiting trajectory which they asymptotically approach, but never reach. STRANGE ATTRACTORS are bounded regions of PHASE SPACE (corresponding to POSITIVE LYAPUNOV CHARACTERISTIC EXPONENTS) having zero MEASURE in the embedding PHASE SPACE and a FRACTAL DIMENSION. Trajectories within a STRANGE ATTRACTOR appear to skip around randomly.

see also BARNSLEY'S FERN, BASIN OF ATTRACTION, CHAOS GAME, FRACTAL DIMENSION, LIMIT CYCLE, LYAPUNOV CHARACTERISTIC EXPONENT, MEASURE, SINK (MAP), STRANGE ATTRACTOR

Auction

A type of sale in which members of a group of buyers offer ever increasing amounts. The bidder making the last bid (for which no higher bid is subsequently made within a specified time limit: "going once, going twice, sold") must then purchase the item in question at this price. Variants of simple bidding are also possible, as in a VICKERY AUCTION.

see also VICKERY AUCTION

Augend

The first of several ADDENDS, or "the one to which the others are added," is sometimes called the augend. Therefore, while a, b, and c are ADDENDS in $a + b + c$, a is the augend.

see also ADDEND, ADDITION

Augmented Amicable Pair

A PAIR of numbers m and n such that

$$\sigma(m) = \sigma(n) = m + n - 1,$$

where $\sigma(m)$ is the DIVISOR FUNCTION. Beck and Najar (1977) found 11 augmented amicable pairs.

see also AMICABLE PAIR, DIVISOR FUNCTION, QUASIAMICABLE PAIR

References

Beck, W. E. and Najar, R. M. "More Reduced Amicable Pairs." *Fib. Quart.* **15**, 331–332, 1977.

Guy, R. K. *Unsolved Problems in Number Theory, 2nd ed.* New York: Springer-Verlag, p. 59, 1994.

Augmented Dodecahedron
see JOHNSON SOLID

Augmented Hexagonal Prism
see JOHNSON SOLID

Augmented Pentagonal Prism
see JOHNSON SOLID

Augmented Polyhedron
A UNIFORM POLYHEDRON with one or more other solids adjoined.

Augmented Sphenocorona
see JOHNSON SOLID

Augmented Triangular Prism
see JOHNSON SOLID

Augmented Tridiminished Icosahedron
see JOHNSON SOLID

Augmented Truncated Cube
see JOHNSON SOLID

Augmented Truncated Dodecahedron
see JOHNSON SOLID

Augmented Truncated Tetrahedron
see JOHNSON SOLID

Aureum Theorema
Gauss's name for the QUADRATIC RECIPROCITY THEOREM.

Aurifeuillean Factorization
A factorization of the form

$$2^{4n+2}+1=(2^{2n+1}-2^{n+1}+1)(2^{2n+1}+2^{n+1}+1). \quad (1)$$

The factorization for $n = 14$ was discovered by Aurifeuille, and the general form was subsequently discovered by Lucas. The large factors are sometimes written as L and M as follows

$$2^{4k-2}+1=(2^{2k-1}-2^k+1)(2^{2k-1}+2^k+1) \quad (2)$$

$$3^{6k-3}+1=(3^{2k-1}+1)(3^{2k-1}-3^k+1)(3^{2k-1}+3^k+1), \quad (3)$$

which can be written

$$2^{2h}+1=L_{2h}M_{2h} \quad (4)$$

$$3^{3h}+1=(3^h+1)L_{3h}M_{3h} \quad (5)$$

$$5^{5h}-1=(5^h-1)L_{5h}M_{5h}, \quad (6)$$

where $h \equiv 2k-1$ and

$$L_{2h}, M_{2h}=2^h+1\mp 2^k \quad (7)$$

$$L_{3h}, M_{3h}=3^h+1\mp 3^k \quad (8)$$

$$L_{5h}, M_{5h}=5^{2h}+3\cdot 5^h+1\mp 5^k(5^h+1). \quad (9)$$

see also GAUSS'S FORMULA

References

Brillhart, J.; Lehmer, D. H.; Selfridge, J.; Wagstaff, S. S. Jr.; and Tuckerman, B. *Factorizations of $b^n \pm 1$, $b = 2, 3, 5, 6, 7, 10, 11, 12$ Up to High Powers, rev. ed.* Providence, RI: Amer. Math. Soc., pp. lxviii–lxxii, 1988.

Wagstaff, S. S. Jr. "Aurifeullian Factorizations and the Period of the Bell Numbers Modulo a Prime." *Math. Comput.* **65**, 383–391, 1996.

Ausdehnungslehre
see EXTERIOR ALGEBRA

Authalic Latitude
An AUXILIARY LATITUDE which gives a SPHERE equal SURFACE AREA relative to an ELLIPSOID. The authalic latitude is defined by

$$\beta=\sin^{-1}\left(\frac{q}{q_p}\right), \quad (1)$$

where

$$q\equiv(1-e^2)\left[\frac{\sin\phi}{1-e^2\sin^2\phi}-\frac{1}{2e}\ln\left(\frac{1-e\sin\phi}{1+e\sin\phi}\right)\right], \quad (2)$$

and q_p is q evaluated at the north pole ($\phi = 90°$). Let R_q be the RADIUS of the SPHERE having the same SURFACE AREA as the ELLIPSOID, then

$$R_q=a\sqrt{\frac{q_p}{2}}. \quad (3)$$

The series for β is

$$\beta=\phi-(\tfrac{1}{3}e^2+\tfrac{31}{180}e^4+\tfrac{59}{560}e^6+\ldots)\sin(2\phi) + (\tfrac{17}{360}e^4+\tfrac{61}{1260}e^6+\ldots)\sin(4\phi) - (\tfrac{383}{45360}e^6+\ldots)\sin(6\phi)+\ldots. \quad (4)$$

The inverse FORMULA is found from

$$\Delta\phi=\frac{(1-e^2\sin^2\phi)^2}{2\cos\phi}\left[\frac{q}{1-e^2}-\frac{\sin\phi}{1-e^2\sin^2\phi}+\frac{1}{2e}\ln\left(\frac{1-e\sin\phi}{1+e\sin\phi}\right)\right], \quad (5)$$

where

$$q=q_p\sin\beta \quad (6)$$

and $\phi_0 = \sin^{-1}(q/2)$. This can be written in series form as

$$\phi = \beta + (\tfrac{1}{3}e^2 + \tfrac{31}{180}e^4 + \tfrac{517}{5040}e^6 + \ldots)\sin(2\beta) + (\tfrac{23}{360}e^4 + \tfrac{251}{3780}e^6 + \ldots)\sin(4\beta) + (\tfrac{761}{45360}e^6 + \ldots)\sin(6\beta) + \ldots. \quad (7)$$

see also LATITUDE

References

Adams, O. S. "Latitude Developments Connected with Geodesy and Cartography with Tables, Including a Table for Lambert Equal-Area Meridional Projections." Spec. Pub. No. 67. U. S. Coast and Geodetic Survey, 1921.

Snyder, J. P. *Map Projections—A Working Manual.* U. S. Geological Survey Professional Paper 1395. Washington, DC: U. S. Government Printing Office, p. 16, 1987.

Autocorrelation

The autocorrelation function is defined by

$$C_f(t) \equiv f \star f = f^*(-t) * f(t) = \int_{-\infty}^{\infty} f^*(\tau)f(t+\tau)\,d\tau, \quad (1)$$

where $*$ denotes CONVOLUTION and $\star$ denotes CROSS-CORRELATION. A finite autocorrelation is given by

$$C_f(\tau) \equiv \langle [y(t) - \bar{y}][y(t+\tau) - \bar{y}] \rangle \quad (2)$$

$$= \lim_{T\to\infty} \int_{-T/2}^{T/2} [y(t) - \bar{y}][y(t+\tau) - \bar{y}]\,dt. \quad (3)$$

If f is a REAL FUNCTION,

$$f^* = f, \quad (4)$$

and an EVEN FUNCTION so that

$$f(-\tau) = f(\tau), \quad (5)$$

then

$$C_f(t) = \int_{-\infty}^{\infty} f(\tau)f(t+\tau)\,d\tau. \quad (6)$$

But let $\tau' \equiv -\tau$, so $d\tau' = -d\tau$, then

$$C_f(t) = \int_{\infty}^{-\infty} f(-\tau)f(t-\tau)\,(-d\tau) = \int_{-\infty}^{\infty} f(-\tau)f(t-\tau)\,d\tau = \int_{-\infty}^{\infty} f(\tau)f(t-\tau)\,d\tau = f * f. \quad (7)$$

The autocorrelation discards phase information, returning only the POWER. It is therefore not reversible.

There is also a somewhat surprising and extremely important relationship between the autocorrelation and the FOURIER TRANSFORM known as the WIENER-KHINTCHINE THEOREM. Let $\mathcal{F}[f(x)] = F(k)$, and F^* denote the COMPLEX CONJUGATE of F, then the FOURIER TRANSFORM of the ABSOLUTE SQUARE of $F(k)$ is given by

$$\mathcal{F}[|F(k)|^2] = \int_{-\infty}^{\infty} f^*(\tau)f(\tau+x)\,d\tau. \quad (8)$$

The autocorrelation is a HERMITIAN OPERATOR since $C_f(-t) = C_f{}^*(t)$. $f \star f$ is MAXIMUM at the ORIGIN. In other words,

$$\int_{-\infty}^{\infty} f(u)f(u+x)\,du \leq \int_{-\infty}^{\infty} f^2(u)\,du. \quad (9)$$

To see this, let ϵ be a REAL NUMBER. Then

$$\int_{-\infty}^{\infty} [f(u) + \epsilon f(u+x)]^2\,du > 0 \quad (10)$$

$$\int_{-\infty}^{\infty} f^2(u)\,du + 2\epsilon \int_{-\infty}^{\infty} f(u)f(u+x)\,du + \epsilon^2 \int_{-\infty}^{\infty} f^2(u+x)\,du > 0 \quad (11)$$

$$\int_{-\infty}^{\infty} f^2(u)\,du + 2\epsilon \int_{-\infty}^{\infty} f(u)f(u+x)\,du + \epsilon^2 \int_{-\infty}^{\infty} f^2(u)\,du > 0. \quad (12)$$

Define

$$a \equiv \int_{-\infty}^{\infty} f^2(u)\,du \quad (13)$$

$$b \equiv 2\int_{-\infty}^{\infty} f(u)f(u+x)\,du. \quad (14)$$

Then plugging into above, we have $a\epsilon^2 + b\epsilon + c > 0$. This QUADRATIC EQUATION does not have any REAL ROOT, so $b^2 - 4ac \leq 0$, i.e., $b/2 \leq a$. It follows that

$$\int_{-\infty}^{\infty} f(u)f(u+x)\,du \leq \int_{-\infty}^{\infty} f^2(u)\,du, \quad (15)$$

with the equality at $x = 0$. This proves that $f \star f$ is MAXIMUM at the ORIGIN.

see also CONVOLUTION, CROSS-CORRELATION, QUANTIZATION EFFICIENCY, WIENER-KHINTCHINE THEOREM

References

Press, W. H.; Flannery, B. P.; Teukolsky, S. A.; and Vetterling, W. T. "Correlation and Autocorrelation Using the FFT." §13.2 in *Numerical Recipes in FORTRAN: The Art of Scientific Computing, 2nd ed.* Cambridge, England: Cambridge University Press, pp. 538–539, 1992.

Automorphic Function

An automorphic function $f(z)$ of a COMPLEX variable z is one which is analytic (except for POLES) in a domain D and which is invariant under a DENUMERABLY INFINITE group of LINEAR FRACTIONAL TRANSFORMATIONS (also known as MÖBIUS TRANSFORMATIONS)

$$z' = \frac{az+b}{cz+d}.$$

Automorphic functions are generalizations of TRIGONOMETRIC FUNCTIONS and ELLIPTIC FUNCTIONS.

see also MODULAR FUNCTION, MÖBIUS TRANSFORMATIONS, ZETA FUCHSIAN

Automorphic Number

A number k such that nk^2 has its last digits equal to k is called n-automorphic. For example, $1 \cdot \underline{5}^2 = 2\underline{5}$ and $1 \cdot \underline{6}^2 = 3\underline{6}$ are 1-automorphic and $2 \cdot \underline{8}^2 = 12\underline{8}$ and $2 \cdot \underline{88}^2 = 154\underline{88}$ are 2-automorphic. de Guerre and Fairbairn (1968) give a history of automorphic numbers.

The first few 1-automorphic numbers are 1, 5, 6, 25, 76, 376, 625, 9376, 90625, ... (Sloane's A003226, Wells 1986, p. 130). There are two 1-automorphic numbers with a given number of digits, one ending in 5 and one in 6 (except that the 1-digit automorphic numbers include 1), and each of these contains the previous number with a digit prepended. Using this fact, it is possible to construct automorphic numbers having more than 25,000 digits (Madachy 1979). The first few 1-automorphic numbers ending with 5 are 5, 25, 625, 0625, 90625, ... (Sloane's A007185), and the first few ending with 6 are 6, 76, 376, 9376, 09376, ... (Sloane's A016090). The 1-automorphic numbers $a(n)$ ending in 5 are IDEMPOTENT (mod 10^n) since

$$[a(n)]^2 \equiv a(n) \pmod{10^n}$$

(Sloane and Plouffe 1995).

The following table gives the 10-digit n-automorphic numbers.

n	n-Automorphic Numbers	Sloane
1	0000000001, 8212890625, 1787109376	—, A007185, A016090
2	0893554688	A030984
3	6666666667, 7262369792, 9404296875	—, A030985, A030986
4	0446777344	A030987
5	3642578125	A030988
6	3631184896	A030989
7	7142857143, 4548984375, 1683872768	A030990, A030991, A030992
8	0223388672	A030993
9	5754123264, 3134765625, 8888888889	A030994, A030995, —

see also IDEMPOTENT, NARCISSISTIC NUMBER, NUMBER PYRAMID, TRIMORPHIC NUMBER

References

Beeler, M.; Gosper, R. W.; and Schroeppel, R. Item 59 in *HAKMEM.* Cambridge, MA: MIT Artificial Intelligence Laboratory, Memo AIM-239, Feb. 1972.

Fairbairn, R. A. "More on Automorphic Numbers." *J. Recr. Math.* **2**, 170–174, 1969.

Fairbairn, R. A. Erratum to "More on Automorphic Numbers." *J. Recr. Math.* **2**, 245, 1969.

de Guerre, V. and Fairbairn, R. A. "Automorphic Numbers." *J. Recr. Math.* **1**, 173–179, 1968.

Hunter, J. A. H. "Two Very Special Numbers." *Fib. Quart.* **2**, 230, 1964.

Hunter, J. A. H. "Some Polyautomorphic Numbers." *J. Recr. Math.* **5**, 27, 1972.

Kraitchik, M. "Automorphic Numbers." §3.8 in *Mathematical Recreations.* New York: W. W. Norton, pp. 77–78, 1942.

Madachy, J. S. *Madachy's Mathematical Recreations.* New York: Dover, pp. 34–54 and 175–176, 1979.

Sloane, N. J. A. Sequences A016090, A003226/M3752, and A007185/M3940 in "An On-Line Version of the Encyclopedia of Integer Sequences."

Wells, D. *The Penguin Dictionary of Curious and Interesting Numbers.* Middlesex: Penguin Books, pp. 171, 178, 191–192, 1986.

Automorphism

An ISOMORPHISM of a system of objects onto itself.

see also ANOSOV AUTOMORPHISM

Automorphism Group

The GROUP of functions from an object G to itself which preserve the structure of the object, denoted $\mathrm{Aut}(G)$. The automorphism group of a GROUP preserves the MULTIPLICATION table, the automorphism group of a GRAPH the INCIDENCE MATRICES, and that of a FIELD the ADDITION and MULTIPLICATION tables.

see also OUTER AUTOMORPHISM GROUP

Autonomous

A differential equation or system of ORDINARY DIFFERENTIAL EQUATIONS is said to be autonomous if it does not explicitly contain the independent variable (usually denoted t). A second-order autonomous differential equation is of the form $F(y, y', y'') = 0$, where $y' \equiv dy/dt \equiv v$. By the CHAIN RULE, y'' can be expressed as

$$y'' = v' = \frac{dv}{dt} = \frac{dv}{dy}\frac{dy}{dt} = \frac{dv}{dy}v.$$

For an autonomous ODE, the solution is independent of the time at which the initial conditions are applied. This means that all particles pass through a given point in phase space. A nonautonomous system of n first-order ODEs can be written as an autonomous system of $n+1$ ODEs by letting $t \equiv x_{n+1}$ and increasing the dimension of the system by 1 by adding the equation

$$\frac{dx_{n+1}}{dt} = 1.$$

Autoregressive Model

see MAXIMUM ENTROPY METHOD

Auxiliary Circle

The CIRCUMCIRCLE of an ELLIPSE, i.e., the CIRCLE whose center corresponds with that of the ELLIPSE and whose RADIUS is equal to the ELLIPSE'S SEMIMAJOR AXIS.

see also CIRCLE, ECCENTRIC ANGLE, ELLIPSE

Auxiliary Latitude

see AUTHALIC LATITUDE, CONFORMAL LATITUDE, GEOCENTRIC LATITUDE, ISOMETRIC LATITUDE, LATITUDE, PARAMETRIC LATITUDE, RECTIFYING LATITUDE, REDUCED LATITUDE

Auxiliary Triangle

see MEDIAL TRIANGLE

Average

see MEAN

Average Absolute Deviation

$$\alpha \equiv \frac{1}{N}\sum_{i=1}^{N}|x_i - \mu| = \langle |x_i - \mu| \rangle .$$

see also ABSOLUTE DEVIATION, DEVIATION, STANDARD DEVIATION, VARIANCE

Average Function

If f is CONTINUOUS on a CLOSED INTERVAL $[a, b]$, then there is at least one number x^* in $[a, b]$ such that

$$\int_a^b f(x)dx = f(x^*)(b-a).$$

The average value of the FUNCTION $(\bar{f})$ on this interval is then given by $f(x^*)$.

see MEAN-VALUE THEOREM

Average Seek Time

see POINT-POINT DISTANCE—1-D

Ax-Kochen Isomorphism Theorem

Let P be the SET of PRIMES, and let $\mathbb{Q}_p$ and $Z_p(t)$ be the FIELDS of p-ADIC NUMBERS and formal POWER series over $Z_p = (0, 1, \ldots, p-1)$. Further, suppose that D is a "nonprincipal maximal filter" on P. Then $\prod_{p\in P} \mathbb{Q}_p/D$ and $\prod_{p\in P} Z_p(t)/D$ are ISOMORPHIC.

see also HYPERREAL NUMBER, NONSTANDARD ANALYSIS

Axial Vector

see PSEUDOVECTOR

Axiom

A PROPOSITION regarded as self-evidently TRUE without PROOF. The word "axiom" is a slightly archaic synonym for POSTULATE. Compare CONJECTURE or HYPOTHESIS, both of which connote apparently TRUE *but not self-evident* statements.

see also ARCHIMEDES' AXIOM, AXIOM OF CHOICE, AXIOMATIC SYSTEM, CANTOR-DEDEKIND AXIOM, CONGRUENCE AXIOMS, CONJECTURE, CONTINUITY AXIOMS, COUNTABLE ADDITIVITY PROBABILITY AXIOM, DEDEKIND'S AXIOM, DIMENSION AXIOM, EILENBERG-STEENROD AXIOMS, EUCLID'S AXIOMS, EXCISION AXIOM, FANO'S AXIOM, FIELD AXIOMS, HAUSDORFF AXIOMS, HILBERT'S AXIOMS, HOMOTOPY AXIOM, INACCESSIBLE CARDINALS AXIOM, INCIDENCE AXIOMS, INDEPENDENCE AXIOM, INDUCTION AXIOM, LAW, LEMMA, LONG EXACT SEQUENCE OF A PAIR AXIOM, ORDERING AXIOMS, PARALLEL AXIOM, PASCH'S AXIOM, PEANO'S AXIOMS, PLAYFAIR'S AXIOM, PORISM, POSTULATE, PROBABILITY AXIOMS, PROCLUS' AXIOM, RULE, T2-SEPARATION AXIOM, THEOREM, ZERMELO'S AXIOM OF CHOICE, ZERMELO-FRAENKEL AXIOMS

Axiom A Diffeomorphism

Let $\phi : M \to M$ be a C^1 DIFFEOMORPHISM on a compact RIEMANNIAN MANIFOLD M. Then ϕ satisfies Axiom A if the NONWANDERING set $\Omega(\phi)$ of ϕ is hyperbolic and the PERIODIC POINTS of ϕ are DENSE in $\Omega(\phi)$. Although it was conjectured that the first of these conditions implies the second, they were shown to be independent in or around 1977. Examples include the ANOSOV DIFFEOMORPHISMS and SMALE HORSESHOE MAP.

In some cases, Axiom A can be replaced by the condition that the DIFFEOMORPHISM is a hyperbolic diffeomorphism on a hyperbolic set (Bowen 1975, Parry and Pollicott 1990).

see also ANOSOV DIFFEOMORPHISM, AXIOM A FLOW, DIFFEOMORPHISM, DYNAMICAL SYSTEM, RIEMANNIAN MANIFOLD, SMALE HORSESHOE MAP

References

Bowen, R. *Equilibrium States and the Ergodic Theory of Anosov Diffeomorphisms.* New York: Springer-Verlag, 1975.

Ott, E. *Chaos in Dynamical Systems.* New York: Cambridge University Press, p. 143, 1993.

Parry, W. and Pollicott, M. "Zeta Functions and the Periodic Orbit Structure of Hyperbolic Dynamics." *Astérisque* No. 187–188, 1990.

Smale, S. "Differentiable Dynamical Systems." *Bull. Amer. Math. Soc.* **73**, 747–817, 1967.

Axiom A Flow

A FLOW defined analogously to the AXIOM A DIFFEOMORPHISM, except that instead of splitting the TANGENT BUNDLE into two invariant sub-BUNDLES, they are split into three (one exponentially contracting, one expanding, and one which is 1-dimensional and tangential to the flow direction).

see also DYNAMICAL SYSTEM

Axiom of Choice

An important and fundamental result in SET THEORY sometimes called ZERMELO'S AXIOM OF CHOICE. It was formulated by Zermelo in 1904 and states that, given any SET of mutually exclusive nonempty SETS, there exists at least one SET that contains exactly one element in common with each of the nonempty SETS.

It is related to HILBERT'S PROBLEM 1B, and was proved to be consistent with other AXIOMS in SET THEORY in 1940 by Gödel. In 1963, Cohen demonstrated that the axiom of choice is independent of the other AXIOMS in Cantorian SET THEORY, so the AXIOM cannot be proved within the system (Boyer and Merzbacher 1991, p. 610).

see also HILBERT'S PROBLEMS, SET THEORY, WELL-ORDERED SET, ZERMELO-FRAENKEL AXIOMS, ZORN'S LEMMA

References

Boyer, C. B. and Merzbacher, U. C. *A History of Mathematics, 2nd ed.* New York: Wiley, 1991.

Cohen, P. J. "The Independence of the Continuum Hypothesis." *Proc. Nat. Acad. Sci. U. S. A.* **50**, 1143–1148, 1963.

Cohen, P. J. "The Independence of the Continuum Hypothesis. II." *Proc. Nat. Acad. Sci. U. S. A.* **51**, 105–110, 1964.

Conway, J. H. and Guy, R. K. *The Book of Numbers.* New York: Springer-Verlag, pp. 274–276, 1996.

Moore, G. H. *Zermelo's Axiom of Choice: Its Origin, Development, and Influence.* New York: Springer-Verlag, 1982.

Axiomatic Set Theory

A version of SET THEORY in which axioms are taken as uninterpreted rather than as formalizations of pre-existing truths.

see also NAIVE SET THEORY, SET THEORY

Axiomatic System

A logical system which possesses an explicitly stated SET of AXIOMS from which THEOREMS can be derived.

see also COMPLETE AXIOMATIC THEORY, CONSISTENCY, MODEL THEORY, THEOREM

Axis

A LINE with respect to which a curve or figure is drawn, measured, rotated, etc. The term is also used to refer to a LINE SEGMENT through a RANGE (Woods 1961).

see also ABSCISSA, ORDINATE, x-AXIS, y-AXIS, z-AXIS

References

Woods, F. S. *Higher Geometry: An Introduction to Advanced Methods in Analytic Geometry.* New York: Dover, p. 8, 1961.

Axonometry

A METHOD for mapping 3-D figures onto the PLANE.

see also CROSS-SECTION, MAP PROJECTION, POHLKE'S THEOREM, PROJECTION, STEREOLOGY

References

Coxeter, H. S. M. *Regular Polytopes, 3rd ed.* New York: Dover, p. 313, 1973.

Hazewinkel, M. (Managing Ed.). *Encyclopaedia of Mathematics: An Updated and Annotated Translation of the Soviet "Mathematical Encyclopaedia."* Dordrecht, Netherlands: Reidel, pp. 322–323, 1988.

Azimuthal Equidistant Projection

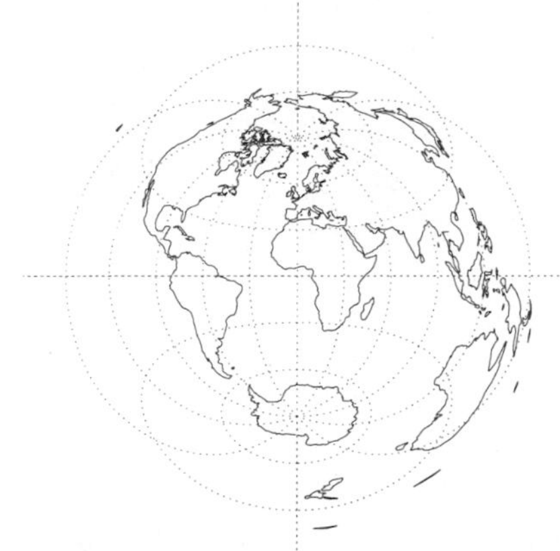

An AZIMUTHAL PROJECTION which is neither equal-AREA nor CONFORMAL. Let ϕ_1 and λ_0 be the LATITUDE and LONGITUDE of the center of the projection, then the transformation equations are given by

$$x = k' \cos\phi \sin(\lambda - \lambda_0) \tag{1}$$

$$y = k'[\cos\phi_1 \sin\phi - \sin\phi_1 \cos\phi \cos(\lambda - \lambda_0)]. \tag{2}$$

Here,

$$k' = \frac{c}{\sin c} \tag{3}$$

and

$$\cos c = \sin\phi_1 \sin\phi + \cos\phi_1 \cos\phi \cos(\lambda - \lambda_0), \tag{4}$$

where c is the angular distance from the center. The inverse FORMULAS are

$$\phi = \sin^{-1}\left(\cos c \sin\phi_1 + \frac{y \sin c \cos\phi_1}{c}\right) \tag{5}$$

and

$$\lambda = \begin{cases} \lambda_0 + \tan^{-1}\left(\frac{x \sin c}{c \cos\phi_1 \cos c - y \sin\phi_1 \sin c}\right) \\ \quad \text{for } \phi_1 \neq \pm 90^\circ \\ \lambda_0 + \tan^{-1}\left(-\frac{x}{y}\right) \\ \quad \text{for } \phi_1 = 90^\circ \\ \lambda_0 + \tan^{-1}\left(\frac{x}{y}\right), \\ \quad \text{for } \phi_1 = -90^\circ, \end{cases} \tag{6}$$

with the angular distance from the center given by

$$c = \sqrt{x^2 + y^2}. \tag{7}$$

References

Snyder, J. P. *Map Projections—A Working Manual.* U. S. Geological Survey Professional Paper 1395. Washington, DC: U. S. Government Printing Office, pp. 191–202, 1987.

Azimuthal Projection

see AZIMUTHAL EQUIDISTANT PROJECTION, LAMBERT AZIMUTHAL EQUAL-AREA PROJECTION, ORTHOGRAPHIC PROJECTION, STEREOGRAPHIC PROJECTION

B

B^*-Algebra

A BANACH ALGEBRA with an ANTIAUTOMORPHIC INVOLUTION $*$ which satisfies

$$x^{**} = x \quad (1)$$
$$x^* y^* = (yx)^* \quad (2)$$
$$x^* + y^* = (x+y)^* \quad (3)$$
$$(cx)^* = \bar{c}x^* \quad (4)$$

and whose NORM satisfies

$$||xx^*|| = ||x||^2. \quad (5)$$

A C^*-ALGEBRA is a special type of B^*-algebra.

see also BANACH ALGEBRA, C^*-ALGEBRA

B_2-Sequence

N.B. A detailed on-line essay by S. Finch was the starting point for this entry.

Also called a SIDON SEQUENCE. An INFINITE SEQUENCE of POSITIVE INTEGERS

$$1 \leq b_1 < b_2 < b_3 < \ldots \quad (1)$$

such that all pairwise sums

$$b_i + b_j \quad (2)$$

for $i \leq j$ are distinct (Guy 1994). An example is 1, 2, 4, 8, 13, 21, 31, 45, 66, 81, ... (Sloane's A005282).

Zhang (1993, 1994) showed that

$$S(B2) \equiv \sup_{\text{all B2 sequences}} \sum_{k=1}^{\infty} \frac{1}{b_k} > 2.1597. \quad (3)$$

The definition can be extended to B_n-sequences (Guy 1994).

see also A-SEQUENCE, MIAN-CHOWLA SEQUENCE

References

Finch, S. "Favorite Mathematical Constants." `http://www.mathsoft.com/asolve/constant/erdos/erdos.html`.

Guy, R. K. "Packing Sums of Pairs," "Three-Subsets with Distinct Sums," and "B_2-Sequences," and B_2-Sequences Formed by the Greedy Algorithm." §C9, C11, E28, and E32 in *Unsolved Problems in Number Theory, 2nd ed.* New York: Springer-Verlag, pp. 115–118, 121–123, 228–229, and 232–233, 1994.

Sloane, N. J. A. Sequence A005282/M1094 in "An On-Line Version of the Encyclopedia of Integer Sequences."

Zhang, Z. X. "A B2-Sequence with Larger Reciprocal Sum." *Math. Comput.* **60**, 835–839, 1993.

Zhang, Z. X. "Finding Finite B2-Sequences with Larger $m - a_m{}^{1/2}$." *Math. Comput.* **63**, 403–414, 1994.

B_p-Theorem

If $O_{p'}(G) = 1$ and if x is a p-element of G, then

$$L_{p'}(C_G(x) \leq E(C_G(x)),$$

where $L_{p'}$ is the p-LAYER.

B-Spline

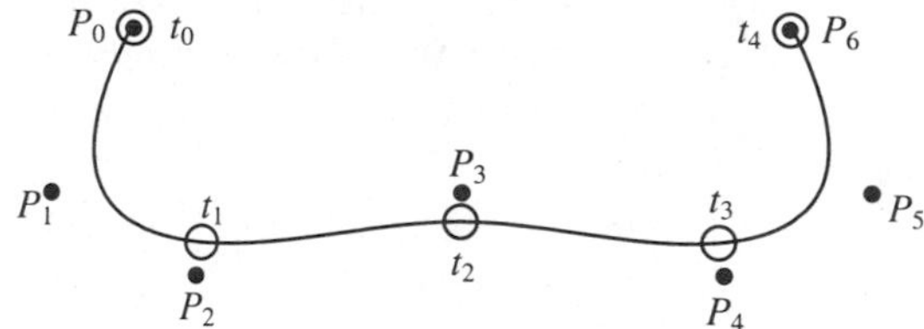

A generalization of the BÉZIER CURVE. Let a vector known as the KNOT VECTOR be defined

$$\mathbf{T} = \{t_0, t_1, \ldots, t_m\}, \quad (1)$$

where $\mathbf{T}$ is a nondecreasing SEQUENCE with $t_i \in [0,1]$, and define control points $\mathbf{P}_0, \ldots, \mathbf{P}_n$. Define the degree as

$$p \equiv m - n - 1. \quad (2)$$

The "knots" $t_{p+1}, \ldots, t_{m-p-1}$ are called INTERNAL KNOTS.

Define the basis functions as

$$N_{i,0}(t) = \begin{cases} 1 & \text{if } t_i \leq t < t_{i+1} \text{ and } t_i < t_{t+1} \\ 0 & \text{otherwise} \end{cases} \quad (3)$$

$$N_{i,p}(t) = \frac{t - t_i}{t_{i+p} - t_i} N_{i,p-1}(t) + \frac{t_{i+p+1} - t}{t_{i+p+1} - t_{i+1}} N_{i+1,p-1}(t). \quad (4)$$

Then the curve defined by

$$\mathbf{C}(t) = \sum_{i=0}^{n} \mathbf{P}_i N_{i,p}(t) \quad (5)$$

is a B-spline. Specific types include the nonperiodic B-spline (first $p+1$ knots equal 0 and last $p+1$ equal to 1) and uniform B-spline (INTERNAL KNOTS are equally spaced). A B-Spline with no INTERNAL KNOTS is a BÉZIER CURVE.

The degree of a B-spline is independent of the number of control points, so a low order can always be maintained for purposes of numerical stability. Also, a curve is $p-k$ times differentiable at a point where k duplicate knot values occur. The knot values determine the extent of the control of the control points.

A nonperiodic B-spline is a B-spline whose first $p+1$ knots are equal to 0 and last $p+1$ knots are equal to 1. A uniform B-spline is a B-spline whose INTERNAL KNOTS are equally spaced.

see also BÉZIER CURVE, NURBS CURVE

B-Tree

B-trees were introduced by Bayer (1972) and McCreight. They are a special m-ary balanced tree used in databases because their structure allows records to be inserted, deleted, and retrieved with guaranteed worst-case performance. An n-node B-tree has height $\mathcal{O}(\lg 2)$, where LG is the LOGARITHM to base 2. The Apple® Macintosh® (Apple Computer, Cupertino, CA) HFS filing system uses B-trees to store disk directories (Benedict 1995). A B-tree satisfies the following properties:

1. The ROOT is either a LEAF (TREE) or has at least two CHILDREN.
2. Each node (except the ROOT and LEAVES) has between $\lceil m/2 \rceil$ and m CHILDREN, where $\lceil x \rceil$ is the CEILING FUNCTION.
3. Each path from the ROOT to a LEAF (TREE) has the same length.

Every 2-3 TREE is a B-tree of order 3. The number of B-trees of order $n = 1, 2, \ldots$ are 0, 1, 1, 1, 2, 2, 3, 4, 5, 8, 14, 23, 32, 43, 63, ... (Ruskey, Sloane's A014535).

see also RED-BLACK TREE

References

Aho, A. V.; Hopcroft, J. E.; and Ullmann, J. D. *Data Structures and Algorithms.* Reading, MA: Addison-Wesley, pp. 369–374, 1987.

Benedict, B. *Using Norton Utilities for the Macintosh.* Indianapolis, IN: Que, pp. B-17–B-33, 1995.

Beyer, R. "Symmetric Binary B-Trees: Data Structures and Maintenance Algorithms." *Acta Informat.* **1**, 290–306, 1972.

Ruskey, F. "Information on B-Trees." http://sue.csc.uvic.ca/~cos/inf/tree/BTrees.html.

Sloane, N. J. A. Sequence A014535 in "An On-Line Version of the Encyclopedia of Integer Sequences."

Baby Monster Group

Also known as FISCHER'S BABY MONSTER GROUP. The SPORADIC GROUP B. It has ORDER

$$2^{41} \cdot 3^{13} \cdot 5^6 \cdot 7^2 \cdot 11 \cdot 13 \cdot 17 \cdot 19 \cdot 23 \cdot 31 \cdot 47.$$

see also MONSTER GROUP

References

Wilson, R. A. "ATLAS of Finite Group Representation." http://for.mat.bham.ac.uk/atlas/BM.html.

BAC-CAB Identity

The VECTOR TRIPLE PRODUCT identity

$$\mathbf{A} \times (\mathbf{B} \times \mathbf{C}) = \mathbf{B}(\mathbf{A} \cdot \mathbf{C}) - \mathbf{C}(\mathbf{A} \cdot \mathbf{B}).$$

This identity can be generalized to n-D

$$\mathbf{a}_2 \times \cdots \times \mathbf{a}_{n-1} \times (\mathbf{b}_1 \times \cdots \times \mathbf{b}_{n-1}) = (-1)^{n+1} \begin{vmatrix} \mathbf{b}_1 & \cdots & \mathbf{b}_{n-1} \\ \mathbf{a}_2 \cdot \mathbf{b}_1 & \cdots & \mathbf{a}_2 \cdot \mathbf{b}_{n-1} \\ \vdots & \ddots & \vdots \\ \mathbf{a}_{n-1} \cdot \mathbf{b}_1 & \cdots & \mathbf{a}_{n-1} \cdot \mathbf{b}_{n-1} \end{vmatrix}.$$

see also LAGRANGE'S IDENTITY

BAC-CAB Rule

see BAC-CAB IDENTITY

Bachelier Function

see BROWN FUNCTION

Bachet's Conjecture

see LAGRANGE'S FOUR-SQUARE THEOREM

Bachet Equation

The DIOPHANTINE EQUATION

$$x^2 + k = y^3,$$

which is also an ELLIPTIC CURVE. The general equation is still the focus of ongoing study.

Backhouse's Constant

Let $P(x)$ be defined as the POWER series whose nth term has a COEFFICIENT equal to the nth PRIME,

$$P(x) \equiv \sum_{k=0}^{\infty} p_k x^k = 1 + 2x + 3x^2 + 5x^3 + 7x^4 + 11x^5 + \ldots,$$

and let $Q(x)$ be defined by

$$Q(x) = \frac{1}{P(x)} = \sum_{k=0}^{\infty} q_k x^k.$$

Then N. Backhouse conjectured that

$$\lim_{n\to\infty} \left| \frac{q_{n+1}}{q_n} \right| = 1.4560749485826896713995953511156\ldots.$$

The constant was subsequently shown to exist by P. Flajolet.

References

Finch, S. "Favorite Mathematical Constants." http://www.mathsoft.com/asolve/constant/backhous/backhous.html.

Bäcklund Transformation

A method for solving classes of nonlinear PARTIAL DIFFERENTIAL EQUATIONS.

see also INVERSE SCATTERING METHOD

References

Infeld, E. and Rowlands, G. *Nonlinear Waves, Solitons, and Chaos.* Cambridge, England: Cambridge University Press, p. 196, 1990.

Miura, R. M. (Ed.) *Bäcklund Transformations, the Inverse Scattering Method, Solitons, and Their Applications.* New York: Springer-Verlag, 1974.

Backtracking

A method of drawing FRACTALS by appropriate numbering of the corresponding tree diagram which does not require storage of intermediate results.

Backus-Gilbert Method

A method which can be used to solve some classes of INTEGRAL EQUATIONS and is especially useful in implementing certain types of data inversion. It has been applied to invert seismic data to obtain density profiles in the Earth.

References

Backus, G. and Gilbert, F. "The Resolving Power of Growth Earth Data." *Geophys. J. Roy. Astron. Soc.* **16**, 169–205, 1968.

Backus, G. E. and Gilbert, F. "Uniqueness in the Inversion of Inaccurate Gross Earth Data." *Phil. Trans. Roy. Soc. London Ser. A* **266**, 123–192, 1970.

Loredo, T. J. and Epstein, R. I. "Analyzing Gamma-Ray Burst Spectral Data." *Astrophys. J.* **336**, 896–919, 1989.

Parker, R. L. "Understanding Inverse Theory." *Ann. Rev. Earth Planet. Sci.* **5**, 35–64, 1977.

Press, W. H.; Flannery, B. P.; Teukolsky, S. A.; and Vetterling, W. T. "Backus-Gilbert Method." §18.6 in *Numerical Recipes in FORTRAN: The Art of Scientific Computing, 2nd ed.* Cambridge, England: Cambridge University Press, pp. 806–809, 1992.

Backward Difference

The backward difference is a FINITE DIFFERENCE defined by

$$\nabla_p \equiv \nabla f_p \equiv f_p - f_{p-1}. \qquad (1)$$

Higher order differences are obtained by repeated operations of the backward difference operator, so

$$\begin{aligned}\nabla_p^2 &= \nabla(\nabla p) = \nabla(f_p - f_{p-1}) = \nabla f_p - \nabla f_{p-1} \qquad (2)\\ &= (f_p - f_{p-1}) - (f_{p-1} - f_{p-2})\\ &= f_p - 2f_{p-1} + f_{p-2} \qquad (3)\end{aligned}$$

In general,

$$\nabla_p^k \equiv \nabla^k f_p \equiv \sum_{m=0}^{k} (-1)^m \binom{k}{m} f_{p-k+m}, \qquad (4)$$

where $\binom{k}{m}$ is a BINOMIAL COEFFICIENT.

NEWTON'S BACKWARD DIFFERENCE FORMULA expresses f_p as the sum of the nth backward differences

$$f_p = f_0 + p\nabla_0 + \tfrac{1}{2!}p(p+1)\nabla_0^2 + \tfrac{1}{3!}p(p+1)(p+2)\nabla_0^3 + \ldots, \qquad (5)$$

where ∇_0^n is the first nth difference computed from the difference table.

see also ADAMS' METHOD, DIFFERENCE EQUATION, DIVIDED DIFFERENCE, FINITE DIFFERENCE, FORWARD DIFFERENCE, NEWTON'S BACKWARD DIFFERENCE FORMULA, RECIPROCAL DIFFERENCE

References

Beyer, W. H. *CRC Standard Mathematical Tables, 28th ed.* Boca Raton, FL: CRC Press, pp. 429 and 433, 1987.

Bader-Deuflhard Method

A generalization of the BULIRSCH-STOER ALGORITHM for solving ORDINARY DIFFERENTIAL EQUATIONS.

References

Bader, G. and Deuflhard, P. "A Semi-Implicit Mid-Point Rule for Stiff Systems of Ordinary Differential Equations." *Numer. Math.* **41**, 373–398, 1983.

Press, W. H.; Flannery, B. P.; Teukolsky, S. A.; and Vetterling, W. T. *Numerical Recipes in FORTRAN: The Art of Scientific Computing, 2nd ed.* Cambridge, England: Cambridge University Press, p. 730, 1992.

Baguenaudier

A PUZZLE involving disentangling a set of rings from a looped double rod (also called CHINESE RINGS). The minimum number of moves needed for n rings is

$$\begin{cases} \frac{1}{3}(2^{n+1} - 2) & n \text{ even} \\ \frac{1}{3}(2^{n+1} - 1) & n \text{ odd.} \end{cases}$$

By simultaneously moving the two end rings, the number of moves can be reduced to

$$\begin{cases} 2^{n-1} - 1 & n \text{ even} \\ 2^{n-1} & n \text{ odd.} \end{cases}$$

The solution of the baguenaudier is intimately related to the theory of GRAY CODES.

References

Dubrovsky, V. "Nesting Puzzles, Part II: Chinese Rings Produce a Chinese Monster." *Quantum* **6**, 61–65 (Mar.) and 58–59 (Apr.), 1996.

Gardner, M. "The Binary Gray Code." In *Knotted Doughnuts and Other Mathematical Entertainments.* New York: W. H. Freeman, pp. 15–17, 1986.

Kraitchik, M. "Chinese Rings." §3.12.3 in *Mathematical Recreations.* New York: W. W. Norton, pp. 89–91, 1942.

Steinhaus, H. *Mathematical Snapshots, 3rd American ed.* New York: Oxford University Press, p. 268, 1983.

Bailey's Method

see LAMBERT'S METHOD

Bailey's Theorem

Let $\Gamma(z)$ be the GAMMA FUNCTION, then

$$\left[\frac{\Gamma(m+\frac{1}{2})}{\Gamma(m)}\right]^2 \underbrace{\left[\frac{1}{m} + \left(\frac{1}{2}\right)^2 \frac{1}{m+1} + \left(\frac{1\cdot 3}{2\cdot 4}\right)^2 \frac{1}{m+2} + \ldots\right]}_{n}$$
$$= \left[\frac{\Gamma(n+\frac{1}{2})}{\Gamma(n)}\right]^2 \underbrace{\left[\frac{1}{n} + \left(\frac{1}{2}\right)^2 \frac{1}{n+1} + \left(\frac{1\cdot 3}{2\cdot 4}\right)^2 \frac{1}{n+2} + \ldots\right]}_{m}.$$

Baire Category Theorem

A nonempty complete METRIC SPACE cannot be represented as the UNION of a COUNTABLE family of nowhere DENSE SUBSETS.

Baire Space

A TOPOLOGICAL SPACE X in which each SUBSET of X of the "first category" has an empty interior. A TOPOLOGICAL SPACE which is HOMEOMORPHIC to a complete METRIC SPACE is a Baire space.

Bairstow's Method

A procedure for finding the quadratic factors for the COMPLEX CONJUGATE ROOTS of a POLYNOMIAL $P(x)$ with REAL COEFFICIENTS.

$$[x-(a+ib)][x-(a-ib)] = x^2+2ax+(a^2+b^2) \equiv x^2+Bx+C. \quad (1)$$

Now write the original POLYNOMIAL as

$$P(x) = (x^2+Bx+C)Q(x)+Rx+S \quad (2)$$

$$R(B+\delta B, C+\delta C) \approx R(B,C)+\frac{\partial R}{\partial B}dB+\frac{\partial R}{\partial C}dC \quad (3)$$

$$S(B+\delta B, C+\delta C) \approx S(B,C)+\frac{\partial S}{\partial B}dB+\frac{\partial S}{\partial C}dC \quad (4)$$

$$\frac{\partial P}{\partial C} = 0 = (x^2+Bx+C)\frac{\partial Q}{\partial C}+Q(x)+\frac{\partial R}{\partial C}+\frac{\partial S}{\partial C} \quad (5)$$

$$-Q(x) = (x^2+Bx+C)\frac{\partial Q}{\partial C}+\frac{\partial R}{\partial C}+\frac{\partial S}{\partial C} \quad (6)$$

$$\frac{\partial P}{\partial B} = 0 = (x^2+Bx+C)\frac{\partial Q}{\partial B}+xQ(x)+\frac{\partial R}{\partial B}+\frac{\partial S}{\partial B} \quad (7)$$

$$-xQ(x) = (x^2+Bx+C)\frac{\partial Q}{\partial B}+\frac{\partial R}{\partial B}+\frac{\partial S}{\partial B}. \quad (8)$$

Now use the 2-D NEWTON'S METHOD to find the simultaneous solutions.

References

Press, W. H.; Flannery, B. P.; Teukolsky, S. A.; and Vetterling, W. T. *Numerical Recipes in C: The Art of Scientific Computing.* Cambridge, England: Cambridge University Press, pp. 277 and 283–284, 1989.

Baker's Dozen

The number 13.

see also 13, DOZEN

Baker's Map

The MAP

$$x_{n+1} = 2\mu x_n, \quad (1)$$

where x is computed modulo 1. A generalized Baker's map can be defined as

$$x_{n+1} = \begin{cases} \lambda_a x_n & y_n < \alpha \\ (1-\lambda_b)+\lambda_b x_n & y_n > \alpha \end{cases} \quad (2)$$

$$y_{n+1} = \begin{cases} \frac{y_n}{\alpha} & y_n < \alpha \\ \frac{y_n-\alpha}{\beta} & y_n > \alpha, \end{cases} \quad (3)$$

where $\beta \equiv 1-\alpha$, $\lambda_a+\lambda_b \leq 1$, and x and y are computed mod 1. The $q=1$ q-DIMENSION is

$$D_1 = 1+\frac{\alpha\ln\left(\frac{1}{\alpha}\right)+\beta\ln\left(\frac{1}{\beta}\right)}{\alpha\ln\left(\frac{1}{\lambda_a}\right)+\beta\ln\left(\frac{1}{\lambda_b}\right)}. \quad (4)$$

If $\lambda_a = \lambda_b$, then the general q-DIMENSION is

$$D_q = 1+\frac{1}{q-1}\frac{\ln(\alpha^q+\beta^q)}{\ln\lambda_a}. \quad (5)$$

References

Lichtenberg, A. and Lieberman, M. *Regular and Stochastic Motion.* New York: Springer-Verlag, p. 60, 1983.

Ott, E. *Chaos in Dynamical Systems.* Cambridge, England: Cambridge University Press, pp. 81–82, 1993.

Rasband, S. N. *Chaotic Dynamics of Nonlinear Systems.* New York: Wiley, p. 32, 1990.

Balanced ANOVA

An ANOVA in which the number of REPLICATES (sets of identical observations) is restricted to be the same for each FACTOR LEVEL (treatment group).

see also ANOVA

Balanced Incomplete Block Design

see BLOCK DESIGN

Ball

The n-ball, denoted $\mathbb{B}^n$, is the interior of a SPHERE $\mathbb{S}^{n-1}$, and sometimes also called the n-DISK. (Although physicists often use the term "SPHERE" to mean the solid ball, mathematicians definitely do not!) Let $\text{Vol}(\mathbb{B}^n)$ denote the volume of an n-D ball of RADIUS r. Then

$$\sum_{n=0}^{\infty} \text{Vol}(B^n) = e^{\pi r^2}[1+\text{erf}(r\sqrt{\pi})],$$

where $\text{erf}(x)$ is the ERF function.

see also ALEXANDER'S HORNED SPHERE, BANACH-TARSKI PARADOX, BING'S THEOREM, BISHOP'S INEQUALITY, BOUNDED, DISK, HYPERSPHERE, SPHERE, WILD POINT

References

Freden, E. Problem 10207. "Summing a Series of Volumes." *Amer. Math. Monthly* **100**, 882, 1993.

Ball Triangle Picking

The determination of the probability for obtaining an OBTUSE TRIANGLE by picking 3 points at random in the unit DISK was generalized by Hall (1982) to the n-D BALL. Buchta (1986) subsequently gave closed form evaluations for Hall's integrals, with the first few solutions being

$$P_2 = \frac{9}{8} - \frac{4}{\pi^2} \approx 0.72$$
$$P_3 = \tfrac{37}{70} \approx 0.53$$
$$P_4 \approx 0.39$$
$$P_5 \approx 0.29.$$

The case P_2 corresponds to the usual DISK case.

see also CUBE TRIANGLE PICKING, OBTUSE TRIANGLE

References

Buchta, C. "A Note on the Volume of a Random Polytope in a Tetrahedron." *Ill. J. Math.* **30**, 653–659, 1986.
Hall, G. R. "Acute Triangles in the n-Ball." *J. Appl. Prob.* **19**, 712–715, 1982.

Ballantine

see BORROMEAN RINGS

Ballieu's Theorem

For any set $\boldsymbol{\mu} = (\mu_1, \mu_2, \ldots, \mu_n)$ of POSITIVE numbers with $\mu_0 = 0$ and

$$M_\mu = \max_{0 \leq k \leq n-1} \frac{\mu_k + \mu_n |b_{n-k}|}{\mu_{k+1}}.$$

Then all the EIGENVALUES λ satisfying $P(\lambda) = 0$, where $P(\lambda)$ is the CHARACTERISTIC POLYNOMIAL, lie on the DISK $|z| \leq M_\mu$.

References

Gradshteyn, I. S. and Ryzhik, I. M. *Tables of Integrals, Series, and Products, 5th ed.* San Diego, CA: Academic Press, p. 1119, 1979.

Ballot Problem

Suppose A and B are candidates for office and there are $2n$ voters, n voting for A and n for B. In how many ways can the ballots be counted so that A is always ahead of or tied with B? The solution is a CATALAN NUMBER C_n.

A related problem also called "the" ballot problem is to let A receive a votes and B b votes with $a > b$. This version of the ballot problem then asks for the probability that A stays ahead of B as the votes are counted (Vardi 1991). The solution is $(a-b)/(a+b)$, as first shown by M. Bertrand (Hilton and Pedersen 1991). Another elegant solution was provided by André (1887) using the so-called ANDRÉ'S REFLECTION METHOD.

The problem can also be generalized (Hilton and Pedersen 1991). Furthermore, the TAK FUNCTION is connected with the ballot problem (Vardi 1991).

see also ANDRÉ'S REFLECTION METHOD, CATALAN NUMBER, TAK FUNCTION

References

André, D. "Solution directe du problème résolu par M. Bertrand." *Comptes Rendus Acad. Sci. Paris* **105**, 436–437, 1887.
Ball, W. W. R. and Coxeter, H. S. M. *Mathematical Recreations and Essays, 13th ed.* New York: Dover, p. 49, 1987.
Carlitz, L. "Solution of Certain Recurrences." *SIAM J. Appl. Math.* **17**, 251–259, 1969.
Comtet, L. *Advanced Combinatorics.* Dordrecht, Netherlands: Reidel, p. 22, 1974.
Feller, W. *An Introduction to Probability Theory and Its Applications, Vol. 1, 3rd ed.* New York: Wiley, pp. 67–97, 1968.
Hilton, P. and Pedersen, J. "The Ballot Problem and Catalan Numbers." *Nieuw Archief voor Wiskunde* **8**, 209–216, 1990.
Hilton, P. and Pedersen, J. "Catalan Numbers, Their Generalization, and Their Uses." *Math. Intel.* **13**, 64–75, 1991.
Kraitchik, M. "The Ballot-Box Problem." §6.13 in *Mathematical Recreations.* New York: W. W. Norton, p. 132, 1942.
Motzkin, T. "Relations Between Hypersurface Cross Ratios, and a Combinatorial Formula for Partitions of a Polygon, for Permanent Preponderance, and for Non-Associative Products." *Bull. Amer. Math. Soc.* **54**, 352–360, 1948.
Vardi, I. *Computational Recreations in Mathematica.* Redwood City, CA: Addison-Wesley, pp. 185–187, 1991.

Banach Algebra

An ALGEBRA A over a FIELD F with a NORM that makes A into a COMPLETE METRIC SPACE, and therefore, a BANACH SPACE. F is frequently taken to be the COMPLEX NUMBERS in order to assure that the SPECTRUM fully characterizes an OPERATOR (i.e., the spectral theorems for normal or compact normal operators do not, in general, hold in the SPECTRUM over the REAL NUMBERS).

see also B^*-ALGEBRA

Banach Fixed Point Theorem

Let f be a contraction mapping from a closed SUBSET F of a BANACH SPACE E into F. Then there exists a unique $z \in F$ such that $f(z) = z$.

see also FIXED POINT THEOREM

References

Debnath, L. and Mikusiński, P. *Introduction to Hilbert Spaces with Applications.* San Diego, CA: Academic Press, 1990.

Banach-Hausdorff-Tarski Paradox

see BANACH-TARSKI PARADOX

Banach Measure

An "AREA" which can be defined for every set—even those without a true geometric AREA—which is rigid and finitely additive.

Banach Space

A normed linear SPACE which is COMPLETE in the norm-determined METRIC. A HILBERT SPACE is always a Banach space, but the converse need not hold.

see also BESOV SPACE, HILBERT SPACE, SCHAUDER FIXED POINT THEOREM

Banach-Steinhaus Theorem

see UNIFORM BOUNDEDNESS PRINCIPLE

Banach-Tarski Paradox

First stated in 1924, this theorem demonstrates that it is possible to dissect a BALL into six pieces which can be reassembled by rigid motions to form two balls of the same size as the original. The number of pieces was subsequently reduced to five. However, the pieces are extremely complicated. A generalization of this theorem is that any two bodies in $\mathbb{R}^3$ which do not extend to infinity and each containing a ball of arbitrary size can be dissected into each other (they are are EQUIDECOMPOSABLE).

References

Stromberg, K. "The Banach-Tarski Paradox." *Amer. Math. Monthly* **86**, 3, 1979.

Wagon, S. *The Banach-Tarski Paradox.* New York: Cambridge University Press, 1993.

Bang's Theorem

The lines drawn to the VERTICES of a face of a TETRAHEDRON from the point of contact of the FACE with the INSPHERE form three ANGLES at the point of contact which are the same three ANGLES in each FACE.

References

Brown, B. H. "Theorem of Bang. Isosceles Tetrahedra." *Amer. Math. Monthly* **33**, 224–226, 1926.

Honsberger, R. *Mathematical Gems II.* Washington, DC: Math. Assoc. Amer., p. 93, 1976.

Bankoff Circle

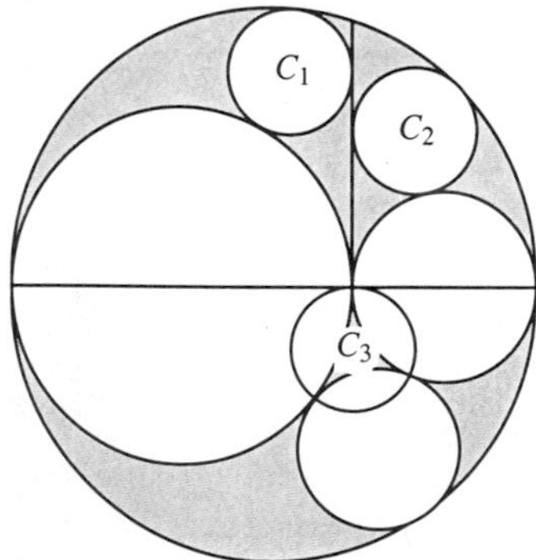

In addition to the ARCHIMEDES' CIRCLES C_1 and C_2 in the ARBELOS figure, there is a third circle C_3 congruent to these two as illustrated in the above figure.

see also ARBELOS

References

Bankoff, L. "Are the Twin Circles of Archimedes Really Twins?" *Math. Mag.* **47**, 214–218, 1974.

Gardner, M. "Mathematical Games: The Diverse Pleasures of Circles that Are Tangent to One Another." *Sci. Amer.* **240**, 18–28, Jan. 1979.

Banzhaf Power Index

The number of ways in which a group of n with weights $\sum_{i=1}^n w_i = 1$ can change a losing coalition (one with $\sum w_i < 1/2$) to a winning one, or vice versa. It was proposed by the lawyer J. F. Banzhaf in 1965.

References

Paulos, J. A. *A Mathematician Reads the Newspaper.* New York: BasicBooks, pp. 9–10, 1995.

Bar (Edge)

The term in rigidity theory for the EDGES of a GRAPH.

see also CONFIGURATION, FRAMEWORK

Bar Polyhex

A POLYHEX consisting of HEXAGONS arranged along a line.

see also BAR POLYIAMOND

References

Gardner, M. *Mathematical Magic Show: More Puzzles, Games, Diversions, Illusions and Other Mathematical Sleight-of-Mind from Scientific American.* New York: Vintage, p. 147, 1978.

Bar Polyiamond

A POLYIAMOND consisting of EQUILATERAL TRIANGLES arranged along a line.

see also BAR POLYHEX

References

Golomb, S. W. *Polyominoes: Puzzles, Patterns, Problems, and Packings, 2nd ed.* Princeton, NJ: Princeton University Press, p. 92, 1994.

Barber Paradox

A man of Seville is shaved by the Barber of Seville IFF the man does not shave himself. Does the barber shave himself? Proposed by Bertrand Russell.

Barbier's Theorem

All CURVES OF CONSTANT WIDTH of width w have the same PERIMETER πw.

Bare Angle Center

The TRIANGLE CENTER with TRIANGLE CENTER FUNCTION

$$\alpha = A.$$

References

Kimberling, C. "Major Centers of Triangles." *Amer. Math. Monthly* **104**, 431–438, 1997.

Barnes G-Function

see G-FUNCTION

Barnes' Lemma

If a CONTOUR in the COMPLEX PLANE is curved such that it separates the increasing and decreasing sequences of POLES, then

$$\frac{1}{2\pi i}\int_{-i\infty}^{i\infty}\Gamma(\alpha+s)\Gamma(\beta+s)\Gamma(\gamma-s)\Gamma(\delta-s)\,ds = \frac{\Gamma(\alpha+\gamma)\Gamma(\alpha+\delta)\Gamma(\beta+\gamma)\Gamma(\beta+\delta)}{\Gamma(\alpha+\beta+\gamma+\delta)},$$

where $\Gamma(z)$ is the GAMMA FUNCTION.

Barnes-Wall Lattice

A lattice which can be constructed from the LEECH LATTICE Λ_{24}.

see also COXETER-TODD LATTICE, LATTICE POINT, LEECH LATTICE

References

Barnes, E. S. and Wall, G. E. "Some Extreme Forms Defined in Terms of Abelian Groups." *J. Austral. Math. Soc.* **1**, 47–63, 1959.

Conway, J. H. and Sloane, N. J. A. "The 16-Dimensional Barnes-Wall Lattice Λ_{16}." §4.10 in *Sphere Packings, Lattices, and Groups, 2nd ed.* New York: Springer-Verlag, pp. 127–129, 1993.

Barnsley's Fern

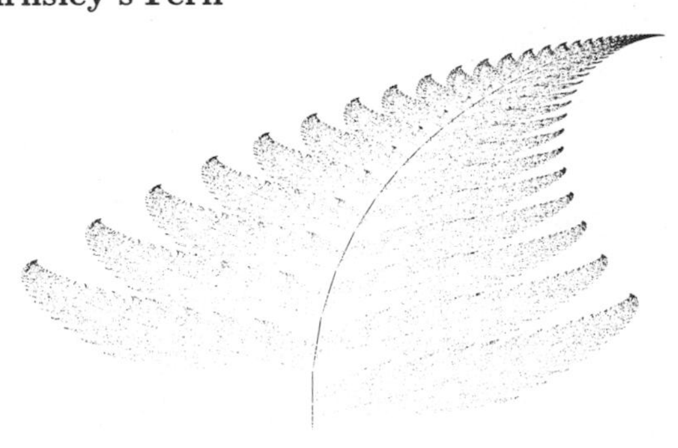

The ATTRACTOR of the ITERATED FUNCTION SYSTEM given by the set of "fern functions"

$$f_1(x,y) = \begin{bmatrix} 0.85 & 0.04 \\ -0.04 & 0.85 \end{bmatrix}\begin{bmatrix} x \\ y \end{bmatrix} + \begin{bmatrix} 0.00 \\ 1.60 \end{bmatrix} \qquad (1)$$

$$f_2(x,y) = \begin{bmatrix} -0.15 & 0.28 \\ 0.26 & 0.24 \end{bmatrix}\begin{bmatrix} x \\ y \end{bmatrix} + \begin{bmatrix} 0.00 \\ 0.44 \end{bmatrix} \qquad (2)$$

$$f_3(x,y) = \begin{bmatrix} 0.20 & -0.26 \\ 0.23 & 0.22 \end{bmatrix}\begin{bmatrix} x \\ y \end{bmatrix} + \begin{bmatrix} 0.00 \\ 1.60 \end{bmatrix} \qquad (3)$$

$$f_4(x,y) = \begin{bmatrix} 0.00 & 0.00 \\ 0.00 & 0.16 \end{bmatrix}\begin{bmatrix} x \\ y \end{bmatrix} \qquad (4)$$

(Barnsley 1993, p. 86; Wagon 1991). These AFFINE TRANSFORMATIONS are contractions. The tip of the fern (which resembles the black spleenwort variety of fern) is the fixed point of f_1, and the tips of the lowest two branches are the images of the main tip under f_2 and f_3 (Wagon 1991).

see also DYNAMICAL SYSTEM, FRACTAL, ITERATED FUNCTION SYSTEM

References

Barnsley, M. *Fractals Everywhere, 2nd ed.* Boston, MA: Academic Press, pp. 86, 90, 102 and Plate 2, 1993.

Gleick, J. *Chaos: Making a New Science.* New York: Penguin Books, p. 238, 1988.

Wagon, S. "Biasing the Chaos Game: Barnsley's Fern." §5.3 in *Mathematica in Action.* New York: W. H. Freeman, pp. 156–163, 1991.

Barrier

A number n is called a barrier of a number-theoretic function $f(m)$ if, for all $m < n$, $m + f(m) \leq n$. Neither the TOTIENT FUNCTION $\phi(n)$ nor the DIVISOR FUNCTION $\sigma(n)$ has barriers.

References

Guy, R. K. *Unsolved Problems in Number Theory, 2nd ed.* New York: Springer-Verlag, pp. 64–65, 1994.

Barth Decic

The Barth decic is a DECIC SURFACE in complex three-dimensional projective space having the maximum possible number of ORDINARY DOUBLE POINTS (345). It is given by the implicit equation

$$\begin{aligned} &8(x^2-\phi^4y^2)(y^2-\phi^4z^2)(z^2-\phi^4x^2) \\ &\quad\times(x^4+y^4+z^4-2x^2y^2-2x^2z^2-2y^2z^2) \\ &+(3+5\phi)(x^2+y^2+z^2-w^2)^2[x^2+y^2+z^2-(2-\phi)w^2]^2w^2 \\ &\qquad = 0, \end{aligned}$$

where ϕ is the GOLDEN MEAN and w is a parameter (Endraß, Nordstrand), taken as $w = 1$ in the above plot. The Barth decic is invariant under the ICOSAHEDRAL GROUP.

see also ALGEBRAIC SURFACE, BARTH SEXTIC, DECIC SURFACE, ORDINARY DOUBLE POINT

References

Barth, W. "Two Projective Surfaces with Many Nodes Admitting the Symmetries of the Icosahedron." *J. Alg. Geom.* **5**, 173–186, 1996.

Endraß, S. "Flächen mit vielen Doppelpunkten." *DMV-Mitteilungen* **4**, 17–20, 4/1995.

Endraß, S. "Barth's Decic." `http://www.mathematik.uni-mainz.de/AlgebraischeGeometrie/docs/Ebarthdecic.shtml`.

Nordstrand, T. "Batch Decic." `http://www.uib.no/people/nfytn/bdectxt.htm`.

Barth Sextic

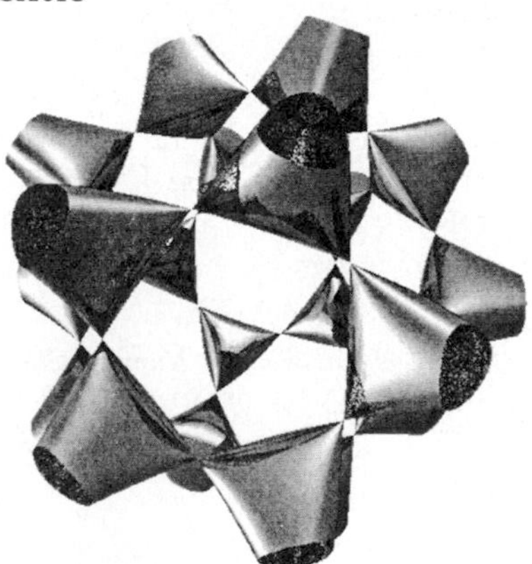

The Barth-sextic is a SEXTIC SURFACE in complex three-dimensional projective space having the maximum possible number of ORDINARY DOUBLE POINTS (65). It is given by the implicit equation

$$\begin{aligned} &4(\phi^2x^2-y^2)(\phi^2y^2-z^2)(\phi^2z^2-x^2) \\ &\qquad -(1+2\phi)(x^2+y^2+z^2-w^2)^2w^2 = 0. \end{aligned}$$

where ϕ is the GOLDEN MEAN, and w is a parameter (Endraß, Nordstrand), taken as $w = 1$ in the above plot. The Barth sextic is invariant under the ICOSAHEDRAL GROUP. Under the map

$$(x, y, z, w) \to (x^2, y^2, z^2, w^2),$$

the surface is the eightfold cover of the CAYLEY CUBIC (Endraß).

see also ALGEBRAIC SURFACE, BARTH DECIC, CAYLEY CUBIC, ORDINARY DOUBLE POINT, SEXTIC SURFACE

References

Barth, W. "Two Projective Surfaces with Many Nodes Admitting the Symmetries of the Icosahedron." *J. Alg. Geom.* **5**, 173–186, 1996.

Endraß, S. "Flächen mit vielen Doppelpunkten." *DMV-Mitteilungen* **4**, 17–20, 4/1995.

Endraß, S. "Barth's Sextic." `http://www.mathematik.uni-mainz.de/AlgebraischeGeometrie/docs/Ebarthsextic.shtml`.

Nordstrand, T. "Barth Sextic." `http://www.uib.no/people/nfytn/sexttxt.htm`.

Bartlett Function

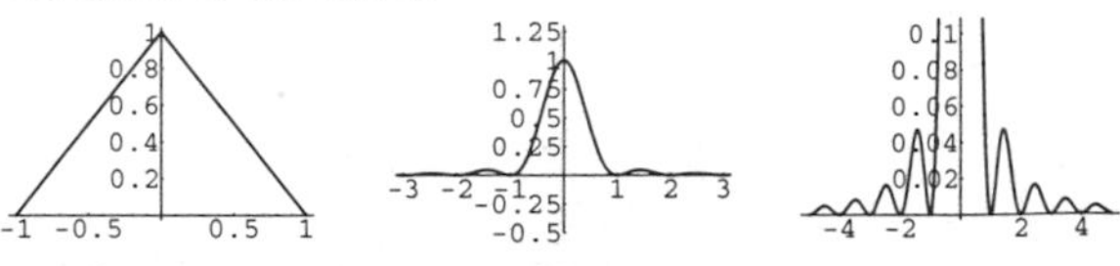

The APODIZATION FUNCTION

$$f(x) = 1 - \frac{|x|}{a} \tag{1}$$

which is a generalization of the one-argument TRIANGLE FUNCTION. Its FULL WIDTH AT HALF MAXIMUM is a. It has INSTRUMENT FUNCTION

$$\begin{aligned} I(x) &= \int_{-a}^{a} e^{-2\pi ikx}\left(1-\frac{|x|}{a}\right)dx \\ &= \int_{-a}^{0} e^{-2\pi ikx}\left(1+\frac{x}{a}\right)dx \\ &\quad + \int_{0}^{a} e^{-2\pi ikx}\left(1-\frac{x}{a}\right)dx. \end{aligned} \tag{2}$$

Letting $x' \equiv -x$ in the first part therefore gives

$$\begin{aligned} \int_{-a}^{0} e^{-2\pi ikx}\left(1+\frac{x}{a}\right)dx &= \int_{a}^{0} e^{2\pi ikx'}\left(1-\frac{x'}{a}\right)(-dx') \\ &= \int_{0}^{a} e^{2\pi ikx}\left(1-\frac{x}{a}\right)dx. \end{aligned} \tag{3}$$

Rewriting (2) using (3) gives

$$\begin{aligned} I(x) &= (e^{2\pi ikx}+e^{-2\pi ikx})\left(1-\frac{x}{a}\right)dx \\ &= 2\int_{0}^{a}\cos(2\pi kx)\left(1-\frac{x}{a}\right)dx. \end{aligned} \tag{4}$$

Integrating the first part and using the integral

$$\int x\cos(bx)\,dx = \frac{1}{b^2}\cos(bx) + \frac{x}{b}\sin(bx) \tag{5}$$

for the second part gives

$$\begin{aligned} I(x) &= 2\left[\frac{\sin(2\pi kx)}{2\pi k}\right. \\ &\quad \left.-\frac{1}{a}\left\{\frac{1}{4\pi^2k^2}\cos(2\pi kx)+\frac{x}{2\pi k}\sin(2\pi kx)\right\}\right]_0^a \\ &= 2\left\{\left[\frac{\sin(2\pi ka)}{2\pi k}-0\right]\right. \\ &\quad \left.-\frac{1}{a}\left[\frac{\cos(2\pi ka)-1}{4\pi^2k^2}+\frac{a\sin(2\pi ka)}{2\pi k}\right]\right\} \\ &= \frac{1}{2\pi^2ak^2}[\cos(2\pi ka)-1] = a\frac{\sin^2(\pi ka)}{\pi^2k^2a^2} \\ &= a\operatorname{sinc}^2(\pi ka), \end{aligned} \tag{6}$$

where $\operatorname{sinc} x$ is the SINC FUNCTION. The peak (in units of a) is 1. The function $I(x)$ is always positive, so there are no NEGATIVE sidelobes. The extrema are given by letting $\beta \equiv \pi ka$ and solving

$$\frac{d}{d\beta}\left(\frac{\sin\beta}{\beta}\right)^2 = 2\frac{\sin\beta}{\beta}\frac{\sin\beta-\beta\cos\beta}{\beta^2} = 0 \tag{7}$$

$$\sin\beta(\sin\beta-\beta\cos\beta) = 0 \tag{8}$$

$$\sin\beta-\beta\cos\beta = 0 \tag{9}$$

$$\tan\beta = \beta. \tag{10}$$

Solving this numerically gives $\beta = 4.49341$ for the first maximum, and the peak POSITIVE sidelobe is 0.047190. The full width at half maximum is given by setting $x \equiv \pi ka$ and solving

$$\operatorname{sinc}^2 x = \tfrac{1}{2} \tag{11}$$

for $x_{1/2}$, yielding

$$x_{1/2} = \pi k_{1/2} a = 1.39156. \tag{12}$$

Therefore, with $L \equiv 2a$,

$$\text{FWHM} = 2k_{1/2} = \frac{0.885895}{a} = \frac{1.77179}{L}. \tag{13}$$

see also APODIZATION FUNCTION, PARZEN APODIZATION FUNCTION, TRIANGLE FUNCTION

References

Bartlett, M. S. "Periodogram Analysis and Continuous Spectra." *Biometrika* **37**, 1–16, 1950.

Barycentric Coordinates

Also known as HOMOGENEOUS COORDINATES or TRILINEAR COORDINATES.

see TRILINEAR COORDINATES

Base Curve

see DIRECTRIX (RULED SURFACE)

Base (Logarithm)

The number used to define a LOGARITHM, which is then written $\log_b$. The symbol $\log x$ is an abbreviation for $\log_{10} x$, $\ln x$ for $\log_e x$ (the NATURAL LOGARITHM), and $\lg x$ for $\log_2 x$.

see also E, LG, LN, LOGARITHM, NAPIERIAN LOGARITHM, NATURAL LOGARITHM

Base (Neighborhood System)

A base for a neighborhood system of a point x is a collection N of OPEN SETS such that x belongs to every member of N, and any OPEN SET containing x also contains a member of N as a SUBSET.

Base (Number)

A REAL NUMBER x can be represented using any INTEGER number b as a base (sometimes also called a RADIX or SCALE). The choice of a base yields to a representation of numbers known as a NUMBER SYSTEM. In base b, the DIGITS 0, 1, ..., $b-1$ are used (where, by convention, for bases larger than 10, the symbols A, B, C, ... are generally used as symbols representing the DECIMAL numbers 10, 11, 12, ...).

Base	Name
2	binary
3	ternary
4	quaternary
5	quinary
6	senary
7	septenary
8	octal
9	nonary
10	decimal
11	undenary
12	duodecimal
16	hexadecimal
20	vigesimal
60	sexagesimal

Let the base b representation of a number x be written

$$(a_n a_{n-1} \ldots a_0 . a_{-1} \ldots)_b, \tag{1}$$

(e.g., 123.456_{10}), then the index of the leading DIGIT needed to represent the number is

$$n \equiv \lfloor \log_b x \rfloor, \tag{2}$$

where $\lfloor x \rfloor$ is the FLOOR FUNCTION. Now, recursively compute the successive DIGITS

$$a_i = \left\lfloor \frac{r_i}{b^i} \right\rfloor, \tag{3}$$

where $r_n \equiv x$ and

$$r_{i-1} = r_i - a_i b^i \tag{4}$$

for $i = n, n-1, \ldots, 1, 0, \ldots$. This gives the base b representation of x. Note that if x is an INTEGER, then i need only run through 0, and that if x has a fractional part, then the expansion may or may not terminate. For example, the HEXADECIMAL representation of 0.1 (which terminates in DECIMAL notation) is the infinite expression $0.19999\ldots_h$.

Some number systems use a mixture of bases for counting. Examples include the Mayan calendar and the old British monetary system (in which ha'pennies, pennies, threepence, sixpence, shillings, half crowns, pounds, and guineas corresponded to units of 1/2, 1, 3, 6, 12, 30, 240, and 252, respectively).

Knuth has considered using TRANSCENDENTAL bases. This leads to some rather unfamiliar results, such as equating π to 1 in "base π," $\pi = 1_\pi$.

see also BINARY, DECIMAL, HEREDITARY REPRESENTATION, HEXADECIMAL, OCTAL, QUATERNARY, SEXAGESIMAL, TERNARY, VIGESIMAL

References

Abramowitz, M. and Stegun, C. A. (Eds.). *Handbook of Mathematical Functions with Formulas, Graphs, and Mathematical Tables, 9th printing.* New York: Dover, p. 28, 1972.

Bogomolny, A. "Base Converter." http://www.cut-the-knot.com/binary.html.

Lauwerier, H. *Fractals: Endlessly Repeated Geometric Figures.* Princeton, NJ: Princeton University Press, pp. 6–11, 1991.

Weisstein, E. W. "Bases." http://www.astro.virginia.edu/~eww6n/math/notebooks/Bases.m.

Base Space

The SPACE B of a FIBER BUNDLE given by the MAP $f: E \to B$, where E is the TOTAL SPACE of the FIBER BUNDLE.

see also FIBER BUNDLE, TOTAL SPACE

Baseball

The numbers 3 and 4 appear prominently in the game of baseball. There are $3 \cdot 3 = 9$ innings in a game, and three strikes are an out. However, 4 balls are needed for a walk. The number of bases can either be regarded as 3 (excluding HOME PLATE) or 4 (including it).

see BASEBALL COVER, HOME PLATE

Baseball Cover

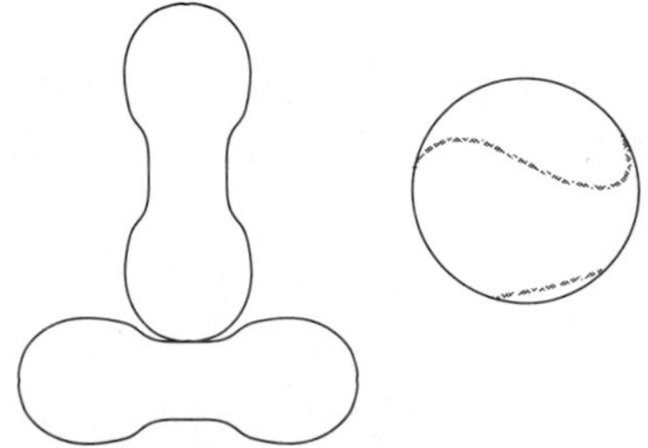

A pair of identical plane regions (mirror symmetric about two perpendicular lines through the center) which can be stitched together to form a baseball (or tennis ball). A baseball has a CIRCUMFERENCE of 9 1/8 inches. The practical consideration of separating the regions far enough to allow the pitcher a good grip requires that the "neck" distance be about 1 3/16 inches. The baseball cover was invented by Elias Drake as a boy in the 1840s. (Thompson's attribution of the current design to trial and error development by C. H. Jackson in the 1860s is apparently unsubstantiated, as discovered by George Bart.)

One way to produce a baseball cover is to draw the regions on a SPHERE, then cut them out. However, it is difficult to produce two identical regions in this manner. Thompson (1996) gives mathematical expressions giving baseball cover curves both in the plane and in 3-D. J. H. Conway has humorously proposed the following "baseball curve conjecture:" no two definitions of "the" baseball curve will give the same answer unless their equivalence was obvious from the start.

see also BASEBALL, HOME PLATE, TENNIS BALL THEOREM, YIN-YANG

References

Thompson, R. B. "Designing a Baseball Cover. 1860's: Patience, Trial, and Error. 1990's: Geometry, Calculus, and Computation." http://www.mathsoft.com/asolve/baseball/baseball.html. Rev. March 5, 1996.

Basin of Attraction

The set of points in the space of system variables such that initial conditions chosen in this set dynamically evolve to a particular ATTRACTOR.

see also WADA BASIN

Basis

A (vector) basis is any SET of n LINEARLY INDEPENDENT VECTORS capable of generating an n-dimensional SUBSPACE of $\mathbb{R}^n$. Given a HYPERPLANE defined by

$$x_1 + x_2 + x_3 + x_4 + x_5 = 0,$$

a basis is found by solving for x_1 in terms of x_2, x_3, x_4, and x_5. Carrying out this procedure,

$$x_1 = -x_2 - x_3 - x_4 - x_5,$$

so

$$\begin{bmatrix} x_1 \\ x_2 \\ x_3 \\ x_4 \\ x_5 \end{bmatrix} = x_2 \begin{bmatrix} -1 \\ 1 \\ 0 \\ 0 \\ 0 \end{bmatrix} + x_3 \begin{bmatrix} -1 \\ 0 \\ 1 \\ 0 \\ 0 \end{bmatrix} + x_4 \begin{bmatrix} -1 \\ 0 \\ 0 \\ 1 \\ 0 \end{bmatrix} + x_5 \begin{bmatrix} -1 \\ 0 \\ 0 \\ 0 \\ 1 \end{bmatrix},$$

and the above VECTOR form an (unnormalized) BASIS. Given a MATRIX A with an orthonormal basis, the MATRIX corresponding to a new basis, expressed in terms of the original $\hat{\mathbf{x}}_1, \ldots, \hat{\mathbf{x}}_n$ is

$$\mathsf{A}' = [\,\mathsf{A}\hat{\mathbf{x}}_1 \quad \ldots \quad \mathsf{A}\hat{\mathbf{x}}_n\,].$$

see also BILINEAR BASIS, MODULAR SYSTEM BASIS, ORTHONORMAL BASIS, TOPOLOGICAL BASIS

Basis Theorem

see HILBERT BASIS THEOREM

Basler Problem

The problem of analytically finding the value of $\zeta(2)$, where ζ is the RIEMANN ZETA FUNCTION.

References

Castellanos, D. "The Ubiquitous Pi. Part I." *Math. Mag.* **61**, 67–98, 1988.

Basset Function

see MODIFIED BESSEL FUNCTION OF THE SECOND KIND

Batch

A set of values of similar meaning obtained in any manner.

References

Tukey, J. W. *Explanatory Data Analysis.* Reading, MA: Addison-Wesley, p. 667, 1977.

Bateman Function

$$k_\nu(x) \equiv \frac{e^{-x}}{\Gamma(1+\frac{1}{2}\nu)} U(-\tfrac{1}{2}\nu, 0, 2x)$$

for $x > 0$, where U is a CONFLUENT HYPERGEOMETRIC FUNCTION OF THE SECOND KIND.

see also CONFLUENT HYPERGEOMETRIC DIFFERENTIAL EQUATION, HYPERGEOMETRIC FUNCTION

Batrachion

A class of CURVE defined at INTEGER values which hops from one value to another. Their name derives from the word *batrachion*, which means "frog-like." Many batrachions are FRACTAL. Examples include the BLANCMANGE FUNCTION, HOFSTADTER-CONWAY \$10,000 SEQUENCE, HOFSTADTER'S Q-SEQUENCE, and MALLOW'S SEQUENCE.

References

Pickover, C. A. "The Crying of Fractal Batrachion 1,489." Ch. 25 in *Keys to Infinity.* New York: W. H. Freeman, pp. 183–191, 1995.

Bauer's Identical Congruence

Let $t(m)$ denote the set of the $\phi(m)$ numbers less than and RELATIVELY PRIME to m, where $\phi(n)$ is the TOTIENT FUNCTION. Define

$$f_m(x) \equiv \prod_{t(m)} (x - t). \qquad (1)$$

A theorem of Lagrange states that

$$f_m(x) \equiv x^{\phi(m)} - 1 \pmod{m}. \qquad (2)$$

This can be generalized as follows. Let p be an ODD PRIME DIVISOR of m and p^a the highest POWER which divides m, then

$$f_m(x) \equiv (x^{p-1} - 1)^{\phi(m)/(p-1)} \pmod{p^a} \qquad (3)$$

and, in particular,

$$f_{p^a}(x) \equiv (x^{p-1} - 1)^{p^{a-1}} \pmod{p^a}. \qquad (4)$$

Furthermore, if $m > 2$ is EVEN and 2^a is the highest POWER of 2 that divides m, then

$$f_m(x) \equiv (x^2 - 1)^{\phi(m)/2} \pmod{2^a} \qquad (5)$$

and, in particular,

$$f_{2^a}(x) \equiv (x^2 - 1)^{2^{a-2}} \pmod{2^a}. \qquad (6)$$

see also LEUDESDORF THEOREM

References

Hardy, G. H. and Wright, E. M. "Bauer's Identical Congruence." §8.5 in *An Introduction to the Theory of Numbers, 5th ed.* Oxford, England: Clarendon Press, pp. 98–100, 1979.

Bauer's Theorem

see BAUER'S IDENTICAL CONGRUENCE

Bauspiel

A construction for the RHOMBIC DODECAHEDRON.

References

Coxeter, H. S. M. *Regular Polytopes, 3rd ed.* New York: Dover, pp. 26 and 50, 1973.

Bayes' Formula

Let A and B_j be SETS. CONDITIONAL PROBABILITY requires that

$$P(A \cap B_j) = P(A)P(B_j|A), \qquad (1)$$

where $\cap$ denotes INTERSECTION ("and"), and also that

$$P(A \cap B_j) = P(B_j \cap A) = P(B_j)P(A|B_j) \qquad (2)$$

and

$$P(B_j \cap A) = P(B_j)P(A|B_j). \quad (3)$$

Since (2) and (3) must be equal,

$$P(A \cap B_j) = P(B_j \cap A). \quad (4)$$

From (2) and (3),

$$P(A \cap B_j) = P(B_j)P(A|B_j). \quad (5)$$

Equating (5) with (2) gives

$$P(A)P(B_j|A) = P(B_j)P(A|B_j), \quad (6)$$

so

$$P(B_j|A) = \frac{P(B_j)P(A|B_j)}{P(A)}. \quad (7)$$

Now, let

$$S \equiv \bigcup_{i=1}^{N} A_i, \quad (8)$$

so A_i is an event is S and $A_i \cap A_j = \varnothing$ for $i \neq j$, then

$$A = A \cap S = A \cap \left(\bigcup_{i=1}^{N} A_i \right) = \bigcup_{i=1}^{N} (A \cap A_i) \quad (9)$$

$$P(A) = P\left(\bigcup_{i=1}^{N} (A \cap A_i) \right) = \sum_{i=1}^{N} P(A \cap A_i). \quad (10)$$

From (5), this becomes

$$P(A) = \sum_{i=1}^{N} P(A_i)P(E|A_i), \quad (11)$$

so

$$P(A_i|A) = \frac{P(A_i)P(A|A_i)}{\sum_{j=1}^{N} P(A_j)P(A|A_j)}. \quad (12)$$

see also CONDITIONAL PROBABILITY, INDEPENDENT STATISTICS

References

Press, W. H.; Flannery, B. P.; Teukolsky, S. A.; and Vetterling, W. T. *Numerical Recipes in FORTRAN: The Art of Scientific Computing, 2nd ed.* Cambridge, England: Cambridge University Press, p. 810, 1992.

Bayes' Theorem

see BAYES' FORMULA

Bayesian Analysis

A statistical procedure which endeavors to estimate parameters of an underlying distribution based on the observed distribution. Begin with a "PRIOR DISTRIBUTION" which may be based on anything, including an assessment of the relative likelihoods of parameters or the results of non-Bayesian observations. In practice, it is common to assume a UNIFORM DISTRIBUTION over the appropriate range of values for the PRIOR DISTRIBUTION.

Given the PRIOR DISTRIBUTION, collect data to obtain the observed distribution. Then calculate the LIKELIHOOD of the observed distribution as a function of parameter values, multiply this likelihood function by the PRIOR DISTRIBUTION, and normalize to obtain a unit probability over all possible values. This is called the POSTERIOR DISTRIBUTION. The MODE of the distribution is then the parameter estimate, and "probability intervals" (the Bayesian analog of CONFIDENCE INTERVALS) can be calculated using the standard procedure. Bayesian analysis is somewhat controversial because the validity of the result depends on how valid the PRIOR DISTRIBUTION is, and this cannot be assessed statistically.

see also MAXIMUM LIKELIHOOD, PRIOR DISTRIBUTION, UNIFORM DISTRIBUTION

References

Hoel, P. G.; Port, S. C.; and Stone, C. J. *Introduction to Statistical Theory.* New York: Houghton Mifflin, pp. 36–42, 1971.

Iversen, G. R. *Bayesian Statistical Inference.* Thousand Oaks, CA: Sage Pub., 1984.

Press, W. H.; Flannery, B. P.; Teukolsky, S. A.; and Vetterling, W. T. *Numerical Recipes in FORTRAN: The Art of Scientific Computing, 2nd ed.* Cambridge, England: Cambridge University Press, pp. 799–806, 1992.

Sivia, D. S. *Data Analysis: A Bayesian Tutorial.* New York: Oxford University Press, 1996.

Bays' Shuffle

A shuffling algorithm used in a class of RANDOM NUMBER generators.

References

Knuth, D. E. §3.2 and 3.3 in *The Art of Computer Programming, Vol. 2: Seminumerical Algorithms, 2nd ed.* Reading, MA: Addison-Wesley, 1981.

Press, W. H.; Flannery, B. P.; Teukolsky, S. A.; and Vetterling, W. T. *Numerical Recipes in FORTRAN: The Art of Scientific Computing, 2nd ed.* Cambridge, England: Cambridge University Press, pp. 270–271, 1992.

Beam Detector

N.B. A detailed on-line essay by S. Finch was the starting point for this entry.

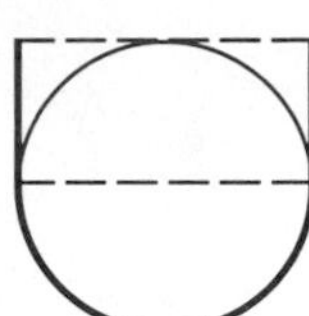

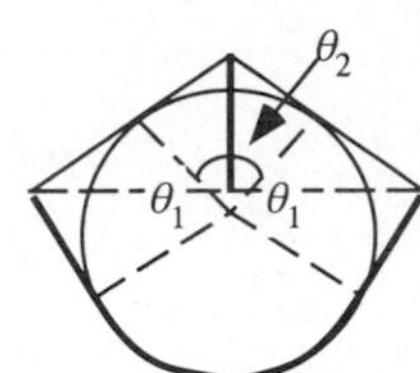

A "beam detector" for a given curve C is defined as a curve (or set of curves) through which every LINE tangent to or intersecting C passes. The shortest 1-arc beam detector, illustrated in the upper left figure, has length $L_1 = \pi + 2$. The shortest known 2-arc beam detector, illustrated in the right figure, has angles

$$\theta_1 \approx 1.286 \text{ rad} \tag{1}$$

$$\theta_2 \approx 1.191 \text{ rad}, \tag{2}$$

given by solving the simultaneous equations

$$2\cos\theta_1 - \sin(\tfrac{1}{2}\theta_2) = 0 \tag{3}$$

$$\tan(\tfrac{1}{2}\theta_1)\cos(\tfrac{1}{2}\theta_2) + \sin(\tfrac{1}{2}\theta_2)[\sec^2(\tfrac{1}{2}\theta_2) + 1] = 2. \tag{4}$$

The corresponding length is

$$L_2 = 2\pi - 2\theta_1 - \theta_2 + 2\tan(\tfrac{1}{2}\theta_1) + \sec(\tfrac{1}{2}\theta_2) - \cos(\tfrac{1}{2}\theta_2) + \tan(\tfrac{1}{2}\theta_1)\sin(\tfrac{1}{2}\theta_2) = 4.8189264563\ldots. \tag{5}$$

A more complicated expression gives the shortest known 3-arc length $L_3 = 4.799891547\ldots$. Finch defines

$$L = \inf_{n\geq 1} L_n \tag{6}$$

as the beam detection constant, or the TRENCH DIGGERS' CONSTANT. It is known that $L \geq \pi$.

References

Croft, H. T.; Falconer, K. J.; and Guy, R. K. §A30 in *Unsolved Problems in Geometry.* New York: Springer-Verlag, 1991.

Faber, V.; Mycielski, J.; and Pedersen, P. "On the Shortest Curve which Meets All Lines which Meet a Circle." *Ann. Polon. Math.* **44**, 249–266, 1984.

Faber, V. and Mycielski, J. "The Shortest Curve that Meets All Lines that Meet a Convex Body." *Amer. Math. Monthly* **93**, 796–801, 1986.

Finch, S. "Favorite Mathematical Constants." `http://www.mathsoft.com/asolve/constant/beam/beam.html`.

Makai, E. "On a Dual of Tarski's Plank Problem." In *Diskrete Geometrie.* 2 Kolloq., Inst. Math. Univ. Salzburg, 127–132, 1980.

Stewart, I. "The Great Drain Robbery." *Sci. Amer.*, 206–207, 106, and 125, Sept. 1995, Dec. 1995, and Feb. 1996.

Bean Curve

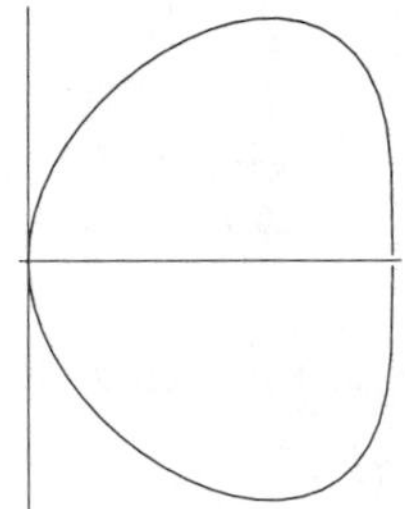

The PLANE CURVE given by the Cartesian equation

$$x^4 + x^2y^2 + y^4 = x(x^2 + y^2).$$

References

Cundy, H. and Rollett, A. *Mathematical Models, 3rd ed.* Stradbroke, England: Tarquin Pub., 1989.

Beast Number

The occult "number of the beast" associated in the Bible with the Antichrist. It has figured in many numerological studies. It is mentioned in Revelation 13:13: "Here is wisdom. Let him that hath understanding count the number of the beast: for it is the number of a man; and his number is 666."

The beast number has several interesting properties which numerologists may find particularly interesting (Keith 1982–83). In particular, the beast number is equal to the sum of the squares of the first 7 PRIMES

$$2^2 + 3^2 + 5^2 + 7^2 + 11^2 + 13^2 + 17^2 = 666, \tag{1}$$

satisfies the identity

$$\phi(666) = 6 \cdot 6 \cdot 6, \tag{2}$$

where ϕ is the TOTIENT FUNCTION, as well as the sum

$$\sum_{i=1}^{6\cdot 6} i = 666. \tag{3}$$

The number 666 is a sum and difference of the first three 6th POWERS,

$$666 = 1^6 - 2^6 + 3^6 \tag{4}$$

(Keith). Another curious identity is that there are exactly two ways to insert "+" signs into the sequence 123456789 to make the sum 666, and exactly one way for the sequence 987654321,

$$666 = 1 + 2 + 3 + 4 + 567 + 89 = 123 + 456 + 78 + 9 \tag{5}$$

$$666 = 9 + 87 + 6 + 543 + 21 \tag{6}$$

(Keith). 666 is a REPDIGIT, and is also a TRIANGULAR NUMBER

$$T_{6\cdot 6} = T_{36} = 666. \tag{7}$$

In fact, it is the largest REPDIGIT TRIANGULAR NUMBER (Bellew and Weger 1975–76). 666 is also a SMITH NUMBER. The first 144 DIGITS of $\pi - 3$, where π is PI, add to 666. In addition $144 = (6+6) \times (6+6)$ (Blatner 1997).

A number of the form 2^i which contains the digits of the beast number "666" is called an APOCALYPTIC NUMBER, and a number having 666 digits is called an APOCALYPSE NUMBER.

see also APOCALYPSE NUMBER, APOCALYPTIC NUMBER, BIMONSTER, MONSTER GROUP

References

Bellew, D. W. and Weger, R. C. "Repdigit Triangular Numbers." *J. Recr. Math.* **8**, 96–97, 1975–76.

Blatner, D. *The Joy of Pi.* New York: Walker, back jacket, 1997.

Castellanos, D. "The Ubiquitous π." *Math. Mag.* **61**, 153–154, 1988.

Hardy, G. H. *A Mathematician's Apology, reprinted with a foreword by C. P. Snow.* New York: Cambridge University Press, p. 96, 1993.

Keith, M. "The Number of the Beast." `http://users.aol.com/s6sj7gt/mike666.htm`.

Keith, M. "The Number 666." *J. Recr. Math.* **15**, 85–87, 1982–1983.

Beatty Sequence

The Beatty sequence is a SPECTRUM SEQUENCE with an IRRATIONAL base. In other words, the Beatty sequence corresponding to an IRRATIONAL NUMBER θ is given by $\lfloor\theta\rfloor$, $\lfloor 2\theta\rfloor$, $\lfloor 3\theta\rfloor$, ..., where $\lfloor x\rfloor$ is the FLOOR FUNCTION. If α and β are POSITIVE IRRATIONAL NUMBERS such that

$$\frac{1}{\alpha}+\frac{1}{\beta}=1,$$

then the Beatty sequences $\lfloor\alpha\rfloor$, $\lfloor 2\alpha\rfloor$, ... and $\lfloor\beta\rfloor$, $\lfloor 2\beta\rfloor$, ... together contain all the POSITIVE INTEGERS without repetition.

References

Gardner, M. *Penrose Tiles and Trapdoor Ciphers... and the Return of Dr. Matrix, reissue ed.* New York: W. H. Freeman, p. 21, 1989.

Graham, R. L.; Lin, S.; and Lin, C.-S. "Spectra of Numbers." *Math. Mag.* **51**, 174–176, 1978.

Guy, R. K. *Unsolved Problems in Number Theory, 2nd ed.* New York: Springer-Verlag, p. 227, 1994.

Sloane, N. J. A. *A Handbook of Integer Sequences.* Boston, MA: Academic Press, pp. 29–30, 1973.

Beauzamy and Dégot's Identity

For P, Q, R, and S POLYNOMIALS in n variables

$$[P\cdot Q, R\cdot S]=\sum_{i_1,\ldots,i_n\geq 0}\frac{A}{i_1!\cdots i_n!},$$

where

$$\begin{aligned}A\equiv{}&[R^{(i_1,\ldots,i_n)}(D_1,\ldots,D_n)Q(x_1,\ldots,x_n)\\&\times P^{(i_1,\ldots,i_n)}(D_1,\ldots,D_n)S(x_1,\ldots,x_n)]\end{aligned}$$

$D_i=\partial/\partial x_i$ is the DIFFERENTIAL OPERATOR, $[X,Y]$ is the BOMBIERI INNER PRODUCT, and

$$P^{(i_1,\ldots,i_n)}=D_1^{i_1}\cdots D_n^{i_n}P.$$

see also REZNIK'S IDENTITY

Bee

A 4-POLYHEX.

References

Gardner, M. *Mathematical Magic Show: More Puzzles, Games, Diversions, Illusions and Other Mathematical Sleight-of-Mind from Scientific American.* New York: Vintage, p. 147, 1978.

Behrens-Fisher Test

see FISHER-BEHRENS PROBLEM

Behrmann Cylindrical Equal-Area Projection

A CYLINDRICAL AREA-PRESERVING projection which uses 30° N as the no-distortion parallel.

References

Dana, P. H. "Map Projections." `http://www.utexas.edu/depts/grg/gcraft/notes/mapproj/mapproj.html`.

Bei

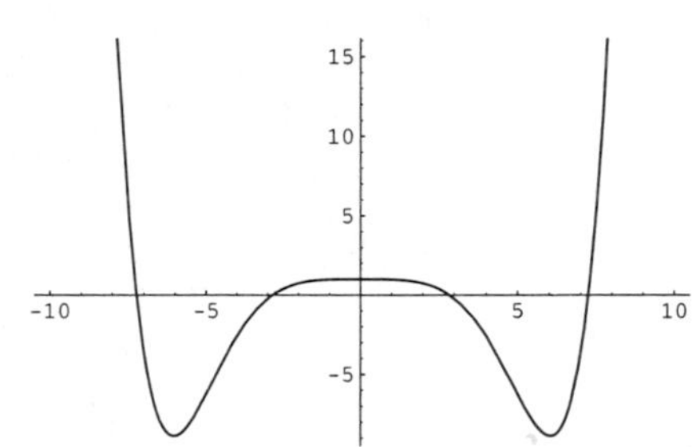

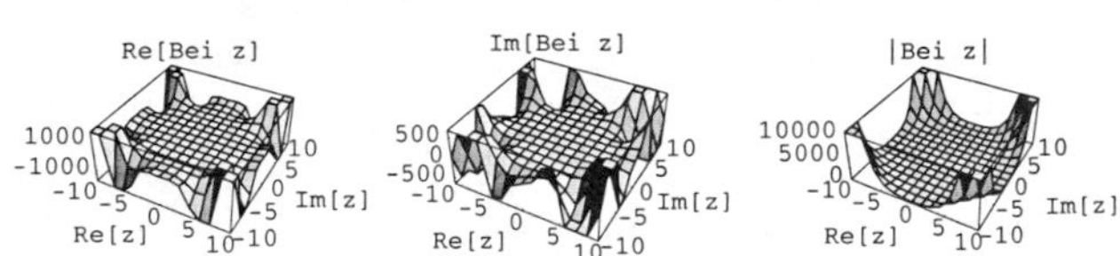

The IMAGINARY PART of

$$J_\nu(xe^{3\pi i/4})=\operatorname{ber}_\nu(x)+i\operatorname{bei}_\nu(x). \quad (1)$$

The special case $\nu=0$ gives

$$J_0(i\sqrt{i}\,x)\equiv\operatorname{ber}(x)+i\operatorname{bei}(x), \quad (2)$$

where $J_0(z)$ is the zeroth order BESSEL FUNCTION OF THE FIRST KIND.

$$\operatorname{bei}(x)\equiv\sum_{n=0}^\infty\frac{(-1)^n\left(\frac{x}{2}\right)^{4n}}{[(2n)!]^2}. \quad (3)$$

see also BER, BESSEL FUNCTION, KEI, KELVIN FUNCTIONS, KER

References

Abramowitz, M. and Stegun, C. A. (Eds.). "Kelvin Functions." §9.9 in *Handbook of Mathematical Functions with Formulas, Graphs, and Mathematical Tables, 9th printing.* New York: Dover, pp. 379–381, 1972.

Spanier, J. and Oldham, K. B. "The Kelvin Functions." Ch. 55 in *An Atlas of Functions.* Washington, DC: Hemisphere, pp. 543–554, 1987.

Bell Curve

see GAUSSIAN DISTRIBUTION, NORMAL DISTRIBUTION

Bell Number

The number of ways a SET of n elements can be PARTITIONED into nonempty SUBSETS is called a BELL NUMBER and is denoted B_n. For example, there are five ways the numbers {1, 2, 3} can be partitioned: {{1}, {2}, {3}}, {{1, 2}, {3}}, {{1, 3}, {2}}, {{1}, {2, 3}}, and {{1, 2, 3}}, so $B_3 = 5$. $B_0 = 1$ and the first few Bell numbers for $n = 1, 2, \ldots$ are 1, 2, 5, 15, 52, 203, 877, 4140, 21147, 115975, ... (Sloane's A000110). Bell numbers are closely related to CATALAN NUMBERS.

The diagram below shows the constructions giving $B_3 = 5$ and $B_4 = 15$, with line segments representing elements in the same SUBSET and dots representing subsets containing a single element (Dickau).

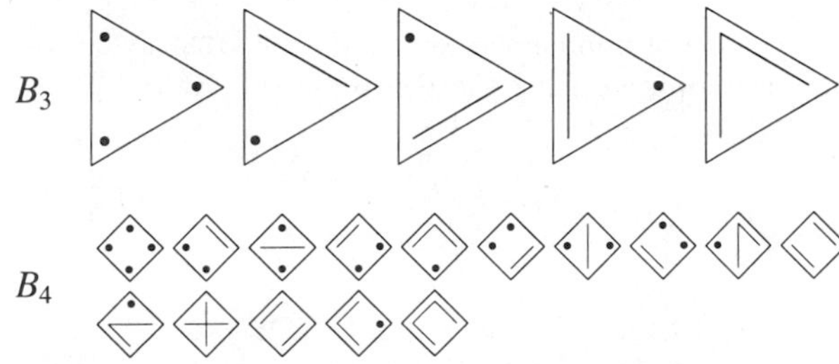

The INTEGERS B_n can be defined by the sum

$$B_n = \sum_{k=1}^{n} \left\{ {n \atop k} \right\}, \tag{1}$$

where $S_n^{(k)} = \left\{ {n \atop k} \right\}$ is a STIRLING NUMBER OF THE SECOND KIND, or by the generating function

$$e^{e^x - 1} = \sum_{n=0}^{\infty} \frac{B_n}{n!} x^n. \tag{2}$$

The Bell numbers can also be generated using the BELL TRIANGLE, using the RECURRENCE RELATION

$$B_{n+1} = \sum_{k=0}^{n} B_k \binom{n}{k}, \tag{3}$$

where $\binom{a}{b}$ is a BINOMIAL COEFFICIENT, or using the formula of Comtet (1974)

$$B_n = \left\lceil e^{-1} \sum_{m=1}^{2n} \frac{m^n}{m!} \right\rceil, \tag{4}$$

where $\lceil x \rceil$ denotes the CEILING FUNCTION.

The Bell number B_n is also equal to $\phi_n(1)$, where $\phi_n(x)$ is a BELL POLYNOMIAL. DOBIŃSKI'S FORMULA gives the nth Bell number

$$B_n = \frac{1}{e} \sum_{k=0}^{\infty} \frac{k^n}{k!}. \tag{5}$$

Lovász (1993) showed that this formula gives the asymptotic limit

$$B_n \sim n^{-1/2} [\lambda(n)]^{n+1/2} e^{\lambda(n)-n-1}, \tag{6}$$

where $\lambda(n)$ is defined implicitly by the equation

$$\lambda(n) \log[\lambda(n)] = n. \tag{7}$$

A variation of DOBIŃSKI'S FORMULA gives

$$B_k = \sum_{m=1}^{n} \frac{m^k}{m!} \sum_{s=0}^{n-m} \frac{(-1)^s}{s} \tag{8}$$

for $1 \leq k \leq n$ (Pitman 1997). de Bruijn (1958) gave the asymptotic formula

$$\frac{\ln B_n}{n} = \ln n - \ln \ln n - 1 + \frac{\ln \ln n}{\ln n} + \frac{1}{lnn} + \frac{1}{2}\left(\frac{\ln \ln n}{\ln n}\right)^2 + \mathcal{O}\left[\frac{\ln \ln n}{(\ln n)^2}\right]. \tag{9}$$

TOUCHARD'S CONGRUENCE states

$$B_{p+k} \equiv B_k + B_{k+1} \pmod{p}, \tag{10}$$

when p is PRIME. The only PRIME Bell numbers for $n \leq 1000$ are B_2, B_3, B_7, B_{13}, B_{42}, and B_{55}. The Bell numbers also have the curious property that

$$\begin{vmatrix} B_0 & B_1 & B_2 & \cdots & B_n \\ B_1 & B_2 & B_3 & \cdots & B_{n+1} \\ \vdots & \vdots & \vdots & \ddots & \vdots \\ B_n & B_{n+1} & B_{n+2} & \cdots & B_{2n} \end{vmatrix} = \prod_{i=1}^{n} n! \tag{11}$$

(Lenard 1986).

see also BELL POLYNOMIAL, BELL TRIANGLE, DOBIŃSKI'S FORMULA, STIRLING NUMBER OF THE SECOND KIND, TOUCHARD'S CONGRUENCE

References
Bell, E. T. "Exponential Numbers." *Amer. Math. Monthly* **41**, 411–419, 1934.
Comtet, L. *Advanced Combinatorics.* Dordrecht, Netherlands: Reidel, 1974.
Conway, J. H. and Guy, R. K. In *The Book of Numbers.* New York: Springer-Verlag, pp. 91–94, 1996.
de Bruijn, N. G. *Asymptotic Methods in Analysis.* New York: Dover, pp. 102–109, 1958.
Dickau, R. M. "Bell Number Diagrams." http://forum.swarthmore.edu/advanced/robertd/bell.html.
Gardner, M. "The Tinkly Temple Bells." Ch. 2 in *Fractal Music, Hypercards, and More Mathematical Recreations from Scientific American Magazine.* New York: W. H. Freeman, 1992.
Gould, H. W. *Bell & Catalan Numbers: Research Bibliography of Two Special Number Sequences, 6th ed.* Morgantown, WV: Math Monongliae, 1985.
Lenard, A. In *Fractal Music, Hypercards, and More Mathematical Recreations from Scientific American Magazine.* (M. Gardner). New York: W. H. Freeman, pp. 35–36, 1992.
Levine, J. and Dalton, R. E. "Minimum Periods, Modulo p, of First Order Bell Exponential Integrals." *Math. Comput.* **16**, 416–423, 1962.
Lovász, L. *Combinatorial Problems and Exercises, 2nd ed.* Amsterdam, Netherlands: North-Holland, 1993.
Pitman, J. "Some Probabilistic Aspects of Set Partitions." *Amer. Math. Monthly* **104**, 201–209, 1997.
Rota, G.-C. "The Number of Partitions of a Set." *Amer. Math. Monthly* **71**, 498–504, 1964.
Sloane, N. J. A. Sequence A000110/M1484 in "An On-Line Version of the Encyclopedia of Integer Sequences."

Bell Polynomial

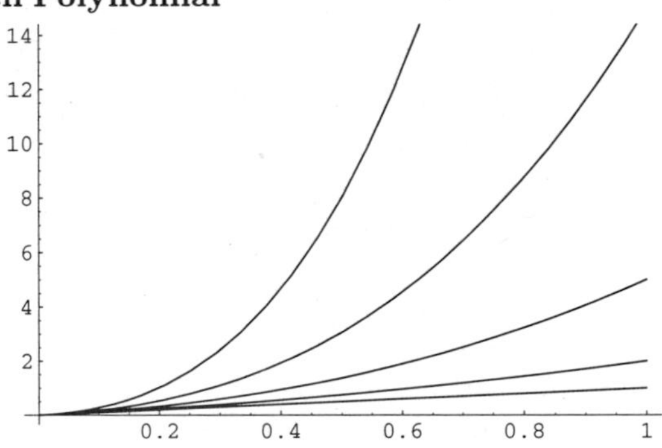

Two different GENERATING FUNCTIONS for the Bell polynomials for $n > 0$ are given by

$$\phi_n(x) \equiv e^{-x} \sum_{k=1}^{\infty} \frac{k^{n-1} x^k}{(k-1)!}$$

or

$$\phi_n(x) = x \sum_{k=1}^{n-1} \binom{n-1}{k-1} \phi_{k-1}(x),$$

where $\binom{n}{k}$ is a BINOMIAL COEFFICIENT.

The Bell polynomials are defined such that $\phi_n(1) = B_n$, where B_n is a BELL NUMBER. The first few Bell polynomials are

$$\begin{aligned}
\phi_0(x) &= 1 \\
\phi_1(x) &= x \\
\phi_2(x) &= x + x^2 \\
\phi_3(x) &= x + 3x^2 + x^3 \\
\phi_4(x) &= x + 7x^2 + 6x^3 + x^4 \\
\phi_5(x) &= x + 15x^2 + 25x^3 + 10x^4 + x^5 \\
\phi_6(x) &= x + 31x^2 + 90x^3 + 65x^4 + 15x^5 + x^6.
\end{aligned}$$

see also BELL NUMBER

References
Bell, E. T. "Exponential Polynomials." *Ann. Math.* **35**, 258–277, 1934.

Bell Triangle

$$\begin{array}{ccccccccccccccc}
1 & 2 & 5 & 15 & 52 & 203 & 877 & \cdots \\
 1 & 3 & 10 & 37 & 151 & 674 & \ddots \\
 & 2 & 7 & 27 & 114 & 523 & \ddots \\
 & & 5 & 20 & 87 & 409 & \ddots \\
 & & & 15 & 67 & 322 & \ddots \\
 & & & & 52 & 255 & \ddots \\
 & & & & & 203 & \ddots \\
 & & & & & & \ddots
\end{array}$$

A triangle of numbers which allow the BELL NUMBERS to be computed using the RECURRENCE RELATION

$$B_{n+1} = \sum_{k=0}^{n} B_k \binom{n}{k}.$$

see also BELL NUMBER, CLARK'S TRIANGLE, LEIBNIZ HARMONIC TRIANGLE, NUMBER TRIANGLE, PASCAL'S TRIANGLE, SEIDEL-ENTRINGER-ARNOLD TRIANGLE

Bellows Conjecture

see FLEXIBLE POLYHEDRON

Beltrami Differential Equation

For a measurable function μ, the Beltrami differential equation is given by

$$f_{z^*} = \mu f_z,$$

where f_z is a PARTIAL DERIVATIVE and z^* denotes the COMPLEX CONJUGATE of z.

see also QUASICONFORMAL MAP

References
Iyanaga, S. and Kawada, Y. (Eds.). *Encyclopedic Dictionary of Mathematics.* Cambridge, MA: MIT Press, p. 1087, 1980.

Beltrami Field

A VECTOR FIELD $\mathbf{u}$ satisfying the vector identity

$$\mathbf{u} \times (\nabla \times \mathbf{u}) = \mathbf{0}$$

where $\mathbf{A} \times \mathbf{B}$ is the CROSS PRODUCT and $\nabla \times \mathbf{A}$ is the CURL is said to be a Beltrami field.

see also DIVERGENCELESS FIELD, IRROTATIONAL FIELD, SOLENOIDAL FIELD

Beltrami Identity

An identity in CALCULUS OF VARIATIONS discovered in 1868 by Beltrami. The EULER-LAGRANGE DIFFERENTIAL EQUATION is

$$\frac{\partial f}{\partial y} - \frac{d}{dx}\left(\frac{\partial f}{\partial y_x}\right) = 0. \tag{1}$$

Now, examine the DERIVATIVE of x

$$\frac{df}{dx} = \frac{\partial f}{\partial y} y_x + \frac{\partial f}{\partial y_x} y_{xx} + \frac{\partial f}{\partial x}. \tag{2}$$

Solving for the $\partial f/\partial y$ term gives

$$\frac{\partial f}{\partial y} y_x = \frac{df}{dx} - \frac{\partial f}{\partial y_x} y_{xx} - \frac{\partial f}{\partial x}. \tag{3}$$

Now, multiplying (1) by y_x gives

$$y_x \frac{\partial f}{\partial y} - y_x \frac{d}{dx}\left(\frac{\partial f}{\partial y_x}\right) = 0. \tag{4}$$

Substituting (3) into (4) then gives

$$\frac{df}{dx} - \frac{\partial f}{\partial y_x} y_{xx} - \frac{\partial f}{\partial x} - y_x \frac{d}{dx}\left(\frac{\partial f}{\partial y_x}\right) = 0 \tag{5}$$

$$-\frac{\partial f}{\partial x} + \frac{d}{dx}\left(f - y_x \frac{\partial f}{\partial y_x}\right) = 0. \tag{6}$$

This form is especially useful if $f_x = 0$, since in that case

$$\frac{d}{dx}\left(f - y_x \frac{\partial f}{\partial y_x}\right) = 0, \tag{7}$$

which immediately gives

$$f - y_x \frac{\partial f}{\partial y_x} = C, \tag{8}$$

where C is a constant of integration.

The Beltrami identity greatly simplifies the solution for the minimal AREA SURFACE OF REVOLUTION about a given axis between two specified points. It also allows straightforward solution of the BRACHISTOCHRONE PROBLEM.

see also BRACHISTOCHRONE PROBLEM, CALCULUS OF VARIATIONS, EULER-LAGRANGE DIFFERENTIAL EQUATION, SURFACE OF REVOLUTION

Bend (Curvature)

Given four mutually tangent circles, their bends are defined as the signed CURVATURES of the CIRCLES. If the contacts are all external, the signs are all taken as POSITIVE, whereas if one circle surrounds the other three, the sign of this circle is taken as NEGATIVE (Coxeter 1969).

see also CURVATURE, DESCARTES CIRCLE THEOREM, SODDY CIRCLES

References

Coxeter, H. S. M. *Introduction to Geometry, 2nd ed.* New York: Wiley, pp. 13–14, 1969.

Bend (Knot)

A KNOT used to join the ends of two ropes together to form a longer length.

References

Owen, P. *Knots.* Philadelphia, PA: Courage, p. 49, 1993.

Benford's Law

Also called the FIRST DIGIT LAW, FIRST DIGIT PHENOMENON, or LEADING DIGIT PHENOMENON. In listings, tables of statistics, etc., the DIGIT 1 tends to occur with PROBABILITY $\sim 30\%$, much greater than the expected 10%. This can be observed, for instance, by examining tables of LOGARITHMS and noting that the first pages are much more worn and smudged than later pages. The table below, taken from Benford (1938), shows the distribution of first digits taken from several disparate sources. Of the 54 million real constants in Plouffe's "Inverse Symbolic Calculator" database, 30% begin with the DIGIT 1.

Title	First Digit									#
	1	2	3	4	5	6	7	8	9	
Rivers, Area	31.0	16.4	10.7	11.3	7.2	8.6	5.5	4.2	5.1	335
Population	33.9	20.4	14.2	8.1	7.2	6.2	4.1	3.7	2.2	3259
Constants	41.3	14.4	4.8	8.6	10.6	5.8	1.0	2.9	10.6	104
Newspapers	30.0	18.0	12.0	10.0	8.0	6.0	6.0	5.0	5.0	100
Specific Heat	24.0	18.4	16.2	14.6	10.6	4.1	3.2	4.8	4.1	1389
Pressure	29.6	18.3	12.8	9.8	8.3	6.4	5.7	4.4	4.7	703
H.P. Lost	30.0	18.4	11.9	10.8	8.1	7.0	5.1	5.1	3.6	690
Mol. Wgt.	26.7	25.2	15.4	10.8	6.7	5.1	4.1	2.8	3.2	1800
Drainage	27.1	23.9	13.8	12.6	8.2	5.0	5.0	2.5	1.9	159
Atomic Wgt.	47.2	18.7	5.5	4.4	6.6	4.4	3.3	4.4	5.5	91
n^{-1}, $\sqrt{n}$	25.7	20.3	9.7	6.8	6.6	6.8	7.2	8.0	8.9	5000
Design	26.8	14.8	14.3	7.5	8.3	8.4	7.0	7.3	5.6	560
Reader's Dig.	33.4	18.5	12.4	7.5	7.1	6.5	5.5	4.9	4.2	308
Cost Data	32.4	18.8	10.1	10.1	9.8	5.5	4.7	5.5	3.1	741
X-Ray Volts	27.9	17.5	14.4	9.0	8.1	7.4	5.1	5.8	4.8	707
Am. League	32.7	17.6	12.6	9.8	7.4	6.4	4.9	5.6	3.0	1458
Blackbody	31.0	17.3	14.1	8.7	6.6	7.0	5.2	4.7	5.4	1165
Addresses	28.9	19.2	12.6	8.8	8.5	6.4	5.6	5.0	5.0	342
n^1, $n^2 \cdots n!$	25.3	16.0	12.0	10.0	8.5	8.8	6.8	7.1	5.5	900
Death Rate	27.0	18.6	15.7	9.4	6.7	6.5	7.2	4.8	4.1	418
Average	30.6	18.5	12.4	9.4	8.0	6.4	5.1	4.9	4.7	1011
Prob. Error	0.8	0.4	0.4	0.3	0.2	0.2	0.2	0.2	0.3	

In fact, the first SIGNIFICANT DIGIT seems to follow a LOGARITHMIC DISTRIBUTION, with

$$P(n) \approx \log(n+1) - \log n$$

for $n = 1, \ldots, 9$. One explanation uses CENTRAL LIMIT-like theorems for the MANTISSAS of random variables under MULTIPLICATION. As the number of variables increases, the density function approaches that of a LOGARITHMIC DISTRIBUTION.

References
Benford, F. "The Law of Anomalous Numbers." *Proc. Amer. Phil. Soc.* **78**, 551–572, 1938.
Boyle, J. "An Application of Fourier Series to the Most Significant Digit Problem." *Amer. Math. Monthly* **101**, 879–886, 1994.
Hill, T. P. "Base-Invariance Implies Benford's Law." *Proc. Amer. Math. Soc.* **12**, 887–895, 1995.
Hill, T. P. "The Significant-Digit Phenomenon." *Amer. Math. Monthly* **102**, 322–327, 1995.
Hill, T. P. "A Statistical Derivation of the Significant-Digit Law." *Stat. Sci.* **10**, 354–363, 1996.
Hill, T. P. "The First Digit Phenomenon." *Amer. Sci.* **86**, 358–363, 1998.
Ley, E. "On the Peculiar Distribution of the U.S. Stock Indices Digits." *Amer. Stat.* **50**, 311–313, 1996.
Newcomb, S. "Note on the Frequency of the Use of Digits in Natural Numbers." *Amer. J. Math.* **4**, 39–40, 1881.
Nigrini, M. "A Taxpayer Compliance Application of Benford's Law." *J. Amer. Tax. Assoc.* **18**, 72–91, 1996.
Plouffe, S. "Inverse Symbolic Calculator." `http://www.cecm.sfu.ca/projects/ISC/`.
Raimi, R. A. "The Peculiar Distribution of First Digits." *Sci. Amer.* **221**, 109–119, Dec. 1969.
Raimi, R. A. "The First Digit Phenomenon." *Amer. Math. Monthly* **83**, 521–538, 1976.

Benham's Wheel

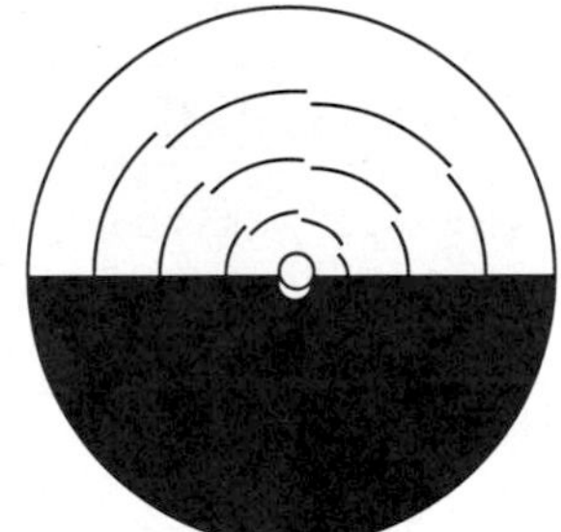

An optical ILLUSION consisting of a spinnable top marked in black with the pattern shown above. When the wheel is spun (especially slowly), the black broken lines appear as green, blue, and red colored bands!

References
Cohen, J. and Gordon, D. A. "The Prevost-Fechner-Benham Subjective Colors." *Psycholog. Bull.* **46**, 97–136, 1949.
Festinger, L.; Allyn, M. R.; and White, C. W. "The Perception of Color with Achromatic Stimulation." *Vision Res.* **11**, 591–612, 1971.
Fineman, M. *The Nature of Visual Illusion.* New York: Dover, pp. 148–151, 1996.
Trolland, T. L. "The Enigma of Color Vision." *Amer. J. Physiology* **2**, 23–48, 1921.

Bennequin's Conjecture

A BRAID with M strands and R components with P positive crossings and N negative crossings satisfies

$$|P - N| \leq 2U + M - R \leq P + N,$$

where U is the UNKNOTTING NUMBER. While the second part of the INEQUALITY was already known to be true (Boileau and Weber, 1983, 1984) at the time the conjecture was proposed, the proof of the entire conjecture was completed using results of Kronheimer and Mrowka on MILNOR'S CONJECTURE (and, independently, using MENASCO'S THEOREM).

see also BRAID, MENASCO'S THEOREM, MILNOR'S CONJECTURE, UNKNOTTING NUMBER

References
Bennequin, D. "L'instanton gordien (d'après P. B. Kronheimer et T. S. Mrowka)." *Astérisque* **216**, 233–277, 1993.
Birman, J. S. and Menasco, W. W. "Studying Links via Closed Braids. II. On a Theorem of Bennequin." *Topology Appl.* **40**, 71–82, 1991.
Boileau, M. and Weber, C. "Le problème de J. Milnor sur le nombre gordien des nœuds algébriques." *Enseign. Math.* **30**, 173–222, 1984.
Boileau, M. and Weber, C. "Le problème de J. Milnor sur le nombre gordien des nœuds algébriques." In *Knots, Braids and Singularities (Plans-sur-Bex, 1982).* Geneva, Switzerland: Monograph. Enseign. Math. Vol. 31, pp. 49–98, 1983.
Cipra, B. *What's Happening in the Mathematical Sciences, Vol. 2.* Providence, RI: Amer. Math. Soc., pp. 8–13, 1994.
Kronheimer, P. B. "The Genus-Minimizing Property of Algebraic Curves." *Bull. Amer. Math. Soc.* **29**, 63–69, 1993.
Kronheimer, P. B. and Mrowka, T. S. "Gauge Theory for Embedded Surfaces. I." *Topology* **32**, 773–826, 1993.
Kronheimer, P. B. and Mrowka, T. S. "Recurrence Relations and Asymptotics for Four-Manifold Invariants." *Bull. Amer. Math. Soc.* **30**, 215–221, 1994.
Menasco, W. W. "The Bennequin-Milnor Unknotting Conjectures." *C. R. Acad. Sci. Paris Sér. I Math.* **318**, 831–836, 1994.

Benson's Formula

An equation for a LATTICE SUM with $n = 3$

$$-b_3(1) = \sum_{i,\,j,\,k=-\infty}^{\infty}{}' \frac{(-1)^{i+j+k+1}}{\sqrt{i^2+j^2+k^2}}$$
$$= 12\pi \sum_{m,\,n=1,\,3,\,\ldots}^{\infty} \operatorname{sech}^2(\tfrac{1}{2}\pi\sqrt{m^2+n^2}).$$

Here, the prime denotes that summation over (0, 0, 0) is excluded. The sum is numerically equal to $-1.74756\ldots$, a value known as "the" MADELUNG CONSTANT.

see also MADELUNG CONSTANTS

References
Borwein, J. M. and Borwein, P. B. *Pi & the AGM: A Study in Analytic Number Theory and Computational Complexity.* New York: Wiley, p. 301, 1987.
Finch, S. "Favorite Mathematical Constants." `http://www.mathsoft.com/asolve/constant/mdlung/mdlung.html`.

Ber

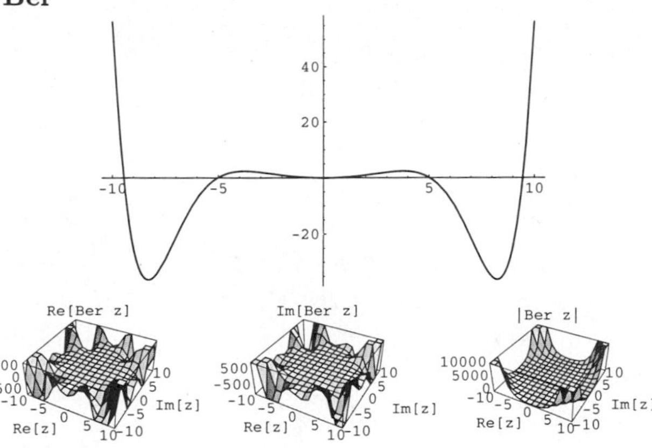

The REAL PART of

$$J_\nu(xe^{3\pi i/4}) = \mathrm{ber}_\nu(x) + i\,\mathrm{bei}_\nu(x). \tag{1}$$

The special case $\nu = 0$ gives

$$J_0(i\sqrt{i}\,x) \equiv \mathrm{ber}(x) + i\,\mathrm{bei}(x), \tag{2}$$

where J_0 is the zeroth order BESSEL FUNCTION OF THE FIRST KIND.

$$\mathrm{ber}(x) \equiv \sum_{n=0}^\infty \frac{(-1)^n \left(\frac{x}{2}\right)^{2+4n}}{[(2n+1)!]^2}. \tag{3}$$

see also BEI, BESSEL FUNCTION, KEI, KELVIN FUNCTIONS, KER

References

Abramowitz, M. and Stegun, C. A. (Eds.). "Kelvin Functions." §9.9 in *Handbook of Mathematical Functions with Formulas, Graphs, and Mathematical Tables, 9th printing.* New York: Dover, pp. 379–381, 1972.

Spanier, J. and Oldham, K. B. "The Kelvin Functions." Ch. 55 in *An Atlas of Functions.* Washington, DC: Hemisphere, pp. 543–554, 1987.

Beraha Constants

The nth Beraha constant is given by

$$\mathrm{Be}_n \equiv 2 + 2\cos\left(\frac{2\pi}{n}\right).$$

The first few are

$$\mathrm{Be}_1 = 4$$
$$\mathrm{Be}_2 = 0$$
$$\mathrm{Be}_3 = 1$$
$$\mathrm{Be}_4 = 2$$
$$\mathrm{Be}_5 = \tfrac{1}{2}(3+\sqrt{5}\,) \approx 2.618$$
$$\mathrm{Be}_6 = 3$$
$$\mathrm{Be}_7 = 2 + 2\cos(\tfrac{2}{7}\pi) \approx 3.247\ldots.$$

They appear to be ROOTS of the CHROMATIC POLYNOMIALS of planar triangular GRAPHS. Be_4 is $\phi + 1$, where ϕ is the GOLDEN RATIO, and Be_7 is the SILVER CONSTANT.

References

Le Lionnais, F. *Les nombres remarquables.* Paris: Hermann, p. 143, 1983.

Berger-Kazdan Comparison Theorem

Let M be a compact n-D MANIFOLD with INJECTIVITY radius $\mathrm{inj}(M)$. Then

$$\mathrm{Vol}(M) \geq \frac{c_n\,\mathrm{inj}(M)}{\pi},$$

with equality IFF M is ISOMETRIC to the standard round SPHERE S^n with RADIUS $\mathrm{inj}(M)$, where $c_n(r)$ is the VOLUME of the standard n-HYPERSPHERE of RADIUS r.

see also BLASCHKE CONJECTURE, HYPERSPHERE, INJECTIVE, ISOMETRY

References

Chavel, I. *Riemannian Geometry: A Modern Introduction.* New York: Cambridge University Press, 1994.

Bergman Kernel

A Bergman kernel is a function of a COMPLEX VARIABLE with the "reproducing kernel" property defined for any DOMAIN in which there exist NONZERO ANALYTIC FUNCTIONS of class $L_2(D)$ with respect to the LEBESGUE MEASURE dV.

References

Hazewinkel, M. (Managing Ed.). *Encyclopaedia of Mathematics: An Updated and Annotated Translation of the Soviet "Mathematical Encyclopaedia."* Dordrecht, Netherlands: Reidel, pp. 356–357, 1988.

Bergman Space

Let G be an open subset of the COMPLEX PLANE $\mathbb{C}$, and let $L_a^2(G)$ denote the collection of all ANALYTIC FUNCTIONS $f : G \to C$ whose MODULUS is square integrable with respect to AREA measure. Then $L_a^2(G)$, sometimes also denoted $A^2(G)$, is called the Bergman space for G. Thus, the Bergman space consists of all the ANALYTIC FUNCTIONS in $L^2(G)$. The Bergman space can also be generalized to $L_a^p(G)$, where $0 < p < \infty$.

Bernoulli Differential Equation

$$\frac{dy}{dx} + p(x)y = q(x)y^n. \tag{1}$$

Let $v \equiv y^{1-n}$ for $n \neq 1$, then

$$\frac{dv}{dx} = (1-n)y^{-n}\frac{dy}{dx}. \tag{2}$$

Rewriting (1) gives

$$y^{-n}\frac{dy}{dx} = q(x) - p(x)y^{1-n} = q(x) - vp(x). \tag{3}$$

Plugging (3) into (2),

$$\frac{dv}{dx} = (1-n)[q(x) - vp(x)]. \tag{4}$$

Now, this is a linear FIRST-ORDER ORDINARY DIFFERENTIAL EQUATION of the form

$$\frac{dv}{dx} + vP(x) = Q(x), \tag{5}$$

where $P(x) \equiv (1-n)p(x)$ and $Q(x) \equiv (1-n)q(x)$. It can therefore be solved analytically using an INTEGRATING FACTOR

$$v = \frac{\int e^{\int P(x)\,dx} Q(x)\,dx + C}{e^{\int P(x)\,dx}} = \frac{(1-n)\int e^{(1-n)\int p(x)\,dx} q(x)\,dx + C}{e^{(1-n)\int p(x)\,dx}}, \tag{6}$$

where C is a constant of integration. If $n = 1$, then equation (1) becomes

$$\frac{dy}{dx} = y(q-p) \tag{7}$$

$$\frac{dy}{y} = (q-p)\,dx \tag{8}$$

$$y = C_2 e^{\int [q(x)-p(x)]\,dx}. \tag{9}$$

The general solution is then, with C_1 and C_2 constants,

$$y = \begin{cases} \left[\frac{(1-n)\int e^{(1-n)\int p(x)\,dx} q(x)\,dx + C_1}{e^{(1-n)\int p(x)\,dx}}\right]^{1/(1-n)} \\ \text{for } n \neq 1 \\ C_2 e^{\int [q(x)-p(x)]\,dx} \\ \text{for } n = 1. \end{cases} \tag{10}$$

Bernoulli Distribution

A DISTRIBUTION given by

$$P(n) = \begin{cases} q \equiv 1-p & \text{for } n = 0 \\ p & \text{for } n = 1 \end{cases} \tag{1}$$

$$= p^n(1-p)^{1-n} \qquad \text{for } n = 0, 1. \tag{2}$$

The distribution of heads and tails in COIN TOSSING is a Bernoulli distribution with $p = q = 1/2$. The GENERATING FUNCTION of the Bernoulli distribution is

$$M(t) = \langle e^{tn} \rangle = \sum_{n=0}^{1} e^{tn} p^n (1-p)^{1-n} = e^0(1-p) + e^t p, \tag{3}$$

so

$$M(t) = (1-p) + pe^t \tag{4}$$

$$M'(t) = pe^t \tag{5}$$

$$M''(t) = pe^t \tag{6}$$

$$M^{(n)}(t) = pe^t, \tag{7}$$

and the MOMENTS about 0 are

$$\mu_1' = \mu = M'(0) = p \tag{8}$$

$$\mu_2' = M''(0) = p \tag{9}$$

$$\mu_n' = M^{(n)}(0) = p. \tag{10}$$

The MOMENTS about the MEAN are

$$\mu_2 = \mu_2' - (\mu_1')^2 = p - p^2 = p(1-p) \tag{11}$$

$$\mu_3 = \mu_3' - 3\mu_2'\mu_1' + 2(\mu_1')^3 = p - 3p^2 + 2p^3 = p(1-p)(1-2p) \tag{12}$$

$$\mu_4 = \mu_4' - 4\mu_3'\mu_1' + 6\mu_2'(\mu_1')^2 - 3(\mu_1')^4 = p - 4p^2 + 6p^3 - 3p^4 = p(1-p)(3p^2 - 3p + 1). \tag{13}$$

The MEAN, VARIANCE, SKEWNESS, and KURTOSIS are then

$$\mu = \mu_1' = p \tag{14}$$

$$\sigma^2 = \mu_2 = p(1-p) \tag{15}$$

$$\gamma_1 = \frac{\mu_3}{\sigma^3} = \frac{p(1-p)(1-2p)}{[p(1-p)]^{3/2}} = \frac{1-2p}{\sqrt{p(1-p)}} \tag{16}$$

$$\gamma_2 = \frac{\mu_4}{\sigma^4} - 3 = \frac{p(1-2p)(2p^2-2p+1)}{p^2(1-p)^2} - 3 = \frac{6p^2 - 6p + 1}{p(1-p)}. \tag{17}$$

To find an estimator for a population mean,

$$\langle p \rangle = \sum_{Np=0}^{N} p \binom{N}{Np} \theta^{Np}(1-\theta)^{Nq} = \theta \sum_{Np=1}^{N} \binom{N-1}{Np-1} \theta^{Np-1}(1-\theta)^{Nq} = \theta[\theta + (1-\theta)]^{N-1} = \theta, \tag{18}$$

so $\langle p \rangle$ is an UNBIASED ESTIMATOR for θ. The probability of Np successes in N trials is then

$$\binom{N}{Np} \theta^{Np}(1-\theta)^{Nq}, \tag{19}$$

where

$$p = \frac{[\text{number of successes}]}{N} \equiv \frac{n}{N}. \tag{20}$$

see also BINOMIAL DISTRIBUTION

Bernoulli Function

see BERNOULLI POLYNOMIAL

Bernoulli Inequality

$$(1+x)^n > 1+nx, \tag{1}$$

where $x \in \mathbb{R} > -1 \neq 0$, $n \in \mathbb{Z} > 1$. This inequality can be proven by taking a MACLAURIN SERIES of $(1+x)^n$,

$$(1+x)^n = 1+nx+\tfrac{1}{2}n(n-1)x^2+\tfrac{1}{6}n(n-1)(n-2)x^3+\ldots. \tag{2}$$

Since the series terminates after a finite number of terms for INTEGRAL n, the Bernoulli inequality for $x > 0$ is obtained by truncating after the first-order term. When $-1 < x < 0$, slightly more finesse is needed. In this case, let $y = |x| = -x > 0$ so that $0 < y < 1$, and take

$$(1-y)^n = 1-ny+\tfrac{1}{2}n(n-1)y^2-\tfrac{1}{6}n(n-1)(n-2)y^3+\ldots. \tag{3}$$

Since each POWER of y multiplies by a number < 1 and since the ABSOLUTE VALUE of the COEFFICIENT of each subsequent term is smaller than the last, it follows that the sum of the third order and subsequent terms is a POSITIVE number. Therefore,

$$(1-y)^n > 1-ny, \tag{4}$$

or

$$(1+x)^n > 1+nx, \quad \text{for } -1 < x < 0, \tag{5}$$

completing the proof of the INEQUALITY over all ranges of parameters.

Bernoulli Lemniscate

see LEMNISCATE

Bernoulli Number

There are two definitions for the Bernoulli numbers. The older one, no longer in widespread use, defines the Bernoulli numbers B_n^* by the equations

$$\frac{x}{e^x-1}+\frac{x}{2}-1 \equiv \sum_{n=1}^{\infty} \frac{(-1)^{n-1}B_n^* x^{2n}}{(2n)!} = \frac{B_1^* x^2}{2!} - \frac{B_2^* x^4}{4!} + \frac{B_3^* x^6}{6!} + \ldots \tag{1}$$

for $|x| < 2\pi$, or

$$1-\frac{x}{2}\cot\left(\frac{x}{2}\right) \equiv \sum_{n=1}^{\infty} \frac{B_n^* x^{2n}}{(2n)!} = \frac{B_1^* x^2}{2!} + \frac{B_2^* x^4}{4!} + \frac{B_3^* x^6}{6!} + \ldots \tag{2}$$

for $|x| < \pi$ (Whittaker and Watson 1990, p. 125). Gradshteyn and Ryzhik (1979) denote these numbers B_n^*, while Bernoulli numbers defined by the newer (National Bureau of Standards) definition are denoted B_n. The B_n^* Bernoulli numbers may be calculated from the integral

$$B_n^* = 4n \int_0^\infty \frac{t^{2n-1}\,dt}{e^{2\pi t}-1}, \tag{3}$$

and analytically from

$$B_n^* = \frac{2(2n)!}{(2\pi)^{2n}} \sum_{p=1}^{\infty} p^{-2n} = \frac{2(2n)!}{(2\pi)^{2n}} \zeta(2n) \tag{4}$$

for $n = 1, 2, \ldots$, where $\zeta(z)$ is the RIEMANN ZETA FUNCTION.

The first few Bernoulli numbers B_n^* are

$$\begin{aligned}
B_1^* &= \tfrac{1}{6}\\
B_2^* &= \tfrac{1}{30}\\
B_3^* &= \tfrac{1}{42}\\
B_4^* &= \tfrac{1}{30}\\
B_5^* &= \tfrac{5}{66}\\
B_6^* &= \tfrac{691}{2,730}\\
B_7^* &= \tfrac{7}{6}\\
B_8^* &= \tfrac{3,617}{510}\\
B_9^* &= \tfrac{43,867}{798}\\
B_{10}^* &= \tfrac{174,611}{330}\\
B_{11}^* &= \tfrac{854,513}{138}.
\end{aligned}$$

Bernoulli numbers defined by the modern definition are denoted B_n and also called "EVEN-index" Bernoulli numbers. These are the Bernoulli numbers returned by the *Mathematica*® (Wolfram Research, Champaign, IL) function `BernoulliB[n]`. These Bernoulli numbers are a superset of the archaic ones B_n^* since

$$B_n \equiv \begin{cases} 1 & \text{for } n=0 \\ -\frac{1}{2} & \text{for } n=1 \\ (-1)^{(n/2)-1} B_{n/2}^* & \text{for } n \text{ even} \\ 0 & \text{for } n \text{ odd.} \end{cases} \tag{5}$$

The B_n can be defined by the identity

$$\frac{x}{e^x-1} \equiv \sum_{n=0}^{\infty} \frac{B_n x^n}{n!}. \tag{6}$$

These relationships can be derived using the generating function

$$F(x,t) = \sum_{n=0}^{\infty} \frac{B_n(x)t^n}{n!}, \tag{7}$$

which converges uniformly for $|t| < 2\pi$ and all x (Castellanos 1988). Taking the partial derivative gives

$$\frac{\partial F(x,t)}{\partial x} = \sum_{n=0}^{\infty} \frac{B_{n-1}(x)t^n}{(n-1)!} = t\sum_{n=0}^{\infty} \frac{B_n(x)t^n}{n!} = tF(x,t). \tag{8}$$

The solution to this differential equation is

$$F(x,t) = T(t)e^{xt}, \tag{9}$$

so integrating gives

$$\begin{aligned}\int_0^1 F(x,t)\,dx &= T(t)\int_0^1 e^{xt}\,dx = T(t)\frac{e^t-1}{t}\\ &= \sum_{n=0}^\infty \frac{t^n}{n!}\int_0^1 B_n(x)\,dx\\ &= 1+\sum_{n=1}^\infty \frac{t^n}{n!}\int_0^1 B_n(x)\,dx = 1 \end{aligned}\tag{10}$$

or

$$\frac{te^{xt}}{e^t-1} = \sum_{n=0}^\infty \frac{B_n(x)t^n}{n!} \tag{11}$$

(Castellanos 1988). Setting $x = 0$ and adding $t/2$ to both sides then gives

$$\tfrac{1}{2}t\coth(\tfrac{1}{2}t) = \sum_{n=0}^\infty \frac{B_{2n}t^{2n}}{(2n)!}. \tag{12}$$

Letting $t = 2ix$ then gives

$$x\cot x = \sum_{n=0}^\infty (-1)^n B_{2n}\frac{2x^{2n}}{(2n)!} \tag{13}$$

for $x \in [-\pi, \pi]$. The Bernoulli numbers may also be calculated from the integral

$$B_n = \frac{n!}{2\pi i}\int \frac{z}{e^z-1}\frac{dz}{z^{n+1}}, \tag{14}$$

or from

$$B_n = \left[\frac{d^n}{dx^n}\frac{x}{e^x-1}\right]_{x=0}. \tag{15}$$

The Bernoulli numbers satisfy the identity

$$\binom{k+1}{1}B_k + \binom{k+1}{2}B_{k-1} + \ldots + \binom{k+1}{k}B_1 + B_0 = 0, \tag{16}$$

where $\binom{n}{k}$ is a BINOMIAL COEFFICIENT. An asymptotic FORMULA is

$$\lim_{n\to\infty} |B_{2n}| \sim 4\sqrt{\pi n}\left(\frac{n}{\pi e}\right)^{2n}. \tag{17}$$

Bernoulli numbers appear in expressions of the form $\sum_{k=1}^n k^p$, where $p = 1, 2, \ldots$. Bernoulli numbers also appear in the series expansions of functions involving $\tan x$, $\cot x$, $\csc x$, $\ln|\sin x|$, $\ln|\cos x|$, $\ln|\tan x|$, $\tanh x$, $\coth x$, and $\operatorname{csch} x$. An analytic solution exists for EVEN orders,

$$B_{2n} = \frac{(-1)^{n-1}2(2n)!}{(2\pi)^2 n}\sum_{p=1}^\infty p^{-2n} = \frac{(-1)^{n-1}2(2n)!}{(2\pi)^{2n}}\zeta(2n) \tag{18}$$

for $n = 1, 2, \ldots$, where $\zeta(2n)$ is the RIEMANN ZETA FUNCTION. Another intimate connection with the RIEMANN ZETA FUNCTION is provided by the identity

$$B_n = (-1)^{n+1}n\zeta(1-n). \tag{19}$$

The DENOMINATOR of B_{2k} is given by the VON STAUDT-CLAUSEN THEOREM

$$\operatorname{denom}(B_{2k}) = \prod_{\substack{p\text{ prime}\\ (p-1)|2k}}^{2k+1} p, \tag{20}$$

which also implies that the DENOMINATOR of B_{2k} is SQUAREFREE (Hardy and Wright 1979). Another curious property is that the fraction part of B_n in DECIMAL has a DECIMAL PERIOD which divides n, and there is a single digit before that period (Conway 1996).

$$\begin{aligned} B_0 &= 1\\ B_1 &= -\tfrac{1}{2}\\ B_2 &= \tfrac{1}{6}\\ B_4 &= -\tfrac{1}{30}\\ B_6 &= \tfrac{1}{42}\\ B_8 &= -\tfrac{1}{30}\\ B_{10} &= \tfrac{5}{66}\\ B_{12} &= -\tfrac{691}{2,730}\\ B_{14} &= \tfrac{7}{6}\\ B_{16} &= -\tfrac{3,617}{510}\\ B_{18} &= \tfrac{43,867}{798}\\ B_{20} &= -\tfrac{174,611}{330}\\ B_{22} &= \tfrac{854,513}{138} \end{aligned}$$

(Sloane's A000367 and A002445). In addition,

$$B_{2n+1} = 0 \tag{21}$$

for $n = 1, 2, \ldots$.

Bernoulli first used the Bernoulli numbers while computing $\sum_{k=1}^n k^p$. He used the property of the FIGURATE NUMBER TRIANGLE that

$$\sum_{i=0}^n a_{ij} = \frac{(n+1)a_{nj}}{j+1}, \tag{22}$$

along with a form for a_{nj} which he derived inductively to compute the sums up to $n = 10$ (Boyer 1968, p. 85). For $p \in \mathbb{Z} > 0$, the sum is given by

$$\sum_{k=1}^{n} k^p = \sum_{k=1}^{n} \frac{(B+n+1)^{[p+1]} - B^{p+1}}{p+1}, \tag{23}$$

where the NOTATION $B^{[k]}$ means the quantity in question is raised to the appropriate POWER k, and all terms of the form B^m are replaced with the corresponding Bernoulli numbers B_m. Written explicitly in terms of a sum of POWERS,

$$\sum_{k=1}^{n} k^p = \frac{B_k p!}{k!(p-k+1)!} n^{p-k+1}. \tag{24}$$

It is also true that the COEFFICIENTS of the terms in such an expansion sum to 1 (which Bernoulli stated without proof). Ramanujan gave a number of curious infinite sum identities involving Bernoulli numbers (Berndt 1994).

G. J. Fee and S. Plouffe have computed $B_{200,000}$, which has $\sim 800,000$ DIGITS (Plouffe). Plouffe and collaborators have also calculated B_n for n up to 72,000.

see also ARGOH'S CONJECTURE, BERNOULLI FUNCTION, BERNOULLI POLYNOMIAL, DEBYE FUNCTIONS, EULER-MACLAURIN INTEGRATION FORMULAS, EULER NUMBER, FIGURATE NUMBER TRIANGLE, GENOCCHI NUMBER, PASCAL'S TRIANGLE, RIEMANN ZETA FUNCTION, VON STAUDT-CLAUSEN THEOREM

References

Abramowitz, M. and Stegun, C. A. (Eds.). "Bernoulli and Euler Polynomials and the Euler-Maclaurin Formula." §23.1 in *Handbook of Mathematical Functions with Formulas, Graphs, and Mathematical Tables, 9th printing.* New York: Dover, pp. 804–806, 1972.

Arfken, G. "Bernoulli Numbers, Euler-Maclaurin Formula." §5.9 in *Mathematical Methods for Physicists, 3rd ed.* Orlando, FL: Academic Press, pp. 327–338, 1985.

Ball, W. W. R. and Coxeter, H. S. M. *Mathematical Recreations and Essays, 13th ed.* New York: Dover, p. 71, 1987.

Berndt, B. C. *Ramanujan's Notebooks, Part IV.* New York: Springer-Verlag, pp. 81–85, 1994.

Boyer, C. B. *A History of Mathematics.* New York: Wiley, 1968.

Castellanos, D. "The Ubiquitous Pi. Part I." *Math. Mag.* **61**, 67–98, 1988.

Conway, J. H. and Guy, R. K. In *The Book of Numbers.* New York: Springer-Verlag, pp. 107–110, 1996.

Gradshteyn, I. S. and Ryzhik, I. M. *Tables of Integrals, Series, and Products, 5th ed.* San Diego, CA: Academic Press, 1980.

Hardy, G. H. and Wright, W. M. *An Introduction to the Theory of Numbers, 5th ed.* Oxford, England: Oxford University Press, pp. 91–93, 1979.

Ireland, K. and Rosen, M. "Bernoulli Numbers." Ch. 15 in *A Classical Introduction to Modern Number Theory, 2nd ed.* New York: Springer-Verlag, pp. 228–248, 1990.

Knuth, D. E. and Buckholtz, T. J. "Computation of Tangent, Euler, and Bernoulli Numbers." *Math. Comput.* **21**, 663–688, 1967.

Plouffe, S. "Plouffe's Inverter: Table of Current Records for the Computation of Constants." `http://lacim.uqam.ca/pi/records.html`.

Ramanujan, S. "Some Properties of Bernoulli's Numbers." *J. Indian Math. Soc.* **3**, 219–234, 1911.

Sloane, N. J. A. Sequences A000367/M4039 and A002445/M4189 in "An On-Line Version of the Encyclopedia of Integer Sequences."

Spanier, J. and Oldham, K. B. "The Bernoulli Numbers, B_n." Ch. 4 in *An Atlas of Functions.* Washington, DC: Hemisphere, pp. 35–38, 1987.

Wagstaff, S. S. Jr. "Ramanujan's Paper on Bernoulli Numbers." *J. Indian Math. Soc.* **45**, 49–65, 1981.

Whittaker, E. T. and Watson, G. N. *A Course in Modern Analysis, 4th ed.* Cambridge, England: Cambridge University Press, 1990.

Bernoulli's Paradox

Suppose the HARMONIC SERIES converges to h:

$$\sum_{k=1}^{\infty} \frac{1}{k} = h.$$

Then rearranging the terms in the sum gives

$$h - 1 = h,$$

which is a contradiction.

References

Boas, R. P. "Some Remarkable Sequences of Integers." Ch. 3 in *Mathematical Plums* (Ed. R. Honsberger). Washington, DC: Math. Assoc. Amer., pp. 39–40, 1979.

Bernoulli Polynomial

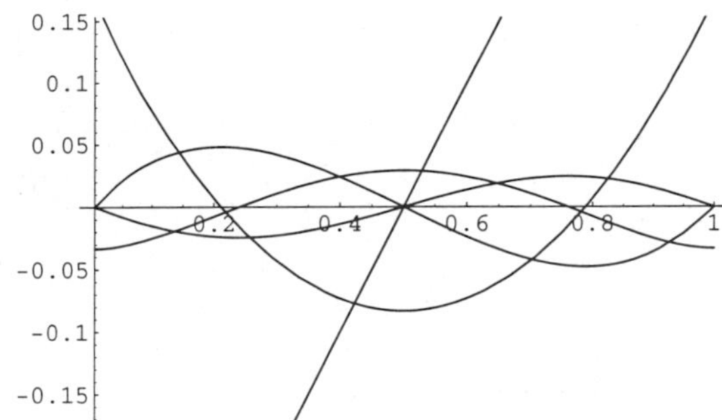

There are two definitions of Bernoulli polynomials in use. The nth Bernoulli polynomial is denoted here by $B_n(x)$, and the archaic Bernoulli polynomial by $B_n^*(x)$. These definitions correspond to the BERNOULLI NUMBERS evaluated at 0,

$$B_n \equiv B_n(0) \tag{1}$$

$$B_n^* \equiv B_n^*(0). \tag{2}$$

They also satisfy

$$B_n(1) = (-1)^n B_n(0) \tag{3}$$

and

$$B_n(1-x) = (-1)^n B_n(x) \tag{4}$$

(Lehmer 1988). The first few Bernoulli POLYNOMIALS are

$$\begin{aligned} B_0(x) &= 1 \\ B_1(x) &= x - \tfrac{1}{2} \\ B_2(x) &= x^2 - x + \tfrac{1}{6} \\ B_3(x) &= x^3 - \tfrac{3}{2}x^2 + \tfrac{1}{2}x \\ B_4(x) &= x^4 - 2x^3 + x^2 - \tfrac{1}{30} \\ B_5(x) &= x^5 - \tfrac{5}{2}x^4 + \tfrac{5}{3}x^3 - \tfrac{1}{6}x \\ B_6(x) &= x^6 - 3x^5 + \tfrac{5}{2}x^4 - \tfrac{1}{2}x^2 + \tfrac{1}{42}. \end{aligned}$$

Bernoulli (1713) defined the POLYNOMIALS in terms of sums of the POWERS of consecutive integers,

$$\sum_{k=0}^{m-1} k^{n-1} = \frac{1}{n}[B_n(m) - B_n(0)]. \tag{5}$$

Euler (1738) gave the Bernoulli POLYNOMIALS $B_n(x)$ in terms of the generating function

$$\frac{te^{tx}}{e^t - 1} \equiv \sum_{n=0}^{\infty} B_n(x)\frac{t^n}{n!}. \tag{6}$$

They satisfy recurrence relation

$$\frac{dB_n}{dx} = nB_{n-1}(x) \tag{7}$$

(Appell 1882), and obey the identity

$$B_n(x) = (B + x)^n, \tag{8}$$

where B^k is interpreted here as $B_k(x)$. Hurwitz gave the FOURIER SERIES

$$B_n(x) = -\frac{n!}{(2\pi i)^n} \sum_{k=-\infty}^{\infty} k^{-n} e^{2\pi i k x}, \tag{9}$$

for $0 < x < 1$, and Raabe (1851) found

$$\frac{1}{m}\sum_{k=0}^{m-1} B_n\left(x + \frac{k}{m}\right) = m^{-n} B_n(mx). \tag{10}$$

A sum identity involving the Bernoulli POLYNOMIALS is

$$\begin{aligned} &\sum_{k=0}^{m} \binom{m}{k} B_k(\alpha) B_{m-k}(\beta) \\ &= -(m-1)B_m(\alpha+\beta) + m(\alpha+\beta-1)B_{m-1}(\alpha+\beta) \end{aligned} \tag{11}$$

for an INTEGER m and arbitrary REAL NUMBERS α and β.

see also BERNOULLI NUMBER, EULER-MACLAURIN INTEGRATION FORMULAS, EULER POLYNOMIAL

References

Abramowitz, M. and Stegun, C. A. (Eds.). "Bernoulli and Euler Polynomials and the Euler-Maclaurin Formula." §23.1 in *Handbook of Mathematical Functions with Formulas, Graphs, and Mathematical Tables, 9th printing.* New York: Dover, pp. 804–806, 1972.

Appell, P. E. "Sur une classe de polynomes." *Annales d'École Normal Superieur, Ser. 2* **9**, 119–144, 1882.

Arfken, G. *Mathematical Methods for Physicists, 3rd ed.* Orlando, FL: Academic Press, p. 330, 1985.

Bernoulli, J. *Ars conjectandi.* Basel, Switzerland, p. 97, 1713. Published posthumously.

Euler, L. "Methodus generalis summandi progressiones." *Comment. Acad. Sci. Petropol.* **6**, 68–97, 1738.

Lehmer, D. H. "A New Approach to Bernoulli Polynomials." *Amer. Math. Monthly.* **95**, 905–911, 1988.

Lucas, E. Ch. 14 in *Théorie des Nombres.* Paris, 1891.

Raabe, J. L. "Zurückführung einiger Summen und bestimmten Integrale auf die Jakob Bernoullische Function." *J. reine angew. Math.* **42**, 348–376, 1851.

Spanier, J. and Oldham, K. B. "The Bernoulli Polynomial $B_n(x)$." Ch. 19 in *An Atlas of Functions.* Washington, DC: Hemisphere, pp. 167–173, 1987.

Bernoulli's Theorem

see WEAK LAW OF LARGE NUMBERS

Bernoulli Trial

An experiment in which s TRIALS are made of an event, with probability p of success in any given TRIAL.

Bernstein-Bézier Curve

see BÉZIER CURVE

Bernstein's Constant

N.B. A detailed on-line essay by S. Finch was the starting point for this entry.

Let $E_n(f)$ be the error of the best uniform approximation to a REAL function $f(x)$ on the INTERVAL $[-1, 1]$ by REAL POLYNOMIALS of degree at most n. If

$$\alpha(x) = |x|, \tag{1}$$

then Bernstein showed that

$$0.267\ldots < \lim_{n\to\infty} 2nE_{2n}(\alpha) < 0.286. \tag{2}$$

He conjectured that the lower limit (β) was $\beta = 1/(2\sqrt{\pi})$. However, this was disproven by Varga and Carpenter (1987) and Varga (1990), who computed

$$\beta = 0.2801694990\ldots. \tag{3}$$

For rational approximations $p(x)/q(x)$ for p and q of degree m and n, D. J. Newman (1964) proved

$$\tfrac{1}{2}e^{-9\sqrt{n}} \leq E_{n,n}(\alpha) \leq 3e^{-\sqrt{n}} \tag{4}$$

for $n \geq 4$. Gonchar (1967) and Bulanov (1975) improved the lower bound to

$$e^{-\pi\sqrt{n+1}} \leq E_{n,n}(\alpha) \leq 3e^{-\sqrt{n}}. \quad (5)$$

Vjacheslavo (1975) proved the existence of POSITIVE constants m and M such that

$$m \leq e^{\pi\sqrt{n}} E_{n,n}(\alpha) < M \quad (6)$$

(Petrushev 1987, pp. 105–106). Varga *et al.* (1993) conjectured and Stahl (1993) proved that

$$\lim_{n\to\infty} e^{\pi\sqrt{2n}} E_{2n,2n}(\alpha) = 8. \quad (7)$$

References

Bulanov, A. P. "Asymptotics for the Best Rational Approximation of the Function Sign x." *Mat. Sbornik* **96**, 171–178, 1975.

Finch, S. "Favorite Mathematical Constants." `http://www.mathsoft.com/asolve/constant/brnstn/brnstn.html`.

Gonchar, A. A. "Estimates for the Growth of Rational Functions and their Applications." *Mat. Sbornik* **72**, 489–503, 1967.

Newman, D. J. "Rational Approximation to $|x|$." *Michigan Math. J.* **11**, 11–14, 1964.

Petrushev, P. P. and Popov, V. A. *Rational Approximation of Real Functions.* New York: Cambridge University Press, 1987.

Stahl, H. "Best Uniform Rational Approximation of $|x|$ on $[-1,1]$." *Russian Acad. Sci. Sb. Math.* **76**, 461–487, 1993.

Varga, R. S. *Scientific Computations on Mathematical Problems and Conjectures.* Philadelphia, PA: SIAM, 1990.

Varga, R. S. and Carpenter, A. J. "On a Conjecture of S. Bernstein in Approximation Theory." *Math. USSR Sbornik* **57**, 547–560, 1987.

Varga, R. S.; Ruttan, A.; and Carpenter, A. J. "Numerical Results on Best Uniform Rational Approximations to $|x|$ on $[-1,+1]$. *Math. USSR Sbornik* **74**, 271–290, 1993.

Vjacheslavo, N. S. "On the Uniform Approximation of $|x|$ by Rational Functions." *Dokl. Akad. Nauk SSSR* **220**, 512–515, 1975.

Bernstein's Inequality

Let P be a POLYNOMIAL of degree n with derivative P'. Then

$$||P'||_\infty \leq n||P||_\infty,$$

where

$$||P||_\infty \equiv \max_{|z|=1} |P(z)|.$$

Bernstein Minimal Surface Theorem

If a MINIMAL SURFACE is given by the equation $z = f(x,y)$ and f has CONTINUOUS first and second PARTIAL DERIVATIVES for all REAL x and y, then f is a PLANE.

References

Hazewinkel, M. (Managing Ed.). *Encyclopaedia of Mathematics: An Updated and Annotated Translation of the Soviet "Mathematical Encyclopaedia."* Dordrecht, Netherlands: Reidel, p. 369, 1988.

Bernstein Polynomial

The POLYNOMIALS defined by

$$B_{i,n}(t) = \binom{n}{i} t^i (1-t)^{n-i},$$

where $\binom{n}{k}$ is a BINOMIAL COEFFICIENT. The Bernstein polynomials of degree n form a basis for the POWER POLYNOMIALS of degree n.

see also BÉZIER CURVE

Bernstein's Polynomial Theorem

If $g(\theta)$ is a trigonometric POLYNOMIAL of degree m satisfying the condition $|g(\theta)| \leq 1$ where θ is arbitrary and real, then $g'(\theta) \leq m$.

References

Szegő, G. *Orthogonal Polynomials, 4th ed.* Providence, RI: Amer. Math. Soc., p. 5, 1975.

Bernstein-Szegő Polynomials

The POLYNOMIALS on the interval $[-1,1]$ associated with the WEIGHT FUNCTIONS

$$w(x) = (1-x^2)^{-1/2}$$
$$w(x) = (1-x^2)^{1/2}$$
$$w(x) = \sqrt{\frac{1-x}{1+x}},$$

also called BERNSTEIN POLYNOMIALS.

References

Szegő, G. *Orthogonal Polynomials, 4th ed.* Providence, RI: Amer. Math. Soc., pp. 31–33, 1975.

Berry-Osseen Inequality

Gives an estimate of the deviation of a DISTRIBUTION FUNCTION as a SUM of independent RANDOM VARIABLES with a NORMAL DISTRIBUTION.

References

Hazewinkel, M. (Managing Ed.). *Encyclopaedia of Mathematics: An Updated and Annotated Translation of the Soviet "Mathematical Encyclopaedia."* Dordrecht, Netherlands: Reidel, p. 369, 1988.

Berry Paradox

There are several versions of the Berry paradox, the original version of which was published by Bertrand Russell and attributed to Oxford University librarian Mr. G. Berry. In one form, the paradox notes that the number "one million, one hundred thousand, one hundred and twenty one" can be named by the description: "the first number not nameable in under ten words." However, this latter expression has only nine words, so the number *can* be named in under ten words, so there is an inconsistency in naming it in this manner!

References

Chaitin, G. J. "The Berry Paradox." *Complexity* **1**, 26–30, 1995.

Bertelsen's Number

An erroneous value of $\pi(10^9)$, where $\pi(x)$ is the PRIME COUNTING FUNCTION. Bertelsen's value of 50,847,478 is 56 lower than the correct value of 50,847,534.

References

Brown, K. S. "Bertelsen's Number." `http://www.seanet.com/~ksbrown/kmath049.htm`.

Bertini's Theorem

The general curve of a system which is LINEARLY INDEPENDENT on a certain number of given irreducible curves will not have a singular point which is not fixed for all the curves of the system.

References

Coolidge, J. L. *A Treatise on Algebraic Plane Curves.* New York: Dover, p. 115, 1959.

Bertrand Curves

Two curves which, at any point, have a common principal NORMAL VECTOR are called Bertrand curves. The product of the TORSIONS of Bertrand curves is a constant.

Bertrand's Paradox

see BERTRAND'S PROBLEM

Bertrand's Postulate

If $n > 3$, there is always at least one PRIME between n and $2n - 2$. Equivalently, if $n > 1$, then there is always at least one PRIME between n and $2n$. It was proved in 1850–51 by Chebyshev, and is therefore sometimes known as CHEBYSHEV'S THEOREM. An elegant proof was later given by Erdős. An extension of this result is that if $n > k$, then there is a number containing a PRIME divisor $> k$ in the sequence $n, n+1, \ldots, n+k-1$. (The case $n = k + 1$ then corresponds to Bertrand's postulate.) This was first proved by Sylvester, independently by Schur, and a simple proof was given by Erdős.

A related problem is to find the least value of θ so that there exists at least one PRIME between n and $n+\mathcal{O}(n^\theta)$ for sufficiently large n (Berndt 1994). The smallest known value is $\theta = 6/11 + \epsilon$ (Lou and Yao 1992).

see also CHOQUET THEORY, DE POLIGNAC'S CONJECTURE, PRIME NUMBER

References

Berndt, B. C. *Ramanujan's Notebooks, Part IV.* New York: Springer-Verlag, p. 135, 1994.

Erdős, P. "Ramanujan and I." In *Proceedings of the International Ramanujan Centenary Conference held at Anna University, Madras, Dec. 21, 1987.* (Ed. K. Alladi). New York: Springer-Verlag, pp. 1–20, 1989.

Lou, S. and Yau, Q. "A Chebyshev's Type of Prime Number Theorem in a Short Interval (II)." *Hardy-Ramanujan J.* **15**, 1–33, 1992.

Bertrand's Problem

What is the PROBABILITY that a CHORD drawn at RANDOM on a CIRCLE of RADIUS r has length $\geq r$? The answer, it turns out, depends on the interpretation of "two points drawn at RANDOM." In the usual interpretation that ANGLES θ_1 and θ_2 are picked at RANDOM on the CIRCUMFERENCE,

$$P = \frac{\pi - \frac{\pi}{3}}{\pi} = \frac{2}{3}.$$

However, if a point is instead placed at RANDOM on a RADIUS of the CIRCLE and a CHORD drawn PERPENDICULAR to it,

$$P = \frac{\frac{\sqrt{3}}{2}r}{r} = \frac{\sqrt{3}}{2}.$$

The latter interpretation is more satisfactory in the sense that the result remains the same for a rotated CIRCLE, a slightly smaller CIRCLE INSCRIBED in the first, or for a CIRCLE of the same size but with its center slightly offset. Jaynes (1983) shows that the interpretation of "RANDOM" as a continuous UNIFORM DISTRIBUTION over the RADIUS is the only one possessing all these three invariances.

References

Bogomolny, A. "Bertrand's Paradox." `http://www.cut-the-knot.com/bertrand.html`.

Jaynes, E. T. *Papers on Probability, Statistics, and Statistical Physics.* Dordrecht, Netherlands: Reidel, 1983.

Pickover, C. A. *Keys to Infinity.* New York: Wiley, pp. 42–45, 1995.

Bertrand's Test

A CONVERGENCE TEST also called DE MORGAN'S AND BERTRAND'S TEST. If the ratio of terms of a SERIES $\{a_n\}_{n=1}^{\infty}$ can be written in the form

$$\frac{a_n}{a_{n+1}} = 1 + \frac{1}{n} + \frac{\rho_n}{n \ln n},$$

then the series converges if $\underline{\lim}_{n\to\infty}\rho_n > 1$ and diverges if $\overline{\lim}_{n\to\infty}\rho_n < 1$, where $\underline{\lim}_{n\to\infty}$ is the LOWER LIMIT and $\overline{\lim}_{n\to\infty}$ is the UPPER LIMIT.

see also KUMMER'S TEST

References

Bromwich, T. J. I'a and MacRobert, T. M. *An Introduction to the Theory of Infinite Series, 3rd ed.* New York: Chelsea, p. 40, 1991.

Bertrand's Theorem

see BERTRAND'S POSTULATE

Besov Space

A type of abstract SPACE which occurs in SPLINE and RATIONAL FUNCTION approximations. The Besov space $B_{p,q}^{\alpha}$ is a complete quasinormed space which is a BANACH SPACE when $1 \leq p,\ q \leq \infty$ (Petrushev and Popov 1987).

References

Bergh, J. and Löfström, J. *Interpolation Spaces.* New York: Springer-Verlag, 1976.

Peetre, J. *New Thoughts on Besov Spaces.* Durham, NC: Duke University Press, 1976.

Petrushev, P. P. and Popov, V. A. "Besov Spaces." §7.2 in *Rational Approximation of Real Functions.* New York: Cambridge University Press, pp. 201–203, 1987.

Triebel, H. *Interpolation Theory, Function Spaces, Differential Operators.* New York: Elsevier, 1978.

Bessel's Correction

The factor $(N-1)/N$ in the relationship between the VARIANCE σ and the EXPECTATION VALUES of the SAMPLE VARIANCE,

$$\langle s^2 \rangle = \frac{N-1}{N}\sigma^2, \tag{1}$$

where

$$s^2 \equiv \langle x^2 \rangle - \langle x \rangle^2. \tag{2}$$

For two samples,

$$\hat{\sigma}^2 = \frac{N_1 {s_1}^2 + N_2 {s_2}^2}{N_1 + N_2 - 2}. \tag{3}$$

see also SAMPLE VARIANCE, VARIANCE

Bessel Differential Equation

$$x^2 \frac{d^2y}{dx^2} + x\frac{dy}{dx} + (x^2 - m^2)y = 0. \tag{1}$$

Equivalently, dividing through by x^2,

$$\frac{d^2y}{dx^2} + \frac{1}{x}\frac{dy}{dx} + \left(1 - \frac{m^2}{x^2}\right)y = 0. \tag{2}$$

The solutions to this equation define the BESSEL FUNCTIONS. The equation has a regular SINGULARITY at 0 and an irregular SINGULARITY at ∞.

A transformed version of the Bessel differential equation given by Bowman (1958) is

$$x^2 \frac{d^2y}{dx^2} + (2p+1)x\frac{dy}{dx} + (a^2x^{2r} + \beta^2)y = 0. \tag{3}$$

The solution is

$$y = x^{-p}\left[C_1 J_{q/r}\left(\frac{\alpha}{r}x^r\right) + C_2 Y_{q/r}\left(\frac{\alpha}{r}x^r\right)\right], \tag{4}$$

where

$$q \equiv \sqrt{p^2 - \beta^2}, \tag{5}$$

J and Y are the BESSEL FUNCTIONS OF THE FIRST and SECOND KINDS, and C_1 and C_2 are constants. Another form is given by letting $y = x^\alpha J_n(\beta x^\gamma)$, $\eta = yx^{-\alpha}$, and $\xi = \beta x^\gamma$ (Bowman 1958, p. 117), then

$$\frac{d^2y}{dx^2} - \frac{2\alpha - 1}{x}\frac{dy}{dx} + \left(\beta^2\gamma^2x^{2\gamma-2} + \frac{\alpha^2 - n^2\gamma^2}{x^2}\right)y = 0. \tag{6}$$

The solution is

$$y = \begin{cases} x^\alpha[AJ_n(\beta x^\gamma) + BY_n(\beta x^\gamma)] & \text{for integral } n \\ AJ_n(\beta x^\gamma) + BJ_{-n}(\beta x^\gamma)] & \text{for nonintegral } n. \end{cases} \tag{7}$$

see also AIRY FUNCTIONS, ANGER FUNCTION, BEI, BER, BESSEL FUNCTION, BOURGET'S HYPOTHESIS, CATALAN INTEGRALS, CYLINDRICAL FUNCTION, DINI EXPANSION, HANKEL FUNCTION, HANKEL'S INTEGRAL, HEMISPHERICAL FUNCTION, KAPTEYN SERIES, LIPSCHITZ'S INTEGRAL, LOMMEL DIFFERENTIAL EQUATION, LOMMEL FUNCTION, LOMMEL'S INTEGRALS, NEUMANN SERIES (BESSEL FUNCTION), PARSEVAL'S INTEGRAL, POISSON INTEGRAL, RAMANUJAN'S INTEGRAL, RICCATI DIFFERENTIAL EQUATION, SONINE'S INTEGRAL, STRUVE FUNCTION, WEBER FUNCTIONS, WEBER'S DISCONTINUOUS INTEGRALS

References

Bowman, F. *Introduction to Bessel Functions.* New York: Dover, 1958.

Morse, P. M. and Feshbach, H. *Methods of Theoretical Physics, Part I.* New York: McGraw-Hill, p. 550, 1953.

Bessel's Finite Difference Formula

An INTERPOLATION formula also sometimes known as

$$f_p = f_0 + p\delta_{1/2} + B_2(\delta_0^2 + \delta_1^2) + B_3\delta_{1/2}^3 + B_4(\delta_0^4 + \delta_1^4) + B_5\delta_{1/2}^5 + \ldots, \quad (1)$$

for $p \in [0, 1]$, where δ is the CENTRAL DIFFERENCE and

$$B_{2n} \equiv \tfrac{1}{2}G_{2n} \equiv \tfrac{1}{2}(E_{2n} + F_{2n}) \quad (2)$$

$$B_{2n+1} \equiv G_{2n+1} - \tfrac{1}{2}G_{2n} \equiv \tfrac{1}{2}(F_{2n} - E_{2n}) \quad (3)$$

$$E_{2n} \equiv G_{2n} - G_{2n+1} \equiv B_{2n} - B_{2n+1} \quad (4)$$

$$F_{2n} \equiv G_{2n+1} \equiv B_{2n} + B_{2n+1}, \quad (5)$$

where G_k are the COEFFICIENTS from GAUSS'S BACKWARD FORMULA and GAUSS'S FORWARD FORMULA and E_k and F_k are the COEFFICIENTS from EVERETT'S FORMULA. The B_ks also satisfy

$$B_{2n}(p) = B_{2n}(q) \quad (6)$$

$$B_{2n+1}(p) = -B_{2n+1}(q), \quad (7)$$

for

$$q \equiv 1 - p. \quad (8)$$

see also EVERETT'S FORMULA

References

Abramowitz, M. and Stegun, C. A. (Eds.). *Handbook of Mathematical Functions with Formulas, Graphs, and Mathematical Tables, 9th printing.* New York: Dover, p. 880, 1972.

Acton, F. S. *Numerical Methods That Work, 2nd printing.* Washington, DC: Math. Assoc. Amer., pp. 90–91, 1990.

Beyer, W. H. *CRC Standard Mathematical Tables, 28th ed.* Boca Raton, FL: CRC Press, p. 433, 1987.

Bessel's First Integral

$$J_n(x) = \frac{1}{\pi}\int_0^\pi \cos(n\theta - x\sin\theta)\,d\theta,$$

where $J_n(x)$ is a BESSEL FUNCTION OF THE FIRST KIND.

Bessel's Formula

see BESSEL'S FINITE DIFFERENCE FORMULA, BESSEL'S INTERPOLATION FORMULA, BESSEL'S STATISTICAL FORMULA

Bessel Function

A function $Z(x)$ defined by the RECURRENCE RELATIONS

$$Z_{m+1} + Z_{m-1} = \frac{2m}{x}Z_m$$

and

$$Z_{m+1} - Z_{m-1} = -2\frac{dZ_m}{dx}.$$

The Bessel functions are more frequently defined as solutions to the DIFFERENTIAL EQUATION

$$x^2\frac{d^2y}{dx^2} + x\frac{dy}{dx} + (x^2 - m^2)y = 0.$$

There are two classes of solution, called the BESSEL FUNCTION OF THE FIRST KIND J and BESSEL FUNCTION OF THE SECOND KIND Y. (A BESSEL FUNCTION OF THE THIRD KIND is a special combination of the first and second kinds.) Several related functions are also defined by slightly modifying the defining equations.

see also BESSEL FUNCTION OF THE FIRST KIND, BESSEL FUNCTION OF THE SECOND KIND, BESSEL FUNCTION OF THE THIRD KIND, CYLINDER FUNCTION, HEMICYLINDRICAL FUNCTION, MODIFIED BESSEL FUNCTION OF THE FIRST KIND, MODIFIED BESSEL FUNCTION OF THE SECOND KIND, SPHERICAL BESSEL FUNCTION OF THE FIRST KIND, SPHERICAL BESSEL FUNCTION OF THE SECOND KIND

References

Abramowitz, M. and Stegun, C. A. (Eds.). "Bessel Functions of Integer Order," "Bessel Functions of Fractional Order," and "Integrals of Bessel Functions." Chs. 9–11 in *Handbook of Mathematical Functions with Formulas, Graphs, and Mathematical Tables, 9th printing.* New York: Dover, pp. 355–389, 435–456, and 480–491, 1972.

Arfken, G. "Bessel Functions." Ch. 11 in *Mathematical Methods for Physicists, 3rd ed.* Orlando, FL: Academic Press, pp. 573–636, 1985.

Bickley, W. G. *Bessel Functions and Formulae.* Cambridge, England: Cambridge University Press, 1957.

Bowman, F. *Introduction to Bessel Functions.* New York: Dover, 1958.

Gray, A. and Matthews, G. B. *A Treatise on Bessel Functions and Their Applications to Physics, 2nd ed.* New York: Dover, 1966.

Luke, Y. L. *Integrals of Bessel Functions.* New York: McGraw-Hill, 1962.

McLachlan, N. W. *Bessel Functions for Engineers, 2nd ed. with corrections.* Oxford, England: Clarendon Press, 1961.

Press, W. H.; Flannery, B. P.; Teukolsky, S. A.; and Vetterling, W. T. "Bessel Functions of Integral Order" and "Bessel Functions of Fractional Order, Airy Functions, Spherical Bessel Functions." §6.5 and 6.7 in *Numerical Recipes in FORTRAN: The Art of Scientific Computing, 2nd ed.* Cambridge, England: Cambridge University Press, pp. 223–229 and 234–245, 1992.

Watson, G. N. *A Treatise on the Theory of Bessel Functions, 2nd ed.* Cambridge, England: Cambridge University Press, 1966.

Bessel Function of the First Kind

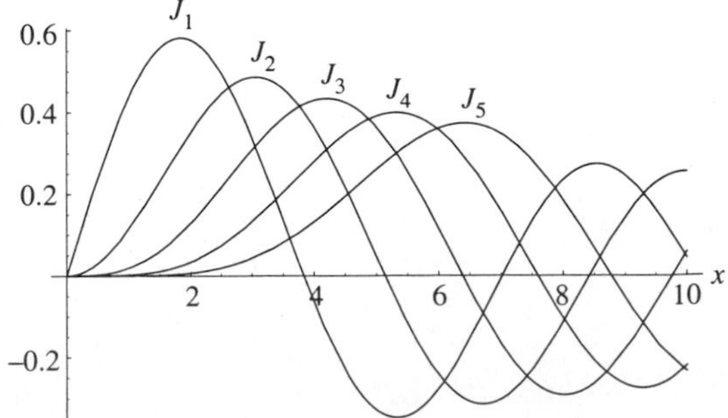

The Bessel functions of the first kind $J_n(x)$ are defined as the solutions to the BESSEL DIFFERENTIAL EQUATION

$$x^2\frac{d^2y}{dx^2}+x\frac{dy}{dx}+(x^2-m^2)y=0 \quad (1)$$

which are nonsingular at the origin. They are sometimes also called CYLINDER FUNCTIONS or CYLINDRICAL HARMONICS. The above plot shows $J_n(x)$ for $n = 1, 2, \ldots, 5$.

To solve the differential equation, apply FROBENIUS METHOD using a series solution of the form

$$y=x^k\sum_{n=0}^{\infty}a_nx^n=\sum_{n=0}^{\infty}a_nx^{n+k}. \quad (2)$$

Plugging into (1) yields

$$x^2\sum_{n=0}^{\infty}(k+n)(k+n-1)a_nx^{k+n-2}+x\sum_{n=0}^{\infty}(k+n)a_nx^{k+n-1}+x^2\sum_{n=0}^{\infty}a_nx^{k+n}-m^2\sum_{n=0}^{\infty}a_nx^{n+k}=0 \quad (3)$$

$$\sum_{n=0}^{\infty}(k+n)(k+n-1)a_nx^{k+n}+\sum_{n=0}^{\infty}(k+n)a_nx^{k+n}+\sum_{n=2}^{\infty}a_{n-2}x^{k+n}-m^2\sum_{n=0}^{\infty}a_nx^{n+k}=0. \quad (4)$$

The INDICIAL EQUATION, obtained by setting $n = 0$, is

$$a_0[k(k-1)+k-m^2]=a_0(k^2-m^2)=0. \quad (5)$$

Since a_0 is defined as the first NONZERO term, $k^2-m^2=0$, so $k=\pm m$. Now, if $k=m$,

$$\sum_{n=0}^{\infty}[(m+n)(m+n-1)+(m+n)-m^2]a_nx^{m+n}+\sum_{n=2}^{\infty}a_{n-2}x^{m+n}=0 \quad (6)$$

$$\sum_{n=0}^{\infty}[(m+n)^2-m^2]a_nx^{m+n}+\sum_{n=2}^{\infty}a_{n-2}x^{m+n}=0 \quad (7)$$

$$\sum_{n=0}^{\infty}n(2m+n)a_nx^{m+n}+\sum_{n=2}^{\infty}a_{n-2}x^{m+n}=0 \quad (8)$$

$$a_1(2m+1)+\sum_{n=2}^{\infty}[a_nn(2m+n)+a_{n-2}]x^{m+n}=0. \quad (9)$$

First, look at the special case $m = -1/2$, then (9) becomes

$$\sum_{n=2}^{\infty}[a_nn(n-1)+a_{n-2}]x^{m+n}=0, \quad (10)$$

so

$$a_n=-\frac{1}{n(n-1)}a_{n-2}. \quad (11)$$

Now let $n \equiv 2l$, where $l = 1, 2, \ldots$.

$$\begin{aligned}a_{2l}&=-\frac{1}{2l(2l-1)}a_{2l-2}\\&=\frac{(-1)^l}{[2l(2l-1)][2(l-1)(2l-3)]\cdots[2\cdot1\cdot1]}a_0\\&=\frac{(-1)^l}{2^ll!(2l-1)!!}a_0,\end{aligned} \quad (12)$$

which, using the identity $2^ll!(2l-1)!!=(2l)!$, gives

$$a_{2l}=\frac{(-1)^l}{(2l)!}a_0. \quad (13)$$

Similarly, letting $n \equiv 2l+1$

$$\begin{aligned}a_{2l+1}&=-\frac{1}{(2l+1)(2l)}a_{2l-1}\\&=\frac{(-1)^l}{[2l(2l+1)][2(l-1)(2l-1)]\cdots[2\cdot1\cdot3][1]}a_1,\end{aligned} \quad (14)$$

which, using the identity $2^ll!(2l+1)!!=(2l+1)!$, gives

$$a_{2l+1}=\frac{(-1)^l}{2^ll!(2l+1)!!}a_1=\frac{(-1)^l}{(2l+1)!}a_1. \quad (15)$$

Plugging back into (2) with $k = m = -1/2$ gives

$$\begin{aligned}y&=x^{-1/2}\sum_{n=0}^{\infty}a_nx^n\\&=x^{-1/2}\left[\sum_{n=1,3,5,\ldots}^{\infty}a_nx^n+\sum_{n=0,2,4,\ldots}^{\infty}a_nx^n\right]\\&=x^{-1/2}\left[\sum_{l=0}^{\infty}a_{2l}x^{2l}+\sum_{l=0}^{\infty}a_{2l+1}x^{2l+1}\right]\\&=x^{-1/2}\left[a_0\sum_{l=0}^{\infty}\frac{(-1)^l}{(2l)!}x^{2l}+a_1\sum_{l=0}^{\infty}\frac{(-1)^l}{(2l+1)!}x^{2l+1}\right]\\&=x^{-1/2}(a_0\cos x+a_1\sin x).\end{aligned} \quad (16)$$

The BESSEL FUNCTIONS of order $\pm1/2$ are therefore defined as

$$J_{-1/2}(x)\equiv\sqrt{\frac{2}{\pi x}}\cos x \quad (17)$$

$$J_{1/2}(x)\equiv\sqrt{\frac{2}{\pi x}}\sin x, \quad (18)$$

so the general solution for $m = \pm 1/2$ is

$$y = a_0' J_{-1/2}(x) + a_1' J_{1/2}(x). \qquad (19)$$

Now, consider a general $m \neq -1/2$. Equation (9) requires

$$a_1(2m+1) = 0 \qquad (20)$$

$$[a_n n(2m+n) + a_{n-2}]x^{m+n} = 0 \qquad (21)$$

for $n = 2, 3, \ldots$, so

$$a_1 = 0 \qquad (22)$$

$$a_n = -\frac{1}{n(2m+n)} a_{n-2} \qquad (23)$$

for $n = 2, 3, \ldots$. Let $n \equiv 2l+1$, where $l = 1, 2, \ldots$, then

$$\begin{aligned} a_{2l+1} &= -\frac{1}{(2l+1)[2(m+1)+1]} a_{2l-1} \\ &= \ldots = f(n,m) a_1 = 0, \end{aligned} \qquad (24)$$

where $f(n,m)$ is the function of l and m obtained by iterating the recursion relationship down to a_1. Now let $n \equiv 2l$, where $l = 1, 2, \ldots$, so

$$\begin{aligned} a_{2l} &= -\frac{1}{2l(2m+2l)} a_{2l-2} = -\frac{1}{4l(m+l)} a_{2l-2} \\ &= \frac{(-1)^l}{[4l(m+l)][4(l-1)(m+l-1)]\cdots[4\cdot(m+1)]} a_0. \end{aligned} \qquad (25)$$

Plugging back into (9),

$$\begin{aligned} y &= \sum_{n=0}^{\infty} a_n x^{n+m} = \sum_{n=1,3,5,\ldots}^{\infty} a_n x^{n+m} + \sum_{n=0,2,4,\ldots}^{\infty} a_n x^{n+m} \\ &= \sum_{l=0}^{\infty} a_{2l+1} x^{2l+m+1} + \sum_{l=0}^{\infty} a_{2l} x^{2l+m} \\ &= a_0 \sum_{l=0}^{\infty} \frac{(-1)^l}{[4l(m+l)][4(l-1)(m+l-1)]\cdots[4\cdot(m+1)]} x^{2l+m} \\ &= a_0 \sum_{l=0}^{\infty} \frac{[(-1)^l m(m-1)\cdots 1] x^{2l+m}}{[4l(m+l)][4(l-1)(m+l-1)]\cdots[m(m-1)\cdots 1]} \\ &= a_0 \sum_{l=0}^{\infty} \frac{(-1)^l m!}{4^l l!(m+l)!} = a_0 \sum_{l=0}^{\infty} \frac{(-1)^l m!}{2^{2l} l!(m+l)!}. \end{aligned} \qquad (26)$$

Now define

$$J_m(x) \equiv \sum_{l=0}^{\infty} \frac{(-1)^l}{2^{2l+m} l!(m+l)!} x^{2l+m}, \qquad (27)$$

where the factorials can be generalized to GAMMA FUNCTIONS for nonintegral m. The above equation then becomes

$$y = a_0 2^m m! J_m(x) = a_0' J_m(x). \qquad (28)$$

Returning to equation (5) and examining the case $k = -m$,

$$a_1(1-2m) + \sum_{n=2}^{\infty} [a_n n(n-2m) + a_{n-2}]x^{n-m} = 0. \qquad (29)$$

However, the sign of m is arbitrary, so the solutions must be the same for $+m$ and $-m$. We are therefore free to replace $-m$ with $-|m|$, so

$$a_1(1+2|m|) + \sum_{n=2}^{\infty} [a_n n(n+2|m|) + a_{n-2}]x^{|m|+n} = 0, \qquad (30)$$

and we obtain the same solutions as before, but with m replaced by $|m|$.

$$J_m(x) = \begin{cases} \sum_{l=0}^{\infty} \frac{(-1)^l}{2^{2l+|m|} l!(|m|+l)!} x^{2l+|m|} & \text{for } |m| \neq -\frac{1}{2} \\ \sqrt{\frac{2}{\pi x}} \cos x & \text{for } m = -\frac{1}{2} \\ \sqrt{\frac{2}{\pi x}} \sin x & \text{for } m = \frac{1}{2}. \end{cases} \qquad (31)$$

We can relate J_m and J_{-m} (when m is an INTEGER) by writing

$$J_{-m}(x) = \sum_{l=0}^{\infty} \frac{(-1)^l}{2^{2l-m} l!(l-m)!} x^{2l-m}. \qquad (32)$$

Now let $l \equiv l' + m$. Then

$$\begin{aligned} J_{-m}(x) &= \sum_{l'+m=0}^{\infty} \frac{(-1)^{l'+m}}{2^{2l'+m}(l'+m)!l!} x^{2l'+m} \\ &= \sum_{l'=-m}^{-1} \frac{(-1)^{l'+m}}{2^{2l'+m} l'!(l'+m)!} x^{2l'+m} \\ &\quad + \sum_{l'=0}^{\infty} \frac{(-1)^{l'+m}}{2^{2l'+m} l'!(l'+m)!} x^{2l'+m}. \end{aligned} \qquad (33)$$

But $l'! = \infty$ for $l' = -m, \ldots, -1$, so the DENOMINATOR is infinite and the terms on the right are zero. We therefore have

$$J_{-m}(x) = \sum_{l=0}^{\infty} \frac{(-1)^{l+m}}{2^{2l+m} l!(l+m)!} x^{2l+m} = (-1)^m J_m(x). \qquad (34)$$

Note that the BESSEL DIFFERENTIAL EQUATION is second-order, so there must be two linearly independent solutions. We have found both only for $|m| = 1/2$. For a general nonintegral order, the independent solutions are J_m and J_{-m}. When m is an INTEGER, the general (real) solution is of the form

$$Z_m \equiv C_1 J_m(x) + C_2 Y_m(x), \qquad (35)$$

where J_m is a Bessel function of the first kind, Y_m (a.k.a. N_m) is the BESSEL FUNCTION OF THE SECOND KIND (a.k.a. NEUMANN FUNCTION or WEBER FUNCTION), and C_1 and C_2 are constants. Complex solutions are given by the HANKEL FUNCTIONS (a.k.a. BESSEL FUNCTIONS OF THE THIRD KIND).

The Bessel functions are ORTHOGONAL in $[0,1]$ with respect to the weight factor x. Except when $2n$ is a NEGATIVE INTEGER,

$$J_m(z) = \frac{z^{-1/2}}{2^{2m+1/2} i^{m+1/2} \Gamma(m+1)} M_{0,m}(2iz), \tag{36}$$

where $\Gamma(x)$ is the GAMMA FUNCTION and $M_{0,m}$ is a WHITTAKER FUNCTION.

In terms of a CONFLUENT HYPERGEOMETRIC FUNCTION OF THE FIRST KIND, the Bessel function is written

$$J_\nu(z) = \frac{(\frac{1}{2}z)^\nu}{\Gamma(\nu+1)} {}_0F_1(\nu+1; -\tfrac{1}{4}z^2). \tag{37}$$

A derivative identity for expressing higher order Bessel functions in terms of $J_0(x)$ is

$$J_n(x) = i^n T_n\left(i\frac{d}{dx}\right) J_0(x), \tag{38}$$

where $T_n(x)$ is a CHEBYSHEV POLYNOMIAL OF THE FIRST KIND. Asymptotic forms for the Bessel functions are

$$J_m(x) \approx \frac{1}{\Gamma(m+1)} \left(\frac{x}{2}\right)^m \tag{39}$$

for $x \ll 1$ and

$$J_m(x) \approx \sqrt{\frac{2}{\pi x}} \cos\left(x - \frac{m\pi}{2} - \frac{\pi}{4}\right) \tag{40}$$

for $x \gg 1$. A derivative identity is

$$\frac{d}{dx}[x^m J_m(x)] = x^m J_{m-1}(x). \tag{41}$$

An integral identity is

$$\int_0^u u' J_0(u')\, du' = u J_1(u). \tag{42}$$

Some sum identities are

$$1 = [J_0(x)]^2 + 2[J_1(x)]^2 + 2[J_2(x)]^2 + \ldots \tag{43}$$

$$1 = J_0(x) + 2J_2(x) + 2J_4(x) + \ldots \tag{44}$$

and the JACOBI-ANGER EXPANSION

$$e^{iz\cos\theta} = \sum_{n=-\infty}^{\infty} i^n J_n(z) e^{in\theta}, \tag{45}$$

which can also be written

$$e^{iz\cos\theta} = J_0(z) + 2\sum_{n=1}^{\infty} i^n J_n(z) \cos(n\theta). \tag{46}$$

The Bessel function addition theorem states

$$J_n(y+z) = \sum_{m=-\infty}^{\infty} J_m(y) J_{n-m}(z). \tag{47}$$

ROOTS of the FUNCTION $J_n(x)$ are given in the following table.

zero	$J_0(x)$	$J_1(x)$	$J_2(x)$	$J_3(x)$	$J_4(x)$	$J_5(x)$
1	2.4048	3.8317	5.1336	6.3802	7.5883	8.7715
2	5.5201	7.0156	8.4172	9.7610	11.0647	12.3386
3	8.6537	10.1735	11.6198	13.0152	14.3725	15.7002
4	11.7915	13.3237	14.7960	16.2235	17.6160	18.9801
5	14.9309	16.4706	17.9598	19.4094	20.8269	22.2178

Let x_n be the nth ROOT of the Bessel function $J_0(x)$, then

$$\sum_{n=1}^{\infty} \frac{1}{x_n J_0(x_n)} = 0.38479\ldots \tag{48}$$

(Le Lionnais 1983).

The ROOTS of its DERIVATIVES are given in the following table.

zero	$J_0'(x)$	$J_1'(x)$	$J_2'(x)$	$J_3'(x)$	$J_4'(x)$	$J_5'(x)$
1	3.8317	1.8412	3.0542	4.2012	5.3175	6.4156
2	7.0156	5.3314	6.7061	8.0152	9.2824	10.5199
3	10.1735	8.5363	9.9695	11.3459	12.6819	13.9872
4	13.3237	11.7060	13.1704	14.5858	15.9641	17.3128
5	16.4706	14.8636	16.3475	17.7887	19.1960	20.5755

Various integrals can be expressed in terms of Bessel functions

$$J_0(z) = \frac{1}{2\pi} \int_0^{2\pi} e^{iz} \cos\phi\, d\phi \tag{49}$$

$$J_n(z) = \frac{1}{\pi} \int_0^{\pi} \cos(z\sin\theta - n\theta)\, d\theta, \tag{50}$$

which is BESSEL'S FIRST INTEGRAL,

$$J_n(z) = \frac{i^{-n}}{\pi} \int_0^{\pi} e^{iz\cos\theta} \cos(n\theta)\, d\theta \tag{51}$$

$$J_n(z) = \frac{1}{2\pi i^n} \int_0^{2\pi} e^{iz\cos\phi} e^{in\phi}\, d\phi \tag{52}$$

for $n = 1, 2, \ldots$,

$$J_n(z) = \frac{2}{\pi} \frac{x^n}{(2m-1)!!} \int_0^{\pi/2} \sin^{2n} u \cos(x\cos u)\, du \tag{53}$$

for $n = 1, 2, \ldots$,

$$J_n(x) = \frac{1}{2\pi i} \int_\gamma e^{(x/2)(z-1/z)} z^{-n-1}\, dz \tag{54}$$

for $n > -1/2$. Integrals involving $J_1(x)$ include

$$\int_0^\infty J_1(x)\,dx = 1 \tag{55}$$

$$\int_0^\infty \left[\frac{J_1(x)}{x}\right]^2 dx = \frac{4}{3\pi} \tag{56}$$

$$\int_0^\infty \left[\frac{J_1(x)}{x}\right]^2 x\,dx = \frac{1}{2}. \tag{57}$$

see also BESSEL FUNCTION OF THE SECOND KIND, DEBYE'S ASYMPTOTIC REPRESENTATION, DIXON-FERRAR FORMULA, HANSEN-BESSEL FORMULA, KAPTEYN SERIES, KNESER-SOMMERFELD FORMULA, MEHLER'S BESSEL FUNCTION FORMULA, NICHOLSON'S FORMULA, POISSON'S BESSEL FUNCTION FORMULA, SCHLÄFLI'S FORMULA, SCHLÖMILCH'S SERIES, SOMMERFELD'S FORMULA, SONINE-SCHAFHEITLIN FORMULA, WATSON'S FORMULA, WATSON-NICHOLSON FORMULA, WEBER'S DISCONTINUOUS INTEGRALS, WEBER'S FORMULA, WEBER-SONINE FORMULA, WEYRICH'S FORMULA

References

Abramowitz, M. and Stegun, C. A. (Eds.). "Bessel Functions J and Y." §9.1 in *Handbook of Mathematical Functions with Formulas, Graphs, and Mathematical Tables, 9th printing.* New York: Dover, pp. 358–364, 1972.

Arfken, G. "Bessel Functions of the First Kind, $J_\nu(x)$" and "Orthogonality." §11.1 and 11.2 in *Mathematical Methods for Physicists, 3rd ed.* Orlando, FL: Academic Press, pp. 573–591 and 591–596, 1985.

Lehmer, D. H. "Arithmetical Periodicities of Bessel Functions." *Ann. Math.* **33**, 143–150, 1932.

Le Lionnais, F. *Les nombres remarquables.* Paris: Hermann, p. 25, 1983.

Morse, P. M. and Feshbach, H. *Methods of Theoretical Physics, Part I.* New York: McGraw-Hill, pp. 619–622, 1953.

Spanier, J. and Oldham, K. B. "The Bessel Coefficients $J_0(x)$ and $J_1(x)$" and "The Bessel Function $J_\nu(x)$." Chs. 52–53 in *An Atlas of Functions.* Washington, DC: Hemisphere, pp. 509–520 and 521–532, 1987.

Bessel Function Fourier Expansion

Let $n \geq 1/2$ and $\alpha_1, \alpha_2, \ldots$ be the POSITIVE ROOTS of $J_n(x) = 0$. An expansion of a function in the interval (0,1) in terms of BESSEL FUNCTIONS OF THE FIRST KIND

$$f(x) = \sum_{l=1}^\infty A_r J_n(x\alpha_r), \tag{1}$$

has COEFFICIENTS found as follows:

$$\int_0^1 xf(x)J_n(x\alpha_l)\,dx = \sum_{r=1}^\infty A_r \int_0^1 xJ_n(x\alpha_r)J_n(x\alpha_l)\,dx. \tag{2}$$

But ORTHOGONALITY of BESSEL FUNCTION ROOTS gives

$$\int_0^1 xJ_n(x\alpha_l)J_n(x\alpha_r)\,dx = \tfrac{1}{2}\delta_{l,r}{J_{n+1}}^2(\alpha_r) \tag{3}$$

(Bowman 1958, p. 108), so

$$\begin{aligned}\int_0^1 xf(x)J_n(x\alpha_l)\,dx &= \tfrac{1}{2}\sum_{r=1}^\infty A_r\delta_{l,r}{J_{n+1}}^2(x\alpha_r)\\ &= \tfrac{1}{2}A_l{J_{n+1}}^2(\alpha_l),\end{aligned} \tag{4}$$

and the COEFFICIENTS are given by

$$A_l = \frac{2}{{J_{n+1}}^2(\alpha_l)}\int_0^1 xf(x)J_n(x\alpha_l)\,dx. \tag{5}$$

References

Bowman, F. *Introduction to Bessel Functions.* New York: Dover, 1958.

Bessel Function of the Second Kind

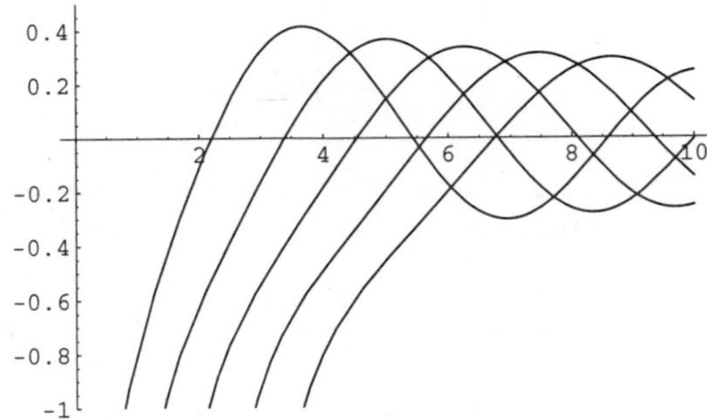

A Bessel function of the second kind $Y_n(x)$ is a solution to the BESSEL DIFFERENTIAL EQUATION which is singular at the origin. Bessel functions of the second kind are also called NEUMANN FUNCTIONS or WEBER FUNCTIONS. The above plot shows $Y_n(x)$ for $n = 1, 2, \ldots, 5$.

Let $v \equiv J_m(x)$ be the first solution and u be the other one (since the BESSEL DIFFERENTIAL EQUATION is second-order, there are two LINEARLY INDEPENDENT solutions). Then

$$xu'' + u' + xu = 0 \tag{1}$$

$$xv'' + v' + xv = 0. \tag{2}$$

Take $v \times (1) - u \times (2)$,

$$x(u''v - uv'') + u'v - uv' = 0 \tag{3}$$

$$\frac{d}{dx}[x(u'v - uv')] = 0, \tag{4}$$

so $x(u'v - uv') = B$, where B is a constant. Divide by xv^2,

$$\frac{u'v - uv'}{v^2} = \frac{d}{dx}\left(\frac{u}{v}\right) = \frac{B}{xv^2} \tag{5}$$

$$\frac{u}{v} = A + B\int\frac{dx}{xv^2}. \tag{6}$$

Rearranging and using $v \equiv J_m(x)$ gives

$$\begin{aligned} u &= AJ_m(x) + BJ_m(x)\int \frac{dx}{x{J_m}^2(x)} \\ &\equiv A'J_m(x) + B'Y_m(x), \end{aligned} \tag{7}$$

where the Bessel function of the second kind is defined by

$$\begin{aligned} Y_m(x) &= \frac{J_m(x)\cos(m\pi) - J_{-m}(x)}{\sin(m\pi)} \\ &= \frac{1}{\pi}\sum_{k=1}^{\infty} \frac{(-1)^k x^{m+2k}}{2^{m+2k}k!(m+k)!}\left[2\ln\left(\frac{x}{2}\right) + 2\gamma - b_{m+k} - b_k\right] \\ &\quad -\frac{1}{\pi}\sum_{k=0}^{m-1} \frac{x^{-m+2k}(m-k-1)!}{2^{-m+2k}k!} \end{aligned} \tag{8}$$

$m = 0, 1, 2, \ldots$, γ is the EULER-MASCHERONI CONSTANT, and

$$b_k \equiv \begin{cases} 0 & k = 0, \\ \sum_{n=0}^{k} \frac{1}{n} & k \neq 0. \end{cases} \tag{9}$$

The function is given by

$$\begin{aligned} Y_n(z) &= \frac{1}{\pi}\int_0^{\pi} \sin(z\sin\theta - n\theta)\, d\theta \\ &\quad -\frac{1}{\pi}\int_0^{\infty} [e^{nt} + e^{-nt}(-1)^n]e^{-z\sinh t}\, dt. \end{aligned} \tag{10}$$

Asymptotic equations are

$$Y_m(x) = \begin{cases} \frac{2}{\pi}\left[\ln(\frac{1}{2}x) + \gamma\right] & m = 0, x \ll 1 \\ -\frac{\Gamma(m)}{\pi}\left(\frac{2}{x}\right)^m & m \neq 0, x \ll 1 \end{cases} \tag{11}$$

$$Y_m(x) = \sqrt{\frac{2}{\pi x}}\sin\left(x - \frac{m\pi}{2} - \frac{\pi}{4}\right) \quad x \gg 1, \tag{12}$$

where $\Gamma(z)$ is a GAMMA FUNCTION.

see also BESSEL FUNCTION OF THE FIRST KIND, BOURGET'S HYPOTHESIS, HANKEL FUNCTION

References

Abramowitz, M. and Stegun, C. A. (Eds.). "Bessel Functions J and Y." §9.1 in *Handbook of Mathematical Functions with Formulas, Graphs, and Mathematical Tables, 9th printing.* New York: Dover, pp. 358–364, 1972.

Arfken, G. "Neumann Functions, Bessel Functions of the Second Kind, $N_\nu(x)$." §11.3 in *Mathematical Methods for Physicists, 3rd ed.* Orlando, FL: Academic Press, pp. 596–604, 1985.

Morse, P. M. and Feshbach, H. *Methods of Theoretical Physics, Part I.* New York: McGraw-Hill, pp. 625–627, 1953.

Spanier, J. and Oldham, K. B. "The Neumann Function $Y_\nu(x)$." Ch. 54 in *An Atlas of Functions.* Washington, DC: Hemisphere, pp. 533–542, 1987.

Bessel Function of the Third Kind

see HANKEL FUNCTION

Bessel's Inequality

If $f(x)$ is piecewise CONTINUOUS and has a general FOURIER SERIES

$$\sum_i a_i\phi_i(x) \tag{1}$$

with WEIGHTING FUNCTION $w(x)$, it must be true that

$$\int \left[f(x) - \sum_i a_i\phi_i(x)\right]^2 w(x)\, dx \geq 0 \tag{2}$$

$$\begin{aligned} \int f^2(x)w(x)\, dx - 2\sum_i a_i \int f(x)\phi_i(x)w(x)\, dx \\ + \sum_i {a_i}^2 \int {\phi_i}^2(x)w(x)\, dx \geq 0. \end{aligned} \tag{3}$$

But the COEFFICIENT of the generalized FOURIER SERIES is given by

$$a_m \equiv \int f(x)\phi_m(x)w(x)\, dx, \tag{4}$$

so

$$\int f^2(x)w(x)\, dx - 2\sum_i {a_i}^2 + \sum_i {a_i}^2 \geq 0 \tag{5}$$

$$\int f^2(x)w(x)\, dx \geq \sum_i {a_i}^2. \tag{6}$$

Equation (6) is an inequality if the functions ϕ_i are not COMPLETE. If they are COMPLETE, then the inequality (2) becomes an equality, so (6) becomes an equality and is known as PARSEVAL'S THEOREM. If $f(x)$ has a simple FOURIER SERIES expansion with COEFFICIENTS a_0, a_1, $\ldots$, a_n and b_1, $\ldots$, b_n, then

$$\tfrac{1}{2}{a_0}^2 + \sum_{k=1}^{\infty}({a_k}^2 + {b_k}^2) \leq \frac{1}{\pi}\int_{-\pi}^{\pi} [f(x)]^2\, dx. \tag{7}$$

The inequality can also be derived from SCHWARZ'S INEQUALITY

$$|\langle f|g\rangle|^2 \leq \langle f|f\rangle \langle g|g\rangle \tag{8}$$

by expanding g in a superposition of EIGENFUNCTIONS of f, $g = \sum_i a_i f_i$. Then

$$\langle f|g\rangle = \sum_i a_i \langle f|f_i\rangle \leq \sum_i a_i \tag{9}$$

$$\begin{aligned} |\langle f|g\rangle|^2 \leq \left|\sum_i a_i\right|^2 = \left(\sum_i a_i\right)\left(\sum_i {a_i}^*\right) \\ = \sum_i a_i {a_i}^* \leq \langle f|f\rangle \langle g|g\rangle. \end{aligned} \tag{10}$$

If y is normalized, then $\langle g|g\rangle = 1$ and

$$\langle f|f\rangle \geq \sum_i a_i a_i^*. \tag{11}$$

see also SCHWARZ'S INEQUALITY, TRIANGLE INEQUALITY

References

Arfken, G. *Mathematical Methods for Physicists, 3rd ed.* Orlando, FL: Academic Press, pp. 526–527, 1985.

Gradshteyn, I. S. and Ryzhik, I. M. *Tables of Integrals, Series, and Products, 5th ed.* San Diego, CA: Academic Press, p. 1102, 1980.

Bessel's Interpolation Formula

see BESSEL'S FINITE DIFFERENCE FORMULA

Bessel Polynomial

see BESSEL FUNCTION

Bessel's Second Integral

see POISSON INTEGRAL

Bessel's Statistical Formula

$$t = \frac{\bar{w} - \omega}{\sigma_w/\sqrt{N}} = \frac{\bar{w} - \omega}{\sqrt{\frac{\sum_{i=1}^n (w_i - \bar{w})^2}{N(N-1)}}}, \tag{1}$$

where

$$\bar{w} \equiv \hat{x}_1 - \bar{x}_2 \tag{2}$$

$$\omega \equiv \mu_{(1)} - \mu_{(2)} \tag{3}$$

$$N \equiv N_1 + N_2. \tag{4}$$

Beta

A financial measure of a fund's sensitivity to market movements which measures the relationship between a fund's excess return over Treasury Bills and the excess return of a benchmark index (which, by definition, has $\beta = 1$). A fund with a beta of β has performed $r = (\beta - 1) \times 100\%$ better (or $|r|$ worse if $r < 0$) than its benchmark index (after deducting the T-bill rate) in up markets and $|r|$ worse (or $|r|$ better if $r < 0$) in down markets.

see also ALPHA, SHARPE RATIO

Beta Distribution

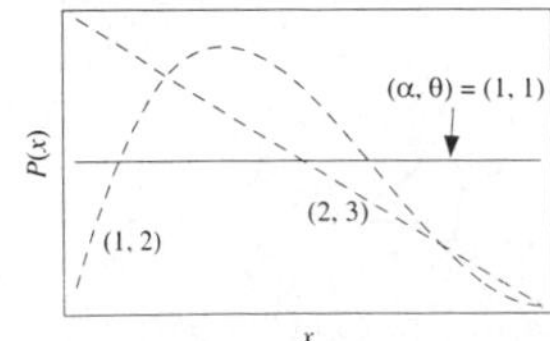

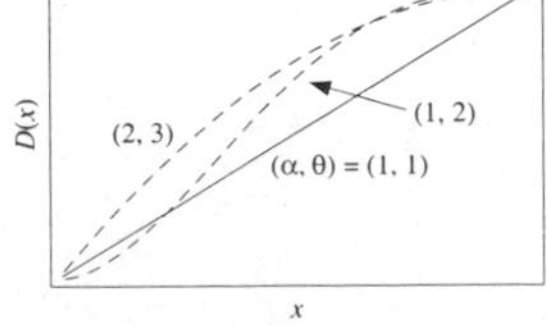

A general type of statistical DISTRIBUTION which is related to the GAMMA DISTRIBUTION. Beta distributions have two free parameters, which are labeled according to one of two notational conventions. The usual definition calls these α and β, and the other uses $\beta' \equiv \beta - 1$ and $\alpha' \equiv \alpha - 1$ (Beyer 1987, p. 534). The above plots are for $(\alpha, \beta) = (1, 1)$ [solid], (1, 2) [dotted], and (2, 3) [dashed]. The probability function $P(x)$ and DISTRIBUTION FUNCTION $D(x)$ are given by

$$P(x) = \frac{(1-x)^{\beta-1}x^{\alpha-1}}{B(\alpha,\beta)} = \frac{\Gamma(\alpha+\beta)}{\Gamma(\alpha)\Gamma(\beta)}(1-x)^{\beta-1}x^{\alpha-1} \tag{1}$$

$$D(x) = I(x; a, b), \tag{2}$$

where $B(a, b)$ is the BETA FUNCTION, $I(x; a, b)$ is the REGULARIZED BETA FUNCTION, and $0 < x < 1$ where α, $\beta > 0$. The distribution is normalized since

$$\int_0^1 P(x)\,dx = \frac{\Gamma(\alpha+\beta)}{\Gamma(\alpha)\Gamma(\beta)} \int_0^1 x^{\alpha-1}(1-x)^{\beta-1}\,dx = \frac{\Gamma(\alpha+\beta)}{\Gamma(\alpha)\Gamma(\beta)} B(\alpha,\beta) = 1. \tag{3}$$

The CHARACTERISTIC FUNCTION is

$$\phi(t) = {}_1F_1(a, a+b, it). \tag{4}$$

The MOMENTS are given by

$$M_r = \int_0^1 (x-\mu)^r\,dx = \frac{\Gamma(\alpha+\beta)\Gamma(\alpha+r)}{\Gamma(\alpha+\beta+r)\Gamma(\alpha)}. \tag{5}$$

The MEAN is

$$\mu = \frac{\Gamma(\alpha+\beta)}{\Gamma(\alpha)\Gamma(\beta)} \int_0^1 x^{\alpha-1}(1-x)^{\beta-1}x\,dx = \frac{\Gamma(\alpha+\beta)}{\Gamma(\alpha)\Gamma(\beta)} B(\alpha+1,\beta) = \frac{\Gamma(\alpha+\beta)}{\Gamma(\alpha)\Gamma(\beta)} \frac{\Gamma(\alpha+1)\Gamma(\beta)}{\Gamma(\alpha+\beta+1)} = \frac{\alpha}{\alpha+\beta}, \tag{6}$$

and the VARIANCE, SKEWNESS, and KURTOSIS are

$$\sigma^2 = \frac{\alpha\beta}{(\alpha+\beta)^2(\alpha+\beta+1)} \tag{7}$$

$$\gamma_1 = \frac{2(\sqrt{\beta}-\sqrt{\alpha})(\sqrt{\alpha}+\sqrt{\beta})\sqrt{1+\alpha+\beta}}{\sqrt{\alpha\beta}\,(\alpha+\beta+2)} \tag{8}$$

$$\gamma_2 = \frac{6(\alpha^2+\alpha^3-4\alpha\beta-2\alpha^2\beta+\beta^2-2\alpha\beta^2+\beta^3)}{\alpha\beta(\alpha+\beta+2)(\alpha+\beta+3)}. \tag{9}$$

The MODE of a variate distributed as $\beta(\alpha, \beta)$ is

$$\hat{x} = \frac{\alpha - 1}{\alpha+\beta-2}. \tag{10}$$

In "normal" form, the distribution is written

$$f(x) = \frac{\Gamma(\alpha+\beta)}{\Gamma(\alpha)\Gamma(\beta)} x^{\alpha-1}(1-x)^{\beta-1} \tag{11}$$

and the MEAN, VARIANCE, SKEWNESS, and KURTOSIS are

$$\mu = \frac{\alpha}{\alpha+\beta} \tag{12}$$

$$\sigma^2 = \frac{\alpha\beta}{(\alpha+\beta)^2(1+\alpha+\beta)} \tag{13}$$

$$\gamma_1 = \frac{2(\sqrt{\alpha}-\sqrt{\beta})(\sqrt{\alpha}+\sqrt{\beta})\sqrt{1+\alpha+\beta}}{\sqrt{\alpha\beta}(\alpha+\beta+2)} \tag{14}$$

$$\gamma_2 = \frac{3(1+\alpha+\beta)(2\alpha^2-2\alpha\beta+\alpha^2\beta+2\beta^2+\alpha\beta^2)}{\alpha\beta(\alpha+\beta+2)(\alpha+\beta+3)}. \tag{15}$$

see also GAMMA DISTRIBUTION

References

Abramowitz, M. and Stegun, C. A. (Eds.). *Handbook of Mathematical Functions with Formulas, Graphs, and Mathematical Tables, 9th printing.* New York: Dover, pp. 944–945, 1972.

Beyer, W. H. *CRC Standard Mathematical Tables, 28th ed.* Boca Raton, FL: CRC Press, pp. 534–535, 1987.

Beta Function

The beta function is the name used by Legendre and Whittaker and Watson (1990) for the EULERIAN INTEGRAL OF THE SECOND KIND. To derive the integral representation of the beta function, write the product of two FACTORIALS as

$$m!n! = \int_0^\infty e^{-u}u^m\,du \int_0^\infty e^{-v}v^n\,dv. \tag{1}$$

Now, let $u \equiv x^2$, $v \equiv y^2$, so

$$m!n! = 4\int^\infty e^{-x^2}x^{2m+1}\,dx \int_0^\infty e^{-y^2}y^{2n+1}\,dy$$

$$= 4\int_{-\infty}^\infty \int_{-\infty}^\infty e^{-(x^2+y^2)}x^{2m+1}y^{2m+1}\,dx\,dy. \tag{2}$$

Transforming to POLAR COORDINATES with $x = r\cos\theta$, $y = r\sin\theta$

$$m!n! = 4\int_0^{\pi/2}\int_0^\infty e^{-r^2}(r\cos\theta)^{2m+1}(r\sin\theta)^{2n+1} r\,dr\,d\theta$$

$$= 4\int_0^\infty e^{-r^2}r^{2m+2n+3}\,dr \int_0^{\pi/2}\cos^{2m+1}\theta\sin^{2n+1}\theta\,d\theta$$

$$= 2(m+n+1)!\int_0^{\pi/2}\cos^{2m+1}\theta\sin^{2n+1}\theta\,d\theta. \tag{3}$$

The beta function is then defined by

$$B(m+1,n+1) = B(n+1,m+1)$$

$$\equiv 2\int_0^{\pi/2}\cos^{2m+1}\theta\sin^{2n+1}\theta\,d\theta = \frac{m!n!}{(m+n+1)!}. \tag{4}$$

Rewriting the arguments,

$$B(p,q) = \frac{\Gamma(p)\Gamma(q)}{\Gamma(p+q)} = \frac{(p-1)!(q-1)!}{(p+q-1)!}. \tag{5}$$

The general trigonometric form is

$$\int_0^{\pi/2}\sin^n x\cos^m x\,dx = \tfrac{1}{2}B(n+\tfrac{1}{2},m+\tfrac{1}{2}). \tag{6}$$

Equation (6) can be transformed to an integral over POLYNOMIALS by letting $u \equiv \cos^2\theta$,

$$B(m+1,n+1) \equiv \frac{m!n!}{(m+n+1)!} = \int_0^1 u^m(1-u)^n\,du \tag{7}$$

$$B(m,n) \equiv \frac{\Gamma(m)\Gamma(n)}{\Gamma(m+n)} = \int_0^1 u^{m-1}(1-u)^{n-1}\,du. \tag{8}$$

To put it in a form which can be used to derive the LEGENDRE DUPLICATION FORMULA, let $x \equiv \sqrt{u}$, so $u = x^2$ and $du = 2x\,dx$, and

$$B(m,n) = \int_0^1 x^{2(m-1)}(1-x^2)^{n-1}(2x\,dx)$$

$$= 2\int_0^1 x^{2m-1}(1-x^2)^{n-1}\,dx. \tag{9}$$

To put it in a form which can be used to develop integral representations of the BESSEL FUNCTIONS and HYPERGEOMETRIC FUNCTION, let $u \equiv x/(1+x)$, so

$$B(m+1,n+1) = \int_0^\infty \frac{u^m\,du}{(1+u)^{m+n+2}}. \tag{10}$$

Various identities can be derived using the GAUSS MULTIPLICATION FORMULA

$$B(np,nq) = \frac{\Gamma(np)\Gamma(nq)}{\Gamma[n(p+q)]}$$

$$= n^{-nq}\frac{B(p,q)B(p+\frac{1}{n},q)\cdots B(p+\frac{n-1}{n},q)}{B(q,q)B(2q,q)\cdots B([n-1]q,q)}. \tag{11}$$

Additional identities include

$$B(p,q+1) = \frac{\Gamma(p)\Gamma(q+1)}{\Gamma(p+q+1)} = \frac{q}{p}\frac{\Gamma(p+1)\Gamma(q)}{\Gamma([p+1]q)}$$

$$= \frac{q}{p}B(p+1,q) \tag{12}$$

$$B(p,q) = B(p+1,q) + B(p,q+1) \tag{13}$$

$$B(p,q+1) = \frac{q}{p+q} B(p,q). \tag{14}$$

If n is a POSITIVE INTEGER, then

$$B(p,n+1) = \frac{1 \cdot 2 \cdots n}{p(p+1)\cdots(p+n)} \tag{15}$$

$$B(p,p)B(p+\tfrac{1}{2}, p+\tfrac{1}{2}) = \frac{\pi}{2^{4p-1}p} \tag{16}$$

$$B(p+q)B(p+q,r) = B(q,r)B(q+r,p). \tag{17}$$

A generalization of the beta function is the incomplete beta function

$$B(t;x,y) \equiv \int_0^t u^{x-1}(1-u)^{y-1}\,du$$
$$= t^p\left[\frac{1}{x} + \frac{1-y}{x+1}t + \ldots + \frac{(1-y)\cdots(n-y)}{n!(x+n)}t^n + \ldots\right]. \tag{18}$$

see also CENTRAL BETA FUNCTION, DIRICHLET INTEGRALS, GAMMA FUNCTION, REGULARIZED BETA FUNCTION

References

Abramowitz, M. and Stegun, C. A. (Eds.). "Beta Function" and "Incomplete Beta Function." §6.2 and 6.6 in *Handbook of Mathematical Functions with Formulas, Graphs, and Mathematical Tables, 9th printing.* New York: Dover, pp. 258 and 263, 1972.

Arfken, G. "The Beta Function." §10.4 in *Mathematical Methods for Physicists, 3rd ed.* Orlando, FL: Academic Press, pp. 560–565, 1985.

Morse, P. M. and Feshbach, H. *Methods of Theoretical Physics, Part I.* New York: McGraw-Hill, p. 425, 1953.

Press, W. H.; Flannery, B. P.; Teukolsky, S. A.; and Vetterling, W. T. "Gamma Function, Beta Function, Factorials, Binomial Coefficients" and "Incomplete Beta Function, Student's Distribution, F-Distribution, Cumulative Binomial Distribution." §6.1 and 6.2 in *Numerical Recipes in FORTRAN: The Art of Scientific Computing, 2nd ed.* Cambridge, England: Cambridge University Press, pp. 206–209 and 219–223, 1992.

Spanier, J. and Oldham, K. B. "The Incomplete Beta Function $B(\nu;\mu;x)$." Ch. 58 in *An Atlas of Functions.* Washington, DC: Hemisphere, pp. 573–580, 1987.

Whittaker, E. T. and Watson, G. N. *A Course of Modern Analysis, 4th ed.* Cambridge, England: Cambridge University Press, 1990.

Beta Function (Exponential)

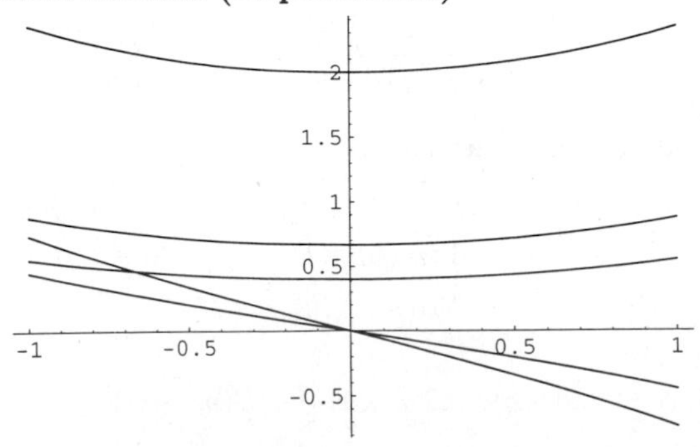

Another "BETA FUNCTION" defined in terms of an integral is the "exponential" beta function, given by

$$\beta_n(z) \equiv \int_{-1}^{1} t^n e^{-zt}\,dt \tag{1}$$
$$= n!z^{-(n+1)}\left[e^z \sum_{k=0}^{n} \frac{(-1)^k z^k}{k!} - e^{-z} \sum_{k=0}^{n} \frac{z^k}{k!}\right]. \tag{2}$$

The exponential beta function satisfies the RECURRENCE RELATION

$$z\beta_n(z) = (-1)^n e^z - e^{-z} + n\beta_{n-1}(z). \tag{3}$$

The first few integral values are

$$\beta_0(z) = \frac{2\sinh z}{z} \tag{4}$$

$$\beta_1(z) = \frac{2(\sinh z - z\cosh z)}{z^2} \tag{5}$$

$$\beta_2(a) = \frac{2(2+z^2)\sinh z - 4z\cosh z}{z^3}. \tag{6}$$

see also ALPHA FUNCTION

Beta Prime Distribution

A distribution with probability function

$$P(x) = \frac{x^{\alpha-1}(1+x)^{-\alpha-\beta}}{B(\alpha,\beta)},$$

where B is a BETA FUNCTION. The MODE of a variate distributed as $\beta'(\alpha,\beta)$ is

$$\hat{x} = \frac{\alpha-1}{\beta+1}.$$

If x is a $\beta'(\alpha,\beta)$ variate, then $1/x$ is a $\beta'(\beta,\alpha)$ variate. If x is a $\beta(\alpha,\beta)$ variate, then $(1-x)/x$ and $x/(1-x)$ are $\beta'(\beta,\alpha)$ and $\beta'(\alpha,\beta)$ variates. If x and y are $\gamma(\alpha_1)$ and $\gamma(\alpha_2)$ variates, then x/y is a $\beta'(\alpha_1,\alpha_2)$ variate. If $x^2/2$ and $y^2/2$ are $\gamma(1/2)$ variates, then $z^2 \equiv (x/y)^2$ is a $\beta'(1/2,1/2)$ variate.

Bethe Lattice

see CAYLEY TREE

Betrothed Numbers

see QUASIAMICABLE PAIR

Betti Group

The free part of the HOMOLOGY GROUP with a domain of COEFFICIENTS in the GROUP of INTEGERS (if this HOMOLOGY GROUP is finitely generated).

References

Hazewinkel, M. (Managing Ed.). *Encyclopaedia of Mathematics: An Updated and Annotated Translation of the Soviet "Mathematical Encyclopaedia."* Dordrecht, Netherlands: Reidel, p. 380, 1988.

Betti Number

Betti numbers are topological objects which were proved to be invariants by Poincaré, and used by him to extend the POLYHEDRAL FORMULA to higher dimensional spaces. The nth Betti number is the rank of the nth HOMOLOGY GROUP. Let p_r be the RANK of the HOMOLOGY GROUP H_r of a TOPOLOGICAL SPACE K. For a closed, orientable surface of GENUS g, the Betti numbers are $p_0 = 1$, $p_1 = 2g$, and $p_2 = 1$. For a nonorientable surface with k CROSS-CAPS, the Betti numbers are $p_0 = 1$, $p_1 = k - 1$, and $p_2 = 0$.

see also EULER CHARACTERISTIC, POINCARÉ DUALITY

Bézier Curve

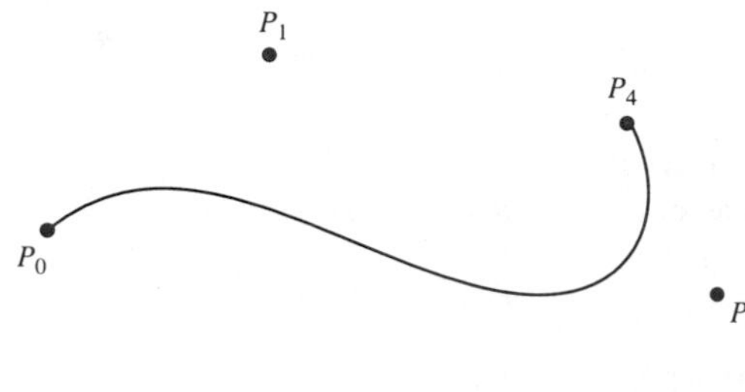

Given a set of n control points, the corresponding Bézier curve (or BERNSTEIN-BÉZIER CURVE) is given by

$$\mathbf{C}(t) = \sum_{i=0}^{n} \mathbf{P}_i B_{i,n}(t),$$

where $B_{i,n}(t)$ is a BERNSTEIN POLYNOMIAL and $t \in [0, 1]$.

A "rational" Bézier curve is defined by

$$\mathbf{C}(t) = \frac{\sum_{i=0}^{n} B_{i,p}(t) w_i \mathbf{P}_i}{\sum_{i=0}^{n} B_{i,p}(t) w_i},$$

where p is the order, $B_{i,p}$ are the BERNSTEIN POLYNOMIALS, $\mathbf{P}_i$ are control points, and the weight w_i of $\mathbf{P}_i$ is the last ordinate of the homogeneous point $\mathbf{P}_i^w$. These curves are closed under perspective transformations, and can represent CONIC SECTIONS exactly.

The Bézier curve always passes through the first and last control points and lies within the CONVEX HULL of the control points. The curve is tangent to $\mathbf{P}_1 - \mathbf{P}_0$ and $\mathbf{P}_n - \mathbf{P}_{n-1}$ at the endpoints. The "variation diminishing property" of these curves is that no line can have more intersections with a Bézier curve than with the curve obtained by joining consecutive points with straight line segments. A desirable property of these curves is that the curve can be translated and rotated by performing these operations on the control points.

Undesirable properties of Bézier curves are their numerical instability for large numbers of control points, and the fact that moving a single control point changes the global shape of the curve. The former is sometimes avoided by smoothly patching together low-order Bézier curves. A generalization of the Bézier curve is the B-SPLINE.

see also B-SPLINE, NURBS CURVE

Bézier Spline

see BÉZIER CURVE, SPLINE

Bezout Numbers

Integers (λ, μ) for a and b such that

$$\lambda a + \mu b = \text{GCD}(a, b).$$

For INTEGERS $a_1, \ldots, a_n$, the Bezout numbers are a set of numbers $k_1, \ldots, k_n$ such that

$$k_1 a_1 + k_2 a_2 + \ldots + k_n a_n = d,$$

where d is the GREATEST COMMON DIVISOR of $a_1, \ldots, a_n$.

see also GREATEST COMMON DIVISOR

Bezout's Theorem

In general, two algebraic curves of degrees m and n intersect in $m \cdot n$ points and cannot meet in more than $m \cdot n$ points unless they have a component in common (i.e., the equations defining them have a common factor). This can also be stated: if P and Q are two POLYNOMIALS with no roots in common, then there exist two other POLYNOMIALS A and B such that $AP + BQ = 1$. Similarly, given N POLYNOMIAL equations of degrees n_1, n_2, $\ldots n_N$ in N variables, there are in general $n_1 n_2 \cdots n_N$ common solutions.

see also POLYNOMIAL

References

Coolidge, J. L. *A Treatise on Algebraic Plane Curves.* New York: Dover, p. 10, 1959.

Bhargava's Theorem

Let the nth composition of a function $f(x)$ be denoted $f^{(n)}(x)$, such that $f^{(0)}(x) = x$ and $f^{(1)}(x) = f(x)$. Denote $f \circ g(x) = f(g(x))$, and define

$$\sum F(a, b, c) = F(a, b, c) + F(b, c, a) + F(c, a, b). \quad (1)$$

Let

$$u \equiv (a, b, c) \quad (2)$$

$$|u| \equiv a + b + c \quad (3)$$

$$||u|| \equiv a^4 + b^4 + c^4, \quad (4)$$

and

$$f(u) = (f_1(u), f_2(u), f_3(u)) \quad (5)$$
$$= (a(b-c), b(c-a), c(a-b)) \quad (6)$$
$$g(u) = (g_1(u), g_2(u), g_3(u))$$
$$= \left(\sum a^2 b, \sum ab^2, 3abc\right). \quad (7)$$

Then if $|u| = 0$,

$$||f^{(m)} \circ g^{(n)}(u)|| = 2(ab+bc+ca)^{2^{m+1}3^n}$$
$$= ||g^{(n)} \circ f^{(m)}(u)||, \quad (8)$$

where $m, n \in \{0, 1, \ldots\}$ and composition is done in terms of components.

see also DIOPHANTINE EQUATION—QUARTIC, FORD'S THEOREM

References

Berndt, B. C. *Ramanujan's Notebooks, Part IV.* New York: Springer-Verlag, pp. 97–100, 1994.
Bhargava, S. "On a Family of Ramanujan's Formulas for Sums of Fourth Powers." *Ganita* **43**, 63–67, 1992.

Bhaskara-Brouckner Algorithm

see SQUARE ROOT

Bi-Connected Component

A maximal SUBGRAPH of an undirected graph such that any two edges in the SUBGRAPH lie on a common simple cycle.

see also STRONGLY CONNECTED COMPONENT

Bianchi Identities

The RIEMANN TENSOR is defined by

$$R_{\lambda\mu\nu\kappa;\eta} = \frac{1}{2}\frac{\partial}{\partial x^\eta}\left(\frac{\partial^2 g_{\lambda\nu}}{\partial x^\kappa \partial x^\mu} - \frac{\partial^2 g_{\mu\nu}}{\partial x^\kappa \partial x^\lambda} - \frac{\partial^2 g_{\lambda\kappa}}{\partial x^\mu \partial x^\nu} + \frac{\partial^2 g_{\mu\kappa}}{\partial x^\nu \partial x^\lambda}\right).$$

Permuting ν, κ, and η (Weinberg 1972, pp. 146–147) gives the Bianchi identities

$$R_{\lambda\mu\nu\kappa;\eta} + R_{\lambda\mu\eta\nu;\kappa} + R_{\lambda\mu\kappa\eta;\nu} = 0.$$

see also BIANCHI IDENTITIES (CONTRACTED), RIEMANN TENSOR

References

Weinberg, S. *Gravitation and Cosmology: Principles and Applications of the General Theory of Relativity.* New York: Wiley, 1972.

Bianchi Identities (Contracted)

CONTRACTING λ with ν in the BIANCHI IDENTITIES

$$R_{\lambda\mu\nu\kappa;\eta} + R_{\lambda\mu\eta\nu;\kappa} + R_{\lambda\mu\kappa\eta;\nu} = 0 \quad (1)$$

gives

$$R_{\mu\kappa;\eta} - R_{\mu\eta;\kappa} + R^\nu{}_{\mu\kappa\eta;\nu} = 0. \quad (2)$$

CONTRACTING again,

$$R_{;\eta} - R^\mu{}_{\eta;\mu} - R^\nu{}_{\eta;\nu} = 0, \quad (3)$$

or

$$(R^\mu{}_\eta - \tfrac{1}{2}\delta^\mu{}_\eta R)_{;\mu} = 0, \quad (4)$$

or

$$(R^{\mu\nu} - \tfrac{1}{2}g^{\mu\nu} R)_{;\mu} = 0. \quad (5)$$

Bias (Estimator)

The bias of an ESTIMATOR $\tilde{\theta}$ is defined as

$$B(\tilde{\theta}) \equiv \langle\tilde{\theta}\rangle - \theta.$$

It is therefore true that

$$\tilde{\theta} - \theta = (\tilde{\theta} - \langle\tilde{\theta}\rangle) + (\langle\tilde{\theta}\rangle - \theta) = (\tilde{\theta} - \langle\tilde{\theta}\rangle) + B(\tilde{\theta}).$$

An ESTIMATOR for which $B = 0$ is said to be UNBIASED.

see also ESTIMATOR, UNBIASED

Bias (Series)

The bias of a SERIES is defined as

$$Q[a_i, a_{i+1}, a_{i+2}] \equiv \frac{a_i a_{i+2} - a_{i+1}{}^2}{a_i a_{i+1} a_{i+2}}.$$

A SERIES is GEOMETRIC IFF $Q = 0$. A SERIES is ARTISTIC IFF the bias is constant.

see also ARTISTIC SERIES, GEOMETRIC SERIES

References

Duffin, R. J. "On Seeing Progressions of Constant Cross Ratio." *Amer. Math. Monthly* **100**, 38–47, 1993.

Biased

An ESTIMATOR which exhibits BIAS.

Biaugmented Pentagonal Prism

see JOHNSON SOLID

Biaugmented Triangular Prism

see JOHNSON SOLID

Biaugmented Truncated Cube

see JOHNSON SOLID

BIBD

see BLOCK DESIGN

Bicentric Polygon

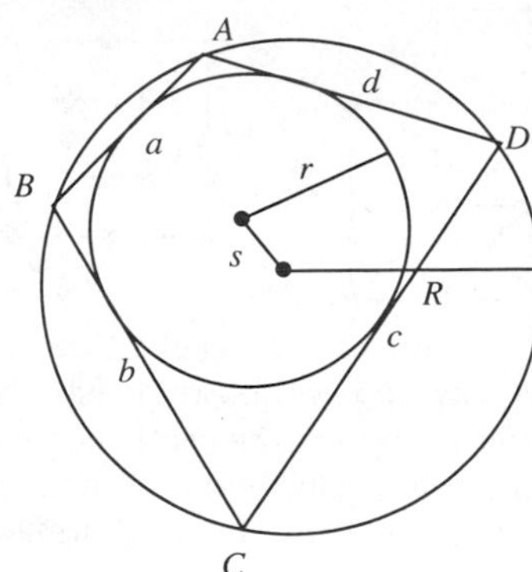

A POLYGON which has both a CIRCUMCIRCLE and an INCIRCLE, both of which touch all VERTICES. All TRIANGLES are bicentric with

$$R^2 - s^2 = 2Rr, \quad (1)$$

where R is the CIRCUMRADIUS, r is the INRADIUS, and s is the separation of centers. In 1798, N. Fuss characterized bicentric POLYGONS of $n = 4$, 5, 6, 7, and 8 sides. For bicentric QUADRILATERALS (FUSS'S PROBLEM), the CIRCLES satisfy

$$2r^2(R^2 - s^2) = (R^2 - s^2)^2 - 4r^2s^2 \quad (2)$$

(Dörrie 1965) and

$$r = \frac{\sqrt{abcd}}{s} \quad (3)$$

$$R = \frac{1}{4}\sqrt{\frac{(ac+bd)(ad+bc)(ab+cd)}{abcd}} \quad (4)$$

(Beyer 1987). In addition,

$$\frac{1}{(R-s)^2} + \frac{1}{(R+s)^2} = \frac{1}{r^2} \quad (5)$$

and

$$a + c = b + d. \quad (6)$$

The AREA of a bicentric quadrilateral is

$$A = \sqrt{abcd}. \quad (7)$$

If the circles permit successive tangents around the INCIRCLE which close the POLYGON for one starting point on the CIRCUMCIRCLE, then they do so for all points on the CIRCUMCIRCLE.

see also PONCELET'S CLOSURE THEOREM

References

Beyer, W. H. (Ed.) *CRC Standard Mathematical Tables, 28th ed.* Boca Raton, FL: CRC Press, p. 124, 1987.

Dörrie, H. "Fuss' Problem of the Chord-Tangent Quadrilateral." §39 in *100 Great Problems of Elementary Mathematics: Their History and Solutions.* New York: Dover, pp. 188–193, 1965.

Bicentric Quadrilateral

A 4-sided BICENTRIC POLYGON, also called a CYCLIC-INSCRIPTABLE QUADRILATERAL.

References

Beyer, W. H. (Ed.) *CRC Standard Mathematical Tables, 28th ed.* Boca Raton, FL: CRC Press, p. 124, 1987.

Bichromatic Graph

A GRAPH with EDGES of two possible "colors," usually identified as red and blue. For a bichromatic graph with R red EDGES and B blue EDGES,

$$R + B \geq 2.$$

see also BLUE-EMPTY GRAPH, EXTREMAL COLORING, EXTREMAL GRAPH, MONOCHROMATIC FORCED TRIANGLE, RAMSEY NUMBER

Bicollared

A SUBSET $X \subset Y$ is said to be bicollared in Y if there exists an embedding $b : X \times [-1,1] \to Y$ such that $b(x,0) = x$ when $x \in X$. The MAP b or its image is then said to be the bicollar.

References

Rolfsen, D. *Knots and Links.* Wilmington, DE: Publish or Perish Press, pp. 34–35, 1976.

Bicorn

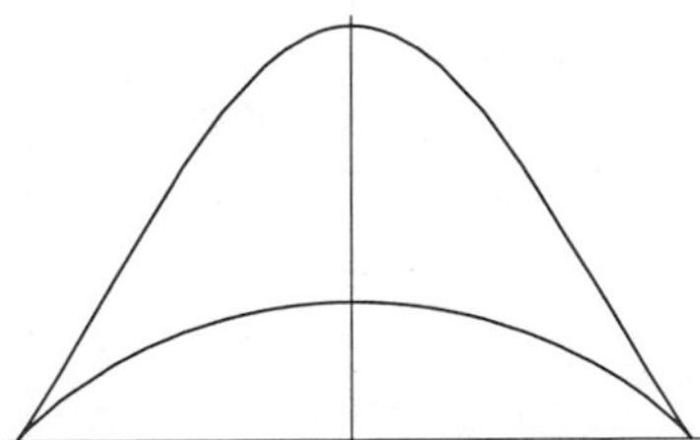

The bicorn is the name of a collection of QUARTIC CURVES studied by Sylvester in 1864 and Cayley in 1867 (MacTutor Archive). The bicorn is given by the parametric equations

$$x = a \sin t$$

$$y = \frac{a \cos^2 t (2 + \cos t)}{3 \sin^2 t}.$$

The graph is similar to that of the COCKED HAT CURVE.

References

Lawrence, J. D. *A Catalog of Special Plane Curves.* New York: Dover, pp. 147–149, 1972.

MacTutor History of Mathematics Archive. "Bicorn." `http://www-groups.dcs.st-and.ac.uk/~history/Curves/Bicorn.html`.

Bicubic Spline

A bicubic spline is a special case of bicubic interpolation which uses an interpolation function of the form

$$y(x_1,x_2)=\sum_{i=1}^{4}\sum_{j=1}^{4}c_{ij}t^{i-1}u^{j-1}$$

$$y_{x_1}(x_1,x_2)=\sum_{i=1}^{4}\sum_{j=1}^{4}(i-1)c_{ij}t^{i-2}u^{j-1}$$

$$y_{x_2}(x_1,x_2)=\sum_{i=1}^{4}\sum_{j=1}^{4}(j-1)c_{ij}t^{i-1}u^{j-2}$$

$$y_{x_1x_2}=\sum_{i=1}^{4}\sum_{j=1}^{4}(i-1)(j-1)c_{ij}t^{i-2}u^{j-2},$$

where c_{ij} are constants and u and t are parameters ranging from 0 to 1. For a bicubic spline, however, the partial derivatives at the grid points are determined globally by 1-D SPLINES.

see also B-SPLINE, SPLINE

References

Press, W. H.; Flannery, B. P.; Teukolsky, S. A.; and Vetterling, W. T. *Numerical Recipes in FORTRAN: The Art of Scientific Computing, 2nd ed.* Cambridge, England: Cambridge University Press, pp. 118–122, 1992.

Bicupola

Two adjoined CUPOLAS.

see also CUPOLA, ELONGATED GYROBICUPOLA, ELONGATED ORTHOBICUPOLA, GYROBICUPOLA, ORTHOBICUPOLA

Bicuspid Curve

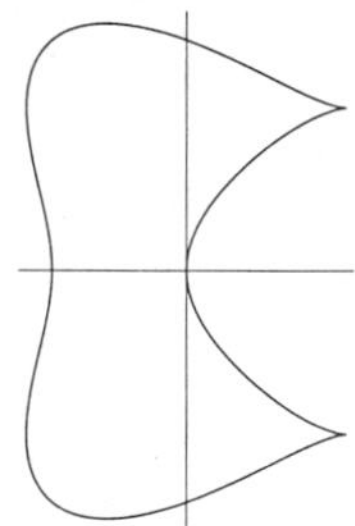

The PLANE CURVE given by the Cartesian equation

$$(x^2-a^2)(x-a)^2+(y^2-a^2)^2=0.$$

Bicylinder

see STEINMETZ SOLID

Bidiakis Cube

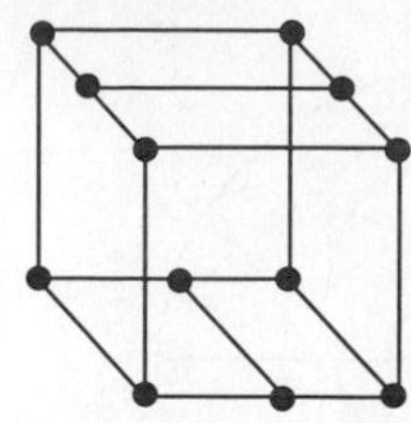

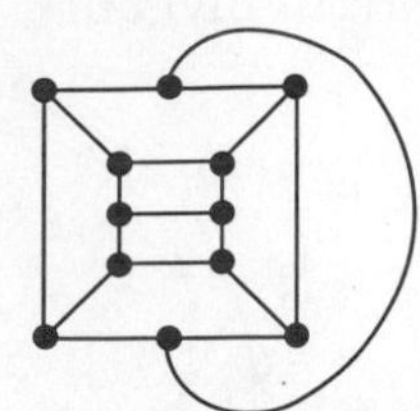

The 12-VERTEX graph consisting of a CUBE in which two opposite faces (say, top and bottom) have edges drawn across them which connect the centers of opposite sides of the faces in such a way that the orientation of the edges added on top and bottom are PERPENDICULAR to each other.

see also BISLIT CUBE, CUBE, CUBICAL GRAPH

Bieberbach Conjecture

The nth COEFFICIENT in the POWER series of a UNIVALENT FUNCTION should be no greater than n. In other words, if

$$f(z)=a_0+a_1z+a_2z^2+\ldots+a_nz^n+\ldots$$

is a conformal transformation of a unit disk on any domain, then$|a_n|\leq n|a_1|$. In more technical terms, "geometric extremality implies metric extremality." The conjecture had been proven for the first six terms (the cases $n=2$, 3, and 4 were done by Bieberbach, Lowner, and Shiffer and Garbedjian, respectively), was known to be false for only a finite number of indices (Hayman 1954), and true for a convex or symmetric domain (Le Lionnais 1983). The general case was proved by Louis de Branges (1985). De Branges proved the MILIN CONJECTURE, which established the ROBERTSON CONJECTURE, which in turn established the Bieberbach conjecture (Stewart 1996).

References

de Branges, L. "A Proof of the Bieberbach Conjecture." *Acta Math.* **154**, 137–152, 1985.
Hayman, W. K. *Multivalent Functions, 2nd ed.* Cambridge, England: Cambridge University Press, 1994.
Hayman, W. K. and Stewart, F. M. "Real Inequalities with Applications to Function Theory." *Proc. Cambridge Phil. Soc.* **50**, 250–260, 1954.
Kazarinoff, N. D. "Special Functions and the Bieberbach Conjecture." *Amer. Math. Monthly* **95**, 689–696, 1988.
Korevaar, J. "Ludwig Bieberbach's Conjecture and its Proof." *Amer. Math. Monthly* **93**, 505–513, 1986.
Le Lionnais, F. *Les nombres remarquables.* Paris: Hermann, p. 53, 1983.
Pederson, R. N. "A Proof of the Bieberbach Conjecture for the Sixth Coefficient." *Arch. Rational Mech. Anal.* **31**, 331–351, 1968/1969.
Pederson, R. and Schiffer, M. "A Proof of the Bieberbach Conjecture for the Fifth Coefficient." *Arch. Rational Mech. Anal.* **45**, 161–193, 1972.
Stewart, I. "The Bieberbach Conjecture." In *From Here to Infinity: A Guide to Today's Mathematics.* Oxford, England: Oxford University Press, pp. 164–166, 1996.

Bienaymé-Chebyshev Inequality

see CHEBYSHEV INEQUALITY

Bifoliate

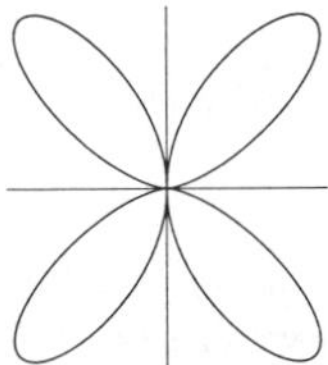

The PLANE CURVE given by the Cartesian equation

$$x^4 + y^4 = 2axy^2.$$

References

Cundy, H. and Rollett, A. *Mathematical Models, 3rd ed.* Stradbroke, England: Tarquin Pub., p. 72, 1989.

Bifolium

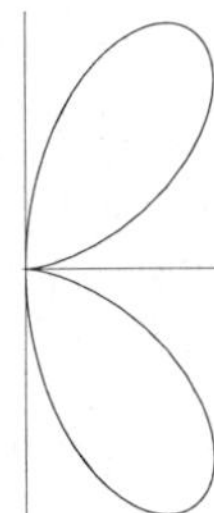

A FOLIUM with $b = 0$. The bifolium is the PEDAL CURVE of the DELTOID, where the PEDAL POINT is the MIDPOINT of one of the three curved sides. The Cartesian equation is

$$(x^2 + y^2)^2 = 4axy^2$$

and the POLAR equation is

$$r = 4a \sin^2 \theta \cos \theta.$$

see also FOLIUM, QUADRIFOLIUM, TRIFOLIUM

References

Lawrence, J. D. *A Catalog of Special Plane Curves.* New York: Dover, pp. 152–153, 1972.

MacTutor History of Mathematics Archive. "Double Folium." `http://www-groups.dcs.st-and.ac.uk/~history/Curves/Double.html`.

Bifurcation

A period doubling, quadrupling, etc., that accompanies the onset of CHAOS. It represents the sudden appearance of a qualitatively different solution for a nonlinear system as some parameter is varied. Bifurcations come in four basic varieties: FLIP BIFURCATION, FOLD BIFURCATION, PITCHFORK BIFURCATION, and TRANSCRITICAL BIFURCATION (Rasband 1990).

see also CODIMENSION, FEIGENBAUM CONSTANT, FEIGENBAUM FUNCTION, FLIP BIFURCATION, HOPF BIFURCATION, LOGISTIC MAP, PERIOD DOUBLING, PITCHFORK BIFURCATION, TANGENT BIFURCATION, TRANSCRITICAL BIFURCATION

References

Guckenheimer, J. and Holmes, P. "Local Bifurcations." Ch. 3 in *Nonlinear Oscillations, Dynamical Systems, and Bifurcations of Vector Fields, 2nd pr., rev. corr.* New York: Springer-Verlag, pp. 117–165, 1983.

Lichtenberg, A. J. and Lieberman, M. A. "Bifurcation Phenomena and Transition to Chaos in Dissipative Systems." Ch. 7 in *Regular and Chaotic Dynamics, 2nd ed.* New York: Springer-Verlag, pp. 457–569, 1992.

Rasband, S. N. "Asymptotic Sets and Bifurcations." §2.4 in *Chaotic Dynamics of Nonlinear Systems.* New York: Wiley, pp. 25–31, 1990.

Wiggins, S. "Local Bifurcations." Ch. 3 in *Introduction to Applied Nonlinear Dynamical Systems and Chaos.* New York: Springer-Verlag, pp. 253–419, 1990.

Bifurcation Theory

The study of the nature and properties of BIFURCATIONS.

see also CHAOS, DYNAMICAL SYSTEM

Bigraph

see BIPARTITE GRAPH

Bigyrate Diminished Rhombicosidodecahedron

see JOHNSON SOLID

Biharmonic Equation

The differential equation obtained by applying the BIHARMONIC OPERATOR and setting to zero.

$$\nabla^4 \phi = 0. \tag{1}$$

In CARTESIAN COORDINATES, the biharmonic equation is

$$\begin{aligned} \nabla^4 \phi &= \nabla^2(\nabla^2)\phi \\ &= \left(\frac{\partial^2}{\partial x^2} + \frac{\partial^2}{\partial y^2} + \frac{\partial^2}{\partial z^2}\right)\left(\frac{\partial^2}{\partial x^2} + \frac{\partial^2}{\partial y^2} + \frac{\partial^2}{\partial z^2}\right)\phi \\ &= \frac{\partial^4 \phi}{\partial x^4} + \frac{\partial^4 \phi}{\partial y^4} + \frac{\partial^4 \phi}{\partial z^4} + 2\frac{\partial^4 \phi}{\partial x^2 \partial y^2} \\ &\quad + 2\frac{\partial^4 \phi}{\partial y^2 \partial z^2} + 2\frac{\partial^4 \phi}{\partial x^2 \partial z^2} = 0. \end{aligned} \tag{2}$$

In POLAR COORDINATES (Kaplan 1984, p. 148)

$$\begin{aligned} \nabla^4 \phi = \phi_{rrrr} &+ \frac{2}{r^2}\phi_{rr\theta\theta} + \frac{1}{r^4}\phi_{\theta\theta\theta\theta} + \frac{2}{r}\phi_{rrr} \\ &- \frac{2}{r^3}\phi_{r\theta\theta} - \frac{1}{r^2}\phi_{rr} + \frac{4}{r^4}\phi_{\theta\theta} + \frac{1}{r^3}\phi_r = 0. \end{aligned} \tag{3}$$

For a radial function $\phi(r)$, the biharmonic equation becomes

$$\begin{aligned}\nabla^4\phi &= \frac{1}{r}\frac{d}{dr}\left\{r\frac{d}{dr}\left[\frac{1}{r}\frac{d}{dr}\left(r\frac{d\phi}{dr}\right)\right]\right\}\\ &= \phi_{rrrr}+\frac{2}{r}\phi_{rrr}-\frac{1}{r^2}\phi_{rr}+\frac{1}{r^3}\phi_r=0. \end{aligned} \tag{4}$$

Writing the inhomogeneous equation as

$$\nabla^4\phi = 64\beta, \tag{5}$$

we have

$$64\beta r\,dr = d\left\{r\frac{d}{dr}\left[\frac{1}{r}\frac{d}{dr}\left(r\frac{d\phi}{dr}\right)\right]\right\} \tag{6}$$

$$32\beta r^2 + C_1 = r\frac{d}{dr}\left[\frac{1}{r}\frac{d}{dr}\left(r\frac{d\phi}{dr}\right)\right] \tag{7}$$

$$\left(32\beta r+\frac{C_1}{r}\right)dr = d\left[\frac{1}{r}\frac{d}{dr}\left(r\frac{d\phi}{dr}\right)\right] \tag{8}$$

$$16\beta r^2 + C_1\ln r + C_2 = \frac{1}{r}\frac{d}{dr}\left(r\frac{d\phi}{dr}\right) \tag{9}$$

$$(16\beta r^3 + C_1 r\ln r + C_2 r)\,dr = d\left(r\frac{d\phi}{dr}\right). \tag{10}$$

Now use

$$\int r\ln r\,dr = \tfrac{1}{2}r^2\ln r - \tfrac{1}{4}r^2 \tag{11}$$

to obtain

$$4\beta r^4 + C_1(\tfrac{1}{2}r^2\ln r - \tfrac{1}{4}r^2) + \tfrac{1}{2}C_2 r^2 + C_3 = r\frac{d\phi}{dr} \tag{12}$$

$$\left(4\beta r^3 + C_1' r\ln r + C_2' r + \frac{C_3}{r}\right)dr = d\phi \tag{13}$$

$$\begin{aligned}\phi(r) &= \beta r^4 + C_1'\left(\tfrac{1}{2}r^2\ln r - \tfrac{1}{4}r^2\right)\\ &\quad + \tfrac{1}{2}C_2' r^2 + C_3\ln r + C_4\\ &= \beta r^4 + ar^2 + b + (cr^2+d)\ln\left(\frac{r}{R}\right). \end{aligned} \tag{14}$$

The homogeneous biharmonic equation can be separated and solved in 2-D BIPOLAR COORDINATES.

References
Kaplan, W. *Advanced Calculus, 4th ed.* Reading, MA: Addison-Wesley, 1991.

Biharmonic Operator

Also known as the BILAPLACIAN.

$$\nabla^4 \equiv (\nabla^2)^2.$$

In n-D space,

$$\nabla^4\left(\frac{1}{r}\right) = \frac{3(15-8n+n^2)}{r^5}.$$

see also BIHARMONIC EQUATION

Bijection

A transformation which is ONE-TO-ONE and ONTO.

see also ONE-TO-ONE, ONTO, PERMUTATION

Bilaplacian

see BIHARMONIC OPERATOR

Bilinear

A function of two variables is bilinear if it is linear with respect to each of its variables. The simplest example is $f(x,y)=xy$.

Bilinear Basis

A bilinear basis is a BASIS, which satisfies the conditions

$$(a\mathbf{x}+b\mathbf{y})\cdot\mathbf{z} = a(\mathbf{x}\cdot\mathbf{z}) + b(\mathbf{y}\cdot\mathbf{z})$$

$$\mathbf{z}\cdot(a\mathbf{x}+b\mathbf{y}) = a(\mathbf{z}\cdot\mathbf{x}) + b(\mathbf{z}\cdot\mathbf{y}).$$

see also BASIS

Billiard Table Problem

Given a billiard table with only corner pockets and sides of INTEGER lengths m and n, a ball sent at a 45° angle from a corner will be pocketed in a corner after $m+n-2$ bounces.

see also ALHAZEN'S BILLIARD PROBLEM, BILLIARDS

Billiards

The game of billiards is played on a RECTANGULAR table (known as a billiard table) upon which balls are placed. One ball (the "cue ball") is then struck with the end of a "cue" stick, causing it to bounce into other balls and REFLECT off the sides of the table. Real billiards can involve spinning the ball so that it does not travel in a straight LINE, but the mathematical study of billiards generally consists of REFLECTIONS in which the reflection and incidence angles are the same. However, strange table shapes such as CIRCLES and ELLIPSES are often considered. Many interesting problems can arise.

For example, ALHAZEN'S BILLIARD PROBLEM seeks to find the point at the edge of a circular "billiards" table at which a cue ball at a given point must be aimed in order to carom once off the edge of the table and strike another ball at a second given point. It was not until 1997 that Neumann proved that the problem is insoluble using a COMPASS and RULER construction.

On an ELLIPTICAL billiard table, the ENVELOPE of a trajectory is a smaller ELLIPSE, a HYPERBOLA, a LINE through the FOCI of the ELLIPSE, or periodic curve (e.g., DIAMOND-shape) (Wagon 1991).

see also ALHAZEN'S BILLIARD PROBLEM, BILLIARD TABLE PROBLEM, REFLECTION PROPERTY

References
Davis, D.; Ewing, C.; He, Z.; and Shen, T. "The Billiards Simulation." http://serendip.brynmawr.edu/chaos/home.html.
Dullin, H. R.; Richter, P.H.; and Wittek, A. "A Two-Parameter Study of the Extent of Chaos in a Billiard System." *Chaos* **6**, 43–58, 1996.
Madachy, J. S. "Bouncing Billiard Balls." In *Madachy's Mathematical Recreations.* New York: Dover, pp. 231–241, 1979.
Neumann, P. Submitted to *Amer. Math. Monthly.*
Pappas, T. "Mathematics of the Billiard Table." *The Joy of Mathematics.* San Carlos, CA: Wide World Publ./Tetra, p. 43, 1989.
Peterson, I. "Billiards in the Round." http://www.sciencenews.org/sn_arc97/3_1_97/mathland.htm.
Wagon, S. "Billiard Paths on Elliptical Tables." §10.2 in *Mathematica in Action.* New York: W. H. Freeman, pp. 330–333, 1991.

Billion

The word billion denotes different numbers in American and British usage. In the American system, one billion equals 10^9. In the British, French, and German systems, one billion equals 10^{12}.

see also LARGE NUMBER, MILLIARD, MILLION, TRILLION

Bilunabirotunda

see JOHNSON SOLID

Bimagic Square

16	41	36	5	27	62	55	18
26	63	54	19	13	44	33	8
1	40	45	12	22	51	58	31
23	50	59	30	4	37	48	9
38	3	10	47	49	24	29	60
52	21	32	57	39	2	11	46
43	14	7	34	64	25	20	53
61	28	17	56	42	15	6	35

If replacing each number by its square in a MAGIC SQUARE produces another MAGIC SQUARE, the square is said to be a bimagic square. The first bimagic square (shown above) has order 8 with magic constant 260 for addition and 11,180 after squaring. Bimagic squares are also called DOUBLY MAGIC SQUARES, and are 2-MULTIMAGIC SQUARES.

see also MAGIC SQUARE, MULTIMAGIC SQUARE, TRIMAGIC SQUARE

References
Ball, W. W. R. and Coxeter, H. S. M. *Mathematical Recreations and Essays, 13th ed.* New York: Dover, p. 212, 1987.
Hunter, J. A. H. and Madachy, J. S. "Mystic Arrays." Ch. 3 in *Mathematical Diversions.* New York: Dover, p. 31, 1975.
Kraitchik, M. "Multimagic Squares." §7.10 in *Mathematical Recreations.* New York: W. W. Norton, pp. 176–178, 1942.

Bimedian

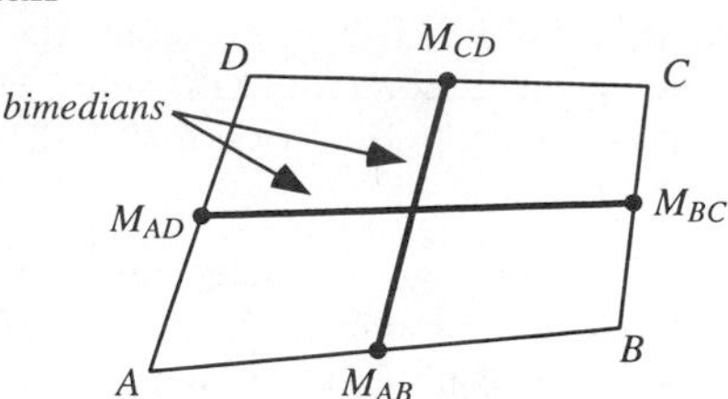

A LINE SEGMENT joining the MIDPOINTS of opposite sides of a QUADRILATERAL.

see also MEDIAN (TRIANGLE), VARIGNON'S THEOREM

Bimodal Distribution

A DISTRIBUTION having two separated peaks.

see also UNIMODAL DISTRIBUTION

Bimonster

The wreathed product of the MONSTER GROUP by Z_2. The bimonster is a quotient of the COXETER GROUP with the following COXETER-DYNKIN DIAGRAM.

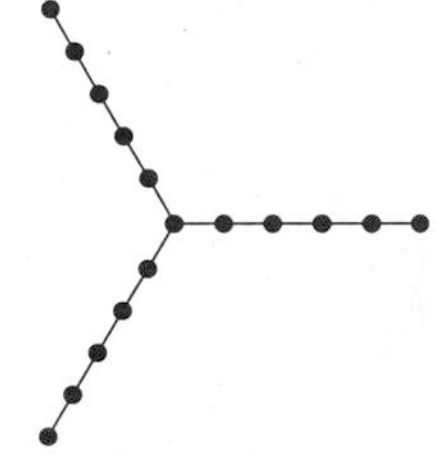

This had been conjectured by Conway, but was proven around 1990 by Ivanov and Norton. If the parameters p, q, r in Coxeter's NOTATION $[3^{p,q,r}]$ are written side by side, the bimonster can be denoted by the BEAST NUMBER 666.

Bin

An interval into which a given data point does or does not fall.

see also HISTOGRAM

Binary

The BASE 2 method of counting in which only the digits 0 and 1 are used. In this BASE, the number 1011 equals $1\cdot 2^0+1\cdot 2+0\cdot 2^2+1\cdot 2^3=11$. This BASE is used in computers, since all numbers can be simply represented as a string of electrically pulsed ons and offs. A NEGATIVE $-n$ is most commonly represented as the complement of the POSITIVE number $n-1$, so $-11=00001011_2$ would be written as the complement of $10=00001010_2$, or 11110101. This allows addition to be carried out with the usual carrying and the left-most digit discarded, so $17-11=6$ gives

$$\begin{array}{rr} 00010001 & 17 \\ \underline{11110101} & \underline{-11} \\ 00000110 & 6 \end{array}$$

The number of times k a given binary number $b_n \cdots b_2 b_1 b_0$ is divisible by 2 is given by the position of the first $b_k = 1$ counting from the right. For example, $12 = 1100$ is divisible by 2 twice, and $13 = 1101$ is divisible by 2 0 times.

Unfortunately, the storage of binary numbers in computers is not entirely standardized. Because computers store information in 8-bit bytes (where a bit is a single binary digit), depending on the "word size" of the machine, numbers requiring more than 8 bits must be stored in multiple bytes. The usual `FORTRAN77` integer size is 4 bytes long. However, a number represented as (byte1 byte2 byte3 byte4) in a VAX would be read and interpreted as (byte4 byte3 byte2 byte1) on a Sun. The situation is even worse for floating point (real) numbers, which are represented in binary as a MANTISSA and CHARACTERISTIC, and worse still for long (8-byte) reals!

Binary multiplication of single bit numbers (0 or 1) is equivalent to the AND operation, as can be seen in the following MULTIPLICATION TABLE.

×	0	1
0	0	0
1	0	1

see also BASE (NUMBER), DECIMAL, HEXADECIMAL, OCTAL, QUATERNARY, TERNARY

References

Lauwerier, H. *Fractals: Endlessly Repeated Geometric Figures.* Princeton, NJ: Princeton University Press, pp. 6–9, 1991.

Pappas, T. "Computers, Counting, & Electricity." *The Joy of Mathematics.* San Carlos, CA: Wide World Publ./Tetra, pp. 24–25, 1989.

Press, W. H.; Flannery, B. P.; Teukolsky, S. A.; and Vetterling, W. T. "Error, Accuracy, and Stability" and "Diagnosing Machine Parameters." §1.2 and §20.1 in *Numerical Recipes in FORTRAN: The Art of Scientific Computing, 2nd ed.* Cambridge, England: Cambridge University Press, pp. 18–21, 276, and 881–886, 1992.

Weisstein, E. W. "Bases." `http://www.astro.virginia.edu/~eww6n/math/notebooks/Bases.m`.

Binary Bracketing

A binary bracketing is a BRACKETING built up entirely of binary operations. The number of binary bracketings of n letters (CATALAN'S PROBLEM) are given by the CATALAN NUMBERS C_{n-1}, where

$$C_n \equiv \frac{1}{n+1}\binom{2n}{n} = \frac{1}{n+1}\frac{(2n)!}{n!^2} = \frac{(2n)!}{(n+1)!n!},$$

where $\binom{2n}{n}$ denotes a BINOMIAL COEFFICIENT and $n!$ is the usual FACTORIAL, as first shown by Catalan in 1838. For example, for the four letters a, b, c, and d there are five possibilities: $((ab)c)d$, $(a(bc))d$, $(ab)(cd)$, $a((bc)d)$, and $a(b(cd))$, written in shorthand as $((xx)x)x$, $(x(xx))x$, $(xx)(xx)$, $x((xx)x)$, and $x(x(xx))$.

see also BRACKETING, CATALAN NUMBER, CATALAN'S PROBLEM

References

Schröder, E. "Vier combinatorische Probleme." *Z. Math. Physik* **15**, 361–376, 1870.

Sloane, N. J. A. Sequences A000108/M1459 in "An On-Line Version of the Encyclopedia of Integer Sequences."

Sloane, N. J. A. and Plouffe, S. Extended entry in *The Encyclopedia of Integer Sequences.* San Diego: Academic Press, 1995.

Stanley, R. P. "Hipparchus, Plutarch, Schröder, and Hough." *Amer. Math. Monthly* **104**, 344–350, 1997.

Binary Operator

An OPERATOR which takes two mathematical objects as input and returns a value is called a binary operator. Binary operators are called compositions by Rosenfeld (1968). Sets possessing a binary multiplication operation include the GROUP, GROUPOID, MONOID, QUASIGROUP, and SEMIGROUP. Sets possessing both a binary multiplication and a binary addition operation include the DIVISION ALGEBRA, FIELD, RING, RINGOID, SEMIRING, and UNIT RING.

see also AND, BOOLEAN ALGEBRA, CLOSURE, DIVISION ALGEBRA, FIELD, GROUP, GROUPOID, MONOID, OPERATOR, OR, MONOID, NOT, QUASIGROUP, RING, RINGOID, SEMIGROUP, SEMIRING, XOR, UNIT RING

References

Rosenfeld, A. *An Introduction to Algebraic Structures.* New York: Holden-Day, 1968.

Binary Quadratic Form

A 2-variable QUADRATIC FORM of the form

$$Q(x,y) = a_{11}x^2 + 2a_{12}xy + a_{22}y^2.$$

see also QUADRATIC FORM, QUADRATIC INVARIANT

Binary Remainder Method

An ALGORITHM for computing a UNIT FRACTION (Stewart 1992).

References

Stewart, I. "The Riddle of the Vanishing Camel." *Sci. Amer.* **266**, 122–124, June 1992.

Binary Tree

A TREE with two BRANCHES at each FORK and with one or two LEAVES at the end of each BRANCH. (This definition corresponds to what is sometimes known as an "extended" binary tree.) The height of a binary tree is the number of levels within the TREE. For a binary tree of height H with n nodes,

$$H \le n \le 2^H - 1.$$

These extremes correspond to a balanced tree (each node except the LEAVES has a left and right CHILD, and all LEAVES are at the same level) and a degenerate tree (each node has only one outgoing BRANCH), respectively. For a search of data organized into a binary tree, the number of search steps $S(n)$ needed to find an item is bounded by

$$\lg n \leq S(n) \leq n.$$

Partial balancing of an arbitrary tree into a so-called AVL binary search tree can improve search speed.

The number of binary trees with n internal nodes is the CATALAN NUMBER C_n (Sloane's A000108), and the number of binary trees of height b is given by Sloane's A001699.

see also B-TREE, QUADTREE, QUATERNARY TREE, RED-BLACK TREE, STERN-BROCOT TREE, WEAKLY BINARY TREE

References

Lucas, J.; Roelants van Baronaigien, D.; and Ruskey, F. "Generating Binary Trees by Rotations." *J. Algorithms* **15**, 343–366, 1993.

Ranum, D. L. "On Some Applications of Fibonacci Numbers." *Amer. Math. Monthly* **102**, 640–645, 1995.

Ruskey, F. "Information on Binary Trees." `http://sue.csc.uvic.ca/~cos/inf/tree/BinaryTrees.html`.

Ruskey, F. and Proskurowski, A. "Generating Binary Trees by Transpositions." *J. Algorithms* **11**, 68–84, 1990.

Sloane, N. J. A. Sequences A000108/M1459 and A001699/M3087 in "An On-Line Version of the Encyclopedia of Integer Sequences."

Binet Forms

The two RECURRENCE SEQUENCES

$$U_n = mU_{n-1} + U_{n-2} \tag{1}$$

$$V_n = mV_{n-1} + V_{n-2} \tag{2}$$

with $U_0 = 0$, $U_1 = 1$ and $V_0 = 2$, $V_1 = m$, can be solved for the individual U_n and V_n. They are given by

$$U_n = \frac{\alpha^n - \beta^n}{\Delta} \tag{3}$$

$$V_n = \alpha^n + \beta^n, \tag{4}$$

where

$$\Delta \equiv \sqrt{m^2+4} \tag{5}$$

$$\alpha \equiv \frac{m+\Delta}{2} \tag{6}$$

$$\beta \equiv \frac{m-\Delta}{2}. \tag{7}$$

A useful related identity is

$$U_{n-1} + U_{n+1} = V_n. \tag{8}$$

BINET'S FORMULA is a special case of the Binet form for U_n corresponding to $m = 1$.

see also FIBONACCI Q-MATRIX

Binet's Formula

A special case of the U_n BINET FORM with $m = 0$, corresponding to the nth FIBONACCI NUMBER,

$$F_n = \frac{(1+\sqrt{5}\,)^n - (1-\sqrt{5}\,)^n}{2^n\sqrt{5}}.$$

It was derived by Binet in 1843, although the result was known to Euler and Daniel Bernoulli more than a century earlier.

see also BINET FORMS, FIBONACCI NUMBER

Bing's Theorem

If M^3 is a closed oriented connected 3-MANIFOLD such that every simple closed curve in M lies interior to a BALL in M, then M is HOMEOMORPHIC with the HYPERSPHERE, $\mathbb{S}^3$.

see also BALL, HYPERSPHERE

References

Bing, R. H. "Necessary and Sufficient Conditions that a 3-Manifold be S^3." *Ann. Math.* **68**, 17–37, 1958.

Rolfsen, D. *Knots and Links.* Wilmington, DE: Publish or Perish Press, pp. 251–257, 1976.

Binomial

A POLYNOMIAL with 2 terms.

see also MONOMIAL, POLYNOMIAL, TRINOMIAL

Binomial Coefficient

The number of ways of picking n unordered outcomes from N possibilities. Also known as a COMBINATION. The binomial coefficients form the rows of PASCAL'S TRIANGLE. The symbols ${}_NC_n$ and

$$\binom{N}{n} \equiv \frac{N!}{(N-n)!n!} \tag{1}$$

are used, where the latter is sometimes known as N CHOOSE n. The number of LATTICE PATHS from the ORIGIN $(0,0)$ to a point (a,b) is the BINOMIAL COEFFICIENT $\binom{a+b}{a}$ (Hilton and Pedersen 1991).

For POSITIVE integer n, the BINOMIAL THEOREM gives

$$(x+a)^n = \sum_{k=0}^{n} \binom{n}{k} x^k a^{n-k}. \tag{2}$$

The FINITE DIFFERENCE analog of this identity is known as the CHU-VANDERMONDE IDENTITY. A similar formula holds for NEGATIVE INTEGRAL n,

$$(x+a)^{-n} = \sum_{k=0}^{\infty} \binom{-n}{k} x^k a^{-n-k}. \tag{3}$$

A general identity is given by

$$\frac{(a+b)^n}{a} = \sum_{j=0}^{n} \binom{n}{j} (a-jc)^{j-1}(b+jc)^{n-j} \tag{4}$$

(Prudnikov *et al.* 1986), which gives the BINOMIAL THEOREM as a special case with $c = 0$.

The binomial coefficients satisfy the identities:

$$\binom{n}{0} = \binom{n}{n} = 1 \tag{5}$$

$$\binom{n}{k} = \binom{n}{n-k} = (-1)^k \binom{k-n-1}{k} \tag{6}$$

$$\binom{n+1}{k} = \binom{n}{k} + \binom{n}{k-1}. \tag{7}$$

Sums of powers include

$$\sum_{k=0}^{n} \binom{n}{k} = 2^n \tag{8}$$

$$\sum_{k=0}^{n} (-1)^k \binom{n}{k} = 0 \tag{9}$$

$$\sum_{k=0}^{n} \binom{n}{k} r^k = (1+r)^n \tag{10}$$

(the BINOMIAL THEOREM), and

$$\sum_{n=0}^{\infty} \binom{2n+s}{n} x^n = {}_2F_1(\tfrac{1}{2}(s+1), \tfrac{1}{2}(s+2); s+1, 4x) = \frac{2^s}{(\sqrt{1-4x}+1)^s \sqrt{1-4x}}, \tag{11}$$

where ${}_2F_1(a, b; c; z)$ is a HYPERGEOMETRIC FUNCTION (Abramowitz and Stegun 1972, p. 555; Graham *et al.* 1994, p. 203). For NONNEGATIVE INTEGERS n and r with $r \leq n+1$,

$$\sum_{k=0}^{n} \frac{(-1)^k}{k+1} \binom{n}{k} \left[\sum_{j=0}^{r-1} (-1)^j \binom{n}{j} (r-j)^{n-k} + \sum_{j=0}^{n-r} (-1)^j \binom{n}{j} (n+1-r-j)^{n-k} \right] = n!. \tag{12}$$

Taking $n = 2r - 1$ gives

$$\sum_{k=0}^{n} \frac{(-1)^k}{k+1} \binom{n}{k} \sum_{j=0}^{r-1} \binom{n}{j} (r-j)^{n-k} = \tfrac{1}{2} n!. \tag{13}$$

Another identity is

$$\sum_{k=0}^{n} \binom{n+k}{k} [x^{n+1}(1-x)^k + (1-x)^{n+1} x^k] = 1 \tag{14}$$

(Beeler *et al.* 1972, Item 42).

RECURRENCE RELATIONS of the sums

$$s_p \equiv \sum_{k=0}^{n} \binom{n}{k}^p \tag{15}$$

are given by

$$2s_1(n) - s_1(n+1) = 0 \tag{16}$$

$$-2(2n+1)s_2(n) + (n+1)s_2(n) = 0 \tag{17}$$

$$-8(n+1)^2 s_3(n) + (-16 - 21n - 7n^2) s_3(n+1) + (n+2)^2 s_3(n+2) = 0 \tag{18}$$

$$-4(n+1)(4n+3)(4n+5)s_4(n) - 2(2n+3)(3n^2+9n+7)s_4(n+1) + (n+2)^3 s_4(n+2) = 0. \tag{19}$$

This sequence for s_3 cannot be expressed as a fixed number of hypergeometric terms (Petkovšek *et al.* 1996, p. 160).

A fascinating series of identities involving binomial coefficients times small powers are

$$\sum_{n=1}^{\infty} \frac{1}{\binom{2n}{n}} = \tfrac{1}{27}(2\pi\sqrt{3}+9) = 0.7363998587\ldots \tag{20}$$

$$\sum_{n=1}^{\infty} \frac{1}{n\binom{2n}{n}} = \tfrac{1}{9}\pi\sqrt{3} = 0.6045997881\ldots \tag{21}$$

$$\sum_{n=1}^{\infty} \frac{1}{n^2\binom{2n}{n}} = \tfrac{1}{3}\zeta(2) = \tfrac{1}{8}\pi^2 \tag{22}$$

$$\sum_{n=1}^{\infty} \frac{1}{n^4\binom{2n}{n}} = \tfrac{17}{36}\zeta(4) = \tfrac{17}{3240}\pi^4 \tag{23}$$

(Comtet 1974, p. 89) and

$$\sum_{n=1}^{\infty} \frac{(-1)^{n-1}}{n^3\binom{2n}{n}} = \tfrac{2}{5}\zeta(3), \tag{24}$$

where $\zeta(z)$ is the RIEMANN ZETA FUNCTION (Le Lionnais 1983, pp. 29, 30, 41, 36, and 35; Guy 1994, p. 257).

As shown by Kummer in 1852, the exact POWER of p dividing $\binom{a+b}{a}$ is equal to

$$\epsilon_0 + \epsilon_1 + \ldots + \epsilon_t, \tag{25}$$

where this is the number of carries in performing the addition of a and b written in base b (Graham *et al.* 1989, Exercise 5.36; Ribenboim 1989; Vardi 1991, p. 68). Kummer's result can also be stated in the form that the exponent of a PRIME p dividing $\binom{n}{m}$ is given by the number of integers $j \geq 0$ for which

$$\text{frac}(m/p^j) > \text{frac}(n/p^j), \qquad (26)$$

where frac(x) denotes the FRACTIONAL PART of x. This inequality may be reduced to the study of the exponential sums $\sum_n \Lambda(n)e(x/n)$, where $\Lambda(n)$ is the MANGOLDT FUNCTION. Estimates of these sums are given by Jutila (1974, 1975), but recent improvements have been made by Granville and Ramare (1996).

R. W. Gosper showed that

$$f(n) = \binom{n-1}{\frac{1}{2}(n-1)} \equiv (-1)^{(n-1)/2} \pmod{n} \qquad (27)$$

for all PRIMES, and conjectured that it holds *only* for PRIMES. This was disproved when Skiena (1990) found it also holds for the COMPOSITE NUMBER $n = 3 \times 11 \times 179$. Vardi (1991, p. 63) subsequently showed that $n = p^2$ is a solution whenever p is a WIEFERICH PRIME and that if $n = p^k$ with $k > 3$ is a solution, then so is $n = p^{k-1}$. This allowed him to show that the only solutions for COMPOSITE $n < 1.3 \times 10^7$ are 5907, 1093^2, and 3511^2, where 1093 and 3511 are WIEFERICH PRIMES.

Consider the binomial coefficients $\binom{2n-1}{n}$, the first few of which are 1, 3, 10, 35, 126, ... (Sloane's A001700). The GENERATING FUNCTION is

$$\frac{1}{2}\left[\frac{1}{\sqrt{1-4x}} - 1\right] = x + 3x^2 + 10x^3 + 35x^4 + \ldots. \qquad (28)$$

These numbers are SQUAREFREE only for $n = 2$, 3, 4, 6, 9, 10, 12, 36, ... (Sloane's A046097), with no others less than $n = 10,000$. Erdős showed that the binomial coefficient $\binom{n}{k}$ is never a POWER of an INTEGER for $n \geq 3$ where $k \neq 0$, 1, $n-1$, and n (Le Lionnais 1983, p. 48).

The binomial coefficients $\binom{n}{\lfloor n/2 \rfloor}$ are called CENTRAL BINOMIAL COEFFICIENTS, where $\lfloor x \rfloor$ is the FLOOR FUNCTION, although the subset of coefficients $\binom{2n}{n}$ is sometimes also given this name. Erdős and Graham (1980, p. 71) conjectured that the CENTRAL BINOMIAL COEFFICIENT $\binom{2n}{n}$ is *never* SQUAREFREE for $n > 4$, and this is sometimes known as the ERDŐS SQUAREFREE CONJECTURE. SÁRKÖZY'S THEOREM (Sárközy 1985) provides a partial solution which states that the BINOMIAL COEFFICIENT $\binom{2n}{n}$ is never SQUAREFREE for all sufficiently large $n \geq n_0$ (Vardi 1991). Granville and Ramare (1996) proved that the *only* SQUAREFREE values are $n = 2$ and 4. Sander (1992) subsequently showed that $\binom{2n \pm d}{n}$ are also never SQUAREFREE for sufficiently large n as long as d is not "too big."

For p, q, and r distinct PRIMES, then the above function satisfies

$$f(pqr)f(p)f(q)f(r) \equiv f(pq)f(pr)p(qr) \pmod{pqr} \qquad (29)$$

(Vardi 1991, p. 66).

The binomial coefficient $\binom{m}{n}$ mod 2 can be computed using the XOR operation n XOR m, making PASCAL'S TRIANGLE mod 2 very easy to construct.

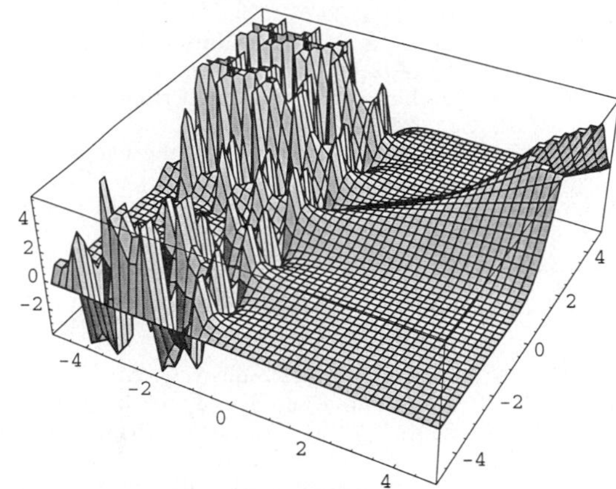

The binomial coefficient "function" can be defined as

$$C(x, y) \equiv \frac{x!}{y!(x-y)!} \qquad (30)$$

(Fowler 1996), shown above. It has a very complicated GRAPH for NEGATIVE x and y which is difficult to render using standard plotting programs.

see also BALLOT PROBLEM, BINOMIAL DISTRIBUTION, BINOMIAL THEOREM, CENTRAL BINOMIAL COEFFICIENT, CHU-VANDERMONDE IDENTITY, COMBINATION, DEFICIENCY, ERDŐS SQUAREFREE CONJECTURE, GAUSSIAN COEFFICIENT, GAUSSIAN POLYNOMIAL, KINGS PROBLEM, MULTINOMIAL COEFFICIENT, PERMUTATION, ROMAN COEFFICIENT, SÁRKÖZY'S THEOREM, STREHL IDENTITY, WOLSTENHOLME'S THEOREM

References

Abramowitz, M. and Stegun, C. A. (Eds.). "Binomial Coefficients." §24.1.1 in *Handbook of Mathematical Functions with Formulas, Graphs, and Mathematical Tables, 9th printing.* New York: Dover, pp. 10 and 822–823, 1972.

Beeler, M.; Gosper, R. W.; and Schroeppel, R. *HAKMEM.* Cambridge, MA: MIT Artificial Intelligence Laboratory, Memo AIM-239, Feb. 1972.

Comtet, L. *Advanced Combinatorics.* Amsterdam, Netherlands: Kluwer, 1974.

Conway, J. H. and Guy, R. K. In *The Book of Numbers.* New York: Springer-Verlag, pp. 66–74, 1996.

Erdős, P.; Graham, R. L.; Nathanson, M. B.; and Jia, X. *Old and New Problems and Results in Combinatorial Number Theory.* New York: Springer-Verlag, 1998.

Fowler, D. "The Binomial Coefficient Function." *Amer. Math. Monthly* **103**, 1–17, 1996.

Graham, R. L.; Knuth, D. E.; and Patashnik, O. "Binomial Coefficients." Ch. 5 in *Concrete Mathematics: A Foundation for Computer Science.* Reading, MA: Addison-Wesley, pp. 153–242, 1990.

Granville, A. and Ramare, O. "Explicit Bounds on Exponential Sums and the Scarcity of Squarefree Binomial Coefficients." *Mathematika* **43**, 73–107, 1996.

Guy, R. K. "Binomial Coefficients," "Largest Divisor of a Binomial Coefficient," and "Series Associated with the ζ-Function." §B31, B33, and F17 in *Unsolved Problems in Number Theory, 2nd ed.* New York: Springer-Verlag, pp. 84–85, 87–89, and 257–258, 1994.
Harborth, H. "Number of Odd Binomial Coefficients." *Not. Amer. Math. Soc.* **23**, 4, 1976.
Hilton, P. and Pedersen, J. "Catalan Numbers, Their Generalization, and Their Uses." *Math. Intel.* **13**, 64–75, 1991.
Jutila, M. "On Numbers with a Large Prime Factor." *J. Indian Math. Soc.* **37**, 43–53, 1973.
Jutila, M. "On Numbers with a Large Prime Factor. II." *J. Indian Math. Soc.* **38**, 125–130, 1974.
Le Lionnais, F. *Les nombres remarquables.* Paris: Hermann, 1983.
Ogilvy, C. S. "The Binomial Coefficients." *Amer. Math. Monthly* **57**, 551–552, 1950.
Petkovšek, M.; Wilf, H. S.; and Zeilberger, D. *A=B.* Wellesley, MA: A. K. Peters, 1996.
Press, W. H.; Flannery, B. P.; Teukolsky, S. A.; and Vetterling, W. T. "Gamma Function, Beta Function, Factorials, Binomial Coefficients." §6.1 in *Numerical Recipes in FORTRAN: The Art of Scientific Computing, 2nd ed.* Cambridge, England: Cambridge University Press, pp. 206–209, 1992.
Prudnikov, A. P.; Marichev, O. I.; and Brychkow, Yu. A. Formula 41 in *Integrals and Series, Vol. 1: Elementary Functions.* Newark, NJ: Gordon & Breach, p. 611, 1986.
Ribenboim, P. *The Book of Prime Number Records, 2nd ed.* New York: Springer-Verlag, pp. 23–24, 1989.
Riordan, J. "Inverse Relations and Combinatorial Identities." *Amer. Math. Monthly* **71**, 485–498, 1964.
Sander, J. W. "On Prime Divisors of Binomial Coefficients." *Bull. London Math. Soc.* **24**, 140–142, 1992.
Sárközy, A. "On the Divisors of Binomial Coefficients, I." *J. Number Th.* **20**, 70–80, 1985.
Skiena, S. *Implementing Discrete Mathematics: Combinatorics and Graph Theory with Mathematica.* Reading, MA: Addison-Wesley, p. 262, 1990.
Sloane, N. J. A. Sequences A046097 and A001700/M2848 in "An On-Line Version of the Encyclopedia of Integer Sequences."
Spanier, J. and Oldham, K. B. "The Binomial Coefficients $\binom{\nu}{m}$." Ch. 6 in *An Atlas of Functions.* Washington, DC: Hemisphere, pp. 43–52, 1987.
Sved, M. "Counting and Recounting." *Math. Intel.* **5**, 21–26, 1983.
Vardi, I. "Application to Binomial Coefficients," "Binomial Coefficients," "A Class of Solutions," "Computing Binomial Coefficients," and "Binomials Modulo and Integer." §2.2, 4.1, 4.2, 4.3, and 4.4 in *Computational Recreations in Mathematica.* Redwood City, CA: Addison-Wesley, pp. 25–28 and 63–71, 1991.
Wolfram, S. "Geometry of Binomial Coefficients." *Amer. Math. Monthly* **91**, 566–571, 1984.

Binomial Distribution

The probability of n successes in N BERNOULLI TRIALS is

$$P(n|N) = \binom{N}{n} p^n (1-p)^{N-n} = \frac{N!}{n!(N-n)!} p^n q^{N-n}. \quad (1)$$

The probability of obtaining more successes than the n observed is

$$P = \sum_{k=n+1}^{N} \binom{N}{k} p^k (1-p)^{N-k} = I_p(n+1, N-N), \quad (2)$$

where

$$I_x(a,b) \equiv \frac{B(x;a,b)}{B(a,b)}, \quad (3)$$

$B(a,b)$ is the BETA FUNCTION, and $B(x;a,b)$ is the incomplete BETA FUNCTION. The CHARACTERISTIC FUNCTION is

$$\phi(t) = (q + pe^{it})^n. \quad (4)$$

The MOMENT-GENERATING FUNCTION M for the distribution is

$$\begin{aligned} M(t) = \langle e^{tn} \rangle &= \sum_{n=0}^{N} e^{tn} \binom{N}{n} p^n q^{N-n} \\ &= \sum_{n=0}^{N} \binom{N}{n} (pe^t)^n (1-p)^{N-n} \\ &= [pe^t + (1-p)]^N \quad (5) \\ M'(t) &= N[pe^t + (1-p)]^{N-1}(pe^t) \quad (6) \\ M''(t) &= N(N-1)[pe^t + (1-p)]^{N-2}(pe^t)^2 \\ &\quad + N[pe^t + (1-p)]^{N-1}(pe^t). \quad (7) \end{aligned}$$

The MEAN is

$$\mu = M'(0) = N(p + 1 - p)p = Np. \quad (8)$$

The MOMENTS about 0 are

$$\mu'_1 = \mu = Np \quad (9)$$

$$\mu'_2 = Np(1 - p + Np) \quad (10)$$

$$\mu'_3 = Np(1 - 3p + 3Np + 2p^2 - 3NP^2 + N^2p^2) \quad (11)$$

$$\begin{aligned} \mu'_4 &= Np(1 - 7p + 7Np + 12p^2 - 18Np^2 + 6N^2p^2 \\ &\quad - 6p^3 + 11Np^3 - 6N^2p^3 + N^3p^3), \quad (12) \end{aligned}$$

so the MOMENTS about the MEAN are

$$\begin{aligned} \mu_2 = \sigma^2 &= [N(N-1)p^2 + Np] - (Np)^2 \\ &= N^2p^2 - Np^2 + Np - N^2p^2 \\ &= Np(1-p) = Npq \quad (13) \\ \mu_3 &= \mu'_3 - 3\mu'_2\mu'_1 + 2(\mu_1)^3 \\ &= Np(1-p)(1-2p) \quad (14) \\ \mu_4 &= \mu'_4 - 4\mu'_3\mu'_1 + 6\mu'_2(\mu'_1)^2 - 3(\mu_1)^4 \\ &= Np(1-p)[3p^2(2-N) + 3p(N-2) + 1]. \quad (15) \end{aligned}$$

The SKEWNESS and KURTOSIS are

$$\gamma_1 = \frac{\mu_3}{\sigma^3} = \frac{Np(1-p)(1-2p)}{[Np(1-p)]^{3/2}}$$
$$= \frac{1-2p}{\sqrt{Np(1-p)}} = \frac{q-p}{\sqrt{Npq}} \tag{16}$$

$$\gamma_2 = \frac{\mu_4}{\sigma^4} - 3 = \frac{6p^2-6p+1}{Np(1-p)} = \frac{1-6pq}{Npq}. \tag{17}$$

An approximation to the Bernoulli distribution for large N can be obtained by expanding about the value $\tilde{n}$ where $P(n)$ is a maximum, i.e., where $dP/dn = 0$. Since the LOGARITHM function is MONOTONIC, we can instead choose to expand the LOGARITHM. Let $n \equiv \tilde{n} + \eta$, then

$$\ln[P(n)] = \ln[P(\tilde{n})] + B_1\eta + \tfrac{1}{2}B_2\eta^2 + \tfrac{1}{3!}B_3\eta^3 + \ldots, \tag{18}$$

where

$$B_k \equiv \left[\frac{d^k \ln[P(n)]}{dn^k}\right]_{n=\tilde{n}}. \tag{19}$$

But we are expanding about the maximum, so, by definition,

$$B_1 = \left[\frac{d\ln[P(n)]}{dn}\right]_{n=\tilde{n}} = 0. \tag{20}$$

This also means that B_2 is negative, so we can write $B_2 = -|B_2|$. Now, taking the LOGARITHM of (1) gives

$$\ln[P(n)] = \ln N! - \ln n! - \ln(N-n)! + n\ln p + (N-n)\ln q. \tag{21}$$

For large n and $N-n$ we can use STIRLING'S APPROXIMATION

$$\ln(n!) \approx n\ln n - n, \tag{22}$$

so

$$\frac{d[\ln(n!)]}{dn} \approx (\ln n + 1) - 1 = \ln n \tag{23}$$
$$\frac{d[\ln(N-n)!]}{dn} \approx \frac{d}{dn}[(N-n)\ln(N-n) - (N-n)]$$
$$= \left[-\ln(N-n) + (N-n)\frac{-1}{N-n} + 1\right]$$
$$= -\ln(N-n), \tag{24}$$

and

$$\frac{d\ln[P(n)]}{dn} \approx -\ln n + \ln(N-n) + \ln p - \ln q. \tag{25}$$

To find $\tilde{n}$, set this expression to 0 and solve for n,

$$\ln\left(\frac{N-\tilde{n}}{\tilde{n}}\frac{p}{q}\right) = 0 \tag{26}$$

$$\frac{N-\tilde{n}}{\tilde{n}}\frac{p}{q} = 1 \tag{27}$$

$$(N-\tilde{n})p = \tilde{n}q \tag{28}$$

$$\tilde{n}(q+p) = \tilde{n} = Np, \tag{29}$$

since $p + q = 1$. We can now find the terms in the expansion

$$B_2 \equiv \left[\frac{d^2\ln[P(n)]}{dn^2}\right]_{n=\tilde{n}} = -\frac{1}{\tilde{n}} - \frac{1}{N-\tilde{n}}$$
$$= -\frac{1}{Np} - \frac{1}{N(1-p)} = -\frac{1}{N}\left(\frac{1}{p}+\frac{1}{q}\right)$$
$$= -\frac{1}{N}\left(\frac{p+q}{pq}\right) = -\frac{1}{Npq} = -\frac{1}{N(1-p)} \tag{30}$$

$$B_3 \equiv \left[\frac{d^3\ln[P(n)]}{dn^3}\right]_{n=\tilde{n}} = \frac{1}{\tilde{n}^2} - \frac{1}{(N-\tilde{n})^2}$$
$$= \frac{1}{N^2p^2} - \frac{1}{N^2q^2} = \frac{q^2-p^2}{N^2p^2q^2}$$
$$= \frac{(1-2p+p^2)-p^2}{N^2p^2(1-p)^2} = \frac{1-2p}{N^2p^2(1-p)^2} \tag{31}$$

$$B_4 \equiv \left[\frac{d^4\ln[P(n)]}{dn^4}\right]_{n=\tilde{n}} = -\frac{2}{\tilde{n}^3} - \frac{2}{(n-\tilde{n})^3}$$
$$= -2\left(\frac{1}{N^3p^3} + \frac{1}{N^3q^3}\right) = \frac{2(p^3+q^3)}{N^3p^3q^3}$$
$$= \frac{2(p^2-pq+q^2)}{N^3p^3q^3}$$
$$= \frac{2[p^2-p(1-p)+(1-2p+p^2)]}{N^3p^3(1-p^3)}$$
$$= \frac{2(3p^2-3p+1)}{N^3p^3(1-p^3)}. \tag{32}$$

Now, treating the distribution as continuous,

$$\lim_{N\to\infty}\sum_{n=0}^{N} P(n) \approx \int P(n)\,dn = \int_{-\infty}^{\infty} P(\tilde{n}+\eta)\,d\eta = 1. \tag{33}$$

Since each term is of order $1/N \sim 1/\sigma^2$ smaller than the previous, we can ignore terms higher than B_2, so

$$P(n) = P(\tilde{n})e^{-|B_2|\eta^2/2}. \tag{34}$$

The probability must be normalized, so

$$\int_{-\infty}^{\infty} P(\tilde{n})e^{-|B_2|\eta^2/2}\,d\eta = P(\tilde{n})\sqrt{\frac{2\pi}{|B_2|}} = 1, \tag{35}$$

and

$$P(n) = \sqrt{\frac{|B_2|}{2\pi}}e^{-|B_2|(n-\tilde{n})^2/2}$$
$$= \frac{1}{\sqrt{2\pi Npq}}\exp\left[-\frac{(n-Np)^2}{2Npq}\right]. \tag{36}$$

Defining $\sigma^2 \equiv 2Npq$,

$$P(n) = \frac{1}{\sigma\sqrt{2\pi}}\exp\left[-\frac{(n-\tilde{n})^2}{2\sigma^2}\right], \tag{37}$$

which is a GAUSSIAN DISTRIBUTION. For $p \ll 1$, a different approximation procedure shows that the binomial distribution approaches the POISSON DISTRIBUTION. The first CUMULANT is

$$\kappa_1 = np, \tag{38}$$

and subsequent CUMULANTS are given by the RECURRENCE RELATION

$$\kappa_{r+1} = pq\frac{d\kappa_r}{dp}. \tag{39}$$

Let x and y be independent binomial RANDOM VARIABLES characterized by parameters n, p and m, p. The CONDITIONAL PROBABILITY of x given that $x + y = k$ is

$$\begin{aligned} P(x = i|x + y = k) &= \frac{P(x = i, x + y = k)}{P(x + y = k)} \\ &= \frac{P(x = i, y = k - i)}{P(x + y = k)} = \frac{P(x = i)P(y = k - i)}{P(x + y = k)} \\ &= \frac{\binom{n}{i}p^i(1-p)^{n-i}\binom{m}{k-i}p^{k-i}(1-p)^{m-(k-i)}}{\binom{n+m}{k}p^k(1-p)^{n+m-k}} \\ &= \frac{\binom{n}{i}\binom{m}{k-i}}{\binom{n+m}{k}}. \end{aligned} \tag{40}$$

Note that this is a HYPERGEOMETRIC DISTRIBUTION!

see also DE MOIVRE-LAPLACE THEOREM, HYPERGEOMETRIC DISTRIBUTION, NEGATIVE BINOMIAL DISTRIBUTION

References

Beyer, W. H. *CRC Standard Mathematical Tables, 28th ed.* Boca Raton, FL: CRC Press, p. 531, 1987.

Press, W. H.; Flannery, B. P.; Teukolsky, S. A.; and Vetterling, W. T. "Incomplete Beta Function, Student's Distribution, F-Distribution, Cumulative Binomial Distribution." §6.2 in *Numerical Recipes in FORTRAN: The Art of Scientific Computing, 2nd ed.* Cambridge, England: Cambridge University Press, pp. 219–223, 1992.

Spiegel, M. R. *Theory and Problems of Probability and Statistics.* New York: McGraw-Hill, p. 108–109, 1992.

Binomial Expansion

see BINOMIAL SERIES

Binomial Formula

see BINOMIAL SERIES, BINOMIAL THEOREM

Binomial Number

A number of the form $a^n \pm b^n$, where a, b, and n are INTEGERS. They can be factored algebraically

$$a^n - b^n = (a-b)(a^{n-1} + a^{n-2}b + \ldots + ab^{n-2} + b^{n-1}) \tag{1}$$

$$a^n + b^n = (a+b)(a^{n-1} - a^{n-2}b + \ldots - ab^{n-2} + b^{n-1}) \tag{2}$$

$$a^{nm} - b^{nm} = (a^m - b^m)[a^{m(n-1)} + a^{m(n-2)}b^m + \ldots + b^{m(n-1)}]. \tag{3}$$

In 1770, Euler proved that if $(a, b) = 1$, then every FACTOR of

$$a^{2^n} + b^{2^n} \tag{4}$$

is either 2 or of the form $2^{n+1}K + 1$. If p and q are PRIMES, then

$$\frac{a^{pq} - 1)(a - 1)}{(a^p - 1)(a^q - 1)} - 1 \tag{5}$$

is DIVISIBLE by every PRIME FACTOR of a^{p-1} not dividing $a^q - 1$.

see also CUNNINGHAM NUMBER, FERMAT NUMBER, MERSENNE NUMBER, RIESEL NUMBER, SIERPIŃSKI NUMBER OF THE SECOND KIND

References

Guy, R. K. "When Does $2^a - 2^b$ Divide $n^a - n^b$." §B47 in *Unsolved Problems in Number Theory, 2nd ed.* New York: Springer-Verlag, p. 102, 1994.

Qi, S and Ming-Zhi, Z. "Pairs where $2^a - a^b$ Divides $n^a - n^b$ for All n." *Proc. Amer. Math. Soc.* **93**, 218–220, 1985.

Schinzel, A. "On Primitive Prime Factors of $a^n - b^n$." *Proc. Cambridge Phil. Soc.* **58**, 555–562, 1962.

Binomial Series

For $|x| < 1$,

$$\begin{aligned} (1 + x)^n &= \sum_{k=0}^{n} \binom{n}{k} x^k \\ &= \binom{n}{0}x^0 + \binom{n}{1}x^1 + \binom{n}{2}x^2 + \ldots \\ &= 1 + \frac{n!}{1!(n-1)!}x + \frac{n!}{(n-2)!2!}x^2 + \ldots \\ &= 1 + nx + \frac{n(n-1)}{2}x^2 + \ldots. \end{aligned}$$

The binomial series also has the CONTINUED FRACTION representation

$$(1 + x)^n = \cfrac{1}{1 - \cfrac{nx}{1 + \cfrac{\frac{1\cdot(1+n)}{1\cdot 2}x}{1 + \cfrac{\frac{1\cdot(1-n)}{2\cdot 3}x}{1 + \cfrac{\frac{2(2+n)}{3\cdot 4}x}{1 + \cfrac{\frac{2(2-n)}{4\cdot 5}x}{1 + \cfrac{\frac{3(3+n)}{5\cdot 6}x}{1 + \ldots}}}}}}}. \tag{5}$$

see also BINOMIAL THEOREM, MULTINOMIAL SERIES, NEGATIVE BINOMIAL SERIES

References

Abramowitz, M. and Stegun, C. A. (Eds.). *Handbook of Mathematical Functions with Formulas, Graphs, and Mathematical Tables, 9th printing.* New York: Dover, pp. 14–15, 1972.

Pappas, T. "Pascal's Triangle, the Fibonacci Sequence & Binomial Formula." *The Joy of Mathematics.* San Carlos, CA: Wide World Publ./Tetra, pp. 40–41, 1989.

Binomial Theorem

The theorem that, for INTEGRAL POSITIVE n,

$$(x+a)^n = \sum_{k=0}^{n} \frac{n!}{k!(n-k)!} x^k a^{n-k} = \sum_{k=0}^{n} \binom{n}{k} x^k a^{n-k},$$

the so-called BINOMIAL SERIES, where $\binom{n}{k}$ are BINOMIAL COEFFICIENTS. The theorem was known for the case $n=2$ by Euclid around 300 BC, and stated in its modern form by Pascal in 1665. Newton (1676) showed that a similar formula (with INFINITE upper limit) holds for NEGATIVE INTEGRAL n,

$$(x+a)^{-n} = \sum_{k=0}^{\infty} \binom{-n}{k} x^k a^{-n-k},$$

the so-called NEGATIVE BINOMIAL SERIES, which converges for $|x| > |a|$.

see also BINOMIAL COEFFICIENT, BINOMIAL SERIES, CAUCHY BINOMIAL THEOREM, CHU-VANDERMONDE IDENTITY, LOGARITHMIC BINOMIAL FORMULA, NEGATIVE BINOMIAL SERIES, q-BINOMIAL THEOREM, RANDOM WALK

References

Abramowitz, M. and Stegun, C. A. (Eds.). *Handbook of Mathematical Functions with Formulas, Graphs, and Mathematical Tables, 9th printing.* New York: Dover, p. 10, 1972.

Arfken, G. *Mathematical Methods for Physicists, 3rd ed.* Orlando, FL: Academic Press, pp. 307–308, 1985.

Conway, J. H. and Guy, R. K. "Choice Numbers Are Binomial Coefficients." In *The Book of Numbers.* New York: Springer-Verlag, pp. 72–74, 1996.

Coolidge, J. L. "The Story of the Binomial Theorem." *Amer. Math. Monthly* **56**, 147–157, 1949.

Courant, R. and Robbins, H. "The Binomial Theorem." §1.6 in *What is Mathematics?: An Elementary Approach to Ideas and Methods, 2nd ed.* Oxford, England: Oxford University Press, pp. 16–18, 1996.

Binomial Triangle

see PASCAL'S TRIANGLE

Binormal Developable

A RULED SURFACE M is said to be a binormal developable of a curve $\mathbf{y}$ if M can be parameterized by $\mathbf{x}(u,v) = \mathbf{y}(u) + v\hat{\mathbf{B}}(u)$, where $\mathbf{B}$ is the BINORMAL VECTOR.

see also NORMAL DEVELOPABLE, TANGENT DEVELOPABLE

References

Gray, A. "Developables." §17.6 in *Modern Differential Geometry of Curves and Surfaces.* Boca Raton, FL: CRC Press, pp. 352–354, 1993.

Binormal Vector

$$\hat{\mathbf{B}} \equiv \hat{\mathbf{T}} \times \hat{\mathbf{N}} \tag{1}$$

$$= \frac{\mathbf{r}' \times \mathbf{r}''}{|\mathbf{r}' \times \mathbf{r}''|}, \tag{2}$$

where the unit TANGENT VECTOR $\mathbf{T}$ and unit "principal" NORMAL VECTOR $\mathbf{N}$ are defined by

$$\hat{\mathbf{T}} \equiv \frac{\mathbf{r}'(s)}{|\mathbf{r}'(s)|} \tag{3}$$

$$\hat{\mathbf{N}} \equiv \frac{\mathbf{r}''(s)}{|\mathbf{r}''(s)|} \tag{4}$$

Here, $\mathbf{r}$ is the RADIUS VECTOR, s is the ARC LENGTH, τ is the TORSION, and κ is the CURVATURE. The binormal vector satisfies the remarkable identity

$$[\dot{\mathbf{B}}, \ddot{\mathbf{B}}, \dddot{\mathbf{B}}] = \tau^5 \frac{d}{ds}\left(\frac{\kappa}{\tau}\right). \tag{5}$$

see also FRENET FORMULAS, NORMAL VECTOR, TANGENT VECTOR

References

Kreyszig, E. "Binormal. Moving Trihedron of a Curve." §13 in *Differential Geometry.* New York: Dover, p. 36–37, 1991.

Bioche's Theorem

If two complementary PLÜCKER CHARACTERISTICS are equal, then each characteristic is equal to its complement except in four cases where the sum of order and class is 9.

References

Coolidge, J. L. *A Treatise on Algebraic Plane Curves.* New York: Dover, p. 101, 1959.

Biotic Potential

see LOGISTIC EQUATION

Bipartite Graph

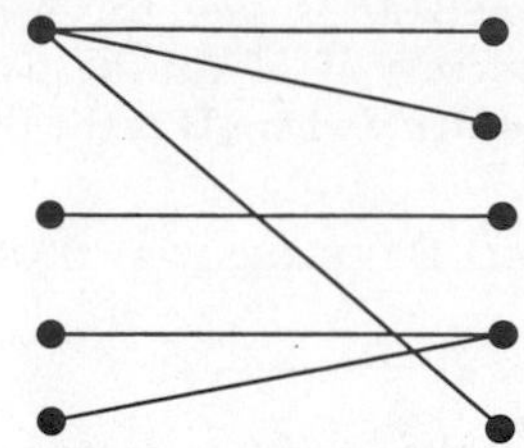

A set of VERTICES decomposed into two disjoint sets such that no two VERTICES within the same set are adjacent. A bigraph is a special case of a k-PARTITE GRAPH with $k = 2$.

see also COMPLETE BIPARTITE GRAPH, k-PARTITE GRAPH, KÖNIG-EGEVÁRY THEOREM

References

Chartrand, G. *Introductory Graph Theory.* New York: Dover, p. 116, 1985.

Saaty, T. L. and Kainen, P. C. *The Four-Color Problem: Assaults and Conquest.* New York: Dover, p. 12, 1986.

Biplanar Double Point

see ISOLATED SINGULARITY

Bipolar Coordinates

Bipolar coordinates are a 2-D system of coordinates. There are two commonly defined types of bipolar coordinates, the first of which is defined by

$$x = \frac{a \sinh v}{\cosh v - \cos u} \tag{1}$$

$$y = \frac{a \sin u}{\cosh v - \cos u}, \tag{2}$$

where $u \in [0, 2\pi)$, $v \in (-\infty, \infty)$. The following identities show that curves of constant u and v are CIRCLES in xy-space.

$$x^2 + (y - a \cot u)^2 = a^2 \csc^2 u \tag{3}$$

$$(x - a \coth v)^2 + y^2 = a^2 \operatorname{csch}^2 v. \tag{4}$$

The SCALE FACTORS are

$$h_u = \frac{a}{\cosh v - \cos u} \tag{5}$$

$$h_v = \frac{a}{\cosh v - \cos u}. \tag{6}$$

The LAPLACIAN is

$$\nabla^2 = \frac{(\cosh v - \cos u)^2}{a^2} \left(\frac{\partial^2}{\partial u^2} + \frac{\partial^2}{\partial v^2} \right). \tag{7}$$

LAPLACE'S EQUATION is separable.

Two-center bipolar coordinates are two coordinates giving the distances from two fixed centers r_1 and r_2, sometimes denoted r and r'. For two-center bipolar coordinates with centers at $(\pm c, 0)$,

$$r_1{}^2 = (x + c)^2 + y^2 \tag{8}$$

$$r_2{}^2 = (x - c)^2 + y^2. \tag{9}$$

Combining (8) and (9) gives

$$r_1{}^2 - r_2{}^2 = 4cx. \tag{10}$$

Solving for CARTESIAN COORDINATES x and y gives

$$x = \frac{r_1{}^2 - r_2{}^2}{4c} \tag{11}$$

$$y = \pm \frac{1}{4c} \sqrt{16c^2 r_1{}^2 - (r_1{}^2 - r_2{}^2 + 4c^2)}. \tag{12}$$

Solving for POLAR COORDINATES gives

$$r = \sqrt{\frac{r_1{}^2 + r_2{}^2 - 2c^2}{2}} \tag{13}$$

$$\theta = \tan^{-1} \left[\sqrt{\frac{8c^2(r_1{}^2 + r_2{}^2 - 2c^2)}{r_1{}^2 - r_2{}^2} - 1} \right]. \tag{14}$$

References

Lockwood, E. H. "Bipolar Coordinates." Ch. 25 in *A Book of Curves.* Cambridge, England: Cambridge University Press, pp. 186–190, 1967.

Bipolar Cylindrical Coordinates

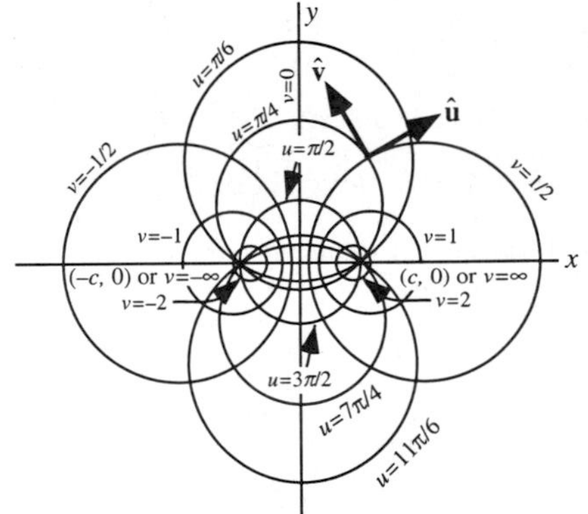

A set of CURVILINEAR COORDINATES defined by

$$x = \frac{a \sinh v}{\cosh v - \cos u} \tag{1}$$

$$y = \frac{a \sin u}{\cosh v - \cos u} \tag{2}$$

$$z = z, \tag{3}$$

where $u \in [0, 2\pi)$, $v \in (-\infty, \infty)$, and $z \in (-\infty, \infty)$. There are several notational conventions, and whereas (u, v, z) is used in this work, Arfken (1970) prefers

(η, ξ, z). The following identities show that curves of constant u and v are CIRCLES in xy-space.

$$x^2 + (y - a\cot u)^2 = a^2 \csc^2 u \quad (4)$$

$$(x - a\coth v)^2 + y^2 = a^2 \operatorname{csch}^2 v. \quad (5)$$

The SCALE FACTORS are

$$h_u = \frac{a}{\cosh v - \cos u} \quad (6)$$

$$h_v = \frac{a}{\cosh v - \cos u} \quad (7)$$

$$h_z = 1. \quad (8)$$

The LAPLACIAN is

$$\nabla^2 = \frac{(\cosh v - \cos u)^2}{a^2}\left(\frac{\partial^2}{\partial u^2} + \frac{\partial^2}{\partial v^2}\right) + \frac{\partial^2}{\partial z^2}. \quad (9)$$

LAPLACE'S EQUATION is not separable in BIPOLAR CYLINDRICAL COORDINATES, but it is in 2-D BIPOLAR COORDINATES.

References
Arfken, G. "Bipolar Coordinates (ξ, η, z)." §2.9 in *Mathematical Methods for Physicists, 2nd ed.* Orlando, FL: Academic Press, pp. 97–102, 1970.

Biprism

Two slant triangular PRISMS fused together.

see also PRISM, SCHMITT-CONWAY BIPRISM

Bipyramid

see DIPYRAMID

Biquadratefree

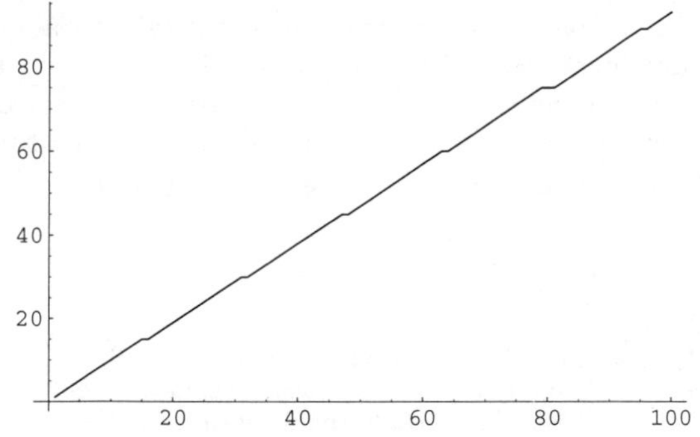

A number is said to be biquadratefree if its PRIME decomposition contains no tripled factors. All PRIMES are therefore trivially biquadratefree. The biquadratefree numbers are 1, 2, 3, 4, 5, 6, 7, 8, 9, 10, 11, 12, 13, 14, 15, 17, ... (Sloane's A046100). The biquadrateful numbers (i.e., those that contain at least one biquadrate) are 16, 32, 48, 64, 80, 81, 96, ... (Sloane's A046101). The number of biquadratefree numbers less than 10, 100, 1000, ... are 10, 93, 925, 9240, 92395, 923939, ..., and their asymptotic density is $1/\zeta(4) = 90/\pi^4 \approx 0.923938$, where $\zeta(n)$ is the RIEMANN ZETA FUNCTION.

see also CUBEFREE, PRIME NUMBER, RIEMANN ZETA FUNCTION, SQUAREFREE

References
Sloane, N. J. A. Sequences A046100 and A046101 in "An On-Line Version of the Encyclopedia of Integer Sequences."

Biquadratic Equation

see QUARTIC EQUATION

Biquadratic Number

A biquadratic number is a fourth POWER, n^4. The first few biquadratic numbers are 1, 16, 81, 256, 625, ... (Sloane's A000583). The minimum number of squares needed to represent the numbers 1, 2, 3, ... are 1, 2, 3, 4, 5, 6, 7, 8, 9, 10, 11, 12, 13, 14, 15, 1, 2, 3, 4, 5, ... (Sloane's A002377), and the number of distinct ways to represent the numbers 1, 2, 3, ... in terms of biquadratic numbers are 1, 1, 1, 1, 1, 1, 1, 1, 1, 1, 1, 1, 1, 1, 2, 2, 2, 2, 2, ... A brute-force algorithm for enumerating the biquadratic permutations of n is repeated application of the GREEDY ALGORITHM.

Every POSITIVE integer is expressible as a SUM of (at most) $g(4) = 19$ biquadratic numbers (WARING'S PROBLEM). Davenport (1939) showed that $G(4) = 16$, meaning that all sufficiently large integers require only 16 biquadratic numbers. The following table gives the first few numbers which require 1, 2, 3, ..., 19 biquadratic numbers to represent them as a sum, with the sequences for 17, 18, and 19 being finite.

#	Sloane	Numbers
1	000290	1, 16, 81, 256, 625, 1296, 2401, 4096, ...
2	003336	2, 17, 32, 82, 97, 162, 257, 272, ...
3	003337	3, 18, 33, 48, 83, 98, 113, 163, ...
4	003338	4, 19, 34, 49, 64, 84, 99, 114, 129, ...
5	003339	5, 20, 35, 50, 65, 80, 85, 100, 115, ...
6	003340	6, 21, 36, 51, 66, 86, 96, 101, 116, ...
7	003341	7, 22, 37, 52, 67, 87, 102, 112, 117, ...
8	003342	8, 23, 38, 53, 68, 88, 103, 118, 128, ...
9	003343	9, 24, 39, 54, 69, 89, 104, 119, 134, ...
10	003344	10, 25, 40, 55, 70, 90, 105, 120, 135, ...
11	003345	11, 26, 41, 56, 71, 91, 106, 121, 136, ...
12	003346	12, 27, 42, 57, 72, 92, 107, 122, 137, ...

The following table gives the numbers which can be represented in n different ways as a sum of k biquadrates.

k	n	Sloane	Numbers
1	1	000290	1, 16, 81, 256, 625, 1296, 2401, 4096, ...
2	2		635318657, 3262811042, 8657437697, ...

The numbers 2, 3, 4, 5, 6, 7, 8, 9, 10, 11, 12, 13, 14, 15, 18, 19, 20, 21, ... (Sloane's A046039) cannot be represented using distinct biquadrates.

see also CUBIC NUMBER, SQUARE NUMBER, WARING'S PROBLEM

References
Davenport, H. "On Waring's Problem for Fourth Powers." *Ann. Math.* **40**, 731–747, 1939.

Biquadratic Reciprocity Theorem

$$x^4 \equiv q \pmod{p}. \qquad (1)$$

This was solved by Gauss using the GAUSSIAN INTEGERS as

$$\left(\frac{\pi}{\sigma}\right)_4 \left(\frac{\sigma}{\pi}\right)_4 = (-1)^{[(N(\pi)-1)/4][(N(\sigma)-1)/4]}, \qquad (2)$$

where π and σ are distinct GAUSSIAN INTEGER PRIMES,

$$N(a+bi) = \sqrt{a^2+b^2} \qquad (3)$$

and N is the norm.

$$\left(\frac{\alpha}{\pi}\right)_4 = \begin{cases} 1 & \text{if } x^4 \equiv \alpha \pmod{\pi} \text{ is solvable} \\ -1, i, \text{ or } -i & \text{otherwise,} \end{cases} \qquad (4)$$

where solvable means solvable in terms of GAUSSIAN INTEGERS.

see also RECIPROCITY THEOREM

Biquaternion

A QUATERNION with COMPLEX coefficients. The ALGEBRA of biquaternions is isomorphic to a full matrix ring over the complex number field (van der Waerden 1985).

see also QUATERNION

References

Clifford, W. K. "Preliminary Sketch of Biquaternions." *Proc. London Math. Soc.* **4**, 381–395, 1873.

Hamilton, W. R. *Lectures on Quaternions: Containing a Systematic Statement of a New Mathematical Method.* Dublin: Hodges and Smith, 1853.

Study, E. "Von den Bewegung und Umlegungen." *Math. Ann.* **39**, 441–566, 1891.

van der Waerden, B. L. *A History of Algebra from al-Khwarizmi to Emmy Noether.* New York: Springer-Verlag, pp. 188–189, 1985.

Birational Transformation

A transformation in which coordinates in two SPACES are expressed rationally in terms of those in another.

see also RIEMANN CURVE THEOREM, WEBER'S THEOREM

Birch Conjecture

see SWINNERTON-DYER CONJECTURE

Birch–Swinnerton-Dyer Conjecture

see SWINNERTON-DYER CONJECTURE

Birkhoff's Ergodic Theorem

Let T be an ergodic ENDOMORPHISM of the PROBABILITY SPACE X and let $f: X \to \mathbb{R}$ be a real-valued MEASURABLE FUNCTION. Then for ALMOST EVERY $x \in X$, we have

$$\frac{1}{n}\sum_{j=1}^n f \circ F^j(x) \to \int f\,dm$$

as $n \to \infty$. To illustrate this, take f to be the characteristic function of some SUBSET A of X so that

$$f(x) = \begin{cases} 1 & \text{if } x \in A \\ 0 & \text{if } x \notin A. \end{cases}$$

The left-hand side of (-1) just says how often the orbit of x (that is, the points x, Tx, T^2x, ...) lies in A, and the right-hand side is just the MEASURE of A. Thus, for an ergodic ENDOMORPHISM, "space-averages = time-averages almost everywhere." Moreover, if T is continuous and uniquely ergodic with BOREL PROBABILITY MEASURE m and f is continuous, then we can replace the ALMOST EVERYWHERE convergence in (-1) to everywhere.

Birotunda

Two adjoined ROTUNDAS.

see also BILUNABIROTUNDA, CUPOLAROTUNDA, ELONGATED GYROCUPOLAROTUNDA, ELONGATED ORTHOCUPOLAROTUNDA, ELONGATED ORTHOBIROTUNDA, GYROCUPOLAROTUNDA, GYROELONGATED ROTUNDA, ORTHOBIROTUNDA, TRIANGULAR HEBESPHENOROTUNDA

Birthday Attack

Birthday attacks are a class of brute-force techniques used in an attempt to solve a class of cryptographic hash function problems. These methods take advantage of functions which, when supplied with a random input, return one of k equally likely values. By repeatedly evaluating the function for different inputs, the same output is expected to be obtained after about $1.2\sqrt{k}$ evaluations.

see also BIRTHDAY PROBLEM

References

RSA Laboratories. "Question 95. What is a Birthday Attack." `http://www.rsa.com/rsalabs/newfaq/q95.html`. "Question 96. How Does the Length of a Hash Value Affect Security?" `http://www.rsa.com/rsalabs/newfaq/q96.html`.

van Oorschot, P. and Wiener, M. "A Known Plaintext Attack on Two-Key Triple Encryption." In *Advances in Cryptology—Eurocrypt '90.* New York: Springer-Verlag, pp. 366–377, 1991.

Yuval, G. "How to Swindle Rabin." *Cryptologia* **3**, 187–189, Jul. 1979.

Birthday Problem

Consider the probability $Q_1(n,d)$ that *no two people* out of a group of n will have matching birthdays out of d equally possible birthdays. Start with an arbitrary person's birthday, then note that the probability that the second person's birthday is different is $(d-1)/d$, that the third person's birthday is different from the first two is $[(d-1)/d][(d-2)/d]$, and so on, up through the nth person. Explicitly,

$$Q_1(n,d) = \frac{d-1}{d}\frac{d-2}{d}\cdots\frac{d-(n-1)}{d} = \frac{(d-1)(d-2)\cdots[d-(n-1)]}{d^n}. \quad (1)$$

But this can be written in terms of FACTORIALS as

$$Q_1(n,d) = \frac{d!}{(d-n)!d^n}, \quad (2)$$

so the probability $P_2(n,365)$ that two people out of a group of n *do have the same birthday* is therefore

$$P_2(n,d) = 1 - Q_1(n,d) = 1 - \frac{d!}{(d-n)!d^n}. \quad (3)$$

If 365-day years have been assumed, i.e., the existence of leap days is ignored, then the number of people needed for there to be at least a 50% chance that two share birthdays is the smallest n such that $P_2(n,365) \geq 1/2$. This is given by $n = 23$, since

$$P_2(23,365) = \frac{38093904702297390785243708291056390518886454060947061}{75091883268515350125426207425223147563269805908203125} \approx 0.507297. \quad (4)$$

The number of people needed to obtain $P_2(n,365) \geq 1/2$ for $n = 1, 2, \ldots$, are 2, 2, 3, 3, 3, 4, 4, 4, 4, 5, ... (Sloane's A033810).

The probability $P_2(n,d)$ can be estimated as

$$P_2(n,d) \approx 1 - e^{-n(n-1)/2d} \quad (5)$$

$$\approx 1 - \left(1 - \frac{n}{2d}\right)^{n-1}, \quad (6)$$

where the latter has error

$$\epsilon < \frac{n^3}{6(d-n+1)^2} \quad (7)$$

(Sayrafiezadeh 1994).

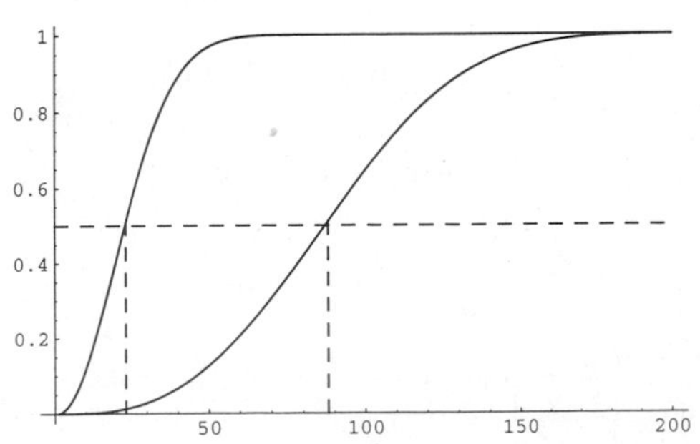

In general, let $Q_i(n,d)$ denote the probability that a birthday is shared by exactly i (and no more) people out of a group of n people. Then the probability that a birthday is shared by k *or more* people is given by

$$P_k(n,d) = 1 - \sum_{i=1}^{k-1} Q_i(n,d). \quad (8)$$

Q_2 can be computed explicitly as

$$Q_2(n,d) = \frac{n!}{d^n}\sum_{i=1}^{\lfloor n/2\rfloor}\frac{1}{2^i}\binom{d}{i}\binom{d-i}{n-2i}$$

$$= \frac{n!}{d^n}\sum_{i=1}^{\lfloor n/2\rfloor}\frac{d!}{2^i i!(n-2i)!(d-n+i)!}$$

$$= \frac{(-1)^n}{d^n}\left[2^{-n/2}\Gamma(1+n)P_n^{(-d)}(\tfrac{1}{2}\sqrt{2}) - \frac{\Gamma(1+d)}{\Gamma(1+d-n)}\right], \quad (9)$$

where $\binom{n}{m}$ is a BINOMIAL COEFFICIENT, $\Gamma(n)$ is a GAMMA FUNCTION, and $P_n^{(\lambda)}(x)$ is an ULTRASPHERICAL POLYNOMIAL. This gives the explicit formula for $P_3(n,d)$ as

$$P_3(n,d) = 1 - Q_1(n,d) - Q_2(n,d) = 1 + \frac{(-1)^{n+1}\Gamma(n+1)P_n^{(-d)}(2^{-1/2})}{2^{n/2}d^n}. \quad (10)$$

$Q_3(n,d)$ cannot be computed in entirely closed form, but a partially reduced form is

$$Q_3(n,d) = \frac{\Gamma(d+1)}{d^n}\left[\frac{(-1)^n F(\frac{9}{8}) - F(-\frac{9}{8})}{\Gamma(d-n+1)} + (-1)^n\Gamma(1+n)\sum_{i=1}^{\lfloor n/3\rfloor}\frac{(-3)^{-i}2^{(i-n)/2}P_{n-3i}^{(i-d)}(\frac{1}{2}\sqrt{2})}{\Gamma(d-i+1)\Gamma(i+1)}\right], \quad (11)$$

where

$$F = F(n,d,a) \equiv 1 - {}_3F_2\left[\begin{matrix}\frac{1}{3}(1-n), \frac{1}{3}(2-n), -\frac{1}{3} \\ \frac{1}{2}(d-n+1), \frac{1}{2}(d-n+2)\end{matrix}; a\right] \quad (12)$$

and ${}_3F_2(a,b,c;d,e;z)$ is a GENERALIZED HYPERGEOMETRIC FUNCTION.

In general, $Q_k(n,d)$ can be computed using the RECURRENCE RELATION

$$Q_k(n,d) = \sum_{i=1}^{\lfloor n/k\rfloor}\left[\frac{n!d!}{d^{ik}i!(k!)^i(n-ik)!(d-i)!} \times \sum_{j=1}^{k-1} Q_j(n-k,d-i)\frac{(d-i)^{n-ik}}{d^{n-ik}}\right] \quad (13)$$

(Finch). However, the time to compute this recursive function grows exponentially with k and so rapidly becomes unwieldy. The minimal number of people to give a 50% probability of having at least n coincident birthdays is 1, 23, 88, 187, 313, 460, 623, 798, 985, 1181, 1385, 1596, 1813, ... (Sloane's A014088; Diaconis and Mosteller 1989).

A good approximation to the number of people n such that $p = P_k(n, d)$ is some given value can given by solving the equation

$$ne^{-n/(dk)} = \left[d^{k-1}k! \ln\left(\frac{1}{1-p}\right)\left(1 - \frac{n}{d(k+1)}\right)\right]^{1/k} \tag{14}$$

for n and taking $\lceil n \rceil$, where $\lceil n \rceil$ is the CEILING FUNCTION (Diaconis and Mosteller 1989). For $p = 0.5$ and $k = 1, 2, 3, \ldots$, this formula gives $n = 1$, 23, 88, 187, 313, 459, 722, 797, 983, 1179, 1382, 1592, 1809, ..., which differ from the true values by from 0 to 4. A much simpler but also poorer approximation for n such that $p = 0.5$ for $k < 20$ is given by

$$n = 47(k - 1.5)^{3/2} \tag{15}$$

(Diaconis and Mosteller 1989), which gives 86, 185, 307, 448, 606, 778, 965, 1164, 1376, 1599, 1832, ... for $k = 3$, 4, ...

The "almost" birthday problem, which asks the number of people needed such that two have a birthday within a day of each other, was considered by Abramson and Moser (1970), who showed that 14 people suffice. An approximation for the minimum number of people needed to get a 50-50 chance that two have a match within k days out of d possible is given by

$$n(k, d) = 1.2\sqrt{\frac{d}{2k+1}} \tag{16}$$

(Sevast'yanov 1972, Diaconis and Mosteller 1989).

see also BIRTHDAY ATTACK, COINCIDENCE, SMALL WORLD PROBLEM

References

Abramson, M. and Moser, W. O. J. "More Birthday Surprises." *Amer. Math. Monthly* **77**, 856–858, 1970.

Ball, W. W. R. and Coxeter, H. S. M. *Mathematical Recreations and Essays, 13th ed.* New York: Dover, pp. 45–46, 1987.

Bloom, D. M. "A Birthday Problem." *Amer. Math. Monthly* **80**, 1141–1142, 1973.

Bogomolny, A. "Coincidence." http://www.cut-the-knot.com/do_you_know/coincidence.html.

Clevenson, M. L. and Watkins, W. "Majorization and the Birthday Inequality." *Math. Mag.* **64**, 183–188, 1991.

Diaconis, P. and Mosteller, F. "Methods of Studying Coincidences." *J. Amer. Statist. Assoc.* **84**, 853–861, 1989.

Feller, W. *An Introduction to Probability Theory and Its Applications, Vol. 1, 3rd ed.* New York: Wiley, pp. 31–32, 1968.

Finch, S. "Puzzle #28 [June 1997]: Coincident Birthdays." http://www.mathsoft.com/mathcad/library/puzzle/soln28/soln28.html.

Gehan, E. A. "Note on the 'Birthday Problem.'" *Amer. Stat.* **22**, 28, Apr. 1968.

Heuer, G. A. "Estimation in a Certain Probability Problem." *Amer. Math. Monthly* **66**, 704–706, 1959.

Hocking, R. L. and Schwertman, N. C. "An Extension of the Birthday Problem to Exactly k Matches." *College Math. J.* **17**, 315–321, 1986.

Hunter, J. A. H. and Madachy, J. S. *Mathematical Diversions.* New York: Dover, pp. 102–103, 1975.

Klamkin, M. S. and Newman, D. J. "Extensions of the Birthday Surprise." *J. Combin. Th.* **3**, 279–282, 1967.

Levin, B. "A Representation for Multinomial Cumulative Distribution Functions." *Ann. Statistics* **9**, 1123–1126, 1981.

McKinney, E. H. "Generalized Birthday Problem." *Amer. Math. Monthly* **73**, 385–387, 1966.

Mises, R. von. "Über Aufteilungs—und Besetzungs-Wahrscheinlichkeiten." *Revue de la Faculté des Sciences de l'Université d'Istanbul, N. S.* **4**, 145–163, 1939. Reprinted in *Selected Papers of Richard von Mises, Vol. 2* (Ed. P. Frank, S. Goldstein, M. Kac, W. Prager, G. Szegő, and G. Birkhoff). Providence, RI: Amer. Math. Soc., pp. 313–334, 1964.

Riesel, H. *Prime Numbers and Computer Methods for Factorization, 2nd ed.* Boston, MA: Birkhäuser, pp. 179–180, 1994.

Sayrafiezadeh, M. "The Birthday Problem Revisited." *Math. Mag.* **67**, 220–223, 1994.

Sevast'yanov, B. A. "Poisson Limit Law for a Scheme of Sums of Dependent Random Variables." *Th. Prob. Appl.* **17**, 695–699, 1972.

Sloane, N. J. A. Sequences A014088 and A033810 in "An On-Line Version of the Encyclopedia of Integer Sequences."

Stewart, I. "What a Coincidence!" *Sci. Amer.* **278**, 95–96, June 1998.

Tesler, L. "Not a Coincidence!" http://www.nomodes.com/coincidence.html.

Bisected Perimeter Point

see NAGEL POINT

Bisection Procedure

Given an interval $[a, b]$, let a_n and b_n be the endpoints at the nth iteration and r_n be the nth approximate solution. Then, the number of iterations required to obtain an error smaller than ϵ is found as follows.

$$b_n - a_n = \frac{1}{2^{n-1}}(b - a) \tag{1}$$

$$r_n \equiv \tfrac{1}{2}(a_n + b_n) \tag{2}$$

$$|r_n - r| \leq \tfrac{1}{2}(b_n - a_n) = 2^{-n}(b - a) < \epsilon \tag{3}$$

$$-n \ln 2 < \ln \epsilon - \ln(b - a), \tag{4}$$

so

$$n > \frac{\ln(b - a) - \ln \epsilon}{\ln 2}. \tag{5}$$

see also ROOT

References

Arfken, G. *Mathematical Methods for Physicists, 3rd ed.* Orlando, FL: Academic Press, pp. 964–965, 1985.

Press, W. H.; Flannery, B. P.; Teukolsky, S. A.; and Vetterling, W. T. "Bracketing and Bisection." §9.1 in *Numerical Recipes in FORTRAN: The Art of Scientific Computing, 2nd ed.* Cambridge, England: Cambridge University Press, pp. 343–347, 1992.

Bisector

Bisection is the division of a given curve or figure into two equal parts (halves).

see also ANGLE BISECTOR, BISECTION PROCEDURE, EXTERIOR ANGLE BISECTOR, HALF, HEMISPHERE, LINE BISECTOR, PERPENDICULAR BISECTOR, TRISECTION

Bishop's Inequality

Let $V(r)$ be the volume of a BALL of radius r in a complete n-D RIEMANNIAN MANIFOLD with RICCI CURVATURE $\geq (n-1)\kappa$. Then $V(r) \geq V_\kappa(r)$, where V_κ is the volume of a BALL in a space having constant SECTIONAL CURVATURE. In addition, if equality holds for some BALL, then this BALL is ISOMETRIC to the BALL of radius r in the space of constant SECTIONAL CURVATURE κ.

References

Chavel, I. *Riemannian Geometry: A Modern Introduction.* New York: Cambridge University Press, 1994.

Bishops Problem

B							
B							B
B							B
B							B
B							B
B							B
B							B
B							

Find the maximum number of bishops $B(n)$ which can be placed on an $n \times n$ CHESSBOARD such that no two attack each other. The answer is $2n-2$ (Dudeney 1970, Madachy 1979), giving the sequence 2, 4, 6, 8, ... (the EVEN NUMBERS) for $n = 2, 3, \ldots$. One maximal solution for $n = 8$ is illustrated above. The number of distinct maximal arrangements of bishops for $n = 1, 2, \ldots$ are 1, 4, 26, 260, 3368, ... (Sloane's A002465). The number of rotationally and reflectively distinct solutions on an $n \times n$ board for $n \geq 2$ is

$$B(n) = \begin{cases} 2^{(n-4)/2}[2^{(n-2)/2}+1] & \text{for } n \text{ even} \\ 2^{(n-3)/2}[2^{(n-3)/2}+1] & \text{for } n \text{ odd} \end{cases}$$

(Dudeney 1970, p. 96; Madachy 1979, p. 45; Pickover 1995). An equivalent formula is

$$B(n) = 2^{n-3} + 2^{\lfloor (n-1)/2 \rfloor - 1},$$

where $\lfloor n \rfloor$ is the FLOOR FUNCTION, giving the sequence for $n = 1, 2, \ldots$ as 1, 1, 2, 3, 6, 10, 20, 36, ... (Sloane's A005418).

			B				
			B				
			B				
			B				
			B				
			B				
			B				
			B				

The minimum number of bishops needed to occupy or attack all squares on an $n \times n$ CHESSBOARD is n, arranged as illustrated above.

see also CHESS, KINGS PROBLEM, KNIGHTS PROBLEM, QUEENS PROBLEM, ROOKS PROBLEM

References

Ahrens, W. *Mathematische Unterhaltungen und Spiele, Vol. 1, 3rd ed.* Leipzig, Germany: Teubner, p. 271, 1921.
Dudeney, H. E. "Bishops—Unguarded" and "Bishops—Guarded." §297 and 298 in *Amusements in Mathematics.* New York: Dover, pp. 88–89, 1970.
Guy, R. K. "The n Queens Problem." §C18 in *Unsolved Problems in Number Theory, 2nd ed.* New York: Springer-Verlag, pp. 133–135, 1994.
Madachy, J. *Madachy's Mathematical Recreations.* New York: Dover, pp. 36–46, 1979.
Pickover, C. A. *Keys to Infinity.* New York: Wiley, pp. 74–75, 1995.
Sloane, N. J. A. Sequences A002465/M3616 and A005418/M0771 in "An On-Line Version of the Encyclopedia of Integer Sequences."

Bislit Cube

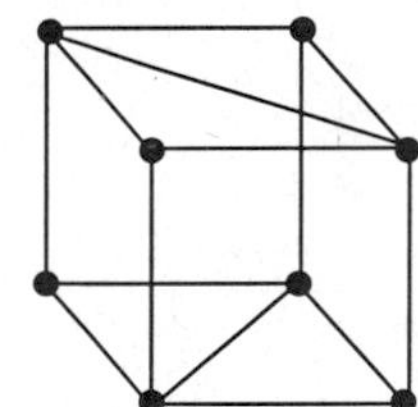
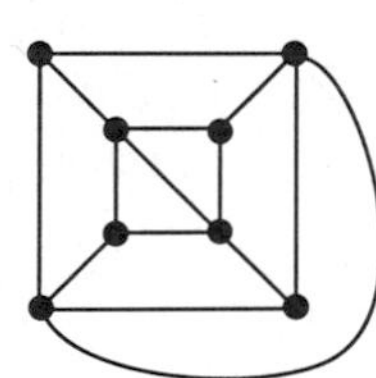

The 8-VERTEX graph consisting of a CUBE in which two opposite faces have DIAGONALS oriented PERPENDICULAR to each other.

see also BIDIAKIS CUBE, CUBE, CUBICAL GRAPH

Bispherical Coordinates

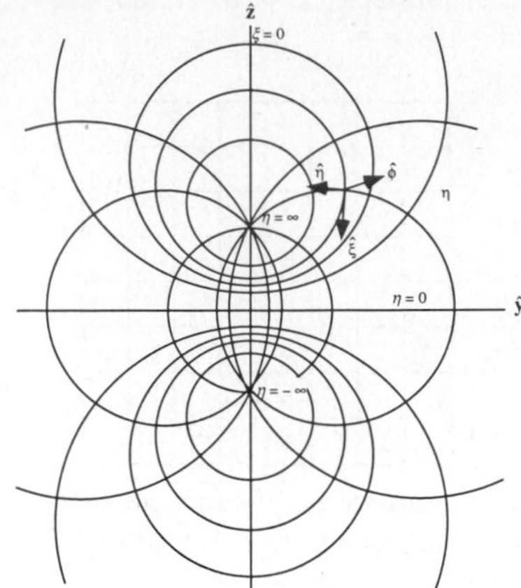

A system of CURVILINEAR COORDINATES defined by

$$x = \frac{a \sin\xi \cos\phi}{\cosh\eta - \cos\xi} \tag{1}$$

$$y = \frac{a \sin\xi \sin\phi}{\cosh\eta - \cos\xi} \tag{2}$$

$$z = \frac{a \sinh\eta}{\cosh\eta - \cos\xi}. \tag{3}$$

The SCALE FACTORS are

$$h_\xi = \frac{a}{\cos\eta - \cos\xi} \tag{4}$$

$$h_\eta = \frac{a}{\cosh\eta - \cos\xi} \tag{5}$$

$$h_\phi = \frac{a \sin\xi}{\cosh\eta - \cos\xi}. \tag{6}$$

The LAPLACIAN is

$$\begin{aligned}\nabla^2 = &\left(\frac{-\cos u \cot^2 u + 3\cosh v \cot^2 u}{\cosh v - \cos u}\right. \\ &\left.+\frac{-3\cosh^2 v \cot u \csc u + \cosh^3 v \csc^2 u}{\cosh v - \cos u}\right)\frac{\partial}{\partial\phi^2} \\ &+(\cos u - \cosh v)\sinh v \frac{\partial}{\partial v} + (\cosh^2 v - \cos u)^2 \frac{\partial^2}{\partial v^2} \\ &+(\cosh v - \cos u)(\cosh v \cot u - \sin u - \cos u \cot u)\frac{\partial}{\partial u} \\ &+(\cosh^2 v - \cos u)^2 \frac{\partial^2}{\partial u^2}.\end{aligned} \tag{7}$$

In bispherical coordinates, LAPLACE'S EQUATION is separable, but the HELMHOLTZ DIFFERENTIAL EQUATION is not.

see also LAPLACE'S EQUATION—BISPHERICAL COORDINATES, TOROIDAL COORDINATES

References

Arfken, G. "Bispherical Coordinates (ξ, η, ϕ)." §2.14 in *Mathematical Methods for Physicists, 2nd ed.* Orlando, FL: Academic Press, pp. 115–117, 1970.

Morse, P. M. and Feshbach, H. *Methods of Theoretical Physics, Part I.* New York: McGraw-Hill, pp. 665–666, 1953.

Bit Complexity

The number of single operations (of ADDITION, SUBTRACTION, and MULTIPLICATION) required to complete an algorithm.

see also STRASSEN FORMULAS

References

Borodin, A. and Munro, I. *The Computational Complexity of Algebraic and Numeric Problems.* New York: American Elsevier, 1975.

Bitangent

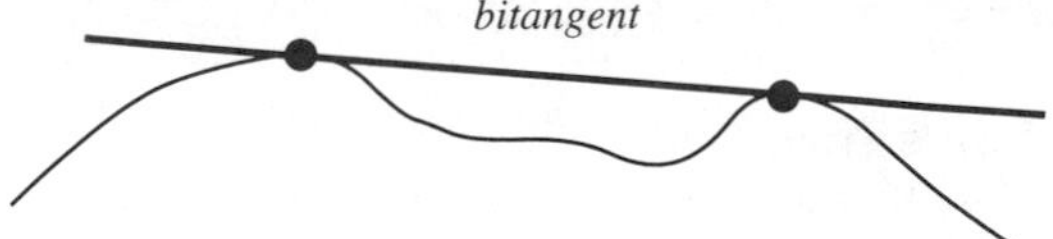

A LINE which is TANGENT to a curve at two distinct points.

see also KLEIN'S EQUATION, PLÜCKER CHARACTERISTICS, SECANT LINE, SOLOMON'S SEAL LINES, TANGENT LINE

Bivariate Distribution

see GAUSSIAN BIVARIATE DISTRIBUTION

Bivector

An antisymmetric TENSOR of second RANK (a.k.a. 2-form).

$$\vec{X} = X_{ab}\omega^a \wedge \omega^b,$$

where $\wedge$ is the WEDGE PRODUCT (or OUTER PRODUCT).

Biweight

see TUKEY'S BIWEIGHT

Black-Scholes Theory

The theory underlying financial derivatives which involves "stochastic calculus" and assumes an uncorrelated LOG NORMAL DISTRIBUTION of continuously varying prices. A simplified "binomial" version of the theory was subsequently developed by Sharpe *et al.* (1995) and Cox *et al.* (1979). It reproduces many results of the full-blown theory, and allows approximation of options for which analytic solutions are not known (Price 1996).

see also GARMAN-KOHLHAGEN FORMULA

References

Black, F. and Scholes, M. S. "The Pricing of Options and Corporate Liabilities." *J. Political Econ.* **81**, 637–659, 1973.

Cox, J. C.; Ross, A.; and Rubenstein, M. "Option Pricing: A Simplified Approach." *J. Financial Economics* **7**, 229–263, 1979.

Price, J. F. "Optional Mathematics is Not Optional." *Not. Amer. Math. Soc.* **43**, 964–971, 1996.

Sharpe, W. F.; Alexander, G. J.; and Bailey, J. V. *Investments, 5th ed.* Englewood Cliffs, NJ: Prentice-Hall, 1995.

Black Spleenwort Fern

see BARNSLEY'S FERN

Blackman Function

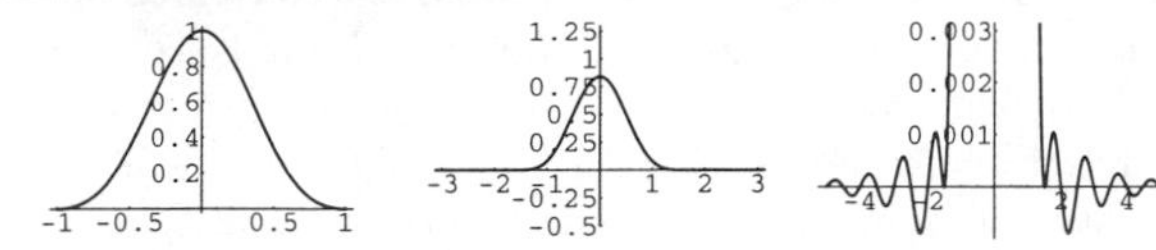

An APODIZATION FUNCTION given by

$$A(x) = 0.42 + 0.5\cos\left(\frac{\pi x}{a}\right) + 0.08\cos\left(\frac{2\pi x}{a}\right). \quad (1)$$

Its FULL WIDTH AT HALF MAXIMUM is $0.810957a$. The APPARATUS FUNCTION is

$$I(k) = \frac{a(0.84 - 0.36a^2k^2 - 2.17\times 10^{-19}a^4k^4)\sin(2\pi ak)}{(1-a^2k^2)(1-4a^2k^2)}. \quad (2)$$

The COEFFICIENTS are approximations to

$$a_0 = \frac{3969}{9304} \quad (3)$$

$$a_1 = \frac{1155}{4652} \quad (4)$$

$$a_2 = \frac{715}{18608}, \quad (5)$$

which would have produced zeros of $I(k)$ at $k = (7/4)a$ and $k = (9/4)a$.

see also APODIZATION FUNCTION

References

Blackman, R. B. and Tukey, J. W. "Particular Pairs of Windows." In *The Measurement of Power Spectra, From the Point of View of Communications Engineering.* New York: Dover, pp. 98–99, 1959.

Blancmange Function

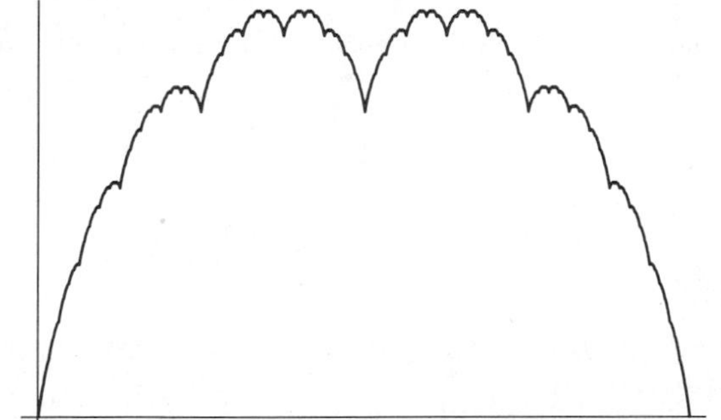

A CONTINUOUS FUNCTION which is nowhere DIFFERENTIABLE. The iterations towards the continuous function are BATRACHIONS resembling the HOFSTADTER-CONWAY \$10,000 SEQUENCE. The first six iterations are illustrated below. The dth iteration contains $N+1$ points, where $N = 2^d$, and can be obtained by setting $b(0) = b(N) = 0$, letting

$$b(m + 2^{n-1}) = 2^n + \tfrac{1}{2}[b(m) + b(m+2^n)],$$

and looping over $n = d$ to 1 by steps of -1 and $m = 0$ to $N-1$ by steps of 2^n.

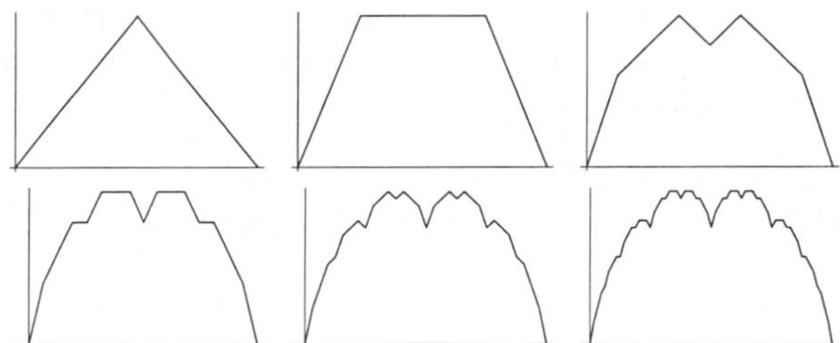

Peitgen and Saupe (1988) refer to this curve as the TAKAGI FRACTAL CURVE.

see also HOFSTADTER-CONWAY \$10,000 SEQUENCE, WEIERSTRAß FUNCTION

References

Dixon, R. *Mathographics.* New York: Dover, pp. 175–176 and 210, 1991.

Peitgen, H.-O. and Saupe, D. (Eds.). "Midpoint Displacement and Systematic Fractals: The Takagi Fractal Curve, Its Kin, and the Related Systems." §A.1.2 in *The Science of Fractal Images.* New York: Springer-Verlag, pp. 246–248, 1988.

Takagi, T. "A Simple Example of the Continuous Function without Derivative." *Proc. Phys. Math. Japan* **1**, 176–177, 1903.

Tall, D. O. "The Blancmange Function, Continuous Everywhere but Differentiable Nowhere." *Math. Gaz.* **66**, 11–22, 1982.

Tall, D. "The Gradient of a Graph." *Math. Teaching* **111**, 48–52, 1985.

Blaschke Conjecture

The only WIEDERSEHEN MANIFOLDS are the standard round spheres. The conjecture has been proven by combining the BERGER-KAZDAN COMPARISON THEOREM with A. Weinstein's results for n EVEN and C. T. Yang's for n ODD.

References

Chavel, I. *Riemannian Geometry: A Modern Introduction.* New York: Cambridge University Press, 1994.

Blaschke's Theorem

A convex planar domain in which the minimal length is > 1 always contains a CIRCLE of RADIUS 1/3.

References

Le Lionnais, F. *Les nombres remarquables.* Paris: Hermann, p. 25, 1983.

Blecksmith-Brillhart-Gerst Theorem

A generalization of SCHRÖTER'S FORMULA.

References

Berndt, B. C. *Ramanujan's Notebooks, Part III.* New York: Springer-Verlag, p. 73, 1985.

Blichfeldt's Lemma

see BLICHFELDT'S THEOREM

Blichfeldt's Theorem

Published in 1914 by Hans Blichfeldt. It states that any bounded planar region with POSITIVE AREA $> A$ placed in any position of the UNIT SQUARE LATTICE can be TRANSLATED so that the number of LATTICE POINTS inside the region will be at least $A + 1$. The theorem can be generalized to n-D.

BLM/Ho Polynomial

A 1-variable unoriented KNOT POLYNOMIAL $Q(x)$. It satisfies

$$Q_{\text{unknot}} = 1 \tag{1}$$

and the SKEIN RELATIONSHIP

$$Q_{L_+} + Q_{L_-} = x(Q_{L_0} + Q_{L_\infty}). \tag{2}$$

It also satisfies

$$Q_{L_1 \# L_2} = Q_{L_1} Q_{L_2}, \tag{3}$$

where # is the KNOT SUM and

$$Q_{L^*} = Q_L, \tag{4}$$

where L^* is the MIRROR IMAGE of L. The BLM/Ho polynomials of MUTANT KNOTS are also identical. Brandt *et al.* (1986) give a number of interesting properties. For any LINK L with ≥ 2 components, $Q_L - 1$ is divisible by $2(x-1)$. If L has c components, then the lowest POWER of x in $Q_L(x)$ is $1-c$, and

$$\lim_{x\to 0} x^{c-1} Q_L(x) = \lim_{(\ell,m)\to(1,0)} (-m)^{c-1} P_L(\ell, m), \tag{5}$$

where P_L is the HOMFLY POLYNOMIAL. Also, the degree of Q_L is less than the CROSSING NUMBER of L. If L is a 2-BRIDGE KNOT, then

$$Q_L(z) = 2z^{-1} V_L(t) V_L(t^{-1} + 1 - 2z^{-1}), \tag{6}$$

where $z \equiv -t - t^{-1}$ (Kanenobu and Sumi 1993).

The POLYNOMIAL was subsequently extended to the 2-variable KAUFFMAN POLYNOMIAL $F(a, z)$, which satisfies

$$Q(x) = F(1, x). \tag{7}$$

Brandt *et al.* (1986) give a listing of Q POLYNOMIALS for KNOTS up to 8 crossings and links up to 6 crossings.

References

Brandt, R. D.; Lickorish, W. B. R.; and Millett, K. C. "A Polynomial Invariant for Unoriented Knots and Links." *Invent. Math.* **84**, 563–573, 1986.

Ho, C. F. "A New Polynomial for Knots and Links—Preliminary Report." *Abstracts Amer. Math. Soc.* **6**, 300, 1985.

Kanenobu, T. and Sumi, T. "Polynomial Invariants of 2-Bridge Knots through 22-Crossings." *Math. Comput.* **60**, 771–778 and S17–S28, 1993.

Stoimenow, A. "Brandt-Lickorish-Millett-Ho Polynomials." `http://www.informatik.hu-berlin.de/~stoimeno/ptab/blmh10.html`.

Weisstein, E. W. "Knots." `http://www.astro.virginia.edu/~eww6n/math/notebooks/Knots.m`.

Bloch Constant

N.B. A detailed on-line essay by S. Finch was the starting point for this entry.

Let F be the set of COMPLEX analytic functions f defined on an open region containing the closure of the unit disk $D = \{z : |z| < 1\}$ satisfying $f(0) = 0$ and $df/dz(0) = 1$. For each f in F, let $b(f)$ be the SUPREMUM of all numbers r such that there is a disk S in D on which f is ONE-TO-ONE and such that $f(S)$ contains a disk of radius r. In 1925, Bloch (Conway 1978) showed that $b(f) \geq 1/72$. Define Bloch's constant by

$$B \equiv \inf\{b(f) : f \in F\}.$$

Ahlfors and Grunsky (1937) derived

$$\begin{aligned} 0.433012701\ldots = \tfrac{1}{4}\sqrt{3} &\leq B \\ &< \frac{1}{\sqrt{1+\sqrt{3}}} \frac{\Gamma(\frac{1}{3})\Gamma(\frac{11}{12})}{\Gamma(\frac{1}{4})} < 0.4718617. \end{aligned}$$

They also conjectured that the upper limit is actually the value of B,

$$\begin{aligned} B &= \frac{1}{\sqrt{1+\sqrt{3}}} \frac{\Gamma(\frac{1}{3})\Gamma(\frac{11}{12})}{\Gamma(\frac{1}{4})} \\ &= \sqrt{\pi}\, 2^{1/4} \frac{\Gamma(\frac{1}{3})}{\Gamma(\frac{1}{4})} \sqrt{\frac{\Gamma(\frac{11}{12})}{\Gamma(\frac{1}{12})}} \\ &= 0.4718617\ldots \end{aligned}$$

(Le Lionnais 1983).

see also LANDAU CONSTANT

References

Conway, J. B. *Functions of One Complex Variable, 2nd ed.* New York: Springer-Verlag, 1989.

Finch, S. "Favorite Mathematical Constants." `http://www.mathsoft.com/asolve/constant/bloch/bloch.html`.

Le Lionnais, F. *Les nombres remarquables.* Paris: Hermann, p. 25, 1983.

Minda, C. D. "Bloch Constants." *J. d'Analyse Math.* **41**, 54–84, 1982.

Bloch-Landau Constant

see LANDAU CONSTANT

Block

see also BLOCK DESIGN, SQUARE POLYOMINO

Block Design

An incidence system (v, k, λ, r, b) in which a set X of v points is partitioned into a family A of b subsets (blocks) in such a way that any two points determine λ blocks, there are k points in each block, and each point is contained in r different blocks. It is also generally required that $k < v$, which is where the "incomplete" comes from in the formal term most often encountered for block designs, BALANCED INCOMPLETE BLOCK DESIGNS (BIBD). The five parameters are not independent, but satisfy the two relations

$$vr = bk \tag{1}$$

$$\lambda(v-1) = r(k-1). \tag{2}$$

A BIBD is therefore commonly written as simply (v, k, λ), since b and r are given in terms of v, k, and λ by

$$b = \frac{v(v-1)\lambda}{k(k-1)} \tag{3}$$

$$r = \frac{\lambda(v-1)}{k-1}. \tag{4}$$

A BIBD is called SYMMETRIC if $b = v$ (or, equivalently, $r = k$).

Writing $X = \{x_i\}_{i=1}^v$ and $A = \{A_j\}_{j=1}^b$, then the INCIDENCE MATRIX of the BIBD is given by the $v \times b$ MATRIX M defined by

$$m_{ij} = \begin{cases} 1 & \text{if } x_i \in A \\ 0 & \text{otherwise.} \end{cases} \tag{5}$$

This matrix satisfies the equation

$$\mathsf{MM}^{\mathrm{T}} = (r-\lambda)\mathsf{I} + \lambda\mathsf{J}, \tag{6}$$

where I is a $v \times v$ IDENTITY MATRIX and J is a $v \times v$ matrix of 1s (Dinitz and Stinson 1992).

Examples of BIBDs are given in the following table.

Block Design	(v, k, λ)
affine plane	$(n^2, n, 1)$
Fano plane	(7, 3, 1))
Hadamard design	symmetric $(4n+3, 2n+1, n)$
projective plane	symmetric $(n^2+n+1, n+1, 1)$
Steiner triple system	$(v, 3, 1)$
unital	$(q^3+1, q+1, 1)$

see also AFFINE PLANE, DESIGN, FANO PLANE, HADAMARD DESIGN, PARALLEL CLASS, PROJECTIVE PLANE, RESOLUTION, RESOLVABLE, STEINER TRIPLE SYSTEM, SYMMETRIC BLOCK DESIGN, UNITAL

References

Dinitz, J. H. and Stinson, D. R. "A Brief Introduction to Design Theory." Ch. 1 in *Contemporary Design Theory: A Collection of Surveys* (Ed. J. H. Dinitz and D. R. Stinson). New York: Wiley, pp. 1–12, 1992.

Ryser, H. J. "The (b, v, r, k, λ)–Configuration." §8.1 in *Combinatorial Mathematics.* Buffalo, NY: Math. Assoc. Amer., pp. 96–102, 1963.

Block Growth

Let $(x_0x_1x_2\ldots)$ be a sequence over a finite ALPHABET A (all the entries are elements of A). Define the block growth function $B(n)$ of a sequence to be the number of ADMISSIBLE words of length n. For example, in the sequence *aabaabaabaabaab*..., the following words are ADMISSIBLE

Length	Admissible Words
1	a, b
2	aa, ab, ba
3	aab, aba, baa
4	$aaba, abaa, baab$

so $B(1) = 2$, $B(2) = 3$, $B(3) = 3$, $B(4) = 3$, and so on. Notice that $B(n) \le B(n+1)$, so the block growth function is always nondecreasing. This is because any ADMISSIBLE word of length n can be extended rightwards to produce an ADMISSIBLE word of length $n+1$. Moreover, suppose $B(n) = B(n+1)$ for some n. Then each admissible word of length n extends to a *unique* ADMISSIBLE word of length $n+1$.

For a SEQUENCE in which each substring of length n uniquely determines the next symbol in the SEQUENCE, there are only finitely many strings of length n, so the process must eventually cycle and the SEQUENCE must be eventually periodic. This gives us the following theorems:

1. If the SEQUENCE is eventually periodic, with least period p, then $B(n)$ is strictly increasing until it reaches p, and $B(n)$ is constant thereafter.
2. If the SEQUENCE is not eventually periodic, then $B(n)$ is strictly increasing and so $B(n) \ge n+1$ for all n. If a SEQUENCE has the property that $B(n) = n+1$ for all n, then it is said to have minimal block growth, and the SEQUENCE is called a STURMIAN SEQUENCE.

The block growth is also called the GROWTH FUNCTION or the COMPLEXITY of a SEQUENCE.

Block Matrix

A square DIAGONAL MATRIX in which the diagonal elements are SQUARE MATRICES of any size (possibly even 1×1), and the off-diagonal elements are 0.

Block (Set)

One of the disjoint SUBSETS making up a SET PARTITION. A block containing n elements is called an n-block. The partitioning of sets into blocks can be denoted using a RESTRICTED GROWTH STRING.

see also BLOCK DESIGN, RESTRICTED GROWTH STRING, SET PARTITION

Blow-Up

A common mechanism which generates SINGULARITIES from smooth initial conditions.

Blue-Empty Coloring

see BLUE-EMPTY GRAPH

Blue-Empty Graph

An EXTREMAL GRAPH in which the forced TRIANGLES are all the same color. Call R the number of red MONOCHROMATIC FORCED TRIANGLES and B the number of blue MONOCHROMATIC FORCED TRIANGLES, then a blue-empty graph is an EXTREMAL GRAPH with $B = 0$. For EVEN n, a blue-empty graph can be achieved by coloring red two COMPLETE SUBGRAPHS of $n/2$ points (the RED NET method). There is no blue-empty coloring for ODD n except for $n = 7$ (Lorden 1962).

see also COMPLETE GRAPH, EXTREMAL GRAPH, MONOCHROMATIC FORCED TRIANGLE, RED NET

References

Lorden, G. "Blue-Empty Chromatic Graphs." *Amer. Math. Monthly* **69**, 114–120, 1962.

Sauvé, L. "On Chromatic Graphs." *Amer. Math. Monthly* **68**, 107–111, 1961.

Board

A subset of $\mathbf{d} \times \mathbf{d}$, where $\mathbf{d} = \{1, 2, \ldots, d\}$.

see also ROOK NUMBER

Boatman's Knot

see CLOVE HITCH

Bochner Identity

For a smooth HARMONIC MAP $u : M \to N$,

$$\Delta(|\nabla u|^2) = |\nabla(du)|^2 + \langle \operatorname{Ric}_M \nabla u, \nabla u \rangle - \langle \operatorname{Riem}_N(u)(\nabla u, \nabla u)\nabla u, \nabla u \rangle\,,$$

where ∇ is the GRADIENT, Ric is the RICCI TENSOR, and Riem is the RIEMANN TENSOR.

References

Eels, J. and Lemaire, L. "A Report on Harmonic Maps." *Bull. London Math. Soc.* **10**, 1–68, 1978.

Bochner's Theorem

Among the continuous functions on $\mathbb{R}^n$, the POSITIVE DEFINITE FUNCTIONS are those functions which are the FOURIER TRANSFORMS of finite measures.

Bode's Rule

$$\int_{x_1}^{x_5} f(x)\,dx = \tfrac{2}{45}h(7f_1 + 32f_2 + 12f_3 + 32f_4 + 7f_5) - \tfrac{8}{945}h^7 f^{(6)}(\xi).$$

see also HARDY'S RULE, NEWTON-COTES FORMULAS, SIMPSON'S 3/8 RULE, SIMPSON'S RULE, TRAPEZOIDAL RULE, WEDDLE'S RULE

References

Abramowitz, M. and Stegun, C. A. (Eds.). *Handbook of Mathematical Functions with Formulas, Graphs, and Mathematical Tables, 9th printing.* New York: Dover, p. 886, 1972.

Bogdanov Map

A 2-D MAP which is conjugate to the HÉNON MAP in its nondissipative limit. It is given by

$$\begin{aligned} x' &= x + y' \\ y' &= y + \epsilon y + kx(x-1) + \mu xy. \end{aligned}$$

see also HÉNON MAP

References

Arrowsmith, D. K.; Cartwright, J. H. E.; Lansbury, A. N.; and Place, C. M. "The Bogdanov Map: Bifurcations, Mode Locking, and Chaos in a Dissipative System." *Int. J. Bifurcation Chaos* **3**, 803–842, 1993.

Bogdanov, R. "Bifurcations of a Limit Cycle for a Family of Vector Fields on the Plane." *Selecta Math. Soviet* **1**, 373–388, 1981.

Bogomolov-Miyaoka-Yau Inequality

Relates invariants of a curve defined over the INTEGERS. If this inequality were proven true, then FERMAT'S LAST THEOREM would follow for sufficiently large exponents. Miyaoka claimed to have proven this inequality in 1988, but the proof contained an error.

see also FERMAT'S LAST THEOREM

References

Cox, D. A. "Introduction to Fermat's Last Theorem." *Amer. Math. Monthly* **101**, 3–14, 1994.

Bohemian Dome

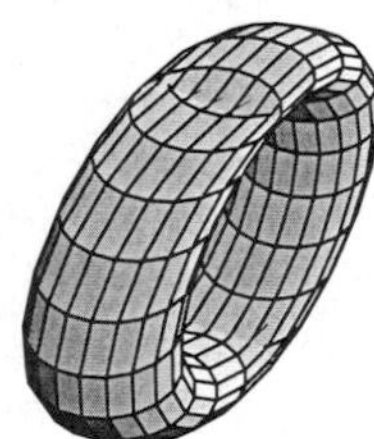

A QUARTIC SURFACE which can be constructed as follows. Given a CIRCLE C and PLANE E PERPENDICULAR to the PLANE of C, move a second CIRCLE K of the same RADIUS as C through space so that its CENTER always lies on C and it remains PARALLEL to E. Then K sweeps out the Bohemian dome. It can be given by the parametric equations

$$\begin{aligned} x &= a\cos u \\ y &= b\cos v + a\sin u \\ z &= c\sin v \end{aligned}$$

where $u, v \in [0, 2\pi)$. In the above plot, $a = 0.5$, $b = 1.5$, and $c = 1$.

see also QUARTIC SURFACE

References
Fischer, G. (Ed.). *Mathematical Models from the Collections of Universities and Museums.* Braunschweig, Germany: Vieweg, pp. 19–20, 1986.
Fischer, G. (Ed.). Plate 50 in *Mathematische Modelle/Mathematical Models, Bildband/Photograph Volume.* Braunschweig, Germany: Vieweg, p. 50, 1986.
Nordstrand, T. "Bohemian Dome." `http://www.uib.no/people/nfytn/bodtxt.htm`.

Bohr-Favard Inequalities

If f has no spectrum in $[-\lambda, \lambda]$, then

$$||f||_\infty \leq \frac{\pi}{2\lambda}||f'||_\infty$$

(Bohr 1935). A related inequality states that if A_k is the class of functions such that

$$f(x) = f(x + 2\pi), f(x), f'(x), \ldots, f^{(k-1)}(x)$$

are absolutely continuous and $\int_0^{2\pi} f(x)\,dx = 0$, then

$$||f||_\infty \leq \frac{4}{\pi}\sum_{\nu=0}^{\infty}\frac{(-1)^{\nu(k+1)}}{(2\nu+1)^{k+1}}||f^{(k)}(x)||_\infty$$

(Northcott 1939). Further, for each value of k, there is always a function $f(x)$ belonging to A_k and not identically zero, for which the above inequality becomes an inequality (Favard 1936). These inequalities are discussed in Mitrinovic *et al.* (1991).

References
Bohr, H. "Ein allgemeiner Satz über die Integration eines trigonometrischen Polynoms." *Prace Matem.-Fiz.* **43**, 1935.
Favard, J. "Application de la formule sommatoire d'Euler à la démonstration de quelques propriétés extrémales des intégrale des fonctions périodiques ou presquepériodiques." *Mat. Tidsskr. B*, 81–94, 1936. [Reviewed in *Zentralblatt f. Math.* **16**, 58–59, 1939.]
Mitrinovic, D. S.; Pecaric, J. E.; and Fink, A. M. *Inequalities Involving Functions and Their Integrals and Derivatives.* Dordrecht, Netherlands: Kluwer, pp. 71–72, 1991.
Northcott, D. G. "Some Inequalities Between Periodic Functions and Their Derivatives." *J. London Math. Soc.* **14**, 198–202, 1939.
Tikhomirov, V. M. "Approximation Theory." In *Analysis II* (Ed. R. V. Gamrelidze). New York: Springer-Verlag, pp. 93–255, 1990.

Bolyai-Gerwein Theorem

see WALLACE-BOLYAI-GERWEIN THEOREM

Bolza Problem

Given the functional

$$U = \int_{t_0}^{t_1} f(y_1, \ldots, y_n; y_1', \ldots, y_n')\,dt + G(y_{10}, \ldots, y_{nr}; y_{11}, \ldots, y_{n1}),$$

find in a class of arcs satisfying p differential and q finite equations

$$\phi_\alpha(y_1, \ldots, y_n; y_1', \ldots, y_n') = 0 \qquad \text{for } \alpha = 1, \ldots, p$$
$$\psi_\beta(y_1, \ldots, y_n) = 0 \qquad \text{for } \beta = 1, \ldots, q$$

as well as the r equations on the endpoints

$$\chi_\gamma(y_{10}, \ldots, y_{nr}; y_{11}, \ldots, y_{n1}) = 0 \qquad \text{for } \gamma = 1, \ldots, r,$$

one which renders U a minimum.

References
Goldstine, H. H. *A History of the Calculus of Variations from the 17th through the 19th Century.* New York: Springer-Verlag, p. 374, 1980.

Bolzano Theorem

see BOLZANO-WEIERSTRAß THEOREM

Bolzano-Weierstraß Theorem

Every BOUNDED infinite set in $\mathbb{R}^n$ has an ACCUMULATION POINT. For $n = 1$, the theorem can be stated as follows: If a SET in a METRIC SPACE, finite-dimensional EUCLIDEAN SPACE, or FIRST-COUNTABLE SPACE has infinitely many members within a finite interval $x \in [a, b]$, then it has at least one LIMIT POINT x such that $x \in [a, b]$. The theorem can be used to prove the INTERMEDIATE VALUE THEOREM.

Bombieri's Inequality

For HOMOGENEOUS POLYNOMIALS P and Q of degree m and n, then

$$[P \cdot Q]_2 \geq \sqrt{\frac{m!n!}{(m+n)!}}[P]_2[Q]_2,$$

where $[P \cdot Q]_2$ is the BOMBIERI NORM. If $m = n$, this becomes

$$[P \cdot Q]_2 \geq [P]_2[Q]_2.$$

see also BEAUZAMY AND DÉGOT'S IDENTITY, REZNIK'S IDENTITY

Bombieri Inner Product

For HOMOGENEOUS POLYNOMIALS P and Q of degree n,

$$[P, Q] \equiv \sum_{i_1, \ldots, i_n \geq 0} (i_1! \cdots i_n!)(a_{i_1, \ldots, i_n} b_{i_1, \ldots, i_n}).$$

Bombieri Norm

For HOMOGENEOUS POLYNOMIALS P of degree m,

$$[P]_2 \equiv \sqrt{[P, P]} = \left(\sum_{|\alpha|=m}\frac{\alpha!}{m!}|a_\alpha|^2\right)^2.$$

see also POLYNOMIAL BAR NORM

Bombieri's Theorem

Define

$$E(x;q,a) \equiv \psi(x;q,a) - \frac{x}{\phi(q)}, \quad (1)$$

where

$$\psi(x;q,a) = \sum_{\substack{n\leq x \\ n\equiv a \pmod q}} \Lambda(n) \quad (2)$$

(Davenport 1980, p. 121), $\Lambda(n)$ is the MANGOLDT FUNCTION, and $\phi(q)$ is the TOTIENT FUNCTION. Now define

$$E(x;q) = \max_{\substack{a \\ (a,q)=1}} |E(x;q,a)| \quad (3)$$

where the sum is over a RELATIVELY PRIME to q, $(a,q)=1$, and

$$E^*(x,q) = \max_{y\leq x} E(y,q). \quad (4)$$

Bombieri's theorem then says that for $A > 0$ fixed,

$$\sum_{q\leq Q} E^*(x,q) \ll \sqrt{x}\, Q(\ln x)^5, \quad (5)$$

provided that $\sqrt{x}\,(\ln x)^{-4} \leq Q \leq \sqrt{x}$.

References

Bombieri, E. "On the Large Sieve." *Mathematika* **12**, 201–225, 1965.

Davenport, H. "Bombieri's Theorem." Ch. 28 in *Multiplicative Number Theory, 2nd ed.* New York: Springer-Verlag, pp. 161–168, 1980.

Bond Percolation

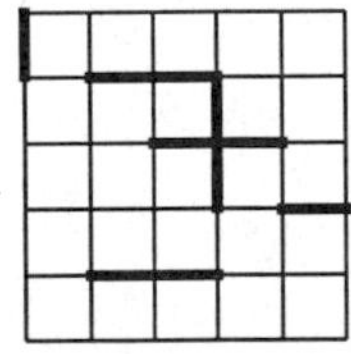
bond percolation

site percolation

A PERCOLATION which considers the lattice edges as the relevant entities (left figure).

see also PERCOLATION THEORY, SITE PERCOLATION

Bonferroni Correction

The Bonferroni correction is a multiple-comparison correction used when several independent STATISTICAL TESTS are being performed simultaneously (since while a given ALPHA VALUE α may be appropriate for each individual comparison, it is not for the set of *all* comparisons). In order to avoid a lot of spurious positives, the ALPHA VALUE needs to be lowered to account for the number of comparisons being performed.

The simplest and most conservative approach is the Bonferroni correction, which sets the ALPHA VALUE for the entire *set* of n comparisons equal to α by taking the ALPHA VALUE for *each* comparison equal to α/n. Explicitly, given n tests T_i for hypotheses H_i $(1 \leq i \leq n)$ under the assumption H_0 that all hypotheses H_i are false, and if the individual test critical values are $\leq \alpha/n$, then the experiment-wide critical value is $\leq \alpha$. In equation form, if

$$P(T_i \text{ passes } |H_0) \leq \frac{\alpha}{n}$$

for $1 \leq i \leq n$, then

$$P(\text{some } T_i \text{ passes } |H_0) \leq \alpha,$$

which follows from BONFERRONI'S INEQUALITY.

Another correction instead uses $1-(1-\alpha)^{1/n}$. While this choice is applicable for two-sided hypotheses, multivariate normal statistics, and positive orthant dependent statistics, it is not, in general, correct (Shaffer 1995).

see also ALPHA VALUE, HYPOTHESIS TESTING, STATISTICAL TEST

References

Bonferroni, C. E. "Il calcolo delle assicurazioni su gruppi di teste." In *Studi in Onore del Professore Salvatore Ortu Carboni.* Rome: Italy, pp. 13–60, 1935.

Bonferroni, C. E. "Teoria statistica delle classi e calcolo delle probabilità." *Pubblicazioni del R Istituto Superiore di Scienze Economiche e Commerciali di Firenze* **8**, 3–62, 1936.

Dewey, M. "Carlo Emilio Bonferroni: Life and Works." `http://www.nottingham.ac.uk/~mhzmd/life.html`.

Miller, R. G. Jr. *Simultaneous Statistical Inference.* New York: Springer-Verlag, 1991.

Perneger, T. V. "What's Wrong with Bonferroni Adjustments." *Brit. Med. J.* **316**, 1236–1238, 1998.

Shaffer, J. P. "Multiple Hypothesis Testing." *Ann. Rev. Psych.* **46**, 561–584, 1995.

Bonferroni's Inequality

Let $P(E_i)$ be the probability that E_i is true, and $P\left(\bigcup_{i=1}^n E_i\right)$ be the probability that E_1, E_2, ..., E_n are all true. Then

$$P\left(\bigcup_{i=1}^n E_i\right) \leq \sum_{i=1}^n P(E_i).$$

Bonferroni Test

see BONFERRONI CORRECTION

Bonne Projection

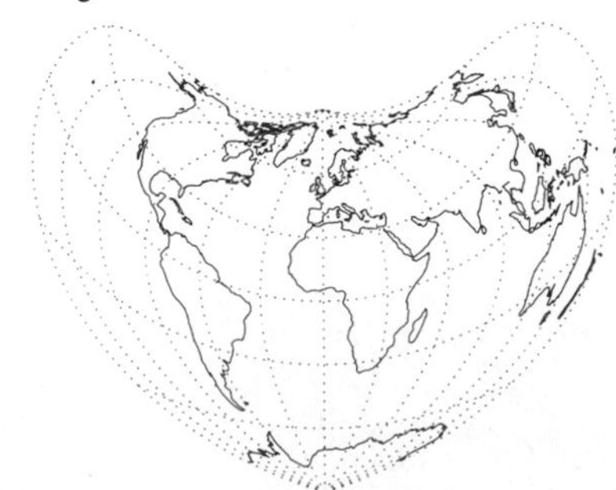

A MAP PROJECTION which resembles the shape of a heart. Let ϕ_1 be the standard parallel and λ_0 the central meridian. Then

$$x = \rho \sin E \quad (1)$$
$$y = R \cot \phi_1 - \rho \cos R, \quad (2)$$

where

$$\rho = \cot \phi_1 + \phi_1 - \phi \quad (3)$$
$$E = \frac{(\lambda - \lambda_0)\cos \phi}{\rho}. \quad (4)$$

The inverse FORMULAS are

$$\phi = \cot \phi_1 + \phi_1 - \rho \quad (5)$$
$$\lambda = \lambda_0 + \frac{\rho}{\cos \phi} \tan^{-1}\left(\frac{x}{\cot \phi_1 - y}\right), \quad (6)$$

where

$$\rho = \pm\sqrt{x^2 + (\cot \phi_1 - y)^2}. \quad (7)$$

References
Snyder, J. P. *Map Projections—A Working Manual.* U. S. Geological Survey Professional Paper 1395. Washington, DC: U. S. Government Printing Office, pp. 138–140, 1987.

Book Stacking Problem

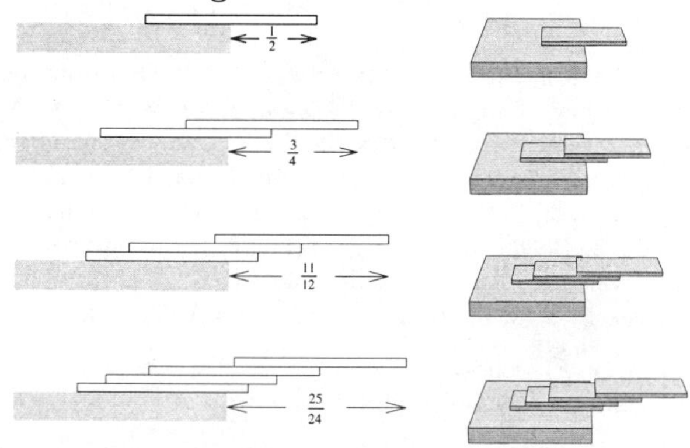

How far can a stack of n books protrude over the edge of a table without the stack falling over? It turns out that the maximum overhang possible d_n for n books (in terms of book lengths) is half the nth partial sum of the HARMONIC SERIES, given explicitly by

$$d_n = \frac{1}{2}\sum_{k=1}^{n}\frac{1}{k} = \tfrac{1}{2}[\gamma + \Psi(1+n)]$$

where $\Psi(z)$ is the DIGAMMA FUNCTION and γ is the EULER-MASCHERONI CONSTANT. The first few values are

$$d_1 = \frac{1}{2} = 0.5$$
$$d_2 = \tfrac{3}{4} = 0.75$$
$$d_3 = \tfrac{11}{12} \approx 0.91667$$
$$d_4 = \tfrac{25}{24} \approx 1.04167,$$

(Sloane's A001008 and A002805).

In order to find the number of stacked books required to obtain d book-lengths of overhang, solve the d_n equation for d, and take the CEILING FUNCTION. For $n = 1, 2, \ldots$ book-lengths of overhang, 4, 31, 227, 1674, 12367, 91380, 675214, 4989191, 36865412, 272400600, ... (Sloane's A014537) books are needed.

References
Dickau, R. M. "The Book-Stacking Problem." http://www.prairienet.org/~pops/BookStacking.html.
Eisner, L. "Leaning Tower of the Physical Review." *Amer. J. Phys.* **27**, 121, 1959.
Gardner, M. *Martin Gardner's Sixth Book of Mathematical Games from Scientific American.* New York: Scribner's, p. 167, 1971.
Graham, R. L.; Knuth, D. E.; and Patashnik, O. *Concrete Mathematics: A Foundation for Computer Science.* Reading, MA: Addison-Wesley, pp. 272–274, 1990.
Johnson, P. B. "Leaning Tower of Lire." *Amer. J. Phys.* **23**, 240, 1955.
Sharp, R. T. "Problem 52." *Pi Mu Epsilon J.* **1**, 322, 1953.
Sharp, R. T. "Problem 52." *Pi Mu Epsilon J.* **2**, 411, 1954.
Sloane, N. J. A. Sequences A014537, A001008/M2885, and A002805/M1589 in "An On-Line Version of the Encyclopedia of Integer Sequences."

Boole's Inequality

$$P\left(\bigcup_{i=1}^{N} E_i\right) \leq \sum_{i=1}^{N} P(E_i).$$

If E_i and E_j are MUTUALLY EXCLUSIVE for all i and j, then the INEQUALITY becomes an equality.

Boolean Algebra

A mathematical object which is similar to a BOOLEAN RING, but which uses the meet and join operators instead of the usual addition and multiplication operators. A Boolean algebra is a set B of elements a, b, ... with BINARY OPERATORS $+$ and $\cdot$ such that

1a. If a and b are in the set B, then $a + b$ is in the set B.

1b. If a and b are in the set B, then $a \cdot b$ is in the set B.

2a. There is an element Z (zero) such that $a + Z = a$ for every element a.

2b. There is an element U (unity) such that $a \cdot U = a$ for every element a.

3a. $a + b = b + a$

3b. $a \cdot b = b \cdot a$

4a. $a + b \cdot c = (a + b)(a + c)$

4b. $a \cdot (b + c) = a \cdot b + a \cdot c$

5. For every element a there is an element a' such that $a + a' = U$ and $a \cdot a' = Z$.

6. There are are least two distinct elements in the set B.

(Bell 1937, p. 444).

In more modern terms, a Boolean algebra is a SET B of elements a, b, ... with the following properties:

1. B has two binary operations, $\wedge$ (WEDGE) and $\vee$ (VEE), which satisfy the IDEMPOTENT laws

$$a \wedge a = a \vee a = a,$$

the COMMUTATIVE laws

$$a \wedge b = b \wedge a$$

$$a \vee b = b \vee a,$$

and the ASSOCIATIVE laws

$$a \wedge (b \wedge c) = (a \wedge b) \wedge c$$

$$a \vee (b \vee c) = (a \vee b) \vee c.$$

2. The operations satisfy the ABSORPTION LAW

$$a \wedge (a \vee b) = a \vee (a \wedge b) = a.$$

3. The operations are mutually distributive

$$a \wedge (b \vee c) = (a \wedge b) \vee (a \wedge c)$$

$$a \vee (b \wedge c) = (a \vee b) \wedge (a \vee c).$$

4. B contains universal bounds O, I which satisfy

$$O \wedge a = O$$

$$O \vee a = a$$

$$I \wedge a = a$$

$$I \vee a = I.$$

5. B has a unary operation $a \to a'$ of complementation which obeys the laws

$$a \wedge a' = O$$

$$a \vee a' = I$$

(Birkhoff and Mac Lane 1965). Under intersection, union, and complement, the subsets of any set I form a Boolean algebra.

Huntington (1933a, b) presented the following basis for Boolean algebra,

1. Commutivity. $x + y = y + x$.
2. Associativity. $(x + y) + z = x + (y + z)$.
3. HUNTINGTON EQUATION. $n(n(x) + y) + n(n(x) + n(y)) = x$.

H. Robbins then conjectured that the HUNTINGTON EQUATION could be replaced with the simpler ROBBINS EQUATION,

$$n(n(x + y) + n(x + n(y))) = x.$$

The ALGEBRA defined by commutivity, associativity, and the ROBBINS EQUATION is called ROBBINS ALGEBRA. Computer theorem proving demonstrated that every ROBBINS ALGEBRA satisfies the second WINKLER CONDITION, from which it follows immediately that all ROBBINS ALGEBRAS are Boolean.

References

Bell, E. T. *Men of Mathematics.* New York: Simon and Schuster, 1986.

Birkhoff, G. and Mac Lane, S. *A Survey of Modern Algebra, 3rd ed.* New York: Macmillian, p. 317, 1965.

Halmos, P. *Lectures on Boolean Algebras.* Princeton, NJ: Van Nostrand, 1963.

Huntington, E. V. "New Sets of Independent Postulates for the Algebra of Logic." *Trans. Amer. Math. Soc.* **35**, 274–304, 1933a.

Huntington, E. V. "Boolean Algebras: A Correction." *Trans. Amer. Math. Soc.* **35**, 557–558, 1933.

McCune, W. "Robbins Algebras are Boolean." `http://www.mcs.anl.gov/~mccune/papers/robbins/`.

Boolean Connective

One of the LOGIC operators AND $\wedge$, OR $\vee$, and NOT $\neg$.

see also QUANTIFIER

Boolean Function

A Boolean function in n variables is a function

$$f(x_1, \ldots, x_n),$$

where each x_i can be 0 or 1 and f is 0 or 1. Determining the number of monotone Boolean functions of n variables is known as DEDEKIND'S PROBLEM. The number of monotonic increasing Boolean functions of n variables is given by 2, 3, 6, 20, 168, 7581, 7828354, ... (Sloane's A000372, Beeler *et al.* 1972, Item 17). The number of inequivalent monotone Boolean functions of n variables is given by 2, 3, 5, 10, 30, ... (Sloane's A003182).

Let $M(n, k)$ denote the number of distinct monotone Boolean functions of n variables with k mincuts. Then

$$M(n, 0) = 1$$
$$M(n, 1) = 2^n$$
$$M(n, 2) = 2^{n-1}(2^n - 1) - 3^n + 2^n$$
$$M(n, 3) = \tfrac{1}{6}(2^n)(2^n - 1)(2^n - 2) - 6^n + 5^n + 4^n - 3^n.$$

References

Beeler, M.; Gosper, R. W.; and Schroeppel, R. *HAKMEM.* Cambridge, MA: MIT Artificial Intelligence Laboratory, Memo AIM-239, Feb. 1972.

Sloane, N. J. A. Sequences A003182/M0729 and A000372/M0817 in "An On-Line Version of the Encyclopedia of Integer Sequences."

Boolean Ring

A RING with a unit element in which every element is IDEMPOTENT.

see also BOOLEAN ALGEBRA

Borchardt-Pfaff Algorithm

see ARCHIMEDES ALGORITHM

Border Square

40	1	2	3	42	41	46
38	31	13	14	32	35	12
39	30	26	21	28	20	11
43	33	27	25	23	17	7
6	16	22	29	24	34	44
5	15	37	36	18	19	45
4	49	48	47	8	9	10

31	13	14	32	35
30	26	21	28	20
33	27	25	23	17
16	22	29	24	34
15	37	36	18	19

26	21	28
27	25	23
22	29	24

A MAGIC SQUARE that remains magic when its border is removed. A nested magic square remains magic after the border is successively removed one ring at a time. An example of a nested magic square is the order 7 square illustrated above (i.e., the order 7, 5, and 3 squares obtained from it are all magic).

see also MAGIC SQUARE

References

Kraitchik, M. "Border Squares." §7.7 in *Mathematical Recreations.* New York: W. W. Norton, pp. 167–170, 1942.

Bordism

A relation between COMPACT boundaryless MANIFOLDS (also called closed MANIFOLDS). Two closed MANIFOLDS are bordant IFF their disjoint union is the boundary of a compact $(n+1)$-MANIFOLD. Roughly, two MANIFOLDS are bordant if together they form the boundary of a MANIFOLD. The word bordism is now used in place of the original term COBORDISM.

References

Budney, R. "The Bordism Project." http://math.cornell.edu/~rybu/bordism/bordism.html.

Bordism Group

There are bordism groups, also called COBORDISM GROUPS or COBORDISM RINGS, and there are singular bordism groups. The bordism groups give a framework for getting a grip on the question, "When is a compact boundaryless MANIFOLD the boundary of another MANIFOLD?" The answer is, precisely when all of its STIEFEL-WHITNEY CLASSES are zero. Singular bordism groups give insight into STEENROD'S REALIZATION PROBLEM: "When can homology classes be realized as the image of fundamental classes of manifolds?" That answer is known, too.

The machinery of the bordism group winds up being important for HOMOTOPY THEORY as well.

References

Budney, R. "The Bordism Project." http://math.cornell.edu/~rybu/bordism/bordism.html.

Borel-Cantelli Lemma

Let $\{A_n\}_{n=0}^\infty$ be a SEQUENCE of events occurring with a certain probability distribution, and let A be the event consisting of the occurrence of a finite number of events A_n, $n = 1, \ldots$. Then if

$$\sum_{n=1}^\infty P(A_n) < \infty,$$

then

$$P(A) = 1.$$

References

Hazewinkel, M. (Managing Ed.). *Encyclopaedia of Mathematics: An Updated and Annotated Translation of the Soviet "Mathematical Encyclopaedia."* Dordrecht, Netherlands: Reidel, pp. 435–436, 1988.

Borel Determinacy Theorem

Let T be a tree defined on a metric over a set of paths such that the distance between paths p and q is $1/n$, where n is the number of nodes shared by p and q. Let A be a Borel set of paths in the topology induced by this metric. Suppose two players play a game by choosing a path down the tree, so that they alternate and each time choose an immediate successor of the previously chosen point. The first player wins if the chosen path is in A. Then one of the players has a winning STRATEGY in this GAME.

see also GAME THEORY, STRATEGY

Borel's Expansion

Let $\phi(t) = \sum_{n=0}^\infty A_n t^n$ be any function for which the integral

$$I(x) \equiv \int_0^\infty e^{-tx} t^p \phi(t)\, dt$$

converges. Then the expansion

$$I(x) = \frac{\Gamma(p+1)}{x^{p+1}} \left[A_0 + (p+1)\frac{A_1}{x} + (p+1)(p+2)\frac{A_2}{x^2} + \ldots\right],$$

where $\Gamma(z)$ is the GAMMA FUNCTION, is usually an ASYMPTOTIC SERIES for $I(x)$.

Borel Measure

If F is the BOREL SIGMA ALGEBRA on some TOPOLOGICAL SPACE, then a MEASURE $m : F \to \mathbb{R}$ is said to be a Borel measure (or BOREL PROBABILITY MEASURE). For a Borel measure, all continuous functions are MEASURABLE.

Borel Probability Measure

see BOREL MEASURE

Borel Set

A DEFINABLE SET derived from the REAL LINE by removing a FINITE number of intervals. Borel sets are measurable and constitute a special type of SIGMA ALGEBRA called a BOREL SIGMA ALGEBRA.

see also STANDARD SPACE

Borel Sigma Algebra

A SIGMA ALGEBRA which is related to the TOPOLOGY of a SET. The Borel *sigma*-algebra is defined to be the SIGMA ALGEBRA generated by the OPEN SETS (or equivalently, by the CLOSED SETS).

see also BOREL MEASURE

Borel Space

A SET equipped with a SIGMA ALGEBRA of SUBSETS.

Borromean Rings

Three mutually interlocked rings named after the Italian Renaissance family who used them on their coat of arms. No two rings are linked, so if one of the rings is cut, all three rings fall apart. They are given the LINK symbol 06^{02}_{03}, and are also called the BALLANTINE. The Borromean rings have BRAID WORD $\sigma_1{}^{-1}\sigma_2\sigma_1{}^{-1}\sigma_2\sigma_1{}^{-1}\sigma_2$ and are also the simplest BRUNNIAN LINK.

References

Cundy, H. and Rollett, A. *Mathematical Models, 3rd ed.* Stradbroke, England: Tarquin Pub., pp. 58–59, 1989.

Gardner, M. *The Unexpected Hanging and Other Mathematical Diversions.* Chicago, IL: University of Chicago Press, 1991.

Jablan, S. "Borromean Triangles." `http://members.tripod.com/~modularity/links.htm`.

Pappas, T. "Trinity of Rings—A Topological Model." *The Joy of Mathematics.* San Carlos, CA: Wide World Publ./Tetra, p. 31, 1989.

Borrow

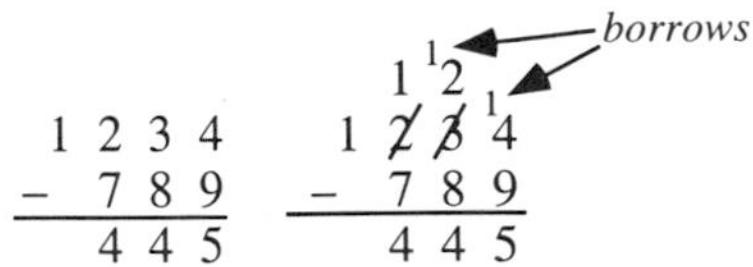

The procedure used in SUBTRACTION to "borrow" 10 from the next higher DIGIT column in order to obtain a POSITIVE DIFFERENCE in the column in question.

see also CARRY

Borsuk's Conjecture

Borsuk conjectured that it is possible to cut an n-D shape of DIAMETER 1 into $n+1$ pieces each with diameter smaller than the original. It is true for $n=2$, 3 and when the boundary is "smooth." However, the minimum number of pieces required has been shown to increase as $\sim 1.1^{\sqrt{n}}$. Since $1.1^{\sqrt{n}} > n+1$ at $n = 9162$, the conjecture becomes false at high dimensions. In fact, the limit has been pushed back to ~ 2000.

see also DIAMETER (GENERAL), KELLER'S CONJECTURE, LEBESGUE MINIMAL PROBLEM

References

Borsuk, K. "Über die Zerlegung einer Euklidischen n-dimensionalen Vollkugel in n Mengen." *Verh. Internat. Math.-Kongr. Zürich* **2**, 192, 1932.

Borsuk, K. "Drei Sätze über die n-dimensionale euklidische Sphäre." *Fund. Math.* **20**, 177–190, 1933.

Cipra, B. "If You Can't See It, Don't Believe It...." *Science* **259**, 26–27, 1993.

Cipra, B. *What's Happening in the Mathematical Sciences, Vol. 1.* Providence, RI: Amer. Math. Soc., pp. 21–25, 1993.

Grünbaum, B. "Borsuk's Problem and Related Questions." In *Convexity, Proceedings of the Seventh Symposium in Pure Mathematics of the American Mathematical Society, Held at the University of Washington, Seattle, June 13–15, 1961.* Providence, RI: Amer. Math. Soc., pp. 271–284, 1963.

Kalai, J. K. G. "A Counterexample to Borsuk's Conjecture." *Bull. Amer. Math. Soc.* **329**, 60–62, 1993. Listernik, L. and Schnirelmann, L. *Topological Methods in Variational Problems.* Moscow, 1930.

Borwein Conjectures

Use the definition of the q-SERIES

$$(a;q)_n \equiv \prod_{j=0}^{n-1}(1-aq^j) \tag{1}$$

and define

$$\begin{bmatrix} N \\ M \end{bmatrix} \equiv \frac{(q^{N-M+1};q)_M}{(q;q)_m}. \tag{2}$$

Then P. Borwein has conjectured that (1) the POLYNOMIALS $A_n(q)$, $B_n(q)$, and $C_n(q)$ defined by

$$(q;q^3)_n(q^2;q^3)_n = A_n(q^3) - qB_n(q^3) - q^2C_n(q^3) \tag{3}$$

have NONNEGATIVE COEFFICIENTS, (2) the POLYNOMIALS $A_n^*(q)$, $B_n^*(q)$, and $C_n^*(q)$ defined by

$$(q;q^3)_n^2(q^2;q^3)_n^2 = A_n^*(q^3) - qB_n^*(q^3) - q^2C_n^*(q^3) \tag{4}$$

have NONNEGATIVE COEFFICIENTS, (3) the POLYNOMIALS $A_n^\star(q)$, $B_n^\star(q)$, $C_n^\star(q)$, $D_n^\star(q)$, and $E_n^\star(q)$ defined by

$$(q;q^5)_n(q^2;q^5)_n(q^3;q^5)_n(q^4;q^5)_n = A_n^\star(q^5) - qB_n^\star(q^5) - q^2C_n^\star(q^5) - q^3D_n^\star(q^5) - q^4E_n^\star(q^5) \tag{5}$$

have NONNEGATIVE COEFFICIENTS, (4) the POLYNOMIALS $A_n^\dagger(m,n,t,q)$, $B_n^\dagger(m,n,t,q)$, and $C_n^\dagger(m,n,t,q)$ defined by

$$(q;q^3)_m(q^2;q^3)_m(zq;q^3)_n(zq^2;q^3)_n = \sum_{t=0}^{2m} z^t[A^\dagger(m,n,t,q^3) - qB^\dagger(m,n,t,q^3) - q^2C^\dagger(m,n,t,q^3)] \quad (6)$$

have NONNEGATIVE COEFFICIENTS, (5) for k ODD and $1 \leq a \leq k/2$, consider the expansion

$$(q^a;q^k)_m(q^{k-a};q^k)_n = \sum_{\nu=(1-k)/2}^{(k-1)/2} (-1)^\nu q^{k(\nu^2+\nu)/2-a\nu} F_\nu(q^k) \quad (7)$$

with

$$F_\nu(q) = \sum_{j=-\infty}^{\infty} (-1)^j q^{j(k^2j+2k\nu+k-2a)/2} \begin{bmatrix} m+n \\ m+\nu+kj \end{bmatrix}, \quad (8)$$

then if a is RELATIVELY PRIME to k and $m = n$, the COEFFICIENTS of $F_\nu(q)$ are NONNEGATIVE, and (6) given $\alpha + \beta < 2K$ and $-K + \beta \leq n - m \leq K - \alpha$, consider

$$G(\alpha,\beta,K;q) = \sum_q (-1)^j q^{j[K(\alpha+\beta)j+K(\alpha+\beta)]/2} \times \begin{bmatrix} m+n \\ m+Kj \end{bmatrix}, \quad (9)$$

the GENERATING FUNCTION for partitions inside an $m \times n$ rectangle with hook difference conditions specified by α, β, and K. Let α and β be POSITIVE RATIONAL NUMBERS and $K > 1$ an INTEGER such that αK and βK are integers. Then if $1 \leq \alpha+\beta \leq 2K-1$ (with strict inequalities for $K = 2$) and $-K + \beta \leq n - m \leq K - \alpha$, then $G(\alpha,\beta,K;q)$ has NONNEGATIVE COEFFICIENTS.

see also q-SERIES

References

Andrews, G. E. *et al.* "Partitions with Prescribed Hook Differences." *Europ. J. Combin.* **8**, 341–350, 1987.

Bressoud, D. M. "The Borwein Conjecture and Partitions with Prescribed Hook Differences." *Electronic J. Combinatorics* **3**, No. 2, R4, 1–14, 1996. http://www.combinatorics.org/Volume_3/volume3_2.html#R4.

Bouligand Dimension

see MINKOWSKI-BOULIGAND DIMENSION

Bound

see GREATEST LOWER BOUND, INFIMUM, LEAST UPPER BOUND, SUPREMUM

Bound Variable

An occurrence of a variable in a LOGIC which is not FREE.

Boundary

The set of points, known as BOUNDARY POINTS, which are members of the CLOSURE of a given set S and the CLOSURE of its complement set. The boundary is sometimes called the FRONTIER.

see also SURGERY

Boundary Conditions

There are several types of boundary conditions commonly encountered in the solution of PARTIAL DIFFERENTIAL EQUATIONS.

1. DIRICHLET BOUNDARY CONDITIONS specify the value of the function on a surface $T = f(\mathbf{r}, t)$.
2. NEUMANN BOUNDARY CONDITIONS specify the normal derivative of the function on a surface,
$$\frac{\partial T}{\partial n} = \hat{\mathbf{n}} \cdot \nabla T = f(\mathbf{r}, y).$$
3. CAUCHY BOUNDARY CONDITIONS specify a weighted average of first and second kinds.
4. ROBIN BOUNDARY CONDITIONS. For an elliptic partial differential equation in a region Ω, Robin boundary conditions specify the sum of αu and the normal derivative of $u = f$ at all points of the boundary of Ω, with α and f being prescribed.

see also BOUNDARY VALUE PROBLEM, DIRICHLET BOUNDARY CONDITIONS, INITIAL VALUE PROBLEM, NEUMANN BOUNDARY CONDITIONS, PARTIAL DIFFERENTIAL EQUATION, ROBIN BOUNDARY CONDITIONS

References

Arfken, G. *Mathematical Methods for Physicists, 3rd ed.* Orlando, FL: Academic Press, pp. 502–504, 1985.

Morse, P. M. and Feshbach, H. "Boundary Conditions and Eigenfunctions." Ch. 6 in *Methods of Theoretical Physics, Part I.* New York: McGraw-Hill, pp. 495–498 and 676–790, 1953.

Boundary Map

The MAP $H_n(X, A) \to H_{n-1}(A)$ appearing in the LONG EXACT SEQUENCE OF A PAIR AXIOM.

see also LONG EXACT SEQUENCE OF A PAIR AXIOM

Boundary Point

A point which is a member of the CLOSURE of a given set S and the CLOSURE of its complement set. If A is a subset of $\mathbb{R}^n$, then a point $\mathbf{x} \in \mathbb{R}^n$ is a boundary point of A if every NEIGHBORHOOD of $\mathbf{x}$ contains at least one point in A and at least one point not in A.

see also BOUNDARY

Boundary Set

A (symmetrical) boundary set of RADIUS r and center $\mathbf{x}_0$ is the set of all points $\mathbf{x}$ such that

$$|\mathbf{x}-\mathbf{x}_0|=r.$$

Let $\mathbf{x}_0$ be the ORIGIN. In $\mathbb{R}^1$, the boundary set is then the pair of points $x=r$ and $x=-r$. In $\mathbb{R}^2$, the boundary set is a CIRCLE. In $\mathbb{R}^3$, the boundary set is a SPHERE.

see also CIRCLE, DISK, OPEN SET, SPHERE

Boundary Value Problem

A boundary value problem is a problem, typically an ORDINARY DIFFERENTIAL EQUATION or a PARTIAL DIFFERENTIAL EQUATION, which has values assigned on the physical boundary of the DOMAIN in which the problem is specified. For example,

$$\begin{cases} \frac{\partial^2 u}{\partial t^2}-\nabla^2 u=f & \text{in } \Omega \\ u(0,t)=u_1 & \text{on } \partial\Omega \\ \frac{\partial u}{\partial t}(0,t)=u_2 & \text{on } \partial\Omega, \end{cases}$$

where $\partial\Omega$ denotes the boundary of Ω, is a boundary problem.

see also BOUNDARY CONDITIONS, INITIAL VALUE PROBLEM

References

Eriksson, K.; Estep, D.; Hansbo, P.; and Johnson, C. *Computational Differential Equations.* Lund: Studentlitteratur, 1996.

Press, W. H.; Flannery, B. P.; Teukolsky, S. A.; and Vetterling, W. T. "Two Point Boundary Value Problems." Ch. 17 in *Numerical Recipes in FORTRAN: The Art of Scientific Computing, 2nd ed.* Cambridge, England: Cambridge University Press, pp. 745–778, 1992.

Bounded

A SET in a METRIC SPACE (X,d) is bounded if it has a FINITE diameter, i.e., there is an $R<\infty$ such that $d(x,y)\leq R$ for all $x,y\in X$. A SET in $\mathbb{R}^n$ is bounded if it is contained inside some BALL ${x_1}^2+\ldots+{x_n}^2\leq R^2$ of FINITE RADIUS R (Adams 1994).

see also BOUND, FINITE

References

Adams, R. A. *Calculus: A Complete Course.* Reading, MA: Addison-Wesley, p. 707, 1994.

Bounded Variation

A FUNCTION $f(x)$ is said to have bounded variation if, over the CLOSED INTERVAL $x\in[a,b]$, there exists an M such that

$$|f(x_i)-f(a)|+|f(x_2)-f(x_1)|+\ldots+|f(b)-f(x_{n-1})|\leq M$$

for all $a<x_1<x_2<\ldots<x_{n-1}<b$.

Bourget Function

$$\begin{aligned} J_{n,k}(z) &= \frac{1}{\pi i}\int t^{-n-1}\left(t+\frac{1}{t}\right)^k \exp\left[\tfrac{1}{2}z\left(t-\frac{1}{t}\right)\right]dt \\ &= \frac{1}{\pi}\int_0^\pi (2\cos\theta)^k\cos(n\theta-z\sin\theta)\,d\theta. \end{aligned}$$

see also BESSEL FUNCTION OF THE FIRST KIND

References

Hazewinkel, M. (Managing Ed.). *Encyclopaedia of Mathematics: An Updated and Annotated Translation of the Soviet "Mathematical Encyclopaedia."* Dordrecht, Netherlands: Reidel, p. 465, 1988.

Bourget's Hypothesis

When n is an INTEGER ≥ 0, then $J_n(z)$ and $J_{n+m}(z)$ have no common zeros other than at $z=0$ for m an INTEGER ≥ 1, where $J_n(z)$ is a BESSEL FUNCTION OF THE FIRST KIND. The theorem has been proved true for m=1 2, 3, and 4.

References

Watson, G. N. *A Treatise on the Theory of Bessel Functions, 2nd ed.* Cambridge, England: Cambridge University Press, 1966.

Boustrophedon Transform

The boustrophedon ("ox-plowing") transform $\mathbf{b}$ of a sequence $\mathbf{a}$ is given by

$$b_n=\sum_{k=0}^n\binom{n}{k}a_kE_{n-k} \tag{1}$$

$$a_n=\sum_{k=0}^n(-1)^{n-k}\binom{n}{k}b_kE_{n-k} \tag{2}$$

for $n\geq 0$, where E_n is a SECANT NUMBER or TANGENT NUMBER defined by

$$\sum_{n=0}^\infty E_n\frac{x^n}{n!}=\sec x+\tan x. \tag{3}$$

The exponential generating functions of $\mathbf{a}$ and $\mathbf{b}$ are related by

$$\mathcal{B}(x)=(\sec x+\tan x)\mathcal{A}(x), \tag{4}$$

where the exponential generating function is defined by

$$\mathcal{A}(x)=\sum_{n=0}^\infty A_n\frac{x^n}{n!}. \tag{5}$$

see also ALTERNATING PERMUTATION, ENTRINGER NUMBER, SECANT NUMBER, SEIDEL-ENTRINGER-ARNOLD TRIANGLE, TANGENT NUMBER

References

Millar, J.; Sloane, N. J. A.; and Young, N. E. "A New Operation on Sequences: The Boustrophedon Transform." *J. Combin. Th. Ser. A* **76**, 44–54, 1996.

Bovinum Problema

see ARCHIMEDES' CATTLE PROBLEM

Bow

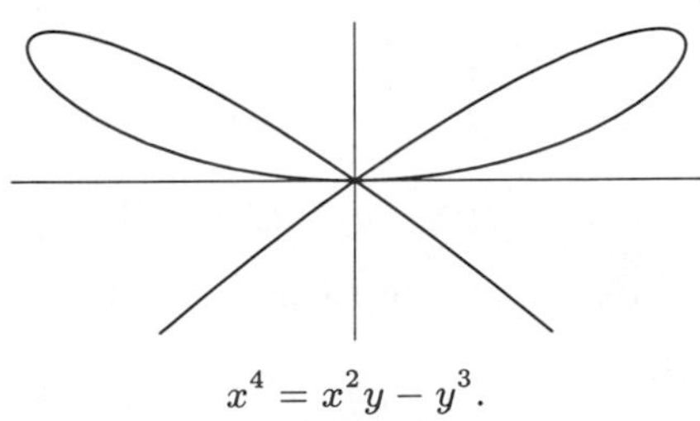

$$x^4 = x^2y - y^3.$$

References

Cundy, H. and Rollett, A. *Mathematical Models, 3rd ed.* Stradbroke, England: Tarquin Pub., p. 72, 1989.

Bowditch Curve

see LISSAJOUS CURVE

Bowley Index

The statistical INDEX

$$P_B \equiv \tfrac{1}{2}(P_L + P_P),$$

where P_L is LASPEYRES' INDEX and P_P is PAASCHE'S INDEX.

see also INDEX

References

Kenney, J. F. and Keeping, E. S. *Mathematics of Statistics, Pt. 1, 3rd ed.* Princeton, NJ: Van Nostrand, p. 66, 1962.

Bowley Skewness

Also known as QUARTILE SKEWNESS COEFFICIENT,

$$\frac{(Q_3 - Q_2) - (Q_2 - Q_1)}{Q_3 - Q_1)} = \frac{Q_1 - 2Q_2 + Q_3}{Q_3 - Q_1},$$

where the Qs denote the INTERQUARTILE RANGES.

see also SKEWNESS

Bowling

Bowling is a game played by rolling a heavy ball down a long narrow track and attempting to knock down ten pins arranged in the form of a TRIANGLE with its vertex oriented towards the bowler. The number 10 is, in fact, the TRIANGULAR NUMBER $T_4 = 4(4+1)/2 = 10$.

Two "bowls" are allowed per "frame." If all the pins are knocked down in the two bowls, the score for that frame is the number of pins knocked down. If some or none of the pins are knocked down on the first bowl, then all the pins knocked down on the second, it is called a "spare," and the number of points tallied is 10 plus the number of pins knocked down on the bowl of the next frame. If all of the pins are knocked down on the first bowl, the number of points tallied is 10 plus the number of pins knocked down on the next two bowls. Ten frames are bowled, unless the last frame is a strike or spare, in which case an additional bowl is awarded.

The maximum number of points possible, corresponding to knocking down all 10 pins on every bowl, is 300.

References

Cooper, C. N. and Kennedy, R. E. "A Generating Function for the Distribution of the Scores of All Possible Bowling Games." In *The Lighter Side of Mathematics* (Ed. R. K. Guy and R. E. Woodrow). Washington, DC: Math. Assoc. Amer., 1994.

Cooper, C. N. and Kennedy, R. E. "Is the Mean Bowling Score Awful?" In *The Lighter Side of Mathematics* (Ed. R. K. Guy and R. E. Woodrow). Washington, DC: Math. Assoc. Amer., 1994.

Box

see CUBOID

Box-and-Whisker Plot

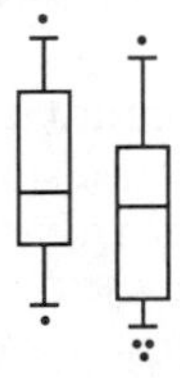

A HISTOGRAM-like method of displaying data invented by J. Tukey (1977). Draw a box with ends at the QUARTILES Q_1 and Q_3. Draw the MEDIAN as a horizontal line in the box. Extend the "whiskers" to the farthest points. For every point that is more than 3/2 times the INTERQUARTILE RANGE from the end of a box, draw a dot on the corresponding top or bottom of the whisker. If two dots have the same value, draw them side by side.

References

Tukey, J. W. *Explanatory Data Analysis.* Reading, MA: Addison-Wesley, pp. 39–41, 1977.

Box Counting Dimension

see CAPACITY DIMENSION

Box Fractal

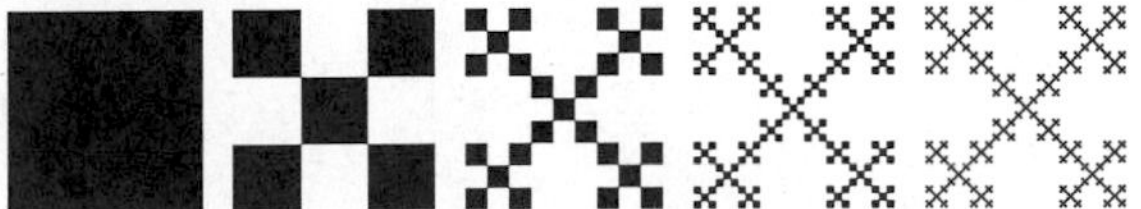

A FRACTAL which can be constructed using STRING REWRITING by creating a matrix with 3 times as many entries as the current matrix using the rules

```
line 1: "*"->"* *"," "->"   "
line 2: "*"->" * "," "->"   "
line 3: "*"->"* *"," "->"   "
```

Let N_n be the number of black boxes, L_n the length of a side of a white box, and A_n the fractional AREA of black boxes after the nth iteration.

$$N_n = 5^n \tag{1}$$
$$L_n = (\tfrac{1}{3})^n = 3^{-n} \tag{2}$$
$$A_n = {L_n}^2 N_n = (\tfrac{5}{9})^n. \tag{3}$$

The CAPACITY DIMENSION is therefore

$$d_{\text{cap}} = -\lim_{n\to\infty} \frac{\ln N_n}{\ln L_n} = -\lim_{n\to\infty} \frac{\ln(5^n)}{\ln(3^{-n})}$$
$$= \frac{\ln 5}{\ln 3} = 1.464973521\ldots. \tag{4}$$

see also CANTOR DUST, SIERPIŃSKI CARPET, SIERPIŃSKI SIEVE

References

Weisstein, E. W. "Fractals." `http://www.astro.virginia.edu/~eww6n/math/notebooks/Fractal.m`.

Box-Muller Transformation

A transformation which transforms from a 2-D continuous UNIFORM DISTRIBUTION to a 2-D GAUSSIAN BIVARIATE DISTRIBUTION (or COMPLEX GAUSSIAN DISTRIBUTION). If x_1 and x_2 are uniformly and independently distributed between 0 and 1, then z_1 and z_2 as defined below have a GAUSSIAN DISTRIBUTION with MEAN $\mu = 0$ and VARIANCE $\sigma^2 = 1$.

$$z_1 = \sqrt{-2\ln x_1}\cos(2\pi x_2) \tag{1}$$
$$z_2 = \sqrt{-2\ln x_1}\sin(2\pi x_2). \tag{2}$$

This can be verified by solving for x_1 and x_2,

$$x_1 = e^{-({z_1}^2+{z_2}^2)/2} \tag{3}$$
$$x_2 = \frac{1}{2\pi}\tan^{-1}\left(\frac{z_2}{z_1}\right). \tag{4}$$

Taking the JACOBIAN yields

$$\frac{\partial(x_1,x_2)}{\partial(z_1,z_2)} = \begin{vmatrix} \frac{\partial x_1}{\partial z_1} & \frac{\partial x_1}{\partial z_2} \\ \frac{\partial x_2}{\partial z_1} & \frac{\partial x_2}{\partial z_2} \end{vmatrix}$$
$$= -\left[\frac{1}{\sqrt{2\pi}}e^{-{z_1}^2/2}\right]\left[\frac{1}{\sqrt{2\pi}}e^{-{z_2}^2/2}\right]. \tag{5}$$

Box-Packing Theorem

The number of "prime" boxes is always finite, where a set of boxes is prime if it cannot be built up from one or more given configurations of boxes.

see also CONWAY PUZZLE, CUBOID, DE BRUIJN'S THEOREM, KLARNER'S THEOREM, SLOTHOUBER-GRAATSMA PUZZLE

References

Honsberger, R. *Mathematical Gems II.* Washington, DC: Math. Assoc. Amer., p. 74, 1976.

Boxcar Function

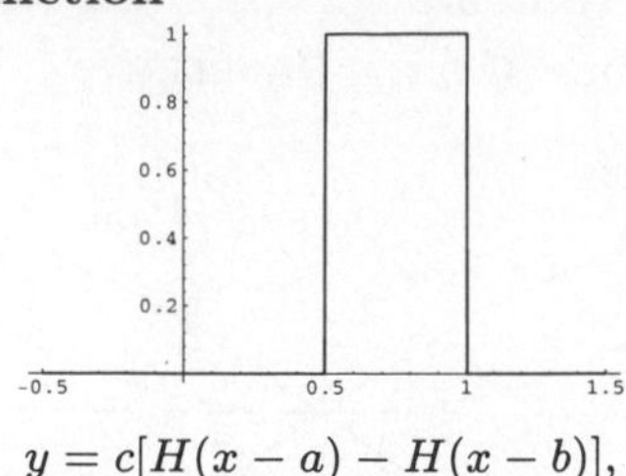

$$y = c[H(x-a) - H(x-b)],$$

where H is the HEAVISIDE STEP FUNCTION.

References

von Seggern, D. *CRC Standard Curves and Surfaces.* Boca Raton, FL: CRC Press, p. 324, 1993.

Boxcars

A roll of two 6s (the highest roll possible) on a pair of 6-sided DICE. The probability of rolling boxcars is 1/36, or 2.777... %.

see also DICE, DOUBLE SIXES, SNAKE EYES

Boy Surface

A NONORIENTABLE SURFACE which is one of the three possible SURFACES obtained by sewing a MÖBIUS STRIP to the edge of a DISK. The other two are the CROSS-CAP and ROMAN SURFACE. The Boy surface is a model of the PROJECTIVE PLANE without singularities and is a SEXTIC SURFACE.

The Boy surface can be described using the general method for NONORIENTABLE SURFACES, but this was not known until the analytic equations were found by Apéry (1986). Based on the fact that it had been proven impossible to describe the surface using quadratic polynomials, Hopf had conjectured that quartic polynomials were also insufficient (Pinkall 1986). Apéry's IMMERSION proved this conjecture wrong, giving the equations explicitly in terms of the standard form for a NONORIENTABLE SURFACE,

$$f_1(x,y,z) = \tfrac{1}{2}[(2x^2-y^2-z^2)(x^2+y^2+z^2) + 2yz(y^2-z^2) + zx(x^2-z^2) + xy(y^2-x^2)] \tag{1}$$
$$f_2(x,y,z) = \tfrac{1}{2}\sqrt{3}[(y^2-z^2)(x^2+y^2+z^2) + zx(z^2-x^2) + xy(y^2-x^2)] \tag{2}$$
$$f_3(x,y,z) = \tfrac{1}{8}(x+y+z)[(x+y+z)^3 + 4(y-x)(z-y)(x-z)]. \tag{3}$$

Plugging in

$$x = \cos u \sin v \quad (4)$$

$$y = \sin u \sin v \quad (5)$$

$$z = \cos v \quad (6)$$

and letting $u \in [0, \pi]$ and $v \in [0, \pi]$ then gives the Boy surface, three views of which are shown above.

The $\mathbb{R}^3$ parameterization can also be written as

$$x = \frac{\sqrt{2}\cos^2 v \cos(2u) + \cos u \sin(2v)}{2 - \sqrt{2}\sin(3u)\sin(2v)} \quad (7)$$

$$y = \frac{\sqrt{2}\cos^2 v \sin(2u) + \cos u \sin(2v)}{2 - \sqrt{2}\sin(3u)\sin(2v)} \quad (8)$$

$$z = \frac{3\cos^2 v}{2 - \sqrt{2}\sin(3u)\sin(2v)} \quad (9)$$

(Nordstrand) for $u \in [-\pi/2, \pi/2]$ and $v \in [0, \pi]$.

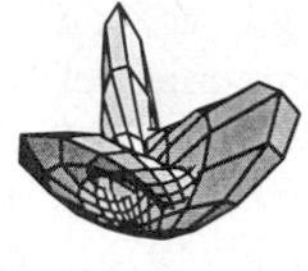

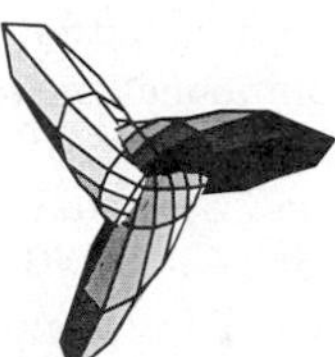

Three views of the surface obtained using this parameterization are shown above.

In fact, a HOMOTOPY (smooth deformation) between the ROMAN SURFACE and Boy surface is given by the equations

$$x(u,v) = \frac{\sqrt{2}\cos(2u)\cos^2 v + \cos u \sin(2v)}{2 - \alpha\sqrt{2}\sin(3u)\sin(2v)} \quad (10)$$

$$y(u,v) = \frac{\sqrt{2}\sin(2u)\cos^2 v - \sin u \sin(2v)}{2 - \alpha\sqrt{2}\sin(3u)\sin(2v)} \quad (11)$$

$$z(u,v) = \frac{3\cos^2 v}{2 - \alpha\sqrt{2}\sin(3u)\sin(2v)} \quad (12)$$

as α varies from 0 to 1, where $\alpha = 0$ corresponds to the ROMAN SURFACE and $\alpha = 1$ to the Boy surface (Wang), shown below.

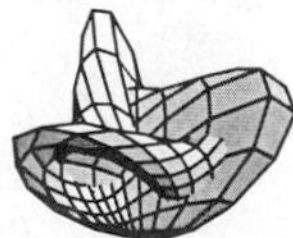
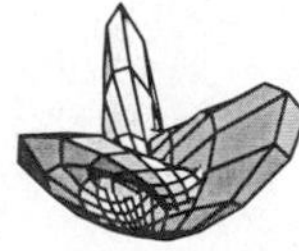

In $\mathbb{R}^4$, the parametric representation is

$$x_0 = 3[(u^2 + v^2 + w^2)(u^2 + v^2) - \sqrt{2}\,vw(3u^2 - v^2)] \quad (13)$$

$$x_1 = \sqrt{2}\,(u^2 + v^2)(u^2 - v^2 + \sqrt{2}\,uw) \quad (14)$$

$$x_2 = \sqrt{2}\,(u^2 + v^2)(2uv - \sqrt{2}\,vw) \quad (15)$$

$$x_3 = 3(u^2 + v^2)^2, \quad (16)$$

and the algebraic equation is

$$\begin{aligned} 64(x_0 - x_3)^3 {x_3}^3 - 48(x_0 - x_3)^2 {x_3}^2 (3{x_1}^2 + 3{x_2}^2 + 2{x_3}^2) \\ +12(x_0 - x_3)x_3[27({x_1}^2 + {x_2}^2)^2 - 24{x_3}^2({x_1}^2 + {x_2}^2) \\ +36\sqrt{2}x_2x_3({x_2}^2 - 3{x_1}^2) + {x_3}^4] \\ +(9{x_1}^2 + 9{x_2}^2 - 2{x_3}^2) \\ \times[-81({x_1}^2 + {x_2}^2)^2 - 72{x_3}^2({x_1}^2 + {x_2}^2) \\ +108\sqrt{2}x_1x_3({x_1}^2 - 3{x_2}^2) + 4{x_3}^4] = 0 \quad (17) \end{aligned}$$

(Apéry 1986). Letting

$$x_0 = 1 \quad (18)$$

$$x_1 = x \quad (19)$$

$$x_2 = y \quad (20)$$

$$x_3 = z \quad (21)$$

gives another version of the surface in $\mathbb{R}^3$.

see also CROSS-CAP, IMMERSION, MÖBIUS STRIP, NONORIENTABLE SURFACE, REAL PROJECTIVE PLANE, ROMAN SURFACE, SEXTIC SURFACE

References

Apéry, F. "The Boy Surface." *Adv. Math.* **61**, 185–266, 1986.

Boy, W. "Über die Curvatura integra und die Topologie geschlossener Flächen." *Math. Ann* **57**, 151–184, 1903.

Brehm, U. "How to Build Minimal Polyhedral Models of the Boy Surface." *Math. Intell.* **12**, 51–56, 1990.

Carter, J. S. "On Generalizing Boy Surface—Constructing a Generator of the 3rd Stable Stem." *Trans. Amer. Math. Soc.* **298**, 103–122, 1986.

Fischer, G. (Ed.). Plates 115–120 in *Mathematische Modelle/Mathematical Models, Bildband/Photograph Volume.* Braunschweig, Germany: Vieweg, pp. 110–115, 1986.

Geometry Center. "Boy's Surface." `http://www.geom.umn.edu/zoo/toptype/pplane/boy/`.

Hilbert, D. and Cohn-Vossen, S. §46–47 in *Geometry and the Imagination.* New York: Chelsea, 1952.

Nordstrand, T. "Boy's Surface." `http://www.uib.no/people/nfytn/boytxt.htm`.

Petit, J.-P. and Souriau, J. "Une représentation analytique de la surface de Boy." *C. R. Acad. Sci. Paris Sér. 1 Math* **293**, 269–272, 1981.

Pinkall, U. *Mathematical Models from the Collections of Universities and Museums* (Ed. G. Fischer). Braunschweig, Germany: Vieweg, pp. 64–65, 1986.

Stewart, I. *Game, Set and Math.* New York: Viking Penguin, 1991.

Wang, P. "Renderings." `http://www.ugcs.caltech.edu/~peterw/portfolio/renderings/`.

Bra

A (COVARIANT) 1-VECTOR denoted $\langle\psi|$. The bra is DUAL to the CONTRAVARIANT KET, denoted $|\psi\rangle$. Taken together, the bra and KET form an ANGLE BRACKET (bra+ket = bracket). The bra is commonly encountered in quantum mechanics.

see also ANGLE BRACKET, BRACKET PRODUCT, COVARIANT VECTOR, DIFFERENTIAL k-FORM, KET, ONE-FORM

Brachistochrone Problem

Find the shape of the CURVE down which a bead sliding from rest and ACCELERATED by gravity will slip (without friction) from one point to another in the least time. This was one of the earliest problems posed in the CALCULUS OF VARIATIONS. The solution, a segment of a CYCLOID, was found by Leibniz, L'Hospital, Newton, and the two Bernoullis.

The time to travel from a point P_1 to another point P_2 is given by the INTEGRAL

$$t_{12} = \int_1^2 \frac{ds}{v}. \tag{1}$$

The VELOCITY at any point is given by a simple application of energy conservation equating kinetic energy to gravitational potential energy,

$$\tfrac{1}{2}mv^2 = mgy, \tag{2}$$

so

$$v = \sqrt{2gy}. \tag{3}$$

Plugging this into (1) then gives

$$t_{12} = \int_1^2 \frac{\sqrt{1+y'^2}}{\sqrt{2gy}}\,dx = \int_1^2 \sqrt{\frac{1+y'^2}{2gy}}\,dx. \tag{4}$$

The function to be varied is thus

$$f = (1+y'^2)^{1/2}(2gy)^{-1/2}. \tag{5}$$

To proceed, one would normally have to apply the full-blown EULER-LAGRANGE DIFFERENTIAL EQUATION

$$\frac{\partial f}{\partial y} - \frac{d}{dx}\left(\frac{\partial f}{\partial y'}\right) = 0. \tag{6}$$

However, the function $f(y, y', x)$ is particularly nice since x does not appear explicitly. Therefore, $\partial f/\partial x = 0$, and we can immediately use the BELTRAMI IDENTITY

$$f - y'\frac{\partial f}{\partial y'} = C. \tag{7}$$

Computing

$$\frac{\partial f}{\partial y'} = y'(1+y'^2)^{-1/2}(2gy)^{-1/2}, \tag{8}$$

subtracting $y'(\partial f/\partial y')$ from f, and simplifying then gives

$$\frac{1}{\sqrt{2gy}\sqrt{1+y'^2}} = C. \tag{9}$$

Squaring both sides and rearranging slightly results in

$$\left[1+\left(\frac{dy}{dx}\right)^2\right]y = \frac{1}{2gC^2} = k^2, \tag{10}$$

where the square of the old constant C has been expressed in terms of a new (POSITIVE) constant k^2. This equation is solved by the parametric equations

$$x = \tfrac{1}{2}k^2(\theta - \sin\theta) \tag{11}$$

$$y = \tfrac{1}{2}k^2(1 - \cos\theta), \tag{12}$$

which are—lo and behold—the equations of a CYCLOID.

If kinetic friction is included, the problem can also be solved analytically, although the solution is significantly messier. In that case, terms corresponding to the normal component of weight and the normal component of the ACCELERATION (present because of path CURVATURE) must be included. Including both terms requires a constrained variational technique (Ashby *et al.* 1975), but including the normal component of weight only gives an elementary solution. The TANGENT and NORMAL VECTORS are

$$\mathbf{T} = \frac{dx}{ds}\hat{\mathbf{x}} + \frac{dy}{ds}\hat{\mathbf{y}} \tag{13}$$

$$\mathbf{N} = -\frac{dy}{ds}\hat{\mathbf{x}} + \frac{dx}{ds}\hat{\mathbf{y}}, \tag{14}$$

gravity and friction are then

$$\mathbf{F}_{\text{gravity}} = mg\hat{\mathbf{y}} \tag{15}$$

$$\mathbf{F}_{\text{friction}} = -\mu(\mathbf{F}_{\text{gravity}}\dot{\mathbf{N}})\mathbf{T} = -\mu mg\frac{dx}{ds}\mathbf{T}, \tag{16}$$

and the components along the curve are

$$\mathbf{F}_{\text{gravity}}\dot{\mathbf{T}} = mg\frac{dy}{ds} \tag{17}$$

$$\mathbf{F}_{\text{friction}}\dot{\mathbf{T}} = -\mu mg\frac{dx}{ds}, \tag{18}$$

so Newton's Second Law gives

$$m\frac{dv}{dt} = mg\frac{dy}{ds} - \mu mg\frac{dx}{ds}. \tag{19}$$

But

$$\frac{dv}{dt} = v\frac{dv}{ds} = \frac{1}{2}\frac{d}{ds}(v^2) \tag{20}$$

$$\tfrac{1}{2}v^2 = g(y - \mu x) \tag{21}$$

$$v = \sqrt{2g(y-\mu x)}, \tag{22}$$

so

$$t=\int\sqrt{\frac{1+(y')^2}{2g(y-\mu x)}}\,dx. \tag{23}$$

Using the EULER-LAGRANGE DIFFERENTIAL EQUATION gives

$$[1+y'^2](1+\mu y')+2(y-\mu x)y''=0. \tag{24}$$

This can be reduced to

$$\frac{1+(y')^2}{(1+\mu y')^2}=\frac{C}{y-\mu x}. \tag{25}$$

Now letting

$$y'=\cot(\tfrac{1}{2}\theta), \tag{26}$$

the solution is

$$x=\tfrac{1}{2}k^2[(\theta-\sin\theta)+\mu(1-\cos\theta)] \tag{27}$$

$$y=\tfrac{1}{2}k^2[(1-\cos\theta)+\mu(\theta+\sin\theta)]. \tag{28}$$

see also CYCLOID, TAUTOCHRONE PROBLEM

References

Ashby, N.; Brittin, W. E.; Love, W. F.; and Wyss, W. "Brachistochrone with Coulomb Friction." *Amer. J. Phys.* **43**, 902–905, 1975.

Haws, L. and Kiser, T. "Exploring the Brachistochrone Problem." *Amer. Math. Monthly* **102**, 328–336, 1995.

Wagon, S. *Mathematica in Action.* New York: W. H. Freeman, pp. 60–66 and 385–389, 1991.

Bracket

see ANGLE BRACKET, BRA, BRACKET POLYNOMIAL, BRACKET PRODUCT, IVERSON BRACKET, KET, LAGRANGE BRACKET, POISSON BRACKET

Bracket Polynomial

A one-variable KNOT POLYNOMIAL related to the JONES POLYNOMIAL. The bracket polynomial, however, is *not* a topological invariant, since it is changed by type I REIDEMEISTER MOVES. However, the SPAN of the bracket polynomial is a knot invariant. The bracket polynomial is occasionally given the grandiose name REGULAR ISOTOPY INVARIANT. It is defined by

$$\langle L\rangle(A,B,d)\equiv\sum_\sigma\langle L|\sigma\rangle d^{||\sigma||}, \tag{1}$$

where A and B are the "splitting variables," σ runs through all "states" of L obtained by SPLITTING the LINK, $\langle L|\sigma\rangle$ is the product of "splitting labels" corresponding to σ, and

$$||\sigma||\equiv N_L-1, \tag{2}$$

where N_L is the number of loops in σ. Letting

$$B=A^{-1} \tag{3}$$

$$d=-A^2-A^{-2} \tag{4}$$

gives a KNOT POLYNOMIAL which is invariant under REGULAR ISOTOPY, and normalizing gives the KAUFFMAN POLYNOMIAL X which is invariant under AMBIENT ISOTOPY. The bracket POLYNOMIAL of the UNKNOT is 1. The bracket POLYNOMIAL of the MIRROR IMAGE K^* is the same as for K but with A replaced by A^{-1}. In terms of the one-variable KAUFFMAN POLYNOMIAL X, the two-variable KAUFFMAN POLYNOMIAL F and the JONES POLYNOMIAL V,

$$X(A)=(-A^3)^{-w(L)}\langle L\rangle, \tag{5}$$

$$\langle L\rangle(A)=F(-A^3,A+A^{-1}) \tag{6}$$

$$\langle L\rangle(A)=V(A^{-4}), \tag{7}$$

where $w(L)$ is the WRITHE of L.

see also SQUARE BRACKET POLYNOMIAL

References

Adams, C. C. *The Knot Book: An Elementary Introduction to the Mathematical Theory of Knots.* New York: W. H. Freeman, pp. 148–155, 1994.

Kauffman, L. "New Invariants in the Theory of Knots." *Amer. Math. Monthly* **95**, 195–242, 1988.

Kauffman, L. *Knots and Physics.* Teaneck, NJ: World Scientific, pp. 26–29, 1991.

Weisstein, E. W. "Knots and Links." `http://www.astro.virginia.edu/~eww6n/math/notebooks/Knots.m`.

Bracket Product

The INNER PRODUCT in an L_2 SPACE represented by an ANGLE BRACKET.

see also ANGLE BRACKET, BRA, KET, L_2 SPACE, ONE-FORM

Bracketing

Take x itself to be a bracketing, then recursively define a bracketing as a sequence $B=(B_1,\ldots,B_k)$ where $k\geq 2$ and each B_i is a bracketing. A bracketing can be represented as a parenthesized string of xs, with parentheses removed from any single letter x for clarity of notation (Stanley 1997). Bracketings built up of binary operations only are called BINARY BRACKETINGS. For example, four letters have 11 possible bracketings:

$$\begin{array}{cccc} xxxx & (xx)xx & x(xx)x & xx(xx) \\ (xxx)x & x(xxx) & ((xx)x)x & (x(xx))x \\ (xx)(xx) & x((xx)x) & x(x(xx)), & \end{array}$$

the last five of which are binary.

The number of bracketings on n letters is given by the GENERATING FUNCTION

$$\tfrac{1}{4}(1+x-\sqrt{1-6x+x^2})=x+x^2+3x^3+11x^4+45x^5$$

(Schröder 1870, Stanley 1997) and the RECURRENCE RELATION

$$s_n=\frac{3(2n-3)s_{n-1}-(n-3)s_{n-2}}{n}$$

(Sloane), giving the sequence for s_n as 1, 1, 3, 11, 45, 197, 903, ... (Sloane's A001003). The numbers are also given by

$$s_n = \sum_{i_1+\ldots+i_k=n} s(i_1)\cdots s(i_k)$$

for $n \geq 2$ (Stanley 1997).

The first PLUTARCH NUMBER 103,049 is equal to s_{10} (Stanley 1997), suggesting that Plutarch's problem of ten compound propositions is equivalent to the number of bracketings. In addition, Plutarch's second number 310,954 is given by $(s_{10} + s_{11})/2 = 310,954$ (Habsieger *et al.* 1998).

see also BINARY BRACKETING, PLUTARCH NUMBERS

References

Habsieger, L.; Kazarian, M.; and Lando, S. "On the Second Number of Plutarch." *Amer. Math. Monthly* **105**, 446, 1998.

Schröder, E. "Vier combinatorische Probleme." *Z. Math. Physik* **15**, 361–376, 1870.

Sloane, N. J. A. Sequence A001003/M2898 in "An On-Line Version of the Encyclopedia of Integer Sequences."

Stanley, R. P. "Hipparchus, Plutarch, Schröder, and Hough." *Amer. Math. Monthly* **104**, 344–350, 1997.

Bradley's Theorem

Let

$$S(\alpha, \beta, m; z) \equiv m\sum_{j=0}^{\infty} \frac{\Gamma(m+j(z+1))\Gamma(\beta+1+jz)}{\Gamma(m+jz+1)\Gamma(\alpha+\beta+1+j(z+1))} \frac{(\alpha)+j}{j!}$$

and α be a NEGATIVE INTEGER. Then

$$S(\alpha, \beta, m; z) = \frac{\Gamma(\beta+1-m)}{\Gamma(\alpha+\beta+1-m)},$$

where $\Gamma(z)$ is the GAMMA FUNCTION.

References

Berndt, B. C. *Ramanujan's Notebooks, Part IV.* New York: Springer-Verlag, pp. 346–348, 1994.

Bradley, D. "On a Claim by Ramanujan about Certain Hypergeometric Series." *Proc. Amer. Math. Soc.* **121**, 1145–1149, 1994.

Brahmagupta's Formula

For a QUADRILATERAL with sides of length a, b, c, and d, the AREA K is given by

$$K = \sqrt{(s-a)(s-b)(s-c)(s-d) - abcd\cos^2[\tfrac{1}{2}(A+B)]}, \quad (1)$$

where

$$s \equiv \tfrac{1}{2}(a+b+c+d) \quad (2)$$

is the SEMIPERIMETER, A is the ANGLE between a and d, and B is the ANGLE between b and c. For a CYCLIC QUADRILATERAL (i.e., a QUADRILATERAL inscribed in a CIRCLE), $A + B = \pi$, so

$$K = \sqrt{(s-a)(s-b)(s-c)(s-d)} \quad (3)$$

$$= \frac{\sqrt{(bc+ad)(ac+bd)(ab+cd)}}{4R}, \quad (4)$$

where R is the RADIUS of the CIRCUMCIRCLE. If the QUADRILATERAL is INSCRIBED in one CIRCLE and CIRCUMSCRIBED on another, then the AREA FORMULA simplifies to

$$K = \sqrt{abcd}. \quad (5)$$

see also BRETSCHNEIDER'S FORMULA, HERON'S FORMULA

References

Coxeter, H. S. M. and Greitzer, S. L. *Geometry Revisited.* Washington, DC: Math. Assoc. Amer., pp. 56–60, 1967.

Johnson, R. A. *Modern Geometry: An Elementary Treatise on the Geometry of the Triangle and the Circle.* Boston, MA: Houghton Mifflin, pp. 81–82, 1929.

Brahmagupta Identity

Let

$$\beta \equiv |B| = x^2 - ty^2,$$

where B is the BRAHMAGUPTA MATRIX, then

$$\det[B(x_1,y_1)B(x_2,y_2)] = \det[B(x_1,y_1)]\det[B(x_2,y_2)] = \beta_1\beta_2.$$

References

Suryanarayan, E. R. "The Brahmagupta Polynomials." *Fib. Quart.* **34**, 30–39, 1996.

Brahmagupta Matrix

$$B(x,y) = \begin{bmatrix} x & y \\ \pm ty & \pm x \end{bmatrix}.$$

It satisfies

$$B(x_1,y_1)B(x_2,y_2) = B(x_1x_2 \pm ty_1y_2, x_1y_2 \pm y_1x_2).$$

Powers of the matrix are defined by

$$B^n = \begin{bmatrix} x & y \\ ty & x \end{bmatrix}^n = \begin{bmatrix} x_n & y_n \\ ty_n & x_n \end{bmatrix} \equiv B_n.$$

The x_n and y_n are called BRAHMAGUPTA POLYNOMIALS. The Brahmagupta matrices can be extended to NEGATIVE INTEGERS

$$B^{-n} = \begin{bmatrix} x & y \\ ty & x \end{bmatrix}^{-n} = \begin{bmatrix} x_{-n} & y_{-n} \\ ty_{-n} & x_{-n} \end{bmatrix} \equiv B_{-n}.$$

see also BRAHMAGUPTA IDENTITY

References

Suryanarayan, E. R. "The Brahmagupta Polynomials." *Fib. Quart.* **34**, 30–39, 1996.

Brahmagupta Polynomial

One of the POLYNOMIALS obtained by taking POWERS of the BRAHMAGUPTA MATRIX. They satisfy the recurrence relation

$$x_{n+1} = xx_n + tyy_n \quad (1)$$

$$y_{n+1} = xy_n + yx_n. \quad (2)$$

A list of many others is given by Suryanarayan (1996). Explicitly,

$$x_n = x^n + t\binom{n}{2}x^{n-2}y^2 + t^2\binom{n}{4}x^{n-4}y^4 + \ldots \quad (3)$$

$$y_n = nx^{n-1}y + t\binom{n}{3}x^{n-3}y^3 + t^2\binom{n}{5}x^{n-5}y^5 + \ldots. \quad (4)$$

The Brahmagupta POLYNOMIALS satisfy

$$\frac{\partial x_n}{\partial x} = \frac{\partial y_n}{\partial y} = nx_{n-1} \quad (5)$$

$$\frac{\partial x_n}{\partial y} = t\frac{\partial y_n}{\partial y} = nty_{n-1}. \quad (6)$$

The first few POLYNOMIALS are

$$x_0 = 0$$
$$x_1 = x$$
$$x_2 = x^2 + ty^2$$
$$x_3 = x^3 + 3txy^2$$
$$x_4 = x^4 + 6tx^2y^2 + t^2y^4$$

and

$$y_0 = 0$$
$$y_1 = y$$
$$y_2 = 2xy$$
$$y_3 = 3x^2y + ty^3$$
$$y_4 = 4x^3y + 4txy^3.$$

Taking $x = y = 1$ and $t = 2$ gives y_n equal to the PELL NUMBERS and x_n equal to half the Pell-Lucas numbers. The Brahmagupta POLYNOMIALS are related to the MORGAN-VOYCE POLYNOMIALS, but the relationship given by Suryanarayan (1996) is incorrect.

References

Suryanarayan, E. R. "The Brahmagupta Polynomials." *Fib. Quart.* **34**, 30–39, 1996.

Brahmagupta's Problem

Solve the PELL EQUATION

$$x^2 - 92y^2 = 1$$

in INTEGERS. The smallest solution is $x = 1151$, $y = 120$.

see also DIOPHANTINE EQUATION, PELL EQUATION

Braid

An intertwining of strings attached to top and bottom "bars" such that each string never "turns back up." In other words, the path of a braid in something that a falling object could trace out if acted upon only by gravity and horizontal forces.

see also BRAID GROUP

References

Christy, J. "Braids." http://www.mathsource.com/cgi-bin/MathSource/Applications/Mathematics/0202-228.

Braid Group

Also called ARTIN BRAID GROUPS. Consider n strings, each oriented vertically from a lower to an upper "bar." If this is the least number of strings needed to make a closed braid representation of a LINK, n is called the BRAID INDEX. Now enumerate the possible braids in a group, denoted B_n. A general n-braid is constructed by iteratively applying the σ_i ($i = 1, \ldots, n-1$) operator, which switches the lower endpoints of the ith and $(i+1)$th strings—keeping the upper endpoints fixed—with the $(i+1)$th string brought *above* the ith string. If the $(i+1)$th string passes *below* the ith string, it is denoted σ_i^{-1}.

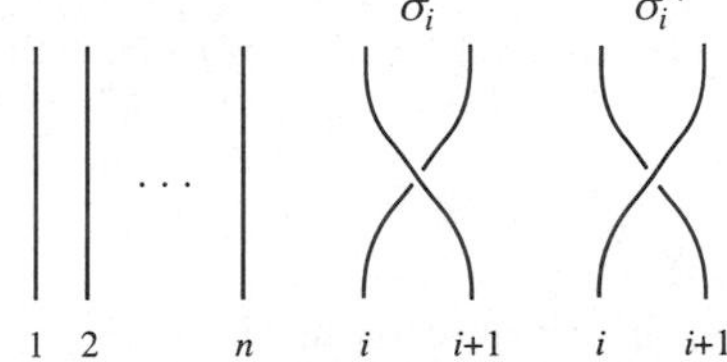

Topological equivalence for different representations of a BRAID WORD $\prod_i \sigma_i$ and $\prod_i \sigma_i'$ is guaranteed by the conditions

$$\begin{cases} \sigma_i\sigma_j = \sigma_j'\sigma_i' & \text{for } |i-j| \geq 2 \\ \sigma_i\sigma_{i+1}\sigma_i = \sigma_{i+1}'\sigma_i'\sigma_{i+1}' & \text{for all } i \end{cases}$$

as first proved by E. Artin. Any n-braid is expressed as a BRAID WORD, e.g., $\sigma_1\sigma_2\sigma_3\sigma_2^{-1}\sigma_1$ is a BRAID WORD for the braid group B_3. When the opposite ends of the braids are connected by nonintersecting lines, KNOTS are formed which are identified by their braid group and BRAID WORD. The BURAU REPRESENTATION gives a matrix representation of the braid groups.

References

Birman, J. S. "Braids, Links, and the Mapping Class Groups." *Ann. Math. Studies*, No. 82. Princeton, NJ: Princeton University Press, 1976.

Birman, J. S. "Recent Developments in Braid and Link Theory." *Math. Intell.* **13**, 52–60, 1991.

Christy, J. "Braids." http://www.mathsource.com/cgi-bin/MathSource/Applications/Mathematics/0202-228.

Jones, V. F. R. "Hecke Algebra Representations of Braid Groups and Link Polynomials." *Ann. Math.* **126**, 335–388, 1987.

Weisstein, E. W. "Knots and Links." http://www.astro.virginia.edu/~eww6n/math/notebooks/Knots.m.

Braid Index

The least number of strings needed to make a closed braid representation of a LINK. The braid index is equal to the least number of SEIFERT CIRCLES in any projection of a KNOT (Yamada 1987). Also, for a nonsplittable LINK with CROSSING NUMBER $c(L)$ and braid index $i(L)$,

$$c(L) \geq 2[i(L) - 1]$$

(Ohyama 1993). Let E be the largest and e the smallest POWER of ℓ in the HOMFLY POLYNOMIAL of an oriented LINK, and i be the braid index. Then the MORTON-FRANKS-WILLIAMS INEQUALITY holds,

$$i \geq \tfrac{1}{2}(E - e) + 1$$

(Franks and Williams 1987). The inequality is sharp for all PRIME KNOTS up to 10 crossings with the exceptions of 09_{042}, 09_{049}, 10_{132}, 10_{150}, and 10_{156}.

References

Franks, J. and Williams, R. F. "Braids and the Jones Polynomial." *Trans. Amer. Math. Soc.* **303**, 97–108, 1987.

Jones, V. F. R. "Hecke Algebra Representations of Braid Groups and Link Polynomials." *Ann. Math.* **126**, 335–388, 1987.

Ohyama, Y. "On the Minimal Crossing Number and the Brad Index of Links." *Canad. J. Math.* **45**, 117–131, 1993.

Yamada, S. "The Minimal Number of Seifert Circles Equals the Braid Index of a Link." *Invent. Math.* **89**, 347–356, 1987.

Braid Word

Any n-braid is expressed as a braid word, e.g., $\sigma_1\sigma_2\sigma_3\sigma_2^{-1}\sigma_1$ is a braid word for the BRAID GROUP B_3. By ALEXANDER'S THEOREM, any LINK is representable by a closed braid, but there is no general procedure for reducing a braid word to its simplest form. However, MARKOV'S THEOREM gives a procedure for identifying different braid words which represent the same LINK.

Let b_+ be the sum of POSITIVE exponents, and b_- the sum of NEGATIVE exponents in the BRAID GROUP B_n. If

$$b_+ - 3b_- - n + 1 > 0,$$

then the closed braid b is not AMPHICHIRAL (Jones 1985).

see also BRAID GROUP

References

Jones, V. F. R. "A Polynomial Invariant for Knots via von Neumann Algebras." *Bull. Amer. Math. Soc.* **12**, 103–111, 1985.

Jones, V. F. R. "Hecke Algebra Representations of Braid Groups and Link Polynomials." *Ann. Math.* **126**, 335–388, 1987.

Braikenridge-Maclaurin Construction

The converse of PASCAL'S THEOREM. Let A_1, B_2, C_1, A_2, and B_1 be the five points on a CONIC. Then the CONIC is the LOCUS of the point

$$C_2 = A_1(z \cdot C_1A_2) \cdot B_1(z \cdot C_1B_2),$$

where z is a line through the point $A_1B_2 \cdot B_1A_2$.

see also PASCAL'S THEOREM

Branch

The segments of a TREE between the points of connection (FORKS).

see also FORK, LEAF (TREE)

Branch Cut

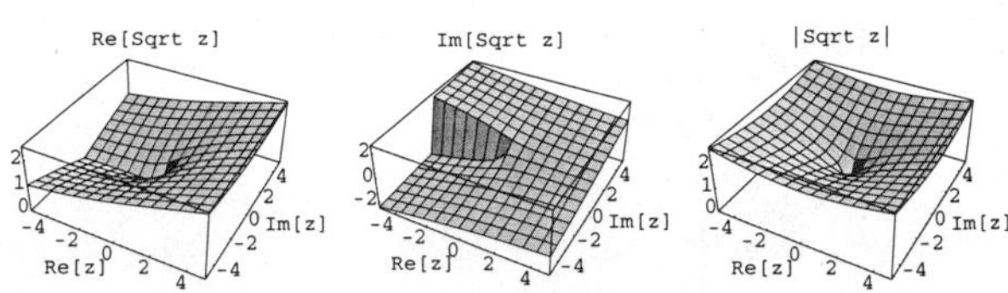

A line in the COMPLEX PLANE across which a FUNCTION is discontinuous.

function	branch cut(s)
$\cos^{-1} z$	$(-\infty, -1)$ and $(1, \infty)$
$\cosh^{-1}$	$(-\infty, 1)$
$\cot^{-1} z$	$(-i, i)$
$\coth^{-1}$	$[-1, 1]$
$\csc^{-1} z$	$(-1, 1)$
csch^{-1}	$(-i, i)$
$\ln z$	$(-\infty, 0]$
$\sec^{-1} z$	$(-1, 1)$
sech^{-1}	$(\infty, 0]$ and $(1, \infty)$
$\sin^{-1} z$	$(-\infty, -1)$ and $(1, \infty)$
$\sinh^{-1}$	$(-i\infty, -i)$ and $(i, i\infty)$
$\sqrt{z}$	$(-\infty, 0)$
$\tan^{-1} z$	$(-i\infty, -i)$ and $(i, i\infty)$
$\tanh^{-1}$	$(-\infty, -1]$ and $[1, \infty)$
$z^n, n \notin \mathbb{Z}$	$(-\infty, 0)$ for $\Re[n] \leq 0$; $(-\infty, 0]$ for $\Re[n] > 0$

see also BRANCH POINT

References

Morse, P. M. and Feshbach, H. *Methods of Theoretical Physics, Part I.* New York: McGraw-Hill, pp. 399–401, 1953.

Branch Line

see BRANCH CUT

Branch Point

An argument at which identical points in the COMPLEX PLANE are mapped to different points. For example, consider

$$f(z) = z^a.$$

Then $f(e^{0i}) = f(1) = 1$, but $f(e^{2\pi i}) = e^{2\pi i a}$, despite the fact that $e^{i0} = e^{2\pi i}$. PINCH POINTS are also called branch points.

see also BRANCH CUT, PINCH POINT

References

Arfken, G. *Mathematical Methods for Physicists, 3rd ed.* Orlando, FL: Academic Press, pp. 397–399, 1985.
Morse, P. M. and Feshbach, H. *Methods of Theoretical Physics, Part I.* New York: McGraw-Hill, pp. 391–392 and 399–401, 1953.

Brauer Chain

A Brauer chain is an ADDITION CHAIN in which each member uses the previous member as a summand. A number n for which a shortest chain exists which is a Brauer chain is called a BRAUER NUMBER.

see also ADDITION CHAIN, BRAUER NUMBER, HANSEN CHAIN

References

Guy, R. K. "Addition Chains. Brauer Chains. Hansen Chains." §C6 in *Unsolved Problems in Number Theory, 2nd ed.* New York: Springer-Verlag, pp. 111–113, 1994.

Brauer Group

The GROUP of classes of finite dimensional central simple ALGEBRAS over k with respect to a certain equivalence.

References

Hazewinkel, M. (Managing Ed.). *Encyclopaedia of Mathematics: An Updated and Annotated Translation of the Soviet "Mathematical Encyclopaedia."* Dordrecht, Netherlands: Reidel, p. 479, 1988.

Brauer Number

A number n for which a shortest chain exists which is a BRAUER CHAIN is called a Brauer number. There are infinitely many non-Brauer numbers.

see also BRAUER CHAIN, HANSEN NUMBER

References

Guy, R. K. "Addition Chains. Brauer Chains. Hansen Chains." §C6 in *Unsolved Problems in Number Theory, 2nd ed.* New York: Springer-Verlag, pp. 111–113, 1994.

Brauer-Severi Variety

An ALGEBRAIC VARIETY over a FIELD K that becomes ISOMORPHIC to a PROJECTIVE SPACE.

References

Hazewinkel, M. (Managing Ed.). *Encyclopaedia of Mathematics: An Updated and Annotated Translation of the Soviet "Mathematical Encyclopaedia."* Dordrecht, Netherlands: Reidel, pp. 480–481, 1988.

Brauer's Theorem

If, in the GERŜGORIN CIRCLE THEOREM for a given m,

$$|a_{jj} - a_{mm}| > \Lambda_j + \Lambda_m$$

for all $j \neq m$, then exactly one EIGENVALUE of A lies in the DISK Γ_m.

References

Gradshteyn, I. S. and Ryzhik, I. M. *Tables of Integrals, Series, and Products, 5th ed.* San Diego, CA: Academic Press, p. 1121, 1979.

Braun's Conjecture

Let $B = \{b_1, b_2, \ldots\}$ be an INFINITE Abelian SEMIGROUP with linear order $b_1 < b_2 < \ldots$ such that b_1 is the unit element and $a < b$ IMPLIES $ac < bc$ for $a, b, c \in B$. Define a MÖBIUS FUNCTION μ on B by $\mu(b_1) = 1$ and

$$\sum_{b_d|b_n} \mu(b_d) = 0$$

for $n = 2, 3, \ldots$. Further suppose that $\mu(b_n) = \mu(n)$ (the true MÖBIUS FUNCTION) for all $n \geq 1$. Then Braun's conjecture states that

$$b_{mn} = b_m b_n$$

for all $m, n \geq 1$.

see also MÖBIUS PROBLEM

References

Flath, A. and Zulauf, A. "Does the Möbius Function Determine Multiplicative Arithmetic?" *Amer. Math. Monthly* **102**, 354–256, 1995.

Breeder

A pair of POSITIVE INTEGERS (a_1, a_2) such that the equations

$$a_1 + a_2 x = \sigma(a_1) = \sigma(a_2)(x+1)$$

have a POSITIVE INTEGER solution x, where $\sigma(n)$ is the DIVISOR FUNCTION. If x is PRIME, then $(a_1, a_2 x)$ is an AMICABLE PAIR (te Riele 1986). (a_1, a_2) is a "special" breeder if

$$a_1 = au$$
$$a_2 = a,$$

where a and u are RELATIVELY PRIME, $(a, u) = 1$. If regular amicable pairs of type $(i, 1)$ with $i \geq 2$ are of the form (au, ap) with p PRIME, then (au, a) are special breeders (te Riele 1986).

References

te Riele, H. J. J. "Computation of All the Amicable Pairs Below 10^{10}." *Math. Comput.* **47**, 361–368 and S9–S35, 1986.

Brelaz's Heuristic Algorithm

An ALGORITHM which can be used to find a good, but not necessarily minimal, EDGE or VERTEX coloring for a GRAPH.

see also CHROMATIC NUMBER

Brent's Factorization Method

A modification of the POLLARD ρ FACTORIZATION METHOD which uses

$$x_{i+1} = {x_i}^2 - c \pmod{n}.$$

References

Brent, R. "An Improved Monte Carlo Factorization Algorithm." *Nordisk Tidskrift for Informationsbehandlung (BIT)* **20**, 176–184, 1980.

Brent's Method

A ROOT-finding ALGORITHM which combines root bracketing, bisection, and INVERSE QUADRATIC INTERPOLATION. It is sometimes known as the VAN WIJNGAARDEN-DEKER-BRENT METHOD.

Brent's method uses a LAGRANGE INTERPOLATING POLYNOMIAL of degree 2. Brent (1973) claims that this method will always converge as long as the values of the function are computable within a given region containing a ROOT. Given three points x_1, x_2, and x_3, Brent's method fits x as a quadratic function of y, then uses the interpolation formula

$$x = \frac{[y-f(x_1)][y-f(x_2)]x_3}{[f(x_3)-f(x_1)][f(x_3)-f(x_2)]} + \frac{[y-f(x_2)][y-f(x_3)]x_1}{[f(x_1)-f(x_2)][f(x_1)-f(x_3)]} + \frac{[y-f(x_3)][y-f(x_1)]x_2}{[f(x_2)-f(x_3)][f(x_2)-f(x_1)]}. \quad (1)$$

Subsequent root estimates are obtained by setting $y = 0$, giving

$$x = b + \frac{P}{Q}, \quad (2)$$

where

$$P = S[R(R-T)(x_3-x_2)-(1-R)(x_2-x_1)] \quad (3)$$

$$Q = (T-1)(R-1)(S-1) \quad (4)$$

with

$$R \equiv \frac{f(x_2)}{f(x_3)} \quad (5)$$

$$S \equiv \frac{f(x_2)}{f(x_1)} \quad (6)$$

$$T \equiv \frac{f(x_1)}{f(x_3)} \quad (7)$$

(Press *et al.* 1992).

References

Brent, R. P. Ch. 3–4 in *Algorithms for Minimization Without Derivatives.* Englewood Cliffs, NJ: Prentice-Hall, 1973.
Forsythe, G. E.; Malcolm, M. A.; and Moler, C. B. §7.2 in *Computer Methods for Mathematical Computations.* Englewood Cliffs, NJ: Prentice-Hall, 1977.
Press, W. H.; Flannery, B. P.; Teukolsky, S. A.; and Vetterling, W. T. "Van Wijngaarden-Dekker-Brent Method." §9.3 in *Numerical Recipes in FORTRAN: The Art of Scientific Computing, 2nd ed.* Cambridge, England: Cambridge University Press, pp. 352–355, 1992.

Brent-Salamin Formula

A formula which uses the ARITHMETIC-GEOMETRIC MEAN to compute PI. It has quadratic convergence and is also called the GAUSS-SALAMIN FORMULA and SALAMIN FORMULA. Let

$$a_{n+1} = \tfrac{1}{2}(a_n + b_n) \quad (1)$$

$$b_{n+1} = \sqrt{a_n b_n} \quad (2)$$

$$c_{n+1} = \tfrac{1}{2}(a_n - b_n) \quad (3)$$

$$d_n \equiv {a_n}^2 - {b_n}^2, \quad (4)$$

and define the initial conditions to be $a_0 = 1$, $b_0 = 1/\sqrt{2}$. Then iterating a_n and b_n gives the ARITHMETIC-GEOMETRIC MEAN, and π is given by

$$\pi = \frac{4[M(1, 2^{-1/2})]^2}{1 - \sum_{j=1}^\infty 2^{j+1} d_j} \quad (5)$$

$$= \frac{4[M(1, 2^{-1/2})]^2}{1 - \sum_{j=1}^\infty 2^{j+1} {c_j}^2}. \quad (6)$$

King (1924) showed that this formula and the LEGENDRE RELATION are equivalent and that either may be derived from the other.

see also ARITHMETIC-GEOMETRIC MEAN, PI

References

Borwein, J. M. and Borwein, P. B. *Pi & the AGM: A Study in Analytic Number Theory and Computational Complexity.* New York: Wiley, pp. 48–51, 1987.
Castellanos, D. "The Ubiquitous Pi. Part II." *Math. Mag.* **61**, 148–163, 1988.
King, L. V. *On the Direct Numerical Calculation of Elliptic Functions and Integrals.* Cambridge, England: Cambridge University Press, 1924.
Lord, N. J. "Recent Calculations of π: The Gauss-Salamin Algorithm." *Math. Gaz.* **76**, 231–242, 1992.
Salamin, E. "Computation of π Using Arithmetic-Geometric Mean." *Math. Comput.* **30**, 565-570, 1976.

Bretschneider's Formula

Given a general QUADRILATERAL with sides of lengths a, b, c, and d (Beyer 1987), the AREA is given by

$$A_{\text{quadrilateral}} = \tfrac{1}{4}\sqrt{4p^2q^2 - (b^2+d^2-a^2-c^2)^2},$$

where p and q are the diagonal lengths.

see also BRAHMAGUPTA'S FORMULA, HERON'S FORMULA

References

Beyer, W. H. (Ed.). *CRC Standard Mathematical Tables, 28th ed.* Boca Raton, FL: CRC Press, p. 123, 1987.

Brianchon Point

The point of CONCURRENCE of the joins of the VERTICES of a TRIANGLE and the points of contact of a CONIC SECTION INSCRIBED in the TRIANGLE. A CONIC INSCRIBED in a TRIANGLE has an equation of the form

$$\frac{f}{u}+\frac{g}{v}+\frac{h}{w}=0,$$

so its Brianchon point has TRILINEAR COORDINATES $(1/f, 1/g, 1/h)$. For KIEPERT'S PARABOLA, the Branchion point has TRIANGLE CENTER FUNCTION

$$\alpha=\frac{1}{a(b^2-c^2)},$$

which is the STEINER POINT.

Brianchon's Theorem

The DUAL of PASCAL'S THEOREM. It states that, given a 6-sided POLYGON CIRCUMSCRIBED on a CONIC SECTION, the lines joining opposite VERTICES (DIAGONALS) meet in a single point.

see also DUALITY PRINCIPLE, PASCAL'S THEOREM

References

Coxeter, H. S. M. and Greitzer, S. L. *Geometry Revisited.* Washington, DC: Math. Assoc. Amer., pp. 77–79, 1967.

Ogilvy, C. S. *Excursions in Geometry.* New York: Dover, p. 110, 1990.

Brick

see EULER BRICK, HARMONIC BRICK, RECTANGULAR PARALLELEPIPED

Bride's Chair

One name for the figure used by Euclid to prove the PYTHAGOREAN THEOREM.

see also PEACOCK'S TAIL, WINDMILL

Bridge Card Game

Bridge is a CARD game played with a normal deck of 52 cards. The number of possible distinct 13-card hands is

$$N=\binom{52}{13}=635{,}013{,}559{,}600.$$

where $\binom{n}{k}$ is a BINOMIAL COEFFICIENT. While the chances of being dealt a hand of 13 CARDS (out of 52) of the same suit are

$$\frac{4}{\binom{52}{13}}=\frac{1}{158{,}753{,}389{,}900},$$

the chance that one of four players will receive a hand of a single suit is

$$\frac{1}{39{,}688{,}347{,}497}.$$

There are special names for specific types of hands. A ten, jack, queen, king, or ace is called an "honor." Getting the three top cards (ace, king, and queen) of three suits and the ace, king, and queen, and jack of the remaining suit is called 13 top honors. Getting all cards of the same suit is called a 13-card suit. Getting 12 cards of same suit with ace high and the 13th card *not* an ace is called 2-card suit, ace high. Getting *no* honors is called a Yarborough.

The probabilities of being dealt 13-card bridge hands of a given type are given below. As usual, for a hand with probability P, the ODDS against being dealt it are $(1/P)-1:1$.

Hand	Exact	Probability
13 top honors	$\frac{4}{N}$	$\frac{1}{158{,}753{,}389{,}900}$
13-card suit	$\frac{4}{N}$	$\frac{1}{158{,}753{,}389{,}900}$
12-card suit, ace high	$\frac{4\cdot 12\cdot 36}{N}$	$\frac{4}{1{,}469{,}938{,}795}$
Yarborough	$\frac{\binom{32}{13}}{N}$	$\frac{5{,}394}{9{,}860{,}459}$
four aces	$\frac{\binom{48}{9}}{N}$	$\frac{11}{4{,}165}$
nine honors	$\frac{\binom{20}{9}\binom{32}{4}}{N}$	$\frac{888{,}212}{93{,}384{,}347}$

Hand	Probability	Odds
13 top honors	6.30×10^{-12}	158,753,389,899:1
13-card suit	6.30×10^{-12}	158,753,389,899:1
12-card suit, ace high	2.72×10^{-9}	367,484,697.8:1
Yarborough	5.47×10^{-4}	1,827.0:1
four aces	2.64×10^{-3}	377.6:1
nine honors	9.51×10^{-3}	104.1:1

see also CARDS, POKER

References

Ball, W. W. R. and Coxeter, H. S. M. *Mathematical Recreations and Essays, 13th ed.* New York: Dover, pp. 48–49, 1987.

Kraitchik, M. "Bridge Hands." §6.3 in *Mathematical Recreations.* New York: W. W. Norton, pp. 119–121, 1942.

Bridge (Graph)

The bridges of a GRAPH are the EDGES whose removal disconnects the GRAPH.

see also ARTICULATION VERTEX

References

Chartrand, G. "Cut-Vertices and Bridges." §2.4 in *Introductory Graph Theory.* New York: Dover, pp. 45–49, 1985.

Bridge Index

A numerical KNOT invariant. For a TAME KNOT K, the bridge index is the least BRIDGE NUMBER of all planar representations of the KNOT. The bridge index of the UNKNOT is defined as 1.

see also BRIDGE NUMBER, CROOKEDNESS

References

Rolfsen, D. *Knots and Links.* Wilmington, DE: Publish or Perish Press, p. 114, 1976.

Schubert, H. "Über eine numerische Knotteninvariante." *Math. Z.* **61**, 245–288, 1954.

Bridge of Königsberg

see KÖNIGSBERG BRIDGE PROBLEM

Bridge Knot

An n-bridge knot is a knot with BRIDGE NUMBER n. The set of 2-bridge knots is identical to the set of rational knots. If L is a 2-BRIDGE KNOT, then the BLM/HO POLYNOMIAL Q and JONES POLYNOMIAL V satisfy

$$Q_L(z) = 2z^{-1}V_L(t)V_L(t^{-1}+1-2z^{-1}),$$

where $z \equiv -t - t^{-1}$ (Kanenobu and Sumi 1993). Kanenobu and Sumi also give a table containing the number of distinct 2-bridge knots of n crossings for $n = 10$ to 22, both not counting and counting MIRROR IMAGES as distinct.

n	K_n	$K_n + K_n^*$
3	0	0
4	0	0
5		
6		
7		
8		
9		
10	45	85
11	91	182
12	176	341
13	352	704
14	693	1365
15	1387	2774
16	2752	5461
17	5504	11008
18	10965	21845
19	21931	43862
20	43776	87381
21	87552	175104
22	174933	349525

References

Kanenobu, T. and Sumi, T. "Polynomial Invariants of 2-Bridge Knots through 22-Crossings." *Math. Comput.* **60**, 771–778 and S17–S28, 1993.

Schubert, H. "Knotten mit zwei Brücken." *Math. Z.* **65**, 133–170, 1956.

Bridge Number

The least number of unknotted arcs lying above the plane in any projection. The knot 05_{05} has bridge number 2. Such knots are called 2-BRIDGE KNOTS. There is a one-to-one correspondence between 2-BRIDGE KNOTS and rational knots. The knot 08_{010} is a 3-bridge knot. A knot with bridge number b is an n-EMBEDDABLE KNOT where $n \leq b$.

see also BRIDGE INDEX

References

Adams, C. C. *The Knot Book: An Elementary Introduction to the Mathematical Theory of Knots.* New York: W. H. Freeman, pp. 64–67, 1994.

Rolfsen, D. *Knots and Links.* Wilmington, DE: Publish or Perish Press, p. 115, 1976.

Brill-Noether Theorem

If the total group of the canonical series is divided into two parts, the difference between the number of points in each part and the double of the dimension of the complete series to which it belongs is the same.

References

Coolidge, J. L. *A Treatise on Algebraic Plane Curves.* New York: Dover, p. 263, 1959.

Bring-Jerrard Quintic Form

A TSCHIRNHAUSEN TRANSFORMATION can be used to algebraically transform a general QUINTIC EQUATION to the form

$$z^5 + c_1 z + c_0 = 0. \quad (1)$$

In practice, the general quintic is first reduced to the PRINCIPAL QUINTIC FORM

$$y^5 + b_2 y^2 + b_1 y + b_0 = 0 \quad (2)$$

before the transformation is done. Then, we require that the sum of the third POWERS of the ROOTS vanishes, so $s_3(y_j) = 0$. We assume that the ROOTS z_i of the Bring-Jerrard quintic are related to the ROOTS y_i of the PRINCIPAL QUINTIC FORM by

$$z_i = \alpha {y_i}^4 + \beta {y_i}^3 + \gamma {y_i}^2 + \delta y_i + \epsilon. \quad (3)$$

In a similar manner to the PRINCIPAL QUINTIC FORM transformation, we can express the COEFFICIENTS c_j in terms of the b_j.

see also BRING QUINTIC FORM, PRINCIPAL QUINTIC FORM, QUINTIC EQUATION

Bring Quintic Form

A TSCHIRNHAUSEN TRANSFORMATION can be used to take a general QUINTIC EQUATION to the form

$$x^5 - x - a = 0,$$

where a may be COMPLEX.

see also BRING-JERRARD QUINTIC FORM, QUINTIC EQUATION

References

Ruppert, W. M. "On the Bring Normal Form of a Quintic in Characteristic 5." *Arch. Math.* **58**, 44–46, 1992.

Brioschi Formula

For a curve with METRIC

$$ds^2 = E\,du^2 + F\,du\,dv + G\,dv^2, \tag{1}$$

where E, F, and G is the first FUNDAMENTAL FORM, the GAUSSIAN CURVATURE is

$$K = \frac{M_1 + M_2}{(EG - F^2)^2}, \tag{2}$$

where

$$M_1 \equiv \begin{vmatrix} -\frac{1}{2}E_{vv} + F_{uv} - \frac{1}{2}G_{uu} & \frac{1}{2}E_u & F_u - \frac{1}{2}E_v \\ F_v - \frac{1}{2}G_u & E & F \\ \frac{1}{2}G_v & F & G \end{vmatrix} \tag{3}$$

$$M_2 \equiv \begin{vmatrix} 0 & \frac{1}{2}E_v & \frac{1}{2}G_u \\ \frac{1}{2}E_v & E & F \\ \frac{1}{2}G_u & F & G \end{vmatrix}, \tag{4}$$

which can also be written

$$K = -\frac{1}{\sqrt{EG}}\left[\frac{\partial}{\partial u}\left(\frac{1}{\sqrt{E}}\frac{\partial\sqrt{G}}{\partial u}\right) + \frac{\partial}{\partial v}\left(\frac{1}{\sqrt{G}}\frac{\partial\sqrt{E}}{\partial v}\right)\right] \tag{5}$$

$$= -\frac{1}{2\sqrt{EG}}\left[\frac{\partial}{\partial u}\left(\frac{G_u}{\sqrt{EG}}\right) + \frac{\partial}{\partial v}\left(\frac{E_v}{\sqrt{EG}}\right)\right]. \tag{6}$$

see also FUNDAMENTAL FORMS, GAUSSIAN CURVATURE

References

Gray, A. *Modern Differential Geometry of Curves and Surfaces.* Boca Raton, FL: CRC Press, pp. 392–393, 1993.

Briot-Bouquet Equation

An ORDINARY DIFFERENTIAL EQUATION of the form

$$x^m y' = f(x, y),$$

where m is a POSITIVE INTEGER, f is ANALYTIC at $x = y = 0$, $f(0,0) = 0$, and $f'_y(0,0) \neq 0$.

References

Hazewinkel, M. (Managing Ed.). *Encyclopaedia of Mathematics: An Updated and Annotated Translation of the Soviet "Mathematical Encyclopaedia."* Dordrecht, Netherlands: Reidel, pp. 481–482, 1988.

Brocard Angle

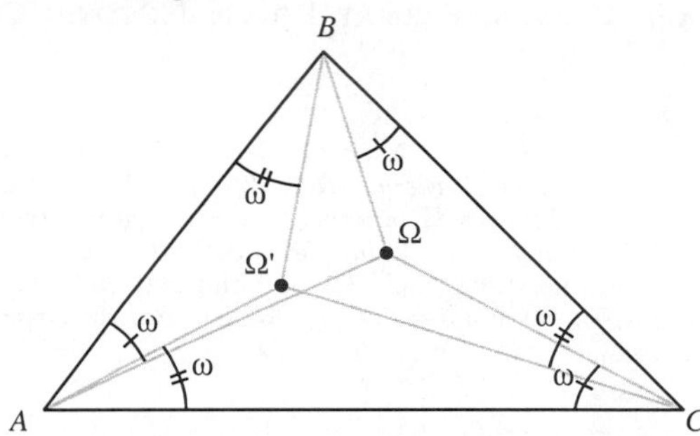

Define the first BROCARD POINT as the interior point Ω of a TRIANGLE for which the ANGLES $\angle\Omega AB$, $\angle\Omega BC$, and $\angle\Omega CA$ are equal. Similarly, define the second BROCARD POINT as the interior point Ω' for which the ANGLES $\angle\Omega' AC$, $\angle\Omega' CB$, and $\angle\Omega' BA$ are equal. Then the ANGLES in both cases are equal, and this angle is called the Brocard angle, denoted ω.

The Brocard angle ω of a TRIANGLE ΔABC is given by the formulas

$$\cot\omega = \cot A + \cot B + \cot C \tag{1}$$

$$= \left(\frac{a^2 + b^2 + c^2}{4\Delta}\right) \tag{2}$$

$$= \frac{1 + \cos\alpha_1\cos\alpha_2\cos\alpha_3}{\sin\alpha_1\sin\alpha_2\sin\alpha_3} \tag{3}$$

$$= \frac{\sin^2\alpha_1 + \sin^2\alpha_2 + \sin^2\alpha_3}{2\sin\alpha_1\sin\alpha_2\sin\alpha_3} \tag{4}$$

$$= \frac{a_1\sin\alpha_1 + a_2\sin\alpha_2 + a_3\sin\alpha_3}{a_1\cos\alpha_1 + a_2\cos\alpha_2 + a_3\cos\alpha_3} \tag{5}$$

$$\csc^2\omega = \csc^2\alpha_1 + \csc^2\alpha_2 + \csc^2\alpha_3 \tag{6}$$

$$\sin\omega = \frac{2\Delta}{\sqrt{{a_1}^2{a_2}^2 + {a_2}^2{a_3}^2 + {a_3}^2{a_1}^2}}, \tag{7}$$

where Δ is the TRIANGLE AREA, A, B, and C are ANGLES, and a, b, and c are side lengths.

If an ANGLE α of a TRIANGLE is given, the maximum possible Brocard angle is given by

$$\cot\omega = \tfrac{3}{2}\tan(\tfrac{1}{2}\alpha) + \tfrac{1}{2}\cos(\tfrac{1}{2}\alpha). \tag{8}$$

Let a TRIANGLE have ANGLES A, B, and C. Then

$$\sin A \sin B \sin C \leq kABC, \tag{9}$$

where

$$k = \left(\frac{3\sqrt{3}}{2\pi}\right)^3 \tag{10}$$

(Le Lionnais 1983). This can be used to prove that

$$8\omega^3 < ABC \tag{11}$$

(Abi-Khuzam 1974).

see also Brocard Circle, Brocard Line, Equi-Brocard Center, Fermat Point, Isogonic Centers

References

Abi-Khuzam, F. "Proof of Yff's Conjecture on the Brocard Angle of a Triangle." *Elem. Math.* **29**, 141–142, 1974.

Johnson, R. A. *Modern Geometry: An Elementary Treatise on the Geometry of the Triangle and the Circle.* Boston, MA: Houghton Mifflin, pp. 263–286 and 289–294, 1929.

Le Lionnais, F. *Les nombres remarquables.* Paris: Hermann, p. 28, 1983.

Brocard Axis

The Line KO passing through the Lemoine Point K and Circumcenter O of a Triangle. The distance $\overline{OK}$ is called the Brocard Diameter. The Brocard axis is Perpendicular to the Lemoine Axis and is the Isogonal Conjugate of Kiepert's Hyperbola. It has equations

$$\sin(B-C)\alpha + \sin(C-A)\beta + \sin(A-B)\gamma = 0$$

$$bc(b^2-c^2)\alpha + ca(c^2-a^2)\beta + ab(a^2-b^2)\gamma = 0.$$

The Lemoine Point, Circumcenter, Isodynamic Points, and Brocard Midpoint all lie along the Brocard axis. Note that the Brocard axis is *not* equivalent to the Brocard Line.

see also Brocard Circle, Brocard Diameter, Brocard Line

Brocard Circle

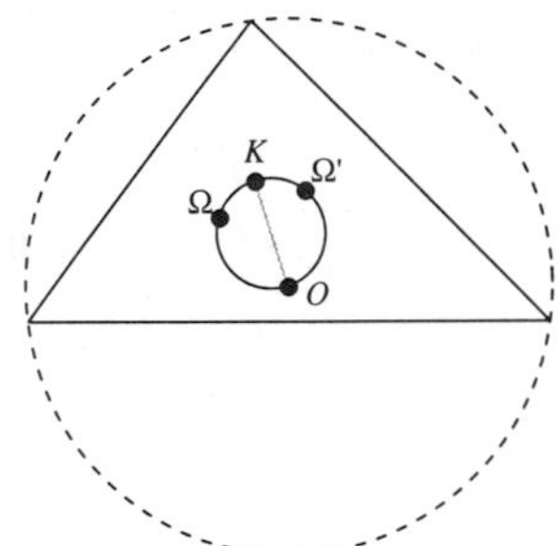

The Circle passing through the first and second Brocard Points Ω and Ω', the Lemoine Point K, and the Circumcenter O of a given Triangle. The Brocard Points Ω and Ω' are symmetrical about the Line $\overleftrightarrow{KO}$, which is called the Brocard Line. The Line Segment $\overline{KO}$ is called the Brocard Diameter, and it has length

$$\overline{OK} = \frac{\overline{O\Omega}}{\cos\omega} = \frac{R\sqrt{1-4\sin^2\omega}}{\cos\omega},$$

where R is the Circumradius and ω is the Brocard Angle. The distance between either of the Brocard Points and the Lemoine Point is

$$\overline{\Omega K} = \overline{\Omega' K} = \overline{\Omega O}\tan\omega.$$

see also Brocard Angle, Brocard Diameter, Brocard Points

References

Johnson, R. A. *Modern Geometry: An Elementary Treatise on the Geometry of the Triangle and the Circle.* Boston, MA: Houghton Mifflin, p. 272, 1929.

Brocard's Conjecture

$$\pi({p_{n+1}}^2) - \pi({p_n}^2) \geq 4$$

for $n \geq 2$ where π is the Prime Counting Function.

see also Andrica's Conjecture

Brocard Diameter

The Line Segment $\overline{KO}$ joining the Lemoine Point K and Circumcenter O of a given Triangle. It is the Diameter of the Triangle's Brocard Circle, and lies along the Brocard Axis. The Brocard diameter has length

$$\overline{OK} = \frac{\overline{O\Omega}}{\cos\omega} = \frac{R\sqrt{1-4\sin^2\omega}}{\cos\omega},$$

where Ω is the first Brocard Point, R is the Circumradius, and ω is the Brocard Angle.

see also Brocard Axis, Brocard Circle, Brocard Line, Brocard Points

Brocard Line

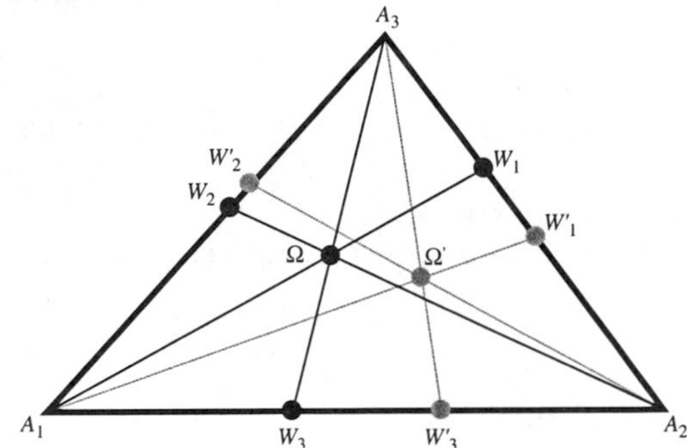

A Line from any of the Vertices A_i of a Triangle to the first Ω or second Ω' Brocard Point. Let the Angle at a Vertex A_i also be denoted A_i, and denote the intersections of $A_1\Omega$ and $A_1\Omega'$ with A_2A_3 as W_1 and W_2. Then the Angles involving these points are

$$\angle A_1\Omega W_3 = A_1 \tag{1}$$

$$\angle W_3\Omega A_2 = A_3 \tag{2}$$

$$\angle A_2\Omega W_1 = A_2. \tag{3}$$

Distances involving the points W_i and W_i' are given by

$$\overline{A_2\Omega} = \frac{a_3}{\sin A_2}\sin\omega \tag{4}$$

$$\frac{\overline{A_2\Omega}}{\overline{A_3\Omega}} = \frac{{a_3}^2}{a_1 a_2} = \frac{\sin(A_3 - \omega)}{\sin\omega} \tag{5}$$

$$\frac{\overline{W_3A_1}}{\overline{W_3A_2}} = \frac{a_2 \sin\omega}{a_1 \sin(A_3 - \omega)} = \left(\frac{a_2}{a_3}\right)^2, \tag{6}$$

where ω is the BROCARD ANGLE (Johnson 1929, pp. 267–268).

The Brocard line, MEDIAN M, and LEMOINE POINT K are concurrent, with $A_1\Omega_1$, A_2K, and A_3M meeting at a point P. Similarly, $A_1\Omega'$, A_2M, and A_3K meet at a point which is the ISOGONAL CONJUGATE point of P (Johnson 1929, pp. 268–269).

see also BROCARD AXIS, BROCARD DIAMETER, BROCARD POINTS, ISOGONAL CONJUGATE, LEMOINE POINT, MEDIAN (TRIANGLE)

References

Johnson, R. A. *Modern Geometry: An Elementary Treatise on the Geometry of the Triangle and the Circle.* Boston, MA: Houghton Mifflin, pp. 263–286, 1929.

Brocard Midpoint

The MIDPOINT of the BROCARD POINTS. It has TRIANGLE CENTER FUNCTION

$$\alpha = a(b^2 + c^2) = \sin(A + \omega),$$

where ω is the BROCARD ANGLE. It lies on the BROCARD AXIS.

References

Kimberling, C. "Central Points and Central Lines in the Plane of a Triangle." *Math. Mag.* **67**, 163–187, 1994.

Brocard Points

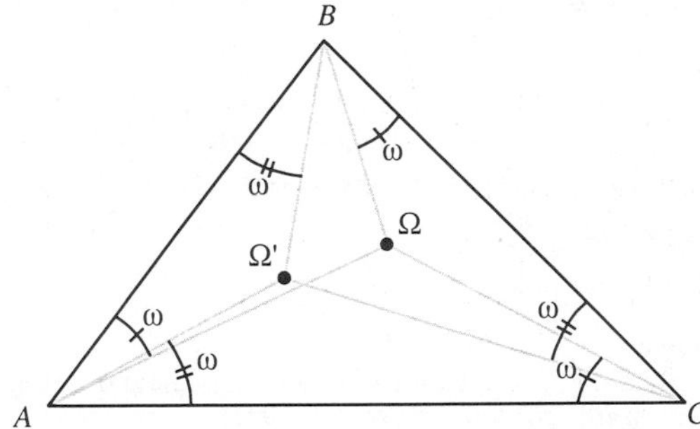

The first Brocard point is the interior point Ω (or τ_1 or Z_1) of a TRIANGLE for which the ANGLES $\angle\Omega AB$, $\angle\Omega BC$, and $\angle\Omega CA$ are equal. The second Brocard point is the interior point Ω' (or τ_2 or Z_2) for which the ANGLES $\angle\Omega' AC$, $\angle\Omega' CB$, and $\angle\Omega' BA$ are equal. The ANGLES in both cases are equal to the BROCARD ANGLE ω,

$$\begin{aligned}\omega &= \angle\Omega AB = \angle\Omega BC = \angle\Omega CA \\ &= \angle\Omega' AC = \angle\Omega' CB = \angle\Omega' BA.\end{aligned}$$

The first two Brocard points are ISOGONAL CONJUGATES (Johnson 1929, p. 266).

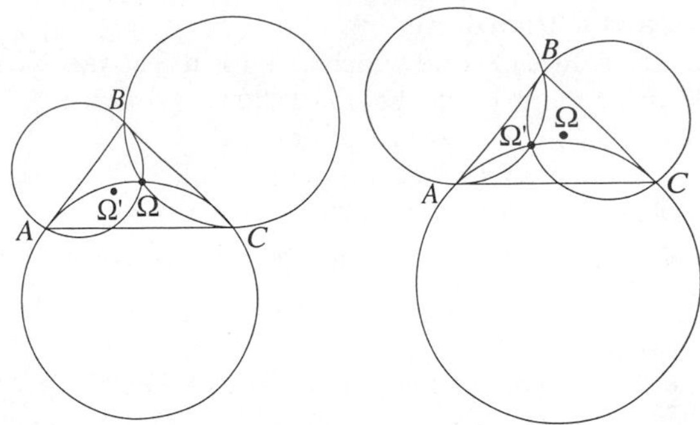

Let C_{BC} be the CIRCLE which passes through the vertices B and C and is TANGENT to the line AC at C, and similarly for C_{AB} and C_{BC}. Then the CIRCLES C_{AB}, C_{BC}, and C_{AC} intersect in the first Brocard point Ω. Similarly, let C'_{BC} be the CIRCLE which passes through the vertices B and C and is TANGENT to the line AB at B, and similarly for C'_{AB} and C'_{AC}. Then the CIRCLES C'_{AB}, C'_{BC}, and C'_{AC} intersect in the second Brocard points Ω' (Johnson 1929, pp. 264–265).

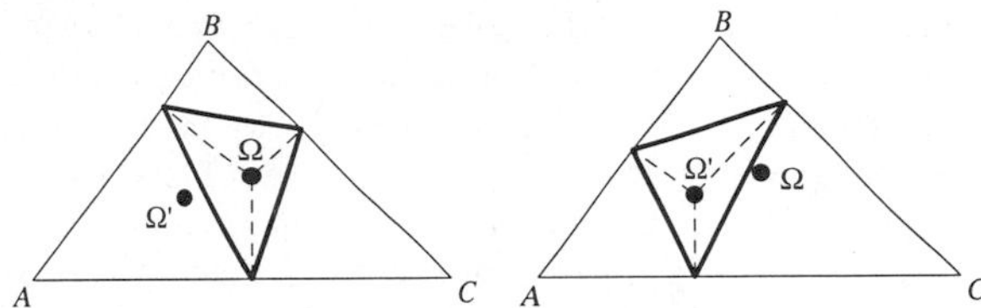

The PEDAL TRIANGLES of Ω and Ω' are congruent, and SIMILAR to the TRIANGLE ΔABC (Johnson 1929, p. 269). Lengths involving the Brocard points include

$$\overline{O\Omega} = \overline{O\Omega'} = R\sqrt{1 - 4\sin^2\omega} \tag{1}$$

$$\overline{\Omega\Omega'} = 2R\sin\omega\sqrt{1 - 4\sin^2\omega}. \tag{2}$$

Brocard's third point is related to a given TRIANGLE by the TRIANGLE CENTER FUNCTION

$$\alpha = a^{-3} \tag{3}$$

(Casey 1893, Kimberling 1994). The third Brocard point Ω'' (or τ_3 or Z_3) is COLLINEAR with the SPIEKER CENTER and the ISOTOMIC CONJUGATE POINT of its TRIANGLE'S INCENTER.

see also BROCARD ANGLE, BROCARD MIDPOINT, EQUI-BROCARD CENTER, YFF POINTS

References

Casey, J. *A Treatise on the Analytical Geometry of the Point, Line, Circle, and Conic Sections, Containing an Account of Its Most Recent Extensions, with Numerous Examples, 2nd ed., rev. enl.* Dublin: Hodges, Figgis, & Co., p. 66, 1893.

Johnson, R. A. *Modern Geometry: An Elementary Treatise on the Geometry of the Triangle and the Circle.* Boston, MA: Houghton Mifflin, pp. 263–286, 1929.

Kimberling, C. "Central Points and Central Lines in the Plane of a Triangle." *Math. Mag.* **67**, 163–187, 1994.

Stroeker, R. J. "Brocard Points, Circulant Matrices, and Descartes' Folium." *Math. Mag.* **61**, 172–187, 1988.

Brocard's Problem

Find the values of n for which $n!+1$ is a SQUARE NUMBER m^2, where $n!$ is the FACTORIAL (Brocard 1876, 1885). The only known solutions are $n = 4$, 5, and 7, and there are no other solutions < 1027. The pairs of numbers (m, n) are called BROWN NUMBERS.

see also BROWN NUMBERS, FACTORIAL, SQUARE NUMBER

References

Brocard, H. Question 166. *Nouv. Corres. Math.* **2**, 287, 1876.

Brocard, H. Question 1532. *Nouv. Ann. Math.* **4**, 391, 1885.

Guy, R. K. *Unsolved Problems in Number Theory, 2nd ed.* New York: Springer-Verlag, p. 193, 1994.

Brocard Triangles

Let the point of intersection of $A_2\Omega$ and $A_3\Omega'$ be B_1, where Ω and Ω' are the BROCARD POINTS, and similarly define B_2 and B_3. $B_1B_2B_3$ is the first Brocard triangle, and is inversely similar to $A_1A_2A_3$. It is inscribed in the BROCARD CIRCLE drawn with OK as the DIAMETER. The triangles $B_1A_2A_3$, $B_2A_3A_1$, and $B_3A_1A_2$ are ISOSCELES TRIANGLES with base angles ω, where ω is the BROCARD ANGLE. The sum of the areas of the ISOSCELES TRIANGLES is Δ, the AREA of TRIANGLE $A_1A_2A_3$. The first Brocard triangle is in perspective with the given TRIANGLE, with A_1B_1, A_2B_2, and A_3B_3 CONCURRENT. The MEDIAN POINT of the first Brocard triangle is the MEDIAN POINT M of the original triangle. The Brocard triangles are in perspective at M.

Let c_1, c_2, and c_3 and c_1', c_2', and c_3' be the CIRCLES intersecting in the BROCARD POINTS Ω and Ω', respectively. Let the two circles c_1 and c_1' tangent at A_1 to A_1A_2 and A_1A_3, and passing respectively through A_3 and A_2, meet again at C_1. The triangle $C_1C_2C_3$ is the second Brocard triangle. Each VERTEX of the second Brocard triangle lies on the second BROCARD CIRCLE.

The two Brocard triangles are in perspective at M.

see also STEINER POINTS, TARRY POINT

References

Johnson, R. A. *Modern Geometry: An Elementary Treatise on the Geometry of the Triangle and the Circle.* Boston, MA: Houghton Mifflin, pp. 277–281, 1929.

Bromwich Integral

The inverse of the LAPLACE TRANSFORM, given by

$$F(t) = \frac{1}{2\pi i}\int_{\gamma-i\infty}^{\gamma+i\infty} e^{st} f(s)\,ds,$$

where γ is a vertical CONTOUR in the COMPLEX PLANE chosen so that all singularities of $f(s)$ are to the left of it.

References

Arfken, G. "Inverse Laplace Transformation." §15.12 in *Mathematical Methods for Physicists, 3rd ed.* Orlando, FL: Academic Press, pp. 853–861, 1985.

Brothers

A PAIR of consecutive numbers.

see also PAIR, SMITH BROTHERS, TWINS

Brouwer Fixed Point Theorem

Any continuous FUNCTION $G : D^n \to D^n$ has a FIXED POINT, where

$$D^n = \{\mathbf{x} \in \mathbb{R}^n : {x_1}^2 + \ldots + {x_n}^2 \leq 1\}$$

is the unit n-BALL.

see also FIXED POINT THEOREM

References

Milnor, J. W. *Topology from the Differentiable Viewpoint.* Princeton, NJ: Princeton University Press, p. 14, 1965.

Browkin's Theorem

For every POSITIVE INTEGER n, there exists a SQUARE in the plane with exactly n LATTICE POINTS in its interior. This was extended by Schinzel and Kulikowski to *all* plane figures of a given shape. The generalization of the SQUARE in 2-D to the CUBE in 3-D was also proved by Browkin.

see also CUBE, SCHINZEL'S THEOREM, SQUARE

References

Honsberger, R. *Mathematical Gems I.* Washington, DC: Math. Assoc. Amer., pp. 121–125, 1973.

Brown's Criterion

A SEQUENCE $\{\nu_i\}$ of nondecreasing POSITIVE INTEGERS is COMPLETE IFF

1. $\nu_1 = 1$.
2. For all $k = 2, 3, \ldots,$

$$s_{k-1} = \nu_1 + \nu_2 + \ldots + \nu_{k-1} \geq \nu_k - 1.$$

A corollary states that a SEQUENCE for which $\nu_1 = 1$ and $\nu_{k+1} \leq 2\nu_k$ is COMPLETE (Honsberger 1985).

see also COMPLETE SEQUENCE

References

Brown, J. L. Jr. "Notes on Complete Sequences of Integers." *Amer. Math. Monthly,* 557–560, 1961.

Honsberger, R. *Mathematical Gems III.* Washington, DC: Math. Assoc. Amer., pp. 123–130, 1985.

Brown Function

For a FRACTAL PROCESS with values $y(t-\Delta t)$ and $y(t+\Delta t)$, the correlation between these two values is given by the Brown function

$$r = 2^{2H-1} - 1,$$

also known as the BACHELIER FUNCTION, LÉVY FUNCTION, or WIENER FUNCTION.

Brown Numbers

Brown numbers are PAIRS (m, n) of INTEGERS satisfying the condition of BROCARD'S PROBLEM, i.e., such that

$$n! + 1 = m^2$$

where $n!$ is the FACTORIAL and m^2 is a SQUARE NUMBER. Only three such PAIRS of numbers are known: (5,4), (11,5), (71,7), and Erdős conjectured that these are the only three such PAIRS. Le Lionnais (1983) points out that there are 3 numbers less than 200,000 for which

$$(n-1)! + 1 \equiv 0 \pmod{n^2},$$

namely 5, 13, and 563.

see also BROCARD'S PROBLEM, FACTORIAL, SQUARE NUMBER

References

Guy, R. K. *Unsolved Problems in Number Theory, 2nd ed.* New York: Springer-Verlag, p. 193, 1994.
Le Lionnais, F. *Les nombres remarquables.* Paris: Hermann, p. 56, 1983.
Pickover, C. A. *Keys to Infinity.* New York: W. H. Freeman, p. 170, 1995.

Broyden's Method

An extension of the secant method of root finding to higher dimensions.

References

Broyden, C. G. "A Class of Methods for Solving Nonlinear Simultaneous Equations." *Math. Comput.* **19**, 577–593, 1965.
Press, W. H.; Flannery, B. P.; Teukolsky, S. A.; and Vetterling, W. T. *Numerical Recipes in FORTRAN: The Art of Scientific Computing, 2nd ed.* Cambridge, England: Cambridge University Press, pp. 382–385, 1992.

Bruck-Ryser-Chowla Theorem

If $n \equiv 1, 2 \pmod 4$, and the SQUAREFREE part of n is divisible by a PRIME $p \equiv 3 \pmod 4$, then no DIFFERENCE SET of ORDER n exists. Equivalently, if a PROJECTIVE PLANE of order n exists, and $n = 1$ or $2 \pmod 4$, then n is the sum of two SQUARES.

Dinitz and Stinson (1992) give the theorem in the following form. If a symmetric (v, k, λ)-BLOCK DESIGN exists, then

1. If v is EVEN, then $k - \lambda$ is a SQUARE NUMBER,
2. If v is ODD, the the DIOPHANTINE EQUATION

$$x^2 = (k-\lambda)y^2 + (-1)^{(v-1)/2}\lambda z^2$$

has a solution in integers, not all of which are 0.

see also BLOCK DESIGN, FISHER'S BLOCK DESIGN INEQUALITY

References

Dinitz, J. H. and Stinson, D. R. "A Brief Introduction to Design Theory." Ch. 1 in *Contemporary Design Theory: A Collection of Surveys* (Ed. J. H. Dinitz and D. R. Stinson). New York: Wiley, pp. 1–12, 1992.
Gordon, D. M. "The Prime Power Conjecture is True for $n < 2,000,000$." *Electronic J. Combinatorics* **1**, R6, 1–7, 1994. http://www.combinatorics.org/Volume_1/volume1.html#R6.
Ryser, H. J. *Combinatorial Mathematics.* Buffalo, NY: Math. Assoc. Amer., 1963.

Bruck-Ryser Theorem

see BRUCK-RYSER-CHOWLA THEOREM

Brun's Constant

The number obtained by adding the reciprocals of the TWIN PRIMES,

$$B \equiv \left(\tfrac{1}{3} + \tfrac{1}{5}\right) + \left(\tfrac{1}{5} + \tfrac{1}{7}\right) + \left(\tfrac{1}{11} + \tfrac{1}{13}\right) + \left(\tfrac{1}{17} + \tfrac{1}{19}\right) + \cdots, \quad (1)$$

By BRUN'S THEOREM, the constant converges to a definite number as $p \to \infty$. Any finite sum underestimates B. Shanks and Wrench (1974) used all the TWIN PRIMES among the first 2 million numbers. Brent (1976) calculated all TWIN PRIMES up to 100 billion and obtained (Ribenboim 1989, p. 146)

$$B \approx 1.90216054, \quad (2)$$

assuming the truth of the first HARDY-LITTLEWOOD CONJECTURE. Using TWIN PRIMES up to 10^{14}, Nicely (1996) obtained

$$B \approx 1.9021605778 \pm 2.1 \times 10^{-9} \quad (3)$$

(Cipra 1995, 1996), in the process discovering a bug in Intel's® Pentium™ microprocessor. The value given by Le Lionnais (1983) is incorrect.

see also TWIN PRIMES, TWIN PRIME CONJECTURE, TWIN PRIMES CONSTANT

References

Ball, W. W. R. and Coxeter, H. S. M. *Mathematical Recreations and Essays, 13th ed.* New York: Dover, p. 64, 1987.
Brent, R. P. "Tables Concerning Irregularities in the Distribution of Primes and Twin Primes Up to 10^{11}." *Math. Comput.* **30**, 379, 1976.
Cipra, B. "How Number Theory Got the Best of the Pentium Chip." *Science* **267**, 175, 1995.
Cipra, B. "Divide and Conquer." *What's Happening in the Mathematical Sciences, 1995–1996, Vol. 3.* Providence, RI: Amer. Math. Soc., pp. 38–47, 1996.
Finch, S. "Favorite Mathematical Constants." http://www.mathsoft.com/asolve/constant/brun/brun.html.
Le Lionnais, F. *Les nombres remarquables.* Paris: Hermann, p. 41, 1983.
Nicely, T. "Enumeration to 10^{14} of the Twin Primes and Brun's Constant." *Virginia J. Sci.* **46**, 195–204, 1996.
Ribenboim, P. *The Book of Prime Number Records, 2nd ed.* New York: Springer-Verlag, 1989.
Shanks, D. and Wrench, J. W. "Brun's Constant." *Math. Comput.* **28**, 293–299, 1974.
Wolf, M. "Generalized Brun's Constants." http://www.ift.uni.wroc.pl/~mwolf/.

Brunn-Minkowski Inequality

The nth root of the CONTENT of the set sum of two sets in Euclidean n-space is greater than or equal to the sum of the nth roots of the CONTENTS of the individual sets.

see also TOMOGRAPHY

References

Cover, T. M. "The Entropy Power Inequality and the Brunn-Minkowski Inequality" §5.10 in In *Open Problems in Communications and Computation.* (Ed. T. M. Cover and B. Gopinath). New York: Springer-Verlag, p. 172, 1987.

Schneider, R. *Convex Bodies: The Brunn-Minkowski Theory.* Cambridge, England: Cambridge University Press, 1993.

Brun's Sum

see BRUN'S CONSTANT

Brun's Theorem

The series producing BRUN'S CONSTANT CONVERGES even if there are an infinite number of TWIN PRIMES. Proved in 1919 by V. Brun.

Brunnian Link

A Brunnian link is a set of n linked loops such that each proper sublink is trivial, so that the removal of any component leaves a set of trivial unlinked UNKNOTS. The BORROMEAN RINGS are the simplest example and have $n = 3$.

see also BORROMEAN RINGS

References

Rolfsen, D. *Knots and Links.* Wilmington, DE: Publish or Perish Press, 1976.

Brute Force Factorization

see DIRECT SEARCH FACTORIZATION

Bubble

A bubble is a MINIMAL SURFACE of the type that is formed by soap film. The simplest bubble is a single SPHERE. More complicated forms occur when multiple bubbles are joined together. Two outstanding problems involving bubbles are to find the arrangements with the smallest PERIMETER (planar problem) or SURFACE AREA (AREA problem) which enclose and separate n given unit areas or volumes in the plane or in space. For $n = 2$, the problems are called the DOUBLE BUBBLE CONJECTURE and the solution to both problems is known to be the DOUBLE BUBBLE.

see also DOUBLE BUBBLE, MINIMAL SURFACE, PLATEAU'S LAWS, PLATEAU'S PROBLEM

References

Morgan, F. "Mathematicians, Including Undergraduates, Look at Soap Bubbles." *Amer. Math. Monthly* **101**, 343–351, 1994.

Pappas, T. "Mathematics & Soap Bubbles." *The Joy of Mathematics.* San Carlos, CA: Wide World Publ./Tetra, p. 219, 1989.

Buchberger's Algorithm

The algorithm for the construction of a GRÖBNER BASIS from an arbitrary ideal basis.

see also GRÖBNER BASIS

References

Becker, T. and Weispfenning, V. *Gröbner Bases: A Computational Approach to Commutative Algebra.* New York: Springer-Verlag, pp. 213–214, 1993.

Buchberger, B. "Theoretical Basis for the Reduction of Polynomials to Canonical Forms." *SIGSAM Bull.* **39**, 19–24, Aug. 1976.

Cox, D.; Little, J.; and O'Shea, D. *Ideals, Varieties, and Algorithms: An Introduction to Algebraic Geometry and Commutative Algebra, 2nd ed.* New York: Springer-Verlag, 1996.

Buckminster Fuller Dome

see GEODESIC DOME

Buffon-Laplace Needle Problem

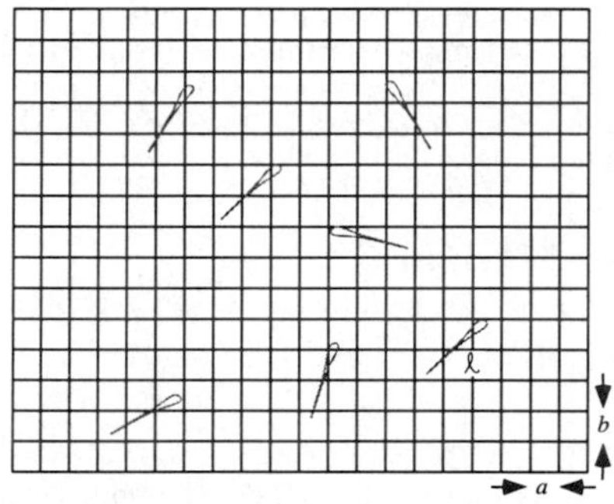

Find the probability $P(\ell, a, b)$ that a needle of length ℓ will land on a line, given a floor with a grid of equally spaced PARALLEL LINES distances a and b apart, with $\ell > a, b$.

$$P(\ell, a, b) = \frac{2\ell(a+b) - \ell^2}{\pi ab}.$$

see also BUFFON'S NEEDLE PROBLEM

Buffon's Needle Problem

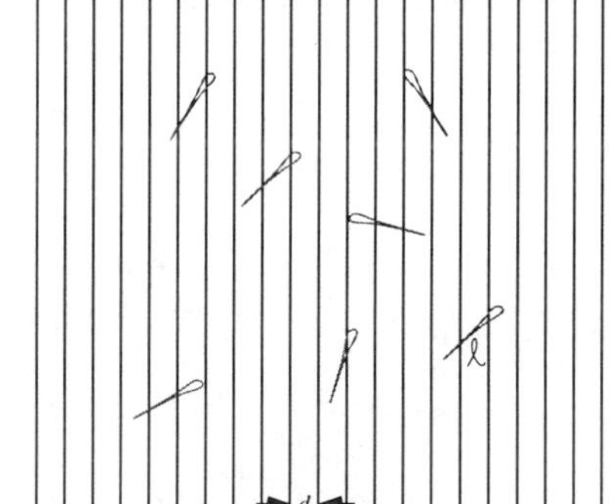

Find the probability $P(\ell, d)$ that a needle of length ℓ will land on a line, given a floor with equally spaced PARALLEL LINES a distance d apart.

$$P(\ell, d) = \int_0^{2\pi} \frac{\ell |\cos\theta|}{d} \frac{d\theta}{2\pi} = \frac{\ell}{2\pi d} 4 \int_0^{\pi/2} \cos\theta \, d\theta$$
$$= \frac{2\ell}{\pi d} [\sin\theta]_0^{\pi/2} = \frac{2\ell}{\pi d}.$$

Several attempts have been made to experimentally determine π by needle-tossing. For a discussion of the relevant statistics and a critical analysis of one of the more accurate (and least believable) needle-tossings, see Badger (1994).

see also BUFFON-LAPLACE NEEDLE PROBLEM

References

Badger, L. "Lazzarini's Lucky Approximation of π." *Math. Mag.* **67**, 83–91, 1994.

Dörrie, H. "Buffon's Needle Problem." §18 in *100 Great Problems of Elementary Mathematics: Their History and Solutions.* New York: Dover, pp. 73–77, 1965.

Kraitchik, M. "The Needle Problem." §6.14 in *Mathematical Recreations.* New York: W. W. Norton, p. 132, 1942.

Wegert, E. and Trefethen, L. N. "From the Buffon Needle Problem to the Kreiss Matrix Theorem." *Amer. Math. Monthly* **101**, 132–139, 1994.

Bulirsch-Stoer Algorithm

An algorithm which finds RATIONAL FUNCTION extrapolations of the form

$$R_{i(i+1)\cdots(i+m)} = \frac{P_\mu(x)}{P_\nu(x)} = \frac{p_0 + p_1 x + \ldots + p_\mu x^\mu}{q_0 + q_1 x + \ldots + q_\nu x^\nu}$$

and can be used in the solution of ORDINARY DIFFERENTIAL EQUATIONS.

References

Bulirsch, R. and Stoer, J. §2.2 in *Introduction to Numerical Analysis.* New York: Springer-Verlag, 1991.

Press, W. H.; Flannery, B. P.; Teukolsky, S. A.; and Vetterling, W. T. "Richardson Extrapolation and the Bulirsch-Stoer Method." §16.4 in *Numerical Recipes in FORTRAN: The Art of Scientific Computing, 2nd ed.* Cambridge, England: Cambridge University Press, pp. 718–725, 1992.

Bullet Nose

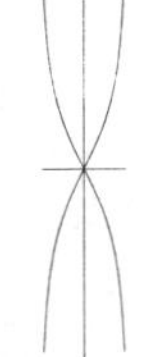

A plane curve with implicit equation

$$\frac{a^2}{x^2} - \frac{b^2}{y^2} = 1. \quad (1)$$

In parametric form,

$$x = a\cos t \quad (2)$$
$$y = b\cot t. \quad (3)$$

The CURVATURE is

$$\kappa = \frac{3ab\cot t\csc t}{(b^2\csc^4 t + a^2\sin^2 t)^{3/2}} \quad (4)$$

and the TANGENTIAL ANGLE is

$$\phi = \tan^{-1}\left(\frac{b\csc^3 t}{a}\right). \quad (5)$$

References

Lawrence, J. D. *A Catalog of Special Plane Curves.* New York: Dover, pp. 127–129, 1972.

Bumping Algorithm

Given a PERMUTATION $\{p_1, p_2, \ldots, p_n\}$ of $\{1, \ldots, n\}$, the bumping algorithm constructs a standard YOUNG TABLEAU by inserting the p_i one by one into an already constructed YOUNG TABLEAU. To apply the bumping algorithm, start with $\{\{p_1\}\}$, which is a YOUNG TABLEAU. If p_1 through p_k have already been inserted, then in order to insert p_{k+1}, start with the first line of the already constructed YOUNG TABLEAU and search for the first element of this line which is greater than p_{k+1}. If there is no such element, append p_{k+1} to the first line and stop. If there is such an element (say, p_p), exchange p_p for p_{k+1}, search the second line using p_p, and so on.

see also YOUNG TABLEAU

References

Skiena, S. *Implementing Discrete Mathematics: Combinatorics and Graph Theory with Mathematica.* Reading, MA: Addison-Wesley, 1990.

Bundle

see FIBER BUNDLE

Burau Representation

Gives a MATRIX representation b_i of a BRAID GROUP in terms of $(n-1)\times(n-1)$ MATRICES. A $-t$ always appears in the (i,i) position.

$$\mathsf{b}_1 = \begin{bmatrix} -t & 0 & 0 & \cdots & 0 \\ -1 & 1 & 0 & \cdots & 0 \\ 0 & 0 & 1 & \cdots & 0 \\ \vdots & \vdots & \vdots & \ddots & \vdots \\ 0 & 0 & 1 & \cdots & 1 \end{bmatrix} \quad (1)$$

$$\mathsf{b}_i = \begin{bmatrix} 1 & \cdots & 0 & 0 & \cdots & 0 \\ \vdots & \ddots & \vdots & \vdots & \ddots & \vdots \\ 0 & \cdots & -t & 0 & \cdots & 0 \\ 0 & \cdots & -t & 0 & \cdots & 0 \\ 0 & \cdots & -1 & 1 & \cdots & 0 \\ 0 & \ddots & 0 & 0 & \ddots & \vdots \\ 0 & \cdots & 0 & 0 & \cdots & 1 \end{bmatrix} \quad (2)$$

$$\mathsf{b}_{n-1} = \begin{bmatrix} 1 & 0 & \cdots & 0 & 0 \\ 0 & 1 & \cdots & 0 & 0 \\ \vdots & \vdots & \ddots & \vdots & \vdots \\ 0 & 0 & \cdots & 0 & -t \\ 0 & 0 & \cdots & 0 & -t \end{bmatrix}. \quad (3)$$

Let Ψ be the MATRIX PRODUCT of BRAID WORDS, then

$$\frac{\det(\mathsf{I} - \Psi)}{1 + t + \cdots + t^{n-1}} = \Delta_L, \quad (4)$$

where Δ_L is the ALEXANDER POLYNOMIAL and det is the DETERMINANT.

References

Burau, W. "Über Zopfgruppen und gleichsinnig verdrilte Verkettungen." *Abh. Math. Sem. Hanischen Univ.* **11**, 171–178, 1936.

Jones, V. "Hecke Algebra Representation of Braid Groups and Link Polynomials." *Ann. Math.* **126**, 335–388, 1987.

Burkhardt Quartic

The VARIETY which is an invariant of degree four and is given by the equation

$$y_0^4 - y_0(y_1^3 + y_2^3 + y_3^3 + y_4^3) + 3y_1y_2y_3y_4 = 0.$$

References

Burkhardt, H. "Untersuchungen aus dem Gebiet der hyperelliptischen Modulfunctionen. II." *Math. Ann.* **38**, 161–224, 1890.

Burkhardt, H. "Untersuchungen aus dem Gebiet der hyperelliptischen Modulfunctionen. III." *Math. Ann.* **40**, 313–343, 1892.

Hunt, B. "The Burkhardt Quartic." Ch. 5 in *The Geometry of Some Special Arithmetic Quotients.* New York: Springer-Verlag, pp. 168–221, 1996.

Burnside's Conjecture

Every non-ABELIAN SIMPLE GROUP has EVEN ORDER.

see also ABELIAN GROUP, SIMPLE GROUP

Burnside's Lemma

Let J be a FINITE GROUP and the image $R(J)$ be a representation which is a HOMEOMORPHISM of J into a PERMUTATION GROUP $S(X)$, where $S(X)$ is the GROUP of all permutations of a SET X. Define the orbits of $R(J)$ as the equivalence classes under $x \sim y$, which is true if there is some permutation p in $R(J)$ such that $p(x) = y$. Define the fixed points of p as the elements x of X for which $p(x) = x$. Then the AVERAGE number of FIXED POINTS of permutations in $R(J)$ is equal to the number of orbits of $R(J)$.

The LEMMA was apparently known by Cauchy (1845) in obscure form and Frobenius (1887) prior to Burnside's (1900) rediscovery. It was subsequently extended and refined by Pólya (1937) for applications in COMBINATORIAL counting problems. In this form, it is known as PÓLYA ENUMERATION THEOREM.

References

Pólya, G. "Kombinatorische Anzahlbestimmungen für Gruppen, Graphen, und chemische Verbindungen." *Acta Math.* **68**, 145–254, 1937.

Burnside Problem

A problem originating with W. Burnside (1902), who wrote, "A still undecided point in the theory of discontinuous groups is whether the ORDER of a GROUP may be not finite, while the order of every operation it contains is finite." This question would now be phrased as "Can a finitely generated group be infinite while every element in the group has finite order?" (Vaughan-Lee 1990). This question was answered by Golod (1964) when he constructed finitely generated infinite p-GROUPS. These GROUPS, however, do not have a finite exponent.

Let F_r be the FREE GROUP of RANK r and let N be the SUBGROUP generated by the set of nth POWERS $\{g^n | g \in F_r\}$. Then N is a normal subgroup of F_r. We define $B(r,n) = F_r/N$ to be the QUOTIENT GROUP. We call $B(r,n)$ the r-generator Burnside group of exponent n. It is the largest r-generator group of exponent n, in the sense that every other such group is a HOMEOMORPHIC image of $B(r,n)$. The Burnside problem is usually stated as: "For which values of r and n is $B(r,n)$ a FINITE GROUP?"

An answer is known for the following values. For $r = 1$, $B(1,n)$ is a CYCLIC GROUP of ORDER n. For $n = 2$, $B(r,2)$ is an elementary ABELIAN 2-group of ORDER 2^n. For $n = 3$, $B(r,3)$ was proved to be finite by Burnside. The ORDER of the $B(r,3)$ groups was established by Levi and van der Waerden (1933), namely 3^a where

$$a \equiv r + \binom{r}{2} + \binom{r}{3}, \quad (1)$$

where $\binom{n}{k}$ is a BINOMIAL COEFFICIENT. For $n = 4$, $B(r,4)$ was proved to be finite by Sanov (1940). Groups of exponent four turn out to be the most complicated for which a POSITIVE solution is known. The precise nilpotency class and derived length are known, as are bounds for the ORDER. For example,

$$|B(2,4)| = 2^{12} \quad (2)$$
$$|B(3,4)| = 2^{69} \quad (3)$$
$$|B(4,4)| = 2^{422} \quad (4)$$
$$|B(5,4)| = 2^{2728}, \quad (5)$$

while for larger values of r the exact value is not yet known. For $n = 6$, $B(r,6)$ was proved to be finite by Hall (1958) with ORDER $2^a 3^b$, where

$$a \equiv 1 + (r-1)3^c \quad (6)$$
$$b \equiv 1 + (r-1)2^r \quad (7)$$
$$c \equiv r + \binom{r}{2} + \binom{r}{3}. \quad (8)$$

No other Burnside groups are known to be finite. On the other hand, for $r > 2$ and $n \geq 665$, with n ODD,

$B(r, n)$ is infinite (Novikov and Adjan 1968). There is a similar fact for $r > 2$ and n a large POWER of 2.

E. Zelmanov was awarded a FIELDS MEDAL in 1994 for his solution of the "restricted" Burnside problem.

see also FREE GROUP

References
Burnside, W. "On an Unsettled Question in the Theory of Discontinuous Groups." *Quart. J. Pure Appl. Math.* **33**, 230–238, 1902.
Golod, E. S. "On Nil-Algebras and Residually Finite p-Groups." *Isv. Akad. Nauk SSSR Ser. Mat.* **28**, 273–276, 1964.
Hall, M. "Solution of the Burnside Problem for Exponent Six." *Ill. J. Math.* **2**, 764–786, 1958.
Levi, F. and van der Waerden, B. L. "Über eine besondere Klasse von Gruppen." *Abh. Math. Sem. Univ. Hamburg* **9**, 154–158, 1933.
Novikov, P. S. and Adjan, S. I. "Infinite Periodic Groups I, II, III." *Izv. Akad. Nauk SSSR Ser. Mat.* **32**, 212–244, 251–524, and 709–731, 1968.
Sanov, I. N. "Solution of Burnside's problem for exponent four." *Leningrad State Univ. Ann. Math. Ser.* **10**, 166–170, 1940.
Vaughan-Lee, M. *The Restricted Burnside Problem, 2nd ed.* New York: Clarendon Press, 1993.

Busemann-Petty Problem

If the section function of a centered convex body in Euclidean n-space ($n \geq 3$) is smaller than that of another such body, is its volume also smaller?

References
Gardner, R. J. "Geometric Tomography." *Not. Amer. Math. Soc.* **42**, 422–429, 1995.

Busy Beaver

A busy beaver is an n-state, 2-symbol, 5-tuple TURING MACHINE which writes the maximum possible number $BB(n)$ of 1s on an initially blank tape before halting. For $n = 0, 1, 2, \ldots, BB(n)$ is given by 0, 1, 4, 6, 13, ≥ 4098, ≥ 136612, ... The busy beaver sequence is also known as RADO'S SIGMA FUNCTION.

see also HALTING PROBLEM, TURING MACHINE

References
Chaitin, G. J. "Computing the Busy Beaver Function." §4.4 in *Open Problems in Communication and Computation* (Ed. T. M. Cover and B. Gopinath). New York: Springer-Verlag, pp. 108–112, 1987.
Dewdney, A. K. "A Computer Trap for the Busy Beaver, the Hardest-Working Turing Machine." *Sci. Amer.* **251**, 19–23, Aug. 1984.
Marxen, H. and Buntrock, J. "Attacking the Busy Beaver 5." *Bull. EATCS* **40**, 247–251, Feb. 1990.
Sloane, N. J. A. Sequence A028444 in "An On-Line Version of the Encyclopedia of Integer Sequences."

Butterfly Catastrophe

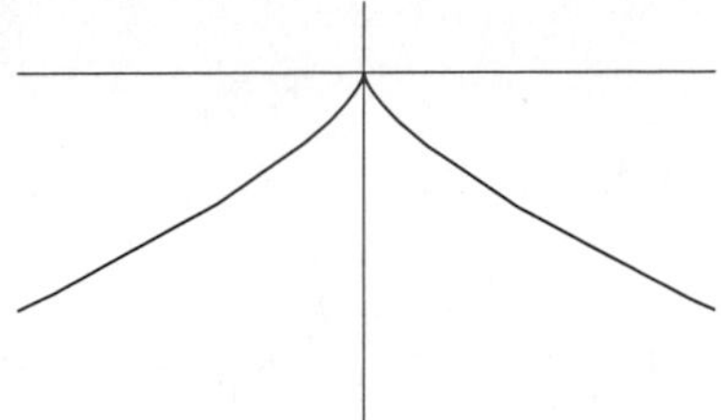

A CATASTROPHE which can occur for four control factors and one behavior axis. The equations

$$x = c(8at^3 + 24t^5)$$
$$y = c(-6at^2 - 15t^4)$$

display such a catastrophe (von Seggern 1993).

References
von Seggern, D. *CRC Standard Curves and Surfaces.* Boca Raton, FL: CRC Press, p. 94, 1993.

Butterfly Curve

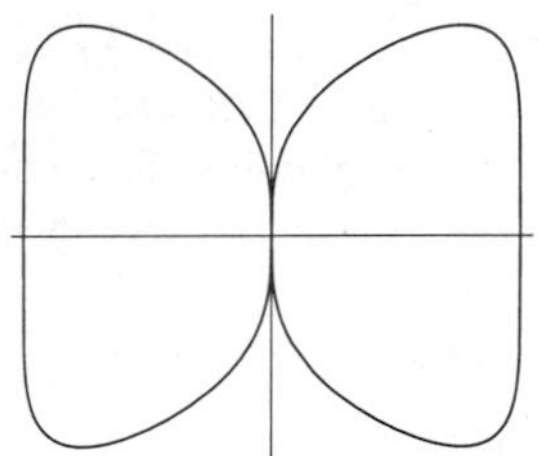

A PLANE CURVE given by the implicit equation

$$y^6 = (x^2 - x^6).$$

see also DUMBBELL CURVE, EIGHT CURVE, PIRIFORM

References
Cundy, H. and Rollett, A. *Mathematical Models, 3rd ed.* Stradbroke, England: Tarquin Pub., p. 72, 1989.

Butterfly Effect

Due to nonlinearities in weather processes, a butterfly flapping its wings in Tahiti can, in theory, produce a tornado in Kansas. This strong dependence of outcomes on very slightly differing initial conditions is a hallmark of the mathematical behavior known as CHAOS.

see also CHAOS, LORENZ SYSTEM

Butterfly Fractal

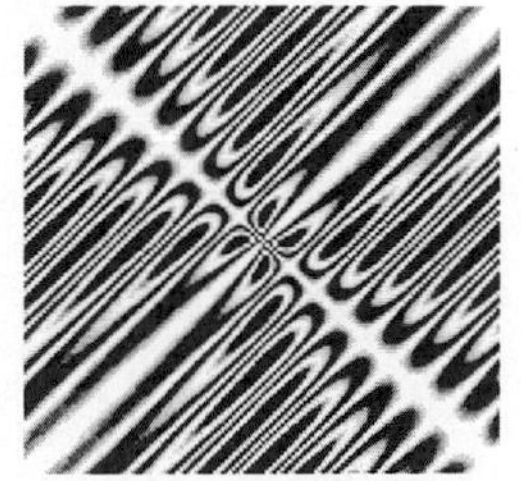

The FRACTAL-like curve generated by the 2-D function

$$f(x, y) = \frac{(x^2 - y^2)\sin\left(\frac{x+y}{a}\right)}{x^2 + y^2}.$$

Butterfly Polyiamond

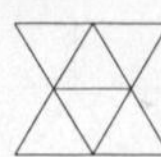

A 6-POLYIAMOND.

References

Golomb, S. W. *Polyominoes: Puzzles, Patterns, Problems, and Packings, 2nd ed.* Princeton, NJ: Princeton University Press, p. 92, 1994.

Butterfly Theorem

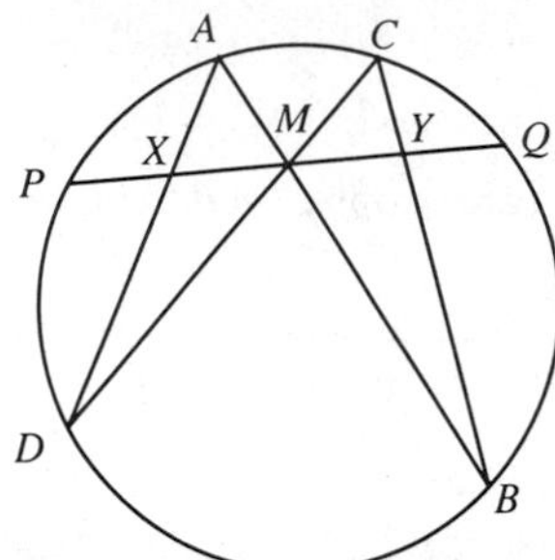

Given a CHORD PQ of a CIRCLE, draw any other two CHORDS AB and CD passing through its MIDPOINT. Call the points where AD and BC meet PQ X and Y. Then M is the MIDPOINT of XY.

see also CHORD, CIRCLE, MIDPOINT

References

Coxeter, H. S. M. and Greitzer, S. L. *Geometry Revisited.* Washington, DC: Math. Assoc. Amer., pp. 45–46, 1967.

C

$\mathbb{C}$

The FIELD of COMPLEX NUMBERS, denoted $\mathbb{C}$.

see also $\mathbb{C}^*$, COMPLEX NUMBER, $\mathbb{I}$, $\mathbb{N}$, $\mathbb{Q}$, $\mathbb{R}$, $\mathbb{Z}$

$\mathbb{C}^*$

The RIEMANN SPHERE $\mathbb{C} \cup \{\infty\}$.

see also $\mathbb{C}$, COMPLEX NUMBER, $\mathbb{Q}$, $\mathbb{R}$, RIEMANN SPHERE, $\mathbb{Z}$

C^*-Algebra

A special type of B^*-ALGEBRA in which the INVOLUTION is the ADJOINT OPERATOR in a HILBERT SPACE.

see also B^*-ALGEBRA, k-THEORY

References

Davidson, K. R. *C^*-Algebras by Example.* Providence, RI: Amer. Math. Soc., 1996.

C-Curve

see LÉVY FRACTAL

C-Determinant

A DETERMINANT appearing in PADÉ APPROXIMANT identities:

$$C_{r/s} = \begin{vmatrix} a_{r-s+1} & a_{r-s+2} & \cdots & a_r \\ \vdots & \vdots & \ddots & \vdots \\ a_r & a_{r+1} & \cdots & a_{r+s-1} \end{vmatrix}.$$

see also PADÉ APPROXIMANT

C-Matrix

Any SYMMETRIC MATRIX ($\mathsf{A}^{\mathrm{T}} = \mathsf{A}$) or SKEW SYMMETRIC MATRIX ($\mathsf{A}^{\mathrm{T}} = -\mathsf{A}$) C_n with diagonal elements 0 and others ± 1 satisfying

$$\mathsf{C}\mathsf{C}^{\mathrm{T}} = (n-1)\mathsf{I},$$

where I is the IDENTITY MATRIX, is known as a C-matrix (Ball and Coxeter 1987). Examples include

$$\mathsf{C}_4 = \begin{bmatrix} 0 & + & + & + \\ - & 0 & - & + \\ - & + & 0 & - \\ - & - & + & 0 \end{bmatrix}$$

$$\mathsf{C}_6 = \begin{bmatrix} 0 & + & + & + & + & + \\ + & 0 & + & - & - & + \\ + & + & 0 & + & + & - \\ + & - & + & 0 & + & - \\ + & - & - & + & 0 & + \\ + & + & - & - & + & 0 \end{bmatrix}.$$

References

Ball, W. W. R. and Coxeter, H. S. M. *Mathematical Recreations and Essays, 13th ed.* New York: Dover, pp. 308–309, 1987.

C-Table

see C-DETERMINANT

Cable Knot

Let K_1 be a TORUS KNOT. Then the SATELLITE KNOT with COMPANION KNOT K_2 is a cable knot on K_2.

References

Adams, C. C. *The Knot Book: An Elementary Introduction to the Mathematical Theory of Knots.* New York: W. H. Freeman, p. 118, 1994.

Rolfsen, D. *Knots and Links.* Wilmington, DE: Publish or Perish Press, pp. 112 and 283, 1976.

Cactus Fractal

A MANDELBROT SET-like FRACTAL obtained by iterating the map

$$z_{n+1} = z_n{}^3 + (z_0 - 1)z_n - z_0.$$

see also FRACTAL, JULIA SET, MANDELBROT SET

Cake Cutting

It is always possible to "fairly" divide a cake among n people using only vertical cuts. Furthermore, it is possible to cut and divide a cake such that each person believes that *everyone* has received $1/n$ of the cake according to his own measure. Finally, if there is some piece on which two people disagree, then there is a way of partitioning and dividing a cake such that each participant believes that he has obtained more than $1/n$ of the cake according to his own measure.

Ignoring the height of the cake, the cake-cutting problem is really a question of fairly dividing a CIRCLE into n equal AREA pieces using cuts in its plane. One method of proving fair cake cutting to always be possible relies on the FROBENIUS-KÖNIG THEOREM.

see also CIRCLE CUTTING, CYLINDER CUTTING, ENVYFREE, FROBENIUS-KÖNIG THEOREM, HAM SANDWICH THEOREM, PANCAKE THEOREM, PIZZA THEOREM, SQUARE CUTTING, TORUS CUTTING

References

Brams, S. J. and Taylor, A. D. "An Envy-Free Cake Division Protocol." *Amer. Math. Monthly* **102**, 9–19, 1995.

Brams, S. J. and Taylor, A. D. *Fair Division: From Cake-Cutting to Dispute Resolution.* New York: Cambridge University Press, 1996.

Dubbins, L. and Spanier, E. "How to Cut a Cake Fairly." *Amer. Math. Monthly* **68**, 1–17, 1961.

Gale, D. "Dividing a Cake." *Math. Intel.* **15**, 50, 1993.

Jones, M. L. "A Note on a Cake Cutting Algorithm of Banach and Knaster." *Amer. Math. Monthly* **104**, 353–355, 1997.

Rebman, K. "How to Get (At Least) a Fair Share of the Cake." In *Mathematical Plums* (Ed. R. Honsberger). Washington, DC: Math. Assoc. Amer., pp. 22–37, 1979.

Steinhaus, H. "Sur la division progmatique." *Ekonometrika (Supp.)* **17**, 315–319, 1949.
Stromquist, W. "How to Cut a Cake Fairly." *Amer. Math. Monthly* **87**, 640–644, 1980.

Cal

see WALSH FUNCTION

Calabi's Triangle

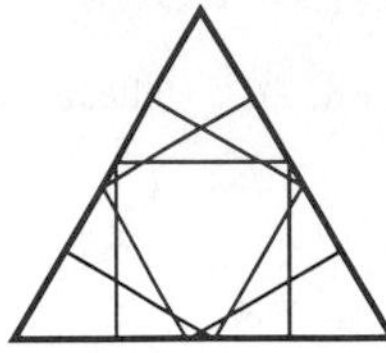
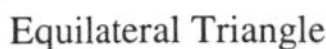
Equilateral Triangle

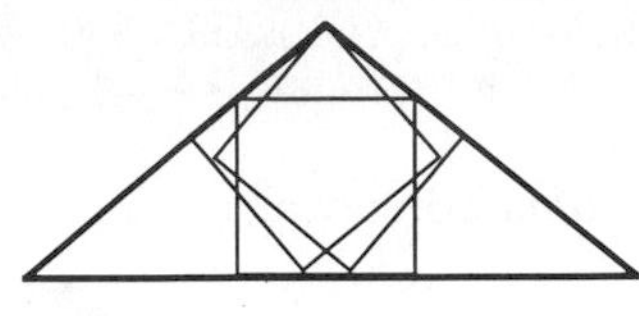
Calabi's Triangle

The one TRIANGLE in addition to the EQUILATERAL TRIANGLE for which the largest inscribed SQUARE can be inscribed in three different ways. The ratio of the sides to that of the base is given by $x = 1.55138752455\ldots$ (Sloane's A046095), where

$$x = \frac{1}{3} + \frac{(-23 + 3i\sqrt{237}\,)^{1/3}}{3 \cdot 2^{2/3}} + \frac{11}{3[2(-23 + 3i\sqrt{237}\,)]^{1/3}}$$

is the largest POSITIVE ROOT of

$$2x^3 - 2x^2 - 3x + 2 = 0,$$

which has CONTINUED FRACTION [1, 1, 1, 4, 2, 1, 2, 1, 5, 2, 1, 3, 1, 1, 390, ...] (Sloane's A046096).

see also GRAHAM'S BIGGEST LITTLE HEXAGON

References
Conway, J. H. and Guy, R. K. "Calabi's Triangle." In *The Book of Numbers.* New York: Springer-Verlag, p. 206, 1996.
Sloane, N. J. A. Sequences A046095 and A046096 in "An On-Line Version of the Encyclopedia of Integer Sequences."

Calabi-Yau Space

A structure into which the 6 extra DIMENSIONS of 10-D string theory curl up.

Calculus

In general, "a" calculus is an abstract theory developed in a purely formal way.

"The" calculus, more properly called ANALYSIS (or REAL ANALYSIS or, in older literature, INFINITESIMAL ANALYSIS) is the branch of mathematics studying the rate of change of quantities (which can be interpreted as SLOPES of curves) and the length, AREA, and VOLUME of objects. The CALCULUS is sometimes divided into DIFFERENTIAL and INTEGRAL CALCULUS, concerned with DERIVATIVES

$$\frac{d}{dx} f(x)$$

and INTEGRALS

$$\int f(x)\,dx,$$

respectively.

While ideas related to calculus had been known for some time (Archimedes' EXHAUSTION METHOD was a form of calculus), it was not until the independent work of Newton and Leibniz that the modern elegant tools and ideas of calculus were developed. Even so, many years elapsed until the subject was put on a mathematically rigorous footing by mathematicians such as Weierstraß.

see also ARC LENGTH, AREA, CALCULUS OF VARIATIONS, CHANGE OF VARIABLES THEOREM, DERIVATIVE, DIFFERENTIAL CALCULUS, ELLIPSOIDAL CALCULUS, EXTENSIONS CALCULUS, FLUENT, FLUXION, FRACTIONAL CALCULUS, FUNCTIONAL CALCULUS, FUNDAMENTAL THEOREMS OF CALCULUS, HEAVISIDE CALCULUS, INTEGRAL, INTEGRAL CALCULUS, JACOBIAN, LAMBDA CALCULUS, KIRBY CALCULUS, MALLIAVIN CALCULUS, PREDICATE CALCULUS, PROPOSITIONAL CALCULUS, SLOPE, TENSOR CALCULUS, UMBRAL CALCULUS, VOLUME

References
Anton, H. *Calculus with Analytic Geometry, 5th ed.* New York: Wiley, 1995.
Apostol, T. M. *Calculus, 2nd ed., Vol. 1: One-Variable Calculus, with an Introduction to Linear Algebra.* Waltham, MA: Blaisdell, 1967.
Apostol, T. M. *Calculus, 2nd ed., Vol. 2: Multi-Variable Calculus and Linear Algebra, with Applications to Differential Equations and Probability.* Waltham, MA: Blaisdell, 1969.
Apostol, T. M. *A Century of Calculus, 2 vols. Pt. 1: 1894–1968. Pt. 2: 1969–1991.* Washington, DC: Math. Assoc. Amer., 1992.
Ayres, F. Jr. and Mendelson, E. *Schaum's Outline of Theory and Problems of Differential and Integral Calculus, 3rd ed.* New York: McGraw-Hill, 1990.
Borden, R. S. *A Course in Advanced Calculus.* New York: Dover, 1998.
Boyer, C. B. *A History of the Calculus and Its Conceptual Development.* New York: Dover, 1989.
Brown, K. S. "Calculus and Differential Equations." `http://www.seanet.com/~ksbrown/icalculu.htm`.
Courant, R. and John, F. *Introduction to Calculus and Analysis, Vol. 1.* New York: Springer-Verlag, 1990.
Courant, R. and John, F. *Introduction to Calculus and Analysis, Vol. 2.* New York: Springer-Verlag, 1990.
Hahn, A. *Basic Calculus: From Archimedes to Newton to Its Role in Science.* New York: Springer-Verlag, 1998.
Kaplan, W. *Advanced Calculus, 4th ed.* Reading, MA: Addison-Wesley, 1992.
Marsden, J. E. and Tromba, A. J. *Vector Calculus, 4th ed.* New York: W. H. Freeman, 1996.
Strang, G. *Calculus.* Wellesley, MA: Wellesley-Cambridge Press, 1991.

Calculus of Variations

A branch of mathematics which is a sort of generalization of CALCULUS. Calculus of variations seeks to find the path, curve, surface, etc., for which a given FUNCTION has a STATIONARY VALUE (which, in physical problems, is usually a MINIMUM or MAXIMUM). Mathematically, this involves finding STATIONARY VALUES of integrals of the form

$$I = \int_b^a f(y, \dot{y}, x)\, dx. \tag{1}$$

I has an extremum only if the EULER-LAGRANGE DIFFERENTIAL EQUATION is satisfied, i.e., if

$$\frac{\partial f}{\partial y} - \frac{d}{dx}\left(\frac{\partial f}{\partial \dot{y}}\right) = 0. \tag{2}$$

The FUNDAMENTAL LEMMA OF CALCULUS OF VARIATIONS states that, if

$$\int_a^b M(x)h(x)\, dx = 0 \tag{3}$$

for all $h(x)$ with CONTINUOUS second PARTIAL DERIVATIVES, then

$$M(x) = 0 \tag{4}$$

on (a, b).

see also BELTRAMI IDENTITY, BOLZA PROBLEM, BRACHISTOCHRONE PROBLEM, CATENARY, ENVELOPE THEOREM, EULER-LAGRANGE DIFFERENTIAL EQUATION, ISOPERIMETRIC PROBLEM, ISOVOLUME PROBLEM, LINDELOF'S THEOREM, PLATEAU'S PROBLEM, POINT-POINT DISTANCE—2-D, POINT-POINT DISTANCE—3-D, ROULETTE, SKEW QUADRILATERAL, SPHERE WITH TUNNEL, UNDULOID, WEIERSTRAß-ERDMAN CORNER CONDITION

References

Arfken, G. "Calculus of Variations." Ch. 17 in *Mathematical Methods for Physicists, 3rd ed.* Orlando, FL: Academic Press, pp. 925–962, 1985.

Bliss, G. A. *Calculus of Variations.* Chicago, IL: Open Court, 1925.

Forsyth, A. R. *Calculus of Variations.* New York: Dover, 1960.

Fox, C. *An Introduction to the Calculus of Variations.* New York: Dover, 1988.

Isenberg, C. *The Science of Soap Films and Soap Bubbles.* New York: Dover, 1992.

Menger, K. "What is the Calculus of Variations and What are Its Applications?" In *The World of Mathematics* (Ed. K. Newman). Redmond, WA: Microsoft Press, pp. 886–890, 1988.

Sagan, H. *Introduction to the Calculus of Variations.* New York: Dover, 1992.

Todhunter, I. *History of the Calculus of Variations During the Nineteenth Century.* New York: Chelsea, 1962.

Weinstock, R. *Calculus of Variations, with Applications to Physics and Engineering.* New York: Dover, 1974.

Calcus

$$1 \text{ calcus} \equiv \tfrac{1}{2304}.$$

see also HALF, QUARTER, SCRUPLE, UNCIA, UNIT FRACTION

Calderón's Formula

$$f(x) = C_\psi \int_{-\infty}^{\infty} \int_{-\infty}^{\infty} \langle f, \psi^{a,b} \rangle\, \psi^{a,b}(x) a^{-2}\, da\, db,$$

where

$$\psi^{a,b}(x) = |a|^{-1/2} \psi\left(\frac{x-b}{a}\right).$$

This result was originally derived using HARMONIC ANALYSIS, but also follows from a WAVELETS viewpoint.

Caliban Puzzle

A puzzle in LOGIC in which one or more facts must be inferred from a set of given facts.

Calvary Cross

see also CROSS

Cameron's Sum-Free Set Constant

A set of POSITIVE INTEGERS S is sum-free if the equation $x + y = z$ has no solutions x, y, $z \in S$. The probability that a random sum-free set S consists entirely of ODD INTEGERS satisfies

$$0.21759 \leq c \leq 0.21862.$$

References

Cameron, P. J. "Cyclic Automorphisms of a Countable Graph and Random Sum-Free Sets." *Graphs and Combinatorics* **1**, 129–135, 1985.

Cameron, P. J. "Portrait of a Typical Sum-Free Set." In *Surveys in Combinatorics 1987* (Ed. C. Whitehead). New York: Cambridge University Press, 13–42, 1987.

Finch, S. "Favorite Mathematical Constants." `http://www.mathsoft.com/asolve/constant/cameron/cameron.html`.

Cancellation

see ANOMALOUS CANCELLATION

Cancellation Law

If $bc \equiv bd \pmod{a}$ and $(b, a) = 1$ (i.e., a and b are RELATIVELY PRIME), then $c \equiv d \pmod{a}$.

see also CONGRUENCE

References

Courant, R. and Robbins, H. *What is Mathematics?: An Elementary Approach to Ideas and Methods, 2nd ed.* Oxford, England: Oxford University Press, p. 36, 1996.

Shanks, D. *Solved and Unsolved Problems in Number Theory, 4th ed.* New York: Chelsea, p. 56, 1993.

Cannonball Problem

Find a way to stack a SQUARE of cannonballs laid out on the ground into a SQUARE PYRAMID (i.e., find a SQUARE NUMBER which is also SQUARE PYRAMIDAL). This corresponds to solving the DIOPHANTINE EQUATION

$$\sum_{i=1}^{k} i^2 = \tfrac{1}{6}k(1+k)(1+2k) = N^2$$

for some pyramid height k. The only solution is $k = 24$, $N = 70$, corresponding to 4900 cannonballs (Ball and Coxeter 1987, Dickson 1952), as conjectured by Lucas (1875, 1876) and proved by Watson (1918).

see also SPHERE PACKING, SQUARE NUMBER, SQUARE PYRAMID, SQUARE PYRAMIDAL NUMBER

References

Ball, W. W. R. and Coxeter, H. S. M. *Mathematical Recreations and Essays, 13th ed.* New York: Dover, p. 59, 1987.

Dickson, L. E. *History of the Theory of Numbers, Vol. 2: Diophantine Analysis.* New York: Chelsea, p. 25, 1952.

Lucas, É. Question 1180. *Nouvelles Ann. Math. Ser. 2* **14**, 336, 1875.

Lucas, É. Solution de Question 1180. *Nouvelles Ann. Math. Ser. 2* **15**, 429–432, 1876.

Ogilvy, C. S. and Anderson, J. T. *Excursions in Number Theory.* New York: Dover, pp. 77 and 152, 1988.

Pappas, T. "Cannon Balls & Pyramids." *The Joy of Mathematics.* San Carlos, CA: Wide World Publ./Tetra, p. 93, 1989.

Watson, G. N. "The Problem of the Square Pyramid." *Messenger. Math.* **48**, 1–22, 1918.

Canonical Form

A clear-cut way of describing every object in a class in a ONE-TO-ONE manner.

see also NORMAL FORM, ONE-TO-ONE

References

Petkovšek, M.; Wilf, H. S.; and Zeilberger, D. *A=B.* Wellesley, MA: A. K. Peters, p. 7, 1996.

Canonical Polyhedron

A POLYHEDRON is said to be canonical if all its EDGES touch a SPHERE and the center of gravity of their contact points is the center of that SPHERE. Each combinatorial type of (GENUS zero) polyhedron contains just one canonical version. The ARCHIMEDEAN SOLIDS and their DUALS are all canonical.

References

Conway, J. H. "Re: polyhedra database." Posting to `geometry.forum` newsgroup, Aug. 31, 1995.

Canonical Transformation

see SYMPLECTIC DIFFEOMORPHISM

Cantor Comb

see CANTOR SET

Cantor-Dedekind Axiom

The points on a line can be put into a ONE-TO-ONE correspondence with the REAL NUMBERS.

see also CARDINAL NUMBER, CONTINUUM HYPOTHESIS, DEDEKIND CUT

Cantor Diagonal Slash

A clever and rather abstract technique used by Georg Cantor to show that the INTEGERS and REALS cannot be put into a ONE-TO-ONE correspondence (i.e., the INFINITY of REAL NUMBERS is "larger" than the INFINITY of INTEGERS). It proceeds by constructing a new member S' of a SET from already known members S by arranging its nth term to differ from the nth term of the nth member of S. The tricky part is that this is done in such a way that the SET including the new member has a larger CARDINALITY than the original SET S.

see also CARDINALITY, CONTINUUM HYPOTHESIS, DENUMERABLE SET

References

Courant, R. and Robbins, H. *What is Mathematics?: An Elementary Approach to Ideas and Methods, 2nd ed.* Oxford, England: Oxford University Press, pp. 81–83, 1996.

Penrose, R. *The Emperor's New Mind: Concerning Computers, Minds, and the Laws of Physics.* Oxford, England: Oxford University Press, pp. 84–85, 1989.

Cantor Dust

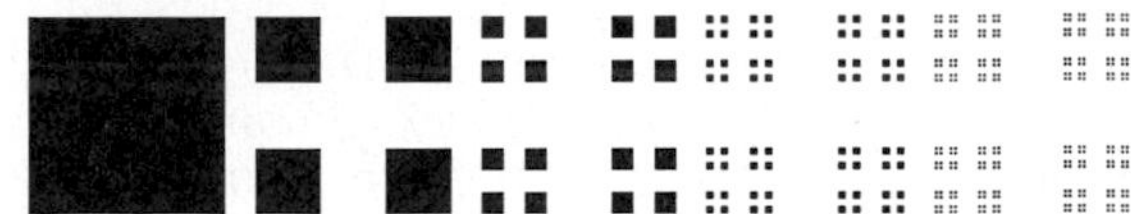

A FRACTAL which can be constructed using STRING REWRITING by creating a matrix three times the size of the current matrix using the rules

```
line 1: "*"->"* *"," "->"   "
line 2: "*"->"   "," "->"   "
line 3: "*"->"* *"," "->"   "
```

Let N_n be the number of black boxes, L_n the length of a side of a white box, and A_n the fractional AREA of black boxes after the nth iteration.

$$N_n = 5^n \tag{1}$$

$$L_n = (\tfrac{1}{3})^n = 3^{-n} \tag{2}$$

$$A_n \doteq {L_n}^2 N_n = (\tfrac{5}{9})^n. \tag{3}$$

The CAPACITY DIMENSION is therefore

$$d_{\text{cap}} = -\lim_{n\to\infty} \frac{\ln N_n}{\ln L_n} = -\lim_{n\to\infty} \frac{\ln(5^n)}{\ln(3^{-n})} = \frac{\ln 5}{\ln 3} = 1.464973521\ldots. \tag{4}$$

see also BOX FRACTAL, SIERPIŃSKI CARPET, SIERPIŃSKI SIEVE

References
Dickau, R. M. "Cantor Dust." http://forum.swarthmore.edu/advanced/robertd/cantor.html.
Ott, E. *Chaos in Dynamical Systems.* New York: Cambridge University Press, pp. 103–104, 1993.
Weisstein, E. W. "Fractals." http://www.astro.virginia.edu/~eww6n/math/notebooks/Fractal.m.

Cantor's Equation

$$\omega^\epsilon = \epsilon,$$

where ω is an ORDINAL NUMBER and ϵ is an INACCESSIBLE CARDINAL.

see also INACCESSIBLE CARDINAL, ORDINAL NUMBER

References
Conway, J. H. and Guy, R. K. *The Book of Numbers.* New York: Springer-Verlag, p. 274, 1996.

Cantor Function

The function whose values are

$$\frac{1}{2}\left(\frac{c_1}{2}+\ldots+\frac{c_{m-1}}{2^{m-1}}+\frac{2}{2^m}\right)$$

for any number between

$$a \equiv \frac{c_1}{3}+\ldots+\frac{c_{m-1}}{3^{m-1}}+\frac{1}{3^m}$$

and

$$b \equiv \frac{c_1}{3}+\ldots+\frac{c_{m-1}}{3^{m-1}}+\frac{2}{3^m}.$$

Chalice (1991) shows that any real-values function $F(x)$ on [0, 1] which is MONOTONE INCREASING and satisfies

1. $F(0) = 0$,
2. $F(x/3) = F(x)/2$,
3. $F(1-x) = 1 - F(x)$

is the Cantor function.

see also CANTOR SET, DEVIL'S STAIRCASE

References
Chalice, D. R. "A Characterization of the Cantor Function." *Amer. Math. Monthly* **98**, 255–258, 1991.
Wagon, S. "The Cantor Function" and "Complex Cantor Sets." §4.2 and 5.1 in *Mathematica in Action.* New York: W. H. Freeman, pp. 102–108 and 143–149, 1991.

Cantor's Paradox

The SET of all SETS is its own POWER SET. Therefore, the CARDINALITY of the SET of all SETS must be bigger than itself.

see also CANTOR'S THEOREM, POWER SET

Cantor Set

The Cantor set (T_∞) is given by taking the interval [0,1] (set T_0), removing the middle third (T_1), removing the middle third of each of the two remaining pieces (T_2), and continuing this procedure ad infinitum. It is therefore the set of points in the INTERVAL [0,1] whose ternary expansions do not contain 1, illustrated below.

This produces the SET of REAL NUMBERS $\{x\}$ such that

$$x = \frac{c_1}{3}+\ldots+\frac{c_n}{3^n}+\ldots, \tag{1}$$

where c_n may equal 0 or 2 for each n. This is an infinite, PERFECT SET. The total length of the LINE SEGMENTS in the nth iteration is

$$\ell_n = \left(\frac{2}{3}\right)^n, \tag{2}$$

and the number of LINE SEGMENTS is $N_n = 2^n$, so the length of each element is

$$\epsilon_n \equiv \frac{\ell}{N} = \left(\frac{1}{3}\right)^n \tag{3}$$

and the CAPACITY DIMENSION is

$$\begin{aligned} d_{\text{cap}} &\equiv -\lim_{\epsilon\to 0^+}\frac{\ln N}{\ln\epsilon} = -\lim_{n\to\infty}\frac{n\ln 2}{-n\ln 3} \\ &= \frac{\ln 2}{\ln 3} = 0.630929\ldots. \end{aligned} \tag{4}$$

The Cantor set is nowhere DENSE, so it has LEBESGUE MEASURE 0.

A general Cantor set is a CLOSED SET consisting entirely of BOUNDARY POINTS. Such sets are UNCOUNTABLE and may have 0 or POSITIVE LEBESGUE MEASURE. The Cantor set is the only totally disconnected, perfect, COMPACT METRIC SPACE up to a HOMEOMORPHISM (Willard 1970).

see also ALEXANDER'S HORNED SPHERE, ANTOINE'S NECKLACE, CANTOR FUNCTION

References
Boas, R. P. Jr. *A Primer of Real Functions.* Washington, DC: Amer. Math. Soc., 1996.
Lauwerier, H. *Fractals: Endlessly Repeated Geometric Figures.* Princeton, NJ: Princeton University Press, pp. 15–20, 1991.
Willard, S. §30.4 in *General Topology.* Reading, MA: Addison-Wesley, 1970.

Cantor Square Fractal

A FRACTAL which can be constructed using STRING REWRITING by creating a matrix three times the size of the current matrix using the rules

```
line 1: "*"->"***"," "->"   "
line 2: "*"->"* *"," "->"   "
line 3: "*"->"***"," "->"   "
```

The first few steps are illustrated above.

The size of the unit element after the nth iteration is

$$L_n = \left(\frac{1}{3}\right)^n$$

and the number of elements is given by the RECURRENCE RELATION

$$N_n = 4N_{n-1} + 5(9^n)$$

where $N_1 \equiv 5$, and the first few numbers of elements are 5, 65, 665, 6305, ... Expanding out gives

$$N_n = 5\sum_{k=0}^{n} 4^{n-k}9^{k-1} = 9^n - 4^n.$$

The CAPACITY DIMENSION is therefore

$$D = -\lim_{n\to\infty}\frac{\ln N_n}{\ln L_n} = -\lim_{n\to\infty}\frac{\ln(9^n - 4^n)}{\ln(3^{-n})}$$
$$= -\lim_{n\to\infty}\frac{\ln(9^n)}{\ln(3^{-n})} = \frac{\ln 9}{\ln 3} = \frac{2\ln 3}{\ln 3} = 2.$$

Since the DIMENSION of the filled part is 2 (i.e., the SQUARE is completely filled), Cantor's square fractal is not a true FRACTAL.

see also BOX FRACTAL, CANTOR DUST

References

Lauwerier, H. *Fractals: Endlessly Repeated Geometric Figures.* Princeton, NJ: Princeton University Press, pp. 82–83, 1991.

Weisstein, E. W. "Fractals." http://www.astro.virginia.edu/~eww6n/math/notebooks/Fractal.m.

Cantor's Theorem

The CARDINAL NUMBER of any set is lower than the CARDINAL NUMBER of the set of all its subsets. A COROLLARY is that there is no highest ℵ (ALEPH).

see also CANTOR'S PARADOX

Cap

see CROSS-CAP, SPHERICAL CAP

Capacity

see TRANSFINITE DIAMETER

Capacity Dimension

A DIMENSION also called the FRACTAL DIMENSION, HAUSDORFF DIMENSION, and HAUSDORFF-BESICOVITCH DIMENSION in which nonintegral values are permitted. Objects whose capacity dimension is different from their TOPOLOGICAL DIMENSION are called FRACTALS. The capacity dimension of a compact METRIC SPACE X is a REAL NUMBER $d_{\rm capacity}$ such that if $n(\epsilon)$ denotes the minimum number of open sets of diameter less than or equal to ϵ, then $n(\epsilon)$ is proportional to ϵ^{-D} as $\epsilon \to 0$. Explicitly,

$$d_{\rm capacity} \equiv -\lim_{\epsilon\to 0^+}\frac{\ln N}{\ln \epsilon}$$

(if the limit exists), where N is the number of elements forming a finite COVER of the relevant METRIC SPACE and ϵ is a bound on the diameter of the sets involved (informally, ϵ is the size of each element used to cover the set, which is taken to to approach 0). If each element of a FRACTAL is equally likely to be visited, then $d_{\rm capacity} = d_{\rm information}$, where $d_{\rm information}$ is the INFORMATION DIMENSION. The capacity dimension satisfies

$$d_{\rm correlation} \leq d_{\rm information} \leq d_{\rm capacity}$$

where $d_{\rm correlation}$ is the CORRELATION DIMENSION, and is conjectured to be equal to the LYAPUNOV DIMENSION.

see also CORRELATION EXPONENT, DIMENSION, HAUSDORFF DIMENSION, KAPLAN-YORKE DIMENSION

References

Nayfeh, A. H. and Balachandran, B. *Applied Nonlinear Dynamics: Analytical, Computational, and Experimental Methods.* New York: Wiley, pp. 538–541, 1995.

Peitgen, H.-O. and Richter, D. H. *The Beauty of Fractals: Images of Complex Dynamical Systems.* New York: Springer-Verlag, 1986.

Wheeden, R. L. and Zygmund, A. *Measure and Integral: An Introduction to Real Analysis.* New York: M. Dekker, 1977.

Carathéodory Derivative

A function f is Carathéodory differentiable at a if there exists a function ϕ which is CONTINUOUS at a such that

$$f(x) - f(a) = \phi(x)(x - a).$$

Every function which is Carathéodory differentiable is also FRÉCHET DIFFERENTIABLE.

see also DERIVATIVE, FRÉCHET DERIVATIVE

Carathéodory's Fundamental Theorem

Each point in the CONVEX HULL of a set S in $\mathbb{R}^n$ is in the convex combination of $n+1$ or fewer points of S.

see also CONVEX HULL, HELLY'S THEOREM

Cardano's Formula

see CUBIC EQUATION

Cardinal Number

In informal usage, a cardinal number is a number used in counting (a COUNTING NUMBER), such as 1, 2, 3, ...

Formally, a cardinal number is a type of number defined in such a way that any method of counting SETS using it gives the same result. (This is not true for the ORDINAL NUMBERS.) In fact, the cardinal numbers are obtained by collecting all ORDINAL NUMBERS which are obtainable by counting a given set. A set has $\aleph_0$ (ALEPH-0) members if it can be put into a ONE-TO-ONE correspondence with the finite ORDINAL NUMBERS.

Two sets are said to have the same cardinal number if all the elements in the sets can be paired off ONE-TO-ONE. An INACCESSIBLE CARDINAL cannot be expressed in terms of a smaller number of smaller cardinals.

see also ALEPH, ALEPH-0 ($\aleph_0$), ALEPH-1 ($\aleph_1$), CANTOR-DEDEKIND AXIOM, CANTOR DIAGONAL SLASH, CONTINUUM, CONTINUUM HYPOTHESIS, EQUIPOLLENT, INACCESSIBLE CARDINALS AXIOM, INFINITY, ORDINAL NUMBER, POWER SET, SURREAL NUMBER, UNCOUNTABLE SET

References

Cantor, G. *Über unendliche, lineare Punktmannigfaltigkeiten, Arbeiten zur Mengenlehre aus dem Jahren 1872–1884.* Leipzig, Germany: Teubner, 1884.

Conway, J. H. and Guy, R. K. "Cardinal Numbers." In *The Book of Numbers.* New York: Springer-Verlag, pp. 277–282, 1996.

Courant, R. and Robbins, H. "Cantor's 'Cardinal Numbers.'" §2.4.3 in *What is Mathematics?: An Elementary Approach to Ideas and Methods, 2nd ed.* Oxford, England: Oxford University Press, pp. 83–86, 1996.

Cardinality

see CARDINAL NUMBER

Cardioid

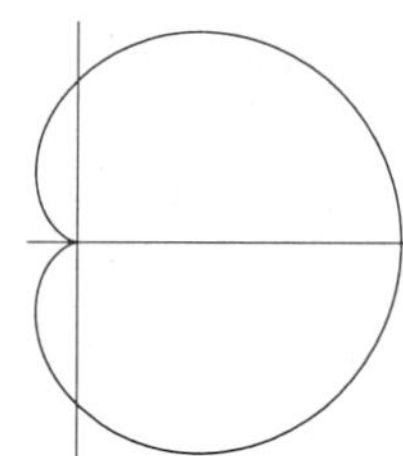

The curve given by the POLAR equation

$$r = a(1 + \cos\theta), \tag{1}$$

sometimes also written

$$r = 2b(1 + \cos\theta), \tag{2}$$

where $b \equiv a/2$, the CARTESIAN equation

$$(x^2 + y^2 - ax)^2 = a^2(x^2 + y^2), \tag{3}$$

and the parametric equations

$$x = a\cos t(1 + \cos t) \tag{4}$$

$$y = a\sin t(1 + \cos t). \tag{5}$$

The cardioid is a degenerate case of the LIMAÇON. It is also a 1-CUSPED EPICYCLOID (with $r = R$) and is the CAUSTIC formed by rays originating at a point on the circumference of a CIRCLE and reflected by the CIRCLE.

The name cardioid was first used by de Castillon in *Philosophical Transactions of the Royal Society* in 1741. Its ARC LENGTH was found by La Hire in 1708. There are exactly three PARALLEL TANGENTS to the cardioid with any given gradient. Also, the TANGENTS at the ends of any CHORD through the CUSP point are at RIGHT ANGLES. The length of any CHORD through the CUSP point is $2a$.

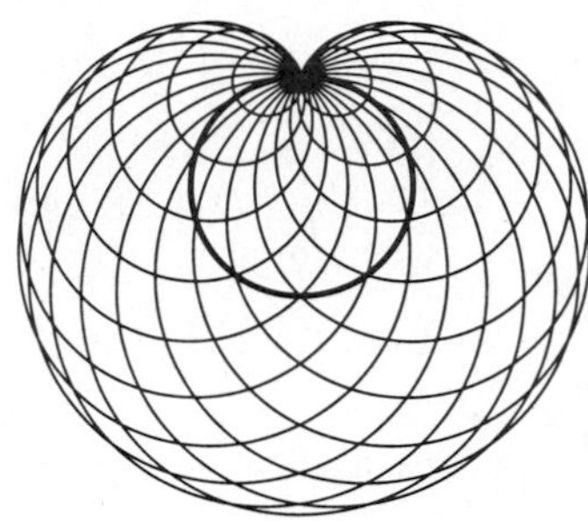

The cardioid may also be generated as follows. Draw a CIRCLE C and fix a point A on it. Now draw a set of CIRCLES centered on the CIRCUMFERENCE of C and passing through A. The ENVELOPE of these CIRCLES is then a cardioid (Pedoe 1995). Let the CIRCLE C be centered at the origin and have RADIUS 1, and let the fixed point be $A = (1, 0)$. Then the RADIUS of a CIRCLE centered at an ANGLE θ from (1, 0) is

$$\begin{aligned} r^2 &= (0 - \cos\theta)^2 + (1 - \sin\theta)^2 \\ &= \cos^2\theta + 1 - 2\sin\theta + \sin^2\theta \\ &= 2(1 - \sin\theta). \end{aligned} \tag{6}$$

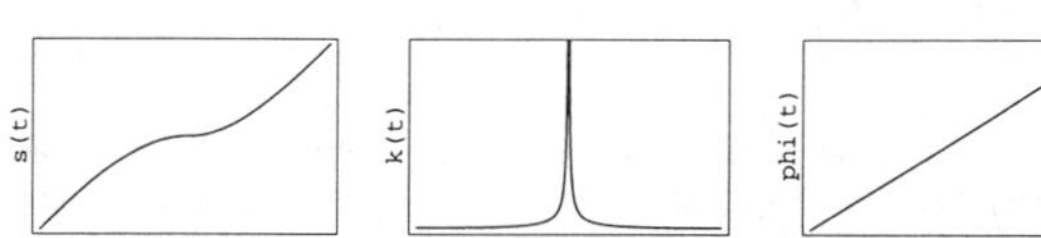

The ARC LENGTH, CURVATURE, and TANGENTIAL ANGLE are

$$s = \int_0^t 2|\cos(\tfrac{1}{2}t)|\,dt = 4a\sin(\tfrac{1}{2}\theta) \tag{7}$$

$$\kappa = \frac{3|\sec(\frac{1}{2}\theta)|}{4a} \tag{8}$$

$$\phi = \tfrac{3}{2}\theta. \tag{9}$$

As usual, care must be taken in the evaluation of $s(t)$ for $t > \pi$. Since (7) comes from an integral involving the

ABSOLUTE VALUE of a function, it must be monotonic increasing. Each QUADRANT can be treated correctly by defining

$$n = \left\lfloor \frac{t}{\pi} \right\rfloor + 1, \qquad (10)$$

where $\lfloor x \rfloor$ is the FLOOR FUNCTION, giving the formula

$$s(t) = (-1)^{1+[n \pmod 2)]} 4\sin(\tfrac{1}{2}t) + 8\left\lfloor \tfrac{1}{2}n \right\rfloor. \qquad (11)$$

The PERIMETER of the curve is

$$\begin{aligned} L &= \int_0^{2\pi} |2a\cos(\tfrac{1}{2}\theta)|\,d\theta = 4a\int_0^{\pi} \cos(\tfrac{1}{2}\theta)\,d\theta \\ &= 4a\int_0^{\pi/2} \cos\phi(2\,d\phi) = 8a\int_0^{\pi/2} \cos\phi\,d\phi \\ &= 8a[\sin\phi]_0^{\pi/2} = 8a. \end{aligned} \qquad (12)$$

The AREA is

$$\begin{aligned} A &= \tfrac{1}{2}\int_0^{2\pi} r^2\,d\theta = \tfrac{1}{2}a^2\int_0^{2\pi} (1 + 2\cos\theta + \cos^2\theta)\,d\theta \\ &= \tfrac{1}{2}a^2\int_0^{2\pi} \{1 + 2\cos\theta + \tfrac{1}{2}[1 + \cos(2\theta)]\}\,d\theta \\ &= \tfrac{1}{2}a^2\int_0^{2\pi} [\tfrac{3}{2} + 2\cos\theta + \tfrac{1}{2}\cos(2\theta)]\,d\theta \\ &= \tfrac{1}{2}a^2[\tfrac{3}{2}\theta + 2\sin\theta + \tfrac{1}{4}\sin(2\theta)]_0^{2\pi} = \tfrac{3}{2}\pi a^2. \end{aligned} \qquad (13)$$

see also CIRCLE, CISSOID, CONCHOID, EQUIANGULAR SPIRAL, LEMNISCATE, LIMAÇON, MANDELBROT SET

References

Gray, A. "Cardioids." §3.3 in *Modern Differential Geometry of Curves and Surfaces.* Boca Raton, FL: CRC Press, pp. 41–42, 1993.

Lawrence, J. D. *A Catalog of Special Plane Curves.* New York: Dover, pp. 118–121, 1972.

Lee, X. "Cardioid." `http://www.best.com/~xah/SpecialPlaneCurves_dir/Cardioid_dir/cardioid.html`.

Lee, X. "Cardioid." `http://www.best.com/~xah/SpecialPlaneCurves_dir/Cardioid_dir/cardioidGG.html`.

Lockwood, E. H. "The Cardioid." Ch. 4 in *A Book of Curves.* Cambridge, England: Cambridge University Press, pp. 34–43, 1967.

MacTutor History of Mathematics Archive. "Cardioid." `http://www-groups.dcs.st-and.ac.uk/~history/Curves/Cardioid.html`.

Pedoe, D. *Circles: A Mathematical View, rev. ed.* Washington, DC: Math. Assoc. Amer., pp. xxvi–xxvii, 1995.

Yates, R. C. "The Cardioid." *Math. Teacher* **52**, 10–14, 1959.

Yates, R. C. "Cardioid." *A Handbook on Curves and Their Properties.* Ann Arbor, MI: J. W. Edwards, pp. 4–7, 1952.

Cardioid Caustic

The CATACAUSTIC of a CARDIOID for a RADIANT POINT at the CUSP is a NEPHROID. The CATACAUSTIC for PARALLEL rays crossing a CIRCLE is a CARDIOID.

Cardioid Evolute

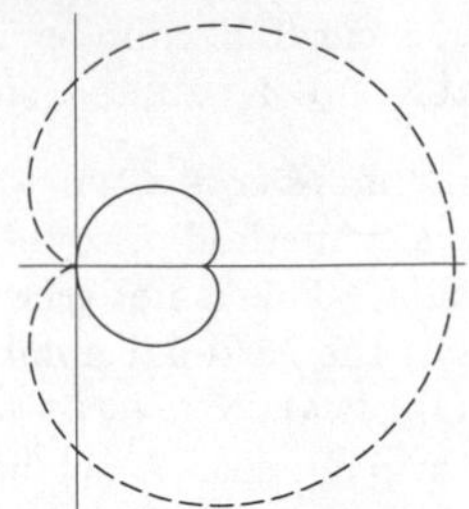

$$\begin{aligned} x &= \tfrac{2}{3}a + \tfrac{1}{3}a\cos\theta(1-\cos\theta) \\ y &= \tfrac{1}{3}a\sin\theta(1-\cos\theta). \end{aligned}$$

This is a mirror-image CARDIOID with $a' = a/3$.

Cardioid Inverse Curve

If the CUSP of the cardioid is taken as the INVERSION CENTER, the cardioid inverts to a PARABOLA.

Cardioid Involute

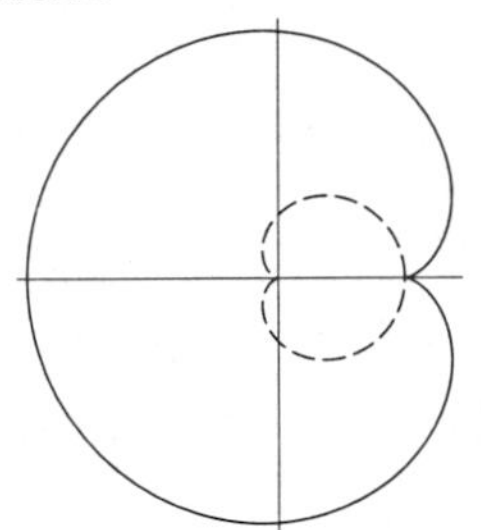

$$\begin{aligned} x &= 2a + 3a\cos\theta(1-\cos\theta) \\ y &= 3a\sin\theta(1-\cos\theta). \end{aligned}$$

This is a mirror-image CARDIOID with $a' = 3a$.

Cardioid Pedal Curve

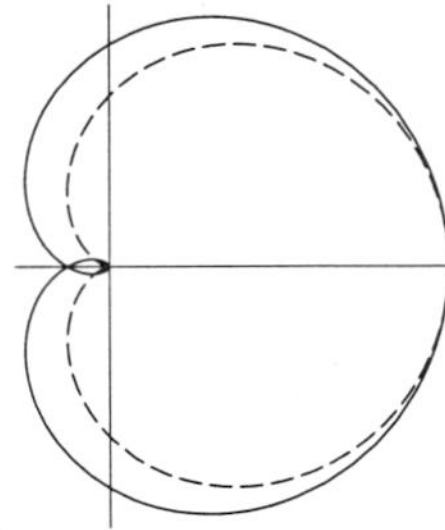

The PEDAL CURVE of the CARDIOID where the PEDAL POINT is the CUSP is CAYLEY'S SEXTIC.

Cards

Cards are a set of n rectangular pieces of cardboard with markings on one side and a uniform pattern on the other. The collection of all cards is called a "deck," and a normal deck of cards consists of 52 cards of four different "suits." The suits are called clubs (♣), diamonds (♢), hearts (♡), and spades (♠). Spades and clubs are colored black, while hearts and diamonds are colored red. The cards of each suit are numbered 1 through 13, where the special terms ace (1), jack (11), queen (12), and king (13) are used instead of numbers 1 and 11–13.

The randomization of the order of cards in a deck is called SHUFFLING. Cards are used in many gambling games (such as POKER), and the investigation of the probabilities of various outcomes in card games was one of the original motivations for the development of modern PROBABILITY theory.

see also BRIDGE CARD GAME, CLOCK SOLITAIRE, COIN, COIN TOSSING, DICE, POKER, SHUFFLE

Carleman's Inequality

Let $\{a_i\}_{i=1}^n$ be a SET of POSITIVE numbers. Then the GEOMETRIC MEAN and ARITHMETIC MEAN satisfy

$$\sum_{i=1}^n (a_1 a_2 \cdots a_i)^{1/i} \leq \frac{e}{n} \sum_{i=1}^n a_i.$$

Here, the constant e is the best possible, in the sense that counterexamples can be constructed for any stricter INEQUALITY which uses a smaller constant.

see also ARITHMETIC MEAN, e, GEOMETRIC MEAN

References

Gradshteyn, I. S. and Ryzhik, I. M. *Tables of Integrals, Series, and Products, 5th ed.* San Diego, CA: Academic Press, p. 1094, 1979.

Hardy, G. H.; Littlewood, J. E.; and Pólya, G. *Inequalities, 2nd ed.* Cambridge, England: Cambridge University Press, pp. 249–250, 1988.

Carlson-Levin Constant

N.B. A detailed on-line essay by S. Finch was the starting point for this entry.

Assume that f is a NONNEGATIVE REAL function on $[0, \infty)$ and that the two integrals

$$\int_0^\infty x^{p-1-\lambda}[f(x)]^p\,dx \tag{1}$$

$$\int_0^\infty x^{q-1+\mu}[f(x)]^q\,dx \tag{2}$$

exist and are FINITE. If $p = q = 2$ and $\lambda = \mu = 1$, Carlson (1934) determined

$$\int_0^\infty f(x)\,dx \leq \sqrt{\pi}\left(\int_0^\infty [f(x)]^2\,dx\right)^{1/4} \times \left(\int_0^\infty x^2[f(x)]^2\,dx\right)^{1/4} \tag{3}$$

and showed that $\sqrt{\pi}$ is the best constant (in the sense that counterexamples can be constructed for any stricter INEQUALITY which uses a smaller constant). For the general case

$$\int_0^\infty f(x)\,dx \leq C\left(\int_0^\infty x^{p-1-\lambda}[f(x)]^p\,dx\right)^s \times \left(\int_0^\infty x^{q-1+\mu}[f(x)]^q\,dx\right)^t, \tag{4}$$

and Levin (1948) showed that the best constant

$$C = \frac{1}{(ps)^s(qt)^t}\left[\frac{\Gamma\left(\frac{s}{\alpha}\right)\Gamma\left(\frac{t}{\alpha}\right)}{(\lambda+\mu)\Gamma\left(\frac{s+t}{\alpha}\right)}\right]^\alpha, \tag{5}$$

where

$$s \equiv \frac{\mu}{p\mu + q\lambda} \tag{6}$$

$$t \equiv \frac{\lambda}{p\mu + q\lambda} \tag{7}$$

$$\alpha \equiv 1 - s - t \tag{8}$$

and $\Gamma(z)$ is the GAMMA FUNCTION.

References

Beckenbach, E. F.; and Bellman, R. *Inequalities.* New York: Springer-Verlag, 1983.

Boas, R. P. Jr. Review of Levin, V. I. "Exact Constants in Inequalities of the Carlson Type." *Math. Rev.* **9**, 415, 1948.

Finch, S. "Favorite Mathematical Constants." `http://www.mathsoft.com/asolve/constant/crlslvn/crlslvn.html`.

Levin, V. I. "Exact Constants in Inequalities of the Carlson Type." *Doklady Akad. Nauk. SSSR (N. S.)* **59**, 635–638, 1948. English review in Boas (1948).

Mitrinovic, D. S.; Pecaric, J. E.; and Fink, A. M. *Inequalities Involving Functions and Their Integrals and Derivatives.* Kluwer, 1991.

Carlson's Theorem

If $f(z)$ is regular and of the form $\mathcal{O}(e^{k|z|})$ where $k < \pi$, for $\Re[z] \geq 0$, and if $f(z) = 0$ for $z = 0, 1, \ldots$, then $f(z)$ is identically zero.

see also GENERALIZED HYPERGEOMETRIC FUNCTION

References

Bailey, W. N. "Carlson's Theorem." §5.3 in *Generalised Hypergeometric Series.* Cambridge, England: Cambridge University Press, pp. 36–40, 1935.

Carlyle Circle

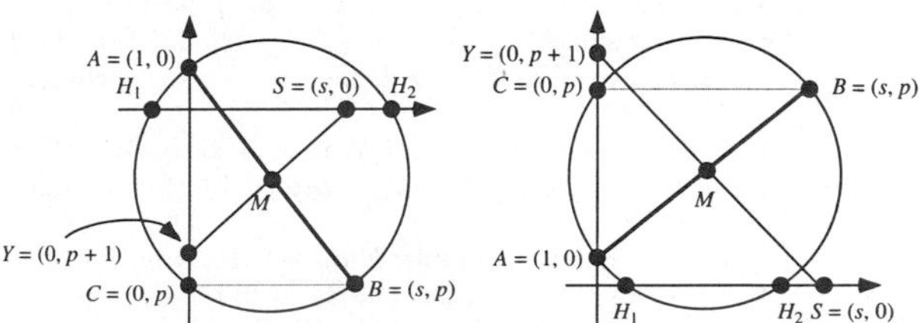

Consider a QUADRATIC EQUATION $x^2 - sx + p = 0$ where s and p denote *signed* lengths. The CIRCLE which has the points $A = (0,1)$ and $B = (s,p)$ as a DIAMETER is then called the Carlyle circle $C_{s,p}$ of the equation. The CENTER of $C_{s,p}$ is then at the MIDPOINT of AB, $M = (s/2, (1+p)/2)$, which is also the MIDPOINT of $S = (s,0)$ and $Y = (0, 1+p)$. Call the points at which $C_{s,p}$ crosses the x-AXIS $H_1 = (x_1, 0)$ and $H_2 = (x_2, 0)$ (with $x_1 \geq x_2$). Then

$$s = x_1 + x_2$$

$$p = x_1 x_2$$

$$(x - x_1)(x - x_2) = x^2 - sx + p,$$

so x_1 and x_2 are the ROOTS of the quadratic equation.

see also 257-GON, 65537-GON, HEPTADECAGON, PENTAGON

References

De Temple, D. W. "Carlyle Circles and the Lemoine Simplicity of Polygonal Constructions." *Amer. Math. Monthly* **98**, 97–108, 1991.

Eves, H. *An Introduction to the History of Mathematics, 6th ed.* Philadelphia, PA: Saunders, 1990.

Leslie, J. *Elements of Geometry and Plane Trigonometry with an Appendix and Very Copious Notes and Illustrations, 4th ed., improved and exp.* Edinburgh: W. & G. Tait, 1820.

Carmichael Condition

A number n satisfies the Carmichael condition IFF $(p-1)|(n/p - 1)$ for all PRIME DIVISORS p of n. This is equivalent to the condition $(p-1)|(n-1)$ for all PRIME DIVISORS p of n.

see also CARMICHAEL NUMBER

References

Borwein, D.; Borwein, J. M.; Borwein, P. B.; and Girgensohn, R. "Giuga's Conjecture on Primality." *Amer. Math. Monthly* **103**, 40–50, 1996.

Carmichael's Conjecture

Carmichael's conjecture asserts that there are an INFINITE number of CARMICHAEL NUMBERS. This was proven by Alford *et al.* (1994).

see also CARMICHAEL NUMBER, CARMICHAEL'S TOTIENT FUNCTION CONJECTURE

References

Alford, W. R.; Granville, A.; and Pomerance, C. "There Are Infinitely Many Carmichael Numbers." *Ann. Math.* **139**, 703–722, 1994.

Cipra, B. *What's Happening in the Mathematical Sciences, Vol. 1.* Providence, RI: Amer. Math. Soc., 1993.

Guy, R. K. "Carmichael's Conjecture." §B39 in *Unsolved Problems in Number Theory, 2nd ed.* New York: Springer-Verlag, p. 94, 1994.

Pomerance, C.; Selfridge, J. L.; and Wagstaff, S. S. Jr. "The Pseudoprimes to $25 \cdot 10^9$." *Math. Comput.* **35**, 1003–1026, 1980.

Ribenboim, P. *The Book of Prime Number Records, 2nd ed.* New York: Springer-Verlag, pp. 29–31, 1989.

Schlafly, A. and Wagon, S. "Carmichael's Conjecture on the Euler Function is Valid Below $10^{10,000,000}$." *Math. Comput.* **63**, 415–419, 1994.

Carmichael Function

$\lambda(n)$ is the LEAST COMMON MULTIPLE (LCM) of all the FACTORS of the TOTIENT FUNCTION $\phi(n)$, except that if $8|n$, then $2^{\alpha-2}$ is a FACTOR instead of $2^{\alpha-1}$.

$$\lambda(n) = \begin{cases} \phi(n) \\ \quad \text{for } n = p^\alpha, p = 2 \text{ and } \alpha \leq 2, \text{ or } p \geq 3 \\ \frac{1}{2}\phi(n) \\ \quad \text{for } n = 2^\alpha \text{ and } \alpha \geq 3 \\ \mathrm{LCM}[\lambda(p_i{}^{\alpha_i})]_i \\ \quad \text{for } n = \prod_i p_i{}^{\alpha_i}. \end{cases}$$

Some special values are

$$\lambda(1) = 1$$
$$\lambda(2) = 1$$
$$\lambda(4) = 2$$
$$\lambda(2^r) = 2^{r-2}$$

for $r \geq 3$, and

$$\lambda(p^r) = \phi(p^r)$$

for p an ODD PRIME and $r \geq 1$. The ORDER of a (mod n) is at most $\lambda(n)$ (Ribenboim 1989). The values of $\lambda(n)$ for the first few n are 1, 1, 2, 2, 4, 2, 6, 4, 10, 2, 12, ... (Sloane's A011773).

see also MODULO MULTIPLICATION GROUP

References

Ribenboim, P. *The Book of Prime Number Records, 2nd ed.* New York: Springer-Verlag, p. 27, 1989.

Riesel, H. "Carmichael's Function." *Prime Numbers and Computer Methods for Factorization, 2nd ed.* Boston, MA: Birkhäuser, pp. 273–275, 1994.

Sloane, N. J. A. Sequence A011773 in "An On-Line Version of the Encyclopedia of Integer Sequences."

Vardi, I. *Computational Recreations in Mathematica.* Redwood City, CA: Addison-Wesley, p. 226, 1991.

Carmichael Number

A Carmichael number is an ODD COMPOSITE NUMBER n which satisfies FERMAT'S LITTLE THEOREM

$$a^{n-1} - 1 \equiv 0 \pmod{n}$$

for *every* choice of a satisfying $(a,n) = 1$ (i.e., a and n are RELATIVELY PRIME) with $1 < a < n$. A Carmichael number is therefore a PSEUDOPRIMES to any base. Carmichael numbers therefore cannot be found to be COMPOSITE using FERMAT'S LITTLE THEOREM. However, if $(a,n) \neq 1$, the congruence of FERMAT'S LITTLE THEOREM is sometimes NONZERO, thus identifying a Carmichael number n as COMPOSITES.

Carmichael numbers are sometimes called ABSOLUTE PSEUDOPRIMES and also satisfy KORSELT'S CRITERION. R. D. Carmichael first noted the existence of such numbers in 1910, computed 15 examples, and conjectured that there were infinitely many (a fact finally proved by Alford *et al.* 1994).

The first few Carmichael numbers are 561, 1105, 1729, 2465, 2821, 6601, 8911, 10585, 15841, 29341, ... (Sloane's A002997). Carmichael numbers have at least three PRIME FACTORS. For Carmichael numbers with exactly three PRIME FACTORS, once one of the PRIMES has been specified, there are only a finite number of Carmichael numbers which can be constructed. Numbers of the form $(6k+1)(12k+1)(18k+1)$ are Carmichael numbers if each of the factors is PRIME (Korselt 1899, Ore 1988, Guy 1994). This can be seen since for

$$N \equiv (6k+1)(12k+1)(18k+1) = 1296k^3+396k^2+36k+1,$$

$N-1$ is a multiple of $36k$ and the LEAST COMMON MULTIPLE of $6k$, $12k$, and $18k$ is $36k$, so $a^{N-1} \equiv 1$ modulo each of the PRIMES $6k+1$, $12k+1$, and $18k+1$, hence $a^{N-1} \equiv 1$ modulo their product. The first few such Carmichael numbers correspond to $k = 1$, 6, 35, 45, 51, 55, 56, ... and are 1729, 294409, 56052361, 118901521, ... (Sloane's A046025). The largest known Carmichael number of this form was found by H. Dubner in 1996 and has 1025 digits.

The smallest Carmichael numbers having 3, 4, ... factors are $561 = 3 \times 11 \times 17$, $41041 = 7 \times 11 \times 13 \times 41$, 825265, 321197185, ... (Sloane's A006931). In total, there are only 43 Carmichael numbers $< 10^6$, 2163 $\leq 2.5 \times 10^{10}$, $105{,}212 \leq 10^{15}$, and $246{,}683 \leq 10^{16}$ (Pinch 1993). Let $C(n)$ denote the number of Carmichael numbers less than n. Then, for sufficiently large n ($n \sim 10^7$ from numerical evidence),

$$C(n) \sim n^{2/7}$$

(Alford *et al.* 1994).

The Carmichael numbers have the following properties:

1. If a PRIME p divides the Carmichael number n, then $n \equiv 1 \pmod{p-1}$ implies that $n \equiv p \pmod{p(p-1)}$.
2. Every Carmichael number is SQUAREFREE.
3. An ODD COMPOSITE SQUAREFREE number n is a Carmichael number IFF n divides the DENOMINATOR of the BERNOULLI NUMBER B_{n-1}.

see also CARMICHAEL CONDITION, PSEUDOPRIME

References

Alford, W. R.; Granville, A.; and Pomerance, C. "There are Infinitely Many Carmichael Numbers." *Ann. Math.* **139**, 703–722, 1994.

Beyer, W. H. *CRC Standard Mathematical Tables, 28th ed.* Boca Raton, FL: CRC Press, p. 87, 1987.

Guy, R. K. "Carmichael Numbers." §A13 in *Unsolved Problems in Number Theory, 2nd ed.* New York: Springer-Verlag, pp. 30–32, 1994.

Korselt, A. "Problème chinois." *L'intermédiaire math.* **6**, 143–143, 1899.

Ore, Ø. *Number Theory and Its History.* New York: Dover, 1988.

Pinch, R. G. E. "The Carmichael Numbers up to 10^{15}." *Math. Comput.* **55**, 381–391, 1993.

Pinch, R. G. E. `ftp://emu.pmms.cam.ac.uk/pub/Carmichael/`.

Pomerance, C.; Selfridge, J. L.; and Wagstaff, S. S. Jr. "The Pseudoprimes to $25 \cdot 10^9$." *Math. Comput.* **35**, 1003–1026, 1980.

Riesel, H. *Prime Numbers and Computer Methods for Factorization, 2nd ed.* Basel: Birkhäuser, pp. 89–90 and 94–95, 1994.

Shanks, D. *Solved and Unsolved Problems in Number Theory, 4th ed.* New York: Chelsea, p. 116, 1993.

Sloane, N. J. A. Sequences A002997/M5462 and A006931/M5463 in "An On-Line Version of the Encyclopedia of Integer Sequences."

Carmichael Sequence

A FINITE, INCREASING SEQUENCE of INTEGERS $\{a_1, \ldots, a_m\}$ such that

$$(a_i - 1) | (a_1 \cdots a_{i-1})$$

for $i = 1, \ldots, m$, where $m|n$ indicates that m DIVIDES n. A Carmichael sequence has exclusive EVEN or ODD elements. There are infinitely many Carmichael sequences for every order.

see also GIUGA SEQUENCE

References

Borwein, D.; Borwein, J. M.; Borwein, P. B.; and Girgensohn, R. "Giuga's Conjecture on Primality." *Amer. Math. Monthly* **103**, 40–50, 1996.

Carmichael's Theorem

If a and n are RELATIVELY PRIME so that the GREATEST COMMON DENOMINATOR $\text{GCD}(a, n) = 1$, then

$$a^{\lambda(n)} \equiv 1 \pmod{n},$$

where λ is the CARMICHAEL FUNCTION.

Carmichael's Totient Function Conjecture

It is thought that the TOTIENT VALENCE FUNCTION $N_\phi(m) \geq 2$ (i.e., the TOTIENT VALENCE FUNCTION never takes the value 1). This assertion is called Carmichael's totient function conjecture and is equivalent to the statement that there exists an $m \neq n$ such that $\phi(n) = \phi(m)$ (Ribenboim 1996, pp. 39–40). Any counterexample to the conjecture must have more than 10,000 DIGITS (Conway and Guy 1996). Recently, the conjecture was reportedly proven by F. Saidak in November, 1997 with a proof short enough to fit on a postcard.

see also TOTIENT FUNCTION, TOTIENT VALENCE FUNCTION

References

Conway, J. H. and Guy, R. K. *The Book of Numbers.* New York: Springer-Verlag, p. 155, 1996.

Ribenboim, P. *The New Book of Prime Number Records.* New York: Springer-Verlag, 1996.

Carnot's Polygon Theorem

If P_1, P_2, ..., are the VERTICES of a finite POLYGON with no "minimal sides" and the side P_iP_j meets a curve in the POINTS P_{ij1} and P_{ij2}, then

$$\frac{\prod_i \overline{P_1P_{12i}} \prod_i \overline{P_2P_{23i}} \cdots \prod_i \overline{P_NP_{N1i}}}{\prod_i \overline{P_NP_{N1i}} \cdots \prod_i \overline{P_2P_{2i1}}} = 1,$$

where $\overline{AB}$ denotes the DISTANCE from POINT A to B.

References

Coolidge, J. L. *A Treatise on Algebraic Plane Curves.* New York: Dover, p. 190, 1959.

Carnot's Theorem

Given any TRIANGLE $A_1A_2A_3$, the signed sum of PERPENDICULAR distances from the CIRCUMCENTER O to the sides is

$$OO_1 + OO_2 + OO_3 = R + r,$$

where r is the INRADIUS and R is the CIRCUMRADIUS. The sign of the distance is chosen to be POSITIVE IFF the entire segment OO_i lies outside the TRIANGLE.

see also JAPANESE TRIANGULATION THEOREM

References

Eves, H. W. *A Survey of Geometry, rev. ed.* Boston, MA: Allyn and Bacon, pp. 256 and 262, 1972.

Honsberger, R. *Mathematical Gems III.* Washington, DC: Math. Assoc. Amer., p. 25, 1985.

Carotid-Kundalini Fractal

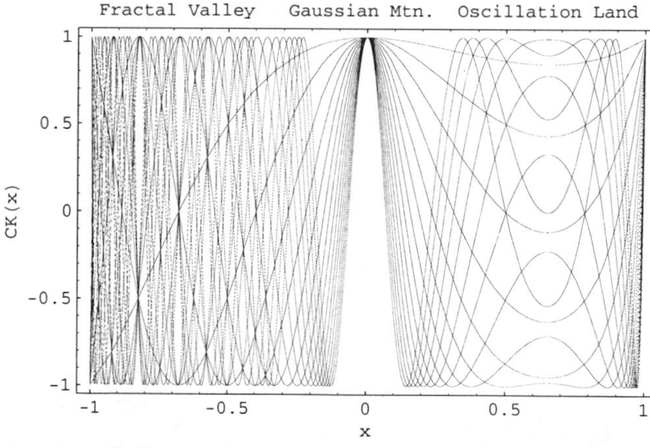

A fractal-like structure is produced for $x < 0$ by superposing plots of CAROTID-KUNDALINI FUNCTIONS CK_n of different orders n. The region $-1 < x < 0$ is called FRACTAL LAND by Pickover (1995), the central region the GAUSSIAN MOUNTAIN RANGE, and the region $x > 0$ OSCILLATION LAND. The plot above shows $n = 1$ to 25. Gaps in FRACTAL LAND occur whenever

$$x \cos^{-1} x = 2\pi \frac{p}{q}$$

for p and q RELATIVELY PRIME INTEGERS. At such points x, the functions assume the $\lceil (q+1)/2 \rceil$ values $\cos(2\pi r/q)$ for $r = 0, 1, \ldots, \lfloor q/2 \rfloor$, where $\lceil z \rceil$ is the CEILING FUNCTION and $\lfloor z \rfloor$ is the FLOOR FUNCTION.

References

Pickover, C. A. "Are Infinite Carotid-Kundalini Functions Fractal?" Ch. 24 in *Keys to Infinity.* New York: W. H. Freeman, pp. 179–181, 1995.

Carotid-Kundalini Function

The FUNCTION given by

$$CK_n(x) \equiv \cos(nx \cos^{-1} x),$$

where n is an INTEGER and $-1 < x < 1$.

see also CAROTID-KUNDALINI FRACTAL

Carry

```
  1 1    ← carries
  1 5 8  ← addend 1
+ 2 4 9  ← addend 2
  4 0 7  ← sum
```

The operating of shifting the leading DIGITS of an ADDITION into the next column to the left when the SUM of that column exceeds a single DIGIT (i.e., 9 in base 10).

see also ADDEND, ADDITION, BORROW

Carrying Capacity

see LOGISTIC GROWTH CURVE

Cartan Matrix

A MATRIX used in the presentation of a LIE ALGEBRA.

References

Jacobson, N. *Lie Algebras.* New York: Dover, p. 121, 1979.

Cartan Relation

The relationship $Sq^i(x \smile y) = \Sigma_{j+k=i} Sq^j(x) \smile Sq^k(y)$ encountered in the definition of the STEENROD ALGEBRA.

Cartan Subgroup

A type of maximal Abelian SUBGROUP.

References

Knapp, A. W. "Group Representations and Harmonic Analysis, Part II." *Not. Amer. Math. Soc.* **43**, 537–549, 1996.

Cartan Torsion Coefficient

The ANTISYMMETRIC parts of the CONNECTION COEFFICIENT ${\Gamma^\lambda}_{\mu\nu}$.

Cartesian Coordinates

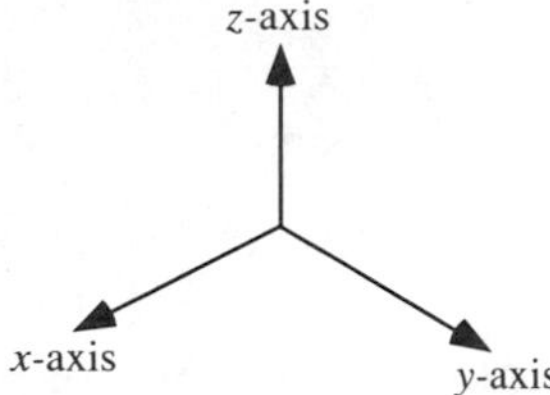

Cartesian coordinates are rectilinear 2-D or 3-D coordinates (and therefore a special case of CURVILINEAR COORDINATES) which are also called RECTANGULAR COORDINATES. The three axes of 3-D Cartesian coordinates, conventionally denoted the x-, y-, and z-AXES (a NOTATION due to Descartes) are chosen to be linear and mutually PERPENDICULAR. In 3-D, the coordinates x, y, and z may lie anywhere in the INTERVAL $(-\infty, \infty)$.

The SCALE FACTORS of Cartesian coordinates are all unity, $h_i = 1$. The LINE ELEMENT is given by

$$d\mathbf{s} = dx\,\hat{\mathbf{x}} + dy\,\hat{\mathbf{y}} + dz\,\hat{\mathbf{z}}, \tag{1}$$

and the VOLUME ELEMENT by

$$dV = dx\,dy\,dz. \tag{2}$$

The GRADIENT has a particularly simple form,

$$\nabla \equiv \hat{\mathbf{x}}\frac{\partial}{\partial x} + \hat{\mathbf{y}}\frac{\partial}{\partial y} + \hat{\mathbf{z}}\frac{\partial}{\partial z}, \tag{3}$$

as does the LAPLACIAN

$$\nabla^2 \equiv \frac{\partial^2}{\partial x^2} + \frac{\partial^2}{\partial y^2} + \frac{\partial^2}{\partial z^2}. \tag{4}$$

The LAPLACIAN is

$$\begin{aligned}\nabla^2\mathbf{F} \equiv \nabla\cdot(\nabla\mathbf{F}) &= \frac{\partial^2\mathbf{F}}{\partial x^2} + \frac{\partial^2\mathbf{F}}{\partial y^2} + \frac{\partial^2\mathbf{F}}{\partial z^2}\\ &= \hat{\mathbf{x}}\left(\frac{\partial^2 F_x}{\partial x^2} + \frac{\partial^2 F_x}{\partial y^2} + \frac{\partial^2 F_x}{\partial z^2}\right)\\ &\quad + \hat{\mathbf{y}}\left(\frac{\partial^2 F_y}{\partial x^2} + \frac{\partial^2 F_y}{\partial y^2} + \frac{\partial^2 F_y}{\partial z^2}\right)\\ &\quad + \hat{\mathbf{z}}\left(\frac{\partial^2 F_z}{\partial x^2} + \frac{\partial^2 F_z}{\partial y^2} + \frac{\partial^2 F_z}{\partial z^2}\right).\end{aligned} \tag{5}$$

The DIVERGENCE is

$$\nabla\cdot\mathbf{F} = \frac{\partial F_x}{\partial x} + \frac{\partial F_y}{\partial y} + \frac{\partial F_z}{\partial z}, \tag{6}$$

and the CURL is

$$\begin{aligned}\nabla\times\mathbf{F} &\equiv \begin{vmatrix}\hat{\mathbf{x}} & \hat{\mathbf{y}} & \hat{\mathbf{z}}\\ \frac{\partial}{\partial x} & \frac{\partial}{\partial y} & \frac{\partial}{\partial z}\\ F_x & F_y & F_z\end{vmatrix}\\ &= \left(\frac{\partial F_z}{\partial y} - \frac{\partial F_y}{\partial z}\right)\hat{\mathbf{x}} + \left(\frac{\partial F_x}{\partial z} - \frac{\partial F_z}{\partial x}\right)\hat{\mathbf{y}}\\ &\quad + \left(\frac{\partial F_y}{\partial x} - \frac{\partial F_x}{\partial y}\right)\hat{\mathbf{z}}.\end{aligned} \tag{7}$$

The GRADIENT of the DIVERGENCE is

$$\begin{aligned}\nabla(\nabla\cdot\mathbf{u}) &= \begin{bmatrix}\frac{\partial}{\partial x}\left(\frac{\partial u_x}{\partial x} + \frac{\partial u_y}{\partial y} + \frac{\partial u_z}{\partial z}\right)\\ \frac{\partial}{\partial y}\left(\frac{\partial u_x}{\partial x} + \frac{\partial u_y}{\partial y} + \frac{\partial u_z}{\partial z}\right)\\ \frac{\partial}{\partial z}\left(\frac{\partial u_x}{\partial x} + \frac{\partial u_y}{\partial y} + \frac{\partial u_z}{\partial z}\right)\end{bmatrix}\\ &= \begin{bmatrix}\frac{\partial}{\partial x}\\ \frac{\partial}{\partial y}\\ \frac{\partial}{\partial z}\end{bmatrix}\left(\frac{\partial u_x}{\partial x} + \frac{\partial u_y}{\partial y} + \frac{\partial u_z}{\partial z}\right).\end{aligned} \tag{8}$$

LAPLACE'S EQUATION is separable in Cartesian coordinates.

see also COORDINATES, HELMHOLTZ DIFFERENTIAL EQUATION—CARTESIAN COORDINATES

References

Arfken, G. "Special Coordinate Systems—Rectangular Cartesian Coordinates." §2.3 in *Mathematical Methods for Physicists, 3rd ed.* Orlando, FL: Academic Press, pp. 94–95, 1985.

Morse, P. M. and Feshbach, H. *Methods of Theoretical Physics, Part I.* New York: McGraw-Hill, p. 656, 1953.

Cartesian Ovals

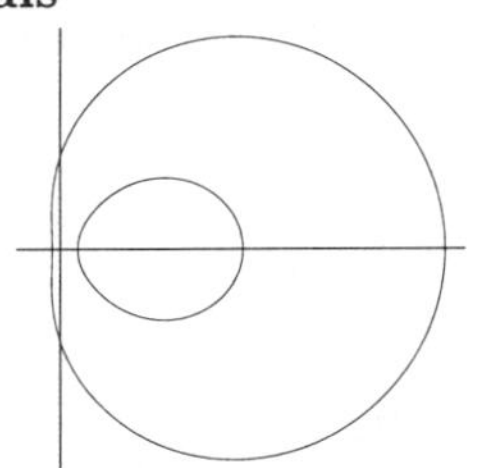

A curve consisting of two ovals which was first studied by Descartes in 1637. It is the locus of a point P whose distances from two FOCI F_1 and F_2 in two-center BIPOLAR COORDINATES satisfy

$$mr \pm nr' = k, \tag{1}$$

where m, n are POSITIVE INTEGERS, k is a POSITIVE real, and r and r' are the distances from F_1 and F_2. If $m = n$, the oval becomes an an ELLIPSE. In CARTESIAN COORDINATES, the Cartesian ovals can be written

$$m\sqrt{(x-a)^2+y^2} + n\sqrt{(x+a)^2+y^2} = k^2 \tag{2}$$

$$\begin{aligned}(x^2+y^2+a^2)(m^2-n^2) - 2ax(m^2+n^2) - k^2\\ = -2n\sqrt{(x+a)^2+y^2},\end{aligned} \tag{3}$$

$$\begin{aligned}&[(m^2-n^2)(x^2+y^2+a^2) - 2ax(m^2+n^2)]^2\\ &= 2(m^2+n^2)(n^2+y^2+a^2) - 4ax(m^2-n^2) - k^2.\end{aligned} \tag{4}$$

Now define

$$b \equiv m^2 - n^2 \tag{5}$$

$$c \equiv m^2 + n^2, \tag{6}$$

and set $a = 1$. Then

$$[b(x^2+y^2)-2cx+b]^2+4bx+k^2-2c = 2c(x^2+y^2). \quad (7)$$

If c' is the distance between F_1 and F_2, and the equation

$$r + mr' = a \quad (8)$$

is used instead, an alternate form is

$$[(1-m^2)(x^2+y^2)+2m^2c'x+a'^2-m^2c'^2]^2 = 4a'^2(x^2+y^2). \quad (9)$$

The curves possess three FOCI. If $m = 1$, one Cartesian oval is a central CONIC, while if $m = a/c$, then the curve is a LIMAÇON and the inside oval touches the outside one. Cartesian ovals are ANALLAGMATIC CURVES.

References

Cundy, H. and Rollett, A. *Mathematical Models, 3rd ed.* Stradbroke, England: Tarquin Pub., p. 35, 1989.

Lawrence, J. D. *A Catalog of Special Plane Curves.* New York: Dover, pp. 155–157, 1972.

Lockwood, E. H. *A Book of Curves.* Cambridge, England: Cambridge University Press, p. 188, 1967.

MacTutor History of Mathematics Archive. "Cartesian Oval." `http://www-groups.dcs.st-and.ac.uk/~history/Curves/Cartesian.html`.

Cartesian Product

see DIRECT PRODUCT (SET)

Cartesian Trident

see TRIDENT OF DESCARTES

Cartography

The study of MAP PROJECTIONS and the making of geographical maps.

see also MAP PROJECTION

Cascade

A $\mathbb{Z}$-ACTION or $\mathbb{N}$-ACTION. A cascade and a single MAP $X \to X$ are essentially the same, but the term "cascade" is preferred by many Russian authors.

see also ACTION, FLOW

Casey's Theorem

Four CIRCLES are TANGENT to a fifth CIRCLE or a straight LINE IFF

$$t_{12}t_{34} \pm t_{13}t_{42} \pm t_{14}t_{23} = 0,$$

where t_{ij} is a common TANGENT to CIRCLES i and j.

see also PURSER'S THEOREM

References

Johnson, R. A. *Modern Geometry: An Elementary Treatise on the Geometry of the Triangle and the Circle.* Boston, MA: Houghton Mifflin, pp. 121–127, 1929.

Casimir Operator

An OPERATOR

$$\Gamma = \sum_{i=1}^{m} e_i^R u^{iR}$$

on a representation R of a LIE ALGEBRA.

References

Jacobson, N. *Lie Algebras.* New York: Dover, p. 78, 1979.

Cassini Ellipses

see CASSINI OVALS

Cassini's Identity

For F_n the nth FIBONACCI NUMBER,

$$F_{n-1}F_{n+1} - F_n{}^2 = (-1)^n.$$

see also FIBONACCI NUMBER

References

Petkovšek, M.; Wilf, H. S.; and Zeilberger, D. *A=B.* Wellesley, MA: A. K. Peters, p. 12, 1996.

Cassini Ovals

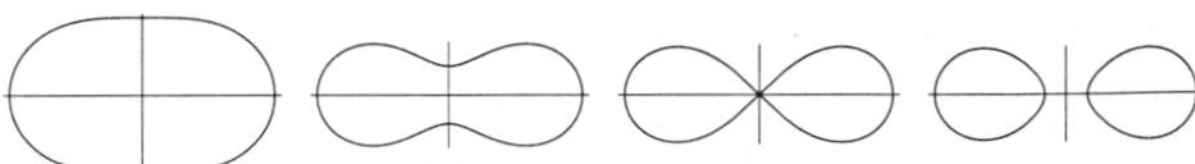

The curves, also called CASSINI ELLIPSES, described by a point such that the product of its distances from two fixed points a distance $2a$ apart is a constant b^2. The shape of the curve depends on b/a. If $a < b$, the curve is a single loop with an OVAL (left figure above) or dog bone (second figure) shape. The case $a = b$ produces a LEMNISCATE (third figure). If $a > b$, then the curve consists of two loops (right figure). The curve was first investigated by Cassini in 1680 when he was studying the relative motions of the Earth and the Sun. Cassini believed that the Sun traveled around the Earth on one of these ovals, with the Earth at one FOCUS of the oval.

Cassini ovals are ANALLAGMATIC CURVES. The Cassini ovals are defined in two-center BIPOLAR COORDINATES by the equation

$$r_1 r_2 = b^2, \quad (1)$$

with the origin at a FOCUS. Even more incredible curves are produced by the locus of a point the product of whose distances from 3 or more fixed points is a constant.

The Cassini ovals have the CARTESIAN equation

$$[(x-a)^2+y^2][(x+a)^2+y^2] = b^4 \quad (2)$$

or the equivalent form

$$(x^2+y^2+a^2)^2 - 4a^2x^2 = b^4 \quad (3)$$

and the polar equation

$$r^4 + a^4 - 2a^2r^2\cos(2\theta) = b^4. \qquad (4)$$

Solving for r^2 using the QUADRATIC EQUATION gives

$$\begin{aligned} r^2 &= \frac{2a^2\cos(2\theta) + \sqrt{4a^4\cos^2(2\theta) - 4(a^4 - b^4)}}{2} \\ &= a^2\cos(2\theta) + \sqrt{a^4\cos^2(2\theta) + b^4 - a^4} \\ &= a^2\cos(2\theta)\sqrt{a^4[\cos^2(2\theta) - 1] + b^4} \\ &= a^2\cos(2\theta) + \sqrt{b^4 - a^4\sin^2(2\theta)} \\ &= a^2\left[\cos(2\theta) + \sqrt{\left(\frac{b}{a}\right)^4 - \sin^2(2\theta)}\right]. \qquad (5) \end{aligned}$$

If $a < b$, the curve has AREA

$$A = \tfrac{1}{2}r^2\,d\theta = 2(\tfrac{1}{2})\int_{-\pi/4}^{\pi/4} r^2\,d\theta = a^2 + b^2E\left(\frac{a^4}{b^4}\right), \qquad (6)$$

where the integral has been done over half the curve and then multiplied by two and $E(x)$ is the complete ELLIPTIC INTEGRAL OF THE SECOND KIND. If $a = b$, the curve becomes

$$r^2 = a^2\left[\cos(2\theta) + \sqrt{1 - \sin^2\theta}\right] = 2a^2\cos(2\theta), \qquad (7)$$

which is a LEMNISCATE having AREA

$$A = 2a^2 \qquad (8)$$

(two loops of a curve $\sqrt{2}$ the linear scale of the usual lemniscate $r^2 = a^2\cos(2\theta)$, which has area $A = a^2/2$ for each loop). If $a > b$, the curve becomes two disjoint ovals with equations

$$r = \pm a\sqrt{\cos(2\theta) \pm \sqrt{\left(\frac{b}{a}\right)^4 - \sin^2(2\theta)}}, \qquad (9)$$

where $\theta \in [-\theta_0, \theta_0]$ and

$$\theta_0 \equiv \tfrac{1}{2}\sin^{-1}\left[\left(\frac{b}{a}\right)^2\right]. \qquad (10)$$

see also CASSINI SURFACE, LEMNISCATE, MANDELBROT SET, OVAL

References

Gray, A. "Cassinian Ovals." §4.2 in *Modern Differential Geometry of Curves and Surfaces.* Boca Raton, FL: CRC Press, pp. 63–65, 1993.

Lawrence, J. D. *A Catalog of Special Plane Curves.* New York: Dover, pp. 153–155, 1972.

Lee, X. "Cassinian Oval." `http://www.best.com/~xah/SpecialPlaneCurves_dir/CassinianOval_dir/cassinianOval.html`.

Lockwood, E. H. *A Book of Curves.* Cambridge, England: Cambridge University Press, pp. 187–188, 1967.

MacTutor History of Mathematics Archive. "Cassinian Ovals." `http://www-groups.dcs.st-and.ac.uk/~history/Curves/Cassinian.html`.

Yates, R. C. "Cassinian Curves." *A Handbook on Curves and Their Properties.* Ann Arbor, MI: J. W. Edwards, pp. 8–11, 1952.

Cassini Projection

A MAP PROJECTION.

$$x = \sin^{-1} B \qquad (1)$$

$$y = \tan^{-1}\left[\frac{\tan\phi}{\cos(\lambda - \lambda_0)}\right], \qquad (2)$$

where

$$B = \cos\phi\sin(\lambda - \lambda_0). \qquad (3)$$

The inverse FORMULAS are

$$\phi = \sin^{-1}(\sin D\cos x) \qquad (4)$$

$$\lambda = \lambda_0 + \tan^{-1}\left(\frac{\tan x}{\cos D}\right), \qquad (5)$$

where

$$D = y + \phi_0. \qquad (6)$$

References

Snyder, J. P. *Map Projections—A Working Manual.* U. S. Geological Survey Professional Paper 1395. Washington, DC: U. S. Government Printing Office, pp. 92–95, 1987.

Cassini Surface

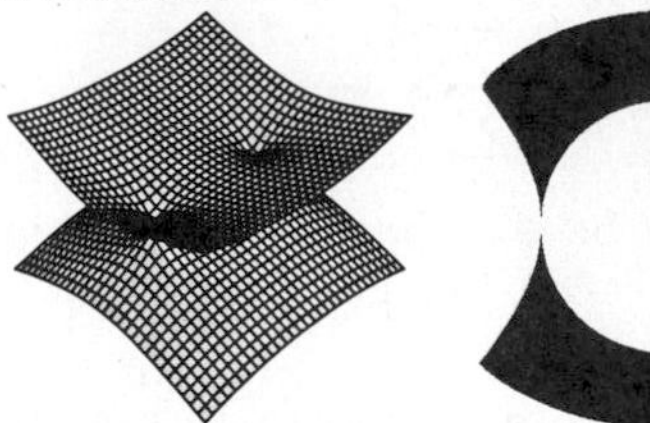

The QUARTIC SURFACE obtained by replacing the constant c in the equation of the CASSINI OVALS

$$[(x-a)^2 + y^2][(x+a)^2 + y^2] = c^2 \qquad (1)$$

by $c = z^2$, obtaining

$$[(x-a)^2 + y^2][(x+a)^2 + y^2] = z^4. \qquad (2)$$

As can be seen by letting $y = 0$ to obtain

$$(x^2 - a^2)^2 = z^4 \qquad (3)$$

$$x^2 + z^2 = a^2, \qquad (4)$$

the intersection of the surface with the $y = 0$ PLANE is a CIRCLE of RADIUS a.

References
Fischer, G. (Ed.). *Mathematical Models from the Collections of Universities and Museums.* Braunschweig, Germany: Vieweg, p. 20, 1986.
Fischer, G. (Ed.). Plate 51 in *Mathematische Modelle/Mathematical Models, Bildband/Photograph Volume.* Braunschweig, Germany: Vieweg, p. 51, 1986.

Castillon's Problem

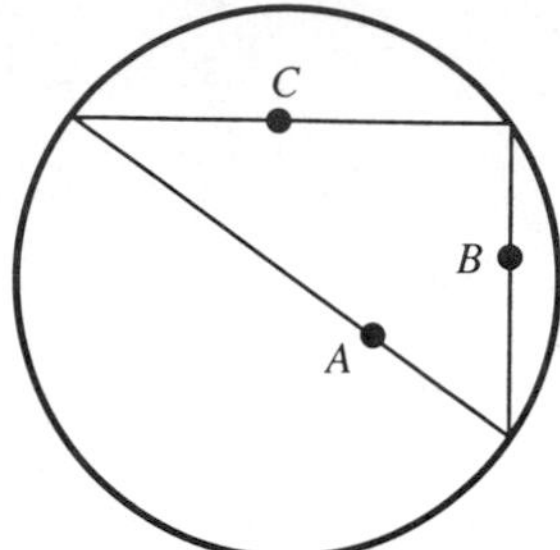

Inscribe a TRIANGLE in a CIRCLE such that the sides of the TRIANGLE pass through three given POINTS A, B, and C.

References
Dörrie, H. "Castillon's Problem." §29 in *100 Great Problems of Elementary Mathematics: Their History and Solutions.* New York: Dover, pp. 144–147, 1965.

Casting Out Nines

An elementary check of a MULTIPLICATION which makes use of the CONGRUENCE $10^n \equiv 1 \pmod 9$ for $n \geq 2$. From this CONGRUENCE, a MULTIPLICATION $ab = c$ must give

$$a \equiv \sum a_i = a^*$$
$$b \equiv \sum b_i = b^*$$
$$c \equiv \sum c_i = c^*,$$

so $ab \equiv a^*b^*$ must be $\equiv c^* \pmod 9$. Casting out nines is sometimes also called "the HINDU CHECK."

References
Conway, J. H. and Guy, R. K. *The Book of Numbers.* New York: Springer-Verlag, pp. 28–29, 1996.

Cat Map

see ARNOLD'S CAT MAP

Catacaustic

The curve which is the ENVELOPE of reflected rays.

Curve	Source	Catacaustic
cardioid	cusp	nephroid
circle	not on circumf.	limaçon
circle	on circumf.	cardioid
circle	point at ∞	nephroid
cissoid of Diocles	focus	cardioid
1 arch of a cycloid	rays $\perp$ axis	2 arches of a cycloid
deltoid	point at ∞	astroid
$\ln x$	rays $\parallel$ axis	catenary
logarithmic spiral	origin	equal logarithmic spiral
parabola	rays $\perp$ axis	Tschirnhausen cubic
quadrifolium	center	astroid
Tschirnhausen cubic	focus	semicubical parabola

see also CAUSTIC, DIACAUSTIC

References
Lawrence, J. D. *A Catalog of Special Plane Curves.* New York: Dover, pp. 60 and 207, 1972.

Catalan's Conjecture

8 and 9 (2^3 and 3^2) are the only consecutive POWERS (excluding 0 and 1), i.e., the only solution to CATALAN'S DIOPHANTINE PROBLEM. Solutions to this problem (CATALAN'S DIOPHANTINE PROBLEM) are equivalent to solving the simultaneous DIOPHANTINE EQUATIONS

$$X^2 - Y^3 = 1$$
$$X^3 - Y^2 = 1.$$

This CONJECTURE has not yet been proved or refuted, although it has been shown to be decidable in a FINITE (but more than astronomical) number of steps. In particular, if n and $n+1$ are POWERS, then $n < \exp\exp\exp\exp 730$ (Guy 1994, p. 155), which follows from R. Tijdeman's proof that there can be only a FINITE number of exceptions should the CONJECTURE not hold.

Hyyrő and Mąkowski proved that there do not exist three consecutive POWERS (Ribenboim 1996), and it is also known that 8 and 9 are the only consecutive CUBIC and SQUARE NUMBERS (in either order).

see also CATALAN'S DIOPHANTINE PROBLEM

References
Guy, R. K. "Difference of Two Power." §D9 in *Unsolved Problems in Number Theory, 2nd ed.* New York: Springer-Verlag, pp. 155–157, 1994.
Ribenboim, P. *Catalan's Conjecture.* Boston, MA: Academic Press, 1994.
Ribenboim, P. "Catalan's Conjecture." *Amer. Math. Monthly* **103**, 529–538, 1996.
Ribenboim, P. "Consecutive Powers." *Expositiones Mathematicae* **2**, 193–221, 1984.

Catalan's Constant

A constant which appears in estimates of combinatorial functions. It is usually denoted K, $\beta(2)$, or G. It is not known if K is IRRATIONAL. Numerically,

$$K = 0.915965594177\ldots \tag{1}$$

(Sloane's A006752). The CONTINUED FRACTION for K is [0, 1, 10, 1, 8, 1, 88, 4, 1, 1, ...] (Sloane's A014538). K can be given analytically by the following expressions,

$$K \equiv \beta(2) \tag{2}$$

$$= \sum_{k=0}^{\infty} \frac{(-1)^k}{(2k+1)^2} = \frac{1}{1^2} - \frac{1}{3^2} + \frac{1}{5^2} + \ldots \tag{3}$$

$$= 1 + \sum_{n=1}^{\infty} \frac{1}{(4n+1)^2} - \frac{1}{9} - \sum_{n=1}^{\infty} \frac{1}{(4n+3)^2} \tag{4}$$

$$= \int_0^1 \frac{\tan^{-1} x\, dx}{x} \tag{5}$$

$$= -\int_0^1 \frac{\ln x\, dx}{1+x^2}, \tag{6}$$

where $\beta(z)$ is the DIRICHLET BETA FUNCTION. In terms of the POLYGAMMA FUNCTION $\Psi_1(x)$,

$$K = \tfrac{1}{16}\Psi_1\left(\tfrac{1}{4}\right) - \tfrac{1}{16}\Psi_1\left(\tfrac{3}{4}\right) \tag{7}$$

$$= \tfrac{1}{80}\Psi_1\left(\tfrac{5}{12}\right) + \tfrac{1}{80}\Psi_1\left(\tfrac{1}{12}\right) - \tfrac{1}{10}\pi^2 \tag{8}$$

$$= \tfrac{1}{32}\Psi_1\left(\tfrac{1}{8}\right) - \tfrac{1}{32}\Psi_1\left(\tfrac{3}{8}\right) - \tfrac{1}{16}\sqrt{2}. \tag{9}$$

Applying CONVERGENCE IMPROVEMENT to (3) gives

$$K = \frac{1}{16}\sum_{m=1}^{\infty}(m+1)\frac{3^m-1}{4^m}\zeta(m+2), \tag{10}$$

where $\zeta(z)$ is the RIEMANN ZETA FUNCTION and the identity

$$\frac{1}{(1-3z)^2} - \frac{1}{(1-z)^2} = \sum_{m=1}^{\infty}(m+1)\frac{3^m-1}{4^m}z^m \tag{11}$$

has been used (Flajolet and Vardi 1996). The Flajolet and Vardi algorithm also gives

$$K = \frac{1}{\sqrt{2}}\prod_{k=1}^{\infty}\left[\left(1-\frac{1}{2^{2^k}}\right)\frac{\zeta(2^k)}{\beta(2^k)}\right]^{1/(2^{k+1})}, \tag{12}$$

where $\beta(z)$ is the DIRICHLET BETA FUNCTION. Glaisher (1913) gave

$$K = 1 - \sum_{n=1}^{\infty}\frac{n\zeta(2n+1)}{16^n} \tag{13}$$

(Vardi 1991, p. 159). W. Gosper used the related FORMULA

$$K = \frac{1}{\sqrt{2}}\left[\frac{1}{\Psi(2)-1}\right]^{2^{1/2}}\prod_{k=2}^{\infty}\left[\frac{1}{-\Psi(2^k)-1}\right]^{1/(2^{k+1})}, \tag{14}$$

where

$$\Psi(m) = \frac{m\psi_{m-1}(\frac{1}{4})}{\pi^m(2^m-1)4^{m-1}B_m}, \tag{15}$$

where B_n is a BERNOULLI NUMBER and $\psi(x)$ is a POLYGAMMA FUNCTION (Finch). The Catalan constant may also be defined by

$$K \equiv \tfrac{1}{2}\int_0^1 K(k)\,dk, \tag{16}$$

where $K(k)$ (not to be confused with Catalan's constant itself, denoted K) is a complete ELLIPTIC INTEGRAL OF THE FIRST KIND.

$$K = \frac{\pi\ln 2}{8} + \sum_{i=1}^{\infty}\frac{a_i}{2^{\lfloor(i+1)/2\rfloor}i^2}, \tag{17}$$

where

$$\{a_i\} = \{\overline{1,1,1,0,-1,-1,-1,0}\} \tag{18}$$

is given by the periodic sequence obtained by appending copies of {1, 1, 1, 0, −1, −1, −1, 0} (in other words, $a_i \equiv a_{[(i-1) \pmod 8)]+1}$ for $i > 8$) and $\lfloor x \rfloor$ is the FLOOR FUNCTION (Nielsen 1909).

see also DIRICHLET BETA FUNCTION

References

Abramowitz, M. and Stegun, C. A. (Eds.). *Handbook of Mathematical Functions with Formulas, Graphs, and Mathematical Tables, 9th printing.* New York: Dover, pp. 807–808, 1972.

Adamchik, V. "32 Representations for Catalan's Constant." `http://www.wolfram.com/~victor/articles/catalan/catalan.html`.

Arfken, G. *Mathematical Methods for Physicists, 3rd ed.* Orlando, FL: Academic Press, pp. 551–552, 1985.

Fee, G. J. "Computation of Catalan's Constant using Ramanujan's Formula." *ISAAC '90. Proc. Internat. Symp. Symbolic Algebraic Comp., Aug. 1990.* Reading, MA: Addison-Wesley, 1990.

Finch, S. "Favorite Mathematical Constants." `http://www.mathsoft.com/asolve/constant/catalan/catalan.html`.

Flajolet, P. and Vardi, I. "Zeta Function Expansions of Classical Constants." Unpublished manuscript. 1996. `http://pauillac.inria.fr/algo/flajolet/Publications/landau.ps`.

Glaisher, J. W. L. "Numerical Values of the Series $1-1/3^n+1/5^n-1/7^n+1/9^n-$&c for $n = 2, 4, 6$." *Messenger Math.* **42**, 35–58, 1913.

Gosper, R. W. "A Calculus of Series Rearrangements." In *Algorithms and Complexity: New Directions and Recent Results* (Ed. J. F. Traub). New York: Academic Press, 1976.

Nielsen, N. *Der Eulersche Dilogarithms.* Leipzig, Germany: Halle, pp. 105 and 151, 1909.

Plouffe, S. "Plouffe's Inverter: Table of Current Records for the Computation of Constants." http://lacim.uqam.ca/pi/records.html.
Sloane, N. J. A. Sequences A014538 and A006752/M4593 in "An On-Line Version of the Encyclopedia of Integer Sequences."
Srivastava, H. M. and Miller, E. A. "A Simple Reducible Case of Double Hypergeometric Series involving Catalan's Constant and Riemann's Zeta Function." *Int. J. Math. Educ. Sci. Technol.* **21**, 375–377, 1990.
Vardi, I. *Computational Recreations in Mathematica.* Reading, MA: Addison-Wesley, p. 159, 1991.
Yang, S. "Some Properties of Catalan's Constant G." *Int. J. Math. Educ. Sci. Technol.* **23**, 549–556, 1992.

Catalan's Diophantine Problem

Find consecutive POWERS, i.e., solutions to

$$a^b - c^d = 1,$$

excluding 0 and 1. CATALAN'S CONJECTURE is that the only solution is $3^2 - 2^3 = 1$, so 8 and 9 (2^3 and 3^2) are the only consecutive POWERS (again excluding 0 and 1).

see also CATALAN'S CONJECTURE

References

Cassels, J. W. S. "On the Equation $a^x - b^y = 1$. II." *Proc. Cambridge Phil. Soc.* **56**, 97–103, 1960.
Inkeri, K. "On Catalan's Problem." *Acta Arith.* **9**, 285–290, 1964.

Catalan Integrals

Special cases of general FORMULAS due to Bessel.

$$J_0(\sqrt{z^2 - y^2}\,) = \frac{1}{\pi}\int_0^\pi e^{y\cos\theta}\cos(z\sin\theta)\,d\theta,$$

where J_0 is a BESSEL FUNCTION OF THE FIRST KIND. Now, let $z \equiv 1 - z'$ and $y \equiv 1 + z'$. Then

$$J_0(2i\sqrt{z}\,) = \frac{1}{\pi}\int_0^\pi e^{(1+z)\cos\theta}\cos[(1-z)\sin\theta]\,d\theta.$$

Catalan Number

The Catalan numbers are an INTEGER SEQUENCE $\{C_n\}$ which appears in TREE enumeration problems of the type, "In how many ways can a regular n-gon be divided into $n - 2$ TRIANGLES if different orientations are counted separately?" (EULER'S POLYGON DIVISION PROBLEM). The solution is the Catalan number C_{n-2} (Dörrie 1965, Honsberger 1973), as graphically illustrated below (Dickau).

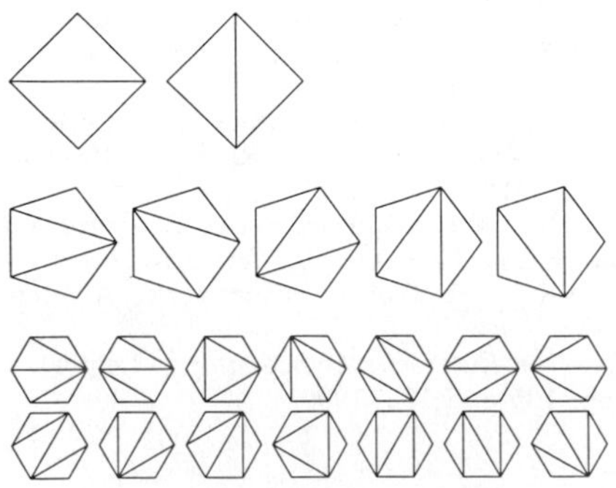

The first few Catalan numbers are 1, 2, 5, 14, 42, 132, 429, 1430, 4862, 16796, ... (Sloane's A000108). The only ODD Catalan numbers are those of the form c_{2^k-1}, and the last DIGIT is five for $k = 9$ to 15. The only PRIME Catalan numbers for $n \leq 2^{15} - 1$ are $C_2 = 2$ and $C_3 = 5$.

The Catalan numbers turn up in many other related types of problems. For instance, the Catalan number C_{n-1} gives the number of BINARY BRACKETINGS of n letters (CATALAN'S PROBLEM). The Catalan numbers also give the solution to the BALLOT PROBLEM, the number of trivalent PLANTED PLANAR TREES (Dickau),

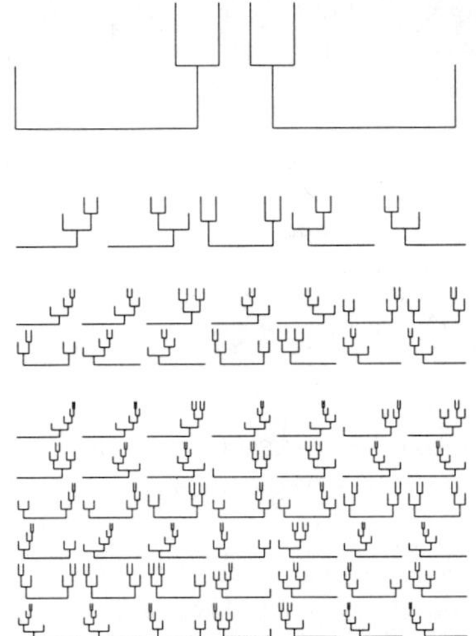

the number of states possible in an n-FLEXAGON, the number of different diagonals possible in a FRIEZE PATTERN with $n + 1$ rows, the number of ways of forming an n-fold exponential, the number of rooted planar binary trees with n internal nodes, the number of rooted plane bushes with n EDGES, the number of extended BINARY TREES with n internal nodes, the number of mountains which can be drawn with n upstrokes and n downstrokes, the number of noncrossing handshakes possible across a round table between n pairs of people (Conway and Guy 1996), and the number of SEQUENCES with NONNEGATIVE PARTIAL SUMS which can be formed from n 1s and n −1s (Bailey 1996, Buraldi 1992)!

An explicit formula for C_n is given by

$$C_n \equiv \frac{1}{n+1}\binom{2n}{n} = \frac{1}{n+1}\frac{(2n)!}{n!^2} = \frac{(2n)!}{(n+1)!n!}, \qquad (1)$$

where $\binom{2n}{n}$ denotes a BINOMIAL COEFFICIENT and $n!$ is the usual FACTORIAL. A RECURRENCE RELATION for C_n is obtained from

$$\begin{aligned}\frac{C_{n+1}}{C_n} &= \frac{(2n+2)!}{(n+2)[(n+1)!]^2}\frac{(n+1)(n!)^2}{(2n)!}\\ &= \frac{(2n+2)(2n+1)(n+1)}{(n+2)(n+1)^2}\\ &= \frac{2(2n+1)(n+1)^2}{(n+1)^2(n+2)} = \frac{2(2n+1)}{n+2},\end{aligned} \qquad (2)$$

so

$$C_{n+1} = \frac{2(2n+1)}{n+2} C_n. \tag{3}$$

Other forms include

$$C_n = \frac{2 \cdot 6 \cdot 10 \cdots (4n-2)}{(n+1)!} \tag{4}$$

$$= \frac{2^n (2n-1)!!}{(n+1)!} \tag{5}$$

$$= \frac{(2n)!}{n!(n+1)!}. \tag{6}$$

SEGNER'S RECURRENCE FORMULA, given by Segner in 1758, gives the solution to EULER'S POLYGON DIVISION PROBLEM

$$E_n = E_2 E_{n-1} + E_3 E_{n-2} + \ldots + E_{n-1} E_2. \tag{7}$$

With $E_1 = E_2 = 1$, the above RECURRENCE RELATION gives the Catalan number $C_{n-2} = E_n$.

The GENERATING FUNCTION for the Catalan numbers is given by

$$\frac{1 - \sqrt{1-4x}}{2x} = \sum_{n=0}^{\infty} C_n x^n = 1 + x + 2x^2 + 5x^3 + \ldots. \tag{8}$$

The asymptotic form for the Catalan numbers is

$$C_k \sim \frac{4^k}{\sqrt{\pi}\, k^{3/2}} \tag{9}$$

(Vardi 1991, Graham *et al.* 1994).

A generalization of the Catalan numbers is defined by

$$_p d_k = \frac{1}{k} \binom{pk}{k-1} = \frac{1}{(p-1)k+1} \binom{pk}{k} \tag{10}$$

for $k \geq 1$ (Klarner 1970, Hilton and Pederson 1991). The usual Catalan numbers $C_k = {}_2d_k$ are a special case with $p = 2$. ${}_pd_k$ gives the number of p-ary TREES with k source-nodes, the number of ways of associating k applications of a given p-ary OPERATOR, the number of ways of dividing a convex POLYGON into k disjoint $(p+1)$-gons with nonintersecting DIAGONALS, and the number of p-GOOD PATHS from $(0, -1)$ to $(k, (p-1)k-1)$ (Hilton and Pederson 1991).

A further generalization is obtained as follows. Let p be an INTEGER > 1, let $P_k = (k, (p-1)k-1)$ with $k \geq 0$, and $q \leq p-1$. Then define ${}_pd_{q0} = 1$ and let ${}_pd_{qk}$ be the number of p-GOOD PATHS from $(1, q-1)$ to P_k (Hilton and Pederson 1991). Formulas for ${}_pd_{qi}$ include the generalized JONAH FORMULA

$$\binom{n-q}{k-1} = \sum_{i=1}^{k} {}_pd_{qi} \binom{n-pi}{k-i} \tag{11}$$

and the explicit formula

$${}_pd_{qk} = \frac{p-q}{pk-q} \binom{pk-q}{k-1}. \tag{12}$$

A RECURRENCE RELATION is given by

$${}_pd_{qk} = \sum_{i,j} {}_pd_{p-r,i}\, {}_pd_{q+r,j} \tag{13}$$

where $i, j, r \geq 1$, $k \geq 1$, $q < p - r$, and $i + j = k + 1$ (Hilton and Pederson 1991).

see also BALLOT PROBLEM, BINARY BRACKETING, BINARY TREE, CATALAN'S PROBLEM, CATALAN'S TRIANGLE, DELANNOY NUMBER, EULER'S POLYGON DIVISION PROBLEM, FLEXAGON, FRIEZE PATTERN, MOTZKIN NUMBER, p-GOOD PATH, PLANTED PLANAR TREE, SCHRÖDER NUMBER, SUPER CATALAN NUMBER

References

Alter, R. "Some Remarks and Results on Catalan Numbers." *Proc. 2nd Louisiana Conf. Comb., Graph Th., and Comput.*, 109–132, 1971.

Alter, R. and Kubota, K. K. "Prime and Prime Power Divisibility of Catalan Numbers." *J. Combin. Th.* **15**, 243–256, 1973.

Bailey, D. F. "Counting Arrangements of 1's and −1's." *Math. Mag.* **69**, 128–131, 1996.

Brualdi, R. A. *Introductory Combinatorics, 3rd ed.* New York: Elsevier, 1997.

Campbell, D. "The Computation of Catalan Numbers." *Math. Mag.* **57**, 195–208, 1984.

Chorneyko, I. Z. and Mohanty, S. G. "On the Enumeration of Certain Sets of Planted Trees." *J. Combin. Th. Ser. B* **18**, 209–221, 1975.

Chu, W. "A New Combinatorial Interpretation for Generalized Catalan Numbers." *Disc. Math.* **65**, 91–94, 1987.

Conway, J. H. and Guy, R. K. In *The Book of Numbers.* New York: Springer-Verlag, pp. 96–106, 1996.

Dershowitz, N. and Zaks, S. "Enumeration of Ordered Trees." *Disc. Math.* **31**, 9–28, 1980.

Dickau, R. M. "Catalan Numbers." `http://forum.swarthmore.edu/advanced/robertd/catalan.html`.

Dörrie, H. "Euler's Problem of Polygon Division." §7 in *100 Great Problems of Elementary Mathematics: Their History and Solutions.* New York: Dover, pp. 21–27, 1965.

Eggleton, R. B. and Guy, R. K. "Catalan Strikes Again! How Likely is a Function to be Convex?" *Math. Mag.* **61**, 211–219, 1988.

Gardner, M. "Catalan Numbers." Ch. 20 in *Time Travel and Other Mathematical Bewilderments.* New York: W. H. Freeman, 1988.

Gardner, M. "Catalan Numbers: An Integer Sequence that Materializes in Unexpected Places." *Sci. Amer.* **234**, 120–125, June 1976.

Gould, H. W. *Bell & Catalan Numbers: Research Bibliography of Two Special Number Sequences, 6th ed.* Morgantown, WV: Math Monongliae, 1985.

Graham, R. L.; Knuth, D. E.; and Patashnik, O. Exercise 9.8 in *Concrete Mathematics: A Foundation for Computer Science, 2nd ed.* Reading, MA: Addison-Wesley, 1994.

Guy, R. K. "Dissecting a Polygon Into Triangles." *Bull. Malayan Math. Soc.* **5**, 57–60, 1958.

Hilton, P. and Pederson, J. "Catalan Numbers, Their Generalization, and Their Uses." *Math. Int.* **13**, 64–75, 1991.

Honsberger, R. *Mathematical Gems I.* Washington, DC: Math. Assoc. Amer., pp. 130–134, 1973.

Honsberger, R. *Mathematical Gems III.* Washington, DC: Math. Assoc. Amer., pp. 146–150, 1985.
Klarner, D. A. "Correspondences Between Plane Trees and Binary Sequences." *J. Comb. Th.* **9**, 401–411, 1970.
Rogers, D. G. "Pascal Triangles, Catalan Numbers and Renewal Arrays." *Disc. Math.* **22**, 301–310, 1978.
Sands, A. D. "On Generalized Catalan Numbers." *Disc. Math.* **21**, 218–221, 1978.
Singmaster, D. "An Elementary Evaluation of the Catalan Numbers." *Amer. Math. Monthly* **85**, 366–368, 1978.
Sloane, N. J. A. *A Handbook of Integer Sequences.* Boston, MA: Academic Press, pp. 18–20, 1973.
Sloane, N. J. A. Sequences A000108/M1459 in "An On-Line Version of the Encyclopedia of Integer Sequences."
Sloane, N. J. A. and Plouffe, S. Extended entry in *The Encyclopedia of Integer Sequences.* San Diego: Academic Press, 1995.
Vardi, I. *Computational Recreations in Mathematica.* Redwood City, CA: Addison-Wesley, pp. 187–188 and 198–199, 1991.
Wells, D. G. *The Penguin Dictionary of Curious and Interesting Numbers.* London: Penguin, pp. 121–122, 1986.

Catalan's Problem

The problem of finding the number of different ways in which a PRODUCT of n different ordered FACTORS can be calculated by pairs (i.e., the number of BINARY BRACKETINGS of n letters). For example, for the four FACTORS a, b, c, and d, there are five possibilities: $((ab)c)d$, $(a(bc))d$, $(ab)(cd)$, $a((bc)d)$, and $a(b(cd))$. The solution was given by Catalan in 1838 as

$$C'_n = \frac{2 \cdot 6 \cdot 10 \cdot (4n-6)}{n!}$$

and is equal to the CATALAN NUMBER $C_{n-1} = C'_n$.

see also BINARY BRACKETING, CATALAN'S DIOPHANTINE PROBLEM, EULER'S POLYGON DIVISION PROBLEM

References

Dörrie, H. *100 Great Problems of Elementary Mathematics: Their History and Solutions.* New York: Dover, p. 23, 1965.

Catalan Solid

The DUAL POLYHEDRA of the ARCHIMEDEAN SOLIDS, given in the following table.

Archimedean Solid	Dual
rhombicosidodecahedron	deltoidal hexecontahedron
small rhombicuboctahedron	deltoidal icositetrahedron
great rhombicuboctahedron	disdyakis dodecahedron
great rhombicosidodecahedron	disdyakis triacontahedron
truncated icosahedron	pentakis dodecahedron
snub dodecahedron (laevo)	pentagonal hexecontahedron (dextro)
snub cube (laevo)	pentagonal icositetrahedron (dextro)
cuboctahedron	rhombic dodecahedron
icosidodecahedron	rhombic triacontahedron
truncated octahedron	tetrakis hexahedron
truncated dodecahedron	triakis icosahedron
truncated cube	triakis octahedron
truncated tetrahedron	triakis tetrahedron

Here are the ARCHIMEDEAN DUALS (Holden 1971, Pearce 1978) displayed in alphabetical order (left to right, then continuing to the next row).

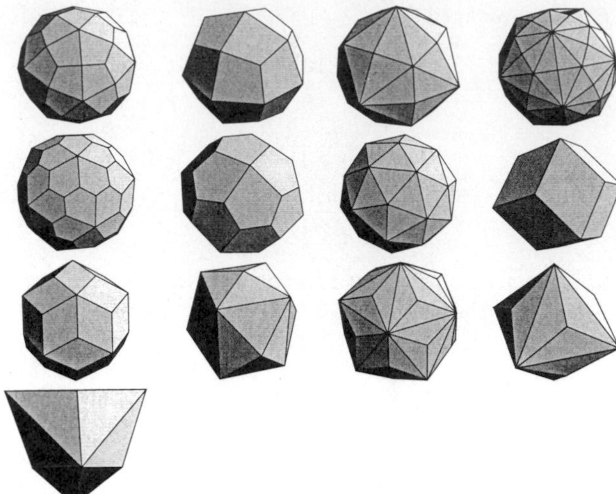

Here are the Archimedean solids paired with the corresponding Catalan solids.

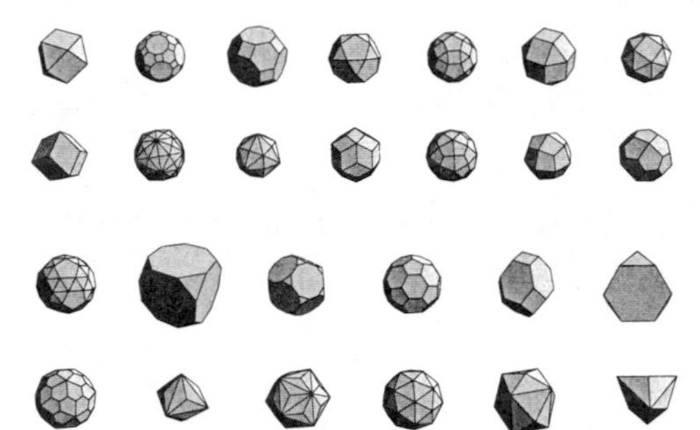

see also ARCHIMEDEAN SOLID, DUAL POLYHEDRON, SEMIREGULAR POLYHEDRON

References

Catalan, E. "Mémoire sur la Théorie des Polyèdres." *J. l'École Polytechnique (Paris)* **41**, 1–71, 1865.
Holden, A. *Shapes, Space, and Symmetry.* New York: Dover, 1991.

Catalan's Surface

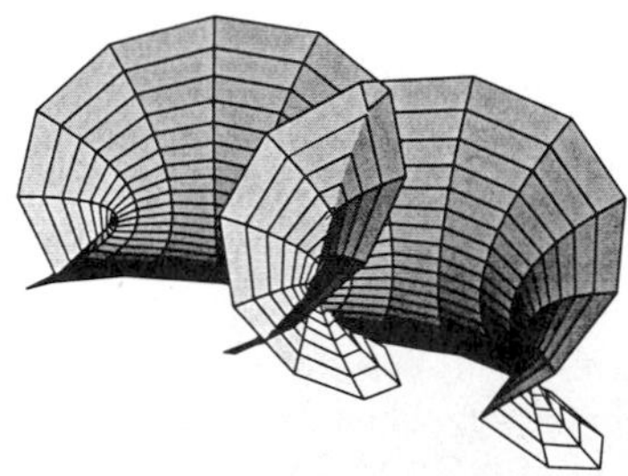

A MINIMAL SURFACE given by the parametric equations

$$x(u,v) = u - \sin u \cosh v \quad (1)$$

$$y(u,v) = 1 - \cos u \cosh v \quad (2)$$

$$z(u,v) = 4\sin(\tfrac{1}{2}u)\sinh(\tfrac{1}{2}v) \quad (3)$$

(Gray 1993), or

$$x(r,\phi) = a\sin(2\phi) - 2a\phi + \tfrac{1}{2}av^2\cos(2\phi) \quad (4)$$
$$y(r,\phi) = -a\cos(2\phi) - \tfrac{1}{2}av^2\cos(2\phi) \quad (5)$$
$$z(r,\phi) = 2av\sin\phi, \quad (6)$$

where

$$v = -r + \frac{1}{r} \quad (7)$$

(do Carmo 1986).

References

Catalan, E. "Mémoir sur les surfaces dont les rayons de courburem en chaque point, sont égaux et des signes contraires." *C. R. Acad. Sci. Paris* **41**, 1019–1023, 1855.

do Carmo, M. P. "Catalan's Surface" §3.5D in *Mathematical Models from the Collections of Universities and Museums* (Ed. G. Fischer). Braunschweig, Germany: Vieweg, pp. 45–46, 1986.

Fischer, G. (Ed.). Plates 94–95 in *Mathematische Modelle/Mathematical Models, Bildband/Photograph Volume.* Braunschweig, Germany: Vieweg, pp. 90–91, 1986.

Gray, A. *Modern Differential Geometry of Curves and Surfaces.*Boca Raton, FL: CRC Press, pp. 448–449, 1993.

Catalan's Triangle

A triangle of numbers with entries given by

$$c_{nm} = \frac{(n+m)!(n-m+1)}{m!(n+1)!}$$

for $0 \leq m \leq n$, where each element is equal to the one above plus the one to the left. Furthermore, the sum of each row is equal to the last element of the next row and also equal to the CATALAN NUMBER C_n.

```
1
1  1
1  2   2
1  3   5   5
1  4   9  14  14
1  5  14  28  42   42
1  6  20  48  90  132  132
```

(Sloane's A009766).

see also BELL TRIANGLE, CLARK'S TRIANGLE, EULER'S TRIANGLE, LEIBNIZ HARMONIC TRIANGLE, NUMBER TRIANGLE, PASCAL'S TRIANGLE, PRIME TRIANGLE, SEIDEL-ENTRINGER-ARNOLD TRIANGLE

References

Sloane, N. J. A. Sequence A009766 in "An On-Line Version of the Encyclopedia of Integer Sequences."

Catalan's Trisectrix

see TSCHIRNHAUSEN CUBIC

Catastrophe

see BUTTERFLY CATASTROPHE, CATASTROPHE THEORY, CUSP CATASTROPHE, ELLIPTIC UMBILIC CATASTROPHE, FOLD CATASTROPHE, HYPERBOLIC UMBILIC CATASTROPHE, PARABOLIC UMBILIC CATASTROPHE, SWALLOWTAIL CATASTROPHE

Catastrophe Theory

Catastrophe theory studies how the qualitative nature of equation solutions depends on the parameters that appear in the equations. Subspecializations include bifurcation theory, nonequilibrium thermodynamics, singularity theory, synergetics, and topological dynamics. For any system that seeks to minimize a function, only seven different local forms of catastrophe "typically" occur for four or fewer variables: (1) FOLD CATASTROPHE, (2) CUSP CATASTROPHE, (3) SWALLOWTAIL CATASTROPHE, (4) BUTTERFLY CATASTROPHE, (5) ELLIPTIC UMBILIC CATASTROPHE, (6) HYPERBOLIC UMBILIC CATASTROPHE, (7) PARABOLIC UMBILIC CATASTROPHE.

More specifically, for any system with fewer than five control factors and fewer than three behavior axes, these are the only seven catastrophes possible. The following tables gives the possible catastrophes as a function of control factors and behavior axes (Goetz).

Control Factors	1 Behavior Axis	2 Behavior Axes
1	fold	—
2	cusp	—
3	swallowtail	hyperbolic umbilic, elliptic umbilic
4	butterfly	parabolic umbilic

References

Arnold, V. I. *Catastrophe Theory, 3rd ed.* Berlin: Springer-Verlag, 1992.

Gilmore, R. *Catastrophe Theory for Scientists and Engineers.* New York: Dover, 1993.

Goetz, P. "Phil's Good Enough Complexity Dictionary." `http://www.cs.buffalo.edu/~goetz/dict.html`.

Saunders, P. T. *An Introduction to Catastrophe Theory.* Cambridge, England: Cambridge University Press, 1980.

Stewart, I. *The Problems of Mathematics, 2nd ed.* Oxford, England: Oxford University Press, p. 211, 1987.

Thom, R. *Structural Stability and Morphogenesis: An Outline of a General Theory of Models.* Reading, MA: Reading, MA: Addison-Wesley, 1993.

Thompson, J. M. T. *Instabilities and Catastrophes in Science and Engineering.* New York: Wiley, 1982.

Woodcock, A. E. R. and Davis, M. *Catastrophe Theory.* New York: E. P. Dutton, 1978.

Zeeman, E. C. *Catastrophe Theory—Selected Papers 1972–1977.* Reading, MA: Addison-Wesley, 1977.

Categorical Game

A GAME in which no draw is possible.

Categorical Variable

A variable which belongs to exactly one of a finite number of CATEGORIES.

Category

A category consists of two things: an OBJECT and a MORPHISM (sometimes called an "arrow"). An OBJECT is some mathematical structure (e.g., a GROUP, VECTOR SPACE, or DIFFERENTIABLE MANIFOLD) and a MORPHISM is a MAP between two OBJECTS. The MORPHISMS are then required to satisfy some fairly natural conditions; for instance, the IDENTITY MAP between any object and itself is always a MORPHISM, and the composition of two MORPHISMS (if defined) is always a MORPHISM.

One usually requires the MORPHISMS to preserve the mathematical structure of the objects. So if the objects are all groups, a good choice for a MORPHISM would be a group HOMOMORPHISM. Similarly, for vector spaces, one would choose linear maps, and for differentiable manifolds, one would choose differentiable maps.

In the category of TOPOLOGICAL SPACES, homomorphisms are usually continuous maps between topological spaces. However, there are also other category structures having TOPOLOGICAL SPACES as objects, but they are not nearly as important as the "standard" category of TOPOLOGICAL SPACES and continuous maps.

see also ABELIAN CATEGORY, ALLEGORY, EILENBERG-STEENROD AXIOMS, GROUPOID, HOLONOMY, LOGOS, MONODROMY, TOPOS

References

Freyd, P. J. and Scedrov, A. *Categories, Allegories.* Amsterdam, Netherlands: North-Holland, 1990.

Category Theory

The branch of mathematics which formalizes a number of algebraic properties of collections of transformations between mathematical objects (such as binary relations, groups, sets, topological spaces, etc.) of the same type, subject to the constraint that the collections contain the identity mapping and are closed with respect to compositions of mappings. The objects studied in category theory are called CATEGORIES.

see also CATEGORY

Catenary

The curve a hanging flexible wire or chain assumes when supported at its ends and acted upon by a uniform gravitational force. The word catenary is derived from the Latin word for "chain." In 1669, Jungius disproved Galileo's claim that the curve of a chain hanging under gravity would be a PARABOLA (MacTutor Archive). The curve is also called the ALYSOID and CHAINETTE. The equation was obtained by Leibniz, Huygens, and Johann Bernoulli in 1691 in response to a challenge by Jakob Bernoulli.

Huygens was the first to use the term catenary in a letter to Leibniz in 1690, and David Gregory wrote a treatise on the catenary in 1690 (MacTutor Archive). If you roll a PARABOLA along a straight line, its FOCUS traces out a catenary. As proved by Euler in 1744, the catenary is also the curve which, when rotated, gives the surface of minimum SURFACE AREA (the CATENOID) for the given bounding CIRCLE.

The CARTESIAN equation for the catenary is given by

$$y = \tfrac{1}{2}a(e^{x/a} + e^{-x/a}) = a\cosh\left(\frac{x}{a}\right), \tag{1}$$

and the CESÀRO EQUATION is

$$(s^2 + a^2)\kappa = -a. \tag{2}$$

The catenary gives the shape of the road over which a regular polygonal "wheel" can travel smoothly. For a regular n-gon, the corresponding catenary is

$$y = -A\cosh\left(\frac{x}{A}\right), \tag{3}$$

where

$$A \equiv R\cos\left(\frac{\pi}{n}\right). \tag{4}$$

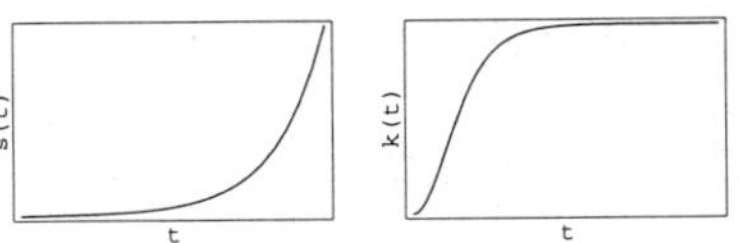

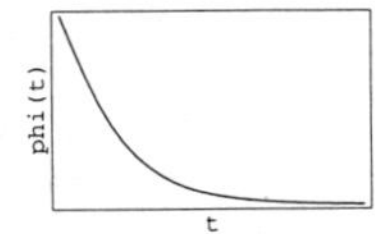

The ARC LENGTH, CURVATURE, and TANGENTIAL ANGLE are

$$s = a\sinh\left(\frac{t}{a}\right), \tag{5}$$

$$\kappa = -\frac{1}{a}\operatorname{sech}^2\left(\frac{t}{a}\right), \tag{6}$$

$$\phi = -2\tan^{-1}\left[\tanh\left(\frac{t}{2a}\right)\right]. \tag{7}$$

The slope is proportional to the ARC LENGTH as measured from the center of symmetry.

see also CALCULUS OF VARIATIONS, CATENOID, LINDELOF'S THEOREM, SURFACE OF REVOLUTION

References

Geometry Center. "The Catenary." http://www.geom.umn.edu/zoo/diffgeom/surfspace/catenoid/catenary.html.

Gray, A. "The Evolute of a Tractrix is a Catenary." §5.3 in *Modern Differential Geometry of Curves and Surfaces.* Boca Raton, FL: CRC Press, pp. 80–81, 1993.

Lawrence, J. D. *A Catalog of Special Plane Curves.* New York: Dover, pp. 195 and 199–200, 1972.

Lockwood, E. H. "The Tractrix and Catenary." Ch. 13 in *A Book of Curves.* Cambridge, England: Cambridge University Press, pp. 118–124, 1967.

MacTutor History of Mathematics Archive. "Catenary." http://www-groups.dcs.st-and.ac.uk/~history/Curves/Catenary.html.

Pappas, T. "The Catenary & the Parabolic Curves." *The Joy of Mathematics.* San Carlos, CA: Wide World Publ./Tetra, p. 34, 1989.

Yates, R. C. "Catenary." *A Handbook on Curves and Their Properties.* Ann Arbor, MI: J. W. Edwards, pp. 12–14, 1952.

Catenary Evolute

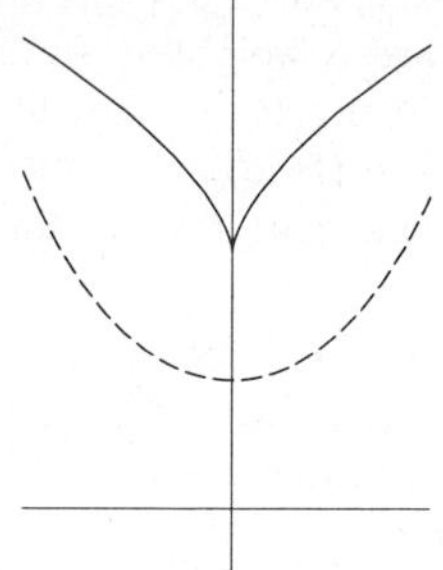

$$x = a[x - \tfrac{1}{2}\sinh(2t)]$$
$$y = 2a\cosh t.$$

Catenary Involute

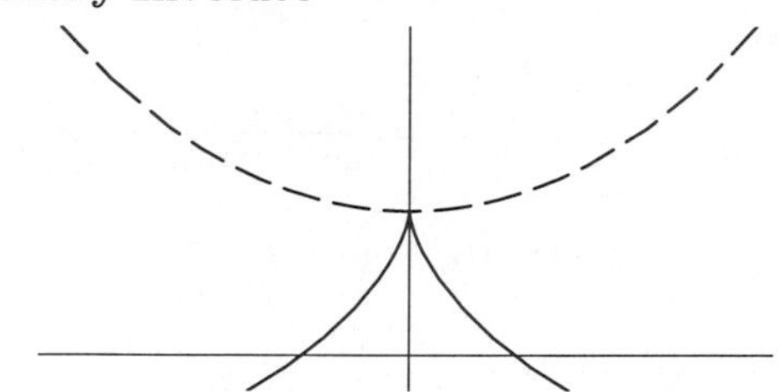

The parametric equation for a CATENARY is

$$\mathbf{r}(t) = a\begin{bmatrix} t \\ \cosh t \end{bmatrix}, \quad (1)$$

so

$$\frac{d\mathbf{r}}{dt} = a\begin{bmatrix} 1 \\ \sinh t \end{bmatrix} \quad (2)$$

$$\left|\frac{d\mathbf{r}}{dt}\right| = a\sqrt{1+\sinh^2 t} = a\cosh t \quad (3)$$

and

$$\hat{\mathbf{T}} = \frac{\frac{d\mathbf{r}}{dt}}{\left|\frac{d\mathbf{r}}{dt}\right|} = \begin{bmatrix} \operatorname{sech} t \\ \tanh t \end{bmatrix} \quad (4)$$

$$ds^2 = |d\mathbf{r}^2| = a^2(1+\sinh^2 t)\,dt^2 = a^2\cosh^2\,dt^2 \quad (5)$$

$$\frac{ds}{dt} = a\cosh t. \quad (6)$$

Therefore,

$$s = a\int \cosh t\,dt = a\sinh t \quad (7)$$

and the equation of the INVOLUTE is

$$x = a(t - \tanh t) \quad (8)$$
$$y = a\operatorname{sech} t. \quad (9)$$

This curve is called a TRACTRIX.

Catenary Radial Curve

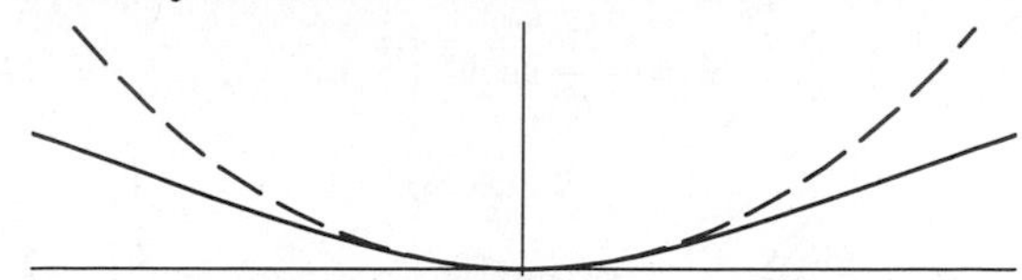

The KAMPYLE OF EUDOXUS.

Catenoid

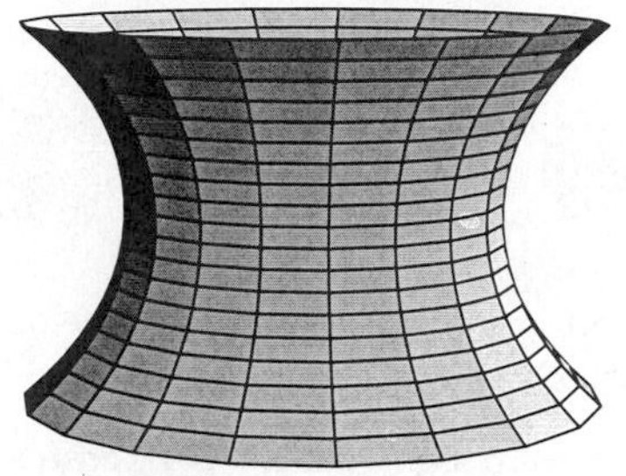

A CATENARY of REVOLUTION. The catenoid and PLANE are the only SURFACES OF REVOLUTION which are also MINIMAL SURFACES. The catenoid can be given by the parametric equations

$$x = c\cosh\left(\frac{v}{c}\right)\cos u \quad (1)$$
$$y = c\cosh\left(\frac{v}{c}\right)\sin u \quad (2)$$
$$z = v, \quad (3)$$

where $u \in [0, 2\pi)$. The differentials are

$$dx = \sinh\left(\frac{v}{c}\right)\cos u\,dv - \cosh\left(\frac{v}{c}\right)\sin u\,du \quad (4)$$
$$dy = \sinh\left(\frac{v}{c}\right)\sin u\,dv + \cosh\left(\frac{v}{c}\right)\cos u\,du \quad (5)$$
$$dz = du, \quad (6)$$

so the LINE ELEMENT is

$$\begin{aligned} ds^2 &= dx^2 + dy^2 + dz^2 \\ &= \left[\sinh^2\left(\frac{v}{c}\right) + 1\right] dv^2 + \cosh^2\left(\frac{v}{c}\right) du^2 \\ &= \cosh^2\left(\frac{v}{c}\right) dv^2 + \cosh^2\left(\frac{v}{c}\right) du^2. \end{aligned} \quad (7)$$

The PRINCIPAL CURVATURES are

$$\kappa_1 = -\frac{1}{c}\operatorname{sech}^2\left(\frac{v}{c}\right) \quad (8)$$
$$\kappa_2 = \frac{1}{c}\operatorname{sech}^2\left(\frac{v}{c}\right). \quad (9)$$

The MEAN CURVATURE of the catenoid is

$$H = 0 \quad (10)$$

and the GAUSSIAN CURVATURE is

$$K = -\frac{1}{c^2}\operatorname{sech}^4\left(\frac{v}{c}\right). \quad (11)$$

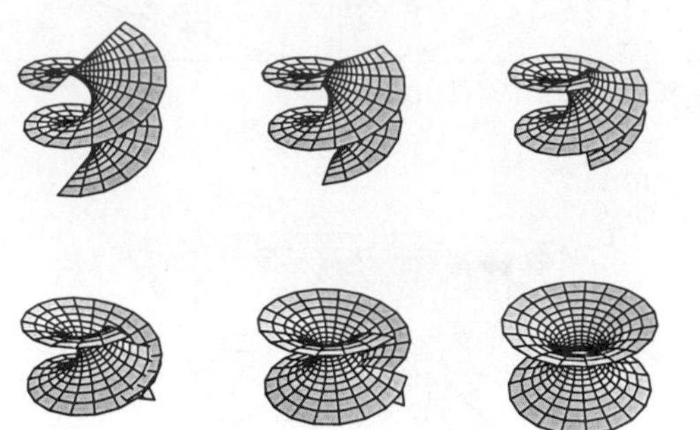

The HELICOID can be continuously deformed into a catenoid with $c = 1$ by the transformation

$$x(u,v) = \cos\alpha \sinh v \sin u + \sin\alpha \cosh v \cos u \quad (12)$$
$$y(u,v) = -\cos\alpha \sinh v \cos u + \sin\alpha \cosh v \sin u \quad (13)$$
$$z(u,v) = u\cos\alpha + v\sin\alpha, \quad (14)$$

where $\alpha = 0$ corresponds to a HELICOID and $\alpha = \pi/2$ to a catenoid.

see also CATENARY, COSTA MINIMAL SURFACE, HELICOID, MINIMAL SURFACE, SURFACE OF REVOLUTION

References
do Carmo, M. P. "The Catenoid." §3.5A in *Mathematical Models from the Collections of Universities and Museums* (Ed. G. Fischer). Braunschweig, Germany: Vieweg, p. 43, 1986.
Fischer, G. (Ed.). Plate 90 in *Mathematische Modelle/Mathematical Models, Bildband/Photograph Volume.* Braunschweig, Germany: Vieweg, p. 86, 1986.
Geometry Center. "The Catenoid." http://www.geom.umn.edu/zoo/diffgeom/surfspace/catenoid/.
Gray, A. "The Catenoid." §18.4 *Modern Differential Geometry of Curves and Surfaces.* Boca Raton, FL: CRC Press, pp. 367–369, 1993.
Meusnier, J. B. "Mémoire sur la courbure des surfaces." *Mém. des savans étrangers* **10** (lu 1776), 477–510, 1785.

Caterpillar Graph

A TREE with every NODE on a central stalk or only one EDGE away from the stalk.

References
Gardner, M. *Wheels, Life, and other Mathematical Amusements.* New York: W. H. Freeman, p. 160, 1983.

Cattle Problem of Archimedes

see ARCHIMEDES' CATTLE PROBLEM

Cauchy Binomial Theorem

$$\sum_{m=0}^{n} y^m q^{m(m+1)/2} \binom{n}{m}_q = \prod_{k=1}^{n}(1 + yq^k),$$

where $\binom{n}{m}_q$ is a GAUSSIAN COEFFICIENT.

see also q-BINOMIAL THEOREM

Cauchy Boundary Conditions

BOUNDARY CONDITIONS of a PARTIAL DIFFERENTIAL EQUATION which are a weighted AVERAGE of DIRICHLET BOUNDARY CONDITIONS (which specify the value of the function on a surface) and NEUMANN BOUNDARY CONDITIONS (which specify the normal derivative of the function on a surface).

see also BOUNDARY CONDITIONS, CAUCHY PROBLEM, DIRICHLET BOUNDARY CONDITIONS, NEUMANN BOUNDARY CONDITIONS

References
Morse, P. M. and Feshbach, H. *Methods of Theoretical Physics, Part I.* New York: McGraw-Hill, pp. 678–679, 1953.

Cauchy's Cosine Integral Formula

$$\int_{-\pi/2}^{\pi/2} \cos^{\mu+\nu-2}\theta e^{i\theta(\mu-\nu+2\xi)}\,d\theta = \frac{\pi\Gamma(\mu+\nu-1)}{2^{\mu+\nu-2}\Gamma(\mu+\xi)\Gamma(\nu-\xi)},$$

where $\Gamma(z)$ is the GAMMA FUNCTION.

Cauchy Criterion

A NECESSARY and SUFFICIENT condition for a SEQUENCE S_i to CONVERGE. The Cauchy criterion is satisfied when, for all $\epsilon > 0$, there is a fixed number N such that $|S_j - S_i| < \epsilon$ for all $i, j > N$.

Cauchy Distribution

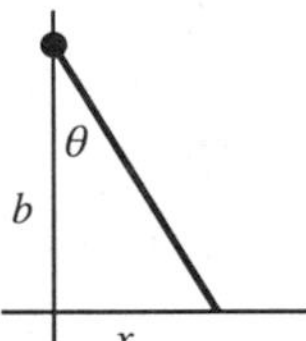

The Cauchy distribution, also called the LORENTZIAN DISTRIBUTION, describes resonance behavior. It also describes the distribution of horizontal distances at which a LINE SEGMENT tilted at a random ANGLE cuts the x-AXIS. Let θ represent the ANGLE that a line, with fixed point of rotation, makes with the vertical axis, as shown above. Then

$$\tan\theta = \frac{x}{b} \quad (1)$$
$$\theta = \tan^{-1}\left(\frac{x}{b}\right) \quad (2)$$
$$d\theta = -\frac{1}{1+\frac{x^2}{b^2}}\frac{dx}{b} = -\frac{b\,dx}{b^2+x^2}, \quad (3)$$

so the distribution of ANGLE θ is given by

$$\frac{d\theta}{\pi} = -\frac{1}{\pi}\frac{b\,dx}{b^2+x^2}. \quad (4)$$

This is normalized over all angles, since

$$\int_{-\pi/2}^{\pi/2} \frac{d\theta}{\pi} = 1 \tag{5}$$

and

$$-\int_{-\infty}^{\infty} \frac{1}{\pi} \frac{b\,dx}{b^2+x^2} = \frac{1}{\pi}\left[\tan^{-1}\left(\frac{b}{x}\right)\right]_{-\infty}^{\infty}$$
$$= \frac{1}{\pi}[\tfrac{1}{2}\pi - (-\tfrac{1}{2}\pi)] = 1. \tag{6}$$

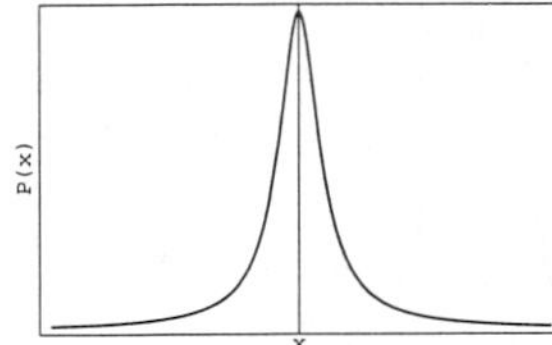

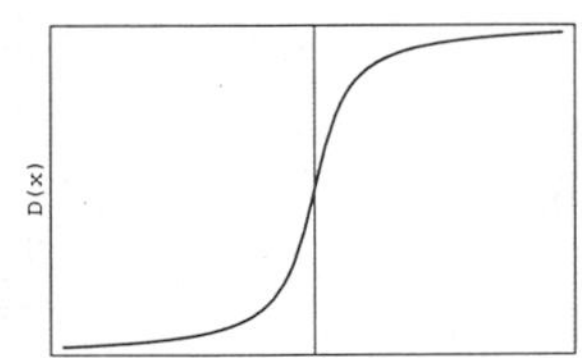

The general Cauchy distribution and its cumulative distribution can be written as

$$P(x) = \frac{1}{\pi} \frac{\frac{1}{2}\Gamma}{(x-\mu)^2 + (\frac{1}{2}\Gamma)^2} \tag{7}$$

$$D(x) = \frac{1}{2} + \frac{1}{\pi}\tan^{-1}\left(\frac{x-\mu}{b}\right), \tag{8}$$

where Γ is the FULL WIDTH AT HALF MAXIMUM ($\Gamma = 2b$ in the above example) and μ is the MEAN ($\mu = 0$ in the above example). The CHARACTERISTIC FUNCTION is

$$\phi(t) = \frac{1}{\pi}\int_{-\infty}^{\infty} \frac{e^{it(\Gamma x/2-\mu)}}{1+x^2}\,dx$$
$$= \frac{e^{-i\mu t}}{\pi}\int_{-\infty}^{\infty} \frac{\cos(\Gamma t x/2)}{1+(\Gamma x/2)^2}\,dx$$
$$= e^{-i\mu t - \Gamma|t|/2}. \tag{9}$$

The MOMENTS are given by

$$\mu_2 = \sigma^2 = \infty \tag{10}$$

$$\mu_3 = \begin{cases} 0 & \text{for } \mu = 0 \\ \infty & \text{for } \mu \neq 0 \end{cases} \tag{11}$$

$$\mu_4 = \infty, \tag{12}$$

and the STANDARD DEVIATION, SKEWNESS, and KURTOSIS by

$$\sigma^2 = \infty \tag{13}$$

$$\gamma_1 = \begin{cases} 0 & \text{for } \mu = 0 \\ \infty & \text{for } \mu \neq 0 \end{cases} \tag{14}$$

$$\gamma_2 = \infty. \tag{15}$$

If X and Y are variates with a NORMAL DISTRIBUTION, then $Z \equiv X/Y$ has a Cauchy distribution with MEAN $\mu = 0$ and full width

$$\Gamma = \frac{2\sigma_y}{\sigma_x}. \tag{16}$$

see also GAUSSIAN DISTRIBUTION, NORMAL DISTRIBUTION

References

Spiegel, M. R. *Theory and Problems of Probability and Statistics.* New York: McGraw-Hill, pp. 114–115, 1992.

Cauchy Equation

see EULER EQUATION

Cauchy's Formula

The GEOMETRIC MEAN is smaller than the ARITHMETIC MEAN,

$$\left(\prod_{i=1}^{N} n_i\right)^{1/N} < \frac{\sum_{i=1}^{N} n_i}{N}.$$

Cauchy Functional Equation

The fifth of HILBERT'S PROBLEMS is a generalization of this equation.

Cauchy-Hadamard Theorem

The RADIUS OF CONVERGENCE of the TAYLOR SERIES

$$a_0 + a_1 z + a_2 z^2 + \ldots$$

is

$$r = \frac{1}{\overline{\lim_{n\to\infty}}(|a_n|)^{1/n}}.$$

see also RADIUS OF CONVERGENCE, TAYLOR SERIES

Cauchy Inequality

A special case of the HÖLDER SUM INEQUALITY with $p = q = 2$,

$$\left(\sum_{k=1}^{n} a_k b_k\right)^2 \leq \left(\sum_{k=1}^{n} a_k{}^2\right)\left(\sum_{k=1}^{n} b_k{}^2\right), \tag{1}$$

where equality holds for $a_k = cb_k$. In 2-D, it becomes

$$(a^2+b^2)(c^2+a^2) \geq (ac+bd)^2. \tag{2}$$

It can be proven by writing

$$\sum_{i=1}^{n}(a_i x + b_i)^2 = \sum_{i=1}^{n} a_i{}^2\left(x + \frac{b_i}{a_i}\right)^2 = 0. \tag{3}$$

If b_i/a_i is a constant c, then $x = -c$. If it is not a constant, then all terms cannot simultaneously vanish for REAL x, so the solution is COMPLEX and can be found using the QUADRATIC EQUATION

$$x = \frac{-2\sum a_i b_i \pm \sqrt{4\left(\sum a_i b_i\right)^2 - 4\sum a_i{}^2 \sum b_i{}^2}}{2\sum a_i{}^2}. \tag{4}$$

In order for this to be COMPLEX, it must be true that

$$\left(\sum_i a_i b_i\right)^2 \leq \left(\sum_i a_i{}^2\right)\left(\sum_i b_i{}^2\right), \qquad (5)$$

with equality when b_i/a_i is a constant. The VECTOR derivation is much simpler,

$$(\mathbf{a}\cdot\mathbf{b})^2 = a^2b^2\cos^2\theta \leq a^2b^2, \qquad (6)$$

where

$$a^2 \equiv \mathbf{a}\cdot\mathbf{a} = \sum_i a_i{}^2, \qquad (7)$$

and similarly for b.

see also CHEBYSHEV INEQUALITY, HÖLDER SUM INEQUALITY

References

Abramowitz, M. and Stegun, C. A. (Eds.). *Handbook of Mathematical Functions with Formulas, Graphs, and Mathematical Tables, 9th printing.* New York: Dover, p. 11, 1972.

Cauchy Integral Formula

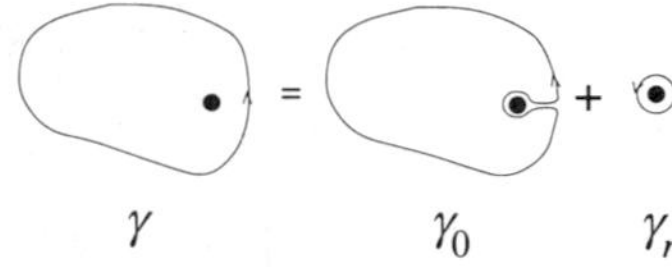

Given a CONTOUR INTEGRAL of the form

$$\int_\gamma \frac{f(z)\,dz}{z-z_0}, \qquad (1)$$

define a path γ_0 as an infinitesimal CIRCLE around the point z_0 (the dot in the above illustration). Define the path γ_r as an arbitrary loop with a cut line (on which the forward and reverse contributions cancel each other out) so as to go around z_0.

The total path is then

$$\gamma = \gamma_0 + \gamma_r, \qquad (2)$$

so

$$\int_\gamma \frac{f(z)\,dz}{z-z_0} = \int_{\gamma_0} \frac{f(z)\,dz}{z-z_0} + \int_{\gamma_r} \frac{f(z)\,dz}{z-z_0}. \qquad (3)$$

From the CAUCHY INTEGRAL THEOREM, the CONTOUR INTEGRAL along any path not enclosing a POLE is 0. Therefore, the first term in the above equation is 0 since γ_0 does not enclose the POLE, and we are left with

$$\int_\gamma \frac{f(z)\,dz}{z-z_0} = \int_{\gamma_r} \frac{f(z)\,dz}{z-z_0}. \qquad (4)$$

Now, let $z \equiv z_0 + re^{i\theta}$, so $dz = ire^{i\theta}\,d\theta$. Then

$$\int_\gamma \frac{f(z)\,dz}{z-z_0} = \int_{\gamma_r} \frac{f(z_0+re^{i\theta})}{re^{i\theta}} ire^{i\theta}\,d\theta = \int_{\gamma_r} f(z_0+re^{i\theta})i\,d\theta. \qquad (5)$$

But we are free to allow the radius r to shrink to 0, so

$$\int_\gamma \frac{f(z)\,dz}{z-z_0} = \lim_{r\to 0}\int_{\gamma_r} f(z_0+re^{i\theta})i\,d\theta = \int_{\gamma_r} f(z_0)i\,d\theta = if(z_0)\int_{\gamma_r} d\theta = 2\pi i f(z_0), \qquad (6)$$

and

$$f(z_0) = \frac{1}{2\pi i}\int_\gamma \frac{f(z)\,dz}{z-z_0}. \qquad (7)$$

If multiple loops are made around the POLE, then equation (7) becomes

$$n(\gamma, z_0)f(z_0) = \frac{1}{2\pi i}\int_\gamma \frac{f(z)\,dz}{z-z_0}, \qquad (8)$$

where $n(\gamma, z_0)$ is the WINDING NUMBER.

A similar formula holds for the derivatives of $f(z)$,

$$\begin{aligned} f'(z_0) &= \lim_{h\to 0}\frac{f(z_0+h)-f(h)}{h} \\ &= \lim_{h\to 0}\frac{1}{2\pi i h}\left(\int_\gamma \frac{f(z)\,dz}{z-z_0-h} - \int_\gamma \frac{f(z)\,dz}{z-z_0}\right) \\ &= \lim_{h\to 0}\frac{1}{2\pi i h}\int_\gamma \frac{f(z)[(z-z_0)-(z-z_0-h)]\,dz}{(z-z_0-h)(z-z_0)} \\ &= \lim_{h\to 0}\frac{1}{2\pi i h}\int_\gamma \frac{hf(z)\,dz}{(z-z_0-h)(z-z_0)} \\ &= \frac{1}{2\pi i}\int_\gamma \frac{f(z)\,dz}{(z-z_0)^2}. \end{aligned} \qquad (9)$$

Iterating again,

$$f''(z_0) = \frac{2}{2\pi i}\int_\gamma \frac{f(z)\,dz}{(z-z_0)^3}. \qquad (10)$$

Continuing the process and adding the WINDING NUMBER n,

$$n(\gamma, z_0)f^{(r)}(z_0) = \frac{r!}{2\pi i}\int_\gamma \frac{f(z)\,dz}{(z-z_0)^{r+1}}. \qquad (11)$$

see also MORERA'S THEOREM

References

Arfken, G. "Cauchy's Integral Formula." §6.4 in *Mathematical Methods for Physicists, 3rd ed.* Orlando, FL: Academic Press, pp. 371–376, 1985.

Morse, P. M. and Feshbach, H. *Methods of Theoretical Physics, Part I.* New York: McGraw-Hill, pp. 367–372, 1953.

Cauchy Integral Test

see INTEGRAL TEST

Cauchy Integral Theorem

If f is continuous and finite on a simply connected region R and has only finitely many points of nondifferentiability in R, then

$$\oint_\gamma f(z)\,dz = 0 \tag{1}$$

for any closed CONTOUR γ completely contained in R. Writing z as

$$z \equiv x + iy \tag{2}$$

and $f(z)$ as

$$f(z) \equiv u + iv \tag{3}$$

then gives

$$\begin{aligned}\oint_\gamma f(z)\,dz &= \int_\gamma (u+iv)(dx + i\,dy)\\ &= \int_\gamma u\,dx - v\,dy + i\int_\gamma v\,dx + u\,dy. \end{aligned} \tag{4}$$

From GREEN'S THEOREM,

$$\int_\gamma f(x,y)\,dx - g(x,y)\,dy = -\iint \left(\frac{\partial g}{\partial x} + \frac{\partial f}{\partial y}\right) dx\,dy, \tag{5}$$

$$\int_\gamma f(x,y)\,dx + g(x,y)\,dy = \iint \left(\frac{\partial g}{\partial x} - \frac{\partial f}{\partial y}\right) dx\,dy \tag{6}$$

so (4) becomes

$$\oint_\gamma f(z)\,dz = -\iint \left(\frac{\partial v}{\partial x} + \frac{\partial u}{\partial y}\right) dx\,dy + i\iint \left(\frac{\partial u}{\partial x} - \frac{\partial v}{\partial y}\right) dx\,dy. \tag{7}$$

But the CAUCHY-RIEMANN EQUATIONS require that

$$\frac{\partial u}{\partial x} = \frac{\partial v}{\partial y} \tag{8}$$

$$\frac{\partial u}{\partial y} = -\frac{\partial v}{\partial x}, \tag{9}$$

so

$$\oint_\gamma f(z)\,dz = 0, \tag{10}$$

Q. E. D.

For a MULTIPLY CONNECTED region,

$$\int_{C_1} f(z)\,dz = \int_{C_2} f(z)\,dz. \tag{11}$$

see also CAUCHY INTEGRAL THEOREM, MORERA'S THEOREM, RESIDUE THEOREM (COMPLEX ANALYSIS)

References

Arfken, G. "Cauchy's Integral Theorem." §6.3 in *Mathematical Methods for Physicists, 3rd ed.* Orlando, FL: Academic Press, pp. 365–371, 1985.

Morse, P. M. and Feshbach, H. *Methods of Theoretical Physics, Part I.* New York: McGraw-Hill, pp. 363–367, 1953.

Cauchy-Kovalevskaya Theorem

The theorem which proves the existence and uniqueness of solutions to the CAUCHY PROBLEM.

see also CAUCHY PROBLEM

Cauchy-Lagrange Identity

$$\begin{aligned}({a_1}^2 + {a_2}^2 + \ldots + {a_n}^2)({b_1}^2 + {b_2}^2 + \ldots + {b_n}^2)\\ = (a_1b_2 - a_2b_1)^2 + (a_1b_3 - a_3b_1)^2 + \ldots\\ +(a_{n-1}b_n - a_nb_{n-1})^2.\end{aligned}$$

From this identity, the n-D CAUCHY INEQUALITY follows.

Cauchy-Maclaurin Theorem

see MACLAURIN-CAUCHY THEOREM

Cauchy Mean Theorem

For numbers > 0, the GEOMETRIC MEAN $<$ the ARITHMETIC MEAN.

Cauchy Principal Value

$$PV\int_{-\infty}^{\infty} f(x)\,dx \equiv \lim_{R\to\infty}\int_{-R}^{R} f(x)\,dx$$

$$PV\int_a^b f(x)\,dx \equiv \lim_{\epsilon\to 0}\left[\int_a^{c-\epsilon} f(x)\,dx + \int_{c+\epsilon}^b f(x)\,dx\right],$$

where $\epsilon > 0$ and $a \leq c \leq b$.

References

Arfken, G. *Mathematical Methods for Physicists, 3rd ed.* Orlando, FL: Academic Press, pp. 401–403, 1985.

Sansone, G. *Orthogonal Functions, rev. English ed.* New York: Dover, p. 158, 1991.

Cauchy Problem

If $f(x,y)$ is an ANALYTIC FUNCTION in a NEIGHBORHOOD of the point (x_0, y_0) (i.e., it can be expanded in a series of NONNEGATIVE INTEGER POWERS of $(x - x_0)$ and $(y - y_0)$), find a solution $y(x)$ of the DIFFERENTIAL EQUATION

$$\frac{dy}{dx} = f(x),$$

with initial conditions $y = y_0$ and $x = x_0$. The existence and uniqueness of the solution were proven by Cauchy and Kovalevskaya in the CAUCHY-KOVALEVSKAYA THEOREM. The Cauchy problem amounts to determining the shape of the boundary and type of equation which yield unique and reasonable solutions for the CAUCHY BOUNDARY CONDITIONS.

see also CAUCHY BOUNDARY CONDITIONS

Cauchy Ratio Test

see RATIO TEST

Cauchy Remainder Form

The remainder of n terms of a TAYLOR SERIES is given by

$$R_n = \frac{(x-c)^{n-1}(x-a)}{(n-1)!} f^{(n)}(c),$$

where $a \leq c \leq x$.

Cauchy-Riemann Equations

Let

$$f(x,y) \equiv u(x,y) + iv(x,y), \tag{1}$$

where

$$z \equiv x + iy, \tag{2}$$

so

$$dz = dx + i\,dy. \tag{3}$$

The total derivative of f with respect to z may then be computed as follows.

$$y = \frac{z-x}{i} \tag{4}$$

$$x = z - iy, \tag{5}$$

so

$$\frac{\partial y}{\partial z} = \frac{1}{i} = -i \tag{6}$$

$$\frac{\partial x}{\partial z} = 1, \tag{7}$$

and

$$\frac{df}{dz} = \frac{\partial f}{\partial x}\frac{\partial x}{\partial z} + \frac{\partial f}{\partial y}\frac{\partial y}{\partial z} = \frac{\partial f}{\partial x} - i\frac{\partial f}{\partial y}. \tag{8}$$

In terms of u and v, (8) becomes

$$\begin{aligned}\frac{df}{dz} &= \left(\frac{\partial u}{\partial x} + i\frac{\partial v}{\partial x}\right) - i\left(\frac{\partial u}{\partial y} + i\frac{\partial v}{\partial y}\right)\\ &= \left(\frac{\partial u}{\partial x} + i\frac{\partial v}{\partial x}\right) + \left(-i\frac{\partial u}{\partial y} + \frac{\partial v}{\partial y}\right).\end{aligned} \tag{9}$$

Along the real, or x-AXIS, $\partial f/\partial y = 0$, so

$$\frac{df}{dz} = \frac{\partial u}{\partial x} + i\frac{\partial v}{\partial x}. \tag{10}$$

Along the imaginary, or y-axis, $\partial f/\partial x = 0$, so

$$\frac{df}{dz} = -i\frac{\partial u}{\partial y} + \frac{\partial v}{\partial y}. \tag{11}$$

If f is COMPLEX DIFFERENTIABLE, then the value of the derivative must be the same for a given dz, regardless of its orientation. Therefore, (10) must equal (11), which requires that

$$\frac{\partial u}{\partial x} = \frac{\partial v}{\partial y} \tag{12}$$

and

$$\frac{\partial v}{\partial x} = -\frac{\partial u}{\partial y}. \tag{13}$$

These are known as the Cauchy-Riemann equations. They lead to the condition

$$\frac{\partial^2 u}{\partial x \partial y} = -\frac{\partial^2 v}{\partial x \partial y}. \tag{14}$$

The Cauchy-Riemann equations may be concisely written as

$$\begin{aligned}\frac{df}{dz^*} &= \frac{\partial f}{\partial x} + i\frac{\partial f}{\partial y} = \left(\frac{\partial u}{\partial x} + i\frac{\partial v}{\partial x}\right) + i\left(\frac{\partial u}{\partial y} + i\frac{\partial v}{\partial y}\right)\\ &= \left(\frac{\partial u}{\partial x} - \frac{\partial v}{\partial y}\right) + i\left(\frac{\partial u}{\partial y} + \frac{\partial v}{\partial x}\right) = 0.\end{aligned} \tag{15}$$

In POLAR COORDINATES,

$$f(re^{i\theta}) \equiv R(r,\theta)e^{i\Theta(r,\theta)}, \tag{16}$$

so the Cauchy-Riemann equations become

$$\frac{\partial R}{\partial r} = \frac{R}{r}\frac{\partial \Theta}{\partial \theta} \tag{17}$$

$$\frac{1}{r}\frac{\partial R}{\partial \theta} = -R\frac{\partial \Theta}{\partial r}. \tag{18}$$

If u and v satisfy the Cauchy-Riemann equations, they also satisfy LAPLACE'S EQUATION in 2-D, since

$$\frac{\partial^2 u}{\partial x^2} + \frac{\partial^2 u}{\partial y^2} = \frac{\partial}{\partial x}\left(\frac{\partial v}{\partial y}\right) + \frac{\partial}{\partial y}\left(-\frac{\partial v}{\partial x}\right) = 0 \tag{19}$$

$$\frac{\partial^2 v}{\partial x^2} + \frac{\partial^2 v}{\partial y^2} = \frac{\partial}{\partial x}\left(-\frac{\partial u}{\partial y}\right) + \frac{\partial}{\partial y}\left(\frac{\partial u}{\partial x}\right) = 0. \tag{20}$$

By picking an arbitrary $f(z)$, solutions can be found which automatically satisfy the Cauchy-Riemann equations and LAPLACE'S EQUATION. This fact is used to find so-called CONFORMAL SOLUTIONS to physical problems involving scalar potentials such as fluid flow and electrostatics.

see also CAUCHY INTEGRAL THEOREM, CONFORMAL SOLUTION, MONOGENIC FUNCTION, POLYGENIC FUNCTION

References

Abramowitz, M. and Stegun, C. A. (Eds.). *Handbook of Mathematical Functions with Formulas, Graphs, and Mathematical Tables, 9th printing.* New York: Dover, p. 17, 1972.

Arfken, G. "Cauchy-Riemann Conditions." §6.2 in *Mathematical Methods for Physicists, 3rd ed.* Orlando, FL: Academic Press, pp. 3560–365, 1985.

Cauchy's Rigidity Theorem

see RIGIDITY THEOREM

Cauchy Root Test

see ROOT TEST

Cauchy-Schwarz Integral Inequality

Let $f(x)$ and $g(x)$ by any two REAL integrable functions of $[a, b]$, then

$$\left[\int_a^b f(x)g(x)\,dx\right]^2 \leq \left[\int_a^b f^2(x)\,dx\right]\left[\int_a^b g^2(x)\,dx\right],$$

with equality IFF $f(x) = kg(x)$ with k real.

References

Gradshteyn, I. S. and Ryzhik, I. M. *Tables of Integrals, Series, and Products, 5th ed.* San Diego, CA: Academic Press, p. 1099, 1993.

Cauchy-Schwarz Sum Inequality

$$|\mathbf{a}\cdot\mathbf{b}| \leq |\mathbf{a}|\,|\mathbf{b}|.$$

$$\left(\sum_{k=1}^n a_k b_k\right)^2 \leq \left(\sum_{k=1}^n {a_k}^2\right)\left(\sum_{k=1}^n {b_k}^2\right).$$

Equality holds IFF the sequences a_1, a_2, ... and b_1, b_2, ... are proportional.

see also FIBONACCI IDENTITY

References

Gradshteyn, I. S. and Ryzhik, I. M. *Tables of Integrals, Series, and Products, 5th ed.* San Diego, CA: Academic Press, p. 1092, 1979.

Cauchy Sequence

A SEQUENCE $a_1, a_2, \ldots$ such that the METRIC $d(a_m, a_n)$ satisfies

$$\lim_{\min(m,n)\to\infty} d(a_m, a_n) = 0.$$

Cauchy sequences in the rationals do not necessarily CONVERGE, but they do CONVERGE in the REALS.

REAL NUMBERS can be defined using either DEDEKIND CUTS or Cauchy sequences.

see also DEDEKIND CUT

Cauchy Test

see RATIO TEST

Caustic

The curve which is the ENVELOPE of reflected (CATACAUSTIC) or refracted (DIACAUSTIC) rays of a given curve for a light source at a given point (known as the RADIANT POINT). The caustic is the EVOLUTE of the ORTHOTOMIC.

References

Lawrence, J. D. *A Catalog of Special Plane Curves.* New York: Dover, p. 60, 1972.

Lee, X. "Caustics." `http://www.best.com/~xah/SpecialPlaneCurves_dir/Caustics_dir/caustics.html`.

Lockwood, E. H. "Caustic Curves." Ch. 24 in *A Book of Curves.* Cambridge, England: Cambridge University Press, pp. 182–185, 1967.

Yates, R. C. "Caustics." *A Handbook on Curves and Their Properties.* Ann Arbor, MI: J. W. Edwards, pp. 15-20, 1952.

Cavalieri's Principle

1. If the lengths of every one-dimensional slice are equal for two regions, then the regions have equal AREAS.
2. If the AREAS of every two-dimensional slice (CROSS-SECTION) are equal for two SOLIDS, then the SOLIDS have equal VOLUMES.

see also CROSS-SECTION, PAPPUS'S CENTROID THEOREM

References

Beyer, W. H. (Ed.) *CRC Standard Mathematical Tables, 28th ed.* Boca Raton, FL: CRC Press, p. 126 and 132, 1987.

Cayley Algebra

The only NONASSOCIATIVE DIVISION ALGEBRA with REAL SCALARS. There is an 8-square identity corresponding to this algebra. The elements of a Cayley algebra are called CAYLEY NUMBERS or OCTONIONS.

References

Kurosh, A. G. *General Algebra.* New York: Chelsea, pp. 226–28, 1963.

Cayley-Bacharach Theorem

Let $X_1, X_2 \subset \mathbb{P}^2$ be CUBIC plane curves meeting in nine points $p_1, \ldots, p_9$. If $X \subset \mathbb{P}^2$ is any CUBIC containing $p_1, \ldots, p_8$, then X contains p_9 as well. It is related to GORENSTEIN RINGS, and is a generalization of PAPPUS'S HEXAGON THEOREM and PASCAL'S THEOREM.

References

Eisenbud, D.; Green, M.; and Harris, J. "Cayley-Bacharach Theorems and Conjectures." *Bull. Amer. Math. Soc.* **33**, 295–324, 1996.

Cayley Cubic

A CUBIC RULED SURFACE (Fischer 1986) in which the director line meets the director CONIC SECTION. Cayley's surface is the unique cubic surface having four ORDINARY DOUBLE POINTS (Hunt), the maximum possible for CUBIC SURFACE (Endraß). The Cayley cubic is invariant under the TETRAHEDRAL GROUP and contains exactly nine lines, six of which connect the four nodes pairwise and the other three of which are coplanar (Endraß).

If the ORDINARY DOUBLE POINTS in projective 3-space are taken as (1, 0, 0, 0), (0, 1, 0, 0), (0, 0, 1, 0), (0, 0, 0, 1), then the equation of the surface in projective coordinates is

$$\frac{1}{x_0} + \frac{1}{x_1} + \frac{1}{x_2} + \frac{1}{x_3} = 0$$

(Hunt). Defining "affine" coordinates with plane at infinity $v = x_0 + x_1 + x_2 + 2x_3$ and

$$x = \frac{x_0}{v}$$
$$y = \frac{x_1}{v}$$
$$z = \frac{x_2}{v}$$

then gives the equation

$$-5(x^2y+x^2z+y^2x+y^2z+z^2y+z^2x)+2(xy+xz+yz) = 0$$

plotted in the left figure above (Hunt). The slightly different form

$$4(x^3 + y^3 + z^3 + w^3) - (x + y + z + w)^3 = 0$$

is given by Endraß which, when rewritten in TETRAHEDRAL COORDINATES, becomes

$$x^2 + y^2 - x^2z + y^2z + z^2 - 1 = 0,$$

plotted in the right figure above.

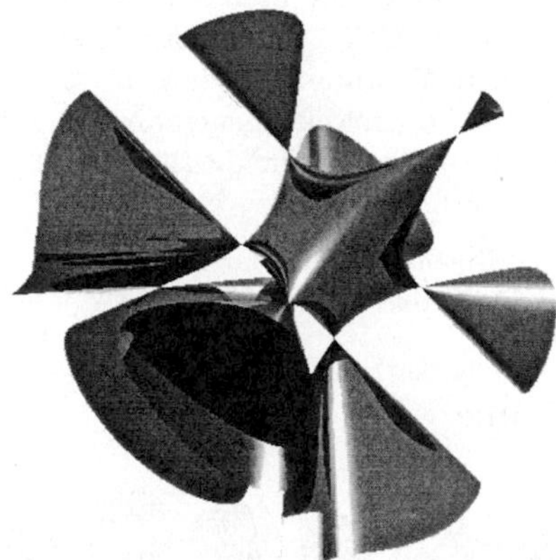

The Hessian of the Cayley cubic is given by

$$0 = x_0{}^2(x_1x_2 + x_1x_3 + x_2x_3) + x_1^2(x_0x_2 + x_0x_3 + x_2x_3) + x_2^2(x_0x_1 + x_0x_3 + x_1x_3) + x_3^2(x_0x_1 + x_0x_2 + x_1x_2).$$

in homogeneous coordinates x_0, x_1, x_2, and x_3. Taking the plane at infinity as $v = 5(x_0 + x_1 + x_2 + 2x_3)/2$ and setting x, y, and z as above gives the equation

$$25[x^3(y+z)+y^3(x+z)+z^3(x+y)]+50(x^2y^2+x^2z^2+y^2z^2) -125(x^2yz+y^2xz+z^2xy)+60xyz-4(xy+xz+yz) = 0,$$

plotted above (Hunt). The Hessian of the Cayley cubic has 14 ORDINARY DOUBLE POINTS, four more than a the general Hessian of a smooth CUBIC SURFACE (Hunt).

References

Endraß, S. "Flächen mit vielen Doppelpunkten." *DMV-Mitteilungen* **4**, 17–20, Apr. 1995.

Endraß, S. "The Cayley Cubic." `http://www.mathematik.uni-mainz.de/AlgebraischeGeometrie/docs/Ecayley.shtml`.

Fischer, G. (Ed.). *Mathematical Models from the Collections of Universities and Museums.* Braunschweig, Germany: Vieweg, p. 14, 1986.

Fischer, G. (Ed.). Plate 33 in *Mathematische Modelle/Mathematical Models, Bildband/Photograph Volume.* Braunschweig, Germany: Vieweg, p. 33, 1986.

Hunt, B. "Algebraic Surfaces." `http://www.mathematik.uni-kl.de/~wwwagag/Galerie.html`.

Hunt, B. *The Geometry of Some Special Arithmetic Quotients.* New York: Springer-Verlag, pp. 115–122, 1996.

Nordstrand, T. "The Cayley Cubic." `http://www.uib.no/people/nfytn/cleytxt.htm`.

Cayley Graph

The representation of a GROUP as a network of directed segments, where the vertices correspond to elements and the segments to multiplication by group generators and their inverses.

see also CAYLEY TREE

References

Grossman, I. and Magnus, W. *Groups and Their Graphs.* New York: Random House, p. 45, 1964.

Cayley's Group Theorem

Every FINITE GROUP of order n can be represented as a PERMUTATION GROUP on n letters, as first proved by Cayley in 1878 (Rotman 1995).

see also FINITE GROUP, PERMUTATION GROUP

References

Rotman, J. J. *An Introduction to the Theory of Groups, 4th ed.* New York: Springer-Verlag, p. 52, 1995.

Cayley-Hamilton Theorem

Given

$$\begin{vmatrix} a_{11} - x & a_{12} & \cdots & a_{1m} \\ a_{21} & a_{22} - x & \cdots & a_{2m} \\ \vdots & \vdots & \ddots & \vdots \\ a_{m1} & a_{m2} & \cdots & a_{mm} - x \end{vmatrix} = x^m + c_{m-1}x^{m-1} + \ldots + c_0, \quad (1)$$

then

$$\mathsf{A}^m + c_{m-1}\mathsf{A}^{m-1} + \ldots + c_0\mathsf{I} = 0, \quad (2)$$

where I is the IDENTITY MATRIX. Cayley verified this identity for $m = 2$ and 3 and postulated that it was true for all m. For $m = 2$, direct verification gives

$$\begin{vmatrix} a - x & b \\ c & d - x \end{vmatrix} = (a - x)(d - x) - bc = x^2 - (a + d)x + (ad - bc) \equiv x^2 + c_1x + c_2 \quad (3)$$

$$A = \begin{bmatrix} a & b \\ c & d \end{bmatrix} \quad (4)$$

$$A^2 = \begin{bmatrix} a & b \\ c & d \end{bmatrix}\begin{bmatrix} a & b \\ c & d \end{bmatrix}$$

$$= \begin{bmatrix} a^2+bc & ab+bd \\ ac+cd & bc+d^2 \end{bmatrix} \quad (5)$$

$$-(a+d)A = \begin{bmatrix} -a^2-ad & -ab-bd \\ -ac-dc & -ad-d^2 \end{bmatrix} \quad (6)$$

$$(ad-bc)I = \begin{bmatrix} ad-bc & 0 \\ 0 & ad-bc \end{bmatrix}, \quad (7)$$

so

$$A^2-(a+d)A+(ad-bc)I = \begin{bmatrix} 0 & 0 \\ 0 & 0 \end{bmatrix}. \quad (8)$$

The Cayley-Hamilton theorem states that a $n \times n$ MATRIX A is annihilated by its CHARACTERISTIC POLYNOMIAL $\det(xI - A)$, which is monic of degree n.

References

Gradshteyn, I. S. and Ryzhik, I. M. *Tables of Integrals, Series, and Products, 5th ed.* San Diego, CA: Academic Press, p. 1117, 1979.

Segercrantz, J. "Improving the Cayley-Hamilton Equation for Low-Rank Transformations." *Amer. Math. Monthly* **99**, 42–44, 1992.

Cayley's Hypergeometric Function Theorem

If

$$(1-z)^{a+b-c}\,{}_2F_1(2a,2b;2c;z) = \sum_{n=0}^{\infty} a_n z^n,$$

then

$${}_2F_1(a,b;c+\tfrac{1}{2};z)\,{}_2F_1(c-a,c-b;c\tfrac{1}{2};z) = \sum_{n=0}^{\infty} \frac{(c)_n}{(c+\frac{1}{2})} a_n z^n,$$

where ${}_2F_1(a,b;c;z)$ is a HYPERGEOMETRIC FUNCTION.

see also HYPERGEOMETRIC FUNCTION

Cayley-Klein Parameters

The parameters α, β, γ, and δ which, like the three EULER ANGLES, provide a way to uniquely characterize the orientation of a solid body. These parameters satisfy the identities

$$\alpha\alpha^* + \gamma\gamma^* = 1 \quad (1)$$

$$\alpha\alpha^* + \beta\beta^* = 1 \quad (2)$$

$$\beta\beta^* + \delta\delta^* = 1 \quad (3)$$

$$\alpha^*\beta + \gamma^*\delta = 0 \quad (4)$$

$$\alpha\delta - \beta\gamma = 1 \quad (5)$$

and

$$\beta = -\gamma^* \quad (6)$$

$$\delta = \alpha^*, \quad (7)$$

where z^* denotes the COMPLEX CONJUGATE. In terms of the EULER ANGLES θ, ϕ, and ψ, the Cayley-Klein parameters are given by

$$\alpha = e^{i(\psi+\phi)/2}\cos(\tfrac{1}{2}\theta) \quad (8)$$

$$\beta = ie^{i(\psi-\phi)/2}\sin(\tfrac{1}{2}\theta) \quad (9)$$

$$\gamma = ie^{-i(\psi-\phi)/2}\sin(\tfrac{1}{2}\theta) \quad (10)$$

$$\delta = e^{-(\psi+\phi)/2}\cos(\tfrac{1}{2}\theta) \quad (11)$$

(Goldstein 1960, p. 155).

The transformation matrix is given in terms of the Cayley-Klein parameters by

$$A = \begin{bmatrix} \frac{1}{2}(\alpha^2-\gamma^2+\delta^2-\beta^2) & \frac{1}{2}i(\gamma^2-\alpha^2+\delta^2-\beta^2) & \gamma\delta-\alpha\beta \\ \frac{1}{2}i(\alpha^2+\gamma^2-\beta^2-\delta^2) & \frac{1}{2}(\alpha^2+\gamma^2+\beta^2+\delta^2) & -i(\alpha\beta+\gamma\delta) \\ \beta\delta-\alpha\gamma & i(\alpha\gamma+\beta\delta) & \alpha\delta+\beta\gamma \end{bmatrix} \quad (12)$$

(Goldstein 1960, p. 153).

The Cayley-Klein parameters may be viewed as parameters of a matrix (denoted Q for its close relationship with QUATERNIONS)

$$Q = \begin{bmatrix} \alpha & \beta \\ \gamma & \delta \end{bmatrix} \quad (13)$$

which characterizes the transformations

$$u' = \alpha u + \beta v \quad (14)$$

$$v' = \gamma u + \delta v. \quad (15)$$

of a linear space having complex axes. This matrix satisfies

$$Q^\dagger Q = QQ^\dagger = I, \quad (16)$$

where I is the IDENTITY MATRIX and $A^\dagger$ the MATRIX TRANSPOSE, as well as

$$|Q|^*|Q| = 1. \quad (17)$$

In terms of the EULER PARAMETERS e_i and the PAULI MATRICES σ_i, the Q-matrix can be written as

$$Q = e_0 I + i(e_1\sigma_1 + e_2\sigma_2 + e_3\sigma_3) \quad (18)$$

(Goldstein 1980, p. 156).

see also EULER ANGLES, EULER PARAMETERS, PAULI MATRICES, QUATERNION

References

Goldstein, H. "The Cayley-Klein Parameters and Related Quantities." §4-5 in *Classical Mechanics, 2nd ed.* Reading, MA: Addison-Wesley, pp. 148–158, 1980.

Cayley-Klein-Hilbert Metric

The METRIC of Felix Klein's model for HYPERBOLIC GEOMETRY,

$$g_{11} = \frac{a^2(1-x_2{}^2)}{(1-x_1{}^2-x_2{}^2)^2}$$
$$g_{12} = \frac{a^2 x_1 x_2}{(1-x_1{}^2-x_2{}^2)^2}$$
$$g_{22} = \frac{a^2(1-x_1{}^2)}{(1-x_1{}^2-x_2{}^2)^2}.$$

see also HYPERBOLIC GEOMETRY

Cayley Number

There are two completely different definitions of Cayley numbers. The first type Cayley numbers is one of the eight elements in a CAYLEY ALGEBRA, also known as an OCTONION. A typical Cayley number is of the form

$$a + bi_0 + ci_1 + di_2 + ei_3 + fi_4 + gi_5 + hi_6,$$

where each of the triples (i_0, i_1, i_3), (i_1, i_2, i_4), (i_2, i_3, i_5), (i_3, i_4, i_6), (i_4, i_5, i_0), (i_5, i_6, i_1), (i_6, i_0, i_2) behaves like the QUATERNIONS (i, j, k). Cayley numbers are *not* ASSOCIATIVE. They have been used in the study of 7- and 8-D space, and a general rotation in 8-D space can be written

$$x' \to ((((((xc_1)c_2)c_3)c_4)c_5)c_6)c_7.$$

The second type of Cayley number is a quantity which describes a DEL PEZZO SURFACE.

see also COMPLEX NUMBER, DEL PEZZO SURFACE, QUATERNION, REAL NUMBER

References

Conway, J. H. and Guy, R. K. "Cayley Numbers." In *The Book of Numbers.* New York: Springer-Verlag, pp. 234–235, 1996.

Okubo, S. *Introduction to Octonion and Other Non-Associative Algebras in Physics.* New York: Cambridge University Press, 1995.

Cayley's Ruled Surface

see CAYLEY CUBIC

Cayley's Sextic

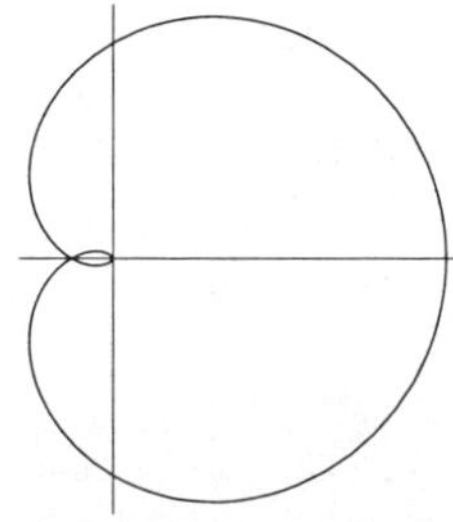

A plane curve discovered by Maclaurin but first studied in detail by Cayley. The name Cayley's sextic is due to R. C. Archibald, who attempted to classify curves in a paper published in Strasbourg in 1900 (MacTutor Archive). Cayley's sextic is given in POLAR COORDINATES by

$$r = a\cos^3(\tfrac{1}{3}\theta), \tag{1}$$

or

$$r = 4b\cos^3(\tfrac{1}{3}\theta), \tag{2}$$

where $b \equiv a/4$. In the latter case, the CARTESIAN equation is

$$4(x^2+y^2-bx)^3 = 27a^2(x^2+y^2)^2. \tag{3}$$

The parametric equations are

$$x(t) = 4a\cos^4(\tfrac{1}{2}t)(2\cos t - 1) \tag{4}$$
$$y(t) = 4a\cos^3(\tfrac{1}{2}t)\sin(\tfrac{3}{2}t). \tag{5}$$

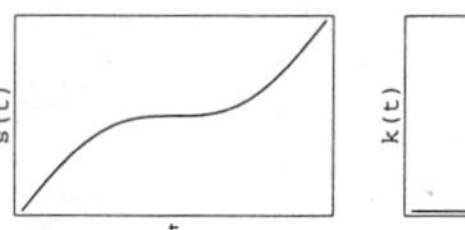

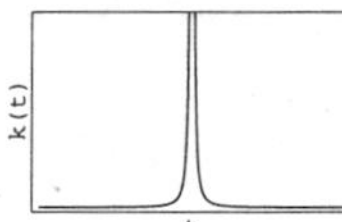

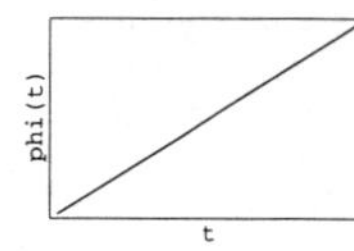

The ARC LENGTH, CURVATURE, and TANGENTIAL ANGLE are

$$s(t) = 3(t + \sin t), \tag{6}$$
$$\kappa(t) = \tfrac{1}{3}\sec^2(\tfrac{1}{2}t), \tag{7}$$
$$\phi(t) = 2t. \tag{8}$$

References

Lawrence, J. D. *A Catalog of Special Plane Curves.* New York: Dover, pp. 178 and 180, 1972.

MacTutor History of Mathematics Archive. "Cayley's Sextic." `http://www-groups.dcs.st-and.ac.uk/~history/Curves/Cayleys.html`.

Cayley's Sextic Evolute

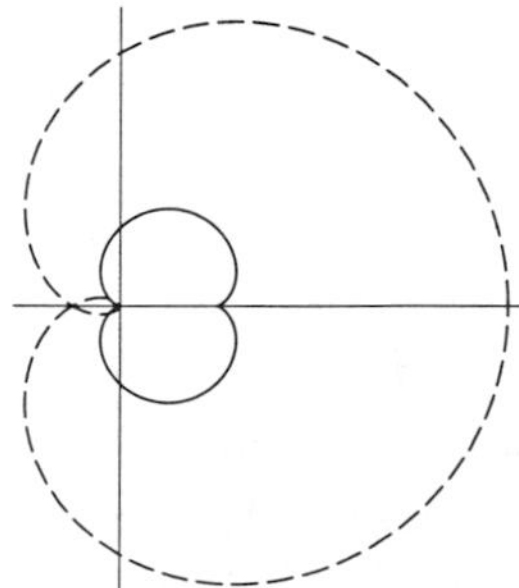

The EVOLUTE of Cayley's sextic is

$$x = \tfrac{1}{8}a + \tfrac{1}{16}a[3\cos(\tfrac{2}{3}t) - \cos(2t)]$$
$$y = \tfrac{1}{16}a[3\sin(\tfrac{2}{3}t) - \sin(2t)],$$

which is a NEPHROID.

Cayley Tree

A TREE in which each NODE has a constant number of branches. The PERCOLATION THRESHOLD for a Cayley tree having z branches is

$$p_c = \frac{1}{z-1}.$$

see also CAYLEY GRAPH

Cayleyian Curve

The ENVELOPE of the lines connecting corresponding points on the JACOBIAN CURVE and STEINERIAN CURVE. The Cayleyian curve of a net of curves of order n has the same GENUS (CURVE) as the JACOBIAN CURVE and STEINERIAN CURVE and, in general, the class $3n(n-1)$.

References

Coolidge, J. L. *A Treatise on Algebraic Plane Curves.* New York: Dover, p. 150, 1959.

Čech Cohomology

The direct limit of the COHOMOLOGY groups with COEFFICIENTS in an ABELIAN GROUP of certain coverings of a TOPOLOGICAL SPACE.

Ceiling Function

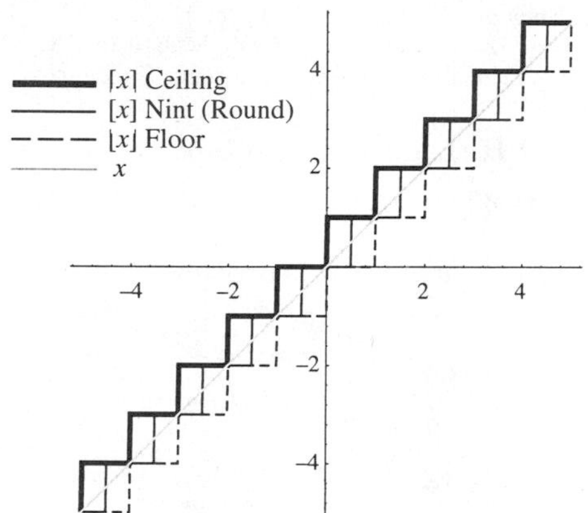

The function $\lceil x \rceil$ which gives the smallest INTEGER $\geq x$, shown as the thick curve in the above plot. Schroeder (1991) calls the ceiling function symbols the "GALLOWS" because of the similarity in appearance to the structure used for hangings. The name and symbol for the ceiling function were coined by K. E. Iverson (Graham *et al.* 1990). It can be implemented as `ceil(x)=-int(-x)`, where `int(x)` is the INTEGER PART of x.

see also FLOOR FUNCTION, INTEGER PART, NINT

References

Graham, R. L.; Knuth, D. E.; and Patashnik, O. "Integer Functions." Ch. 3 in *Concrete Mathematics: A Foundation for Computer Science.* Reading, MA: Addison-Wesley, pp. 67–101, 1990.

Iverson, K. E. *A Programming Language.* New York: Wiley, p. 12, 1962.

Schroeder, M. *Fractals, Chaos, Power Laws: Minutes from an Infinite Paradise.* New York: W. H. Freeman, p. 57, 1991.

Cell

A finite regular POLYTOPE.

see also 16-CELL, 24-CELL, 120-CELL, 600-CELL

Cellular Automaton

A grid (possibly 1-D) of cells which evolves according to a set of rules based on the states of surrounding cells. von Neumann was one of the first people to consider such a model, and incorporated a cellular model into his "universal constructor." von Neumann proved that an automaton consisting of cells with four orthogonal neighbors and 29 possible states would be capable of simulating a TURING MACHINE for some configuration of about 200,000 cells (Gardner 1983, p. 227).

1-D automata are called "elementary" and are represented by a row of pixels with states either 0 or 1. These can be represented with an 8-bit binary number, as shown by Stephen Wolfram. Wolfram further restricted the number from $2^8 = 256$ to 32 by requiring certain symmetry conditions.

The most well-known cellular automaton is Conway's game of LIFE, popularized in Martin Gardner's *Scientific American* columns. Although the computation of successive LIFE generations was originally done by hand, the computer revolution soon arrived and allowed more extensive patterns to be studied and propagated.

see LIFE, LANGTON'S ANT

References

Adami, C. *Artificial Life.* Cambridge, MA: MIT Press, 1998.

Buchi, J. R. and Siefkes, D. (Eds.). *Finite Automata, Their Algebras and Grammars: Towards a Theory of Formal Expressions.* New York: Springer-Verlag, 1989.

Burks, A. W. (Ed.). *Essays on Cellular Automata.* Urbana-Champaign, IL: University of Illinois Press, 1970.

Cipra, B. "Cellular Automata Offer New Outlook on Life, the Universe, and Everything." In *What's Happening in the Mathematical Sciences, 1995–1996, Vol. 3.* Providence, RI: Amer. Math. Soc., pp. 70–81, 1996.

Dewdney, A. K. *The Armchair Universe: An Exploration of Computer Worlds.* New York: W. H. Freeman, 1988.

Gardner, M. "The Game of Life, Parts I–III." Chs. 20–22 in *Wheels, Life, and Other Mathematical Amusements.* New York: W. H. Freeman, pp. 219 and 222, 1983.

Gutowitz, H. (Ed.). *Cellular Automata: Theory and Experiment.* Cambridge, MA: MIT Press, 1991.

Levy, S. *Artificial Life: A Report from the Frontier Where Computers Meet Biology.* New York: Vintage, 1993.

Martin, O.; Odlyzko, A.; and Wolfram, S. "Algebraic Aspects of Cellular Automata." *Communications in Mathematical Physics* **93**, 219–258, 1984.

McIntosh, H. V. "Cellular Automata." `http://www.cs.cinvestav.mx/mcintosh/cellular.html`.

Preston, K. Jr. and Duff, M. J. B. *Modern Cellular Automata: Theory and Applications.* New York: Plenum, 1985.

Sigmund, K. *Games of Life: Explorations in Ecology, Evolution and Behaviour.* New York: Penguin, 1995.

Sloane, N. J. A. Sequences A006977/M2497 in "An On-Line Version of the Encyclopedia of Integer Sequences."

Sloane, N. J. A. and Plouffe, S. Extended entry in *The Encyclopedia of Integer Sequences.* San Diego: Academic Press, 1995.

Toffoli, T. and Margolus, N. *Cellular Automata Machines: A New Environment for Modeling.* Cambridge, MA: MIT Press, 1987.
Wolfram, S. "Statistical Mechanics of Cellular Automata." *Rev. Mod. Phys.* **55**, 601–644, 1983.
Wolfram, S. (Ed.). *Theory and Application of Cellular Automata.* Reading, MA: Addison-Wesley, 1986.
Wolfram, S. *Cellular Automata and Complexity: Collected Papers.* Reading, MA: Addison-Wesley, 1994.
Wuensche, A. and Lesser, M. *The Global Dynamics of Cellular Automata: An Atlas of Basin of Attraction Fields of One-Dimensional Cellular Automata.* Reading, MA: Addison-Wesley, 1992.

Cellular Space

A HAUSDORFF SPACE which has the structure of a so-called CW-COMPLEX.

Center

A special POINT which usually has some symmetric placement with respect to points on a curve or in a SOLID. The center of a CIRCLE is equidistant from all points on the CIRCLE and is the intersection of any two distinct DIAMETERS. The same holds true for the center of a SPHERE.

see also CENTER (GROUP), CENTER OF MASS, CIRCUMCENTER, CURVATURE CENTER, ELLIPSE, EQUIBROCARD CENTER, EXCENTER, HOMOTHETIC CENTER, INCENTER, INVERSION CENTER, ISOGONIC CENTERS, MAJOR TRIANGLE CENTER, NINE-POINT CENTER, ORTHOCENTER, PERSPECTIVE CENTER, POINT, RADICAL CENTER, SIMILITUDE CENTER, SPHERE, SPIEKER CENTER, TAYLOR CENTER, TRIANGLE CENTER, TRIANGLE CENTER FUNCTION, YFF CENTER OF CONGRUENCE

Center Function

see TRIANGLE CENTER FUNCTION

Center of Gravity

see CENTER OF MASS

Center (Group)

The center of a GROUP is the set of elements which commute with every member of the GROUP. It is equal to the intersection of the CENTRALIZERS of the GROUP elements.

see also ISOCLINIC GROUPS, NILPOTENT GROUP

Center of Mass

see CENTROID (GEOMETRIC)

Centered Cube Number

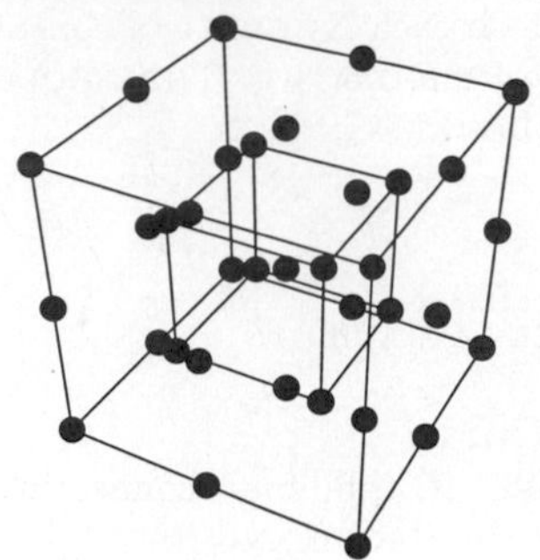

A FIGURATE NUMBER of the form,

$$CCub_n = n^3 + (n-1)^3 = (2n-1)(n^2-n+1).$$

The first few are 1, 9, 35, 91, 189, 341, ... (Sloane's A005898). The GENERATING FUNCTION for the centered cube numbers is

$$\frac{x(x^3+5x^2+5x+1)}{(x-1)^4} = x + 9x^2 + 35x^3 + 91x^4 + \ldots.$$

see also CUBIC NUMBER

References

Conway, J. H. and Guy, R. K. *The Book of Numbers.* New York: Springer-Verlag, p. 51, 1996.
Sloane, N. J. A. Sequence A005898/M4616 in "An On-Line Version of the Encyclopedia of Integer Sequences."

Centered Hexagonal Number

see HEX NUMBER

Centered Pentagonal Number

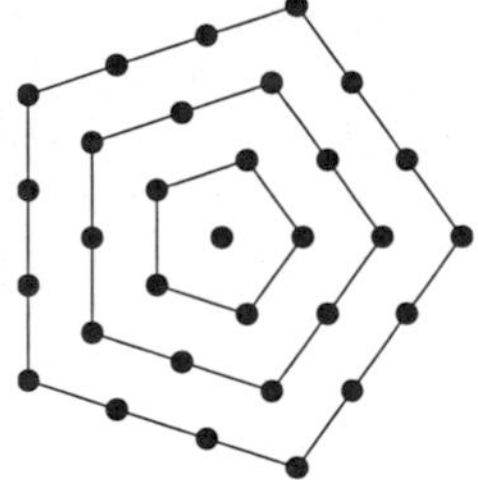

A CENTERED POLYGONAL NUMBER consisting of a central dot with five dots around it, and then additional dots in the gaps between adjacent dots. The general term is $(5n^2-5n+2)/2$, and the first few such numbers are 1, 6, 16, 31, 51, 76, ... (Sloane's A005891). The GENERATING FUNCTION of the centered pentagonal numbers is

$$\frac{x(x^2+3x+1)}{(x-1)^3} = x + 6x^2 + 16x^3 + 31x^4 + \ldots.$$

see also CENTERED SQUARE NUMBER, CENTERED TRIANGULAR NUMBER

References

Sloane, N. J. A. Sequence A005891/M4112 in "An On-Line Version of the Encyclopedia of Integer Sequences."

Centered Polygonal Number

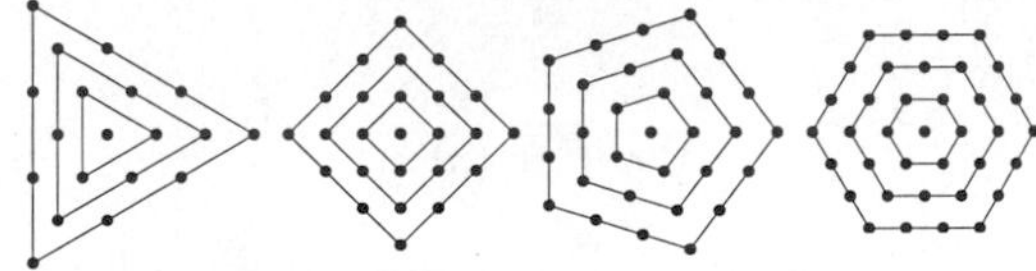

A FIGURATE NUMBER in which layers of POLYGONS are drawn centered about a point instead of with the point at a VERTEX.

see also CENTERED PENTAGONAL NUMBER, CENTERED SQUARE NUMBER, CENTERED TRIANGULAR NUMBER

References

Sloane, N. J. A. and Plouffe, S. Extended entry for sequence M3826 in *The Encyclopedia of Integer Sequences.* San Diego, CA: Academic Press, 1995.

Centered Square Number

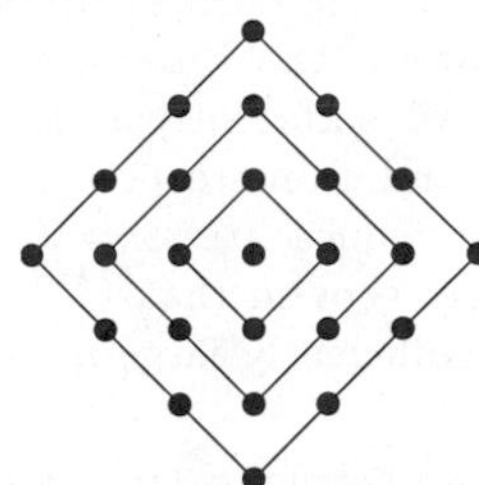

A CENTERED POLYGONAL NUMBER consisting of a central dot with four dots around it, and then additional dots in the gaps between adjacent dots. The general term is $n^2 + (n-1)^2$, and the first few such numbers are 1, 5, 13, 25, 41, ... (Sloane's A001844). Centered square numbers are the sum of two consecutive SQUARE NUMBERS and are congruent to 1 (mod 4). The GENERATING FUNCTION giving the centered square numbers is

$$\frac{x(x+1)^2}{(1-x)^3} = x + 5x^2 + 13x^3 + 25x^4 + \ldots.$$

see also CENTERED PENTAGONAL NUMBER, CENTERED POLYGONAL NUMBER, CENTERED TRIANGULAR NUMBER, SQUARE NUMBER

References

Conway, J. H. and Guy, R. K. *The Book of Numbers.* New York: Springer-Verlag, p. 41, 1996.
Sloane, N. J. A. Sequence A001844/M3826 in "An On-Line Version of the Encyclopedia of Integer Sequences."

Centered Triangular Number

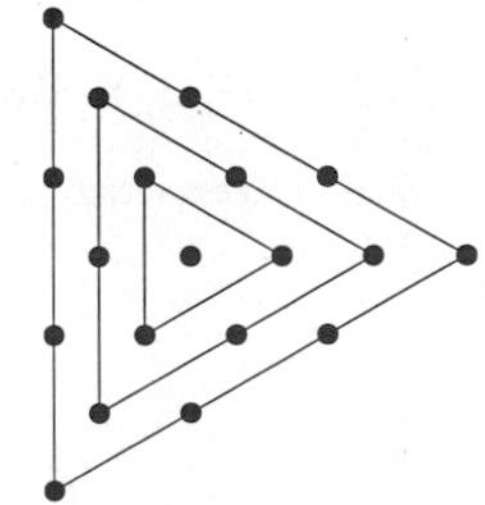

A CENTERED POLYGONAL NUMBER consisting of a central dot with three dots around it, and then additional dots in the gaps between adjacent dots. The general term is $(3n^2 - 3n + 2)/2$, and the first few such numbers are 1, 4, 10, 19, 31, 46, 64, ... (Sloane's A005448). The GENERATING FUNCTION giving the centered triangular numbers is

$$\frac{x(x^2+x+1)}{(1-x)^3} = x + 4x^2 + 10x^3 + 19x^4 + \ldots.$$

see also CENTERED PENTAGONAL NUMBER, CENTERED SQUARE NUMBER

References

Sloane, N. J. A. Sequence A005448/M3378 in "An On-Line Version of the Encyclopedia of Integer Sequences."

Centillion

In the American system, 10^{303}.

see also LARGE NUMBER

Central Angle

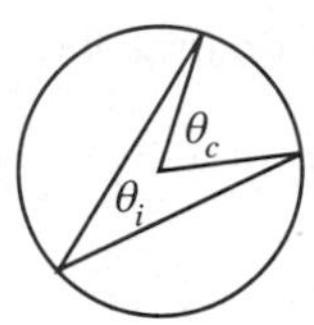

An ANGLE having its VERTEX at a CIRCLE's center which is formed by two points on the CIRCLE'S CIRCUMFERENCE. For angles with the same endpoints,

$$\theta_c = 2\theta_i,$$

where θ_i is the INSCRIBED ANGLE.

References

Pedoe, D. *Circles: A Mathematical View, rev. ed.* Washington, DC: Math. Assoc. Amer., pp. xxi–xxii, 1995.

Central Beta Function

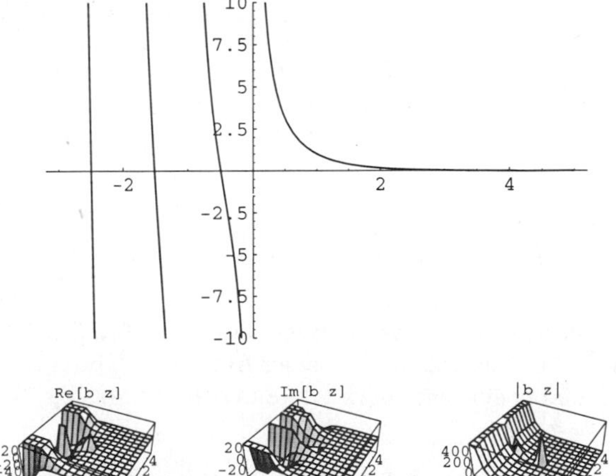

The central beta function is defined by

$$\beta(p) \equiv B(p,p), \tag{1}$$

where $B(p,q)$ is the BETA FUNCTION. It satisfies the identities

$$\beta(p) = 2^{1-2p}B(p,\tfrac{1}{2}) \tag{2}$$
$$= 2^{1-2p}\cos(\pi p)B(\tfrac{1}{2}-p,p) \tag{3}$$
$$= \int_0^1 \frac{t^p\,dt}{(1+t)^{2p}} \tag{4}$$
$$= \frac{2}{p}\prod_{n=1}^{\infty}\frac{n(n+2p)}{(n+p)(n+p)}. \tag{5}$$

With $p = 1/2$, the latter gives the WALLIS FORMULA. When $p = a/b$,

$$b\beta(a/b) = 2^{1-2a/b}J(a,b), \tag{6}$$

where

$$J(a,b) \equiv \int_0^1 \frac{t^{a-1}\,dt}{\sqrt{1-t^b}}. \tag{7}$$

The central beta function satisfies

$$(2+4x)\beta(1+x) = x\beta(x) \tag{8}$$

$$(1-2x)\beta(1-x)\beta(x) = 2\pi\cot(\pi x) \tag{9}$$

$$\beta(\tfrac{1}{2}-x) = 2^{4x-1}\tan(\pi x)\beta(x) \tag{10}$$

$$\beta(x)\beta(x+\tfrac{1}{2}) = 2^{4x+1}\pi\beta(2x)\beta(2x+\tfrac{1}{2}). \tag{11}$$

For p an ODD POSITIVE INTEGER, the central beta function satisfies the identity

$$\beta(px) = \frac{1}{\sqrt{p}}\prod_{k=1}^{(p-1)/2}\frac{2x+\frac{2k-1}{p}}{2\pi}\prod_{k=0}^{p-1}\beta\left(x+\frac{k}{p}\right). \tag{12}$$

see also BETA FUNCTION, REGULARIZED BETA FUNCTION

References

Borwein, J. M. and Zucker, I. J. "Elliptic Integral Evaluation of the Gamma Function at Rational Values of Small Denominators." *IMA J. Numerical Analysis* **12**, 519–526, 1992.

Central Binomial Coefficient

The nth central binomial coefficient is defined as $\binom{n}{\lfloor n/2\rfloor}$, where $\binom{n}{k}$ is a BINOMIAL COEFFICIENT and $\lfloor n\rfloor$ is the FLOOR FUNCTION. The first few values are 1, 2, 3, 6, 10, 20, 35, 70, 126, 252, ... (Sloane's A001405). The central binomial coefficients have GENERATING FUNCTION

$$\frac{1-4x^2-\sqrt{1-4x^2}}{2(2x^3-x^2)} = 1+2x+3x^2+6x^3+10x^4+\ldots.$$

The central binomial coefficients are SQUAREFREE only for $n = 1, 2, 3, 4, 5, 7, 8, 11, 17, 19, 23, 71, \ldots$ (Sloane's A046098), with no others less than 1500.

The above coefficients are a superset of the alternative "central" binomial coefficients

$$\binom{2n}{n} = \frac{(2n)!}{(n!)^2},$$

which have GENERATING FUNCTION

$$\frac{1}{\sqrt{1-4x}} = 1+2x+6x^2+20x^3+70x^4+\ldots.$$

The first few values are 2, 6, 20, 70, 252, 924, 3432, 12870, 48620, 184756, ... (Sloane's A000984).

Erdős and Graham (1980, p. 71) conjectured that the central binomial coefficient $\binom{2n}{n}$ is *never* SQUAREFREE for $n > 4$, and this is sometimes known as the ERDŐS SQUAREFREE CONJECTURE. SÁRKÖZY'S THEOREM (Sárközy 1985) provides a partial solution which states that the BINOMIAL COEFFICIENT $\binom{2n}{n}$ is never SQUAREFREE for all sufficiently large $n \geq n_0$ (Vardi 1991). Granville and Ramare (1996) proved that the *only* SQUAREFREE values are $n = 2$ and 4. Sander (1992) subsequently showed that $\binom{2n\pm d}{n}$ are also never SQUAREFREE for sufficiently large n as long as d is not "too big."

see also BINOMIAL COEFFICIENT, CENTRAL TRINOMIAL COEFFICIENT, ERDŐS SQUAREFREE CONJECTURE, SÁRKÖZY'S THEOREM, QUOTA SYSTEM

References

Granville, A. and Ramare, O. "Explicit Bounds on Exponential Sums and the Scarcity of Squarefree Binomial Coefficients." *Mathematika* **43**, 73–107, 1996.

Sander, J. W. "On Prime Divisors of Binomial Coefficients." *Bull. London Math. Soc.* **24**, 140–142, 1992.

Sárközy, A. "On Divisors of Binomial Coefficients. I." *J. Number Th.* **20**, 70–80, 1985.

Sloane, N. J. A. Sequences A046098, A000984/M1645, and A001405/M0769 in "An On-Line Version of the Encyclopedia of Integer Sequences."

Vardi, I. "Application to Binomial Coefficients," "Binomial Coefficients," "A Class of Solutions," "Computing Binomial Coefficients," and "Binomials Modulo and Integer." §2.2, 4.1, 4.2, 4.3, and 4.4 in *Computational Recreations in Mathematica.* Redwood City, CA: Addison-Wesley, pp. 25–28 and 63–71, 1991.

Central Conic

An ELLIPSE or HYPERBOLA.

see also CONIC SECTION

References

Coxeter, H. S. M. and Greitzer, S. L. *Geometry Revisited.* Washington, DC: Math. Assoc. Amer., pp. 146–150, 1967.

Ogilvy, C. S. *Excursions in Geometry.* New York: Dover, p. 77, 1990.

Central Difference

The central difference for a function tabulated at equal intervals f_i is defined by

$$\delta(f_{n+1/2}) = \delta_{n+1/2} = \delta^1_{n+1/2} \equiv f_{n+1} - f_n. \quad (1)$$

Higher order differences may be computed for EVEN and ODD powers,

$$\delta^{2k}_{n+1/2} = \sum_{j=0}^{2k} (-1)^j \binom{2k}{j} f_{n+k-j} \quad (2)$$

$$\delta^{2k+1}_{n+1/2} = \sum_{j=0}^{2k+1} (-1)^j \binom{2k+1}{j} f_{n+k+1-j}. \quad (3)$$

see also BACKWARD DIFFERENCE, DIVIDED DIFFERENCE, FORWARD DIFFERENCE

References

Abramowitz, M. and Stegun, C. A. (Eds.). "Differences." §25.1 in *Handbook of Mathematical Functions with Formulas, Graphs, and Mathematical Tables, 9th printing.* New York: Dover, pp. 877–878, 1972.

Central Limit Theorem

Let $x_1, x_2, \ldots, x_N$ be a set of N INDEPENDENT random variates and each x_i have an *arbitrary* probability distribution $P(x_1, \ldots, x_N)$ with MEAN μ_i and a finite VARIANCE $\sigma_i{}^2$. Then the normal form variate

$$X_{\text{norm}} \equiv \frac{\sum_{i=1}^N x_i - \sum_{i=1}^N \mu_i}{\sqrt{\sum_{i=1}^N \sigma_i{}^2}} \quad (1)$$

has a limiting distribution which is NORMAL (GAUSSIAN) with MEAN $\mu = 0$ and VARIANCE $\sigma^2 = 1$. If conversion to normal form is not performed, then the variate

$$X \equiv \frac{1}{N}\sum_{i=1}^N x_i \quad (2)$$

is NORMALLY DISTRIBUTED with $\mu_X = \mu_x$ and $\sigma_X = \sigma_x/\sqrt{N}$. To prove this, consider the INVERSE FOURIER TRANSFORM of $P_X(f)$.

$$\begin{aligned}\mathcal{F}^{-1}[P_X(f)] &\equiv \int_{-\infty}^{\infty} e^{2\pi i f X} p(X)\,dX \\ &= \int_{-\infty}^{\infty} \sum_{n=0}^{\infty} \frac{(2\pi i f X)^n}{n!} p(X)\,dX \\ &= \sum_{n=0}^{\infty} \frac{(2\pi i f)^n}{n!} \int_{-\infty}^{\infty} X^n p(X)\,dx \\ &= \sum_{n=0}^{\infty} \frac{(2\pi i f)^n}{n!} \langle x \rangle^n. \end{aligned} \quad (3)$$

Now write

$$\begin{aligned}\langle X^n \rangle &= \langle N^{-n}(x_1 + x_2 + \ldots + x_N)^n \rangle \\ &= \int_{-\infty}^{\infty} N^{-n}(x_1 + \ldots + x_N)^n p(x_1)\cdots p(x_N)\,dx_1 \cdots dx_N, \end{aligned} \quad (4)$$

so we have

$$\begin{aligned}\mathcal{F}^{-1}[P_X(f)] &= \sum_{n=0}^{\infty} \frac{(2\pi i f)^n}{n!} \langle X^n \rangle \\ &= \sum_{n=0}^{\infty} \frac{(2\pi i f)^n}{n!} \int_{-\infty}^{\infty} N^{-n}(x_1 + \ldots + x_N)^n \\ &\qquad \times p(x_1)\cdots p(x_N)\,dx_1 \cdots dx_N \\ &= \int_{-\infty}^{\infty} \sum_{n=0}^{\infty} \left[\frac{2\pi i f(x_1 + \ldots + x_N)}{N}\right]^n \frac{1}{n!} \\ &\qquad \times p(x_1)\cdots p(x_N)\,dx_1 \cdots dx_N \\ &= \int_{-\infty}^{\infty} e^{2\pi i f(x_1 + \ldots + x_N)/N} p(x_1)\cdots p(x_N)\,dx_1 \cdots dx_N \\ &= \left[\int_{-\infty}^{\infty} e^{2\pi i f x_1/N} p(x_1)\,dx_1\right] \\ &\qquad \times \cdots \times \left[\int_{-\infty}^{\infty} e^{2\pi i f x_N/N} p(x_N)\,dx_N\right] \\ &= \left[\int_{-\infty}^{\infty} e^{2\pi i f x/N} p(x)\,dx\right]^N \\ &= \left\{\int_{-\infty}^{\infty} \left[1 + \left(\frac{2\pi i f}{N}\right) x + \frac{1}{2}\left(\frac{2\pi i f}{N}\right)^2 x^2 + \ldots\right] p(x)\,dx\right\}^N \\ &= \left[\int_{-\infty}^{\infty} p(x)\,dx + \frac{2\pi i f}{N}\int_{-\infty}^{\infty} x p(x)\,dx \right. \\ &\qquad \left. - \frac{(2\pi f)^2}{2N^2}\int_{-\infty}^{\infty} x^2 p(x)\,dx + \mathcal{O}(N^{-3})\right]^N \\ &= \left[1 + \frac{2\pi i f}{N}\langle x \rangle - \frac{(2\pi f)^2}{2N^2}\langle x^2 \rangle + \mathcal{O}(N^{-3})\right]^N \\ &= \exp\left\{N \ln\left[1 + \frac{2\pi i f}{N}\langle x \rangle - \frac{(2\pi f)^2}{2N^2}\langle x^2 \rangle + \mathcal{O}(N^{-3})\right]\right\}. \end{aligned} \quad (5)$$

Now expand

$$\ln(1+x) = x - \tfrac{1}{2}x^2 + \tfrac{1}{3}x^3 + \ldots, \quad (6)$$

so

$$\begin{aligned}&\mathcal{F}^{-1}[P_X(f)] \\ &\quad \approx \exp\left\{N\left[\frac{2\pi i f}{N}\langle x \rangle - \frac{(2\pi f)^2}{2N^2}\langle x^2 \rangle \right.\right. \\ &\qquad \left.\left. + \frac{1}{2}\frac{(2\pi i f)^2}{N^2}\langle x \rangle^2 + \mathcal{O}(N^{-3})\right]\right\} \\ &\quad = \exp\left[2\pi i f\langle x \rangle - \frac{(2\pi f)^2(\langle x^2 \rangle - \langle x \rangle^2)}{2N} + \mathcal{O}(N^{-2})\right] \\ &\quad \approx \exp\left[2\pi i f \mu_x - \frac{(2\pi f)^2 \sigma_x{}^2}{2N}\right], \end{aligned} \quad (7)$$

since

$$\mu_x \equiv \langle x \rangle \quad (8)$$

$$\sigma_x^{\ 2} \equiv \langle x^2 \rangle - \langle x \rangle^2. \quad (9)$$

Taking the FOURIER TRANSFORM,

$$\begin{aligned} P_X &\equiv \int_{-\infty}^{\infty} e^{-2\pi i f x} \mathcal{F}^{-1}[P_X(f)]\,df \\ &= \int_{-\infty}^{\infty} e^{2\pi i f(\mu_x - x) - (2\pi f)^2 \sigma_x^{\ 2}/2N}\,df. \quad (10) \end{aligned}$$

This is of the form

$$\int_{-\infty}^{\infty} e^{iaf - bf^2}\,df, \quad (11)$$

where $a \equiv 2\pi(\mu_x - x)$ and $b \equiv (2\pi\sigma_x)^2/2N$. But, from Abramowitz and Stegun (1972, p. 302, equation 7.4.6),

$$\int_{-\infty}^{\infty} e^{iaf - bf^2}\,df = e^{-a^2/4b}\sqrt{\frac{\pi}{b}}. \quad (12)$$

Therefore,

$$\begin{aligned} P_X &= \sqrt{\frac{\pi}{\frac{(2\pi\sigma_x)^2}{2N}}} \exp\left\{\frac{-[2\pi(\mu_x - x)]^2}{4\frac{(2\pi\sigma_x)^2}{2N}}\right\} \\ &= \sqrt{\frac{2\pi N}{4\pi^2\sigma_x^{\ 2}}} \exp\left[-\frac{4\pi^2(\mu_x - x)^2 2N}{4\cdot 4\pi^2\sigma_x^{\ 2}}\right] \\ &= \frac{\sqrt{N}}{\sigma_x\sqrt{2\pi}} e^{-(\mu_x - x)^2 N/2\sigma_x^{\ 2}}. \quad (13) \end{aligned}$$

But $\sigma_X = \sigma_x/\sqrt{N}$ and $\mu_X = \mu_x$, so

$$P_X = \frac{1}{\sigma_X\sqrt{2\pi}} e^{-(\mu_X - x)^2/2\sigma_X^{\ 2}}. \quad (14)$$

The "fuzzy" central limit theorem says that data which are influenced by many small and unrelated random effects are approximately NORMALLY DISTRIBUTED.

see also LINDEBERG CONDITION, LINDEBERG-FELLER CENTRAL LIMIT THEOREM, LYAPUNOV CONDITION

References

Abramowitz, M. and Stegun, C. A. (Eds.). *Handbook of Mathematical Functions with Formulas, Graphs, and Mathematical Tables, 9th printing.* New York: Dover, 1972.

Spiegel, M. R. *Theory and Problems of Probability and Statistics.* New York: McGraw-Hill, pp. 112–113, 1992.

Zabell, S. L. "Alan Turing and the Central Limit Theorem." *Amer. Math. Monthly* **102**, 483–494, 1995.

Central Trinomial Coefficient

The nth central binomial coefficient is defined as the coefficient of x^n in the expansion of $(1+x+x^2)^n$. The first few are 1, 3, 7, 19, 51, 141, 393, ... (Sloane's A002426). This sequence cannot be expressed as a fixed number of hypergeometric terms (Petkovšek *et al.* 1996, p. 160). The GENERATING FUNCTION is given by

$$f(x) = \frac{1}{\sqrt{(1+x)(1-3x)}} = 1 + x + 3x^2 + 7x^3 + \ldots.$$

see also CENTRAL BINOMIAL COEFFICIENT

References

Petkovšek, M.; Wilf, H. S.; and Zeilberger, D. *A=B.* Wellesley, MA: A. K. Peters, 1996.

Sloane, N. J. A. Sequence A002426/M2673 in "An On-Line Version of the Encyclopedia of Integer Sequences."

Centralizer

The centralizer of a FINITE non-ABELIAN SIMPLE GROUP G is an element z of order 2 such that

$$C_G(z) = \{g \in G : gz = zg\}.$$

see also CENTER (GROUP), NORMALIZER

Centrode

$$\mathbf{C} \equiv \tau\mathbf{T} + \kappa\mathbf{B},$$

where τ is the TORSION, κ is the CURVATURE, **T** is the TANGENT VECTOR, and **B** is the BINORMAL VECTOR.

Centroid (Function)

By analogy with the GEOMETRIC CENTROID, the centroid of an arbitrary function $f(x)$ is defined as

$$\langle x \rangle = \frac{\int_{-\infty}^{\infty} x f(x)\,dx}{\int_{-\infty}^{\infty} f(x)\,dx}.$$

References

Bracewell, R. *The Fourier Transform and Its Applications.* New York: McGraw-Hill, pp. 139–140 and 156, 1965.

Centroid (Geometric)

The CENTER OF MASS of a 2-D planar LAMINA or a 3-D solid. The mass of a LAMINA with surface density function $\sigma(x, y)$ is

$$M = \iint \sigma(x, y)\,dA. \quad (1)$$

The coordinates of the centroid (also called the CENTER OF GRAVITY) are

$$\bar{x} = \frac{\iint x\sigma(x, y)\,dA}{M} \quad (2)$$

$$\bar{y} = \frac{\iint y\sigma(x,y)\,dA}{M}. \tag{3}$$

The centroids of several common laminas along the non-symmetrical axis are summarized in the following table.

Figure	$\bar{y}$
parabolic segment	$\frac{3}{5}h$
semicircle	$\frac{4r}{3\pi}$

In 3-D, the mass of a solid with density function $\rho(x,y,z)$ is

$$M = \iiint \rho(x,y,z)\,dV, \tag{4}$$

and the coordinates of the center of mass are

$$\bar{x} = \frac{\iiint x\rho(x,y,z)\,dV}{M} \tag{5}$$

$$\bar{y} = \frac{\iiint y\rho(x,y,z)\,dV}{M} \tag{6}$$

$$\bar{z} = \frac{\iiint z\rho(x,y,z)\,dV}{M}. \tag{7}$$

Figure	$\bar{z}$
cone	$\frac{1}{4}h$
conical frustum	$\frac{h(R_1{}^2+2R_1R_2+3R_2{}^2)}{4(R_1{}^2+R_1R_2+R_2{}^2)}$
hemisphere	$\frac{3}{8}R$
paraboloid	$\frac{2}{3}h$
pyramid	$\frac{1}{4}h$

see also PAPPUS'S CENTROID THEOREM

References

Beyer, W. H. *CRC Standard Mathematical Tables, 28th ed.* Boca Raton, FL: CRC Press, p. 132, 1987.

McLean, W. G. and Nelson, E. W. "First Moments and Centroids." Ch. 9 in *Schaum's Outline of Theory and Problems of Engineering Mechanics: Statics and Dynamics, 4th ed.* New York: McGraw-Hill, pp. 134–162, 1988.

Centroid (Orthocentric System)

The centroid of the four points constituting an ORTHOCENTRIC SYSTEM is the center of the common NINE-POINT CIRCLE (Johnson 1929, p. 249). This fact automatically guarantees that the centroid of the INCENTER and EXCENTERS of a TRIANGLE is located at the CIRCUMCENTER.

References

Johnson, R. A. *Modern Geometry: An Elementary Treatise on the Geometry of the Triangle and the Circle.* Boston, MA: Houghton Mifflin, 1929.

Centroid (Triangle)

The centroid (CENTER OF MASS) of the VERTICES of a TRIANGLE is the point M (or G) of intersection of the TRIANGLE'S three MEDIANS, also called the MEDIAN POINT (Johnson 1929, p. 249). The centroid is always in the interior of the TRIANGLE, and has TRILINEAR COORDINATES

$$\frac{1}{a} : \frac{1}{b} : \frac{1}{c}, \tag{1}$$

or

$$\csc A : \csc B : \csc C. \tag{2}$$

If the sides of a TRIANGLE are divided so that

$$\frac{\overline{A_2P_1}}{\overline{P_1A_3}} = \frac{\overline{A_3P_2}}{\overline{P_2A_1}} = \frac{\overline{A_1P_2}}{\overline{P_3A_2}} = \frac{p}{q}, \tag{3}$$

the centroid of the TRIANGLE $\Delta P_1P_2P_3$ is M (Johnson 1929, p. 250).

Pick an interior point X. The TRIANGLES BXC, CXA, and AXB have equal areas IFF X corresponds to the centroid. The centroid is located one third of the way from each VERTEX to the MIDPOINT of the opposite side. Each median divides the triangle into two equal areas; all the medians together divide it into six equal parts, and the lines from the MEDIAN POINT to the VERTICES divide the whole into three equivalent TRIANGLES. In general, for any line in the plane of a TRIANGLE ABC,

$$d = \tfrac{1}{3}(d_A + d_B + d_C), \tag{4}$$

where d, d_A, d_B, and d_C are the distances from the centroid and VERTICES to the line. A TRIANGLE will balance at the centroid, and along any line passing through the centroid. The TRILINEAR POLAR of the centroid is called the LEMOINE AXIS. The PERPENDICULARS from the centroid are proportional to $s_i{}^{-1}$,

$$a_1p_2 = a_2p_2 = a_3p_3 = \tfrac{2}{3}\Delta, \tag{5}$$

where Δ is the AREA of the TRIANGLE. Let P be an arbitrary point, the VERTICES be A_1, A_2, and A_3, and the centroid M. Then

$$\overline{PA_1}^2 + \overline{PA_2}^2 + \overline{PA_3}^2 = \overline{MA_1}^2 + \overline{MA_2}^2 + \overline{MA_3}^2 + 3\overline{PM}^2. \tag{6}$$

If O is the CIRCUMCENTER of the triangle's centroid, then

$$\overline{OM}^2 = R^2 - \tfrac{1}{9}(a^2 + b^2 + c^2). \tag{7}$$

The centroid lies on the EULER LINE.

The centroid of the PERIMETER of a TRIANGLE is the triangle's SPIEKER CENTER (Johnson 1929, p. 249).

see also CIRCUMCENTER, EULER LINE, EXMEDIAN POINT, INCENTER, ORTHOCENTER

References

Carr, G. S. *Formulas and Theorems in Pure Mathematics, 2nd ed.* New York: Chelsea, p. 622, 1970.

Coxeter, H. S. M. and Greitzer, S. L. *Geometry Revisited.* Washington, DC: Math. Assoc. Amer., p. 7, 1967.
Dixon, R. *Mathographics.* New York: Dover, pp. 55–57, 1991.
Johnson, R. A. *Modern Geometry: An Elementary Treatise on the Geometry of the Triangle and the Circle.* Boston, MA: Houghton Mifflin, pp. 173–176 and 249, 1929.
Kimberling, C. "Central Points and Central Lines in the Plane of a Triangle." *Math. Mag.* **67**, 163–187, 1994.
Kimberling, C. "Centroid." `http://www.evansville.edu/~ck6/tcenters/class/centroid.html`.

Certificate of Compositeness

see COMPOSITENESS CERTIFICATE

Certificate of Primality

see PRIMALITY CERTIFICATE

Cesàro Equation

An INTRINSIC EQUATION which expresses a curve in terms of its ARC LENGTH s and RADIUS OF CURVATURE R (or equivalently, the CURVATURE κ).

see also ARC LENGTH, INTRINSIC EQUATION, NATURAL EQUATION, RADIUS OF CURVATURE, WHEWELL EQUATION

References

Yates, R. C. "Intrinsic Equations." *A Handbook on Curves and Their Properties.* Ann Arbor, MI: J. W. Edwards, pp. 123–126, 1952.

Cesàro Fractal

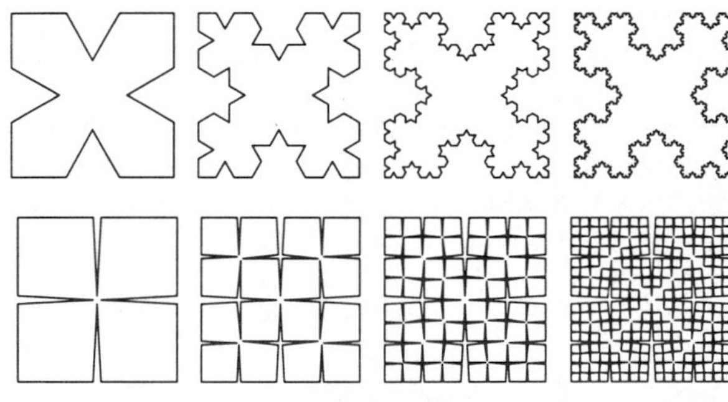

A FRACTAL also known as the TORN SQUARE FRACTAL. The base curves and motifs for the two fractals illustrated above are show below.

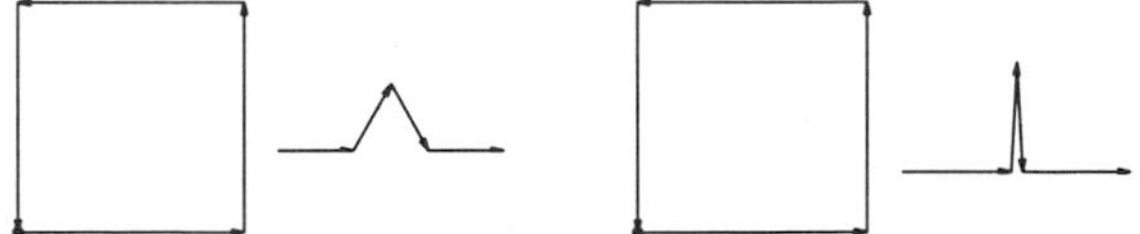

see also FRACTAL, KOCH SNOWFLAKE

References

Lauwerier, H. *Fractals: Endlessly Repeated Geometric Figures.* Princeton, NJ: Princeton University Press, p. 43, 1991.
Pappas, T. *The Joy of Mathematics.* San Carlos, CA: Wide World Publ./Tetra, p. 79, 1989.
Weisstein, E. W. "Fractals." `http://www.astro.virginia.edu/~eww6n/math/notebooks/Fractal.m`.

Cesàro Mean

see FEJES TÓTH'S INTEGRAL

Ceva's Theorem

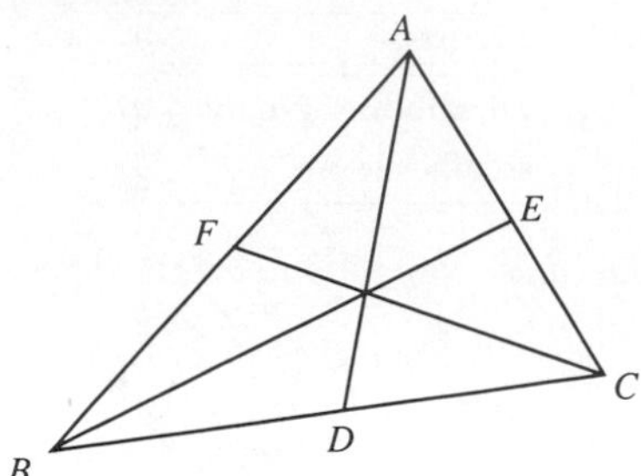

Given a TRIANGLE with VERTICES A, B, and C and points along the sides D, E, and F, a NECESSARY and SUFFICIENT condition for the CEVIANS AD, BE, and CF to be CONCURRENT (intersect in a single point) is that

$$BD \cdot CE \cdot AF = DC \cdot EA \cdot FB. \tag{1}$$

Let $P = [V_1, \ldots, V_n]$ be an arbitrary n-gon, C a given point, and k a POSITIVE INTEGER such that $1 \leq k \leq n/2$. For $i = 1, \ldots, n$, let W_i be the intersection of the lines CV_i and $V_{i-k}V_{i+k}$, then

$$\prod_{i=1}^{n} \left[\frac{V_{i-k}W_i}{W_i V_{i+k}}\right] = 1. \tag{2}$$

Here, $AB||CD$ and

$$\left[\frac{AB}{CD}\right] \tag{3}$$

is the RATIO of the lengths $[A, B]$ and $[C, D]$ with a plus or minus sign depending on whether these segments have the same or opposite directions (Grünbaum and Shepard 1995).

Another form of the theorem is that three CONCURRENT lines from the VERTICES of a TRIANGLE divide the opposite sides in such fashion that the product of three nonadjacent segments equals the product of the other three (Johnson 1929, p. 147).

see also HOEHN'S THEOREM, MENELAUS' THEOREM

References

Beyer, W. H. (Ed.) *CRC Standard Mathematical Tables, 28th ed.* Boca Raton, FL: CRC Press, p. 122, 1987.
Coxeter, H. S. M. and Greitzer, S. L. *Geometry Revisited.* Washington, DC: Math. Assoc. Amer., pp. 4–5, 1967.
Grünbaum, B. and Shepard, G. C. "Ceva, Menelaus, and the Area Principle." *Math. Mag.* **68**, 254–268, 1995.
Johnson, R. A. *Modern Geometry: An Elementary Treatise on the Geometry of the Triangle and the Circle.* Boston, MA: Houghton Mifflin, pp. 145–151, 1929.
Pedoe, D. *Circles: A Mathematical View, rev. ed.* Washington, DC: Math. Assoc. Amer., p. xx, 1995.

Cevian

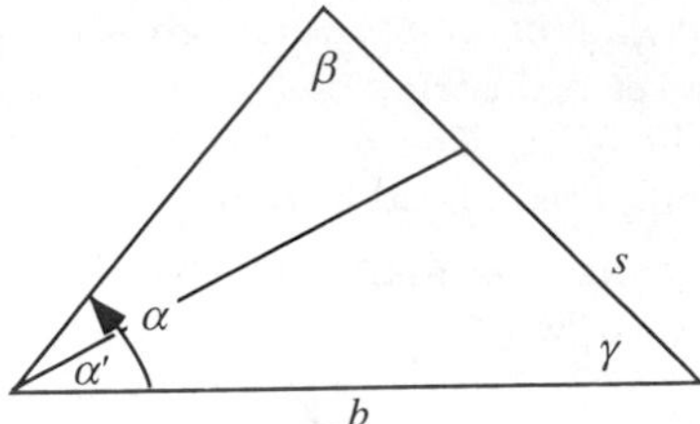

A line segment which joins a VERTEX of a TRIANGLE with a point on the opposite side (or its extension). In the above figure,

$$s = \frac{b \sin \alpha'}{\sin(\gamma + \alpha')}.$$

References

Thébault, V. "On the Cevians of a Triangle." *Amer. Math. Monthly* **60**, 167–173, 1953.

Cevian Conjugate Point

see ISOTOMIC CONJUGATE POINT

Cevian Transform

Vandeghen's (1965) name for the transformation taking points to their ISOTOMIC CONJUGATE POINTS.

see also ISOTOMIC CONJUGATE POINT

References

Vandeghen, A. "Some Remarks on the Isogonal and Cevian Transforms. Alignments of Remarkable Points of a Triangle." *Amer. Math. Monthly* **72**, 1091–1094, 1965.

Cevian Triangle

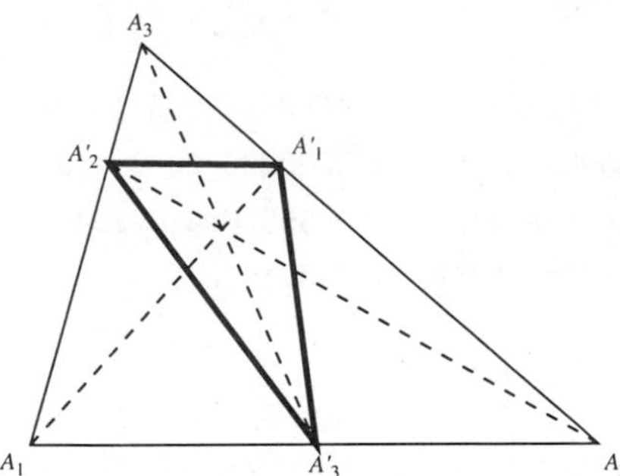

Given a center $\alpha : \beta : \gamma$, the cevian triangle is defined as that with VERTICES $0 : \beta : \gamma$, $\alpha : 0 : \gamma$, and $\alpha : \beta : 0$. If $A'B'C'$ is the CEVIAN TRIANGLE of X and $A''B''C''$ is the ANTICEVIAN TRIANGLE, then X and A'' are HARMONIC CONJUGATE POINTS with respect to A and A'.

see also ANTICEVIAN TRIANGLE

Chain

Let P be a finite PARTIALLY ORDERED SET. A chain in P is a set of pairwise comparable elements (i.e., a TOTALLY ORDERED subset). The WIDTH of P is the maximum CARDINALITY of an ANTICHAIN in P. For a PARTIAL ORDER, the size of the longest CHAIN is called the WIDTH.

see also ADDITION CHAIN, ANTICHAIN, BRAUER CHAIN, CHAIN (GRAPH), DILWORTH'S LEMMA, HANSEN CHAIN

Chain Fraction

see CONTINUED FRACTION

Chain (Graph)

A chain of a GRAPH is a SEQUENCE $\{x_1, x_2, \ldots, x_n\}$ such that (x_1, x_2), (x_2, x_3), ..., (x_{n-1}, x_n) are EDGES of the GRAPH.

Chain Rule

If $g(x)$ is DIFFERENTIABLE at the point x and $f(x)$ is DIFFERENTIABLE at the point $g(x)$, then $f \circ g$ is DIFFERENTIABLE at x. Furthermore, let $y = f(g(x))$ and $u = g(x)$, then

$$\frac{dy}{dx} = \frac{dy}{du} \cdot \frac{du}{dx}. \tag{1}$$

There are a number of related results which also go under the name of "chain rules." For example, if $z = f(x, y)$, $x = g(t)$, and $y = h(t)$, then

$$\frac{dz}{dt} = \frac{\partial z}{\partial x}\frac{dx}{dt} + \frac{\partial z}{\partial y}\frac{dy}{dt}. \tag{2}$$

The "general" chain rule applies to two sets of functions

$$y_1 = f_1(u_1, \ldots, u_p)$$

$$\vdots (3)$$

$$y_m = f_m(u_1, \ldots, u_p)$$

and

$$u_1 = g_1(x_1, \ldots, x_n)$$

$$\vdots (4)$$

$$u_p = g_p(x_1, \ldots, x_n).$$

Defining the $m \times n$ JACOBI MATRIX by

$$\left(\frac{\partial y_i}{\partial x_j}\right) = \begin{bmatrix} \frac{\partial y_1}{\partial x_1} & \frac{\partial y_1}{\partial x_2} & \cdots & \frac{\partial y_1}{\partial x_n} \\ \vdots & \vdots & \ddots & \vdots \\ \frac{\partial y_m}{\partial x_1} & \frac{\partial y_m}{\partial x_2} & \cdots & \frac{\partial y_m}{\partial x_n} \end{bmatrix}, \tag{5}$$

and similarly for $(\partial y_i/\partial u_j)$ and $(\partial u_i/\partial x_j)$ then gives

$$\left(\frac{\partial y_i}{\partial x_j}\right) = \left(\frac{\partial y_i}{\partial u_j}\right)\left(\frac{\partial u_i}{\partial x_j}\right). \tag{6}$$

In differential form, this becomes

$$dy_1 = \left(\frac{\partial y_1}{\partial u_1}\frac{\partial u_1}{\partial x_1} + \ldots + \frac{\partial y_1}{\partial u_p}\frac{\partial u_p}{\partial x_1}\right) dx_1 + \left(\frac{\partial y_1}{\partial u_1}\frac{\partial u_1}{\partial x_2} + \ldots + \frac{\partial y_1}{\partial u_p}\frac{\partial u_p}{\partial x_2}\right) dx_2 + \ldots \quad (7)$$

(Kaplan 1984).

see also DERIVATIVE, JACOBIAN, POWER RULE, PRODUCT RULE

References
Anton, H. *Calculus with Analytic Geometry, 2nd ed.* New York: Wiley, p. 165, 1984.
Kaplan, W. "Derivatives and Differentials of Composite Functions" and "The General Chain Rule." §2.8 and 2.9 in *Advanced Calculus, 3rd ed.* Reading, MA: Addison-Wesley, pp. 101–105 and 106–110, 1984.

Chained Arrow Notation

A NOTATION which generalizes ARROW NOTATION and is defined as

$$\underbrace{a \uparrow \cdots \uparrow b}_{c} \equiv a \to b \to c.$$

see also ARROW NOTATION

References
Conway, J. H. and Guy, R. K. *The Book of Numbers.* New York: Springer-Verlag, p. 61, 1996.

Chainette

see CATENARY

Chair

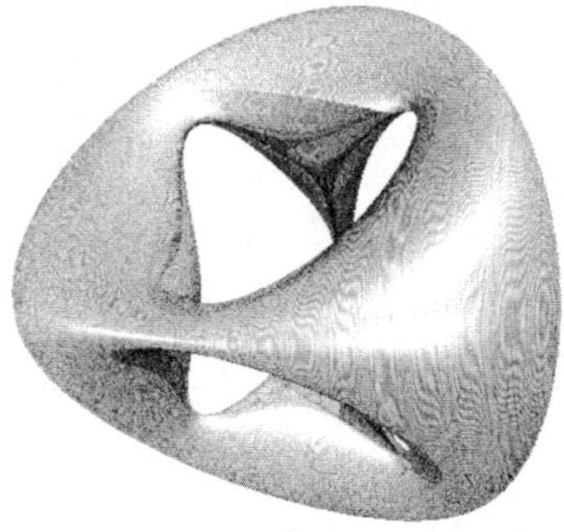

A SURFACE with tetrahedral symmetry which, according to Nordstrand, looks like an inflatable chair from the 1970s. It is given by the implicit equation

$$(x^2+y^2+z^2-ak^2)^2-b[(z-k)^2-2x^2][(z+k)^2-2y^2]=0.$$

see also BRIDE'S CHAIR

References
Nordstrand, T. "Chair." http://www.uib.no/people/nfytn/chairtxt.htm.

Chaitin's Constant

An IRRATIONAL NUMBER Ω which gives the probability that for any set of instructions, a UNIVERSAL TURING MACHINE will halt. The digits in Ω are random and cannot be computed ahead of time.

see also HALTING PROBLEM, TURING MACHINE, UNIVERSAL TURING MACHINE

References
Finch, S. "Favorite Mathematical Constants." http://www.mathsoft.com/asolve/constant/chaitin/chaitin.html.
Gardner, M. "The Random Number Ω Bids Fair to Hold the Mysteries of the Universe." *Sci. Amer.* **241**, 20–34, Nov. 1979.
Gardner, M. "Chaitin's Omega." Ch. 21 in *Fractal Music, Hypercards, and More Mathematical Recreations from Scientific American Magazine.* New York: W. H. Freeman, 1992.
Kobayashi, K. "Sigma(N)O-Complete Properties of Programs and Lartin-Lof Randomness." *Information Proc. Let.* **46**, 37–42, 1993.

Chaitin's Number

see CHAITIN'S CONSTANT

Chaitin's Omega

see CHAITIN'S CONSTANT

Champernowne Constant

Champernowne's number 0.1234567891011... (Sloane's A033307) is the decimal obtained by concatenating the POSITIVE INTEGERS. It is NORMAL in base 10. In 1961, Mahler showed it to also be TRANSCENDENTAL.

The CONTINUED FRACTION of the Champernowne constant is [0, 8, 9, 1, 149083, 1, 1, 1, 4, 1, 1, 1, 3, 4, 1, 1, 1, 15,

$$\begin{aligned}&457540111391031076483646628242956118599603939\cdots\\&710457555000662004393090262659256314937953207\cdots\\&747128656313864120937550355209460718308998457\cdots\\&5801469863148833592141783010987,\end{aligned}$$

6, 1, 1, 21, 1, 9, 1, 1, 2, 3, 1, 7, 2, 1, 83, 1, 156, 4, 58, 8, 54, ...] (Sloane's A030167). The next term of the CONTINUED FRACTION is huge, having 2504 digits. In fact, the coefficients eventually become unbounded, making the continued fraction difficult to calculate for too many more terms. Large terms greater than 10^5 occur at positions 5, 19, 41, 102, 163, 247, 358, 460, ... and have 6, 166, 2504, 140, 33102, 109, 2468, 136, ... digits (Plouffe). Interestingly, the COPELAND-ERDŐS CONSTANT, which is the decimal obtained by concatenating the PRIMES, has a well-behaved CONTINUED FRACTION which does not show the "large term" phenomenon.

see also COPELAND-ERDŐS CONSTANT, SMARANDACHE SEQUENCES

References

Champernowne, D. G. "The Construction of Decimals Normal in the Scale of Ten." *J. London Math. Soc.* **8**, 1933.

Finch, S. "Favorite Mathematical Constants." `http://www.mathsoft.com/asolve/constant/cntfrc/cntfrc.html`.

Sloane, N. J. A. Sequences A030167 and A033307 in "An On-Line Version of the Encyclopedia of Integer Sequences."

Change of Variables Theorem

A theorem which effectively describes how lengths, areas, volumes, and generalized n-dimensional volumes (CONTENTS) are distorted by DIFFERENTIABLE FUNCTIONS. In particular, the change of variables theorem reduces the whole problem of figuring out the distortion of the content to understanding the infinitesimal distortion, i.e., the distortion of the DERIVATIVE (a linear MAP), which is given by the linear MAP's DETERMINANT. So $f : \mathbb{R}^n \to \mathbb{R}^n$ is an AREA-PRESERVING linear MAP IFF $|\det(f)| = 1$, and in more generality, if S is any subset of $\mathbb{R}^n$, the CONTENT of its image is given by $|\det(f)|$ times the CONTENT of the original. The change of variables theorem takes this infinitesimal knowledge, and applies CALCULUS by breaking up the DOMAIN into small pieces and adds up the change in AREA, bit by bit.

The change of variable formula persists to the generality of DIFFERENTIAL FORMS on MANIFOLDS, giving the formula

$$\int_M (f^*\omega) = \int_W (\omega)$$

under the conditions that M and W are compact connected oriented MANIFOLDS with nonempty boundaries, $f : M \to W$ is a smooth map which is an orientation-preserving DIFFEOMORPHISM of the boundaries.

In 2-D, the explicit statement of the theorem is

$$\int_R f(x,y)\,dx\,dy = \int_{R^*} f[x(u,v),y(u,v)] \left|\frac{\partial(x,y)}{\partial(u,v)}\right| du\,dv$$

and in 3-D, it is

$$\int_R f(x,y,z)\,dx\,dy\,dz = \int_{R^*} f[x(u,v,w),y(u,v,w),z(u,v,w)] \left|\frac{\partial(x,y,z)}{\partial(u,v,w)}\right| du\,dv\,dw,$$

where $R = f(R^*)$ is the image of the original region R^*,

$$\left|\frac{\partial(x,y,z)}{\partial(u,v,w)}\right|$$

is the JACOBIAN, and f is a global orientation-preserving DIFFEOMORPHISM of R and R^* (which are open subsets of $\mathbb{R}^n$).

The change of variables theorem is a simple consequence of the CURL THEOREM and a little DE RHAM COHOMOLOGY. The generalization to n-D requires no additional assumptions other than the regularity conditions on the boundary.

see also IMPLICIT FUNCTION THEOREM, JACOBIAN

References

Kaplan, W. "Change of Variables in Integrals." §4.6 in *Advanced Calculus, 3rd ed.* Reading, MA: Addison-Wesley, pp. 238–245, 1984.

Chaos

A DYNAMICAL SYSTEM is chaotic if it

1. Has a DENSE collection of points with periodic orbits,
2. Is sensitive to the initial condition of the system (so that initially nearby points can evolve quickly into very different states), and
3. Is TOPOLOGICALLY TRANSITIVE.

Chaotic systems exhibit irregular, unpredictable behavior (the BUTTERFLY EFFECT). The boundary between linear and chaotic behavior is characterized by PERIOD DOUBLING, following by quadrupling, etc.

An example of a simple physical system which displays chaotic behavior is the motion of a magnetic pendulum over a plane containing two or more attractive magnets. The magnet over which the pendulum ultimately comes to rest (due to frictional damping) is highly dependent on the starting position and velocity of the pendulum (Dickau). Another such system is a double pendulum (a pendulum with another pendulum attached to its end).

see also ACCUMULATION POINT, ATTRACTOR, BASIN OF ATTRACTION, BUTTERFLY EFFECT, CHAOS GAME, FEIGENBAUM CONSTANT, FRACTAL DIMENSION, GINGERBREADMAN MAP, HÉNON-HEILES EQUATION, HÉNON MAP, LIMIT CYCLE, LOGISTIC EQUATION, LYAPUNOV CHARACTERISTIC EXPONENT, PERIOD THREE THEOREM, PHASE SPACE, QUANTUM CHAOS, RESONANCE OVERLAP METHOD, ŠARKOVSKII'S THEOREM, SHADOWING THEOREM, SINK (MAP), STRANGE ATTRACTOR

References

Bai-Lin, H. *Chaos.* Singapore: World Scientific, 1984.

Baker, G. L. and Gollub, J. B. *Chaotic Dynamics: An Introduction, 2nd ed.* Cambridge: Cambridge University Press, 1996.

Cvitanovic, P. *Universality in Chaos: A Reprint Selection, 2nd ed.* Bristol: Adam Hilger, 1989.

Dickau, R. M. "Magnetic Pendulum." `http://forum.swarthmore.edu/advanced/robertd/magneticpendulum.html`.

Drazin, P. G. *Nonlinear Systems.* Cambridge, England: Cambridge University Press, 1992.

Field, M. and Golubitsky, M. *Symmetry in Chaos: A Search for Pattern in Mathematics, Art and Nature.* Oxford, England: Oxford University Press, 1992.

Gleick, J. *Chaos: Making a New Science.* New York: Penguin, 1988.

Guckenheimer, J. and Holmes, P. *Nonlinear Oscillations, Dynamical Systems, and Bifurcations of Vector Fields, 3rd ed.* New York: Springer-Verlag, 1997.
Lichtenberg, A. and Lieberman, M. *Regular and Stochastic Motion, 2nd ed.* New York: Springer-Verlag, 1994.
Lorenz, E. N. *The Essence of Chaos.* Seattle, WA: University of Washington Press, 1996.
Ott, E. *Chaos in Dynamical Systems.* New York: Cambridge University Press, 1993.
Ott, E.; Sauer, T.; and Yorke, J. A. *Coping with Chaos: Analysis of Chaotic Data and the Exploitation of Chaotic Systems.* New York: Wiley, 1994.
Peitgen, H.-O.; Jürgens, H.; and Saupe, D. *Chaos and Fractals: New Frontiers of Science.* New York: Springer-Verlag, 1992.
Poon, L. "Chaos at Maryland." `http://www-chaos.umd.edu`.
Rasband, S. N. *Chaotic Dynamics of Nonlinear Systems.* New York: Wiley, 1990.
Strogatz, S. H. *Nonlinear Dynamics and Chaos, with Applications to Physics, Biology, Chemistry, and Engineering.* Reading, MA: Addison-Wesley, 1994.
Tabor, M. *Chaos and Integrability in Nonlinear Dynamics: An Introduction.* New York: Wiley, 1989.
Tufillaro, N.; Abbott, T. R.; and Reilly, J. *An Experimental Approach to Nonlinear Dynamics and Chaos.* Redwood City, CA: Addison-Wesley, 1992.
Wiggins, S. *Global Bifurcations and Chaos: Analytical Methods.* New York: Springer-Verlag, 1988.
Wiggins, S. *Introduction to Applied Nonlinear Dynamical Systems and Chaos.* New York: Springer-Verlag, 1990.

Chaos Game

Pick a point at random inside a regular n-gon. Then draw the next point a fraction r of the distance between it and a VERTEX picked at random. Continue the process (after throwing out the first few points). The result of this "chaos game" is sometimes, but not always, a FRACTAL. The case $(n,r) = (4,1/2)$ gives the interior of a SQUARE with all points visited with equal probability.

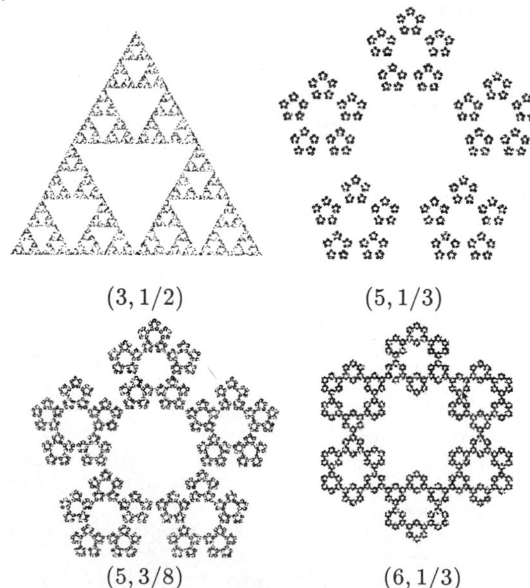

$(3,1/2)$ $(5,1/3)$ $(5,3/8)$ $(6,1/3)$

see also BARNSLEY'S FERN

References

Barnsley, M. F. and Rising, H. *Fractals Everywhere, 2nd ed.* Boston, MA: Academic Press, 1993.
Dickau, R. M. "The Chaos Game." `http://forum.swarthmore.edu/advanced/robertd/chaos_game.html`.
Wagon, S. *Mathematica in Action.* New York: W. H. Freeman, pp. 149–163, 1991.
Weisstein, E. W. "Fractals." `http://www.astro.virginia.edu/~eww6n/math/notebooks/Fractal.m`.

Character (Group)

The GROUP THEORY term for what is known to physicists as the TRACE. All members of the same CONJUGACY CLASS in the same representation have the same character. Members of other CONJUGACY CLASSES may also have the same character, however. An (abstract) GROUP can be uniquely identified by a listing of the characters of its various representations, known as a CHARACTER TABLE. Some of the SCHÖNFLIES SYMBOLS denote different sets of symmetry operations but correspond to the same abstract GROUP and so have the same CHARACTER TABLES.

Character (Multiplicative)

A continuous HOMEOMORPHISM of a GROUP into the NONZERO COMPLEX NUMBERS. A multiplicative character ω gives a REPRESENTATION on the 1-D SPACE $\mathbb{C}$ of COMPLEX NUMBERS, where the REPRESENTATION action by $g \in G$ is multiplication by $\omega(g)$. A multiplicative character is UNITARY if it has ABSOLUTE VALUE 1 everywhere.

References

Knapp, A. W. "Group Representations and Harmonic Analysis, Part II." *Not. Amer. Math. Soc.* **43**, 537–549, 1996.

Character (Number Theory)

A number theoretic function $\chi_k(n)$ for POSITIVE integral n is a character modulo k if

$$\chi_k(1) = 1$$
$$\chi_k(n) = \chi_k(n+k)$$
$$\chi_k(m)\chi_k(n) = \chi_k(mn)$$

for all m, n, and

$$\chi_k(n) = 0$$

if $(k,n) \neq 1$. χ_k can only assume values which are $\phi(k)$ ROOTS OF UNITY, where ϕ is the TOTIENT FUNCTION.

see also DIRICHLET L-SERIES

Character Table

C_1	E
A	1

C_s	E	σ_h		
A	1	1	x, y, R_z	x^2, y^2, z^2, xy
B	1	-1	z, R_x, R_y	yz, xz

C_i	E	i		
A_g	1	1	R_x, R_y, R_z	$x^2, y^2, z^2, xy, xz, yz$
A_u	1	−1	x, y, z	

C_2	E	C_2		
A	1	1	z, R_z	x^2, y^2, z^2, xy
B	1	−1	x, y, R_x, R_y	yz, xz

C_3	E	C_3	${C_3}^2$		$\varepsilon = \exp(2\pi i/3)$
A	1	1	1	z, R_z	x^2, y^2, z^2, xy
E	$\left\{\begin{matrix}1 \\ 1\end{matrix}\right.$	$\begin{matrix}\varepsilon \\ \varepsilon^*\end{matrix}$	$\left.\begin{matrix}\varepsilon^* \\ \varepsilon\end{matrix}\right\}$	$(x, y)(R_x, R_y)$	$(x^2 - y^2, xy)(yz, xz)$

C_4	E	C_3	C_2	${C_4}^3$		
A	1	1	1	1	z, R_z	$x^2 + y^2, z^2$
B	1	−1	1	−1		$x^2 - y^2, xy$
E	$\left\{\begin{matrix}1 \\ 1\end{matrix}\right.$	$\begin{matrix}i \\ -i\end{matrix}$	$\begin{matrix}-1 \\ 1\end{matrix}$	$\left.\begin{matrix}-i \\ i\end{matrix}\right\}$	$(x, y)(R_x, R_y)$	(yz, xz)

C_5	E	C_5	${C_5}^2$	${C_5}^3$	${C_5}^4$		$\varepsilon = \exp(2\pi i/5)$
A	1	1	1	1	1	z, R_z	$x^2 + y^2, z^2$
E_1	$\left\{\begin{matrix}1 \\ 1\end{matrix}\right.$	$\begin{matrix}\varepsilon \\ \varepsilon^*\end{matrix}$	$\begin{matrix}\varepsilon^2 \\ \varepsilon^{2*}\end{matrix}$	$\begin{matrix}\varepsilon^{2*} \\ \varepsilon^2\end{matrix}$	$\left.\begin{matrix}\varepsilon^* \\ \varepsilon\end{matrix}\right\}$	$(x, y)(R_x, R_y)$	(yz, xz)
E_2	$\left\{\begin{matrix}1 \\ 1\end{matrix}\right.$	$\begin{matrix}\varepsilon^2 \\ \varepsilon^{2*}\end{matrix}$	$\begin{matrix}\varepsilon^* \\ \varepsilon\end{matrix}$	$\begin{matrix}\varepsilon \\ \varepsilon^*\end{matrix}$	$\left.\begin{matrix}\varepsilon^{2*} \\ \varepsilon^2\end{matrix}\right\}$		$(x^2 - y^2, xy)$

C_6	E	C_6	C_3	C_2	${C_3}^2$	${C_6}^5$		$\varepsilon = \exp(2\pi i/6)$
A	1	1	1	1	1	1	z, R_z	$x^2 + y^2, z^2$
B	1	−1	1	−1	1	−1		
E_1	$\left\{\begin{matrix}1 \\ 1\end{matrix}\right.$	$\begin{matrix}\varepsilon \\ \varepsilon^*\end{matrix}$	$\begin{matrix}-\varepsilon^* \\ -\varepsilon\end{matrix}$	$\begin{matrix}-1 \\ -1\end{matrix}$	$\begin{matrix}-\varepsilon \\ -\varepsilon^*\end{matrix}$	$\left.\begin{matrix}\varepsilon^* \\ \varepsilon\end{matrix}\right\}$	(x, y) (R_x, R_y)	(yz, xz)
E_2	$\left\{\begin{matrix}1 \\ 1\end{matrix}\right.$	$\begin{matrix}-\varepsilon \\ -\varepsilon^*\end{matrix}$	$\begin{matrix}-\varepsilon \\ -\varepsilon^*\end{matrix}$	$\begin{matrix}1 \\ 1\end{matrix}$	$\begin{matrix}-\varepsilon^* \\ -\varepsilon\end{matrix}$	$\left.\begin{matrix}-\varepsilon \\ \varepsilon^*\end{matrix}\right\}$		$(x^2 - y^2, xy)$

D_2	E	$C_2(z)$	$C_2(y)$	$C_2(x)$		
A_1	1	1	1	1		$x^2 + y^2, z^2$
B_1	1	1	−1	−1	z, R_z	xy
B_2	1	−1	1	−1	y, R_y	xz
B_3	1	−1	−1	1	z, R_z	yz

D_3	E	$2C_3$	$3C_2$		
A_1	1	1	1		$x^2 + y^2, z^2$
A_2	1	1	−1	z, R_z	xy
E	2	−1	0	$(x, y)(R_x, R_y)$	$(x^2 - y^2, xy)(xz, yz)$

D_4	E	$2C_4$	C_2	$2C_2'$	$2C_2''$		
A_1	1	1	1	1	1		$x^2 + y^2, z^2$
A_2	1	1	1	−1	−1	z, R_z	
B_1	1	−1	1	1	−1		$x^2 - y^2$
B_2	1	−1	1	−1	1		xy
E	2	0	−2	0	0	$(x, y)(R_x, R_y)$	(xz, yz)

D_5	E	$2C_5$	$2{C_5}^2$	$5C_2$		
A_1	1	1	1	1		$x^2 + y^2, z^2$
B_1	1	1	1	−1	z, R_z	
B_2	2	2 cos 72°	2 cos 144°	0	$(x, y)(R_x, R_y)$	(xz, yz)
B_3	2	2 cos 144°	2 cos 72°	0		$(x^2 - y^2, xy)$

D_6	E	$2C_6$	$2C_3$	C_2	$3C_2'$	$3C_2''$		
A_1	1	1	1	1	1	1		$x^2 + y^2, z^2$
A_2	1	1	1	1	−1	−1	z, R_z	
B_1	1	−1	1	−1	1	−1		
B_2	1	−1	1	−1	−1	1	$(x, y)(R_x, R_y)$	
E_1	2	1	−1	−2	0	0		(xz, yz)
E_2	2	−1	−1	2	0	0		$(x^2 - y^2, xy)$

C_{2v}	E	C_2	$\sigma_v(xz)$	$\sigma_v'(yz)$		
A_1	1	1	1	1	z	x^2, y^2, z^2
A_2	1	1	−1	−1	R_z	xy
B_1	1	−1	1	−1	x, R_y	xz
B_2	1	−1	−1	1	y, R_x	yz

C_{3v}	E	$2C_3$	$3\sigma_v$		
A_1	1	1	1	z	$x^2 + y^2, z^2$
A_2	1	1	−1	R_z	
E	2	−1	0	$(x, y)(R_x, R_y)$	$(x^2 - y^2, xy)(xz, yz)$

C_{4v}	E	$2C_4$	C_2	$2\sigma_v$	$2\sigma_d$		
A_1	1	1	1	1	1	z	$x^2 + y^2, z^2$
A_2	1	1	1	−1	−1	R_z	
B_1	1	−1	1	1	−1		$x^2 - y^2$
B_2	1	−1	1	−1	1		xy
E	2	0	−2	0	0	$(x, y)(R_x, R_y)$	(xz, yz)

C_{5v}	E	$2C_5$	$2{C_5}^2$	$5\sigma_v$		
A_1	1	1	1	1	z	$x^2 + y^2, z^2$
B_1	1	1	1	−1	R_z	
B_2	2	2 cos 72°	2 cos 144°	0	$(x, y)(R_x, R_y)$	(xz, yz)
B_3	2	2 cos 144°	2 cos 72°	0		$(x^2 - y^2, xy)$

C_{6v}	E	$2C_6$	$2C_3$	C_2	$3\sigma_v$	$3\sigma_d$		
A_1	1	1	1	1	1	1	z	$x^2 + y^2, z^2$
A_2	1	1	1	1	−1	−1	R_z	
B_1	1	−1	1	−1	1	−1		
B_2	1	−1	1	−1	−1	1		
E_1	2	1	−1	−2	0	0	$(x, y)(R_x, R_y)$	(xz, yz)
E_2	2	−1	−1	2	0	0		$(x^2 - y^2, xy)$

$C_{\infty v}$	E	${C_\infty}^\Phi$	...	$\infty\sigma_v$		
$A_1 \equiv \Sigma^+$	1	1	...	1	z	$x^2 + y^2, z^2$
$A_2 \equiv \Sigma^-$	1	1	...	−1	R_z	
$E_1 \equiv \Pi$	2	2 cos Φ	...	0	$(x, y); (R_x, R_y)$	(xz, yz)
$E_2 \equiv \Delta$	2	2 cos 2Φ	...	0		$(x^2 - y^2, xy)$
$E_3 \equiv \Phi$	2	2 cos 3Φ	...	0		
⋮	⋮	⋮	⋱	⋮		

References

Bishop, D. M. "Character Tables." Appendix 1 in *Group Theory and Chemistry.* New York: Dover, pp. 279–288, 1993.

Cotton, F. A. *Chemical Applications of Group Theory, 3rd ed.* New York: Wiley, 1990.

Iyanaga, S. and Kawada, Y. (Eds.). "Characters of Finite Groups." Appendix B, Table 5 in *Encyclopedic Dictionary of Mathematics.* Cambridge, MA: MIT Press, pp. 1496–1503, 1980.

Characteristic Class

Characteristic classes are COHOMOLOGY classes in the BASE SPACE of a VECTOR BUNDLE, defined through OBSTRUCTION theory, which are (perhaps partial) obstructions to the existence of k everywhere linearly independent vector FIELDS on the VECTOR BUNDLE. The most common examples of characteristic classes are the CHERN, PONTRYAGIN, and STIEFEL-WHITNEY CLASSES.

Characteristic (Elliptic Integral)

A parameter n used to specify an ELLIPTIC INTEGRAL OF THE THIRD KIND.

see also AMPLITUDE, ELLIPTIC INTEGRAL, MODULAR ANGLE, MODULUS (ELLIPTIC INTEGRAL), NOME, PARAMETER

References

Abramowitz, M. and Stegun, C. A. (Eds.). *Handbook of Mathematical Functions with Formulas, Graphs, and Mathematical Tables, 9th printing.* New York: Dover, p. 590, 1972.

Characteristic Equation

The equation which is solved to find a MATRIX's EIGENVALUES, also called the CHARACTERISTIC POLYNOMIAL. Given a 2×2 system of equations with MATRIX

$$\mathsf{M} \equiv \begin{bmatrix} a & b \\ c & d \end{bmatrix}, \tag{1}$$

the MATRIX EQUATION is

$$\begin{bmatrix} a & b \\ c & d \end{bmatrix} \begin{bmatrix} x \\ y \end{bmatrix} = t \begin{bmatrix} x \\ y \end{bmatrix}, \tag{2}$$

which can be rewritten

$$\begin{bmatrix} a-t & b \\ c & d-t \end{bmatrix} \begin{bmatrix} x \\ y \end{bmatrix} = t \begin{bmatrix} 0 \\ 0 \end{bmatrix}. \tag{3}$$

M can have no MATRIX INVERSE, since otherwise

$$\begin{bmatrix} x \\ y \end{bmatrix} = \mathsf{M}^{-1} \begin{bmatrix} 0 \\ 0 \end{bmatrix} = \begin{bmatrix} 0 \\ 0 \end{bmatrix}, \tag{4}$$

which contradicts our ability to pick arbitrary x and y. Therefore, M has no inverse, so its DETERMINANT is 0. This gives the characteristic equation

$$\begin{vmatrix} a-t & b \\ c & d-t \end{vmatrix} = 0, \tag{5}$$

where $|\mathsf{A}|$ denotes the DETERMINANT of A. For a general $k \times k$ MATRIX

$$\begin{bmatrix} a_{11} & a_{12} & \cdots & a_{1k} \\ a_{21} & a_{22} & \cdots & a_{2k} \\ \vdots & \vdots & \ddots & \vdots \\ a_{k1} & a_{k2} & \cdots & a_{kk} \end{bmatrix}, \tag{6}$$

the characteristic equation is

$$\begin{vmatrix} a_{11}-t & a_{12} & \cdots & a_{1k} \\ a_{21} & a_{22}-t & \cdots & a_{2k} \\ \vdots & \vdots & \ddots & \vdots \\ a_{k1} & a_{k2} & \cdots & a_{kk}-t \end{vmatrix} = 0. \tag{7}$$

see also BALLIEU'S THEOREM, CAYLEY-HAMILTON THEOREM, PARODI'S THEOREM, ROUTH-HURWITZ THEOREM

References

Gradshteyn, I. S. and Ryzhik, I. M. *Tables of Integrals, Series, and Products, 5th ed.* San Diego, CA: Academic Press, pp. 1117–1119, 1979.

Characteristic (Euler)

see EULER CHARACTERISTIC

Characteristic Factor

A characteristic factor is a factor in a particular factorization of the TOTIENT FUNCTION $\phi(n)$ such that the product of characteristic factors gives the representation of a corresponding abstract GROUP as a DIRECT PRODUCT. By computing the characteristic factors, any ABELIAN GROUP can be expressed as a DIRECT PRODUCT of CYCLIC SUBGROUPS, for example, $Z_2 \otimes Z_4$ or $Z_2 \otimes Z_2 \otimes Z_2$. There is a simple algorithm for determining the characteristic factors of MODULO MULTIPLICATION GROUPS.

see also CYCLIC GROUP, DIRECT PRODUCT (GROUP), MODULO MULTIPLICATION GROUP, TOTIENT FUNCTION

References

Shanks, D. *Solved and Unsolved Problems in Number Theory, 4th ed.* New York: Chelsea, p. 94, 1993.

Characteristic (Field)

For a FIELD K with multiplicative identity 1, consider the numbers $2 = 1+1$, $3 = 1+1+1$, $4 = 1+1+1+1$, etc. Either these numbers are all different, in which case we say that K has characteristic 0, or two of them will be equal. In this case, it is straightforward to show that, for some number p, we have $\underbrace{1+1+\ldots+1}_{p \text{ times}} = 0$. If p is chosen to be as small as possible, then p will be a PRIME, and we say that K has characteristic p. The FIELDS $\mathbb{Q}$, $\mathbb{R}$, $\mathbb{C}$, and the p-ADIC NUMBERS $\mathbb{Q}_p$ have characteristic 0. For p a PRIME, the GALOIS FIELD $\mathrm{GF}(p^n)$ has characteristic p.

If H is a SUBFIELD of K, then H and K have the same characteristic.

see also FIELD, SUBFIELD

Characteristic Function

The characteristic function $\phi(t)$ is defined as the FOURIER TRANSFORM of the PROBABILITY DENSITY FUNCTION,

$$\begin{aligned}\phi(t) = \mathcal{F}[P(x)] &= \int_{-\infty}^{\infty} e^{itx} P(x)\,dx && (1)\\ &= \int_{-\infty}^{\infty} P(x)\,dx + it\int_{-\infty}^{\infty} xP(x)\,dx\\ &\quad + \tfrac{1}{2}(it)^2 \int_{-\infty}^{\infty} x^2 P(x)\,dx + \ldots && (2)\\ &= \sum_{k=0}^{\infty} \frac{(it)^k}{k!}\mu'_k && (3)\\ &= 1 + it\mu'_1 - \tfrac{1}{2}t^2\mu'_2 - \tfrac{1}{3!}it^3\mu'_3 + \tfrac{1}{4!}t^4\mu'_4 + \ldots, && (4)\end{aligned}$$

where μ'_n (sometimes also denoted ν_n) is the nth MOMENT about 0 and $\mu'_0 \equiv 1$. The characteristic function can therefore be used to generate MOMENTS about 0,

$$\phi^{(n)}(0) \equiv \left[\frac{d^n\phi}{dt^n}\right]_{t=0} = i^n \mu'_n \qquad (5)$$

or the CUMULANTS κ_n,

$$\ln \phi(t) \equiv \sum_{n=0}^{\infty} \kappa_n \frac{(it)^n}{n!}. \qquad (6)$$

A DISTRIBUTION is not uniquely specified by its MOMENTS, but is uniquely specified by its characteristic function.

see also CUMULANT, MOMENT, MOMENT-GENERATING FUNCTION, PROBABILITY DENSITY FUNCTION

References

Abramowitz, M. and Stegun, C. A. (Eds.). *Handbook of Mathematical Functions with Formulas, Graphs, and Mathematical Tables, 9th printing.* New York: Dover, p. 928, 1972.

Kenney, J. F. and Keeping, E. S. "Moment-Generating and Characteristic Functions," "Some Examples of Moment-Generating Functions," and "Uniqueness Theorem for Characteristic Functions." §4.6–4.8 in *Mathematics of Statistics, Pt. 2, 2nd ed.* Princeton, NJ: Van Nostrand, pp. 72–77, 1951.

Characteristic (Partial Differential Equation)

Paths in a 2-D plane used to transform PARTIAL DIFFERENTIAL EQUATIONS into systems of ORDINARY DIFFERENTIAL EQUATIONS. They were invented by Riemann. For an example of the use of characteristics, consider the equation

$$u_t - 6uu_x = 0.$$

Now let $u(s) = u(x(s), t(s))$. Since

$$\frac{du}{ds} = \frac{dx}{ds}u_x + \frac{dt}{ds}u_t,$$

it follows that $dt/ds = 1$, $dx/ds = -6u$, and $du/ds = 0$. Integrating gives $t(s) = s$, $x(s) = -6su_0(x)$, and $u(s) = u_0(x)$, where the constants of integration are 0 and $u_0(x) = u(x, 0)$.

Characteristic Polynomial

The expanded form of the CHARACTERISTIC EQUATION.

$$\det(x\mathsf{I} - \mathsf{A}),$$

where A is an $n \times n$ MATRIX and I is the IDENTITY MATRIX.

see also CAYLEY-HAMILTON THEOREM

Characteristic (Real Number)

For a REAL NUMBER x, $\lfloor x \rfloor = \mathtt{int(x)}$ is called the characteristic. Here, $\lfloor x \rfloor$ is the FLOOR FUNCTION.

see also MANTISSA, SCIENTIFIC NOTATION

Charlier's Check

A check which can be used to verify correct computation of MOMENTS.

Chasles-Cayley-Brill Formula

The number of coincidences of a (ν, ν') correspondence of value γ on a curve of GENUS p is given by

$$\nu + \nu' + 2p\gamma.$$

see also ZEUTHEN'S THEOREM

References

Coolidge, J. L. *A Treatise on Algebraic Plane Curves.* New York: Dover, p. 129, 1959.

Chasles's Contact Theorem

If a one-parameter family of curves has index N and class M, the number tangent to a curve of order n_1 and class m_1 in general position is

$$m_1 N + n_1 M.$$

References

Coolidge, J. L. *A Treatise on Algebraic Plane Curves.* New York: Dover, p. 436, 1959.

Chasles's Polars Theorem

If the TRILINEAR POLARS of the VERTICES of a TRIANGLE are distinct from the respectively opposite sides, they meet the sides in three COLLINEAR points.

see also COLLINEAR, TRIANGLE, TRILINEAR POLAR

Chasles's Theorem

If two projective PENCILS of curves of orders n and n' have no common curve, the LOCUS of the intersections of corresponding curves of the two is a curve of order $n+n'$ through all the centers of either PENCIL. Conversely, if a curve of order $n+n'$ contains all centers of a PENCIL of order n to the multiplicity demanded by NOETHER'S FUNDAMENTAL THEOREM, then it is the LOCUS of the intersections of corresponding curves of this PENCIL and one of order n' projective therewith.

see also NOETHER'S FUNDAMENTAL THEOREM, PENCIL

References

Coolidge, J. L. *A Treatise on Algebraic Plane Curves.* New York: Dover, p. 33, 1959.

Chebyshev Approximation Formula

Using a CHEBYSHEV POLYNOMIAL OF THE FIRST KIND T, define

$$c_j \equiv \frac{2}{N}\sum_{k=1}^{N} f(x_k)T_j(x_k)$$
$$= \frac{2}{N}\sum_{k=1}^{N} f\left[\cos\left\{\frac{\pi(k-\frac{1}{2})}{N}\right\}\right]\cos\left\{\frac{\pi j(k-\frac{1}{2})}{N}\right\}.$$

Then

$$f(x) \approx \sum_{k=0}^{N-1} c_k T_k(x) - \tfrac{1}{2}c_0.$$

It is exact for the N zeros of $T_N(x)$. This type of approximation is important because, when truncated, the error is spread smoothly over $[-1,1]$. The Chebyshev approximation formula is very close to the MINIMAX POLYNOMIAL.

References

Press, W. H.; Flannery, B. P.; Teukolsky, S. A.; and Vetterling, W. T. "Chebyshev Approximation," "Derivatives or Integrals of a Chebyshev-Approximated Function," and "Polynomial Approximation from Chebyshev Coefficients." §5.8, 5.9, and 5.10 in *Numerical Recipes in FORTRAN: The Art of Scientific Computing, 2nd ed.* Cambridge, England: Cambridge University Press, pp. 184–188, 189–190, and 191–192, 1992.

Chebyshev Constants

N.B. A detailed on-line essay by S. Finch was the starting point for this entry.

The constants

$$\lambda_{m,n} = \inf_{r\in R_{m,n}} \sup_{x\geq 0} |e^{-x} - r(x)|,$$

where

$$r(x) = \frac{p(x)}{q(x)},$$

p and q are mth and nth order POLYNOMIALS, and $R_{m,n}$ is the set all RATIONAL FUNCTIONS with REAL coefficients.

see also ONE-NINTH CONSTANT, RATIONAL FUNCTION

References

Finch, S. "Favorite Mathematical Constants." http://www.mathsoft.com/asolve/constant/onenin/onenin.html.

Petrushev, P. P. and Popov, V. A. *Rational Approximation of Real Functions.* New York: Cambridge University Press, 1987.

Varga, R. S. *Scientific Computations on Mathematical Problems and Conjectures.* Philadelphia, PA: SIAM, 1990. Philadelphia, PA: SIAM, 1990.

Chebyshev Deviation

$$\max_{a\leq x\leq b}\{|f(x)-\rho(x)|w(x)\}.$$

References

Szegő, G. *Orthogonal Polynomials, 4th ed.* Providence, RI: Amer. Math. Soc., p. 41, 1975.

Chebyshev Differential Equation

$$(1-x^2)\frac{d^2y}{dx^2} - x\frac{dy}{dx} + m^2y = 0 \quad (1)$$

for $|x|<1$. The Chebyshev differential equation has regular SINGULARITIES at -1, 1, and ∞. It can be solved by series solution using the expansions

$$y = \sum_{n=0}^{\infty} a_n x^n \quad (2)$$
$$y' = \sum_{n=0}^{\infty} na_n x^{n-1} = \sum_{n=1}^{\infty} na_n x^{n-1}$$
$$= \sum_{n=0}^{\infty}(n+1)a_{n+1}x^n \quad (3)$$
$$y'' = \sum_{n=0}^{\infty}(n+1)na_{n+1}x^{n-1} = \sum_{n=1}^{\infty}(n+1)na_{n+1}x^{n-1}$$
$$= \sum_{n=0}^{\infty}(n+2)(n+1)a_{n+2}x^n. \quad (4)$$

Now, plug (2-4) into the original equation (1) to obtain

$$(1-x^2)\sum_{n=0}^{\infty}(n+2)(n+1)a_{n+2}x^n$$
$$-x\sum_{n=0}^{\infty}(n+1)n_{n+1}x^n + m^2\sum_{n=0}^{\infty}a_nx^n = 0 \quad (5)$$

$$\sum_{n=0}^{\infty}(n+2)(n+1)a_{n+2}x^n - \sum_{n=0}^{\infty}(n+2)(n+1)a_{n+2}x^{n+2}$$
$$-\sum_{n=0}^{\infty}(n+1)a_{n+1}x^{n+1} + m^2\sum_{n=0}^{\infty}a_nx^n = 0 \quad (6)$$

$$\sum_{n=0}^{\infty}(n+2)(n+1)a_{n+2}x^n - \sum_{n=2}^{\infty}n(n-1)a_nx^{n+2} - \sum_{n=1}^{\infty}na_nx^n + m^2\sum_{n=0}^{\infty}a_nx^n = 0 \quad (7)$$

$$2\cdot 1a_2 + 3\cdot 2a_3x - 1\cdot ax + m^2a_0 + m^2a_1x + \sum_{n=2}^{\infty}[(n+2)(n+1)a_{n+2} - n(n-1)a_n - na_n + m^2a_n]x^n = 0 \quad (8)$$

$$(2a_2 + m^2a_0) + [(m^2-1)a_1 + 6a_3]x + \sum_{n=2}^{\infty}[(n+2)(n+1)a_{n+2} + (m^2-n^2)a_n]x^n = 0, \quad (9)$$

so

$$2a_2 + m^2a_0 = 0 \quad (10)$$

$$(m^2-1)a_1 + 6a_3 = 0 \quad (11)$$

$$a_{n+2} = \frac{n^2-m^2}{(n+1)(n+2)}a_n \quad \text{for } n = 2, 3, \ldots. \quad (12)$$

The first two are special cases of the third, so the general recurrence relation is

$$a_{n+2} = \frac{n^2-m^2}{(n+1)(n+2)}a_n \quad \text{for } n = 0, 1, \ldots. \quad (13)$$

From this, we obtain for the EVEN COEFFICIENTS

$$a_2 = -\tfrac{1}{2}m^2a_0 \quad (14)$$

$$a_4 = \frac{2^2-m^2}{3\cdot 4}a_2 = \frac{(2^2-m^2)(-m^2)}{1\cdot 2\cdot 3\cdot 4}a_0 \quad (15)$$

$$a_{2n} = \frac{[(2n)^2-m^2][(2n-2)^2-m^2]\cdots[-m^2]}{(2n)!}a_0, \quad (16)$$

and for the ODD COEFFICIENTS

$$a_3 = \frac{1-m^2}{6}a_0 \quad (17)$$

$$a_5 = \frac{3^2-m^2}{4\cdot 5}a_3 = \frac{(3^2-m^2)(1^2-m^2)}{5!}a_1 \quad (18)$$

$$a_{2n-1} = \frac{[(2n-1)^2-m^2][(2n-3)^2-m^2]\cdots[1^2-m^2]}{(2n+1)!}a_1. \quad (19)$$

So the general solution is

$$y = a_0\left[1 + \sum_{k=2,4,\ldots}^{\infty}\frac{[k^2-m^2][(k-2)^2-m^2]\cdots[-m^2]}{k!}x^k\right] + a_1\left[x + \sum_{k=3,5,\ldots}^{\infty}\frac{[(k-2)^2-m^2][(k-2)^2-m^2]\cdots[1^2-m^2]}{k!}x^k\right] \quad (20)$$

If n is EVEN, then y_1 terminates and is a POLYNOMIAL solution, whereas if n is ODD, then y_2 terminates and is a POLYNOMIAL solution. The POLYNOMIAL solutions defined here are known as CHEBYSHEV POLYNOMIALS OF THE FIRST KIND. The definition of the CHEBYSHEV POLYNOMIAL OF THE SECOND KIND gives a similar, but distinct, recurrence relation

$$a'_{n+2} = \frac{(n+1)^2-m^2}{(n+2)(n+3)}a'_n \quad \text{for } n = 0, 1, \ldots. \quad (21)$$

Chebyshev Function

$$\theta(x) \equiv \sum_{p\leq x}\ln p,$$

where the sum is over PRIMES p, so

$$\lim_{x\to\infty}\frac{x}{\theta(x)} = 1.$$

Chebyshev-Gauss Quadrature

Also called CHEBYSHEV QUADRATURE. A GAUSSIAN QUADRATURE over the interval $[-1,1]$ with WEIGHTING FUNCTION $W(x) = 1/\sqrt{1-x^2}$. The ABSCISSAS for quadrature order n are given by the roots of the CHEBYSHEV POLYNOMIAL OF THE FIRST KIND $T_n(x)$, which occur symmetrically about 0. The WEIGHTS are

$$w_i = -\frac{A_{n+1}\gamma_n}{A_nT'_n(x_i)T_{n+1}(x_i)} = \frac{A_n}{A_{n-1}}\frac{\gamma_{n-1}}{T_{n-1}(x_i)T'_n(x_i)}, \quad (1)$$

where A_n is the COEFFICIENT of x^n in $T_n(x)$. For HERMITE POLYNOMIALS,

$$A_n = 2^{n-1}, \quad (2)$$

so

$$\frac{A_{n+1}}{A_n} = 2. \quad (3)$$

Additionally,

$$\gamma_n = \tfrac{1}{2}\pi, \quad (4)$$

so

$$w_i = -\frac{\pi}{T_{n+1}(x_i)T'_n(x_i)}. \quad (5)$$

Since

$$T_n(x) = \cos(n\cos^{-1}x), \quad (6)$$

the ABSCISSAS are given explicitly by

$$x_i = \cos\left[\frac{(2i-1)\pi}{2n}\right]. \quad (7)$$

Since

$$T'_n(x_i) = \frac{(-1)^{i+1}n}{\alpha_i} \quad (8)$$

$$T_{n+1}(x_i) = (-1)^i\sin\alpha_i, \quad (9)$$

where

$$\alpha_i = \frac{(2i-1)\pi}{2n}, \quad (10)$$

all the WEIGHTS are

$$w_i = \frac{\pi}{n}. \quad (11)$$

The explicit FORMULA is then

$$\int_{-1}^{1} \frac{f(x)\,dx}{\sqrt{1-x^2}} = \frac{\pi}{n}\sum_{k=1}^{n} f\left[\cos\left(\frac{2k-1}{2n}\pi\right)\right] + \frac{2\pi}{2^{2n}(2n)!} f^{(2n)}(\xi). \quad (12)$$

n	x_i	w_i
2	±0.707107	1.5708
3	0	1.0472
	±0.866025	1.0472
4	±0.382683	0.785398
	±0.92388	0.785398
5	0	0.628319
	±0.587785	0.628319
	±0.951057	0.628319

References

Hildebrand, F. B. *Introduction to Numerical Analysis.* New York: McGraw-Hill, pp. 330–331, 1956.

Chebyshev Inequality

Apply MARKOV'S INEQUALITY with $a \equiv k^2$ to obtain

$$P[(x-\mu)^2 \geq k^2] \leq \frac{\langle (x-\mu)^2 \rangle}{k^2} = \frac{\sigma^2}{k^2}. \quad (1)$$

Therefore, if a RANDOM VARIABLE x has a finite MEAN μ and finite VARIANCE σ^2, then $\forall\ k \geq 0$,

$$P(|x-\mu| \geq k) \leq \frac{\sigma^2}{k^2} \quad (2)$$

$$P(|x-\mu| \geq k\sigma) \leq \frac{1}{k^2}. \quad (3)$$

References

Abramowitz, M. and Stegun, C. A. (Eds.). *Handbook of Mathematical Functions with Formulas, Graphs, and Mathematical Tables, 9th printing.* New York: Dover, p. 11, 1972.

Chebyshev Integral

$$\int x^p(1-x)^q\,dx.$$

Chebyshev Integral Inequality

$$\int_a^b f_1(x)\,dx \int_a^b f_2(x)\,dx \cdots \int_a^b f_n(x)\,dx \leq (b-a)^{n-1}\int_a^b f(x_1)f(x_2)\cdots f_n(x)\,dx,$$

where f_1, f_2, ..., f_n are NONNEGATIVE integrable functions on $[a,b]$ which are monotonic increasing or decreasing.

References

Gradshteyn, I. S. and Ryzhik, I. M. *Tables of Integrals, Series, and Products, 5th ed.* San Diego, CA: Academic Press, p. 1092, 1979.

Chebyshev Phenomenon

see PRIME QUADRATIC EFFECT

Chebyshev Polynomial of the First Kind

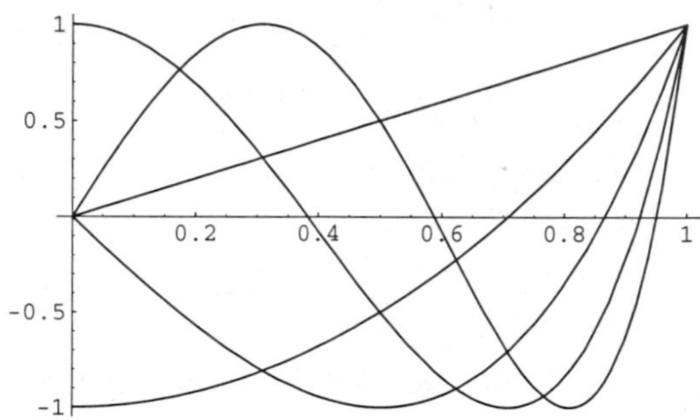

A set of ORTHOGONAL POLYNOMIALS defined as the solutions to the CHEBYSHEV DIFFERENTIAL EQUATION and denoted $T_n(x)$. They are used as an approximation to a LEAST SQUARES FIT, and are a special case of the ULTRASPHERICAL POLYNOMIAL with $\alpha = 0$. The Chebyshev polynomials of the first kind $T_n(x)$ are illustrated above for $x \in [0,1]$ and $n = 1, 2, \ldots, 5$.

The Chebyshev polynomials of the first kind can be obtained from the generating functions

$$g_1(t,x) \equiv \frac{1-t^2}{1-2xt+t^2} = T_0(x) + 2\sum_{n=1}^{\infty} T_n(x)t^n \quad (1)$$

and

$$g_2(t,x) \equiv \frac{1-xt}{1-2xt+t^2} = \sum_{n=0}^{\infty} T_n(x)t^n \quad (2)$$

for $|x| \leq 1$ and $|t| < 1$ (Beeler *et al.* 1972, Item 15). (A closely related GENERATING FUNCTION is the basis for the definition of CHEBYSHEV POLYNOMIAL OF THE SECOND KIND.) They are normalized such that $T_n(1) = 1$. They can also be written

$$T_n(x) = \frac{n}{2}\sum_{r=0}^{\lfloor n/2 \rfloor} \frac{(-1)^r}{n-r}\binom{n-r}{r}(2x)^{n-2r}, \quad (3)$$

or in terms of a DETERMINANT

$$T_n = \begin{vmatrix} x & 1 & 0 & 0 & \cdots & 0 & 0 \\ 1 & 2x & 1 & 0 & \cdots & 0 & 0 \\ 0 & 1 & 2x & 1 & \cdots & 0 & 0 \\ 0 & 0 & 1 & 2x & \cdots & 0 & 0 \\ \vdots & \vdots & \vdots & \vdots & \ddots & \vdots & \vdots \\ 0 & 0 & 0 & 0 & \cdots & 1 & 2x \end{vmatrix}. \tag{4}$$

In closed form,

$$T_n(x) = \cos(n\cos^{-1}x) = \sum_{m=0}^{\lfloor n/2 \rfloor} \binom{n}{2m} x^{n-2m}(x^2-1)^m, \tag{5}$$

where $\binom{n}{k}$ is a BINOMIAL COEFFICIENT and $\lfloor x \rfloor$ is the FLOOR FUNCTION. Therefore, zeros occur when

$$x = \cos\left[\frac{\pi(k-\frac{1}{2})}{n}\right] \tag{6}$$

for $k = 1, 2, \ldots, n$. Extrema occur for

$$x = \cos\left(\frac{\pi k}{n}\right), \tag{7}$$

where $k = 0, 1, \ldots, n$. At maximum, $T_n(x) = 1$, and at minimum, $T_n(x) = -1$. The Chebyshev POLYNOMIALS are ORTHONORMAL with respect to the WEIGHTING FUNCTION $(1-x^2)^{-1/2}$

$$\int_{-1}^{1} \frac{T_m(x)T_n(x)\,dx}{\sqrt{1-x^2}} = \begin{cases} \frac{1}{2}\pi\delta_{nm} & \text{for } m \neq 0,\ n \neq 0 \\ \pi & \text{for } m = n = 0, \end{cases} \tag{8}$$

where δ_{mn} is the KRONECKER DELTA. Chebyshev polynomials of the first kind satisfy the additional discrete identity

$$\sum_{k=1}^{m} T_i(x_k)T_j(x_k) = \begin{cases} \frac{1}{2}m\delta_{ij} & \text{for } i \neq 0,\ j \neq 0 \\ m & \text{for } i = j = 0, \end{cases} \tag{9}$$

where x_k for $k = 1, \ldots, m$ are the m zeros of $T_m(x)$. They also satisfy the RECURRENCE RELATIONS

$$T_{n+1}(x) = 2xT_n(x) - T_{n-1}(x) \tag{10}$$

$$T_{n+1}(x) = xT_n(x) - \sqrt{(1-x^2)\{1-[T_n(x)]^2\}} \tag{11}$$

for $n \geq 1$. They have a COMPLEX integral representation

$$T_n(x) = \frac{1}{4\pi i}\int_\gamma \frac{(1-z^2)z^{-n-1}\,dz}{1-2xz+z^2} \tag{12}$$

and a Rodrigues representation

$$T_n(x) = \frac{(-1)^n\sqrt{\pi}\,(1-x^2)^{1/2}}{2n(n-\frac{1}{2})!}\frac{d^n}{dx^n}[(1-x^2)^{n-1/2}]. \tag{13}$$

Using a FAST FIBONACCI TRANSFORM with multiplication law

$$(A,B)(C,D) = (AD+BC+2xAC, BD-AC) \tag{14}$$

gives

$$(T_{n+1}(x), -T_n(x)) = (T_1(x), -T_0(x))(1,0)^n. \tag{15}$$

Using GRAM-SCHMIDT ORTHONORMALIZATION in the range (−1,1) with WEIGHTING FUNCTION $(1-x^2)^{(-1/2)}$ gives

$$p_0(x) = 1 \tag{16}$$

$$\begin{aligned} p_1(x) &= \left[x - \frac{\int_{-1}^1 x(1-x^2)^{-1/2}\,dx}{\int_{-1}^1 (1-x^2)^{-1/2}\,dx}\right] \\ &= x - \frac{[-(1-x^2)^{1/2}]_{-1}^1}{[\sin^{-1}x]_{-1}^1} = x \end{aligned} \tag{17}$$

$$\begin{aligned} p_2(x) &= \left[x - \frac{\int_{-1}^1 x^3(1-x^2)^{-1/2}\,dx}{\int_{-1}^1 x^2(1-x^2)^{-1/2}\,dx}\right]x \\ &\quad - \left[\frac{\int_{-1}^1 x^2(1-x^2)^{-1/2}\,dx}{\int_{-1}^1 (1-x^2)^{-1/2}\,dx}\right]\cdot 1 \\ &= [x-0]x - \frac{\frac{\pi}{2}}{\pi} = x^2 - \frac{1}{2}, \end{aligned} \tag{18}$$

etc. Normalizing such that $T_n(1) = 1$ gives

$$\begin{aligned} T_0(x) &= 1 \\ T_1(x) &= x \\ T_2(x) &= 2x^2 - 1 \\ T_3(x) &= 4x^3 - 3x \\ T_4(x) &= 8x^4 - 8x^2 + 1 \\ T_5(x) &= 16x^5 - 20x^3 + 5x \\ T_6(x) &= 32x^6 - 48x^4 + 18x^2 - 1. \end{aligned}$$

The Chebyshev polynomial of the first kind is related to the BESSEL FUNCTION OF THE FIRST KIND $J_n(x)$ and MODIFIED BESSEL FUNCTION OF THE FIRST KIND $I_n(x)$ by the relations

$$J_n(x) = i^n T_n\left(i\frac{d}{dx}\right)J_0(x) \tag{19}$$

$$I_n(x) = T_n\left(\frac{d}{dx}\right)I_0(x). \tag{20}$$

Letting $x \equiv \cos\theta$ allows the Chebyshev polynomials of the first kind to be written as

$$T_n(x) = \cos(n\theta) = \cos(n\cos^{-1}x). \tag{21}$$

The second linearly dependent solution to the transformed differential equation

$$\frac{d^2T_n}{d\theta^2} + n^2T_n = 0 \tag{22}$$

is then given by

$$V_n(x) = \sin(n\theta) = \sin(n\cos^{-1}x), \tag{23}$$

which can also be written

$$V_n(x) = \sqrt{1-x^2}\, U_{n-1}(x), \tag{24}$$

where U_n is a CHEBYSHEV POLYNOMIAL OF THE SECOND KIND. Note that $V_n(x)$ is therefore *not* a POLYNOMIAL.

The POLYNOMIAL

$$x^n - 2^{1-n}T_n(x) \tag{25}$$

(of degree $n-2$) is the POLYNOMIAL of degree $< n$ which stays closest to x^n in the interval $(-1,1)$. The maximum deviation is 2^{1-n} at the $n+1$ points where

$$x = \cos\left(\frac{k\pi}{n}\right), \tag{26}$$

for $k = 0, 1, \ldots, n$ (Beeler *et al.* 1972, Item 15).

see also CHEBYSHEV APPROXIMATION FORMULA, CHEBYSHEV POLYNOMIAL OF THE SECOND KIND

References

Abramowitz, M. and Stegun, C. A. (Eds.). "Orthogonal Polynomials." Ch. 22 in *Handbook of Mathematical Functions with Formulas, Graphs, and Mathematical Tables, 9th printing.* New York: Dover, pp. 771–802, 1972.

Arfken, G. "Chebyshev (Tschebyscheff) Polynomials" and "Chebyshev Polynomials—Numerical Applications." §13.3 and 13.4 in *Mathematical Methods for Physicists, 3rd ed.* Orlando, FL: Academic Press, pp. 731–748, 1985.

Beeler, M.; Gosper, R. W.; and Schroeppel, R. *HAKMEM.* Cambridge, MA: MIT Artificial Intelligence Laboratory, Memo AIM-239, Feb. 1972.

Iyanaga, S. and Kawada, Y. (Eds.). "Čebyšev (Tschebyscheff) Polynomials." Appendix A, Table 20.II in *Encyclopedic Dictionary of Mathematics.* Cambridge, MA: MIT Press, pp. 1478–1479, 1980.

Rivlin, T. J. *Chebyshev Polynomials.* New York: Wiley, 1990.

Spanier, J. and Oldham, K. B. "The Chebyshev Polynomials $T_n(x)$ and $U_n(x)$." Ch. 22 in *An Atlas of Functions.* Washington, DC: Hemisphere, pp. 193–207, 1987.

Chebyshev Polynomial of the Second Kind

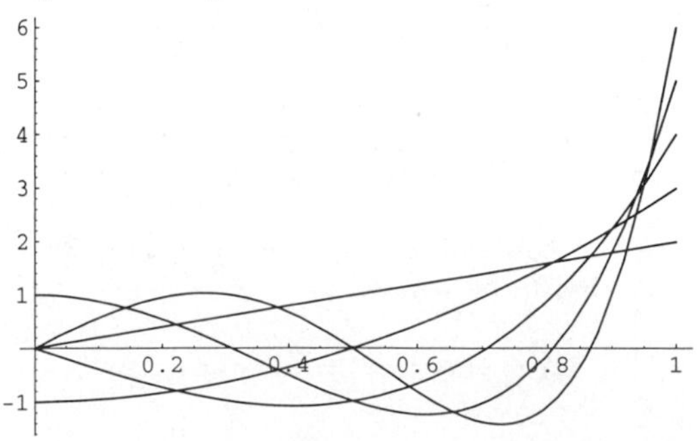

A modified set of Chebyshev POLYNOMIALS defined by a slightly different GENERATING FUNCTION. Used to develop four-dimensional SPHERICAL HARMONICS in angular momentum theory. They are also a special case of the ULTRASPHERICAL POLYNOMIAL with $\alpha = 1$. The Chebyshev polynomials of the second kind $U_n(x)$ are illustrated above for $x \in [0,1]$ and $n = 1, 2, \ldots, 5$.

The defining GENERATING FUNCTION of the Chebyshev polynomials of the second kind is

$$g_2(t,x) = \frac{1}{1-2xt+t^2} = \sum_{n=0}^{\infty} U_n(x)t^n \tag{1}$$

for $|x| < 1$ and $|t| < 1$. To see the relationship to a CHEBYSHEV POLYNOMIAL OF THE FIRST KIND (T), take $\partial g/\partial t$,

$$\begin{aligned}\frac{\partial g}{\partial t} &= -(1-2xt+t^2)^{-2}(-2x+2t)\\ &= 2(t-x)(1-2xt+t^2)^{-2}\\ &= \sum_{n=0}^{\infty} nU_n(x)t^{n-1}.\end{aligned} \tag{2}$$

Multiply (2) by t,

$$(2t^2-2xt)(1-2xt+t^2)^{-2} = \sum_{n=0}^{\infty} nU_n(x)t^n \tag{3}$$

and take (3)−(2),

$$\begin{aligned}\frac{(2t^2-2tx)-(1-2xt+t^2)}{(1-2xt+t^2)^2} &= \frac{t^2-1}{(1-2xt+t)^2}\\ &= \sum_{n=0}^{\infty}(n-1)U_n(x)t^n.\end{aligned} \tag{4}$$

The Rodrigues representation is

$$U_n(x) = \frac{(-1)^n(n+1)\sqrt{\pi}}{2^{n+1}(n+\frac{1}{2})!(1-x^2)^{1/2}}\frac{d^n}{dx^n}[(1-x^2)^{n+1/2}]. \tag{5}$$

The polynomials can also be written

$$\begin{aligned}U_n(x) &= \sum_{r=0}^{\lfloor n/2\rfloor}(-1)^r\binom{n-r}{r}(2x)^{n-2r}\\ &= \sum_{m=0}^{\lceil n/2\rceil}\binom{n+1}{2m+1}x^{n-2m}(x^2-1)^m,\end{aligned} \tag{6}$$

where $\lfloor x\rfloor$ is the FLOOR FUNCTION and $\lceil x\rceil$ is the CEILING FUNCTION, or in terms of a DETERMINANT

$$U_n = \begin{vmatrix} 2x & 1 & 0 & 0 & \cdots & 0 & 0\\ 0 & 2x & 1 & 0 & \cdots & 0 & 0\\ 0 & 1 & 2x & 1 & \cdots & 0 & 0\\ \vdots & \vdots & \ddots & \ddots & \ddots & \ddots & \vdots\\ 0 & 0 & 0 & 0 & \cdots & 1 & 2x\end{vmatrix}. \tag{7}$$

The first few POLYNOMIALS are

$$U_0(x) = 1$$
$$U_1(x) = 2x$$
$$U_2(x) = 4x^2 - 1$$
$$U_3(x) = 8x^3 - 4x$$
$$U_4(x) = 16x^4 - 12x^2 + 1$$
$$U_5(x) = 32x^5 - 32x^3 + 6x$$
$$U_6(x) = 64x^6 - 80x^4 + 24x^2 - 1.$$

Letting $x \equiv \cos\theta$ allows the Chebyshev polynomials of the second kind to be written as

$$U_n(x) = \frac{\sin[(n+1)\theta]}{\sin\theta}. \tag{8}$$

The second linearly dependent solution to the transformed differential equation is then given by

$$W_n(x) = \frac{\cos[(n+1)\theta]}{\sin\theta}, \tag{9}$$

which can also be written

$$W_n(x) = (1-x^2)^{-1/2}T_{n+1}(x), \tag{10}$$

where T_n is a CHEBYSHEV POLYNOMIAL OF THE FIRST KIND. Note that $W_n(x)$ is therefore *not* a POLYNOMIAL.

see also CHEBYSHEV APPROXIMATION FORMULA, CHEBYSHEV POLYNOMIAL OF THE FIRST KIND, ULTRASPHERICAL POLYNOMIAL

References

Abramowitz, M. and Stegun, C. A. (Eds.). "Orthogonal Polynomials." Ch. 22 in *Handbook of Mathematical Functions with Formulas, Graphs, and Mathematical Tables, 9th printing.* New York: Dover, pp. 771–802, 1972.

Arfken, G. "Chebyshev (Tschebyscheff) Polynomials" and "Chebyshev Polynomials—Numerical Applications." §13.3 and 13.4 in *Mathematical Methods for Physicists, 3rd ed.* Orlando, FL: Academic Press, pp. 731–748, 1985.

Rivlin, T. J. *Chebyshev Polynomials.* New York: Wiley, 1990.

Spanier, J. and Oldham, K. B. "The Chebyshev Polynomials $T_n(x)$ and $U_n(x)$." Ch. 22 in *An Atlas of Functions.* Washington, DC: Hemisphere, pp. 193–207, 1987.

Chebyshev Quadrature

A GAUSSIAN QUADRATURE-like FORMULA for numerical estimation of integrals. It uses WEIGHTING FUNCTION $W(x) = 1$ in the interval $[-1, 1]$ and forces all the weights to be equal. The general FORMULA is

$$\int_{-1}^{1} f(x)\,dx = \frac{2}{n}\sum_{i=1}^{n} f(x_i).$$

The ABSCISSAS are found by taking terms up to y^n in the MACLAURIN SERIES of

$$s_n(y) = \exp\left\{\tfrac{1}{2}n\left[-2 + \ln(1-y)\left(1-\frac{1}{y}\right) + \ln(1+y)\left(1+\frac{1}{y}\right)\right]\right\},$$

and then defining

$$G_n(x) \equiv x^n s_n\left(\frac{1}{x}\right).$$

The ROOTS of $G_n(x)$ then give the ABSCISSAS. The first few values are

$$G_0(x) = 1$$
$$G_1(x) = x$$
$$G_2(x) = \tfrac{1}{3}(3x^2 - 1)$$
$$G_3(x) = \tfrac{1}{2}(2x^3 - x)$$
$$G_4(x) = \tfrac{1}{45}(45x^4 - 30x^2 + 1)$$
$$G_5(x) = \tfrac{1}{72}(72x^5 - 60x^3 + 7x)$$
$$G_6(x) = \tfrac{1}{105}(105x^6 - 105x^4 + 21x^2 - 1)$$
$$G_7(x) = \tfrac{1}{6480}(6480x^7 - 7560x^5 + 2142x^3 - 149x)$$
$$G_8(x) = \tfrac{1}{42525}(42525x^8 - 56700x^6 + 20790x^4 - 2220x^2 - 43)$$
$$G_9(x) = \tfrac{1}{22400}(22400x^9 - 33600x^7 + 15120x^5 - 2280x^3 + 53x).$$

Because the ROOTS are all REAL for $n \leq 7$ and $n = 9$ only (Hildebrand 1956), these are the only permissible orders for Chebyshev quadrature. The error term is

$$E_n = \begin{cases} c_n \frac{f^{(n+1)}(\xi)}{(n+1)!} & n \text{ odd} \\ c_n \frac{f^{(n+2)}(\xi)}{(n+2)!} & n \text{ even,} \end{cases}$$

where

$$c_n = \begin{cases} \int_{-1}^{1} xG_n(x)\,dx & n \text{ odd} \\ \int_{-1}^{1} x^2G_n(x)\,dx & n \text{ even.} \end{cases}$$

The first few values of c_n are 2/3, 8/45, 1/15, 32/945, 13/756, and 16/1575 (Hildebrand 1956). Beyer (1987) gives abscissas up to $n = 7$ and Hildebrand (1956) up to $n = 9$.

n	x_i
2	± 0.57735
3	0
	± 0.707107
4	± 0.187592
	± 0.794654
5	0
	± 0.374541
	± 0.832497
6	± 0.266635
	± 0.422519
	± 0.866247
7	0
	± 0.323912
	± 0.529657
	± 0.883862
9	0
	± 0.167906
	± 0.528762
	± 0.601019
	± 0.911589

The ABSCISSAS and weights can be computed analytically for small n.

n	x_i
2	$\pm\frac{1}{3}\sqrt{3}$
3	0
	$\pm\frac{1}{2}\sqrt{2}$
4	$\pm\sqrt{\frac{\sqrt{5}-2}{3\sqrt{5}}}$
	$\pm\sqrt{\frac{\sqrt{5}+2}{3\sqrt{5}}}$
5	0
	$\pm\frac{1}{2}\sqrt{\frac{5-\sqrt{11}}{3}}$
	$\pm\frac{1}{2}\sqrt{\frac{5+\sqrt{11}}{3}}$

see also CHEBYSHEV QUADRATURE, LOBATTO QUADRATURE

References

Beyer, W. H. *CRC Standard Mathematical Tables, 28th ed.* Boca Raton, FL: CRC Press, p. 466, 1987.

Hildebrand, F. B. *Introduction to Numerical Analysis.* New York: McGraw-Hill, pp. 345–351, 1956.

Chebyshev-Radau Quadrature

A GAUSSIAN QUADRATURE-like FORMULA over the interval $[-1, 1]$ which has WEIGHTING FUNCTION $W(x) = x$. The general FORMULA is

$$\int_{-1}^{1} x f(x)\,dx = \sum_{i=1}^{n} w_i[f(x_i) - f(-x_i)].$$

n	x_i	w_i
1	0.7745967	0.4303315
2	0.5002990	0.2393715
	0.8922365	0.2393715
3	0.4429861	0.1599145
	0.7121545	0.1599145
	0.9293066	0.1599145
4	0.3549416	0.1223363
	0.6433097	0.1223363
	0.7783202	0.1223363
	0.9481574	0.1223363

References

Beyer, W. H. *CRC Standard Mathematical Tables, 28th ed.* Boca Raton, FL: CRC Press, p. 466, 1987.

Chebyshev Sum Inequality

If

$$a_1 \geq a_2 \geq \ldots \geq a_n$$

$$b_1 \geq b_2 \geq \ldots \geq b_n,$$

then

$$n\sum_{k=1}^{n} a_k b_k \geq \left(\sum_{k=1}^{n} a_k\right)\left(\sum_{k=1}^{n} b_k\right).$$

This is true for *any* distribution.

see also CAUCHY INEQUALITY, HÖLDER SUM INEQUALITY

References

Gradshteyn, I. S. and Ryzhik, I. M. *Tables of Integrals, Series, and Products, 5th ed.* San Diego, CA: Academic Press, p. 1092, 1979.

Hardy, G. H.; Littlewood, J. E.; and Pólya, G. *Inequalities, 2nd ed.* Cambridge, England: Cambridge University Press, pp. 43–44, 1988.

Chebyshev-Sylvester Constant

In 1891, Chebyshev and Sylvester showed that for sufficiently large x, there exists at least one prime number p satisfying

$$x < p < (1 + \alpha)x,$$

where $\alpha = 0.092\ldots$. Since the PRIME NUMBER THEOREM shows the above inequality is true for all $\alpha > 0$ for sufficiently large x, this constant is only of historical interest.

References

Le Lionnais, F. *Les nombres remarquables.* Paris: Hermann, p. 22, 1983.

Chebyshev's Theorem

see BERTRAND'S POSTULATE

Checker-Jumping Problem

Seeks the minimum number of checkers placed on a board required to allow pieces to move by a sequence of horizontal or vertical jumps (removing the piece jumped over) n rows beyond the forward-most initial checker. The first few cases are 2, 4, 8, 20. It is, however, impossible to reach level 5.

References

Honsberger, R. *Mathematical Gems II.* Washington, DC: Math. Assoc. Amer., pp. 23–28, 1976.

Checkerboard

see CHESSBOARD

Checkers

Beeler *et al.* (1972, Item 93) estimated that there are about 10^{12} possible positions. However, this disagrees with the estimate of Jon Schaeffer of 5×10^{20} plausible positions, with 10^{18} reachable under the rules of the game. Because "solving" checkers may require only the SQUARE ROOT of the number of positions in the search space (i.e., 10^9), so there is hope that some day checkers may be solved (i.e., it may be possible to guarantee a win for the first player to move before the game is even started; Dubuque 1996).

Depending on how they are counted, the number of EULERIAN CIRCUITS on an $n \times n$ checkerboard are either 1, 40, 793, 12800, 193721, ... (Sloane's A006240) or 1, 13, 108, 793, 5611, 39312, ... (Sloane's A006239).

see also CHECKERBOARD, CHECKER-JUMPING PROBLEM

References

Beeler, M.; Gosper, R. W.; and Schroeppel, R. *HAKMEM.* Cambridge, MA: MIT Artificial Intelligence Laboratory, Memo AIM-239, Feb. 1972.

Dubuque, W. "Re: number of legal chess positions." math-fun@cs.arizona.edu posting, Aug 15, 1996.

Kraitchik, M. "Chess and Checkers" and "Checkers (Draughts)." §12.1.1 and 12.1.10 in *Mathematical Recreations.* New York: W. W. Norton, pp. 267–276 and 284–287, 1942.

Schaeffer, J. *One Jump Ahead: Challenging Human Supremacy in Checkers.* New York: Springer-Verlag, 1997.

Sloane, N. J. A. Sequences A006239/M4909 and A006240/M5271 in "An On-Line Version of the Encyclopedia of Integer Sequences."

Checksum

A sum of the digits in a given transmission modulo some number. The simplest form of checksum is a parity bit appended on to 7-bit numbers (e.g., ASCII characters) such that the total number of 1s is always EVEN ("even parity") or ODD ("odd parity"). A significantly more sophisticated checksum is the CYCLIC REDUNDANCY CHECK (or CRC), which is based on the algebra of polynomials over the integers (mod 2). It is substantially more reliable in detecting transmission errors, and is one common error-checking protocol used in modems.

see also CYCLIC REDUNDANCY CHECK, ERROR-CORRECTING CODE

References

Press, W. H.; Flannery, B. P.; Teukolsky, S. A.; and Vetterling, W. T. "Cyclic Redundancy and Other Checksums." Ch. 20.3 in *Numerical Recipes in FORTRAN: The Art of Scientific Computing, 2nd ed.* Cambridge, England: Cambridge University Press, pp. 888–895, 1992.

Cheeger's Finiteness Theorem

Consider the set of compact n-RIEMANNIAN MANIFOLDS M with diameter$(M) \leq d$, Volume$(M) \geq V$, and $|\mathcal{K}| \leq \kappa$ where κ is the SECTIONAL CURVATURE. Then there is a bound on the number of DIFFEOMORPHISMS classes of this set in terms of the constants n, d, V, and κ.

References

Chavel, I. *Riemannian Geometry: A Modern Introduction.* New York: Cambridge University Press, 1994.

Chefalo Knot

A fake KNOT created by tying a SQUARE KNOT, then looping one end twice through the KNOT such that when both ends are pulled, the KNOT vanishes.

Chen's Theorem

Every "large" EVEN INTEGER may be written as $2n = p + m$ where p is a PRIME and $m \in P_2$ is the SET of SEMIPRIMES (i.e., 2-ALMOST PRIMES).

see also ALMOST PRIME, PRIME NUMBER, SEMIPRIME

References

Rivera, C. "Problems & Puzzles (Conjectures): Chen's Conjecture." `http://www.sci.net.mx/~crivera/ppp/conj_002.htm`.

Chern Class

A GADGET defined for COMPLEX VECTOR BUNDLES. The Chern classes of a COMPLEX MANIFOLD are the Chern classes of its TANGENT BUNDLE. The ith Chern class is an OBSTRUCTION to the existence of $(n - i + 1)$ everywhere COMPLEX linearly independent VECTOR FIELDS on that VECTOR BUNDLE. The ith Chern class is in the $(2i)$th cohomology group of the base SPACE.

see also OBSTRUCTION, PONTRYAGIN CLASS, STIEFEL-WHITNEY CLASS

Chern Number

The Chern number is defined in terms of the CHERN CLASS of a MANIFOLD as follows. For any collection CHERN CLASSES such that their cup product has the same DIMENSION as the MANIFOLD, this cup product can be evaluated on the MANIFOLD'S FUNDAMENTAL CLASS. The resulting number is called the Chern number for that combination of Chern classes. The most important aspect of Chern numbers is that they are COBORDISM invariant.

see also PONTRYAGIN NUMBER, STIEFEL-WHITNEY NUMBER

Chernoff Face

A way to display n variables on a 2-D surface. For instance, let x be eyebrow slant, y be eye size, z be nose length, etc.

References

Gonick, L. and Smith, W. *The Cartoon Guide to Statistics.* New York: Harper Perennial, p. 212, 1993.

Chess

Chess is a game played on an 8×8 board, called a CHESSBOARD, of alternating black and white squares. Pieces with different types of allowed moves are placed on the board, a set of black pieces in the first two rows and a set of white pieces in the last two rows. The pieces are called the bishop (2), king (1), knight (2), pawn (8), queen (1), and rook (2). The object of the game is to capture the opponent's king. It is believed that chess was played in India as early as the sixth century AD.

In a game of 40 moves, the number of possible board positions is at least 10^{120} according to Peterson (1996). However, this value does not agree with the 10^{40} possible positions given by Beeler *et al.* (1972, Item 95). This value was obtained by estimating the number of pawn positions (in the no-captures situation, this is 15^8), times all pieces in all positions, dividing by 2 for each of the (rook, knight) which are interchangeable, dividing by 2 for each pair of bishops (since half the positions will have the bishops on the same color squares). There are more positions with one or two captures, since the pawns can then switch columns (Schroeppel 1996). Shannon (1950) gave the value

$$P(40) \approx \frac{64!}{32!(8!)^2(2!)^6} \approx 10^{43}.$$

The number of chess games which end in exactly n plies (including games that mate in fewer than n plies) for $n = 1, 2, 3, \ldots$ are 20, 400, 8902, 197742, 4897256, 119060679, 3195913043, ... (K. Thompson, Sloane's A007545). Rex Stout's fictional detective Nero Wolfe quotes the number of possible games after ten moves as follows: "Wolfe grunted. One hundred and sixty-nine million, five hundred and eighteen thousand, eight hundred and twenty-nine followed by twenty-one ciphers. The number of ways the first ten moves, both sides, may be played" (Stout 1983). The number of chess positions after n moves for $n = 1, 2, \ldots$ are 20, 400, 5362, 71852, 809896?, 9132484?, ... (Schwarzkopf 1994, Sloane's A019319).

Cunningham (1889) incorrectly found 197,299 games and 71,782 positions after the fourth move. C. Flye St. Marie was the first to find the correct number of positions after four moves: 71,852. Dawson (1946) gives the source as *Intermediare des Mathematiques* (1895), but K. Fabel writes that Flye St. Marie corrected the number 71,870 (which he found in 1895) to 71,852 in 1903. The history of the determination of the chess sequences is discussed in Schwarzkopf (1994).

Two problems in recreational mathematics ask

1. How many pieces of a given type can be placed on a CHESSBOARD without any two attacking.
2. What is the smallest number of pieces needed to occupy or attack every square.

The answers are given in the following table (Madachy 1979).

Piece	Max.	Min.
bishops	14	8
kings	16	9
knights	32	12
queens	8	5
rooks	8	8

see also BISHOPS PROBLEM, CHECKERBOARD, CHECKERS, FAIRY CHESS, GO, GOMORY'S THEOREM, HARD HEXAGON ENTROPY CONSTANT, KINGS PROBLEM, KNIGHT'S TOUR, MAGIC TOUR, QUEENS PROBLEM, ROOKS PROBLEM, TOUR

References

Ball, W. W. R. and Coxeter, H. S. M. *Mathematical Recreations and Essays, 13th ed.* New York: Dover, pp. 124–127, 1987.

Beeler, M.; Gosper, R. W.; and Schroeppel, R. *HAKMEM.* Cambridge, MA: MIT Artificial Intelligence Laboratory, Memo AIM-239, Feb. 1972.

Dawson, T. R. "A Surprise Correction." *The Fairy Chess Review* **6**, 44, 1946.

Dickins, A. "A Guide to Fairy Chess." p. 28, 1967/1969/1971.

Dudeney, H. E. "Chessboard Problems." *Amusements in Mathematics.* New York: Dover, pp. 84–109, 1970.

Fabel, K. "Nüsse." *Die Schwalbe* **84**, 196, 1934.

Fabel, K. "Weihnachtsnüsse." *Die Schwalbe* **190**, 97, 1947.

Fabel, K. "Weihnachtsnüsse." *Die Schwalbe* **195**, 14, 1948.

Fabel, K. "Eröffnungen." *Am Rande des Schachbretts*, 34–35, 1947.

Fabel, K. "Die ersten Schritte." *Rund um das Schachbrett*, 107–109, 1955.

Fabel, K. "Eröffnungen." *Schach und Zahl* **8**, 1966/1971.

Hunter, J. A. H. and Madachy, J. S. *Mathematical Diversions.* New York: Dover, pp. 86–89, 1975.

Kraitchik, M. "Chess and Checkers." §12.1.1 in *Mathematical Recreations.* New York: W. W. Norton, pp. 267–276, 1942.

Madachy, J. S. "Chessboard Placement Problems." Ch. 2 in *Madachy's Mathematical Recreations.* New York: Dover, pp. 34–54, 1979.

Peterson, I. "The Soul of a Chess Machine: Lessons Learned from a Contest Pitting Man Against Computer." *Sci. News* **149**, 200–201, Mar. 30, 1996.

Petković, M. *Mathematics and Chess.* New York: Dover, 1997.

Schroeppel, R. "Reprise: Number of legal chess positions." tech-news@cs.arizona.edu posting, Aug. 18, 1996.

Schwarzkopf, B. "Die ersten Züge." *Problemkiste*, 142–143, No. 92, Apr. 1994.

Shannon, C. "Programming a Computer for Playing Chess." *Phil. Mag.* **41**, 256–275, 1950.

Sloane, N. J. A. Sequences A019319 and A007545/M5100 in "An On-Line Version of the Encyclopedia of Integer Sequences."

Stout, R. "Gambit." In *Seven Complete Nero Wolfe Novels.* New York: Avenic Books, p. 475, 1983.

Chessboard

A board containing 8×8 squares alternating in color between black and white on which the game of CHESS is played. The checkerboard is identical to the chessboard except that chess's black and white squares are colored red and white in CHECKERS. It is impossible to cover a chessboard from which two opposite corners have been removed with DOMINOES.

see also CHECKERS, CHESS, DOMINO, GOMORY'S THEOREM, WHEAT AND CHESSBOARD PROBLEM

References

Pappas, T. "The Checkerboard." *The Joy of Mathematics.* San Carlos, CA: Wide World Publ./Tetra, pp. 136 and 232, 1989.

Chevalley Groups

Finite SIMPLE GROUPS of LIE-TYPE. They include four families of linear SIMPLE GROUPS: $PSL(n,q)$, $PSU(n,q)$, $PSp(2n,q)$, or $P\Omega^\epsilon(n,q)$.

see also TWISTED CHEVALLEY GROUPS

References

Wilson, R. A. "ATLAS of Finite Group Representation." `http://for.mat.bham.ac.uk/atlas#chev`.

Chevalley's Theorem

Let $f(x)$ be a member of a FINITE FIELD $F[x_1, x_2, \ldots, x_n]$ and suppose $f(0,0,\ldots,0) = 0$ and n is greater than the degree of f, then f has at least two zeros in $A^n(F)$.

References

Chevalley, C. "Démonstration d'une hypothèse de M. Artin." *Abhand. Math. Sem. Hamburg* **11**, 73–75, 1936.

Ireland, K. and Rosen, M. "Chevalley's Theorem." §10.2 in *A Classical Introduction to Modern Number Theory, 2nd ed.* New York: Springer-Verlag, pp. 143–144, 1990.

Chevron

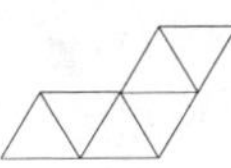

A 6-POLYIAMOND.

References

Golomb, S. W. *Polyominoes: Puzzles, Patterns, Problems, and Packings, 2nd ed.* Princeton, NJ: Princeton University Press, p. 92, 1994.

Chi

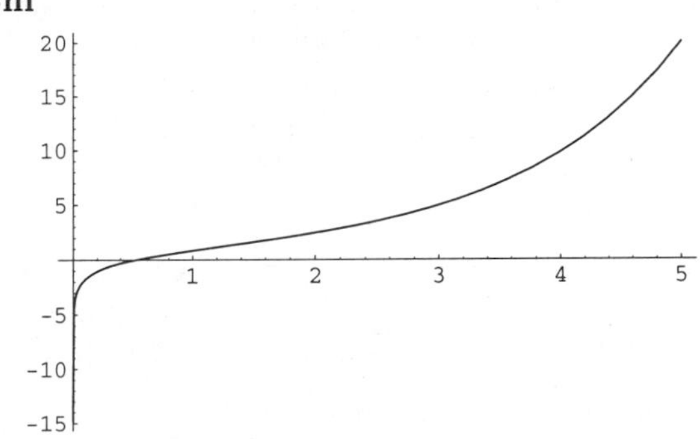

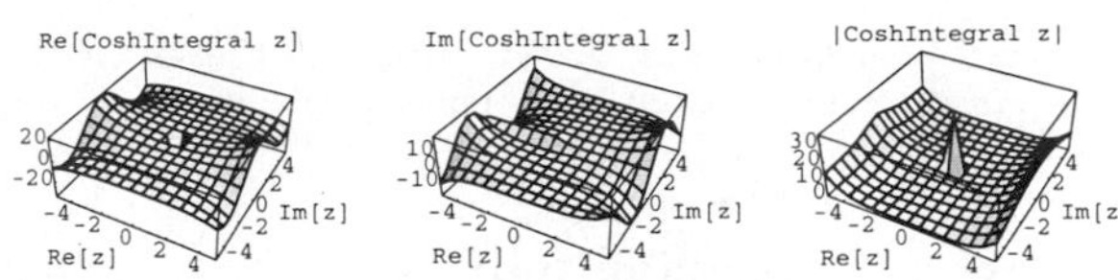

$$\mathrm{Chi}(z) = \gamma + \ln z + \int_0^z \frac{\cosh t - 1}{t}\,dt,$$

where γ is the EULER-MASCHERONI CONSTANT. The function is given by the *Mathematica*® (Wolfram Research, Champaign, IL) command `CoshIntegral[z]`.

see also COSINE INTEGRAL, SHI, SINE INTEGRAL

References

Abramowitz, M. and Stegun, C. A. (Eds.). "Sine and Cosine Integrals." §5.2 in *Handbook of Mathematical Functions with Formulas, Graphs, and Mathematical Tables, 9th printing.* New York: Dover, pp. 231–233, 1972.

Chi Distribution

The probability density function and cumulative distribution function are

$$P_n(x) = \frac{2^{1-n/2} x^{n-1} e^{-x^2/2}}{\Gamma(\tfrac{1}{2}n)} \tag{1}$$

$$D_n(x) = Q(\tfrac{1}{2}n, \tfrac{1}{2}x^2), \tag{2}$$

where Q is the REGULARIZED GAMMA FUNCTION.

$$\mu = \frac{\sqrt{2}\,\Gamma(\tfrac{1}{2}(n+1))}{\Gamma(\tfrac{1}{2}n)} \tag{3}$$

$$\sigma^2 = \frac{2[\Gamma(\tfrac{1}{2}n)\Gamma(1+\tfrac{1}{2}n) - \Gamma^2(\tfrac{1}{2}(n+1))]}{\Gamma^2(\tfrac{1}{2}n)} \tag{4}$$

$$\gamma_1 = \frac{2\Gamma^3(\tfrac{1}{2}(n+1)) - 3\Gamma(\tfrac{1}{2}n)\Gamma(\tfrac{1}{2}(n+1))\Gamma(1+\tfrac{1}{2}n)}{[\Gamma(\tfrac{1}{2}n)\Gamma(1+\tfrac{1}{2}n) - \Gamma^2(\tfrac{1}{2}(n+1))]^{3/2}}$$

$$+\frac{\Gamma^2(\frac{1}{2}n)\Gamma(\frac{3+n}{2})}{[\Gamma(\frac{1}{2}n)\Gamma(1+\frac{1}{2}n)-\Gamma^2(\frac{1}{2}(n+1))]^{3/2}} \tag{5}$$

$$\gamma_2 = \frac{-3\Gamma^4(\frac{1}{2}(n+1))+6\Gamma(\frac{1}{2}n)+\Gamma^2(\frac{1}{2}(n+1))\Gamma(1+\frac{1}{2}n)}{[\Gamma(\frac{1}{2}n)\Gamma(\frac{2+n}{2})-\Gamma^2(\frac{1}{2}(n+1))]^2}$$
$$+\frac{-4\Gamma^2(\frac{1}{2}n)\Gamma(\frac{1}{2}(n+1))\Gamma(\frac{3+n}{2})+\Gamma^3(\frac{1}{2}n)\Gamma(\frac{4+n}{2})}{[\Gamma(\frac{1}{2}n)\Gamma(\frac{2+n}{2})-\Gamma^2(\frac{1}{2}(n+1))]^2}, \tag{6}$$

where μ is the MEAN, σ^2 the VARIANCE, γ_1 the SKEWNESS, and γ_2 the KURTOSIS. For $n = 1$, the χ distribution is a HALF-NORMAL DISTRIBUTION with $\theta = 1$. For $n = 2$, it is a RAYLEIGH DISTRIBUTION with $\sigma = 1$.

see also CHI-SQUARED DISTRIBUTION, HALF-NORMAL DISTRIBUTION, RAYLEIGH DISTRIBUTION

Chi Inequality

The inequality

$$(j+1)a_j + a_i \geq (j+1)i,$$

which is satisfied by all A-SEQUENCES.

References

Levine, E. and O'Sullivan, J. "An Upper Estimate for the Reciprocal Sum of a Sum-Free Sequence." *Acta Arith.* **34**, 9–24, 1977.

Chi-Squared Distribution

A χ^2 distribution is a GAMMA DISTRIBUTION with $\theta \equiv 2$ and $\alpha \equiv r/2$, where r is the number of DEGREES OF FREEDOM. If Y_i have NORMAL INDEPENDENT distributions with MEAN 0 and VARIANCE 1, then

$$\chi^2 \equiv \sum_{i=1}^{n} {Y_i}^2 \tag{1}$$

is distributed as χ^2 with n DEGREES OF FREEDOM. If ${\chi_i}^2$ are independently distributed according to a χ^2 distribution with $n_1, n_2, \ldots, n_k$ DEGREES OF FREEDOM, then

$$\sum_{j=1}^{k} {\chi_j}^2 \tag{2}$$

is distributed according to χ^2 with $n \equiv \sum_{j=1}^{n} n_j$ DEGREES OF FREEDOM.

$$P_n(x) = \begin{cases} \frac{x^{r/2-1}e^{-x/2}}{\Gamma(\frac{1}{2}r)2^{r/2}} & \text{for } 0 \leq x < \infty \\ 0 & \text{for } x < 0. \end{cases} \tag{3}$$

The cumulative distribution function is then

$$D_n(\chi^2) = \int_0^{\chi^2} \frac{t^{r/2-1}e^{-t/2}\,dt}{\Gamma(\frac{1}{2}r)2^{r/2}}$$
$$= \frac{\gamma(\frac{1}{2}n, \frac{1}{2}\chi^2)}{\Gamma(\frac{1}{2}n)} = P(\tfrac{1}{2}n, \tfrac{1}{2}\chi^2), \tag{4}$$

where $P(a, z)$ is a REGULARIZED GAMMA FUNCTION. The CONFIDENCE INTERVALS can be found by finding the value of x for which $D_n(x)$ equals a given value. The MOMENT-GENERATING FUNCTION of the χ^2 distribution is

$$M(t) = (1-2t)^{-r/2} \tag{5}$$
$$R(t) \equiv \ln M(t) = -\tfrac{1}{2}r\ln(1-2t) \tag{6}$$
$$R'(t) = \frac{r}{1-2t} \tag{7}$$
$$R''(t) = \frac{2r}{(1-2t)^2}, \tag{8}$$

so

$$\mu = R'(0) = r \tag{9}$$
$$\sigma^2 = R''(0) = 2r \tag{10}$$
$$\gamma_1 = 2\sqrt{\frac{2}{r}} \tag{11}$$
$$\gamma_2 = \frac{12}{r}. \tag{12}$$

The nth MOMENT about zero for a distribution with n DEGREES OF FREEDOM is

$$m_n' = 2^n \frac{\Gamma(n+\frac{1}{2}r)}{\Gamma(\frac{1}{2}r)} = r(r+2)\cdots(r+2n-2), \tag{13}$$

and the moments about the MEAN are

$$\mu_2 = 2r \tag{14}$$
$$\mu_3 = 8r \tag{15}$$
$$\mu_4 = 12n^2 + 48n. \tag{16}$$

The nth CUMULANT is

$$\kappa_n = 2^n\Gamma(n)(\tfrac{1}{2}r) = 2^{n-1}(n-1)!r. \tag{17}$$

The MOMENT-GENERATING FUNCTION is

$$M(t) = e^{rt/\sqrt{2r}}\left(1-\frac{2t}{\sqrt{2r}}\right)^{-r/2}$$
$$= \left[e^{t\sqrt{2/r}}\left(1-\sqrt{\frac{2}{r}}\,t\right)\right]^{-r/2}$$
$$= \left[1-\frac{t^2}{r}-\frac{1}{3}\left(\frac{2}{r}\right)^{3/2}t^3-\ldots\right]^{-r/2}. \tag{18}$$

As $r \to \infty$,

$$\lim_{r\to\infty} M(t) = e^{t^2/2}, \tag{19}$$

so for large r,

$$\sqrt{2\chi^2} = \sqrt{\sum_i \frac{(x_i-\mu_i)^2}{{\sigma_i}^2}} \tag{20}$$

is approximately a GAUSSIAN DISTRIBUTION with MEAN $\sqrt{2r}$ and VARIANCE $\sigma^2 = 1$. Fisher showed that

$$\frac{\chi^2 - r}{\sqrt{2r-1}} \tag{21}$$

is an improved estimate for moderate r. Wilson and Hilferty showed that

$$\left(\frac{\chi^2}{r}\right)^{1/3} \tag{22}$$

is a nearly GAUSSIAN DISTRIBUTION with MEAN $\mu = 1 - 2/(9r)$ and VARIANCE $\sigma^2 = 2/(9r)$.

In a GAUSSIAN DISTRIBUTION,

$$P(x)\,dx = \frac{1}{\sigma\sqrt{2\pi}} e^{-(x-\mu)^2/2\sigma^2}\,dx, \tag{23}$$

let

$$z \equiv (x-\mu)^2/\sigma^2. \tag{24}$$

Then

$$dz = \frac{2(x-\mu)}{\sigma^2}\,dx = \frac{2\sqrt{z}}{\sigma}\,dx \tag{25}$$

so

$$dx = \frac{\sigma}{2\sqrt{z}}\,dz. \tag{26}$$

But

$$P(z)\,dz = 2P(x)\,dx, \tag{27}$$

so

$$P(x)\,dx = 2\,\frac{1}{\sigma\sqrt{2\pi}} e^{-z/2}\,dz = \frac{1}{\sigma\sqrt{\pi}} e^{-z/2}\,dz. \tag{28}$$

This is a χ^2 distribution with $r = 1$, since

$$P(z)\,dz = \frac{z^{1/2-1}e^{-z/2}}{\Gamma(\frac{1}{2})2^{1/2}}\,dz = \frac{x^{-1/2}e^{-1/2}}{\sqrt{2\pi}}\,dz. \tag{29}$$

If X_i are independent variates with a NORMAL DISTRIBUTION having MEANS μ_i and VARIANCES $\sigma_i{}^2$ for $i = 1, \ldots, n$, then

$$\tfrac{1}{2}\chi^2 \equiv \sum_{i=1}^{n} \frac{(x_i - \mu_i)^2}{2\sigma_i{}^2} \tag{30}$$

is a GAMMA DISTRIBUTION variate with $\alpha = n/2$,

$$P(\tfrac{1}{2}\chi^2)\,d(\tfrac{1}{2}\chi^2) = \frac{1}{\Gamma(\frac{1}{2}n)} e^{-\chi^2/2}(\tfrac{1}{2}\chi^2)^{(n/2)-1}\,d(\tfrac{1}{2}\chi^2). \tag{31}$$

The noncentral chi-squared distribution is given by

$$P(x) = 2^{-n/2}e^{-(\lambda+x)/2}x^{n/2-1}F(\tfrac{1}{2}n, \tfrac{1}{4}\lambda x), \tag{32}$$

where

$$F(a,z) \equiv \frac{{}_0F_1(;a;z)}{\Gamma(a)}, \tag{33}$$

${}_0F_1$ is the CONFLUENT HYPERGEOMETRIC LIMIT FUNCTION and Γ is the GAMMA FUNCTION. The MEAN, VARIANCE, SKEWNESS, and KURTOSIS are

$$\mu = \lambda + n \tag{34}$$

$$\sigma^2 = 2(2\lambda + n) \tag{35}$$

$$\gamma_1 = \frac{2\sqrt{2}\,(3\lambda+n)}{(2\lambda+n)^{3/2}} \tag{36}$$

$$\gamma_2 = \frac{12(4\lambda+n)}{(2\lambda+n)^2}. \tag{37}$$

see also CHI DISTRIBUTION, SNEDECOR'S F-DISTRIBUTION

References

Abramowitz, M. and Stegun, C. A. (Eds.). *Handbook of Mathematical Functions with Formulas, Graphs, and Mathematical Tables, 9th printing.* New York: Dover, pp. 940–943, 1972.

Beyer, W. H. *CRC Standard Mathematical Tables, 28th ed.* Boca Raton, FL: CRC Press, p. 535, 1987.

Press, W. H.; Flannery, B. P.; Teukolsky, S. A.; and Vetterling, W. T. "Incomplete Gamma Function, Error Function, Chi-Square Probability Function, Cumulative Poisson Function." §6.2 in *Numerical Recipes in FORTRAN: The Art of Scientific Computing, 2nd ed.* Cambridge, England: Cambridge University Press, pp. 209–214, 1992.

Spiegel, M. R. *Theory and Problems of Probability and Statistics.* New York: McGraw-Hill, pp. 115–116, 1992.

Chi-Squared Test

Let the probabilities of various classes in a distribution be $p_1, p_2, \ldots, p_k$. The expected frequency

$$\chi_s{}^2 = \sum_{i=1}^{k} \frac{(m_i - Np_i)^2}{Np_i}$$

is a measure of the deviation of a sample from expectation. Karl Pearson proved that the limiting distribution of $\chi_s{}^2$ is χ^2 (Kenney and Keeping 1951, pp. 114–116).

$$\begin{aligned}
\Pr(\chi^2 \geq \chi_s{}^2) &= \int_{\chi_s{}^2}^{\infty} f(\chi^2)\,d(\chi^2) \\
&= \frac{1}{2}\int_{\chi_s{}^2}^{\infty} \frac{\left(\frac{\chi^2}{2}\right)^{(k-3)/2}}{\Gamma\left(\frac{k-1}{2}\right)} e^{-\chi^2/2}\,d(\chi^2) \\
&= 1 - \frac{\Gamma\left(\frac{1}{2}\chi_s{}^2, \frac{k-1}{2}\right)}{\Gamma\left(\frac{k-1}{2}\right)} \\
&= 1 - I\left(\frac{\chi_s{}^2}{\sqrt{2(k-1)}}, \frac{k-3}{2}\right),
\end{aligned}$$

where $I(x,n)$ is PEARSON'S FUNCTION. There are some subtleties involved in using the χ^2 test to fit curves (Kenney and Keeping 1951, pp. 118–119).

When fitting a one-parameter solution using χ^2, the best-fit parameter value can be found by calculating χ^2

at three points, plotting against the parameter values of these points, then finding the minimum of a PARABOLA fit through the points (Cuzzi 1972, pp. 162–168).

References
Cuzzi, J. *The Subsurface Nature of Mercury and Mars from Thermal Microwave Emission.* Ph.D. Thesis. Pasadena, CA: California Institute of Technology, 1972.
Kenney, J. F. and Keeping, E. S. *Mathematics of Statistics, Pt. 2, 2nd ed.* Princeton, NJ: Van Nostrand, 1951.

Child

A node which is one EDGE further away from a given EDGE in a ROOTED TREE.

see also ROOT (TREE), ROOTED TREE, SIBLING

Chinese Hypothesis

A PRIME p always satisfies the condition that $2^p - 2$ is divisible by p. However, this condition is not true *exclusively* for PRIME (e.g., $2^{341} - 2$ is divisible by $341 = 11 \cdot 31$). COMPOSITE NUMBERS n (such as 341) for which $2^n - 2$ is divisible by n are called POULET NUMBERS, and are a special class of FERMAT PSEUDOPRIMES. The Chinese hypothesis is a special case of FERMAT'S LITTLE THEOREM.

see also CARMICHAEL NUMBER, EULER'S THEOREM, FERMAT'S LITTLE THEOREM, FERMAT PSEUDOPRIME, POULET NUMBER, PSEUDOPRIME

References
Shanks, D. *Solved and Unsolved Problems in Number Theory, 4th ed.* New York: Chelsea, pp. 19–20, 1993.

Chinese Remainder Theorem

Let r and s be POSITIVE INTEGERS which are RELATIVELY PRIME and let a and b be any two INTEGERS. Then there is an INTEGER N such that

$$N \equiv a \pmod{r} \tag{1}$$

and

$$N \equiv b \pmod{s}. \tag{2}$$

Moreover, N is uniquely determined modulo rs. An equivalent statement is that if $(r, s) = 1$, then every pair of RESIDUE CLASSES modulo r and s corresponds to a simple RESIDUE CLASS modulo rs.

The theorem can also be generalized as follows. Given a set of simultaneous CONGRUENCES

$$x \equiv a_i \pmod{m_i} \tag{3}$$

for $i = 1, \ldots, r$ and for which the m_i are pairwise RELATIVELY PRIME, the solution of the set of CONGRUENCES is

$$x = a_1 b_1 \frac{M}{m_1} + \ldots + a_r b_r \frac{M}{m_r} \pmod{M}, \tag{4}$$

where

$$M \equiv m_1 m_2 \cdots m_r \tag{5}$$

and the b_i are determined from

$$b_i \frac{M}{m_i} \equiv 1 \pmod{m_i}. \tag{6}$$

References
Ireland, K. and Rosen, M. "The Chinese Remainder Theorem." §3.4 in *A Classical Introduction to Modern Number Theory, 2nd ed.* New York: Springer-Verlag, pp. 34–38, 1990.
Uspensky, J. V. and Heaslet, M. A. *Elementary Number Theory.* New York: McGraw-Hill, pp. 189–191, 1939.
Wagon, S. "The Chinese Remainder Theorem." §8.4 in *Mathematica in Action.* New York: W. H. Freeman, pp. 260–263, 1991.

Chinese Rings

see BAGUENAUDIER

Chiral

Having forms of different HANDEDNESS which are not mirror-symmetric.

see also DISYMMETRIC, ENANTIOMER, HANDEDNESS, MIRROR IMAGE, REFLEXIBLE

Choice Axiom

see AXIOM OF CHOICE

Choice Number

see COMBINATION

Cholesky Decomposition

Given a symmetric POSITIVE DEFINITE MATRIX A, the Cholesky decomposition is an upper TRIANGULAR MATRIX U such that

$$\mathsf{A} = \mathsf{U}^{\mathrm{T}} \mathsf{U}.$$

see also LU DECOMPOSITION, QR DECOMPOSITION

References
Nash, J. C. "The Choleski Decomposition." Ch. 7 in *Compact Numerical Methods for Computers: Linear Algebra and Function Minimisation, 2nd ed.* Bristol, England: Adam Hilger, pp. 84–93, 1990.
Press, W. H.; Flannery, B. P.; Teukolsky, S. A.; and Vetterling, W. T. "Cholesky Decomposition." §2.9 in *Numerical Recipes in FORTRAN: The Art of Scientific Computing, 2nd ed.* Cambridge, England: Cambridge University Press, pp. 89–91, 1992.

Choose

An alternative term for a BINOMIAL COEFFICIENT, in which $\binom{n}{k}$ is read as "n choose k." R. K. Guy suggested this pronunciation around 1950, when the notations ${}^nC_r$ and ${}_nC_r$ were commonly used. Leo Moser liked the pronunciation and he and others spread it around. It got the final seal of approval from Donald Knuth when he incorporated it into the TeX mathematical typesetting language as `{n\choose k}`.

Choquet Theory

Erdős proved that there exist at least one PRIME of the form $4k+1$ and at least one PRIME of the form $4k+3$ between n and $2n$ for all $n > 6$.

see also EQUINUMEROUS, PRIME NUMBER

Chord

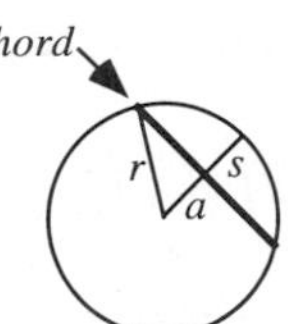

The LINE SEGMENT joining two points on a curve. The term is often used to describe a LINE SEGMENT whose ends lie on a CIRCLE. In the above figure, r is the RADIUS of the CIRCLE, a is called the APOTHEM, and s the SAGITTA.

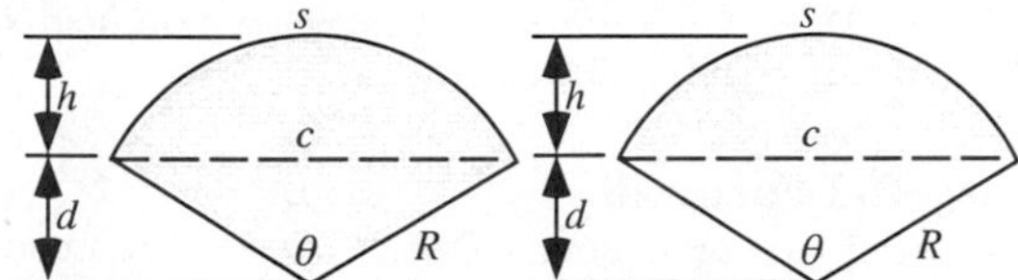

The shaded region in the left figure is called a SECTOR, and the shaded region in the right figure is called a SEGMENT.

All ANGLES inscribed in a CIRCLE and subtended by the same chord are equal. The converse is also true: The LOCUS of all points from which a given segment subtends equal ANGLES is a CIRCLE.

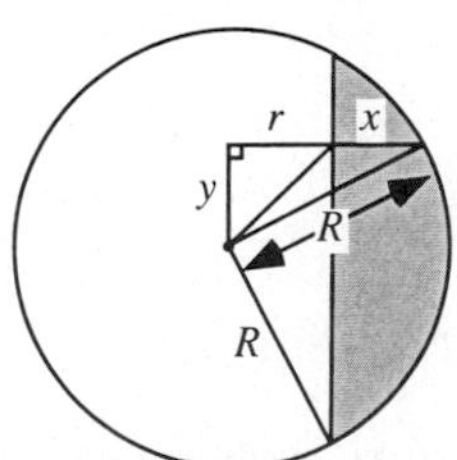

Let a CIRCLE of RADIUS R have a CHORD at distance r. The AREA enclosed by the CHORD, shown as the shaded region in the above figure, is then

$$A = 2\int_0^{\sqrt{R^2-r^2}} x(y)\,dy. \tag{1}$$

But

$$y^2 + (r+x)^2 = R^2, \tag{2}$$

so

$$x(y) = \sqrt{R^2-y^2} - r \tag{3}$$

and

$$\begin{aligned} A &= 2\int_0^{\sqrt{R^2-r^2}} (\sqrt{R^2-y^2} - r)\,dy \\ &= \left[y\sqrt{R^2-y^2} + R^2\tan^{-1}\left(\frac{y}{\sqrt{R^2-y^2}}\right) \right. \\ &\quad \left. - 2ry \right]_0^{\sqrt{R^2-r^2}} \\ &= r\sqrt{R^2-r^2} + R^2\tan^{-1}\left[\left(\frac{R}{r}\right)^2 - 1\right] - 2r\sqrt{R^2-r^2} \\ &= R^2\tan^{-1}\left[\left(\frac{R}{r}\right)^2 - 1\right] - r\sqrt{R^2-r^2}. \end{aligned} \tag{4}$$

Checking the limits, when $r = R$, $A = 0$ and when $r \to 0$,

$$A = \tfrac{1}{2}\pi R^2, \tag{5}$$

see also ANNULUS, APOTHEM, BERTRAND'S PROBLEM, CONCENTRIC CIRCLES, RADIUS, SAGITTA, SECTOR, SEGMENT

Chordal

see RADICAL AXIS

Chordal Theorem

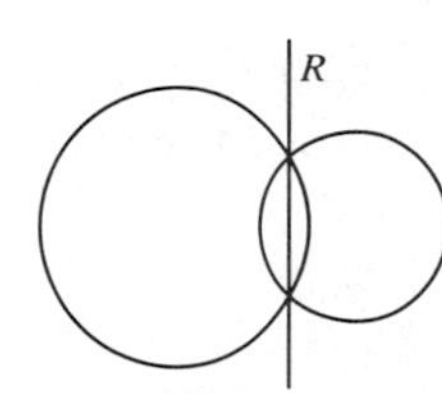

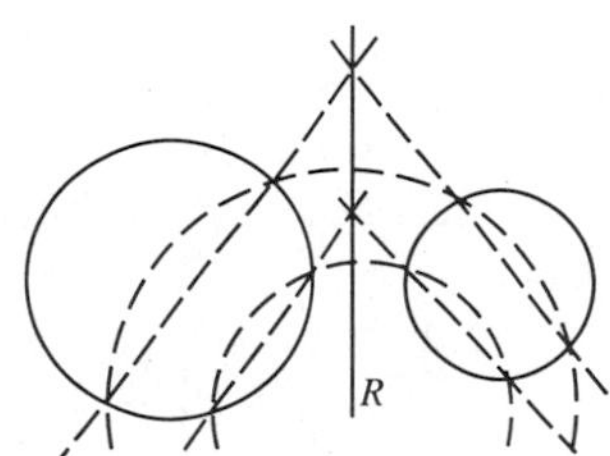

The LOCUS of the point at which two given CIRCLES possess the same POWER is a straight line PERPENDICULAR to the line joining the MIDPOINTS of the CIRCLE and is known as the chordal (or RADICAL AXIS) of the two CIRCLES.

References

Dörrie, H. *100 Great Problems of Elementary Mathematics: Their History and Solutions.* New York: Dover, p. 153, 1965.

Chow Coordinates

A generalization of GRASSMANN COORDINATES to m-D varieties of degree d in P^n, where P^n is an n-D projective space. To define the Chow coordinates, take the intersection of a m-D VARIETY Z of degree d by an $(n-m)$-D SUBSPACE U of P^n. Then the coordinates of the d points of intersection are algebraic functions of the GRASSMANN COORDINATES of U, and by taking a symmetric function of the algebraic functions, a hHOMOGENEOUS POLYNOMIAL known as the Chow form of Z is obtained. The Chow coordinates are then the COEFFICIENTS of the Chow form. Chow coordinates can generate the smallest field of definition of a divisor.

References
Chow, W.-L. and van der Waerden., B. L. "Zur algebraische Geometrie IX." *Math. Ann.* **113**, 692–704, 1937.
Wilson, W. S.; Chern, S. S.; Abhyankar, S. S.; Lang, S.; and Igusa, J.-I. "Wei-Liang Chow." *Not. Amer. Math. Soc.* **43**, 1117–1124, 1996.

Chow Ring

The intersection product for classes of rational equivalence between cycles on an ALGEBRAIC VARIETY.

References
Chow, W.-L. "On Equivalence Classes of Cycles in an Algebraic Variety." *Ann. Math.* **64**, 450–479, 1956.
Wilson, W. S.; Chern, S. S.; Abhyankar, S. S.; Lang, S.; and Igusa, J.-I. "Wei-Liang Chow." *Not. Amer. Math. Soc.* **43**, 1117–1124, 1996.

Chow Variety

The set $C_{n,m,d}$ of all m-D varieties of degree d in an n-D projective space P^n into an M-D projective space P^M.

References
Wilson, W. S.; Chern, S. S.; Abhyankar, S. S.; Lang, S.; and Igusa, J.-I. "Wei-Liang Chow." *Not. Amer. Math. Soc.* **43**, 1117–1124, 1996.

Christoffel-Darboux Formula

For three consecutive ORTHOGONAL POLYNOMIALS

$$p_n(x) = (A_n x + B_n)p_{n-1}x - C_n p_{n-2}(x) \quad (1)$$

for $n = 2, 3, \ldots$, where $A_n > 0$, B_n, and $C_n > 0$ are constants. Denoting the highest COEFFICIENT of $p_n(x)$ by k_n,

$$A_n = \frac{k_n}{k_{n-1}} \quad (2)$$

$$C_n = \frac{A_n}{A_{n-1}} = \frac{k_n k_{n-2}}{{k_{n-1}}^2}. \quad (3)$$

Then

$$p_0(x)p_0(y) + \ldots + p_n(x)p_n(y) = \frac{k_n}{k_{n+1}} \frac{p_{n+1}(x)p_n(y) - p_n(x)p_{n+1}(y)}{x-y}. \quad (4)$$

In the special case of $x = y$, (4) gives

$$[p_0(x)]^2 + \ldots + [p_n(x)]^2 = \frac{k_n}{k_{n+1}}[p'_{n+1}(x)p_n(x) - p'_n(x)p_{n+1}(x)]. \quad (5)$$

References
Abramowitz, M. and Stegun, C. A. (Eds.). *Handbook of Mathematical Functions with Formulas, Graphs, and Mathematical Tables, 9th printing.* New York: Dover, p. 785, 1972.
Szegő, G. *Orthogonal Polynomials, 4th ed.* Providence, RI: Amer. Math. Soc., pp. 42–44, 1975.

Christoffel-Darboux Identity

$$\sum_{k=0}^{\infty} \frac{\phi_k(x)\phi_k(y)}{\gamma_k} = \frac{\phi_{m+1}(x)\phi_m(y) - \phi_m(x)\phi_{m+1}(y)}{a_m \gamma_m (x-y)}, \quad (1)$$

where $\phi_k(x)$ are ORTHOGONAL POLYNOMIALS with WEIGHTING FUNCTION $W(x)$,

$$\gamma_m \equiv \int [\phi_m(x)]^2 W(x)\, dx, \quad (2)$$

and

$$a_k \equiv \frac{A_{k+1}}{A_k} \quad (3)$$

where A_k is the COEFFICIENT of x^k in $\phi_k(x)$.

References
Hildebrand, F. B. *Introduction to Numerical Analysis.* New York: McGraw-Hill, p. 322, 1956.

Christoffel Formula

Let $\{p_n(x)\}$ be orthogonal POLYNOMIALS associated with the distribution $d\alpha(x)$ on the interval $[a,b]$. Also let

$$\rho \equiv c(x - x_1)(x - x_2)\cdots(x - x_l)$$

(for $c \neq 0$) be a POLYNOMIAL of order l which is NONNEGATIVE in this interval. Then the orthogonal POLYNOMIALS $\{q(x)\}$ associated with the distribution $\rho(x)\, d\alpha(x)$ can be represented in terms of the POLYNOMIALS $p_n(x)$ as

$$\rho(x)q_n(x) = \begin{vmatrix} p_n(x) & p_{n+1}(x) & \cdots & p_{n+l}(x) \\ p_n(x_1) & p_{n+1}(x_1) & \cdots & p_{n+l}(x_1) \\ \vdots & \vdots & \ddots & \vdots \\ p_n(x_l) & p_{n+1}(x_l) & \cdots & p_{n+l}(x_l) \end{vmatrix}.$$

In the case of a zero x_k of multiplicity $m > 1$, we replace the corresponding rows by the derivatives of order 0, 1, 2, ..., $m-1$ of the POLYNOMIALS $p_n(x_l), \ldots, p_{n+l}(x_l)$ at $x = x_k$.

References
Szegő, G. *Orthogonal Polynomials, 4th ed.* Providence, RI: Amer. Math. Soc., pp. 29–30, 1975.

Christoffel Number

One of the quantities λ_i appearing in the GAUSS-JACOBI MECHANICAL QUADRATURE. They satisfy

$$\lambda_1 + \lambda_2 + \ldots + \lambda_n = \int_a^b d\alpha(x) = \alpha(b) - \alpha(a) \quad (1)$$

and are given by

$$\lambda_\nu = \int_a^b \left[\frac{p_n(x)}{p_n'(x_\nu)(x-x_\nu)}\right]^2 d\alpha(x) \tag{2}$$

$$\lambda_\nu = -\frac{k_{n+1}}{k_n}\frac{1}{p_{n+1}(x_\nu)p_n'(x_\nu)} \tag{3}$$

$$= \frac{k_n}{k_{n-1}}\frac{1}{p_{n-1}(x_\nu)P_n'(x_\nu)} \tag{4}$$

$$(\lambda_\nu)^{-1} = [p_0(x_\nu)]^2 + \ldots + [p_n(x_\nu)]^2, \tag{5}$$

where k_n is the higher COEFFICIENT of $p_n(x)$.

References
Szegő, G. *Orthogonal Polynomials, 4th ed.* Providence, RI: Amer. Math. Soc., pp. 47–48, 1975.

Christoffel Symbol of the First Kind

Variously denoted $[ij,k]$, $\left[{i \atop k}{j \atop }\right]$, Γ_{abc}, or $\{ab,c\}$.

$$[ij,k] = \begin{bmatrix} i & j \\ & k \end{bmatrix} \equiv g_{mk}\Gamma^m_{ij} = g_{mk}\vec{e}^{\,m}\cdot\frac{\partial\vec{e}_i}{\partial q^j} = \vec{e}_k\cdot\frac{\partial\vec{e}_i}{\partial q^j}, \tag{1}$$

where g_{mk} is the METRIC TENSOR and

$$\vec{e}_i \equiv \frac{\partial\vec{r}}{\partial q^i} = h_i\hat{e}_i. \tag{2}$$

But

$$\frac{\partial g_{ij}}{\partial q^k} = \frac{\partial}{\partial q^k}(\vec{e}_i\cdot\vec{e}_j) = \frac{\partial\vec{e}_i}{\partial q^k}\cdot\vec{e}_j + \vec{e}_i\cdot\frac{\partial\vec{e}_j}{\partial q^k}$$
$$= [ik,j] + [jk,i], \tag{3}$$

so

$$[ab,c] = \tfrac{1}{2}(g_{ac,b} + g_{bc,a} - g_{ab,c}). \tag{4}$$

References
Arfken, G. *Mathematical Methods for Physicists, 3rd ed.* Orlando, FL: Academic Press, pp. 160–167, 1985.

Christoffel Symbol of the Second Kind

Variously denoted $\left\{{m \atop i\ \ j}\right\}$ or Γ^m_{ij}.

$$\left\{\begin{matrix} & m & \\ i & & j\end{matrix}\right\} \equiv \Gamma^m_{ij} = \vec{e}^{\,m}\cdot\frac{\partial\vec{e}_i}{\partial q^j} = g^{km}[ij,k]$$
$$= \frac{1}{2}g^{km}\left(\frac{\partial g_{ik}}{\partial q^j} + \frac{\partial g_{jk}}{\partial q^i} - \frac{\partial g_{ij}}{\partial q^k}\right), \tag{1}$$

where Γ^m_{ij} is a CONNECTION COEFFICIENT and $\{bc,d\}$ is a CHRISTOFFEL SYMBOL OF THE FIRST KIND.

$$\left\{\begin{matrix} & a & \\ b & & c\end{matrix}\right\} = g_{ad}\{bc,d\}. \tag{2}$$

The Christoffel symbols are given in terms of the first FUNDAMENTAL FORM E, F, and G by

$$\Gamma^1_{11} = \frac{GE_u - 2FF_u + FE_v}{2(EG-F^2)} \tag{3}$$

$$\Gamma^1_{12} = \frac{GE_v - FG_u}{2(EG-F^2)} \tag{4}$$

$$\Gamma^1_{22} = \frac{2GF_v - GG_u - FG_v}{2(EG-F^2)} \tag{5}$$

$$\Gamma^2_{11} = \frac{2EF_u - EE_v - FE_u}{2(EG-F^2)} \tag{6}$$

$$\Gamma^2_{12} = \frac{EG_u - FE_v}{2(EG-F^2)} \tag{7}$$

$$\Gamma^2_{22} = \frac{EG_v - 2FF_v + FG_u}{2(EG-F^2)}, \tag{8}$$

and $\Gamma^1_{21} = \Gamma^1_{12}$ and $\Gamma^2_{21} = \Gamma^2_{12}$. If $F = 0$, the Christoffel symbols of the second kind simplify to

$$\Gamma^1_{11} = \frac{E_u}{2E} \tag{9}$$

$$\Gamma^1_{12} = \frac{E_v}{2E} \tag{10}$$

$$\Gamma^1_{22} = -\frac{G_u}{2E} \tag{11}$$

$$\Gamma^2_{11} = -\frac{E_v}{2G} \tag{12}$$

$$\Gamma^2_{12} = \frac{G_u}{2G} \tag{13}$$

$$\Gamma^2_{22} = \frac{G_v}{2G} \tag{14}$$

(Gray 1993).

The following relationships hold between the Christoffel symbols of the second kind and coefficients of the first FUNDAMENTAL FORM,

$$\Gamma^1_{11}E + \Gamma^2_{11}F = \tfrac{1}{2}E_u \tag{15}$$

$$\Gamma^1_{12}E + \Gamma^2_{12}F = \tfrac{1}{2}E_v \tag{16}$$

$$\Gamma^1_{22}E + \Gamma^2_{22}F = F_v - \tfrac{1}{2}G_u \tag{17}$$

$$\Gamma^1_{11}F + \Gamma^2_{11}G = F_u - \tfrac{1}{2}E_v \tag{18}$$

$$\Gamma^1_{12}F + \Gamma^2_{12}G = \tfrac{1}{2}G_u \tag{19}$$

$$\Gamma^1_{22}F + \Gamma^2_{22}G = \tfrac{1}{2}G_v \tag{20}$$

$$\Gamma^1_{11} + \Gamma^2_{12} = (\ln\sqrt{EG-F^2}\,)_u \tag{21}$$

$$\Gamma^1_{12} + \Gamma^2_{22} = (\ln\sqrt{EG-F^2}\,)_v \tag{22}$$

(Gray 1993).

For a surface given in MONGE'S FORM $z = F(x,y)$,

$$\Gamma^k_{ij} = \frac{z_{ij}z_k}{1 + z_1{}^2 + z_2{}^2}. \tag{23}$$

see also CHRISTOFFEL SYMBOL OF THE FIRST KIND, CONNECTION COEFFICIENT, GAUSS EQUATIONS

References
Arfken, G. *Mathematical Methods for Physicists, 3rd ed.* Orlando, FL: Academic Press, pp. 160–167, 1985.
Gray, A. "Christoffel Symbols." §20.3 in *Modern Differential Geometry of Curves and Surfaces.* Boca Raton, FL: CRC Press, pp. 397–400, 1993.
Morse, P. M. and Feshbach, H. *Methods of Theoretical Physics, Part I.* New York: McGraw-Hill, pp. 47–48, 1953.

Chromatic Number

The fewest number of colors $\gamma(G)$ necessary to color a GRAPH or surface. The chromatic number of a surface of GENUS g is given by the HEAWOOD CONJECTURE,

$$\gamma(g) = \left\lfloor \tfrac{1}{2}(7 + \sqrt{48g+1}) \right\rfloor,$$

where $\lfloor x \rfloor$ is the FLOOR FUNCTION. $\gamma(g)$ is sometimes also denoted $\chi(g)$. For $g = 0, 1, \ldots$, the first few values of $\chi(g)$ are 4, 7, 8, 9, 10, 11, 12, 12, 13, 13, 14, 15, 15, 16, ... (Sloane's A000934).

The fewest number of colors necessary to color each EDGE of a GRAPH so that no two EDGES incident on the same VERTEX have the same color is called the "EDGE chromatic number."

see also BRELAZ'S HEURISTIC ALGORITHM, CHROMATIC POLYNOMIAL, EDGE-COLORING, EULER CHARACTERISTIC, HEAWOOD CONJECTURE, MAP COLORING, TORUS COLORING

References
Chartrand, G. "A Scheduling Problem: An Introduction to Chromatic Numbers." §9.2 in *Introductory Graph Theory.* New York: Dover, pp. 202–209, 1985.
Eppstein, D. "The Chromatic Number of the Plane." http://www.ics.uci.edu/~eppstein/junkyard/plane-color/.
Sloane, N. J. A. Sequence A000934/M3292 in "An On-Line Version of the Encyclopedia of Integer Sequences."

Chromatic Polynomial

A POLYNOMIAL $P(z)$ of a graph g which counts the number of ways to color g with exactly z colors. Tutte (1970) showed that the chromatic POLYNOMIALS of planar triangular graphs possess a ROOT close to $\phi^2 = 2.618033\ldots$, where ϕ is the GOLDEN MEAN. More precisely, if n is the number of VERTICES of G, then

$$P_G(\phi^2) \leq \phi^{5-n}$$

(Le Lionnais 1983).

References
Le Lionnais, F. *Les nombres remarquables.* Paris: Hermann, p. 46, 1983.
Tutte, W. T. "On Chromatic Polynomials and the Golden Ratio." *J. Combin. Th.* **9**, 289–296, 1970.

Chu Identity

see CHU-VANDERMONDE IDENTITY

Chu Space

A Chu space is a binary relation from a SET A to an antiset X which is defined as a SET which transforms via converse functions.

References
Stanford Concurrency Group. "Guide to Papers on Chu Spaces." http://boole.stanford.edu/chuguide.html.

Chu-Vandermonde Identity

$$(x+a)_n = \sum_{k=0}^{\infty} \binom{n}{k} (a)_k (x)_{n-k}$$

where $\binom{n}{k}$ is a BINOMIAL COEFFICIENT and $(a)_n \equiv a(a-1)\cdots(a-n+1)$ is the POCHHAMMER SYMBOL. A special case gives the identity

$$\sum_{l=0}^{\max(k,n)} \binom{m}{k-l}\binom{n}{l} = \binom{m+n}{k}.$$

see also BINOMIAL THEOREM, UMBRAL CALCULUS

References
Petkovšek, M.; Wilf, H. S.; and Zeilberger, D. *A=B.* Wellesley, MA: A. K. Peters, pp. 130 and 181–182, 1996.

Church's Theorem

No decision procedure exists for ARITHMETIC.

Church's Thesis

see CHURCH-TURING THESIS

Church-Turing Thesis

The TURING MACHINE concept defines what is meant mathematically by an algorithmic procedure. Stated another way, a function f is effectively COMPUTABLE IFF it can be computed by a TURING MACHINE.

see also ALGORITHM, COMPUTABLE FUNCTION, TURING MACHINE

References
Penrose, R. *The Emperor's New Mind: Concerning Computers, Minds, and the Laws of Physics.* Oxford, England: Oxford University Press, pp. 47–49, 1989.

Chvátal's Art Gallery Theorem

see ART GALLERY THEOREM

Chvátal's Theorem

Let the GRAPH G have VERTICES with VALENCES $d_1 \leq \ldots \leq d_m$. If for every $i < n/2$ we have either $d_i \geq i+1$ or $d_{n-i} \geq n-i$, then the GRAPH is HAMILTONIAN.

ci

see COSINE INTEGRAL

Ci

see COSINE INTEGRAL

Cigarettes

It is possible to place 7 cigarettes in such a way that each touches the other if $l/d > 7\sqrt{3}/2$ (Gardner 1959, p. 115).

References

Gardner, M. *The Scientific American Book of Mathematical Puzzles & Diversions.* New York: Simon and Schuster, 1959.

Cin

see COSINE INTEGRAL

Circle

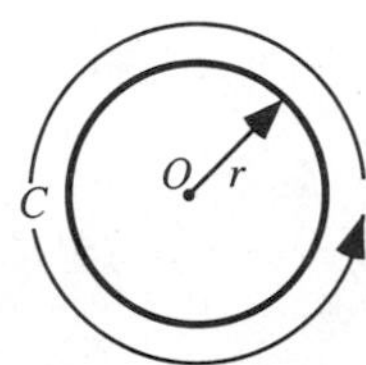

A circle is the set of points equidistant from a given point O. The distance r from the CENTER is called the RADIUS, and the point O is called the CENTER. Twice the RADIUS is known as the DIAMETER $d = 2r$. The PERIMETER C of a circle is called the CIRCUMFERENCE, and is given by

$$C = \pi d = 2\pi r. \tag{1}$$

The circle is a CONIC SECTION obtained by the intersection of a CONE with a PLANE PERPENDICULAR to the CONE's symmetry axis. A circle is the degenerate case of an ELLIPSE with equal semimajor and semiminor axes (i.e., with ECCENTRICITY 0). The interior of a circle is called a DISK. The generalization of a circle to 3-D is called a SPHERE, and to n-D for $n \geq 4$ a HYPERSPHERE.

The region of intersection of two circles is called a LENS. The region of intersection of three symmetrically placed circles (as in a VENN DIAGRAM), in the special case of the center of each being located at the intersection of the other two, is called a REULEAUX TRIANGLE.

The parametric equations for a circle of RADIUS a are

$$x = a\cos t \tag{2}$$
$$y = a\sin t. \tag{3}$$

For a body moving uniformly around the circle,

$$x' = -a\sin t \tag{4}$$
$$y' = a\cos t, \tag{5}$$

and

$$x'' = -a\cos t \tag{6}$$
$$y'' = -a\sin t. \tag{7}$$

When normalized, the former gives the equation for the unit TANGENT VECTOR of the circle, $(-\sin t, \cos t)$. The circle can also be parameterized by the rational functions

$$x = \frac{1-t^2}{t(1+t)} \tag{8}$$
$$y = \frac{2t}{1+t^2}, \tag{9}$$

but an ELLIPTIC CURVE cannot. The following plots show a sequence of NORMAL and TANGENT VECTORS for the circle.

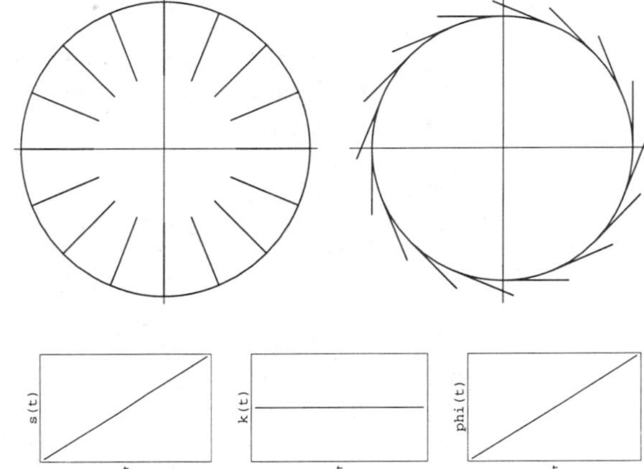

The ARC LENGTH s, CURVATURE κ, and TANGENTIAL ANGLE ϕ of the circle are

$$s(t) = \int ds = \int \sqrt{x'^2 + y'^2}\, dt = at \tag{10}$$
$$\kappa(t) = \frac{x'y'' - y'x''}{(x'^2 + y'^2)^{3/2}} = \frac{1}{a} \tag{11}$$
$$\phi(t) = \int \kappa(t)\, dt = \frac{t}{a}. \tag{12}$$

The CESÀRO EQUATION is

$$\kappa = \frac{1}{a}. \tag{13}$$

In POLAR COORDINATES, the equation of the circle has a particularly simple form.

$$r = a \tag{14}$$

is a circle of RADIUS a centered at ORIGIN,

$$r = 2a\cos\theta \tag{15}$$

is circle of RADIUS a centered at $(a, 0)$, and

$$r = 2a\sin\theta \tag{16}$$

is a circle of RADIUS a centered on $(0, a)$. In CARTESIAN COORDINATES, the equation of a circle of RADIUS a centered on (x_0, y_0) is

$$(x - x_0)^2 + (y - y_0)^2 = a^2. \tag{17}$$

In PEDAL COORDINATES with the PEDAL POINT at the center, the equation is

$$pa = r^2. \tag{18}$$

The circle having P_1P_2 as a diameter is given by

$$(x - x_1)(x - x_2) + (y - y_1)(y - y_2) = 0. \tag{19}$$

The equation of a circle passing through the three points (x_i, y_i) for $i = 1, 2, 3$ (the CIRCUMCIRCLE of the TRIANGLE determined by the points) is

$$\begin{vmatrix} x^2+y^2 & x & y & 1 \\ {x_1}^2+{y_1}^2 & x_1 & y_1 & 1 \\ {x_2}^2+{y_2}^2 & x_2 & y_2 & 1 \\ {x_3}^2+{y_3}^2 & x_3 & y_3 & 1 \end{vmatrix} = 0. \tag{20}$$

The CENTER and RADIUS of this circle can be identified by assigning coefficients of a QUADRATIC CURVE

$$ax^2 + cy^2 + dx + ey + f = 0, \tag{21}$$

where $a = c$ and $b = 0$ (since there is no xy cross term). COMPLETING THE SQUARE gives

$$a\left(x + \frac{d}{2a}\right)^2 + a\left(y + \frac{e}{2a}\right)^2 + f - \frac{d^2+e^2}{4a} = 0. \tag{22}$$

The CENTER can then be identified as

$$x_0 = -\frac{d}{2a} \tag{23}$$

$$y_0 = -\frac{e}{2a} \tag{24}$$

and the RADIUS as

$$r = \sqrt{\frac{d^2+e^2}{4a^2} - \frac{f}{a}}, \tag{25}$$

where

$$a = \begin{vmatrix} x_1 & y_1 & 1 \\ x_2 & y_2 & 1 \\ x_3 & y_3 & 1 \end{vmatrix} \tag{26}$$

$$d = -\begin{vmatrix} {x_1}^2+{y_1}^2 & y_1 & 1 \\ {x_2}^2+{y_2}^2 & y_2 & 1 \\ {x_3}^2+{y_3}^2 & y_3 & 1 \end{vmatrix} \tag{27}$$

$$e = \begin{vmatrix} {x_1}^2+{y_1}^2 & x_1 & 1 \\ {x_2}^2+{y_2}^2 & x_2 & 1 \\ {x_3}^2+{y_3}^2 & x_3 & 1 \end{vmatrix} \tag{28}$$

$$f = -\begin{vmatrix} {x_1}^2+{y_1}^2 & x_1 & y_1 \\ {x_2}^2+{y_2}^2 & x_2 & y_2 \\ {x_3}^2+{y_3}^2 & x_3 & y_3 \end{vmatrix}. \tag{29}$$

Four or more points which lie on a circle are said to be CONCYCLIC. Three points are trivially concyclic since three noncollinear points determine a circle.

The CIRCUMFERENCE-to-DIAMETER ratio C/d for a circle is constant as the size of the circle is changed (as it must be since scaling a plane figure by a factor s increases its PERIMETER by s), and d also scales by s. This ratio is denoted π (PI), and has been proved TRANSCENDENTAL. With d the DIAMETER and r the RADIUS,

$$C = \pi d = 2\pi r. \tag{30}$$

Knowing C/d, we can then compute the AREA of the circle either geometrically or using CALCULUS. From CALCULUS,

$$A = \int_0^{2\pi} d\theta \int_0^r r\,dr = (2\pi)(\tfrac{1}{2}r^2) = \pi r^2. \tag{31}$$

Now for a few geometrical derivations. Using concentric strips, we have

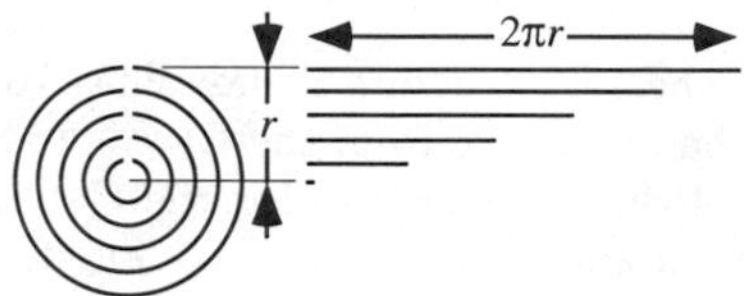

As the number of strips increases to infinity, we are left with a TRIANGLE on the right, so

$$A = \tfrac{1}{2}(2\pi r)r = \pi r^2. \tag{32}$$

This derivation was first recorded by Archimedes in *Measurement of a Circle* (ca. 225 BC). If we cut the circle instead into wedges,

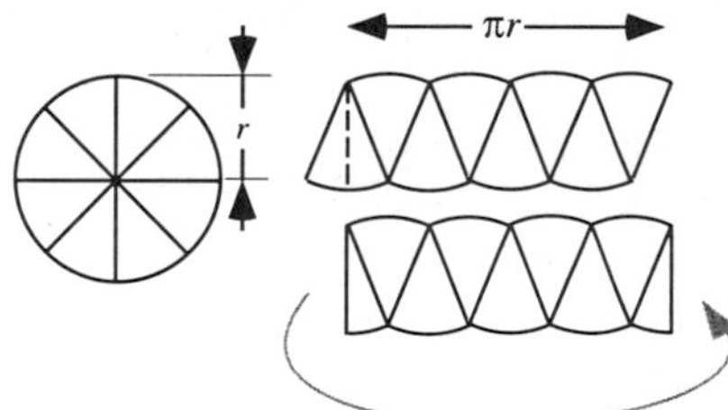

As the number of wedges increases to infinity, we are left with a RECTANGLE, so

$$A = (\pi r)r = \pi r^2. \tag{33}$$

see also ARC, BLASCHKE'S THEOREM, BRAHMAGUPTA'S FORMULA, BROCARD CIRCLE, CASEY'S THEOREM, CHORD, CIRCUMCIRCLE, CIRCUMFERENCE, CLIFFORD'S CIRCLE THEOREM, CLOSED DISK, CONCENTRIC CIRCLES, COSINE CIRCLE, COTES CIRCLE PROPERTY, DIAMETER, DISK, DROZ-FARNY CIRCLES, EULER TRIANGLE FORMULA, EXCIRCLE, FEUERBACH'S THEOREM,

FIVE DISKS PROBLEM, FLOWER OF LIFE, FORD CIRCLE, FUHRMANN CIRCLE, GERŜGORIN CIRCLE THEOREM, HOPF CIRCLE, INCIRCLE, INVERSIVE DISTANCE, JOHNSON CIRCLE, KINNEY'S SET, LEMOINE CIRCLE, LENS, MAGIC CIRCLES, MALFATTI CIRCLES, MCCAY CIRCLE, MIDCIRCLE, MONGE'S THEOREM, MOSER'S CIRCLE PROBLEM, NEUBERG CIRCLES, NINE-POINT CIRCLE, OPEN DISK, P-CIRCLE, PARRY CIRCLE, PI, POLAR CIRCLE, POWER (CIRCLE), PRIME CIRCLE, PTOLEMY'S THEOREM, PURSER'S THEOREM, RADICAL AXIS, RADIUS, REULEAUX TRIANGLE, SEED OF LIFE, SEIFERT CIRCLE, SEMICIRCLE, SODDY CIRCLES, SPHERE, TAYLOR CIRCLE, TRIANGLE INSCRIBING IN A CIRCLE, TRIPLICATE-RATIO CIRCLE, TUCKER CIRCLES, UNIT CIRCLE, VENN DIAGRAM, VILLARCEAU CIRCLES, YIN-YANG

References

Beyer, W. H. *CRC Standard Mathematical Tables, 28th ed.* Boca Raton, FL: CRC Press, pp. 125 and 197, 1987.

Casey, J. "The Circle." Ch. 3 in *A Treatise on the Analytical Geometry of the Point, Line, Circle, and Conic Sections, Containing an Account of Its Most Recent Extensions, with Numerous Examples, 2nd ed., rev. enl.* Dublin: Hodges, Figgis, & Co., pp. 96–150, 1893.

Courant, R. and Robbins, H. *What is Mathematics?: An Elementary Approach to Ideas and Methods, 2nd ed.* Oxford, England: Oxford University Press, pp. 74–75, 1996.

Dunham, W. "Archimedes' Determination of Circular Area." Ch. 4 in *Journey Through Genius: The Great Theorems of Mathematics.* New York: Wiley, pp. 84–112, 1990.

Eppstein, D. "Circles and Spheres." `http://www.ics.uci.edu/~eppstein/junkyard/sphere.html`.

Lawrence, J. D. *A Catalog of Special Plane Curves.* New York: Dover, pp. 65–66, 1972.

MacTutor History of Mathematics Archive. "Circle." `http://www-groups.dcs.st-and.ac.uk/~history/Curves/Circle.html`.

Pappas, T. "Infinity & the Circle" and "Japanese Calculus." *The Joy of Mathematics.* San Carlos, CA: Wide World Publ./Tetra, pp. 68 and 139, 1989.

Pedoe, D. *Circles: A Mathematical View, rev. ed.* Washington, DC: Math. Assoc. Amer., 1995.

Yates, R. C. "The Circle." *A Handbook on Curves and Their Properties.* Ann Arbor, MI: J. W. Edwards, pp. 21–25, 1952.

Circles-and-Squares Fractal

A FRACTAL produced by iteration of the equation

$$z_{n+1} = {z_n}^2 \pmod{m}$$

which results in a MØIRÉ-like pattern.

see also FRACTAL, MØIRÉ PATTERN

Circle Caustic

Consider a point light source located at a point $(\mu, 0)$. The CATACAUSTIC of a unit CIRCLE for the light at $\mu = \infty$ is the NEPHROID

$$x = \tfrac{1}{4}[3\cos t - \cos(3t)] \tag{1}$$
$$y = \tfrac{1}{4}[3\sin t - \sin(3t)]. \tag{2}$$

The CATACAUSTIC for the light at a finite distance $\mu > 1$ is the curve

$$x = \frac{\mu(1 - 3\mu\cos t + 2\mu\cos^3 t)}{-(1 + 2\mu^2) + 3\mu\cos t} \tag{3}$$
$$y = \frac{2\mu^2 \sin^3 t}{1 + 2\mu^2 - 3\mu\cos t}, \tag{4}$$

and for the light on the CIRCUMFERENCE of the CIRCLE $\mu = 1$ is the CARDIOID

$$x = \tfrac{2}{3}\cos t(1 + \cos t) - \tfrac{1}{3} \tag{5}$$
$$y = \tfrac{2}{3}\sin t(1 + \cos t). \tag{6}$$

If the point is inside the circle, the catacaustic is a discontinuous two-part curve. These four cases are illustrated below.

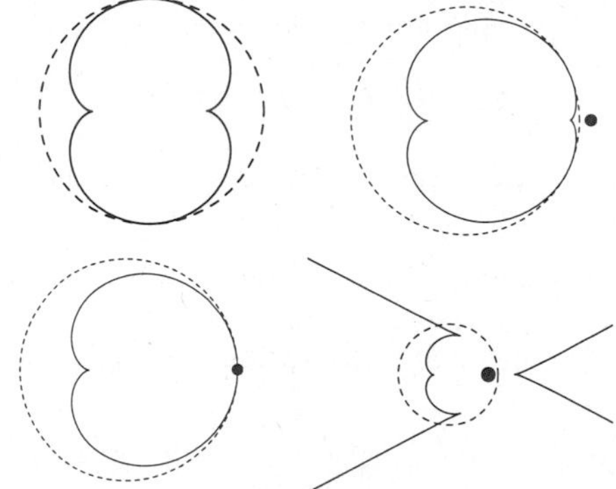

The CATACAUSTIC for PARALLEL rays crossing a CIRCLE is a CARDIOID.

see also CATACAUSTIC, CAUSTIC

Circle-Circle Intersection

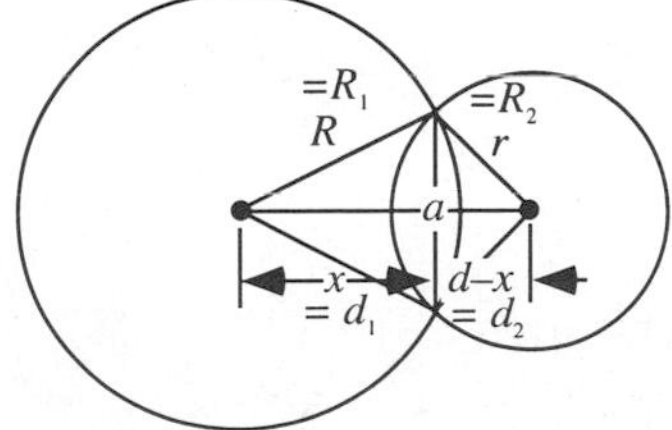

Let two CIRCLES of RADII R and r and centered at $(0, 0)$ and $(d, 0)$ intersect in a LENS-shaped region. The equations of the two circles are

$$x^2 + y^2 = R^2 \tag{1}$$
$$(x - d)^2 + y^2 = r^2. \tag{2}$$

Combining (1) and (2) gives

$$(x-d)^2+(R^2-x^2)=r^2. \tag{3}$$

Multiplying through and rearranging gives

$$x^2-2dx+d^2-x^2=r^2-R^2. \tag{4}$$

Solving for x results in

$$x=\frac{d^2-r^2+R^2}{2d}. \tag{5}$$

The line connecting the cusps of the LENS therefore has half-length given by plugging x back in to obtain

$$\begin{aligned} y^2 &= R^2-x^2=R^2-\left(\frac{d^2-r^2+R^2}{2d}\right)^2 \\ &= \frac{4d^2R^2-(d^2-r^2+R^2)^2}{4d^2}, \end{aligned} \tag{6}$$

giving a length of

$$\begin{aligned} a &= \frac{1}{d}\sqrt{4d^2R^2-(d^2-r^2+R^2)^2} \\ &= \frac{1}{d}[(-d+r-R)(-d-r+R) \\ &\quad \times [(-d+r+R)(d+r+R)]^{1/2}. \end{aligned} \tag{7}$$

This same formulation applies directly to the SPHERE-SPHERE INTERSECTION problem.

To find the AREA of the asymmetric "LENS" in which the CIRCLES intersect, simply use the formula for the circular SEGMENT of radius $R'$and triangular height d'

$$A(R',d')=R'^2\cos^{-1}\left(\frac{d'}{R'}\right)-d'\sqrt{R'^2-d'^2} \tag{8}$$

twice, one for each half of the "LENS." Noting that the heights of the two segment triangles are

$$d_1=x=\frac{d^2-r^2+R^2}{2d} \tag{9}$$

$$d_2=d-x=\frac{d^2+r^2-R^2}{2d}. \tag{10}$$

The result is

$$\begin{aligned} A &= A(R_1,d_1)+A(R_2,d_2) \\ &= r^2\cos^{-1}\left(\frac{d^2+r^2-R^2}{2dr}\right) \\ &\quad +R^2\cos^{-1}\left(\frac{d^2+R^2-r^2}{2dR}\right) \\ &\quad -\tfrac{1}{2}\sqrt{(d-r-R)(d+r-R)(d-r+R)(d+r+R)}. \end{aligned} \tag{11}$$

The limiting cases of this expression can be checked to give 0 when $d=R+r$ and

$$A=2R^2\cos^{-1}\left(\frac{d}{2R}\right)-\tfrac{1}{2}d\sqrt{4R^2-d^2} \tag{12}$$

$$=2A(\tfrac{1}{2}d,R) \tag{13}$$

when $r=R$, as expected. In order for half the area of two UNIT DISKS ($R=1$) to overlap, set $A=\pi R^2/2=\pi/2$ in the above equation

$$\tfrac{1}{2}\pi=2\cos^{-1}(\tfrac{1}{2}d)-\tfrac{1}{2}d\sqrt{4-d^2} \tag{14}$$

and solve numerically, yielding $d\approx 0.807946$.

see also LENS, SEGMENT, SPHERE-SPHERE INTERSECTION

Circle Cutting

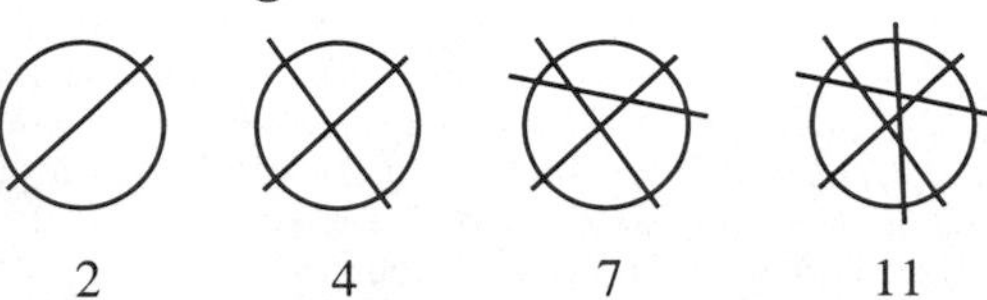

Determining the maximum number of pieces in which it is possible to divide a CIRCLE for a given number of cuts is called the circle cutting, or sometimes PANCAKE CUTTING, problem. The minimum number is always $n+1$, where n is the number of cuts, and it is always possible to obtain any number of pieces between the minimum and maximum. The first cut creates 2 regions, and the nth cut creates n new regions, so

$$f(1)=2 \tag{1}$$

$$f(2)=2+f(1) \tag{2}$$

$$f(n)=n+f(n-1). \tag{3}$$

Therefore,

$$\begin{aligned} f(n) &= n+[(n-1)+f(n-2)] \\ &= n+(n-1)+\ldots+2+f(1)=\sum_{k=2}^{n}kf(1) \\ &= \sum_{k=1}^{n}k-1+f(1)=\tfrac{1}{2}n(n+1)-1+2 \\ &= \tfrac{1}{2}(n^2+n+2). \end{aligned} \tag{4}$$

Evaluating for $n=1, 2, \ldots$ gives 2, 4, 7, 11, 16, 22, ... (Sloane's A000124).

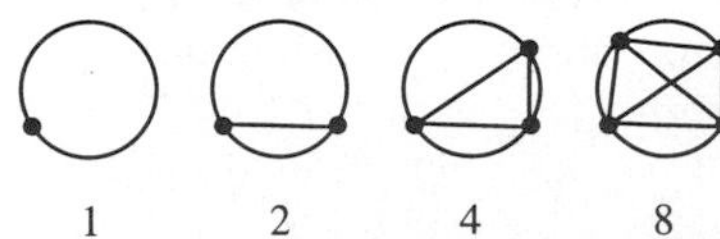

A related problem, sometimes called MOSER'S CIRCLE PROBLEM, is to find the number of pieces into which a CIRCLE is divided if n points on its CIRCUMFERENCE are joined by CHORDS with no three CONCURRENT. The answer is

$$g(n) = \binom{n}{4} + \binom{n}{2} + 1 \qquad (5)$$

$$= \tfrac{1}{24}(n^4 - 6n^3 + 23n^2 - 18n + 24), \qquad (6)$$

(Yaglom and Yaglom 1987, Guy 1988, Conway and Guy 1996, Noy 1996), where $\binom{n}{m}$ is a BINOMIAL COEFFICIENT. The first few values are 1, 2, 4, 8, 16, 31, 57, 99, 163, 256, ... (Sloane's A000127). This sequence and problem are an example of the danger in making assumptions based on limited trials. While the series starts off like 2^{n-1}, it begins differing from this GEOMETRIC SERIES at $n = 6$.

see also CAKE CUTTING, CYLINDER CUTTING, HAM SANDWICH THEOREM, PANCAKE THEOREM, PIZZA THEOREM, SQUARE CUTTING, TORUS CUTTING

References

Conway, J. H. and Guy, R. K. "How Many Regions." In *The Book of Numbers.* New York: Springer-Verlag, pp. 76–79, 1996.

Guy, R. K. "The Strong Law of Small Numbers." *Amer. Math. Monthly* **95**, 697–712, 1988.

Noy, M. "A Short Solution of a Problem in Combinatorial Geometry." *Math. Mag.* **69**, 52–53, 1996.

Sloane, N. J. A. Sequences A000124/M1041 and A000127/M1119 in "An On-Line Version of the Encyclopedia of Integer Sequences."

Yaglom, A. M. and Yaglom, I. M. Problem 47. *Challenging Mathematical Problems with Elementary Solutions, Vol. 1.* New York: Dover, 1987.

Circle Evolute

$$x = \cos t \qquad x' = -\sin t \qquad x'' = -\cos t \qquad (1)$$

$$y = \sin t \qquad y' = \cos t \qquad y'' = -\sin t, \qquad (2)$$

so the RADIUS OF CURVATURE is

$$R = \frac{(x'^2 + y'^2)^{3/2}}{y''x' - x''y'}$$

$$= \frac{(\sin^2 t + \cos^2 t)^{3/2}}{(-\sin t)(-\sin t) - (-\cos t)\cos t} = 1, \qquad (3)$$

and the TANGENT VECTOR is

$$\hat{\mathbf{T}} = \begin{bmatrix} -\sin t \\ \cos t \end{bmatrix}. \qquad (4)$$

Therefore,

$$\cos\tau = \hat{\mathbf{T}} \cdot \hat{\mathbf{x}} = -\sin t \qquad (5)$$

$$\sin\tau = \hat{\mathbf{T}} \cdot \hat{\mathbf{y}} = \cos t, \qquad (6)$$

so

$$\xi(t) = x - R\sin\tau = \cos t - 1 \cdot \cos t = 0 \qquad (7)$$

$$\eta(t) = y + R\cos\tau = \sin t + 1 \cdot (-\sin t) = 0, \qquad (8)$$

and the EVOLUTE degenerates to a POINT at the ORIGIN.

see also CIRCLE INVOLUTE

References

Gray, A. *Modern Differential Geometry of Curves and Surfaces.* Boca Raton, FL: CRC Press, p. 77, 1993.

Lauwerier, H. *Fractals: Endlessly Repeated Geometric Figures.* Princeton, NJ: Princeton University Press, pp. 55–59, 1991.

Circle Inscribing

If r is the RADIUS of a CIRCLE inscribed in a RIGHT TRIANGLE with sides a and b and HYPOTENUSE c, then

$$r = \tfrac{1}{2}(a + b - c).$$

see INSCRIBED, POLYGON

Circle Involute

First studied by Huygens when he was considering clocks without pendula for use on ships at sea. He used the circle involute in his first pendulum clock in an attempt to force the pendulum to swing in the path of a CYCLOID.

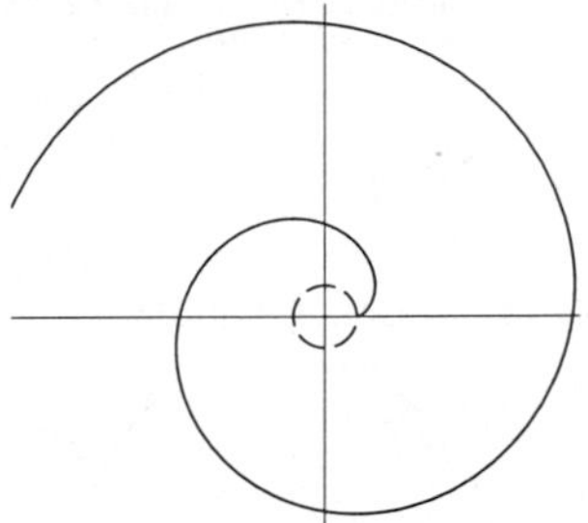

For a CIRCLE with $a = 1$, the parametric equations of the circle and their derivatives are given by

$$x = \cos t \qquad x' = -\sin t \qquad x'' = -\cos t \qquad (1)$$

$$y = \sin t \qquad y' = \cos t \qquad y'' = -\sin t. \qquad (2)$$

The TANGENT VECTOR is

$$\hat{\mathbf{T}} = \begin{bmatrix} -\sin t \\ \cos t \end{bmatrix} \qquad (3)$$

and the ARC LENGTH along the circle is

$$s = \int \sqrt{x'^2 + y'^2}\, dt = \int dt = t, \qquad (4)$$

so the involute is given by

$$\mathbf{r}_i = \mathbf{r} - s\hat{\mathbf{T}} = \begin{bmatrix} \cos t \\ \sin t \end{bmatrix} - t \begin{bmatrix} -\sin t \\ \cos t \end{bmatrix} = \begin{bmatrix} \cos t + t\sin t \\ \sin t - t\cos t \end{bmatrix}, \qquad (5)$$

or

$$x = a(\cos t + t\sin t) \qquad (6)$$

$$y = a(\sin t - t\cos t). \qquad (7)$$

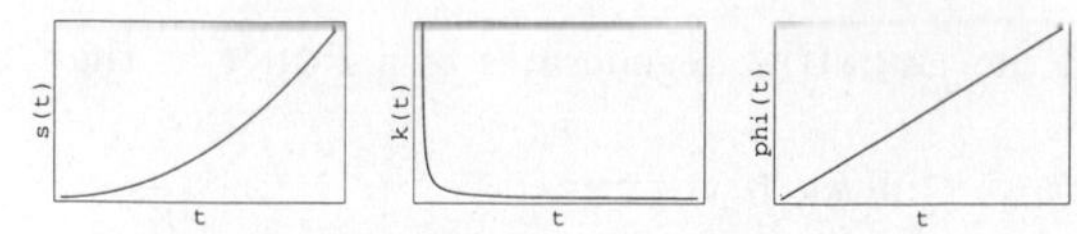

The ARC LENGTH, CURVATURE, and TANGENTIAL ANGLE are

$$s = \int ds = \int \sqrt{x'^2 + y'^2}\, dt = \tfrac{1}{2}at^2 \quad (8)$$

$$\kappa = \frac{1}{at} \quad (9)$$

$$\phi = t. \quad (10)$$

The CESÀRO EQUATION is

$$\kappa = \frac{1}{\sqrt{as}}. \quad (11)$$

see also CIRCLE, CIRCLE EVOLUTE, ELLIPSE INVOLUTE, INVOLUTE

References
Gray, A. *Modern Differential Geometry of Curves and Surfaces.* Boca Raton, FL: CRC Press, p. 83, 1993.
Lawrence, J. D. *A Catalog of Special Plane Curves.* New York: Dover, pp. 190–191, 1972.
MacTutor History of Mathematics Archive. "Involute of a Circle." `http://www-groups.dcs.st-and.ac.uk/~history/Curves/Involute.html`.

Circle Involute Pedal Curve

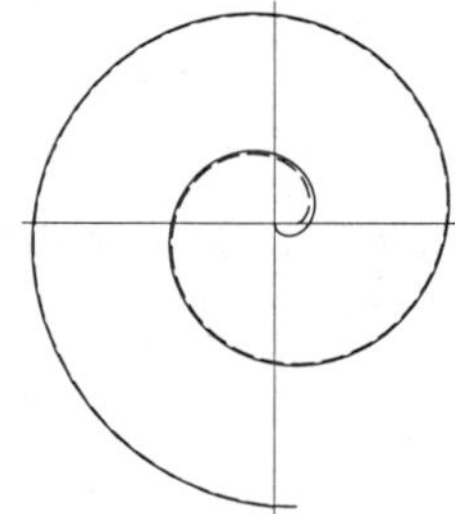

The PEDAL CURVE of CIRCLE INVOLUTE

$$f = \cos t + t \sin t$$
$$g = \sin t - t \cos t$$

with the center as the PEDAL POINT is the ARCHIMEDES' SPIRAL

$$x = t \sin t$$
$$y = -t \cos t.$$

Circle Lattice Points

For every POSITIVE INTEGER n, there exists a CIRCLE which contains exactly n lattice points in its interior. H. Steinhaus proved that for every POSITIVE INTEGER n, there exists a CIRCLE of AREA n which contains exactly n lattice points in its interior.

SCHINZEL'S THEOREM shows that for every POSITIVE INTEGER n, there exists a CIRCLE in the PLANE having exactly n LATTICE POINTS on its CIRCUMFERENCE. The theorem also explicitly identifies such "SCHINZEL CIRCLES" as

$$\begin{cases} (x - \frac{1}{2})^2 + y^2 = \frac{1}{4}5^{k-1} & \text{for } n = 2k \\ (x - \frac{1}{3})^2 + y^2 = \frac{1}{9}5^{2k} & \text{for } n = 2k+1. \end{cases} \quad (1)$$

Note, however, that these solutions do not necessarily have the smallest possible RADIUS. For example, while the SCHINZEL CIRCLE centered at (1/3, 0) and with RADIUS 625/3 has nine lattice points on its CIRCUMFERENCE, so does the CIRCLE centered at (1/3, 0) with RADIUS 65/3.

Let r be the smallest INTEGER RADIUS of a CIRCLE centered at the ORIGIN (0, 0) with $L(r)$ LATTICE POINTS. In order to find the number of lattice points of the CIRCLE, it is only necessary to find the number in the first octant, i.e., those with $0 \leq y \leq \lfloor r/\sqrt{2} \rfloor$, where $\lfloor z \rfloor$ is the FLOOR FUNCTION. Calling this $N(r)$, then for $r \geq 1$, $L(r) = 8N(r) - 4$, so $L(r) \equiv 4 \pmod 8$. The multiplication by eight counts all octants, and the subtraction by four eliminates points on the axes which the multiplication counts twice. (Since $\sqrt{2}$ is IRRATIONAL, the MIDPOINT of a are is never a LATTICE POINT.)

GAUSS'S CIRCLE PROBLEM asks for the number of lattice points *within* a CIRCLE of RADIUS r

$$N(r) = 1 + 4\lfloor r \rfloor + 4 \sum_{i=1}^{\lfloor r \rfloor} \left\lfloor \sqrt{r^2 - i^2} \right\rfloor. \quad (2)$$

Gauss showed that

$$N(r) = \pi r^2 + E(r), \quad (3)$$

where

$$|E(r)| \leq 2\sqrt{2}\,\pi r. \quad (4)$$

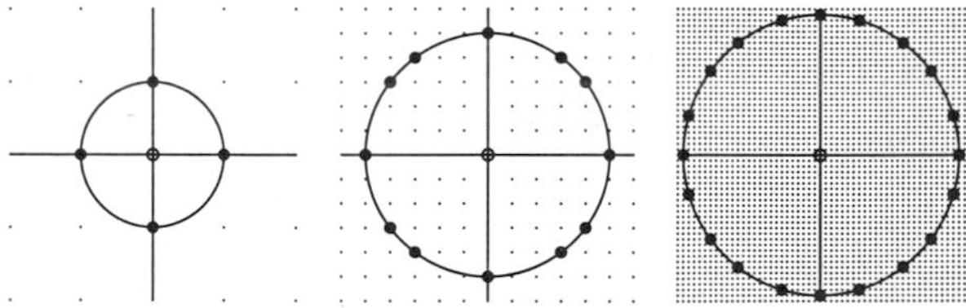

The number of lattice points on the CIRCUMFERENCE of circles centered at (0, 0) with radii 0, 1, 2, ... are 1, 4, 4, 4, 4, 12, 4, 4, 4, 4, 12, 4, 4, ... (Sloane's A046109). The following table gives the smallest RADIUS $r \leq 111{,}000$ for a circle centered at (0, 0) having a given number of LATTICE POINTS $L(r)$. Note that the high water mark radii are always multiples of five.

$L(r)$	r
1	0
4	1
12	5
20	25
28	125
36	65
44	3,125
52	15,625
60	325
68	$\leq 390,625$
76	$\leq 1,953,125$
84	1,625
92	$\leq 48,828,125$
100	4,225
108	1,105
132	40,625
140	21,125
180	5,525
252	27,625
300	71,825
324	32,045

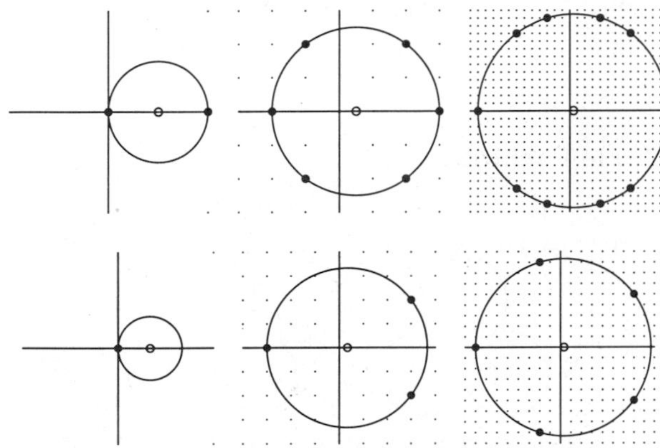

If the CIRCLE is instead centered at (1/2, 0), then the CIRCLES of RADII 1/2, 3/2, 5/2, ... have 2, 2, 6, 2, 2, 2, 6, 6, 6, 2, 2, 2, 10, 2, ... (Sloane's A046110) on their CIRCUMFERENCES. If the CIRCLE is instead centered at (1/3, 0), then the number of lattice points on the CIRCUMFERENCE of the CIRCLES of RADIUS 1/3, 2/3, 4/3, 5/3, 7/3, 8/3, ... are 1, 1, 1, 3, 1, 1, 3, 1, 3, 1, 1, 3, 1, 3, 1, 1, 5, 3, ... (Sloane's A046111).

Let

1. a_n be the RADIUS of the CIRCLE centered at (0, 0) having $8n+4$ lattice points on its CIRCUMFERENCE,
2. $b_n/2$ be the RADIUS of the CIRCLE centered at (1/2, 0) having $4n+2$ lattice points on its CIRCUMFERENCE,
3. $c_n/3$ be the RADIUS of CIRCLE centered at (1/3, 0) having $2n+1$ lattice points on its CIRCUMFERENCE.

Then the sequences $\{a_n\}$, $\{b_n\}$, and $\{c_n\}$ are equal, with the exception that $b_n = 0$ if $2|n$ and $c_n = 0$ if $3|n$. However, the sequences of *smallest* radii having the above numbers of lattice points are equal in the three cases and given by 1, 5, 25, 125, 65, 3125, 15625, 325, ... (Sloane's A046112).

KULIKOWSKI'S THEOREM states that for every POSITIVE INTEGER n, there exists a 3-D SPHERE which has exactly n LATTICE POINTS on its surface. The SPHERE is given by the equation

$$(x-a)^2 + (y-b)^2 + (z-\sqrt{2})^2 = c^2 + 2,$$

where a and b are the coordinates of the center of the so-called SCHINZEL CIRCLE and c is its RADIUS (Honsberger 1973).

see also CIRCLE, CIRCUMFERENCE, GAUSS'S CIRCLE PROBLEM, KULIKOWSKI'S THEOREM, LATTICE POINT, SCHINZEL CIRCLE, SCHINZEL'S THEOREM

References

Honsberger, R. "Circles, Squares, and Lattice Points." Ch. 11 in *Mathematical Gems I.* Washington, DC: Math. Assoc. Amer., pp. 117–127, 1973.

Kulikowski, T. "Sur l'existence d'une sphère passant par un nombre donné aux coordonnées entières." *L'Enseignement Math. Ser. 2* **5**, 89–90, 1959.

Schinzel, A. "Sur l'existence d'un cercle passant par un nombre donné de points aux coordonnées entières." *L'Enseignement Math. Ser. 2* **4**, 71–72, 1958.

Sierpiński, W. "Sur quelques problèmes concernant les points aux coordonnées entières." *L'Enseignement Math. Ser. 2* **4**, 25–31, 1958.

Sierpiński, W. "Sur un problème de H. Steinhaus concernant les ensembles de points sur le plan." *Fund. Math.* **46**, 191–194, 1959.

Sierpiński, W. *A Selection of Problems in the Theory of Numbers.* New York: Pergamon Press, 1964.

Weisstein, E. W. "Circle Lattice Points." `http://www.astro.virginia.edu/~eww6n/math/notebooks/CircleLatticePoints.m`.

Circle Lattice Theorem

see GAUSS'S CIRCLE PROBLEM

Circle Map

A 1-D MAP which maps a CIRCLE onto itself

$$\theta_{n+1} = \theta_n + \Omega - \frac{K}{2\pi}\sin(2\pi\theta_n), \tag{1}$$

where θ_{n+1} is computed mod 1. Note that the circle map has two parameters: Ω and K. Ω can be interpreted as an externally applied frequency, and K as a strength of nonlinearity. The 1-D JACOBIAN is

$$\frac{\partial\theta_{n+1}}{\partial\theta_n} = 1 - K\cos(2\pi\theta_n), \tag{2}$$

so the circle map is not AREA-PRESERVING. It is related to the STANDARD MAP

$$I_{n+1} = I_n + \frac{K}{2\pi}\sin(2\pi\theta_n) \tag{3}$$

$$\theta_{n+1} = \theta_n + I_{n+1}, \tag{4}$$

for I and θ computed mod 1. Writing θ_{n+1} as

$$\theta_{n+1} = \theta_n + I_n + \frac{K}{2\pi}\sin(2\pi\theta_n) \qquad (5)$$

gives the circle map with $I_n = \Omega$ and $K = -K$. The unperturbed circle map has the form

$$\theta_{n+1} = \theta_n + \Omega. \qquad (6)$$

If Ω is RATIONAL, then it is known as the map WINDING NUMBER, defined by

$$\Omega = W \equiv \frac{p}{q}, \qquad (7)$$

and implies a periodic trajectory, since θ_n will return to the same point (at most) every q ORBITS. If Ω is IRRATIONAL, then the motion is quasiperiodic. If K is NONZERO, then the motion may be periodic in some finite region surrounding each RATIONAL Ω. This execution of periodic motion in response to an IRRATIONAL forcing is known as MODE LOCKING.

If a plot is made of K vs. Ω with the regions of periodic MODE-LOCKED parameter space plotted around RATIONAL Ω values (WINDING NUMBERS), then the regions are seen to widen upward from 0 at $K = 0$ to some finite width at $K = 1$. The region surrounding each RATIONAL NUMBER is known as an ARNOLD TONGUE. At $K = 0$, the ARNOLD TONGUES are an isolated set of MEASURE zero. At $K = 1$, they form a CANTOR SET of DIMENSION $d \approx 0.08700$. For $K > 1$, the tongues overlap, and the circle map becomes noninvertible. The circle map has a FEIGENBAUM CONSTANT

$$\delta \equiv \lim_{n\to\infty} \frac{\theta_n - \theta_{n-1}}{\theta_{n+1} - \theta_n} = 2.833. \qquad (8)$$

see also ARNOLD TONGUE, DEVIL'S STAIRCASE, MODE LOCKING, WINDING NUMBER (MAP)

Circle Method

see PARTITION FUNCTION P

Circle Negative Pedal Curve

The NEGATIVE PEDAL CURVE of a circle is an ELLIPSE if the PEDAL POINT is inside the CIRCLE, and a HYPERBOLA if the PEDAL POINT is outside the CIRCLE.

Circle Notation

A NOTATION for LARGE NUMBERS due to Steinhaus (1983) in which $\textcircled{n}$ is defined in terms of STEINHAUS-MOSER NOTATION as n in n SQUARES. The particular number known as the MEGA is then defined as follows.

$$\textcircled{2} = \boxed{4} = \boxed{\triangle\!\!\!\triangle 2} = \boxed{\triangle 2^2} = \boxed{4^4} = \boxed{256}$$

see also MEGA, MEGISTRON, STEINHAUS-MOSER NOTATION

References

Steinhaus, H. *Mathematical Snapshots, 3rd American ed.* New York: Oxford University Press, pp. 28–29, 1983.

Circle Order

A POSET P is a circle order if it is ISOMORPHIC to a SET of DISKS ordered by containment.

see also ISOMORPHIC POSETS, PARTIALLY ORDERED SET

Circle Orthotomic

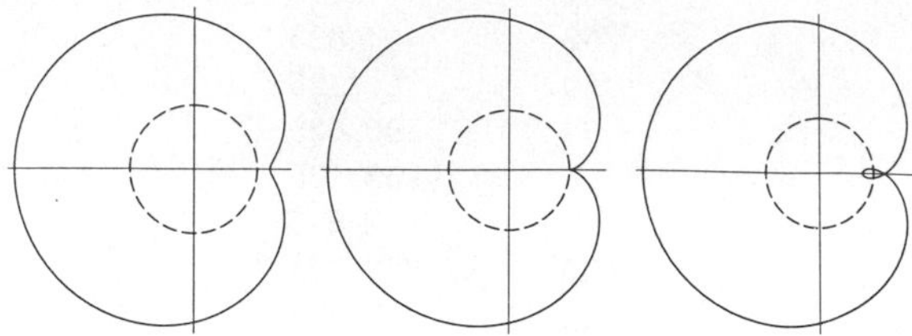

The ORTHOTOMIC of the CIRCLE represented by

$$x = \cos t \qquad (1)$$
$$y = \sin t \qquad (2)$$

with a source at (x, y) is

$$x = x\cos(2t) - y\sin(2t) + 2\sin t \qquad (3)$$
$$y = -x\sin(2t) - y\cos(2t) + 2\cos t. \qquad (4)$$

Circle Packing

The densest packing of spheres in the PLANE is the hexagonal lattice of the bee's honeycomb (illustrated above), which has a PACKING DENSITY of

$$\eta = \frac{\pi}{2\sqrt{3}} = 0.9068996821\ldots.$$

Gauss proved that the hexagonal lattice is the densest plane *lattice* packing, and in 1940, L. Fejes Tóth proved that the hexagonal lattice is indeed the densest of *all* possible plane packings.

Solutions for the smallest diameter CIRCLES into which n UNIT CIRCLES can be packed have been proved optimal for $n = 1$ through 10 (Kravitz 1967). The best known results are summarized in the following table.

n	d exact	d approx.
1	1	1.00000
2	2	2.00000
3	$1+\frac{2}{3}\sqrt{3}$	2.15470...
4	$1+\sqrt{2}$	2.41421...
5	$1+\sqrt{2(1+1/\sqrt{5})}$	2.70130...
6	3	3.00000
7	3	3.00000
8	$1+\csc(\pi/7)$	3.30476...
9	$1+\sqrt{2(2+\sqrt{2})}$	3.61312...
10		3.82...
11		
12		4.02...

For CIRCLE packing inside a SQUARE, proofs are known only for $n = 1$ to 9.

n	d exact	d approx.
1	1	1.000
2		0.58...
3		0.500...
4	$\frac{1}{2}$	0.500
5		0.41...
6		0.37...
7		0.348...
8		0.341...
9	$\frac{1}{3}$	0.333...
10		0.148204...

The smallest SQUARE into which two UNIT CIRCLES, one of which is split into two pieces by a chord, can be packed is not known (Goldberg 1968, Ogilvy 1990).

see also HYPERSPHERE PACKING, MALFATTI'S RIGHT TRIANGLE PROBLEM, MERGELYAN-WESLER THEOREM, SPHERE PACKING

References

Conway, J. H. and Sloane, N. J. A. *Sphere Packings, Lattices, and Groups, 2nd ed.* New York: Springer-Verlag, 1992.
Eppstein, D. "Covering and Packing." http://www.ics.uci.edu/~eppstein/junkyard/cover.html.
Folkman, J. H. and Graham, R. "A Packing Inequality for Compact Convex Subsets of the Plane." *Canad. Math. Bull.* **12**, 745–752, 1969.
Gardner, M. "Mathematical Games: The Diverse Pleasures of Circles that Are Tangent to One Another." *Sci. Amer.* **240**, 18–28, Jan. 1979.
Gardner, M. "Tangent Circles." Ch. 10 in *Fractal Music, Hypercards, and More Mathematical Recreations from Scientific American Magazine.* New York: W. H. Freeman, 1992.
Goldberg, M. "Problem E1924." *Amer. Math. Monthly* **75**, 195, 1968.
Goldberg, M. "The Packing of Equal Circles in a Square." *Math. Mag.* **43**, 24–30, 1970.
Goldberg, M. "Packing of 14, 16, 17, and 20 Circles in a Circle." *Math. Mag.* **44**, 134–139, 1971.
Graham, R. L. and Luboachevsky, B. D. "Repeated Patterns of Dense Packings of Equal Disks in a Square." *Electronic J. Combinatorics* **3**, R16, 1–17, 1996. http://www.combinatorics.org/Volume_3/volume3.html#R16.
Kravitz, S. "Packing Cylinders into Cylindrical Containers." *Math. Mag.* **40**, 65–70, 1967.
McCaughan, F. "Circle Packings." http://www.pmms.cam.ac.uk/~gjm11/cpacking/info.html.
Molland, M. and Payan, Charles. "A Better Packing of Ten Equal Circles in a Square." *Discrete Math.* **84**, 303–305, 1990.
Ogilvy, C. S. *Excursions in Geometry.* New York: Dover, p. 145, 1990.
Reis, G. E. "Dense Packing of Equal Circle within a Circle." *Math. Mag.* **48**, 33–37, 1975.
Schaer, J. "The Densest Packing of Nine Circles in a Square." *Can. Math. Bul.* **8**, 273–277, 1965.
Schaer, J. "The Densest Packing of Ten Equal Circles in a Square." *Math. Mag.* **44**, 139–140, 1971.
Valette, G. "A Better Packing of Ten Equal Circles in a Square." *Discrete Math.* **76**, 57–59, 1989.

Circle Pedal Curve

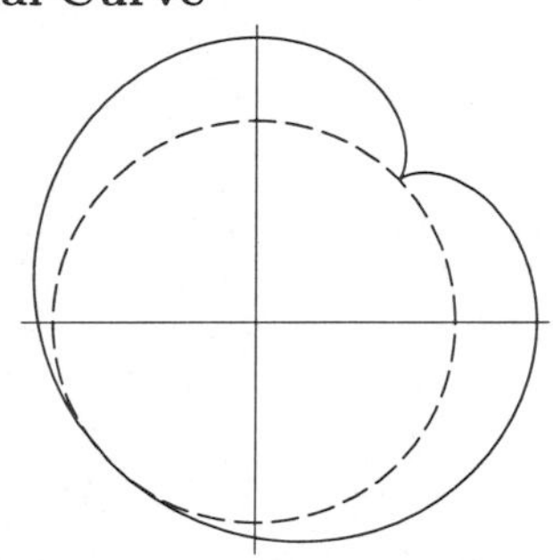

The PEDAL CURVE of a CIRCLE is a CARDIOID if the PEDAL POINT is taken on the CIRCUMFERENCE,

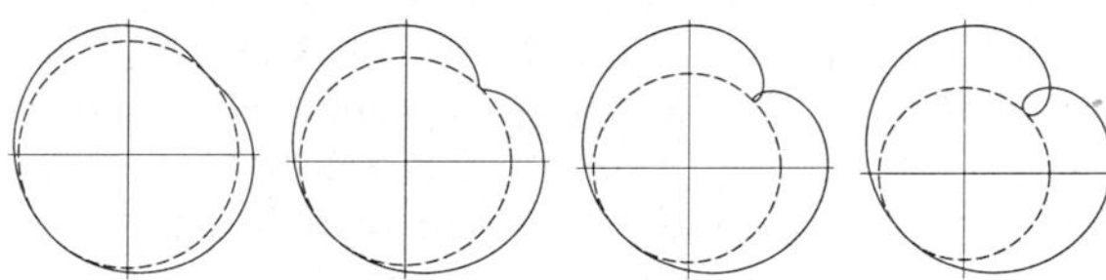

and otherwise a LIMAÇON.

Circle-Point Midpoint Theorem

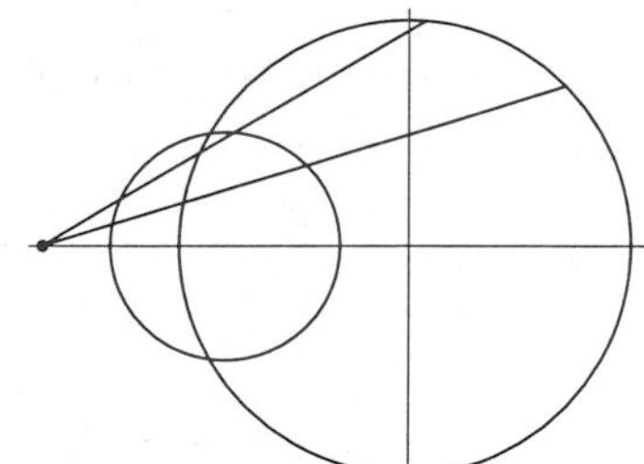

Taking the locus of MIDPOINTS from a fixed point to a circle of radius r results in a circle of radius $r/2$. This follows trivially from

$$\begin{aligned}\mathbf{r}(\theta) &= \begin{bmatrix}-x\\0\end{bmatrix} + \frac{1}{2}\left(\begin{bmatrix}r\cos\theta\\r\sin\theta\end{bmatrix} - \begin{bmatrix}-x\\0\end{bmatrix}\right)\\ &= \begin{bmatrix}\frac{1}{2}r\cos\theta - \frac{1}{2}x\\ \frac{1}{2}\sin\theta\end{bmatrix}.\end{aligned}$$

References

Johnson, R. A. *Modern Geometry: An Elementary Treatise on the Geometry of the Triangle and the Circle.* Boston, MA: Houghton Mifflin, p. 17, 1929.

Circle Radial Curve

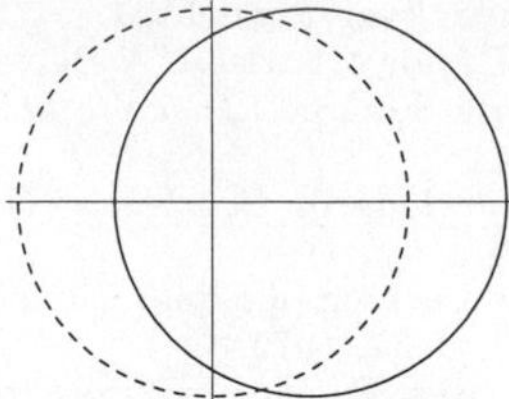

The RADIAL CURVE of a unit CIRCLE from a RADIAL POINT $(x, 0)$ is another CIRCLE with parametric equations

$$x(t) = x - \cos t$$
$$y(t) = -\sin t.$$

Circle Squaring

Construct a SQUARE equal in AREA to a CIRCLE using only a STRAIGHTEDGE and COMPASS. This was one of the three GEOMETRIC PROBLEMS OF ANTIQUITY, and was perhaps first attempted by Anaxagoras. It was finally proved to be an impossible problem when PI was proven to be TRANSCENDENTAL by Lindemann in 1882.

However, approximations to circle squaring are given by constructing lengths close to $\pi = 3.1415926\ldots$. Ramanujan (1913–14) and Olds (1963) give geometric constructions for $355/113 = 3.1415929\ldots$. Gardner (1966, pp. 92–93) gives a geometric construction for $3 + 16/113 = 3.1415929\ldots$. Dixon (1991) gives constructions for $6/5(1+\phi) = 3.141640\ldots$ and $\sqrt{4+[3-\tan(30°)]} = 3.141533\ldots$.

While the circle cannot be squared in EUCLIDEAN SPACE, it *can* in GAUSS-BOLYAI-LOBACHEVSKY SPACE (Gray 1989).

see also GEOMETRIC CONSTRUCTION, QUADRATURE, SQUARING

References

Conway, J. H. and Guy, R. K. *The Book of Numbers.* New York: Springer-Verlag, pp. 190–191, 1996.

Dixon, R. *Mathographics.* New York: Dover, pp. 44–49 and 52–53, 1991.

Dunham, W. "Hippocrates' Quadrature of the Lune." Ch. 1 in *Journey Through Genius: The Great Theorems of Mathematics.* New York: Wiley, pp. 20–26, 1990.

Gardner, M. "The Transcendental Number Pi." Ch. 8 in *Martin Gardner's New Mathematical Diversions from Scientific American.* New York: Simon and Schuster, 1966.

Gray, J. *Ideas of Space.* Oxford, England: Oxford University Press, 1989.

Meyers, L. F. "Update on William Wernick's 'Triangle Constructions with Three Located Points.'" *Math. Mag.* **69**, 46–49, 1996.

Olds, C. D. *Continued Fractions.* New York: Random House, pp. 59–60, 1963.

Ramanujan, S. "Modular Equations and Approximations to π." *Quart. J. Pure. Appl. Math.* **45**, 350–372, 1913–1914.

Circle Strophoid

The STROPHOID of a CIRCLE with pole at the center and fixed point on the CIRCUMFERENCE is a FREETH'S NEPHROID.

Circle Tangents

There are four CIRCLES that touch all the sides of a given TRIANGLE. These are all touched by the CIRCLE through the intersection of the ANGLE BISECTORS of the TRIANGLE, known as the NINE-POINT CIRCLE.

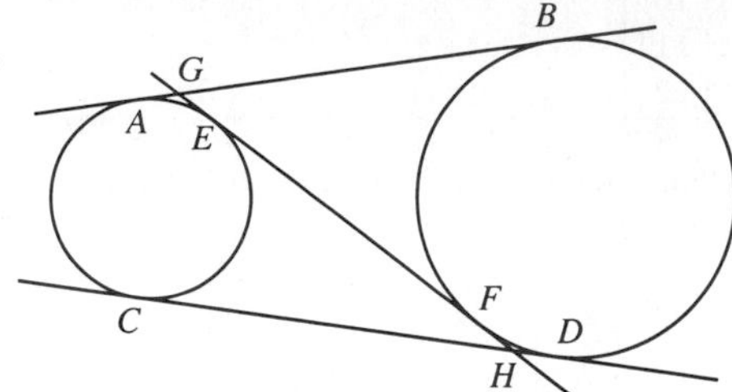

Given the above figure, $GE = FH$, since

$$\begin{aligned} AB &= AG + GB = GE + GF = GE + (GE + EF) \\ &= 2G + EF \\ CD &= CH + HD = EH + FH = FH + (FH + EF) \\ &= EF + 2FH. \end{aligned}$$

Because $AB = CD$, it follows that $GE = FH$.

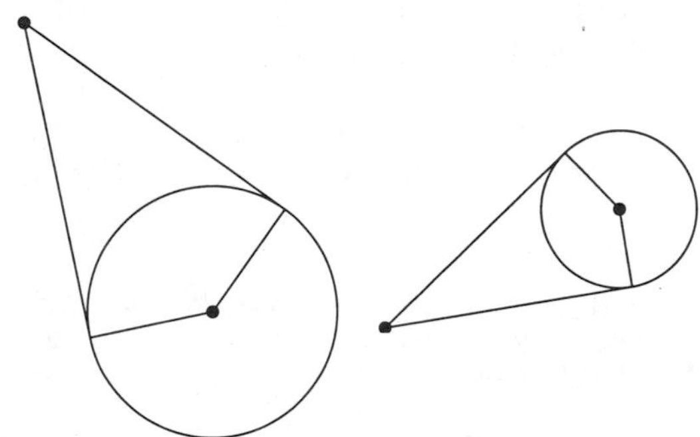

The line tangent to a CIRCLE of RADIUS a centered at (x, y)

$$x' = x + a\cos t$$
$$y' = y + a\sin t$$

through $(0, 0)$ can be found by solving the equation

$$\begin{bmatrix} x + a\cos t \\ y + a\sin t \end{bmatrix} \cdot \begin{bmatrix} a\cos t \\ a\sin t \end{bmatrix} = 0,$$

giving

$$t = \pm\cos^{-1}\left(\frac{-ax \pm y\sqrt{x^2+y^2-a^2}}{x^2+y^2}\right).$$

Two of these four solutions give tangent lines, as illustrated above.

see also KISSING CIRCLES PROBLEM, MIQUEL POINT, MONGE'S PROBLEM, PEDAL CIRCLE, TANGENT LINE, TRIANGLE

References

Dixon, R. *Mathographics.* New York: Dover, p. 21, 1991.

Honsberger, R. *More Mathematical Morsels.* Washington, DC: Math. Assoc. Amer., pp. 4–5, 1991.

Circuit

see CYCLE (GRAPH)

Circuit Rank

Also known as the CYCLOMATIC NUMBER. The circuit rank is the smallest number of EDGES γ which must be removed from a GRAPH of N EDGES and n nodes such that no CIRCUIT remains.

$$\gamma = N - n + 1.$$

Circulant Determinant

Gradshteyn and Ryzhik (1970) define circulants by

$$\begin{vmatrix} x_1 & x_2 & x_3 & \cdots & x_n \\ x_n & x_1 & x_2 & \cdots & x_{n-1} \\ x_{n-1} & x_n & x_1 & \cdots & x_{n-2} \\ \vdots & \vdots & \vdots & \ddots & \vdots \\ x_2 & x_3 & x_4 & \cdots & x_1 \end{vmatrix} = \prod_{j=1} (x_1 + x_2\omega_j + x_3{\omega_j}^2 + \ldots + x_n{\omega_j}^{n-1}), \quad (1)$$

where ω_j is the nth ROOT OF UNITY. The second-order circulant determinant is

$$\begin{vmatrix} x_1 & x_2 \\ x_2 & x_1 \end{vmatrix} = (x_1 + x_2)(x_1 - x_2), \quad (2)$$

and the third order is

$$\begin{vmatrix} x_1 & x_2 & x_3 \\ x_3 & x_1 & x_2 \\ x_2 & x_3 & x_1 \end{vmatrix} = (x_1 + x_2 + x_3)(x_1 + \omega x_2 + \omega^2 x_3)(x_1 + \omega^2 x_2 + \omega x_3), \quad (3)$$

where ω and ω^2 are the COMPLEX CUBE ROOTS of UNITY.

The EIGENVALUES λ of the corresponding $n \times n$ circulant matrix are

$$\lambda_j = x_1 + x_2\omega_j + x_3{\omega_j}^2 + \ldots + x_n{\omega_j}^{n-1}. \quad (4)$$

see also CIRCULANT MATRIX

References

Gradshteyn, I. S. and Ryzhik, I. M. *Tables of Integrals, Series, and Products, 5th ed.* San Diego, CA: Academic Press, pp. 1111–1112, 1979.

Vardi, I. *Computational Recreations in Mathematica.* Reading, MA: Addison-Wesley, p. 114, 1991.

Circulant Graph

A GRAPH of n VERTICES in which the ith VERTEX is adjacent to the $(i + j)$th and $(i - j)$th VERTICES for each j in a list l.

Circulant Matrix

An $n \times n$ MATRIX C defined as follows,

$$\mathsf{C} = \begin{bmatrix} 1 & \binom{n}{1} & \binom{n}{2} & \cdots & \binom{n}{n-1} \\ \binom{n}{n-1} & 1 & \binom{n}{1} & \cdots & \binom{n}{n-2} \\ \vdots & \vdots & \vdots & \ddots & \vdots \\ \binom{n}{1} & \binom{n}{2} & \binom{n}{3} & \cdots & 1 \end{bmatrix}$$

$$\mathsf{C} = \prod_{j=0}^{n-1} [(1 + \omega_j)^n - 1],$$

where $\omega_0 \equiv 1$, ω_1, ..., ω_{n-1} are the nth ROOTS OF UNITY. Circulant matrices are examples of LATIN SQUARES.

see also CIRCULANT DETERMINANT

References

Davis, P. J. *Circulant Matrices, 2nd ed.* New York: Chelsea, 1994.

Stroeker, R. J. "Brocard Points, Circulant Matrices, and Descartes' Folium." *Math. Mag.* **61**, 172–187, 1988.

Vardi, I. *Computational Recreations in Mathematica.* Reading, MA: Addison-Wesley, p. 114, 1991.

Circular Cylindrical Coordinates

see CYLINDRICAL COORDINATES

Circular Functions

The functions describing the horizontal and vertical positions of a point on a CIRCLE as a function of ANGLE (COSINE and SINE) and those functions derived from them:

$$\cot x \equiv \frac{1}{\tan x} = \frac{\cos x}{\sin x} \quad (1)$$

$$\csc x \equiv \frac{1}{\sin x} \quad (2)$$

$$\sec x \equiv \frac{1}{\cos x} \quad (3)$$

$$\tan x \equiv \frac{\sin x}{\cos x}. \quad (4)$$

The study of circular functions is called TRIGONOMETRY.

see also COSECANT, COSINE, COTANGENT, ELLIPTIC FUNCTION, GENERALIZED HYPERBOLIC FUNCTIONS, HYPERBOLIC FUNCTIONS, SECANT, SINE, TANGENT, TRIGONOMETRY

References

Abramowitz, M. and Stegun, C. A. (Eds.). "Circular Functions." §4.3 in *Handbook of Mathematical Functions with Formulas, Graphs, and Mathematical Tables, 9th printing.* New York: Dover, pp. 71–79, 1972.

Circular Permutation

The number of ways to arrange n distinct objects along a CIRCLE is

$$P_n = (n-1)!.$$

The number is $(n-1)!$ instead of the usual FACTORIAL $n!$ since all CYCLIC PERMUTATIONS of objects are equivalent because the CIRCLE can be rotated.

see also PERMUTATION, PRIME CIRCLE

Circumcenter

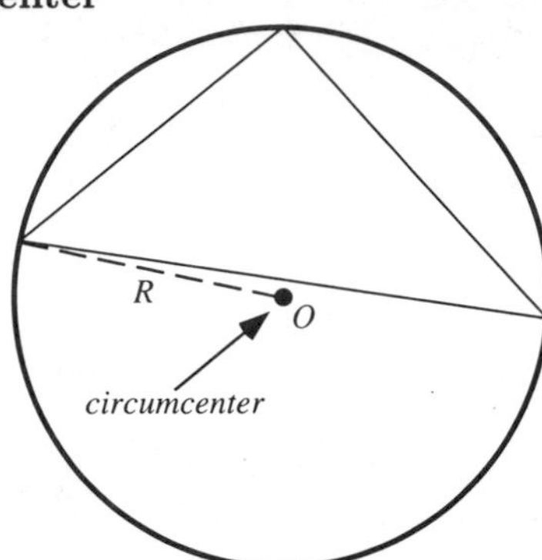

The center O of a TRIANGLE'S CIRCUMCIRCLE. It can be found as the intersection of the PERPENDICULAR BISECTORS. If the TRIANGLE is ACUTE, the circumcenter is in the interior of the TRIANGLE. In a RIGHT TRIANGLE, the circumcenter is the MIDPOINT of the HYPOTENUSE.

$$\overline{OO_1} + \overline{OO_2} + \overline{OO_3} = R + r, \tag{1}$$

where O_i are the MIDPOINTS of sides A_i, R is the CIRCUMRADIUS, and r is the INRADIUS (Johnson 1929, p. 190). The TRILINEAR COORDINATES of the circumcenter are

$$\cos A : \cos B : \cos C, \tag{2}$$

and the exact trilinears are therefore

$$R\cos A : R\cos B : R\cos C. \tag{3}$$

The AREAL COORDINATES are

$$(\tfrac{1}{2}a\cot A, \tfrac{1}{2}b\cot B, \tfrac{1}{2}c\cot C). \tag{4}$$

The distance between the INCENTER and circumcenter is $\sqrt{R(R-2r)}$. Given an interior point, the distances to the VERTICES are equal IFF this point is the circumcenter. It lies on the BROCARD AXIS.

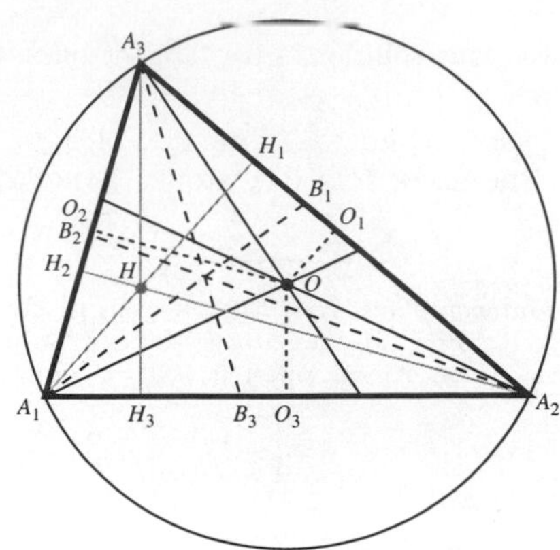

The circumcenter O and ORTHOCENTER H are ISOGONAL CONJUGATES.

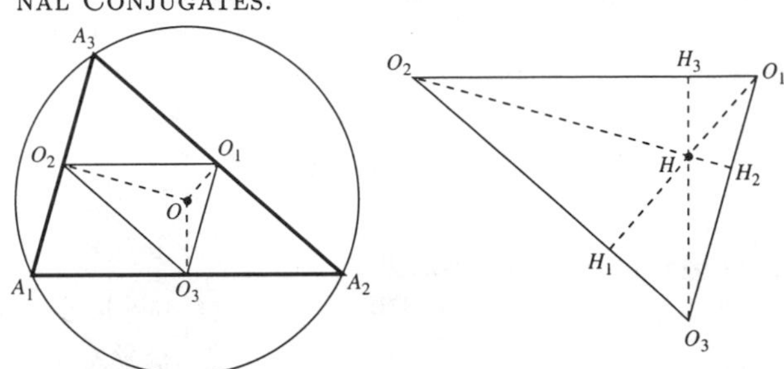

The ORTHOCENTER H of the PEDAL TRIANGLE $\Delta O_1O_2O_3$ formed by the CIRCUMCENTER O concurs with the circumcenter O itself, as illustrated above. The circumcenter also lies on the EULER LINE.

see also BROCARD DIAMETER, CARNOT'S THEOREM, CENTROID (TRIANGLE), CIRCLE, EULER LINE, INCENTER, ORTHOCENTER

References

Carr, G. S. *Formulas and Theorems in Pure Mathematics, 2nd ed.* New York: Chelsea, p. 623, 1970.

Dixon, R. *Mathographics.* New York: Dover, p. 55, 1991.

Eppstein, D. "Circumcenters of Triangles." `http://www.ics.uci.edu/~eppstein/junkyard/circumcenter.html`.

Johnson, R. A. *Modern Geometry: An Elementary Treatise on the Geometry of the Triangle and the Circle.* Boston, MA: Houghton Mifflin, 1929.

Kimberling, C. "Central Points and Central Lines in the Plane of a Triangle." *Math. Mag.* **67**, 163–187, 1994.

Kimberling, C. "Circumcenter." `http://www.evansville.edu/~ck6/tcenters/class/ccenter.html`.

Circumcircle

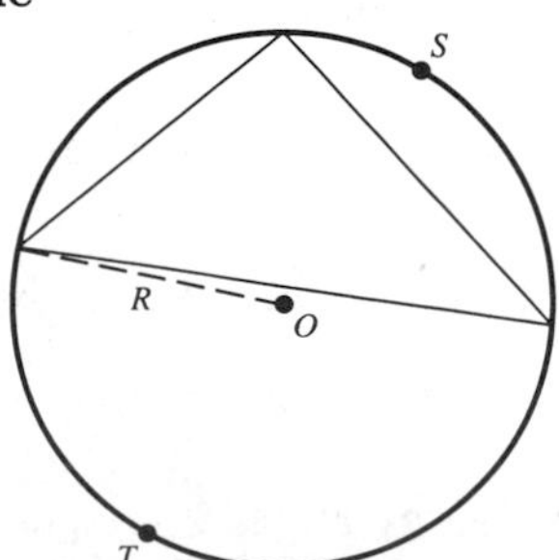

A TRIANGLE'S circumscribed circle. Its center O is called the CIRCUMCENTER, and its RADIUS R the CIRCUMRADIUS. The circumcircle can be specified using TRILINEAR COORDINATES as

$$\beta\gamma a + \gamma\alpha b + \alpha\beta c = 0. \tag{1}$$

The STEINER POINT S and TARRY POINT T lie on the circumcircle.

A GEOMETRIC CONSTRUCTION for the circumcircle is given by Pedoe (1995, pp. xii–xiii). The equation for the circumcircle of the TRIANGLE with VERTICES (x_i, y_i) for $i = 1, 2, 3$ is

$$\begin{vmatrix} x^2+y^2 & x & y & 1 \\ {x_1}^2+{y_1}^2 & x_1 & y_1 & 1 \\ {x_2}^2+{y_2}^2 & x_2 & y_2 & 1 \\ {x_3}^2+{y_3}^2 & x_3 & y_3 & 1 \end{vmatrix} = 0. \tag{2}$$

Expanding the DETERMINANT,

$$a(x^2+y^2)+2dx+2fy+g=0, \tag{3}$$

where

$$a = \begin{vmatrix} x_1 & y_1 & 1 \\ x_2 & y_2 & 1 \\ x_3 & y_3 & 1 \end{vmatrix} \tag{4}$$

$$d = -\tfrac{1}{2}\begin{vmatrix} {x_1}^2+{y_1}^2 & y_1 & 1 \\ {x_2}^2+{y_2}^2 & y_2 & 1 \\ {x_3}^2+{y_3}^2 & y_3 & 1 \end{vmatrix} \tag{5}$$

$$f = \tfrac{1}{2}\begin{vmatrix} {x_1}^2+{y_1}^2 & x_1 & 1 \\ {x_2}^2+{y_2}^2 & x_2 & 1 \\ {x_3}^2+{y_3}^2 & x_3 & 1 \end{vmatrix} \tag{6}$$

$$g = -\begin{vmatrix} {x_1}^2+{y_1}^2 & x_1 & y_1 \\ {x_2}^2+{y_2}^2 & x_2 & y_2 \\ {x_3}^2+{y_3}^2 & x_3 & y_3 \end{vmatrix}. \tag{7}$$

COMPLETING THE SQUARE gives

$$a\left(x+\frac{d}{a}\right)^2 + a\left(y+\frac{f}{a}\right)^2 - \frac{d^2}{a} - \frac{f^2}{a} + g = 0 \tag{8}$$

which is a CIRCLE of the form

$$(x-x_0)^2+(y-y_0)^2=r^2, \tag{9}$$

with CIRCUMCENTER

$$x_0 = -\frac{d}{a} \tag{10}$$

$$y_0 = -\frac{f}{a} \tag{11}$$

and CIRCUMRADIUS

$$r = \sqrt{\frac{f^2+d^2}{a^2} - \frac{g}{a}}. \tag{12}$$

see also CIRCLE, CIRCUMCENTER, CIRCUMRADIUS, EXCIRCLE, INCIRCLE, PARRY POINT, PURSER'S THEOREM, STEINER POINTS, TARRY POINT

References

Pedoe, D. *Circles: A Mathematical View, rev. ed.* Washington, DC: Math. Assoc. Amer., 1995.

Circumference

The PERIMETER of a CIRCLE. For RADIUS r or DIAMETER $d = 2r$,

$$C = 2\pi r = \pi d,$$

where π is PI.

see also CIRCLE, DIAMETER, PERIMETER, PI, RADIUS

Circuminscribed

Given two closed curves, the circuminscribed curve is simultaneously INSCRIBED in the outer one and CIRCUMSCRIBED on the inner one.

see also PONCELET'S CLOSURE THEOREM

Circumradius

The radius of a TRIANGLE'S CIRCUMCIRCLE or of a POLYHEDRON'S CIRCUMSPHERE, denoted R. For a TRIANGLE,

$$R = \frac{abc}{\sqrt{(a+b+c)(b+c-a)(c+a-b)(a+b-c)}}, \tag{1}$$

where the side lengths of the TRIANGLE are a, b, and c.

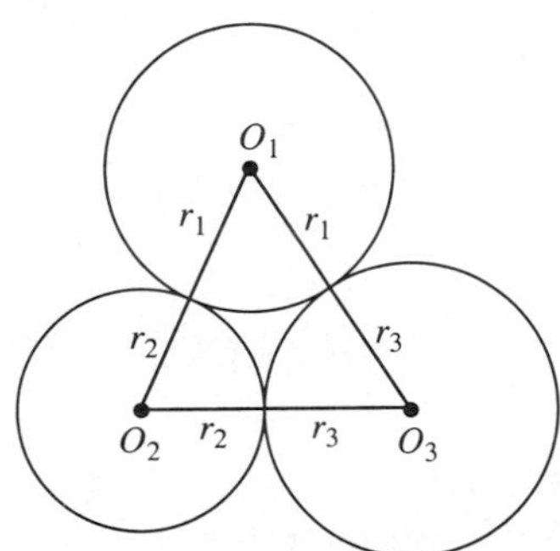

This equation can also be expressed in terms of the RADII of the three mutually tangent CIRCLES centered at the TRIANGLE'S VERTICES. Relabeling the diagram for the SODDY CIRCLES with VERTICES O_1, O_2, and O_3 and the radii r_1, r_2, and r_3, and using

$$a = r_1 + r_2 \tag{2}$$

$$b = r_2 + r_3 \tag{3}$$

$$c = r_1 + r_3 \tag{4}$$

then gives

$$R = \frac{(r_1+r_2)(r_1+r_3)(r_2+r_3)}{4\sqrt{r_1r_2r_3(r_1+r_2+r_3)}}. \tag{5}$$

If O is the CIRCUMCENTER and M is the triangle CENTROID, then

$$\overline{OM}^2 = R^2 - \tfrac{1}{9}(a^2+b^2+c^2). \tag{6}$$

$$Rr = \frac{a_1a_2a_3}{4s} \tag{7}$$

$$\cos\alpha_1 + \cos\alpha_2 + \cos\alpha_3 = 1 + \frac{r}{R} \tag{8}$$

$$r = 2R\cos\alpha_1 \cos\alpha_2 \cos\alpha_3 \tag{9}$$

$${a_1}^2 + {a_2}^2 + {a_3}^2 = 4rR + 8R^2 \tag{10}$$

(Johnson 1929, pp. 189–191). Let d be the distance between INRADIUS r and circumradius R, $d = \overline{rR}$. Then

$$R^2 - d^2 = 2Rr \tag{11}$$

$$\frac{1}{R-d} + \frac{1}{R+d} = \frac{1}{r} \tag{12}$$

(Mackay 1886–87). These and many other identities are given in Johnson (1929, pp. 186–190).

For an ARCHIMEDEAN SOLID, expressing the circumradius in terms of the INRADIUS r and MIDRADIUS ρ gives

$$R = \tfrac{1}{2}(r + \sqrt{r^2 + a^2}\,) \tag{13}$$

$$= \sqrt{\rho^2 + \tfrac{1}{4}a^2} \tag{14}$$

for an ARCHIMEDEAN SOLID.

see also CARNOT'S THEOREM, CIRCUMCIRCLE, CIRCUMSPHERE

References

Johnson, R. A. *Modern Geometry: An Elementary Treatise on the Geometry of the Triangle and the Circle.* Boston, MA: Houghton Mifflin, 1929.

Mackay, J. S. "Historical Notes on a Geometrical Theorem and its Developments [18th Century]." *Proc. Edinburgh Math. Soc.* **5**, 62–78, 1886–1887.

Circumscribed

A geometric figure which touches only the VERTICES (or other extremities) of another figure.

see also CIRCUMCENTER, CIRCUMCIRCLE, CIRCUMINSCRIBED, CIRCUMRADIUS, INSCRIBED

Circumsphere

A SPHERE circumscribed in a given solid. Its radius is called the CIRCUMRADIUS.

see also INSPHERE

Cis

$$\operatorname{Cis} x \equiv e^{ix} = \cos x + i \sin x.$$

Cissoid

Given two curves C_1 and C_2 and a fixed point O, let a line from O cut C at Q and C at R. Then the LOCUS of a point P such that $OP = QR$ is the cissoid. The word cissoid means "ivy shaped."

Curve 1	Curve 2	Pole	Cissoid
line	parallel line	any point	line
line	circle	center	conchoid of Nicomedes
circle	tangent line	on C	oblique cissoid
circle	tangent line	on C opp. tangent	cissoid of Diocles
circle	radial line	on C	strophoid
circle	concentric circle	center	circle
circle	same circle	$(a\sqrt{2}, 0)$	lemniscate

see also CISSOID OF DIOCLES

References

Lawrence, J. D. *A Catalog of Special Plane Curves.* New York: Dover, pp. 53–56 and 205, 1972.

Lee, X. "Cissoid." http://www.best.com/~xah/SpecialPlaneCurves_dir/Cissoid_dir/cissoid.html.

Lockwood, E. H. "Cissoids." Ch. 15 in *A Book of Curves.* Cambridge, England: Cambridge University Press, pp. 130–133, 1967.

Yates, R. C. "Cissoid." *A Handbook on Curves and Their Properties.* Ann Arbor, MI: J. W. Edwards, pp. 26–30, 1952.

Cissoid of Diocles

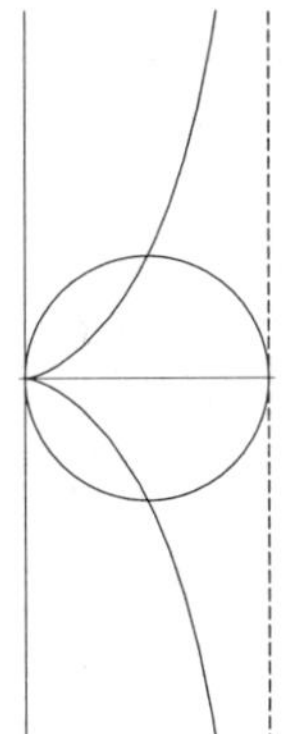

A curve invented by Diocles in about 180 BC in connection with his attempt to duplicate the cube by geometrical methods. The name "cissoid" first appears in the work of Geminus about 100 years later. Fermat and Roberval constructed the tangent in 1634. Huygens and Wallis found, in 1658, that the AREA between the curve and its asymptote was $3a$ (MacTutor Archive). From a given point there are either one or three TANGENTS to the cissoid.

Given an origin O and a point P on the curve, let S be the point where the extension of the line OP intersects the line $x = 2a$ and R be the intersection of the CIRCLE of RADIUS a and center $(a, 0)$ with the extension of OP. Then the cissoid of Diocles is the curve which satisfies $OP = RS$.

The cissoid of Diocles is the ROULETTE of the VERTEX of a PARABOLA rolling on an equal PARABOLA. Newton gave a method of drawing the cissoid of Diocles using two line segments of equal length at RIGHT ANGLES. If they are moved so that one line always passes through a fixed point and the end of the other line segment slides along a straight line, then the MIDPOINT of the sliding line segment traces out a cissoid of Diocles.

The cissoid of Diocles is given by the parametric equations

$$x = 2a\sin^2\theta \tag{1}$$

$$y = \frac{2a\sin^3\theta}{\cos\theta}. \tag{2}$$

Converting these to POLAR COORDINATES gives

$$r^2 = x^2 + y^2 = 4a^2\left(\sin^4\theta + \frac{\sin^6\theta}{\cos^2\theta}\right)$$

$$= 4a^2\sin^4\theta(1+\tan^2\theta) = 4a^2\sin^4\theta\sec^2\theta, \tag{3}$$

so

$$r = 2a\sin^2\theta\sec\theta = 2a\sin\theta\tan\theta. \tag{4}$$

In CARTESIAN COORDINATES,

$$\frac{x^3}{2a-x} = \frac{8a^3\sin^6\theta}{2a-2a\sin^2\theta} = 4a^2\frac{\sin^6\theta}{1-\sin^2\theta}$$

$$= 4a^2\frac{\sin^6\theta}{\cos^2\theta} = y^2. \tag{5}$$

An equivalent form is

$$x(x^2+y^2) = 2ay^2. \tag{6}$$

Using the alternative parametric form

$$x(t) = \frac{2at^2}{1+t^2} \tag{7}$$

$$y(t) = \frac{2at^3}{1+t^2} \tag{8}$$

(Gray 1993), gives the CURVATURE as

$$\kappa(t) = \frac{3}{a|t|(t^2+4)^{3/2}}. \tag{9}$$

References

Gray, A. "The Cissoid of Diocles." §3.4 in *Modern Differential Geometry of Curves and Surfaces.*Boca Raton, FL: CRC Press, pp. 43–46, 1993.

Lawrence, J. D. *A Catalog of Special Plane Curves.* New York: Dover, pp. 98–100, 1972.

Lee, X. "Cissoid of Diocles." `http://www.best.com/~xah/SpecialPlaneCurves_dir/CissoidOfDiocles_dir/cissoidOfDiocles.html`.

Lockwood, E. H. *A Book of Curves.* Cambridge, England: Cambridge University Press, pp. 130–133, 1967.

MacTutor History of Mathematics Archive. "Cissoid of Diocles." `http://www-groups.dcs.st-and.ac.uk/~history/Curves/Cissoid.html`.

Yates, R. C. "Cissoid." *A Handbook on Curves and Their Properties.* Ann Arbor, MI: J. W. Edwards, pp. 26–30, 1952.

Cissoid of Diocles Caustic

The CAUSTIC of the cissoid where the RADIANT POINT is taken as $(8a, 0)$ is a CARDIOID.

Cissoid of Diocles Inverse Curve

If the cusp of the CISSOID OF DIOCLES is taken as the INVERSION CENTER, then the cissoid inverts to a PARABOLA.

Cissoid of Diocles Pedal Curve

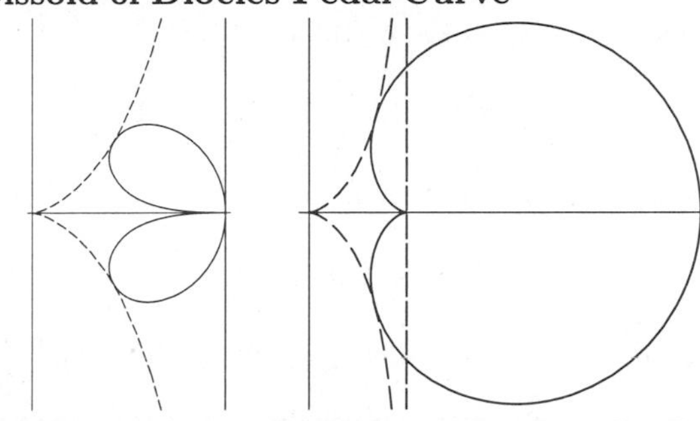

The PEDAL CURVE of the cissoid, when the PEDAL POINT is on the axis beyond the ASYMPTOTE at a distance from the cusp which is four times that of the ASYMPTOTE is a CARDIOID.

Clairaut's Differential Equation

$$y = x\frac{dy}{dx} + f\left(\frac{dy}{dx}\right)$$

or

$$y = px + f(p),$$

where f is a FUNCTION of one variable and $p \equiv dy/dx$. The general solution is $y = cx + f(c)$. The singular solution ENVELOPES are $x = -f'(c)$ and $y = f(c) - cf'(c)$.

see also D'ALEMBERT'S EQUATION

References

Boyer, C. B. *A History of Mathematics.* New York: Wiley, p. 494, 1968.

Clarity

The RATIO of a measure of the size of a "fit" to the size of a "residual."

References

Tukey, J. W. *Explanatory Data Analysis.* Reading, MA: Addison-Wesley, p. 667, 1977.

Clark's Triangle

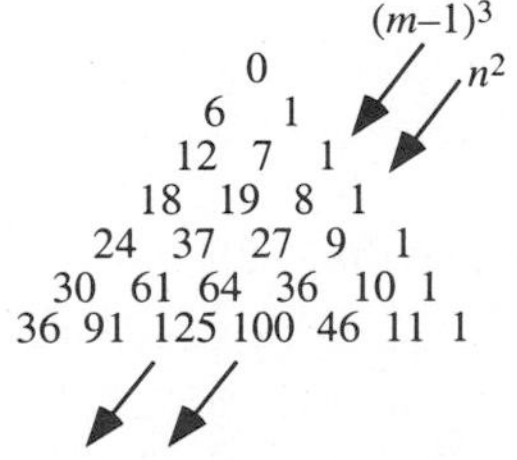

A NUMBER TRIANGLE created by setting the VERTEX equal to 0, filling one diagonal with 1s, the other diagonal with multiples of an INTEGER f, and filling in the remaining entries by summing the elements on either side from one row above. Call the first column $n = 0$ and the last column $m = n$ so that

$$c(m,0) = fm \quad (1)$$

$$c(m,m) = 1, \quad (2)$$

then use the RECURRENCE RELATION

$$c(m,n) = c(m-1,n-1) + c(m-1,n) \quad (3)$$

to compute the rest of the entries. For $n = 1$, we have

$$c(m,1) = c(m-1,0) + c(m-1,1) \quad (4)$$

$$c(m,1) - c(m-1,1) = c(m-1,0) = f(m-1). \quad (5)$$

For arbitrary m, the value can be computed by SUMMING this RECURRENCE,

$$c(m,1) = f\left(\sum_{k=1}^{m-1} k\right) + 1 = \tfrac{1}{2}fm(m-1) + 1. \quad (6)$$

Now, for $n = 2$ we have

$$c(m,2) = c(m-1,1) + c(m-1,2) \quad (7)$$

$$c(m,2) - c(m-1,2) = c(m-1,1) = \tfrac{1}{2}f(m-1)m + 1, \quad (8)$$

so SUMMING the RECURRENCE gives

$$\begin{aligned} c(m,2) &= \sum_{k=1}^{m}[\tfrac{1}{2}fk(k-1)+1] = \sum_{k=1}^{m}(\tfrac{1}{2}fk^2 - \tfrac{1}{2}fk + 1) \\ &= \tfrac{1}{2}f[\tfrac{1}{6}m(m+1)(2m+1)] - \tfrac{1}{2}f[\tfrac{1}{2}m(m+1)] + m \\ &= \tfrac{1}{6}(m-1)(fm^2 - 2fm + 6). \quad (9) \end{aligned}$$

Similarly, for $n = 3$ we have

$$\begin{aligned} c(m,3) - c(m-1,3) &= c(m-1,2) \\ &= \tfrac{1}{6}fm^3 - fm^2 + (\tfrac{11}{6}f + 1)m - (f+2). \quad (10) \end{aligned}$$

Taking the SUM,

$$c(m,3) = \sum_{k=2}^{m} \tfrac{1}{6}fk^3 - fk^2 + (\tfrac{11}{6}f+1)k - (f+2). \quad (11)$$

Evaluating the SUM gives

$$c(m,3) = \tfrac{1}{24}(m-1)(m-2)(fm^2 - 3fm + 12). \quad (12)$$

So far, this has just been relatively boring ALGEBRA. But the amazing part is that if $f = 6$ is chosen as the INTEGER, then $c(m,2)$ and $c(m,3)$ simplify to

$$\begin{aligned} c(m,2) &= \tfrac{1}{6}(m-1)(6m^2 - 12m + 6) \\ &= (m-1)^3 \quad (13) \\ c(m,3) &= \tfrac{1}{4}(m-1)^2(m-2)^2, \quad (14) \end{aligned}$$

which are consecutive CUBES $(m-1)^3$ and nonconsecutive SQUARES $n^2 = [(m-1)(m-2)/2]^2$.

see also BELL TRIANGLE, CATALAN'S TRIANGLE, EULER'S TRIANGLE, LEIBNIZ HARMONIC TRIANGLE, NUMBER TRIANGLE, PASCAL'S TRIANGLE, SEIDEL-ENTRINGER-ARNOLD TRIANGLE, SUM

References

Clark, J. E. "Clark's Triangle." *Math. Student* **26**, No. 2, p. 4, Nov. 1978.

Class

see CHARACTERISTIC CLASS, CLASS INTERVAL, CLASS (MULTIPLY PERFECT NUMBER), CLASS NUMBER, CLASS (SET), CONJUGACY CLASS

Class (Group)

see CONJUGACY CLASS

Class Interval

The constant bin size in a HISTOGRAM.

see also SHEPPARD'S CORRECTION

Class (Map)

A MAP $u : \mathbb{R}^n \to \mathbb{R}^n$ from a DOMAIN G is called a map of class C^r if each component of

$$u(x) = (u_1(x_1, \ldots, x_n), \ldots, u_m(x_1, \ldots, x_n))$$

is of class C^r ($0 \leq r \leq \infty$ or $r = \omega$) in G, where C^d denotes a continuous function which is differentiable d times.

Class (Multiply Perfect Number)

The number k in the expression $s(n) = kn$ for a MULTIPLY PERFECT NUMBER is called its class.

Class Number

For any IDEAL I, there is an IDEAL I_i such that

$$II_i = z, \quad (1)$$

where z is a PRINCIPAL IDEAL, (i.e., an IDEAL of rank 1). Moreover, there is a finite list of ideals I_i such that this equation may be satisfied for every I. The size of this list is known as the class number. When the class number is 1, the RING corresponding to a given IDEAL has unique factorization and, in a sense, the class number is a measure of the failure of unique factorization in the original number ring.

A finite series giving exactly the class number of a RING is known as a CLASS NUMBER FORMULA. A CLASS NUMBER FORMULA is known for the full ring of cyclotomic integers, as well as for any subring of the cyclotomic integers. Finding the class number is a computationally difficult problem.

Let $h(d)$ denote the class number of a quadratic ring, corresponding to the BINARY QUADRATIC FORM

$$ax^2 + bxy + cy^2, \tag{2}$$

with DISCRIMINANT

$$d \equiv b^2 - 4ac. \tag{3}$$

Then the class number $h(d)$ for DISCRIMINANT d gives the number of possible factorizations of $ax^2 + bxy + cy^2$ in the QUADRATIC FIELD $\mathbb{Q}(\sqrt{d})$. Here, the factors are of the form $x + y\sqrt{d}$, with x and y half INTEGERS.

Some fairly sophisticated mathematics shows that the class number for discriminant d can be given by the CLASS NUMBER FORMULA

$$h(d) \equiv \begin{cases} -\frac{1}{2\ln\eta}\sum_{r=1}^{d-1}(d|r)\ln\sin\left(\frac{\pi r}{d}\right) & \text{for } d > 0 \\ -\frac{w(d)}{2|d|}\sum_{r=1}^{|d|-1}(d|r)r & \text{for } d < 0, \end{cases} \tag{4}$$

where $(d|r)$ is the KRONECKER SYMBOL, $\eta(d)$ is the FUNDAMENTAL UNIT, $w(d)$ is the number of substitutions which leave the BINARY QUADRATIC FORM unchanged

$$w(d) = \begin{cases} 6 & \text{for } d = -3 \\ 4 & \text{for } d = -4 \\ 2 & \text{otherwise,} \end{cases} \tag{5}$$

and the sums are taken over all terms where the KRONECKER SYMBOL is defined (Cohn 1980). The class number for $d > 0$ can also be written

$$\eta^{2h(d)} = \prod_{r=1}^{d-1} \sin^{-(d|r)}\left(\frac{\pi r}{d}\right) \tag{6}$$

for $d > 0$, where the PRODUCT is taken over terms for which the KRONECKER SYMBOL is defined.

The class number is related to the DIRICHLET L-SERIES by

$$h(d) = \frac{L_d(1)}{\kappa(d)}, \tag{7}$$

where $\kappa(d)$ is the DIRICHLET STRUCTURE CONSTANT.

Wagner (1996) shows that class number $h(-d)$ satisfies the INEQUALITY

$$h(-d) > \frac{1}{55}\prod_{p|d}^{*}\left(1 - \frac{\lfloor 2\sqrt{p}\rfloor}{p+1}\right)\ln d, \tag{8}$$

for $-d < 0$, where $\lfloor x \rfloor$ is the FLOOR FUNCTION, the product is over PRIMES dividing d, and the $*$ indicates that the GREATEST PRIME FACTOR of d is omitted from the product.

The *Mathematica*® (Wolfram Research, Champaign, IL) function `NumberTheory`NumberTheoryFunctions`` `ClassNumber[n]` gives the class number $h(d)$ for d a NEGATIVE SQUAREFREE number of the form $4k + 1$.

GAUSS'S CLASS NUMBER PROBLEM asks to determine a complete list of fundamental DISCRIMINANTS $-d$ such that the CLASS NUMBER is given by $h(-d) = m$ for a given m. This problem has been solved for $n \leq 7$ and ODD $n \leq 23$. Gauss conjectured that the class number $h(-d)$ of an IMAGINARY quadratic field with DISCRIMINANT $-d$ tends to infinity with d, an assertion now known as GAUSS'S CLASS NUMBER CONJECTURE.

The discriminants d having $h(-d) = 1, 2, 3, 4, 5, \ldots$ are Sloane's A014602 (Cohen 1993, p. 229; Cox 1997, p. 271), Sloane's A014603 (Cohen 1993, p. 229), Sloane's A006203 (Cohen 1993, p. 504), Sloane's A013658 (Cohen 1993, p. 229), Sloane's A046002, Sloane's A046003, ... The complete set of negative discriminants having class numbers 1–5 and ODD 7–23 are known. Buell (1977) gives the smallest and largest fundamental class numbers for $d < 4{,}000{,}000$, partitioned into EVEN discriminants, discriminants 1 (mod 8), and discriminants 5 (mod 8). Arno *et al.* (1993) give complete lists of values of d with $h(-d) = k$ for ODD $k = 5, 7, 9, \ldots, 23$. Wagner gives complete lists of values for $k = 5$, 6, and 7.

Lists of NEGATIVE discriminants corresponding to IMAGINARY QUADRATIC FIELDS $\mathbb{Q}(\sqrt{-d(n)})$ having small class numbers $h(-d)$ are given in the table below. In the table, N is the number of "fundamental" values of $-d$ with a given class number $h(-d)$, where "fundamental" means that $-d$ is not divisible by any SQUARE NUMBER s^2 such that $h(-d/s^2) < h(-d)$. For example, although $h(-63) = 2$, -63 is not a fundamental discriminant since $63 = 3^2 \cdot 7$ and $h(-63/3^2) = h(-7) = 1 < h(-63)$. EVEN values $8 \leq h(-d) \leq 18$ have been computed by Weisstein. The number of negative discriminants having class number 1, 2, 3, ... are 9, 18, 16, 54, 25, 51, 31, ... (Sloane's A046125). The largest negative discriminants having class numbers 1, 2, 3, ... are 163, 427, 907, 1555, 2683, ... (Sloane's A038552).

The following table lists the numbers with small class numbers ≤ 11. Lists including larger class numbers are given by Weisstein.

$h(-d)$	N	d
1	9	3, 4, 7, 8, 11, 19, 43, 67, 163
2	18	15, 20, 24, 35, 40, 51, 52, 88, 91, 115, 123, 148, 187, 232, 235, 267, 403, 427
3	16	23, 31, 59, 83, 107, 139, 211, 283, 307, 331, 379, 499, 547, 643, 883, 907

$h(-d)$	N	d
4	54	39, 55, 56, 68, 84, 120, 132, 136, 155, 168, 184, 195, 203, 219, 228, 259, 280, 291, 292, 312, 323, 328, 340, 355, 372, 388, 408, 435, 483, 520, 532, 555, 568, 595, 627, 667, 708, 715, 723, 760, 763, 772, 795, 955, 1003, 1012, 1027, 1227, 1243, 1387, 1411, 1435, 1507, 1555
5	25	47, 79, 103, 127, 131, 179, 227, 347, 443, 523, 571, 619, 683, 691, 739, 787, 947, 1051, 1123, 1723, 1747, 1867, 2203, 2347, 2683
6	51	87, 104, 116, 152, 212, 244, 247, 339, 411, 424, 436, 451, 472, 515, 628, 707, 771, 808, 835, 843, 856, 1048, 1059, 1099, 1108, 1147, 1192, 1203, 1219, 1267, 1315, 1347, 1363, 1432, 1563, 1588, 1603, 1843, 1915, 1963, 2227, 2283, 2443, 2515, 2563, 2787, 2923, 3235, 3427, 3523, 3763
7	31	71, 151, 223, 251, 463, 467, 487, 587, 811, 827, 859, 1163, 1171, 1483, 1523, 1627, 1787, 1987, 2011, 2083, 2179, 2251, 2467, 2707, 3019, 3067, 3187, 3907, 4603, 5107, 5923
8	131	95, 111, 164, 183, 248, 260, 264, 276, 295, 299, 308, 371, 376, 395, 420, 452, 456, 548, 552, 564, 579, 580, 583, 616, 632, 651, 660, 712, 820, 840, 852, 868, 904, 915, 939, 952, 979, 987, 995, 1032, 1043, 1060, 1092, 1128, 1131, 1155, 1195, 1204, 1240, 1252, 1288, 1299, 1320, 1339, 1348, 1380, 1428, 1443, 1528, 1540, 1635, 1651, 1659, 1672, 1731, 1752, 1768, 1771, 1780, 1795, 1803, 1828, 1848, 1864, 1912, 1939, 1947, 1992, 1995, 2020, 2035, 2059, 2067, 2139, 2163, 2212, 2248, 2307, 2308, 2323, 2392, 2395, 2419, 2451, 2587, 2611, 2632, 2667, 2715, 2755, 2788, 2827, 2947, 2968, 2995, 3003, 3172, 3243, 3315, 3355, 3403, 3448, 3507, 3595, 3787, 3883, 3963, 4123, 4195, 4267, 4323, 4387, 4747, 4843, 4867, 5083, 5467, 5587, 5707, 5947, 6307
9	34	199, 367, 419, 491, 563, 823, 1087, 1187, 1291, 1423, 1579, 2003, 2803, 3163, 3259, 3307, 3547, 3643, 4027, 4243, 4363, 4483, 4723, 4987, 5443, 6043, 6427, 6763, 6883, 7723, 8563, 8803, 9067, 10627
10	87	119, 143, 159, 296, 303, 319, 344, 415, 488, 611, 635, 664, 699, 724, 779, 788, 803, 851, 872, 916, 923, 1115, 1268, 1384, 1492, 1576, 1643, 1684, 1688, 1707, 1779, 1819, 1835, 1891, 1923, 2152, 2164, 2363, 2452, 2643, 2776, 2836, 2899, 3028, 3091, 3139, 3147, 3291, 3412, 3508, 3635, 3667, 3683, 3811, 3859, 3928, 4083, 4227, 4372, 4435, 4579, 4627, 4852, 4915, 5131, 5163, 5272, 5515, 5611, 5667, 5803, 6115, 6259, 6403, 6667, 7123, 7363, 7387, 7435, 7483, 7627, 8227, 8947, 9307, 10147, 10483, 13843
11	41	167, 271, 659, 967, 1283, 1303, 1307, 1459, 1531, 1699, 2027, 2267, 2539, 2731, 2851, 2971, 3203, 3347, 3499, 3739, 3931, 4051, 5179, 5683, 6163, 6547, 7027, 7507, 7603, 7867, 8443, 9283, 9403, 9643, 9787, 10987, 13003, 13267, 14107, 14683, 15667

The table below gives lists of POSITIVE fundamental discriminants d having small class numbers $h(d)$, corresponding to REAL quadratic fields. All POSITIVE SQUAREFREE values of $d \leq 97$ (for which the KRONECKER SYMBOL is defined) are included.

$h(d)$	d
1	5, 13, 17, 21, 29, 37, 41, 53, 57, 61, 69, 73, 77
2	65

The POSITIVE d for which $h(d) = 1$ is given by Sloane's A014539.

see also CLASS NUMBER FORMULA, DIRICHLET L-SERIES, DISCRIMINANT (BINARY QUADRATIC FORM), GAUSS'S CLASS NUMBER CONJECTURE, GAUSS'S CLASS NUMBER PROBLEM, HEEGNER NUMBER, IDEAL, j-FUNCTION

References

Arno, S. "The Imaginary Quadratic Fields of Class Number 4." *Acta Arith.* **40**, 321–334, 1992.

Arno, S.; Robinson, M. L.; and Wheeler, F. S. "Imaginary Quadratic Fields with Small Odd Class Number." `http://www.math.uiuc.edu/Algebraic-Number-Theory/0009/`.

Buell, D. A. "Small Class Numbers and Extreme Values of L-Functions of Quadratic Fields." *Math. Comput.* **139**, 786–796, 1977.

Cohen, H. *A Course in Computational Algebraic Number Theory.* New York: Springer-Verlag, 1993.

Cohn, H. *Advanced Number Theory.* New York: Dover, pp. 163 and 234, 1980.

Cox, D. A. *Primes of the Form x^2+ny^2: Fermat, Class Field Theory and Complex Multiplication.* New York: Wiley, 1997.

Davenport, H. "Dirichlet's Class Number Formula." Ch. 6 in *Multiplicative Number Theory, 2nd ed.* New York: Springer-Verlag, pp. 43–53, 1980.

Iyanaga, S. and Kawada, Y. (Eds.). "Class Numbers of Algebraic Number Fields." Appendix B, Table 4 in *Encyclopedic Dictionary of Mathematics.* Cambridge, MA: MIT Press, pp. 1494–1496, 1980.

Montgomery, H. and Weinberger, P. "Notes on Small Class Numbers." *Acta. Arith.* **24**, 529–542, 1974.

Sloane, N. J. A. Sequences A014539, A038552, A046125, and A003657/M2332 in "An On-Line Version of the Encyclopedia of Integer Sequences."

Stark, H. M. "A Complete Determination of the Complex Quadratic Fields of Class Number One." *Michigan Math. J.* **14**, 1–27, 1967.
Stark, H. M. "On Complex Quadratic Fields with Class Number Two." *Math. Comput.* **29**, 289–302, 1975.
Wagner, C. "Class Number 5, 6, and 7." *Math. Comput.* **65**, 785–800, 1996.
Weisstein, E. W. "Class Numbers." `http://www.astro.virginia.edu/~eww6n/math/notebooks/ClassNumbers.m`.

Class Number Formula

A class number formula is a finite series giving exactly the CLASS NUMBER of a RING. For a RING of quadratic integers, the class number is denoted $h(d)$, where d is the discriminant. A class number formula is known for the full ring of cyclotomic integers, as well as for any subring of the cyclotomic integers. This formula includes the quadratic case as well as many cubic and higher-order rings.

see also CLASS NUMBER

Class Representative

A set of class representatives is a SUBSET of X which contains exactly one element from each EQUIVALENCE CLASS.

Class (Set)

A class is a special kind of SET invented to get around RUSSELL'S PARADOX while retaining the arbitrary criteria for membership which leads to difficulty for SETS. The members of classes are SETS, but it is possible to have the class C of "all SETS which are not members of themselves" without producing a paradox (since C is a proper class (and not a SET), it is not a candidate for membership in C).

see also AGGREGATE, RUSSELL'S PARADOX, SET

Classical Groups

The four following types of GROUPS,

1. LINEAR GROUPS,
2. ORTHOGONAL GROUPS,
3. SYMPLECTIC GROUPS, and
4. UNITARY GROUPS,

which were studied before more exotic types of groups (such as the SPORADIC GROUPS) were discovered.

see also GROUP, LINEAR GROUP, ORTHOGONAL GROUP, SYMPLECTIC GROUP, UNITARY GROUP

Classification

The classification of a collection of objects generally means that a list has been constructed with exactly one member from each ISOMORPHISM type among the objects, and that tools and techniques can effectively be used to identify any combinatorially given object with its unique representative in the list. Examples of mathematical objects which have been classified include the finite SIMPLE GROUPS and 2-MANIFOLDS but not, for example, KNOTS.

Classification Theorem

The classification theorem of FINITE SIMPLE GROUPS, also known as the ENORMOUS THEOREM, which states that the FINITE SIMPLE GROUPS can be classified completely into

1. CYCLIC GROUPS $\mathbb{Z}_p$ of PRIME ORDER,
2. ALTERNATING GROUPS A_n of degree at least five,
3. LIE-TYPE CHEVALLEY GROUPS $PSL(n,q)$, $PSU(n,q)$, $PsP(2n,q)$, and $P\Omega^\epsilon(n,q)$,
4. LIE-TYPE (TWISTED CHEVALLEY GROUPS or the TITS GROUP) ${}^3D_4(q)$, $E_6(q)$, $E_7(q)$, $E_8(q)$, $F_4(q)$, ${}^2F_4(2^n)'$, $G_2(q)$, ${}^2G_2(3^n)$, ${}^2B(2^n)$,
5. SPORADIC GROUPS M_{11}, M_{12}, M_{22}, M_{23}, M_{24}, $J_2 = HJ$, Suz, HS, McL, Co_3, Co_2, Co_1, He, Fi_{22}, Fi_{23}, Fi'_{24}, HN, Th, B, M, J_1, $O'N$, J_3, Ly, Ru, J_4.

The "PROOF" of this theorem is spread throughout the mathematical literature and is estimated to be approximately 15,000 pages in length.

see also FINITE GROUP, GROUP, j-FUNCTION, SIMPLE GROUP

References

Cartwright, M. "Ten Thousand Pages to Prove Simplicity." *New Scientist* **109**, 26–30, 1985.
Cipra, B. "Are Group Theorists Simpleminded?" *What's Happening in the Mathematical Sciences, 1995–1996, Vol. 3.* Providence, RI: Amer. Math. Soc., pp. 82–99, 1996.
Cipra, B. "Slimming an Outsized Theorem." *Science* **267**, 794–795, 1995.
Gorenstein, D. "The Enormous Theorem." *Sci. Amer.* **253**, 104–115, Dec. 1985.
Solomon, R. "On Finite Simple Groups and Their Classification." *Not. Amer. Math. Soc.* **42**, 231–239, 1995.

Clausen Formula

Clausen's ${}_4F_3$ identity

$$ {}_4F_3\begin{bmatrix} a & b & c & d \\ e & f & g & \end{bmatrix}; 1\Big] = \frac{(2a)_{|d|}(a+b)_{|d|}(2b)_{|d|}}{(2a+2b)_{|d|}a_{|d|}b_{|d|}}, $$

holds for $a+b+c-d=1/2$, $e=a+b+1/2$, $a+f=d+1=b+g$, d a nonpositive integer, and $(a)_n$ is the POCHHAMMER SYMBOL (Petkovšek *et al.* 1996).

Another identity ascribed to Clausen which involves the HYPERGEOMETRIC FUNCTION ${}_2F_1(a,b;c;z)$ and the GENERALIZED HYPERGEOMETRIC FUNCTION ${}_3F_2(a,b,c;d,e;z)$ is given by

$$ \left[{}_2F_1\left(\begin{matrix} a, b \\ a+b+\frac{1}{2} \end{matrix}; x\right)\right]^2 = {}_3F_2\left(\begin{matrix} 2a, a+b, 2b \\ a+b+\frac{1}{2}, 2a+2b \end{matrix}; x\right). $$

see also GENERALIZED HYPERGEOMETRIC FUNCTION, HYPERGEOMETRIC FUNCTION

References

Petkovšek, M.; Wilf, H. S.; and Zeilberger, D. *A=B*. Wellesley, MA: A. K. Peters, pp. 43 and 127, 1996.

Clausen Function

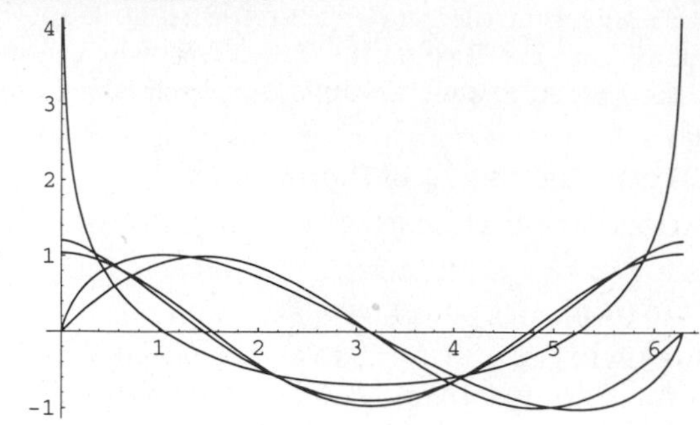

Define

$$S_n(x) \equiv \sum_{k=1}^{\infty} \frac{\sin(kx)}{k^n} \tag{1}$$

$$C_n(x) \equiv \sum_{k=1}^{\infty} \frac{\cos(kx)}{k^n}, \tag{2}$$

and write

$$\mathrm{Cl}_n(x) \equiv \begin{cases} S_n(x) = \sum_{k=1}^{\infty} \frac{\sin(kx)}{k^n} & n \text{ even} \\ C_n(x) = \sum_{k=1}^{\infty} \frac{\cos(kx)}{k^n} & n \text{ odd.} \end{cases} \tag{3}$$

Then the Clausen function $\mathrm{Cl}_n(x)$ can be given symbolically in terms of the POLYLOGARITHM as

$$\mathrm{Cl}_n(x) = \begin{cases} \frac{1}{2}i[\mathrm{Li}_n(e^{-ix}) - \mathrm{Li}_n(e^{ix})] & n \text{ even} \\ \frac{1}{2}[\mathrm{Li}_n(e^{-ix}) + \mathrm{Li}_n(e^{ix})] & n \text{ odd.} \end{cases}$$

For $n = 1$, the function takes on the special form

$$\mathrm{Cl}_1(x) = C_1(x) = -\ln|2\sin(\tfrac{1}{2}x)| \tag{4}$$

and for $n = 2$, it becomes CLAUSEN'S INTEGRAL

$$\mathrm{Cl}_2(x) = S_2(x) = -\int_0^x \ln[2\sin(\tfrac{1}{2}t)]\,dt. \tag{5}$$

The symbolic sums of opposite parity are summable symbolically, and the first few are given by

$$C_2(x) = \tfrac{1}{6}\pi^2 - \tfrac{1}{2}\pi x + \tfrac{1}{4}x^2 \tag{6}$$

$$C_4(x) = \tfrac{1}{90} - \tfrac{1}{12}\pi^2 x^2 + \tfrac{1}{12}\pi x^3 - \tfrac{1}{48}x^4 \tag{7}$$

$$S_1(x) = \tfrac{1}{2}(\pi - x) \tag{8}$$

$$S_3(x) = \tfrac{1}{6}\pi^2 x - \tfrac{1}{4}\pi x^2 + \tfrac{1}{12}x^3 \tag{9}$$

$$S_5(x) = \tfrac{1}{90}\pi^4 x - \tfrac{1}{36}\pi^2 x^3 + \tfrac{1}{48}\pi x^4 - \tfrac{1}{240}x^5 \tag{10}$$

for $0 \le x \le 2\pi$ (Abramowitz and Stegun 1972).

see also CLAUSEN'S INTEGRAL, POLYGAMMA FUNCTION, POLYLOGARITHM

References

Abramowitz, M. and Stegun, C. A. (Eds.). "Clausen's Integral and Related Summations" §27.8 in *Handbook of Mathematical Functions with Formulas, Graphs, and Mathematical Tables, 9th printing.* New York: Dover, pp. 1005–1006, 1972.

Arfken, G. *Mathematical Methods for Physicists, 3rd ed.* Orlando, FL: Academic Press, p. 783, 1985.

Clausen, R. "Über die Zerlegung reeller gebrochener Funktionen." *J. reine angew. Math.* **8**, 298–300, 1832.

Grosjean, C. C. "Formulae Concerning the Computation of the Clausen Integral $\mathrm{Cl}_2(\alpha)$." *J. Comput. Appl. Math.* **11**, 331–342, 1984.

Jolley, L. B. W. *Summation of Series.* London: Chapman, 1925.

Wheelon, A. D. *A Short Table of Summable Series.* Report No. SM-14642. Santa Monica, CA: Douglas Aircraft Co., 1953.

Clausen's Integral

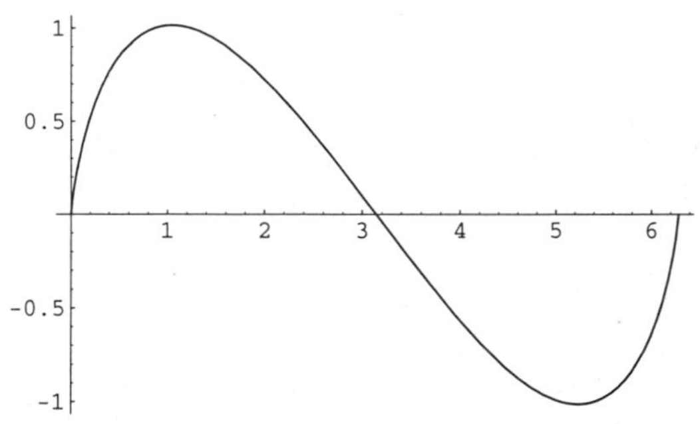

The CLAUSEN FUNCTION

$$\mathrm{Cl}_2(\theta) = -\int_0^\theta \ln[2\sin(\tfrac{1}{2}t)]\,dt.$$

see also CLAUSEN FUNCTION

References

Abramowitz, M. and Stegun, C. A. (Eds.). *Handbook of Mathematical Functions with Formulas, Graphs, and Mathematical Tables, 9th printing.* New York: Dover, pp. 1005–1006, 1972.

Ashour, A. and Sabri, A. "Tabulation of the Function $\psi(\theta) = \sum_{n=1}^{\infty} \frac{\sin(n\theta)}{n^2}$." *Math. Tables Aids Comp.* **10**, 54 and 57–65, 1956.

Clausen, R. "Über die Zerlegung reeller gebrochener Funktionen." *J. reine angew. Math.* **8**, 298–300, 1832.

CLEAN Algorithm

An iterative algorithm which DECONVOLVES a sampling function (the "DIRTY BEAM") from an observed brightness ("DIRTY MAP") of a radio source. This algorithm is of fundamental importance in radio astronomy, where it is used to create images of astronomical sources which are observed using arrays of radio telescopes ("synthesis imaging"). As a result of the algorithm's importance to synthesis imaging, a great deal of effort has gone into optimizing and adjusting the ALGORITHM. CLEAN is a nonlinear algorithm, since linear DECONVOLUTION algorithms such as WIENER FILTERING and inverse filtering are inapplicable to applications with invisible distributions (i.e., incomplete sampling of the spatial frequency plane) such as map obtained in synthesis imaging.

The basic CLEAN method was developed by Högbom (1974). It was originally designed for point sources, but it has been found to work well for extended sources as well when given a reasonable starting model. The Högbom CLEAN constructs discrete approximations I_n to the CLEAN MAP in the (ξ, η) plane from the CONVOLUTION equation

$$b' * I = I', \tag{1}$$

where b' is the DIRTY BEAM, I' is the DIRTY MAP (both in the (ξ, η) PLANE), and $f * g$ denotes a CONVOLUTION.

The CLEAN algorithm starts with an initial approximation $I_0 = 0$. At the nth iteration, it then searches for the largest value in the residual map

$$I_n = I' - b' * I_{n-1}. \tag{2}$$

A DELTA FUNCTION is then centered at the location of the largest residual flux and given an amplitude μ (the so-called "LOOP GAIN") times this value. An antenna's response to the DELTA FUNCTION, the DIRTY BEAM, is then subtracted from I_{n-1} to yield I_n. Iteration continues until a specified iteration limit N is reached, or until the peak residual or ROOT-MEAN-SQUARE residual decreases to some level. The resulting final map is denoted I_N, and the position of each DELTA FUNCTION is saved in a "CLEAN component" table in the CLEAN MAP file. At the point where component subtraction is stopped, it is assumed that the residual brightness distribution consists mainly of NOISE.

To diminish high spatial frequency features which may be spuriously extrapolated from the measured data, each CLEAN component is convolved with the so-called CLEAN BEAM b, which is simply a suitably smoothed version of the sampling function ("DIRTY BEAM"). Usually, a GAUSSIAN is used. A good CLEAN BEAM should:

1. Have a unity FOURIER TRANSFORM inside the sampled region of (u, v) space,
2. Have a FOURIER TRANSFORM which tends to 0 outside the sampled (u, v) region as quickly as possible, and
3. Not have any effects produced by NEGATIVE sidelobes larger than the NOISE level.

A CLEAN MAP is produced when the final residual map is added to the the approximate solution,

$$[\text{clean map}] = I_N * b + [I' - b' * I_N] \tag{3}$$

in order to include the NOISE.

CLEAN will always converge to one (of possibly many) solutions if the following three conditions are satisfied (Schwarz 1978):

1. The beam must be symmetric.
2. The FOURIER TRANSFORM of the DIRTY BEAM is NONNEGATIVE (positive definite or positive semidefinite).
3. There must be no spatial frequencies present in the dirty image which are not also present in the DIRTY BEAM.

These conditions are almost always satisfied in practice. If the number of CLEAN components does not exceed the number of independent (u, v) points, CLEAN converges to a solution which is the least squares fit of the FOURIER TRANSFORMS of the DELTA FUNCTION components to the measured visibility (Thompson *et al.* 1986, p. 347). Schwarz claims that the CLEAN algorithm is equivalent to a least squares fitting of cosine and sine parts in the (u, v) plane of the visibility data. Schwab has produced a NOISE analysis of the CLEAN algorithm in the case of least squares minimization of a noiseless image which involves an $N \times M$ MATRIX. However, no NOISE analysis has been performed for a REAL image.

Poor modulation of short spacings results in an underestimation of the flux, which is manifested in a bowl of negative surface brightness surrounding an object. Providing an estimate of the "zero spacing" flux (the total flux of the source, which cannot be directly measured by an interferometer) can considerably reduce this effect. Modulations or stripes can occur at spatial frequencies corresponding to undersampled parts of the (u, v) plane. This can result in a golf ball-like mottling for disk sources such as planets, or a corrugated pattern of parallel lines of peaks and troughs ("stripes"). A more accurate model can be used to suppress the "golf ball" modulations, but may not eliminate the corrugations. A tapering function which deemphasizes data near $(u, v) = (0, 0)$ can also be used. Stripes can sometimes be eliminated using the Cornwell smoothness-stabilized CLEAN (a.k.a. Prussian helmet algorithm; Thompson *et al.* 1986). CLEANing part way, then restarting the CLEAN also seems to eliminate the stripes, although this fact is more disturbing than reassuring. Stability the the CLEAN algorithm is discussed by Tan (1986).

In order to CLEAN a map of a given dimension, it is necessary to have a beam pattern twice as large so a point source can be subtracted from any point in the map. Because the CLEAN algorithm uses a FAST FOURIER TRANSFORM, the size must also be a POWER of 2.

There are many variants of the basic Högbom CLEAN which extend the method to achieve greater speed and produce more realistic maps. Alternate nonlinear DECONVOLUTION methods, such as the MAXIMUM ENTROPY METHOD, may also be used, but are generally slower than the CLEAN technique. The Astronomical Image Processing Software (AIPS) of the National Radio Astronomical Observatory includes 2-D

DECONVOLUTION algorithms in the tasks **DCONV** and **UVMAP**. Among the variants of the basic Högbom CLEAN are Clark, Cornwell smoothness stabilized (Prussian helmet), Cotton-Schwab, Gerchberg-Saxton (Fienup), Steer, Steer-Dewdney-Ito, and van Cittert iteration.

In the Clark (1980) modification, CLEAN picks out only the largest residual points, and subtracts approximate point source responses in the (ξ, η) plane during minor (Högbom CLEAN) cycles. It only occasionally (during major cycles) computes the full I_n residual map by subtracting the identified point source responses in the (u, v) plane using a FAST FOURIER TRANSFORM for the CONVOLUTION. The ALGORITHM then returns to a minor cycle. This algorithm modifies the Högbom method to take advantage of the array processor (although it also works without one). It is therefore a factor of 2–10 faster than the simple Högbom routine. It is implemented as the AIPS task **APCLN**.

The Cornwell smoothness stabilized variant was developed because, when dealing with two-dimensional extended structures, CLEAN can produce artifacts in the form of low-level high frequency stripes running through the brighter structure. These stripes derive from poor interpolations into unsampled or poorly sampled regions of the (u, v) plane. When dealing with quasi-one-dimensional sources (i.e., jets), the artifacts resemble knots (which may not be so readily recognized as spurious). **APCLN** can invoke a modification of CLEAN that is intended to bias it toward generating smoother solutions to the deconvolution problem while preserving the requirement that the transform of the CLEAN components list fits the data. The mechanism for introducing this bias is the addition to the DIRTY BEAM of a DELTA FUNCTION (or "spike") of small amplitude (**PHAT**) while searching for the CLEAN components. The beam used for the deconvolution resembles the helmet worn by German military officers in World War I, hence the name "Prussian helmet" CLEAN.

The theory underlying the Cornwell smoothness stabilized algorithm is given by Cornwell (1982, 1983), where it is described as the smoothness stabilized CLEAN. It is implemented in the AIPS tasks **APCLN** and **MX**. The spike performs a NEGATIVE feedback into the dirty image, thus suppressing features not required by the data. Spike heights of a few percent and lower than usual loop gains are usually needed. Also according to the **MX** documentation,

$$\mathtt{PHAT} \approx \frac{(\text{noise})^2}{2(\text{signal})^2} = \frac{1}{2(\text{SNR})^2}.$$

Unfortunately, the addition of a Prussian helmet generally has "limited success," so resorting to another deconvolution method such as the MAXIMUM ENTROPY METHOD is sometimes required.

The Cotton-Schwab uses the Clark method, but the major cycle subtractions of CLEAN components are performed on ungridded visibility data. The Cotton-Schwab technique is often faster than the Clark variant. It is also capable of including the w baseline term, thus removing distortions from noncoplanar baselines. It is often faster than the Clark method. The Cotton-Schwab technique is implemented as the AIPS task **MX**.

The Gerchberg-Saxton variant, also called the Fienup variant, is a technique originally introduced for solving the phase problem in electron microscopy. It was subsequently adapted for visibility amplitude measurements only. A Gerchberg-Saxton map is constrained to be NONZERO, and positive. Data and image plane constraints are imposed alternately while transforming to and from the image plane. If the boxes to CLEAN are chosen to surround the source snugly, then the algorithm will converge faster and will have more chance of finding a unique image. The algorithm is slow, but should be comparable to the Clark technique (**APCLN**) if the map contains many picture elements. However, the resolution is data dependent and varies across the map. It is implemented as the AIPS task **APGS** (Pearson 1984).

The Steer variant is a modification of the Clark variant (Cornwell 1982). It is slow, but should be comparable to the Clark algorithm if the map contains many picture elements. The algorithm used in the program is due to David Steer. The principle is similar to Barry Clark's CLEAN except that in the minor cycle only points above the (trim level)×(peak in the residual map) are selected. In the major cycle these are removed using a FAST FOURIER TRANSFORM. If boxes are chosen to surround the source snugly, then the algorithm will converge faster and will have more chance of finding a unique image. It is implemented in AIPS as the experimental program **STEER** and as the Steer-Dewdney-Ito variant combined with the Clark algorithm as **SDCLN**.

The Steer-Dewdney-Ito variant is similar to the Clark variant, but the components are taken as all pixels having residual flux greater than a cutoff value times the current peak residual. This method should avoid the "ripples" produced by the standard CLEAN on extended emission. The AIPS task **SDCLN** does an AP-based CLEAN of the the Clark type, but differs from **APCLN** in that it offers the option to switch to the Steer-Dewdney-Ito method.

Finally, van Cittert iteration consists of two steps:

1. Estimate a correction to add to the current map estimate by multiplying the residuals by some weight. In the classical van Cittert algorithm, this weight is a constant, where as in CLEAN the weight is zero everywhere except at the peak of the residuals.
2. Add the step to the current estimate, and subtract the estimate, convolved with the DIRTY BEAM, from the residuals.

Though it is a simple algorithm, it works well (if slowly) for cases where the DIRTY BEAM is positive semidefinite (as it is in astronomical observations). The basic idea is that the DIRTY MAP is a reasonably good estimate of the deconvolved map. The different iterations vary only in the weight to apply to each residual in determining the correction step. van Cittert iteration is implemented as the AIPS task APVC, which is a rather experimental and ad hoc procedure. In some limiting cases, it reduces to the standard CLEAN algorithm (though it would be impractically slow).

see also CLEAN BEAM, CLEAN MAP, DIRTY BEAM, DIRTY MAP

References

Christiansen, W. N. and Högbom, J. A. *Radiotelescopes, 2nd ed.* Cambridge, England: Cambridge University Press, pp. 214–216, 1985.

Clark, B. G. "An Efficient Implementation of the Algorithm 'CLEAN'." *Astron. Astrophys.* **89**, 377–378, 1980.

Cornwell, T. J. "Can CLEAN be Improved?" VLA Scientific Memorandum No. 141, 1982.

Cornwell, T. J. "Image Restoration (and the CLEAN Technique)." Lecture 9. NRAO VLA Workshop on Synthesis Mapping, p. 113, 1982.

Cornwell, T. J. "A Method of Stabilizing the CLEAN Algorithm." *Astron. Astrophys.* **121**, 281–285, 1983.

Cornwell, T. and Braun, R. "Deconvolution." Ch. 8 in *Synthesis Imaging in Radio Astronomy: Third NRAO Summer School, 1988* (Ed. R. A. Perley, F. R. Schwab, and A. H. Bridle). San Francisco, CA: Astronomical Society of the Pacific, pp. 178–179, 1989.

Högbom, J. A. "Aperture Synthesis with a Non-Regular Distribution of Interferometric Baselines." *Astron. Astrophys. Supp.* **15**, 417–426, 1974.

National Radio Astronomical Observatory. Astronomical Image Processing Software (AIPS) software package. APCLN, MX, and UVMAP tasks.

Pearson, T. J. and Readhead, A. C. S. "Image Formation by Self-Calibration in Radio Astronomy." *Ann. Rev. Astron. Astrophys.* **22**, 97–130, 1984.

Schwarz, U. J. "Mathematical-Statistical Description of the Iterative Beam Removing Technique (Method CLEAN)." *Astron. Astrophys.* **65**, 345–356, 1978.

Tan, S. M. "An Analysis of the Properties of CLEAN and Smoothness Stabilized CLEAN—Some Warnings." *Mon. Not. Royal Astron. Soc.* **220**, 971–1001, 1986.

Thompson, A. R.; Moran, J. M.; and Swenson, G. W. Jr. *Interferometry and Synthesis in Radio Astronomy.* New York: Wiley, p. 348, 1986.

CLEAN Beam

An ELLIPTICAL GAUSSIAN fit to the DIRTY BEAM in order to remove sidelobes. The CLEAN beam is convolved with the final CLEAN iteration to diminish spurious high spatial frequencies.

see also CLEAN ALGORITHM, CLEAN MAP, DECONVOLUTION, DIRTY BEAM, DIRTY MAP

CLEAN Map

The deconvolved map extracted from a finitely sampled DIRTY MAP by the CLEAN ALGORITHM, MAXIMUM ENTROPY METHOD, or any other DECONVOLUTION procedure.

see also CLEAN ALGORITHM, CLEAN BEAM, DECONVOLUTION, DIRTY BEAM, DIRTY MAP

Clebsch-Aronhold Notation

A notation used to describe curves. The fundamental principle of Clebsch-Aronhold notation states that if each of a number of forms be replaced by a POWER of a linear form in the same number of variables equal to the order of the given form, and if a sufficient number of equivalent symbols are introduced by the ARONHOLD PROCESS so that no actual COEFFICIENT appears except to the first degree, then every identical relation holding for the new specialized forms holds for the general ones.

References

Coolidge, J. L. *A Treatise on Algebraic Plane Curves.* New York: Dover, p. 79, 1959.

Clebsch Diagonal Cubic

A CUBIC ALGEBRAIC SURFACE given by the equation

$$x_0{}^3 + x_1{}^3 + x_2{}^3 + x_3{}^3 + x_4{}^3 = 0, \tag{1}$$

with the added constraint

$$x_0 + x_1 + x_2 + x_3 + x_4 = 0. \tag{2}$$

The implicit equation obtained by taking the plane at infinity as $x_0 + x_1 + x_2 + x_3/2$ is

$$\begin{aligned}81(x^3+y^3+z^3) - 189(x^2y+x^2z+y^2x+y^2z+z^2x+z^2y)\\ +54xyz + 126(xy+xz+yz) - 9(x^2+y^2+z^2)\\ -9(x+y+z)+1 = 0\end{aligned} \tag{3}$$

(Hunt, Nordstrand). On Clebsch's diagonal surface, all 27 of the complex lines (SOLOMON'S SEAL LINES) present on a general smooth CUBIC SURFACE are real. In addition, there are 10 points on the surface where 3 of the 27 lines meet. These points are called ECKARDT POINTS (Fischer 1986, Hunt), and the Clebsch diagonal surface is the unique CUBIC SURFACE containing 10 such points (Hunt).

If one of the variables describing Clebsch's diagonal surface is dropped, leaving the equations

$$x_0{}^3 + x_1{}^3 + x_2{}^3 + x_3{}^3 = 0, \tag{4}$$

$$x_0 + x_1 + x_2 + x_3 = 0, \tag{5}$$

the equations degenerate into two intersecting PLANES given by the equation

$$(x+y)(x+z)(y+z) = 0. \tag{6}$$

see also CUBIC SURFACE, ECKARDT POINT

References

Fischer, G. (Ed.). *Mathematical Models from the Collections of Universities and Museums.* Braunschweig, Germany: Vieweg, pp. 9–11, 1986.

Fischer, G. (Ed.). Plates 10–12 in *Mathematische Modelle/Mathematical Models, Bildband/Photograph Volume.* Braunschweig, Germany: Vieweg, pp. 13–15, 1986.

Hunt, B. *The Geometry of Some Special Arithmetic Quotients.* New York: Springer-Verlag, pp. 122–128, 1996.

Nordstrand, T. "Clebsch Diagonal Surface." http://www.uib.no/people/nfytn/clebtxt.htm.

Clebsch-Gordon Coefficient

A mathematical symbol used to integrate products of three SPHERICAL HARMONICS. Clebsch-Gordon coefficients commonly arise in applications involving the addition of angular momentum in quantum mechanics. If products of more than three SPHERICAL HARMONICS are desired, then a generalization known as WIGNER $6j$-SYMBOLS or WIGNER $9j$-SYMBOLS is used. The Clebsch-Gordon coefficients are written

$$C^j_{m_1 m_2} = (j_1 j_2 m_1 m_2 | j_1 j_2 j m) \tag{1}$$

and are defined by

$$\Psi_{JM} = \sum_{M=M_1+M_2} C^J_{M_1 M_2} \Psi_{M_1 M_2}, \tag{2}$$

where $J \equiv J_1 + J_2$. The Clebsch-Gordon coefficients are sometimes expressed using the related RACAH V-COEFFICIENTS

$$V(j_1 j_2 j; m_1 m_2 m) \tag{3}$$

or WIGNER $3j$-SYMBOLS. Connections among the three are

$$(j_1 j_2 m_1 m_2 | j_1 j_2 m) = (-1)^{-j_1+j_2-m} \sqrt{2j+1} \begin{pmatrix} j_1 & j_2 & j \\ m_1 & m_2 & -m \end{pmatrix} \tag{4}$$

$$(j_1 j_2 m_1 m_2 | j_1 j_2 j m) = (-1)^{j+m} \sqrt{2j+1} V(j_1 j_2 j; m_1 m_2 - m) \tag{5}$$

$$V(j_1 j_2 j; m_1 m_2 m) = (-1)^{-j_1+j_2+j} \begin{pmatrix} j_1 & j_2 & j_1 \\ m_2 & m_1 & m_2 \end{pmatrix}. \tag{6}$$

They have the symmetry

$$(j_1 j_2 m_1 m_2 | j_1 j_2 j m) = (-1)^{j_1+j_2-j} (j_2 j_1 m_2 m_1 | j_2 j_1 j m), \tag{7}$$

and obey the orthogonality relationships

$$\sum_{j,m} (j_1 j_2 m_1 m_2 | j_1 j_2 j m)(j_1 j_2 j m | j_1 j_2 m_1' m_2') = \delta_{m_1 m_1'} \delta_{m_2 m_2'} \tag{8}$$

$$\sum_{m_1, m_2} (j_1 j_2 m_1 m_2 | j_1 j_2 j m)(j_1 j_2 j' m' | j_1 j_2 m_1 m_2) = \delta_{jj'} \delta_{mm'}. \tag{9}$$

see also RACAH V-COEFFICIENT, RACAH W-COEFFICIENT, WIGNER $3j$-SYMBOL, WIGNER $6j$-SYMBOL, WIGNER $9j$-SYMBOL

References

Abramowitz, M. and Stegun, C. A. (Eds.). "Vector-Addition Coefficients." §27.9 in *Handbook of Mathematical Functions with Formulas, Graphs, and Mathematical Tables, 9th printing.* New York: Dover, pp. 1006–1010, 1972.

Cohen-Tannoudji, C.; Diu, B.; and Laloë, F. "Clebsch-Gordon Coefficients." Complement B_X in *Quantum Mechanics, Vol. 2.* New York: Wiley, pp. 1035–1047, 1977.

Condon, E. U. and Shortley, G. §3.6–3.14 in *The Theory of Atomic Spectra.* Cambridge, England: Cambridge University Press, pp. 56–78, 1951.

Fano, U. and Fano, L. *Basic Physics of Atoms and Molecules.* New York: Wiley, p. 240, 1959.

Messiah, A. "Clebsch-Gordon (C.-G.) Coefficients and '3j' Symbols." Appendix C.I in *Quantum Mechanics, Vol. 2.* Amsterdam, Netherlands: North-Holland, pp. 1054–1060, 1962.

Shore, B. W. and Menzel, D. H. "Coupling and Clebsch-Gordon Coefficients." §6.2 in *Principles of Atomic Spectra.* New York: Wiley, pp. 268–276, 1968.

Sobel'man, I. I. "Angular Momenta." Ch. 4 in *Atomic Spectra and Radiative Transitions, 2nd ed.* Berlin: Springer-Verlag, 1992.

Clement Matrix

see KAC MATRIX

Clenshaw Recurrence Formula

The downward Clenshaw recurrence formula evaluates a sum of products of indexed COEFFICIENTS by functions which obey a recurrence relation. If

$$f(x) = \sum_{k=0}^{N} c_k F_k(x)$$

and

$$F_{n+1}(x) = \alpha(n,x) F_n(x) + \beta(n,x) F_{n-1}(x),$$

where the c_ks are known, then define

$$y_{N+2} = y_{N+1} = 0$$
$$y_k = \alpha(k,x)y_{k+1} + \beta(k+1,x)y_{k+2} + c_k$$

for $k = N, N-1, \ldots$ and solve backwards to obtain y_2 and y_1.

$$c_k = y_k - \alpha(k,x)y_{k+1} - \beta(k+1,x)y_{k+2}$$

$$\begin{aligned} f(x) &= \sum_{k=0}^{N} c_k F_k(x) \\ &= c_0F_0(x) + [y_1 - \alpha(1,x)y_2 - \beta(2,x)y_3]F_1(x) \\ &\quad +[y_2 - \alpha(2,x)y_3 - \beta(3,x)y_4]F_2(x) \\ &\quad +[y_3 - \alpha(3,x)y_4 - \beta(4,x)y_5]F_3(x) \\ &\quad +[y_4 - \alpha(4,x)y_5 - \beta(5,x)y_6]F_4(x) + \ldots \\ &= c_0F_0(x) + y_1F_1(x) + y_2[F_2(x) - \alpha(1,x)F_1(x)] \\ &\quad + y_3[F_3(x) - \alpha(2,x)F_2(x) - \beta(2,x)] \\ &\quad + y_4[F_4(x) - \alpha(3,x)F_3(x) - \beta(3,x)] + \ldots \\ &= c_0F_0(x) + y_2[\{\alpha(1,x)F_1(x) + \beta(1,x)F_0(x)\} \\ &\quad - \alpha(1,x)F_1(x)] + y_1F_1(x) \\ &= c_0F_0(x) + y_1F_1(x) + \beta(1,x)F_0(x)y_2. \end{aligned}$$

The upward Clenshaw recurrence formula is

$$y_{-2} = y_{-1} = 0$$

$$y_k = \frac{1}{\beta(k+1,x)}[y_{k-2} - \alpha(k,x)y_{k-1} - c_k]$$

for $k = 0, 1, \ldots, N-1$.

$$f(x) = c_NF_N(x) - \beta(N,x)F_{N-1}(x)y_{N-1} - F_N(x)y_{N-2}.$$

References

Press, W. H.; Flannery, B. P.; Teukolsky, S. A.; and Vetterling, W. T. "Recurrence Relations and Clenshaw's Recurrence Formula." §5.5 in *Numerical Recipes in FORTRAN: The Art of Scientific Computing, 2nd ed.* Cambridge, England: Cambridge University Press, pp. 172–178, 1992.

Cliff Random Number Generator

A RANDOM NUMBER generator produced by iterating

$$X_{n+1} = |100 \ln X_n \pmod 1|$$

for a SEED $X_0 = 0.1$. This simple generator passes the NOISE SPHERE test for randomness by showing no structure.

see also RANDOM NUMBER, SEED

References

Pickover, C. A. "Computers, Randomness, Mind, and Infinity." Ch. 31 in *Keys to Infinity.* New York: W. H. Freeman, pp. 233–247, 1995.

Clifford Algebra

Let V be an n-D linear SPACE over a FIELD K, and let Q be a QUADRATIC FORM on V. A Clifford algebra is then defined over the $T(V)/I(Q)$, where $T(V)$ is the tensor algebra over V and I is a particular IDEAL of $T(V)$.

References

Iyanaga, S. and Kawada, Y. (Eds.). "Clifford Algebras." §64 in *Encyclopedic Dictionary of Mathematics.* Cambridge, MA: MIT Press, pp. 220–222, 1980.

Lounesto, P. "Counterexamples to Theorems Published and Proved in Recent Literature on Clifford Algebras, Spinors, Spin Groups, and the Exterior Algebra." `http://www.hit.fi/~lounesto/counterexamples.htm`.

Clifford's Circle Theorem

Let C_1, C_2, C_3, and C_4 be four CIRCLES of GENERAL POSITION through a point P. Let P_{ij} be the second intersection of the CIRCLES C_i and C_j. Let C_{ijk} be the CIRCLE $P_{ij}P_{ik}P_{jk}$. Then the four CIRCLES P_{234}, P_{134}, P_{124}, and P_{123} all pass through the point P_{1234}. Similarly, let C_5 be a fifth CIRCLE through P. Then the five points P_{2345}, P_{1345}, P_{1245}, P_{1235} and P_{1234} all lie on one CIRCLE C_{12345}. And so on.

see also CIRCLE, COX'S THEOREM

Clifford's Curve Theorem

The dimension of a special series can never exceed half its order.

References

Coolidge, J. L. *A Treatise on Algebraic Plane Curves.* New York: Dover, p. 263, 1959.

Clique

In a GRAPH of N VERTICES, a subset of pairwise adjacent VERTICES is known as a clique. A clique is a fully connected subgraph of a given graph. The problem of finding the size of a clique for a given GRAPH is an NP-COMPLETE PROBLEM. The number of graphs on n nodes having 3 cliques are 0, 0, 1, 4, 12, 31, 67, ... (Sloane's A005289).

see also CLIQUE NUMBER, MAXIMUM CLIQUE PROBLEM, RAMSEY NUMBER, TURÁN'S THEOREM

References

Sloane, N. J. A. Sequence A005289/M3440 in "An On-Line Version of the Encyclopedia of Integer Sequences."

Clique Number

The number of VERTICES in the largest CLIQUE of G, denoted $\omega(G)$. For an arbitrary GRAPH,

$$\omega(G) \geq \sum_{i=1}^{n} \frac{1}{n-d_i},$$

where d_i is the DEGREE of VERTEX i.

References

Aigner, M. "Turán's Graph Theorem." *Amer. Math. Monthly* **102**, 808–816, 1995.

Clock Solitaire

A solitaire game played with CARDS. The chance of winning is 1/13, and the AVERAGE number of CARDS turned up is 42.4.

References

Gardner, M. *Mathematical Magic Show: More Puzzles, Games, Diversions, Illusions and Other Mathematical Sleight-of-Mind from Scientific American.* New York: Vintage, pp. 244–247, 1978.

Close Packing

see SPHERE PACKING

Closed Curve

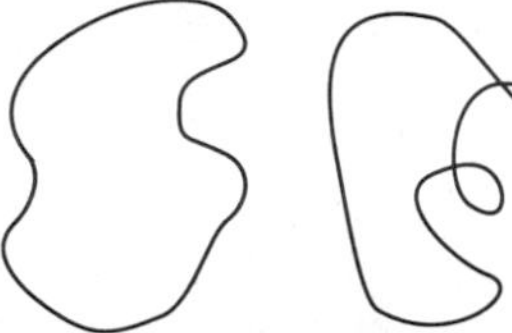

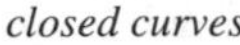

closed curves *open curves*

A CURVE with no endpoints which completely encloses an AREA. A closed curve is formally defined as the continuous IMAGE of a CLOSED SET.

see also SIMPLE CURVE

Closed Curve Problem

Find NECESSARY and SUFFICIENT conditions that determine when the integral curve of two periodic functions $\kappa(s)$ and $\tau(s)$ with the same period L is a CLOSED CURVE.

Closed Disk

An n-D closed disk of RADIUS r is the collection of points of distance $\leq r$ from a fixed point in EUCLIDEAN n-space.

see also DISK, OPEN DISK

Closed Form

A discrete FUNCTION $A(n,k)$ is called closed form (or sometimes "hypergeometric") in two variables if the ratios $A(n+1,k)/A(n,k)$ and $A(n,k+1)/A(n,k)$ are both RATIONAL FUNCTIONS. A pair of closed form functions (F,G) is said to be a WILF-ZEILBERGER PAIR if

$$F(n+1,k) - F(n,k) = G(n,k+1) - G(n,k).$$

see also RATIONAL FUNCTION, WILF-ZEILBERGER PAIR

References

Petkovšek, M.; Wilf, H. S.; and Zeilberger, D. *A=B*. Wellesley, MA: A. K. Peters, p. 141, 1996.

Zeilberger, D. "Closed Form (Pun Intended!)." *Contemporary Math.* **143**, 579–607, 1993.

Closed Graph Theorem

A linear OPERATOR between two BANACH SPACES is continuous IFF it has a "closed" GRAPH.

see also BANACH SPACE

References

Zeidler, E. *Applied Functional Analysis: Applications to Mathematical Physics.* New York: Springer-Verlag, 1995.

Closed Interval

An INTERVAL which includes its LIMIT POINTS. If the endpoints of the interval are FINITE numbers a and b, then the INTERVAL is denoted $[a,b]$. If one of the endpoints is $\pm\infty$, then the interval still contains all of its LIMIT POINTS, so $[a,\infty)$ and $(-\infty,b]$ are also closed intervals.

see also HALF-CLOSED INTERVAL, OPEN INTERVAL

Closed Set

There are several equivalent definitions of a closed SET. A SET S is closed if

1. The COMPLEMENT of S is an OPEN SET,
2. S is its own CLOSURE,
3. Sequences/nets/filters in S which converge do so within S,
4. Every point outside S has a NEIGHBORHOOD disjoint from S.

The POINT-SET TOPOLOGICAL definition of a closed set is a set which contains all of its LIMIT POINTS. Therefore, a closed set C is one for which, whatever point x is picked outside of C, x can always be isolated in some OPEN SET which doesn't touch C.

see also CLOSED INTERVAL

Closure

A SET S and a BINARY OPERATOR $*$ are said to exhibit closure if applying the BINARY OPERATOR to two elements S returns a value which is itself a member of S.

The term "closure" is also used to refer to a "closed" version of a given set. The closure of a SET can be defined in several equivalent ways, including

1. The SET plus its LIMIT POINTS, also called "boundary" points, the union of which is also called the "frontier,"
2. The unique smallest CLOSED SET containing the given SET,
3. The COMPLEMENT of the interior of the COMPLEMENT of the set,
4. The collection of all points such that every NEIGHBORHOOD of them intersects the original SET in a nonempty SET.

In topologies where the T2-SEPARATION AXIOM is assumed, the closure of a finite SET S is S itself.

see also BINARY OPERATOR, EXISTENTIAL CLOSURE, REFLEXIVE CLOSURE, TIGHT CLOSURE, TRANSITIVE CLOSURE

Clothoid

see also CORNU SPIRAL

Clove Hitch

A HITCH also called the BOATMAN'S KNOT or PEG KNOT.

References

Owen, P. *Knots.* Philadelphia, PA: Courage, pp. 24–27, 1993.

Clump

see RUN

Cluster

Given a lattice, a cluster is a group of filled cells which are all connected to their neighbors vertically or horizontally.

see also CLUSTER PERIMETER, PERCOLATION THEORY, s-CLUSTER, s-RUN

References

Stauffer, D. and Aharony, A. *Introduction to Percolation Theory, 2nd ed.* London: Taylor & Francis, 1992.

Cluster Perimeter

The number of empty neighbors of a CLUSTER.

see also PERIMETER POLYNOMIAL

Coanalytic Set

A DEFINABLE SET which is the complement of an ANALYTIC SET.

see also ANALYTIC SET

Coastline Paradox

Determining the length of a country's coastline is not as simple as it first appears, as first considered by L. F. Richardson (1881–1953). In fact, the answer depends on the length of the RULER you use for the measurements. A shorter RULER measures more of the sinuosity of bays and inlets than a larger one, so the estimated length continues to increase as the RULER length decreases.

In fact, a coastline is an example of a FRACTAL, and plotting the length of the RULER versus the measured length of the coastline on a log-log plot gives a straight line, the slope of which is the FRACTAL DIMENSION of the coastline (and will be a number between 1 and 2).

References

Lauwerier, H. *Fractals: Endlessly Repeated Geometric Figures.* Princeton, NJ: Princeton University Press, pp. 29–31, 1991.

Coates-Wiles Theorem

In 1976, Coates and Wiles showed that ELLIPTIC CURVES with COMPLEX MULTIPLICATION having an infinite number of solutions have L-functions which are zero at the relevant fixed point. This is a special case of the SWINNERTON-DYER CONJECTURE.

References

Cipra, B. "Fermat Prover Points to Next Challenges." *Science* **271**, 1668–1669, 1996.

Coaxal Circles

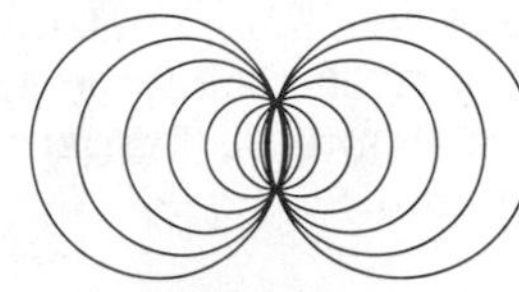

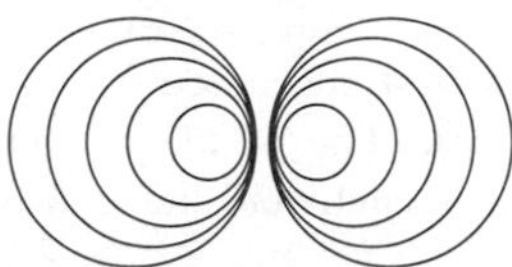

CIRCLES which share a RADICAL LINE with a given circle are said to be coaxal. The centers of coaxal circles are COLLINEAR. It is possible to combine the two types of coaxal systems illustrated above such that the sets are orthogonal.

see also CIRCLE, COAXALOID SYSTEM, GAUSS-BODENMILLER THEOREM, RADICAL LINE

References

Coxeter, H. S. M. and Greitzer, S. L. *Geometry Revisited.* Washington, DC: Math. Assoc. Amer., pp. 35–36 and 122, 1967.

Dixon, R. *Mathographics.* New York: Dover, pp. 68–72, 1991.

Johnson, R. A. *Modern Geometry: An Elementary Treatise on the Geometry of the Triangle and the Circle.* Boston, MA: Houghton Mifflin, pp. 34–37, 199, and 279, 1929.

Coaxal System

A system of COAXAL CIRCLES.

Coaxaloid System

A system of circles obtained by multiplying each RADIUS in a COAXAL SYSTEM by a constant.

References

Johnson, R. A. *Modern Geometry: An Elementary Treatise on the Geometry of the Triangle and the Circle.* Boston, MA: Houghton Mifflin, pp. 276–277, 1929.

Cobordant Manifold

Two open MANIFOLDS M and M' are cobordant if there exists a MANIFOLD with boundary W^{n+1} such that an acceptable restrictive relationship holds.

see also COBORDISM, h-COBORDISM THEOREM, MORSE THEORY

Cobordism

see BORDISM, h-COBORDISM

Cobordism Group
see BORDISM GROUP

Cobordism Ring
see BORDISM GROUP

Cochleoid

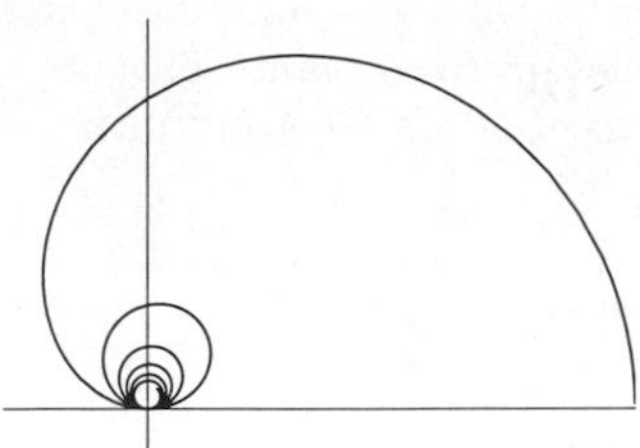

The cochleoid, whose name means "snail-form" in Latin, was first discussed by J. Peck in 1700 (MacTutor Archive). The points of contact of PARALLEL TANGENTS to the cochleoid lie on a STROPHOID.

In POLAR COORDINATES,

$$r = \frac{a\sin\theta}{\theta}. \tag{1}$$

In CARTESIAN COORDINATES,

$$(x^2 + y^2)\tan^{-1}\left(\frac{y}{x}\right) = ay. \tag{2}$$

The CURVATURE is

$$\kappa = \frac{2\sqrt{2}\,\theta^3[2\theta - \sin(2\theta)]}{[1 + 2\theta^2 - \cos(2\theta) - 2\theta\sin(2\theta)]^{3/2}}. \tag{3}$$

see also QUADRATRIX OF HIPPIAS

References
Lawrence, J. D. *A Catalog of Special Plane Curves.* New York: Dover, pp. 192 and 196, 1972.
MacTutor History of Mathematics Archive. "Cochleoid." `http://www-groups.dcs.st-and.ac.uk/~history/Curves/Cochleoid.html`.

Cochleoid Inverse Curve

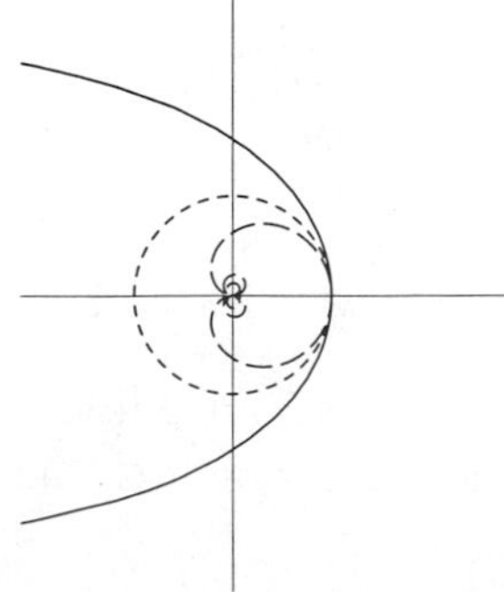

The INVERSE CURVE of the COCHLEOID

$$r = \frac{\sin\theta}{\theta} \tag{1}$$

with INVERSION CENTER at the ORIGIN and inversion radius k is the QUADRATRIX OF HIPPIAS.

$$x = kt\cot\theta \tag{2}$$

$$y = kt. \tag{3}$$

Cochloid
see CONCHOID OF NICOMEDES

Cochran's Theorem
The converse of FISHER'S THEOREM.

Cocked Hat Curve

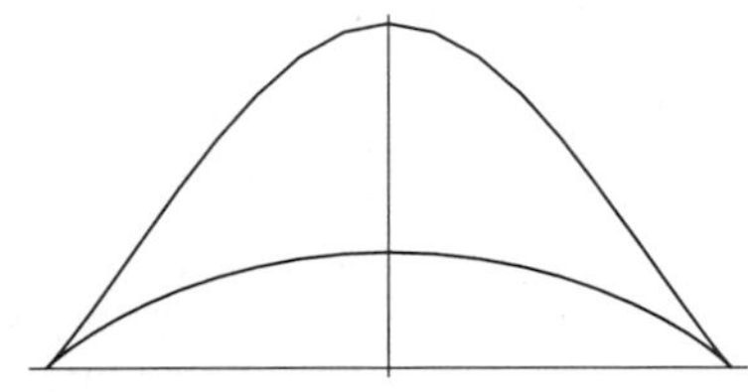

The PLANE CURVE

$$(x^2 + 2ay - a^2)^2 = y^2(a^2 - x^2),$$

which is similar to the BICORN.

References
Cundy, H. and Rollett, A. *Mathematical Models, 3rd ed.* Stradbroke, England: Tarquin Pub., p. 72, 1989.

Cocktail Party Graph

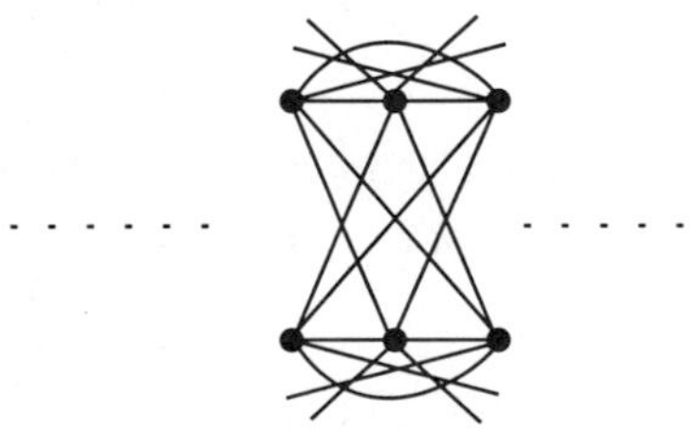

A GRAPH consisting of two rows of paired nodes in which all nodes but the paired ones are connected with an EDGE. It is the complement of the LADDER GRAPH.

Coconut
see MONKEY AND COCONUT PROBLEM

Codazzi Equations
see MAINARDI-CODAZZI EQUATIONS

Code
A code is a set of n-tuples of elements ("WORDS") taken from an ALPHABET.

see also ALPHABET, CODING THEORY, ENCODING, ERROR-CORRECTING CODE, GRAY CODE, HUFFMAN CODING, ISBN, LINEAR CODE, WORD

Codimension

The minimum number of parameters needed to fully describe all possible behaviors near a nonstructurally stable element.

see also BIFURCATION

Coding Theory

Coding theory, sometimes called ALGEBRAIC CODING THEORY, deals with the design of ERROR-CORRECTING CODES for the reliable transmission of information across noisy channels. It makes use of classical and modern algebraic techniques involving FINITE FIELDS, GROUP THEORY, and polynomial algebra. It has connections with other areas of DISCRETE MATHEMATICS, especially NUMBER THEORY and the theory of experimental designs.

see also ENCODING, ERROR-CORRECTING CODE, GALOIS FIELD, HADAMARD MATRIX

References

Alexander, B. "At the Dawn of the Theory of Codes." *Math. Intel.* **15**, 20–26, 1993.

Golomb, S. W.; Peile, R. E.; and Scholtz, R. A. *Basic Concepts in Information Theory and Coding: The Adventures of Secret Agent 00111.* New York: Plenum, 1994.

Humphreys, O. F. and Prest, M. Y. *Numbers, Groups, and Codes.* New York: Cambridge University Press, 1990.

MacWilliams, F. J. and Sloane, N. J. A. *The Theory of Error-Correcting Codes.* New York: Elsevier, 1978.

Roman, S. *Coding and Information Theory.* New York: Springer-Verlag, 1992.

Coefficient

A multiplicative factor (usually indexed) such as one of the constants a_i in the POLYNOMIAL $a_n x^n + a_{n-1}x^{n-1} + \ldots + a_2x^2 + a_1x + a_0$.

see also BINOMIAL COEFFICIENT, CARTAN TORSION COEFFICIENT, CENTRAL BINOMIAL COEFFICIENT, CLEBSCH-GORDON COEFFICIENT, COEFFICIENT FIELD, COMMUTATION COEFFICIENT, CONNECTION COEFFICIENT, CORRELATION COEFFICIENT, CROSS-CORRELATION COEFFICIENT, EXCESS COEFFICIENT, GAUSSIAN COEFFICIENT, LAGRANGIAN COEFFICIENT, MULTINOMIAL COEFFICIENT, PEARSON'S SKEWNESS COEFFICIENTS, PRODUCT-MOMENT COEFFICIENT OF CORRELATION, QUARTILE SKEWNESS COEFFICIENT, QUARTILE VARIATION COEFFICIENT, RACAH V-COEFFICIENT, RACAH W-COEFFICIENT, REGRESSION COEFFICIENT, ROMAN COEFFICIENT, TRIANGLE COEFFICIENT, UNDETERMINED COEFFICIENTS METHOD, VARIATION COEFFICIENT

Coefficient Field

Let V be a VECTOR SPACE over a FIELD K, and let A be a nonempty SET. For an appropriately defined AFFINE SPACE A, K is called the COEFFICIENT field.

Coercive Functional

A bilinear FUNCTIONAL ϕ on a normed SPACE E is called coercive (or sometimes ELLIPTIC) if there exists a POSITIVE constant K such that

$$\phi(x,x) \geq K\|x\|^2$$

for all $x \in E$.

see also LAX-MILGRAM THEOREM

References

Debnath, L. and Mikusiński, P. *Introduction to Hilbert Spaces with Applications.* San Diego, CA: Academic Press, 1990.

Cofactor

The MINOR of a DETERMINANT is another DETERMINANT $|\mathsf{C}|$ formed by omitting the ith row and jth column of the original DETERMINANT $|\mathsf{M}|$.

$$C_{ij} \equiv (-1)^{i+j} a_i M_{ij}.$$

see also DETERMINANT EXPANSION BY MINORS, MINOR

Cohen-Kung Theorem

Guarantees that the trajectory of LANGTON'S ANT is unbounded.

Cohomology

Cohomology is an invariant of a TOPOLOGICAL SPACE, formally "dual" to HOMOLOGY, and so it detects "holes" in a SPACE. Cohomology has more algebraic structure than HOMOLOGY, making it into a graded ring (multiplication given by "cup product"), whereas HOMOLOGY is just a graded ABELIAN GROUP invariant of a SPACE.

A generalized homology or cohomology theory must satisfy all of the EILENBERG-STEENROD AXIOMS with the exception of the dimension axiom.

see also ALEKSANDROV-ČECH COHOMOLOGY, ALEXANDER-SPANIER COHOMOLOGY, ČECH COHOMOLOGY, DE RHAM COHOMOLOGY, HOMOLOGY (TOPOLOGY)

Cohomotopy Group

Cohomotopy groups are similar to HOMOTOPY GROUPS. A cohomotopy group is a GROUP related to the HOMOTOPY classes of MAPS from a SPACE X into a SPHERE $\mathbb{S}^n$.

see also HOMOTOPY GROUP

Coin

A flat disk which acts as a two-sided DIE.

see BERNOULLI TRIAL, CARDS, COIN PARADOX, COIN TOSSING, DICE, FELLER'S COIN-TOSSING CONSTANTS, FOUR COINS PROBLEM, GAMBLER'S RUIN

References

Brooke, M. *Fun for the Money.* New York: Scribner's, 1963.

Coin Flipping

see COIN TOSSING

Coin Paradox

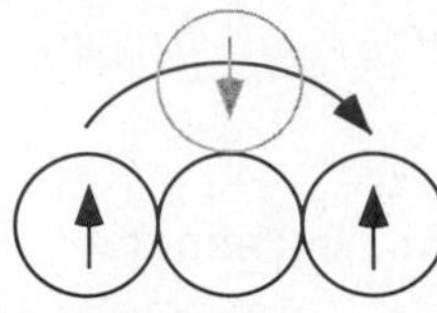

After a half rotation of the coin on the left around the central coin (of the same RADIUS), the coin undergoes a complete rotation.

References

Pappas, T. "The Coin Paradox." *The Joy of Mathematics.* San Carlos, CA: Wide World Publ./Tetra, p. 220, 1989.

Coin Problem

Let there be $n \geq 2$ INTEGERS $0 < a_1 < \ldots < a_n$ with $(a_1, a_2, \ldots, a_n) = 1$ (all RELATIVELY PRIME). For large enough $N = \sum_{i=1}^n a_i x_i$, there is a solution in NONNEGATIVE INTEGERS x_i. The greatest $N = g(a_1, a_2, \ldots a_n)$ for which there is no solution is called the coin problem. Sylvester showed

$$g(a_1, a_2) = (a_1 - 1)(a_2 - 1) - 1,$$

and an explicit solution is known for $n = 3$, but no closed form solution is known for larger N.

References

Guy, R. K. "The Money-Changing Problem." §C7 in *Unsolved Problems in Number Theory, 2nd ed.* New York: Springer-Verlag, pp. 113–114, 1994.

Coin Tossing

An idealized coin consists of a circular disk of zero thickness which, when thrown in the air and allowed to fall, will rest with either side face up ("heads" H or "tails" T) with equal probability. A coin is therefore a two-sided DIE. A coin toss corresponds to a BERNOULLI DISTRIBUTION with $p = 1/2$. Despite slight differences between the sides and NONZERO thickness of actual coins, the distribution of their tosses makes a good approximation to a $p = 1/2$ BERNOULLI DISTRIBUTION.

There are, however, some rather counterintuitive properties of coin tossing. For example, it is twice as likely that the triple TTH will be encountered before THT than after it, and three times as likely that THH will precede HTT. Furthermore, it is six times as likely that HTT will be the first of HTT, TTH, and TTT to occur (Honsberger 1979). More amazingly still, *spinning* a penny instead of tossing it results in heads only about 30% of the time (Paulos 1995).

Let $w(n)$ be the probability that no RUN of three consecutive heads appears in n independent tosses of a COIN. The following table gives the first few values of $w(n)$.

n	$w(n)$
0	1
1	1
2	1
3	$\frac{7}{8}$
4	$\frac{13}{16}$
5	$\frac{3}{4}$

Feller (1968, pp. 278–279) proved that

$$\lim_{n\to\infty} w(n)\alpha^{n+1} = \beta, \tag{1}$$

where

$$\alpha = \tfrac{1}{3}[(136 + 24\sqrt{33})^{1/3} - 8(136 + 24\sqrt{33})^{-1/3} - 2] = 1.087378025\ldots \tag{2}$$

and

$$\beta = \frac{2-\alpha}{4-3\alpha} = 1.236839845\ldots. \tag{3}$$

The corresponding constants for a RUN of $k > 1$ heads are α_k, the smallest POSITIVE ROOT of

$$1 - x + (\tfrac{1}{2}x)^{k+1} = 0, \tag{4}$$

and

$$\beta_k = \frac{2-\alpha}{k+1-k\alpha_k}. \tag{5}$$

These are modified for unfair coins with $P(H) = p$ and $P(T) = q = 1-p$ to α_k', the smallest POSITIVE ROOT of

$$1 - x + qp^k x^{k+1} = 0, \tag{6}$$

and

$$\beta_k' = \frac{1-p\alpha_k'}{(k+1-k\alpha_k')p} \tag{7}$$

(Feller 1968, pp. 322–325).

see also BERNOULLI DISTRIBUTION, CARDS, COIN, DICE, GAMBLER'S RUIN, MARTINGALE, RUN, SAINT PETERSBURG PARADOX

References

Feller, W. *An Introduction to Probability Theory and Its Application, Vol. 1, 3rd ed.* New York: Wiley, 1968.

Finch, S. "Favorite Mathematical Constants." http://www.mathsoft.com/asolve/constant/feller/feller.html.

Ford, J. "How Random is a Coin Toss?" *Physics Today* **36**, 40–47, 1983.

Honsberger, R. "Some Surprises in Probability." Ch. 5 in *Mathematical Plums* (Ed. R. Honsberger). Washington, DC: Math. Assoc. Amer., pp. 100–103, 1979.

Keller, J. B. "The Probability of Heads." *Amer. Math. Monthly* **93**, 191–197, 1986.

Paulos, J. A. *A Mathematician Reads the Newspaper.* New York: BasicBooks, p. 75, 1995.

Peterson, I. *Islands of Truth: A Mathematical Mystery Cruise.* New York: W. H. Freeman, pp. 238–239, 1990.

Spencer, J. "Combinatorics by Coin Flipping." *Coll. Math. J.*, **17**, 407–412, 1986.

Coincidence

A coincidence is a surprising concurrence of events, perceived as meaningfully related, with no apparent causal connection (Diaconis and Mosteller 1989).

see also BIRTHDAY PROBLEM, LAW OF TRULY LARGE NUMBERS, ODDS, PROBABILITY, RANDOM NUMBER

References

Bogomolny, A. "Coincidence." http://www.cut-the-knot.com/do_you_know/coincidence.html.

Falk, R. "On Coincidences." *Skeptical Inquirer* **6**, 18–31, 1981–82.

Falk, R. "The Judgment of Coincidences: Mine Versus Yours." *Amer. J. Psych.* **102**, 477–493, 1989.

Falk, R. and MacGregor, D. "The Surprisingness of Coincidences." In *Analysing and Aiding Decision Processes* (Ed. P. Humphreys, O. Svenson, and A. Vári). New York: Elsevier, pp. 489–502, 1984.

Diaconis, P. and Mosteller, F. "Methods of Studying Coincidences." *J. Amer. Statist. Assoc.* **84**, 853–861, 1989.

Jung, C. G. *Synchronicity: An Acausal Connecting Principle.* Princeton, NJ: Princeton University Press, 1973.

Kammerer, P. *Das Gesetz der Serie: Eine Lehre von den Wiederholungen im Lebens—und im Weltgeschehen.* Stuttgart, Germany: Deutsche Verlags-Anstahlt, 1919.

Stewart, I. "What a Coincidence!" *Sci. Amer.* **278**, 95–96, June 1998.

Colatitude

The polar angle on a SPHERE measured from the North Pole instead of the equator. The angle ϕ in SPHERICAL COORDINATES is the COLATITUDE. It is related to the LATITUDE δ by $\phi = 90° - \delta$.

see also LATITUDE, LONGITUDE, SPHERICAL COORDINATES

Colinear

see COLLINEAR

Collatz Problem

A problem posed by L. Collatz in 1937, also called the $3x+1$ MAPPING, HASSE'S ALGORITHM, KAKUTANI'S PROBLEM, SYRACUSE ALGORITHM, SYRACUSE PROBLEM, THWAITES CONJECTURE, and ULAM'S PROBLEM (Lagarias 1985). Thwaites (1996) has offered a £1000 reward for resolving the CONJECTURE. Let n be an INTEGER. Then the Collatz problem asks if iterating

$$f(n) = \begin{cases} \frac{1}{2}n & \text{for } n \text{ even} \\ 3n+1 & \text{for } n \text{ odd} \end{cases} \tag{1}$$

always returns to 1 for POSITIVE n. This question has been tested and found to be true for all numbers $< 5.6 \times 10^{13}$ (Leavens and Vermeulen 1992), and more recently, 10^{15} (Vardi 1991, p. 129). The members of the SEQUENCE produced by the Collatz are sometimes known as HAILSTONE NUMBERS. Because of the difficulty in solving this problem, Erdős commented that "mathematics is not yet ready for such problems" (Lagarias 1985). If NEGATIVE numbers are included, there are four known cycles (excluding the trivial 0 cycle): (4, 2, 1), (−2, −1), (−5, −7, −10), and (−17, −25, −37, −55, −82, −41, −61, −91, −136, −68, −34). The number of tripling steps needed to reach 1 for $n = 1, 2, \ldots$ are 0, 0, 2, 0, 1, 2, 5, 0, 6, ... (Sloane's A006667).

The Collatz problem was modified by Terras (1976, 1979), who asked if iterating

$$T(x) = \begin{cases} \frac{1}{2}x & \text{for } x \text{ even} \\ \frac{1}{2}(3x+1) & \text{for } x \text{ odd} \end{cases} \tag{2}$$

always returns to 1. If NEGATIVE numbers are included, there are 4 known cycles: (1, 2), (−1), (−5, −7, −10), and (−17, −25, −37, −55, −82, −41, −61, −91, −136, −68, −34). It is a special case of the "generalized Collatz problem" with $d = 2$, $m_0 = 1$, $m_1 = 3$, $r_0 = 0$, and $r_1 = -1$. Terras (1976, 1979) also proved that the set of INTEGERS $S_k \equiv \{n : n \text{ has stopping time} \leq k\}$ has a limiting asymptotic density $F(k)$, so the limit

$$F(k) = \lim_{x\to\infty} \frac{1}{x}, \tag{3}$$

for $\{n : n \leq x \text{ and } \sigma(n) \leq k\}$ exists. Furthermore, $F(k) \to 1$ as $k \to \infty$, so almost all INTEGERS have a finite stopping time. Finally, for all $k \geq 1$,

$$1 - F(k) = \lim_{x\to\infty} \frac{1}{x} \leq 2^{-\eta k}, \tag{4}$$

where

$$\eta = 1 - H(\theta) = 0.05004\ldots \tag{5}$$

$$H(x) = -x\lg x - (1-x)\lg(1-x) \tag{6}$$

$$\theta = \frac{1}{\lg 3} \tag{7}$$

(Lagarias 1985).

Conway proved that the original Collatz problem has no nontrivial cycles of length < 400. Lagarias (1985) showed that there are no nontrivial cycles with length $< 275,000$. Conway (1972) also proved that Collatz-type problems can be formally UNDECIDABLE.

A generalization of the COLLATZ PROBLEM lets $d \geq 2$ be a POSITIVE INTEGER and $m_0, \ldots, m_{d-1}$ be NONZERO INTEGERS. Also let $r_i \in \mathbb{Z}$ satisfy

$$r_i \equiv i m_i \pmod{d}. \tag{8}$$

Then

$$T(x) = \frac{m_i x - r_i}{d} \tag{9}$$

for $x \equiv i \pmod{d}$ defines a generalized Collatz mapping. An equivalent form is

$$T(x) = \left\lfloor \frac{m_i x}{d} \right\rfloor + X_i \tag{10}$$

for $x \equiv i \pmod{d}$ where $X_0, \ldots, X_{d-1}$ are INTEGERS and $\lfloor r \rfloor$ is the FLOOR FUNCTION. The problem is connected with ERGODIC THEORY and MARKOV CHAINS (Matthews 1995). Matthews (1995) obtained the following table for the mapping

$$T_k(x) = \begin{cases} \frac{1}{2}x & \text{for } x \equiv 0 \pmod{2} \\ \frac{1}{2}(3x+k) & \text{for } x \equiv 1 \pmod{2}, \end{cases} \quad (11)$$

where $k = T_{5^k}$.

k	# Cycles	Max. Cycle Length
0	5	27
1	10	34
2	13	118
3	17	118
4	19	118
5	21	165
6	23	433

Matthews and Watts (1984) proposed the following conjectures.

1. If $|m_0 \cdots m_{d-1}| < d^d$, then all trajectories $\{T^K(n)\}$ for $n \in \mathbb{Z}$ eventually cycle.
2. If $|m_0 \cdots m_{d-1}| > d^d$, then almost all trajectories $\{T^K(n)\}$ for $n \in \mathbb{Z}$ are divergent, except for an exceptional set of INTEGERS n satisfying

$$\#\{n \in S | -X \le n < X\} = o(X).$$

3. The number of cycles is finite.
4. If the trajectory $\{T^K(n)\}$ for $n \in \mathbb{Z}$ is not eventually cyclic, then the iterates are uniformly distribution mod d^α for each $\alpha \ge 1$, with

$$\lim_{N\to\infty} \frac{1}{N+1} \operatorname{card}\{K \le N | T^K(n) \equiv j \pmod{d^\alpha}\} = d^{-\alpha} \quad (12)$$

for $0 \le j \le d^\alpha - 1$.

Matthews believes that the map

$$T(x) = \begin{cases} 7x+3 & \text{for } x \equiv 0 \pmod{3} \\ \frac{1}{3}(7x+2) & \text{for } x \equiv 1 \pmod{3} \\ \frac{1}{3}(x-2) & \text{for } x \equiv 2 \pmod{3} \end{cases} \quad (13)$$

will either reach 0 (mod 3) or will enter one of the cycles (-1) or $(-2, -4)$, and offers a \$100 (Australian?) prize for a proof.

see also HAILSTONE NUMBER

References

Applegate, D. and Lagarias, J. C. "Density Bounds for the $3x+1$ Problem 1. Tree-Search Method." *Math. Comput.* **64**, 411–426, 1995.

Applegate, D. and Lagarias, J. C. "Density Bounds for the $3x+1$ Problem 2. Krasikov Inequalities." *Math. Comput.* **64**, 427–438, 1995.

Beeler, M.; Gosper, R. W.; and Schroeppel, R. *HAKMEM.* Cambridge, MA: MIT Artificial Intelligence Laboratory, Memo AIM-239, Feb. 1972.

Burckel, S. "Functional Equations Associated with Congruential Functions." *Theor. Comp. Sci.* **123**, 397–406, 1994.

Conway, J. H. "Unpredictable Iterations." *Proc. 1972 Number Th. Conf.*, University of Colorado, Boulder, Colorado, pp. 49–52, 1972.

Crandall, R. "On the '$3x+1$' Problem." *Math. Comput.* **32**, 1281–1292, 1978.

Everett, C. "Iteration of the Number Theoretic Function $f(2n) = n$, $f(2n+1) = f(3n+2)$." *Adv. Math.* **25**, 42–45, 1977.

Guy, R. K. "Collatz's Sequence." §E16 in *Unsolved Problems in Number Theory, 2nd ed.* New York: Springer-Verlag, pp. 215–218, 1994.

Lagarias, J. C. "The $3x+1$ Problem and Its Generalizations." *Amer. Math. Monthly* **92**, 3–23, 1985. http://www.cecm.sfu.ca/organics/papers/lagarias/.

Leavens, G. T. and Vermeulen, M. "$3x+1$ Search Programs." *Comput. Math. Appl.* **24**, 79–99, 1992.

Matthews, K. R. "The Generalized $3x+1$ Mapping." http://www.maths.uq.oz.au/~krm/survey.dvi. Rev. Sept. 10, 1995.

Matthews, K. R. "A Generalized $3x+1$ Conjecture." [\$100 Reward for a Proof.] ftp://www.maths.uq.edu.au/pub/krm/gnubc/challenge.

Matthews, K. R. and Watts, A. M. "A Generalization of Hasses's Generalization of the Syracuse Algorithm." *Acta Arith.* **43**, 167–175, 1984.

Sloane, N. J. A. Sequence A006667/M0019 in "An On-Line Version of the Encyclopedia of Integer Sequences."

Terras, R. "A Stopping Time Problem on the Positive Integers." *Acta Arith.* **30**, 241–252, 1976.

Terras, R. "On the Existence of a Density." *Acta Arith.* **35**, 101–102, 1979.

Thwaites, B. "Two Conjectures, or How to win £1100." *Math. Gaz.* **80**, 35–36, 1996.

Vardi, I. "The $3x+1$ Problem." Ch. 7 in *Computational Recreations in Mathematica.* Redwood City, CA: Addison-Wesley, pp. 129–137, 1991.

Collinear

Three or more points P_1, P_2, P_3, ..., are said to be collinear if they lie on a single straight LINE L. (Two points are always collinear.) This will be true IFF the ratios of distances satisfy

$$x_2 - x_1 : y_2 - y_1 : z_2 - z_1 = x_3 - x_1 : y_3 - y_1 : z_3 - z_1.$$

Two points are trivially collinear since two points determine a LINE.

see also CONCYCLIC, DIRECTED ANGLE, N-CLUSTER, SYLVESTER'S LINE PROBLEM

Collineation

A transformation of the plane which transforms COLLINEAR points into COLLINEAR points. A projective collineation transforms every 1-D form projectively, and a perspective collineation is a collineation which leaves all lines through a point and points through a line invariant. In an ELATION, the center and axis are incident; in a HOMOLOGY they are not. For further discussion, see Coxeter (1969, p. 248).

see also AFFINITY, CORRELATION, ELATION, EQUI-AFFINITY, HOMOLOGY (GEOMETRY), PERSPECTIVE COLLINEATION, PROJECTIVE COLLINEATION

References

Coxeter, H. S. M. "Collineations and Correlations." §14.6 in *Introduction to Geometry, 2nd ed.* New York: Wiley, pp. 247–251, 1969.

Cologarithm

The LOGARITHM of the RECIPROCAL of a number, equal to the NEGATIVE of the LOGARITHM of the number itself,

$$\operatorname{colog} x \equiv \log\left(\frac{1}{x}\right) = -\log x.$$

see also ANTILOGARITHM, LOGARITHM

Colon Product

Let **AB** and **CD** be DYADS. Their colon product is defined by

$$\mathbf{AB}:\mathbf{CD} \equiv \mathbf{C}\cdot\mathbf{AB}\cdot\mathbf{D} = (\mathbf{A}\cdot\mathbf{C})(\mathbf{B}\cdot\mathbf{D}).$$

Colorable

Color each segment of a KNOT DIAGRAM using one of three colors. If

1. at any crossing, either the colors are all different or all the same, and
2. at least two colors are used,

then a KNOT is said to be colorable (or more specifically, THREE-COLORABLE). Colorability is invariant under REIDEMEISTER MOVES, and can be generalized. For instance, for five colors 0, 1, 2, 3, and 4, a KNOT is five-colorable if

1. at any crossing, three segments meet. If the overpass is numbered a and the two underpasses B and C, then $2a \equiv b + c \pmod 5$, and
2. at least two colors are used.

Colorability cannot alway distinguish HANDEDNESS. For instance, three-colorability can distinguish the mirror images of the TREFOIL KNOT but not the FIGURE-OF-EIGHT KNOT. Five-colorability, on the other hand, distinguishes the MIRROR IMAGES of the FIGURE-OF-EIGHT KNOT but not the TREFOIL KNOT.

see also COLORING, THREE-COLORABLE

Coloring

A coloring of plane regions, LINK segments, etc., is an assignment of a distinct labelling (which could be a number, letter, color, etc.) to each component. Coloring problems generally involve TOPOLOGICAL considerations (i.e., they depend on the abstract study of the arrangement of objects), and theorems about colorings, such as the famous FOUR-COLOR THEOREM, can be extremely difficult to prove.

see also COLORABLE, EDGE-COLORING, FOUR-COLOR THEOREM, k-COLORING, POLYHEDRON COLORING, SIX-COLOR THEOREM, THREE-COLORABLE, VERTEX COLORING

References

Eppstein, D. "Coloring." `http://www.ics.uci.edu/~eppstein/junkyard/color.html`.

Saaty, T. L. and Kainen, P. C. *The Four-Color Problem: Assaults and Conquest.* New York: Dover, 1986.

Columbian Number

see SELF NUMBER

Colunar Triangle

Given a SCHWARZ TRIANGLE $(p\ q\ r)$, replacing each VERTEX with its antipodes gives the three colunar SPHERICAL TRIANGLES

$$(p\ q'\ r'), (p'\ q\ r'), (p'\ q'\ r),$$

where

$$\frac{1}{p} + \frac{1}{p'} = 1$$
$$\frac{1}{q} + \frac{1}{q'} = 1$$
$$\frac{1}{r} + \frac{1}{r'} = 1.$$

see also SCHWARZ TRIANGLE, SPHERICAL TRIANGLE

References

Coxeter, H. S. M. *Regular Polytopes, 3rd ed.* New York: Dover, p. 112, 1973.

Comb Function

see SHAH FUNCTION

Combination

The number of ways of picking r unordered outcomes from n possibilities. Also known as the BINOMIAL COEFFICIENT or CHOICE NUMBER and read "n choose r."

$${}_nC_r \equiv \binom{n}{r} \equiv \frac{n!}{r!(n-r)!},$$

where $n!$ is a FACTORIAL.

see also BINOMIAL COEFFICIENT, DERANGEMENT, FACTORIAL, PERMUTATION, SUBFACTORIAL

References

Conway, J. H. and Guy, R. K. "Choice Numbers." In *The Book of Numbers.* New York: Springer-Verlag, pp. 67–68, 1996.

Ruskey, F. "Information on Combinations of a Set." `http://sue.csc.uvic.ca/~cos/inf/comb/CombinationsInfo.html`.

Combination Lock

Let a combination of n buttons be a SEQUENCE of disjoint nonempty SUBSETS of the SET $\{1, 2, \ldots, n\}$. If the number of possible combinations is denoted a_n, then a_n satisfies the RECURRENCE RELATION

$$a_n = \sum_{i=0}^{n-1} \binom{n}{n-i} a_i, \tag{1}$$

with $a_0 = 1$. This can also be written

$$a_n = \frac{d^n}{dx^n} \left(\frac{1}{2-e^x} \right) \bigg|_{x=0} = \tfrac{1}{2} \sum_{k=0}^{\infty} \frac{k^n}{2^k}, \tag{2}$$

where the definition $0^0 = 1$ has been used. Furthermore,

$$a_n = \sum_{k=1}^{n} A_{n,k} 2^{n-k} = \sum_{k=1}^{n} A_{n,k} 2^{k-1}, \tag{3}$$

where $A_{n,k}$ are EULERIAN NUMBERS. In terms of the STIRLING NUMBERS OF THE SECOND KIND $s(n,k)$,

$$a_n = \sum_{k=1}^{n} k! s(n,k). \tag{4}$$

a_n can also be given in closed form as

$$a_n = \tfrac{1}{2} \operatorname{Li}_{-n}(\tfrac{1}{2}), \tag{5}$$

where $\operatorname{Li}_n(z)$ is the POLYLOGARITHM. The first few values of a_n for $n = 1, 2, \ldots$ are 1, 3, 13, 75, 541, 4683, 47293, 545835, 7087261, 102247563, ... (Sloane's A000670).

The quantity

$$b_n \equiv \frac{a_n}{n!} \tag{6}$$

satisfies the inequality

$$\frac{1}{2(\ln 2)^n} \leq b_n \leq \frac{1}{(\ln 2)^n}. \tag{7}$$

References

Sloane, N. J. A. Sequence A000670/M2952 in "An On-Line Version of the Encyclopedia of Integer Sequences."

Velleman, D. J. and Call, G. S. "Permutations and Combination Locks." *Math. Mag.* **68**, 243–253, 1995.

Combinatorial Species

see SPECIES

Combinatorial Topology

Combinatorial topology is a special type of ALGEBRAIC TOPOLOGY that uses COMBINATORIAL methods. For example, SIMPLICIAL HOMOLOGY is a combinatorial construction in ALGEBRAIC TOPOLOGY, so it belongs to combinatorial topology.

see also ALGEBRAIC TOPOLOGY, SIMPLICIAL HOMOLOGY, TOPOLOGY

Combinatorics

The branch of mathematics studying the enumeration, combination, and permutation of sets of elements and the mathematical relations which characterize these properties.

see also ANTICHAIN, CHAIN, DILWORTH'S LEMMA, DIVERSITY CONDITION, ERDŐS-SZEKERES THEOREM, INCLUSION-EXCLUSION PRINCIPLE, KIRKMAN'S SCHOOLGIRL PROBLEM, KIRKMAN TRIPLE SYSTEM, LENGTH (PARTIAL ORDER), PARTIAL ORDER, PIGEONHOLE PRINCIPLE, RAMSEY'S THEOREM, SCHRÖDER-BERNSTEIN THEOREM, SCHUR'S LEMMA, SPERNER'S THEOREM, TOTAL ORDER, VAN DER WAERDEN'S THEOREM, WIDTH (PARTIAL ORDER)

References

Abramowitz, M. and Stegun, C. A. (Eds.). "Combinatorial Analysis." Ch. 24 in *Handbook of Mathematical Functions with Formulas, Graphs, and Mathematical Tables, 9th printing.* New York: Dover, pp. 821–8827, 1972.

Aigner, M. *Combinatorial Theory.* New York: Springer-Verlag, 1997.

Bellman, R. and Hall, M. *Combinatorial Analysis.* Amer. Math. Soc., 1979.

Biggs, N. L. "The Roots of Combinatorics." *Historia Mathematica* **6**, 109–136, 1979.

Bose, R. C. and Manvel, B. *Introduction to Combinatorial Theory.* New York: Wiley, 1984.

Brown, K. S. "Combinatorics." `http://www.seanet.com/~ksbrown/icombina.htm`.

Cameron, P. J. *Combinatorics: Topics, Techniques, Algorithms.* New York: Cambridge University Press, 1994.

Cohen, D. *Basic Techniques of Combinatorial Theory.* New York: Wiley, 1978.

Cohen, D. E. *Combinatorial Group Theory: A Topological Approach.* New York: Cambridge University Press, 1989.

Colbourn, C. J. and Dinitz, J. H. *CRC Handbook of Combinatorial Designs.* Boca Raton, FL: CRC Press, 1996.

Comtet, L. *Advanced Combinatorics.* Dordrecht, Netherlands: Reidel, 1974.

Coolsaet, K. "Index of Combinatorial Objects." `http://www.hogent.be/~kc/ico/`.

Dinitz, J. H. and Stinson, D. R. (Eds.). *Contemporary Design Theory: A Collection of Surveys.* New York: Wiley, 1992.

Electronic Journal of Combinatorics. `http://www.combinatorics.org/previous_volumes.html`.

Eppstein, D. "Combinatorial Geometry." `http://www.ics.uci.edu/~eppstein/junkyard/combinatorial.html`.

Erickson, M. J. *Introduction to Combinatorics.* New York: Wiley, 1996.

Fields, J. "On-Line Dictionary of Combinatorics." `http://math.uic.edu/~fields/dic/`.

Godsil, C. D. "Problems in Algebraic Combinatorics." *Electronic J. Combinatorics* **2**, F1, 1–20, 1995. `http://www.combinatorics.org/Volume_2/volume2.html#F1`.

Graham, R. L.; Grötschel, M.; and Lovász, L. (Eds.). *Handbook of Combinatorics, 2 vols.* Cambridge, MA: MIT Press, 1996.

Graham, R. L.; Knuth, D. E.; and Patashnik, O. *Concrete Mathematics: A Foundation for Computer Science, 2nd ed.* Reading, MA: Addison-Wesley, 1994.

Hall, M. Jr. *Combinatorial Theory, 2nd ed.* New York: Wiley, 1986.

Knuth, D. E. (Ed.). *Stable Marriage and Its Relation to Other Combinatorial Problems.* Providence, RI: Amer. Math. Soc., 1997.

Kučera, L. *Combinatorial Algorithms.* Bristol, England: Adam Hilger, 1989.
Liu, C. L. *Introduction to Combinatorial Mathematics.* New York: McGraw-Hill, 1968.
MacMahon, P. A. *Combinatory Analysis.* New York: Chelsea, 1960.
Nijenhuis, A. and Wilf, H. *Combinatorial Algorithms for Computers and Calculators, 2nd ed.* New York: Academic Press, 1978.
Riordan, J. *Combinatorial Identities, reprint ed. with corrections.* Huntington, NY: Krieger, 1979.
Riordan, J. *An Introduction to Combinatorial Analysis.* New York: Wiley, 1980.
Roberts, F. S. *Applied Combinatorics.* Englewood Cliffs, NJ: Prentice-Hall, 1984.
Rota, G.-C. (Ed.). *Studies in Combinatorics.* Providence, RI: Math. Assoc. Amer., 1978.
Ruskey, F. "The (Combinatorial) Object Server." http://sue.csc.uvic.ca/~cos.
Ryser, H. J. *Combinatorial Mathematics.* Buffalo, NY: Math. Assoc. Amer., 1963.
Skiena, S. S. *Implementing Discrete Mathematics: Combinatorics and Graph Theory with Mathematica.* Reading, MA: Addison-Wesley, 1990.
Sloane, N. J. A. "An On-Line Version of the Encyclopedia of Integer Sequences." http://www.research.att.com/~njas/sequences/eisonline.html.
Sloane, N. J. A. and Plouffe, S. *The Encyclopedia of Integer Sequences.* San Diego, CA: Academic Press, 1995.
Street, A. P. and Wallis, W. D. *Combinatorial Theory: An Introduction.* Winnipeg, Manitoba: Charles Babbage Research Center, 1977.
Tucker, A. *Applied Combinatorics, 3rd ed.* New York: Wiley, 1995.
van Lint, J. H. and Wilson, R. M. *A Course in Combinatorics.* New York: Cambridge University Press, 1992.
Wilf, H. S. *Combinatorial Algorithms: An Update.* Philadelphia, PA: SIAM, 1989.

Comma Derivative

$$A_{,k} \equiv \frac{\partial A}{\partial x^k} \equiv \partial_k A$$
$$A^k_{,k} \equiv \frac{1}{g_k}\frac{\partial A^k}{\partial x^k} \equiv \partial_k A^k.$$

see also COVARIANT DERIVATIVE, SEMICOLON DERIVATIVE

Comma of Didymus

The musical interval by which four fifths exceed a seventeenth (i.e., two octaves and a major third),

$$\frac{(\frac{3}{2})^4}{2^2(\frac{5}{4})} = \frac{3^4}{2^4 \cdot 5} = \frac{81}{80} = 1.0125,$$

also called a SYNTONIC COMMA.

see also COMMA OF PYTHAGORAS, DIESIS, SCHISMA

Comma of Pythagoras

The musical interval by which twelve fifths exceed seven octaves,

$$\frac{(\frac{3}{2})^{12}}{2^7} = \frac{3^{12}}{2^{19}} = \frac{531441}{524288} = 1.013643265.$$

Successive CONTINUED FRACTION CONVERGENTS to $\log 2/\log(3/2)$ give increasingly close approximations m/n of m fifths by n octaves as 1, 2, 5/3, 12/7, 41/24, 53/31, 306/179, 665/389, ... (Sloane's A005664 and A046102; Jeans 1968, p. 188), shown in **bold** in the table below. All near-equalities of m fifths and n octaves having

$$R \equiv \frac{(\frac{3}{2})^m}{2^n} = \frac{3^m}{2^{m+n}}$$

with $|R-1| < 0.02$ are given in the following table.

m	n	Ratio	m	n	Ratio
12	**7**	1.013643265	265	155	1.010495356
41	**24**	0.9886025477	294	172	0.9855324037
53	**31**	1.002090314	**306**	**179**	0.9989782832
65	38	1.015762098	318	186	1.012607608
94	55	0.9906690375	347	203	0.9875924759
106	62	1.004184997	359	210	1.001066462
118	69	1.017885359	371	217	1.014724276
147	86	0.9927398469	400	234	0.9896568543
159	93	1.006284059	412	241	1.003159005
188	110	0.9814251419	424	248	1.016845369
200	117	0.994814985	453	265	0.9917255479
212	124	1.008387509	465	272	1.005255922
241	141	0.9834766286	477	279	1.018970895
253	148	0.9968944607	494	289	0.9804224033

see also COMMA OF DIDYMUS, DIESIS, SCHISMA

References

Conway, J. H. and Guy, R. K. *The Book of Numbers.* New York: Springer-Verlag, p. 257, 1995.
Guy, R. K. "Small Differences Between Powers of 2 and 3." §F23 in *Unsolved Problems in Number Theory, 2nd ed.* New York: Springer-Verlag, p. 261, 1994.
Sloane, N. J. A. Sequences A005664 and A046102 in "An On-Line Version of the Encyclopedia of Integer Sequences."

Common Cycloid

see CYCLOID

Common Residue

The value of b, where $a \equiv b \pmod{m}$, taken to be NONNEGATIVE and smaller than m.

see also MINIMAL RESIDUE, RESIDUE (CONGRUENCE)

Commutation Coefficient

A coefficient which gives the difference between partial derivatives of two coordinates with respect to the other coordinate,

$$c^{\mu}{}_{\alpha\beta}\vec{e}_\mu = [\vec{e}_\alpha, \vec{e}_\beta] = \nabla_\alpha \vec{e}_\beta - \nabla_\beta \vec{e}_\alpha.$$

see also CONNECTION COEFFICIENT

Commutative

Let A denote an $\mathbb{R}$-algebra, so that A is a VECTOR SPACE over R and

$$A \times A \to A$$
$$(x, y) \mapsto x \cdot y.$$

Now define

$$Z \equiv \{x \in a : x \cdot y \text{ for some } y \in A \neq 0\},$$

where $0 \in Z$. An ASSOCIATIVE $\mathbb{R}$-algebra is commutative if $x \cdot y = y \cdot x$ for all $x, y \in A$. Similarly, a RING is commutative if the MULTIPLICATION operation is commutative, and a LIE ALGEBRA is commutative if the COMMUTATOR $[A, B]$ is 0 for every A and B in the LIE ALGEBRA.

see also ABELIAN, ASSOCIATIVE, TRANSITIVE

References

Finch, S. "Zero Structures in Real Algebras." http://www.mathsoft.com/asolve/zerodiv/zerodiv.html.

MacDonald, I. G. and Atiyah, M. F. *Introduction to Commutative Algebra.* Reading, MA: Addison-Wesley, 1969.

Commutative Algebra

An ALGEBRA in which the $+$ operators and $\times$ are COMMUTATIVE.

see also ALGEBRAIC GEOMETRY, GRÖBNER BASIS

References

MacDonald, I. G. and Atiyah, M. F. *Introduction to Commutative Algebra.* Reading, MA: Addison-Wesley, 1969.

Cox, D.; Little, J.; and O'Shea, D. *Ideals, Varieties, and Algorithms: An Introduction to Algebraic Geometry and Commutative Algebra, 2nd ed.* New York: Springer-Verlag, 1996.

Samuel, P. and Zariski, O. *Commutative Algebra, Vol. 2.* New York: Springer-Verlag, 1997.

Commutator

Let $\tilde{A}$, $\tilde{B}$, ... be OPERATORS. Then the commutator of $\tilde{A}$ and $\tilde{B}$ is defined as

$$[\tilde{A}, \tilde{B}] \equiv \tilde{A}\tilde{B} - \tilde{B}\tilde{A}. \qquad (1)$$

Let a, b, ... be constants. Identities include

$$[f(x), x] = 0 \qquad (2)$$
$$[\tilde{A}, \tilde{A}] = 0 \qquad (3)$$
$$[\tilde{A}, \tilde{B}] = -[\tilde{B}, \tilde{A}] \qquad (4)$$
$$[\tilde{A}, \tilde{B}\tilde{C}] = [\tilde{A}, \tilde{B}]\tilde{C} + \tilde{B}[\tilde{A}, \tilde{C}] \qquad (5)$$
$$[\tilde{A}\tilde{B}, \tilde{C}] = [\tilde{A}, \tilde{C}]\tilde{B} + \tilde{A}[\tilde{B}, \tilde{C}] \qquad (6)$$
$$[a + \tilde{A}, b + \tilde{B}] = [\tilde{A}, \tilde{B}] \qquad (7)$$
$$[\tilde{A} + \tilde{B}, \tilde{C} + \tilde{D}] = [\tilde{A}, \tilde{C}] + [\tilde{A}, \tilde{D}] + [\tilde{B}, \tilde{C}] + [\tilde{B}, \tilde{D}]. \qquad (8)$$

The commutator can be interpreted as the "infinitesimal" of the commutator of a LIE GROUP.

Let A and B be TENSORS. Then

$$[A, B] \equiv \nabla_A B - \nabla_B A. \qquad (9)$$

see also ANTICOMMUTATOR, JACOBI IDENTITIES

Compact Group

If the parameters of a LIE GROUP vary over a CLOSED INTERVAL, the GROUP is compact. Every representation of a compact group is equivalent to a UNITARY representation.

Compact Manifold

A MANIFOLD which can be "charted" with finitely many EUCLIDEAN SPACE charts. The CIRCLE is the only compact 1-D MANIFOLD. The SPHERE and n-TORUS are the only compact 2-D MANIFOLDS. It is an open question if the known compact MANIFOLDS in 3-D are complete, and it is not even known what a complete list in 4-D should look like. The following terse table therefore summarizes current knowledge about the number of compact manifolds $N(D)$ of D dimensions.

D	$N(D)$
1	1
2	2

see also TYCHONOF COMPACTNESS THEOREM

Compact Set

The SET S is compact if, from any SEQUENCE of elements X_1, X_2, ... of S, a subsequence can always be extracted which tends to some limit element X of S. Compact sets are therefore closed and bounded.

Compact Space

A TOPOLOGICAL SPACE is compact if every open cover of X has a finite subcover. In other words, if X is the union of a family of open sets, there is a finite subfamily whose union is X. A subset A of a TOPOLOGICAL SPACE X is compact if it is compact as a TOPOLOGICAL SPACE with the relative topology (i.e., every family of open sets of X whose union contains A has a finite subfamily whose union contains A).

Compact Surface

A surface with a finite number of TRIANGLES in its TRIANGULATION. The SPHERE and TORUS are compact, but the PLANE and TORUS minus a DISK are not.

Compactness Theorem

Inside a BALL B in $\mathbb{R}^3$,

$$\{\text{rectifiable currents } S \text{ in } BL \text{ AREA } S \leq c, \quad \text{length } \partial S \leq c\}$$

is compact under the FLAT NORM.

References

Morgan, F. "What Is a Surface?" *Amer. Math. Monthly* **103**, 369–376, 1996.

Companion Knot

Let K_1 be a knot inside a TORUS. Now knot the TORUS in the shape of a second knot (called the companion knot) K_2. Then the new knot resulting from K_1 is called the SATELLITE KNOT K_3.

References

Adams, C. C. *The Knot Book: An Elementary Introduction to the Mathematical Theory of Knots.* New York: W. H. Freeman, pp. 115–118, 1994.

Comparability Graph

The comparability graph of a POSET $P = (X, \leq)$ is the GRAPH with vertex set X for which vertices x and y are adjacent IFF either $x \leq y$ or $y \leq x$ in P.

see also INTERVAL GRAPH, PARTIALLY ORDERED SET

Comparison Test

Let $\sum a_k$ and $\sum b_k$ be a SERIES with POSITIVE terms and suppose $a_1 \leq b_1$, $a_2 \leq b_2$, ...

1. If the bigger series CONVERGES, then the smaller series also CONVERGES.
2. If the smaller series DIVERGES, then the bigger series also DIVERGES.

see also CONVERGENCE TESTS

References

Arfken, G. *Mathematical Methods for Physicists, 3rd ed.* Orlando, FL: Academic Press, pp. 280–281, 1985.

Compass

A tool with two arms joined at their ends which can be used to draw CIRCLES. In GEOMETRIC CONSTRUCTIONS, the classical Greek rules stipulate that the compass cannot be used to mark off distances, so it must "collapse" whenever one of its arms is removed from the page. This results in significant complication in the complexity of GEOMETRIC CONSTRUCTIONS.

see also CONSTRUCTIBLE POLYGON, GEOMETRIC CONSTRUCTION, GEOMETROGRAPHY, MASCHERONI CONSTRUCTION, PLANE GEOMETRY, POLYGON, PONCELET-STEINER THEOREM, RULER, SIMPLICITY, STEINER CONSTRUCTION, STRAIGHTEDGE

References

Dixon, R. "Compass Drawings." Ch. 1 in *Mathographics.* New York: Dover, pp. 1–78, 1991.

Compatible

Let $||\mathsf{A}||$ be the MATRIX NORM associated with the MATRIX A and $||\mathbf{x}||$ be the VECTOR NORM associated with a VECTOR $\mathbf{x}$. Let the product $\mathsf{A}\mathbf{x}$ be defined, then $||\mathsf{A}||$ and $||\mathbf{x}||$ are said to be compatible if

$$||\mathsf{A}\mathbf{x}|| \leq ||\mathsf{A}||\,||\mathbf{x}||.$$

References

Gradshteyn, I. S. and Ryzhik, I. M. *Tables of Integrals, Series, and Products, 5th ed.* San Diego, CA: Academic Press, p. 1115, 1980.

Complement Graph

The complement GRAPH $\bar{G}$ of G has the same VERTICES as G but contains precisely those two-element SUBSETS which are not in G.

Complement Knot

see KNOT COMPLEMENT

Complement Set

Given a set S with a subset E, the complement of E is defined as

$$E' \equiv \{F : F \in S, F \notin E\}. \tag{1}$$

If $E = S$, then

$$E' \equiv S' = \varnothing, \tag{2}$$

where $\varnothing$ is the EMPTY SET. Given a single SET, the second PROBABILITY AXIOM gives

$$1 = P(S) = P(E \cup E'). \tag{3}$$

Using the fact that $E \cap E' = \varnothing$,

$$1 = P(E) + P(E') \tag{4}$$

$$P(E') = 1 - P(E). \tag{5}$$

This demonstrates that

$$P(S') = P(\varnothing) = 1 - P(S) = 1 - 1 = 0. \tag{6}$$

Given two SETS,

$$P(E \cap F') = P(E) - P(E \cap F) \tag{7}$$

$$P(E' \cap F') = 1 - P(E) - P(F) + P(E \cap F). \tag{8}$$

Complementary Angle

Two ANGLES α and $\pi/2 - \alpha$ are said to be complementary.

see also ANGLE, SUPPLEMENTARY ANGLE

Complete

see COMPLETE AXIOMATIC THEORY, COMPLETE BIGRAPH, COMPLETE FUNCTIONS, COMPLETE GRAPH, COMPLETE QUADRANGLE, COMPLETE QUADRILATERAL, COMPLETE SEQUENCE, COMPLETE SPACE, COMPLETENESS PROPERTY, WEAKLY COMPLETE SEQUENCE

Complete Axiomatic Theory

An axiomatic theory (such as a GEOMETRY) is said to be complete if each valid statement in the theory is capable of being proven true or false.

see also CONSISTENCY

Complete Bigraph

see COMPLETE BIPARTITE GRAPH

Complete Bipartite Graph

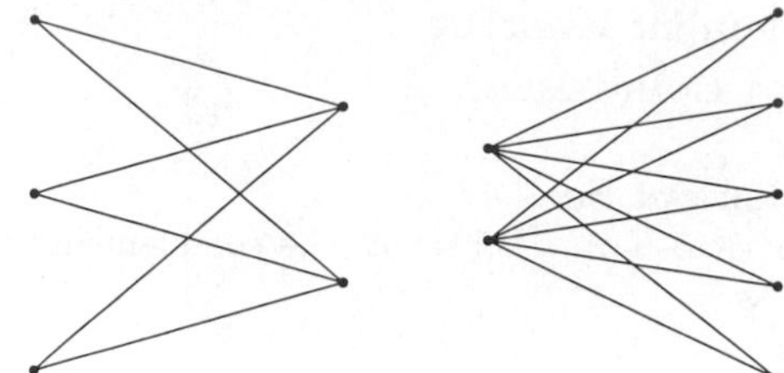

A BIPARTITE GRAPH (i.e., a set of VERTICES decomposed into two disjoint sets such that there are no two VERTICES within the same set are adjacent) such that every pair of VERTICES in the two sets are adjacent. If there are p and q VERTICES in the two sets, the complete bipartite graph (sometimes also called a COMPLETE BIGRAPH) is denoted $K_{p,q}$. The above figures show $K_{3,2}$ and $K_{2,5}$.

see also BIPARTITE GRAPH, COMPLETE GRAPH, COMPLETE k-PARTITE GRAPH, k-PARTITE GRAPH, THOMASSEN GRAPH, UTILITY GRAPH

References
Saaty, T. L. and Kainen, P. C. *The Four-Color Problem: Assaults and Conquest.* New York: Dover, p. 12, 1986.

Complete Functions

A set of ORTHONORMAL FUNCTIONS $\phi_n(x)$ is termed complete in the CLOSED INTERVAL $x \in [a, b]$ if, for every piecewise CONTINUOUS FUNCTION $f(x)$ in the interval, the minimum square error

$$E_n \equiv \|f - (c_1\phi_1 + \ldots + c_n\phi_n)\|^2$$

(where $\|$ denotes the NORM) converges to zero as n becomes infinite. Symbolically, a set of functions is complete if

$$\lim_{m\to\infty} \int_a^b \left[f(x) - \sum_{n=0}^{m} a_n\phi_n(x) \right]^2 w(x)\,dx = 0,$$

where $w(x)$ is a WEIGHTING FUNCTION and the above is a LEBESGUE INTEGRAL.

see also BESSEL'S INEQUALITY, HILBERT SPACE

References
Arfken, G. "Completeness of Eigenfunctions." §9.4 in *Mathematical Methods for Physicists, 3rd ed.* Orlando, FL: Academic Press, pp. 523–538, 1985.

Complete Graph

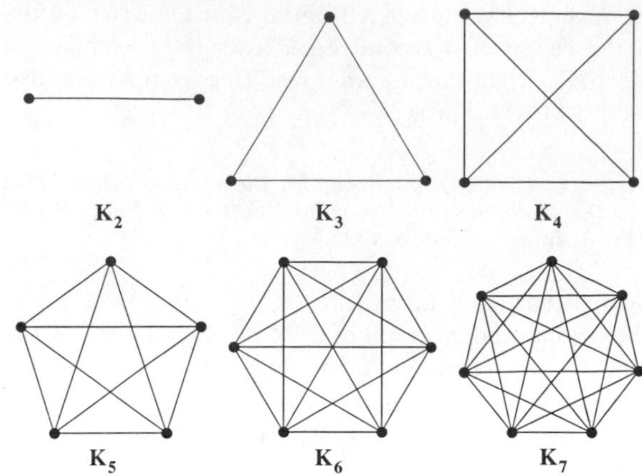

A GRAPH in which each pair of VERTICES is connected by an EDGE. The complete graph with n VERTICES is denoted K_n. In older literature, complete GRAPHS are called UNIVERSAL GRAPHS.

K_4 is the TETRAHEDRAL GRAPH and is therefore a PLANAR GRAPH. K_5 is nonplanar. Conway and Gordon (1983) proved that every embedding of K_6 is INTRINSICALLY LINKED with at least one pair of linked triangles. They also showed that any embedding of K_7 contains a knotted HAMILTONIAN CYCLE.

The number of EDGES in K_v is $v(v-1)/2$, and the GENUS is $(v-3)(v-4)/12$ for $v \geq 3$. The number of distinct variations for K_n (GRAPHS which cannot be transformed into each other without passing nodes through an EDGE or another node) for $n = 1, 2, \ldots$ are 1, 1, 1, 1, 1, 1, 6, 3, 411, 37, ... The ADJACENCY MATRIX of the complete graph takes the particularly simple form of all 1s with 0s on the diagonal.

It is not known in general if a set of TREES with $1, 2, \ldots, n-1$ EDGES can always be packed into K_n. However, if the choice of TREES is restricted to either the path or star from each family, then the packing can always be done (Zaks and Liu 1977, Honsberger 1985).

References
Chartrand, G. *Introductory Graph Theory.* New York: Dover, pp. 29–30, 1985.
Conway, J. H. and Gordon, C. M. "Knots and Links in Spatial Graphs." *J. Graph Th.* **7**, 445–453, 1983.
Honsberger, R. *Mathematical Gems III.* Washington, DC: Math. Assoc. Amer., pp. 60–63, 1985.
Saaty, T. L. and Kainen, P. C. *The Four-Color Problem: Assaults and Conquest.* New York: Dover, p. 12, 1986.
Zaks, S. and Liu, C. L. "Decomposition of Graphs into Trees." *Proc. Eighth Southeastern Conference on Combinatorics, Graph Theory, and Computing.* pp. 643–654, 1977.

Complete k-Partite Graph

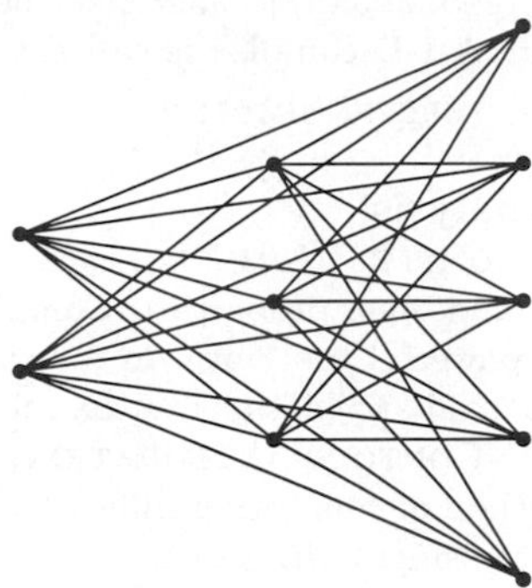

A k-PARTITE GRAPH (i.e., a set of VERTICES decomposed into k disjoint sets such that no two VERTICES within the same set are adjacent) such that every pair of VERTICES in the k sets are adjacent. If there are p, q, ..., r VERTICES in the k sets, the complete k-partite graph is denoted $K_{p,q,\ldots,r}$. The above figure shows $K_{2,3,5}$.

see also COMPLETE GRAPH, COMPLETE k-PARTITE GRAPH, k-PARTITE GRAPH

References

Saaty, T. L. and Kainen, P. C. *The Four-Color Problem: Assaults and Conquest.* New York: Dover, p. 12, 1986.

Complete Metric Space

A complete metric space is a METRIC SPACE in which every CAUCHY SEQUENCE is CONVERGENT. Examples include the REAL NUMBERS with the usual metric and the p-ADIC NUMBERS.

Complete Permutation

see DERANGEMENT

Complete Quadrangle

If the four points making up a QUADRILATERAL are joined pairwise by six distinct lines, a figure known as a complete quadrangle results. Note that a complete quadrilateral is defined differently from a COMPLETE QUADRANGLE.

The midpoints of the sides of any complete quadrangle and the three diagonal points all lie on a CONIC known as the NINE-POINT CONIC. If it is an ORTHOCENTRIC QUADRILATERAL, the CONIC reduces to a CIRCLE. The ORTHOCENTERS of the four TRIANGLES of a complete quadrangle are COLLINEAR on the RADICAL LINE of the CIRCLES on the diameters of a QUADRILATERAL.

see also COMPLETE QUADRANGLE, PTOLEMY'S THEOREM

References

Coxeter, H. S. M. *Introduction to Geometry, 2nd ed.* New York: Wiley, pp. 230–231, 1969.

Demir, H. "The Compleat [sic] Cyclic Quadrilateral." *Amer. Math. Monthly* **79**, 777–778, 1972.

Johnson, R. A. *Modern Geometry: An Elementary Treatise on the Geometry of the Triangle and the Circle.* Boston, MA: Houghton Mifflin, pp. 61–62, 1929.

Ogilvy, C. S. *Excursions in Geometry.* New York: Dover, pp. 101–104, 1990.

Complete Quadrilateral

The figure determined by four lines and their six points of intersection (Johnson 1929, pp. 61–62). Note that this is different from a COMPLETE QUADRANGLE. The midpoints of the diagonals of a complete quadrilateral are COLLINEAR (Johnson 1929, pp. 152–153).

A theorem due to Steiner (Mention 1862, Johnson 1929, Steiner 1971) states that in a complete quadrilateral, the bisectors of angles are CONCURRENT at 16 points which are the incenters and EXCENTERS of the four TRIANGLES. Furthermore, these points are the intersections of two sets of four CIRCLES each of which is a member of a conjugate coaxal system. The axes of these systems intersect at the point common to the CIRCUMCIRCLES of the quadrilateral.

see also COMPLETE QUADRANGLE, GAUSS-BODENMILLER THEOREM, POLAR CIRCLE

References

Coxeter, H. S. M. *Introduction to Geometry, 2nd ed.* New York: Wiley, pp. 230–231, 1969.

Johnson, R. A. *Modern Geometry: An Elementary Treatise on the Geometry of the Triangle and the Circle.* Boston, MA: Houghton Mifflin, pp. 61–62, 149, 152–153, and 255–256, 1929.

Mention, M. J. "Démonstration d'un Théorème de M. Steiner." *Nouv. Ann. Math., 2nd Ser.* **1**, 16–20, 1862.

Mention, M. J. "Démonstration d'un Théorème de M. Steiner." *Nouv. Ann. Math., 2nd Ser.* **1**, 65–67, 1862.

Steiner, J. *Gesammelte Werke, 2nd ed, Vol. 1.* New York: Chelsea, p. 223, 1971.

Complete Residue System

A set of numbers a_0, a_1, ..., a_{m-1} (mod m) form a complete set of residues if they satisfy

$$a_i \equiv i \pmod{m}$$

for $i = 0, 1, \ldots, m-1$. In other words, a complete system of residues is formed by a base and a modulus if the residues r_i in $b^i \equiv r_i \pmod{m}$ for $i = 1, \ldots, m-1$ run through the values $1, 2, \ldots, m-1$.

see also HAUPT-EXPONENT

Complete Sequence

A SEQUENCE of numbers $V = \{\nu_n\}$ is complete if every POSITIVE INTEGER n is the sum of some subsequence of V, i.e., there exist $a_i = 0$ or 1 such that

$$n = \sum_{i=1}^{\infty} a_i \nu_i$$

(Honsberger 1985, pp. 123–126). The FIBONACCI NUMBERS are complete. In fact, dropping one number still leaves a complete sequence, although dropping two numbers does not (Honsberger 1985, pp. 123 and 126). The SEQUENCE of PRIMES with the element $\{1\}$ prepended,

$$\{1, 2, 3, 5, 7, 11, 13, 17, 19, 23, \ldots\}$$

is complete, even if any number of PRIMES each > 7 are dropped, as long as the dropped terms do not include two consecutive PRIMES (Honsberger 1985, pp. 127–128). This is a consequence of BERTRAND'S POSTULATE.

see also BERTRAND'S POSTULATE, BROWN'S CRITERION, FIBONACCI DUAL THEOREM, GREEDY ALGORITHM, WEAKLY COMPLETE SEQUENCE, ZECKENDORF'S THEOREM

References

Brown, J. L. Jr. "Unique Representations of Integers as Sums of Distinct Lucas Numbers." *Fib. Quart.* **7**, 243–252, 1969.

Hoggatt, V. E. Jr.; Cox, N.; and Bicknell, M. "A Primer for Fibonacci Numbers. XII." *Fib. Quart.* **11**, 317–331, 1973.

Honsberger, R. *Mathematical Gems III.* Washington, DC: Math. Assoc. Amer., 1985.

Complete Space

A SPACE of COMPLETE FUNCTIONS.

see also COMPLETE METRIC SPACE

Completely Regular Graph

A POLYHEDRAL GRAPH is completely regular if the DUAL GRAPH is also REGULAR. There are only five types. Let ρ be the number of EDGES at each node, ρ^* the number of EDGES at each node of the DUAL GRAPH, V the number of VERTICES, E the number of EDGES, and F the number of faces in the PLATONIC SOLID corresponding to the given graph. The following table summarizes the completely regular graphs.

Type	ρ	ρ^*	V	E	F
Tetrahedral	3	3	4	6	4
Cubical	3	4	8	12	6
Dodecahedral	3	5	20	39	12
Octahedral	4	3	6	12	8
Icosahedral	5	3	12	30	20

Completeness Property

All lengths can be expressed as REAL NUMBERS.

Completing the Square

The conversion of an equation of the form $ax^2 + bx + c$ to the form

$$a\left(x + \frac{b}{2a}\right)^2 + \left(c - \frac{b^2}{4a}\right),$$

which, defining $B \equiv b/2a$ and $C \equiv c - b^2/4a$, simplifies to

$$a(x + B)^2 + C.$$

Complex

A finite SET of SIMPLEXES such that no two have a common point. A 1-D complex is called a GRAPH.

see also CW-COMPLEX, SIMPLICIAL COMPLEX

Complex Analysis

The study of COMPLEX NUMBERS, their DERIVATIVES, manipulation, and other properties. Complex analysis is an extremely powerful tool with an unexpectedly large number of practical applications to the solution of physical problems. CONTOUR INTEGRATION, for example, provides a method of computing difficult INTEGRALS by investigating the singularities of the function in regions of the COMPLEX PLANE near and between the limits of integration.

The most fundamental result of complex analysis is the CAUCHY-RIEMANN EQUATIONS, which give the conditions a FUNCTION must satisfy in order for a complex generalization of the DERIVATIVE, the so-called COMPLEX DERIVATIVE, to exist. When the COMPLEX DERIVATIVE is defined "everywhere," the function is said to be ANALYTIC. A single example of the unexpected power of complex analysis is PICARD'S THEOREM, which states that an ANALYTIC FUNCTION assumes every COMPLEX NUMBER, with possibly one exception, infinitely often in any NEIGHBORHOOD of an ESSENTIAL SINGULARITY!

see also ANALYTIC CONTINUATION, BRANCH CUT, BRANCH POINT, CAUCHY INTEGRAL FORMULA, CAUCHY INTEGRAL THEOREM, CAUCHY PRINCIPAL VALUE, CAUCHY-RIEMANN EQUATIONS, COMPLEX NUMBER, CONFORMAL MAP, CONTOUR INTEGRATION, DE MOIVRE'S IDENTITY, EULER FORMULA, INSIDE-OUTSIDE THEOREM, JORDAN'S LEMMA, LAURENT SERIES, LIOUVILLE'S CONFORMALITY THEOREM, MONOGENIC FUNCTION, MORERA'S THEOREM, PERMANENCE OF ALGEBRAIC FORM, PICARD'S THEOREM, POLE, POLYGENIC FUNCTION, RESIDUE (COMPLEX ANALYSIS)

References

Arfken, G. "Functions of a Complex Variable I: Analytic Properties, Mapping" and "Functions of a Complex Variable II: Calculus of Residues." Chs. 6–7 in *Mathematical Methods for Physicists, 3rd ed.* Orlando, FL: Academic Press, pp. 352–395 and 396–436, 1985.

Boas, R. P. *Invitation to Complex Analysis.* New York: Random House, 1987.

Churchill, R. V. and Brown, J. W. *Complex Variables and Applications, 6th ed.* New York: McGraw-Hill, 1995.

Conway, J. B. *Functions of One Complex Variable, 2nd ed.* New York: Springer-Verlag, 1995.

Forsyth, A. R. *Theory of Functions of a Complex Variable, 3rd ed.* Cambridge, England: Cambridge University Press, 1918.

Lang, S. *Complex Analysis, 3rd ed.* New York: Springer-Verlag, 1993.

Morse, P. M. and Feshbach, H. "Functions of a Complex Variable" and "Tabulation of Properties of Functions of Complex Variables." Ch. 4 in *Methods of Theoretical Physics, Part I.* New York: McGraw-Hill, pp. 348–491 and 480–485, 1953.

Complex Conjugate

The complex conjugate of a COMPLEX NUMBER $z \equiv a+bi$ is defined to be $z^* \equiv a-bi$. The complex conjugate is ASSOCIATIVE, $(z_1+z_2)^* = z_1{}^* + z_2{}^*$, since

$$\begin{aligned}(a_1+b_1i)^*+(a_2+b_2i)^* &= a_1-ib_1+a_2-ib_2\\ &= (a_1-ib_1)+(a_2-ib_2)\\ &= (a_1+b_1)^*+(a_2+b_2)^*,\end{aligned}$$

and DISTRIBUTIVE, $(z_1z_2)^* = z_1{}^*z_2{}^*$, since

$$\begin{aligned}[(a_1+b_1i)(a_2+b_2i)]^* &= [(a_1a_2-b_1b_2)+i(a_1b_2+a_2b_1)]^*\\ &= (a_1a_2-b_1b_2)-i(a_1b_2+a_2b_1)\\ &= (a_1-ib_1)(a_2-ib_2)\\ &= (a_1+ib_1)^*(a_2+ib_2)^*.\end{aligned}$$

References

Abramowitz, M. and Stegun, C. A. (Eds.). *Handbook of Mathematical Functions with Formulas, Graphs, and Mathematical Tables, 9th printing.* New York: Dover, p. 16, 1972.

Complex Derivative

A DERIVATIVE of a COMPLEX function, which must satisfy the CAUCHY-RIEMANN EQUATIONS in order to be COMPLEX DIFFERENTIABLE.

see also CAUCHY-RIEMANN EQUATIONS, COMPLEX DIFFERENTIABLE, DERIVATIVE

Complex Differentiable

If the CAUCHY-RIEMANN EQUATIONS are satisfied for a function $f(x) = u(x) + iv(x)$ and the PARTIAL DERIVATIVES of $u(x)$ and $v(x)$ are CONTINUOUS, then the COMPLEX DERIVATIVE df/dz exists.

see also ANALYTIC FUNCTION, CAUCHY-RIEMANN EQUATIONS, COMPLEX DERIVATIVE, PSEUDOANALYTIC FUNCTION

Complex Function

A FUNCTION whose RANGE is in the COMPLEX NUMBERS is said to be a complex function.

see also REAL FUNCTION, SCALAR FUNCTION, VECTOR FUNCTION

Complex Matrix

A MATRIX whose elements may contain COMPLEX NUMBERS. The MATRIX PRODUCT of two 2×2 complex matrices is given by

$$\begin{aligned}&\begin{bmatrix} x_{11}+y_{11}i & x_{12}+y_{12}i\\ x_{21}+y_{21}i & x_{22}+y_{22}i\end{bmatrix}\begin{bmatrix} u_{11}+v_{11}i & u_{12}+v_{12}i\\ u_{21}+v_{21}i & u_{22}+v_{22}i\end{bmatrix}\\ &\qquad = \begin{bmatrix} R_{11} & R_{12}\\ R_{21} & R_{22}\end{bmatrix} + i\begin{bmatrix} I_{11} & I_{12}\\ I_{21} & I_{22}\end{bmatrix},\end{aligned}$$

where

$$\begin{aligned}R_{11} &= u_{11}x_{11}+u_{21}x_{12}-v_{11}y_{11}-v_{21}y_{12}\\ R_{12} &= u_{12}x_{11}+u_{22}x_{12}-v_{12}y_{11}-v_{22}y_{12}\\ R_{21} &= u_{11}x_{21}+u_{21}x_{22}-v_{11}y_{21}-v_{21}y_{22}\\ R_{22} &= u_{12}x_{21}+u_{22}x_{22}-v_{12}y_{21}-v_{22}y_{22}\\ I_{11} &= v_{11}x_{11}+v_{21}x_{12}+u_{11}y_{11}+u_{21}y_{12}\\ I_{12} &= v_{12}x_{11}+v_{22}x_{12}+u_{12}y_{11}+u_{22}y_{12}\\ I_{21} &= v_{11}x_{21}+v_{21}x_{22}+u_{11}y_{21}+u_{21}y_{22}\\ I_{22} &= v_{12}x_{21}+v_{22}x_{22}+u_{12}y_{21}+u_{22}y_{22}.\end{aligned}$$

see also REAL MATRIX

Complex Multiplication

Two COMPLEX NUMBERS $x = a+ib$ and $y = c+id$ are multiplied as follows:

$$\begin{aligned}xy = (a+ib)(c+id) &= ac+ibc+iad-bd\\ &= (ac-bd)+i(ad+bc).\end{aligned}$$

However, the multiplication can be carried out using only three REAL multiplications, ac, bd, and $(a+b)(c+d)$ as

$$\begin{aligned}\Re[(a+ib)(c+id)] &= ac-bd\\ \Im[(a+ib)(c+id)] &= (a+b)(c+d)-ac-bd.\end{aligned}$$

Complex multiplication has a special meaning for ELLIPTIC CURVES.

see also COMPLEX NUMBER, ELLIPTIC CURVE, IMAGINARY PART, MULTIPLICATION, REAL PART

References

Cox, D. A. *Primes of the Form x^2+ny^2: Fermat, Class Field Theory and Complex Multiplication.* New York: Wiley, 1997.

Complex Number

The complex numbers are the FIELD $\mathbb{C}$ of numbers of the form $x+iy$, where x and y are REAL NUMBERS and i is the IMAGINARY NUMBER equal to $\sqrt{-1}$. When a single letter $z = x+iy$ is used to denote a complex number, it is sometimes called an "AFFIX." The FIELD of complex numbers includes the FIELD of REAL NUMBERS as a SUBFIELD.

Through the EULER FORMULA, a complex number

$$z = x+iy \tag{1}$$

may be written in "PHASOR" form

$$z = |z|(\cos\theta + i\sin\theta) = |z|e^{i\theta}. \tag{2}$$

Here, $|z|$ is known as the MODULUS and θ is known as the ARGUMENT or PHASE. The ABSOLUTE SQUARE of

z is defined by $|z|^2 = zz^*$, and the argument may be computed from

$$\arg(z) = \theta = \tan^{-1}\left(\frac{y}{x}\right). \quad (3)$$

DE MOIVRE'S IDENTITY relates POWERS of complex numbers

$$z^n = |z|^n[\cos(n\theta) + i\sin(n\theta)]. \quad (4)$$

Finally, the REAL $\Re(z)$ and IMAGINARY PARTS $\Im(z)$ are given by

$$\Re(z) = \tfrac{1}{2}(z + z^*) \quad (5)$$

$$\Im(z) = \frac{z - z^*}{2i} = -\tfrac{1}{2}i(z - z^*) = \tfrac{1}{2}i(z^* - z). \quad (6)$$

The POWERS of complex numbers can be written in closed form as follows:

$$z^n = \left[x^n - \binom{n}{2}x^{n-2}y^2 + \binom{n}{4}x^{n-4}y^4 - \ldots\right] + i\left[\binom{n}{1}x^{n-1}y - \binom{n}{3}x^{n-3}y^3 + \ldots\right]. \quad (7)$$

The first few are explicitly

$$z^2 = (x^2 - y^2) + i(2xy) \quad (8)$$

$$z^3 = (x^3 - 3xy^2) + i(3x^2y - y^3) \quad (9)$$

$$z^4 = (x^4 - 6x^2y^2 + y^4) + i(4x^3y - 4xy^3) \quad (10)$$

$$z^5 = (x^5 - 10x^3y^2 + 5xy^4) + i(5x^4y - 10x^2y^3 + y^5) \quad (11)$$

(Abramowitz and Stegun 1972).

see also ABSOLUTE SQUARE, ARGUMENT (COMPLEX NUMBER), COMPLEX PLANE, I, IMAGINARY NUMBER, MODULUS, PHASE, PHASOR, REAL NUMBER, SURREAL NUMBER

References

Abramowitz, M. and Stegun, C. A. (Eds.). *Handbook of Mathematical Functions with Formulas, Graphs, and Mathematical Tables, 9th printing.* New York: Dover, pp. 16–17, 1972.

Arfken, G. *Mathematical Methods for Physicists, 3rd ed.* Orlando, FL: Academic Press, pp. 353–357, 1985.

Courant, R. and Robbins, H. "Complex Numbers." §2.5 in *What is Mathematics?: An Elementary Approach to Ideas and Methods, 2nd ed.* Oxford, England: Oxford University Press, pp. 88–103, 1996.

Morse, P. M. and Feshbach, H. "Complex Numbers and Variables." §4.1 in *Methods of Theoretical Physics, Part I.* New York: McGraw-Hill, pp. 349–356, 1953.

Press, W. H.; Flannery, B. P.; Teukolsky, S. A.; and Vetterling, W. T. "Complex Arithmetic." §5.4 in *Numerical Recipes in FORTRAN: The Art of Scientific Computing, 2nd ed.* Cambridge, England: Cambridge University Press, pp. 171–172, 1992.

Complex Plane

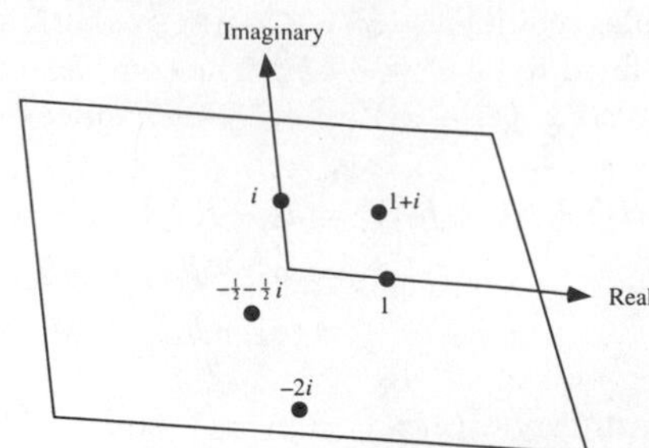

The plane of COMPLEX NUMBERS spanned by the vectors 1 and i, where i is the IMAGINARY NUMBER. Every COMPLEX NUMBER corresponds to a unique POINT in the complex plane. The LINE in the plane with $i = 0$ is the REAL LINE. The complex plane is sometimes called the ARGAND PLANE or GAUSS PLANE, and a plot of COMPLEX NUMBERS in the plane is sometimes called an ARGAND DIAGRAM.

see also AFFINE COMPLEX PLANE, ARGAND DIAGRAM, ARGAND PLANE, BERGMAN SPACE, COMPLEX PROJECTIVE PLANE

References

Courant, R. and Robbins, H. "The Geometric Interpretation of Complex Numbers." §5.2 in *What is Mathematics?: An Elementary Approach to Ideas and Methods, 2nd ed.* Oxford, England: Oxford University Press, pp. 92–97, 1996.

Complex Projective Plane

The set $\mathbb{P}^2$ is the set of all EQUIVALENCE CLASSES $[a, b, c]$ of ordered triples $(a, b, c) \in \mathbb{C}^3\backslash(0, 0, 0)$ under the equivalence relation $(a, b, c) \sim (a', b', c')$ if $(a, b, c) = (\lambda a', \lambda b', \lambda c')$ for some NONZERO COMPLEX NUMBER λ.

Complex Representation

see PHASOR

Complex Structure

The complex structure of a point $\mathbf{x} = x_1, x_2$ in the PLANE is defined by the linear MAP $J : \mathbb{R}^2 \to \mathbb{R}^2$

$$J(x_1, x_2) = (-x_2, x_1),$$

and corresponds to a clockwise rotation by $\pi/2$. This map satisfies

$$J^2 = -I$$

$$(J\mathbf{x}) \cdot (J\mathbf{y}) = \mathbf{x} \cdot \mathbf{y}$$

$$(J\mathbf{x}) \cdot \mathbf{x} = 0,$$

where I is the IDENTITY MAP.

More generally, if V is a 2-D VECTOR SPACE, a linear map $J : V \to V$ such that $J^2 = -I$ is called a complex structure on V. If $V = \mathbb{R}^2$, this collapses to the previous definition.

References

Gray, A. *Modern Differential Geometry of Curves and Surfaces.*Boca Raton, FL: CRC Press, pp. 3 and 229, 1993.

Complexity (Number)

The number of 1s needed to represent an INTEGER using only additions, multiplications, and parentheses are called the integer's complexity. For example,

$$\begin{aligned} 1 &= 1 \\ 2 &= 1+1 \\ 3 &= 1+1+1 \\ 4 &= (1+1)(1+1) = 1+1+1+1 \\ 5 &= (1+1)(1+1)+1 = 1+1+1+1+1 \\ 6 &= (1+1)(1+1+1) \\ 7 &= (1+1)(1+1+1)+1 \\ 8 &= (1+1)(1+1)(1+1) \\ 9 &= (1+1+1)(1+1+1) \\ 10 &= (1+1+1)(1+1+1)+1 \\ &= (1+1)(1+1+1+1+1) \end{aligned}$$

So, for the first few n, the complexity is 1, 2, 3, 4, 5, 5, 6, 6, 6, 7, 8, 7, 8, ... (Sloane's A005245).

References

Guy, R. K. "Expressing Numbers Using Just Ones." §F26 in *Unsolved Problems in Number Theory, 2nd ed.* New York: Springer-Verlag, p. 263, 1994.

Guy, R. K. "Some Suspiciously Simple Sequences." *Amer. Math. Monthly* **93**, 186–190, 1986.

Guy, R. K. "Monthly Unsolved Problems, 1969–1987." *Amer. Math. Monthly* **94**, 961–970, 1987.

Guy, R. K. "Unsolved Problems Come of Age." *Amer. Math. Monthly* **96**, 903–909, 1989.

Rawsthorne, D. A. "How Many 1's are Needed?" *Fib. Quart.* **27**, 14–17, 1989.

Sloane, N. J. A. Sequence A005245/M0457 in "An On-Line Version of the Encyclopedia of Integer Sequences."

Complexity (Sequence)

see BLOCK GROWTH

Complexity Theory

Divides problems into "easy" and "hard" categories. A problem is easy and assigned to the P-PROBLEM (POLYNOMIAL time) class if the number of steps needed to solve it is bounded by some POWER of the problem's size. A problem is hard and assigned to the NP-PROBLEM (nondeterministic POLYNOMIAL time) class if the number of steps is not bounded and may grow exponentially.

However, if a solution is known to an NP-PROBLEM, it can be reduced to a single period verification. A problem is NP-COMPLETE if an ALGORITHM for solving it can be translated into one for solving any other NP-PROBLEM. Examples of NP-COMPLETE PROBLEMS include the HAMILTONIAN CYCLE and TRAVELING SALESMAN PROBLEMS. LINEAR PROGRAMMING, thought to be an NP-PROBLEM, was shown to actually be a P-PROBLEM by L. Khachian in 1979. It is not known if all apparently NP-PROBLEMS are actually P-PROBLEMS.

see also BIT COMPLEXITY, NP-COMPLETE PROBLEM, NP-PROBLEM, P-PROBLEM

References

Bridges, D. S. *Computability.* New York: Springer-Verlag, 1994.

Brookshear, J. G. *Theory of Computation: Formal Languages, Automata, and Complexity.* Redwood City, CA: Benjamin/Cummings, 1989.

Cooper, S. B.; Slaman, T. A.; and Wainer, S. S. (Eds.). *Computability, Enumerability, Unsolvability: Directions in Recursion Theory.* New York: Cambridge University Press, 1996.

Garey, M. R. and Johnson, D. S. *Computers and Intractability: A Guide to the Theory of NP-Completeness.* New York: W. H. Freeman, 1983.

Goetz, P. "Phil Goetz's Complexity Dictionary." `http://www.cs.buffalo.edu/~goetz/dict.html`.

Hopcroft, J. E. and Ullman, J. D. *Introduction to Automated Theory, Languages, and Computation.* Reading, MA: Addison-Wesley, 1979.

Lewis, H. R. and Papadimitriou, C. H. *Elements of the Theory of Computation, 2nd ed.* Englewood Cliffs, NJ: Prentice-Hall, 1997.

Sudkamp, T. A. *Language and Machines: An Introduction to the Theory of Computer Science, 2nd ed.* Reading, MA: Addison-Wesley, 1996.

Welsh, D. J. A. *Complexity: Knots, Colourings and Counting.* New York: Cambridge University Press, 1993.

Component

A GROUP L is a component of H if L is a QUASISIMPLE GROUP which is a SUBNORMAL SUBGROUP of H.

see also GROUP, QUASISIMPLE GROUP, SUBGROUP, SUBNORMAL

Composite Knot

A KNOT which is not a PRIME KNOT. Composite knots are special cases of SATELLITE KNOTS.

see also KNOT, PRIME KNOT, SATELLITE KNOT

Composite Number

A POSITIVE INTEGER which is not PRIME (i.e., which has FACTORS other than 1 and itself).

A composite number C can always be written as a PRODUCT in at least two ways (since $1 \cdot C$ is always possible). Call these two products

$$C = ab = cd, \qquad (1)$$

then it is obviously the case that $C|ab$ (C divides ab). Set

$$c = mn, \qquad (2)$$

where m is the part of C which divides a, and n the part of C which divides n. Then there are p and q such that

$$a = mp \qquad (3)$$

$$b = nq. \qquad (4)$$

Solving $ab = cd$ for d gives

$$d = \frac{ab}{c} = \frac{(mp)(nq)}{mn} = pq. \qquad (5)$$

It then follows that

$$\begin{aligned} S &\equiv a^2 + b^2 + c^2 + d^2 \\ &= m^2p^2 + n^2q^2 + m^2n^2 + p^2q^2 \\ &= (m^2 + q^2)(n^2 + p^2). \end{aligned} \qquad (6)$$

It therefore follows that $a^2+b^2+c^2+d^2$ is never PRIME! In fact, the more general result that

$$S \equiv a^k + b^k + c^k + d^k \qquad (7)$$

is never PRIME for k an INTEGER ≥ 0 also holds (Honsberger 1991).

There are infinitely many integers of the form $\lfloor (3/2)^n \rfloor$ and $\lfloor (4/3)^n \rfloor$ which are composite, where $\lfloor x \rfloor$ is the FLOOR FUNCTION (Forman and Shapiro, 1967; Guy 1994, p. 220). The first few composite $\lfloor (3/2)^n \rfloor$ occur for $n = 8$, 9, 10, 11, 12, 13, 14, 15, 16, 17, 18, 19, 20, 23, ..., and the the few composite $\lfloor (4/3)^n \rfloor$ occur for $n = 5$, 8, 13, 14, 15, 16, 17, 18, 19, 20, 21, 22, ...

see also AMENABLE NUMBER, GRIMM'S CONJECTURE, HIGHLY COMPOSITE NUMBER, PRIME FACTORIZATION PRIME GAPS, PRIME NUMBER

References

Forman, W. and Shapiro, H. N. "An Arithmetic Property of Certain Rational Powers." *Comm. Pure Appl. Math.* **20**, 561–573, 1967.

Guy, R. K. *Unsolved Problems in Number Theory, 2nd ed.* New York: Springer-Verlag, 1994.

Honsberger, R. *More Mathematical Morsels.* Washington, DC: Math. Assoc. Amer., pp. 19–20, 1991.

Sloane, N. J. A. Sequence A002808/M3272 in "An On-Line Version of the Encyclopedia of Integer Sequences."

Composite Runs

see PRIME GAPS

Compositeness Certificate

A compositeness certificate is a piece of information which guarantees that a given number p is COMPOSITE. Possible certificates consist of a FACTOR of a number (which, in general, is much quicker to check by direct division than to determine initially), or of the determination that either

$$a^{p-1} \not\equiv 1 \pmod{p},$$

(i.e., p violates FERMAT'S LITTLE THEOREM), or

$$a \neq -1, 1 \text{ and } a^2 \equiv 1 \pmod{p}.$$

A quantity a satisfying either property is said to be a WITNESS to p's compositeness.

see also ADLEMAN-POMERANCE-RUMELY PRIMALITY TEST, FERMAT'S LITTLE THEOREM, MILLER'S PRIMALITY TEST, PRIMALITY CERTIFICATE, WITNESS

Compositeness Test

A test which always identifies PRIME numbers correctly, but may incorrectly identify a COMPOSITE NUMBER as a PRIME.

see also PRIMALITY TEST

Composition

The combination of two FUNCTIONS to form a single new OPERATOR. The composition of two functions f and g is denoted $f \circ g$ and is defined by

$$f \circ g = f(g(x))$$

when f and g are both functions of x.

An operation called composition is also defined on BINARY QUADRATIC FORMS. For two numbers represented by two forms, the product can then be represented by the composition. For example, the composition of the forms $2x^2 + 15y^2$ and $3x^2 + 10y^2$ is given by $6x^2 + 5y^2$, and in this case, the product of 17 and 13 would be represented as $(6 \cdot 36 + 5 \cdot 1 = 221)$. There are several algorithms for computing binary quadratic form composition, which is the basis for some factoring methods.

see also ADEM RELATIONS, BINARY OPERATOR, BINARY QUADRATIC FORM

Composition Series

Every FINITE GROUP G of order greater than one possesses a finite series of SUBGROUPS, called a composition series, such that

$$I \subset H_s \subset \ldots \subset H_2 \subset H_1 \subset G,$$

where H_{i+1} is a maximal subgroup of H_i. The QUOTIENT GROUPS G/H_1, H_1/H_2, ..., H_{s-1}/H_s, H_s are called composition quotient groups.

see also FINITE GROUP, JORDAN-HÖLDER THEOREM, QUOTIENT GROUP, SUBGROUP

References

Lomont, J. S. *Applications of Finite Groups.* New York: Dover, p. 26, 1993.

Composition Theorem

Let

$$Q(x,y) \equiv x^2 + y^2.$$

Then

$$Q(x,y)Q(x',y') = Q(xx' - yy', x'y + xy'),$$

since

$$\begin{aligned} (x^2 + y^2)(x'^2 + y'^2) &= (xx' - yy')^2 + (xy' + x'y)^2 \\ &= x^2x'^2 + y^2y'^2 + x'^2y^2 + x^2y'^2. \end{aligned}$$

see also GENUS THEOREM

Compound Interest

Let P be the PRINCIPAL (initial investment), r be the annual compounded rate, $i^{(n)}$ the "nominal rate," n be the number of times INTEREST is compounded per year (i.e., the year is divided into n CONVERSION PERIODS), and t be the number of years (the "term"). The INTEREST rate per CONVERSION PERIOD is then

$$r \equiv \frac{i^{(n)}}{n}. \qquad (1)$$

If interest is compounded n times at an annual rate of r (where, for example, 10% corresponds to $r = 0.10$), then the effective rate over $1/n$ the time (what an investor would earn if he did not redeposit his interest after each compounding) is

$$(1+r)^{1/n}. \qquad (2)$$

The total amount of holdings A after a time t when interest is re-invested is then

$$A = P\left(1+\frac{i^{(n)}}{n}\right)^{nt} = P(1+r)^{nt}. \qquad (3)$$

Note that even if interest is compounded continuously, the return is still finite since

$$\lim_{n\to\infty}\left(1+\frac{1}{n}\right)^n = e, \qquad (4)$$

where e is the base of the NATURAL LOGARITHM.

The time required for a given PRINCIPAL to double (assuming $n = 1$ CONVERSION PERIOD) is given by solving

$$2P = P(1+r)^t, \qquad (5)$$

or

$$t = \frac{\ln 2}{\ln(1+r)}, \qquad (6)$$

where LN is the NATURAL LOGARITHM. This function can be approximated by the so-called RULE OF 72:

$$t \approx \frac{0.72}{r}. \qquad (7)$$

see also e, INTEREST, LN, NATURAL LOGARITHM, PRINCIPAL, RULE OF 72, SIMPLE INTEREST

References

Kellison, S. G. *The Theory of Interest, 2nd ed.* Burr Ridge, IL: Richard D. Irwin, pp. 14–16, 1991.

Milanfar, P. "A Persian Folk Method of Figuring Interest." *Math. Mag.* **69**, 376, 1996.

Compound Polyhedron

see POLYHEDRON COMPOUND

Computability

see COMPLEXITY THEORY

Computable Function

Any computable function can be incorporated into a PROGRAM using while-loops (i.e., "while something is true, do something else"). For-loops (which have a fixed iteration limit) are a special case of while-loops, so computable functions could also be coded using a combination of for- and while-loops. The ACKERMANN FUNCTION is the simplest example of a well-defined TOTAL FUNCTION which is computable but not PRIMITIVE RECURSIVE, providing a counterexample to the belief in the early 1900s that every computable function was also primitive recursive (Dötzel 1991).

see also ACKERMANN FUNCTION, CHURCH'S THESIS, COMPUTABLE NUMBER, PRIMITIVE RECURSIVE FUNCTION, TURING MACHINE

References

Dötzel, G. "A Function to End All Functions." *Algorithm: Recreational Programming* **2**, 16–17, 1991.

Computable Number

A number which can be computed to any number of DIGITS desired by a TURING MACHINE. Surprisingly, most IRRATIONALS are not computable numbers!

References

Penrose, R. *The Emperor's New Mind: Concerning Computers, Minds, and the Laws of Physics.* Oxford, England: Oxford University Press, 1989.

Computational Complexity

see COMPLEXITY THEORY

Concatenated Number Sequences

see CONSECUTIVE NUMBER SEQUENCES

Concatenation

The concatenation of two strings a and b is the string ab formed by joining a and b. Thus the concatenation of the strings "book" and "case" is the string "bookcase". The concatenation of two strings a and b is often denoted ab, $a||b$, or (in *Mathematica*® (Wolfram Research, Champaign, IL) $a <> b$. Concatenation is an associative operation, so that the concatenation of three or more strings, for example abc, $abcd$, etc., is well-defined.

The concatenation of two or more numbers is the number formed by concatenating their numerals. For example, the concatenation of 1, 234, and 5678 is 12345678. The value of the result depends on the numeric base, which is typically understood from context.

The formula for the concatenation of numbers p and q in base b is

$$p||q = pb^{l(q)} + q,$$

where

$$l(q) = \lfloor \log_b q \rfloor + 1$$

is the LENGTH of q in base b and $\lfloor x \rfloor$ is the FLOOR FUNCTION.

see also CONSECUTIVE NUMBER SEQUENCES, LENGTH (NUMBER), SMARANDACHE SEQUENCES

Concave

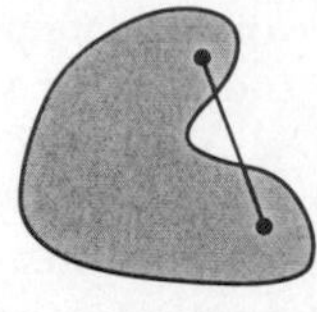

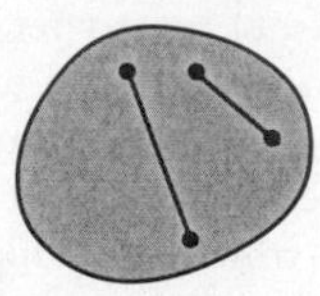

concave convex

A SET in $\mathbb{R}^d$ is concave if it *does not* contain all the LINE SEGMENTS connecting any pair of its points. If the SET *does* contain all the LINE SEGMENTS, it is called CONVEX.

see also CONNECTED SET, CONVEX FUNCTION, CONVEX HULL, CONVEX OPTIMIZATION THEORY, CONVEX POLYGON, DELAUNAY TRIANGULATION, SIMPLY CONNECTED

Concave Function

A function $f(x)$ is said to be concave on an interval $[a, b]$ if, for any points x_1 and x_2 in $[a, b]$, the function $-f(x)$ is CONVEX on that interval. If the second DERIVATIVE of f

$$f''(x) > 0,$$

on an open interval (a, b) (where $f''(x)$ is the second DERIVATIVE), then f is concave up on the interval. If

$$f''(x) < 0$$

on the interval, then f is concave down on it.

References

Gradshteyn, I. S. and Ryzhik, I. M. *Tables of Integrals, Series, and Products, 5th ed.* San Diego, CA: Academic Press, p. 1100, 1980.

Concentrated

Let μ be a POSITIVE MEASURE on a SIGMA ALGEBRA M, and let λ be an arbitrary (real or complex) MEASURE on M. If there is a SET $A \in M$ such that $\lambda(E) = \lambda(A \cap E)$ for every $E \in M$, then *lambda* is said to be concentrated on A. This is equivalent to requiring that $\lambda(E) = 0$ whenever $E \cap A = \varnothing$.

see also ABSOLUTELY CONTINUOUS, MUTUALLY SINGULAR

References

Rudin, W. *Functional Analysis.* New York: McGraw-Hill, p. 121, 1991.

Concentric

Two geometric figures are said to be concentric if their CENTERS coincide. The region between two concentric CIRCLES is called an ANNULUS.

see also ANNULUS, CONCENTRIC CIRCLES, CONCYCLIC, ECCENTRIC

Concentric Circles

The region between two CONCENTRIC circles of different RADII is called an ANNULUS.

Given two concentric circles with RADII R and $2R$, what is the probability that a chord chosen at random from the outer circle will cut across the inner circle? Depending on how the "random" CHORD is chosen, 1/2, 1/3, or 1/4 could all be correct answers.

1. Picking any two points on the outer circle and connecting them gives 1/3.
2. Picking any random point on a diagonal and then picking the CHORD that perpendicularly bisects it gives 1/2.
3. Picking any point on the large circle, drawing a line to the center, and then drawing the perpendicularly bisected CHORD gives 1/4.

So some care is obviously needed in specifying what is meant by "random" in this problem.

Given an arbitrary CHORD BB' to the larger of two concentric CIRCLES centered on O, the distance between inner and outer intersections is equal on both sides ($AB = A'B'$). To prove this, take the PERPENDICULAR to BB' passing through O and crossing at P. By symmetry, it must be true that PA and PA' are equal. Similarly, PB and PB' must be equal. Therefore, $PB - PA = AB$ equals $PB' - PA' = A'B'$. Incidentally, this is also true for HOMEOIDS, but the proof is nontrivial.

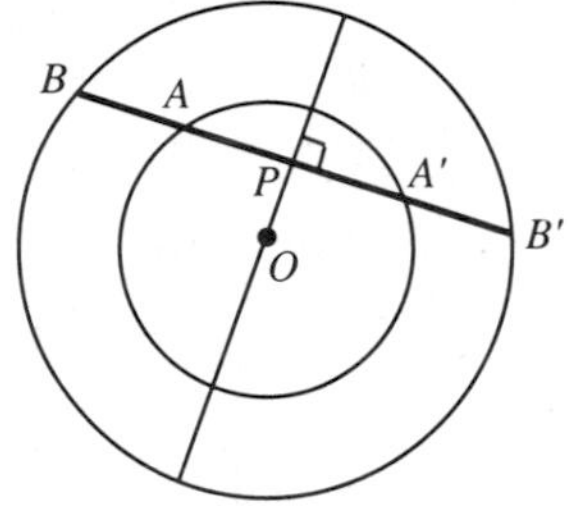

see also ANNULUS

Concho-Spiral

The SPACE CURVE with parametric equations

$$r = \mu^u a$$
$$\theta = u$$
$$z = \mu^u c.$$

see also CONICAL SPIRAL, SPIRAL

Conchoid

A curve whose name means "shell form." Let C be a curve and O a fixed point. Let P and P' be points on a line from O to C meeting it at Q, where $P'Q = QP = k$, with k a given constant. For example, if C is a CIRCLE and O is on C, then the conchoid is a LIMAÇON, while in the special case that k is the DIAMETER of C, then the conchoid is a CARDIOID. The equation for a parametrically represented curve $(f(t), g(t))$ with $O = (x_0, y_0)$ is

$$x = f \pm \frac{k(f - x_0)}{\sqrt{(f - x_0)^2 + (g - y_0)^2}}$$
$$y = g \pm \frac{k(g - y_0)}{\sqrt{(f - x_0)^2 + (g - y_0)^2}}.$$

see also CONCHO-SPIRAL, CONCHOID OF DE SLUZE, CONCHOID OF NICOMEDES, CONICAL SPIRAL, DÜRER'S CONCHOID

References
Lawrence, J. D. *A Catalog of Special Plane Curves.* New York: Dover, pp. 49–51, 1972.
Lee, X. "Conchoid." `http://www.best.com/~xah/SpecialPlaneCurves_dir/Conchoid_dir/conchoid.html`.
Lockwood, E. H. "Conchoids." Ch. 14 in *A Book of Curves.* Cambridge, England: Cambridge University Press, pp. 126–129, 1967.
Yates, R. C. "Conchoid." *A Handbook on Curves and Their Properties.* Ann Arbor, MI: J. W. Edwards, pp. 31–33, 1952.

Conchoid of de Sluze

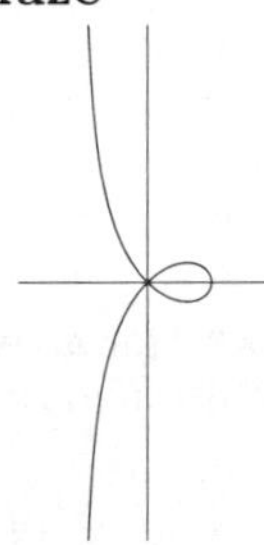

A curve first constructed by René de Sluze in 1662. In CARTESIAN COORDINATES,

$$a(x - a)(x^2 + y^2) = k^2 x^2,$$

and in POLAR COORDINATES,

$$r = \frac{k^2 \cos\theta}{a} + a \sec\theta.$$

The above curve has $k^2/a = 1$, $a = -0.5$.

Conchoid of Nicomedes

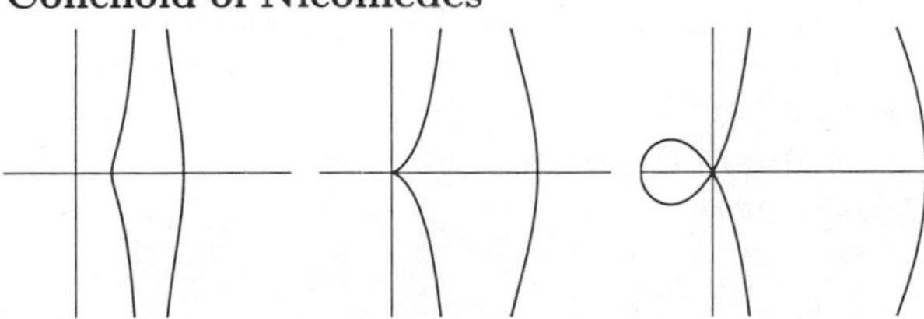

A curve studied by the Greek mathematician Nicomedes in about 200 BC, also called the COCHLOID. It is the LOCUS of points a fixed distance away from a line as measured along a line from the FOCUS point (MacTutor Archive). Nicomedes recognized the three distinct forms seen in this family. This curve was a favorite with 17th century mathematicians and could be used to solve the problems of CUBE DUPLICATION and ANGLE TRISECTION.

In POLAR COORDINATES,

$$r = b + a \sec\theta. \tag{1}$$

In CARTESIAN COORDINATES,

$$(x - a)^2(x^2 + y^2) = b^2 x^2. \tag{2}$$

The conchoid has $x = a$ as an asymptote and the AREA between either branch and the ASYMPTOTE is infinite. The AREA of the loop is

$$A = a\sqrt{b^2 - a^2} - 2ab \ln\left(\frac{b + \sqrt{b^2 - a^2}}{a}\right) + b^2 \cos^{-1}\left(\frac{a}{b}\right). \tag{3}$$

see also CONCHOID

References
Lawrence, J. D. *A Catalog of Special Plane Curves.* New York: Dover, pp. 135–139, 1972.
Lee, X. "Conchoid of Nicomedes." `http://www.best.com/~xah/SpecialPlaneCurves_dir/ConchoidOfNicomedes_dir/conchoidOfNicomedes.html`.
MacTutor History of Mathematics Archive. "Conchoid." `http://www-groups.dcs.st-and.ac.uk/~history/Curves/Conchoid.html`.
Pappas, T. "Conchoid of Nicomedes." *The Joy of Mathematics.* San Carlos, CA: Wide World Publ./Tetra, pp. 94–95, 1989.
Yates, R. C. "Conchoid." *A Handbook on Curves and Their Properties.* Ann Arbor, MI: J. W. Edwards, pp. 31–33, 1952.

Concordant Form

A concordant form is an integer TRIPLE (a, b, N) where

$$\begin{cases} a^2 + b^2 = c^2 \\ a^2 + Nb^2 = d^2, \end{cases}$$

with c and d integers. Examples include

$$\begin{cases} 14663^2 + 111384^2 = 112345^2 \\ 14663^2 + 47 \cdot 111384^2 = 763751^2 \end{cases}$$
$$\begin{cases} 1141^2 + 13260^2 = 13309^2 \\ 1141^2 + 53 \cdot 13260^2 = 96541^2 \end{cases}$$
$$\begin{cases} 2873161^2 + 2401080^2 = 3744361^2 \\ 2873161^2 + 83 \cdot 2401080^2 = 22062761^2. \end{cases}$$

Dickson (1962) states that C. H. Brooks and S. Watson found in *The Ladies' and Gentlemen's Diary* (1857) that $x^2 + y^2$ and $x^2 + Ny^2$ can be simultaneously squares for $N < 100$ only for 1, 7, 10, 11, 17, 20, 22, 23, 24, 27, 30, 31, 34, 41, 42, 45, 49, 50, 52, 57, 58, 59, 60, 61, 68, 71, 72, 74, 76, 77, 79, 82, 85, 86, 90, 92, 93, 94, 97,

99, and 100 (which evidently omits 47, 53, and 83 from above). The list of concordant primes less than 1000 is now complete with the possible exception of the 16 primes 103, 131, 191, 223, 271, 311, 431, 439, 443, 593, 607, 641, 743, 821, 929, and 971 (Brown).

see also CONGRUUM

References

Brown, K. S. "Concordant Forms." http://www.seanet.com/~ksbrown/kmath286.htm.

Dickson, L. E. *History of the Theory of Numbers, Vol. 1: Divisibility and Primality.* New York: Chelsea, p. 475, 1952.

Concur

Two or more lines which intersect in a POINT are said to concur.

see also CONCURRENT

Concurrent

Two or more LINES are said to be concurrent if they intersect in a single point. Two LINES concur if their TRILINEAR COORDINATES satisfy

$$\begin{vmatrix} l_1 & m_1 & n_1 \\ l_2 & m_2 & n_2 \\ l_3 & m_3 & n_3 \end{vmatrix} = 0. \tag{1}$$

Three LINES concur if their TRILINEAR COORDINATES satisfy

$$l_1\alpha + m_1\beta + n_1\gamma = 0 \tag{2}$$
$$l_2\alpha + m_2\beta + n_2\gamma = 0 \tag{3}$$
$$l_3\alpha + m_3\beta + n_3\gamma = 0, \tag{4}$$

in which case the point is

$$m_2n_3 - n_2m_3 : n_2l_3 - l_2n_3 : l_2m_3 - m_2l_3. \tag{5}$$

Three lines

$$A_1x + B_1y + C_1 = 0 \tag{6}$$
$$A_2x + B_2y + C_2 = 0 \tag{7}$$
$$A_3x + B_3y + C_3 = 0. \tag{8}$$

are concurrent if their COEFFICIENTS satisfy

$$\begin{vmatrix} A_1 & B_1 & C_1 \\ A_2 & B_2 & C_2 \\ A_3 & B_3 & C_3 \end{vmatrix} = 0. \tag{9}$$

see also CONCYCLIC, POINT

Concyclic

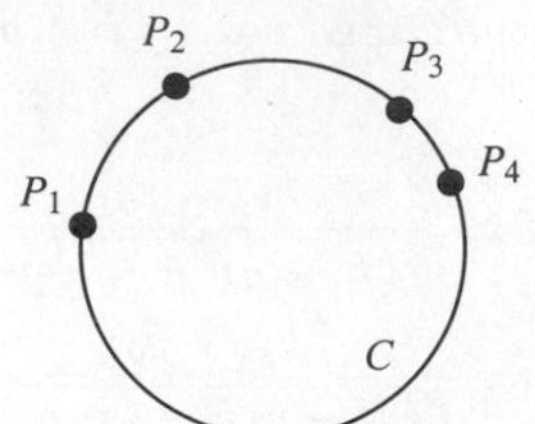

Four or more points P_1, P_2, P_3, P_4, ... which lie on a CIRCLE C are said to be concyclic. Three points are trivially concyclic since three noncollinear points determine a CIRCLE. The number of the n^2 LATTICE POINTS $x, y \in [1, n]$ which can be picked with no four concyclic is $\mathcal{O}(n^{2/3} - \epsilon)$ (Guy 1994).

A theorem states that if any four consecutive points of a POLYGON are not concyclic, then its AREA can be increased by making them concyclic. This fact arises in some PROOFS that the solution to the ISOPERIMETRIC PROBLEM is the CIRCLE.

see also CIRCLE, COLLINEAR, CONCENTRIC, CYCLIC HEXAGON, CYCLIC PENTAGON, CYCLIC QUADRILATERAL, ECCENTRIC, N-CLUSTER

References

Guy, R. K. "Lattice Points, No Four on a Circle." §F3 in *Unsolved Problems in Number Theory, 2nd ed.* New York: Springer-Verlag, p. 241, 1994.

Condition

A requirement NECESSARY for a given statement or theorem to hold. Also called a CRITERION.

see also BOUNDARY CONDITIONS, CARMICHAEL CONDITION, CAUCHY BOUNDARY CONDITIONS, CONDITION NUMBER, DIRICHLET BOUNDARY CONDITIONS, DIVERSITY CONDITION, FELLER-LÉVY CONDITION, HÖLDER CONDITION, LICHNEROWICZ CONDITIONS, LINDEBERG CONDITION, LIPSCHITZ CONDITION, LYAPUNOV CONDITION, NEUMANN BOUNDARY CONDITIONS, ROBERTSON CONDITION, ROBIN BOUNDARY CONDITIONS, TAYLOR'S CONDITION, TRIANGLE CONDITION, WEIERSTRAß-ERDMAN CORNER CONDITION, WINKLER CONDITIONS

Condition Number

The ratio of the largest to smallest SINGULAR VALUE of a system. A system is said to be singular if the condition number is INFINITE, and ill-conditioned if it is too large.

Conditional Convergence

If the SERIES

$$\sum_{n=0}^{\infty} u_n$$

CONVERGES, but

$$\sum_{n=0}^{\infty} |u_n|$$

does not, where $|x|$ is the ABSOLUTE VALUE, then the SERIES is said to be conditionally CONVERGENT.

see also ABSOLUTE CONVERGENCE, CONVERGENCE TESTS, RIEMANN SERIES THEOREM, SERIES

Conditional Probability

The conditional probability of A given that B has occurred, denoted $P(A|B)$, equals

$$P(A|B) = \frac{P(A \cap B)}{P(B)}, \tag{1}$$

which can be proven directly using a VENN DIAGRAM. Multiplying through, this becomes

$$P(A|B)P(B) = P(A \cap B), \tag{2}$$

which can be generalized to

$$P(A \cup B \cup C) = P(A)P(B|A)P(C|A \cup B). \tag{3}$$

Rearranging (1) gives

$$P(B|A) = \frac{P(B \cap A)}{P(A)}. \tag{4}$$

Solving (4) for $P(B \cap A) = P(A \cap B)$ and plugging in to (1) gives

$$P(A|B) = \frac{P(A)P(B|A)}{P(B)}. \tag{5}$$

see also BAYES' FORMULA

Condom Problem

see GLOVE PROBLEM

Condon-Shortley Phase

The $(-1)^m$ phase factor in some definitions of the SPHERICAL HARMONICS and associated LEGENDRE POLYNOMIALS. Using the Condon-Shortley convention gives

$$Y_n^m(\theta, \phi) = (-1)^m \sqrt{\frac{2n+1}{4\pi} \frac{(n-m)!}{(n+m)!}} P_n^m(\cos\theta) e^{im\phi}.$$

see also LEGENDRE POLYNOMIAL, SPHERICAL HARMONIC

References

Arfken, G. *Mathematical Methods for Physicists, 3rd ed.* Orlando, FL: Academic Press, pp. 682 and 692, 1985.

Condon, E. U. and Shortley, G. *The Theory of Atomic Spectra.* Cambridge, England: Cambridge University Press, 1951.

Shore, B. W. and Menzel, D. H. *Principles of Atomic Spectra.* New York: Wiley, p. 158, 1968.

Conductor

see j-CONDUCTOR

Cone

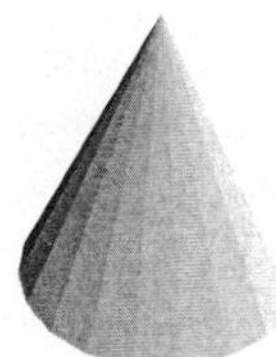

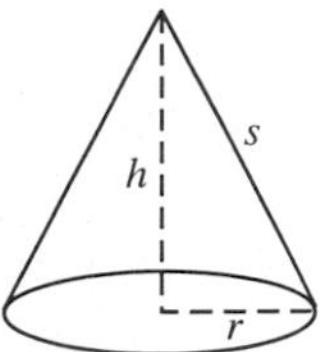

A cone is a PYRAMID with a circular CROSS-SECTION. A right cone is a cone with its vertex above the center of its base. A right cone of height h can be described by the parametric equations

$$x = r(h - z)\cos\theta \tag{1}$$
$$y = r(h - z)\sin\theta \tag{2}$$
$$z = z \tag{3}$$

for $z \in [0, h]$ and $\theta \in [0, 2\pi)$. The VOLUME of a cone is therefore

$$V = \tfrac{1}{3} A_b h, \tag{4}$$

where A_b is the base AREA and h is the height. If the base is circular, then

$$V = \tfrac{1}{3}\pi r^2 h. \tag{5}$$

This amazing fact was first discovered by Eudoxus, and other proofs were subsequently found by Archimedes in *On the Sphere and Cylinder* (ca. 225 BC) and Euclid in Proposition XII.10 of his *Elements* (Dunham 1990).

The CENTROID can be obtained by setting $R_2 = 0$ in the equation for the centroid of the CONICAL FRUSTUM,

$$\bar{z} = \frac{\langle z \rangle}{V} = \frac{h({R_1}^2 + R_1 R_2 + {R_2}^2)}{4({R_1}^2 + 2R_1 R_2 + 3{R_2}^2)}, \tag{6}$$

(Beyer 1987, p. 133) yielding

$$\bar{z} = \tfrac{1}{4} h. \tag{7}$$

For a right circular cone, the SLANT HEIGHT s is

$$s = \sqrt{r^2 + h^2} \tag{8}$$

and the surface AREA (not including the base) is

$$S = \pi r s = \pi r \sqrt{r^2 + h^2}. \tag{9}$$

In discussions of CONIC SECTIONS, the word cone is often used to refer to two similar cones placed apex to apex. This allows the HYPERBOLA to be defined as the intersection of a PLANE with both NAPPES (pieces) of the cone.

The LOCUS of the apex of a variable cone containing an ELLIPSE fixed in 3-space is a HYPERBOLA through the FOCI of the ELLIPSE. In addition, the LOCUS of the apex of a cone containing that HYPERBOLA is the original ELLIPSE. Furthermore, the ECCENTRICITIES of the ELLIPSE and HYPERBOLA are reciprocals.

see also CONIC SECTION, CONICAL FRUSTUM, CYLINDER, NAPPE, PYRAMID, SPHERE

References
Beyer, W. H. (Ed.) *CRC Standard Mathematical Tables, 28th ed.* Boca Raton, FL: CRC Press, pp. 129 and 133, 1987.
Dunham, W. *Journey Through Genius: The Great Theorems of Mathematics.* New York: Wiley, pp. 76–77, 1990.
Yates, R. C. "Cones." *A Handbook on Curves and Their Properties.* Ann Arbor, MI: J. W. Edwards, pp. 34–35, 1952.

Cone Graph

A GRAPH $C_n + \overline{K_m}$, where C_n is a CYCLIC GRAPH and K_m is a COMPLETE GRAPH.

Cone Net

The mapping of a grid of regularly ruled squares onto a CONE with no overlap or misalignment. Cone nets are possible for vertex angles of 90°, 180°, and 270°, and are beautifully illustrated by Steinhaus (1983).

References
Steinhaus, H. *Mathematical Snapshots, 3rd American ed.* New York: Oxford University Press, pp. 224–228, 1983.

Cone (Space)

The JOIN of a TOPOLOGICAL SPACE X and a point P, $C(X) = X * P$.

References
Rolfsen, D. *Knots and Links.* Wilmington, DE: Publish or Perish Press, p. 6, 1976.

Cone-Sphere Intersection

Let a CONE of opening parameter c and vertex at $(0,0,0)$ intersect a SPHERE of RADIUS r centered at (x_0, y_0, z_0), with the CONE oriented such that its axis does not pass through the center of the SPHERE. Then the equations of the curve of intersection are

$$\frac{x^2+y^2}{c^2} = z^2 \tag{1}$$

$$(x-x_0)^2 + (y-y_0)^2 + (z-z_0)^2 = r^2. \tag{2}$$

Combining (1) and (2) gives

$$(x-x_0)^2+(y-y_0)^2+\frac{x^2+y^2}{c^2}-\frac{2z_0}{c}\sqrt{x^2+y^2}+{z_0}^2 = r^2 \tag{3}$$

$$x^2\left(1+\frac{1}{c^2}\right) - 2x_0x + y^2\left(1+\frac{1}{c^2}\right) - 2y_0y + ({x_0}^2+{y_0}^2+{z_0}^2-r^2) - \frac{2z_0}{c}\sqrt{x^2+y^2} = 0. \tag{4}$$

Therefore, x and y are connected by a complicated QUARTIC EQUATION, and x, y, and z by a QUADRATIC EQUATION.

If the CONE-SPHERE intersection is *on-axis* so that a CONE of opening parameter c and vertex at $(0,0,z_0)$ is oriented with its AXIS along a radial of the SPHERE of radius r centered at $(0,0,0)$, then the equations of the curve of intersection are

$$(z-z_0)^2 = \frac{x^2+y^2}{c^2} \tag{5}$$

$$x^2+y^2+z^2 = r^2. \tag{6}$$

Combining (5) and (6) gives

$$c^2(z-z_0)^2 + z2 = r^2 \tag{7}$$

$$c^2(z^2 - 2z_0z + {z_0}^2) + z^2 = r^2 \tag{8}$$

$$z^2(c^2+1) - 2c^2z_0z + ({z_0}^2c^2 - r^2) = 0. \tag{9}$$

Using the QUADRATIC EQUATION gives

$$\begin{aligned} z &= \frac{2c^2z_0 \pm \sqrt{4c^4{z_0}^2 - 4(c^2+1)({z_0}^2c^2-r^2)}}{2(c^2+1)} \\ &= \frac{c^2z_0 \pm \sqrt{c^2(r^2-{z_0}^2)+r^2}}{c^2+1}. \end{aligned} \tag{10}$$

So the curve of intersection is planar. Plugging (10) into (5) shows that the curve is actually a CIRCLE, with RADIUS given by

$$a = \sqrt{r^2 - z^2}. \tag{11}$$

Confidence Interval

The probability that a measurement will fall within a given CLOSED INTERVAL $[a,b]$. For a continuous distribution,

$$\mathrm{CI}(a,b) \equiv \int_b^a P(x)\,dx, \tag{1}$$

where $P(x)$ is the PROBABILITY DISTRIBUTION FUNCTION. Usually, the confidence interval of interest is symmetrically placed around the mean, so

$$\mathrm{CI}(x) \equiv \mathrm{CI}(\mu - x, \mu + x) = \int_{\mu-x}^{\mu+x} P(x)\,dx, \tag{2}$$

where μ is the MEAN. For a GAUSSIAN DISTRIBUTION, the probability that a measurement falls within $n\sigma$ of the mean μ is

$$\begin{aligned} \mathrm{CI}(n\sigma) &\equiv \frac{1}{\sigma\sqrt{2\pi}} \int_{\mu-n\sigma}^{\mu+n\sigma} e^{-(x-\mu)^2/2\sigma^2}\,dx \\ &= \frac{2}{\sigma\sqrt{2\pi}} \int_0^{\mu+n\sigma} e^{-(x-\mu)^2/2\sigma^2}\,dx. \end{aligned} \quad (3)$$

Now let $u \equiv (x-\mu)/\sqrt{2}\,\sigma$, so $du = dx/\sqrt{2}\,\sigma$. Then

$$\begin{aligned} \mathrm{CI}(n\sigma) &= \frac{2}{\sigma\sqrt{2\pi}}\sqrt{2}\,\sigma \int_0^{n/\sqrt{2}} e^{-u^2}\,du \\ &= \frac{2}{\sqrt{\pi}} \int_0^{n/\sqrt{2}} e^{-u^2}\,du = \mathrm{erf}\left(\frac{n}{\sqrt{2}}\right), \end{aligned} \quad (4)$$

where $\mathrm{erf}(x)$ is the so-called ERF function. The variate value producing a confidence interval CI is often denoted x_{CI}, so

$$x_{\mathrm{CI}} = \sqrt{2}\,\mathrm{erf}^{-1}(\mathrm{CI}). \quad (5)$$

range	CI
σ	0.6826895
2σ	0.9544997
3σ	0.9973002
4σ	0.9999366
5σ	0.9999994

To find the standard deviation range corresponding to a given confidence interval, solve (4) for n.

$$n = \sqrt{2}\,\mathrm{erf}^{-1}(\mathrm{CI}) \quad (6)$$

CI	range
0.800	$\pm 1.28155\sigma$
0.900	$\pm 1.64485\sigma$
0.950	$\pm 1.95996\sigma$
0.990	$\pm 2.57583\sigma$
0.995	$\pm 2.80703\sigma$
0.999	$\pm 3.29053\sigma$

Configuration

A finite collection of points $p = (p_1, \ldots, p_n)$, $p_i \in \mathbb{R}^d$, where $\mathbb{R}^d$ is a EUCLIDEAN SPACE.

see also BAR (EDGE), EUCLIDEAN SPACE, FRAMEWORK, RIGID

Confluent Hypergeometric Differential Equation

$$xy'' + (b-x)y' - ay = 0, \quad (1)$$

where $y' \equiv dy/dx$ and with boundary conditions

$${}_1F_1(a;b;0) = 1 \quad (2)$$

$$\left[\frac{\partial}{\partial x}\,{}_1F_1(a;b;x)\right]_{x=0} = \frac{a}{b}. \quad (3)$$

The equation has a REGULAR SINGULAR POINT at 0 and an irregular singularity at ∞. The solutions are called CONFLUENT HYPERGEOMETRIC FUNCTION OF THE FIRST or SECOND KINDS. Solutions of the first kind are denoted ${}_1F_1(a;b;x)$ or $M(a,b,x)$.

see also HYPERGEOMETRIC DIFFERENTIAL EQUATION, WHITTAKER DIFFERENTIAL EQUATION

References

Abramowitz, M. and Stegun, C. A. (Eds.). *Handbook of Mathematical Functions with Formulas, Graphs, and Mathematical Tables, 9th printing.* New York: Dover, p. 504, 1972.

Arfken, G. "Confluent Hypergeometric Functions." §13.6 in *Mathematical Methods for Physicists, 3rd ed.* Orlando, FL: Academic Press, pp. 753–758, 1985.

Morse, P. M. and Feshbach, H. *Methods of Theoretical Physics, Part I.* New York: McGraw-Hill, pp. 551–555, 1953.

Confluent Hypergeometric Function

see CONFLUENT HYPERGEOMETRIC FUNCTION OF THE FIRST KIND, CONFLUENT HYPERGEOMETRIC FUNCTION OF THE SECOND KIND

Confluent Hypergeometric Function of the First Kind

The confluent hypergeometric function a degenerate form the HYPERGEOMETRIC FUNCTION ${}_2F_1(a,b;c;z)$ which arises as a solution the the CONFLUENT HYPERGEOMETRIC DIFFERENTIAL EQUATION. It is commonly denoted ${}_1F_1(a;b;z)$, $M(a,b,z)$, or $\Phi(a;b;z)$, and is also known as KUMMER'S FUNCTION of the first kind. An alternate form of the solution to the Confluent Hypergeometric Differential Equation is known as the WHITTAKER FUNCTION.

The confluent hypergeometric function has a HYPERGEOMETRIC SERIES given by

$${}_1F_1(a;b;z) = 1 + \frac{a}{b}z + \frac{a(a+1)}{b(b+1)}\frac{z^2}{2!} + \ldots = \sum_{k=0}^{\infty} \frac{(a)_k}{(b)_k}\frac{z^k}{k!}, \quad (1)$$

where $(a)_k$ and $(b)_k$ are POCHHAMMER SYMBOLS. If a and b are INTEGERS, $a < 0$, and either $b > 0$ or $b < a$, then the series yields a POLYNOMIAL with a finite number of terms. If b is an INTEGER ≤ 0, then ${}_1F_1(a;b;z)$ is undefined. The confluent hypergeometric function also has an integral representation

$${}_1F_1(a;b;z) = \frac{\Gamma(b)}{\Gamma(b-a)\Gamma(a)} \int_0^1 e^{zt} t^{a-1} (1-t)^{b-a-1}\,dt \quad (2)$$

(Abramowitz and Stegun 1972, p. 505).

BESSEL FUNCTIONS, the ERROR FUNCTION, the incomplete GAMMA FUNCTION, HERMITE POLYNOMIAL, LAGUERRE POLYNOMIAL, as well as other are all special cases of this function (Abramowitz and Stegun 1972, p. 509).

KUMMER'S SECOND FORMULA gives

$$ {}_1F_1(\tfrac{1}{2}+m;2m+1;z) = M_{0,m}(z) = z^{m+1/2} \times \left[1+\sum_{p=1}^{\infty} \frac{z^{2p}}{2^{4p}p!(m+1)(m+2)\cdots(m+p)}\right], \quad (3) $$

where ${}_1F_1$ is the CONFLUENT HYPERGEOMETRIC FUNCTION and $m \neq -1/2, -1, -3/2, \ldots$.

see also CONFLUENT HYPERGEOMETRIC DIFFERENTIAL EQUATION, CONFLUENT HYPERGEOMETRIC FUNCTION OF THE SECOND KIND, CONFLUENT HYPERGEOMETRIC LIMIT FUNCTION, GENERALIZED HYPERGEOMETRIC FUNCTION, HEINE HYPERGEOMETRIC SERIES, HYPERGEOMETRIC FUNCTION, HYPERGEOMETRIC SERIES, KUMMER'S FORMULAS, WEBER-SONINE FORMULA, WHITTAKER FUNCTION

References

Abramowitz, M. and Stegun, C. A. (Eds.). "Confluent Hypergeometric Functions." Ch. 13 in *Handbook of Mathematical Functions with Formulas, Graphs, and Mathematical Tables, 9th printing.* New York: Dover, pp. 503–515, 1972.

Arfken, G. "Confluent Hypergeometric Functions." §13.6 in *Mathematical Methods for Physicists, 3rd ed.* Orlando, FL: Academic Press, pp. 753–758, 1985.

Iyanaga, S. and Kawada, Y. (Eds.). "Hypergeometric Function of Confluent Type." Appendix A, Table 19.I in *Encyclopedic Dictionary of Mathematics.* Cambridge, MA: MIT Press, p. 1469, 1980.

Morse, P. M. and Feshbach, H. *Methods of Theoretical Physics, Part I.* New York: McGraw-Hill, pp. 551–554 and 604–605, 1953.

Slater, L. J. *Confluent Hypergeometric Functions.* Cambridge, England: Cambridge University Press, 1960.

Spanier, J. and Oldham, K. B. "The Kummer Function $M(a;c;x)$." Ch. 47 in *An Atlas of Functions.* Washington, DC: Hemisphere, pp. 459–469, 1987.

Confluent Hypergeometric Function of the Second Kind

Gives the second linearly independent solution to the CONFLUENT HYPERGEOMETRIC DIFFERENTIAL EQUATION. It is also known as the KUMMER'S FUNCTION of the second kind, the TRICOMI FUNCTION, or the GORDON FUNCTION. It is denoted $U(a,b,z)$ and has an integral representation

$$ U(a,b,z) = \frac{1}{\Gamma(a)}\int_0^\infty e^{-zt}t^{a-1}(1+t)^{b-a-1}\,dt $$

(Abramowitz and Stegun 1972, p. 505). The WHITTAKER FUNCTIONS give an alternative form of the solution. For small z, the function behaves as z^{1-b}.

see also BATEMAN FUNCTION, CONFLUENT HYPERGEOMETRIC FUNCTION OF THE FIRST KIND, CONFLUENT HYPERGEOMETRIC LIMIT FUNCTION, COULOMB WAVE FUNCTION, CUNNINGHAM FUNCTION, GORDON FUNCTION, HYPERGEOMETRIC FUNCTION, POISSON-CHARLIER POLYNOMIAL, TORONTO FUNCTION, WEBER FUNCTIONS, WHITTAKER FUNCTION

References

Abramowitz, M. and Stegun, C. A. (Eds.). "Confluent Hypergeometric Functions." Ch. 13 in *Handbook of Mathematical Functions with Formulas, Graphs, and Mathematical Tables, 9th printing.* New York: Dover, pp. 503–515, 1972.

Arfken, G. "Confluent Hypergeometric Functions." §13.6 in *Mathematical Methods for Physicists, 3rd ed.* Orlando, FL: Academic Press, pp. 753–758, 1985.

Morse, P. M. and Feshbach, H. *Methods of Theoretical Physics, Part I.* New York: McGraw-Hill, pp. 671–672, 1953.

Spanier, J. and Oldham, K. B. "The Tricomi Function $U(a;c;x)$." Ch. 48 in *An Atlas of Functions.* Washington, DC: Hemisphere, pp. 471–477, 1987.

Confluent Hypergeometric Limit Function

$$ {}_0F_1(;a;z) \equiv \lim_{q\to\infty} {}_1F_1\left(q;a;\frac{z}{q}\right). \quad (1) $$

It has a series expansion

$$ {}_0F_1(;a;z) = \sum_{n=0}^{\infty} \frac{z^n}{(a)_n n!} \quad (2) $$

and satisfies

$$ z\frac{d^2y}{dz^2} + a\frac{dy}{dz} - y = 0. \quad (3) $$

A BESSEL FUNCTION OF THE FIRST KIND can be expressed in terms of this function by

$$ J_n(x) = \frac{(\frac{1}{2}x)^n}{n!}\,{}_0F_1(;n+1;-\tfrac{1}{4}x^2) \quad (4) $$

(Petkovšek *et al.* 1996).

see also CONFLUENT HYPERGEOMETRIC FUNCTION, GENERALIZED HYPERGEOMETRIC FUNCTION, HYPERGEOMETRIC FUNCTION

References

Petkovšek, M.; Wilf, H. S.; and Zeilberger, D. *A=B.* Wellesley, MA: A. K. Peters, p. 38, 1996.

Confocal Conics

Confocal conics are CONIC SECTIONS sharing a common FOCUS. Any two confocal CENTRAL CONICS are orthogonal (Ogilvy 1990, p. 77).

see also CONIC SECTION, FOCUS

References

Ogilvy, C. S. *Excursions in Geometry.* New York: Dover, pp. 77–78, 1990.

Confocal Ellipsoidal Coordinates

The confocal ellipsoidal coordinates (called simply ellipsoidal coordinates by Morse and Feshbach 1953) are given by the equations

$$\frac{x^2}{a^2+\xi}+\frac{y^2}{b^2+\xi}+\frac{z^2}{c^2+\xi}=1 \quad (1)$$

$$\frac{x^2}{a^2+\eta}+\frac{y^2}{b^2+\eta}+\frac{z^2}{c^2+\eta}=1 \quad (2)$$

$$\frac{x^2}{a^2+\zeta}+\frac{y^2}{b^2+\zeta}+\frac{z^2}{c^2+\zeta}=1, \quad (3)$$

where $-c^2 < \xi < \infty$, $-b^2 < \eta < -c^2$, and $-a^2 < \zeta < -b^2$. Surfaces of constant ξ are confocal ELLIPSOIDS, surfaces of constant η are one-sheeted HYPERBOLOIDS, and surfaces of constant ζ are two-sheeted HYPERBOLOIDS. For every (x, y, z), there is a unique set of ellipsoidal coordinates. However, (ξ, η, ζ) specifies eight points symmetrically located in octants. Solving for x, y, and z gives

$$x^2=\frac{(a^2+\xi)(a^2+\eta)(a^2+\zeta)}{(b^2-a^2)(c^2-a^2)} \quad (4)$$

$$y^2=\frac{(b^2+\xi)(b^2+\eta)(b^2+\zeta)}{(a^2-b^2)(c^2-b^2)} \quad (5)$$

$$z^2=\frac{(c^2+\xi)(c^2+\eta)(c^2+\zeta)}{(a^2-c^2)(b^2-c^2)}. \quad (6)$$

The LAPLACIAN is

$$\nabla^2\Psi=(\eta-\zeta)f(\xi)\frac{\partial}{\partial\xi}\left[f(\xi)\frac{\partial\Psi}{\partial\xi}\right]$$
$$+(\zeta-\xi)f(\eta)\frac{\partial}{\partial\eta}\left[f(\eta)\frac{\partial\Psi}{\partial\eta}\right]+(\xi-\eta)f(\zeta)\frac{\partial}{\partial\zeta}\left[f(\zeta)\frac{\partial\Psi}{\partial\zeta}\right], \quad (7)$$

where

$$f(x)\equiv\sqrt{(x+a^2)(x+b^2)(x+c^2)}. \quad (8)$$

Another definition is

$$\frac{x^2}{a^2-\lambda}+\frac{y^2}{b^2-\lambda}+\frac{z^2}{c^2-\lambda}=1 \quad (9)$$

$$\frac{x^2}{a^2-\mu}+\frac{y^2}{b^2-\mu}+\frac{z^2}{c^2-\mu}=1 \quad (10)$$

$$\frac{x^2}{a^2-\nu}+\frac{y^2}{b^2-\nu}+\frac{z^2}{c^2-\nu}=1, \quad (11)$$

where

$$\lambda<c^2<\mu<b^2<\nu<a^2 \quad (12)$$

(Arfken 1970, pp. 117–118). Byerly (1959, p. 251) uses a slightly different definition in which the Greek variables are replaced by their squares, and $a = 0$. Equation (9) represents an ELLIPSOID, (10) represents a one-sheeted HYPERBOLOID, and (11) represents a two-sheeted HYPERBOLOID. In terms of CARTESIAN COORDINATES,

$$x^2=\frac{(a^2-\lambda)(a^2-\mu)(a^2-\nu)}{(a^2-b^2)(a^2-c^2)} \quad (13)$$

$$y^2=\frac{(b^2-\lambda)(b^2-\mu)(b^2-\nu)}{(b^2-a^2)(b^2-c^2)} \quad (14)$$

$$z^2=\frac{(c^2-\lambda)(c^2-\mu)(c^2-\nu)}{(c^2-a^2)(c^2-b^2)}. \quad (15)$$

The SCALE FACTORS are

$$h_\lambda=\sqrt{\frac{(\mu-\lambda)(\nu-\lambda)}{4(a^2-\lambda)(b^2-\lambda)(c^2-\lambda)}} \quad (16)$$

$$h_\mu=\sqrt{\frac{(\nu-\mu)(\lambda-\mu)}{4(a^2-\mu)(b^2-\mu)(c^2-\mu)}} \quad (17)$$

$$h_\nu=\sqrt{\frac{(\lambda-\nu)(\mu-\nu)}{4(a^2-\nu)(b^2-\nu)(c^2-\nu)}}. \quad (18)$$

The LAPLACIAN is

$$\begin{aligned}\nabla^2 &= 2\frac{a^2b^2+a^2c^2+b^2c^2-2\nu(a^2+b^2+c^2)+3\nu^2}{(\mu-\nu)(\nu-\lambda)}\frac{\partial}{\partial\nu}\\ &+\frac{4(a^2-\nu)(b^2-\nu)(c^2-\nu)}{(\mu-\nu)(\nu-\lambda)}\frac{\partial^2}{\partial\nu^2}\\ &+2\frac{a^2b^2+a^2c^2+b^2c^2-2\mu(a^2+b^2+c^2)+3\mu^2}{(\nu-\mu)(\mu-\lambda)}\frac{\partial}{\partial\mu}\\ &+\frac{4(a^2-\mu)(b^2-\mu)(c^2-\mu)}{(\mu-\lambda)(\nu-\mu)}\frac{\partial^2}{\partial\mu^2}\\ &+2\frac{-(a^2b^2+a^2c^2+b^2c^2)+2\lambda(a^2+b^2+c^2)-3\lambda^2}{(\mu-\lambda)(\nu-\lambda)}\frac{\partial}{\partial\lambda}\\ &+\frac{4(a^2-\lambda)(b^2-\lambda)(c^2-\lambda)}{(\mu-\lambda)(\nu-\lambda)}\frac{\partial^2}{\partial\lambda^2}.\end{aligned} \quad (19)$$

Using the NOTATION of Byerly (1959, pp. 252–253), this can be reduced to

$$\nabla^2=(\mu^2-\nu^2)\frac{\partial^2}{\partial\alpha^2}+(\lambda^2-\nu^2)\frac{\partial^2}{\partial\beta^2}+(\lambda^2-\mu^2)\frac{\partial^2}{\partial\gamma^2}, \quad (20)$$

where

$$\alpha=c\int_c^\lambda\frac{d\lambda}{\sqrt{(\lambda^2-b^2)(\lambda^2-c^2)}}$$
$$=F\left(\frac{b}{c},\frac{\pi}{2}\right)-F\left(\frac{b}{c},\sin^{-1}\left(\frac{c}{\lambda}\right)\right) \quad (21)$$

$$\beta=c\int_b^\mu\frac{d\mu}{\sqrt{(c^2-\mu^2)(\mu^2-b^2)}}$$
$$=F\left[\sqrt{1-\frac{b^2}{c^2}},\sin^{-1}\left(\sqrt{\frac{1-\frac{b^2}{\mu^2}}{1-\frac{b^2}{c^2}}}\right)\right] \quad (22)$$

$$\gamma=c\int_0^\nu\frac{d\nu}{\sqrt{(b^2-\nu^2)(c^2-\nu^2)}}$$
$$=F\left(\frac{b}{c},\sin^{-1}\left(\frac{\nu}{b}\right)\right). \quad (23)$$

Here, F is an ELLIPTIC INTEGRAL OF THE FIRST KIND. In terms of α, β, and γ,

$$\lambda = c\,\mathrm{dc}\left(\alpha, \frac{b}{c}\right) \tag{24}$$

$$\mu = b\,\mathrm{nd}\left(\beta, \sqrt{1-\frac{b^2}{c^2}}\right) \tag{25}$$

$$\nu = b\,\mathrm{sn}\left(\gamma, \frac{b}{c}\right), \tag{26}$$

where dc, nd and sn are JACOBI ELLIPTIC FUNCTIONS. The HELMHOLTZ DIFFERENTIAL EQUATION is separable in confocal ellipsoidal coordinates.

see also HELMHOLTZ DIFFERENTIAL EQUATION—CONFOCAL ELLIPSOIDAL COORDINATES

References

Abramowitz, M. and Stegun, C. A. (Eds.). "Definition of Elliptical Coordinates." §21.1 in *Handbook of Mathematical Functions with Formulas, Graphs, and Mathematical Tables, 9th printing.* New York: Dover, p. 752, 1972.

Arfken, G. "Confocal Ellipsoidal Coordinates (ξ_1, ξ_2, ξ_3)." §2.15 in *Mathematical Methods for Physicists, 2nd ed.* New York: Academic Press, pp. 117–118, 1970.

Byerly, W. E. *An Elementary Treatise on Fourier's Series, and Spherical, Cylindrical, and Ellipsoidal Harmonics, with Applications to Problems in Mathematical Physics.* New York: Dover, 1959.

Morse, P. M. and Feshbach, H. *Methods of Theoretical Physics, Part I.* New York: McGraw-Hill, p. 663, 1953.

Confocal Parabolic Coordinates

see CONFOCAL PARABOLOIDAL COORDINATES

Confocal Paraboloidal Coordinates

$$\frac{x^2}{a^2-\lambda} + \frac{y^2}{b^2-\lambda} = z - \lambda \tag{1}$$

$$\frac{x^2}{a^2-\mu} + \frac{y^2}{b^2-\mu} = z - \mu \tag{2}$$

$$\frac{x^2}{a^2-\nu} + \frac{y^2}{b^2-\nu} = z - \nu, \tag{3}$$

where $\lambda \in (-\infty, b^2)$, $\mu \in (b^2, a^2)$, and $\nu \in (a^2, \infty)$.

$$x^2 = \frac{(a^2-\lambda)(a^2-\mu)(a^2-\nu)}{(b^2-a^2)} \tag{4}$$

$$y^2 = \frac{(b^2-\lambda)(b^2-\mu)(b^2-\nu)}{(a^2-b^2)} \tag{5}$$

$$z = \lambda + \mu + \nu - a^2 - b^2. \tag{6}$$

The SCALE FACTORS are

$$h_\lambda = \sqrt{\frac{(\mu-\lambda)(\nu-\lambda)}{4(a^2-\lambda)(b^2-\lambda)}} \tag{7}$$

$$h_\mu = \sqrt{\frac{(\nu-\mu)(\lambda-\mu)}{4(a^2-\mu)(b^2-\mu)}} \tag{8}$$

$$h_\nu = \sqrt{\frac{(\lambda-\nu)(\mu-\nu)}{16(a^2-\nu)(b^2-\nu)}}. \tag{9}$$

The LAPLACIAN is

$$\nabla^2 = \frac{2(a^2+b^2-2\nu)}{(\mu-\nu)(\nu-\lambda)}\frac{\partial}{\partial\nu} + \frac{4(a^2-\nu)(\nu-b^2)}{(\mu-\nu)(\nu-\lambda)}\frac{\partial^2}{\nu^2}$$
$$+\frac{2(a^2+b^2-2\mu)}{(\mu-\lambda)(\nu-\mu)}\frac{\partial}{\partial\mu} + \frac{4(a^2-\mu)(\mu-b^2)}{(\mu-\lambda)(\nu-\mu)}\frac{\partial^2}{\partial\mu^2}$$
$$+\frac{2(2\lambda-a^2-b^2)}{(\mu-\lambda)(\nu-\lambda)}\frac{\partial}{\partial\lambda} + \frac{4(\lambda-a^2)(\lambda-b^2)}{(\mu-\lambda)(\nu-\lambda)}\frac{\partial^2}{\partial\lambda^2}. \tag{10}$$

The HELMHOLTZ DIFFERENTIAL EQUATION is SEPARABLE.

see also HELMHOLTZ DIFFERENTIAL EQUATION—CONFOCAL PARABOLOIDAL COORDINATES

References

Arfken, G. "Confocal Parabolic Coordinates (ξ_1, ξ_2, ξ_3)." §2.17 in *Mathematical Methods for Physicists, 2nd ed.* Orlando, FL: Academic Press, pp. 119–120, 1970.

Morse, P. M. and Feshbach, H. *Methods of Theoretical Physics, Part I.* New York: McGraw-Hill, p. 664, 1953.

Conformal Latitude

An AUXILIARY LATITUDE defined by

$$\chi \equiv 2\tan^{-1}\left\{\tan(\tfrac{1}{4}\pi + \tfrac{1}{2}\phi)\left[\frac{1-e\sin\phi}{1+e\sin\phi}\right]^{e/2}\right\} - \tfrac{1}{2}\pi$$
$$= 2\tan^{-1}\left[\frac{1+\sin\phi}{1-\sin\phi}\left(\frac{1-e\sin\phi}{1+e\sin\phi}\right)^e\right]^{1/2} - \tfrac{1}{2}\pi$$
$$= \phi - (\tfrac{1}{2}e^2 + \tfrac{5}{24}e^4 + \tfrac{3}{32}e^6 + \tfrac{281}{5760}e^8 + \ldots)\sin(2\phi)$$
$$+ (\tfrac{5}{48}e^4 + \tfrac{7}{80}e^6 + \tfrac{697}{11520}e^8 + \ldots)\sin(4\phi)$$
$$- (\tfrac{13}{480}e^6 + \tfrac{461}{13440} + \ldots)\sin(6\phi)$$
$$+ (\tfrac{1237}{161280}e^8 + \ldots)\sin(8\phi) + \ldots.$$

The inverse is obtained by iterating the equation

$$\phi = 2\tan^{-1}\left[\tan(\tfrac{1}{4}\pi + \tfrac{1}{2}\chi)\left(\frac{1+e\sin\phi}{1-e\sin\phi}\right)^{e/2}\right] - \tfrac{1}{2}\pi$$

using $\phi = \chi$ as the first trial. A series form is

$$\phi = \chi + (\tfrac{1}{2}e^2 + \tfrac{5}{24}e^4 + \tfrac{1}{12}e^6 + \tfrac{13}{360}e^8 + \ldots)\sin(2\chi)$$
$$+ (\tfrac{7}{48}e^4 + \tfrac{29}{240}e^6 + \tfrac{811}{11520}e^9 + \ldots)\sin(4\chi)$$
$$+ (\tfrac{7}{120}e^6 + \tfrac{81}{1120}e^8 + \ldots)\sin(6\chi)$$
$$+ (\tfrac{4279}{161280}e^8 + \ldots)\sin(8\chi) + \ldots$$

The conformal latitude was called the ISOMETRIC LATITUDE by Adams (1921), but this term is now used to refer to a different quantity.

see also AUXILIARY LATITUDE, LATITUDE

References

Adams, O. S. "Latitude Developments Connected with Geodesy and Cartography with Tables, Including a Table for Lambert Equal-Area Meridianal Projections." Spec. Pub. No. 67. U. S. Coast and Geodetic Survey, pp. 18 and 84–85, 1921.

Snyder, J. P. *Map Projections—A Working Manual.* U. S. Geological Survey Professional Paper 1395. Washington, DC: U. S. Government Printing Office, pp. 15–16, 1987.

Conformal Map

A TRANSFORMATION which preserves ANGLES is known as conformal. For a transformation to be conformal, it must be an ANALYTIC FUNCTION and have a NONZERO DERIVATIVE. Let θ and ϕ be the tangents to the curves γ and $f(\gamma)$ at z_0 and w_0,

$$w - w_0 \equiv f(z) - f(z_0) = \frac{f(z) - f(z_0)}{z - z_0}(z - z_0) \quad (1)$$

$$\arg(w - w_0) = \arg\left[\frac{f(z) - f(z_0)}{z - z_0}\right] + \arg(z - z_0). \quad (2)$$

Then as $w \to w_0$ and $z \to z_0$,

$$\phi = \arg f'(z_0) + \theta \quad (3)$$

$$|w| = |f'(z_0)|\,|z|. \quad (4)$$

see also ANALYTIC FUNCTION, HARMONIC FUNCTION, MÖBIUS TRANSFORMATION, QUASICONFORMAL MAP, SIMILAR

References

Arfken, G. "Conformal Mapping." §6.7 in *Mathematical Methods for Physicists, 3rd ed.* Orlando, FL: Academic Press, pp. 392–394, 1985.

Bergman, S. *The Kernel Function and Conformal Mapping.* New York: Amer. Math. Soc., 1950.

Katznelson, Y. *An Introduction to Harmonic Analysis.* New York: Dover, 1976.

Morse, P. M. and Feshbach, H. "Conformal Mapping." §4.7 in *Methods of Theoretical Physics, Part I.* New York: McGraw-Hill, pp. 358–362 and 443–453, 1953.

Nehari, Z. *Conformal Map.* New York: Dover, 1982.

Conformal Solution

By letting $w \equiv f(z)$, the REAL and IMAGINARY PARTS of w must satisfy the CAUCHY-RIEMANN EQUATIONS and LAPLACE'S EQUATION, so they automatically provide a scalar POTENTIAL and a so-called stream function. If a physical problem can be found for which the solution is valid, we obtain a solution—which may have been very difficult to obtain directly—by working backwards. Let

$$Az^n = Ar^n e^{in\theta}, \quad (1)$$

the REAL and IMAGINARY PARTS then give

$$\phi = Ar^n \cos(n\theta) \quad (2)$$

$$\psi = Ar^n \sin(n\theta). \quad (3)$$

For $n = -2$,

$$\phi = \frac{A}{r^2}\cos(2\theta) \quad (4)$$

$$\psi = -\frac{A}{r^2}\sin(2\theta), \quad (5)$$

which is a double system of LEMNISCATES (Lamb 1945, p. 69). For $n = -1$,

$$\phi = \frac{A}{r}\cos\theta \quad (6)$$

$$\psi = -\frac{A}{r}\sin\theta. \quad (7)$$

This solution consists of two systems of CIRCLES, and ϕ is the POTENTIAL FUNCTION for two PARALLEL opposite charged line charges (Feynman *et al.* 1989, §7-5; Lamb 1945, p. 69). For $n = 1/2$,

$$\phi = Ar^{1/2}\cos\left(\frac{\theta}{2}\right) = A\sqrt{\frac{\sqrt{x^2+y^2}+x}{2}} \quad (8)$$

$$\psi = Ar^{1/2}\sin\left(\frac{\theta}{2}\right) = A\sqrt{\frac{\sqrt{x^2+y^2}-x}{2}}. \quad (9)$$

ϕ gives the field near the edge of a thin plate (Feynman *et al.* 1989, §7-5). For $n = 1$,

$$\phi = Ar\cos\theta = Ax \quad (10)$$

$$\psi = Ar\sin\theta = Ay. \quad (11)$$

This is two straight lines (Lamb 1945, p. 68). For $n = 3/2$,

$$w = Ar^{3/2}e^{3i\theta/2}. \quad (12)$$

ϕ gives the field near the outside of a rectangular corner (Feynman *et al.* 1989, §7-5). For $n = 2$,

$$w = A(x+iy)^2 = A[(x^2-y^2)+2ixy] \quad (13)$$

$$\phi = A(x^2-y^2) = Ar^2\cos(2\theta) \quad (14)$$

$$\psi = 2Axy = Ar^2\sin(2\theta). \quad (15)$$

These are two PERPENDICULAR HYPERBOLAS, and ϕ is the POTENTIAL FUNCTION near the middle of two point charges or the field on the opening side of a charged RIGHT ANGLE conductor (Feynman 1989, §7-3).

see also CAUCHY-RIEMANN EQUATIONS, CONFORMAL MAP, LAPLACE'S EQUATION

References

Feynman, R. P.; Leighton, R. B.; and Sands, M. *The Feynman Lectures on Physics, Vol. 1.* Redwood City, CA: Addison-Wesley, 1989.

Lamb, H. *Hydrodynamics, 6th ed.* New York: Dover, 1945.

Conformal Tensor

see WEYL TENSOR

Conformal Transformation

see CONFORMAL MAP

Congruence

If $b-c$ is integrally divisible by a, then b and c are said to be congruent with MODULUS a. This is written mathematically as $b \equiv c \pmod{a}$. If $b-c$ is *not* divisible by a, then we say $b \not\equiv c \pmod{a}$. The (mod a) is sometimes omitted when the MODULUS a is understood for a given computation, so care must be taken not to confuse the symbol $\equiv$ with that for an EQUIVALENCE. The quantity b is called the RESIDUE or REMAINDER. The COMMON RESIDUE is taken to be NONNEGATIVE and smaller than m, and the MINIMAL RESIDUE is b or $b-m$, whichever is smaller in ABSOLUTE VALUE. In many computer languages (such as **FORTRAN** or *Mathematic*®), the COMMON RESIDUE of $c \pmod{a}$ is written `mod(c,a)`.

Congruence arithmetic is perhaps most familiar as a generalization of the arithmetic of the clock: 40 minutes past the hour plus 35 minutes gives $40+35 \equiv 15 \pmod{60}$, or 15 minutes past the hour, and 10 o'clock a.m. plus five hours gives $10+5 \equiv 3 \pmod{12}$, or 3 o'clock p.m. Congruences satisfy a number of important properties, and are extremely useful in many areas of NUMBER THEORY. Using congruences, simple DIVISIBILITY TESTS to check whether a given number is divisible by another number can sometimes be derived. For example, if the sum of a number's digits is divisible by 3 (9), then the original number is divisible by 3 (9).

Congruences also have their limitations. For example, if $a \equiv b$ and $c \equiv d \pmod{n}$, then it follows that $a^x \equiv b^x$, but usually not that $x^c \equiv x^d$ or $a^c \equiv b^d$. In addition, by "rolling over," congruences discard absolute information. For example, knowing the number of minutes past the hour is useful, but knowing the hour the minutes are past is often more useful still.

Let $a \equiv a' \pmod{m}$ and $b \equiv b' \pmod{m}$, then important properties of congruences include the following, where $\Rightarrow$ means "IMPLIES":

1. Equivalence: $a \equiv b \pmod{0} \Rightarrow a = b$.
2. Determination: either $a \equiv b \pmod{m}$ or $a \not\equiv b$ (mom m).
3. Reflexivity: $a \equiv a \pmod{m}$.
4. Symmetry: $a \equiv b \pmod{m} \Rightarrow b \equiv a \pmod{m}$.
5. Transitivity: $a \equiv b \pmod{m}$ and $b \equiv c \pmod{m} \Rightarrow a \equiv c \pmod{m}$.
6. $a+b \equiv a'+b' \pmod{m}$.
7. $a-b \equiv a'-b' \pmod{m}$.
8. $ab \equiv a'b' \pmod{m}$.
9. $a \equiv b \pmod{m} \Rightarrow ka \equiv kb \pmod{m}$.
10. $a \equiv b \pmod{m} \Rightarrow a^n \equiv b^n \pmod{m}$.
11. $a \equiv b \pmod{m_1}$ and $a \equiv b \pmod{m_2} \Rightarrow a \equiv b \pmod{[m_1, m_2]}$, where $[m_1, m_2]$ is the LEAST COMMON MULTIPLE.
12. $ak \equiv bk \pmod{m} \Rightarrow a \equiv b \left(\bmod \frac{m}{(k,m)}\right)$, where (k,m) is the GREATEST COMMON DIVISOR.
13. If $a \equiv b \pmod{m}$, then $P(a) \equiv P(b) \pmod{m}$, for $P(x)$ a POLYNOMIAL.

Properties (6–8) can be proved simply by defining

$$a \equiv a' + rd \tag{1}$$

$$b \equiv b' + sd, \tag{2}$$

where r and s are INTEGERS. Then

$$a+b = a'+b'+(r+s)d \tag{3}$$

$$a-b = a'-b'+(r-s)d \tag{4}$$

$$ab = a'b' + (a's+b'r+rsd)d, \tag{5}$$

so the properties are true.

Congruences also apply to FRACTIONS. For example, note that (mod 7)

$$2 \times 4 \equiv 1 \qquad 3 \times 3 \equiv 2 \qquad 6 \times 6 \equiv 1 \pmod{7}, \tag{6}$$

so

$$\tfrac{1}{2} \equiv 4 \qquad \tfrac{1}{4} \equiv 2 \qquad \tfrac{2}{3} \equiv 3 \qquad \tfrac{1}{6} \equiv 6 \pmod{7}. \tag{7}$$

To find p/q mod m, use an ALGORITHM similar to the GREEDY ALGORITHM. Let $q_0 \equiv q$ and find

$$p_0 = \left\lceil \frac{m}{q_0} \right\rceil, \tag{8}$$

where $\lceil x \rceil$ is the CEILING FUNCTION, then compute

$$q_1 \equiv q_0 p_0 \pmod{m}. \tag{9}$$

Iterate until $q_n = 1$, then

$$\frac{p}{q} \equiv p \prod_{i=0}^{n-1} p_i \pmod{m}. \tag{10}$$

This method always works for m PRIME, and sometimes even for m COMPOSITE. However, for a COMPOSITE m, the method can fail by reaching 0 (Conway and Guy 1996).

A LINEAR CONGRUENCE

$$ax \equiv b \pmod{m} \tag{11}$$

is solvable IFF the congruence

$$b \equiv 0 \pmod{(a,m)} \tag{12}$$

is solvable, where $d \equiv (a,m)$ is the GREATEST COMMON DIVISOR, in which case the solutions are x_0, $x_0 + m/d$, $x_0 + 2m/d$, ..., $x_0 + (d-1)m/d$, where $x_0 < m/d$. If $d = 1$, then there is only one solution.

A general QUADRATIC CONGRUENCE

$$a_2x^2 + a_1x + a_0 \equiv 0 \pmod{n} \tag{13}$$

can be reduced to the congruence

$$x^2 \equiv q \pmod{p} \tag{14}$$

and can be solved using EXCLUDENTS. Solution of the general polynomial congruence

$$a_mx^m + \ldots + a_2x^2 + a_1x + a_0 \equiv 0 \pmod{n} \tag{15}$$

is intractable. Any polynomial congruence will give congruent results when congruent values are substituted.

Two simultaneous congruences

$$x \equiv a \pmod{m} \tag{16}$$

$$x \equiv b \pmod{n} \tag{17}$$

are solvable only when $x \equiv b \pmod{(m,n)}$, and the single solution is

$$x \equiv x_0 \pmod{[m,n]}, \tag{18}$$

where $x_0 < m/d$.

see also CANCELLATION LAW, CHINESE REMAINDER THEOREM, COMMON RESIDUE, CONGRUENCE AXIOMS, DIVISIBILITY TESTS, GREATEST COMMON DIVISOR, LEAST COMMON MULTIPLE, MINIMAL RESIDUE, MODULUS (CONGRUENCE), QUADRATIC RECIPROCITY LAW, RESIDUE (CONGRUENCE)

References

Conway, J. H. and Guy, R. K. "Arithmetic Modulo *p*." In *The Book of Numbers.* New York: Springer-Verlag, pp. 130–132, 1996.

Courant, R. and Robbins, H. "Congruences." §2 in Supplement to Ch. 1 in *What is Mathematics?: An Elementary Approach to Ideas and Methods, 2nd ed.* Oxford, England: Oxford University Press, pp. 31–40, 1996.

Shanks, D. *Solved and Unsolved Problems in Number Theory, 4th ed.* New York: Chelsea, p. 55, 1993.

Weisstein, E. W. "Fractional Congruences." http://www.astro.virginia.edu/~eww6n/math/notebooks/ModFraction.m.

Congruence Axioms

The five of HILBERT'S AXIOMS which concern geometric equivalence.

see also CONGRUENCE AXIOMS, CONTINUITY AXIOMS, HILBERT'S AXIOMS, INCIDENCE AXIOMS, ORDERING AXIOMS, PARALLEL POSTULATE

References

Hilbert, D. *The Foundations of Geometry, 2nd ed.* Chicago, IL: Open Court, 1980.

Iyanaga, S. and Kawada, Y. (Eds.). "Hilbert's System of Axioms." §163B in *Encyclopedic Dictionary of Mathematics.* Cambridge, MA: MIT Press, pp. 544–545, 1980.

Congruence (Geometric)

Two geometric figures are said to be congruent if they are equivalent to within a ROTATION. This relationship is written $A \cong B$. (Unfortunately, this symbol is also used to denote ISOMORPHIC GROUPS.)

see also SIMILAR

Congruence Transformation

A transformation of the form $g = \mathsf{D}^{\mathrm{T}}\eta\mathsf{D}$, where $\det(\mathsf{D}) \neq 0$ and $\det(\mathsf{D})$ is the DETERMINANT.

see also SYLVESTER'S INERTIA LAW

Congruent

A number a is said to be congruent to b modulo m if $m|a-b$ (m DIVIDES $a-b$).

Congruent Incircles Point

The point Y for which TRIANGLES BYC, CYA, and AYB have congruent INCIRCLES. It is a special case of an ELKIES POINT.

References

Kimberling, C. "Central Points and Central Lines in the Plane of a Triangle." *Math. Mag.* **67**, 163–187, 1994.

Congruent Isoscelizers Point

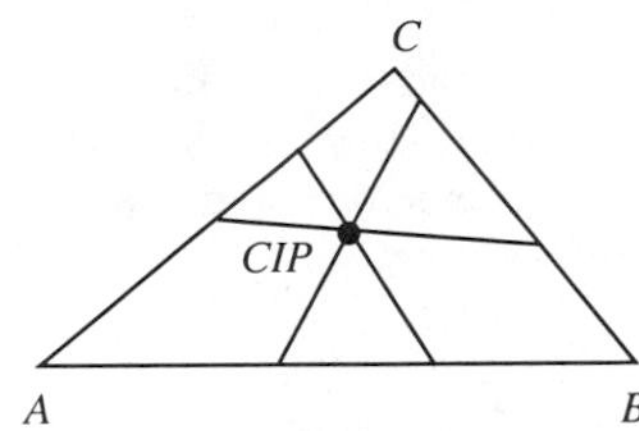

In 1989, P. Yff proved there is a unique configuration of ISOSCELIZERS for a given TRIANGLE such that all three have the same length. Furthermore, these ISOSCELIZERS meet in a point called the congruent isoscelizers point, which has TRIANGLE CENTER FUNCTION

$$\alpha = \cos(\tfrac{1}{2}B) + \cos(\tfrac{1}{2}C) - \cos(\tfrac{1}{2}A).$$

see also CONGRUENT ISOSCELIZERS POINT, ISOSCELIZER

References

Kimberling, C. "Congruent Isoscelizers Point." http://www.evansville.edu/~ck6/tcenters/recent/conisos.html.

Congruent Numbers

A set of numbers (a, x, y, t) such that

$$\begin{cases} x^2 + ay^2 = z^2 \\ x^2 - ay^2 = t^2. \end{cases}$$

They are a generalization of the CONGRUUM PROBLEM, which is the case $y = 1$. For $a = 101$, the smallest solution is

$$\begin{aligned} x &= 2015242462949760001961 \\ y &= 118171431852779451900 \\ z &= 2339148435306225006961 \\ t &= 1628124370727269996961. \end{aligned}$$

see also CONGRUUM

References

Guy, R. K. "Congruent Number." §D76 in *Unsolved Problems in Number Theory, 2nd ed.* New York: Springer-Verlag, pp. 195–197, 1994.

Congruum

A number h which satisfies the conditions of the CONGRUUM PROBLEM:

$$x^2 + h = a^2$$

and

$$x^2 - h = b^2.$$

see also CONCORDANT FORM, CONGRUUM PROBLEM

Congruum Problem

Find a SQUARE NUMBER x^2 such that, when a given number h is added or subtracted, new SQUARE NUMBERS are obtained so that

$$x^2 + h = a^2 \tag{1}$$

and

$$x^2 - h = b^2. \tag{2}$$

This problem was posed by the mathematicians Théodore and Jean de Palerma in a mathematical tournament organized by Frederick II in Pisa in 1225. The solution (Ore 1988, pp. 188–191) is

$$x = m^2 + n^2 \tag{3}$$

$$h = 4mn(m^2 - n^2), \tag{4}$$

where m and n are INTEGERS. Fibonacci proved that all numbers h (the CONGRUA) are divisible by 24. FERMAT'S RIGHT TRIANGLE THEOREM is equivalent to the result that a congruum cannot be a SQUARE NUMBER. A table for small m and n is given in Ore (1988, p. 191), and a larger one (for $h \leq 1000$) by Lagrange (1977).

m	n	h	x
2	1	24	5
3	1	96	10
3	2	120	13
4	1	240	17
4	3	336	25

see also CONCORDANT FORM, CONGRUENT NUMBERS, SQUARE NUMBER

References

Alter, R. and Curtz, T. B. "A Note on Congruent Numbers." *Math. Comput.* **28**, 303–305, 1974.

Alter, R.; Curtz, T. B.; and Kubota, K. K. "Remarks and Results on Congruent Numbers." In *Proc. Third Southeastern Conference on Combinatorics, Graph Theory, and Computing, 1972, Boca Raton, FL.* Boca Raton, FL: Florida Atlantic University, pp. 27–35, 1972.

Bastien, L. "Nombres congruents." *Interméd. des Math.* **22**, 231–232, 1915.

Gérardin, A. "Nombres congruents." *Interméd. des Math.* **22**, 52–53, 1915.

Lagrange, J. "Construction d'une table de nombres congruents." *Calculateurs en Math., Bull. Soc. math. France.*, Mémoire 49–50, 125–130, 1977.

Ore, Ø. *Number Theory and Its History.* New York: Dover, 1988.

Conic

see CONIC SECTION

Conic Constant

$$K \equiv -e^2,$$

where e is the ECCENTRICITY of a CONIC SECTION.

see also CONIC SECTION, ECCENTRICITY

Conic Double Point

see ISOLATED SINGULARITY

Conic Equidistant Projection

A MAP PROJECTION with transformation equations

$$x = \rho \sin\theta \tag{1}$$

$$y = \rho_0 - \rho\cos\theta, \tag{2}$$

where

$$\rho = (G - \phi) \tag{3}$$

$$\theta = n(\lambda - \lambda_0) \tag{4}$$

$$\rho_0 = (G - \theta_0) \tag{5}$$

$$G = \frac{\cos\phi_1}{n} + \phi_1 \tag{6}$$

$$n = \frac{\cos\phi_1 - \cos\phi_2}{\phi_2 - \phi_1}. \tag{7}$$

The inverse FORMULAS are given by

$$\phi = G - \rho \tag{8}$$

$$\lambda = \lambda_0 + \frac{\theta}{n}, \tag{9}$$

where

$$\rho = \operatorname{sgn}(n)\sqrt{x^2 + (\rho_0 - y)^2} \tag{10}$$

$$\theta = \tan^{-1}\left(\frac{x}{\rho_0 - y}\right). \tag{11}$$

Conic Projection

see ALBERS EQUAL-AREA CONIC PROJECTION, CONIC EQUIDISTANT PROJECTION, LAMBERT AZIMUTHAL EQUAL-AREA PROJECTION, POLYCONIC PROJECTION

Conic Section

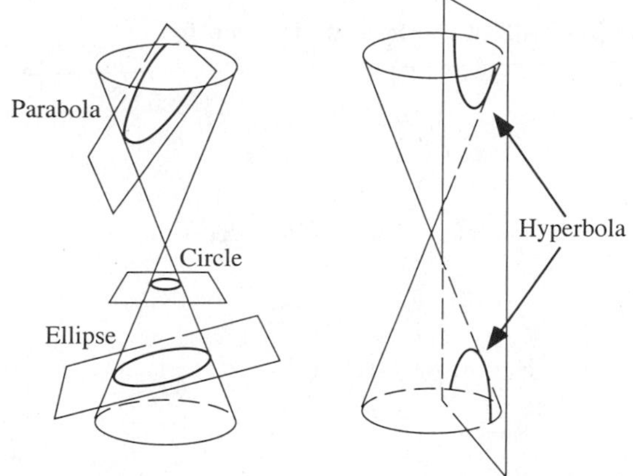

The conic sections are the nondegenerate curves generated by the intersections of a PLANE with one or two NAPPES of a CONE. For a PLANE parallel to a CROSS-SECTION, a CIRCLE is produced. The closed curve produced by the intersection of a single NAPPE with an inclined PLANE is an ELLIPSE or PARABOLA. The curve produced by a PLANE intersecting both NAPPES is a HYPERBOLA. The ELLIPSE and HYPERBOLA are known as CENTRAL CONICS.

Because of this simple geometric interpretation, the conic sections were studied by the Greeks long before their application to inverse square law orbits was known. Apollonius wrote the classic ancient work on the subject entitled *On Conics*. Kepler was the first to notice that planetary orbits were ELLIPSES, and Newton was then able to derive the shape of orbits mathematically using CALCULUS, under the assumption that gravitational force goes as the inverse square of distance. Depending on the energy of the orbiting body, orbit shapes which are any of the four types of conic sections are possible.

A conic section may more formally be defined as the locus of a point P that moves in the PLANE of a fixed point F called the FOCUS and a fixed line d called the DIRECTRIX (with F not on d) such that the ratio of the distance of P from F to its distance from d is a constant e called the ECCENTRICITY. For a FOCUS $(0,0)$ and DIRECTRIX $x = -a$, the equation is

$$x^2 + y^2 = e^2(x+a)^2.$$

If $e = 1$, the conic is a PARABOLA, if $e < 1$, the conic is an ELLIPSE, and if $e > 1$, it is a HYPERBOLA.

In standard form, a conic section is written

$$y^2 = 2Rx - (1 - e^2)x^2,$$

where R is the RADIUS OF CURVATURE and e is the ECCENTRICITY. Five points in a plane determine a conic (Le Lionnais 1983, p. 56).

see also BRIANCHON'S THEOREM, CENTRAL CONIC, CIRCLE, CONE, ECCENTRICITY, ELLIPSE, FERMAT CONIC, HYPERBOLA, NAPPE, PARABOLA, PASCAL'S THEOREM, QUADRATIC CURVE, SEYDEWITZ'S THEOREM, SKEW CONIC, STEINER'S THEOREM

References

Besant, W. H. *Conic Sections, Treated Geometrically, 8th ed. rev.* Cambridge, England: Deighton, Bell, 1890.

Casey, J. "Special Relations of Conic Sections" and "Invariant Theory of Conics." Chs. 9 and 15 in *A Treatise on the Analytical Geometry of the Point, Line, Circle, and Conic Sections, Containing an Account of Its Most Recent Extensions, with Numerous Examples, 2nd ed., rev. enl.* Dublin: Hodges, Figgis, & Co., pp. 307–332 and 462–545, 1893.

Coolidge, J. L. *A History of the Conic Sections and Quadric Surfaces.* New York: Dover, 1968.

Coxeter, H. S. M. and Greitzer, S. L. *Geometry Revisited.* Washington, DC: Math. Assoc. Amer., pp. 138–141, 1967.

Downs, J. W. *Conic Sections.* Dale Seymour Pub., 1993.

Iyanaga, S. and Kawada, Y. (Eds.). "Conic Sections." §80 in *Encyclopedic Dictionary of Mathematics.* Cambridge, MA: MIT Press, pp. 271–276, 1980.

Le Lionnais, F. *Les nombres remarquables.* Paris: Hermann, p. 56, 1983.

Lee, X. "Conic Sections." `http://www.best.com/~xah/SpecialPlaneCurves_dir/ConicSections_dir/conicSections.html`.

Ogilvy, C. S. "The Conic Sections." Ch. 6 in *Excursions in Geometry.* New York: Dover, pp. 73–85, 1990.

Pappas, T. "Conic Sections." *The Joy of Mathematics.* San Carlos, CA: Wide World Publ./Tetra, pp. 196–197, 1989.

Salmon, G. *Conic Sections, 6th ed.* New York: Chelsea, 1954.

Smith, C. *Geometric Conics.* London: MacMillan, 1894.

Sommerville, D. M. Y. *Analytical Conics, 3rd ed.* London: G. Bell and Sons, 1961.

Yates, R. C. "Conics." *A Handbook on Curves and Their Properties.* Ann Arbor, MI: J. W. Edwards, pp. 36–56, 1952.

Conic Section Tangent

Given a CONIC SECTION

$$x^2 + y^2 + 2gx + 2fy + c = 0,$$

the tangent at (x_1, y_1) is given by the equation

$$xx_1 + yy_1 + g(x + x_1) + f(y + y_1) + c = 0.$$

Conical Coordinates

Arfken (1970) and Morse and Feshbach (1953) use slightly different definitions of these coordinates. The system used in *Mathematica*® (Wolfram Research, Inc., Champaign, Illinois) is

$$x = \frac{\lambda\mu\nu}{ab} \tag{1}$$

$$y = \frac{\lambda}{a}\sqrt{\frac{(\mu^2-a^2)(\nu^2-a^2)}{a^2-b^2}} \tag{2}$$

$$z = \frac{\lambda}{b}\sqrt{\frac{(\mu^2-b^2)(\nu^2-b^2)}{b^2-a^2}}, \tag{3}$$

where $b^2 > \mu^2 > c^2 > \nu^2$. The NOTATION of Byerly replaces λ with r, and a and b with b and c. The above equations give

$$x^2+y^2+z^2=\lambda^2 \tag{4}$$

$$\frac{x^2}{\mu^2}+\frac{y^2}{\mu^2-a^2}+\frac{z^2}{\mu^2-b^2}=0 \tag{5}$$

$$\frac{x^2}{\nu^2}+\frac{y^2}{\nu^2-a^2}+\frac{z^2}{\nu^2-b^2}=0. \tag{6}$$

The SCALE FACTORS are

$$h_\lambda = 1 \tag{7}$$

$$h_\mu = \sqrt{\frac{\lambda^2(\mu^2-\nu^2)}{(\mu^2-a^2)(b^2-\mu^2)}} \tag{8}$$

$$h_\nu = \sqrt{\frac{\lambda^2(\mu^2-\nu^2)}{(\nu^2-a^2)(\nu^2-b^2)}}. \tag{9}$$

The LAPLACIAN is

$$\begin{aligned}\nabla^2 = {} & \frac{\nu(2\nu^2-a^2-b^2)}{(\mu-\nu)(\mu+\nu)\lambda^2}\frac{\partial}{\partial\nu} \\ & +\frac{(a-\nu)(a+\nu)(\nu-b)(\nu+b)}{(\nu-\mu)(\nu+\mu)\lambda^2}\frac{\partial^2}{\partial\nu^2} \\ & +\frac{\mu(2\mu^2-a^2-b^2)}{(\nu-\mu)(\nu+\mu)\lambda^2}\frac{\partial}{\partial\mu} \\ & +\frac{(\mu-b)(\mu+b)(\mu-a)(\mu+a)}{(\nu-\mu)(\nu+\mu)\lambda^2}\frac{\partial^2}{\partial\mu^2} \\ & +\frac{2}{\lambda}\frac{\partial}{\partial\lambda}+\frac{\partial^2}{\partial\lambda^2}.\end{aligned} \tag{10}$$

The HELMHOLTZ DIFFERENTIAL EQUATION is separable in conical coordinates.

see also HELMHOLTZ DIFFERENTIAL EQUATION—CONICAL COORDINATES

References

Arfken, G. "Conical Coordinates (ξ_1, ξ_2, ξ_3)." §2.16 in *Mathematical Methods for Physicists, 2nd ed.* Orlando, FL: Academic Press, pp. 118–119, 1970.

Byerly, W. E. *An Elementary Treatise on Fourier's Series, and Spherical, Cylindrical, and Ellipsoidal Harmonics, with Applications to Problems in Mathematical Physics.* New York: Dover, p. 263, 1959.

Morse, P. M. and Feshbach, H. *Methods of Theoretical Physics, Part I.* New York: McGraw-Hill, p. 659, 1953.

Spence, R. D. "Angular Momentum in Sphero-Conal Coordinates." *Amer. J. Phys.* **27**, 329–335, 1959.

Conical Frustum

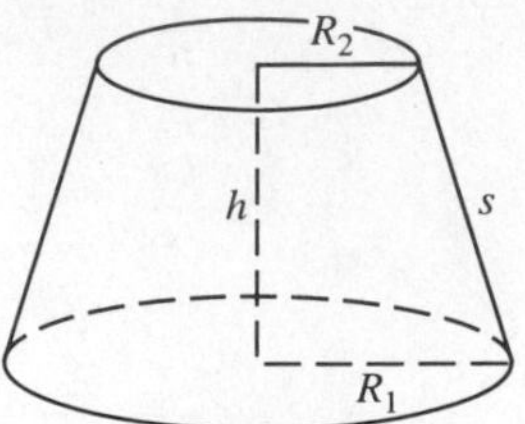

A conical frustum is a FRUSTUM created by slicing the top off a CONE (with the cut made parallel to the base). For a right circular CONE, let s be the slant height and R_1 and R_2 the top and bottom RADII. Then

$$s = \sqrt{(R_1-R_2)^2+h^2}. \tag{1}$$

The SURFACE AREA, not including the top and bottom CIRCLES, is

$$A = \pi(R_1+R_2)s = \pi(R_1+R_2)\sqrt{(R_1-R_2)^2+h^2}. \tag{2}$$

The VOLUME of the frustum is given by

$$V = \pi\int_0^h [r(z)]^2\,dz. \tag{3}$$

But

$$r(z) = R_1+(R_2-R_1)\frac{z}{h}, \tag{4}$$

so

$$\begin{aligned} V &= \pi\int_0^h \left[R_1+(R_2-R_1)\frac{z}{h}\right]^2 dz \\ &= \tfrac{1}{3}\pi h({R_1}^2+R_1R_2+{R_2}^2). \end{aligned} \tag{5}$$

This formula can be generalized to any PYRAMID by letting A_i be the base AREAS of the top and bottom of the frustum. Then the VOLUME can be written as

$$V = \tfrac{1}{3}h(A_1+A_2+\sqrt{A_1A_2}\,). \tag{6}$$

The weighted mean of z over the frustum is

$$\langle z\rangle = \pi\int_0^h z[r(z)]^2\,dz = \tfrac{1}{12}h^2({R_1}^2+2R_1R_2+3{R_2}^2). \tag{7}$$

The CENTROID is then given by

$$\bar{z} = \frac{\langle z\rangle}{V} = \frac{h({R_1}^2+R_1R_2+{R_2}^2)}{4({R_1}^2+2R_1R_2+3{R_2}^2)} \tag{8}$$

(Beyer 1987, p. 133). The special case of the CONE is given by taking $R_2=0$, yielding $\bar{z}=h/4$.

see also CONE, FRUSTUM, PYRAMIDAL FRUSTUM, SPHERICAL SEGMENT

References

Beyer, W. H. (Ed.) *CRC Standard Mathematical Tables, 28th ed.* Boca Raton, FL: CRC Press, pp. 129–130 and 133, 1987.

Conical Function

Functions which can be expressed in terms of LEGENDRE FUNCTIONS OF THE FIRST and SECOND KINDS. See Abramowitz and Stegun (1972, p. 337).

$$\begin{aligned} P^{\mu}_{-1/2+ip}(\cos\theta) &= 1 + \frac{4p^2+1^2}{2^2}\sin^2(\tfrac{1}{2}\theta) \\ &\quad + \frac{(4p^2+1^2)(4p^2+3^2)}{2^2 4^2}\sin^4(\tfrac{1}{2}\theta) + \ldots \\ &= \frac{2}{\pi}\int_0^\theta \frac{\cosh(pt)\,dt}{\sqrt{2(\cos t - \cos\theta)}} \\ Q^{\mu}_{-1/2\mp ip}(\cos\theta) &= \pm i\sinh(p\pi)\int_0^\infty \frac{\cos(pt)\,dt}{\sqrt{2(\cosh t + \cos\theta)}} \\ &\quad + \int_0^\infty \frac{\cosh(pt)\,dt}{\sqrt{2(\cos t - \cos\theta)}}. \end{aligned}$$

see also TOROIDAL FUNCTION

References

Abramowitz, M. and Stegun, C. A. (Eds.). "Conical Functions." §8.12 in *Handbook of Mathematical Functions with Formulas, Graphs, and Mathematical Tables, 9th printing.* New York: Dover, p. 337, 1972.

Iyanaga, S. and Kawada, Y. (Eds.). *Encyclopedic Dictionary of Mathematics.* Cambridge, MA: MIT Press, p. 1464, 1980.

Conical Spiral

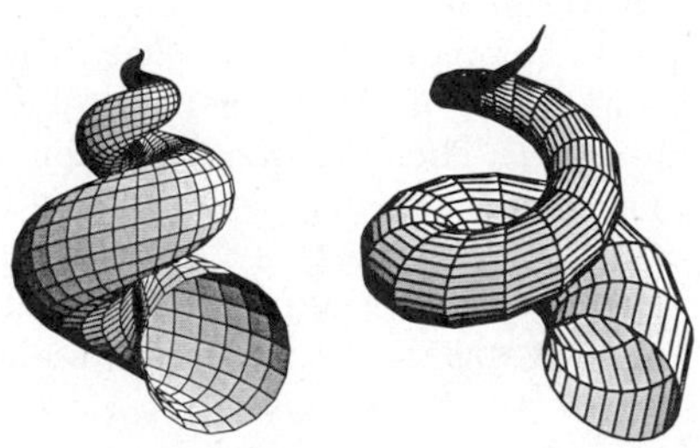

A surface modeled after the shape of a SEASHELL. One parameterization (left figure) is given by

$$x = 2[1 - e^{u/(6\pi)}]\cos u \cos^2(\tfrac{1}{2}v) \quad (1)$$

$$y = 2[-1 + e^{u/(6\pi)}]\cos^2(\tfrac{1}{2}v)\sin u \quad (2)$$

$$z = 1 - e^{u/(3\pi)} - \sin v + e^{u/(6\pi)}\sin v, \quad (3)$$

where $v \in [0, 2\pi)$, and $u \in [0, 6\pi)$ (Wolfram). Nordstrand gives the parameterization

$$x = \left[\left(1 - \frac{v}{2\pi}\right)(1+\cos u) + c\right]\cos(nv) \quad (4)$$

$$x = \left[\left(1 - \frac{v}{2\pi}\right)(1+\cos u) + c\right]\sin(nv) \quad (5)$$

$$z = \frac{bv}{2\pi} + a\sin u\left(1 - \frac{v}{2\pi}\right) \quad (6)$$

for $u, v \in [0, 2\pi]$ (right figure with $a = 0.2$, $b = 1$, $c = 0.1$, and $n = 2$).

References

Gray, A. "Sea Shells." §11.6 in *Modern Differential Geometry of Curves and Surfaces.* Boca Raton, FL: CRC Press, pp. 223–223, 1993.

Nordstrand, T. "Conic Spiral or Seashell." http://www.uib.no/people/nfytn/shelltxt.htm.

Wolfram Research "Mathematica Version 2.0 Graphics Gallery." http://www.mathsource.com/cgi-bin/MathSource/Applications/Graphics/3D/0207-155.

Conical Wedge

The SURFACE also called the CONOCUNEUS OF WALLIS and given by the parametric equation

$$\begin{aligned} x &= u\cos v \\ y &= u\sin v \\ z &= c(1 - 2\cos^2 v). \end{aligned}$$

see also CYLINDRICAL WEDGE, WEDGE

References

von Seggern, D. *CRC Standard Curves and Surfaces.* Boca Raton, FL: CRC Press, p. 302, 1993.

Conjecture

A proposition which is consistent with known data, but has neither been verified nor shown to be false. It is synonymous with HYPOTHESIS.

see also ABC CONJECTURE, ABHYANKAR'S CONJECTURE, ABLOWITZ-RAMANI-SEGUR CONJECTURE, ANDRICA'S CONJECTURE, ANNULUS CONJECTURE, ARGOH'S CONJECTURE, ARTIN'S CONJECTURE, AXIOM, BACHET'S CONJECTURE, BENNEQUIN'S CONJECTURE, BIEBERBACH CONJECTURE, BIRCH CONJECTURE, BLASCHKE CONJECTURE, BORSUK'S CONJECTURE, BORWEIN CONJECTURES, BRAUN'S CONJECTURE, BROCARD'S CONJECTURE, BURNSIDE'S CONJECTURE, CARMICHAEL'S CONJECTURE, CATALAN'S CONJECTURE, CRAMÉR CONJECTURE, DE POLIGNAC'S CONJECTURE, DIESIS, DODECAHEDRAL CONJECTURE, DOUBLE BUBBLE CONJECTURE, EBERHART'S CONJECTURE, EULER'S CONJECTURE, EULER POWER CONJECTURE, EULER QUARTIC CONJECTURE, FEIT-THOMPSON CONJECTURE, FERMAT'S CONJECTURE, FLYPING CONJECTURE, GILBREATH'S CONJECTURE, GIUGA'S CONJECTURE, GOLDBACH CONJECTURE, GRIMM'S CONJECTURE, GUY'S CONJECTURE, HARDY-LITTLEWOOD CONJECTURES, HASSE'S CONJECTURE, HEAWOOD CONJECTURE, HYPOTHESIS, JACOBIAN CONJECTURE, KAPLAN-YORKE CONJECTURE, KELLER'S CONJECTURE, KELVIN'S CONJECTURE, KEPLER CONJECTURE, KREISEL CONJECTURE, KUMMER'S CONJECTURE, LEMMA, LOCAL DENSITY CONJECTURE, MERTENS CONJECTURE, MILIN CONJECTURE, MILNOR'S CONJECTURE, MORDELL CONJECTURE, NETTO'S CONJECTURE, NIRENBERG'S CONJECTURE, ORE'S CONJECTURE, PADÉ CONJECTURE,

PALINDROMIC NUMBER CONJECTURE, PILLAI'S CONJECTURE, POINCARÉ CONJECTURE, PÓLYA CONJECTURE, PORISM, PRIME k-TUPLES CONJECTURE, PRIME PATTERNS CONJECTURE, PRIME POWER CONJECTURE, PROOF, QUILLEN-LICHTENBAUM CONJECTURE, RAMANUJAN-PETERSSON CONJECTURE, ROBERTSON CONJECTURE, SAFAREVICH CONJECTURE, SAUSAGE CONJECTURE, SCHANUEL'S CONJECTURE, SCHISMA, SCHOLZ CONJECTURE, SEIFERT CONJECTURE, SELFRIDGE'S CONJECTURE, SHANKS' CONJECTURE, SMITH CONJECTURE, SWINNERTON-DYER CONJECTURE, SZPIRO'S CONJECTURE, TAIT'S HAMILTONIAN GRAPH CONJECTURE, TAIT'S KNOT CONJECTURES, TANIYAMA-SHIMURA CONJECTURE, TAU CONJECTURE, THEOREM, THURSTON'S GEOMETRIZATION CONJECTURE, THWAITES CONJECTURE, VOJTA'S CONJECTURE, WANG'S CONJECTURE, WARING'S PRIME CONJECTURE, WARING'S SUM CONJECTURE, ZARANKIEWICZ'S CONJECTURE

References

Rivera, C. "Problems & Puzzles (Conjectures)." http://www.sci.net.mx/~crivera/ppp/conjectures.htm.

Conjugacy Class

A complete set of mutually conjugate GROUP elements. Each element in a GROUP belongs to exactly one class, and the identity ($I = 1$) element is always in its own class. The ORDERS of all classes must be integral FACTORS of the ORDER of the GROUP. From the last two statements, a GROUP of PRIME order has one class for each element. More generally, in an ABELIAN GROUP, each element is in a conjugacy class by itself. Two operations belong to the same class when one may be replaced by the other in a new COORDINATE SYSTEM which is accessible by a symmetry operation (Cotton 1990, p. 52). These sets correspond directly to the sets of equivalent operation.

Let G be a FINITE GROUP of ORDER $|G|$. If $|G|$ is ODD, then

$$|G| \equiv s \pmod{16}$$

(Burnside 1955, p. 295). Furthermore, if every PRIME p_i DIVIDING $|G|$ satisfies $p_i \equiv 1 \pmod 4$, then

$$|G| \equiv s \pmod{32}$$

(Burnside 1955, p. 320). Poonen (1995) showed that if every PRIME p_i DIVIDING $|G|$ satisfies $p_i \equiv 1 \pmod m$ for $m \geq 2$, then

$$|G| \equiv s \left(\text{mod } 2m^2\right).$$

References

Burnside, W. *Theory of Groups of Finite Order, 2nd ed.* New York: Dover, 1955.

Cotton, F. A. *Chemical Applications of Group Theory, 3rd ed.* New York: Wiley, 1990.

Poonen, B. "Congruences Relating the Order of a Group to the Number of Conjugacy Classes." *Amer. Math. Monthly* **102**, 440–442, 1995.

Conjugate Element

Given a GROUP with elements A and X, there must be an element B which is a SIMILARITY TRANSFORMATION of $A, B = X^{-1}AX$ so A and B are conjugate with respect to X. Conjugate elements have the following properties:

1. Every element is conjugate with itself.
2. If A is conjugate with B with respect to X, then B is conjugate to A with respect to X.
3. If A is conjugate with B and C, then B and C are conjugate with each other.

see also CONJUGACY CLASS, CONJUGATE SUBGROUP

Conjugate Gradient Method

An ALGORITHM for calculating the GRADIENT $\nabla f(\mathbf{P})$ of a function at an n-D point $\mathbf{P}$. It is more robust than the simpler STEEPEST DESCENT METHOD.

References

Press, W. H.; Flannery, B. P.; Teukolsky, S. A.; and Vetterling, W. T. *Numerical Recipes in FORTRAN: The Art of Scientific Computing, 2nd ed.* Cambridge, England: Cambridge University Press, pp. 413–417, 1992.

Conjugate Points

see HARMONIC CONJUGATE POINTS, ISOGONAL CONJUGATE, ISOTOMIC CONJUGATE POINT

Conjugate Subgroup

A SUBGROUP H of an original GROUP G has elements h_i. Let x be a fixed element of the original GROUP G which is not a member of H. Then the transformation xh_ix^{-1}, ($i = 1, 2, \ldots$) generates a conjugate SUBGROUP xHx^{-1}. If, for all x, $xHx^{-1} = H$, then H is a SELF-CONJUGATE (also called INVARIANT or NORMAL) SUBGROUP. All SUBGROUPS of an ABELIAN GROUP are invariant.

Conjugation

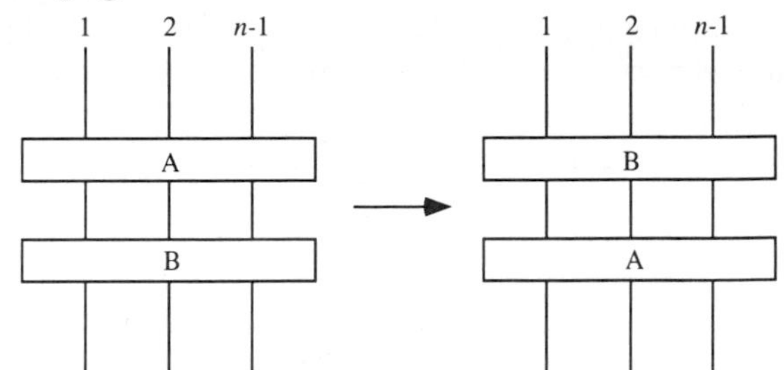

A type I MARKOV MOVE.

see also MARKOV MOVES, STABILIZATION

Conjunction

A product of ANDs, denoted

$$\bigwedge_{k+1}^{n} A_k.$$

see also AND, DISJUNCTION

Connected Graph

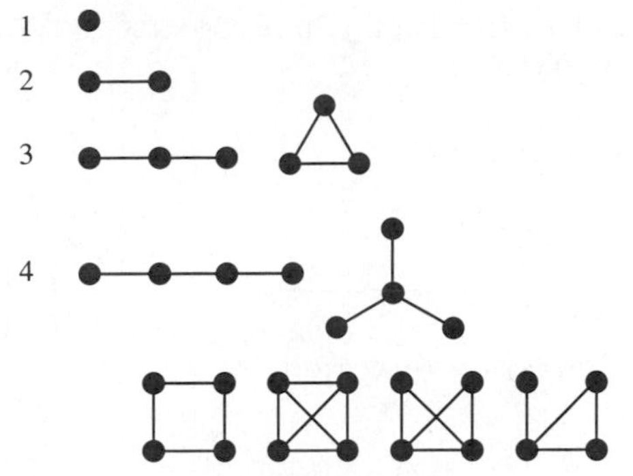

A GRAPH which is connected (as a TOPOLOGICAL SPACE), i.e., there is a path from any point to any other point in the GRAPH. The number of n-VERTEX (unlabeled) connected graphs for $n = 1, 2, \ldots$ are 1, 1, 2, 6, 21, 112, 853, 11117, ... (Sloane's A001349).

References
Chartrand, G. "Connected Graphs." §2.3 in *Introductory Graph Theory*. New York: Dover, pp. 41–45, 1985.
Sloane, N. J. A. Sequence A001349/M1657 in "An On-Line Version of the Encyclopedia of Integer Sequences."

Connected Set

A connected set is a SET which cannot be partitioned into two nonempty SUBSETS which are open in the relative topology induced on the SET. Equivalently, it is a SET which cannot be partitioned into two nonempty SUBSETS such that each SUBSET has no points in common with the closure of the other.

The REAL NUMBERS are a connected set.

see also CLOSED SET, EMPTY SET, OPEN SET, SET, SUBSET

Connected Space

A SPACE D is connected if any two points in D can be connected by a curve lying wholly within D. A SPACE is 0-connected (a.k.a. PATHWISE-CONNECTED) if every MAP from a 0-SPHERE to the SPACE extends continuously to the 1-DISK. Since the 0-SPHERE is the two endpoints of an interval (1-DISK), every two points have a path between them. A space is 1-connected (a.k.a. SIMPLY CONNECTED) if it is 0-connected and if every MAP from the 1-SPHERE to it extends continuously to a MAP from the 2-DISK. In other words, every loop in the SPACE is contractible. A SPACE is n-MULTIPLY CONNECTED if it is $(n-1)$-connected and if every MAP from the n-SPHERE into it extends continuously over the $(n+1)$-DISK.

A theorem of Whitehead says that a SPACE is infinitely connected IFF it is contractible.

see also CONNECTIVITY, LOCALLY PATHWISE-CONNECTED SPACE, MULTIPLY CONNECTED, PATHWISE-CONNECTED, SIMPLY CONNECTED

Connected Sum

The connected sum $M_1\#M_2$ of n-manifolds M_1 and M_2 is formed by deleting the interiors of n-BALLS B_i^n in M_i^n and attaching the resulting punctured MANIFOLDS $M_i - \dot{B}_i$ to each other by a HOMEOMORPHISM $h: \partial B_2 \to \partial B_1$, so

$$M_1\#M_2 = (M_1 - \dot{B}_1)\bigcup_h (M_2 - \dot{B}_2).$$

B_i is required to be interior to M_i and ∂B_i bicollared in M_i to ensure that the connected sum is a MANIFOLD.

The connected sum of two KNOTS is called a KNOT SUM.

see also KNOT SUM

References
Rolfsen, D. *Knots and Links.* Wilmington, DE: Publish or Perish Press, p. 39, 1976.

Connected Sum Decomposition

Every COMPACT 3-MANIFOLD is the CONNECTED SUM of a unique collection of PRIME 3-MANIFOLDS.

see also JACO-SHALEN-JOHANNSON TORUS DECOMPOSITION

Connection

see CONNECTION COEFFICIENT, GAUSS-MANIN CONNECTION

Connection Coefficient

A quantity also known as a CHRISTOFFEL SYMBOL OF THE SECOND KIND. Connection COEFFICIENTS are defined by

$$\Gamma^{\vec{e}_\alpha}_{\vec{e}_\beta \vec{e}_\gamma} \equiv \vec{e}^{\,\alpha} \cdot (\nabla_{\vec{e}_\gamma} \vec{e}_\beta) \tag{1}$$

(long form) or

$$\Gamma^\alpha_{\beta\gamma} \equiv \vec{e}^{\,\alpha} \cdot (\nabla_\gamma \vec{e}_\beta), \tag{2}$$

(abbreviated form), and satisfy

$$\nabla_{\vec{e}_\gamma} \vec{e}_\beta = \Gamma^{\vec{e}_\alpha}_{\vec{e}_\beta \vec{e}_\gamma} \vec{e}_\alpha \tag{3}$$

(long form) and

$$\nabla_\gamma \vec{e}_\beta = \Gamma^\alpha_{\beta\gamma} \vec{e}_\alpha \tag{4}$$

(abbreviated form).

Connection COEFFICIENTS are not TENSORS, but have TENSOR-like CONTRAVARIANT and COVARIANT indices. A fully COVARIANT connection COEFFICIENT is given by

$$\Gamma_{\alpha\beta\gamma} \equiv \tfrac{1}{2}(g_{\alpha\beta,\gamma} + g_{\alpha\gamma,\beta} + c_{\alpha\beta\gamma} + c_{\alpha\gamma\beta} - c_{\beta\gamma\alpha}), \tag{5}$$

where the gs are the METRIC TENSORS, the cs are COMMUTATION COEFFICIENTS, and the commas indicate the

COMMA DERIVATIVE. In an ORTHONORMAL BASIS, $g_{\alpha\beta,\gamma} = 0$ and $g_{\mu\gamma} = \delta_{\mu\gamma}$, so

$$\Gamma_{\alpha\beta\gamma} = \Gamma^{\mu}_{\alpha\beta} g_{\mu\gamma} = \Gamma_{\alpha\beta}{}^{\mu} = \tfrac{1}{2}(c_{\alpha\beta\gamma} + c_{\alpha\gamma\beta} - c_{\beta\gamma\alpha}) \quad (6)$$

and

$$\Gamma_{ijk} = 0 \quad \text{for } i \neq j \neq k \quad (7)$$

$$\Gamma_{iik} = -\frac{1}{2}\frac{\partial g_{ii}}{\partial x^k} \quad \text{for } i \neq k \quad (8)$$

$$\Gamma_{iji} = \Gamma_{jii} = \frac{1}{2}\frac{\partial g_{ii}}{\partial x^j} \quad (9)$$

$$\Gamma^k_{ij} = 0 \quad \text{for } i \neq j \neq k \quad (10)$$

$$\Gamma^k_{ii} = -\frac{1}{2g_{kk}}\frac{\partial g_{ii}}{\partial x^k} \quad \text{for } i \neq k \quad (11)$$

$$\Gamma^i_{ij} = \Gamma^i_{ji} = \frac{1}{2g_{ii}}\frac{\partial g_{ii}}{\partial x^j} = \frac{1}{2}\frac{\partial \ln g_{ii}}{\partial x^j}. \quad (12)$$

For TENSORS of RANK 3, the connection COEFFICIENTS may be concisely summarized in MATRIX form:

$$\Gamma^\theta \equiv \begin{bmatrix} \Gamma^\theta_{rr} & \Gamma^\theta_{r\theta} & \Gamma^\theta_{r\phi} \\ \Gamma^\theta_{\theta r} & \Gamma^\theta_{\theta\theta} & \Gamma^\theta_{\theta\phi} \\ \Gamma^\theta_{\phi r} & \Gamma^\theta_{\phi\theta} & \Gamma^\theta_{\phi\phi} \end{bmatrix}. \quad (13)$$

Connection COEFFICIENTS arise in the computation of GEODESICS. The GEODESIC EQUATION of free motion is

$$d\tau^2 = -\eta_{\alpha\beta}\, d\xi^\alpha\, d\xi^\beta, \quad (14)$$

or

$$\frac{d^2\xi^\alpha}{d\tau^2} = 0. \quad (15)$$

Expanding,

$$\frac{d}{d\tau}\left(\frac{\partial \xi^\alpha}{\partial x^\mu}\frac{dx^\mu}{d\tau}\right) = \frac{\partial \xi^\alpha}{\partial x^\mu}\frac{d^2x^\mu}{d\tau^2} + \frac{\partial^2 \xi^\alpha}{\partial x^\mu \partial x^\nu}\frac{dx^\mu}{d\tau}\frac{dx^\nu}{d\tau} = 0 \quad (16)$$

$$\frac{\partial \xi^\alpha}{\partial x^\mu}\frac{d^2x^\mu}{d\tau^2}\frac{\partial x^\lambda}{\partial \xi^\alpha} + \frac{\partial^2 \xi^\alpha}{\partial x^\mu \partial x^\nu}\frac{dx^\mu}{d\tau}\frac{dx^\nu}{d\tau}\frac{\partial x^\lambda}{\partial \xi^\alpha} = 0. \quad (17)$$

But

$$\frac{\partial \xi^\alpha}{\partial x^\nu}\frac{\partial x^\lambda}{\partial \xi^\alpha} = \delta^\lambda_\mu, \quad (18)$$

so

$$\delta^\lambda_\mu \frac{d^2x^\mu}{d\tau^2} + \left[\frac{\partial^2 \xi^\alpha}{\partial x^\mu \partial x^\nu}\frac{\partial x^\lambda}{\partial \xi^\alpha}\right]\frac{dx^\mu}{d\tau}\frac{dx^\nu}{d\tau} = \frac{d^2x^\lambda}{d\tau^2} + \Gamma^\lambda_{\mu\nu}\frac{dx^\mu}{d\tau}\frac{dx^\nu}{d\tau}, \quad (19)$$

where

$$\Gamma^\lambda_{\mu\nu} \equiv \frac{\partial^2 \xi^\alpha}{\partial x^\mu \partial x^\nu}\frac{\partial x^\lambda}{\partial \xi^\alpha}. \quad (20)$$

see also CARTAN TORSION COEFFICIENT, CHRISTOFFEL SYMBOL OF THE FIRST KIND, CHRISTOFFEL SYMBOL OF THE SECOND KIND, COMMA DERIVATIVE, COMMUTATION COEFFICIENT, CURVILINEAR COORDINATES, SEMICOLON DERIVATIVE, TENSOR

Connectivity

see CONNECTED SPACE, EDGE CONNECTIVITY, VERTEX CONNECTIVITY

Connes Function

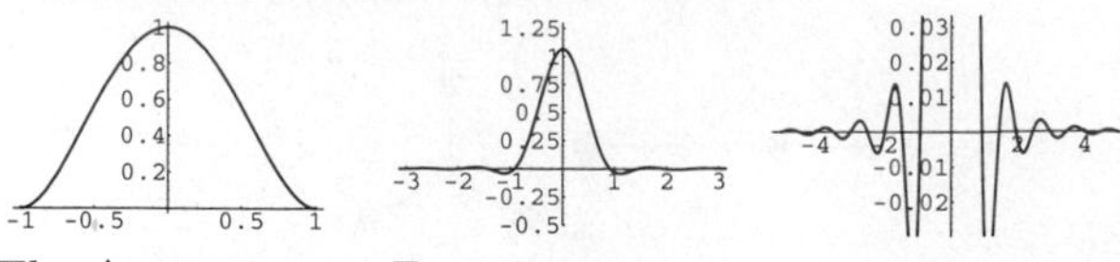

The APODIZATION FUNCTION

$$A(x) = \left(1 - \frac{x^2}{a^2}\right)^2.$$

Its FULL WIDTH AT HALF MAXIMUM is $\sqrt{4 - 2\sqrt{2}}\,a$, and its INSTRUMENT FUNCTION is

$$I(x) = 8a\sqrt{2\pi}\,\frac{J_{5/2}(2\pi ka)}{(2\pi ka)^{5/2}},$$

where $J_n(z)$ is a BESSEL FUNCTION OF THE FIRST KIND.

see also APODIZATION FUNCTION

Conocuneus of Wallis

see CONICAL WEDGE

Conoid

see PLÜCKER'S CONOID, RIGHT CONOID

Consecutive Number Sequences

Consecutive number sequences are sequences constructed by concatenating numbers of a given type. Many of these sequences were considered by Smarandache, so they are sometimes known as SMARANDACHE SEQUENCES.

The nth term of the consecutive integer sequence consists of the concatenation of the first n POSITIVE integers: 1, 12, 123, 1234, ... (Sloane's A007908; Smarandache 1993, Dumitrescu and Seleacu 1994, sequence 1; Mudge 1995; Stephen 1998). This sequence gives the digits of the CHAMPERNOWNE CONSTANT and contains no PRIMES in the first 4,470 terms (Weisstein). This is roughly consistent with simple arguments based on the distribution of prime which suggest that only a single prime is expected in the first 15,000 or so terms. The number of digits of the n term can be computed by noticing the pattern in the following table, where $d = \lfloor \log_{10} n \rfloor + 1$ is the number of digits in n.

d	n Range	Digits
1	1–9	n
2	10–99	$9 + 2(n-9)$
3	100–999	$9 + 90 \cdot 2 + 3(n-99)$
4	1000–9999	$9 + 90 \cdot 2 + 900 \cdot 3 + 4(n-999)$

Therefore, the number of digits $D(n)$ in the nth term can be written

$$D(n) = d(n+1-10^{d-1}) + \sum_{k=1}^{d-1} 9k \cdot 10^{k-1}$$
$$= (n+1)d - \frac{10^d - 1}{9},$$

where the second term is the REPUNIT R_d.

The nth term of the reverse integer sequence consists of the concatenation of the first n POSITIVE integers written backwards: 1, 21, 321, 4321, ... (Sloane's A000422; Smarandache 1993, Dumitrescu and Seleacu 1994, Stephen 1998). The only PRIME in the first 3,576 terms (Weisstein) of this sequence is the 82nd term 828180...321 (Stephen 1998), which has 155 digits. This is roughly consistent with simple arguments based on the distribution of prime which suggest that a single prime is expected in the first 15,000 or so terms. The terms of the reverse integer sequence have the same number of digits as do the consecutive integer sequence.

The concatenation of the first n PRIMES gives 2, 23, 235, 2357, 235711, ... (Sloane's A019518; Smith 1996, Mudge 1997). This sequence converges to the digits of the COPELAND-ERDŐS CONSTANT and is PRIME for terms 1, 2, 4, 128, 174, 342, 435, 1429, ... (Sloane's A046035; Ibstedt 1998, pp. 78–79), with no others less than 2,305 (Weisstein).

The concatenation of the first n ODD NUMBERS gives 1, 13, 135, 1357, 13579, ... (Sloane's A019519; Smith 1996, Marimutha 1997, Mudge 1997). This sequence is PRIME for terms 2, 10, 16, 34, 49, 2570, ... (Sloane's A046036; Weisstein, Ibstedt 1998, pp. 75–76), with no others less than 2,650 (Weisstein). The 2570th term, given by 1 3 5 7...5137 5139, has 9725 digits and was discovered by Weisstein in Aug. 1998.

The concatenation of the first n EVEN NUMBERS gives 2, 24, 246, 2468, 246810, ... (Sloane's A019520; Smith 1996; Marimutha 1997; Mudge 1997; Ibstedt 1998, pp. 77–78).

The concatenation of the first n SQUARE NUMBERS gives 1, 14, 149, 14916, ... (Sloane's A019521; Marimutha 1997). The only PRIME in the first 2,090 terms is the third term, 149, (Weisstein).

The concatenation of the first n CUBIC NUMBERS gives 1, 18, 1827, 182764, ... (Sloane's A019522; Marimutha 1997). There are no PRIMES in the first 1,830 terms (Weisstein).

see also CHAMPERNOWNE CONSTANT, CONCATENATION, COPELAND-ERDŐS CONSTANT, CUBIC NUMBER, DEMLO NUMBER, EVEN NUMBER, ODD NUMBER, SMARANDACHE SEQUENCES, SQUARE NUMBER

References
Dumitrescu, C. and Seleacu, V. (Ed.). *Some Notions and Questions in Number Theory.* Glendale, AZ: Erhus University Press, 1994.
Ibstedt, H. "Smarandache Concatenated Sequences." Ch. 5 in *Computer Analysis of Number Sequences.* Lupton, AZ: American Research Press, pp. 75–79, 1998.
Marimutha, H. "Smarandache Concatenate Type Sequences." *Bull. Pure Appl. Sci.* **16E**, 225–226, 1997.
Mudge, M. "Top of the Class." *Personal Computer World,* 674–675, June 1995.
Mudge, M. "Not Numerology but Numeralogy!" *Personal Computer World,* 279–280, 1997.
Smarandache, F. *Only Problems, Not Solutions!, 4th ed.* Phoenix, AZ: Xiquan, 1993.
Smith, S. "A Set of Conjectures on Smarandache Sequences." *Bull. Pure Appl. Sci.* **15E**, 101–107, 1996.
Stephen, R. W. "Factors and Primes in Two Smarandache Sequences." *Smarandache Notions J.* **9**, 4–10, 1998. http://www.tmt.de/~stephen/sm.ps.gz.

Conservation of Number Principle

A generalization of Poncelet's PERMANENCE OF MATHEMATICAL RELATIONS PRINCIPLE made by H. Schubert in 1874–79. The conservation of number principle asserts that the number of solutions of any determinate algebraic problem in any number of parameters under variation of the parameters is invariant in such a manner that no solutions become INFINITE. Schubert called the application of this technique the CALCULUS OF ENUMERATIVE GEOMETRY.

see also DUALITY PRINCIPLE, HILBERT'S PROBLEMS, PERMANENCE OF MATHEMATICAL RELATIONS PRINCIPLE

References
Bell, E. T. *The Development of Mathematics, 2nd ed.* New York: McGraw-Hill, p. 340, 1945.

Conservative Field

The following conditions are equivalent for a conservative VECTOR FIELD:

1. For any oriented simple closed curve C, the LINE INTEGRAL $\oint_C \mathbf{F} \cdot d\mathbf{s} = 0$.
2. For any two oriented simple curves C_1 and C_2 with the same endpoints, $\int_{C_1} \mathbf{F} \cdot d\mathbf{s} = \int_{C_2} \mathbf{F} \cdot d\mathbf{s}$.
3. There exists a SCALAR POTENTIAL FUNCTION f such that $\mathbf{F} = \nabla f$, where ∇ is the GRADIENT.
4. The CURL $\nabla \times \mathbf{F} = \mathbf{0}$.

see also CURL, GRADIENT, LINE INTEGRAL, POTENTIAL FUNCTION, VECTOR FIELD

Consistency

The absence of contradiction (i.e., the ability to prove that a statement and its NEGATIVE are both true) in an AXIOMATIC THEORY is known as consistency.

see also COMPLETE AXIOMATIC THEORY, CONSISTENCY STRENGTH

Consistency Strength

If the CONSISTENCY of one of two propositions implies the CONSISTENCY of the other, the first is said to have greater consistency strength.

Constant

Any REAL NUMBER which is "significant" (or interesting) in some way. In this work, the term "constant" is generally reserved for REAL nonintegral numbers of interest, while "NUMBER" is reserved for interesting INTEGERS (e.g., BRUN'S CONSTANT, but BEAST NUMBER).

Certain constants are known to many DECIMAL DIGITS and recur throughout many diverse areas of mathematics, often in unexpected and surprising places (e.g., PI, e, and to some extent, the EULER-MASCHERONI CONSTANT γ). Other constants are more specialized and may be known to only a few DIGITS. S. Plouffe maintains a site about the computation and identification of numerical constants. Plouffe's site also contains a page giving the largest number of DIGITS computed for the most common constants. S. Finch maintains a delightful, more expository site containing detailed essays and references on constants both common and obscure.

see also ABUNDANT NUMBER, ALLADI-GRINSTEAD CONSTANT, APÉRY'S CONSTANT, ARCHIMEDES' CONSTANT, ARTIN'S CONSTANT, BACKHOUSE'S CONSTANT, BERAHA CONSTANTS, BERNSTEIN'S CONSTANT, BLOCH CONSTANT, BRUN'S CONSTANT, CAMERON'S SUM-FREE SET CONSTANT, CARLSON-LEVIN CONSTANT, CATALAN'S CONSTANT, CHAITIN'S CONSTANT, CHAMPERNOWNE CONSTANT, CHEBYSHEV CONSTANTS, CHEBYSHEV-SYLVESTER CONSTANT, COMMA OF DIDYMUS, COMMA OF PYTHAGORAS, CONIC CONSTANT, CONSTANT FUNCTION, CONSTANT PROBLEM, CONTINUED FRACTION CONSTANT, CONWAY'S CONSTANT, COPELAND-ERDŐS CONSTANT, COPSON-DE BRUIJN CONSTANT, DE BRUIJN-NEWMAN CONSTANT, DELIAN CONSTANT, DIESIS, DU BOIS RAYMOND CONSTANTS, e, ELLISON–MENDÈS-FRANCE CONSTANT, ERDŐS RECIPROCAL SUM CONSTANTS, EULER-MASCHERONI CONSTANT, EXTREME VALUE DISTRIBUTION, FAVARD CONSTANTS, FELLER'S COIN-TOSSING CONSTANTS, FRANSÉN-ROBINSON CONSTANT, FREIMAN'S CONSTANT, GAUSS'S CIRCLE PROBLEM, GAUSS'S CONSTANT, GAUSS-KUZMIN-WIRSING CONSTANT, GELFOND-SCHNEIDER CONSTANT, GEOMETRIC PROBABILITY CONSTANTS, GIBBS CONSTANT, GLAISHER-KINKELIN CONSTANT, GOLDEN MEAN, GOLOMB CONSTANT, GOLOMB-DICKMAN CONSTANT, GOMPERTZ CONSTANT, GROSSMAN'S CONSTANT, GROTHENDIECK'S MAJORANT, HADAMARD-VALLÉE POUSSIN CONSTANTS, HAFNER-SARNAK-MCCURLEY CONSTANT, HALPHEN CONSTANT, HARD SQUARE ENTROPY CONSTANT, HARDY-LITTLEWOOD CONSTANTS, HERMITE CONSTANTS, HILBERT'S CONSTANTS, INFINITE PRODUCT, ITERATED EXPONENTIAL CONSTANTS, KHINTCHINE'S CONSTANT, KHINTCHINE-LÉVY CONSTANT, KOEBE'S CONSTANT, KOLMOGOROV CONSTANT, LAL'S CONSTANT, LANDAU CONSTANT, LANDAU-KOLMOGOROV CONSTANTS, LANDAU-RAMANUJAN CONSTANT, LEBESGUE CONSTANTS (FOURIER SERIES), LEBESGUE CONSTANTS (LAGRANGE INTERPOLATION), LEGENDRE'S CONSTANT, LEHMER'S CONSTANT, LENGYEL'S CONSTANT, LÉVY CONSTANT, LINNIK'S CONSTANT, LIOUVILLE'S CONSTANT, LIOUVILLE-ROTH CONSTANT, LUDOLPH'S CONSTANT, MADELUNG CONSTANTS, MAGIC CONSTANT, MAGIC GEOMETRIC CONSTANTS, MASSER-GRAMAIN CONSTANT, MERTENS CONSTANT, MILLS' CONSTANT, MOVING SOFA CONSTANT, NAPIER'S CONSTANT, NIELSEN-RAMANUJAN CONSTANTS, NIVEN'S CONSTANT, OMEGA CONSTANT, ONE-NINTH CONSTANT, OTTER'S TREE ENUMERATION CONSTANTS, PARITY CONSTANT, PI, PISOT-VIJAYARAGHAVAN CONSTANTS, PLASTIC CONSTANT, PLOUFFE'S CONSTANT, POLYGON CIRCUMSCRIBING CONSTANT, POLYGON INSCRIBING CONSTANT, PORTER'S CONSTANT, PYTHAGORAS'S CONSTANT, QUADRATIC RECURRENCE, QUADTREE, RABBIT CONSTANT, RAMANUJAN CONSTANT, RANDOM WALK, RÉNYI'S PARKING CONSTANTS, ROBBIN CONSTANT, SALEM CONSTANTS, SELF-AVOIDING WALK, SHAH-WILSON CONSTANT, SHALLIT CONSTANT, SHAPIRO'S CYCLIC SUM CONSTANT, SIERPIŃSKI CONSTANT, SILVER CONSTANT, SILVERMAN CONSTANT, SMARANDACHE CONSTANTS, SOLDNER'S CONSTANT, SPHERE PACKING, STIELTJES CONSTANTS, STOLARSKY-HARBORTH CONSTANT, SYLVESTER'S SEQUENCE, THUE CONSTANT, THUE-MORSE CONSTANT, TOTIENT FUNCTION CONSTANTS, TRAVELING SALESMAN CONSTANTS, TREE SEARCHING, TWIN PRIMES CONSTANT, VARGA'S CONSTANT, W2-CONSTANT, WEIERSTRAß CONSTANT, WHITNEY-MIKHLIN EXTENSION CONSTANTS, WILBRAHAM-GIBBS CONSTANT, WIRTINGER-SOBOLEV ISOPERIMETRIC CONSTANTS

References

Borwein, J. and Borwein, P. *A Dictionary of Real Numbers.* London: Chapman & Hall, 1990.

Finch, S. "Favorite Mathematical Constants." `http://www.mathsoft.com/asolve/constant/constant.html`.

Le Lionnais, F. *Les nombres remarquables.* Paris: Hermann, 1983.

Plouffe, S. "Inverse Symbolic Calculator Table of Constants." `http://www.cecm.sfu.ca/projects/ISC/I_d.html`.

Plouffe, S. "Plouffe's Inverter." `http://www.lacim.uqam.ca/pi/`.

Plouffe, S. "Plouffe's Inverter: Table of Current Records for the Computation of Constants." `http://lacim.uqam.ca/pi/records.html`.

Wells, D. W. *The Penguin Dictionary of Curious and Interesting Numbers.* Harmondsworth, England: Penguin Books, 1986.

Constant Function

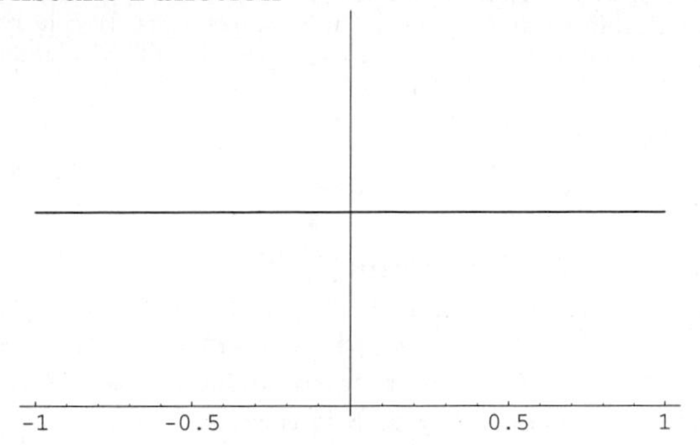

A FUNCTION $f(x) = c$ which does not change as its parameters vary. The GRAPH of a 1-D constant FUNCTION is a straight LINE. The DERIVATIVE of a constant FUNCTION c is

$$\frac{d}{dx}c = 0, \tag{1}$$

and the INTEGRAL is

$$\int c\,dx = cx. \tag{2}$$

The FOURIER TRANSFORM of the constant function $f(x) = 1$ is given by

$$\mathcal{F}[1] = \int_{-\infty}^{\infty} e^{-2\pi ikx}\,dx = \delta(k), \tag{3}$$

where $\delta(k)$ is the DELTA FUNCTION.

see also FOURIER TRANSFORM—1

References

Spanier, J. and Oldham, K. B. "The Constant Function *c*." Ch. 1 in *An Atlas of Functions.* Washington, DC: Hemisphere, pp. 11–14, 1987.

Constant Precession Curve

see CURVE OF CONSTANT PRECESSION

Constant Problem

Given an expression involving known constants, integration in finite terms, computation of limits, etc., determine if the expression is equal to ZERO. The constant problem is a very difficult unsolved problem in TRANSCENDENTAL NUMBER theory. However, it is known that the problem is UNDECIDABLE if the expression involves oscillatory functions such as SINE. However, the FERGUSON-FORCADE ALGORITHM is a practical algorithm for determining if there exist integers a_i for given real numbers x_i such that

$$a_1x_1 + a_2x_2 + \ldots + a_nx_n = 0,$$

or else establish bounds within which no relation can exist (Bailey 1988).

see also FERGUSON-FORCADE ALGORITHM, INTEGER RELATION, SCHANUEL'S CONJECTURE

References

Bailey, D. H. "Numerical Results on the Transcendence of Constants Involving π, e, and Euler's Constant." *Math. Comput.* **50**, 275–281, 1988.

Sackell, J. "Zero-Equivalence in Function Fields Defined by Algebraic Differential Equations." *Trans. Amer. Math. Soc.* **336**, 151–171, 1993.

Constant Width Curve

see CURVE OF CONSTANT WIDTH

Constructible Number

A number which can be represented by a FINITE number of ADDITIONS, SUBTRACTIONS, MULTIPLICATIONS, DIVISIONS, and FINITE SQUARE ROOT extractions of integers. Such numbers correspond to LINE SEGMENTS which can be constructed using only STRAIGHTEDGE and COMPASS.

All RATIONAL NUMBERS are constructible, and all constructible numbers are ALGEBRAIC NUMBERS (Courant and Robbins 1996, p. 133). If a CUBIC EQUATION with rational coefficients has no rational root, then none of its roots is constructible (Courant and Robbins, p. 136).

In particular, let F_0 be the FIELD of RATIONAL NUMBERS. Now construct an extension field F_1 of constructible numbers by the adjunction of $\sqrt{k_0}$, where k_0 is in F_0, but $\sqrt{k_0}$ is not, consisting of all numbers of the form $a_0 + b_0\sqrt{k_0}$, where $a_0, b_0 \in F_0$. Next, construct an extension field F_2 of F_1 by the adjunction of $\sqrt{k_1}$, defined as the numbers $a_1 + b_1\sqrt{k_1}$, where $a_1, b_1 \in F_1$, and k_1 is a number in F_1 for which $\sqrt{k_1}$ does not lie in F_1. Continue the process n times. Then constructible numbers are precisely those which can be reached by such a sequence of extension fields F_n, where n is a measure of the "complexity" of the construction (Courant and Robbins 1996).

see also ALGEBRAIC NUMBER, COMPASS, CONSTRUCTIBLE POLYGON, EUCLIDEAN NUMBER, RATIONAL NUMBER, STRAIGHTEDGE

References

Courant, R. and Robbins, H. "Constructible Numbers and Number Fields." §3.2 in *What is Mathematics?: An Elementary Approach to Ideas and Methods, 2nd ed.* Oxford, England: Oxford University Press, pp. 127–134, 1996.

Constructible Polygon

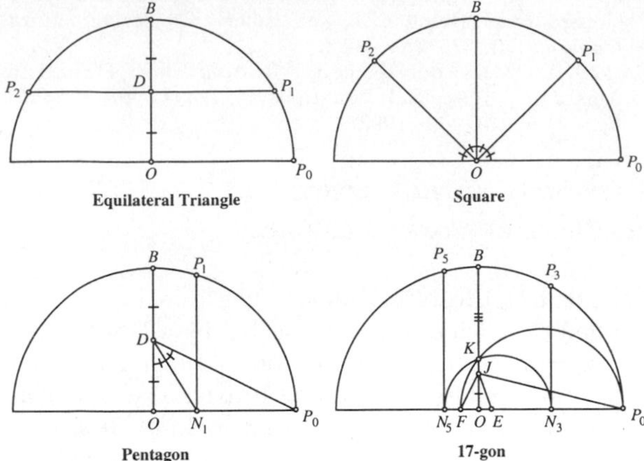

Equilateral Triangle Square Pentagon 17-gon

COMPASS and STRAIGHTEDGE constructions dating back to Euclid were capable of inscribing regular polygons of 3, 4, 5, 6, 8, 10, 12, 16, 20, 24, 32, 40, 48, 64, ..., sides. However, this listing is not a complete enumeration of "constructible" polygons. A regular n-gon ($n \geq 3$) can be constructed by STRAIGHTEDGE and COMPASS IFF

$$n = 2^k p_1 p_2 \cdots p_s,$$

where k is in INTEGER ≥ 0 and the p_i are distinct FERMAT PRIMES. FERMAT NUMBERS are of the form

$$F_m = 2^{2^m} + 1,$$

where m is an INTEGER ≥ 0. The only known PRIMES of this form are 3, 5, 17, 257, and 65537. The fact that this condition was SUFFICIENT was first proved by Gauss in 1796 when he was 19 years old. That this condition was also NECESSARY was not explicitly proven by Gauss, and the first proof of this fact is credited to Wantzel (1836).

see also COMPASS, CONSTRUCTIBLE NUMBER, GEOMETRIC CONSTRUCTION, GEOMETROGRAPHY, HEPTADECAGON, HEXAGON, OCTAGON, PENTAGON, POLYGON, SQUARE, STRAIGHTEDGE, TRIANGLE

References

Ball, W. W. R. and Coxeter, H. S. M. *Mathematical Recreations and Essays, 13th ed.* New York: Dover, pp. 94–96, 1987.

Courant, R. and Robbins, H. *What is Mathematics?: An Elementary Approach to Ideas and Methods, 2nd ed.* Oxford, England: Oxford University Press, p. 119, 1996.

De Temple, D. W. "Carlyle Circles and the Lemoine Simplicity of Polygonal Constructions." *Amer. Math. Monthly* **98**, 97–108, 1991.

Dixon, R. "Compass Drawings." Ch. 1 in *Mathographics.* New York: Dover, pp. 1–78, 1991.

Gauss, C. F. §365 and 366 in *Disquisitiones Arithmeticae.* Leipzig, Germany, 1801. Translated by A. A. Clarke. New Haven, CT: Yale University Press, 1965.

Kazarinoff, N. D. "On Who First Proved the Impossibility of Constructing Certain Regular Polygons with Ruler and Compass Alone." *Amer. Math. Monthly* **75**, 647–648, 1968.

Ogilvy, C. S. *Excursions in Geometry.* New York: Dover, pp. 137–138, 1990.

Wantzel, P. L. "Recherches sur les moyens de reconnaître si un Problème de Géométrie peut se résoudre avec la règle et le compas." *J. Math. pures appliq.* **1**, 366–372, 1836.

Construction

see GEOMETRIC CONSTRUCTION

Constructive Dilemma

A formal argument in LOGIC in which it is stated that (1) $P \Rightarrow Q$ and $R \Rightarrow S$ (where $\Rightarrow$ means "IMPLIES"), and (2) either P or R is true, from which two statements it follows that either Q or S is true.

see also DESTRUCTIVE DILEMMA, DILEMMA

Contact Angle

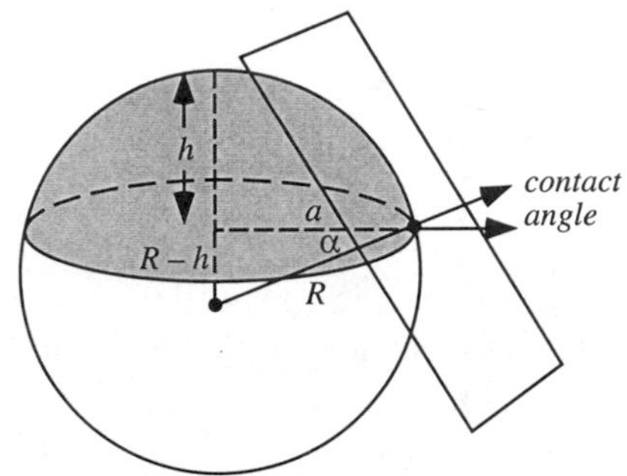

The ANGLE α between the normal vector of a SPHERE (or other geometric object) at a point where a PLANE is tangent to it and the normal vector of the plane. In the above figure,

$$\alpha = \cos^{-1}\left(\frac{a}{R}\right)$$
$$= \sin^{-1}\left(\frac{R-h}{R}\right).$$

see also SPHERICAL CAP

Contact Number

see KISSING NUMBER

Contact Triangle

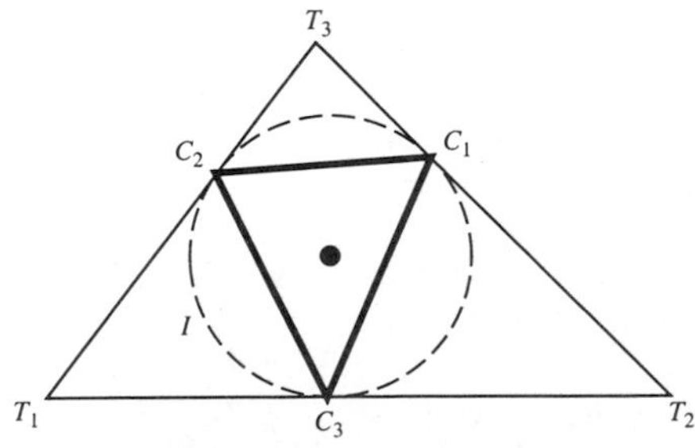

The TRIANGLE formed by the points of intersection of a TRIANGLE T's INCIRCLE with T. This is the PEDAL TRIANGLE of T with the INCENTER as the PEDAL POINT (c.f., TANGENTIAL TRIANGLE). The contact triangle and TANGENTIAL TRIANGLE are perspective from the GERGONNE POINT.

see also GERGONNE POINT, PEDAL TRIANGLE, TANGENTIAL TRIANGLE

References

Oldknow, A. "The Euler-Gergonne-Soddy Triangle of a Triangle." *Amer. Math. Monthly* **103**, 319–329, 1996.

Content

The generalized VOLUME for an n-D object (the "HYPERVOLUME").

see also VOLUME

Contiguous Function

A HYPERGEOMETRIC FUNCTION in which one parameter changes by $+1$ or -1 is said to be contiguous. There are 26 functions contiguous to ${}_2F_1(a,b,c;x)$ taking one pair at a time. There are 325 taking two or more pairs at a time. See Abramowitz and Stegun (1972, pp. 557–558).

see also HYPERGEOMETRIC FUNCTION

References

Abramowitz, M. and Stegun, C. A. (Eds.). *Handbook of Mathematical Functions with Formulas, Graphs, and Mathematical Tables, 9th printing.* New York: Dover, 1972.

Continued Fraction

A "general" continued fraction representation of a REAL NUMBER x is of the form

$$x = b_0 + \cfrac{a_1}{b_1 + \cfrac{a_2}{b_2 + \cfrac{a_3}{b_3 + \ldots}}}, \tag{1}$$

which can be written

$$x = b_0 + \frac{a_1}{b_1+}\frac{a_2}{b_2+}\cdots. \tag{2}$$

The SIMPLE CONTINUED FRACTION representation of x (which is usually what is meant when the term "continued fraction" is used without qualification) of a number is given by

$$x = a_0 + \cfrac{1}{a_1 + \cfrac{1}{a_2 + \cfrac{1}{a_3 + \ldots}}}, \tag{3}$$

which can be written in a compact abbreviated NOTATION as

$$x = [a_0, a_1, a_2, a_3, \ldots]. \tag{4}$$

Here,

$$a_0 = \lfloor x \rfloor \tag{5}$$

is the integral part of σ (where $\lfloor x \rfloor$ is the FLOOR FUNCTION),

$$a_1 = \left\lfloor \frac{1}{x - a_0} \right\rfloor \tag{6}$$

is the integral part of the RECIPROCAL of $x - a_0$, a_2 is the integral part of the reciprocal of the remainder, etc. The quantities a_i are called PARTIAL QUOTIENTS. An archaic word for a continued fraction is ANTHYPHAIRETIC RATIO.

Continued fractions provide, in some sense, a series of "best" estimates for an IRRATIONAL NUMBER. Functions can also be written as continued fractions, providing a series of better and better rational approximations. Continued fractions have also proved useful in the proof of certain properties of numbers such as e and π (PI). Because irrationals which are square roots of RATIONAL NUMBERS have periodic continued fractions, an exact representation for a tabulated numerical value (i.e., 1.414... for PYTHAGORAS'S CONSTANT, $\sqrt{2}$) can sometimes be found.

Continued fractions are also useful for finding near commensurabilities between events with different periods. For example, the Metonic cycle used for calendrical purposes by the Greeks consists of 235 lunar months which very nearly equal 19 solar years, and 235/19 is the sixth CONVERGENT of the ratio of the lunar phase (synodic) period and solar period (365.2425/29.53059). Continued fractions can also be used to calculate gear ratios, and were used for this purpose by the ancient Greeks (Guy 1990).

If only the first few terms of a continued fraction are kept, the result is called a CONVERGENT. Let P_n/Q_n be convergents of a nonsimple continued fraction. Then

$$P_{-1} \equiv 1 \qquad Q_{-1} \equiv 0 \tag{7}$$

$$P_0 \equiv b_0 \qquad Q_0 \equiv 1 \tag{8}$$

and subsequent terms are calculated from the RECURRENCE RELATIONS

$$P_j = b_j P_{j-1} + a_j P_{j-2} \tag{9}$$

$$Q_j = b_j Q_{j-1} + a_j Q_{j-2} \tag{10}$$

for $j = 1, 2, \ldots, n$. It is also true that

$$P_n Q_{n-1} - P_{n-1} Q_n = (-1)^{n-1} \prod_{k=1}^{n} a_k. \tag{11}$$

The error in approximating a number by a given CONVERGENT is roughly the MULTIPLICATIVE INVERSE of the square of the DENOMINATOR of the first neglected term.

A *finite simple* continued fraction representation terminates after a finite number of terms. To "round" a continued fraction, truncate the last term unless it is ± 1, in which case it should be added to the previous term (Beeler *et al.* 1972, Item 101A). To take one over a continued fraction, add (or possibly delete) an initial 0 term. To negate, take the NEGATIVE of all terms, optionally using the identity

$$[-a,-b,-c,-d,\ldots]=[-a-1,1,b-1,c,d,\ldots]. \quad (12)$$

A particularly beautiful identity involving the terms of the continued fraction is

$$\frac{[a_0,a_1,\ldots,a_n]}{[a_0,a_1,\ldots,a_{n-1}]}=\frac{[a_n,a_{n-1},\ldots,a_1,a_0]}{[a_n,a_{n-1},\ldots,a_1]}. \quad (13)$$

Finite simple fractions represent rational numbers and all rational numbers are represented by finite continued fractions. There are two possible representations for a finite simple fraction:

$$[a_1,\ldots,a_n]=\begin{cases}[a_1,\ldots,a_{n-1},a_n-1,1] & \text{for } a_n>1\\ [a_1,\ldots,a_{n-2},a_{n-1}+1] & \text{for } a_n=1.\end{cases} \quad (14)$$

On the other hand, an infinite simple fraction represents a unique IRRATIONAL NUMBER, and each IRRATIONAL NUMBER has a unique infinite continued fraction.

Consider the CONVERGENTS p_n/q_n of a *simple* continued fraction, and define

$$p_{-1}\equiv 0 \qquad q_{-1}\equiv 1 \quad (15)$$

$$p_0\equiv 1 \qquad q_0\equiv 0 \quad (16)$$

$$p_1\equiv a_1 \qquad q_1\equiv 1. \quad (17)$$

Then subsequent terms can be calculated from the RECURRENCE RELATIONS

$$p_i=a_ip_{i-1}+p_{i-2} \quad (18)$$

$$q_i=a_iq_{i-1}+q_{i-2}. \quad (19)$$

The CONTINUED FRACTION FUNDAMENTAL RECURRENCE RELATION for *simple* continued fractions is

$$p_nq_{n-1}-p_{n-1}q_n=(-1)^n. \quad (20)$$

It is also true that if $a_1\neq 0$,

$$\frac{p_n}{p_{n-1}}=[a_n,a_{n-1},\ldots,a_1] \quad (21)$$

$$\frac{q_n}{q_{n-1}}=[a_n,\ldots,a_2]. \quad (22)$$

Furthermore,

$$\frac{p_n}{q_n}=\frac{p_{n+1}-p_{n-1}}{q_{n+1}-q_{n-1}} \quad (23)$$

$$p_n=(n-1)p_{n-1}+(n-1)p_{n-2}+(n-2)p_{n-3} +\ldots+3p_2+2p_1+p_1+1. \quad (24)$$

Also, if $p/q>1$ and

$$\frac{p}{q}=[a_1,a_2,\ldots,a_n], \quad (25)$$

then

$$\frac{q}{p}=[0,a_1,\ldots,a_n]. \quad (26)$$

Similarly, if $p/q<1$ so

$$\frac{p}{q}=[0,a_1,\ldots,a_n], \quad (27)$$

then

$$\frac{q}{p}=[a_1,\ldots,a_n]. \quad (28)$$

The convergents also satisfy

$$c_n-c_{n-1}=\frac{(-1)^n}{q_nq_{n-1}} \quad (29)$$

$$c_n-c_{n-2}=\frac{a_n(-1)^{n-1}}{q_nq_{n-2}}. \quad (30)$$

The ODD convergents c_{2n+1} of an infinite simple continued fraction form an INCREASING SEQUENCE, and the EVEN convergents c_{2n} form a DECREASING SEQUENCE (so any ODD convergent is less than any EVEN convergent). Summarizing,

$$c_1<c_3<c_5<\cdots<c_{2n+1}<\cdots <c_{2n}<\cdots<c_6<c_4<c_2. \quad (31)$$

Furthermore, each convergent for $n\geq 3$ lies between the two preceding ones. Each convergent is nearer to the value of the infinite continued fraction than the previous one. Let p_n/q_n be the nth continued fraction representation. Then

$$\frac{1}{(a_{n+1}+2){q_n}^2}<\left|\sigma-\frac{p_n}{q_n}\right|<\frac{1}{a_{n+1}{q_n}^2}. \quad (32)$$

The SQUARE ROOT of a SQUAREFREE INTEGER has a periodic continued fraction of the form

$$\sqrt{n}=[a_1,\overline{a_2,\ldots,a_n,2a_1}] \quad (33)$$

(Rose 1994, p. 130). Furthermore, if D is not a SQUARE NUMBER, then the terms of the continued fraction of $\sqrt{D}$ satisfy

$$0<a_n<2\sqrt{D}. \quad (34)$$

In particular,

$$[\overline{a}]=\frac{a+\sqrt{a^2+4}}{2} \quad (35)$$

$$[1,\overline{a}]=\frac{-1+\sqrt{1+4a}}{2} \quad (36)$$

$$[a,\overline{2a}]=\sqrt{a^2+1} \quad (37)$$

$$[\overline{ac,a}]=\frac{b+\sqrt{b^2+4c}}{2} \quad (38)$$

$$\overline{[a_1,\ldots,a_n]}$$
$$= \frac{-(q_{n-1}-p_n)+\sqrt{(q_{n-1}-p_n)^2+4q_np_{n-1}}}{2q_n} \quad (39)$$

$$[a_1,\overline{b_1,\ldots,b_n}] = a_1 + \frac{1}{\overline{[b_1,\ldots,b_n]}} \quad (40)$$

$$\overline{[b_1,\ldots,b_n]} = \frac{\overline{[b_1,\ldots,b_n]}p_n+p_{n-1}}{\overline{[b_1,\ldots,b_n]}q_n+q_{n-1}}. \quad (41)$$

The first follows from

$$\alpha = n + \cfrac{1}{n+\cfrac{1}{n+\cfrac{1}{n+\ldots}}}$$
$$= n + \cfrac{1}{n+\left(\cfrac{1}{n+\cfrac{1}{n+\ldots}}\right)}. \quad (42)$$

Therefore,

$$\alpha - n = \cfrac{1}{n+\cfrac{1}{n+\cfrac{1}{n+\ldots}}}, \quad (43)$$

so plugging (43) into (42) gives

$$\alpha = n + \frac{1}{n+(\alpha-n)} = n + \frac{1}{\alpha}. \quad (44)$$

Expanding

$$\alpha^2 - n\alpha - 1 = 0, \quad (45)$$

and solving using the QUADRATIC FORMULA gives

$$\alpha = \frac{n+\sqrt{n^2+4}}{2}. \quad (46)$$

The analog of this treatment in the general case gives

$$\alpha = \frac{\alpha p_n + p_{n-1}}{\alpha q_n + q_{n-1}}. \quad (47)$$

The following table gives the repeating simple continued fractions for the square roots of the first few integers (excluding the trivial SQUARE NUMBERS).

N	$\alpha_{\sqrt{N}}$	N	$\alpha_{\sqrt{N}}$
2	$[1,\overline{2}]$	22	$[4,\overline{1,2,4,2,1,8}]$
3	$[1,\overline{1,2}]$	23	$[4,\overline{1,3,1,8}]$
5	$[2,\overline{4}]$	24	$[4,\overline{1,8}]$
6	$[2,\overline{2,4}]$	26	$[5,\overline{10}]$
7	$[2,\overline{1,1,1,4}]$	27	$[5,\overline{5,10}]$
8	$[2,\overline{1,4}]$	28	$[5,\overline{3,2,3,10}]$
10	$[3,\overline{6}]$	29	$[5,\overline{2,1,1,2,10}]$
11	$[3,\overline{3,6}]$	30	$[5,\overline{2,10}]$
12	$[3,\overline{2,6}]$	31	$[5,\overline{1,1,3,5,3,1,1,10}]$
13	$[3,\overline{1,1,1,1,6}]$	32	$[5,\overline{1,1,1,10}]$
14	$[3,\overline{1,2,1,6}]$	33	$[5,\overline{1,2,1,10}]$
15	$[3,\overline{1,6}]$	34	$[5,\overline{1,4,1,10}]$
17	$[4,\overline{8}]$	35	$[5,\overline{1,10}]$
18	$[4,\overline{4,8}]$	37	$[6,\overline{12}]$
19	$[4,\overline{2,1,3,1,2,8}]$	38	$[6,\overline{6,12}]$
20	$[4,\overline{2,8}]$	39	$[6,\overline{4,12}]$
21	$[4,\overline{1,1,2,1,1,8}]$	40	$[6,\overline{3,12}]$

The periods of the continued fractions of the square roots of the first few nonsquare integers 2, 3, 5, 6, 7, 8, 10, 11, 12, 13, ... (Sloane's A000037) are 1, 2, 1, 2, 4, 2, 1, 2, 2, 5, ... (Sloane's A013943; Williams 1981, Jacobson *et al.* 1995). An upper bound for the length is roughly $\mathcal{O}(\ln D\sqrt{D})$.

An even stronger result is that a continued fraction is periodic IFF it is a ROOT of a quadratic POLYNOMIAL. Calling the portion of a number x remaining after a given convergent the "tail," it must be true that the relationship between the number x and terms in its tail is of the form

$$x = \frac{ax+b}{cd+d}, \quad (48)$$

which can only lead to a QUADRATIC EQUATION.

LOGARITHMS $\log_{b_0} b_1$ can be computed by defining b_2, ... and the POSITIVE INTEGER n_1, ... such that

$$b_1^{n_1} < b_0 < b_1^{n_1+1} \qquad b_2 = \frac{b_0}{b_1^{n_1}} \quad (49)$$

$$b_2^{n_2} < b_1 < b_2^{n_2+1} \qquad b_3 = \frac{b_1}{b_2^{n_2}} \quad (50)$$

and so on. Then

$$\log_{b_0} b_1 = [n_1, n_2, n_3, \ldots]. \quad (51)$$

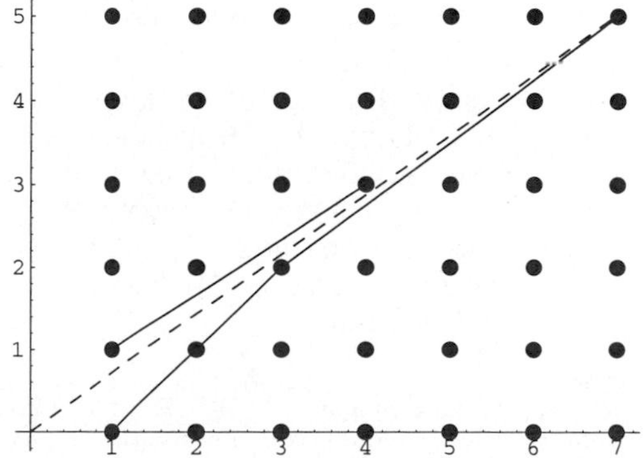

A geometric interpretation for a reduced FRACTION y/x consists of a string through a LATTICE of points with ends at $(1,0)$ and (x,y) (Klein 1907, 1932; Steinhaus 1983; Ball and Coxeter 1987, pp. 86–87; Davenport 1992). This interpretation is closely related to a similar one for the GREATEST COMMON DIVISOR. The pegs it presses against (x_i, y_i) give alternate CONVERGENTS y_i/x_i, while the other CONVERGENTS are obtained from the pegs it presses against with the initial end at $(0,1)$. The above plot is for $e-2$, which has convergents 0, 1, 2/3, 3/4, 5/7, ...

Let the continued fraction for x be written $[a_1, a_2, \ldots, a_n]$. Then the limiting value is *almost always* KHINTCHINE'S CONSTANT

$$K \equiv \lim_{n\to\infty} (a_1 a_2 \cdots a_n)^{1/n} = 2.68545\ldots. \tag{52}$$

Continued fractions can be used to express the POSITIVE ROOTS of any POLYNOMIAL equation. Continued fractions can also be used to solve linear DIOPHANTINE EQUATIONS and the PELL EQUATION. Euler showed that if a CONVERGENT SERIES can be written in the form

$$c_1 + c_1c_2 + c_1c_2c_3 + \ldots, \tag{53}$$

then it is equal to the continued fraction

$$\cfrac{c_1}{1 - \cfrac{c_2}{1 + c_2 - \cfrac{c_3}{1 + c_3 - \ldots}}}. \tag{54}$$

Gosper has invented an ALGORITHM for performing analytic ADDITION, SUBTRACTION, MULTIPLICATION, and DIVISION using continued fractions. It requires keeping track of eight INTEGERS which are conceptually arranged at the VERTICES of a CUBE. The ALGORITHM has not, however, appeared in print (Gosper 1996).

An algorithm for computing the continued fraction for $(ax+b)/(cx+d)$ from the continued fraction for x is given by Beeler *et al.* (1972, Item 101), Knuth (1981, Exercise 4.5.3.15, pp. 360 and 601), and Fowler (1991). (In line 9 of Knuth's solution, $X_k \leftarrow \lfloor A/C \rfloor$ should be replaced by $X_k \leftarrow \min(\lfloor A/C \rfloor, \lfloor (A+B)/(C+D) \rfloor)$.) Beeler *et al.* (1972) and Knuth (1981) also mention the bivariate case $(axy+bx+cy+d)/(Axy+Bx+Cy+D)$.

see also GAUSSIAN BRACKETS, HURWITZ'S IRRATIONAL NUMBER THEOREM, KHINTCHINE'S CONSTANT, LAGRANGE'S CONTINUED FRACTION THEOREM, LAMÉ'S THEOREM, LÉVY CONSTANT, PADÉ APPROXIMANT, PARTIAL QUOTIENT, PI, QUADRATIC IRRATIONAL NUMBER, QUOTIENT-DIFFERENCE ALGORITHM, SEGRE'S THEOREM

References

Abramowitz, M. and Stegun, C. A. (Eds.). *Handbook of Mathematical Functions with Formulas, Graphs, and Mathematical Tables, 9th printing.* New York: Dover, p. 19, 1972.

Acton, F. S. "Power Series, Continued Fractions, and Rational Approximations." Ch. 11 in *Numerical Methods That Work, 2nd printing.* Washington, DC: Math. Assoc. Amer., 1990.

Ball, W. W. R. and Coxeter, H. S. M. *Mathematical Recreations and Essays, 13th ed.* New York: Dover, pp. 54–57 and 86–87, 1987.

Beeler, M.; Gosper, R. W.; and Schroeppel, R. *HAKMEM.* Cambridge, MA: MIT Artificial Intelligence Laboratory, Memo AIM-239, pp. 36–44, Feb. 1972.

Beskin, N. M. *Fascinating Fractions.* Moscow: Mir Publishers, 1980.

Brezinski, C. *History of Continued Fractions and Padé Approximants.* New York: Springer-Verlag, 1980.

Conway, J. H. and Guy, R. K. "Continued Fractions." In *The Book of Numbers.* New York: Springer-Verlag, pp. 176–179, 1996.

Courant, R. and Robbins, H. "Continued Fractions. Diophantine Equations." §2.4 in Supplement to Ch. 1 in *What is Mathematics?: An Elementary Approach to Ideas and Methods, 2nd ed.* Oxford, England: Oxford University Press, pp. 49–51, 1996.

Davenport, H. §IV.12 in *The Higher Arithmetic: An Introduction to the Theory of Numbers, 6th ed.* New York: Cambridge University Press, 1992.

Euler, L. *Introduction to Analysis of the Infinite, Book I.* New York: Springer-Verlag, 1980.

Fowler, D. H. *The Mathematics of Plato's Academy.* Oxford, England: Oxford University Press, 1991.

Guy, R. K. "Continued Fractions" §F20 in *Unsolved Problems in Number Theory, 2nd ed.* New York: Springer-Verlag, p. 259, 1994.

Jacobson, M. J. Jr.; Lukes, R. F.; and Williams, H. C. "An Investigation of Bounds for the Regulator of Quadratic Fields." *Experiment. Math.* **4**, 211–225, 1995.

Khinchin, A. Ya. *Continued Fractions.* New York: Dover, 1997.

Kimberling, C. "Continued Fractions." `http://www.evansville.edu/~ck6/integer/contfr.html`.

Klein, F. *Ausgewählte Kapitel der Zahlentheorie.* Germany: Teubner, 1907.

Klein, F. *Elementary Number Theory.* New York, p. 44, 1932.

Kline, M. *Mathematical Thought from Ancient to Modern Times.* New York: Oxford University Press, 1972.

Knuth, D. E. *The Art of Computer Programming, Vol. 2: Seminumerical Algorithms, 2nd ed.* Reading, MA: Addison-Wesley, p. 316, 1981.

Moore, C. D. *An Introduction to Continued Fractions.* Washington, DC: National Council of Teachers of Mathematics, 1964.

Olds, C. D. *Continued Fractions.* New York: Random House, 1963.

Pettofrezzo, A. J. and Bykrit, D. R. *Elements of Number Theory.* Englewood Cliffs, NJ: Prentice-Hall, 1970.

Press, W. H.; Flannery, B. P.; Teukolsky, S. A.; and Vetterling, W. T. "Evaluation of Continued Fractions." §5.2 in *Numerical Recipes in FORTRAN: The Art of Scientific Computing, 2nd ed.* Cambridge, England: Cambridge University Press, pp. 163–167, 1992.

Rose, H. E. *A Course in Number Theory, 2nd ed.* Oxford, England: Oxford University Press, 1994.

Rosen, K. H. *Elementary Number Theory and Its Applications.* New York: Addison-Wesley, 1980.

Sloane, N. J. A. Sequences A013943 and A000037/M0613 in "An On-Line Version of the Encyclopedia of Integer Sequences."

Steinhaus, H. *Mathematical Snapshots, 3rd American ed.* New York: Oxford University Press, pp. 39–42, 1983.

Van Tuyl, A. L. "Continued Fractions." `http://www.calvin.edu/academic/math/confrac/`.
Wagon, S. "Continued Fractions." §8.5 in *Mathematica in Action*. New York: W. H. Freeman, pp. 263–271, 1991.
Wall, H. S. *Analytic Theory of Continued Fractions*. New York: Chelsea, 1948.
Williams, H. C. "A Numerical Investigation into the Length of the Period of the Continued Fraction Expansion of $\sqrt{D}$." *Math. Comp.* **36**, 593–601, 1981.

Continued Fraction Constant

A continued fraction with partial quotients which increase in ARITHMETIC PROGRESSION is

$$[A+D, A+2D, A+3D, \ldots] = \frac{I_{A/D}\left(\frac{2}{D}\right)}{I_{1+A/D}\left(\frac{2}{D}\right)},$$

where $I_n(x)$ is a MODIFIED BESSEL FUNCTION OF THE FIRST KIND (Beeler *et al.* 1972, Item 99). A special case is

$$C = 0 + \cfrac{1}{1+\cfrac{1}{2+\cfrac{1}{3+\cfrac{1}{4+\cfrac{1}{5+\ldots}}}}},$$

which has the value

$$C = \frac{I_1(2)}{I_0(2)} = 0.697774658\ldots$$

(Lehmer 1973, Rabinowitz 1990).

see also e, GOLDEN MEAN, MODIFIED BESSEL FUNCTION OF THE FIRST KIND, PI, RABBIT CONSTANT, THUE-MORSE CONSTANT

References

Beeler, M.; Gosper, R. W.; and Schroeppel, R. *HAKMEM*. Cambridge, MA: MIT Artificial Intelligence Laboratory, Memo AIM-239, Feb. 1972.
Finch, S. "Favorite Mathematical Constants." `http://www.mathsoft.com/asolve/constant/cntfrc/cntfrc.html`.
Guy, R. K. "Review: The Mathematics of Plato's Academy." *Amer. Math. Monthly* **97**, 440–443, 1990.
Lehmer, D. H. "Continued Fractions Containing Arithmetic Progressions." *Scripta Math.* **29**, 17–24, 1973.
Rabinowitz, S. Problem E3264. "Asymptotic Estimates from Convergents of a Continued Fraction." *Amer. Math. Monthly* **97**, 157–159, 1990.

Continued Fraction Factorization Algorithm

A PRIME FACTORIZATION ALGORITHM which uses RESIDUES produced in the CONTINUED FRACTION of $\sqrt{mN}$ for some suitably chosen m to obtain a SQUARE NUMBER. The ALGORITHM solves

$$x^2 \equiv y^2 \pmod{n}$$

by finding an m for which $m^2 \pmod{n}$ has the smallest upper bound. The method requires (by conjecture) about $\exp(\sqrt{2\log n \log\log n})$ steps, and was the fastest PRIME FACTORIZATION ALGORITHM in use before the QUADRATIC SIEVE FACTORIZATION METHOD, which eliminates the 2 under the SQUARE ROOT (Pomerance 1996), was developed.

see also EXPONENT VECTOR, PRIME FACTORIZATION ALGORITHMS

References

Morrison, M. A. and Brillhart, J. "A Method of Factoring and the Factorization of F_7." *Math. Comput.* **29**, 183–205, 1975.
Pomerance, C. "A Tale of Two Sieves." *Not. Amer. Math. Soc.* **43**, 1473–1485, 1996.

Continued Fraction Fundamental Recurrence Relation

For a SIMPLE CONTINUED FRACTION $\sigma = [a_0, a_1, \ldots]$ with CONVERGENTS p_n/q_n, the fundamental RECURRENCE RELATION is given by

$$p_n q_{n-1} - p_{n-1} q_n = (-1)^n.$$

References

Olds, C. D. *Continued Fractions*. New York: Random House, p. 27, 1963.

Continued Fraction Map

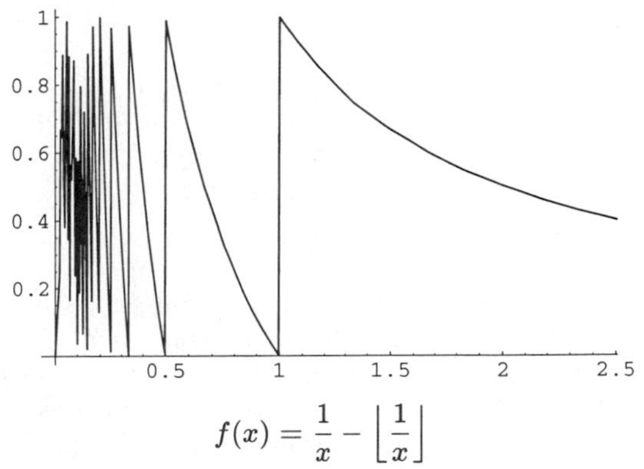

$$f(x) = \frac{1}{x} - \left\lfloor \frac{1}{x} \right\rfloor$$

for $x \in [0,1]$, where $\lfloor x \rfloor$ is the FLOOR FUNCTION. The INVARIANT DENSITY of the map is

$$\rho(y) = \frac{1}{(1+y)\ln 2}.$$

References

Beck, C. and Schlögl, F. *Thermodynamics of Chaotic Systems*. Cambridge, England: Cambridge University Press, pp. 194–195, 1995.

Continued Fraction Unit Fraction Algorithm

An algorithm for computing a UNIT FRACTION, called the FAREY SEQUENCE method by Bleicher (1972).

References

Bleicher, M. N. "A New Algorithm for the Expansion of Continued Fractions." *J. Number Th.* **4**, 342–382, 1972.

Continued Square Root

Expressions of the form

$$\lim_{k\to\infty} x_0 + \sqrt{x_1 + \sqrt{x_2 + \sqrt{\ldots + x_k}}}.$$

Herschfeld (1935) proved that a continued square root of REAL NONNEGATIVE terms converges IFF $(x_n)^{2^{-n}}$ is bounded. He extended this result to arbitrary POWERS (which include continued square roots and CONTINUED FRACTIONS as well), which is known as HERSCHFELD'S CONVERGENCE THEOREM.

see also CONTINUED FRACTION, HERSCHFELD'S CONVERGENCE THEOREM, SQUARE ROOT

References

Herschfeld, A. "On Infinite Radicals." *Amer. Math. Monthly* **42**, 419–429, 1935.

Pólya, G. and Szegő, G. *Problems and Theorems in Analysis, Vol. 1.* New York: Springer-Verlag, 1997.

Sizer, W. S. "Continued Roots." *Math. Mag.* **59**, 23–27, 1986.

Continued Vector Product

see VECTOR TRIPLE PRODUCT

Continuity

The property of being CONTINUOUS.

see also CONTINUITY AXIOMS, CONTINUITY CORRECTION, CONTINUITY PRINCIPLE, CONTINUOUS DISTRIBUTION, CONTINUOUS FUNCTION, CONTINUOUS SPACE, FUNDAMENTAL CONTINUITY THEOREM

Continuity Axioms

"The" continuity axiom is an additional AXIOM which must be added to those of Euclid's *Elements* in order to guarantee that two equal CIRCLES of RADIUS r intersect each other if the separation of their centers is less than $2r$ (Dunham 1990). The continuity *axioms* are the three of HILBERT'S AXIOMS which concern geometric equivalence.

ARCHIMEDES' LEMMA is sometimes also known as "the continuity axiom."

see also CONGRUENCE AXIOMS, HILBERT'S AXIOMS, INCIDENCE AXIOMS, ORDERING AXIOMS, PARALLEL POSTULATE

References

Dunham, W. *Journey Through Genius: The Great Theorems of Mathematics.* New York: Wiley, p. 38, 1990.

Hilbert, D. *The Foundations of Geometry.* Chicago, IL: Open Court, 1980.

Iyanaga, S. and Kawada, Y. (Eds.). "Hilbert's System of Axioms." §163B in *Encyclopedic Dictionary of Mathematics.* Cambridge, MA: MIT Press, pp. 544–545, 1980.

Continuity Correction

A correction to a discrete BINOMIAL DISTRIBUTION to approximate a continuous distribution.

$$P(a \leq X \leq b) \approx P\left(\frac{a - \frac{1}{2} - np}{\sqrt{np(1-p)}} \leq z \leq \frac{b - \frac{1}{2} - np}{\sqrt{np(1-p)}}\right),$$

where

$$z \equiv \frac{(x - \mu)}{\sigma}$$

is a continuous variate with a NORMAL DISTRIBUTION and X is a variate of a BINOMIAL DISTRIBUTION.

see also BINOMIAL DISTRIBUTION, NORMAL DISTRIBUTION

Continuity Principle

see PERMANENCE OF MATHEMATICAL RELATIONS PRINCIPLE

Continuous

A general mathematical property obeyed by mathematical objects in which all elements are within a NEIGHBORHOOD of nearby points.

see also ABSOLUTELY CONTINUOUS, CONTINUOUS DISTRIBUTION, CONTINUOUS FUNCTION, CONTINUOUS SPACE, JUMP

Continuous Distribution

A DISTRIBUTION for which the variables may take on a continuous range of values. Abramowitz and Stegun (1972, p. 930) give a table of the parameters of most common discrete distributions.

see also BETA DISTRIBUTION, BIVARIATE DISTRIBUTION, CAUCHY DISTRIBUTION, CHI DISTRIBUTION, CHI-SQUARED DISTRIBUTION, CORRELATION COEFFICIENT, DISCRETE DISTRIBUTION, DOUBLE EXPONENTIAL DISTRIBUTION, EQUALLY LIKELY OUTCOMES DISTRIBUTION, EXPONENTIAL DISTRIBUTION, EXTREME VALUE DISTRIBUTION, F-DISTRIBUTION, FERMI-DIRAC DISTRIBUTION, FISHER'S z-DISTRIBUTION, FISHER-TIPPETT DISTRIBUTION, GAMMA DISTRIBUTION, GAUSSIAN DISTRIBUTION, HALF-NORMAL DISTRIBUTION, LAPLACE DISTRIBUTION, LATTICE DISTRIBUTION, LÉVY DISTRIBUTION, LOGARITHMIC DISTRIBUTION, LOG-SERIES DISTRIBUTION, LOGISTIC DISTRIBUTION, LORENTZIAN DISTRIBUTION, MAXWELL DISTRIBUTION, NORMAL DISTRIBUTION, PARETO DISTRIBUTION, PASCAL DISTRIBUTION, PEARSON TYPE III DISTRIBUTION, POISSON DISTRIBUTION, PÓLYA DISTRIBUTION, RATIO DISTRIBUTION, RAYLEIGH DISTRIBUTION, RICE DISTRIBUTION, SNEDECOR'S F-DISTRIBUTION, STUDENT'S t-DISTRIBUTION, STUDENT'S z-DISTRIBUTION, UNIFORM DISTRIBUTION, WEIBULL DISTRIBUTION

References
Abramowitz, M. and Stegun, C. A. (Eds.). *Handbook of Mathematical Functions with Formulas, Graphs, and Mathematical Tables, 9th printing.* New York: Dover, pp. 927 and 930, 1972.

Continuous Function

A continuous function is a FUNCTION $f: X \to Y$ where the pre-image of every OPEN SET in Y is OPEN in X. A function $f(x)$ in a single variable x is said to be continuous at point x_0 if

1. $f(x_0)$ is defined, so x_0 is the DOMAIN of f.
2. $\lim_{x\to x_0} f(x)$ exists.
3. $\lim_{x\to x_0} f(x) = f(x_0)$,

where lim denotes a LIMIT. If f is DIFFERENTIABLE at point x_0, then it is also continuous at x_0. If f and g are continuous at x_0, then

1. $f+g$ is continuous at x_0.
2. $f-g$ is continuous at x_0.
3. $f\times g$ is continuous at x_0.
4. $f \div g$ is continuous at x_0 if $g(x_0)\neq 0$ and is discontinuous at x_0 if $g(x_0)=0$.
5. $f\circ g$ is continuous, where $\circ$ denotes using g as the argument to f.

see also CRITICAL POINT, DIFFERENTIABLE, LIMIT, NEIGHBORHOOD, STATIONARY POINT

Continuous Space

A TOPOLOGICAL SPACE.

see also NET

Continuum

The nondenumerable set of REAL NUMBERS, denoted C. It satisfies

$$\aleph_0 + C = C \tag{1}$$

and

$$C^r = C, \tag{2}$$

where $\aleph_0$ is $\aleph_0$ (ALEPH-0). It is also true that

$$\aleph_0^{\aleph_0} = C. \tag{3}$$

However,

$$C^C = F \tag{4}$$

is a SET larger than the continuum. Paradoxically, there are exactly as many points C on a LINE (or LINE SEGMENT) as in a PLANE, a 3-D SPACE, or finite HYPERSPACE, since all these SETS can be put into a ONE-TO-ONE correspondence with each other.

The CONTINUUM HYPOTHESIS, first proposed by Georg Cantor, holds that the CARDINAL NUMBER of the continuum is the same as that of $\aleph_1$. The surprising truth is that this proposition is UNDECIDABLE, since neither it nor its converse contradicts the tenets of SET THEORY.

see also ALEPH-0 ($\aleph_0$), ALEPH-1 ($\aleph_1$), CONTINUUM HYPOTHESIS, DENUMERABLE SET

Continuum Hypothesis

The proposal originally made by Georg Cantor that there is no infinite SET with a CARDINAL NUMBER between that of the "small" infinite SET of INTEGERS $\aleph_0$ and the "large" infinite set of REAL NUMBERS C (the "CONTINUUM"). Symbolically, the continuum hypothesis is that $\aleph_1 = C$. Gödel showed that no contradiction would arise if the continuum hypothesis were added to conventional ZERMELO-FRAENKEL SET THEORY. However, using a technique called FORCING, Paul Cohen (1963, 1964) proved that no contradiction would arise if the *negation* of the continuum hypothesis was added to SET THEORY. Together, Gödel's and Cohen's results established that the validity of the continuum hypothesis depends on the version of SET THEORY being used, and is therefore UNDECIDABLE (assuming the ZERMELO-FRAENKEL AXIOMS together with the AXIOM OF CHOICE).

Conway and Guy (1996) give a generalized version of the Continuum Hypothesis which is also UNDECIDABLE: is $2^{\aleph_\alpha} = \aleph_{\alpha+1}$ for every α?

see also ALEPH-0 ($\aleph_0$), ALEPH-1 ($\aleph_1$), AXIOM OF CHOICE, CARDINAL NUMBER, CONTINUUM, DENUMERABLE SET, FORCING, HILBERT'S PROBLEMS, LEBESGUE MEASURABILITY PROBLEM, UNDECIDABLE, ZERMELO-FRAENKEL AXIOMS, ZERMELO-FRAENKEL SET THEORY

References
Cohen, P. J. "The Independence of the Continuum Hypothesis." *Proc. Nat. Acad. Sci. U. S. A.* **50**, 1143–1148, 1963.
Cohen, P. J. "The Independence of the Continuum Hypothesis. II." *Proc. Nat. Acad. Sci. U. S. A.* **51**, 105–110, 1964.
Cohen, P. J. *Set Theory and the Continuum Hypothesis.* New York: W. A. Benjamin, 1966.
Conway, J. H. and Guy, R. K. *The Book of Numbers.* New York: Springer-Verlag, p. 282, 1996.
Gödel, K. *The Consistency of the Continuum-Hypothesis.* Princeton, NJ: Princeton University Press, 1940.
McGough, N. "The Continuum Hypothesis." `http://www.jazzie.com/ii/math/ch` and `http://www.best.com/~ii/math/ch/`.

Contour

A path in the COMPLEX PLANE over which CONTOUR INTEGRATION is performed.

see also CONTOUR INTEGRATION

Contour Integral

see CONTOUR INTEGRATION

Contour Integration

Let $P(x)$ and $Q(x)$ be POLYNOMIALS of DEGREES n and m with COEFFICIENTS $b_n, \ldots, b_0$ and $c_m, \ldots, c_0$. Take the contour in the upper half-plane, replace x by z, and write $z \equiv Re^{i\theta}$. Then

$$\int_{-\infty}^{\infty} \frac{P(z)\,dz}{Q(z)} = \lim_{R\to\infty} \int_{-R}^{R} \frac{P(z)\,dz}{Q(z)}. \tag{1}$$

Define a path γ_R which is straight along the REAL axis from $-R$ to R and makes a circular arc to connect the two ends in the upper half of the COMPLEX PLANE. The RESIDUE THEOREM then gives

$$\lim_{R\to\infty}\int_{\gamma_R}\frac{P(z)\,dz}{Q(z)} = \lim_{R\to\infty}\int_{-R}^{R}\frac{P(z)\,dz}{Q(z)} + \lim_{R\to\infty}\int_0^{\pi}\frac{P(Re^{i\theta})}{Q(Re^{i\theta})}iRe^{i\theta}\,d\theta = 2\pi i\sum_{\Im[z]>0}\operatorname{Res}\left[\frac{P(z)}{Q(z)}\right], \quad (2)$$

where Res denotes the RESIDUES. Solving,

$$\lim_{R\to\infty}\int_{-R}^{R}\frac{P(z)\,dz}{Q(z)} = 2\pi i\sum_{\Im[z]>0}\operatorname{Res}\frac{P(z)}{Q(z)} - \lim_{R\to\infty}\int_0^{\pi}\frac{P(Re^{i\theta})}{Q(Re^{i\theta})}iRe^{i\theta}\,d\theta. \quad (3)$$

Define

$$\begin{aligned} I_r &\equiv \lim_{R\to\infty}\int_0^{\pi}\frac{P(Re^{i\theta})}{Q(Re^{i\theta})}iRe^{i\theta}\,d\theta \\ &= \lim_{R\to\infty}\int_0^{\pi}\frac{b_n(Re^{i\theta})^n + b_{n-1}(Re^{i\theta})^{n-1}+\ldots+b_0}{c_m(Re^{i\theta})^m + c_{m-1}(Re^{i\theta})^{m-1}+\ldots+c_0}iR\,d\theta \\ &= \lim_{R\to\infty}\int_0^{\pi}\frac{b_n}{c_m}(Re^{i\theta})^{n-m}iR\,d\theta \\ &= \lim_{R\to\infty}\int_0^{\pi}\frac{b_n}{c_m}R^{n+1-m}i(e^{i\theta})^{n-m}\,d\theta \quad (4) \end{aligned}$$

and set

$$\epsilon \equiv -(n+1-m), \quad (5)$$

then equation (4) becomes

$$I_r \equiv \lim_{R\to\infty}\frac{i}{R^{\epsilon}}\frac{b_n}{c_m}\int_0^{\pi}e^{i(n-m)\theta}\,d\theta. \quad (6)$$

Now,

$$\lim_{R\to\infty} R^{-\epsilon} = 0 \quad (7)$$

for $\epsilon > 0$. That means that for $-n-1+m \geq 1$, or $m \geq n+2$, $I_R = 0$, so

$$\int_{-\infty}^{\infty}\frac{P(z)\,dz}{Q(z)} = 2\pi i\sum_{\Im[z]>0}\operatorname{Res}\left[\frac{P(z)}{Q(z)}\right] \quad (8)$$

for $m \geq n+2$. Apply JORDAN'S LEMMA with $f(x) \equiv P(x)/Q(x)$. We must have

$$\lim_{x\to\infty} f(x) = 0, \quad (9)$$

so we require $m \geq n+1$. Then

$$\int_{-\infty}^{\infty}\frac{P(z)}{Q(z)}e^{iaz}\,dz = 2\pi i\sum_{\Im[z]>0}\operatorname{Res}\left[\frac{P(z)}{Q(z)}e^{iaz}\right] \quad (10)$$

for $m \geq n+1$.

Since this must hold separately for REAL and IMAGINARY PARTS, this result can be extended to

$$\int_{-\infty}^{\infty}\frac{P(x)}{Q(x)}\cos(ax)\,dx = 2\pi\Re\left\{\sum_{\Im[z]>0}\operatorname{Res}\left[\frac{P(z)}{Q(z)}e^{iaz}\right]\right\} \quad (11)$$

$$\int_{-\infty}^{\infty}\frac{P(x)}{Q(x)}\sin(ax)\,dx = 2\pi\Im\left\{\sum_{\Im[z]>0}\operatorname{Res}\left[\frac{P(z)}{Q(z)}e^{iaz}\right]\right\}. \quad (12)$$

It is also true that

$$\int_{-\infty}^{\infty}\frac{P(z)}{Q(z)}\ln(az)\,dz = 0. \quad (13)$$

see also CAUCHY INTEGRAL FORMULA, CAUCHY INTEGRAL THEOREM, INSIDE-OUTSIDE THEOREM, JORDAN'S LEMMA, RESIDUE (COMPLEX ANALYSIS), SINE INTEGRAL

References

Morse, P. M. and Feshbach, H. *Methods of Theoretical Physics, Part I.* New York: McGraw-Hill, pp. 353–356, 1953.

Contracted Cycloid

see CURTATE CYCLOID

Contraction

see DILATION

Contraction (Graph)

The merging of nodes in a GRAPH by eliminating segments between two nodes.

Contraction (Tensor)

The contraction of a TENSOR is obtained by setting unlike indices equal and summing according to the EINSTEIN SUMMATION convention. Contraction reduces the RANK of a TENSOR by 2. For a second RANK TENSOR,

$$\operatorname{contr}(B'^i_j) \equiv B'^i_i$$

$$B'^i_i = \frac{\partial x'_i}{\partial x_k}\frac{\partial x_l}{\partial x'_i}B^k_l = \frac{\partial x_l}{\partial x_k}B^k_l = \delta^l_k B^k_l = B^k_k.$$

Therefore, the contraction is invariant, and must be a SCALAR. In fact, this SCALAR is known as the TRACE of a MATRIX in MATRIX theory.

References

Arfken, G. "Contraction, Direct Product." §3.2 in *Mathematical Methods for Physicists, 3rd ed.* Orlando, FL: Academic Press, pp. 124–126, 1985.

Contradiction Law

No A is not-A.

see also NOT

Contravariant Tensor

A contravariant tensor is a TENSOR having specific transformation properties (c.f., a COVARIANT TENSOR). To examine the transformation properties of a contravariant tensor, first consider a TENSOR of RANK 1 (a VECTOR)

$$d\mathbf{r} = dx_1\hat{\mathbf{x}}_1 + dx_2\hat{\mathbf{x}}_2 + dx_3\hat{\mathbf{x}}_3, \quad (1)$$

for which

$$dx'_i = \frac{\partial x'_i}{\partial x_j} dx_j. \quad (2)$$

Now let $A_i \equiv dx_i$, then any set of quantities A_j which transform according to

$$A'_i = \frac{\partial x'_i}{\partial x_j} A_j, \quad (3)$$

or, defining

$$a_{ij} \equiv \frac{\partial x'_i}{\partial x_j}, \quad (4)$$

according to

$$A'_i = a_{ij} A_j \quad (5)$$

is a contravariant tensor. Contravariant tensors are indicated with raised indices, i.e., a^μ.

COVARIANT TENSORS are a type of TENSOR with differing transformation properties, denoted a_ν. However, in 3-D CARTESIAN COORDINATES,

$$\frac{\partial x_j}{\partial x'_i} = \frac{\partial x'_i}{\partial x_j} \equiv a_{ij} \quad (6)$$

for $i, j = 1, 2, 3$, meaning that contravariant and covariant tensors are equivalent. The two types of tensors do differ in higher dimensions, however. Contravariant FOUR-VECTORS satisfy

$$a^\mu = \Lambda^\mu_\nu a^\nu, \quad (7)$$

where Λ is a LORENTZ TENSOR.

To turn a COVARIANT TENSOR into a contravariant tensor, use the METRIC TENSOR $g^{\mu\nu}$ to write

$$a^\mu \equiv g^{\mu\nu} a_\nu. \quad (8)$$

Covariant and contravariant indices can be used simultaneously in a MIXED TENSOR.

see also COVARIANT TENSOR, FOUR-VECTOR, LORENTZ TENSOR, METRIC TENSOR, MIXED TENSOR, TENSOR

References

Arfken, G. "Noncartesian Tensors, Covariant Differentiation." §3.8 in *Mathematical Methods for Physicists, 3rd ed.* Orlando, FL: Academic Press, pp. 158–164, 1985.

Morse, P. M. and Feshbach, H. *Methods of Theoretical Physics, Part I.* New York: McGraw-Hill, pp. 44–46, 1953.

Contravariant Vector

A CONTRAVARIANT TENSOR of RANK 1.

see also CONTRAVARIANT TENSOR, VECTOR

Control Theory

The mathematical study of how to manipulate the parameters affecting the behavior of a system to produce the desired or optimal outcome.

see also KALMAN FILTER, LINEAR ALGEBRA, PONTRYAGIN MAXIMUM PRINCIPLE

References

Zabczyk, J. *Mathematical Control Theory: An Introduction.* Boston, MA: Birkhäuser, 1993.

Convective Acceleration

The acceleration of an element of fluid, given by the CONVECTIVE DERIVATIVE of the VELOCITY $\mathbf{v}$,

$$\frac{D\mathbf{v}}{Dt} = \frac{\partial \mathbf{v}}{\partial t} + \mathbf{v} \cdot \nabla \mathbf{v},$$

where ∇ is the GRADIENT operator.

see also ACCELERATION, CONVECTIVE DERIVATIVE, CONVECTIVE OPERATOR

References

Batchelor, G K. *An Introduction to Fluid Dynamics.* Cambridge, England: Cambridge University Press, p. 73, 1977.

Convective Derivative

A DERIVATIVE taken with respect to a moving coordinate system, also called a LAGRANGIAN DERIVATIVE. It is given by

$$\frac{D}{Dt} = \frac{\partial}{\partial t} + \mathbf{v} \cdot \nabla,$$

where ∇ is the GRADIENT operator and $\mathbf{v}$ is the VELOCITY of the fluid. This type of derivative is especially useful in the study of fluid mechanics. When applied to $\mathbf{v}$,

$$\frac{D\mathbf{v}}{Dt} = \frac{\partial \mathbf{v}}{\partial t} + (\nabla \times \mathbf{v}) \times \mathbf{v} + \nabla(\tfrac{1}{2}\mathbf{v}^2).$$

see also CONVECTIVE OPERATOR, DERIVATIVE, VELOCITY

References

Batchelor, G K. *An Introduction to Fluid Dynamics.* Cambridge, England: Cambridge University Press, p. 73, 1977.

Convective Operator

Defined for a VECTOR FIELD $\mathbf{A}$ by $(\mathbf{A} \cdot \nabla)$, where ∇ is the GRADIENT operator.

Applied in arbitrary orthogonal 3-D coordinates to a VECTOR FIELD $\mathbf{B}$, the convective operator becomes

$$[(\mathbf{A} \cdot \nabla)\mathbf{B}]_j = \sum_{k=1}^{3} \left[\frac{A_k}{h_k} \frac{\partial B_j}{\partial q_k} + \frac{B_k}{h_k h_j} \left(A_j \frac{\partial h_j}{\partial q_k} - A_k \frac{\partial h_k}{\partial q_j} \right) \right], \quad (1)$$

where the h_is are related to the METRIC TENSORS by $h_i = \sqrt{g_{ii}}$. In CARTESIAN COORDINATES,

$$(\mathbf{A}\cdot\nabla)\mathbf{B} = \left(A_x\frac{\partial B_x}{\partial x} + A_y\frac{\partial B_x}{\partial y} + A_z\frac{\partial B_x}{\partial z}\right)\hat{\mathbf{x}} + \left(A_x\frac{\partial B_y}{\partial x} + A_y\frac{\partial B_y}{\partial y} + A_z\frac{\partial B_y}{\partial z}\right)\hat{\mathbf{y}} + \left(A_x\frac{\partial B_z}{\partial x} + A_y\frac{\partial B_z}{\partial y} + A_z\frac{\partial B_z}{\partial z}\right)\hat{\mathbf{z}}. \quad (2)$$

In CYLINDRICAL COORDINATES,

$$(\mathbf{A}\cdot\nabla)\mathbf{B} = \left(A_r\frac{\partial B_r}{\partial r} + \frac{A_\phi}{r}\frac{\partial B_r}{\partial \phi} + A_z\frac{\partial B_r}{\partial z} - \frac{A_\phi B_\phi}{r}\right)\hat{\mathbf{r}} + \left(A_r\frac{\partial B_\phi}{\partial r} + \frac{A_\phi}{r}\frac{\partial B_\phi}{\partial \phi} + A_z\frac{\partial B_\phi}{\partial z} + \frac{A_\phi B_r}{r}\right)\hat{\boldsymbol{\phi}} + \left(A_r\frac{\partial B_z}{\partial r} + \frac{A_\phi}{r}\frac{\partial B_z}{\partial \phi} + A_z\frac{\partial B_z}{\partial z}\right)\hat{\mathbf{z}}. \quad (3)$$

In SPHERICAL COORDINATES,

$$(\mathbf{A}\cdot\nabla)\mathbf{B} = \left(A_r\frac{\partial B_r}{\partial r} + \frac{A_\theta}{r}\frac{\partial B_r}{\partial \theta} + \frac{A_\phi}{r\sin\theta}\frac{\partial B_r}{\partial \phi} - \frac{A_\theta B_\theta + A_\phi B_\phi}{r}\right)\hat{\mathbf{r}} + \left(A_r\frac{\partial B_\theta}{\partial r} + \frac{A_\theta}{r}\frac{\partial B_\theta}{\partial \theta} + \frac{A_\phi}{r\sin\theta}\frac{\partial B_\theta}{\partial \phi} + \frac{A_\theta B_r}{r} - \frac{A_\phi B_\phi\cot\theta}{r}\right)\hat{\boldsymbol{\theta}} + \left(A_r\frac{\partial B_\phi}{\partial r} + \frac{A_\theta}{r}\frac{\partial B_\phi}{\partial \theta} + \frac{A_\phi}{r\sin\theta}\frac{\partial B_\phi}{\partial \phi} + \frac{A_\phi B_r}{r} + \frac{A_\phi B_\theta\cot\theta}{r}\right)\hat{\boldsymbol{\phi}}. \quad (4)$$

see also CONVECTIVE ACCELERATION, CONVECTIVE DERIVATIVE, CURVILINEAR COORDINATES, GRADIENT

Convergence Acceleration

see CONVERGENCE IMPROVEMENT

Convergence Improvement

The improvement of the convergence properties of a SERIES, also called CONVERGENCE ACCELERATION, such that a SERIES reaches its limit to within some accuracy with fewer terms than required before. Convergence improvement can be effected by forming a linear combination with a SERIES whose sum is known. Useful sums include

$$\sum_{n=1}^{\infty}\frac{1}{n(n+1)} = 1 \quad (1)$$

$$\sum_{n=1}^{\infty}\frac{1}{n(n+1)(n+2)} = \frac{1}{4} \quad (2)$$

$$\sum_{n=1}^{\infty}\frac{1}{n(n+1)(n+2)(n+3)} = \frac{1}{18} \quad (3)$$

$$\sum_{n=1}^{\infty}\frac{1}{n(n+1)\cdots(n+p)} = \frac{1}{p\cdot p!}. \quad (4)$$

Kummer's transformation takes a convergent series

$$s = \sum_{k=0}^{\infty} a_k \quad (5)$$

and another convergent series

$$c = \sum_{k=0}^{\infty} c_k \quad (6)$$

with known c such that

$$\lim_{k\to\infty}\frac{a_k}{c_k} = \lambda \neq 0. \quad (7)$$

Then a series with more rapid convergence to the same value is given by

$$s = \lambda c + \sum_{k=0}^{\infty}\left(1 - \lambda\frac{c_k}{a_k}\right)a_k \quad (8)$$

(Abramowitz and Stegun 1972).

EULER'S TRANSFORM takes a convergent alternating series

$$\sum_{k=0}^{\infty}(-1)^k a_k = a_0 - a_1 + a_2 - \ldots \quad (9)$$

into a series with more rapid convergence to the same value to

$$s = \sum_{k=0}^{\infty}\frac{(-1)^k\Delta^k a_0}{2^{k+1}}, \quad (10)$$

where

$$\Delta^k a_0 = \sum_{m=0}^{k} \equiv (-1)^m\binom{k}{m}a_{k-m} \quad (11)$$

(Abramowitz and Stegun 1972; Beeler *et al.* 1972, Item 120).

Given a series of the form

$$S = \sum_{n=1}^{\infty} f\left(\frac{1}{n}\right), \quad (12)$$

where $f(z)$ is an ANALYTIC at 0 and on the closed unit DISK, and

$$f(z)|_{z\to 0} = \mathcal{O}(z^2), \quad (13)$$

then the series can be rearranged to

$$S = \sum_{n=1}^{\infty}\sum_{m=2}^{\infty} f_m\left(\frac{1}{n}\right)^m = \sum_{m=2}^{\infty}\sum_{n=1}^{\infty} f_m\left(\frac{1}{n}\right)^m = \sum_{m=2}^{\infty} f_m\zeta(m), \quad (14)$$

where

$$f(z) = \sum_{m=2}^{\infty} f_m z^m \qquad (15)$$

is the MACLAURIN SERIES of f and $\zeta(z)$ is the RIEMANN ZETA FUNCTION (Flajolet and Vardi 1996). The transformed series exhibits geometric convergence. Similarly, if $f(z)$ is ANALYTIC in $|z| \leq 1/n_0$ for some POSITIVE INTEGER n_0, then

$$S = \sum_{n=1}^{n_0-1} f\left(\frac{1}{n}\right) + \sum_{m=2}^{\infty} f_m \left[\zeta(m) - \frac{1}{1^m} - \ldots - \frac{1}{(n_0-1)^m}\right], \qquad (16)$$

which converges geometrically (Flajolet and Vardi 1996). (16) can also be used to further accelerate the convergence of series (14).

see also EULER'S TRANSFORM, WILF-ZEILBERGER PAIR

References

Abramowitz, M. and Stegun, C. A. (Eds.). *Handbook of Mathematical Functions with Formulas, Graphs, and Mathematical Tables, 9th printing.* New York: Dover, p. 16, 1972.

Arfken, G. *Mathematical Methods for Physicists, 3rd ed.* Orlando, FL: Academic Press, pp. 288–289, 1985.

Beeler, M.; Gosper, R. W.; and Schroeppel, R. *HAKMEM.* Cambridge, MA: MIT Artificial Intelligence Laboratory, Memo AIM-239, Feb. 1972.

Flajolet, P. and Vardi, I. "Zeta Function Expansions of Classical Constants." Unpublished manuscript. 1996. `http://pauillac.inria.fr/algo/flajolet/Publications/landau.ps`.

Convergence Tests

A test to determine if a given SERIES CONVERGES or DIVERGES.

see also ABEL'S UNIFORM CONVERGENCE TEST, BERTRAND'S TEST, D'ALEMBERT RATIO TEST, DIVERGENCE TESTS, ERMAKOFF'S TEST, GAUSS'S TEST, INTEGRAL TEST, KUMMER'S TEST, RAABE'S TEST, RATIO TEST, RIEMANN SERIES THEOREM, ROOT TEST

References

Arfken, G. "Convergence Tests." §5.2 in *Mathematical Methods for Physicists, 3rd ed.* Orlando, FL: Academic Press, pp. 280–293, 1985.

Bromwich, T. J. I'a and MacRobert, T. M. *An Introduction to the Theory of Infinite Series, 3rd ed.* New York: Chelsea, pp. 55–57, 1991.

Convergent

The RATIONAL NUMBER obtained by keeping only a limited number of terms in a CONTINUED FRACTION is called a convergent. For example, in the SIMPLE CONTINUED FRACTION for the GOLDEN RATIO,

$$\phi = 1 + \cfrac{1}{1 + \cfrac{1}{1 + \ldots}},$$

the convergents are

$$1, 1 + \frac{1}{1} = \frac{3}{2}, 1 + \frac{1}{1 + \frac{1}{1}} = \frac{5}{3}, \ldots.$$

The word convergent is also used to describe a CONVERGENT SEQUENCE or CONVERGENT SERIES.

see also CONTINUED FRACTION, CONVERGENT SEQUENCE, CONVERGENT SERIES, PARTIAL QUOTIENT, SIMPLE CONTINUED FRACTION

Convergent Sequence

A SEQUENCE S_n converges to the limit S

$$\lim_{n\to\infty} S_n = S$$

if, for any $\epsilon > 0$, there exists an N such that $|S_n - S| < \epsilon$ for $n > N$. If S_n does not converge, it is said to DIVERGE. Every bounded MONOTONIC SEQUENCE converges. Every unbounded SEQUENCE diverges. This condition can also be written as

$$\overline{\lim_{n\to\infty}} S_n = \underline{\lim_{n\to\infty}} S_n = S.$$

see also CONDITIONAL CONVERGENCE, STRONG CONVERGENCE, WEAK CONVERGENCE

Convergent Series

The infinite SERIES $\sum_{n=1}^{\infty} a_n$ is convergent if the SEQUENCE of partial sums

$$S_n = \sum_{k=1}^{n} a_k$$

is convergent. Conversely, a SERIES is divergent if the SEQUENCE of partial sums is divergent. If $\sum u_k$ and $\sum v_k$ are convergent SERIES, then $\sum(u_k + v_k)$ and $\sum(u_k - v_k)$ are convergent. If $c \neq 0$, then $\sum u_k$ and $c\sum u_k$ both converge or both diverge. Convergence and divergence are unaffected by deleting a finite number of terms from the beginning of a series. Constant terms in the denominator of a sequence can usually be deleted without affecting convergence. All but the highest POWER terms in POLYNOMIALS can usually be deleted in both NUMERATOR and DENOMINATOR of a SERIES without affecting convergence. If a SERIES converges absolutely, then it converges.

see also CONVERGENCE TESTS, RADIUS OF CONVERGENCE

References

Bromwich, T. J. I'a. and MacRobert, T. M. *An Introduction to the Theory of Infinite Series, 3rd ed.* New York: Chelsea, 1991.

Conversion Period

The period of time between INTEREST payments.

see also COMPOUND INTEREST, INTEREST, SIMPLE INTEREST

Convex

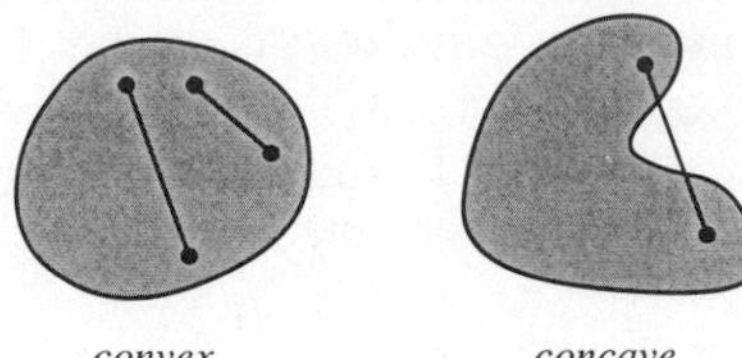

convex *concave*

A SET in EUCLIDEAN SPACE $\mathbb{R}^d$ is convex if it contains *all* the LINE SEGMENTS connecting any pair of its points. If the SET does not contain all the LINE SEGMENTS, it is called CONCAVE.

see also CONNECTED SET, CONVEX FUNCTION, CONVEX HULL, CONVEX OPTIMIZATION THEORY, CONVEX POLYGON, DELAUNAY TRIANGULATION, MINKOWSKI CONVEX BODY THEOREM, SIMPLY CONNECTED

References

Croft, H. T.; Falconer, K. J.; and Guy, R. K. "Convexity." Ch. A in *Unsolved Problems in Geometry.* New York: Springer-Verlag, pp. 6–47, 1994.

Convex Function

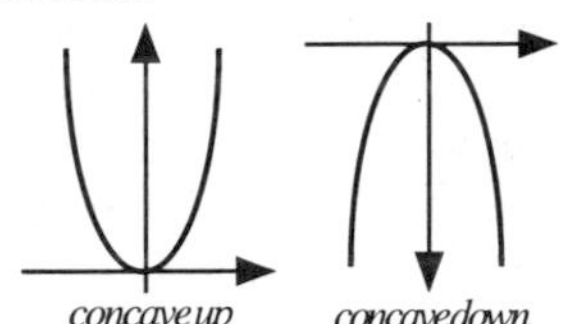

concave up *concave down*

A function whose value at the MIDPOINT of every INTERVAL in its DOMAIN does not exceed the AVERAGE of its values at the ends of the INTERVAL. In other words, a function $f(x)$ is convex on an INTERVAL $[a, b]$ if for any two points x_1 and x_2 in $[a, b]$,

$$f[\tfrac{1}{2}(x_1 + x_2)] \leq \tfrac{1}{2}[f(x_1) + f(x_2)].$$

If $f(x)$ has a second DERIVATIVE in $[a, b]$, then a NECESSARY and SUFFICIENT condition for it to be convex on that INTERVAL is that the second DERIVATIVE $f''(x) > 0$ for all x in $[a, b]$.

see also CONCAVE FUNCTION, LOGARITHMICALLY CONVEX FUNCTION

References

Eggleton, R. B. and Guy, R. K. "Catalan Strikes Again! How Likely is a Function to be Convex?" *Math. Mag.* **61**, 211–219, 1988.

Gradshteyn, I. S. and Ryzhik, I. M. *Tables of Integrals, Series, and Products, 5th ed.* San Diego, CA: Academic Press, p. 1100, 1980.

Convex Hull

The convex hull of a set of points S is the INTERSECTION of all convex sets containing S. For N points $p_1, \ldots, p_N$, the convex hull C is then given by the expression

$$C \equiv \left\{ \sum_{j=1}^{N} \lambda_j p_j : \lambda_j \geq 0 \text{ for all } j \text{ and } \sum_{j=1}^{N} \lambda_j = 1 \right\}.$$

see also CARATHÉODORY'S FUNDAMENTAL THEOREM, CROSS POLYTOPE, GROEMER PACKING, GROEMER THEOREM, SAUSAGE CONJECTURE, SYLVESTER'S FOUR-POINT PROBLEM

References

Santaló, L. A. *Integral Geometry and Geometric Probability.* Reading, MA: Addison-Wesley, 1976.

Convex Optimization Theory

The problem of maximizing a linear function over a convex polyhedron, also known as OPERATIONS RESEARCH or OPTIMIZATION THEORY. The general problem of convex optimization is to find the minimum of a convex (or quasiconvex) function f on a FINITE-dimensional convex body A. Methods of solution include Levin's algorithm and the method of circumscribed ELLIPSOIDS, also called the Nemirovsky-Yudin-Shor method.

References

Tokhomirov, V. M. "The Evolution of Methods of Convex Optimization." *Amer. Math. Monthly* **103**, 65–71, 1996.

Convex Polygon

A POLYGON is CONVEX if it contains *all* the LINE SEGMENTS connecting any pair of its points. Let $f(n)$ be the smallest number such that when W is a set of more than $f(n)$ points in GENERAL POSITION (with no three points COLLINEAR) in the plane, all of the VERTICES of some convex n-gon are contained in W. The answers for $n = 2$, 3, and 4 are 2, 4, and 8. It is conjectured that $f(n) = 2^{n-2}$, but only proven that

$$2^{n-2} \leq f(n) \leq \binom{2n-4}{n-2},$$

where $\binom{n}{k}$ is a BINOMIAL COEFFICIENT.

Convex Polyhedron

A POLYHEDRON for which a line connecting any two (noncoplanar) points on the surface always lies in the interior of the polyhedron. The 92 convex polyhedra having only REGULAR POLYGONS as faces are called the JOHNSON SOLIDS, which include the PLATONIC SOLIDS and ARCHIMEDEAN SOLIDS. No method is known for computing the VOLUME of a general convex polyhedron (Ogilvy 1990, p. 173).

see also ARCHIMEDEAN SOLID, DELTAHEDRON, JOHNSON SOLID, KEPLER-POINSOT SOLID, PLATONIC SOLID, REGULAR POLYGON

References

Ogilvy, C. S. *Excursions in Geometry.* New York: Dover, 1990.

Convolution

A convolution is an integral which expresses the amount of overlap of one function $g(t)$ as it is shifted over another function $f(t)$. It therefore "blends" one function with another. For example, in synthesis imaging, the measured DIRTY MAP is a convolution of the "true" CLEAN MAP with the DIRTY BEAM (the FOURIER TRANSFORM of the sampling distribution). The convolution is sometimes also known by its German name, FALTUNG ("folding"). A convolution over a finite range $[0, t]$ is given by

$$f(t) * g(t) \equiv \int_0^t f(\tau)g(t-\tau)\,d\tau, \tag{1}$$

where the symbol $f*g$ (occasionally also written as $f \otimes g$) denotes convolution of f and g. Convolution is more often taken over an infinite range,

$$f(t)*g(t) \equiv \int_{-\infty}^{\infty} f(\tau)g(t-\tau)\,d\tau = \int_{-\infty}^{\infty} g(\tau)f(t-\tau)\,d\tau. \tag{2}$$

Let f, g, and h be arbitrary functions and a a constant. Convolution has the following properties:

$$f * g = g * f \tag{3}$$

$$f * (g * h) = (f * g) * h \tag{4}$$

$$f * (g + h) = (f * g) + (f * h) \tag{5}$$

$$a(f * g) = (af) * g = f * (ag). \tag{6}$$

The INTEGRAL identity

$$\int_a^x \int_a^x f(t)\,dt\,dx = \int_a^x (x-t)f(t)\,dt \tag{7}$$

also gives a convolution. Taking the DERIVATIVE of a convolution gives

$$\frac{d}{dx}(f*g) = \frac{df}{dx} * g = f * \frac{dg}{dx}. \tag{8}$$

The AREA under a convolution is the product of areas under the factors,

$$\begin{aligned}\int_{-\infty}^{\infty} (f*g)\,dx &= \int_{-\infty}^{\infty}\left[\int_{-\infty}^{\infty} f(u)g(x-u)\,du\right] dx \\ &= \int_{-\infty}^{\infty} f(u)\left[\int_{-\infty}^{\infty} g(x-u)\,dx\right] du \\ &= \left[\int_{-\infty}^{\infty} f(u)\,du\right]\left[\int_{-\infty}^{\infty} g(x)\,dx\right].\end{aligned} \tag{9}$$

The horizontal CENTROIDS add

$$\int_{-\infty}^{\infty} \langle x(f*g)\rangle\,dx = \langle xf\rangle + \langle xg\rangle, \tag{10}$$

as do the VARIANCES

$$\int_{-\infty}^{\infty} \left\langle x^2(f*g)\right\rangle dx = \left\langle x^2 f\right\rangle + \left\langle x^2 g\right\rangle, \tag{11}$$

where

$$\langle x^n f\rangle \equiv \frac{\int_{-\infty}^{\infty} x^n f(x)\,dx}{\int_{-\infty}^{\infty} f(x)\,dx}. \tag{12}$$

see also AUTOCORRELATION, CONVOLUTION THEOREM, CROSS-CORRELATION, WIENER-KHINTCHINE THEOREM

References

Bracewell, R. "Convolution." Ch. 3 in *The Fourier Transform and Its Applications.* New York: McGraw-Hill, pp. 25–50, 1965.

Hirschman, I. I. and Widder, D. V. *The Convolution Transform.* Princeton, NJ: Princeton University Press, 1955.

Morse, P. M. and Feshbach, H. *Methods of Theoretical Physics, Part I.* New York: McGraw-Hill, pp. 464–465, 1953.

Press, W. H.; Flannery, B. P.; Teukolsky, S. A.; and Vetterling, W. T. "Convolution and Deconvolution Using the FFT." §13.1 in *Numerical Recipes in FORTRAN: The Art of Scientific Computing, 2nd ed.* Cambridge, England: Cambridge University Press, pp. 531–537, 1992.

Convolution Theorem

Let $f(t)$ and $f(t)$ be arbitrary functions of time t with FOURIER TRANSFORMS. Take

$$f(t) = \mathcal{F}^{-1}[F(\nu)] = \int_{-\infty}^{\infty} F(\nu)e^{2\pi i\nu t}\,d\nu \tag{1}$$

$$g(t) = \mathcal{F}^{-1}[G(\nu)] = \int_{-\infty}^{\infty} G(\nu)e^{2\pi i\nu t}\,d\nu, \tag{2}$$

where $\mathcal{F}^{-1}$ denotes the inverse FOURIER TRANSFORM (where the transform pair is defined to have constants $A = 1$ and $B = -2\pi$). Then the CONVOLUTION is

$$\begin{aligned}f*g &\equiv \int_{-\infty}^{\infty} g(t')f(t-t')dt' \\ &= \int_{-\infty}^{\infty} g(t')\left[\int_{-\infty}^{\infty} F(\nu)e^{2\pi i\nu(t-t')}\,d\nu\right] dt'.\end{aligned} \tag{3}$$

Interchange the order of integration,

$$\begin{aligned}f*g &= \int_{-\infty}^{\infty} F(\nu)\left[\int_{-\infty}^{\infty} g(t')e^{-2\pi i\nu t'}\,dt'\right] e^{2\pi i\nu t}\,d\nu \\ &= \int_{-\infty}^{\infty} F(\nu)G(\nu)e^{2\pi i\nu t}\,d\nu \\ &= \mathcal{F}^{-1}[F(\nu)G(\nu)].\end{aligned} \tag{4}$$

So, applying a FOURIER TRANSFORM to each side, we have

$$\mathcal{F}[f*g] = \mathcal{F}[f]\mathcal{F}[g]. \tag{5}$$

The convolution theorem also takes the alternate forms

$$\mathcal{F}[fg] = \mathcal{F}[f] * \mathcal{F}[g] \quad (6)$$

$$\mathcal{F}(\mathcal{F}[f]\mathcal{F}[g]) = f * g \quad (7)$$

$$\mathcal{F}(\mathcal{F}[f] * \mathcal{F}[g]) = fg. \quad (8)$$

see also AUTOCORRELATION, CONVOLUTION, FOURIER TRANSFORM, WIENER-KHINTCHINE THEOREM

References

Arfken, G. "Convolution Theorem." §15.5 in *Mathematical Methods for Physicists, 3rd ed.* Orlando, FL: Academic Press, pp. 810–814, 1985.

Bracewell, R. "Convolution Theorem." *The Fourier Transform and Its Applications.* New York: McGraw-Hill, pp. 108–112, 1965.

Conway-Alexander Polynomial

see ALEXANDER POLYNOMIAL

Conway's Constant

The constant

$$\lambda = 1.303577269034296\ldots$$

(Sloane's A014715) giving the asymptotic rate of growth $C\lambda^k$ of the number of DIGITS in the kth term of the LOOK AND SAY SEQUENCE. λ is given by the largest ROOT of the POLYNOMIAL

$$\begin{aligned}0 &= x^{71}\\ &- x^{69} - 2x^{68} - x^{67} + 2x^{66} + 2x^{65} + x^{64} - x^{63} - x^{62} - x^{61}\\ &- x^{60} - x^{59} + 2x^{58} + 5x^{57} + 3x^{56} - 2x^{55} - 10x^{54}\\ &- 3x^{53} - 2x^{52} + 6x^{51} + 6x^{50} + x^{49} + 9x^{48} - 3x^{47}\\ &- 7x^{46} - 8x^{45} - 8x^{44} + 10x^{43} + 6x^{42} + 8x^{41} - 4x^{40}\\ &- 12x^{39} + 7x^{38} - 7x^{37} + 7x^{36} + x^{35} - 3x^{34} + 10x^{33}\\ &+ x^{32} - 6x^{31} - 2x^{30} - 10x^{29} - 3x^{28} + 2x^{27} + 9x^{26}\\ &- 3x^{25} + 14x^{24} - 8x^{23} - 7x^{21} + 9x^{20} - 3x^{19} - 4x^{18}\\ &- 10x^{17} - 7x^{16} + 12x^{15} + 7x^{14} + 2x^{13} - 12x^{12}\\ &- 4x^{11} - 2x^{10} - 5x^{9} + x^{7} - 7x^{6}\\ &+ 7x^{5} - 4x^{4} + 12x^{3} - 6x^{2} + 3x - 6.\end{aligned}$$

The POLYNOMIAL given in Conway (1987, p. 188) contains a misprint. The CONTINUED FRACTION for λ is 1, 3, 3, 2, 2, 54, 5, 2, 1, 16, 1, 30, 1, 1, 1, 2, 2, 1, 14, 1, ... (Sloane's A014967).

see also CONWAY SEQUENCE, COSMOLOGICAL THEOREM, LOOK AND SAY SEQUENCE

References

Conway, J. H. "The Weird and Wonderful Chemistry of Audioactive Decay." §5.11 in *Open Problems in Communications and Computation* (Ed. T. M. Cover and B. Gopinath). New York: Springer-Verlag, pp. 173–188, 1987.

Conway, J. H. and Guy, R. K. "The Look and Say Sequence." In *The Book of Numbers.* New York: Springer-Verlag, pp. 208–209, 1996.

Finch, S. "Favorite Mathematical Constants." http://www.mathsoft.com/asolve/constant/cnwy/cnwy.html.

Sloane, N. J. A. Sequence A014967 in "An On-Line Version of the Encyclopedia of Integer Sequences."

Vardi, I. *Computational Recreations in Mathematica.* Reading, MA: Addison-Wesley, pp. 13–14, 1991.

Conway's Game of Life

see LIFE

Conway Groups

The AUTOMORPHISM GROUP Co_1 of the LEECH LATTICE modulo a center of order two is called "the" Conway group. There are 15 exceptional CONJUGACY CLASSES of the Conway group. This group, combined with the GROUPS Co_2 and Co_3 obtained similarly from the LEECH LATTICE by stabilization of the 1-D and 2-D sublattices, are collectively called Conway groups. The Conway groups are SPORADIC GROUPS.

see also LEECH LATTICE, SPORADIC GROUP

References

Wilson, R. A. "ATLAS of Finite Group Representation." http://for.mat.bham.ac.uk/atlas/Co1.html, Co2.html, Co3.html.

Conway's Knot

The KNOT with BRAID WORD

$${\sigma_2}^3\sigma_1{\sigma_3}^{-1}{\sigma_2}^{-2}\sigma_1{\sigma_2}^{-1}\sigma_1{\sigma_3}^{-1}.$$

The JONES POLYNOMIAL of Conway's knot is

$$t^{-4}(-1 + 2t - 2t^2 + 2t^3 + t^6 - 2t^7 + 2t^8 - 2t^9 + t^{10}),$$

the *same* as for the KINOSHITA-TERASAKA KNOT.

Conway's Knot Notation

A concise NOTATION based on the concept of the TANGLE used by Conway (1967) to enumerate KNOTS up to 11 crossings. An ALGEBRAIC KNOT containing no NEGATIVE signs in its Conway knot NOTATION is an ALTERNATING KNOT.

References

Conway, J. H. "An Enumeration of Knots and Links, and Some of Their Algebraic Properties." In *Computation Problems in Abstract Algebra* (Ed. J. Leech). Oxford, England: Pergamon Press, pp. 329–358, 1967.

Conway's Life

see LIFE

Conway Notation

see CONWAY'S KNOT NOTATION, CONWAY POLYHEDRON NOTATION

Conway Polyhedron Notation

A NOTATION for POLYHEDRA which begins by specifying a "seed" polyhedron using a capital letter. The PLATONIC SOLIDS are denoted T (TETRAHEDRON), O (OCTAHEDRON), C (CUBE), I (ICOSAHEDRON), and D (DODECAHEDRON), according to their first letter. Other polyhedra include the PRISMS, Pn, ANTIPRISMS, An, and PYRAMIDS, Yn, where $n \geq 3$ specifies the number of sides of the polyhedron's base.

Operations to be performed on the polyhedron are then specified with lower-case letters preceding the capital letter.

see also POLYHEDRON, SCHLÄFLI SYMBOL, WYTHOFF SYMBOL

References

Hart, G. "Conway Notation for Polyhedra." http://www.li.net/~george/virtual-polyhedra/conway_notation.html.

Conway Polynomial

see ALEXANDER POLYNOMIAL

Conway Puzzle

Construct a $5 \times 5 \times 5$ cube from 13 $1 \times 2 \times 4$ blocks, 1 $2 \times 2 \times 2$ block, 1 $1 \times 2 \times 2$ and 3 $1 \times 1 \times 3$ blocks.

see also BOX-PACKING THEOREM, CUBE DISSECTION, DE BRUIJN'S THEOREM, KLARNER'S THEOREM, POLYCUBE, SLOTHOUBER-GRAATSMA PUZZLE

References

Honsberger, R. *Mathematical Gems II.* Washington, DC: Math. Assoc. Amer., pp. 77–80, 1976.

Conway Sequence

The LOOK AND SAY SEQUENCE generated from a starting DIGIT of 3, as given by Vardi (1991).

see also CONWAY'S CONSTANT, LOOK AND SAY SEQUENCE

References

Vardi, I. *Computational Recreations in Mathematica.* Reading, MA: Addison-Wesley, pp. 13–14, 1991.

Conway Sphere

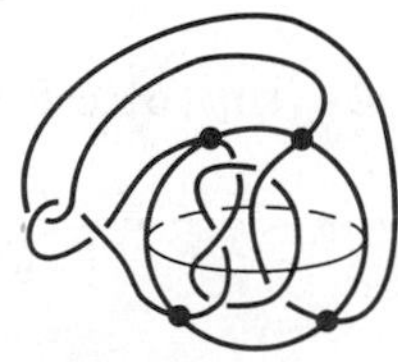

A sphere with four punctures occurring where a KNOT passes through the surface.

References

Adams, C. C. *The Knot Book: An Elementary Introduction to the Mathematical Theory of Knots.* New York: W. H. Freeman, p. 94, 1994.

Coordinate Geometry

see ANALYTIC GEOMETRY

Coordinate System

A system of COORDINATES.

Coordinates

A set of n variables which fix a geometric object. If the coordinates are distances measured along PERPENDICULAR axes, they are known as CARTESIAN COORDINATES. The study of GEOMETRY using one or more coordinate systems is known as ANALYTIC GEOMETRY.

see also AREAL COORDINATES, BARYCENTRIC COORDINATES, BIPOLAR COORDINATES, BIPOLAR CYLINDRICAL COORDINATES, BISPHERICAL COORDINATES, CARTESIAN COORDINATES, CHOW COORDINATES, CIRCULAR CYLINDRICAL COORDINATES, CONFOCAL ELLIPSOIDAL COORDINATES, CONFOCAL PARABOLOIDAL COORDINATES, CONICAL COORDINATES, CURVILINEAR COORDINATES, CYCLIDIC COORDINATES, CYLINDRICAL COORDINATES, ELLIPSOIDAL COORDINATES, ELLIPTIC CYLINDRICAL COORDINATES, GAUSSIAN COORDINATE SYSTEM, GRASSMANN COORDINATES, HARMONIC COORDINATES, HOMOGENEOUS COORDINATES, OBLATE SPHEROIDAL COORDINATES, ORTHOCENTRIC COORDINATES, PARABOLIC COORDINATES, PARABOLIC CYLINDRICAL COORDINATES, PARABOLOIDAL COORDINATES, PEDAL COORDINATES, POLAR COORDINATES, PROLATE SPHEROIDAL COORDINATES, QUADRIPLANAR COORDINATES, RECTANGULAR COORDINATES, SPHERICAL COORDINATES, TOROIDAL COORDINATES, TRILINEAR COORDINATES

References

Arfken, G. "Coordinate Systems." Ch. 2 in *Mathematical Methods for Physicists, 3rd ed.* Orlando, FL: Academic Press, pp. 85–117, 1985.

Woods, F. S. *Higher Geometry: An Introduction to Advanced Methods in Analytic Geometry.* New York: Dover, p. 1, 1961.

Coordination Number

see KISSING NUMBER

Copeland-Erdős Constant

The decimal 0.23571113171923... (Sloane's A033308) obtained by concatenating the PRIMES: 2, 23, 235, 2357, 235711, ... (Sloane's A033308; one of the SMARANDACHE SEQUENCES). In 1945, Copeland and Erdős showed that it is a NORMAL NUMBER. The first few digits of the CONTINUED FRACTION of the Copeland-Erdős are 0, 4, 4, 8, 16, 18, 5, 1, ... (Sloane's A030168). The positions of the first occurrence of n in the CONTINUED FRACTION are 8, 16, 20, 2, 7, 15, 12, 4, 17, 254, ... (Sloane's A033309). The incrementally largest terms are 1, 27, 154, 1601, 2135, ... (Sloane's A033310), which occur at positions 2, 5, 11, 19, 1801, ... (Sloane's A033311).

see also CHAMPERNOWNE CONSTANT, PRIME NUMBER

References

Sloane, N. J. A. Sequences A030168, A033308, A033309, A033310, and A033311 in "An On-Line Version of the Encyclopedia of Integer Sequences."

Coplanar

Three noncollinear points determine a plane and so are trivially coplanar. Four points are coplanar IFF the volume of the TETRAHEDRON defined by them is 0,

$$\begin{vmatrix} x_1 & y_1 & z_1 & 0 \\ x_2 & y_2 & z_2 & 0 \\ x_3 & y_3 & z_3 & 0 \\ x_4 & y_4 & z_4 & 0 \end{vmatrix}.$$

Coprime

see RELATIVELY PRIME

Copson-de Bruijn Constant

see DE BRUIJN CONSTANT

Copson's Inequality

Let $\{a_n\}$ be a NONNEGATIVE SEQUENCE and $f(x)$ a NONNEGATIVE integrable function. Define

$$A_n = \sum_{k=1}^{n} a_k \tag{1}$$

$$B_n = \sum_{k=n}^{\infty} a_k \tag{2}$$

and

$$F(x) = \int_0^x f(t)\,dt \tag{3}$$

$$G(x) = \int_x^\infty f(t)\,dt, \tag{4}$$

and take $0 < p < 1$. For integrals,

$$\int_0^\infty \left[\frac{G(x)}{x}\right]^p dx > \left(\frac{p}{p-1}\right)^p \int_0^\infty [f(x)]^p\,dx \tag{5}$$

(unless f is identically 0). For sums,

$$\left(1+\frac{1}{p-1}\right) B_1{}^p + \sum_{n=2}^{\infty}\left(\frac{B_n}{n}\right)^p > \left(\frac{p}{p-1}\right)^p \sum_{n=1}^{\infty} a_n{}^p \tag{6}$$

(unless all $a_n = 0$).

References

Beesack, P. R. "On Some Integral Inequalities of E. T. Copson." In *General Inequalities 2* (Ed. E. F. Beckenbach). Basel: Birkhäuser, 1980.

Copson, E. T. "Some Integral Inequalities." *Proc. Royal Soc. Edinburgh* **75A**, 157–164, 1975–1976.

Hardy, G. H.; Littlewood, J. E.; and Pólya, G. Theorems 326–327, 337–338, and 345 in *Inequalities*. Cambridge, England: Cambridge University Press, 1934.

Mitrinovic, D. S.; Pecaric, J. E.; and Fink, A. M. *Inequalities Involving Functions and Their Integrals and Derivatives.* Dordrecht, Netherlands: Kluwer, 1991.

Copula

A function that joins univariate distribution functions to form multivariate distribution functions. A 2-D copula is a function $C : I^2 \to I$ such that

$$C(0,t) = C(t,0) = 0$$

and

$$C(1,t) = C(t,1) = t$$

for all $t \in I$, and

$$C(u_2,v_2) - C(u_1,v_2) - C(u_2,v_1) + C(u_1,v_1) \geq 0$$

for all $u_1, u_2, v_1, v-2 \in I$ such that $u_1 \leq u_2$ and $v_1 \leq v-2$.

see also SKLAR'S THEOREM

Cork Plug

A 3-D SOLID which can stopper a SQUARE, TRIANGULAR, or CIRCULAR HOLE. There is an infinite family of such shapes. The one with smallest VOLUME has TRIANGULAR CROSS-SECTIONS and $V = \pi r^3$; that with the largest VOLUME is made using two cuts from the top diameter to the EDGE and has VOLUME $V = 4\pi r^3/3$.

see also STEREOLOGY, TRIP-LET

Corkscrew Surface

A surface also called the TWISTED SPHERE.

References

Gray, A. *Modern Differential Geometry of Curves and Surfaces.*Boca Raton, FL: CRC Press, pp. 493–494, 1993.

Cornish-Fisher Asymptotic Expansion

$$y \approx m + \sigma w,$$

where

$$\begin{aligned} w = x &+ [\gamma_1 h_1(x)] + [\gamma_2 h_2(x) + \gamma_1{}^2 h_{11}(x)] \\ &+ [\gamma_3 h_3(x) + \gamma_1\gamma_2 h_{12}(x) + \gamma_1{}^3 h_{111}(x)] \\ &+ [\gamma_4 h_4(x) + \gamma_2{}^2 h_{22}(x) + \gamma_1\gamma_3 h_{13}(x) \\ &+ \gamma_1{}^2\gamma_2 h_{112}(x) + \gamma_1{}^4 h_{1111}(x)] + \ldots, \end{aligned}$$

where

$$\begin{aligned}
h_1(x) &= \tfrac{1}{6}\,\mathrm{He}_2(x)\\
h_2(x) &= \tfrac{1}{24}\,\mathrm{He}_3(x)\\
h_{11}(x) &= -\tfrac{1}{36}[2\,\mathrm{He}_3(x)+\mathrm{He}_1(x)]\\
h_3(x) &= \tfrac{1}{120}\,\mathrm{He}_4(x)\\
h_{12}(x) &= -\tfrac{1}{24}[\mathrm{He}_4(x)+\mathrm{He}_2(x)]\\
h_{111}(x) &= \tfrac{1}{324}[12\,\mathrm{He}_4(x)+19\,\mathrm{He}_2(x)]\\
h_4(x) &= \tfrac{1}{720}\,\mathrm{He}_5(x)\\
h_{22}(x) &= -\tfrac{1}{384}[3\,\mathrm{He}_5(x)+6\,\mathrm{He}_3(x)+2\,\mathrm{He}_1(x)]\\
h_{13}(x) &= -\tfrac{1}{180}[2\,\mathrm{He}_5+3\,\mathrm{He}_3(x)]\\
h_{112}(x) &= \tfrac{1}{288}[14\,\mathrm{He}_5(x)+37\,\mathrm{He}_3(x)+8\,\mathrm{He}_1(x)]\\
h_{1111}(x) &= -\tfrac{1}{7776}[252\,\mathrm{He}_5(x)+832\,\mathrm{He}_3(x)+227\,\mathrm{He}_1(x)].
\end{aligned}$$

see also EDGEWORTH SERIES, GRAM-CHARLIER SERIES

References

Abramowitz, M. and Stegun, C. A. (Eds.). *Handbook of Mathematical Functions with Formulas, Graphs, and Mathematical Tables, 9th printing.* New York: Dover, p. 935, 1972.

Cornu Spiral

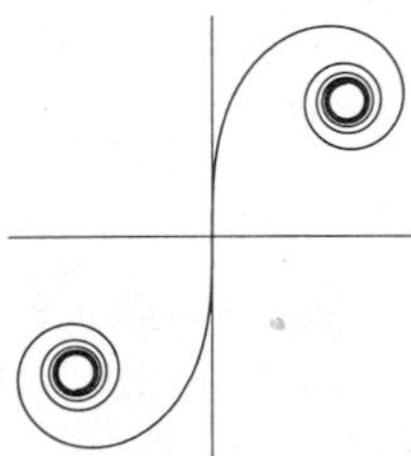

A plot in the COMPLEX PLANE of the points

$$B(z) = C(t) + iS(t) = \int_0^t e^{i\pi x^2/2}\,dx, \tag{1}$$

where $C(z)$ and $S(z)$ are the FRESNEL INTEGRALS. The Cornu spiral is also known as the CLOTHOID or EULER'S SPIRAL. A Cornu spiral describes diffraction from the edge of a half-plane.

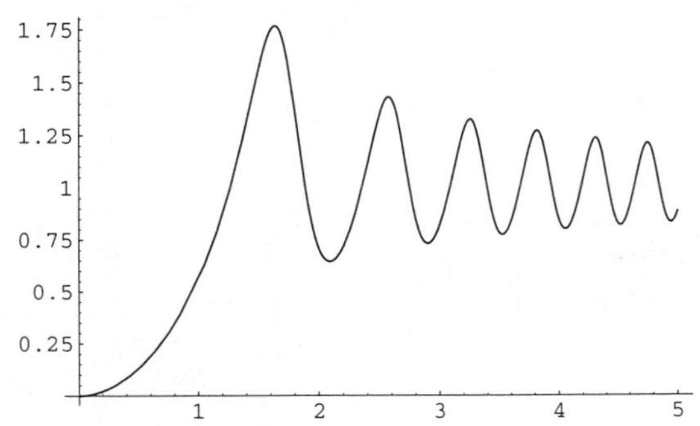

The SLOPE of the Cornu spiral

$$m(t) = \frac{S(t)}{C(t)} \tag{2}$$

is plotted above.

The SLOPE of the curve's TANGENT VECTOR (above right figure) is

$$m_T(t) = \frac{S'(t)}{C'(t)} = \tan(\tfrac{1}{2}\pi t^2), \tag{3}$$

plotted below.

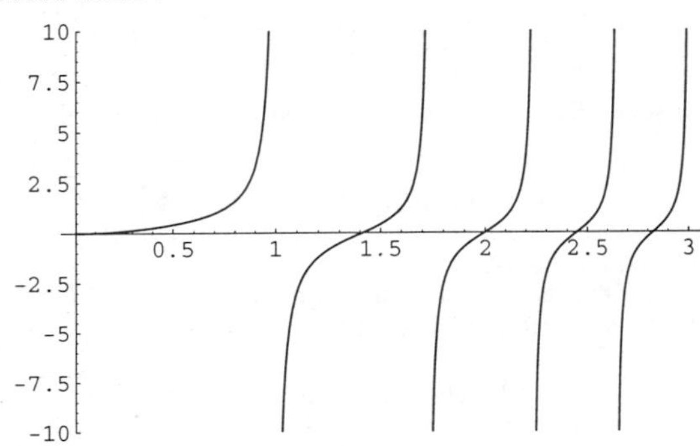

The CESÀRO EQUATION for a Cornu spiral is $\rho = c^2/s$, where ρ is the RADIUS OF CURVATURE and s the ARC LENGTH. The TORSION is $\tau = 0$.

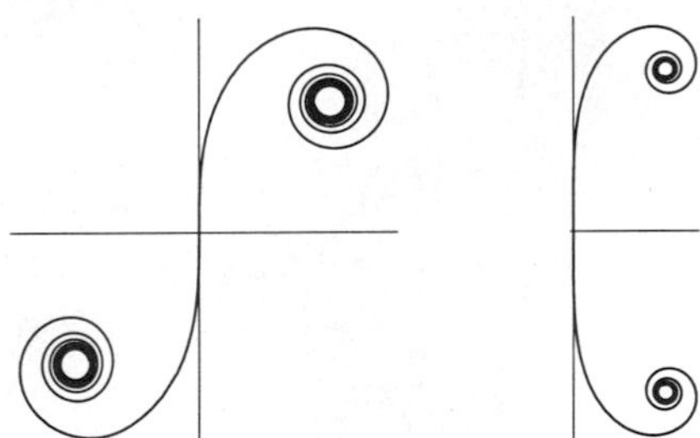

Gray (1993) defines a generalization of the Cornu spiral given by parametric equations

$$x(t) = a\int_0^t \sin\left(\frac{u^{n+1}}{n+1}\right)du \tag{4}$$

$$y(t) = a\int_0^t \cos\left(\frac{u^{n+1}}{n+1}\right)du. \tag{5}$$

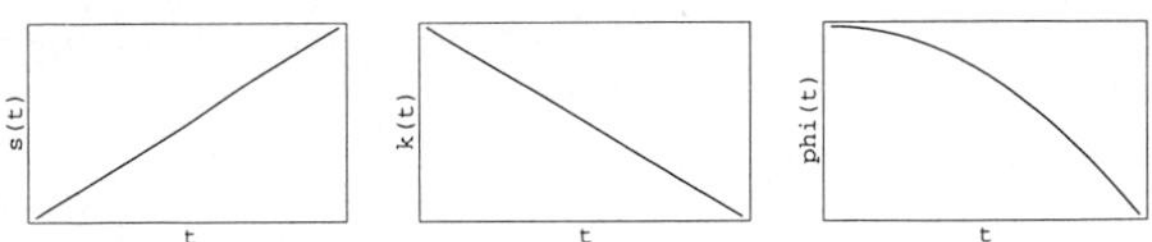

The ARC LENGTH, CURVATURE, and TANGENTIAL ANGLE of this curve are

$$s(t) = at \tag{3}$$

$$\kappa(t) = -\frac{t^n}{a} \tag{4}$$

$$\phi(t) = -\frac{t^{n+1}}{n+1}. \tag{5}$$

The CESÀRO EQUATION is

$$\kappa = -\frac{a}{s^n}. \tag{6}$$

Dillen (1990) describes a class of "polynomial spirals" for which the CURVATURE is a polynomial function of the ARC LENGTH. These spirals are a further generalization of the Cornu spiral.

see also FRESNEL INTEGRALS, NIELSEN'S SPIRAL

References

Dillen, F. "The Classification of Hypersurfaces of a Euclidean Space with Parallel Higher Fundamental Form." *Math. Z.* **203**, 635–643, 1990.

Gray, A. "Clothoids." §3.6 in *Modern Differential Geometry of Curves and Surfaces.* Boca Raton, FL: CRC Press, pp. 50–52, 1993.

Lawrence, J. D. *A Catalog of Special Plane Curves.* New York: Dover, pp. 190–191, 1972.

Cornucopia

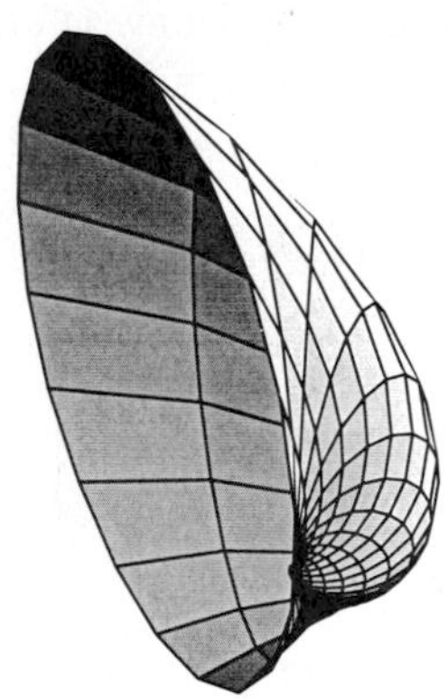

The SURFACE given by the parametric equations

$$\begin{aligned} x &= e^{bv}\cos v + e^{av}\cos u\cos v \\ y &= e^{bv}\sin v + e^{av}\cos u\sin v \\ z &= e^{av}\sin u. \end{aligned}$$

References

von Seggern, D. *CRC Standard Curves and Surfaces.* Boca Raton, FL: CRC Press, p. 304, 1993.

Corollary

An immediate consequence of a result already proved. Corollaries usually state more complicated THEOREMS in a language simpler to use and apply.

see also LEMMA, PORISM, THEOREM

Corona (Polyhedron)

see AUGMENTED SPHENOCORONA, HEBESPHENOMEGACORONA, SPHENOCORONA, SPHENOMEGACORONA

Corona (Tiling)

The first corona of a TILE is the set of all tiles that have a common boundary point with that tile (including the original tile itself). The second corona is the set of tiles that share a point with something in the first corona, and so on.

References

Eppstein, D. "Heesch's Problem." http://www.ics.uci.edu/~eppstein/junkyard/heesch.

Correlation

see AUTOCORRELATION, CORRELATION COEFFICIENT, CORRELATION (GEOMETRIC), CORRELATION (STATISTICAL), CROSS-CORRELATION

Correlation Coefficient

The correlation coefficient is a quantity which gives the quality of a LEAST SQUARES FITTING to the original data. To define the correlation coefficient, first consider the sum of squared values ss_{xx}, ss_{xy}, and ss_{yy} of a set of n data points (x_i, y_i) about their respective means,

$$\begin{aligned} \mathrm{ss}_{xx} &\equiv \Sigma(x_i-\bar{x})^2 = \Sigma x^2 - 2\bar{x}\Sigma x + \Sigma\bar{x}^2 \\ &= \Sigma x^2 - 2n\bar{x}^2 + n\bar{x}^2 = \Sigma x^2 - n\bar{x}^2 \qquad (1) \\ \mathrm{ss}_{yy} &\equiv \Sigma(y_i-\bar{y})^2 = \Sigma y^2 - 2\bar{y}\Sigma y + \Sigma\bar{y}^2 \\ &= \Sigma y^2 - 2n\bar{y}^2 + n\bar{y}^2 = \Sigma y^2 - n\bar{y}^2 \qquad (2) \\ \mathrm{ss}_{xy} &\equiv \Sigma(x_i-\bar{x})(y_i-\bar{y}) = \Sigma(x_iy_i - \bar{x}y_i - x_i\bar{y} + \bar{x}\bar{y}) \\ &= \Sigma xy - n\bar{x}\bar{y} - n\bar{x}\bar{y} + n\bar{x}\bar{y} = \Sigma xy - n\bar{x}\bar{y}. \qquad (3) \end{aligned}$$

For linear LEAST SQUARES FITTING, the COEFFICIENT b in

$$y = a + bx \tag{4}$$

is given by

$$b = \frac{n\sum xy - \sum x\sum y}{n\sum x^2 - \left(\sum x\right)^2} = \frac{\mathrm{ss}_{xy}}{\mathrm{ss}_{xx}}, \tag{5}$$

and the COEFFICIENT b' in

$$x = a' + b'y \tag{6}$$

is given by

$$b' = \frac{n\sum xy - \sum x\sum y}{n\sum y^2 - \left(\sum y\right)^2}. \tag{7}$$

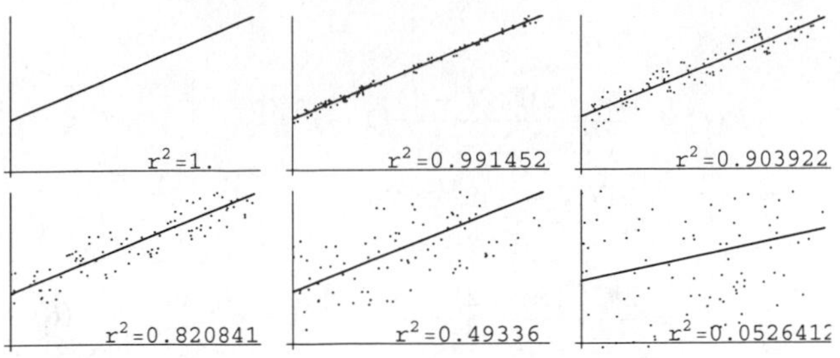

The correlation coefficient r^2 (sometimes also denoted R^2) is then defined by

$$r \equiv \sqrt{bb'} = \frac{n\sum xy - \sum x \sum y}{\sqrt{\left[n\sum x^2 - \left(\sum x\right)^2\right]\left[n\sum y^2 - \left(\sum y\right)^2\right]}}, \tag{8}$$

which can be written more simply as

$$r^2 = \frac{{\text{ss}_{xy}}^2}{\text{ss}_{xx}\text{ss}_{yy}}. \tag{9}$$

The correlation coefficient is also known as the PRODUCT-MOMENT COEFFICIENT OF CORRELATION or PEARSON'S CORRELATION. The correlation coefficients for linear fits to increasingly noise data are shown above.

The correlation coefficient has an important physical interpretation. To see this, define

$$A \equiv (\Sigma x^2 - n\bar{x}^2)^{-1} \tag{10}$$

and denote the "expected" value for y_i as $\hat{y}_i$. Sums of $\hat{y}_i$ are then

$$\begin{aligned}\hat{y}_i &= a + bx_i = \bar{y} - b\bar{x} + bx_i = \bar{x} + b(x_i - \bar{x})\\ &= A(\bar{y}\Sigma x^2 - \bar{x}\Sigma xy + x_i\Sigma xy - n\bar{x}\bar{y}x_i)\\ &= A[\bar{y}\Sigma x^2 + (x_i - \bar{x})\Sigma xy - n\bar{x}\bar{y}x_i] \end{aligned} \tag{11}$$

$$\Sigma\hat{y}_i = A(n\bar{y}\Sigma x^2 - n^2\bar{x}^2\bar{y}) \tag{12}$$

$$\begin{aligned}\Sigma\hat{y}_i^2 &= A^2[n\bar{y}^2(\Sigma x^2)^2 - n^2\bar{x}^2\bar{y}^2(\Sigma x^2)\\ &\quad - 2n\bar{x}\bar{y}(\Sigma xy)(\Sigma x^2) + 2n^2\bar{x}^3\bar{y}(\Sigma xy)\\ &\quad + (\Sigma x^2)(\Sigma xy)^2 - n\bar{x}^2(\Sigma xy)]\end{aligned} \tag{13}$$

$$\begin{aligned}\Sigma y_i\hat{y}_i &= A\Sigma[y_i\bar{y}\Sigma x^2 + y_i(x_i - \bar{x})\Sigma xy - n\bar{x}\bar{y}x_iy_i]\\ &= A[n\bar{y}^2\Sigma x^2 + (\Sigma xy)^2 - n\bar{x}\bar{y}\Sigma xy - n\bar{x}\bar{y}(\Sigma xy)]\\ &= A[n\bar{y}^2\Sigma x^2 + (\Sigma xy)^2 - 2n\bar{x}\bar{y}\Sigma xy].\end{aligned} \tag{14}$$

The sum of squared residuals is then

$$\begin{aligned}\text{SSR} &\equiv \Sigma(\hat{y}_i - \bar{y})^2 = \Sigma(\hat{y}_i^2 - 2\bar{y}\hat{y}_i + \bar{y}^2)\\ &= A^2(\Sigma xy - n\bar{x}\bar{y})^2(\Sigma x^2 - n\bar{x}^2) = \frac{(\Sigma xy - n\bar{x}\bar{y})^2}{\Sigma x^2 - n\bar{x}^2}\\ &= b\,\text{ss}_{xy} = \frac{{\text{ss}_{xy}}^2}{\text{ss}_{xx}} = \text{ss}_{yy}r^2 = b^2\text{ss}_{xx},\end{aligned} \tag{15}$$

and the sum of squared errors is

$$\begin{aligned}\text{SSE} &\equiv \Sigma(y_i - \hat{y}_i)^2 = \Sigma(y_i - \bar{y} + b\bar{x} - bx_i)^2\\ &= \Sigma[y_i - \bar{y} - b(x_i - \bar{x})]^2\\ &= \Sigma(y_i - \bar{y})^2 + b^2\Sigma(x_i - \bar{x})^2 - 2b\Sigma(x_i - \bar{x})(y_i - \bar{y})\\ &= \text{ss}_{yy} + b^2\text{ss}_{xx} - 2b\text{ss}_{xy}.\end{aligned} \tag{16}$$

But

$$b = \frac{\text{ss}_{xy}}{\text{ss}_{xx}} \tag{17}$$

$$r^2 = \frac{{\text{ss}_{xy}}^2}{\text{ss}_{xx}\text{ss}_{yy}}, \tag{18}$$

so

$$\begin{aligned}\text{SSE} &= \text{ss}_{yy} + \frac{{\text{ss}_{xy}}^2}{{\text{ss}_{xx}}^2}\text{ss}_{xx} - 2\frac{\text{ss}_{xy}}{\text{ss}_{xx}}\text{ss}_{xy}\\ &= \text{ss}_{yy} - \frac{{\text{ss}_{xy}}^2}{\text{ss}_{xx}}\end{aligned} \tag{19}$$

$$\begin{aligned}&= \text{ss}_{yy}\left(1 - \frac{{\text{ss}_{xy}}^2}{{\text{ss}_{xx}}^2}\right) = \text{ss}_{yy}(1 - r^2)\\ &= {s_y}^2 - {s_{\hat{y}}}^2,\end{aligned} \tag{20}$$

and

$$\text{SSE} + \text{SSR} = \text{ss}_{yy}(1 - r^2) + \text{ss}_{yy}r^2 = \text{ss}_{yy}. \tag{21}$$

The square of the correlation coefficient r^2 is therefore given by

$$r^2 \equiv \frac{\text{SSR}}{\text{ss}_{yy}} = \frac{{\text{ss}_{xy}}^2}{\text{ss}_{xx}\text{ss}_{yy}} = \frac{(\Sigma xy - n\bar{x}\bar{y})^2}{(\Sigma x^2 - n\bar{x}^2)(\Sigma y^2 - n\bar{y}^2)}. \tag{22}$$

In other words, r^2 is the proportion of ss_{yy} which is accounted for by the regression.

If there is complete correlation, then the lines obtained by solving for best-fit (a, b) and (a', b') coincide (since all data points lie on them), so solving (6) for y and equating to (4) gives

$$y = -\frac{a'}{b'} + \frac{x}{b'} = a + bx. \tag{23}$$

Therefore, $a = -a'/b'$ and $b = 1/b'$, giving

$$r^2 = bb' = 1. \tag{24}$$

The correlation coefficient is independent of both origin and scale, so

$$r(u, v) = r(x, y), \tag{25}$$

where

$$u \equiv \frac{x - x_0}{h} \tag{26}$$

$$v \equiv \frac{y - y_0}{h}. \tag{27}$$

see also CORRELATION INDEX, CORRELATION COEFFICIENT—GAUSSIAN BIVARIATE DISTRIBUTION, CORRELATION RATIO, LEAST SQUARES FITTING, REGRESSION COEFFICIENT

References
Acton, F. S. *Analysis of Straight-Line Data.* New York: Dover, 1966.
Kenney, J. F. and Keeping, E. S. "Linear Regression and Correlation." Ch. 15 in *Mathematics of Statistics, Pt. 1, 3rd ed.* Princeton, NJ: Van Nostrand, pp. 252–285, 1962.
Gonick, L. and Smith, W. *The Cartoon Guide to Statistics.* New York: Harper Perennial, 1993.
Press, W. H.; Flannery, B. P.; Teukolsky, S. A.; and Vetterling, W. T. "Linear Correlation." §14.5 in *Numerical Recipes in FORTRAN: The Art of Scientific Computing, 2nd ed.* Cambridge, England: Cambridge University Press, pp. 630–633, 1992.

Correlation Coefficient—Gaussian Bivariate Distribution

For a GAUSSIAN BIVARIATE DISTRIBUTION, the distribution of correlation COEFFICIENTS is given by

$$\begin{aligned}P(r) &= \frac{1}{\pi}(N-2)(1-r^2)^{(N-4)/2}(1-\rho^2)^{(N-1)/2} \\ &\quad\times\int_0^\infty \frac{d\beta}{(\cosh\beta-\rho r)^{N-1}} \\ &= \frac{1}{\pi}(N-2)(1-r^2)^{(N-4)/2}(1-\rho^2)^{(N-1)/2}\sqrt{\frac{\pi}{2}}\frac{\Gamma(N-1)}{\Gamma(N-\frac{1}{2})} \\ &\quad\times(1-\rho r)^{-(N-3/2)}{}_2F_1(\tfrac{1}{2},\tfrac{1}{2},\tfrac{2N-1}{2};\tfrac{\rho r+1}{2}) \\ &= \frac{(N-2)\Gamma(N-1)(1-\rho^2)^{(N-1)/2}(1-r^2)^{(N-4)/2}}{\sqrt{2\pi}\,\Gamma(N-\frac{1}{2})(1-\rho r)^{N-3/2}} \\ &\quad\times\left[1+\frac{1}{4}\frac{\rho r+1}{2N-1}+\frac{9}{16}\frac{(\rho r+1)^2}{(2N-1)(2N+1)}+\ldots\right], \end{aligned} \tag{1}$$

where ρ is the population correlation COEFFICIENT, ${}_2F_1(a,b;c;x)$ is a HYPERGEOMETRIC FUNCTION, and $\Gamma(z)$ is the GAMMA FUNCTION (Kenney and Keeping 1951, pp. 217–221). The MOMENTS are

$$\langle r\rangle = \rho - \frac{\rho(1-\rho^2)}{2n} \tag{2}$$

$$\operatorname{var}(r) = \frac{(1-\rho^2)^2}{n}\left(1+\frac{11\rho^2}{2n}+\cdots\right) \tag{3}$$

$$\gamma_1 = \frac{6\rho}{\sqrt{n}}\left(1+\frac{77\rho^2-30}{12n}+\cdots\right)$$

$$\gamma_2 = \frac{6}{n}(12\rho^2-1)+\ldots, \tag{4}$$

where $n \equiv N-1$. If the variates are uncorrelated, then $\rho = 0$ and

$$\begin{aligned}{}_2F_1(\tfrac{1}{2},\tfrac{1}{2},\tfrac{2N-1}{2};\tfrac{\rho r+1}{2}) &= {}_2F_1(\tfrac{1}{2},\tfrac{1}{2},\tfrac{2N-1}{2};\tfrac{1}{2}) \\ &= \frac{\Gamma(N-\frac{1}{2})2^{3/2-N}\sqrt{\pi}}{[\Gamma(\frac{N}{2})]^2},\end{aligned} \tag{5}$$

so

$$\begin{aligned}P(r) &= \frac{(N-2)\Gamma(N-1)}{\sqrt{2\pi}\,\Gamma(N-\frac{1}{2})}(1-r^2)^{(N-4)/2} \\ &\quad\times\frac{\Gamma(N-\frac{1}{2})2^{3/2-N}\sqrt{\pi}}{[\Gamma(\frac{N}{2})]^2} \\ &= \frac{2^{1-N}(N-2)\Gamma(N-1)}{[\Gamma(\frac{N}{2})]^2}(1-r^2)^{(N-4/2)}.\end{aligned} \tag{6}$$

But from the LEGENDRE DUPLICATION FORMULA,

$$\sqrt{\pi}\,\Gamma(N-1) = 2^{N-2}\Gamma(\tfrac{N}{2})\Gamma(\tfrac{N-1}{2}), \tag{7}$$

so

$$\begin{aligned}P(r) &= \frac{(2^{1-N})(2^{N-2})(N-2)\Gamma(\frac{N}{2})\Gamma(\frac{N-1}{2})}{\sqrt{\pi}\,[\Gamma(\frac{N}{2})]^2}(1-r^2)^{(N-4)/2} \\ &= \frac{(N-2)\Gamma(\frac{N-1}{2})}{2\sqrt{\pi}\,\Gamma(\frac{N}{2})}(1-r^2)^{(N-4)/2} \\ &= \frac{1}{\sqrt{\pi}}\frac{\frac{\nu}{2}\Gamma(\frac{\nu+1}{2})}{\Gamma(\frac{\nu}{2}+1)}(1-r^2)^{(\nu-2)/2} \\ &= \frac{1}{\sqrt{\pi}}\frac{\Gamma(\frac{\nu+1}{2})}{\Gamma(\frac{\nu}{2})}(1-r^2)^{(\nu-2)/2}.\end{aligned} \tag{8}$$

The uncorrelated case can be derived more simply by letting β be the true slope, so that $\eta = \alpha+\beta x$. Then

$$t \equiv (b-\beta)\frac{s_x}{s_y}\sqrt{\frac{N-2}{1-r^2}} = \frac{(b-\beta)r}{b}\sqrt{\frac{N-2}{1-r^2}} \tag{9}$$

is distributed as STUDENT'S t with $\nu \equiv N-2$ DEGREES OF FREEDOM. Let the population regression COEFFICIENT ρ be 0, then $\beta = 0$, so

$$t = r\sqrt{\frac{\nu}{1-r^2}}, \tag{10}$$

and the distribution is

$$P(t)\,dt = \frac{1}{\sqrt{\nu\pi}}\frac{\Gamma(\frac{\nu+1}{2})}{\Gamma(\frac{\nu}{2})\left(1+\frac{t^2}{\nu}\right)^{(\nu+1)/2}}\,dt. \tag{11}$$

Plugging in for t and using

$$\begin{aligned}dt &= \sqrt{\nu}\left[\frac{\sqrt{1-r^2}-r(\frac{1}{2})(-2r)(1-r^2)^{-1/2}}{1-r^2}\right]dr \\ &= \sqrt{\frac{\nu}{1-r^2}}\left(\frac{1-r^2+r^2}{1-r^2}\right)dr \\ &= \sqrt{\frac{\nu}{(1-r)^3}}\,dr\end{aligned} \tag{12}$$

gives

$$P(t)\,dt = \frac{1}{\sqrt{\nu\pi}} \frac{\Gamma(\frac{\nu+1}{2})}{\Gamma(\frac{\nu}{2})\left[1+\frac{r^2\nu}{(1-r^2)\nu}\right]^{(\nu+1)/2}} \sqrt{\frac{\nu}{(1-r)^3}}\,dr$$
$$= \frac{(1-r^2)^{-3/2}}{\sqrt{\pi}} \frac{\Gamma(\frac{\nu+1}{2})}{\Gamma(\frac{\nu}{2})\left(\frac{1}{1-r^2}\right)^{(\nu+1)/2}}\,dr$$
$$= \frac{1}{\sqrt{\pi}} \frac{\Gamma(\frac{\nu+1}{2})}{\Gamma(\frac{\nu}{2})}(1-r^2)^{-3/2}(1-r^2)^{(\nu+1)/2}\,dr$$
$$= \frac{1}{\sqrt{\pi}} \frac{\Gamma(\frac{\nu+1}{2})}{\Gamma(\frac{\nu}{2})}(1-r^2)^{(\nu-2)/2}\,dr, \tag{13}$$

so

$$P(r) = \frac{1}{\sqrt{\pi}} \frac{\Gamma\left(\frac{\nu+1}{2}\right)}{\Gamma\left(\frac{\nu}{2}\right)}(1-r^2)^{(\nu-2)/2} \tag{14}$$

as before. See Bevington (1969, pp. 122–123) or Pugh and Winslow (1966, §12-8). If we are interested instead in the probability that a correlation COEFFICIENT would be obtained $\geq |r|$, where r is the observed COEFFICIENT, then

$$P_c(r,N) = 2\int_{|r|}^{1} P(r',N)\,dr' = 1 - 2\int_0^{|r|} P(r',N)\,dr'$$
$$= 1 - \frac{2}{\sqrt{\pi}} \frac{\Gamma(\frac{\nu+1}{2})}{\Gamma(\frac{\nu}{2})} \int_0^{|r|} (1-r^2)^{(\nu-2)/2}\,dr. \tag{15}$$

Let $I \equiv \frac{1}{2}(\nu-2)$. For EVEN ν, the exponent I is an INTEGER so, by the BINOMIAL THEOREM,

$$(1-r^2)^I = \sum_{k=0}^{I} \binom{I}{k}(-r^2)^k \tag{16}$$

and

$$P_c(r) = 1 - \frac{2}{\sqrt{\pi}} \frac{\Gamma(\frac{\nu+1}{2})}{\Gamma(\frac{\nu}{2})}(-1)^k \frac{I!}{(I-k)!k!} \int_0^{|r|} \sum_{k=0}^{I} r'^{2k}\,dr'$$
$$= 1 - \frac{2}{\sqrt{\pi}} \frac{\Gamma(\frac{\nu+1}{2})}{\Gamma(\frac{\nu}{2})} \sum_{k=0}^{I} \left[(-1)^k \frac{I!}{(I-k)!k!} \frac{|r|^{2k+1}}{2k+1}\right]. \tag{17}$$

For ODD ν, the integral is

$$P_c(r) = 1 - 2\int_0^{|r|} P(r')\,dr'$$
$$= 1 - \frac{2}{\sqrt{\pi}} \frac{\Gamma(\frac{\nu+1}{2})}{\Gamma(\frac{\nu}{2})} \int_0^{|r|} (\sqrt{1-r^2}\,)^{\nu-2}\,dr. \tag{18}$$

Let $r \equiv \sin x$ so $dr = \cos x\,dx$, then

$$P_c(r) = 1 - \frac{2}{\sqrt{\pi}} \frac{\Gamma[(\frac{\nu+1}{2})]}{\Gamma(\frac{\nu}{2})} \int_0^{\sin^{-1}|r|} \cos^{\nu-2} x \cos x\,dx$$
$$= 1 - \frac{2}{\sqrt{\pi}} \frac{\Gamma(\frac{\nu+1}{2})}{\Gamma(\frac{\nu}{2})} + \int_0^{\sin^{-1}|r|} \cos^{\nu-1} x\,dx. \tag{19}$$

But ν is ODD, so $\nu - 1 \equiv 2n$ is EVEN. Therefore

$$\frac{2}{\sqrt{\pi}} \frac{\Gamma(\frac{\nu+1}{2})}{\Gamma(\frac{\nu}{2})} = \frac{2}{\sqrt{\pi}} \frac{\Gamma(n+1)}{\Gamma(n+\frac{1}{2})} = \frac{2}{\sqrt{\pi}} \frac{n!}{\frac{(2n-1)!!\sqrt{\pi}}{2^n}}$$
$$= \frac{2}{\pi} \frac{2^n n!}{(2n-1)!!} = \frac{2}{\pi} \frac{(2n)!!}{(2n-1)!!}. \tag{20}$$

Combining with the result from the COSINE INTEGRAL gives

$$P_c(r) = 1 - \frac{2}{\pi} \frac{(2n)!!(2n-1)!!}{(2n-1)!!(2n)!!}$$
$$\times \left[\sin x \sum_{k=0}^{n-1} \frac{(2k)!!}{(2k+1)!!} \cos^{2k+1} x + x\right]_0^{\sin^{-1}|r|}. \tag{21}$$

Use

$$\cos^{2k-1} x = (1-r^2)^{(2k-1)/2} = (1-r^2)^{(k-1/2)}, \tag{22}$$

and define $J \equiv n - 1 = (\nu-3)/2$, then

$$P_c(r)$$
$$= 1 - \frac{2}{\pi}\left[\sin^{-1}|r| + |r| \sum_{k=0}^{J} \frac{(2k)!!}{(2k+1)!!}(1-r^2)^{k+1/2}\right]. \tag{23}$$

(In Bevington 1969, this is given incorrectly.) Combining the correct solutions

$$P_c(r) = \begin{cases} 1 - \frac{2}{\sqrt{\pi}} \frac{\Gamma[(\nu+1)/2]}{\Gamma(\nu/2)} \sum_{k=0}^{I} \left[(-1)^k \frac{I!}{(I-k)!k!} \frac{|r|^{2k+1}}{2k+1}\right] \\ \text{for } \nu \text{ even} \\ 1 - \frac{2}{\pi}\left[\sin^{-1}|r| + |r| \sum_{k=0}^{J} \frac{(2k)!!}{(2k+1)!!}(1-r^2)^{k+1/2}\right] \\ \text{for } \nu \text{ odd} \end{cases} \tag{24}$$

If $\rho \neq 0$, a skew distribution is obtained, but the variable z defined by

$$z \equiv \tanh^{-1} r \tag{25}$$

is approximately normal with

$$\mu_z = \tanh^{-1} \rho \tag{26}$$
$${\sigma_z}^2 = \frac{1}{N-3} \tag{27}$$

(Kenney and Keeping 1962, p. 266).

Let b_j be the slope of a best-fit line, then the multiple correlation COEFFICIENT is

$$R^2 \equiv \sum_{j=1}^{n} \left(b_j \frac{{s_{jy}}^2}{{s_y}^2}\right) = \sum_{j=1}^{n} \left(b_j \frac{s_j}{s_y} r_{jy}\right), \tag{28}$$

where s_{jy} is the sample VARIANCE.

On the surface of a SPHERE,

$$r \equiv \frac{\int fg\,d\Omega}{\int f\,d\Omega \int g\,d\Omega}, \tag{29}$$

where $d\Omega$ is a differential SOLID ANGLE. This definition guarantees that $-1 < r < 1$. If f and g are expanded in REAL SPHERICAL HARMONICS,

$$f(\theta,\phi) \equiv \sum_{l=0}^{\infty}\sum_{m=0}^{l}[C_l^m Y_l^{m\,c}(\theta,\phi)\sin(m\phi) + S_l^m Y_l^{m\,s}(\theta,\phi)] \tag{30}$$

$$g(\theta,\phi) \equiv \sum_{l=0}^{\infty}\sum_{m=0}^{l}[A_l^m Y_l^{m\,c}(\theta,\phi)\sin(m\phi) + B_l^m Y_l^{m\,s}(\theta,\phi)]. \tag{31}$$

Then

$$r_l = \frac{\sum_{m=0}^{l}(C_l^m A_l^m + S_l^m B_l^m)}{\sqrt{\sum_{m=0}^{l}(C_l^{m\,2} + S_l^{m\,2})}\sqrt{\sum_{m=0}^{l}(A_l^{m\,2} + B_l^{m\,2})}}. \tag{32}$$

The confidence levels are then given by

$$\begin{aligned}
G_1(r) &= r \\
G_2(r) &= r(1 + \tfrac{1}{2}s^2) = \tfrac{1}{2}r(3 - r^2) \\
G_3(r) &= r[1 + \tfrac{1}{2}s^2(1 + \tfrac{3}{4}s^2)] = \tfrac{1}{8}r(15 - 10r^2 + 3r^4) \\
G_4(r) &= r\{1 + \tfrac{1}{2}s^2[1 + \tfrac{3}{4}s^2(1 + \tfrac{5}{6}s^2)]\} \\
&= \tfrac{1}{16}r(35 - 35r^2 + 21r^4 - 5r^6),
\end{aligned}$$

where

$$s \equiv \sqrt{1 - r^2} \tag{33}$$

(Eckhardt 1984).

see also FISHER'S z'-TRANSFORMATION, SPEARMAN RANK CORRELATION, SPHERICAL HARMONIC

References

Bevington, P. R. *Data Reduction and Error Analysis for the Physical Sciences.* New York: McGraw-Hill, 1969.

Eckhardt, D. H. "Correlations Between Global Features of Terrestrial Fields." *Math. Geology* **16**, 155–171, 1984.

Kenney, J. F. and Keeping, E. S. *Mathematics of Statistics, Pt. 1, 3rd ed.* Princeton, NJ: Van Nostrand, 1962.

Kenney, J. F. and Keeping, E. S. *Mathematics of Statistics, Pt. 2, 2nd ed.* Princeton, NJ: Van Nostrand, 1951.

Pugh, E. M. and Winslow, G. H. *The Analysis of Physical Measurements.* Reading, MA: Addison-Wesley, 1966.

Correlation Dimension

Define the correlation integral as

$$C(\epsilon) \equiv \lim_{n\to\infty}\frac{1}{N^2}\sum_{\substack{i,j=1\\ i\neq j}}^{\infty} H(\epsilon - ||x_i - x_j||), \tag{1}$$

where H is the HEAVISIDE STEP FUNCTION. When the below limit exists, the correlation dimension is then defined as

$$D_2 \equiv d_{\rm cor} \equiv \lim_{\epsilon,\epsilon'\to 0^+}\frac{\ln\left[\frac{C(\epsilon)}{C(\epsilon')}\right]}{\ln\left(\frac{\epsilon}{\epsilon'}\right)}. \tag{2}$$

If ν is the CORRELATION EXPONENT, then

$$\lim_{\epsilon\to 0}\nu \to D_2. \tag{3}$$

It satisfies

$$d_{\rm cor} \leq d_{\rm inf} \leq d_{\rm cap} \stackrel{?}{=} d_{\rm Lya}. \tag{4}$$

To estimate the correlation dimension of an M-dimensional system with accuracy $(1-Q)$ requires $N_{\rm min}$ data points, where

$$N_{\rm min} \geq \left[\frac{R(2-Q)}{2(1-Q)}\right]^M, \tag{5}$$

where $R \geq 1$ is the length of the "plateau region." If an ATTRACTOR exists, then an estimate of D_2 saturates above some M given by

$$M \geq 2D + 1, \tag{6}$$

which is sometimes known as the fractal Whitney embedding prevalence theorem.

see also CORRELATION EXPONENT, q-DIMENSION

References

Nayfeh, A. H. and Balachandran, B. *Applied Nonlinear Dynamics: Analytical, Computational, and Experimental Methods.* New York: Wiley, pp. 547–548, 1995.

Correlation Exponent

A measure ν of a STRANGE ATTRACTOR which allows the presence of CHAOS to be distinguished from random noise. It is related to the CAPACITY DIMENSION D and INFORMATION DIMENSION σ, satisfying

$$\nu \leq \sigma \leq D. \tag{1}$$

It satisfies

$$\nu \leq D_{\rm KY}, \tag{2}$$

where $D_{\rm KY}$ is the KAPLAN-YORKE DIMENSION. As the cell size goes to zero,

$$\lim_{\epsilon\to 0}\nu \to D_2, \tag{3}$$

where D_2 is the CORRELATION DIMENSION.

References

Grassberger, P. and Procaccia, I. "Measuring the Strangeness of Strange Attractors." *Physica D* **9**, 189–208, 1983.

Correlation (Geometric)

A point-to-line and line-to-point TRANSFORMATION which transforms points A into lines a' and lines b into points B' such that a' passes through B' IFF A' lies on b.

see also POLARITY

Correlation Index

$$r_c \equiv \frac{s_{y\hat{y}}}{s_y s_{\hat{y}}}$$

$$r_c{}^2 = \frac{s_{\hat{y}}{}^2}{s_y{}^2} = 1 - \frac{\text{SSE}}{s_y{}^2}.$$

see also CORRELATION COEFFICIENT

Correlation Integral

Consider a set of points $\mathbf{X}_i$ on an ATTRACTOR, then the correlation integral is

$$C(l) \equiv \lim_{N\to\infty} \frac{1}{N^2} f,$$

where f is the number of pairs (i, j) whose distance $|\mathbf{X}_i - \mathbf{X}_j| < l$. For small l,

$$C(l) \sim l^\nu,$$

where ν is the CORRELATION EXPONENT.

References

Grassberger, P. and Procaccia, I. "Measuring the Strangeness of Strange Attractors." *Physica D* **9**, 189–208, 1983.

Correlation Ratio

Let there be N_i observations of the ith phenomenon, where $i = 1, \ldots, p$ and

$$N \equiv \sum N_i \tag{1}$$

$$\bar{y}_i \equiv \frac{1}{N_i} \sum_\alpha y_{i\alpha} \tag{2}$$

$$\bar{y} \equiv \frac{1}{N} \sum_i \sum_\alpha y_{i\alpha}. \tag{3}$$

Then

$$E_{yx}{}^2 \equiv \frac{\sum_i N_i(\bar{y}_i - \bar{y})^2}{\sum_i \sum_\alpha (y_{i\alpha} - \bar{y})^2}. \tag{4}$$

Let η_{yx} be the population correlation ratio. If $N_i = N_j$ for $i \neq j$, then

$$f(E^2) = \frac{e^{-\lambda}(E^2)^{a-1}(1-E^2)^{b-1}{}_1F_1(a, b; \lambda E^2)}{B(a, b)}, \tag{5}$$

where

$$\lambda \equiv \frac{N\eta^2}{2(1-\eta^2)} \tag{6}$$

$$a \equiv \frac{n_1}{2} \tag{7}$$

$$b \equiv \frac{n_2}{2}, \tag{8}$$

and ${}_1F_1(a, b; z)$ is the CONFLUENT HYPERGEOMETRIC LIMIT FUNCTION. If $\lambda = 0$, then

$$f(E^2) = \beta(a, b) \tag{9}$$

(Kenney and Keeping 1951, pp. 323–324).

see also CORRELATION COEFFICIENT, REGRESSION COEFFICIENT

References

Kenney, J. F. and Keeping, E. S. *Mathematics of Statistics, Pt. 2, 2nd ed.* Princeton, NJ: Van Nostrand, 1951.

Correlation (Statistical)

For two variables x and y,

$$\text{cor}(x, y) \equiv \frac{\text{cov}(x, y)}{\sigma_x \sigma_y}, \tag{1}$$

where σ_x denotes STANDARD DEVIATION and $\text{cov}(x, y)$ is the COVARIANCE of these two variables. For the general case of variables x_i and x_j, where $i, j = 1, 2, \ldots, n$,

$$\text{cor}(x_i, x_j) = \frac{\text{cov}(x_i, x_j)}{\sqrt{V_{ii} V_{jj}}}, \tag{2}$$

where V_{ii} are elements of the COVARIANCE MATRIX. In general, a correlation gives the strength of the relationship between variables. The variance of any quantity is alway NONNEGATIVE by definition, so

$$\text{var}\left(\frac{x}{\sigma_x} + \frac{y}{\sigma_y}\right) \geq 0. \tag{3}$$

From a property of VARIANCES, the sum can be expanded

$$\text{var}\left(\frac{x}{\sigma_x}\right) + \text{var}\left(\frac{y}{\sigma_y}\right) + 2\,\text{cov}\left(\frac{x}{\sigma_x}, \frac{y}{\sigma_y}\right) \geq 0 \tag{4}$$

$$\frac{1}{\sigma_x{}^2}\,\text{var}(x) + \frac{1}{\sigma_y{}^2}\,\text{var}(y) + \frac{2}{\sigma_x\sigma_y}\,\text{cov}(x, y) \geq 0 \tag{5}$$

$$1 + 1 + \frac{2}{\sigma_x\sigma_y}\,\text{cov}(x, y) = 2 + \frac{2}{\sigma_x\sigma_y}\,\text{cov}(x, y) \geq 0. \tag{6}$$

Therefore,

$$\text{cor}(x, y) = \frac{\text{cov}(x, y)}{\sigma_x\sigma_y} \geq -1. \tag{7}$$

Similarly,

$$\text{var}\left(\frac{x}{\sigma_x}\right) - \left(\frac{y}{\sigma_y}\right) \geq 0 \tag{8}$$

$$\text{var}\left(\frac{x}{\sigma_x}\right) + \text{var}\left(-\frac{y}{\sigma_y}\right) + 2\,\text{cov}\left(\frac{x}{\sigma_x}, -\frac{y}{\sigma_y}\right) \geq 0 \tag{9}$$

$$\frac{1}{{\sigma_x}^2}\,\text{var}(x) + \frac{1}{{\sigma_y}^2}\,\text{var}(y) - \frac{2}{\sigma_x\sigma_y}\,\text{cov}(x,y) \geq 0 \tag{10}$$

$$1 + 1 - \frac{2}{\sigma_x\sigma_y}\,\text{cov}(x,y) = 2 - \frac{2}{\sigma_x\sigma_y}\,\text{cov}(x,y) \geq 0. \tag{11}$$

Therefore,

$$\text{cor}(x,y) = \frac{\text{cov}(x,y)}{\sigma_x\sigma_y} \leq 1, \tag{12}$$

so $-1 \leq \text{cor}(x,y) \leq 1$. For a linear combination of two variables,

$$\begin{aligned}\text{var}(y - bx) &= \text{var}(y) + \text{var}(-bx) + 2\,\text{cov}(y, -bx)\\ &= \text{var}(y) + b^2\,\text{var}(x) - 2b\,\text{cov}(x,y)\\ &= {\sigma_y}^2 + {\sigma_x}^2 - 2b\,\text{cov}(x,y).\end{aligned} \tag{13}$$

Examine the cases where $\text{cor}(x,y) = \pm 1$,

$$\text{cor}(x,y) \equiv \frac{\text{cov}(x,y)}{\sigma_x\sigma_y} = \pm 1 \tag{14}$$

$$\text{var}(y - bx) = b^2{\sigma_x}^2 + {\sigma_y}^2 \mp 2b\sigma_x\sigma_y = (b\sigma_x \mp \sigma_y)^2. \tag{15}$$

The VARIANCE will be zero if $b \equiv \pm\sigma_y/\sigma_x$, which requires that the argument of the VARIANCE is a constant. Therefore, $y - bx = a$, so $y = a + bx$. If $\text{cor}(x,y) = \pm 1$, y is either perfectly correlated ($b > 0$) or perfectly anticorrelated ($b < 0$) with x.

see also COVARIANCE, COVARIANCE MATRIX, VARIANCE

Cosecant

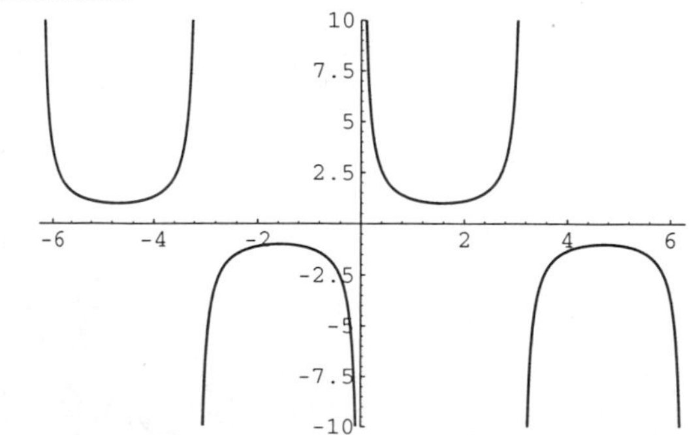

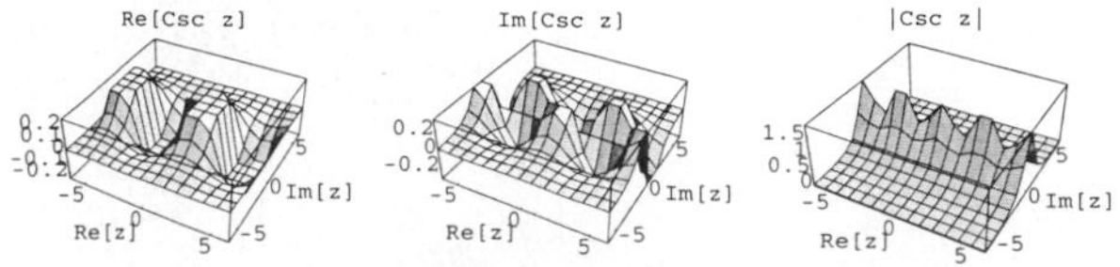

The function defined by $\csc x \equiv 1/\sin x$, where $\sin x$ is the SINE. The MACLAURIN SERIES of the cosecant function is

$$\begin{aligned}\csc x = \frac{1}{x} + \tfrac{1}{6}x + \tfrac{7}{360}x^3 + \tfrac{31}{15120}x^5 + \ldots\\ + \frac{(-1)^{n+1}2(2^{2n-1}-1)B_{2n}}{(2n)!}x^{2n-1} + \ldots,\end{aligned}$$

where B_{2n} is a BERNOULLI NUMBER.

see also INVERSE COSECANT, SECANT, SINE

References

Abramowitz, M. and Stegun, C. A. (Eds.). "Circular Functions." §4.3 in *Handbook of Mathematical Functions with Formulas, Graphs, and Mathematical Tables, 9th printing.* New York: Dover, pp. 71–79, 1972.

Spanier, J. and Oldham, K. B. "The Secant sec(x) and Cosecant csc(x) Functions." Ch. 33 in *An Atlas of Functions.* Washington, DC: Hemisphere, pp. 311–318, 1987.

Coset

Consider a countable SUBGROUP H with ELEMENTS h_i and an element x not in H, then

$$xh_i \tag{1}$$

$$h_ix \tag{2}$$

for $i = 1, 2, \ldots$ are left and right cosets of the SUBGROUP H with respect to x. The coset of a SUBGROUP has the same number of ELEMENTS as the SUBGROUP. The ORDER of any SUBGROUP is a divisor of the ORDER of the GROUP. The original GROUP can be represented by

$$G = H + x_1H + x_2H + \ldots. \tag{3}$$

For G a not necessarily FINITE GROUP with H a SUBGROUP of G, define an EQUIVALENCE RELATION $x \sim y$ if $x = hy$ for some h in H. Then the EQUIVALENCE CLASSES are the left (or right, depending on convention) cosets of H in G, namely the sets

$$\{x \in G : x = ha \text{ for some } h \text{ in } H\}, \tag{4}$$

where a is an element of G.

see also EQUIVALENCE CLASS, GROUP, SUBGROUP

Cosh

see HYPERBOLIC COSINE

Cosine

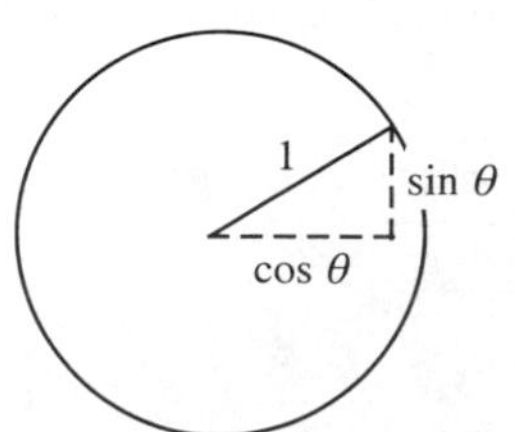

Let θ be an ANGLE measured counterclockwise from the x-axis along the arc of the unit CIRCLE. Then $\cos\theta$ is the horizontal coordinate of the arc endpoint. As a result of this definition, the cosine function is periodic with period 2π.

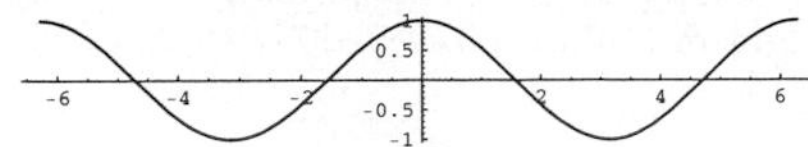

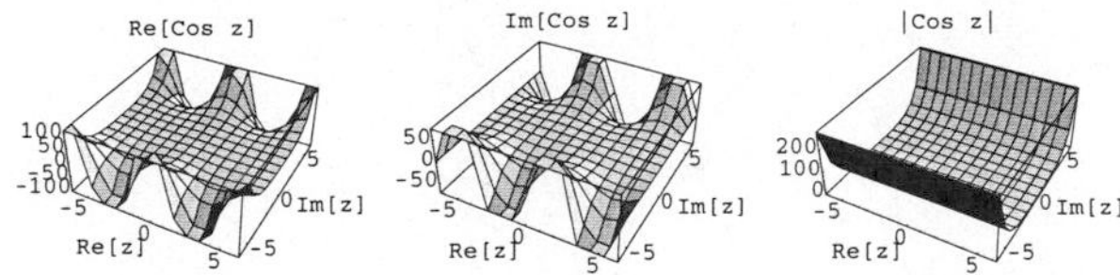

The cosine function has a FIXED POINT at 0.739085.

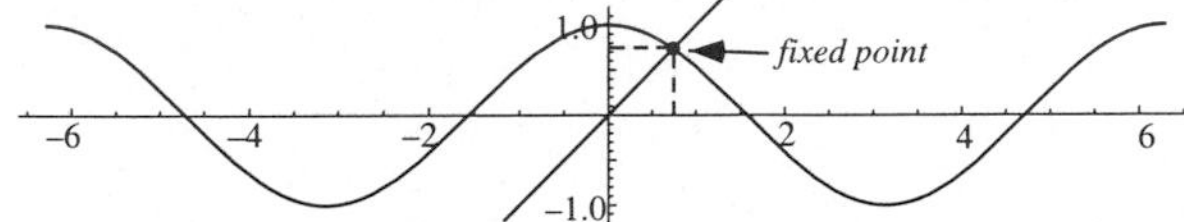

The cosine function can be defined algebraically using the infinite sum

$$\cos x \equiv \sum_{n=0}^{\infty} \frac{(-1)^n x^{2n}}{(2n)!} = 1 - \frac{x^2}{2!} + \frac{x^4}{4!} - \frac{x^6}{6!} + \ldots, \quad (1)$$

or the INFINITE PRODUCT

$$\cos x = \prod_{n=1}^{\infty} \left[1 - \frac{4x^2}{\pi^2(2n-1)^2}\right]. \quad (2)$$

A close approximation to $\cos(x)$ for $x \in [0, \pi/2]$ is

$$\cos\left(\frac{\pi}{2}x\right) \approx 1 - \frac{x^2}{x + (1-x)\sqrt{\frac{2-x}{3}}} \quad (3)$$

(Hardy 1959). The difference between $\cos x$ and Hardy's approximation is plotted below.

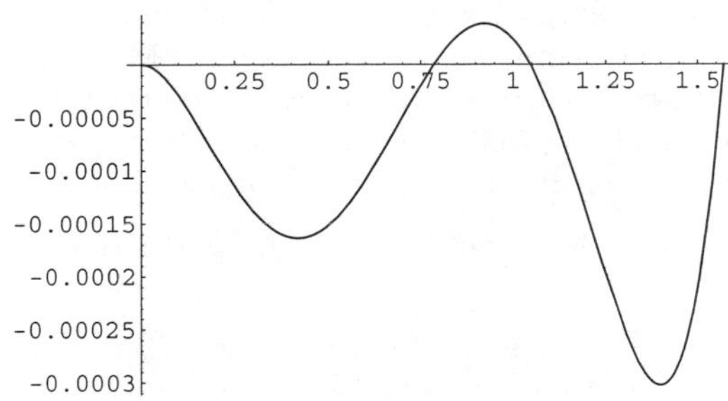

The FOURIER TRANSFORM of $\cos(2\pi k_0 x)$ is given by

$$\begin{aligned}\mathcal{F}[\cos(2\pi k_0 x)] &= \int_{-\infty}^{\infty} e^{-2\pi i k x} \cos(2\pi k_0 x)\, dx \\ &= \tfrac{1}{2}[\delta(k - k_0) + \delta(k + k_0)], \quad (4)\end{aligned}$$

where $\delta(k)$ is the DELTA FUNCTION.

The cosine sum rule gives an expansion of the COSINE function of a multiple ANGLE in terms of a sum of POWERS of sines and cosines,

$$\begin{aligned}\cos(n\theta) &= 2\cos\theta\cos[(n-1)\theta] - \cos[(n-2)\theta] \\ &= \cos^n\theta - \binom{n}{2}\cos^{n-2}\theta\sin^2\theta \\ &\quad + \binom{n}{4}\cos^{n-4}\theta\sin^4\theta - \ldots. \quad (5)\end{aligned}$$

Summing the COSINE of a multiple angle from $n = 0$ to $N - 1$ can be done in closed form using

$$\sum_{n=0}^{N-1} \cos(nx) = \Re\left[\sum_{n=0}^{N-1} e^{inx}\right]. \quad (6)$$

The EXPONENTIAL SUM FORMULAS give

$$\begin{aligned}\sum_{n=0}^{N-1} \cos(nx) &= \Re\left[\frac{\sin(\frac{1}{2}Nx)}{\sin(\frac{1}{2}x)} e^{i(N-1)x/2}\right] \\ &= \frac{\sin(\frac{1}{2}Nx)}{\sin(\frac{1}{2}x)} \cos[\tfrac{1}{2}x(N-1)]. \quad (7)\end{aligned}$$

Similarly,

$$\sum_{n=0}^{\infty} p^n \cos(nx) = \Re\left[\sum_{n=0}^{\infty} p^n e^{inx}\right], \quad (8)$$

where $|p| < 1$. The EXPONENTIAL SUM FORMULA gives

$$\begin{aligned}\sum_{n=0}^{\infty} p^n \cos(nx) &= \Re\left[\frac{1 - pe^{-ix}}{1 - 2p\cos x + p^2}\right] \\ &= \frac{1 - p\cos x}{1 - 2p\cos x + p^2}. \quad (9)\end{aligned}$$

Cvijović and Klinowski (1995) note that the following series

$$C_\nu(\alpha) = \sum_{k=0}^{\infty} \frac{\cos(2k+1)\alpha}{(2k+1)^\nu} \quad (10)$$

has closed form for $\nu = 2n$,

$$C_{2n}(\alpha) = \frac{(-1)^n}{4(2n-1)!} \pi^{2n} E_{2n-1}\left(\frac{\alpha}{\pi}\right), \quad (11)$$

where $E_n(x)$ is an EULER POLYNOMIAL.

see also EULER POLYNOMIAL, EXPONENTIAL SUM FORMULAS, FOURIER TRANSFORM—COSINE, HYPERBOLIC COSINE, SINE, TANGENT, TRIGONOMETRIC FUNCTIONS

References

Abramowitz, M. and Stegun, C. A. (Eds.). "Circular Functions." §4.3 in *Handbook of Mathematical Functions with*

Formulas, Graphs, and Mathematical Tables, 9th printing. New York: Dover, pp. 71–79, 1972.
Hardy, G. H. *Ramanujan: Twelve Lectures on Subjects Suggested by His Life and Work, 3rd ed.* New York: Chelsea, p. 68, 1959.
Cvijović, D. and Klinowski, J. "Closed-Form Summation of Some Trigonometric Series." *Math. Comput.* **64**, 205–210, 1995.
Hansen, E. R. *A Table of Series and Products.* Englewood Cliffs, NJ: Prentice-Hall, 1975.
Project Mathematics! *Sines and Cosines, Parts I–III.* Videotapes (28, 30, and 30 minutes). California Institute of Technology. Available from the Math. Assoc. Amer.
Spanier, J. and Oldham, K. B. "The Sine sin(x) and Cosine cos(x) Functions." Ch. 32 in *An Atlas of Functions.* Washington, DC: Hemisphere, pp. 295–310, 1987.

Cosine Apodization Function

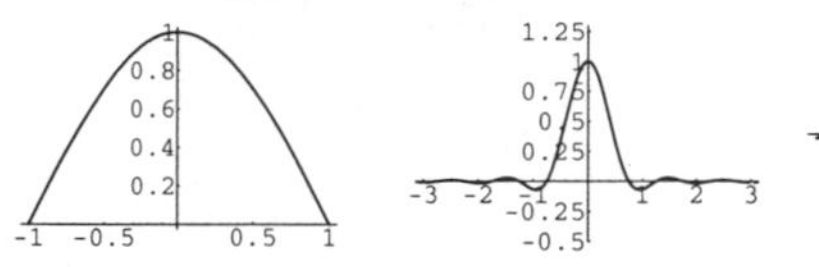

The APODIZATION FUNCTION

$$A(x) = \cos\left(\frac{\pi x}{2a}\right).$$

Its FULL WIDTH AT HALF MAXIMUM is $4a/3$. Its INSTRUMENT FUNCTION is

$$I(k) = \frac{4a\cos(2\pi ak)}{\pi(1-16a^2k^2)}.$$

see also APODIZATION FUNCTION

Cosine Circle

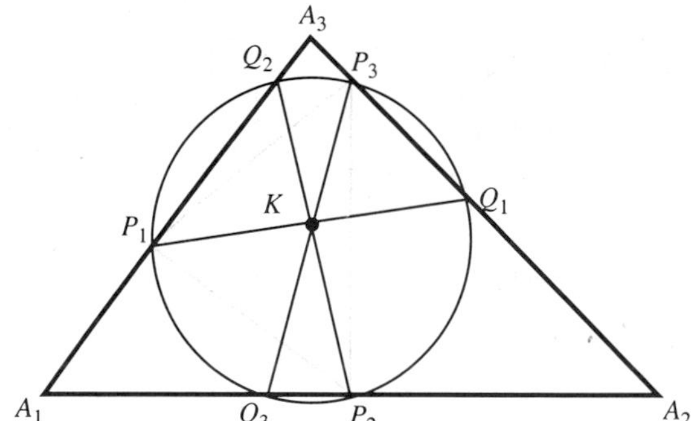

Also called the second LEMOINE CIRCLE. Draw lines through the LEMOINE POINT K and PARALLEL to the sides of the TRIANGLES. The points where the antiparallel lines intersect the sides then lie on a CIRCLE known as the cosine circle with center at K. The CHORDS P_2Q_3, P_3Q_1, and P_1Q_2 are proportional to the COSINES of the ANGLES of $\Delta A_1A_2A_3$, giving the circle its name.

TRIANGLES $P_1P_2P_3$ and $\Delta A_1A_2A_3$ are directly similar, and TRIANGLES $\Delta Q_1Q_2Q_3$ and $A_1A_2A_3$ are similar. The MIQUEL POINT of $\Delta P_1P_2P_3$ is at the BROCARD POINT Ω of $\Delta P_1P_2P_3$.

see also BROCARD POINTS, LEMOINE CIRCLE, MIQUEL POINT, TUCKER CIRCLES

References

Johnson, R. A. *Modern Geometry: An Elementary Treatise on the Geometry of the Triangle and the Circle.* Boston, MA: Houghton Mifflin, pp. 271–273, 1929.

Cosine Integral

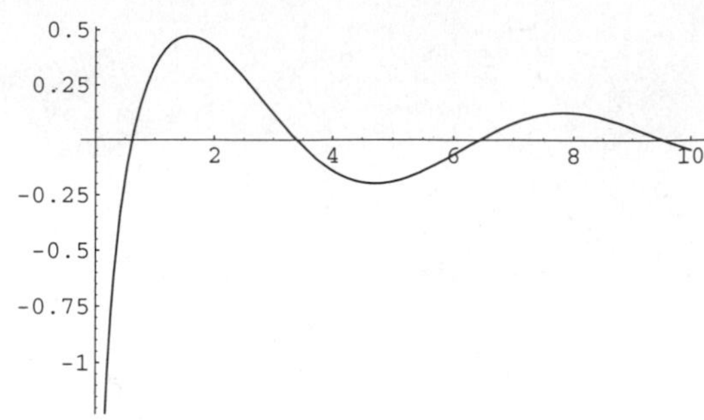

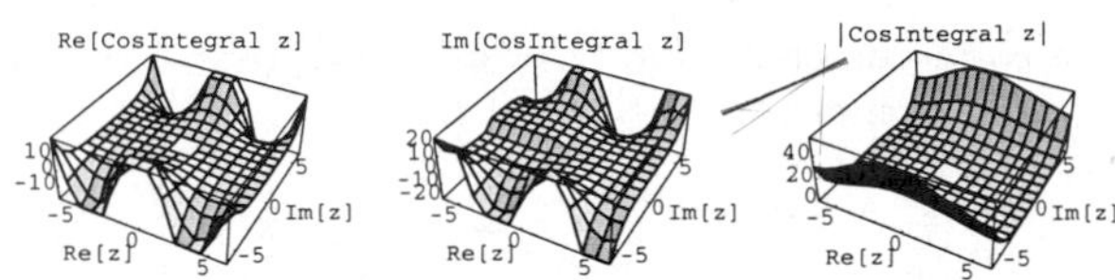

There are (at least) three types of "cosine integrals," denoted ci(x), Ci(x), and Cin(x):

$$\mathrm{ci}(x) \equiv -\int_x^\infty \frac{\cos t\,dt}{t} \quad (1)$$

$$= \tfrac{1}{2}[\mathrm{ei}(ix)+\mathrm{ei}(-ix)] \quad (2)$$

$$= -\tfrac{1}{2}[\mathrm{E}_1(ix)+\mathrm{E}_1(-ix)], \quad (3)$$

$$\mathrm{Ci}(x) \equiv \gamma + \ln z + \int_0^z \frac{\cos t - 1}{t}\,dt \quad (4)$$

$$\mathrm{Cin}(x) \equiv \int_0^z \frac{(1-\cos t)\,dt}{t} \quad (5)$$

$$= -\,\mathrm{Ci}(x) + \ln x + \gamma. \quad (6)$$

Here, ei(x) is the EXPONENTIAL INTEGRAL, $\mathrm{E}_n(x)$ is the E_n-FUNCTION, and γ is the EULER-MASCHERONI CONSTANT. ci(x) is the function returned by the *Mathematica*® (Wolfram Research, Champaign, IL) command `CosIntegral[x]` and displayed above.

To compute the integral of an EVEN power times a cosine,

$$I \equiv \int x^{2n}\cos(mx)\,dx, \quad (7)$$

use INTEGRATION BY PARTS. Let

$$u = x^{2n} \qquad dv = \cos(mx)\,dx \quad (8)$$

$$du = 2nx^{2n-1}\,dx \qquad v = \frac{1}{m}\sin(mx), \quad (9)$$

so

$$I = \frac{1}{m}x^{2n}\sin(mx) - \frac{2n}{m}\int x^{2n-1}\sin(mx)\,dx. \tag{10}$$

Using INTEGRATION BY PARTS again,

$$u = x^{2n-1} \qquad dv = \sin(mx)\,dx \tag{11}$$

$$du = (2n-1)x^{2n-2}\,dx \qquad v = -\frac{1}{m}\cos(mx), \tag{12}$$

and

$$\begin{aligned}\int x^{2n}\cos(mx)\,dx &= \frac{1}{m}x^{2n}\sin(mx) - \frac{2n}{m}\left[-\frac{1}{m}x^{2n-1}\cos(mx)\right.\\ &\quad\left.+\frac{2n-1}{m}\int x^{2n-2}\cos(mx)\,dx\right]\\ &= \frac{1}{m}x^{2n}\sin(mx) + \frac{2n}{m^2}x^{2n-1}\cos(mx)\\ &\quad-\frac{(2n)(2n-1)}{m^2}\int x^{2n-2}\cos(mx)\,dx\\ &= \frac{1}{m}x^{2n}\sin(mx) + \frac{2n}{m^2}x^{2n-1}\cos(mx)\\ &\quad+\ldots+\frac{(2n)!}{m^{2n}}\int x^0\cos(mx)\,dx\\ &= \frac{1}{m}x^{2n}\sin(mx) + \frac{2n}{m^2}x^{2n-1}\cos(mx)\\ &\quad+\ldots+\frac{(2n)!}{m^{2n+1}}\sin(mx)\end{aligned}$$

$$= \sin(mx)\sum_{k=0}^{n}(-1)^{k+1}\frac{(2n)!}{(2n-2k)!m^{2k+1}}x^{2n-2k} + \cos(mx)\sum_{k=1}^{n}(-1)^{k+1}\frac{(2n)!}{(2k-2n-1)!m^{2k}}x^{2n-2k+1}. \tag{13}$$

Letting $k' \equiv n-k$,

$$\begin{aligned}\int x^{2n}\cos(mx)\,dx &= \sin(mx)\sum_{k=0}^{n}(-1)^{n-k+1}\frac{(2n)!}{(2k)!m^{2n-2k+1}}x^{2k}\\ &\quad+\cos(mx)\sum_{k=0}^{n-1}(-1)^{n-k+1}\frac{(2n)!}{(2k-1)!m^{2n-2k}}x^{2k+1}\\ &= (-1)^{n+1}(2n)!\left[\sin(mx)\sum_{k=0}^{n}\frac{(-1)^k}{(2k)!m^{2n-2k+1}}x^{2k}\right.\\ &\quad\left.+\cos(mx)\sum_{k=1}^{n}\frac{(-1)^{k+1}}{(2k-3)!m^{2n-2k+2}}x^{2k-1}\right].\end{aligned} \tag{14}$$

To find a closed form for an integral power of a cosine function,

$$I \equiv \int \cos^m x\,dx, \tag{15}$$

perform an INTEGRATION BY PARTS so that

$$u = \cos^{m-1}x \qquad dv = \cos x\,dx \tag{16}$$

$$du = -(m-1)\cos^{m-2}x\sin x\,dx \qquad v = \sin x. \tag{17}$$

Therefore

$$\begin{aligned}I &= \sin x\cos^{m-1}x + (m-1)\int\cos^{m-2}x\sin^2 x\,dx\\ &= \sin x\cos^{m-1}x\\ &\quad+(m-1)\left[\int\cos^{m-2}x\,dx - \int\cos^m x\,dx\right]\\ &= \sin x\cos^{m-1}x + (m-1)\left[\int\cos^{m-2}x\,dx - I\right],\end{aligned} \tag{18}$$

so

$$I[1+(m-1)] = \sin x\cos^{m-1}x + (m-1)\int\cos^{m-2}x\,dx \tag{19}$$

$$I = \int\cos^m x\,dx = \frac{\sin x\cos^{m-1}x}{m} + \frac{m-1}{m}\int\cos^{m-2}x\,dx. \tag{20}$$

Now, if m is EVEN so $m \equiv 2n$, then

$$\begin{aligned}\int\cos^{2n}x\,dx &= \frac{\sin x\cos^{2n-1}x}{2n} + \frac{2n-1}{2n}\int\cos^{2n-2}x\,dx\\ &= \frac{\sin x\cos^{2n-1}x}{2n} + \frac{2n-1}{n}\left[\frac{\sin x\cos^{2n-3}x}{2n-2}\right.\\ &\quad\left.+\frac{2n-3}{2n-2}\int\cos^{2n-4}x\,dx\right]\\ &= \sin x\left[\frac{1}{2n}\cos^{2n-1}x + \frac{2n-1}{(2n)(2n-2)}\cos^{2n-3}x\right]\\ &\quad+\frac{(2n-1)(2n-3)}{(2n)(2n-2)}\int\cos^{2n-4}x\,dx\\ &= \sin x\left[\frac{1}{2n}\cos^{2n-1}x\right.\\ &\quad\left.+\frac{2n-1}{(2n)(2n-2)}\cos^{2n-3}x+\ldots\right]\\ &\quad+\frac{(2n-1)(2n-3)\cdots 1}{(2n)(2n-2)\cdots 2}\int\cos^0 x\,dx\\ &= \sin x\sum_{k=1}^{n}\frac{(2n-2k)!!}{(2n)!!}\frac{(2n-1)!!}{(2n-2k+1)!!}\cos^{2n-2k+1}x\\ &\quad+\frac{(2n-1)!!}{(2n)!!}x.\end{aligned} \tag{21}$$

Now let $k' \equiv n-k+1$, so $n-k=k'-1$,

$$\int \cos^{2n} x\,dx$$
$$= \sin x \sum_{k=1}^{n} \frac{(2k-2)!!}{(2n)!!}\frac{(2n-1)!!}{(2k-1)!!}\cos^{2k-1} x + \frac{(2n-1)!!}{(2n)!!} x$$
$$= \frac{(2n-1)!!}{(2n)!!}\left[\sin x \sum_{k=0}^{n-1} \frac{(2k)!!}{(2k+1)!!}\cos^{2k+1} x + x\right]. \qquad (22)$$

Now if m is ODD so $m \equiv 2n+1$, then

$$\int \cos^{2n+1} x\,dx$$
$$= \frac{\sin x \cos^{2n} x}{2n+1} + \frac{2n}{2n+1}\int \cos^{2n-1} x\,dx$$
$$= \frac{\sin x \cos^{2n} x}{2n+1} + \frac{2n}{2n+1}\left[\frac{\sin x \cos^{2n-2} x}{2n-1} + \frac{2n-2}{2n-1}\int \cos^{2n-3} x\,dx\right]$$
$$= \sin x\left[\frac{1}{2n+1}\cos^{2n} x + \frac{2n}{(2n+1)(2n-1)}\cos^{2n-2} x\right] + \frac{(2n)(2n-2)}{(2n+1)(2n-1)}\int \cos^{2n-3} x\,dx$$
$$= \sin x\left[\frac{1}{2n+1}\cos^{2n} x + \frac{2n}{(2n+1)(2n-1)}\cos^{2n-2} x + \ldots\right] + \frac{(2n)(2n-2)\cdots 2}{(2n+1)(2n-1)\cdots 3}\int \cos x\,dx$$
$$= \sin x \sum_{k=0}^{n} \frac{(2n-2k-1)!!}{(2n+1)!!}\frac{(2n)!!}{(2n-2k)!!}\cos^{2n-2k} x. \qquad (23)$$

Now let $k' \equiv n-k$,

$$\int \cos^{2n} x\,dx = \frac{(2n)!!}{(2n+1)!!}\sin x \sum_{k=0}^{n} \frac{(2k-1)!!}{(2k)!!}\cos^{2k} x. \qquad (24)$$

The general result is then

$$\int \cos^m x\,dx = \begin{cases} \frac{(2n-1)!!}{(2n)!!}\left[\sin x \sum_{k=0}^{n-1} \frac{(2k)!!}{(2k+1)!!}\cos^{2k+1} x + x\right] \\ \quad \text{for } m = 2n \\ \frac{(2n)!!}{(2n+1)!!}\sin x \sum_{k=0}^{n} \frac{(2k-1)!!}{(2k)!!}\cos^{2k} x \\ \quad \text{for } m = 2n+1. \end{cases} \qquad (25)$$

The infinite integral of a cosine times a Gaussian can also be done in closed form,

$$\int_{-\infty}^{\infty} e^{-ax^2}\cos(kx)\,dx = \sqrt{\frac{\pi}{a}}\,e^{-k^2/4a}. \qquad (26)$$

see also CHI, DAMPED EXPONENTIAL COSINE INTEGRAL, NIELSEN'S SPIRAL, SHI, SICI SPIRAL, SINE INTEGRAL

References

Abramowitz, M. and Stegun, C. A. (Eds.). "Sine and Cosine Integrals." §5.2 in *Handbook of Mathematical Functions with Formulas, Graphs, and Mathematical Tables, 9th printing.* New York: Dover, pp. 231–233, 1972.

Arfken, G. *Mathematical Methods for Physicists, 3rd ed.* Orlando, FL: Academic Press, pp. 342–343, 1985.

Press, W. H.; Flannery, B. P.; Teukolsky, S. A.; and Vetterling, W. T. "Fresnel Integrals, Cosine and Sine Integrals." §6.79 in *Numerical Recipes in FORTRAN: The Art of Scientific Computing, 2nd ed.* Cambridge, England: Cambridge University Press, pp. 248–252, 1992.

Spanier, J. and Oldham, K. B. "The Cosine and Sine Integrals." Ch. 38 in *An Atlas of Functions.* Washington, DC: Hemisphere, pp. 361–372, 1987.

Cosines Law

see LAW OF COSINES

Cosmic Figure

see PLATONIC SOLID

Cosmological Theorem

There exists an INTEGER N such that every string in the LOOK AND SAY SEQUENCE "decays" in at most N days to a compound of "common" and "transuranic elements."

The table below gives the periodic table of atoms associated with the LOOK AND SAY SEQUENCE as named by Conway (1987). The "abundance" is the average number of occurrences for long strings out of every million atoms. The asymptotic abundances are zero for transuranic elements, and 27.246... for arsenic (As), the next rarest element. The most common element is hydrogen (H), having an abundance of 91,970.383.... The starting element is U, represented by the string "3," and subsequent terms are those giving a description of the current term: one three (13); one one, one three (1113); three ones, one three (3113), etc.

Abundance	n	E_n	E_n is the derivate of E_{n+1}
102.56285249	92	U	3
9883.5986392	91	Pa	12
7581.9047125	90	Th	1113
6926.9352045	89	Ac	3113
5313.7894999	88	Ra	132113
4076.3134078	87	Fr	1113122113
3127.0209328	86	Rn	311311222113
2398.7998311	85	At	Ho.1322113
1840.1669683	84	Po	1113222113
1411.6286100	83	Bi	3113322113
1082.8883285	82	Pb	Pm.123222113
830.70513293	81	Tl	111213322113
637.25039755	80	Hg	31121123222113
488.84742982	79	Au	132112211213322113
375.00456738	78	Pt	111312212221121123222113
287.67344775	77	Ir	3113112211322112211213322113
220.68001229	76	Os	1321132122211322212221121123222113
169.28801808	75	Re	111312211312113221133211322112211213322113
315.56655252	74	W	Ge.Ca.312211322212221121123222113
242.07736666	73	Ta	13112221133211322112211213322113
2669.0970363	72	Hf	11132.Pa.H.Ca.W
2047.5173200	71	Lu	311312
1570.6911808	70	Yb	1321131112
1204.9083841	69	Tm	11131221133112
1098.5955997	68	Er	311311222.Ca.Co
47987.529438	67	Ho	1321132.Pm
36812.186418	66	Dy	111312211312
28239.358949	65	Tb	3113112221131112
21662.972821	64	Gd	Ho.13221133112
20085.668709	63	Eu	1113222.Ca.Co
15408.115182	62	Sm	311332
29820.456167	61	Pm	132.Ca.Zn
22875.863883	60	Nd	111312
17548.529287	59	Pr	31131112
13461.825166	58	Ce	1321133112
10326.833312	57	La	11131.H.Ca.Co
7921.9188284	56	Ba	311311
6077.0611889	55	Cs	13211321
4661.8342720	54	Xe	11131221131211
3576.1856107	53	I	311311222113111221
2743.3629718	52	Te	Ho.1322113312211
2104.4881933	51	Sb	Eu.Ca.3112221
1614.3946687	50	Sn	Pm.13211
1238.4341972	49	In	11131221
950.02745646	48	Cd	3113112211
728.78492056	47	Ag	132113212221
559.06537946	46	Pd	111312211312113211
428.87015041	45	Rh	311311222113111221131221
328.99480576	44	Ru	Ho.132211331222113112211
386.07704943	43	Tc	Eu.Ca.311322113212221
296.16736852	42	Mo	13211322211312113211
227.19586752	41	Nb	1113122113322113111221131221
174.28645997	40	Zr	Er.12322211331222113112211
133.69860315	39	Y	1112133.H.Ca.Tc
102.56285249	38	Sr	3112112.U
78.678000089	37	Rb	1321122112
60.355455682	36	Kr	11131221222112
46.299868152	35	Br	3113112211322112

Abundance	n	E_n	E_n is the derivate of E_{n+1}
35.517547944	34	Se	13211321222113222112
27.246216076	33	As	1113122113121132211332112
1887.4372276	32	Ge	31131122211311122113222.Na
1447.8905642	31	Ga	Ho.13221133122211332
23571.391336	30	Zn	Eu.Ca.Ac.H.Ca.312
18082.082203	29	Cu	131112
13871.123200	28	Ni	11133112
45645.877256	27	Co	Zn.32112
35015.858546	26	Fe	13122112
26861.360180	25	Mn	111311222112
20605.882611	24	Cr	31132.Si
15807.181592	23	V	13211312
12126.002783	22	Ti	11131221131112
9302.0974443	21	Sc	3113112221133112
56072.543129	20	Ca	Ho.Pa.H.12.Co
43014.360913	19	K	1112
32997.170122	18	Ar	3112
25312.784218	17	Cl	132112
19417.939250	16	S	1113122112
14895.886658	15	P	311311222112
32032.812960	14	Si	Ho.1322112
24573.006696	13	Al	1113222112
18850.441228	12	Mg	3113322112
14481.448773	11	Na	Pm.123222112
11109.006696	10	Ne	111213322112
8521.9396539	9	F	31121123222112
6537.3490750	8	O	132112211213322112
5014.9302464	7	N	111312212221121123222112
3847.0525419	6	C	3113112211322112211213322112
2951.1503716	5	B	1321132122211322212221121123222112
2263.8860325	4	Be	111312211312113221133211322112211213322112
4220.0665982	3	Li	Ge.Ca.312211322212221121123222122
3237.2968588	2	He	13112221133211322112211213322112
91790.383216	1	H	Hf.Pa.22.Ca.Li

see also CONWAY'S CONSTANT, LOOK AND SAY SEQUENCE

References

Conway, J. H. "The Weird and Wonderful Chemistry of Audioactive Decay." §5.11 in *Open Problems in Communication and Computation* (Ed. T. M. Cover and B. Gopinath). New York: Springer-Verlag, pp. 173–188, 1987.

Conway, J. H. "The Weird and Wonderful Chemistry of Audioactive Decay." *Eureka,* 5–18, 1985.

Ekhad, S. B. and Zeilberger, D. "Proof of Conway's Lost Cosmological Theorem." *Electronic Research Announcement of the Amer. Math. Soc.* **3**, 78–82, 1997. `http://www.mathtemple.edu/~zeilberg/mamarim/mamarimhtml/horton.html`.

Costa-Hoffman-Meeks Minimal Surface

see COSTA MINIMAL SURFACE

Costa Minimal Surface

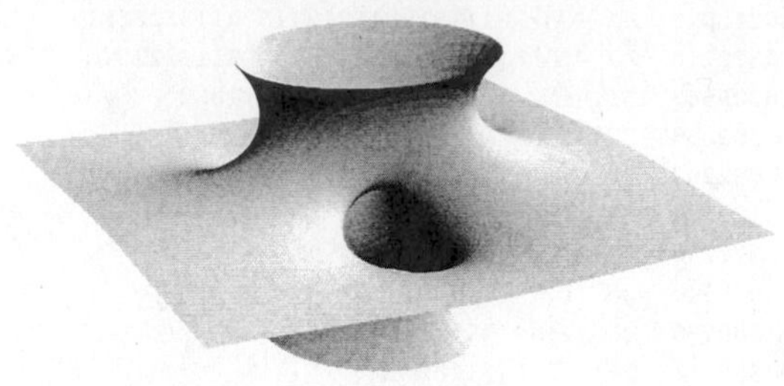

A complete embedded MINIMAL SURFACE of finite topology. It has no BOUNDARY and does not intersect itself. It can be represented parametrically by

$$x = \tfrac{1}{2}\Re\left\{-\zeta(u+iv) + \pi u + \frac{\pi^2}{4e_1} + \frac{\pi}{2e_1}[\zeta(u+iv-\tfrac{1}{2}) - \zeta(u+iv-\tfrac{1}{2}i)]\right\}$$

$$y = \tfrac{1}{2}\Re\left\{-i\zeta(u+iv) + \pi v + \frac{\pi^2}{4e_1} - \frac{\pi}{2e_1}[i\zeta(u+iv-\tfrac{1}{2}) - i\zeta(u+iv-\tfrac{1}{2}i)]\right\}$$

$$z = \tfrac{1}{4}\sqrt{2\pi}\ln\left|\frac{\wp(u+iv)-e_1}{\wp(u+iv)+e_1}\right|,$$

where $\zeta(z)$ is the WEIERSTRAß ZETA FUNCTION, $\wp(g_2, g_3; z)$ is the WEIERSTRAß ELLIPTIC FUNCTION, $c = 189.07272$, $e_1 = 6.87519$, and the invariants are given by $g_2 = c$ and $g_3 = 0$.

References

Costa, A. "Examples of a Complete Minimal Immersion in R^3 of Genus One and Three Embedded Ends." *Bil. Soc. Bras. Mat.* **15**, 47–54, 1984.

do Carmo, M. P. *Mathematical Models from the Collections of Universities and Museums* (Ed. G. Fischer). Braunschweig, Germany: Vieweg, p. 43, 1986.

Gray, A. *Modern Differential Geometry of Curves and Surfaces.* Boca Raton, FL: CRC Press, 1993.

Gray, A. Images of the Costa surface. `ftp://bianchi.umd.edu/pub/COSTAPS/`.

Nordstrand, T. "Costa-Hoffman-Meeks Minimal Surface." `http://www.uib.no/people/nfytn/costatxt.htm`.

Peterson, I. "The Song in the Stone: Developing the Art of Telecarving a Minimal Surface." *Sci. News* **149**, 110–111, Feb. 17, 1996.

Cosymmedian Triangles

Extend the SYMMEDIAN LINES of a TRIANGLE $\Delta A_1A_2A_3$ to meet the CIRCUMCIRCLE at P_1, P_2, P_3. Then the LEMOINE POINT K of $\Delta A_1A_2A_3$ is also the LEMOINE POINT of $\Delta P_1P_2P_3$. The TRIANGLES $\Delta A_1A_2A_3$ and $\Delta P_1P_2P_3$ are cosymmedian triangles, and have the same BROCARD CIRCLE, second BROCARD TRIANGLE, BROCARD ANGLE, BROCARD POINTS, and CIRCUMCIRCLE.

see also BROCARD ANGLE, BROCARD CIRCLE, BROCARD POINTS, BROCARD TRIANGLES, CIRCUMCIRCLE, LEMOINE POINT, SYMMEDIAN LINE

Cotangent

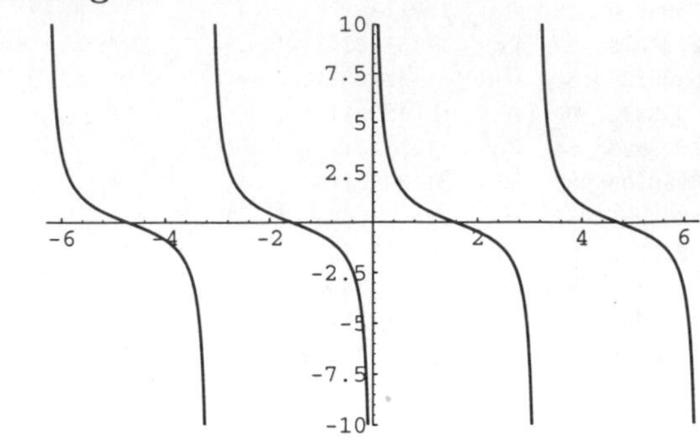

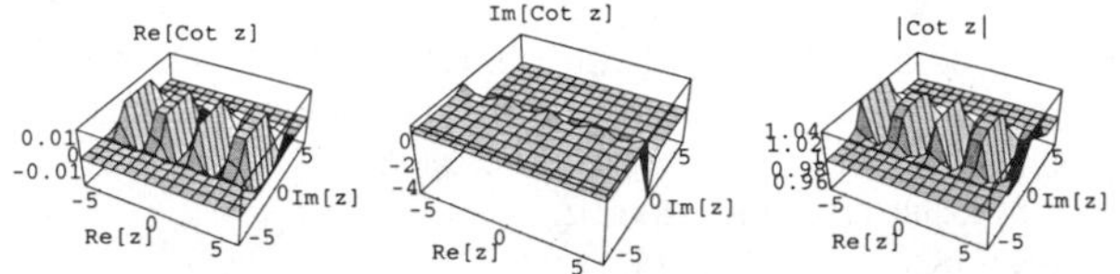

The function defined by $\cot x \equiv 1/\tan x$, where $\tan x$ is the TANGENT. The MACLAURIN SERIES for $\cot x$ is

$$\cot x = \frac{1}{x} - \tfrac{1}{3}x - \tfrac{1}{45}x^3 - \tfrac{2}{945}x^5 - \tfrac{1}{4725}x^7 - \ldots - \frac{(-1)^{n+1}2^{2n}B_{2n}}{(2n)!} - \ldots,$$

where B_n is a BERNOULLI NUMBER.

$$\pi\cot(\pi x) = \frac{1}{x} + 2x\sum_{n=1}^{\infty}\frac{1}{x^2-n^2}.$$

It is known that, for $n \geq 3$, $\cot(\pi/n)$ is rational only for $n = 4$.

see also HYPERBOLIC COTANGENT, INVERSE COTANGENT, LEHMER'S CONSTANT, TANGENT

References

Abramowitz, M. and Stegun, C. A. (Eds.). "Circular Functions." §4.3 in *Handbook of Mathematical Functions with Formulas, Graphs, and Mathematical Tables, 9th printing.* New York: Dover, pp. 71–79, 1972.

Spanier, J. and Oldham, K. B. "The Tangent $\tan(x)$ and Cotangent $\cot(x)$ Functions." Ch. 34 in *An Atlas of Functions.* Washington, DC: Hemisphere, pp. 319–330, 1987.

Cotangent Bundle

The cotangent bundle of a MANIFOLD is similar to the TANGENT BUNDLE, except that it is the set (x, f) where $x \in M$ and f is a dual vector in the TANGENT SPACE to $x \in M$. The cotangent bundle is denoted by T^*M.

see also TANGENT BUNDLE

Cotes Circle Property

$$x^{2n}+1=\left[x^2-2x\cos\left(\frac{\pi}{2n}\right)+1\right]\times\left[x^2-2x\cos\left(\frac{3\pi}{2n}\right)+1\right]\times\cdots\times\left[x^2-2x\cos\left(\frac{(2n-1)\pi}{2n}\right)+1\right].$$

Cotes Number

The numbers $\lambda_{\nu n}$ in the GAUSSIAN QUADRATURE formula

$$Q_n(f)=\sum_{\nu=1}^{n}\lambda_{\nu n}f(x_{\nu n}).$$

see also GAUSSIAN QUADRATURE

References
Cajori, F. *A History of Mathematical Notations, Vols. 1-2.* New York: Dover, p. 42, 1993.

Cotes' Spiral

The planar orbit of a particle under a r^{-3} force field. It is an EPISPIRAL.

Coth

see HYPERBOLIC COTANGENT.

Coulomb Wave Function

A special case of the CONFLUENT HYPERGEOMETRIC FUNCTION OF THE FIRST KIND. It gives the solution to the radial Schrödinger equation in the Coulomb potential $(1/r)$ of a point nucleus

$$\frac{d^2W}{d\rho^2}+\left[1-\frac{2\eta}{\rho}-\frac{L(L+1)}{\rho^2}\right]W=0. \quad (1)$$

The complete solution is

$$W=C_1F_L(\eta,\rho)+C_2G_L(\eta,\rho). \quad (2)$$

The Coulomb function of the first kind is

$$F_L(\eta,\rho)=C_L(\eta)\rho^{L+1}e^{-i\rho}{}_1F_1(L+1-i\eta;2L+2;2i\rho), \quad (3)$$

where

$$C_L(\eta)\equiv\frac{2^Le^{-\pi\eta/2}|\Gamma(L+1+i\eta)|}{\Gamma(2L+2)}, \quad (4)$$

${}_1F_1(a;b;z)$ is the CONFLUENT HYPERGEOMETRIC FUNCTION, $\Gamma(z)$ is the GAMMA FUNCTION, and the Coulomb function of the second kind is

$$G_L(\eta,\rho)=\frac{2\eta}{C_0{}^2(\eta)}F_L(\eta,\rho)\left[\ln(2\rho)+\frac{q_L(\eta)}{p_L(\eta)}\right]+\frac{1}{(2L+1)C_L(\eta)}\rho^{-L}\sum_{K=-L}^{\infty}a_k^L(\eta)\rho^{K+L}, \quad (5)$$

where q_L, p_L, and a_k^L are defined in Abramowitz and Stegun (1972, p. 538).

References
Abramowitz, M. and Stegun, C. A. (Eds.). "Coulomb Wave Functions." Ch. 14 in *Handbook of Mathematical Functions with Formulas, Graphs, and Mathematical Tables, 9th printing.* New York: Dover, pp. 537–544, 1972.
Morse, P. M. and Feshbach, H. *Methods of Theoretical Physics, Part I.* New York: McGraw-Hill, pp. 631–633, 1953.

Count

The largest n such that $|z_n|<4$ in a MANDELBROT SET. Points of different count are often assigned different colors.

Countable Additivity Probability Axiom

For a COUNTABLE SET of n disjoint events E_1, E_2, ..., E_n

$$P\left(\bigcup_{i=1}^{n}E_i\right)=\sum_{i=1}^{n}P(E_i).$$

see also COUNTABLE SET

Countable Set

A SET which is either FINITE or COUNTABLY INFINITE.

see also ALEPH-0, ALEPH-1, COUNTABLY INFINITE SET, FINITE, INFINITE, UNCOUNTABLY INFINITE SET

Countable Space

see FIRST-COUNTABLE SPACE

Countably Infinite Set

Any SET which can be put in a ONE-TO-ONE correspondence with the NATURAL NUMBERS (or INTEGERS), and so has CARDINAL NUMBER $\aleph_0$. Examples of countable sets include the INTEGERS and ALGEBRAIC NUMBERS. Georg Cantor showed that the number of REAL NUMBERS is rigorously larger than a countably infinite set, and the postulate that this number, the "CONTINUUM," is equal to $\aleph_1$ is called the CONTINUUM HYPOTHESIS.

see also ALEPH-0, ALEPH-1, CANTOR DIAGONAL SLASH, CARDINAL NUMBER, CONTINUUM HYPOTHESIS, COUNTABLE SET,

Counting Generalized Principle

If r experiments are performed with n_i possible outcomes for each experiment $i=1,2,\ldots,r$, then there are a total of $\prod_{i=1}^{r}n_i$ possible outcomes.

Counting Number

A POSITIVE INTEGER: 1, 2, 3, 4, ... (Sloane's A000027), also called a NATURAL NUMBER. However, 0 is sometimes also included in the list of counting numbers. Due to lack of standard terminology, the following terms are recommended in preference to "counting number," "NATURAL NUMBER," and "WHOLE NUMBER."

Set	Name	Symbol
$\ldots, -2, -1, 0, 1, 2, \ldots$	integers	$\mathbb{Z}$
1, 2, 3, 4, ...	positive integers	$\mathbb{Z}^+$
0, 1, 2, 3, 4 ...	nonnegative integers	$\mathbb{Z}^*$
$-1, -2, -3, -4, \ldots$	negative integers	$\mathbb{Z}^-$

see also NATURAL NUMBER, WHOLE NUMBER, $\mathbb{Z}$, $\mathbb{Z}^-$, $\mathbb{Z}^+$, $\mathbb{Z}^*$

References

Sloane, N. J. A. Sequence A000027/M0472 in "An On-Line Version of the Encyclopedia of Integer Sequences."

Coupon Collector's Problem

Let n objects be picked repeatedly with probability p_i that object i is picked on a given try, with

$$\sum_i p_i = 1.$$

Find the earliest time at which all n objects have been picked at least once.

References

Hildebrand, M. V. "The Birthday Problem." *Amer. Math. Monthly* **100**, 643, 1993.

Covariance

Given n sets of variates denoted $\{x_1\}, \ldots, \{x_n\}$, a quantity called the COVARIANCE MATRIX is defined by

$$\begin{aligned} V_{ij} &= \operatorname{cov}(x_i, x_j) & (1)\\ &\equiv \langle (x_i - \mu_i)(x_j - \mu_j)\rangle & (2)\\ &= \langle x_i x_j\rangle - \langle x_i\rangle \langle x_j\rangle, & (3)\end{aligned}$$

where $\mu_i = \langle x_i\rangle$ and $\mu_j = \langle x_j\rangle$ are the MEANS of x_i and x_j, respectively. An individual element V_{ij} of the COVARIANCE MATRIX is called the covariance of the two variates x_i and x_j, and provides a measure of how strongly correlated these variables are. In fact, the derived quantity

$$\operatorname{cor}(x_i, x_j) \equiv \frac{\operatorname{cov}(x_i, x_j)}{\sigma_i \sigma_j}, \qquad (4)$$

where σ_i, σ_j are the STANDARD DEVIATIONS, is called the CORRELATION of x_i and x_j. Note that if x_i and x_j are taken from the same set of variates (say, x), then

$$\operatorname{cov}(x, x) = \langle x^2\rangle - \langle x\rangle^2 = \operatorname{var}(x), \qquad (5)$$

giving the usual VARIANCE $\operatorname{var}(x)$. The covariance is also symmetric since

$$\operatorname{cov}(x, y) = \operatorname{cov}(y, x). \qquad (6)$$

For two variables, the covariance is related to the VARIANCE by

$$\operatorname{var}(x + y) = \operatorname{var}(x) + \operatorname{var}(y) + 2\operatorname{cov}(x, y). \qquad (7)$$

For two independent variates $x = x_i$ and $y = x_j$,

$$\operatorname{cov}(x, y) = \langle xy\rangle - \mu_x\mu_y = \langle x\rangle \langle y\rangle - \mu_x\mu_y = 0, \qquad (8)$$

so the covariance is zero. However, if the variables are correlated in some way, then their covariance will be NONZERO. In fact, if $\operatorname{cov}(x, y) > 0$, then y tends to increase as x increases. If $\operatorname{cov}(x, y) < 0$, then y tends to decrease as x increases.

The covariance obeys the identity

$$\begin{aligned} \operatorname{cov}(x + z, y) &= \langle (x + z)y - \langle x + z\rangle \langle y\rangle\rangle\\ &= \langle xy\rangle + \langle zy\rangle - (\langle x\rangle + \langle z\rangle)\langle y\rangle\\ &= \langle xy\rangle - \langle x\rangle \langle y\rangle + \langle zy\rangle - \langle z\rangle \langle y\rangle\\ &= \operatorname{cov}(x, y) + \operatorname{cov}(z, y). & (9)\end{aligned}$$

By induction, it therefore follows that

$$\operatorname{cov}\left(\sum_{i=1}^n x_i, y\right) = \sum_{i=1}^n \operatorname{cov}(x_i, y) \qquad (10)$$

$$\begin{aligned} \operatorname{cov}\left(\sum_{i=1}^n x_i, \sum_{j=1}^m y_j\right) &= \sum_{i=1}^n \operatorname{cov}\left(x_i, \sum_{j=1}^m y_j\right) & (11)\\ &= \sum_{i=1}^n \operatorname{cov}\left(\sum_{j=1}^m y_j, x_i\right) & (12)\\ &= \sum_{i=1}^n \sum_{j=1}^m \operatorname{cov}(y_j, x_i) & (13)\\ &= \sum_{i=1}^n \sum_{j=1}^m \operatorname{cov}(x_i, y_j). & (14)\end{aligned}$$

see also CORRELATION (STATISTICAL), COVARIANCE MATRIX, VARIANCE

Covariance Matrix

Given n sets of variates denoted $\{x_1\}, \ldots, \{x_n\}$, the first-order covariance matrix is defined by

$$V_{ij} = \operatorname{cov}(x_i, x_j) \equiv \langle (x_i - \mu_i)(x_j - \mu_j)\rangle,$$

where μ_i is the MEAN. Higher order matrices are given by

$$V_{ij}^{mn} = \langle (x_i - \mu_i)^m (x_j - \mu_j)^n\rangle.$$

An individual matrix element $V_{ij} = \operatorname{cov}(x_i, x_j)$ is called the COVARIANCE of x_i and x_j.

see also CORRELATION (STATISTICAL), COVARIANCE, VARIANCE

Covariant Derivative

The covariant derivative of a TENSOR A^α (also called the SEMICOLON DERIVATIVE since its symbol is a semicolon)

$$A^\alpha{}_{;\alpha} = \nabla \cdot \mathbf{A} = A^k_{,k} + \Gamma^k_{jk} A^j, \tag{1}$$

and of A_j is

$$A_{j;k} = \frac{1}{g^{kk}} \frac{\partial A_j}{\partial x_k} - \Gamma^i_{jk} A_i, \tag{2}$$

where Γ is a CONNECTION COEFFICIENT.

see also CONNECTION COEFFICIENT, COVARIANT TENSOR, DIVERGENCE

References

Morse, P. M. and Feshbach, H. *Methods of Theoretical Physics, Part I.* New York: McGraw-Hill, pp. 48–50, 1953.

Covariant Tensor

A covariant tensor is a TENSOR having specific transformation properties (c.f., a CONTRAVARIANT TENSOR). To examine the transformation properties of a covariant tensor, first consider the GRADIENT

$$\nabla \phi \equiv \frac{\partial \phi}{\partial x_1} \hat{\mathbf{x}}_1 + \frac{\partial \phi}{\partial x_2} \hat{\mathbf{x}}_2 + \frac{\partial \phi}{\partial x_3} \hat{\mathbf{x}}_3, \tag{1}$$

for which

$$\frac{\partial \phi'}{\partial x'_i} = \frac{\partial \phi}{\partial x_j} \frac{\partial x_j}{\partial x'_i}, \tag{2}$$

where $\phi(x_1, x_2, x_3) = \phi'(x'_1, x'_2, x'_3)$. Now let

$$A_i \equiv \frac{\partial \phi}{\partial x_i}, \tag{3}$$

then any set of quantities A_j which transform according to

$$A'_i = \frac{\partial x_j}{\partial x'_j} A'_j \tag{4}$$

or, defining

$$a_{ij} \equiv \frac{\partial x_j}{\partial x'_i}, \tag{5}$$

according to

$$A_i = a_{ij} A'_j \tag{6}$$

is a covariant tensor. Covariant tensors are indicated with lowered indices, i.e., a_μ.

CONTRAVARIANT TENSORS are a type of TENSOR with differing transformation properties, denoted a^ν. However, in 3-D CARTESIAN COORDINATES,

$$\frac{\partial x_j}{\partial x'_i} = \frac{\partial x'_i}{\partial x_j} \equiv a_{ij} \tag{7}$$

for $i, j = 1, 2, 3$, meaning that contravariant and covariant tensors are equivalent. The two types of tensors do differ in higher dimensions, however. Covariant FOUR-VECTORS satisfy

$$a_\mu = \Lambda^\nu_\mu a_\nu, \tag{8}$$

where Λ is a LORENTZ TENSOR.

To turn a CONTRAVARIANT TENSOR into a covariant tensor, use the METRIC TENSOR $g_{\mu\nu}$ to write

$$a_\mu \equiv g_{\mu\nu} a^\nu. \tag{9}$$

Covariant and contravariant indices can be used simultaneously in a MIXED TENSOR.

see also CONTRAVARIANT TENSOR, FOUR-VECTOR, LORENTZ TENSOR, METRIC TENSOR, MIXED TENSOR, TENSOR

References

Arfken, G. "Noncartesian Tensors, Covariant Differentiation." §3.8 in *Mathematical Methods for Physicists, 3rd ed.* Orlando, FL: Academic Press, pp. 158–164, 1985.

Morse, P. M. and Feshbach, H. *Methods of Theoretical Physics, Part I.* New York: McGraw-Hill, pp. 44–46, 1953.

Covariant Vector

A COVARIANT TENSOR of RANK 1.

Cover

A group C of SUBSETS of X whose UNION contains the given set X ($\cup\{S : S \in C\} = X$) is called a cover (or a COVERING). A MINIMAL COVER is a cover for which removal of one member destroys the covering property. There are various types of specialized covers, including proper covers, antichain covers, minimal covers, k-covers, and k^*-covers. The number of possible covers for a set of N elements is

$$|C(N)| = \frac{1}{2} \sum_{k=0}^{N} (-1)^k \binom{N}{k} 2^{2^{N-l}},$$

the first few of which are 1, 5, 109, 32297, 2147321017, 9223372023970362989, ... (Sloane's A003465). The number of proper covers for a set of N elements is

$$\begin{aligned} |C'(N)| &= |C(N)| - \tfrac{1}{4} 2^{2^N} \\ &= \frac{1}{2} \sum_{k=0}^{N} (-1)^k \binom{N}{k} 2^{2^{N-l}} - \frac{2^{2^N}}{4}, \end{aligned}$$

the first few of which are 0, 1, 45, 15913, 1073579193, ... (Sloane's A007537).

see also MINIMAL COVER

References

Eppstein, D. "Covering and Packing." `http://www.ics.uci.edu/~eppstein/junkyard/cover.html`.

Macula, A. J. "Covers of a Finite Set." *Math. Mag.* **67**, 141–144, 1994.

Sloane, N. J. A. Sequences A003465/M4024 and A007537/M5287 in "An On-Line Version of the Encyclopedia of Integer Sequences."

Cover Relation

The transitive reflexive reduction of a PARTIAL ORDER. An element z of a POSET $(X, \leq)$ covers another element x provided that there exists no third element y in the poset for which $x \leq y \leq z$. In this case, z is called an "upper cover" of x and x a "lower cover" of z.

Covering

see COVER

Covering Dimension

see LEBESGUE COVERING DIMENSION

Covering System

A system of congruences a_i mod n_i with $1 \leq i \leq k$ is called a covering system if every INTEGER y satisfies $y \equiv a_i \pmod{n}$ for at least one value of i.

see also EXACT COVERING SYSTEM

References

Guy, R. K. "Covering Systems of Congruences." §F13 in *Unsolved Problems in Number Theory, 2nd ed.* New York: Springer-Verlag, pp. 251–253, 1994.

Coversine

$$\operatorname{covers} A \equiv 1 - \sin A,$$

where $\sin A$ is the SINE.

see also EXSECANT, HAVERSINE, SINE, VERSINE

References

Abramowitz, M. and Stegun, C. A. (Eds.). *Handbook of Mathematical Functions with Formulas, Graphs, and Mathematical Tables, 9th printing.* New York: Dover, p. 78, 1972.

Cox's Theorem

Let $\sigma_1, \ldots, \sigma_4$ be four PLANES in GENERAL POSITION through a point P and let P_{ij} be a point on the LINE $\sigma_i \cdot \sigma_j$. Let σ_{ijk} denote the PLANE $P_{ij}P_{ik}P_{jk}$. Then the four PLANES σ_{234}, σ_{134}, σ_{124}, σ_{123} all pass through one point P_{1234}. Similarly, let $\sigma_1, \ldots, \sigma_5$ be five PLANES in GENERAL POSITION through P. Then the five points P_{2345}, P_{1345}, P_{1245}, P_{1235}, and P_{1234} all lie in one PLANE. And so on.

see also CLIFFORD'S CIRCLE THEOREM

Coxeter Diagram

see COXETER-DYNKIN DIAGRAM

Coxeter-Dynkin Diagram

A labeled graph whose nodes are indexed by the generators of a COXETER GROUP having (P_i, P_j) as an EDGE labeled by M_{ij} whenever $M_{ij} > 2$, where M_{ij} is an element of the COXETER MATRIX. Coxeter-Dynkin diagrams are used to visualize COXETER GROUPS. A Coxeter-Dynkin diagram is associated with each RATIONAL DOUBLE POINT (Fischer 1986).

see also COXETER GROUP, DYNKIN DIAGRAM, RATIONAL DOUBLE POINT

References

Arnold, V. I. "Critical Points of Smooth Functions." *Proc. Int. Congr. Math.* **1**, 19–39, 1974.
Fischer, G. (Ed.). *Mathematical Models from the Collections of Universities and Museums.* Braunschweig, Germany: Vieweg, pp. 12–13, 1986.

Coxeter Graph

see COXETER-DYNKIN DIAGRAM

Coxeter Group

A group generated by the elements P_i for $i = 1, \ldots, n$ subject to

$$(P_iP_j)^{M_{ij}} = 1,$$

where M_{ij} are the elements of a COXETER MATRIX. Coxeter used the NOTATION $[3^{p,q,r}]$ for the Coxeter group generated by the nodes of a Y-shaped COXETER-DYNKIN DIAGRAM whose three arms have p, q, and r EDGES. A Coxeter group of this form is finite IFF

$$\frac{1}{p+q} + \frac{1}{q+1} + \frac{1}{r+1} > 1.$$

see also BIMONSTER

References

Arnold, V. I. "Snake Calculus and Combinatorics of Bernoulli, Euler, and Springer Numbers for Coxeter Groups." *Russian Math. Surveys* **47**, 3–45, 1992.

Coxeter's Loxodromic Sequence of Tangent Circles

An infinite sequence of CIRCLES such that every four consecutive CIRCLES are mutually tangent, and the CIRCLES' RADII $\ldots$, R_{-n}, $\ldots$, R_{-1}, R_0, R_1, R_2, R_3, R_4, $\ldots$, R_n, $R_n + 1$, $\ldots$, are in GEOMETRIC PROGRESSION with ratio

$$k \equiv \frac{R_{n+1}}{R_n} = \phi + \sqrt{\phi},$$

where ϕ is the GOLDEN RATIO (Gardner 1979ab). Coxeter (1968) generalized the sequence to SPHERES.

see also ARBELOS, GOLDEN RATIO, HEXLET, PAPPUS CHAIN, STEINER CHAIN

References

Coxeter, D. "Coxeter on 'Firmament.'" `http://www.bangor.ac.uk/SculMath/image/donald.htm`.
Coxeter, H. S. M. "Loxodromic Sequences of Tangent Spheres." *Aequationes Math.* **1**, 112–117, 1968.

Gardner, M. "Mathematical Games: The Diverse Pleasures of Circles that Are Tangent to One Another." *Sci. Amer.* **240**, 18–28, Jan. 1979a.
Gardner, M. "Mathematical Games: How to be a Psychic, Even if You are a Horse or Some Other Animal." *Sci. Amer.* **240**, 18–25, May 1979b.

Coxeter Matrix

An $n \times n$ SQUARE MATRIX M with

$$M_{ii} = 1$$
$$M_{ij} = M_{ji} > 1$$

for all $i, j = 1, \ldots, n$.

see also COXETER GROUP

Coxeter-Todd Lattice

The complex LATTICE Λ_6^ω corresponding to real lattice K_{12} having the densest HYPERSPHERE PACKING (KISSING NUMBER) in 12-D. The associated AUTOMORPHISM GROUP G_0 was discovered by Mitchell (1914). The order of G_0 is given by

$$|\operatorname{Aut}(\Lambda_6^\omega)| = 2^9 \cdot 3^7 \cdot 5 \cdot 7 = 39,191,040.$$

The order of the AUTOMORPHISM GROUP of K_{12} is given by

$$|\operatorname{Aut}(K_{12})| = 2^{10} \cdot 3^7 \cdot 5 \cdot 7$$

(Conway and Sloane 1983).

see also BARNES-WALL LATTICE, LEECH LATTICE

References

Conway, J. H. and Sloane, N. J. A. "The Coxeter-Todd Lattice, the Mitchell Group and Related Sphere Packings." *Math. Proc. Camb. Phil. Soc.* **93**, 421–440, 1983.
Conway, J. H. and Sloane, N. J. A. "The 12-Dimensional Coxeter-Todd Lattice K_{12}." §4.9 in *Sphere Packings, Lattices, and Groups, 2nd ed.* New York: Springer-Verlag, pp. 127–129, 1993.
Coxeter, H. S. M. and Todd, J. A. "As Extreme Duodenary Form." *Canad. J. Math.* **5**, 384–392, 1953.
Mitchell, H. H. "Determination of All Primitive Collineation Groups in More than Four Variables." *Amer. J. Math.* **36**, 1–12, 1914.
Todd, J. A. "The Characters of a Collineation Group in Five Dimensions." *Proc. Roy. Soc. London Ser. A* **200**, 320–336, 1950.

Cramér Conjecture

An unproven CONJECTURE that

$$\overline{\lim_{n\to\infty}} \frac{p_{n+1} - p_n}{(\ln p_n)^2} = 1,$$

where p_n is the nth PRIME.

References

Cramér, H. "On the Order of Magnitude of the Difference Between Consecutive Prime Numbers." *Acta Arith.* **2**, 23–46, 1936.
Guy, R. K. *Unsolved Problems in Number Theory, 2nd ed.* New York: Springer-Verlag, p. 7, 1994.
Riesel, H. "The Cramér Conjecture." *Prime Numbers and Computer Methods for Factorization, 2nd ed.* Boston, MA: Birkhäuser, pp. 79–82, 1994.
Rivera, C. "Problems & Puzzles (Conjectures): Cramer's Conjecture." http://www.sci.net.mx/~crivera/ppp/conj_007.htm.

Cramér-Euler Paradox

A curve of order n is generally determined by $n(n+3)/2$ points. So a CONIC SECTION is determined by five points and a CUBIC CURVE should require nine. But the MACLAURIN-BEZOUT THEOREM says that two curves of degree n intersect in n^2 points, so two CUBICS intersect in nine points. This means that $n(n+3)/2$ points do not always uniquely determine a single curve of order n. The paradox was publicized by Stirling, and explained by Plücker.

see also CUBIC CURVE, MACLAURIN-BEZOUT THEOREM

Cramer's Rule

Given a set of linear equations

$$\begin{cases} a_1x + b_1y + c_1z = d_1 \\ a_2x + b_2y + c_2z = d_2 \\ a_3x + b_3y + c_3z = d_3, \end{cases} \tag{1}$$

consider the DETERMINANT

$$D \equiv \begin{vmatrix} a_1 & b_1 & c_1 \\ a_2 & b_2 & c_2 \\ a_3 & b_3 & c_3 \end{vmatrix}. \tag{2}$$

Now multiply D by x, and use the property of DETERMINANTS that MULTIPLICATION by a constant is equivalent to MULTIPLICATION of each entry in a given row by that constant

$$x \begin{vmatrix} a_1 & b_1 & c_1 \\ a_2 & b_2 & c_2 \\ a_3 & b_3 & c_3 \end{vmatrix} = \begin{vmatrix} a_1x & b_1 & c_1 \\ a_2x & b_2 & c_2 \\ a_3x & b_3 & c_3 \end{vmatrix}. \tag{3}$$

Another property of DETERMINANTS enables us to add a constant times any column to any column and obtain the same DETERMINANT, so add y times column 2 and z times column 3 to column 1,

$$xD = \begin{vmatrix} a_1x + b_1y + c_1z & b_1 & c_1 \\ a_2x + b_2y + c_2z & b_2 & c_2 \\ a_3x + b_3y + c_3z & b_3 & c_3 \end{vmatrix} = \begin{vmatrix} d_1 & b_1 & c_1 \\ d_2 & b_2 & c_2 \\ d_3 & b_3 & c_3 \end{vmatrix}. \tag{4}$$

If $\mathbf{d} = \mathbf{0}$, then (4) reduces to $xD = 0$, so the system has nondegenerate solutions (i.e., solutions other than (0, 0, 0)) only if $D = 0$ (in which case there is a family of solutions). If $\mathbf{d} \neq \mathbf{0}$ and $D = 0$, the system has no unique solution. If instead $\mathbf{d} \neq \mathbf{0}$ and $D \neq 0$, then solutions are given by

$$x = \frac{\begin{vmatrix} d_1 & b_1 & c_1 \\ d_2 & b_2 & c_2 \\ d_3 & b_3 & c_3 \end{vmatrix}}{D}, \tag{5}$$

and similarly for

$$y = \frac{\begin{vmatrix} a_1 & d_1 & c_1 \\ a_2 & d_2 & c_2 \\ a_3 & d_3 & c_3 \end{vmatrix}}{D} \tag{6}$$

$$z = \frac{\begin{vmatrix} a_1 & b_1 & d_1 \\ a_2 & b_2 & d_2 \\ a_3 & b_3 & d_3 \end{vmatrix}}{D}. \tag{7}$$

This procedure can be generalized to a set of n equations so, given a system of n linear equations

$$\begin{bmatrix} a_{11} & a_{12} & \cdots & a_{1n} \\ \vdots & \vdots & \ddots & \vdots \\ a_{1n1} & a_{n2} & \cdots & a_{nn} \end{bmatrix} \begin{bmatrix} x_1 \\ \vdots \\ x_n \end{bmatrix} = \begin{bmatrix} d_1 \\ \vdots \\ d_n \end{bmatrix}, \tag{8}$$

let

$$D \equiv \begin{vmatrix} a_{11} & a_{12} & \cdots & a_{1n} \\ \vdots & \vdots & \ddots & \vdots \\ a_{1n1} & a_{n2} & \cdots & a_{nn} \end{vmatrix}. \tag{9}$$

If $\mathbf{d} = \mathbf{0}$, then nondegenerate solutions exist only if $D = 0$. If $\mathbf{d} \neq \mathbf{0}$ and $D = 0$, the system has no unique solution. Otherwise, compute

$$D_k \equiv \begin{vmatrix} a_{11} & \cdots & a_{1(k-1)} & d_1 & a_{1(k+1)} & \cdots & a_{1n} \\ \vdots & \ddots & \vdots & \vdots & \vdots & \ddots & \vdots \\ a_{n1} & \cdots & a_{n(k-1)} & d_n & a_{n(k+1)} & \cdots & a_{nn} \end{vmatrix}. \tag{10}$$

Then $x_k = D_k/D$ for $1 \leq k \leq n$. In the 3-D case, the VECTOR analog of Cramer's rule is

$$(\mathbf{A}\times\mathbf{B})\times(\mathbf{C}\times\mathbf{D}) = (\mathbf{A}\cdot\mathbf{B}\times\mathbf{D})\mathbf{C}-(\mathbf{A}\cdot\mathbf{B}\times\mathbf{C})\mathbf{D}. \tag{11}$$

see also DETERMINANT, LINEAR ALGEBRA, MATRIX, SYSTEM OF EQUATIONS, VECTOR

Cramér's Theorem

If X and Y are INDEPENDENT variates and $X + Y$ is a GAUSSIAN DISTRIBUTION, then both X and Y must have GAUSSIAN DISTRIBUTIONS. This was proved by Cramér in 1936.

Craps

A game played with two DICE. If the total is 7 or 11 (a "natural"), the thrower wins and retains the DICE for another throw. If the total is 2, 3, or 12 ("craps"), the thrower loses but retains the DICE. If the total is any other number (called the thrower's "point"), the thrower must continue throwing and roll the "point" value again before throwing a 7. If he succeeds, he wins and retains the DICE, but if a 7 appears first, the player loses and passes the dice. The probability of winning is $244/495 \approx 0.493$ (Kraitchik 1942).

References

Kenney, J. F. and Keeping, E. S. *Mathematics of Statistics, Pt. 2, 2nd ed.* Princeton, NJ: Van Nostrand, pp. 12–13, 1951.

Kraitchik, M. "Craps." §6.5 in *Mathematical Recreations.* New York: W. W. Norton, pp. 123–126, 1942.

CRC

see CYCLIC REDUNDANCY CHECK

Creative Telescoping

see TELESCOPING SUM, ZEILBERGER'S ALGORITHM

Cremona Transformation

An entire Cremona transformation is a BIRATIONAL TRANSFORMATION of the PLANE. Cremona transformations are MAPS of the form

$$x_{i+1} = f(x_i, y_i)$$
$$y_{i+1} = g(x_i, y_i),$$

in which f and g are POLYNOMIALS. A quadratic Cremona transformation is always factorable.

see also NOETHER'S TRANSFORMATION THEOREM

References

Coolidge, J. L. *A Treatise on Algebraic Plane Curves.* New York: Dover, pp. 203–204, 1959.

Cribbage

Cribbage is a game in which each of two players is dealt a hand of six CARDS. Each player then discards two of his six cards to a four-card "crib" which alternates between players. After the discard, the top card in the remaining deck is turned up. Cards are then alternating played out by the two players, with points being scored for pairs, runs, cumulative total of 15 and 31, and playing the last possible card ("go") not giving a total over 31. All face cards are counted as 10 for the purpose of playing out, but the normal values of Jack = 11, Queen = 12, King = 13 are used to determine runs. Aces are always low (ace = 1). After all cards have been played, each player counts the four cards in his hand taken in conjunction with the single top card. Points are awarded for pairs, flushes, runs, and combinations of cards giving 15. A Jack having the same suit as a top card is awarded an additional point for "nobbs." The crib is then also counted and scored. The winner is the first person to "peg" a certain score, as recorded on a "cribbage board."

The best possible score in a hand is 29, corresponding to three 5s and a Jack with a top 5 the same suit as the Jack. Hands with scores of 25, 26, and 27 are not possible.

see also BRIDGE CARD GAME, CARDS, POKER

Criss-Cross Method

A standard form of the LINEAR PROGRAMMING problem of maximizing a linear function over a CONVEX POLYHEDRON is to maximize $\mathbf{c} \cdot \mathbf{x}$ subject to $\mathsf{m}\mathbf{x} \leq \mathbf{b}$ and $\mathbf{x} \geq \mathbf{0}$, where m is a given $s \times d$ matrix, $\mathbf{c}$ and $\mathbf{b}$ are given d-vector and s-vectors, respectively. The Criss-cross method always finds a VERTEX solution if an optimal solution exists.

see also CONVEX POLYHEDRON, LINEAR PROGRAMMING, VERTEX (POLYHEDRON)

Criterion

A requirement Necessary for a given statement or theorem to hold. Also called a Condition.

see also Brown's Criterion, Cauchy Criterion, Euler's Criterion, Gauss's Criterion, Korselt's Criterion, Leibniz Criterion, Pocklington's Criterion, Vandiver's Criteria, Weyl's Criterion

Critical Line

The Line $\Re(s) = 1/2$ in the Complex Plane on which the Riemann Hypothesis asserts that all nontrivial (Complex) Roots of the Riemann Zeta Function $\zeta(s)$ lie. Although it is known that an Infinite number of zeros lie on the critical line and that these comprise *at least* 40% of all zeros, the Riemann Hypothesis is still unproven.

see also Riemann Hypothesis, Riemann Zeta Function

References
Vardi, I. *Computational Recreations in Mathematica.* Reading, MA: Addison-Wesley, p. 142, 1991.

Critical Point

A Function $y = f(x)$ has critical points at all points x_0 where $f'(x_0) = 0$ or $f(x)$ is not Differentiable. A Function $z = f(x, y)$ has critical points where the Gradient $\nabla f = \mathbf{0}$ or $\partial f/\partial x$ or the Partial Derivative $\partial f/\partial y$ is not defined.

see also Fixed Point, Inflection Point, Only Critical Point in Town Test, Stationary Point

Critical Strip

see Critical Line

Crook

A 6-Polyiamond.

References
Golomb, S. W. *Polyominoes: Puzzles, Patterns, Problems, and Packings, 2nd ed.* Princeton, NJ: Princeton University Press, p. 92, 1994.

Crookedness

Let a Knot K be parameterized by a Vector Function $\mathbf{v}(t)$ with $t \in \mathbb{S}^1$, and let $\mathbf{w}$ be a fixed Unit Vector in $\mathbb{R}^3$. Count the number of Relative Minima of the projection function $\mathbf{w}\cdot\mathbf{v}(t)$. Then the Minimum such number over all directions $\mathbf{w}$ and all K of the given type is called the crookedness $\mu(K)$. Milnor (1950) showed that $2\pi\mu(K)$ is the Infimum of the total curvature of K. For any Tame Knot K in $\mathbb{R}^3$, $\mu(K) = b(K)$ where $b(K)$ is the Bridge Index.

see also Bridge Index

References
Milnor, J. W. "On the Total Curvature of Knots." *Ann. Math.* **52**, 248–257, 1950.
Rolfsen, D. *Knots and Links.* Wilmington, DE: Publish or Perish Press, p. 115, 1976.

Cross

In general, a cross is a figure formed by two intersecting Line Segments. In Linear Algebra, a cross is defined as a set of n mutually Perpendicular pairs of Vectors of equal magnitude from a fixed origin in Euclidean n-Space.

The word "cross" is also used to denote the operation of the Cross Product, so $\mathbf{a} \times \mathbf{b}$ would be pronounced "**a** cross **b**."

see also Cross Product, Dot, Eutactic Star, Gaullist Cross, Greek Cross, Latin Cross, Maltese Cross, Papal Cross, Saint Andrew's Cross, Saint Anthony's Cross, Star

Cross-Cap

The self-intersection of a one-sided Surface. It can be described as a circular Hole which, when entered, exits from its opposite point (from a topological viewpoint, both singular points on the cross-cap are equivalent). The cross-cap has a segment of double points which terminates at two "Pinch Points" known as Whitney Singularities.

The cross-cap can be generated using the general method for Nonorientable Surfaces using the polynomial function

$$\mathbf{f}(x, y, z) = (xz, yz, \tfrac{1}{2}(z^2 - x^2)) \tag{1}$$

(Pinkall 1986). Transforming to Spherical Coordinates gives

$$x(u, v) = \tfrac{1}{2}\cos u \sin(2v) \tag{2}$$
$$y(u, v) = \tfrac{1}{2}\sin u \sin(2v) \tag{3}$$
$$z(u, v) = \tfrac{1}{2}(\cos^2 v - \cos^2 u \sin^2 v) \tag{4}$$

for $u \in [0, 2\pi)$ and $v \in [0, \pi/2]$. To make the equations slightly simpler, all three equations are normally multiplied by a factor of 2 to clear the arbitrary scaling constant. Three views of the cross-cap generated using this equation are shown above. Note that the middle one looks suspiciously like Maeder's Owl Minimal Surface.

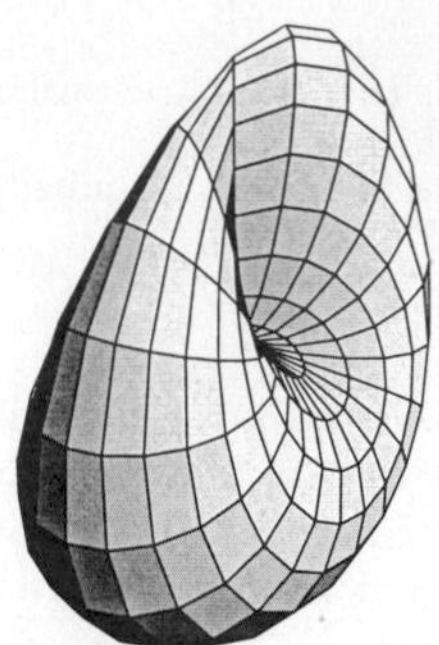

Another representation is

$$\mathbf{f}(x,y,z) = (yz, 2xy, x^2 - y^2), \quad (5)$$

(Gray 1993), giving parametric equations

$$x = \tfrac{1}{2}\sin u \sin(2v) \quad (6)$$

$$y = \sin(2u)\sin^2 v \quad (7)$$

$$z = \cos(2u)\sin^2 v, \quad (8)$$

(Geometry Center) where, for aesthetic reasons, the y- and z-coordinates have been multiplied by 2 to produce a squashed, but topologically equivalent, surface. Nordstrand gives the implicit equation

$$4x^2(x^2+y^2+z^2+z) + y^2(y^2+z^2-1) = 0 \quad (9)$$

which can be solved for z to yield

$$z = \frac{-2x^2 \pm \sqrt{(y^2+2x^2)(1-4x^2-y^2)}}{4x^2+y^2}. \quad (10)$$

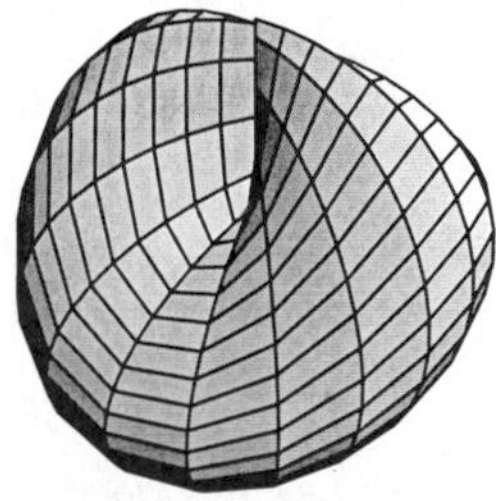

Taking the inversion of a cross-cap such that $(0,0,-1/2)$ is sent to ∞ gives a CYLINDROID, shown above (Pinkall 1986).

The cross-cap is one of the three possible SURFACES obtained by sewing a MÖBIUS STRIP to the edge of a DISK. The other two are the BOY SURFACE and ROMAN SURFACE.

see also BOY SURFACE, MÖBIUS STRIP, NONORIENTABLE SURFACE, PROJECTIVE PLANE, ROMAN SURFACE

References

Fischer, G. (Ed.). Plate 107 in *Mathematische Modelle/Mathematical Models, Bildband/Photograph Volume.* Braunschweig, Germany: Vieweg, p. 108, 1986.

Geometry Center. "The Crosscap." http://www.geom.umn.edu/zoo/toptype/pplane/cap/.

Pinkall, U. *Mathematical Models from the Collections of Universities and Museums* (Ed. G. Fischer). Braunschweig, Germany: Vieweg, p. 64, 1986.

Cross-Correlation

Let $\star$ denote cross-correlation. Then the cross-correlation of two functions $f(t)$ and $g(t)$ of a real variable t is defined by

$$f \star g \equiv f^*(-t) * g(t), \quad (1)$$

where $*$ denotes CONVOLUTION and $f^*(t)$ is the COMPLEX CONJUGATE of $f(t)$. The CONVOLUTION is defined by

$$f(t) * g(t) = \int_{-\infty}^{\infty} f(\tau)g(t-\tau)d\tau, \quad (2)$$

therefore

$$f \star g \equiv \int_{-\infty}^{\infty} f^*(-\tau)g(t-\tau)\,d\tau. \quad (3)$$

Let $\tau' \equiv -\tau$, so $d\tau' = -d\tau$ and

$$f \star g = \int_{\infty}^{-\infty} f^*(\tau')g(t+\tau')(-d\tau')$$

$$= \int_{-\infty}^{\infty} f^*(\tau)g(t+\tau)\,d\tau. \quad (4)$$

The cross-correlation satisfies the identity

$$(g \star h) \star (g \star h) = (g \star g) \star (h \star h). \quad (5)$$

If f or g is EVEN, then

$$f \star g = f * g, \quad (6)$$

where $*$ denotes CONVOLUTION.

see also AUTOCORRELATION, CONVOLUTION, CROSS-CORRELATION THEOREM

Cross-Correlation Coefficient

The COEFFICIENT ρ in a GAUSSIAN BIVARIATE DISTRIBUTION.

Cross-Correlation Theorem

Let $f \star g$ denote the CROSS-CORRELATION of functions $f(t)$ and $g(t)$. Then

$$f \star g = \int_{-\infty}^{\infty} f^*(\tau)g(t+\tau)\,d\tau$$

$$= \int_{-\infty}^{\infty}\left[\int_{-\infty}^{\infty} F^*(\nu)e^{2\pi i\nu\tau}\,d\nu \int_{-\infty}^{\infty} G(\nu')e^{-2\pi i\nu'(t+\tau)}\,d\nu'\right]d\tau$$

$$= \int_{-\infty}^{\infty}\int_{-\infty}^{\infty}\int_{-\infty}^{\infty} F^*(\nu)G(\nu')e^{-2\pi i\tau(\nu'-\nu)}e^{-2\pi i\nu' t}\,d\tau\,d\nu\,d\nu'$$

$$= \int_{-\infty}^{\infty}\int_{-\infty}^{\infty} F^*(\nu)G(\nu')e^{-2\pi i\nu' t}\left[\int_{-\infty}^{\infty} e^{-2\pi i\tau(\nu'-\nu)}\,d\tau\right]d\nu\,d\nu'$$

$$= \int_{-\infty}^{\infty}\int_{-\infty}^{\infty} F^*(\nu)G(\nu')e^{-2\pi i\nu' t}\delta(\nu'-\nu)\,d\nu'\,d\nu$$

$$= \int_{-\infty}^{\infty} F^*(\nu)G(\nu)e^{-2\pi i\nu t}\,d\nu$$

$$= \mathcal{F}[F^*(\nu)G(\nu)], \quad (1)$$

where $\mathcal{F}$ denotes the FOURIER TRANSFORM and

$$f(t) \equiv \mathcal{F}[F(\nu)] = \int_{-\infty}^{\infty} F(\nu)e^{-2\pi i\nu t}\,dt \tag{2}$$

$$g(t) \equiv \mathcal{F}[G(\nu)] = \int_{-\infty}^{\infty} G(\nu)e^{-2\pi i\nu t}\,dt. \tag{3}$$

Applying a FOURIER TRANSFORM on each side gives the cross-correlation theorem,

$$f \star g = \mathcal{F}[F^*(\nu)G(\nu)]. \tag{4}$$

If $F = G$, then the cross-correlation theorem reduces to the WIENER-KHINTCHINE THEOREM.

see also FOURIER TRANSFORM, WIENER-KHINTCHINE THEOREM

Cross Curve

see CRUCIFORM

Cross Fractal

see CANTOR SQUARE FRACTAL

Cross of Lorraine

see GAULLIST CROSS

Cross Polytope

A regular POLYTOPE in n-D (generally assumed to satisfy $n \geq 5$) corresponding to the CONVEX HULL of the points formed by permuting the coordinates ($\pm$ 1, 0, 0, ..., 0). It is denoted β_n and has SCHLÄFLI SYMBOL $\{3^{n-2}, 4\}$. In 3-D, the cross polytope is the OCTAHEDRON.

see also MEASURE POLYTOPE, SIMPLEX

Cross Product

For VECTORS **u** and **v**,

$$\mathbf{u}\times\mathbf{v} = \hat{\mathbf{x}}(u_yv_z - u_zv_y) - \hat{\mathbf{y}}(u_xv_z - u_zv_x) + \hat{\mathbf{z}}(u_xv_y - u_yv_x). \tag{1}$$

This can be written in a shorthand NOTATION which takes the form of a DETERMINANT

$$\mathbf{u}\times\mathbf{v} = \begin{vmatrix} \hat{\mathbf{x}} & \hat{\mathbf{y}} & \hat{\mathbf{z}} \\ u_x & u_y & u_z \\ v_x & v_y & v_z \end{vmatrix}. \tag{2}$$

It is also true that

$$|\mathbf{u}\times\mathbf{v}| = |\mathbf{u}|\,|\mathbf{v}|\sin\theta, \tag{3}$$

$$= |\mathbf{u}|\,|\mathbf{v}|\sqrt{1-(\hat{\mathbf{u}}\cdot\hat{\mathbf{v}})^2}, \tag{4}$$

where θ is the angle between **u** and **v**, given by the DOT PRODUCT

$$\cos\theta \equiv \hat{\mathbf{u}}\cdot\hat{\mathbf{v}}. \tag{5}$$

Identities involving the cross product include

$$\frac{d}{dt}[\mathbf{r}_1(t)\times\mathbf{r}_2(t)] = \mathbf{r}_1(t)\times\frac{d\mathbf{r}_2}{dt} + \frac{d\mathbf{r}_1}{dt}\times\mathbf{r}_2(t) \tag{6}$$

$$\mathbf{A}\times\mathbf{B} = -\mathbf{B}\times\mathbf{A} \tag{7}$$

$$\mathbf{A}\times(\mathbf{B}+\mathbf{C}) = \mathbf{A}\times\mathbf{B} + \mathbf{A}\times\mathbf{C} \tag{8}$$

$$(t\mathbf{A})\times\mathbf{B} = t(\mathbf{A}\times\mathbf{B}). \tag{9}$$

For a proof that $\mathbf{A}\times\mathbf{B}$ is a PSEUDOVECTOR, see Arfken (1985, pp. 22–23). In TENSOR notation,

$$\mathbf{A}\times\mathbf{B} = \epsilon_{ijk}A^jB^k, \tag{10}$$

where ϵ_{ijk} is the LEVI-CIVITA TENSOR.

see also DOT PRODUCT, SCALAR TRIPLE PRODUCT

References

Arfken, G. "Vector or Cross Product." §1.4 in *Mathematical Methods for Physicists, 3rd ed.* Orlando, FL: Academic Press, pp. 18–26, 1985.

Cross-Ratio

$$[a,b,c,d] \equiv \frac{(a-b)(c-d)}{(a-d)(c-b)}. \tag{1}$$

For a MÖBIUS TRANSFORMATION f,

$$[a,b,c,d] = [f(a), f(b), f(c), f(d)]. \tag{2}$$

There are six different values which the cross-ratio may take, depending on the order in which the points are chosen. Let $\lambda \equiv [a,b,c,d]$. Possible values of the cross-ratio are then λ, $1-\lambda$, $1/\lambda$, $(\lambda-1)/\lambda$, $1/(1-\lambda)$, and $\lambda/(\lambda-1)$.

Given lines a, b, c, and d which intersect in a point O, let the lines be cut by a line l, and denote the points of intersection of l with each line by A, B, C, and D. Let the distance between points A and B be denoted AB, etc. Then the cross-ratio

$$[AB, CD] \equiv \frac{(AB)(CD)}{(BC)(AD)} \tag{3}$$

is the same for any position of the l (Coxeter and Greitzer 1967). Note that the definitions $(AB/AD)/(BC/CD)$ and $(CA/CB)/(DA/DB)$ are used instead by Kline (1990) and Courant and Robbins (1966), respectively. The identity

$$[AD, BC] + [AB, DC] = 1 \tag{4}$$

holds IFF $AC//BD$, where $//$ denotes SEPARATION.

The cross-ratio of four points on a radial line of an INVERSION CIRCLE is preserved under INVERSION (Ogilvy 1990, p. 40).

see also MÖBIUS TRANSFORMATION, SEPARATION

References
Courant, R. and Robbins, H. *What is Mathematics?: An Elementary Approach to Ideas and Methods, 2nd ed.* Oxford, England: Oxford University Press, 1996.
Coxeter, H. S. M. and Greitzer, S. L. *Geometry Revisited.* Washington, DC: Math. Assoc. Amer., pp. 107–108, 1967.
Kline, M. *Mathematical Thought from Ancient to Modern Times, Vol. 1.* Oxford, England: Oxford University Press, 1990.
Ogilvy, C. S. *Excursions in Geometry.* New York: Dover, pp. 39–41, 1990.

Cross-Section

The cross-section of a SOLID is a LAMINA obtained by its intersection with a PLANE. The cross-section of an object therefore represents an infinitesimal "slice" of a solid, and may be different depending on the orientation of the slicing plane. While the cross-section of a SPHERE is always a DISK, the cross-section of a CUBE may be a SQUARE, HEXAGON, or other shape.

see also AXONOMETRY, CAVALIERI'S PRINCIPLE, LAMINA, PLANE, PROJECTION, RADON TRANSFORM, STEREOLOGY

Crossed Ladders Problem

Given two crossed LADDERS resting against two buildings, what is the distance between the buildings? Let the height at which they cross be c and the lengths of the LADDERS a and b. The height at which b touches the building k is then obtained by solving

$$k^4 - 2ck^3 + k^2(a^2 - b^2) - 2ck(a^2 - b^2) + c^2(a^2 - b^2) = 0.$$

Call the horizontal distance from the top of a to the crossing u, and the distance from the top of b, v. Call the height at which a touches the building h. There are solutions in which a, b, h, k, u, and v are all INTEGERS. One is $a = 119$, $b = 70$, $c = 30$, and $u + v = 56$.

see also LADDER

References
Gardner, M. *Mathematical Circus: More Puzzles, Games, Paradoxes and Other Mathematical Entertainments from Scientific American.* New York: Knopf, pp. 62–64, 1979.

Crossed Trough

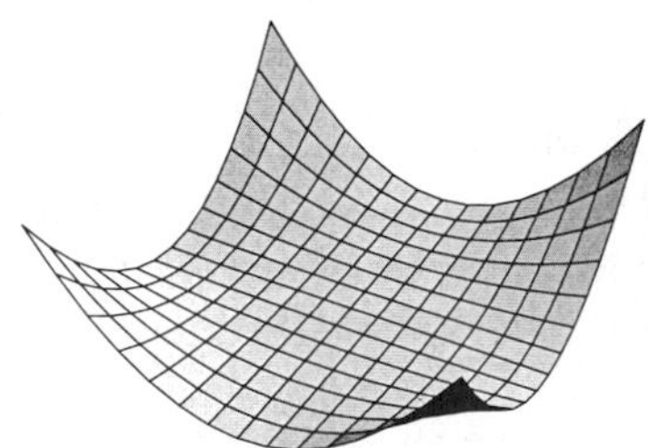

The SURFACE

$$z = cx^2y^2.$$

see also MONKEY SADDLE

References
von Seggern, D. *CRC Standard Curves and Surfaces.* Boca Raton, FL: CRC Press, p. 286, 1993.

Crossing Number (Graph)

Given a "good" GRAPH (i.e., one for which all intersecting EDGES intersect in a single point and arise from four distinct VERTICES), the crossing number is the minimum possible number of crossings with which the GRAPH can be drawn. A GRAPH with crossing number 0 is a PLANAR GRAPH. Garey and Johnson (1983) showed that determining the crossing number is an NP-COMPLETE PROBLEM. GUY'S CONJECTURE suggests that the crossing number for the COMPLETE GRAPH K_n is

$$\frac{1}{4}\left\lfloor\frac{n}{2}\right\rfloor\left\lfloor\frac{n-1}{2}\right\rfloor\left\lfloor\frac{n-2}{2}\right\rfloor\left\lfloor\frac{n-3}{2}\right\rfloor, \tag{1}$$

which can be rewritten

$$\begin{cases} \frac{1}{64}n(n-2)^2(n-4) & \text{for } n \text{ even} \\ \frac{1}{64}(n-1)^2(n-3)^2 & \text{for } n \text{ odd.} \end{cases} \tag{2}$$

The first few predicted and known values are given in the following table (Sloane's A000241).

Order	Predicted	Known
1	0	0
2	0	0
3	0	0
4	0	0
5	1	1
6	3	3
7	9	9
8	18	18
9	36	36
10	60	60
11	100	
12	150	
13	225	
14	315	
15	441	
16	588	

ZARANKIEWICZ'S CONJECTURE asserts that the crossing number for a COMPLETE BIGRAPH is

$$\left\lfloor\frac{n}{2}\right\rfloor\left\lfloor\frac{n-1}{2}\right\rfloor\left\lfloor\frac{m}{2}\right\rfloor\left\lfloor\frac{m-1}{2}\right\rfloor. \tag{3}$$

It has been checked up to $m, n = 7$, and Zarankiewicz has shown that, in general, the FORMULA provides an upper bound to the actual number. The table below gives known results. When the number is not known exactly, the prediction of ZARANKIEWICZ'S CONJECTURE is given in parentheses.

	1	2	3	4	5	6	7
1	0	0	0	0	0	0	0
2		0	0	0	0	0	0
3			1	2	4	6	9
4				4	8	12	18
5					16	24	36
6						36	54
7							77, 79, or (81)

Consider the crossing number for a rectilinear GRAPH G which may have only straight EDGES, denoted $\bar{\nu}(G)$. For a COMPLETE GRAPH of order $n \geq 10$, the rectilinear crossing number is always larger than the general graph crossing number. The first few values for COMPLETE GRAPHS are 0, 0, 0, 0, 1, 3, 9, 19, 36, 61 or 62, ... (Sloane's A014540). The $n = 10$ lower limit is from Singer (1986), who proved that

$$\bar{\nu}(K_n) \leq \tfrac{1}{312}(5n^4 - 39n^3 + 91n^2 - 57n). \quad (4)$$

Jensen (1971) has shown that

$$\bar{\nu}(K_n) \leq \tfrac{7}{432}n^4 + \mathcal{O}(n^3). \quad (5)$$

Consider the crossing number for a toroidal GRAPH. For a COMPLETE GRAPH, the first few are 0, 0, 0, 0, 0, 0, 0, 4, 9, 23, 42, 70, 105, 154, 226, 326, ... (Sloane's A014543). The toroidal crossing numbers for a COMPLETE BIGRAPH are given in the following table.

	1	2	3	4	5	6	7
1	0	0	0	0	0	0	
2		0	0	0	0	0	
3			0	0	0	0	
4					2		
5					5	8	
6						12	
7							

see also GUY'S CONJECTURE, ZARANKIEWICZ'S CONJECTURE

References

Gardner, M. "Crossing Numbers." Ch. 11 in *Knotted Doughnuts and Other Mathematical Entertainments.* New York: W. H. Freeman, pp. 133–144, 1986.

Garey, M. R. and Johnson, D. S. "Crossing Number is NP-Complete." *SIAM J. Alg. Discr. Meth.* **4**, 312–316, 1983.

Guy, R. K. "Latest Results on Crossing Numbers." In *Recent Trends in Graph Theory, Proc. New York City Graph Theory Conference, 1st, 1970.* (Ed. New York City Graph Theory Conference Staff). New York: Springer-Verlag, 1971.

Guy, R. K. and Jenkyns, T. "The Toroidal Crossing Number of $K_{m,n}$." *J. Comb. Th.* **6**, 235–250, 1969.

Guy, R. K.; Jenkyns, T.; and Schaer, J. "Toroidal Crossing Number of the Complete Graph." *J. Comb. Th.* **4**, 376–390, 1968.

Jensen, H. F. "An Upper Bound for the Rectilinear Crossing Number of the Complete Graph." *J. Comb. Th. Ser. B* **10**, 212–216, 1971.

Kleitman, D. J. "The Crossing Number of $K_{5,n}$." *J. Comb. Th.* **9**, 315–323, 1970.

Singer, D. Unpublished manuscript "The Rectilinear Crossing Number of Certain Graphs," 1971. Quoted in Gardner, M. *Knotted Doughnuts and Other Mathematical Entertainments.* New York: W. H. Freeman, 1986.

Sloane, N. J. A. Sequences A014540, A014543, and A000241/M2772 in "An On-Line Version of the Encyclopedia of Integer Sequences."

Tutte, W. T. "Toward a Theory of Crossing Numbers." *J. Comb. Th.* **8**, 45–53, 1970.

Crossing Number (Link)

The least number of crossings that occur in any projection of a LINK. In general, it is difficult to find the crossing number of a given LINK.

References

Adams, C. C. *The Knot Book: An Elementary Introduction to the Mathematical Theory of Knots.* New York: W. H. Freeman, pp. 67–69, 1994.

Crout's Method

A ROOT finding technique used in LU DECOMPOSITION. It solves the N^2 equations

$$\begin{array}{ll} i < j & \alpha_{i1}\beta_{1j} + \alpha_{i2}\beta_{2j} + \ldots + \alpha_{ii}\beta_{ij} = a_{ij} \\ i = j & \alpha_{i1}\beta_{1j} + \alpha_{i2}\beta_{2j} + \ldots + \alpha_{ii}\beta_{jj} = a_{ij} \\ i > j & \alpha_{i1}\beta_{1j} + \alpha_{i2}\beta_{2j} + \ldots + \alpha_{ij}\beta_{jj} = a_{ij} \end{array}$$

for the $N^2 + N$ unknowns α_{ij} and β_{ij}.

see also LU DECOMPOSITION

References

Press, W. H.; Flannery, B. P.; Teukolsky, S. A.; and Vetterling, W. T. *Numerical Recipes in FORTRAN: The Art of Scientific Computing, 2nd ed.* Cambridge, England: Cambridge University Press, pp. 36–38, 1992.

Crowd

A group of SOCIABLE NUMBERS of order 3.

Crown

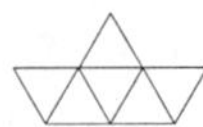

A 6-POLYIAMOND.

References

Golomb, S. W. *Polyominoes: Puzzles, Patterns, Problems, and Packings, 2nd ed.* Princeton, NJ: Princeton University Press, p. 92, 1994.

Crucial Point

The HOMOTHETIC CENTER of the ORTHIC TRIANGLE and the triangular hull of the three EXCIRCLES. It has TRIANGLE CENTER FUNCTION

$$\alpha = \tan A = \sin(2B) + \sin(2C) - \sin(2A).$$

References

Kimberling, C. "Central Points and Central Lines in the Plane of a Triangle." *Math. Mag.* **67**, 163–187, 1994.

Lyness, R. and Veldkamp, G. R. Problem 682 and Solution. *Crux Math.* **9**, 23–24, 1983.

Cruciform

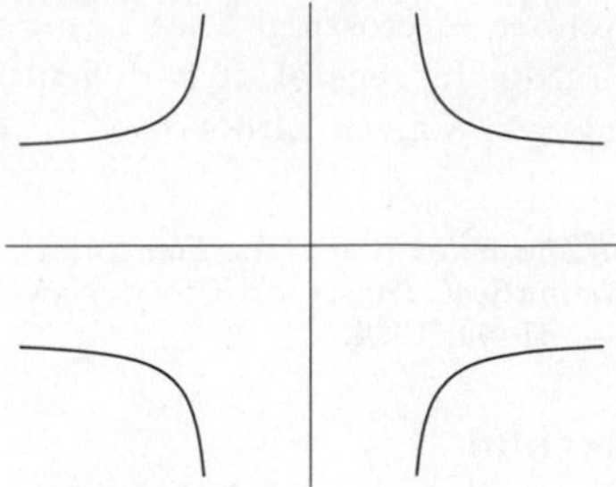

A plane curve also called the CROSS CURVE and POLICEMAN ON POINT DUTY CURVE (Cundy and Rollett 1989). It is given by the equation

$$x^2y^2 - a^2x^2 - a^2y^2 = 0, \tag{1}$$

which is equivalent to

$$1 - \frac{a^2}{y^2} - \frac{a^2}{x^2} = 0 \tag{2}$$

$$\frac{a^2}{x^2} + \frac{b^2}{y^2} = 1, \tag{3}$$

or, rewriting,

$$y^2 = \frac{a^2x^2}{x^2 - a^2}. \tag{4}$$

In parametric form,

$$x = a \sec t \tag{5}$$

$$y = b \csc t. \tag{6}$$

The CURVATURE is

$$\kappa = \frac{3ab \csc^2 t \sec^2 t}{(b^2 \cos^2 t \csc^2 t + a^2 \sec^2 t \tan^2 t)^{3/2}}. \tag{7}$$

References
Cundy, H. and Rollett, A. *Mathematical Models, 3rd ed.* Stradbroke, England: Tarquin Pub., p. 71, 1989.
Lawrence, J. D. *A Catalog of Special Plane Curves.* New York: Dover, pp. 127 and 130–131, 1972.

Crunode

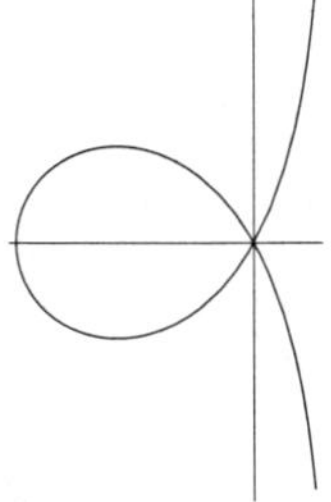

A point where a curve intersects itself so that two branches of the curve have distinct tangent lines. The MACLAURIN TRISECTRIX, shown above, has a crunode at the origin.

see also ACNODE, SPINODE, TACNODE

Cryptarithm

see CRYPTARITHMETIC

Cryptarithmetic

A number PUZZLE in which a group of arithmetical operations has some or all of its DIGITS replaced by letters or symbols, and where the original DIGITS must be found. In such a puzzle, each letter represents a unique digit.

see also ALPHAMETIC, DIGIMETIC, SKELETON DIVISION

References
Bogomolny, A. "Cryptarithms." http://www.cut-the-knot.com/st_crypto.html.
Brooke, M. *One Hundred & Fifty Puzzles in Crypt-Arithmetic.* New York: Dover, 1963.
Kraitchik, M. "Cryptarithmetic." §3.11 in *Mathematical Recreations.* New York: W. W. Norton, pp. 79–83, 1942.
Marks, R. W. *The New Mathematics Dictionary and Handbook.* New York: Bantam Books, 1964.

Cryptography

The science and mathematics of encoding and decoding information.

see also CRYPTARITHM, KNAPSACK PROBLEM, PUBLIC-KEY CRYPTOGRAPHY

References
Davies, D. W. *The Security of Data in Networks.* Los Angeles, CA: IEEE Computer Soc., 1981.
Diffie, W. and Hellman, M. "New Directions in Cryptography." *IEEE Trans. Info. Th.* **22**, 644–654, 1976.
Honsberger, R. "Four Clever Schemes in Cryptography." Ch. 10 in *Mathematical Gems III.* Washington, DC: Math. Assoc. Amer., pp. 151–173, 1985.
Simmons, G. J. "Cryptology, The Mathematics of Secure Communications." *Math. Intel.* **1**, 233–246, 1979.

Crystallography Restriction

If a discrete GROUP of displacements in the plane has more than one center of rotation, then the only rotations that can occur are by 2, 3, 4, and 6. This can be shown as follows. It must be true that the sum of the interior angles divided by the number of sides is a divisor of 360°.

$$\frac{180°(n-2)}{n} = \frac{360°}{m},$$

where m is an INTEGER. Therefore, symmetry will be possible only for

$$\frac{2n}{n-2} = m,$$

where m is an INTEGER. This will hold for 1-, 2-, 3-, 4-, and 6-fold symmetry. That it does not hold for $n > 6$ is seen by noting that $n = 6$ corresponds to $m = 3$. The $m = 2$ case requires that $n = n - 2$ (impossible), and the $m = 1$ case requires that $n = -2$ (also impossible).

see also POINT GROUPS, SYMMETRY

Császár Polyhedron

A POLYHEDRON topologically equivalent to a TORUS discovered in the late 1940s. It has 7 VERTICES, 14 faces, and 21 EDGES, and is the DUAL POLYHEDRON of the SZILASSI POLYHEDRON. Its SKELETON is ISOMORPHIC to the COMPLETE GRAPH K_7.

see also SZILASSI POLYHEDRON, TOROIDAL POLYHEDRON

References

Császár, Á. "A Polyhedron without Diagonals." *Acta Sci. Math.* **13**, 140–142, 1949–1950.

Gardner, M. "The Császár Polyhedron." Ch. 11 in *Time Travel and Other Mathematical Bewilderments.* New York: W. H. Freeman, 1988.

Gardner, M. *Fractal Music, Hypercards, and More: Mathematical Recreations from Scientific American Magazine.* New York: W. H. Freeman, pp. 118–120, 1992.

Hart, G. "Toroidal Polyhedra." `http://www.li.net/~george/virtual-polyhedra/toroidal.html`.

Csch

see HYPERBOLIC COSECANT

Cube

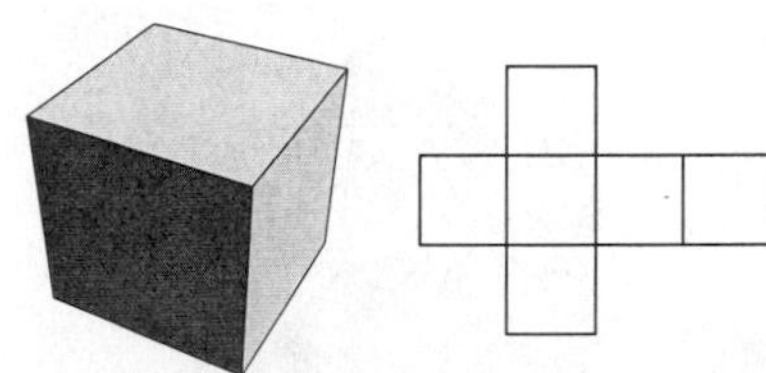

The three-dimensional PLATONIC SOLID (P_3) which is also called the HEXAHEDRON. The cube is composed of six SQUARE faces 6{4} which meet each other at RIGHT ANGLES, and has 8 VERTICES and 12 EDGES. It is described by the SCHLÄFLI SYMBOL {4,3}. It is a ZONOHEDRON. It is also the UNIFORM POLYHEDRON U_6 with WYTHOFF SYMBOL $3 \mid 2\,4$. It has the O_h OCTAHEDRAL GROUP of symmetries. The DUAL POLYHEDRON of the cube is the OCTAHEDRON.

Because the VOLUME of a cube of side length n is given by n^3, a number of the form n^3 is called a CUBIC NUMBER (or sometimes simply "a cube"). Similarly, the operation of taking a number to the third POWER is called CUBING.

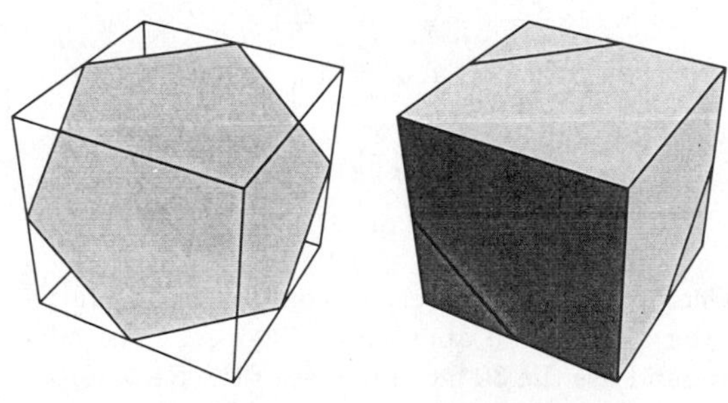

The cube cannot be STELLATED. A PLANE passing through the MIDPOINTS of opposite sides (perpendicular to a C_3 axis) cuts the cube in a regular HEXAGONAL CROSS-SECTION (Gardner 1960; Steinhaus 1983, p. 170; Cundy and Rollett 1989, p. 157; Holden 1991, pp. 22–23). Since there are four such axes, there are four possibly hexagonal cross-sections. If the vertices of the cube are $(\pm1, \pm1 \pm 1)$, then the vertices of the inscribed HEXAGON are $(0,-1,-1)$, $(1,0,-1)$, $(1,1,0)$, $(0,1,1)$, $(-1,0,1)$, and $(-1,-1,0)$. The largest SQUARE which will fit inside a cube of side a has each corner a distance 1/4 from a corner of a cube. The resulting SQUARE has side length $3\sqrt{2}\,a/4$, and the cube containing that side is called PRINCE RUPERT'S CUBE.

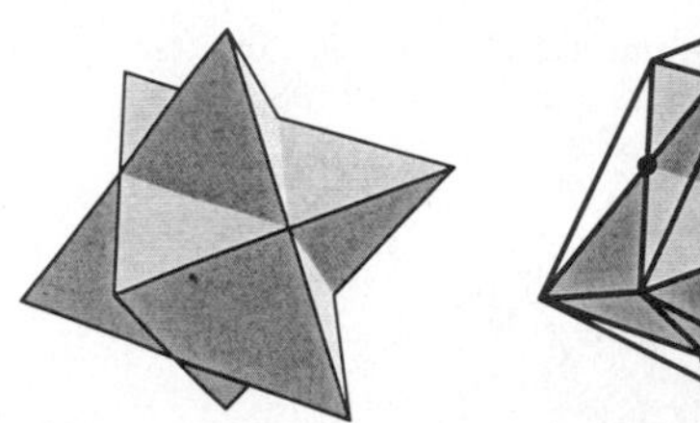

The solid formed by the faces having the sides of the STELLA OCTANGULA (left figure) as DIAGONALS is a cube (right figure; Ball and Coxeter 1987).

The VERTICES of a cube of side length 2 with face-centered axes are given by $(\pm1, \pm1, \pm1)$. If the cube is oriented with a space diagonal along the z-axis, the coordinates are $(0, 0, \sqrt{3})$, $(0, 2\sqrt{2/3}, 1/\sqrt{3})$, $(\sqrt{2}, \sqrt{2/3}, -1/\sqrt{3})$, $(\sqrt{2}, -\sqrt{2/3}, 1/\sqrt{3})$, $(0, -2\sqrt{2/3}, -1/\sqrt{3})$, $(-\sqrt{2}, -\sqrt{2/3}, 1/\sqrt{3})$, $(-\sqrt{2}, \sqrt{2/3}, -1/\sqrt{3})$, and the negatives of these vectors. A FACETED version is the GREAT CUBICUBOCTAHEDRON.

A cube of side length 1 has INRADIUS, MIDRADIUS, and CIRCUMRADIUS of

$$r = \tfrac{1}{2} = 0.5 \qquad (1)$$

$$\rho = \tfrac{1}{2}\sqrt{2} \approx 0.70710 \qquad (2)$$

$$R = \tfrac{1}{2}\sqrt{3} \approx 0.86602. \qquad (3)$$

The cube has a DIHEDRAL ANGLE of

$$\alpha = \tfrac{1}{2}\pi. \qquad (4)$$

The SURFACE AREA and VOLUME of the cube are

$$S = 6a^2 \qquad (5)$$

$$V = a^3. \qquad (6)$$

see also AUGMENTED TRUNCATED CUBE, BIAUGMENTED TRUNCATED CUBE, BIDIAKIS CUBE, BISLIT CUBE, BROWKIN'S THEOREM, CUBE DISSECTION, CUBE DOVETAILING PROBLEM, CUBE DUPLICATION, CUBIC NUMBER, CUBICAL GRAPH, HADWIGER PROBLEM, HYPERCUBE, KELLER'S CONJECTURE, PRINCE

RUPERT'S CUBE, RUBIK'S CUBE, SOMA CUBE, STELLA OCTANGULA, TESSERACT

References

Beyer, W. H. (Ed.) *CRC Standard Mathematical Tables, 28th ed.* Boca Raton, FL: CRC Press, p. 127, 1987.

Cundy, H. and Rollett, A. "Hexagonal Section of a Cube." §3.15.1 in *Mathematical Models, 3rd ed.* Stradbroke, England: Tarquin Pub., p. 157, 1989.

Davie, T. "The Cube (Hexahedron)." `http://www.dcs.st-and.ac.uk/~d/mathrecs/polyhedra/cube.html`.

Eppstein, D. "Rectilinear Geometry." `http://www.ics.uci.edu/~eppstein/junkyard/rect.html`.

Gardner, M. "Mathematical Games: More About the Shapes that Can Be Made with Complex Dominoes." *Sci. Amer.* **203**, 186–198, Nov. 1960.

Holden, A. *Shapes, Space, and Symmetry.* New York: Dover, 1991.

Steinhaus, H. *Mathematical Snapshots, 3rd American ed.* New York: Oxford University Press, 1983.

Cube 2-Compound

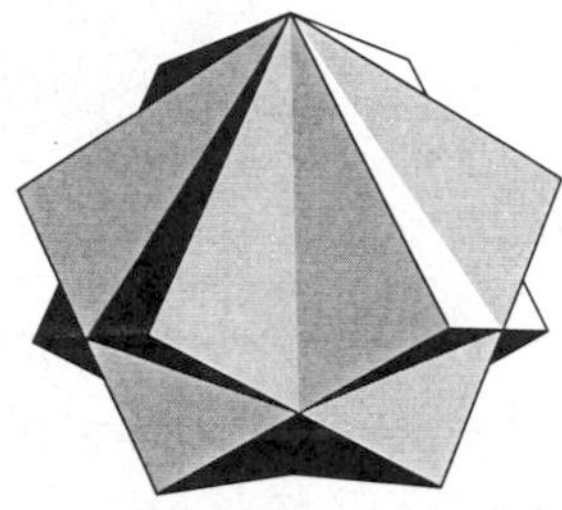

A POLYHEDRON COMPOUND obtained by allowing two CUBES to share opposite VERTICES, then rotating one a sixth of a turn (Holden 1971, p. 34).

see also CUBE, CUBE 3-COMPOUND, CUBE 4-COMPOUND, CUBE 5-COMPOUND, POLYHEDRON COMPOUND

References

Holden, A. *Shapes, Space, and Symmetry.* New York: Dover, 1991.

Cube 3-Compound

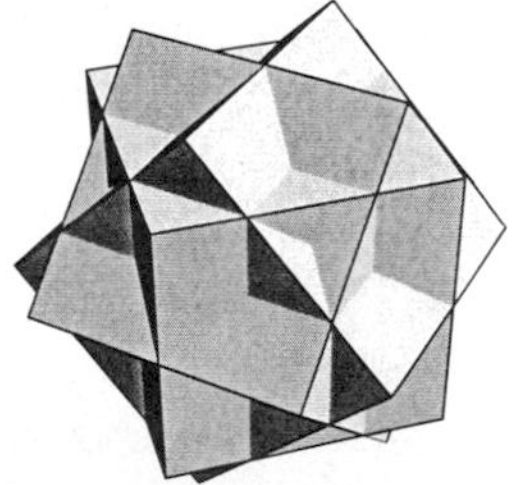

A compound with the symmetry of the CUBE which arises by joining three CUBES such that each shares two C_2 axes (Holden 1971, p. 35).

see also CUBE, CUBE 2-COMPOUND, CUBE 4-COMPOUND, CUBE 5-COMPOUND, POLYHEDRON COMPOUND

References

Holden, A. *Shapes, Space, and Symmetry.* New York: Dover, 1991.

Cube 4-Compound

A compound with the symmetry of the CUBE which arises by joining four CUBES such that each C_3 axis falls along the C_3 axis of one of the other CUBES (Holden 1971, p. 35).

see also CUBE, CUBE 2-COMPOUND, CUBE 3-COMPOUND, CUBE 5-COMPOUND, POLYHEDRON COMPOUND

References

Holden, A. *Shapes, Space, and Symmetry.* New York: Dover, 1991.

Cube 5-Compound

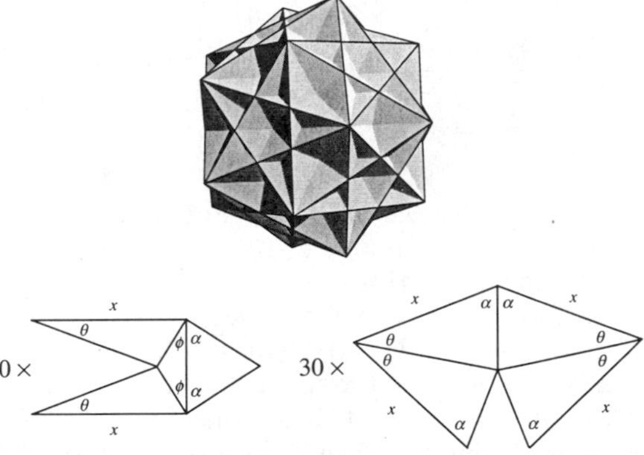

A POLYHEDRON COMPOUND consisting of the arrangement of five CUBES in the VERTICES of a DODECAHEDRON. In the above figure, let a be the length of a CUBE EDGE. Then

$$x = \tfrac{1}{2}a(3-\sqrt{5})$$

$$\theta = \tan^{-1}\left(\frac{3-\sqrt{5}}{2}\right) \approx 20^\circ 54'$$

$$\phi = \tan^{-1}\left(\frac{\sqrt{5}-1}{2}\right) \approx 31^\circ 43'$$

$$\psi = 90^\circ - \phi \approx 58^\circ 17'$$

$$\alpha = 90^\circ - \theta \approx 69^\circ 6'.$$

The compound is most easily constructed using pieces like the ones in the above line diagram. The cube 5-compound has the 30 facial planes of the RHOMBIC TRIACONTAHEDRON (Ball and Coxeter 1987).

see also CUBE, CUBE 2-COMPOUND, CUBE 3-COMPOUND, CUBE 4-COMPOUND, DODECAHEDRON,

POLYHEDRON COMPOUND, RHOMBIC TRIACONTAHEDRON

References

Ball, W. W. R. and Coxeter, H. S. M. *Mathematical Recreations and Essays, 13th ed.* New York: Dover, pp. 135 and 137, 1987.

Cundy, H. and Rollett, A. *Mathematical Models, 3rd ed.* Stradbroke, England: Tarquin Pub., pp. 135–136, 1989.

Gardner, M. *Fractal Music, Hypercards, and More: Mathematical Recreations from Scientific American Magazine.* New York: W. H. Freeman, pp. 297–298, 1992.

Honsberger, R. *Mathematical Gems II.* Washington, DC: Math. Assoc. Amer., pp. 75–80, 1976.

Hunter, J. A. H. and Madachy, J. S. *Mathematical Diversions.* New York: Dover, pp. 69–70, 1975.

Sloane, N. J. A. Sequence A014544 in "An On-Line Version of the Encyclopedia of Integer Sequences."

Cube Dissection

A CUBE can be divided into n subcubes for only $n = 1$, 8, 15, 20, 22, 27, 29, 34, 36, 38, 39, 41, 43, 45, 46, and $n \geq 48$ (Sloane's A014544).

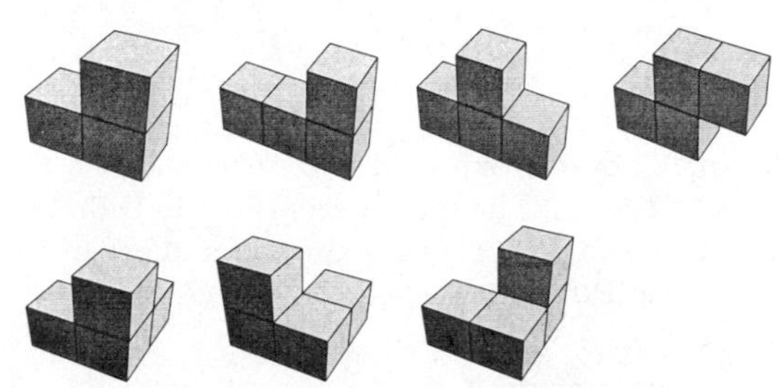

The seven pieces used to construct the $3 \times 3 \times 3$ cube dissection known as the SOMA CUBE are one 3-POLYCUBE and six 4-POLYCUBES ($1 \cdot 3 + 6 \cdot 4 = 27$), illustrated above.

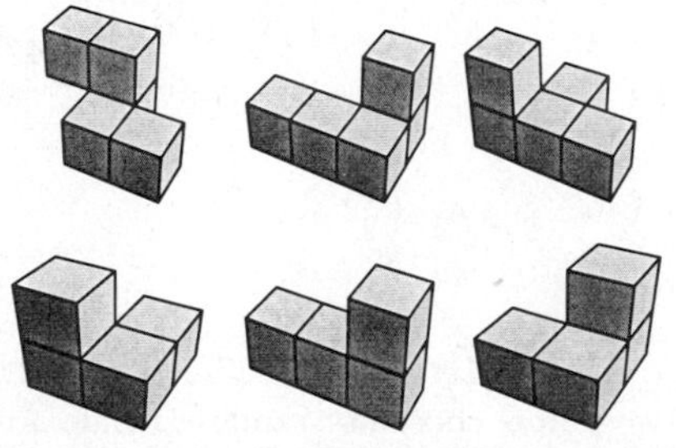

Another $3 \times 3 \times 3$ cube dissection due to Steinhaus uses three 5-POLYCUBES and three 4-POLYCUBES ($3 \cdot 5 + 3 \cdot 4 = 27$), illustrated above.

It is possible to cut a 1×3 RECTANGLE into two identical pieces which will form a CUBE (without overlapping) when folded and joined. In fact, an INFINITE number of solutions to this problem were discovered by C. L. Baker (Hunter and Madachy 1975).

see also CONWAY PUZZLE, DISSECTION, HADWIGER PROBLEM, POLYCUBE, SLOTHOUBER-GRAATSMA PUZZLE, SOMA CUBE

References

Ball, W. W. R. and Coxeter, H. S. M. *Mathematical Recreations and Essays, 13th ed.* New York: Dover, pp. 112–113, 1987.

Cundy, H. and Rollett, A. *Mathematical Models, 3rd ed.* Stradbroke, England: Tarquin Pub., pp. 203–205, 1989.

Gardner, M. "Block Packing." Ch. 18 in *Time Travel and Other Mathematical Bewilderments.* New York: W. H. Freeman, pp. 227–239, 1988.

Cube Dovetailing Problem

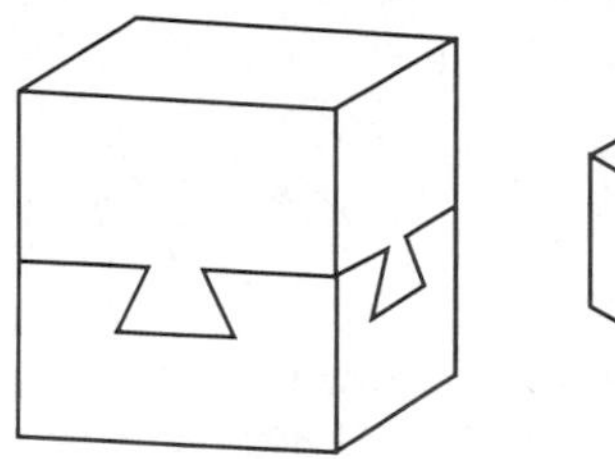

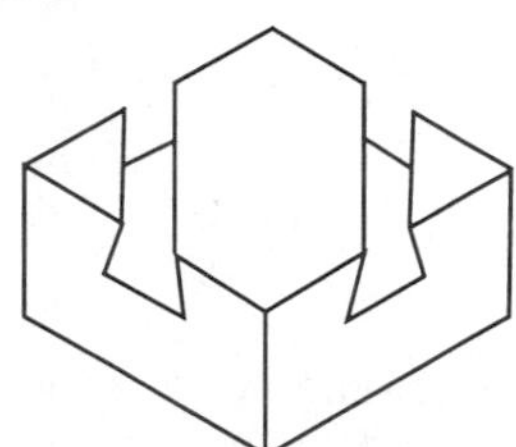

Given the figure on the left (without looking at the solution on the right), determine how to disengage the two slotted CUBE halves without cutting, breaking, or distorting.

References

Dudeney, H. E. *Amusements in Mathematics.* New York: Dover, pp. 145 and 249, 1958.

Ogilvy, C. S. *Excursions in Mathematics.* New York: Dover, pp. 57, 59, and 143, 1994.

Cube Duplication

Also called the DELIAN PROBLEM or DUPLICATION OF THE CUBE. A classical problem of antiquity which, given the EDGE of a CUBE, requires a second CUBE to be constructed having double the VOLUME of the first using only a STRAIGHTEDGE and COMPASS.

Under these restrictions, the problem cannot be solved because the DELIAN CONSTANT $2^{1/3}$ (the required RATIO of sides of the original CUBE and that to be constructed) is not a EUCLIDEAN NUMBER. The problem can be solved, however, using a NEUSIS CONSTRUCTION.

see also ALHAZEN'S BILLIARD PROBLEM, COMPASS, CUBE, DELIAN CONSTANT, GEOMETRIC PROBLEMS OF ANTIQUITY, NEUSIS CONSTRUCTION, STRAIGHTEDGE

References

Ball, W. W. R. and Coxeter, H. S. M. *Mathematical Recreations and Essays, 13th ed.* New York: Dover, pp. 93–94, 1987.

Conway, J. H. and Guy, R. K. *The Book of Numbers.* New York: Springer-Verlag, pp. 190–191, 1996.

Courant, R. and Robbins, H. "Doubling the Cube" and "A Classical Construction for Doubling the Cube." §3.3.1 and 3.5.1 in *What is Mathematics?: An Elementary Approach to Ideas and Methods, 2nd ed.* Oxford, England: Oxford University Press, pp. 134–135 and 146, 1996.

Dörrie, H. "The Delian Cube-Doubling Problem." §35 in *100 Great Problems of Elementary Mathematics: Their History and Solutions.* New York: Dover, pp. 170–172, 1965.

Cube-Octahedron Compound

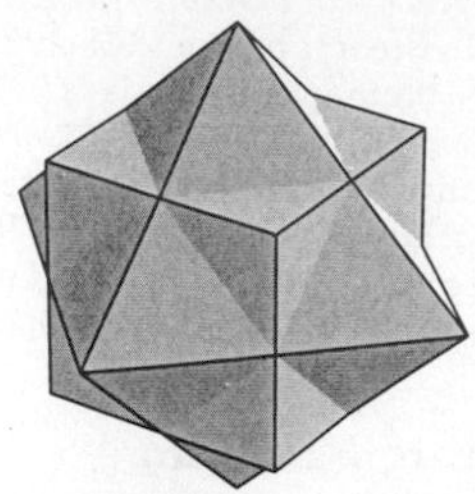

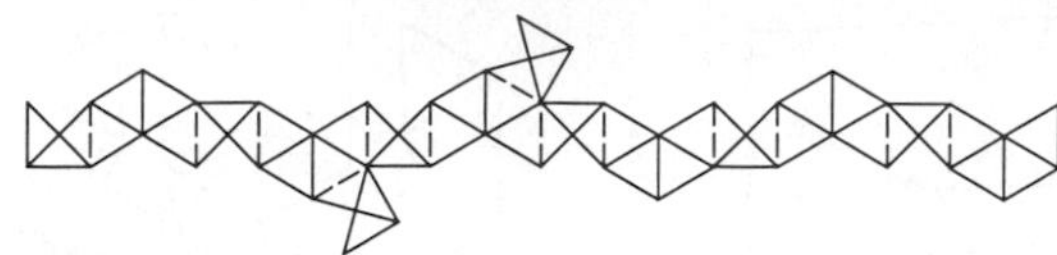

A POLYHEDRON COMPOUND composed of a CUBE and its DUAL POLYHEDRON, the OCTAHEDRON. The 14 vertices are given by (±1, ±1, ±1), (±2, 0, 0), (0, ±2, 0), (0, 0, ±2).

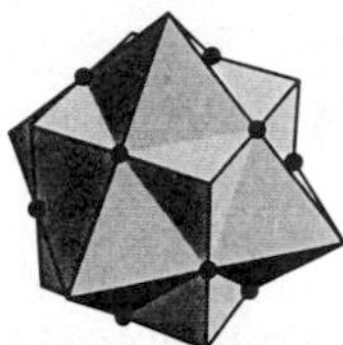 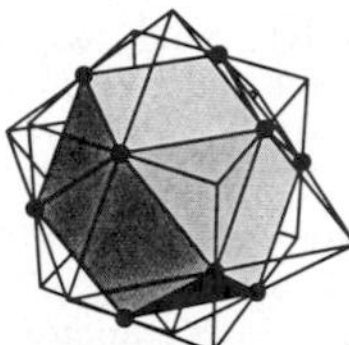 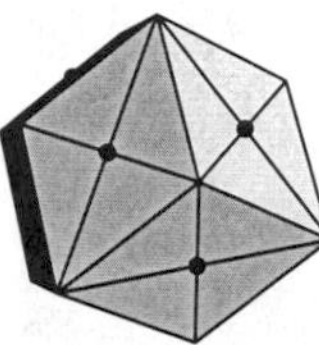

The solid common to both the CUBE and OCTAHEDRON (left figure) in a cube-octahedron compound is a CUBOCTAHEDRON (middle figure). The edges intersecting in the points plotted above are the diagonals of RHOMBUSES, and the 12 RHOMBUSES form a RHOMBIC DODECAHEDRON (right figure; Ball and Coxeter 1987).

see also CUBE, CUBOCTAHEDRON, OCTAHEDRON, POLYHEDRON COMPOUND

References

Ball, W. W. R. and Coxeter, H. S. M. *Mathematical Recreations and Essays, 13th ed.* New York: Dover, p. 137, 1987.

Coxeter, H. S. M. *Introduction to Geometry, 2nd ed.* New York: Wiley, p. 158, 1969.

Cundy, H. and Rollett, A. *Mathematical Models, 3rd ed.* Stradbroke, England: Tarquin Pub., p. 130, 1989.

Cube Point Picking

N.B. A detailed on-line essay by S. Finch was the starting point for this entry.

Let two points be picked randomly from a unit n-D HYPERCUBE. The expected distance between the points $\Delta(N)$ is then

$$\Delta(1) = \tfrac{1}{3}$$
$$\Delta(2) = \tfrac{1}{15}[\sqrt{2}+2+5\ln(1+\sqrt{2})] = 0.521405433\ldots$$
$$\Delta(3) = \tfrac{1}{105}[4+17\sqrt{2}-6\sqrt{3}+21\ln(1+\sqrt{2}) + 42\ln(2+\sqrt{3})-7\pi] = 0.661707182\ldots$$
$$\Delta(4) = 0.77766\ldots$$
$$\Delta(5) = 0.87852\ldots$$
$$\Delta(6) = 0.96895\ldots$$
$$\Delta(7) = 1.05159\ldots$$
$$\Delta(8) = 1.12817\ldots.$$

The function $\Delta(n)$ satisfies

$$\tfrac{1}{3}n^{1/2} \leq \Delta(n) \leq (\tfrac{1}{6}n)^{1/2}\sqrt{\frac{1}{3}\left[1+2\left(1-\frac{3}{5n}\right)^{1/2}\right]}$$

(Anderssen *et al.* 1976).

Pick N points $p_1, \ldots, p_N$ randomly in a unit n-cube. Let C be the CONVEX HULL, so

$$C \equiv \left\{\sum_{j=1}^N \lambda_j p_j : \lambda_j \geq 0 \text{ for all } j \text{ and } \sum_{j=1}^N \lambda_j = 1\right\}.$$

Let $V(n,N)$ be the expected n-D VOLUME (the CONTENT) of C, $S(n,N)$ be the expected $(n-1)$-D SURFACE AREA of C, and $P(n,N)$ the expected number of VERTICES on the POLYGONAL boundary of C. Then

$$\lim_{N\to\infty} \frac{N[1-V(2,N)]}{\ln N} = \frac{8}{3}$$
$$\lim_{N\to\infty} \sqrt{N}[4-S(2,N)] = \sqrt{2\pi}\left[2-\int_0^1(\sqrt{1+t^2}-1)t^{-3/2}\,dt\right] = 4.2472965\ldots,$$

and

$$\lim_{N\to\infty} P(2,N) - \tfrac{8}{3}\ln N = \tfrac{8}{3}(\gamma - \ln 2) = -0.309150708\ldots$$

(Rényi and Sulanke 1963, 1964). The average DISTANCE between two points chosen at random inside a unit cube is

$$\tfrac{1}{105}(4+17\sqrt{2}-6\sqrt{3}+21\ln(1+\sqrt{2})+42\ln(2+\sqrt{3})-7\pi)$$

(Robbins 1978, Le Lionnais 1983).

Pick n points on a CUBE, and space them as far apart as possible. The best value known for the minimum straight LINE distance between any two points is given in the following table.

n	$d(n)$
5	1.1180339887498
6	1.0606601482100
7	1
8	1
9	0.86602540378463
10	0.74999998333331
11	0.70961617562351
12	0.70710678118660
13	0.70710678118660
14	0.70710678118660
15	0.625

see also CUBE TRIANGLE PICKING, DISCREPANCY THEOREM, POINT PICKING

References

Anderssen, R. S.; Brent, R. P.; Daley, D. J.; and Moran, A. P. "Concerning $\int_0^1 \cdots \int_0^1 \sqrt{x_1{}^2 + \ldots + x_k{}^2}\, dx_1 \cdots dx_k$ and a Taylor Series Method." *SIAM J. Appl. Math.* **30**, 22–30, 1976.

Bolis, T. S. Solution to Problem E2629. "Average Distance Between Two Points in a Box." *Amer. Math. Monthly* **85**, 277–278, 1978.

Finch, S. "Favorite Mathematical Constants." `http://www.mathsoft.com/asolve/constant/geom/geom.html`.

Ghosh, B. "Random Distances within a Rectangle and Between Two Rectangles." *Bull. Calcutta Math. Soc.* **43**, 17–24, 1951.

Holshouser, A. L.; King, L. R.; and Klein, B. G. Solution to Problem E3217, "Minimum Average Distance Between Points in a Rectangle." *Amer. Math. Monthly* **96**, 64–65, 1989.

Le Lionnais, F. *Les nombres remarquables.* Paris: Hermann, p. 30, 1983.

Rényi, A. and Sulanke, R. "Über die konvexe Hülle von n zufällig gewählten Punkten, I." *Z. Wahrscheinlichkeits* **2**, 75–84, 1963.

Rényi, A. and Sulanke, R. "Über die konvexe Hülle von n zufällig gewählten Punkten, II." *Z. Wahrscheinlichkeits* **3**, 138–147, 1964.

Robbins, D. "Average Distance Between Two Points in a Box." *Amer. Math. Monthly* **85**, 278, 1978.

Santaló, L. A. *Integral Geometry and Geometric Probability.* Reading, MA: Addison-Wesley, 1976.

Cube Power

A number raised to the third POWER. x^3 is read as "x cubed."

see also CUBIC NUMBER

Cube Root

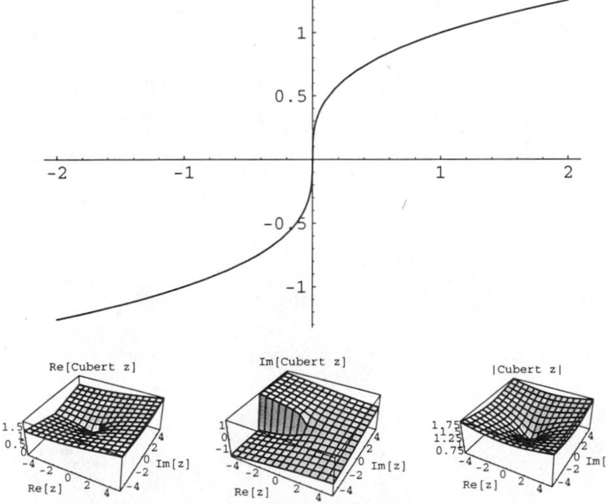

Given a number z, the cube root of z, denoted $\sqrt[3]{z}$ or $z^{1/3}$ (z to the 1/3 POWER), is a number a such that $a^3 = z$. There are three (not necessarily distinct) cube roots for any number.

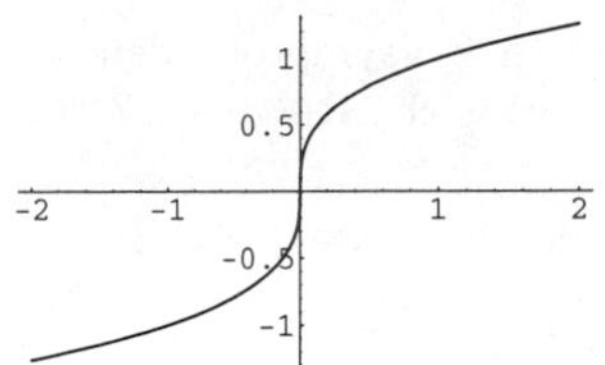

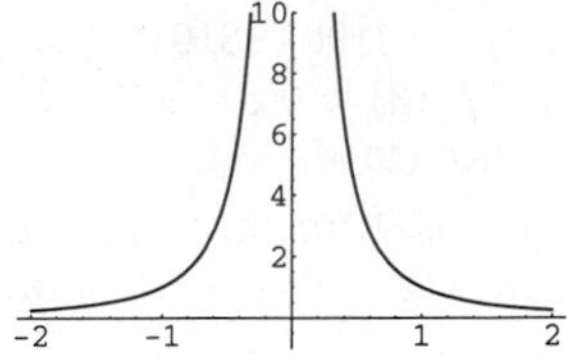

For real arguments, the cube root is an INCREASING FUNCTION, although the usual derivative test cannot be used to establish this fact at the ORIGIN since the the derivative approaches infinity there (as illustrated above).

see also CUBE DUPLICATION, CUBED, DELIAN CONSTANT, GEOMETRIC PROBLEMS OF ANTIQUITY, k-MATRIX, SQUARE ROOT

Cube Triangle Picking

Pick 3 points at random in the unit n-HYPERCUBE. Denote the probability that the three points form an OBTUSE TRIANGLE $\Pi(n)$. Langford (1969) proved

$$\Pi(2) = \tfrac{97}{150} + \tfrac{1}{40}\pi = 0.725206483\ldots.$$

see also BALL TRIANGLE PICKING, CUBE POINT PICKING

References

Finch, S. "Favorite Mathematical Constants." `http://www.mathsoft.com/asolve/constant/geom/geom.html`.

Langford, E. "The Probability that a Random Triangle is Obtuse." *Biometrika* **56**, 689–690, 1969.

Santaló, L. A. *Integral Geometry and Geometric Probability.* Reading, MA: Addison-Wesley, 1976.

Cubed

A number to the POWER 3 is said to be cubed, so that x^3 is called "x cubed."

see also CUBE ROOT, SQUARED

Cubefree

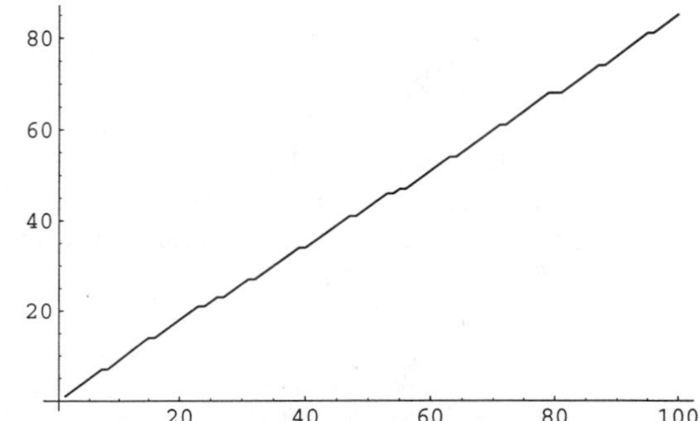

A number is said to be cubefree if its PRIME decomposition contains no tripled factors. All PRIMES are therefore trivially cubefree. The cubefree numbers are 1, 2, 3, 4, 5, 6, 7, 9, 10, 11, 12, 13, 14, 15, 17, ... (Sloane's A004709). The cubeful numbers (i.e., those that contain at least one cube) are 8, 16, 24, 27, 32, 40, 48, 54, ... (Sloane's A046099). The number of cubefree numbers less than 10, 100, 1000, ... are 9, 85, 833,

8319, 83190, 831910, ..., and their asymptotic density is $1/\zeta(3) \approx 0.831907$, where $\zeta(n)$ is the RIEMANN ZETA FUNCTION.

see also BIQUADRATEFREE, PRIME NUMBER, RIEMANN ZETA FUNCTION, SQUAREFREE

References

Sloane, N. J. A. Sequences A004709 and A046099 in "An On-Line Version of the Encyclopedia of Integer Sequences."

Cubic Curve

A cubic curve is an ALGEBRAIC CURVE of degree 3. An algebraic curve over a FIELD K is an equation $f(X,Y)=0$, where $f(X,Y)$ is a POLYNOMIAL in X and Y with COEFFICIENTS in K, and the degree of f is the MAXIMUM degree of each of its terms (MONOMIALS).

Newton showed that all cubics can be generated by the projection of the five divergent cubic parabolas. Newton's classification of cubic curves appeared in the chapter "Curves" in *Lexicon Technicum* by John Harris published in London in 1710. Newton also classified all cubics into 72 types, missing six of them. In addition, he showed that any cubic can be obtained by a suitable projection of the ELLIPTIC CURVE

$$y^2 = ax^3 + bx^2 + cx + d, \tag{1}$$

where the projection is a BIRATIONAL TRANSFORMATION, and the general cubic can also be written as

$$y^2 = x^3 + ax + b. \tag{2}$$

Newton's first class is equations of the form

$$xy^2 + ey = ax^3 + bx^2 + cx + d. \tag{3}$$

This is the hardest case and includes the SERPENTINE CURVE as one of the subcases. The third class was

$$ay^2 = x(x^2 - 2bx + c), \tag{4}$$

which is called NEWTON'S DIVERGING PARABOLAS. Newton's 66th curve was the TRIDENT OF NEWTON. Newton's classification of cubics was criticized by Euler because it lacked generality. Plücker later gave a more detailed classification with 219 types.

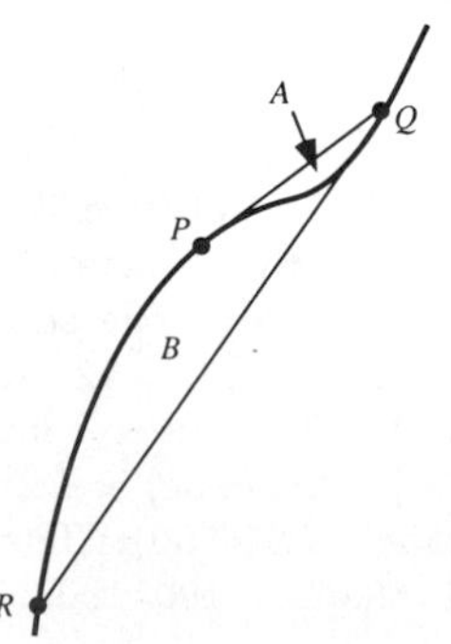

Pick a point P, and draw the tangent to the curve at P. Call the point where this tangent intersects the curve Q. Draw another tangent and call the point of intersection with the curve R. Every curve of third degree has the property that, with the areas in the above labeled figure,

$$B = 16A \tag{5}$$

(Honsberger 1991).

see also CAYLEY-BACHARACH THEOREM, CUBIC EQUATION

References

Honsberger, R. *More Mathematical Morsels.* Washington, DC: Math. Assoc. Amer., pp. 114–118, 1991.

Newton, I. *Mathematical Works, Vol. 2.* New York: Johnson Reprint Corp., pp. 135–161, 1967.

Wall, C. T. C. "Affine Cubic Functions III." *Math. Proc. Cambridge Phil. Soc.* **87**, 1–14, 1980.

Westfall, R. S. *Never at Rest: A Biography of Isaac Newton.* New York: Cambridge University Press, 1988.

Yates, R. C. "Cubic Parabola." *A Handbook on Curves and Their Properties.* Ann Arbor, MI: J. W. Edwards, pp. 56–59, 1952.

Cubic Equation

A cubic equation is a POLYNOMIAL equation of degree three. Given a general cubic equation

$$z^3 + a_2z^2 + a_1z + a_0 = 0 \tag{1}$$

(the COEFFICIENT a_3 of z^3 may be taken as 1 without loss of generality by dividing the entire equation through by a_3), first attempt to eliminate the a_2 term by making a substitution of the form

$$z \equiv x - \lambda. \tag{2}$$

Then

$$(x-\lambda)^3 + a_2(x-\lambda)^2 + a_1(x-\lambda) + a_0 = 0 \tag{3}$$

$$(x^3 - 3\lambda x^2 + 3\lambda^2 x - \lambda^3) + a_2(x^2 - 2\lambda x + \lambda^2) + a_1(x-\lambda) + a_0 = 0 \tag{4}$$

$$x^3 + x^2(a_2 - 3\lambda) + x(a_1 - 2a_2\lambda + 3\lambda^2) + (a_0 - a_1\lambda + a_2\lambda^2 - \lambda^3) = 0. \tag{5}$$

The x^2 is eliminated by letting $\lambda = a_2/3$, so

$$z \equiv x - \tfrac{1}{3}a_2. \tag{6}$$

Then

$$z^3 = (x - \tfrac{1}{3}a_2)^3 = x^3 - a_2x^2 + \tfrac{1}{3}a_2{}^2x - \tfrac{1}{27}a_2{}^3 \tag{7}$$

$$a_2z^2 = a_2(x - \tfrac{1}{3}a_2)^2 = a_2x^2 - \tfrac{2}{3}a_2{}^2x + \tfrac{1}{9}a_2{}^3 \tag{8}$$

$$a_1z = a_1(x - \tfrac{1}{3}a_2) = a_1x - \tfrac{1}{3}a_1a_2, \tag{9}$$

so equation (1) becomes

$$x^3 + (-a_2 + a_2)x^2 + (\tfrac{1}{3}a_2{}^2 - \tfrac{2}{3}a_2{}^2 + a_1)x - (\tfrac{1}{27}a_2{}^3 - \tfrac{1}{9}a_2{}^3 + \tfrac{1}{3}a_1a_2 - a_0) = 0 \tag{10}$$

$$x^3 + (a_1 - \tfrac{1}{3}a_2{}^2)x - (\tfrac{1}{3}a_1a_2 - \tfrac{2}{27}a_2{}^3 - a_0) = 0 \quad (11)$$

$$x^3 + 3 \cdot \frac{3a_1 - a_2{}^2}{9}x - 2 \cdot \frac{9a_1a_2 - 27a_0 - 2a_2{}^3}{54} = 0. \quad (12)$$

Defining

$$p \equiv \frac{3a_1 - a_2{}^2}{3} \quad (13)$$

$$q \equiv \frac{9a_1a_2 - 27a_0 - 2a_2{}^3}{27} \quad (14)$$

then allows (12) to be written in the standard form

$$x^3 + px = q. \quad (15)$$

The simplest way to proceed is to make VIETA'S SUBSTITUTION

$$x = w - \frac{p}{3w}, \quad (16)$$

which reduces the cubic to the equation

$$w^3 - \frac{p^3}{27w^3} - q = 0, \quad (17)$$

which is easily turned into a QUADRATIC EQUATION in w^3 by multiplying through by w^3 to obtain

$$(w^3)^2 - q(w^3) - \tfrac{1}{27}p^3 = 0 \quad (18)$$

(Birkhoff and Mac Lane 1965, p. 106). The result from the QUADRATIC EQUATION is

$$w^3 = \tfrac{1}{2}\left(q \pm \sqrt{q^2 + \tfrac{4}{27}p^3}\right) = \tfrac{1}{2}q \pm \sqrt{\tfrac{1}{4}q^2 + \tfrac{1}{27}p^3}$$
$$= R \pm \sqrt{R^2 + Q^3}, \quad (19)$$

where Q and R are are sometimes more useful to deal with than are p and q. There are therefore six solutions for w (two corresponding to each sign for each ROOT of w^3). Plugging w back in to (17) gives three pairs of solutions, but each pair is equal, so there are three solutions to the cubic equation.

Equation (12) may also be explicitly factored by attempting to pull out a term of the form $(x - B)$ from the cubic equation, leaving behind a quadratic equation which can then be factored using the QUADRATIC FORMULA. This process is equivalent to making Vieta's substitution, but does a slightly better job of motivating Vieta's "magic" substitution, and also at producing the explicit formulas for the solutions. First, define the intermediate variables

$$Q \equiv \frac{3a_1 - a_2{}^2}{9} \quad (20)$$

$$R \equiv \frac{9a_2a_1 - 27a_0 - 2a_2{}^3}{54} \quad (21)$$

(which are identical to p and q up to a constant factor). The general cubic equation (12) then becomes

$$x^3 + 3Qx - 2R = 0. \quad (22)$$

Let B and C be, for the moment, arbitrary constants. An identity satisfied by PERFECT CUBIC equations is that

$$x^3 - B^3 = (x - B)(x^2 + Bx + B^2). \quad (23)$$

The general cubic would therefore be directly factorable if it did not have an x term (i.e., if $Q = 0$). However, since in general $Q \neq 0$, add a multiple of $(x - B)$—say $C(x-B)$—to both sides of (23) to give the slightly messy identity

$$(x^3 - B^3) + C(x - B) = (x - B)(x^2 + Bx + B^2 + C) = 0, \quad (24)$$

which, after regrouping terms, is

$$x^3 + Cx - (B^3 + BC) = (x - B)[x^2 + Bx + (B^2 + C)] = 0. \quad (25)$$

We would now like to match the COEFFICIENTS C and $-(B^3 + BC)$ with those of equation (22), so we must have

$$C = 3Q \quad (26)$$

$$B^3 + BC = 2R. \quad (27)$$

Plugging the former into the latter then gives

$$B^3 + 3QB = 2R. \quad (28)$$

Therefore, if we can find a value of B satisfying the above identity, we have factored a linear term from the cubic, thus reducing it to a QUADRATIC EQUATION. The trial solution accomplishing this miracle turns out to be the symmetrical expression

$$B = [R + \sqrt{Q^3 + R^2}\,]^{1/3} + [R - \sqrt{Q^3 + R^2}\,]^{1/3}. \quad (29)$$

Taking the second and third POWERS of B gives

$$B^2 = [R + \sqrt{Q^3 + R^2}]^{2/3} + 2[R^2 - (Q^3 + R^2)]^{1/3}$$
$$+ [R - \sqrt{Q^3 + R^2}]^{2/3}$$
$$= [R + \sqrt{Q^3 + R^2}]^{2/3} + [R - \sqrt{Q^3 + R^2}]^{2/3} - 2Q \quad (30)$$

$$B^3 = -2QB + \left\{[R + \sqrt{Q^3 + R^2}]^{1/3} + [R - \sqrt{Q^3 + R^2}]^{1/3}\right\}$$
$$\times \left\{[R + \sqrt{Q^3 + R^2}]^{2/3} + [R - \sqrt{Q^3 + R^2}]^{2/3}\right\}$$
$$= [R + \sqrt{Q^3 + R^2}] + [R - \sqrt{Q^3 + R^2}]$$
$$+ [R - \sqrt{Q^3 + R^2}]^{1/3}[R - \sqrt{Q^3 + R^2}]^{2/3}$$
$$+ [R - \sqrt{Q^3 + R^2}]^{2/3}[R - \sqrt{Q^3 + R^2}]^{1/3} - 2QB$$
$$= -2QB + 2R + [R^2 - (Q^3 + R^2)]^{1/3}$$
$$\times \left[\left(R + \sqrt{Q^3 + R^2}\right)^{1/3} + \left(R - \sqrt{Q^3 - R^2}\right)^{1/3}\right]$$
$$= -2QB + 2R - QB = -3QB + 2R. \quad (31)$$

Plugging B^3 and B into the left side of (28) gives

$$(-3QB + 2R) + 3QB = 2R, \quad (32)$$

so we have indeed found the factor $(x - B)$ of (22), and we need now only factor the quadratic part. Plugging $C = 3Q$ into the quadratic part of (25) and solving the resulting

$$x^2 + Bx + (B^2 + 3Q) = 0 \quad (33)$$

then gives the solutions

$$\begin{aligned} x &= \tfrac{1}{2}[-B \pm \sqrt{B^2 - 4(B^2 + 3Q)}\,] \\ &= -\tfrac{1}{2}B \pm \tfrac{1}{2}\sqrt{-3B^2 - 12Q} \\ &= -\tfrac{1}{2}B \pm \tfrac{1}{2}\sqrt{3}\,i\sqrt{B^2 + 4Q}\,. \quad (34) \end{aligned}$$

These can be simplified by defining

$$\begin{aligned} A &\equiv [R + \sqrt{Q^3 + R^2}\,]^{1/3} - [R - \sqrt{Q^3 + R^2}\,]^{1/3} \quad (35) \\ A^2 &= [R + \sqrt{Q^3 + R^2}\,]^{2/3} - 2[R^2 - (Q^3 + R^2)]^{1/3} \\ &\quad + [R - \sqrt{Q^3 + R^2}\,]^{2/3} \\ &= [R + \sqrt{Q^3 + R^2}\,]^{2/3} + [R - \sqrt{Q^3 + R^2}\,]^{2/3} + 2Q \\ &= B^2 + 4Q, \quad (36) \end{aligned}$$

so that the solutions to the quadratic part can be written

$$x = -\tfrac{1}{2}B \pm \tfrac{1}{2}\sqrt{3}\,iA. \quad (37)$$

Defining

$$\begin{aligned} D &\equiv Q^3 + R^2 \quad (38) \\ S &\equiv \sqrt[3]{R + \sqrt{D}} \quad (39) \\ T &\equiv \sqrt[3]{R - \sqrt{D}}, \quad (40) \end{aligned}$$

where D is the DISCRIMINANT (which is defined slightly differently, including the opposite SIGN, by Birkhoff and Mac Lane 1965) then gives very simple expressions for A and B, namely

$$\begin{aligned} B &= S + T \quad (41) \\ A &= S - T. \quad (42) \end{aligned}$$

Therefore, at last, the ROOTS of the original equation in z are then given by

$$\begin{aligned} z_1 &= -\tfrac{1}{3}a_2 + (S + T) \quad (43) \\ z_2 &= -\tfrac{1}{3}a_2 - \tfrac{1}{2}(S + T) + \tfrac{1}{2}i\sqrt{3}\,(S - T) \quad (44) \\ z_3 &= -\tfrac{1}{3}a_2 - \tfrac{1}{2}(S + T) - \tfrac{1}{2}i\sqrt{3}\,(S - T), \quad (45) \end{aligned}$$

with a_2 the COEFFICIENT of z^2 in the original equation, and S and T as defined above. These three equations giving the three ROOTS of the cubic equation are sometimes known as CARDANO'S FORMULA. Note that if the equation is in the standard form of Vieta

$$x^3 + px = q, \quad (46)$$

in the variable x, then $a_2 = 0$, $a_1 = p$, and $a_0 = -q$, and the intermediate variables have the simple form (c.f. Beyer 1987)

$$\begin{aligned} Q &= \tfrac{1}{3}p \quad (47) \\ R &= \tfrac{1}{2}q \quad (48) \\ D &\equiv Q^3 + R^2 = \left(\frac{p}{3}\right)^3 + \left(\frac{q}{2}\right)^2. \quad (49) \end{aligned}$$

The equation for z_1 in CARDANO'S FORMULA does not have an i appearing in it explicitly while z_2 and z_3 do, but this does not say anything about the number of REAL and COMPLEX ROOTS (since S and T are themselves, in general, COMPLEX). However, determining which ROOTS are REAL and which are COMPLEX can be accomplished by noting that if the DISCRIMINANT $D > 0$, one ROOT is REAL and two are COMPLEX CONJUGATES; if $D = 0$, all ROOTS are REAL and at least two are equal; and if $D < 0$, all ROOTS are REAL and unequal. If $D < 0$, define

$$\theta \equiv \cos^{-1}\left(\frac{R}{\sqrt{-Q^3}}\right). \quad (50)$$

Then the REAL solutions are of the form

$$\begin{aligned} z_1 &= 2\sqrt{-Q}\cos\left(\frac{\theta}{3}\right) - \tfrac{1}{3}a_2 \quad (51) \\ z_2 &= 2\sqrt{-Q}\cos\left(\frac{\theta + 2\pi}{3}\right) - \tfrac{1}{3}a_2 \quad (52) \\ z_3 &= 2\sqrt{-Q}\cos\left(\frac{\theta + 4\pi}{3}\right) - \tfrac{1}{3}a_2. \quad (53) \end{aligned}$$

This procedure can be generalized to find the REAL ROOTS for any equation in the standard form (46) by using the identity

$$\sin^3\theta - \tfrac{3}{4}\sin\theta + \tfrac{1}{4}\sin(3\theta) = 0 \quad (54)$$

(Dickson 1914) and setting

$$x \equiv \sqrt{\frac{4|p|}{3}}\,y \quad (55)$$

(Birkhoff and Mac Lane 1965, pp. 90–91), then

$$\left(\frac{4|p|}{3}\right)^{3/2} y^3 + p\sqrt{\frac{4|p|}{3}}\,y = q \quad (56)$$

$$y^3 + \tfrac{3}{4}\frac{p}{|p|}y = \left(\frac{3}{4|p|}\right)^{3/2} q \quad (57)$$

$$4y^3 + 3\,\mathrm{sgn}(p)y = \tfrac{1}{2}q\left(\frac{3}{|p|}\right)^{3/2} \equiv C. \tag{58}$$

If $p > 0$, then use

$$\sinh(3\theta) = 4\sinh^3\theta + 3\sinh\theta \tag{59}$$

to obtain

$$y = \sinh(\tfrac{1}{3}\sinh^{-1}C). \tag{60}$$

If $p < 0$ and $|C| \geq 1$, use

$$\cosh(3\theta) = 4\cosh^3\theta - 3\cosh\theta, \tag{61}$$

and if $p < 0$ and $|C| \leq 1$, use

$$\cos(3\theta) = 4\cos^3\theta - 3\cos\theta, \tag{62}$$

to obtain

$$y = \begin{cases} \cosh(\frac{1}{3}\cosh^{-1}C) & \text{for } C \geq 1 \\ -\cosh(\frac{1}{3}\cosh^{-1}|C|) & \text{for } C \leq -1 \\ \cos(\frac{1}{3}\cos^{-1}C) \text{ [three solutions]} & \text{for } |C| < 1. \end{cases} \tag{63}$$

The solutions to the original equation are then

$$x_i = 2\sqrt{\frac{|p|}{3}}\,y_i - \tfrac{1}{3}a_2. \tag{64}$$

An alternate approach to solving the cubic equation is to use LAGRANGE RESOLVENTS. Let $\omega \equiv e^{2\pi i/3}$, define

$$(1, x_1) = x_1 + x_2 + x_3 \tag{65}$$

$$(\omega, x_1) = x_1 + \omega x_2 + \omega^2 x_3 \tag{66}$$

$$(\omega^2, x_1) = x_1 + \omega^2 x_2 + \omega x_3, \tag{67}$$

where x_i are the ROOTS of

$$x^3 + px + q = 0, \tag{68}$$

and consider the equation

$$[x - (u_1 + u_2)][x - (\omega u_1 + \omega^2 u_2)][x - (\omega^2 u_1 + \omega u_2)] = 0, \tag{69}$$

where u_1 and u_2 are COMPLEX NUMBERS. The ROOTS are then

$$x_j = \omega^j u_1 + \omega^{2j} u_2 \tag{70}$$

for $j = 0, 1, 2$. Multiplying through gives

$$x^3 - 3u_1u_2x - (u_1{}^3 + u_2{}^3) = 0, \tag{71}$$

or

$$x^3 + px + q = 0, \tag{72}$$

where

$$u_1{}^3 + u_2{}^3 = -q \tag{73}$$

$$u_1{}^3 u_2{}^3 = -\left(\frac{1}{3}\right)^3. \tag{74}$$

The solutions satisfy NEWTON'S IDENTITIES

$$z_1 + z_2 + z_3 = -a_2 \tag{75}$$

$$z_1z_2 + z_2z_3 + z_1z_3 = a_1 \tag{76}$$

$$z_1z_2z_3 = -a_0. \tag{77}$$

In standard form, $a_2 = 0$, $a_1 = p$, and $a_0 = -q$, so we have the identities

$$p = z_1z_2 - z_3{}^2 \tag{78}$$

$$(z_1 - z_2)^2 = -(4p - 3z_3{}^2) \tag{79}$$

$$z_1{}^2 + z_2{}^2 + z_3{}^2 = -2p. \tag{80}$$

Some curious identities involving the roots of a cubic equation due to Ramanujan are given by Berndt (1994).

see also QUADRATIC EQUATION, QUARTIC EQUATION, QUINTIC EQUATION, SEXTIC EQUATION

References

Abramowitz, M. and Stegun, C. A. (Eds.). *Handbook of Mathematical Functions with Formulas, Graphs, and Mathematical Tables, 9th printing.* New York: Dover, p. 17, 1972.

Berger, M. §16.4.1–16.4.11.1 in *Geometry I.* New York: Springer-Verlag, 1994.

Berndt, B. C. *Ramanujan's Notebooks, Part IV.* New York: Springer-Verlag, pp. 22–23, 1994.

Beyer, W. H. *CRC Standard Mathematical Tables, 28th ed.* Boca Raton, FL: CRC Press, pp. 9–11, 1987.

Birkhoff, G. and Mac Lane, S. *A Survey of Modern Algebra, 3rd ed.* New York: Macmillan, pp. 90–91, 106–107, and 414–417, 1965.

Dickson, L. E. "A New Solution of the Cubic Equation." *Amer. Math. Monthly* **5**, 38–39, 1898.

Dickson, L. E. *Elementary Theory of Equations.* New York: Wiley, pp. 36–37, 1914.

Dunham, W. "Cardano and the Solution of the Cubic." Ch. 6 in *Journey Through Genius: The Great Theorems of Mathematics.* New York: Wiley, pp. 133–154, 1990.

Ehrlich, G. §4.16 in *Fundamental Concepts of Abstract Algebra.* Boston, MA: PWS-Kent, 1991.

Jones, J. "Omar Khayyám and a Geometric Solution of the Cubic." `http://jwilson.coe.uga.edu/emt669/Student.Folders/Jones.June/omar/omarpaper.html`.

Kennedy, E. C. "A Note on the Roots of a Cubic." *Amer. Math. Monthly* **40**, 411–412, 1933.

King, R. B. *Beyond the Quartic Equation.* Boston, MA: Birkhäuser, 1996.

Press, W. H.; Flannery, B. P.; Teukolsky, S. A.; and Vetterling, W. T. "Quadratic and Cubic Equations." §5.6 in *Numerical Recipes in FORTRAN: The Art of Scientific Computing, 2nd ed.* Cambridge, England: Cambridge University Press, pp. 178–180, 1992.

Spanier, J. and Oldham, K. B. "The Cubic Function $x^3 + ax^2 + bx + c$ and Higher Polynomials." Ch. 17 in *An Atlas of Functions.* Washington, DC: Hemisphere, pp. 131–147, 1987.

van der Waerden, B. L. §64 in *Algebra.* New York: Frederick Ungar, 1970.

Cubic Number

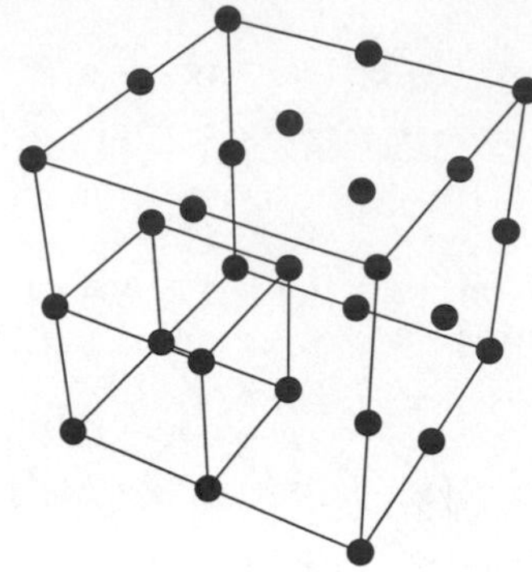

A FIGURATE NUMBER of the form n^3, for n a POSITIVE INTEGER. The first few are 1, 8, 27, 64, ... (Sloane's A000578). The GENERATING FUNCTION giving the cubic numbers is

$$\frac{x(x^2+4x+1)}{(x-1)^4} = x + 8x^2 + 27x^3 + \ldots. \quad (1)$$

The HEX PYRAMIDAL NUMBERS are equivalent to the cubic numbers (Conway and Guy 1996).

The number of *positive* cubes needed to represent the numbers 1, 2, 3, ... are 1, 2, 3, 4, 5, 6, 7, 1, 2, 3, 4, 5, 6, 7, 8, 2, ... (Sloane's A02376), and the number of distinct ways to represent the numbers 1, 2, 3, ... in terms of positive cubes are 1, 1, 1, 1, 1, 1, 1, 2, 2, 2, 2, 2, 2, 2, 2, 3, 3, 3, 3, 3, 3, 3, 3, 4, 4, 4, 5, 5, 5, 5, ... (Sloane's A003108). In the early twentieth century, Dickson, Pillai, and Niven proved that every POSITIVE INTEGER is the sum of not more than nine CUBES (so $g(3) = 9$ in WARING'S PROBLEM).

In 1939, Dickson proved that the only INTEGERS requiring nine CUBES are 23 and 239. Wieferich proved that only 15 INTEGERS require eight CUBES: 15, 22, 50, 114, 167, 175, 186, 212, 213, 238, 303, 364, 420, 428, and 454 (Sloane's A018889). The quantity $G(3)$ in WARING'S PROBLEM therefore satisfies $G(3) \leq 7$, and the largest number known requiring seven cubes is 8042. The following table gives the first few numbers which require at least $N = 1, 2, 3, \ldots, 9$ (positive) cubes to represent them as a sum.

N	Sloane	Numbers
1	000578	1, 8, 27, 64, 125, 216, 343, 512, ...
2	003325	2, 9, 16, 28, 35, 54, 65, 72, 91, ...
3	003072	3, 10, 17, 24, 29, 36, 43, 55, 62, ...
4	003327	4, 11, 18, 25, 30, 32, 37, 44, 51, ...
5	003328	5, 12, 19, 26, 31, 33, 38, 40, 45, ...
6		6, 13, 20, 34, 39, 41, 46, 48, 53, ...
7	018890	7, 14, 21, 42, 47, 49, 61, 77, ...
8	018889	15, 22, 50, 114, 167, 175, 186, ...,
9	—	23, 239

There is a finite set of numbers which cannot be expressed as the sum of *distinct* cubes: 2, 3, 4, 5, 6, 7, 10, 11, 12, 13, 14, 15, 16, 17, 18, 19, 20, 21, 22, 23, 24, 25, 26, ... (Sloane's A001476). The following table gives the numbers which can be represented in W different ways as a sum of N positive cubes. For example,

$$157 = 4^3 + 4^3 + 3^3 + 1^3 + 1^3 = 5^3 + 2^3 + 2^3 + 2^3 + 2^3 \quad (2)$$

can be represented in $W = 2$ ways by $N = 5$ cubes. The smallest number representable in $W = 2$ ways as a sum of $N = 2$ cubes,

$$1729 = 1^3 + 12^3 = 9^3 + 10^3, \quad (3)$$

is called the HARDY-RAMANUJAN NUMBER and has special significance in the history of mathematics as a result of a story told by Hardy about Ramanujan. Sloane's A001235 is defined as the sequence of numbers which are the sum of cubes in two *or more* ways, and so appears identical in the first few terms.

N	W	Sloane	Numbers
1	1	000578	1, 8, 27, 64, 125, 216, 343, 512, ...
2	1	025403	2, 9, 16, 28, 35, 54, 65, 72, 91, ...
2	2		1729, 4104, 13832, 20683, 32832, ...
2	3	003825	87539319, 119824488, 143604279, ...
2	4	003826	6963472309248, 12625136269928, ...
2	5		48988659276962496, ...
2	6		8230545258248091551205888, ...
3	1	025395	3, 10, 17, 24, 29, 36, 43, 55, 62, ...

It is believed to be possible to express any number as a SUM of four (positive or negative) cubes, although this has not been proved for numbers of the form $9n \pm 4$. In fact, all numbers *not* of the form $9n \pm 4$ are known to be expressible as the SUM of *three* (positive or negative) cubes except 30, 33, 42, 52, 74, 110, 114, 156, 165, 195, 290, 318, 366, 390, 420, 435, 444, 452, 462, 478, 501, 530, 534, 564, 579, 588, 600, 606, 609, 618, 627, 633, 732, 735, 758, 767, 786, 789, 795, 830, 834, 861, 894, 903, 906, 912, 921, 933, 948, 964, 969, and 975 (Guy 1994, p. 151).

The following table gives the possible residues (mod n) for cubic numbers for $n = 1$ to 20, as well as the number of distinct residues $s(n)$.

n	$s(n)$	$x^3 \pmod n$
2	2	0, 1
3	3	0, 1, 2
4	3	0, 1, 3
5	5	0, 1, 2, 3, 4
6	6	0, 1, 2, 3, 4, 5
7	3	0, 1, 6
8	5	0, 1, 3, 5, 7
9	3	0, 1, 8
10	10	0, 1, 2, 3, 4, 5, 6, 7, 8, 9
11	11	0, 1, 2, 3, 4, 5, 6, 7, 8, 9, 10
12	9	0, 1, 3, 4, 5, 7, 8, 9, 11
13	5	0, 1, 5, 8, 12
14	6	0, 1, 6, 7, 8, 13
15	15	0, 1, 2, 3, 4, 5, 6, 7, 8, 9, 10, 11, 12, 13, 14
16	10	0, 1, 3, 5, 7, 8, 9, 11, 13, 15
17	17	0, 1, 2, 3, 4, 5, 6, 7, 8, 9, 10, 11, 12, 13, 14, 15, 16
18	6	0, 1, 8, 9, 10, 17
19	7	0, 1, 7, 8, 11, 12, 18
20	15	0, 1, 3, 4, 5, 7, 8, 9, 11, 12, 13, 15, 16, 17, 19

Dudeney found two RATIONAL NUMBERS other than 1 and 2 whose cubes sum to 9,

$$\frac{415280564497}{348671682660} \text{ and } \frac{676702467503}{348671682660}. \tag{4}$$

The problem of finding two RATIONAL NUMBERS whose cubes sum to six was "proved" impossible by Legendre. However, Dudeney found the simple solutions 17/21 and 37/21.

The only three consecutive INTEGERS whose cubes sum to a cube are given by the DIOPHANTINE EQUATION

$$3^3 + 4^3 + 5^3 = 6^3. \tag{5}$$

CATALAN'S CONJECTURE states that 8 and 9 (2^3 and 3^2) are the only consecutive POWERS (excluding 0 and 1), i.e., the only solution to CATALAN'S DIOPHANTINE PROBLEM. This CONJECTURE has not yet been proved or refuted, although R. Tijdeman has proved that there can be only a finite number of exceptions should the CONJECTURE not hold. It is also known that 8 and 9 are the only consecutive cubic and SQUARE NUMBERS (in either order).

There are six POSITIVE INTEGERS equal to the sum of the DIGITS of their cubes: 1, 8, 17, 18, 26, and 27 (Moret Blanc 1879). There are four POSITIVE INTEGERS equal to the sums of the cubes of their digits:

$$153 = 1^3 + 5^3 + 3^3 \tag{6}$$

$$370 = 3^3 + 7^3 + 0^3 \tag{7}$$

$$371 = 3^3 + 7^3 + 1^3 \tag{8}$$

$$407 = 4^3 + 0^3 + 7^3 \tag{9}$$

(Ball and Coxeter 1987). There are two SQUARE NUMBERS of the form $n^3 - 4$: $4 = 2^3 - 4$ and $121 = 5^3 - 4$ (Le Lionnais 1983). A cube cannot be the concatenation of two cubes, since if c^3 is the concatenation of a^3 and b^3, then $c^3 = 10^k a^3 + b^3$, where k is the number of digits in b^3. After shifting any powers of 1000 in 10^k into a^3, the original problem is equivalent to finding a solution to one of the DIOPHANTINE EQUATIONS

$$c^3 - b^3 = a^3 \tag{10}$$

$$c^3 - b^3 = 10a^3 \tag{11}$$

$$c^3 - b^3 = 100a^3. \tag{12}$$

None of these have solutions in integers, as proved independently by Sylvester, Lucas, and Pepin (Dickson 1966, pp. 572–578).

see also BIQUADRATIC NUMBER, CENTERED CUBE NUMBER, CLARK'S TRIANGLE, DIOPHANTINE EQUATION—CUBIC, HARDY-RAMANUJAN NUMBER, PARTITION, SQUARE NUMBER

References

Ball, W. W. R. and Coxeter, H. S. M. *Mathematical Recreations and Essays, 13th ed.* New York: Dover, p. 14, 1987.

Conway, J. H. and Guy, R. K. *The Book of Numbers.* New York: Springer-Verlag, pp. 42–44, 1996.

Davenport, H. "On Waring's Problem for Cubes." *Acta Math.* **71**, 123–143, 1939.

Dickson, L. E. *History of the Theory of Numbers, Vol. 2: Diophantine Analysis.* New York: Chelsea, 1966.

Guy, R. K. "Sum of Four Cubes." §D5 in *Unsolved Problems in Number Theory, 2nd ed.* New York: Springer-Verlag, pp. 151–152, 1994.

Le Lionnais, F. *Les nombres remarquables.* Paris: Hermann, p. 53, 1983.

Sloane, N. J. A. Sequences A000578/M4499, A02376/M0466, and A003108/M0209 in "An On-Line Version of the Encyclopedia of Integer Sequences."

Cubic Reciprocity Theorem

A RECIPROCITY THEOREM for the case $n = 3$ solved by Gauss using "INTEGERS" of the form $a + b\rho$, when ρ is a root if $x^2 + x + 1 = 0$ and a, b are INTEGERS.

see also RECIPROCITY THEOREM

References

Ireland, K. and Rosen, M. "Cubic and Biquadratic Reciprocity." Ch. 9 in *A Classical Introduction to Modern Number Theory, 2nd ed.* New York: Springer-Verlag, pp. 108–137, 1990.

Cubic Spline

A cubic spline is a SPLINE constructed of piecewise third-order POLYNOMIALS which pass through a set of control points. The second DERIVATIVE of each POLYNOMIAL is zero at the endpoints.

References

Burden, R. L.; Faires, J. D.; and Reynolds, A. C. *Numerical Analysis, 6th ed.* Boston, MA: Brooks/Cole, pp. 120–121, 1997.

Press, W. H.; Flannery, B. P.; Teukolsky, S. A.; and Vetterling, W. T. "Cubic Spline Interpolation." §3.3 in *Numerical Recipes in FORTRAN: The Art of Scientific Computing, 2nd ed.* Cambridge, England: Cambridge University Press, pp. 107–110, 1992.

Cubic Surface

An ALGEBRAIC SURFACE of ORDER 3. Schläfli and Cayley classified the singular cubic surfaces. On the general cubic, there exists a curious geometrical structure called DOUBLE SIXES, and also a particular arrangement of 27 (possibly complex) lines, as discovered by Schläfli (Salmon 1965, Fischer 1986) and sometimes called SOLOMON'S SEAL LINES. A nonregular cubic surface can contain 3, 7, 15, or 27 real lines (Segre 1942, Le Lionnais 1983). The CLEBSCH DIAGONAL CUBIC contains all possible 27. The maximum number of ORDINARY DOUBLE POINTS on a cubic surface is four, and the unique cubic surface having four ORDINARY DOUBLE POINTS is the CAYLEY CUBIC.

Schoutte (1910) showed that the 27 lines can be put into a ONE-TO-ONE correspondence with the vertices of a particular POLYTOPE in 6-D space in such a manner that all incidence relations between the lines are mirrored in the connectivity of the POLYTOPE and conversely (Du Val 1931). A similar correspondence can be made between the 28 bitangents of the general plane QUARTIC CURVE and a 7-D POLYTOPE (Coxeter 1928) and between the tritangent planes of the canonical curve of genus 4 and an 8-D POLYTOPE (Du Val 1933).

A smooth cubic surface contains 45 TRITANGENTS (Hunt). The Hessian of smooth cubic surface contains at least 10 ORDINARY DOUBLE POINTS, although the Hessian of the CAYLEY CUBIC contains 14 (Hunt).

see also CAYLEY CUBIC, CLEBSCH DIAGONAL CUBIC, DOUBLE SIXES, ECKARDT POINT, ISOLATED SINGULARITY, NORDSTRAND'S WEIRD SURFACE, SOLOMON'S SEAL LINES, TRITANGENT

References

Bruce, J. and Wall, C. T. C. "On the Classification of Cubic Surfaces." *J. London Math. Soc.* **19**, 245–256, 1979.

Cayley, A. "A Memoir on Cubic Surfaces." *Phil. Trans. Roy. Soc.* **159**, 231–326, 1869.

Coxeter, H. S. M. "The Pure Archimedean Polytopes in Six and Seven Dimensions." *Proc. Cambridge Phil. Soc.* **24**, 7–9, 1928.

Du Val, P. "On the Directrices of a Set of Points in a Plane." *Proc. London Math. Soc. Ser. 2* **35**, 23–74, 1933.

Fischer, G. (Ed.). *Mathematical Models from the Collections of Universities and Museums.* Braunschweig, Germany: Vieweg, pp. 9–14, 1986.

Fladt, K. and Baur, A. *Analytische Geometrie spezieler Flächen und Raumkurven.* Braunschweig, Germany: Vieweg, pp. 248–255, 1975.

Hunt, B. "Algebraic Surfaces." `http://www.mathematik.uni-kl.de/~wwwagag/Galerie.html`.

Hunt, B. "The 27 Lines on a Cubic Surface" and "Cubic Surfaces." Ch. 4 and Appendix B.4 in *The Geometry of Some Special Arithmetic Quotients.* New York: Springer-Verlag, pp. 108–167 and 302–310, 1996.

Le Lionnais, F. *Les nombres remarquables.* Paris: Hermann, p. 49, 1983.

Rodenberg, C. "Zur Classification der Flächen dritter Ordnung." *Math. Ann.* **14**, 46–110, 1878.

Salmon, G. *Analytic Geometry of Three Dimensions.* New York: Chelsea, 1965.

Schläfli, L. "On the Distribution of Surface of Third Order into Species." *Phil. Trans. Roy. Soc.* **153**, 193–247, 1864.

Schoutte, P. H. "On the Relation Between the Vertices of a Definite Sixdimensional Polytope and the Lines of a Cubic Surface." *Proc. Roy. Acad. Amsterdam* **13**, 375–383, 1910.

Segre, B. *The Nonsingular Cubic Surface.* Oxford, England: Clarendon Press, 1942.

Cubical Conic Section

see CUBICAL ELLIPSE, CUBICAL HYPERBOLA, CUBICAL PARABOLA, SKEW CONIC

Cubical Ellipse

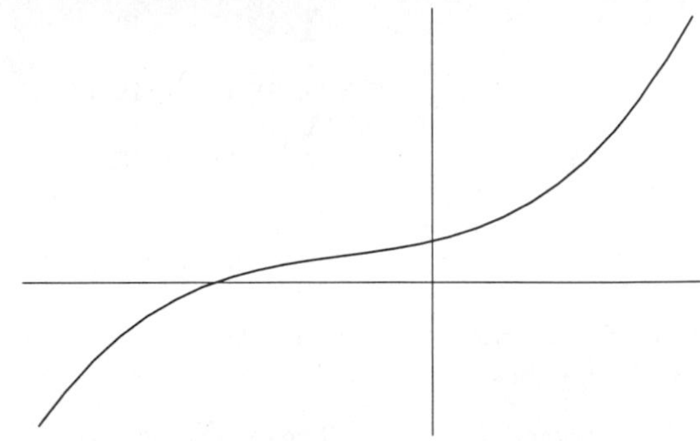

An equation of the form

$$y = ax^3 + bx^2 + cx + d$$

where only one ROOT is real.

see also CUBICAL CONIC SECTION, CUBICAL HYPERBOLA, CUBICAL PARABOLA, CUBICAL PARABOLIC HYPERBOLA, ELLIPSE, SKEW CONIC

Cubical Graph

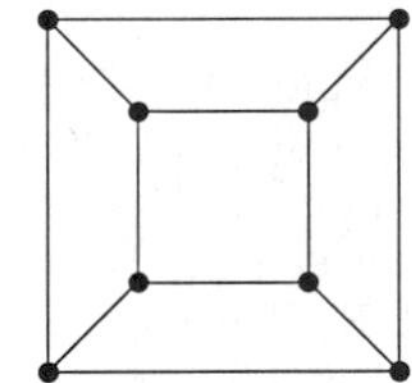

An 8-vertex POLYHEDRAL GRAPH.

see also BIDIAKIS CUBE, BISLIT CUBE, DODECAHEDRAL GRAPH, ICOSAHEDRAL GRAPH, OCTAHEDRAL GRAPH, TETRAHEDRAL GRAPH

Cubical Hyperbola

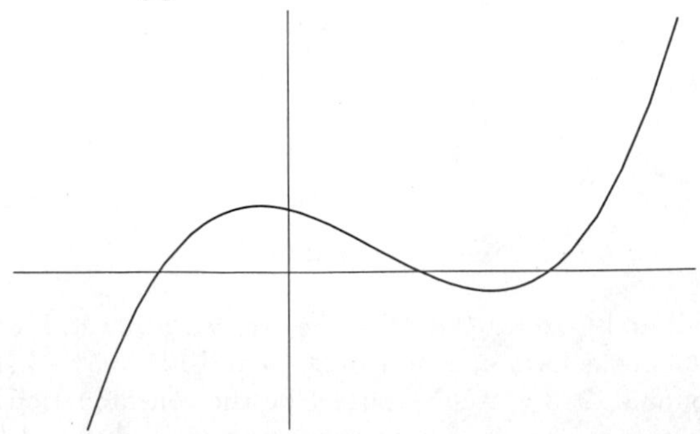

An equation of the form

$$y = ax^3 + bx^2 + cx + d,$$

where the three ROOTS are REAL and distinct, i.e.,

$$y = a(x - r_1)(x - r_2)(x - r_3)$$
$$= a[x^3 - (r_1 + r_2 + r_3)x^2 + (r_1r_2 + r_1r_3 + r_2r_3)x - r_1r_2r_3].$$

see also CUBICAL CONIC SECTION, CUBICAL ELLIPSE, CUBICAL HYPERBOLA, CUBICAL PARABOLA, HYPERBOLA

Cubical Parabola

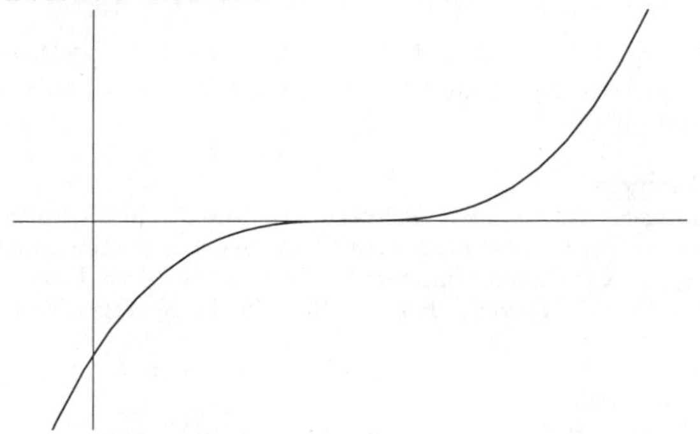

An equation of the form

$$y = ax^3 + bx^2 + cx + d,$$

where the three ROOTS of the equation coincide (and are therefore real), i.e.,

$$y = a(x - r)^3 = a(x^3 - 3rx^2 - 3r^2x - r^3).$$

see also CUBICAL CONIC SECTION, CUBICAL ELLIPSE, CUBICAL HYPERBOLA, CUBICAL PARABOLIC HYPERBOLA, PARABOLA, SEMICUBICAL PARABOLA

Cubical Parabolic Hyperbola

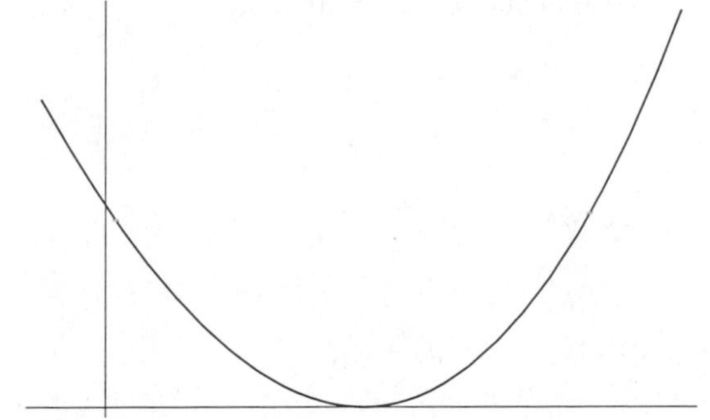

An equation of the form

$$y = ax^3 + bx^2 + cx + d,$$

where two of the ROOTS of the equation coincide (and all three are therefore real), i.e.,

$$y = a(x - r_1)^2(x - r_2)$$
$$= a[x^3 - (2r_1 + r_2)x^2 + r_1(r_1 + 2r_2)x - {r_1}^2r_2].$$

see also CUBICAL CONIC SECTION, CUBICAL ELLIPSE, CUBICAL HYPERBOLA, CUBICAL PARABOLA, HYPERBOLA

Cubicuboctahedron

see GREAT CUBICUBOCTAHEDRON, SMALL CUBICUBOCTAHEDRON

Cubique d'Agnesi

see WITCH OF AGNESI

Cubitruncated Cuboctahedron

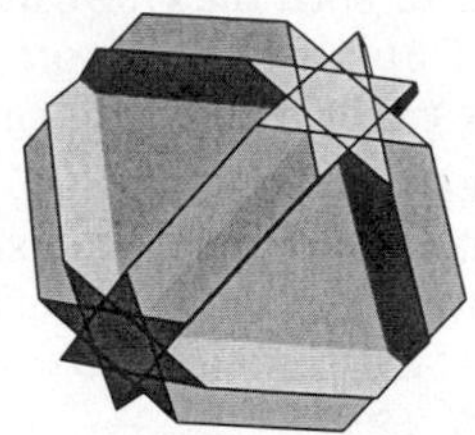

The UNIFORM POLYHEDRON U_{16} whose DUAL is the TETRADYAKIS HEXAHEDRON. It has WYTHOFF SYMBOL $3\,\frac{4}{3}\,4\,|$. Its faces are $8\{6\} + 6\{8\} + 6\{\frac{8}{3}\}$. It is a FACETED OCTAHEDRON. The CIRCUMRADIUS for a cubitruncated cuboctahedron of unit edge length is

$$R = \tfrac{1}{2}\sqrt{7}.$$

References

Wenninger, M. J. *Polyhedron Models.* Cambridge, England: Cambridge University Press, pp. 113–114, 1971.

Cuboctahedron

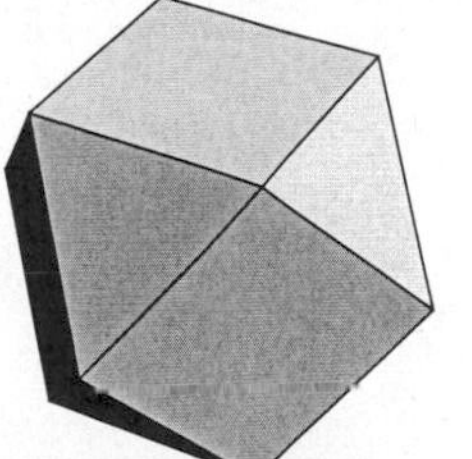

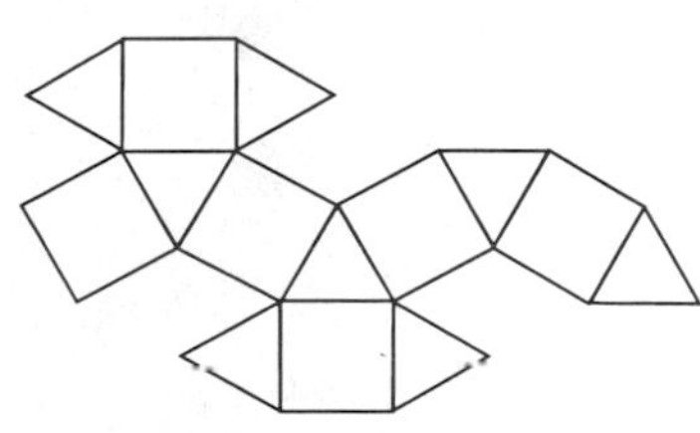

An ARCHIMEDEAN SOLID (also called the DYMAXION or HEPTAPARALLELOHEDRON) whose DUAL is the RHOMBIC DODECAHEDRON. It is one of the two convex QUASIREGULAR POLYHEDRA and has SCHLÄFLI SYMBOL $\left\{ {3 \atop 4} \right\}$. It is also UNIFORM POLYHEDRON U_7 and has WYTHOFF SYMBOL $2\,|\,3\,4$. Its faces are $\{3\} + 6\{4\}$. It has the O_h OCTAHEDRAL GROUP of symmetries.

The VERTICES of a cuboctahedron with EDGE length of $\sqrt{2}$ are $(0, \pm 1, \pm 1)$, $(\pm 1, 0, \pm 1)$, and $(\pm 1, \pm 1, 0)$. The INRADIUS, MIDRADIUS, and CIRCUMRADIUS for $a = 1$ are

$$r = \tfrac{3}{4} = 0.75$$
$$\rho = \tfrac{1}{2}\sqrt{3} \approx 0.86602$$
$$R = 1.$$

FACETED versions include the CUBOHEMIOCTAHEDRON and OCTAHEMIOCTAHEDRON.

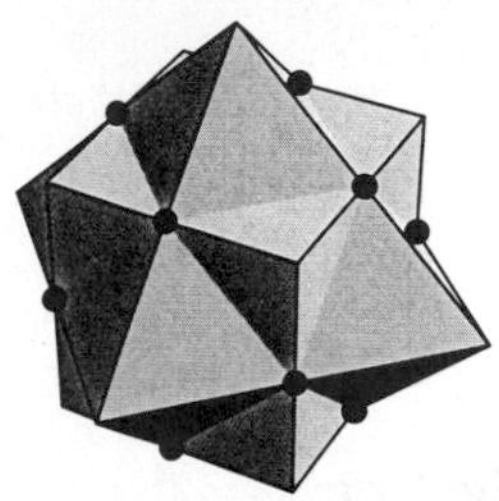
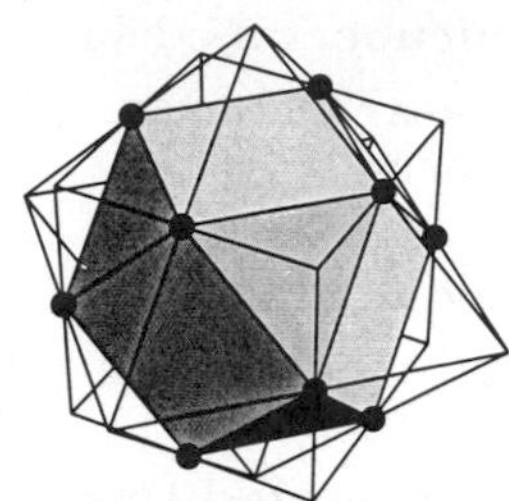

The solid common to both the CUBE and OCTAHEDRON (left figure) in a CUBE-OCTAHEDRON COMPOUND is a CUBOCTAHEDRON (right figure; Ball and Coxeter 1987).

see also ARCHIMEDEAN SOLID, CUBE, CUBE-OCTAHEDRON COMPOUND, CUBOHEMIOCTAHEDRON, OCTAHEDRON, OCTAHEMIOCTAHEDRON, QUASIREGULAR POLYHEDRON, RHOMBIC DODECAHEDRON, RHOMBUS

References

Ball, W. W. R. and Coxeter, H. S. M. *Mathematical Recreations and Essays, 13th ed.* New York: Dover, p. 137, 1987.

Ghyka, M. *The Geometry of Art and Life.* New York: Dover, p. 54, 1977.

Cuboctatruncated Cuboctahedron

see CUBITRUNCATED CUBOCTAHEDRON

Cubocycloid

see ASTROID

Cubohemioctahedron

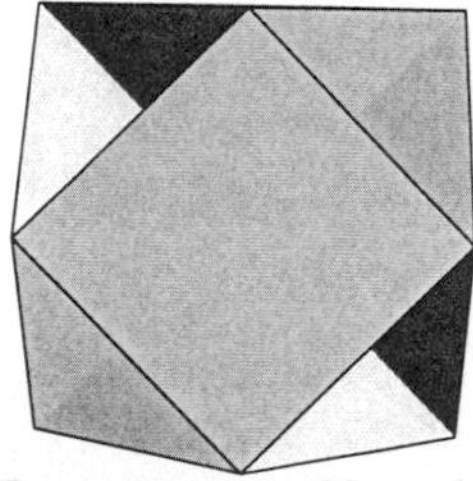

The UNIFORM POLYHEDRON U_{15} whose DUAL is the HEXAHEMIOCTAHEDRON. It has WYTHOFF SYMBOL $\frac{4}{3}\,4\,|\,3$. Its faces are $4\{6\} + 6\{4\}$. It is a FACETED version of the CUBOCTAHEDRON. Its CIRCUMRADIUS for unit edge length is

$$R = 1.$$

References

Wenninger, M. J. *Polyhedron Models.* Cambridge, England: Cambridge University Press, pp. 121–122, 1971.

Cuboid

A rectangular PARALLELEPIPED.

see also EULER BRICK, PARALLELEPIPED, SPIDER AND FLY PROBLEM

Cullen Number

A number of the form

$$C_n = 2^n n + 1.$$

The first few are 3, 9, 25, 65, 161, 385, ... (Sloane's A002064). The only Cullen numbers C_n for $n < 300{,}000$ which are PRIME are for $n = 1$, 141, 4713, 5795, 6611, 18496, 32292, 32469, 59656, 90825, 262419, ... (Sloane's A005849; Ballinger). Cullen numbers are DIVISIBLE by $p = 2n - 1$ if p is a PRIME of the form $8k \pm 3$.

see also CUNNINGHAM NUMBER, FERMAT NUMBER, SIERPIŃSKI NUMBER OF THE FIRST KIND, WOODALL NUMBER

References

Ballinger, R. "Cullen Primes: Definition and Status." `http://ballingerr.xray.ufl.edu/proths/cullen.html`.

Guy, R. K. "Cullen Numbers." §B20 in *Unsolved Problems in Number Theory, 2nd ed.* New York: Springer-Verlag, p. 77, 1994.

Keller, W. "New Cullen Primes." *Math. Comput.* **64**, 1733–1741, 1995.

Leyland, P. `ftp://sable.ox.ac.uk/pub/math/factors/cullen`.

Ribenboim, P. *The New Book of Prime Number Records.* New York: Springer-Verlag, pp. 360–361, 1996.

Sloane, N. J. A. Sequences A002064/M2795 and A005849/M5401 in "An On-Line Version of the Encyclopedia of Integer Sequences."

Cumulant

Let $\phi(t)$ be the CHARACTERISTIC FUNCTION, defined as the FOURIER TRANSFORM of the PROBABILITY DENSITY FUNCTION,

$$\phi(t) = \mathcal{F}[P(x)] = \int_{-\infty}^{\infty} e^{itx} P(x)\,dx. \tag{1}$$

Then the cumulants κ_n are defined by

$$\ln \phi(t) \equiv \sum_{n=0}^{\infty} \kappa_n \frac{(it)^n}{n!}. \tag{2}$$

Taking the MACLAURIN SERIES gives

$$\begin{aligned}\ln \phi(t) &= (it)\mu_1' + \tfrac{1}{2}(it)^2(\mu_2' - {\mu_1'}^2) \\ &+ \tfrac{1}{3!}(it)^3(2{\mu_1'}^3 - 3\mu_1'\mu_2' + \mu_3') \\ &+ \tfrac{1}{4!}(it)^4(-6{\mu_1'}^4 + 12{\mu_1'}^2\mu_2' - 3{\mu_2'}^2 - 4\mu_1'\mu_3' + \mu_4') \\ &+ \tfrac{1}{5!}(it)^5[-24{\mu_1'}^5 + 60{\mu_1'}^3\mu_2' + 20{\mu_1'}^2\mu_3' + 10\mu_2'\mu_3' \\ &+ 5\mu_1'(6{\mu_2'}^2 - \mu_4') + \mu_5'] + \ldots,\end{aligned} \tag{3}$$

where μ'_n are MOMENTS about 0, so

$$\kappa_1 = \mu'_1 \tag{4}$$
$$\kappa_2 = \mu'_2 - {\mu'_1}^2 \tag{5}$$
$$\kappa_3 = 2{\mu'_1}^3 - 3\mu'_1\mu'_2 + \mu'_3 \tag{6}$$
$$\kappa_4 = -6{\mu'_1}^4 + 12{\mu'_1}^2\mu'_2 - 3{\mu'_2}^2 - 4\mu'_1\mu'_3 + \mu'_4 \tag{7}$$
$$\kappa_5 = -24{\mu'_1}^5 + 60{\mu'_1}^3\mu'_2 + 20{\mu'_1}^2\mu'_3 + 10\mu'_2\mu'_3 + 5\mu'_1(6{\mu'_2}^2 - \mu'_4) + \mu'_5. \tag{8}$$

In terms of the MOMENTS μ_n about the MEAN,

$$\kappa_1 = \mu \tag{9}$$
$$\kappa_2 = \mu_2 = \sigma^2 \tag{10}$$
$$\kappa_3 = \mu_3 \tag{11}$$
$$\kappa_4 = \mu_4 - 3{\mu_2}^2 \tag{12}$$
$$\kappa_5 = \mu_5 - 10\mu_2\mu_3, \tag{13}$$

where μ is the MEAN and $\sigma^2 \equiv \mu_2$ is the VARIANCE.

The k-STATISTICS are UNBIASED ESTIMATORS of the cumulants.

see also CHARACTERISTIC FUNCTION, CUMULANT-GENERATING FUNCTION, k-STATISTIC, KURTOSIS, MEAN, MOMENT, SHEPPARD'S CORRECTION, SKEWNESS, VARIANCE

References

Abramowitz, M. and Stegun, C. A. (Eds.). *Handbook of Mathematical Functions with Formulas, Graphs, and Mathematical Tables, 9th printing.* New York: Dover, p. 928, 1972.

Kenney, J. F. and Keeping, E. S. "Cumulants and the Cumulant-Generating Function," "Additive Property of Cumulants," and "Sheppard's Correction." §4.10–4.12 in *Mathematics of Statistics, Pt. 2, 2nd ed.* Princeton, NJ: Van Nostrand, pp. 77–82, 1951.

Cumulant-Generating Function

Let $M(h)$ be the MOMENT-GENERATING FUNCTION. Then

$$K(h) \equiv \ln M(h) = \kappa_1 h + \tfrac{1}{2!}h^2\kappa_2 + \tfrac{1}{3!}h^3\kappa_3 + \ldots.$$

If

$$L = \sum_{j=1}^{M} c_j x_j$$

is a function of N independent variables, the cumulant generating function for L is then

$$K(h) = \sum_{j=1}^{N} K_j(c_j h).$$

see also CUMULANT, MOMENT-GENERATING FUNCTION

References

Abramowitz, M. and Stegun, C. A. (Eds.). *Handbook of Mathematical Functions with Formulas, Graphs, and Mathematical Tables, 9th printing.* New York: Dover, p. 928, 1972.

Kenney, J. F. and Keeping, E. S. "Cumulants and the Cumulant-Generating Function" and "Additive Property of Cumulants." §4.10–4.11 in *Mathematics of Statistics, Pt. 2, 2nd ed.* Princeton, NJ: Van Nostrand, pp. 77–80, 1951.

Cumulative Distribution Function

see DISTRIBUTION FUNCTION

Cundy and Rollett's Egg

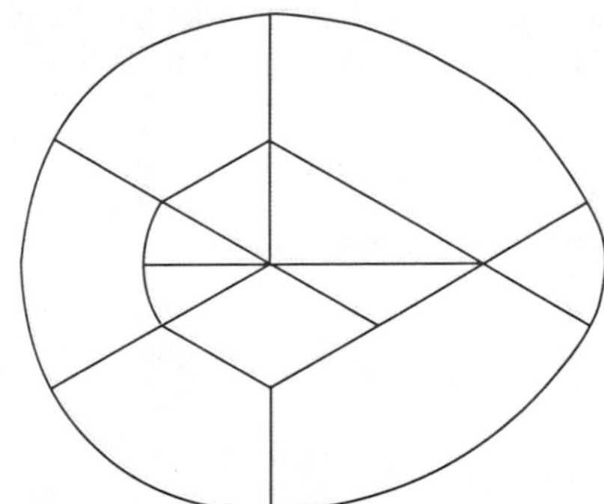

An OVAL dissected into pieces which are to used to create pictures. The resulting figures resemble those constructed out of TANGRAMS.

see also DISSECTION, EGG, OVAL, TANGRAM

References

Cundy, H. and Rollett, A. *Mathematical Models, 3rd ed.* Stradbroke, England: Tarquin Pub., pp. 19–21, 1989.

Dixon, R. *Mathographics.* New York: Dover, p. 11, 1991.

Cunningham Chain

A SEQUENCE of PRIMES $q_1 < q_2 < \ldots < q_k$ is a Cunningham chain of the first kind (second kind) of length k if $q_{i+1} = 2q_i + 1$ ($q_{i+1} = 2q_i - 1$) for $i = 1, \ldots, k-1$. Cunningham PRIMES of the first kind are SOPHIE GERMAIN PRIMES.

The two largest known Cunningham chains (of the first kind) of length three are $(384205437 \cdot 2^{4000} - 1,\ 384205437 \cdot 2^{4001} - 1,\ 384205437 \cdot 2^{4002} - 1)$ and $(651358155 \cdot 2^{3291} - 1,\ 651358155 \cdot 2^{3292} - 1,\ 651358155 \cdot 2^{3293} - 1)$, both discovered by W. Roonguthai in 1998.

see also PRIME ARITHMETIC PROGRESSION, PRIME CLUSTER

References

Guy, R. K. "Cunningham Chains." §A7 in *Unsolved Problems in Number Theory, 2nd ed.* New York: Springer-Verlag, pp. 18–19, 1994.

Ribenboim, P. *The New Book of Prime Number Records.* New York: Springer-Verlag, p. 333, 1996.

Roonguthai, W. "Yves Gallot's Proth and Cunningham Chains." `http://ksc9.th.com/warut/cunningham.html`.

Cunningham Function

Sometimes also called the PEARSON-CUNNINGHAM FUNCTION. It can be expressed using WHITTAKER FUNCTIONS (Whittaker and Watson 1990, p. 353).

$$\omega_{n,m}(x) \equiv \frac{e^{\pi i(m/2-n)+x}}{\Gamma(1+n-\frac{1}{2}m)} U(\tfrac{1}{2}m-n, 1+m, x),$$

where U is a CONFLUENT HYPERGEOMETRIC FUNCTION OF THE SECOND KIND (Abramowitz and Stegun 1972, p. 510).

see also CONFLUENT HYPERGEOMETRIC FUNCTION OF THE SECOND KIND, WHITTAKER FUNCTION

References

Abramowitz, M. and Stegun, C. A. (Eds.). *Handbook of Mathematical Functions with Formulas, Graphs, and Mathematical Tables, 9th printing.* New York: Dover, 1972.

Whittaker, E. T. and Watson, G. N. *A Course in Modern Analysis, 4th ed.* Cambridge, England: Cambridge University Press, 1990.

Cunningham Number

A BINOMIAL NUMBER of the form $C^{\pm}(b,n) \equiv b^n \pm 1$. Bases b^k which are themselves powers need not be considered since they correspond to $(b^k)^n \pm 1 = b^{kn} \pm 1$. PRIME NUMBERS of the form $C^{\pm}(b,n)$ are very rare.

A NECESSARY (but not SUFFICIENT) condition for $C^+(2,n) = 2^n + 1$ to be PRIME is that n be of the form $n = 2^m$. Numbers of the form $F_m = C^+(2, 2^m) = 2^{2^m} + 1$ are called FERMAT NUMBERS, and the only known PRIMES occur for $C^+(2,1) = 3$, $C^+(2,2) = 5$, $C^+(2,4) = 17$, $C^+(2,8) = 257$, and $C^+(2,16) = 65537$ (i.e., $n = 0$, 1, 2, 3, 4). The only other PRIMES $C^+(b,n)$ for nontrivial $b \leq 11$ and $2 \leq n \leq 1000$ are $C^+(6,2) = 37$, $C^+(6,4) = 1297$, and $C^+(10,2) = 101$.

PRIMES of the form $C^-(b,n)$ are also very rare. The MERSENNE NUMBERS $M_n = C^-(2,n) = 2^n - 1$ are known to be prime only for 37 values, the first few of which are $n = 2$, 3, 5, 7, 13, 17, 19, ... (Sloane's A000043). There are no other PRIMES $C^-(b,n)$ for nontrivial $b \leq 20$ and $2 \leq n \leq 1000$.

In 1925, Cunningham and Woodall (1925) gathered together all that was known about the PRIMALITY and factorization of the numbers $C^{\pm}(b,n)$ and published a small book of tables. These tables collected from scattered sources the known prime factors for the bases 2 and 10 and also presented the authors' results of 30 years' work with these and other bases.

Since 1925, many people have worked on filling in these tables. D. H. Lehmer, a well-known mathematician who died in 1991, was for many years a leader of these efforts. Lehmer was a mathematician who was at the forefront of computing as modern electronic computers became a reality. He was also known as the inventor of some ingenious pre-electronic computing devices specifically designed for factoring numbers.

Updated factorizations were published in Brillhart *et al.* (1988). The current archive of Cunningham number factorizations for $b = 1, \ldots, \pm 12$ is kept on `ftp://sable.ox.ac.uk/pub/math/cunningham`. The tables have been extended by Brent and te Riele (1992) to $b = 13, \ldots, 100$ with $m < 255$ for $b < 30$ and $m < 100$ for $b \geq 30$. All numbers with exponent 58 and smaller, and all composites with ≤ 90 digits have now been factored.

see also BINOMIAL NUMBER, CULLEN NUMBER, FERMAT NUMBER, MERSENNE NUMBER, REPUNIT, RIESEL NUMBER, SIERPIŃSKI NUMBER OF THE FIRST KIND, WOODALL NUMBER

References

Brent, R. P. and te Riele, H. J. J. "Factorizations of $a^n \pm 1$, $13 \leq a < 100$." Report NM-R9212, *Centrum voor Wiskunde en Informatica.* Amsterdam, June 1992. The text is available electronically at `ftp://sable.ox.ac.uk/pub/math/factors/BMtR_13-99.dvi`, and the files at `BMtR_13-99`. Updates are given in `BMtR_13-99_update1` (94-09-01) and `BMtR_13-99_update2` (95-06-01).

Brillhart, J.; Lehmer, D. H.; Selfridge, J.; Wagstaff, S. S. Jr.; and Tuckerman, B. *Factorizations of $b^n \pm 1$, $b = 2, 3, 5, 6, 7, 10, 11, 12$ Up to High Powers, rev. ed.* Providence, RI: Amer. Math. Soc., 1988. Updates are available electronically from `ftp://sable.ox.ac.uk/pub/math/cunningham/`.

Cunningham, A. J. C. and Woodall, H. J. *Factorisation of $y^n \mp 1$, $y = 2, 3, 5, 6, 7, 10, 11, 12$ Up to High Powers (n).* London: Hodgson, 1925.

Mudge, M. "Not Numerology but Numeralogy!" *Personal Computer World,* 279–280, 1997.

Ribenboim, P. "Numbers $k \times 2^n \pm 1$." §5.7 in *The New Book of Prime Number Records.* New York: Springer-Verlag, pp. 355–360, 1996.

Sloane, N. J. A. Sequence A000043/M0672 in "An On-Line Version of the Encyclopedia of Integer Sequences."

Cunningham Project

see CUNNINGHAM NUMBER

Cupola

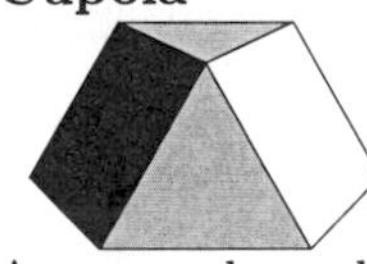
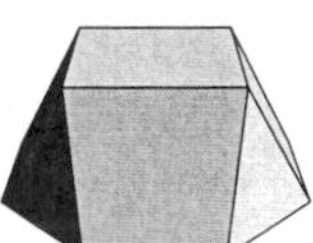
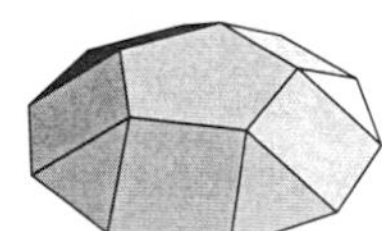

An n-gonal cupola Q_n (possible for only $n = 3$, 4, 5) is a POLYHEDRON having n TRIANGULAR and n SQUARE faces separating an $\{n\}$ and a $\{2n\}$ REGULAR POLYGON. The coordinates of the base VERTICES are

$$\left(R\cos\left[\frac{\pi(2k+1)}{2n}\right], R\sin\left[\frac{\pi(2k+1)}{2n}\right], 0\right), \tag{1}$$

and the coordinates of the top VERTICES are

$$\left(r\cos\left[\frac{2k\pi}{n}\right], r\sin\left[\frac{2k\pi}{n}\right], z\right), \tag{2}$$

where R and r are the CIRCUMRADII of the base and top

$$R = \tfrac{1}{2}a\csc\left(\frac{\pi}{2n}\right) \quad (3)$$

$$r = \tfrac{1}{2}a\csc\left(\frac{\pi}{n}\right), \quad (4)$$

and z is the height, obtained by letting $k = 0$ in the equations (1) and (2) to obtain the coordinates of neighboring bottom and top VERTICES,

$$\mathbf{b} = \begin{bmatrix} R\cos\left(\frac{\pi}{2n}\right) \\ R\sin\left(\frac{\pi}{2n}\right) \\ 0 \end{bmatrix} \quad (5)$$

$$\mathbf{t} = \begin{bmatrix} r \\ 0 \\ z \end{bmatrix}. \quad (6)$$

Since all side lengths are a,

$$|\mathbf{b} - \mathbf{t}|^2 = a^2. \quad (7)$$

Solving for z then gives

$$\left[R\cos\left(\frac{\pi}{2n}\right) - r\right]^2 + R^2\sin^2\left(\frac{\pi}{2n}\right) + z^2 = a^2 \quad (8)$$

$$z^2 + R^2 + r^2 - 2rR\cos\left(\frac{\pi}{2n}\right) = a^2 \quad (9)$$

$$z = \sqrt{a^2 - 2rR\cos\left(\frac{\pi}{2n}\right) - r^2 - R^2} = a\sqrt{1 - \tfrac{1}{4}\csc^2\left(\frac{\pi}{n}\right)}. \quad (10)$$

see also BICUPOLA, ELONGATED CUPOLA, GYROELONGATED CUPOLA, PENTAGONAL CUPOLA, SQUARE CUPOLA, TRIANGULAR CUPOLA

References
Johnson, N. W. "Convex Polyhedra with Regular Faces." *Canad. J. Math.* **18**, 169–200, 1966.

Cupolarotunda

A CUPOLA adjoined to a ROTUNDA.

see also GYROCUPOLAROTUNDA, ORTHOCUPOLAROTUNDA

Curl

The curl of a TENSOR field is given by

$$(\nabla \times A)^\alpha = \epsilon^{\alpha\mu\nu}A_{\nu;\mu}, \quad (1)$$

where ϵ_{ijk} is the LEVI-CIVITA TENSOR and ";" is the COVARIANT DERIVATIVE. For a VECTOR FIELD, the curl is denoted

$$\operatorname{curl}(\mathbf{F}) \equiv \nabla \times \mathbf{F}, \quad (2)$$

and $\nabla \times \mathbf{F}$ is normal to the PLANE in which the "circulation" is MAXIMUM. Its magnitude is the limiting value of circulation per unit AREA,

$$(\nabla \times \mathbf{F}) \cdot \hat{\mathbf{n}} \equiv \lim_{A \to 0} \frac{\oint_C \mathbf{F} \cdot d\mathbf{s}}{A}. \quad (3)$$

Let

$$\mathbf{F} \equiv F_1\hat{\mathbf{u}}_1 + F_2\hat{\mathbf{u}}_2 + F_3\hat{\mathbf{u}}_3 \quad (4)$$

and

$$h_i \equiv \left|\frac{\partial \mathbf{r}}{\partial u_i}\right|, \quad (5)$$

then

$$\nabla \times \mathbf{F} \equiv \frac{1}{h_1h_2h_3}\begin{vmatrix} h_1\hat{\mathbf{u}}_1 & h_2\hat{\mathbf{u}}_2 & h_3\hat{\mathbf{u}}_3 \\ \frac{\partial}{\partial u_1} & \frac{\partial}{\partial u_2} & \frac{\partial}{\partial u_3} \\ h_1F_1 & h_2F_2 & h_3F_3 \end{vmatrix}$$
$$= \frac{1}{h_2h_3}\left[\frac{\partial}{\partial u_2}(h_3F_3) - \frac{\partial}{\partial u_3}(h_2F_2)\right]\hat{\mathbf{u}}_1 + \frac{1}{h_1h_3}\left[\frac{\partial}{\partial u_3}(h_1F_1) - \frac{\partial}{\partial u_1}(h_3F_3)\right]\hat{\mathbf{u}}_2 + \frac{1}{h_1h_2}\left[\frac{\partial}{\partial u_1}(h_2F_2) - \frac{\partial}{\partial u_2}(h_1F_1)\right]\hat{\mathbf{u}}_3. \quad (6)$$

Special cases of the curl formulas above can be given for CURVILINEAR COORDINATES.

see also CURL THEOREM, DIVERGENCE, GRADIENT, VECTOR DERIVATIVE

References
Arfken, G. "Curl, $\nabla\times$." §1.8 in *Mathematical Methods for Physicists, 3rd ed.* Orlando, FL: Academic Press, pp. 42–47, 1985.

Curl Theorem

A special case of STOKES' THEOREM in which F is a VECTOR FIELD and M is an oriented, compact embedded 2-MANIFOLD with boundary in $\mathbb{R}^3$, given by

$$\int_S (\nabla \times \mathbf{F}) \cdot d\mathbf{a} = \int_{\partial S} \mathbf{F} \cdot d\mathbf{s}. \quad (1)$$

There are also alternate forms. If

$$\mathbf{F} \equiv \mathbf{c}F, \quad (2)$$

then

$$\int_S d\mathbf{a} \times \nabla F = \int_C F d\mathbf{s}. \quad (3)$$

and if

$$\mathbf{F} \equiv \mathbf{c} \times \mathbf{P}, \quad (4)$$

then

$$\int_S (d\mathbf{a} \times \nabla) \times \mathbf{P} = \int_C d\mathbf{s} \times \mathbf{P}. \quad (5)$$

see also CHANGE OF VARIABLES THEOREM, CURL, STOKES' THEOREM

References
Arfken, G. "Stokes's Theorem." §1.12 in *Mathematical Methods for Physicists, 3rd ed.* Orlando, FL: Academic Press, pp. 61–64, 1985.

Curlicue Fractal

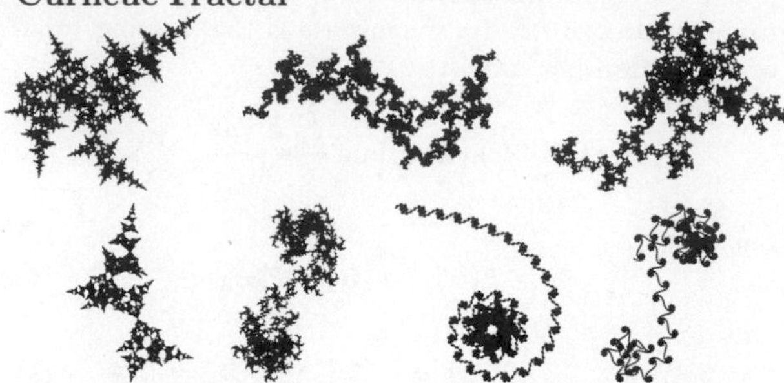

The curlicue fractal is a figure obtained by the following procedure. Let s be an IRRATIONAL NUMBER. Begin with a line segment of unit length, which makes an ANGLE $\phi_0 \equiv 0$ to the horizontal. Then define θ_n iteratively by

$$\theta_{n+1} = (\theta_n + 2\pi s) \pmod{2\pi},$$

with $\theta_0 = 0$. To the end of the previous line segment, draw a line segment of unit length which makes an angle

$$\phi_{n+1} = \theta_n + \phi_n \pmod{2\pi},$$

to the horizontal (Pickover 1995). The result is a FRACTAL, and the above figures correspond to the curlicue fractals with 10,000 points for the GOLDEN RATIO ϕ, $\ln 2$, e, $\sqrt{2}$, the EULER-MASCHERONI CONSTANT γ, π, and FEIGENBAUM CONSTANT δ.

The TEMPERATURE of these curves is given in the following table.

Constant	Temperature
golden ratio ϕ	46
$\ln 2$	51
e	58
$\sqrt{2}$	58
Euler-Mascheroni constant γ	63
π	90
Feigenbaum constant δ	92

References

Berry, M. and Goldberg, J. "Renormalization of Curlicues." *Nonlinearity* **1**, 1–26, 1988.

Moore, R. and van der Poorten, A. "On the Thermodynamics of Curves and Other Curlicues." *McQuarie Univ. Math. Rep.* 89-0031, April 1989.

Pickover, C. A. "The Fractal Golden Curlicue is Cool." Ch. 21 in *Keys to Infinity.* New York: W. H. Freeman, pp. 163–167, 1995.

Pickover, C. A. *Mazes for the Mind: Computers and the Unexpected.* New York: St. Martin's Press, 1993.

Sedgewick, R. *Algorithms.* Reading, MA: Addison-Wesley, 1988.

Stewart, I. *Another Fine Math You've Got Me Into....* New York: W. H. Freeman, 1992.

Current

A linear FUNCTIONAL on a smooth differential form.

see also FLAT NORM, INTEGRAL CURRENT, RECTIFIABLE CURRENT

Curtate Cycloid

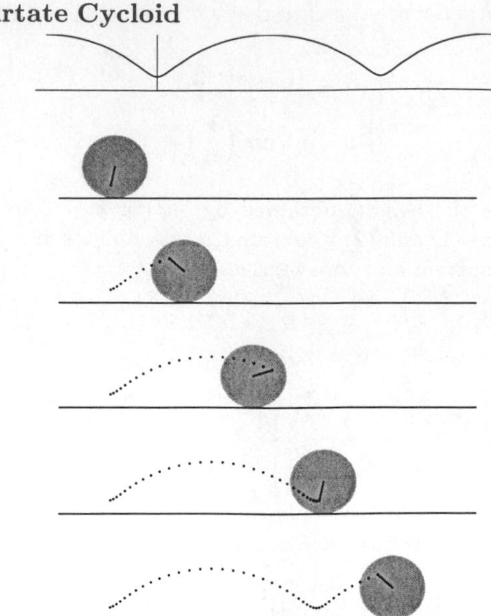

The path traced out by a fixed point at a RADIUS $b < a$, where a is the RADIUS of a rolling CIRCLE, sometimes also called a CONTRACTED CYCLOID.

$$x = a\phi - b\sin\phi \quad (1)$$

$$y = a - b\cos\phi. \quad (2)$$

The ARC LENGTH from $\phi = 0$ is

$$s = 2(a+b)E(u), \quad (3)$$

where

$$\sin(\tfrac{1}{2}\phi) = \operatorname{sn} u \quad (4)$$

$$k^2 = \frac{4ab}{(a+c)^2}, \quad (5)$$

and $E(u)$ is a complete ELLIPTIC INTEGRAL OF THE SECOND KIND and $\operatorname{sn} u$ is a JACOBI ELLIPTIC FUNCTION.

see also CYCLOID, PROLATE CYCLOID

References

Cundy, H. and Rollett, A. *Mathematical Models, 3rd ed.* Stradbroke, England: Tarquin Pub., 1989.

Wagon, S. *Mathematica in Action.* New York: W. H. Freeman, pp. 46–50, 1991.

Curtate Cycloid Evolute

The EVOLUTE of the CURTATE CYCLOID

$$x = a\phi - b\sin\phi \quad (1)$$

$$y = a - b\cos\phi. \quad (2)$$

is given by

$$x = \frac{a[-2b\phi + 2a\phi\cos\phi - 2a\sin\phi + b\sin(2\phi)]}{2(a\cos\phi - b)} \quad (3)$$

$$y = \frac{a(a - b\cos\phi)^2}{b(a\cos\phi - b)}. \quad (4)$$

Curvature

In general, there are two important types of curvature: EXTRINSIC CURVATURE and INTRINSIC CURVATURE. The EXTRINSIC CURVATURE of curves in 2- and 3-space was the first type of curvature to be studied historically, culminating in the FRENET FORMULAS, which describe a SPACE CURVE entirely in terms of its "curvature," TORSION, and the initial starting point and direction.

After the curvature of 2- and 3-D curves was studied, attention turned to the curvature of surfaces in 3-space. The main curvatures which emerged from this scrutiny are the MEAN CURVATURE, GAUSSIAN CURVATURE, and the WEINGARTEN MAP. MEAN CURVATURE was the most important for applications at the time and was the most studied, but Gauss was the first to recognize the importance of the GAUSSIAN CURVATURE.

Because GAUSSIAN CURVATURE is "intrinsic," it is detectable to 2-dimensional "inhabitants" of the surface, whereas MEAN CURVATURE and the WEINGARTEN MAP are not detectable to someone who can't study the 3-dimensional space surrounding the surface on which he resides. The importance of GAUSSIAN CURVATURE to an inhabitant is that it controls the surface AREA of SPHERES around the inhabitant.

Riemann and many others generalized the concept of curvature to SECTIONAL CURVATURE, SCALAR CURVATURE, the RIEMANN TENSOR, RICCI CURVATURE, and a host of other INTRINSIC and EXTRINSIC CURVATURES. General curvatures no longer need to be numbers, and can take the form of a MAP, GROUP, GROUPOID, tensor field, etc.

The simplest form of curvature and that usually first encountered in CALCULUS is an EXTRINSIC CURVATURE. In 2-D, let a PLANE CURVE be given by CARTESIAN parametric equations $x = x(t)$ and $y = y(t)$. Then the curvature κ is defined by

$$\kappa \equiv \frac{d\phi}{ds} = \frac{\frac{d\phi}{dt}}{\frac{ds}{dt}} = \frac{\frac{d\phi}{dt}}{\sqrt{\left(\frac{dx}{dt}\right)^2 + \left(\frac{dy}{dt}\right)^2}} = \frac{\frac{d\phi}{dt}}{\sqrt{x'^2 + y'^2}}, \quad (1)$$

where ϕ is the POLAR ANGLE and s is the ARC LENGTH. As can readily be seen from the definition, curvature therefore has units of inverse distance. The $d\phi/dt$ derivative in the above equation can be eliminated by using the identity

$$\tan\phi = \frac{dy}{dx} = \frac{dy/dt}{dx/dt} = \frac{y'}{x'}, \quad (2)$$

so

$$\frac{d}{dt}(\tan\phi) = \sec^2\phi \frac{d\phi}{dt} = \frac{x'y'' - y'x''}{x'^2} \quad (3)$$

and

$$\begin{aligned}\frac{d\phi}{dt} &= \frac{1}{1+\tan^2\phi}\frac{x'y''-y'x''}{x'^2} \\ &= \frac{1}{\frac{y'^2}{x'^2}}\frac{x'y''-y'x''}{x'^2} = \frac{x'y''-y'x''}{x'^2+y'^2}. \end{aligned} \quad (4)$$

Combining (2) and (4) gives

$$\kappa = \frac{x'y''-y'x''}{(x'^2+y'^2)^{3/2}}. \quad (5)$$

For a 2-D curve written in the form $y = f(x)$, the equation of curvature becomes

$$\kappa = \frac{\frac{d^2y}{dx^2}}{\left[1+\left(\frac{dy}{dx}\right)^2\right]^{3/2}}. \quad (6)$$

If the 2-D curve is instead parameterized in POLAR COORDINATES, then

$$\kappa = \frac{r^2 + 2{r_\theta}^2 - r r_{\theta\theta}}{(r^2 + {r_\theta}^2)^{3/2}}, \quad (7)$$

where $r_\theta \equiv \partial r/\partial\theta$ (Gray 1993). In PEDAL COORDINATES, the curvature is given by

$$\kappa = \frac{1}{r}\frac{dp}{dr}. \quad (8)$$

The curvature for a 2-D curve given implicitly by $g(x,y) = 0$ is given by

$$\kappa = \frac{g_{xx}{g_y}^2 - 2g_{xy}g_xg_y + g_{yy}{g_x}^2}{({g_x}^2+{g_y}^2)^{3/2}} \quad (9)$$

(Gray 1993).

Now consider a parameterized SPACE CURVE $\mathbf{r}(t)$ in 3-D for which the TANGENT VECTOR $\hat{\mathbf{T}}$ is defined as

$$\hat{\mathbf{T}} \equiv \frac{\frac{d\mathbf{r}}{dt}}{\left|\frac{d\mathbf{r}}{dt}\right|} = \frac{\frac{d\mathbf{r}}{dt}}{\frac{ds}{dt}}. \quad (10)$$

Therefore,

$$\frac{d\mathbf{r}}{dt} = \frac{ds}{dt}\hat{\mathbf{T}} \quad (11)$$

$$\frac{d^2\mathbf{r}}{dt^2} = \frac{d^2s}{dt^2}\hat{\mathbf{T}} + \frac{ds}{dt}\frac{d\hat{\mathbf{T}}}{dt} = \frac{ds^2}{dt^2}\hat{\mathbf{T}} + \kappa\hat{\mathbf{N}}\left(\frac{ds}{dt}\right)^2, \quad (12)$$

where $\hat{\mathbf{N}}$ is the NORMAL VECTOR. But

$$\begin{aligned}\frac{d\mathbf{r}}{dt} \times \frac{d^2\mathbf{r}}{dt^2} &= \frac{ds}{dt}\frac{d^2s}{dt^2}(\hat{\mathbf{T}}\times\hat{\mathbf{T}}) + \kappa\left(\frac{ds}{dt}\right)^3(\hat{\mathbf{T}}\times\hat{\mathbf{N}}) \\ &= \kappa\left(\frac{ds}{dt}\right)^3(\hat{\mathbf{T}}\times\hat{\mathbf{N}}) \end{aligned} \quad (13)$$

$$\left|\frac{d\mathbf{r}}{dt} \times \frac{d^2\mathbf{r}}{dt^2}\right| = \kappa \left(\frac{ds}{dt}\right)^3 = \kappa \left|\frac{d\mathbf{r}}{dt}\right|^3, \tag{14}$$

so

$$\kappa = \left|\frac{d\hat{\mathbf{T}}}{ds}\right| = \frac{\left|\frac{d\mathbf{r}}{dt} \times \frac{d^2\mathbf{r}}{dt^2}\right|}{\left|\frac{d\mathbf{r}}{dt}\right|^3}. \tag{15}$$

The curvature of a 2-D curve is related to the RADIUS OF CURVATURE of the curve's OSCULATING CIRCLE. Consider a CIRCLE specified parametrically by

$$x = a \cos t \tag{16}$$

$$y = a \sin t \tag{17}$$

which is tangent to the curve at a given point. The curvature is then

$$\kappa = \frac{x'y'' - y'x''}{(x'^2 + y'^2)^{3/2}} = \frac{a^2}{a^3} = \frac{1}{a}, \tag{18}$$

or one over the RADIUS OF CURVATURE. The curvature of a CIRCLE can also be repeated in vector notation. For the CIRCLE with $0 \leq t < 2\pi$, the ARC LENGTH is

$$\begin{aligned} s(t) &= \int_0^t \sqrt{\left(\frac{dx}{dt}\right)^2 + \left(\frac{dy}{dt}\right)^2}\, dt \\ &= \int_0^t \sqrt{a^2 \cos^2 t + a^2 \sin^2 t}\, dt = at, \end{aligned} \tag{19}$$

so $t = s/a$ and the equations of the CIRCLE can be rewritten as

$$x = a \cos\left(\frac{s}{a}\right) \tag{20}$$

$$y = a \sin\left(\frac{s}{a}\right). \tag{21}$$

The POSITION VECTOR is then given by

$$\mathbf{r}(s) = a \cos\left(\frac{s}{a}\right) \hat{\mathbf{x}} + a \sin\left(\frac{s}{a}\right) \hat{\mathbf{y}}, \tag{22}$$

and the TANGENT VECTOR is

$$\hat{\mathbf{T}} = \frac{d\mathbf{r}}{ds} = -\sin\left(\frac{s}{a}\right) \hat{\mathbf{x}} + \cos\left(\frac{s}{a}\right) \hat{\mathbf{y}}, \tag{23}$$

so the curvature is related to the RADIUS OF CURVATURE a by

$$\begin{aligned} \kappa &= \left|\frac{d\hat{\mathbf{T}}}{ds}\right| = \left|-\frac{1}{a}\cos\left(\frac{s}{a}\right)\hat{\mathbf{x}} - \frac{1}{a}\sin\left(\frac{s}{a}\right)\hat{\mathbf{y}}\right| \\ &= \sqrt{\frac{\cos^2\left(\frac{s}{a}\right) + \sin^2\left(\frac{s}{a}\right)}{a^2}} = \frac{1}{a}, \end{aligned} \tag{24}$$

as expected.

Four very important derivative relations in differential geometry related to the FRENET FORMULAS are

$$\dot{\mathbf{r}} = \mathbf{T} \tag{25}$$

$$\ddot{\mathbf{r}} = \kappa\mathbf{N} \tag{26}$$

$$\dddot{\mathbf{r}} = \dot{\kappa}\mathbf{N} + \kappa(\tau\mathbf{B} - \kappa\mathbf{T}) \tag{27}$$

$$[\dot{\mathbf{r}}, \ddot{\mathbf{r}}, \dddot{\mathbf{r}}] = \kappa^2\tau, \tag{28}$$

where $\mathbf{T}$ is the TANGENT VECTOR, $\mathbf{N}$ is the NORMAL VECTOR, $\mathbf{B}$ is the BINORMAL VECTOR, and τ is the TORSION (Coxeter 1969, p. 322).

The curvature at a point on a surface takes on a variety of values as the PLANE through the normal varies. As κ varies, it achieves a minimum and a maximum (which are in perpendicular directions) known as the PRINCIPAL CURVATURES. As shown in Coxeter (1969, pp. 352–353),

$$\kappa^2 - \sum b_i^i \kappa + \det(b_i^j) = 0 \tag{29}$$

$$\kappa^2 - 2H\kappa + K = 0, \tag{30}$$

where K is the GAUSSIAN CURVATURE, H is the MEAN CURVATURE, and det denotes the DETERMINANT.

The curvature κ is sometimes called the FIRST CURVATURE and the TORSION τ the SECOND CURVATURE. In addition, a THIRD CURVATURE (sometimes called TOTAL CURVATURE)

$$\sqrt{ds_T{}^2 + ds_B{}^2} \tag{31}$$

is also defined. A signed version of the curvature of a CIRCLE appearing in the DESCARTES CIRCLE THEOREM for the radius of the fourth of four mutually tangent circles is called the BEND.

see also BEND (CURVATURE), CURVATURE CENTER, CURVATURE SCALAR, EXTRINSIC CURVATURE, FIRST CURVATURE, FOUR-VERTEX THEOREM, GAUSSIAN CURVATURE, INTRINSIC CURVATURE, LANCRET EQUATION, LINE OF CURVATURE, MEAN CURVATURE, NORMAL CURVATURE, PRINCIPAL CURVATURES, RADIUS OF CURVATURE, RICCI CURVATURE, RIEMANN TENSOR, SECOND CURVATURE, SECTIONAL CURVATURE, SODDY CIRCLES, THIRD CURVATURE, TORSION (DIFFERENTIAL GEOMETRY), WEINGARTEN MAP

References

Coxeter, H. S. M. *Introduction to Geometry, 2nd ed.* New York: Wiley, 1969.

Fischer, G. (Ed.). Plates 79–85 in *Mathematische Modelle/Mathematical Models, Bildband/Photograph Volume.* Braunschweig, Germany: Vieweg, pp. 74–81, 1986.

Gray, A. "Curvature of Curves in the Plane," "Drawing Plane Curves with Assigned Curvature," and "Drawing Space Curves with Assigned Curvature." §1.5, 6.4, and 7.8 in *Modern Differential Geometry of Curves and Surfaces.* Boca Raton, FL: CRC Press, pp. 11–13, 68–69, 113–118, and 145–147, 1993.

Kreyszig, E. "Principal Normal, Curvature, Osculating Circle." §12 in *Differential Geometry.* New York: Dover, pp. 34–36, 1991.

Yates, R. C. "Curvature." *A Handbook on Curves and Their Properties.* Ann Arbor, MI: J. W. Edwards, pp. 60–64, 1952.

Curvature Center

The point on the POSITIVE RAY of the NORMAL VECTOR at a distance $\rho(s)$, where ρ is the RADIUS OF CURVATURE. It is given by

$$\mathbf{z} = \mathbf{x} + \rho\mathbf{N} = \mathbf{x} + \rho^2 \frac{\mathbf{T}}{ds}, \tag{1}$$

where $\mathbf{N}$ is the NORMAL VECTOR and $\mathbf{T}$ is the TANGENT VECTOR. It can be written in terms of $\mathbf{x}$ explicitly as

$$\mathbf{z} = \mathbf{x} + \frac{\mathbf{x}''(\mathbf{x}'\cdot\mathbf{x}')^2 - \mathbf{x}'(\mathbf{x}'\cdot\mathbf{x}')(\mathbf{x}'\cdot\mathbf{x}'')}{(\mathbf{x}'\cdot\mathbf{x}')(\mathbf{x}''\cdot\mathbf{x}'') - (\mathbf{x}'\cdot\mathbf{x}'')^2}. \tag{2}$$

For a CURVE represented parametrically by $(f(t), g(t))$,

$$\alpha = f - \frac{(f'^2 - g'^2)g'}{f'g'' - f''g'} \tag{3}$$

$$\beta = g + \frac{(f'^2 - g'^2)f'}{f'g'' - f''g'}. \tag{4}$$

References

Gray, A. *Modern Differential Geometry of Curves and Surfaces.* Boca Raton, FL: CRC Press, 1993.

Curvature Scalar

The curvature scalar is given by

$$R \equiv g^{\mu\kappa} R_{\mu\kappa},$$

where $g^{\mu\kappa}$ is the METRIC TENSOR and $R_{\mu\kappa}$ is the RICCI TENSOR.

see also CURVATURE, GAUSSIAN CURVATURE, MEAN CURVATURE, METRIC TENSOR, RADIUS OF CURVATURE, RICCI TENSOR, RIEMANN-CHRISTOFFEL TENSOR

Curvature Vector

$$\mathbf{K} \equiv \frac{d\mathbf{T}}{ds},$$

where $\mathbf{T}$ is the TANGENT VECTOR defined by

$$\mathbf{T} \equiv \frac{\frac{d\mathbf{x}}{ds}}{\left|\frac{d\mathbf{x}}{ds}\right|}.$$

Curve

A CONTINUOUS MAP from a 1-D SPACE to an n-D SPACE. Loosely speaking, the word "curve" is often used to mean the GRAPH of a 2- or 3-D curve. The simplest curves can be represented parametrically in n-D SPACE as

$$x_1 = f_1(t)$$
$$x_2 = f_2(t)$$
$$\vdots$$
$$x_n = f_n(t).$$

Other simple curves can be simply defined only implicitly, i.e., in the form

$$f(x_1, x_2, \ldots) = 0.$$

see also ARCHIMEDEAN SPIRAL, ASTROID, ASYMPTOTIC CURVE, BASEBALL COVER, BATRACHION, BICORN, BIFOLIUM, BOW, BULLET NOSE, BUTTERFLY CURVE, CARDIOID, CASSINI OVALS, CATALAN'S TRISECTRIX, CATENARY, CAUSTIC, CAYLEY'S SEXTIC, CESÀRO EQUATION, CIRCLE, CIRCLE INVOLUTE, CISSOID, CISSOID OF DIOCLES, COCHLEOID, CONCHOID, CONCHOID OF NICOMEDES, CROSS CURVE, CRUCIFORM, CUBICAL PARABOLA, CURVE OF CONSTANT PRECESSION, CURVE OF CONSTANT WIDTH, CURTATE CYCLOID, CYCLOID, DELTA CURVE, DELTOID, DEVIL'S CURVE, DEVIL ON TWO STICKS, DUMBBELL CURVE, DÜRER'S CONCHOID, EIGHT CURVE, ELECTRIC MOTOR CURVE, ELLIPSE, ELLIPSE INVOLUTE, ELLIPTIC CURVE, ENVELOPE, EPICYCLOID, EQUIPOTENTIAL CURVE, EUDOXUS'S KAMPYLE, EVOLUTE, EXPONENTIAL RAMP, FERMAT CONIC, FOLIUM OF DESCARTES, FREETH'S NEPHROID, FREY CURVE, GAUSSIAN FUNCTION, GERONO LEMNISCATE, GLISSETTE, GUDERMANNIAN FUNCTION, GUTSCHOVEN'S CURVE, HIPPOPEDE, HORSE FETTER, HYPERBOLA, HYPERELLIPSE, HYPOCYCLOID, HYPOELLIPSE, INVOLUTE, ISOPTIC CURVE, KAPPA CURVE, KERATOID CUSP, KNOT CURVE, LAMÉ CURVE, LEMNISCATE, L'HOSPITAL'S CUBIC, LIMAÇON, LINKS CURVE, LISSAJOUS CURVE, LITUUS, LOGARITHMIC SPIRAL, MACLAURIN TRISECTRIX, MALTESE CROSS, MILL, NATURAL EQUATION, NEGATIVE PEDAL CURVE, NEPHROID, NIELSEN'S SPIRAL, ORTHOPTIC CURVE, PARABOLA, PEAR CURVE, PEAR-SHAPED CURVE, PEARLS OF SLUZE, PEDAL CURVE, PEG TOP, PIRIFORM, PLATEAU CURVES, POLICEMAN ON POINT DUTY CURVE, PROLATE CYCLOID, PURSUIT CURVE, QUADRATRIX OF HIPPIAS, RADIAL CURVE, RHODONEA, ROSE, ROULETTE, SEMICUBICAL PARABOLA, SERPENTINE CURVE, SICI SPIRAL, SIGMOID CURVE, SINUSOIDAL SPIRAL, SPACE CURVE, STROPHOID, SUPERELLIPSE, SWASTIKA, SWEEP SIGNAL, TALBOT'S CURVE, TEARDROP CURVE, TRACTRIX, TRIDENT, TRIDENT OF DESCARTES, TRIDENT OF NEWTON, TROCHOID, TSCHIRNHAUSEN CUBIC, VERSIERA, WATT'S CURVE, WHEWELL EQUATION, WITCH OF AGNESI

References

Cundy, H. and Rollett, A. *Mathematical Models, 3rd ed.* Stradbroke, England: Tarquin Pub., pp. 71–75, 1989.

"Geometry." *The New Encyclopædia Britannica, 15th ed.* **19**, pp. 946–951, 1990.

Gray, A. "Famous Plane Curves." Ch. 3 in *Modern Differential Geometry of Curves and Surfaces.* Boca Raton, FL: CRC Press, pp. 37–55, 1993.

Lawrence, J. D. *A Catalog of Special Plane Curves.* New York: Dover, 1972.

Lee, X. "A Catalog of Special Plane Curves." `http://www.best.com/~xah/SpecialPlaneCurves_dir/specialPlaneCurves.html`.

Lockwood, E. H. *A Book of Curves.* Cambridge, England: Cambridge University Press, 1961.
MacTutor History of Mathematics Archive. http://www-groups.dcs.st-and.ac.uk/~history/Curves/Curves.html.
Oakley, C. O. *Analytic Geometry.* New York: Barnes and Noble, 1957.
Shikin, E. V. *Handbook and Atlas of Curves.* Boca Raton, FL: CRC Press, 1995.
Smith, P. F.; Gale, A. S.; and Neelley, J. H. *New Analytic Geometry, Alternate Edition.* Boston, MA: Ginn and Company, 1938.
von Seggern, D. *CRC Standard Curves and Surfaces.* Boca Raton, FL: CRC Press, 1993.
Walker, R. J. *Algebraic Curves.* New York: Springer-Verlag, 1978.
Weisstein, E. W. "Plane Curves." http://www.astro.virginia.edu/~eww6n/math/notebooks/Curves.m.
Yates, R. C. *A Handbook on Curves and Their Properties.* Ann Arbor, MI: J. W. Edwards, 1947.
Yates, R. C. *The Trisection Problem.* Reston, VA: National Council of Teachers of Mathematics, 1971.
Zwillinger, D. (Ed.). "Algebraic Curves." §8.1 in *CRC Standard Mathematical Tables and Formulae, 3rd ed.* Boca Raton, FL: CRC Press, 1996. http://www.geom.umn.edu/docs/reference/CRC-formulas/node33.html.

Curve of Constant Breadth

see CURVE OF CONSTANT WIDTH

Curve of Constant Precession

A curve whose CENTRODE revolves about a fixed axis with constant ANGLE and SPEED when the curve is traversed with unit SPEED. The TANGENT INDICATRIX of a curve of constant precession is a SPHERICAL HELIX. An ARC LENGTH parameterization of a curve of constant precession with NATURAL EQUATIONS

$$\kappa(s) = -\omega \sin(\mu s) \tag{1}$$
$$\tau(s) = \omega \cos(\mu s) \tag{2}$$

is

$$x(s) = \frac{\alpha+\mu}{2\alpha}\frac{\sin[(\alpha-\mu)s]}{\alpha-\mu} - \frac{\alpha-\mu}{2\alpha}\frac{\sin[(\alpha+\mu)s]}{\alpha+\mu} \tag{3}$$
$$y(s) = -\frac{\alpha+\mu}{2\alpha}\frac{\cos[(\alpha-\mu)s]}{\alpha-\mu} + \frac{\alpha-\mu}{2\alpha}\frac{\cos[(\alpha+\mu)s]}{\alpha+\mu} \tag{4}$$
$$z(s) = \frac{\omega}{\mu\alpha}\sin(\mu s), \tag{5}$$

where

$$\alpha \equiv \sqrt{\omega^2+\mu^2} \tag{6}$$

and ω, and μ are constant. This curve lies on a circular one-sheeted HYPERBOLOID

$$x^2+y^2-\frac{\mu^2}{\omega^2}z^2 = \frac{4\mu^2}{\omega^4}. \tag{7}$$

The curve is closed IFF μ/α is RATIONAL.

References
Scofield, P. D. "Curves of Constant Precession." *Amer. Math. Monthly* **102**, 531–537, 1995.

Curve of Constant Slope

see GENERALIZED HELIX

Curve of Constant Width

Curves which, when rotated in a square, make contact with all four sides. The "width" of a closed convex curve is defined as the distance between parallel lines bounding it ("supporting lines"). Every curve of constant width is convex. Curves of constant width have the same "width" regardless of their orientation between the parallel lines. In fact, they also share the same PERIMETER (BARBIER'S THEOREM). Examples include the CIRCLE (with largest AREA), and REULEAUX TRIANGLE (with smallest AREA) but there are an infinite number. A curve of constant width can be used in a special drill chuck to cut square "HOLES."

A generalization gives solids of constant width. These do not have the same surface AREA for a given width, but their shadows are curves of constant width with the *same* width!

see also DELTA CURVE, KAKEYA NEEDLE PROBLEM, REULEAUX TRIANGLE

References
Bogomolny, A. "Shapes of Constant Width." http://www.cut-the-knot.com/do_you_know/cwidth.html.
Böhm, J. "Convex Bodies of Constant Width." Ch. 4 in *Mathematical Models from the Collections of Universities and Museums* (Ed. G. Fischer). Braunschweig, Germany: Vieweg, pp. 96–100, 1986.
Fischer, G. (Ed.). Plates 98–102 in *Mathematische Modelle/Mathematical Models, Bildband/Photograph Volume.* Braunschweig, Germany: Vieweg, pp. 89 and 96, 1986.
Gardner, M. Ch. 18 in *The Unexpected Hanging and Other Mathematical Diversions.* Chicago, IL: Chicago University Press, 1991.
Goldberg, M. "Circular-Arc Rotors in Regular Polygons." *Amer. Math. Monthly* **55**, 393–402, 1948.
Kelly, P. *Convex Figures.* New York: Harcourt Brace, 1995.
Rademacher, H. and Toeplitz, O. *The Enjoyment of Mathematics: Selections from Mathematics for the Amateur.* Princeton, NJ: Princeton University Press, 1957.
Yaglom, I. M. and Boltyanski, V. G. *Convex Figures.* New York: Holt, Rinehart, and Winston, 1961.

Curvilinear Coordinates

A general METRIC $g_{\mu\nu}$ has a LINE ELEMENT

$$ds^2 = g_{\mu\nu}du^\mu du^\nu, \tag{1}$$

where EINSTEIN SUMMATION is being used. Curvilinear coordinates are defined as those with a diagonal METRIC so that

$$g_{\mu\nu} \equiv \delta^\mu_\nu {h_\mu}^2, \tag{2}$$

where δ^μ_ν is the KRONECKER DELTA. Curvilinear coordinates therefore have a simple LINE ELEMENT

$$ds^2 = \delta^\mu_\nu {h_\mu}^2 du^\mu du^\nu = {h_\mu}^2 du^{\mu 2}, \tag{3}$$

which is just the PYTHAGOREAN THEOREM, so the differential VECTOR is

$$d\mathbf{r} = h_\mu du_\mu \hat{\mathbf{u}}_\mu, \tag{4}$$

or

$$d\mathbf{r} = \frac{\partial \mathbf{r}}{\partial u_1} du_1 + \frac{\partial \mathbf{r}}{\partial u_2} du_2 + \frac{\partial \mathbf{r}}{\partial u_3} du_3, \tag{5}$$

where the SCALE FACTORS are

$$h_i \equiv \left| \frac{\partial \mathbf{r}}{\partial u_i} \right| \tag{6}$$

and

$$\hat{\mathbf{u}}_i \equiv \frac{\frac{\partial \mathbf{r}}{\partial u_i}}{\left|\frac{\partial \mathbf{r}}{\partial u_i}\right|} = \frac{1}{h_i} \frac{\partial \mathbf{r}}{\partial u_i}. \tag{7}$$

Equation (5) may therefore be re-expressed as

$$d\mathbf{r} = h_1 du_1 \hat{\mathbf{u}}_1 + h_2 du_2 \hat{\mathbf{u}}_2 + h_3 du_3 \hat{\mathbf{u}}_3. \tag{8}$$

The GRADIENT is

$$\mathrm{grad}(\phi) \equiv \nabla\phi = \frac{1}{h_1}\frac{\partial \phi}{\partial u_1}\hat{\mathbf{u}}_1 + \frac{1}{h_2}\frac{\partial \phi}{\partial u_2}\hat{\mathbf{u}}_2 + \frac{1}{h_3}\frac{\partial \phi}{\partial u_3}\hat{\mathbf{u}}_3, \tag{9}$$

the DIVERGENCE is

$$\begin{aligned}\mathrm{div}(F) \equiv \nabla\cdot\mathbf{F} \equiv \frac{1}{h_1h_2h_3}\left[\frac{\partial}{\partial u_1}(h_2h_3F_1)\right.\\ \left. + \frac{\partial}{\partial u_2}(h_3h_1F_2) + \frac{\partial}{\partial u_3}(h_1h_2F_3)\right],\end{aligned} \tag{10}$$

and the CURL is

$$\begin{aligned}\nabla\times\mathbf{F} &\equiv \frac{1}{h_1h_2h_3}\begin{vmatrix} h_1\hat{\mathbf{u}}_1 & h_2\hat{\mathbf{u}}_2 & h_3\hat{\mathbf{u}}_3 \\ \frac{\partial}{\partial u_1} & \frac{\partial}{\partial u_2} & \frac{\partial}{\partial u_3} \\ h_1F_1 & h_2F_2 & h_3F_3 \end{vmatrix} \\ &= \frac{1}{h_2h_3}\left[\frac{\partial}{\partial u_2}(h_3F_3) - \frac{\partial}{\partial u_3}(h_2F_2)\right]\hat{\mathbf{u}}_1 \\ &\quad + \frac{1}{h_1h_3}\left[\frac{\partial}{\partial u_3}(h_1F_1) - \frac{\partial}{\partial u_1}(h_3F_3)\right]\hat{\mathbf{u}}_2 \\ &\quad + \frac{1}{h_1h_2}\left[\frac{\partial}{\partial u_1}(h_2F_2) - \frac{\partial}{\partial u_2}(h_1F_1)\right]\hat{\mathbf{u}}_3.\end{aligned} \tag{11}$$

Orthogonal curvilinear coordinates satisfy the additional constraint that

$$\hat{\mathbf{u}}_i \cdot \hat{\mathbf{u}}_j = \delta_{ij}. \tag{12}$$

Therefore, the LINE ELEMENT is

$$ds^2 = d\mathbf{r}\cdot d\mathbf{r} = {h_1}^2{du_1}^2 + {h_2}^2{du_2}^2 + {h_3}^2{du_3}^2 \tag{13}$$

and the VOLUME ELEMENT is

$$\begin{aligned} dV &= |(h_1\hat{\mathbf{u}}_1\,du_1)\cdot(h_2\hat{\mathbf{u}}_2\,du_2)\times(h_3\hat{\mathbf{u}}_3\,du_3)| \\ &= h_1h_2h_3\,du_1\,du_2\,du_3 \\ &= \left|\frac{\partial r}{\partial u_1}\cdot\frac{\partial r}{\partial u_2}\times\frac{\partial r}{\partial u_3}\right| du_1\,du_2\,du_3 \\ &= \begin{vmatrix} \frac{\partial x}{\partial u_1} & \frac{\partial x}{\partial u_2} & \frac{\partial x}{\partial u_3} \\ \frac{\partial y}{\partial u_1} & \frac{\partial y}{\partial u_2} & \frac{\partial y}{\partial u_3} \\ \frac{\partial z}{\partial u_1} & \frac{\partial z}{\partial u_2} & \frac{\partial z}{\partial u_3} \end{vmatrix} du_1\,du_2\,du_3 \\ &= \left|\frac{\partial(x,y,z)}{\partial(u_1,u_2,u_3)}\right| du_1\,du_2\,du_3, \end{aligned} \tag{14}$$

where the latter is the JACOBIAN.

Orthogonal curvilinear coordinate systems include BIPOLAR CYLINDRICAL COORDINATES, BISPHERICAL COORDINATES, CARTESIAN COORDINATES, CONFOCAL ELLIPSOIDAL COORDINATES, CONFOCAL PARABOLOIDAL COORDINATES, CONICAL COORDINATES, CYCLIDIC COORDINATES, CYLINDRICAL COORDINATES, ELLIPSOIDAL COORDINATES, ELLIPTIC CYLINDRICAL COORDINATES, OBLATE SPHEROIDAL COORDINATES, PARABOLIC COORDINATES, PARABOLIC CYLINDRICAL COORDINATES, PARABOLOIDAL COORDINATES, POLAR COORDINATES, PROLATE SPHEROIDAL COORDINATES, SPHERICAL COORDINATES, and TOROIDAL COORDINATES. These are degenerate cases of the CONFOCAL ELLIPSOIDAL COORDINATES.

see also CHANGE OF VARIABLES THEOREM, CURL, DIVERGENCE, GRADIENT, JACOBIAN, LAPLACIAN

References

Arfken, G. "Curvilinear Coordinates" and "Differential Vector Operators." §2.1 and 2.2 in *Mathematical Methods for Physicists, 3rd ed.* Orlando, FL: Academic Press, pp. 86–90 and 90–94, 1985.

Gradshteyn, I. S. and Ryzhik, I. M. *Tables of Integrals, Series, and Products, 5th ed.* San Diego, CA: Academic Press, pp. 1084–1088, 1980.

Morse, P. M. and Feshbach, H. "Curvilinear Coordinates" and "Table of Properties of Curvilinear Coordinates." §1.3 in *Methods of Theoretical Physics, Part I.* New York: McGraw-Hill, pp. 21–31 and 115–117, 1953.

Cushion

The QUARTIC SURFACE resembling a squashed round cushion on a barroom stool and given by the equation

$$\begin{aligned} z^2x^2 - z^4 - 2zx^2 + 2z^3 + x^2 - z^2 \\ -(x^2-z)^2 - y^4 - 2x^2y^2 - y^2z^2 + 2y^2z + y^2 = 0. \end{aligned}$$

see also QUARTIC SURFACE

References
Nordstrand, T. "Surfaces." http://www.uib.no/people/nfytn/surfaces.htm.

Cusp

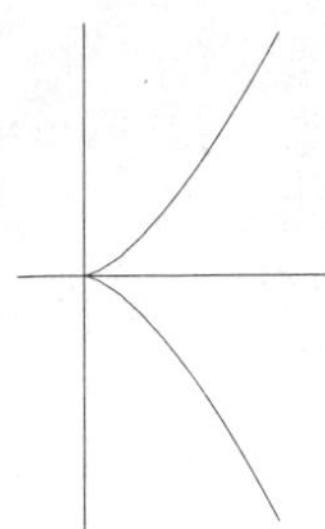

A function $f(x)$ has a cusp (also called a SPINODE) at a point x_0 if $f(x)$ is CONTINUOUS at x_0 and

$$\lim_{x \to x_0} f'(x) = \infty$$

from one side while

$$\lim_{x \to x_0} f'(x) = -\infty$$

from the other side, so the curve is CONTINUOUS but the DERIVATIVE is not. A cusp is a type of DOUBLE POINT. The above plot shows the curve $x^3 - y^2 = 0$, which has a cusp at the ORIGIN.

see also DOUBLE CUSP, DOUBLE POINT, ORDINARY DOUBLE POINT, RAMPHOID CUSP, SALIENT POINT

References
Walker, R. J. *Algebraic Curves.* New York: Springer-Verlag, pp. 57–58, 1978.

Cusp Catastrophe

A CATASTROPHE which can occur for two control factors and one behavior axis. The equation $y = x^{2/3}$ has a cusp catastrophe.

see also CATASTROPHE

References
von Seggern, D. *CRC Standard Curves and Surfaces.* Boca Raton, FL: CRC Press, p. 28, 1993.

Cusp Form

A cusp form on $\Gamma_0(N)$, the group of INTEGER matrices with determinant 1 which are upper triangular mod N, is an ANALYTIC FUNCTION on the upper half-plane consisting of the COMPLEX NUMBERS with POSITIVE IMAGINARY PART. Weight n cusp forms satisfy

$$f\left(\frac{az+b}{cz+d}\right) = (cz+d)^n f(z)$$

for all matrices

$$\begin{bmatrix} a & b \\ c & d \end{bmatrix} \in \Gamma_0(N).$$

see also MODULAR FORM

Cusp Map

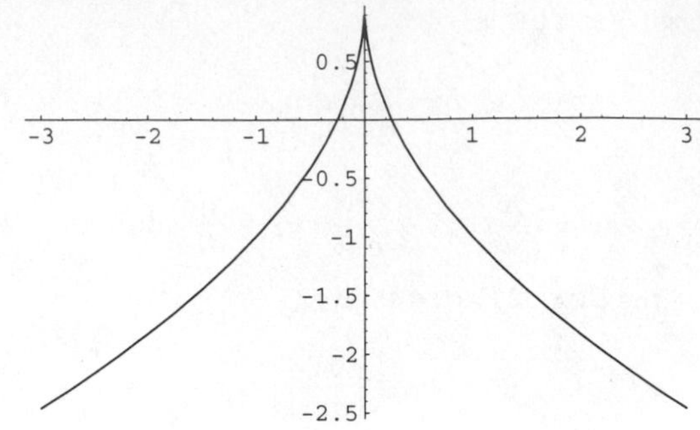

The function

$$f(x) = 1 - 2|x|^{1/2}$$

for $x \in [-1, 1]$. The INVARIANT DENSITY is

$$\rho(y) = \tfrac{1}{2}(1 - y).$$

References
Beck, C. and Schlögl, F. *Thermodynamics of Chaotic Systems.* Cambridge, England: Cambridge University Press, p. 195, 1995.

Cusp Point

see CUSP

Cut-Vertex

see ARTICULATION VERTEX

Cutting

see ARRANGEMENT, CAKE CUTTING, CIRCLE CUTTING, CYLINDER CUTTING, PANCAKE CUTTING, PIE CUTTING, SQUARE CUTTING, TORUS CUTTING

CW-Approximation Theorem

If X is any SPACE, then there is a CW-COMPLEX Y and a MAP $f : Y \to X$ inducing ISOMORPHISMS on all HOMOTOPY, HOMOLOGY, and COHOMOLOGY groups.

CW-Complex

A CW-complex is a homotopy-theoretic generalization of the notion of a SIMPLICIAL COMPLEX. A CW-complex is any SPACE X which can be built by starting off with a discrete collection of points called X^0, then attaching 1-D DISKS D^1 to X^0 along their boundaries S^0, writing X^1 for the object obtained by attaching the D^1s to X^0, then attaching 2-D DISKS D^2 to X^1 along their boundaries S^1, writing X^2 for the new SPACE, and so on, giving spaces X^n for every n. A CW-complex is any SPACE that has this sort of decomposition into SUBSPACES X^n built up in such a hierarchical fashion (so the X^ns must exhaust all of X). In particular, X^n may be built from X^{n-1} by attaching infinitely many n-DISKS, and the attaching MAPS $S^{n-1} \to X^{n-1}$ may be any continuous MAPS.

The main importance of CW-complexes is that, for the sake of HOMOTOPY, HOMOLOGY, and COHOMOLOGY groups, every SPACE is a CW-complex. This is called the CW-APPROXIMATION THEOREM. Another is WHITEHEAD'S THEOREM, which says that MAPS between CW-complexes that induce ISOMORPHISMS on all HOMOTOPY GROUPS are actually HOMOTOPY equivalences.

see also COHOMOLOGY, CW-APPROXIMATION THEOREM, HOMOLOGY GROUP, HOMOTOPY GROUP, SIMPLICIAL COMPLEX, SPACE, SUBSPACE, WHITEHEAD'S THEOREM

Cycle (Circle)

A CIRCLE with an arrow indicating a direction.

Cycle (Graph)

A subset of the EDGE-set of a graph that forms a CHAIN (GRAPH), the first node of which is also the last (also called a CIRCUIT).

see also CYCLIC GRAPH, HAMILTONIAN CYCLE, WALK

Cycle Graph

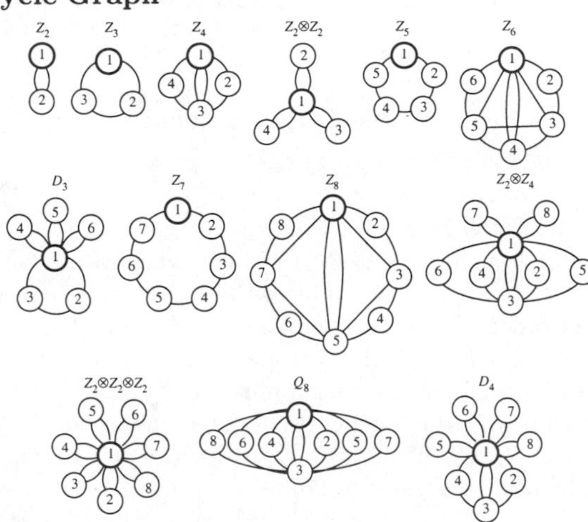

A cycle graph is a GRAPH which shows cycles of a GROUP as well as the connectivity between the cycles. Several examples are shown above. For Z_4, the group elements A_i satisfy ${A_i}^4 = 1$, where 1 is the IDENTITY ELEMENT, and two elements satisfy ${A_1}^2 = {A_3}^2 = 1$.

For a CYCLIC GROUP of COMPOSITE ORDER n (e.g., Z_4, Z_6, Z_8), the degenerate subcycles corresponding to factors dividing n are often not shown explicitly since their presence is implied.

see also CHARACTERISTIC FACTOR, CYCLIC GROUP

References

Shanks, D. *Solved and Unsolved Problems in Number Theory, 4th ed.* New York: Chelsea, pp. 83–98, 1993.

Cycle (Map)

An n-cycle is a finite sequence of points $Y_0, \ldots, Y_{n-1}$ such that, under a MAP G,

$$\begin{aligned} Y_1 &= G(Y_0) \\ Y_2 &= G(Y_1) \\ Y_{n-1} &= G(Y_{n-2}) \\ Y_0 &= G(Y_{n-1}). \end{aligned}$$

In other words, it is a periodic trajectory which comes back to the same point after n iterations of the cycle. Every point Y_j of the cycle satisfies $Y_j = G^n(Y_j)$ and is therefore a FIXED POINT of the mapping G^n. A fixed point of G is simply a CYCLE of period 1.

Cycle (Permutation)

A SUBSET of a PERMUTATION whose elements trade places with one another. A cycle decomposition of a PERMUTATION can therefore be viewed as a CLASS of a PERMUTATION GROUP. For example, in the PERMUTATION GROUP {4, 2, 1, 3}, {1, 3, 4} is a 3-cycle ($1 \to 3$, $3 \to 4$, and $4 \to 1$) and {2} is a 1-cycle ($2 \to 2$). Every PERMUTATION GROUP on n symbols can be uniquely expressed as a product of disjoint cycles. The cyclic decomposition of a PERMUTATION can be computed in *Mathematica®* (Wolfram Research, Champaign, IL) with the function `ToCycles` and the PERMUTATION corresponding to a cyclic decomposition can be computed with `FromCycles`. According to Vardi (1991), the Mathematica code for `ToCycles` is one of the most obscure ever written.

To find the number $N(m,n)$ of m cycles in a PERMUTATION GROUP of order n, take

$$N(n,m) = (-1)^{n-m} S_1(n,m),$$

where S_1 is the STIRLING NUMBER OF THE FIRST KIND.

see also GOLOMB-DICKMAN CONSTANT, PERMUTATION, PERMUTATION GROUP, SUBSET

References

Skiena, S. *Implementing Discrete Mathematics: Combinatorics and Graph Theory with Mathematica.* Reading, MA: Addison-Wesley, p. 20, 1990.
Vardi, I. *Computational Recreations in Mathematica.* Redwood City, CA: Addison-Wesley, p. 223, 1991.

Cyclic Graph

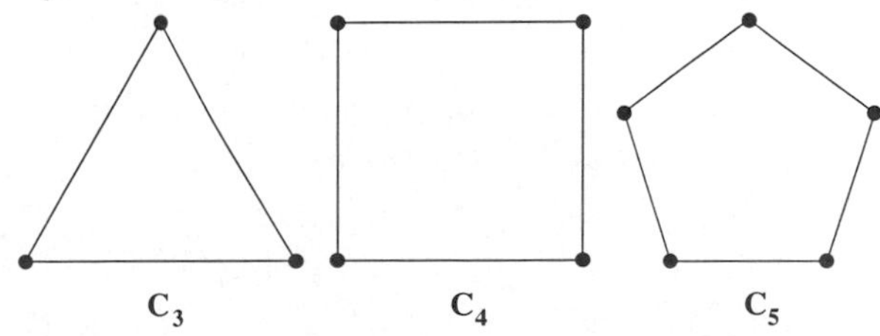

A GRAPH of n nodes and n edges such that node i is connected to the two adjacent nodes $i+1$ and $i-1$ (mod n), where the nodes are numbered $0, 1, \ldots, n-1$.

see also CYCLE (GRAPH), CYCLE GRAPH, STAR GRAPH, WHEEL GRAPH

Cyclic Group

A cyclic group Z_n of ORDER n is a GROUP defined by the element X (the GENERATOR) and its n POWERS up to

$$X^n = I,$$

where I is the IDENTITY ELEMENT. Cyclic groups are both ABELIAN and SIMPLE. There exists a unique cyclic group of every order $n \geq 2$, so cyclic groups of the same order are always isomorphic (Shanks 1993, p. 74), and all GROUPS of PRIME ORDER are cyclic.

Examples of cyclic groups include Z_2, Z_3, Z_4, and the MODULO MULTIPLICATION GROUPS M_m such that $m = 2$, 4, p^n, or $2p^n$, for p an ODD PRIME and $n \geq 1$ (Shanks 1993, p. 92). By computing the CHARACTERISTIC FACTORS, any ABELIAN GROUP can be expressed as a DIRECT PRODUCT of cyclic SUBGROUPS, for example, $Z_2 \otimes Z_4$ or $Z_2 \otimes Z_2 \otimes Z_2$.

see also ABELIAN GROUP, CHARACTERISTIC FACTOR, FINITE GROUP—Z_2, FINITE GROUP—Z_3, FINITE GROUP—Z_2, FINITE GROUP—Z_5, FINITE GROUP—Z_6, MODULO MULTIPLICATION GROUP, SIMPLE GROUP

References

Lomont, J. S. "Cyclic Groups." §3.10.A in *Applications of Finite Groups.* New York: Dover, p. 78, 1987.

Shanks, D. *Solved and Unsolved Problems in Number Theory, 4th ed.* New York: Chelsea, 1993.

Cyclic Hexagon

A hexagon (not necessarily regular) on whose VERTICES a CIRCLE may be CIRCUMSCRIBED. Let

$$\sigma_i \equiv \sum_{i,j,\ldots,n=1} {a_i}^2 {a_j}^2 \cdots {a_n}^2, \tag{1}$$

where the sum runs over all distinct permutations of the SQUARES of the six side lengths, so

$$\sigma_1 = {a_1}^2 + {a_2}^2 + {a_3}^2 + {a_4}^2 + {a_5}^2 + {a_6}^2 \tag{2}$$

$$\begin{aligned}\sigma_2 = {} & {a_1}^2{a_2}^2 + {a_1}^2{a_3}^2 + {a_1}^2{a_4}^2 + {a_1}^2{a_5}^2 + {a_1}^2{a_6}^2 \\ & + {a_2}^2{a_3}^2 + {a_2}^2{a_4}^2 + {a_2}^2{a_5}^2 + {a_2}^2{a_6}^2 \\ & + {a_3}^2{a_4}^2 + {a_3}^2{a_5}^2 + {a_3}^2{a_6}^2 \\ & + {a_4}^2{a_5}^2 + {a_4}^2{a_6}^2 + {a_5}^2{a_6}^2\end{aligned} \tag{3}$$

$$\begin{aligned}\sigma_3 = {} & {a_1}^2{a_2}^2{a_3}^2 + {a_1}^2{a_2}^2{a_4}^2 + {a_1}^2{a_2}^2{a_5}^2 + {a_1}^2{a_2}^2{a_6}^2 \\ & + {a_2}^2{a_3}^2{a_4}^2 + {a_2}^2{a_3}^2{a_5}^2 + {a_2}^2{a_3}^2{a_6}^2 \\ & + {a_3}^2{a_4}^2{a_5}^2 + {a_3}^2{a_4}^2{a_6}^2 + {a_4}^2{a_5}^2{a_6}^2\end{aligned} \tag{4}$$

$$\begin{aligned}\sigma_4 = {} & {a_1}^2{a_2}^2{a_3}^2{a_4}^2 + {a_1}^2{a_2}^2{a_3}^2{a_5}^2 + {a_1}^2{a_2}^2{a_3}^2{a_6}^2 \\ & + {a_1}^2{a_3}^2{a_4}^2{a_5}^2 + {a_1}^2{a_3}^2{a_4}^2{a_6}^2 \\ & + {a_1}^2{a_3}^2{a_5}^2{a_6}^2 + {a_1}^2{a_4}^2{a_5}^2{a_6}^2 \\ & + {a_2}^2{a_3}^2{a_4}^2{a_5}^2 + {a_2}^2{a_3}^2{a_4}^2{a_6}^2 + {a_2}^2{a_3}^2{a_5}^2{a_6}^2 \\ & + {a_2}^2{a_4}^2{a_5}^2{a_6}^2 + {a_3}^2{a_4}^2{a_5}^2{a_6}^2\end{aligned} \tag{5}$$

$$\begin{aligned}\sigma_5 = {} & {a_1}^2{a_2}^2{a_3}^2{a_4}^2{a_5}^2 + {a_1}^2{a_2}^2{a_3}^2{a_4}^2{a_6}^2 \\ & + {a_1}^2{a_2}^2{a_3}^2{a_5}^2{a_6}^2 + {a_1}^2{a_2}^2{a_4}^2{a_5}^2{a_6}^2 \\ & + {a_1}^2{a_3}^2{a_4}^2{a_5}^2{a_6}^2 + {a_2}^2{a_3}^2{a_4}^2{a_5}^2{a_6}^2\end{aligned} \tag{6}$$

$$\sigma_6 = {a_1}^2{a_2}^2{a_3}^2{a_4}^2{a_5}^2{a_6}^2. \tag{7}$$

Then define

$$t_2 = u - 4\sigma_2 + {\sigma_1}^2 \tag{8}$$

$$t_3 = 8\sigma_3 + \sigma_1 t_2 - 16\sqrt{\sigma_6} \tag{9}$$

$$t_4 = {t_2}^2 - 64\sigma_4 + 64\sigma_1\sqrt{\sigma_6} \tag{10}$$

$$t_5 = 128\sigma_5 + 32t_2\sqrt{\sigma_6} \tag{11}$$

$$u = 16K^2. \tag{12}$$

The AREA of the hexagon then satisfies

$$u{t_4}^3 + {t_3}^2{t_4}^2 - 16{t_3}^3t_5 - 18ut_3t_4t_5 - 27u^2{t_5}^2 = 0, \tag{13}$$

or this equation with $\sqrt{\sigma_6}$ replaced by $-\sqrt{\sigma_6}$, a seventh order POLYNOMIAL in u. This is $1/(4u^2)$ times the DISCRIMINANT of the CUBIC EQUATION

$$z^3 + 2t_3z^2 - ut_4z + 2y^2t_5. \tag{14}$$

see also CONCYCLIC, CYCLIC PENTAGON, CYCLIC POLYGON, FUHRMANN'S THEOREM

References

Robbins, D. P. "Areas of Polygons Inscribed in a Circle." *Discr. Comput. Geom.* **12**, 223–236, 1994.

Robbins, D. P. "Areas of Polygons Inscribed in a Circle." *Amer. Math. Monthly* **102**, 523–530, 1995.

Cyclic-Inscriptable Quadrilateral

see BICENTRIC QUADRILATERAL

Cyclic Number

A number having $n-1$ DIGITS which, when MULTIPLIED by 1, 2, 3, ..., $n-1$, produces the same digits in a different order. Cyclic numbers are generated by the UNIT FRACTIONS $1/n$ which have maximal period DECIMAL EXPANSIONS (which means n must be PRIME). The first few numbers which generate cyclic numbers are 7, 17, 19, 23, 29, 47, 59, 61, 97, ... (Sloane's A001913). A much larger generator is 17389.

It has been conjectured, but not yet proven, that an INFINITE number of cyclic numbers exist. In fact, the FRACTION of PRIMES which generate cyclic numbers seems to be approximately 3/8. See Yates (1973) for a table of PRIME period lengths for PRIMES $< 1,370,471$. When a cyclic number is multiplied by its generator, the result is a string of 9s. This is a special case of MIDY'S THEOREM.

07 = 0.142857
17 = 0.0588235294117647
19 = 0.052631578947368421
23 = 0.0434782608695652173913
29 = 0.0344827586206896551724137931
47 = 0.021276595744680851063829787234042553191···
···4893617
59 = 0.016949152542372881355932203389830508474···
···5762711864406779661
61 = 0.016393442622950819672131147540983606557···

$\cdots 377049180327868852459$

$97 = 0.010309278350515463917525773195876288659\cdots$

$\cdots 793814432989690721649484536082474226804 12\cdots$

$\cdots 3711340206185567$

see also DECIMAL EXPANSION, MIDY'S THEOREM

References

Gardner, M. Ch. 10 in *Mathematical Circus: More Puzzles, Games, Paradoxes and Other Mathematical Entertainments from Scientific American.* New York: Knopf, 1979.

Guttman, S. "On Cyclic Numbers." *Amer. Math. Monthly* **44**, 159–166, 1934.

Kraitchik, M. "Cyclic Numbers." §3.7 in *Mathematical Recreations.* New York: W. W. Norton, pp. 75–76, 1942.

Rao, K. S. "A Note on the Recurring Period of the Reciprocal of an Odd Number." *Amer. Math. Monthly* **62**, 484–487, 1955.

Sloane, N. J. A. Sequence A001913/M4353 in "An On-Line Version of the Encyclopedia of Integer Sequences."

Yates, S. *Primes with Given Period Length.* Trondheim, Norway: Universitetsforlaget, 1973.

Cyclic Pentagon

A cyclic pentagon is a not necessarily regular PENTAGON on whose VERTICES a CIRCLE may be CIRCUMSCRIBED. Let

$$\sigma_i \equiv \sum_{i,j,\ldots,n=1} {a_i}^2 {a_j}^2 \cdots {a_n}^2, \quad (1)$$

where the SUM runs over all distinct PERMUTATIONS of the SQUARES of the 5 side lengths, so

$$\sigma_1 = {a_1}^2 + {a_2}^2 + {a_3}^2 + {a_4}^2 + {a_5}^2 \quad (2)$$

$$\sigma_2 = {a_1}^2{a_2}^2 + {a_1}^2{a_3}^2 + {a_1}^2{a_4}^2 + {a_1}^2{a_5}^2 + {a_2}^2{a_3}^2 + {a_2}^2{a_4}^2 + {a_2}^2{a_5}^2 + {a_3}^2{a_4}^2 + {a_3}^2{a_5}^2 + {a_4}^2{a_5}^2 \quad (3)$$

$$\sigma_3 = {a_1}^2{a_2}^2{a_3}^2 + {a_1}^2{a_2}^2{a_4}^2 + {a_1}^2{a_2}^2{a_5}^2 + {a_2}^2{a_3}^2{a_4}^2 + {a_2}^2{a_3}^2{a_5}^2 + {a_3}^2{a_4}^2{a_5}^2 \quad (4)$$

$$\sigma_4 = {a_1}^2{a_2}^2{a_3}^2{a_4}^2 + {a_1}^2{a_2}^2{a_3}^2{a_5}^2 + {a_1}^2{a_3}^2{a_4}^2{a_5}^2 + {a_2}^2{a_3}^2{a_4}^2{a_5}^2 \quad (5)$$

$$\sigma_5 = {a_1}^2{a_2}^2{a_3}^2{a_4}^2{a_5}^2. \quad (6)$$

Then define

$$t_2 = u - 4\sigma_2 + {\sigma_1}^2 \quad (7)$$

$$t_3 = 8\sigma_3 + \sigma_1 t_2 \quad (8)$$

$$t_4 = -64\sigma_4 + {t_2}^2 \quad (9)$$

$$t_5 = 128\sigma_5 \quad (10)$$

$$u = 16K^2. \quad (11)$$

The AREA of the pentagon then satisfies

$$u{t_4}^3 + {t_3}^2{t_4}^2 - 16{t_3}^3 t_5 - 18ut_3t_4t_5 - 27u^2{t_5}^2 = 0, \quad (12)$$

a seventh order POLYNOMIAL in u. This is $1/(4u^2)$ times the DISCRIMINANT of the CUBIC EQUATION

$$z^3 + 2t_3z^2 - ut_4z + 2y^2t_5. \quad (13)$$

see also CONCYCLIC, CYCLIC HEXAGON, CYCLIC POLYGON

References

Robbins, D. P. "Areas of Polygons Inscribed in a Circle." *Discr. Comput. Geom.* **12**, 223–236, 1994.

Robbins, D. P. "Areas of Polygons Inscribed in a Circle." *Amer. Math. Monthly* **102**, 523–530, 1995.

Cyclic Permutation

A PERMUTATION which shifts all elements of a SET by a fixed offset, with the elements shifted off the end inserted back at the beginning. For a SET with elements a_0, a_1, $\ldots$, a_{n-1}, this can be written $a_i \to a_{i+k \pmod{n}}$ for a shift of k.

see also PERMUTATION

Cyclic Polygon

A cyclic polygon is a POLYGON with VERTICES upon which a CIRCLE can be CIRCUMSCRIBED. Since every TRIANGLE has a CIRCUMCIRCLE, every TRIANGLE is cyclic. It is conjectured that for a cyclic polygon of $2m+1$ sides, $16K^2$ (where K is the AREA) satisfies a MONIC POLYNOMIAL of degree Δ_m, where

$$\Delta_m \equiv \sum_{k=0}^{m-1} (m-k)\binom{2m+1}{k} \quad (1)$$

$$= \frac{1}{2}\left[(2m+1)\binom{2m}{m} - 2^{2m}\right] \quad (2)$$

(Robbins 1995). It is also conjectured that a cyclic polygon with $2m+2$ sides satisfies one of two POLYNOMIALS of degree Δ_m. The first few values of Δ_m are 1, 7, 38, 187, 874, ... (Sloane's A000531).

For TRIANGLES ($n = 3 = 2 \cdot 1 + 1$), the POLYNOMIAL is HERON'S FORMULA, which may be written

$$16K^2 = 2a^2b^2 + 2a^2c^2 + 2b^2c^2 - a^4 - b^4 - c^4, \quad (3)$$

and which is of order $\Delta_1 = 1$ in $16K^2$. For a CYCLIC QUADRILATERAL, the POLYNOMIAL is BRAHMAGUPTA'S FORMULA, which may be written

$$16K^2 = -a^4 + 2a^2b^2 - b^4 + 2a^2c^2 + 2b^2c^2 - c^4 + 8abcd + 2a^2d^2 + 2b^2d^2 + 2c^2d^2 - d^4, \quad (4)$$

which is of order $\Delta_1 = 1$ in $16K^2$. Robbins (1995) gives the corresponding FORMULAS for the CYCLIC PENTAGON and CYCLIC HEXAGON.

see also CONCYCLIC, CYCLIC HEXAGON, CYCLIC PENTAGON, CYCLIC QUADRANGLE, CYCLIC QUADRILATERAL

References

Robbins, D. P. "Areas of Polygons Inscribed in a Circle." *Discr. Comput. Geom.* **12**, 223–236, 1994.

Robbins, D. P. "Areas of Polygons Inscribed in a Circle." *Amer. Math. Monthly* **102**, 523–530, 1995.

Sloane, N. J. A. Sequence A000531 in "An On-Line Version of the Encyclopedia of Integer Sequences."

Cyclic Quadrangle

Let A_1, A_2, A_3, and A_4 be four POINTS on a CIRCLE, and H_1, H_2, H_3, H_4 the ORTHOCENTERS of TRIANGLES $\Delta A_2A_3A_4$, etc. If, from the eight POINTS, four with different subscripts are chosen such that three are from one set and the fourth from the other, these POINTS form an ORTHOCENTRIC SYSTEM. There are eight such systems, which are analogous to the six sets of ORTHOCENTRIC SYSTEMS obtained using the feet of the ANGLE BISECTORS, ORTHOCENTER, and VERTICES of a generic TRIANGLE.

On the other hand, if all the POINTS are chosen from one set, or two from each set, with all different subscripts, the four POINTS lie on a CIRCLE. There are four pairs of such CIRCLES, and eight POINTS lie by fours on eight equal CIRCLES.

The SIMSON LINE of A_4 with regard to TRIANGLE $\Delta A_1A_2A_3$ is the same as that of H_4 with regard to the TRIANGLE $\Delta H_1A_2A_3$.

see also ANGLE BISECTOR, CONCYCLIC, CYCLIC POLYGON, CYCLIC QUADRILATERAL, ORTHOCENTRIC SYSTEM

References

Johnson, R. A. *Modern Geometry: An Elementary Treatise on the Geometry of the Triangle and the Circle.* Boston, MA: Houghton Mifflin, pp. 251–253, 1929.

Cyclic Quadrilateral

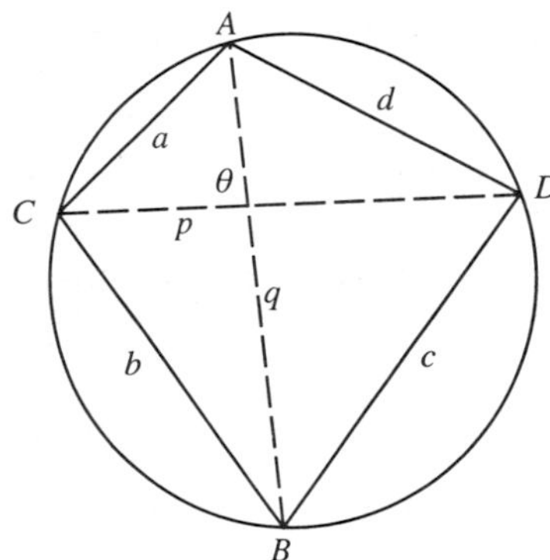

A QUADRILATERAL for which a CIRCLE can be circumscribed so that it touches each VERTEX. The AREA is then given by a special case of BRETSCHNEIDER'S FORMULA. Let the sides have lengths a, b, c, and d, let s be the SEMIPERIMETER

$$s \equiv \tfrac{1}{2}(a+b+c+d), \tag{1}$$

and let R be the CIRCUMRADIUS. Then

$$A = \sqrt{(s-a)(s-b)(s-c)(s-d)} \tag{2}$$

$$= \frac{\sqrt{(ac+bd)(ad+bc)(ab+cd)}}{4R}. \tag{3}$$

Solving for the CIRCUMRADIUS gives

$$R = \tfrac{1}{4}\sqrt{\frac{(ac+bd)(ad+bc)(ab+cd)}{(s-a)(s-b)(s-c)(s-d)}}. \tag{4}$$

The DIAGONALS of a cyclic quadrilateral have lengths

$$p = \sqrt{\frac{(ab+cd)(ac+bd)}{ad+bc}} \tag{5}$$

$$q = \sqrt{\frac{(ac+bd)(ad+bc)}{ab+cd}}, \tag{6}$$

so that $pq = ac+bd$. In general, there are three essentially distinct cyclic quadrilaterals (modulo ROTATION and REFLECTION) whose edges are permutations of the lengths a, b, c, and d. Of the six corresponding DIAGONAL lengths, three are distinct. In addition to p and q, there is therefore a "third" DIAGONAL which can be denoted r. It is given by the equation

$$r = \sqrt{\frac{(ad+bc)(ab+cd)}{ac+bd}}. \tag{7}$$

This allows the AREA formula to be written in the particularly beautiful and simple form

$$A = \frac{pqr}{4R}. \tag{8}$$

The DIAGONALS are sometimes also denoted p, q, and r.

The AREA of a cyclic quadrilateral is the MAXIMUM possible for any QUADRILATERAL with the given side lengths. Also, the opposite ANGLES of a cyclic quadrilateral sum to π RADIANS (Dunham 1990).

A cyclic quadrilateral with RATIONAL sides a, b, c, and d, DIAGONALS p and q, CIRCUMRADIUS R, and AREA A is given by $a = 25$, $b = 33$, $c = 39$, $d = 65$, $p = 60$, $q = 52$, $R = 65/2$, and $A = 1344$.

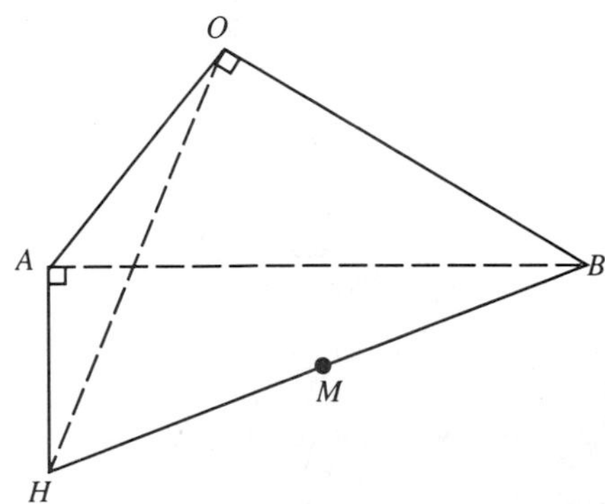

Let $AHBO$ be a QUADRILATERAL such that the angles $\angle HAB$ and $\angle HOB$ are RIGHT ANGLES, then $AHBO$ is a cyclic quadrilateral (Dunham 1990). This is a COROLLARY of the theorem that, in a RIGHT TRIANGLE, the MIDPOINT of the HYPOTENUSE is equidistant from the three VERTICES. Since M is the MIDPOINT of both RIGHT TRIANGLES ΔAHB and ΔBOH, it is equidistant from all four VERTICES, so a CIRCLE centered at M may be drawn through them. This theorem is one of the building blocks of Heron's derivation of HERON'S FORMULA.

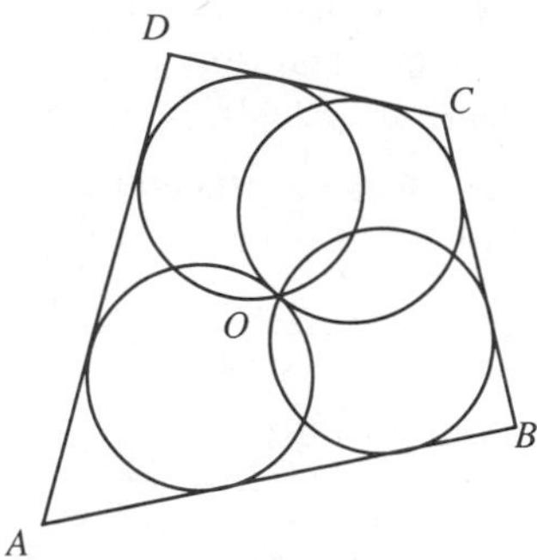

Place four equal CIRCLES so that they intersect in a point. The quadrilateral $ABCD$ is then a cyclic quadrilateral (Honsberger 1991). For a CONVEX cyclic quadrilateral Q, consider the set of CONVEX cyclic quadrilaterals $Q_{||}$ whose sides are PARALLEL to Q. Then the $Q_{||}$ of maximal AREA is the one whose DIAGONALS are PERPENDICULAR (Gürel 1996).

see also BRETSCHNEIDER'S FORMULA, CONCYCLIC, CYCLIC POLYGON, CYCLIC QUADRANGLE, EULER BRICK, HERON'S FORMULA, PTOLEMY'S THEOREM, QUADRILATERAL

References

Beyer, W. H. *CRC Standard Mathematical Tables, 28th ed.* Boca Raton, FL: CRC Press, p. 123, 1987.

Dunham, W. *Journey Through Genius: The Great Theorems of Mathematics.* New York: Wiley, p. 121, 1990.

Gürel, E. Solution to Problem 1472. "Maximal Area of Quadrilaterals." *Math. Mag.* **69**, 149, 1996.

Honsberger, R. *More Mathematical Morsels.* Washington, DC: Math. Assoc. Amer., pp. 36–37, 1991.

Cyclic Redundancy Check

A sophisticated CHECKSUM (often abbreviated CRC), which is based on the algebra of polynomials over the integers (mod 2). It is substantially more reliable in detecting transmission errors, and is one common error-checking protocol used in modems.

see also CHECKSUM, ERROR-CORRECTING CODE

References

Press, W. H.; Flannery, B. P.; Teukolsky, S. A.; and Vetterling, W. T. "Cyclic Redundancy and Other Checksums." Ch. 20.3 in *Numerical Recipes in FORTRAN: The Art of Scientific Computing, 2nd ed.* Cambridge, England: Cambridge University Press, pp. 888–895, 1992.

Cyclid

see CYCLIDE

Cyclide

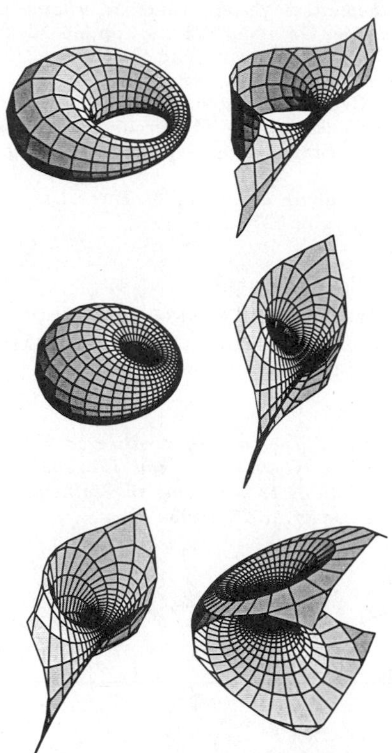

A pair of focal conics which are the envelopes of two one-parameter families of spheres, sometimes also called a CYCLID. The cyclide is a QUARTIC SURFACE, and the lines of curvature on a cyclide are all straight lines or circular arcs (Pinkall 1986). The STANDARD TORI and their inversions in a SPHERE S centered at a point $\mathbf{x}_0$ and of RADIUS r, given by

$$I(\mathbf{x}_0, r) = \mathbf{x}_0 + \frac{\mathbf{x} - \mathbf{x}_0 r^2}{|\mathbf{x} - \mathbf{x}_0|^2},$$

are both cyclides (Pinkall 1986). Illustrated above are RING CYCLIDES, HORN CYCLIDES, and SPINDLE CYCLIDES. The figures on the right correspond to $\mathbf{x}_0$ lying on the torus itself, and are called the PARABOLIC RING CYCLIDE, PARABOLIC HORN CYCLIDE, and PARABOLIC SPINDLE CYCLIDE, respectively.

see also CYCLIDIC COORDINATES, HORN CYCLIDE, PARABOLIC HORN CYCLIDE, PARABOLIC RING CYCLIDE, RING CYCLIDE, SPINDLE CYCLIDE, STANDARD TORI

References

Bierschneider-Jakobs, A. "Cyclides." `http://www.mi.uni-erlangen.de/~biersch/cyclides.html`.

Byerly, W. E. *An Elementary Treatise on Fourier's Series, and Spherical, Cylindrical, and Ellipsoidal Harmonics, with Applications to Problems in Mathematical Physics.* New York: Dover, p. 273, 1959.

Eisenhart, L. P. "Cyclides of Dupin." §133 in *A Treatise on the Differential Geometry of Curves and Surfaces.* New York: Dover, pp. 312–314, 1960.

Fischer, G. (Ed.). Plates 71–77 in *Mathematische Modelle/Mathematical Models, Bildband/Photograph Volume.* Braunschweig, Germany: Vieweg, pp. 66–72, 1986.
Nordstrand, T. "Dupin Cyclide." `http://www.uib.no/people/nfytn/dupintxt.htm`.
Pinkall, U. "Cyclides of Dupin." §3.3 in *Mathematical Models from the Collections of Universities and Museums* (Ed. G. Fischer). Braunschweig, Germany: Vieweg, pp. 28–30, 1986.
Salmon, G. *Analytic Geometry of Three Dimensions.* New York: Chelsea, p. 527, 1979.

Cyclidic Coordinates

A general system of CURVILINEAR COORDINATES based on the CYCLIDE in which LAPLACE'S EQUATION is SEPARABLE.

References
Byerly, W. E. *An Elementary Treatise on Fourier's Series, and Spherical, Cylindrical, and Ellipsoidal Harmonics, with Applications to Problems in Mathematical Physics.* New York: Dover, p. 273, 1959.

Cycloid

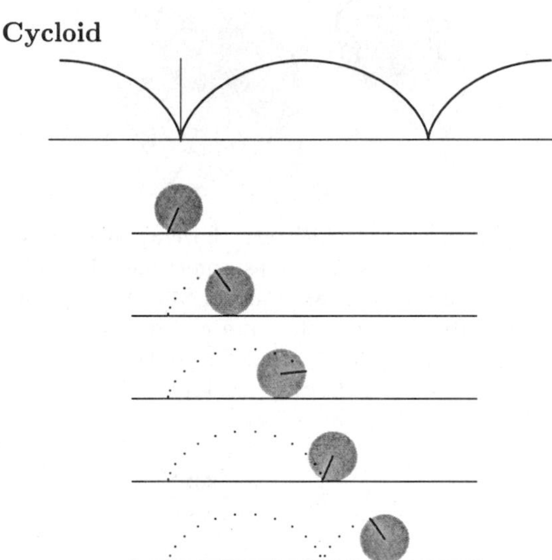

The cycloid is the locus of a point on the rim of a CIRCLE of RADIUS a rolling along a straight LINE. It was studied and named by Galileo in 1599. Galileo attempted to find the AREA by weighing pieces of metal cut into the shape of the cycloid. Torricelli, Fermat, and Descartes all found the AREA. The cycloid was also studied by Roberval in 1634, Wren in 1658, Huygens in 1673, and Johann Bernoulli in 1696. Roberval and Wren found the ARC LENGTH (MacTutor Archive). Gear teeth were also made out of cycloids, as first proposed by Desargues in the 1630s (Cundy and Rollett 1989).

In 1696, Johann Bernoulli challenged other mathematicians to find the curve which solves the BRACHISTOCHRONE PROBLEM, knowing the solution to be a cycloid. Leibniz, Newton, Jakob Bernoulli and L'Hospital all solved Bernoulli's challenge. The cycloid also solves the TAUTOCHRONE PROBLEM. Because of the frequency with which it provoked quarrels among mathematicians in the 17th century, the cycloid became known as the "Helen of Geometers" (Boyer 1968, p. 389).

The cycloid is the CATACAUSTIC of a CIRCLE for a RADIANT POINT on the circumference, as shown by Jakob and Johann Bernoulli in 1692. The CAUSTIC of the cycloid when the rays are parallel to the y-axis is a cycloid with twice as many arches. The RADIAL CURVE of a CYCLOID is a CIRCLE. The EVOLUTE and INVOLUTE of a cycloid are identical cycloids.

If the cycloid has a CUSP at the ORIGIN, its equation in CARTESIAN COORDINATES is

$$x = a\cos^{-1}\left(\frac{a-y}{a}\right) \mp \sqrt{2ay-y^2}. \tag{1}$$

In parametric form, this becomes

$$x = a(t-\sin t) \tag{2}$$
$$y = a(1-\cos t). \tag{3}$$

If the cycloid is upside-down with a cusp at $(0, a)$, (2) and (3) become

$$x = 2a\sin^{-1}\left(\frac{y}{2a}\right) + \sqrt{2ay-y^2} \tag{4}$$

or

$$x = a(t+\sin t) \tag{5}$$
$$y = a(1-\cos t) \tag{6}$$

(sign of $\sin t$ flipped for x).

The DERIVATIVES of the parametric representation (2) and (3) are

$$x' = a(1-\cos t) \tag{7}$$
$$y' = a\sin t \tag{8}$$

$$\begin{aligned}\frac{dy}{dx} &= \frac{y'}{x'} = \frac{a\sin t}{a(1-\cos t)} = \frac{\sin t}{1-\cos t}\\ &= \frac{2\sin(\frac{1}{2}t)\cos(\frac{1}{2}t)}{2\sin^2(\frac{1}{2}t)} = \cot(\tfrac{1}{2}t).\end{aligned} \tag{9}$$

The squares of the derivatives are

$$x'^2 = a^2(1-2\cos t+\cos^2 t) \tag{10}$$
$$y'^2 = a^2\sin^2 t, \tag{11}$$

so the ARC LENGTH of a single cycle is

$$\begin{aligned}L &= \int ds = \int_0^{2\pi} \sqrt{x'^2+y'^2}\,dt\\ &= a\int_0^{2\pi}\sqrt{(1-2\cos t+\cos^2 t)+\sin^2 t}\,dt\\ &= a\sqrt{2}\int_0^{2\pi}\sqrt{1-\cos t}\,dt = 2a\int_0^{2\pi}\sqrt{\frac{1-\cos t}{2}}\,dt\\ &= 2a\int_0^{2\pi}\left|\sin(\tfrac{1}{2}t)\right|\,dt.\end{aligned} \tag{12}$$

Now let $u \equiv t/2$ so $du = dt/2$. Then

$$\begin{aligned} L &= 4a \int_0^\pi \sin u\, du = 4a[-\cos u]_0^\pi \\ &= -4a[(-1) - 1] = 8a. \end{aligned} \tag{13}$$

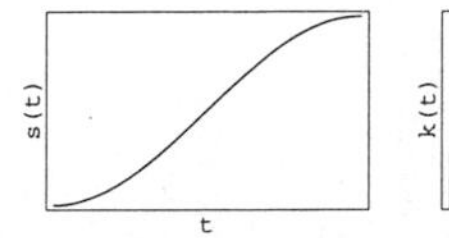

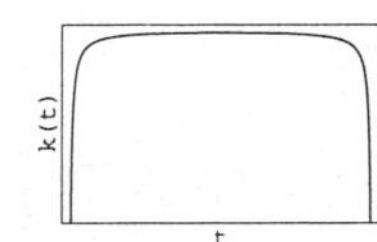

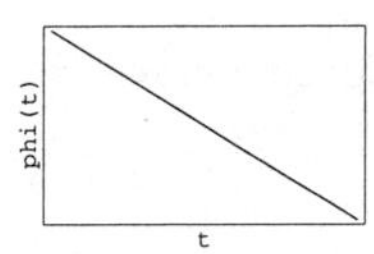

The ARC LENGTH, CURVATURE, and TANGENTIAL ANGLE are

$$s = 8a \sin^2(\tfrac{1}{4}t) \tag{14}$$
$$\kappa = -\tfrac{1}{4}a \csc(\tfrac{1}{2}t) \tag{15}$$
$$\phi = -\tfrac{1}{2}at. \tag{16}$$

The AREA under a single cycle is

$$\begin{aligned} A &= \int_0^{2\pi} y\,dx = a^2 \int_0^{2\pi} (1 - \cos\phi)(1 - \cos\phi)\, d\phi \\ &= a^2 \int_0^{2\pi} (1 - \cos\phi)^2\, d\phi \\ &= a^2 \int_0^{2\pi} (1 - 2\cos\phi + \cos^2\phi)\, d\phi \\ &= a^2 \int_0^{2\pi} \{1 - 2\cos\phi + \tfrac{1}{2}[1 + \cos(2\phi)]\}\, d\phi \\ &= a^2 \int_0^{2\pi} [\tfrac{3}{2} - 2\cos\phi + \tfrac{1}{2}\cos(2\phi)]\, d\phi \\ &= a^2[\tfrac{3}{2}\phi - 2\sin\phi + \tfrac{1}{4}\sin(2\phi)]_0^{2\pi} \\ &= a^2 \tfrac{3}{2} 2\pi = 3\pi a^2. \end{aligned} \tag{17}$$

The NORMAL is

$$\hat{\mathbf{T}} = \frac{1}{\sqrt{2 - 2\cos t}} \begin{bmatrix} 1 - \cos t \\ \sin t \end{bmatrix}. \tag{18}$$

see also CURTATE CYCLOID, CYCLIDE, CYCLOID EVOLUTE, CYCLOID INVOLUTE, EPICYCLOID, HYPOCYCLOID, PROLATE CYCLOID, TROCHOID

References

Bogomolny, A. "Cycloids." http://www.cut-the-knot.com/pythagoras/cycloids.html.

Boyer, C. B. *A History of Mathematics.* New York: Wiley, 1968.

Cundy, H. and Rollett, A. *Mathematical Models, 3rd ed.* Stradbroke, England: Tarquin Pub., 1989.

Gray, A. "Cycloids." §3.1 in *Modern Differential Geometry of Curves and Surfaces.* Boca Raton, FL: CRC Press, pp. 37–39, 1993.

Lawrence, J. D. *A Catalog of Special Plane Curves.* New York: Dover, pp. 192 and 197, 1972.

Lee, X. "Cycloid." http://www.best.com/~xah/SpecialPlaneCurves_dir/Cycloid_dir/cycloid.html.

Lockwood, E. H. "The Cycloid." Ch. 9 in *A Book of Curves.* Cambridge, England: Cambridge University Press, pp. 80–89, 1967.

MacTutor History of Mathematics Archive. "Cycloid." http://www-groups.dcs.st-and.ac.uk/~history/Curves/Cycloid.html.

Muterspaugh, J.; Driver, T.; and Dick, J. E. "The Cycloid and Tautochronism." http://ezinfo.ucs.indiana.edu/~jedick/project/intro.html.

Pappas, T. "The Cycloid—The Helen of Geometry." *The Joy of Mathematics.* San Carlos, CA: Wide World Publ./Tetra, pp. 6–8, 1989.

Wagon, S. "Rolling Circles." Ch. 2 in *Mathematica in Action.* New York: W. H. Freeman, pp. 39–66, 1991.

Yates, R. C. "Cycloid." *A Handbook on Curves and Their Properties.* Ann Arbor, MI: J. W. Edwards, pp. 65–70, 1952.

Cycloid Evolute

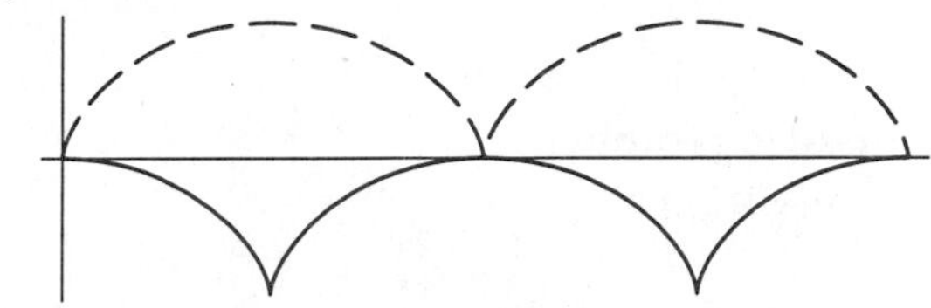

The EVOLUTE of the CYCLOID

$$x(t) = a(t - \sin t)$$
$$y(t) = a(1 - \cos t)$$

is given by

$$x(t) = a(t + \sin t)$$
$$y(t) = a(\cos t - 1).$$

As can be seen in the above figure, the EVOLUTE is simply a shifted copy of the original CYCLOID, so the CYCLOID is its own EVOLUTE.

Cycloid Involute

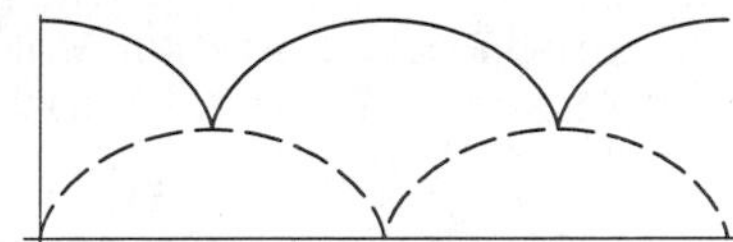

The INVOLUTE of the CYCLOID

$$x(t) = a(t - \sin t)$$
$$y(t) = a(1 - \cos t)$$

is given by

$$x(t) = a(t + \sin t)$$
$$y(t) = a(3 + \cos t).$$

As can be seen in the above figure, the INVOLUTE is simply a shifted copy of the original CYCLOID, so the CYCLOID is its own INVOLUTE!

Cycloid Radial Curve

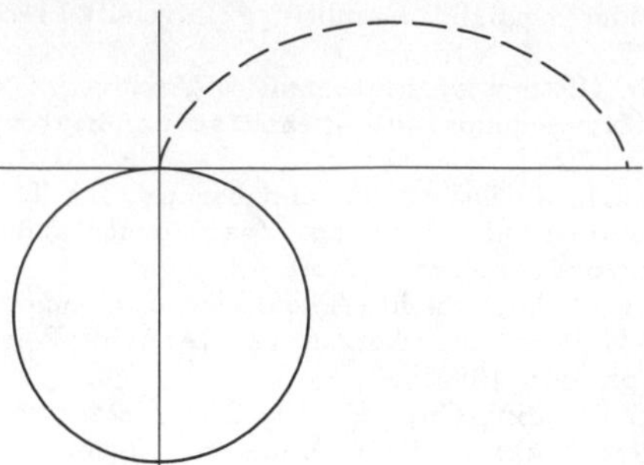

The RADIAL CURVE of the CYCLOID is the CIRCLE

$$x = x_0 + 2a\sin\phi$$
$$y = -2a + y_0 + 2a\cos\phi.$$

Cyclomatic Number

see CIRCUIT RANK

Cyclotomic Equation

The equation

$$x^p = 1,$$

where solutions $\zeta_k = e^{2\pi ik/p}$ are the ROOTS OF UNITY sometimes called DE MOIVRE NUMBERS. Gauss showed that the cyclotomic equation can be reduced to solving a series of QUADRATIC EQUATIONS whenever p is a FERMAT PRIME. Wantzel (1836) subsequently showed that this condition is not only SUFFICIENT, but also NECESSARY. An "irreducible" cyclotomic equation is an expression of the form

$$\frac{x^p - 1}{x - 1} = x^{p-1} + x^{p-2} + \ldots + 1 = 0,$$

where p is PRIME. Its ROOTS z_i satisfy $|z_i| = 1$.

see also CYCLOTOMIC POLYNOMIAL, DE MOIVRE NUMBER, POLYGON, PRIMITIVE ROOT OF UNITY

References

Courant, R. and Robbins, H. *What is Mathematics?: An Elementary Approach to Ideas and Methods, 2nd ed.* Oxford, England: Oxford University Press, pp. 99–100, 1996.

Scott, C. A. "The Binomial Equation $x^p - 1 = 0$." *Amer. J. Math.* **8**, 261–264, 1886.

Wantzel, P. L. "Recherches sur les moyens de reconnaître si un Problème de Géométrie peut se résoudre avec la règle et le compas." *J. Math. pures appliq.* **1**, 366–372, 1836.

Cyclotomic Factorization

$$z^p - y^p = (z - y)(z - \zeta y)\cdots(z - \zeta^{p-1}y),$$

where $\zeta \equiv e^{2\pi i/p}$ (a DE MOIVRE NUMBER) and p is a PRIME.

Cyclotomic Field

The smallest field containing $m \in \mathbb{Z} \geq 1$ with ζ a PRIME ROOT OF UNITY is denoted $\mathbb{R}_m(\zeta)$.

$$x^p + y^p = \prod_{k=1}^{p}(x + \zeta^k y).$$

Specific cases are

$$\mathbb{R}_3 = \mathbb{Q}(\sqrt{-3})$$
$$\mathbb{R}_4 = \mathbb{Q}(\sqrt{-1})$$
$$\mathbb{R}_6 = \mathbb{Q}(\sqrt{-3}),$$

where $\mathbb{Q}$ denotes a QUADRATIC FIELD.

Cyclotomic Integer

A number of the form

$$a_0 + a_1\zeta + \ldots + a_{p-1}\zeta^{p-1},$$

where

$$\zeta \equiv e^{2\pi i/p}$$

is a DE MOIVRE NUMBER and p is a PRIME number. Unique factorizations of cyclotomic INTEGERS fail for $p > 23$.

Cyclotomic Invariant

Let p be an ODD PRIME and F_n the CYCLOTOMIC FIELD of p^{n+1}th ROOTS of unity over the rational FIELD. Now let $p^{e(n)}$ be the POWER of p which divides the CLASS NUMBER h_n of F_n. Then there exist INTEGERS $\mu_p, \lambda_p \geq 0$ and ν_p such that

$$e(n) = \mu_p p^n + \lambda_p n + \nu_p$$

for all sufficiently large n. For REGULAR PRIMES, $\mu_p = \lambda_p = \nu_p = 0$.

References

Johnson, W. "Irregular Primes and Cyclotomic Invariants." *Math. Comput.* **29**, 113–120, 1975.

Cyclotomic Number

see DE MOIVRE NUMBER, SYLVESTER CYCLOTOMIC NUMBER

Cyclotomic Polynomial

A polynomial given by

$$\Phi_d(x) = \prod_{k=1}^{d}(x - \zeta_k), \qquad (1)$$

where ζ_i are the primitive dth ROOTS OF UNITY in $\mathbb{C}$ given by $\zeta_k = e^{2\pi ik/d}$. The numbers ζ_k are sometimes called DE MOIVRE NUMBERS. $\Phi_d(x)$ is an irreducible

POLYNOMIAL in $\mathbb{Z}[x]$ with degree $\phi(d)$, where ϕ is the TOTIENT FUNCTION. For d PRIME,

$$\Phi_p = \sum_{k=0}^{p-1} x^k, \tag{2}$$

i.e., the coefficients are all 1. Φ_{105} has coefficients of -2 for x^7 and x^{41}, making it the first cyclotomic polynomial to have a coefficient other than ± 1 and 0. This is true because 105 is the first number to have three distinct ODD PRIME factors, i.e., $105 = 3 \cdot 5 \cdot 7$ (McClellan and Rader 1979, Schroeder 1997). Migotti (1883) showed that COEFFICIENTS of Φ_{pq} for p and q distinct PRIMES can be only 0, ± 1. Lam and Leung (1996) considered

$$\Phi_{pq} \equiv \sum_{k=0}^{pq-1} a_k x^k \tag{3}$$

for p, q PRIME. Write the TOTIENT FUNCTION as

$$\phi(pq) = (p-1)(q-1) = rp + sq \tag{4}$$

and let

$$0 \leq k \leq (p-1)(q-1), \tag{5}$$

then

1. $a_k = 1$ IFF $k = ip + jq$ for some $i \in [0, r]$ and $j \in [0, s]$,
2. $a_k = -1$ IFF $k + pq = ip + jq$ for $i \in [r+1, q-1]$ and $j \in [s+1, p-1]$,
3. otherwise $a_k = 0$.

The number of terms having $a_k = 1$ is $(r+1)(s+1)$, and the number of terms having $a_k = -1$ is $(p-s-1)(q-r-1)$. Furthermore, assume $q > p$, then the middle COEFFICIENT of Φ_{pq} is $(-1)^r$.

The LOGARITHM of the cyclotomic polynomial

$$\Phi_n(x) = \prod_{d|n} (1 - x^{n/d})^{\mu(d)} \tag{6}$$

is the MÖBIUS INVERSION FORMULA (Vardi 1991, p. 225).

The first few cyclotomic POLYNOMIALS are

$$\begin{aligned}
\Phi_1(x) &= x - 1 \\
\Phi_2(x) &= x + 1 \\
\Phi_3(x) &= x^2 + x + 1 \\
\Phi_4(x) &= x^2 + 1 \\
\Phi_5(x) &= x^4 + x^3 + x^2 + x + 1 \\
\Phi_6(x) &= x^2 - x + 1 \\
\Phi_7(x) &= x^6 + x^5 + x^4 + x^3 + x^2 + x + 1 \\
\Phi_8(x) &= x^4 + 1 \\
\Phi_9(x) &= x^6 + x^3 + 1 \\
\Phi_{10}(x) &= x^4 - x^3 + x^2 - x + 1.
\end{aligned}$$

The smallest values of n for which Φ_n has one or more coefficients ± 1, ± 2, ± 3, ... are 0, 105, 385, 1365, 1785, 2805, 3135, 6545, 6545, 10465, 10465, 10465, 10465, 10465, 11305, ... (Sloane's A013594).

The POLYNOMIAL $x^n - 1$ can be factored as

$$x^n - 1 = \prod_{d|n} \Phi_d(x), \tag{7}$$

where $\Phi_d(x)$ is a CYCLOTOMIC POLYNOMIAL. Furthermore,

$$x^n + 1 = \frac{x^{2n-1}}{x^n - 1} = \frac{\prod_{d|2n} \Phi_d(x)}{\prod_{d|n} \Phi_d(x)} = \prod_{d|m} \Phi_{2^t d}(x). \tag{8}$$

The COEFFICIENTS of the inverse of the cyclotomic POLYNOMIAL

$$\begin{aligned}
\frac{1}{1+x+x^2} &= 1 - x + x^3 - x^4 + x^6 - x^7 + x^9 - x^{10} + \ldots \\
&\equiv \sum_{n=0}^{\infty} c_n x^n
\end{aligned} \tag{9}$$

can also be computed from

$$c_n = 1 - 2\left\lfloor \tfrac{1}{3}(n+2) \right\rfloor + \left\lfloor \tfrac{1}{3}(n+1) \right\rfloor + \left\lfloor \tfrac{1}{3}n \right\rfloor, \tag{10}$$

where $\lfloor x \rfloor$ is the FLOOR FUNCTION.

see also AURIFEUILLEAN FACTORIZATION, MÖBIUS INVERSION FORMULA

References

Beiter, M. "The Midterm Coefficient of the Cyclotomic Polynomial $F_{pq}(x)$." *Amer. Math. Monthly* **71**, 769–770, 1964.

Beiter, M. "Magnitude of the Coefficients of the Cyclotomic Polynomial F_{pq}." *Amer. Math. Monthly* **75**, 370–372, 1968.

Bloom, D. M. "On the Coefficients of the Cyclotomic Polynomials." *Amer. Math. Monthly* **75**, 372–377, 1968.

Carlitz, L. "The Number of Terms in the Cyclotomic Polynomial $F_{pq}(x)$." *Amer. Math. Monthly* **73**, 979–981, 1966.

Conway, J. H. and Guy, R. K. *The Book of Numbers.* New York: Springer-Verlag, 1996.

de Bruijn, N. G. "On the Factorization of Cyclic Groups." *Indag. Math.* **15**, 370–377, 1953.

Lam, T. Y. and Leung, K. H. "On the Cyclotomic Polynomial $\Phi_{pq}(X)$." *Amer. Math. Monthly* **103**, 562–564, 1996.

Lehmer, E. "On the Magnitude of Coefficients of the Cyclotomic Polynomials." *Bull. Amer. Math. Soc.* **42**, 389–392, 1936.

McClellan, J. H. and Rader, C. *Number Theory in Digital Signal Processing.* Englewood Cliffs, NJ: Prentice-Hall, 1979.

Migotti, A. "Zur Theorie der Kreisteilungsgleichung." *Sitzber. Math.-Naturwiss. Classe der Kaiser. Akad. der Wiss., Wien* **87**, 7–14, 1883.

Schroeder, M. R. *Number Theory in Science and Communication, with Applications in Cryptography, Physics, Digital Information, Computing, and Self-Similarity, 3rd ed.* New York: Springer-Verlag, p. 245, 1997.

Sloane, N. J. A. Sequence A013594 in "An On-Line Version of the Encyclopedia of Integer Sequences."

Vardi, I. *Computational Recreations in Mathematica.* Redwood City, CA: Addison-Wesley, pp. 8 and 224–225, 1991.

Cylinder

A cylinder is a solid of circular CROSS-SECTION in which the centers of the CIRCLES all lie on a single LINE. The cylinder was extensively studied by Archimedes in his 2-volume work *On the Sphere and Cylinder* in ca. 225 BC.

A cylinder is called a right cylinder if it is "straight" in the sense that its cross-sections lie directly on top of each other; otherwise, the cylinder is called oblique. The surface of a cylinder of height h and RADIUS r can be described parametrically by

$$x = r\cos\theta \quad (1)$$
$$y = r\sin\theta \quad (2)$$
$$z = z, \quad (3)$$

for $z \in [0, h]$ and $\theta \in [0, 2\pi)$. These are the basis for CYLINDRICAL COORDINATES. The SURFACE AREA (of the sides) and VOLUME of the cylinder of height h and RADIUS r are

$$S = 2\pi rh \quad (4)$$
$$V = \pi r^2 h. \quad (5)$$

Therefore, if top and bottom caps are added, the volume-to-surface area ratio for a cylindrical container is

$$\frac{V}{S} = \frac{\pi r^2 h}{2\pi rh + 2\pi r^2} = \frac{1}{2}\left(\frac{1}{r} + \frac{1}{h}\right)^{-1}, \quad (6)$$

which is related to the HARMONIC MEAN of the radius r and height h.

see also CONE, CYLINDER-SPHERE INTERSECTION, CYLINDRICAL SEGMENT, ELLIPTIC CYLINDER, GENERALIZED CYLINDER, SPHERE, STEINMETZ SOLID, VIVIANI'S CURVE

References

Beyer, W. H. (Ed.) *CRC Standard Mathematical Tables, 28th ed.* Boca Raton, FL: CRC Press, p. 129, 1987.

Cylinder Cutting

The maximum number of pieces into which a cylinder can be divided by n oblique cuts is given by

$$f(n) = \binom{n+1}{3} + n + 1 = \tfrac{1}{6}(n+2)(n+3),$$

where $\binom{a}{b}$ is a BINOMIAL COEFFICIENT. This problem is sometimes also called CAKE CUTTING or PIE CUTTING. For $n = 1, 2, \ldots$ cuts, the maximum number of pieces is 2, 4, 8, 15, 26, 42, ... (Sloane's A000125).

see also CIRCLE CUTTING, HAM SANDWICH THEOREM, PANCAKE THEOREM, TORUS CUTTING

References

Bogomolny, A. "Can You Cut a Cake into 8 Pieces with Three Movements." http://www.cut-the-knot.com/do_you_know/cake.html.

Sloane, N. J. A. Sequence A000125/M1100 in "An On-Line Version of the Encyclopedia of Integer Sequences."

Cylinder-Cylinder Intersection

see STEINMETZ SOLID

Cylinder Function

The cylinder function is defined as

$$C(x,y) \equiv \begin{cases} 1 & \text{for } \sqrt{x^2+y^2} \leq a \\ 0 & \text{for } \sqrt{x^2+y^2} > a. \end{cases} \quad (1)$$

The BESSEL FUNCTIONS are sometimes also called cylinder functions. To find the FOURIER TRANSFORM of the cylinder function, let

$$k_x = k\cos\alpha \quad (2)$$
$$k_y = k\sin\alpha \quad (3)$$

$$x = r\cos\theta \quad (4)$$
$$y = r\sin\theta. \quad (5)$$

Then

$$\begin{aligned} F(k,a) &= \mathcal{F}(C(x,y)) \\ &= \int_0^{2\pi}\int_0^a e^{i(k\cos\alpha r\cos\theta + k\sin\alpha r\sin\theta)} r\,dr\,d\theta \\ &= \int_0^{2\pi}\int_0^a e^{ikr\cos(\theta-\alpha)} r\,dr\,d\theta. \end{aligned} \quad (6)$$

Let $b = \theta - \alpha$, so $db = d\theta$. Then

$$\begin{aligned} F(k,a) &= \int_{-\alpha}^{2\pi-\alpha}\int_0^a e^{ikr\cos b} r\,dr\,d\theta \\ &= \int_0^{2\pi}\int_0^a e^{ikr\cos b} r\,dr\,d\theta \\ &= 2\pi\int_0^a J_0(kr) r\,dr, \end{aligned} \quad (7)$$

where J_0 is a zeroth order BESSEL FUNCTION OF THE FIRST KIND. Let $u \equiv kr$, so $du = k\,dr$, then

$$\begin{aligned} F(k,a) &= \frac{2\pi}{k^2}\int_0^{ka} J_0(u) u\,du = \frac{2\pi}{k^2}[uJ_1(u)]_0^{ka} \\ &= \frac{2\pi a}{k} J_1(ka) = 2\pi a^2 \frac{J_1(ka)}{ka}. \end{aligned} \quad (8)$$

As defined by Watson (1966), a "cylinder function" is any function which satisfies the RECURRENCE RELATIONS

$$\mathcal{C}_{\nu-1}(z) + \mathcal{C}_{\nu+1}(z) = \frac{2\nu}{z}\mathcal{C}_\nu(z) \tag{9}$$

$$\mathcal{C}_{\nu-1}(z) - \mathcal{C}_{\nu+1}(z) = 2\mathcal{C}'_\nu(z). \tag{10}$$

This class of functions can be expressed in terms of BESSEL FUNCTIONS.

see also BESSEL FUNCTION OF THE FIRST KIND, CYLINDER FUNCTION, CYLINDRICAL FUNCTION, HEMISPHERICAL FUNCTION

References

Watson, G. N. *A Treatise on the Theory of Bessel Functions, 2nd ed.* Cambridge, England: Cambridge University Press, 1966.

Cylinder-Sphere Intersection

see VIVIANI'S CURVE

Cylindrical Coordinates

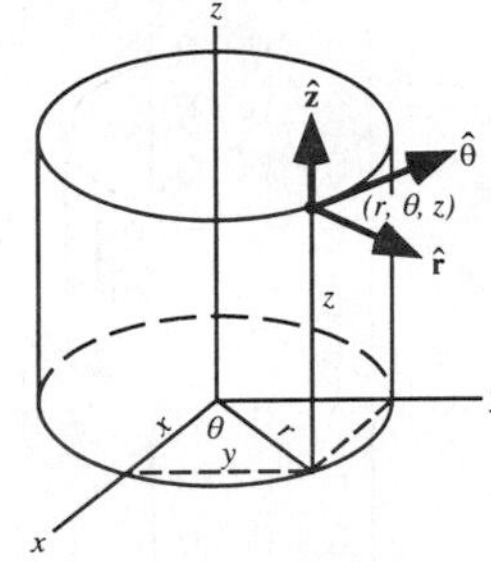

Cylindrical coordinates are a generalization of 2-D POLAR COORDINATES to 3-D by superposing a height (z) axis. Unfortunately, there are a number of different notations used for the other two coordinates. Either r or ρ is used to refer to the radial coordinate and either ϕ or θ to the azimuthal coordinates. Arfken (1985), for instance, uses (ρ, ϕ, z), while Beyer (1987) uses (r, θ, z). In this work, the NOTATION (r, θ, z) is used.

$$r = \sqrt{x^2 + y^2} \tag{1}$$

$$\theta = \tan^{-1}\left(\frac{y}{x}\right) \tag{2}$$

$$z = z, \tag{3}$$

where $r \in [0, \infty)$, $\theta \in [0, 2\pi)$, and $z \in (-\infty, \infty)$. In terms of x, y, and z

$$x = r\cos\theta \tag{4}$$

$$y = r\sin\theta \tag{5}$$

$$z = z. \tag{6}$$

Morse and Feshbach (1953) define the cylindrical coordinates by

$$x = \xi_1\xi_2 \tag{7}$$

$$y = \xi_1\sqrt{1 - {\xi_2}^2} \tag{8}$$

$$z = \xi_3, \tag{9}$$

where $\xi_1 = r$ and $\xi_2 = \cos\theta$. The METRIC elements of the cylindrical coordinates are

$$g_{rr} = 1 \tag{10}$$

$$g_{\theta\theta} = r^2 \tag{11}$$

$$g_{zz} = 1, \tag{12}$$

so the SCALE FACTORS are

$$g_r = 1 \tag{13}$$

$$g_\theta = r \tag{14}$$

$$g_z = 1. \tag{15}$$

The LINE ELEMENT is

$$ds = dr\,\hat{\mathbf{r}} + r\,d\theta\,\hat{\boldsymbol{\theta}} + dz\,\hat{\mathbf{z}}, \tag{16}$$

and the VOLUME ELEMENT is

$$dV = r\,dr\,d\theta\,dz. \tag{17}$$

The JACOBIAN is

$$\left|\frac{\partial(x, y, z)}{\partial(r, \theta, z)}\right| = r. \tag{18}$$

A CARTESIAN VECTOR is given in CYLINDRICAL COORDINATES by

$$\mathbf{r} = \begin{bmatrix} r\cos\theta \\ r\sin\theta \\ z \end{bmatrix}. \tag{19}$$

To find the UNIT VECTORS,

$$\hat{\mathbf{r}} \equiv \frac{\frac{d\mathbf{r}}{dr}}{\left|\frac{d\mathbf{r}}{dr}\right|} = \begin{bmatrix} \cos\theta \\ \sin\theta \\ 0 \end{bmatrix} \tag{20}$$

$$\hat{\boldsymbol{\theta}} \equiv \frac{\frac{d\mathbf{r}}{d\theta}}{\left|\frac{d\mathbf{r}}{d\theta}\right|} = \begin{bmatrix} -\sin\theta \\ \cos\theta \\ 0 \end{bmatrix} \tag{21}$$

$$\hat{\mathbf{z}} \equiv \frac{\frac{d\mathbf{r}}{dz}}{\left|\frac{d\mathbf{r}}{dz}\right|} = \begin{bmatrix} 0 \\ 0 \\ 1 \end{bmatrix}. \tag{22}$$

Derivatives of unit VECTORS with respect to the coordinates are

$$\frac{\partial\hat{\mathbf{r}}}{\partial r} = \mathbf{0} \tag{23}$$

$$\frac{\partial\hat{\mathbf{r}}}{\partial\theta} = \begin{bmatrix} -\sin\theta \\ \cos\theta \\ 0 \end{bmatrix} = \hat{\boldsymbol{\theta}} \tag{24}$$

$$\frac{\partial\hat{\mathbf{r}}}{\partial z} = \mathbf{0} \tag{25}$$

$$\frac{\partial\hat{\boldsymbol{\theta}}}{\partial r} = \mathbf{0} \tag{26}$$

$$\frac{\partial\hat{\boldsymbol{\theta}}}{\partial\theta} = \begin{bmatrix} -\cos\theta \\ -\sin\theta \\ 0 \end{bmatrix} = -\hat{\mathbf{r}} \tag{27}$$

$$\frac{\partial\hat{\boldsymbol{\theta}}}{\partial z} = \mathbf{0} \tag{28}$$

$$\frac{\partial\hat{\mathbf{z}}}{\partial r} = \mathbf{0} \tag{29}$$

$$\frac{\partial\hat{\mathbf{z}}}{\partial\theta} = \mathbf{0} \tag{30}$$

$$\frac{\partial\hat{\mathbf{z}}}{\partial z} = \mathbf{0}. \tag{31}$$

The GRADIENT of a VECTOR FIELD in cylindrical coordinates is given by

$$\nabla \equiv \hat{\mathbf{r}}\frac{\partial}{\partial r} + \hat{\boldsymbol{\theta}}\frac{1}{r}\frac{\partial}{\partial \theta} + \hat{\mathbf{z}}\frac{\partial}{\partial z}, \tag{32}$$

so the GRADIENT components become

$$\nabla_r \hat{\mathbf{r}} = \mathbf{0} \tag{33}$$
$$\nabla_\theta \hat{\mathbf{r}} = \frac{1}{r}\hat{\boldsymbol{\theta}} \tag{34}$$
$$\nabla_z \hat{\mathbf{r}} = \mathbf{0} \tag{35}$$
$$\nabla_r \hat{\boldsymbol{\theta}} = \mathbf{0} \tag{36}$$
$$\nabla_\theta \hat{\boldsymbol{\theta}} = -\frac{1}{r}\hat{\mathbf{r}} \tag{37}$$
$$\nabla_z \hat{\boldsymbol{\theta}} = \mathbf{0} \tag{38}$$
$$\nabla_r \hat{\mathbf{z}} = \mathbf{0} \tag{39}$$
$$\nabla_\theta \hat{\mathbf{z}} = \mathbf{0} \tag{40}$$
$$\nabla_z \hat{\mathbf{z}} = \mathbf{0}. \tag{41}$$

Now, since the CONNECTION COEFFICIENTS are defined by

$$\Gamma^i_{jk} = \hat{\mathbf{x}}_i \cdot (\nabla_k \hat{\mathbf{x}}_j), \tag{42}$$

$$\Gamma^r = \begin{bmatrix} 0 & 0 & 0 \\ 0 & -\frac{1}{r} & 0 \\ 0 & 0 & 0 \end{bmatrix} \tag{43}$$

$$\Gamma^\theta = \begin{bmatrix} 0 & \frac{1}{r} & 0 \\ 0 & 0 & 0 \\ 0 & 0 & 0 \end{bmatrix} \tag{44}$$

$$\Gamma^z = \begin{bmatrix} 0 & 0 & 0 \\ 0 & 0 & 0 \\ 0 & 0 & 0 \end{bmatrix}. \tag{45}$$

The COVARIANT DERIVATIVES, given by

$$A_{j;k} = \frac{1}{g^{kk}}\frac{\partial A_j}{\partial x_k} - \Gamma^i_{jk} A_i, \tag{46}$$

are

$$A_{r;r} = \frac{\partial A_r}{\partial r} - \Gamma^i_{rr} A_i = \frac{\partial A_r}{\partial r} \tag{47}$$

$$A_{r;\theta} = \frac{1}{r}\frac{\partial A_r}{\partial \theta} - \Gamma^i_{r\theta} A_i = \frac{1}{r}\frac{\partial A_\theta}{\partial r} - \Gamma^\theta_{r\theta} A_\theta$$
$$= \frac{1}{r}\frac{\partial A_r}{\partial \theta} - \frac{A_\theta}{r} \tag{48}$$

$$A_{r;z} = \frac{\partial A_r}{\partial z} - \Gamma^i_{rz} A_i = \frac{\partial A_r}{\partial z} \tag{49}$$

$$A_{\theta;r} = \frac{\partial A_\theta}{\partial r}\Gamma^i_{\theta r} A_i = \frac{\partial A_\theta}{\partial r} \tag{50}$$

$$A_{\theta;\theta} = \frac{1}{r}\frac{\partial A_\theta}{\partial \theta} - \Gamma^i_{\theta\theta} A_i = \frac{1}{r}\frac{\partial A_\theta}{\partial \theta} - \Gamma^r_{\theta\theta} A_r$$
$$= \frac{1}{r}\frac{\partial A_\theta}{\partial \theta} + \frac{A_r}{r} \tag{51}$$

$$A_{\theta;z} = \frac{\partial A_\theta}{\partial z} - \Gamma^i_{\theta z} A_i = \frac{\partial A_\theta}{\partial z} \tag{52}$$

$$A_{z;r} = \frac{\partial A_z}{\partial r} - \Gamma^i_{zr} A_i = \frac{\partial A_z}{\partial r} \tag{53}$$

$$A_{z;\theta} = \frac{1}{r}\frac{\partial A_z}{\partial \theta} - \Gamma^i_{z\theta} A_i = \frac{1}{r}\frac{\partial A_z}{\partial \theta} \tag{54}$$

$$A_{z;z} = \frac{\partial A_z}{\partial z} - \Gamma^i_{zz} A_i = \frac{\partial A_z}{\partial z}. \tag{55}$$

CROSS PRODUCTS of the coordinate axes are

$$\hat{\mathbf{r}} \times \hat{\mathbf{z}} = -\hat{\boldsymbol{\theta}} \tag{56}$$
$$\hat{\boldsymbol{\theta}} \times \hat{\mathbf{z}} = \hat{\mathbf{r}} \tag{57}$$
$$\hat{\mathbf{r}} \times \hat{\boldsymbol{\theta}} = \hat{\mathbf{z}}. \tag{58}$$

The COMMUTATION COEFFICIENTS are given by

$$c_{\alpha\beta}^{\mu} \vec{e}_\mu = [\vec{e}_\alpha, \vec{e}_\beta] = \nabla_\alpha \vec{e}_\beta - \nabla_\beta \vec{e}_\alpha, \tag{59}$$

But

$$[\hat{\mathbf{r}}, \hat{\mathbf{r}}] = [\hat{\boldsymbol{\theta}}, \hat{\boldsymbol{\theta}}] = [\hat{\boldsymbol{\phi}}, \hat{\boldsymbol{\phi}}] = \mathbf{0}, \tag{60}$$

so $c_{rr}^{\alpha} = c_{\theta\theta}^{\alpha} = c_{\phi\phi}^{\alpha} = 0$, where $\alpha = r, \theta, \phi$. Also

$$[\hat{\mathbf{r}}, \hat{\boldsymbol{\theta}}] = -[\hat{\boldsymbol{\theta}}, \hat{\mathbf{r}}] = \nabla_r \hat{\boldsymbol{\theta}} - \nabla_\theta \hat{\mathbf{r}} = 0 - \frac{1}{r}\hat{\boldsymbol{\theta}} = -\frac{1}{r}\hat{\boldsymbol{\theta}}, \tag{61}$$

so $c_{r\theta}^{\theta} = -c_{\theta r}^{\theta} = -\frac{1}{r}$, $c_{r\theta}^{r} = c_{r\theta}^{\phi} = 0$. Finally,

$$[\hat{\mathbf{r}}, \hat{\boldsymbol{\phi}}] = [\hat{\boldsymbol{\theta}}, \hat{\boldsymbol{\phi}}] = 0. \tag{62}$$

Summarizing,

$$c^r = \begin{bmatrix} 0 & 0 & 0 \\ 0 & 0 & 0 \\ 0 & 0 & 0 \end{bmatrix} \tag{63}$$

$$c^\theta = \begin{bmatrix} 0 & -\frac{1}{r} & 0 \\ \frac{1}{r} & 0 & 0 \\ 0 & 0 & 0 \end{bmatrix} \tag{64}$$

$$c^\phi = \begin{bmatrix} 0 & 0 & 0 \\ 0 & 0 & 0 \\ 0 & 0 & 0 \end{bmatrix}. \tag{65}$$

Time DERIVATIVES of the VECTOR are

$$\dot{\mathbf{r}} = \begin{bmatrix} \cos\theta\,\dot{r} - r\sin\theta\,\dot{\theta} \\ \sin\theta\,\dot{r} + r\cos\theta\,\dot{\theta} \\ \dot{z} \end{bmatrix} = \dot{r}\,\hat{\mathbf{r}} + r\dot{\theta}\,\hat{\boldsymbol{\theta}} + \dot{z}\,\hat{\mathbf{z}} \tag{66}$$

$$\ddot{\mathbf{r}} = \begin{bmatrix} -\sin\theta\,\dot{r}\dot{\theta} + \cos\theta\,\ddot{r} - \sin\theta\,\dot{r}\dot{\theta} - r\cos\theta\,\dot{\theta}^2 - r\sin\theta\,\ddot{\theta} \\ \cos\theta\,\dot{r}\dot{\theta} + \sin\theta\,\ddot{r} + \cos\theta\,\dot{r}\dot{\theta} - r\sin\theta\,\dot{\theta}^2 + r\cos\theta\,\ddot{\theta} \\ \ddot{z} \end{bmatrix}$$

$$= \begin{bmatrix} -2\sin\theta\,\dot{r}\dot{\theta} + \cos\theta\,\ddot{r} - r\cos\theta\,\dot{\theta}^2 - r\sin\theta\,\ddot{\theta} \\ 2\cos\theta\,\dot{r}\dot{\theta} + \sin\theta\,\ddot{r} - r\sin\theta\,\dot{\theta}^2 + r\cos\theta\,\ddot{\theta} \\ \ddot{z} \end{bmatrix}$$

$$= (\ddot{r} - r\dot{\theta}^2)\hat{\mathbf{r}} + (2\dot{r}\dot{\theta} + r\ddot{\theta})\hat{\boldsymbol{\theta}} + \ddot{z}\,\hat{\mathbf{z}}. \tag{67}$$

SPEED is given by

$$v \equiv |\dot{\mathbf{r}}| = \sqrt{\dot{r}^2 + r^2\dot{\theta}^2 + \dot{z}^2}. \tag{68}$$

Time derivatives of the unit VECTORS are

$$\dot{\hat{\mathbf{r}}} = \begin{bmatrix} -\sin\theta\,\dot{\theta} \\ \cos\theta\,\dot{\theta} \\ 0 \end{bmatrix} = \dot{\theta}\,\hat{\boldsymbol{\theta}} \tag{69}$$

$$\dot{\hat{\boldsymbol{\theta}}} = \begin{bmatrix} -\cos\theta\,\dot{\theta} \\ -\sin\theta\,\dot{\theta} \\ 0 \end{bmatrix} = -\dot{\theta}\,\hat{\mathbf{r}} \tag{70}$$

$$\dot{\hat{\mathbf{z}}} = \begin{bmatrix} 0 \\ 0 \\ 0 \end{bmatrix} = \mathbf{0}. \tag{71}$$

CROSS PRODUCTS of the axes are

$$\hat{\mathbf{r}} \times \hat{\mathbf{z}} = -\hat{\boldsymbol{\theta}} \tag{72}$$
$$\hat{\boldsymbol{\theta}} \times \hat{\mathbf{z}} = \hat{\mathbf{r}} \tag{73}$$
$$\hat{\mathbf{r}} \times \hat{\boldsymbol{\theta}} = \hat{\mathbf{z}}. \tag{74}$$

The CONVECTIVE DERIVATIVE is

$$\frac{D\dot{\mathbf{r}}}{Dt} \equiv \left(\frac{\partial}{\partial t} + \dot{\mathbf{r}} \cdot \nabla\right) \dot{\mathbf{r}} = \frac{\partial \dot{\mathbf{r}}}{\partial t} + \dot{\mathbf{r}} \cdot \nabla \dot{\mathbf{r}}. \tag{75}$$

To rewrite this, use the identity

$$\nabla(\mathbf{A}\cdot\mathbf{B}) = \mathbf{A}\times(\nabla\times\mathbf{B}) + \mathbf{B}\times(\nabla\times\mathbf{A}) + (\mathbf{A}\cdot\nabla)\mathbf{B} + (\mathbf{B}\cdot\nabla)\mathbf{A} \tag{76}$$

and set $\mathbf{A} = \mathbf{B}$, to obtain

$$\nabla(\mathbf{A}\cdot\mathbf{A}) = 2\mathbf{A} \times (\nabla \times \mathbf{A}) + 2(\mathbf{A}\cdot\nabla)\mathbf{A}, \tag{77}$$

so

$$(\mathbf{A}\cdot\nabla)\mathbf{A} = \nabla(\tfrac{1}{2}\mathbf{A}^2) - \mathbf{A} \times (\nabla \times \mathbf{A}). \tag{78}$$

Then

$$\frac{D\dot{\mathbf{r}}}{Dt} = \ddot{\mathbf{r}} + \nabla(\tfrac{1}{2}\dot{\mathbf{r}}^2) - \dot{\mathbf{r}}\times(\nabla\times\dot{\mathbf{r}}) = \ddot{\mathbf{r}} + (\nabla\times\dot{\mathbf{r}})\times\dot{\mathbf{r}} + \nabla(\tfrac{1}{2}\dot{\mathbf{r}}^2). \tag{79}$$

The CURL in the above expression gives

$$\nabla \times \dot{\mathbf{r}} = \frac{1}{r}\frac{\partial}{\partial r}(r^2\dot{\theta})\hat{\mathbf{z}} = 2\dot{\theta}\hat{\mathbf{z}}, \tag{80}$$

so

$$\begin{aligned} -\dot{\mathbf{r}} \times (\nabla \times \dot{\mathbf{r}}) &= -2\dot{\theta}(\dot{r}\hat{\mathbf{r}} \times \hat{\mathbf{z}} + r\dot{\theta}\hat{\boldsymbol{\theta}} \times \hat{\mathbf{z}}) \\ &= -2\dot{\theta}(-\dot{r}\hat{\boldsymbol{\theta}} + r\dot{\theta}\hat{\mathbf{r}}) = 2\dot{r}\dot{\theta}\hat{\boldsymbol{\theta}} - 2r\dot{\theta}^2\hat{\mathbf{r}}. \end{aligned} \tag{81}$$

We expect the gradient term to vanish since SPEED does not depend on position. Check this using the identity $\nabla(f^2) = 2f\nabla f$,

$$\nabla(\tfrac{1}{2}\dot{\mathbf{r}}^2) = \tfrac{1}{2}\nabla(\dot{r}^2 + r^2\dot{\theta}^2 + \dot{z}^2) = \dot{r}\nabla\dot{r} + r\dot{\theta}\nabla(r\dot{\theta}) + \dot{z}\nabla\dot{z}. \tag{82}$$

Examining this term by term,

$$\dot{r}\nabla\dot{r} = \dot{r}\frac{\partial}{\partial t}\nabla r = \dot{r}\frac{\partial}{\partial t}\hat{\mathbf{r}} = \dot{r}\dot{\hat{\mathbf{r}}} = \dot{r}\dot{\theta}\hat{\boldsymbol{\theta}} \tag{83}$$

$$\begin{aligned} r\dot{\theta}\nabla(r\dot{\theta}) &= r\dot{\theta}\left[r\frac{\partial}{\partial t}\nabla\theta + \dot{\theta}\nabla r\right] = r\dot{\theta}\left[r\frac{\partial}{\partial t}\left(\frac{1}{r}\hat{\boldsymbol{\theta}}\right) + \dot{\theta}\hat{\mathbf{r}}\right] \\ &= r\dot{\theta}\left[r\left(-\frac{1}{r^2}\dot{r}\hat{\boldsymbol{\theta}} + \frac{1}{r}\dot{\hat{\boldsymbol{\theta}}}\right) + \dot{\theta}\hat{\mathbf{r}}\right] \\ &= -\dot{\theta}\dot{r}\hat{\boldsymbol{\theta}} + r\dot{\theta}(-\dot{\theta}\hat{\mathbf{r}}) + r\dot{\theta}^2\hat{\mathbf{r}} = -\dot{\theta}\dot{r}\hat{\boldsymbol{\theta}} \end{aligned} \tag{84}$$

$$\dot{z}\nabla\dot{z} = \dot{z}\frac{\partial}{\partial t}\nabla z = \dot{z}\frac{\partial}{\partial t}\hat{\mathbf{z}} = \dot{z}\dot{\hat{\mathbf{z}}} = \mathbf{0}, \tag{85}$$

so, as expected,

$$\nabla(\tfrac{1}{2}\dot{\mathbf{r}}^2) = \mathbf{0}. \tag{86}$$

We have already computed $\ddot{\mathbf{r}}$, so combining all three pieces gives

$$\begin{aligned} \frac{D\dot{\mathbf{r}}}{Dt} &= (\ddot{r} - r\dot{\theta}^2 - 2r\dot{\theta}^2)\hat{\mathbf{r}} + (2\dot{r}\dot{\theta} + 2\dot{r}\dot{\theta} + r\ddot{\theta})\hat{\boldsymbol{\theta}} + \ddot{z}\hat{\mathbf{z}} \\ &= (\ddot{r} - 3r\dot{\theta}^2)\hat{\mathbf{r}} + (4\dot{r}\dot{\theta} + r\ddot{\theta})\hat{\boldsymbol{\theta}} + \ddot{z}\hat{\mathbf{z}}. \end{aligned} \tag{87}$$

The DIVERGENCE is

$$\begin{aligned} \nabla \cdot A = A^r_{;r} &= A^r_{,r} + \left(\Gamma^r_{rr}A^t + \Gamma^r_{\theta r}A^\theta + \Gamma^r_{zr}A^z\right) + A^\theta_{,\theta} \\ &\quad + \left(\Gamma^\theta_{r\theta}A^r + \Gamma^\theta_{\theta\theta}A^\theta + \Gamma^\theta_{z\theta}A^z\right) \\ &\quad + A^z_{,z} + \left(\Gamma^z_{rz}A^r + \Gamma^z_{\theta z}A^\theta + \Gamma^z_{zz}A^z\right) \\ &= A^r_{,r} + A^\theta_{,\theta} + A^z_{,z} + (0+0+0) + \left(\frac{1}{r} + 0 + 0\right) \\ &\quad + (0+0+0) \\ &= \frac{1}{g_r}\frac{\partial}{\partial r}A^r + \frac{1}{g_\theta}\frac{\partial}{\partial\theta}A^\theta + \frac{1}{g_z}\frac{\partial}{\partial z}A^z + \frac{1}{r}A^r \\ &= \left(\frac{\partial}{\partial r} + \frac{1}{r}\right)A^r + \frac{1}{r}\frac{\partial}{\partial\theta}A^\theta + \frac{\partial}{\partial z}A^z, \end{aligned} \tag{88}$$

or, in VECTOR notation

$$\nabla \cdot \mathbf{F} = \frac{1}{r}\frac{\partial}{\partial r}(rF_r) + \frac{1}{r}\frac{\partial F_\theta}{\partial\theta} + \frac{\partial F_z}{\partial z}. \tag{89}$$

The CROSS PRODUCT is

$$\begin{aligned} \nabla \times \mathbf{F} &= \left(\frac{1}{r}\frac{\partial F_z}{\partial\theta} - \frac{\partial F_\theta}{\partial z}\right)\hat{\mathbf{r}} + \left(\frac{\partial F_r}{\partial z} - \frac{\partial F_z}{\partial r}\right)\hat{\boldsymbol{\theta}} \\ &\quad + \frac{1}{r}\left[\frac{\partial}{\partial r}(rF_\theta) - \frac{\partial F_r}{\partial\theta}\right]\hat{\mathbf{z}}, \end{aligned} \tag{90}$$

and the LAPLACIAN is

$$\begin{aligned} \nabla^2 f &\equiv \frac{1}{r}\frac{\partial}{\partial r}\left(r\frac{\partial f}{\partial r}\right) + \frac{1}{r^2}\frac{\partial^2 f}{\partial\theta^2} + \frac{\partial^2 f}{\partial z^2} \\ &= \frac{\partial^2 f}{\partial r^2} + \frac{1}{r}\frac{\partial f}{\partial r} + \frac{1}{r^2}\frac{\partial^2 f}{\partial\theta^2} + \frac{\partial^2 f}{\partial z^2}. \end{aligned} \tag{91}$$

The vector LAPLACIAN is

$$\nabla^2\mathbf{v} = \begin{bmatrix} \frac{\partial^2 v_r}{\partial r^2} + \frac{1}{r^2}\frac{\partial^2 v_r}{\partial\phi^2} + \frac{\partial^2 v_r}{z^2} + \frac{1}{r}\frac{\partial v_r}{\partial r} - \frac{2}{r^2}\frac{\partial v_\phi}{\partial\phi} - \frac{v_r}{r^2} \\ \frac{\partial^2}{\partial r^2} + \frac{1}{r^2}\frac{\partial^2 v_\phi}{\partial\phi^2} + \frac{\partial^2 v_\phi}{\partial z^2} + \frac{1}{r}\frac{\partial v_\phi}{\partial r} + \frac{2}{r^2}\frac{\partial v_r}{\partial\phi} - \frac{v_\phi}{r^2} \\ \frac{\partial^2 v_z}{\partial r^2} + \frac{1}{r^2}\frac{\partial^2 v_z}{\partial\phi^2} + \frac{\partial^2 v_z}{\partial z^2} + \frac{1}{r}\frac{\partial v_z}{\partial r} \end{bmatrix}. \tag{92}$$

The HELMHOLTZ DIFFERENTIAL EQUATION is separable in cylindrical coordinates and has STÄCKEL DETERMINANT $S = 1$ (for r, θ, z) or $S = 1/(1 - \xi_2{}^2)$ (for Morse and Feshbach's ξ_1, ξ_2, ξ_3).

see also ELLIPTIC CYLINDRICAL COORDINATES, HELMHOLTZ DIFFERENTIAL EQUATION—CIRCULAR CYLINDRICAL COORDINATES, POLAR COORDINATES, SPHERICAL COORDINATES

References

Arfken, G. "Circular Cylindrical Coordinates." §2.4 in *Mathematical Methods for Physicists, 3rd ed.* Orlando, FL: Academic Press, pp. 95–101, 1985.

Beyer, W. H. *CRC Standard Mathematical Tables, 28th ed.* Boca Raton, FL: CRC Press, p. 212, 1987.

Morse, P. M. and Feshbach, H. *Methods of Theoretical Physics, Part I.* New York: McGraw-Hill, p. 657, 1953.

Cylindrical Equal-Area Projection

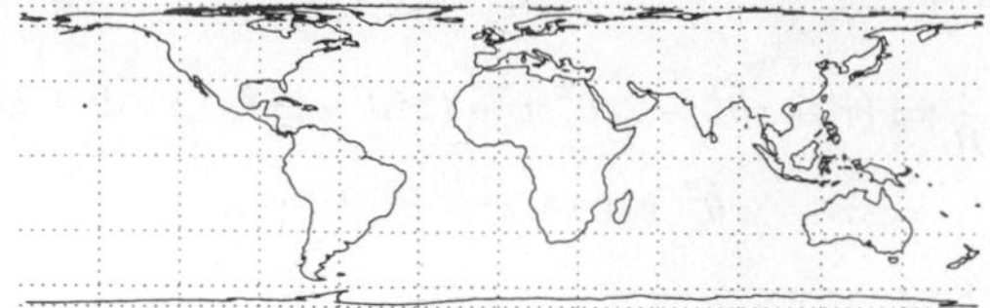

The MAP PROJECTION having transformation equations,

$$x = (\lambda - \lambda_0)\cos\phi_s \tag{1}$$

$$y = \frac{\sin\phi}{\cos\phi_s} \tag{2}$$

for the normal aspect, and inverse transformation equations

$$\phi = \sin^{-1}(y\cos\phi_s) \tag{3}$$

$$\lambda = \frac{x}{\cos\phi_s} + \lambda_0. \tag{4}$$

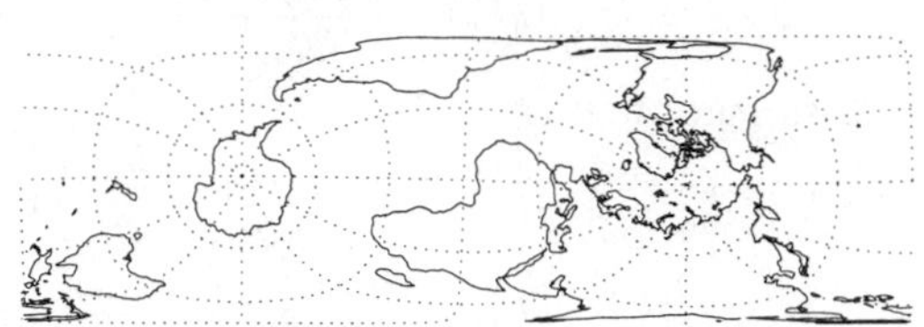

An oblique form of the cylindrical equal-area projection is given by the equations

$$\lambda_p = \tan^{-1}\left(\frac{\cos\phi_1\sin\phi_2\cos\lambda_1 - \sin\phi_1\cos\phi_2\cos\lambda_2}{\sin\phi_1\cos\phi_2\sin\lambda_2 - \cos\phi_1\sin\phi_2\sin\lambda_1}\right) \tag{5}$$

$$\phi_p = \tan^{-1}\left[-\frac{\cos(\lambda_p - \lambda_1)}{\tan\phi_1}\right], \tag{6}$$

and the inverse FORMULAS are

$$\phi = \sin^{-1}(y\sin\phi_p + \sqrt{1-y^2}\,\cos\phi_p\sin x) \tag{7}$$

$$\lambda = \lambda_0 + \tan^{-1}\left(\frac{\sqrt{1-y^2}\,\sin\phi_p\sin x - y\cos\phi_p}{\sqrt{1-y^2}\,\cos x}\right). \tag{8}$$

A transverse form of the cylindrical equal-area projection is given by the equations

$$x = \cos\phi\sin(\lambda - \lambda_0) \tag{9}$$

$$y = \tan^{-1}\left[\frac{\tan\phi}{\cos(\lambda - \lambda_0)}\right] - \phi_0, \tag{10}$$

and the inverse FORMULAS are

$$\phi = \sin^{-1}[\sqrt{1-x^2}\sin(y+\phi_0)] \tag{11}$$

$$\lambda = \lambda_0 + \tan^{-1}\left[\frac{x}{\sqrt{1-x^2}}\cos(y+\phi_0)\right]. \tag{12}$$

References

Snyder, J. P. *Map Projections—A Working Manual.* U. S. Geological Survey Professional Paper 1395. Washington, DC: U. S. Government Printing Office, pp. 76–85, 1987.

Cylindrical Equidistant Projection

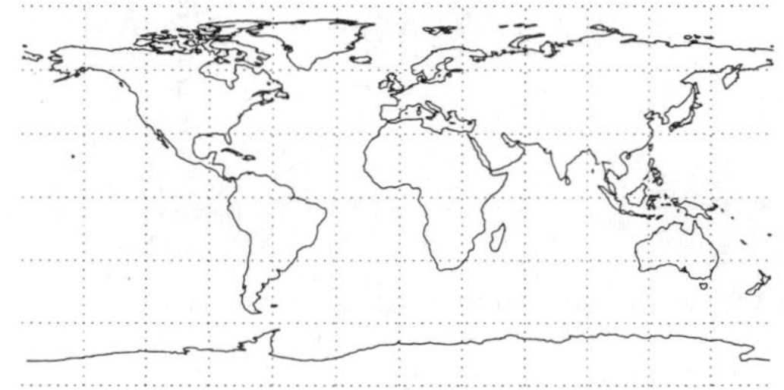

The MAP PROJECTION having transformation equations

$$x = (\lambda - \lambda_0)\cos\phi_1 \tag{1}$$

$$y = \phi, \tag{2}$$

and the inverse FORMULAS are

$$\phi = y \tag{3}$$

$$\lambda = \lambda_0 + \frac{x}{\cos\phi_1}. \tag{4}$$

References

Snyder, J. P. *Map Projections—A Working Manual.* U. S. Geological Survey Professional Paper 1395. Washington, DC: U. S. Government Printing Office, pp. 90–91, 1987.

Cylindrical Function

$$R_m(x,y) \equiv \frac{J'_m(x)Y'_m(y) - J'_m(y)Y'_m(x)}{J_m(x)Y'_m(y) - J'_m(y)Y_m(x)}$$

$$S_m(x,y) \equiv \frac{J'_m(x)Y_m(y) - J_m(y)Y'_m(x)}{J_m(x)Y_m(y) - J_m(y)Y_m(x)}.$$

see also CYLINDER FUNCTION, HEMISPHERICAL FUNCTION

Cylindrical Harmonics

see BESSEL FUNCTION OF THE FIRST KIND

Cylindrical Hoof

see CYLINDRICAL WEDGE

Cylindrical Projection

see BEHRMANN CYLINDRICAL EQUAL-AREA PROJECTION, CYLINDRICAL EQUAL-AREA PROJECTION, CYLINDRICAL EQUIDISTANT PROJECTION, GALL'S STEREOGRAPHIC PROJECTION, MERCATOR PROJECTION, MILLER CYLINDRICAL PROJECTION, PETERS PROJECTION, PSEUDOCYLINDRICAL PROJECTION

Cylindrical Segment

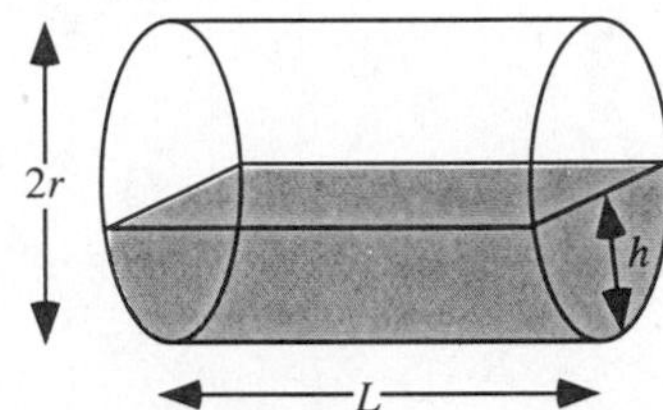

The solid portion of a CYLINDER below a cutting PLANE which is oriented PARALLEL to the CYLINDER's axis of symmetry. For a CYLINDER of RADIUS r and length L, the VOLUME of the cylindrical segment is given by multiplying the AREA of a circular SEGMENT of height h by L,

$$V = Lr^2 \cos^{-1}\left(\frac{r-h}{r}\right) - (r-h)L\sqrt{2rh-h^2}.$$

see also CYLINDRICAL WEDGE, SECTOR, SEGMENT, SPHERICAL SEGMENT

Cylindrical Wedge

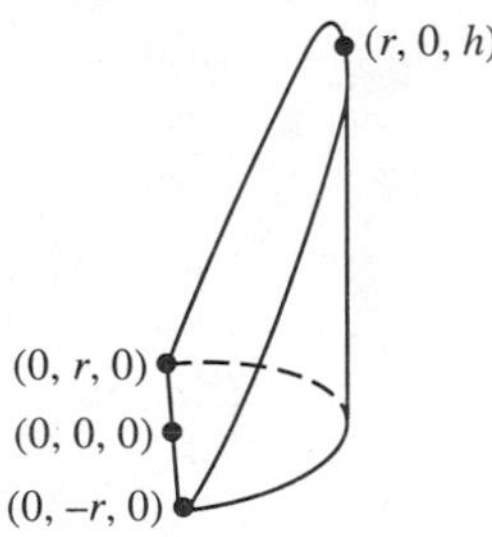

The solid cut from a CYLINDER by a tilted PLANE passing through a DIAMETER of the base. It is also called a CYLINDRICAL HOOF. Let the height of the wedge be h and the radius of the CYLINDER from which it is cut r. Then plugging the points $(0, -r, 0)$, $(0, r, 0)$, and $(r, 0, h)$ into the 3-point equation for a PLANE gives the equation for the plane as

$$hx - rz = 0. \tag{1}$$

Combining with the equation of the CIRCLE which describes the curved part remaining of the cylinder (and writing $t = x$) then gives the parametric equations of the "tongue" of the wedge as

$$x = t \tag{2}$$
$$y = \pm\sqrt{r^2 - t^2} \tag{3}$$
$$z = \frac{ht}{r} \tag{4}$$

for $t \in [0, r]$. To examine the form of the tongue, it needs to be rotated into a convenient plane. This can be accomplished by first rotating the plane of the curve by 90° about the x-AXIS using the ROTATION MATRIX $\mathsf{R}_x(90°)$ and then by the ANGLE

$$\theta = \tan^{-1}\left(\frac{h}{r}\right) \tag{5}$$

above the z-AXIS. The transformed plane now rests in the xz-plane and has parametric equations

$$x = \frac{t\sqrt{h^2 + r^2}}{r} \tag{6}$$
$$z = \pm\sqrt{r^2 - t^2} \tag{7}$$

and is shown below.

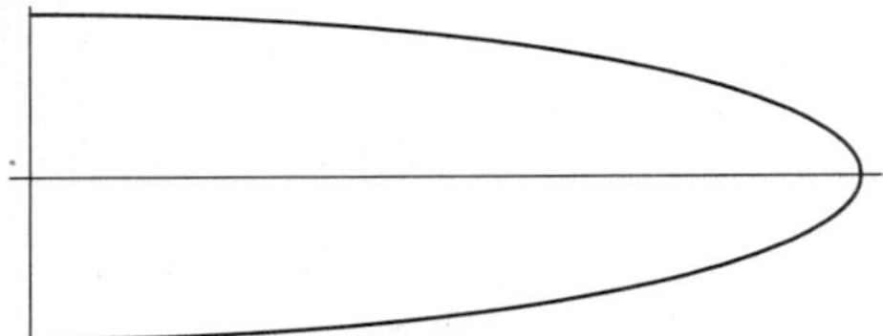

The length of the tongue (measured down its middle) is obtained by plugging $t = r$ into the above equation for x, which becomes

$$L = \sqrt{h^2 + r^2} \tag{8}$$

(and which follows immediately from the PYTHAGOREAN THEOREM). The VOLUME of the wedge is given by

$$V = \tfrac{2}{3} r^2 h. \tag{9}$$

see also CONICAL WEDGE, CYLINDRICAL SEGMENT

Cylindroid

see PLÜCKER'S CONOID

D

d'Alembert's Equation

The ORDINARY DIFFERENTIAL EQUATION

$$y = xf(y') + g(y'),$$

where $y' \equiv dy/dx$ and f and g are given functions.

d'Alembert Ratio Test

see RATIO TEST

d'Alembert's Solution

A method of solving the 1-D WAVE EQUATION.

see also WAVE EQUATION

d'Alembert's Theorem

If three CIRCLES A, B, and C are taken in pairs, the external similarity points of the three pairs lie on a straight line. Similarly, the external similarity point of one pair and the two internal similarity points of the other two pairs lie upon a straight line, forming a similarity axis of the three CIRCLES.

References

Dörrie, H. *100 Great Problems of Elementary Mathematics: Their History and Solutions.* New York: Dover, p. 155, 1965.

d'Alembertian Operator

Written in the NOTATION of PARTIAL DERIVATIVES,

$$\Box^2 \equiv \nabla^2 - \frac{1}{c^2}\frac{\partial^2}{\partial t^2},$$

where c is the speed of light. Writing in TENSOR notation

$$\Box^2 \phi \equiv (g^{\lambda\kappa}\phi_{;\lambda})_{;\kappa} = g^{\lambda\kappa}\frac{\partial^2 \phi}{\partial x^\lambda \partial x^\kappa} - \Gamma^\lambda \frac{\partial \phi}{\partial x^\lambda}.$$

see also HARMONIC COORDINATES

d-Analog

The d-analog of INFINITY FACTORIAL is given by

$$[\infty!]_d = \prod_{n=3}^{\infty}\left(1 - \frac{2^d}{n^d}\right).$$

This INFINITE PRODUCT can be evaluated in closed form for small POSITIVE integral $d \geq 2$.

see also q-ANALOG

References

Finch, S. "Favorite Mathematical Constants." `http://www.mathsoft.com/asolve/constant/infprd/infprd.html.`

D-Number

A NATURAL NUMBER $n > 3$ such that

$$n|(a^{n-2} - a)$$

whenever $(a, n) = 1$ (a and n are RELATIVELY PRIME) and $a \leq n$. There are an infinite number of such numbers, the first few being 9, 15, 21, 33, 39, 51, ... (Sloane's A033553).

see also KNÖDEL NUMBERS

References

Makowski, A. "Generalization of Morrow's D-Numbers." *Simon Stevin* **36**, 71, 1962/1963.

Sloane, N. J. A. Sequence A033553 in "An On-Line Version of the Encyclopedia of Integer Sequences."

D-Statistic

see KOLMOGOROV-SMIRNOV TEST

D-Triangle

Let the circles c_2 and c_3' used in the construction of the BROCARD POINTS which are tangent to A_2A_3 at A_2 and A_3, respectively, meet again at D_1. The points $D_1D_2D_3$ then define the D-triangle. The VERTICES of the D-triangle lie on the respective APOLLONIUS CIRCLES.

see also APOLLONIUS CIRCLES, BROCARD POINTS

References

Johnson, R. A. *Modern Geometry: An Elementary Treatise on the Geometry of the Triangle and the Circle.* Boston, MA: Houghton Mifflin, pp. 284–285, 296 and 307, 1929.

Daisy

A figure resembling a daisy or sunflower in which copies of a geometric figure of increasing size are placed at regular intervals along a spiral. The resulting figure appears to have multiple spirals spreading out from the center.

see also PHYLLOTAXIS, SPIRAL, SWIRL, WHIRL

References

Dixon, R. "On Drawing a Daisy." §5.1 in *Mathographics.* New York: Dover, pp. 122–143, 1991.

Damped Exponential Cosine Integral

$$\int_0^\infty e^{-\omega T}\cos(\omega t)\,d\omega. \quad (1)$$

Integrate by parts with

$$u \equiv e^{-\omega T} \qquad dv = \cos(\omega t)\,d\omega \quad (2)$$

$$du \equiv -Te^{-\omega T}\,d\omega \qquad v = \frac{1}{t}\sin(\omega t), \quad (3)$$

so

$$\int e^{-\omega T}\cos(\omega t)\,d\omega$$

$$= \frac{1}{t}e^{-\omega t}\sin(\omega t) + \frac{T}{t}\int e^{-\omega T}\sin(\omega t)\,d\omega. \quad (4)$$

Now integrate

$$\int e^{-\omega T}\sin(\omega t)\,d\omega \quad (5)$$

by parts. Let

$$v = e^{-\omega T} \qquad dv = \sin(\omega t)\,d\omega \quad (6)$$

$$du = -Te^{-\omega T}\,d\omega \qquad v = -\frac{1}{t}\cos(\omega t), \quad (7)$$

so

$$\int e^{-\omega t}\sin(\omega t)\,d\omega = -\frac{1}{t}\cos(\omega t) - \frac{T}{t}\int e^{-\omega T}\cos(\omega t)\,d\omega \quad (8)$$

and

$$\int e^{\omega T}\cos(\omega t)\,d\omega = \frac{1}{t}e^{-\omega t}\sin(\omega t)$$

$$-\frac{T}{t^2}e^{-\omega t}\cos(\omega t) - \frac{T^2}{t^2}\int e^{-\omega T}\cos(\omega t)\,d\omega \quad (9)$$

$$\left(1+\frac{T^2}{t^2}\right)\int e^{-\omega T}\cos(\omega t)\,d\omega$$

$$= e^{-\omega T}\left[\frac{1}{t}\sin(\omega t) - \frac{T}{t^2}\cos(\omega t)\right] \quad (10)$$

$$\frac{t^2+T^2}{t^2}\int e^{-\omega T}\cos(\omega t)\,d\omega$$

$$= \frac{e^{-\omega t}}{t^2}[t\sin(\omega T) - T\cos(\omega t)] \quad (11)$$

$$\int e^{-\omega T}\cos(\omega t)\,d\omega = \frac{e^{-\omega T}}{t^2+T^2}[t\sin(\omega t) - T\cos(\omega T)]. \quad (12)$$

Therefore,

$$\int_0^\infty e^{-\omega T}\cos(\omega t)\,d\omega = 0 + \frac{T}{t^2+T^2} = \frac{T}{t^2+T^2}. \quad (13)$$

see also COSINE INTEGRAL, FOURIER TRANSFORM—LORENTZIAN FUNCTION, LORENTZIAN FUNCTION

Dandelin Spheres

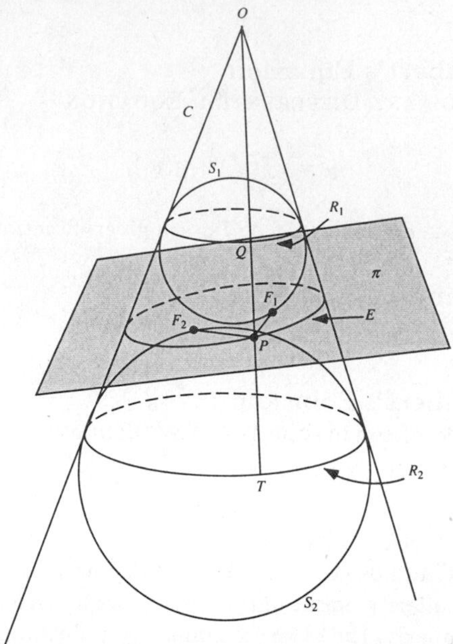

The inner and outer SPHERES TANGENT internally to a CONE and also to a PLANE intersecting the CONE are called Dandelin spheres.

The SPHERES can be used to show that the intersection of the PLANE with the CONE is an ELLIPSE. Let π be a PLANE intersecting a right circular CONE with vertex O in the curve E. Call the SPHERES TANGENT to the CONE and the PLANE S_1 and S_2, and the CIRCLES on which the CIRCLES are TANGENT to the CONE R_1 and R_2. Pick a line along the CONE which intersects R_1 at Q, E at P, and R_2 at T. Call the points on the PLANE where the CIRCLES are TANGENT F_1 and F_2. Because intersecting tangents have the same length,

$$F_1P = QP$$

$$F_2P = TP.$$

Therefore,

$$PF_1 + PF_2 = QP + PT = QT,$$

which is a constant independent of P, so E is an ELLIPSE with $a = QT/2$.

see also CONE, SPHERE

References

Honsberger, R. "Kepler's Conics." Ch. 9 in *Mathematical Plums* (Ed. R. Honsberger). Washington, DC: Math. Assoc. Amer., p. 170, 1979.

Honsberger, R. *More Mathematical Morsels.* Washington, DC: Math. Assoc. Amer., pp. 40–44, 1991.

Ogilvy, C. S. *Excursions in Geometry.* New York: Dover, pp. 80–81, 1990.

Ogilvy, C. S. *Excursions in Mathematics.* New York: Dover, pp. 68–69, 1994.

Danielson-Lanczos Lemma

The DISCRETE FOURIER TRANSFORM of length N (where N is EVEN) can be rewritten as the sum of two DISCRETE FOURIER TRANSFORMS, each of length $N/2$. One is formed from the EVEN numbered points; the other from the ODD numbered points. Denote the kth point of the DISCRETE FOURIER TRANSFORM by F_n. Then

$$\begin{aligned} F_n &= \sum_{k=0}^{N-1} f_k e^{-2\pi ink/N} \\ &= \sum_{k=0}^{N/2-1} e^{-2\pi ikn/(N/2)} f_{2k} + W^n \sum_{k=0}^{N/2-1} e^{-2\pi ikn/(N/2)} f_{2k+1} \\ &= F_n^e + W_n F_n^o, \end{aligned}$$

where $W \equiv e^{-2\pi i/N}$ and $n = 0, \ldots, N$. This procedure can be applied recursively to break up the $N/2$ even and ODD points to their $N/4$ EVEN and ODD points. If N is a POWER of 2, this procedure breaks up the original transform into $\lg N$ transforms of length 1. Each transform of an individual point has $F_n^{eeo\cdots} = f_k$ for some k. By reversing the patterns of evens and odds, then letting $e = 0$ and $o = 1$, the value of k in BINARY is produced. This is the basis for the FAST FOURIER TRANSFORM.

see also DISCRETE FOURIER TRANSFORM, FAST FOURIER TRANSFORM, FOURIER TRANSFORM

References
Press, W. H.; Flannery, B. P.; Teukolsky, S. A.; and Vetterling, W. T. *Numerical Recipes in C: The Art of Scientific Computing.* Cambridge, England: Cambridge University Press, pp. 407–411, 1989.

Darboux Integral

A variant of the RIEMANN INTEGRAL defined when the UPPER and LOWER INTEGRALS, taken as limits of the LOWER SUM

$$L(f;\phi;N) = \sum_{r=1}^{n} m(f;\delta_r) - \phi(x_{r-1})$$

and UPPER SUM

$$U(f;\phi;N) = \sum_{r=1}^{n} M(f;\delta_r) - \phi(x_{r-1}),$$

are equal. Here, $f(x)$ is a REAL FUNCTION, $\phi(x)$ is a monotonic increasing function with respect to which the sum is taken, $m(f;S)$ denotes the lower bound of $f(x)$ over the interval S, and $M(f;S)$ denotes the upper bound.

see also LOWER INTEGRAL, LOWER SUM, RIEMANN INTEGRAL, UPPER INTEGRAL, UPPER SUM

References
Kestelman, H. *Modern Theories of Integration, 2nd rev. ed.* New York: Dover, p. 250, 1960.

Darboux-Stieltjes Integral

see DARBOUX INTEGRAL

Darboux Vector

The rotation VECTOR of the TRIHEDRON of a curve with CURVATURE $\kappa \neq 0$ when a point moves along a curve with unit SPEED. It is given by

$$\mathbf{D} = \tau\mathbf{T} + \kappa\mathbf{B}, \tag{1}$$

where τ is the TORSION, $\mathbf{T}$ the TANGENT VECTOR, and $\mathbf{B}$ the BINORMAL VECTOR. The Darboux vector field satisfies

$$\dot{\mathbf{T}} = \mathbf{D} \times \mathbf{T} \tag{2}$$

$$\dot{\mathbf{N}} = \mathbf{D} \times \mathbf{N} \tag{3}$$

$$\dot{\mathbf{B}} = \mathbf{D} \times \mathbf{B}. \tag{4}$$

see also BINORMAL VECTOR, CURVATURE, TANGENT VECTOR, TORSION (DIFFERENTIAL GEOMETRY)

References
Gray, A. *Modern Differential Geometry of Curves and Surfaces.* Boca Raton, FL: CRC Press, p. 151, 1993.

Darling's Products

A generalization of the HYPERGEOMETRIC FUNCTION identity

$$\begin{aligned} &{}_2F_1(\alpha,\beta;\gamma;z)\,{}_2F_1(1-\alpha,1-\beta;2-\gamma;z) \\ &= {}_2F_1(\alpha+1-\gamma,\beta+1-\gamma;2-\gamma;z)\,{}_2F_1(\gamma-\alpha,\gamma-\beta;\gamma;z) \end{aligned} \tag{1}$$

to the GENERALIZED HYPERGEOMETRIC FUNCTION ${}_3F_2(a,b,c;d,e;x)$. Darling's products are

$$\begin{aligned} &{}_3F_2\begin{bmatrix}\alpha,\beta,\gamma;z\\ \delta,\epsilon\end{bmatrix} {}_3F_2\begin{bmatrix}1-\alpha,1-\beta,1-\gamma;z\\ 2-\delta,2-\epsilon\end{bmatrix} \\ &= \frac{\epsilon-1}{\epsilon-\delta}\,{}_3F_2\begin{bmatrix}\alpha+1-\delta,\beta+1-\delta,\gamma+1-\delta;z\\ 2-\delta,\epsilon+1-\delta\end{bmatrix} \\ &\quad \times{}_3F_2\begin{bmatrix}\delta-\alpha,\delta-\beta,\delta-\gamma;z\\ \delta,\delta+1-\epsilon\end{bmatrix} \\ &+\frac{\delta-1}{\delta-\epsilon}\,{}_3F_2\begin{bmatrix}\alpha+1-\epsilon,\beta+1-\epsilon,\gamma+1-\epsilon;z\\ 2-\epsilon,\delta+1-\epsilon\end{bmatrix} \\ &\quad \times{}_3F_2\begin{bmatrix}\epsilon-\alpha,\epsilon-\beta,\epsilon-\gamma;z\\ \epsilon,\epsilon+1-\delta\end{bmatrix} \end{aligned} \tag{2}$$

and

$$\begin{aligned} &(1-z)^{\alpha+\beta+\gamma-\delta-\epsilon}\,{}_3F_2\begin{bmatrix}\alpha,\beta,\gamma;z\\ \delta,\epsilon\end{bmatrix} \\ &= \frac{\epsilon-1}{\epsilon-\delta}\,{}_3F_2\begin{bmatrix}\delta-\alpha,\delta-\beta,\delta-\gamma;z\\ \delta,\delta+1-\epsilon\end{bmatrix} \\ &\quad \times{}_3F_2\begin{bmatrix}\epsilon-\alpha,\epsilon-\beta,\epsilon-\gamma;z\\ \epsilon-1,\epsilon+1-\delta\end{bmatrix} \\ &+\frac{\delta-1}{\delta-\epsilon}\,{}_3F_2\begin{bmatrix}\epsilon-\alpha,\epsilon-\beta,\epsilon-\gamma;z\\ \epsilon,\epsilon+1-\delta\end{bmatrix} \\ &\quad \times{}_3F_2\begin{bmatrix}\delta-\alpha,\delta-\beta,\delta-\gamma;z\\ \delta-1,\delta+1-\epsilon\end{bmatrix}, \end{aligned} \tag{3}$$

which reduce to (1) when $\gamma = \epsilon \to \infty$.

References
Bailey, W. N. "Darling's Theorems of Products." §10.3 in *Generalised Hypergeometric Series.* Cambridge, England: Cambridge University Press, pp. 88–92, 1935.

Dart

see PENROSE TILES

Darwin-de Sitter Spheroid

A SURFACE OF REVOLUTION of the form

$$r(\phi) = a[1 - e\sin^2\phi - (\tfrac{3}{8}e^2 + k)\sin^2(2\phi)],$$

where k is a second-order correction to the figure of a rotating fluid.

see also OBLATE SPHEROID, PROLATE SPHEROID, SPHEROID

References
Zharkov, V. N. and Trubitsyn, V. P. *Physics of Planetary Interiors.* Tucson, AZ: Pachart Publ. House, 1978.

Darwin's Expansions

Series expansions of the PARABOLIC CYLINDER FUNCTION $U(a, x)$ and $W(a, x)$. The formulas can be found in Abramowitz and Stegun (1972).

References
Abramowitz, M. and Stegun, C. A. (Eds.). *Handbook of Mathematical Functions with Formulas, Graphs, and Mathematical Tables, 9th printing.* New York: Dover, pp. 689–690 and 694–695, 1972.

Data Structure

A formal structure for the organization of information. Examples of data structures include the LIST, QUEUE, STACK, and TREE.

Database

A database can be roughly defined as a structure consisting of

1. A collection of information (the data),
2. A collection of queries that can be submitted, and
3. A collection of algorithms by which the structure responds to queries, searches the data, and returns the results.

References
Petkovšek, M.; Wilf, H. S.; and Zeilberger, D. *A=B.* Wellesley, MA: A. K. Peters, p. 48, 1996.

Daubechies Wavelet Filter

A WAVELET used for filtering signals. Daubechies (1988, p. 980) has tabulated the numerical values up to order $p = 10$.

References
Daubechies, I. "Orthonormal Bases of Compactly Supported Wavelets." *Comm. Pure Appl. Math.* **41**, 909–996, 1988.
Press, W. H.; Flannery, B. P.; Teukolsky, S. A.; and Vetterling, W. T. "Interpolation and Extrapolation." Ch. 3 in *Numerical Recipes in FORTRAN: The Art of Scientific Computing, 2nd ed.* Cambridge, England: Cambridge University Press, pp. 584–586, 1992.

Davenport-Schinzel Sequence

Form a sequence from an ALPHABET of letters $[1, n]$ such that there are no consecutive letters and no alternating subsequences of length greater than d. Then the sequence is a Davenport-Schinzel sequence if it has maximal length $N_d(n)$. The value of $N_1(n)$ is the trivial sequence of 1s: 1, 1, 1, ... (Sloane's A000012). The values of $N_2(n)$ are the POSITIVE INTEGERS 1, 2, 3, 4, ... (Sloane's A000027). The values of $N_3(n)$ are the ODD INTEGERS 1, 3, 5, 7, ... (Sloane's A005408). The first nontrivial Davenport-Schinzel sequence $N_4(n)$ is given by 1, 4, 8, 12, 17, 22, 27, 32, ... (Sloane's A002004). Additional sequences are given by Guy (1994, p. 221) and Sloane.

References
Davenport, H. and Schinzel, A. "A Combinatorial Problem Connected with Differential Equations." *Amer. J. Math.* **87**, 684–690, 1965.
Guy, R. K. "Davenport-Schinzel Sequences." §E20 in *Unsolved Problems in Number Theory, 2nd ed.* New York: Springer-Verlag, pp. 220–222, 1994.
Roselle, D. P. and Stanton, R. G. "Results of Davenport-Schinzel Sequences." In *Proc. Louisiana Conference on Combinatorics, Graph Theory, and Computing. Louisiana State University, Baton Rouge, March 1–5, 1970* (Ed. R. C. Mullin, K. B. Reid, and D. P. Roselle). Winnipeg, Manitoba: Utilitas Mathematica, pp. 249–267, 1960.
Sharir, M. and Agarwal, P. *Davenport-Schinzel Sequences and Their Geometric Applications.* New York: Cambridge University Press, 1995.
Sloane, N. J. A. Sequences A000012/M0003, A000027/M0472, A002004/M3328, and A005408/M2400 in "An On-Line Version of the Encyclopedia of Integer Sequences."

Dawson's Integral

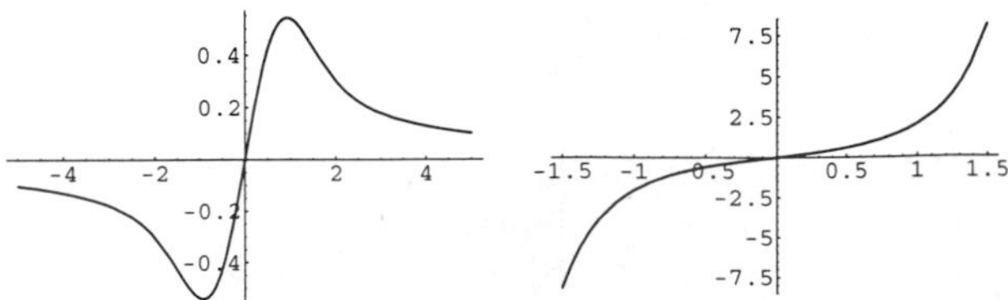

An INTEGRAL which arises in computation of the Voigt lineshape:

$$D(x) \equiv e^{-x^2} \int_0^x e^{y^2}\, dy. \tag{1}$$

It is sometimes generalized such that

$$D_{\pm}(x) \equiv e^{\mp x^2} \int_0^x e^{\pm y^2}\,dy, \tag{2}$$

giving

$$D_{+}(x) = \tfrac{1}{2}\sqrt{\pi}\,e^{-x^2}\operatorname{erfi}(x) \tag{3}$$

$$D_{-}(x) = \tfrac{1}{2}\sqrt{\pi}\,e^{x^2}\operatorname{erf}(x), \tag{4}$$

where $\operatorname{erf}(z)$ is the ERF function and $\operatorname{erfi}(z)$ is the imaginary error function ERFI. $D_{+}(x)$ is illustrated in the left figure above, and $D_{-}(x)$ in the right figure. D_{+} has a maximum at $D'_{+}(x) = 0$, or

$$1 - \sqrt{\pi}\,e^{-x^2}x^2\operatorname{erfi}(x) = 0, \tag{5}$$

giving

$$D_{+}(0.9241388730) = 0.5410442246, \tag{6}$$

and an inflection at $D''_{+}(x) = 0$, or

$$-2x + \sqrt{\pi}\,e^{-x^2}(2x^2 - 1)\operatorname{erfi}(x) = 0, \tag{7}$$

giving

$$D_{+}(1.5019752683) = 0.4276866160. \tag{8}$$

References

Abramowitz, M. and Stegun, C. A. (Eds.). *Handbook of Mathematical Functions with Formulas, Graphs, and Mathematical Tables, 9th printing.* New York: Dover, p. 298, 1972.

Press, W. H.; Flannery, B. P.; Teukolsky, S. A.; and Vetterling, W. T. "Dawson's Integrals." §6.10 in *Numerical Recipes in FORTRAN: The Art of Scientific Computing, 2nd ed.* Cambridge, England: Cambridge University Press, pp. 252–254, 1992.

Spanier, J. and Oldham, K. B. "Dawson's Integral." Ch. 42 in *An Atlas of Functions.* Washington, DC: Hemisphere, pp. 405–410, 1987.

Day of Week

see FRIDAY THE THIRTEENTH, WEEKDAY

de Bruijn Constant

Also called the COPSON-DE BRUIJN CONSTANT. It is defined by

$$\sum_{n=1}^{\infty} a_n \leq c \sum_{n=1}^{\infty} \sqrt{\frac{a_n{}^2 + a_{n+1}{}^2 + a_{n+2}{}^2 + \ldots}{n}}.$$

where

$$c = 1.0164957714\ldots.$$

References

Copson, E. T. "Note on Series of Positive Terms." *J. London Math. Soc.* **2**, 9–12, 1927.

Copson, E. T. "Note on Series of Positive Terms." *J. London Math. Soc.* **3**, 49–51, 1928.

de Bruijn, N. G. *Asymptotic Methods in Analysis.* New York: Dover, 1981.

Finch, S. "Favorite Mathematical Constants." `http://www.mathsoft.com/asolve/constant/copson/copson.html`.

de Bruijn Diagram

see DE BRUIJN GRAPH

de Bruijn Graph

A graph whose nodes are sequences of symbols from some ALPHABET and whose edges indicate the sequences which might overlap.

References

Golomb, S. W. *Shift Register Sequences.* San Francisco, CA: Holden-Day, 1967.

Ralston, A. "de Bruijn Sequences—A Model Example of the Interaction of Discrete Mathematics and Computer Science." *Math. Mag.* **55**, 131–143, 1982.

de Bruijn-Newman Constant

N.B. A detailed on-line essay by S. Finch was the starting point for this entry.

Let Ξ be the XI FUNCTION defined by

$$\Xi(iz) = \tfrac{1}{2}(z^2 - \tfrac{1}{4})\pi^{-z/2-\frac{1}{4}}\Gamma(\tfrac{1}{2}z + \tfrac{1}{4})\zeta(z + \tfrac{1}{2}). \tag{1}$$

$\Xi(z/2)/8$ can be viewed as the FOURIER TRANSFORM of the signal

$$\Phi(t) = \sum_{n=1}^{\infty}(2\pi^2 n^4 e^{9t} - 3\pi n^2 e^{5t})e^{-\pi n^2 e^{4t}} \tag{2}$$

for $t \in \mathbb{R} \geq 0$. Then denote the FOURIER TRANSFORM of $\Phi(t)e^{\lambda t^2}$ as $H(\lambda, z)$,

$$\mathcal{F}[\Phi(t)e^{\lambda t^2}] = H(\lambda, z). \tag{3}$$

de Bruijn (1950) proved that H has only REAL zeros for $\lambda \geq 1/2$. C. M. Newman (1976) proved that there exists a constant Λ such that H has only REAL zeros IFF $\lambda \geq \Lambda$. The best current lower bound (Csordas *et al.* 1993, 1994) is $\Lambda > -5.895 \times 10^{-9}$. The RIEMANN HYPOTHESIS is equivalent to the conjecture that $\Lambda \leq 0$.

References

Csordas, G.; Odlyzko, A.; Smith, W.; and Varga, R. S. "A New Lehmer Pair of Zeros and a New Lower Bound for the de Bruijn-Newman Constant." *Elec. Trans. Numer. Analysis* **1**, 104–111, 1993.

Csordas, G.; Smith, W.; and Varga, R. S. "Lehmer Pairs of Zeros, the de Bruijn-Newman Constant and the Riemann Hypothesis." *Constr. Approx.* **10**, 107–129, 1994.

de Bruijn, N. G. "The Roots of Trigonometric Integrals." *Duke Math. J.* **17**, 197–226, 1950.

Finch, S. "Favorite Mathematical Constants." `http://www.mathsoft.com/asolve/constant/dbnwm/dbnwm.html`.

Newman, C. M. "Fourier Transforms with only Real Zeros." *Proc. Amer. Math. Soc.* **61**, 245–251, 1976.

de Bruijn Sequence

The shortest sequence such that every string of length n on the ALPHABET a occurs as a contiguous subrange of the sequence described by a. Every de Bruijn sequence corresponds to an EULERIAN CYCLE on a "DE BRUIJN GRAPH." Surprisingly, it turns out that the lexicographic sequence of LYNDON WORDS of lengths DIVISIBLE by n gives the lexicographically smallest de Bruijn sequence (Ruskey).

References

Ruskey, F. "Information on Necklaces, Lyndon Words, de Bruijn Sequences." `http://sue.csc.uvic.ca/~cos/inf/neck/NecklaceInfo.html`.

de Bruijn's Theorem

A box can be packed with a HARMONIC BRICK $a \times ab \times abc$ IFF the box has dimensions $ap \times abq \times abcr$ for some natural numbers p, q, r (i.e., the box is a multiple of the brick).

see also BOX-PACKING THEOREM, CONWAY PUZZLE, DE BRUIJN'S THEOREM, KLARNER'S THEOREM

References

Honsberger, R. *Mathematical Gems II.* Washington, DC: Math. Assoc. Amer., pp. 69–72, 1976.

de Jonquières Theorem

The total number of groups of a g_N^r consisting in a point of multiplicity k_1, one of multiplicity k_2, ..., one of multiplicity k_ρ, where

$$\sum k_i = N \tag{1}$$

$$\sum (k_i - 1) = r, \tag{2}$$

and where α_1 points have one multiplicity, α_2 another, etc., and

$$\Pi = k_1 k_2 \cdots k_\rho \tag{3}$$

is

$$\frac{\Pi p(p-1)\cdots(p-\rho)}{\alpha_1!\alpha_2!\cdots}\left[\frac{\Pi}{p-\rho} - \frac{\sum_i \frac{\partial \Pi}{\partial k_i}}{p-\rho+1} + \frac{\sum_{ij} \frac{\partial^2 \Pi}{\partial k_i \partial k_j}}{p-\rho+2} + \ldots\right]. \tag{4}$$

References

Coolidge, J. L. *A Treatise on Algebraic Plane Curves.* New York: Dover, p. 288, 1959.

de Jonquières Transformation

A transformation which is of the same type as its inverse. A de Jonquières transformation is always factorable.

References

Coolidge, J. L. *A Treatise on Algebraic Plane Curves.* New York: Dover, pp. 203–204, 1959.

de la Loubere's Method

A method for constructing MAGIC SQUARES of ODD order, also called the SIAMESE METHOD.

see also MAGIC SQUARE

de Longchamps Point

The reflection of the ORTHOCENTER about the CIRCUMCENTER. This point is also the ORTHOCENTER of the ANTICOMPLEMENTARY TRIANGLE. It has TRIANGLE CENTER FUNCTION

$$\alpha = \cos A - \cos B \cos C.$$

It lies on the EULER LINE.

References

Altshiller-Court, N. "On the de Longchamps Circle of the Triangle." *Amer. Math. Monthly* **33**, 368–375, 1926.
Kimberling, C. "Central Points and Central Lines in the Plane of a Triangle." *Math. Mag.* **67**, 163–187, 1994.
Vandeghen, A. "Soddy's Circles and the de Longchamps Point of a Triangle." *Amer. Math. Monthly* **71**, 176–179, 1964.

de Mere's Problem

The probability of getting at least one "6" in four rolls of a single 6-sided DIE is

$$1 - \left(\tfrac{5}{6}\right)^4 = 0.518\ldots, \tag{1}$$

which is slightly higher than the probability of at least one double 6 in 24 throws,

$$1 - \left(\tfrac{35}{36}\right)^{24} = 0.491\ldots. \tag{2}$$

de Mere suspected that (1) was higher than (2). He posed the question to Pascal, who solved the problem and proved de Mere correct.

see also DICE

References

Kraitchik, M. "A Dice Problem." §6.2 in *Mathematical Recreations.* New York: W. W. Norton, pp. 118–119, 1942.

de Moivre's Identity

$$e^{i(n\theta)} = (e^{i\theta})^n. \tag{1}$$

From the EULER FORMULA it follows that

$$\cos(n\theta) + i\sin(n\theta) = (\cos\theta + i\sin\theta)^n. \tag{2}$$

A similar identity holds for the HYPERBOLIC FUNCTIONS,

$$(\cosh z + \sinh z)^n = \cosh(nz) + \sinh(nz). \tag{3}$$

References

Arfken, G. *Mathematical Methods for Physicists, 3rd ed.* Orlando, FL: Academic Press, pp. 356–357, 1985.
Courant, R. and Robbins, H. *What is Mathematics?: An Elementary Approach to Ideas and Methods, 2nd ed.* Oxford, England: Oxford University Press, pp. 96–100, 1996.

de Moivre Number

A solution $\zeta_k = e^{2\pi ik/d}$ to the CYCLOTOMIC EQUATION

$$x^d = 1.$$

The de Moivre numbers give the coordinates in the COMPLEX PLANE of the VERTICES of a regular POLYGON with d sides and unit RADIUS.

n	de Moivre Numbers
2	± 1
3	$1, \frac{1}{2}(-1 \pm i\sqrt{3})$
4	$\pm 1, \pm i$
5	$1, \frac{1}{4}\left(-1+\sqrt{5} \pm (1+\sqrt{5})\sqrt{\frac{5-5\sqrt{5}}{2}}\, i\right),$
	$-\frac{1+\sqrt{5}}{4} \pm \frac{\sqrt{5-\sqrt{5}}}{2\sqrt{2}}\, i$
6	$\pm 1, \pm\frac{1}{2}(\pm 1 + i\sqrt{3})$

see also CYCLOTOMIC EQUATION, CYCLOTOMIC POLYNOMIAL, EUCLIDEAN NUMBER

References

Conway, J. H. and Guy, R. K. *The Book of Numbers.* New York: Springer-Verlag, 1996.

de Moivre-Laplace Theorem

The sum of those terms of the BINOMIAL SERIES of $(p+q)^s$ for which the number of successes x falls between d_1 and d_2 is approximately

$$Q \approx \frac{1}{\sqrt{2\pi}} \int_{t_1}^{t_2} e^{-t^2/2}\, dt, \tag{1}$$

where

$$t_1 \equiv \frac{d_1 - \frac{1}{2} - sp}{\sigma} \tag{2}$$

$$t_2 \equiv \frac{d_2 + \frac{1}{2}s - sp}{\sigma} \tag{3}$$

$$\sigma \equiv \sqrt{spq}. \tag{4}$$

Uspensky (1937) has shown that

$$Q = \frac{1}{\sqrt{2\pi}} \int_{t_1}^{t_2} e^{-t^2/2}\, dt + \frac{q-p}{6\sigma}\left[(1-t^2)\frac{1}{2\pi}e^{-t^2/2}\right]_{t_1}^{t_2} + \Omega, \tag{5}$$

where

$$|\Omega| < \frac{0.12 + 0.18|p-q|}{\sigma^2} + e^{-3\sigma/2} \tag{6}$$

for $\sigma \geq 5$.

A COROLLARY states that the probability that x successes in s trials will differ from the expected value sp by more than d is

$$P_\delta \approx 1 - 2\int_0^\delta \phi(t)\, dt, \tag{7}$$

where

$$\delta \equiv \frac{d + \frac{1}{2}}{\sigma}. \tag{8}$$

Uspensky (1937) showed that

$$Q_{\delta_1} \equiv P(|x - sp| \leq d) = 2\int_0^{\delta_1} \phi(t)\, dt + \frac{1 - \theta_1 - \theta_2}{\sigma}\phi(\delta_1) + \Omega_1, \tag{9}$$

where

$$\delta_1 \equiv \frac{d}{\delta} \tag{10}$$

$$\theta_1 \equiv (sq + d) - \lfloor sq + d \rfloor \tag{11}$$

$$\theta_2 \equiv (sp + d) - \lfloor sp + d \rfloor \tag{12}$$

and

$$|\Omega_1| < \frac{0.20 + 0.25|p-q|}{\sigma^2} + e^{-3\sigma/2}, \tag{13}$$

for $\sigma \geq 5$.

References

Uspensky, J. V. *Introduction to Mathematical Probability.* New York: McGraw-Hill, 1937.

de Moivre's Quintic

$$x^5 + ax^3 + \tfrac{1}{5}a^2x + b = 0.$$

see also QUINTIC EQUATION

de Morgan's and Bertrand's Test

see BERTRAND'S TEST

de Morgan's Duality Law

For every proposition involving logical addition and multiplication ("or" and "and"), there is a corresponding proposition in which the words "addition" and "multiplication" are interchanged.

de Morgan's Laws

Let $\cup$ represent "or", $\cap$ represent "and", and $'$ represent "not." Then, for two logical units E and F,

$$(E \cup F)' = E' \cap F'$$

$$(E \cap F)' = E' \cup F'.$$

de Polignac's Conjecture

Every EVEN NUMBER is the difference of two consecutive PRIMES in infinitely many ways. If true, taking the difference 2, this conjecture implies that there are infinitely many TWIN PRIMES (Ball and Coxeter 1987). The CONJECTURE has never been proven true or refuted.

see also EVEN NUMBER, TWIN PRIMES

References

Ball, W. W. R. and Coxeter, H. S. M. *Mathematical Recreations and Essays, 13th ed.* New York: Dover, p. 64, 1987.

de Polignac, A. "Six propositions arithmologiques déduites de crible d'Ératosthène." *Nouv. Ann. Math.* **8**, 423–429, 1849.

de Rham Cohomology

de Rham cohomology is a formal set-up for the analytic problem: If you have a DIFFERENTIAL k-FORM ω on a MANIFOLD M, is it the EXTERIOR DERIVATIVE of another DIFFERENTIAL k-FORM ω'? Formally, if $\omega = d\omega'$ then $d\omega = 0$. This is more commonly stated as $d \circ d = 0$, meaning that if ω is to be the EXTERIOR DERIVATIVE of a DIFFERENTIAL k-FORM, a NECESSARY condition that ω must satisfy is that its EXTERIOR DERIVATIVE is zero.

de Rham cohomology gives a formalism that aims to answer the question, "Are all differential k-forms on a MANIFOLD with zero EXTERIOR DERIVATIVE the EXTERIOR DERIVATIVES of $(k+1)$-forms?" In particular, the kth de Rham cohomology vector space is defined to be the space of all k-forms with EXTERIOR DERIVATIVE 0, modulo the space of all boundaries of $(k+1)$-forms. This is the trivial VECTOR SPACE IFF the answer to our question is yes.

The fundamental result about de Rham cohomology is that it is a topological invariant of the MANIFOLD, namely: the kth de Rham cohomology VECTOR SPACE of a MANIFOLD M is canonically isomorphic to the ALEXANDER-SPANIER COHOMOLOGY VECTOR SPACE $H^k(M;\mathbb{R})$ (also called cohomology with compact support). In the case that M is compact, ALEXANDER-SPANIER COHOMOLOGY is exactly singular cohomology.

see also ALEXANDER-SPANIER COHOMOLOGY, CHANGE OF VARIABLES THEOREM, DIFFERENTIAL k-FORM, EXTERIOR DERIVATIVE, VECTOR SPACE

de Sluze Conchoid

see CONCHOID OF DE SLUZE

de Sluze Pearls

see PEARLS OF SLUZE

Debye's Asymptotic Representation

An asymptotic expansion for a HANKEL FUNCTION OF THE FIRST KIND

$$H_\nu^{(1)}(x) \sim \frac{1}{\sqrt{\pi}} \exp\{ix[\cos\alpha + (\alpha - \pi/2)\sin\alpha]\}$$
$$\times \left[\frac{e^{i\pi/4}}{X} + (\tfrac{1}{8} + \tfrac{5}{24}\tan^2\alpha)\frac{3e^{3\pi i/4}}{2X^3}\right.$$
$$\left. + (\tfrac{3}{128} + \tfrac{77}{576}\tan^\alpha + \tfrac{385}{3456}\tan^4\alpha)\frac{3\cdot e^{5\pi i/4}}{2^2X^5} + \ldots\right],$$

where

$$\frac{\nu}{x} = \sin\alpha,$$

$$1 - \frac{\nu}{x} > \frac{3}{x}\nu^{1/2},$$

and

$$X \equiv \sqrt{-x\cos(\tfrac{1}{2}\alpha)}.$$

see also HANKEL FUNCTION OF THE FIRST KIND

References

Iyanaga, S. and Kawada, Y. (Eds.). *Encyclopedic Dictionary of Mathematics.* Cambridge, MA: MIT Press, p. 1475, 1980.

Debye Functions

$$\int_0^x \frac{t^n\,dt}{e^t - 1} = x^n \left[\frac{1}{n} - \frac{x}{2(n+1)} + \sum_{k=1}^\infty \frac{B_{2k}x^{2k}}{(2k+n)(2k!)}\right], \tag{1}$$

where $|x| < 2\pi$ and B_n are BERNOULLI NUMBERS.

$$\int_x^\infty \frac{t^n\,dt}{e^t-1} = \sum_{k=1}^\infty e^{-kx}\left[\frac{x^n}{k} + \frac{nx^{n-1}}{k^2} + \frac{n(n-1)x^{n-2}}{k^3} + \ldots + \frac{n!}{k^{n+1}}\right], \tag{2}$$

where $x > 0$. The sum of these two integrals is

$$\int_0^\infty \frac{t^n\,dt}{e^t-1} = n!\zeta(n+1), \tag{3}$$

where $\zeta(z)$ is the RIEMANN ZETA FUNCTION.

References

Abramowitz, M. and Stegun, C. A. (Eds.). "Debye Functions." §27.1 in *Handbook of Mathematical Functions with Formulas, Graphs, and Mathematical Tables, 9th printing.* New York: Dover, pp. 998, 1972.

Decagon

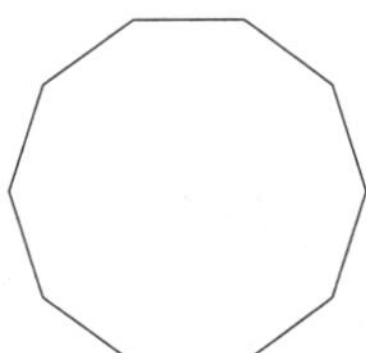

The constructible regular 10-sided POLYGON with SCHLÄFLI SYMBOL $\{10\}$. The INRADIUS r, CIRCUMRADIUS R, and AREA can be computed directly from the formulas for a general regular POLYGON with side length s and $n = 10$ sides,

$$r = \tfrac{1}{2}s\cot\left(\frac{\pi}{10}\right) = \tfrac{1}{2}\sqrt{25 - 10\sqrt{5}}\,s \tag{1}$$

$$R = \tfrac{1}{2}s\csc\left(\frac{\pi}{10}\right) = \tfrac{1}{2}(1+\sqrt{5})s = \phi s \tag{2}$$

$$A = \tfrac{1}{4}ns^2\cot\left(\frac{\pi}{10}\right) = \tfrac{5}{2}\sqrt{5+2\sqrt{5}}\,s^2. \tag{3}$$

Here, ϕ is the GOLDEN MEAN.

see also DECAGRAM, DODECAGON, TRIGONOMETRY VALUES—$\pi/10$, UNDECAGON

References

Dixon, R. *Mathographics.* New York: Dover, p. 18, 1991.

Decagonal Number

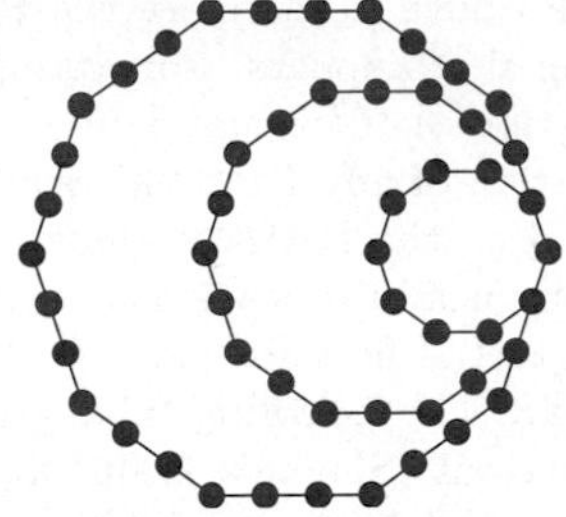

A FIGURATE NUMBER of the form $4n^2 - 3n$. The first few are 1, 10, 27, 52, 85, ... (Sloane's A001107). The GENERATING FUNCTION giving the decagonal numbers is

$$\frac{x(7x+1)}{(1-x)^3} = x + 10x^2 + 27x^3 + 52x^4 + \ldots.$$

References

Sloane, N. J. A. Sequence A001107/M4690 in "An On-Line Version of the Encyclopedia of Integer Sequences."

Decagram

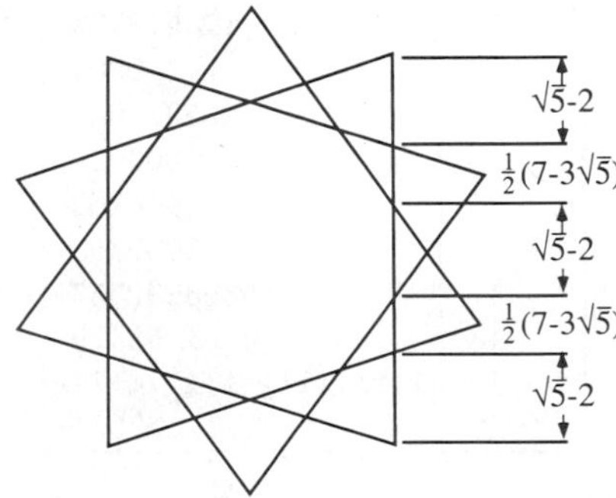

The STAR POLYGON $\left\{ {10 \atop 3} \right\}$.

see also DECAGON, STAR POLYGON

Decic Surface

A SURFACE which can be represented implicitly by a POLYNOMIAL of degree 10 in x, y, and z. An example is the BARTH DECIC.

see also BARTH DECIC, CUBIC SURFACE, QUADRATIC SURFACE, QUARTIC SURFACE

Decidable

A "theory" in LOGIC is decidable if there is an ALGORITHM that will decide on input ϕ whether or not ϕ is a SENTENCE true of the FIELD of REAL NUMBERS $\mathbb{R}$.

see also CHURCH'S THESIS, GÖDEL'S COMPLETENESS THEOREM, GÖDEL'S INCOMPLETENESS THEOREM, KREISEL CONJECTURE, TARSKI'S THEOREM, UNDECIDABLE, UNIVERSAL STATEMENT

References

Kemeny, J. G. "Undecidable Problems of Elementary Number Theory." *Math. Ann.* **135**, 160–169, 1958.

Decillion

In the American system, 10^{33}.

see also LARGE NUMBER

Decimal

The base 10 notational system for representing REAL NUMBERS.

see also 10, BASE (NUMBER), BINARY, HEXADECIMAL, OCTAL

References

Pappas, T. "The Evolution of Base Ten." *The Joy of Mathematics.* San Carlos, CA: Wide World Publ./Tetra, pp. 2–3, 1989.

Weisstein, E. W. "Bases." `http://www.astro.virginia.edu/~eww6n/math/notebooks/Bases.m`.

Decimal Expansion

The decimal expansion of a number is its representation in base 10. For example, the decimal expansion of 25^2 is 625, of π is 3.14159..., and of 1/9 is 0.1111....

If $r \equiv p/q$ has a finite decimal expansion, then

$$\begin{aligned} r &= \frac{a_1}{10} + \frac{a_2}{10^2} + \ldots + \frac{a_n}{10^n} \\ &= \frac{a_1 10^{n-1} + a_2 10^{n-2} + \ldots + a_n}{10^n} \\ &= \frac{a_1 10^{n-1} + a_2 10^{n-2} + \ldots + a_n}{2^n 5^n}. \end{aligned} \quad (1)$$

FACTORING possible common multiples gives

$$r = \frac{p}{2^\alpha 5^\beta}, \quad (2)$$

where $p \not\equiv 0 \pmod{2, 5}$. Therefore, the numbers with finite decimal expansions are fractions of this form. The number of decimals is given by $\max(\alpha, \beta)$. Numbers which have a finite decimal expansion are called REGULAR NUMBERS.

Any NONREGULAR fraction m/n is periodic, and has a period $\lambda(n)$ independent of m, which is at most $n-1$ DIGITS long. If n is RELATIVELY PRIME to 10, then the period of m/n is a divisor of $\phi(n)$ and has at most $\phi(n)$ DIGITS, where ϕ is the TOTIENT FUNCTION. When a rational number m/n with $(m,n) = 1$ is expanded, the period begins after s terms and has length t, where s and t are the smallest numbers satisfying

$$10^2 \equiv 10^{s+t} \pmod{n}. \quad (3)$$

When $n \not\equiv 0 \pmod{2, 5}$, $s = 0$, and this becomes a purely periodic decimal with

$$10^t \equiv 1 \pmod{n}. \quad (4)$$

As an example, consider $n = 84$.

$$\begin{array}{llll} 10^0 \equiv 1 & 10^1 \equiv 10 & 10^2 \equiv 16 & 10^3 \equiv -8 \\ 10^4 \equiv 4 & 10^5 \equiv 40 & 10^6 \equiv -20 & 10^7 \equiv -32, \\ 10^8 \equiv 16 \end{array}$$

so $s = 2$, $t = 6$. The decimal representation is $1/84 = 0.01\overline{190476}$. When the DENOMINATOR of a fraction m/n has the form $n = n_0 2^\alpha 5^\beta$ with $(n_0, 10) = 1$, then the period begins after $\max(\alpha, \beta)$ terms and the length of the period is the exponent to which 10 belongs (mod n_0), i.e., the number x such that $10^x \equiv 1 \pmod{n_0}$. If q is PRIME and $\lambda(q)$ is EVEN, then breaking the repeating DIGITS into two equal halves and adding gives all 9s. For example, $1/7 = 0.\overline{142857}$, and $142 + 857 = 999$. For $1/q$ with a PRIME DENOMINATOR other than 2 or 5, all cycles n/q have the same length (Conway and Guy 1996).

If n is a PRIME and 10 is a PRIMITIVE ROOT of n, then the period $\lambda(n)$ of the repeating decimal $1/n$ is given by

$$\lambda(n) = \phi(n), \tag{5}$$

where $\phi(n)$ is the TOTIENT FUNCTION. Furthermore, the decimal expansions for p/n, with $p = 1, 2, \ldots, n-1$ have periods of length $n-1$ and differ only by a cyclic permutation. Such numbers are called LONG PRIMES by Conway and Guy (1996). An equivalent definition is that

$$10^i \equiv 1 \pmod{n} \tag{6}$$

for $i = n - 1$ and no i less than this. In other words, a NECESSARY (but not SUFFICIENT) condition is that the number $9R_{n-1}$ (where R_n is a REPUNIT) is DIVISIBLE by n, which means that R_n is DIVISIBLE by n.

The first few numbers with maximal decimal expansions, called FULL REPTEND PRIMES, are 7, 17, 19, 23, 29, 47, 59, 61, 97, 109, 113, 131, 149, 167, ... (Sloane's A001913). The decimals corresponding to these are called CYCLIC NUMBERS. No general method is known for finding FULL REPTEND PRIMES. Artin conjectured that ARTIN'S CONSTANT $C = 0.3739558136\ldots$ is the fraction of PRIMES p for with $1/p$ has decimal maximal period (Conway and Guy 1996). D. Lehmer has generalized this conjecture to other bases, obtaining values which are small rational multiples of C.

To find DENOMINATORS with short periods, note that

$$\begin{aligned}
10^1 - 1 &= 3^2 \\
10^2 - 1 &= 3^2 \cdot 11 \\
10^3 - 1 &= 3^3 \cdot 37 \\
10^4 - 1 &= 3^2 \cdot 11 \cdot 101 \\
10^5 - 1 &= 3^2 \cdot 41 \cdot 271 \\
10^6 - 1 &= 3^3 \cdot 7 \cdot 11 \cdot 13 \cdot 37 \\
10^7 - 1 &= 3^2 \cdot 239 \cdot 4649 \\
10^8 - 1 &= 3^2 \cdot 11 \cdot 73 \cdot 101 \cdot 137 \\
10^9 - 1 &= 3^4 \cdot 37 \cdot 333667 \\
10^{10} - 1 &= 3^2 \cdot 11 \cdot 41 \cdot 271 \cdot 9091 \\
10^{11} - 1 &= 3^2 \cdot 21649 \cdot 513239 \\
10^{12} - 1 &= 3^3 \cdot 7 \cdot 11 \cdot 13 \cdot 37 \cdot 101 \cdot 9901.
\end{aligned}$$

The period of a fraction with DENOMINATOR equal to a PRIME FACTOR above is therefore the POWER of 10 in which the factor first appears. For example, 37 appears in the factorization of $10^3 - 1$ and $10^9 - 1$, so its period is 3. Multiplication of any FACTOR by a $2^\alpha 5^\beta$ still gives the same period as the FACTOR alone. A DENOMINATOR obtained by a multiplication of two FACTORS has a period equal to the first POWER of 10 in which both FACTORS appear. The following table gives the PRIMES having small periods (Sloane's A046106, A046107, and A046108; Ogilvy and Anderson 1988).

period	primes
1	3
2	11
3	37
4	101
5	41, 271
6	7, 13
7	239, 4649
8	73, 137
9	333667
10	9091
11	21649, 513239
12	9901
13	53, 79, 265371653
14	909091
15	31, 2906161
16	17, 5882353
17	2071723, 5363222357
18	19, 52579
19	1111111111111111111
20	3541, 27961

A table of the periods e of small PRIMES other than the special $p = 5$, for which the decimal expansion is not periodic, follows (Sloane's A002371).

p	e	p	e	p	e
3	1	31	15	67	33
7	6	37	3	71	35
11	2	41	5	73	8
13	6	43	21	79	13
17	16	47	46	83	41
19	18	53	13	89	44
23	22	59	58	97	96
29	28	61	60	101	4

Shanks (1873ab) computed the periods for all PRIMES up to 120,000 and published those up to 29,989.

see also FRACTION, MIDY'S THEOREM, REPEATING DECIMAL

References

Conway, J. H. and Guy, R. K. "Fractions Cycle into Decimals." In *The Book of Numbers.* New York: Springer-Verlag, pp. 157–163 and 166–171, 1996.

Das, R. C. "On Bose Numbers." *Amer. Math. Monthly* **56**, 87–89, 1949.

Dickson, L. E. *History of the Theory of Numbers, Vol. 1: Divisibility and Primality.* New York: Chelsea, pp. 159–179, 1952.

Lehmer, D. H. "A Note on Primitive Roots." *Scripta Math.* **26**, 117–119, 1963.
Ogilvy, C. S. and Anderson, J. T. *Excursions in Number Theory.* New York: Dover, p. 60, 1988.
Rademacher, H. and Toeplitz, O. *The Enjoyment of Mathematics: Selections from Mathematics for the Amateur.* Princeton, NJ: Princeton University Press, pp. 147–163, 1957.
Rao, K. S. "A Note on the Recurring Period of the Reciprocal of an Odd Number." *Amer. Math. Monthly* **62**, 484–487, 1955.
Shanks, W. "On the Number of Figures in the Period of the Reciprocal of Every Prime Number Below 20,000." *Proc. Roy. Soc. London* **22**, 200, 1873a.
Shanks, W. "On the Number of Figures in the Period of the Reciprocal of Every Prime Number Between 20,000 and 30,000." *Proc. Roy. Soc. London* **22**, 384, 1873b.
Shiller, J. K. "A Theorem in the Decimal Representation of Rationals." *Amer. Math. Monthly* **66**, 797–798, 1959.
Sloane, N. J. A. Sequences A001913/M4353 and A002371/M4050 in "An On-Line Version of the Encyclopedia of Integer Sequences."

Decimal Period

see DECIMAL EXPANSION

Decision Problem

Does there exist an ALGORITHM for deciding whether or not a specific mathematical assertion does or does not have a proof? The decision problem is also known as the ENTSCHEIDUNGSPROBLEM (which, not so coincidentally, is German for "decision problem"). Using the concept of the TURING MACHINE, Turing showed the answer to be NEGATIVE for elementary NUMBER THEORY. J. Robinson and Tarski showed the decision problem is undecidable for arbitrary FIELDS.

Decision Theory

A branch of GAME THEORY dealing with strategies to maximize the outcome of a given process in the face of uncertain conditions.

see also NEWCOMB'S PARADOX, OPERATIONS RESEARCH, PRISONER'S DILEMMA

Decomposition

A rewriting of a given quantity (e.g., a MATRIX) in terms of a combination of "simpler" quantities.

see also CHOLESKY DECOMPOSITION, CONNECTED SUM DECOMPOSITION, JACO-SHALEN-JOHANNSON TORUS DECOMPOSITION, LU DECOMPOSITION, QR DECOMPOSITION, SINGULAR VALUE DECOMPOSITION

Deconvolution

The inversion of a CONVOLUTION equation, i.e., the solution for f of an equation of the form

$$f * g = h + \epsilon,$$

given g and h, where ϵ is the NOISE and $*$ denotes the CONVOLUTION. Deconvolution is ill-posed and will usually not have a unique solution even in the absence of NOISE.

Linear deconvolution ALGORITHMS include INVERSE FILTERING and WIENER FILTERING. Nonlinear ALGORITHMS include the CLEAN ALGORITHM, MAXIMUM ENTROPY METHOD, and LUCY.

see also CLEAN ALGORITHM, CONVOLUTION, LUCY, MAXIMUM ENTROPY METHOD, WIENER FILTER

References

Cornwell, T. and Braun, R. "Deconvolution." Ch. 8 in *Synthesis Imaging in Radio Astronomy: Third NRAO Summer School, 1988* (Ed. R. A. Perley, F. R. Schwab, and A. H. Bridle). San Francisco, CA: Astronomical Society of the Pacific, pp. 167–183, 1989.
Press, W. H.; Flannery, B. P.; Teukolsky, S. A.; and Vetterling, W. T. "Convolution and Deconvolution Using the FFT." §13.1 in *Numerical Recipes in FORTRAN: The Art of Scientific Computing, 2nd ed.* Cambridge, England: Cambridge University Press, pp. 531–537, 1992.

Decreasing Function

A function $f(x)$ decreases on an INTERVAL I if $f(b) < f(a)$ for all $b > a$, where $a, b \in I$. Conversely, a function $f(x)$ increases on an INTERVAL I if $f(b) > f(a)$ for all $b > a$ with $a, b \in I$.

If the DERIVATIVE $f'(x)$ of a CONTINUOUS FUNCTION $f(x)$ satisfies $f'(x) < 0$ on an OPEN INTERVAL (a, b), then $f(x)$ is decreasing on (a, b). However, a function may decrease on an interval without having a derivative defined at all points. For example, the function $-x^{1/3}$ is decreasing everywhere, including the origin $x = 0$, despite the fact that the DERIVATIVE is not defined at that point.

see also DERIVATIVE, INCREASING FUNCTION, NONDECREASING FUNCTION, NONINCREASING FUNCTION

Decreasing Sequence

A SEQUENCE $\{a_1, a_2, \ldots\}$ for which $a_1 \geq a_2 \geq \ldots$.

see also INCREASING SEQUENCE

Decreasing Series

A SERIES $s_1, s_2, \ldots$ for which $s_1 \geq s_2 \geq \ldots$.

Dedekind's Axiom

For every partition of all the points on a line into two nonempty SETS such that no point of either lies between two points of the other, there is a point of one SET which lies between every other point of that SET and every point of the other SET.

Dedekind Cut

A set partition of the RATIONAL NUMBERS into two nonempty subsets S_1 and S_2 such that all members of S_1 are less than those of S_2 and such that S_1 has no greatest member. REAL NUMBERS can be defined using either Dedekind cuts or CAUCHY SEQUENCES.

see also CANTOR-DEDEKIND AXIOM, CAUCHY SEQUENCE

References

Courant, R. and Robbins, H. "Alternative Methods of Defining Irrational Numbers. Dedekind Cuts." §2.2.6 in *What is Mathematics?: An Elementary Approach to Ideas and Methods, 2nd ed.* Oxford, England: Oxford University Press, pp. 71–72, 1996.

Dedekind Eta Function

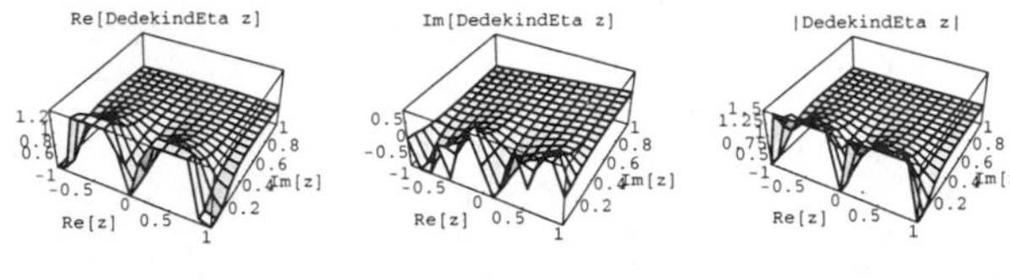

Let

$$q = e^{2\pi iz}, \quad (1)$$

then the Dedekind eta function is defined by

$$\eta(z) \equiv q^{1/24} \prod_{n=1}^{\infty} (1 - q^n), \quad (2)$$

which can be written as

$$\eta(z) = q^{1/24} \left\{ 1 + \sum_{n=1}^{\infty} (-1)^n [q^{n(3n-1)/2} + q^{n(3n+1)/2}] \right\} \quad (3)$$

(Weber 1902, pp. 85 and 112; Atkin and Morain 1993). η is a MODULAR FORM. Letting $\zeta_{24} = 2^{2\pi i/24}$ be a ROOT OF UNITY, $\eta(z)$ satisfies

$$\eta(z+1) = \zeta_{24}\eta(z) \quad (4)$$

$$\eta\left(-\frac{1}{z}\right) = -\sqrt{zi}\,\eta(z) \quad (5)$$

(Weber 1902, p. 113; Atkin and Morain 1993).

see also DIRICHLET ETA FUNCTION, THETA FUNCTION, WEBER FUNCTIONS

References

Atkin, A. O. L. and Morain, F. "Elliptic Curves and Primality Proving." *Math. Comput.* **61**, 29–68, 1993.
Weber, H. *Lehrbuch der Algebra, Vols. I-II.* New York: Chelsea, 1902.

Dedekind Function

$$\psi(n) = n \prod_{\substack{\text{distinct prime}\\ \text{factors } p \text{ of } n}} (1 + p^{-1}),$$

where the PRODUCT is over the distinct PRIME FACTORS of n. The first few values are 1, 3, 4, 6, 6, 12, 8, 12, 12, 18, ... (Sloane's A001615).

see also DEDEKIND ETA FUNCTION, EULER PRODUCT, TOTIENT FUNCTION

References

Cox, D. A. *Primes of the Form x^2+ny^2: Fermat, Class Field Theory and Complex Multiplication.* New York: Wiley, p. 228, 1997.
Guy, R. K. *Unsolved Problems in Number Theory, 2nd ed.* New York: Springer-Verlag, p. 96, 1994.
Sloane, N. J. A. Sequence A001615/M2315 in "An On-Line Version of the Encyclopedia of Integer Sequences."

Dedekind's Problem

The determination of the number of monotone BOOLEAN FUNCTIONS of n variables is called Dedekind's problem.

Dedekind Ring

A abstract commutative RING in which every NONZERO IDEAL is a unique product of PRIME IDEALS.

Dedekind Sum

Given RELATIVELY PRIME INTEGERS p and q, the Dedekind sum is defined by

$$s(p,q) \equiv \sum_{i=1}^{q} \left(\left(\frac{i}{q}\right)\right) \left(\left(\frac{pi}{q}\right)\right), \quad (1)$$

where

$$((x)) \equiv \begin{cases} x - \lfloor x \rfloor - \frac{1}{2} & x \in \mathbb{Z} \\ 0 & x \notin \mathbb{Z}. \end{cases} \quad (2)$$

Dedekind sums obey 2-term

$$s(p,q) + s(q,p) = -\frac{1}{4} + \frac{1}{12}\left(\frac{p}{q} + \frac{q}{p} + \frac{1}{pq}\right), \quad (3)$$

and 3-term

$$s(bc',a)+s(ca',b)+s(ab',c) = -\frac{1}{4}+\frac{1}{12}\left(\frac{a}{bc} + \frac{b}{ca} + \frac{c}{ab}\right) \quad (4)$$

reciprocity laws, where a, b, c are pairwise COPRIME and

$$aa' \equiv 1 \pmod{b} \quad (5)$$

$$bb' \equiv 1 \pmod{c} \quad (6)$$

$$cc' \equiv 1 \pmod{a}. \quad (7)$$

Let p, q, u, $v \in \mathbb{N}$ with $(p,q) = (u,v) = 1$ (i.e., are pairwise RELATIVELY PRIME), then the Dedekind sums also satisfy

$$s(p,q) + s(u,v) = s(pu' - qv', pv + qu) - \tfrac{1}{4} + \frac{1}{12}\left(\frac{q}{vt} + \frac{v}{tq} + \frac{t}{qv}\right), \quad (8)$$

where $t = pv + qu$, and u', v' are any INTEGERS such that $uu' + vv' = 1$ (Pommersheim 1993).

References

Pommersheim, J. "Toric Varieties, Lattice Points, and Dedekind Sums." *Math. Ann.* **295**, 1–24, 1993.

Deducible

If q is logically deducible from p, this is written $p \vdash q$.

Deep Theorem

Qualitatively, a deep theorem is a theorem whose proof is long, complicated, difficult, or appears to involve branches of mathematics which are not obviously related to the theorem itself (Shanks 1993). Shanks (1993) cites the QUADRATIC RECIPROCITY THEOREM as an example of a deep theorem.

see also THEOREM

References

Shanks, D. "Is the Quadratic Reciprocity Law a Deep Theorem?" §2.25 in *Solved and Unsolved Problems in Number Theory, 4th ed.* New York: Chelsea, pp. 64–66, 1993.

Defective Matrix

A MATRIX whose EIGENVECTORS are not COMPLETE.

Defective Number

see DEFICIENT NUMBER

Deficiency

The deficiency of a BINOMIAL COEFFICIENT $\binom{n+k}{k}$ with $k \leq n$ as the number of i for which $b_i = 1$, where

$$n + i = a_i b_i,$$

$1 \leq i \leq k$, the PRIME factors of b_i are $> k$, and $\prod a_i = k!$, where $k1$ is the FACTORIAL.

see also ABUNDANCE

References

Guy, R. K. *Unsolved Problems in Number Theory, 2nd ed.* New York: Springer-Verlag, pp. 84–85, 1994.

Deficient Number

Numbers which are not PERFECT and for which

$$s(N) \equiv \sigma(N) - N < N,$$

or equivalently

$$\sigma(n) < 2n,$$

where $\sigma(N)$ is the DIVISOR FUNCTION. Deficient numbers are sometimes called DEFECTIVE NUMBERS (Singh 1997). PRIMES, POWERS of PRIMES, and any divisors of a PERFECT or deficient number are all deficient. The first few deficient numbers are 1, 2, 3, 4, 5, 7, 8, 9, 10, 11, 13, 14, 15, 16, 17, 19, 21, 22, 23, ... (Sloane's A002855).

see also ABUNDANT NUMBER, LEAST DEFICIENT NUMBER, PERFECT NUMBER

References

Dickson, L. E. *History of the Theory of Numbers, Vol. 1: Divisibility and Primality.* New York: Chelsea, pp. 3–33, 1952.

Guy, R. K. *Unsolved Problems in Number Theory, 2nd ed.* New York: Springer-Verlag, p. 45, 1994.

Singh, S. *Fermat's Enigma: The Epic Quest to Solve the World's Greatest Mathematical Problem.* New York: Walker, p. 11, 1997.

Sloane, N. J. A. Sequence A002855/M0514 in "An On-Line Version of the Encyclopedia of Integer Sequences."

Definable Set

An ANALYTIC, BOREL, or COANALYTIC SET.

Defined

If A and B are equal by definition (i.e., A is defined as B), then this is written symbolically as $A \equiv B$ or $A := B$.

Definite Integral

An INTEGRAL

$$\int_a^b f(x)\,dx$$

with upper and lower limits. The first FUNDAMENTAL THEOREM OF CALCULUS allows definite integrals to be computed in terms of INDEFINITE INTEGRALS, since if F is the INDEFINITE INTEGRAL for $f(x)$, then

$$\int_a^b f(x)\,dx = F(b) - F(a).$$

see also CALCULUS, FUNDAMENTAL THEOREMS OF CALCULUS, INDEFINITE INTEGRAL, INTEGRAL

Degenerate

A limiting case in which a class of object changes its nature so as to belong to another, usually simpler, class. For example, the POINT is a degenerate case of the CIRCLE as the RADIUS approaches 0, and the CIRCLE is a degenerate form of an ELLIPSE as the ECCENTRICITY approaches 0. Another example is the two identical ROOTS of the second-order POLYNOMIAL $(x-1)^2$. Since the n ROOTS of an nth degree POLYNOMIAL are usually distinct, ROOTS which coincide are said to be degenerate. Degenerate cases often require special treatment in numerical and analytical solutions. For example, a simple search for both ROOTS of the above equation would find only a single one: 1

The word degenerate also has several very specific and technical meanings in different branches of mathematics.

References

Arfken, G. *Mathematical Methods for Physicists, 3rd ed.* Orlando, FL: Academic Press, pp. 513–514, 1985.

Degree

The word "degree" has many meanings in mathematics. The most common meaning is the unit of ANGLE measure defined such that an entire rotation is 360°. This unit harks back to the Babylonians, who used a base 60 number system. 360° likely arises from the Babylonian year, which was composed of 360 days (12 months of 30 days each). The degree is subdivided into 60 MINUTES per degree, and 60 SECONDS per MINUTE.

see also ARC MINUTE, ARC SECOND, DEGREE OF FREEDOM, DEGREE (MAP), DEGREE (POLYNOMIAL), DEGREE (VERTEX), INDEGREE, LOCAL DEGREE, OUTDEGREE

Degree (Algebraic Surface)

see ORDER (ALGEBRAIC SURFACE)

Degree of Freedom

The number of degrees of freedom in a problem, distribution, etc., is the number of parameters which may be independently varied.

see also LIKELIHOOD RATIO

Degree (Map)

Let $f : M \mapsto N$ be a MAP between two compact, connected, oriented n-D MANIFOLDS without boundary. Then f induces a HOMEOMORPHISM f_* from the HOMOLOGY GROUPS $H_n(M)$ to $H_n(N)$, both canonically isomorphic to the INTEGERS, and so f_* can be thought of as a HOMEOMORPHISM of the INTEGERS. The INTEGER $d(f)$ to which the number 1 gets sent is called the degree of the MAP f.

There is an easy way to compute $d(f)$ if the MANIFOLDS involved are smooth. Let $x \in \mathbb{N}$, and approximate f by a smooth map HOMOTOPIC to f such that x is a "regular value" of f (which exist and are everywhere by SARD'S THEOREM). By the IMPLICIT FUNCTION THEOREM, each point in $f^{-1}(x)$ has a NEIGHBORHOOD such that f restricted to it is a DIFFEOMORPHISM. If the DIFFEOMORPHISM is orientation preserving, assign it the number $+1$, and if it is orientation reversing, assign it the number -1. Add up all the numbers for all the points in $f^{-1}(x)$, and that is the $d(f)$, the degree of f. One reason why the degree of a map is important is because it is a HOMOTOPY invariant. A sharper result states that two self-maps of the n-sphere are homotopic IFF they have the same degree. This is equivalent to the result that the nth HOMOTOPY GROUP of the n-SPHERE is the set $\mathbb{Z}$ of INTEGERS. The ISOMORPHISM is given by taking the degree of any representation.

One important application of the degree concept is that homotopy classes of maps from n-spheres to n-spheres are classified by their degree (there is exactly one homotopy class of maps for every INTEGER n, and n is the degree of those maps).

Degree (Polynomial)

see ORDER (POLYNOMIAL)

Degree Sequence

Given an (undirected) GRAPH, a degree sequence is a monotonic nonincreasing sequence of the degrees of its VERTICES. A degree sequence is said to be k-connected if there exists some k-CONNECTED GRAPH corresponding to the degree sequence. For example, while the degree sequence {1, 2, 1} is 1- but not 2-connected, {2, 2, 2} is 2-connected. The number of degree sequences for $n = 1, 2, \ldots$ is given by 1, 2, 4, 11, 31, 102, ... (Sloane's A004251).

see also GRAPHICAL PARTITION

References

Ruskey, F. "Information on Degree Sequences." http://sue.csc.uvic.ca/~cos/inf/nump/DegreeSequences.html.

Ruskey, F.; Cohen, R.; Eades, P.; and Scott, A. "Alley CATs in Search of Good Homes." *Congres. Numer.* **102**, 97–110, 1994.

Sloane, N. J. A. Sequence A004251/M1250 in "An On-Line Version of the Encyclopedia of Integer Sequences."

Degree (Vertex)

see VERTEX DEGREE

Dehn Invariant

An invariant defined using the angles of a 3-D POLYHEDRON. It remains constant under solid DISSECTION and reassembly. However, solids with the same volume can have different Dehn invariants. Two POLYHEDRA can be dissected into each other only if they have the same volume and the same Dehn invariant.

see also DISSECTION, EHRHART POLYNOMIAL

Dehn's Lemma

If you have an embedding of a 1-SPHERE in a 3-MANIFOLD which exists continuously over the 2-DISK, then it also extends over the DISK as an embedding. It was proposed by Dehn in 1910, but a correct proof was not obtained until the work of Papakyriakopoulos (1957ab).

References

Hempel, J. *3-Manifolds.* Princeton, NJ: Princeton University Press, 1976.

Papakyriakopoulos, C. D. "On Dehn's Lemma and the Asphericity of Knots." *Proc. Nat. Acad. Sci. USA* **43**, 169–172, 1957a.

Papakyriakopoulos, C. D. "On Dehn's Lemma and the Asphericity of Knots." *Ann. Math.* **66**, 1–26, 1957.

Rolfsen, D. *Knots and Links.* Wilmington, DE: Publish or Perish Press, pp. 100–101, 1976.

Dehn Surgery

The operation of drilling a tubular NEIGHBORHOOD of a KNOT K in $\mathbb{S}^3$ and then gluing in a solid TORUS so that its meridian curve goes to a (p,q)-curve on the TORUS boundary of the KNOT exterior. Every compact connected 3-MANIFOLD comes from Dehn surgery on a LINK in $\mathbb{S}^3$.

see also KIRBY CALCULUS

References

Adams, C. C. *The Knot Book: An Elementary Introduction to the Mathematical Theory of Knots.* New York: W. H. Freeman, p. 260, 1994.

Del

see GRADIENT

Del Pezzo Surface

A SURFACE which is related to CAYLEY NUMBERS.

References

Coxeter, H. S. M. *Regular Polytopes, 3rd ed.* New York: Dover, p. 211, 1973.

Hunt, B. "Del Pezzo Surfaces." §4.1.4 in *The Geometry of Some Special Arithmetic Quotients.* New York: Springer-Verlag, pp. 128–129, 1996.

Delannoy Number

The Delannoy numbers are defined by

$$D(a,b) = D(a-1,b) + D(a,b-1) + D(a-1,b-1),$$

where $D(0,0) = 1$. They are the number of lattice paths from $(0,0)$ to (b,a) in which only east (1, 0), north (0, 1), and northeast (1, 1) steps are allowed (i.e, $\rightarrow$, $\uparrow$, and $\nearrow$).

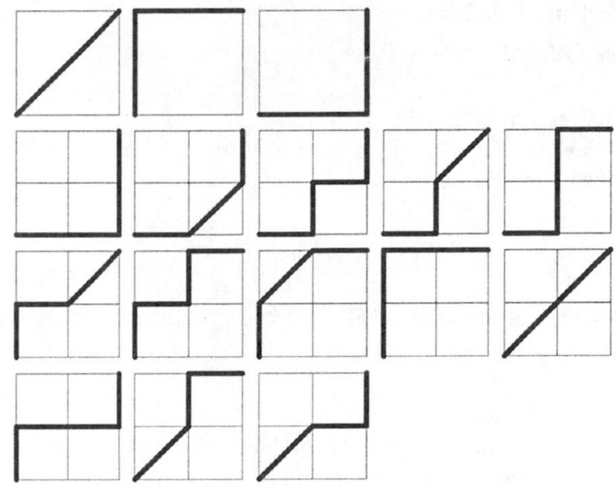

For $n \equiv a = b$, the Delannoy numbers are the number of "king walks"

$$D(n,n) = P_n(3),$$

where $P_n(x)$ is a LEGENDRE POLYNOMIAL (Moser 1955, Vardi 1991). Another expression is

$$D(n,n) = \sum_{k=0}^{n} \binom{n}{k} \binom{n+k}{k} = {}_2F_1(-n, n+1; 1, -1),$$

where $\binom{a}{b}$ is a BINOMIAL COEFFICIENT and ${}_2F_1(a,b;c;z)$ is a HYPERGEOMETRIC FUNCTION. The values of $D(n,n)$ for $n = 1, 2, \ldots$ are 3, 13, 63, 321, 1683, 8989, 48639, ... (Sloane's A001850).

The SCHRÖDER NUMBERS bear the same relation to the Delannoy numbers as the CATALAN NUMBERS do to the BINOMIAL COEFFICIENTS.

see also BINOMIAL COEFFICIENT, CATALAN NUMBER, MOTZKIN NUMBER, SCHRÖDER NUMBER

References

Sloane, N. J. A. Sequence A001850/M2942 in "An On-Line Version of the Encyclopedia of Integer Sequences."

Delaunay Triangulation

The NERVE of the cells in a VORONOI DIAGRAM, which is the triangular of the CONVEX HULL of the points in the diagram. The Delaunay triangulation of a VORONOI DIAGRAM in $\mathbb{R}^2$ is the diagram's planar dual.

see also TRIANGULATION

Delian Constant

The number $2^{1/3}$ (the CUBE ROOT of 2) which is to be constructed in the CUBE DUPLICATION problem. This number is not a EUCLIDEAN NUMBER although it is an ALGEBRAIC of third degree.

References

Conway, J. H. and Guy, R. K. "Three Greek Problems." In *The Book of Numbers.* New York: Springer-Verlag, pp. 192–194, 1996.

Delian Problem

see CUBE DUPLICATION

Delta Amplitude

Given an AMPLITUDE ϕ and a MODULUS m in an ELLIPTIC INTEGRAL,

$$\Delta(\phi) \equiv \sqrt{1 - m\sin^2\phi}.$$

see also AMPLITUDE, ELLIPTIC INTEGRAL, MODULUS (ELLIPTIC INTEGRAL)

Delta Curve

A curve which can be turned continuously inside an EQUILATERAL TRIANGLE. There are an infinite number of delta curves, but the simplest are the CIRCLE and lens-shaped Δ-biangle. All the Δ curves of height h have the same PERIMETER $2\pi h/3$. Also, at each position of a Δ curve turning in an EQUILATERAL TRIANGLE, the perpendiculars to the sides at the points of contact are CONCURRENT at the instantaneous center of rotation.

see also REULEAUX TRIANGLE

References

Honsberger, R. *Mathematical Gems I.* Washington, DC: Math. Assoc. Amer., pp. 56–59, 1973.

Delta Function

Defined as the limit of a class of DELTA SEQUENCES. Sometimes called the IMPULSE SYMBOL. The most commonly used (equivalent) definitions are

$$\delta(x) \equiv \lim_{n\to\infty} \frac{1}{2\pi} \frac{\sin[(n+\frac{1}{2})x]}{\sin(\frac{1}{2}x)} \tag{1}$$

(the so-called DIRICHLET KERNEL) and

$$\begin{aligned}\delta(x) &\equiv \lim_{n\to\infty} \frac{\sin(nx)}{\pi x} &(2)\\ &= \frac{1}{2\pi}\int_{-\infty}^{\infty} e^{-ikx}\,dk &(3)\\ &= \frac{1}{2\pi}\int_{-\infty}^{\infty} e^{-ikx}\,dk &(4)\\ &= \mathcal{F}[1], &(5)\end{aligned}$$

where $\mathcal{F}$ is the FOURIER TRANSFORM. Some identities include

$$\delta(x-a) = 0 \tag{6}$$

for $x \neq a$,

$$\int_{a-\epsilon}^{a+\epsilon} \delta(x-a)\,dx = 1, \tag{7}$$

where ϵ is any POSITIVE number, and

$$\int_{-\infty}^{\infty} f(x)\delta(x-a)\,dx = f(a) \tag{8}$$

$$\int_{-\infty}^{\infty} f(x)\delta'(x-a)dx = -f'(a) \tag{9}$$

$$x\int f(x)\delta(x-x_0)\,dx = x_0\int f(x)\delta(x-x_0)\,dx \tag{10}$$

$$\delta' * f = \int_{-\infty}^{\infty} \delta'(a-x)f(x)\,dx = f'(x) \tag{11}$$

$$\int_{-\infty}^{\infty} |\delta'(x)|\,dx = \infty \tag{12}$$

$$x^2\delta'(x) = 0 \tag{13}$$

$$\delta'(-x) = -\delta'(x) \tag{14}$$

$$x\delta'(x) = -\delta(x). \tag{15}$$

(15) can be established using INTEGRATION BY PARTS as follows:

$$\begin{aligned}\int f(x)x\delta'(x)\,dx &= -\int \delta(x)\frac{d}{dx}[xf(x)]\,dx\\ &= -\int \delta[f(x) + xf'(x)]\,dx\\ &= -\int f(x)\delta(x)\,dx. &(16)\end{aligned}$$

Additional identities are

$$\delta(ax) = \frac{1}{a}\delta(x) \tag{17}$$

$$\delta(x^2 - a^2) = \frac{1}{2a}[\delta(x+a) + \delta(x-a)] \tag{18}$$

$$\delta[g(x)] = \sum_i \frac{\delta(x-x_i)}{|g'(x_i)|}, \tag{19}$$

where the x_is are the ROOTS of g. For example, examine

$$\delta(x^2 + x - 2) = \delta[(x-1)(x+2)]. \tag{20}$$

Then $g'(x) = 2x+1$, so $g'(x_1) = g'(1) = 3$ and $g'(x_2) = g'(-2) = -3$, and we have

$$\delta(x^2 + x - 2) = \tfrac{1}{3}\delta(x-1) + \tfrac{1}{3}\delta(x+2). \tag{21}$$

A FOURIER SERIES expansion of $\delta(x-a)$ gives

$$a_n = \frac{1}{\pi}\int_{-\pi}^{\pi} \delta(x-a)\cos(nx)\,dx = \frac{1}{\pi}\cos(na) \tag{22}$$

$$b_n = \frac{1}{\pi}\int_{-\pi}^{\pi} \delta(x-a)\sin(nx)\,dx = \frac{1}{\pi}\sin(na), \tag{23}$$

so

$$\begin{aligned}\delta(x-a) &= \frac{1}{2\pi} + \frac{1}{\pi}\sum_{n=1}^{\infty}[\cos(na)\cos(nx) + \sin(na)\sin(nx)]\\ &= \frac{1}{2\pi} + \frac{1}{\pi}\sum_{n=1}^{\infty}\cos[n(x-a)]. &(24)\end{aligned}$$

The FOURIER TRANSFORM of the delta function is

$$\mathcal{F}[\delta(x-x_0)] = \int_{-\infty}^{\infty} e^{-2\pi ikx}\delta(x-x_0)\,dx = e^{-2\pi ikx_0}. \tag{25}$$

Delta functions can also be defined in 2-D, so that in 2-D CARTESIAN COORDINATES

$$\delta^2(x-x_0, y-y_0) = \delta(x-x_0)\delta(y-y_0), \tag{26}$$

and in 3-D, so that in 3-D CARTESIAN COORDINATES

$$\delta^3(x-x_0, y-y_0, z-z_0) = \delta(x-x_0)\delta(y-y_0)\delta(z-z_0), \tag{27}$$

in CYLINDRICAL COORDINATES

$$\delta^3(r,\theta,z) = \frac{\delta(r)\delta(z)}{\pi r}, \tag{28}$$

and in SPHERICAL COORDINATES,

$$\delta^3(r,\theta,\phi) = \frac{\delta(r)}{2\pi r^2}. \tag{29}$$

A series expansion in CYLINDRICAL COORDINATES gives

$$\begin{aligned}\delta^3(\mathbf{r}_1-\mathbf{r}_2) &= \frac{1}{r_1}\delta(r_1-r_2)\delta(\phi_1-\phi_2)\delta(z_1-z_2)\\ &= \frac{1}{r_1}\delta(r_1-r_2)\frac{1}{2\pi}\sum_{m=-\infty}^{\infty} e^{im(\phi_1-\phi_2)}\frac{1}{2\pi}\int_{-\infty}^{\infty} e^{ik(z_1-z_2)}\,dk.\end{aligned} \tag{30}$$

The delta function also obeys the so-called SIFTING PROPERTY

$$\int f(\mathbf{y})\delta(\mathbf{x}-\mathbf{y})\,d\mathbf{y} = f(\mathbf{x}). \tag{31}$$

see also DELTA SEQUENCE, DOUBLET FUNCTION, FOURIER TRANSFORM—DELTA FUNCTION

References

Arfken, G. *Mathematical Methods for Physicists, 3rd ed.* Orlando, FL: Academic Press, pp. 481–485, 1985.

Spanier, J. and Oldham, K. B. "The Dirac Delta Function $\delta(x-a)$." Ch. 10 in *An Atlas of Functions.* Washington, DC: Hemisphere, pp. 79–82, 1987.

Delta Sequence

A SEQUENCE of strongly peaked functions for which

$$\lim_{n\to\infty}\int_{-\infty}^{\infty} \delta_n(x)f(x)\,dx = f(n) \tag{1}$$

so that in the limit as $n\to\infty$, the sequences become DELTA FUNCTIONS. Examples include

$$\delta_n(x) = \begin{cases}0 & x<-\frac{1}{2n}\\ n & -\frac{1}{2n}<x<\frac{1}{2n}\\ 0 & x>\frac{1}{2n}\end{cases} \tag{2}$$

$$= \frac{n}{\sqrt{\pi}}e^{-n^2x^2} \tag{3}$$

$$= \frac{n}{\pi}\operatorname{sinc}(ax) \equiv \frac{\sin(nx)}{\pi x} \tag{4}$$

$$= \frac{1}{\pi x}\frac{e^{inx}-e^{-inx}}{2i} \tag{5}$$

$$= \frac{1}{2\pi ix}\left[e^{ixt}\right]_{-n}^{n} \tag{6}$$

$$= \frac{1}{2\pi}\int_{-n}^{n} e^{ixt}\,dt \tag{7}$$

$$= \frac{1}{2\pi}\frac{\sin\left[\left(n+\frac{1}{2}\right)x\right]}{\sin\left(\frac{1}{2}x\right)}, \tag{8}$$

where (8) is known as the DIRICHLET KERNEL.

Delta Variation

see VARIATION

Deltahedron

A semiregular POLYHEDRON whose faces are all EQUILATERAL TRIANGLES. There are an infinite number of deltahedra, but only eight convex ones (Freudenthal and van der Waerden 1947). They have 4, 6, 8, 10, 12, 14, 16, and 20 faces. These are summarized in the table below, and illustrated in the following figures.

n	Name
4	tetrahedron
6	triangular dipyramid
8	octahedron
10	pentagonal dipyramid
12	snub disphenoid
14	triaugmented triangular prism
16	gyroelongated square dipyramid
20	icosahedron

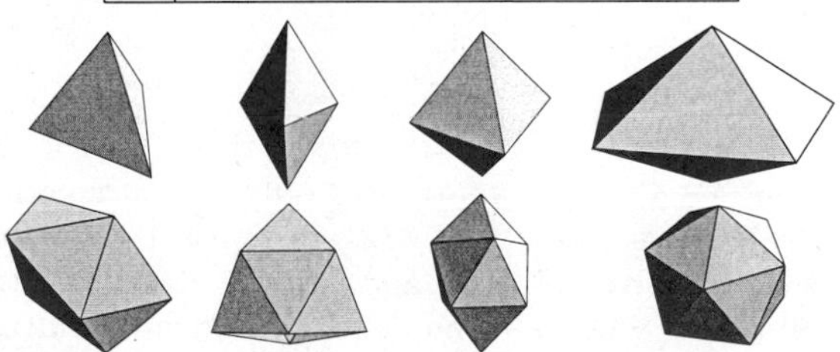

The STELLA OCTANGULA is a concave deltahedron with 24 sides:

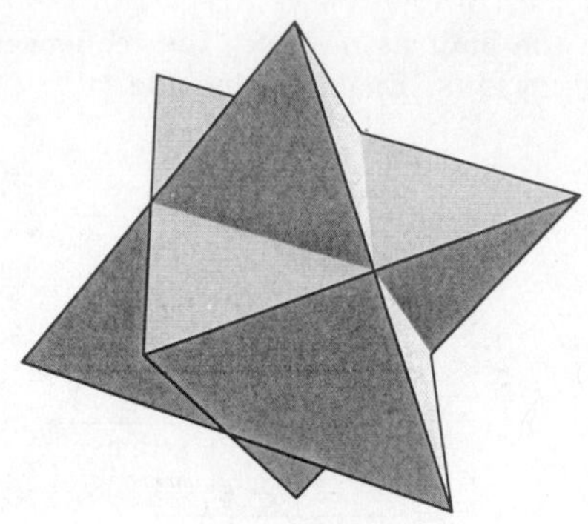

Another with 60 faces is a "caved in" DODECAHEDRON which is ICOSAHEDRON STELLATION I_{20}.

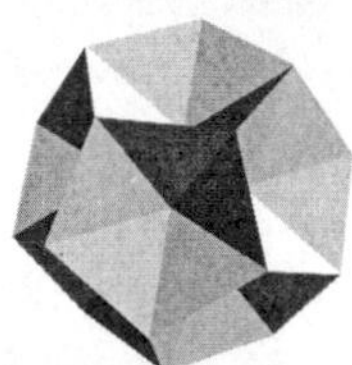

Cundy (1952) identifies 17 concave deltahedra with two kinds of VERTICES.

see also GYROELONGATED SQUARE DIPYRAMID, ICOSAHEDRON, OCTAHEDRON, PENTAGONAL DIPYRAMID, SNUB DISPHENOID TETRAHEDRON, TRIANGULAR DIPYRAMID, TRIAUGMENTED TRIANGULAR PRISM

References

Cundy, H. M. "Deltahedra." *Math. Gaz.* **36**, 263–266, 1952.

Freudenthal, H. and van der Waerden, B. L. "On an Assertion of Euclid." *Simon Stevin* **25**, 115–121, 1947.

Gardner, M. *Fractal Music, Hypercards, and More: Mathematical Recreations from Scientific American Magazine.* New York: W. H. Freeman, pp. 40, 53, and 58–60, 1992.

Pugh, A. *Polyhedra: A Visual Approach.* Berkeley, CA: University of California Press, pp. 35–36, 1976.

Deltoid

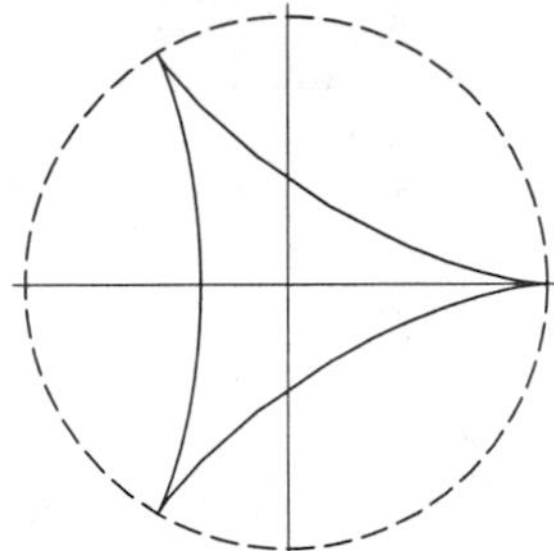

A 3-cusped HYPOCYCLOID, also called a TRICUSPOID, which has $n \equiv a/b = 3$ or $3/2$, where a is the RADIUS of the large fixed CIRCLE and b is the RADIUS of the small rolling CIRCLE. The deltoid was first considered by Euler in 1745 in connection with an optical problem. It was also investigated by Steiner in 1856 and is sometimes called STEINER'S HYPOCYCLOID (MacTutor Archive). The equation of the deltoid is obtained by setting $n = 3$ in the equation of the HYPOCYCLOID, yielding the parametric equations

$$x = [\tfrac{2}{3}\cos\phi - \tfrac{1}{3}\cos(2\phi)]a = 2b\cos\phi + b\cos(2\phi) \quad (1)$$
$$y = [\tfrac{2}{3}\sin\phi + \tfrac{1}{3}\sin(2\phi)]a = 2b\sin\phi - b\sin(2\phi). \quad (2)$$

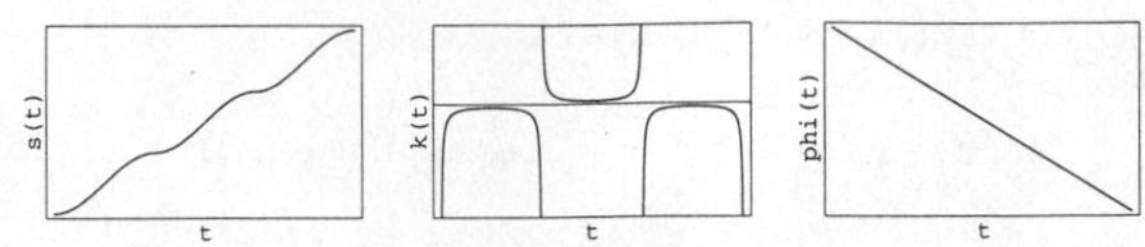

The ARC LENGTH, CURVATURE, and TANGENTIAL ANGLE are

$$s(t) = 4\int_0^t |\sin(\tfrac{3}{2}t')|\,dt' = \tfrac{16}{3}\sin^2(\tfrac{3}{4}t) \quad (3)$$
$$\kappa(t) = -\tfrac{1}{8}\csc(\tfrac{3}{2}t) \quad (4)$$
$$\phi(t) = -\tfrac{1}{2}t. \quad (5)$$

As usual, care must be taken in the evaluation of $s(t)$ for $t > 2\pi/3$. Since the form given above comes from an integral involving the ABSOLUTE VALUE of a function, it must be monotonic increasing. Each branch can be treated correctly by defining

$$n = \left\lfloor \frac{3t}{2\pi} \right\rfloor + 1, \quad (6)$$

where $\lfloor x \rfloor$ is the FLOOR FUNCTION, giving the formula

$$s(t) = (-1)^{1+[n \pmod 2]}\tfrac{16}{3}\sin^2(\tfrac{3}{4}t) + \tfrac{32}{3}\left\lfloor \tfrac{1}{2}n \right\rfloor. \quad (7)$$

The total ARC LENGTH is computed from the general HYPOCYCLOID equation

$$s_n = \frac{8a(n-1)}{n}. \quad (8)$$

With $n = 3$, this gives

$$s_3 = \tfrac{16}{3}a. \quad (9)$$

The AREA is given by

$$A_n = \frac{(n-1)(n-2)}{n^2}\pi a^2 \quad (10)$$

with $n = 3$,

$$A_3 = \tfrac{2}{9}\pi a^2. \quad (11)$$

The length of the tangent to the tricuspoid, measured between the two points P, Q in which it cuts the curve again, is constant and equal to $4a$. If you draw TANGENTS at P and Q, they are at RIGHT ANGLES.

References

Gray, A. *Modern Differential Geometry of Curves and Surfaces.* Boca Raton, FL: CRC Press, p. 53, 1993.

Lawrence, J. D. *A Catalog of Special Plane Curves.* New York: Dover, pp. 131–135, 1972.

Lee, X. "Deltoid." http://www.best.com/~xah/SpecialPlaneCurves_dir/Deltoid_dir/deltoid.html.
Lockwood, E. H. "The Deltoid." Ch. 8 in *A Book of Curves*. Cambridge, England: Cambridge University Press, pp. 72–79, 1967.
Macbeth, A. M. "The Deltoid, I." *Eureka* **10**, 20–23, 1948.
Macbeth, A. M. "The Deltoid, II." *Eureka* **11**, 26–29, 1949.
Macbeth, A. M. "The Deltoid, III." *Eureka* **12**, 5–6, 1950.
MacTutor History of Mathematics Archive. "Tricuspoid." http://www-groups.dcs.st-and.ac.uk/~history/Curves/Tricuspoid.html.
Yates, R. C. "Deltoid." *A Handbook on Curves and Their Properties*. Ann Arbor, MI: J. W. Edwards, pp. 71–74, 1952.

Deltoid Caustic

The caustic of the DELTOID when the rays are PARALLEL in any direction is an ASTROID.

Deltoid Evolute

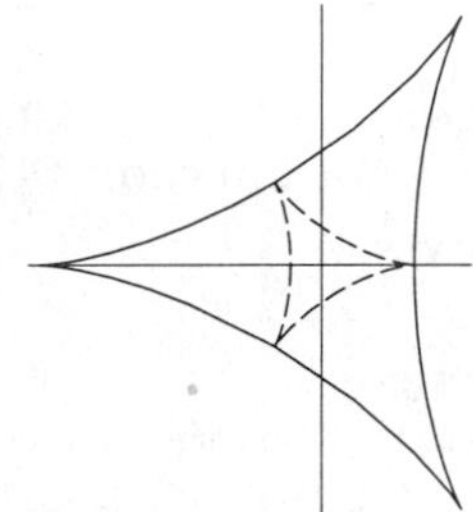

A HYPOCYCLOID EVOLUTE for $n = 3$ is another DELTOID scaled by a factor $n/(n-2) = 3/1 = 3$ and rotated $1/(2 \cdot 3) = 1/6$ of a turn.

Deltoid Involute

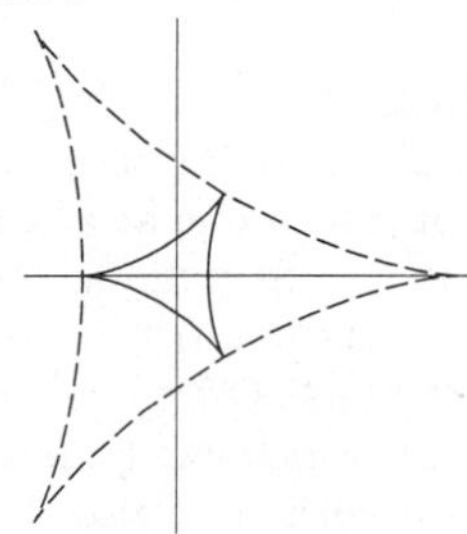

A HYPOCYCLOID INVOLUTE for $n = 3$ is another DELTOID scaled by a factor $(n-2)/n = 1/3$ and rotated $1/(2 \cdot 3) = 1/6$ of a turn.

Deltoid Pedal Curve

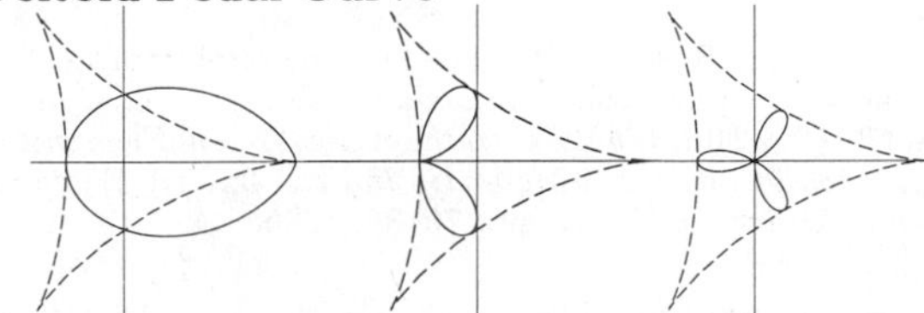

The PEDAL CURVE for a DELTOID with the PEDAL POINT at the CUSP is a FOLIUM. For the PEDAL POINT at the CUSP (NEGATIVE x-intercept), it is a BIFOLIUM. At the center, or anywhere on the inscribed EQUILATERAL TRIANGLE, it is a TRIFOLIUM.

Deltoid Radial Curve

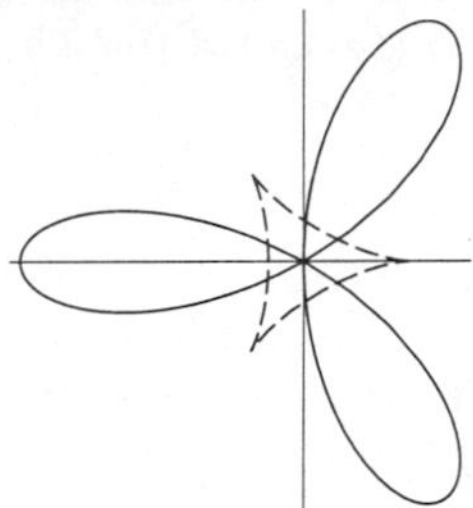

The TRIFOLIUM

$$x = x_0 + 4a\cos\phi - 4a\cos(2\phi)$$
$$y = y_0 + 4a\sin\phi + 4a\sin(2\phi).$$

Deltoidal Hexecontahedron

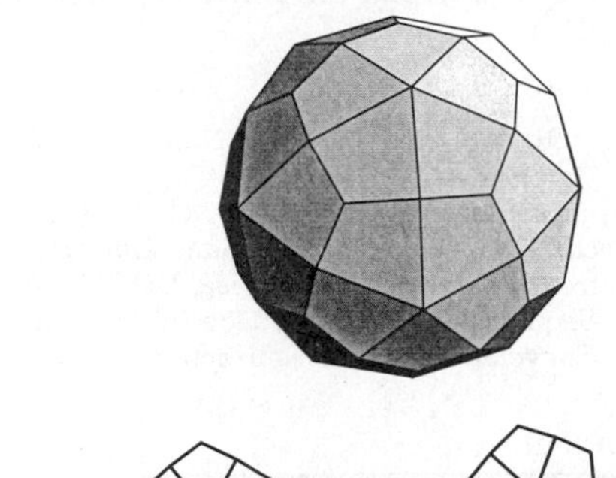

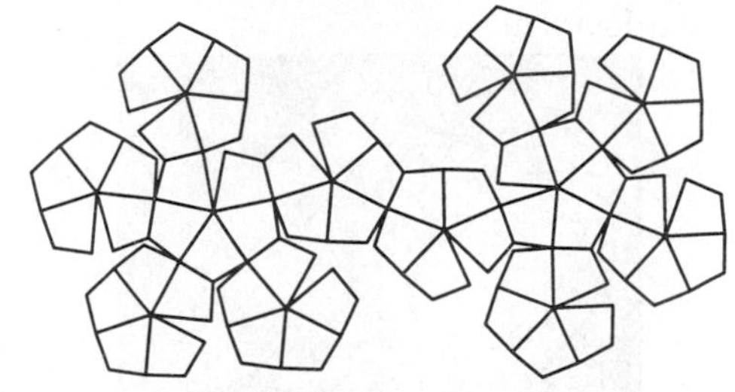

The DUAL POLYHEDRON of the RHOMBICOSIDODECAHEDRON.

Deltoidal Icositetrahedron

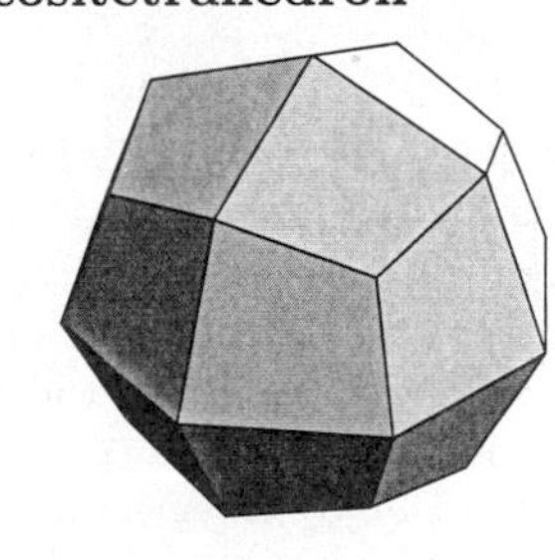

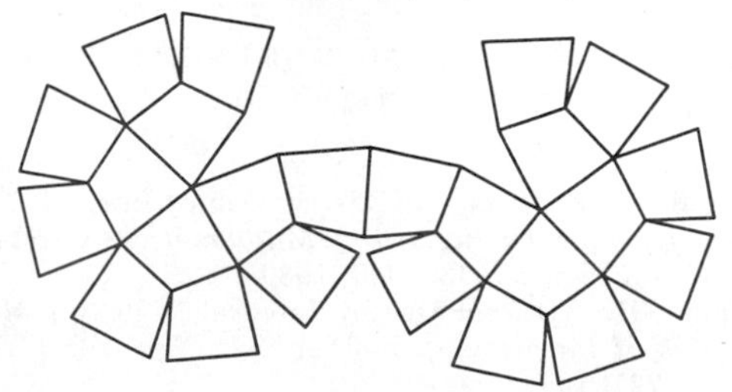

The DUAL POLYHEDRON of the SMALL RHOMBICUBOCTAHEDRON. It is also called the TRAPEZOIDAL ICOSITETRAHEDRON.

Demlo Number

The initially PALINDROMIC NUMBERS 1, 121, 12321, 1234321, 123454321, ... (Sloane's A002477). For the first through ninth terms, the sequence is given by the GENERATING FUNCTION

$$-\frac{10x+1}{(x-1)(10x-1)(100x-1)} = 1 + 121x + 12321x^2 + 1234321x^3 + \ldots$$

(Plouffe 1992, Sloane and Plouffe 1995). The definition of this sequence is slightly ambiguous from the tenth term on.

see also CONSECUTIVE NUMBER SEQUENCES, PALINDROMIC NUMBER

References

Kaprekar, D. R. "On Wonderful Demlo Numbers." *Math. Student* **6**, 68–70, 1938.

Plouffe, S. "Approximations de Séries Génératrices et quelques conjectures." Montréal, Canada: Université du Québec à Montréal, Mémoire de Maîtrise, UQAM, 1992.

Sloane, N. J. A. Sequence A002477/M5386 in "An On-Line Version of the Encyclopedia of Integer Sequences."

Dendrite Fractal

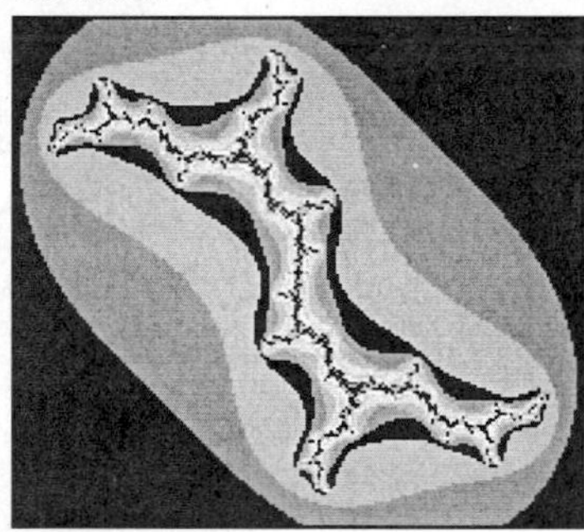

A JULIA SET with $c = i$.

Denjoy Integral

A type of INTEGRAL which is an extension of both the RIEMANN INTEGRAL and the LEBESGUE INTEGRAL. The original Denjoy integral is now called a Denjoy integral "in the restricted sense," and a more general type is now called a Denjoy integral "in the wider sense." The independently discovered PERON INTEGRAL turns out to be equivalent to the Denjoy integral "in the restricted sense."

see also INTEGRAL, LEBESGUE INTEGRAL, PERON INTEGRAL, RIEMANN INTEGRAL

References

Iyanaga, S. and Kawada, Y. (Eds.). "Denjoy Integrals." §103 in *Encyclopedic Dictionary of Mathematics.* Cambridge, MA: MIT Press, pp. 337–340, 1980.

Kestelman, H. "General Denjoy Integral." §9.2 in *Modern Theories of Integration, 2nd rev. ed.* New York: Dover, pp. 217–227, 1960.

Denominator

The number q in a FRACTION p/q.

see also FRACTION, NUMERATOR, RATIO, RATIONAL NUMBER

Dense

A set A in a FIRST-COUNTABLE SPACE is dense in B if $B = A \cup L$, where L is the limit of sequences of elements of A. For example, the rational numbers are dense in the reals. In general, a SUBSET A of X is dense if its CLOSURE $\text{cl}(A) = X$.

see also CLOSURE, DENSITY, DERIVED SET, PERFECT SET

Density

see DENSITY (POLYGON), DENSITY (SEQUENCE), NATURAL DENSITY

Density (Polygon)

The number q in a STAR POLYGON $\{\frac{p}{q}\}$.

see also STAR POLYGON

Density (Sequence)

Let a SEQUENCE $\{a_i\}_{i=1}^\infty$ be strictly increasing and composed of NONNEGATIVE INTEGERS. Call $A(x)$ the number of terms not exceeding x. Then the density is given by $\lim_{x\to\infty} A(x)/x$ if the LIMIT exists.

References

Guy, R. K. *Unsolved Problems in Number Theory, 2nd ed.* New York: Springer-Verlag, p. 199, 1994.

Denumerable Set

A SET is denumerable if a prescription can be given for identifying its members one at a time. Such a set is said to have CARDINAL NUMBER $\aleph_0$. Examples of denumerable sets include ALGEBRAIC NUMBERS, INTEGERS, and RATIONAL NUMBERS. Once one denumerable set S is given, any other set which can be put into a ONE-TO-ONE correspondence with S is also denumerable. Examples of nondenumerable sets include the REAL, COMPLEX, IRRATIONAL, and TRANSCENDENTAL NUMBERS.

see also ALEPH-0, ALEPH-1, CANTOR DIAGONAL SLASH, CONTINUUM, HILBERT HOTEL

References

Courant, R. and Robbins, H. "The Denumerability of the Rational Number and the Non-Denumerability of the Continuum." §2.4.2 in *What is Mathematics?: An Elementary Approach to Ideas and Methods, 2nd ed.* Oxford, England: Oxford University Press, pp. 79–83, 1996.

Denumerably Infinite

see DENUMERABLE SET

Depth (Graph)

The depth $E(G)$ of a GRAPH G is the minimum number of PLANAR GRAPHS P_i needed such that the union $\cup_i P_i = G$.

see also PLANAR GRAPH

Depth (Size)

The depth of a box is the horizontal DISTANCE from front to back (usually not necessarily defined to be smaller than the WIDTH, the horizontal DISTANCE from side to side).

see also HEIGHT, WIDTH (SIZE)

Depth (Statistics)

The smallest RANK (either up or down) of a set of data.

References

Tukey, J. W. *Exploratory Data Analysis.* Reading, MA: Addison-Wesley, p. 30, 1977.

Depth (Tree)

The depth of a RESOLVING TREE is the number of levels of links, not including the top. The depth of the link is the minimal depth for any RESOLVING TREE of that link. The only links of length 0 are the trivial links. A KNOT of length 1 is always a trivial KNOT and links of depth one are always HOPF LINKS, possibly with a few additional trivial components (Bleiler and Scharlemann). The LINKS of depth two have also been classified (Thompson and Scharlemann).

References

Adams, C. C. *The Knot Book: An Elementary Introduction to the Mathematical Theory of Knots.* New York: W. H. Freeman, p. 169, 1994.

Derangement

A PERMUTATION of n ordered objects in which none of the objects appears in its natural place. The function giving this quantity is the SUBFACTORIAL $!n$, defined by

$$!n \equiv n! \sum_{k=0}^{n} \frac{(-1)^k}{k!} \tag{1}$$

or

$$!n \equiv \left[\frac{n!}{e}\right], \tag{2}$$

where $k!$ is the usual FACTORIAL and $[x]$ is the NINT function. These are also called RENCONTRES NUMBERS (named after rencontres solitaire), or COMPLETE PERMUTATIONS, or derangements. The number of derangements $!n = d(n)$ of length n satisfy the RECURRENCE RELATIONS

$$d(n) = (n-1)[d(n-1) + d(n-2)] \tag{3}$$

and

$$d(n) = nd(n-1) + (-1)^n, \tag{4}$$

with $d(1) = 0$ and $d(2) = 1$. The first few are 0, 1, 2, 9, 44, 265, 1854, ... (Sloane's A000166). This sequence cannot be expressed as a fixed number of hypergeometric terms (Petkovšek *et al.* 1996, pp. 157–160).

see also MARRIED COUPLES PROBLEM, PERMUTATION, ROOT, SUBFACTORIAL

References

Aitken, A. C. *Determinants and Matrices.* Westport, CT: Greenwood Pub., p. 135, 1983.

Ball, W. W. R. and Coxeter, H. S. M. *Mathematical Recreations and Essays, 13th ed.* New York: Dover, pp. 46–47, 1987.

Coolidge, J. L. *An Introduction to Mathematical Probability.* Oxford, England: Oxford University Press, p. 24, 1925.

Courant, R. and Robbins, H. *What is Mathematics?: An Elementary Approach to Ideas and Methods, 2nd ed.* Oxford, England: Oxford University Press, pp. 115–116, 1996.

de Montmort, P. R. *Essai d'analyse sur les jeux de hasard.* Paris, p. 132, 1713.

Dickau, R. M. "Derangements." http://forum.swarthmore.edu/advanced/robertd/derangements.html.

Durell, C. V. and Robson, A. *Advanced Algebra.* London, p. 459, 1937.

Petkovšek, M.; Wilf, H. S.; and Zeilberger, D. *A=B.* Wellesley, MA: A. K. Peters, 1996.

Roberts, F. S. *Applied Combinatorics.* Englewood Cliffs, NJ: Prentice-Hall, 1984.

Ruskey, F. "Information on Derangements." http://sue.csc.uvic.ca/~cos/inf/perm/Derangements.html.

Sloane, N. J. A. Sequence A000166/M1937 in "An On-Line Version of the Encyclopedia of Integer Sequences."

Stanley, R. P. *Enumerative Combinatorics, Vol. 1.* New York: Cambridge University Press, p. 67, 1986.

Vardi, I. *Computational Recreations in Mathematica.* Reading, MA: Addison-Wesley, p. 123, 1991.

Derivative

The derivative of a FUNCTION represents an infinitesimal change in the function with respect to whatever parameters it may have. The "simple" derivative of a function f with respect to x is denoted either $f'(x)$ or $\frac{df}{dx}$ (and often written in-line as df/dx). When derivatives are taken with respect to time, they are often denoted using Newton's FLUXION notation, $\frac{dx}{dt} = \dot{x}$. The derivative of a function $f(x)$ with respect to the variable x is defined as

$$f'(x) \equiv \lim_{h \to 0} \frac{f(x+h) - f(x)}{h}. \tag{1}$$

Note that in order for the limit to exist, both $\lim_{h \to 0^+}$ and $\lim_{h \to 0^-}$ must exist and be equal, so the FUNCTION must be continuous. However, continuity is a NECESSARY but *not* SUFFICIENT condition for differentiability. Since some DISCONTINUOUS functions can be integrated, in a sense there are "more" functions which can be integrated than differentiated. In a letter to Stieltjes, Hermite wrote, "I recoil with dismay and horror at this lamentable plague of functions which do not have derivatives."

A 3-D generalization of the derivative to an arbitrary direction is known as the DIRECTIONAL DERIVATIVE.

In general, derivatives are mathematical objects which exist between smooth functions on manifolds. In this formalism, derivatives are usually assembled into "TANGENT MAPS."

Simple derivatives of some simple functions follow.

$$\frac{d}{dx}x^n = nx^{n-1} \tag{2}$$

$$\frac{d}{dx}\ln|x| = \frac{1}{x} \tag{3}$$

$$\frac{d}{dx}\sin x = \cos x \tag{4}$$

$$\frac{d}{dx}\cos x = -\sin x \tag{5}$$

$$\begin{aligned}\frac{d}{dx}\tan x &= \frac{d}{dx}\left(\frac{\sin x}{\cos x}\right) = \frac{\cos x\cos x - \sin x(-\sin x)}{\cos^2 x}\\ &= \frac{1}{\cos^2 x} = \sec^2 x\end{aligned} \tag{6}$$

$$\begin{aligned}\frac{d}{dx}\csc x &= \frac{d}{dx}(\sin x)^{-1} = -(\sin x)^{-2}\cos x = -\frac{\cos x}{\sin^2 x}\\ &= -\csc x\cot x\end{aligned} \tag{7}$$

$$\begin{aligned}\frac{d}{dx}\sec x &= \frac{d}{dx}(\cos x)^{-1} = -(\cos x)^{-2}(-\sin x) = \frac{\sin x}{\cos^2 x}\\ &= \sec x\tan x\end{aligned} \tag{8}$$

$$\begin{aligned}\frac{d}{dx}\cot x &= \frac{d}{dx}\left(\frac{\cos x}{\sin x}\right) = \frac{\sin x(-\sin x) - \cos x\cos x}{\cos^2 x}\\ &= -\frac{1}{\cos^2 x} = -\csc^2 x\end{aligned} \tag{9}$$

$$\frac{d}{dx}e^x = e^x \tag{10}$$

$$\begin{aligned}\frac{d}{dx}a^x &= \frac{d}{dx}e^{\ln a^x} = \frac{d}{dx}e^{x\ln a}\\ &= (\ln a)e^{x\ln a} = (\ln a)a^x\end{aligned} \tag{11}$$

$$\frac{d}{dx}\sin^{-1}x = \frac{1}{\sqrt{1-x^2}} \tag{12}$$

$$\frac{d}{dx}\cos^{-1}x = -\frac{1}{\sqrt{1-x^2}} \tag{13}$$

$$\frac{d}{dx}\tan^{-1}x = \frac{1}{1+x^2} \tag{14}$$

$$\frac{d}{dx}\cot^{-1}x = -\frac{1}{1+x^2} \tag{15}$$

$$\frac{d}{dx}\sec^{-1}x = \frac{1}{x\sqrt{x^2-1}} \tag{16}$$

$$\frac{d}{dx}\csc^{-1}x = -\frac{1}{x\sqrt{x^2-1}} \tag{17}$$

$$\frac{d}{dx}\sinh x = \cosh x \tag{18}$$

$$\frac{d}{dx}\cosh x = \sinh x \tag{19}$$

$$\frac{d}{dx}\tanh x = \operatorname{sech}^2 x \tag{20}$$

$$\frac{d}{dx}\coth x = -\operatorname{csch}^2 x \tag{21}$$

$$\frac{d}{dx}\operatorname{sech} x = -\operatorname{sech} x\tanh x \tag{22}$$

$$\frac{d}{dx}\operatorname{csch} x = -\operatorname{csch} x\coth x \tag{23}$$

$$\frac{d}{dx}\operatorname{sn} x = \operatorname{cn} x\operatorname{dn} x \tag{24}$$

$$\frac{d}{dx}\operatorname{cn} x = -\operatorname{sn} x\operatorname{dn} x \tag{25}$$

$$\frac{d}{dx}\operatorname{dn} x = -k^2\operatorname{sn} x\operatorname{cn} x. \tag{26}$$

Derivatives of sums are equal to the sum of derivatives so that

$$[f(x) + \ldots + h(x)]' = f'(x) + \ldots + h'(x). \tag{27}$$

In addition, if c is a constant,

$$\frac{d}{dx}[cf(x)] = cf'(x). \tag{28}$$

Furthermore,

$$\frac{d}{dx}[f(x)g(x)] = f(x)g'(x) + f'(x)g(x), \tag{29}$$

where f' denotes the DERIVATIVE of f with respect to x. This derivative rule can be applied iteratively to yield derivate rules for products of three or more functions, for example,

$$\begin{aligned}[fgh]' &= (fg)h' + (fg)'h = fgh' + (fg' + f'g)h\\ &= f'gh + fg'h + fgh'.\end{aligned} \tag{30}$$

Other rules involving derivatives include the CHAIN RULE, POWER RULE, PRODUCT RULE, and QUOTIENT RULE. Miscellaneous other derivative identities include

$$\frac{dy}{dx} = \frac{\frac{dy}{dt}}{\frac{dx}{dt}} \tag{31}$$

$$\frac{dy}{dx} = \frac{1}{\frac{dx}{dy}}. \tag{32}$$

If $F(x,y) = C$, where C is a constant, then

$$dF = \frac{\partial F}{\partial y}dy + \frac{\partial F}{\partial x}dx = 0, \tag{33}$$

so

$$\frac{dy}{dx} = -\frac{\frac{\partial F}{\partial x}}{\frac{\partial F}{\partial y}}. \tag{34}$$

A vector derivative of a vector function

$$\mathbf{X}(t) \equiv \begin{bmatrix} x_1(t)\\ x_2(t)\\ \vdots\\ x_k(t)\end{bmatrix} \tag{35}$$

can be defined by

$$\frac{d\mathbf{X}}{dt} = \begin{bmatrix} \frac{dx_1}{dt} \\ \frac{dx_2}{dt} \\ \vdots \\ \frac{dx_k}{dt} \end{bmatrix}. \tag{36}$$

see also BLANCMANGE FUNCTION, CARATHÉODORY DERIVATIVE, COMMA DERIVATIVE, CONVECTIVE DERIVATIVE, COVARIANT DERIVATIVE, DIRECTIONAL DERIVATIVE, EULER-LAGRANGE DERIVATIVE, FLUXION, FRACTIONAL CALCULUS, FRÉCHET DERIVATIVE, LAGRANGIAN DERIVATIVE, LIE DERIVATIVE, POWER RULE, SCHWARZIAN DERIVATIVE, SEMICOLON DERIVATIVE, WEIERSTRAß FUNCTION

References

Abramowitz, M. and Stegun, C. A. (Eds.). *Handbook of Mathematical Functions with Formulas, Graphs, and Mathematical Tables, 9th printing.* New York: Dover, p. 11, 1972.

Anton, H. *Calculus with Analytic Geometry, 5th ed.* New York: Wiley, 1987.

Beyer, W. H. "Derivatives." *CRC Standard Mathematical Tables, 28th ed.* Boca Raton, FL: CRC Press, pp. 229–232, 1987.

Derivative Test

see FIRST DERIVATIVE TEST, SECOND DERIVATIVE TEST

Derived Set

The LIMIT POINTS of a SET P, denoted P'.

see also DENSE, LIMIT POINT, PERFECT SET

Dervish

A QUINTIC SURFACE having the maximum possible number of ORDINARY DOUBLE POINTS (31), which was constructed by W. Barth in 1994 (Endraß). The implicit equation of the surface is

$$64(x-w)[x^4 - 4x^3w - 10x^2y^2 - 4x^2w^2 + 16xw^3 - 20xy^2w + 5y^4 + 16w^4 - 20y^2w^2] - 5\sqrt{5-\sqrt{5}}\,(2z - \sqrt{5-\sqrt{5}}\,w) \times [4(x^2+y^2+z^2) + (1+3\sqrt{5})w^2]^2,$$

where w is a parameter (Endraß). The surface can also be described by the equation

$$aF + q = 0, \tag{1}$$

where

$$F = h_1h_2h_3h_4h_5, \tag{2}$$

$$h_1 = x - z \tag{3}$$

$$h_2 = \cos\left(\frac{2\pi}{5}\right)x - \sin\left(\frac{2\pi}{5}\right)y - z \tag{4}$$

$$h_3 = \cos\left(\frac{4\pi}{5}\right)x - \sin\left(\frac{4\pi}{5}\right)y - z \tag{5}$$

$$h_4 = \cos\left(\frac{6\pi}{5}\right)x - \sin\left(\frac{6\pi}{5}\right)y - z \tag{6}$$

$$h_5 = \cos\left(\frac{8\pi}{5}\right)x - \sin\left(\frac{8\pi}{5}\right)y - z \tag{7}$$

$$q = (1-cz)(x^2+y^2-1+rz^2)^2, \tag{8}$$

and

$$r = \tfrac{1}{4}(1+\sqrt{5}) \tag{9}$$

$$a = -\frac{8}{5}\left(1+\frac{1}{\sqrt{5}}\right)\sqrt{5-\sqrt{5}} \tag{10}$$

$$c = \tfrac{1}{2}\sqrt{5-\sqrt{5}} \tag{11}$$

(Nordstrand).

The dervish is invariant under the GROUP D_5 and contains exactly 15 lines. Five of these are the intersection of the surface with a D_5-invariant cone containing 16 nodes, five are the intersection of the surface with a D_5-invariant plane containing 10 nodes, and the last five are the intersection of the surface with a second D_5-invariant plane containing no nodes (Endraß).

References

Endraß, S. "Togliatti Surfaces." `http://www.mathematik.uni-mainz.de/AlgebraischeGeometrie/docs/Etogliatti.shtml`.

Endraß, S. "Flächen mit vielen Doppelpunkten." *DMV-Mitteilungen* **4**, 17–20, 4/1995.

Endraß, S. *Symmetrische Fläche mit vielen gewöhnlichen Doppelpunkten.* Ph.D. thesis. Erlangen, Germany, 1996.

Nordstrand, T. "Dervish." `http://www.uib.no/people/nfytn/dervtxt.htm`.

Desargues' Theorem

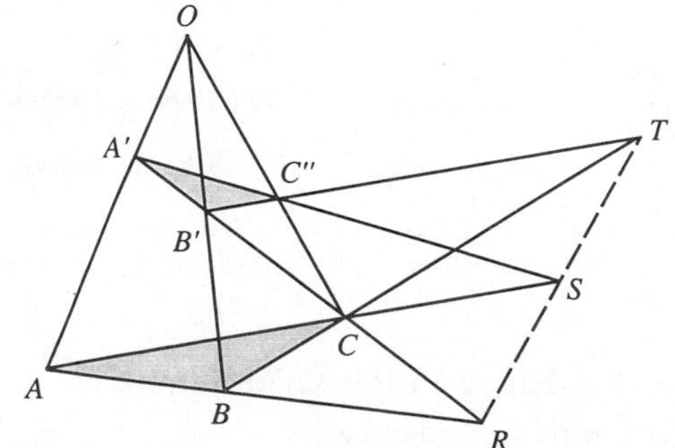

If the three straight LINES joining the corresponding VERTICES of two TRIANGLES ABC and $A'B'C'$ all meet in a point (the PERSPECTIVE CENTER), then the three intersections of pairs of corresponding sides lie on a straight LINE (the PERSPECTIVE AXIS). Equivalently, if two TRIANGLES are PERSPECTIVE from a POINT, they are PERSPECTIVE from a LINE.

Desargues' theorem is essentially its own dual according to the DUALITY PRINCIPLE of PROJECTIVE GEOMETRY.

see also DUALITY PRINCIPLE, PAPPUS'S HEXAGON THEOREM, PASCAL LINE, PASCAL'S THEOREM, PERSPECTIVE AXIS, PERSPECTIVE CENTER, PERSPECTIVE TRIANGLES

References
Coxeter, H. S. M. and Greitzer, S. L. *Geometry Revisited.* Washington, DC: Math. Assoc. Amer., pp. 70–72, 1967.
Ogilvy, C. S. *Excursions in Geometry.* New York: Dover, pp. 89–92, 1990.

Descartes Circle Theorem

A special case of APOLLONIUS' PROBLEM requiring the determination of a CIRCLE touching three mutually tangent CIRCLES (also called the KISSING CIRCLES PROBLEM). There are two solutions: a small circle surrounded by the three original CIRCLES, and a large circle surrounding the original three. Frederick Soddy gave the FORMULA for finding the RADIUS of the so-called inner and outer SODDY CIRCLES given the RADII of the other three. The relationship is

$$2({\kappa_1}^2 + {\kappa_2}^2 + {\kappa_3}^2 + {\kappa_4}^2) = (\kappa_1 + \kappa_2 + \kappa_3 + \kappa_4)^2,$$

where κ_i are the CURVATURES of the CIRCLES. Here, the NEGATIVE solution corresponds to the outer SODDY CIRCLE and the POSITIVE solution to the inner SODDY CIRCLE. This formula was known to Descartes and Viète (Boyer and Merzbach 1991, p. 159), but Soddy extended it to SPHERES. In n-D space, $n+2$ mutually touching n-SPHERES can always be found, and the relationship of their CURVATURES is

$$n\left(\sum_{i=1}^{n+2} {\kappa_i}^2\right) = \left(\sum_{i=1}^{n+2} \kappa_i\right)^2.$$

see also APOLLONIUS' PROBLEM, FOUR COINS PROBLEM, SODDY CIRCLES, SPHERE PACKING

References
Boyer, C. B. and Merzbach, U. C. *A History of Mathematics, 2nd ed.* New York: Wiley, 1991.
Coxeter, H. S. M. *Introduction to Geometry, 2nd ed.* New York: Wiley, pp. 13–16, 1969.
Wilker, J. B. "Four Proofs of a Generalization of the Descartes Circle Theorem." *Amer. Math. Monthly* **76**, 278–282, 1969.

Descartes-Euler Polyhedral Formula

see POLYHEDRAL FORMULA

Descartes Folium

see FOLIUM OF DESCARTES

Descartes' Formula

see DESCARTES TOTAL ANGULAR DEFECT

Descartes Ovals

see CARTESIAN OVALS

Descartes' Sign Rule

A method of determining the maximum number of POSITIVE and NEGATIVE REAL ROOTS of a POLYNOMIAL.

For POSITIVE ROOTS, start with the SIGN of COEFFICIENT of the lowest (or highest) POWER. Count the number of SIGN changes n as you proceed from the lowest to the highest POWER (ignoring POWERS which do not appear). Then n is the *maximum* number of POSITIVE ROOTS. Furthermore, the number of allowable ROOTS is n, $n-2$, $n-4$, ... For example, consider the POLYNOMIAL

$$f(x) = x^7 + x^6 - x^4 - x^3 - x^2 + x - 1.$$

Since there are three SIGN changes, there are a maximum of three possible POSITIVE ROOTS.

For NEGATIVE ROOTS, starting with a POLYNOMIAL $f(x)$, write a new POLYNOMIAL $g(x)$ with the SIGNS of all ODD POWERS reversed, while leaving the SIGNS of the EVEN POWERS unchanged. Then proceed as before to count the number of SIGN changes n. Then n is the *maximum* number of NEGATIVE ROOTS. For example, consider the POLYNOMIAL

$$f(x) = x^7 + x^6 - x^4 - x^3 - x^2 + x - 1,$$

and compute the new POLYNOMIAL

$$g(x) = -x^7 + x^6 - x^4 + x^3 - x^2 - x - 1.$$

There are four SIGN changes, so there are a maximum of four NEGATIVE ROOTS.

see also BOUND, STURM FUNCTION

References
Anderson, B.; Jackson, J.; and Sitharam, M. "Descartes' Rule of Signs Revisited." *Amer. Math. Monthly* **105**, 447–451, 1998.
Hall, H. S. and Knight, S. R. *Higher Algebra: A Sequel to Elementary Algebra for Schools.* London: Macmillan, pp. 459–460, 1950.
Struik, D. J. (Ed.). *A Source Book in Mathematics 1200–1800.* Princeton, NJ: Princeton University Press, pp. 89–93, 1986.

Descartes Total Angular Defect

The total angular defect is the sum of the ANGULAR DEFECTS over all VERTICES of a POLYHEDRON, where the ANGULAR DEFECT δ at a given VERTEX is the difference between the sum of face angles and 2π. For any convex POLYHEDRON, the Descartes total angular defect is

$$\Delta = \sum_i \delta_i = 4\pi. \tag{1}$$

This is equivalent to the POLYHEDRAL FORMULA for a closed rectilinear surface, which satisfies

$$\Delta = 2\pi(V - E + F). \tag{2}$$

A POLYHEDRON with N_0 equivalent VERTICES is called a PLATONIC SOLID and can be assigned a SCHLÄFLI SYMBOL $\{p, q\}$. It then satisfies

$$N_0 = \frac{4\pi}{\delta} \tag{3}$$

and

$$\delta = 2\pi - q\left(1 - \frac{2}{p}\right)\pi, \tag{4}$$

so

$$N_0 = \frac{4p}{2p + 2q - pq}. \tag{5}$$

see also ANGULAR DEFECT, PLATONIC SOLID, POLYHEDRAL FORMULA, POLYHEDRON

Descriptive Set Theory

The study of DEFINABLE SETS and functions in POLISH SPACES.

References

Becker, H. and Kechris, A. S. *The Descriptive Set Theory of Polish Group Actions.* New York: Cambridge University Press, 1996.

Design

A formal description of the constraints on the possible configurations of an experiment which is subject to given conditions. A design is sometimes called an EXPERIMENTAL DESIGN.

see also BLOCK DESIGN, COMBINATORICS, DESIGN THEORY, HADAMARD DESIGN, HOWELL DESIGN, SPHERICAL DESIGN, SYMMETRIC BLOCK DESIGN, TRANSVERSAL DESIGN

Design Theory

The study of DESIGNS and, in particular, NECESSARY and SUFFICIENT conditions for the existence of a BLOCK DESIGN.

see also BRUCK-RYSER-CHOWLA THEOREM, FISHER'S BLOCK DESIGN INEQUALITY

References

Assmus, E. F. Jr. and Key, J. D. *Designs and Their Codes.* New York: Cambridge University Press, 1993.

Colbourn, C. J. and Dinitz, J. H. *CRC Handbook of Combinatorial Designs.* Boca Raton, FL: CRC Press, 1996.

Dinitz, J. H. and Stinson, D. R. (Eds.). "A Brief Introduction to Design Theory." Ch. 1 in *Contemporary Design Theory: A Collection of Surveys.* New York: Wiley, pp. 1–12, 1992.

Desmic Surface

Let Δ_1, Δ_2, and Δ_3 be tetrahedra in projective 3-space $\mathbb{P}^3$. Then the tetrahedra are said to be desmically related if there exist constants α, β, and γ such that

$$\alpha\Delta_1 + \beta\Delta_2 + \gamma\Delta_3 = 0.$$

A desmic surface is then defined as a QUARTIC SURFACE which can be written as

$$a\Delta_1 + b\Delta_2 + c\Delta_3 = 0$$

for desmically related tetrahedra Δ_1, Δ_2, and Δ_3. Desmic surfaces have 12 ORDINARY DOUBLE POINTS, which are the vertices of three tetrahedra in 3-space (Hunt).

see also QUARTIC SURFACE

References

Hunt, B. "Desmic Surfaces." §B.5.2 in *The Geometry of Some Special Arithmetic Quotients.* New York: Springer-Verlag, pp. 311–315, 1996.

Jessop, C. §13 in *Quartic Surfaces with Singular Points.* Cambridge, England: Cambridge University Press, 1916.

Destructive Dilemma

A formal argument in LOGIC in which it is stated that

1. $P \Rightarrow Q$ and $R \Rightarrow S$ (where $\Rightarrow$ means "IMPLIES"), and
2. Either not-Q or not-S is true, from which two statements it follows that either not-P or not-R is true.

see also CONSTRUCTIVE DILEMMA, DILEMMA

Determinant

Determinants are mathematical objects which are very useful in the analysis and solution of systems of linear equations. As shown in CRAMER'S RULE, a nonhomogeneous system of linear equations has a nontrivial solution IFF the determinant of the system's MATRIX is NONZERO (so that the MATRIX is nonsingular). A 2×2 determinant is defined to be

$$\det\begin{bmatrix} a & b \\ c & d \end{bmatrix} \equiv \begin{vmatrix} a & b \\ c & d \end{vmatrix} \equiv ad - bc. \tag{1}$$

A $k \times k$ determinant can be expanded by MINORS to obtain

$$\begin{vmatrix} a_{11} & a_{12} & a_{13} & \cdots & a_{1k} \\ a_{21} & a_{22} & a_{23} & \cdots & a_{2k} \\ \vdots & \vdots & \vdots & \ddots & \vdots \\ a_{k1} & a_{k2} & a_{k3} & \cdots & a_{kk} \end{vmatrix} = a_{11} \begin{vmatrix} a_{22} & a_{23} & \cdots & a_{2k} \\ \vdots & \vdots & \ddots & \vdots \\ a_{k2} & a_{k3} & \cdots & a_{kk} \end{vmatrix} - a_{12} \begin{vmatrix} a_{21} & a_{23} & \cdots & a_{2k} \\ \vdots & \vdots & \ddots & \vdots \\ a_{k1} & a_{k3} & \cdots & a_{kk} \end{vmatrix} + \ldots \pm a_{1k} \begin{vmatrix} a_{21} & a_{22} & \cdots & a_{2(k-1)} \\ \vdots & \vdots & \ddots & \vdots \\ a_{k1} & a_{k2} & \cdots & a_{k(k-1)} \end{vmatrix}. \quad (2)$$

A general determinant for a MATRIX A has a value

$$|\mathsf{A}| = \sum_i a_{ij} a^{ij}, \quad (3)$$

with no implied summation over i and where a^{ij} is the COFACTOR of a_{ij} defined by

$$a^{ij} \equiv (-1)^{i+j} C_{ij}. \quad (4)$$

Here, C is the $(n-1) \times (n-1)$ MATRIX formed by eliminating row i and column j from A, i.e., by DETERMINANT EXPANSION BY MINORS.

Given an $n \times n$ determinant, the additive inverse is

$$|-\mathsf{A}| = (-1)^n |\mathsf{A}|. \quad (5)$$

Determinants are also DISTRIBUTIVE, so

$$|\mathsf{AB}| = |\mathsf{A}|\,|\mathsf{B}|. \quad (6)$$

This means that the determinant of a MATRIX INVERSE can be found as follows:

$$|\mathsf{I}| = |\mathsf{AA}^{-1}| = |\mathsf{A}|\,|\mathsf{A}^{-1}| = 1, \quad (7)$$

where I is the IDENTITY MATRIX, so

$$|\mathsf{A}| = \frac{1}{|\mathsf{A}^{-1}|}. \quad (8)$$

Determinants are MULTILINEAR in rows and columns, since

$$\begin{vmatrix} a_1 & a_2 & a_3 \\ a_4 & a_5 & a_6 \\ a_7 & a_8 & a_9 \end{vmatrix} = \begin{vmatrix} a_1 & 0 & 0 \\ a_4 & a_5 & a_6 \\ a_7 & a_8 & a_9 \end{vmatrix} + \begin{vmatrix} 0 & a_2 & 0 \\ a_4 & a_5 & a_6 \\ a_7 & a_8 & a_9 \end{vmatrix} + \begin{vmatrix} 0 & 0 & a_3 \\ a_4 & a_5 & a_6 \\ a_7 & a_8 & a_9 \end{vmatrix} \quad (9)$$

and

$$\begin{vmatrix} a_1 & a_2 & a_3 \\ a_4 & a_5 & a_6 \\ a_7 & a_8 & a_9 \end{vmatrix} = \begin{vmatrix} a_1 & a_2 & a_3 \\ 0 & a_5 & a_6 \\ 0 & a_8 & a_9 \end{vmatrix} + \begin{vmatrix} 0 & a_2 & a_3 \\ a_4 & a_5 & a_6 \\ 0 & a_8 & a_9 \end{vmatrix} + \begin{vmatrix} 0 & a_2 & a_3 \\ 0 & a_5 & a_6 \\ a_7 & a_8 & a_9 \end{vmatrix}. \quad (10)$$

The determinant of the SIMILARITY TRANSFORMATION of a matrix is equal to the determinant of the original MATRIX

$$|\mathsf{BAB}^{-1}| = |\mathsf{B}|\,|\mathsf{A}|\,|\mathsf{B}^{-1}| = |\mathsf{B}|\,|\mathsf{A}| \frac{1}{|\mathsf{B}^{-1}|} = |\mathsf{A}|. \quad (11)$$

The determinant of a similarity transformation minus a multiple of the unit MATRIX is given by

$$\begin{aligned} |\mathsf{B}^{-1}\mathsf{AB} - \lambda\mathsf{I}| &= |\mathsf{B}^{-1}\mathsf{AB} - \mathsf{B}^{-1}\lambda\mathsf{IB}| = |\mathsf{B}^{-1}(\mathsf{A} - \lambda\mathsf{I})\mathsf{B}| \\ &= |\mathsf{B}^{-1}|\,|\mathsf{A} - \lambda\mathsf{I}|\,|\mathsf{B}| = |\mathsf{A} - \lambda\mathsf{I}|. \end{aligned} \quad (12)$$

The determinant of a MATRIX TRANSPOSE equals the determinant of the original MATRIX,

$$|\mathsf{A}| = |\mathsf{A}^{\mathrm{T}}|, \quad (13)$$

and the determinant of a COMPLEX CONJUGATE is equal to the COMPLEX CONJUGATE of the determinant

$$|\mathsf{A}^*| = |\mathsf{A}|^*. \quad (14)$$

Let ϵ be a small number. Then

$$|\mathsf{I} + \epsilon\mathsf{A}| = 1 + \epsilon\,\mathrm{Tr}(\mathsf{A}) + \mathcal{O}(\epsilon^2), \quad (15)$$

where $\mathrm{Tr}(\mathsf{A})$ is the trace of A. The determinant takes on a particularly simple form for a TRIANGULAR MATRIX

$$\begin{vmatrix} a_{11} & a_{21} & \cdots & a_{k1} \\ 0 & a_{22} & \cdots & a_{k2} \\ \vdots & \vdots & \ddots & \vdots \\ 0 & 0 & \cdots & a_{kk} \end{vmatrix} = \prod_{n=1}^{k} a_{nn}. \quad (16)$$

Important properties of the determinant include the following.

1. Switching two rows or columns changes the sign.
2. Scalars can be factored out from rows and columns.
3. Multiples of rows and columns can be added together without changing the determinant's value.
4. Scalar multiplication of a row by a constant c multiplies the determinant by c.
5. A determinant with a row or column of zeros has value 0.
6. Any determinant with two rows or columns equal has value 0.

Property 1 can be established by induction. For a 2×2 MATRIX, the determinant is

$$\begin{vmatrix} a_1 & b_1 \\ a_2 & b_2 \end{vmatrix} = a_1b_2 - b_1a_2 = -(b_1a_2 - a_1b_2) = -\begin{vmatrix} b_1 & a_1 \\ b_2 & a_2 \end{vmatrix} \tag{17}$$

For a 3×3 MATRIX, the determinant is

$$\begin{aligned}\begin{vmatrix} a_1 & b_1 & c_1 \\ a_2 & b_2 & c_2 \\ a_3 & b_3 & c_3 \end{vmatrix} &= a_1\begin{vmatrix} b_2 & c_2 \\ b_3 & c_3 \end{vmatrix} - b_1\begin{vmatrix} a_2 & c_2 \\ a_3 & c_3 \end{vmatrix} + c_1\begin{vmatrix} a_2 & b_2 \\ a_3 & b_3 \end{vmatrix} \\ &= -\left(a_1\begin{vmatrix} c_2 & b_2 \\ c_3 & b_3 \end{vmatrix} + b_1\begin{vmatrix} c_2 & a_2 \\ c_3 & a_3 \end{vmatrix} - c_1\begin{vmatrix} a_2 & b_2 \\ a_3 & b_3 \end{vmatrix}\right) \\ &= -\begin{vmatrix} a_1 & c_1 & b_1 \\ a_2 & c_2 & b_2 \\ a_3 & c_3 & b_3 \end{vmatrix} \\ &= -\left(-a_1\begin{vmatrix} b_2 & c_2 \\ b_3 & c_3 \end{vmatrix} + b_1\begin{vmatrix} a_2 & c_2 \\ a_3 & c_3 \end{vmatrix} + c_1\begin{vmatrix} b_2 & a_2 \\ b_3 & a_3 \end{vmatrix}\right) \\ &= -\begin{vmatrix} b_1 & a_1 & c_1 \\ b_2 & a_2 & c_2 \\ b_3 & a_3 & c_3 \end{vmatrix} \\ &= -\left(-a_1\begin{vmatrix} c_2 & b_2 \\ c_3 & b_3 \end{vmatrix} - b_1\begin{vmatrix} a_2 & c_2 \\ a_3 & c_3 \end{vmatrix} + c_1\begin{vmatrix} b_2 & a_2 \\ b_3 & a_3 \end{vmatrix}\right) \\ &= -\begin{vmatrix} c_1 & b_1 & a_1 \\ c_2 & b_2 & a_2 \\ c_3 & b_3 & a_3 \end{vmatrix}.\end{aligned} \tag{18}$$

Property 2 follows likewise. For 2×2 and 3×3 matrices,

$$\begin{vmatrix} ka_1 & b_1 \\ ka_2 & b_2 \end{vmatrix} = k(a_1b_2) - k(b_1a_2) = k\begin{vmatrix} a_1 & b_1 \\ a_2 & b_2 \end{vmatrix} \tag{19}$$

and

$$\begin{aligned}\begin{vmatrix} ka_1 & b_1 & c_1 \\ ka_2 & b_2 & c_2 \\ ka_3 & b_3 & c_3 \end{vmatrix} &= ka_1\begin{vmatrix} b_2 & c_2 \\ b_3 & c_3 \end{vmatrix} - b_1\begin{vmatrix} ka_2 & c_2 \\ ka_3 & c_3 \end{vmatrix} \\ &\quad + c_1\begin{vmatrix} ka_2 & b_2 \\ ka_3 & b_3 \end{vmatrix} = k\begin{vmatrix} a_1 & b_1 & c_1 \\ a_2 & b_2 & c_2 \\ a_3 & b_3 & c_3 \end{vmatrix}.\end{aligned} \tag{20}$$

Property 3 follows from the identity

$$\begin{aligned}\begin{vmatrix} a_1+kb_1 & b_1 & c_1 \\ a_2+kb_2 & b_2 & c_2 \\ a_3+kb_3 & b_3 & c_3 \end{vmatrix} &= (a_1+kb_1)\begin{vmatrix} b_2 & c_2 \\ b_3 & c_3 \end{vmatrix} \\ &\quad - b_1\begin{vmatrix} a+kb_2 & c_2 \\ a_3+kb_3 & c_3 \end{vmatrix} + c_1\begin{vmatrix} a_2+kb_2 & b_2 \\ a_3+kb_3 & b_3 \end{vmatrix}.\end{aligned} \tag{21}$$

If a_{ij} is an $n \times n$ MATRIX with a_{ij} REAL NUMBERS, then $\det[a_{ij}]$ has the interpretation as the oriented n-dimensional CONTENT of the PARALLELEPIPED spanned by the column vectors $[a_{i,1}], \ldots, [a_{i,n}]$ in $\mathbb{R}^n$. Here, "oriented" means that, up to a change of $+$ or $-$ SIGN, the number is the n-dimensional CONTENT, but the SIGN depends on the "orientation" of the column vectors involved. If they agree with the standard orientation, there is a $+$ SIGN; if not, there is a $-$ SIGN. The PARALLELEPIPED spanned by the n-D vectors $\mathbf{v}_1$ through $\mathbf{v}_i$ is the collection of points

$$t_1\mathbf{v}_1 + \ldots + t_i\mathbf{v}_i, \tag{22}$$

where t_j is a REAL NUMBER in the CLOSED INTERVAL [0,1].

There are an infinite number of 3×3 determinants with no 0 or ± 1 entries having unity determinant. One parametric family is

$$\begin{vmatrix} -8n^2-8n & 2n+1 & 4n \\ -4n^2-4n & n+1 & 2n+1 \\ -4n^2-4n-1 & n & 2n-1 \end{vmatrix}. \tag{23}$$

Specific examples having small entries include

$$\begin{vmatrix} 2 & 3 & 2 \\ 4 & 2 & 3 \\ 9 & 6 & 7 \end{vmatrix}, \begin{vmatrix} 2 & 3 & 5 \\ 3 & 2 & 3 \\ 9 & 5 & 7 \end{vmatrix}, \begin{vmatrix} 2 & 3 & 6 \\ 3 & 2 & 3 \\ 17 & 11 & 16 \end{vmatrix}, \ldots \tag{24}$$

(Guy 1989, 1994).

see also CIRCULANT DETERMINANT, COFACTOR, HESSIAN DETERMINANT, HYPERDETERMINANT, IMMANANT, JACOBIAN, KNOT DETERMINANT, MATRIX, MINOR, PERMANENT, VANDERMONDE DETERMINANT, WRONSKIAN

References

Arfken, G. "Determinants." §4.1 in *Mathematical Methods for Physicists, 3rd ed.* Orlando, FL: Academic Press, pp. 168–176, 1985.

Guy, R. K. "Unsolved Problems Come of Age." *Amer. Math. Monthly* **96**, 903–909, 1989.

Guy, R. K. "A Determinant of Value One." §F28 in *Unsolved Problems in Number Theory, 2nd ed.* New York: Springer-Verlag, pp. 265–266, 1994.

Determinant (Binary Quadratic Form)

The determinant of a BINARY QUADRATIC FORM

$$Au^2 + 2Buv + Cv^2$$

is

$$D \equiv B^2 - AC.$$

It is equal to 1/4 of the corresponding DISCRIMINANT.

Determinant Expansion by Minors

Also known as LAPLACIAN DETERMINANT EXPANSION BY MINORS. Let $|\mathsf{M}|$ denote the DETERMINANT of a MATRIX M, then

$$|\mathsf{M}| = \sum_{i=1}^{k} (-1)^{i+j} a_i M_{ij},$$

where M_{ij} is called a MINOR,

$$|\mathsf{M}| = \sum_{i=1}^{k} a_i C_{ij},$$

where C_{ij} is called a COFACTOR.

see also COFACTOR, DETERMINANT

References

Arfken, G. *Mathematical Methods for Physicists, 3rd ed.* Orlando, FL: Academic Press, pp. 169–170, 1985.

Determinant (Knot)

see KNOT DETERMINANT

Determinant Theorem

Given a MATRIX m, the following are equivalent:

1. $|\mathsf{m}| \neq 0$.
2. The columns of m are linearly independent.
3. The rows of m are linearly independent.
4. $\text{Range}(\mathsf{m}) = \mathbb{R}^n$.
5. $\text{Null}(\mathsf{m}) = \{0\}$.
6. m has a MATRIX INVERSE.

see also DETERMINANT, MATRIX INVERSE, NULLSPACE, RANGE (IMAGE)

Developable Surface

A surface on which the GAUSSIAN CURVATURE K is everywhere 0.

see also BINORMAL DEVELOPABLE, NORMAL DEVELOPABLE, SYNCLASTIC, TANGENT DEVELOPABLE

Deviation

The DIFFERENCE of a quantity from some fixed value, usually the "correct" or "expected" one.

see ABSOLUTE DEVIATION, AVERAGE ABSOLUTE DEVIATION, DIFFERENCE, DISPERSION (STATISTICS), MEAN DEVIATION, SIGNED DEVIATION, STANDARD DEVIATION

Devil's Curve

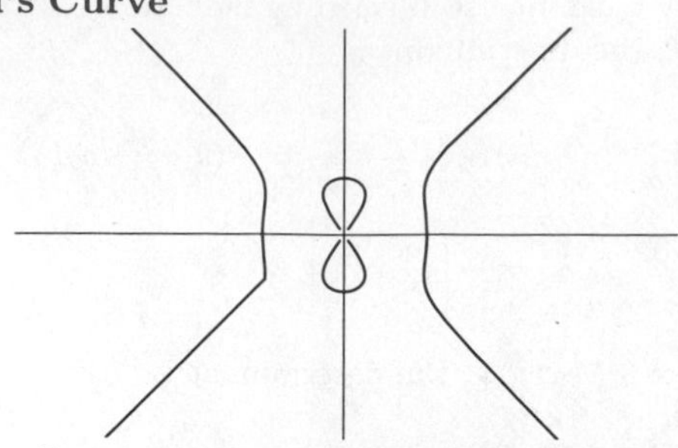

The devil's curve was studied by G. Cramer in 1750 and Lacroix in 1810 (MacTutor Archive). It appeared in *Nouvelles Annales* in 1858. The Cartesian equation is

$$y^4 - a^2 y^2 = x^4 - b^2 x^2, \tag{1}$$

equivalent to

$$y^2(y^2 - a^2) = x^2(x^2 - b^2), \tag{2}$$

the polar equation is

$$r^2(\sin^2\theta - \cos^2\theta) = a^2 \sin^2\theta - b^2 \cos^2\theta, \tag{3}$$

and the parametric equations are

$$x = \cos t \sqrt{\frac{a^2 \sin^2 t - b^2 \cos^2 t}{\sin^2 t - \cos^2 t}} \tag{4}$$

$$y = \sin t \sqrt{\frac{a^2 \sin^2 t - b^2 \cos^2 t}{\sin^2 t - \cos^2 t}}. \tag{5}$$

A special case of the Devil's curve is the so-called ELECTRIC MOTOR CURVE:

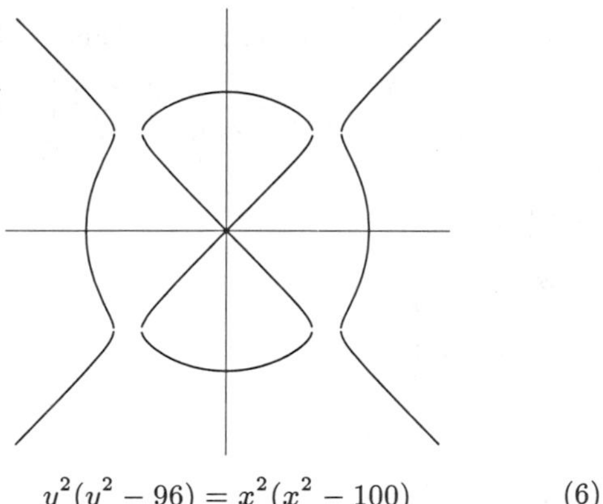

$$y^2(y^2 - 96) = x^2(x^2 - 100) \tag{6}$$

(Cundy and Rollett 1989).

see also ELECTRIC MOTOR CURVE

References

Cundy, H. and Rollett, A. *Mathematical Models, 3rd ed.* Stradbroke, England: Tarquin Pub., p. 71, 1989.

Gray, A. *Modern Differential Geometry of Curves and Surfaces.* Boca Raton, FL: CRC Press, p. 71, 1993.

Lawrence, J. D. *A Catalog of Special Plane Curves.* New York: Dover, pp. 151–152, 1972.

MacTutor History of Mathematics Archive. "Devil's Curve." `http://www-groups.dcs.st-and.ac.uk/~history/Curves/Devils.html`.

Devil's Staircase

A plot of the WINDING NUMBER W resulting from MODE LOCKING as a function of Ω for the CIRCLE MAP with $K = 1$. At each value of Ω, the WINDING NUMBER is some RATIONAL NUMBER. The result is a monotonic increasing "staircase" for which the simplest RATIONAL NUMBERS have the largest steps. For $K = 1$, the MEASURE of quasiperiodic states (Ω IRRATIONAL) on the Ω-axis has become zero, and the measure of MODE-LOCKED state has become 1. The DIMENSION of the Devil's staircase $\approx 0.8700 \pm 3.7 \times 10^{-4}$.

see also CANTOR FUNCTION

References

Mandelbrot, B. B. *The Fractal Geometry of Nature.* New York: W. H. Freeman, 1983.

Ott, E. *Chaos in Dynamical Systems.* New York: Cambridge University Press, 1993.

Rasband, S. N. *Chaotic Dynamics of Nonlinear Systems.* New York: Wiley, p. 132, 1990.

Devil on Two Sticks

see DEVIL'S CURVE

Diabolical Cube

A 6-piece POLYCUBE DISSECTION of the 3×3 CUBE.

see also CUBE DISSECTION, SOMA CUBE

References

Gardner, M. "Polycubes." Ch. 3 in *Knotted Doughnuts and Other Mathematical Entertainments.* New York: W. H. Freeman, pp. 29–30, 1986.

Diabolical Square

see PANMAGIC SQUARE

Diabolo

A 2-POLYABOLO.

Diacaustic

The ENVELOPE of refracted rays for a given curve.

see also CATACAUSTIC, CAUSTIC

References

Lawrence, J. D. *A Catalog of Special Plane Curves.* New York: Dover, p. 60, 1972.

Diagonal Matrix

A diagonal matrix is a MATRIX A of the form

$$a_{ij} = c_i \delta_{ij}, \tag{1}$$

where δ is the KRONECKER DELTA, c_i are constants, and there is no summation over indices. The general diagonal matrix is therefore SQUARE and of the form

$$\begin{bmatrix} c_1 & 0 & \cdots & 0 \\ 0 & c_2 & \cdots & 0 \\ \vdots & \vdots & \ddots & \vdots \\ 0 & 0 & \cdots & c_n \end{bmatrix}. \tag{2}$$

Given a MATRIX equation of the form

$$\begin{bmatrix} a_{11} & \cdots & a_{1n} \\ \vdots & \ddots & \vdots \\ a_{n1} & \cdots & a_{nn} \end{bmatrix} \begin{bmatrix} \lambda_1 & \cdots & 0 \\ \vdots & \ddots & \vdots \\ 0 & \cdots & \lambda_n \end{bmatrix} = \begin{bmatrix} \lambda_1 & \cdots & 0 \\ \vdots & \ddots & \vdots \\ 0 & \cdots & \lambda_n \end{bmatrix} \begin{bmatrix} a_{11} & \cdots & a_{1n} \\ \vdots & \ddots & \vdots \\ a_{n1} & \cdots & a_{nn} \end{bmatrix}, \tag{3}$$

multiply through to obtain

$$\begin{bmatrix} a_{11}\lambda_1 & \cdots & a_{1n}\lambda_n \\ \vdots & \ddots & \vdots \\ a_{n1}\lambda_1 & \cdots & a_{nn}\lambda_n \end{bmatrix} = \begin{bmatrix} a_{11}\lambda_1 & \cdots & a_{1n}\lambda_1 \\ \vdots & \ddots & \vdots \\ a_{n1}\lambda_n & \cdots & a_{nn}\lambda_n \end{bmatrix}. \tag{4}$$

Since in general, $\lambda_i \neq \lambda_j$ for $i \neq j$, this can be true only if off-diagonal components vanish. Therefore, A must be diagonal.

Given a diagonal matrix T,

$$\mathsf{T}^n = \begin{bmatrix} t_1 & 0 & \cdots & 0 \\ 0 & t_2 & \cdots & 0 \\ \vdots & \vdots & \ddots & \vdots \\ 0 & 0 & \cdots & t_k \end{bmatrix}^n = \begin{bmatrix} {t_1}^n & 0 & \cdots & 0 \\ 0 & {t_2}^n & \cdots & 0 \\ \vdots & \vdots & \ddots & \vdots \\ 0 & 0 & \cdots & {t_k}^n \end{bmatrix}. \tag{5}$$

see also MATRIX, TRIANGULAR MATRIX, TRIDIAGONAL MATRIX

References

Arfken, G. *Mathematical Methods for Physicists, 3rd ed.* Orlando, FL: Academic Press, pp. 181–184 and 217–229, 1985.

Diagonal Metric

A METRIC g_{ij} which is zero for $i \neq j$.

see also METRIC

Diagonal (Polygon)

A LINE SEGMENT connecting two nonadjacent VERTICES of a POLYGON. The number of ways a fixed convex n-gon can be divided into TRIANGLES by nonintersecting diagonals is C_{n-2} (with C_{n-3} diagonals), where C_n is a CATALAN NUMBER. This is EULER'S POLYGON DIVISION PROBLEM. Counting the number of regions determined by drawing the diagonals of a regular n-gon is a more difficult problem, as is determining the number of n-tuples of CONCURRENT diagonals (Beller *et al.* 1972, Item 2).

The number of regions which the diagonals of a CONVEX POLYGON divide its center if no three are concurrent in its interior is

$$N = \binom{n}{4} + \binom{n-1}{2} = \tfrac{1}{24}(n-1)(n-2)(n^2 - 3n + 12).$$

The first few values are 0, 0, 1, 4, 11, 25, 50, 91, 154, 246, ... (Sloane's A006522).

see also CATALAN NUMBER, DIAGONAL (POLYHEDRON), EULER'S POLYGON DIVISION PROBLEM, POLYGON, VERTEX (POLYGON)

References

Beeler, M.; Gosper, R. W.; and Schroeppel, R. *HAKMEM.* Cambridge, MA: MIT Artificial Intelligence Laboratory, Memo AIM-239, Feb. 1972.

Sloane, N. J. A. Sequence A006522/M3413 in "An On-Line Version of the Encyclopedia of Integer Sequences."

Diagonal (Polyhedron)

A LINE SEGMENT connecting two nonadjacent sides of a POLYHEDRON. The only simple POLYHEDRON with no diagonals is the TETRAHEDRON. The only known TOROIDAL POLYHEDRON with no diagonals is the CSÁSZÁR POLYHEDRON.

see also DIAGONAL (POLYGON), EULER BRICK, POLYHEDRON, SPACE DIAGONAL

Diagonal Ramsey Number

A RAMSEY NUMBER of the form $R(k,k;2)$.

see also RAMSEY NUMBER

Diagonal Slash

see CANTOR DIAGONAL SLASH

Diagonal (Solidus)

see SOLIDUS

Diagonalization

see MATRIX DIAGONALIZATION

Diagonals Problem

see EULER BRICK

Diagram

A schematic mathematical illustration showing the relationships between or properties of mathematical objects.

see also ALTERNATING KNOT DIAGRAM, ARGAND DIAGRAM, COXETER-DYNKIN DIAGRAM, DE BRUIJN DIAGRAM, DYNKIN DIAGRAM, FERRERS DIAGRAM, HASSE DIAGRAM, HEEGAARD DIAGRAM, KNOT DIAGRAM, LINK DIAGRAM, STEM-AND-LEAF DIAGRAM, VENN DIAGRAM, VORONOI DIAGRAM, YOUNG DIAGRAM

Diameter

The diameter of a CIRCLE is the DISTANCE from a point on the CIRCLE to point π RADIANS away. If r is the RADIUS, $d = 2r$.

see also BROCARD DIAMETER, CIRCUMFERENCE, DIAMETER (GENERAL), DIAMETER (GRAPH), PI, RADIUS, TRANSFINITE DIAMETER

Diameter (General)

The farthest DISTANCE between two points on the boundary of a closed figure.

see also BORSUK'S CONJECTURE

References

Eppstein, D. "Width, Diameter, and Geometric Inequalities." http://www.ics.uci.edu/~eppstein/junkyard/diam.html.

Diameter (Graph)

The length of the "longest shortest path" between two VERTICES of a GRAPH. In other words, a graph's diameter is the largest number of vertices which must be traversed in order to travel from one vertex to another when paths which backtrack, detour, or loop are excluded from consideration.

Diamond

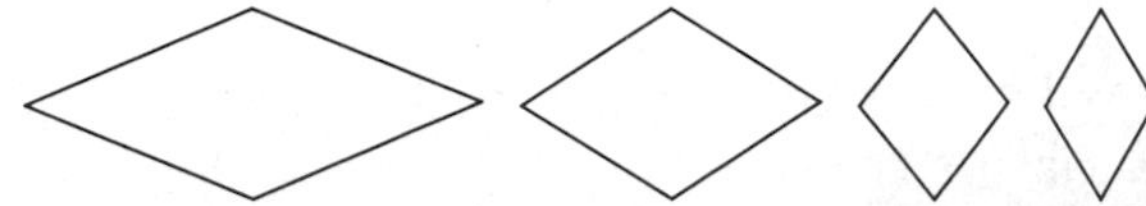

A convex QUADRILATERAL having sides of equal length and PERPENDICULAR PLANES of symmetry passing through opposite pairs of VERTICES. The LOZENGE is a special case of a diamond.

see also KITE, LOZENGE, PARALLELOGRAM, QUADRILATERAL, RHOMBUS

Dice

A die (plural "dice") is a SOLID with markings on each of its faces. The faces are usually all the same shape, making PLATONIC SOLIDS and ARCHIMEDEAN SOLID DUALS the obvious choices. The die can be "rolled" by throwing it in the air and allowing it to come to rest on one of its faces. Dice are used in many games of chance as a way of picking RANDOM NUMBERS on which to bet, and are used in board or roll-playing games to determine the number of spaces to move, results of a conflict, etc. A COIN can be viewed as a degenerate 2-sided case of a die.

The most common type of die is a six-sided CUBE with the numbers 1–6 placed on the faces. The value of the roll is indicated by the number of "spots" showing on the top. For the six-sided die, opposite faces are arranged to always sum to seven. This gives two possible MIRROR IMAGE arrangements in which the numbers 1, 2, and 3 may be arranged in a clockwise or counterclockwise order about a corner. Commercial dice may, in fact, have either orientation. The illustrations below show 6-sided dice with counterclockwise and clockwise arrangements, respectively.

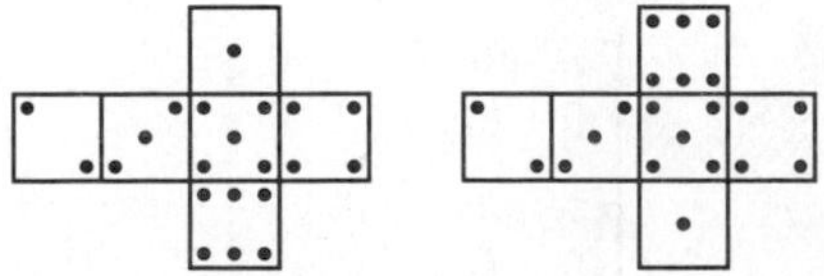

The CUBE has the nice property that there is an upward-pointing face opposite the bottom face from which the value of the "roll" can easily be read. This would not be true, for instance, for a TETRAHEDRAL die, which would have to be picked up and turned over to reveal the number underneath (although it could be determined by noting which number 1–4 was *not* visible on one of the upper three faces). The arrangement of spots ⁙ corresponding to a roll of 5 on a six-sided die is called the QUINCUNX. There are also special names for certain rolls of two six-sided dice: two 1s are called SNAKE EYES and two 6s are called BOXCARS.

Shapes of dice other than the usual 6-sided CUBE are commercially available from companies such as Dice & Games, Ltd.®

Diaconis and Keller (1989) show that there exist "fair" dice other than the usual PLATONIC SOLIDS and duals of the ARCHIMEDEAN SOLIDS, where a fair die is one for which its symmetry group acts transitively on its faces. However, they did not explicitly provide any examples.

The probability of obtaining p points (a roll of p) on n s-sided dice can be computed as follows. The number of ways in which p can be obtained is the COEFFICIENT of x^p in

$$f(x) = (x + x^2 + \ldots + x^s)^n, \tag{1}$$

since each possible arrangement contributes one term. $f(x)$ can be written as a MULTINOMIAL SERIES

$$f(x) = x^n \left(\sum_{i=0}^{s-1} x^i\right)^n = x^n \left(\frac{1-x^s}{1-x}\right)^n, \tag{2}$$

so the desired number c is the COEFFICIENT of x^p in

$$x^n(1-x^s)^n(1-x)^{-n}. \tag{3}$$

Expanding,

$$x^n \sum_{k=0}^{n} (-1)^k \binom{n}{k} x^{sk} \sum_{l=0}^{\infty} \binom{n+l-1}{l} x^l, \tag{4}$$

so in order to get the COEFFICIENT of x^p, include all terms with

$$p = n + sk + l. \tag{5}$$

c is therefore

$$c = \sum_{k=0}^{n} (-1)^k \binom{n}{k} \binom{p-sk-1}{p-sk-n}. \tag{6}$$

But $p - sk - n > 0$ only when $k < (p-n)/s$, so the other terms do not contribute. Furthermore,

$$\binom{p-sk-1}{p-sk-n} = \binom{p-sk-1}{n-1}, \tag{7}$$

so

$$c = \sum_{k=0}^{\lfloor (p-n)/s \rfloor} (-1)^k \binom{n}{k} \binom{p-sk-1}{n-1}, \tag{8}$$

where $\lfloor x \rfloor$ is the FLOOR FUNCTION, and

$$P(p,n,s) = \frac{1}{s^n} \sum_{k=0}^{\lfloor (p-n)/s \rfloor} (-1)^k \binom{n}{k} \binom{p-sk-1}{n-1}. \tag{9}$$

Consider now $s = 6$. For $n = 2$ six-sided dice,

$$k_{\max} \equiv \left\lfloor \frac{p-2}{6} \right\rfloor = \begin{cases} 0 & \text{for } 2 \le p \le 7 \\ 1 & \text{for } 12 \le p \le 8, \end{cases} \tag{10}$$

and

$$\begin{aligned} P(p,2,6) &= \frac{1}{6^2} \sum_{k=0}^{k_{\max}} (-1)^k \binom{2}{k} \binom{p-6k-1}{1} \\ &= \frac{1}{6^2} \sum_{k=0}^{k_{\max}} (-1)^k \frac{2!}{k!(2-k)!} (p-6k-1) \\ &= \frac{1}{36} \sum_{k=0}^{k_{\max}} (1-2k)(k+1)(p-6k-1) \\ &= \frac{1}{36} \begin{cases} p-1 & \text{for } 2 \le p \le 7 \\ 13-p & \text{for } 8 \le p \le 12 \end{cases} \\ &= \frac{6 - |p-7|}{36} \quad \text{for } 2 \le p \le 12. \end{aligned} \tag{11}$$

The most common roll is therefore seen to be a 7, with probability $6/36 = 1/6$, and the least common rolls are 2 and 12, both with probability $1/36$.

For $n = 3$ six-sided dice,

$$k_{\max} = \left\lfloor \frac{p-3}{6} \right\rfloor = \begin{cases} 0 & \text{for } 3 \le p \le 8 \\ 1 & \text{for } 9 \le p \le 14 \\ 2 & \text{for } 15 \le p \le 18, \end{cases} \tag{12}$$

and

$$\begin{aligned} &P(p,3,6) \\ &= \frac{1}{6^3} \sum_{k=0}^{k_{\max}} (-1)^k \binom{3}{k} \binom{p-6k-1}{2} \\ &= \frac{1}{6^3} \sum_{k=0}^{k_{\max}} (-1)^k \frac{3!}{k!(3-k)!} \frac{(p-6k-1)(p-6k-2)}{2} \\ &= \frac{1}{216} \begin{cases} \frac{(p-1)(p-2)}{2} & \text{for } 3 \le p \le 8 \\ \frac{(p-1)(p-2)}{2} - 3\frac{(p-7)(p-8)}{2} & \text{for } 9 \le p \le 14 \\ \frac{(p-1)(p-2)}{2} - 3\frac{(p-7)(p-8)}{2} + 3\frac{(p-13)(p-14)}{2} & \text{for } 15 \le p \le 18 \end{cases} \\ &= \frac{1}{216} \begin{cases} \frac{1}{2}(p-1)(p-2) & \text{for } 3 \le p \le 8 \\ -p^2 + 21p - 83 & \text{for } 9 \le p \le 14 \\ \frac{1}{2}(19-p)(20-p) & \text{for } 15 \le p \le 18. \end{cases} \end{aligned} \tag{13}$$

For three six-sided dice, the most common rolls are 10 and 11, both with probability $1/8$; and the least common rolls are 3 and 18, both with probability $1/216$.

For four six-sided dice, the most common roll is 14, with probability 73/648; and the least common rolls are 4 and 24, both with probability 1/1296.

In general, the likeliest roll p_L for n s-sided dice is given by

$$p_L(n,s) = \left\lfloor \tfrac{1}{2}n(s+1) \right\rfloor, \qquad (14)$$

which can be written explicitly as

$$p_L(n,s) = \begin{cases} \frac{1}{2}n(s+1) & \text{for } n \text{ even} \\ \frac{1}{2}[n(s+1)-1] & \text{for } n \text{ odd, } s \text{ even} \\ \frac{1}{2}n(s+1) & \text{for } n \text{ odd, } s \text{ odd.} \end{cases} \qquad (15)$$

For 6-sided dice, the likeliest rolls are given by

$$p_L(n,6) = \left\lfloor \tfrac{7}{2}n \right\rfloor = \begin{cases} \frac{7}{2}n & \text{for } n \text{ even} \\ \frac{1}{2}(7n-1) & \text{for } n \text{ odd, } s \text{ even} \\ \frac{7}{2}n & \text{for } n \text{ odd, } s \text{ odd,} \end{cases} \qquad (16)$$

or 7, 10, 14, 17, 21, 24, 28, 31, 35, ... for $n = 2, 3, \ldots$ (Sloane's A030123) dice. The probabilities corresponding to the most likely rolls can be computed by plugging $p = p_L$ into the general formula together with

$$k_L(n,s) = \begin{cases} \frac{1}{2}n & \text{for } n \text{ even} \\ \left\lfloor \frac{n(s-1)-1}{2s} \right\rfloor & \text{for } n \text{ odd, } s \text{ even} \\ \left\lfloor \frac{n(s-1)}{2s} \right\rfloor & \text{for } n \text{ odd, } s \text{ odd.} \end{cases} \qquad (17)$$

Unfortunately, $P(p_L, n, s)$ does not have a simple closed-form expression in terms of s and n. However, the probabilities of obtaining the likeliest roll totals can be found explicitly for a particular s. For n 6-sided dice, the probabilities are 1/6, 1/8, 73/648, 65/648, 361/3888, 24017/279936, 7553/93312, ... for $n = 2, 3, \ldots$.

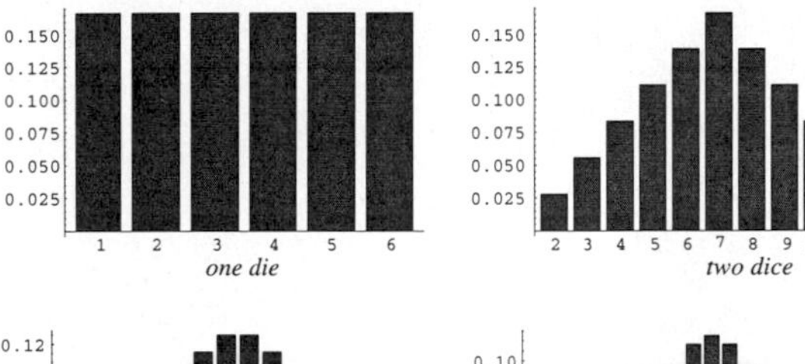

one die   two dice

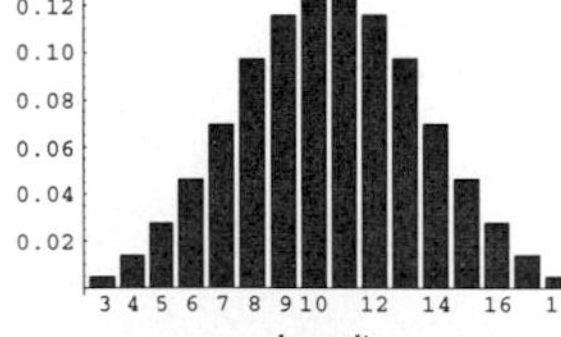

three dice

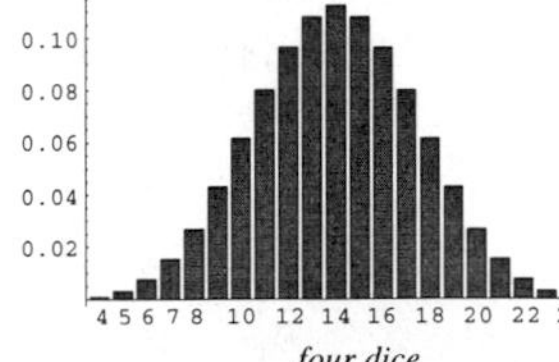

four dice

The probabilities for obtaining a given total using n 6-sided dice are shown above for n = 1, 2, 3, and 4 dice. They can be seen to approach a GAUSSIAN DISTRIBUTION as the number of dice is increased.

see also BOXCARS, COIN TOSSING, CRAPS, DE MERE'S PROBLEM, EFRON'S DICE, POKER, QUINCUNX, SICHERMAN DICE, SNAKE EYES

References

Diaconis, P. and Keller, J. B. "Fair Dice." *Amer. Math. Monthly* **96**, 337–339, 1989.

Dice & Games, Ltd. "Dice & Games Hobby Games Accessories." `http://www.dice.co.uk/hob.htm`.

Gardner, M. "Dice." Ch. 18 in *Mathematical Magic Show: More Puzzles, Games, Diversions, Illusions and Other Mathematical Sleight-of-Mind from Scientific American.* New York: Vintage, pp. 251–262, 1978.

Robertson, L. C.; Shortt, R. M.; Landry, S. G. "Dice with Fair Sums." *Amer. Math. Monthly* **95**, 316–328, 1988.

Sloane, N. J. A. Sequence A030123 in "An On-Line Version of the Encyclopedia of Integer Sequences."

Dichroic Polynomial

A POLYNOMIAL $Z_G(q,v)$ in two variables for abstract GRAPHS. A GRAPH with one VERTEX has $Z = q$. Adding a VERTEX not attached by any EDGES multiplies the Z by q. Picking a particular EDGE of a GRAPH G, the POLYNOMIAL for G is defined by adding the POLYNOMIAL of the GRAPH with that EDGE deleted to v times the POLYNOMIAL of the graph with that EDGE collapsed to a point. Setting $v = -1$ gives the number of distinct VERTEX colorings of the GRAPH. The dichroic POLYNOMIAL of a PLANAR GRAPH can be expressed as the SQUARE BRACKET POLYNOMIAL of the corresponding ALTERNATING LINK by

$$Z_G(q,v) = q^{N/2} B_{L(G)},$$

where N is the number of VERTICES in G. Dichroic POLYNOMIALS for some simple GRAPHS are

$$\begin{aligned} Z_{K_1} &= q \\ Z_{K_2} &= q^2 + vq \\ Z_{K_3} &= q^3 + 3vq^2 + 3v^2q + v^3q. \end{aligned}$$

References

Adams, C. C. *The Knot Book: An Elementary Introduction to the Mathematical Theory of Knots.* New York: W. H. Freeman, pp. 231–235, 1994.

Dido's Problem

Find the figure bounded by a line which has the maximum AREA for a given PERIMETER. The solution is a SEMICIRCLE.

see also ISOPERIMETRIC PROBLEM, ISOVOLUME PROBLEM, PERIMETER, SEMICIRCLE

Diesis

The musical interval by which an octave exceeds three major thirds,

$$\frac{2}{(\frac{5}{4})^3} = \frac{2^7}{5^3} = \frac{128}{125} = 1.024.$$

Taking CONTINUED FRACTION CONVERGENTS of $\log(5/4)/\log(2)$ gives the increasing accurate approximations m/n of m octaves and n major thirds: 1/3,

9/28, 19/59, 47/146, 207/643, 1289/4004, ... (Sloane's A046103 and A046104). Other near equalities of m octaves and n major thirds having

$$R \equiv \frac{2^m}{(\frac{5}{4})^n} = \frac{2^{m+2n}}{5^n}$$

with $|R-1| < 0.02$ are given in the following table.

m	n	Ratio	m	n	Ratio
9	**28**	0.9903520314	104	323	1.012011267
10	31	1.01412048	113	351	1.002247414
18	56	0.9807971462	122	379	0.9925777621
19	**59**	1.004336278	123	382	1.016399628
28	87	0.9946464728	131	407	0.983001403
29	90	1.018517988	132	410	1.006593437
37	115	0.9850501549	141	438	0.9968818549
38	118	1.008691359	150	466	0.9872639701
47	**146**	0.9989595361	151	469	1.010958305
56	174	0.9893216059	160	497	1.001204611
57	177	1.013065324	169	525	0.9915450208
66	205	1.003291302	170	528	1.015342101
75	233	0.9936115791	178	553	0.9819786256
76	236	1.017458257	179	556	1.005546113
84	261	0.9840252458	188	584	0.9958446353
85	264	1.007641852	189	587	1.019744907
94	292	0.9979201548	197	612	0.9862367575
103	320	0.9882922525	198	615	1.00990644

see also Comma of Didymus, Comma of Pythagoras, Schisma

References

Sloane, N. J. A. Sequences A046103 and A046104 in "An On-Line Version of the Encyclopedia of Integer Sequences."

Diffeomorphism

A diffeomorphism is a Map between Manifolds which is Differentiable and has a Differentiable inverse.

see also Anosov Diffeomorphism, Axiom A Diffeomorphism, Symplectic Diffeomorphism, Tangent Map

Difference

The difference of two numbers n_1 and n_2 is $n_1 - n_2$, where the Minus sign denotes Subtraction.

see also Backward Difference, Finite Difference, Forward Difference

Difference Equation

A difference equation is the discrete analogue of a Differential Equation. A difference equation involves a Function with Integer-valued arguments $f(n)$ in a form like

$$f(n) - f(n-1) = g(n), \tag{1}$$

where g is some Function. The above equation is the discrete analog of the first-order Ordinary Differential Equation

$$f'(x) = g(x). \tag{2}$$

Examples of difference equations often arise in Dynamical Systems. Examples include the iteration involved in the Mandelbrot and Julia Set definitions,

$$f(n+1) = f(n)^2 + c, \tag{3}$$

with c a constant, as well as the Logistic Equation

$$f(n+1) = rf(n)[1-f(n)], \tag{4}$$

with r a constant.

see also Finite Difference, Recurrence Relation

References

Batchelder, P. M. *An Introduction to Linear Difference Equations.* New York: Dover, 1967.
Bellman, R. E. and Cooke, K. L. *Differential-Difference Equations.* New York: Academic Press, 1963.
Beyer, W. H. "Finite Differences." *CRC Standard Mathematical Tables, 28th ed.* Boca Raton, FL: CRC Press, pp. 429–460, 1988.
Brand, L. *Differential and Difference Equations.* New York: Wiley, 1966.
Goldberg, S. *Introduction to Difference Equations, with Illustrative Examples from Economics, Psychology, and Sociology.* New York: Dover, 1986.
Levy, H. and Lessman, F. *Finite Difference Equations.* New York: Dover, 1992.
Richtmyer, R. D. and Morton, K. W. *Difference Methods for Initial-Value Problems, 2nd ed.* New York: Interscience Publishers, 1967.

Difference Operator

see Backward Difference, Forward Difference

Difference Quotient

$$\Delta_h f(x) \equiv \frac{f(x+h) - f(x)}{h} = \frac{\Delta f}{h}.$$

It gives the slope of the Secant Line passing through $f(x)$ and $f(x+h)$. In the limit $n \to 0$, the difference quotient becomes the Partial Derivative

$$\lim_{h \to 0} \Delta_{x(h)} f(x,y) = \frac{\partial f}{\partial x}.$$

Difference Set

Let G be a Group of Order h and D be a set of k elements of G. If the set of differences $d_i - d_j$ contains every Nonzero element of G exactly λ times, then D is a (h, k, λ)-difference set in G of Order $n = k - \lambda$. If $\lambda = 1$, the difference set is called planar. The quadratic residues in the Galois Field $GF(11)$ form a difference set. If there is a difference set of size k in a group G, then $2\binom{k}{2}$ must be a multiple of $|G| - 1$, where $\binom{k}{2}$ is a Binomial Coefficient.

see also Bruck-Ryser-Chowla Theorem, First Multiplier Theorem, Prime Power Conjecture

References

Gordon, D. M. "The Prime Power Conjecture is True for $n < 2,000,000$." *Electronic J. Combinatorics* **1**, R6, 1–7, 1994. http://www.combinatorics.org/Volume_1/volume1.html#R6.

Difference of Successes

If x_1/n_1 and x_2/n_2 are the observed proportions from standard NORMALLY DISTRIBUTED samples with proportion of success θ, then the probability that

$$w \equiv \frac{x_1}{n_1} - \frac{x_2}{n_2} \quad (1)$$

will be as great as observed is

$$P_\delta = 1 - 2\int_0^{|\delta|} \phi(t)\,dt, \quad (2)$$

where

$$\delta \equiv \frac{w}{\sigma_w} \quad (3)$$

$$\sigma_w \equiv \sqrt{\hat\theta(1-\hat\theta)\left(\frac{1}{n_1}+\frac{1}{n_2}\right)} \quad (4)$$

$$\hat\theta \equiv \frac{x_1+x_2}{n_1+n_2}. \quad (5)$$

Here, $\hat\theta$ is the UNBIASED ESTIMATOR. The SKEWNESS and KURTOSIS of this distribution are

$${\gamma_1}^2 = \frac{(n_1-n_2)^2}{n_1n_2(n_1+n_2)} \frac{1-4\hat\theta(1-\hat\theta)}{\hat\theta(1-\hat\theta)} \quad (6)$$

$$\gamma_2 = \frac{{n_1}^2 - n_1n_2 + {n_2}^2}{n_1n_2(n_1+n_2)} \frac{1-6\hat\theta(1-\hat\theta)}{\hat\theta(1-\hat\theta)}. \quad (7)$$

Difference Table

A table made by subtracting adjacent entries in a sequence, then repeating the process with those numbers.

see also FINITE DIFFERENCE, QUOTIENT-DIFFERENCE TABLE

Different

Two quantities are said to be different (or "unequal") if they are not EQUAL.

The term "different" also has a technical usage related to MODULES. Let a MODULE M in an INTEGRAL DOMAIN D_1 for $R(\sqrt{D})$ be expressed using a two-element basis as

$$M = [\xi_1, \xi_2],$$

where ξ_1 and ξ_2 are in D_1. Then the different of the MODULE is defined as

$$\Delta = \Delta(M) = \begin{vmatrix} \xi_1 & \xi_2 \\ \xi_1' & \xi_2' \end{vmatrix} = \xi_1\xi_2' - \xi_1'\xi_2.$$

The different $\Delta \neq 0$ IFF ξ_1 and ξ_2 are linearly independent. The DISCRIMINANT is defined as the square of the different.

see also DISCRIMINANT (MODULE), EQUAL, MODULE

References

Cohn, H. *Advanced Number Theory.* New York: Dover, pp. 72–73, 1980.

Different Prime Factors

see DISTINCT PRIME FACTORS

Differentiable

A FUNCTION is said to be differentiable at a point if its DERIVATIVE exists at that point. Let $z = x + iy$ and $f(z) = u(x,y) + iv(x,y)$ on some region G containing the point z_0. If $f(z)$ satisfies the CAUCHY-RIEMANN EQUATIONS and has continuous first PARTIAL DERIVATIVES at z_0, then $f'(z_0)$ exists and is given by

$$f'(z_0) = \lim_{z\to z_0} \frac{f(z)-f(z_0)}{z-z_0},$$

and the function is said to be COMPLEX DIFFERENTIABLE. Amazingly, there exist CONTINUOUS FUNCTIONS which are nowhere differentiable. Two examples are the BLANCMANGE FUNCTION and WEIERSTRAß FUNCTION.

see also BLANCMANGE FUNCTION, CAUCHY-RIEMANN EQUATIONS, COMPLEX DIFFERENTIABLE, CONTINUOUS FUNCTION, DERIVATIVE, PARTIAL DERIVATIVE, WEIERSTRAß FUNCTION

Differentiable Manifold

see SMOOTH MANIFOLD

Differential

A DIFFERENTIAL 1-FORM.

see also EXACT DIFFERENTIAL, INEXACT DIFFERENTIAL

Differential Calculus

That portion of "the" CALCULUS dealing with DERIVATIVES.

see also INTEGRAL CALCULUS

Differential Equation

An equation which involves the DERIVATIVES of a function as well as the function itself. If PARTIAL DERIVATIVES are involved, the equation is called a PARTIAL DIFFERENTIAL EQUATION; if only ordinary DERIVATIVES are present, the equation is called an ORDINARY DIFFERENTIAL EQUATION. Differential equations play an extremely important and useful role in applied math, engineering, and physics, and much mathematical and numerical machinery has been developed for the solution of differential equations.

see also INTEGRAL EQUATION, ORDINARY DIFFERENTIAL EQUATION, PARTIAL DIFFERENTIAL EQUATION

References

Arfken, G. "Differential Equations." Ch. 8 in *Mathematical Methods for Physicists, 3rd ed.* Orlando, FL: Academic Press, pp. 437–496, 1985.

Dormand, J. R. *Numerical Methods for Differential Equations: A Computational Approach.* Boca Raton, FL: CRC Press, 1996.

Differential Form

see DIFFERENTIAL k-FORM

Differential Geometry

Differential geometry is the study of RIEMANNIAN MANIFOLDS. Differential geometry deals with metrical notions on MANIFOLDS, while DIFFERENTIAL TOPOLOGY deals with those nonmetrical notions of MANIFOLDS.

see also DIFFERENTIAL TOPOLOGY

References

Eisenhart, L. P. *A Treatise on the Differential Geometry of Curves and Surfaces.* New York: Dover, 1960.

Gray, A. *Modern Differential Geometry of Curves and Surfaces.* Boca Raton, FL: CRC Press, 1993.

Kreyszig, E. *Differential Geometry.* New York: Dover, 1991.

Lipschutz, M. M. *Theory and Problems of Differential Geometry.* New York: McGraw-Hill, 1969.

Spivak, M. *A Comprehensive Introduction to Differential Geometry, 2nd ed, 5 vols.* Berkeley, CA: Publish or Perish Press, 1979.

Struik, D. J. *Lectures on Classical Differential Geometry.* New York: Dover, 1988.

Weatherburn, C. E. *Differential Geometry of Three Dimensions, 2 vols.* Cambridge, England: Cambridge University Press, 1961.

Differential k-Form

A differential k-form is a TENSOR of RANK k which is antisymmetric under exchange of any pair of indices. The number of algebraically independent components in n-D is $\binom{n}{p}$, where this is a BINOMIAL COEFFICIENT. In particular, a 1-form (often simply called a "differential") is a quantity

$$\omega^1 = b_1\,dx_1 + b_2\,dx_2, \tag{1}$$

where $b_1 = b_1(x_1, x_2)$ and $b_2 = b_2(x_1, x_2)$ are the components of a COVARIANT TENSOR. Changing variables from $\mathbf{x}$ to $\mathbf{y}$ gives

$$\omega^1 = \sum_{i=1}^{2} b_i\,dx_i = \sum_{i=1}^{2}\sum_{j=1}^{2} b_i \frac{\partial x_i}{\partial y_j} = \sum_{j=1}^{2} \bar{b}_j\,dy_j, \tag{2}$$

where

$$\bar{b}_j \equiv \sum_{i=1}^{2} b_i \frac{\partial x_i}{\partial y_j}, \tag{3}$$

which is the covariant transformation law. 2-forms can be constructed from the WEDGE PRODUCT of 1-forms. Let

$$\theta_1 = b_1\,dx_1 + b_2\,dx_2 \tag{4}$$

$$\theta_2 \equiv c_1\,dx_1 + c_2\,dx_2, \tag{5}$$

then $\theta_1 \wedge \theta_2$ is a 2-form denoted ω^2. Changing variables $x_1(y_1, y_2)$ to $x_2(y_1, y_2)$ gives

$$dx_1 = \frac{\partial x_1}{\partial y_1} dy_1 + \frac{\partial x_1}{\partial y_2} dy_2 \tag{6}$$

$$dx_2 = \frac{\partial x_2}{\partial y_1} dy_1 + \frac{\partial x_2}{\partial y_2} dy_2, \tag{7}$$

so

$$\begin{aligned} dx_1 \wedge dx_2 &= \left(\frac{\partial x_1}{\partial y_1}\frac{\partial x_2}{\partial y_2} - \frac{\partial x_1}{\partial y_2}\frac{\partial x_2}{\partial y_1}\right) dy_1 \wedge dy_2 \\ &= \frac{\partial(x_1, x_2)}{\partial(y_1, y_2)} dy_1 \wedge dy_2. \end{aligned} \tag{8}$$

Similarly, a 4-form can be constructed from WEDGE PRODUCTS of two 2-forms or four 1-forms

$$\omega^4 = {\omega_1}^2 \wedge {\omega_2}^2 = ({\omega_1}^1 \wedge {\omega_2}^1) \wedge ({\omega_3}^1 \wedge {\omega_4}^1). \tag{9}$$

see also ANGLE BRACKET, BRA, EXTERIOR DERIVATIVE, KET, ONE-FORM, SYMPLECTIC FORM, WEDGE PRODUCT

References

Weintraub, S. H. *Differential Forms: A Complement to Vector Calculus.* San Diego, CA: Academic Press, 1996.

Differential Operator

The OPERATOR representing the computation of a DERIVATIVE,

$$\tilde{D} \equiv \frac{d}{dx}.$$

The second derivative is then denoted $\tilde{D}^2$, the third $\tilde{D}^3$, etc. The INTEGRAL is denoted $\tilde{D}^{-1}$.

see also CONVECTIVE DERIVATIVE, DERIVATIVE, FRACTIONAL DERIVATIVE, GRADIENT

Differential Structure

see EXOTIC R4, EXOTIC SPHERE

Differential Topology

The motivating force of TOPOLOGY, consisting of the study of smooth (differentiable) MANIFOLDS. Differential topology deals with nonmetrical notions of MANIFOLDS, while DIFFERENTIAL GEOMETRY deals with metrical notions of MANIFOLDS.

see also DIFFERENTIAL GEOMETRY

References

Dieudonné, J. *A History of Algebraic and Differential Topology: 1900–1960.* Boston, MA: Birkhäuser, 1989.

Munkres, J. R. *Elementary Differential Topology.* Princeton, NJ: Princeton University Press, 1963.

Differentiation

The computation of a DERIVATIVE.

see also CALCULUS, DERIVATIVE, INTEGRAL, INTEGRATION

Digamma Function

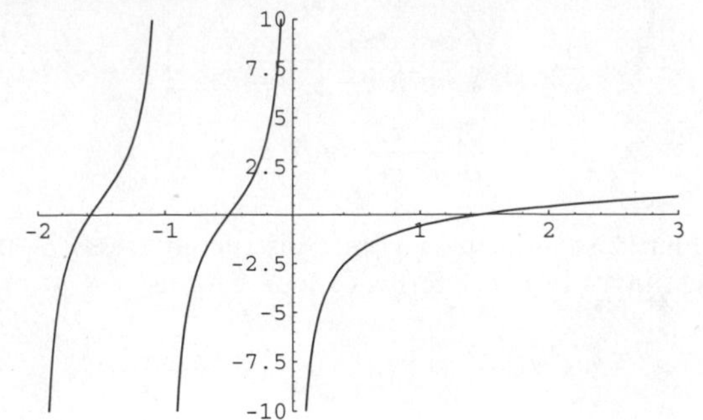

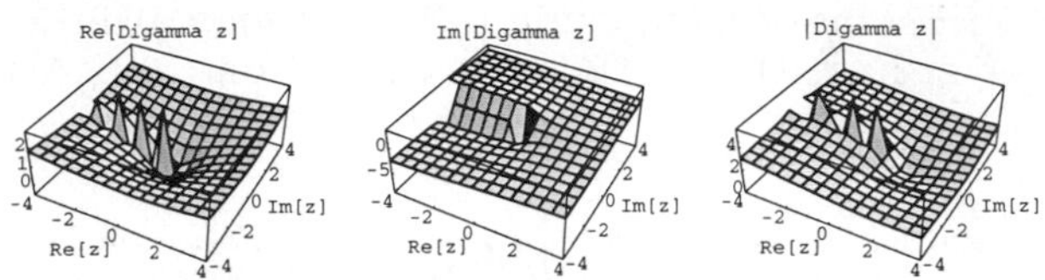

Two notations are used for the digamma function. The $\Psi(z)$ digamma function is defined by

$$\Psi(z) \equiv \frac{d}{dz} \ln \Gamma(z) = \frac{\Gamma'(z)}{\Gamma(z)}, \tag{1}$$

where Γ is the GAMMA FUNCTION, and is the function returned by the function `PolyGamma[z]` in *Mathematica*® (Wolfram Research, Champaign, IL). The F digamma function is defined by

$$F(z) \equiv \frac{d}{dz} \ln z! \tag{2}$$

and is equal to

$$F(z) = \Psi(z+1). \tag{3}$$

From a series expansion of the FACTORIAL function,

$$\begin{aligned} F(z) = \frac{d}{dz} \lim_{n\to\infty} [&\ln n! + z \ln n \\ &- \ln(z+1) - \ln(z+2) - \ldots - \ln(z+n)] \end{aligned} \tag{4}$$

$$= \lim_{n\to\infty} \left(\ln n - \frac{1}{z+1} - \frac{1}{z+2} - \ldots - \frac{1}{z+n} \right) \tag{5}$$

$$= -\gamma - \sum_{n=1}^{\infty} \left(\frac{1}{z+n} - \frac{1}{n} \right) \tag{6}$$

$$= -\gamma + \sum_{n=1}^{\infty} \frac{z}{n(n+z)} \tag{7}$$

$$= \ln z + \frac{1}{2z} - \sum_{n=1}^{\infty} \frac{B_{2n}}{2nz^{2n}}, \tag{8}$$

where γ is the EULER-MASCHERONI CONSTANT and B_{2n} are BERNOULLI NUMBERS.

The nth DERIVATIVE of $\Psi(z)$ is called the POLYGAMMA FUNCTION and is denoted $\psi_n(z)$. Since the digamma function is the zeroth derivative of $\Psi(z)$ (i.e., the function itself), it is also denoted $\psi_0(z)$.

The digamma function satisfies

$$\Psi(z) = \int_0^\infty \left(\frac{e^{-t}}{t} - \frac{e^{-zt}}{1-e^{-t}} \right) dt. \tag{9}$$

For integral $z \equiv n$,

$$\Psi(n) = -\gamma + \sum_{k=1}^{n-1} \frac{1}{k} = -\gamma + H_{n-1}, \tag{10}$$

where γ is the EULER-MASCHERONI CONSTANT and H_n is a HARMONIC NUMBER. Other identities include

$$\frac{d\Psi}{dz} = \sum_{n=0}^{\infty} \frac{1}{(z+n)^2} \tag{11}$$

$$\Psi(1-z) - \Psi(z) = \pi \cot(\pi z) \tag{12}$$

$$\Psi(z+1) = \Psi(z) + \frac{1}{z} \tag{13}$$

$$\Psi(2z) = \tfrac{1}{2}\Psi(z) + \tfrac{1}{2}\Psi(z+\tfrac{1}{2}) + \ln 2. \tag{14}$$

Special values are

$$\Psi(\tfrac{1}{2}) = -\gamma - 2\ln 2 \tag{15}$$

$$\Psi(1) = -\gamma. \tag{16}$$

At integral values,

$$\psi_0(n+1) = -\gamma + \sum_{k=1}^{n} \frac{1}{k}, \tag{17}$$

and at half-integral values,

$$\psi_0(\tfrac{1}{2} \pm n) = -\ln(4\gamma) + 2\sum_{k=1}^{n} \frac{1}{2k-1}. \tag{18}$$

At rational arguments, $\psi_0(p/q)$ is given by the explicit equation

$$\begin{aligned} \psi_0\left(\frac{p}{q}\right) &= -\gamma - \ln(2q) - \tfrac{1}{2}\pi \cot\left(\frac{p}{q}\pi\right) \\ &+2 \sum_{k=1}^{\lceil q/2 \rceil - 1} \cos\left(\frac{2\pi pk}{q}\right) \ln\left[\sin\left(\frac{\pi k}{q}\right)\right] \end{aligned} \tag{19}$$

for $0 < p < q$ (Knuth 1973). These give the special values

$$\psi_0(\tfrac{1}{2}) = -\gamma - 2\ln 2 \tag{20}$$

$$\psi_0(\tfrac{1}{3}) = \tfrac{1}{6}(-6\gamma - \pi\sqrt{3} - 9\ln 3) \tag{21}$$

$$\psi_0(\tfrac{2}{3}) = \tfrac{1}{6}(-6\gamma + \pi\sqrt{3} - 9\ln 3) \tag{22}$$

$$\psi_0(\tfrac{1}{4}) = -\gamma - \tfrac{1}{2}\pi - 3\ln 2 \tag{23}$$

$$\psi_0(\tfrac{3}{4}) = \tfrac{1}{2}(-2\gamma + \pi - 6\ln 2) \tag{24}$$

$$\psi_0(1) = -\gamma, \tag{25}$$

where γ is the EULER-MASCHERONI CONSTANT. Sums and differences of $\psi_1(r/s)$ for small integral r and s can be expressed in terms of CATALAN'S CONSTANT and π.

see also GAMMA FUNCTION, HARMONIC NUMBER, HURWITZ ZETA FUNCTION, POLYGAMMA FUNCTION

References

Abramowitz, M. and Stegun, C. A. (Eds.). "Psi (Digamma) Function." §6.3 in *Handbook of Mathematical Functions with Formulas, Graphs, and Mathematical Tables, 9th printing.* New York: Dover, pp. 258–259, 1972.

Arfken, G. "Digamma and Polygamma Functions." §10.2 in *Mathematical Methods for Physicists, 3rd ed.* Orlando, FL: Academic Press, pp. 549–555, 1985.

Knuth, D. E. *The Art of Computer Programming, Vol. 1: Fundamental Algorithms, 2nd ed.* Reading, MA: Addison-Wesley, p. 94, 1973.

Spanier, J. and Oldham, K. B. "The Digamma Function $\psi(x)$." Ch. 44 in *An Atlas of Functions.* Washington, DC: Hemisphere, pp. 423–434, 1987.

Digimetic

A CRYPTARITHM in which DIGITS are used to represent other DIGITS.

Digit

The number of digits D in an INTEGER n is the number of numbers in some base (usually 10) required to represent it. The numbers 1 to 9 are therefore single digits, while the numbers 10 to 99 are double digits. Terms such as "double-digit inflation" are occasionally encountered, although this particular usage has thankfully not been needed in the U.S. for some time. The number of (base 10) digits in a number n can be calculated as

$$D = \lfloor \log_{10} n + 1 \rfloor,$$

where $\lfloor x \rfloor$ is the FLOOR FUNCTION.

see also 196-ALGORITHM, ADDITIVE PERSISTENCE, DIGITADITION, DIGITAL ROOT, FACTORION, FIGURES, LENGTH (NUMBER), MULTIPLICATIVE PERSISTENCE, NARCISSISTIC NUMBER, SCIENTIFIC NOTATION, SIGNIFICANT DIGITS, SMITH NUMBER

Digitadition

Start with an INTEGER n, known as the GENERATOR. Add the SUM of the GENERATOR's digits to the GENERATOR to obtain the digitadition n'. A number can have more than one GENERATOR. If a number has no GENERATOR, it is called a SELF NUMBER. The sum of all numbers in a digitadition series is given by the last term minus the first plus the sum of the DIGITS of the last.

If the digitadition process is performed on n' to yield *its* digitadition n'', on n'' to yield n''', etc., a single-digit number, known as the DIGITAL ROOT of n, is eventually obtained. The digital roots of the first few integers are 1, 2, 3, 4, 5, 6, 7, 8, 9, 1, 2, 3, 4, 5, 6, 7, 9, 1, ... (Sloane's A010888).

If the process is generalized so that the kth (instead of first) powers of the digits of a number are repeatedly added, a periodic sequence of numbers is eventually obtained for any given starting number n. If the original number n is equal to the sum of the kth powers of its digits, it is called a NARCISSISTIC NUMBER. If the original number is the smallest number in the eventually periodic sequence of numbers in the repeated k-digitaditions, it is called a RECURRING DIGITAL INVARIANT. Both NARCISSISTIC NUMBERS and RECURRING DIGITAL INVARIANTS are relatively rare.

The only possible periods for repeated 2-digitaditions are 1 and 8, and the periods of the first few positive integers are 1, 8, 8, 8, 8, 8, 1, 8, 8, 1, ... The possible periods p for n-digitaditions are summarized in the following table, together with digitaditions for the first few integers and the corresponding sequence numbers.

n	Sloane	ps	n-Digitaditions
2	031176	1, 8	1, 8, 8, 8, 8, 8, 1, 8, 8, ...
3	031178	1, 2, 3	1, 1, 1, 3, 1, 1, 1, 1, 1, ...
4	031182	1, 2, 7	1, 7, 7, 7, 7, 7, 7, 7, 7, ...
5	031186	1, 2, 4, 6, 10, 12, 22, 28	1, 12, 22, 4, 10, 22, 28, 10, 22, 1, ...
6	031195	1, 2, 3, 4, 10, 30	1, 10, 30, 30, 30, 10, 10, 10, 3, 1, 10, ...
7	031200	1, 2, 3, 6, 12, 14, 21, 27, 30, 56, 92	1, 92, 14, 30, 92, 56, 6, 92, 56, 1, 92, 27, ...
8	031211	1, 25, 154	1, 25, 154, 154, 154, 154, 25, 154, 154, 1, 25, ...
9	031212	1, 2, 3, 4, 8, 10, 19, 24, 28, 30, 80, 93	1, 30, 93, 1, 19, 80, 4, 30, 80, 1, 30, 93, 4, 10, ...
10	031212	1, 6, 7, 17, 81, 123	1, 30, 93, 1, 19, 80, 4, 30, 80, 1, 30, 93, 4, 10, ...

The numbers having period-1 2-digitaded sequences are also called HAPPY NUMBERS. The first few numbers having period p n-digitaditions are summarized in the following table, together with their sequence numbers.

n	p	Sloane	Members
2	1	007770	1, 7, 10, 13, 19, 23, 28, 31, 32, ...
2	8	031177	2, 3, 4, 5, 6, 8, 9, 11, 12, 14, 15, ...
3	1	031179	1, 2, 3, 5, 6, 7, 8, 9, 10, 11, 12, ...
3	2	031180	49, 94, 136, 163, 199, 244, 316, ...
3	3	031181	4, 13, 16, 22, 25, 28, 31, 40, 46, ...
4	1	031183	1, 10, 12, 17, 21, 46, 64, 71, 100, ...
4	2	031184	66, 127, 172, 217, 228, 271, 282, ...
4	7	031185	2, 3, 4, 5, 6, 7, 8, 9, 11, 13, 14, ...
5	1	031187	1, 10, 100, 145, 154, 247, 274, ...
5	2	031188	133, 139, 193, 199, 226, 262, ...
5	4	031189	4, 37, 40, 55, 73, 124, 142, ...
5	6	031190	16, 61, 106, 160, 601, 610, 778, ...
5	10	031191	5, 8, 17, 26, 35, 44, 47, 50, 53, ...
5	12	031192	2, 11, 14, 20, 23, 29, 32, 38, 41, ...
5	22	031193	3, 6, 9, 12, 15, 18, 21, 24, 27, ...
5	28	031194	7, 13, 19, 22, 25, 28, 31, 34, 43, ...
6	1	011557	1, 10, 100, 1000, 10000, 100000, ...
6	2	031357	3468, 3486, 3648, 3684, 3846, ...
6	3	031196	9, 13, 31, 37, 39, 49, 57, 73, 75, ...
6	4	031197	255, 466, 525, 552, 646, 664, ...
6	10	031198	2, 6, 7, 8, 11, 12, 14, 15, 17, 19, ...
6	30	031199	3, 4, 5, 16, 18, 22, 29, 30, 33, ...
7	1	031201	1, 10, 100, 1000, 1259, 1295, ...
7	2	031202	22, 202, 220, 256, 265, 526, 562, ...
7	3	031203	124, 142, 148, 184, 214, 241, 259, ...
7	6		7, 70, 700, 7000, 70000, 700000, ...
7	12	031204	17, 26, 47, 59, 62, 71, 74, 77, 89, ...
7	14	031205	3, 30, 111, 156, 165, 249, 294, ...
7	21	031206	19, 34, 43, 91, 109, 127, 172, 190, ...
7	27	031207	12, 18, 21, 24, 39, 42, 45, 54, 78, ...
7	30	031208	4, 13, 16, 25, 28, 31, 37, 40, 46, ...
7	56	031209	6, 9, 15, 27, 33, 36, 48, 51, 57, ...
7	92	031210	2, 5, 8, 11, 14, 20, 23, 29, 32, 35, ...
8	1		1, 10, 14, 17, 29, 37, 41, 71, 73, ...
8	25		2, 7, 11, 15, 16, 20, 23, 27, 32, ...
8	154		3, 4, 5, 6, 8, 9, 12, 13, 18, 19, ...
9	1		1, 4, 10, 40, 100, 400, 1000, 1111, ...
9	2		127, 172, 217, 235, 253, 271, 325, ...
9	3		444, 4044, 4404, 4440, 4558, ...
9	4		7, 13, 31, 67, 70, 76, 103, 130, ...
9	8		22, 28, 34, 37, 43, 55, 58, 73, 79, ...
9	10		14, 38, 41, 44, 83, 104, 128, 140, ...
9	19		5, 26, 50, 62, 89, 98, 155, 206, ...
9	24		16, 61, 106, 160, 337, 373, 445, ...
9	28		19, 25, 46, 49, 52, 64, 91, 94, ...
9	30		2, 8, 11, 17, 20, 23, 29, 32, 35, ...
9	80		6, 9, 15, 18, 24, 33, 42, 48, 51, ...
9	93		3, 12, 21, 27, 30, 36, 39, 45, 54, ...
10	1	011557	1, 10, 100, 1000, 10000, 100000, ...
10	6		266, 626, 662, 1159, 1195, 1519, ...
10	7		46, 58, 64, 85, 122, 123, 132, ...
10	17		2, 4, 5, 11, 13, 20, 31, 38, 40, ...
10	81		17, 18, 37, 71, 73, 81, 107, 108, ...
10	123		3, 6, 7, 8, 9, 12, 14, 15, 16, 19, ...

see also 196-ALGORITHM, ADDITIVE PERSISTENCE, DIGIT, DIGITAL ROOT, MULTIPLICATIVE PERSISTENCE, NARCISSISTIC NUMBER, RECURRING DIGITAL INVARIANT

Digital Root

Consider the process of taking a number, adding its DIGITS, then adding the DIGITS of numbers derived from it, etc., until the remaining number has only one DIGIT. The number of additions required to obtain a single DIGIT from a number n is called the ADDITIVE PERSISTENCE of n, and the DIGIT obtained is called the digital root of n.

For example, the sequence obtained from the starting number 9876 is (9876, 30, 3), so 9876 has an ADDITIVE PERSISTENCE of 2 and a digital root of 3. The digital roots of the first few integers are 1, 2, 3, 4, 5, 6, 7, 8, 9, 1, 2, 3, 4, 5, 6, 7, 9, 1, ... (Sloane's A010888). The digital root of an INTEGER n can therefore be computed without actually performing the iteration using the simple congruence formula

$$\begin{cases} n \pmod 9 & n \not\equiv 0 \pmod 9 \\ 9 & n \equiv 0 \pmod 9. \end{cases}$$

see also ADDITIVE PERSISTENCE, DIGITADITION, KAPREKAR NUMBER, MULTIPLICATIVE DIGITAL ROOT, MULTIPLICATIVE PERSISTENCE, NARCISSISTIC NUMBER, RECURRING DIGITAL INVARIANT, SELF NUMBER

References

Sloane, N. J. A. Sequences A010888 and A007612/M1114 in "An On-Line Version of the Encyclopedia of Integer Sequences."

Digon

The DEGENERATE POLYGON (corresponding to a LINE SEGMENT) with SCHLÄFLI SYMBOL {2}.

see also LINE SEGMENT, POLYGON, TRIGONOMETRY VALUES—$\pi/2$

Digraph

see DIRECTED GRAPH

Dihedral Angle

The ANGLE between two PLANES. The dihedral angle between the planes

$$A_1x + B_1y + C_1z + D_1 = 0 \quad (1)$$

$$A_2x + B_2y + C_2z + D_2 = 0 \quad (2)$$

is

$$\cos\theta = \frac{A_1A_2 + B_1B_2 + C_1C_2}{\sqrt{{A_1}^2 + {B_1}^2 + {C_1}^2}\sqrt{{A_2}^2 + {B_2}^2 + {C_2}^2}}. \quad (3)$$

see also ANGLE, PLANE, VERTEX ANGLE

Dihedral Group

A GROUP of symmetries for an n-sided REGULAR POLYGON, denoted D_n. The ORDER of D_n is $2n$.

see also FINITE GROUP—D_3, FINITE GROUP—D_4

References

Arfken, G. "Dihedral Groups, D_n." *Mathematical Methods for Physicists, 3rd ed.* Orlando, FL: Academic Press, p. 248, 1985.

Lomont, J. S. "Dihedral Groups." §3.10.B in *Applications of Finite Groups.* New York: Dover, pp. 78–80, 1987.

Dijkstra's Algorithm

An ALGORITHM for finding the shortest path between two VERTICES.

see also FLOYD'S ALGORITHM

Dijkstra Tree

The shortest path-spanning TREE from a VERTEX of a GRAPH.

Dilation

An AFFINE TRANSFORMATION in which the scale is reduced. A dilation is also known as a CONTRACTION or HOMOTHECY. Any dilation which is not a simple translation has a unique FIXED POINT. The opposite of a dilation is an EXPANSION.

see also AFFINE TRANSFORMATION, EXPANSION, HOMOTHECY

References

Coxeter, H. S. M. and Greitzer, S. L. *Geometry Revisited.* Washington, DC: Math. Assoc. Amer., pp. 94–95, 1967.

Dilemma

Informally, a situation in which a decision must be made from several alternatives, none of which is obviously the optimal one. In formal LOGIC, a dilemma is a specific type of argument using two conditional statements which may take the form of a CONSTRUCTIVE DILEMMA or a DESTRUCTIVE DILEMMA.

see also CONSTRUCTIVE DILEMMA, DESTRUCTIVE DILEMMA, MONTY HALL PROBLEM, PARADOX, PRISONER'S DILEMMA

Dilogarithm

A special case of the POLYLOGARITHM $\mathrm{Li}_n(z)$ for $n = 2$. It is denoted $\mathrm{Li}_2(z)$, or sometimes $L_2(z)$, and is defined by the sum

$$\mathrm{Li}_2(z) = \sum_{k=1}^{\infty} \frac{z^k}{k^2}$$

or the integral

$$\mathrm{Li}_2(z) \equiv \int_z^0 \frac{\ln(1-t)\,dt}{t}.$$

There are several remarkable identities involving the POLYLOGARITHM function.

see also ABEL'S FUNCTIONAL EQUATION, POLYLOGARITHM, SPENCE'S INTEGRAL

References

Abramowitz, M. and Stegun, C. A. (Eds.). "Dilogarithm." §27.7 in *Handbook of Mathematical Functions with Formulas, Graphs, and Mathematical Tables, 9th printing.* New York: Dover, pp. 1004–1005, 1972.

Dilworth's Lemma

The WIDTH of a set P is equal to the minimum number of CHAINS needed to COVER P. Equivalently, if a set P of $ab + 1$ elements is PARTIALLY ORDERED, then P contains a CHAIN of size $a + 1$ or an ANTICHAIN of size $b + 1$. Letting N be the CARDINALITY of P, W the WIDTH, and L the LENGTH, this last statement says $N \leq LW$. Dilworth's lemma is a generalization of the ERDŐS-SZEKERES THEOREM. RAMSEY'S THEOREM generalizes Dilworth's Lemma.

see also COMBINATORICS, ERDŐS-SZEKERES THEOREM, RAMSEY'S THEOREM

Dilworth's Theorem

see DILWORTH'S LEMMA

Dimension

The notion of dimension is important in mathematics because it gives a precise parameterization of the conceptual or visual complexity of any geometric object. In fact, the concept can even be applied to abstract objects which cannot be directly visualized. For example, the notion of time can be considered as one-dimensional, since it can be thought of as consisting of only "now," "before" and "after." Since "before" and "after," regardless of how far back or how far into the future they are, are extensions, time is like a line, a 1-dimensional object.

To see how lower and higher dimensions relate to each other, take any geometric object (like a POINT, LINE, CIRCLE, PLANE, etc.), and "drag" it in an opposing direction (drag a POINT to trace out a LINE, a LINE to trace out a box, a CIRCLE to trace out a CYLINDER, a DISK to a solid CYLINDER, etc.). The result is an object which is qualitatively "larger" than the previous object, "qualitative" in the sense that, regardless of how you drag the original object, you always trace out an object of the same "qualitative size." The POINT could be made into a straight LINE, a CIRCLE, a HELIX, or some other CURVE, but all of these objects are qualitatively of the same dimension. The notion of dimension was invented for the purpose of measuring this "qualitative" topological property.

Making things a bit more formal, finite collections of objects (e.g., points in space) are considered 0-dimensional. Objects that are "dragged" versions of 0-dimensional objects are then called 1-dimensional. Similarly, objects which are dragged 1-dimensional objects are 2-dimensional, and so on. Dimension is formalized in mathematics as the intrinsic dimension of a TOPOLOGICAL SPACE. This dimension is called the LEBESGUE COVERING DIMENSION (also known simply as the TOPOLOGICAL DIMENSION). The archetypal example is EUCLIDEAN n-space $\mathbb{R}^n$, which has topological dimension n. The basic ideas leading up to this result (including the DIMENSION INVARIANCE THEOREM, DOMAIN INVARIANCE THEOREM, and LEBESGUE COVERING DIMENSION) were developed by Poincaré, Brouwer, Lebesgue, Urysohn, and Menger.

There are several branchings and extensions of the notion of topological dimension. Implicit in the notion of the LEBESGUE COVERING DIMENSION is that dimension, in a sense, is a measure of how an object fills space. If it takes up a lot of room, it is higher dimensional, and if it takes up less room, it is lower dimensional. HAUSDORFF DIMENSION (also called FRACTAL DIMENSION) is a fine tuning of this definition that allows notions of objects with dimensions other than INTEGERS. FRACTALS are objects whose HAUSDORFF DIMENSION is different from their TOPOLOGICAL DIMENSION.

The concept of dimension is also used in ALGEBRA, primarily as the dimension of a VECTOR SPACE over a FIELD. This usage stems from the fact that VECTOR SPACES over the reals were the first VECTOR SPACES to be studied, and for them, their topological dimension can be calculated by purely algebraic means as the CARDINALITY of a maximal linearly independent subset. In particular, the dimension of a SUBSPACE of $\mathbb{R}^n$ is equal to the number of LINEARLY INDEPENDENT VECTORS needed to generate it (i.e., the number of VECTORS in its BASIS). Given a transformation A of $\mathbb{R}^n$,

$$\dim[\text{Range}(A)] + \dim[\text{Null}(A)] = \dim(\mathbb{R}^n).$$

see also CAPACITY DIMENSION, CODIMENSION, CORRELATION DIMENSION, EXTERIOR DIMENSION, FRACTAL DIMENSION, HAUSDORFF DIMENSION, HAUSDORFF-BESICOVITCH DIMENSION, KAPLAN-YORKE DIMENSION, KRULL DIMENSION, LEBESGUE COVERING DIMENSION, LEBESGUE DIMENSION, LYAPUNOV DIMENSION, POSET DIMENSION, q-DIMENSION, SIMILARITY DIMENSION, TOPOLOGICAL DIMENSION

References
Abbott, E. A. *Flatland: A Romance of Many Dimensions.* New York: Dover, 1992.
Hinton, C. H. *The Fourth Dimension.* Pomeroy, WA: Health Research, 1993.
Manning, H. *The Fourth Dimension Simply Explained.* Magnolia, MA: Peter Smith, 1990.
Manning, H. *Geometry of Four Dimensions.* New York: Dover, 1956.
Neville, E. H. *The Fourth Dimension.* Cambridge, England: Cambridge University Press, 1921.
Rucker, R. von Bitter. *The Fourth Dimension: A Guided Tour of the Higher Universes.* Boston, MA: Houghton Mifflin, 1984.
Sommerville, D. M. Y. *An Introduction to the Geometry of n Dimensions.* New York: Dover, 1958.

Dimension Axiom

One of the EILENBERG-STEENROD AXIOMS. Let X be a single point space. $H_n(X) = 0$ unless $n = 0$, in which case $H_0(X) = G$ where G are some GROUPS. The H_0 are called the COEFFICIENTS of the HOMOLOGY THEORY $H(\cdot)$.

see also EILENBERG-STEENROD AXIOMS, HOMOLOGY (TOPOLOGY)

Dimension Invariance Theorem

$\mathbb{R}^n$ is HOMEOMORPHIC to $\mathbb{R}^m$ IFF $n = m$. This theorem was first proved by Brouwer.

see also DOMAIN INVARIANCE THEOREM

Dimensionality Theorem

For a finite GROUP of h elements with an n_ith dimensional ith irreducible representation,

$$\sum_i {n_i}^2 = h.$$

Diminished Polyhedron

A UNIFORM POLYHEDRON with pieces removed.

Diminished Rhombicosidodecahedron

see JOHNSON SOLID

Dini Expansion

An expansion based on the ROOTS of

$$x^{-n}[xJ_n'(x) + HJ_n(x)] = 0,$$

where $J_n(x)$ is a BESSEL FUNCTION OF THE FIRST KIND, is called a Dini expansion.

see also BESSEL FUNCTION FOURIER EXPANSION

References
Bowman, F. *Introduction to Bessel Functions.* New York: Dover, p. 109, 1958.

Dini's Surface

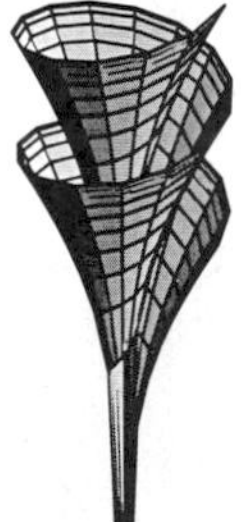

A surface of constant NEGATIVE CURVATURE obtained by twisting a PSEUDOSPHERE and given by the parametric equations

$$x = a\cos u \sin v \tag{1}$$

$$y = a\sin u \sin v \tag{2}$$

$$z = a\{\cos v + \ln[\tan(\tfrac{1}{2}v)]\} + bu. \tag{3}$$

The above figure corresponds to $a = 1$, $b = 0.2$, $u \in [0, 4\pi]$, and $v \in (0, 2]$.

see also PSEUDOSPHERE

References

Geometry Center. "Dini's Surface." `http://www.geom.umn.edu/zoo/diffgeom/surfspace/dini/`.

Gray, A. *Modern Differential Geometry of Curves and Surfaces.* Boca Raton, FL: CRC Press, pp. 494–495, 1993.

Nordstrand, T. "Dini's Surface." `http://www.uib.no/people/nfytn/dintxt.htm`.

Dini's Test

A test for the convergence of FOURIER SERIES. Let

$$\phi_x(t) \equiv f(x+t) + f(x-t) - 2f(x),$$

then if

$$\int_0^\pi \frac{|\phi_x(t)|\,dt}{t}$$

is FINITE, the FOURIER SERIES converges to $f(x)$ at x.

see also FOURIER SERIES

References

Sansone, G. *Orthogonal Functions, rev. English ed.* New York: Dover, pp. 65–68, 1991.

Dinitz Problem

Given any assignment of n-element sets to the n^2 locations of a square $n \times n$ array, is it always possible to find a PARTIAL LATIN SQUARE? The fact that such a PARTIAL LATIN SQUARE can always be found for a 2×2 array can be proven analytically, and techniques were developed which also proved the existence for 4×4 and 6×6 arrays. However, the general problem eluded solution until it was answered in the affirmative by Galvin in 1993 using results of Janssen (1993ab) and F. Maffray.

see also PARTIAL LATIN SQUARE

References

Chetwynd, A. and Häggkvist, R. "A Note on List-Colorings." *J. Graph Th.* **13**, 87–95, 1989.

Cipra, B. "Quite Easily Done." In *What's Happening in the Mathematical Sciences* **2**, pp. 41–46, 1994.

Erdős, P.; Rubin, A.; and Taylor, H. "Choosability in Graphs." *Congr. Numer.* **26**, 125–157, 1979.

Häggkvist, R. "Towards a Solution of the Dinitz Problem?" *Disc. Math.* **75**, 247–251, 1989.

Janssen, J. C. M. "The Dinitz Problem Solved for Rectangles." *Bull. Amer. Math. Soc.* **29**, 243–249, 1993a.

Janssen, J. C. M. *Even and Odd Latin Squares.* Ph.D. thesis. Lehigh University, 1993b.

Kahn, J. "Recent Results on Some Not-So-Recent Hypergraph Matching and Covering Problems." *Proceedings of the Conference on Extremal Problems for Finite Sets.* Visegràd, Hungary, 1991.

Kahn, J. "Coloring Nearly-Disjoint Hypergraphs with $n + o(n)$ Colors." *J. Combin. Th. Ser. A* **59**, 31–39, 1992.

Diocles's Cissoid

see CISSOID OF DIOCLES

Diophantine Equation

An equation in which only INTEGER solutions are allowed. HILBERT'S 10TH PROBLEM asked if a technique for solving a general Diophantine existed. A general method exists for the solution of first degree Diophantine equations. However, the impossibility of obtaining a general solution was proven by Julia Robinson and Martin Davis in 1970, following proof of the result that the equation $n = F_{2m}$ (where F_{2m} is a FIBONACCI NUMBER) is Diophantine by Yuri Matijasevič (Matijasevič 1970, Davis 1973, Davis and Hersh 1973, Matijasevič 1993).

No general method is known for quadratic or higher Diophantine equations. Jones and Matijasevič (1982) proved that no ALGORITHMS can exist to determine if an arbitrary Diophantine equation in nine variables has solutions. Ogilvy and Anderson (1988) give a number of Diophantine equations with known and unknown solutions.

D. Wilson has compiled a list of the smallest nth POWERS which are the sums of n *distinct* smaller nth POWERS. The first few are 3, 5, 6, 15, 12, 25, 40, ... (Sloane's A030052):

$$\begin{aligned}
3^1 &= 1^1 + 2^1\\
5^2 &= 3^2 + 4^2\\
6^3 &= 3^3 + 4^3 + 5^3\\
15^4 &= 4^4 + 6^4 + 8^4 + 9^4 + 14^4\\
12^5 &= 4^5 + 5^5 + 6^5 + 7^5 + 9^5 + 11^5\\
25^6 &= 1^6 + 2^6 + 3^6 + 5^6 + 6^6 + 7^6 + 8^6 + 9^6 + 10^6\\
&\quad + 12^6 + 13^6 + 15^6 + 16^6 + 17^6 + 18^6 + 23^6\\
40^7 &= 1^7 + 3^7 + 5^7 + 9^7 + 12^7 + 14^7 + 16^7 + 17^7\\
&\quad + 18^7 + 20^7 + 21^7 + 22^7 + 25^7 + 28^7 + 39^7\\
84^8 &= 1^8 + 2^8 + 3^8 + 5^8 + 7^8 + 9^8 + 10^8 + 11^8\\
&\quad + 12^8 + 13^8 + 14^8 + 15^8 + 16^8 + 17^8 + 18^8\\
&\quad + 19^8 + 21^8 + 23^8 + 24^8 + 25^8 + 26^8 + 27^8\\
&\quad + 29^8 + 32^8 + 33^8 + 35^8 + 37^8 + 38^8 + 39^8\\
&\quad + 41^8 + 42^8 + 43^8 + 45^8 + 46^8 + 47^8 + 48^8\\
&\quad + 49^8 + 51^8 + 52^8 + 53^8 + 57^8 + 58^8 + 59^8\\
&\quad + 61^8 + 63^8 + 69^8 + 73^8\\
47^9 &= 1^9 + 2^9 + 4^9 + 7^9 + 11^9 + 14^9 + 15^9 + 18^9\\
&\quad + 26^9 + 27^9 + 30^9 + 31^9 + 32^9 + 33^9\\
&\quad + 36^9 + 38^9 + 39^9 + 43^9\\
63^{10} &= 1^{10} + 2^{10} + 4^{10} + 5^{10} + 6^{10} + 8^{10} + 12^{10}\\
&\quad + 15^{10} + 16^{10} + 17^{10} + 20^{10} + 21^{10} + 25^{10}\\
&\quad + 26^{10} + 27^{10} + 28^{10} + 30^{10} + 36^{10} + 37^{10}\\
&\quad + 38^{10} + 40^{10} + 51^{10} + 62^{10}.
\end{aligned}$$

see also ABC CONJECTURE, ARCHIMEDES' CATTLE PROBLEM, BACHET EQUATION, BRAHMAGUPTA'S

PROBLEM, CANNONBALL PROBLEM, CATALAN'S PROBLEM, DIOPHANTINE EQUATION—LINEAR, DIOPHANTINE EQUATION—QUADRATIC, DIOPHANTINE EQUATION—CUBIC, DIOPHANTINE EQUATION—QUARTIC, DIOPHANTINE EQUATION—5TH POWERS, DIOPHANTINE EQUATION—6TH POWERS, DIOPHANTINE EQUATION—7TH POWERS, DIOPHANTINE EQUATION—8TH POWERS, DIOPHANTINE EQUATION—9TH POWERS, DIOPHANTINE EQUATION—10TH POWERS, DIOPHANTINE EQUATION—nTH POWERS, DIOPHANTUS PROPERTY, EULER BRICK, EULER QUARTIC CONJECTURE, FERMAT'S LAST THEOREM, FERMAT SUM THEOREM, GENUS THEOREM, HURWITZ EQUATION, MARKOV NUMBER, MONKEY AND COCONUT PROBLEM, MULTIGRADE EQUATION, p-ADIC NUMBER, PELL EQUATION, PYTHAGOREAN QUADRUPLE, PYTHAGOREAN TRIPLE

References

Beiler, A. H. *Recreations in the Theory of Numbers: The Queen of Mathematics Entertains.* New York: Dover, 1966.

Carmichael, R. D. *The Theory of Numbers, and Diophantine Analysis.* New York: Dover, 1959.

Chen, S. "Equal Sums of Like Powers: On the Integer Solution of the Diophantine System." `http://www.nease.net/~chin/eslp/`.

Chen, S. "References." `http://www.nease.net/~chin/eslp/referenc.htm`

Davis, M. "Hilbert's Tenth Problem is Unsolvable." *Amer. Math. Monthly* **80**, 233–269, 1973.

Davis, M. and Hersh, R. "Hilbert's 10th Problem." *Sci. Amer.*, pp. 84–91, Nov. 1973.

Dörrie, H. "The Fermat-Gauss Impossibility Theorem." §21 in *100 Great Problems of Elementary Mathematics: Their History and Solutions.* New York: Dover, pp. 96–104, 1965.

Guy, R. K. "Diophantine Equations." Ch. D in *Unsolved Problems in Number Theory, 2nd ed.* New York: Springer-Verlag, pp. 139–198, 1994.

Hardy, G. H. and Wright, E. M. *An Introduction to the Theory of Numbers, 5th ed.* Oxford, England: Clarendon Press, 1979.

Hunter, J. A. H. and Madachy, J. S. "Diophantos and All That." Ch. 6 in *Mathematical Diversions.* New York: Dover, pp. 52–64, 1975.

Ireland, K. and Rosen, M. "Diophantine Equations." Ch. 17 in *A Classical Introduction to Modern Number Theory, 2nd ed.* New York: Springer-Verlag, pp. 269–296, 1990.

Jones, J. P. and Matijasevič, Yu. V. "Exponential Diophantine Representation of Recursively Enumerable Sets." *Proceedings of the Herbrand Symposium, Marseilles, 1981.* Amsterdam, Netherlands: North-Holland, pp. 159–177, 1982.

Lang, S. *Introduction to Diophantine Approximations, 2nd ed.* New York: Springer-Verlag, 1995.

Matijasevič, Yu. V. "Solution to of the Tenth Problem of Hilbert." *Mat. Lapok* **21**, 83–87, 1970.

Matijasevič, Yu. V. *Hilbert's Tenth Problem.* Cambridge, MA: MIT Press, 1993.

Mordell, L. J. *Diophantine Equations.* New York: Academic Press, 1969.

Nagel, T. *Introduction to Number Theory.* New York: Wiley, 1951.

Ogilvy, C. S. and Anderson, J. T. "Diophantine Equations." Ch. 6 in *Excursions in Number Theory.* New York: Dover, pp. 65–83, 1988.

Sloane, N. J. A. Sequence A030052 in "An On-Line Version of the Encyclopedia of Integer Sequences."

Diophantine Equation—5th Powers

The 2-1 fifth-order Diophantine equation

$$A^5 + B^5 = C^5 \qquad (1)$$

is a special case of FERMAT'S LAST THEOREM with $n = 5$, and so has no solution. No solutions to the 2-2 equation

$$A^5 + B^5 = C^5 + D^5 \qquad (2)$$

are known, despite the fact that sums up to 1.02×10^{26} have been checked (Guy 1994, p. 140), improving on the results on Lander *et al.* (1967), who checked up to 2.8×10^{14}. (In fact, no solutions are known for POWERS of 6 or 7 either.)

No solutions to the 3-1 equation

$$A^5 + B^5 + C^5 = D^5 \qquad (3)$$

are known (Lander *et al.* 1967), nor are any 3-2 solutions up to 8×10^{12} (Lander *et al.* 1967).

Parametric solutions are known for the 3-3 (Guy 1994, pp. 140 and 142). Swinnerton-Dyer (1952) gave two parametric solutions to the 3-3 equation but, forty years later, W. Gosper discovered that the second scheme has an unfixable bug. The smallest primitive 3-3 solutions are

$$24^5 + 28^5 + 67^5 = 3^5 + 54^5 + 62^5 \qquad (4)$$
$$18^5 + 44^5 + 66^5 = 13^5 + 51^5 + 64^5 \qquad (5)$$
$$21^5 + 43^5 + 76^5 = 8^5 + 62^5 + 68^5 \qquad (6)$$
$$56^5 + 67^5 + 83^5 = 53^5 + 72^5 + 81^5 \qquad (7)$$
$$49^5 + 75^5 + 107^5 = 39^5 + 92^5 + 100^5 \qquad (8)$$

(Moessner 1939, Moessner 1948, Lander *et al.* 1967).

For 4 fifth POWERS, we have the 4-1 equation

$$27^5 + 84^5 + 110^5 + 133^5 = 144^5 \qquad (9)$$

(Lander and Parkin 1967, Lander *et al.* 1967), but it is not known if there is a parametric solution (Guy 1994, p. 140). Sastry's (1934) 5-1 solution gives some 4-2 solutions. The smallest primitive 4-2 solutions are

$$4^5 + 10^5 + 20^5 + 28^5 = 3^5 + 29^5 \qquad (10)$$
$$5^5 + 13^5 + 25^5 + 37^5 = 12^5 + 38^5 \qquad (11)$$
$$26^5 + 29^5 + 35^5 + 50^5 = 28^5 + 52^5 \qquad (12)$$
$$5^5 + 25^5 + 62^5 + 63^5 = 61^5 + 64^5 \qquad (13)$$
$$6^5 + 50^5 + 53^5 + 82^5 = 16^5 + 85^5 \qquad (14)$$
$$56^5 + 63^5 + 72^5 + 86^5 = 31^5 + 96^5 \qquad (15)$$
$$44^5 + 58^5 + 67^5 + 94^5 = 14^5 + 99^5 \qquad (16)$$
$$11^5 + 13^5 + 37^5 + 99^5 = 63^5 + 97^5 \qquad (17)$$
$$48^5 + 57^5 + 76^5 + 100^5 = 25^5 + 106^5 \qquad (18)$$
$$58^5 + 76^5 + 79^5 + 102^5 = 54^5 + 111^5 \qquad (19)$$

(Rao 1934, Moessner 1948, Lander *et al.* 1967).

A two-parameter solution to the 4-3 equation was given by Xeroudakes and Moessner (1958). Gloden (1949) also gave a parametric solution. The smallest solution is

$$1^5 + 8^5 + 14^5 + 27^5 = 3^5 + 22^5 + 25^5 \quad (20)$$

(Rao 1934, Lander *et al.* 1967). Several parametric solutions to the 4-4 equation were found by Xeroudakes and Moessner (1958). The smallest 4-4 solution is

$$5^5 + 6^5 + 6^5 + 8^5 = 4^5 + 7^5 + 7^5 + 7^5 \quad (21)$$

(Rao 1934, Lander *et al.* 1967). The first 4-4-4 equation is

$$\begin{aligned} 3^5 + 48^5 + 52^5 + 61^5 &= 13^5 + 36^5 + 51^5 + 64^5 \\ &= 18^5 + 36^5 + 44^5 + 66^5 \quad (22) \end{aligned}$$

(Lander *et al.* 1967).

Sastry (1934) found a 2-parameter solution for 5-1 equations

$$\begin{aligned} (75v^5 - u^5)^5 + (u^5 + 25v^5)^5 + (u^5 - 25v^5)^5 \\ +(10u^3v^2)^5 + (50uv^4)^5 = (u^5 + 75v^5)^5 \quad (23) \end{aligned}$$

(quoted in Lander and Parkin 1967), and Lander and Parkin (1967) found the smallest numerical solutions. Lander *et al.* (1967) give a list of the smallest solutions, the first few being

$$\begin{aligned} 19^5 + 43^5 + 46^5 + 47^5 + 67^5 &= 72^5 \quad (24) \\ 21^5 + 23^5 + 37^5 + 79^5 + 84^5 &= 94^5 \quad (25) \\ 7^5 + 43^5 + 57^5 + 80^5 + 100^5 &= 107^5 \quad (26) \\ 8^5 + 120^5 + 191^5 + 259^5 + 347^5 &= 365^5 \quad (27) \\ 79^5 + 202^5 + 258^5 + 261^5 + 395^5 &= 415^5 \quad (28) \\ 4^5 + 26^5 + 139^5 + 296^5 + 412^5 &= 427^5 \quad (29) \\ 31^5 + 105^5 + 139^5 + 314^5 + 416^5 &= 435^5 \quad (30) \\ 54^5 + 91^5 + 101^5 + 404^5 + 430^5 &= 480^5 \quad (31) \\ 19^5 + 201^5 + 347^5 + 388^5 + 448^5 &= 503^5 \quad (32) \\ 159^5 + 172^5 + 200^5 + 356^5 + 513^5 &= 530^5 \quad (33) \\ 218^5 + 276^5 + 385^5 + 409^5 + 495^5 &= 553^5 \quad (34) \\ 2^5 + 298^5 + 351^5 + 474^5 + 500^5 &= 575^5 \quad (35) \end{aligned}$$

(Lander and Parkin 1967, Lander *et al.* 1967).

The smallest primitive 5-2 solutions are

$$\begin{aligned} 4^5 + 5^5 + 7^5 + 16^5 + 21^5 &= 1^5 + 22^5 \quad (36) \\ 9^5 + 11^5 + 14^5 + 18^5 + 30^5 &= 23^5 + 29^5 \quad (37) \\ 10^5 + 14^5 + 26^5 + 31^5 + 33^5 &= 16^5 + 38^5 \quad (38) \\ 4^5 + 22^5 + 29^5 + 35^5 + 36^5 &= 24^5 + 42^5 \quad (39) \\ 8^5 + 15^5 + 17^5 + 19^5 + 45^5 &= 30^5 + 44^5 \quad (40) \\ 5^5 + 6^5 + 26^5 + 27^5 + 44^5 &= 36^5 + 42^5 \quad (41) \end{aligned}$$

(Rao 1934, Lander *et al.* 1967).

The 6-1 equation has solutions

$$\begin{aligned} 4^5 + 5^5 + 6^5 + 7^5 + 9^5 + 11^5 &= 12^5 \quad (42) \\ 5^5 + 10^5 + 11^5 + 16^5 + 19^5 + 29^5 &= 30^5 \quad (43) \\ 15^5 + 16^5 + 17^5 + 22^5 + 24^5 + 28^5 &= 32^5 \quad (44) \\ 13^5 + 18^5 + 23^5 + 31^5 + 36^5 + 66^5 &= 67^5 \quad (45) \\ 7^5 + 20^5 + 29^5 + 31^5 + 34^5 + 66^5 &= 67^5 \quad (46) \\ 22^5 + 35^5 + 48^5 + 58^5 + 61^5 + 64^5 &= 78^5 \quad (47) \\ 4^5 + 13^5 + 19^5 + 20^5 + 67^5 + 96^5 &= 99^5 \quad (48) \\ 6^5 + 17^5 + 60^5 + 64^5 + 73^5 + 89^5 &= 99^5 \quad (49) \end{aligned}$$

(Martin 1887, 1888, Lander and Parkin 1967, Lander *et al.* 1967).

The smallest 7-1 solution is

$$1^5 + 7^5 + 8^5 + 14^5 + 15^5 + 18^5 + 20^5 = 23^5 \quad (50)$$

(Lander *et al.* 1967).

References

Berndt, B. C. *Ramanujan's Notebooks, Part IV.* New York: Springer-Verlag, p. 95, 1994.

Gloden, A. "Über mehrgeradige Gleichungen." *Arch. Math.* **1**, 482–483, 1949.

Guy, R. K. "Sums of Like Powers. Euler's Conjecture." §D1 in *Unsolved Problems in Number Theory, 2nd ed.* New York: Springer-Verlag, pp. 139–144, 1994.

Lander, L. J. and Parkin, T. R. "A Counterexample to Euler's Sum of Powers Conjecture." *Math. Comput.* **21**, 101–103, 1967.

Lander, L. J.; Parkin, T. R.; and Selfridge, J. L. "A Survey of Equal Sums of Like Powers." *Math. Comput.* **21**, 446–459, 1967.

Martin, A. "Methods of Finding *n*th-Power Numbers Whose Sum is an *n*th Power; With Examples." *Bull. Philos. Soc. Washington* **10**, 107–110, 1887.

Martin, A. *Smithsonian Misc. Coll.* **33**, 1888.

Martin, A. "About Fifth-Power Numbers whose Sum is a Fifth Power." *Math. Mag.* **2**, 201–208, 1896.

Moessner, A. "Einige numerische Identitäten." *Proc. Indian Acad. Sci. Sect. A* **10**, 296–306, 1939.

Moessner, A. "Alcune richerche di teoria dei numeri e problemi diofantei." *Bol. Soc. Mat. Mexicana* **2**, 36–39, 1948.

Rao, K. S. "On Sums of Fifth Powers." *J. London Math. Soc.* **9**, 170–171, 1934.

Sastry, S. "On Sums of Powers." *J. London Math. Soc.* **9**, 242–246, 1934.

Swinnerton-Dyer, H. P. F. "A Solution of $A^5 + B^5 + C^5 = D^5 + E^5 + F^5$." *Proc. Cambridge Phil. Soc.* **48**, 516–518, 1952.

Xeroudakes, G. and Moessner, A. "On Equal Sums of Like Powers." *Proc. Indian Acad. Sci. Sect. A* **48**, 245–255, 1958.

Diophantine Equation—6th Powers

The 2-1 equation

$$A^6 + B^6 = C^6 \quad (1)$$

Diophantine Equation—6th Powers

is a special case of FERMAT'S LAST THEOREM with $n = 6$, and so has no solution. Ekl (1996) has searched and found no solutions to the 2-2

$$A^6 + B^6 = C^6 + D^6 \quad (2)$$

with sums less than 7.25×10^{26}.

No solutions are known to the 3-1 or 3-2 equations. However, parametric solutions are known for the 3-3 equation

$$A^6 + B^6 + C^6 = D^6 + E^6 + F^6 \quad (3)$$

(Guy 1994, pp. 140 and 142). Known solutions are

$$3^6 + 19^6 + 22^6 = 10^6 + 15^6 + 23^6 \quad (4)$$
$$36^6 + 37^6 + 67^6 = 15^6 + 52^6 + 65^6 \quad (5)$$
$$33^6 + 47^6 + 74^6 = 23^6 + 54^6 + 73^6 \quad (6)$$
$$32^6 + 43^6 + 81^6 = 3^6 + 55^6 + 80^6 \quad (7)$$
$$37^6 + 50^6 + 81^6 = 11^6 + 65^6 + 78^6 \quad (8)$$
$$25^6 + 62^6 + 138^6 = 82^6 + 92^6 + 135^6 \quad (9)$$
$$51^6 + 113^6 + 136^6 = 40^6 + 125^6 + 129^6 \quad (10)$$
$$71^6 + 92^6 + 147^6 = 1^6 + 132^6 + 133^6 \quad (11)$$
$$111^6 + 121^6 + 230^6 = 26^6 + 169^6 + 225^6 \quad (12)$$
$$75^6 + 142^6 + 245^6 = 14^6 + 163^6 + 243^6 \quad (13)$$

(Rao 1934, Lander *et al.* 1967).

No solutions are known to the 4-1 or 4-2 equations. The smallest primitive 4-3 solutions are

$$41^6 + 58^6 + 73^6 = 15^6 + 32^6 + 65^6 + 70^6 \quad (14)$$
$$61^6 + 62^6 + 85^6 = 52^6 + 56^6 + 69^6 + 83^6 \quad (15)$$
$$61^6 + 74^6 + 85^6 = 26^6 + 56^6 + 71^6 + 87^6 \quad (16)$$
$$11^6 + 88^6 + 90^6 = 21^6 + 74^6 + 78^6 + 92^6 \quad (17)$$
$$26^6 + 83^6 + 95^6 = 23^6 + 24^6 + 28^6 + 101^6 \quad (18)$$

(Lander *et al.* 1967). Moessner (1947) gave three parametric solutions to the 4-4 equation. The smallest 4-4 solution is

$$2^6 + 2^6 + 9^6 + 9^6 = 3^6 + 5^6 + 6^6 + 10^6 \quad (19)$$

(Rao 1934, Lander *et al.* 1967). The smallest 4-4-4 solution is

$$1^6 + 34^6 + 49^6 + 111^6 = 7^6 + 43^6 + 69^6 + 110^6$$
$$= 18^6 + 25^6 + 77^6 + 109^6 \quad (20)$$

(Lander *et al.* 1967).

No n-1 solutions are known for $n \leq 6$ (Lander *et al.* 1967). No solution to the 5-1 equation is known (Guy 1994, p. 140) or the 5-2 equation.

No solutions are known to the 6-1 or 6-2 equations.

The smallest 7-1 solution is

$$74^6 + 234^6 + 402^6 + 474^6 + 702^6 + 894^6 + 1077^6 = 1141^6 \quad (21)$$

(Lander *et al.* 1967). The smallest 7-2 solution is

$$18^6 + 22^6 + 36^6 + 58^6 + 69^6 + 78^6 + 78^6 = 56^6 + 91^6 \quad (22)$$

(Lander *et al.* 1967).

The smallest primitive 8-1 solutions are

$$8^6 + 12^6 + 30^6 + 78^6 + 102^6 + 138^6 + 165^6 + 246^6 = 251^6 \quad (23)$$
$$48^6 + 111^6 + 156^6 + 186^6 + 188^6 + 228^6 + 240^6 + 426^6 = 431^6 \quad (24)$$
$$93^6 + 93^6 + 195^6 + 197^6 + 303^6 + 303^6 + 303^6 + 411^6 = 440^6 \quad (25)$$
$$219^6 + 255^6 + 261^6 + 267^6 + 289^6 + 351^6 + 351^6 + 351^6 = 440^6 \quad (26)$$
$$12^6 + 66^6 + 138^6 + 174^6 + 212^6 + 288^6 + 306^6 + 441^6 = 455^6 \quad (27)$$
$$12^6 + 48^6 + 222^6 + 236^6 + 333^6 + 384^6 + 390^6 + 426^6 = 493^6 \quad (28)$$
$$66^6 + 78^6 + 144^6 + 228^6 + 256^6 + 288^6 + 435^6 + 444^6 = 499^6 \quad (29)$$
$$16^6 + 24^6 + 60^6 + 156^6 + 204^6 + 276^6 + 330^6 + 492^6 = 502^6 \quad (30)$$
$$61^6 + 96^6 + 156^6 + 228^6 + 276^6 + 318^6 + 354^6 + 534^6 = 547^6 \quad (31)$$
$$170^6 + 177^6 + 276^6 + 312^6 + 312^6 + 408^6 + 450^6 + 498^6 = 559^6 \quad (32)$$
$$60^6 + 102^6 + 126^6 + 261^6 + 270^6 + 338^6 + 354^6 + 570^6 = 581^6 \quad (33)$$
$$57^6 + 146^6 + 150^6 + 360^6 + 390^6 + 402^6 + 444^6 + 528^6 = 583^6 \quad (34)$$
$$33^6 + 72^6 + 122^6 + 192^6 + 204^6 + 390^6 + 534^6 + 534^6 = 607^6 \quad (35)$$
$$12^6 + 90^6 + 114^6 + 114^6 + 273^6 + 306^6 + 492^6 + 592^6 = 623^6 \quad (36)$$

(Lander *et al.* 1967). The smallest 8-2 solution is

$$8^6 + 10^6 + 12^6 + 15^6 + 24^6 + 30^6 + 33^6 + 36^6 = 35^6 + 37^6 \quad (37)$$

(Lander *et al.* 1967).

The smallest 9-1 solution is

$$1^6 + 17^6 + 19^6 + 22^6 + 31^6 + 37^6 + 37^6 + 41^6 + 49^6 = 54^6 \quad (38)$$

(Lander *et al.* 1967). The smallest 9-2 solution is

$$1^6+5^6+5^6+7^6+13^6+13^6+13^6+17^6+19^6=6^6+21^6 \quad (39)$$

(Lander *et al.* 1967).

The smallest 10-1 solution is

$$2^6+4^6+7^6+14^6+16^6+26^6+26^6+30^6+32^6+32^6=39^6 \quad (40)$$

(Lander *et al.* 1967). The smallest 10-2 solution is

$$1^6+1^6+1^6+4^6+4^6+7^6+9^6+11^6+11^6+11^6=12^6+12^6 \quad (41)$$

(Lander *et al.* 1967).

The smallest 11-1 solution is

$$2^6+5^6+5^6+5^6+7^6+7^6+9^6+9^6+10^6+14^6+17^6=18^6 \quad (42)$$

(Lander *et al.* 1967).

There is also at least one 16-1 identity,

$$1^6+2^6+4^6+5^6+6^6+7^6+9^6+12^6+13^6+15^6 \\ +16^6+18^6+20^6+21^6+22^6+23^6=28^6 \quad (43)$$

(Martin 1893). Moessner (1959) gave solutions for 16-1, 18-1, 20-1, and 23-1.

References

Ekl, R. L. "Equal Sums of Four Seventh Powers." *Math. Comput.* **65**, 1755–1756, 1996.

Guy, R. K. "Sums of Like Powers. Euler's Conjecture." §D1 in *Unsolved Problems in Number Theory, 2nd ed.* New York: Springer-Verlag, pp. 139–144, 1994.

Lander, L. J.; Parkin, T. R.; and Selfridge, J. L. "A Survey of Equal Sums of Like Powers." *Math. Comput.* **21**, 446–459, 1967.

Martin, A. "On Powers of Numbers Whose Sum is the Same Power of Some Number." *Quart. J. Math.* **26**, 225–227, 1893.

Moessner, A. "On Equal Sums of Like Powers." *Math. Student* **15**, 83–88, 1947.

Moessner, A. "Einige zahlentheoretische Untersuchungen und diophantische Probleme." *Glasnik Mat.-Fiz. Astron. Drustvo Mat. Fiz. Hrvatske Ser. 2* **14**, 177–182, 1959.

Rao, S. K. "On Sums of Sixth Powers." *J. London Math. Soc.* **9**, 172–173, 1934.

Diophantine Equation—7th Powers

The 2-1 equation

$$A^7+B^7=C^7 \quad (1)$$

is a special case of FERMAT'S LAST THEOREM with $n=7$, and so has no solution. No solutions to the 2-2 equation

$$A^7+B^7=C^7+D^7 \quad (2)$$

are known

No solutions to the 3-1 or 3-2 equations are known, neither are solutions to the 3-3 equation

$$A^7+B^7+C^7=D^7+E^7+F^7 \quad (3)$$

(Ekl 1996).

No 4-1, 4-2, or 4-3 solutions are known. Guy (1994, p. 140) asked if a 4-4 equation exists for 7th POWERS. An affirmative answer was provided by (Ekl 1996),

$$149^7+123^7+14^7+10^7=146^7+129^7+90^7+15^7 \quad (4)$$

$$194^7+150^7+105^7+23^7=192^7+152^7+132^7+38^7. \quad (5)$$

A 4-5 solution is known.

No 5-1, 5-2, or 5-3 solutions are known. Numerical solutions to the 5-4 equation are given by Gloden (1948). The smallest 5-4 solution is

$$3^7+11^7+26^7+29^7+52^7=12^7+16^7+43^7+50^7 \quad (6)$$

(Lander *et al.* 1967). Gloden (1949) gives parametric solutions to the 5-5 equation. The first few 5-5 solutions are

$$8^7+8^7+13^7+16^7+19^7 \\ =2^7+12^7+15^7+17^7+18^7 \quad (7)$$

$$4^7+8^7+14^7+16^7+23^7 \\ =7^7+7^7+9^7+20^7+22^7 \quad (8)$$

$$11^7+12^7+18^7+21^7+26^7 \\ =9^7+10^7+22^7+23^7+24^7 \quad (9)$$

$$6^7+12^7+20^7+22^7+27^7 \\ =10^7+13^7+13^7+25^7+26^7 \quad (10)$$

$$3^7+13^7+17^7+24^7+38^7 \\ =14^7+26^7+32^7+32^7+33^7 \quad (11)$$

(Lander *et al.* 1967).

No 6-1, 6-2, or 6-3 solutions are known. A parametric solution to the 6-6 equation was given by Sastry and Rai (1948). The smallest is

$$2^7+3^7+6^7+6^7+10^7+13^7=1^7+1^7+7^7+7^7+12^7+12^7 \quad (12)$$

(Lander *et al.* 1967).

There are no known solutions to the 7-1 equation (Guy 1994, p. 140). A 7^2-2 solution is

$$2^7+26^7 \\ =4^7+8^7+13^7+14^7+14^7+16^7+18^7+22^7+23^7+23^7 \\ =7^7+7^7+9^7+13^7+14^7+18^7+20^7+22^7+22^7+23^7 \quad (13)$$

(Lander *et al.* 1967). The smallest 7-3 solution is

$$7^7+7^7+12^7+16^7+27^7+28^7+31^7 = 26^7+30^7+30^7 \quad (14)$$

(Lander *et al.* 1967).

The smallest 8-1 solution is

$$12^7+35^7+53^7+58^7+64^7+83^7+85^7+90^7 = 102^7 \quad (15)$$

(Lander *et al.* 1967). The smallest 8-2 solution is

$$5^7+6^7+7^7+15^7+15^7+20^7+28^7+31^7 = 10^7+33^7 \quad (16)$$

(Lander *et al.* 1967).

The smallest 9-1 solution is

$$6^7+14^7+20^7+22^7+27^7+33^7+41^7+50^7+59^7 = 62^7 \quad (17)$$

(Lander *et al.* 1967).

References

Ekl, R. L. "Equal Sums of Four Seventh Powers." *Math. Comput.* **65**, 1755–1756, 1996.

Gloden, A. "Zwei Parameterlösungen einer mehrgeradigen Gleichung." *Arch. Math.* **1**, 480–482, 1949.

Guy, R. K. "Sums of Like Powers. Euler's Conjecture." §D1 in *Unsolved Problems in Number Theory, 2nd ed.* New York: Springer-Verlag, pp. 139–144, 1994.

Lander, L. J.; Parkin, T. R.; and Selfridge, J. L. "A Survey of Equal Sums of Like Powers." *Math. Comput.* **21**, 446–459, 1967.

Sastry, S. and Rai, T. "On Equal Sums of Like Powers." *Math. Student* **16**, 18–19, 1948.

Diophantine Equation—8th Powers

The 2-1 equation

$$A^8 + B^8 = C^8 \quad (1)$$

is a special case of FERMAT'S LAST THEOREM with $n = 8$, and so has no solution. No 2-2 solutions are known.

No 3-1, 3-2, or 3-3 solutions are known.

No 4-1, 4-2, 4-3, or 4-4 solutions are known.

No 5-1, 5-2, 5-3, or 5-4 solutions are known, but Letac (1942) found a solution to the 5-5 equation. The smallest 5-5 solution is

$$1^8+10^8+11^8+20^8+43^8 = 5^8+28^8+32^8+35^8+41^8 \quad (2)$$

(Lander *et al.* 1967).

No 6-1, 6-2, 6-3, or 6-4 solutions are known. Moessner and Gloden (1944) found solutions to the 6-6 equation. The smallest 6-6 solution is

$$3^8+6^8+8^8+10^8+15^8+23^8 = 5^8+9^8+9^8+12^8+20^8+22^8 \quad (3)$$

(Lander *et al.* 1967).

No 7-1, 7-2, or 7-3 solutions are known. The smallest 7-4 solution is

$$7^8+9^8+16^8+22^8+22^8+28^8+34^8 = 6^8+11^8+20^8+35^8 \quad (4)$$

(Lander *et al.* 1967). Moessner and Gloden (1944) found solutions to the 7-6 equation. Parametric solutions to the 7-7 equation were given by Moessner (1947) and Gloden (1948). The smallest 7-7 solution is

$$1^8+3^8+5^8+6^8+6^8+8^8+13^8 = 4^8+7^8+9^8+9^8+10^8+11^8+12^8 \quad (5)$$

(Lander *et al.* 1967).

No 8-1 or 8-2 solutions are known. The smallest 8-3 solution is

$$6^8+12^8+16^8+16^8+38^8+38^8+40^8+47^8 = 8^8+17^8+50^8 \quad (6)$$

(Lander *et al.* 1967). Sastry (1934) used the smallest 17-1 solution to give a parametric 8-8 solution. The smallest 8-8 solution is

$$1^8+3^8+7^8+7^8+7^8+10^8+10^8+12^8 = 4^8+5^8+5^8+6^8+6^8+11^8+11^8+11^8 \quad (7)$$

(Lander *et al.* 1967).

No solutions to the 9-1 equation is known. The smallest 9-2 solution is

$$2^8+7^8+8^8+16^8+17^8+20^8+20^8+24^8+24^8 = 11^8+27^8 \quad (8)$$

(Lander *et al.* 1967). Letac (1942) found solutions to the 9-9 equation.

No solutions to the 10-1 equation are known.

The smallest 11-1 solution is

$$14^8+18^8+2\cdot 44^8+66^8+70^8+92^8 +93^8+96^8+106^8+112^8 = 125^8 \quad (9)$$

(Lander *et al.* 1967).

The smallest 12-1 solution is

$$2\cdot 8^8+10^8+3\cdot 24^8+26^8+30^8 +34^8+44^8+52^8+63^8 = 65^8 \quad (10)$$

(Lander *et al.* 1967).

The general identity

$$(2^{8k+4}+1)^8 = (2^{8k+4}-1)^8+(2^{7k+4})^8 +(2^{k+1})^8+7[(2^{5k+3})^8+(2^{3k+2})^8] \quad (11)$$

gives a solution to the 17-1 equation (Lander *et al.* 1967).

References

Gloden, A. "Parametric Solutions of Two Multi-Degreed Equalities." *Amer. Math. Monthly* **55**, 86–88, 1948.

Lander, L. J.; Parkin, T. R.; and Selfridge, J. L. "A Survey of Equal Sums of Like Powers." *Math. Comput.* **21**, 446–459, 1967.

Letac, A. *Gazetta Mathematica* **48**, 68–69, 1942.

Moessner, A. "On Equal Sums of Like Powers." *Math. Student* **15**, 83–88, 1947.

Moessner, A. and Gloden, A. "Einige Zahlentheoretische Untersuchungen und Resultante." *Bull. Sci. École Polytech. de Timisoara* **11**, 196–219, 1944.

Sastry, S. "On Sums of Powers." *J. London Math. Soc.* **9**, 242–246, 1934.

Diophantine Equation—9th Powers

The 2-1 equation

$$A^9 + B^9 = C^9 \quad (1)$$

is a special case of FERMAT'S LAST THEOREM with $n = 9$, and so has no solution. There is no known 2-2 solution.

There are no known 3-1, 3-2, or 3-3 solutions.

There are no known 4-1, 4-2, 4-3, or 4-4 solutions.

There are no known 5-1, 5-2, 5-3, 5-4, or 5-5 solutions.

There are no known 6-1, 6-2, 6-3, 6-4, or 6-5 solutions. The smallest 6-6 solution is

$$1^9 + 13^9 + 13^9 + 14^9 + 18^9 + 23^9 = 5^9 + 9^9 + 10^9 + 15^9 + 21^9 + 22^9 \quad (2)$$

(Lander *et al.* 1967).

There are no known 7-1, 7-2, 7-3, 7-4, or 7-5 solutions.

There are no known 8-1, 8-2, 8-3, 8-4, or 8-5 solutions.

There are no known 9-1, 9-2, 9-3, 9-4, or 9-5 solutions.

There are no known 10-1, 10-2, or 10-3 solutions. The smallest 10-4 solution is

$$2^9 + 6^9 + 6^9 + 9^9 + 10^9 + 11^9 + 14^9 + 18^9 + 2 \cdot 19^9 = 5^9 + 12^9 + 16^9 + 21^9 \quad (3)$$

(Lander *et al.* 1967). No 10-5 solution is known. Moessner (1947) gives a parametric solution to the 10-10 equation.

There are no known 11-1 or 11-2 solutions. The smallest 11-3 solution is

$$2^9 + 3^9 + 6^9 + 7^9 + 9^9 + 9^9 + 19^9 + 19^9 + 21^9 + 25^9 + 29^9 = 13^9 + 16^9 + 30^9 \quad (4)$$

(Lander *et al.* 1967). The smallest 11-5 solution is

$$3^9 + 5^9 + 5^9 + 9^9 + 9^9 + 12^9 + 15^9 + 15^9 + 16^9 + 21^9 + 21^9 = 7^9 + 8^9 + 14^9 + 20^9 + 22^9 \quad (5)$$

(Lander *et al.* 1967). Palamá (1953) gave a solution to the 11-11 equation.

There is no known 12-1 solution. The smallest 12-2 solution is

$$4 \cdot 2^9 + 2 \cdot 3^9 + 4^9 + 7^9 + 16^9 + 17^9 + 2 \cdot 19^9 = 15^9 + 21^9 \quad (6)$$

(Lander *et al.* 1967).

There are no known 13-1 or 14-1 solutions. The smallest 15-1 solution is

$$2^9 + 2^9 + 4^9 + 6^9 + 6^9 + 7^9 + 9^9 + 9^9 + 10^9 + 15^9 + 18^9 + 21^9 + 21^9 + 23^9 + 23^9 = 26^9 \quad (7)$$

(Lander *et al.* 1967).

References

Lander, L. J.; Parkin, T. R.; and Selfridge, J. L. "A Survey of Equal Sums of Like Powers." *Math. Comput.* **21**, 446–459, 1967.

Moessner, A. "On Equal Sums of Like Powers." *Math. Student* **15**, 83–88, 1947.

Palamá, G. "Diophantine Systems of the Type $\sum_{i=1}^p a_i{}^k = \sum_{i=1}^p b_i{}^k$ ($k = 1, 2, \ldots, n, n+2, n+4, \ldots, n+2r$)." *Scripta Math.* **19**, 132–134, 1953.

Diophantine Equation—10th Powers

The 2-1 equation

$$A^{10} + B^{10} = C^{10} \quad (1)$$

is a special case of FERMAT'S LAST THEOREM with $n = 10$, and so has no solution. The smallest values for which n-1, n-2, etc., have solutions are 23, 19, 24, 23, 16, 27, and 7, corresponding to

$$5 \cdot 1^{10} + 2^{10} + 3^{10} + 6^{10} + 6 \cdot 7^{10} + 4 \cdot 9^{10} + 10^{10} + 2 \cdot 12^{10} + 13^{10} + 14^{10} = 15^{10} \quad (2)$$

$$5 \cdot 2^{10} + 5^{10} + 6^{10} + 10^{10} + 6 \cdot 11^{10} + 2 \cdot 12^{10} + 3 \cdot 15^{10} = 9^{10} + 17^{10} \quad (3)$$

$$1^{10} + 2^{10} + 3^{10} + 10 \cdot 4^{10} + 7^{10} + 7 \cdot 8^{10} + 10^{10} + 12^{10} + 16^{10} = 11^{10} + 2 \cdot 15^{10} \quad (4)$$

$$5 \cdot 1^{10} + 2 \cdot 2^{10} + 3 \cdot 3^{10} + 4^{10} + 4 \cdot 6^{10} + 3 \cdot 7^{10} + 8^{10} + 2 \cdot 10^{10} + 2 \cdot 14^{10} + 15^{10} = 3 \cdot 11^{10} + 16^{10} \quad (5)$$

$$4 \cdot 1^{10} + 2^{10} + 2 \cdot 4^{10} + 6^{10} + 2 \cdot 12^{10} + 5 \cdot 13^{10} + 15^{10} = 2 \cdot 3^{10} + 8^{10} + 14^{10} + 16^{10} \quad (6)$$

$$1^{10}+4\cdot 3^{10}+2\cdot 4^{10}+2\cdot 5^{10}+7\cdot 6^{10}$$
$$+9\cdot 7^{10}+10^{10}+13^{10}=2\cdot 2^{10}+8^{10}+11^{10}+2\cdot 12^{10} \quad (7)$$

$$1^{10}+28^{10}+31^{10}+32^{10}+55^{10}+61^{10}+68^{10}$$
$$=17^{10}+20^{10}+23^{10}+44^{10}+49^{10}+64^{10}+67^{10} \quad (8)$$

(Lander *et al.* 1967).

References
Lander, L. J.; Parkin, T. R.; and Selfridge, J. L. "A Survey of Equal Sums of Like Powers." *Math. Comput.* **21**, 446–459, 1967.

Diophantine Equation—Cubic

The 2-1 equation

$$A^3+B^3=C^3 \quad (1)$$

is a case of FERMAT'S LAST THEOREM with $n=3$. In fact, this particular case was known not to have any solutions long before the general validity of FERMAT'S LAST THEOREM was established. The 2-2 equation

$$A^3+B^3=C^3+D^3 \quad (2)$$

has a known parametric solution (Dickson 1966, pp. 550–554; Guy 1994, p. 140), and 10 solutions with sum $< 10^5$,

$$1729=1^3+12^3=9^3+10^3 \quad (3)$$
$$4104=2^3+16^3=9^3+15^3 \quad (4)$$
$$13832=2^3+24^3=18^3+20^3 \quad (5)$$
$$20683=10^3+27^3=19^3+24^3 \quad (6)$$
$$32832=4^3+32^3=18^3+30^3 \quad (7)$$
$$39312=2^3+34^3=15^3+33^3 \quad (8)$$
$$40033=9^3+34^3=16^3+33^3 \quad (9)$$
$$46683=3^3+36^3=16^3+33^3 \quad (10)$$
$$64232=17^3+39^3=26^3+36^3 \quad (11)$$
$$65728=12^3+40^3=31^3+33^3 \quad (12)$$

(Sloane's A001235; Moreau 1898). The first number (Madachy 1979, pp. 124 and 141) in this sequence, the so-called HARDY-RAMANUJAN NUMBER, is associated with a story told about Ramanujan by G. H. Hardy, but was known as early as 1657 (Berndt and Bhargava 1993). The smallest number representable in n ways as a sum of cubes is called the nth TAXICAB NUMBER.

Ramanujan gave a general solution to the 2-2 equation as

$$(\alpha+\lambda^2\gamma)^3+(\lambda\beta+\gamma)^3=(\lambda\alpha+\gamma)^3+(\beta+\lambda^2\gamma)^3 \quad (13)$$

where

$$\alpha^2+\alpha\beta+\beta^2=3\lambda\gamma^2 \quad (14)$$

(Berndt 1994, p. 107). Another form due to Ramanujan is

$$(A^2+7AB-9B^2)^3+(2A^2-4AB+12B^2)^3$$
$$=(2A^2+10B^2)^3+(A^2-9AB-B^2)^3. \quad (15)$$

Hardy and Wright (1979, Theorem 412) prove that there are numbers that are expressible as the sum of two cubes in n ways for any n (Guy 1994, pp. 140–141). The proof is constructive, providing a method for computing such numbers: given RATIONALS NUMBERS r and s, compute

$$t=\frac{r(r^3+2s^3)}{r^3-s^3} \quad (16)$$
$$u=\frac{s(2r^3+s^3)}{r^3-s^3} \quad (17)$$
$$v=\frac{t(t^3-2u^3)}{t^3+u^3} \quad (18)$$
$$w=\frac{u(2t^3-u^3)}{t^3+u^3}. \quad (19)$$

Then

$$r^3+s^3=t^3-u^3=v^3+w^3 \quad (20)$$

The DENOMINATORS can now be cleared to produce an integer solution. If r/s is picked to be large enough, the v and w will be POSITIVE. If r/s is still larger, the v/w will be large enough for v and w to be used as the inputs to produce a third pair, etc. However, the resulting integers may be quite large, even for $n=2$. E.g., starting with $3^3+1^3=28$, the algorithm finds

$$28=\left(\tfrac{28340511}{21446828}\right)^3+\left(\tfrac{63284705}{21446828}\right)^3, \quad (21)$$

giving

$$28\cdot 21446828^3=(3\cdot 21446828)^3+21446828^3 \quad (22)$$
$$=28340511^3+63284705^3. \quad (23)$$

The numbers representable in three ways as a sum of two cubes (a 2-2-2 equation) are

$$87539319=167^3+436^3=228^3+423^3=255^3+414^3 \quad (24)$$
$$119824488=11^3+493^3=90^3+492^3=346^3+428^3 \quad (25)$$
$$143604279=111^3+522^3=359^3+460^3=408^3+423^3 \quad (26)$$
$$175959000=70^3+560^3=198^3+552^3=315^3+525^3 \quad (27)$$
$$327763000=300^3+670^3=339^3+661^3=510^3+580^3 \quad (28)$$

(Guy 1994, Sloane's A003825). Wilson (1997) found 32 numbers representable in four ways as the sum of two cubes (a 2-2-2-2 equation). The first is

$$\begin{aligned}6963472309248 &= 2421^2 + 19083^2 = 5436^2 + 18948^2\\ &= 102020^3 + 18072^2 = 13322^3 + 15530^3. \quad (29)\end{aligned}$$

The smallest known numbers so representable are 6963472309248, 12625136269928, 21131226514944, 26059452841000, ... (Sloane's A003826). Wilson also found six five-way sums,

$$\begin{aligned}48988659276962496 &= 38787^3 + 365757^3\\ &= 107839^3 + 362753^3\\ &= 205292^3 + 342952^3\\ &= 221424^3 + 336588^3\\ &= 231518^3 + 331954^3 \quad (30)\end{aligned}$$

$$\begin{aligned}490593422681271000 &= 48369^3 + 788631^3\\ &= 233775^3 + 781785^3\\ &= 285120^3 + 776070^3\\ &= 543145^3 + 691295^3\\ &= 579240^3 + 666630^3 \quad (31)\end{aligned}$$

$$\begin{aligned}6355491080314102272 &= 103113^3 + 1852215^3\\ &= 580488^3 + 1833120^3\\ &= 788724^3 + 1803372^3\\ &= 1150792^3 + 1690544^3\\ &= 1462050^3 + 1478238^3 \quad (32)\end{aligned}$$

$$\begin{aligned}27365551142421413376 &= 167751^3 + 3013305^3\\ &= 265392^3 + 3012792^3\\ &= 944376^3 + 2982240^3\\ &= 1283148^3 + 2933844^3\\ &= 1872184^3 + 2750288^3 \quad (33)\end{aligned}$$

$$\begin{aligned}1199962860219870469632 &= 591543^3 + 10625865^3\\ &= 935856^3 + 10624056^3\\ &= 3330168^3 + 10516320^3\\ &= 6601912^3 + 9698384^3\\ &= 8387550^3 + 8480418^3 \quad (34)\end{aligned}$$

$$\begin{aligned}111549833098123426841016 &= 1074073^3 + 48137999^3\\ &= 8787870^3 + 48040356^3\\ &= 13950972^3 + 47744382^3\\ &= 24450192^3 + 45936462^3\\ &= 33784478^3 + 41791204^3, \quad (35)\end{aligned}$$

and a single six-way sum

$$\begin{aligned}&8230545258248091551205888\\ &= 11239317^3 + 201891435^3\\ &= 17781264^3 + 201857064^3\\ &= 63273192^3 + 199810080^3\\ &= 85970916^3 + 196567548^3\\ &= 125436328^3 + 184269296^3\\ &= 159363450^3 + 161127942^3. \quad (36)\end{aligned}$$

The first rational solution to the 3-1 equation

$$A^3 + B^3 + C^3 = D^3 \quad (37)$$

was found by Euler and Vieta (Dickson 1966, pp. 550–554). Hardy and Wright (1979, pp. 199–201) give a solution which can be based on the identities

$$\begin{aligned}a^3(a^3+b^3)^3 &= b^3(a^3+b^3)^3 + a^3(a^3-2b^3)^3\\ &\quad + b^3(2a^3-b^3)^3 \quad (38)\\ a^3(a^3+2b^3)^3 &= a^3(a^3-b^3)^3 + b^3(a^3-b^3)^3\\ &\quad + b^3(2a^3+b^3)^3. \quad (39)\end{aligned}$$

This is equivalent to the general 2-2 solution found by Ramanujan (Berndt 1994, pp. 54 and 107). The smallest integral solutions are

$$\begin{aligned}3^3 + 4^3 + 5^3 &= 6^3 \quad (40)\\ 1^3 + 6^3 + 8^3 &= 9^3 \quad (41)\\ 7^3 + 14^3 + 17^3 &= 20^3 \quad (42)\\ 11^3 + 15^3 + 27^3 &= 29^3 \quad (43)\\ 28^3 + 53^3 + 75^3 &= 84^3 \quad (44)\\ 26^3 + 55^3 + 78^3 &= 87^3 \quad (45)\\ 33^3 + 70^3 + 92^3 &= 105^3 \quad (46)\end{aligned}$$

(Beeler *et al.* 1972; Madachy 1979, pp. 124 and 141). Other general solutions have been found by Binet (1841) and Schwering (1902), although Ramanujan's formulation is the simplest. No general solution giving *all* POSITIVE integral solutions is known (Dickson 1966, pp. 550–561).

4-1 equations include

$$\begin{aligned}11^3 + 12^3 + 13^3 + 14^3 &= 20^3 \quad (47)\\ 5^3 + 7^3 + 9^3 + 10^3 &= 13^3. \quad (48)\end{aligned}$$

A solution to the 4-4 equation is

$$2^3 + 3^3 + 10^3 + 11^3 = 1^3 + 5^3 + 8^3 + 12^3 \quad (49)$$

(Madachy 1979, pp. 118 and 133).

5-1 equations

$$\begin{aligned}1^3 + 3^3 + 4^3 + 5^3 + 8^3 &= 9^3 \quad (50)\\ 3^3 + 4^3 + 5^3 + 8^3 + 10^3 &= 12^3, \quad (51)\end{aligned}$$

and a 6-1 equation is given by

$$1^3 + 5^3 + 6^3 + 7^3 + 8^3 + 10^3 = 13^3. \quad (52)$$

A 6-6 equation also exists:

$$1^3 + 2^3 + 4^3 + 8^3 + 9^3 + 12^3 = 3^3 + 5^3 + 6^3 + 7^3 + 10^3 + 11^3 \quad (53)$$

(Madachy 1979, p. 142).

Euler gave the general solution to

$$A^3 + B^3 = C^2 \tag{54}$$

as

$$A = 3n^3 + 6n^2 - n \tag{55}$$
$$B = -3n^3 + 6n^2 + n \tag{56}$$
$$C = 6n^2(3n^2 + 1). \tag{57}$$

see also CANNONBALL PROBLEM, HARDY-RAMANUJAN NUMBER, SUPER-3 NUMBER, TAXICAB NUMBER, TRIMORPHIC NUMBER

References

Beeler, M.; Gosper, R. W.; and Schroeppel, R. Item 58 in *HAKMEM*. Cambridge, MA: MIT Artificial Intelligence Laboratory, Memo AIM-239, Feb. 1972.

Berndt, B. C. *Ramanujan's Notebooks, Part IV*. New York: Springer-Verlag, 1994.

Berndt, B. C. and Bhargava, S. "Ramanujan—For Lowbrows." *Amer. Math. Monthly* **100**, 645–656, 1993.

Binet, J. P. M. "Note sur une question relative à la théorie des nombres." *C. R. Acad. Sci. (Paris)* **12**, 248–250, 1841.

Dickson, L. E. *History of the Theory of Numbers, Vol. 2: Diophantine Analysis*. New York: Chelsea, 1966.

Guy, R. K. "Sums of Like Powers. Euler's Conjecture." §D1 in *Unsolved Problems in Number Theory, 2nd ed.* New York: Springer-Verlag, pp. 139–144, 1994.

Hardy, G. H. *Ramanujan: Twelve Lectures on Subjects Suggested by His Life and Work, 3rd ed.* New York: Chelsea, p. 68, 1959.

Hardy, G. H. and Wright, E. M. *An Introduction to the Theory of Numbers, 5th ed.* Oxford, England: Clarendon Press, 1979.

Madachy, J. S. *Madachy's Mathematical Recreations*. New York: Dover, 1979.

Moreau, C. "Plus petit nombre égal à la somme de deux cubes de deux façons." *L'Intermediaire Math.* **5**, 66, 1898.

Schwering, K. "Vereinfachte Lösungen des Eulerschen Aufgabe: $x^3 + y^3 + z^3 + v^3 = 0$." *Arch. Math. Phys.* **2**, 280–284, 1902.

Shanks, D. *Solved and Unsolved Problems in Number Theory, 4th ed.* New York: Chelsea, p. 157, 1993.

Sloane, N. J. A. Sequences A001235 and A003825 in "An On-Line Version of the Encyclopedia of Integer Sequences."

Wilson, D. Personal communication, Apr. 17, 1997.

Diophantine Equation—Linear

A linear Diophantine equation (in two variables) is an equation of the general form

$$ax + by = c, \tag{1}$$

where solutions are sought with a, b, and c INTEGERS. Such equations can be solved completely, and the first known solution was constructed by Brahmagupta. Consider the equation

$$ax + by = 1. \tag{2}$$

Now use a variation of the EUCLIDEAN ALGORITHM, letting $a = r_1$ and $b = r_2$

$$r_1 = q_1 r_2 + r_3 \tag{3}$$
$$r_2 = q_2 r_3 + r_4 \tag{4}$$
$$r_{n-3} = q_{n-3} r_{n-2} + r_{n-1} \tag{5}$$
$$r_{n-2} = q_{n-2} r_{n-1} + 1. \tag{6}$$

Starting from the bottom gives

$$1 = r_{n-2} - q_{n-2} r_{n-1} \tag{7}$$
$$r_{n-1} = r_{n-3} - q_{n-3} r_{n-2}, \tag{8}$$

so

$$\begin{aligned} 1 &= r_{n-2} - q_{n-2}(r_{n-3} - q_{n-3} r_{n-2}) \\ &= -q_{n-2} r_{n-3} + (1 - q_{n-2} q_{n-3}) r_{n-2}. \end{aligned} \tag{9}$$

Continue this procedure all the way back to the top.

Take as an example the equation

$$1027x + 712y = 1. \tag{10}$$

Proceed as follows.

$$\begin{array}{rl|rl} 1027 = 712\cdot 1 + 315 & | & 1 = -165\cdot 1027 + 238\cdot 712 & \uparrow \\ 712 = 315\cdot 2 + 82 & | & 1 = 73\cdot 712 - 165\cdot 315 & | \\ 315 = 82\cdot 3 + 69 & | & 1 = -19\cdot 315 + 73\cdot 82 & | \\ 82 = 69\cdot 1 + 13 & | & 1 = 16\cdot 82 - 19\cdot 69 & | \\ 69 = 13\cdot 5 + 4 & | & 1 = -3\cdot 69 + 16\cdot 13 & | \\ 13 = 4\cdot 3 + 1 & \downarrow & 1 = 1\cdot 13 - 3\cdot 4 & | \\ & & 1 = 0\cdot 4 + 1\cdot 1, & | \end{array}$$

The solution is therefore $x = -165$, $y = 238$. The above procedure can be simplified by noting that the two leftmost columns are offset by one entry and alternate signs, as they must since

$$1 = -A_{i+1} r_i + A_i r_{i+1} \tag{11}$$
$$r_{i+1} = r_{i-1} - r_i q_{i-1} \tag{12}$$
$$1 = A_i r_{i-1} - (A_i q_{i-1} + A_{i+1}), \tag{13}$$

so the COEFFICIENTS of r_{i-1} and r_{i+1} are the same and

$$A_{i-1} = -(A_i q_{i-1} + A_{i+1}). \tag{14}$$

Repeating the above example using this information therefore gives

$$\begin{array}{rl|lll} 1027 = 712\cdot 1 + 315 & | & (-) & 165\cdot 1 + 73 = 238 & \uparrow \\ 712 = 315\cdot 2 + 82 & | & (+) & 73\cdot 2 + 19 = 165 & | \\ 315 = 82\cdot 3 + 69 & | & (-) & 19\cdot 3 + 16 = 73 & | \\ 82 = 69\cdot 1 + 13 & | & (+) & 16\cdot 1 + 3 = 19 & | \\ 69 = 13\cdot 5 + 4 & | & (-) & 3\cdot 5 + 1 = 16 & | \\ 13 = 4\cdot 3 + 1 & \downarrow & (+) & 1\cdot 3 + 0 = 3 & | \\ & & (-) & 0\cdot 1 + 1 = 1 & | \end{array}$$

and we recover the above solution.

Call the solutions to

$$ax + by = 1 \tag{15}$$

x_0 and y_0. If the signs in front of ax or by are NEGATIVE, then solve the above equation and take the signs of the solutions from the following table:

equation	x	y
$ax + by = 1$	x_0	y_0
$ax - by = 1$	x_0	$-y_0$
$-ax + by = 1$	$-x_0$	y_0
$-ax - by = 1$	$-x_0$	$-y_0$

In fact, the solution to the equation

$$ax - by = 1 \tag{16}$$

is equivalent to finding the CONTINUED FRACTION for a/b, with a and b RELATIVELY PRIME (Olds 1963). If there are n terms in the fraction, take the $(n-1)$th convergent p_{n-1}/q_{n-1}. But

$$p_n q_{n-1} - p_{n-1} q_n = (-1)^n, \tag{17}$$

so one solution is $x_0 = (-1)^n q_{n-1}$, $y_0 = (-1)^n p_{n-1}$, with a general solution

$$x = x_0 + kb \tag{18}$$

$$y = y_0 + ka \tag{19}$$

with k an arbitrary INTEGER. The solution in terms of smallest POSITIVE INTEGERS is given by choosing an appropriate k.

Now consider the general first-order equation of the form

$$ax + by = c. \tag{20}$$

The GREATEST COMMON DIVISOR $d \equiv \text{GCD}(a, b)$ can be divided through yielding

$$a'x + b'y = c', \tag{21}$$

where $a' \equiv a/d$, $b' \equiv b/d$, and $c' \equiv c/d$. If $d \nmid c$, then c' is not an INTEGER and the equation cannot have a solution in INTEGERS. A necessary and sufficient condition for the general first-order equation to have solutions in INTEGERS is therefore that $d|c$. If this is the case, then solve

$$a'x + b'y = 1 \tag{22}$$

and multiply the solutions by c', since

$$a'(c'x) + b'(c'y) = c'. \tag{23}$$

References

Courant, R. and Robbins, H. "Continued Fractions. Diophantine Equations." §2.4 in Supplement to Ch. 1 in *What is Mathematics?: An Elementary Approach to Ideas and Methods, 2nd ed.* Oxford, England: Oxford University Press, pp. 49–51, 1996.

Dickson, L. E. "Linear Diophantine Equations and Congruences." Ch. 2 in *History of the Theory of Numbers, Vol. 2: Diophantine Analysis.* New York: Chelsea, pp. 41–99, 1952.

Olds, C. D. Ch. 2 in *Continued Fractions.* New York: Random House, 1963.

Diophantine Equation—*n*th Powers

The 2-1 equation

$$A^n + B^n = C^n \tag{1}$$

is a special case of FERMAT'S LAST THEOREM and so has no solutions for $n \geq 3$. Lander *et al.* (1967) give a table showing the smallest n for which a solution to

$$x_1{}^k + x_2{}^k + \ldots + x_m{}^k = y_1{}^k + y_2{}^k + \ldots + y_n{}^k,$$

with $1 \leq m \leq n$ is known.

	k								
m	2	3	4	5	6	7	8	9	10
1	2	3	3	4	7	8	11	15	23
2	2	2	2	4	7	8	9	12	19
3				3	3	7	8	11	24
4						4	7	10	23
5						5	5	11	16
6								6	27
7									7

Take the results from the RAMANUJAN 6-10-8 IDENTITY that for $ad = bc$, with

$$F_{2m}(a,b,c,d) = (a+b+c)^{2m} + (b+c+d)^{2m} - (c+d+a)^{2m} - (d+a+b)^{2m} + (a-d)^{2m} - (b-c)^{2m} \tag{2}$$

and

$$f_{2m}(x,y) = (1+x+y)^{2m} + (x+y+xy)^{2m} - (y+xy+1)^{2m} - (xy+1+x)^{2m} + (1-xy)^{2m} - (x-y)^{2m}, \tag{3}$$

then

$$F_{2m}(a,b,c,d) = a^{2m} f_{2m}(x,y). \tag{4}$$

Using

$$f_2(x,y) = 0 \tag{5}$$

$$f_4(x,y) = 0 \tag{6}$$

now gives

$$(a+b+c)^n + (b+c+d)^n + (a-d)^n = (c+d+a)^n + (d+a+b)^n + (b-c)^n \tag{7}$$

for $n = 2$ or 4.

see also RAMANUJAN 6-10-8 IDENTITY

References

Berndt, B. C. *Ramanujan's Notebooks, Part IV.* New York: Springer-Verlag, p. 101, 1994.

Berndt, B. C. and Bhargava, S. "Ramanujan—For Lowbrows." *Amer. Math. Monthly* **100**, 644–656, 1993.

Dickson, L. E. *History of the Theory of Numbers, Vol. 2: Diophantine Analysis.* New York: Chelsea, pp. 653–657, 1966.

Gloden, A. *Mehrgradige Gleichungen.* Groningen, Netherlands: P. Noordhoff, 1944.
Guy, R. K. *Unsolved Problems in Number Theory, 2nd ed.* New York: Springer-Verlag, 1994.
Lander, L. J.; Parkin, T. R.; and Selfridge, J. L. "A Survey of Equal Sums of Like Powers." *Math. Comput.* **21**, 446–459, 1967.
Reznick, B. "Sums of Even Powers of Real Linear Forms." *Mem. Amer. Math. Soc.* No. 463, **96**. Providence, RI: Amer. Math. Soc., 1992.

Diophantine Equation—Quadratic

An equation of the form

$$x^2 - Dy^2 = 1, \tag{1}$$

where D is an INTEGER is called a PELL EQUATION. Pell equations, as well as the analogous equation with a minus sign on the right, can be solved by finding the CONTINUED FRACTION for $\sqrt{D}$. (The trivial solution $x = 1$, $y = 0$ is ignored in all subsequent discussion.) Let p_n/q_n denote the nth CONVERGENT $[a_1, a_2, \ldots, a_n]$, then we are looking for a convergent which obeys the identity

$$p_n{}^2 - Dq_n{}^2 = (-1)^n, \tag{2}$$

which turns out to always be possible since the CONTINUED FRACTION of a QUADRATIC SURD always becomes periodic at some term a_{r+1}, where $a_{r+1} = 2a_1$, i.e.,

$$\sqrt{D} = [a_1, \overline{a_2, \ldots, a_r, 2a_1}\,]. \tag{3}$$

Writing $n = rk$ gives

$$p_{rk}{}^2 - Dq_{rk}{}^2 = (-1)^{rk}, \tag{4}$$

for k a POSITIVE INTEGER. If r is ODD, solutions to

$$x^2 - Dy^2 = \pm 1 \tag{5}$$

can be obtained if k is chosen to be EVEN or ODD, but if r is EVEN, there are no values of k which can make the exponent ODD.

If r is EVEN, then $(-1)^r$ is POSITIVE and the solution in terms of smallest INTEGERS is $x = p_r$ and $y = q_r$, where p_r/q_r is the rth CONVERGENT. If r is ODD, then $(-1)^r$ is NEGATIVE, but we can take $k = 2$ in this case, to obtain

$$p_{2r}{}^2 - Dq_{2r}{}^2 = 1, \tag{6}$$

so the solution in smallest INTEGERS is $x = p_{2r}$, $y = q_{2r}$. Summarizing,

$$(x, y) = \begin{cases} (p_r, q_r) & \text{for } r \text{ even} \\ (p_{2r}, p_{2r}) & \text{for } r \text{ odd.} \end{cases} \tag{7}$$

The more complicated equation

$$x^2 - Dy^2 = \pm c \tag{8}$$

can also be solved for certain values of c and D, but the procedure is more complicated (Chrystal 1961). However, if a single solution to the above equation is known, other solutions can be found. Let p and q be solutions to (8), and r and s solutions to the "unit" form". Then

$$(p^2 - Dq^2)(r^2 - Ds^2) = \pm c \tag{9}$$

$$(pr \pm Dqs)^2 - D(ps \pm qr)^2 = \pm c. \tag{10}$$

Call a Diophantine equation consisting of finding m POWERS equal to a sum of n equal POWERS an "$m - n$ equation." The 2-1 equation

$$A^2 = B^2 + C^2, \tag{11}$$

which corresponds to finding a PYTHAGOREAN TRIPLE (A, B, C) has a well-known general solution (Dickson 1966, pp. 165-170). To solve the equation, note that every PRIME of the form $4x + 1$ can be expressed as the sum of two RELATIVELY PRIME squares in exactly one way. To find in how many ways a general number m can be expressed as a sum of two squares, factor it as follows

$$m = 2^{a_0} p_1{}^{2a_1} \cdots p_n{}^{2a_n} q_1{}^{b_1} \cdots q_r{}^{b_r}, \tag{12}$$

where the ps are primes of the form $4x - 1$ and the qs are primes of the form $x + 1$. If the as are integral, then define

$$B \equiv (2b_1 + 1)(2b_2 + 1) \cdots (2b_r + 1) - 1. \tag{13}$$

Then m is a sum of two *unequal* squares in

$$N(m) = \begin{cases} 0 \\ \quad \text{for any } a_i \text{ half-integral} \\ \frac{1}{2}(b_1 + 1)(b_2 + 1) \cdots (b_r + 1) \\ \quad \text{for all } a_i \text{ integral, } B \text{ odd} \\ \frac{1}{2}(b_1 + 1)(b_2 + 1) \cdots (b_r + 1) - \frac{1}{2} \\ \quad \text{for all } a_i \text{ integral, } B \text{ even.} \end{cases} \tag{14}$$

If zero is counted as a square, both POSITIVE and NEGATIVE numbers are included, and the order of the two squares is distinguished, Jacobi showed that the number of ways a number can be written as the sum of two squares is four times the excess of the number of DIVISORS of the form $4x + 1$ over the number of DIVISORS of the form $4x - 1$.

A set of INTEGERS satisfying the 3-1 equation

$$A^2 + B^2 + C^2 = D^2 \tag{15}$$

is called a PYTHAGOREAN QUADRUPLE. Parametric solutions to the 2-2 equation

$$A^2 + B^2 = C^2 + D^2 \tag{16}$$

are known (Dickson 1966; Guy 1994, p. 140).

Solutions to an equation of the form

$$(A^2+B^2)(C^2+D^2)=E^2+F^2 \quad (17)$$

are given by the FIBONACCI IDENTITY

$$(a^2+b^2)(c^2+d^2)=(ac\pm bd)^2+(bc\mp ad)^2\equiv e^2+f^2. \quad (18)$$

Another similar identity is the EULER FOUR-SQUARE IDENTITY

$$(a_1{}^2+a_2{}^2)(b_1{}^2+b_2{}^2)(c_1{}^2+c_2{}^2)(d_1{}^2+d_2{}^2) = e_1{}^2+e_2{}^2+e_3{}^2+e_4{}^2 \quad (19)$$

$$\begin{aligned}(a_1{}^2+a_2{}^2+a_3{}^2+a_4{}^2)(b_1{}^2+b_2{}^2+b_3{}^2+b_4{}^2)\\ =(a_1b_1-a_2b_2-a_3b_3-a_4b_4)^2\\ +(a_1b_2+a_2b_1+a_3b_4-a_4b_3)^2\\ +(a_1b_3-a_2b_4+a_3b_1+a_4b_2)^2\\ +(a_1b_4+a_2b_3-a_3b_2+a_4b_1)^2.\end{aligned} \quad (20)$$

Degen's eight-square identity holds for eight squares, but no other number, as proved by Cayley. The two-square identity underlies much of TRIGONOMETRY, the four-square identity some of QUATERNIONS, and the eight-square identity, the CAYLEY ALGEBRA (a noncommutative nonassociative algebra; Bell 1945).

RAMANUJAN'S SQUARE EQUATION

$$2^n-7=x^2 \quad (21)$$

has been proved to have only solutions $n = 3, 4, 5, 7$, and 15 (Beeler *et al.* 1972, Item 31).

see also ALGEBRA, CANNONBALL PROBLEM, CONTINUED FRACTION, FERMAT DIFFERENCE EQUATION, LAGRANGE NUMBER (DIOPHANTINE EQUATION), PELL EQUATION, PYTHAGOREAN QUADRUPLE, PYTHAGOREAN TRIPLE, QUADRATIC RESIDUE

References

Beeler, M.; Gosper, R. W.; and Schroeppel, R. *HAKMEM.* Cambridge, MA: MIT Artificial Intelligence Laboratory, Memo AIM-239, Feb. 1972.

Beiler, A. H. "The Pellian." Ch. 22 in *Recreations in the Theory of Numbers: The Queen of Mathematics Entertains.* New York: Dover, pp. 248–268, 1966.

Bell, E. T. *The Development of Mathematics, 2nd ed.* New York: McGraw-Hill, p. 159, 1945.

Chrystal, G. *Textbook of Algebra, 2 vols.* New York: Chelsea, 1961.

Degan, C. F. *Canon Pellianus.* Copenhagen, Denmark, 1817.

Dickson, L. E. "Number of Representations as a Sum of 5, 6, 7, or 8 Squares." Ch. 13 in *Studies in the Theory of Numbers.* Chicago, IL: University of Chicago Press, 1930.

Dickson, L. E. *History of the Theory of Numbers, Vol. 2: Diophantine Analysis.* New York: Chelsea, 1966.

Guy, R. K. *Unsolved Problems in Number Theory, 2nd ed.* New York: Springer-Verlag, 1994.

Lam, T. Y. *The Algebraic Theory of Quadratic Forms.* Reading, MA: W. A. Benjamin, 1973.

Rajwade, A. R. *Squares.* Cambridge, England: Cambridge University Press, 1993.

Scharlau, W. *Quadratic and Hermitian Forms.* Berlin: Springer-Verlag, 1985.

Shapiro, D. B. "Products of Sums and Squares." *Expo. Math.* **2**, 235–261, 1984.

Smarandache, F. "Un metodo de resolucion de la ecuacion diofantica." *Gaz. Math.* **1**, 151–157, 1988.

Smarandache, F. " Method to Solve the Diophantine Equation $ax^2 - by^2 + c = 0$." In *Collected Papers, Vol. 1.* Bucharest, Romania: Tempus, 1996.

Taussky, O. "Sums of Squares." *Amer. Math. Monthly* **77**, 805–830, 1970.

Whitford, E. E. *Pell Equation.* New York: Columbia University Press, 1912.

Diophantine Equation—Quartic

Call an equation involving quartics m-n if a sum of m quartics is equal to a sum of n fourth POWERS. The 2-1 equation

$$A^4+B^4=C^4 \quad (1)$$

is a case of FERMAT'S LAST THEOREM with $n = 4$ and therefore has no solutions. In fact, the equations

$$A^4\pm B^4=C^2 \quad (2)$$

also have no solutions in INTEGERS.

Parametric solutions to the 2-2 equation

$$A^4+B^4=C^4+D^4 \quad (3)$$

are known (Euler 1802; Gérardin 1917; Guy 1994, pp. 140–141). A few specific solutions are

$$59^4+158^4=133^4+134^4=635{,}318{,}657 \quad (4)$$
$$7^4+239^4=157^4+227^4=3{,}262{,}811{,}042 \quad (5)$$
$$193^4+292^4=256^4+257^4=8{,}657{,}437{,}697 \quad (6)$$
$$298^4+497^4=271^4+502^4=68{,}899{,}596{,}497 \quad (7)$$
$$514^4+359^4=103^4+542^4=86{,}409{,}838{,}577 \quad (8)$$
$$222^4+631^4=503^4+558^4=160{,}961{,}094{,}577 \quad (9)$$
$$21^4+717^4=471^4+681^4=264{,}287{,}694{,}402 \quad (10)$$
$$76^4+1203^4=653^4+1176^4=2{,}094{,}447{,}251{,}857 \quad (11)$$
$$997^4+1342^4=878^4+1381^4=4{,}231{,}525{,}221{,}377 \quad (12)$$
$$27^4+2379^4=577^4+728^4=32{,}031{,}536{,}780{,}322 \quad (13)$$

(Sloane's A001235; Richmond 1920, Leech 1957), the smallest of which is due to Euler. Lander *et al.* (1967) give a list of 25 primitive 2-2 solutions. General (but incomplete) solutions are given by

$$x=a+b \quad (14)$$
$$y=c-d \quad (15)$$
$$u=a-b \quad (16)$$
$$v=c+d, \quad (17)$$

Diophantine Equation—Quartic

where

$$a = n(m^2 + n^2)(-m^4 + 18m^2n^2 - n^4) \quad (18)$$
$$b = 2m(m^6 + 10m^4n^2 + m^2n^4 + 4n^6) \quad (19)$$
$$c = 2n(4m^6 + m^4n^2 + 10m^2n^4 + n^6) \quad (20)$$
$$d = m(m^2 + n^2)(-m^4 + 18m^2n^2 - n^4) \quad (21)$$

(Hardy and Wright 1979).

In 1772, Euler proposed that the 3-1 equation

$$A^4 + B^4 + C^4 = D^4 \quad (22)$$

had no solutions in INTEGERS (Lander *et al.* 1967). This assertion is known as the EULER QUARTIC CONJECTURE. Ward (1948) showed there were no solutions for $D \leq 10,000$, which was subsequently improved to $D \leq 220,000$ by Lander *et al.* (1967). However, the EULER QUARTIC CONJECTURE was disproved in 1987 by Noam D. Elkies, who, using a geometric construction, found

$$2{,}682{,}440^4 + 15{,}365{,}639^4 + 18{,}796{,}760^4 = 20{,}615{,}673^4 \quad (23)$$

and showed that infinitely many solutions existed (Guy 1994, p. 140). In 1988, Roger Frye found

$$95{,}800^4 + 217{,}519^4 + 414{,}560^4 = 422{,}481^4 \quad (24)$$

and proved that there are no solutions in smaller INTEGERS (Guy 1994, p. 140). Another solution was found by Allan MacLeod in 1997,

$$638{,}523{,}249^4 = 630{,}662{,}624^4 + 275{,}156{,}240^4 + 219{,}076{,}465^4. \quad (25)$$

It is not known if there is a parametric solution.

In contrast, there are many solutions to the 3-1 equation

$$A^4 + B^4 + C^4 = 2D^4 \quad (26)$$

(see below).

Parametric solutions to the 3-2 equation

$$A^4 + B^4 = C^4 + D^4 + E^4 \quad (27)$$

are known (Gérardin 1910, Ferrari 1913). The smallest 3-2 solution is

$$3^4 + 5^4 + 8^4 = 7^4 + 7^4 \quad (28)$$

(Lander *et al.* 1967).

Ramanujan gave the 3-3 equations

$$2^4 + 4^4 + 7^4 = 3^4 + 6^4 + 6^4 \quad (29)$$
$$3^4 + 7^4 + 8^4 = 1^4 + 2^4 + 9^4 \quad (30)$$
$$6^4 + 9^4 + 12^4 = 2^4 + 2^4 + 13^4 \quad (31)$$

(Berndt 1994, p. 101). Similar examples can be found in Martin (1896). Parametric solutions were given by Gérardin (1911).

Ramanujan also gave the general expression

$$3^4 + (2x^4 - 1)^4 + (4x^5 + x)^4 = (4x^4 + 1)^4 + (6x^4 - 3)^4 + (4x^5 - 5x)^4 \quad (32)$$

(Berndt 1994, p. 106). Dickson (1966, pp. 653–655) cites several FORMULAS giving solutions to the 3-3 equation, and Haldeman (1904) gives a general FORMULA.

The 4-1 equation

$$A^4 + B^4 + C^4 + D^4 = E^4 \quad (33)$$

has solutions

$$30^4 + 120^4 + 272^4 + 315^4 = 353^4 \quad (34)$$
$$240^4 + 340^4 + 430^4 + 599^4 = 651^4 \quad (35)$$
$$435^4 + 710^4 + 1384^4 + 2420^4 = 2487^4 \quad (36)$$
$$1130^4 + 1190^4 + 1432^4 + 2365^4 = 2501^4 \quad (37)$$
$$850^4 + 1010^4 + 1546^4 + 2745^4 = 2829^4 \quad (38)$$
$$2270^4 + 2345^4 + 2460^4 + 3152^4 = 3723^4 \quad (39)$$
$$350^4 + 1652^4 + 3230^4 + 3395^4 = 3973^4 \quad (40)$$
$$205^4 + 1060^4 + 2650^4 + 4094^4 = 4267^4 \quad (41)$$
$$1394^4 + 1750^4 + 3545^4 + 3670^4 = 4333^4 \quad (42)$$
$$699^4 + 700^4 + 2840^4 + 4250^4 = 4449^4 \quad (43)$$
$$380^4 + 1660^4 + 1880^4 + 4907^4 = 4949^4 \quad (44)$$
$$1000^4 + 1120^4 + 3233^4 + 5080^4 = 5281^4 \quad (45)$$
$$410^4 + 1412^4 + 3910^4 + 5055^4 = 5463^4 \quad (46)$$
$$955^4 + 1770^4 + 2634^4 + 5400^4 = 5491^4 \quad (47)$$
$$30^4 + 1680^4 + 3043^4 + 5400^4 = 5543^4 \quad (48)$$
$$1354^4 + 1810^4 + 4355^4 + 5150^4 = 5729^4 \quad (49)$$
$$542^4 + 2770^4 + 4280^4 + 5695^4 = 6167^4 \quad (50)$$
$$50^4 + 885^4 + 5000^4 + 5984^4 = 6609^4 \quad (51)$$
$$1490^4 + 3468^4 + 4790^4 + 6185^4 = 6801^4 \quad (52)$$
$$1390^4 + 2850^4 + 5365^4 + 6368^4 = 7101^4 \quad (53)$$
$$160^4 + 1345^4 + 2790^4 + 7166^4 = 7209^4 \quad (54)$$
$$800^4 + 3052^4 + 5440^4 + 6635^4 = 7339^4 \quad (55)$$
$$2230^4 + 3196^4 + 5620^4 + 6995^4 = 7703^4 \quad (56)$$

(Norrie 1911, Patterson 1942, Leech 1958, Brudno 1964, Lander *et al.* 1967), but it is not known if there is a parametric solution (Guy 1994, p. 139).

Ramanujan gave the 4-2 equation

$$3^4 + 9^4 = 5^4 + 5^4 + 6^4 + 6^4, \quad (57)$$

and the 4-3 identities

$$2^4+2^4+7^4=4^4+4^4+5^4+6^4 \quad (58)$$

$$3^4+9^4+14^4=7^4+8^4+10^4+13^4 \quad (59)$$

$$7^4+10^4+13^4=5^4+5^4+6^4+14^4 \quad (60)$$

(Berndt 1994, p. 101). Haldeman (1904) gives general FORMULAS for 4-2 and 4-3 equations.

There are an infinite number of solutions to the 5-1 equation

$$A^4+B^4+C^4+D^4+E^4=F^4. \quad (61)$$

Some of the smallest are

$$2^4+2^4+3^4+4^2+4^2=5^4 \quad (62)$$

$$4^4+6^4+8^4+9^4+14^4=15^4 \quad (63)$$

$$4^4+21^4+22^4+26^4+28^4=35^4 \quad (64)$$

$$1^4+2^4+12^4+24^4+44^4=45^4 \quad (65)$$

$$1^4+8^4+12^4+32^4+64^4=65^4 \quad (66)$$

$$2^4+39^4+44^4+46^4+52^4=65^4 \quad (67)$$

$$22^4+52^4+57^4+74^4+76^4=95^4 \quad (68)$$

$$22^4+28^4+63^4+72^4+94^4=105^4 \quad (69)$$

(Berndt 1994). Berndt and Bhargava (1993) and Berndt (1994, pp. 94–96) give Ramanujan's solutions for arbitrary s, t, m, and n,

$$(8s^2+40st-24t^2)^4+(6s^2-44st-18t^2)^4 +(14s^2-4st-42t^2)^4+(9s^2+27t^2)^4+(4s^2+12t^2)^4 =(15s^2+45t^2)^4, \quad (70)$$

and

$$(4m^2-12n^2)^4+(3m^2+9n^2)^4+(2m^2-12mn-6n^2)^4 +(4m^2+12n^2)^4+(2m^2+12mn-6n^2)^4=(5m^2+15n^2)^4. \quad (71)$$

These are also given by Dickson (1966, p. 649), and two general FORMULAS are given by Beiler (1966, p. 290). Other solutions are given by Fauquembergue (1898), Haldeman (1904), and Martin (1910).

Ramanujan gave

$$2(ab+ac+bc)^2=a^4+b^4+c^4 \quad (72)$$

$$2(ab+ac+bc)^4=a^4(b-c)^4+b^4(c-a)^4+c^4(a-b)^4 \quad (73)$$

$$2(ab+ac+bc)^6=(a^2b+b^2c+c^2a)^4 +(ab^2+bc^2+ca^2)^4+(3abc)^4 \quad (74)$$

$$2(ab+ac+bc)^8=(a^3+2abc)^4(b-c)^4 +(b^3+2abc)^4(c-a)^4+(c^3+2abc)^4(a-b)^4, \quad (75)$$

where

$$a+b+c=0 \quad (76)$$

(Berndt 1994, pp. 96–97). FORMULA (73) is equivalent to FERRARI'S IDENTITY

$$(a^2+2ac-2bc-b^2)^4+(b^2-2ab-2ac-c^2)^4 +(c^2+2ab+2bc-a^2)^4=2(a^2+b^2+c^2-ab+ac+bc)^4. \quad (77)$$

BHARGAVA'S THEOREM is a general identity which gives the above equations as a special case, and may have been the route by which Ramanujan proceeded. Another identity due to Ramanujan is

$$(a+b+c)^4+(b+c+d)^4+(a-d)^4 =(c+d+a)^4+(d+a+b)^4+(b-c)^4, \quad (78)$$

where $a/b=c/d$, and 4 may also be replaced by 2 (Ramanujan 1957, Hirschhorn 1998).

V. Kyrtatas noticed that $a=3$, $b=7$, $c=20$, $d=25$, $e=38$, and $f=39$ satisfy

$$\frac{a^4+b^4+c^4}{d^4+e^4+f^4}=\frac{a+b+c}{d+e+f} \quad (79)$$

and asks if there are any other distinct integer solutions.

The first few numbers n which are a sum of two or more fourth POWERS ($m-1$ equations) are 353, 651, 2487, 2501, 2829, ... (Sloane's A003294). The only number of the form

$$4x^4+y^4 \quad (80)$$

which is PRIME is 5 (Baudran 1885, Le Lionnais 1983).

see also BHARGAVA'S THEOREM, FORD'S THEOREM

References

Barbette, E. *Les sommes de p-iémes puissances distinctes égales à une p-iéme puissance.* Doctoral Dissertation, Liege, Belgium. Paris: Gauthier-Villars, 1910.

Beiler, A. H. *Recreations in the Theory of Numbers: The Queen of Mathematics Entertains.* New York: Dover, 1966.

Berndt, B. C. *Ramanujan's Notebooks, Part IV.* New York: Springer-Verlag, 1994.

Berndt, B. C. and Bhargava, S. "Ramanujan—For Lowbrows." *Am. Math. Monthly* **100**, 645–656, 1993.

Bhargava, S. "On a Family of Ramanujan's Formulas for Sums of Fourth Powers." *Ganita* **43**, 63–67, 1992.

Brudno, S. "A Further Example of $A^4+B^4+C^4+D^4=E^4$." *Proc. Cambridge Phil. Soc.* **60**, 1027–1028, 1964.

Dickson, L. E. *History of the Theory of Numbers, Vol. 2: Diophantine Analysis.* New York: Chelsea, 1966.

Euler, L. *Nova Acta Acad. Petrop. as annos 1795–1796* **13**, 45, 1802.

Fauquembergue, E. *L'intermédiaire des Math.* **5**, 33, 1898.

Ferrari, F. *L'intermédiaire des Math.* **20**, 105–106, 1913.

Guy, R. K. "Sums of Like Powers. Euler's Conjecture" and "Some Quartic Equations." §D1 and D23 in *Unsolved Problems in Number Theory, 2nd ed.* New York: Springer-Verlag, pp. 139–144 and 192–193, 1994.

Haldeman, C. B. "On Biquadrate Numbers." *Math. Mag.* **2**, 285–296, 1904.
Hardy, G. H. and Wright, E. M. §13.7 in *An Introduction to the Theory of Numbers, 5th ed.* Oxford, England: Clarendon Press, 1979.
Hirschhorn, M. D. "Two or Three Identities of Ramanujan." *Amer. Math. Monthly* **105**, 52–55, 1998.
Lander, L. J.; Parkin, T. R.; and Selfridge, J. L. "A Survey of Equal Sums of Like Powers." *Math. Comput.* **21**, 446–459, 1967.
Le Lionnais, F. *Les nombres remarquables.* Paris: Hermann, p. 56, 1983.
Leech, J. "Some Solutions of Diophantine Equations." *Proc. Cambridge Phil. Soc.* **53**, 778–780, 1957.
Leech, J. "On $A^4+B^4+C^4+D^4=E^4$." *Proc. Cambridge Phil. Soc.* **54**, 554–555, 1958.
Martin, A. "About Biquadrate Numbers whose Sum is a Biquadrate." *Math. Mag.* **2**, 173–184, 1896.
Martin, A. "About Biquadrate Numbers whose Sum is a Biquadrate—II." *Math. Mag.* **2**, 325–352, 1904.
Norrie, R. *University of St. Andrews 500th Anniversary Memorial Volume.* Edinburgh, Scotland: pp. 87–89, 1911.
Patterson, J. O. "A Note on the Diophantine Problem of Finding Four Biquadrates whose Sum is a Biquadrate." *Bull. Amer. Math. Soc.* **48**, 736–737, 1942.
Ramanujan, S. *Notebooks.* New York: Springer-Verlag, pp. 385–386, 1987.
Richmond, H. W. "On Integers Which Satisfy the Equation $t^3 \pm x^3 \pm y^3 \pm z^3 = 0$." *Trans. Cambridge Phil. Soc.* **22**, 389–403, 1920.
Sloane, N. J. A. Sequences A001235 and A003294/M5446 in "An On-Line Version of the Encyclopedia of Integer Sequences."
Ward, M. "Euler's Problem on Sums of Three Fourth Powers." *Duke Math. J.* **15**, 827–837, 1948.

Diophantine Quadruple

see DIOPHANTINE SET

Diophantine Set

A set S of POSITIVE integers is said to be Diophantine IFF there exists a POLYNOMIAL Q with integral coefficients in $m \geq 1$ indeterminates such that

$$S = \{Q(x_1, \ldots, x_m) \geq 1 : x_1 \geq 1, \ldots, x_m \geq 1\}.$$

It has been proved that the set of PRIME numbers is a Diophantine set.

References

Ribenboim, P. *The New Book of Prime Number Records.* New York: Springer-Verlag, pp. 189–192, 1995.

Diophantus Property

A set of POSITIVE INTEGERS $S = \{a_1, \ldots, a_m\}$ satisfies the Diophantus property $D(n)$ of order n if, for all $i, j = 1, \ldots, m$ with $i \neq j$,

$$a_i a_j + n = {b_{ij}}^2, \tag{1}$$

where n and b_{ij} are INTEGERS. The set S is called a Diophantine n-tuple. Fermat found the first $D(1)$ quadruple: $\{1, 3, 8, 120\}$. General $D(1)$ quadruples are

$$\{F_{2n}, F_{2n+2}, F_{2n+4}, 4F_{2n+1}F_{2n+2}F_{2n+3},\} \tag{2}$$

where F_n are FIBONACCI NUMBERS, and

$$\{n, n+2, 4n+4, 4(n+1)(2n+1)(2n+3)\}. \tag{3}$$

The quadruplet

$$\{2F_{n-1}, 2F_{n+1}, 2{F_n}^3F_{n+1}F_{n+2}, 2F_{n+1}F_{n+2}F_{n+3}(2{F_{n+1}}^2 - {F_n}^2)\} \tag{4}$$

is $D({F_n}^2)$ (Dujella 1996). Dujella (1993) showed there exist no Diophantine quadruples $D(4k+2)$.

References

Aleksandriiskii, D. *Arifmetika i kniga o mnogougol'nyh chislakh.* Moscow: Nauka, 1974.
Brown, E. "Sets in Which $xy+k$ is Always a Square." *Math. Comput.* **45**, 613–620, 1985.
Davenport, H. and Baker, A. "The Equations $3x^2 - 2 = y^2$ and $8x^2 - 7 = z^2$." *Quart. J. Math. (Oxford) Ser. 2* **20**, 129–137, 1969.
Dujella, A. "Generalization of a Problem of Diophantus." *Acta Arithm.* **65**, 15–27, 1993.
Dujella, A. "Diophantine Quadruples for Squares of Fibonacci and Lucas Numbers." *Portugaliae Math.* **52**, 305–318, 1995.
Dujella, A. "Generalized Fibonacci Numbers and the Problem of Diophantus." *Fib. Quart.* **34**, 164–175, 1996.
Hoggatt, V. E. Jr. and Bergum, G. E. "A Problem of Fermat and the Fibonacci Sequence." *Fib. Quart.* **15**, 323–330, 1977.
Jones, B. W. "A Variation of a Problem of Davenport and Diophantus." *Quart. J. Math. (Oxford) Ser. (2)* **27**, 349–353, 1976.

Diophantus' Riddle

"Diophantus' youth lasts 1/6 of his life. He grew a beard after 1/12 more of his life. After 1/7 more of his life, Diophantus married. Five years later, he had a son. The son lived exactly half as long as his father, and Diophantus died just four years after his son's death. All of this totals the years Diophantus lived."

Let D be the number of years Diophantus lived, and let S be the number of years his son lived. Then the above word problem gives the two equations

$$D = (\tfrac{1}{6} + \tfrac{1}{12} + \tfrac{1}{7})D + 5 + S + 4$$
$$S = \tfrac{1}{2}D.$$

Solving this simultaneously gives $S = 42$ as the age of the son and $D = 84$ as the age of Diophantus.

References

Pappas, T. "Diophantus' Riddle." *The Joy of Mathematics.* San Carlos, CA: Wide World Publ./Tetra, pp. 123 and 232, 1989.

Dipyramid

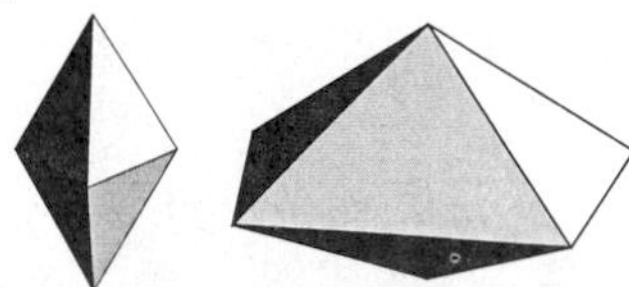

Two PYRAMIDS symmetrically placed base-to-base, also called a BIPYRAMID. They are the DUALS of the Archimedean PRISMS.

see also ELONGATED DIPYRAMID, PENTAGONAL DIPYRAMID, PRISM, PYRAMID, TRAPEZOHEDRON, TRIANGULAR DIPYRAMID, TRIGONAL DIPYRAMID

References
Cundy, H. and Rollett, A. *Mathematical Models, 3rd ed.* Stradbroke, England: Tarquin Pub., p. 117, 1989.

Dirac Delta Function

see DELTA FUNCTION

Dirac Matrices

Define the 4×4 matrices

$$\sigma_i = \mathsf{I} \otimes \sigma_{i,\,\text{Pauli}} \tag{1}$$

$$\rho_i = \sigma_{i,\,\text{Pauli}} \otimes \mathsf{I}, \tag{2}$$

where $\sigma_{i,\,\text{Pauli}}$ are the PAULI MATRICES, I is the IDENTITY MATRIX, $i = 1, 2, 3$, and $\mathsf{A} \otimes \mathsf{B}$ is the matrix DIRECT PRODUCT. Explicitly,

$$\mathsf{I} = \begin{bmatrix} 1 & 0 & 0 & 0 \\ 0 & 1 & 0 & 0 \\ 0 & 0 & 1 & 0 \\ 0 & 0 & 0 & 1 \end{bmatrix} \tag{3}$$

$$\sigma_1 = \begin{bmatrix} 0 & 1 & 0 & 0 \\ 1 & 0 & 0 & 0 \\ 0 & 0 & 0 & 1 \\ 0 & 0 & 1 & 0 \end{bmatrix} \tag{4}$$

$$\sigma_2 = \begin{bmatrix} 0 & -i & 0 & 0 \\ i & 0 & 0 & 0 \\ 0 & 0 & 0 & -i \\ 0 & 0 & i & 0 \end{bmatrix} \tag{5}$$

$$\sigma_3 = \begin{bmatrix} 1 & 0 & 0 & 0 \\ 0 & -1 & 0 & 0 \\ 0 & 0 & 1 & 0 \\ 0 & 0 & 0 & -1 \end{bmatrix} \tag{6}$$

$$\rho_1 = \begin{bmatrix} 0 & 0 & 1 & 0 \\ 0 & 0 & 0 & 1 \\ 1 & 0 & 0 & 0 \\ 0 & 1 & 0 & 0 \end{bmatrix} \tag{7}$$

$$\rho_2 = \begin{bmatrix} 0 & 0 & -i & 0 \\ 0 & 0 & 0 & -i \\ i & 0 & 0 & 0 \\ 0 & i & 0 & 0 \end{bmatrix} \tag{8}$$

$$\rho_3 = \begin{bmatrix} 1 & 0 & 0 & 0 \\ 0 & 1 & 0 & 0 \\ 0 & 0 & -1 & 0 \\ 0 & 0 & 0 & -1 \end{bmatrix}. \tag{9}$$

These matrices satisfy the anticommutation identities

$$\sigma_i\sigma_j + \sigma_j\sigma_i = 2\delta_{ij}\mathsf{I} \tag{10}$$

$$\rho_i\rho_j + \rho_j\rho_i = 2\delta_{ij}\mathsf{I}, \tag{11}$$

where δ_{ij} is the KRONECKER DELTA, the commutation identity

$$[\sigma_i, \rho_j] = \sigma_i\rho_j - \rho_j\sigma_i = 0, \tag{12}$$

and are cyclic under permutations of indices

$$\sigma_i\sigma_j = i\sigma_k \tag{13}$$

$$\rho_i\rho_j = i\rho_k. \tag{14}$$

A total of 16 Dirac matrices can be defined via

$$\mathsf{E}_{ij} = \rho_i\sigma_j \tag{15}$$

for $i, j = 0, 1, 2, 3$ and where $\sigma_0 = \rho_0 \equiv \mathsf{I}$. These matrix satisfy

1. $|\mathsf{E}_{ij}| = 1$, where $|\mathsf{A}|$ is the DETERMINANT,
2. $\mathsf{E}_{ij}^2 = \mathsf{I}$,
3. $\mathsf{E}_{ij} = \mathsf{E}_{ij}^\dagger$, making them Hermitian, and therefore unitary,
4. $\mathrm{tr}(\mathsf{E}_{ij}) = 0$, except $\mathrm{tr}(\mathsf{E}_{00}) = 4$,
5. Any two E_{ij} multiplied together yield a Dirac matrix to within a multiplicative factor of $-i$ or $\pm i$,
6. The E_{ij} are linearly independent,
7. The E_{ij} form a complete set, i.e., any 4×4 constant matrix may be written as

$$\mathsf{A} = \sum_{i,j=0}^{3} c_{ij}\mathsf{E}_{ij}, \tag{16}$$

where the c_{ij} are real or complex and are given by

$$c_{mn} = \tfrac{1}{4}\,\mathrm{tr}(\mathsf{A}\mathsf{E}_{mn}) \tag{17}$$

(Arfken 1985).

Dirac's original matrices were written α_i and were defined by

$$\alpha_i = \mathsf{E}_{1i} = \rho_1\sigma_i \tag{18}$$

$$\alpha_4 = \mathsf{E}_{30} = \rho_3, \tag{19}$$

for $i = 1, 2, 3$, giving

$$\alpha_1 = \mathsf{E}_{1i} = \begin{bmatrix} 0 & 0 & 0 & 1 \\ 0 & 0 & 1 & 0 \\ 0 & 1 & 0 & 0 \\ 1 & 0 & 0 & 0 \end{bmatrix} \tag{20}$$

$$\alpha_2 = \mathsf{E}_{2i} = \begin{bmatrix} 0 & 0 & 0 & -i \\ 0 & 0 & i & 0 \\ 0 & -i & 0 & 0 \\ i & 0 & 0 & 0 \end{bmatrix} \tag{21}$$

$$\alpha_3 = \mathsf{E}_{3i} = \begin{bmatrix} 0 & 0 & 1 & 0 \\ 0 & 0 & 0 & -1 \\ 1 & 0 & 0 & 0 \\ 0 & -1 & 0 & 0 \end{bmatrix} \tag{22}$$

$$\alpha_4 = \mathsf{E}_{30} = \begin{bmatrix} 1 & 0 & 0 & 0 \\ 0 & 1 & 0 & 0 \\ 0 & 0 & -1 & 0 \\ 0 & 0 & 0 & -1 \end{bmatrix}. \tag{23}$$

The additional matrix

$$\alpha_5 = \mathsf{E}_{20} = \rho_2 = \begin{bmatrix} 0 & 0 & -i & 0 \\ 0 & 0 & 0 & -i \\ i & 0 & 0 & 0 \\ 0 & i & 0 & 0 \end{bmatrix} \quad (24)$$

is sometimes defined. Other sets of Dirac matrices are sometimes defined as

$$y_i = \mathsf{E}_{2i} \quad (25)$$

$$y_4 = \mathsf{E}_{30} \quad (26)$$

$$y_5 = -\mathsf{E}_{10} \quad (27)$$

and

$$\delta_i = \mathsf{E}_{3i} \quad (28)$$

for $i = 1, 2, 3$ (Arfken 1985) and

$$\gamma_i = \begin{bmatrix} 0 & \sigma_i \\ -\sigma_i & 0 \end{bmatrix} \quad (29)$$

$$\gamma_4 = \begin{bmatrix} \mathsf{I} & 0 \\ 2\mathsf{I} & -\mathsf{I} \end{bmatrix} \quad (30)$$

for $i = 1, 2, 3$ (Goldstein 1980).

Any of the 15 Dirac matrices (excluding the identity matrix) commute with eight Dirac matrices and anticommute with the other eight. Let $\mathsf{M} \equiv \frac{1}{2}(1 + \mathsf{E}_{ij})$, then

$$\mathsf{M}^2 = \mathsf{M}. \quad (31)$$

In addition

$$\begin{bmatrix} \alpha_1 \\ \alpha_2 \\ \alpha_3 \end{bmatrix} \times \begin{bmatrix} \alpha_1 \\ \alpha_2 \\ \alpha_3 \end{bmatrix} = 2i\sigma. \quad (32)$$

The products of α_i and y_i satisfy

$$\alpha_1\alpha_2\alpha_3\alpha_4\alpha_5 = 1 \quad (33)$$

$$y_1 y_2 y_3 y_4 y_5 = 1. \quad (34)$$

The 16 Dirac matrices form six anticommuting sets of five matrices each:

1. $\alpha_1, \alpha_2, \alpha_3, \alpha_4, \alpha_5$,
2. y_1, y_2, y_3, y_4, y_5,
3. $\delta_1, \delta_2, \delta_3, \rho_1, \rho_2$,
4. $\alpha_1, y_1, \delta_1, \sigma_2, \sigma_3$,
5. $\alpha_2, y_2, \delta_2, \sigma_1, \sigma_3$,
6. $\alpha_3, y_3, \delta_3, \sigma_1, \sigma_2$.

see also PAULI MATRICES

References

Arfken, G. *Mathematical Methods for Physicists, 3rd ed.* Orlando, FL: Academic Press, pp. 211–213, 1985.

Goldstein, H. *Classical Mechanics, 2nd ed.* Reading, MA: Addison-Wesley, p. 580, 1980.

Dirac's Theorem

A GRAPH with $n \geq 3$ VERTICES in which each VERTEX has VALENCY $\geq n/2$ has a HAMILTONIAN CIRCUIT.

see also HAMILTONIAN CIRCUIT

Direct Product (Group)

The expression of a GROUP as a product of SUBGROUPS. The CHARACTERS of the representations of a direct product are equal to the products of the CHARACTERS of the representations based on the individual sets of functions. For R_1 and R_2,

$$\chi(R_1 \otimes R_2) = \chi(R_1)\chi(R_2).$$

The representation of a direct product Γ_{AB} will contain the totally symmetric representation only if the irreducible Γ_A equals the irreducible Γ_B.

Direct Product (Matrix)

Given two $n \times m$ MATRICES, their direct product $\mathsf{C} = \mathsf{A} \otimes \mathsf{B}$ is an $(mn) \times (nm)$ MATRIX with elements defined by

$$C_{\alpha\beta} = \mathsf{A}_{ij}\mathsf{B}_{kl}, \quad (1)$$

where

$$\alpha \equiv n(i-1) + k \quad (2)$$

$$\beta \equiv n(j-1) + l. \quad (3)$$

For a 2×2 MATRIX,

$$\mathsf{A} \otimes \mathsf{B} = \begin{bmatrix} a_{11}\mathsf{B} & a_{12}\mathsf{B} \\ a_{21}\mathsf{B} & a_{22}\mathsf{B} \end{bmatrix} \quad (4)$$

$$= \begin{bmatrix} a_{11}b_{11} & a_{11}b_{12} & a_{12}b_{11} & a_{12}b_{12} \\ a_{11}b_{21} & a_{11}b_{22} & a_{12}b_{21} & a_{12}b_{22} \\ a_{21}b_{11} & a_{21}b_{12} & a_{22}b_{11} & a_{22}b_{12} \\ a_{21}b_{21} & a_{21}b_{22} & a_{22}b_{21} & a_{22}b_{22} \end{bmatrix}. \quad (5)$$

Direct Product (Set)

The direct product of two sets A and B is defined to be the set of all points (a, b) where $a \in A$ and $b \in B$. The direct product is denoted $A \times B$ or $A \otimes B$ and is also called the CARTESIAN PRODUCT, since it originated in Descartes' formulation of analytic geometry. In the Cartesian view, points in the plane are specified by their vertical and horizontal coordinates, with points on a line being specified by just one coordinate. The main examples of direct products are EUCLIDEAN 3-space ($\mathbb{R} \otimes \mathbb{R} \otimes \mathbb{R}$, where $\mathbb{R}$ are the REAL NUMBERS), and the plane ($\mathbb{R} \times \mathbb{R}$).

Direct Product (Tensor)

For a first-RANK TENSOR (i.e., a VECTOR),

$$a'_i b'^j \equiv \frac{\partial x_k}{\partial x'_i} a_k \frac{\partial x'_j}{\partial x_l} b^l = \frac{\partial x_k}{\partial x'_i} \frac{\partial x'_j}{\partial x_l}(a_k b^l), \qquad (1)$$

which is a second-RANK TENSOR. The CONTRACTION of a direct product of first-RANK TENSORS is the SCALAR

$$\mathrm{contr}(a'_i b'^j) = a'_i b'^i = a_k b^k. \qquad (2)$$

For a second-RANK TENSOR,

$$A^i_j B_{kl} = C^{ikl}_j \qquad (3)$$

$$C^{ikl'}_j = \frac{\partial x'_i}{\partial x_m}\frac{\partial x_n}{\partial x'_j}\frac{\partial x'_k}{\partial x_p}\frac{\partial x'_l}{\partial x_q} C^{mpq}_n. \qquad (4)$$

For a general TENSOR, the direct product of two TENSORS is a TENSOR of RANK equal to the sum of the two initial RANKS. The direct product is ASSOCIATIVE, but not COMMUTATIVE.

References

Arfken, G. "Contraction, Direct Product." §3.2 in *Mathematical Methods for Physicists, 3rd ed.* Orlando, FL: Academic Press, pp. 124–126, 1985.

Direct Search Factorization

Direct search factorization is the simplest PRIME FACTORIZATION ALGORITHM. It consists of searching for factors of a number by systematically performing TRIAL DIVISIONS, usually using a sequence of increasing numbers. Multiples of small PRIMES are commonly excluded to reduce the number of trial DIVISORS, but just including them is sometimes faster than the time required to exclude them. This approach is very inefficient, and can be used only with fairly small numbers.

When using this method on a number n, only DIVISORS up to $\lfloor\sqrt{n}\rfloor$ (where $\lfloor x \rfloor$ is the FLOOR FUNCTION) need to be tested. This is true since if all INTEGERS less than this had been tried, then

$$\frac{n}{\lfloor\sqrt{n}\rfloor + 1} < \sqrt{n}. \qquad (1)$$

In other words, all possible FACTORS have had their COFACTORS already tested. It is also true that, when the smallest PRIME FACTOR p of n is $> \sqrt[3]{n}$, then its COFACTOR m (such that $n = pm$) must be PRIME. To prove this, suppose that the smallest p is $> \sqrt[3]{n}$. If $m = ab$, then the smallest value a and b could assume is p. But then

$$n = pm = pab = p^3 > n, \qquad (2)$$

which cannot be true. Therefore, m must be PRIME, so

$$n = p_1 p_2. \qquad (3)$$

see also PRIME FACTORIZATION ALGORITHMS, TRIAL DIVISION

Direct Sum (Module)

The direct sum of two MODULES V and W over the same RING R is given by $V \otimes W$ with MODULE operations defined by

$$r \cdot (v, w) = (rv, rw)$$

$$(v, w) \oplus (y, z) = (v + y, w + z).$$

The direct sum of an arbitrary family of MODULES over the same RING is also defined. If J is the indexing set for the family of MODULES, then the direct sum is represented by the collection of functions with finite support from J to the union of all these MODULES such that the function sends $j \in J$ to an element in the MODULE indexed by j.

The dimension of a direct sum is the product of the dimensions of the quantities summed. The significant property of the direct sum is that it is the coproduct in the category of MODULES. This general definition gives as a consequence the definition of the direct sum of ABELIAN GROUPS (since they are MODULES over the INTEGERS) and the direct sum of VECTOR SPACES (since they are MODULES over a FIELD).

Directed Angle

The symbol $\measuredangle ABC$ denotes the directed angle from AB to BC, which is the signed angle through which AB must be rotated about B to coincide with BC. Four points $ABCD$ lie on a CIRCLE (i.e., are CONCYCLIC) IFF $\measuredangle ABC = \measuredangle ADC$. It is also true that

$$\measuredangle l_1 l_2 + \measuredangle l_2 l_1 = 0° \text{ or } 180°.$$

Three points A, B, and C are COLLINEAR IFF $\measuredangle ABC = 0$. For any four points, A, B, C, and D,

$$\measuredangle ABC + \measuredangle CDA = \measuredangle BAD + \measuredangle DCB.$$

see also ANGLE, COLLINEAR, CONCYCLIC, MIQUEL EQUATION

References

Johnson, R. A. *Modern Geometry: An Elementary Treatise on the Geometry of the Triangle and the Circle.* Boston, MA: Houghton Mifflin, pp. 11–15, 1929.

Directed Graph

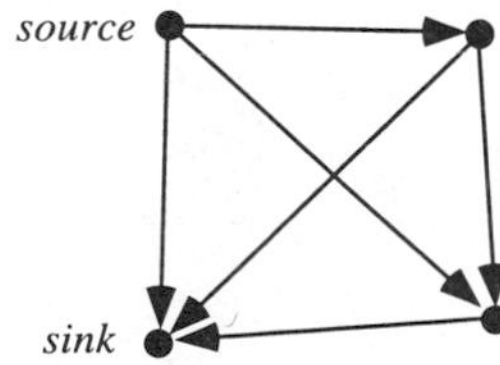

A GRAPH in which each EDGE is replaced by a directed EDGE, also called a DIGRAPH or REFLEXIVE GRAPH. A COMPLETE directed graph is called a TOURNAMENT. If G is an undirected connected GRAPH, then one can always direct the circuit EDGES of G and leave the SEPARATING EDGES undirected so that there is a directed path from any node to another. Such a GRAPH is said to be transitive if the adjacency relation is transitive. The number of directed graphs of n nodes for $n = 1$, 2, ... are 1, 1, 3, 16, 218, 9608, ... (Sloane's A000273).

see also ARBORESCENCE, CAYLEY GRAPH, INDEGREE, NETWORK, OUTDEGREE, SINK (DIRECTED GRAPH), SOURCE, TOURNAMENT

References

Sloane, N. J. A. Sequence A000273/M3032 in "An On-Line Version of the Encyclopedia of Integer Sequences."

Direction Cosine

Let a be the ANGLE between $\mathbf{v}$ and $\mathbf{x}$, b the ANGLE between $\mathbf{v}$ and $\mathbf{y}$, and c the ANGLE between $\mathbf{v}$ and $\mathbf{z}$. Then the direction cosines are equivalent to the (x, y, z) coordinates of a UNIT VECTOR $\hat{\mathbf{v}}$,

$$\alpha \equiv \cos a \equiv \frac{\mathbf{v} \cdot \hat{\mathbf{x}}}{|\mathbf{v}|} \tag{1}$$

$$\beta \equiv \cos b \equiv \frac{\mathbf{v} \cdot \hat{\mathbf{y}}}{|\mathbf{v}|} \tag{2}$$

$$\gamma \equiv \cos c \equiv \frac{\mathbf{v} \cdot \hat{\mathbf{z}}}{|\mathbf{v}|}. \tag{3}$$

From these definitions, it follows that

$$\alpha^2 + \beta^2 + \gamma^2 = 1. \tag{4}$$

To find the JACOBIAN when performing integrals over direction cosines, use

$$\theta = \sin^{-1}\left(\sqrt{\alpha^2 + \beta^2}\right) \tag{5}$$

$$\phi = \tan^{-1}\left(\frac{\beta}{\alpha}\right) \tag{6}$$

$$\gamma = \sqrt{1 - \alpha^2 - \beta^2}. \tag{7}$$

The JACOBIAN is

$$\left|\frac{\partial(\theta, \phi)}{\partial(\alpha, \beta)}\right| = \begin{vmatrix} \frac{\partial\theta}{\partial\alpha} & \frac{\partial\theta}{\partial\beta} \\ \frac{\partial\phi}{\partial\alpha} & \frac{\partial\phi}{\partial\beta} \end{vmatrix}. \tag{8}$$

Using

$$\frac{d}{dx}(\sin^{-1} x) = \frac{1}{\sqrt{1 - x^2}} \tag{9}$$

$$\frac{d}{dx}(\tan^{-1} x) = \frac{1}{1 + x^2}, \tag{10}$$

$$\begin{aligned} \left|\frac{\partial(\theta, \phi)}{\partial(\alpha, \beta)}\right| &= \begin{vmatrix} \frac{\frac{1}{2}(\alpha^2+\beta^2)^{-1/2}2\alpha}{\sqrt{1-\alpha^2-\beta^2}} & \frac{\frac{1}{2}(\alpha^2+\beta^2)^{-1/2}2\beta}{\sqrt{1-\alpha^2-\beta^2}} \\ \frac{-\alpha^{-2}\beta}{1+\frac{\beta^2}{\alpha^2}} & \frac{\alpha^{-1}}{1+\frac{\beta^2}{\alpha^2}} \end{vmatrix} \\ &= \frac{1}{\sqrt{1 - \alpha^2 - \beta^2}} \frac{(\alpha^2 + \beta^2)^{-1/2}}{1 + \frac{\beta^2}{\alpha^2}} \left(1 + \frac{\beta^2}{\alpha^2}\right) \\ &= \frac{1}{\sqrt{(\alpha^2 + \beta^2)(1 - \alpha^2 - \beta^2)}}, \end{aligned} \tag{11}$$

so

$$\begin{aligned} d\Omega = \sin\theta \, d\phi \, d\theta &= \sqrt{\alpha^2 + \beta^2} \left|\frac{\partial(\theta, \phi)}{\partial(\alpha, \beta)}\right| d\alpha \, d\beta \\ &= \frac{d\alpha \, d\beta}{\sqrt{1 - \alpha^2 - \beta^2}} = \frac{d\alpha \, d\beta}{\gamma}. \end{aligned} \tag{12}$$

Direction cosines can also be defined between two sets of CARTESIAN COORDINATES,

$$\alpha_1 \equiv \hat{\mathbf{x}}' \cdot \hat{\mathbf{x}} \tag{13}$$

$$\alpha_2 \equiv \hat{\mathbf{x}}' \cdot \hat{\mathbf{y}} \tag{14}$$

$$\alpha_3 \equiv \hat{\mathbf{x}}' \cdot \hat{\mathbf{z}} \tag{15}$$

$$\beta_1 \equiv \hat{\mathbf{y}}' \cdot \hat{\mathbf{x}} \tag{16}$$

$$\beta_2 \equiv \hat{\mathbf{y}}' \cdot \hat{\mathbf{y}} \tag{17}$$

$$\beta_3 \equiv \hat{\mathbf{y}}' \cdot \hat{\mathbf{z}} \tag{18}$$

$$\gamma_1 \equiv \hat{\mathbf{z}}' \cdot \hat{\mathbf{x}} \tag{19}$$

$$\gamma_2 \equiv \hat{\mathbf{z}}' \cdot \hat{\mathbf{y}} \tag{20}$$

$$\gamma_3 \equiv \hat{\mathbf{z}}' \cdot \hat{\mathbf{z}}. \tag{21}$$

Projections of the unprimed coordinates onto the primed coordinates yield

$$\hat{\mathbf{x}}' = (\hat{\mathbf{x}}' \cdot \hat{\mathbf{x}})\hat{\mathbf{x}} + (\hat{\mathbf{x}}' \cdot \hat{\mathbf{y}})\hat{\mathbf{y}} + (\hat{\mathbf{x}}' \cdot \hat{\mathbf{z}})\hat{\mathbf{z}} = \alpha_1\hat{\mathbf{x}} + \alpha_2\hat{\mathbf{y}} + \alpha_3\hat{\mathbf{z}} \tag{22}$$

$$\hat{\mathbf{y}}' = (\hat{\mathbf{y}}' \cdot \hat{\mathbf{x}})\hat{\mathbf{x}} + (\hat{\mathbf{y}}' \cdot \hat{\mathbf{y}})\hat{\mathbf{y}} + (\hat{\mathbf{y}}' \cdot \hat{\mathbf{z}})\hat{\mathbf{z}} = \beta_1\hat{\mathbf{x}} + \beta_2\hat{\mathbf{y}} + \beta_3\hat{\mathbf{z}} \tag{23}$$

$$\hat{\mathbf{z}}' = (\hat{\mathbf{z}}' \cdot \hat{\mathbf{x}})\hat{\mathbf{x}} + (\hat{\mathbf{z}}' \cdot \hat{\mathbf{x}})\hat{\mathbf{y}} + (\hat{\mathbf{z}}' \cdot \hat{\mathbf{z}})\hat{\mathbf{z}} = \gamma_1\hat{\mathbf{x}} + \gamma_2\hat{\mathbf{y}} + \gamma_3\hat{\mathbf{z}}, \tag{24}$$

and

$$x' = \mathbf{r} \cdot \hat{\mathbf{x}}' = \alpha_1 x + \alpha_2 y + \alpha_3 z \tag{25}$$

$$y' = \mathbf{r} \cdot \hat{\mathbf{y}}' = \beta_1 x + \beta_2 y + \beta_3 z \tag{26}$$

$$z' = \mathbf{r} \cdot \hat{\mathbf{z}}' = \gamma_1 x + \gamma_2 y + \gamma_3 z. \tag{27}$$

Projections of the primed coordinates onto the unprimed coordinates yield

$$\begin{aligned} \hat{\mathbf{x}} &= (\hat{\mathbf{x}} \cdot \hat{\mathbf{x}}')\hat{\mathbf{x}}' + (\hat{\mathbf{x}} \cdot \hat{\mathbf{y}}')\hat{\mathbf{y}}' + (\hat{\mathbf{x}} \cdot \hat{\mathbf{z}}')\hat{\mathbf{z}}' \\ &= \alpha_1\hat{\mathbf{x}}' + \beta_1\hat{\mathbf{y}}' + \gamma_1\hat{\mathbf{z}}' \end{aligned} \tag{28}$$

$$\begin{aligned} \hat{\mathbf{y}} &= (\hat{\mathbf{y}} \cdot \hat{\mathbf{x}}')\hat{\mathbf{x}}' + (\hat{\mathbf{y}} \cdot \hat{\mathbf{y}}')\hat{\mathbf{y}}' + (\hat{\mathbf{y}} \cdot \hat{\mathbf{z}}')\hat{\mathbf{z}}' \\ &= \alpha_2\hat{\mathbf{x}}' + \beta_2\hat{\mathbf{y}}' + \gamma_2\hat{\mathbf{z}}' \end{aligned} \tag{29}$$

$$\begin{aligned} \hat{\mathbf{z}} &= (\hat{\mathbf{z}} \cdot \hat{\mathbf{x}}')\hat{\mathbf{x}}' + (\hat{\mathbf{z}} \cdot \hat{\mathbf{x}}')\hat{\mathbf{y}}' + (\hat{\mathbf{z}} \cdot \hat{\mathbf{z}}')\hat{\mathbf{z}}' \\ &= \alpha_3\hat{\mathbf{x}}' + \beta_3\hat{\mathbf{y}}' + \gamma_3\hat{\mathbf{z}}', \end{aligned} \tag{30}$$

and

$$x = \mathbf{r} \cdot \hat{\mathbf{x}} = \alpha_1 x + \beta_1 y + \gamma_1 z \tag{31}$$

$$y = \mathbf{r} \cdot \hat{\mathbf{y}} = \alpha_2 x + \beta_2 y + \gamma_2 z \tag{32}$$

$$z = \mathbf{r} \cdot \hat{\mathbf{z}} = \alpha_3 x + \beta_3 y + \gamma_3 z. \quad (33)$$

Using the orthogonality of the coordinate system, it must be true that

$$\hat{\mathbf{x}} \cdot \hat{\mathbf{y}} = \hat{\mathbf{y}} \cdot \hat{\mathbf{z}} = \hat{\mathbf{z}} \cdot \hat{\mathbf{x}} = 0 \quad (34)$$

$$\hat{\mathbf{x}} \cdot \hat{\mathbf{x}} = \hat{\mathbf{y}} \cdot \hat{\mathbf{y}} = \hat{\mathbf{z}} \cdot \hat{\mathbf{z}} = 1, \quad (35)$$

giving the identities

$$\alpha_l \alpha_m + \beta_l \beta_m + \gamma_l \gamma_m = 0 \quad (36)$$

for $l, m = 1, 2, 3$ and $l \neq m$, and

$$\alpha_l{}^2 + \beta_l{}^2 + \gamma_l{}^2 = 1 \quad (37)$$

for $l = 1, 2, 3$. These two identities may be combined into the single identity

$$\alpha_l \alpha_m + \beta_l \beta_m + \gamma_l \gamma_m = \delta_{lm}, \quad (38)$$

where δ_{lm} is the KRONECKER DELTA.

Directional Derivative

$$\nabla_{\mathbf{u}} f \equiv \nabla f \cdot \frac{\mathbf{u}}{|\mathbf{u}|} \propto \lim_{h \to 0} \frac{f(\mathbf{x} + h\mathbf{u}) - f(\mathbf{x})}{h}. \quad (1)$$

$\nabla_{\mathbf{u}} f(x_0, y_0, z_0)$ is the rate at which the function $w = f(x, y, z)$ changes at (x_0, y_0, z_0) in the direction $\mathbf{u}$. Let $\mathbf{u}$ be a UNIT VECTOR in CARTESIAN COORDINATES, so

$$|\mathbf{u}| = \sqrt{u_x{}^2 + u_y{}^2 + u_z{}^2} = 1, \quad (2)$$

then

$$\nabla_{\mathbf{u}} f = \frac{\partial f}{\partial x} u_x + \frac{\partial f}{\partial y} u_y + \frac{\partial f}{\partial z} u_z. \quad (3)$$

The directional derivative is often written in the notation

$$\frac{d}{ds} \equiv \hat{\mathbf{s}} \cdot \nabla = s_x \frac{\partial}{\partial x} + s_y \frac{\partial}{\partial y} + s_z \frac{\partial}{\partial z}. \quad (4)$$

Directly Similar

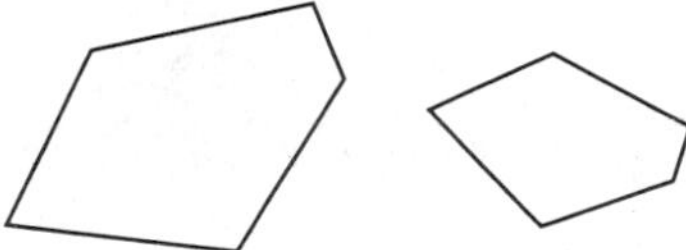

directly similar

Two figures are said to be SIMILAR when all corresponding ANGLES are equal, and are directly similar when all corresponding ANGLES are equal and described in the same rotational sense.

see also FUNDAMENTAL THEOREM OF DIRECTLY SIMILAR FIGURES, INVERSELY SIMILAR, SIMILAR

Director Curve

The curve $\mathbf{d}(u)$ in the RULED SURFACE parameterization

$$\mathbf{x}(u, v) = \mathbf{b}(u) + v\mathbf{d}(u).$$

see also DIRECTRIX (RULED SURFACE), RULED SURFACE, RULING

References

Gray, A. *Modern Differential Geometry of Curves and Surfaces.* Boca Raton, FL: CRC Press, p. 333, 1993.

Directrix (Conic Section)

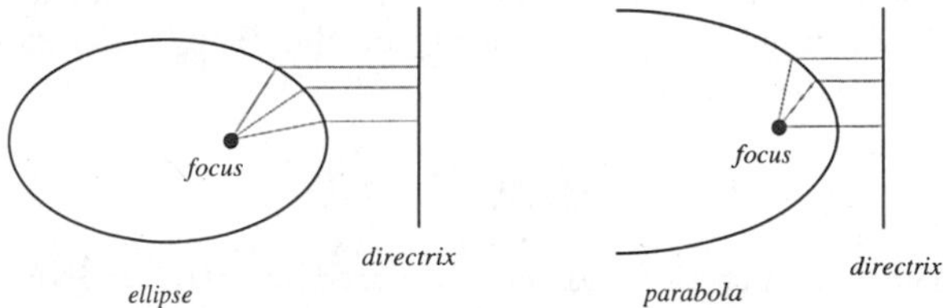

The LINE which, together with the point known as the FOCUS, serves to define a CONIC SECTION.

see also CONIC SECTION, ELLIPSE, FOCUS, HYPERBOLA, PARABOLA

References

Coxeter, H. S. M. *Introduction to Geometry, 2nd ed.* New York: Wiley, pp. 115–116, 1969.
Coxeter, H. S. M. and Greitzer, S. L. *Geometry Revisited.* Washington, DC: Math. Assoc. Amer., pp. 141–144, 1967.

Directrix (Graph)

A CYCLE.

Directrix (Ruled Surface)

The curve $\mathbf{b}(u)$ in the RULED SURFACE parameterization

$$\mathbf{x}(u, v) = \mathbf{b}(u) + v\mathbf{d}(u)$$

is called the directrix (or BASE CURVE).

see also DIRECTOR CURVE, RULED SURFACE

References

Gray, A. *Modern Differential Geometry of Curves and Surfaces.* Boca Raton, FL: CRC Press, p. 333, 1993.

Dirichlet Beta Function

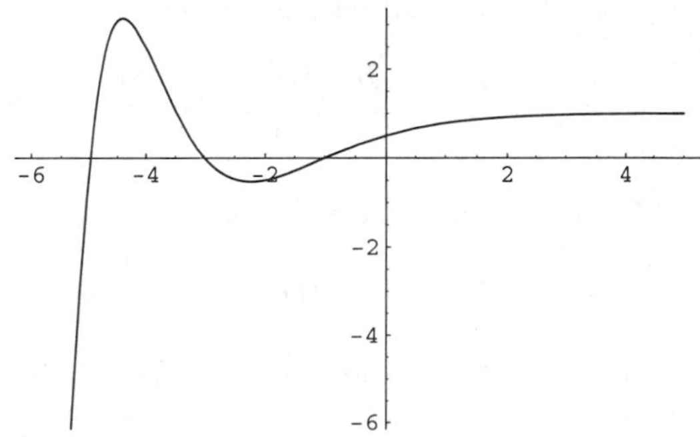

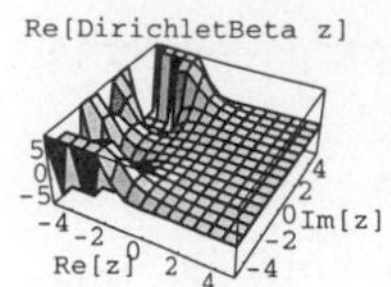

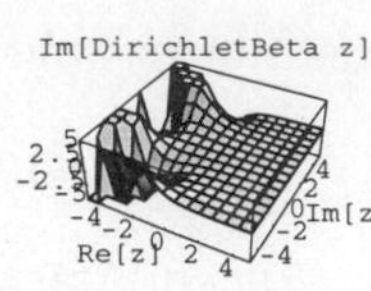

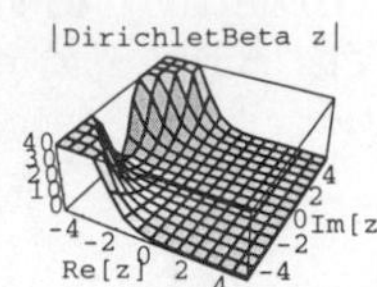

$$\beta(x) \equiv \sum_{n=0}^{\infty} (-1)^n (2n+1)^{-x} \tag{1}$$

$$\beta(x) = 2^{-x} \Phi(-1, x, \tfrac{1}{2}), \tag{2}$$

where Φ is the LERCH TRANSCENDENT. The beta function can be written in terms of the HURWITZ ZETA FUNCTION $\zeta(x, a)$ by

$$\beta(x) = \frac{1}{4^x} [\zeta(x, \tfrac{1}{4}) - \zeta(x, \tfrac{3}{4})]. \tag{3}$$

The beta function can be evaluated directly for POSITIVE ODD x as

$$\beta(2k+1) = \frac{(-1)^k E_{2k}}{2(2k)!} (\tfrac{1}{2}\pi)^{2k+1}, \tag{4}$$

where E_n is an EULER NUMBER. The beta function can be defined over the whole COMPLEX PLANE using ANALYTIC CONTINUATION,

$$\beta(1-z) = \left(\frac{2}{\pi}\right)^z \sin(\tfrac{1}{2}\pi z)\Gamma(z)\beta(z), \tag{5}$$

where $\Gamma(z)$ is the GAMMA FUNCTION.

Particular values for β are

$$\beta(1) = \tfrac{1}{4}\pi \tag{6}$$

$$\beta(2) \equiv K \tag{7}$$

$$\beta(3) = \tfrac{1}{32}\pi^3, \tag{8}$$

where K is CATALAN'S CONSTANT.

see also CATALAN'S CONSTANT, DIRICHLET ETA FUNCTION, DIRICHLET LAMBDA FUNCTION, HURWITZ ZETA FUNCTION, LERCH TRANSCENDENT, RIEMANN ZETA FUNCTION, ZETA FUNCTION

References

Abramowitz, M. and Stegun, C. A. (Eds.). *Handbook of Mathematical Functions with Formulas, Graphs, and Mathematical Tables, 9th printing.* New York: Dover, pp. 807–808, 1972.

Spanier, J. and Oldham, K. B. "The Zeta Numbers and Related Functions." Ch. 3 in *An Atlas of Functions.* Washington, DC: Hemisphere, pp. 25–33, 1987.

Dirichlet Boundary Conditions

PARTIAL DIFFERENTIAL EQUATION BOUNDARY CONDITIONS which give the value of the function on a surface, e.g., $T = f(\mathbf{r}, t)$.

see also BOUNDARY CONDITIONS, CAUCHY BOUNDARY CONDITIONS

References

Morse, P. M. and Feshbach, H. *Methods of Theoretical Physics, Part I.* New York: McGraw-Hill, p. 679, 1953.

Dirichlet's Box Principle

A.k.a. the PIGEONHOLE PRINCIPLE. Given n boxes and $m > n$ objects, at least one box must contain more than one object. This statement has important applications in number theory and was first stated by Dirichlet in 1834.

see also FUBINI PRINCIPLE

References

Chartrand, G. *Introductory Graph Theory.* New York: Dover, p. 38, 1985.

Shanks, D. *Solved and Unsolved Problems in Number Theory, 4th ed.* New York: Chelsea, pp. 161, 1993.

Dirichlet's Boxing-In Principle

see DIRICHLET'S BOX PRINCIPLE

Dirichlet Conditions

see DIRICHLET BOUNDARY CONDITIONS, DIRICHLET FOURIER SERIES CONDITIONS

Dirichlet Divisor Problem

Let $d(n) = \nu(n) = \sigma_0(n)$ be the number of DIVISORS of n (including n itself). For a PRIME p, $\nu(p) = 2$. In general,

$$\sum_{k=1}^{n} \nu(k) = n \ln n + (2\gamma - 1)n + \mathcal{O}(n^\theta),$$

where γ is the EULER-MASCHERONI CONSTANT. Dirichlet originally gave $\theta \approx 1/2$. As of 1988, this had been reduced to $\theta \approx 7/22$.

see also DIVISOR FUNCTION

Dirichlet Energy

Let h be a real-valued HARMONIC FUNCTION on a bounded DOMAIN Ω, then the Dirichlet energy is defined as $\int_\Omega |\nabla h|^2 \, dx$, where ∇ is the GRADIENT.

see also ENERGY

Dirichlet Eta Function

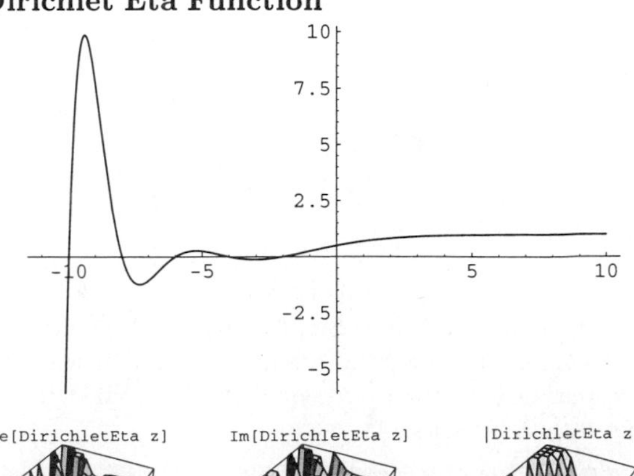

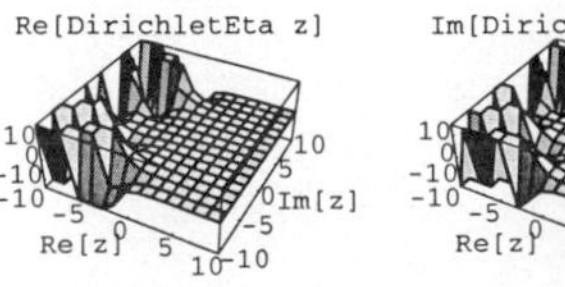

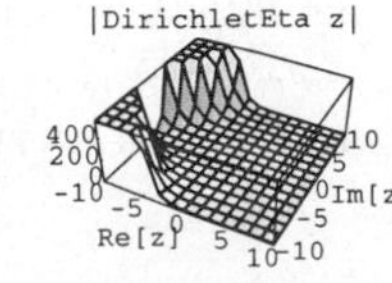

$$\eta(x) \equiv \sum_{n=1}^{\infty} (-1)^{n-1} n^{-x} = (1 - 2^{1-x})\zeta(x), \qquad (1)$$

where $n = 1, 2, \ldots$, and $\zeta(x)$ is the RIEMANN ZETA FUNCTION. Particular values are given in Abramowitz and Stegun (1972, p. 811). The eta function is related to the RIEMANN ZETA FUNCTION and DIRICHLET LAMBDA FUNCTION by

$$\frac{\zeta(\nu)}{2^\nu} = \frac{\lambda(\nu)}{2^\nu - 1} = \frac{\eta(\nu)}{2^\nu - 2} \qquad (2)$$

and

$$\zeta(\nu) + \eta(\nu) = 2\lambda(\nu) \qquad (3)$$

(Spanier and Oldham 1987). The value $\eta(1)$ may be computed by noting that the MACLAURIN SERIES for $\ln(1 + x)$ for $-1 < x \leq 1$ is

$$\ln(1 + x) = x - \tfrac{1}{2}x^2 + \tfrac{1}{3}x^3 - \tfrac{1}{4}x^4 + \ldots. \qquad (4)$$

Therefore,

$$\begin{aligned} \ln 2 = \ln(1 + 1) &= 1 - \tfrac{1}{2} + \tfrac{1}{3} - \tfrac{1}{4} + \ldots \\ &= \sum_{n=1}^{\infty} \frac{(-1)^{n-1}}{n} = \eta(1). \end{aligned} \qquad (5)$$

Values for EVEN INTEGERS are related to the analytical values of the RIEMANN ZETA FUNCTION. $\eta(0)$ is defined to be $\frac{1}{2}$.

$$\begin{aligned} \eta(0) &= \tfrac{1}{2} \\ \eta(1) &= \ln 2 \\ \eta(2) &= \frac{\pi^2}{12} \\ \eta(3) &= 0.90154\ldots \\ \eta(4) &= \frac{7\pi^4}{720}. \end{aligned}$$

see also DEDEKIND ETA FUNCTION, DIRICHLET BETA FUNCTION, DIRICHLET LAMBDA FUNCTION, RIEMANN ZETA FUNCTION, ZETA FUNCTION

References

Abramowitz, M. and Stegun, C. A. (Eds.). *Handbook of Mathematical Functions with Formulas, Graphs, and Mathematical Tables, 9th printing.* New York: Dover, pp. 807–808, 1972.

Spanier, J. and Oldham, K. B. "The Zeta Numbers and Related Functions." Ch. 3 in *An Atlas of Functions.* Washington, DC: Hemisphere, pp. 25–33, 1987.

Dirichlet's Formula

If g is continuous and $\mu, \nu > 0$, then

$$\int_0^t (t - \xi)^{\mu-1}\, d\xi \int_0^\xi (\xi - x)^{\nu-1} g(\xi, x)\, dx = \int_0^t dx \int_x^t (t - \xi)^{\mu-1} (\xi - x)^{\nu-1} g(\xi, x)\, d\xi.$$

Dirichlet Fourier Series Conditions

A piecewise regular function which

1. Has a finite number of finite discontinuities and
2. Has a finite number of extrema

can be expanded in a FOURIER SERIES which converges to the function at continuous points and the mean of the POSITIVE and NEGATIVE limits at points of discontinuity.

see also FOURIER SERIES

Dirichlet Function

Let c and $d \neq c$ be REAL NUMBERS (usually taken as $c = 1$ and $d = 0$). The Dirichlet function is defined by

$$D(x) = \begin{cases} c & \text{for } x \text{ rational} \\ d & \text{for } x \text{ irrational.} \end{cases}$$

The function is CONTINUOUS at IRRATIONAL x and discontinuous at RATIONAL points. The function can be written analytically as

$$D(x) = \lim_{m,n \to \infty} \cos[(m!\pi x)^n].$$

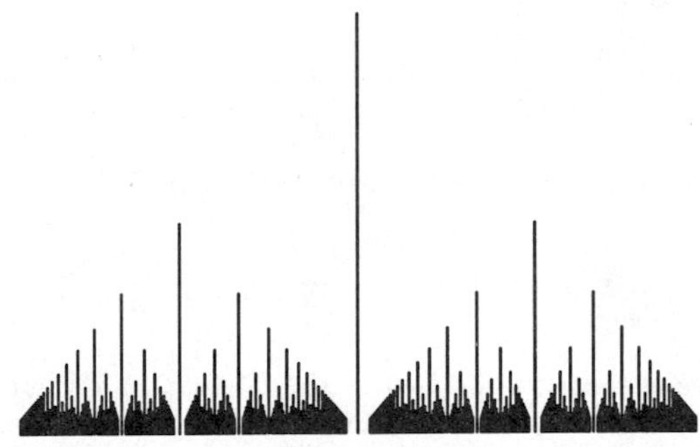

Because the Dirichlet function cannot be plotted without producing a solid blend of lines, a modified version can be defined as

$$D_M(x) = \begin{cases} 0 & \text{for } x \text{ rational} \\ b & \text{for } x = a/b \text{ with } a/b \text{ a reduced fraction} \end{cases}$$

(Dixon 1991), illustrated above.

see also CONTINUOUS FUNCTION, IRRATIONAL NUMBER, RATIONAL NUMBER

References

Dixon, R. *Mathographics.* New York: Dover, pp. 177 and 184–186, 1991.

Tall, D. "The Gradient of a Graph." *Math. Teaching* **111**, 48–52, 1985.

Dirichlet Integrals

There are several types of integrals which go under the name of a "Dirichlet integral." The integral

$$D[u] = \int_\Omega |\nabla u|^2\, dV \qquad (1)$$

appears in DIRICHLET'S PRINCIPLE.

The integral

$$\frac{1}{2\pi}\int_{-\pi}^{\pi} f(x)\frac{\sin[(n+\frac{1}{2})x]}{\sin(\frac{1}{2}x)}\,dx, \tag{2}$$

where the kernel is the DIRICHLET KERNEL, gives the nth partial sum of the FOURIER SERIES.

Another integral is denoted

$$\delta_k \equiv \frac{1}{\pi}\int_{-\infty}^{\infty} \frac{\sin\alpha_k\rho_k}{\rho_k}e^{i\rho_k\gamma_k}\,d\rho_k = \begin{cases} 0 & \text{for } |\gamma_k| > \alpha_k \\ 1 & \text{for } |\gamma_k| < \alpha_k \end{cases} \tag{3}$$

for $k = 1, \ldots, n$.

There are two types of Dirichlet integrals which are denoted using the letters C, D, I, and J. The type 1 Dirichlet integrals are denoted I, J, and IJ, and the type 2 Dirichlet integrals are denoted C, D, and CD.

The type 1 integrals are given by

$$\begin{aligned} I &\equiv \iint\cdots\int f(t_1+t_2+\ldots+t_n) \\ &\quad\times {t_1}^{\alpha_1-1}{t_2}^{\alpha_2-1}\cdots{t_n}^{\alpha_n-1}\,dt_1\,dt_2\,dt_n \\ &= \frac{\Gamma(\alpha_1)\Gamma(\alpha_2)\cdots\Gamma(\alpha_n)}{\Gamma\left(\sum_n \alpha_n\right)}\int_0^1 f(\tau)\tau^{\left(\sum_n \alpha\right)-1}\,d\tau, \end{aligned} \tag{4}$$

where $\Gamma(z)$ is the GAMMA FUNCTION. In the case $n = 2$,

$$I = \iint_T x^p y^q\,dx\,dy = \frac{p!q!}{(p+q+2)!} = \frac{B(p+1,q+1)}{p+q+2}, \tag{5}$$

where the integration is over the TRIANGLE T bounded by the x-axis, y-axis, and line $x + y = 1$ and $B(x, y)$ is the BETA FUNCTION.

The type 2 integrals are given for b-D vectors $\mathbf{a}$ and $\mathbf{r}$, and $0 \leq c \leq b$,

$$C_{\mathbf{a}}^{(b)}(\mathbf{r},m) = \frac{\Gamma(m+R)}{\Gamma(m)\prod_{i=1}^b \Gamma(r_i)} \times\int_0^{a_1}\cdots\int_0^{a_b}\frac{\prod_{i=1}^b {x_i}^{r_i-1}\,dx_i}{\left(1+\sum_{i=1}^b x_i\right)^{m+R}} \tag{6}$$

$$D_{\mathbf{a}}^{(b)}(\mathbf{r},m) = \frac{\Gamma(m+R)}{\Gamma(m)\prod_{i=1}^b \Gamma(r_i)} \times\int_{a_1}^{\infty}\cdots\int_{a_k}^{\infty}\frac{\prod_{i=1}^b {x_i}^{r_i-1}\,dx_i}{\left(1+\sum_{i=1}^b x_i\right)^{m+R}} \tag{7}$$

$$CD_{\mathbf{a}}^{(c,d-c)}(\mathbf{r},m) = \frac{\Gamma(m+R)}{\Gamma(m)\prod_{i=1}^b \Gamma(r_i)} \times\int_0^{a_c}\int_{a_{c+1}}^{\infty}\int_{a_b}^{\infty}\frac{\prod_{i=1}^b {x_i}^{r_i-1}\,dx_i}{\left(1+\sum_{i=1}^b x_i\right)^{m+R}}, \tag{8}$$

where

$$R \equiv \sum_{i=1}^k r_i \tag{9}$$

$$a_i \equiv \frac{p_i}{1-\sum_{i=1}^k p_i}, \tag{10}$$

and p_i are the cell probabilities. For equal probabilities, $a_i = 1$. The Dirichlet D integral can be expanded as a MULTINOMIAL SERIES as

$$\begin{aligned} D_{\mathbf{a}}^{(b)}(\mathbf{r},m) &= \frac{1}{\left(1+\sum_{i=1}^b\right)^m} \\ &\quad\times\sum_{x_1<r_1}\cdots\sum_{x_b<r_b}\binom{m-1+\sum_{a=1}^b x_i}{m-1, x_1, \ldots, x_b} \\ &\quad\times\prod_{i=1}^b\left(\frac{a_i}{1+\sum_{k=1}^b a_k}\right)^{x_i}. \end{aligned} \tag{11}$$

For small b, C and D can be expressed analytically either partially or fully for general arguments and $a_i = 1$.

$$C_1^{(1)}(r_2;r_1) = \frac{\Gamma(r_1+r_2)\,{}_2F_1(r_2,r_1+r_2;1+r_2;-1)}{r_2\Gamma(r_1)\Gamma(r_2)} \tag{12}$$

$$\begin{aligned} C_1^{(2)}(r_2,r_3;r_1) &= \frac{\Gamma(r_1+r_2+r_3)}{r_2\Gamma(r_1)\Gamma(r_2)\Gamma(r_3)} \\ &\quad\times\int_0^1 {}_2F_1\,y^{r_3-1}(1+y)^{-(r_1+r_2+r_3)}\,dy, \end{aligned} \tag{13}$$

where

$${}_2F_1 \equiv {}_2F_1(r_2,r_1+r_2+r_3;1+r_2,-(1+y)^{-1}) \tag{14}$$

is a HYPERGEOMETRIC FUNCTION.

$$D_1^{(1)}(r_2;r_1) = \frac{\Gamma(r_1+r_2)\,{}_2F_1(r_1,r_1+r_2;1+r_1;-1)}{r_1\Gamma(r_1)\Gamma(r_2)} \tag{15}$$

$$\begin{aligned} D_1^{(2)}(r_2,r_3;r_1) &= \frac{\Gamma(r_1+r_2+r_3)}{(r_1+r_3)\Gamma(r_1)\Gamma(r_2)\Gamma(r_3)} \\ &\quad\times\int_1^{\infty} {}_2F_1\,y^{r_3-1}\,dy, \end{aligned} \tag{16}$$

where

$${}_2F_1 \equiv {}_2F_1(r_1+r_3,r_1+r_2+r_3;1+r_1+r_3;-1-y). \tag{17}$$

References

Sobel, M.; Uppuluri, R. R.; and Frankowski, K. *Selected Tables in Mathematical Statistics, Vol. 4: Dirichlet Distribution—Type 1.* Providence, RI: Amer. Math. Soc., 1977.

Sobel, M.; Uppuluri, R. R.; and Frankowski, K. *Selected Tables in Mathematical Statistics, Vol. 9: Dirichlet Integrals of Type 2 and Their Applications.* Providence, RI: Amer. Math. Soc., 1985.

Weisstein, E. W. "Dirichlet Integrals." `http://www.astro.virginia.edu/~eww6n/math/notebooks/DirichletIntegrals.m`.

Dirichlet Kernel

The Dirichlet kernel D_n^M is obtained by integrating the CHARACTER $e^{i\langle\xi,x\rangle}$ over the BALL $|\xi| \leq M$,

$$D_n^M = -\frac{1}{2\pi r}\frac{d}{dr}D_{n-2}^M.$$

The Dirichlet kernel of a DELTA SEQUENCE is given by

$$\delta_n(x) \equiv \frac{1}{2\pi}\frac{\sin[(n+\frac{1}{2})x]}{\sin(\frac{1}{2}x)}.$$

The integral of this kernal is called the DIRICHLET INTEGRAL $D[u]$.

see also DELTA SEQUENCE, DIRICHLET INTEGRALS, DIRICHLET'S LEMMA

Dirichlet L-Series

Series of the form

$$L_k(s,\chi) \equiv \sum_{n=1}^{\infty} \chi_k(n) n^{-s}, \tag{1}$$

where the CHARACTER (NUMBER THEORY) $\chi_k(n)$ is an INTEGER function with period m. These series appear in number theory (they were used, for instance, to prove DIRICHLET'S THEOREM) and can be written as sums of LERCH TRANSCENDENTS with z a POWER of $e^{2\pi i/m}$. The DIRICHLET ETA FUNCTION

$$\eta(s) \equiv \sum_{n=1}^{\infty} \frac{(-1)^{n+1}}{n^s} = (1-2^{1-s})\zeta(s) \tag{2}$$

(for $s \neq 1$) and DIRICHLET BETA FUNCTION

$$L_{-4}(s) = \beta(s) \equiv \sum_{n=0}^{\infty} \frac{(-1)^n}{(2n+1)^s} \tag{3}$$

and RIEMANN ZETA FUNCTION

$$L_{+1}(s) = \zeta(s) \tag{4}$$

are Dirichlet series (Borwein and Borwein 1987, p. 289). χ_k is called primitive if the CONDUCTOR $f(\chi) = k$. Otherwise, χ_k is imprimitive. A primitive L-series modulo k is then defined as one for which $\chi_k(n)$ is primitive. All imprimitive L-series can be expressed in terms of primitive L-series.

Let $P = 1$ or $P = \prod_{i=1}^{t} p_i$, where p_i are distinct ODD PRIMES. Then there are three possible types of primitive L-series with REAL COEFFICIENTS. The requirement of REAL COEFFICIENTS restricts the CHARACTER to $\chi_k(n) = \pm 1$ for all k and n. The three type are then

1. If $k = P$ (e.g., $k = 1, 3, 5, \ldots$) or $k = 4P$ (e.g., $k = 4, 12, 20$, dots), there is exactly one primitive L-series.
2. If $k = 8P$ (e.g., $k = 8, 24, \ldots$), there are two primitive L-series.
3. If $k = 2P, Pp_i$, or $2^\alpha P$ where $\alpha > 3$ (e.g., $k = 2, 6, 9, \ldots$), there are no primitive L-series

(Zucker and Robertson 1976). All primitive L-series are algebraically independent and divide into two types according to

$$\chi_k(k-1) = \pm 1. \tag{5}$$

Primitive L-series of these types are denoted $L_\pm$. For a primitive L-series with REAL CHARACTER (NUMBER THEORY), if $k = P$, then

$$L_k = \begin{cases} L_{-k} & \text{if } P \equiv 3 \pmod 4 \\ L_k & \text{if } P \equiv 1 \pmod 4. \end{cases} \tag{6}$$

If $k = 4P$, then

$$L_k = \begin{cases} L_{-k} & \text{if } P \equiv 1 \pmod 4 \\ L_k & \text{if } P \equiv 3 \pmod 4, \end{cases} \tag{7}$$

and if $k = 8P$, then there is a primitive function of each type (Zucker and Robertson 1976).

The first few primitive NEGATIVE L-series are L_{-3}, L_{-4}, L_{-7}, L_{-8}, L_{-11}, L_{-15}, L_{-19}, L_{-20}, L_{-23}, L_{-24}, L_{-31}, L_{-35}, L_{-39}, L_{-40}, L_{-43}, L_{-47}, L_{-51}, L_{-52}, L_{-55}, L_{-56}, L_{-59}, L_{-67}, L_{-68}, L_{-71}, L_{-79}, L_{-83}, L_{-84}, L_{-87}, L_{-88}, L_{-91}, L_{-95}, ... (Sloane's A003657), corresponding to the negated discriminants of imaginary quadratic fields. The first few primitive POSITIVE L-series are L_{+1}, L_{+5}, L_{+8}, L_{+12}, L_{+13}, L_{+17}, L_{+21}, L_{+24}, L_{+28}, L_{+29}, L_{+33}, L_{+37}, L_{+40}, L_{+41}, L_{+44}, L_{+53}, L_{+56}, L_{+57}, L_{+60}, L_{+61}, L_{+65}, L_{+69}, L_{+73}, L_{+76}, L_{+77}, L_{+85}, L_{+88}, L_{+89}, L_{+92}, L_{+93}, L_{+97}, ... (Sloane's A046113).

The KRONECKER SYMBOL is a REAL CHARACTER modulo k, and is in fact essentially the only type of REAL primitive CHARACTER (Ayoub 1963). Therefore,

$$L_{+d}(s) = \sum_{n=1}^{\infty} (d|n) n^{-s} \tag{8}$$

$$L_{-d}(s) = \sum_{n=1}^{\infty} (-d|n) n^{-s}, \tag{9}$$

where $(d|n)$ is the KRONECKER SYMBOL. The functional equations for $L_\pm$ are

$$L_{-k}(s) = 2^s \pi^{s-1} k^{-s+1/2} \Gamma(1-s) \cos(\tfrac{1}{2}s\pi) L_{-k}(1-s) \tag{10}$$

$$L_{+k}(s) = 2^s \pi^{s-1} k^{-s+1/2} \Gamma(1-s) \sin(\tfrac{1}{2}s\pi) L_{+k}(1-s). \tag{11}$$

For m a POSITIVE INTEGER

$$L_{+k}(-2m) = 0 \quad (12)$$
$$L_{-k}(1-2m) = 0 \quad (13)$$
$$L_{+k}(2m) = Rk^{-1/2}\pi^{2m} \quad (14)$$
$$L_{-k}(2m-1) = R'k^{-1/2}\pi^{2m-1} \quad (15)$$
$$L_{+k}(1-2m) = \frac{(-1)^m(2m-1)!R}{(2k)^{2m-1}} \quad (16)$$
$$L_{-k}(-2k) = \frac{(-1)^m R'(2m)!}{(2k)^{2m}}, \quad (17)$$

where R and R' are RATIONAL NUMBERS. $L_{+k}(1)$ can be expressed in terms of transcendentals by

$$L_d(1) = h(d)\kappa(d), \quad (18)$$

where $h(d)$ is the CLASS NUMBER and $\kappa(d)$ is the DIRICHLET STRUCTURE CONSTANT. Some specific values of primitive L-series are

$$L_{-15}(1) = \frac{2\pi}{\sqrt{15}}$$
$$L_{-11}(1) = \frac{\pi}{\sqrt{11}}$$
$$L_{-8}(1) = \frac{\pi}{2\sqrt{2}}$$
$$L_{-7}(1) = \frac{\pi}{\sqrt{7}}$$
$$L_{-4}(1) = \tfrac{1}{4}\pi$$
$$L_{-3}(1) = \frac{\pi}{3\sqrt{3}}$$
$$L_{+5}(1) = \frac{2}{\sqrt{5}}\ln\left(\frac{1+\sqrt{5}}{2}\right)$$
$$L_{+8}(1) =$$
$$L_{+12}(1) = \frac{\ln(2+\sqrt{3})}{\sqrt{3}}$$
$$L_{+13}(1) = \frac{2}{\sqrt{13}}\ln\left(\frac{3+\sqrt{13}}{2}\right)$$
$$L_{+17}(1) = \frac{2}{\sqrt{17}}\ln(4+\sqrt{17})$$
$$L_{+21}(1) = \frac{2}{\sqrt{21}}\ln\left(\frac{5+\sqrt{21}}{2}\right)$$
$$L_{+24}(1) = \frac{\ln(5+2\sqrt{6})}{\sqrt{6}}.$$

No general forms are known for $L_{-k}(2m)$ and $L_{+k}(2m-1)$ in terms of known transcendentals. For example,

$$L_{-4}(2) = \beta(2) \equiv K, \quad (19)$$

where K is defined as CATALAN'S CONSTANT.

see also DIRICHLET BETA FUNCTION, DIRICHLET ETA FUNCTION

References

Ayoub, R. G. *An Introduction to the Analytic Theory of Numbers.* Providence, RI: Amer. Math. Soc., 1963.

Borwein, J. M. and Borwein, P. B. *Pi & the AGM: A Study in Analytic Number Theory and Computational Complexity.* New York: Wiley, 1987.

Buell, D. A. "Small Class Numbers and Extreme Values of L-Functions of Quadratic Fields." *Math. Comput.* **139**, 786–796, 1977.

Ireland, K. and Rosen, M. "Dirichlet L-Functions." Ch. 16 in *A Classical Introduction to Modern Number Theory, 2nd ed.* New York: Springer-Verlag, pp. 249–268, 1990.

Sloane, N. J. A. Sequences A046113 and A003657/M2332 in "An On-Line Version of the Encyclopedia of Integer Sequences."

Weisstein, E. W. "Class Numbers." http://www.astro.virginia.edu/~eww6n/math/notebooks/ClassNumbers.m.

Zucker, I. J. and Robertson, M. M. "Some Properties of Dirichlet L-Series." *J. Phys. A: Math. Gen.* **9**, 1207–1214, 1976.

Dirichlet Lambda Function

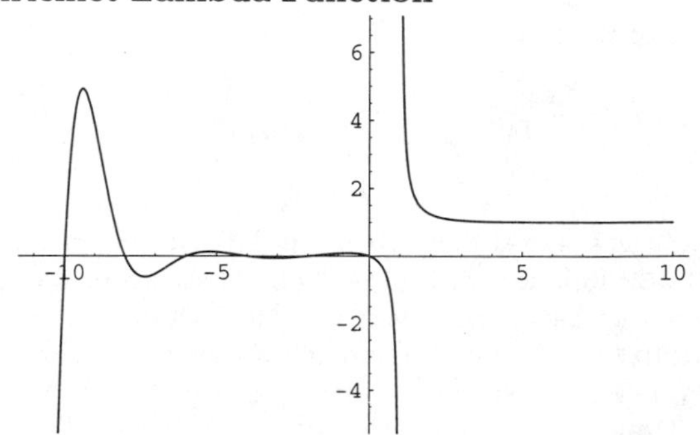

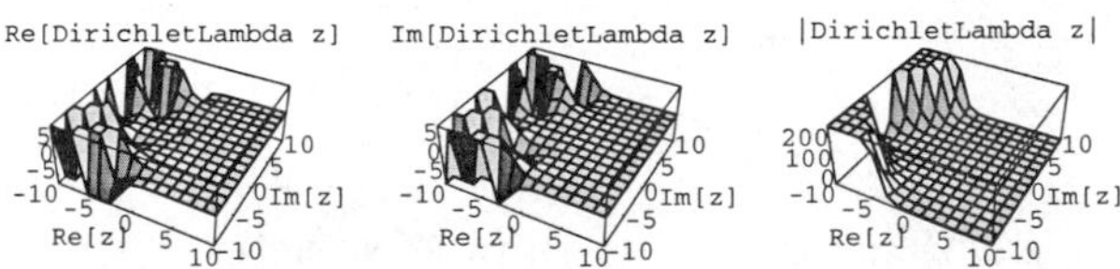

$$\lambda(x) \equiv \sum_{n=0}^{\infty}(2n+1)^{-x} = (1-2^{-x})\zeta(x) \quad (1)$$

for $x = 2, 3, \ldots$, where $\zeta(x)$ is the RIEMANN ZETA FUNCTION. The function is undefined at $x = 1$. It can be computed in closed form where $\zeta(x)$ can, that is for EVEN POSITIVE n. It is related to the RIEMANN ZETA FUNCTION and DIRICHLET ETA FUNCTION by

$$\frac{\zeta(\nu)}{2^\nu} = \frac{\lambda(\nu)}{2^\nu-1} = \frac{\eta(\nu)}{2^\nu-2} \quad (2)$$

and

$$\zeta(\nu) + \eta(\nu) = 2\lambda(\nu) \quad (3)$$

(Spanier and Oldham 1987). Special values of $\lambda(n)$ include

$$\lambda(2) = \frac{\pi^2}{8} \quad (4)$$
$$\lambda(4) = \frac{\pi^4}{96}. \quad (5)$$

see also DIRICHLET BETA FUNCTION, DIRICHLET ETA FUNCTION, RIEMANN ZETA FUNCTION, ZETA FUNCTION

References
Abramowitz, M. and Stegun, C. A. (Eds.). *Handbook of Mathematical Functions with Formulas, Graphs, and Mathematical Tables, 9th printing.* New York: Dover, pp. 807–808, 1972.
Spanier, J. and Oldham, K. B. "The Zeta Numbers and Related Functions." Ch. 3 in *An Atlas of Functions.* Washington, DC: Hemisphere, pp. 25–33, 1987.

Dirichlet's Lemma

$$\int_0^\pi \frac{\sin[(n+\frac{1}{2})x]}{2\sin(\frac{1}{2}x)}\,dx = \tfrac{1}{2}\pi,$$

where the KERNEL is the DIRICHLET KERNEL.

References
Cohn, H. *Advanced Number Theory.* New York: Dover, p. 37, 1980.
Gradshteyn, I. S. and Ryzhik, I. M. *Tables of Integrals, Series, and Products, 5th ed.* San Diego, CA: Academic Press, p. 1101, 1979.

Dirichlet's Principle

Also known as THOMSON'S PRINCIPLE. There exists a function u that minimizes the functional

$$D[u] = \int_\Omega |\nabla u|^2\,dV$$

(called the DIRICHLET INTEGRAL) for $\Omega \subset \mathbb{R}^2$ or $\mathbb{R}^3$ among all the functions $u \in C^{(1)}(\Omega) \cap C^{(0)}(\bar{\Omega})$ which take on given values f on the boundary $\partial\Omega$ of Ω, and that function u satisfies $\nabla^2 = 0$ in Ω, $u|_{\partial\Omega} = f$, $u \in C^{(2)}(\Omega) \cap C^{(0)}(\bar{\Omega})$. Weierstraß showed that Dirichlet's argument contained a subtle fallacy. As a result, it can be claimed only that there exists a lower bound to which $D[u]$ comes arbitrarily close without being forced to actually reach it. Kneser, however, obtained a valid proof of Dirichlet's principle.

see also DIRICHLET'S BOX PRINCIPLE, DIRICHLET INTEGRALS

Dirichlet Region

see VORONOI POLYGON

Dirichlet Series

A sum $\sum a_n e^{\lambda_n z}$, where a_n and z are COMPLEX and λ_n is REAL and MONOTONIC increasing.

see also DIRICHLET L-SERIES

Dirichlet Structure Constant

$$\kappa(d) = \begin{cases} \frac{2\ln\eta(d)}{\sqrt{d}} & \text{for } d > 0 \\ \frac{2\pi}{w(d)\sqrt{|d|}} & \text{for } d < 0, \end{cases}$$

where $\eta(d)$ is the FUNDAMENTAL UNIT and $w(d)$ is the number of substitutions which leave the binary quadratic form unchanged

$$w(d) = \begin{cases} 6 & \text{for } d = -3 \\ 4 & \text{for } d = -4 \\ 2 & \text{otherwise.} \end{cases}$$

see also CLASS NUMBER, DIRICHLET L-SERIES

References
Weisstein, E. W. "Class Numbers." http://www.astro.virginia.edu/~eww6n/math/notebooks/ClassNumbers.m.

Dirichlet Tessellation

see VORONOI DIAGRAM

Dirichlet's Test

Let

$$\left|\sum_{n=1}^{p} a_n\right| < K,$$

where K is independent of p. Then if $f_n \geq f_{n+1} > 0$ and

$$\lim_{n\to\infty} f_n = 0,$$

it follows that

$$\sum_{n=1}^{\infty} a_n f_n$$

CONVERGES.

see also CONVERGENCE TESTS

Dirichlet's Theorem

Given an ARITHMETIC SERIES of terms $an+b$, for $n = 1, 2, \ldots$, the series contains an infinite number of PRIMES if a and b are RELATIVELY PRIME, i.e., $(a,b) = 1$. Dirichlet proved this theorem using DIRICHLET L-SERIES.

see also PRIME ARITHMETIC PROGRESSION, PRIME PATTERNS CONJECTURE, RELATIVELY PRIME, SIERPIŃSKI'S PRIME SEQUENCE THEOREM

References
Courant, R. and Robbins, H. "Primes in Arithmetical Progressions." §1.2b in Supplement to Ch. 1 in *What is Mathematics?: An Elementary Approach to Ideas and Methods, 2nd ed.* Oxford, England: Oxford University Press, pp. 26–27, 1996.
Shanks, D. *Solved and Unsolved Problems in Number Theory, 4th ed.* New York: Chelsea, pp. 22–23, 1993.

Dirty Beam

The FOURIER TRANSFORM of the (u,v) sampling distribution in synthesis imaging

$$b' = \mathcal{F}^{-1}(S), \quad (1)$$

also called the SYNTHESIZED BEAM. It is called a "beam" by way of analogy with the DIRTY MAP

$$\begin{aligned} I' &= \mathcal{F}^{-1}(VS) = \mathcal{F}^{-1}[V] * \mathcal{F}^{-1}[S] \\ &= I * \mathcal{F}^{-1}(S) \equiv I * b', \end{aligned} \quad (2)$$

where $*$ denotes CONVOLUTION. Here, I' is the intensity which would be observed for an extended source by an antenna with response pattern b_1,

$$I' = b_1(\theta'') * I(\theta''). \quad (3)$$

The dirty beam is often a complicated function. In order to avoid introducing any high spatial frequency features when CLEANing, an elliptical Gaussian is usually fit to the dirty beam, producing a CLEAN BEAM which is CONVOLVED with the final iteration.

see also CLEAN ALGORITHM, CLEAN MAP, DIRTY MAP

Dirty Map

From the van Cittert-Zernicke theorem, the relationship between observed visibility function $V(u,v)$ and source brightness $I(\xi,\eta)$ in synthesis imaging is given by

$$\begin{aligned} I(\xi,\eta) &= \int_{-\infty}^{\infty}\int_{-\infty}^{\infty} V(u,v)e^{2\pi i(\xi u+\eta v)}\,du\,dv \\ &= \mathcal{F}^{-1}[V(u,v)]. \end{aligned} \quad (1)$$

But the visibility function is sampled only at discrete points $S(u,v)$ (finite sampling), so only an approximation to I, called the "dirty map" and denoted I', is measured. It is given by

$$\begin{aligned} I'(\xi,\eta) &= \int_{-\infty}^{\infty}\int_{-\infty}^{\infty} S(u,v)V(u,v)e^{2\pi i(\xi u+\eta v)}\,du\,dv \\ &= \mathcal{F}^{-1}[VS], \end{aligned} \quad (2)$$

where $S(u,v)$ is the sampling function and $V(u,v)$ is the observed visibility function. Let $*$ denote CONVOLUTION and rearrange the CONVOLUTION THEOREM,

$$\mathcal{F}[f * g] = \mathcal{F}[f]\mathcal{F}[g] \quad (3)$$

into the form

$$\mathcal{F}[\mathcal{F}^{-1}[f] * \mathcal{F}^{-1}[g]] = fg, \quad (4)$$

from which it follows that

$$\mathcal{F}^{-1}[f] * \mathcal{F}^{-1}[g] = \mathcal{F}^{-1}[fg]. \quad (5)$$

Now note that

$$I = \mathcal{F}^{-1}[V] \quad (6)$$

is the CLEAN MAP, and define the "DIRTY BEAM" as the inverse FOURIER TRANSFORM of the sampling function,

$$b' \equiv \mathcal{F}^{-1}[S]. \quad (7)$$

The dirty map is then given by

$$I' = \mathcal{F}^{-1}[VS] = \mathcal{F}^{-1}[V] * \mathcal{F}^{-1}[S] = I * b'. \quad (8)$$

In order to deconvolve the desired CLEAN MAP I from the measured dirty map I' and the known DIRTY BEAM b', the CLEAN ALGORITHM is often used.

see also CLEAN ALGORITHM, CLEAN MAP, DIRTY BEAM

Disc

see DISK

Disconnected Form

A FORM which is the sum of two FORMS involving separate sets of variables.

Disconnectivity

Disconnectivities are mathematical entities which stand in the way of a SPACE being contractible (i.e., shrunk to a point, where the shrinking takes place inside the SPACE itself). When dealing with TOPOLOGICAL SPACES, a disconnectivity is interpreted as a "HOLE" in the space. Disconnectivities in SPACE are studied through the EXTENSION PROBLEM or the LIFTING PROBLEM.

see also EXTENSION PROBLEM, HOLE, LIFTING PROBLEM

Discontinuity

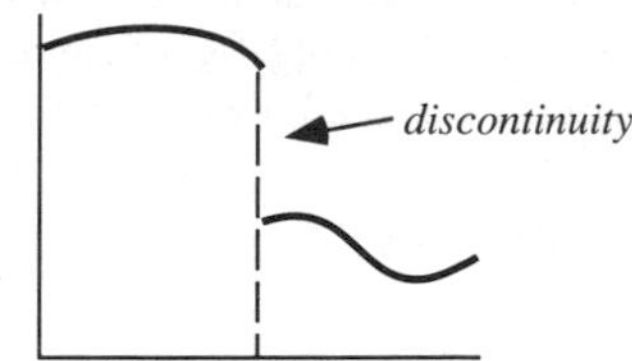

A point at which a mathematical object is DISCONTINUOUS.

Discontinuous

Not CONTINUOUS. A point at which a function is discontinuous is called a DISCONTINUITY, or sometimes a JUMP.

References

Yates, R. C. "Functions with Discontinuous Properties." *A Handbook on Curves and Their Properties.* Ann Arbor, MI: J. W. Edwards, pp. 100–107, 1952.

Discordant Permutation

see MARRIED COUPLES PROBLEM

Discrepancy Theorem

Let $s_1, s_2, \ldots$ be an infinite series of real numbers lying between 0 and 1. Then corresponding to any arbitrarily large K, there exists a positive integer n and two subintervals of equal length such that the number of s_ν with $\nu = 1, 2, \ldots, n$ which lie in one of the subintervals differs from the number of such s_ν that lie in the other subinterval by more than K (van der Corput 1935ab, van Aardenne-Ehrenfest 1945, 1949, Roth 1954).

This statement can be refined as follows. Let N be a large integer and $s_1, s_2, \ldots, s_N$ be a sequence of N real numbers lying between 0 and 1. Then for any integer $1 \leq n \leq N$ and any real number α satisfying $0 < \alpha < 1$, let $D_n(\alpha)$ denote the number of s_ν with $\nu = 1, 2, \ldots, n$ that satisfy $0 \leq s_\nu < \alpha$. Then there exist n and α such that

$$|D_n(\alpha) - n\alpha| > c_1 \frac{\ln\ln N}{\ln\ln\ln N}$$

where c_1 is a positive constant.

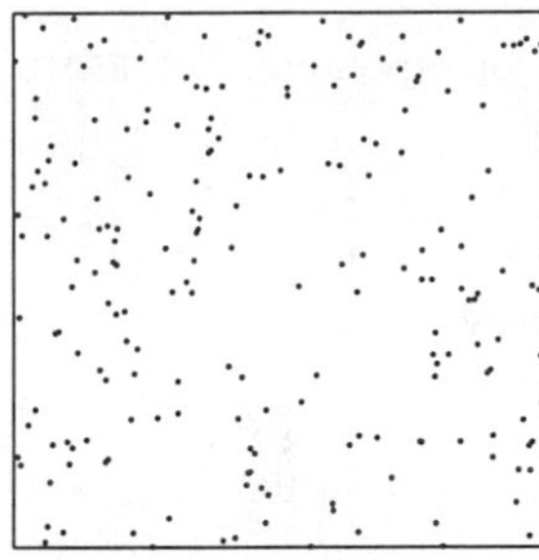

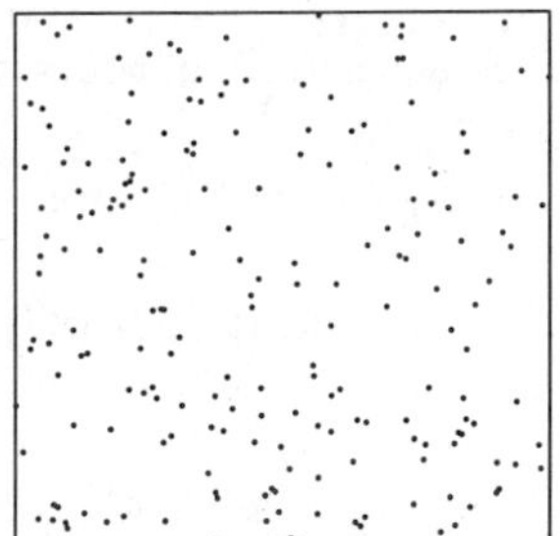

This result can be further strengthened, which is most easily done by reformulating the problem. Let $N > 1$ be an integer and $P_1, P_2, \ldots, P_N$ be N (not necessarily distinct) points in the square $0 \leq x \leq 1$, $0 \leq y \leq 1$. Then

$$\int_0^1 \int_0^1 [S(x,y) - Nxy]^2 \, dx \, dy > c_2 \ln N,$$

where c_2 is a positive constant and $S(u,v)$ is the number of points in the rectangle $0 \leq x < u$, $0 \leq y < v$ (Roth 1954). Therefore,

$$|S(x,y) - Nxy| > c_3 \sqrt{\ln N},$$

and the original result can be stated as the fact that there exist n and α such that

$$|D_n(\alpha) - n\alpha| > c_4 \sqrt{\ln N}.$$

The randomly distributed points shown in the above squares have $|S(x,y) - Nxy|^2 = 6.40$ and 9.11, respectively.

Similarly, the discrepancy of a set of N points in a unit d-HYPERCUBE satisfies

$$|S(x,y) - Nxy| > c(\ln N)^{(d-1)/2}$$

(Roth 1954, 1976, 1979, 1980).

see also 18-POINT PROBLEM, CUBE POINT PICKING

References

Berlekamp, E. R. and Graham, R. L. "Irregularities in the Distributions of Finite Sequences." *J. Number Th.* **2**, 152–161, 1970.

Roth, K. F. "On Irregularities of Distribution." *Mathematika* **1**, 73–79, 1954.

Roth, K. F. "On Irregularities of Distribution. II." *Comm. Pure Appl. Math.* **29**, 739–744, 1976.

Roth, K. F. "On Irregularities of Distribution. III." *Acta Arith.* **35**, 373–384, 1979.

Roth, K. F. "On Irregularities of Distribution. IV." *Acta Arith.* **37**, 67–75, 1980

van Aardenne-Ehrenfest, T. "Proof of the Impossibility of a Just Distribution of an Infinite Sequence Over an Interval." *Proc. Kon. Ned. Akad. Wetensch.* **48**, 3–8, 1945.

van Aardenne-Ehrenfest, T. *Proc. Kon. Ned. Akad. Wetensch.* **52**, 734–739, 1949.

van der Corput, J. G. *Proc. Kon. Ned. Akad. Wetensch.* **38**, 813–821, 1935a.

van der Corput, J. G. *Proc. Kon. Ned. Akad. Wetensch.* **38**, 1058–1066, 1935b.

Discrete Distribution

A DISTRIBUTION whose variables can take on only discrete values. Abramowitz and Stegun (1972, p. 929) give a table of the parameters of most common discrete distributions.

see also BERNOULLI DISTRIBUTION, BINOMIAL DISTRIBUTION, CONTINUOUS DISTRIBUTION, DISTRIBUTION, GEOMETRIC DISTRIBUTION, HYPERGEOMETRIC DISTRIBUTION, NEGATIVE BINOMIAL DISTRIBUTION, POISSON DISTRIBUTION, PROBABILITY, STATISTICS, UNIFORM DISTRIBUTION

References

Abramowitz, M. and Stegun, C. A. (Eds.). *Handbook of Mathematical Functions with Formulas, Graphs, and Mathematical Tables, 9th printing.* New York: Dover, pp. 927 and 929, 1972.

Discrete Fourier Transform

The FOURIER TRANSFORM is defined as

$$f(\nu) = \mathcal{F}[f(t)] = \int_{-\infty}^{\infty} f(t) e^{-2\pi i \nu t} \, dt. \tag{1}$$

Now consider generalization to the case of a discrete function, $f(t) \to f(t_k)$ by letting $f_k \equiv f(t_k)$, where $t_k \equiv k\Delta$, with $k = 0, \ldots, N-1$. Choose the frequency step such that

$$\nu_n = \frac{n}{N\Delta}, \tag{2}$$

with $n = -N/2, \ldots, 0, \ldots, N/2$. There are $N+1$ values of n, so there is one relationship between the frequency components. Writing this out as per Press *et al.* (1989)

$$\mathcal{F}[f(t)] = \sum_{k=0}^{N-1} f_k e^{-2\pi i (n/N\Delta) k\Delta} \Delta = \Delta \sum_{k=0}^{N-1} f_k e^{-2\pi i n k/N}, \tag{3}$$

and

$$F_n \equiv \sum_{k=0}^{N-1} f_k e^{-2\pi ink/N}. \quad (4)$$

The inverse transform is

$$f_k = \frac{1}{N}\sum_{n=0}^{N-1} F_n e^{2\pi ink/N}. \quad (5)$$

Note that $F_{-n} = F_{N-n}$, $n = 1, 2, \ldots$, so an alternate formulation is

$$\nu_n = \frac{n}{N\Delta}, \quad (6)$$

where the NEGATIVE frequencies $-\nu_c < \nu < 0$ have $N/2+1 \leq n \leq N-1$, POSITIVE frequencies $0 < \nu < \nu_c$ have $1 \leq n \leq N/2-1$, with zero frequency $n = 0$. $n = N/2$ corresponds to both $\nu = \nu_c$ and $\nu = -\nu_c$. The discrete Fourier transform can be computed using a FAST FOURIER TRANSFORM.

The discrete Fourier transform is a special case of the z-TRANSFORM.

see also FAST FOURIER TRANSFORM, FOURIER TRANSFORM, HARTLEY TRANSFORM, WINOGRAD TRANSFORM, z-TRANSFORM

References

Arfken, G. "Discrete Orthogonality—Discrete Fourier Transform." §14.6 in *Mathematical Methods for Physicists, 3rd ed.* Orlando, FL: Academic Press, pp. 787–792, 1985.

Press, W. H.; Flannery, B. P.; Teukolsky, S. A.; and Vetterling, W. T. "Fourier Transform of Discretely Sampled Data." §12.1 in *Numerical Recipes in C: The Art of Scientific Computing.* Cambridge, England: Cambridge University Press, pp. 494–498, 1989.

Discrete Mathematics

The branch of mathematics dealing with objects which can assume only certain "discrete" values. Discrete objects can be characterized by INTEGERS (or RATIONAL NUMBERS), whereas continuous objects require REAL NUMBERS. The study of how discrete objects combine with one another and the probabilities of various outcomes is known as COMBINATORICS.

see also COMBINATORICS

References

Balakrishnan, V. K. *Introductory Discrete Mathematics.* New York: Dover, 1997.

Bobrow, L. S. and Arbib, M. A. *Discrete Mathematics: Applied Algebra for Computer and Information Science.* Philadelphia, PA: Saunders, 1974.

Dossey, J. A.; Otto, A. D.; Spence, L.; and Eynden, C. V. *Discrete Mathematics, 3rd ed.* Reading, MA: Addison-Wesley, 1997.

Skiena, S. S. *Implementing Discrete Mathematics.* Reading, MA: Addison-Wesley, 1990.

Discrete Set

A finite SET or an infinitely COUNTABLE SET of elements.

Discrete Uniform Distribution

see EQUALLY LIKELY OUTCOMES DISTRIBUTION

Discriminant

A discriminant is a quantity (usually invariant under certain classes of transformations) which characterizes certain properties of a quantity's ROOTS. The concept of the discriminant is used for BINARY QUADRATIC FORMS, ELLIPTIC CURVES, METRICS, MODULES, POLYNOMIALS, QUADRATIC CURVES, QUADRATIC FIELDS, QUADRATIC FORMS, and in the SECOND DERIVATIVE TEST.

Discriminant (Binary Quadratic Form)

The discriminant of a BINARY QUADRATIC FORM

$$au^2 + buv + cv^2$$

is defined by

$$d \equiv b^2 - 4ac.$$

It is equal to four times the corresponding DETERMINANT.

see also CLASS NUMBER

Discriminant (Elliptic Curve)

An ELLIPTIC CURVE is of the form

$$y^2 = x^3 + a_2x^2 + a_1x + a_0.$$

Let the ROOTS of y^2 be r_1, r_2, and r_3. The discriminant is then defined as

$$\Delta = k(r_1 - r_2)^2(r_1 - r_3)^2(r_2 - r_3)^2.$$

see also FREY CURVE, MINIMAL DISCRIMINANT

Discriminant (Metric)

Given a METRIC $g_{\alpha\beta}$, the discriminant is defined by

$$g \equiv \det(g_{\alpha\beta}) = \begin{vmatrix} g_{11} & g_{12} \\ g_{21} & g_{22} \end{vmatrix} = g_{11}g_{22} - (g_{12})^2. \quad (1)$$

Let g be the discriminant and $\bar{g}$ the transformed discriminant, then

$$\bar{g} = D^2 g \quad (2)$$

$$g = \bar{D}^2 \bar{g}, \quad (3)$$

where

$$D \equiv \frac{\partial(u^1, u^2)}{\partial(\bar{u}^1, \bar{u}^2)} = \begin{vmatrix} \frac{\partial u^1}{\partial \bar{u}^1} & \frac{\partial u^1}{\partial \bar{u}^2} \\ \frac{\partial u^2}{\partial \bar{u}^1} & \frac{\partial u^2}{\partial \bar{u}^2} \end{vmatrix} \quad (4)$$

$$\bar{D} \equiv \frac{\partial(\bar{u}^1, \bar{u}^2)}{\partial(u^1, u^2)} = \begin{vmatrix} \frac{\partial \bar{u}^1}{\partial u^1} & \frac{\partial \bar{u}^1}{\partial u^2} \\ \frac{\partial \bar{u}^2}{\partial u^1} & \frac{\partial \bar{u}^2}{\partial u^2} \end{vmatrix}. \quad (5)$$

Discriminant (Module)

Let a MODULE M in an INTEGRAL DOMAIN D_1 for $R(\sqrt{D})$ be expressed using a two-element basis as

$$M = [\xi_1, \xi_2],$$

where ξ_1 and ξ_2 are in D_1. Then the DIFFERENT of the MODULE is defined as

$$\Delta = \Delta(M) = \begin{vmatrix} \xi_1 & \xi_2 \\ \xi_1' & \xi_2' \end{vmatrix} = \xi_1\xi_2' - \xi_1'\xi_2$$

and the discriminant is defined as the square of the DIFFERENT (Cohn 1980).

For IMAGINARY QUADRATIC FIELDS $\mathbb{Q}(\sqrt{n})$ (with $n < 0$), the discriminants are given in the following table.

-1	-2^2	-33	$-2^2 \cdot 3 \cdot 11$	-67	-67
-2	-2^3	-34	$-2^3 \cdot 17$	-69	$-2^2 \cdot 3 \cdot 23$
-3	-3	-35	$-5 \cdot 7$	-70	$-2^3 \cdot 5 \cdot 7$
-5	$-2^2 \cdot 5$	-37	$-2^2 \cdot 37$	-71	-71
-6	$-2^3 \cdot 3$	-39	$-3 \cdot 13$	-73	$-2^2 \cdot 73$
-7	-7	-41	$-2^2 \cdot 41$	-74	$-2^3 \cdot 37$
-10	$-2^3 \cdot 5$	-42	$-2^3 \cdot 3 \cdot 7$	-77	$-2^2 \cdot 7 \cdot 11$
-11	-11	-43	-43	-78	$-2^3 \cdot 3 \cdot 13$
-13	$-2^2 \cdot 13$	-46	$-2^3 \cdot 23$	-79	-79
-14	$-2^3 \cdot 7$	-47	-47	-82	$-2^3 \cdot 41$
-15	$-3 \cdot 5$	-51	$-3 \cdot 17$	-83	-83
-17	$-2^2 \cdot 17$	-53	$-2^2 \cdot 53$	-85	$-2^2 \cdot 5 \cdot 17$
-19	-19	-55	$-5 \cdot 11$	-86	$-2^3 \cdot 43$
-21	$-2^2 \cdot 3 \cdot 7$	-57	$-2^2 \cdot 3 \cdot 19$	-87	$-3 \cdot 29$
-22	$-2^3 \cdot 11$	-58	$-2^3 \cdot 29$	-89	$-2^2 \cdot 89$
-23	-23	-59	-59	-91	$-7 \cdot 13$
-26	$-2^3 \cdot 13$	-61	$-2^2 \cdot 61$	-93	$-2^2 \cdot 3 \cdot 31$
-29	$-2^2 \cdot 29$	-62	$-2^3 \cdot 31$	-94	$-2^3 \cdot 47$
-30	$-2^3 \cdot 3 \cdot 5$	-65	$-2^2 \cdot 5 \cdot 13$	-95	$-5 \cdot 19$
-31	-31	-66	$-2^3 \cdot 3 \cdot 11$	-97	$-2^2 \cdot 97$

The discriminants of REAL QUADRATIC FIELDS $\mathbb{Q}(\sqrt{n})$ ($n > 0$) are given in the following table.

2	2^3	34	$2^3 \cdot 17$	67	$67 \cdot 2^2$
3	$3 \cdot 2^2$	35	$7 \cdot 2^2 \cdot 5$	69	$3 \cdot 23$
5	5	37	37	70	$7 \cdot 2^3 \cdot 5$
6	$3 \cdot 2^3$	38	$19 \cdot 2^3$	71	$71 \cdot 2^2$
7	$7 \cdot 2^2$	39	$3 \cdot 2^2 \cdot 13$	73	73
10	$2^3 \cdot 5$	41	41	74	$2^3 \cdot 37$
11	$11 \cdot 2^2$	42	$3 \cdot 2^3 \cdot 7$	77	$7 \cdot 11$
13	13	43	$43 \cdot 2^2$	78	$3 \cdot 2^3 \cdot 13$
14	$7 \cdot 2^3$	46	$23 \cdot 2^3$	79	$79 \cdot 2^2$
15	$3 \cdot 2^2 \cdot 5$	47	$47 \cdot 2^2$	82	$2^3 \cdot 41$
17	17	51	$3 \cdot 2^2 \cdot 17$	83	$83 \cdot 2^2$
19	$19 \cdot 2^2$	53	53	85	$5 \cdot 17$
21	$3 \cdot 7$	55	$11 \cdot 2^2 \cdot 5$	86	$43 \cdot 2^3$
22	$11 \cdot 2^3$	57	$3 \cdot 19$	87	$3 \cdot 2^2 \cdot 13$
23	$23 \cdot 2^2$	58	$2^3 \cdot 29$	89	89
26	$2^3 \cdot 13$	59	$59 \cdot 2^2$	91	$7 \cdot 2^2 \cdot 13$
29	29	61	61	93	$3 \cdot 31$
30	$3 \cdot 2^3 \cdot 5$	62	$31 \cdot 2^3$	94	$47 \cdot 2^3$
31	$31 \cdot 2^2$	65	$5 \cdot 13$	95	$19 \cdot 2^2 \cdot 5$
33	$3 \cdot 11$	66	$3 \cdot 2^3 \cdot 11$	97	97

see also DIFFERENT, FUNDAMENTAL DISCRIMINANT, MODULE

References

Cohn, H. *Advanced Number Theory.* New York: Dover, pp. 72–73 and 261–274, 1980.

Discriminant (Polynomial)

The PRODUCT of the SQUARES of the differences of the POLYNOMIAL ROOTS x_i. For a POLYNOMIAL of degree n,

$$D_n \equiv \prod_{\substack{i,j \\ i<j}}^{n} (x_i - x_j)^2. \tag{1}$$

The discriminant of the QUADRATIC EQUATION

$$ax^2 + bx + c = 0 \tag{2}$$

is usually taken as

$$D = b^2 - 4ac. \tag{3}$$

However, using the general definition of the POLYNOMIAL DISCRIMINANT gives

$$D \equiv \prod_{i<j} (z_i - z_j)^2 = \frac{b^2 - 4ac}{a^2}, \tag{4}$$

where z_i are the ROOTS.

The discriminant of the CUBIC EQUATION

$$z^3 + a_2 z^2 + a_1 z + a_0 = 0 \tag{5}$$

is commonly defined as

$$D \equiv Q^3 + R^2, \tag{6}$$

where

$$Q \equiv \frac{3a_1 - a_2{}^2}{9} \tag{7}$$

$$R \equiv \frac{9a_2 a_1 - 27a_0 - 2a_2{}^3}{54}. \tag{8}$$

However, using the general definition of the polynomial discriminant for the standard form CUBIC EQUATION

$$z^3 + pz = q \tag{9}$$

gives

$$D \equiv \prod_{i<j} (z_i - z_j)^2 = P^2 = -4p^3 - 27q^2, \tag{10}$$

where z_i are the ROOTS and

$$P = (z_1 - z_2)(z_2 - z_3)(z_1 - z_3). \tag{11}$$

The discriminant of a QUARTIC EQUATION

$$x^4 + a_3x^3 + a_2x^2 + a_1x + a_0 = 0 \quad (12)$$

is

$$-27a_1^4 + 18a_3a_2a_1^3 - 4a_3^3a_1^3 - 4a_2^3a_1^2 + a_3^2a_2^2a_1^2$$
$$+a_0(144a_2a_1^2 - 6a_3^2a_1^2 - 80a_3a_2^2a_1 + 18a_3^3a_2a_1 + 16a_2^4$$
$$-4a_3^2a_2^3) + a_0^2(-192a_3a_1 - 128a_2^2 + 144a_3^2a_2 - 27a_3^4)$$
$$-256a_0^3 \quad (13)$$

(Beeler *et al.* 1972, Item 4).

see also RESULTANT

References

Beeler, M.; Gosper, R. W.; and Schroeppel, R. *HAKMEM.* Cambridge, MA: MIT Artificial Intelligence Laboratory, Memo AIM-239, Feb. 1972.

Discriminant (Quadratic Curve)

Given a general QUADRATIC CURVE

$$Ax^2 + Bxy + Cy^2 + Dx + Ey + F = 0, \quad (1)$$

the quantity X is known as the discriminant, where

$$X \equiv B^2 - 4AC, \quad (2)$$

and is invariant under ROTATION. Using the COEFFICIENTS from QUADRATIC EQUATIONS for a rotation by an angle θ,

$$A' = \tfrac{1}{2}A[1+\cos(2\theta)] + \tfrac{1}{2}B\sin(2\theta) + \tfrac{1}{2}C[1-\cos(2\theta)]$$
$$= \frac{A+C}{2} + \frac{B}{2}\sin(2\theta) + \frac{A-C}{2}\cos(2\theta) \quad (3)$$
$$B' = G\cos\left(2\theta + \delta - \frac{\pi}{2}\right) = G\sin(2\theta+\delta) \quad (4)$$
$$C' = \tfrac{1}{2}A[1-\cos(2\theta)] - \tfrac{1}{2}B\sin(2\theta) + \tfrac{1}{2})C[1+\cos(2\theta)]$$
$$= \frac{A+C}{2} - \frac{B}{2}\sin(2\theta) + \frac{C-A}{2}\cos(2\theta). \quad (5)$$

Now let

$$G \equiv \sqrt{B^2 + (A-C)^2} \quad (6)$$
$$\delta \equiv \tan^{-1}\left(\frac{B}{C-A}\right) \quad (7)$$
$$\delta_2 \equiv \tan^{-1}\left(\frac{A-C}{B}\right) = -\cot^{-1}\left(\frac{B}{C-A}\right), \quad (8)$$

and use

$$\cot^{-1}(x) = \tfrac{1}{2}\pi - \tan^{-1}(x) \quad (9)$$
$$\delta_2 = \delta - \tfrac{1}{2}\pi \quad (10)$$

to rewrite the primed variables

$$A' = \frac{A+C}{2} + \tfrac{1}{2}G\cos(2\theta+\delta) \quad (11)$$
$$B' = B\cos(2\theta) + (C-A)\sin(2\theta) = G\cos(2\theta+\delta_2) \quad (12)$$
$$C' = \frac{A+C}{2} - \tfrac{1}{2}G\cos(2\theta+\delta). \quad (13)$$

From (11) and (13), it follows that

$$4A'C' = (A+C)^2 - G^2\cos(2\theta+\delta). \quad (14)$$

Combining with (12) yields, for an arbitrary θ

$$X \equiv B'^2 - 4A'C'$$
$$= G^2\sin^2(2\theta+\delta) + G^2\cos^2(2\theta+\delta) - (A+C)^2$$
$$= G^2 - (A+C)^2 = B^2 + (A-C)^2 - (A+C)^2$$
$$= B^2 - 4AC, \quad (15)$$

which is therefore invariant under rotation. This invariant therefore provides a useful shortcut to determining the shape represented by a QUADRATIC CURVE. Choosing θ to make $B' = 0$ (see QUADRATIC EQUATION), the curve takes on the form

$$A'x^2 + C'y^2 + D'x + E'y + F = 0. \quad (16)$$

COMPLETING THE SQUARE and defining new variables gives

$$A'x'^2 + C'y'^2 = H. \quad (17)$$

Without loss of generality, take the sign of H to be positive. The discriminant is

$$X = B'^2 - 4A'C' = -4A'C'. \quad (18)$$

Now, if $-4A'C' < 0$, then A' and C' both have the same sign, and the equation has the general form of an ELLIPSE (if A' and B' are positive). If $-4A'C' > 0$, then A' and C' have opposite signs, and the equation has the general form of a HYPERBOLA. If $-4A'C' = 0$, then either A' or C' is zero, and the equation has the general form of a PARABOLA (if the NONZERO A' or C' is positive). Since the discriminant is invariant, these conclusions will also hold for an arbitrary choice of θ, so they also hold when $-4A'C'$ is replaced by the original $B^2 - 4AC$. The general result is

1. If $B^2 - 4AC < 0$, the equation represents an ELLIPSE, a CIRCLE (degenerate ELLIPSE), a POINT (degenerate CIRCLE), or has no graph.
2. If $B^2 - 4AC > 0$, the equation represents a HYPERBOLA or pair of intersecting lines (degenerate HYPERBOLA).
3. If $B^2 - 4AC = 0$, the equation represents a PARABOLA, a LINE (degenerate PARABOLA), a pair of PARALLEL lines (degenerate PARABOLA), or has no graph.

Discriminant (Quadratic Form)

see DISCRIMINANT (BINARY QUADRATIC FORM)

Discriminant (Second Derivative Test)

$$D \equiv f_{xx}f_{yy} - f_{xy}f_{yx} = f_{xx}f_{yy} - {f_{xy}}^2,$$

where f_{ij} are PARTIAL DERIVATIVES.

see also SECOND DERIVATIVE TEST

Disdyakis Dodecahedron

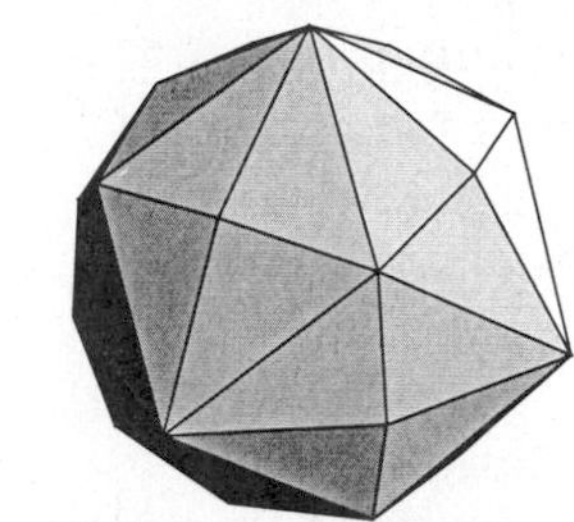

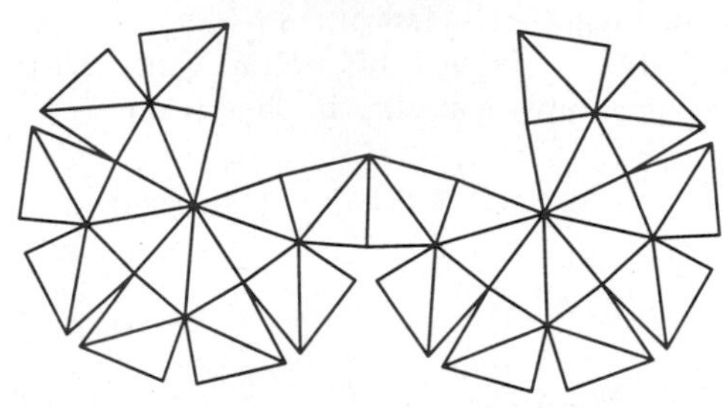

The DUAL POLYHEDRON of the ARCHIMEDEAN GREAT RHOMBICUBOCTAHEDRON, also called the HEXAKIS OCTAHEDRON.

see also GREAT DISDYAKIS DODECAHEDRON

Disdyakis Triacontahedron

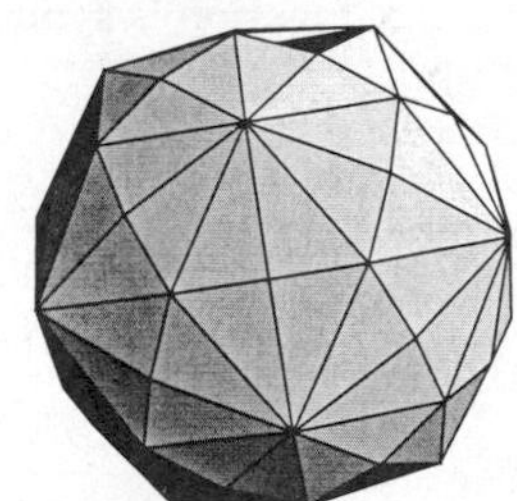

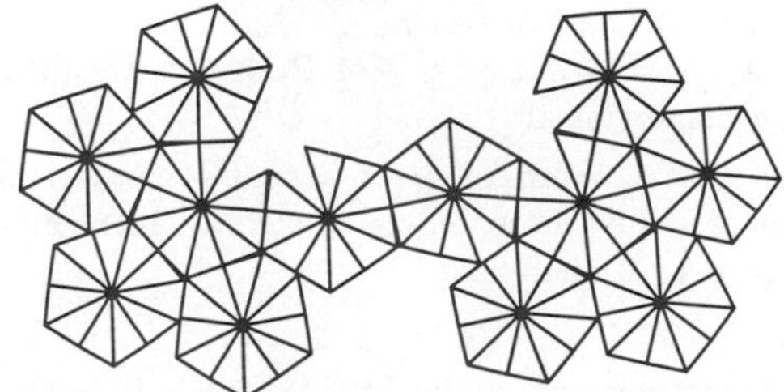

The DUAL POLYHEDRON of the ARCHIMEDEAN GREAT RHOMBICOSIDODECAHEDRON. It is also called the HEXAKIS ICOSAHEDRON.

Disjoint

see MUTUALLY EXCLUSIVE

Disjunction

A product of ORs, denoted

$$\bigvee_{k+1}^{n} A_k.$$

see also CONJUNCTION, OR

Disjunctive Game

see NIM-HEAP

Disk

An n-D disk (or DISC) of RADIUS r is the collection of points of distance $\leq r$ (CLOSED DISK) or $< r$ (OPEN DISK) from a fixed point in EUCLIDEAN n-space. A disk is the SHADOW of a BALL on a PLANE PERPENDICULAR to the BALL-RADIANT POINT line.

The n-disk for $n \geq 3$ is called a BALL, and the boundary of the n-disk is a $(n-1)$-HYPERSPHERE. The standard n-disk, denoted $\mathbb{D}^n$ (or $\mathbb{B}^n$), has its center at the ORIGIN and has RADIUS $r = 1$.

see also BALL, CLOSED DISK, DISK COVERING PROBLEM, FIVE DISKS PROBLEM, HYPERSPHERE, MERGELYAN-WESLER THEOREM, OPEN DISK, POLYDISK, SPHERE, UNIT DISK

Disk Covering Problem

N.B. A detailed on-line essay by S. Finch was the starting point for this entry.

Given a UNIT DISK, find the smallest RADIUS $r(n)$ required for n equal disks to completely cover the UNIT DISK. For a symmetrical arrangement with $n = 5$ (the FIVE DISKS PROBLEM), $r(5) = \phi - 1 = 1/\phi = 0.6180340\ldots$, where ϕ is the GOLDEN RATIO. However, the radius can be reduced in the general disk covering problem where symmetry is not required. The first few such values are

$$\begin{aligned} r(1) &= 1 \\ r(2) &= 1 \\ r(3) &= \tfrac{1}{2}\sqrt{3} \\ r(4) &= \tfrac{1}{2}\sqrt{2} \\ r(5) &= 0.609382864\ldots \\ r(6) &= 0.555 \\ r(7) &= \tfrac{1}{2} \\ r(8) &= 0.437 \\ r(9) &= 0.422 \\ r(10) &= 0.398. \end{aligned}$$

Here, values for $n = 6, 8, 9, 10$ were obtained using computer experimentation by Zahn (1962). The value $r(5)$ is equal to $\cos(\theta+\phi/2)$, where θ and ϕ are solutions to

$$2\sin\theta - \sin(\theta + \tfrac{1}{2}\phi + \psi) - \sin(\psi - \theta - \tfrac{1}{2}\phi) = 0 \quad (1)$$

$$2\sin\phi - \sin(\theta + \tfrac{1}{2}\phi + \chi) - \sin(\chi - \theta - \tfrac{1}{2}\phi) = 0 \quad (2)$$

$$2\sin\theta + \sin(\chi + \theta) - \sin(\chi - \theta) - \sin(\psi + \phi) - \sin(\psi - \phi) - 2\sin(\psi - 2\theta) = 0 \quad (3)$$

$$\cos(2\psi - \chi + \phi) - \cos(2\psi + \chi - \phi) - 2\cos\chi + \cos(2\psi + \chi - 2\theta) + \cos(2\psi - \chi - 2\theta) = 0 \quad (4)$$

(Neville 1915). It is also given by $1/x$, where x is the largest real root of

$$a(y)x^6 - b(y)x^5 + c(y)x^4 - d(y)x^3 + e(y)x^2 - f(y)x + g(y) = 0 \quad (5)$$

maximized over all y, subject to the constraints

$$\sqrt{2} < x < 2y + 1 \quad (6)$$

$$-1 < y < 1, \quad (7)$$

and with

$$a(y) = 80y^2 + 64y \quad (8)$$

$$b(y) = 416y^3 + 384y^2 + 64y \quad (9)$$

$$c(y) = 848y^4 + 928y^3 + 352y^2 + 32y \quad (10)$$

$$d(y) = 768y^5 + 992y^4 + 736y^3 + 288y^2 + 96y$$

$$e(y) = 256y^6 + 384y^5 + 592y^4 + 480y^3 + 336y^2 + 96y + 16 \quad (11)$$

$$f(y) = 128y^5 + 192y^4 + 256y^3 + 160y^2 + 96y + 32 \quad (12)$$

$$g(y) = 64y^2 + 64y + 16 \quad (13)$$

(Bezdek 1983, 1984).

Letting $N(\epsilon)$ be the smallest number of DISKS of RADIUS ϵ needed to cover a disk D, the limit of the ratio of the AREA of D to the AREA of the disks is given by

$$\lim_{\epsilon\to 0^+} \frac{1}{\epsilon^2 N(\epsilon)} = \frac{3\sqrt{3}}{2\pi} \quad (14)$$

(Kershner 1939, Verblunsky 1949).

see also FIVE DISKS PROBLEM

References

Ball, W. W. R. and Coxeter, H. S. M. "The Five-Disc Problem." In *Mathematical Recreations and Essays, 13th ed.* New York: Dover, pp. 97–99, 1987.

Bezdek, K. "Über einige Kreisüberdeckungen." *Beiträge Algebra Geom.* **14**, 7–13, 1983.

Bezdek, K. "Über einige optimale Konfigurationen von Kreisen." *Ann. Univ. Sci. Budapest Eötvös Sect. Math.* **27**, 141–151, 1984.

Finch, S. "Favorite Mathematical Constants." `http://www.mathsoft.com/asolve/constant/circle/circle.html`.

Kershner, R. "The Number of Circles Covering a Set." *Amer. J. Math.* **61**, 665–671, 1939.

Neville, E. H. "On the Solution of Numerical Functional Equations, Illustrated by an Account of a Popular Puzzle and of its Solution." *Proc. London Math. Soc.* **14**, 308–326, 1915.

Verblunsky, S. "On the Least Number of Unit Circles which Can Cover a Square." *J. London Math. Soc.* **24**, 164–170, 1949.

Zahn, C. T. "Black Box Maximization of Circular Coverage." *J. Res. Nat. Bur. Stand. B* **66**, 181–216, 1962.

Disk Lattice Points

see GAUSS'S CIRCLE PROBLEM

Dispersion Numbers

see MAGIC GEOMETRIC CONSTANTS

Dispersion Relation

Any pair of equations giving the REAL PART of a function as an integral of its IMAGINARY PART and the IMAGINARY PART as an integral of its REAL PART. Dispersion relationships imply causality in physics. Let

$$f(x_0) \equiv u(x_0) + iv(x_0), \quad (1)$$

then

$$u(x_0) = \frac{1}{\pi} PV \int_{-\infty}^{\infty} \frac{v(x)\,dx}{x - x_0} \quad (2)$$

$$v(x_0) = -\frac{1}{\pi} PV \int_{-\infty}^{\infty} \frac{u(x)\,dx}{x - x_0}, \quad (3)$$

where PV denotes the CAUCHY PRINCIPAL VALUE and $u(x_0)$ and $v(x_0)$ are HILBERT TRANSFORMS of each other. If the COMPLEX function is symmetric such that $f(-x) = f^*(x)$, then

$$u(x_0) = \frac{2}{\pi} PV \int_0^{\infty} \frac{xv(x)\,dx}{x^2 - {x_0}^2} \quad (4)$$

$$v(x_0) = -\frac{2}{\pi} PV \int_0^{\infty} \frac{xu(x)\,dx}{x^2 - {x_0}^2}. \quad (5)$$

Dispersion (Sequence)

An array $B = b_{ij}$, $i, j \geq 1$ of POSITIVE INTEGERS is called a dispersion if

1. The first column of B is a strictly increasing sequence, and there exists a strictly increasing sequence $\{s_k\}$ such that
2. $b_{12} = s_1 \geq 2$,
3. The complement of the SET $\{b_{i1} : i \geq 1\}$ is the SET $\{s_k\}$,
4. $b_{ij} = s_{b_{i,j-1}}$ for all $j \geq 3$ for $i = 1$ and for all $g \geq 2$ for all $i \geq 2$.

If an array $B = b_{ij}$ is a dispersion, then it is an INTERSPERSION.

see also INTERSPERSION

References

Kimberling, C. "Interspersions and Dispersions." *Proc. Amer. Math. Soc.* **117**, 313–321, 1993.

Dispersion (Statistics)

$$(\Delta u)^2{}_i \equiv (u_i - \bar{u})^2.$$

see also ABSOLUTE DEVIATION, SIGNED DEVIATION, VARIANCE

Disphenocingulum

see JOHNSON SOLID

Disphenoid

A TETRAHEDRON with identical ISOSCELES or SCALENE faces.

Dissection

Any two rectilinear figures with equal AREA can be dissected into a finite number of pieces to form each other. This is the WALLACE-BOLYAI-GERWEIN THEOREM. For minimal dissections of a TRIANGLE, PENTAGON, and OCTAGON into a SQUARE, see Stewart (1987, pp. 169–170) and Ball and Coxeter (1987, pp. 89–91). The TRIANGLE to SQUARE dissection (HABERDASHER'S PROBLEM) is particularly interesting because it can be built from hinged pieces which can be folded and unfolded to yield the two shapes (Gardner 1961; Stewart 1987, p. 169; Pappas 1989).

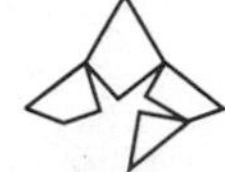
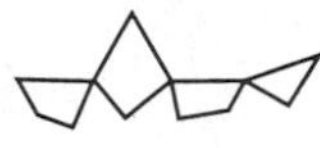

Laczkovich (1988) proved that the CIRCLE can be squared in a finite number of dissections ($\sim 10^{50}$). Furthermore, any shape whose boundary is composed of smoothly curving pieces can be dissected into a SQUARE.

The situation becomes considerably more difficult moving from 2-D to 3-D. In general, a POLYHEDRON cannot be dissected into other POLYHEDRA of a specified type. A CUBE *can* be dissected into n^3 CUBES, where n is any INTEGER. In 1900, Dehn proved that not every PRISM cannot be dissected into a TETRAHEDRON (Lenhard 1962, Ball and Coxeter 1987) The third of HILBERT'S PROBLEMS asks for the determination of two TETRAHEDRA which cannot be decomposed into congruent TETRAHEDRA directly or by adjoining congruent TETRAHEDRA. Max Dehn showed this could not be done in 1902, and W. F. Kagon obtained the same result independently in 1903. A quantity growing out of Dehn's work which can be used to analyze the possibility of performing a given solid dissection is the DEHN INVARIANT.

The table below is an updated version of the one given in Gardner (1991, p. 50). Many of the improvements are due to G. Theobald (Frederickson 1997). The minimum number of pieces known to dissect a regular n-gon (where n is a number in the first column) into a k-gon (where k is a number is the bottom row) is read off by the intersection of the corresponding row and column. In the table, $\{n\}$ denotes a regular n-gon, GR a GOLDEN RECTANGLE, GC a GREEK CROSS, LC a LATIN CROSS, MC a MALTESE CROSS, SW a SWASTIKA, {5/2} a five-point star (solid PENTAGRAM), {6/2} a six-point star (i.e., HEXAGRAM or solid STAR OF DAVID), and {8/3} the solid OCTAGRAM.

	{3}	{4}	{5}	{6}	{7}	{8}	{9}	{10}	{12}
{4}	4								
{5}	6	6							
{6}	5	5	7						
{7}	8	7	9	8					
{8}	7	5	9	8	11				
{9}	8	9	12	11	14	13			
{10}	7	7	10	9	11	10	13		
{12}	8	6	10	6	11	10	14	12	
GR	4	3	6	5	7	6	9	6	7
GC	5	4	7	7	9	9	12	10	6
LC	5	5	8	6	8	8	11	10	7
MC		7		14					
SW		6		12					
{5/2}	7	7	9	9	11	10	14	6	12
{6/2}	5	5	8	6	9	8	11	9	9
{8/3}	8	8	9	9	12	6	13	12	12

	GR	GC	LC	MC	SW	{5/2}	{6/2}
GC	5						
LC	5	7					
MC		8					
SW		8	9				
{5/2}	7	12	10	10			
{6/2}	5	8	8			11	
{8/3}	7	10	11			13	10

The best-known dissections of one regular convex n-gon into another are shown for $n = 3, 4, 5, 6, 7, 8, 9, 10$, and 12 in the following illustrations due to Theobald.

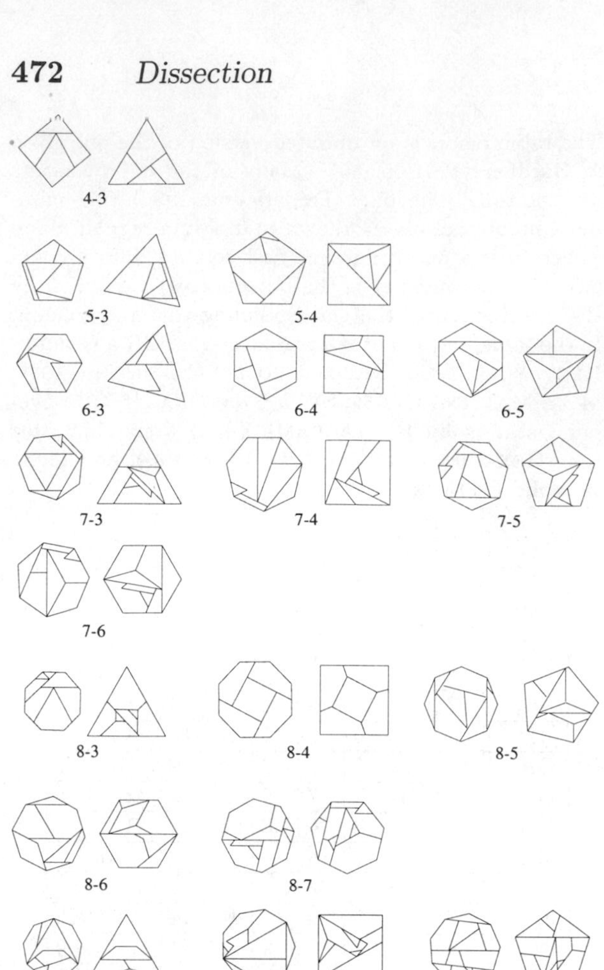

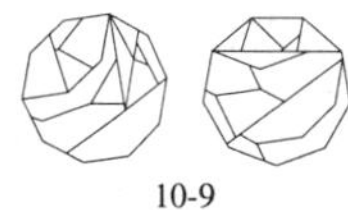

12-4 12-5

12-6 12-7 12-8

12-9 12-10

The best-known dissections of regular concave polygons are illustrated below for {5/2}, {6/2}, and {8/3} (Theobald).

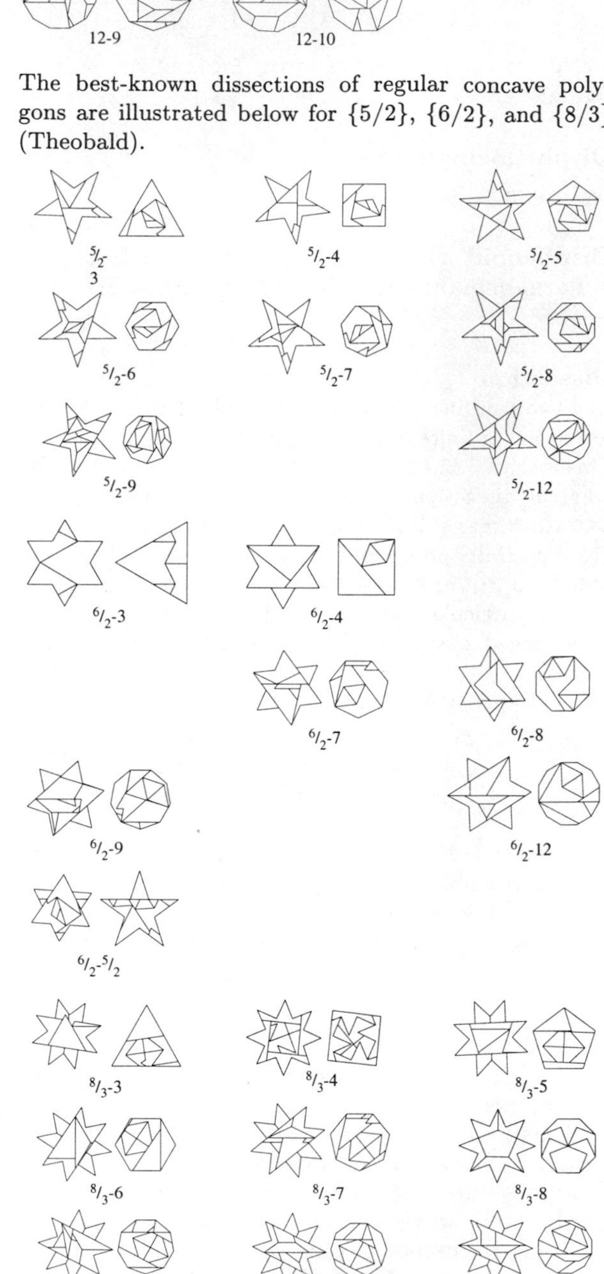

The best-known dissections of various crosses are illustrated below (Theobald).

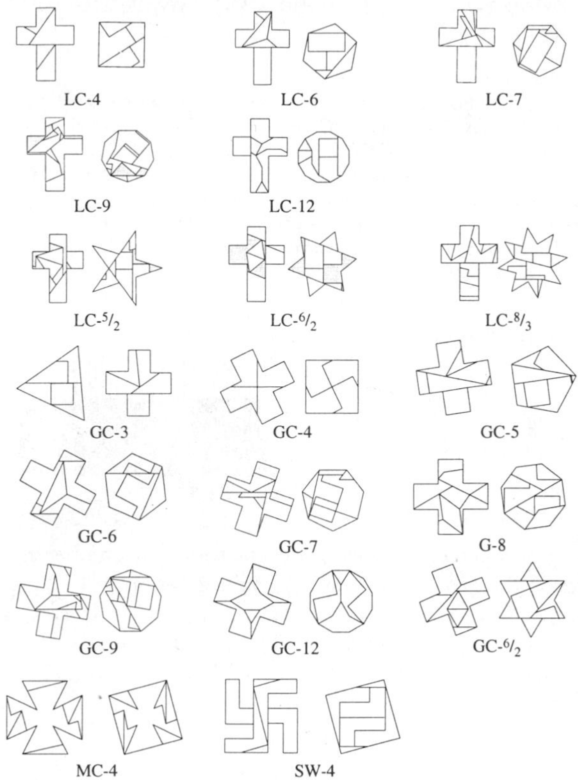

The best-known dissections of the GOLDEN RECTANGLE are illustrated below (Theobald).

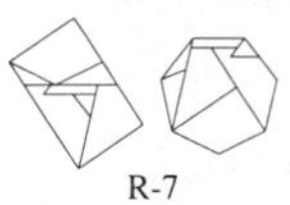
R-7

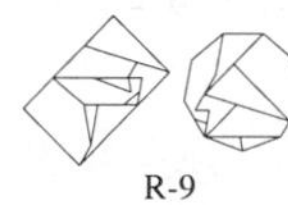
R-9

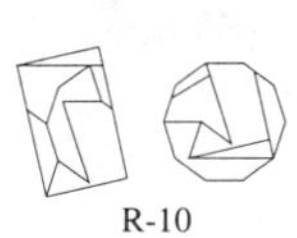
R-10

see also BANACH-TARSKI PARADOX, CUNDY AND ROLLETT'S EGG, DECAGON, DEHN INVARIANT, DIABOLICAL CUBE, DISSECTION PUZZLES, DODECAGON, EHRHART POLYNOMIAL, EQUIDECOMPOSABLE, EQUILATERAL TRIANGLE, GOLDEN RECTANGLE, HEPTAGON HEXAGON, HEXAGRAM, HILBERT'S PROBLEMS, LATIN CROSS, MALTESE CROSS, NONAGON, OCTAGON, OCTAGRAM, PENTAGON, PENTAGRAM, POLYHEDRON DISSECTION, PYTHAGOREAN SQUARE PUZZLE, PYTHAGOREAN THEOREM, REP-TILE, SOMA CUBE, SQUARE, STAR OF LAKSHMI, SWASTIKA, T-PUZZLE, TANGRAM, WALLACE-BOLYAI-GERWEIN THEOREM

References

Ball, W. W. R. and Coxeter, H. S. M. *Mathematical Recreations and Essays, 13th ed.* New York: Dover, pp. 87–94, 1987.

Coffin, S. T. *The Puzzling World of Polyhedral Dissections.* New York: Oxford University Press, 1990.

Cundy, H. and Rollett, A. Ch. 2 in *Mathematical Models, 3rd ed.* Stradbroke, England: Tarquin Pub., 1989.

Eppstein, D. "Dissection." `http://www.ics.uci.edu/~eppstein/junkyard/dissect.html`.

Eppstein, D. "Dissection Tiling." `http://www.ics.uci.edu/~eppstein/junkyard/distile`.

Eriksson, K. "Splitting a Polygon into Two Congruent Pieces." *Amer. Math. Monthly* **103**, 393–400, 1996.

Frederickson, G. *Dissections: Plane and Fancy.* New York: Cambridge University Press, 1997.

Gardner, M. *The Second Scientific American Book of Mathematical Puzzles & Diversions: A New Selection.* New York: Simon and Schuster, 1961.

Gardner, M. "Paper Cutting." Ch. 5 in *Martin Gardner's New Mathematical Diversions from Scientific American.* New York: Simon and Schuster, 1966.

Gardner, M. *The Unexpected Hanging and Other Mathematical Diversions.* Chicago, IL: Chicago University Press, 1991.

Hunter, J. A. H. and Madachy, J. S. *Mathematical Diversions.* New York: Dover, pp. 65–67, 1975.

Kraitchik, M. "Dissection of Plane Figures." §8.1 in *Mathematical Recreations.* New York: W. W. Norton, pp. 193–198, 1942.

Laczkovich, M. "Von Neumann's Paradox with Translation." *Fund. Math.* **131**, 1–12, 1988.

Lenhard, H.-C. "Über fünf neue Tetraeder, die einem Würfel äquivalent sind." *Elemente Math.* **17**, 108–109, 1962.

Lindgren, H. "Geometric Dissections." *Austral. Math. Teacher* **7**, 7–10, 1951.

Lindgren, H. "Geometric Dissections." *Austral. Math. Teacher* **9**, 17–21, 1953.

Lindgren, H. "Going One Better in Geometric Dissections." *Math. Gaz.* **45**, 94–97, 1961.

Lindgren, H. *Recreational Problems in Geometric Dissection and How to Solve Them.* New York: Dover, 1972.

Madachy, J. S. "Geometric Dissection." Ch. 1 in *Madachy's Mathematical Recreations.* New York: Dover, pp. 15–33, 1979.

Pappas, T. "A Triangle to a Square." *The Joy of Mathematics.* San Carlos, CA: Wide World Publ./Tetra, pp. 9 and 230, 1989.

Stewart, I. *The Problems of Mathematics, 2nd ed.* Oxford, England: Oxford University Press, 1987.

Dissection Puzzles

A puzzle in which one object is to be converted to another by making a finite number of cuts and reassembling it. The cuts are often, but not always, restricted to straight lines. Sometimes, a given puzzle is precut and is to be re-assembled into two or more given shapes.

see also CUNDY AND ROLLETT'S EGG, PYTHAGOREAN SQUARE PUZZLE, T-PUZZLE, TANGRAM

Dissipative System

A system in which the phase space volume contracts along a trajectory. This means that the generalized DIVERGENCE is less than zero,

$$\frac{\partial f_i}{\partial x_i} < 0,$$

where EINSTEIN SUMMATION has been used.

Distance

Let $\gamma(t)$ be a smooth curve in a MANIFOLD M from x to y with $\gamma(0) = x$ and $\gamma(1) = y$. Then $\gamma'(t) \in T_{\gamma(t)}$, where

T_x is the TANGENT SPACE of M at x. The LENGTH of γ with respect to the Riemannian structure is given by

$$\int_0^1 \|\gamma'(t)\|_{\gamma(t)}\, dt, \tag{1}$$

and the distance $d(x,y)$ between x and y is the shortest distance between x and y given by

$$d(x,y) = \inf_{\gamma:x \text{ to } y} \int \|\gamma'(t)\|_{\gamma(t)}\, dt. \tag{2}$$

In order to specify the relative distances of $n > 1$ points in the plane, $1+2(n-2) = 2n-3$ coordinates are needed, since the first can always be taken as (0, 0) and the second as $(x, 0)$, which defines the x-AXIS. The remaining $n-2$ points need two coordinates each. However, the total number of distances is

$$\binom{n}{2} = \frac{n!}{2!(n-2)!} = \tfrac{1}{2}n(n-1), \tag{3}$$

where $\binom{n}{k}$ is a BINOMIAL COEFFICIENT. The distances between $n > 1$ points are therefore subject to m relationships, where

$$m \equiv \tfrac{1}{2}n(n-1) - (2n-3) = \tfrac{1}{2}(n-2)(n-3). \tag{4}$$

For $n = 1, 2, \ldots$, this gives 0, 0, 0, 1, 3, 6, 10, 15, 21, 28, ... (Sloane's A000217) relationships, and the number of relationships between n points is the TRIANGULAR NUMBER T_{n-3}.

Although there are no relationships for $n = 2$ and $n = 3$ points, for $n = 4$ (a QUADRILATERAL), there is one (Weinberg 1972):

$$\begin{aligned} 0 = {} & {d_{12}}^4{d_{34}}^2 + {d_{13}}^4{d_{24}}^2 + {d_{14}}^4{d_{23}}^2 + {d_{23}}^4{d_{14}}^2 \\ & + d_{24}^4 d_{13}^2 + d_{34}^4 d_{12}^2 \\ & + d_{12}^2 d_{23}^2 d_{31}^2 + d_{12}^2 d_{24}^2 d_{41}^2 + d_{13}^2 d_{34}^2 d_{41}^2 \\ & + d_{23}^2 d_{34}^2 d_{42}^2 - d_{12}^2 d_{23}^2 d_{34}^2 - d_{13}^2 d_{32}^2 d_{24}^2 \\ & - d_{12}^2 d_{24}^2 d_{43}^2 - d_{14}^2 d_{42}^2 d_{23}^2 - d_{13}^2 d_{34}^2 d_{42}^2 \\ & - d_{14}^2 d_{43}^2 d_{32}^2 - d_{23}^2 d_{31}^2 d_{14}^2 - d_{21}^2 d_{13}^2 d_{34}^2 \\ & - d_{24}^2 d_{41}^2 d_{13}^2 - d_{21}^2 d_{14}^2 d_{43}^2 - d_{31}^2 d_{12}^2 d_{24}^2 \\ & - d_{32}^2 d_{21}^2 d_{14}^2. \end{aligned} \tag{5}$$

This equation can be derived by writing

$$d_{ij} \equiv \sqrt{(x_i - x_j)^2 + (y_i - y_j)^2} \tag{6}$$

and eliminating x_i and y_j from the equations for d_{12}, d_{13}, d_{14}, d_{23}, d_{24}, and d_{34}.

see also ARC LENGTH, CUBE POINT PICKING, EXPANSIVE, LENGTH (CURVE), METRIC, PLANAR DISTANCE, POINT-LINE DISTANCE—2-D, POINT-LINE DISTANCE—3-D, POINT-PLANE DISTANCE, POINT-POINT DISTANCE—1-D, POINT-POINT DISTANCE—2-D, POINT-POINT DISTANCE—3-D, SPACE DISTANCE, SPHERE

References

Gray, A. "The Intuitive Idea of Distance on a Surface." §13.1 in *Modern Differential Geometry of Curves and Surfaces.* Boca Raton, FL: CRC Press, pp. 251–255, 1993.

Sloane, N. J. A. Sequence A000217/M2535 in "An On-Line Version of the Encyclopedia of Integer Sequences."

Weinberg, S. *Gravitation and Cosmology: Principles and Applications of the General Theory of Relativity.* New York: Wiley, p. 7, 1972.

Distinct Prime Factors

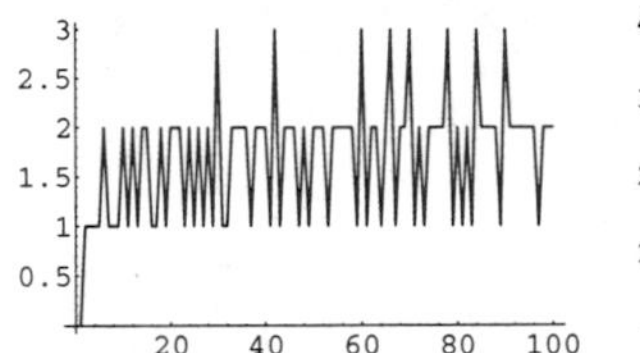

The number of distinct prime factors of a number n is denoted $\omega(n)$. The first few values for $n = 1, 2, \ldots$ are 0, 1, 1, 1, 1, 2, 1, 1, 1, 2, 1, 2, 1, 2, 2, 1, 1, 2, 1, 2, ... (Sloane's A001221). The first few values of the SUMMATORY FUNCTION

$$\sum_{k=2}^{n} \omega(k)$$

are 1, 2, 3, 4, 6, 7, 8, 9, 11, 12, 14, 15, 17, 19, 20, 21, ... (Sloane's A013939), and the asymptotic value is

$$\sum_{k=2}^{n} \omega(k) = n \ln\ln n + B_1 n + o(n),$$

where B_1 is MERTENS CONSTANT. In addition,

$$\sum_{k=2}^{n} [\omega(k)]^2 = n(\ln\ln n)^2 + \mathcal{O}(n \ln\ln n).$$

see also DIVISOR FUNCTION, GREATEST PRIME FACTOR, HARDY-RAMANUJAN THEOREM, HETEROGENEOUS NUMBERS, LEAST PRIME FACTOR, MERTENS CONSTANT, PRIME FACTORS

References

Hardy, G. H. and Wright, E. M. "The Number of Prime Factors of n" and "The Normal Order of $\omega(n)$ and $\Omega(n)$." §22.10 and 22.11 in *An Introduction to the Theory of Numbers, 5th ed.* Oxford, England: Clarendon Press, pp. 354–358, 1979.

Sloane, N. J. A. Sequences A013939 and A001221/M0056 in "An On-Line Version of the Encyclopedia of Integer Sequences."

Distribution

The distribution of a variable is a description of the relative numbers of times each possible outcome will occur in a number of trials. The function describing the distribution is called the PROBABILITY FUNCTION, and the function describing the probability that a given value or any value smaller than it will occur is called the DISTRIBUTION FUNCTION.

Formally, a distribution can be defined as a normalized MEASURE, and the distribution of a RANDOM VARIABLE x is the MEASURE P_x on $\mathbb{S}'$ defined by setting

$$P_x(A') = P\{s \in S : x(s) \in A'\},$$

where $(S, \mathbb{S}, P)$ is a PROBABILITY SPACE, $(S, \mathbb{S})$ is a MEASURABLE SPACE, and P a MEASURE on $\mathbb{S}$ with $P(S) = 1$.

see also CONTINUOUS DISTRIBUTION, DISCRETE DISTRIBUTION, DISTRIBUTION FUNCTION, MEASURABLE SPACE, MEASURE, PROBABILITY, PROBABILITY DENSITY FUNCTION, RANDOM VARIABLE, STATISTICS

References

Doob, J. L. "The Development of Rigor in Mathematical Probability (1900–1950)." *Amer. Math. Monthly* **103**, 586–595, 1996.

Distribution Function

The distribution function $D(x)$, sometimes also called the PROBABILITY DISTRIBUTION FUNCTION, describes the probability that a trial X takes on a value less than or equal to a number x. The distribution function is therefore related to a continuous PROBABILITY DENSITY FUNCTION $P(x)$ by

$$D(x) = P(X \leq x) \equiv \int_{-\infty}^{x} P(x')\,dx', \tag{1}$$

so $P(x)$ (when it exists), is simply the derivative of the distribution function

$$P(x) = D'(x) = [P(x')]_{-\infty}^{x} = P(x) - P(-\infty). \tag{2}$$

Similarly, the distribution function is related to a discrete probability $P(x)$ by

$$D(x) = P(X \leq x) = \sum_{X \leq x} P(x). \tag{3}$$

In general, there exist distributions which are neither continuous nor discrete.

A JOINT DISTRIBUTION FUNCTION can be defined if outcomes are dependent on two parameters:

$$D(x, y) \equiv P(X \leq x, Y \leq y) \tag{4}$$

$$D_x(x) \equiv D(x, \infty) \tag{5}$$

$$D_y(y) \equiv D(\infty, y). \tag{6}$$

Similarly, a multiple distribution function can be defined if outcomes depend on n parameters:

$$D(a_1, \ldots, a_n) \equiv P(x_1 \leq a_1, \ldots, x_n \leq a_n). \tag{7}$$

Given a continuous $P(x)$, assume you wish to generate numbers distributed as $P(x)$ using a random number generator. If the random number generator yields a uniformly distributed value y_i in [0,1] for each trial i, then compute

$$D(x) \equiv \int^{x} P(x')\,dx'. \tag{8}$$

The FORMULA connecting y_i with a variable distributed as $P(x)$ is then

$$x_i = D^{-1}(y_i), \tag{9}$$

where $D^{-1}(x)$ is the inverse function of $D(x)$. For example, if $P(x)$ were a GAUSSIAN DISTRIBUTION so that

$$D(x) = \frac{1}{2}\left[1 + \operatorname{erf}\left(\frac{x-\mu}{\sigma\sqrt{2}}\right)\right], \tag{10}$$

then

$$x_i = \sigma\sqrt{2}\operatorname{erf}^{-1}(2y_i - 1) + \mu. \tag{11}$$

If $P(x) = Cx^n$ for $x \in (x_{\min}, x_{\max})$, then normalization gives

$$\int_{x_{\min}}^{x_{\max}} P(x)\,dx = C\frac{[x^{n+1}]_{x_{\min}}^{x_{\max}}}{n+1} = 1, \tag{12}$$

so

$$C = \frac{n+1}{{x_{\max}}^{n+1} - {x_{\min}}^{n+1}}. \tag{13}$$

Let y be a uniformly distributed variate on [0, 1]. Then

$$\begin{aligned} D(x) &= \int_{x_{\min}}^{x} P(x)\,dx = C\int_{x_{\min}}^{x} x^n\,dx \\ &= \frac{C}{n+1}(x^{n+1} - {x_{\min}}^{n+1}) \equiv y, \end{aligned} \tag{14}$$

and the variate given by

$$\begin{aligned} x &= \left(\frac{n+1}{C}y + {x_{\min}}^{n+1}\right)^{1/(n+1)} \\ &= [({x_{\max}}^{n+1} - {x_{\min}}^{n+1})y + {x_{\min}}^{n+1}]^{1/(n+1)} \end{aligned} \tag{15}$$

is distributed as $P(x)$.

A distribution with constant VARIANCE of y for all values of x is known as a HOMOSCEDASTIC distribution. The method of finding the value at which the distribution is a maximum is known as the MAXIMUM LIKELIHOOD method.

see also BERNOULLI DISTRIBUTION, BETA DISTRIBUTION, BINOMIAL DISTRIBUTION, BIVARIATE DISTRIBUTION, CAUCHY DISTRIBUTION, CHI DISTRIBUTION, CHI-SQUARED DISTRIBUTION, CORNISH-FISHER

ASYMPTOTIC EXPANSION, CORRELATION COEFFICIENT, DISTRIBUTION, DOUBLE EXPONENTIAL DISTRIBUTION, EQUALLY LIKELY OUTCOMES DISTRIBUTION, EXPONENTIAL DISTRIBUTION, EXTREME VALUE DISTRIBUTION, F-DISTRIBUTION, FERMI-DIRAC DISTRIBUTION, FISHER'S z-DISTRIBUTION, FISHER-TIPPETT DISTRIBUTION, GAMMA DISTRIBUTION, GAUSSIAN DISTRIBUTION, GEOMETRIC DISTRIBUTION, HALF-NORMAL DISTRIBUTION, HYPERGEOMETRIC DISTRIBUTION, JOINT DISTRIBUTION FUNCTION, LAPLACE DISTRIBUTION, LATTICE DISTRIBUTION, LÉVY DISTRIBUTION, LOGARITHMIC DISTRIBUTION, LOG-SERIES DISTRIBUTION, LOGISTIC DISTRIBUTION, LORENTZIAN DISTRIBUTION, MAXWELL DISTRIBUTION, NEGATIVE BINOMIAL DISTRIBUTION, NORMAL DISTRIBUTION, PARETO DISTRIBUTION, PASCAL DISTRIBUTION, PEARSON TYPE III DISTRIBUTION, POISSON DISTRIBUTION, PÓLYA DISTRIBUTION, RATIO DISTRIBUTION, RAYLEIGH DISTRIBUTION, RICE DISTRIBUTION, SNEDECOR'S F-DISTRIBUTION, STUDENT'S t-DISTRIBUTION, STUDENT'S z-DISTRIBUTION, UNIFORM DISTRIBUTION, WEIBULL DISTRIBUTION

References

Abramowitz, M. and Stegun, C. A. (Eds.). "Probability Functions." Ch. 26 in *Handbook of Mathematical Functions with Formulas, Graphs, and Mathematical Tables, 9th printing.* New York: Dover, pp. 925–964, 1972.

Iyanaga, S. and Kawada, Y. (Eds.). "Distribution of Typical Random Variables." Appendix A, Table 22 in *Encyclopedic Dictionary of Mathematics.* Cambridge, MA: MIT Press, pp. 1483–1486, 1980.

Distribution (Functional)

A functional distribution, also called a GENERALIZED FUNCTION, is a generalization of the concept of a function. Functional distributions are defined as continuous linear FUNCTIONALS over a SPACE of infinitely differentiable functions such that all continuous functions have SCHWARZIAN DERIVATIVES which are themselves distributions. The most commonly encountered functional distribution is the DELTA FUNCTION.

see also DELTA FUNCTION, GENERALIZED FUNCTION, SCHWARZIAN DERIVATIVE

References

Friedlander, F. G. *Introduction to the Theory of Distributions.* Cambridge, England: Cambridge University Press, 1982.

Gel'fand, I. M. and Shilov, G. E. *Generalized Functions, Vol. 1: Properties and Operations.* New York: Harcourt Brace, 1977.

Gel'fand, I. M. and Shilov, G. E. *Generalized Functions, Vol. 2: Spaces of Fundamental and Generalized Functions.* New York: Harcourt Brace, 1977.

Gel'fand, I. M. and Shilov, G. E. *Generalized Functions, Vol. 3: Theory of Differential Equations.* New York: Harcourt Brace, 1977.

Gel'fand, I. M. and Vilenkin, N. Ya. *Generalized Functions, Vol. 4: Applications of Harmonic Analysis.* New York: Harcourt Brace, 1977.

Gel'fand, I. M.; Graev, M. I.; and Vilenkin, N. Ya. *Generalized Functions, Vol. 5: Integral Geometry and Representation Theory.* New York: Harcourt Brace, 1977.

Griffel, D. H. *Applied Functional Analysis.* Englewood Cliffs, NJ: Prentice-Hall, 1984.

Halperin, I. and Schwartz, L. *Introduction to the Theory of Distributions, Based on the Lectures Given by Laurent Schwarz.* Toronto, Canada: University of Toronto Press, 1952.

Lighthill, M. J. *Introduction to Fourier Analysis and Generalised Functions.* Cambridge, England: Cambridge University Press, 1958.

Richards, I. and Young, H. *The Theory of Distributions: A Nontechnical Introduction.* New York: Cambridge University Press, 1995.

Rudin, W. *Functional Analysis, 2nd ed.* New York: McGraw-Hill, 1991.

Strichartz, R. *Fourier Transforms and Distribution Theory.* Boca Raton, FL: CRC Press, 1993.

Zemanian, A. H. *Distribution Theory and Transform Analysis: An Introduction to Generalized Functions, with Applications.* New York: Dover, 1987.

Distribution Parameter

The distribution parameter of a NONCYLINDRICAL RULED SURFACE parameterized by

$$\mathbf{x}(u,v) = \boldsymbol{\sigma}(u) + v\boldsymbol{\delta}(u), \tag{1}$$

where $\boldsymbol{\sigma}$ is the STRICTION CURVE and $\boldsymbol{\delta}$ the DIRECTOR CURVE, is the function p defined by

$$p = \frac{\det(\boldsymbol{\sigma}'\boldsymbol{\delta}\boldsymbol{\delta}')}{\boldsymbol{\delta}'\cdot\boldsymbol{\delta}'}. \tag{2}$$

The GAUSSIAN CURVATURE of a RULED SURFACE is given in terms of its distribution parameter by

$$K = -\frac{[p(u)]^2}{\{[p(u)]^2 + v^2\}^2}. \tag{3}$$

see also NONCYLINDRICAL RULED SURFACE, STRICTION CURVE

References

Gray, A. *Modern Differential Geometry of Curves and Surfaces.* Boca Raton, FL: CRC Press, pp. 347–348, 1993.

Distribution (Statistical)

The set of probabilities for each possible event.

see DISTRIBUTION FUNCTION

Distributive

Elements of an ALGEBRA which obey the identity

$$A(B+C) = AB + AC$$

are said to be distributive over the operation +.

see also ASSOCIATIVE, COMMUTATIVE, TRANSITIVE

Distributive Lattice

A LATTICE which satisfies the identities

$$(x \wedge y) \vee (x \wedge z) = x \wedge (y \vee z)$$

$$(x \vee y) \wedge (x \vee z) = x \vee (y \wedge z)$$

is said to be distributive.

see also LATTICE, MODULAR LATTICE

References

Grätzer, G. *Lattice Theory: First Concepts and Distributive Lattices.* San Francisco, CA: W. H. Freeman, pp. 35–36, 1971.

Disymmetric

An object which is not superimposable on its MIRROR IMAGE is said to be disymmetric. All asymmetric objects are disymmetric, and an object with no IMPROPER ROTATION (rotoinversion) axis must also be disymmetric.

Ditrigonal Dodecadodecahedron

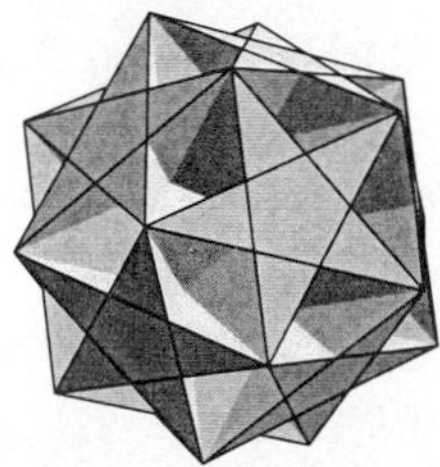

The UNIFORM POLYHEDRON U_{41}, also called the DITRIGONAL DODECAHEDRON, whose DUAL POLYHEDRON is the MEDIAL TRIAMBIC ICOSAHEDRON. It has WYTHOFF SYMBOL $3 \mid \frac{5}{3}\, 5$. Its faces are $12\{\frac{5}{2}\} + 12\{5\}$. It is a FACETED version of the SMALL DITRIGONAL ICOSIDODECAHEDRON. The CIRCUMRADIUS for unit edge length is

$$R = \tfrac{1}{2}\sqrt{3}.$$

References

Wenninger, M. J. *Polyhedron Models.* Cambridge, England: Cambridge University Press, pp. 123–124, 1989.

Ditrigonal Dodecahedron

see DITRIGONAL DODECADODECAHEDRON

Divergence

The divergence of a VECTOR FIELD $\mathbf{F}$ is given by

$$\operatorname{div}(\mathbf{F}) \equiv \nabla \cdot \mathbf{F} \equiv \lim_{V \to 0} \frac{\oint_S \mathbf{F} \cdot d\mathbf{a}}{V}. \quad (1)$$

Define

$$\mathbf{F} \equiv F_1 \hat{\mathbf{u}}_1 + F_2 \hat{\mathbf{u}}_2 + F_3 \hat{\mathbf{u}}_3. \quad (2)$$

Then in arbitrary orthogonal CURVILINEAR COORDINATES,

$$\operatorname{div}(F) \equiv \nabla \cdot \mathbf{F} \equiv \frac{1}{h_1 h_2 h_3} \left[\frac{\partial}{\partial u_1}(h_2 h_3 F_1) + \frac{\partial}{\partial u_2}(h_3 h_1 F_2) + \frac{\partial}{\partial u_3}(h_1 h_2 F_3) \right]. \quad (3)$$

If $\nabla \cdot \mathbf{F} = 0$, then the field is said to be a DIVERGENCELESS FIELD. For divergence in individual coordinate systems, see CURVILINEAR COORDINATES.

$$\nabla \cdot \frac{\mathsf{A}\mathbf{x}}{|\mathbf{x}|} = \frac{\operatorname{Tr}(\mathsf{A})}{|\mathbf{x}|} - \frac{\mathbf{x}^{\mathrm{T}}(\mathsf{A}\mathbf{x})}{|\mathbf{x}|^3}. \quad (4)$$

The divergence of a TENSOR A is

$$\nabla \cdot A \equiv A^{\alpha}_{;\alpha} = A^k_{,k} + \Gamma^k_{jk} A^j, \quad (5)$$

where ; is the COVARIANT DERIVATIVE. Expanding the terms gives

$$\begin{aligned} A^{\alpha}_{;\alpha} = A^{\alpha}_{,\alpha} &+ (\Gamma^{\alpha}_{\alpha\alpha} A^{\alpha} + \Gamma^{\alpha}_{\beta\alpha} A^{\beta} + \Gamma^{\alpha}_{\gamma\alpha} A^{\gamma}) \\ &+ A^{\beta}_{,\beta} + \left(\Gamma^{\beta}_{\alpha\beta} A^{\alpha} + \Gamma^{\beta}_{\beta\beta} A^{\beta} + \Gamma^{\beta}_{\gamma\beta} A^{\gamma}\right) \\ &+ A^{\gamma}_{,\gamma} + \left(\Gamma^{\gamma}_{\alpha\gamma} A^{\alpha} + \Gamma^{\gamma}_{\beta\gamma} A^{\beta} + \Gamma^{\gamma}_{\gamma\gamma} A^{\gamma}\right). \end{aligned} \quad (6)$$

see also CURL, CURL THEOREM, GRADIENT, GREEN'S THEOREM, DIVERGENCE THEOREM, VECTOR DERIVATIVE

References

Arfken, G. "Divergence, $\nabla\cdot$." §1.7 in *Mathematical Methods for Physicists, 3rd ed.* Orlando, FL: Academic Press, pp. 37–42, 1985.

Divergence Tests

If

$$\lim_{k \to \infty} u_k \neq 0,$$

then the series $\{u_n\}$ diverges.

see also CONVERGENCE TESTS, CONVERGENT SERIES, DINI'S TEST, SERIES

Divergence Theorem

A.k.a. GAUSS'S THEOREM. Let V be a region in space with boundary ∂V. Then

$$\int_V (\nabla \cdot \mathbf{F})\, dV = \int_{\partial V} \mathbf{F} \cdot d\mathbf{a}. \quad (1)$$

Let S be a region in the plane with boundary ∂S.

$$\int_S \nabla \cdot \mathbf{F}\, dA = \int_{\partial S} \mathbf{F} \cdot \mathbf{n}\, ds. \quad (2)$$

If the VECTOR FIELD $\mathbf{F}$ satisfies certain constraints, simplified forms can be used. If $\mathbf{F}(x, y, z) = v(x, y, z)\mathbf{c}$ where $\mathbf{c}$ is a constant vector $\neq \mathbf{0}$, then

$$\int_S \mathbf{F} \cdot d\mathbf{a} = \mathbf{c} \cdot \int_S v\, d\mathbf{a}. \quad (3)$$

But

$$\nabla \cdot (f\mathbf{v}) = (\nabla f) \cdot \mathbf{v} + f(\nabla \cdot \mathbf{v}), \tag{4}$$

so

$$\int_V \nabla \cdot (\mathbf{c}v)\, dV = \mathbf{c} \cdot \int_V (\nabla v + v\nabla \cdot \mathbf{c})\, dV = \mathbf{c} \cdot \int_V \nabla v\, dV \tag{5}$$

$$\mathbf{c} \cdot \left(\int_S v\, d\mathbf{a} - \int_V \nabla v\, dV \right) = 0. \tag{6}$$

But $\mathbf{c} \neq \mathbf{0}$, and $\mathbf{c} \cdot \mathbf{f}(v)$ must vary with v so that $\mathbf{c} \cdot \mathbf{f}(v)$ cannot always equal zero. Therefore,

$$\int_S v\, d\mathbf{a} = \int_V \nabla v\, dV. \tag{7}$$

If $\mathbf{F}(x,y,z) = \mathbf{c} \times P(x,y,z)$, where $\mathbf{c}$ is a constant vector $\neq \mathbf{0}$, then

$$\int_S d\mathbf{a} \times \mathbf{P} = \int_V \nabla \times \mathbf{P}\, dV. \tag{8}$$

see also CURL THEOREM, GRADIENT, GREEN'S THEOREM

References

Arfken, G. "Gauss's Theorem." §1.11 in *Mathematical Methods for Physicists, 3rd ed.* Orlando, FL: Academic Press, pp. 57–61, 1985.

Divergenceless Field

A divergenceless field, also called a SOLENOIDAL FIELD, is a FIELD for which $\nabla \cdot \mathbf{F} \equiv \mathbf{0}$. Therefore, there exists a $\mathbf{G}$ such that $\mathbf{F} = \nabla \times \mathbf{G}$. Furthermore, $\mathbf{F}$ can be written as

$$\mathbf{F} = \nabla \times (T\mathbf{r}) + \nabla^2(S\mathbf{r}) \equiv \mathbf{T} + \mathbf{S},$$

where

$$\mathbf{T} \equiv \nabla \times (T\mathbf{r}) = -\mathbf{r} \times (\nabla T)$$

$$\mathbf{S} \equiv \nabla^2(S\mathbf{r}) = \nabla\left[\frac{\partial}{\partial r}(rS)\right] - \mathbf{r}\nabla^2 S.$$

Following Lamb, $\mathbf{T}$ and $\mathbf{S}$ are called TOROIDAL FIELD and POLOIDAL FIELD.

see also BELTRAMI FIELD, IRROTATIONAL FIELD, POLOIDAL FIELD, SOLENOIDAL FIELD, TOROIDAL FIELD

Divergent Sequence

A divergent sequence is a SEQUENCE for which the LIMIT exists but is not CONVERGENT.

see also CONVERGENT SEQUENCE, DIVERGENT SERIES

Divergent Series

A SERIES which is not CONVERGENT. Series may diverge by marching off to infinity or by oscillating.

see also CONVERGENT SERIES, DIVERGENT SEQUENCE

References

Bromwich, T. J. I'a and MacRobert, T. M. *An Introduction to the Theory of Infinite Series, 3rd ed.* New York: Chelsea, 1991.

Diversity Condition

For any group of k men out of N, there must be at least k jobs for which they are collectively qualified.

Divide

To divide is to perform the operation of DIVISION, i.e., to see how many time a DIVISOR d goes into another number n. n divided by d is written n/d or $n \div d$. The result need not be an INTEGER, but if it is, some additional terminology is used. $d|n$ is read "d divides n" and means that d is a PROPER DIVISOR of n. In this case, n is said to be DIVISIBLE by d. Clearly, $1|n$ and $n|n$. By convention, $n|0$ for every n except 0 (Hardy and Wright 1979). The "divided" operation satisfies

$$b|a \text{ and } c|b \Rightarrow c|a$$

$$b|a \Rightarrow bc|ac$$

$$c|a \text{ and } c|b \Rightarrow c|(ma+nb).$$

$d' \nmid n$ is read "d' does not divide n" and means that d' is not a PROPER DIVISOR of n. $a^k||b$ means a^k divides b exactly.

see also CONGRUENCE, DIVISIBLE, DIVISION, DIVISOR

References

Hardy, G. H. and Wright, E. M. *An Introduction to the Theory of Numbers, 5th ed.* Oxford, England: Clarendon Press, p. 1, 1979.

Divided Difference

The divided difference $f[x_1, x_2, \ldots, x_n]$ on n points x_1, x_2, ..., x_n of a function $f(x)$ is defined by $f[x_1] \equiv f(x_1)$ and

$$f[x_1, x_2, \ldots, x_n] = \frac{f[x_1, \ldots, x_n] - f[x_2, \ldots, x_n]}{x_1 - x_n} \tag{1}$$

for $n \geq 2$. The first few differences are

$$[x_0, x_1] = \frac{f_0 - f_1}{x_0 - x_1} \tag{2}$$

$$[x_0, x_1, x_2] = \frac{[x_0, x_1] - [x_1, x_2]}{x_0 - x_2} \tag{3}$$

$$[x_0, x_1, \ldots, x_n] = \frac{[x_0, \ldots, x_{n-1}] - [x_1, \ldots, x_n]}{x_0 - x_n}. \tag{4}$$

Defining

$$\pi_n(x) \equiv (x - x_0)(x - x_1) \cdots (x - x_n) \tag{5}$$

and taking the DERIVATIVE

$$\pi'_n(x_k) = (x_k - x_0) \cdots (x_k - x_{k-1})(x_k - x_{k+1}) \cdots (x_k - x_n) \tag{6}$$

gives the identity

$$[x_0, x_1, \ldots, x_n] = \sum_{k=0}^{n} \frac{f_k}{\pi'_n(x_k)}. \tag{7}$$

Consider the following question: does the property

$$f[x_1, x_2, \ldots, x_n] = h(x_1 + x_2 + \ldots + x_n) \qquad (8)$$

for $n \geq 2$ and $h(x)$ a given function guarantee that $f(x)$ is a POLYNOMIAL of degree $\leq n$? Aczél (1985) showed that the answer is "yes" for $n = 2$, and Bailey (1992) showed it to be true for $n = 3$ with differentiable $f(x)$. Schwaiger (1994) and Andersen (1996) subsequently showed the answer to be "yes" for all $n \geq 3$ with restrictions on $f(x)$ or $h(x)$.

see also NEWTON'S DIVIDED DIFFERENCE INTERPOLATION FORMULA, RECIPROCAL DIFFERENCE

References

Abramowitz, M. and Stegun, C. A. (Eds.). *Handbook of Mathematical Functions with Formulas, Graphs, and Mathematical Tables, 9th printing.* New York: Dover, pp. 877–878, 1972.

Aczél, J. "A Mean Value Property of the Derivative of Quadratic Polynomials—Without Mean Values and Derivatives." *Math. Mag.* **58**, 42–45, 1985.

Andersen, K. M. "A Characterization of Polynomials." *Math. Mag.* **69**, 137–142, 1996.

Bailey, D. F. "A Mean-Value Property of Cubic Polynomials—Without Mean Values." *Math. Mag.* **65**, 123–124, 1992.

Beyer, W. H. (Ed.) *CRC Standard Mathematical Tables, 28th ed.* Boca Raton, FL: CRC Press, pp. 439–440, 1987.

Schwaiger, J. "On a Characterization of Polynomials by Divided Differences." *Aequationes Math.* **48**, 317–323, 1994.

Divine Proportion

see GOLDEN RATIO

Divisibility Tests

Write a decimal number a out digit by digit in the form $a_n \ldots a_3 a_2 a_1 a_0$. It is always true that $10^0 = 1 \equiv 1$ for any base.

2 $10^1 \equiv 0$, so $10^n \equiv 0$ for $n \geq 1$. Therefore, if the last digit a_0 is divisible by 2 (i.e., is EVEN), then so is a.

3 $10^1 \equiv 1$, $10^2 \equiv 1$, ..., $10^n \equiv 1$. Therefore, if $\sum_{i=1}^n a_i$ is divisible by 3, so is a.

4 $10^1 \equiv 2$, $10^2 \equiv 0$, $\ldots 10^n \equiv 0$. So if the last two digits are divisible by 4, more specifically if $r \equiv a_0 + 2a_1$ is, then so is a.

5 $10^1 \equiv 0$, so $10^n \equiv 0$ for $n \geq 1$. Therefore, if the last digit a_0 is divisible by 5 (i.e., is 5 or 0), then so is a_0.

6 $10^1 \equiv -2$, $10^2 \equiv -2$, so $10^n \equiv -2$. Therefore, if $r \equiv a_0 - 2\sum_{i=1}^n a_i$ is divisible by 6, so is a. If a is divisible by 3 and is EVEN, it is also divisible by 6.

7 $10^1 \equiv 3$, $10^2 \equiv 2$, $10^3 \equiv -1$, $10^4 \equiv -3$, $10^5 \equiv -2$, $10^6 \equiv 1$, and the sequence then repeats. Therefore, if $r \equiv (a_0 + 3a_1 + 2a_2 - a_3 - 3a_4 - 2a_5) + (a_6 + 3a_7 + \ldots) + \ldots$ is divisible by 7, so is a.

8 $10^1 \equiv 2$, $10^2 \equiv 4$, $10^3 \equiv 0$, ..., $10^n \equiv 0$. Therefore, if the last three digits are divisible by 8, more specifically if $r \equiv a_0 + 2a_1 + 4a_2$ is, then so is a.

9 $10^1 \equiv 1$, $10^2 \equiv 1$, ..., $10^3 \equiv 1$. Therefore, if $\sum_{i=1}^n a_i$ is divisible by 9, so is a.

10 $10^1 \equiv 0$, so if the last digit is 0, then a is divisible by 10.

11 $10^1 \equiv -1$, $10^2 \equiv 1$, $10^3 \equiv -1$, $10^4 \equiv 1$, ... Therefore, if $r \equiv a_0 - a_1 + a_2 - a_3 + \ldots$ is divisible by 11, then so is a.

12 $10^1 \equiv -2$, $10^2 \equiv 4$, $10^3 \equiv 4$, ... Therefore, if $r \equiv a_0 - 2a_1 + 4(a_2 + a_3 + \ldots)$ is divisible by 12, then so is a. Divisibility by 12 can also be checked by seeing if a is divisible by 3 and 4.

13 $10^1 \equiv -3$, $10^2 \equiv -4$, $10^3 \equiv -1$, $10^4 \equiv 3$, $10^5 \equiv 4$, $10^6 \equiv 1$, and the pattern repeats. Therefore, if $r \equiv (a_0 - 3a_1 - 4a_2 - a_3 + 3a_4 + 4a_5) + (a_6 - 3a_7 + \ldots) + \ldots$ is divisible by 13, so is a.

For additional tests for 13, see Gardner (1991).

References

Dickson, L. E. *History of the Theory of Numbers, Vol. 1: Divisibility and Primality.* New York: Chelsea, pp. 337–346, 1952.

Gardner, M. Ch. 14 in *The Unexpected Hanging and Other Mathematical Diversions.* Chicago, IL: Chicago University Press, 1991.

Divisible

A number n is said to be divisible by d if d is a PROPER DIVISOR of n. The sum of any n consecutive INTEGERS is divisible by $n!$, where $n!$ is the FACTORIAL.

see also DIVIDE, DIVISOR, DIVISOR FUNCTION

References

Guy, R. K. "Divisibility." Ch. B in *Unsolved Problems in Number Theory, 2nd ed.* New York: Springer-Verlag, pp. 44–104, 1994.

Division

Taking the RATIO x/y of two numbers x and y, also written $x \div y$. Here, y is called the DIVISOR. The symbol "/" is called a SOLIDUS (or DIAGONAL), and the symbol "÷" is called the OBELUS. Division in which the fractional (remainder) is discarded is called INTEGER DIVISION, and is sometimes denoted using a backslash, \.

see also ADDITION, DIVIDE, INTEGER DIVISION, LONG DIVISION, MULTIPLICATION, OBELUS, ODDS, RATIO, SKELETON DIVISION, SOLIDUS, SUBTRACTION, TRIAL DIVISION

Division Algebra

A division algebra, also called a DIVISION RING or SKEW FIELD, is a RING in which every NONZERO element has a multiplicative inverse, but multiplication is not COMMUTATIVE. Explicitly, a division algebra is a set together with two BINARY OPERATORS $S(+, *)$ satisfying the following conditions:

1. Additive associativity: For all $a, b, c \in S$, $(a+b)+c = a+(b+c)$,
2. Additive commutativity: For all $a, b \in S$, $a+b = b+a$,
3. Additive identity: There exists an element $0 \in S$ such that for all $a \in S$, $0+a = a+0 = a$,
4. Additive inverse: For every $a \in S$ there exists a $-a \in S$ such that $a+(-a) = (-a)+a = 0$,
5. Multiplicative associativity: For all $a, b, c \in S$, $(a*b)*c = a*(b*c)$,
6. Multiplicative identity: There exists an element $1 \in S$ not equal to 0 such that for all $a \in S$, $1*a = a*1 = a$,
7. Multiplicative inverse: For every $a \in S$ not equal to 0, there exists $a^{-1} \in S$, $a*a^{-1} = a^{-1}*a = 1$,
8. Left and right distributivity: For all $a, b, c \in S$, $a*(b+c) = (a*b)+(a*c)$ and $(b+c)*a = (b*a)+(c*a)$.

Thus a division algebra $(S, +, *)$ is a UNIT RING for which $(S - \{0\}, *)$ is a GROUP. A division algebra must contain at least two elements. A COMMUTATIVE division algebra is called a FIELD.

In 1878 and 1880, Frobenius and Peirce proved that the only associative REAL division algebras are real numbers, COMPLEX NUMBERS, and QUATERNIONS. The CAYLEY ALGEBRA is the only NONASSOCIATIVE DIVISION ALGEBRA. Hurwitz (1898) proved that the ALGEBRAS of REAL NUMBERS, COMPLEX NUMBERS, QUATERNIONS, and CAYLEY NUMBERS are the only ones where multiplication by unit "vectors" is distance-preserving. Adams (1956) proved that n-D vectors form an ALGEBRA in which division (except by 0) is always possible only for $n = 1$, 2, 4, and 8.

see also CAYLEY NUMBER, FIELD, GROUP, NONASSOCIATIVE ALGEBRA, QUATERNION, UNIT RING

References

Dickson, L. E. *Algebras and Their Arithmetics.* Chicago, IL: University of Chicago Press, 1923.

Dixon, G. M. *Division Algebras: Octonions, Quaternions, Complex Numbers and the Algebraic Design of Physics.* Dordrecht, Netherlands: Kluwer, 1994.

Herstein, I. N. *Topics in Algebra, 2nd ed.* New York: Wiley, pp. 326–329, 1975.

Hurwitz, A. "Ueber die Composition der quadratischen Formen von beliebig vielen Variabeln." *Nachr. Gesell. Wiss. Göttingen, Math.-Phys. Klasse,* 309–316, 1898.

Kurosh, A. G. *General Algebra.* New York: Chelsea, pp. 221–243, 1963.

Petro, J. "Real Division Algebras of Dimension > 1 contain $\mathbb{C}$." *Amer. Math. Monthly* **94**, 445–449, 1987.

Division Lemma

When ac is DIVISIBLE by a number b that is RELATIVELY PRIME to a, then c must be DIVISIBLE by b.

Division Ring

see DIVISION ALGEBRA

Divisor

A divisor of a number N is a number d which DIVIDES N, also called a FACTOR. The total number of divisors for a given number N can be found as follows. Write a number in terms of its PRIME FACTORIZATION

$$N = {p_1}^{\alpha_1} {p_2}^{\alpha_2} \cdots {p_r}^{\alpha_r}. \quad (1)$$

For any divisor d of N, $N = dd'$ where

$$d = {p_1}^{\delta_1} {p_2}^{\delta_2} \cdots {p_r}^{\delta_r}, \quad (2)$$

so

$$d' = {p_1}^{\alpha_1-\delta_1} {p_2}^{\alpha_2-\delta_2} \cdots {p_r}^{\alpha_r-\delta_r}. \quad (3)$$

Now, $\delta_1 = 0, 1, \ldots, \alpha_1$, so there are $\alpha_1 + 1$ possible values. Similarly, for δ_n, there are $\alpha_n + 1$ possible values, so the total number of divisors $\nu(N)$ of N is given by

$$\nu(N) = \prod_{n=1}^{r} (\alpha_n + 1). \quad (4)$$

The function $\nu(N)$ is also sometimes denoted $d(N)$ or $\sigma_0(N)$. The product of divisors can be found by writing the number N in terms of all possible products

$$N = \begin{cases} d^{(1)} d'^{(1)} \\ \vdots \\ d^{(\nu)} d'^{(\nu)} \end{cases}, \quad (5)$$

so

$$\begin{aligned} N^{\nu(N)} &= [d^{(1)} \cdots d^{(\nu)}][d'^{(1)} d'^{(\nu)}] \\ &= \prod_{i=1}^{\nu} d_i \prod_{i=1}^{\nu} d_i' = \left(\prod d\right)^2, \end{aligned} \quad (6)$$

and

$$\prod d = N^{\nu(N)/2}. \quad (7)$$

The GEOMETRIC MEAN of divisors is

$$G \equiv \left(\prod d\right)^{\nu(N)/2} = [N^{\nu(n)/2}]^{1/\nu(N)} = \sqrt{N}. \quad (8)$$

The sum of the divisors can be found as follows. Let $N \equiv ab$ with $a \neq b$ and $(a, b) = 1$. For any divisor d of N, $d = a_i b_i$, where a_i is a divisor of a and b_i is a divisor of b. The divisors of a are 1, a_1, a_2, ..., and a. The divisors of b are 1, b_1, b_2, ..., b. The sums of the divisors are then

$$\sigma(a) = 1 + a_1 + a_2 + \ldots + a \quad (9)$$

$$\sigma(b) = 1 + b_1 + b_2 + \ldots + b. \quad (10)$$

For a given a_i,

$$a_i(1 + b_1 + b_2 + \ldots + b) = a_i \sigma(b). \quad (11)$$

Summing over all a_i,

$$(1 + a_1 + a_2 + \ldots + a)\sigma(b) = \sigma(a)\sigma(b), \qquad (12)$$

so $\sigma(N) = \sigma(ab) = \sigma(a)\sigma(b)$. Splitting a and b into prime factors,

$$\sigma(N) = \sigma({p_1}^{\alpha_1})\sigma({p_2}^{\alpha_2})\cdots\sigma({p_r}^{\alpha_r}). \qquad (13)$$

For a prime POWER ${p_i}^{\alpha_i}$, the divisors are 1, p_i, ${p_i}^2$, ..., ${p_i}^{\alpha_i}$, so

$$\sigma({p_i}^{\alpha_i}) = 1 + p_i + {p_i}^2 + \ldots + {p_i}^{\alpha_i} = \frac{{p_i}^{\alpha_i+1} - 1}{p_i - 1}. \qquad (14)$$

For N, therefore,

$$\sigma(N) = \prod_{i=1}^{r} \frac{{p_i}^{\alpha_i+1} - 1}{p_i - 1}. \qquad (15)$$

For the special case of N a PRIME, (15) simplifies to

$$\sigma(p) = \frac{p^2 - 1}{p - 1} = p + 1. \qquad (16)$$

For N a POWER of two, (15) simplifies to

$$\sigma(2^\alpha) = \frac{2^{\alpha+1} - 1}{2 - 1} = 2^{\alpha+1} - 1. \qquad (17)$$

The ARITHMETIC MEAN is

$$A(N) \equiv \frac{\sigma(N)}{\nu(N)}. \qquad (18)$$

The HARMONIC MEAN is

$$\frac{1}{H} \equiv \frac{1}{n}\left(\sum \frac{1}{d}\right). \qquad (19)$$

But $N = dd'$, so $\frac{1}{\nu(N)} = \frac{d'}{N}$ and

$$\sum \frac{1}{d} = \frac{1}{N}\sum d' = \frac{1}{N}\sum d = \frac{\sigma(N)}{N}, \qquad (20)$$

and we have

$$\frac{1}{H(N)} = \frac{1}{N(\nu)}\frac{\sigma(N)}{N} = \frac{A(N)}{N} \qquad (21)$$

$$N = A(N)H(N). \qquad (22)$$

Given three INTEGERS chosen at random, the probability that no common factor will divide them all is

$$[\zeta(3)]^{-1} \approx 1.202^{-1} = 0.832\ldots, \qquad (23)$$

where $\zeta(3)$ is APÉRY'S CONSTANT.

Let $f(n)$ be the number of elements in the greatest subset of $[1, n]$ such that none of its elements are divisible by two others. For n sufficiently large,

$$0.6725\ldots \leq \frac{f(n)}{n} \leq 0.673\ldots \qquad (24)$$

(Le Lionnais 1983, Lebensold 1976/1977).

see also ALIQUANT DIVISOR, ALIQUOT DIVISOR, ALIQUOT SEQUENCE, DIRICHLET DIVISOR PROBLEM, DIVISOR FUNCTION, e-DIVISOR, EXPONENTIAL DIVISOR, GREATEST COMMON DIVISOR, INFINARY DIVISOR, k-ARY DIVISOR, PERFECT NUMBER, PROPER DIVISOR, UNITARY DIVISOR

References

Guy, R. K. "Solutions of $d(n) = d(n+1)$." §B18 in *Unsolved Problems in Number Theory, 2nd ed.* New York: Springer-Verlag, pp. 73–75, 1994.

Le Lionnais, F. *Les nombres remarquables.* Paris: Hermann, p. 43, 1983.

Lebensold, K. "A Divisibility Problem." *Studies Appl. Math.* **56**, 291–294, 1976/1977.

Divisor Function

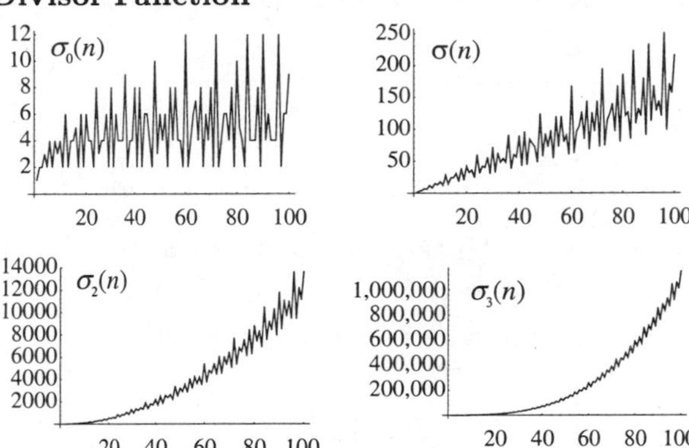

$\sigma_k(n)$ is defined as the sum of the kth POWERS of the DIVISORS of n. The function $\sigma_0(n)$ gives the total number of DIVISORS of n and is often denoted $d(n)$, $\nu(n)$, $\tau(n)$, or $\Omega(n)$ (Hardy and Wright 1979, pp. 354–355). The first few values of $\sigma_0(n)$ are 1, 2, 2, 3, 2, 4, 2, 4, 3, 4, 2, 6, ... (Sloane's A000005). The function $\sigma_1(n)$ is equal to the sum of DIVISORS of n and is often denoted $\sigma(n)$. The first few values of $\sigma(n)$ are 1, 3, 4, 7, 6, 12, 8, 15, 13, 18, ... (Sloane's A000203). The first few values of $\sigma_2(n)$ are 1, 5, 10, 21, 26, 50, 50, 85, 91, 130, ... (Sloane's A001157). The first few values of $\sigma_3(n)$ are 1, 9, 28, 73, 126, 252, 344, 585, 757, 1134, ... (Sloane's A001158).

The sum of the DIVISORS of n excluding n itself (i.e., the PROPER DIVISORS of n) is called the RESTRICTED DIVISOR FUNCTION and is denoted $s(n)$. The first few values are 0, 1, 1, 3, 1, 6, 1, 7, 4, 8, 1, 16, ... (Sloane's A001065).

As an illustrative example, consider the number 140, which has DIVISORS $d_i = 1, 2, 4, 5, 7, 10, 14, 20, 28, 35, 70$, and 140 (for a total of $N = 12$ of them). Therefore,

$$d(140) = N = 12 \tag{1}$$

$$\sigma(140) = \sum_i^N d_i = 336 \tag{2}$$

$$\sigma_2(140) = \sum_i^N d_i{}^2 = 27,300 \tag{3}$$

$$\sigma_3(140) = \sum_i^N d_i{}^3 = 3,164,112. \tag{4}$$

The $\sigma(n)$ function has the series expansion

$$\sigma(n) = \tfrac{1}{6}\pi^2 n \left[1 + \frac{(-1)^n}{2^2} + \frac{2\cos(\frac{2}{3}n\pi)}{3^2} + \frac{2\cos(\frac{1}{2}n\pi)}{4^2} + \frac{2[\cos(\frac{2}{5}n\pi) + \cos(\frac{4}{5}n\pi)]}{5^2} + \ldots\right] \tag{5}$$

(Hardy 1959). It also satisfies the INEQUALITY

$$\frac{\sigma(n)}{n\ln\ln n} \leq e^\gamma + \frac{2(1-\sqrt{2}) + \gamma - \ln(4\pi)}{\sqrt{\ln n}\,\ln\ln n} + \mathcal{O}\left(\frac{1}{\sqrt{\ln n}\,(\ln\ln n)^2}\right), \tag{6}$$

where γ is the EULER-MASCHERONI CONSTANT (Robin 1984, Erdős 1989).

Let a number n have PRIME factorization

$$n = \prod_{j=1}^r p_j{}^{\alpha_j}, \tag{7}$$

then

$$\sigma(n) = \prod_{j=1}^r \frac{p_j{}^{\alpha_j+1} - 1}{p_j - 1} \tag{8}$$

(Berndt 1985). GRONWALL'S THEOREM states that

$$\overline{\lim_{n\to\infty}} \frac{\sigma(n)}{n\ln\ln n} = e^\gamma, \tag{9}$$

where γ is the EULER-MASCHERONI CONSTANT.

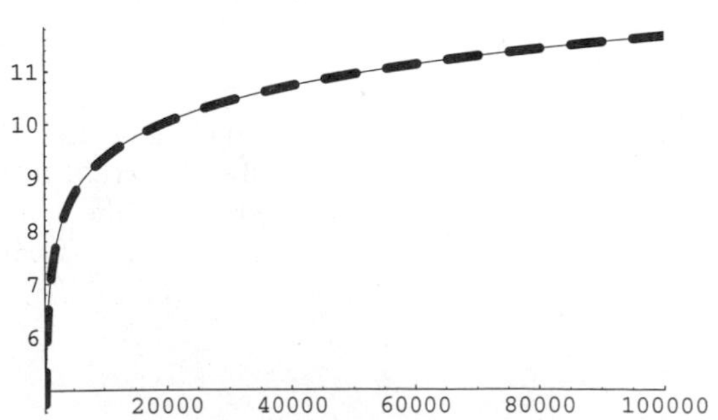

In general,

$$\sigma_k(n) \equiv \sum_{d|n} d^k. \tag{10}$$

In 1838, Dirichlet showed that the average number of DIVISORS of all numbers from 1 to n is asymptotic to

$$\frac{\sum_{i=1}^n \sigma_0(i)}{n} \sim \ln n + 2\gamma - 1 \tag{11}$$

(Conway and Guy 1996), as illustrated above, where the thin solid curve plots the actual values and the thick dashed curve plots the asymptotic function.

A curious identity derived using MODULAR FORM theory is given by

$$\sigma_7(n) = \sigma_3(n) + 120\sum_{k=1}^{n-1} \sigma_3(k)\sigma_3(n-k). \tag{12}$$

The asymptotic SUMMATORY FUNCTION of $\sigma_0(n) = \Omega(n)$ is given by

$$\sum_{k=2}^n \Omega(k) = n\ln\ln n + B_2 + o(n), \tag{13}$$

where

$$B_2 = \gamma + \sum_{p\ \text{prime}} \left[\ln(1-p^{-1}) + \frac{1}{p-1}\right] \approx 1.034653 \tag{14}$$

(Hardy and Wright 1979, p. 355). This is related to the DIRICHLET DIVISOR PROBLEM. The SUMMATORY FUNCTIONS for σ_a with $a > 1$ are

$$\sum_{k=1}^n \sigma_a(k) = \frac{\zeta(a+1)}{a+1} n^{a+1} + \mathcal{O}(n^a). \tag{15}$$

For $a = 1$,

$$\sum_{k=1}^n \sigma_1(k) = \frac{\pi^2}{12} n^2 + \mathcal{O}(n\ln n). \tag{16}$$

The divisor function is ODD IFF n is a SQUARE NUMBER or twice a SQUARE NUMBER. The divisor function satisfies the CONGRUENCE

$$n\sigma(n) \equiv 2 \pmod{\phi(n)}, \tag{17}$$

for all PRIMES and no COMPOSITE NUMBERS with the exception of 4, 6, and 22 (Subbarao 1974). $\tau(n)$ is PRIME whenever $\sigma(n)$ is (Honsberger 1991). Factorizations of $\sigma(p^a)$ for PRIME p are given by Sorli.

see also DIRICHLET DIVISOR PROBLEM, DIVISOR, FACTOR, GREATEST PRIME FACTOR, GRONWALL'S THEOREM, LEAST PRIME FACTOR, MULTIPLY PERFECT

NUMBER, ORE'S CONJECTURE, PERFECT NUMBER, $r(n)$, RESTRICTED DIVISOR FUNCTION, SILVERMAN CONSTANT, TAU FUNCTION, TOTIENT FUNCTION, TOTIENT VALENCE FUNCTION, TWIN PEAKS

References

Abramowitz, M. and Stegun, C. A. (Eds.). "Divisor Functions." §24.3.3 in *Handbook of Mathematical Functions with Formulas, Graphs, and Mathematical Tables, 9th printing.* New York: Dover, p. 827, 1972.

Berndt, B. C. *Ramanujan's Notebooks: Part I.* New York: Springer-Verlag, p. 94, 1985.

Conway, J. H. and Guy, R. K. *The Book of Numbers.* New York: Springer-Verlag, pp. 260–261, 1996.

Dickson, L. E. *History of the Theory of Numbers, Vol. 1: Divisibility and Primality.* New York: Chelsea, pp. 279–325, 1952.

Dirichlet, G. L. "Sur l'usage des séries infinies dans la théorie des nombres." *J. reine angew. Math.* **18**, 259–274, 1838.

Erdős, P. "Ramanujan and I." In *Proceedings of the International Ramanujan Centenary Conference held at Anna University, Madras, Dec. 21, 1987.* (Ed. K. Alladi). New York: Springer-Verlag, pp. 1–20, 1989.

Guy, R. K. "Solutions of $m\sigma(m) = n\sigma(n)$," "Analogs with $d(n)$, $\sigma_k(n)$," "Solutions of $\sigma(n) = \sigma(n+1)$," and "Solutions of $\sigma(q) + \sigma(r) = \sigma(q+r)$." §B11, B12, B13 and B15 in *Unsolved Problems in Number Theory, 2nd ed.* New York: Springer-Verlag, pp. 67–70, 1994.

Hardy, G. H. *Ramanujan: Twelve Lectures on Subjects Suggested by His Life and Work, 3rd ed.* New York: Chelsea, p. 141, 1959.

Hardy, G. H. and Weight, E. M. *An Introduction to the Theory of Numbers, 5th ed.* Oxford, England: Oxford University Press, pp. 354–355, 1979.

Honsberger, R. *More Mathematical Morsels.* Washington, DC: Math. Assoc. Amer., pp. 250–251, 1991.

Robin, G. "Grandes valeurs de la fonction somme des diviseurs et hypothese de Riemann." *J. Math. Pures Appl.* **63**, 187–213, 1984.

Sloane, N. J. A. Sequences A000005/M0246, A000203/M2329, A001065/M2226, A001157/M3799, A001158/M4605 in "An On-Line Version of the Encyclopedia of Integer Sequences."

Sorli, R. "Factorization Tables." `http://www.maths.uts.edu.au/staff/ron/fact/fact.html`.

Subbarao, M. V. "On Two Congruences for Primality." *Pacific J. Math.* **52**, 261–268, 1974.

Divisor Theory

A generalization by Kronecker of Kummer's theory of PRIME IDEAL factors. A divisor on a full subcategory C of mod(A) is an additive mapping χ on C with values in a SEMIGROUP of IDEALS on A.

see also IDEAL, IDEAL NUMBER, PRIME IDEAL, SEMIGROUP

References

Edwards, H. M. *Divisor Theory.* Boston, MA: Birkhäuser, 1989.

Vasconcelos, W. V. *Divisor Theory in Module Categories.* Amsterdam, Netherlands: North-Holland, pp. 63–64, 1974.

Dixon's Factorization Method

In order to find INTEGERS x and y such that

$$x^2 \equiv y^2 \pmod{n} \tag{1}$$

(a modified form of FERMAT'S FACTORIZATION METHOD), in which case there is a 50% chance that GCD$(n, x-y)$ is a FACTOR of n, choose a RANDOM INTEGER r_i, compute

$$g(r_i) \equiv {r_i}^2 \pmod{n}, \tag{2}$$

and try to factor $g(r_i)$. If $g(r_i)$ is not easily factorable (up to some small trial divisor d), try another r_i. In practice, the trial rs are usually taken to be $\lfloor\sqrt{n}\rfloor + k$, with $k = 1, 2, \ldots$, which allows the QUADRATIC SIEVE FACTORIZATION METHOD to be used. Continue finding and factoring $g(r_i)$s until $N \equiv \pi d$ are found, where π is the PRIME COUNTING FUNCTION. Now for each $g(r_i)$, write

$$g(r_i) = {p_{1i}}^{a_{1i}} {p_{2i}}^{a_{2i}} \cdots {p_{Ni}}^{a_{Ni}}, \tag{3}$$

and form the EXPONENT VECTOR

$$\mathbf{v}(r_i) = \begin{bmatrix} a_{1i} \\ a_{2i} \\ \vdots \\ a_{Ni} \end{bmatrix}. \tag{4}$$

Now, if a_{ki} are even for any k, then $g(r_i)$ is a SQUARE NUMBER and we have found a solution to (1). If not, look for a linear combination $\sum_i c_i \mathbf{v}(r_i)$ such that the elements are all even, i.e.,

$$c_1 \begin{bmatrix} a_{11} \\ a_{21} \\ \vdots \\ a_{N1} \end{bmatrix} + c_2 \begin{bmatrix} a_{12} \\ a_{22} \\ \vdots \\ a_{N2} \end{bmatrix} + \ldots + c_N \begin{bmatrix} a_{1N} \\ a_{2N} \\ \vdots \\ a_{NN} \end{bmatrix} = \begin{bmatrix} 0 \\ 0 \\ \vdots \\ 0 \end{bmatrix} \pmod{2} \tag{5}$$

$$\begin{bmatrix} a_{11} & a_{12} & \cdots & a_{1N} \\ a_{21} & a_{22} & \cdots & a_{2N} \\ \vdots & \vdots & \ddots & \vdots \\ a_{N1} & a_{N2} & \cdots & a_{NN} \end{bmatrix} \begin{bmatrix} c_1 \\ c_2 \\ \vdots \\ c_N \end{bmatrix} = \begin{bmatrix} 0 \\ 0 \\ \vdots \\ 0 \end{bmatrix} \pmod{2}. \tag{6}$$

Since this must be solved only mod 2, the problem can be simplified by replacing the a_{ij}s with

$$b_{ij} = \begin{cases} 0 & \text{for } a_{ij} \text{ even} \\ 1 & \text{for } a_{ij} \text{ odd.} \end{cases} \tag{7}$$

GAUSSIAN ELIMINATION can then be used to solve

$$\mathbf{bc} = \mathbf{z} \tag{8}$$

for $\mathbf{c}$, where $\mathbf{z}$ is a VECTOR equal to $\mathbf{0}$ (mod 2). Once $\mathbf{c}$ is known, then we have

$$\prod_k g(r_k) \equiv \prod_k {r_k}^2 \pmod{n}, \tag{9}$$

where the products are taken over all k for which $c_k = 1$. Both sides are PERFECT SQUARES, so we have a 50% chance that this yields a nontrivial factor of n. If it does not, then we proceed to a different $\mathbf{z}$ and repeat the procedure. There is no guarantee that this method will yield a factor, but in practice it produces factors faster than any method using trial divisors. It is especially amenable to parallel processing, since each processor can work on a different value of r.

References

Bressoud, D. M. *Factorization and Prime Testing.* New York: Springer-Verlag, pp. 102–104, 1989.
Dixon, J. D. "Asymptotically Fast Factorization of Integers." *Math. Comput.* **36**, 255–260, 1981.
Lenstra, A. K. and Lenstra, H. W. Jr. "Algorithms in Number Theory." In *Handbook of Theoretical Computer Science, Volume A: Algorithms and Complexity* (Ed. J. van Leeuwen). New York: Elsevier, pp. 673–715, 1990.
Pomerance, C. "A Tale of Two Sieves." *Not. Amer. Math. Soc.* **43**, 1473–1485, 1996.

Dixon-Ferrar Formula

Let $J_\nu(z)$ be a BESSEL FUNCTION OF THE FIRST KIND, $Y_\nu(z)$ a BESSEL FUNCTION OF THE SECOND KIND, and $K_\nu(z)$ a MODIFIED BESSEL FUNCTION OF THE FIRST KIND. Also let $\Re[z] > 0$ and $|\Re[z]| < 1/2$. Then

$$J_\nu^2(z) + Y_\nu^2(z) = \frac{8\cos(\nu\pi)}{\pi^2}\int_0^\infty K_{2\nu}(2z\sinh t)\,dt.$$

see also NICHOLSON'S FORMULA, WATSON'S FORMULA

References

Gradshteyn, I. S. and Ryzhik, I. M. Eqn. 6.518 in *Tables of Integrals, Series, and Products, 5th ed.* San Diego, CA: Academic Press, p. 671, 1979.
Iyanaga, S. and Kawada, Y. (Eds.). *Encyclopedic Dictionary of Mathematics.* Cambridge, MA: MIT Press, p. 1476, 1980.

Dixon's Random Squares Factorization Method

see DIXON'S FACTORIZATION METHOD

Dixon's Theorem

$$\begin{aligned}&{}_3F_2\left[\begin{matrix} n, -x, -y \\ x+n+1, y+n+1\end{matrix}\right]\\ &= \Gamma(x+n+1)\Gamma(y+n+1)\Gamma(\tfrac{1}{2}n+1)\Gamma(x+y+\tfrac{1}{2}n+1)\\ &\times\Gamma(n+1)\Gamma(x+y+n+1)\Gamma(x+\tfrac{1}{2}n+1)\Gamma(y+\tfrac{1}{2}n+1),\end{aligned}$$

where ${}_3F_2(a,b,c;d,e;z)$ is a GENERALIZED HYPERGEOMETRIC FUNCTION and $\Gamma(z)$ is the GAMMA FUNCTION. It can be derived from the DOUGALL-RAMANUJAN IDENTITY. It can be written more symmetrically as

$${}_3F_2(a,b,c;d,e;1) = \frac{(\frac{1}{2}a)!(a-b)!(a-c)!(\frac{1}{2}a-b-c)!}{a!(\frac{1}{2}a-b)!(\frac{1}{2}a-c)!(a-b-c)!},$$

where $1 + a/2 - b - c$ has a positive REAL PART, $d = a - b + 1$, and $e = a - c + 1$. The identity can also be written as the beautiful symmetric sum

$$\sum_k (-1)^k \binom{a+b}{a+k}\binom{a+c}{c+k}\binom{b+c}{b+k} = \frac{(a+b+c)!}{a!b!c!}$$

(Petkovšek 1996).

see also DOUGALL-RAMANUJAN IDENTITY, GENERALIZED HYPERGEOMETRIC FUNCTION

References

Bailey, W. N. *Generalised Hypergeometric Series.* Cambridge, England: Cambridge University Press, 1935.
Cartier, P. and Foata, D. *Problèmes combinatoires de commutation et réarrangements.* New York: Springer-Verlag, 1969.
Knuth, D. E. *The Art of Computer Programming, Vol. 1: Fundamental Algorithms, 2nd ed.* Reading, MA: Addison-Wesley, 1973.
Petkovšek, M.; Wilf, H. S.; and Zeilberger, D. *A=B.* Wellesley, MA: A. K. Peters, p. 43, 1996.
Zeilberger, D. and Bressoud, D. "A Proof of Andrew's q-Dyson Conjecture." *Disc. Math.* **54**, 201–224, 1985.

Dobiński's Formula

Gives the nth BELL NUMBER,

$$B_n = \frac{1}{e}\sum_{k=0}^\infty \frac{k^n}{k!}. \tag{1}$$

It can be derived by dividing the formula for a STIRLING NUMBER OF THE SECOND KIND by $m!$, yielding

$$\frac{m^n}{m!} = \sum_{k=1}^m \left\{ n \atop k \right\} \frac{1}{(m-k)!}. \tag{2}$$

Then

$$\sum_{m=1}^\infty \frac{m^n}{m!}\lambda^m = \left(\sum_{k=1}^n \left\{ n \atop k \right\}\lambda^k\right)\left(\sum_{k=0}^\infty \frac{\lambda^j}{j!}\right), \tag{3}$$

and

$$\sum_{k=1}^n \left\{ n \atop k \right\}\lambda^k = e^{-\lambda}\sum_{m=1}^\infty \frac{m^n}{m!}\lambda^m. \tag{4}$$

Now setting $\lambda = 1$ gives the identity (Dobiński 1877; Rota 1964; Berge 1971, p. 44; Comtet 1974, p. 211; Roman 1984, p. 66; Lupas 1988; Wilf 1990, p. 106; Chen and Yeh 1994; Pitman 1997).

References

Berge, C. *Principles of Combinatorics.* New York: Academic Press, 1971.
Chen, B. and Yeh, Y.-N. "Some Explanations of Dobinski's Formula." *Studies Appl. Math.* **92**, 191–199, 1994.
Comtet, L. *Advanced Combinatorics.* Boston, MA: Reidel, 1974.
Dobiński, G. "Summierung der Reihe $\sum n^m/n!$ für $m = 1, 2, 3, 4, 5, \ldots$." *Grunert Archiv (Arch. Math. Phys.)* **61**, 333–336, 1877.

Foata, D. *La série génératrice exponentielle dans les problèmes d'énumération.* Vol. 54 of *Séminaire de Mathématiques supérieures.* Montréal, Canada: Presses de l'Université de Montréal, 1974.

Lupas, A. "Dobiński-Type Formula for Binomial Polynomials." *Stud. Univ. Babes-Bolyai Math.* **33**, 30–44, 1988.

Pitman, J. "Some Probabilistic Aspects of Set Partitions." *Amer. Math. Monthly* **104**, 201–209, 1997.

Roman, S. *The Umbral Calculus.* New York: Academic Press, 1984.

Rota, G.-C. "The Number of Partitions of a Set." *Amer. Math. Monthly* **71**, 498–504, 1964.

Wilf, H. *Generatingfunctionology, 2nd ed.* San Diego, CA: Academic Press, 1990.

Dodecadodecahedron

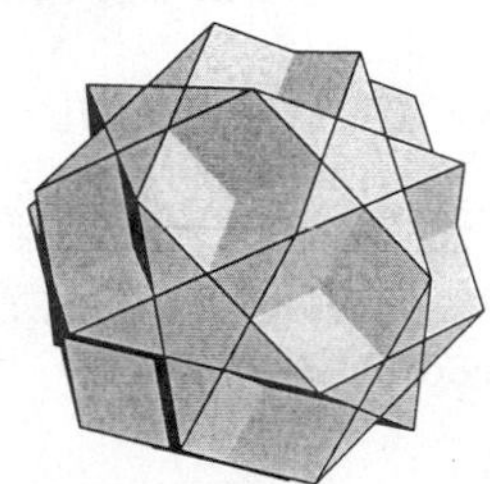

The UNIFORM POLYHEDRON U_{36} whose DUAL POLYHEDRON is the MEDIAL RHOMBIC TRIACONTAHEDRON. The solid is also called the GREAT DODECADODECAHEDRON, and its DUAL POLYHEDRON is also called the SMALL STELLATED TRIACONTAHEDRON. It can be obtained by TRUNCATING a GREAT DODECAHEDRON or FACETING a ICOSIDODECAHEDRON with PENTAGONS and covering remaining open spaces with PENTAGRAMS (Holden 1991, p. 103). A FACETED version is the GREAT DODECAHEMICOSAHEDRON. The dodecadodecahedron is an ARCHIMEDEAN SOLID STELLATION. The dodecadodecahedron has SCHLÄFLI SYMBOL $\{\frac{5}{2}, 5\}$ and WYTHOFF SYMBOL $2 \mid \frac{5}{2}\, 5$. Its faces are $12\{\frac{5}{2}\} + 12\{5\}$, and its CIRCUMRADIUS for unit edge length is

$$R = 1.$$

References

Cundy, H. and Rollett, A. *Mathematical Models, 3rd ed.* Stradbroke, England: Tarquin Pub., p. 123, 1989.

Holden, A. *Shapes, Space, and Symmetry.* New York: Dover, 1991.

Wenninger, M. J. *Polyhedron Models.* Cambridge, England: Cambridge University Press, p. 112, 1989.

Dodecagon

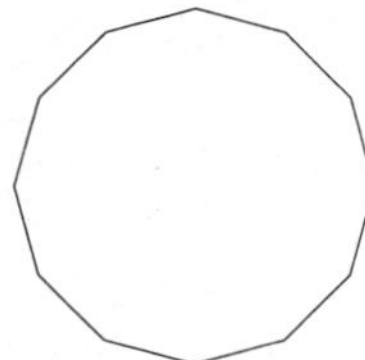

The constructible regular 12-sided POLYGON with SCHLÄFLI SYMBOL $\{12\}$. The INRADIUS r, CIRCUMRADIUS R, and AREA A can be computed directly from the formulas for a general regular POLYGON with side length s and $n = 12$ sides,

$$r = \tfrac{1}{2} s \cot\left(\frac{\pi}{12}\right) = \tfrac{1}{2}(2 + \sqrt{3})s \qquad (1)$$

$$R = \tfrac{1}{2} s \cot\left(\frac{\pi}{12}\right) = \tfrac{1}{2}(\sqrt{2} + \sqrt{6})s \qquad (2)$$

$$A = \tfrac{1}{4} n s^2 \cot\left(\frac{\pi}{12}\right) = 3(2 + \sqrt{3})s^2. \qquad (3)$$

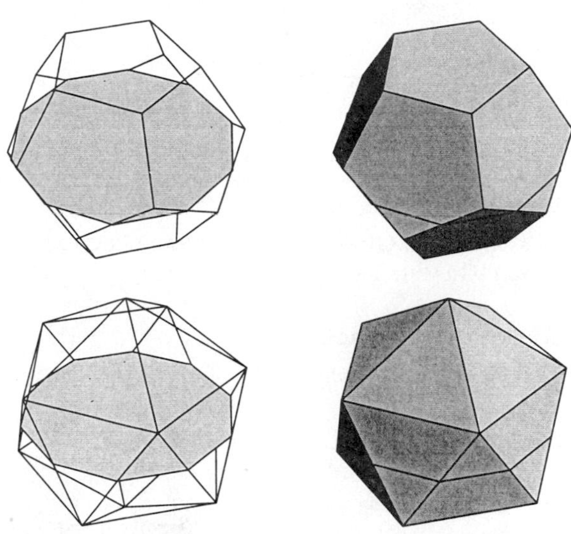

A PLANE PERPENDICULAR to a C_5 axis of a DODECAHEDRON or ICOSAHEDRON cuts the solid in a regular DECAGONAL CROSS-SECTION (Holden 1991, pp. 24–25).

The GREEK, LATIN, and MALTESE CROSSES are all irregular dodecagons.

see also DECAGON, DODECAGRAM, DODECAHEDRON, GREEK CROSS, LATIN CROSS, MALTESE CROSS, TRIGONOMETRY VALUES—$\pi/12$, UNDECAGON

References

Holden, A. *Shapes, Space, and Symmetry.* New York: Dover, 1991.

Dodecagram

The STAR POLYGON $\left\{\begin{smallmatrix}12\\5\end{smallmatrix}\right\}$.

see also STAR POLYGON, TRIGONOMETRY VALUES—$\pi/12$

Dodecahedral Conjecture

In any unit SPHERE PACKING, the volume of any VORONOI CELL around any sphere is at least as large as a regular DODECAHEDRON of INRADIUS 1. If true, this would provide a bound on the densest possible sphere packing greater than any currently known. It would not, however, be sufficient to establish the KEPLER CONJECTURE.

Dodecahedral Graph

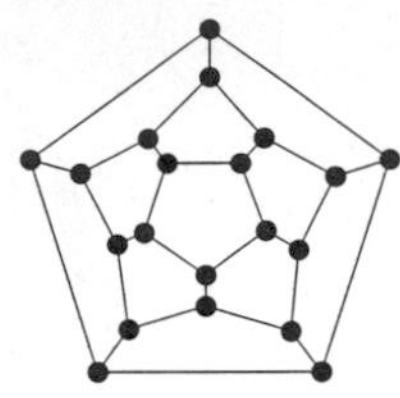

A POLYHEDRAL GRAPH.

see also CUBICAL GRAPH, ICOSAHEDRAL GRAPH, OCTAHEDRAL GRAPH, TETRAHEDRAL GRAPH

Dodecahedral Space

see POINCARÉ MANIFOLD

Dodecahedron

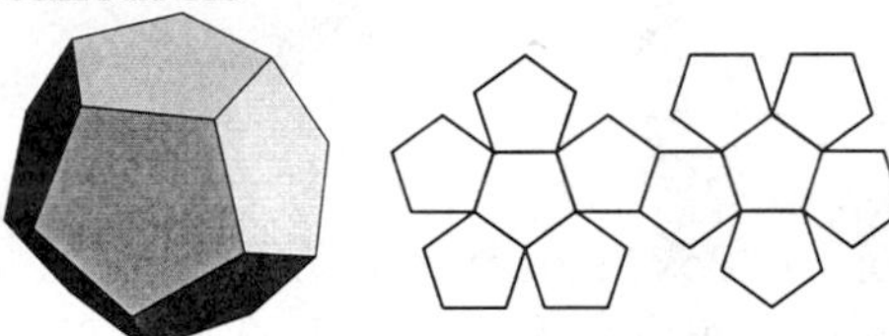

The regular dodecahedron is the PLATONIC SOLID (P_4) composed of 20 VERTICES, 30 EDGES, and 12 PENTAGONAL FACES. It is given by the symbol 12{5}, the SCHLÄFLISYMBOL {5,3}. It is also UNIFORM POLYHEDRON U_{23} and has WYTHOFF SYMBOL 3 | 2 5. The dodecahedron has the ICOSAHEDRAL GROUP I_h of symmetries.

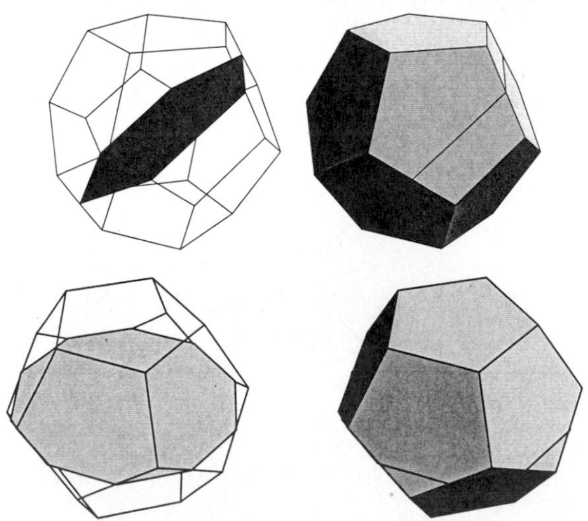

A PLANE PERPENDICULAR to a C_3 axis of a dodecahedron cuts the solid in a regular HEXAGONAL CROSS-SECTION (Holden 1991, p. 27). A PLANE PERPENDICULAR to a C_5 axis of a dodecahedron cuts the solid in a regular DECAGONAL CROSS-SECTION (Holden 1991, p. 24).

The DUAL POLYHEDRON of the dodecahedron is the ICOSAHEDRON.

When the dodecahedron with edge length $\sqrt{10-2\sqrt{5}}$ is oriented with two opposite faces parallel to the xy-PLANE, the vertices of the top and bottom faces lie at $z=\pm(\phi+1)$ and the other VERTICES lie at $z=\pm(\phi-1)$, where ϕ is the GOLDEN RATIO. The explicit coordinates are

$$\pm\left(2\cos(\tfrac{2}{5}\pi i), 2\sin(\tfrac{2}{5}\pi i), \phi+1\right) \tag{1}$$

$$\pm\left(2\phi\cos(\tfrac{2}{5}\pi i), 2\phi\sin(\tfrac{2}{5}\pi i), \phi-1\right) \tag{2}$$

with $i=0, 1, \ldots, 4$, where ϕ is the GOLDEN RATIO. Explicitly, these coordinates are

$$\mathbf{x}_{10}^{\pm} = \pm(2, 0, \tfrac{1}{2}(3+\sqrt{5})) \tag{3}$$

$$\mathbf{x}_{11}^{\pm} = \pm(\tfrac{1}{2}(\sqrt{5}-1), \tfrac{1}{2}\sqrt{10+2\sqrt{5}}, \tfrac{1}{2}(3+\sqrt{5})) \tag{4}$$

$$\mathbf{x}_{12}^{\pm} = \pm(-\tfrac{1}{2}(1+\sqrt{5}), \tfrac{1}{2}\sqrt{10-2\sqrt{5}}, \tfrac{1}{2}(3+\sqrt{5})) \tag{5}$$

$$\mathbf{x}_{13}^{\pm} = \pm(-\tfrac{1}{2}(1+\sqrt{5}), -\tfrac{1}{2}\sqrt{10-2\sqrt{5}}, \tfrac{1}{2}(3+\sqrt{5})) \tag{6}$$

$$\mathbf{x}_{14}^{\pm} = \pm(\tfrac{1}{2}(\sqrt{5}-1), -\tfrac{1}{2}\sqrt{10+2\sqrt{5}}, \tfrac{1}{2}(3+\sqrt{5})) \tag{7}$$

$$\mathbf{x}_{20}^{\pm} = \pm(1+\sqrt{5}, 0, \tfrac{1}{2}(\sqrt{5}-1)) \tag{8}$$

$$\mathbf{x}_{21}^{\pm} = \pm(1, \sqrt{5+2\sqrt{5}}, \tfrac{1}{2}(\sqrt{5}-1)) \tag{9}$$

$$\mathbf{x}_{22}^{\pm} = \pm(-\tfrac{1}{2}(3+\sqrt{5}), \tfrac{1}{2}\sqrt{10+2\sqrt{5}}, \tfrac{1}{2}(\sqrt{5}-1)) \tag{10}$$

$$\mathbf{x}_{23}^{\pm} = \pm(-\tfrac{1}{2}(3+\sqrt{5}), -\tfrac{1}{2}\sqrt{10+2\sqrt{5}}, \tfrac{1}{2}(\sqrt{5}-1)) \tag{11}$$

$$\mathbf{x}_{24}^{\pm} = \pm(1, -\sqrt{5+2\sqrt{5}}, \tfrac{1}{2}(\sqrt{5}-1)), \tag{12}$$

where $\mathbf{x}_{1i}^{+}$ are the top vertices, $\mathbf{x}_{2i}^{+}$ are the vertices above the mid-plane, $\mathbf{x}_{2i}^{-}$ are the vertices below the mid-plane, and $\mathbf{x}_{2i}^{-}$ are the bottom vertices. The VERTICES of a dodecahedron can be given in a simple form for a dodecahedron of side length $a=\sqrt{5}-1$ by $(0, \pm\phi^{-1}, \pm\phi)$, $(\pm\phi, 0, \pm\phi^{-1})$, $(\pm\phi^{-1}, \pm\phi, 0)$, and $(\pm 1, \pm 1, \pm 1)$.

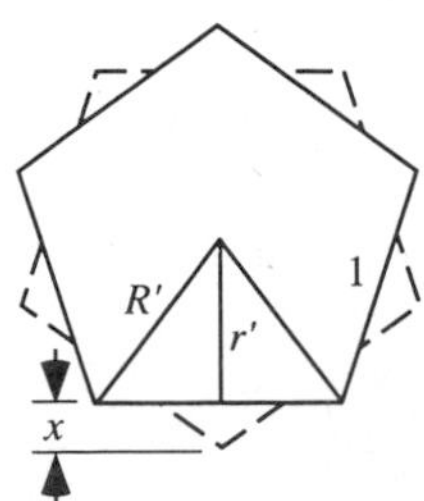

For a dodecahedron of unit edge length $a = 1$, the CIRCUMRADIUS R' and INRADIUS r' of a PENTAGONAL FACE are

$$R' = \tfrac{1}{10}\sqrt{50+10\sqrt{5}} \quad (13)$$

$$r' = \tfrac{1}{10}\sqrt{25+10\sqrt{5}}. \quad (14)$$

The SAGITTA x is then given by

$$x \equiv R' - r' = \tfrac{1}{10}\sqrt{125-10\sqrt{5}}. \quad (15)$$

Now consider the following figure.

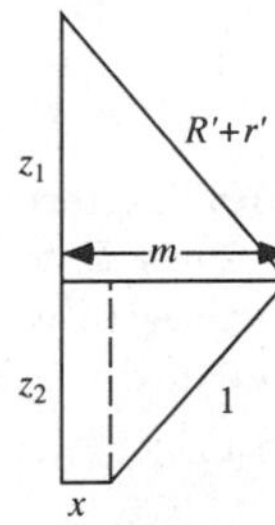

Using the PYTHAGOREAN THEOREM on the figure then gives

$$z_1{}^2 + m^2 = (R'+r)^2 \quad (16)$$

$$z_2{}^2 + (m-x)^2 = 1 \quad (17)$$

$$\left(\frac{z_1+z_2}{2}\right)^2 + R'^2 = \left(\frac{z_1-z_2}{2}\right)^2 + (m+r')^2. \quad (18)$$

Equation (18) can be written

$$z_1 z_2 + r^2 = (m+r')^2. \quad (19)$$

Solving (16), (17), and (19) simultaneously gives

$$m = r' = \tfrac{1}{10}\sqrt{25+10\sqrt{5}} \quad (20)$$

$$z_1 = 2r' = \tfrac{1}{5}\sqrt{25+10\sqrt{5}} \quad (21)$$

$$z_2 = R' = \tfrac{1}{10}\sqrt{50+10\sqrt{5}}. \quad (22)$$

The INRADIUS of the dodecahedron is then given by

$$r = \tfrac{1}{2}(z_1+z_2), \quad (23)$$

so

$$r^2 = \tfrac{1}{4}\left(\tfrac{1}{10}\sqrt{50+10\sqrt{5}} + \tfrac{1}{5}\sqrt{25+10\sqrt{5}}\right)^2 = \tfrac{1}{40}(25+11\sqrt{5}), \quad (24)$$

and

$$r = \sqrt{\frac{25+11\sqrt{5}}{40}} = \tfrac{1}{20}\sqrt{250+110\sqrt{5}} = 1.11351\ldots. \quad (25)$$

Now,

$$R^2 = R'^2 + r^2 = [\tfrac{1}{100}(50+10\sqrt{5}) + \tfrac{1}{400}(250+110\sqrt{5})] = \tfrac{3}{8}(3+\sqrt{5}), \quad (26)$$

and the CIRCUMRADIUS is

$$R = a\sqrt{\tfrac{3}{8}(3+\sqrt{5})} = \tfrac{1}{4}(\sqrt{15}+\sqrt{3}) = 1.40125\ldots. \quad (27)$$

The INTERRADIUS is given by

$$\rho^2 = r'^2 + r^2 = [\tfrac{1}{100}(25+10\sqrt{5}) + \tfrac{1}{400}(250+110\sqrt{5})] = \tfrac{1}{8}(7+3\sqrt{5}), \quad (28)$$

so

$$\rho = \tfrac{1}{4}(3+\sqrt{5}) = 1.30901\ldots. \quad (29)$$

The AREA of a single FACE is the AREA of a PENTAGON,

$$A = \tfrac{1}{4}\sqrt{25+10\sqrt{5}}. \quad (30)$$

The VOLUME of the dodecahedron can be computed by summing the volume of the 12 constituent PENTAGONAL PYRAMIDS,

$$V = 12(\tfrac{1}{3}Ar) = 12(\tfrac{1}{3})(\tfrac{1}{4}\sqrt{25+10\sqrt{5}})(\tfrac{1}{20}\sqrt{250+110\sqrt{5}}) = \tfrac{1}{20}(75+35\sqrt{5}) = \tfrac{1}{4}(15+7\sqrt{5}). \quad (31)$$

Apollonius showed that the VOLUME V and SURFACE AREA A of the dodecahedron and its DUAL the ICOSAHEDRON are related by

$$\frac{V_{\text{icosahedron}}}{V_{\text{dodecahedron}}} = \frac{A_{\text{icosahedron}}}{A_{\text{dodecahedron}}}. \quad (32)$$

The HEXAGONAL SCALENOHEDRON is an irregular dodecahedron.

see also AUGMENTED DODECAHEDRON, AUGMENTED TRUNCATED DODECAHEDRON, DODECAGON, DODECAHEDRON-ICOSAHEDRON COMPOUND, ELONGATED DODECAHEDRON, GREAT DODECAHEDRON, GREAT STELLATED DODECAHEDRON, HYPERBOLIC DODECAHEDRON, ICOSAHEDRON, METABIAUGMENTED DODECAHEDRON, METABIAUGMENTED TRUNCATED DODECAHEDRON, PARABIAUGMENTED DODECAHEDRON, PARABIAUGMENTED TRUNCATED DODECAHEDRON, PYRITOHEDRON, RHOMBIC DODECAHEDRON, SMALL STELLATED DODECAHEDRON, TRIAUGMENTED DODECAHEDRON, TRIAUGMENTED TRUNCATED DODECAHEDRON, TRIGONAL DODECAHEDRON, TRIGONOMETRY VALUES—$\pi/5$ TRUNCATED DODECAHEDRON

References

Cundy, H. and Rollett, A. *Mathematical Models, 3rd ed.* Stradbroke, England: Tarquin Pub., 1989.

Davie, T. "The Dodecahedron." `http://www.dcs.st-and.ac.uk/~ad/mathrecs/polyhedra/dodecahedron.html`.

Holden, A. *Shapes, Space, and Symmetry.* New York: Dover, 1991.

Dodecahedron-Icosahedron Compound

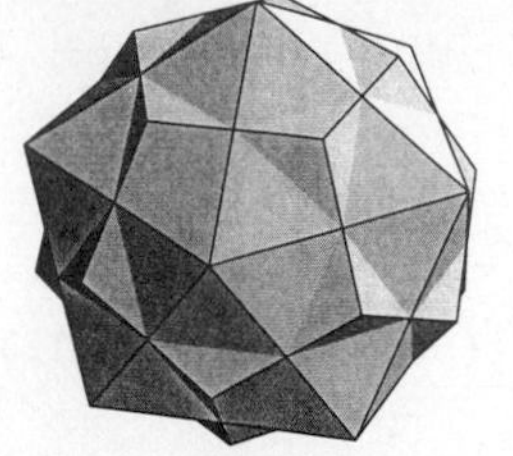

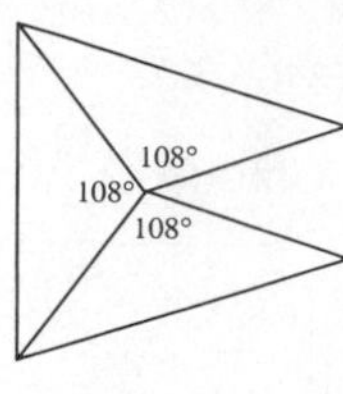

A POLYHEDRON COMPOUND of a DODECAHEDRON and ICOSAHEDRON which is most easily constructed by adding 20 triangular PYRAMIDS, constructed as above, to an ICOSAHEDRON. In the compound, the DODECAHEDRON and ICOSAHEDRON are rotated $\pi/5$ radians with respect to each other, and the ratio of the ICOSAHEDRON to DODECAHEDRON edges lengths are the GOLDEN RATIO ϕ.

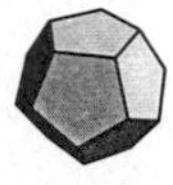 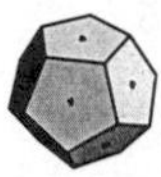   

 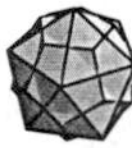  

The above figure shows compounds composed of a DODECAHEDRON of unit edge length and ICOSAHEDRA having edge lengths varying from $\sqrt{5}/2$ (inscribed in the dodecahedron) to 2 (circumscribed about the dodecahedron).

The intersecting edges of the compound form the DIAGONALS of 30 RHOMBUSES comprising the TRIACONTAHEDRON, which is the the DUAL POLYHEDRON of the ICOSIDODECAHEDRON (Ball and Coxeter 1987). The dodecahedron-icosahedron is the first STELLATION of the ICOSIDODECAHEDRON.

see also DODECAHEDRON, ICOSAHEDRON, ICOSIDODECAHEDRON, POLYHEDRON COMPOUND

References
Cundy, H. and Rollett, A. *Mathematical Models, 2nd ed.* Stradbroke, England: Tarquin Pub., p. 131, 1989.
Wenninger, M. J. *Polyhedron Models.* Cambridge, England: Cambridge University Press, p. 76, 1989.

Dodecahedron Stellations

The dodecahedron has three STELLATIONS: the GREAT DODECAHEDRON, GREAT STELLATED DODECAHEDRON, and SMALL STELLATED DODECAHEDRON. The only STELLATIONS of PLATONIC SOLIDS which are UNIFORM POLYHEDRA are these three and one ICOSAHEDRON STELLATION. Bulatov has produced 270 stellations of a deformed dodecahedron.

see also ICOSAHEDRON STELLATIONS, STELLATED POLYHEDRON, STELLATION

References
Bulatov, V.v "270 Stellations of Deformed Dodecahedron." http://www.physics.orst.edu/~bulatov/polyhedra/dodeca270/.

Dodecahedron 2-Compound

A compound of two dodecahedra with the symmetry of the CUBE arises by combining the two dodecahedra rotated 90° with respect to each other about a common C_2 axis (Holden 1991, p. 37).

see also POLYHEDRON COMPOUND

References
Holden, A. *Shapes, Space, and Symmetry.* New York: Dover, 1991.

Domain

A connected OPEN SET. The term domain is also used to describe the set of values D for which a FUNCTION is defined. The set of values to which D is sent by the function (MAP) is then called the RANGE.

see also MAP, ONE-TO-ONE, ONTO, RANGE (IMAGE), REINHARDT DOMAIN

Domain Invariance Theorem

The Invariance of Domain Theorem is that if $f : A \to \mathbb{R}^n$ is a ONE-TO-ONE continuous MAP from A, a compact subset of $\mathbb{R}^n$, then the interior of A is mapped to the interior of $f(A)$.

see also DIMENSION INVARIANCE THEOREM

Dome

see BOHEMIAN DOME, GEODESIC DOME, HEMISPHERE, SPHERICAL CAP, TORISPHERICAL DOME, VAULT

Dominance

The dominance RELATION on a SET of points in EUCLIDEAN n-space is the INTERSECTION of the n coordinate-wise orderings. A point p dominates a point q provided that every coordinate of p is at least as large as the corresponding coordinate of q.

The dominance orders in $\mathbb{R}^n$ are precisely the POSETS of DIMENSION at most n.

see also PARTIALLY ORDERED SET, REALIZER

Domino

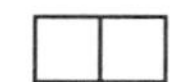

The unique 2-POLYOMINO consisting of two equal squares connected along a complete EDGE.

The FIBONACCI NUMBER F_{n+1} gives the number of ways for 2×1 dominoes to cover a $2 \times n$ CHECKERBOARD, as illustrated in the following diagrams (Dickau).

  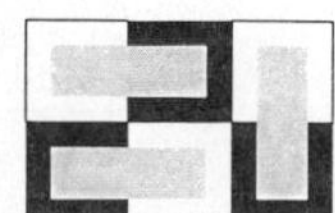

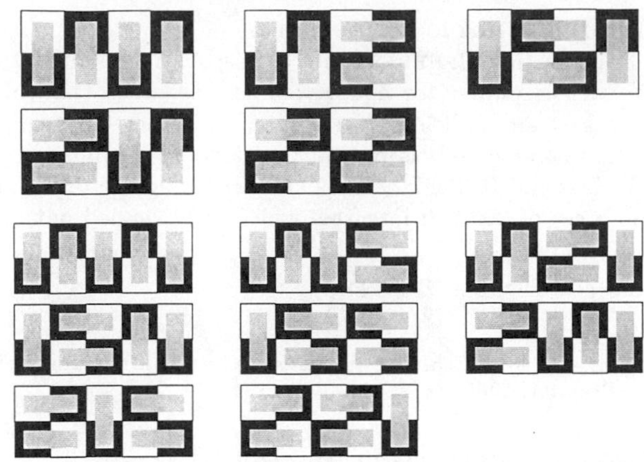

see also FIBONACCI NUMBER, GOMORY'S THEOREM, HEXOMINO, PENTOMINO, POLYOMINO, TETROMINO, TRIOMINO

References

Dickau, R. M. "Fibonacci Numbers." http://www.prairienet.org/~pops/fibboard.html.

Gardner, M. "Polyominoes." Ch. 13 in *The Scientific American Book of Mathematical Puzzles & Diversions.* New York: Simon and Schuster, pp. 124–140, 1959.

Kraitchik, M. "Dominoes." §12.1.22 in *Mathematical Recreations.* New York: W. W. Norton, pp. 298–302, 1942.

Lei, A. "Domino." http://www.cs.ust.hk/~philipl/omino/domino.html.

Madachy, J. S. "Domino Recreations." *Madachy's Mathematical Recreations.* New York: Dover, pp. 209–219, 1979.

Domino Problem

see WANG'S CONJECTURE

Donaldson Invariants

Distinguish between smooth MANIFOLDS in 4-D.

Donkin's Theorem

The product of three translations along the directed sides of a TRIANGLE through twice the lengths of these sides is the identity.

Donut

see TORUS

Doob's Theorem

A theorem proved by Doob (1942) which states that any random process which is both GAUSSIAN and MARKOV has the following forms for its correlation function, spectral density, and probability densities:

$$C_y(\tau) = {\sigma_y}^2 e^{-\tau/\tau_r}$$

$$G_y(f) = \frac{4{\tau_r}^{-1}{\sigma_y}^2}{(2\pi f)^2 + {\tau_r}^{-2}}$$

$$p_1(y) = \frac{1}{\sqrt{2\pi{\sigma_y}^2}} e^{-(y-\bar{y})^2/2{\sigma_y}^2}$$

$$p_2(y_1|y_2,\tau) = \frac{1}{\sqrt{2\pi(1-e^{-2\tau/\tau_r}){\sigma_y}^2}} \times \exp\left\{-\frac{[(y_2-\bar{y}) - e^{-\tau/\tau_r}(y_1-\bar{y})]^2}{2(1-e^{-2\tau/\tau_r}){\sigma_y}^2}\right\},$$

where $\bar{y}$ is the MEAN, σ_y the STANDARD DEVIATION, and τ_r the relaxation time.

References

Doob, J. L. "Topics in the Theory of Markov Chains." *Trans. Amer. Math. Soc.* **52**, 37–64, 1942.

Dot

The "dot" $\cdot$ has several meanings in mathematics, including MULTIPLICATION ($a \cdot b$ is pronounced "a times b"), computation of a DOT PRODUCT ($\mathbf{a}\cdot\mathbf{b}$ is pronounced "**a** dot **b**"), or computation of a time DERIVATIVE ($\dot{a}$ is pronounced "a dot").

see also DERIVATIVE, DOT PRODUCT, TIMES

Dot Product

The dot product can be defined by

$$\mathbf{X}\cdot\mathbf{Y} = |\mathbf{X}|\,|\mathbf{Y}|\cos\theta, \tag{1}$$

where θ is the angle between the vectors. It follows immediately that $\mathbf{X}\cdot\mathbf{Y} = 0$ if $\mathbf{X}$ is PERPENDICULAR to $\mathbf{Y}$. The dot product is also called the INNER PRODUCT and written $\langle a, b\rangle$. By writing

$$A_x = A\cos\theta_A \qquad B_x = B\cos\theta_B \tag{2}$$

$$A_y = A\sin\theta_A \qquad B_y = B\sin\theta_B, \tag{3}$$

it follows that (1) yields

$$\begin{aligned}\mathbf{A}\cdot\mathbf{B} &= AB\cos(\theta_A - \theta_B)\\ &= AB(\cos\theta_A\cos\theta_B + \sin\theta_A\sin\theta_B)\\ &= A\cos\theta_A B\cos\theta_B + A\sin\theta_A B\sin\theta_B\\ &= A_xB_x + A_yB_y.\end{aligned} \tag{4}$$

So, in general,

$$\mathbf{X}\cdot\mathbf{Y} = x_1y_1 + \ldots + x_ny_n. \tag{5}$$

The dot product is COMMUTATIVE

$$\mathbf{X}\cdot\mathbf{Y} = \mathbf{Y}\cdot\mathbf{X}, \tag{6}$$

ASSOCIATIVE

$$(r\mathbf{X})\cdot\mathbf{Y} = r(\mathbf{X}\cdot\mathbf{Y}), \tag{7}$$

and DISTRIBUTIVE

$$\mathbf{X}\cdot(\mathbf{Y}+\mathbf{Z}) = \mathbf{X}\cdot\mathbf{Y} + \mathbf{X}\cdot\mathbf{Z}. \tag{8}$$

The DERIVATIVE of a dot product of VECTORS is

$$\frac{d}{dt}[\mathbf{r}_1(t)\cdot\mathbf{r}_2(t)] = \mathbf{r}_1(t)\cdot\frac{d\mathbf{r}_2}{dt} + \frac{d\mathbf{r}_1}{dt}\cdot\mathbf{r}_2(t). \tag{9}$$

The dot product is invariant under rotations

$$\begin{aligned}\mathbf{A}' \cdot \mathbf{B}' &= A_i' B_i' = a_{ij} A_j a_{ik} B_k = (a_{ij} a_{ik}) A_j B_k \\ &= \delta_{jk} A_j B_k = A_j B_j = \mathbf{A} \cdot \mathbf{B}, \end{aligned} \tag{10}$$

where EINSTEIN SUMMATION has been used.

The dot product is also defined for TENSORS A and B by

$$A \cdot B \equiv A^\alpha B_\alpha. \tag{11}$$

see also CROSS PRODUCT, INNER PRODUCT, OUTER PRODUCT, WEDGE PRODUCT

References

Arfken, G. "Scalar or Dot Product." §1.3 in *Mathematical Methods for Physicists, 3rd ed.* Orlando, FL: Academic Press, pp. 13–18, 1985.

Douady's Rabbit Fractal

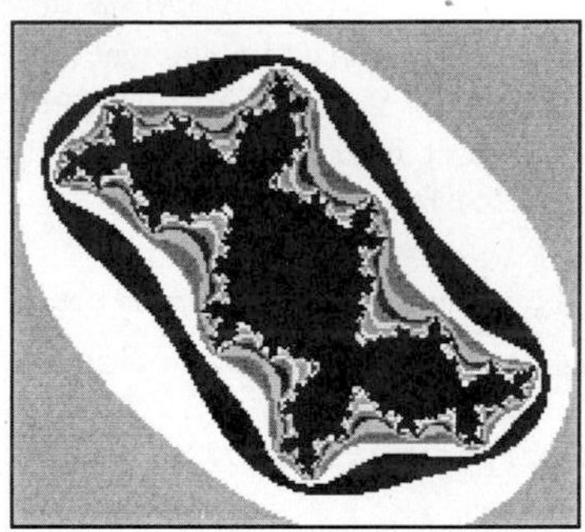

A JULIA SET with $c = -0.123 + 0.745i$, also known as the DRAGON FRACTAL.

see also SAN MARCO FRACTAL, SIEGEL DISK FRACTAL

References

Wagon, S. *Mathematica in Action.* New York: W. H. Freeman, p. 176, 1991.

Double Bubble

The planar double bubble (three circular arcs meeting in two points at equal 120° ANGLES) has the minimum PERIMETER for enclosing two equal areas (Foisy 1993, Morgan 1995).

see also APPLE, BUBBLE, DOUBLE BUBBLE CONJECTURE, SPHERE-SPHERE INTERSECTION

References

Campbell, P. J. (Ed.). Reviews. *Math. Mag.* **68**, 321, 1995.

Foisy, J.; Alfaro, M.; Brock, J.; Hodges, N.; and Zimba, J. "The Standard Double Soap Bubble in $\mathbb{R}^2$ Uniquely Minimizes Perimeter." *Pacific J. Math.* **159**, 47–59, 1993.

Morgan, F. "The Double Bubble Conjecture." *FOCUS* **15**, 6–7, 1995.

Peterson, I. "Toil and Trouble over Double Bubbles." *Sci. News* **148**, 101, Aug. 12, 1995.

Double Bubble Conjecture

Two partial SPHERES with a separating boundary (which is planar for equal volumes) separate two volumes of air with less AREA than any other boundary. The planar case was proved true for equal volumes by J. Hass and R. Schlafy in 1995 by reducing the problem to a set of 200,260 integrals which they carried out on an ordinary PC.

see also DOUBLE BUBBLE

References

Haas, J. and Schlafy, R. "Double Bubbles Minimize." Preprint, 1995.

Double Contraction Relation

A TENSOR t is said to satisfy the double contraction relation when

$$t_{ij}^{m*} t_{ij}^n = \delta_{mn}.$$

This equation is satisfied by

$$\begin{aligned}\hat{t}^0 &= \frac{2\hat{\mathbf{z}}\hat{\mathbf{z}} - \hat{\mathbf{x}}\hat{\mathbf{x}} - \hat{\mathbf{y}}\hat{\mathbf{y}}}{\sqrt{6}} \\ \hat{t}^{\pm 1} &= \mp\tfrac{1}{2}(\hat{\mathbf{x}}\hat{\mathbf{z}} + \hat{\mathbf{z}}\hat{\mathbf{x}}) - \tfrac{1}{2}i(\hat{\mathbf{y}}\hat{\mathbf{z}} - \hat{\mathbf{z}}\hat{\mathbf{y}}) \\ \hat{t}^{\pm 2} &= \mp\tfrac{1}{2}(\hat{\mathbf{x}}\hat{\mathbf{x}} + \hat{\mathbf{y}}\hat{\mathbf{y}}) - \tfrac{1}{2}i(\hat{\mathbf{x}}\hat{\mathbf{y}} - \hat{\mathbf{y}}\hat{\mathbf{x}}),\end{aligned}$$

where the hat denotes zero trace, symmetric unit TENSORS. These TENSORS are used to define the SPHERICAL HARMONIC TENSOR.

see also SPHERICAL HARMONIC TENSOR, TENSOR

References

Arfken, G. "Alternating Series." *Mathematical Methods for Physicists, 3rd ed.* Orlando, FL: Academic Press, p. 140, 1985.

Double Cusp

see DOUBLE POINT

Double Exponential Distribution

see FISHER-TIPPETT DISTRIBUTION, LAPLACE DISTRIBUTION

Double Exponential Integration

An excellent NUMERICAL INTEGRATION technique used by *Maple V R4®* (Waterloo Maple Inc.) for numerical computation of integrals.

see also INTEGRAL, INTEGRATION, NUMERICAL INTEGRATION

References

Davis, P. J. and Rabinowitz, P. *Methods of Numerical Integration, 2nd ed.* New York: Academic Press, p. 214, 1984.

Di Marco, G.; Favati, P.; Lotti, G.; and Romani, F. "Asymptotic Behaviour of Automatic Quadrature." *J. Complexity* **10**, 296–340, 1994.

Mori, M. *Developments in the Double Exponential Formula for Numerical Integration. Proceedings of the International Congress of Mathematicians, Kyoto 1990.* New York: Springer-Verlag, pp. 1585–1594, 1991.

Mori, M. and Ooura, T. "Double Exponential Formulas for Fourier Type Integrals with a Divergent Integrand." In *Contributions in Numerical Mathematics* (Ed. R. P. Agarwal). World Scientific Series in Applicable Analysis, Vol. 2, pp. 301–308, 1993.
Ooura, T. and Mori, M. "The Double Exponential Formula for Oscillatory Functions over the Half Infinite Interval." *J. Comput. Appl. Math.* **38**, 353–360, 1991.
Takahasi, H. and Mori, M. "Double Exponential Formulas for Numerical Integration." *Pub. RIMS Kyoto Univ.* **9**, 721–741, 1974.
Toda, H. and Ono, H. "Some Remarks for Efficient Usage of the Double Exponential Formulas." *Kokyuroku RIMS Kyoto Univ.* **339**, 74–109, 1978.

Double Factorial

The double factorial is a generalization of the usual FACTORIAL $n!$ defined by

$$n!! \equiv \begin{cases} n\cdot(n-2)\ldots 5\cdot 3\cdot 1 & n \text{ odd} \\ n\cdot(n-2)\ldots 6\cdot 4\cdot 2 & n \text{ even} \\ 1 & n=-1,0. \end{cases} \quad (1)$$

For $n = 0, 1, 2, \ldots$, the first few values are 1, 1, 2, 3, 8, 15, 48, 105, 384, ... (Sloane's A006882).

There are many identities relating double factorials to FACTORIALS. Since

$$\begin{aligned}&(2n+1)!!2^n n!\\ &= [(2n+1)(2n-1)\cdots 1][2n][2(n-1)][2(n-2)]\cdots 2(1)\\ &= [(2n+1)(2n-1)\cdots 1][2n(2n-2)(2n-4)\cdots 2]\\ &= (2n+1)(2n)(2n-1)(2n-2)(2n-3)(2n-4)\cdots 2(1)\\ &= (2n+1)!, \end{aligned} \quad (2)$$

it follows that $(2n+1)!! = \frac{(2n+1)!}{2^n n!}$. Since

$$\begin{aligned}(2n)!! &= (2n)(2n-2)(2n-4)\cdots 2\\ &= [2(n)][2(n-1)][2(n-2)]\cdots 2 = 2^n n!, \end{aligned} \quad (3)$$

it follows that $(2n)!! = 2^n n!$. Since

$$\begin{aligned}&(2n-1)!!2^n n!\\ &= [(2n-1)(2n-3)\cdots 1][2n][2(n-1)][2(n-2)]\cdots 2(1)\\ &= (2n-1)(2n-3)\cdots 1][2n(2n-2)(2n-4)\cdots 2]\\ &= 2n(2n-1)(2n-2)(2n-3)(2n-4)\cdots 2(1)\\ &= (2n)!, \end{aligned} \quad (4)$$

it follows that

$$(2n-1)!! = \frac{(2n)!}{2^n n!}. \quad (5)$$

Similarly, for $n = 0, 1, \ldots$,

$$(-2n-1)!! = \frac{(-1)^n}{(2n-1)!!} = \frac{(-1)n2^n n!}{(2n)!}. \quad (6)$$

For n ODD,

$$\begin{aligned}\frac{n!}{n!!} &= \frac{n(n-1)(n-2)\cdots(1)}{n(n-2)(n-4)\cdots(1)}\\ &= (n-1)(n-3)\cdots(1) = (n-1)!!. \end{aligned} \quad (7)$$

For n EVEN,

$$\begin{aligned}\frac{n!}{n!!} &= \frac{n(n-1)(n-2)\cdots(2)}{n(n-2)(n-4)\cdots(2)}\\ &= (n-1)(n-3)\cdots(2) = (n-1)!!. \end{aligned} \quad (8)$$

Therefore, for any n,

$$\frac{n!}{n!!} = (n-1)!! \quad (9)$$

$$n! = n!!(n-1)!!. \quad (10)$$

The FACTORIAL may be further generalized to the MULTIFACTORIAL

see also FACTORIAL, MULTIFACTORIAL

References
Sloane, N. J. A. Sequence A006882/M0876 in "An On-Line Version of the Encyclopedia of Integer Sequences."

Double Folium

see BIFOLIUM

Double-Free Set

A SET of POSITIVE integers is double-free if, for any integer x, the SET $\{x, 2x\} \not\subset S$ (or equivalently, if $x \in S$ IMPLIES $2x \notin S$). Define

$$r(n) = \max\{S : S \subset \{1, 2, \ldots, n\} \text{ is double-free}\}.$$

Then an asymptotic formula is

$$r(n) \sim \tfrac{2}{3}n + \mathcal{O}(\ln n)$$

(Wang 1989).

see also TRIPLE-FREE SET

References
Finch, S. "Favorite Mathematical Constants." `http://www.mathsoft.com/asolve/constant/triple/triple.html`.
Wang, E. T. H. "On Double-Free Sets of Integers." *Ars Combin.* **28**, 97–100, 1989.

Double Gamma Function

see DIGAMMA FUNCTION

Double Point

A point traced out twice as a closed curve is traversed. The maximum number of double points for a nondegenerate QUARTIC CURVE is three. An ORDINARY DOUBLE POINT is called a NODE.

Arnold (1994) gives pictures of spherical and PLANE CURVES with up to five double points, as well as other curves.

see also BIPLANAR DOUBLE POINT, CONIC DOUBLE POINT, CRUNODE, CUSP, ELLIPTIC CONE POINT, GAUSS'S DOUBLE POINT THEOREM, NODE (ALGEBRAIC CURVE), ORDINARY DOUBLE POINT, QUADRUPLE POINT RATIONAL DOUBLE POINT, SPINODE, TACNODE, TRIPLE POINT, UNIPLANAR DOUBLE POINT

References

Aicardi, F. Appendix to "Plane Curves, Their Invariants, Perestroikas, and Classifications." In *Singularities & Bifurcations* (V. I. Arnold). Providence, RI: Amer. Math. Soc., pp. 80–91, 1994.

Fischer, G. (Ed.). *Mathematical Models from the Collections of Universities and Museums.* Braunschweig, Germany: Vieweg, pp. 12–13, 1986.

Double Sixes

Two sextuples of SKEW LINES on the general CUBIC SURFACE such that each line of one is SKEW to one LINE in the other set. Discovered by Schläfli.

see also BOXCARS, CUBIC SURFACE, SOLOMON'S SEAL LINES

References

Fischer, G. (Ed.). *Mathematical Models from the Collections of Universities and Museums.* Braunschweig, Germany: Vieweg, p. 11, 1986.

Double Sum

A nested sum over two variables. Identities involving double sums include the following:

$$\sum_{p=0}^{\infty}\sum_{q=0}^{p} a_{q,p-q} = \sum_{m=0}^{\infty}\sum_{n=0}^{\infty} a_{n,m} = \sum_{r=0}^{\infty}\sum_{s=0}^{\lfloor r/2\rfloor} a_{s,r-2s}, \quad (1)$$

where

$$\lfloor r/2\rfloor = \begin{cases} \frac{1}{2}r & r \text{ even} \\ \frac{1}{2}(r-1) & r \text{ odd} \end{cases} \quad (2)$$

is the FLOOR FUNCTION, and

$$\sum_{i=1}^{n}\sum_{j=1}^{n} x_i x_j = n^2 \left\langle x^2 \right\rangle. \quad (3)$$

Consider the sum

$$S(a,b,c;s) = \sum_{(m,n)\neq(0,0)} (am^2+bmn+cn^2)^{-s} \quad (4)$$

over binary QUADRATIC FORMS. If S can be decomposed into a linear sum of products of DIRICHLET L-SERIES, it is said to be solvable. The related sums

$$S_1(a,b,c;s) = \sum_{(m,n)\neq(0,0)} (-1)^m (am^2+bmn+cn^2)^{-s} \quad (5)$$

$$S_2(a,b,c;s) = \sum_{(m,n)\neq(0,0)} (-1)^n (am^2+bmn+cn^2)^{-s} \quad (6)$$

$$S_{1,2}(a,b,c;s) = \sum_{(m,n)\neq(0,0)} (-1)^{m+n} (am^2+bmn+cn^2)^{-s} \quad (7)$$

can also be defined, which gives rise to such impressive FORMULAS as

$$S_1(1,0,58;1) = -\frac{\pi\ln(27+5\sqrt{29})}{\sqrt{58}}. \quad (8)$$

A complete table of the principal solutions of all solvable $S(a,b,c;s)$ is given in Glasser and Zucker (1980, pp. 126–131).

see also EULER SUM

References

Glasser, M. L. and Zucker, I. J. "Lattice Sums in Theoretical Chemistry." *Theoretical Chemistry: Advances and Perspectives, Vol. 5.* New York: Academic Press, 1980.

Zucker, I. J. and Robertson, M. M. "A Systematic Approach to the Evaluation of $\sum_{(m,n\neq 0,0)}(am^2+bmn+cn^2)^{-s}$." *J. Phys. A: Math. Gen.* **9**, 1215–1225, 1976.

Doublet Function

$$y = \delta'(x-a),$$

where $\delta(x)$ is the DELTA FUNCTION.

see also DELTA FUNCTION

References

von Seggern, D. *CRC Standard Curves and Surfaces.* Boca Raton, FL: CRC Press, p. 324, 1993.

Doubly Even Number

An even number N for which $N \equiv 0 \pmod 4$. The first few POSITIVE doubly even numbers are 4, 8, 12, 16, ... (Sloane's A008586).

see also EVEN FUNCTION, ODD NUMBER, SINGLY EVEN NUMBER

References

Sloane, N. J. A. Sequence A008586 in "An On-Line Version of the Encyclopedia of Integer Sequences."

Doubly Magic Square

see BIMAGIC SQUARE

Dougall-Ramanujan Identity

Discovered by Ramanujan around 1910. From Hardy (1959, pp. 102–103),

$$\sum_{n=0}^{\infty}(-1)^n(s+2n)\frac{s^{(n)}}{1^{(n)}}\frac{(x+y+z+u+2s+1)^{(n)}}{(x+y+z+u-s)_{(n)}} \times \prod_{x,y,z,u}\frac{x_{(n)}}{(x+s+1)^{(n)}}$$
$$= \frac{s}{\Gamma(s+1)\Gamma(x+y+z+u+s+1)} \times \prod_{x,y,z,u}\frac{\Gamma(x+s+1)\Gamma(y+z+u+s+1)}{\Gamma(z+u+s+1)}, \quad (1)$$

where

$$a^{(n)} \equiv a(a+1)\cdots(a+n-1) \quad (2)$$

$$a_{(n)} \equiv a(a-1)\cdots(a-n+1) \quad (3)$$

(here, the POCHHAMMER SYMBOL has been written $a^{(n)}$). This can be rewritten as

$${}_7F_6\left[\begin{matrix} s, 1+\frac{1}{2}s, -x-y, -z, -u, x-y+z+u+2s+1 \\ \frac{1}{2}s, x+s+1, y+s+1, z+s+1, u+s+1, \\ -x-y-z-u-s \end{matrix}; 1\right]$$
$$= \frac{1}{\Gamma(s+1)\Gamma(x+y+z+u+s+1)} \times \prod_{x,t,z,u}\frac{\Gamma(x+s+1)\Gamma(y+z+u+s+1)}{\Gamma(z+u+s+1)}. \quad (4)$$

In a more symmetric form, if $n = 2a_1+1 = a_2+a_3+a_4+a_5$, $a_6 = 1+a_1/2$, $a_7 = -n$, and $b_i = 1+a_1-a_{i+1}$ for $i = 1, 2, \ldots, 6$, then

$${}_7F_6\left[\begin{matrix} a_1, a_2, a_3, a_4, a_5, a_6, a_7 \\ b_1, b_2, b_3, b_4, b_5, b_6 \end{matrix}; 1\right]$$
$$\frac{(a_1+1)_n(a_1-a_2-a_3+1)_n}{(a_1-a_2+1)_n(a_1-a_3+1)_n} \times \frac{(a_1-a_2-a_4+1)_n(a_1-a_3-a_4+1)_n}{(a_1-a_4+1)_n(a_1-a_2-a_3-a_4+1)_n}, \quad (5)$$

where $(a)_n$ is the POCHHAMMER SYMBOL (Petkovšek *et al.* 1996).

The identity is a special case of JACKSON'S IDENTITY.

see also DIXON'S THEOREM, DOUGALL'S THEOREM, GENERALIZED HYPERGEOMETRIC FUNCTION, HYPERGEOMETRIC FUNCTION, JACKSON'S IDENTITY, SAALSCHÜTZ'S THEOREM

References

Dixon, A. C. "Summation of a Certain Series." *Proc. London Math. Soc.* **35**, 285–289, 1903.

Hardy, G. H. *Ramanujan: Twelve Lectures on Subjects Suggested by His Life and Work, 3rd ed.* New York: Chelsea, 1959.

Petkovšek, M.; Wilf, H. S.; and Zeilberger, D. *A=B.* Wellesley, MA: A. K. Peters, pp. 43, 126–127, and 183–184, 1996.

Dougall's Theorem

$${}_5F_4\left[\begin{matrix} \frac{1}{2}n+1, n, -x, -y, -z \\ \frac{1}{2}n, x+n+1, y+n+1, z+n+1 \end{matrix}\right] = \frac{\Gamma(x+n+1)\Gamma(y+n+1)\Gamma(z+n+1)\Gamma(x+y+z+n+1)}{\Gamma(n+1)\Gamma(x+y+n+1)\Gamma(y+z+n+1)\Gamma(x+z+n+1)},$$

where ${}_5F_4(a,b,c,d,e;f,g,h,i;z)$ is a GENERALIZED HYPERGEOMETRIC FUNCTION and $\Gamma(z)$ is the GAMMA FUNCTION.

see also DOUGALL-RAMANUJAN IDENTITY, GENERALIZED HYPERGEOMETRIC FUNCTION

Doughnut

see TORUS

Douglas-Neumann Theorem

If the lines joining corresponding points of two directly similar figures are divided proportionally, then the LOCUS of the points of the division will be a figure directly similar to the given figures.

References

Eves, H. "Solution to Problem E521." *Amer. Math. Monthly* **50**, 64, 1943.

Musselman, J. R. "Problem E521." *Amer. Math. Monthly* **49**, 335, 1942.

Dovetailing Problem

see CUBE DOVETAILING PROBLEM

Dowker Notation

A simple way to describe a knot projection. The advantage of this notation is that it enables a KNOT DIAGRAM to be drawn quickly.

For an oriented ALTERNATING KNOT with n crossings, begin at an arbitrary crossing and label it 1. Now follow the undergoing strand to the next crossing, and denote it 2. Continue around the knot following the same strand until each crossing has been numbered twice. Each crossing will have one even number and one odd number, with the numbers running from 1 to $2n$.

Now write out the ODD NUMBERS 1, 3, ..., $2n-1$ in a row, and underneath write the even crossing number corresponding to each number. The Dowker NOTATION is this bottom row of numbers. When the sequence of even numbers can be broken into two permutations of consecutive sequences (such as $\{4,6,2\}$ $\{10,12,8\}$), the knot is composite and is not uniquely determined by the Dowker notation. Otherwise, the knot is prime and the NOTATION uniquely defines a single knot (for amphichiral knots) or corresponds to a single knot or its MIRROR IMAGE (for chiral knots).

For general nonalternating knots, the procedure is modified slightly by making the sign of the even numbers

POSITIVE if the crossing is on the top strand, and NEGATIVE if it is on the bottom strand.

These data are available only for knots, but not for links, from Berkeley's gopher site.

References

Adams, C. C. *The Knot Book: An Elementary Introduction to the Mathematical Theory of Knots.* New York: W. H. Freeman, pp. 35–40, 1994.
Dowker, C. H. and Thistlethwaite, M. B. "Classification of Knot Projections." *Topol. Appl.* **16**, 19–31, 1983.

Down Arrow Notation

An inverse of the up ARROW NOTATION defined by

$$e \downarrow n = \ln n$$
$$e \downarrow\downarrow n = \ln^* n$$
$$e \downarrow\downarrow\downarrow n = \ln^{**} n,$$

where $\ln^* n$ is the number of times the NATURAL LOGARITHM must be iterated to obtain a value $\leq e$.

see also ARROW NOTATION

References

Vardi, I. *Computational Recreations in Mathematica.* Redwood City, CA: Addison-Wesley, pp. 12 and 231–232, 1991.

Dozen

12.

see also BAKER'S DOZEN, GROSS

Dragon Curve

Nonintersecting curves which can be iterated to yield more and more sinuosity. They can be constructed by taking a path around a set of dots, representing a left turn by 1 and a right turn by 0. The first-order curve is then denoted 1. For higher order curves, add a 1 to the end, then copy the string of digits preceding it to the end but switching its center digit. For example, the second-order curve is generated as follows: (1)1 → (1)1(0) → 110, and the third as: (110)1 → (110)1(100) → 1101100. Continuing gives 110110011100100... (Sloane's A014577). The OCTAL representation sequence is 1, 6, 154, 66344, ... (Sloane's A003460). The dragon curves of orders 1 to 9 are illustrated below.

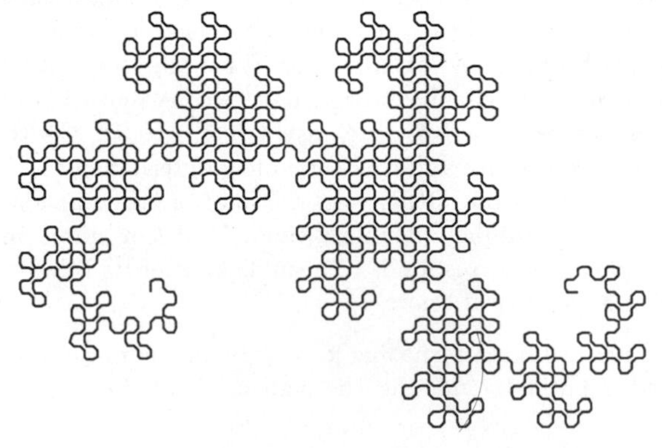

This procedure is equivalent to drawing a RIGHT ANGLE and subsequently replacing each RIGHT ANGLE with another smaller RIGHT ANGLE (Gardner 1978). In fact, the dragon curve can be written as a LINDENMAYER SYSTEM with initial string "FX", STRING REWRITING rules "X" -> "X+YF+", "Y" -> "-FX-Y", and angle 90°.

see also LINDENMAYER SYSTEM, PEANO CURVE

References

Dickau, R. M. "Two-Dimensional L-Systems." http://forum.swarthmore.edu/advanced/robertd/lsys2d.html.
Dixon, R. *Mathographics.* New York: Dover, pp. 180–181, 1991.
Dubrovsky, V. "Nesting Puzzles, Part I: Moving Oriental Towers." *Quantum* **6**, 53–57 (Jan.) and 49–51 (Feb.), 1996.
Dubrovsky, V. "Nesting Puzzles, Part II: Chinese Rings Produce a Chinese Monster." *Quantum* **6**, 61–65 (Mar.) and 58–59 (Apr.), 1996.
Gardner, M. *Mathematical Magic Show: More Puzzles, Games, Diversions, Illusions and Other Mathematical Sleight-of-Mind from Scientific American.* New York: Vintage, pp. 207–209 and 215–220, 1978.
Lauwerier, H. *Fractals: Endlessly Repeated Geometric Figures.* Princeton, NJ: Princeton University Press, pp. 48–53, 1991.
Peitgen, H.-O. and Saupe, D. (Eds.). *The Science of Fractal Images.* New York: Springer-Verlag, p. 284, 1988.
Sloane, N. J. A. Sequences A014577 and A003460/M4300 in "An On-Line Version of the Encyclopedia of Integer Sequences."
Vasilyev, N. and Gutenmacher, V. "Dragon Curves." *Quantum* **6**, 5–10, 1995.

Dragon Fractal

see DOUADY'S RABBIT FRACTAL

Draughts

see CHECKERS

Drinfeld's Symmetric Space

A set of points which do not lie on any of a certain class of HYPERPLANES.

References

Teitelbaum, J. "The Geometry of p-adic Symmetric Spaces." *Not. Amer. Math. Soc.* **42**, 1120–1126, 1995.

Droz-Farny Circles

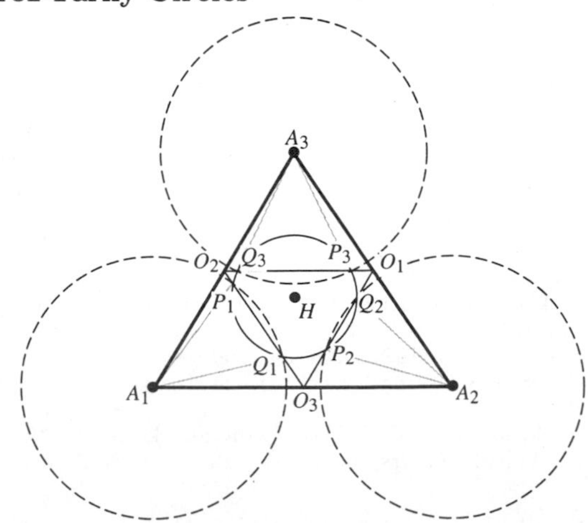

Draw a CIRCLE with center H which cuts the lines O_2O_3, O_3O_1, and O_1O_2 (where O_i are the MIDPOINTS) at P_1, Q_1; P_2, Q_2; and P_3, Q_3 respectively, then

$$\overline{A_1P_1} = \overline{A_2P_2} = \overline{A_3P_3} = \overline{A_1Q_1} = \overline{A_2Q_2} = \overline{A_3Q_3}.$$

Conversely, if equal CIRCLES are drawn about the VERTICES of a TRIANGLE, they cut the lines joining the MIDPOINTS of the corresponding sides in six points. These points lie on a CIRCLE whose center is the ORTHOCENTER. If r is the RADIUS of the equal CIRCLES centered on the vertices A_1, A_2, and A_3, and R_0 is the RADIUS of the CIRCLE about H, then

$$R_1{}^2 = 4R^2 + r^2 - \tfrac{1}{2}(a_1{}^2 + a_2{}^2 + a_3{}^2).$$

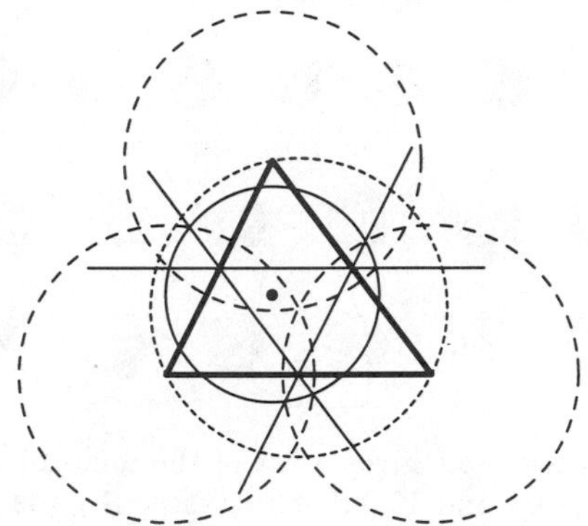

If the circles equal to the CIRCUMCIRCLE are drawn about the VERTICES of a triangle, they cut the lines joining midpoints of the adjacent sides in points of a CIRCLE R_2 with center H and RADIUS

$$R_2{}^2 = 5R^2 - \tfrac{1}{2}(a_1{}^2 + a_2{}^2 + a_3{}^2).$$

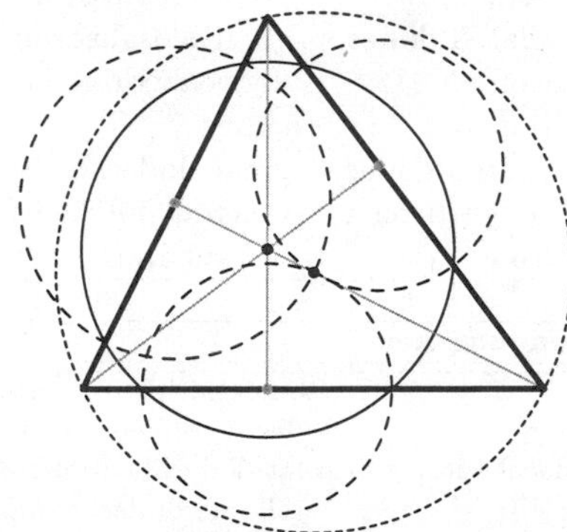

It is equivalent to the circle obtained by drawing circles with centers at the feet of the altitudes and passing through the CIRCUMCENTER. These circles cut the corresponding sides in six points on a circle R_2' whose center is H.

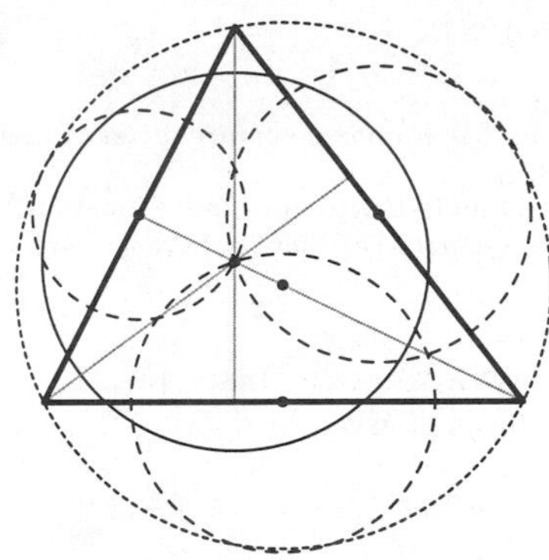

Furthermore, the circles about the midpoints of the sides and passing though H cut the sides in six points lying on another equivalent circle R_2'' whose center is O. In summary, the second Droz-Farny circle passes through 12 notable points, two on each of the sides and two on each of the lines joining midpoints of the sides.

References

Goormaghtigh, R. "Droz-Farny's Theorem." *Scripta Math.* **16**, 268–271, 1950.

Johnson, R. A. *Modern Geometry: An Elementary Treatise on the Geometry of the Triangle and the Circle.* Boston, MA: Houghton Mifflin, pp. 256–258, 1929.

Drum

see ISOSPECTRAL MANIFOLDS

Du Bois Raymond Constants

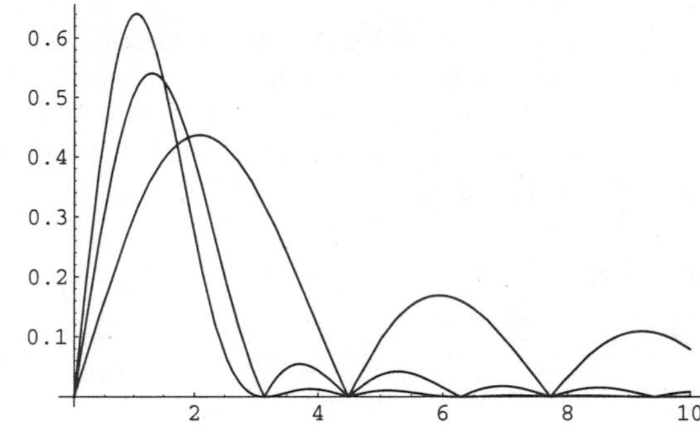

The constants C_n defined by

$$C_n \equiv \int_0^\infty \left| \frac{d}{dt} \left(\frac{\sin t}{t} \right)^n \right| dt - 1$$

which are difficult to compute numerically. The first few are

$$C_1 \approx 455$$
$$C_2 \approx 0.1945$$
$$C_3 \approx 0.028254$$
$$C_4 \approx 0.00524054.$$

Rather surprisingly, the second Du Bois Raymond constant is given analytically by

$$C_2 = \tfrac{1}{2}(e^2 - 7) = 0.1945280494\ldots$$

(Le Lionnais 1983).

References
Le Lionnais, F. *Les nombres remarquables.* Paris: Hermann, p. 23, 1983.
Plouffe, S. "Dubois-Raymond 2nd Constant." http://lacim.uqam.ca/piDATA/dubois.txt.

Dual Basis

Given a CONTRAVARIANT BASIS $\{\vec{e}_1, \ldots, \vec{e}_n\}$, its dual COVARIANT basis is given by

$$\vec{e}^{\,\alpha} \cdot \vec{e}_\beta = g(\vec{e}^{\,\alpha}, \vec{e}_\beta) = \delta^\alpha_\beta,$$

where g is the METRIC and δ^α_β is the mixed KRONECKER DELTA. In EUCLIDEAN SPACE with an ORTHONORMAL BASIS,

$$\vec{e}^{\,j} = \vec{e}_j,$$

so the BASIS and its dual are the same.

Dual Bivector

A dual BIVECTOR is defined by

$$\tilde{X}_{ab} \equiv \tfrac{1}{2}\epsilon_{abcd}X^{cd},$$

and a self-dual BIVECTOR by

$$X^*_{ab} \equiv X_{ab} + i\tilde{X}_{ab}.$$

Dual Graph

The dual graph G^* of a POLYHEDRAL GRAPH G has VERTICES each of which corresponds to a face of G and each of whose faces corresponds to a VERTEX of G. Two nodes in G^* are connected by an EDGE if the corresponding faces in G have a boundary EDGE in common.

Dual Map

see PULLBACK MAP

Dual Polyhedron

By the DUALITY PRINCIPLE, for every POLYHEDRON, there exists another POLYHEDRON in which faces and VERTICES occupy complementary locations. This POLYHEDRON is known as the dual, or RECIPROCAL. The dual polyhedron of a PLATONIC SOLID or ARCHIMEDEAN SOLID can be drawn by constructing EDGES tangent to the RECIPROCATING SPHERE (a.k.a. MIDSPHERE and INTERSPHERE) which are PERPENDICULAR to the original EDGES.

The dual of a general solid can be computed by connecting the midpoints of the sides surrounding each VERTEX, and constructing the corresponding tangent POLYGON. (The tangent polygon is the polygon which is tangent to the CIRCUMCIRCLE of the POLYGON produced by connecting the MIDPOINT on the sides surrounding the given VERTEX.) The process is illustrated below for the PLATONIC SOLIDS. The POLYHEDRON COMPOUNDS consisting of a POLYHEDRON and its dual are generally very attractive, and are also illustrated below for the PLATONIC SOLIDS.

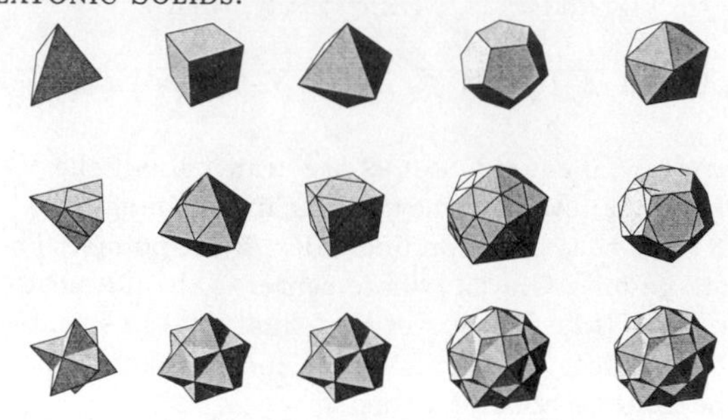

The ARCHIMEDEAN SOLIDS and their duals are illustrated below.

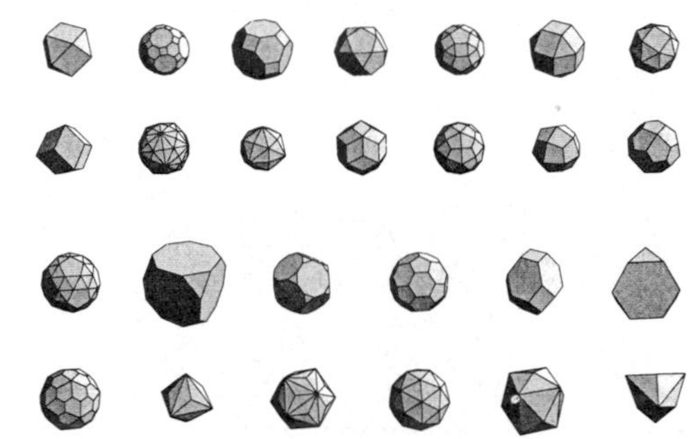

The following table gives a list of the duals of the PLATONIC SOLIDS and KEPLER-POINSOT SOLIDS together with the names of the POLYHEDRON-dual COMPOUNDS.

Polyhedron	Dual
Császár polyhedron	Szilassi polyhedron
cube	octahedron
cuboctahedron	rhombic dodecahedron
dodecahedron	icosahedron
great dodecahedron	small stellated dodec.
great icosahedron	great stellated dodec.
great stellated dodec.	great icosahedron
icosahedron	dodecahedron
octahedron	cube
small stellated dodec.	great dodecahedron
Szilassi polyhedron	Császár polyhedron
tetrahedron	tetrahedron

polyhedron	compound
cube	cube-octahedron compound
dodecahedron	dodec.-icosahedron compound
great dodecahedron	great dodecahedron-small stellated dodec. compound
great icosahedron	great icosahedron-great stellated dodec. compound
great stellated dodec.	great icosahedron-great stellated dodec. compound
icosahedron	dodec.-icosahedron compound
octahedron	cube-octahedron compound
small stellated dodec.	great dodec.-small stellated dodec. compound
tetrahedron	stella octangula

see also DUALITY PRINCIPLE, POLYHEDRON COMPOUND, RECIPROCATING SPHERE

References

Weisstein, E. W. "Polyhedron Duals." http://www.astro.virginia.edu/~eww6n/math/notebooks/Duals.m.

Wenninger, M. *Dual Models*. Cambridge, England: Cambridge University Press, 1983.

Dual Scalar

Given a third RANK TENSOR,

$$V_{ijk} \equiv \det [\mathbf{A} \quad \mathbf{B} \quad \mathbf{C}],$$

where det is the DETERMINANT, the dual scalar is defined as

$$V \equiv \frac{1}{3!}\epsilon_{ijk}V_{ijk},$$

where ϵ_{ijk} is the LEVI-CIVITA TENSOR.

see also DUAL TENSOR, LEVI-CIVITA TENSOR

Dual Solid

see DUAL POLYHEDRON

Dual Tensor

Given an antisymmetric second RANK TENSOR C_{ij}, a dual pseudotensor C_i is defined by

$$C_i \equiv \tfrac{1}{2}\epsilon_{ijk}C_{jk}, \tag{1}$$

where

$$C_i \equiv \begin{bmatrix} C_{23} \\ C_{31} \\ C_{12} \end{bmatrix} \tag{2}$$

$$C_{jk} \equiv \begin{bmatrix} 0 & C_{12} & -C_{31} \\ -C_{12} & 0 & C_{23} \\ C_{31} & -C_{23} & 0 \end{bmatrix}. \tag{3}$$

see also DUAL SCALAR

References

Arfken, G. "Pseudotensors, Dual Tensors." §3.4 in *Mathematical Methods for Physicists, 3rd ed.* Orlando, FL: Academic Press, pp. 128–137, 1985.

Dual Voting

A term in SOCIAL CHOICE THEORY meaning each alternative receives equal weight for a single vote.

see also ANONYMOUS, MONOTONIC VOTING

Duality Principle

All the propositions in PROJECTIVE GEOMETRY occur in dual pairs which have the property that, starting from either proposition of a pair, the other can be immediately inferred by interchanging the parts played by the words "point" and "line." A similar duality exists for RECIPROCATION (Casey 1893).

see also BRIANCHON'S THEOREM, CONSERVATION OF NUMBER PRINCIPLE, DESARGUES' THEOREM, DUAL POLYHEDRON, PAPPUS'S HEXAGON THEOREM, PASCAL'S THEOREM, PERMANENCE OF MATHEMATICAL RELATIONS PRINCIPLE, PROJECTIVE GEOMETRY, RECIPROCATION

References

Casey, J. "Theory of Duality and Reciprocal Polars." Ch. 13 in *A Treatise on the Analytical Geometry of the Point, Line, Circle, and Conic Sections, Containing an Account of Its Most Recent Extensions, with Numerous Examples, 2nd ed., rev. enl.* Dublin: Hodges, Figgis, & Co., pp. 382–392, 1893.

Ogilvy, C. S. *Excursions in Geometry.* New York: Dover, pp. 107–110, 1990.

Duality Theorem

Dual pairs of LINEAR PROGRAMS are in "strong duality" if both are possible. The theorem was first conceived by John von Neumann. The first written proof was an Air Force report by George Dantzig, but credit is usually given to Tucker, Kuhn, and Gale.

see also LINEAR PROGRAMMING

Duffing Differential Equation

The most general forced form of the Duffing equation is

$$\ddot{x} + \delta\dot{x} + (\beta x^3 \pm {\omega_0}^2 x) = A\sin(\omega t + \phi). \tag{1}$$

If there is no forcing, the right side vanishes, leaving

$$\ddot{x} + \delta\dot{x} + (\beta x^3 \pm {\omega_0}^2 x) = 0. \tag{2}$$

If $\delta = 0$ and we take the plus sign,

$$\ddot{x} + {\omega_0}^2 x + \beta x^3 = 0. \tag{3}$$

This equation can display chaotic behavior. For $\beta > 0$, the equation represents a "hard spring," and for $\beta < 0$, it represents a "soft spring." If $\beta < 0$, the phase portrait curves are closed. Returning to (1), take $\beta = 1$, $\omega_0 = 1$, $A = 0$, and use the minus sign. Then the equation is

$$\ddot{x} + \delta\dot{x} + (x^3 - x) = 0 \tag{4}$$

(Ott 1993, p. 3). This can be written as a system of first-order ordinary differential equations by writing

$$\dot{x} = y, \tag{5}$$

$$\dot{y} = x - x^3 - \delta y. \tag{6}$$

The fixed points of these differential equations

$$\dot{x} = y = 0, \tag{7}$$

so $y = 0$, and

$$\dot{y} = x - x^3 - \delta y = x(1 - x^2) - 0 \tag{8}$$

giving $x = 0, \pm 1$. Differentiating,

$$\ddot{x} = \dot{y} = x - x^3 - \delta y \tag{9}$$

$$\ddot{y} = (1 - 3x^2)\dot{x} - \delta \dot{y} \tag{10}$$

$$\begin{bmatrix} \ddot{x} \\ \ddot{y} \end{bmatrix} = \begin{bmatrix} 0 & 1 \\ 1 - 3x^2 & -\delta \end{bmatrix} \begin{bmatrix} \dot{x} \\ \dot{y} \end{bmatrix}. \tag{11}$$

Examine the stability of the point (0,0):

$$\begin{vmatrix} 0 - \lambda & 1 \\ 1 & -\delta - \lambda \end{vmatrix} = \lambda(\lambda + \delta) - 1 = \lambda^2 + \lambda\delta - 1 = 0 \tag{12}$$

$$\lambda_\pm^{(0,0)} = \tfrac{1}{2}(-\delta \pm \sqrt{\delta^2 + 4}\,). \tag{13}$$

But $\delta^2 \geq 0$, so $\lambda_\pm^{(0,0)}$ is real. Since $\sqrt{\delta^2 + 4} > |\delta|$, there will always be one POSITIVE ROOT, so this fixed point is unstable. Now look at (± 1, 0).

$$\begin{vmatrix} 0 - \lambda & 1 \\ -2 & -\delta - \lambda \end{vmatrix} = \lambda(\lambda + \delta) + 2 = \lambda^2 + \lambda\delta + 2 = 0 \tag{14}$$

$$\lambda_\pm^{(\pm 1,0)} = \tfrac{1}{2}(-\delta \pm \sqrt{\delta^2 - 8}\,). \tag{15}$$

For $\delta > 0$, $\Re[\lambda_\pm^{(\pm 1,0)}] < 0$, so the point is asymptotically stable. If $\delta = 0$, $\lambda_\pm^{(\pm 1,0)} = \pm i\sqrt{2}$, so the point is linearly stable. If $\delta \in (-2\sqrt{2}, 0)$, the radical gives an IMAGINARY PART and the REAL PART is > 0, so the point is unstable. If $\delta = -2\sqrt{2}$, $\lambda_\pm^{(\pm 1,0)} = \sqrt{2}$, which has a POSITIVE REAL ROOT, so the point is unstable. If $\delta < -2\sqrt{2}$, then $|\delta| < \sqrt{\delta^2 - 8}$, so both ROOTS are POSITIVE and the point is unstable. Summarizing,

$$\begin{cases} \text{asymptotically stable} & \delta > 0 \\ \text{linearly stable (superstable)} & \delta = 0 \\ \text{unstable} & \delta < 0. \end{cases} \tag{16}$$

Now specialize to the case $\delta = 0$, which can be integrated by quadratures. In this case, the equations become

$$\dot{x} = y \tag{17}$$

$$\dot{y} = x - x^3. \tag{18}$$

Differentiating (17) and plugging in (18) gives

$$\ddot{x} = \dot{y} = x - x^3. \tag{19}$$

Multiplying both sides by $\dot{x}$ gives

$$\ddot{x}\dot{x} - \dot{x}x + \dot{x}x^3 = 0 \tag{20}$$

$$\frac{d}{dt}(\tfrac{1}{2}\dot{x}^2 - \tfrac{1}{2}x^2 + \tfrac{1}{4}x^4) = 0, \tag{21}$$

so we have an invariant of motion h,

$$h \equiv \tfrac{1}{2}\dot{x}^2 - \tfrac{1}{2}x^2 + \tfrac{1}{4}x^4. \tag{22}$$

Solving for $\dot{x}^2$ gives

$$\dot{x}^2 = \left(\frac{dx}{dt}\right)^2 = 2h + x^2 - \tfrac{1}{2}x^4 \tag{23}$$

$$\frac{dx}{dt} = \sqrt{2h + x^2 + \tfrac{1}{2}x^2}, \tag{24}$$

so

$$t = \int dt = \int \frac{dx}{\sqrt{2h + x^2 + \frac{1}{2}x^2}}. \tag{25}$$

Note that the invariant of motion h satisfies

$$\dot{x} = \frac{\partial h}{\partial \dot{x}} = \frac{\partial h}{\partial y} \tag{26}$$

$$\frac{\partial h}{\partial x} = -x + x^3 = -\dot{y}, \tag{27}$$

so the equations of the Duffing oscillator are given by the HAMILTONIAN SYSTEM

$$\begin{cases} \dot{x} = \frac{\partial h}{\partial y} \\ \dot{y} = -\frac{\partial h}{\partial x}. \end{cases} \tag{28}$$

References

Ott, E. *Chaos in Dynamical Systems.* New York: Cambridge University Press, 1993.

Duhamel's Convolution Principle

Can be used to invert a LAPLACE TRANSFORM.

Dumbbell Curve

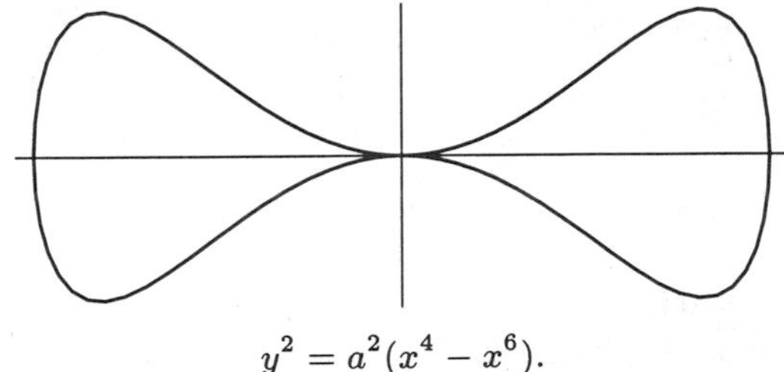

$$y^2 = a^2(x^4 - x^6).$$

see also BUTTERFLY CURVE, EIGHT CURVE, PIRIFORM

References

Cundy, H. and Rollett, A. *Mathematical Models, 3rd ed.* Stradbroke, England: Tarquin Pub., p. 72, 1989.

Duodecillion

In the American system, 10^{39}.

see also LARGE NUMBER

Dupin's Cyclide

see CYCLIDE

Dupin's Indicatrix

A pair of conics obtained by expanding an equation in MONGE'S FORM $z = F(x, y)$ in a MACLAURIN SERIES

$$\begin{aligned} z &= z(0,0) + z_1 x + z_2 y \\ &\quad + \tfrac{1}{2}(z_{11}x^2 + 2z_{12}xy + z_{22}y^2) + \ldots \\ &= \tfrac{1}{2}(b_{11}x^2 + 2b_{12}xy + b_{22}y^2). \end{aligned}$$

This gives the equation

$$b_{11}x^2 + 2b_{12}xy + b_{22}y^2 = \pm 1.$$

Amazingly, the radius of the indicatrix in any direction is equal to the SQUARE ROOT of the RADIUS OF CURVATURE in that direction (Coxeter 1969).

References

Coxeter, H. S. M. "Dupin's Indicatrix" §19.8 in *Introduction to Geometry, 2nd ed.* New York: Wiley, pp. 363–365, 1969.

Dupin's Theorem

In three mutually orthogonal systems of the surfaces, the LINES OF CURVATURE on any surface in one of the systems are its intersections with the surfaces of the other two systems.

Duplication of the Cube

see CUBE DUPLICATION

Duplication Formula

see LEGENDRE DUPLICATION FORMULA

Durand's Rule

The NEWTON-COTES FORMULA

$$\begin{aligned} &\int_{x_1}^{x_n} f(x)\,dx \\ &\quad = h(\tfrac{2}{5}f_1 + \tfrac{11}{10}f_2 + f_3 + \ldots + f_{n-2} + \tfrac{11}{10}f_{n-1} + \tfrac{2}{5}f_n). \end{aligned}$$

see also BODE'S RULE, HARDY'S RULE, NEWTON-COTES FORMULAS, SIMPSON'S 3/8 RULE, SIMPSON'S RULE, TRAPEZOIDAL RULE, WEDDLE'S RULE

References

Beyer, W. H. (Ed.) *CRC Standard Mathematical Tables, 28th ed.* Boca Raton, FL: CRC Press, p. 127, 1987.

Dürer's Conchoid

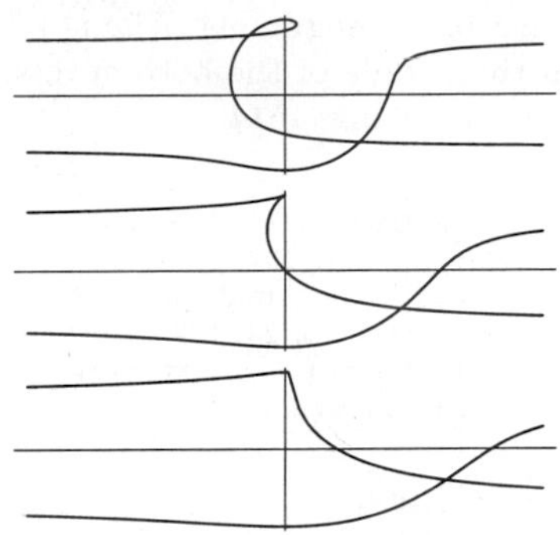

These curves appear in Dürer's work *Instruction in Measurement with Compasses and Straight Edge* (1525) and arose in investigations of perspective. Dürer constructed the curve by drawing lines QRP and $P'QR$ of length 16 units through $Q(q, 0)$ and $R(r, 0)$, where $q+r = 13$. The locus of P and P' is the curve, although Dürer found only one of the two branches of the curve.

The ENVELOPE of the lines QRP and $P'QR$ is a PARABOLA, and the curve is therefore a GLISSETTE of a point on a line segment sliding between a PARABOLA and one of its TANGENTS.

Dürer called the curve "Muschellini," which means CONCHOID. However, it is not a true CONCHOID and so is sometimes called DÜRER'S SHELL CURVE. The Cartesian equation is

$$\begin{aligned} 2y^2(x^2+y^2) - 2by^2(x+y) + (b^2 - 3a^2)y^2 - a^2x^2 \\ + 2a^2b(x+y) + a^2(a^2 - b^2) = 0. \end{aligned}$$

The above curves are for $(a, b) = (3, 1)$, $(3, 3)$, $(3, 5)$. There are a number of interesting special cases. If $b = 0$, the curve becomes two coincident straight lines $x = 0$. For $a = 0$, the curve becomes the line pair $x = b/2$, $x = -b/2$, together with the CIRCLE $x + y = b$. If $a = b/2$, the curve has a CUSP at $(-2a, a)$.

References

Lawrence, J. D. *A Catalog of Special Plane Curves.* New York: Dover, pp. 157–159, 1972.

Lockwood, E. H. *A Book of Curves.* Cambridge, England: Cambridge University Press, p. 163, 1967.

MacTutor History of Mathematics Archive. "Dürer's Shell Curves." `http://www-groups.dcs.st-and.ac.uk/~history/Curves/Durers.html`.

Dürer's Magic Square

16	3	2	13
5	10	11	8
9	6	7	12
4	15	14	1

Dürer's magic square is a MAGIC SQUARE with MAGIC CONSTANT 34 used in an engraving entitled *Melencolia I* by Albrecht Dürer (The British Museum). The engraving shows a disorganized jumble of scientific equipment lying unused while an intellectual sits absorbed in thought. Dürer's magic square is located in the upper left-hand corner of the engraving. The numbers 15 and 14 appear in the middle of the bottom row, indicating the date of the engraving, 1514.

References

Boyer, C. D. and Merzbach, U. C. *A History of Mathematics.* New York: Wiley, pp. 296–297, 1991.

Hunter, J. A. H. and Madachy, J. S. *Mathematical Diversions.* New York: Dover, p. 24, 1975.

Rivera, C. "Melancholia." `http://www.sci.net.mx/~crivera/melancholia.htm`.

Dürer's Shell Curve

see DÜRER'S CONCHOID

Durfee Polynomial

Let $F(n)$ be a family of PARTITIONS of n and let $F(n,d)$ denote the set of PARTITIONS in $F(n)$ with DURFEE SQUARE of size d. The Durfee polynomial of $F(n)$ is then defined as the polynomial

$$P_{F,n} = \sum |F(n,d)| y^d,$$

where $0 \leq d \leq \sqrt{n}$.

see also DURFEE SQUARE, PARTITION

References

Canfield, E. R.; Corteel, S.; and Savage, C. D. "Durfee Polynomials." *Electronic J. Combinatorics* **5**, No. 1, R32, 1–21, 1998. `http://www.combinatorics.org/Volume_5/v5i1toc.html#R32`.

Durfee Square

The length of the largest-sized SQUARE contained within the FERRERS DIAGRAM of a PARTITION.

see also DURFEE POLYNOMIAL, FERRERS DIAGRAM, PARTITION

Dvoretzky's Theorem

Each centered convex body of sufficiently high dimension has an "almost spherical" k-dimensional central section.

Dyad

Dyads extend VECTORS to provide an alternative description to second RANK TENSORS. A dyad $D(\mathbf{A}, \mathbf{B})$ of a pair of VECTORS $\mathbf{A}$ and $\mathbf{B}$ is defined by $D(\mathbf{A}, \mathbf{B}) \equiv \mathbf{AB}$. The DOT PRODUCT is defined by

$$\mathbf{A} \cdot \mathbf{BC} \equiv (\mathbf{A} \cdot \mathbf{B})\mathbf{C}$$

$$\mathbf{AB} \cdot \mathbf{C} \equiv \mathbf{A}(\mathbf{B} \cdot \mathbf{C}),$$

and the COLON PRODUCT by

$$\mathbf{AB} : \mathbf{CD} \equiv \mathbf{C} \cdot \mathbf{AB} \cdot \mathbf{D} = (\mathbf{A} \cdot \mathbf{C})(\mathbf{B} \cdot \mathbf{D}).$$

References

Morse, P. M. and Feshbach, H. "Dyadics and Other Vector Operators." §1.6 in *Methods of Theoretical Physics, Part I.* New York: McGraw-Hill, pp. 54–92, 1953.

Dyadic

A linear POLYNOMIAL of DYADS $\mathbf{AB} + \mathbf{CD} + \ldots$ consisting of nine components A_{ij} which transform as

$$(A_{ij})' = \sum_{m,n} \frac{h_m h_n}{h_i' h_j'} \frac{\partial x_m}{\partial x_i'} \frac{\partial x_n}{\partial x_j'} A_{mn} \quad (1)$$

$$= \sum_{m,n} \frac{h_i' h_j'}{h_m h_n} \frac{\partial x_i'}{\partial x_m} \frac{\partial x_j'}{\partial x_n} A_{mn} \quad (2)$$

$$= \sum_{m,n} \frac{h_i' h_n}{h_m h_j'} \frac{\partial x_i'}{\partial x_m} \frac{\partial x_n}{\partial x_j'} A_{mn}. \quad (3)$$

Dyadics are often represented by Gothic capital letters. The use of dyadics is nearly archaic since TENSORS perform the same function but are notationally simpler.

A unit dyadic is also called the IDEMFACTOR and is defined such that

$$\mathbf{I} \cdot \mathbf{A} \equiv \mathbf{A}. \quad (4)$$

In CARTESIAN COORDINATES,

$$\mathbf{I} = \hat{\mathbf{x}}\hat{\mathbf{x}} + \hat{\mathbf{y}}\hat{\mathbf{y}} + \hat{\mathbf{z}}\hat{\mathbf{z}}, \quad (5)$$

and in SPHERICAL COORDINATES

$$\mathbf{I} = \nabla \mathbf{r}. \quad (6)$$

see also DYAD, TETRADIC

References

Arfken, G. "Dyadics." §3.5 in *Mathematical Methods for Physicists, 3rd ed.* Orlando, FL: Academic Press, pp. 137–140, 1985.

Morse, P. M. and Feshbach, H. "Dyadics and Other Vector Operators." §1.6 in *Methods of Theoretical Physics, Part I.* New York: McGraw-Hill, pp. 54–92, 1953.

Dyck's Theorem

see VON DYCK'S THEOREM

Dye's Theorem

For any two ergodic measure-preserving transformations on nonatomic PROBABILITY SPACES, there is an ISOMORPHISM between the two PROBABILITY SPACES carrying orbits onto orbits.

Dymaxion

Buckminster Fuller's term for the CUBOCTAHEDRON.

see also CUBOCTAHEDRON, MECON

Dynamical System

A means of describing how one state develops into another state over the course of time. Technically, a dynamical system is a smooth action of the reals or the INTEGERS on another object (usually a MANIFOLD). When the reals are acting, the system is called a continuous dynamical system, and when the INTEGERS are acting, the system is called a discrete dynamical system. If f is any CONTINUOUS FUNCTION, then the evolution of a variable x can be given by the formula

$$x_{n+1} = f(x_n). \tag{1}$$

This equation can also be viewed as a difference equation

$$x_{n+1} - x_n = f(x_n) - x_n, \tag{2}$$

so defining

$$g(x) \equiv f(x) - x \tag{3}$$

gives

$$x_{n+1} - x_n = g(x_n) * 1, \tag{4}$$

which can be read "as n changes by 1 unit, x changes by $g(x)$." This is the discrete analog of the DIFFERENTIAL EQUATION

$$x'(n) = g(x(n)). \tag{5}$$

see also ANOSOV DIFFEOMORPHISM, ANOSOV FLOW, AXIOM A DIFFEOMORPHISM, AXIOM A FLOW, BIFURCATION THEORY, CHAOS, ERGODIC THEORY, GEODESIC FLOW

References

Aoki, N. and Hiraide, K. *Topological Theory of Dynamical Systems.* Amsterdam, Netherlands: North-Holland, 1994.

Golubitsky, M. *Introduction to Applied Nonlinear Dynamical Systems and Chaos.* New York: Springer-Verlag, 1997.

Guckenheimer, J. and Holmes, P. *Nonlinear Oscillations, Dynamical Systems, and Bifurcations of Vector Fields, 3rd ed.* New York: Springer-Verlag, 1997.

Lichtenberg, A. and Lieberman, M. *Regular and Stochastic Motion, 2nd ed.* New York: Springer-Verlag, 1994.

Ott, E. *Chaos in Dynamical Systems.* New York: Cambridge University Press, 1993.

Rasband, S. N. *Chaotic Dynamics of Nonlinear Systems.* New York: Wiley, 1990.

Strogatz, S. H. *Nonlinear Dynamics and Chaos, with Applications to Physics, Biology, Chemistry, and Engineering.* 1994.

Tabor, M. *Chaos and Integrability in Nonlinear Dynamics: An Introduction.* New York: Wiley, 1989.

Dynkin Diagram

A diagram used to describe CHEVALLEY GROUPS.

see also COXETER-DYNKIN DIAGRAM

References

Jacobson, N. *Lie Algebras.* New York: Dover, p. 128, 1979.

E

e

The base of the NATURAL LOGARITHM, named in honor of Euler. It appears in many mathematical contexts involving LIMITS and DERIVATIVES, and can be defined by

$$e \equiv \lim_{x\to\infty}\left(1+\frac{1}{x}\right)^x, \tag{1}$$

or by the infinite sum

$$e = \sum_{k=0}^{\infty}\frac{1}{k!}. \tag{2}$$

The numerical value of e is

$$e = 2.718281828459045235360287471352662497757\ldots \tag{3}$$

(Sloane's A001113).

Euler proved that e is IRRATIONAL, and Liouville proved in 1844 that e does not satisfy any QUADRATIC EQUATION with integral COEFFICIENTS. Hermite proved e to be TRANSCENDENTAL in 1873. It is not known if $\pi + e$ or π/e is IRRATIONAL. However, it is known that $\pi + e$ and π/e do not satisfy any POLYNOMIAL equation of degree ≤ 8 with INTEGER COEFFICIENTS of average size 10^9 (Bailey 1988, Borwein *et al.* 1989).

The special case of the EULER FORMULA

$$e^{ix} = \cos x + i\sin x \tag{4}$$

with $x = \pi$ gives the beautiful identity

$$e^{i\pi} + 1 = 0, \tag{5}$$

an equation connecting the fundamental numbers i, PI, e, 1, and 0 (ZERO).

Some CONTINUED FRACTION representations of e include

$$e = 2 + \cfrac{1}{1+\cfrac{1}{2+\cfrac{2}{3+\cfrac{3}{\ddots}}}} \tag{6}$$

$$= [2,1,2,1,1,4,1,1,6,\ldots] \tag{7}$$

(Sloane's A003417) and

$$\frac{e-1}{e+1} = [2,6,10,14,\ldots] \tag{8}$$

$$e-1 = [1,1,2,1,1,4,1,1,6,\ldots] \tag{9}$$

$$\tfrac{1}{2}(e-1) = [0,1,6,10,14,\ldots] \tag{10}$$

$$\sqrt{e} = [1,1,1,1,5,1,1,1,9,1,\ldots]. \tag{11}$$

The first few convergents of the CONTINUED FRACTION are 3, 8/3, 11/4, 19/7, 87/32, 106/39, 193/71, ... (Sloane's A007676 and A007677).

Using the RECURRENCE RELATION

$$a_n = n(a_{n-1}+1) \tag{12}$$

with $a_1 = a^{-1}$, compute

$$\prod_{n=1}^{\infty}(1+{a_n}^{-1}). \tag{13}$$

The result is e^a. Gosper gives the unusual equation connecting π and e,

$$\sum_{n=1}^{\infty}\frac{1}{n^2}\cos\left(\frac{9}{n\pi+\sqrt{n^2\pi^2-9}}\right) = -\frac{\pi^2}{12e^3} = -0.040948222\ldots. \tag{14}$$

Rabinowitz and Wagon (1995) give an ALGORITHM for computing digits of e based on earlier DIGITS, but a much simpler SPIGOT ALGORITHM was found by Sales (1968). Around 1966, MIT hacker Eric Jensen wrote a very concise program (requiring less than a page of assembly language) that computed e by converting from factorial base to decimal.

Let $p(n)$ be the probability that a random ONE-TO-ONE function on the INTEGERS 1, ..., n has at least one FIXED POINT. Then

$$\lim_{n\to\infty} p(n) = \sum_{k=1}^{\infty}\frac{(-1)^{k+1}}{k!} = 1-\frac{1}{e} = 0.6321205588\ldots. \tag{15}$$

STIRLING'S FORMULA gives

$$\lim_{n\to\infty}\frac{(n!)^{1/n}}{n} = \frac{1}{e}. \tag{16}$$

Castellanos (1988) gives several curious approximations to e,

$$e \approx 2 + \frac{54^2+41^2}{80^2} \tag{17}$$

$$\approx (\pi^4+\pi^5)^{1/6} \tag{18}$$

$$\approx \frac{271801}{99990} \tag{19}$$

$$\approx \left(150 - \frac{87^3+12^5}{83^3}\right)^{1/5} \tag{20}$$

$$\approx 4 - \frac{300^4-100^4-1291^2+9^2}{91^5} \tag{21}$$

$$\approx \left(1097 - \frac{55^5+311^3-11^3}{68^5}\right)^{1/7}, \tag{22}$$

which are good to 6, 7, 9, 10, 12, and 15 digits respectively.

Examples of e MNEMONICS (Gardner 1959, 1991) include:

> "By omnibus I traveled to Brooklyn" (6 digits).
>
> "To disrupt a playroom is commonly a practice of children" (10 digits).
>
> "It enables a numskull to memorize a quantity of numerals" (10 digits).
>
> "I'm forming a mnemonic to remember a function in analysis" (10 digits).
>
> "He repeats: I shouldn't be tippling, I shouldn't be toppling here!" (11 digits).
>
> "In showing a painting to probably a critical or venomous lady, anger dominates. O take guard, or she raves and shouts" (21 digits). Here, the word "O" stands for the number 0.

A much more extensive mnemonic giving 40 digits is

> "We present a mnemonic to memorize a constant so exciting that Euler exclaimed: '!' when first it was found, yes, loudly '!'. My students perhaps will compute e, use power or Taylor series, an easy summation formula, obvious, clear, elegant!"

(Barel 1995). In the latter, 0s are represented with "!". A list of e mnemonics in several languages is maintained by A. P. Hatzipolakis.

Scanning the decimal expansion of e until all n-digit numbers have occurred, the last appearing is 6, 12, 548, 1769, 92994, 513311, ... (Sloane's A032511). These end at positions 21, 372, 8092, 102128, 1061613, 12108841, ...

see also CARLEMAN'S INEQUALITY, COMPOUND INTEREST, DE MOIVRE'S IDENTITY, EULER FORMULA, EXPONENTIAL FUNCTION, HERMITE-LINDEMANN THEOREM, NATURAL LOGARITHM

References

Bailey, D. H. "Numerical Results on the Transcendence of Constants Involving π, e, and Euler's Constant." *Math. Comput.* **50**, 275–281, 1988.

Barel, Z. "A Mnemonic for e." *Math. Mag.* **68**, 253, 1995.

Borwein, J. M.; Borwein, P. B.; and Bailey, D. H. "Ramanujan, Modular Equations, and Approximations to Pi or How to Compute One Billion Digits of Pi." *Amer. Math. Monthly* **96**, 201–219, 1989.

Castellanos, D. "The Ubiquitous Pi. Part I." *Math. Mag.* **61**, 67–98, 1988.

Conway, J. H. and Guy, R. K. *The Book of Numbers.* New York: Springer-Verlag, pp. 201 and 250–254, 1996.

Finch, S. "Favorite Mathematical Constants." `http://www.mathsoft.com/asolve/constant/e/e.html`.

Gardner, M. "Memorizing Numbers." Ch. 11 in *The Scientific American Book of Mathematical Puzzles and Diversions.* New York: Simon and Schuster, pp. 103 and 109, 1959.

Gardner, M. Ch. 3 in *The Unexpected Hanging and Other Mathematical Diversions.* Chicago, IL: Chicago University Press, p. 40, 1991.

Hatzipolakis, A. P. "PiPhilology." `http://users.hol.gr/~xpolakis/piphil.html`.

Hermite, C. "Sur la fonction exponentielle." *C. R. Acad. Sci. Paris* **77**, 18–24, 74–79, and 226–233, 1873.

Le Lionnais, F. *Les nombres remarquables.* Paris: Hermann, p. 47, 1983.

Maor, E. *e: The Story of a Number.* Princeton, NJ: Princeton University Press, 1994.

Minkus, J. "A Continued Fraction." Problem 10327. *Amer. Math. Monthly* **103**, 605–606, 1996.

Mitchell, U. G. and Strain, M. "The Number e." *Osiris* **1**, 476–496, 1936.

Olds, C. D. "The Simple Continued Fraction Expression of e." *Amer. Math. Monthly* **77**, 968–974, 1970.

Plouffe, S. "Plouffe's Inverter: Table of Current Records for the Computation of Constants." `http://lacim.uqam.ca/pi/records.html`.

Rabinowitz, S. and Wagon, S. "A Spigot Algorithm for the Digits of π." *Amer. Math. Monthly* **102**, 195–203, 1995.

Sales, A. H. J. "The Calculation of e to Many Significant Digits." *Computer J.* **11**, 229–230, 1968.

Sloane, N. J. A. Sequences A032511, A001113/M1727, A003417/M0088, A007676/M0869, and A007677/M2343 in "An On-Line Version of the Encyclopedia of Integer Sequences."

e-Divisor

d is called an e-divisor (or EXPONENTIAL DIVISOR) of

$$n = {p_1}^{a_1} {p_2}^{a_2} \cdots {p_r}^{a_r}$$

if $d|n$ and

$$d = {p_1}^{b_1} {p_2}^{b_2} \cdots {p_r}^{b_r}$$

where $b_j|a_j$ with $1 \leq j \leq r$.

see also e-PERFECT NUMBER

References

Guy, R. K. "Exponential-Perfect Numbers." §B17 in *Unsolved Problems in Number Theory, 2nd ed.* New York: Springer-Verlag, pp. 73, 1994.

Straus, E. G. and Subbarao, M. V. "On Exponential Divisors." *Duke Math. J.* **41**, 465–471, 1974.

E_n-Function

The $E_n(x)$ function is defined by the integral

$$E_n(x) \equiv \int_1^\infty \frac{e^{-xt}\,dt}{t^n} \tag{1}$$

and is given by the *Mathematica*® (Wolfram Research, Champaign, IL) function `ExpIntegralE[n,x]`. Defining $t \equiv \eta^{-1}$ so that $dt = -\eta^{-2} d\eta$,

$$E_n(x) = \int_0^1 e^{-x/\eta} \eta^{n-2}\,d\eta \tag{2}$$

$$E_n(0) = \frac{1}{n-1}. \tag{3}$$

The function satisfies the RECURRENCE RELATIONS

$$E'_n(x) = -E_{n-1}(x) \tag{4}$$

$$n\,E_{n+1}(x) = e^{-x} - x\,E_n(x). \tag{5}$$

Equation (4) can be derived from

$$\mathrm{E}_n(x) = \int_1^\infty \frac{e^{-tx}}{t^n}\,dt \tag{6}$$

$$\begin{aligned}\mathrm{E}'_n(x) &= \frac{d}{dx}\int_1^\infty \frac{e^{-tx}}{t^n}\,dt = \int_1^\infty \frac{d}{dx}\left(\frac{e^{-tx}}{t^n}\right)dt \\ &= -\int_1^\infty t\frac{e^{-tx}}{t^n}\,dt \\ &= -\int_1^\infty \frac{e^{-tx}}{t^{n-1}}\,dt = -\mathrm{E}_{n-1}(x),\end{aligned} \tag{7}$$

and (5) using integrating by parts, letting

$$u = \frac{1}{t^n} \qquad dv = e^{-tx}\,dt \tag{8}$$

$$du = -\frac{n}{t^{n+1}}\,dt \qquad v = -\frac{e^{-tx}}{x}, \tag{9}$$

$$\begin{aligned}\mathrm{E}_n(x) &= \int u\,dv = uv - \int v\,du \\ &= -\frac{e^{-tx}}{xt^n} - \frac{n}{x}\int_1^\infty \frac{e^{-tx}\,dx}{t^{n+1}} \\ &= x\int_1^\infty \frac{e^{-tx}\,dt}{t^n} \\ &= -\left[\frac{1}{e^{tx}t^n}\right]_1^\infty - n\int_1^\infty \frac{e^{-tx}\,dx}{t^{n+1}} \\ &= x\,\mathrm{E}_n(x) = e^{-x} - n\,\mathrm{E}_{n+1}(x).\end{aligned} \tag{10}$$

Solving (10) for $n\,\mathrm{E}_n(x)$ gives (5). An asymptotic expansion gives

$$(n-1)!\,\mathrm{E}_n(x) = (-x)^{n-1}\,\mathrm{E}_1(x) + e^{-x}\sum_{s=0}^{n} -2(n-s-2)!(-x)^s, \tag{11}$$

so

$$\mathrm{E}_n(x) = \frac{e^{-x}}{x}\left[1 - \frac{n}{x} + \frac{n(n+1)}{x^2} + \ldots\right]. \tag{12}$$

The special case $n = 1$ gives

$$\mathrm{E}_1(x) \equiv -\,\mathrm{ei}\,(-x) = \int_1^\infty \frac{e^{-tx}\,dt}{t} = \int_x^\infty \frac{e^{-u}\,du}{u}, \tag{13}$$

where $\mathrm{ei}(x)$ is the EXPONENTIAL INTEGRAL, which is also equal to

$$\mathrm{E}_1(x) = -\gamma - \ln x - \sum_{n=1}^{\infty} \frac{(-1)^n x^n}{n!n}, \tag{14}$$

where γ is the EULER-MASCHERONI CONSTANT.

$$\mathrm{E}_1(0) = \infty \tag{15}$$

$$\mathrm{E}_1(ix) = -\,\mathrm{ci}(x) + i\,\mathrm{si}(x), \tag{16}$$

where $\mathrm{ci}(x)$ and $\mathrm{si}(x)$ are the COSINE INTEGRAL and SINE INTEGRAL.

see also COSINE INTEGRAL, E_t-FUNCTION, EXPONENTIAL INTEGRAL, GOMPERTZ CONSTANT, SINE INTEGRAL

References

Abramowitz, M. and Stegun, C. A. (Eds.). "Exponential Integral and Related Functions." Ch. 5 in *Handbook of Mathematical Functions with Formulas, Graphs, and Mathematical Tables, 9th printing.* New York: Dover, pp. 227–233, 1972.

Press, W. H.; Flannery, B. P.; Teukolsky, S. A.; and Vetterling, W. T. "Exponential Integrals." §6.3 in *Numerical Recipes in FORTRAN: The Art of Scientific Computing, 2nd ed.* Cambridge, England: Cambridge University Press, pp. 215–219, 1992.

Spanier, J. and Oldham, K. B. "The Exponential Integral Ei(x) and Related Functions." Ch. 37 in *An Atlas of Functions.* Washington, DC: Hemisphere, pp. 351–360, 1987.

E_t-Function

A function which arises in FRACTIONAL CALCULUS.

$$E_t(\nu, a) = \frac{1}{\Gamma(\nu)}e^{at}\int_0^t x^{\nu-1}e^{-ax}\,dx = t^\nu e^{at}\gamma(\nu, at), \tag{1}$$

where γ is the incomplete GAMMA FUNCTION and Γ the complete GAMMA FUNCTION. The E_t function satisfies the RECURRENCE RELATION

$$E_t(\nu, a) = aE_t(\nu+1, a) + \frac{t^\nu}{\Gamma(\nu+1)}. \tag{2}$$

A special value is

$$E_t(0, a) = e^{at}. \tag{3}$$

see also E_n-FUNCTION

References

Abramowitz, M. and Stegun, C. A. (Eds.). "Exponential Integral and Related Functions." Ch. 5 in *Handbook of Mathematical Functions with Formulas, Graphs, and Mathematical Tables, 9th printing.* New York: Dover, pp. 227–233, 1972.

e-Multiperfect Number

A number n is called a k e-perfect number if $\sigma_e(n) = kn$, where $\sigma_e(n)$ is the SUM of the e-DIVISORS of n.

see also e-DIVISOR, e-PERFECT NUMBER

References

Guy, R. K. "Exponential-Perfect Numbers." §B17 in *Unsolved Problems in Number Theory, 2nd ed.* New York: Springer-Verlag, pp. 73, 1994.

e-Perfect Number

A number n is called an e-perfect number if $\sigma_e(n) = 2n$, where $\sigma_e(n)$ is the SUM of the e-DIVISORS of n. If m is SQUAREFREE, then $\sigma_e(m) = m$. As a result, if n is e-perfect and m is SQUAREFREE with $m \perp b$, then mn is e-perfect. There are no ODD e-perfect numbers.

see also e-DIVISOR

References

Guy, R. K. "Exponential-Perfect Numbers." §B17 in *Unsolved Problems in Number Theory, 2nd ed.* New York: Springer-Verlag, pp. 73, 1994.

Subbarao, M. V. and Suryanarayan, D. "Exponential Perfect and Unitary Perfect Numbers." *Not. Amer. Math. Soc.* **18**, 798, 1971.

Ear

A PRINCIPAL VERTEX x_i of a SIMPLE POLYGON P is called an ear if the diagonal $[x_{i-1}, x_{i+1}]$ that bridges x_i lies entirely in P. Two ears x_i and x_j are said to overlap if

$$\operatorname{int}[x_{i-1}, x_i, x_{i+1}] \cap \operatorname{int}[x_{j-1}, x_j, x_{j+1}] = \varnothing.$$

The TWO-EARS THEOREM states that, except for TRIANGLES, every SIMPLE POLYGON has at least two nonoverlapping ears.

see also ANTHROPOMORPHIC POLYGON, MOUTH, TWO-EARS THEOREM

References

Meisters, G. H. "Polygons Have Ears." *Amer. Math. Monthly* **82**, 648–751, 1975.

Meisters, G. H. "Principal Vertices, Exposed Points, and Ears." *Amer. Math. Monthly* **87**, 284–285, 1980.

Toussaint, G. "Anthropomorphic Polygons." *Amer. Math. Monthly* **122**, 31–35, 1991.

Early Election Results

Let Jones and Smith be the only two contestants in an election that will end in a deadlock when all votes for Jones (J) and Smith (S) are counted. What is the EXPECTATION VALUE of $X_k \equiv |S - J|$ after k votes are counted? The solution is

$$\langle X_k \rangle = \frac{2N\binom{N-1}{\lfloor k/2 \rfloor}\binom{N-1}{\lfloor k/2 \rfloor - 1}}{\binom{2N}{k}} = \begin{cases} \frac{k(2N-k)}{2N}\binom{N}{k/2}^2\binom{2N}{k}^{-1} & \text{for } k \text{ even} \\ \frac{k(2N-k+1)}{2N}\binom{N}{(k-1)/2}^2\binom{2N}{k-1}^{-1} & \text{for } k \text{ odd.} \end{cases}$$

References

Handelsman, M. B. Solution to Problem 10248. "Early Returns in a Tied Election." *Amer. Math. Monthly* **102**, 554–556, 1995.

Eban Number

The sequence of numbers whose names (in English) do not contain the letter "e" (i.e., "e" is "banned"). The first few eban numbers are 2, 4, 6, 30, 32, 34, 36, 40, 42, 44, 46, 50, 52, 54, 56, 60, 62, 64, 66, 2000, 2002, 2004, ... (Sloane's A006933); i.e., two, four, six, thirty, etc.

References

Sloane, N. J. A. Sequence A006933/M1030 in "An On-Line Version of the Encyclopedia of Integer Sequences."

Eberhart's Conjecture

If q_n is the nth prime such that M_{q_n} is a MERSENNE PRIME, then

$$q_n \sim (3/2)^n.$$

It was modified by Wagstaff (1983) to yield

$$q_n \sim (2^{e^{-\gamma}})^n,$$

where γ is the EULER-MASCHERONI CONSTANT.

References

Ribenboim, P. *The Book of Prime Number Records, 2nd ed.* New York: Springer-Verlag, pp. 332–333, 1989.

Wagstaff, S. S. "Divisors of Mersenne Numbers." *Math. Comput.* **40**, 385–397, 1983.

Eccentric

Not CONCENTRIC.

see also CONCENTRIC, CONCYCLIC

Eccentric Angle

The angle θ measured from the CENTER of an ELLIPSE to a point on the ELLIPSE.

see also ECCENTRICITY, ELLIPSE

Eccentric Anomaly

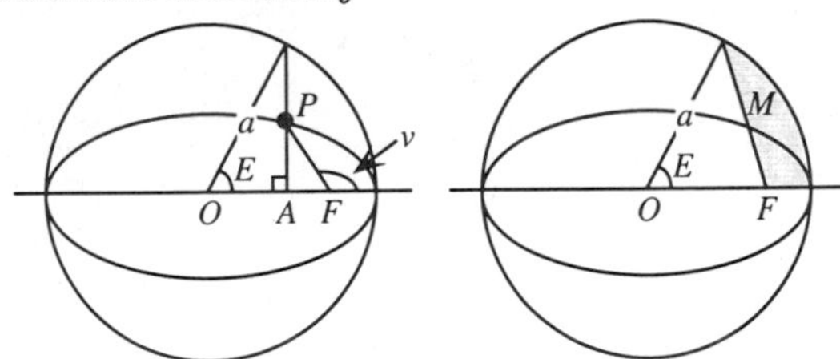

The ANGLE obtained by drawing the AUXILIARY CIRCLE of an ELLIPSE with center O and FOCUS F, and drawing a LINE PERPENDICULAR to the SEMIMAJOR AXIS and intersecting it at A. The ANGLE E is then defined as illustrated above. Then for an ELLIPSE with ECCENTRICITY e,

$$AF = OF - AO = ae - a\cos E. \quad (1)$$

But the distance AF is also given in terms of the distance from the FOCUS $r = FP$ and the SUPPLEMENT of the ANGLE from the SEMIMAJOR AXIS v by

$$AF = r\cos(\pi - v) = -r\cos v. \quad (2)$$

Equating these two expressions gives

$$r = \frac{a(\cos E - e)}{\cos v}, \quad (3)$$

which can be solved for $\cos v$ to obtain

$$\cos v = \frac{a(\cos E - e)}{r}. \quad (4)$$

To get E in terms of r, plug (4) into the equation of the ELLIPSE

$$r = \frac{a(1-e^2)}{1+e\cos v} \quad (5)$$

$$r(1+e\cos v) = a(1-e^2) \quad (6)$$

$$r\left(1+\frac{ae\cos E}{r} - \frac{e^2}{r}\right) = r + ae\cos E - e^2 = a(1-e^2) \quad (7)$$

$$r = a(1-e^2) - ea\cos E + e^2 a = a(1 - e\cos E). \quad (8)$$

Differentiating gives

$$\dot{r} = ae\dot{E}\sin E. \quad (9)$$

The eccentric anomaly is a very useful concept in orbital mechanics, where it is related to the so-called mean anomaly M by KEPLER'S EQUATION

$$M = E - e\sin E. \quad (10)$$

M can also be interpreted as the AREA of the shaded region in the above figure (Finch).

see also ECCENTRICITY, ELLIPSE, KEPLER'S EQUATION

References
Danby, J. M. *Fundamentals of Celestial Mechanics, 2nd ed., rev. ed.* Richmond, VA: Willmann-Bell, 1988.
Finch, S. "Favorite Mathematical Constants." `http://www.mathsoft.com/asolve/constant/lpc/lpc.html`.

Eccentricity

A quantity defined for a CONIC SECTION which can be given in terms of SEMIMAJOR and SEMIMINOR AXES for an ELLIPSE. For an ELLIPSE with SEMIMAJOR AXIS a and SEMIMINOR AXIS b,

$$e \equiv \sqrt{1 - \frac{b^2}{a^2}}.$$

The eccentricity can be interpreted as the fraction of the distance to the semimajor axis at which the FOCUS lies,

$$e = \frac{c}{a},$$

where c is the distance from the center of the CONIC SECTION to the FOCUS. The table below gives the type of CONIC SECTION corresponding to various ranges of eccentricity e.

e	Curve
$e = 0$	circle
$0 < e < 1$	ellipse
$e = 1$	parabola
$e > 1$	hyperbola

see also CIRCLE, CONIC SECTION, ECCENTRIC ANOMALY, ELLIPSE, FLATTENING, HYPERBOLA, OBLATENESS, PARABOLA, SEMIMAJOR AXIS, SEMIMINOR AXIS

Eccentricity (Graph)

The length of the longest shortest path from a VERTEX in a GRAPH.

see also DIAMETER (GRAPH)

Echidnahedron

ICOSAHEDRON STELLATION #4.

References
Wenninger, M. J. *Polyhedron Models.* Cambridge, England: Cambridge University Press, p. 65, 1971.

Eckardt Point

On the CLEBSCH DIAGONAL CUBIC, all 27 of the complex lines present on a general smooth CUBIC SURFACE are real. In addition, there are 10 points on the surface where three of the 27 lines meet. These points are called Eckardt points (Fischer 1986).

see also CLEBSCH DIAGONAL CUBIC, CUBIC SURFACE

References
Fischer, G. (Ed.). *Mathematical Models from the Collections of Universities and Museums.* Braunschweig, Germany: Vieweg, p. 11, 1986.

Eckert IV Projection

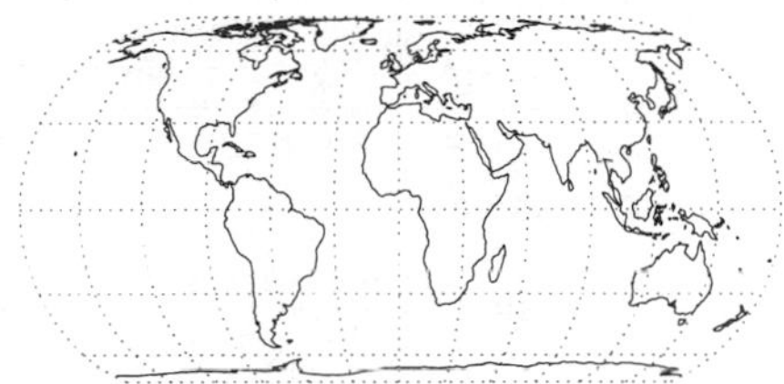

The equations are

$$x = \frac{2}{\sqrt{\pi(4+\pi)}}(\lambda - \lambda_0)(1+\cos\theta) \quad (1)$$

$$y = 2\sqrt{\frac{\pi}{4+\pi}}\sin\theta, \quad (2)$$

where θ is the solution to

$$\theta + \sin\theta\cos\theta + 2\sin\theta = (2 + \tfrac{1}{2}\pi)\sin\phi. \quad (3)$$

This can be solved iteratively using NEWTON'S METHOD with $\theta_0 = \phi/2$ to obtain

$$\Delta\theta = -\frac{\theta + \sin\theta\cos\theta + 2\sin\theta - (2 - \frac{1}{2}\pi)\sin\phi}{2\cos\theta(1+\cos\theta)}. \quad (4)$$

The inverse FORMULAS are

$$\phi = \sin^{-1}\left(\frac{\theta + \sin\theta\cos\theta + 2\sin\theta}{2 + \frac{1}{2}\pi}\right) \quad (5)$$

$$\lambda = \lambda_0 + \frac{\pi\sqrt{4+\pi}\,x}{1+\cos\theta}, \quad (6)$$

where

$$\theta = \sin^{-1}\left(\frac{y}{2}\sqrt{\frac{4+\pi}{\pi}}\right). \quad (7)$$

References

Snyder, J. P. *Map Projections—A Working Manual.* U. S. Geological Survey Professional Paper 1395. Washington, DC: U. S. Government Printing Office, pp. 253–258, 1987.

Eckert VI Projection

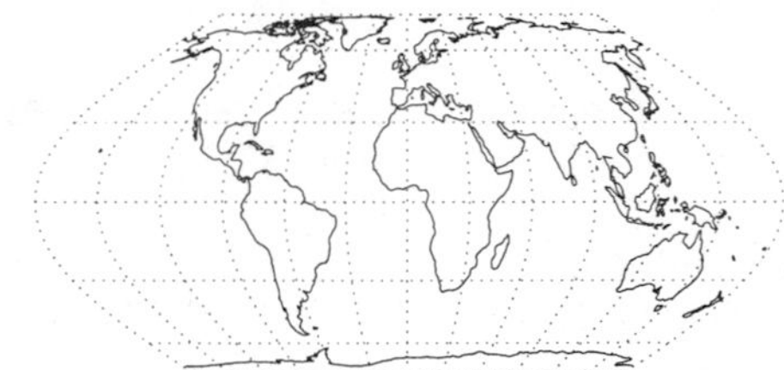

The equations are

$$x = \frac{(\lambda - \lambda_0)(1+\cos\theta)}{\sqrt{2+\pi}} \quad (1)$$

$$y = \frac{2\theta}{\sqrt{2+\pi}}, \quad (2)$$

where θ is the solution to

$$\theta + \sin\theta = (1 + \tfrac{1}{2}\pi)\sin\phi. \quad (3)$$

This can be solved iteratively using NEWTON'S METHOD with $\theta_0 = \phi$ to obtain

$$\Delta\theta = -\frac{\theta + \sin\theta - (1 + \frac{1}{2}\pi)\sin\phi}{1+\cos\theta}. \quad (4)$$

The inverse FORMULAS are

$$\phi = \sin^{-1}\left(\frac{\theta + \sin\theta}{1 + \frac{1}{2}\pi}\right) \quad (5)$$

$$\lambda = \lambda_0 + \frac{\sqrt{2+\pi}\,x}{1+\cos\theta}, \quad (6)$$

where

$$\theta = \tfrac{1}{2}\sqrt{2+\pi}\,y. \quad (7)$$

References

Snyder, J. P. *Map Projections—A Working Manual.* U. S. Geological Survey Professional Paper 1395. Washington, DC: U. S. Government Printing Office, pp. 253–258, 1987.

Economized Rational Approximation

A PADÉ APPROXIMATION perturbed with a CHEBYSHEV POLYNOMIAL OF THE FIRST KIND to reduce the leading COEFFICIENT in the ERROR.

Eddington Number

$$136 \cdot 2^{256} \approx 1.575 \times 10^{79}.$$

According to Eddington, the exact number of protons in the universe, where 136 was the RECIPROCAL of the fine structure constant as best as it could be measured in his time.

see also LARGE NUMBER

Edge-Coloring

An edge-coloring of a GRAPH G is a coloring of the edges of G such that adjacent edges (or the edges bounding different regions) receive different colors. BRELAZ'S HEURISTIC ALGORITHM can be used to find a good, but not necessarily minimal, edge-coloring.

see also BRELAZ'S HEURISTIC ALGORITHM, CHROMATIC NUMBER, k-COLORING

References

Saaty, T. L. and Kainen, P. C. *The Four-Color Problem: Assaults and Conquest.* New York: Dover, p. 13, 1986.

Edge Connectivity

The minimum number of EDGES whose deletion from a GRAPH disconnects it.

see also VERTEX CONNECTIVITY

Edge (Graph)

For an undirected GRAPH, an unordered pair of nodes which specify the line connecting them. For a DIRECTED GRAPH, the edge is an ordered pair of nodes.

see also EDGE NUMBER, NULL GRAPH, TAIT COLORING, TAIT CYCLE, VERTEX (GRAPH)

Edge Number

The number of EDGES in a GRAPH, denoted $|E|$.

see also EDGE (GRAPH)

Edge (Polygon)

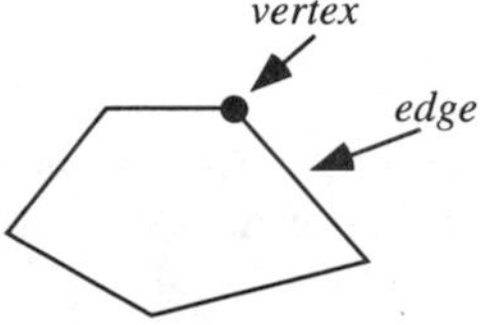

A LINE SEGMENT on the boundary of a FACE, also called a SIDE.

see also EDGE (POLYHEDRON), VERTEX (POLYGON)

Edge (Polyhedron)

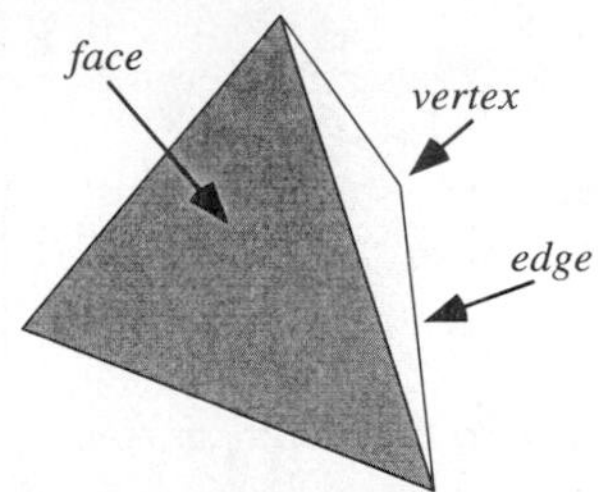

A LINE SEGMENT where two FACES of a POLYHEDRON meet, also called a SIDE.

see also EDGE (POLYGON), VERTEX (POLYHEDRON)

Edge (Polytope)

A 1-D LINE SEGMENT where two 2-D FACES of an n-D POLYTOPE meet, also called a SIDE.

see also EDGE (POLYGON), EDGE (POLYHEDRON)

Edgeworth Series

Approximate a distribution in terms of a NORMAL DISTRIBUTION. Let

$$\phi(t) \equiv \frac{1}{\sqrt{2\pi}} e^{-t^2/2},$$

then

$$f(t) = \phi(t) + \tfrac{1}{3!}\gamma_1 \phi^{(3)}(t) + \left[\frac{\gamma_2}{4!}\phi^{(4)}(t) + \frac{10{\gamma_1}^2}{6!}\phi^{(6)}(t)\right] + \ldots.$$

see also CORNISH-FISHER ASYMPTOTIC EXPANSION, GRAM-CHARLIER SERIES

References

Abramowitz, M. and Stegun, C. A. (Eds.). *Handbook of Mathematical Functions with Formulas, Graphs, and Mathematical Tables, 9th printing.* New York: Dover, p. 935, 1972.

Kenney, J. F. and Keeping, E. S. *Mathematics of Statistics, Pt. 2, 2nd ed.* Princeton, NJ: Van Nostrand, p. 108, 1951.

Edmonds' Map

A nonreflexible regular map of GENUS 7 with eight VERTICES, 28 EDGES, and eight HEPTAGONAL faces.

Efron's Dice

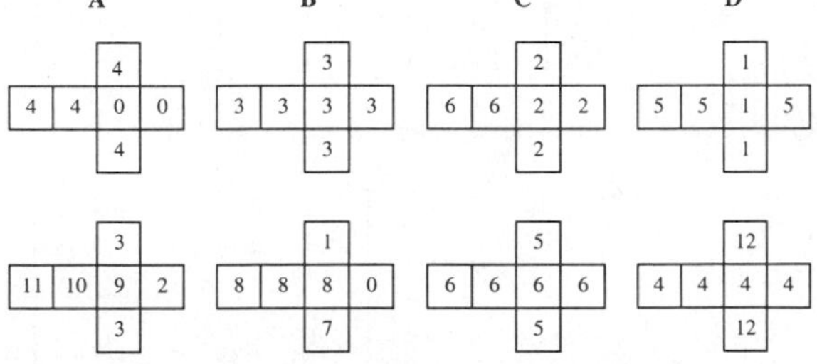

A set of four nontransitive DICE such that the probabilities of A winning against B, B against C, C against D, and D against A are all 2:1. A set in which ties may occur, in which case the DICE are rolled again, which gives ODDS of 11:6 is

Die	Top	Row	Bottom
A	1	10, 2, 9, 3	11
B	0	8, 1, 7, 8	9
C	5	6, 7, 5, 7	6
D	5	4, 12, 3, 4	11

see also DICE, SICHERMAN DICE

References

Gardner, M. "Mathematical Games: The Paradox of the Nontransitive Dice and the Elusive Principle of Indifference." *Sci. Amer.* **223**, 110–114, Dec. 1970.

Honsberger, R. "Some Surprises in Probability." Ch. 5 in *Mathematical Plums* (Ed. R. Honsberger). Washington, DC: Math. Assoc. Amer., pp. 94–97, 1979.

Egg

An OVAL with one end more pointed than the other.

see also ELLIPSE, MOSS'S EGG, OVAL, OVOID, THOM'S EGGS

Egyptian Fraction

see UNIT FRACTION

Ehrhart Polynomial

Let Δ denote an integral convex POLYTOPE of DIMENSION n in a lattice M, and let $l_\Delta(k)$ denote the number of LATTICE POINTS in Δ dilated by a factor of the integer k,

$$l_\Delta(k) = \#(k\Delta \cap M) \tag{1}$$

for $k \in \mathbb{Z}^+$. Then l_Δ is a polynomial function in k of degree n with rational coefficients

$$l_\Delta(k) = a_n k^n + a_{n-1}k^{n-1} + \ldots + a_0 \tag{2}$$

called the Ehrhart polynomial (Ehrhart 1967, Pommersheim 1993). Specific coefficients have important geometric interpretations.

1. a_n is the CONTENT of Δ.
2. a_{n-1} is half the sum of the CONTENTS of the $(n-1)$-D faces of Δ.
3. $a_0 = 1$.

Let $S_2(\Delta)$ denote the sum of the lattice lengths of the edges of Δ, then the case $n = 2$ corresponds to PICK'S THEOREM,

$$l_\Delta(k) = \mathrm{Vol}(\Delta)k^2 + \tfrac{1}{2}S_2(\Delta) + 1. \tag{3}$$

Let $S_3(\Delta)$ denote the sum of the lattice volumes of the 2-D faces of Δ, then the case $n = 3$ gives

$$l_\Delta(k) = \mathrm{Vol}(\Delta)k^3 + \tfrac{1}{2}S_3(\Delta)k^2 + a_1 k + 1, \tag{4}$$

where a rather complicated expression is given by Pommersheim (1993), since a_1 can unfortunately *not* be interpreted in terms of the edges of Δ. The Ehrhart polynomial of the tetrahedron with vertices at (0, 0, 0), (a, 0, 0), (0, b, 0), (0, 0, c) is

$$\begin{aligned} l_\Delta(k) &= \tfrac{1}{6}abck^3 + \tfrac{1}{4}(ab+ac+bc+d)k^2 \\ &+ \left[\frac{1}{12}\left(\frac{ac}{b}+\frac{bc}{a}+\frac{ab}{c}+\frac{d^2}{abc}\right)\right. \\ &+\tfrac{1}{4}(a+b+c+A+B+C) - As\left(\frac{bc}{d},\frac{aA}{d}\right) \\ &\left.-Bs\left(\frac{ac}{d},\frac{bB}{d}\right) - Cs\left(\frac{ab}{d},\frac{cC}{d}\right)\right] k+1, \end{aligned} \quad (5)$$

where $s(x,y)$ is a DEDEKIND SUM, $A = \gcd(b,c)$, $B = \gcd(a,c)$, $C = \gcd(a,b)$ (here, gcd is the GREATEST COMMON DENOMINATOR), and $d = ABC$ (Pommersheim 1993).

see also DEHN INVARIANT, PICK'S THEOREM

References

Ehrhart, E. "Sur une problème de géométrie diophantine linéaire." *J. Reine angew. Math.* **227**, 1–29, 1967.

MacDonald, I. G. "The Volume of a Lattice Polyhedron." *Proc. Camb. Phil. Soc.* **59**, 719–726, 1963.

McMullen, P. "Valuations and Euler-Type Relations on Certain Classes of Convex Polytopes." *Proc. London Math. Soc.* **35**, 113–135, 1977.

Pommersheim, J. "Toric Varieties, Lattices Points, and Dedekind Sums." *Math. Ann.* **295**, 1–24, 1993.

Reeve, J. E. "On the Volume of Lattice Polyhedra." *Proc. London Math. Soc.* **7**, 378–395, 1957.

Reeve, J. E. "A Further Note on the Volume of Lattice Polyhedra." *Proc. London Math. Soc.* **34**, 57–62, 1959.

Ei

see EXPONENTIAL INTEGRAL, E_n-FUNCTION

Eigenfunction

If $\tilde{L}$ is a linear OPERATOR on a FUNCTION SPACE, then f is an eigenfunction for $\tilde{L}$ and λ is the associated EIGENVALUE whenever $\tilde{L}f = \lambda f$.

see also EIGENVALUE, EIGENVECTOR

Eigenvalue

Let A be a linear transformation represented by a MATRIX A. If there is a VECTOR $\mathbf{X} \in \mathbb{R}^n \neq \mathbf{0}$ such that

$$\mathsf{A}\mathbf{X} = \lambda\mathbf{X} \quad (1)$$

for some SCALAR λ, then λ is the eigenvalue of A with corresponding (right) EIGENVECTOR $\mathbf{X}$. Letting A be a $k \times k$ MATRIX,

$$\begin{bmatrix} a_{11} & a_{12} & \cdots & a_{1k} \\ a_{21} & a_{22} & \cdots & a_{2k} \\ \vdots & \vdots & \ddots & \vdots \\ a_{k1} & a_{k2} & \cdots & a_{kk} \end{bmatrix} \quad (2)$$

with eigenvalue λ, then the corresponding EIGENVECTORS satisfy

$$\begin{bmatrix} a_{11} & a_{12} & \cdots & a_{1k} \\ a_{21} & a_{22} & \cdots & a_{2k} \\ \vdots & \vdots & \ddots & \vdots \\ a_{k1} & a_{k2} & \cdots & a_{kk} \end{bmatrix} \begin{bmatrix} x_1 \\ x_2 \\ \vdots \\ x_k \end{bmatrix} = \lambda \begin{bmatrix} x_1 \\ x_2 \\ \vdots \\ x_k \end{bmatrix}, \quad (3)$$

which is equivalent to the homogeneous system

$$\begin{bmatrix} a_{11}-\lambda & a_{12} & \cdots & a_{1k} \\ a_{21} & a_{22}-\lambda & \cdots & a_{2k} \\ \vdots & \vdots & \ddots & \vdots \\ a_{k1} & a_{k2} & \cdots & a_{kk}-\lambda \end{bmatrix} \begin{bmatrix} x_1 \\ x_2 \\ \vdots \\ x_k \end{bmatrix} = \begin{bmatrix} 0 \\ 0 \\ \vdots \\ 0 \end{bmatrix}. \quad (4)$$

Equation (4) can be written compactly as

$$(\mathsf{A} - \lambda\mathsf{I})\mathbf{X} = \mathbf{0}, \quad (5)$$

where I is the IDENTITY MATRIX.

As shown in CRAMER'S RULE, a system of linear equations has nontrivial solutions only if the DETERMINANT vanishes, so we obtain the CHARACTERISTIC EQUATION

$$|\mathsf{A} - \lambda\mathsf{I}| = 0. \quad (6)$$

If all k λs are different, then plugging these back in gives $k-1$ independent equations for the k components of each corresponding EIGENVECTOR. The EIGENVECTORS will then be orthogonal and the system is said to be nondegenerate. If the eigenvalues are n-fold DEGENERATE, then the system is said to be degenerate and the EIGENVECTORS are not linearly independent. In such cases, the additional constraint that the EIGENVECTORS be orthogonal,

$$\mathbf{X}_i \cdot \mathbf{X}_j = X_i X_j \delta_{ij}, \quad (7)$$

where δ_{ij} is the KRONECKER DELTA, can be applied to yield n additional constraints, thus allowing solution for the EIGENVECTORS.

Assume A has nondegenerate eigenvalues $\lambda_1, \lambda_2, \ldots, \lambda_n$ and corresponding linearly independent EIGENVECTORS $\mathbf{X}_1, \mathbf{X}_2, \ldots, \mathbf{X}_k$ which can be denoted

$$\begin{bmatrix} x_{11} \\ x_{12} \\ \vdots \\ x_{1k} \end{bmatrix}, \begin{bmatrix} x_{21} \\ x_{22} \\ \vdots \\ x_{2k} \end{bmatrix}, \ldots \begin{bmatrix} x_{k1} \\ x_{k2} \\ \vdots \\ x_{kk} \end{bmatrix}. \quad (8)$$

Define the matrices composed of eigenvectors

$$\mathsf{P} \equiv [\,\mathbf{X}_1 \quad \mathbf{X}_2 \quad \cdots \quad \mathbf{X}_k\,] = \begin{bmatrix} x_{11} & x_{21} & \cdots & x_{k1} \\ x_{12} & x_{22} & \cdots & x_{k2} \\ \vdots & \vdots & \ddots & \vdots \\ x_{1k} & x_{2k} & \cdots & x_{kk} \end{bmatrix} \quad (9)$$

and eigenvalues

$$\mathsf{D} \equiv \begin{bmatrix} \lambda_1 & 0 & \cdots & 0 \\ 0 & \lambda_2 & \cdots & 0 \\ \vdots & \vdots & \ddots & \vdots \\ 0 & 0 & \cdots & \lambda_k \end{bmatrix}, \tag{10}$$

where D is a DIAGONAL MATRIX. Then

$$\begin{aligned} \mathsf{AP} &= \mathsf{A}[\mathbf{X}_1 \quad \mathbf{X}_2 \quad \cdots \quad \mathbf{X}_k] \\ &= [\mathsf{A}\mathbf{X}_1 \quad \mathsf{A}\mathbf{X}_2 \quad \cdots \quad \mathsf{A}\mathbf{X}_k] \\ &= [\lambda_1\mathbf{X}_1 \quad \lambda_2\mathbf{X}_2 \quad \cdots \quad \lambda_k\mathbf{X}_k] \\ &= \begin{bmatrix} \lambda_1 x_{11} & \lambda_2 x_{21} & \cdots & \lambda_k x_{k1} \\ \lambda_1 x_{12} & \lambda_2 x_{22} & \cdots & \lambda_k x_{k2} \\ \vdots & \vdots & \ddots & \vdots \\ \lambda_1 x_{1k} & \lambda_2 x_{2k} & \cdots & \lambda_k x_{kk} \end{bmatrix} \\ &= \begin{bmatrix} x_{11} & x_{21} & \cdots & x_{k1} \\ x_{12} & x_{22} & \cdots & x_{k2} \\ \vdots & \vdots & \ddots & \vdots \\ x_{1k} & x_{2k} & \cdots & x_{kk} \end{bmatrix} \begin{bmatrix} \lambda_1 & 0 & \cdots & 0 \\ 0 & \lambda_2 & \cdots & 0 \\ \vdots & \vdots & \ddots & \vdots \\ 0 & 0 & \cdots & \lambda_k \end{bmatrix} \\ &= \mathsf{PD}, \end{aligned} \tag{11}$$

so

$$\mathsf{A} = \mathsf{PDP}^{-1}. \tag{12}$$

Furthermore,

$$\begin{aligned} \mathsf{A}^2 &= (\mathsf{PDP}^{-1})(\mathsf{PDP}^{-1}) = \mathsf{PD}(\mathsf{P}^{-1}\mathsf{P})\mathsf{DP}^{-1} \\ &= \mathsf{PD}^2\mathsf{P}^{-1}. \end{aligned} \tag{13}$$

By induction, it follows that for $n > 0$,

$$\mathsf{A}^n = \mathsf{PD}^n\mathsf{P}^{-1}. \tag{14}$$

The inverse of A is

$$\mathsf{A}^{-1} = (\mathsf{PDP}^{-1})^{-1} = \mathsf{PD}^{-1}\mathsf{P}^{-1}, \tag{15}$$

where the inverse of the DIAGONAL MATRIX D is trivially given by

$$\mathsf{D}^{-1} = \frac{1}{k} \begin{bmatrix} {\lambda_1}^{-1} & 0 & \cdots & 0 \\ 0 & {\lambda_2}^{-1} & \cdots & 0 \\ \vdots & \vdots & \ddots & \vdots \\ 0 & 0 & \cdots & {\lambda_k}^{-1} \end{bmatrix}. \tag{16}$$

Equation (14) therefore holds for both POSITIVE and NEGATIVE n.

A further remarkable result involving the matrices P and D follows from the definition

$$\begin{aligned} e^{\mathsf{A}} &\equiv \sum_{n=0}^{\infty} \frac{\mathsf{A}^n}{n!} = \sum_{n=0}^{\infty} \frac{\mathsf{PD}^n\mathsf{P}^{-1}}{n!} \\ &= \mathsf{P}\left(\frac{\sum_{n=0}^{\infty} \mathsf{D}^n}{n!}\right)\mathsf{P}^{-1} = \mathsf{P}e^{\mathsf{D}}\mathsf{P}^{-1}. \end{aligned} \tag{17}$$

Since D is a DIAGONAL MATRIX,

$$\begin{aligned} e^{\mathsf{D}} &= \sum_{n=0}^{\infty} \frac{\mathsf{D}^n}{n!} = \sum_{n=0}^{\infty} \frac{1}{n!} \begin{bmatrix} {\lambda_1}^n & 0 & \cdots & 0 \\ 0 & {\lambda_2}^n & \cdots & 0 \\ \vdots & \vdots & \ddots & \vdots \\ 0 & 0 & \cdots & {\lambda_k}^n \end{bmatrix} \\ &= \begin{bmatrix} \sum_{i=0}^{\infty} {\lambda_1}^i & 0 & \cdots & 0 \\ 0 & \sum_{i=0}^{\infty} {\lambda_2}^i & \cdots & 0 \\ \vdots & \vdots & \ddots & \vdots \\ 0 & 0 & \cdots & \sum_{i=0}^{\infty} {\lambda_k}^i \end{bmatrix} \\ &= \begin{bmatrix} e^{\lambda_1} & 0 & \cdots & 0 \\ 0 & e^{\lambda_2} & \cdots & 0 \\ \vdots & \vdots & \ddots & \vdots \\ 0 & 0 & \cdots & e^{\lambda_k} \end{bmatrix}, \end{aligned} \tag{18}$$

e^{D} can be found using

$$\mathsf{D}^n = \begin{bmatrix} {\lambda_1}^n & 0 & \cdots & 0 \\ 0 & {\lambda_2}^n & \cdots & 0 \\ \vdots & \vdots & \ddots & \vdots \\ 0 & 0 & \cdots & {\lambda_k}^n \end{bmatrix}. \tag{19}$$

Assume we know the eigenvalue for

$$\mathsf{A}\mathbf{X} = \lambda\mathbf{X}. \tag{20}$$

Adding a constant times the IDENTITY MATRIX to A,

$$(\mathsf{A} + c\,\mathsf{I})\mathbf{X} = (\lambda + c)\mathbf{X} \equiv \lambda'\mathbf{X}, \tag{21}$$

so the new eigenvalues equal the old plus c. Multiplying A by a constant c

$$(c\,\mathsf{A})\mathbf{X} = c\,(\lambda\mathbf{X}) \equiv \lambda'\mathbf{X}, \tag{22}$$

so the new eigenvalues are the old multiplied by c.

Now consider a SIMILARITY TRANSFORMATION of A. Let $|\mathsf{A}|$ be the DETERMINANT of A, then

$$\begin{aligned} |\mathsf{Z}^{-1}\mathsf{AZ} - \lambda\mathsf{I}| &= |\mathsf{Z}^{-1}(\mathsf{A} - \lambda\mathsf{I})\mathsf{Z}| \\ &= |\mathsf{Z}|\,|\mathsf{A} - \lambda\mathsf{I}|\,|\mathsf{Z}^{-1}| = |\mathsf{A} - \lambda\mathsf{I}|, \end{aligned} \tag{23}$$

so the eigenvalues are the same as for A.

see also BRAUER'S THEOREM, CONDITION NUMBER, EIGENFUNCTION, EIGENVECTOR, FROBENIUS THEOREM, GERŠGORIN CIRCLE THEOREM, LYAPUNOV'S FIRST THEOREM, LYAPUNOV'S SECOND THEOREM, OSTROWSKI'S THEOREM, PERRON'S THEOREM, PERRON-FROBENIUS THEOREM, POINCARÉ SEPARATION THEOREM, RANDOM MATRIX, SCHUR'S INEQUALITIES, STURMIAN SEPARATION THEOREM, SYLVESTER'S INERTIA LAW, WIELANDT'S THEOREM

References

Arfken, G. "Eigenvectors, Eigenvalues." §4.7 in *Mathematical Methods for Physicists, 3rd ed.* Orlando, FL: Academic Press, pp. 229–237, 1985.

Nash, J. C. "The Algebraic Eigenvalue Problem." Ch. 9 in *Compact Numerical Methods for Computers: Linear Algebra and Function Minimisation, 2nd ed.* Bristol, England: Adam Hilger, pp. 102–118, 1990.

Press, W. H.; Flannery, B. P.; Teukolsky, S. A.; and Vetterling, W. T. "Eigensystems." Ch. 11 in *Numerical Recipes in FORTRAN: The Art of Scientific Computing, 2nd ed.* Cambridge, England: Cambridge University Press, pp. 449–489, 1992.

Eigenvector

A right eigenvector satisfies

$$\mathsf{A}\mathbf{X} = \lambda \mathbf{X}, \tag{1}$$

where $\mathbf{X}$ is a column VECTOR. The right EIGENVALUES therefore satisfy

$$|\mathsf{A} - \lambda \mathsf{I}| = 0. \tag{2}$$

A left eigenvector satisfies

$$\mathbf{X}\mathsf{A} = \lambda \mathbf{X}, \tag{3}$$

where $\mathbf{X}$ is a row VECTOR, so

$$(\mathbf{X}\mathsf{A})^{\mathrm{T}} = \lambda_L \mathbf{X}^{\mathrm{T}} \tag{4}$$

$$\mathsf{A}^{\mathrm{T}} \mathbf{X}^{\mathrm{T}} = \lambda_L \mathbf{X}^{\mathrm{T}}, \tag{5}$$

where $\mathbf{X}^{\mathrm{T}}$ is the transpose of $\mathbf{X}$. The left EIGENVALUES satisfy

$$|\mathsf{A}^{\mathrm{T}} - \lambda_L \mathsf{I}| = |\mathsf{A}^{\mathrm{T}} - \lambda_L \mathsf{I}^{\mathrm{T}}| = |(\mathsf{A} - \lambda_L \mathsf{I})^{\mathrm{T}}| = |(\mathsf{A} - \lambda_L \mathsf{I})|, \tag{6}$$

(since $|\mathsf{A}| = |\mathsf{A}^{\mathrm{T}}|$) where $|\mathsf{A}|$ is the DETERMINANT of A. But this is the same equation satisfied by the right EIGENVALUES, so the left and right EIGENVALUES are the same. Let $\mathbf{X}_R$ be a MATRIX formed by the columns of the right eigenvectors and $\mathbf{X}_L$ be a MATRIX formed by the rows of the left eigenvectors. Let

$$\mathsf{D} \equiv \begin{bmatrix} \lambda_1 & \cdots & 0 \\ \vdots & \ddots & \vdots \\ 0 & \cdots & \lambda_n \end{bmatrix}. \tag{7}$$

Then

$$\mathsf{A}\mathbf{X}_R = \mathbf{X}_R \mathsf{D} \qquad \mathbf{X}_L \mathsf{A} = \mathsf{D}\mathbf{X}_L \tag{8}$$

$$\mathbf{X}_L \mathsf{A} \mathbf{X}_R = \mathbf{X}_L \mathbf{X}_R \mathsf{D} \qquad \mathbf{X}_L \mathsf{A} \mathbf{X}_R = \mathsf{D} \mathbf{X}_L \mathbf{X}_R, \tag{9}$$

so

$$\mathbf{X}_L \mathbf{X}_R \mathsf{D} = \mathsf{D} \mathbf{X}_L \mathbf{X}_R. \tag{10}$$

But this equation is of the form $\mathsf{CD} = \mathsf{DC}$ where D is a DIAGONAL MATRIX, so it must be true that $\mathsf{C} \equiv \mathbf{X}_L \mathbf{X}_R$ is also diagonal. In particular, if A is a SYMMETRIC MATRIX, then the left and right eigenvectors are transposes of each other. If A is a SELF-ADJOINT MATRIX, then the left and right eigenvectors are conjugate HERMITIAN MATRICES.

Given a 3×3 MATRIX A with eigenvectors $\mathbf{x}_1$, $\mathbf{x}_2$, and $\mathbf{x}_3$ and corresponding EIGENVALUES λ_1, λ_2, and λ_3, then an arbitrary VECTOR $\mathbf{y}$ can be written

$$\mathbf{y} = b_1 \mathbf{x}_1 + b_2 \mathbf{x}_2 + b_3 \mathbf{x}_3. \tag{11}$$

Applying the MATRIX A,

$$\begin{aligned} \mathsf{A}\mathbf{y} &= b_1 \mathsf{A}\mathbf{x}_1 + b_2 \mathsf{A}\mathbf{x}_2 + b_3 \mathsf{A}\mathbf{x}_3 \\ &= \lambda_1 \left(b_1 \mathbf{x}_1 + \frac{\lambda_2}{\lambda_1} b_2 \mathbf{x}_2 + \frac{\lambda_3}{\lambda_1} b_3 \mathbf{x}_3 \right), \end{aligned} \tag{12}$$

so

$$\mathsf{A}^n \mathbf{y} = \lambda_1{}^n \left[b_1 \mathbf{x}_1 + \left(\frac{\lambda_2}{\lambda_1}\right)^n b_2 \mathbf{x}_2 + \left(\frac{\lambda_3}{\lambda_1}\right)^n b_3 \mathbf{x}_3 \right]. \tag{13}$$

If $\lambda_1 > \lambda_2, \lambda_3$, it therefore follows that

$$\lim_{n\to\infty} \mathsf{A}^n \mathbf{y} = \lambda_1{}^n b_1 \mathbf{x}_1, \tag{14}$$

so repeated application of the matrix to an arbitrary vector results in a vector proportional to the EIGENVECTOR having the largest EIGENVALUE.

see also EIGENFUNCTION, EIGENVALUE

References

Arfken, G. "Eigenvectors, Eigenvalues." §4.7 in *Mathematical Methods for Physicists, 3rd ed.* Orlando, FL: Academic Press, pp. 229–237, 1985.

Press, W. H.; Flannery, B. P.; Teukolsky, S. A.; and Vetterling, W. T. "Eigensystems." Ch. 11 in *Numerical Recipes in FORTRAN: The Art of Scientific Computing, 2nd ed.* Cambridge, England: Cambridge University Press, pp. 449–489, 1992.

Eight Curve

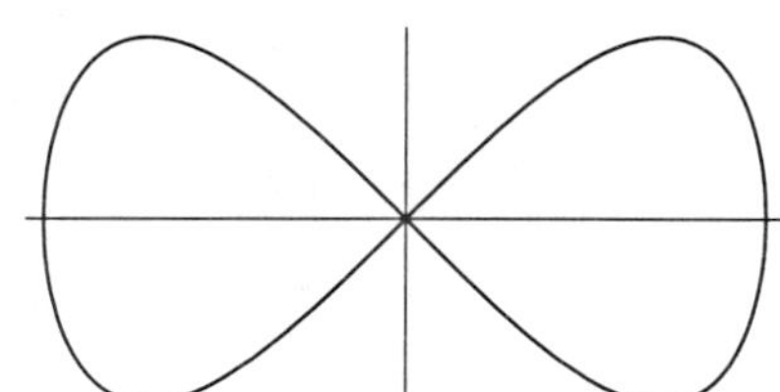

A curve also known as the GERONO LEMNISCATE. It is given by CARTESIAN COORDINATES

$$x^4 = a^2(x^2 - y^2), \tag{1}$$

POLAR COORDINATES,

$$r^2 = a^2 \sec^4\theta \cos(2\theta), \tag{2}$$

and parametric equations

$$x = a \sin t \tag{3}$$

$$y = a \sin t \cos t. \tag{4}$$

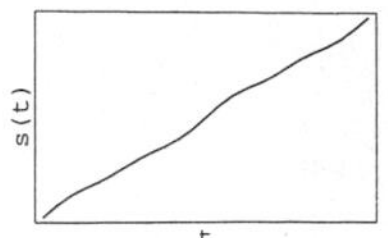

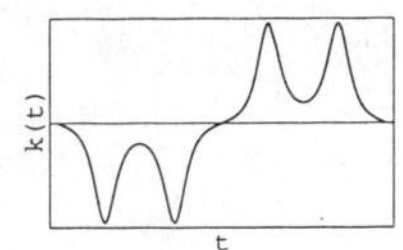

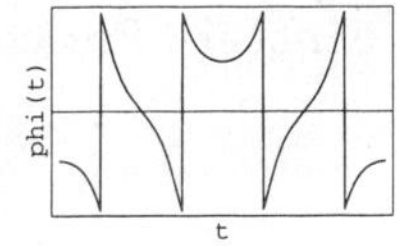

The CURVATURE and TANGENTIAL ANGLE are

$$\kappa(t) = -\frac{3\sin t + \sin(3t)}{2[\cos^2 t + \cos^2(2t)]^{3/2}} \tag{5}$$

$$\phi(t) = -\tan^{-1}[\cos t \sec(2t)]. \tag{6}$$

see also BUTTERFLY CURVE, DUMBBELL CURVE, EIGHT SURFACE, PIRIFORM

References

Cundy, H. and Rollett, A. *Mathematical Models, 3rd ed.* Stradbroke, England: Tarquin Pub., p. 71, 1989.

Lawrence, J. D. *A Catalog of Special Plane Curves.* New York: Dover, pp. 124–126, 1972.

Lee, X. "Lemniscate of Gerono." `http://www.best.com/~xah/SpecialPlaneCurves_dir/LemniscateOfGerono_dir/lemniscateOfGerono.html`.

MacTutor History of Mathematics Archive. "Eight Curve." `http://www-groups.dcs.st-and.ac.uk/~history/Curves/Eight.html`.

Eight-Point Circle Theorem

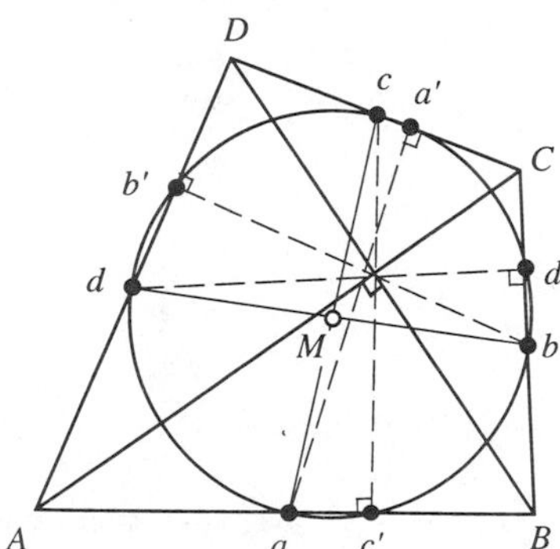

Let $ABCD$ be a QUADRILATERAL with PERPENDICULAR DIAGONALS. The MIDPOINTS of the sides (a, b, c, and d) determine a PARALLELOGRAM (the VARIGNON PARALLELOGRAM) with sides PARALLEL to the DIAGONALS. The eight-point circle passes through the four MIDPOINTS and the four feet of the PERPENDICULARS from the opposite sides a', b', c', and d'.

see also FEUERBACH'S THEOREM

References

Brand, L. "The Eight-Point Circle and the Nine-Point Circle." *Amer. Math. Monthly* **51**, 84–85, 1944.

Honsberger, R. *Mathematical Gems II.* Washington, DC: Math. Assoc. Amer., pp. 11–13, 1976.

Eight Surface

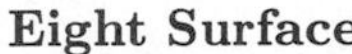

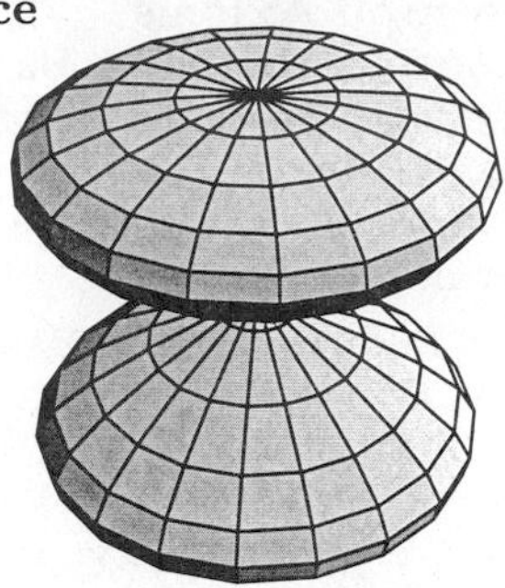

The SURFACE OF REVOLUTION given by the parametric equations

$$x(u,v) = \cos u \sin(2v) \tag{1}$$

$$y(u,v) = \sin u \sin(2v) \tag{2}$$

$$z(u,v) = \sin v \tag{3}$$

for $u \in [0, 2\pi)$ and $v \in [-\pi/2, \pi/2]$.

see also EIGHT CURVE

References

Gray, A. *Modern Differential Geometry of Curves and Surfaces.* Boca Raton, FL: CRC Press, pp. 209–210 and 224, 1993.

Eikonal Equation

$$\sum_{i=1}^{n} \left(\frac{\partial u}{\partial x_i}\right)^2 = 1.$$

Eilenberg-Mac Lane Space

For any ABELIAN GROUP G and any NATURAL NUMBER n, there is a unique SPACE (up to HOMOTOPY type) such that all HOMOTOPY GROUPS except for the nth are trivial (including the 0th HOMOTOPY GROUPS, meaning the SPACE is path-connected), and the nth HOMOTOPY GROUP is ISOMORPHIC to the GROUP G. In the case where $n = 1$, the GROUP G can be non-ABELIAN as well.

Eilenberg-Mac Lane spaces have many important applications. One of them is that every TOPOLOGICAL SPACE has the HOMOTOPY type of an iterated FIBRATION of Eilenberg-Mac Lane spaces (called a POSTNIKOV SYSTEM). In addition, there is a spectral sequence relating the COHOMOLOGY of Eilenberg-Mac Lane spaces to the HOMOTOPY GROUPS of SPHERES.

Eilenberg-Mac Lane-Steenrod-Milnor Axioms

see EILENBERG-STEENROD AXIOMS

Eilenberg-Steenrod Axioms

A family of FUNCTORS $H_n(\cdot)$ from the CATEGORY of pairs of TOPOLOGICAL SPACES and continuous maps, to the CATEGORY of ABELIAN GROUPS and group homomorphisms satisfies the Eilenberg-Steenrod axioms if the following conditions hold.

1. LONG EXACT SEQUENCE OF A PAIR AXIOM. For every pair (X, A), there is a natural long exact sequence
$$\ldots \to H_n(A) \to H_n(X) \to H_n(X,A) \to H_{n-1}(A) \to \ldots,$$
where the MAP $H_n(A) \to H_n(X)$ is induced by the INCLUSION MAP $A \to X$ and $H_n(X) \to H_n(X, A)$ is induced by the INCLUSION MAP $(X, \phi) \to (X, A)$. The MAP $H_n(X, A) \to H_{n-1}(A)$ is called the BOUNDARY MAP.
2. HOMOTOPY AXIOM. If $f : (X, A) \to (Y, B)$ is homotopic to $g : (X, A) \to (Y, B)$, then their INDUCED MAPS $f_* : H_n(X, A) \to H_n(Y, B)$ and $g_* : H_n(X, A) \to H_n(Y, B)$ are the same.
3. EXCISION AXIOM. If X is a SPACE with SUBSPACES A and U such that the CLOSURE of A is contained in the interior of U, then the INCLUSION MAP $(X\ U, A\ U) \to (X, A)$ induces an isomorphism $H_n(X\ U, A\ U) \to H_n(X, A)$.
4. DIMENSION AXIOM. Let X be a single point space. $H_n(X) = 0$ unless $n = 0$, in which case $H_0(X) = G$ where G are some GROUPS. The H_0 are called the COEFFICIENTS of the HOMOLOGY theory $H(\cdot)$.

These are the axioms for a generalized homology theory. For a cohomology theory, instead of requiring that $H(\cdot)$ be a FUNCTOR, it is required to be a co-functor (meaning the INDUCED MAP points in the opposite direction). With that modification, the axioms are essentially the same (except that all the induced maps point backwards).

see also ALEKSANDROV-ČECH COHOMOLOGY

Ein Function

$$\mathrm{Ein}(z) \equiv \int_0^z \frac{(1 - e^{-t})\, dt}{t} = \mathrm{E}_1(z) + \ln z + \gamma,$$

where γ is the EULER-MASCHERONI CONSTANT and E_1 is the E_n-FUNCTION with $n = 1$.

see also E_n-FUNCTION

Einstein Functions

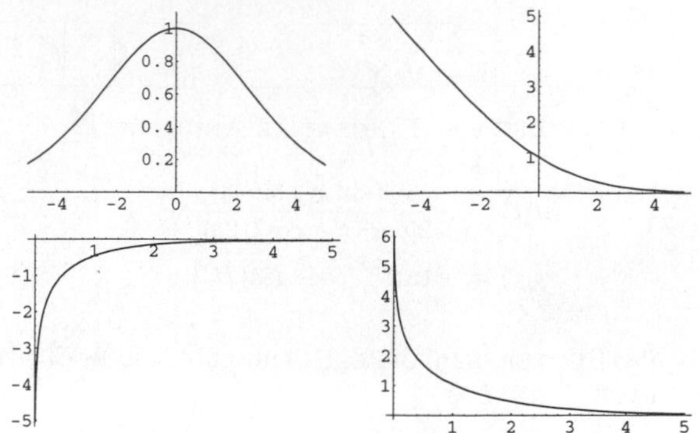

The functions $x^2e^x/(e^x - 1)^2$, $x/(e^x - 1)$, $\ln(1 - e^{-x})$, and $x/(e^x - 1) - \ln(1 - e^{-x})$.

References

Abramowitz, M. and Stegun, C. A. (Eds.). "Debye Functions." §27.1 in *Handbook of Mathematical Functions with Formulas, Graphs, and Mathematical Tables, 9th printing.* New York: Dover, pp. 999–1000, 1972.

Einstein Summation

The implicit convention that repeated indices are summed over so that, for example,

$$a_i a_i \equiv \sum_i a_i a_i.$$

Eisenstein Integer

The numbers $a + b\omega$, where

$$\omega \equiv \tfrac{1}{2}(-1 + i\sqrt{3})$$

is one of the ROOTS of $z^3 = 1$, the others being 1 and

$$\omega^2 = \tfrac{1}{2}(-1 - i\sqrt{3}).$$

Eisenstein integers are members of the QUADRATIC FIELD $\mathbb{Q}(\sqrt{-3})$, and the COMPLEX NUMBERS $\mathbb{Z}[\omega]$. Every Eisenstein integer has a unique factorization. Specifically, any NONZERO Eisenstein integer is uniquely the product of POWERS of -1, ω, and the "positive" EISENSTEIN PRIMES (Conway and Guy 1996). Every Eisenstein integer is within a distance $|n|/\sqrt{3}$ of some multiple of a given Eisenstein integer n.

Dörrie (1965) uses the alternative notation

$$J \equiv \tfrac{1}{2}(1 + i\sqrt{3}) \tag{1}$$
$$O \equiv \tfrac{1}{2}(1 - i\sqrt{3}). \tag{2}$$

for $-\omega^2$ and $-\omega$, and calls numbers of the form $aJ + bO$ G-NUMBERS. O and J satisfy

$$J + O = 1 \tag{3}$$
$$JO = 1 \tag{4}$$
$$J^2 + O = 0 \tag{5}$$
$$O^2 + J = 0 \tag{6}$$
$$J^3 = -1 \tag{7}$$
$$O^3 = -1. \tag{8}$$

The sum, difference, and products of G numbers are also G numbers. The norm of a G number is

$$N(aJ + bO) = a^2 + b^2 - ab. \tag{9}$$

The analog of FERMAT'S THEOREM for Eisenstein integers is that a PRIME NUMBER p can be written in the form

$$a^2 - ab + b^2 = (a + b\omega)(a + b\omega^2)$$

IFF $3 \nmid p + 1$. These are precisely the PRIMES of the form $3m^2 + n^2$ (Conway and Guy 1996).

see also EISENSTEIN PRIME, EISENSTEIN UNIT, GAUSSIAN INTEGER, INTEGER

References
Conway, J. H. and Guy, R. K. *The Book of Numbers.* New York: Springer-Verlag, pp. 220–223, 1996.
Cox, D. A. §4A in *Primes of the Form $x^2 + ny^2$: Fermat, Class Field Theory and Complex Multiplication.* New York: Wiley, 1989.
Dörrie, H. "The Fermat-Gauss Impossibility Theorem." §21 in *100 Great Problems of Elementary Mathematics: Their History and Solutions.* New York: Dover, pp. 96–104, 1965.
Guy, R. K. "Gaussian Primes. Eisenstein-Jacobi Primes." §A16 in *Unsolved Problems in Number Theory, 2nd ed.* New York: Springer-Verlag, pp. 33–36, 1994.
Riesel, H. Appendix 4 in *Prime Numbers and Computer Methods for Factorization, 2nd ed.* Boston, MA: Birkhäuser, 1994.
Wagon, S. "Eisenstein Primes." *Mathematica in Action.* New York: W. H. Freeman, pp. 278–279, 1991.

Eisenstein-Jacobi Integer

see EISENSTEIN INTEGER

Eisenstein Prime

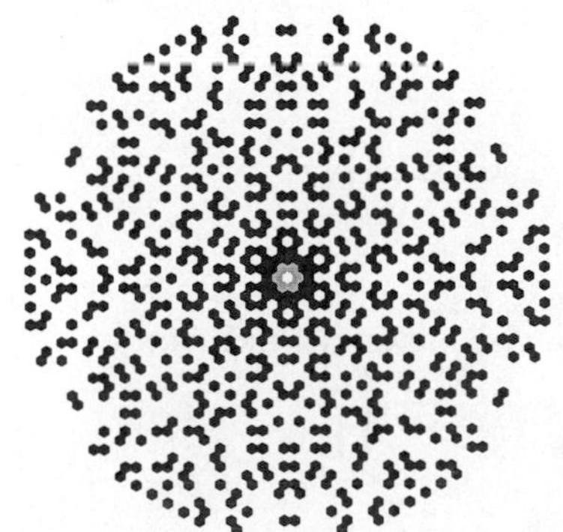

Let ω be the CUBE ROOT of unity $(-1 + i\sqrt{3})/2$. Then the Eisenstein primes are

1. Ordinary PRIMES CONGRUENT to 2 (mod 3),
2. $1 - \omega$ is prime in $\mathbb{Z}[\omega]$,
3. Any ordinary PRIME CONGRUENT to 1 (mod 3) factors as $\alpha\alpha^*$, where each of α and α^* are primes in $\mathbb{Z}[\omega]$ and α and α^* are not "associates" of each other (where associates are equivalent modulo multiplication by an EISENSTEIN UNIT).

References
Cox, D. A. §4A in *Primes of the Form $x^2 + ny^2$: Fermat, Class Field Theory and Complex Multiplication.* New York: Wiley, 1989.
Guy, R. K. "Gaussian Primes. Eisenstein-Jacobi Primes." §A16 in *Unsolved Problems in Number Theory, 2nd ed.* New York: Springer-Verlag, pp. 33–36, 1994.
Wagon, S. "Eisenstein Primes." *Mathematica in Action.* New York: W. H. Freeman, pp. 278–279, 1991.

Eisenstein Series

$$E_r(t) = \sum_{m,n}{}' \frac{1}{(mt+n)^{2r}},$$

where the sum Σ' excludes $m = n = 0$, $\Im[t] > 0$, and r is an INTEGER > 2. The Eisenstein series satisfies the remarkable property

$$E_r\left(\frac{at+b}{ct+d}\right) = (ct+d)^{2r} E_r(t).$$

see also RAMANUJAN-EISENSTEIN SERIES

Eisenstein Unit

The Eisenstein units are the EISENSTEIN INTEGERS ± 1, $\pm\omega$, $\pm\omega^2$, where

$$\omega = \tfrac{1}{2}(-1 + i\sqrt{3})$$
$$\omega^2 = \tfrac{1}{2}(-1 - i\sqrt{3}).$$

see also EISENSTEIN INTEGER, EISENSTEIN PRIME

References
Conway, J. H. and Guy, R. K. *The Book of Numbers.* New York: Springer-Verlag, pp. 220–223, 1996.

Elastica

The elastica formed by bent rods and considered in physics can be generalized to curves in a RIEMANNIAN MANIFOLD which are a CRITICAL POINT for

$$F^\lambda(\gamma) = \int_\gamma (\kappa^2 + \lambda),$$

where κ is the GEODESIC CURVATURE of γ, λ is a REAL NUMBER, and γ is closed or satisfies some specified boundary condition. The curvature of an elastica must satisfy

$$0 = 2\kappa''(s) + \kappa^3(s) + 2\kappa(s)G(s) - \lambda\kappa(s),$$

where κ is the signed curvature of γ, $G(s)$ is the GAUSSIAN CURVATURE of the oriented Riemannian surface M along γ, κ'' is the second derivative of κ with respect to s, and λ is a constant.

References

Barros, M. and Garay, O. J. "Free Elastic Parallels in a Surface of Revolution." *Amer. Math. Monthly* **103**, 149–156, 1996.

Bryant, R. and Griffiths, P. "Reduction for Constrained Variational Problems and $\int(k^2/s)\,ds$." *Amer. J. Math.* **108**, 525–570, 1986.

Langer, J. and Singer, D. A. "Knotted Elastic Curves in R^3." *J. London Math. Soc.* **30**, 512–520, 1984.

Langer, J. and Singer, D. A. "The Total Squared of Closed Curves." *J. Diff. Geom.* **20**, 1–22, 1984.

Elation

A perspective COLLINEATION in which the center and axis are incident.

see also HOMOLOGY (GEOMETRY)

Elder's Theorem

A generalization of STANLEY'S THEOREM. It states that the total number of occurrences of an INTEGER k among all unordered PARTITIONS of n is equal to the number of occasions that a part occurs k or more times in a PARTITION, where a PARTITION which contains r parts that each occur k or more times contributes r to the sum in question.

see also STANLEY'S THEOREM

References

Honsberger, R. *Mathematical Gems III.* Washington, DC: Math. Assoc. Amer, pp. 8–9, 1985.

Election

see EARLY ELECTION RESULTS, VOTING

Electric Motor Curve

see DEVIL'S CURVE

Element

If x is a member of a set A, then x is said to be an element of A, written $x \in A$. If x is not an element of A, this is written $x \notin A$. The term element also refers to a particular member of a GROUP, or entry in a MATRIX.

Elementary Function

A function built up of compositions of the EXPONENTIAL FUNCTION and the TRIGONOMETRIC FUNCTIONS and their inverses by ADDITION, MULTIPLICATION, DIVISION, root extractions (the ELEMENTARY OPERATIONS) under repeated compositions. Not all functions are elementary. For example, the NORMAL DISTRIBUTION FUNCTION

$$\Phi(x) \equiv \frac{1}{\sqrt{2\pi}} \int_0^x e^{-t^2/2}\,dt$$

is a notorious example of a nonelementary function. Nonelementary functions are called TRANSCENDENTAL FUNCTIONS.

see also ALGEBRAIC FUNCTION, ELEMENTARY OPERATION, ELEMENTARY SYMMETRIC FUNCTION, TRANSCENDENTAL FUNCTION

References

Shanks, D. *Solved and Unsolved Problems in Number Theory, 4th ed.* New York: Chelsea, p. 145, 1993.

Elementary Matrix

The elementary MATRICES are the PERMUTATION MATRIX p_{ij} and the SHEAR MATRIX e_{ij}^s.

Elementary Operation

One of the operations of ADDITION, SUBTRACTION, MULTIPLICATION, DIVISION, and root extraction.

see also ALGEBRAIC FUNCTION, ELEMENTARY FUNCTION

Elementary Symmetric Function

The elementary symmetric functions Π_n on n variables $\{x_1, \ldots, x_n\}$ are defined by

$$\Pi_1 = \sum_{1\leq i\leq n} x_i \quad (1)$$

$$\Pi_2 = \sum_{1\leq i<j\leq n} x_i x_j \quad (2)$$

$$\Pi_3 = \sum_{1\leq i<j<k\leq n} x_i x_j x_k \quad (3)$$

$$\Pi_4 = \sum_{1\leq i<j<k<l\leq n} x_i x_j x_k x_l \quad (4)$$

$$\vdots$$

$$\Pi_n = \prod_{i\leq i\leq n} x_i. \quad (5)$$

Alternatively, Π_j can be defined as the coefficient of x^{n-j} in the GENERATING FUNCTION

$$\prod_{1\leq i\leq n} (x + x_i). \quad (6)$$

The elementary symmetric functions satisfy the relationships

$$\sum_{i=1}^{n} x_i{}^2 = \Pi_1{}^2 - 2\Pi_2 \quad (7)$$

$$\sum_{i=1}^{n} x_i{}^3 = \Pi_1{}^3 - 3\Pi_1\Pi_2 + 3\Pi_3 \quad (8)$$

$$\sum_{i=1}^{n} x_i{}^4 = \Pi_1{}^4 - 4\Pi_1{}^2\Pi_2 + 2\Pi_2{}^2 + 4\Pi_1\Pi_3 - 4\Pi_4 \quad (9)$$

(Beeler *et al.* 1972, Item 6).

see also FUNDAMENTAL THEOREM OF SYMMETRIC FUNCTIONS, NEWTON'S RELATIONS, SYMMETRIC FUNCTION

References

Beeler, M.; Gosper, R. W.; and Schroeppel, R. *HAKMEM.* Cambridge, MA: MIT Artificial Intelligence Laboratory, Memo AIM-239, Feb. 1972.

Elements

The classic treatise in geometry written by Euclid and used as a textbook for more than 1,000 years in western Europe. *The Elements*, which went through more than 2,000 editions and consisted of 465 propositions, are divided into 13 "books" (an archaic word for "chapters").

Book	Contents
1	triangles
2	rectangles
3	Circles
4	polygons
5	proportion
6	similarity
7–10	number theory
11	solid geometry
12	pyramids
13	platonic solids

The elements started with 23 definitions, five POSTULATES, and five "common notions," and systematically built the rest of plane and solid geometry upon this foundation. The five EUCLID'S POSTULATES are

1. It is possible to draw a straight LINE from any POINT to another POINT.
2. It is possible to produce a finite straight LINE continuously in a straight LINE.
3. It is possible to describe a CIRCLE with any CENTER and RADIUS.
4. All RIGHT ANGLES are equal to one another.
5. If a straight LINE falling on two straight LINES makes the interior ANGLES on the same side less than two RIGHT ANGLES, the straight LINES (if extended indefinitely) meet on the side on which the ANGLES which are less than two RIGHT ANGLES lie.

(Dunham 1990). Euclid's fifth postulate is known as the PARALLEL POSTULATE. After more than two millennia of study, this POSTULATE was found to be independent of the others. In fact, equally valid NON-EUCLIDEAN GEOMETRIES were found to be possible by changing the assumption of this POSTULATE. Unfortunately, Euclid's postulates were not rigorously complete and left a large number of gaps. Hilbert needed a total of 20 postulates to construct a logically complete geometry.

see also PARALLEL POSTULATE

References

Casey, J. *A Sequel to the First Six Books of the Elements of Euclid, 6th ed.* Dublin: Hodges, Figgis, & Co., 1892.
Dixon, R. *Mathographics.* New York: Dover, pp. 26–27, 1991.
Dunham, W. *Journey Through Genius: The Great Theorems of Mathematics.* New York: Wiley, pp. 30–83, 1990.
Heath, T. L. *The Thirteen Books of the Elements, 2nd ed., Vol. 1: Books I and II.* New York: Dover, 1956.
Heath, T. L. *The Thirteen Books of the Elements, 2nd ed., Vol. 2: Books III–IX.* New York: Dover, 1956.
Heath, T. L. *The Thirteen Books of the Elements, 2nd ed., Vol. 3: Books X–XIII.* New York: Dover, 1956.
Joyce, D. E. "Euclid's Elements." `http://aleph0.clarku.edu/~djoyce/java/elements/elements.html`

Elevator Paradox

A fact noticed by physicist G. Gamow when he had an office on the second floor and physicist M. Stern had an office on the sixth floor of a seven-story building (Gamow and Stern 1958, Gardner 1986). Gamow noticed that about 5/6 of the time, the first elevator to stop on his floor was going down, whereas about the same fraction of time, the first elevator to stop on the sixth floor was going up. This actually makes perfect sense, since 5 of the 6 floors 1, 3, 4, 5, 6, 7 are above the second, and 5 of the 6 floors 1, 2, 3, 4, 5, 7 are below the sixth. However, the situation takes some unexpected turns if more than one elevator is involved, as discussed by Gardner (1986).

References

Gamow, G. and Stern, M. *Puzzle Math.* New York: Viking, 1958.
Gardner, M. "Elevators." Ch. 10 in *Knotted Doughnuts and Other Mathematical Entertainments.* New York: W. H. Freeman, pp. 123–132, 1986.

Elkies Point

Given POSITIVE numbers s_a, s_b, and s_c, the Elkies point is the unique point Y in the interior of a TRIANGLE ΔABC such that the respective INRADII r_a, r_b, r_c of the TRIANGLES ΔBYC, ΔCYA, and ΔAYB satisfy $r_a : r_b : r_c = s_a : s_b : s_c$.

see also CONGRUENT INCIRCLES POINT, INRADIUS

References

Kimberling, C. "Central Points and Central Lines in the Plane of a Triangle." *Math. Mag.* **67**, 163–187, 1994.
Kimberling, C. and Elkies, N. "Problem 1238 and Solution." *Math. Mag.* **60**, 116–117, 1987.

Ellipse

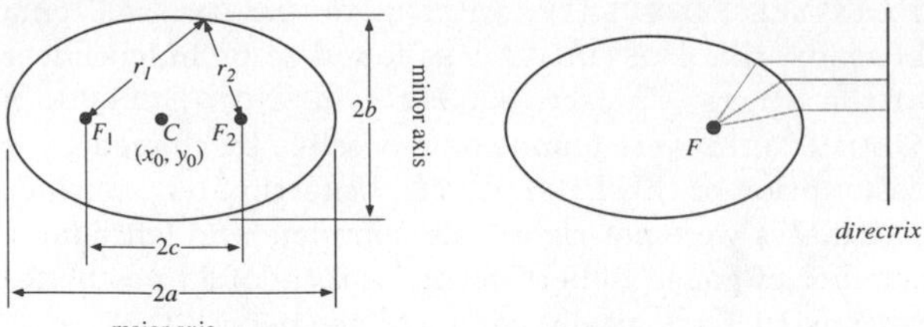

A curve which is the LOCUS of all points in the PLANE the SUM of whose distances r_1 and r_2 from two fixed points F_1 and F_2 (the FOCI) separated by a distance of $2c$ is a given POSITIVE constant $2a$ (left figure). This results in the two-center BIPOLAR COORDINATE equation

$$r_1 + r_2 = 2a, \tag{1}$$

where a is the SEMIMAJOR AXIS and the ORIGIN of the coordinate system is at one of the FOCI. The ellipse can also be defined as the LOCUS of points whose distance from the FOCUS is proportional to the horizontal distance from a vertical line known as the DIRECTRIX (right figure).

The ellipse was first studied by Menaechmus, investigated by Euclid, and named by Apollonius. The FOCUS and DIRECTRIX of an ellipse were considered by Pappus. In 1602, Kepler believed that the orbit of Mars was OVAL; he later discovered that it was an ellipse with the Sun at one FOCUS. In fact, Kepler introduced the word "FOCUS" and published his discovery in 1609. In 1705 Halley showed that the comet which is now named after him moved in an elliptical orbit around the Sun (MacTutor Archive).

A ray passing through a FOCUS will pass through the other focus after a single bounce. Reflections not passing through a FOCUS will be tangent to a confocal HYPERBOLA or ELLIPSE, depending on whether the ray passes between the FOCI or not. Let an ellipse lie along the x-AXIS and find the equation of the figure (1) where r_1 and r_2 are at $(-c, 0)$ and $(c, 0)$. In CARTESIAN COORDINATES,

$$\sqrt{(x+c)^2 + y^2} + \sqrt{(x-c)^2 + y^2} = 2a. \tag{2}$$

Bring the second term to the right side and square both sides,

$$(x+c)^2 + y^2 = 4a^2 - 4a\sqrt{(x-c)^2 + y^2} + (x-c)^2 + y^2. \tag{3}$$

Now solve for the SQUARE ROOT term and simplify

$$\begin{aligned}\sqrt{(x-c)^2 + y^2} \\ = -\frac{1}{4a}(x^2 + 2xc + c^2 + y^2 - 4a^2 - x^2 + 2xc - c^2 - y^2) \\ = -\frac{1}{4a}(4xc - 4a^2) = a - \frac{c}{a}x.\end{aligned} \tag{4}$$

Square one final time to clear the remaining SQUARE ROOT,

$$x^2 - 2xc + c^2 + y^2 = a^2 - 2cx + \frac{c^2}{a^2}x^2. \tag{5}$$

Grouping the x terms then gives

$$x^2 \frac{a^2 - c^2}{a^2} + y^2 = a^2 - c^2, \tag{6}$$

which can be written in the simple form

$$\frac{x^2}{a^2} + \frac{y^2}{a^2 - c^2} = 1. \tag{7}$$

Defining a new constant

$$b^2 \equiv a^2 - c^2 \tag{8}$$

puts the equation in the particularly simple form

$$\frac{x^2}{a^2} + \frac{y^2}{b^2} = 1. \tag{9}$$

The parameter b is called the SEMIMINOR AXIS by analogy with the parameter a, which is called the SEMIMAJOR AXIS. The fact that b as defined above is actually the SEMIMINOR AXIS is easily shown by letting r_1 and r_2 be equal. Then two RIGHT TRIANGLES are produced, each with HYPOTENUSE a, base c, and height $b \equiv \sqrt{a^2 - c^2}$. Since the largest distance along the MINOR AXIS will be achieved at this point, b is indeed the SEMIMINOR AXIS.

If, instead of being centered at (0, 0), the CENTER of the ellipse is at (x_0, y_0), equation (9) becomes

$$\frac{(x - x_0)^2}{a^2} + \frac{(y - y_0)^2}{b^2} = 1. \tag{10}$$

As can be seen from the CARTESIAN EQUATION for the ellipse, the curve can also be given by a simple parametric form analogous to that of a CIRCLE, but with the x and y coordinates having different scalings,

$$x = a \cos t \tag{11}$$

$$y = b \sin t. \tag{12}$$

The unit TANGENT VECTOR of the ellipse so parameterized is

$$x_T(t) = -\frac{a \sin t}{\sqrt{b^2 \cos^2 t + a^2 \sin^2 t}} \tag{13}$$

$$y_T(t) = \frac{b \cos t}{\sqrt{b^2 \cos^2 t + a^2 \sin^2 t}}. \tag{14}$$

A sequence of NORMAL and TANGENT VECTORS are plotted below for the ellipse.

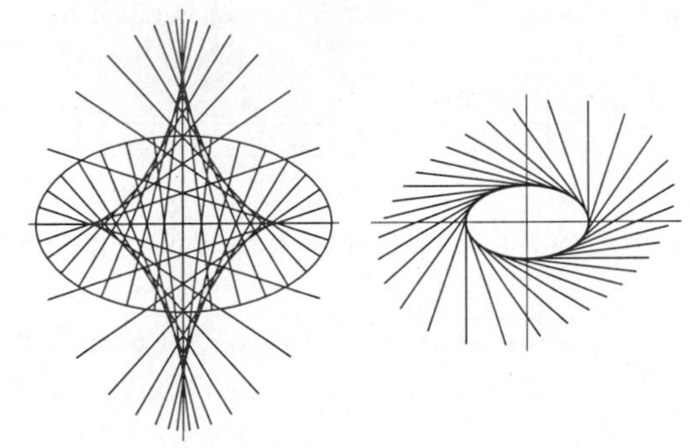

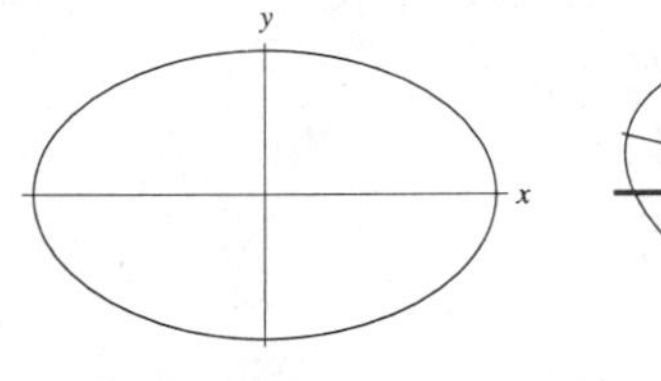

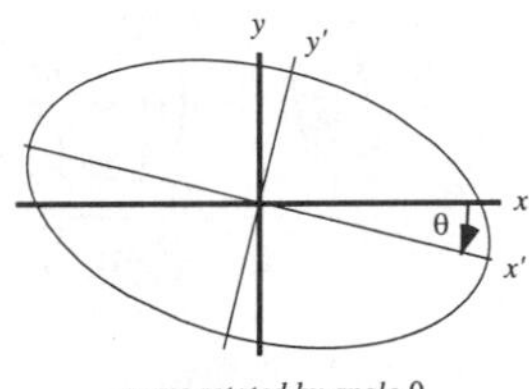

curve rotated by angle θ

For an ellipse centered at the ORIGIN but inclined at an arbitrary ANGLE θ to the x-AXIS, the parametric equations are

$$\begin{bmatrix} x \\ y \end{bmatrix} = \begin{bmatrix} \cos\theta & \sin\theta \\ -\sin\theta & \cos\theta \end{bmatrix} \begin{bmatrix} a\cos t \\ b\sin t \end{bmatrix}$$
$$= \begin{bmatrix} a\cos\theta\cos t + b\sin\theta\sin t \\ -a\sin\theta\cos t + b\cos\theta\sin t \end{bmatrix}. \tag{15}$$

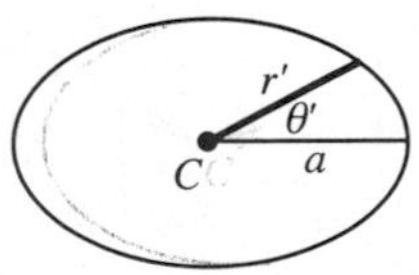

In POLAR COORDINATES, the ANGLE θ' measured from the *center* of the ellipse is called the ECCENTRIC ANGLE. Writing r' for the distance of a point from the ellipse center, the equation in POLAR COORDINATES is just given by the usual

$$x = r'\cos\theta' \tag{16}$$
$$y = r'\sin\theta'. \tag{17}$$

Here, the coordinates θ' and r' are written with primes to distinguish them from the more common polar coordinates for an ellipse which are centered on a *focus*. Plugging the polar equations into the Cartesian equation (9) and solving for r'^2 gives

$$r'^2 = \frac{b^2a^2}{b^2\cos^2\theta' + a^2\sin^2\theta'}. \tag{18}$$

Define a new constant $0 \leq e < 1$ called the ECCENTRICITY (where $e = 0$ is the case of a CIRCLE) to replace b

$$e \equiv \sqrt{1 - \frac{b^2}{a^2}}, \tag{19}$$

from which it also follows from (8) that

$$a^2e^2 = a^2 - b^2 \equiv c^2 \tag{20}$$
$$c = ae \tag{21}$$
$$b^2 = a^2(1 - e^2). \tag{22}$$

Therefore (18) can be written as

$$r'^2 = \frac{a^2(1-e^2)}{1 - e^2\cos^2\theta'} \tag{23}$$

$$r' = a\sqrt{\frac{1-e^2}{1-e^2\cos^2\theta'}}. \tag{24}$$

If $e \ll 1$, then

$$r' = a\{1 - \tfrac{1}{2}e^2\sin^2\theta' - \tfrac{1}{16}e^4[5 + 3\cos(2\theta')]\sin^2\theta' + \ldots\}, \tag{25}$$

so

$$\frac{\Delta r'}{a} \equiv \frac{a - r'}{a} \approx \tfrac{1}{2}e^2\sin^2\theta'. \tag{26}$$

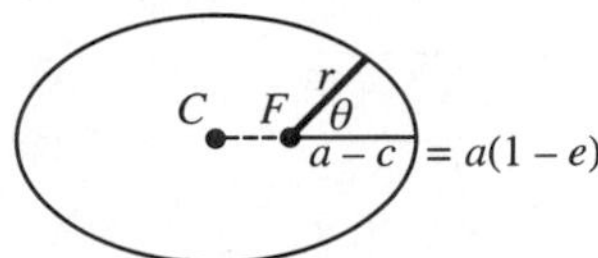

If r and θ are measured from a FOCUS instead of from the center, as they commonly are in orbital mechanics, then the equations of the ellipse are

$$x = c + r\cos\theta \tag{27}$$
$$y = r\sin\theta, \tag{28}$$

and (9) becomes

$$\frac{(c + r\cos\theta)^2}{a^2} + \frac{r^2\sin^2\theta}{b^2} = 1. \tag{29}$$

Clearing the DENOMINATORS gives

$$b^2(c^2 + 2cr\cos\theta + r^2\cos^2\theta) + a^2r^2\sin^2\theta = a^2b^2 \tag{30}$$

$$b^2c^2 + 2rcb^2\cos\theta + b^2r^2\cos^2\theta + a^2r^2 - a^2r^2\cos^2\theta = a^2b^2. \tag{31}$$

Plugging in (21) and (22) to re-express b and c in terms of a and e,

$$a^2(1-e^2)a^2e^2 + 2aea^2(1-e^2)r\cos\theta + a^2(1-e^2)r^2\cos^2\theta$$
$$+ a^2r^2 - a^2r^2\cos^2\theta = a^2[a^2(1-e^2)]. \tag{32}$$

Simplifying,

$$-r^2 + [er\cos\theta - a(1-e^2)]^2 = 0 \tag{33}$$

$$r = \pm[er\cos\theta - a(1-e^2)]. \tag{34}$$

The sign can be determined by requiring that r must be POSITIVE. When $e = 0$, (34) becomes $r = \pm(-a)$, but since a is always POSITIVE, we must take the NEGATIVE sign, so (34) becomes

$$r = a(1-e^2) - er\cos\theta \tag{35}$$

$$r(1+e\cos\theta) = a(1-e^2) \tag{36}$$

$$r = \frac{a(1-e^2)}{1+e\cos\theta}. \tag{37}$$

The distance from a FOCUS to a point with horizontal coordinate x is found from

$$\cos\theta = \frac{c+x}{r}. \tag{38}$$

Plugging this into (37) yields

$$r + e(c+x) = a(1-e^2) \tag{39}$$

$$r = a(1-e^2) - e(c+x). \tag{40}$$

Summarizing relationships among the parameters characterizing an ellipse,

$$b = a\sqrt{1-e^2} = \sqrt{a^2-c^2} \tag{41}$$

$$c = \sqrt{a^2-b^2} = ae \tag{42}$$

$$e = \sqrt{1-\frac{b^2}{a^2}} = \frac{c}{a}. \tag{43}$$

The ECCENTRICITY can therefore be interpreted as the position of the FOCUS as a fraction of the SEMIMAJOR AXIS.

In PEDAL COORDINATES with the PEDAL POINT at the FOCUS, the equation of the ellipse is

$$\frac{b^2}{p^2} = \frac{2a}{r} - 1. \tag{44}$$

To find the RADIUS OF CURVATURE, return to the parametric coordinates centered at the center of the ellipse and compute the first and second derivatives,

$$x' = -a\sin t \tag{45}$$

$$y' = b\cos t \tag{46}$$

$$x'' = -a\cos t \tag{47}$$

$$y'' = -b\sin t. \tag{48}$$

Therefore,

$$\begin{aligned} R &= \frac{(x'^2+y'^2)^{3/2}}{x'y''-x''y'} \\ &= \frac{(a^2\sin^2 t+b^2\cos^2 t)^{3/2}}{-a\sin t(-b\sin t)-(a\cos t)(b\cos t)} \\ &= \frac{(a^2\sin^2 t+b^2\cos^2 t)^{3/2}}{ab(\sin^2 t+\cos^2 t)} \\ &= \frac{(a^2\sin^2 t+b^2\cos^2 t)^{3/2}}{ab}. \end{aligned} \tag{49}$$

Similarly, the unit TANGENT VECTOR is given by

$$\hat{\mathbf{T}} = \begin{bmatrix} -a\sin t \\ b\cos t \end{bmatrix} \frac{1}{\sqrt{a^2\sin^2 t+b^2\cos^2 t}}. \tag{50}$$

The ARC LENGTH of the ellipse can be computed using

$$\begin{aligned} s &= \int \sqrt{x'^2+y'^2}\,dt = \int \sqrt{a^2\cos^2 t+b^2\sin^2 t}\,dt \\ &= a\int \sqrt{(1-\sin^2 t)+\frac{b^2}{a^2}\sin^2 t}\,dt \\ &= a\int \sqrt{1-\left(1-\frac{b^2}{a^2}\right)\sin^2 t}\,dt \\ &= a\int \sqrt{1-e^2\sin^2 t}\,dt = aE(t,e), \end{aligned} \tag{51}$$

where E is an incomplete ELLIPTIC INTEGRAL OF THE SECOND KIND. Again, note that t is a parameter which does not have a direct interpretation in terms of an ANGLE. However, the relationship between the polar angle from the ellipse center θ and the parameter t follows from

$$\theta = \tan^{-1}\left(\frac{y}{x}\right) = \tan^{-1}\left(\frac{b}{a}\tan t\right). \tag{52}$$

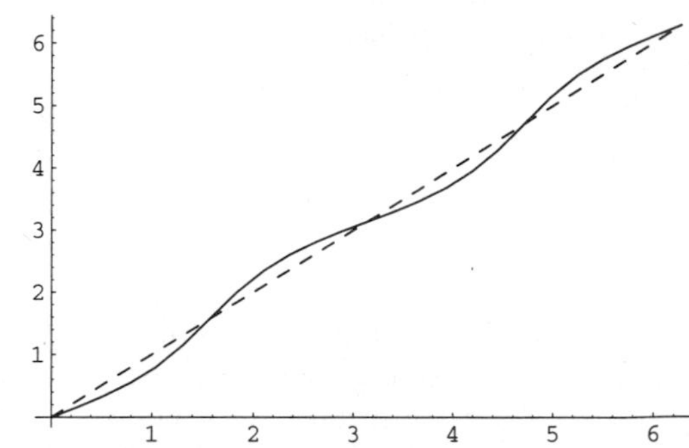

This function is illustrated above with θ shown as the solid curve and t as the dashed, with $b/a = 0.6$. Care must be taken to make sure that the correct branch of the INVERSE TANGENT function is used. As can be seen, θ weaves back and forth around t, with crossings occurring at multiples of $\pi/2$.

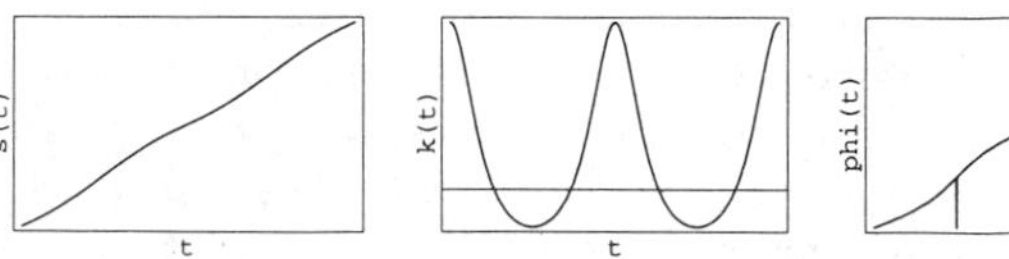

The CURVATURE and TANGENTIAL ANGLE of the ellipse are given by

$$\kappa = \frac{ab}{(b^2\cos^2 t+a^2\sin^2 t)^{3/2}} \tag{53}$$

$$\phi = -\tan^{-1}\left(\frac{b}{a}\cos t\right). \tag{54}$$

The entire PERIMETER p of the ellipse is given by setting $t = 2\pi$ (corresponding to $\theta = 2\pi$), which is equivalent to four times the length of one of the ellipse's QUADRANTS,

$$p = aE(2\pi, e) = 4aE(\tfrac{1}{2}\pi, e) = 4aE(e), \tag{55}$$

where $E(e)$ is a complete ELLIPTIC INTEGRAL OF THE SECOND KIND with MODULUS k. The PERIMETER can be computed numerically by the rapidly converging GAUSS-KUMMER SERIES

$$\begin{aligned} p &= \pi(a+b)\sum_{n=0}^{\infty}\binom{\frac{1}{2}}{n}^2 h^{2n} \\ &= \pi(a+b)(1+\tfrac{1}{4}h^2+\tfrac{1}{64}h^4+\tfrac{1}{256}h^6+\ldots), \end{aligned} \tag{56}$$

where

$$h \equiv \frac{a-b}{a+b} \tag{57}$$

and $\binom{n}{k}$ is a BINOMIAL COEFFICIENT. Approximations to the PERIMETER include

$$\begin{aligned} p &\approx \pi\sqrt{2(a^2+b^2)} && (58) \\ &\approx \pi[3(a+b) - \sqrt{(a+3b)(3a+b)}] && (59) \\ &\approx \pi(a+b)\left(1+\frac{3t}{10+\sqrt{4-3t}}\right), && (60) \end{aligned}$$

where the last two are due to Ramanujan (1913-14),

$$t \equiv \left(\frac{a-b}{a+b}\right)^2, \tag{61}$$

and (60) is accurate to within $\sim 3\cdot 2^{-17}t^5$.

The maximum and minimum distances from the FOCUS are called the APOAPSIS and PERIAPSIS, and are given by

$$\begin{aligned} r_+ &= r_{\text{apoapsis}} = a(1+e) && (62) \\ r_- &= \text{periapsis} = a(1-e). && (63) \end{aligned}$$

The AREA of an ellipse may be found by direct INTEGRATION

$$\begin{aligned} A &= \int_{-a}^{a}\int_{-b\sqrt{a^2-x^2}/a}^{b\sqrt{a^2-x^2}/a} dy\,dx = \int_{-a}^{a}\frac{2b}{a}\sqrt{a^2-x^2}\,dx \\ &= \frac{2b}{a}\left\{\frac{1}{2}\left[x\sqrt{a^2-x^2}+a^2\sin^{-1}\left(\frac{x}{|a|}\right)\right]\right\}_{x=-a}^{a} \\ &= ab[\sin^{-1}1 - \sin^{-1}(-1)] = ab\left[\frac{\pi}{2}-\left(\frac{\pi}{2}\right)\right] = \pi ab. \end{aligned} \tag{64}$$

The AREA can also be computed more simply by making the change of coordinates $x' \equiv (b/a)x$ and $y' \equiv y$ from the elliptical region R to the new region R'. Then the equation becomes

$$\frac{1}{a^2}\left(\frac{a}{b}x'\right)^2 + \frac{y'^2}{b^2} = 1, \tag{65}$$

or $x'^2 + y'^2 = b^2$, so R' is a CIRCLE of RADIUS b. Since

$$\frac{\partial x}{\partial x'} = \left(\frac{\partial x'}{\partial x}\right)^{-1} = \left(\frac{b}{a}\right)^{-1} = \frac{a}{b}, \tag{66}$$

the JACOBIAN is

$$\left|\frac{\partial(x,y)}{\partial(x',y')}\right| = \begin{vmatrix} \frac{\partial x}{\partial x'} & \frac{\partial y'}{\partial x'} \\ \frac{\partial x}{\partial y'} & \frac{\partial y}{\partial y'} \end{vmatrix} = \begin{vmatrix} \frac{a}{b} & 0 \\ 0 & 1 \end{vmatrix} = \frac{a}{b}. \tag{67}$$

The AREA is therefore

$$\begin{aligned} \iint_R dx\,dy &= \iint_{R'} \left|\frac{\partial(x,y)}{\partial(x',y')}\right| dx'\,dy' \\ &= \frac{a}{b}\iint_{R'} dx'dy' = \frac{a}{b}(\pi b^2) = \pi ab, \end{aligned} \tag{68}$$

as before. The AREA of an arbitrary ellipse given by the QUADRATIC EQUATION

$$ax^2 + bxy + cy^2 = 1 \tag{69}$$

is

$$A = \frac{2\pi}{\sqrt{4ac-b^2}}. \tag{70}$$

The AREA of an ELLIPSE with semiaxes a and b with respect to a PEDAL POINT P is

$$A = \tfrac{1}{2}\pi(a^2+b^2+|OP|^2). \tag{71}$$

The ellipse INSCRIBED in a given TRIANGLE and tangent at its MIDPOINTS is called the MIDPOINT ELLIPSE. The LOCUS of the centers of the ellipses INSCRIBED in a TRIANGLE is the interior of the MEDIAL TRIANGLE. Newton gave the solution to inscribing an ellipse in a convex QUADRILATERAL (Dörrie 1965, p. 217). The centers of the ellipses INSCRIBED in a QUADRILATERAL all lie on the straight line segment joining the MIDPOINTS of the DIAGONALS (Chakerian 1979, pp. 136–139).

The AREA of an ellipse with BARYCENTRIC COORDINATES (α, β, γ) INSCRIBED in a TRIANGLE of unit AREA is

$$\Delta = \pi\sqrt{(1-2\alpha)(1-2\beta)(1-2\gamma)}. \tag{72}$$

(Chakerian 1979, pp. 142–145).

The LOCUS of the apex of a variable CONE containing an ellipse fixed in 3-space is a HYPERBOLA through the FOCI of the ellipse. In addition, the LOCUS of the apex of a CONE containing that HYPERBOLA is the original ellipse. Furthermore, the ECCENTRICITIES of the ellipse and HYPERBOLA are reciprocals. The LOCUS of centers of a PAPPUS CHAIN of CIRCLES is an ellipse. Surprisingly, the locus of the end of a garage door mounted on rollers along a vertical track but extending beyond the track is a quadrant of an ellipse (the envelopes of positions is an ASTROID).

see also CIRCLE, CONIC SECTION, ECCENTRIC ANOMALY, ECCENTRICITY, ELLIPTIC CONE, ELLIPTIC CURVE, ELLIPTIC CYLINDER, HYPERBOLA, MIDPOINT ELLIPSE, PARABOLA, PARABOLOID, QUADRATIC CURVE, REFLECTION PROPERTY, SALMON'S THEOREM, STEINER'S ELLIPSE

References

Beyer, W. H. *CRC Standard Mathematical Tables, 28th ed.* Boca Raton, FL: CRC Press, pp. 126 and 198–199, 1987.

Casey, J. "The Ellipse." Ch. 6 in *A Treatise on the Analytical Geometry of the Point, Line, Circle, and Conic Sections, Containing an Account of Its Most Recent Extensions, with Numerous Examples, 2nd ed., rev. enl.* Dublin: Hodges, Figgis, & Co., pp. 201–249, 1893.

Chakerian, G. D. "A Distorted View of Geometry." Ch. 7 in *Mathematical Plums* (Ed. R. Honsberger). Washington, DC: Math. Assoc. Amer., 1979.

Courant, R. and Robbins, H. *What is Mathematics?: An Elementary Approach to Ideas and Methods, 2nd ed.* Oxford, England: Oxford University Press, p. 75, 1996.

Dörrie, H. *100 Great Problems of Elementary Mathematics: Their History and Solutions.* New York: Dover, 1965.

Lawrence, J. D. *A Catalog of Special Plane Curves.* New York: Dover, pp. 72–78, 1972.

Lee, X. "Ellipse." `http://www.best.com/~xah/SpecialPlaneCurves_dir/Ellipse_dir/ellipse.html`.

Lockwood, E. H. "The Ellipse." Ch. 2 in *A Book of Curves.* Cambridge, England: Cambridge University Press, pp. 13–24, 1967.

MacTutor History of Mathematics Archive. "Ellipse." `http://www-groups.dcs.st-and.ac.uk/~history/Curves/Ellipse.html`.

Ramanujan, S. "Modular Equations and Approximations to π." *Quart. J. Pure. Appl. Math.* **45**, 350–372, 1913–1914.

Ellipse Caustic Curve

For an ELLIPSE given by

$$x = r\cos t \qquad (1)$$

$$y = \sin t \qquad (2)$$

with light source at $(x, 0)$, the CAUSTIC is

$$x = \frac{N_x}{D_x} \qquad (3)$$

$$y = \frac{N_y}{D_y}, \qquad (4)$$

where

$$\begin{aligned} N_x = {} & 2rx(3-5r^2) + (-6r^2+6r^4-3x^2+9r^2x^2)\cos t \\ & + 6rx(1-r^2)\cos(2t) \\ & + (-2r^2+2r^4-x^2-r^2x^2)\cos(3t) \end{aligned} \qquad (5)$$

$$\begin{aligned} D_x = {} & 2r(1+2r^2+4x^2) + 3x(1-5r^2)\cos t \\ & + (6r+6r^3)\cos(2t) + x(1-r^2)\cos(3t) \end{aligned} \qquad (6)$$

$$N_y = 8r(-1+r^2-x^2)\sin^3 t \qquad (7)$$

$$\begin{aligned} D_y = {} & 2r(-1-r^2-4x^2) + 3(-x+5r^2)\cos t \\ & + 6r(1-r^2)\cos(2t) + x(-1+r^2)\cos(3t). \end{aligned} \qquad (8)$$

At $(\infty, 0)$,

$$x = \frac{\cos t[-1+5r^2-\cos(2t)(1+r^2)]}{4r} \qquad (9)$$

$$y = \sin^3 t. \qquad (10)$$

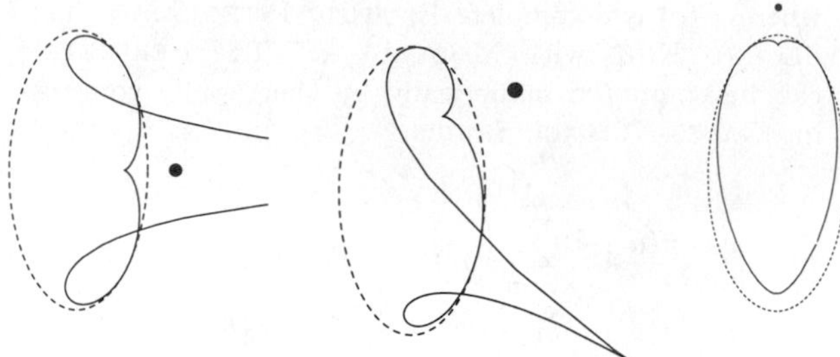

Ellipse Envelope

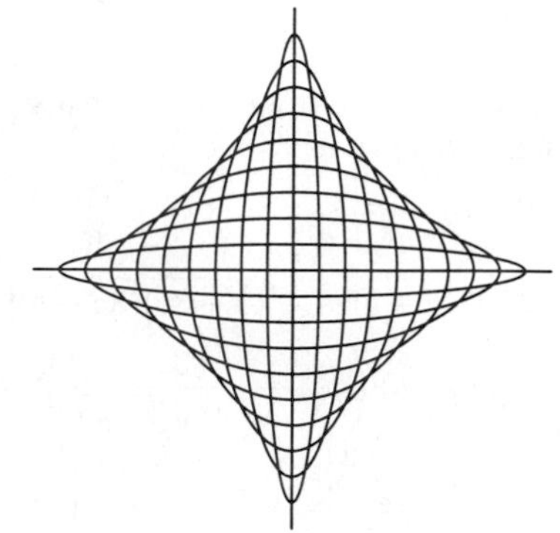

Consider the family of ELLIPSES

$$\frac{x^2}{c^2} + \frac{y^2}{(1-c)^2} - 1 = 0 \qquad (1)$$

for $c \in [0, 1]$. The PARTIAL DERIVATIVE with respect to c is

$$-\frac{2x^2}{c^3} + \frac{2y^2}{(1-c)^3} = 0 \qquad (2)$$

$$\frac{x^2}{c^3} - \frac{y^2}{(1-c)^3} = 0. \qquad (3)$$

Combining (1) and (3) gives the set of equations

$$\begin{bmatrix} \frac{1}{c^2} & \frac{1}{(1-c)^2} \\ \frac{1}{c^3} & -\frac{1}{(1-c)^3} \end{bmatrix} \begin{bmatrix} x^2 \\ y^2 \end{bmatrix} = \begin{bmatrix} 1 \\ 0 \end{bmatrix} \qquad (4)$$

$$\begin{aligned} \begin{bmatrix} x^2 \\ y^2 \end{bmatrix} &= \frac{1}{\Delta} \begin{bmatrix} -\frac{1}{(1-c)^3} & -\frac{1}{(1-c)^2} \\ -\frac{1}{c^3} & \frac{1}{c^2} \end{bmatrix} \begin{bmatrix} 1 \\ 0 \end{bmatrix} \\ &= \frac{1}{\Delta} \begin{bmatrix} -\frac{1}{(1-c)^3} \\ -\frac{1}{c^3} \end{bmatrix}, \end{aligned} \qquad (5)$$

where the DISCRIMINANT is

$$\Delta = -\frac{1}{c^2(1-c)^3} - \frac{1}{c^3(1-c)^2} = -\frac{1}{c^3(1-c)^3}, \qquad (6)$$

so (5) becomes

$$\begin{bmatrix} x^2 \\ y^2 \end{bmatrix} = \begin{bmatrix} c^3 \\ (1-c)^3 \end{bmatrix}. \tag{7}$$

Eliminating c then gives

$$x^{2/3} + y^{2/3} = 1, \tag{8}$$

which is the equation of the ASTROID. If the curve is instead represented parametrically, then

$$x = c\cos t \tag{9}$$

$$y = (1-c)\sin t. \tag{10}$$

Solving

$$\frac{\partial x}{\partial t}\frac{\partial y}{\partial c} - \frac{\partial x}{\partial c}\frac{\partial y}{\partial t} = (-c\sin t)(-\sin t) - (\cos t)[(1-c)\cos t] = c(\sin^2 t + \cos^2 t) - \cos^2 t = c - \cos^2 t = 0 \tag{11}$$

for c gives

$$c = \cos^2 t, \tag{12}$$

so substituting this back into (9) and (10) gives

$$x = (\cos^2 t)\cos t = \cos^3 t \tag{13}$$

$$y = (1-\cos^2 t)\sin t = \sin^3 t, \tag{14}$$

the parametric equations of the ASTROID.

see also ASTROID, ELLIPSE, ENVELOPE

Ellipse Evolute

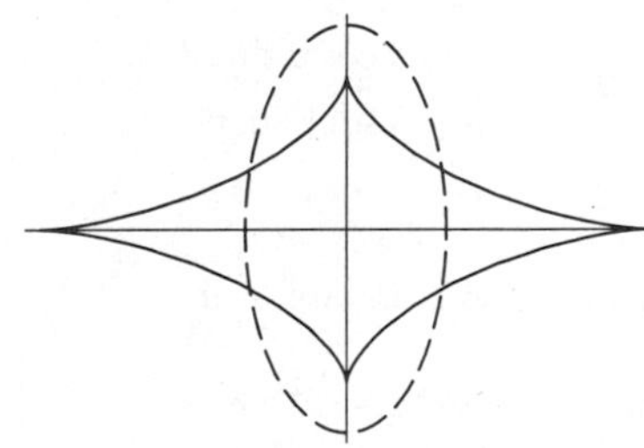

The EVOLUTE of an ELLIPSE is given by the parametric equations

$$x = \frac{a^2 - b^2}{a}\cos^3 t \tag{1}$$

$$y = \frac{b^2 - a^2}{b}\sin^3 t, \tag{2}$$

which can be combined and written

$$(ax)^{2/3} + (by)^{2/3} = [(a^2-b^2)\cos^3 t]^{2/3} + [(b^2-a^2)\sin^3 t]^{2/3} = (a^2-b^2)^{2/3}(\sin^2 t + \cos^2 t) = (a^2-b^2)^{2/3} = c^{4/3}, \tag{3}$$

which is a stretched ASTROID called the LAMÉ CURVE. From a point inside the EVOLUTE, four NORMALS can be drawn to the ellipse, but from a point outside, only two NORMALS can be drawn.

see also ASTROID, ELLIPSE, EVOLUTE, LAMÉ CURVE

References

Gray, A. *Modern Differential Geometry of Curves and Surfaces.* Boca Raton, FL: CRC Press, p. 77, 1993.

Ellipse Involute

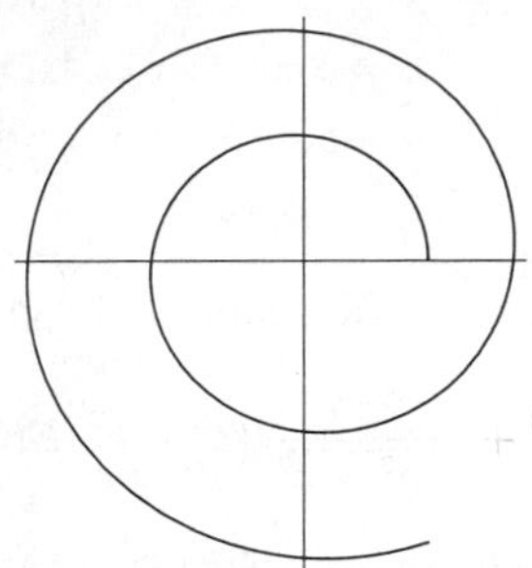

From ELLIPSE, the TANGENT VECTOR is

$$\mathbf{T} = \begin{bmatrix} -a\sin t \\ b\cos t \end{bmatrix}, \tag{1}$$

and the ARC LENGTH is

$$s = a\int \sqrt{1 - e^2\sin^2 t}\,dt = aE(t,e), \tag{2}$$

where $E(t,e)$ is an incomplete ELLIPTIC INTEGRAL OF THE SECOND KIND. Therefore,

$$\mathbf{r}_i = \mathbf{r} - s\hat{\mathbf{T}} = \begin{bmatrix} a\cos t \\ b\sin t \end{bmatrix} - aeE(t,e)\begin{bmatrix} -a\sin t \\ b\cos t \end{bmatrix} \tag{3}$$

$$= \begin{bmatrix} a\{\cos t + aeE(t,e)\sin t\} \\ b\{\sin t - aeE(t,e)\cos t\}. \end{bmatrix} \tag{4}$$

Ellipse Pedal Curve

The pedal curve of an ellipse with a FOCUS as the PEDAL POINT is a CIRCLE.

Ellipsoid

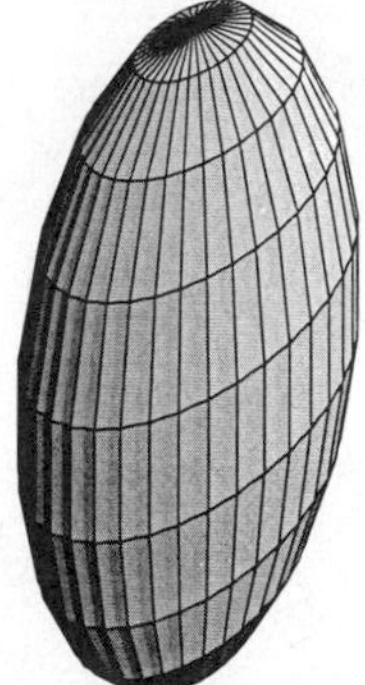

A QUADRATIC SURFACE which is given in CARTESIAN COORDINATES by

$$\frac{x^2}{a^2} + \frac{y^2}{b^2} + \frac{z^2}{c^2} = 1, \tag{1}$$

where the semi-axes are of lengths a, b, and c. In SPHERICAL COORDINATES, this becomes

$$\frac{r^2\cos^2\theta\sin^2\phi}{a^2} + \frac{r^2\sin^2\theta\sin^2\phi}{b^2} + \frac{r^2\cos^2\phi}{c^2} = 1. \tag{2}$$

The parametric equations are

$$x = a\cos\theta\sin\phi \quad (3)$$
$$y = b\sin\theta\sin\phi \quad (4)$$
$$z = c\cos\phi. \quad (5)$$

The SURFACE AREA (Bowman 1961, pp. 31–32) is

$$S = 2\pi c^2 + \frac{2\pi b}{\sqrt{a^2-c^2}}[(a^2-c^2)E(\theta)+c^2\theta], \quad (6)$$

where $E(\theta)$ is a COMPLETE ELLIPTIC INTEGRAL OF THE SECOND KIND,

$${e_1}^2 \equiv \frac{a^2-c^2}{a^2} \quad (7)$$
$${e_2}^2 \equiv \frac{b^2-c^2}{b^2} \quad (8)$$
$$k \equiv \frac{e_2}{a_1}, \quad (9)$$

and θ is given by inverting the expression

$$e_1 = \operatorname{sn}(\theta, k), \quad (10)$$

where $\operatorname{sn}(\theta, k)$ is a JACOBI ELLIPTIC FUNCTION. The VOLUME of an ellipsoid is

$$V = \tfrac{4}{3}\pi abc. \quad (11)$$

If two axes are the same, the figure is called a SPHEROID (depending on whether $c < a$ or $c > a$, an OBLATE SPHEROID or PROLATE SPHEROID, respectively), and if all three are the same, it is a SPHERE.

A different parameterization of the ellipsoid is the so-called stereographic ellipsoid, given by the parametric equations

$$x(u,v) = \frac{a(1-u^2-v^2)}{1+u^2+v^2} \quad (12)$$
$$y(u,v) = \frac{2bu}{1+u^2+v^2} \quad (13)$$
$$z(u,v) = \frac{2cv}{1+u^2+v^2}. \quad (14)$$

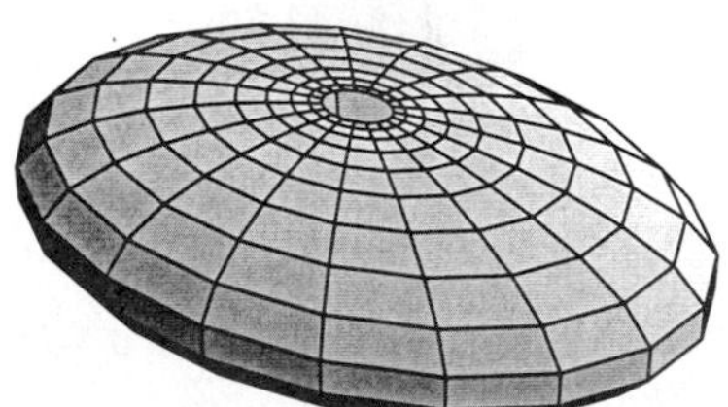

A third parameterization is the Mercator parameterization

$$x(u,v) = a\operatorname{sech} v\cos u \quad (15)$$
$$y(u,v) = b\operatorname{sech} v\sin u \quad (16)$$
$$z(u,v) = c\tanh v \quad (17)$$

(Gray 1993).

The SUPPORT FUNCTION of the ellipsoid is

$$h = \left(\frac{x^2}{a^4}+\frac{y^2}{b^4}+\frac{z^2}{c^4}\right)^{-1/2}, \quad (18)$$

and the GAUSSIAN CURVATURE is

$$K = \frac{h^4}{a^2b^2c^2} \quad (19)$$

(Gray 1993, p. 296).

see also CONVEX OPTIMIZATION THEORY, OBLATE SPHEROID, PROLATE SPHEROID, SPHERE, SPHEROID

References

Beyer, W. H. *CRC Standard Mathematical Tables, 28th ed.* Boca Raton, FL: CRC Press, p. 131, 1987.

Bowman, F. *Introduction to Elliptic Functions, with Applications.* New York: Dover, 1961.

Fischer, G. (Ed.). Plate 65 in *Mathematische Modelle/Mathematical Models, Bildband/Photograph Volume.* Braunschweig, Germany: Vieweg, p. 60, 1986.

Gray, A. "The Ellipsoid" and "The Stereographic Ellipsoid." §11.2 and 11.3 in *Modern Differential Geometry of Curves and Surfaces.* Boca Raton, FL: CRC Press, pp. 215–217, and 296, 1993.

Ellipsoid Geodesic

An ELLIPSOID can be specified parametrically by

$$x = a\cos u\sin v \quad (1)$$
$$y = b\sin u\sin v \quad (2)$$
$$z = c\cos v. \quad (3)$$

The GEODESIC parameters are then

$$P = \sin^2 v(b^2\cos^2 u + a^2\sin^2 u) \quad (4)$$
$$Q = \tfrac{1}{4}(b^2-a^2)\sin(2u)\sin(2v) \quad (5)$$
$$R = \cos^2 v(a^2\cos^2 u + b^2\sin^2 u) + c^2\sin^2 v. \quad (6)$$

When the coordinates of a point are on the QUADRIC

$$\frac{x^2}{a}+\frac{y^2}{b}+\frac{z^2}{c} = 1 \quad (7)$$

and expressed in terms of the parameters p and q of the confocal quadrics passing through that point (in other words, having $a+p$, $b+p$, $c+p$, and $a+q$, $b+q$, $c+q$ for the squares of their semimajor axes), then the equation of a GEODESIC can be expressed in the form

$$\frac{q\,dq}{\sqrt{q(a+q)(b+q)(c+q)(\theta+q)}} \pm \frac{p\,dp}{\sqrt{p(a+p)(b+p)(c+p)(\theta+p)}} = 0, \quad (8)$$

with θ an arbitrary constant, and the ARC LENGTH element ds is given by

$$-2\frac{ds}{pq} = \frac{dq}{\sqrt{q(a+q)(b+q)(c+q)(\theta+q)}} \pm \frac{dp}{\sqrt{p(a+p)(b+p)(c+p)(\theta+p)}}, \quad (9)$$

where upper and lower signs are taken together.

see also OBLATE SPHEROID GEODESIC, SPHERE GEODESIC

References

Eisenhart, L. P. *A Treatise on the Differential Geometry of Curves and Surfaces.* New York: Dover, pp. 236–241, 1960.

Forsyth, A. R. *Calculus of Variations.* New York: Dover, p. 447, 1960.

Ellipsoidal Calculus

Ellipsoidal calculus is a method for solving problems in control and estimation theory having unknown but bounded errors in terms of sets of approximating ellipsoidal-value functions. Ellipsoidal calculus has been especially useful in the study of LINEAR PROGRAMMING.

References

Kurzhanski, A. B. and Vályi, I. *Ellipsoidal Calculus for Estimation and Control.* Boston, MA: Birkhäuser, 1996.

Ellipsoidal Coordinates

see CONFOCAL ELLIPSOIDAL COORDINATES

Ellipsoidal Harmonic

see ELLIPSOIDAL HARMONIC OF THE FIRST KIND, ELLIPSOIDAL HARMONIC OF THE SECOND KIND

Ellipsoidal Harmonic of the First Kind

The first solution to LAMÉ'S DIFFERENTIAL EQUATION, denoted $E_n^m(x)$ for $m = 1, \ldots, 2n+1$. They are also called LAMÉ FUNCTIONS. The product of two ellipsoidal harmonics of the first kind is a SPHERICAL HARMONIC. Whittaker and Watson (1990, pp. 536–537) write

$$\Theta_p = \frac{x^2}{a^2+\theta_p} + \frac{y^2}{b^2+\theta_p} + \frac{z^2}{c^2+\theta_p} - 1 \quad (1)$$

$$\Pi(\Theta) \equiv \Theta_1\Theta_2\cdots\Theta_m, \quad (2)$$

and give various types of ellipsoidal harmonics and their highest degree terms as

1. $\Pi(\Theta) : 2m$
2. $x\Pi(\Theta), y\Pi(\Theta), z\Pi(\Theta) : 2m+1$
3. $yz\Pi(\Theta), zx\Pi(\Theta), xy\Pi(\Theta) : 2m+2$
4. $xyz\Pi(\Theta) : 2m+3$.

A Lamé function of degree n may be expressed as

$$(\theta+a^2)^{\kappa_1}(\theta+b^2)^{\kappa_2}(\theta+c^2)^{\kappa_3}\prod_{p=1}^{m}(\theta-\theta_p), \quad (3)$$

where $\kappa_i = 0$ or $1/2$, θ_i are REAL and unequal to each other and to $-a^2$, $-b^2$, and $-c^2$, and

$$\tfrac{1}{2}n = m + \kappa_1 + \kappa_2 + \kappa_3. \quad (4)$$

Byerly (1959) uses the RECURRENCE RELATIONS to explicitly compute some ellipsoidal harmonics, which he denotes by $K(x)$, $L(x)$, $M(x)$, and $N(x)$,

$$\begin{aligned}
K_0(x) &= 1\\
L_0(x) &= 0\\
M_0(x) &= 0\\
N_0(x) &= 0\\
K_1(x) &= x\\
L_1(x) &= \sqrt{x^2-b^2}\\
M_1(x) &= \sqrt{x^2-c^2}\\
N_1(x) &= 0\\
K_2^{p_1}(x) &= x^2 - \tfrac{1}{3}[b^2+c^2-\sqrt{(b^2+c^2)^2-3b^2c^2}\,]\\
K_2^{p_2}(x) &= x^2 - \tfrac{1}{3}[b^2+c^2+\sqrt{(b^2+c^2)^2-3b^2c^2}\,]\\
L_2(x) &= x\sqrt{x^2-b^2}\\
M_2(x) &= x\sqrt{x^2-c^2}\\
N_2(x) &= \sqrt{(x^2-b^2)(x^2-c^2)}\\
K_3^{p_1}(x) &= x^3 - \tfrac{1}{5}x[2(b^2+c^2)\\
&\quad - \sqrt{4(b^2+c^2)^2-15b^2c^2}\,]\\
K_3^{p_2}(x) &= x^3 - \tfrac{1}{5}x[2(b^2+c^2)\\
&\quad + \sqrt{4(b^2+c^2)^2-15b^2c^2}\,]\\
L_3^{q_1}(x) &= \sqrt{x^2-b^2}[x^2 - \tfrac{1}{5}(b^2+2c^2\\
&\quad - \sqrt{(b^2+2c^2)^2-5b^2c^2}\,)]\\
L_3^{q_2}(x) &= \sqrt{x^2-b^2}[x^2 - \tfrac{1}{5}(b^2+2c^2\\
&\quad + \sqrt{(b^2+2c^2)^2-5b^2c^2}\,)]\\
M_3^{q_1}(x) &= \sqrt{x^2-c^2}[x^2 - \tfrac{1}{5}(2b^2+c^2\\
&\quad - \sqrt{(2b^2+c^2)^2-5b^2c^2}\,)]\\
M_3^{q_2}(x) &= \sqrt{x^2-c^2}[x^2 - \tfrac{1}{5}(2b^2+c^2\\
&\quad + \sqrt{(2b^2+c^2)^2-5b^2c^2}\,)]\\
M_3^{q_3}(x) &= x\sqrt{(x^2-b^2)(x^2-c^2)}
\end{aligned}$$

see also ELLIPSOIDAL HARMONIC OF THE SECOND KIND, STIELTJES' THEOREM

References

Byerly, W. E. *An Elementary Treatise on Fourier's Series, and Spherical, Cylindrical, and Ellipsoidal Harmonics,*

with Applications to Problems in Mathematical Physics. New York: Dover, pp. 254–258, 1959.
Whittaker, E. T. and Watson, G. N. *A Course in Modern Analysis, 4th ed.* Cambridge, England: Cambridge University Press, 1990.

Ellipsoidal Harmonic of the Second Kind

Given by

$$F_m^p(x) = (2m+1)E_m^p(x) \times \int_x^\infty \frac{dx}{(x^2-b^2)(x^2-c^2)[E_m^p(x)]^2}.$$

Elliptic Alpha Function

Elliptic alpha functions relate the complete ELLIPTIC INTEGRALS OF THE FIRST $K(k_r)$ and SECOND KINDS $E(k_r)$ at ELLIPTIC INTEGRAL SINGULAR VALUES k_r according to

$$\alpha(r) = \frac{E'(k_r)}{K(k_r)} - \frac{\pi}{4[K(k_r)]^2} \tag{1}$$

$$= \frac{\pi}{4[K(k_r)]^2} + \sqrt{r} - \frac{E(k_r)\sqrt{r}}{K(k_r)} \tag{2}$$

$$= \frac{\pi^{-1} - 4\sqrt{r}\,q\frac{d\vartheta_4(q)}{dq}\frac{1}{\vartheta_4(q)}}{\vartheta_3{}^4(q)}, \tag{3}$$

where $\vartheta_3(q)$ is a THETA FUNCTION and

$$k_r = \lambda^*(r) \tag{4}$$

$$q = e^{-\pi\sqrt{r}}, \tag{5}$$

and $\lambda^*(r)$ is the ELLIPTIC LAMBDA FUNCTION. The elliptic alpha function is related to the ELLIPTIC DELTA FUNCTION by

$$\alpha(r) = \tfrac{1}{2}[\sqrt{r} - \delta(r)]. \tag{6}$$

It satisfies

$$\alpha(4r) = (1+k_r)^2\alpha(r) - 2\sqrt{r}\,k_r, \tag{7}$$

and has the limit

$$\lim_{r\to\infty}\left[\alpha(r) - \frac{1}{\pi}\right] \approx 8\left(\sqrt{r} - \frac{1}{\pi}\right)e^{-\pi\sqrt{r}} \tag{8}$$

(Borwein *et al.* 1989). A few specific values (Borwein and Borwein 1987, p. 172) are

$$\alpha(1) = \tfrac{1}{2}$$
$$\alpha(2) = \sqrt{2} - 1$$
$$\alpha(3) = \tfrac{1}{2}(\sqrt{3} - 1)$$
$$\alpha(4) = 2(\sqrt{2} - 1)^2$$
$$\alpha(5) = \tfrac{1}{2}\left(\sqrt{5} - \sqrt{2\sqrt{5} - 2}\,\right)$$
$$\alpha(6) = 5\sqrt{6} + 6\sqrt{3} - 8\sqrt{2} - 11$$
$$\alpha(7) = \tfrac{1}{2}(\sqrt{7} - 2)$$
$$\alpha(8) = 2(10 + 7\sqrt{2})\left(1 - \sqrt{\sqrt{8} - 2}\,\right)^2$$
$$\alpha(9) = \tfrac{1}{2}[3 - 3^{3/4}\sqrt{2}(\sqrt{3} - 1)]$$
$$\alpha(10) = -103 + 72\sqrt{2} - 46\sqrt{5} + 33\sqrt{10}$$
$$\alpha(12) = 264 + 154\sqrt{3} - 188\sqrt{2} - 108\sqrt{6}$$
$$\alpha(13) = \tfrac{1}{2}\left(\sqrt{13} - \sqrt{74\sqrt{13} - 258}\,\right)$$
$$\alpha(15) = \tfrac{1}{2}(\sqrt{15} - \sqrt{5} - 1)$$
$$\alpha(16) = \frac{4(\sqrt{8} - 1)}{(2^{1/4} + 1)^4}$$
$$\alpha(18) = -3057 + 2163\sqrt{2} + 1764\sqrt{3} - 1248\sqrt{6}$$
$$\alpha(22) = -12479 - 8824\sqrt{2} + 3762\sqrt{11} + 2661\sqrt{22}$$
$$\alpha(25) = \tfrac{5}{2}[1 - 25^{1/4}(7 - 3\sqrt{5})]$$
$$\alpha(27) = 3[\tfrac{1}{2}(\sqrt{3} + 1) - 2^{1/3}]$$
$$\alpha(30) = \tfrac{1}{2}\{\sqrt{30} - (2 + \sqrt{5})^2(3 + \sqrt{10})^2 \times \left(-6 - 5\sqrt{2} - 3\sqrt{5} - 2\sqrt{10} + \sqrt{6}\sqrt{57 + 40\sqrt{2}}\,\right) \times [56 + 38\sqrt{2} + \sqrt{30}(2 + \sqrt{5})(3 + \sqrt{10})]\}$$
$$\alpha(37) = \tfrac{1}{2}\left[\sqrt{37} - (171 - 25\sqrt{37})\sqrt{\sqrt{37} - 6}\,\right]$$
$$\alpha(49) = \tfrac{7}{2} - \sqrt{7[\sqrt{2}\,7^{3/4}(33011 + 12477\sqrt{7}) - 21(9567 + 3616\sqrt{7})]}$$
$$\alpha(46) = \tfrac{1}{2}\left[\sqrt{46} + \left(18 + 13\sqrt{2} + \sqrt{661 + 468\sqrt{2}}\,\right)^2 \times \left(18 + 13\sqrt{2} - 3\sqrt{2}\sqrt{147 + 104\sqrt{2}} + \sqrt{661 + 468\sqrt{2}}\,\right) \times \left(200 + 14\sqrt{2} + 26\sqrt{23} + 18\sqrt{46} + \sqrt{46}\sqrt{661 + 468\sqrt{2}}\,\right)\right]$$
$$\alpha(58) = [\tfrac{1}{2}(\sqrt{29} + 5)]^6(99\sqrt{29} - 444)(99\sqrt{2} - 70 - 13\sqrt{29})$$
$$= 3(-40768961 + 28828008\sqrt{2} - 7570606\sqrt{29} + 5353227\sqrt{58})$$
$$\alpha(64) = \frac{8[2(\sqrt{8} - 1) - (2^{1/4} - 1)^4]}{\left(\sqrt{\sqrt{2} + 1} + 2^{5/8}\right)^4}.$$

J. Borwein has written an ALGORITHM which uses lattice basis reduction to provide algebraic values for $\alpha(n)$.

see also ELLIPTIC INTEGRAL OF THE FIRST KIND, ELLIPTIC INTEGRAL OF THE SECOND KIND, ELLIPTIC INTEGRAL SINGULAR VALUE, ELLIPTIC LAMBDA FUNCTION

References

Borwein, J. M. and Borwein, P. B. *Pi & the AGM: A Study in Analytic Number Theory and Computational Complexity.* New York: Wiley, 1987.
Borwein, J. M.; Borwein, P. B.; and Bailey, D. H. "Ramanujan, Modular Equations, and Approximations to Pi, or How to Compute One Billion Digits of Pi." *Amer. Math. Monthly* **96**, 201–219, 1989.

Weisstein, E. W. "Elliptic Singular Values." http://www.astro.virginia.edu/~eww6n/math/notebooks/EllipticSingular.m.

Elliptic Cone

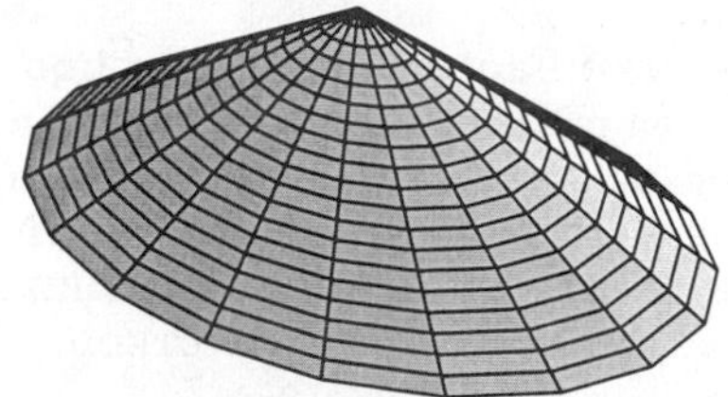

A CONE with ELLIPTICAL CROSS-SECTION. The parametric equations for an elliptic cone of height h, SEMIMAJOR AXIS a, and SEMIMINOR AXIS b are

$$\begin{aligned} x &= (h-z)a\cos\theta \\ y &= (h-z)b\sin\theta \\ z &= z, \end{aligned}$$

where $\theta \in [0, 2\pi)$ and $z \in [0, h]$.

see also CONE, ELLIPTIC CYLINDER, ELLIPTIC PARABOLOID, HYPERBOLIC PARABOLOID

References

Fischer, G. (Ed.). Plate 68 in *Mathematische Modelle/Mathematical Models, Bildband/Photograph Volume.* Braunschweig, Germany: Vieweg, p. 63, 1986.

Elliptic Cone Point

see ISOLATED SINGULARITY

Elliptic Curve

Informally, an elliptic curve is a type of CUBIC CURVE whose solutions are confined to a region of space which is topologically equivalent to a TORUS. Formally, an elliptic curve over a FIELD K is a nonsingular CUBIC CURVE in two variables, $f(X,Y) = 0$, with a K-rational point (which may be a point at infinity). The FIELD K is usually taken to be the COMPLEX NUMBERS $\mathbb{C}$, REALS $\mathbb{R}$, RATIONALS $\mathbb{Q}$, algebraic extensions of $\mathbb{Q}$, p-ADIC NUMBERS $\mathbb{Q}_p$, or a FINITE FIELD.

By an appropriate change of variables, a general elliptic curve over a FIELD of CHARACTERISTIC $\neq 2, 3$

$$\begin{aligned} Ax^3 + Bx^2y + Cxy^2 + Dy^3 + Ex^2 \\ +Fxy + Gy^2 + Hx + Iy + J = 0, \quad (1) \end{aligned}$$

where A, B, ..., are elements of K, can be written in the form

$$y^2 = x^3 + ax + b, \quad (2)$$

where the right side of (2) has no repeated factors. If K has CHARACTERISTIC three, then the best that can be done is to transform the curve into

$$y^2 = x^3 + ax^2 + bx + c \quad (3)$$

(the x^2 term cannot be eliminated). If K has CHARACTERISTIC two, then the situation is even worse. A general form into which an elliptic curve over any K can be transformed is called the WEIERSTRAß FORM, and is given by

$$y^2 + ay = x^3 + bx^2 + cxy + dx + e, \quad (4)$$

where a, b, c, d, and e are elements of K. Luckily, $\mathbb{Q}$, $\mathbb{R}$, and $\mathbb{C}$ all have CHARACTERISTIC zero.

Whereas CONIC SECTIONS can be parameterized by the rational functions, elliptic curves cannot. The simplest parameterization functions are ELLIPTIC FUNCTIONS. ABELIAN VARIETIES can be viewed as generalizations of elliptic curves.

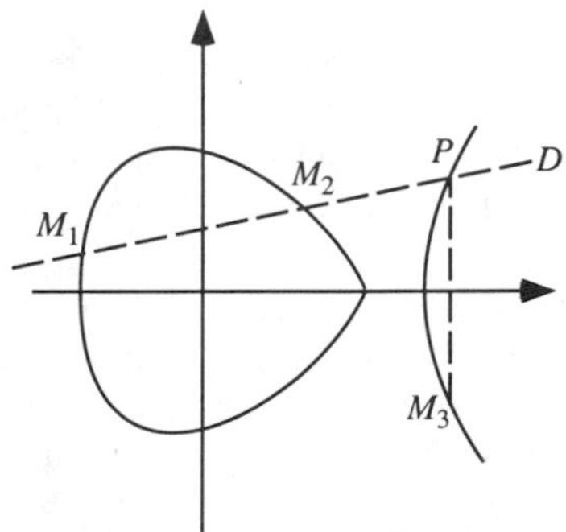

If the underlying FIELD of an elliptic curve is algebraically closed, then a straight line cuts an elliptic curve at three points (counting multiple roots at points of tangency). If two are known, it is possible to compute the third. If two of the intersection points are K-RATIONAL, then so is the third. Let (x_1, y_1) and (x_2, y_2) be two points on an elliptic curve E with DISCRIMINANT

$$\Delta_E = -16(4a^3 + 27b^2) \quad (5)$$

satisfying

$$\Delta_E \neq 0. \quad (6)$$

A related quantity known as the j-INVARIANT of E is defined as

$$j(E) \equiv \frac{2^8 3^3 a^3}{4a^3 + 27b^2}. \quad (7)$$

Now define

$$\lambda = \begin{cases} \frac{y_1 - y_2}{x_1 - x_2} & \text{for } x_1 \neq x_2 \\ \frac{3x_1{}^2 + a}{2y_1} & \text{for } x_1 = x_2. \end{cases} \quad (8)$$

Then the coordinates of the third point are

$$\begin{aligned} x_3 &= \lambda^2 - x_1 - x_2 \quad &(9) \\ y_3 &= \lambda(x_3 - x_1) + y_1. \quad &(10) \end{aligned}$$

For elliptic curves over $\mathbb{Q}$, Mordell proved that there are a finite number of integral solutions. The MORDELL-WEIL THEOREM says that the GROUP of RATIONAL

POINTS of an elliptic curve over $\mathbb{Q}$ is finitely generated. Let the ROOTS of y^2 be r_1, r_2, and r_3. The discriminant is then

$$\Delta = k(r_1 - r_2)^2(r_1 - r_3)^2(r_2 - r_3)^2. \tag{11}$$

The amazing TANIYAMA-SHIMURA CONJECTURE states that all rational elliptic curves are also modular. This fact is far from obvious, and despite the fact that the conjecture was proposed in 1955, it was not proved until 1995. Even so, Wiles' proof surprised most mathematicians, who had believed the conjecture unassailable. As a side benefit, Wiles' proof of the TANIYAMA-SHIMURA CONJECTURE also laid to rest the famous and thorny problem which had baffled mathematicians for hundreds of years, FERMAT'S LAST THEOREM.

Curves with small CONDUCTORS are listed in Swinnerton-Dyer (1975) and Cremona (1997). Methods for computing integral points (points with integral coordinates) are given in Gebel *et al.* and Stroeker and Tzanakis (1994).

see also ELLIPTIC CURVE GROUP LAW, FERMAT'S LAST THEOREM, FREY CURVE, j-INVARIANT, MINIMAL DISCRIMINANT, MORDELL-WEIL THEOREM, OCHOA CURVE, RIBET'S THEOREM, SIEGEL'S THEOREM, SWINNERTON-DYER CONJECTURE, TANIYAMA-SHIMURA CONJECTURE, WEIERSTRAß FORM

References

Atkin, A. O. L. and Morain, F. "Elliptic Curves and Primality Proving." *Math. Comput.* **61**, 29–68, 1993.

Cassels, J. W. S. *Lectures on Elliptic Curves.* New York: Cambridge University Press, 1991.

Cremona, J. E. *Algorithms for Modular Elliptic Curves, 2nd ed.* Cambridge, England: Cambridge University Press, 1997.

Cremona, J. E. "Elliptic Curve Data." ftp://euclid.ex.ac.uk/pub/cremona/data/.

Du Val, P. *Elliptic Functions and Elliptic Curves.* Cambridge: Cambridge University Press, 1973.

Gebel, J.; Pethő, A.; and Zimmer, H. G. "Computing Integral Points on Elliptic Curves." *Acta Arith.* **68**, 171–192, 1994.

Ireland, K. and Rosen, M. "Elliptic Curves." Ch. 18 in *A Classical Introduction to Modern Number Theory, 2nd ed.* New York: Springer-Verlag, pp. 297–318, 1990.

Katz, N. M. and Mazur, B. *Arithmetic Moduli of Elliptic Curves.* Princeton, NJ: Princeton University Press, 1985.

Knapp, A. W. *Elliptic Curves.* Princeton, NJ: Princeton University Press, 1992.

Koblitz, N. *Introduction to Elliptic Curves and Modular Forms.* New York: Springer-Verlag, 1993.

Lang, S. *Elliptic Curves: Diophantine Analysis.* Berlin: Springer-Verlag, 1978.

Silverman, J. H. *The Arithmetic of Elliptic Curves.* New York: Springer-Verlag, 1986.

Silverman, J. H. *The Arithmetic of Elliptic Curves II.* New York: Springer-Verlag, 1994.

Silverman, J. H. and Tate, J. T. *Rational Points on Elliptic Curves.* New York: Springer-Verlag, 1992.

Stroeker, R. J. and Tzanakis, N. "Solving Elliptic Diophantine Equations by Estimating Linear Forms in Elliptic Logarithms." *Acta Arith.* **67**, 177–196, 1994.

Swinnerton-Dyer, H. P. F. "Correction to: 'On 1-adic Representations and Congruences for Coefficients of Modular Forms.'" In *Modular Functions of One Variable, Vol. 4, Proc. Internat. Summer School for Theoret. Phys., Univ. Antwerp, Antwerp, RUCA, July-Aug. 1972.* Berlin: Springer-Verlag, 1975.

Elliptic Curve Factorization Method

A factorization method, abbreviated ECM, which computes a large multiple of a point on a random ELLIPTIC CURVE modulo the number to be factored N. It tends to be faster than the POLLARD ρ FACTORIZATION and POLLARD $p-1$ FACTORIZATION METHOD.

see also ATKIN-GOLDWASSER-KILIAN-MORAIN CERTIFICATE, ELLIPTIC CURVE PRIMALITY PROVING, ELLIPTIC PSEUDOPRIME

References

Atkin, A. O. L. and Morain, F. "Finding Suitable Curves for the Elliptic Curve Method of Factorization." *Math. Comput.* **60**, 399–405, 1993.

Brent, R. P. "Some Integer Factorization Algorithms Using Elliptic Curves." *Austral. Comp. Sci. Comm.* **8**, 149–163, 1986.

Brent, R. P. "Parallel Algorithms for Integer Factorisation." In *Number Theory and Cryptography* (Ed. J. H. Loxton). New York: Cambridge University Press, 26–37, 1990. ftp://nimbus.anu.edu.au/pub/Brent/115.dvi.Z.

Brillhart, J.; Lehmer, D. H.; Selfridge, J.; Wagstaff, S. S. Jr.; and Tuckerman, B. *Factorizations of $b^n \pm 1$, $b = 2, 3, 5, 6, 7, 10, 11, 12$ Up to High Powers, rev. ed.* Providence, RI: Amer. Math. Soc., p. lxxxiii, 1988.

Eldershaw, C. and Brent, R. P. "Factorization of Large Integers on Some Vector and Parallel Computers." ftp://nimbus.anu.edu.au/pub/Brent/156tr.dvi.Z.

Lenstra, A. K. and Lenstra, H. W. Jr. "Algorithms in Number Theory." In *Handbook of Theoretical Computer Science, Volume A: Algorithms and Complexity* (Ed. J. van Leeuwen). Elsevier, pp. 673–715, 1990.

Lenstra, H. W. Jr. "Factoring Integers with Elliptic Curves." *Ann. Math.* **126**, 649–673, 1987.

Montgomery, P. L. "Speeding the Pollard and Elliptic Curve Methods of Factorization." *Math. Comput.* **48**, 243–264, 1987.

Elliptic Curve Group Law

The GROUP of an ELLIPTIC CURVE which has been transformed to the form

$$y^2 = x^3 + ax + b$$

is the set of K-RATIONAL POINTS, including the single POINT AT INFINITY. The group law (addition) is defined as follows: Take 2 K-RATIONAL POINTS P and Q. Now 'draw' a straight line through them and compute the third point of intersection R (also a K-RATIONAL POINT). Then

$$P + Q + R = 0$$

gives the identity point at infinity. Now find the inverse of R, which can be done by setting $R = (a, b)$ giving $-R = (a, -b)$.

This remarkable result is only a special case of a more general procedure. Essentially, the reason is that this type of ELLIPTIC CURVE has a single point at infinity which is an inflection point (the line at infinity meets the curve at a single point at infinity, so it must be an intersection of multiplicity three).

Elliptic Curve Primality Proving

A class of algorithm, abbreviated ECPP, which provides certificates of primality using sophisticated results from the theory of ELLIPTIC CURVES. A detailed description and list of references are given by Atkin and Morain (1990, 1993).

Adleman and Huang (1987) designed an independent algorithm using elliptic curves of genus two.

see also ATKIN-GOLDWASSER-KILIAN-MORAIN CERTIFICATE, ELLIPTIC CURVE FACTORIZATION METHOD, ELLIPTIC PSEUDOPRIME

References

Adleman, L. M. and Huang, M. A. "Recognizing Primes in Random Polynomial Time." In *Proc. 19th STOC, New York City, May 25–27, 1986.* New York: ACM Press, pp. 462–469, 1987.

Atkin, A. O. L. Lecture notes of a conference, Boulder, CO, Aug. 1986.

Atkin, A. O. L. and Morain, F. "Elliptic Curves and Primality Proving." Res. Rep. 1256, INRIA, June 1990.

Atkin, A. O. L. and Morain, F. "Elliptic Curves and Primality Proving." *Math. Comput.* **61**, 29–68, 1993.

Bosma, W. "Primality Testing Using Elliptic Curves." Techn. Rep. 85–12, Math. Inst., Univ. Amsterdam, 1985.

Chudnovsky, D. V. and Chudnovsky, G. V. "Sequences of Numbers Generated by Addition in Formal Groups and New Primality and Factorization Tests." Res. Rep. RC 11262, IBM, Yorktown Heights, NY, 1985.

Cohen, H. *Cryptographie, factorisation et primalité: l'utilisation des courbes elliptiques.* Paris: C. R. J. Soc. Math. France, Jan. 1987.

Kaltofen, E.; Valente, R.; and Yui, N. "An Improved Las Vegas Primality Test." Res. Rep. 89–12, Rensselaer Polytechnic Inst., Troy, NY, May 1989.

Elliptic Cylinder

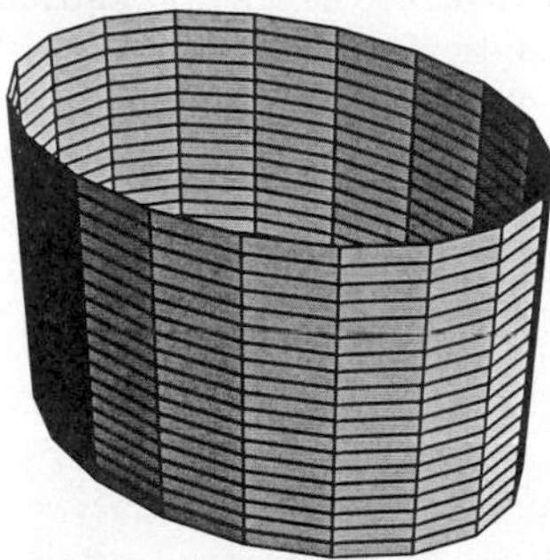

A CYLINDER with ELLIPTICAL CROSS-SECTION. The parametric equations for an elliptic cylinder of height h, SEMIMAJOR AXIS a, and SEMIMINOR AXIS b are

$$x = a\cos\theta$$
$$y = b\sin\theta$$
$$z = z,$$

where $\theta \in [0, 2\pi)$ and $z \in [0, h]$.

see also CONE, CYLINDER, ELLIPTIC CONE, ELLIPTIC PARABOLOID

Elliptic Cylindrical Coordinates

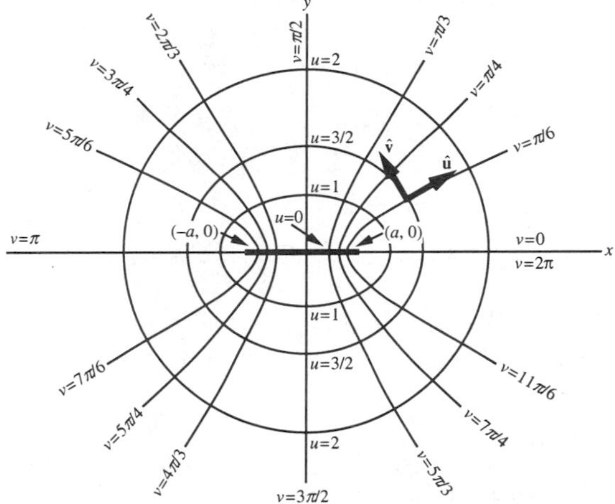

The v coordinates are the asymptotic angle of confocal PARABOLA segments symmetrical about the x axis. The u coordinates are confocal ELLIPSES centered on the origin.

$$x = a\cosh u\cos v \tag{1}$$
$$y = a\sinh u\sin v \tag{2}$$
$$z = z, \tag{3}$$

where $u \in [0, \infty)$, $v \in [0, 2\pi)$, and $z \in (-\infty, \infty)$. They are related to CARTESIAN COORDINATES by

$$\frac{x^2}{a^2\cosh^2 u} + \frac{y^2}{a^2\sinh^2 u} = 1 \tag{4}$$

$$\frac{x^2}{a^2\cos^2 v} - \frac{y^2}{a^2\sin^2 v} = 1. \tag{5}$$

The SCALE FACTORS are

$$h_1 = a\sqrt{\cosh^2 u\sin^2 v + \sinh^2 u\cos^2 v} \tag{6}$$
$$= a\sqrt{\frac{\cosh(2u) - \cos(2v)}{2}} \tag{7}$$
$$= a\sqrt{\sinh^2 u + \sin^2 v} \tag{8}$$
$$h_2 = a\sqrt{\sinh^2 u\sin^2 v + \sinh^2 u\cos^2 v} \tag{9}$$
$$= a\sqrt{\frac{\cosh(2u) - \cos(2v)}{2}} \tag{10}$$
$$= a\sqrt{\sinh^2 u + \sin^2 v} \tag{11}$$
$$h_3 = 1. \tag{12}$$

The LAPLACIAN is

$$\nabla^2 = \frac{1}{a^2(\sinh^2 u + \sin^2 v)}\left(\frac{\partial^2}{\partial u^2} + \frac{\partial^2}{\partial v^2}\right) + \frac{\partial^2}{\partial z^2}. \tag{13}$$

Let

$$q_1 = \cosh u \quad (14)$$

$$q_2 = \cos v \quad (15)$$

$$q_3 = z. \quad (16)$$

Then the new SCALE FACTORS are

$$h_{q_1} = a\sqrt{\frac{{q_1}^2 - {q_2}^2}{{q_1}^2 - 1}} \quad (17)$$

$$h_{q_2} = a\sqrt{\frac{{q_1}^2 - {q_2}^2}{1 - {q_1}^2}} \quad (18)$$

$$h_{q_3} = 1. \quad (19)$$

The HELMHOLTZ DIFFERENTIAL EQUATION is SEPARABLE.

see also CYLINDRICAL COORDINATES, HELMHOLTZ DIFFERENTIAL EQUATION—ELLIPTIC CYLINDRICAL COORDINATES

References

Arfken, G. "Elliptic Cylindrical Coordinates (u, v, z)." §2.7 in *Mathematical Methods for Physicists, 2nd ed.* Orlando, FL: Academic Press, pp. 95–97, 1970.

Morse, P. M. and Feshbach, H. *Methods of Theoretical Physics, Part I.* New York: McGraw-Hill, p. 657, 1953.

Elliptic Delta Function

$$\delta(r) = \sqrt{r} - 2\alpha(r),$$

where α is the ELLIPTIC ALPHA FUNCTION.

see also ELLIPTIC ALPHA FUNCTION, ELLIPTIC INTEGRAL SINGULAR VALUE

References

Borwein, J. M. and Borwein, P. B. *Pi & the AGM: A Study in Analytic Number Theory and Computational Complexity.* New York: Wiley, 1987.

Weisstein, E. W. "Elliptic Singular Values." http://www.astro.virginia.edu/~eww6n/math/notebooks/EllipticSingular.m.

Elliptic Exponential Function

The inverse of the ELLIPTIC LOGARITHM

$$\operatorname{eln}(x) \equiv \int_x^\infty \frac{dt}{\sqrt{t^3 + at^2 + bt}}.$$

It is doubly periodic in the COMPLEX PLANE.

Elliptic Fixed Point (Differential Equations)

A FIXED POINT for which the STABILITY MATRIX is purely IMAGINARY, $\lambda_\pm = \pm i\omega$ (for $\omega > 0$).

see also DIFFERENTIAL EQUATION, FIXED POINT, HYPERBOLIC FIXED POINT (DIFFERENTIAL EQUATIONS), PARABOLIC FIXED POINT, STABLE IMPROPER NODE, STABLE NODE, STABLE SPIRAL POINT, STABLE STAR, UNSTABLE IMPROPER NODE, UNSTABLE NODE, UNSTABLE SPIRAL POINT, UNSTABLE STAR

References

Tabor, M. "Classification of Fixed Points." §1.4.b in *Chaos and Integrability in Nonlinear Dynamics: An Introduction.* New York: Wiley, pp. 22–25, 1989.

Elliptic Fixed Point (Map)

A FIXED POINT of a LINEAR TRANSFORMATION (MAP) for which the rescaled variables satisfy

$$(\delta - \alpha)^2 + 4\beta\gamma < 0.$$

see also HYPERBOLIC FIXED POINT (MAP), LINEAR TRANSFORMATION, PARABOLIC FIXED POINT

Elliptic Function

A doubly periodic function with periods $2\omega_1$ and $2\omega_2$ such that

$$f(z + 2\omega_1) = f(z + 2\omega_2) = f(z), \quad (1)$$

which is ANALYTIC and has no singularities except for POLES in the finite part of the COMPLEX PLANE. The ratio ω_1/ω_2 must not be purely real. If this ratio is real, the function reduces to a singly periodic function if it is rational and a constant if the ratio is irrational (Jacobi, 1835). ω_1 and ω_2 are labeled such that $\Im(\omega_2/\omega_1) > 0$. A "cell" of an elliptic function is defined as a parallelogram region in the COMPLEX PLANE in which the function is not multi-valued. Properties obeyed by elliptic functions include

1. The number of POLES in a cell is finite.
2. The number of ROOTS in a cell is finite.
3. The sum of RESIDUES in any cell is 0.
4. LIOUVILLE'S ELLIPTIC FUNCTION THEOREM: An elliptic function with no POLES in a cell is a constant.
5. The number of zeros of $f(z) - c$ (the "order") equals the number of POLES of $f(z)$.
6. The simplest elliptic function has order two, since a function of order one would have a simple irreducible POLE, which would need to have a NONZERO residue. By property (3), this is impossible.
7. Elliptic functions with a single POLE of order 2 with RESIDUE 0 are called WEIERSTRAß ELLIPTIC FUNCTIONS. Elliptic functions with two simple POLES having residues a_0 and $-a_0$ are called JACOBI ELLIPTIC FUNCTIONS.
8. Any elliptic function is expressible in terms of either WEIERSTRAß ELLIPTIC FUNCTION or JACOBI ELLIPTIC FUNCTIONS.
9. The sum of the AFFIXES of ROOTS equals the sum of the AFFIXES of the POLES.
10. An algebraic relationship exists between any two elliptic functions with the same periods.

The elliptic functions are inversions of the ELLIPTIC INTEGRALS. The two standard forms of these functions are known as JACOBI ELLIPTIC FUNCTIONS and WEIERSTRAß ELLIPTIC FUNCTIONS. JACOBI ELLIPTIC FUNCTIONS arise as solutions to differential equations of the form

$$\frac{d^2x}{dt^2} = A + Bx + Cx^2 + Dx^3, \quad (2)$$

and WEIERSTRAß ELLIPTIC FUNCTIONS arise as solutions to differential equations of the form

$$\frac{d^2x}{dt^2} = A + Bx + Cx^2. \tag{3}$$

see also ELLIPTIC CURVE, ELLIPTIC INTEGRAL, JACOBI ELLIPTIC FUNCTIONS, LIOUVILLE'S ELLIPTIC FUNCTION THEOREM, MODULAR FORM, MODULAR FUNCTION, NEVILLE THETA FUNCTION, THETA FUNCTION, WEIERSTRAß ELLIPTIC FUNCTIONS

References

Akhiezer, N. I. *Elements of the Theory of Elliptic Functions.* Providence, RI: Amer. Math. Soc., 1990.

Bellman, R. E. *A Brief Introduction to Theta Functions.* New York: Holt, Rinehart and Winston, 1961.

Borwein, J. M. and Borwein, P. B. *Pi & the AGM: A Study in Analytic Number Theory and Computational Complexity.* New York: Wiley, 1987.

Bowman, F. *Introduction to Elliptic Functions, with Applications.* New York: Dover, 1961.

Byrd, P. F. and Friedman, M. D. *Handbook of Elliptic Integrals for Engineers and Scientists, 2nd ed., rev.* Berlin: Springer-Verlag, 1971.

Cayley, A. *An Elementary Treatise on Elliptic Functions, 2nd ed.* London: G. Bell, 1895.

Chandrasekharan, K. *Elliptic Functions.* Berlin: Springer-Verlag, 1985.

Du Val, P. *Elliptic Functions and Elliptic Curves.* Cambridge, England: Cambridge University Press, 1973.

Dutta, M. and Debnath, L. *Elements of the Theory of Elliptic and Associated Functions with Applications.* Calcutta, India: World Press, 1965.

Eagle, A. *The Elliptic Functions as They Should Be: An Account, with Applications, of the Functions in a New Canonical Form.* Cambridge, England: Galloway and Porter, 1958.

Greenhill, A. G. *The Applications of Elliptic Functions.* London: Macmillan, 1892.

Hancock, H. *Lectures on the Theory of Elliptic Functions.* New York: Wiley, 1910.

Jacobi, C. G. J. *Fundamentia Nova Theoriae Functionum Ellipticarum.* Regiomonti, Sumtibus fratrum Borntraeger, 1829.

King, L. V. *On the Direct Numerical Calculation of Elliptic Functions and Integrals.* Cambridge, England: Cambridge University Press, 1924.

Lang, S. *Elliptic Functions, 2nd ed.* New York: Springer-Verlag, 1987.

Lawden, D. F. *Elliptic Functions and Applications.* New York: Springer Verlag, 1989.

Morse, P. M. and Feshbach, H. *Methods of Theoretical Physics, Part I.* New York: McGraw-Hill, pp. 427 and 433–434, 1953.

Murty, M. R. (Ed.). *Theta Functions.* Providence, RI: Amer. Math. Soc., 1993.

Neville, E. H. *Jacobian Elliptic Functions, 2nd ed.* Oxford, England: Clarendon Press, 1951.

Petkovšek, M.; Wilf, H. S.; and Zeilberger, D. "Elliptic Function Identities." §1.8 in *A=B.* Wellesley, MA: A. K. Peters, pp. 13–15, 1996.

Whittaker, E. T. and Watson, G. N. Chs. 20–22 in *A Course of Modern Analysis, 4th ed.* Cambridge, England: University Press, 1943.

Elliptic Functional

see COERCIVE FUNCTIONAL

Elliptic Geometry

A constant curvature NON-EUCLIDEAN GEOMETRY which replaces the PARALLEL POSTULATE with the statement "through any point in the plane, there exist no lines PARALLEL to a given line." Elliptic geometry is sometimes also called RIEMANNIAN GEOMETRY. It can be visualized as the surface of a SPHERE on which "lines" are taken as GREAT CIRCLES. In elliptic geometry, the sum of angles of a TRIANGLE is $> 180°$.

see also EUCLIDEAN GEOMETRY, HYPERBOLIC GEOMETRY, NON-EUCLIDEAN GEOMETRY

Elliptic Group Modulo p

$E(a,b)/p$ denotes the elliptic GROUP modulo p whose elements are 1 and ∞ together with the pairs of INTEGERS (x,y) with $0 \leq x, y < p$ satisfying

$$y^2 \equiv x^3 + ax + b \pmod{p} \tag{1}$$

with a and b INTEGERS such that

$$4a^3 + 27b^2 \not\equiv 0 \pmod{p}. \tag{2}$$

Given (x_1, y_1), define

$$(x_i, y_i) \equiv (x_1, y_1)^i \pmod{p}. \tag{3}$$

The ORDER h of $E(a,b)/p$ is given by

$$h = 1 + \sum_{x=1}^{p} \left[\left(\frac{x^3 + ax + b}{p} \right) + 1 \right], \tag{4}$$

where $(x^3 + ax + b/p)$ is the LEGENDRE SYMBOL, although this FORMULA quickly becomes impractical. However, it has been proven that

$$p + 1 - 2\sqrt{p} \leq h(E(a,b)/p) \leq p + 1 + 2\sqrt{p}. \tag{5}$$

Furthermore, for p a PRIME > 3 and and INTEGER n in the above interval, there exists a and b such that

$$h(E(a,b)/p) = n, \tag{6}$$

and the orders of elliptic GROUPS mod p are nearly uniformly distributed in the interval.

Elliptic Helicoid

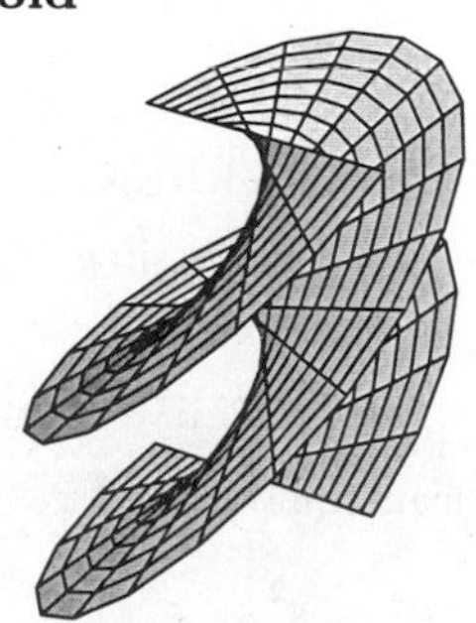

A generalization of the HELICOID to the parametric equations

$$x(u,v) = av\cos u$$
$$y(u,v) = bv\sin u$$
$$z(u,v) = cu.$$

see also HELICOID

References

Gray, A. *Modern Differential Geometry of Curves and Surfaces.* Boca Raton, FL: CRC Press, p. 264, 1993.

Elliptic Hyperboloid

The elliptic hyperboloid is the generalization of the HYPERBOLOID to three distinct semimajor axes. The elliptic hyperboloid of one sheet is a RULED SURFACE and has Cartesian equation

$$\frac{x^2}{a^2} + \frac{y^2}{b^2} - \frac{z^2}{c^2} = 1, \tag{1}$$

and parametric equations

$$x(u,v) = a\sqrt{1+u^2}\cos v \tag{2}$$
$$y(u,v) = b\sqrt{1+u^2}\sin v \tag{3}$$
$$z(u,v) = cu \tag{4}$$

for $v \in [0, 2\pi)$, or

$$x(u,v) = a(\cos u \mp v\sin u) \tag{5}$$
$$y(u,v) = b(\sin u \pm v\cos u) \tag{6}$$
$$z(u,v) = \pm cv, \tag{7}$$

or

$$x(u,v) = a\cosh v\cos u \tag{8}$$
$$y(u,v) = b\cosh v\sin u \tag{9}$$
$$z(u,v) = c\sinh v. \tag{10}$$

The two-sheeted elliptic hyperboloid oriented along the z-AXIS has Cartesian equation

$$\frac{x^2}{a^2} + \frac{y^2}{a^2} - \frac{z^2}{c^2} = -1, \tag{11}$$

and parametric equations

$$x = a\sinh u\cos v \tag{12}$$
$$y = b\sinh u\sin v \tag{13}$$
$$z = c \pm \cosh u. \tag{14}$$

The two-sheeted elliptic hyperboloid oriented along the x-AXIS has Cartesian equation

$$\frac{x^2}{a^2} - \frac{y^2}{a^2} - \frac{z^2}{c^2} = 1 \tag{15}$$

and parametric equations

$$x = a\cosh u\cosh v \tag{16}$$
$$y = b\sinh u\cosh v \tag{17}$$
$$z = c\sinh v. \tag{18}$$

see also HYPERBOLOID, RULED SURFACE

References

Gray, A. *Modern Differential Geometry of Curves and Surfaces.* Boca Raton, FL: CRC Press, pp. 296–297, 1993.

Elliptic Integral

An elliptic integral is an INTEGRAL of the form

$$\int \frac{A(x) + B(x)\sqrt{S(x)}}{C(x) + D(x)\sqrt{S(x)}}\,dx, \tag{1}$$

or

$$\int \frac{A(x)\,dx}{B(x)\sqrt{S(x)}}, \tag{2}$$

where A, B, C, and D are POLYNOMIALS in x and S is a POLYNOMIAL of degree 3 or 4. Another form is

$$\int R(w,x)\,dx, \tag{3}$$

where R is a RATIONAL FUNCTION of x and y, w^2 is a function of x CUBIC or QUADRATIC in x, $R(w,x)$ contains at least one ODD POWER of w, and w^2 has no repeated factors.

Elliptic integrals can be viewed as generalizations of the TRIGONOMETRIC FUNCTIONS and provide solutions to a wider class of problems. For instance, while the ARC LENGTH of a CIRCLE is given as a simple function of the parameter, computing the ARC LENGTH of an ELLIPSE requires an elliptic integral. Similarly, the position of a pendulum is given by a TRIGONOMETRIC FUNCTION as a function of time for small angle oscillations, but the full solution for arbitrarily large displacements requires the use of elliptic integrals. Many other problems in electromagnetism and gravitation are solved by elliptic integrals.

A very useful class of functions known as ELLIPTIC FUNCTIONS is obtained by inverting elliptic integrals (by analogy with the inverse trigonometric functions). ELLIPTIC FUNCTIONS (among which the JACOBI ELLIPTIC FUNCTIONS and WEIERSTRASS ELLIPTIC FUNCTION are the two most common forms) provide a powerful tool for analyzing many deep problems in NUMBER THEORY, as well as other areas of mathematics.

All elliptic integrals can be written in terms of three "standard" types. To see this, write

$$R(w,x) \equiv \frac{P(w,x)}{Q(w,x)} = \frac{wP(w,x)Q(-w,x)}{wQ(w,x)Q(-w,x)}. \tag{4}$$

But since $w^2 = f(x)$,

$$Q(w,x)Q(-w,x) \equiv Q_1(w,x) = Q_1(w,x), \quad (5)$$

then

$$wP(w,x)Q(-w,x) = A + Bx + Cw + Dx^2 + Ewx + Fw^2 + Gw^2x + Hw^3x$$
$$= (A + Bx + Dx^2 + Fw^2 + Gw^2x) + w(c + Ex + Hw^2x + \ldots)$$
$$= P_1(x) + wP_2(x), \quad (6)$$

so

$$R(w,x) = \frac{P_1(x) + wP_2(x)}{wQ_1(w)} = \frac{R_1(x)}{w} + R_2(x). \quad (7)$$

But any function $\int R_2(x)\,dx$ can be evaluated in terms of elementary functions, so the only portion that need be considered is

$$\frac{\int R_1(x)}{w}\,dx. \quad (8)$$

Now, any quartic can be expressed as S_1S_2 where

$$S_1 \equiv a_1x^2 + 2b_1x + c_1 \quad (9)$$
$$S_2 \equiv a_2x^2 + 2b_2x + c_2. \quad (10)$$

The COEFFICIENTS here are real, since pairs of COMPLEX ROOTS are COMPLEX CONJUGATES

$$[x - (R + Ii)][x - (R - Ii)]$$
$$= x^2 + x(-R + Ii - R - Ii) + (R^2 - I^2 i)$$
$$= x^2 - 2Rx + (R^2 + I^2). \quad (11)$$

If all four ROOTS are real, they must be arranged so as not to interleave (Whittaker and Watson 1990, p. 514). Now define a quantity λ such that $S_1 + \lambda S_2$

$$(a_1 - \lambda a_2)x^2 - (2b_1 - 2b_2\lambda)x + (c_1 - \lambda c_2) \quad (12)$$

is a SQUARE NUMBER and

$$2\sqrt{(a_1 - \lambda a_2)(c_1 - \lambda_2)} = 2(b_1 - b_2\lambda) \quad (13)$$

$$(a_1 - \lambda a_2)(c_1 - \lambda c_2) - (b_1 - \lambda b_2)^2 = 0. \quad (14)$$

Call the ROOTS of this equation λ_1 and λ_2, then

$$S_1 - \lambda_1 S_2 = \left[\sqrt{(a_1 - \lambda_1 a_2)x} + \sqrt{c_1 - \lambda c_2}\right]^2$$
$$= (a_1 - \lambda_1 a_2)\left(x + \sqrt{\frac{c_1 - \lambda_1 c_2}{a_1 - \lambda_1 a_2}}\right)$$
$$\equiv (a_1 - \lambda_1 a_2)(x - \alpha)^2 \quad (15)$$

$$S_1 - \lambda_2 S_2 = \left[\sqrt{(a_1 - \lambda_1 a_2)x} + \sqrt{c_1 - \lambda c_2}\right]^2$$
$$= (a_1 - \lambda_1 a_2)\left(x + \sqrt{\frac{c_1 - \lambda_2 c_2}{a_1 - \lambda_2 a_2}}\right)$$
$$\equiv (a_1 - \lambda_2 a_2)(x - \beta)^2. \quad (16)$$

Taking (15)-(16) and $\lambda_2(1) - \lambda_1(2)$ gives

$$S_2(\lambda_2 - \lambda_1) = (a_1 - \lambda_1 a_2)(x - \alpha)^2 - (a_1 - \lambda_2 a_2)(x - \beta)^2 \quad (17)$$

$$S_1(\lambda_2 - \lambda_1) = \lambda_2(a_1 - \lambda_1 a_2)(x - \alpha)^2 - \lambda_1(a_1 - \lambda_2 a_2)(x - \beta^2). \quad (18)$$

Solving gives

$$S_1 = \frac{a_1 - \lambda_1 a_2}{\lambda_2 - \lambda_1}(x - \alpha)^2 - \frac{a_1 - \lambda_2 a_2}{\lambda_2 - \lambda_1}(x - \beta)^2$$
$$\equiv A_1(x - \alpha)^2 + B_1(x - \beta)^2 \quad (19)$$

$$S_2 = \frac{\lambda_2(a_1 - \lambda_1 a_2)}{\lambda_2 - \lambda_1}(x - \alpha)^2 - \frac{\lambda_1(a_1 - \lambda_2 a_2)}{\lambda_2 - \lambda_1}(x - \beta)^2$$
$$\equiv A_2(x - \alpha)^2 + B_2(x - \beta)^2, \quad (20)$$

so we have

$$w^2 = S_1S_2$$
$$= [A_1(x - \alpha)^2 + B_1(x - \beta)^2][A^2(x - \alpha)^2 + B^2(x - \beta)^2]. \quad (21)$$

Now let

$$t \equiv \frac{x - \alpha}{x - \beta} \quad (22)$$

$$dy = [(x - \beta)^{-1} - (x - \alpha)(x - \beta)^{-2}]\,dx$$
$$= \frac{(x - \beta) - (x - \alpha)}{(x - \beta)^2}\,dx$$
$$= \frac{\alpha - \beta}{(x - \beta)^2}\,dx, \quad (23)$$

so

$$w^2 = (x - \beta)^4\left[A_1\left(\frac{x - \alpha}{x - \beta}\right)^2 + B_1\right] \times \left[A_2\left(\frac{x - \alpha}{x - \beta}\right) + B_2\right]$$
$$= (x - \beta)^4(A_1t^2 + B_1)(A_2t^2 + B_2), \quad (24)$$

and

$$w = (x - \beta)^2\sqrt{(A_1t^2 + B_1)(A_2t^2 + B_2)} \quad (25)$$

$$\frac{dx}{w} = \left[\frac{(x - \beta)^2}{\alpha - \beta}\,dt\right]\frac{1}{(x - \beta)^2\sqrt{(A_1t^2 + B_1)(A_2t^2 + B_2)}}$$
$$= \frac{dt}{(\alpha - \beta)\sqrt{(A_1t^2 + B_1)(A_2t^2 + B_2)}}. \quad (26)$$

Now let

$$R_3(t) \equiv \frac{R_1(x)}{\alpha - \beta}, \quad (27)$$

so

$$\int \frac{R_1(x)\,dx}{w} = \int \frac{R_3(t)\,dt}{\sqrt{(A_1t^2 + B_1)(A_2t^2 + B_2)}}. \quad (28)$$

Rewriting the EVEN and ODD parts

$$R_3(t) + R_3(-t) \equiv 2R_4(t^2) \quad (29)$$
$$R_3(t) - R_3(-t) \equiv 2tR_5(t^2), \quad (30)$$

gives

$$R_3(t) \equiv \tfrac{1}{2}(R_{\rm even} - R_{\rm odd}) = R_4(t^2) + tR_5(t^2), \quad (31)$$

so we have

$$\int \frac{R_1(x)\,dx}{w} = \int \frac{R_4(t^2)\,dt}{\sqrt{(A_1t^2+B_1)(A_2t^2+B_2)}} + \int \frac{R_5(t^2)t\,dt}{\sqrt{(A_1t^2+B_1)(A_2t^2+B_2)}}. \quad (32)$$

Letting

$$u \equiv t^2 \quad (33)$$
$$du = 2t\,dt \quad (34)$$

reduces the second integral to

$$\frac{1}{2}\int \frac{R_5(u)\,du}{\sqrt{(A_1u+B_1)(A_2u+B_2)}}, \quad (35)$$

which can be evaluated using elementary functions. The first integral can then be reduced by INTEGRATION BY PARTS to one of the three Legendre elliptic integrals (also called Legendre-Jacobi ELLIPTIC INTEGRALS), known as incomplete elliptic integrals of the first, second, and third kind, denoted $F(\phi, k)$, $E(\phi, k)$, and $\Pi(n; \phi, k)$, respectively (von Kármán and Biot 1940, Whittaker and Watson 1990, p. 515). If $\phi = \pi/2$, then the integrals are called complete elliptic integrals and are denoted $K(k)$, $E(k)$, $\Pi(n; k)$.

Incomplete elliptic integrals are denoted using a MODULUS k, PARAMETER $m \equiv k^2$, or MODULAR ANGLE $\alpha \equiv \sin^{-1} k$. An elliptic integral is written $I(\phi|m)$ when the PARAMETER is used, $I(\phi, k)$ when the MODULUS is used, and $I(\phi\backslash\alpha)$ when the MODULAR ANGLE is used. Complete elliptic integrals are defined when $\phi = \pi/2$ and can be expressed using the expansion

$$(1-k^2\sin^2\theta)^{-1/2} = \sum_{n=0}^{\infty} \frac{(2n-1)!!}{(2n)!!} k^{2n} \sin^{2n}\theta. \quad (36)$$

An elliptic integral in standard form

$$\int_a^x \frac{dx}{\sqrt{f(x)}}, \quad (37)$$

where

$$f(x) = a_4x^4 + a_3x^3 + a_2x^2 + a_1x + a_0, \quad (38)$$

can be computed analytically (Whittaker and Watson 1990, p. 453) in terms of the WEIERSTRAß ELLIPTIC FUNCTION with invariants

$$g_2 = a_0a_4 - 4a_1a_3 + 3a_2^2 \quad (39)$$
$$g_3 = a_0a_2a_4 - 2a_1a_2a_3 - a_4a_1^2 - a_3^2a_0. \quad (40)$$

If $a \equiv x_0$ is a root of $f(x) = 0$, then the solution is

$$x = x_0 + \tfrac{1}{4}f'(x_0)[\wp(z; g_2, g_3) - \tfrac{1}{24}f''(x_0)]^{-1}. \quad (41)$$

For an arbitrary lower bound,

$$x = a + \frac{\sqrt{f(a)}\wp'(z) + \frac{1}{2}f'(a)[\wp(z) - \frac{1}{24}f''(a)] + \frac{1}{24}f(a)f'''(a)}{2[\wp(z) - \frac{1}{24}f''(a)]^2 - \frac{1}{48}f(a)f^{(a)}(a)}, \quad (42)$$

where $\wp(z) \equiv \wp(z; g_2, g_3)$ is a WEIERSTRAß ELLIPTIC FUNCTION.

A generalized elliptic integral can be defined by the function

$$T(a,b) \equiv \frac{2}{\pi}\int_0^{\pi/2} \frac{d\theta}{\sqrt{a^2\cos^2\theta + b^2\sin^2\theta}} \quad (43)$$
$$= \frac{2}{\pi}\int_0^{\pi/2} \int \frac{d\theta}{\cos\theta\sqrt{a^2 + b^2\tan^2\theta}} \quad (44)$$

(Borwein and Borwein 1987). Now let

$$t \equiv b\tan\theta \quad (45)$$
$$dt = b\sec^2\theta\,d\theta. \quad (46)$$

But

$$\sec\theta = \sqrt{1+\tan^2\theta}, \quad (47)$$

so

$$dt = \frac{b}{\cos\theta}\sec\theta\,d\theta = \frac{b}{\cos\theta}\sqrt{1+\tan^2\theta}\,d\theta = \frac{b}{\cos\theta}\sqrt{1+\left(\frac{t}{b}\right)^2}\,d\theta = \frac{d\theta}{\cos\theta}\sqrt{b^2+t^2}, \quad (48)$$

and

$$\frac{d\theta}{\cos\theta} = \frac{dt}{\sqrt{b^2+t^2}}, \quad (49)$$

and the equation becomes

$$T(a,b) = \frac{2}{\pi}\int_0^\infty \frac{dt}{\sqrt{(a^2+t^2)(b^2+t^2)}} = \frac{1}{\pi}\int_{-\infty}^\infty \frac{dt}{\sqrt{(a^2+t^2)(b^2+t^2)}}. \quad (50)$$

Now we make the further substitution $u \equiv \frac{1}{2}(t - ab/t)$. The differential becomes

$$du = \tfrac{1}{2}(1 + ab/t^2)\,dt, \qquad (51)$$

but $2u = t - ab/t$, so

$$2u/t = 1 - ab/t^2 \qquad (52)$$

$$ab/t^2 = 1 - 2u/t \qquad (53)$$

and

$$1 + ab/t^2 = 2 - 2u/t = 2(1 - u/t). \qquad (54)$$

However, the left side is always positive, so

$$1 + ab/t^2 = 2 - 2u/t = 2|1 - u/t| \qquad (55)$$

and the differential is

$$dt = \frac{du}{\left|1 - \frac{u}{t}\right|}. \qquad (56)$$

We need to take some care with the limits of integration. Write (50) as

$$\int_{-\infty}^{\infty} f(t)\,dt = \int_{-\infty}^{0^-} f(t)\,dt + \int_{0^+}^{\infty} f(t)\,dt. \qquad (57)$$

Now change the limits to those appropriate for the u integration

$$\int_{-\infty}^{\infty} g(u)\,du + \int_{-\infty}^{\infty} g(u)\,du = 2\int_{-\infty}^{\infty} g(u)\,du, \qquad (58)$$

so we have picked up a factor of 2 which must be included. Using this fact and plugging (56) in (50) therefore gives

$$T(a,b) = \frac{2}{\pi}\int_{-\infty}^{\infty} \frac{du}{\left|1 - \frac{u}{t}\right| \sqrt{a^2b^2 + (a^2 + b^2)t^2 + t^4}}. \qquad (59)$$

Now note that

$$u^2 = \frac{t^4 - 2abt^2 + a^2b^2}{4t^2} \qquad (60)$$

$$4u^2t^2 = t^4 - 2abt^2 + 2abt^2 \qquad (61)$$

$$a^2b^2 + t^4 = 4u^2t^2 + 2abt^2. \qquad (62)$$

Plug (62) into (59) to obtain

$$\begin{aligned} T(a,b) &= \frac{2}{\pi}\int_{-\infty}^{\infty} \frac{du}{\left|1 - \frac{u}{t}\right| \sqrt{4u^2t^2 + 2abt^2 + (a^2+b^2)t^2}} \\ &= \frac{2}{\pi}\int_{-\infty}^{\infty} \frac{du}{|t-u|\sqrt{4u^2 + (a+b)^2}}. \end{aligned} \qquad (63)$$

But

$$2ut = t^2 - ab \qquad (64)$$

$$t^2 - 2ut - ab = 0 \qquad (65)$$

$$t = \tfrac{1}{2}(2u \pm \sqrt{4u^2 + 4ab}) = u \pm \sqrt{u^2 + ab}, \qquad (66)$$

so

$$t - u = \pm\sqrt{u^2 + ab}, \qquad (67)$$

and (63) becomes

$$\begin{aligned} T(a,b) &= \frac{2}{\pi}\int_{-\infty}^{\infty} \frac{du}{\sqrt{[4u^2 + (a+b)^2](u^2 + ab)}} \\ &= \frac{1}{\pi}\int_{-\infty}^{\infty} \frac{du}{\sqrt{\left[u^2 + \left(\frac{a+b}{2}\right)^2\right](u^2 + ab)}}. \end{aligned} \qquad (68)$$

We have therefore demonstrated that

$$T(a,b) = T(\tfrac{1}{2}(a+b), \sqrt{ab}). \qquad (69)$$

We can thus iterate

$$a_{i+1} = \tfrac{1}{2}(a_i + b_i) \qquad (70)$$

$$b_{i+1} = \sqrt{a_i b_i}, \qquad (71)$$

as many times as we wish, without changing the value of the integral. But this iteration is the same as and therefore converges to the ARITHMETIC-GEOMETRIC MEAN, so the iteration terminates at $a_i = b_i = M(a_0, b_0)$, and we have

$$\begin{aligned} T(a_0,b_0) &= T(M(a_0,b_0), M(a_0,b_0)) \\ &= \frac{1}{\pi}\int_{-\infty}^{\infty} \frac{dt}{M^2(a_0,b_0) + t^2} \\ &= \frac{1}{\pi M(a_0,b_0)}\left[\tan^{-1}\left(\frac{t}{M(a_0,b_0)}\right)\right]_{-\infty}^{\infty} \\ &= \frac{1}{\pi M(a_0,b_0)}\left[\frac{\pi}{2} - \left(\frac{\pi}{2}\right)\right] \\ &= \frac{1}{M(a_0,b_0)}. \end{aligned} \qquad (72)$$

Complete elliptic integrals arise in finding the arc length of an ELLIPSE and the period of a pendulum. They also arise in a natural way from the theory of THETA FUNCTIONS. Complete elliptic integrals can be computed using a procedure involving the ARITHMETIC-GEOMETRIC MEAN. Note that

$$\begin{aligned} T(a,b) &\equiv \frac{2}{\pi}\int_0^{\pi/2} \frac{d\theta}{\sqrt{a^2\cos^2\theta + b^2\sin^2\theta}} \\ &= \frac{2}{\pi}\int_0^{\pi/2} \frac{d\theta}{a\sqrt{\cos^2\theta + \left(\frac{b}{a}\right)^2\sin^2\theta}} \\ &= \frac{2}{a\pi}\int_0^{\pi/2} \frac{d\theta}{\sqrt{1 - \left(1 - \frac{b^2}{a^2}\right)^2\sin^2\theta}}. \end{aligned} \qquad (73)$$

So we have

$$T(a,b) = \frac{2}{a\pi} K\left(1 - \frac{b^2}{a^2}\right) = \frac{1}{M(a,b)}, \quad (74)$$

where $K(k)$ is the complete ELLIPTIC INTEGRAL OF THE FIRST KIND. We are free to let $a \equiv a_0 \equiv 1$ and $b \equiv b_0 \equiv k'$, so

$$\frac{2}{\pi} K(\sqrt{1 - k'^2}) = \frac{2}{\pi} K(k) = \frac{1}{M(1,k')}, \quad (75)$$

since $k \equiv \sqrt{1 - k'^2}$, so

$$K(k) = \frac{\pi}{2M(1,k')}. \quad (76)$$

But the ARITHMETIC-GEOMETRIC MEAN is defined by

$$a_i = \tfrac{1}{2}(a_{i-1} + b_{i-1}) \quad (77)$$

$$b_i = \sqrt{a_{i-1} b_{i-1}} \quad (78)$$

$$c_i = \begin{cases} \tfrac{1}{2}(a_{i-1} - b_{i-1}) & i > 0 \\ \sqrt{{a_0}^2 - {b_0}^2} & i = 0, \end{cases} \quad (79)$$

where

$$c_{n-1} = \tfrac{1}{2} a_n - b_n = \frac{{c_n}^2}{4a_{n+1}} \leq \frac{{c_n}^2}{4M(a_0, b_0)}, \quad (80)$$

so we have

$$K(k) = \frac{\pi}{2a_N}, \quad (81)$$

where a_N is the value to which a_n converges. Similarly, taking instead $a_0' = 1$ and $b_0' = k$ gives

$$K'(k) = \frac{\pi}{2a_N'}. \quad (82)$$

Borwein and Borwein (1987) also show that defining

$$U(a,b) \equiv \frac{\pi}{2} \int_0^{\pi/2} \sqrt{a^2 \cos^2 + b^2 \sin^2 \theta}\, d\theta = aE'\left(\frac{b}{a}\right) \quad (83)$$

leads to

$$2U(a_{n+1}, b_{n+1}) - U(a_n, b_n) = a_n b_n T(a_n, b_n), \quad (84)$$

so

$$\frac{K(k) - E(k)}{K(k)} = \tfrac{1}{2}({c_0}^2 + 2{c_1}^2 + 2^2 {c_2}^2 + \ldots + 2^n {c_n}^2) \quad (85)$$

for $a_0 \equiv 1$ and $b_0 \equiv k'$, and

$$\frac{K'(k) - E'(k)}{K'(k)} = \tfrac{1}{2}({c_0'}^2 + 2{c_1'}^2 + 2^2 {c_2'}^2 + \ldots + 2^n {c_n'}^2). \quad (86)$$

The elliptic integrals satisfy a large number of identities. The complementary functions and moduli are defined by

$$K'(k) \equiv K(\sqrt{1 - k^2}) = K(k'). \quad (87)$$

Use the identity of generalized elliptic integrals

$$T(a,b) = T(\tfrac{1}{2}(a+b), \sqrt{ab}) \quad (88)$$

to write

$$\begin{aligned} \frac{1}{a} K\left(\sqrt{1 - \frac{b^2}{a^2}}\right) &= \frac{2}{a+b} K\left(\sqrt{1 - \frac{4ab}{(a+b)^2}}\right) \\ &= \frac{2}{a+b} K\left(\sqrt{\frac{a^2 + b^2 - 2ab}{(a+b)^2}}\right) \\ &= \frac{2}{a+b} K\left(\frac{a-b}{a+b}\right) \end{aligned} \quad (89)$$

$$K\left(\sqrt{1 - \frac{b^2}{a^2}}\right) = \frac{2}{1 + \frac{b}{a}} K\left(\frac{1 - \frac{b}{a}}{1 + \frac{b}{a}}\right). \quad (90)$$

Define

$$k' \equiv \frac{b}{a}, \quad (91)$$

and use

$$k \equiv \sqrt{1 - k'^2}, \quad (92)$$

so

$$K(k) = \frac{2}{1 + k'} K\left(\frac{1 - k'}{1 + k'}\right). \quad (93)$$

Now letting $l \equiv (1 - k')/(1 + k')$ gives

$$l(1 + k') = 1 - k' \Rightarrow k'(l + 1) = 1 - l \quad (94)$$

$$k' = \frac{1 - l}{1 + l} \quad (95)$$

$$\begin{aligned} k &= \sqrt{1 - k'^2} = \sqrt{1 - \left(\frac{1-l}{1+l}\right)^2} \\ &= \sqrt{\frac{(1+l)^2 - (1-l)^2}{(1+l)^2}} = \frac{2\sqrt{l}}{1+l}, \end{aligned} \quad (96)$$

and

$$\begin{aligned} \tfrac{1}{2}(1 + k') &= \frac{1}{2}\left(1 + \frac{1-l}{1+l}\right) = \frac{1}{2}\left[\frac{(1+l) + (1-l)}{1+l}\right] \\ &= \frac{1}{1+l}. \end{aligned} \quad (97)$$

Writing k instead of l,

$$K(k) = \frac{1}{k+1} K\left(\frac{2\sqrt{k}}{1+k}\right). \quad (98)$$

Similarly, from Borwein and Borwein (1987),

$$E(k) = \frac{1+k}{2} E\left(\frac{2\sqrt{k}}{1+k}\right) + \frac{k'^2}{2} K(k) \tag{99}$$

$$E(k) = (1+k')E\left(\frac{1-k'}{1+k'}\right) - k'K(k). \tag{100}$$

Expressions in terms of the complementary function can be derived from interchanging the moduli and their complements in (93), (98), (99), and (100).

$$\begin{aligned} K'(k) = K(k') &= \frac{2}{1+k} K\left(\frac{1-k}{1+k}\right) \\ &= \frac{2}{1+k} K'\left(\sqrt{1-\left(\frac{1-k}{1+k}\right)^2}\right) \\ &= \frac{2}{1+k} K'\left(\frac{2\sqrt{k}}{1+k}\right) \end{aligned} \tag{101}$$

$$K'(k) = \frac{1}{1+k'} K\left(\frac{2\sqrt{k'}}{1+k'}\right) = \frac{1}{1+k'} K'\left(\frac{1-k'}{1+k'}\right), \tag{102}$$

and

$$E'(k) = (1+k)E'\left(\frac{2\sqrt{k}}{1+k}\right) - kK'(k) \tag{103}$$

$$E'(k) = \left(\frac{1+k'}{2}\right) E'\left(\frac{1-k'}{1+k'}\right) + \frac{k^2}{2} K'(k). \tag{104}$$

Taking the ratios

$$\frac{K'(k)}{K(k)} = 2 \frac{K'\left(\frac{2\sqrt{k}}{1+k}\right)}{K\left(\frac{2\sqrt{k}}{1+k}\right)} = \frac{1}{2} \frac{K'\left(\frac{1-k'}{1+k'}\right)}{K\left(\frac{1-k'}{1+k'}\right)} \tag{105}$$

gives the MODULAR EQUATION of degree 2. It is also true that

$$K(x) = \frac{4}{(1+\sqrt{x'})^2} K\left(\left[\frac{1-\sqrt[4]{1-x^4}}{1+\sqrt[4]{1-x^4}}\right]^2\right). \tag{106}$$

see also ABELIAN INTEGRAL, AMPLITUDE, ARGUMENT (ELLIPTIC INTEGRAL), CHARACTERISTIC (ELLIPTIC INTEGRAL), DELTA AMPLITUDE, ELLIPTIC FUNCTION, ELLIPTIC INTEGRAL OF THE FIRST KIND, ELLIPTIC INTEGRAL OF THE SECOND KIND, ELLIPTIC INTEGRAL OF THE THIRD KIND, ELLIPTIC INTEGRAL SINGULAR VALUE, HEUMAN LAMBDA FUNCTION, JACOBI ZETA FUNCTION, MODULAR ANGLE, MODULUS (ELLIPTIC INTEGRAL), NOME, PARAMETER

References

Abramowitz, M. and Stegun, C. A. (Eds.). "Elliptic Integrals." Ch. 17 in *Handbook of Mathematical Functions with Formulas, Graphs, and Mathematical Tables, 9th printing.* New York: Dover, pp. 587–607, 1972.

Arfken, G. "Elliptic Integrals." §5.8 in *Mathematical Methods for Physicists, 3rd ed.* Orlando, FL: Academic Press, pp. 321–327, 1985.

Borwein, J. M. and Borwein, P. B. *Pi & the AGM: A Study in Analytic Number Theory and Computational Complexity.* New York: Wiley, 1987.

Hancock, H. *Elliptic Integrals.* New York: Wiley, 1917.

King, L. V. *The Direct Numerical Calculation of Elliptic Functions and Integrals.* London: Cambridge University Press, 1924.

Press, W. H.; Flannery, B. P.; Teukolsky, S. A.; and Vetterling, W. T. "Elliptic Integrals and Jacobi Elliptic Functions." §6.11 in *Numerical Recipes in FORTRAN: The Art of Scientific Computing, 2nd ed.* Cambridge, England: Cambridge University Press, pp. 254–263, 1992.

Prudnikov, A. P.; Brychkov, Yu. A.; and Marichev, O. I. *Integrals and Series, Vol. 1: Elementary Functions.* New York: Gordon & Breach, 1986.

Timofeev, A. F. *Integration of Functions.* Moscow and Leningrad: GTTI, 1948.

von Kármán, T. and Biot, M. A. *Mathematical Methods in Engineering: An Introduction to the Mathematical Treatment of Engineering Problems.* New York: McGraw-Hill, p. 121, 1940.

Whittaker, E. T. and Watson, G. N. *A Course in Modern Analysis, 4th ed.* Cambridge, England: Cambridge University Press, 1990.

Elliptic Integral of the First Kind

Let the MODULUS k satisfy $0 < k^2 < 1$. (This may also be written in terms of the PARAMETER $m \equiv k^2$ or MODULAR ANGLE $\alpha \equiv \sin^{-1} k$.) The incomplete elliptic integral of the first kind is then defined as

$$F(\phi, k) = \int_0^\phi \frac{d\theta}{\sqrt{1-k^2\sin^2\theta}}. \tag{1}$$

Let

$$t \equiv \sin\theta \tag{2}$$

$$dt = \cos\theta\, d\theta = \sqrt{1-t^2}\, d\theta \tag{3}$$

$$\begin{aligned} F(\phi, k) &= \int_0^{\sin\phi} \frac{1}{\sqrt{1-k^2t^2}} \frac{dt}{\sqrt{1-t^2}} \\ &= \int_0^{\sin\phi} \frac{dt}{\sqrt{(1-k^2t^2)(1-t^2)}}. \end{aligned} \tag{4}$$

Let

$$v \equiv \tan\theta \tag{5}$$

$$dv \equiv \sec^2\theta\, d\theta = (1+v^2)\, d\theta, \tag{6}$$

so the integral can also be written as

$$\begin{aligned} F(\phi, k) &= \int_0^{\tan\phi} \frac{1}{\sqrt{1-k^2\frac{v^2}{1+u^2}}} \frac{du}{1+v^2} \\ &= \int_0^{\tan\phi} \frac{dv}{\sqrt{1+v^2}\sqrt{(1+v^2)-k^2v^2}} \qquad (7) \\ &= \int_0^{\tan\phi} \frac{dv}{\sqrt{(1+v^2)(1+k'v^2)}}, \qquad (8) \end{aligned}$$

where $k'^2 \equiv 1 - k^2$ is the complementary MODULUS. The integral

$$I = \frac{1}{\sqrt{2}} \int_0^{\theta_0} \frac{d\theta}{\sqrt{\cos\theta - \cos\theta_0}}, \tag{9}$$

which arises in computing the period of a pendulum, is also an elliptic integral of the first kind. Use

$$\cos\theta = 1 - 2\sin^2(\tfrac{1}{2}\theta) \tag{10}$$

$$\sin(\tfrac{1}{2}\theta) = \sqrt{\frac{1-\cos\theta}{2}} \tag{11}$$

to write

$$\begin{aligned}\sqrt{\cos\theta - \cos\theta_0} &= \sqrt{1 - 2\sin^2(\tfrac{1}{2}\theta) - \cos\theta_0} \\ &= \sqrt{1-\cos\theta_0}\sqrt{1 - \frac{2}{1-\cos\theta_0}\sin^2(\tfrac{1}{2}\theta)} \\ &= \sqrt{2}\sin(\tfrac{1}{2}\theta_0)\sqrt{1 - \csc^2(\tfrac{1}{2}\theta_0)\sin^2(\tfrac{1}{2}\theta)},\end{aligned} \tag{12}$$

so

$$I = \frac{1}{2}\int_0^{\theta_0} \frac{d\theta}{\sin(\tfrac{1}{2}\theta_0)\sqrt{1-\csc^2(\tfrac{1}{2}\theta_0)\sin^2(\tfrac{1}{2}\theta)}}. \tag{13}$$

Now let

$$\sin(\tfrac{1}{2}\theta) = \sin(\tfrac{1}{2}\theta_0)\sin\phi, \tag{14}$$

so the angle θ is transformed to

$$\phi = \sin^{-1}\left(\frac{\sin\theta}{\theta_0}\right), \tag{15}$$

which ranges from 0 to $\pi/2$ as θ varies from 0 to θ_0. Taking the differential gives

$$\tfrac{1}{2}\cos(\tfrac{1}{2}\theta)\,d\theta = \sin(\tfrac{1}{2}\theta_0)\cos\phi\,d\phi, \tag{16}$$

or

$$\tfrac{1}{2}\sqrt{1-\sin^2(\tfrac{1}{2}\theta_0)\sin^2\phi}\,d\theta = \sin(\tfrac{1}{2}\theta_0)\cos\phi\,d\phi. \tag{17}$$

Plugging this in gives

$$\begin{aligned}I &= \int_0^{\pi/2} \frac{1}{\sqrt{1-\sin^2(\tfrac{1}{2}\theta_0)\sin^2\phi}} \frac{\sin(\tfrac{1}{2}\theta_0)\cos\phi\,d\phi}{\sin(\tfrac{1}{2}\theta_0)\sqrt{1-\sin^2\phi}} \\ &= \int_0^{\pi/2} \frac{d\phi}{\sqrt{1-\sin^2(\tfrac{1}{2}\theta_0)\sin^2\phi}} = K(\sin(\tfrac{1}{2}\theta_0)),\end{aligned} \tag{18}$$

so

$$I = \frac{1}{\sqrt{2}}\int_0^{\theta_0} \frac{d\theta}{\sqrt{\cos\theta - \cos\theta_0}} = K(\sin(\tfrac{1}{2}\theta_0)). \tag{19}$$

Making the slightly different substitution $\phi = \theta/2$, so $d\theta = 2\,d\phi$ leads to an equivalent, but more complicated expression involving an *incomplete* elliptic function of the first kind,

$$\begin{aligned}I &= 2\frac{1}{\sqrt{2}}\frac{1}{\sqrt{2}}\csc(\tfrac{1}{2}\theta_0)\int_0^{\theta_0} \frac{d\phi}{\sqrt{1-\csc^2(\tfrac{1}{2}\theta_0)\sin^2\phi}} \\ &= \csc(\tfrac{1}{2}\theta_0)F(\tfrac{1}{2}\theta_0, \csc(\tfrac{1}{2}\theta_0)).\end{aligned} \tag{20}$$

Therefore, we have proven the identity

$$\csc x F(x, \csc x) = K(\sin x). \tag{21}$$

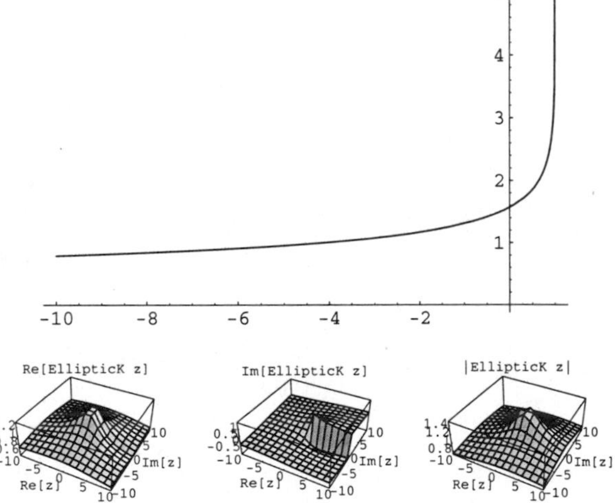

The complete elliptic integral of the first kind, illustrated above as a function of $m = k^2$, is defined by

$$\begin{aligned}K(k) &\equiv F(\tfrac{1}{2}\pi, k) && (22)\\ &= \sum_{n=0}^\infty \frac{(2n-1)!!}{(2n)!!}k^{2n}\int_0^{2\pi}\sin^{2n}\theta\,d\theta && (23)\\ &= \tfrac{1}{2}\pi{\vartheta_3}^2(q) && (24)\\ &= \sum_{n=0}^\infty \frac{(2n-1)!!}{(2n)!!}k^{2n}\frac{\pi}{2}\frac{(2n-1)!!}{(2n)!!} \\ &= \frac{\pi}{2}\sum_{n=0}^\infty\left[\frac{(2n-1)!!}{(2n)!!}\right]^2 k^{2n} && (25)\\ &= \tfrac{1}{2}\pi\,{}_2F_1(\tfrac{1}{2}, \tfrac{1}{2}, 1; k^2) && (26)\\ &= \frac{\pi}{2\sqrt{1-k^2}}P_{-1/2}\left(\frac{1+k^2}{1-k^2}\right), && (27)\end{aligned}$$

where

$$q = e^{-\pi K'(k)/K(k)} \tag{28}$$

is the NOME (for $|q| < 1$), ${}_2F_1(a, b; c; x)$ is the HYPERGEOMETRIC FUNCTION, and $P_n(x)$ is a LEGENDRE POLYNOMIAL. $K(k)$ satisfies the LEGENDRE RELATION

$$E(k)K'(k) + E'(k)K(k) - K(k)K'(k) = \tfrac{1}{2}\pi, \tag{29}$$

where $E(k)$ and $K(k)$ are complete elliptic integrals of the first and SECOND KINDS, and $E'(k)$ and $K'(k)$ are the complementary integrals. The modulus k is often suppressed for conciseness, so that $E(k)$ and $K(k)$ are often simply written E and K, respectively.

The DERIVATIVE of $K(k)$ is

$$\frac{dK}{dk} \equiv \int_0^1 \frac{dt}{\sqrt{(1-t^2)(1-k'^2t^2)}} = \frac{E(k)}{k(1-k^2)} - \frac{K(k)}{k} \tag{30}$$

$$\frac{d}{dk}\left(kk'^2\frac{dK}{dk}\right) = kK, \tag{31}$$

so

$$E = k(1-k^2)\left(\frac{dK}{dk}+\frac{K}{k}\right) = (1-k^2)\left(k\frac{dK}{dk}+k\right) \tag{32}$$

(Whittaker and Watson 1990, pp. 499 and 521).

see also AMPLITUDE, CHARACTERISTIC (ELLIPTIC INTEGRAL), ELLIPTIC INTEGRAL SINGULAR VALUE, GAUSS'S TRANSFORMATION, LANDEN'S TRANSFORMATION, LEGENDRE RELATION, MODULAR ANGLE, MODULUS (ELLIPTIC INTEGRAL), PARAMETER

References

Abramowitz, M. and Stegun, C. A. (Eds.). "Elliptic Integrals." Ch. 17 in *Handbook of Mathematical Functions with Formulas, Graphs, and Mathematical Tables, 9th printing.* New York: Dover, pp. 587–607, 1972.

Spanier, J. and Oldham, K. B. "The Complete Elliptic Integrals $K(p)$ and $E(p)$" and "The Incomplete Elliptic Integrals $F(p;\phi)$ and $E(p;\phi)$." Chs. 61–62 in *An Atlas of Functions.* Washington, DC: Hemisphere, pp. 609–633, 1987.

Whittaker, E. T. and Watson, G. N. *A Course in Modern Analysis, 4th ed.* Cambridge, England: Cambridge University Press, 1990.

Elliptic Integral of the Second Kind

Let the MODULUS k satisfy $0 < k^2 < 1$. (This may also be written in terms of the PARAMETER $m \equiv k^2$ or MODULAR ANGLE $\alpha \equiv \sin^{-1} k$.) The incomplete elliptic integral of the second kind is then defined as

$$E(\phi,k) \equiv \int_0^\phi \sqrt{1-k^2\sin^2\theta}\,d\theta. \tag{1}$$

A generalization replacing $\sin\theta$ with $\sinh\theta$ gives

$$-iE(i\phi,-k) = \int_0^\phi \sqrt{1-k^2\sinh^2\theta}\,d\theta. \tag{2}$$

To place the elliptic integral of the second kind in a slightly different form, let

$$t \equiv \sin\theta \tag{3}$$

$$dt = \cos\theta\,d\theta = \sqrt{1-t^2}\,d\theta, \tag{4}$$

so the elliptic integral can also be written as

$$\begin{aligned} E(\phi,k) &= \int_0^{\sin\phi} \sqrt{1-k^2t^2}\,\frac{dt}{\sqrt{1-t^2}} \\ &= \int_0^{\sin\phi} \sqrt{\frac{1-k^2t^2}{1-t^2}}\,dt. \end{aligned} \tag{5}$$

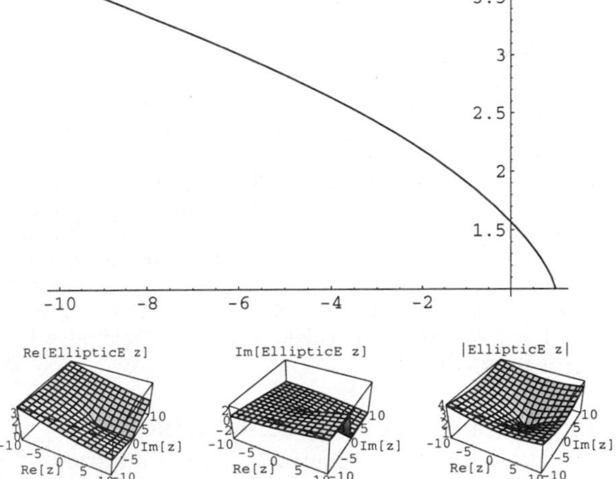

The complete elliptic integral of the second kind, illustrated above as a function of the PARAMETER m, is defined by

$$\begin{aligned} E(k) &\equiv E(\tfrac{1}{2}\pi,k) &(6)\\ &= \frac{\pi}{2}\left\{1-\sum_{n=1}^{\infty}\left[\frac{(2n-1)!!}{(2n)!!)}\right]^2\frac{k^{2n}}{2n-1}\right\} &(7)\\ &= \tfrac{1}{2}\pi\,{}_2F_1(-\tfrac{1}{2},\tfrac{1}{2},1;k^2) &(8)\\ &= \int_0^K \operatorname{dn}^2 u\,du, &(9) \end{aligned}$$

where ${}_2F_1(a,b;c;x)$ is the HYPERGEOMETRIC FUNCTION and $\operatorname{dn} u$ is a JACOBI ELLIPTIC FUNCTION. The complete elliptic integral of the second kind satisfies the LEGENDRE RELATION

$$E(k)K'(k) + E'(k)K(k) - K(k)K'(k) = \tfrac{1}{2}\pi, \tag{10}$$

where E and K are complete ELLIPTIC INTEGRALS OF THE FIRST and second kinds, and E' and K' are the complementary integrals. The DERIVATIVE is

$$\frac{dE}{dk} = \frac{E(k)-K(k)}{k} \tag{11}$$

(Whittaker and Watson 1990, p. 521). If k_r is a singular value (i.e.,

$$k_r = \lambda^*(r), \tag{12}$$

where λ^* is the ELLIPTIC LAMBDA FUNCTION), and $K(k_r)$ and the ELLIPTIC ALPHA FUNCTION $\alpha(r)$ are also known, then

$$E(k) = \frac{K(k)}{\sqrt{r}}\left[\frac{\pi}{3[K(k)]^2} - \alpha(r)\right] + K(k). \tag{13}$$

see also ELLIPTIC INTEGRAL OF THE FIRST KIND, ELLIPTIC INTEGRAL OF THE THIRD KIND, ELLIPTIC INTEGRAL SINGULAR VALUE

References

Abramowitz, M. and Stegun, C. A. (Eds.). "Elliptic Integrals." Ch. 17 in *Handbook of Mathematical Functions with Formulas, Graphs, and Mathematical Tables, 9th printing.* New York: Dover, pp. 587–607, 1972.

Spanier, J. and Oldham, K. B. "The Complete Elliptic Integrals $K(p)$ and $E(p)$" and "The Incomplete Elliptic Integrals $F(p;\phi)$ and $E(p;\phi)$." Chs. 61 and 62 in *An Atlas of Functions.* Washington, DC: Hemisphere, pp. 609–633, 1987.

Whittaker, E. T. and Watson, G. N. *A Course in Modern Analysis, 4th ed.* Cambridge, England: Cambridge University Press, 1990.

Elliptic Integral of the Third Kind

Let $0 < k^2 < 1$. The incomplete elliptic integral of the third kind is then defined as

$$\Pi(n;\phi,k) = \int_0^\phi \frac{d\theta}{(1-n\sin^2\theta)\sqrt{1-k^2\sin^2\theta}} \tag{1}$$

$$= \int_0^{\sin\phi} \frac{dt}{(1-nt^2)\sqrt{(1-t^2)(1-k^2t^2)}}, \tag{2}$$

where n is a constant known as the CHARACTERISTIC.

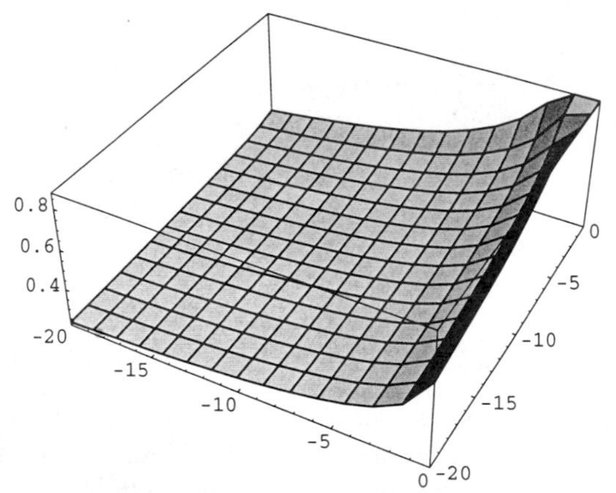

The complete elliptic integral of the second kind

$$\Pi(n|m) = \Pi(n; \tfrac{1}{2}\pi|m) \tag{3}$$

is illustrated above.

see also ELLIPTIC INTEGRAL OF THE FIRST KIND, ELLIPTIC INTEGRAL OF THE SECOND KIND, ELLIPTIC INTEGRAL SINGULAR VALUE

References

Abramowitz, M. and Stegun, C. A. (Eds.). "Elliptic Integrals" and "Elliptic Integrals of the Third Kind." Ch. 17 and §17.7 in *Handbook of Mathematical Functions with Formulas, Graphs, and Mathematical Tables, 9th printing.* New York: Dover, pp. 587–607, 1972.

Elliptic Integral Singular Value

When the MODULUS k has a singular value, the complete elliptic integrals may be computed in analytic form in terms of GAMMA FUNCTIONS. Abel (quoted in Whittaker and Watson 1990, p. 525) proved that whenever

$$\frac{K'(k)}{K(k)} = \frac{a+b\sqrt{n}}{c+d\sqrt{n}}, \tag{1}$$

where a, b, c, d, and n are INTEGERS, $K(k)$ is a complete ELLIPTIC INTEGRAL OF THE FIRST KIND, and $K'(k) \equiv K(\sqrt{1-k^2})$ is the complementary complete ELLIPTIC INTEGRAL OF THE FIRST KIND, then the MODULUS k is the ROOT of an algebraic equation with INTEGER COEFFICIENTS.

A MODULUS k_r such that

$$\frac{K'(k_r)}{K(k_r)} = \sqrt{r}, \tag{2}$$

is called a singular value of the elliptic integral. The ELLIPTIC LAMBDA FUNCTION $\lambda^*(r)$ gives the value of k_r. Selberg and Chowla (1967) showed that $K(\lambda^*(r))$ and $E(\lambda^*(r))$ are expressible in terms of a finite number of GAMMA FUNCTIONS. The complete ELLIPTIC INTEGRALS OF THE SECOND KIND $E(k_r)$ and $E'(k_r)$ can be expressed in terms of $K(k_r)$ and $K'(k_r)$ with the aid of the ELLIPTIC ALPHA FUNCTION $\alpha(r)$.

The following table gives the values of $K(k_r)$ for small integral r in terms of GAMMA FUNCTIONS.

$$K(k_1) = \frac{\Gamma^2(\frac{1}{4})}{4\sqrt{\pi}}$$

$$K(k_2) = \frac{\sqrt{\sqrt{2}+1}\,\Gamma(\frac{1}{8})\Gamma(\frac{3}{8})}{2^{13/4}\sqrt{\pi}}$$

$$K(k_3) = \frac{3^{1/4}\Gamma^3(\frac{1}{3})}{2^{7/3}\pi}$$

$$K(k_4) = \frac{(\sqrt{2}+1)\Gamma^2(\frac{1}{4})}{2^{7/2}\sqrt{\pi}}$$

$$K(k_5) = (\sqrt{5}+2)^{1/4}\sqrt{\frac{\Gamma(\frac{1}{20})\Gamma(\frac{3}{20})\Gamma(\frac{7}{20})\Gamma(\frac{9}{20})}{160\pi}}$$

$$K(k_6) = \sqrt{(\sqrt{2}-1)(\sqrt{3}+\sqrt{2})(2+\sqrt{3})} \times \sqrt{\frac{\Gamma(\frac{1}{24})\Gamma(\frac{5}{24})\Gamma(\frac{7}{24})\Gamma(\frac{11}{24})}{384\pi}}$$

$$K(k_7) = \frac{\Gamma(\frac{1}{7})\Gamma(\frac{2}{7})\Gamma(\frac{4}{7})}{7^{1/4}\cdot 4\pi}$$

$$K(k_8) = \sqrt{\frac{2\sqrt{2}+\sqrt{1+5\sqrt{2}}}{4\sqrt{2}}}\,\frac{(\sqrt{2}+1)^{1/4}\Gamma(\frac{1}{8})\Gamma(\frac{3}{8})}{8\sqrt{\pi}}$$

$$K(k_9) = \frac{3^{1/4}\sqrt{2+\sqrt{3}}}{12\sqrt{\pi}\Gamma^2(\frac{1}{4})}$$

$$K(k_{10}) = \sqrt{(2+3\sqrt{2}+\sqrt{5})} \times \sqrt{\frac{\Gamma(\frac{1}{40})\Gamma(\frac{7}{40})\Gamma(\frac{9}{40})\Gamma(\frac{11}{40})\Gamma(\frac{13}{40})\Gamma(\frac{19}{40})\Gamma(\frac{23}{40})\Gamma(\frac{37}{40})}{2560\pi^3}}$$

$$K(k_{11}) = [2+(17+3\sqrt{33})^{1/3}-(3\sqrt{33}-17)^{1/3}]^2 \times \frac{\Gamma(\frac{1}{11})\Gamma(\frac{3}{11})\Gamma(\frac{4}{11})\Gamma(\frac{5}{11})\Gamma(\frac{9}{11})}{11^{1/4}144\pi^2}$$

$$K(k_{12}) = \frac{3^{1/4}(\sqrt{2}+1)(\sqrt{3}+\sqrt{2})\sqrt{2-\sqrt{3}}\Gamma^3(\frac{1}{3})}{2^{13/3}\pi}$$

$$K(k_{13}) = \frac{(18+5\sqrt{13})^{1/4}}{\sqrt{6656\pi^5}} \times \sqrt{\Gamma(\tfrac{1}{52})\Gamma(\tfrac{7}{52})\Gamma(\tfrac{9}{52})\Gamma(\tfrac{11}{52})\Gamma(\tfrac{15}{52})\Gamma(\tfrac{17}{52})} \times \sqrt{\Gamma(\tfrac{19}{52})\Gamma(\tfrac{25}{52})\Gamma(\tfrac{29}{52})\Gamma(\tfrac{31}{52})\Gamma(\tfrac{47}{52})\Gamma(\tfrac{49}{52})}$$

$$K(k_{15}) = \sqrt{\frac{(\sqrt{5}+1)\Gamma(\frac{1}{15})\Gamma(\frac{2}{15})\Gamma(\frac{4}{15})\Gamma(\frac{8}{15})}{240\pi}}$$

$$K(k_{16}) = \frac{(2^{1/4}+1)^2\Gamma^2(\frac{1}{4})}{2^{9/2}\sqrt{\pi}}$$

$$K(k_{17}) = C_1\left[\frac{\Gamma(\frac{1}{68})\Gamma(\frac{3}{68})\Gamma(\frac{7}{68})\Gamma(\frac{11}{68})\Gamma(\frac{13}{68})}{\Gamma(\frac{5}{68})\Gamma(\frac{15}{68})\Gamma(\frac{19}{68})\Gamma(\frac{29}{68})}\right]^{1/4} \times [\Gamma(\tfrac{21}{68})\Gamma(\tfrac{25}{68})\Gamma(\tfrac{27}{68})\Gamma(\tfrac{31}{68})\Gamma(\tfrac{33}{68})]^{1/4}$$

$$K(k_{25}) = \frac{\sqrt{5}+2}{20}\frac{\Gamma^2(\frac{1}{4})}{\sqrt{\pi}},$$

where $\Gamma(z)$ is the GAMMA FUNCTION and C_1 is an algebraic number (Borwein and Borwein 1987, p. 298).

Borwein and Zucker (1992) give amazing expressions for singular values of complete elliptic integrals in terms of CENTRAL BETA FUNCTIONS

$$\beta(p) \equiv B(p,p). \tag{3}$$

Furthermore, they show that $K(k_n)$ is *always* expressible in terms of these functions for $n \equiv 1, 2 \pmod 4$. In such cases, the Γ functions appearing in the expression are of the form $\Gamma(t/4n)$ where $1 \leq t \leq (2n-1)$ and $(t, 4n) = 1$. The terms in the numerator depend on the sign of the KRONECKER SYMBOL $\{t/4n\}$. Values for the first few n are

$$K(k_1) = 2^{-2}\beta(\tfrac{1}{4})$$

$$K(k_2) = 2^{-13/4}\beta(\tfrac{1}{8})$$

$$K(k_3) = 2^{-4/3}3^{-1/4}\beta(\tfrac{1}{3}) = 2^{-5/3}3^{-3/4}\beta(\tfrac{1}{6})$$

$$K(k_5) = 2^{-33/20}5^{-5/8}(11+5\sqrt{5})^{1/4}\sin(\tfrac{1}{20}\pi)\beta(\tfrac{1}{2}) = 2^{-29/20}5^{-3/8}(1+\sqrt{5})^{1/4}\sin(\tfrac{3}{20}\pi)\beta(\tfrac{3}{20})$$

$$K(k_6) = 2^{-47/12}3^{-3/4}(\sqrt{2}-1)(\sqrt{3}+1)\beta(\tfrac{1}{24}) = 2^{-43/12}3^{-1/4}(\sqrt{3}-1)\beta(\tfrac{5}{24})$$

$$K(k_7) = 2\cdot 7^{-3/4}\sin(\tfrac{1}{7}\pi)\sin(\tfrac{2}{7}\pi)B(\tfrac{1}{7},\tfrac{2}{7}) = 2^{-2/7}7^{-1/4}\frac{\beta(\frac{1}{7})\beta(\frac{2}{7})}{\beta(\frac{1}{14})}$$

$$K(k_{10}) = 2^{-61/20}5^{-1/4}(\sqrt{5}-2)^{1/2}(\sqrt{10}+3)\frac{\beta(\frac{1}{8})\beta(\frac{7}{40})}{\beta(\frac{1}{3}40)} = 2^{-15/4}5^{-3/4}(\sqrt{5}-2)^{1/2}\frac{\beta(\frac{1}{40})\beta(\frac{1}{9}40)}{\beta(\frac{3}{8})}$$

$$K(k_{11}) = R\cdot 2^{-7/11}\sin(\tfrac{1}{11}\pi)\sin(\tfrac{3}{11}\pi)B(\tfrac{1}{22},\tfrac{3}{22})$$

$$K(k_{13}) = 2^{-3}13^{-5/8}(5\sqrt{13}+18)^{1/4} \times [\tan(\tfrac{1}{52}\pi)\tan(\tfrac{3}{52}\pi)\tan(\tfrac{9}{52}\pi)]^{1/2}\frac{\beta(\frac{1}{52})\beta(\frac{9}{52})}{\beta(\frac{23}{52})}$$

$$K(k_{14}) = \sqrt{\sqrt{4\sqrt{2}+2}+\sqrt{2}+\sqrt{2\sqrt{2}-1}}\cdot 2^{-13/4}7^{-3/8} \times \left[\frac{\tan(\frac{5}{56}\pi)\tan(\frac{13}{56}\pi)}{\tan(\frac{11}{56}\pi)}\right]^{1/4}\sqrt{\frac{\beta(\frac{5}{56})\beta(\frac{13}{56})\beta(\frac{1}{8})}{\beta(\frac{11}{56})}}$$

$$K(k_{15}) = 2^{-1}3^{-3/4}5^{-7/12}B(\tfrac{1}{15},\tfrac{4}{15}) = \frac{2^{-2}3^{-3/4}5^{-3/4}(\sqrt{5}-1)\beta(\frac{1}{15})\beta(\frac{4}{15})}{\beta(\frac{1}{3})}$$

$$K(k_{17}) = C_2\left[\frac{\beta(\frac{1}{68})\beta(\frac{3}{68}(\beta(\frac{7}{68})\beta(\frac{9}{68})\beta(\frac{11}{68})\beta(\frac{13}{68})}{\beta(\frac{5}{68})\beta(\frac{15}{68})}\right]^{1/4},$$

where R is the REAL ROOT of

$$x^3 - 4x = 4 = 0 \tag{4}$$

and C_2 is an algebraic number (Borwein and Zucker 1992). Note that $K(k_{11})$ is the only value in the above list which cannot be expressed in terms of CENTRAL BETA FUNCTIONS.

Using the ELLIPTIC ALPHA FUNCTION, the ELLIPTIC INTEGRALS OF THE SECOND KIND can also be found from

$$E = \frac{\pi}{4\sqrt{r}K} + \left[1 - \frac{\alpha(r)}{\sqrt{r}}\right]K \tag{5}$$

$$E' = \frac{\pi}{4K} + \alpha(r)K, \tag{6}$$

and by definition,

$$K' = K\sqrt{n}. \tag{7}$$

see also CENTRAL BETA FUNCTION, ELLIPTIC ALPHA FUNCTION, ELLIPTIC DELTA FUNCTION, ELLIPTIC INTEGRAL OF THE FIRST KIND, ELLIPTIC INTEGRAL OF THE SECOND KIND, ELLIPTIC LAMBDA FUNCTION, GAMMA FUNCTION, MODULUS (ELLIPTIC INTEGRAL)

References

Abel, N. *J. für Math* **3**, 184, 1881. Reprinted in Abel, N. H. *Oeuvres Completes* (Ed. L. Sylow and S. Lie). New York: Johnson Reprint Corp., p. 377, 1988.

Borwein, J. M. and Borwein, P. B. *Pi & the AGM: A Study in Analytic Number Theory and Computational Complexity.* New York: Wiley, pp. 139 and 298, 1987.

Borwein, J. M. and Zucker, I. J. "Elliptic Integral Evaluation of the Gamma Function at Rational Values of Small

Denominator." *IMA J. Numerical Analysis* **12**, 519–526, 1992.

Bowman, F. *Introduction to Elliptic Functions, with Applications.* New York: Dover, pp. 75, 95, and 98, 1961.

Glasser, M. L. and Wood, V. E. "A Closed Form Evaluation of the Elliptic Integral." *Math. Comput.* **22**, 535–536, 1971.

Selberg, A. and Chowla, S. "On Epstein's Zeta-Function." *J. Reine. Angew. Math.* **227**, 86–110, 1967.

Weisstein, E. W. "Elliptic Singular Values." `http://www.astro.virginia.edu/~eww6n/math/notebooks/EllipticSingular.m`.

Whittaker, E. T. and Watson, G. N. *A Course in Modern Analysis, 4th ed.* Cambridge, England: Cambridge University Press, pp. 524–528, 1990.

Wrigge, S. "An Elliptic Integral Identity." *Math. Comput.* **27**, 837–840, 1973.

Zucker, I. J. "The Evaluation in Terms of Γ-Functions of the Periods of Elliptic Curves Admitting Complex Multiplication." *Math. Proc. Cambridge Phil. Soc.* **82**, 111–118, 1977.

Elliptic Integral Singular Value—k_1

The first SINGULAR VALUE k_1, corresponding to

$$K'(k_1) = K(k_1), \tag{1}$$

is given by

$$k_1 = \frac{1}{\sqrt{2}} \tag{2}$$

$$k_1' = \frac{1}{\sqrt{2}}. \tag{3}$$

As shown in LEMNISCATE FUNCTION,

$$K\left(\frac{1}{\sqrt{2}}\right) \equiv \int_0^1 \frac{dt}{\sqrt{(1-t^2)\left(1-\frac{1}{2}t^2\right)}} = \sqrt{2}\int_0^1 \frac{dt}{\sqrt{1-t^4}}. \tag{4}$$

Let

$$u \equiv t^4 \tag{5}$$

$$du = 4t^3\,dt = 4u^{3/4}\,dt \tag{6}$$

$$dt = \tfrac{1}{4}u^{-3/4}\,du, \tag{7}$$

then

$$K\left(\frac{1}{\sqrt{2}}\right) = \frac{\sqrt{2}}{4}\int_0^1 u^{-3/4}(1-u)^{-1/2}\,du = \frac{\sqrt{2}}{4}B(\tfrac{1}{4},\tfrac{1}{2}) = \frac{\Gamma(\frac{1}{4})\Gamma(\frac{1}{2})}{\Gamma(\frac{3}{4})}\frac{\sqrt{2}}{4}, \tag{8}$$

where $B(a,b)$ is the BETA FUNCTION and $\Gamma(z)$ is the GAMMA FUNCTION. Now use

$$\Gamma(\tfrac{1}{2}) = \sqrt{\pi} \tag{9}$$

and

$$\frac{1}{\Gamma(1-x)} = \frac{\sin(\pi x)}{\pi}\Gamma(x), \tag{10}$$

so

$$\frac{1}{\Gamma\left(\frac{3}{4}\right)} = \frac{1}{\Gamma\left(1-\frac{1}{4}\right)} = \frac{\sin\left(\frac{\pi}{4}\right)}{\pi}\Gamma(\tfrac{1}{4}) = \frac{1}{\pi\sqrt{2}}\Gamma(\tfrac{1}{4}). \tag{11}$$

Therefore,

$$K\left(\frac{1}{\sqrt{2}}\right) = \frac{\Gamma^2(\frac{1}{4})\sqrt{\pi}\sqrt{2}}{4\pi\sqrt{2}} = \frac{\Gamma^2(\frac{1}{4})}{4\sqrt{\pi}}. \tag{12}$$

Now consider

$$E\left(\frac{1}{\sqrt{2}}\right) \equiv \int_0^1 \sqrt{\frac{1-\frac{1}{2}t^2}{1-t^2}}\,dt. \tag{13}$$

Let

$$t^2 \equiv 1-u^2 \tag{14}$$

$$2t\,dt = -2u\,du \tag{15}$$

$$dt = -\frac{1}{t}u\,du = u(1-u^2)^{-1/2}\,du, \tag{16}$$

so

$$E\left(\frac{1}{\sqrt{2}}\right) = \int_0^1 \sqrt{\frac{1-\frac{1}{2}(1-u^2)}{1-(1-u^2)}}\,u(1-u^2)^{-1/2}\,du = \int_0^1 \frac{\sqrt{\frac{1}{2}(1+u^2)}}{u}u(1-u^2)^{-1/2}\,du = \frac{1}{\sqrt{2}}\int_0^1 \sqrt{\frac{1+u^2}{1-u^2}}\,du. \tag{17}$$

Now note that

$$\left(\frac{1}{\sqrt{1-u^4}}+\frac{u^2}{\sqrt{1-u^4}}\right)^2 = \frac{(1+u^2)^2}{1-u^4} = \frac{(1+u^2)^2}{(1+u^2)(1-u^2)} = \frac{1+u^2}{1-u^2}, \tag{18}$$

so

$$E\left(\frac{1}{\sqrt{2}}\right) = \frac{1}{\sqrt{2}}\int_0^1 \sqrt{\frac{1+u^2}{1-u^2}}\,du = \frac{1}{\sqrt{2}}\int_0^1 \left(\frac{1}{\sqrt{1-u^4}}+\frac{u^2}{\sqrt{1-u^4}}\right)du = \frac{1}{2}K\left(\frac{1}{\sqrt{2}}\right)+\frac{1}{\sqrt{2}}\int_0^1 \frac{u^2\,du}{\sqrt{1-u^4}}. \tag{19}$$

Now let

$$t \equiv u^4 \tag{20}$$

$$dt = 4u^3\,du, \tag{21}$$

so

$$\int_0^1 \frac{u^2\,du}{\sqrt{1-u^4}} = \frac{1}{4}\int_0^1 t^{1/2}t^{-3/4}(1-t)^{-1/2}\,dt$$
$$= \tfrac{1}{4}\int_0^1 t^{-1/4}(1-t)^{-1/2}\,dt$$
$$= \tfrac{1}{4}B(\tfrac{3}{4},\tfrac{1}{2}) = \frac{\Gamma(\frac{3}{4})\Gamma(\frac{1}{2})}{4\Gamma(\frac{5}{4})}. \qquad (22)$$

But

$$[\Gamma(\tfrac{5}{4})]^{-1} = [\tfrac{1}{4}\Gamma(\tfrac{1}{4})]^{-1} \qquad (23)$$
$$\Gamma(\tfrac{3}{4}) = \pi\sqrt{2}\,[\Gamma(\tfrac{1}{4})]^{-1} \qquad (24)$$
$$\Gamma(\tfrac{1}{2}) = \sqrt{\pi}, \qquad (25)$$

so

$$\int_0^1 \frac{u^2\,du}{\sqrt{1-u^4}} = \frac{1}{4}\frac{\pi\sqrt{2}\cdot 4\sqrt{\pi}}{\Gamma^2(\frac{1}{4})} = \frac{\sqrt{2}\pi^{3/2}}{\Gamma^2(\frac{1}{4})} \qquad (26)$$

$$E\left(\frac{1}{\sqrt{2}}\right) = \tfrac{1}{2}K + \frac{\pi^{3/2}}{\Gamma^2(\frac{1}{4})} = \frac{\Gamma^2(\frac{1}{4})}{8\sqrt{\pi}} + \frac{\pi^{3/2}}{\Gamma^2(\frac{1}{4})}$$
$$= \frac{1}{4}\sqrt{\frac{\pi}{2}}\left[\frac{\Gamma(\frac{1}{4})}{\Gamma(\frac{3}{4})} + \frac{\Gamma(\frac{3}{4})}{\Gamma(\frac{5}{4})}\right]. \qquad (27)$$

Summarizing (12) and (27) gives

$$K\left(\frac{1}{\sqrt{2}}\right) = \frac{\Gamma^2(\frac{1}{4})}{4\sqrt{\pi}}$$
$$K'\left(\frac{1}{\sqrt{2}}\right) = \frac{\Gamma^2(\frac{1}{4})}{4\sqrt{\pi}}$$
$$E\left(\frac{1}{\sqrt{2}}\right) = \frac{\Gamma^2(\frac{1}{4})}{8\sqrt{\pi}} + \frac{\pi^{3/2}}{\Gamma^2(\frac{1}{4})}$$
$$E'\left(\frac{1}{\sqrt{2}}\right) = \frac{\Gamma^2(\frac{1}{4})}{8\sqrt{\pi}} + \frac{\pi^{3/2}}{\Gamma^2(\frac{1}{4})}.$$

Elliptic Integral Singular Value—k_2

The second SINGULAR VALUE k_2, corresponding to

$$K'(k_2) = \sqrt{2}\,K(k_2), \qquad (1)$$

is given by

$$k_2 = \tan\left(\frac{\pi}{8}\right) = \sqrt{2}-1, \qquad (2)$$
$$k_2' = \sqrt{2}\,(\sqrt{2}-1). \qquad (3)$$

For this modulus,

$$E(\sqrt{2}-1) = \frac{1}{4}\sqrt{\frac{\pi}{4}}\left[\frac{\Gamma(\frac{1}{8})}{\Gamma(\frac{5}{8})} + \frac{\Gamma(\frac{5}{8})}{\Gamma(\frac{9}{8})}\right]. \qquad (4)$$

Elliptic Integral Singular Value—k_3

The third SINGULAR VALUE k_3, corresponding to

$$K'(k_3) = \sqrt{3}\,K(k_3), \qquad (1)$$

is given by

$$k_3 = \sin\left(\frac{\pi}{12}\right) = \tfrac{1}{4}(\sqrt{6}-\sqrt{2}). \qquad (2)$$

As shown by Legendre,

$$K(k_3) = \frac{\sqrt{\pi}}{2\cdot 3^{3/4}}\frac{\Gamma(\frac{1}{6})}{\Gamma(\frac{2}{3})} \qquad (3)$$

(Whittaker and Watson 1990, p. 525). In addition,

$$E(k_3) = \frac{\pi}{4\sqrt{3}}\frac{1}{K} + \frac{\sqrt{3}+1}{2\sqrt{3}}K$$
$$= \frac{1}{4}\left(\frac{\pi}{\sqrt{3}}\right)^{1/2}\left[\left(1+\frac{1}{\sqrt{3}}\right)\frac{\Gamma(\frac{1}{3})}{\Gamma(\frac{5}{6})} + \frac{2\Gamma(\frac{5}{6})}{\Gamma(\frac{1}{3})}\right], \qquad (4)$$

and

$$E'(k_3) = \frac{\pi\sqrt{3}}{4}\frac{1}{K'(k_3)} + \frac{\sqrt{3}-1}{2\sqrt{3}}K'(k_3). \qquad (5)$$

Summarizing,

$$K[\tfrac{1}{4}(\sqrt{6}-\sqrt{2})] = \frac{\sqrt{\pi}}{2\cdot 3^{3/4}}\frac{\Gamma(\frac{1}{6})}{\Gamma(\frac{2}{3})} \qquad (6)$$
$$K'[\tfrac{1}{4}(\sqrt{6}-\sqrt{2})] = \sqrt{3}K = \frac{\sqrt{\pi}}{2\cdot 3^{1/4}}\frac{\Gamma(\frac{1}{6})}{\Gamma(\frac{2}{3})} \qquad (7)$$
$$E[\tfrac{1}{4}(\sqrt{6}-\sqrt{2})]$$
$$= \frac{1}{4}\left(\frac{\pi}{\sqrt{3}}\right)^{1/2}\left[\left(1+\frac{1}{\sqrt{3}}\right)\frac{\Gamma(\frac{1}{3})}{\Gamma(\frac{5}{6})} + \frac{2\Gamma(\frac{5}{6})}{\Gamma(\frac{1}{3})}\right] \qquad (8)$$
$$E'[\tfrac{1}{4}(\sqrt{6}-\sqrt{2})]$$
$$= \frac{\sqrt{\pi}}{2}\left[3^{3/4}\frac{\Gamma(\frac{2}{3})}{\Gamma(\frac{1}{6})} + \frac{\sqrt{3}-1}{2\cdot 3^{3/4}}\frac{\Gamma(\frac{1}{6})}{\Gamma(\frac{2}{3})}\right]. \qquad (9)$$

(Whittaker and Watson 1990).

see also THETA FUNCTION

References

Ramanujan, S. "Modular Equations and Approximations to π." *Quart. J. Pure. Appl. Math.* **45**, 350–372, 1913–1914.

Whittaker, E. T. and Watson, G. N. *A Course in Modern Analysis, 4th ed.* Cambridge, England: Cambridge University Press, pp. 525–527 and 535, 1990.

Elliptic Lambda Function

The λ GROUP is the SUBGROUP of the GAMMA GROUP with a and d ODD; b and c EVEN. The function

$$\lambda(t) \equiv \lambda(q) \equiv k^2(q) = \left[\frac{\vartheta_2(q)}{\vartheta_3(q)}\right]^4, \tag{1}$$

where

$$q \equiv e^{i\pi t} \tag{2}$$

is a λ-MODULAR FUNCTION and ϑ_i are THETA FUNCTIONS.

$\lambda^*(r)$ gives the value of the MODULUS k_r for which the complementary and normal complete ELLIPTIC INTEGRALS OF THE FIRST KIND are related by

$$\frac{K'(k_r)}{K(k_r)} = \sqrt{r}. \tag{3}$$

It can be computed from

$$\lambda^*(r) \equiv k(q) = \frac{\vartheta_2{}^2(q)}{\vartheta_3{}^2(q)}, \tag{4}$$

where

$$q \equiv e^{-\pi\sqrt{r}}, \tag{5}$$

and ϑ_i is a THETA FUNCTION.

From the definition of the lambda function,

$$\lambda^*(r') = \lambda^*\left(\frac{1}{r}\right) = \lambda^{*\prime}(r). \tag{6}$$

For all rational r, $K(\lambda^*(r))$ and $E(\lambda^*(r))$ are expressible in terms of a finite number of GAMMA FUNCTIONS (Selberg and Chowla 1967). $\lambda^*(r)$ is related to the RAMANUJAN g- AND G-FUNCTIONS by

$$\lambda^*(n) = \tfrac{1}{2}(\sqrt{1+G_n^{-12}} - \sqrt{1-G_n^{-12}}) \tag{7}$$

$$\lambda^*(n) = g_n^6(\sqrt{g_n^{12}+g_n^{-12}} - g_n^6). \tag{8}$$

Special values are

$$\lambda^*(\tfrac{2}{29}) = (13\sqrt{58}-99)(\sqrt{2}+1)^6$$
$$\lambda^*(\tfrac{2}{5}) = (\sqrt{10}-3)(\sqrt{2}+1)^2$$
$$\lambda^*(\tfrac{2}{3}) = (2-\sqrt{3})(\sqrt{2}+\sqrt{3})$$
$$\lambda^*(\tfrac{3}{4}) = (\sqrt{3}-\sqrt{2})^2(\sqrt{2}+1)^2$$
$$\lambda^*(1) = \frac{1}{\sqrt{2}}$$
$$\lambda^*(2) = \sqrt{2}-1$$
$$\lambda^*(3) = \tfrac{1}{4}\sqrt{2}(\sqrt{3}-1)$$
$$\lambda^*(4) = 3-2\sqrt{2}$$
$$\lambda^*(5) = \tfrac{1}{2}\left(\sqrt{\sqrt{5}-1} - \sqrt{3-\sqrt{5}}\right)$$
$$\lambda^*(6) = (2-\sqrt{3})(\sqrt{3}-\sqrt{2})$$
$$\lambda^*(7) = \tfrac{1}{8}\sqrt{2}(3-\sqrt{7})$$
$$\lambda^*(8) = \left(\sqrt{2}+1-\sqrt{2\sqrt{2}+2}\right)^2$$
$$\lambda^*(9) = \tfrac{1}{2}(\sqrt{2}-3^{1/4})(\sqrt{3}-1)$$
$$\lambda^*(10) = (\sqrt{10}-3)(\sqrt{2}-1)^2$$
$$\lambda^*(11) = \tfrac{1}{12}\sqrt{6}\left(\sqrt{1+2x_{11}-4x_{11}{}^{-1}} - \sqrt{11+2x_{11}-4x_{11}{}^{-1}}\right)$$
$$\lambda^*(12) = (\sqrt{3}-\sqrt{2})^2(\sqrt{2}-1)^2 = 15-10\sqrt{2}+8\sqrt{3}-6\sqrt{6}$$
$$\lambda^*(13) = \tfrac{1}{2}(\sqrt{5\sqrt{13}-17} - \sqrt{19-5\sqrt{13}})$$
$$\lambda^*(14) = -11-8\sqrt{2}-2(\sqrt{2}+2)\sqrt{5+4\sqrt{2}} + \sqrt{11+8\sqrt{2}}(2+2\sqrt{2}+\sqrt{2}\sqrt{5+4\sqrt{2}})$$
$$\lambda^*(15) = \tfrac{1}{16}\sqrt{2}(3-\sqrt{5})(\sqrt{5}-\sqrt{3})(2-\sqrt{3})$$
$$\lambda^*(16) = \frac{(2^{1/4}-1)^2}{(2^{1/4}+1)^2}$$
$$\lambda^*(17) = \tfrac{1}{4}\sqrt{2}(\sqrt{42+10\sqrt{17}} - 13\sqrt{-3+\sqrt{17}}\sqrt{5+\sqrt{17}} - 3\sqrt{17}\sqrt{-3+\sqrt{17}}\sqrt{5+\sqrt{17}} - \sqrt{-38-10\sqrt{17}+13\sqrt{-3+\sqrt{17}}\sqrt{5+\sqrt{17}} + 3\sqrt{17}\sqrt{-3+\sqrt{17}}\sqrt{5+\sqrt{17}}})$$
$$\lambda^*(18) = (\sqrt{2}-1)^3(2-\sqrt{3})^2$$
$$\lambda^*(22) = (3\sqrt{11}-7\sqrt{2})(10-3\sqrt{11})$$
$$\lambda^*(30) = (\sqrt{3}-\sqrt{2})^2(2-\sqrt{3})(\sqrt{6}-\sqrt{5})(4-\sqrt{15})$$
$$\lambda^*(34) = (\sqrt{2}-1)^2(3\sqrt{2}-\sqrt{17}) \times(\sqrt{297+72\sqrt{17}} - \sqrt{296+72\sqrt{17}})$$
$$\lambda^*(42) = (\sqrt{2}-1)^2(2-\sqrt{3})^2(\sqrt{7}-\sqrt{6})(8-3\sqrt{7})$$
$$\lambda^*(58) = (13\sqrt{58}-99)(\sqrt{2}-1)^6$$
$$\lambda^*(210) = (\sqrt{2}-1)^2(2-\sqrt{3})(\sqrt{7}-\sqrt{6})^2(8-3\sqrt{7}) \times(\sqrt{10}-3)^2(4-\sqrt{15})^2(\sqrt{15}-\sqrt{14})(6-\sqrt{35}),$$

where

$$x_{11} \equiv (17+3\sqrt{33})^{1/3}.$$

In addition,

$$\lambda^*(1') = \frac{1}{\sqrt{2}}$$
$$\lambda^*(2') = \sqrt{2\sqrt{2}-2}$$
$$\lambda^*(3') = \tfrac{1}{4}\sqrt{2}(\sqrt{3}+1)$$
$$\lambda^*(4') = 2^{1/4}(2\sqrt{2}-2)$$
$$\lambda^*(5') = \tfrac{1}{2}\left(\sqrt{\sqrt{5}-1} + \sqrt{3-\sqrt{5}}\right)$$
$$\lambda^*(7') = \tfrac{1}{8}\sqrt{2}(3+\sqrt{7})$$
$$\lambda^*(9') = \tfrac{1}{2}(\sqrt{2}+3^{1/4})(\sqrt{3}-1)$$
$$\lambda^*(12') = 2\sqrt{-208+147\sqrt{2}-120\sqrt{3}+85\sqrt{6}}.$$

see also ELLIPTIC ALPHA FUNCTION, ELLIPTIC INTEGRAL OF THE FIRST KIND, MODULUS (ELLIPTIC INTEGRAL), RAMANUJAN g- AND G-FUNCTIONS, THETA FUNCTION

References
Borwein, J. M. and Borwein, P. B. *Pi & the AGM: A Study in Analytic Number Theory and Computational Complexity.* New York: Wiley, pp. 139 and 298, 1987.
Bowman, F. *Introduction to Elliptic Functions, with Applications.* New York: Dover, pp. 75, 95, and 98, 1961.
Selberg, A. and Chowla, S. "On Epstein's Zeta-Function." *J. Reine. Angew. Math.* **227**, 86–110, 1967.
Watson, G. N. "Some Singular Moduli (1)." *Quart. J. Math.* **3**, 81–98, 1932.

Elliptic Logarithm

A generalization of integrals of the form

$$\int_\infty^x \frac{dt}{\sqrt{t^2+at}},$$

which can be expressed in terms of logarithmic and inverse trigonometric functions to

$$\operatorname{eln}(x) \equiv \int_x^\infty \frac{dt}{\sqrt{t^3+at^2+bt}}.$$

The inverse of the elliptic logarithm is the ELLIPTIC EXPONENTIAL FUNCTION.

Elliptic Modular Function

$$\varphi(z) = \left[\frac{\vartheta_2{}^4(0,z)}{\vartheta_3{}^4(0,z)}\right]^{1/8},$$

where ϑ is a THETA FUNCTION. A special case is

$$\varphi(-e^{-\pi\sqrt{3}}) = (4\sqrt{3}-7)^{1/8}.$$

see also MODULAR FUNCTION

Elliptic Paraboloid

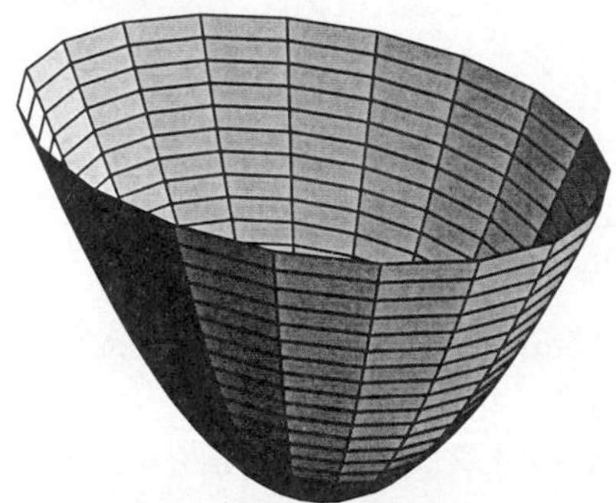

A QUADRATIC SURFACE which has ELLIPTICAL CROSS-SECTION. The elliptic paraboloid of height h, SEMIMAJOR AXIS a, and SEMIMINOR AXIS b can be specified parametrically by

$$x = a\sqrt{u}\cos v$$
$$y = b\sqrt{u}\sin v$$
$$z = u.$$

for $v \in [0, 2\pi)$ and $u \in [0, h]$.

see also ELLIPTIC CONE, ELLIPTIC CYLINDER, PARABOLOID

References
Fischer, G. (Ed.). Plate 66 in *Mathematische Modelle/Mathematical Models, Bildband/Photograph Volume.* Braunschweig, Germany: Vieweg, p. 61, 1986.

Elliptic Partial Differential Equation

A second-order PARTIAL DIFFERENTIAL EQUATION, i.e., one of the form

$$Au_{xx} + 2Bu_{xy} + Cu_{yy} + Du_x + Eu_y + F = 0, \quad (1)$$

is called elliptic if the MATRIX

$$\mathsf{Z} \equiv \begin{bmatrix} A & B \\ B & C \end{bmatrix} \quad (2)$$

is POSITIVE DEFINITE. LAPLACE'S EQUATION and POISSON'S EQUATION are examples of elliptic partial differential equations. For an elliptic partial differential equation, BOUNDARY CONDITIONS are used to give the constraint $u(x,y) = g(x,y)$ on $\partial\Omega$, where

$$u_{xx} + u_{yy} = f(u_x, u_y, u, x, y) \quad (3)$$

holds in Ω.

see also HYPERBOLIC PARTIAL DIFFERENTIAL EQUATION, PARABOLIC PARTIAL DIFFERENTIAL EQUATION, PARTIAL DIFFERENTIAL EQUATION

Elliptic Plane

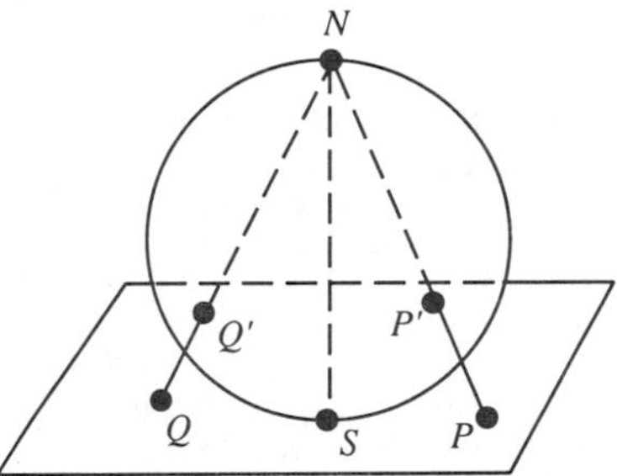

The REAL PROJECTIVE PLANE with elliptic METRIC where the distance between two points P and Q is defined as the RADIAN ANGLE between the projection of the points on the surface of a SPHERE (which is tangent to the plane at a point S) from the ANTIPODE N of the tangent point.

References
Coxeter, H. S. M. *Introduction to Geometry, 2nd ed.* New York: Wiley, p. 94, 1969.

Elliptic Point

A point $\mathbf{p}$ on a REGULAR SURFACE $M \in \mathbb{R}^3$ is said to be elliptic if the GAUSSIAN CURVATURE $K(\mathbf{p}) > 0$ or equivalently, the PRINCIPAL CURVATURES κ_1 and κ_2 have the same sign.

see also ANTICLASTIC, ELLIPTIC FIXED POINT (DIFFERENTIAL EQUATIONS), ELLIPTIC FIXED POINT (MAP), GAUSSIAN CURVATURE, HYPERBOLIC POINT, PARABOLIC POINT, PLANAR POINT, SYNCLASTIC

References

Gray, A. *Modern Differential Geometry of Curves and Surfaces.* Boca Raton, FL: CRC Press, p. 280, 1993.

Elliptical Projection

see MOLLWEIDE PROJECTION

Elliptic Pseudoprime

Let E be an ELLIPTIC CURVE defined over the FIELD of RATIONAL NUMBERS $\mathbb{Q}(\sqrt{-d})$ having equation

$$y^2 = x^3 + ax + b$$

with a and b INTEGERS. Let P be a point on E with integer coordinates and having infinite order in the additive group of rational points of E, and let n be a COMPOSITE NATURAL NUMBER such that $(-d/n) = -1$, where $(-d/n)$ is the JACOBI SYMBOL. Then if

$$(n+1)P \equiv 0 \pmod{n},$$

n is called an elliptic pseudoprime for (E, P).

see also ATKIN-GOLDWASSER-KILIAN-MORAIN CERTIFICATE, ELLIPTIC CURVE PRIMALITY PROVING, STRONG ELLIPTIC PSEUDOPRIME

References

Balasubramanian, R. and Murty, M. R. "Elliptic Pseudoprimes. II." Submitted.

Gordon, D. M. "The Number of Elliptic Pseudoprimes." *Math. Comput.* **52**, 231–245, 1989.

Gordon, D. M. "Pseudoprimes on Elliptic Curves." In *Théorie des nombres* (Ed. J. M. DeKoninck and C. Levesque). Berlin: de Gruyter, pp. 290–305, 1989.

Miyamoto, I. and Murty, M. R. "Elliptic Pseudoprimes." *Math. Comput.* **53**, 415–430, 1989.

Ribenboim, P. *The New Book of Prime Number Records, 3rd ed.* New York: Springer-Verlag, pp. 132–134, 1996.

Elliptic Rotation

Leaves the CIRCLE

$$x^2 + y^2 = 1$$

invariant.

$$x' = x\cos\theta - y\sin\theta$$
$$y' = x\sin\theta + y\sin\theta.$$

see also EQUIAFFINITY

Elliptic Theta Function

see NEVILLE THETA FUNCTION, THETA FUNCTION

Elliptic Torus

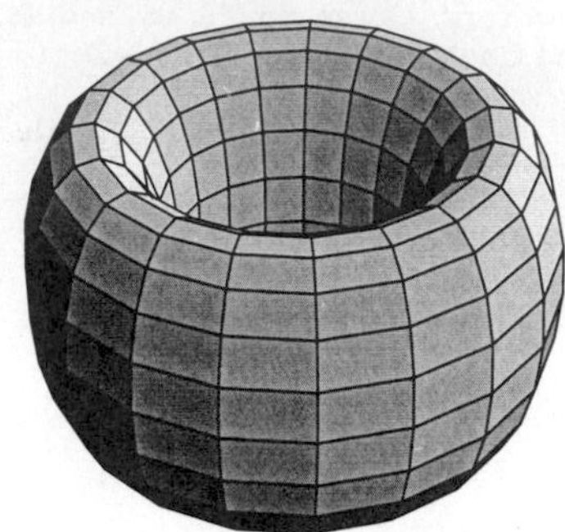

A generalization of the ring TORUS produced by stretching or compressing in the z direction. It is given by the parametric equations

$$x(u,v) = (a + b\cos v)\cos u$$
$$y(u,v) = (a + b\cos v)\sin u$$
$$z(u,v) = c\sin v.$$

see also TORUS

References

Gray, A. "Tori." §11.4 in *Modern Differential Geometry of Curves and Surfaces.* Boca Raton, FL: CRC Press, pp. 218–220, 1993.

Elliptic Umbilic Catastrophe

A CATASTROPHE which can occur for three control factors and two behavior axes.

see also HYPERBOLIC UMBILIC CATASTROPHE

Ellipticity

Given a SPHEROID with equatorial radius a and polar radius c,

$$e \equiv \begin{cases} \sqrt{\frac{a^2-c^2}{a^2}} & a > c \text{ (oblate spheroid)} \\ \sqrt{\frac{c^2-a^2}{a^2}}. & a < c \text{ (prolate spheroid)} \end{cases}$$

see also FLATTENING, OBLATE SPHEROID, PROLATE SPHEROID, SPHEROID

Ellison–Mendès-France Constant

$$\sum_{n\le x} \frac{1}{n} \ln\left(\frac{x}{n}\right) = \tfrac{1}{2}(\ln x)^2 + \gamma \ln x + D + \mathcal{O}(x^{-1}),$$

where γ is the EULER-MASCHERONI CONSTANT, and

$$D = 2.723\ldots$$

is the Ellision-Mendès-France constant.

References

Ellison, W. J. and Mendès-France, M. *Les nombres premiers.* Paris: Hermann, 1975.

Le Lionnais, F. *Les nombres remarquables.* Paris: Hermann, p. 47, 1983.

Elongated Cupola

A n-gonal CUPOLA adjoined to a $2n$-gonal PRISM.

see also ELONGATED PENTAGONAL CUPOLA, ELONGATED SQUARE CUPOLA, ELONGATED TRIANGULAR CUPOLA

Elongated Dipyramid

see also ELONGATED PENTAGONAL DIPYRAMID, ELONGATED SQUARE DIPYRAMID, ELONGATED TRIANGULAR DIPYRAMID

Elongated Dodecahedron

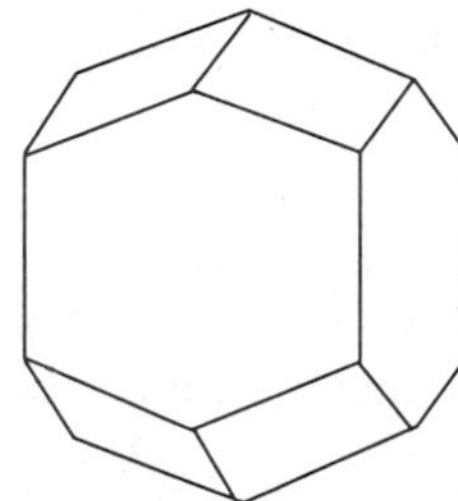

A SPACE-FILLING POLYHEDRON and PARALLELOHEDRON.

References

Coxeter, H. S. M. *Regular Polytopes, 3rd ed.* New York: Dover, pp. 29–30 and 257, 1973.

Elongated Gyrobicupola

see ELONGATED PENTAGONAL GYROBICUPOLA, ELONGATED SQUARE GYROBICUPOLA, ELONGATED TRIANGULAR GYROBICUPOLA

Elongated Gyrocupolarotunda

see ELONGATED PENTAGONAL GYROCUPOLAROTUNDA

Elongated Orthobicupola

see ELONGATED PENTAGONAL ORTHOBICUPOLA, ELONGATED TRIANGULAR ORTHOBICUPOLA

Elongated Orthobirotunda

see ELONGATED PENTAGONAL ORTHOBIROTUNDA

Elongated Orthocupolarotunda

see ELONGATED PENTAGONAL ORTHOCUPOLAROTUNDA

Elongated Pentagonal Cupola

see JOHNSON SOLID

Elongated Pentagonal Dipyramid

see JOHNSON SOLID

Elongated Pentagonal Gyrobicupola

see JOHNSON SOLID

Elongated Pentagonal Gyrobirotunda

see JOHNSON SOLID

Elongated Pentagonal Gyrocupolarotunda

see JOHNSON SOLID

Elongated Pentagonal Orthobicupola

see JOHNSON SOLID

Elongated Pentagonal Orthobirotunda

see JOHNSON SOLID

Elongated Pentagonal Orthocupolarotunda

see JOHNSON SOLID

Elongated Pentagonal Pyramid

see JOHNSON SOLID

Elongated Pentagonal Rotunda

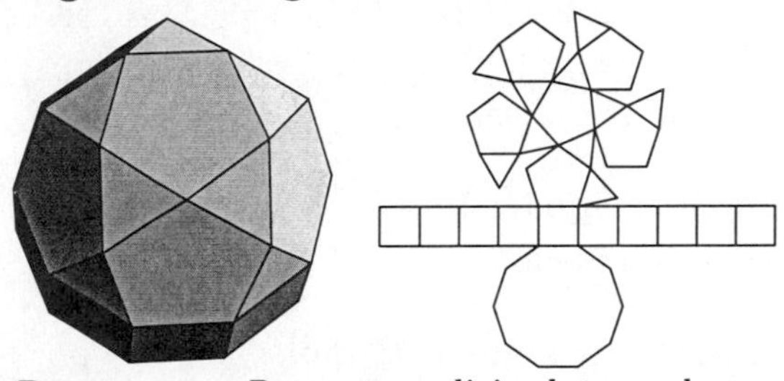

A PENTAGONAL ROTUNDA adjoined to a decagonal PRISM which is JOHNSON SOLID J_{21}.

Elongated Pyramid

An n-gonal PYRAMID adjoined to an n-gonal PRISM.

see also ELONGATED PENTAGONAL PYRAMID, ELONGATED SQUARE PYRAMID, ELONGATED TRIANGULAR PYRAMID, GYROELONGATED PYRAMID

Elongated Rotunda

see ELONGATED PENTAGONAL ROTUNDA

Elongated Square Cupola

see JOHNSON SOLID

Elongated Square Dipyramid

see JOHNSON SOLID

Elongated Square Gyrobicupola

A nonuniform POLYHEDRON obtained by rotating the bottom third of a SMALL RHOMBICUBOCTAHEDRON (Ball and Coxeter 1987, p. 137). It is also called MILLER'S SOLID, the MILLER-AŠKINUZE SOLID, or the PSEUDORHOMBICUBOCTAHEDRON, and is JOHNSON SOLID J_{37}.

see also SMALL RHOMBICUBOCTAHEDRON

References

Aškinuze, V. G. "O čisle polupravil'nyh mnogogrannikov." *Math. Prosvešč.* **1**, 107–118, 1957.

Ball, W. W. R. and Coxeter, H. S. M. *Mathematical Recreations and Essays, 13th ed.* New York: Dover, pp. 137–138, 1987.

Cromwell, P. R. *Polyhedra.* New York: Cambridge University Press, pp. 91–92, 1997.

Elongated Square Pyramid

see JOHNSON SOLID

Elongated Triangular Cupola

see JOHNSON SOLID

Elongated Triangular Dipyramid

see JOHNSON SOLID

Elongated Triangular Gyrobicupola

see JOHNSON SOLID

Elongated Triangular Orthobicupola

see JOHNSON SOLID

Elongated Triangular Pyramid

see JOHNSON SOLID

Elsasser Function

$$E(y,u) \equiv \int_{-1/2}^{1/2} \exp\left[-\frac{2\pi yu\sinh(2\pi y)}{\cosh(2\pi y)-\cos(2\pi x)}\right] dx.$$

Embeddable Knot

A KNOT K is an n-embeddable knot if it can be placed on a GENUS n standard embedded surface without crossings, but K cannot be placed on any standardly embedded surface of lower GENUS without crossings. Any KNOT is an n-embeddable knot for some n. The FIGURE-OF-EIGHT KNOT is a 2-EMBEDDABLE KNOT. A knot with BRIDGE NUMBER b is an n-embeddable knot where $n \leq b$.

see also TUNNEL NUMBER

Embedding

see EXTRINSIC CURVATURE, HYPERBOLOID EMBEDDING, INJECTION, SPHERE EMBEDDING

Empty Set

The SET containing no elements, denoted $\varnothing$. Strangely, the empty set is both OPEN and CLOSED for any SET X and TOPOLOGY. A GROUPOID, SEMIGROUP, QUASIGROUP, RINGOID, and SEMIRING can be empty. A MONOID, GROUP, and RINGS must have at least one element, while DIVISION RINGS and FIELDS must have at least two elements.

References

Conway, J. H. and Guy, R. K. *The Book of Numbers.* New York: Springer-Verlag, p. 266, 1996.

Enantiomer

Two objects which are MIRROR IMAGES of each other are called enantiomers. The term enantiomer is synonymous with ENANTIOMORPH.

see also AMPHICHIRAL KNOT, CHIRAL, DISYMMETRIC, HANDEDNESS, MIRROR IMAGE, REFLEXIBLE

References

Ball, W. W. R. and Coxeter, H. S. M. "Polyhedra." Ch. 5 in *Mathematical Recreations and Essays, 13th ed.* New York: Dover, pp. 130–161, 1987.

Enantiomorph

see ENANTIOMER

Encoding

An encoding is a way of representing a number or expression in terms of another (usually simpler) one. However, multiple expressions can also be encoded as a single expression, as in, for example,

$$(a,b) \equiv \tfrac{1}{2}[(a+b)^2+3a+b]$$

which encodes a and b uniquely as a single number.

a	b	(a,b)
0	0	0
0	1	1
1	0	2
0	2	3
1	2	4
2	0	5

see also CODE, CODING THEORY

Endogenous Variable

An economic variable which is independent of the relationships determining the equilibrium levels, but nonetheless affects the equilibrium.

see also EXOGENOUS VARIABLE

References

Iyanaga, S. and Kawada, Y. (Eds.). *Encyclopedic Dictionary of Mathematics.* Cambridge, MA: MIT Press, p. 458, 1980.

Endomorphism

A SURJECTIVE MORPHISM from an object to itself. In ERGODIC THEORY, let X be a SET, F a SIGMA ALGEBRA on X and m a PROBABILITY MEASURE. A MAP $T : X \to X$ is called an endomorphism or MEASURE-PRESERVING TRANSFORMATION if

1. T is SURJECTIVE,
2. T is MEASURABLE,
3. $m(T^{-1}A) = m(A)$ for all $A \in F$.

An endomorphism is called ERGODIC if it is true that $T^{-1}A = A$ IMPLIES $m(A) = 0$ or 1, where $T^{-1}A = \{x \in X : T(x) \in A\}$.

see also MEASURABLE FUNCTION, MEASURE-PRESERVING TRANSFORMATION, MORPHISM, SIGMA ALGEBRA, SURJECTIVE

Endraß Octic

Endraß surfaces are a pair of OCTIC SURFACES which have 168 ORDINARY DOUBLE POINTS. This is the maximum number known to exist for an OCTIC SURFACE, although the rigorous upper bound is 174. The equations of the surfaces $X_8^\pm$ are

$$\begin{aligned} &64(x^2 - w^2)(y^2 - w^2)[(x+y)^2 - 2w^2] \\ &\quad [(x-y)^2 - 2w^2] - \{-4(1 \pm \sqrt{2})(x^2+y^2)^2 \\ &\quad +[8(2 \pm \sqrt{2})z^2 + 2(2 \pm 7\sqrt{2})w^2](x^2+y^2) \\ &\qquad -16z^4 + 8(1 \mp 2\sqrt{2})z^2w^2 - (1+12\sqrt{2})w^4\}^2 = 0, \end{aligned}$$

where w is a parameter taken as $w = 1$ in the above plots. All ORDINARY DOUBLE POINTS of X_8^+ are real, while 24 of those in X_8^- are complex. The surfaces were discovered in a 5-D family of octics with 112 nodes, and are invariant under the GROUP $D_8 \otimes Z_2$.

see also OCTIC SURFACE

References

Endraß, S. "Octics with 168 Nodes." `http://www.mathematik.uni-mainz.de/AlgebraischeGeometrie/docs/Eendrassoctic.shtml`.

Endraß, S. "Flächen mit vielen Doppelpunkten." *DMV-Mitteilungen* **4**, 17–20, 4/1995.

Endraß, S. "A Proctive Surface of Degree Eight with 168 Nodes." *J. Algebraic Geom.* **6**, 325–334, 1997.

Energy

The term energy has an important physical meaning in physics and is an extremely useful concept. A much more abstract mathematical generalization is defined as follows. Let Ω be a SPACE with MEASURE $\mu \geq 0$ and let $\Phi(P, Q)$ be a real function on the PRODUCT SPACE $\Omega \times \Omega$. When

$$\begin{aligned} (\mu, nu) &= \iint \Phi(P,Q)\, d\mu(Q)\, d\nu(P) \\ &= \int \Phi(P,\mu)\, d\nu(P) \end{aligned}$$

exists for measures $\mu, \nu \geq 0$, (μ, ν) is called the MUTUAL ENERGY and (μ, μ) is called the ENERGY.

see also DIRICHLET ENERGY, MUTUAL ENERGY

References

Iyanaga, S. and Kawada, Y. (Eds.). "General Potential." §335.B in *Encyclopedic Dictionary of Mathematics.* Cambridge, MA: MIT Press, p. 1038, 1980.

Engel's Theorem

A finite-dimensional LIE ALGEBRA all of whose elements are ad-NILPOTENT is itself a NILPOTENT LIE ALGEBRA.

Enneacontagon

A 90-sided POLYGON.

Enneacontahedron

A ZONOHEDRON constructed from the 10 diameters of the DODECAHEDRON which has 90 faces, 30 of which are RHOMBS of one type and the other 60 of which are RHOMBS of another. The enneacontahedron somewhat resembles a figure of Sharp.

see also DODECAHEDRON, RHOMB, ZONOHEDRON

References

Ball, W. W. R. and Coxeter, H. S. M. *Mathematical Recreations and Essays, 13th ed.* New York: Dover, pp. 142–143, 1987.

Sharp, A. *Geometry Improv'd.* London, p. 87, 1717.

Enneadecagon

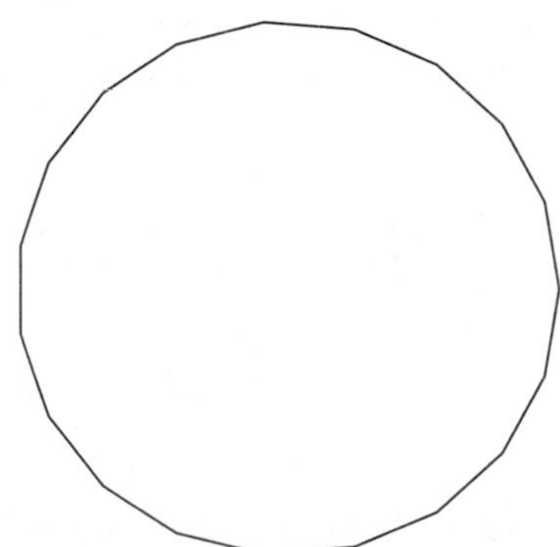

A 19-sided POLYGON, sometimes also called the ENNEAKAIDECAGON.

Enneagon

see NONAGON

Enneagonal Number

see NONAGONAL NUMBER

Enneakaidecagon

see ENNEADECAGON

Enneper's Surfaces

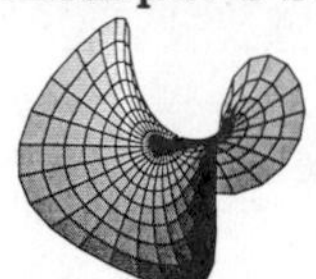 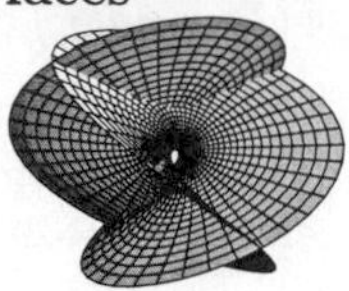 

The Enneper surfaces are a three-parameter family of surfaces with constant curvature. In general, they are described by elliptic functions. However, special cases which can be specified parametrically using ELEMENTARY FUNCTION include the KUEN SURFACE, REMBS' SURFACES, and SIEVERT'S SURFACE. The surfaces shown above can be generated using the ENNEPER-WEIERSTRAß PARAMETERIZATION with

$$f(\zeta) = 1 \tag{1}$$

$$g(\zeta) = \zeta. \tag{2}$$

Letting $z = re^{i\phi}$ and taking the REAL PART give

$$x = \Re[re^{i\phi} - \tfrac{1}{3}r^3 e^{3i\phi}] \tag{3}$$

$$y = \Re[ire^{i\phi} + \tfrac{1}{3}ir^3 e^{3i\phi}] \tag{4}$$

$$z = \Re[r^2 e^{2i\phi}], \tag{5}$$

where $r \in [0, 1]$ and $\phi \in [-\pi, \pi)$. Letting $z = u + iv$ instead gives the figure on the right,

$$x = u - \tfrac{1}{3}u^3 + uv^2 \tag{6}$$

$$y = -v - u^2 v + \tfrac{1}{3}v^3 \tag{7}$$

$$z = u^2 - v^2 \tag{8}$$

(do Carmo 1986, Gray 1993, Nordstrand). This surface has a HOLE in its middle. Nordstrand gives the implicit form

$$\left(\frac{y^2 - x^2}{2z} + \tfrac{2}{9}z^2 + \tfrac{2}{3}\right)^3 - 6\left[\frac{(y^2 - x^2)}{4z} - \tfrac{1}{4}(x^2 + y^2 + \tfrac{8}{9}z^2) + \tfrac{2}{9}\right]^2 = 0. \tag{9}$$

References

Dickson, S. "Minimal Surfaces." *Mathematica J.* **1**, 38–40, 1990.

do Carmo, M. P. "Enneper's Surface." §3.5C in *Mathematical Models from the Collections of Universities and Museums* (Ed. G. Fischer). Braunschweig, Germany: Vieweg, p. 43, 1986.

Enneper, A. "Analytisch-geometrische Untersuchungen." *Nachr. Königl. Gesell. Wissensch. Georg-Augustus-Univ. Göttingen* **12**, 258–277, 1868.

Fischer, G. (Ed.). Plate 92 in *Mathematische Modelle/Mathematical Models, Bildband/Photograph Volume.* Braunschweig, Germany: Vieweg, p. 88, 1986.

Gray, A. *Modern Differential Geometry of Curves and Surfaces.* Boca Raton, FL: CRC Press, p. 265, 1993.

Maeder, R. *The Mathematica Programmer.* San Diego, CA: Academic Press, pp. 150–151, 1994.

Nordstrand, T. "Enneper's Minimal Surface." http://www.uib.no/people/nfytn/enntxt.htm.

Reckziegel, H. "Enneper's Surfaces." §3.4.4 in *Mathematical Models from the Collections of Universities and Museums* (Ed. G. Fischer). Braunschweig, Germany: Vieweg, pp. 37–39, 1986.

Wolfram Research "Mathematica Version 2.0 Graphics Gallery." http://www.mathsource.com/cgi-bin/MathSource/Applications/Graphics/3D/0207-155.

Enneper-Weierstraß Parameterization

Gives a parameterization of a MINIMAL SURFACE.

$$\Re\int\begin{bmatrix} f(1-g^2) \\ if(1+g^2) \\ 2fg \end{bmatrix} d\zeta.$$

see also MINIMAL SURFACE

References

Dickson, S. "Minimal Surfaces." *Mathematica J.* **1**, 38–40, 1990.

do Carmo, M. P. *Mathematical Models from the Collections of Universities and Museums* (Ed. G. Fischer). Braunschweig, Germany: Vieweg, p. 41, 1986.

Weierstraß, K. "Über die Flächen deren mittlere Krümmung überall gleich null ist." *Monatsber. Berliner Akad.*, 612–625, 1866.

Enormous Theorem

see CLASSIFICATION THEOREM

Enriques Surfaces

An Enriques surface X is a smooth compact complex surface having irregularity $q(X) = 0$ and nontrivial canonical sheaf K_X such that $K_X^2 = O_X$ (Endraß). Such surfaces cannot be embedded in projective 3-space, but there nonetheless exist transformations onto singular surfaces in projective 3-space. There exists a family of such transformed surfaces of degree six which passes through each edge of a TETRAHEDRON twice. A subfamily with tetrahedral symmetry is given by the two-parameter (r, c) family of surfaces

$$f_r x_0 x_1 x_2 x_3 + c({x_0}^2 {x_1}^2 {x_2}^2 + {x_0}^2 {x_1}^2 {x_3}^2 + {x_0}^2 {x_2}^2 {x_3}^2 + {x_1}^2 {x_2}^2 {x_3}^2 = 0$$

and the polynomial f_r is a sphere with radius r,

$$f_r = (3 - r)({x_0}^2 + {x_1}^2 + {x_2}^2 + {x_3}^2) - 2(1 + r)(x_0 x_1 + x_0 x_2 + x_0 x_3 + x_1 x_2 + x_1 x_3 + x_2 x_3)$$

(Endraß).

References

Angermüller, G. and Barth, W. "Elliptic Fibres on Enriques Surfaces." *Compos. Math.* **47**, 317–332, 1982.

Barth, W. and Peters, C. "Automorphisms of Enriques Surfaces." *Invent. Math.* **73**, 383–411, 1983.

Barth, W. P.; Peters, C. A.; and van de Ven, A. A. *Compact Complex Surfaces.* New York: Springer-Verlag, 1984.

Barth, W. "Lectures on K3- and Enriques Surfaces." In *Algebraic Geometry, Sitges (Barcelona) 1983, Proceedings of a Conference Held in Sitges (Barcelona), Spain, October 5–12, 1983* (Ed. E. Casas-Alvero, G. E. Welters, and S. Xambó-Descamps). New York: Springer-Verlag, pp. 21–57, 1983.

Endraß, S. "Enriques Surfaces." `http://www.mathematik.uni-mainz.de/Algebraische Geometrie/docs/enriques.shtml`.

Enriques, F. *Le superficie algebriche.* Bologna, Italy: Zanichelli, 1949.

Enriques, F. "Sulla classificazione." *Atti Accad. Naz. Lincei* **5**, 1914.

Hunt, B. *The Geometry of Some Special Arithmetic Quotients.* New York: Springer-Verlag, p. 317, 1996.

Entire Function

If a function is ANALYTIC on $\mathbb{C}^*$, where $\mathbb{C}^*$ denotes the extended COMPLEX PLANE, then it is said to be entire.

see also ANALYTIC FUNCTION, HOLOMORPHIC FUNCTION, MEROMORPHIC

Entringer Number

The Entringer numbers $E(n,k)$ are the number of PERMUTATIONS of $\{1,2,\ldots,n+1\}$, starting with $k+1$, which, after initially falling, alternately fall then rise. The Entringer numbers are given by

$$E(0,0) = 1$$
$$E(n,0) = 0$$

together with the RECURRENCE RELATION

$$E(n,k) = E(n,k+1) + E(n-1,n-k).$$

The numbers $E(n) = E(n,n)$ are the SECANT and TANGENT NUMBERS given by the MACLAURIN SERIES

$$\sec x + \tan x = A_0 + A_1 x + A_2 \frac{x^2}{2!} + A_3 \frac{x^3}{3!} + A_4 \frac{x^4}{4!} + A_5 \frac{x^5}{5!} + \ldots.$$

see also ALTERNATING PERMUTATION, BOUSTROPHEDON TRANSFORM, EULER ZIGZAG NUMBER, PERMUTATION, SECANT NUMBER, SEIDEL-ENTRINGER-ARNOLD TRIANGLE, TANGENT NUMBER, ZAG NUMBER, ZIG NUMBER

References

Entringer, R. C. "A Combinatorial Interpretation of the Euler and Bernoulli Numbers." *Nieuw. Arch. Wisk.* **14**, 241–246, 1966.

Millar, J.; Sloane, N. J. A.; and Young, N. E. "A New Operation on Sequences: The Boustrophedon Transform." *J. Combin. Th. Ser. A* **76**, 44–54, 1996.

Poupard, C. "De nouvelles significations enumeratives des nombres d'Entringer." *Disc. Math.* **38**, 265–271, 1982.

Entropy

In physics, the word entropy has important physical implications as the amount of "disorder" of a system. In mathematics, a more abstract definition is used. The (Shannon) entropy of a variable X is defined as

$$H(X) \equiv -\sum_x p(x)\ln[p(x)],$$

where $p(x)$ is the probability that X is in the state x, and $p\ln p$ is defined as 0 if $p=0$. The joint entropy of variables $X_1, \ldots, X_n$ is then defined by

$$H(X_1,\ldots,X_n) \equiv -\sum_{x_1}\cdots\sum_{x_n} p(x_1,\ldots,x_n)\ln[p(x_1,\ldots,x_n)].$$

see also KOLMOGOROV ENTROPY, KOLMOGOROV-SINAI ENTROPY, MAXIMUM ENTROPY METHOD, METRIC ENTROPY, ORNSTEIN'S THEOREM, REDUNDANCY, SHANNON ENTROPY, TOPOLOGICAL ENTROPY

References

Ott, E. "Entropies." §4.5 in *Chaos in Dynamical Systems.* New York: Cambridge University Press, pp. 138–144, 1993.

Entscheidungsproblem

see DECISION PROBLEM

Enumerative Geometry

Schubert's application of the CONSERVATION OF NUMBER PRINCIPLE.

see also CONSERVATION OF NUMBER PRINCIPLE, DUALITY PRINCIPLE, HILBERT'S PROBLEMS, PERMANENCE OF MATHEMATICAL RELATIONS PRINCIPLE

References

Bell, E. T. *The Development of Mathematics, 2nd ed.* New York: McGraw-Hill, p. 340, 1945.

Envelope

The envelope of a one-parameter family of curves given implicitly by

$$U(x,y,c) = 0, \tag{1}$$

or in parametric form by $(f(t,c), g(t,c))$, is a curve which touches every member of the family. For a curve represented by $(f(t,c), g(t,c))$, the envelope is found by solving

$$0 = \frac{\partial f}{\partial t}\frac{\partial g}{\partial c} - \frac{\partial f}{\partial c}\frac{\partial g}{\partial t}. \tag{2}$$

For a curve represented implicitly, the envelope is given by simultaneously solving

$$\frac{\partial U}{\partial c} = 0 \tag{3}$$

$$U(x,y,c) = 0. \tag{4}$$

see also ASTROID, CARDIOID, CATACAUSTIC, CAUSTIC, CAYLEYIAN CURVE, DÜRER'S CONCHOID, ELLIPSE ENVELOPE, ENVELOPE THEOREM, EVOLUTE, GLISSETTE, HEDGEHOG, KIEPERT'S PARABOLA, LINDELOF'S THEOREM, NEGATIVE PEDAL CURVE

References

Lawrence, J. D. *A Catalog of Special Plane Curves.* New York: Dover, pp. 33–34, 1972.

Lee, X. "Envelope." http://www.best.com/~xah/SpecialPlaneCurves_dir/Envelope_dir/envelope.html.

Yates, R. C. "Envelopes." *A Handbook on Curves and Their Properties.* Ann Arbor, MI: J. W. Edwards, pp. 75–80, 1952.

Envelope Theorem

Relates EVOLUTES to single paths in the CALCULUS OF VARIATIONS. Proved in the general case by Darboux and Zermelo (1894) and Kneser (1898). It states: "When a single parameter family of external paths from a fixed point O has an ENVELOPE, the integral from the fixed point to any point A on the ENVELOPE equals the integral from the fixed point to any second point B on the ENVELOPE plus the integral along the envelope to the first point on the ENVELOPE, $J_{OA} = J_{OB} + J_{BA}$."

References

Kimball, W. S. *Calculus of Variations by Parallel Displacement.* London: Butterworth, p. 292, 1952.

Envyfree

An agreement in which all parties feel as if they have received the best deal.

Epicycloid

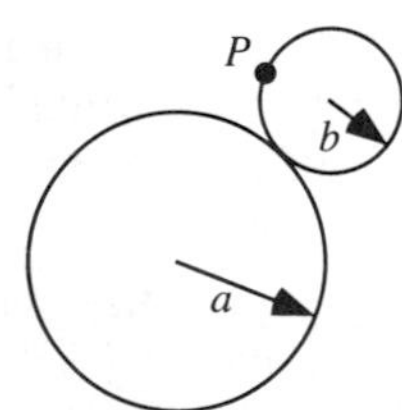

The path traced out by a point P on the EDGE of a CIRCLE of RADIUS b rolling on the outside of a CIRCLE of RADIUS a.

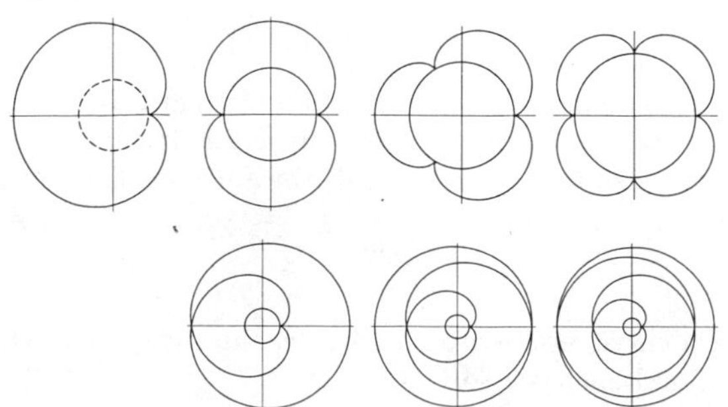

It is given by the equations

$$x = (a+b)\cos\phi - b\cos\left(\frac{a+b}{b}\phi\right) \tag{1}$$

$$y = (a+b)\sin\phi - b\sin\left(\frac{a+b}{b}\phi\right) \tag{2}$$

$$x^2 = (a+b)^2\cos^2\phi - 2b(a+b)\cos\phi\cos\left(\frac{a+b}{b}\phi\right) + b^2\cos^2\left(\frac{a+b}{b}\phi\right) \tag{3}$$

$$y^2 = (a+b)^2\sin^2\phi - 2b(a+b)\sin\phi\sin\left(\frac{a+b}{b}\phi\right) + b^2\sin^2\left(\frac{a+b}{b}\phi\right) \tag{4}$$

$$r^2 = x^2 + y^2 = (a+b)^2 + b^2 - 2b(a+b)\left\{\cos\left[\left(\frac{a}{b}+1\right)\phi\right]\cos\phi + \sin\left[\left(\frac{a}{b}+1\right)\phi\right]\sin\phi\right\}. \tag{5}$$

But

$$\cos\alpha\cos\beta + \sin\alpha\sin\beta = \cos(\alpha-\beta), \tag{6}$$

so

$$r^2 = (a+b)^2 + b^2 - 2b(a+b)\cos\left[\left(\frac{a}{b}+1\right)\phi - \phi\right] = (a+b)^2 + b^2 - 2b(a+b)\cos\left(\frac{a}{b}\phi\right). \tag{7}$$

Note that ϕ is the parameter here, *not* the polar angle. The polar angle from the center is

$$\tan\theta = \frac{y}{x} = \frac{(a+b)\sin\phi - b\sin\left(\frac{a+b}{b}\phi\right)}{(a+b)\cos\phi - b\cos\left(\frac{a+b}{b}\phi\right)}. \tag{8}$$

To get n CUSPS in the epicycloid, $b = a/n$, because then n rotations of b bring the point on the edge back to its starting position.

$$\begin{aligned} r^2 &= a^2\left[\left(1+\frac{1}{n}\right)^2 + \left(\frac{1}{n}\right)^2 - 2\left(\frac{1}{n}\right)\left(1+\frac{1}{n}\right)\cos(n\phi)\right] \\ &= a^2\left[1 + \frac{2}{n} + \frac{1}{n^2} + \frac{1}{n^2} - \left(\frac{2}{n}\right)\left(\frac{n+1}{n}\right)\cos(n\phi)\right] \\ &= a^2\left[\frac{n^2+2n+2}{n^2} - \frac{2(n+1)}{n^2}\cos(n\phi)\right] \\ &= \frac{a^2}{n^2}\left[(n^2+2n+2) - 2(n+1)\cos(n\phi)\right], \end{aligned} \tag{9}$$

so

$$\tan\theta = \frac{a\left(\frac{n+1}{n}\right)\sin\phi - \frac{a}{n}\sin[(n+1)\phi]}{a\left(\frac{n+1}{n}\right)\cos\phi - \frac{a}{n}\cos[(n+1)\phi]} = \frac{(n+1)\sin\phi - \sin[(n+1)\phi]}{(n+1)\cos\phi - \cos[(n+1)\phi]}. \tag{10}$$

An epicycloid with one cusp is called a CARDIOID, one with two cusps is called a NEPHROID, and one with five cusps is called a RANUNCULOID.

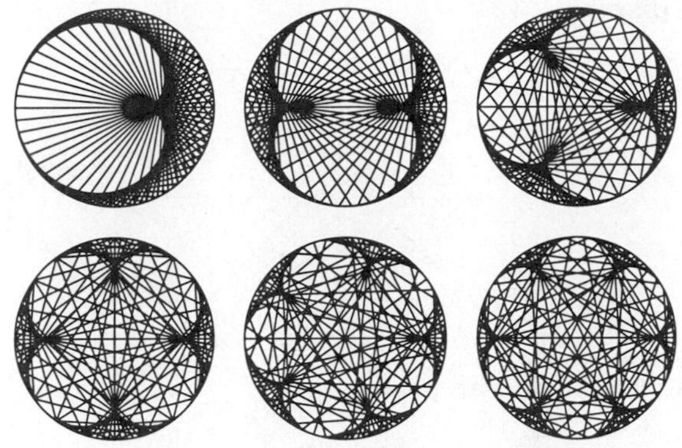

n-epicycloids can also be constructed by beginning with the DIAMETER of a CIRCLE, offsetting one end by a series of steps while at the same time offsetting the other end by steps n times as large. After traveling around the CIRCLE once, an n-cusped epicycloid is produced, as illustrated above (Madachy 1979).

Epicycloids have TORSION

$$\tau = 0 \tag{11}$$

and satisfy

$$\frac{s^2}{a^2} + \frac{\rho^2}{b^2} = 1, \tag{12}$$

where ρ is the RADIUS OF CURVATURE ($1/\kappa$).

see also CARDIOID, CYCLIDE, CYCLOID, EPICYCLOID—1-CUSPED, HYPOCYCLOID, NEPHROID, RANUNCULOID

References

Bogomolny, A. "Cycloids." http://www.cut-the-knot.com/pythagoras/cycloids.html.

Lawrence, J. D. *A Catalog of Special Plane Curves.* New York: Dover, pp. 160–164 and 169, 1972.

Lee, X. "Epicycloid and Hypocycloid." http://www.best.com/~xah/SpecialPlaneCurves_dir/EpiHypocycloid_dir/epiHypocycloid.html.

MacTutor History of Mathematics Archive. "Epicycloid." http://www-groups.dcs.st-and.ac.uk/~history/Curves/Epicycloid.html.

Madachy, J. S. *Madachy's Mathematical Recreations.* New York: Dover, pp. 219–225, 1979.

Wagon, S. *Mathematica in Action.* New York: W. H. Freeman, pp. 50–52, 1991.

Yates, R. C. "Epi- and Hypo-Cycloids." *A Handbook on Curves and Their Properties.* Ann Arbor, MI: J. W. Edwards, pp. 81–85, 1952.

Epicycloid—1-Cusped

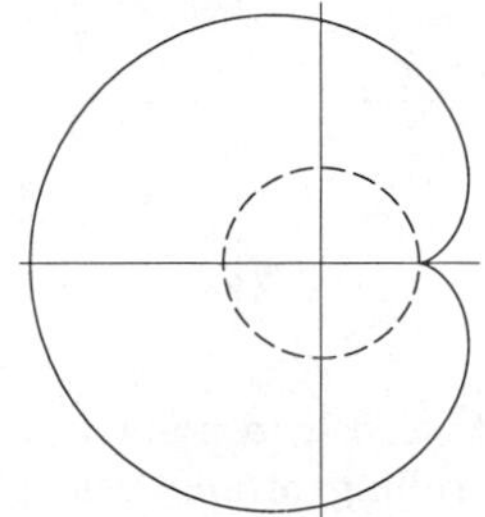

A 1-cusped epicycloid has $b = a$, so $n = 1$. The radius measured from the center of the large circle for a 1-cusped epicycloid is given by EPICYCLOID equation (9) with $n = 1$ so

$$\begin{aligned} r^2 &= \frac{a^2}{n^2}\left[(n^2 + 2n + 2) - 2(n+1)\cos(n\phi)\right] \\ &= a^2[(1^2 + 2\cdot 1 + 2) - 2(1+1)\cos(1\cdot\phi)] \\ &= a^2(5 - 4\cos\phi) \end{aligned} \tag{1}$$

$$r = a\sqrt{5 - 4\cos\phi}, \tag{2}$$

and

$$\tan\theta = \frac{2\sin\phi - \sin(2\phi)}{2\cos\phi - \cos(2\phi)}. \tag{3}$$

The 1-cusped epicycloid is just an offset CARDIOID.

Epicycloid—2-Cusped

see NEPHROID

Epicycloid Evolute

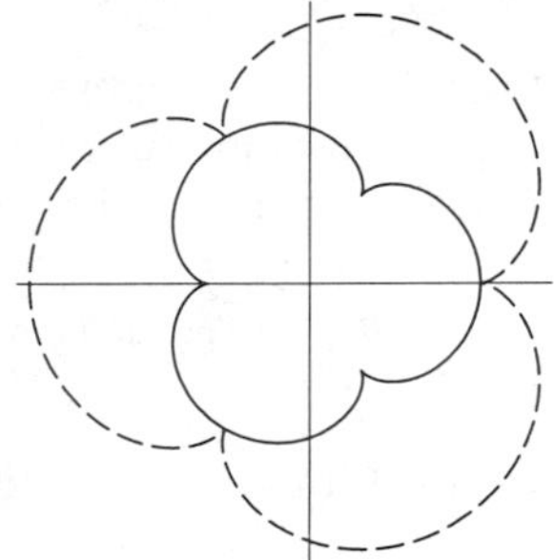

The EVOLUTE of the EPICYCLOID

$$\begin{aligned} x &= (a+b)\cos t - b\cos\left[\left(\frac{a+b}{b}\right)t\right] \\ y &= (a+b)\sin t - b\sin\left[\left(\frac{a+b}{b}\right)t\right] \end{aligned}$$

is another EPICYCLOID given by

$$\begin{aligned} x &= \frac{a}{a+2b}\left\{(a+b)\cos t + b\cos\left[\left(\frac{a+b}{b}\right)t\right]\right\} \\ y &= \frac{a}{a+2b}\left\{(a+b)\sin t + b\cos\left[\left(\frac{a+b}{b}\right)t\right]\right\}. \end{aligned}$$

Epicycloid Involute

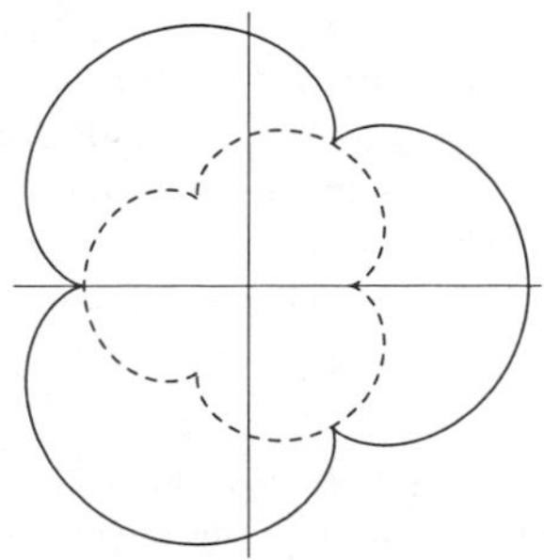

The INVOLUTE of the EPICYCLOID

$$x = (a+b)\cos t - b\cos\left[\left(\frac{a+b}{b}\right)t\right]$$

$$y = (a+b)\sin t - b\sin\left[\left(\frac{a+b}{b}\right)t\right]$$

is another EPICYCLOID given by

$$x = \frac{a+2b}{a}\left\{(a+b)\cos t + b\cos\left[\left(\frac{a+b}{b}\right)t\right]\right\}$$

$$y = \frac{a+2b}{a}\left\{(a+b)\sin t + b\cos\left[\left(\frac{a+b}{b}\right)t\right]\right\}.$$

Epicycloid Pedal Curve

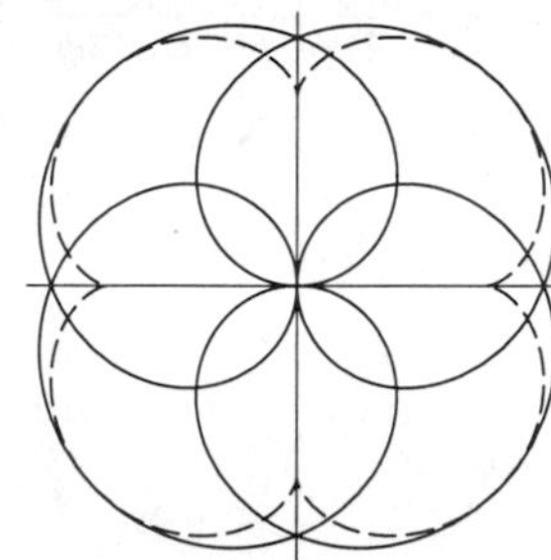

The PEDAL CURVE of an EPICYCLOID with PEDAL POINT at the center, shown for an epicycloid with four cusps, is not a ROSE as claimed by Lawrence (1972).

References

Lawrence, J. D. *A Catalog of Special Plane Curves.* New York: Dover, p. 204, 1972.

Epicycloid Radial Curve

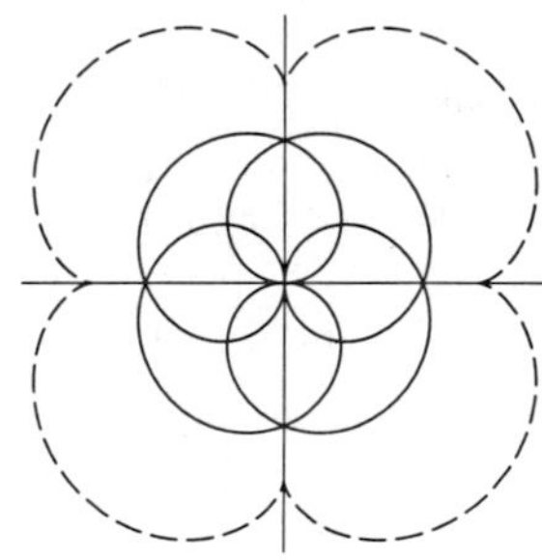

The RADIAL CURVE of an EPICYCLOID is shown above for an epicycloid with four cusps. It is not a ROSE, as claimed by Lawrence (1972).

References

Lawrence, J. D. *A Catalog of Special Plane Curves.* New York: Dover, p. 202, 1972.

Epimenides Paradox

A PARADOX, also called the LIAR'S PARADOX, attributed to the philosopher Epimenides in the sixth century BC. "All Cretans are liers... One of their own poets has said so." A sharper version of the paradox is the EUBULIDES PARADOX, "This statement is false."

see also EUBULIDES PARADOX, SOCRATES' PARADOX

References

Hofstadter, D. R. *Gödel, Escher, Bach: An Eternal Golden Braid.* New York: Vintage Books, p. 17, 1989.

Epimorphism

A SURJECTIVE MORPHISM.

Epispiral

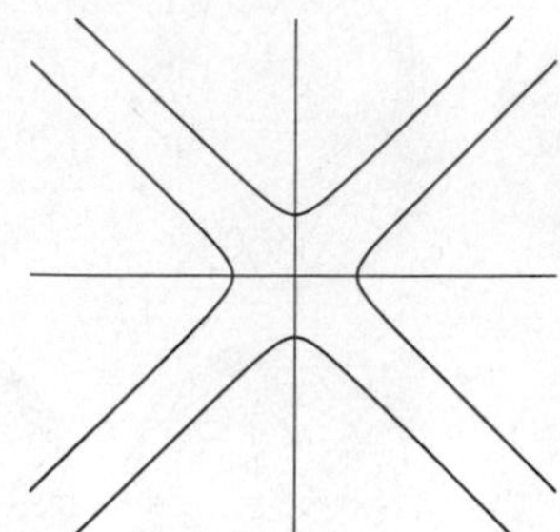

A plane curve with polar equation

$$r = \frac{a}{\cos(n\theta)}.$$

There are n sections if n is ODD and $2n$ if n is EVEN.

References

Lawrence, J. D. *A Catalog of Special Plane Curves.* New York: Dover, pp. 192–193, 1972.

Epispiral Inverse Curve

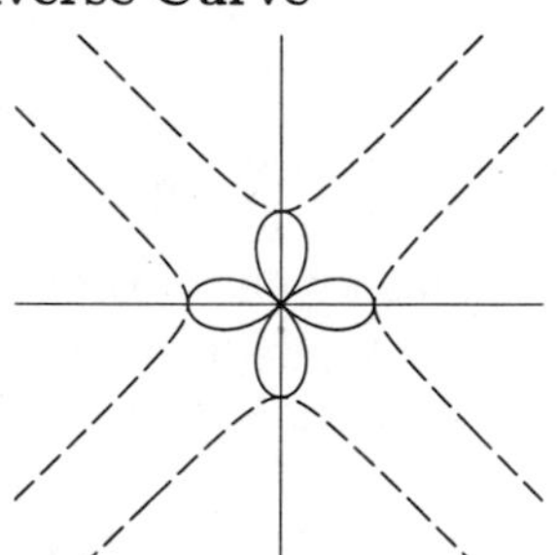

The INVERSE CURVE of the EPISPIRAL

$$r = a\sec(nt)$$

with INVERSION CENTER at the origin and inversion radius k is the ROSE

$$r = \frac{k\cos(nt)}{a}.$$

Epitrochoid

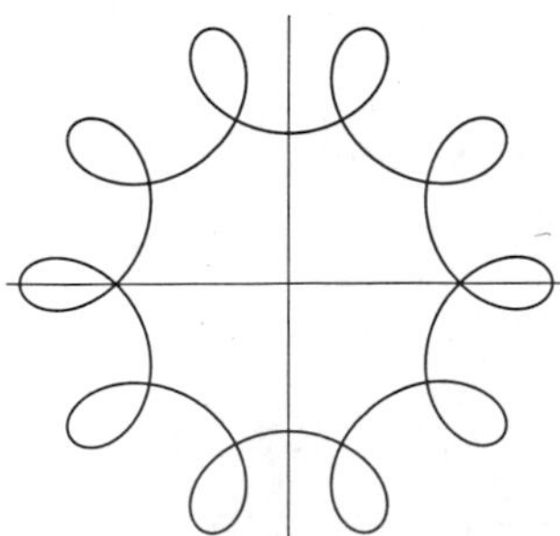

The ROULETTE traced by a point P attached to a CIRCLE of radius b rolling around the outside of a fixed

CIRCLE of radius a. These curves were studied by Dürer (1525), Desargues (1640), Huygens (1679), Leibniz, Newton (1686), L'Hospital (1690), Jakob Bernoulli (1690), la Hire (1694), Johann Bernoulli (1695), Daniel Bernoulli (1725), Euler (1745, 1781). An epitrochoid appears in Dürer's work *Instruction in Measurement with Compasses and Straight Edge* (1525). He called epitrochoids SPIDER LINES because the lines he used to construct the curves looked like a spider.

The parametric equations for an epitrochoid are

$$x = m\cos t - h\cos\left(\frac{m}{b}t\right)$$
$$y = m\sin t - h\sin\left(\frac{m}{b}t\right),$$

where $m \equiv a+b$ and h is the distance from P to the center of the rolling CIRCLE. Special cases include the LIMAÇON with $a = b$, the CIRCLE with $a = 0$, and the EPICYCLOID with $h = b$.

see also EPICYCLOID, HYPOTROCHOID, SPIROGRAPH

References
Lawrence, J. D. *A Catalog of Special Plane Curves.* New York: Dover, pp. 168–170, 1972.
Lee, X. "Epitrochoid." http://www.best.com/~xah/SpecialPlaneCurves_dir/Epitrochoid_dir/epitrochoid.html.
Lee, X. "Epitrochoid and Hypotrochoid Movie Gallery." http://www.best.com/~xah/SpecialPlaneCurves_dir/EpiHypoTMovieGallery_dir/epiHypoTMovieGallery.html.

Epitrochoid Evolute

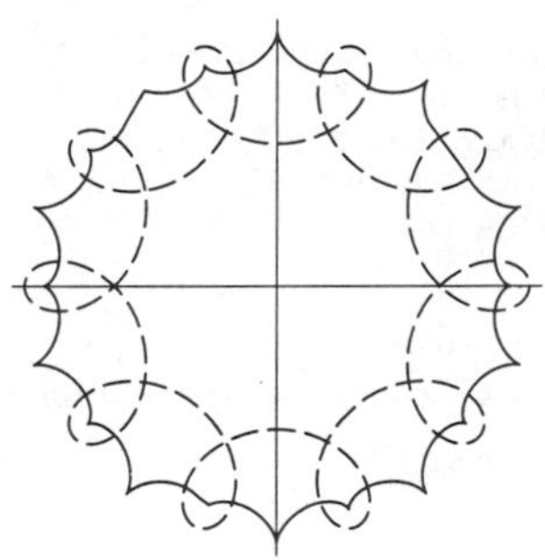

Epsilon

In mathematics, a small POSITIVE INFINITESIMAL quantity whose LIMIT is usually taken to be 0. The late mathematician P. Erdős also used the term "epsilons" to refer to children.

Epsilon-Neighborhood

see NEIGHBORHOOD

Epstein Zeta Function

$$Z\begin{vmatrix}\mathbf{g}\\\mathbf{h}\end{vmatrix}(q;s) = \sum_{\mathbf{l}} \frac{e^{-2\pi i\mathbf{h}\cdot\mathbf{l}}}{[q(\mathbf{l}+\mathbf{g})]^{s/2}},$$

where $\mathbf{g}$ and $\mathbf{h}$ are arbitrary VECTORS, the SUM runs over a d-dimensional LATTICE, and $\mathbf{l} = -\mathbf{g}$ is omitted if $\mathbf{g}$ is a lattice VECTOR.

see also ZETA FUNCTION

References
Glasser, M. L. and Zucker, I. J. "Lattice Sums in Theoretical Chemistry." *Theoretical Chemistry: Advances and Perspectives, Vol. 5.* New York: Academic Press, pp. 69–70, 1980.
Shanks, D. "Calculation and Applications of Epstein Zeta Functions." *Math. Comput.* **29**, 271–287, 1975.

Equal

Two quantities are said to be equal if they are, in some well-defined sense, equivalent. Equality of quantities a and b is written $a = b$.

A symbol with three horizontal line segments ($\equiv$) resembling the equals sign is used to denote both equality by definition (e.g., $A \equiv B$ means A is DEFINED to be equal to B) and CONGRUENCE (e.g., $13 \equiv 12 \pmod 1$) means 13 divided by 12 leaves a REMAINDER of 1—a fact known to all readers of analog clocks).

see also CONGRUENCE, DEFINED, DIFFERENT, EQUAL BY DEFINITION, EQUALITY, EQUIVALENT, ISOMORPHISM

Equal by Definition

see DEFINED

Equal Detour Point

The center of an outer SODDY CIRCLE. It has TRIANGLE CENTER FUNCTION

$$\alpha = 1 + \frac{2\Delta}{a(b+c-a)} = \sec(\tfrac{1}{2}A)\cos(\tfrac{1}{2}B)\cos(\tfrac{1}{2}C) + 1.$$

Given a point Y not between A and B, a detour of length

$$|AY| + |YB| - |AB|$$

is made walking from A to B via Y, the point is of equal detour if the three detours from one side to another via Y are equal. If ABC has no ANGLE $> 2\sin^{-1}(4/5)$, then the point given by the above TRILINEAR COORDINATES is the unique equal detour point. Otherwise, the ISOPERIMETRIC POINT is also equal detour.

References
Kimberling, C. "Central Points and Central Lines in the Plane of a Triangle." *Math. Mag.* **67**, 163–187, 1994.
Kimberling, C. "Isoperimetric Point and Equal Detour Point." http://www.evansville.edu/~ck6/tcenters/recent/isoper.html.
Veldkamp, G. R. "The Isoperimetric Point and the Point(s) of Equal Detour." *Amer. Math. Monthly* **92**, 546–558, 1985.

Equal Parallelians Point

The point of intersection of the three LINE SEGMENTS, each parallel to one side of a TRIANGLE and touching the other two, such that all three segments are of the same length. The TRILINEAR COORDINATES are

$$bc(ca+ab-bc) : ca(ab+bc-ca) : ab(bc+ca-ab).$$

References

Kimberling, C. "Equal Parallelians Point." `http://www.evansville.edu/~ck6/tcenters/recent/eqparal.html`.

Equality

A mathematical statement of the equivalence of two quantities. The equality "A is equal to B" is written $A = B$.

see also EQUAL, INEQUALITY

Equally Likely Outcomes Distribution

Let there be a set S with N elements, each of them having the same probability. Then

$$\begin{aligned} P(S) &= P\left(\bigcup_{i=1}^N E_i\right) = \sum_{i=1}^N P(E_i) \\ &= P(E_i)\sum_{i=1}^N 1 = NP(E_i). \end{aligned}$$

Using $P(S) \equiv 1$ gives

$$P(E_i) = \frac{1}{N}.$$

see also UNIFORM DISTRIBUTION

Equi-Brocard Center

The point Y for which the TRIANGLES BYC, CYA, and AYB have equal BROCARD ANGLES.

References

Kimberling, C. "Central Points and Central Lines in the Plane of a Triangle." *Math. Mag.* **67**, 163–187, 1994.

Equiaffinity

An AREA-preserving AFFINITY. Equiaffinities include the ELLIPTIC ROTATION, HYPERBOLIC ROTATION, HYPERBOLIC ROTATION (CROSSED), and PARABOLIC ROTATION.

Equiangular Spiral

see LOGARITHMIC SPIRAL

Equianharmonic Case

The case of the WEIERSTRAß ELLIPTIC FUNCTION with invariants $g_2 = 0$ and $g_3 = 1$.

see also LEMNISCATE CASE, PSEUDOLEMNISCATE CASE

References

Abramowitz, M. and Stegun, C. A. (Eds.). "Equianharmonic Case ($g_2 = 0$, $g_3 = 1$)." §18.13 in *Handbook of Mathematical Functions with Formulas, Graphs, and Mathematical Tables, 9th printing.* New York: Dover, p. 652, 1972.

Equichordal Point

A point P for which all the CHORDS passing through P are of the same length. It satisfies

$$px + py = [\text{const}],$$

where p is the CHORD length. It is an open question whether a plane convex region can have two equichordal points.

see also EQUICHORDAL PROBLEM, EQUIPRODUCT POINT, EQUIRECIPROCAL POINT

Equichordal Problem

Is there a planar body bounded by a simple closed curve and star-shaped with respect to two interior points p and q whose point X-rays at p and q are both constant? Rychlik (1997) has answered the question in the negative.

see also EQUICHORDAL POINT

References

Rychlik, M. "The Equichordal Point Problem." *Elec. Res. Announcements Amer. Math. Soc.* **2**, 108–123, 1996.
Rychlik, M. "A Complete Solution to the Equichordal Problem of Fujiwara, Blaschke, Rothe, and Weitzenböck." *Invent. Math.* **129**, 141–212, 1997.

Equidecomposable

The ability of two plane or space regions to be DISSECTED into each other.

Equidistance Postulate

PARALLEL lines are everywhere equidistant. This POSTULATE is equivalent to the PARALLEL AXIOM.

References

Dunham, W. "Hippocrates' Quadrature of the Lune." Ch. 1 in *Journey Through Genius: The Great Theorems of Mathematics.* New York: Wiley, p. 54, 1990.

Equidistant Cylindrical Projection

see CYLINDRICAL EQUIDISTANT PROJECTION

Equidistributed Sequence

A sequence of REAL NUMBERS $\{x_n\}$ is equidistributed if the probability of finding x_n in any subinterval is proportional to the subinterval length.

see also WEYL'S CRITERION

References

Kuipers, L. and Niederreiter, H. *Uniform Distribution of Sequences.* New York: Wiley, 1974.

Pólya, G. and Szegő, G. *Problems and Theorems in Analysis I.* New York: Springer-Verlag, p. 88, 1972.

Vardi, I. *Computational Recreations in Mathematica.* Reading, MA: Addison-Wesley, pp. 155–156, 1991.

Equilateral Hyperbola

see RECTANGULAR HYPERBOLA

Equilateral Triangle

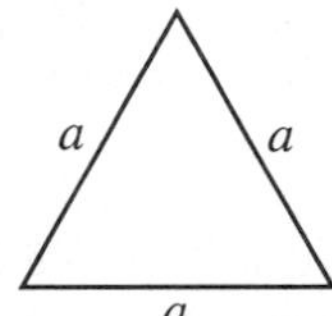

An equilateral triangle is a TRIANGLE with all three sides of equal length s. An equilateral triangle also has three equal $60°$ ANGLES.

An equilateral triangle can be constructed by TRISECTING all three ANGLES of any TRIANGLE (MORLEY'S THEOREM). NAPOLEON'S THEOREM states that if three equilateral triangles are drawn on the LEGS of any TRIANGLE (either all drawn inwards or outwards) and the centers of these triangles are connected, the result is another equilateral triangle.

Given the distances of a point from the three corners of an equilateral triangle, a, b, and c, the length of a side s is given by

$$3(a^4+b^4+c^4+s^4)=(a^2+b^2+c^2+s^2)^2 \qquad (1)$$

(Gardner 1977, pp. 56–57 and 63). There are infinitely many solutions for which a, b, and c are INTEGERS. In these cases, one of a, b, c, and s is DIVISIBLE by 3, one by 5, one by 7, and one by 8 (Guy 1994, p. 183).

The ALTITUDE h of an equilateral triangle is

$$h=\tfrac{1}{2}\sqrt{3}\,s, \qquad (2)$$

where s is the side length, so the AREA is

$$A=\tfrac{1}{2}sh=\tfrac{1}{4}\sqrt{3}\,s^2. \qquad (3)$$

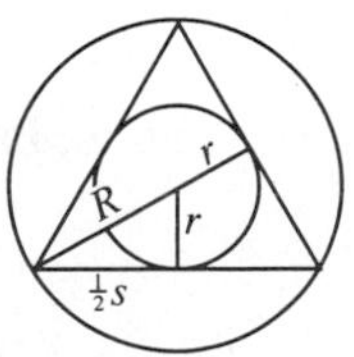

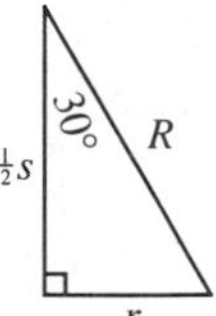

The INRADIUS r, CIRCUMRADIUS R, and AREA A can be computed directly from the formulas for a general regular POLYGON with side length s and $n=3$ sides,

$$r=\tfrac{1}{2}s\cot\left(\frac{\pi}{3}\right)=\tfrac{1}{2}s\tan\left(\frac{\pi}{6}\right)=\tfrac{1}{6}\sqrt{3}\,s \qquad (4)$$

$$R=\tfrac{1}{2}s\csc\left(\frac{\pi}{3}\right)=\tfrac{1}{2}s\sec\left(\frac{\pi}{6}\right)=\tfrac{1}{3}\sqrt{3}\,s \qquad (5)$$

$$A=\tfrac{1}{4}ns^2\cot\left(\frac{\pi}{3}\right)=\tfrac{1}{4}\sqrt{3}\,s^2. \qquad (6)$$

The AREAS of the INCIRCLE and CIRCUMCIRCLE are

$$A_r=\pi r^2=\tfrac{1}{12}\pi s^2 \qquad (7)$$

$$A_R=\pi R^2=\tfrac{1}{3}\pi s^2. \qquad (8)$$

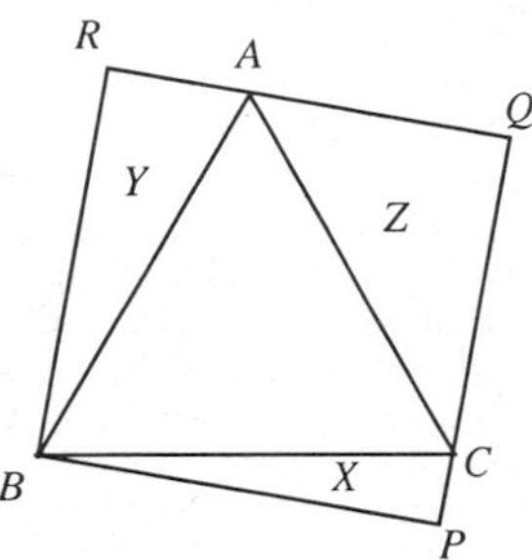

Let any RECTANGLE be circumscribed about an EQUILATERAL TRIANGLE. Then

$$X+Y=Z, \qquad (9)$$

where X, Y, and Z are the AREAS of the triangles in the figure (Honsberger 1985).

Begin with an arbitrary TRIANGLE and find the EXCENTRAL TRIANGLE. Then find the EXCENTRAL TRIANGLE of that triangle, and so on. Then the resulting triangle approaches an equilateral triangle. The only RATIONAL TRIANGLE is the equilateral triangle (Conway and Guy 1996). A POLYHEDRON composed of only equilateral triangles is known as a DELTAHEDRON.

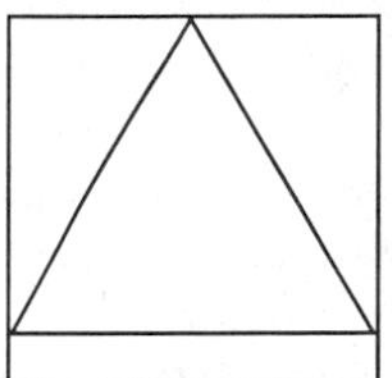

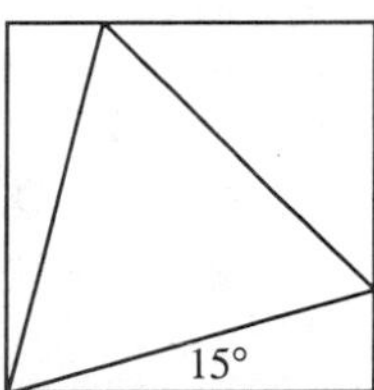

The largest equilateral triangle which can be inscribed in a UNIT SQUARE (left) has side length and area

$$s=1 \qquad (10)$$

$$A=\tfrac{1}{4}\sqrt{3}. \qquad (11)$$

The smallest equilateral triangle which can be inscribed (right) is oriented at an angle of $15°$ and has side length and area

$$s=\sec(15°)=\sqrt{6}-\sqrt{2} \qquad (12)$$

$$A=2\sqrt{3}-3 \qquad (13)$$

(Madachy 1979).

see also ACUTE TRIANGLE, DELTAHEDRON, EQUILIC QUADRILATERAL, FERMAT POINT, GYROELONGATED SQUARE DIPYRAMID, ICOSAHEDRON, ISOGONIC CENTERS, ISOSCELES TRIANGLE, MORLEY'S THEOREM, OCTAHEDRON, PENTAGONAL DIPYRAMID, RIGHT TRIANGLE, SCALENE TRIANGLE, SNUB DISPHENOID, TETRAHEDRON, TRIANGLE, TRIANGULAR DIPYRAMID, TRIAUGMENTED TRIANGULAR PRISM, VIVIANI'S THEOREM

References

Beyer, W. H. (Ed.) *CRC Standard Mathematical Tables, 28th ed.* Boca Raton, FL: CRC Press, p. 121, 1987.
Conway, J. H. and Guy, R. K. "The Only Rational Triangle." In *The Book of Numbers.* New York: Springer-Verlag, pp. 201 and 228–239, 1996.
Dixon, R. *Mathographics.* New York: Dover, p. 33, 1991.
Gardner, M. *Mathematical Carnival: A New Round-Up of Tantalizers and Puzzles from Scientific American.* New York: Vintage Books, 1977.
Guy, R. K. "Rational Distances from the Corners of a Square." §D19 in *Unsolved Problems in Number Theory, 2nd ed.* New York: Springer-Verlag, pp. 181–185, 1994.
Honsberger, R. "Equilateral Triangles." Ch. 3 in *Mathematical Gems I.* Washington, DC: Math. Assoc. Amer., 1973.
Honsberger, R. *Mathematical Gems III.* Washington, DC: Math. Assoc. Amer., pp. 19–21, 1985.
Madachy, J. S. *Madachy's Mathematical Recreations.* New York: Dover, pp. 115 and 129–131, 1979.

Equilibrium Point

An equilibrium point in GAME THEORY is a set of strategies $\{\hat{x}_1, \ldots, \hat{x}_n\}$ such that the ith payoff function $K_i(\mathbf{x})$ is larger or equal for any other ith strategy, i.e.,

$$K_i(\hat{x}_1, \ldots, \hat{x}_n) \geq K_i(\hat{x}_1, \ldots, \hat{x}_{i-1}, x_i, \hat{x}_{i+1}, \ldots, \hat{x}_n).$$

see NASH EQUILIBRIUM

Equilic Quadrilateral

A QUADRILATERAL in which a pair of opposite sides have the same length and are inclined at 60° to each other (or equivalently, satisfy $\langle A\rangle + \langle B\rangle = 120°$). Some interesting theorems hold for such quadrilaterals. Let $ABCD$ be an equilic quadrilateral with $AD = BC$ and $\langle A\rangle + \langle B\rangle = 120°$. Then

1. The MIDPOINTS P, Q, and R of the diagonals and the side CD always determine an EQUILATERAL TRIANGLE.
2. If EQUILATERAL TRIANGLE PCD is drawn outwardly on CD, then ΔPAB is also an EQUILATERAL TRIANGLE.
3. If EQUILATERAL TRIANGLES are drawn on AC, DC, and DB away from AB, then the three new VERTICES P, Q, and R are COLLINEAR.

See Honsberger (1985) for additional theorems.

References

Garfunkel, J. "The Equilic Quadrilateral." *Pi Mu Epsilon J.*, 317–329, Fall 1981.
Honsberger, R. *Mathematical Gems III.* Washington, DC: Math. Assoc. Amer., pp. 32–35, 1985.

Equinumerous

Let A and B be two classes of POSITIVE integers. Let $A(n)$ be the number of integers in A which are less than or equal to n, and let $B(n)$ be the number of integers in B which are less than or equal to n. Then if

$$A(n) \sim B(n),$$

A and B are said to be equinumerous.

The four classes of PRIMES $8k+1$, $8k+3$, $8k+5$, $8k+7$ are equinumerous. Similarly, since $8k+1$ and $8k+5$ are both of the form $4k+1$, and $8k+3$ and $8k+7$ are both of the form $4k+3$, $4k+1$ and $4k+3$ are also equinumerous.

see also BERTRAND'S POSTULATE, CHOQUET THEORY, PRIME COUNTING FUNCTION

References

Shanks, D. *Solved and Unsolved Problems in Number Theory, 4th ed.* New York: Chelsea, pp. 21–22 and 31–32, 1993.

Equipollent

Two statements in LOGIC are said to be equipollent if they are deducible from each other. Two SETS with the same CARDINAL NUMBER are also said to be equipollent. The term EQUIPOTENT is sometimes used instead of equipollent.

Equipotent

see EQUIPOLLENT

Equipotential Curve

A curve in 2-D on which the value of a function $f(x,y)$ is a constant. Other synonymous terms are ISARITHM and ISOPLETH.

see also LEMNISCATE

Equiproduct Point

A point, such as interior points of a disk, such that

$$(px)(py) = [\text{const}],$$

where p is the CHORD length.

see also EQUICHORDAL POINT, EQUIRECIPROCAL POINT

Equireciprocal Point

A point, such as the FOCI of an ELLIPSE, which satisfies

$$\frac{1}{px} + \frac{1}{py} = [\text{const}],$$

where p is the CHORD length.

see also EQUICHORDAL POINT, EQUIPRODUCT POINT

Equirectangular Projection

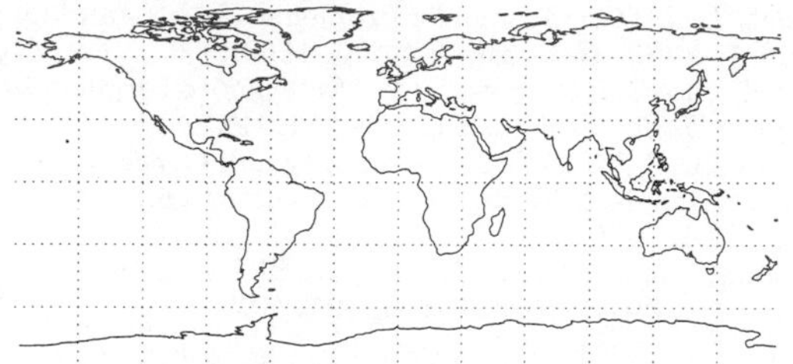

A MAP PROJECTION, also called a RECTANGULAR PROJECTION, in which the horizontal coordinate is the longitude and the vertical coordinate is the latitude.

Equiripple

A distribution of ERROR such that the ERROR remaining is always given approximately by the last term dropped.

Equitangential Curve

see TRACTRIX

Equivalence Class

An equivalence class is defined as a SUBSET of the form $\{x \in X : xRa\}$, where a is an element of X and the NOTATION "xRy" is used to mean that there is an EQUIVALENCE RELATION between x and y. It can be shown that any two equivalence classes are either equal or disjoint, hence the collection of equivalence classes forms a partition of X. For all $a, b \in X$, we have aRb IFF a and b belong to the same equivalence class.

A set of CLASS REPRESENTATIVES is a SUBSET of X which contains EXACTLY ONE element from each equivalence class.

For n a POSITIVE INTEGER, and a, b INTEGERS, consider the CONGRUENCE $a \equiv b \pmod{n}$, then the equivalence classes are the sets $\{\ldots, -2n, -n, 0, n, 2n, \ldots\}$, $\{\ldots, 1-2n, 1-n, 1, 1+n, 1+2n, \ldots\}$ etc. The standard CLASS REPRESENTATIVES are taken to be $0, 1, 2, \ldots, n-1$.

see also CONGRUENCE, COSET

References

Shanks, D. *Solved and Unsolved Problems in Number Theory, 4th ed.* New York: Chelsea, pp. 56–57, 1993.

Equivalence Problem

see METRIC EQUIVALENCE PROBLEM

Equivalence Relation

An equivalence relation on a set X is a SUBSET of $X \times X$, i.e., a collection R of ordered pairs of elements of X, satisfying certain properties. Write "xRy" to mean (x, y) is an element of R, and we say "x is related to y," then the properties are

1. Reflexive: aRa for all $a \in X$,
2. Symmetric: aRb IMPLIES bRa for all $a, b \in X$
3. Transitive: aRb and bRc imply aRc for all $a, b, c \in X$,

where these three properties are completely independent. Other notations are often used to indicate a relation, e.g., $a \equiv b$ or $a \sim b$.

see also EQUIVALENCE CLASS, TEICHMÜLLER SPACE

References

Stewart, I. and Tall, D. *The Foundations of Mathematics.* Oxford, England: Oxford University Press, 1977.

Equivalent

If $A \Rightarrow B$ and $B \Rightarrow A$ (i.e, $A \Rightarrow B \wedge B \Rightarrow A$, where $\Rightarrow$ denotes IMPLIES), then A and B are said to be equivalent, a relationship which is written symbolically as $A \Leftrightarrow B$ or $A \rightleftharpoons B$. However, if A and B are "equivalent by definition" (i.e., A is DEFINED to be B), this is written $A \equiv B$, a notation which conflicts with that for a CONGRUENCE.

see also DEFINED, IFF, IMPLIES

Equivalent Matrix

An $m \times n$ MATRIX A is said to be equivalent to another $m \times n$ MATRIX B IFF

$$\mathsf{B} = \mathsf{PAQ}$$

for P and Q any $m \times n$ and $n \times n$ MATRICES, respectively.

References

Gradshteyn, I. S. and Ryzhik, I. M. *Tables of Integrals, Series, and Products, 5th ed.* San Diego, CA: Academic Press, p. 1103, 1979.

Eratosthenes Sieve

1	2	3	~~4~~ (2)	5	~~6~~ (2)	7	~~8~~ (2)	9	~~10~~ (2)
11	~~12~~ (2)	13	~~14~~ (2)	15	~~16~~ (2)	17	~~18~~ (2)	19	~~20~~ (2)
21	~~22~~ (2)	23	~~24~~ (2)	25	~~26~~ (2)	27	~~28~~ (2)	29	~~30~~ (2)
31	~~32~~ (2)	33	~~34~~ (2)	35	~~36~~ (2)	37	~~38~~ (2)	39	~~40~~ (2)
41	~~42~~ (2)	43	~~44~~ (2)	45	~~46~~ (2)	47	~~48~~ (2)	49	~~50~~ (2)

1	2	3	~~4~~ (2)	5	~~6~~ (23)	7	~~8~~ (2)	~~9~~ (3)	~~10~~ (2)
11	~~12~~ (23)	13	~~14~~ (2)	~~15~~ (3)	~~16~~ (2)	17	~~18~~ (23)	19	~~20~~ (2)
~~21~~ (3)	~~22~~ (2)	23	~~24~~ (23)	25	~~26~~ (2)	~~27~~ (3)	~~28~~ (2)	29	~~30~~ (23)
31	~~32~~ (2)	~~33~~ (3)	~~34~~ (2)	35	~~36~~ (23)	37	~~38~~ (2)	~~39~~ (3)	~~40~~ (2)
41	~~42~~ (23)	43	~~44~~ (2)	~~45~~ (3)	~~46~~ (2)	47	~~48~~ (23)	49	~~50~~ (2)

1	2	3	~~4~~ (2)	5	~~6~~ (23)	7	~~8~~ (2)	~~9~~ (3)	~~10~~ (25)
11	~~12~~ (23)	13	~~14~~ (2)	~~15~~ (35)	~~16~~ (2)	17	~~18~~ (23)	19	~~20~~ (25)
~~21~~ (3)	~~22~~ (2)	23	~~24~~ (23)	~~25~~ (5)	~~26~~ (2)	~~27~~ (3)	~~28~~ (2)	29	~~30~~ (235)
31	~~32~~ (2)	~~33~~ (3)	~~34~~ (2)	~~35~~ (5)	~~36~~ (23)	37	~~38~~ (2)	~~39~~ (3)	~~40~~ (25)
41	~~42~~ (23)	43	~~44~~ (2)	~~45~~ (35)	~~46~~ (2)	47	~~48~~ (23)	49	~~50~~ (25)

1	2	3	~~4~~ (2)	5	~~6~~ (23)	7	~~8~~ (2)	~~9~~ (3)	~~10~~ (25)
11	~~12~~ (23)	13	~~14~~ (27)	~~15~~ (35)	~~16~~ (2)	17	~~18~~ (23)	19	~~20~~ (25)
~~21~~ (37)	~~22~~ (2)	23	~~24~~ (23)	~~25~~ (5)	~~26~~ (2)	~~27~~ (3)	~~28~~ (27)	29	~~30~~ (235)
31	~~32~~ (2)	~~33~~ (3)	~~34~~ (2)	~~35~~ (57)	~~36~~ (23)	37	~~38~~ (2)	~~39~~ (3)	~~40~~ (25)
41	~~42~~ (237)	43	~~44~~ (2)	~~45~~ (35)	~~46~~ (2)	47	~~48~~ (23)	~~49~~ (7)	~~50~~ (25)

An ALGORITHM for making tables of PRIMES. Sequentially write down the INTEGERS from 2 to the highest number n you wish to include in the table. Cross out all numbers > 2 which are divisible by 2 (every second number). Find the smallest remaining number > 2. It is 3. So cross out all numbers > 3 which are divisible by 3 (every third number). Find the smallest remaining number > 3. It is 5. So cross out all numbers > 5 which are divisible by 5 (every fifth number).

Continue until you have crossed out all numbers divisible by $\lfloor\sqrt{n}\rfloor$, where $\lfloor x \rfloor$ is the FLOOR FUNCTION. The numbers remaining are PRIME. This procedure is illustrated in the above diagram which sieves up to 50, and therefore crosses out PRIMES up to $\lfloor\sqrt{50}\rfloor = 7$. If the procedure is then continued up to n, then the number of cross-outs gives the number of distinct PRIME factors of each number.

References

Conway, J. H. and Guy, R. K. *The Book of Numbers.* New York: Springer-Verlag, pp. 127–130, 1996.
Ribenboim, P. *The New Book of Prime Number Records.* New York: Springer-Verlag, pp. 20–21, 1996.

Erdős-Anning Theorem

If an infinite number of points in the PLANE are all separated by INTEGER distances, then all the points lie on a straight LINE.

Erdős-Kac Theorem

A deeper result than the HARDY-RAMANUJAN THEOREM. Let $N(x,a,b)$ be the number of INTEGERS in $[3,x]$ such that inequality

$$a \leq \frac{\omega(n) - \ln\ln n}{\sqrt{\ln\ln n}} \leq b$$

holds, where $\omega(n)$ is the number of different PRIME factors of n. Then

$$\lim_{x\to\infty} N(x,a,b) = \frac{(x+o(x))}{\sqrt{2\pi}} \int_a^b e^{-t^2/2}\,dt.$$

The theorem is discussed in Kac (1959).

References

Kac, M. *Statistical Independence in Probability, Analysis and Number Theory.* New York: Wiley, 1959.
Riesel, H. "The Erdős-Kac Theorem." *Prime Numbers and Computer Methods for Factorization, 2nd ed.* Boston, MA: Birkhäuser, pp. 158–159, 1994.

Erdős-Mordell Theorem

If O is any point inside a TRIANGLE ΔABC, and P, Q, and R are the feet of the perpendiculars from O upon the respective sides BC, CA, and AB, then

$$OA + OB + OC \geq 2(OP + OQ + OR).$$

Oppenheim (1961) and Mordell (1962) also showed that

$$OA \times OB \times OC \geq (OQ + OR)(OR + OP)(OP + OQ).$$

References

Bankoff, L. "An Elementary Proof of the Erdős-Mordell Theorem." *Amer. Math. Monthly* **65**, 521, 1958.
Brabant, H. "The Erdős-Mordell Inequality Again." *Nieuw Tijdschr. Wisk.* **46**, 87, 1958/1959.
Casey, J. *A Sequel to the First Six Books of the Elements of Euclid, 6th ed.* Dublin: Hodges, Figgis, & Co., p. 253, 1892.
Coxeter, H. S. M. *Introduction to Geometry, 2nd ed.* New York: Wiley, p. 9, 1969.
Erdős, P. "Problem 3740." *Amer. Math. Monthly* **42**, 396, 1935.
Fejes-Tóth, L. *Lagerungen in der Ebene auf der Kugel und im Raum.* Berlin: Springer, 1953.
Mordell, L. J. "On Geometric Problems of Erdős and Oppenheim." *Math. Gaz.* **46**, 213–215, 1962.
Mordell, L. J. and Barrow, D. F. "Solution to Problem 3740." *Amer. Math. Monthly* **44**, 252–254, 1937.
Oppenheim, A. "The Erdős Inequality and Other Inequalities for a Triangle." *Amer. Math. Monthly* **68**, 226–230 and 349, 1961.
Veldkamp, G. R. "The Erdős-Mordell Inequality." *Nieuw Tijdschr. Wisk.* **45**, 193–196, 1957/1958.

Erdős Number

An author's Erdős number is 1 if he has co-authored a paper with Erdős, 2 if he has co-authored a paper with someone who has co-authored a paper with Erdős, etc.

References

Grossman, J. and Ion, P. "The Erdős Number Project." `http://www.acs.oakland.edu/~grossman/erdoshp.html`.

Erdős Reciprocal Sum Constants

see A-SEQUENCE, $B2$-SEQUENCE, NONAVERAGING SEQUENCE

Erdős-Selfridge Function

The Erdős-Selfridge function $g(k)$ is defined as the least integer bigger than $k+1$ such that all prime factors of $\binom{g(k)}{k}$ exceed k (Ecklund *et al.* 1974). The best lower bound known is

$$g(k) \geq \exp\left(c\frac{\ln^3 k}{\ln\ln k}^{1/2}\right)$$

(Granville and Ramare 1996). Scheidler and Williams (1992) tabulated $g(k)$ up to $k = 140$, and Lukes *et al.* (1997) tabulated $g(k)$ for $135 \leq k \leq 200$. The values for $n = 2, 3, \ldots$ are 4, 7, 7, 23, 62, 143, 44, 159, 46, 47, 174, 2239, ... (Sloane's A046105).

see also BINOMIAL COEFFICIENT, LEAST PRIME FACTOR

References

Ecklund, E. F. Jr.; Erdős, P.; and Selfridge, J. L. "A New Function Associated with the prime factors of $\binom{n}{k}$. *Math. Comput.* **28**, 647–649, 1974.
Erdős, P.; Lacampagne, C. B.; and Selfridge, J. L. "Estimates of the Least Prime Factor of a Binomial Coefficient." *Math. Comput.* **61**, 215–224, 1993.
Granville, A. and Ramare, O. "Explicit Bounds on Exponential Sums and the Scarcity of Squarefree Binomial Coefficients." *Mathematika* **43**, 73–107, 1996.
Lukes, R. F.; Scheidler, R.; and Williams, H. C. "Further Tabulation of the Erdős-Selfridge Function." *Math. Comput.* **66**, 1709–1717, 1997.
Scheidler, R. and Williams, H. C. "A Method of Tabulating the Number-Theoretic Function $g(k)$." *Math. Comput.* **59**, 251–257, 1992.
Sloane, N. J. A. Sequence A046105 in "An On-Line Version of the Encyclopedia of Integer Sequences."

Erdős Squarefree Conjecture

The CENTRAL BINOMIAL COEFFICIENT $\binom{2n}{n}$ is never SQUAREFREE for $n > 4$. This was proved true for all sufficiently large n by SÁRKÖZY'S THEOREM. Goetgheluck (1988) proved the CONJECTURE true for $4 < n \leq 2^{42205184}$ and Vardi (1991) for $4 < n < 2^{774840978}$. The conjecture was proved true in its entirely by Granville and Ramare (1996).

see also CENTRAL BINOMIAL COEFFICIENT

References

Erdős, P. and Graham, R. L. *Old and New Problems and Results in Combinatorial Number Theory.* Geneva, Switzerland: L'Enseignement Mathématique Université de Genève, Vol. 28, p. 71, 1980.

Goetgheluck, P. "Prime Divisors of Binomial Coefficients." *Math. Comput.* **51**, 325–329, 1988.

Granville, A. and Ramare, O. "Explicit Bounds on Exponential Sums and the Scarcity of Squarefree Binomial Coefficients." *Mathematika* **43**, 73–107, 1996.

Sander, J. W. "On Prime Divisors of Binomial Coefficients." *Bull. London Math. Soc.* **24**, 140–142, 1992.

Sander, J. W. "A Story of Binomial Coefficients and Primes." *Amer. Math. Monthly* **102**, 802–807, 1995.

Sárközy, A. "On Divisors of Binomial Coefficients. I." *J. Number Th.* **20**, 70–80, 1985.

Vardi, I. "Applications to Binomial Coefficients." *Computational Recreations in Mathematica.* Reading, MA: Addison-Wesley, pp. 25–28, 1991.

Erdős-Szekeres Theorem

Suppose $a, b \in \mathbb{N}$, $n = ab + 1$, and $x_1, \ldots, x_n$ is a sequence of n REAL NUMBERS. Then this sequence contains a MONOTONIC increasing (decreasing) subsequence of $a + 1$ terms or a MONOTONIC decreasing (increasing) subsequence of $b + 1$ terms. DILWORTH'S LEMMA is a generalization of this theorem.

see also COMBINATORICS

Erf

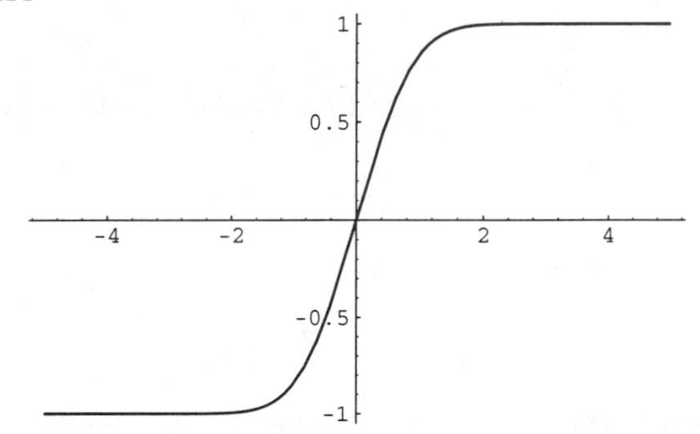

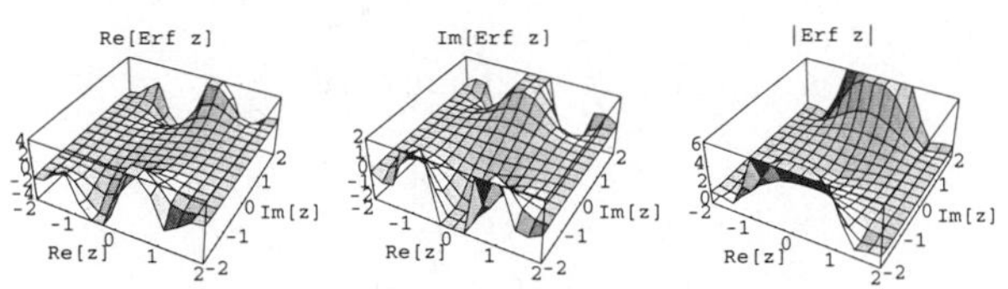

The "error function" encountered in integrating the GAUSSIAN DISTRIBUTION.

$$\begin{aligned} \operatorname{erf}(z) &\equiv \frac{2}{\sqrt{\pi}} \int_0^z e^{-t^2}\, dt && (1)\\ &= 1 - \operatorname{erfc}(z) && (2)\\ &= \sqrt{\pi}\, \gamma(\tfrac{1}{2}, z^2), && (3) \end{aligned}$$

where ERFC is the complementary error function and $\gamma(x, a)$ is the incomplete GAMMA FUNCTION. It can also be defined as a MACLAURIN SERIES

$$\operatorname{erf}(z) = \frac{2}{\sqrt{\pi}} \sum_{n=0}^{\infty} \frac{(-1)^n z^{2n+1}}{n!(2n+1)}. \tag{4}$$

Erf has the values

$$\operatorname{erf}(0) = 0 \tag{5}$$

$$\operatorname{erf}(\infty) = 1. \tag{6}$$

It is an ODD FUNCTION

$$\operatorname{erf}(-z) = -\operatorname{erf}(z), \tag{7}$$

and satisfies

$$\operatorname{erf}(z) + \operatorname{erfc}(z) = 1. \tag{8}$$

Erf may be expressed in terms of a CONFLUENT HYPERGEOMETRIC FUNCTION OF THE FIRST KIND M as

$$\operatorname{erf}(z) = \frac{2z}{\sqrt{\pi}} M(\tfrac{1}{2}, \tfrac{3}{2}, -z^2) = \frac{2z}{\sqrt{\pi}} e^{-z^2} M(1, \tfrac{3}{2}, z^2). \tag{9}$$

Erf is bounded by

$$\frac{1}{x + \sqrt{x^2 + 2}} < e^{x^2} \int_x^\infty e^{-t^2}\, dt \leq \frac{1}{x + \sqrt{x^2 + \frac{4}{\pi}}}. \tag{10}$$

Its DERIVATIVE is

$$\frac{d^n}{dz^n} \operatorname{erf}(z) = (-1)^{n-1} \frac{2}{\sqrt{\pi}} H_n(z) e^{-z^2}, \tag{11}$$

where H_n is a HERMITE POLYNOMIAL. The first DERIVATIVE is

$$\frac{d}{dz} \operatorname{erf}(z) = \frac{2}{\sqrt{\pi}} e^{-z^2/2}, \tag{12}$$

and the integral is

$$\int \operatorname{erf}(z)\, dz = z \operatorname{erf}(z) + \frac{e^{-z^2}}{\sqrt{\pi}}. \tag{13}$$

For $x \ll 1$, erf may be computed from

$$\begin{aligned}\operatorname{erf}(x) &= \frac{2}{\sqrt{\pi}}\int_0^x e^{-t^2}\,dt & (14)\\ &= \frac{2}{\sqrt{\pi}}\int_0^x \sum_{k=0}^{\infty}\frac{(-t^2)^k}{k!}\,dt\\ &= \frac{2}{\sqrt{\pi}}\int_0^x \sum_{k=0}^{\infty}\frac{(-1)^k t^{2k}}{k!}\,dt\\ &= \frac{2}{\sqrt{\pi}}\sum_{k=0}^{\infty}\frac{x^{2k+1}(-1)^k}{k!(2k+1)} & (15)\\ &= \frac{2}{\sqrt{\pi}}(x - \tfrac{1}{3}x^3 + \tfrac{1}{10}x^5 - \tfrac{1}{42}x^7 + \tfrac{1}{216}x^9\\ &\quad - \tfrac{1}{1320}x^{11} + \ldots) & (16)\\ &= \frac{2}{\sqrt{\pi}}e^{-x^2}x\left[1 + \frac{2x^2}{1\cdot 3} + \frac{(2x)^2}{1\cdot 3\cdot 5} + \ldots\right] & (17)\end{aligned}$$

(Acton 1990). For $x \gg 1$,

$$\begin{aligned}\operatorname{erf}(x) &= \frac{2}{\sqrt{\pi}}\left(\int_0^\infty e^{-t^2}\,dt - \int_x^\infty e^{-t^2}\,dt\right)\\ &= 1 - \frac{2}{\sqrt{\pi}}\int_x^\infty e^{-t^2}\,dt. & (18)\end{aligned}$$

Using INTEGRATION BY PARTS gives

$$\begin{aligned}\int_x^\infty e^{-t^2}\,dt &= -\frac{1}{2}\int_x^\infty \frac{1}{x}d(e^{-x^2})\\ &= -\frac{1}{2}\left[\frac{e^{-t^2}}{t}\right]_x^\infty - \frac{1}{2}\int_x^\infty \frac{e^{-t^2}\,dt}{t^2}\\ &= \frac{e^{-x^2}}{2x} + \frac{1}{4}\int_x^\infty \frac{1}{t^3}d(e^{-t^2})\\ &= \frac{e^{-x^2}}{2x} - \frac{e^{-x^2}}{4x^3} - \ldots, & (19)\end{aligned}$$

so

$$\operatorname{erf}(x) = 1 - \frac{e^{-x^2}}{\sqrt{\pi}\,x}\left(1 - \frac{1}{2x^2} - \ldots\right) \quad (20)$$

and continuing the procedure gives the ASYMPTOTIC SERIES

$$\begin{aligned}\operatorname{erf}(x) = 1 - \frac{e^{-x^2}}{\sqrt{\pi}}(x^{-1} - \tfrac{1}{2}x^{-3}\\ + \tfrac{3}{4}x^{-5} - \tfrac{15}{8}x^{-7} + \tfrac{105}{16}x^{-9} + \ldots). \quad (21)\end{aligned}$$

A COMPLEX generalization of erf is defined as

$$\begin{aligned}w(z) &\equiv e^{-z^2}\operatorname{erfc}(-iz) & (22)\\ &= e^{-z^2}\left(1 + \frac{2i}{\sqrt{\pi}} + \frac{2i}{\sqrt{\pi}}\int_0^z e^{t^2}\,dt\right) & (23)\\ &= \frac{i}{\pi}\int_{-\infty}^\infty \frac{e^{-t^2}\,dt}{z-t} = \frac{2iz}{\pi}\int_0^\infty \frac{e^{-t^2}\,dt}{z^2-t^2}. & (24)\end{aligned}$$

see also DAWSON'S INTEGRAL, ERFC, ERFI, GAUSSIAN INTEGRAL, NORMAL DISTRIBUTION FUNCTION, PROBABILITY INTEGRAL

References

Abramowitz, M. and Stegun, C. A. (Eds.). "Error Function" and "Repeated Integrals of the Error Function." §7.1–7.2 in *Handbook of Mathematical Functions with Formulas, Graphs, and Mathematical Tables, 9th printing.* New York: Dover, pp. 297–300, 1972.

Acton, F. S. *Numerical Methods That Work, 2nd printing.* Washington, DC: Math. Assoc. Amer., p. 16, 1990.

Arfken, G. *Mathematical Methods for Physicists, 3rd ed.* Orlando, FL: Academic Press, pp. 568–569, 1985.

Spanier, J. and Oldham, K. B. "The Error Function erf(x) and Its Complement erfc(x)." Ch. 40 in *An Atlas of Functions.* Washington, DC: Hemisphere, pp. 385–393, 1987.

Erfc

The "complementary error function"

$$\begin{aligned}\operatorname{erfc}(x) &\equiv \frac{2}{\sqrt{\pi}}\int_x^\infty e^{-t^2}\,dt & (1)\\ &= 1 - \operatorname{erf}(x) & (2)\\ &= \sqrt{\pi}\,\gamma(\tfrac{1}{2}, z^2), & (3)\end{aligned}$$

where γ is the incomplete GAMMA FUNCTION. It has the values

$$\operatorname{erfc}(0) = 1 \quad (4)$$

$$\operatorname{erfc}(\infty) = 0 \quad (5)$$

$$\operatorname{erfc}(-x) = 2 - \operatorname{erfc}(x) \quad (6)$$

$$\int_0^\infty \operatorname{erfc}(x)\,dx = \frac{1}{\sqrt{\pi}} \quad (7)$$

$$\int_0^\infty \operatorname{erfc}^2(x)\,dx = \frac{2-\sqrt{2}}{\sqrt{\pi}}. \quad (8)$$

A generalization is obtained from the differential equation

$$\frac{d^2y}{dz^2} + 2z\frac{dy}{dz} - 2ny = 0. \quad (9)$$

The general solution is then

$$y = A\operatorname{erfci}_n(z) + B\operatorname{erfci}_n(-z), \quad (10)$$

where $\operatorname{erfci}_n(z)$ is the erfc integral. For integral $n \geq 1$,

$$\begin{aligned}\operatorname{erfci}_n(z) &= \underbrace{\int\cdots\int}_{n}\operatorname{erfc}(z)\,dz & (11)\\ &= \frac{2}{\sqrt{\pi}}\int_0^\infty \frac{(t-z)^2}{n!}e^{-t^2}\,dt. & (12).\end{aligned}$$

The definition can be extended to $n = -1$ and 0 using

$$\mathrm{erfci}_{-1}(z) = \frac{2}{\sqrt{\pi}} e^{-z^2} \tag{13}$$

$$\mathrm{erfci}_0(z) = \mathrm{erfc}(z). \tag{14}$$

see also ERF, ERFI

References

Abramowitz, M. and Stegun, C. A. (Eds.). "Repeated Integrals of the Error Function." §7.2 in *Handbook of Mathematical Functions with Formulas, Graphs, and Mathematical Tables, 9th printing.* New York: Dover, pp. 299–300, 1972.

Press, W. H.; Flannery, B. P.; Teukolsky, S. A.; and Vetterling, W. T. "Incomplete Gamma Function, Error Function, Chi-Square Probability Function, Cumulative Poisson Function." §6.2 in *Numerical Recipes in FORTRAN: The Art of Scientific Computing, 2nd ed.* Cambridge, England: Cambridge University Press, pp. 209–214, 1992.

Spanier, J. and Oldham, K. B. "The Error Function $\mathrm{erf}(x)$ and Its Complement $\mathrm{erfc}(x)$" and "The $\exp(x)$ and $\mathrm{erfc}(\sqrt{x})$ and Related Functions." Chs. 40 and 41 in *An Atlas of Functions.* Washington, DC: Hemisphere, pp. 385–393 and 395–403, 1987.

Erfi

$$\mathrm{erfi}(z) \equiv -i\,\mathrm{erf}(iz).$$

see also ERF, ERFC

Ergodic Measure

An ENDOMORPHISM is called ergodic if it is true that $T^{-1}A = A$ IMPLIES $m(A) = 0$ or 1, where $T^{-1}A = \{x \in X : T(x) \in A\}$. Examples of ergodic endomorphisms include the MAP $X \to 2x \bmod 1$ on the unit interval with LEBESGUE MEASURE, certain AUTOMORPHISMS of the TORUS, and "Bernoulli shifts" (and more generally "Markov shifts").

Given a MAP T and a SIGMA ALGEBRA, there may be many ergodic measures. If there is only one ergodic measure, then T is called uniquely ergodic. An example of a uniquely ergodic transformation is the MAP $x \mapsto x + a \bmod 1$ on the unit interval when a is irrational. Here, the unique ergodic measure is LEBESGUE MEASURE.

Ergodic Theory

Ergodic theory can be described as the statistical and qualitative behavior of measurable group and semigroup actions on MEASURE SPACES. The GROUP is most commonly $\mathbb{N}$, $\mathbb{R}$, $\mathbb{R}^+$, and $\mathbb{Z}$.

Ergodic theory had its origins in the work of Boltzmann in statistical mechanics. Its mathematical origins are due to von Neumann, Birkhoff, and Koopman in the 1930s. It has since grown to be a huge subject and has applications not only to statistical mechanics, but also to number theory, differential geometry, functional analysis, etc. There are also many internal problems (e.g., ergodic theory being applied to ergodic theory) which are interesting.

see also AMBROSE-KAKUTANI THEOREM, BIRKHOFF'S ERGODIC THEOREM, DYE'S THEOREM, DYNAMICAL SYSTEM, HOPF'S THEOREM, ORNSTEIN'S THEOREM

References

Billingsley, P. *Ergodic Theory and Information.* New York: Wiley, 1965.

Cornfeld, I.; Fomin, S.; and Sinai, Ya. G. *Ergodic Theory.* New York: Springer-Verlag, 1982.

Katok, A. and Hasselblatt, B. *An Introduction to the Modern Theory of Dynamical Systems.* Cambridge, England: Cambridge University Press, 1996.

Nadkarni, M. G. *Basic Ergodic Theory.* India: Hindustan Book Agency, 1995.

Parry, W. *Topics in Ergodic Theory.* Cambridge, England: Cambridge University Press, 1982.

Smorodinsky, M. *Ergodic Theory, Entropy.* Berlin: Springer-Verlag, 1971.

Walters, P. *Ergodic Theory: Introductory Lectures.* New York: Springer-Verlag, 1975.

Ergodic Transformation

A transformation which has only trivial invariant SUBSETS is said to be invariant.

Erlanger Program

A program initiated by F. Klein in an 1872 lecture to describe geometric structures in terms of their group AUTOMORPHISMS.

References

Klein, F. "Vergleichende Betrachtungen über neuere geometrische Forschungen." 1872.

Yaglom, I. M. *Felix Klein and Sophus Lie: Evolution of the Idea of Symmetry in the Nineteenth Century.* Boston, MA: Birkhäuser, 1988.

Ermakoff's Test

The series $\sum f(n)$ for a monotonic nonincreasing $f(x)$ is convergent if

$$\overline{\lim_{x\to\infty}} \frac{e^x f(e^x)}{f(x)} < 1$$

and divergent if

$$\underline{\lim_{x\to\infty}} \frac{e^x f(e^x)}{f(x)} > 1.$$

References

Bromwich, T. J. I'a and MacRobert, T. M. *An Introduction to the Theory of Infinite Series, 3rd ed.* New York: Chelsea, p. 43, 1991.

Error

The difference between a quantity and its estimated or measured quantity.

see also ABSOLUTE ERROR, PERCENTAGE ERROR, RELATIVE ERROR

Error-Correcting Code

An error-correcting code is an algorithm for expressing a sequence of numbers such that any errors which are introduced can be detected and corrected (within certain limitations) based on the remaining numbers. The study of error-correcting codes and the associated mathematics is known as CODING THEORY.

Error detection is much simpler than error correction, and one or more "check" digits are commonly embedded in credit card numbers in order to detect mistakes. Early space probes like Mariner used a type of error-correcting code called a block code, and more recent space probes use convolution codes. Error-correcting codes are also used in CD players, high speed modems, and cellular phones. Modems use error detection when they compute CHECKSUMS, which are sums of the digits in a given transmission modulo some number. The ISBN used to identify books also incorporates a check DIGIT.

A powerful check for 13 DIGIT numbers consists of the following. Write the number as a string of DIGITS $a_1a_2a_3\cdots a_{13}$. Take $a_1+a_3+\ldots+a_{13}$ and double. Now add the number of DIGITS in ODD positions which are >4 to this number. Now add $a_2+a_4+\ldots+a_{12}$. The check number is then the number required to bring the last DIGIT to 0. This scheme detects all single DIGIT errors and all TRANSPOSITIONS of adject DIGITS except 0 and 9.

see also CHECKSUM, CODING THEORY, GALOIS FIELD, HADAMARD MATRIX, ISBN

References

Conway, J. H. and Sloane, N. J. A. "Error-Correcting Codes." §3.2 in *Sphere Packings, Lattices, and Groups, 2nd ed.* New York: Springer-Verlag, pp. 75–88, 1993.

Gallian, J. "How Computers Can Read and Correct ID Numbers." *Math Horizons*, pp. 14–15, Winter 1993.

Guy, R. K. *Unsolved Problems in Number Theory, 2nd ed.* New York: Springer-Verlag, pp. 119–121, 1994.

MacWilliams, F. J. and Sloane, N. J. A. *The Theory of Error-Correcting Codes.* Amsterdam, Netherlands: North-Holland, 1977.

Error Curve

see GAUSSIAN FUNCTION

Error Function

see ERF, ERFC

Error Function Distribution

A NORMAL DISTRIBUTION with MEAN 0.

$$P(x) = \frac{h}{\sqrt{\pi}} e^{-h^2x^2}. \tag{1}$$

The CHARACTERISTIC FUNCTION is

$$\phi(t) = e^{-t^2/(4h^2)}. \tag{2}$$

The MEAN, VARIANCE, SKEWNESS, and KURTOSIS are

$$\mu = 0 \tag{3}$$

$$\sigma^2 = \frac{1}{2h^2} \tag{4}$$

$$\gamma_1 = 0 \tag{5}$$

$$\gamma_2 = 0. \tag{6}$$

The CUMULANTS are

$$\kappa_1 = 0 \tag{7}$$

$$\kappa_2 = \frac{1}{2h^2} \tag{8}$$

$$\kappa_n = 0 \tag{9}$$

for $n \geq 3$.

Error Propagation

Given a FORMULA $y = f(x)$ with an ABSOLUTE ERROR in x of dx, the ABSOLUTE ERROR is dy. The RELATIVE ERROR is dy/y. If $x = f(u,v)$, then

$$x_i - \bar{x} = (u_i - \bar{u})\frac{\partial x}{\partial u} + (v_i - \bar{v})\frac{\partial x}{\partial v} + \ldots, \tag{1}$$

so

$$\begin{aligned} {\sigma_x}^2 &\equiv \frac{1}{N-1}\sum_i^N (x_i - \bar{x})^2 \\ &= \frac{1}{N-1}\sum_i^N \left[(u_i-\bar{u})^2\left(\frac{\partial x}{\partial u}\right)^2 + (v_i-\bar{v})^2\left(\frac{\partial x}{\partial v}\right)^2 \right. \\ &\quad \left. + 2(u_i-\bar{u})(v_i-\bar{v})\left(\frac{\partial x}{\partial u}\right)\left(\frac{\partial x}{\partial v}\right) + \ldots\right]. \end{aligned} \tag{2}$$

The definitions of VARIANCE and COVARIANCE then give

$${\sigma_u}^2 \equiv \frac{1}{N-1}\sum_{i=1}^N (u_i - \bar{u})^2 \tag{3}$$

$${\sigma_v}^2 \equiv \frac{1}{N-1}\sum_{i=1}^N (v_i - \bar{v})^2 \tag{4}$$

$${\sigma_{uv}}^2 \equiv \frac{1}{N-1}\sum_{i=1}^N (u_i - \bar{u})(v_i - \bar{v}), \tag{5}$$

so

$$\begin{aligned} {\sigma_x}^2 &= {\sigma_u}^2\left(\frac{\partial x}{\partial u}\right)^2 + {\sigma_v}^2\left(\frac{\partial x}{\partial v}\right)^2 \\ &\quad + 2\sigma_{uv}\left(\frac{\partial x}{\partial u}\right)\left(\frac{\partial x}{\partial v}\right) + \ldots. \end{aligned} \tag{6}$$

If u and v are uncorrelated, then $\sigma_{uv} = 0$ so

$${\sigma_x}^2 = {\sigma_u}^2\left(\frac{\partial x}{\partial u}\right)^2 + {\sigma_v}^2. \tag{7}$$

Now consider addition of quantities with errors. For $x = au \pm bv$, $\partial x/\partial u = a$ and $\partial x/\partial v = \pm b$, so

$$\sigma_x{}^2 = a^2\sigma_u{}^2 + b^2\sigma_v{}^2 \pm 2ab\sigma_{uv}{}^2. \quad (8)$$

For division of quantities with $x = \pm au/v$, $\partial x/\partial u = \pm a/v$ and $\partial x/\partial v = \mp au/v^2$, so

$$\sigma_x{}^2 = \frac{a^2}{v^2}\sigma_u{}^2 + \frac{a^2u^2}{\sigma_v{}^4} - 2\frac{a}{v}\frac{au}{v^2}\sigma_{uv}{}^2. \quad (9)$$

$$\left(\frac{\sigma_x}{x}\right)^2 = \frac{a^2}{v^2}\frac{v^2}{a^2u^2}\sigma_u{}^2 + \frac{a^2u^2}{v^4}\frac{v^2}{a^2u^2} - 2\left(\frac{a}{v}\right)\left(\frac{au}{v^2}\right)\sigma_{uv}{}^2$$
$$= \left(\frac{\sigma_u}{u}\right)^2 + \left(\frac{\sigma_v}{v}\right)^2 - 2\left(\frac{\sigma_{uv}}{u}\right)\left(\frac{\sigma_{uv}}{v}\right). \quad (10)$$

For exponentiation of quantities with

$$x = a^{\pm bu} = (e^{\ln a})^{\pm bu} = e^{\pm b(\ln a)u}, \quad (11)$$

$$\frac{\partial x}{\partial u} = \pm b(\ln a)e^{\pm b\ln au} = \pm b(\ln a)x, \quad (12)$$

so

$$\sigma_x = \sigma_u b(\ln a)x \quad (13)$$

$$\frac{\sigma_x}{x} = b\ln a\sigma_u. \quad (14)$$

If $a = e$, then

$$\frac{\sigma_x}{x} = b\sigma_u. \quad (15)$$

For LOGARITHMS of quantities with $x = a\ln(\pm bu)$, $\partial x/\partial u = a(\pm b)/(\pm bu) = a/u$, so

$$\sigma_x{}^2 = \sigma_u{}^2\left(\frac{a^2}{u^2}\right) \quad (16)$$

$$\sigma_x = a\frac{\sigma_u}{u}. \quad (17)$$

For multiplication with $x = \pm auv$, $\partial x/\partial u = \pm av$ and $\partial x/\partial v = \pm au$, so

$$\sigma_x{}^2 = a^2v^2\sigma_u{}^2 + a^2u^2\sigma_v{}^2 + 2a^2uv\sigma_{uv}{}^2 \quad (18)$$

$$\left(\frac{\sigma_x}{x}\right)^2 = \frac{a^2v^2}{a^2u^2v^2}\sigma_u{}^2 + \frac{a^2u^2}{a^2u^2v^2}\sigma_v{}^2 + \frac{2a^2uv}{a^2u^2v^2}\sigma_{uv}{}^2$$
$$= \left(\frac{\sigma_u}{u}\right)^2 + \left(\frac{\sigma_v}{v}\right)^2 + 2\left(\frac{\sigma_{uv}}{u}\right)\left(\frac{\sigma_{uv}}{v}\right). \quad (19)$$

For POWERS, with $x = au^{\pm b}$, $\partial x/\partial u = \pm abu^{\pm b-1} = \pm bx/u$, so

$$\sigma_x{}^2 = \sigma_u{}^2\frac{b^2x^2}{u^2} \quad (20)$$

$$\frac{\sigma_x}{x} = b\frac{\sigma_u}{u}. \quad (21)$$

see also ABSOLUTE ERROR, PERCENTAGE ERROR, RELATIVE ERROR

References

Abramowitz, M. and Stegun, C. A. (Eds.). *Handbook of Mathematical Functions with Formulas, Graphs, and Mathematical Tables, 9th printing.* New York: Dover, p. 14, 1972.

Bevington, P. R. *Data Reduction and Error Analysis for the Physical Sciences.* New York: McGraw-Hill, pp. 58–64, 1969.

Escher's Map

$$f(z) \mapsto z^{(1+\cos\beta+i\sin\beta)/2}.$$

Escribed Circle

see EXCIRCLE

Essential Singularity

A SINGULARITY a for which $f(z)(z-a)^n$ is not DIFFERENTIABLE for any INTEGER $n > 0$.

see also PICARD'S THEOREM, WEIERSTRAß-CASORATI THEOREM

Estimate

An estimate is an educated guess for an unknown quantity or outcome based on known information. The making of estimates is an important part of statistics, since care is needed to provide as accurate an estimate as possible using as little input data as possible. Often, an estimate for the uncertainty ΔE of an estimate E can also be determined statistically. A rule that tells how to calculate an estimate based on the measurements contained in a sample is called an ESTIMATOR.

see also BIAS (ESTIMATOR), ERROR, ESTIMATOR

References

Iyanaga, S. and Kawada, Y. (Eds.). "Statistical Estimation and Statistical Hypothesis Testing." Appendix A, Table 23 in *Encyclopedic Dictionary of Mathematics.* Cambridge, MA: MIT Press, pp. 1486–1489, 1980.

Estimator

An estimator is a rule that tells how to calculate an ESTIMATE based on the measurements contained in a sample. For example, the "sample MEAN" AVERAGE $\bar{x}$ is an estimator for the population MEAN μ.

The mean square error of an estimator $\tilde{\theta}$ is defined by

$$\text{MSE} \equiv \left\langle(\tilde{\theta} - \theta)^2\right\rangle.$$

Let B be the BIAS, then

$$\text{MSE} = \left\langle[(\tilde{\theta} - \langle\tilde{\theta}\rangle) + B(\tilde{\theta})]^2\right\rangle$$
$$= \left\langle(\tilde{\theta} - \langle\tilde{\theta}\rangle)^2\right\rangle + B^2(\tilde{\theta}) \equiv V(\tilde{\theta}) + B^2(\tilde{\theta}),$$

where V is the estimator VARIANCE.

see also BIAS (ESTIMATOR), ERROR, ESTIMATE, k-STATISTIC

Eta Function

see DEDEKIND ETA FUNCTION, DIRICHLET ETA FUNCTION, THETA FUNCTION

Ethiopian Multiplication

see RUSSIAN MULTIPLICATION

Etruscan Venus Surface

A 3-D shadow of a 4-D KLEIN BOTTLE.

see also IDA SURFACE

References

Peterson, I. *Islands of Truth: A Mathematical Mystery Cruise.* New York: W. H. Freeman, pp. 42–44, 1990.

Eubulides Paradox

The PARADOX "This statement is false," stated in the fourth century BC. It is a sharper version of the EPIMENIDES PARADOX, "All Cretans are liers... One of their own poets has said so."

see also EPIMENIDES PARADOX, SOCRATES' PARADOX

References

Hofstadter, D. R. *Gödel, Escher, Bach: An Eternal Golden Braid.* New York: Vintage Books, p. 17, 1989.

Euclid's Axioms

see EUCLID'S POSTULATES

Euclid's Elements

see ELEMENTS

Euclid's Fifth Postulate

see EUCLID'S POSTULATES

Euclid Number

The nth Euclid number is defined by

$$E_n \equiv 1 + \prod_{i=1}^{n} p_i,$$

where p_i is the ith PRIME. The first few E_n are 3, 7, 31, 211, 2311, 30031, 510511, 9699691, 223092871, 6469693231, ... (Sloane's A006862). The largest factor of E_n are 3, 7, 31, 211, 2311, 509, 277, 27953, ... (Sloane's A002585). The n of the first few PRIME Euclid numbers E_n are 1, 2, 3, 4, 5, 11, 75, 171, 172, 384, 457, 616, 643, ... (Sloane's A014545) up to a search limit of 700. It is not known if there are an INFINITE number of PRIME Euclid numbers (Guy 1994, Ribenboim 1996).

see also SMARANDACHE SEQUENCES

References

Guy, R. K. *Unsolved Problems in Number Theory, 2nd ed.* New York: Springer-Verlag, 1994.

Ribenboim, P. *The New Book of Prime Number Records.* New York: Springer-Verlag, 1996.

Sloane, N. J. A. Sequences A014544, A006862/M2698, and A002585/M2697 in "An On-Line Version of the Encyclopedia of Integer Sequences."

Wagon, S. *Mathematica in Action.* New York: W. H. Freeman, pp. 35–37, 1991.

Euclid's Postulates

1. A straight LINE SEGMENT can be drawn joining any two points.
2. Any straight LINE SEGMENT can be extended indefinitely in a straight LINE.
3. Given any straight LINE SEGMENT, a CIRCLE can be drawn having the segment as RADIUS and one endpoint as center.
4. All RIGHT ANGLES are congruent.
5. If two lines are drawn which intersect a third in such a way that the sum of the inner angles on one side is less than two RIGHT ANGLES, then the two lines inevitably must intersect each other on that side if extended far enough. This postulate is equivalent to what is known as the PARALLEL POSTULATE.

Euclid's fifth postulate cannot be proven as a theorem, although this was attempted by many people. Euclid himself used only the first four postulates ("ABSOLUTE GEOMETRY") for the first 28 propositions of the *Elements*, but was forced to invoke the PARALLEL POSTULATE on the 29th. In 1823, Janos Bolyai and Nicolai Lobachevsky independently realized that entirely self-consistent "NON-EUCLIDEAN GEOMETRIES" could be created in which the parallel postulate *did not hold.* (Gauss had also discovered but suppressed the existence of non-Euclidean geometries.)

see also ABSOLUTE GEOMETRY, CIRCLE, *Elements*, LINE SEGMENT, NON-EUCLIDEAN GEOMETRY, PARALLEL POSTULATE, PASCH'S THEOREM, RIGHT ANGLE

References

Hofstadter, D. R. *Gödel, Escher, Bach: An Eternal Golden Braid.* New York: Vintage Books, pp. 88–92, 1989.

Euclid's Principle

see EUCLID'S THEOREMS

Euclid's Theorems

A theorem sometimes called "Euclid's First Theorem" or EUCLID'S PRINCIPLE states that if p is a PRIME and $p|ab$, then $p|a$ or $p|b$ (where | means DIVIDES). A COROLLARY is that $p|a^n \Rightarrow p|a$ (Conway and Guy 1996). The FUNDAMENTAL THEOREM OF ARITHMETIC is another COROLLARY (Hardy and Wright 1979).

Euclid's Second Theorem states that the number of PRIMES is INFINITE. This theorem, also called the INFINITUDE OF PRIMES theorem, was proved by Euclid in Proposition IX.20 of the *Elements*. Ribenboim (1989) gives nine (and a half) proofs of this theorem. Euclid's elegant proof proceeds as follows. Given a finite sequence of consecutive PRIMES 2, 3, 5, ..., p, the number

$$N = 2 \cdot 3 \cdot 5 \cdots p + 1, \tag{1}$$

known as the ith EUCLID NUMBER when $p = p_i$ is the ith PRIME, is either a new PRIME or the product of PRIMES.

If N is a PRIME, then it must be greater than the previous PRIMES, since one plus the product of PRIMES must be greater than each PRIME composing the product. Now, if N is a product of PRIMES, then at least one of the PRIMES must be greater than p. This can be shown as follows. If N is COMPOSITE and not greater than p, then one of its factors (say F) must be one of the PRIMES in the sequence, 2, 3, 5, ..., p. It therefore DIVIDES the product $2 \cdot 3 \cdot 5 \cdots p$. However, since it is a factor of N, it also DIVIDES N. But a number which DIVIDES two numbers a and $b < a$ also DIVIDES their difference $a - b$, so F must also divide

$$N-(2\cdot3\cdot5\cdots p) = (2\cdot3\cdot5\cdots p+1)-(2\cdot3\cdot5\cdots p) = 1. \quad (2)$$

However, in order to divide 1, F must be 1, which is contrary to the assumption that it is a PRIME in the sequence 2, 3, 5, ... It therefore follows that if N is composite, it has at least one factor greater than p. Since N is either a PRIME greater than p or contains a factor greater than p, a PRIME larger than the largest in the finite sequence can always be found, so there are an infinite number of PRIMES. Hardy (1967) remarks that this proof is "as fresh and significant as when it was discovered" so that "two thousand years have not written a wrinkle" on it.

A similar argument shows that $p! \pm 1$ is PRIME, and

$$1 \cdot 3 \cdot 5 \cdot 7 \cdots p + 1 \quad (3)$$

must be either PRIME or be divisible by a PRIME $> p$. Kummer used a variation of this proof, which is also a proof by contradiction. It assumes that there exist only a finite number of PRIMES $N = p_1, p_2, \ldots, p_r$. Now consider $N - 1$. It must be a product of PRIMES, so it has a PRIME divisor p_i in common with N. Therefore, $p_i | N - (N-1) = 1$ which is nonsense, so we have proved the initial assumption is wrong by contradiction.

It is also true that there are runs of COMPOSITE NUMBERS which are arbitrarily long. This can be seen by defining

$$n \equiv j! = \prod_{i=1}^{j} i, \quad (4)$$

where $j!$ is a FACTORIAL. Then the $j - 1$ consecutive numbers $n + 2, n + 3, \ldots, n + j$ are COMPOSITE, since

$$n+2 = (1 \cdot 2 \cdots j) + 2 = 2(1 \cdot 3 \cdot 4 \cdots n + 1) \quad (5)$$
$$n+3 = (1 \cdot 2 \cdots j) + 3 = 3(1 \cdot 2 \cdot 4 \cdot 5 \cdots n + 1) \quad (6)$$
$$n+j = (1 \cdot 2 \cdots j) + j = j[1 \cdot 2 \cdots (j-1) + 1]. \quad (7)$$

Guy (1981, 1988) points out that while $p_1 p_2 \cdots p_n + 1$ is not necessarily PRIME, letting q be the next PRIME after $p_1 p_2 \cdots p_n + 1$, the number $q - p_1 p_2 \cdots p_n + 1$ is almost always a PRIME, although it has not been proven that this must *always* be the case.

see also DIVIDE, EUCLID NUMBER, PRIME NUMBER

References

Ball, W. W. R. and Coxeter, H. S. M. *Mathematical Recreations and Essays, 13th ed.* New York: Dover, p. 60, 1987.

Conway, J. H. and Guy, R. K. "There are Always New Primes!" In *The Book of Numbers.* New York: Springer-Verlag, pp. 133–134, 1996.

Cosgrave, J. B. "A Remark on Euclid's Proof of the Infinitude of Primes." *Amer. Math. Monthly* **96**, 339–341, 1989.

Courant, R. and Robbins, H. *What is Mathematics?: An Elementary Approach to Ideas and Methods, 2nd ed.* Oxford, England: Oxford University Press, p. 22, 1996.

Dunham, W. "Great Theorem: The Infinitude of Primes." *Journey Through Genius: The Great Theorems of Mathematics.* New York: Wiley, pp. 73–75, 1990.

Guy, R. K. §A12 in *Unsolved Problems in Number Theory.* New York: Springer-Verlag, 1981.

Guy, R. K. "The Strong Law of Small Numbers." *Amer. Math. Monthly* **95**, 697–712, 1988.

Hardy, G. H. *A Mathematician's Apology.* Cambridge, England: Cambridge University Press, 1992.

Ribenboim, P. *The Book of Prime Number Records, 2nd ed.* New York: Springer-Verlag, pp. 3–12, 1989.

Euclidean Algorithm

An ALGORITHM for finding the GREATEST COMMON DIVISOR of two numbers a and b, also called Euclid's algorithm. It is an example of a P-PROBLEM whose time complexity is bounded by a quadratic function of the length of the input values (Banach and Shallit). Let $a = bq + r$, then find a number u which DIVIDES both a and b (so that $a = su$ and $b = tu$), then u also DIVIDES r since

$$r = a - bq = su - qtu = (s - qt)u. \quad (1)$$

Similarly, find a number v which DIVIDES b and r (so that $b = s'v$ and $r = t'v$), then v DIVIDES a since

$$a = bq + r = s'vq + t'v = (s'q + t')v. \quad (2)$$

Therefore, every common DIVISOR of a and b is a common DIVISOR of b and r, so the procedure can be iterated as follows

$$a = bq_1 + r_1 \quad (3)$$
$$b = q_2 r_1 + r_2 \quad (4)$$
$$r_1 = q_3 r_2 + r_3 \quad (5)$$
$$r_{n-2} = q_n r_{n-1} + r_n \quad (6)$$
$$r_{n-1} = q_{n+1} r_n, \quad (7)$$

where r_n is $\text{GCD}(a, b) \equiv (a, b)$. Lamé showed that the number of steps needed to arrive at the GREATEST COMMON DIVISOR for two numbers less than N is

$$\text{steps} \leq \frac{\log_{10} N}{\log_{10} \phi} + \frac{\log_{10} \sqrt{5}}{\log_{10} \phi} \quad (8)$$

where ϕ is the GOLDEN MEAN, or ≤ 5 times the number of digits in the smaller number. Numerically, Lamé's expression evaluates to

$$\text{steps} \leq 4.785 \log_{10} N + 1.6723. \quad (9)$$

As shown by LAMÉ'S THEOREM, the worst case occurs when the ALGORITHM is applied to two consecutive FIBONACCI NUMBERS. Heilbronn showed that the average number of steps is $12 \ln 2/\pi^2 \log_{10} n = 0.843 \log_{10} n$ for all pairs (n, b) with $b < n$. Kronecker showed that the shortest application of the ALGORITHM uses least absolute remainders. The QUOTIENTS obtained are distributed as shown in the following table (Wagon 1991).

Quotient	%
1	41.5
2	17.0
3	9.3

For details, see Uspensky and Heaslet (1939) or Knuth (1973). Let $T(m, n)$ be the number of divisions required to compute $\text{GCD}(m, n)$ using the Euclidean algorithm, and define $T(m, 0) = 0$ if $m \geq 0$. Then

$$T(m,n) = \begin{cases} 1 + T(n, m \bmod n) & \text{for } m \geq n \\ 1 + T(n, m) & \text{for } m < n. \end{cases} \quad (10)$$

Define the functions

$$T(n) = \frac{1}{n} \sum_{0 \leq m < n} T(m, n) \quad (11)$$

$$\tau(n) = \frac{1}{\phi(n)} \sum_{\substack{0 \leq m < n \\ \text{GCD}(m,n)=1}} T(m, n) \quad (12)$$

$$A(N) = \frac{1}{N^2} \sum_{\substack{1 \leq m \leq N \\ 1 \leq n \leq N}} T(m, n), \quad (13)$$

where ϕ is the TOTIENT FUNCTION, $T(n)$ is the average number of divisions when n is fixed and m chosen at random, $\tau(n)$ is the average number of divisions when n is fixed and m is a random number coprime to n, and $A(N)$ is the average number of divisions when m and n are both chosen at random in $[1, N]$. Norton (1990) showed that

$$T(n) = \frac{12 \ln 2}{\pi^2} \left[\ln n - \sum_{d|n} \frac{\Lambda(d)}{d} \right] + C + \frac{1}{n} \sum_{d|n} \phi(d) \mathcal{O}(d^{-1/6+\epsilon}), \quad (14)$$

where Λ is the VON MANGOLDT FUNCTION and C is PORTER'S CONSTANT. Porter (1975) showed that

$$\tau(n) = \frac{12 \ln 2}{\pi^2} \ln n + C + \mathcal{O}(n^{-1/6} + \epsilon), \quad (15)$$

and Norton (1990) proved that

$$A(N) = \frac{12 \ln 2}{\pi^2} \left[\ln N - \tfrac{1}{2} + \frac{6}{\pi^2} \zeta'(2) \right] + C - \tfrac{1}{2} + \mathcal{O}(N^{-1/6+\epsilon}). \quad (16)$$

There exist 22 QUADRATIC FIELDS in which there is a Euclidean algorithm (Inkeri 1947).

see also FERGUSON-FORCADE ALGORITHM

References

Bach, E. and Shallit, J. *Algorithmic Number Theory, Vol. 1: Efficient Algorithms.* Cambridge, MA: MIT Press, 1996.

Courant, R. and Robbins, H. "The Euclidean Algorithm." §2.4 in Supplement to Ch. 1 in *What is Mathematics?: An Elementary Approach to Ideas and Methods, 2nd ed.* Oxford, England: Oxford University Press, pp. 42–51, 1996.

Dunham, W. *Journey Through Genius: The Great Theorems of Mathematics.* New York: Wiley, pp. 69–70, 1990.

Finch, S. "Favorite Mathematical Constants." `http://www.mathsoft.com/asolve/constant/porter/porter.html`.

Inkeri, K. "Über den Euklidischen Algorithmus in quadratischen Zahlkörpern." *Ann. Acad. Sci. Fennicae. Ser. A. I. Math.-Phys.* **1947**, 1–35, 1947.

Knuth, D. E. *The Art of Computer Programming, Vol. 1: Fundamental Algorithms, 2nd ed.* Reading, MA: Addison-Wesley, 1973.

Knuth, D. E. *The Art of Computer Programming, Vol. 2: Seminumerical Algorithms, 2nd ed.* Reading, MA: Addison-Wesley, 1981.

Norton, G. H. "On the Asymptotic Analysis of the Euclidean Algorithm." *J. Symb. Comput.* **10**, 53–58, 1990.

Porter, J. W. "On a Theorem of Heilbronn." *Mathematika* **22**, 20–28, 1975.

Uspensky, J. V. and Heaslet, M. A. *Elementary Number Theory.* New York: McGraw-Hill, 1939.

Wagon, S. "The Ancient and Modern Euclidean Algorithm" and "The Extended Euclidean Algorithm." §8.1 and 8.2 in *Mathematica in Action.* New York: W. H. Freeman, pp. 247–252 and 252–256, 1991.

Euclidean Construction

see GEOMETRIC CONSTRUCTION

Euclidean Geometry

A GEOMETRY in which EUCLID'S FIFTH POSTULATE holds, sometimes also called PARABOLIC GEOMETRY. 2-D Euclidean geometry is called PLANE GEOMETRY, and 3-D Euclidean geometry is called SOLID GEOMETRY. Hilbert proved the CONSISTENCY of Euclidean geometry.

see also ELLIPTIC GEOMETRY, GEOMETRIC CONSTRUCTION, GEOMETRY, HYPERBOLIC GEOMETRY, NON-EUCLIDEAN GEOMETRY, PLANE GEOMETRY

References

Altshiller-Court, N. *College Geometry: A Second Course in Plane Geometry for Colleges and Normal Schools, 2nd ed., rev. enl.* New York: Barnes and Noble, 1952.

Casey, J. *A Treatise on the Analytical Geometry of the Point, Line, Circle, and Conic Sections, Containing an Account of Its Most Recent Extensions with Numerous Examples, 2nd rev. enl. ed.* Dublin: Hodges, Figgis, & Co., 1893.

Coxeter, H. S. M. and Greitzer, S. L. *Geometry Revisited.* Washington, DC: Math. Assoc. Amer., 1967
Coxeter, H. S. M. *Introduction to Geometry, 2nd ed.* New York: Wiley, 1969.
Gallatly, W. *The Modern Geometry of the Triangle, 2nd ed.* London: Hodgson, 1913.
Heath, T. L. *The Thirteen Books of the Elements, 2nd ed., Vol. 1: Books I and II.* New York: Dover, 1956.
Heath, T. L. *The Thirteen Books of the Elements, 2nd ed., Vol. 2: Books III–IX.* New York: Dover, 1956.
Heath, T. L. *The Thirteen Books of the Elements, 2nd ed., Vol. 3: Books X–XIII.* New York: Dover, 1956.
Honsberger, R. *Episodes in Nineteenth and Twentieth Century Euclidean Geometry.* Washington, DC: Math. Assoc. Amer., 1995.
Johnson, R. A. *Modern Geometry: An Elementary Treatise on the Geometry of the Triangle and the Circle.* Boston, MA: Houghton Mifflin, 1929.
Johnson, R. A. *Advanced Euclidean Geometry.* New York: Dover, 1960.
Klee, V. "Some Unsolved Problems in Plane Geometry." *Math. Mag.* **52**, 131–145, 1979.
Klee, V. and Wagon, S. *Old and New Unsolved Problems in Plane Geometry and Number Theory, rev. ed.* Washington, DC: Math. Assoc. Amer., 1991.

Euclidean Group

The GROUP of ROTATIONS and TRANSLATIONS.

see also ROTATION, TRANSLATION

References

Lomont, J. S. *Applications of Finite Groups.* New York: Dover, 1987.

Euclidean Metric

The FUNCTION $f : \mathbb{R}^n \times \mathbb{R}^n \to \mathbb{R}$ that assigns to any two VECTORS $(x_1, \ldots, x_n)$ and $(y_1, \ldots, y_n)$ the number

$$\sqrt{(x_1 - y_1)^2 + \ldots + (x_n - y_n)^2},$$

and so gives the "standard" distance between any two VECTORS in $\mathbb{R}^n$.

Euclidean Motion

A Euclidean motion of $\mathbb{R}^n$ is an AFFINE TRANSFORMATION whose linear part is an ORTHOGONAL TRANSFORMATION.

see also RIGID MOTION

References

Gray, A. *Modern Differential Geometry of Curves and Surfaces.* Boca Raton, FL: CRC Press, p. 105, 1993.

Euclidean Norm

see L_2-NORM

Euclidean Number

A Euclidean number is a number which can be obtained by repeatedly solving the QUADRATIC EQUATION. Euclidean numbers, together with the RATIONAL NUMBERS, can be constructed using classical GEOMETRIC CONSTRUCTIONS. However, the cases for which the values of the TRIGONOMETRIC FUNCTIONS SINE, COSINE, TANGENT, etc., can be written in closed form involving square roots of REAL NUMBERS are much more restricted.

see also ALGEBRAIC INTEGER, ALGEBRAIC NUMBER, CONSTRUCTIBLE NUMBER, RADICAL INTEGER

References

Conway, J. H. and Guy, R. K. "Three Greek Problems." In *The Book of Numbers.* New York: Springer-Verlag, pp. 192–194, 1996.

Euclidean Plane

The 2-D EUCLIDEAN SPACE denoted $\mathbb{R}^2$.

see also COMPLEX PLANE, EUCLIDEAN SPACE

Euclidean Space

Euclidean n-space is the SPACE of all n-tuples of REAL NUMBERS, $(x_1, x_2, \ldots, x_n)$ and is denoted $\mathbb{R}^n$. $\mathbb{R}^n$ is a VECTOR SPACE and has LEBESGUE COVERING DIMENSION n. Elements of $\mathbb{R}^n$ are called n-VECTORS. $\mathbb{R}^1 = \mathbb{R}$ is the set of REAL NUMBERS (i.e., the REAL LINE), and $\mathbb{R}^2$ is called the EUCLIDEAN PLANE. In Euclidean space, COVARIANT and CONTRAVARIANT quantities are equivalent so $\vec{e}^j = \vec{e}_j$.

see also EUCLIDEAN PLANE, REAL LINE, VECTOR

References

Gray, A. "Euclidean Spaces." §1.1 in *Modern Differential Geometry of Curves and Surfaces.* Boca Raton, FL: CRC Press, pp. 2–4, 1993.

Eudoxus's Kampyle

see KAMPYLE OF EUDOXUS

Euler's $6n + 1$ Theorem

Every PRIME of the form $6n + 1$ can be written in the form $x^2 + 3y^2$.

Euler's Addition Theorem

Let $g(x) \equiv (1 - x^2)(1 - k^2x^2)$. Then

$$\int_0^a \frac{dx}{\sqrt{g(x)}} + \int_0^b \frac{dx}{\sqrt{g(x)}} = \int_0^c \frac{dx}{\sqrt{g(x)}},$$

where

$$c \equiv \frac{b\sqrt{g(a)} + a\sqrt{g(b)}}{\sqrt{1 - k^2a^2b^2}}.$$

Euler Angles

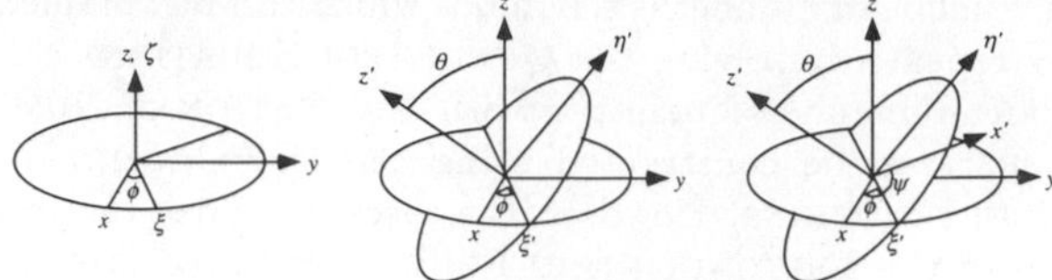

According to EULER'S ROTATION THEOREM, any ROTATION may be described using three ANGLES. If the ROTATIONS are written in terms of ROTATION MATRICES B, C, and D, then a general ROTATION A can be written as

$$\mathsf{A} = \mathsf{BCD}. \tag{1}$$

The three angles giving the three rotation matrices are called Euler angles. There are several conventions for Euler angles, depending on the axes about which the rotations are carried out. Write the MATRIX A as

$$\mathsf{A} \equiv \begin{bmatrix} a_{11} & a_{12} & a_{13} \\ a_{21} & a_{22} & a_{23} \\ a_{11} & a_{12} & a_{13} \end{bmatrix}. \tag{2}$$

In the so-called "x-convention," illustrated above,

$$\mathsf{D} \equiv \begin{bmatrix} \cos\phi & \sin\phi & 0 \\ -\sin\phi & \cos\phi & 0 \\ 0 & 0 & 1 \end{bmatrix} \tag{3}$$

$$\mathsf{C} \equiv \begin{bmatrix} 1 & 0 & 0 \\ 0 & \cos\theta & \sin\theta \\ 0 & -\sin\theta & \cos\theta \end{bmatrix} \tag{4}$$

$$\mathsf{B} \equiv \begin{bmatrix} \cos\psi & \sin\psi & 0 \\ -\sin\psi & \cos\psi & 0 \\ 0 & 0 & 1 \end{bmatrix}, \tag{5}$$

so

$$\begin{aligned} a_{11} &= \cos\psi\cos\phi - \cos\theta\sin\phi\sin\psi \\ a_{12} &= \cos\psi\sin\phi + \cos\theta\cos\phi\sin\psi \\ a_{13} &= \sin\psi\sin\theta \\ a_{21} &= -\sin\psi\cos\phi - \cos\theta\sin\phi\cos\psi \\ a_{22} &= -\sin\psi\sin\phi + \cos\theta\cos\phi\cos\psi \\ a_{23} &= \cos\psi\sin\theta \\ a_{31} &= \sin\theta\sin\phi \\ a_{32} &= -\sin\theta\cos\phi \\ a_{33} &= \cos\theta \end{aligned}$$

To obtain the components of the ANGULAR VELOCITY $\boldsymbol{\omega}$ in the body axes, note that for a MATRIX

$$\mathsf{A} \equiv [\,\mathbf{A}_1 \quad \mathbf{A}_2 \quad \mathbf{A}_3\,], \tag{6}$$

it is true that

$$\begin{bmatrix} a_{11} & a_{12} & a_{13} \\ a_{21} & a_{22} & a_{23} \\ a_{31} & a_{32} & a_{33} \end{bmatrix} \begin{bmatrix} \omega_x \\ \omega_y \\ \omega_z \end{bmatrix} = \begin{bmatrix} a_{11}\omega_x + a_{12}\omega_y + a_{13}\omega_z \\ a_{21}\omega_x + a_{22}\omega_y + a_{23}\omega_z \\ a_{31}\omega_x + a_{32}\omega_y + a_{33}\omega_z \end{bmatrix} \tag{7}$$

$$= \mathbf{A}_1\omega_x + \mathbf{A}_2\omega_y + \mathbf{A}_3\omega_z. \tag{8}$$

Now, ω_z corresponds to rotation about the ϕ axis, so look at the ω_z component of $\mathsf{A}\boldsymbol{\omega}$,

$$\boldsymbol{\omega}_\phi = \mathbf{A}_1\omega_z = \begin{bmatrix} \sin\psi\sin\theta \\ \cos\psi\sin\theta \\ \cos\theta \end{bmatrix} \dot{\phi}. \tag{9}$$

The line of nodes corresponds to a rotation by θ about the ξ-axis, so look at the ω_ξ component of $\mathsf{B}\boldsymbol{\omega}$,

$$\boldsymbol{\omega}_\theta = \mathbf{B}_1\omega_\xi = \mathbf{B}_1\dot{\theta} = \begin{bmatrix} \cos\psi \\ -\sin\psi \\ 0 \end{bmatrix} \dot{\theta}. \tag{10}$$

Similarly, to find rotation by ψ about the remaining axis, look at the ω_ψ component of $\mathsf{B}\boldsymbol{\omega}$,

$$\boldsymbol{\omega}_\psi = \mathbf{B}_3\omega_\psi = \mathbf{B}_3\dot{\psi} = \begin{bmatrix} 0 \\ 0 \\ 1 \end{bmatrix} \dot{\psi}. \tag{11}$$

Combining the pieces gives

$$\boldsymbol{\omega} = \begin{bmatrix} \sin\psi\sin\theta\dot{\phi} + \cos\psi\dot{\theta} \\ \cos\psi\sin\theta\dot{\phi} - \sin\psi \\ \cos\theta\dot{\phi} + \dot{\psi}. \end{bmatrix} \tag{12}$$

For more details, see Goldstein (1980, p. 176) and Landau and Lifschitz (1976, p. 111).

The x-convention Euler angles are given in terms of the CAYLEY-KLEIN PARAMETERS by

$$\begin{aligned} \phi &= -2i\ln\left(\pm\frac{\alpha^{1/2}\gamma^{1/4}}{\beta^{1/4}(1+\beta\gamma)^{1/4}}\right), \\ &\quad -2i\ln\left(\pm\frac{i\alpha^{1/2}\gamma^{1/4}}{\beta^{1/4}(1+\beta\gamma)^{1/4}}\right) \end{aligned} \tag{13}$$

$$\begin{aligned} \psi &= -2i\ln\left(\pm\frac{\alpha^{1/2}\beta^{1/4}}{\gamma^{1/4}(1+\beta\gamma)^{1/4}}\right), \\ &\quad -2i\ln\left(\pm\frac{i\alpha^{1/2}\beta^{1/4}}{\gamma^{1/4}(1+\beta\gamma)^{1/4}}\right) \end{aligned} \tag{14}$$

$$\theta = \pm 2\cos^{-1}(\pm\sqrt{1+\beta\gamma}). \tag{15}$$

In the "y-convention,"

$$\phi_x \equiv \phi_y + \tfrac{1}{2}\pi \tag{16}$$

$$\psi_x \equiv \psi_y - \tfrac{1}{2}\pi. \tag{17}$$

Therefore,

$$\sin\phi_x = \cos\phi_y \tag{18}$$

$$\cos\phi_x = -\sin\phi_y \tag{19}$$

$$\sin\psi_x = -\cos\psi_y \tag{20}$$

$$\cos\psi_x = \sin\psi_y \tag{21}$$

$$\mathsf{D} \equiv \begin{bmatrix} -\sin\phi & \cos\phi & 0 \\ -\cos\phi & -\sin\phi & 0 \\ 0 & 0 & 1 \end{bmatrix} \tag{22}$$

$$\mathsf{C} \equiv \begin{bmatrix} 1 & 0 & 0 \\ 0 & \cos\theta & \sin\theta \\ 0 & -\sin\theta & \cos\theta \end{bmatrix} \tag{23}$$

$$\mathsf{B} \equiv \begin{bmatrix} \sin\psi & -\cos\psi & 0 \\ \cos\psi & \sin\psi & 0 \\ 0 & 0 & 1 \end{bmatrix} \tag{24}$$

and A is given by

$$\begin{aligned} a_{11} &= -\sin\psi\sin\phi + \cos\theta\cos\phi\cos\psi \\ a_{12} &= \sin\psi\cos\phi + \cos\theta\sin\phi\cos\psi \\ a_{13} &= -\cos\psi\sin\theta \\ a_{21} &= -\cos\psi\sin\phi - \cos\theta\cos\phi\sin\psi \\ a_{22} &= \cos\psi\cos\phi - \cos\theta\sin\phi\sin\psi \\ a_{23} &= \sin\psi\sin\theta \\ a_{31} &= \sin\theta\cos\phi \\ a_{32} &= \sin\theta\sin\phi \\ a_{33} &= \cos\theta. \end{aligned}$$

In the "xyz" (pitch-roll-yaw) convention, θ is pitch, ψ is roll, and ϕ is yaw.

$$\mathsf{D} \equiv \begin{bmatrix} \cos\phi & \sin\phi & 0 \\ -\sin\phi & \cos\phi & 0 \\ 0 & 0 & 1 \end{bmatrix} \tag{25}$$

$$\mathsf{C} \equiv \begin{bmatrix} \cos\theta & 0 & -\sin\theta \\ 0 & 1 & 0 \\ \sin\theta & 0 & \cos\theta \end{bmatrix} \tag{26}$$

$$\mathsf{B} \equiv \begin{bmatrix} 1 & 0 & 0 \\ 0 & \cos\psi & \sin\psi \\ 0 & -\sin\psi & \cos\psi \end{bmatrix} \tag{27}$$

and A is given by

$$\begin{aligned} a_{11} &= \cos\theta\cos\phi \\ a_{12} &= \cos\theta\sin\phi \\ a_{13} &= -\sin\theta \\ a_{21} &= \sin\psi\sin\theta\cos\phi - \cos\psi\sin\phi \\ a_{22} &= \sin\psi\sin\theta\sin\phi + \cos\psi\cos\phi \\ a_{23} &= \cos\theta\sin\psi \\ a_{31} &= \cos\psi\sin\theta\cos\phi + \sin\psi\sin\phi \\ a_{32} &= \cos\psi\sin\theta\sin\phi - \sin\psi\cos\phi \\ a_{33} &= \cos\theta\cos\psi. \end{aligned}$$

A set of parameters sometimes used instead of angles are the EULER PARAMETERS e_0, e_1, e_2 and e_3, defined by

$$e_0 \equiv \cos\left(\frac{\phi}{2}\right) \tag{28}$$

$$\mathbf{e} \equiv \begin{bmatrix} e_1 \\ e_2 \\ e_3 \end{bmatrix} = \hat{\mathbf{n}}\sin\left(\frac{\phi}{2}\right). \tag{29}$$

Using EULER PARAMETERS (which are QUATERNIONS), an arbitrary ROTATION MATRIX can be described by

$$\begin{aligned} a_{11} &= {e_0}^2 + {e_1}^2 - {e_2}^2 - {e_3}^2 \\ a_{12} &= 2(e_1e_2 + e_0e_3) \\ a_{13} &= 2(e_1e_3 - e_0e_2) \\ a_{21} &= 2(e_1e_2 - e_0e_3) \\ a_{22} &= {e_0}^2 - {e_1}^2 + {e_2}^2 - {e_3}^2 \\ a_{23} &= 2(e_2e_3 + e_0e_1) \\ a_{31} &= 2(e_1e_3 + e_0e_2) \\ a_{32} &= 2(e_2e_3 - e_0e_1) \\ a_{33} &= {e_0}^2 - {e_1}^2 - {e_2}^2 + {e_3}^2 \end{aligned}$$

(Goldstein 1960, p. 153).

If the coordinates of two pairs of n points $\mathbf{x}_i$ and $\mathbf{x}'_i$ are known, one rotated with respect to the other, then the Euler rotation matrix can be obtained in a straightforward manner using LEAST SQUARES FITTING. Write the points as arrays of vectors, so

$$[\mathbf{x}'_1 \quad \cdots \quad \mathbf{x}'_n] = \mathsf{A}[\mathbf{x}_1 \quad \cdots \quad \mathbf{x}_n]. \tag{30}$$

Writing the arrays of vectors as matrices gives

$$\mathsf{X}' = \mathsf{AX} \tag{31}$$

$$\mathsf{X}'\mathsf{X}^{\mathrm{T}} = \mathsf{AXX}^{\mathrm{T}}, \tag{32}$$

and solving for A gives

$$\mathsf{A} = \mathsf{X}'\mathsf{X}^{\mathrm{T}}(\mathsf{XX}^{\mathrm{T}})^{-1}. \tag{33}$$

However, we want the angles θ, ϕ, and ψ, not their combinations contained in the MATRIX A. Therefore, write the 3×3 MATRIX

$$\mathsf{A} = \begin{bmatrix} f_1(\theta,\phi,\psi) & f_2(\theta,\phi,\psi) & f_3(\theta,\phi,\psi) \\ f_4(\theta,\phi,\psi) & f_5(\theta,\phi,\psi) & f_6(\theta,\phi,\psi) \\ f_7(\theta,\phi,\psi) & f_7(\theta,\phi,\psi) & f_9(\theta,\phi,\psi) \end{bmatrix} \tag{34}$$

as a 1×9 VECTOR

$$\mathbf{f} = \begin{bmatrix} f_1(\theta,\phi,\psi) \\ \vdots \\ f_9(\theta,\phi,\psi) \end{bmatrix}. \tag{35}$$

Now set up the matrices

$$\begin{bmatrix} \left.\frac{\partial f_1}{\partial\theta}\right|_{\theta_i,\phi_i,\psi_i} & \left.\frac{\partial f_1}{\partial\phi}\right|_{\theta_i,\phi_i,\psi_i} & \left.\frac{\partial f_1}{\partial\psi}\right|_{\theta_i,\phi_i,\psi_i} \\ \vdots & \vdots & \vdots \\ \left.\frac{\partial f_9}{\partial\theta}\right|_{\theta_i,\phi_i,\psi_i} & \left.\frac{\partial f_9}{\partial\phi}\right|_{\theta_i,\phi_i,\psi_i} & \left.\frac{\partial f_9}{\partial\psi}\right|_{\theta_i,\phi_i,\psi_i} \end{bmatrix} \begin{bmatrix} d\theta \\ d\phi \\ d\psi \end{bmatrix} = d\mathbf{f}. \tag{36}$$

Using NONLINEAR LEAST SQUARES FITTING then gives solutions which converge to (θ, ϕ, ψ).

see also CAYLEY-KLEIN PARAMETERS, EULER PARAMETERS, EULER'S ROTATION THEOREM, INFINITESIMAL ROTATION, QUATERNION, ROTATION, ROTATION MATRIX

References

Arfken, G. *Mathematical Methods for Physicists, 3rd ed.* Orlando, FL: Academic Press, pp. 198–200, 1985.
Goldstein, H. "The Euler Angles" and "Euler Angles in Alternate Conventions." §4-4 and Appendix B in *Classical Mechanics, 2nd ed.* Reading, MA: Addison-Wesley, pp. 143–148 and 606–610, 1980.
Landau, L. D. and Lifschitz, E. M. *Mechanics, 3rd ed.* Oxford, England: Pergamon Press, 1976.

Euler-Bernoulli Triangle

see SEIDEL-ENTRINGER-ARNOLD TRIANGLE

Euler Brick

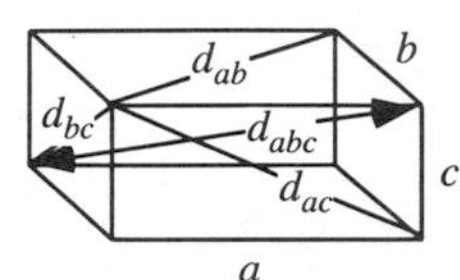

A RECTANGULAR PARALLELEPIPED ("BRICK") with integral edges $a > b > c$ and face diagonals d_{ij} given by

$$d_{ab} = \sqrt{a^2 + b^2} \tag{1}$$

$$d_{ac} = \sqrt{a^2 + c^2} \tag{2}$$

$$d_{bc} = \sqrt{b^2 + c^2}. \tag{3}$$

The problem is also called the BRICK, DIAGONALS PROBLEM, PERFECT BOX, PERFECT CUBOID, or RATIONAL CUBOID problem.

Euler found the smallest solution, which has sides $a = 240$, $b = 117$, and $c = 44$ and face DIAGONALS $d_{ab} = 267$, $d_{ac} = 244$, and $d_{bc} = 125$. Kraitchik gave 257 cuboids with the ODD edge less than 1 million (Guy 1994, p. 174). F. Helenius has compiled a list of the 5003 smallest (measured by the longest edge) Euler bricks. The first few are (240, 117, 44), (275, 252, 240), (693, 480, 140), (720, 132, 85), (792, 231, 160), ... (Sloane's A031173, A031174, and A031175). Parametric solutions for Euler bricks are also known.

No solution is known in which the oblique SPACE DIAGONAL

$$d_{abc} = \sqrt{a^2 + b^2 + c^2} \tag{4}$$

is also an INTEGER. If such a brick exists, the smallest side must be at least 1,281,000,000 (R. Rathbun 1996). Such a solution is equivalent to solving the DIOPHANTINE EQUATIONS

$$A^2 + B^2 = C^2 \tag{5}$$

$$A^2 + D^2 = E^2 \tag{6}$$

$$B^2 + D^2 = F^2 \tag{7}$$

$$B^2 + E^2 = G^2. \tag{8}$$

A solution with integral SPACE DIAGONAL and two out of three face diagonals is $a = 672$, $b = 153$, and $c = 104$, giving $d_{ab} = 3\sqrt{52777}$, $d_{ac} = 680$, $d_{bc} = 185$, and $d_{abc} = 697$. A solution giving integral space and face diagonals with only a single nonintegral EDGE is $a = 18720$, $b = \sqrt{211773121}$, and $c = 7800$, giving $d_{ab} = 23711$, $d_{ac} = 20280$, $d_{bc} = 16511$, and $d_{abc} = 24961$.

see also CUBOID, CYCLIC QUADRILATERAL, DIAGONAL (POLYHEDRON), PARALLELEPIPED, PYTHAGOREAN QUADRUPLE

References

Guy, R. K. "Is There a Perfect Cuboid? Four Squares whose Sums in Pairs are Square. Four Squares whose Differences are Square." §D18 in *Unsolved Problems in Number Theory, 2nd ed.* New York: Springer-Verlag, pp. 173–181, 1994.
Helenius, F. First 1000 Primitive Euler Bricks. `notebooks/EulerBricks.dat`.
Leech, J. "The Rational Cuboid Revisited." *Amer. Math. Monthly* **84**, 518–533, 1977. Erratum in *Amer. Math. Monthly* **85**, 472, 1978.
Sloane, N. J. A. Sequences A031173, A031174, and A031175 in "An On-Line Version of the Encyclopedia of Integer Sequences."
Rathbun, R. L. Personal communication, 1996.
Spohn, W. G. "On the Integral Cuboid." *Amer. Math. Monthly* **79**, 57–59, 1972.
Spohn, W. G. "On the Derived Cuboid." *Canad. Math. Bull.* **17**, 575–577, 1974.
Wells, D. G. *The Penguin Dictionary of Curious and Interesting Numbers.* London: Penguin, p. 127, 1986.

Euler Chain

A CHAIN (GRAPH) whose EDGES consist of all graph EDGES.

Euler Characteristic

Let a closed surface have GENUS g. Then the POLYHEDRAL FORMULA becomes the POINCARÉ FORMULA

$$\chi \equiv V - E + F = 2 - 2g, \tag{1}$$

where χ is the Euler characteristic, sometimes also known as the EULER-POINCARÉ CHARACTERISTIC. In terms of the INTEGRAL CURVATURE of the surface K,

$$\iint K\,da = 2\pi\chi. \tag{2}$$

The Euler characteristic is sometimes also called the EULER NUMBER. It can also be expressed as

$$\chi = p_0 - p_1 + p_2, \tag{3}$$

where p_i is the ith BETTI NUMBER of the space.

see also CHROMATIC NUMBER, MAP COLORING

Euler's Circle

see NINE-POINT CIRCLE

Euler's Conjecture

$$g(k) = 2^k + \left\lfloor \left(\frac{3}{2}\right)^k \right\rfloor - 2,$$

where $g(k)$ is the quantity appearing in WARING'S PROBLEM, and $\lfloor x \rfloor$ is the FLOOR FUNCTION.

see also WARING'S PROBLEM

Euler Constant

see e, EULER-MASCHERONI CONSTANT, MACLAURIN-CAUCHY THEOREM

Euler's Criterion

Let $p = 2m + 1$ be an ODD PRIME and a a POSITIVE INTEGER with $p \nmid a$. Then

$$a^m \equiv 1 \pmod{p} \tag{1}$$

IFF there exists an INTEGER t such that

$$p \equiv t^2 \pmod{p}. \tag{2}$$

In other words,

$$a^{(p-1)/2} \equiv \frac{a}{p} \pmod{p}, \tag{3}$$

where (a/p) is the LEGENDRE SYMBOL.

see also QUADRATIC RESIDUE

References

Rosen, K. H. Ch. 9 in *Elementary Number Theory and Its Applications, 3rd ed.* Reading, MA: Addison-Wesley, 1993.

Shanks, D. *Solved and Unsolved Problems in Number Theory, 4th ed.* New York: Chelsea, pp. 33–37, 1993.

Wagon, S. *Mathematica in Action.* New York: W. H. Freeman, p. 293, 1991.

Euler Curvature Formula

$$\kappa = \kappa_1 \cos^2\theta + \kappa_2 \sin^2\theta,$$

where κ is the normal CURVATURE in a direction making an ANGLE θ with the first principle direction.

Euler Differential Equation

The general nonhomogeneous equation is

$$x^2 \frac{d^2y}{dx^2} + \alpha x \frac{dy}{dx} + \beta y = S(x). \tag{1}$$

The homogeneous equation is

$$x^2 y'' + \alpha x y' + \beta y = 0 \tag{2}$$

$$y'' + \frac{\alpha}{x} y' + \frac{\beta}{x^2} y = 0. \tag{3}$$

Now attempt to convert the equation from

$$y'' + p(x)y' + q(x)y = 0 \tag{4}$$

to one with constant COEFFICIENTS

$$\frac{d^2y}{dz^2} + A\frac{dy}{dz} + By = 0 \tag{5}$$

by using the standard transformation for linear SECOND-ORDER ORDINARY DIFFERENTIAL EQUATIONS. Comparing (3) and (5), the functions $p(x)$ and $q(x)$ are

$$p(x) \equiv \frac{\alpha}{x} = \alpha x^{-1} \tag{6}$$

$$q(x) \equiv \frac{\beta}{x^2} = \beta x^{-2}. \tag{7}$$

Let $B \equiv \beta$ and define

$$\begin{aligned} z &\equiv B^{-1/2} \int \sqrt{q(x)}\, dx = \beta^{-1/2} \int \sqrt{\beta x^{-2}}\, dx \\ &= \int x^{-1}\, dx = \ln x. \end{aligned} \tag{8}$$

Then A is given by

$$\begin{aligned} A &\equiv \frac{q'(x) + 2p(x)q(x)}{2[q(x)]^{3/2}} B^{1/2} \\ &= \frac{-2\beta x^{-3} + 2(\alpha x^{-1})(\beta x^{-2})}{2(\beta x^{-2})^{3/2}} \beta^{1/2} \\ &= \alpha - 1, \end{aligned} \tag{9}$$

which is a constant. Therefore, the equation becomes a second-order ODE with constant COEFFICIENTS

$$\frac{d^2y}{dz^2} + (\alpha - 1)\frac{dy}{dz} + \beta y = 0. \tag{10}$$

Define

$$\begin{aligned} r_1 &\equiv \tfrac{1}{2}\left(-A + \sqrt{A^2 - 4B}\right) \\ &= \tfrac{1}{2}\left[1 - \alpha + \sqrt{(\alpha-1)^2 - 4\beta}\right] \end{aligned} \tag{11}$$

$$\begin{aligned} r_2 &\equiv \tfrac{1}{2}\left(-A - \sqrt{A^2 - 4B}\right) \\ &= \tfrac{1}{2}\left[1 - \alpha - \sqrt{(\alpha-1)^2 - 4\beta}\right] \end{aligned} \tag{12}$$

and

$$a \equiv \tfrac{1}{2}(1 - \alpha) \tag{13}$$

$$b \equiv \tfrac{1}{2}\sqrt{4\beta - (\alpha - 1)^2}. \tag{14}$$

The solutions are

$$y = \begin{cases} c_1 e^{r_1 z} + c_2 e^{r_2 z} & (\alpha-1)^2 > 4\beta \\ (c_1 + c_2 z)e^{az} & (\alpha-1)^2 = 4\beta \\ e^{az}[c_1 \cos(bz) + c_2 \sin(bz)] & (\alpha-1)^2 < 4\beta. \end{cases} \tag{15}$$

In terms of the original variable x,

$$y = \begin{cases} c_1 |x|^{r_1} + c_2 |x|^{r_2} & (\alpha-1)^2 > 4\beta \\ (c_1 + c_2 \ln|x|)|x|^a & (\alpha-1)^2 = 4\beta \\ |x|^a[c_1 \cos(b \ln|x|) + c_2 \sin(b \ln|x|)] & (\alpha-1)^2 < 4\beta. \end{cases} \tag{16}$$

Euler's Displacement Theorem

The general displacement of a rigid body (or coordinate frame) with one point fixed is a ROTATION about some axis. Furthermore, a ROTATION may be described in any basis using three ANGLES.

see also EUCLIDEAN MOTION, EULER ANGLES, RIGID MOTION, ROTATION

Euler's Distribution Theorem

For signed distances,

$$\overline{AB}\cdot\overline{CD}+\overline{AC}\cdot\overline{DB}+\overline{AD}\cdot\overline{BC}=0,$$

since

$$(b-a)(d-c)+(c-a)(b-d)+(d-a)(c-b)=0.$$

References

Johnson, R. A. *Modern Geometry: An Elementary Treatise on the Geometry of the Triangle and the Circle.* Boston, MA: Houghton Mifflin, p. 3, 1929.

Euler Equation

see also EULER DIFFERENTIAL EQUATION, EULER FORMULA, EULER-LAGRANGE DIFFERENTIAL EQUATION

Euler's Factorization Method

Works by expressing N as a QUADRATIC FORM in two different ways. Then

$$N=a^2+b^2=c^2+d^2, \tag{1}$$

so

$$a^2-c^2=d^2-b^2 \tag{2}$$

$$(a-c)(a+c)=(d-b)(d+b). \tag{3}$$

Let k be the GREATEST COMMON DIVISOR of $a-c$ and $d-b$ so

$$a-c=kl \tag{4}$$

$$d-b=km \tag{5}$$

$$(l,m)=1, \tag{6}$$

(where (l,m) denotes the GREATEST COMMON DIVISOR of l and m), and

$$l(a+c)=m(d+b). \tag{7}$$

But since $(l,m)=1$, $m|a+c$ and

$$a+c=mn, \tag{8}$$

which gives

$$b+d=ln, \tag{9}$$

so we have

$$\begin{aligned}[(\tfrac{1}{2}k)^2+(\tfrac{1}{2}n)^2](l^2+m^2)&=\tfrac{1}{4}(k^2+n^2)(l^2+m^2)\\ &=\tfrac{1}{4}[(kn)^2+(kl)^2+(nm)^2+(nl)^2]\\ &=\tfrac{1}{4}[(d-b)^2+(a-c)^2+(a+c)^2+(d+b)^2]\\ &=\tfrac{1}{4}(2a^2+2b^2+2c^2+2d^2)\\ &=\tfrac{1}{4}(2N+2N)=N.\end{aligned} \tag{10}$$

see also PRIME FACTORIZATION ALGORITHMS

Euler's Finite Difference Transformation

A transformation for the acceleration of the convergence of slowly converging ALTERNATING SERIES,

$$\sum_{k=0}^{\infty}(-1)^k a_k=\sum_{n=0}^{\infty}\frac{\Delta^k a_0}{2^{n+1}}.$$

References

Iyanaga, S. and Kawada, Y. (Eds.). *Encyclopedic Dictionary of Mathematics.* Cambridge, MA: MIT Press, p. 1163, 1980.

Euler Formula

The Euler formula states

$$e^{ix}=\cos x+i\sin x, \tag{1}$$

where i is the IMAGINARY NUMBER. Note that the EULER POLYHEDRAL FORMULA is sometimes also called the Euler formula, as is the EULER CURVATURE FORMULA. The equivalent expression

$$ix=\ln(\cos x+i\sin x) \tag{2}$$

had previously been published by Cotes (1714). The special case of the formula with $x=\pi$ gives the beautiful identity

$$e^{i\pi}+1=0, \tag{3}$$

an equation connecting the fundamental numbers i, PI, e, 1, and 0 (ZERO).

The Euler formula can be demonstrated using a series expansion

$$\begin{aligned}e^{ix}&=\sum_{n=0}^{\infty}\frac{(ix)^n}{n!}\\ &=\sum_{n=0}^{\infty}\frac{(-1)^n x^{2n}}{(2n)!}+i\sum_{n=1}^{\infty}\frac{(-1)^{n-1}x^{2n-1}}{(2n-1)!}\\ &=\cos x+i\sin x.\end{aligned} \tag{4}$$

It can also be proven using a COMPLEX integral. Let

$$z\equiv\cos\theta+i\sin\theta \tag{5}$$

$$dz=(-\sin\theta+i\cos\theta)\,d\theta=i(\cos\theta+i\sin\theta)\,d\theta=iz\,d\theta \tag{6}$$

$$\int\frac{dz}{z}=\int i\,d\theta \tag{7}$$

$$\ln z=i\theta, \tag{8}$$

so

$$z=e^{i\theta}\equiv\cos\theta+i\sin\theta. \tag{9}$$

see also DE MOIVRE'S IDENTITY, EULER POLYHEDRAL FORMULA

References
Castellanos, D. "The Ubiquitous Pi. Part I." *Math. Mag.* **61**, 67–98, 1988.
Conway, J. H. and Guy, R. K. "Euler's Wonderful Relation." *The Book of Numbers.* New York: Springer-Verlag, pp. 254–256, 1996.
Cotes, R. *Philosophical Transactions* **29**, 32, 1714.
Euler, L. *Miscellanea Berolinensia* **7**, 179, 1743.
Euler, L. *Introductio in Analysin Infinitorum, Vol. 1.* Lausanne, p. 104, 1748.

Euler Four-Square Identity

The amazing polynomial identity

$$\begin{aligned}(a_1{}^2 + a_2{}^2 + a_3{}^2 + a_4{}^2)(b_1{}^2 + b_2{}^2 + b_3{}^2 + b_4{}^2) \\ = (a_1b_1 - a_2b_2 - a_3b_3 - a_4b_4)^2 \\ +(a_1b_2 + a_2b_1 + a_3b_4 - a_4b_3)^2 \\ +(a_1b_3 - a_2b_4 + a_3b_1 + a_4b_2)^2 \\ +(a_1b_4 + a_2b_3 - a_3b_2 + a_4b_1)^2,\end{aligned}$$

communicated by Euler in a letter to Goldbach on April 15, 1705. The identity also follows from the fact that the norm of the product of two QUATERNIONS is the product of the norms (Conway and Guy 1996).

see also FIBONACCI IDENTITY, LAGRANGE'S FOUR-SQUARE THEOREM

References
Conway, J. H. and Guy, R. K. *The Book of Numbers.* New York: Springer-Verlag, p. 232, 1996.
Petkovšek, M.; Wilf, H. S.; and Zeilberger, D. *A=B.* Wellesley, MA: A. K. Peters, p. 8, 1996.

Euler's Graeco-Roman Squares Conjecture

Euler conjectured that there do not exist GRAECO-ROMAN SQUARES (now known as EULER SQUARES) of order $n = 4k + 2$ for $k = 1, 2, \ldots$. Such squares were found to exist in 1959, refuting the CONJECTURE.

see also EULER SQUARE, LATIN SQUARE

Euler Graph

A GRAPH containing an EULERIAN CIRCUIT. An undirected GRAPH is Eulerian IFF every VERTEX has EVEN DEGREE. A DIRECTED GRAPH is Eulerian IFF every VERTEX has equal INDEGREE and OUTDEGREE. A planar BIPARTITE GRAPH is DUAL to a planar Euler graph and vice versa. The number of Euler graphs with n nodes are 1, 1, 2, 3, 7, 16, 54, 243, ... (Sloane's A002854).

References
Sloane, N. J. A. Sequence A002854/M0846 in "An On-Line Version of the Encyclopedia of Integer Sequences."

Euler's Homogeneous Function Theorem

Let $f(x,y)$ be a HOMOGENEOUS FUNCTION of order n so that

$$f(tx,ty) = t^n(x,y). \tag{1}$$

Then define $x' \equiv xt$ and $y' \equiv yt$. Then

$$\begin{aligned}nt^{n-1}f(x,y) &= \frac{\partial f}{\partial x'}\frac{\partial x'}{\partial t} + \frac{\partial f}{\partial y'}\frac{\partial y'}{\partial t} \\ &= x\frac{\partial f}{\partial x'} + y\frac{\partial f}{\partial y'} = x\frac{\partial f}{\partial (xt)} + y\frac{\partial f}{\partial (yt)}. \end{aligned}\tag{2}$$

Let $t = 1$, then

$$x\frac{\partial f}{\partial x} + y\frac{\partial f}{\partial y} = nf(x,y). \tag{3}$$

This can be generalized to an arbitrary number of variables

$$x_i\frac{\partial f}{\partial x_i} = nf(\mathbf{x}), \tag{4}$$

where EINSTEIN SUMMATION has been used.

Euler's Hypergeometric Transformations

$${}_2F_1(a,b;c;z) = \int_0^1 \frac{t^{b-1}(1-t)^{c-b-1}}{(1-tz)^a}\,dt, \tag{1}$$

where ${}_2F_1(a,b;c;z)$ is a HYPERGEOMETRIC FUNCTION. The solution can be written using the Euler's transformations

$$t \to t \tag{2}$$
$$t \to 1-t \tag{3}$$
$$t \to (1-z-tz)^{-1} \tag{4}$$
$$t \to \frac{1-t}{1-tz} \tag{5}$$

in the equivalent forms

$$\begin{aligned}{}_2F_1(a,b;c;z) &= (1-z)^{-a}\,{}_2F_1(a,c-b;c;z/(z-1)) &(6)\\ &= (1-z)^{-b}\,{}_2F_1(c-a,b;c;z/(z-1)) &(7)\\ &= (1-z)^{c-a-b}\,{}_2F_1(c-a,c-b;c;z). &(8)\end{aligned}$$

see also HYPERGEOMETRIC FUNCTION

References
Morse, P. M. and Feshbach, H. *Methods of Theoretical Physics, Part I.* New York: McGraw-Hill, pp. 585–591, 1953.

Euler Identity

For $|z| < 1$,

$$\prod_{p=1}^{\infty}(1+z^p) = \prod_{q=1}^{\infty}(1-z^{2q-1})^{-1}.$$

see also JACOBI TRIPLE PRODUCT, q-SERIES

Euler's Idoneal Number

see IDONEAL NUMBER

Euler Integral

Euler integration was defined by Schanuel and subsequently explored by Rota, Chen, and Klain. The Euler integral of a FUNCTION $f : \mathbb{R} \to \mathbb{R}$ (assumed to be piecewise-constant with finitely many discontinuities) is the sum of

$$f(x) - \tfrac{1}{2}[f(x_+) + f(x_-)]$$

over the finitely many discontinuities of f. The n-D Euler integral can be defined for classes of functions $\mathbb{R}^n \to \mathbb{R}$. Euler integration is additive, so the Euler integral of $f+g$ equals the sum of the Euler integrals of f and g.

see also EULER MEASURE

Euler-Jacobi Pseudoprime

An Euler-Jacobi pseudoprime is a number n such that

$$2^{(n-1)/2} \equiv \frac{2}{n} \pmod{n}.$$

The first few are 561, 1105, 1729, 1905, 2047, 2465, ... (Sloane's A006971).

see also PSEUDOPRIME

References

Sloane, N. J. A. Sequence A006971/M5461 in "An On-Line Version of the Encyclopedia of Integer Sequences."

Euler L-Function

A special case of the ARTIN L-FUNCTION for the POLYNOMIAL x^2+1. It is given by

$$L(s) = \prod_{p \text{ odd prime}} \frac{1}{1-\chi^-(p)p^{-s}},$$

where

$$\begin{aligned} \chi^-(p) &\equiv \begin{cases} 1 & \text{for } p \equiv 1 \pmod 4 \\ -1 & \text{for } p \equiv 3 \pmod 4 \end{cases} \\ &= \left(\frac{-1}{p}\right), \end{aligned}$$

where $(-1/p)$ is a LEGENDRE SYMBOL.

References

Knapp, A. W. "Group Representations and Harmonic Analysis, Part II." *Not. Amer. Math. Soc.* **43**, 537–549, 1996.

Euler-Lagrange Differential Equation

A fundamental equation of CALCULUS OF VARIATIONS which states that if J is defined by an INTEGRAL of the form

$$J = \int f(x, y, \dot{y})\, dx, \tag{1}$$

where

$$\dot{y} \equiv \frac{dy}{dt}, \tag{2}$$

then J has a STATIONARY VALUE if the Euler-Lagrange differential equation

$$\frac{\partial f}{\partial y} - \frac{d}{dt}\left(\frac{\partial f}{\partial \dot{y}}\right) = 0 \tag{3}$$

is satisfied. If time DERIVATIVE NOTATION is replaced instead by space variable notation, the equation becomes

$$\frac{\partial f}{\partial y} - \frac{d}{dx}\frac{\partial f}{\partial y_x} = 0. \tag{4}$$

In many physical problems, f_x (the PARTIAL DERIVATIVE of f with respect to x) turns out to be 0, in which case a manipulation of the Euler-Lagrange differential equation reduces to the greatly simplified and partially integrated form known as the BELTRAMI IDENTITY,

$$f - y_x \frac{\partial f}{\partial y_x} = C. \tag{5}$$

For three independent variables (Arfken 1985, pp. 924–944), the equation generalizes to

$$\frac{\partial f}{\partial u} - \frac{\partial}{\partial x}\frac{\partial f}{\partial u_x} - \frac{\partial}{\partial y}\frac{\partial f}{\partial u_y} - \frac{\partial}{\partial z}\frac{\partial f}{\partial u_z} = 0. \tag{6}$$

Problems in the CALCULUS OF VARIATIONS often can be solved by solution of the appropriate Euler-Lagrange equation.

To derive the Euler-Lagrange differential equation, examine

$$\begin{aligned} \delta J \equiv \delta \int L(q, \dot{q}, t)\, dt &= \int \left(\frac{\partial L}{\partial q}\delta q + \frac{\partial L}{\partial \dot{q}}\delta \dot{q}\right) dt \\ &= \int \left[\frac{\partial L}{\partial q}\,\delta q + \frac{\partial L}{\partial \dot{q}}\frac{d(\delta q)}{dt}\right] dt, \end{aligned} \tag{7}$$

since $\delta\dot{q} = d(\delta q)/dt$. Now, integrate the second term by PARTS using

$$u = \frac{\partial L}{\partial \dot{q}} \qquad dv = d(\delta q) \tag{8}$$

$$du = \frac{d}{dt}\left(\frac{\partial L}{\partial \dot{q}}\right) dt \qquad v = \delta q, \tag{9}$$

so

$$\int \frac{\partial L}{\partial \dot{q}} \frac{d(\delta q)}{dt}\, dt = \int \frac{\partial L}{\partial \dot{q}}\, d(\delta q)$$
$$= \left[\frac{\partial L}{\partial \dot{q}} \delta q\right]_{t_1}^{t_2} - \int_{t_1}^{t_2} \left(\frac{d}{dt}\frac{\partial L}{\partial \dot{q}}\, dt\right) \delta q. \quad (10)$$

Combining (7) and (10) then gives

$$\delta J = \left[\frac{\partial L}{\partial \dot{q}} \delta q\right]_{t_1}^{t_2} + \int_{t_1}^{t_2} \left(\frac{\partial L}{\partial q} - \frac{d}{dt}\frac{\partial L}{\partial \dot{q}}\right) \delta q\, dt. \quad (11)$$

But we are varying the path only, not the endpoints, so $\delta q(t_1) = \delta q(t_2) = 0$ and (11) becomes

$$\delta J = \int_{t_1}^{t_2} \left(\frac{\partial L}{\partial q} - \frac{d}{dt}\frac{\partial L}{\partial \dot{q}}\right) \delta q\, dt. \quad (12)$$

We are finding the STATIONARY VALUES such that $\delta J = 0$. These must vanish for any small change δq, which gives from (12),

$$\frac{\partial L}{\partial q} - \frac{d}{dt}\left(\frac{\partial L}{\partial \dot{q}}\right) = 0. \quad (13)$$

This is the Euler-Lagrange differential equation.

The variation in J can also be written in terms of the parameter κ as

$$\delta J = \int [f(x, y + \kappa v, \dot{y} + \kappa \dot{v}) - f(x, y, \dot{y})]\, dt$$
$$= \kappa I_1 + \tfrac{1}{2}\kappa^2 I_2 + \tfrac{1}{6}\kappa^3 I_3 + \tfrac{1}{24}\kappa^4 I_4 + \ldots, \quad (14)$$

where

$$v = \delta y \quad (15)$$
$$\dot{v} = \delta \dot{y} \quad (16)$$

and the first, second, etc., variations are

$$I_1 = \int (v f_y + \dot{v} f_{\dot{y}})\, dt \quad (17)$$
$$I_2 = \int (v^2 f_{yy} + 2v\dot{v} f_{y\dot{y}} + \dot{v}^2 f_{\dot{y}\dot{y}})\, dt \quad (18)$$
$$I_3 = \int (v^3 f_{yyy} + 3v^2\dot{v} f_{yy\dot{y}} + 3v\dot{v}^2 f_{y\dot{y}\dot{y}} + \dot{v}^3 f_{\dot{y}\dot{y}\dot{y}})\, dt \quad (19)$$
$$I_4 = \int (v^4 f_{yyyy} + 4v^3\dot{v} f_{yyy\dot{y}} + 6v^2\dot{v}^2 f_{yy\dot{y}\dot{y}} + 4v\dot{v}^3 f_{y\dot{y}\dot{y}\dot{y}} + \dot{v}^4 f_{\dot{y}\dot{y}\dot{y}\dot{y}})\, dt. \quad (20)$$

The second variation can be re-expressed using

$$\frac{d}{dt}(v^2\lambda) = v^2\dot{\lambda} + 2v\dot{v}\lambda, \quad (21)$$

so

$$I_2 + [v^2\lambda]_2^1 = \int_1^2 [v^2(f_{yy} + \dot{\lambda}) + 2v\dot{v}(f_{y\dot{y}} + \lambda) + \dot{v}^2 f_{\dot{y}\dot{y}}]\, dt. \quad (22)$$

But

$$[v^2\lambda]_2^1 = 0. \quad (23)$$

Now choose λ such that

$$f_{\dot{y}\dot{y}}(f_{yy} + \dot{\lambda}) = (f_{y\dot{y}} + \lambda)^2 \quad (24)$$

and z such that

$$f_{y\dot{y}} + \lambda = -\frac{f_{\dot{y}\dot{y}}}{z}\frac{dz}{dt} \quad (25)$$

so that z satisfies

$$f_{\dot{y}\dot{y}}\ddot{z} + \dot{f}_{\dot{y}\dot{y}}\dot{z} - (f_{yy} - \dot{f}_{y\dot{y}})z = 0. \quad (26)$$

It then follows that

$$I_2 = \int f_{\dot{y}\dot{y}}\left(\dot{v} + \frac{f_{y\dot{y}} + \lambda}{f_{\dot{y}\dot{y}} v}\right)^2 dt = \int f_{\dot{y}\dot{y}}\left(\dot{v} - \frac{v}{z}\frac{dz}{dt}\right)^2. \quad (27)$$

see also BELTRAMI IDENTITY, BRACHISTOCHRONE PROBLEM, CALCULUS OF VARIATIONS, EULER-LAGRANGE DERIVATIVE

References

Arfken, G. *Mathematical Methods for Physicists, 3rd ed.* Orlando, FL: Academic Press, 1985.

Forsyth, A. R. *Calculus of Variations.* New York: Dover, pp. 17–20 and 29, 1960.

Morse, P. M. and Feshbach, H. "The Variational Integral and the Euler Equations." §3.1 in *Methods of Theoretical Physics, Part I.* New York: McGraw-Hill, pp. 276–280, 1953.

Euler-Lagrange Derivative

The derivative

$$\frac{\delta L}{\delta q} \equiv \frac{\partial L}{\partial q} - \frac{d}{dt}\left(\frac{\partial L}{\partial \dot{q}}\right)$$

appearing in the EULER-LAGRANGE DIFFERENTIAL EQUATION.

Euler Line

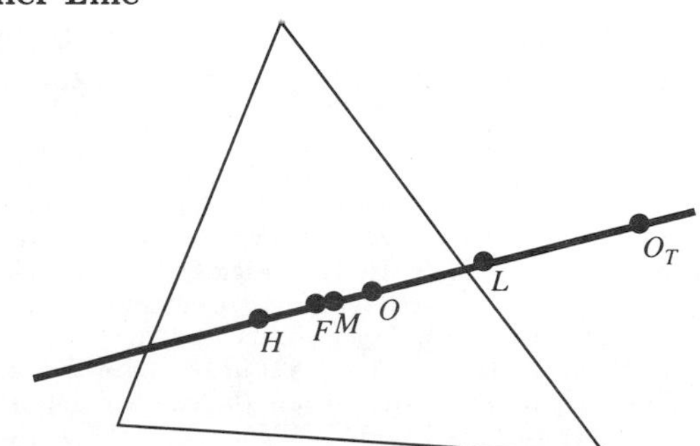

The line on which the ORTHOCENTER H, CENTROID M, CIRCUMCENTER O, DE LONGCHAMPS POINT L, NINE-POINT CENTER F, and the TANGENTIAL TRIANGLE CIRCUMCIRCLE O_T of a TRIANGLE lie. The INCENTER lies on the Euler line only if the TRIANGLE is an ISOSCELES TRIANGLE. The Euler line consists of all points with TRILINEAR COORDINATES $\alpha:\beta:\gamma$ which satisfy

$$\begin{vmatrix} \alpha & \beta & \gamma \\ \cos A & \cos B & \cos C \\ \cos B\cos C & \cos C\cos A & \cos A\cos B \end{vmatrix} = 0, \qquad (1)$$

which simplifies to

$$\alpha\cos A(\cos^2 B-\cos^2 C)+\beta\cos B(\cos^2 C-\cos^2 A) +\gamma\cos C(\cos^2 A-\cos^2 B)=0. \qquad (2)$$

This can also be written

$$\alpha\sin(2A)\sin(B-C)+\beta\sin(2B)\sin(C-A) +\gamma\sin(2C)\sin(A-B)=0. \qquad (3)$$

The Euler line may also be given parametrically by

$$P(\lambda)=O+\lambda H \qquad (4)$$

(Oldknow 1996).

λ	Center
-2	point at infinity
-1	de Longchamps point L
0	circumcenter O
1	centroid G
2	nine-point center F
∞	orthocenter H

The ORTHOCENTER is twice as far from the CENTROID as is the CIRCUMCENTER. The CIRCUMCENTER O, NINE-POINT CENTER F, CENTROID G, and ORTHOCENTER H form a HARMONIC RANGE.

The Euler line intersects the SODDY LINE in the DE LONGCHAMPS POINT, and the GERGONNE LINE in the EVANS POINT. The ISOTOMIC CONJUGATE of the Euler line is called JERABEK'S HYPERBOLA (Casey 1893, Vandeghen 1965).

see also CENTROID (TRIANGLE), CIRCUMCENTER, EVANS POINT, GERGONNE LINE, JERABEK'S HYPERBOLA, DE LONGCHAMPS POINT, NINE-POINT CENTER, ORTHOCENTER, SODDY LINE, TANGENTIAL TRIANGLE

References
Casey, J. *A Treatise on the Analytical Geometry of the Point, Line, Circle, and Conic Sections, Containing an Account of Its Most Recent Extensions with Numerous Examples, 2nd rev. enl. ed.* Dublin: Hodges, Figgis, & Co., 1893.
Coxeter, H. S. M. and Greitzer, S. L. *Geometry Revisited.* Washington, DC: Math. Assoc. Amer., pp. 18–20, 1967.
Dörrie, H. "Euler's Straight Line." §27 in *100 Great Problems of Elementary Mathematics: Their History and Solutions.* New York: Dover, pp. 141–142, 1965.
Ogilvy, C. S. *Excursions in Geometry.* New York: Dover, pp. 117–119, 1990.
Oldknow, A. "The Euler-Gergonne-Soddy Triangle of a Triangle." *Amer. Math. Monthly* **103**, 319–329, 1996.
Vandeghen, A. "Some Remarks on the Isogonal and Cevian Transforms. Alignments of Remarkable Points of a Triangle." *Amer. Math. Monthly* **72**, 1091–1094, 1965.

Euler-Lucas Pseudoprime

Let $U(P,Q)$ and $V(P,Q)$ be LUCAS SEQUENCES generated by P and Q, and define

$$D\equiv P^2-4Q.$$

Then

$$\begin{cases} U_{(n-(D/n))/2}\equiv 0 \pmod{n} & \text{when } (Q/n)=1 \\ V_{(n-(D/n))/2}\equiv D \pmod{n} & \text{when } (Q/n)=-1, \end{cases}$$

where (Q/n) is the LEGENDRE SYMBOL. An ODD COMPOSITE NUMBER n such that $(n,QD)=1$ (i.e., n and QD are RELATIVELY PRIME) is called an Euler-Lucas pseudoprime with parameters (P,Q).

see also PSEUDOPRIME, STRONG LUCAS PSEUDOPRIME

References
Ribenboim, P. "Euler-Lucas Pseudoprimes (elpsp(P,Q)) and Strong Lucas Pseudoprimes (slpsp(P,Q))." §2.X.C in *The New Book of Prime Number Records.* New York: Springer-Verlag, pp. 130–131, 1996.

Euler's Machin-Like Formula

The MACHIN-LIKE FORMULA

$$\tfrac{1}{4}\pi=\tan^{-1}(\tfrac{1}{2})+\tan^{-1}(\tfrac{1}{3}).$$

The other 2-term MACHIN-LIKE FORMULAS are HERMANN'S FORMULA, HUTTON'S FORMULA, and MACHIN'S FORMULA.

see also INVERSE TANGENT

Euler-Maclaurin Integration Formulas

The first Euler-Maclaurin integration formula is

$$\int_0^1 f(x)\,dx=\tfrac{1}{2}[f(1)+f(0)] -\sum_{p=1}^{q}\frac{1}{(2p)!}B_{2p}[f^{(2p-1)}(1)-f^{(2p-1)}(0)] +\frac{1}{(2q)!}\int_0^1 f^{(2q)}(x)B_{2q}(x)\,dx, \qquad (1)$$

where B_n are BERNOULLI NUMBERS. SUMS may be converted to INTEGRALS by inverting the FORMULA to obtain

$$\sum_{m=1}^{n} f(m)=\int_1^n f(x)\,dx+\tfrac{1}{2}[f(1)+f(n)] +\frac{B_2}{2!}[f'(n)-f'(1)]+\ldots. \qquad (2)$$

For a more general case when $f(x)$ is tabulated at n values $f_1, f_2, \ldots, f_n$,

$$\int_{x_1}^{x_n} f(x)\,dx = h[\tfrac{1}{2}f_1 + f_2 + f_3 + \ldots + f_{n-1} + \tfrac{1}{2}f_n] - \sum_{k=1}^{\infty} \frac{B_{2k}h^{2k}}{(2k)!}[f_n{}^{(2k-1)} - f_1{}^{(2k-1)}]. \quad (3)$$

The Euler-Maclaurin formula is implemented in *Mathematica*® (Wolfram Research, Champaign, IL) as the function `NSum` with option `Method->Integrate`.

The second Euler-Maclaurin integration formula is used when $f(x)$ is tabulated at n values $f_{3/2}, f_{5/2}, \ldots, f_{n-1/2}$:

$$\int_{x_1}^{x_n} f(x)\,dx = h[f_{3/2} + f_{5/2} + f_{7/2} + \ldots + f_{n-3/2} + f_{n-1/2}] - \sum_{k=1}^{\infty} \frac{B_{2k}h^{2k}}{(2k)!}(1-2^{-2k+1})[f_n{}^{(2k-1)} - f_1{}^{(2k-1)}]. \quad (4)$$

see also SUM, WYNN'S EPSILON METHOD

References

Abramowitz, M. and Stegun, C. A. (Eds.). *Handbook of Mathematical Functions with Formulas, Graphs, and Mathematical Tables, 9th printing.* New York: Dover, pp. 16 and 806, 1972.

Arfken, G. "Bernoulli Numbers, Euler-Maclaurin Formula." §5.9 in *Mathematical Methods for Physicists, 3rd ed.* Orlando, FL: Academic Press, pp. 327–338, 1985.

Borwein, J. M.; Borwein, P. B.; and Dilcher, K. "Pi, Euler Numbers, and Asymptotic Expansions." *Amer. Math. Monthly* **96**, 681–687, 1989.

Vardi, I. "The Euler-Maclaurin Formula." §8.3 in *Computational Recreations in Mathematica.* Reading, MA: Addison-Wesley, pp. 159–163, 1991.

Euler-Mascheroni Constant

The Euler-Mascheroni constant is denoted γ (or sometimes C) and has the numerical value

$$\gamma \approx 0.577215664901532860606512090082402431042\ldots \quad (1)$$

(Sloane's A001620). The CONTINUED FRACTION of the Euler-Mascheroni constant is [0, 1, 1, 2, 1, 2, 1, 4, 3, 13, 5, 1, 1, 8, 1, 2, 4, 1, 1, 40, ...] (Sloane's A002852). The first few CONVERGENTS are 1, 1/2, 3/5, 4/7, 11/19, 15/26, 71/123, 228/395, 3035/5258, 15403/26685, ... (Sloane's A046114 and A046115). The positions at which the digits 1, 2, ... first occur in the CONTINUED FRACTION are 2, 4, 9, 8, 11, 69, 24, 14, 139, 52, 22, ... (Sloane's A033149). The sequence of largest terms in the CONTINUED FRACTION is 1, 2, 4, 13, 40, 49, 65, 399, 2076, ... (Sloane's A033091), which occur at positions 2, 4, 8, 10, 20, 31, 34, 40, 529, ... (Sloane's A033092).

It is not known if this constant is IRRATIONAL, let alone TRANSCENDENTAL. However, Conway and Guy (1996) are "prepared to bet that it is transcendental," although they do not expect a proof to be achieved within their lifetimes.

The Euler-Mascheroni constant arises in many integrals

$$\gamma \equiv -\int_0^\infty e^{-x}\ln x\,dx \quad (2)$$

$$= \int_0^\infty \left(\frac{1}{1-e^{-x}} - \frac{1}{x}\right) e^{-x}\,dx \quad (3)$$

$$= \int_0^\infty \frac{1}{x}\left(\frac{1}{1+x} - e^{-x}\right) dx. \quad (4)$$

and sums

$$\gamma \equiv 1 + \sum_{k=2}^{\infty}\left[\frac{1}{k} + \ln\left(\frac{k-1}{k}\right)\right] \quad (5)$$

$$= \lim_{m\to\infty}\left(\sum_{n=1}^{m}\frac{1}{n} - \ln m\right) \quad (6)$$

$$= \sum_{n=2}^{\infty}(-1)^n\frac{\zeta(n)}{n} \quad (7)$$

$$= \ln\left(\frac{4}{\pi}\right) - \sum_{n=1}^{\infty}\frac{(-1)^n\zeta(n+1)}{2^n(n+1)} \quad (8)$$

$$= \lim_{n\to\infty}\left[\sum_{k=1}^{n}k^{-1} - \ln n - \frac{1}{2n} + \sum_{k=1}^{n}\frac{B_{2k}}{(2k)n^{2k}}\right], \quad (9)$$

where $\zeta(z)$ is the RIEMANN ZETA FUNCTION and B_n are the BERNOULLI NUMBERS. It is also given by the EULER PRODUCT

$$e^\gamma = \lim_{n\to\infty}\frac{1}{\ln n}\prod_{i=1}^{n}\frac{1}{1-\frac{1}{p_i}}, \quad (10)$$

where the product is over PRIMES p. Another connection with the PRIMES was provided by Dirichlet's 1838 proof that the average number of DIVISORS of all numbers from 1 to n is asymptotic to

$$\frac{\sum_{i=1}^{n}\sigma_0(i)}{n} \sim \ln n + 2\gamma - 1 \quad (11)$$

(Conway and Guy 1996). de la Vallée Poussin (1898) proved that, if a large number n is divided by all PRIMES $\leq n$, then the average amount by which the QUOTIENT is less than the next whole number is γ.

INFINITE PRODUCTS involving γ also arise from the G-FUNCTION with POSITIVE INTEGER n. The cases $G(2)$ and $G(3)$ give

$$\prod_{n=1}^{\infty} e^{-1+1/(2n)}\left(1+\frac{1}{n}\right)^n = \frac{e^{1+\gamma/2}}{\sqrt{2\pi}} \tag{12}$$

$$\prod_{n=1}^{\infty} e^{-2+2/n}\left(1+\frac{2}{n}\right)^n = \frac{e^{3+2\gamma}}{2\pi}. \tag{13}$$

The Euler-Mascheroni constant is also given by the limits

$$\gamma = \lim_{s\to 1}\frac{\zeta(s)-1}{s-1} \tag{14}$$

$$= -\Gamma'(1) \tag{15}$$

$$= \lim_{x\to\infty}\left[x-\Gamma\left(\frac{1}{x}\right)\right] \tag{16}$$

(Le Lionnais 1983).

The difference between the nth convergent in (6) and γ is given by

$$\sum_{k=1}^{n}\frac{1}{k} - \ln n - \gamma = \int_n^\infty \frac{x-\lfloor x\rfloor}{x^2}\,dx, \tag{17}$$

where $\lfloor x\rfloor$ is the FLOOR FUNCTION, and satisfies the INEQUALITY

$$\frac{1}{2(n+1)} < \sum_{k=1}^{n}\frac{1}{k} - \ln n - \gamma < \frac{1}{2n} \tag{18}$$

(Young 1991). A series with accelerated convergence is

$$\gamma = \tfrac{3}{2} - \ln 2 - \sum_{m=2}^{\infty}(-1)^m\frac{m-1}{m}[\zeta(m)-1] \tag{19}$$

(Flajolet and Vardi 1996). Another series is

$$\gamma = \sum_{n=1}^{\infty}(-1)^n\frac{\lfloor \lg n\rfloor}{n} \tag{20}$$

(Vacca 1910, Gerst 1969), where LG is the LOGARITHM to base 2. The convergence of this series can be greatly improved using Euler's CONVERGENCE IMPROVEMENT transformation to

$$\gamma = \sum_{k=1}^{\infty}2^{-(k+1)}\sum_{j=0}^{k-1}\frac{1}{\binom{2^{k-j}+j}{j}}, \tag{21}$$

where $\binom{a}{b}$ is a BINOMIAL COEFFICIENT (Beeler *et al.* 1972, Item 120, with $k-j$ replacing the undefined i). Bailey (1988) gives

$$\gamma = \frac{2^n}{e^{2^n}}\sum_{m=0}^{\infty}\frac{2^{mn}}{(m+1)!}\sum_{t=0}^{m}\frac{1}{t+1} - n\ln 2 + \mathcal{O}\left(\frac{1}{2^n e^{2^n}}\right), \tag{22}$$

which is an improvement over Sweeney (1963).

The symbol γ is sometimes also used for

$$\gamma' \equiv e^\gamma \approx 1.781072 \tag{23}$$

(Gradshteyn and Ryzhik 1979, p. xxvii).

Odena (1982–1983) gave the strange approximation

$$(0.11111111)^{1/4} = 0.577350\ldots, \tag{24}$$

and Castellanos (1988) gave

$$(\tfrac{7}{83})^{2/9} = 0.57721521\ldots \tag{25}$$

$$\left(\frac{520^6+22}{52^4}\right)^{1/6} = 0.5772156634\ldots \tag{26}$$

$$\left(\frac{80^3+92}{61^4}\right)^{1/6} = 0.57721566457\ldots \tag{27}$$

$$\frac{990^3-55^3-79^2-4^2}{70^5} = 0.5772156649015295\ldots. \tag{28}$$

No quadratically converging algorithm for computing γ is known (Bailey 1988). 7,000,000 digits of γ have been computed as of Feb. 1998 (Plouffe).

see also EULER PRODUCT, MERTENS THEOREM, STIELTJES CONSTANTS

References

Bailey, D. H. "Numerical Results on the Transcendence of Constants Involving π, e, and Euler's Constant." *Math. Comput.* **50**, 275–281, 1988.

Beeler, M.; Gosper, R. W.; and Schroeppel, R. *HAKMEM.* Cambridge, MA: MIT Artificial Intelligence Laboratory, Memo AIM-239, Feb. 1972.

Brent, R. P. "Computation of the Regular Continued Fraction for Euler's Constant." *Math. Comput.* **31**, 771–777, 1977.

Brent, R. P. and McMillan, E. M. "Some New Algorithms for High-Precision Computation of Euler's Constant." *Math. Comput.* **34**, 305–312, 1980.

Castellanos, D. "The Ubiquitous Pi. Part I." *Math. Mag.* **61**, 67–98, 1988.

Conway, J. H. and Guy, R. K. "The Euler-Mascheroni Number." In *The Book of Numbers.* New York: Springer-Verlag, pp. 260–261, 1996.

de la Vallée Poussin, C.-J. Untitled communication. *Annales de la Soc. Sci. Bruxelles* **22**, 84–90, 1898.

DeTemple, D. W. "A Quicker Convergence to Euler's Constant." *Amer. Math. Monthly* **100**, 468–470, 1993.

Dirichlet, G. L. *J. für Math.* **18**, 273, 1838.

Finch, S. "Favorite Mathematical Constants." `http://www.mathsoft.com/asolve/constant/euler/euler.html`.

Flajolet, P. and Vardi, I. "Zeta Function Expansions of Classical Constants." Unpublished manuscript, 1996. `http://pauillac.inria.fr/algo/flajolet/Publications/landau.ps`.

Gerst, I. "Some Series for Euler's Constant." *Amer. Math. Monthly* **76**, 273–275, 1969.

Glaisher, J. W. L. "On the History of Euler's Constant." *Messenger of Math.* **1**, 25–30, 1872.

Gradshteyn, I. S. and Ryzhik, I. M. *Tables of Integrals, Series, and Products, 5th ed.* San Diego, CA: Academic Press, 1979.
Knuth, D. E. "Euler's Constant to 1271 Places." *Math. Comput.* **16**, 275–281, 1962.
Le Lionnais, F. *Les nombres remarquables.* Paris: Hermann, p. 28, 1983.
Plouffe, S. "Plouffe's Inverter: Table of Current Records for the Computation of Constants." `http://lacim.uqam.ca/pi/records.html`.
Sloane, N. J. A. Sequences A033091, A033092, A046114, A046115, A001620/M3755, and A002852/M0097 in "An On-Line Version of the Encyclopedia of Integer Sequences."
Sweeney, D. W. "On the Computation of Euler's Constant." *Math. Comput.* **17**, 170–178, 1963.
Vacca, G. "A New Series for the Eulerian Constant." *Quart. J. Pure Appl. Math.* **41**, 363–368, 1910.
Young, R. M. "Euler's Constant." *Math. Gaz.* **75**, 187–190, 1991.

Euler-Mascheroni Integrals

Define

$$I_n \equiv (-1)^n \int_0^\infty (\ln z)^n e^{-z}\,dz, \qquad (1)$$

then

$$I_0 = \int_0^\infty e^{-z}\,dz = [-e^{-z}]_0^\infty = (0+1) = 1 \qquad (2)$$

$$I_1 = -\int_0^\infty (\ln z)e^{-z}\,dz = \gamma \qquad (3)$$

$$I_2 = \gamma^2 + \tfrac{1}{6}\pi^2 \qquad (4)$$

$$I_3 = \gamma^3 + \tfrac{1}{2}\gamma\pi^2 + 2\zeta(3) \qquad (5)$$

$$I_4 = \gamma^4 + \gamma^2\pi^2 - \tfrac{3}{20}\pi^4 + 8\gamma\zeta(3), \qquad (6)$$

where γ is the EULER-MASCHERONI CONSTANT and $\zeta(3)$ is APÉRY'S CONSTANT.

Euler Measure

Define the Euler measure of a polyhedral set as the EULER INTEGRAL of its indicator function. It is easy to show by induction that the Euler measure of a closed bounded convex POLYHEDRON is always 1 (independent of dimension), while the Euler measure of a d-D relative-open bounded convex POLYHEDRON is $(-1)^d$.

Euler Number

The Euler numbers, also called the SECANT NUMBERS or ZIG NUMBERS, are defined for $|x| < \pi/2$ by

$$\operatorname{sech} x - 1 \equiv -\frac{E_1^* x^2}{2!} + \frac{E_2^* x^4}{4!} - \frac{E_3^* x^6}{6!} + \ldots \qquad (1)$$

$$\sec x - 1 \equiv \frac{E_1^* x^2}{2!} + \frac{E_2^* x^4}{4!} + \frac{E_3^* x^6}{6!} + \ldots, \qquad (2)$$

where sech is the HYPERBOLIC SECANT and sec is the SECANT. Euler numbers give the number of ODD ALTERNATING PERMUTATIONS and are related to GENOCCHI NUMBERS. The base e of the NATURAL LOGARITHM is sometimes known as Euler's number.

Some values of the Euler numbers are

$$\begin{aligned}
E_1^* &= 1\\
E_2^* &= 5\\
E_3^* &= 61\\
E_4^* &= 1{,}385\\
E_5^* &= 50{,}521\\
E_6^* &= 2{,}702{,}765\\
E_7^* &= 199{,}360{,}981\\
E_8^* &= 19{,}391{,}512{,}145\\
E_9^* &= 2{,}404{,}879{,}675{,}441\\
E_{10}^* &= 370{,}371{,}188{,}237{,}525\\
E_{11}^* &= 69{,}348{,}874{,}393{,}137{,}901\\
E_{12}^* &= 15{,}514{,}534{,}163{,}557{,}086{,}905
\end{aligned}$$

(Sloane's A000364). The first few PRIME Euler numbers E_n occur for $n = 2$, 3, 19, 227, 255, ... (Sloane's A014547) up to a search limit of $n = 1415$.

The slightly different convention defined by

$$E_{2n} = (-1)^n E_n^* \qquad (3)$$

$$E_{2n+1} = 0 \qquad (4)$$

is frequently used. These are, for example, the Euler numbers computed by the *Mathematica*® (Wolfram Research, Champaign, IL) function `EulerE[n]`. This definition has the particularly simple series definition

$$\operatorname{sech} x - 1 \equiv \sum_{k=0}^{\infty} \frac{E_k x^k}{k!} \qquad (5)$$

and is equivalent to

$$E_n = 2^n E_n(\tfrac{1}{2}), \qquad (6)$$

where $E_n(x)$ is an EULER POLYNOMIAL.

To confuse matters further, the EULER CHARACTERISTIC is sometimes also called the "Euler number."

see also BERNOULLI NUMBER, EULERIAN NUMBER, EULER POLYNOMIAL, EULER ZIGZAG NUMBER, GENOCCHI NUMBER

References

Abramowitz, M. and Stegun, C. A. (Eds.). "Bernoulli and Euler Polynomials and the Euler-Maclaurin Formula." §23.1 in *Handbook of Mathematical Functions with Formulas, Graphs, and Mathematical Tables, 9th printing.* New York: Dover, pp. 804–806, 1972.
Conway, J. H. and Guy, R. K. In *The Book of Numbers.* New York: Springer-Verlag, pp. 110–111, 1996.
Guy, R. K. "Euler Numbers." §B45 in *Unsolved Problems in Number Theory, 2nd ed.* New York: Springer-Verlag, p. 101, 1994.
Knuth, D. E. and Buckholtz, T. J. "Computation of Tangent, Euler, and Bernoulli Numbers." *Math. Comput.* **21**, 663–688, 1967.
Sloane, N. J. A. Sequences A014547 and A000364/M4019 in "An On-Line Version of the Encyclopedia of Integer Sequences."
Spanier, J. and Oldham, K. B. "The Euler Numbers, E_n." Ch. 5 in *An Atlas of Functions.* Washington, DC: Hemisphere, pp. 39–42, 1987.

Euler Parameters

The four parameters e_0, e_1, e_2, and e_3 describing a finite rotation about an arbitrary axis. The Euler parameters are defined by

$$e_0 \equiv \cos\left(\frac{\phi}{2}\right) \tag{1}$$

$$\mathbf{e} \equiv \begin{bmatrix} e_1 \\ e_2 \\ e_3 \end{bmatrix} = \hat{\mathbf{n}} \sin\left(\frac{\phi}{2}\right), \tag{2}$$

and are a QUATERNION in scalar-vector representation

$$(e_0, \mathbf{e}) = e_0 + e_1 x + e_2 j + e_3 k. \tag{3}$$

Because EULER'S ROTATION THEOREM states that an arbitrary rotation may be described by only three parameters, a relationship must exist between these four quantities

$$e_0^{\ 2} + \mathbf{e} \cdot \mathbf{e} = e_0^{\ 2} + e_1^{\ 2} + e_2^{\ 2} + e_3^{\ 2} = 1 \tag{4}$$

(Goldstein 1980, p. 153). The rotation angle is then related to the Euler parameters by

$$\cos\phi = 2e_0^{\ 2} - 1 = e_0^{\ 2} - \mathbf{e}\cdot\mathbf{e} = e_0^{\ 2} - e_1^{\ 2} - e_2^{\ 2} - e_3^{\ 2} \tag{5}$$

$$\hat{\mathbf{n}} \sin\phi = 2\mathbf{e}e_0. \tag{6}$$

The Euler parameters may be given in terms of the EULER ANGLES by

$$e_0 = \cos[\tfrac{1}{2}(\phi + \psi)] \cos(\tfrac{1}{2}\theta) \tag{7}$$

$$e_1 = \sin[\tfrac{1}{2}(\phi - \psi)] \sin(\tfrac{1}{2}\theta) \tag{8}$$

$$e_2 = \cos[\tfrac{1}{2}(\phi - \psi)] \sin(\tfrac{1}{2}\theta) \tag{9}$$

$$e_3 = \sin[\tfrac{1}{2}(\phi + \psi)] \cos(\tfrac{1}{2}\theta) \tag{10}$$

(Goldstein 1980, p. 155).

Using the Euler parameters, the ROTATION FORMULA becomes

$$\mathbf{r}' = \mathbf{r}(e_0^{\ 2} - e_1^{\ 2} - e_2^{\ 2} - e_3^{\ 2}) + 2\mathbf{e}(\mathbf{e}\cdot\mathbf{r}) + (\mathbf{r}\times\hat{\mathbf{n}})\sin\phi, \tag{11}$$

and the ROTATION MATRIX becomes

$$\begin{bmatrix} x' \\ y' \\ z' \end{bmatrix} = \mathsf{A} \begin{bmatrix} x \\ y \\ z \end{bmatrix}, \tag{12}$$

where the elements of the matrix are

$$a_{ij} = \delta_{ij}(e_0^{\ 2} - e_k e_k) + 2e_i e_j + 2\epsilon_{ijk} e_0 e_k. \tag{13}$$

Here, EINSTEIN SUMMATION has been used, δ_{ij} is the KRONECKER DELTA, and ϵ_{ijk} is the PERMUTATION SYMBOL. Written out explicitly, the matrix elements are

$$a_{11} = e_0^{\ 2} + e_1^{\ 2} - e_2^{\ 2} - e_3^{\ 2} \tag{14}$$

$$a_{12} = 2(e_1 e_2 + e_0 e_3) \tag{15}$$

$$a_{13} = 2(e_1 e_3 - e_0 e_2) \tag{16}$$

$$a_{21} = 2(e_1 e_2 - e_0 e_3) \tag{17}$$

$$a_{22} = e_0^{\ 2} - e_1^{\ 2} + e_2^{\ 2} - e_3^{\ 2} \tag{18}$$

$$a_{23} = 2(e_2 e_3 + e_0 e_1) \tag{19}$$

$$a_{31} = 2(e_1 e_3 + e_0 e_2) \tag{20}$$

$$a_{32} = 2(e_2 e_3 - e_0 e_1) \tag{21}$$

$$a_{33} = e_0^{\ 2} - e_1^{\ 2} - e_2^{\ 2} + e_3^{\ 2}. \tag{22}$$

see also EULER ANGLES, QUATERNION, ROTATION MATRIX

References

Arfken, G. *Mathematical Methods for Physicists, 3rd ed.* Orlando, FL: Academic Press, pp. 198–200, 1985.

Goldstein, H. *Classical Mechanics, 2nd ed.* Reading, MA: Addison-Wesley, 1980.

Landau, L. D. and Lifschitz, E. M. *Mechanics, 3rd ed.* Oxford, England: Pergamon Press, 1976.

Euler's Pentagonal Number Theorem

$$\prod_{n=1}^{\infty}(1 - x^n) = \sum_{n=-\infty}^{\infty} (-1)^n x^{n(3n+1)/2}, \tag{1}$$

where $n(3n+1)/2$ are generalized PENTAGONAL NUMBERS. Related equalities are

$$\prod_{k=1}^{\infty}(1 - x^k t) = \sum_{n=0}^{\infty} \frac{(-1)^n x^{n(n+1)/2} t^n}{\prod_{k=1}^{n}(1 - x^k)} \tag{2}$$

$$\prod_{k=1}^{\infty}(1 - x^k t)^{-1} = \sum_{n=0}^{\infty} \frac{t^n}{\prod_{k=1}^{n}(1 - x^k)}. \tag{3}$$

see also PARTITION FUNCTION P, PENTAGONAL NUMBER

Euler's Phi Function

see TOTIENT FUNCTION

Euler-Poincaré Characteristic

see EULER CHARACTERISTIC

Euler's Polygon Division Problem

The problem of finding in how many ways E_n a PLANE convex POLYGON of n sides can be divided into TRIANGLES by diagonals. Euler first proposed it to Christian Goldbach in 1751, and the solution is the CATALAN NUMBER $E_n = C_{n-2}$.

see also CATALAN NUMBER, CATALAN'S PROBLEM

References

Guy, R. K. "Dissecting a Polygon Into Triangles." *Bull. Malayan Math. Soc.* **5**, 57–60, 1958.

Euler Polyhedral Formula

see POLYHEDRAL FORMULA

Euler Polynomial

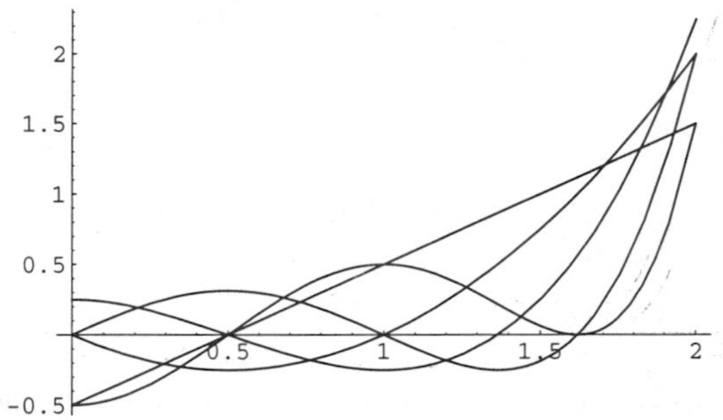

A POLYNOMIAL $E_n(x)$ given by the sum

$$\frac{2e^{xt}}{e^t+1} \equiv \sum_{n=0}^{\infty} E_n(x) \frac{t^n}{n!}. \quad (1)$$

Euler polynomials are related to the BERNOULLI NUMBERS by

$$E_{n-1}(x) = \frac{2^n}{n}\left[B_n\left(\frac{x+1}{2}\right) - B_n\left(\frac{x}{2}\right)\right] \quad (2)$$

$$= \frac{2}{n}\left[B_n(x) - 2^n B_n\left(\frac{x}{2}\right)\right] \quad (3)$$

$$E_{n-2}(x) = 2\binom{n}{2}^{-1} \sum_{k=0}^{n-2} \binom{n}{k}[(2^{n-k}-1)B_{n-k}B_k(x)], \quad (4)$$

where $\binom{n}{k}$ is a BINOMIAL COEFFICIENT. Setting $x = 1/2$ and normalizing by 2^n gives the EULER NUMBER

$$E_n = 2^n E_n(\tfrac{1}{2}). \quad (5)$$

Call $E'_n = E_n(0)$, then the first few terms are $-1/2$, 0, 1/4, $-1/2$, 0, 17/8, 0, 31/2, 0, ... The terms are the same but with the SIGNS reversed if $x = 1$. These values can be computed using the double sum

$$E_n(0) = 2^{-n} \sum_{j=1}^{n} \left[(-1)^{j+n+1} j^k \sum_{k=0}^{n-j} \binom{n+1}{k}\right]. \quad (6)$$

The BERNOULLI NUMBERS B_n for $n > 1$ can be expressed in terms of the E'_n by

$$B_n = -\frac{nE'_{n-1}}{2(2^n-1)}. \quad (7)$$

see also BERNOULLI POLYNOMIAL, EULER NUMBER, GENOCCHI NUMBER

References

Abramowitz, M. and Stegun, C. A. (Eds.). "Bernoulli and Euler Polynomials and the Euler-Maclaurin Formula." §23.1 in *Handbook of Mathematical Functions with Formulas, Graphs, and Mathematical Tables, 9th printing.* New York: Dover, pp. 804–806, 1972.

Gradshteyn, I. S. and Ryzhik, I. M. *Tables of Integrals, Series, and Products, 5th ed.* San Diego, CA: Academic Press, 1979.

Spanier, J. and Oldham, K. B. "The Euler Polynomials $E_n(x)$." Ch. 20 in *An Atlas of Functions.* Washington, DC: Hemisphere, pp. 175–181, 1987.

Euler Polynomial Identity

see EULER FOUR-SQUARE IDENTITY

Euler Power Conjecture

see EULER'S SUM OF POWERS CONJECTURE

Euler Product

For $\sigma > 1$,

$$\zeta(\sigma) \equiv \sum_{n=1}^{\infty} \frac{1}{n^\sigma} = \prod_p \frac{1}{1-\frac{1}{p^\sigma}},$$

where $\zeta(z)$ is the RIEMANN ZETA FUNCTION.

$$e^\gamma = \lim_{n\to\infty} \frac{1}{\ln n} \prod_{i=1}^{n} \frac{1}{1-\frac{1}{p_i}},$$

where the product is over PRIMES p, where γ is the EULER-MASCHERONI CONSTANT.

see also DEDEKIND FUNCTION

Euler Pseudoprime

Euler pseudoprimes to a base a are ODD COMPOSITE numbers such that $(a, n) = 1$ and the JACOBI SYMBOL satisfies

$$\left(\frac{a}{n}\right) \equiv a^{(n-1)/2} \pmod{n}.$$

No ODD COMPOSITE number is an Euler pseudoprime for all bases a RELATIVELY PRIME to it. This class includes some CARMICHAEL NUMBERS and all STRONG PSEUDOPRIMES to base a. An Euler pseudoprime is pseudoprime to at most 1/2 of all possible bases less than itself. The first few Euler pseudoprimes are 341, 561, 1105, 1729, 1905, 2047, ... (Sloane's A006970).

see also PSEUDOPRIME, STRONG PSEUDOPRIME

References

Guy, R. K. "Pseudoprimes. Euler Pseudoprimes. Strong Pseudoprimes." §A12 in *Unsolved Problems in Number Theory, 2nd ed.* New York: Springer-Verlag, pp. 27–30, 1994.

Sloane, N. J. A. Sequence A006970/M5442 in "An On-Line Version of the Encyclopedia of Integer Sequences."

Euler's Quadratic Residue Theorem

A number D that possesses no common divisor with a prime number p is either a QUADRATIC RESIDUE or nonresidue of p, depending whether $D^{(p-1)/2}$ is congruent mod p to ± 1.

Euler Quartic Conjecture

Euler conjectured that there are no POSITIVE INTEGER solutions to the quartic DIOPHANTINE EQUATION

$$A^4 + B^4 = C^4 + D^4.$$

This conjecture was disproved by N. D. Elkies in 1988, who found an infinite class of solutions.

see also DIOPHANTINE EQUATION—QUARTIC

References

Berndt, B. C. and Bhargava, S. "Ramanujan—For Low-brows." *Amer. Math. Monthly* **100**, 644–656, 1993.

Guy, R. K. *Unsolved Problems in Number Theory, 2nd ed.* New York: Springer-Verlag, pp. 139–140, 1994.

Lander, L. J.; Parkin, T. R.; and Selfridge, J. L. "A Survey of Equal Sums of Like Powers." *Math. Comput.* **21**, 446–459, 1967.

Ward, M. "Euler's Problem on Sums of Three Fourth Powers." *Duke Math. J.* **15**, 827–837, 1948.

Euler's Rotation Theorem

An arbitrary ROTATION may be described by only three parameters.

see also EULER ANGLES, EULER PARAMETERS, ROTATION MATRIX

Euler's Rule

The numbers $2^n pq$ and $2^n r$ are AMICABLE NUMBERS if the three INTEGERS

$$p \equiv 2^m(2^{n-m}+1)-1$$
$$q \equiv 2^m(2^{n-m}+1)-1$$
$$r \equiv 2^{n+m}(2^{n-m}+1)^2-1$$

are all PRIME numbers for some POSITIVE INTEGER m satisfying $1 \leq m \leq n-1$ (Dickson 1952, p. 42). However, there are exotic AMICABLE NUMBERS which do not satisfy Euler's rule, so it is a SUFFICIENT but not NECESSARY condition for amicability.

see also AMICABLE NUMBERS

References

Dickson, L. E. *History of the Theory of Numbers, Vol. 1: Divisibility and Primality.* New York: Chelsea, 1952.

Euler's Series Transformation

Accelerates the rate of CONVERGENCE for an ALTERNATING SERIES

$$S = \sum_{s=0}^{\infty}(-1)^s u_s$$
$$= u_0 - u_1 + u_2 - \ldots - u_{n-1} + \sum_{s=0}^{\infty} \frac{(-1)^2}{2^{s+1}}[\Delta^s u_n] \quad (1)$$

for n EVEN and Δ the FORWARD DIFFERENCE operator

$$\Delta^k u_n \equiv \sum_{m=0}^{k}(-1)^m \binom{k}{m} u_{n+k-m}, \quad (2)$$

where $\binom{k}{m}$ are BINOMIAL COEFFICIENTS. The POSITIVE terms in the series can be converted to an ALTERNATING SERIES using

$$\sum_{r=1}^{\infty} v_r = \sum_{r=1}^{\infty}(-1)^{r-1}w_r, \quad (3)$$

where

$$w_r \equiv v_r + 2v_{2r} + 4v_{4r} + 8v_{8r} + \ldots. \quad (4)$$

see also ALTERNATING SERIES

References

Abramowitz, M. and Stegun, C. A. (Eds.). *Handbook of Mathematical Functions with Formulas, Graphs, and Mathematical Tables, 9th printing.* New York: Dover, p. 16, 1972.

Euler's Spiral

see CORNU SPIRAL

Euler Square

A square ARRAY made by combining n objects of two types such that the first and second elements form LATIN SQUARES. Euler squares are also known as GRAECO-LATIN SQUARES, GRAECO-ROMAN SQUARES, or LATIN-GRAECO SQUARES. For many years, Euler squares were known to exist for $n = 3$, 4, and for every ODD n except $n = 3k$. EULER'S GRAECO-ROMAN SQUARES CONJECTURE maintained that there do not exist Euler squares of order $n = 4k + 2$ for $k = 1, 2, \ldots$. However, such squares were found to exist in 1959, refuting the CONJECTURE.

see also LATIN RECTANGLE, LATIN SQUARE, ROOM SQUARE

References

Beezer, R. "Graeco-Latin Squares." `http://buzzard.ups.edu/squares.html`.

Kraitchik, M. "Euler (Graeco-Latin) Squares." §7.12 in *Mathematical Recreations.* New York: W. W. Norton, pp. 179–182, 1942.

Euler Sum

In response to a letter from Goldbach, Euler considered DOUBLE SUMS of the form

$$s(m,n) = \sum_{k=1}^{\infty}\left(1+\frac{1}{2}+\ldots+\frac{1}{k}\right)^m (k+1)^{-n} \quad (1)$$
$$= \sum_{k=1}^{\infty}[\gamma + \psi_0(k+1)]^m (k+1)^{-n} \quad (2)$$

with $m \geq 1$ and $n \geq 2$ and where γ is the EULER-MASCHERONI CONSTANT and $\Psi(x) = \psi_0(x)$ is the DIGAMMA FUNCTION. Euler found explicit formulas in terms of the RIEMANN ZETA FUNCTION for $s(1,n)$ with $n \geq 2$, and E. Au-Yeung numerically discovered

$$\sum_{k=1}^{\infty}\left(1+\frac{1}{2}+\ldots+\frac{1}{k}\right)^2 k^{-2} = \tfrac{17}{4}\zeta(4), \qquad (3)$$

where $\zeta(z)$ is the RIEMANN ZETA FUNCTION, which was subsequently rigorously proven true (Borwein and Borwein 1995). Sums involving k^{-n} can be re-expressed in terms of sums the form $(k+1)^{-n}$ via

$$\begin{aligned}\sum_{k=1}^{\infty}&\left(1+\frac{1}{2^m}+\ldots+\frac{1}{k^m}\right)k^{-n}\\ &=\sum_{k=0}^{\infty}\left[1+\frac{2}{2^m}+\ldots+\frac{1}{(k+1)^m}\right](k+1)^{-n}\\ &=\sum_{k=1}^{\infty}\left(1+\frac{1}{2^m}+\ldots+\frac{1}{k^m}\right)(k+1)^{-n}+\sum_{k=1}^{\infty}k^{-(m+n)}\\ &\equiv \sigma_h(m,n)+\zeta(m+n) \qquad (4)\end{aligned}$$

and

$$\begin{aligned}\sum_{k=1}^{\infty}&\left(1+\frac{1}{2}+\ldots+\frac{1}{k}\right)^2 k^{-n}\\ &= s_h(2,n)+2s_h(1,n+1)+\zeta(n+2), \qquad (5)\end{aligned}$$

where σ_h is defined below.

Bailey *et al.* (1994) subsequently considered sums of the forms

$$s_h(m,n)=\sum_{k=1}^{\infty}\left(1+\frac{1}{2}+\ldots+\frac{1}{k}\right)^m (k+1)^{-n} \qquad (6)$$

$$s_a(m,n)=\sum_{k=1}^{\infty}\left[1-\frac{1}{2}+\ldots+\frac{(-1)^{k+1}}{k}\right]^m (k+1)^{-n} \qquad (7)$$

$$a_h(m,n)=\sum_{k=1}^{\infty}\left(1+\frac{1}{2}+\ldots+\frac{1}{k}\right)^m (-1)^{k+1}(k+1)^{-n} \qquad (8)$$

$$a_a(m,n)=\sum_{k=1}^{\infty}\left(1-\frac{1}{2}+\ldots+\frac{(-1)^{k+1}}{k}\right)^m (-1)^{k+1}(k+1)^{-n} \qquad (9)$$

$$\sigma_h(m,n)=\sum_{k=1}^{\infty}\left(1+\frac{1}{2^m}+\ldots+\frac{1}{k^m}\right)(k+1)^{-n} \qquad (10)$$

$$\sigma_a(m,n)=\sum_{k=1}^{\infty}\left(1-\frac{1}{2^m}+\ldots+\frac{(-1)^{k+1}}{k^m}\right)(k+1)^{-n} \qquad (11)$$

$$\alpha_h(m,n)=\sum_{k=1}^{\infty}\left(1+\frac{1}{2^m}+\ldots+\frac{1}{k^m}\right)(-1)^{k+1}(k+1)^{-n} \qquad (12)$$

$$\alpha_a(m,n)=\sum_{k=1}^{\infty}\left(1-\frac{1}{2^m}+\ldots+\frac{(-1)^{k+1}}{k^m}\right)(-1)^{k+1}(k+1)^{-n}, \qquad (13)$$

where s_h and s_a have the special forms

$$s_h=\sum_{k=1}^{\infty}[\gamma+\psi_0(n+1)]^m k+1^{-n} \qquad (14)$$

$$\begin{aligned}a_a=\sum_{k=1}^{\infty}&\{\ln 2+\tfrac{1}{2}(-1)^n\\ &\times[\psi_0(\tfrac{1}{2}n+\tfrac{1}{2})-\psi_0(\tfrac{1}{2}n+1)]\}^m(k+1)^{-m}. \qquad (15)\end{aligned}$$

Analytic single or double sums over $\zeta(z)$ can be constructed for

$$s_h(1,n)=\tfrac{1}{2}n\zeta(n+1)-\tfrac{1}{2}\sum_{k=1}^{n-2}\zeta(n-k)\zeta(k+1) \qquad (16)$$

$$\begin{aligned}s_h(2,n)&=\tfrac{1}{3}n(n+1)\zeta(n+2)+\zeta(2)\zeta(n)\\ &-\tfrac{1}{2}n\sum_{k=0}^{n-2}\zeta(n-k)\zeta(k+2)\\ &+\tfrac{1}{3}\sum_{k=2}^{n-2}\zeta(n-k)\sum_{j=1}^{k-1}\zeta(j+1)\zeta(k+1-j)+\sigma_h(2,n) \qquad (17)\end{aligned}$$

$$\begin{aligned}s_h(2,2n-1)&=\tfrac{1}{6}(2n^2-7n-3)\zeta(2n+1)+\zeta(2)\zeta(2n-1)\\ &-\tfrac{1}{2}\sum_{k=1}^{n-2}(2k-1)\zeta(2n-1-2k)\zeta(2k+2)\\ &+\tfrac{1}{3}\sum_{k=1}^{n-2}\zeta(2k+1)\sum_{j=1}^{n-2-k}\zeta(2j+1)\zeta(2n-1-2k-2j) \qquad (18)\end{aligned}$$

$$\sigma_h(1,n)=s_h(1,n) \qquad (19)$$

$$\begin{aligned}\sigma_h(2,2n-1)&=-\tfrac{1}{2}(2n^2+n+1)\zeta(2n+1)+\zeta(2)\zeta(2n-1)\\ &+\sum_{k=1}^{n-1}2k\zeta(k+1)\zeta(2n-2k) \qquad (20)\end{aligned}$$

$$\begin{aligned}\sigma_h(m\text{ even},n\text{ odd})&=\tfrac{1}{2}\left[\binom{m+n}{m}-1\right]\zeta(m+n)+\zeta(m)\zeta(n)\\ &-\sum_{j=1}^{m+n}\left[\binom{2j-2}{m-1}+\binom{2j-2}{n-1}\right]\\ &\times\zeta(2j-1)\zeta(m+n-2j+1) \qquad (21)\end{aligned}$$

$$\begin{aligned}\sigma_h(m\text{ odd},n\text{ even})&=-\tfrac{1}{2}\left[\binom{m+n}{m}+1\right]\zeta(m+n)\\ &+\sum_{k=1}^{m+n}\left[\binom{2j-2}{m-1}+\binom{2j-2}{n-1}\right]\\ &\times\zeta(2j-1)\zeta(m+n-2j+1), \qquad (22)\end{aligned}$$

where $\binom{n}{m}$ is a BINOMIAL COEFFICIENT. Explicit formulas inferred using the PSLQ ALGORITHM include

$$s_h(2,2)=\tfrac{3}{2}\zeta(4)+\tfrac{1}{2}[\zeta(2)]^2 \qquad (23)$$

$$=\tfrac{11}{360}\pi^4 \qquad (24)$$

$$s_h(2,4) = \tfrac{2}{3}\zeta(6) - \tfrac{1}{3}\zeta(2)\zeta(4) + \tfrac{1}{3}[\zeta(2)]^3 - [\zeta(3)]^2 \quad (25)$$
$$= \tfrac{37}{22680}\pi^6 - [\zeta(3)]^2 \quad (26)$$
$$s_h(3,2) = \tfrac{15}{2}\zeta(5) + \zeta(2)\zeta(3) \quad (27)$$
$$s_h(3,3) = -\tfrac{33}{16}\zeta(6) + 2[\zeta(3)]^2 \quad (28)$$
$$s_h(3,4) = \tfrac{119}{16}\zeta(7) - \tfrac{33}{4}\zeta(3)\zeta(4) + 2\zeta(2)\zeta(5) \quad (29)$$
$$s_h(3,6) = \tfrac{197}{24}\zeta(9) - \tfrac{33}{4}\zeta(4)\zeta(5) - \tfrac{37}{8}\zeta(3)\zeta(6) + [\zeta(3)]^3 + 3\zeta(2)\zeta(7) \quad (30)$$
$$s_h(4,2) = \tfrac{859}{24}\zeta(6) + 3[\zeta(3)]^2 \quad (31)$$
$$s_h(4,3) = -\tfrac{109}{8}\zeta(7) + \tfrac{37}{2}\zeta(3)\zeta(4) - 5\zeta(2)\zeta(5) \quad (32)$$
$$s_h(4,5) = -\tfrac{29}{2}\zeta(9) + \tfrac{37}{2}\zeta(4)\zeta(5) + \tfrac{33}{4}\zeta(3)\zeta(6) - \tfrac{8}{3}[\zeta(3)]^3 - 7\zeta(2)\zeta(7) \quad (33)$$
$$s_h(5,2) = \tfrac{1855}{16}\zeta(7) + 33\zeta(3)\zeta(4) + \tfrac{57}{2}\zeta(2)\zeta(5) \quad (34)$$
$$s_h(5,4) = \tfrac{890}{9}\zeta(9) + 66\zeta(4)\zeta(5) - \tfrac{4295}{24}\zeta(3)\zeta(6) - 5[\zeta(3)]^3 + \tfrac{265}{8}\zeta(2)\zeta(7) \quad (35)$$
$$s_h(6,3) = -\tfrac{3073}{12}\zeta(9) - 243\zeta(4)\zeta(5) + \tfrac{2097}{4}\zeta(3)\zeta(6) + \tfrac{67}{3}[\zeta(3)]^3 - \tfrac{651}{8}\zeta(2)\zeta(7) \quad (36)$$
$$s_h(7,2) = \tfrac{134701}{36}\zeta(9) + \tfrac{15697}{8}\zeta(4)\zeta(5) + \tfrac{29555}{24}\zeta(3)\zeta(6) + 56[\zeta(3)]^3 + \tfrac{3287}{4}\zeta(2)\zeta(7), \quad (37)$$

$$s_a(2,2) = 6\,\mathrm{Li}_4(\tfrac{1}{2}) + \tfrac{1}{4}(\ln 2)^4 - \tfrac{29}{8}\zeta(4) + \tfrac{3}{2}\zeta(2)(\ln 2)^2 \quad (38)$$
$$s_a(2,3) = 4\,\mathrm{Li}_5(\tfrac{1}{2}) - \tfrac{1}{30}(\ln 2)^5 - \tfrac{17}{32}\zeta(5) - \tfrac{11}{8}\zeta(4)\ln 2 + \tfrac{7}{4}\zeta(3)(\ln 2)^2 + \tfrac{1}{3}\zeta(2)(\ln 2)^3 - \tfrac{3}{4}\zeta(2)\zeta(3), \quad (39)$$
$$s_a(3,2) = -24\,\mathrm{Li}_5(\tfrac{1}{2}) + 6\ln 2\,\mathrm{Li}_4(\tfrac{1}{2}) + \tfrac{9}{20}(\ln 2)^5 + \tfrac{659}{32}\zeta(5) - \tfrac{285}{16}\zeta(4)\ln 2 + \tfrac{5}{2}\zeta(2)(\ln 2)^3 + \tfrac{1}{2}\zeta(2)\zeta(3), \quad (40)$$

$$a_h(2,2) = -2\,\mathrm{Li}_4(\tfrac{1}{2}) - \tfrac{1}{12}(\ln 2)^4 + \tfrac{99}{48}\zeta(4) - \tfrac{7}{4}\zeta(3)\ln 2 + \tfrac{1}{2}\zeta(2)(\ln 2)^2 \quad (41)$$
$$a_h(2,3) = -4\,\mathrm{Li}_5(\tfrac{1}{2}) - 4(\ln 2)\,\mathrm{Li}_4(\tfrac{1}{2}) - \tfrac{2}{15}(\ln 2)^5 + \tfrac{107}{32}\zeta(5) - \tfrac{7}{4}\zeta(3)(\ln 2)^2 + \tfrac{2}{3}\zeta(2)(\ln 2)^3 + \tfrac{3}{8}\zeta(2)\zeta(3) \quad (42)$$
$$a_h(3,2) = 6\,\mathrm{Li}_5(\tfrac{1}{2}) + 6(\ln 2)\,\mathrm{Li}_4(\tfrac{1}{2}) + \tfrac{1}{5}(\ln 2)^5 - \tfrac{33}{8}\zeta(5) + \tfrac{21}{8}\zeta(3)(\ln 2)^2 - \zeta(2)(\ln 2)^3 - \tfrac{15}{16}\zeta(2)\zeta(3), \quad (43)$$

and

$$a_a(2,2) = -4\,\mathrm{Li}_4(\tfrac{1}{2}) - \tfrac{1}{6}(\ln 2)^4 + \tfrac{37}{16}\zeta(4) + \tfrac{7}{4}\zeta(3)(\ln 2) - 2\zeta(2)(\ln 2)^2 \quad (44)$$
$$a_a(2,3) = 4(\ln 2)\,\mathrm{Li}_4(\tfrac{1}{2}) + \tfrac{1}{6}(\ln 2)^5 - \tfrac{79}{32}\zeta(5) + \tfrac{11}{8}\zeta(4)(\ln 2) - \zeta(2)(\ln 2)^3 + \tfrac{3}{8}\zeta(2)\zeta(3) \quad (45)$$
$$a_a(3,2) = 30\,\mathrm{Li}_5(\tfrac{1}{2}) - \tfrac{1}{4}(\ln 2)^5 - \tfrac{1813}{64}\zeta(5) + \tfrac{285}{16}\zeta(4)(\ln 2) + \tfrac{21}{8}\zeta(3)(\ln 2)^2 - \tfrac{7}{2}\zeta(2)(\ln 2)^3 + \tfrac{3}{4}\zeta(2)\zeta(3), \quad (46)$$

where Li_n is a POLYLOGARITHM, and $\zeta(z)$ is the RIEMANN ZETA FUNCTION (Bailey and Plouffe). Of these, only $s_h(3,2)$, $s_h(3,3)$ and the identities for $s_a(m,n)$, $a_h(m,n)$ and $a_a(m,n)$ have been rigorously established.

References

Bailey, D. and Plouffe, S. "Recognizing Numerical Constants." http://www.cecm.sfu.ca/organics/papers/bailey/.

Bailey, D. H.; Borwein, J. M.; and Girgensohn, R. "Experimental Evaluation of Euler Sums." *Exper. Math.* **3**, 17–30, 1994.

Berndt, B. C. *Ramanujan's Notebooks: Part I.* New York: Springer-Verlag, 1985.

Borwein, D. and Borwein, J. M. "On an Intriguing Integral and Some Series Related to $\zeta(4)$." *Proc. Amer. Math. Soc.* **123**, 1191–1198, 1995.

Borwein, D.; Borwein, J. M.; and Girgensohn, R. "Explicit Evaluation of Euler Sums." *Proc. Edinburgh Math. Soc.* **38**, 277–294, 1995.

de Doelder, P. J. "On Some Series Containing $\Psi(x) - \Psi(y)$ and $(\Psi(x) - \Psi(y))^2$ for Certain Values of x and y." *J. Comp. Appl. Math.* **37**, 125–141, 1991.

Euler's Sum of Powers Conjecture

Euler conjectured that at least n nth POWERS are required for $n > 2$ to provide a sum that is itself an nth POWER. The conjecture was disproved by Lander and Parkin (1967) with the counterexample

$$27^5 + 84^5 + 110^5 + 133^5 = 144^5.$$

see also DIOPHANTINE EQUATION

References

Lander, L. J. and Parkin, T. R. "A Counterexample to Euler's Sum of Powers Conjecture." *Math. Comput.* **21**, 101–103, 1967.

Euler's Theorem

A generalization of FERMAT'S LITTLE THEOREM. Euler published a proof of the following more general theorem in 1736. Let $\phi(n)$ denote the TOTIENT FUNCTION. Then

$$a^{\phi(n)} \equiv 1 \pmod{n}$$

for all a RELATIVELY PRIME to n.

see also CHINESE HYPOTHESIS, EULER'S DISPLACEMENT THEOREM, EULER'S DISTRIBUTION THEOREM, FERMAT'S LITTLE THEOREM, TOTIENT FUNCTION

References

Shanks, D. *Solved and Unsolved Problems in Number Theory, 4th ed.* New York: Chelsea, p. 21 and 23–25, 1993.

Euler Totient Function

see TOTIENT FUNCTION

Euler's Totient Rule

The number of bases mod p in which $1/p$ has cycle length l is the same as the number of FRACTIONS $0/(p-1)$, $1/(p-1)$, ..., $(p-2)/(p-1)$ which have least DENOMINATOR l.

see also TOTIENT FUNCTION

References

Conway, J. H. and Guy, R. K. *The Book of Numbers.* New York: Springer-Verlag, pp. 167–168, 1996.

Euler's Transform

A technique for SERIES CONVERGENCE IMPROVEMENT which takes a convergent alternating series

$$\sum_{k=0}^{\infty}(-1)^k a_k = a_0 - a_1 + a_2 - \ldots \quad (1)$$

into a series with more rapid convergence to the same value to

$$s = \sum_{k=0}^{\infty} \frac{(-1)^k \Delta^k a_0}{2^{k+1}}, \quad (2)$$

where the FORWARD DIFFERENCE is defined by

$$\Delta^k a_0 = \sum_{m=0}^{k} \equiv (-1)^m \binom{k}{m} a_{k-m} \quad (3)$$

(Abramowitz and Stegun 1972; Beeler *et al.* 1972, Item 120).

see also FORWARD DIFFERENCE

References

Abramowitz, M. and Stegun, C. A. (Eds.). *Handbook of Mathematical Functions with Formulas, Graphs, and Mathematical Tables, 9th printing.* New York: Dover, p. 16, 1972.

Beeler, M.; Gosper, R. W.; and Schroeppel, R. *HAKMEM.* Cambridge, MA: MIT Artificial Intelligence Laboratory, Memo AIM-239, Feb. 1972.

Euler Transformation

see EULER'S FINITE DIFFERENCE TRANSFORMATION, EULER'S HYPERGEOMETRIC TRANSFORMATIONS, EULER'S TRANSFORM

Euler's Triangle

The triangle of numbers $A_{n,k}$ given by

$$A_{n,1} = A_{n,n} = 1$$

and the RECURRENCE RELATION

$$A_{n+1,k} = kA_{n,k} + (n+2-k)A_{n,k-1}$$

for $k \in [2, n]$, where $A_{n,k}$ are EULERIAN NUMBERS.

$$\begin{array}{c} 1 \\ 1 \quad 1 \\ 1 \quad 4 \quad 1 \\ 1 \quad 11 \quad 11 \quad 1 \\ 1 \quad 26 \quad 66 \quad 26 \quad 1 \\ 1 \quad 57 \quad 302 \quad 302 \quad 57 \quad 1 \\ 1 \; 120 \; 1191 \; 2416 \; 1191 \; 120 \; 1. \end{array}$$

The numbers 1, 1, 1, 1, 4, 1, 1, 11, 11, 1, ... are Sloane's A008292. Amazingly, the Z-TRANSFORMS of t^n

$$\frac{(z-1)^n}{T^n z} Z[t^n] = \frac{(1-z)^n}{T^n z} \lim_{x \to 0} \frac{\partial^n}{\partial x^n}\left(\frac{z}{z - e^{-xT}}\right)$$

are generators for Euler's triangle.

see also CLARK'S TRIANGLE, EULERIAN NUMBER, LEIBNIZ HARMONIC TRIANGLE, NUMBER TRIANGLE, PASCAL'S TRIANGLE, SEIDEL-ENTRINGER-ARNOLD TRIANGLE, Z-TRANSFORM

References

Sloane, N. J. A. Sequence A008292 in "An On-Line Version of the Encyclopedia of Integer Sequences."

Euler Triangle Formula

Let O and I be the CIRCUMCENTER and INCENTER of a TRIANGLE with CIRCUMRADIUS R and INRADIUS r. Let d be the distance between O and I. Then

$$d^2 = R^2 - 2rR.$$

Euler Walk

see EULERIAN TRAIL

Euler Zigzag Number

The number of ALTERNATING PERMUTATIONS for n elements is sometimes called an Euler zigzag number. Denote the number of ALTERNATING PERMUTATIONS on n elements for which the first element is k by $E(n,k)$. Then $E(1,1)$ and

$$E(n,k) = \begin{cases} 0 & \text{for } k \geq n \text{ or } k < 1 \\ E(n,k+1) \\ \quad +E(n-1,n-k) & \text{otherwise.} \end{cases}$$

see also ALTERNATING PERMUTATION, ENTRINGER NUMBER, SECANT NUMBER, TANGENT NUMBER

References

Ruskey, F. "Information of Alternating Permutations." http://sue.csc.uvic.ca/~cos/inf/perm/Alternating.html.

Sloane, N. J. A. Sequence A000111/M1492 in "An On-Line Version of the Encyclopedia of Integer Sequences."

Eulerian Circuit

An EULERIAN TRAIL which starts and ends at the same VERTEX. In other words, it is a CYCLE which uses each EDGE exactly once. The term EULERIAN CYCLE is also used synonymously with Eulerian circuit. For technical reasons, Eulerian circuits are easier to study mathematically than are HAMILTONIAN CIRCUITS. As a generalization of the KÖNIGSBERG BRIDGE PROBLEM, Euler showed (without proof) that a CONNECTED GRAPH has an Eulerian circuit IFF it has no VERTICES of ODD DEGREE.

see also EULER GRAPH, HAMILTONIAN CIRCUIT

Eulerian Cycle

see EULERIAN CIRCUIT

Eulerian Integral of the First Kind

Legendre and Whittaker and Watson's (1990) term for the BETA FUNCTION.

References

Whittaker, E. T. and Watson, G. N. *A Course in Modern Analysis, 4th ed.* Cambridge, England: Cambridge University Press, 1990.

Eulerian Integral of the Second Kind

$$\Pi(z,n) \equiv \int_0^n \left(1-\frac{t}{n}\right)^n t^{z-1}\,dt = n^z \int_0^1 (1-\tau)^n \tau^{z-1}\,d\tau$$
$$= \frac{n!}{z(z+1)\cdots(z+n)} n^z.$$

Eulerian Number

The number of PERMUTATIONS of length n with $k \leq n$ RUNS, denoted $\left\langle {n \atop k} \right\rangle$, $A_{n,k}$, or $A(n,k)$. The Eulerian numbers are given explicitly by the sum

$$\left\langle {n \atop k} \right\rangle = \sum_{j=0}^{k} (-1)^j \binom{n+1}{j} (k-j)^n. \tag{1}$$

Making the definition

$$b_{n,1} = 1 \tag{2}$$
$$b_{1,n} = 1 \tag{3}$$

together with the RECURRENCE RELATION

$$b_{n,k} = n b_{n,k-1} + k b_{n-1,k} \tag{4}$$

for $n > k$ then gives

$$\left\langle {n \atop k} \right\rangle = b_{k,n-k+1}. \tag{5}$$

The arrangement of the numbers into a triangle gives EULER'S TRIANGLE, whose entries are 1, 1, 1, 1, 4, 1, 1, 11, 11, 1, ... (Sloane's A008292). Therefore, they represent a sort of generalization of the BINOMIAL COEFFICIENTS where the defining RECURRENCE RELATION weights the sum of neighbors by their row and column numbers, respectively.

The Eulerian numbers satisfy

$$\sum_{k=1}^{n} \left\langle {n \atop k} \right\rangle = n!. \tag{6}$$

Eulerian numbers also arise in the surprising context of integrating the SINC FUNCTION, and also in sums of the form

$$\sum_{k=1}^{\infty} k^n r^k = \mathrm{Li}_{-n}(r) = \frac{r}{(1-r)^{n+1}} \sum_{i=1}^{n} \left\langle {n \atop i} \right\rangle r^{n-i}, \tag{7}$$

where $\mathrm{Li}_m(z)$ is the POLYLOGARITHM function.

see also COMBINATION LOCK, EULER NUMBER, EULER'S TRIANGLE, EULER ZIGZAG NUMBER, POLYLOGARITHM, SINC FUNCTION, WORPITZKY'S IDENTITY, Z-TRANSFORM

References

Carlitz, L. "Eulerian Numbers and Polynomials." *Math. Mag.* **32**, 247–260, 1959.
Foata, D. and Schützenberger, M.-P. *Théorie Géométrique des Polynômes Eulériens.* Berlin: Springer-Verlag, 1970.
Kimber, A. C. "Eulerian Numbers." Supplement to *Encyclopedia of Statistical Sciences.* (Eds. S. Kotz, N. L. Johnson, and C. B. Read). New York: Wiley, pp. 59–60, 1989.
Salama, I. A. and Kupper, L. L. "A Geometric Interpretation for the Eulerian Numbers." *Amer. Math. Monthly* **93**, 51–52, 1986.
Sloane, N. J. A. Sequence A008292 in "An On-Line Version of the Encyclopedia of Integer Sequences."

Eulerian Trail

A WALK on the EDGES of a GRAPH which uses each EDGE exactly once. A CONNECTED GRAPH has an Eulerian trail IFF it has at most two VERTICES of ODD DEGREE.

see also EULERIAN CIRCUIT

Eutactic Star

An orthogonal projection of a CROSS onto a 3-D SUBSPACE. It is said to be normalized if the CROSS vectors are all of unit length.

see also HADWIGER'S PRINCIPAL THEOREM

Evans Point

The intersection of the GERGONNE LINE and the EULER LINE. It does not appear to have a simple parametric representation.

References

Oldknow, A. "The Euler-Gergonne-Soddy Triangle of a Triangle." *Amer. Math. Monthly* **103**, 319–329, 1996.

Eve

see APPLE, ROOT, SNAKE, SNAKE EYES, SNAKE OIL METHOD, SNAKE POLYIAMOND

Even Function

A function $f(x)$ such that $f(x) = f(-x)$. An even function times an ODD FUNCTION is odd.

Even Number

An INTEGER of the form $N = 2n$, where n is an INTEGER. The even numbers are therefore $\ldots, -4, -2, 0, 2, 4, 6, 8, 10, \ldots$ (Sloane's A005843). Since the even numbers are integrally divisible by two, $N \equiv 0 \pmod 2$ for even N. An even number N for which $N \equiv 2 \pmod 4$ is called a SINGLY EVEN NUMBER, and an even number N for which $N \equiv 0 \pmod 4$ is called a DOUBLY EVEN NUMBER. An integer which is not even is called an ODD NUMBER. The GENERATING FUNCTION of the even numbers is

$$\frac{2x}{(x-1)^2} = 2x + 4x^2 + 6x^3 + 8x^4 + \ldots.$$

see also DOUBLY EVEN NUMBER, EVEN FUNCTION, ODD NUMBER, SINGLY EVEN NUMBER

References

Sloane, N. J. A. Sequence A005843/M0985 in "An On-Line Version of the Encyclopedia of Integer Sequences."

Eventually Periodic

A PERIODIC SEQUENCE such as $\{1, 1, 1, 2, 1, 2, 1, 2, 1, 2, 1, 1, 2, 1, \ldots\}$ which is periodic from some point onwards.

see also PERIODIC SEQUENCE

Everett's Formula

$$f_p = (1-p)f_0 + pf_1 + E_2\delta_0^2 + F_2\delta_1^2 + E_4\delta_0^4 + F_4\delta_1^4 + E_6\delta_0^6 + F_6\delta_1^6 + \ldots, \quad (1)$$

for $p \in [0, 1]$, where δ is the CENTRAL DIFFERENCE and

$$E_{2n} \equiv G_{2n} - G_{2n+1} \equiv B_{2n} - B_{2n+1} \quad (2)$$

$$F_{2n} \equiv G_{2n+1} \equiv B_{2n} + B_{2n+1}, \quad (3)$$

where G_k are the COEFFICIENTS from GAUSS'S BACKWARD FORMULA and GAUSS'S FORWARD FORMULA and B_k are the COEFFICIENTS from BESSEL'S FINITE DIFFERENCE FORMULA. The E_ks and F_ks also satisfy

$$E_{2n}(p) = F_{2n}(q) \quad (4)$$

$$F_{2n}(p) = E_{2n}(q), \quad (5)$$

for

$$q \equiv 1 - p. \quad (6)$$

see also BESSEL'S FINITE DIFFERENCE FORMULA

References

Abramowitz, M. and Stegun, C. A. (Eds.). *Handbook of Mathematical Functions with Formulas, Graphs, and Mathematical Tables, 9th printing.* New York: Dover, pp. 880–881, 1972.

Acton, F. S. *Numerical Methods That Work, 2nd printing.* Washington, DC: Math. Assoc. Amer., pp. 92–93, 1990.

Beyer, W. H. *CRC Standard Mathematical Tables, 28th ed.* Boca Raton, FL: CRC Press, p. 433, 1987.

Everett Interpolation

see EVERETT'S FORMULA

Eversion

A curve on the unit sphere S^2 is an eversion if it has no corners or cusps (but it may be self-intersecting). These properties are guaranteed by requiring that the curve's velocity never vanishes. A mapping $\boldsymbol{\sigma} : S^1 \to S^2$ forms an immersion of the CIRCLE into the SPHERE IFF, for all $\theta \in \mathbb{R}$,

$$\left|\frac{d}{d\theta}[\boldsymbol{\sigma}(e^{i\theta})]\right| > 0.$$

Smale (1958) showed it is possible to turn a SPHERE inside out (SPHERE EVERSION) using eversion.

see also SPHERE EVERSION

References

Smale, S. "A Classification of Immersions of the Two-Sphere." *Trans. Amer. Math. Soc.* **90**, 281–290, 1958.

Evolute

An evolute is the locus of centers of curvature (the envelope) of a plane curve's normals. The original curve is then said to be the INVOLUTE of its evolute. Given a plane curve represented parametrically by $(f(t), g(t))$, the equation of the evolute is given by

$$x = f - R\sin\tau \quad (1)$$

$$y = g + R\cos\tau, \quad (2)$$

where (x, y) are the coordinates of the running point, R is the RADIUS OF CURVATURE

$$R = \frac{(f'^2 + g'^2)^{3/2}}{f'g'' - f''g'}, \quad (3)$$

and τ is the angle between the unit TANGENT VECTOR

$$\hat{\mathbf{T}} = \frac{\mathbf{x}'}{|\mathbf{x}'|} = \frac{1}{\sqrt{f'^2 + g'^2}}\begin{bmatrix} f' \\ g' \end{bmatrix} \quad (4)$$

and the x-AXIS,

$$\cos\tau = \hat{\mathbf{T}} \cdot \hat{\mathbf{x}} \quad (5)$$

$$\sin\tau = \hat{\mathbf{T}} \cdot \hat{\mathbf{y}}. \quad (6)$$

Combining gives

$$x = f - \frac{(f'^2 + g'^2)g'}{f'g'' - f''g'} \quad (7)$$

$$y = g + \frac{(f'^2 + g'^2)g'}{f'g'' - f''g'}. \quad (8)$$

The definition of the evolute of a curve is independent of parameterization for any differentiable function (Gray 1993). If E is the evolute of a curve I, then I is said to be the INVOLUTE of E. The centers of the OSCULATING CIRCLES to a curve form the evolute to that curve (Gray 1993, p. 90).

The following table lists the evolutes of some common curves.

Curve	Evolute
astroid	astroid 2 times as large
cardioid	cardioid 1/3 as large
cayley's sextic	nephroid
circle	point (0, 0)
cycloid	equal cycloid
deltoid	deltoid 3 times as large
ellipse	Lamé curve
epicycloid	enlarged epicycloid
hypocycloid	similar hypocycloid
limaçon	circle catacaustic for a point source
logarithmic spiral	equal logarithmic spiral
nephroid	nephroid 1/2 as large
parabola	Neile's parabola
tractrix	catenary

see also INVOLUTE, OSCULATING CIRCLE

References

Cayley, A. "On Evolutes of Parallel Curves." *Quart. J. Pure Appl. Math.* **11**, 183–199, 1871.

Dixon, R. "String Drawings." Ch. 2 in *Mathographics.* New York: Dover, pp. 75–78, 1991.

Gray, A. "Evolutes." §5.1 in *Modern Differential Geometry of Curves and Surfaces.* Boca Raton, FL: CRC Press, pp. 76–80, 1993.

Jeffrey, H. M. "On the Evolutes of Cubic Curves." *Quart. J. Pure Appl. Math.* **11**, 78–81 and 145–155, 1871.

Lawrence, J. D. *A Catalog of Special Plane Curves.* New York: Dover, pp. 40 and 202, 1972.

Lee, X. "Evolute." `http://www.best.com/~xah/SpecialPlaneCurves_dir/Evolute_dir/evolute.html`.

Lockwood, E. H. "Evolutes and Involutes." Ch. 21 in *A Book of Curves.* Cambridge, England: Cambridge University Press, pp. 166–171, 1967.

Yates, R. C. "Evolutes." *A Handbook on Curves and Their Properties.* Ann Arbor, MI: J. W. Edwards, pp. 86–92, 1952.

Exact Covering System

A system of congruences $a_i \bmod n_i$ with $1 \leq i \leq k$ is called a COVERING SYSTEM if every INTEGER y satisfies $y \equiv a_i \pmod{n}$ for at least one value of i. A covering system in which each integer is covered by just one congruence is called an exact covering system.

see also COVERING SYSTEM

References

Guy, R. K. "Exact Covering Systems." §F14 in *Unsolved Problems in Number Theory, 2nd ed.* New York: Springer-Verlag, pp. 253–256, 1994.

Exact Differential

A differential of the form

$$df = P(x,y)\,dx + Q(x,y)\,dy \quad (1)$$

is exact (also called a TOTAL DIFFERENTIAL) if $\int df$ is path-independent. This will be true if

$$df = \frac{\partial f}{\partial x}\,dx + \frac{\partial f}{\partial y}\,dy, \quad (2)$$

so P and Q must be of the form

$$P(x,y) = \frac{\partial f}{\partial x} \qquad Q(x,y) = \frac{\partial f}{\partial y}. \quad (3)$$

But

$$\frac{\partial P}{\partial y} = \frac{\partial^2 f}{\partial y \partial x} \quad (4)$$

$$\frac{\partial Q}{\partial x} = \frac{\partial^2 f}{\partial x \partial y}, \quad (5)$$

so

$$\frac{\partial P}{\partial y} = \frac{\partial Q}{\partial x}. \quad (6)$$

see also PFAFFIAN FORM, INEXACT DIFFERENTIAL

Exact Period

see LEAST PERIOD

Exact Trilinear Coordinates

The TRILINEAR COORDINATES $\alpha : \beta : \gamma$ of a point P relative to a TRIANGLE are PROPORTIONAL to the directed distances $a' : b' : c'$ from P to the side lines (i.e, $a' : b' : c' = k\alpha : b' = k\beta : k\gamma$). Letting k be the constant of proportionality,

$$k \equiv \frac{2\Delta}{a\alpha + b\beta + c\gamma},$$

where Δ is the AREA of ΔABC and a, b, and c are the lengths of its sides. When the trilinears are chosen so that $k = 1$, the coordinates are known as exact trilinear coordinates.

see also TRILINEAR COORDINATES

Exactly One

"Exactly one" means "one and only one," sometimes also referred to as "JUST ONE." J. H. Conway has also humorously suggested "onee" (one and only one) by analogy with IFF (if and only if), "twoo" (two and only two), and "threee" (three and only three). This refinement is sometimes needed in formal mathematical discourse because, for example, if you have two apples, you also have one apple, but you do not have *exactly one* apple.

In 2-valued LOGIC, exactly one is equivalent to the exclusive or operator XOR,

$$P(E)\ \mathrm{XOR}\ P(F) = P(E) + P(F) - 2P(E \cap F).$$

see also IFF, PRECISELY UNLESS, XOR

Exactly When

see IFF

Excenter

The center J_i of an EXCIRCLE. There are three excenters for a given TRIANGLE, denoted J_1, J_2, J_3. The INCENTER I and excenters J_i of a TRIANGLE are an ORTHOCENTRIC SYSTEM.

$$\overline{OI}^2 + \overline{OJ_1}^2 + \overline{OJ_2}^2 + \overline{OJ_3}^2 = 12R^2,$$

where O is the CIRCUMCENTER, J_i are the excenters, and R is the CIRCUMRADIUS (Johnson 1929, p. 190). Denote the MIDPOINTS of the original TRIANGLE M_1, M_2, and M_3. Then the lines J_1M_1, J_2M_2, and J_3M_3 intersect in a point known as the MITTENPUNKT.

see also CENTROID (ORTHOCENTRIC SYSTEM), EXCENTER-EXCENTER CIRCLE, EXCENTRAL TRIANGLE, EXCIRCLE, INCENTER, MITTENPUNKT

References

Dixon, R. *Mathographics.* New York: Dover, pp. 58–59, 1991.
Johnson, R. A. *Modern Geometry: An Elementary Treatise on the Geometry of the Triangle and the Circle.* Boston, MA: Houghton Mifflin, 1929.

Excenter-Excenter Circle

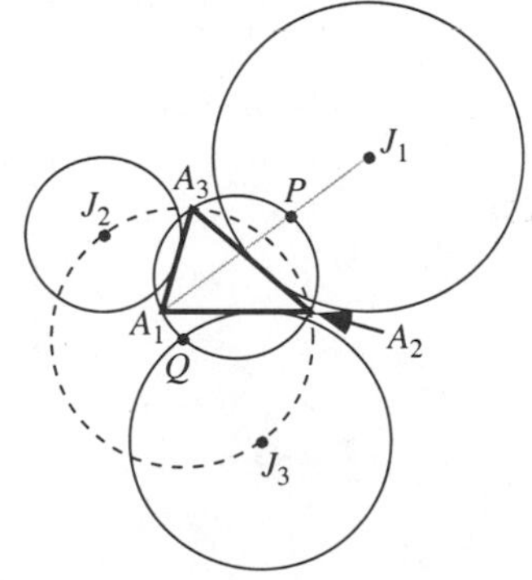

Given a TRIANGLE $\Delta A_1A_2A_3$, the points A_1, I, and J_1 lie on a line, where I is the INCENTER and J_1 is the EXCENTER corresponding to A_1. Furthermore, the circle with J_2J_3 as the diameter has Q as its center, where P is the intersection of A_1J_1 with the CIRCUMCIRCLE of $A_1A_2A_3$ and Q is the point opposite P on the CIRCUMCIRCLE. The circle with diameter J_2J_3 also passes through A_2 and A_3 and has radius

$$r = \tfrac{1}{2}a_1 \csc(\tfrac{1}{2}\alpha_1) = 2R\cos(\tfrac{1}{2}\alpha_1).$$

It arises because the points I, J_1, J_2, and J_3 form an ORTHOCENTRIC SYSTEM.

see also EXCENTER, INCENTER-EXCENTER CIRCLE, ORTHOCENTRIC SYSTEM

References

Johnson, R. A. *Modern Geometry: An Elementary Treatise on the Geometry of the Triangle and the Circle.* Boston, MA: Houghton Mifflin, pp. 185–186, 1929.

Excentral Triangle

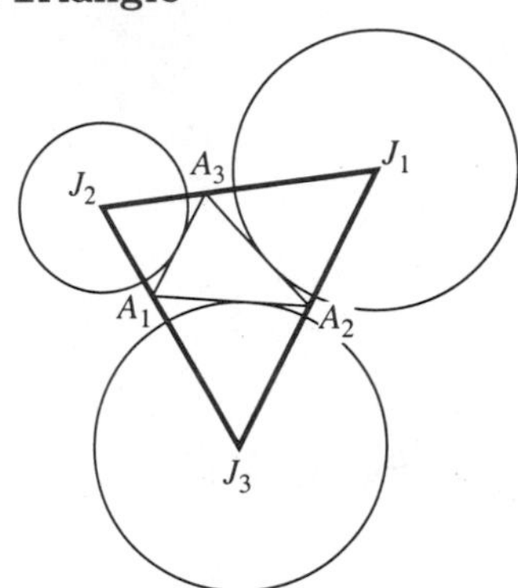

The TRIANGLE $J = \Delta J_1J_2J_3$ with VERTICES corresponding to the EXCENTERS of a given TRIANGLE A, also called the TRITANGENT TRIANGLE.

Beginning with an arbitrary TRIANGLE A, find the excentral triangle J. Then find the excentral triangle $J'$of that TRIANGLE, and so on. Then the resulting TRIANGLE $J^{(\infty)}$ approaches an EQUILATERAL TRIANGLE.

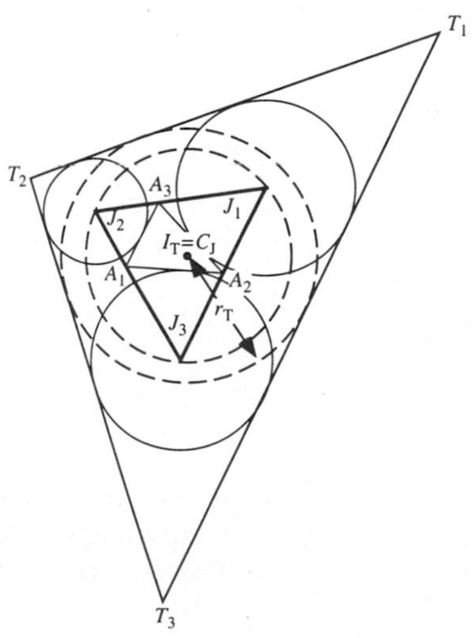

Call T the TRIANGLE tangent externally to the EXCIRCLES of A. Then the INCENTER I_T of K coincides with the CIRCUMCENTER C_J of TRIANGLE $\Delta J_1 J_2 J_3$, where J_i are the EXCENTERS of A. The INRADIUS r_T of the INCIRCLE of T is

$$r_T = 2R + r = \tfrac{1}{2}(r + r_1 + r_2 + r_3),$$

where R is the CIRCUMRADIUS of A, r is the INRADIUS, and r_i are the EXRADII (Johnson 1929, p. 192).

see also EXCENTER, EXCENTER-EXCENTER CIRCLE, EXCIRCLE, MITTENPUNKT

References

Johnson, R. A. *Modern Geometry: An Elementary Treatise on the Geometry of the Triangle and the Circle.* Boston, MA: Houghton Mifflin, 1929.

Excess

see KURTOSIS

Excess Coefficient

see KURTOSIS

Excessive Number

see ABUNDANT NUMBER

Excircle

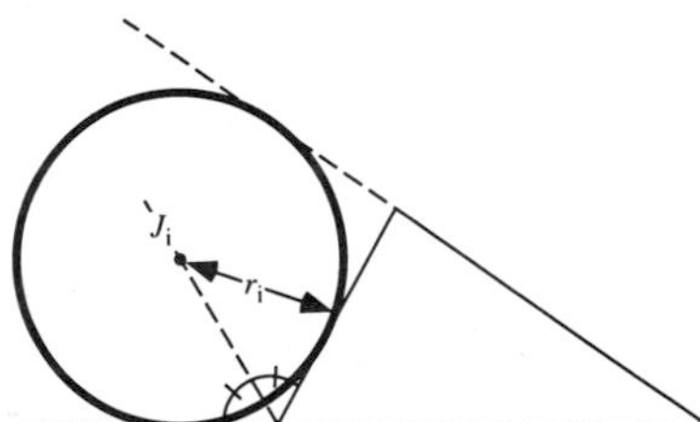

Given a TRIANGLE, extend two nonadjacent sides. The CIRCLE tangent to these two lines and to the other side of the TRIANGLE is called an ESCRIBED CIRCLE, or excircle. The CENTER J_i of the excircle is called the EXCENTER and lies on the external ANGLE BISECTOR of the opposite ANGLE. Every TRIANGLE has three excircles, and the TRILINEAR COORDINATES of the EXCENTERS are $-1 : 1 : 1$, $1 : -1 : 1$, and $1 : 1 : -1$. The RADIUS r_i of the excircle i is called its EXRADIUS.

Given a TRIANGLE with INRADIUS r, let h_i be the ALTITUDES of the excircles, and r_i their RADII (the EXRADII). Then

$$\frac{1}{h_1} + \frac{1}{h_2} + \frac{1}{h_3} = \frac{1}{r_1} + \frac{1}{r_2} + \frac{1}{r_3} = \frac{1}{r}$$

(Johnson 1929, p. 189).

see also EXCENTER, EXCENTER-EXCENTER CIRCLE, EXCENTRAL TRIANGLE, FEUERBACH'S THEOREM, NAGEL POINT, TRIANGLE TRANSFORMATION PRINCIPLE

References

Coxeter, H. S. M. and Greitzer, S. L. *Geometry Revisited.* Washington, DC: Math. Assoc. Amer., pp. 11-13, 1967.

Johnson, R. A. *Modern Geometry: An Elementary Treatise on the Geometry of the Triangle and the Circle.* Boston, MA: Houghton Mifflin, pp. 176–177 and 182–194, 1929.

Excision Axiom

One of the EILENBERG-STEENROD AXIOMS which states that, if X is a SPACE with SUBSPACES A and U such that the CLOSURE of A is contained in the interior of U, then the INCLUSION MAP $(X\ U, A\ U) \to (X, A)$ induces an isomorphism $H_n(X\ U, A\ U) \to H_n(X, A)$.

Excluded Middle Law

A law in (2-valued) LOGIC which states there is no third alternative to TRUTH or FALSEHOOD. In other words, every statement must be either A or not-A. This fact no longer holds in THREE-VALUED LOGIC or FUZZY LOGIC.

Excludent

A method which can be used to solve any QUADRATIC CONGRUENCE. This technique relies on the fact that solving

$$x^2 \equiv b \pmod{p}$$

is equivalent to finding a value y such that

$$b + py = x^2.$$

Pick a few small moduli m. If y mod m does not make $b+py$ a quadratic residue of m, then this value of y may be excluded. Furthermore, values of $y > p/4$ are never necessary.

Excludent Factorization Method

Also known as the difference of squares. It was first used by Fermat and improved by Gauss. Gauss looked for INTEGERS x and y satisfying

$$y^2 \equiv x^2 - N \pmod{E}$$

for various moduli E. This allowed the exclusion of many potential factors. This method works best when factors are of approximately the same size, so it is sometimes better to attempt mN for some suitably chosen value of m.

see also PRIME FACTORIZATION ALGORITHMS

Exclusive Or

see XOR

Exeter Point

Define A' to be the point (other than the VERTEX A) where the MEDIAN through A meets the CIRCUMCIRCLE of ABC, and define B' and C' similarly. Then the Exeter point is the PERSPECTIVE CENTER of the TRIANGLE $A'B'C'$ and the TANGENTIAL TRIANGLE. It has TRIANGLE CENTER FUNCTION

$$\alpha = a(b^4 + c^4 - a^4).$$

References

Kimberling, C. "Central Points and Central Lines in the Plane of a Triangle." *Math. Mag.* **67**, 163–187, 1994.

Kimberling, C. "Exeter Point." `http://www.evansville.edu/~ck6/tcenters/recent/exeter.html`.

Kimberling, C. and Lossers, O. P. "Problem 6557 and Solution." *Amer. Math. Monthly* **97**, 535–537, 1990.

Exhaustion Method

The method of exhaustion was a INTEGRAL-like limiting process used by Archimedes to compute the AREA and VOLUME of 2-D LAMINA and 3-D SOLIDS.

see also INTEGRAL, LIMIT

Existence

If at least one solution can be determined for a given problem, a solution to that problem is said to exist. Frequently, mathematicians seek to prove the existence of solutions and then investigate their UNIQUENESS.

see also EXISTS, UNIQUE

Existential Closure

A class of processes which attempt to round off a domain and simplify its theory by adjoining elements.

see also MODEL COMPLETION

References

Kenneth, M. "Domain Extension and the Philosophy of Mathematics." *J. Philos.* **86**, 553–562, 1989.

Exists

If there exists an A, this is written $\exists A$. Similarly, A does not exit is written $\nexists A$.

see also EXISTENCE, FOR ALL, QUANTIFIER

Exmedian

The line through the VERTEX of a TRIANGLE which is PARALLEL to the opposite side.

References

Johnson, R. A. *Modern Geometry: An Elementary Treatise on the Geometry of the Triangle and the Circle.* Boston, MA: Houghton Mifflin, p. 176, 1929.

Exmedian Point

The point of intersection of two EXMEDIANS.

References

Johnson, R. A. *Modern Geometry: An Elementary Treatise on the Geometry of the Triangle and the Circle.* Boston, MA: Houghton Mifflin, p. 176, 1929.

Exogenous Variable

An economic variable that is related to other economic variables and determines their equilibrium levels.

see also ENDOGENOUS VARIABLE

References

Iyanaga, S. and Kawada, Y. (Eds.). *Encyclopedic Dictionary of Mathematics.* Cambridge, MA: MIT Press, p. 458, 1980.

Exotic $\mathbb{R}^4$

Donaldson (1983) showed there exists an exotic smooth DIFFERENTIAL STRUCTURE on $\mathbb{R}^4$. Donaldson's result has been extended to there being precisely a CONTINUUM of nondiffeomorphic DIFFERENTIAL STRUCTURES on $\mathbb{R}^4$.

see also EXOTIC SPHERE

References

Donaldson, S. K. "Self-Dual Connections and the Topology of Smooth 4-Manifold." *Bull. Amer. Math. Soc.* **8**, 81–83, 1983.

Monastyrsky, M. *Modern Mathematics in the Light of the Fields Medals.* Wellesley, MA: A. K. Peters, 1997.

Exotic Sphere

Milnor (1963) found more than one smooth structure on the 7-D HYPERSPHERE. Generalizations have subsequently been found in other dimensions. Using SURGERY theory, it is possible to relate the number of DIFFEOMORPHISM classes of exotic spheres to higher homotopy groups of spheres (Kosinski 1992). Kervaire and Milnor (1963) computed a list of the number $N(d)$ of distinct (up to DIFFEOMORPHISM) DIFFERENTIAL STRUCTURES on spheres indexed by the DIMENSION d of the sphere. For $d = 1, 2, \ldots$, assuming the POINCARÉ CONJECTURE, they are 1, 1, 1, ≥ 1, 1, 1, 28, 2, 8, 6, 992, 1, 3, 2, 16256, 2, 16, 16, ... (Sloane's A001676). The status of $d = 4$ is still unresolved: at least one exotic structure exists, but it is not known if others do as well.

The only exotic Euclidean spaces are a CONTINUUM of EXOTIC $\mathbb{R}^4$ structures.

see also EXOTIC $\mathbb{R}^4$, HYPERSPHERE

References

Kervaire, M. A. and Milnor, J. W. "Groups of Homotopy Spheres: I." *Ann. Math.* **77**, 504–537, 1963.

Kosinski, A. A. §X.6 in *Differential Manifolds.* Boston, MA: Academic Press, 1992.

Milnor, J. "Topological Manifolds and Smooth Manifolds." *Proc. Internat. Congr. Mathematicians (Stockholm, 1962)* Djursholm: Inst. Mittag-Leffler, pp. 132–138, 1963.

Milnor, J. W. and Stasheff, J. D. *Characteristic Classes.* Princeton, NJ: Princeton University Press, 1973.

Monastyrsky, M. *Modern Mathematics in the Light of the Fields Medals.* Wellesley, MA: A. K. Peters, 1997.

Novikov, S. P. (Ed.). *Topology I.* New York: Springer-Verlag, 1996.

Sloane, N. J. A. Sequence A001676/M5197 in "An On-Line Version of the Encyclopedia of Integer Sequences."

Exp

see EXPONENTIAL FUNCTION

Expansion

An AFFINE TRANSFORMATION in which the scale is increased. It is the opposite of a DILATION (CONTRACTION).

see also DILATION

Expansive

Let ϕ be a MAP. Then ϕ is expansive if the DISTANCE $d(\phi^n x, \phi^n y) < \delta$ for all $n \in \mathbb{Z}$, then $x = y$. Equivalently, ϕ is expansive if the orbits of two points x and y are always very close.

Expectation Value

For one discrete variable,

$$\langle f(x) \rangle = \sum_x P(x). \tag{1}$$

For one continuous variable,

$$\langle f(x) \rangle = \int f(x)P(x)\,dx. \tag{2}$$

The expectation value satisfies

$$\langle ax + by \rangle = a\langle x \rangle + b\langle y \rangle \tag{3}$$

$$\langle a \rangle = a \tag{4}$$

$$\left\langle \sum x \right\rangle = \sum \langle x \rangle. \tag{5}$$

For multiple discrete variables

$$\langle f(x_1, \ldots, x_n) \rangle = \sum_{x_1, \ldots, x_n} P(x_1, \ldots, x_n). \tag{6}$$

For multiple continuous variables

$$\langle f(x_1, \ldots, x_n) \rangle = \int f(x_1, \ldots, x_n) P(x_1, \ldots, x_n)\,dx_1 \cdots dx_n. \tag{7}$$

The (multiple) expectation value satisfies

$$\begin{aligned} \langle (x - \mu_x)(y - \mu_y) \rangle &= \langle xy - \mu_x y - \mu_y x + \mu_x \mu_y \rangle \\ &= \langle xy \rangle - \mu_x \mu_y - \mu_y \mu_x + \mu_x \mu_y \\ &= \langle xy \rangle - \langle x \rangle \langle y \rangle, \end{aligned} \tag{8}$$

where μ_i is the MEAN for the variable i.

see also MEAN

Experimental Design

see DESIGN

Exploration Problem

see JEEP PROBLEM

Exponent

The POWER p in an expression a^p.

Exponent Laws

The laws governing the combination of EXPONENTS (POWERS) are

$$x^m \cdot x^n = x^{m+n} \tag{1}$$

$$\frac{x^m}{x^n} = x^{m-n} \tag{2}$$

$$(x^m)^n = x^{mn} \tag{3}$$

$$(xy)^m = x^m y^m \tag{4}$$

$$\left(\frac{x}{y}\right)^n = \frac{x^n}{y^n} \tag{5}$$

$$x^{-n} = \frac{1}{x^n} \tag{6}$$

$$\left(\frac{x}{y}\right)^{-n} = \left(\frac{y}{x}\right)^n, \tag{7}$$

where quantities in the DENOMINATOR are taken to be nonzero. Special cases include

$$x^1 = x \tag{8}$$

and

$$x^0 = 1 \tag{9}$$

for $x \neq 0$. The definition $0^0 = 1$ is sometimes used to simplify formulas, but it should be kept in mind that this equality is a definition and not a fundamental mathematical truth.

see also EXPONENT, POWER

Exponent Vector

Let p_i denote the ith PRIME, and write

$$m = \prod_i p_i{}^{v_i}.$$

Then the exponent vector is $\mathbf{v}(m) = (v_1, v_2, \ldots)$.

see also DIXON'S FACTORIZATION METHOD

References

Pomerance, C. "A Tale of Two Sieves." *Not. Amer. Math. Soc.* **43**, 1473–1485, 1996.

Exponential Digital Invariant

see NARCISSISTIC NUMBER

Exponential Distribution

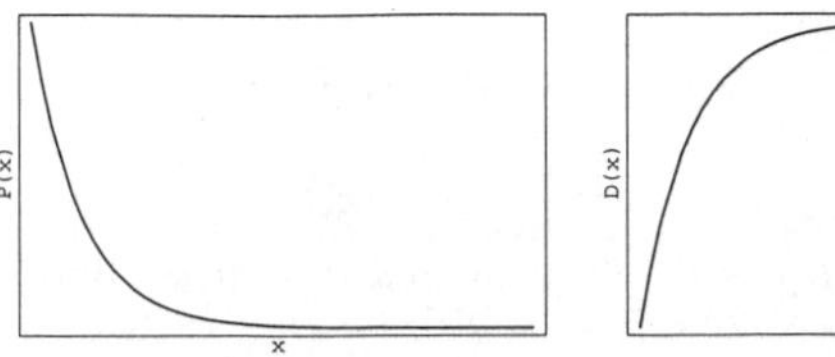

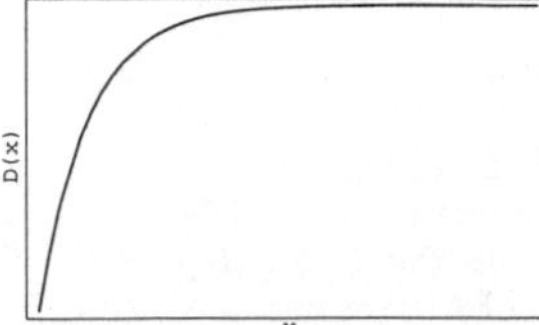

Given a POISSON DISTRIBUTION with rate of change λ, the distribution of waiting times between successive changes (with $k = 0$) is

$$D(x) \equiv P(X \leq x) = 1 - P(X > x) = 1 - \frac{(\lambda x)^0 e^{-\lambda x}}{0!} = 1 - e^{-\lambda x} \quad (1)$$

$$P(x) = D'(x) = \lambda e^{-\lambda x}, \quad (2)$$

which is normalized since

$$\int_0^\infty P(x)\,dx = \lambda \int_0^\infty e^{-\lambda x}\,dx = -[e^{-\lambda x}]_0^\infty = -(0-1) = 1. \quad (3)$$

This is the only MEMORYLESS RANDOM DISTRIBUTION. Define the MEAN waiting time between successive changes as $\theta \equiv \lambda^{-1}$. Then

$$P(x) = \begin{cases} \frac{1}{\theta} e^{-x/\theta} & x \geq 0 \\ 0 & x < 0. \end{cases} \quad (4)$$

The MOMENT-GENERATING FUNCTION is

$$M(t) = \int_0^\infty e^{tx} \left(\frac{1}{\theta}\right) e^{-x/\theta}\,dx = \frac{1}{\theta} \int_0^\infty e^{-(1-\theta t)x/\theta}\,dx = \left[\frac{e^{-(1-\theta t)x/\theta}}{1-\theta t}\right]_0^\infty = \frac{1}{1-\theta t} \quad (5)$$

$$M'(t) = \frac{\theta}{(1-\theta t)^2} \quad (6)$$

$$M''(t) = \frac{2\theta^2}{(1-\theta t)^3}, \quad (7)$$

so

$$R(t) \equiv \ln M(t) = -\ln(1-\theta t) \quad (8)$$

$$R'(t) = \frac{\theta}{1-\theta t} \quad (9)$$

$$R''(t) = \frac{\theta^2}{(1-\theta t)^2} \quad (10)$$

$$\mu = R'(0) = \theta \quad (11)$$

$$\sigma^2 = R''(0) = \theta^2. \quad (12)$$

The SKEWNESS and KURTOSIS are given by

$$\gamma_1 = 2 \quad (13)$$

$$\gamma_2 = 6. \quad (14)$$

The MEAN and VARIANCE can also be computed directly

$$\langle x \rangle \equiv \int_0^\infty P(x)\,dx = \frac{1}{s} \int_0^\infty x e^{-x/s}\,dx. \quad (15)$$

Use the integral

$$\int x e^{ax}\,dx = \frac{e^{ax}}{a^2}(ax-1) \quad (16)$$

to obtain

$$\langle x \rangle = \frac{1}{s} \left[\frac{e^{-x/s}}{(-\frac{1}{s})^2} \left\{\left(-\frac{1}{s}\right) x - 1\right\}\right]_0^\infty = -s\left[e^{-x/s}\left(1 + \frac{x}{s}\right)\right]_0^\infty = -s(0-1) = s. \quad (17)$$

Now, to find

$$\langle x^2 \rangle = \frac{1}{s} \int_0^\infty x^2 e^{-x/s}\,dx, \quad (18)$$

use the integral

$$\int x^2 e^{-x/s}\,dx = \frac{e^{ax}}{a^3}(2 - 2ax + a^2x^2) \quad (19)$$

$$\langle x^2 \rangle = \frac{1}{s} \left[\frac{e^{-x/s}}{(-\frac{1}{s})^3} \left(2 + \frac{2}{s}x + \frac{1}{s^2}x^2\right)\right]_0^\infty = -s^2(0-2) = 2s^2, \quad (20)$$

giving

$$\sigma^2 \equiv \langle x^2 \rangle - \langle x \rangle^2 = 2s^2 - s^2 = s^2 \quad (21)$$

$$\sigma \equiv \sqrt{\operatorname{var}(x)} = s. \quad (22)$$

If a generalized exponential probability function is defined by

$$P_{(\alpha,\beta)}(x) = \frac{1}{\beta} e^{-(x-\alpha)/\beta}, \quad (23)$$

then the CHARACTERISTIC FUNCTION is

$$\phi(t) = \frac{e^{i\alpha t}}{1 - i\beta t}, \quad (24)$$

and the MEAN, VARIANCE, SKEWNESS, and KURTOSIS are

$$\mu = \alpha + \beta \quad (25)$$

$$\sigma^2 = \beta^2 \quad (26)$$

$$\gamma_1 = 2 \quad (27)$$

$$\gamma_2 = 6. \quad (28)$$

see also DOUBLE EXPONENTIAL DISTRIBUTION

References

Balakrishnan, N. and Basu, A. P. *The Exponential Distribution: Theory, Methods, and Applications.* New York: Gordon and Breach, 1996.

Beyer, W. H. *CRC Standard Mathematical Tables, 28th ed.* Boca Raton, FL: CRC Press, pp. 534–535, 1987.

Spiegel, M. R. *Theory and Problems of Probability and Statistics.* New York: McGraw-Hill, p. 119, 1992.

Exponential Divisor

see e-DIVISOR

Exponential Function

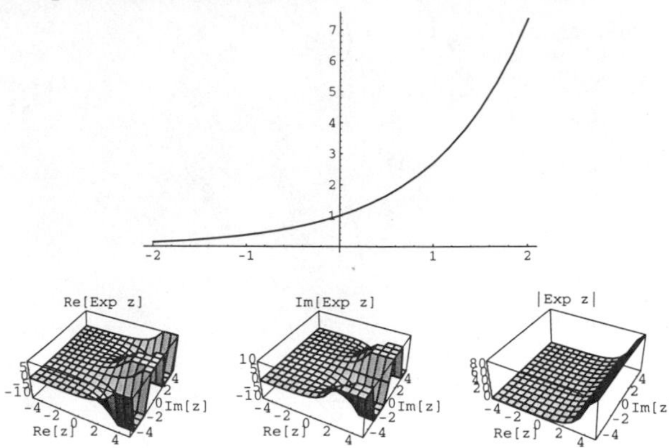

The exponential function is defined by

$$\exp(x) \equiv e^x, \tag{1}$$

where e is the constant 2.718.... It satisfies the identity

$$\exp(x + y) = \exp(x)\exp(y). \tag{2}$$

If $z \equiv x + iy$,

$$e^z = e^{x+iy} = e^x e^{iy} = e^x(\cos y + i \sin y). \tag{3}$$

If

$$a + bi = e^{x+iy}, \tag{4}$$

then

$$y = \tan^{-1}\left(\frac{b}{a}\right) \tag{5}$$

$$\begin{aligned} x &= \ln\left\{b \csc\left[\tan^{-1}\left(\frac{b}{a}\right)\right]\right\} \\ &= \ln\left\{a \sec\left[\tan^{-1}\left(\frac{b}{a}\right)\right]\right\}. \end{aligned} \tag{6}$$

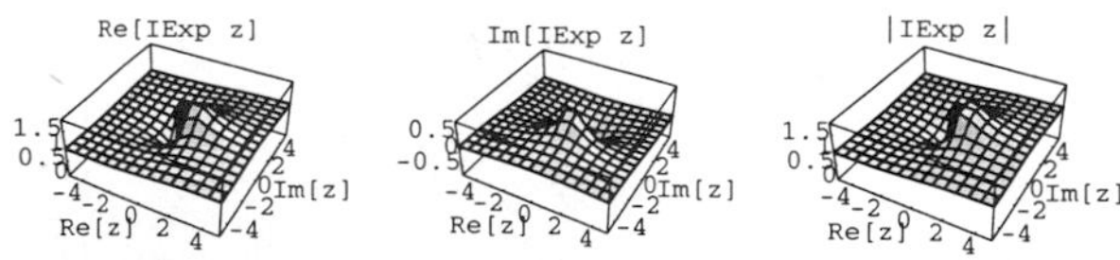

The above plot shows the function $e^{1/z}$.

see also EULER FORMULA, EXPONENTIAL RAMP, FOURIER TRANSFORM—EXPONENTIAL FUNCTION, SIGMOID FUNCTION

References

Abramowitz, M. and Stegun, C. A. (Eds.). "Exponential Function." §4.2 in *Handbook of Mathematical Functions with Formulas, Graphs, and Mathematical Tables, 9th printing.* New York: Dover, p. 69–71, 1972.

Fischer, G. (Ed.). Plates 127–128 in *Mathematische Modelle/Mathematical Models, Bildband/Photograph Volume.* Braunschweig, Germany: Vieweg, pp. 124–125, 1986.

Spanier, J. and Oldham, K. B. "The Exponential Function $\exp(bx + c)$" and "Exponentials of Powers $\exp(-ax^\nu)$." Chs. 26–27 in *An Atlas of Functions.* Washington, DC: Hemisphere, pp. 233–261, 1987.

Yates, R. C. "Exponential Curves." *A Handbook on Curves and Their Properties.* Ann Arbor, MI: J. W. Edwards, pp. 86–97, 1952.

Exponential Function (Truncated)

see EXPONENTIAL SUM FUNCTION

Exponential Inequality

For $c < 1$,

$$x^c < 1 + c(x - 1).$$

For $c > 1$,

$$x^c > 1 + c(x - 1).$$

Exponential Integral

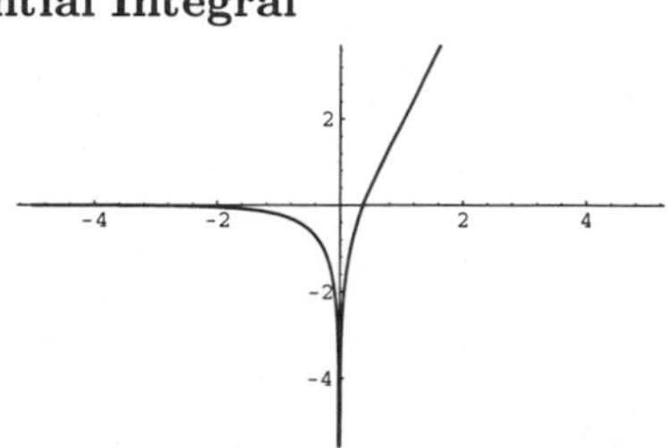

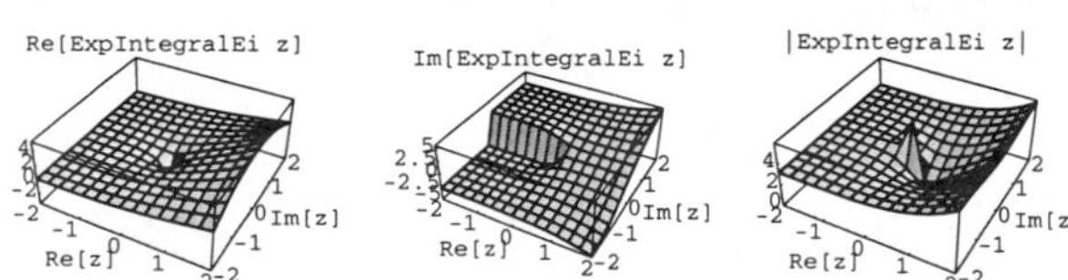

Let $E_1(x)$ be the E_n-FUNCTION with $n = 1$,

$$E_1(x) \equiv \int_1^\infty \frac{e^{-tx}\,dt}{t} = \int_x^\infty \frac{e^{-u}\,du}{u}. \tag{1}$$

Then define the exponential integral $\mathrm{ei}(x)$ by

$$E_1(x) = -\,\mathrm{ei}(-x), \tag{2}$$

where the retention of the $-\,\mathrm{ei}(-x)$ NOTATION is a historical artifact. Then $\mathrm{ei}(x)$ is given by the integral

$$\mathrm{ei}(x) = -\int_{-x}^\infty \frac{e^{-t}\,dt}{t}. \tag{3}$$

This function is given by the *Mathematica*® (Wolfram Research, Champaign, IL) function **ExpIntegralEi[x]**. The exponential integral can also be written

$$\mathrm{ei}(ix) = \mathrm{ci}(x) + i\,\mathrm{si}(x), \tag{4}$$

where $\mathrm{ci}(x)$ and $\mathrm{si}(x)$ are COSINE and SINE INTEGRAL.

The real ROOT of the exponential integral occurs at 0.37250741078..., which is not known to be expressible in terms of other standard constants. The quantity $-e\,\mathrm{ei}(-1) = 0.596347362\ldots$ is known as the GOMPERTZ CONSTANT.

see also COSINE INTEGRAL, E_n-FUNCTION, GOMPERTZ CONSTANT, SINE INTEGRAL

References

Arfken, G. *Mathematical Methods for Physicists, 3rd ed.* Orlando, FL: Academic Press, pp. 566–568, 1985.

Morse, P. M. and Feshbach, H. *Methods of Theoretical Physics, Part I.* New York: McGraw-Hill, pp. 434–435, 1953.

Press, W. H.; Flannery, B. P.; Teukolsky, S. A.; and Vetterling, W. T. "Exponential Integrals." §6.3 in *Numerical Recipes in FORTRAN: The Art of Scientific Computing, 2nd ed.* Cambridge, England: Cambridge University Press, pp. 215–219, 1992.

Spanier, J. and Oldham, K. B. "The Exponential Integral Ei(x) and Related Functions." Ch. 37 in *An Atlas of Functions.* Washington, DC: Hemisphere, pp. 351–360, 1987.

Exponential Map

On a LIE GROUP, exp is a MAP from the LIE ALGEBRA to its LIE GROUP. If you think of the LIE ALGEBRA as the TANGENT SPACE to the identity of the LIE GROUP, $\exp(v)$ is defined to be $h(1)$, where h is the unique LIE GROUP HOMEOMORPHISM from the REAL NUMBERS to the LIE GROUP such that its velocity at time 0 is v.

On a RIEMANNIAN MANIFOLD, exp is a MAP from the TANGENT BUNDLE of the MANIFOLD to the MANIFOLD, and $\exp(v)$ is defined to be $h(1)$, where h is the unique GEODESIC traveling through the base-point of v such that its velocity at time 0 is v.

The three notions of exp (exp from COMPLEX ANALYSIS, exp from LIE GROUPS, and exp from Riemannian geometry) are all linked together, the strongest link being between the LIE GROUPS and Riemannian geometry definition. If G is a compact LIE GROUP, it admits a left and right invariant RIEMANNIAN METRIC. With respect to that metric, the two exp maps agree on their common domain. In other words, one-parameter subgroups are geodesics. In the case of the MANIFOLD $\mathbb{S}^1$, the CIRCLE, if we think of the tangent space to 1 as being the IMAGINARY axis (y-AXIS) in the COMPLEX PLANE, then

$$\begin{aligned}\exp_{\text{Riemannian geometry}}(v) &= \exp_{\text{Lie Groups}}(v)\\ &= \exp_{\text{complex analysis}}(v),\end{aligned}$$

and so the three concepts of the exponential all agree in this case.

see also EXPONENTIAL FUNCTION

Exponential Matrix

see MATRIX EXPONENTIAL

Exponential Ramp

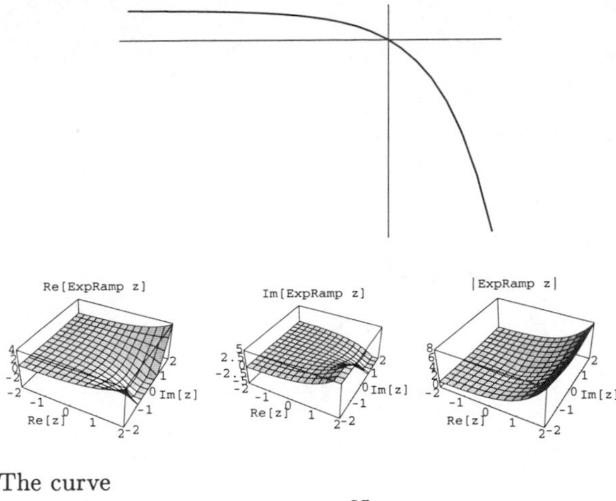

The curve

$$y = 1 - e^{ax}.$$

see also EXPONENTIAL FUNCTION, SIGMOID FUNCTION

References

von Seggern, D. *CRC Standard Curves and Surfaces.* Boca Raton, FL: CRC Press, p. 158, 1993.

Exponential Sum Formulas

$$\begin{aligned}\sum_{n=0}^{N-1} e^{iNx} &= \frac{1-e^{iNx}}{1-e^{ix}} = \frac{-e^{iNx/2}\left(e^{-iNx/2}-e^{iNx/2}\right)}{-e^{ix/2}\left(e^{-ix/2}-e^{ix/2}\right)}\\ &= \frac{\sin(\frac{1}{2}Nx)}{\sin(\frac{1}{2}x)} e^{ix(N-1)/2}, \end{aligned} \tag{1}$$

where

$$\sum_{n=0}^{N-1} r^n = \frac{1-r^n}{1-r} \tag{2}$$

has been used. Similarly,

$$\begin{aligned}\sum_{n=0}^{N-1} p^n e^{inx} &= \frac{1-p^N e^{iNx}}{1-pe^{ix}} = \frac{(1-p^N e^{iNx})(1-pe^{-ix})}{(1-pe^{ix})(1-pe^{-ix})}\\ &= \frac{1-p^N e^{iNx} - pe^{-ix} + p^{N+1} e^{ix(N-1)}}{1-p(e^{ix}+e^{-ix})+p^2}\\ &= \frac{p^{N+1}e^{ix(N-1)} - p^N e^{iNx} + 1 - pe^{-ix}}{1-2p\cos x + p^2}. \end{aligned} \tag{3}$$

This gives

$$\sum_{n=0}^{\infty} p^n e^{inx} = \lim_{N\to\infty} \sum_{n=0}^{N-1} p^n e^{inx} = \frac{1-pe^{-ix}}{1-2p\cos x + p^2}. \tag{4}$$

By looking at the REAL and IMAGINARY PARTS of these FORMULAS, sums involving sines and cosines can be obtained.

Exponential Sum Function

$$\text{es}_n(x) \equiv \exp_n(x) \equiv \sum_{m=0}^{n} \frac{x^m}{m!}.$$

see also GAMMA FUNCTION

Exradius

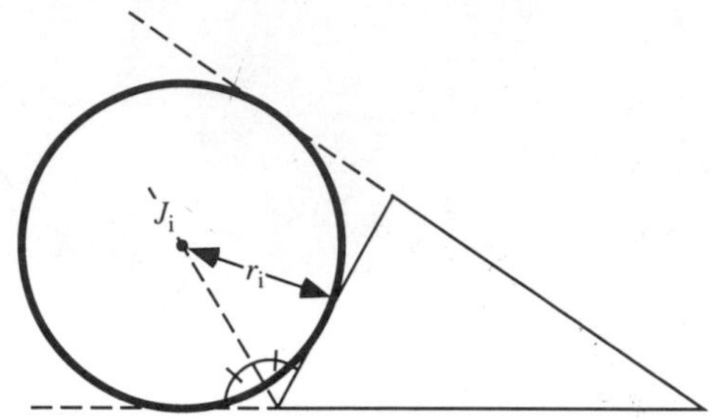

The RADIUS of an EXCIRCLE. Let a TRIANGLE have exradius r_a (sometimes denoted ρ_a), opposite side of length a, AREA Δ, and SEMIPERIMETER s. Then

$$r_a{}^2 = \left(\frac{\Delta}{s-a}\right)^2 \tag{1}$$

$$= \frac{s(s-c)(s-b)}{s-a} \tag{2}$$

$$= 4R\sin(\tfrac{1}{2}\alpha_1)\cos(\tfrac{1}{2}\alpha_2)\cos(\tfrac{1}{2}\alpha_3) \tag{3}$$

(Johnson 1929, p. 189) where R is the CIRCUMRADIUS. Let r be the INRADIUS, then

$$4R = r_a + r_b + r_c - r \tag{4}$$

$$\frac{1}{r_a} + \frac{1}{r_b} + \frac{1}{r_c} = \frac{1}{r} \tag{5}$$

$$rr_ar_br_c = \Delta^2. \tag{6}$$

Some fascinating FORMULAS due to Feuerbach are

$$r_2r_3 + r_3r_1 + r_1r_3 = s^2 \tag{7}$$

$$r(r_2r_3 + r_3r_1 + r_1r_2) = s\Delta = r_1r_2r_3 \tag{8}$$

$$r(r_1 + r_2 + r_3) = a_2a_3 + a_3a_1 + a_1a_2 - s^2 \tag{9}$$

$$rr_1 + rr_2 + rr_3 + r_1r_2 + r_2r_3 + r_3r_1 = a_2a_3 + a_3a_1 + a_1a_2 \tag{10}$$

$$r_2r_3 + r_3r_1 + r_1r_2 - rr_1 - rr_2 - rr_3 - \tfrac{1}{2}(a_1{}^2 + a_2{}^2 + a_3{}^2) \tag{11}$$

(Johnson 1929, pp. 190–191).

see also CIRCLE, CIRCUMRADIUS, EXCIRCLE, INRADIUS, RADIUS

References

Johnson, R. A. *Modern Geometry: An Elementary Treatise on the Geometry of the Triangle and the Circle.* Boston, MA: Houghton Mifflin, 1929.

Mackay, J. S. "Formulas Connected with the Radii of the Incircle and Excircles of a Triangle." *Proc. Edinburgh Math. Soc.* **12**, 86–105.

Mackay, J. S. "Formulas Connected with the Radii of the Incircle and Excircles of a Triangle." *Proc. Edinburgh Math. Soc.* **13**, 103–104.

Exsecant

$$\text{exsec}\, x \equiv \sec x - 1,$$

where $\sec x$ is the SECANT.

see also COVERSINE, HAVERSINE, SECANT, VERSINE

References

Abramowitz, M. and Stegun, C. A. (Eds.). *Handbook of Mathematical Functions with Formulas, Graphs, and Mathematical Tables, 9th printing.* New York: Dover, p. 78, 1972.

Extended Cycloid

see PROLATE CYCLOID

Extended Goldbach Conjecture

see GOLDBACH CONJECTURE

Extended Greatest Common Divisor

see GREATEST COMMON DIVISOR

Extended Mean-Value Theorem

Let the functions f and g be DIFFERENTIABLE on the OPEN INTERVAL (a,b) and CONTINUOUS on the CLOSED INTERVAL $[a,b]$. If $g'(x) \neq 0$ for any $x \in (a,b)$, then there is at least one point $c \in (a,b)$ such that

$$\frac{f'(c)}{g'(c)} = \frac{f(b)-f(a)}{g(b)-g(a)}.$$

see also MEAN-VALUE THEOREM

Extended Riemann Hypothesis

The first quadratic nonresidue mod p of a number is always less than $2(\ln p)^2$.

see also RIEMANN HYPOTHESIS

References

Bach, E. *Analytic Methods in the Analysis and Design of Number-Theoretic Algorithms.* Cambridge, MA: MIT Press, 1985.

Wagon, S. *Mathematica in Action.* New York: W. H. Freeman, p. 295, 1991.

Extension

The definition of a SET by enumerating its members. An extensional definition can always be reduced to an INTENTIONAL one.

see also INTENSION

References

Russell, B. "Definition of Number." *Introduction to Mathematical Philosophy.* New York: Simon and Schuster, 1971.

Extension Problem

Given a SUBSPACE A of a SPACE X and a MAP from A to a SPACE Y, is it possible to extend that MAP to a MAP from X to Y?

see also LIFTING PROBLEM

Extensions Calculus

see EXTERIOR ALGEBRA

Extent

The RADIUS of the smallest CIRCLE centered at one of the points of an N-CLUSTER, which contains all the points in the N-CLUSTER.

see also N-CLUSTER

Exterior

That portion of a region lying "outside" a specified boundary.

see also INTERIOR

Exterior Algebra

The ALGEBRA of the EXTERIOR PRODUCT, also called an ALTERNATING ALGEBRA or GRASSMANN ALGEBRA. The study of exterior algebra is also called AUSDEHNUNGSLEHRE and EXTENSIONS CALCULUS. Exterior algebras are GRADED ALGEBRAS.

In particular, the exterior algebra of a VECTOR SPACE is the DIRECT SUM over k in the natural numbers of the VECTOR SPACES of alternating k-forms on that VECTOR SPACE. The product on this algebra is then the wedge product of forms. The exterior algebra for a VECTOR SPACE V is constructed by forming monomials u, $v \wedge w$, $x \wedge y \wedge z$, etc., where u, v, w, x, y, and z are vectors in V and $\wedge$ is asymmetric multiplication. The sums formed from linear combinations of the MONOMIALS are the elements of an exterior algebra.

References

Forder, H. G. *The Calculus of Extension.* Cambridge, England: Cambridge University Press, 1941.

Lounesto, P. "Counterexamples to Theorems Published and Proved in Recent Literature on Clifford Algebras, Spinors, Spin Groups, and the Exterior Algebra." `http://www.hit.fi/~lounesto/counterexamples.htm`.

Exterior Angle Bisector

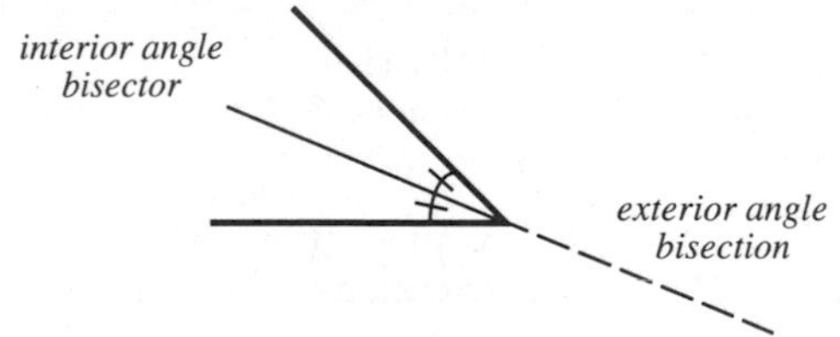

The exterior bisector of an ANGLE is the LINE or LINE SEGMENT which cuts it into two equal ANGLES on the opposite "side" as the ANGLE.

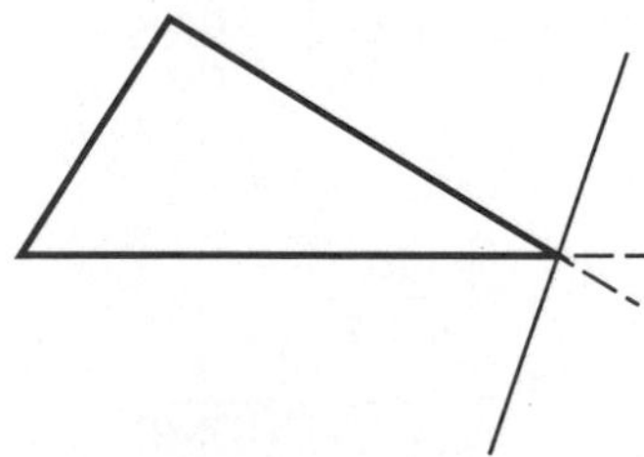

For a TRIANGLE, the exterior angle bisector bisects the SUPPLEMENTARY ANGLE at a given VERTEX. It also divides the opposite side externally in the ratio of adjacent sides.

see also ANGLE BISECTOR, ISODYNAMIC POINTS

Exterior Angle Theorem

In any TRIANGLE, if one of the sides is extended, the exterior angle is greater than both the interior and opposite angles.

References

Dunham, W. *Journey Through Genius: The Great Theorems of Mathematics.* New York: Wiley, p. 41, 1990.

Exterior Derivative

Consider a DIFFERENTIAL k-FORM

$$\omega^1 = b_1\,dx_1 + b_2\,dx_2. \tag{1}$$

Then its exterior derivative is

$$d\omega^1 = db_1 \wedge dx_1 + db_2 \wedge dx_2, \tag{2}$$

where $\wedge$ is the WEDGE PRODUCT. Similarly, consider

$$\omega^1 = b_1(x_1, x_2)\,dx_1 + b_2(x_1, x_2)\,dx_2. \tag{3}$$

Then

$$\begin{aligned} d\omega^1 &= db_1 \wedge dx_1 + db_2 \wedge dx_2 \\ &= \left(\frac{\partial b_1}{\partial x_1} dx_1 + \frac{\partial b_1}{\partial x_2} dx_2\right) \wedge dx_1 \\ &\quad + \left(\frac{\partial b_2}{\partial x_1} dx_1 + \frac{\partial b_2}{\partial x_2} dx_2\right) \wedge dx_2. \end{aligned} \tag{4}$$

Denote the exterior derivative by

$$Dt \equiv \frac{\partial}{\partial x} \wedge t. \tag{5}$$

Then for a 0-form t,

$$(Dt)_\mu \equiv \frac{\partial t}{\partial x^\mu}, \tag{6}$$

for a 1-form t,

$$(Dt)_{\mu\nu} \equiv \frac{1}{2}\left(\frac{\partial t_\nu}{\partial x^\mu} - \frac{\partial t_\mu}{\partial x^\nu}\right), \tag{7}$$

and for a 2-form t,

$$(Dt)_{ijk} \equiv \tfrac{1}{3}\epsilon_{ijk}\left(\frac{\partial t_{23}}{\partial x^1} + \frac{\partial t_{31}}{\partial x^2} + \frac{\partial t_{12}}{\partial x^3}\right), \tag{8}$$

where ϵ_{ijk} is the PERMUTATION TENSOR.

The second exterior derivative is

$$D^2 t = \frac{\partial}{\partial x} \wedge \left(\frac{\partial}{\partial x} \wedge t\right) = \left(\frac{\partial}{\partial x} \wedge \frac{\partial}{\partial x}\right) \wedge t = 0, \tag{9}$$

which is known as POINCARÉ'S LEMMA.

see also DIFFERENTIAL k-FORM, POINCARÉ'S LEMMA, WEDGE PRODUCT

Exterior Dimension

A type of DIMENSION which can be used to characterize FAT FRACTALS.

see also FAT FRACTAL

References

Grebogi, C.; McDonald, S. W.; Ott, E.; and Yorke, J. A. "Exterior Dimension of Fat Fractals." *Phys. Let. A* **110**, 1–4, 1985.

Grebogi, C.; McDonald, S. W.; Ott, E.; and Yorke, J. A. Erratum to "Exterior Dimension of Fat Fractals." *Phys. Let. A* **113**, 495, 1986.

Ott, E. *Chaos in Dynamical Systems.* New York: Cambridge University Press, p. 98, 1993.

Exterior Product

see WEDGE PRODUCT

Exterior Snowflake

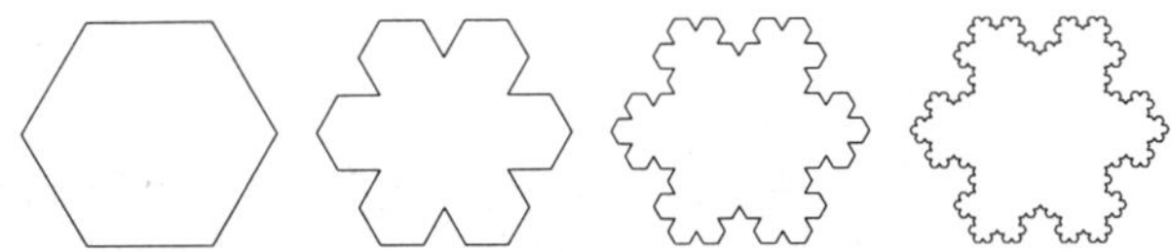

A FRACTAL.

see also FLOWSNAKE FRACTAL, KOCH ANTISNOWFLAKE, KOCH SNOWFLAKE, PENTAFLAKE

References

Wagon, S. *Mathematica in Action.* New York: W. H. Freeman, pp. 193–195, 1991.

Weisstein, E. W. "Fractals." http://www.astro.virginia.edu/~eww6n/math/notebooks/Fractal.m.

Extra Strong Lucas Pseudoprime

Given the LUCAS SEQUENCE $U_n(b,-1)$ and $V_n(b,-1)$, define $\Delta = b^2 - 4$. Then an extra strong Lucas pseudoprime to the base b is a COMPOSITE NUMBER $n = 2^r s + (\Delta/n)$, where s is ODD and $(n, 2\Delta) = 1$ such that either $U_s \equiv 0 \pmod{n}$ and $V_s \equiv \pm 2 \pmod{n}$, or $V_{2^t s} \equiv 0 \pmod{n}$ for some t with $0 \leq t < r-1$. An extra strong Lucas pseudoprime is a STRONG LUCAS PSEUDOPRIME with parameters $(b,-1)$. COMPOSITE n are extra strong pseudoprimes for at most 1/8 of possible bases (Grantham 1997).

see also LUCAS PSEUDOPRIME, STRONG LUCAS PSEUDOPRIME

References

Grantham, J. "Frobenius Pseudoprimes." http://www.clark.net/pub/grantham/pseudo/pseudo.ps

Grantham, J. "A Frobenius Probable Prime Test with High Confidence." 1997. http://www.clark.net/pub/grantham/pseudo/pseudo2.ps

Jones, J. P. and Mo, Z. "A New Primality Test Using Lucas Sequences." Preprint.

Extrapolation

see RICHARDSON EXTRAPOLATION

Extremal Coloring

see EXTREMAL GRAPH

Extremal Graph

A two-coloring of a COMPLETE GRAPH K_n of n nodes which contains exactly the number $N \equiv (R+B)_{\min}$ of MONOCHROMATIC FORCED TRIANGLES and no more (i.e., a minimum of $R+B$ where R and B are the numbers of red and blue TRIANGLES). Goodman (1959) showed that for an extremal graph,

$$N(n) = \begin{cases} \frac{1}{3}m(m-1)(m-2) & \text{for } n = 2m \\ \frac{1}{3}2m(m-1)(4m+1) & \text{for } n = 4m+1 \\ \frac{1}{3}2m(m+1)(4m-1) & \text{for } n = 4m+3. \end{cases}$$

This is sometimes known as GOODMAN'S FORMULA. Schwenk (1972) rewrote it in the form

$$N(n) = \binom{n}{3} - \left\lfloor \tfrac{1}{2}n \left\lfloor \tfrac{1}{4}(n-1)^2 \right\rfloor \right\rfloor,$$

sometimes known as SCHWENK'S FORMULA, where $\lfloor x \rfloor$ is the FLOOR FUNCTION. The first few values of $N(n)$ for $n = 1, 2, \ldots$ are 0, 0, 0, 0, 0, 2, 4, 8, 12, 20, 28, 40, 52, 70, 88, ... (Sloane's A014557).

see also BICHROMATIC GRAPH, BLUE-EMPTY GRAPH, GOODMAN'S FORMULA, MONOCHROMATIC FORCED TRIANGLE, SCHWENK'S FORMULA

References

Goodman, A. W. "On Sets of Acquaintances and Strangers at Any Party." *Amer. Math. Monthly* **66**, 778–783, 1959.

Schwenk, A. J. "Acquaintance Party Problem." *Amer. Math. Monthly* **79**, 1113–1117, 1972.

Sloane, N. J. A. Sequence A014557 in "An On-Line Version of the Encyclopedia of Integer Sequences."

Extremals

A field of extremals is a plane region which is SIMPLY CONNECTED by a one-parameter family of extremals. The concept was invented by Weierstraß.

Extreme and Mean Ratio

see GOLDEN MEAN

Extreme Value Distribution

N.B. A detailed on-line essay by S. Finch was the starting point for this entry.

Let M_n denote the "extreme" (i.e., largest) ORDER STATISTIC $X^{(n)}$ for a distribution of n elements X_i taken from a continuous UNIFORM DISTRIBUTION. Then the distribution of the M_n is

$$P(M_n < x) = \begin{cases} 0 & \text{if } x < 0 \\ x^n & \text{if } 0 \leq x \leq 1 \\ 1 & \text{if } x > 1, \end{cases} \tag{1}$$

and the MEAN and VARIANCE are

$$\mu = \frac{n}{n+1} \tag{2}$$

$$\sigma^2 = \frac{n}{(n+1)^2(n+2)}. \tag{3}$$

If X_i are taken from a STANDARD NORMAL DISTRIBUTION, then its cumulative distribution is

$$F(x) = \frac{1}{\sqrt{2\pi}} \int_{-\infty}^{x} e^{-t^2/2}\,dt = \tfrac{1}{2} + \Phi(x), \tag{4}$$

where $\Phi(x)$ is the NORMAL DISTRIBUTION FUNCTION. The probability distribution of M_n is then

$$P(M_n < x) = [F(x)]^n = \frac{n}{\sqrt{2\pi}} \int_{-\infty}^{x} [F(t)]^{n-1} e^{-t^2/2}\,dt. \tag{5}$$

The MEAN $\mu(n)$ and VARIANCE $\sigma^2(n)$ are expressible in closed form for small n,

$$\mu(1) = 0 \tag{6}$$

$$\mu(2) = \frac{1}{\sqrt{\pi}} \tag{7}$$

$$\mu(3) = \frac{3}{2\sqrt{\pi}} \tag{8}$$

$$\mu(4) = \frac{3}{2\sqrt{\pi}} \left[1 + \frac{2}{\pi}\sin^{-1}(\tfrac{1}{3})\right] \tag{9}$$

$$\mu(5) = \frac{5}{4\sqrt{\pi}} \left[1 + \frac{6}{\pi}\sin^{-1}(\tfrac{1}{3})\right] \tag{10}$$

and

$$\sigma^2(1) = 1 \tag{11}$$

$$\sigma^2(2) = 1 - \frac{1}{\pi} \tag{12}$$

$$\sigma^2(3) = \frac{4\pi - 9 + 2\sqrt{3}}{4\pi} \tag{13}$$

$$\sigma^2(4) = 1 + \frac{\sqrt{3}}{\pi} - [\mu(4)]^2 \tag{14}$$

$$\sigma^2(5) = 1 + \frac{5\sqrt{3}}{4\pi} + \frac{5\sqrt{3}}{2\pi^2}\sin^{-1}(\tfrac{1}{4}) - [\mu(5)]^2. \tag{15}$$

No exact expression is known for $\mu(6)$ or $\sigma^2(6)$, but there is an equation connecting them

$$[\mu(6)]^2 + \sigma^2(6) = 1 + \frac{5\sqrt{3}}{4\pi} + \frac{15\sqrt{3}}{2\pi^2}\sin^{-1}(\tfrac{1}{4}). \tag{16}$$

An analog to the CENTRAL LIMIT THEOREM states that the asymptotic normalized distribution of M_n satisfies one of the three distributions

$$P(y) = \exp(-e^{-y}) \tag{17}$$

$$P(y) = \begin{cases} 0 & \text{if } y \leq 0 \\ \exp(-y^{-a}) & \end{cases} \tag{18}$$

$$P(y) = \begin{cases} \exp[-(-y)^a] & \text{if } y \leq 0 \\ 1 & \text{if } y > 0, \end{cases} \tag{19}$$

also known as GUMBEL, Fréchet, and WEIBULL DISTRIBUTIONS, respectively.

see also FISHER-TIPPETT DISTRIBUTION, ORDER STATISTIC

References

Balakrishnan, N. and Cohen, A. C. *Order Statistics and Inference.* New York: Academic Press, 1991.

David, H. A. *Order Statistics, 2nd ed.* New York: Wiley, 1981.

Finch, S. "Favorite Mathematical Constants." `http://www.mathsoft.com/asolve/constant/extval/extval.html`.

Gibbons, J. D. *Nonparametric Statistical Inference.* New York: McGraw-Hill, 1971.

Extreme Value Theorem

If a function f is continuous on a closed interval $[a, b]$, then f has both a MAXIMUM and a MINIMUM on $[a, b]$. If f has an extreme value on an open interval (a, b), then the extreme value occurs at a CRITICAL POINT. This theorem is sometimes also called the WEIERSTRAß EXTREME VALUE THEOREM.

Extremum

A MAXIMUM or MINIMUM. An extremum may be LOCAL (a.k.a. a RELATIVE EXTREMUM; an extremum in a given region which is not the overall MAXIMUM or MINIMUM) or GLOBAL. Functions with many extrema can be very difficult to GRAPH. Notorious examples include the functions $\cos(1/x)$ and $\sin(1/x)$ near $x = 0$

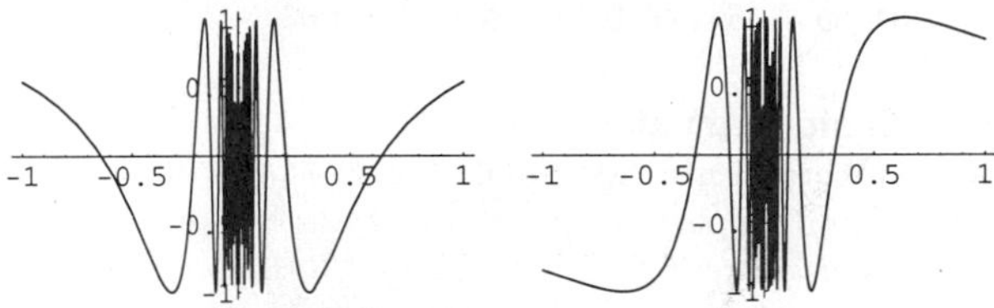

and $\sin(e^{2x+9})$ near 0 and 1.

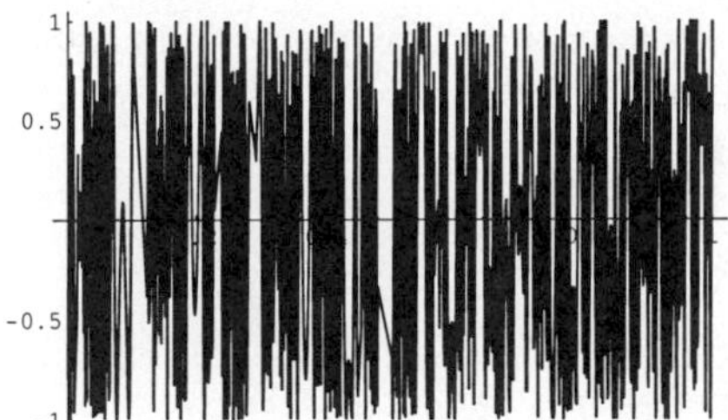

The latter has

$$\left\lfloor \frac{e^{11}}{\pi} - \frac{1}{2} \right\rfloor - \left\lceil \frac{e^9}{\pi} - \frac{1}{2} \right\rceil + 1 = 19058 - 2579 + 1 = 16480$$

extrema in the CLOSED INTERVAL [0,1] (Mulcahy 1996).

see also GLOBAL EXTREMUM, GLOBAL MAXIMUM, GLOBAL MINIMUM, KUHN-TUCKER THEOREM, LAGRANGE MULTIPLIER, LOCAL EXTREMUM, LOCAL MAXIMUM, LOCAL MINIMUM, MAXIMUM, MINIMUM

References

Abramowitz, M. and Stegun, C. A. (Eds.). *Handbook of Mathematical Functions with Formulas, Graphs, and Mathematical Tables, 9th printing.* New York: Dover, p. 14, 1972.

Mulcahy, C. "Plotting and Scheming with Wavelets." *Math. Mag.* **69**, 323–343, 1996.

Tikhomirov, V. M. *Stories About Maxima and Minima.* Providence, RI: Amer. Math. Soc., 1991.

Extremum Test

Consider a function $f(x)$ in 1-D. If $f(x)$ has a relative extremum at x_0, then either $f'(x_0) = 0$ or f is not DIFFERENTIABLE at x_0. Either the first or second DERIVATIVE tests may be used to locate relative extrema of the first kind.

A NECESSARY condition for $f(x)$ to have a MINIMUM (MAXIMUM) at x_0 is

$$f'(x_0) = 0,$$

and

$$f''(x_0) \geq 0 \qquad (f''(x_0) \leq 0).$$

A SUFFICIENT condition is $f'(x_0) = 0$ and $f''(x_0) > 0$ ($f''(x_0) < 0$). Let $f'(x_0) = 0$, $f''(x_0) = 0$, ..., $f^{(n)}(x_0) = 0$, but $f^{(n+1)}(x_0) \neq 0$. Then $f(x)$ has a RELATIVE MAXIMUM at x_0 if n is ODD and $f^{(n+1)}(x_0) < 0$, and $f(x)$ has a RELATIVE MINIMUM at x_0 if n is ODD and $f^{(n+1)}(x_0) > 0$. There is a SADDLE POINT at x_0 if n is EVEN.

see also EXTREMUM, FIRST DERIVATIVE TEST, RELATIVE MAXIMUM, RELATIVE MINIMUM, SADDLE POINT (FUNCTION), SECOND DERIVATIVE TEST

Extrinsic Curvature

A curvature of a SUBMANIFOLD of a MANIFOLD which depends on its particular EMBEDDING. Examples of extrinsic curvature include the CURVATURE and TORSION of curves in 3-space, or the mean curvature of surfaces in 3-space.

see also CURVATURE, INTRINSIC CURVATURE, MEAN CURVATURE

F

F-Distribution

Arises in the testing of whether two observed samples have the same VARIANCE. Let ${\chi_m}^2$ and ${\chi_n}^2$ be independent variates distributed as CHI-SQUARED with m and n DEGREES OF FREEDOM. Define a statistic $F_{n,m}$ as the ratio of the dispersions of the two distributions

$$F_{n,m} \equiv \frac{{\chi_n}^2/n}{{\chi_m}^2/m}. \tag{1}$$

This statistic then has an F-distribution with probability function and cumulative distribution

$$F_{n,m}(x) = \frac{\Gamma(\frac{n+m}{2})n^{n/2}m^{m/2}}{\Gamma(\frac{n}{2})\Gamma(\frac{m}{2})} \frac{x^{n/2-1}}{(m+nx)^{(n+m)/2}} \tag{2}$$

$$= \frac{m^{m/2}n^{n/2}x^{n/2-1}}{(m+nx)^{(n+m)/2}B(\frac{1}{2}n,\frac{1}{2}m)} \tag{3}$$

$$= I(1;\tfrac{1}{2}m;\tfrac{1}{2}n) - I\left(\frac{m}{m+nx};\tfrac{1}{2}m;\tfrac{1}{2}n\right), \tag{4}$$

where $\Gamma(z)$ is the GAMMA FUNCTION, $B(a,b)$ is the BETA FUNCTION, and $I(a,b;x)$ is the REGULARIZED BETA FUNCTION. The MEAN, VARIANCE, SKEWNESS and KURTOSIS are

$$\mu = \frac{m}{m-2} \tag{5}$$

$$\sigma^2 = \frac{2m^2(m+n-2)}{n(m-2)^2(m-4)} \tag{6}$$

$$\gamma_1 = \frac{2(m+2n-2)}{m-6}\sqrt{\frac{2(m-4)}{n(m+n-2)}} \tag{7}$$

$$\gamma_2 = \frac{12(-16+20m-8m^2+m^3+44n)}{n(m-6)(m-8)(n+m-2)} + \frac{12(-32mn+5m^2n-22n^2+5mn^2)}{n(m-6)(m-8)(n+m-2)}. \tag{8}$$

The probability that F would be as large as it is if the first distribution has a smaller variance than the second is denoted $Q(F_{n,m})$.

The noncentral F-distribution is given by

$$P(x) = e^{-\lambda/2+(\lambda n_1 x)/[2(n_2+n_1 x)]} \times {n_1}^{n_1/2}{n_2}^{n_2/2}x^{n_1/2-1}(n_2+n_1x)^{-(n_1+n_2)/2} \times \frac{\Gamma(\frac{1}{2}n_1)\Gamma(1+\frac{1}{2}n_2)L_{n_2/2}^{n_1/2-1}\left(-\frac{\lambda n_1 x}{2(n_2+n_1x)}\right)}{B(\frac{1}{2}n_1,\frac{1}{2}n_2)\Gamma[\frac{1}{2}(n_1+n_2)]}, \tag{9}$$

where $\Gamma(z)$ is the GAMMA FUNCTION, $B(\alpha,\beta)$ is the BETA FUNCTION, and $L_m^n(z)$ is an associated LAGUERRE POLYNOMIAL.

see also BETA FUNCTION, GAMMA FUNCTION, REGULARIZED BETA FUNCTION, SNEDECOR'S *F*-DISTRIBUTION

References

Abramowitz, M. and Stegun, C. A. (Eds.). *Handbook of Mathematical Functions with Formulas, Graphs, and Mathematical Tables, 9th printing.* New York: Dover, pp. 946–949, 1972.

Press, W. H.; Flannery, B. P.; Teukolsky, S. A.; and Vetterling, W. T. "Incomplete Beta Function, Student's Distribution, F-Distribution, Cumulative Binomial Distribution." §6.2 in *Numerical Recipes in FORTRAN: The Art of Scientific Computing, 2nd ed.* Cambridge, England: Cambridge University Press, pp. 219–223, 1992.

Spiegel, M. R. *Theory and Problems of Probability and Statistics.* New York: McGraw-Hill, pp. 117–118, 1992.

F-Polynomial

see KAUFFMAN POLYNOMIAL F

F-Ratio

The RATIO of two independent estimates of the VARIANCE of a NORMAL DISTRIBUTION.

see also *F*-DISTRIBUTION, NORMAL DISTRIBUTION, VARIANCE

F-Ratio Distribution

see *F*-DISTRIBUTION

Fabry Imbedding

A representation of a PLANAR GRAPH as a planar straight line graph such that no two EDGES cross.

Face

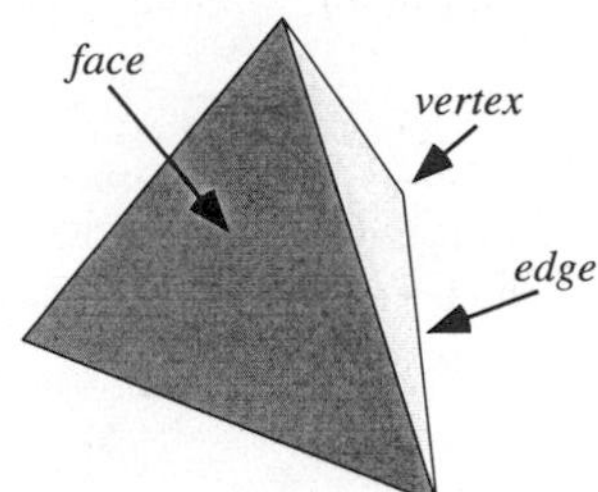

The intersection of an n-D POLYTOPE with a tangent HYPERPLANE. 0-D faces are known as VERTICES (nodes), 1-D faces as EDGES, $(n-2)$-D faces as RIDGES, and $(n-1)$-D faces as FACETS.

see also EDGE (POLYHEDRON), FACET, POLYTOPE, RIDGE, VERTEX (POLYHEDRON)

Facet

An $(n-1)$-D FACE of an n-D POLYTOPE. A procedure for generating facets is known as FACETING.

Faceting

Using a set of corners of a SOLID that lie in a plane to form the VERTICES of a new POLYGON is called faceting. Such POLYGONS may outline new FACES that join to enclose a new SOLID, even if the sides of the POLYGONS do not fall along EDGES of the original SOLID.

References

Holden, A. *Shapes, Space, and Symmetry.* New York: Columbia University Press, p. 94, 1971.

Factor

A factor is a portion of a quantity, usually an INTEGER or POLYNOMIAL. The determination of factors is called FACTORIZATION (or sometimes "FACTORING"). It is usually desired to break factors down into the smallest possible pieces so that no factor is itself factorable. For INTEGERS, the determination of factors is called PRIME FACTORIZATION. For large quantities, the determination of all factors is usually very difficult except in exceptional circumstances.

see also DIVISOR, FACTORIZATION, GREATEST PRIME FACTOR, LEAST PRIME FACTOR, PRIME FACTORIZATION ALGORITHMS

Factor Base

The primes with LEGENDRE SYMBOL $(n/p) = 1$ (less than $N = \pi(d)$ for trial divisor d) which need be considered when using the QUADRATIC SIEVE FACTORIZATION METHOD.

see also DIXON'S FACTORIZATION METHOD

References

Morrison, M. A. and Brillhart, J. "A Method of Factoring and the Factorization of F_7." *Math. Comput.* **29**, 183–205, 1975.

Factor (Graph)

A 1-factor of a GRAPH with n VERTICES is a set of $n/2$ separate EDGES which collectively contain all n of the VERTICES of G among their endpoints.

Factor Group

see QUOTIENT GROUP

Factor Level

A grouping of statistics.

Factor Ring

see QUOTIENT RING

Factor Space

see QUOTIENT SPACE

Factorial

The factorial $n!$ is defined for a POSITIVE INTEGER n as

$$n! \equiv \begin{cases} n\cdot(n-1)\cdots 2\cdot 1 & n=1,2,\ldots \\ 1 & n=0. \end{cases} \tag{1}$$

The first few factorials for $n = 0, 1, 2, \ldots$ are 1, 1, 2, 6, 24, 120, ... (Sloane's A000142). An older NOTATION for the factorial is $\underline{|n}$ (Dudeney 1970, Gardner 1978, Conway and Guy 1996).

As n grows large, factorials begin acquiring tails of trailing ZEROS. To calculate the number of trailing ZEROS for $n!$, use

$$Z = \sum_{k=1}^{k_{\max}} \left\lfloor \frac{n}{5^k} \right\rfloor, \tag{2}$$

where

$$k_{\max} \equiv \left\lfloor \frac{\ln n}{\ln 5} \right\rfloor \tag{3}$$

and $\lfloor x \rfloor$ is the FLOOR FUNCTION (Gardner 1978, p. 63; Ogilvy and Anderson 1988, pp. 112–114). For $n = 1, 2, \ldots$, the number of trailing zeros are 0, 0, 0, 0, 1, 1, 1, 1, 1, 2, 2, 2, 2, 2, 3, 3, ... (Sloane's A027868). This is a special application of the general result that the POWER of a PRIME p dividing $n!$ is

$$\epsilon_p(n) = \sum_{r\geq 0} \left\lfloor \frac{n}{p^r} \right\rfloor \tag{4}$$

(Graham *et al.* 1994, Vardi 1991). Stated another way, the exact POWER of a PRIME p which divides $n!$ is

$$\frac{n - \text{sum of digits of the base-}p\text{ representation of }n}{p-1}. \tag{5}$$

By noting that

$$n! = \Gamma(n+1), \tag{6}$$

where $\Gamma(n)$ is the GAMMA FUNCTION for INTEGERS n, the definition can be generalized to COMPLEX values

$$z! \equiv \Gamma(z+1) \equiv \int_0^\infty e^{-t} t^z \, dt. \tag{7}$$

This defines $z!$ for all COMPLEX values of z, except when z is a NEGATIVE INTEGER, in which case $z! = \infty$. Using the identities for GAMMA FUNCTIONS, the values of $(\frac{1}{2}n)!$ (half integral values) can be written explicitly

$$(-\tfrac{1}{2})! = \sqrt{\pi} \tag{8}$$

$$(\tfrac{1}{2})! = \tfrac{1}{2}\sqrt{\pi} \tag{9}$$

$$(n-\tfrac{1}{2})! = \frac{\sqrt{\pi}}{2^n}(2n-1)!! \tag{10}$$

$$(n+\tfrac{1}{2})! = \frac{\sqrt{\pi}}{2^{n+1}}(2n+1)!!, \tag{11}$$

where $n!!$ is a DOUBLE FACTORIAL.

For INTEGERS s and n with $s < n$,

$$\frac{(s-n)!}{(2s-2n)!} = \frac{(-1)^{n-s}(2n-2s)!}{(n-s)!}. \quad (12)$$

The LOGARITHM of $z!$ is frequently encountered

$$\ln(z!) = \frac{1}{2}\ln\left[\frac{\pi z}{\sin(\pi z)}\right] - \gamma - \sum_{n=1}^{\infty}\frac{\zeta(2n+1)}{2n+1}z^{2n+1} \quad (13)$$

$$= \frac{1}{2}\ln\left[\frac{\pi z}{\sin(\pi z)}\right] - \frac{1}{2}\ln\left(\frac{1+z}{1-z}\right) + (1-\gamma)z - \sum_{n=1}^{\infty}[\zeta(2n+1)-1]\frac{z^{2n+1}}{2n+1} \quad (14)$$

$$= \ln\left[\lim_{n\to\infty}\frac{n!}{(z+1)(z+2)\cdots(z+n)}n^z\right] \quad (15)$$

$$= \lim_{n\to\infty}[\ln(n!) + z\ln n - \ln(z+1) - \ln(z+2) - \ldots - \ln(z+n)] \quad (16)$$

$$= \sum_{n=1}^{\infty}\frac{z^n}{n!}F_{n-1}(0) \quad (17)$$

$$= -\gamma z + \sum_{n=2}^{\infty}(-1)^n\frac{z^n}{n}\zeta(n) \quad (18)$$

$$= -\ln(1+z) + z(1-\gamma) + \sum_{n=2}^{\infty}(-1)^n[\zeta(n)-1]\frac{z^n}{n}, \quad (19)$$

where γ is the EULER-MASCHERONI CONSTANT, ζ is the RIEMANN ZETA FUNCTION, and F_n is the POLYGAMMA FUNCTION. The factorial can be expanded in a series

$$z! = \sqrt{2\pi}\,z^{z+1/2}e^{-z}(1 + \tfrac{1}{12}z^{-1} + \tfrac{1}{288}z^{-2} - \tfrac{139}{51840}z^{-3} + \ldots). \quad (20)$$

STIRLING'S SERIES gives the series expansion for $\ln(z!)$,

$$\ln(z!) = \tfrac{1}{2}\ln(2\pi) + \left(z+\tfrac{1}{2}\right)\ln z - z + \frac{B_2}{2z} + \ldots + \frac{B_{2n}}{2n(2n-1)z^{2n-1}} + \ldots$$

$$= \tfrac{1}{2}\ln(2\pi) + \left(z+\tfrac{1}{2}\right)\ln z - z + \tfrac{1}{12}z^{-1} - \tfrac{1}{360}z^{-3} + \tfrac{1}{1260}z^{-5} - \ldots, \quad (21)$$

where B_n is a BERNOULLI NUMBER.

Identities satisfied by sums of factorials include

$$\sum_{k=0}^{\infty}\frac{1}{k!} = e = 2.718281828\ldots \quad (22)$$

$$\sum_{k=0}^{\infty}\frac{(-1)^k}{k!} = e^{-1} = 0.3678794412\ldots \quad (23)$$

$$\sum_{k=0}^{\infty}\frac{1}{(k!)^2} = I_0(2) = 2.279585302\ldots \quad (24)$$

$$\sum_{k=0}^{\infty}\frac{(-1)^k}{(k!)^2} = J_0(2) = 0.2238907791\ldots \quad (25)$$

$$\sum_{k=0}^{\infty}\frac{1}{(2k)!} = \cosh 1 = 1.543080635\ldots \quad (26)$$

$$\sum_{k=0}^{\infty}\frac{(-1)^k}{(2k)!} = \cos 1 = 0.5403023059\ldots \quad (27)$$

$$\sum_{k=0}^{\infty}\frac{1}{(2k+1)!} = \sinh 1 = 1.175201194\ldots \quad (28)$$

$$\sum_{k=0}^{\infty}\frac{(-1)^k}{(2k+1)!} = \sin 1 = 0.8414709848\ldots \quad (29)$$

(Spanier and Oldham 1987), where I_0 is a MODIFIED BESSEL FUNCTION OF THE FIRST KIND, J_0 is a BESSEL FUNCTION OF THE FIRST KIND, cosh is the HYPERBOLIC COSINE, cos is the COSINE, sinh is the HYPERBOLIC SINE, and sin is the SINE.

Let h be the exponent of the greatest POWER of a PRIME p dividing $n!$. Then

$$h = \sum_{\substack{i=1\\ p^i\le n}}\left\lfloor\frac{n}{p^i}\right\rfloor. \quad (30)$$

Let g be the number of 1s in the BINARY representation of n. Then

$$g + h = n \quad (31)$$

(Honsberger 1976). In general, as discovered by Legendre in 1808, the POWER m of the PRIME p dividing $n!$ is given by

$$m = \sum_{k=0}^{\infty}\left\lfloor\frac{n}{p^k}\right\rfloor = \frac{n-(n_0+n_1+\ldots+n_N)}{p-1}, \quad (32)$$

where the INTEGERS $n_1, \ldots, n_N$ are the digits of n in base p (Ribenboim 1989).

The sum-of-factorials function is defined by

$$\Sigma(n) \equiv \sum_{k=1}^{n}k!$$

$$= \frac{-e + \mathrm{ei}(1) + \pi i + \mathrm{E}_{2n+1}(-1)\Gamma(n+2)}{e}, \quad (33)$$

$$= \frac{-e + \mathrm{ei}(1) + \Re[\mathrm{E}_{2n+1}(-1)]\Gamma(n+2)}{e}, \quad (34)$$

where $\mathrm{ei}(1) \approx 1.89512$ is the EXPONENTIAL INTEGRAL, E_n is the E_n-FUNCTION, and i is the IMAGINARY NUMBER. The first few values are 1, 3, 9, 33, 153, 873, 5913, 46233, 409113, ... (Sloane's A007489). $\Sigma(n)$ cannot be written as a hypergeometric term plus a constant (Petkovšek *et al.* 1996). However the sum

$$\Sigma'(n) \equiv \sum_{k=1}^{n} kk! = (n+1)! - 1 \tag{35}$$

has a simple form, with the first few values being 1, 5, 23, 119, 719, 5039, ... (Sloane's A033312).

The numbers $n!+1$ are prime for $n = 1$, 2, 3, 11, 27, 37, 41, 73, 77, 116, 154, ... (Sloane's A002981), and the numbers $n!-1$ are prime for $n = 3$, 4, 6, 7, 12, 14, 30, 32, 33, 38, 94, 166, ... (Sloane's A002982). In general, the power-product sequences (Mudge 1997) are given by $S_k^{\pm}(n) = (n!)^k \pm 1$. The first few terms of $S_2^+(n)$ are 2, 5, 37, 577, 14401, 518401, ... (Sloane's A020549), and $S_2^+(n)$ is PRIME for $n = 1$, 2, 3, 4, 5, 9, 10, 11, 13, 24, 65, 76, ... (Sloane's A046029). The first few terms of $S_2^-(n)$ are 0, 3, 35, 575, 14399, 518399, ... (Sloane's A046030), but $S_2^-(n)$ is PRIME for only $n = 2$ since $S_2^-(n) = (n!)^2 - 1 = (n!+1)(n!-1)$ for $n > 2$. The first few terms of $S_3^-(n)$ are 0, 7, 215, 13823, 1727999, ..., and the first few terms of $S_3^+(n)$ are 2, 9, 217, 13825, 1728001, ... (Sloane's A19514).

There are only four INTEGERS equal to the sum of the factorials of their digits. Such numbers are called FACTORIONS. While no factorial is a SQUARE NUMBER, D. Hoey listed sums $< 10^{12}$ of distinct factorials which give SQUARE NUMBERS, and J. McCranie gave the one additional sum less than $21! = 5.1 \times 10^{19}$:

$$\begin{aligned}
0!+1!+2! &= 2^2\\
1!+2!+3! &= 3^2\\
1!+4! &= 5^2\\
1!+5! &= 11^2\\
4!+5! &= 12^2\\
1!+2!+3!+6! &= 27^2\\
1!+5!+6! &= 29^2\\
1!+7! &= 71^2\\
4!+5!+7! &= 72^2\\
1!+2!+3!+7!+8! &= 213^2\\
1!+4!+5!+6!+7!+8! &= 215^2\\
1!+2!+3!+6!+9! &= 603^2\\
1!+4!+8!+9! &= 635^2\\
1!+2!+3!+6!+7!+8!+10! &= 1917^2
\end{aligned}$$

$$\begin{aligned}
&1!+2!+3!+7!+8!+9!+10!+11!+12!\\
&\qquad +13!+14!+15! = 1183893^2
\end{aligned}$$

(Sloane's A014597). The first few values for which the alternating SUM

$$\sum_{i=1}^{n} (-1)^{n-i} i! \tag{36}$$

is PRIME are 3, 4, 5, 6, 7, 8, 41, 59, 61, 105, 160, ... (Sloane's A014615, Guy 1994, p. 100). The only known factorials which are products of factorial in an ARITHMETIC SEQUENCE are

$$\begin{aligned}
0!1! &= 1!\\
1!2! &= 2!\\
0!1!2! &= 2!\\
6!7! &= 10!\\
1!3!5! &= 6!\\
1!3!5!7! &= 10!
\end{aligned}$$

(Madachy 1979).

There are no identities of the form

$$n! = a_1!a_2!\cdots a_r! \tag{37}$$

for $r \geq 2$ with $a_i \geq a_j \geq 2$ for $i < j$ for $n \leq 18160$ except

$$9! = 7!3!3!2! \tag{38}$$

$$10! = 7!6! = 7!5!3! \tag{39}$$

$$16! = 14!5!2! \tag{40}$$

(Guy 1994, p. 80).

There are three numbers less than 200,000 for which

$$(n-1)! + 1 \equiv 0 \pmod{n^2}, \tag{41}$$

namely 5, 13, and 563 (Le Lionnais 1983). BROWN NUMBERS are pairs (m, n) of INTEGERS satisfying the condition of BROCARD'S PROBLEM, i.e., such that

$$n! + 1 = m^2. \tag{42}$$

Only three such numbers are known: (5, 4), (11, 5), (71, 7). Erdős conjectured that these are the only three such pairs (Guy 1994, p. 193).

see also ALLADI-GRINSTEAD CONSTANT, BROCARD'S PROBLEM, BROWN NUMBERS, DOUBLE FACTORIAL, FACTORIAL PRIME, FACTORION, GAMMA FUNCTION, HYPERFACTORIAL, MULTIFACTORIAL, POCHHAMMER SYMBOL, PRIMORIAL, ROMAN FACTORIAL, STIRLING'S SERIES, SUBFACTORIAL, SUPERFACTORIAL

References

Conway, J. H. and Guy, R. K. "Factorial Numbers." In *The Book of Numbers*. New York: Springer-Verlag, pp. 65–66, 1996.

Dudeney, H. E. *Amusements in Mathematics*. New York: Dover, p. 96, 1970.

Gardner, M. "Factorial Oddities." Ch. 4 in *Mathematical Magic Show: More Puzzles, Games, Diversions, Illusions and Other Mathematical Sleight-of-Mind from Scientific American.* New York: Vintage, pp. 50–65, 1978.
Graham, R. L.; Knuth, D. E.; and Patashnik, O. "Factorial Factors." §4.4 in *Concrete Mathematics: A Foundation for Computer Science.* Reading, MA: Addison-Wesley, pp. 111—115, 1990.
Guy, R. K. "Equal Products of Factorials," "Alternating Sums of Factorials," and "Equations Involving Factorial *n*." §B23, B43, and D25 in *Unsolved Problems in Number Theory, 2nd ed.* New York: Springer-Verlag, pp. 80, 100, and 193–194, 1994.
Honsberger, R. *Mathematical Gems II.* Washington, DC: Math. Assoc. Amer., p. 2, 1976.
Le Lionnais, F. *Les nombres remarquables.* Paris: Hermann, p. 56, 1983.
Leyland, P. ftp://sable.ox.ac.uk/pub/math/factors/factorial-.Z and ftp://sable.ox.ac.uk/pub/math/factors/factorial+.Z.
Madachy, J. S. *Madachy's Mathematical Recreations.* New York: Dover, p. 174, 1979.
Mudge, M. "Not Numerology but Numeralogy!" *Personal Computer World,* 279–280, 1997.
Ogilvy, C. S. and Anderson, J. T. *Excursions in Number Theory.* New York: Dover, 1988.
Petkovšek, M.; Wilf, H. S.; and Zeilberger, D. *A=B.* Wellesley, MA: A. K. Peters, p. 86, 1996.
Press, W. H.; Flannery, B. P.; Teukolsky, S. A.; and Vetterling, W. T. "Gamma Function, Beta Function, Factorials, Binomial Coefficients." §6.1 in *Numerical Recipes in FORTRAN: The Art of Scientific Computing, 2nd ed.* Cambridge, England: Cambridge University Press, pp. 206–209, 1992.
Ribenboim, P. *The Book of Prime Number Records, 2nd ed.* New York: Springer-Verlag, pp. 22–24, 1989.
Sloane, N. J. A. Sequences A014615, A014597, A033312, A020549, A000142/M1675, and A007489/M2818 in "An On-Line Version of the Encyclopedia of Integer Sequences."
Spanier, J. and Oldham, K. B. "The Factorial Function *n*! and Its Reciprocal." Ch. 2 in *An Atlas of Functions.* Washington, DC: Hemisphere, pp. 19–33, 1987.
Vardi, I. *Computational Recreations in Mathematica.* Reading, MA: Addison-Wesley, p. 67, 1991.

Factorial Moment

$$\nu_{(r)} \equiv \sum_x x^{(r)} f(x),$$

where

$$x^{(r)} \equiv x(x-1)\cdots(x-r+1).$$

Factorial Number

see FACTORIAL

Factorial Prime

A PRIME of the form $n! \pm 1$. $n!+1$ is PRIME for 1, 2, 3, 11, 27, 37, 41, 73, 77, 116, 154, 320, 340, 399, 427, 872, 1477, ... (Sloane's A002981) up to a search limit 4850. $n!-1$ is PRIME for 3, 4, 6, 7, 12, 14, 30, 32, 33, 38, 94, 116, 324, 379, 469, 546, 974, 1963, 3507, 3610, ... (Sloane's A002982) up to a search limit of 4850.

References
Borning, A. "Some Results for $k!+1$ and $2\cdot 3\cdot 5\cdot p+1$." *Math. Comput.* **26**, 567–570, 1972.
Buhler, J. P.; Crandall, R. E.; and Penk, M. A. "Primes of the Form $M!+1$ and $2\cdot3\cdot5\cdots p+1$." *Math. Comput.* **38**, 639–643, 1982.
Caldwell, C. K. "On the Primality of $N!\pm1$ and $2\cdot3\cdot5\cdots p\pm 1$." *Math. Comput.* **64**, 889–890, 1995.
Dubner, H. "Factorial and Primorial Primes." *J. Rec. Math.* **19**, 197–203, 1987.
Guy, R. K. *Unsolved Problems in Number Theory, 2nd ed.* New York: Springer-Verlag, p. 7, 1994.
Sloane, N. J. A. Sequences A002981/M0908 and A002982/M2321 in "An On-Line Version of the Encyclopedia of Integer Sequences."
Temper, M. "On the Primality of $k!+1$ and $\cdot3\cdot5\cdots p+1$." *Math. Comput.* **34**, 303–304, 1980.

Factorial Sum

Sums with unity NUMERATOR and FACTORIALS in the DENOMINATOR which can be expressed analytically include

$$\sum_{i=1}^n \frac{1}{(n+i-k)!(n-i)!} = \frac{{}_2F_1(1,-n;1+n-k;-1)-1}{\Gamma(1+n)\Gamma(1+n-k)} \quad (1)$$

$$\sum_{i=1}^n \frac{1}{(n+i-1)!(n-i)!} = \frac{n\sqrt{\pi}}{2\Gamma(\frac{1}{2}+n)\Gamma(1+n)} \quad (2)$$

$$\sum_{i=1}^n \frac{1}{(n+i)!(n-i)!} = \frac{\sqrt{\pi}}{2\Gamma(\frac{1}{2}+n)\Gamma(1+n)} - \frac{1}{2\Gamma^2(1+n)} \quad (3)$$

$$\sum_{i=1}^n \frac{1}{(n+i+1)!(n-i)!} = \frac{\sqrt{\pi}}{2\Gamma(\frac{3}{2}+n)\Gamma(1+n)} - \frac{1}{2\Gamma(1+n)\Gamma(2+n)}, \quad (4)$$

where ${}_2F_1(a,b;c;z)$ is a HYPERGEOMETRIC FUNCTION and $\Gamma(z)$ is a GAMMA FUNCTION.

Sums with i in the NUMERATOR having analytic solutions include

$$\sum_{i=1}^n \frac{i}{(n+i-k)!(n-i)!} = \frac{n\,{}_2F_1(2,1-n;2-k+n;-1)}{(1-k+n)\Gamma(1+n)\Gamma(1-k+n)} \quad (5)$$

$$\sum_{i=1}^n \frac{i}{(n+i-1)!(n-i)!} = \frac{1}{2\Gamma(n)}\left[\frac{\sqrt{\pi}}{2\Gamma(\frac{1}{2}+n)} + \frac{n}{\Gamma(1+n)}\right] \quad (6)$$

$$\sum_{i=1}^n \frac{i}{(n+i)!(n-i)!} = \frac{n}{2\Gamma^2(1+n)} \quad (7)$$

$$\sum_{i=1}^{n} \frac{i}{(n+i+1)!(n-i)!} = \frac{1}{2\Gamma(1+n)}\left[\frac{1}{\Gamma(2+n)} - \frac{(n^2+3n+2)\sqrt{\pi}}{2\Gamma(\frac{3}{2}+n)}\right]. \quad (8)$$

A sum with i^2 in the NUMERATOR is

$$\sum_{i=1}^{n} \frac{i^2}{(n+i-k)!(n-i)!} = \frac{n}{(1-k+n)(2-k+n)\Gamma(1+n)\Gamma(1-k+n)} \times[(2-k+n)\,{}_2F_1(2,1-n;2-k+n;-1) + 2(n-1)\,{}_2F_1(3,2-n;3-k+n;-1)], \quad (9)$$

where ${}_2F_1(a,b;c;z)$ is the HYPERGEOMETRIC FUNCTION.

Sums of factorial POWERS include

$$\sum_{n=0}^{\infty} \frac{(n!)^2}{(2n)!} = \frac{4}{3} + \frac{2\pi}{9\sqrt{3}} \quad (10)$$

$$\sum_{n=0}^{\infty} \frac{(n!)^3}{(3n)!} = \int_0^1 [P(t) + Q(t)\cos^{-1} R(t)]\,dt, \quad (11)$$

where

$$P(t) = \frac{2(8+7t^2-7t^3)}{(4-t^2+t^3)^2} \quad (12)$$

$$Q(t) = \frac{4t(1-t)(5+t^2-t^3)}{(4-t^2+t^3)^2\sqrt{(1-t)(4-t^2+t^3)}} \quad (13)$$

$$R(t) = 1 - \tfrac{1}{2}(t^2-t^3) \quad (14)$$

(Beeler *et al.* 1972, Item 116).

References

Beeler, M.; Gosper, R. W.; and Schroeppel, R. *HAKMEM.* Cambridge, MA: MIT Artificial Intelligence Laboratory, Memo AIM-239, Feb. 1972.

Factoring

see FACTORIZATION

Factorion

A factorion is an INTEGER which is equal to the sum of FACTORIALS of its digits. There are exactly four such numbers:

$$1 = 1! \quad (1)$$

$$2 = 2! \quad (2)$$

$$145 = 1! + 4! + 5! \quad (3)$$

$$40,585 = 4! + 0! + 5! + 8! + 5! \quad (4)$$

(Gardner 1978, Madachy 1979, Pickover 1995). The factorion of an n-digit number cannot exceed $n \cdot 9!$ digits.

see also FACTORIAL

References

Gardner, M. "Factorial Oddities." Ch. 4 in *Mathematical Magic Show: More Puzzles, Games, Diversions, Illusions and Other Mathematical Sleight-of-Mind from Scientific American.* New York: Vintage, pp. 61 and 64, 1978.

Madachy, J. S. *Madachy's Mathematical Recreations.* New York: Dover, p. 167, 1979.

Pickover, C. A. "The Loneliness of the Factorions." Ch. 22 in *Keys to Infinity.* New York: W. H. Freeman, pp. 169–171 and 319–320, 1995.

Factorization

The finding of FACTORS (DIVISORS) of a given INTEGER, POLYNOMIAL, etc. Factorization is also called FACTORING.

see also FACTOR, PRIME FACTORIZATION ALGORITHMS

Fagnano's Point

The point of coincidence of P and P' in FAGNANO'S PROBLEM.

Fagnano's Problem

In a given ACUTE-angled TRIANGLE ΔABC, INSCRIBE another TRIANGLE whose PERIMETER is as small as possible. The answer is the PEDAL TRIANGLE of ΔABC.

References

Coxeter, H. S. M. and Greitzer, S. L. *Geometry Revisited.* Washington, DC: Math. Assoc. Amer., pp. 88–89, 1967.

Fagnano's Theorem

If $P(x,y)$ and $P(x',y')$ are two points on an ELLIPSE

$$\frac{x^2}{a^2} + \frac{y^2}{b^2} = 1, \quad (1)$$

with ECCENTRIC ANGLES ϕ and ϕ' such that

$$\tan\phi\tan\phi' = \frac{b}{a} \quad (2)$$

and $A = P(a,0)$ and $B = P(0,b)$. Then

$$\operatorname{arc} BP + \operatorname{arc} BP' = \frac{e^2 xx'}{a}. \quad (3)$$

This follows from the identity

$$E(u,k) + E(v,k) - E(k) = k^2 \operatorname{sn}(u,k)\operatorname{sn}(v,k), \quad (4)$$

where $E(u,k)$ is an incomplete ELLIPTIC INTEGRAL OF THE SECOND KIND, $E(k)$ is a complete ELLIPTIC INTEGRAL OF THE SECOND KIND, and $\operatorname{sn}(v,k)$ is a JACOBI ELLIPTIC FUNCTION. If P and P' coincide, the point where they coincide is called FAGNANO'S POINT.

Fair Game

A GAME which is not biased toward any player.

see also GAME, MARTINGALE

Fairy Chess

A variation of CHESS involving a change in the form of the board, the rules of play, or the pieces used. For example, the normal rules of chess can be used but with a cylindrical or MÖBIUS STRIP connection of the edges.

see also CHESS

References

Kraitchik, M. "Fairy Chess." §12.2 in *Mathematical Recreations.* New York: W. W. Norton, pp. 276–279, 1942.

Fallacy

A fallacy is an incorrect result arrived at by apparently correct, though actually specious reasoning. The most common example of a mathematical fallacy is the "proof" that $1 = 2$ as follows. Let $a = b$, then

$$ab = a^2$$

$$ab - b^2 = a^2 - b^2$$

$$b(a-b) = (a+b)(a-b)$$

$$b = a+b$$

$$b = 2b$$

$$1 = 2.$$

The incorrect step is division by $a-b$ (equal to 0), which is invalid. Ball and Coxeter (1987) give other such examples in the areas of both arithmetic and geometry.

References

Ball, W. W. R. and Coxeter, H. S. M. *Mathematical Recreations and Essays, 13th ed.* New York: Dover, pp. 41–45 and 76–84, 1987.

Pappas, T. "Geometric Fallacy & the Fibonacci Sequence." *The Joy of Mathematics.* San Carlos, CA: Wide World Publ./Tetra, p. 191, 1989.

False

A statement which is rigorously not TRUE. Regular two-valued LOGIC allows statements to be only TRUE or false, but FUZZY LOGIC treats "truth" as a continuum which can have a value between 0 and 1.

see also ALETHIC, FUZZY LOGIC, LOGIC, TRUE, TRUTH TABLE, UNDECIDABLE

False Position Method

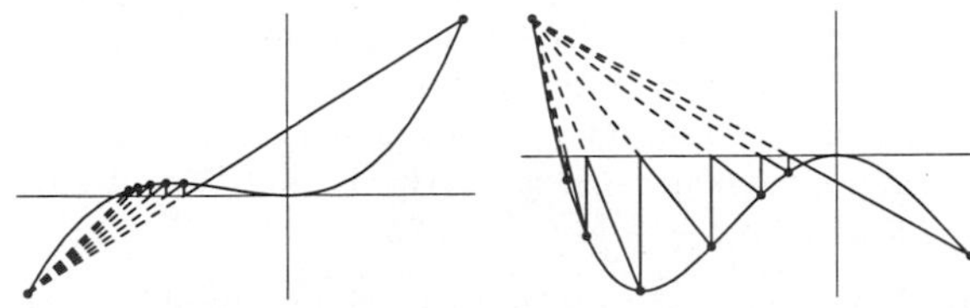

An ALGORITHM for finding ROOTS which uses the point where the linear approximation crosses the axis as the next iteration and keeps the same initial point for each iteration. Using the two-point form of the line

$$y - y_1 = \frac{f(x_{n-1}) - f(x_1)}{x_{n-1} - x_1}(x_n - x_1)$$

with $y = 0$, using $y_1 = f(x_1)$, and solving for x_n therefore gives the iteration

$$x_n = x_1 - \frac{x_{n-1} - x_1}{f(x_{n-1}) - f(x_1)} f(x_1).$$

see also BRENT'S METHOD, RIDDERS' METHOD, SECANT METHOD

References

Abramowitz, M. and Stegun, C. A. (Eds.). *Handbook of Mathematical Functions with Formulas, Graphs, and Mathematical Tables, 9th printing.* New York: Dover, p. 18, 1972.

Press, W. H.; Flannery, B. P.; Teukolsky, S. A.; and Vetterling, W. T. "Secant Method, False Position Method, and Ridders' Method." §9.2 in *Numerical Recipes in FORTRAN: The Art of Scientific Computing, 2nd ed.* Cambridge, England: Cambridge University Press, pp. 347–352, 1992.

Faltung (Form)

Let A and B be bilinear forms

$$A = A(x,y) = \sum\sum a_{ij}x_iy_i$$

$$B = B(x,y) = \sum\sum b_{ij}x_iy_i$$

and suppose that A and B are bounded in $[p,p']$ with bounds M and N. Then

$$F = F(A,B) = \sum\sum f_{ij}x_iy_j,$$

where the series

$$f_{ij} = \sum_k a_{ik}b_{kj}$$

is absolutely convergent, is called the faltung of A and B. F is bounded in $[p,p']$, and its bound does not exceed MN.

References

Hardy, G. H.; Littlewood, J. E.; and Pólya, G. *Inequalities, 2nd ed.* Cambridge, England: Cambridge University Press, pp. 210–211, 1988.

Faltung (Function)

see CONVOLUTION

Fan

A SPREAD in which each node has a FINITE number of children.

see also SPREAD (TREE)

Fano's Axiom

The three diagonal points of a COMPLETE QUADRILATERAL are never COLLINEAR.

Fano Plane

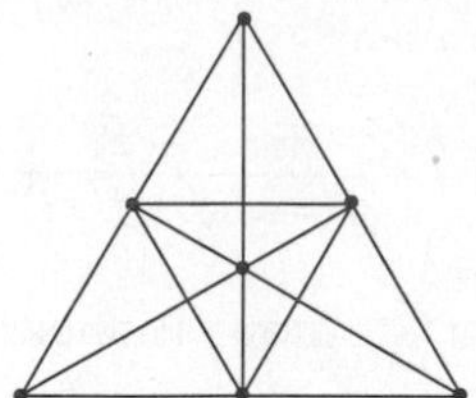

The 2-D PROJECTIVE PLANE over $GF(2)$ ("of order two"), illustrated above. It is a BLOCK DESIGN with $\nu = 7$, $k = 3$, $\lambda = 1$, $r = 3$, and $b = 7$, and is also the STEINER TRIPLE SYSTEM $S(7)$.

The Fano plane also solves the TRANSYLVANIA LOTTERY, which picks three numbers from the INTEGERS 1–14. Using two Fano planes we can guarantee matching two by playing just 14 times as follows. Label the VERTICES of one Fano plane by the INTEGERS 1–7, the other plane by the INTEGERS 8–14. The 14 tickets to play are the 14 lines of the two planes. Then if (a, b, c) is the winning ticket, at least two of a, b, c are either in the interval [1, 7] or [8, 14]. These two numbers are on exactly one line of the corresponding plane, so one of our tickets matches them.

The Lehmers (1974) found an application of the Fano plane for factoring INTEGERS via QUADRATIC FORMS. Here, the triples of forms used form the lines of the PROJECTIVE GEOMETRY on seven points, whose planes are Fano configurations corresponding to pairs of residue classes mod 24 (Lehmer and Lehmer 1974, Guy 1975, Shanks 1985). The group of AUTOMORPHISMS (incidence-preserving BIJECTIONS) of the Fano plane is the SIMPLE GROUP of ORDER 168 (Klein 1870).

see also DESIGN, PROJECTIVE PLANE, STEINER TRIPLE SYSTEM, TRANSYLVANIA LOTTERY

References

Guy, R. "How to Factor a Number." *Proc. Fifth Manitoba Conf. on Numerical Math.*, 49–89, 1975.

Lehmer, D. H. and Lehmer, E. "A New Factorization Technique Using Quadratic Forms." *Math. Comput.* **28**, 625–635, 1974.

Shanks, D. *Solved and Unsolved Problems in Number Theory, 3rd ed.* New York: Chelsea, pp. 202 and 238, 1985.

Far Out

A word used by Tukey to describe data points which are outside the outer FENCES.

References

Tukey, J. W. *Explanatory Data Analysis.* Reading, MA: Addison-Wesley, p. 44, 1977.

Far-Out Point

For a TRIANGLE with side lengths a, b, and c, the far-out point has TRIANGLE CENTER FUNCTION

$$\alpha = a(b^4 + c^4 - a^4 - b^2c^2).$$

As $a : b : c$ approaches $1 : 1 : 1$, this point moves out along the EULER LINE to infinity.

References

Kimberling, C. "Central Points and Central Lines in the Plane of a Triangle." *Math. Mag.* **67**, 163–187, 1994.

Kimberling, C.; Lyness, R. C.; and Veldkamp, G. R. "Problem 1195 and Solution." *Crux Math.* **14**, 177–179, 1988.

Farey Sequence

The Farey sequence F_n for any POSITIVE INTEGER n is the set of irreducible RATIONAL NUMBERS a/b with $0 \leq a \leq b \leq n$ and $(a, b) = 1$ arranged in increasing order.

$$F_1 = \{\tfrac{0}{1}, \tfrac{1}{1}\} \tag{1}$$

$$F_2 = \{\tfrac{0}{1}, \tfrac{1}{2}, \tfrac{1}{1}\} \tag{2}$$

$$F_3 = \{\tfrac{0}{1}, \tfrac{1}{3}, \tfrac{1}{2}, \tfrac{2}{3}, \tfrac{1}{1}\} \tag{3}$$

$$F_4 = \{\tfrac{0}{1}, \tfrac{1}{4}, \tfrac{1}{3}, \tfrac{1}{2}, \tfrac{2}{3}, \tfrac{3}{4}, \tfrac{1}{1}\} \tag{4}$$

$$F_5 = \{\tfrac{0}{1}, \tfrac{1}{5}, \tfrac{1}{4}, \tfrac{1}{3}, \tfrac{2}{5}, \tfrac{1}{2}, \tfrac{3}{5}, \tfrac{2}{3}, \tfrac{3}{4}, \tfrac{4}{5}, \tfrac{1}{1}\}. \tag{5}$$

There is always an ODD number of terms, and the middle term is always 1/2. Let p/q, p'/q', and p''/q'' be three successive terms in a Farey series. Then

$$qp' - pq' = 1 \tag{6}$$

$$\frac{p'}{q'} = \frac{p + p''}{q + q''}. \tag{7}$$

These two statements are actually equivalent.

The number of terms $N(n)$ in the Farey sequence for the INTEGER n is

$$N(n) = 1 + \sum_{k=1}^{n} \phi(k) = 1 + \Phi(n), \tag{8}$$

where $\phi(k)$ is the TOTIENT FUNCTION and $\Phi(n)$ is the SUMMATORY FUNCTION of $\phi(k)$, giving 2, 3, 5, 7, 11, 13, 19, ... (Sloane's A005728). The asymptotic limit for the function $N(n)$ is

$$N(n) \sim \frac{3n^2}{\pi^2} = 0.3039635509n^2 \tag{9}$$

(Vardi 1991, p. 155). For a method of computing a successive sequence from an existing one of n terms, insert the MEDIANT fraction $(a + b)/(c + d)$ between terms a/c and b/d when $c + d \leq n$ (Hardy and Wright 1979, pp. 25–26; Conway and Guy 1996).

FORD CIRCLES provide a method of visualizing the Farey sequence. The Farey sequence F_n defines a subtree of the STERN-BROCOT TREE obtained by pruning unwanted branches (Graham *et al.* 1994).

see also FORD CIRCLE, MEDIANT, RANK (SEQUENCE), STERN-BROCOT TREE

References

Beiler, A. H. "Farey Tails." Ch. 16 in *Recreations in the Theory of Numbers: The Queen of Mathematics Entertains.* New York: Dover, 1966.

Conway, J. H. and Guy, R. K. "Farey Fractions and Ford Circles." *The Book of Numbers.* New York: Springer-Verlag, pp. 152–154 and 156, 1996.

Dickson, L. E. *History of the Theory of Numbers, Vol. 1: Divisibility and Primality.* New York: Chelsea, pp. 155–158, 1952.

Farey, J. "On a Curious Property of Vulgar Fractions." *London, Edinburgh and Dublin Phil. Mag.* **47**, 385, 1816.

Graham, R. L.; Knuth, D. E.; and Patashnik, O. *Concrete Mathematics: A Foundation for Computer Science, 2nd ed.* Reading, MA: Addison-Wesley, pp. 118–119, 1994.

Guy, R. K. "Mahler's Generalization of Farey Series." §F27 in *Unsolved Problems in Number Theory, 2nd ed.* New York: Springer-Verlag, pp. 263–265, 1994.

Hardy, G. H. and Wright, E. M. *An Introduction to the Theory of Numbers, 5th ed.* Oxford, England: Clarendon Press, 1979.

Sloane, N. J. A. Sequences A005728/M0661, A006842/M0041, and A006843/M0081 in "An On-Line Version of the Encyclopedia of Integer Sequences."

Sylvester, J. J. "On the Number of Fractions Contained in Any Farey Series of Which the Limiting Number is Given." *London, Edinburgh and Dublin Phil. Mag. (5th Series)* **15**, 251, 1883.

Vardi, I. *Computational Recreations in Mathematica.* Reading, MA: Addison-Wesley, p. 155, 1991.

Weisstein, E. W. "Plane Geometry." `http://www.astro.virginia.edu/~eww6n/math/notebooks/PlaneGeometry.m`.

Farey Series

see FAREY SEQUENCE

Farkas's Lemma

The INEQUALITY $\langle f_0, x\rangle \leq 0$ follows from

$$\langle f_1, x\rangle \leq 0, \ldots, \langle f_n, x\rangle \leq 0$$

IFF there exist NONNEGATIVE numbers $\lambda_1, \ldots, \lambda_n$ with

$$\sum_{k=1}^{n} \lambda_k f_k = f_0.$$

This LEMMA is used in the proof of the KUHN-TUCKER THEOREM.

see also KUHN-TUCKER THEOREM, LAGRANGE MULTIPLIER

Faro Shuffle

see RIFFLE SHUFFLE

Fast Fibonacci Transform

For a general second-order recurrence equation

$$f_{n+1} = xf_n + yf_{n-1}, \tag{1}$$

define a multiplication rule on ordered pairs by

$$(A, B)(C, D) = (AD + BC + xAC, BD + yAC). \tag{2}$$

The inverse is then given by

$$(A, B)^{-1} = \frac{(-A, xA + B)}{B^2 + xAB - yA^2}, \tag{3}$$

and we have the identity

$$(f_1, yf_0)(1, 0)^n = (f_{n+1}, yf_n) \tag{4}$$

(Beeler *et al.* 1972, Item 12).

References

Beeler, M.; Gosper, R. W.; and Schroeppel, R. *HAKMEM.* Cambridge, MA: MIT Artificial Intelligence Laboratory, Memo AIM-239, Feb. 1972.

Fast Fourier Transform

The fast Fourier transform (FFT) is a DISCRETE FOURIER TRANSFORM ALGORITHM which reduces the number of computations needed for N points from $2N^2$ to $2N \lg N$, where LG is the base-2 LOGARITHM. If the function to be transformed is not harmonically related to the sampling frequency, the response of an FFT looks like a SINC FUNCTION (although the integrated POWER is still correct). ALIASING (LEAKAGE) can be reduced by APODIZATION using a TAPERING FUNCTION. However, ALIASING reduction is at the expense of broadening the spectral response.

FFTs were first discussed by Cooley and Tukey (1965), although Gauss had actually described the critical factorization step as early as 1805 (Gergkand 1969, Strang 1993). A DISCRETE FOURIER TRANSFORM can be computed using an FFT by means of the DANIELSON-LANCZOS LEMMA if the number of points N is a POWER of two. If the number of points N is not a POWER of two, a transform can be performed on sets of points corresponding to the prime factors of N which is slightly degraded in speed. An efficient real Fourier transform algorithm or a fast HARTLEY TRANSFORM (Bracewell 1965) gives a further increase in speed by approximately a factor of two. Base-4 and base-8 fast Fourier transforms use optimized code, and can be 20–30% faster than base-2 fast Fourier transforms. PRIME factorization is slow when the factors are large, but discrete Fourier transforms can be made fast for $N = 2$, 3, 4, 5, 7, 8, 11, 13, and 16 using the WINOGRAD TRANSFORM ALGORITHM (Press *et al.* 1992, pp. 412–413, Arndt).

Fast Fourier transform algorithms generally fall into two classes: decimation in time, and decimation in frequency. The Cooley-Tukey FFT ALGORITHM first rearranges the input elements in bit-reversed order, then builds the output transform (decimation in time). The basic idea is to break up a transform of length N into two transforms of length $N/2$ using the identity

$$\sum_{n=0}^{N-1} a_n e^{-2\pi ink/N} = \sum_{n=0}^{N/2-1} a_{2n} e^{-2\pi i(2n)k/N} + \sum_{n=0}^{N/2-1} a_{2n+1} e^{-2\pi i(2n+1)k/N}$$

$$= \sum_{n=0}^{N/2-1} a_n^{\text{even}} e^{-2\pi ink/(N/2)} + e^{-2\pi ik/N} \sum_{n=0}^{N/2-1} a_n^{\text{odd}} e^{-2\pi ink/(N/2)},$$

sometimes called the DANIELSON-LANCZOS LEMMA. The easiest way to visualize this procedure is perhaps via the FOURIER MATRIX.

The Sande-Tukey ALGORITHM (Stoer and Burlisch 1980) first transforms, then rearranges the output values (decimation in frequency).

see also DANIELSON-LANCZOS LEMMA, DISCRETE FOURIER TRANSFORM, FOURIER MATRIX, FOURIER TRANSFORM, HARTLEY TRANSFORM, NUMBER THEORETIC TRANSFORM, WINOGRAD TRANSFORM

References

Arndt, J. "FFT Code and Related Stuff." `http://www.jjj.de/fxt/`.

Bell Laboratories. "Netlib FFTPack." `http://netlib.bell-labs.com/netlib/fftpack/`.

Blahut, R. E. *Fast Algorithms for Digital Signal Processing.* New York: Addison-Wesley, 1984.

Bracewell, R. *The Fourier Transform and Its Applications.* New York: McGraw-Hill, 1965.

Brigham, E. O. *The Fast Fourier Transform and Applications.* Englewood Cliffs, NJ: Prentice Hall, 1988.

Cooley, J. W. and Tukey, O. W. "An Algorithm for the Machine Calculation of Complex Fourier Series." *Math. Comput.* **19**, 297–301, 1965.

Duhamel, P. and Vetterli, M. "Fast Fourier Transforms: A Tutorial Review." *Signal Processing* **19**, 259–299, 1990.

Gergkand, G. D. "A Guided Tour of the Fast Fourier Transform." *IEEE Spectrum,* pp. 41–52, July 1969.

Lipson, J. D. *Elements of Algebra and Algebraic Computing.* Reading, MA: Addison-Wesley, 1981.

Nussbaumer, H. J. *Fast Fourier Transform and Convolution Algorithms, 2nd ed.* New York: Springer-Verlag, 1982.

Papoulis, A. *The Fourier Integral and its Applications.* New York: McGraw-Hill, 1962.

Press, W. H.; Flannery, B. P.; Teukolsky, S. A.; and Vetterling, W. T. "Fast Fourier Transform." Ch. 12 in *Numerical Recipes in FORTRAN: The Art of Scientific Computing, 2nd ed.* Cambridge, England: Cambridge University Press, pp. 490–529, 1992.

Stoer, J. and Burlisch, R. *Introduction to Numerical Analysis.* New York: Springer-Verlag, 1980.

Strang, G. "Wavelet Transforms Versus Fourier Transforms." *Bull. Amer. Math. Soc.* **28**, 288–305, 1993.

Van Loan, C. *Computational Frameworks for the Fast Fourier Transform.* Philadelphia, PA: SIAM, 1992.

Walker, J. S. *Fast Fourier Transform, 2nd ed.* Boca Raton, FL: CRC Press, 1996.

Fat Fractal

A CANTOR SET with LEBESGUE MEASURE greater than 0.

see also CANTOR SET, EXTERIOR DERIVATIVE, FRACTAL, LEBESGUE MEASURE

References

Ott, E. "Fat Fractals." §3.9 in *Chaos in Dynamical Systems.* New York: Cambridge University Press, pp. 97–100, 1993.

Fatou Dust

see FATOU SET

Fatou's Lemma

If a SEQUENCE $\{f_n\}$ of NONNEGATIVE measurable functions is defined on a measurable set E, then

$$\int_E \liminf_{n\to\infty} f_n\,d\mu \leq \liminf_{n\to\infty} \int_E f_n\,d\mu.$$

References

Zeidler, E. *Applied Functional Analysis: Applications to Mathematical Physics.* New York: Springer-Verlag, 1995.

Fatou Set

A set consisting of the complementary set of complex numbers to a JULIA SET.

see also JULIA SET

References

Schroeder, M. *Fractals, Chaos, Power Laws.* New York: W. H. Freeman, p. 39, 1991.

Fatou's Theorems

Let $f(\theta)$ be LEBESGUE INTEGRABLE and let

$$f(r,\theta) = \frac{1}{2\pi}\int_{-\pi}^{\pi} f(t)\frac{1-r^2}{1-2r\cos(t-\theta)+r^2}\,dt \tag{1}$$

be the corresponding POISSON INTEGRAL. Then ALMOST EVERYWHERE in $-\pi \leq \theta \leq \pi$,

$$\lim_{r\to 0^-} f(r,\theta) = f(\theta). \tag{2}$$

Let

$$F(z) = c_0 + c_1 z + c_2 z^2 + \ldots + c_n z^n + \ldots \tag{3}$$

be regular for $|z| < 1$, and let the integral

$$\frac{1}{2\pi}\int_{-\pi}^{\pi} |F(re^{i\theta})|^2\,d\theta \tag{4}$$

be bounded for $r < 1$. This condition is equivalent to the convergence of

$$|c_0|^2 + |c_1|^2 + \ldots + |c_n|^2 + \ldots. \tag{5}$$

Then almost everywhere in $-\pi \leq \theta \leq \pi$,

$$\lim_{r\to 0^-} F(re^{i\theta}) = F(e^{i\theta}). \qquad (6)$$

Furthermore, $F(e^{i\theta})$ is measurable, $|F(e^{i\theta})|^2$ is LEBESGUE INTEGRABLE, and the FOURIER SERIES of $F(e^{i\theta})$ is given by writing $z = e^{i\theta}$.

References
Szegő, G. *Orthogonal Polynomials, 4th ed.* Providence, RI: Amer. Math. Soc., p. 274, 1975.

Faulhaber's Formula

In a 1631 edition of *Academiae Algebrae*, J. Faulhaber published the general formula for the SUM of pth POWERS of the first n POSITIVE INTEGERS,

$$\sum_{k=1}^n k^p = \frac{1}{p+1}\sum_{i=1}^{p+1}(-1)^{\delta_{ip}}\binom{p+1}{i}B_{p+1-i}n^i, \qquad (1)$$

where δ_{ip} is the KRONECKER DELTA, $\binom{n}{i}$ is a BINOMIAL COEFFICIENT, and B_i is the ith BERNOULLI NUMBER. Computing the sums for $p = 1, \ldots, 10$ gives

$$\sum_{k=1}^n k = \tfrac{1}{2}(n^2+n) \qquad (2)$$

$$\sum_{k=1}^n k^2 = \tfrac{1}{6}(2n^3+3n^2+n) \qquad (3)$$

$$\sum_{k=1}^n k^3 = \tfrac{1}{4}(n^4+2n^3+n^2) \qquad (4)$$

$$\sum_{k=1}^n k^4 = \tfrac{1}{30}(6n^5+15n^4+10n^3-n) \qquad (5)$$

$$\sum_{k=1}^n k^5 = \tfrac{1}{12}(2n^6+6n^5+5n^4-n^2) \qquad (6)$$

$$\sum_{k=1}^n k^6 = \tfrac{1}{42}(6n^7+21n^6+21n^5-7n^3+n) \qquad (7)$$

$$\sum_{k=1}^n k^7 = \tfrac{1}{24}(3n^8+12n^7+14n^6-7n^4+2n^2) \qquad (8)$$

$$\sum_{k=1}^n k^8 = \tfrac{1}{90}(10n^9+45n^8+60n^7-42n^5 + 20n^3-3n) \qquad (9)$$

$$\sum_{k=1}^n k^9 = \tfrac{1}{20}(2n^{10}+10n^9+15n^8-14n^6 + 10n^4-3n^2) \qquad (10)$$

$$\sum_{k=1}^n k^{10} = \tfrac{1}{66}(6n^{11}+33n^{10}+55n^9-66n^7 + 66n^5-33n^3+5n). \qquad (11)$$

see also POWER, SUM

References
Conway, J. H. and Guy, R. K. *The Book of Numbers.* New York: Springer-Verlag, p. 106, 1996.

Favard Constants

N.B. A detailed on-line essay by S. Finch was the starting point for this entry.

Let $T_n(x)$ be an arbitrary trigonometric POLYNOMIAL

$$T_n(x) = \tfrac{1}{2}a_0 + \left\{\sum_{k=1}^n [a_k\cos(kx) + b_k\sin(kx)]\right\},$$

where the COEFFICIENTS are real. Let the rth derivative of $T_n(x)$ be bounded in $[-1,1]$, then there exists a POLYNOMIAL $T_n(x)$ for which

$$|f(x) - T_n(x)| \leq \frac{K_r}{(n+1)^r},$$

for all x, where K_r is the rth Favard constant, which is the smallest constant possible.

$$K_r = \frac{4}{\pi}\sum_{k=0}^\infty \left[\frac{(-1)^k}{2k+1}\right]^{r+1}.$$

These can be expressed by

$$K_r = \begin{cases} \frac{4}{\pi}\lambda(r+1) & \text{for } r \text{ odd} \\ \frac{4}{\pi}\beta(r+1) & \text{for } r \text{ even,} \end{cases}$$

where λ is the DIRICHLET LAMBDA FUNCTION and β is the DIRICHLET BETA FUNCTION. Explicitly,

$$K_0 = 1$$
$$K_1 = \tfrac{1}{2}\pi$$
$$K_2 = \tfrac{1}{8}\pi^2$$
$$K_3 = \tfrac{1}{24}\pi^3.$$

References
Finch, S. "Favorite Mathematical Constants." `http://www.mathsoft.com/asolve/constant/favard/favard.html`.
Kolmogorov, A. N. "Zur Grössenordnung des Restgliedes Fourierscher reihen differenzierbarer Funktionen." *Ann. Math.* **36**, 521–526, 1935.
Zygmund, A. G. *Trigonometric Series, Vols. 1–2, 2nd ed.* New York: Cambridge University Press, 1959.

Feigenbaum Constant

A universal constant for functions approaching CHAOS via period doubling. It was discovered by Feigenbaum in 1975 and demonstrated rigorously by Lanford (1982) and Collet and Eckmann (1979, 1980). The Feigenbaum constant δ characterizes the geometric approach of the bifurcation parameter to its limiting value. Let μ_k be the point at which a period 2^k cycle becomes unstable.

Denote the converged value by μ_∞. Assuming geometric convergence, the difference between this value and μ_k is denoted

$$\lim_{k\to\infty} \mu_\infty - \mu_k = \frac{\Gamma}{\delta^k}, \tag{1}$$

where Γ is a constant and δ is a constant > 1. Solving for δ gives

$$\delta \equiv \lim_{n\to\infty} \frac{\mu_{n+1} - \mu_n}{\mu_{n+2} - \mu_{n+1}} \tag{2}$$

(Rasband 1990, p. 23). For the LOGISTIC EQUATION,

$$\delta = 4.669216091\ldots \tag{3}$$

$$\Gamma = 2.637\ldots \tag{4}$$

$$\mu_\infty = 3.5699456\ldots. \tag{5}$$

Amazingly, the Feigenbaum constant $\delta \approx 4.669$ is "universal" (i.e., the same) for all 1-D MAPS $f(x)$ if $f(x)$ has a single locally quadratic MAXIMUM. More specifically, the Feigenbaum constant is universal for 1-D MAPS if the SCHWARZIAN DERIVATIVE

$$D_{\text{Schwarzian}} \equiv \frac{f'''(x)}{f'(x)} - \frac{3}{2}\left[\frac{f''(x)}{f'(x)}\right]^2 \tag{6}$$

is NEGATIVE in the bounded interval (Tabor 1989, p. 220). Examples of maps which are universal include the HÉNON MAP, LOGISTIC MAP, LORENZ SYSTEM, Navier-Stokes truncations, and sine map $x_{n+1} = a\sin(\pi x_n)$. The value of the Feigenbaum constant can be computed explicitly using functional group renormalization theory. The universal constant also occurs in phase transitions in physics and, curiously, is very nearly equal to

$$\pi + \tan^{-1}(e^\pi) = 4.669201932\ldots. \tag{7}$$

The CIRCLE MAP is *not* universal, and has a Feigenbaum constant of $\delta \approx 2.833$. For an AREA-PRESERVING 2-D MAP with

$$x_{n+1} = f(x_n, y_n) \tag{8}$$

$$y_{n+1} = g(x_n, y_n), \tag{9}$$

the Feigenbaum constant is $\delta = 0.7210978\ldots$ (Tabor 1989, p. 225). For a function of the form

$$f(x) = 1 - a|x|^n \tag{10}$$

with a and n constant and n an INTEGER, the Feigenbaum constant for various n is given in the following table (Briggs 1991, Briggs *et al.* 1991), which updates the values in Tabor (1989, p. 225).

n	δ
2	5.9679
4	7.2846
6	8.3494
8	9.2962

An additional constant α, defined as the separation of adjacent elements of PERIOD DOUBLED ATTRACTORS from one double to the next, has a value

$$\lim_{n\to\infty} \frac{d_n}{d_{n+1}} \equiv -\alpha = -2.502907875\ldots \tag{11}$$

for "universal" maps (Rasband 1990, p. 37). This value may be approximated from functional group renormalization theory to the zeroth order by

$$1 - \alpha^{-1} = \frac{1-\alpha^{-2}}{[1-\alpha^{-2}(1-\alpha^{-1})]^2}, \tag{12}$$

which, when the QUINTIC EQUATION is numerically solved, gives $\alpha = -2.48634\ldots$, only 0.7% off from the actual value (Feigenbaum 1988).

see also ATTRACTOR, BIFURCATION, FEIGENBAUM FUNCTION, LINEAR STABILITY, LOGISTIC MAP, PERIOD DOUBLING

References

Briggs, K. "A Precise Calculation of the Feigenbaum Constants." *Math. Comput.* **57**, 435–439, 1991.

Briggs, K.; Quispel, G.; and Thompson, C. "Feigenvalues for Mandelsets." *J. Phys. A: Math. Gen.* **24** 3363–3368, 1991.

Briggs, K.; Quispel, G.; and Thompson, C. "Feigenvalues for Mandelsets." `http://epidem13.plantsci.cam.ac.uk/~kbriggs/`.

Collett, P. and Eckmann, J.-P. "Properties of Continuous Maps of the Interval to Itself." *Mathematical Problems in Theoretical Physics* (Ed. K. Osterwalder). New York: Springer-Verlag, 1979.

Collett, P. and Eckmann, J.-P. *Iterated Maps on the Interval as Dynamical Systems.* Boston, MA: Birkhäuser, 1980.

Eckmann, J.-P. and Wittwer, P. *Computer Methods and Borel Summability Applied to Feigenbaum's Equations.* New York: Springer-Verlag, 1985.

Feigenbaum, M. J. "Presentation Functions, Fixed Points, and a Theory of Scaling Function Dynamics." *J. Stat. Phys.* **52**, 527–569, 1988.

Finch, S. "Favorite Mathematical Constants." `http://www.mathsoft.com/asolve/constant/fgnbaum/fgnbaum.html`.

Finch, S. "Generalized Feigenbaum Constants." `http://www.mathsoft.com/asolve/constant/fgnbaum/general.html`.

Lanford, O. E. "A Computer-Assisted Proof of the Feigenbaum Conjectures." *Bull. Amer. Math. Soc.* **6**, 427–434, 1982.

Rasband, S. N. *Chaotic Dynamics of Nonlinear Systems.* New York: Wiley, 1990.

Stephenson, J. W. and Wang, Y. "Numerical Solution of Feigenbaum's Equation." *Appl. Math. Notes* **15**, 68–78, 1990.

Stephenson, J. W. and Wang, Y. "Relationships Between the Solutions of Feigenbaum's Equations." *Appl. Math. Let.* **4**, 37–39, 1991.

Tabor, M. *Chaos and Integrability in Nonlinear Dynamics: An Introduction.* New York: Wiley, 1989.

Feigenbaum Function

Consider an arbitrary 1-D MAP

$$x_{n+1} = F(x_n) \tag{1}$$

at the onset of CHAOS. After a suitable rescaling, the Feigenbaum function

$$g(x) = \lim_{n\to\infty} \frac{1}{F^{(2^n)}(0)} F^{(2^n)}(xF^{(2^n)}(0)) \tag{2}$$

is obtained. This function satisfies

$$g(g(x)) = -\frac{1}{\alpha} g(\alpha x), \tag{3}$$

with $\alpha = 2.50290\ldots$, a quantity related to the FEIGENBAUM CONSTANT.

see also BIFURCATION, CHAOS, FEIGENBAUM CONSTANT

References

Grassberger, P. and Procaccia, I. "Measuring the Strangeness of Strange Attractors." *Physica D* **9**, 189–208, 1983.

Feit-Thompson Conjecture

Concerns PRIMES p and q for which $p^q - 1$ and $q^p - 1$ have a common factor. The only (p, q) pair with both values less than 400,000 is (17, 3313), with a common factor 112,643.

References

Wells, D. G. *The Penguin Dictionary of Curious and Interesting Numbers.* London: Penguin, p. 17, 1986.

Feit-Thompson Theorem

Every FINITE SIMPLE GROUP (which is not CYCLIC) has EVEN ORDER, and the ORDER of every FINITE SIMPLE noncommutative group is DOUBLY EVEN, i.e., divisible by 4 (Feit and Thompson 1963).

see also BURNSIDE PROBLEM, FINITE GROUP, ORDER (GROUP), SIMPLE GROUP

References

Feit, W. and Thompson, J. G. "Solvability of Groups of Odd Order." *Pacific J. Math.* **13**, 775–1029, 1963.

Fejes Tóth's Integral

$$\frac{1}{2\pi(n+1)} \int_{-\pi}^{\pi} f(x) \left\{ \frac{\sin[\frac{1}{2}(n+1)x]}{\sin(\frac{1}{2}x)} \right\}^2 dx$$

gives the nth CESÀRO MEAN of the FOURIER SERIES of $f(x)$.

References

Szegő, G. *Orthogonal Polynomials, 4th ed.* Providence, RI: Amer. Math. Soc., p. 12, 1975.

Fejes Tóth's Problem

How can n points be distributed on a UNIT SPHERE such that they maximize the minimum distance between any pair of points? In 1943, Fejes Tóth proved that for N points, there always exist two points whose distance d is

$$d \le \sqrt{4 - \csc^2\left[\frac{\pi N}{6(N-2)}\right]},$$

and that the limit is exact for $N = 3$, 4, 6, and 12.

For two points, the points should be at opposite ends of a DIAMETER. For four points, they should be placed at the VERTICES of an inscribed TETRAHEDRON. There is no best solution for five points since the distance cannot be reduced below that for six points. For six points, they should be placed at the VERTICES of an inscribed OCTAHEDRON. For seven points, the best solution is four equilateral spherical triangles with angles of 80°. For eight points, the best dispersal is *not* the VERTICES of the inscribed CUBE, but of a square ANTIPRISM with equal EDGES. The solution for nine points is eight equilateral spherical triangles with angles of $\cos^{-1}(1/4)$. For 12 points, the solution is an inscribed ICOSAHEDRON.

The general problem has not been solved.

see also THOMSON PROBLEM

References

Ogilvy, C. S. *Excursions in Mathematics.* New York: Dover, p. 99, 1994.

Ogilvy, C. S. Solved by L. Moser. "Minimal Configuration of Five Points on a Sphere." Problem E946. *Amer. Math. Monthly* **58**, 592, 1951.

Schütte, K. and van der Waerden, B. L. "Auf welcher Kügel haben 5, 6, 7, 8 oder 9 Pünkte mit Mindestabstand Eins Platz?" *Math. Ann.* **123**, 96–124, 1951.

Whyte, L. L. "Unique Arrangement of Points on a Sphere." *Amer. Math. Monthly* **59**, 606–611, 1952.

Feller's Coin-Tossing Constants

see COIN TOSSING

Feller-Lévy Condition

Given a sequence of independent random variates X_1, X_2, ..., if ${\sigma_k}^2 = \mathrm{var}(X_k)$ and

$${\rho_n}^2 \equiv \max_{k\le n}\left(\frac{{\sigma_k}^2}{{s_n}^2}\right),$$

then

$$\lim_{n\to\infty} {\rho_n}^2 = 0.$$

This means that if the LINDEBERG CONDITION holds for the sequence of variates X_1, ..., then the VARIANCE of an individual term in the sum S_n of X_k is asymptotically negligible. For such sequences, the LINDEBERG CONDITION is NECESSARY as well as SUFFICIENT for the LINDEBERG-FELLER CENTRAL LIMIT THEOREM to hold.

References

Zabell, S. L. "Alan Turing and the Central Limit Theorem." *Amer. Math. Monthly* **102**, 483–494, 1995.

Fence

Values one STEP outside the HINGES are called inner fences, and values two steps outside the HINGES are called outer fences. Tukey calls values outside the outer fences FAR OUT.

see also ADJACENT VALUE

References

Tukey, J. W. *Explanatory Data Analysis.* Reading, MA: Addison-Wesley, p. 44, 1977.

Fence Poset

A PARTIAL ORDER defined by $(i-1, j)$, $(i+1, j)$ for ODD i.

see also PARTIAL ORDER

References

Ruskey, F. "Information on Ideals of Partially Ordered Sets." http://sue.csc.uvic.ca/~cos/inf/pose/Ideals.html.

Ferguson-Forcade Algorithm

A practical algorithm for determining if there exist integers a_i for given real numbers x_i such that

$$a_1x_1 + a_2x_2 + \ldots + a_nx_n = 0,$$

or else establish bounds within which no such INTEGER RELATION can exist (Ferguson and Forcade 1979). A nonrecursive variant of the original algorithm was subsequently devised by Ferguson (1987). The Ferguson-Forcade algorithm has shown that there are no algebraic equations of degree ≤ 8 with integer coefficients having Euclidean norms below certain bounds for e/π, $e+\pi$, $\ln\pi$, γ, e^γ, γ/e, γ/π, and $\ln\gamma$, where e is the base for the NATURAL LOGARITHM, π is PI, and γ is the EULER-MASCHERONI CONSTANT (Bailey 1988).

Constant	Bound
e/π	6.1030×10^{14}
$e+\pi$	2.2753×10^{18}
$\ln\pi$	8.7697×10^{9}
γ	3.5739×10^{9}
e^γ	1.6176×10^{17}
γ/e	1.8440×10^{11}
γ/π	6.5403×10^{9}
$\ln\gamma$	2.6881×10^{10}

see also CONSTANT PROBLEM, EUCLIDEAN ALGORITHM, INTEGER RELATION, PSLQ ALGORITHM

References

Bailey, D. H. "Numerical Results on the Transcendence of Constants Involving π, e, and Euler's Constant." *Math. Comput.* **50**, 275–281, 1988.
Ferguson, H. R. P. "A Short Proof of the Existence of Vector Euclidean Algorithms." *Proc. Amer. Math. Soc.* **97**, 8–10, 1986.
Ferguson, H. R. P. "A Non-Inductive GL(n, Z) Algorithm that Constructs Linear Relations for n Z-Linearly Dependent Real Numbers." *J. Algorithms* **8**, 131–145, 1987.
Ferguson, H. R. P. and Forcade, R. W. "Generalization of the Euclidean Algorithm for Real Numbers to All Dimensions Higher than Two." *Bull. Amer. Math. Soc.* **1**, 912–914, 1979.

Fermat $4n+1$ Theorem

Every PRIME of the form $4n+1$ is a sum of two SQUARE NUMBERS in one unique way (up to the order of SUMMANDS). The theorem was stated by Fermat, but the first published proof was by Euler.

see also SIERPIŃSKI'S PRIME SEQUENCE THEOREM, SQUARE NUMBER

References

Hardy, G. H. and Wright, E. M. "Some Notation." Th. 251 in *An Introduction to the Theory of Numbers, 5th ed.* Oxford, England: Clarendon Press, 1979.

Fermat's Algorithm

see FERMAT'S FACTORIZATION METHOD

Fermat Compositeness Test

Uses FERMAT'S LITTLE THEOREM

Fermat's Congruence

see FERMAT'S LITTLE THEOREM

Fermat Conic

A PLANE CURVE of the form $y = x^n$. For $n > 0$, the curve is a generalized PARABOLA; for $n < 0$ it is a generalized HYPERBOLA.

see also CONIC SECTION, HYPERBOLA, PARABOLA

Fermat's Conjecture

see FERMAT'S LAST THEOREM

Fermat Difference Equation

see PELL EQUATION

Fermat Diophantine Equation

see FERMAT DIFFERENCE EQUATION

Fermat Equation

The DIOPHANTINE EQUATION

$$x^n + y^n = z^n.$$

The assertion that this equation has no nontrivial solutions for $n > 2$ is called FERMAT'S LAST THEOREM.

see also FERMAT'S LAST THEOREM

Fermat-Euler Theorem

see FERMAT'S LITTLE THEOREM

Fermat's Factorization Method

Given a number n, look for INTEGERS x and y such that $n = x^2 - y^2$. Then

$$n = (x - y)(x + y) \tag{1}$$

and n is factored. Any ODD NUMBER can be represented in this form since then $n = ab$, a and b are ODD, and

$$a = x + y \tag{2}$$
$$b = x - y. \tag{3}$$

Adding and subtracting,

$$a + b = 2x \tag{4}$$
$$a - b = 2y, \tag{5}$$

so solving for x and y gives

$$x = \tfrac{1}{2}(a + b) \tag{6}$$
$$y = \tfrac{1}{2}(a - b). \tag{7}$$

Therefore,

$$x^2 - y^2 = \tfrac{1}{4}[(a + b)^2 - (a - b)^2] = ab. \tag{8}$$

As the first trial for x, try $x_1 \lceil \sqrt{n} \rceil$, where $\lceil x \rceil$ is the CEILING FUNCTION. Then check if

$$\Delta x_1 = {x_1}^2 - n \tag{9}$$

is a SQUARE NUMBER. There are only 22 combinations of the last two digits which a SQUARE NUMBER can assume, so most combinations can be eliminated. If Δx_1 is not a SQUARE NUMBER, then try

$$x_2 = x_1 + 1, \tag{10}$$

so

$$\begin{aligned}\Delta x_2 &= {x_2}^2 - n = (x_1 + 1)^2 - n = {x_1}^2 + 2x_1 + 1 - n \\ &= \Delta x_1 + 2x_1 + 1.\end{aligned} \tag{11}$$

Continue with

$$\begin{aligned}\Delta x_3 &= {x_3}^2 - n = (x_2 + 1)^2 - n = {x_2}^2 + 2x_2 + 1 - n \\ &= \Delta x_2 + 2x_2 + 1 = \Delta x_2 + 2x_1 + 3,\end{aligned} \tag{12}$$

so subsequent differences are obtained simply by adding two.

Maurice Kraitchik sped up the ALGORITHM by looking for x and y satisfying

$$x^2 \equiv y^2 \pmod{n}, \tag{13}$$

i.e., $n|(x^2 - y^2)$. This congruence has uninteresting solutions $x \equiv \pm y \pmod{n}$ and interesting solutions $x \not\equiv \pm y \pmod{n}$. It turns out that if n is ODD and DIVISIBLE by at least two different PRIMES, then at least half of the solutions to $x^2 \equiv y^2 \pmod{n}$ with xy COPRIME to n are interesting. For such solutions, $(n, x - y)$ is neither n nor 1 and is therefore a nontrivial factor of n (Pomerance 1996). This ALGORITHM can be used to prove primality, but is not practical. In 1931, Lehmer and Powers discovered how to search for such pairs using CONTINUED FRACTIONS. This method was improved by Morrison and Brillhart (1975) into the CONTINUED FRACTION FACTORIZATION ALGORITHM, which was the fastest ALGORITHM in use before the QUADRATIC SIEVE FACTORIZATION METHOD was developed.

see also PRIME FACTORIZATION ALGORITHMS, SMOOTH NUMBER

References

Lehmer, D. H. and Powers, R. E. "On Factoring Large Numbers." *Bull. Amer. Math. Soc.* **37**, 770–776, 1931.

Morrison, M. A. and Brillhart, J. "A Method of Factoring and the Factorization of F_7." *Math. Comput.* **29**, 183–205, 1975.

Pomerance, C. "A Tale of Two Sieves." *Not. Amer. Math. Soc.* **43**, 1473–1485, 1996.

Fermat's Last Theorem

A theorem first proposed by Fermat in the form of a note scribbled in the margin of his copy of the ancient Greek text *Arithmetica* by Diophantus. The scribbled note was discovered posthumously, and the original is now lost. However, a copy was preserved in a book published by Fermat's son. In the note, Fermat claimed to have discovered a proof that the DIOPHANTINE EQUATION $x^n + y^n = z^n$ has no INTEGER solutions for $n > 2$.

The full text of Fermat's statement, written in Latin, reads "Cubum autem in duos cubos, aut quadratoquadratum in duos quadratoquadratos & generaliter nullam in infinitum ultra quadratum potestatem in duos eiusdem nominis fas est diuidere cuius rei demonstrationem mirabilem sane detexi. Hanc marginis exiguitas non caperet." In translation, "It is impossible for a cube to be the sum of two cubes, a fourth power to be the sum of two fourth powers, or in general for any number that is a power greater than the second to be the sum of two like powers. I have discovered a truly marvelous demonstration of this proposition that this margin is too narrow to contain."

As a result of Fermat's marginal note, the proposition that the DIOPHANTINE EQUATION

$$x^n + y^n = z^n, \tag{1}$$

where x, y, z, and n are INTEGERS, has no NONZERO solutions for $n > 2$ has come to be known as Fermat's Last Theorem. It was called a "THEOREM" on the strength of Fermat's statement, despite the fact that no other mathematician was able to prove it for hundreds of years.

Note that the restriction $n > 2$ is obviously necessary since there are a number of elementary formulas for generating an infinite number of PYTHAGOREAN TRIPLES (x, y, z) satisfying the equation for $n = 2$,

$$x^2 + y^2 = z^2. \tag{2}$$

A first attempt to solve the equation can be made by attempting to factor the equation, giving

$$(z^{n/2} + y^{n/2})(z^{n/2} - y^{n/2}) = x^n. \tag{3}$$

Since the product is an exact POWER,

$$\begin{cases} z^{n/2} + y^{n/2} = 2^{n-1}p^n \\ z^{n/2} - y^{n/2} = 2q^n \end{cases} \text{or} \quad \begin{cases} z^{n/2} + y^{n/2} = 2p^n \\ z^{n/2} - y^{n/2} = 2^{n-1}q^n. \end{cases} \tag{4}$$

Solving for y and z gives

$$\begin{cases} z^{n/2} = 2^{n-2}p^n + q^n \\ y^{n/2} = 2^{n-2}p^n - q^n \end{cases} \text{or} \quad \begin{cases} z^{n/2} = p^n + 2^{n-2}q^n \\ y^{n/2} = p^n - 2^{n-2}q^n, \end{cases} \tag{5}$$

which give

$$\begin{cases} z = (2^{n-2}p^n + q^n)^{2/n} \\ y = (2^{n-2}p^n - q^n)^{2/n} \end{cases} \text{or} \quad \begin{cases} z = (p^n + 2^{n-2}q^n)^{2/n} \\ y = (p^n - 2^{n-2}q^n)^{2/n}. \end{cases} \tag{6}$$

However, since solutions to these equations in RATIONAL NUMBERS are no easier to find than solutions to the original equation, this approach unfortunately does not provide any additional insight.

It is sufficient to prove Fermat's Last Theorem by considering PRIME POWERS only, since the arguments can otherwise be written

$$(x^m)^p + (y^m)^p = (z^m)^p, \tag{7}$$

so redefining the arguments gives

$$x^p + y^p = z^p. \tag{8}$$

The so-called "first case" of the theorem is for exponents which are RELATIVELY PRIME to x, y, and z ($p \nmid x, y, z$) and was considered by Wieferich. Sophie Germain proved the first case of Fermat's Last Theorem for any ODD PRIME p when $2p+1$ is also a PRIME. Legendre subsequently proved that if p is a PRIME such that $4p+1$, $8p+1$, $10p+1$, $14p+1$, or $16p+1$ is also a PRIME, then the first case of Fermat's Last Theorem holds for p. This established Fermat's Last Theorem for $p < 100$. In 1849, Kummer proved it for all REGULAR PRIMES and COMPOSITE NUMBERS of which they are factors (Vandiver 1929, Ball and Coxeter 1987).

Kummer's attack led to the theory of IDEALS, and Vandiver developed VANDIVER'S CRITERIA for deciding if a given IRREGULAR PRIME satisfies the theorem. Genocchi (1852) proved that the first case is true for p if $(p, p-3)$ is not an IRREGULAR PAIR. In 1858, Kummer showed that the first case is true if either $(p, p-3)$ or $(p, p-5)$ is an IRREGULAR PAIR, which was subsequently extended to include $(p, p-7)$ and $(p, p-9)$ by Mirimanoff (1905). Wieferich (1909) proved that if the equation is solved in integers RELATIVELY PRIME to an ODD PRIME p, then

$$2^{p-1} \equiv 1 \pmod{p^2}. \tag{9}$$

(Ball and Coxeter 1987). Such numbers are called WIEFERICH PRIMES. Mirimanoff (1909) subsequently showed that

$$3^{p-1} \equiv 1 \pmod{p^2} \tag{10}$$

must also hold for solutions RELATIVELY PRIME to an ODD PRIME p, which excludes the first two WIEFERICH PRIMES 1093 and 3511. Vandiver (1914) showed

$$5^{p-1} \equiv 1 \pmod{p^2}, \tag{11}$$

and Frobenius extended this to

$$11^{p-1}, 17^{p-1} \equiv 1 \pmod{p^2}. \tag{12}$$

It has also been shown that if p were a PRIME of the form $6x - 1$, then

$$7^{p-1}, 13^{p-1}, 19^{p-1} \equiv 1 \pmod{p^2}, \tag{13}$$

which raised the smallest possible p in the "first case" to 253,747,889 by 1941 (Rosser 1941). Granville and Monagan (1988) showed if there exists a PRIME p satisfying Fermat's Last Theorem, then

$$q^{p-1} \equiv 1 \pmod{p^2} \tag{14}$$

for $q = 5, 7, 11, \ldots, 71$. This establishes that the first case is true for all PRIME exponents up to 714,591,416,091,398 (Vardi 1991).

The "second case" of Fermat's Last Theorem (for $p|x, y, z$) proved harder than the first case.

Euler proved the general case of the theorem for $n = 3$, Fermat $n = 4$, Dirichlet and Lagrange $n = 5$. In 1832, Dirichlet established the case $n = 14$. The $n = 7$ case was proved by Lamé (1839), using the identity

$$\begin{aligned}(X+Y+Z)^7 &- (X^7+Y^7+Z^7) \\ &= 7(X+Y)(X+Z)(Y+Z) \\ &\quad \times[(X^2+Y^2+Z^2+XY+XZ+YZ)^2 \\ &\quad + XYZ(X+Y+Z)].\end{aligned} \tag{15}$$

Although some errors were present in this proof, these were subsequently fixed by Lebesgue (1840). Much additional progress was made over the next 150 years, but no completely general result had been obtained. Buoyed by false confidence after his proof that PI is TRANSCENDENTAL, the mathematician Lindemann proceeded to publish several proofs of Fermat's Last Theorem, all of them invalid (Bell 1937, pp. 464–465). A prize of 100,000 German marks (known as the Wolfskel Prize) was also offered for the first valid proof (Ball and Coxeter 1987, p. 72).

A recent false alarm for a general proof was raised by Y. Miyaoka (Cipra 1988) whose proof, however, turned out to be flawed. Other attempted proofs among both professional and amateur mathematicians are discussed by vos Savant (1993), although vos Savant erroneously claims that work on the problem by Wiles (discussed below) is invalid. By the time 1993 rolled around, the general case of Fermat's Last Theorem had been shown to be true for all exponents up to 4×10^6 (Cipra 1993). However, given that a proof of Fermat's Last Theorem requires truth for *all* exponents, proof for any finite number of exponents does not constitute any significant progress towards a proof of the general theorem (although the fact that no counterexamples were found for this many cases is highly suggestive).

In 1993, a bombshell was dropped. In that year, the general theorem was partially proven by Andrew Wiles (Cipra 1993, Stewart 1993) by proving the SEMISTABLE case of the TANIYAMA-SHIMURA CONJECTURE. Unfortunately, several holes were discovered in the proof shortly thereafter when Wiles' approach via the TANIYAMA-SHIMURA CONJECTURE became hung up on properties of the SELMER GROUP using a tool called an "Euler system." However, the difficulty was circumvented by Wiles and R. Taylor in late 1994 (Cipra 1994, 1995ab) and published in Taylor and Wiles (1995) and Wiles (1995). Wiles' proof succeeds by (1) replacing ELLIPTIC CURVES with Galois representations, (2) reducing the problem to a CLASS NUMBER FORMULA, (3) proving that FORMULA, and (4) tying up loose ends that arise because the formalisms fail in the simplest degenerate cases (Cipra 1995a).

The proof of Fermat's Last Theorem marks the end of a mathematical era. Since virtually all of the tools which were eventually brought to bear on the problem had yet to be invented in the time of Fermat, it is interesting to speculate about whether he actually was in possession of an elementary proof of the theorem. Judging by the temerity with which the problem resisted attack for so long, Fermat's alleged proof seems likely to have been illusionary.

see also ABC CONJECTURE, BOGOMOLOV-MIYAOKA-YAU INEQUALITY, MORDELL CONJECTURE, PYTHAGOREAN TRIPLE, RIBET'S THEOREM, SELMER GROUP, SOPHIE GERMAIN PRIME, SZPIRO'S CONJECTURE, TANIYAMA-SHIMURA CONJECTURE, VOJTA'S CONJECTURE, WARING FORMULA

References

Ball, W. W. R. and Coxeter, H. S. M. *Mathematical Recreations and Essays, 13th ed.* New York: Dover, pp. 69–73, 1987.

Beiler, A. H. "The Stone Wall." Ch. 24 in *Recreations in the Theory of Numbers: The Queen of Mathematics Entertains.* New York: Dover, 1966.

Bell, E. T. *Men of Mathematics.* New York: Simon and Schuster, 1937.

Bell, E. T. *The Last Problem.* New York: Simon and Schuster, 1961.

Cipra, B. A. "Fermat Theorem Proved." *Science* **239**, 1373, 1988.

Cipra, B. A. "Mathematics—Fermat's Last Theorem Finally Yields." *Science* **261**, 32–33, 1993.

Cipra, B. A. "Is the Fix in on Fermat's Last Theorem?" *Science* **266**, 725, 1994.

Cipra, B. A. "Fermat's Theorem—At Last." *What's Happening in the Mathematical Sciences, 1995–1996, Vol. 3.* Providence, RI: Amer. Math. Soc., pp. 2–14, 1996.

Cipra, B. A. "Princeton Mathematician Looks Back on Fermat Proof." *Science* **268**, 1133–1134, 1995b.

Courant, R. and Robbins, H. "Pythagorean Numbers and Fermat's Last Theorem." §2.3 in Supplement to Ch. 1 in *What is Mathematics?: An Elementary Approach to Ideas and Methods, 2nd ed.* Oxford, England: Oxford University Press, pp. 40–42, 1996.

Cox, D. A. "Introduction to Fermat's Last Theorem." *Amer. Math. Monthly* **101**, 3–14, 1994.

Dickson, L. E. "Fermat's Last Theorem, $ax^r + by^s = cz^t$, and the Congruence $x^n + y^n \equiv z^n \pmod{p}$." Ch. 26 in *History of the Theory of Numbers, Vol. 2: Diophantine Analysis.* New York: Chelsea, pp. 731–776, 1952.

Edwards, H. M. *Fermat's Last Theorem: A Genetic Introduction to Algebraic Number Theory.* New York: Springer-Verlag, 1977.

Edwards, H. M. "Fermat's Last Theorem." *Sci. Amer.*, Oct. 1978.

Granville, A. "Review of BBC's Horizon Program, 'Fermat's Last Theorem'." *Not. Amer. Math. Soc.* **44**, 26–28, 1997.

Granville, A. and Monagan, M. B. "The First Case of Fermat's Last Theorem is True for All Prime Exponents up to 714,591,416,091,389." *Trans. Amer. Math. Soc.* **306**, 329–359, 1988.

Guy, R. K. "The Fermat Problem." §D2 in *Unsolved Problems in Number Theory, 2nd ed.* New York: Springer-Verlag, pp. 144–146, 1994.

Hanson, A. "Fermat Project." `http://www.cica.indiana.edu/projects/Fermat/`.

Kolata, G. "Andrew Wiles: A Math Whiz Battles 350-Year-Old Puzzle." *New York Times*, June 29, 1993.

Lynch, J. "Fermat's Last Theorem." BBC Horizon television documentary. `http://www.bbc.co.uk/horizon/fermat.shtml`.

Lynch, J. (Producer and Writer). "The Proof." NOVA television episode. 52 mins. Broadcast by the U. S. Public Broadcasting System on Oct. 28, 1997.

Mirimanoff, D. "Sur le dernier théorème de Fermat et le critérium de wiefer." *Enseiggnement Math.* **11**, 455–459, 1909.

Mordell, L. J. *Fermat's Last Theorem.* New York: Chelsea, 1956.

Murty, V. K. (Ed.). *Fermat's Last Theorem: Proceedings of the Fields Institute for Research in Mathematical Sciences on Fermat's Last Theorem, Held 1993–1994 Toronto, Ontario, Canada.* Providence, RI: Amer. Math. Soc., 1995.

Osserman, R. (Ed.). *Fermat's Last Theorem. The Theorem and Its Proof: An Exploration of Issues and Ideas.* 98 min. videotape and 56 pp. book. 1994.

Ribenboim, P. *Lectures on Fermat's Last Theorem.* New York: Springer-Verlag, 1979.

Ribet, K. A. and Hayes, B. "Fermat's Last Theorem and Modern Arithmetic." *Amer. Sci.* **82**, 144–156, March/April 1994.

Ribet, K. A. and Hayes, B. Correction to "Fermat's Last Theorem and Modern Arithmetic." *Amer. Sci.* **82**, 205, May/June 1994.

Rosser, B. "On the First Case of Fermat's Last Theorem." *Bull. Amer. Math. Soc.* **45**, 636–640, 1939.

Rosser, B. "A New Lower Bound for the Exponent in the First Case of Fermat's Last Theorem." *Bull. Amer. Math. Soc.* **46**, 299–304, 1940.

Rosser, B. "An Additional Criterion for the First Case of Fermat's Last Theorem." *Bull. Amer. Math. Soc.* **47**, 109–110, 1941.

Shanks, D. *Solved and Unsolved Problems in Number Theory, 4th ed.* New York: Chelsea, pp. 144–149, 1993.

Singh, S. *Fermat's Enigma: The Quest to Solve the World's Greatest Mathematical Problem.* New York: Walker & Co., 1997.

Stewart, I. "Fermat's Last Time-Trip." *Sci. Amer.* **269**, 112–115, 1993.

Taylor, R. and Wiles, A. "Ring-Theoretic Properties of Certain Hecke Algebras." *Ann. Math.* **141**, 553–572, 1995.

van der Poorten, A. *Notes on Fermat's Last Theorem.* New York: Wiley, 1996.

Vandiver, H. S. "On Fermat's Last Theorem." *Trans. Amer. Math. Soc.* **31**, 613–642, 1929.

Vandiver, H. S. *Fermat's Last Theorem and Related Topics in Number Theory.* Ann Arbor, MI: 1935.

Vandiver, H. S. "Fermat's Last Theorem: Its History and the Nature of the Known Results Concerning It." *Amer. Math. Monthly,* **53**, 555–578, 1946.

Vardi, I. *Computational Recreations in Mathematica.* Reading, MA: Addison-Wesley, pp. 59–61, 1991.

vos Savant, M. *The World's Most Famous Math Problem.* New York: St. Martin's Press, 1993.

Wieferich, A. "Zum letzten Fermat'schen Theorem." *J. reine angew. Math.* **136**, 293–302, 1909.

Wiles, A. "Modular Elliptic-Curves and Fermat's Last Theorem." *Ann. Math.* **141**, 443–551, 1995.

Fermat's Lesser Theorem

see FERMAT'S LITTLE THEOREM

Fermat's Little Theorem

If p is a PRIME number and a a NATURAL NUMBER, then

$$a^p \equiv a \pmod{p}. \tag{1}$$

Furthermore, if $p \nmid a$ (p does not divide a), then there exists some smallest exponent d such that

$$a^d - 1 \equiv 0 \pmod{p} \tag{2}$$

and d divides $p-1$. Hence,

$$a^{p-1} - 1 \equiv 0 \pmod{p}. \tag{3}$$

This is a generalization of the CHINESE HYPOTHESIS and a special case of EULER'S THEOREM. It is sometimes called FERMAT'S PRIMALITY TEST and is a NECESSARY but not SUFFICIENT test for primality. Although it was presumably proved (but suppressed) by Fermat, the first proof was published by Euler in 1749.

The theorem is easily proved using mathematical INDUCTION. Suppose $p|a^p - a$. Then examine

$$(a+1)^p - (a+1). \tag{4}$$

From the BINOMIAL THEOREM,

$$(a+1)^p = a^p + \binom{p}{1}a^{p-1} + \binom{p}{2}a^{p-2} + \ldots + \binom{p}{p-1}a + 1. \tag{5}$$

Rewriting,

$$(a+1)^p - a^p - 1 = \binom{p}{1}a^{p-1} + \binom{p}{2}a^{p-2} + \ldots + \binom{p}{p-1}a. \tag{6}$$

But p divides the right side, so it also divides the left side. Combining with the induction hypothesis gives that p divides the sum

$$[(a+1)^p - a^p - 1] + (a^p - a) = (a+1)^p - (a+1), \tag{7}$$

as assumed, so the hypothesis is true for any a. The theorem is sometimes called FERMAT'S SIMPLE THEOREM. WILSON'S THEOREM follows as a COROLLARY of Fermat's Little Theorem.

Fermat's little theorem shows that, if p is PRIME, there does not exists a base $a < p$ with $(a,p) = 1$ such that $a^{p-1} - 1$ possesses a nonzero residue modulo p. If such base a exists, p is therefore guaranteed to be composite. However, the lack of a nonzero residue in Fermat's little theorem does *not* guarantee that p is PRIME. The property of unambiguously certifying composite numbers while passing some PRIMES make Fermat's little theorem a COMPOSITENESS TEST which is sometimes called the FERMAT COMPOSITENESS TEST. COMPOSITE NUMBERS known as FERMAT PSEUDOPRIMES (or sometimes simply "PSEUDOPRIMES") have zero residue for some as and so are not identified as composite. Worse still, there exist numbers known as CARMICHAEL NUMBERS (the smallest of which is 561) which give zero residue for *any* choice of the base a RELATIVELY PRIME to p. However, FERMAT'S LITTLE THEOREM CONVERSE provides a criterion for certifying the primality of a number.

A number satisfying Fermat's little theorem for some nontrivial base and which is not known to be composite is called a PROBABLE PRIME. A table of the smallest PSEUDOPRIMES P for the first 100 bases a follows (Sloane's A007535).

a	P	a	P	a	P	a	P	a	P
2	341	22	69	42	205	62	63	82	91
3	91	23	33	43	77	63	341	83	105
4	15	24	25	44	45	64	65	84	85
5	124	25	28	45	76	65	133	85	129
6	35	26	27	46	133	66	91	86	87
7	25	27	65	47	65	67	85	87	91
8	9	28	87	48	49	68	69	88	91
9	28	29	35	49	66	69	85	89	99
10	33	30	49	50	51	70	169	90	91
11	15	31	49	51	65	71	105	91	115
12	65	32	33	52	85	72	85	92	93
13	21	33	85	53	65	73	111	93	301
14	15	34	35	54	55	74	75	94	95
15	341	35	51	55	63	75	91	95	141
16	51	36	91	56	57	76	77	96	133
17	45	37	45	57	65	77	95	97	105
18	25	38	39	58	95	78	341	98	99
19	45	39	95	59	87	79	91	99	145
20	21	40	91	60	341	80	81	100	259
21	55	41	105	61	91	81	85		

see also Binomial Theorem, Carmichael Number, Chinese Hypothesis, Composite Number, Compositeness Test, Euler's Theorem, Fermat's Little Theorem Converse, Fermat Pseudoprime, Modulo Multiplication Group, Pratt Certificate, Primality Test, Prime Number, Pseudoprime, Relatively Prime, Totient Function, Wieferich Prime, Wilson's Theorem, Witness

References

Ball, W. W. R. and Coxeter, H. S. M. *Mathematical Recreations and Essays, 13th ed.* New York: Dover, p. 61, 1987.
Conway, J. H. and Guy, R. K. *The Book of Numbers.* New York: Springer-Verlag, pp. 141–142, 1996.
Courant, R. and Robbins, H. "Fermat's Theorem." §2.2 in Supplement to Ch. 1 in *What is Mathematics?: An Elementary Approach to Ideas and Methods, 2nd ed.* Oxford, England: Oxford University Press, pp. 37–38, 1996.
Shanks, D. *Solved and Unsolved Problems in Number Theory, 4th ed.* New York: Chelsea, p. 20, 1993.
Sloane, N. J. A. Sequence A007535/M5440 in "An On-Line Version of the Encyclopedia of Integer Sequences."

Fermat's Little Theorem Converse

The converse of Fermat's Little Theorem is also known as Lehmer's Theorem. It states that, if an Integer x is Prime to m and $x^{m-1} \equiv 1 \pmod m$ and there is no Integer $e < m-1$ for which $x^e \equiv 1 \pmod m$, then m is Prime. Here, x is called a Witness to the primality of m. This theorem is the basis for the Pratt Primality Certificate.

see also Fermat's Little Theorem, Pratt Certificate, Primality Certificate, Witness

References

Riesel, H. *Prime Numbers and Computer Methods for Factorization, 2nd ed.* Boston, MA: Birkhäuser, p. 96, 1994.
Wagon, S. *Mathematica in Action.* New York: W. H. Freeman, pp. 278–279, 1991.

Fermat-Lucas Number

A number of the form $2^n + 1$ obtained by setting $x = 1$ in a Fermat-Lucas Polynomial. The first few are 3, 5, 9, 17, 33, ... (Sloane's A000051).

see also Fermat Number (Lucas)

References

Sloane, N. J. A. Sequence A000051/M0717 in "An On-Line Version of the Encyclopedia of Integer Sequences."

Fermat Number

A Binomial Number of the form $F_n = 2^{2^n} + 1$. The first few for $n = 0, 1, 2, \ldots$ are 3, 5, 17, 257, 65537, 4294967297, ... (Sloane's A000215). The number of Digits for a Fermat number is

$$\begin{aligned} D(n) &= \lfloor [\log(2^{2^n} + 1)] + 1 \rfloor \approx \lfloor \log(2^{2^n}) + 1 \rfloor \\ &= \lfloor 2^n \log 2 + 1 \rfloor . \end{aligned} \tag{1}$$

Being a Fermat number is the Necessary (but not Sufficient) form a number

$$N_n \equiv 2^n + 1 \tag{2}$$

must have in order to be Prime. This can be seen by noting that if $N_n = 2^n + 1$ is to be Prime, then n cannot have any Odd factors b or else N_n would be a factorable number of the form

$$\begin{aligned} 2^n + 1 &= (2^a)^b + 1 \\ &= (2^a + 1)[2^{a(b-1)} - 2^{a(b-2)} + 2^{a(b-3)} - \ldots + 1]. \end{aligned} \tag{3}$$

Therefore, for a Prime N_n, n must be a Power of 2.

Fermat conjectured in 1650 that every Fermat number is Prime, but only Composite Fermat numbers F_n are known for $n \geq 5$. Eisenstein (1844) proposed as a problem the proof that there are an infinite number of Fermat primes (Ribenboim 1996, p. 88), but this has not yet been achieved. An anonymous writer proposed that numbers of the form $2^2 + 1$, $2^{2^2} + 1$, $2^{2^{2^2}} + 1$ were Prime. However, this conjecture was refuted when Selfridge (1953) showed that

$$F_{16} = 2^{2^{16}} + 1 = 2^{2^{2^{2^2}}} + 1 \tag{4}$$

is Composite (Ribenboim 1996, p. 88). Numbers of the form $a^{2^n} + b^{2^n}$ are called generalized Fermat numbers (Ribenboim 1996, pp. 359–360).

Fermat numbers satisfy the Recurrence Relation

$$F_m = F_0 F_1 \cdots F_{m-1} + 2. \tag{5}$$

F_n can be shown to be Prime iff it satisfies Pépin's Test

$$3^{(F_n - 1)/2} \equiv -1 \pmod{F_n}. \tag{6}$$

PÉPIN'S THEOREM

$$3^{2^{2^n-1}} \equiv -1 \pmod{F_n} \qquad (7)$$

is also NECESSARY and SUFFICIENT.

In 1770, Euler showed that any FACTOR of F_n must have the form

$$2^{n+1}K+1, \qquad (8)$$

where K is a POSITIVE INTEGER. In 1878, Lucas increased the exponent of 2 by one, showing that FACTORS of Fermat numbers must be of the form

$$2^{n+2}L+1. \qquad (9)$$

If

$$F = p_1 p_2 \cdots p_r \qquad (10)$$

is the factored part of $F_n = FC$ (where C is the cofactor to be tested for primality), compute

$$A \equiv 3^{F_n-1} \pmod{F_n} \qquad (11)$$
$$B \equiv 3^{F-1} \pmod{F_n} \qquad (12)$$
$$R \equiv A - B \pmod{C}. \qquad (13)$$

Then if $R \equiv 0$, the cofactor is a PROBABLE PRIME to the base 3^F; otherwise C is COMPOSITE.

In order for a POLYGON to be circumscribed about a CIRCLE (i.e., a CONSTRUCTIBLE POLYGON), it must have a number of sides N given by

$$N = 2^k F_0 \cdots F_n, \qquad (14)$$

where the F_n are *distinct* Fermat primes. This is equivalent to the statement that the trigonometric functions $\sin(k\pi/N)$, $\cos(k\pi/N)$, etc., can be computed in terms of finite numbers of additions, multiplications, and square root extractions iff N is of the above form. The only known Fermat PRIMES are

$$F_0 = 3$$
$$F_1 = 5$$
$$F_2 = 17$$
$$F_3 = 257$$
$$F_4 = 65537$$

and it seems unlikely that any more exist.

Factoring Fermat numbers is extremely difficult as a result of their large size. In fact, only F_5 to F_{11} have been complete factored, as summarized in the following table. Written out explicitly, the complete factorizations are

$$F_5 = 641 \cdot 6700417$$
$$F_6 = 274177 \cdot 67280421310721$$
$$F_7 = 59649589127497217 \cdot 5704689200685129054721$$
$$F_8 = 1238926361552897 \cdot 93461639715357977769163\cdots$$
$$\cdots 558199606896584051237541638188580280321$$
$$F_9 = 2424833 \cdot 7455602825647884208337395736200\cdots$$
$$\cdots 54918783366342657 \cdot P99$$
$$F_{10} = 45592577 \cdot 6487031809 \cdot 4659775785220018\cdots$$
$$\cdots 4326456074307677819 2897 \cdot P252$$
$$F_{11} = 319489 \cdot 974849 \cdot 167988556341760475137$$
$$\cdot 3560841906445833920513 \cdot P564.$$

Here, the final large PRIME is not explicitly given since it can be computed by dividing F_n by the other given factors.

F_n	Digits	Facts.	Digits	Reference
5	10	2	3, 7	Euler 1732
6	20	2	6, 14	Landry 1880
7	39	2	7, 22	Morrison and Brillhart 1975
8	78	2	16, 62	Brent and Pollard 1981
9	155	3	7, 49, 99	Manasse and Lenstra (In Cipra 1993)
10	309	4	8, 10, 40, 252	Brent 1995
11	617	5	6, 6, 21, 22, 564	Brent 1988

Tables of known factors of Fermat numbers are given by Keller (1983), Brillhart *et al.* (1988), Young and Buell (1988), Riesel (1994), and Pomerance (1996). Young and Buell (1988) discovered that F_{20} is COMPOSITE, and Crandall *et al.* (1995) that F_{22} is COMPOSITE. A current list of the known factors of Fermat numbers is maintained by Keller, and reproduced in the form of a *Mathematica®* notebook by Weisstein. In these tables, since all factors are of the form $k2^n+1$, the known factors are expressed in the concise form (k, n). The number of factors for Fermat numbers F_n for $n = 0, 1, 2, \ldots$ are 1, 1, 1, 1, 1, 2, 2, 2, 2, 3, 4, 5, ...

see also CULLEN NUMBER, PÉPIN'S TEST, PÉPIN'S THEOREM, POCKLINGTON'S THEOREM, POLYGON, PROTH'S THEOREM, SELFRIDGE-HURWITZ RESIDUE, WOODALL NUMBER

References

Ball, W. W. R. and Coxeter, H. S. M. *Mathematical Recreations and Essays, 13th ed.* New York: Dover, pp. 68–69 and 94–95, 1987.

Brent, R. P. "Factorization of the Eighth Fermat Number." *Amer. Math. Soc. Abstracts* **1**, 565, 1980.

Brent, R. P. "Factorisation of F10." `http://cslab.anu.edu.au/~rpb/F10.html`.

Brent, R. P "Factorization of the Tenth and Eleventh Fermat Numbers." Submitted to *Math. Comput.* `ftp://nimbus.anu.edu.au/pub/Brent/rpb161tr.dvi.Z`.

Brent, R. P. and Pollard, J. M. "Factorization of the Eighth Fermat Number." *Math. Comput.* **36**, 627–630, 1981.

Brillhart, J.; Lehmer, D. H.; Selfridge, J.; Wagstaff, S. S. Jr.; and Tuckerman, B. *Factorizations of* $b^n \pm 1$, $b = 2, 3, 5, 6, 7, 10, 11, 12$ *Up to High Powers, rev. ed.* Providence, RI: Amer. Math. Soc., pp. lxxxvii and 2–3 of Update 2.2, 1988.

Cipra, B. "Big Number Breakdown." *Science* **248**, 1608, 1990.

Conway, J. H. and Guy, R. K. "Fermat's Numbers." In *The Book of Numbers.* New York: Springer-Verlag, pp. 137–141, 1996.

Cormack, G. V. and Williams, H. C. "Some Very Large Primes of the Form $k \cdot 2^m + 1$." *Math. Comput.* **35**, 1419–1421, 1980.

Courant, R. and Robbins, H. *What is Mathematics?: An Elementary Approach to Ideas and Methods, 2nd ed.* Oxford, England: Oxford University Press, pp. 25–26 and 119, 1996.

Crandall, R.; Doenias, J.; Norrie, C.; and Young, J. "The Twenty-Second Fermat Number is Composite." *Math. Comput.* **64**, 863–868, 1995.

Dickson, L. E. "Fermat Numbers $F_n = 2^{2^n} + 1$." Ch. 15 in *History of the Theory of Numbers, Vol. 1: Divisibility and Primality.* New York: Chelsea, pp. 375–380, 1952.

Dixon, R. *Mathographics.* New York: Dover, p. 53, 1991.

Euler, L. "Observationes de theoremate quodam Fermatiano aliisque ad numeros primos spectantibus." *Acad. Sci. Petropol.* **6**, 103–107, ad annos 1732–33 (1738). In *Leonhardi Euleri Opera Omnia,* Ser. I, Vol. II. Leipzig: Teubner, pp. 1–5, 1915.

Gostin, G. B. "A Factor of F_{17}." *Math. Comput.* **35**, 975–976, 1980.

Gostin, G. B. "New Factors of Fermat Numbers." *Math. Comput.* **64**, 393–395, 1995.

Gostin, G. B. and McLaughlin, P. B. Jr. "Six New Factors of Fermat Numbers." *Math. Comput.* **38**, 645–649, 1982.

Guy, R. K. "Mersenne Primes. Repunits. Fermat Numbers. Primes of Shape $k \cdot 2^n + 2$." §A3 in *Unsolved Problems in Number Theory, 2nd ed.* New York: Springer-Verlag, pp. 8–13, 1994.

Hallyburton, J. C. Jr. and Brillhart, J. "Two New Factors of Fermat Numbers." *Math. Comput.* **29**, 109–112, 1975.

Hardy, G. H. and Wright, E. M. *An Introduction to the Theory of Numbers, 5th ed.* Oxford, England: Clarendon Press, pp. 14–15, 1979.

Keller, W. "Factor of Fermat Numbers and Large Primes of the Form $k \cdot 2^n + 1$." *Math. Comput.* **41**, 661–673, 1983.

Keller, W. "Factors of Fermat Numbers and Large Primes of the Form $k \cdot 2^n + 1$, II." In prep.

Keller, W. "Prime Factors $k \cdot 2^n + 1$ of Fermat Numbers F_m and Complete Factoring Status." `http://ballingerr.xray.ufl.edu/proths/fermat.html`.

Kraitchik, M. "Fermat Numbers." §3.6 in *Mathematical Recreations.* New York: W. W. Norton, pp. 73–75, 1942.

Landry, F. "Note sur la décomposition du nombre $2^{64} + 1$ (Extrait)." *C. R. Acad. Sci. Paris,* **91**, 138, 1880.

Lenstra, A. K.; Lenstra, H. W. Jr.; Manasse, M. S.; and Pollard, J. M. "The Factorization of the Ninth Fermat Number." *Math. Comput.* **61**, 319–349, 1993.

Morrison, M. A. and Brillhart, J. "A Method of Factoring and the Factorization of F_7." *Math. Comput.* **29**, 183–205, 1975.

Pomerance, C. "A Tale of Two Sieves." *Not. Amer. Math. Soc.* **43**, 1473–1485, 1996.

Ribenboim, P. "Fermat Numbers" and "Numbers $k \times 2^n \pm 1$." §2.6 and 5.7 in *The New Book of Prime Number Records.* New York: Springer-Verlag, pp. 83–90 and 355–360, 1996.

Riesel, H. *Prime Numbers and Computer Methods for Factorization, 2nd ed.* Basel: Birkhäuser, pp. 384–388, 1994.

Robinson, R. M. "A Report on Primes of the Form $k \cdot 2^n + 1$ and on Factors of Fermat Numbers." *Proc. Amer. Math. Soc.* **9**, 673–681, 1958.

Selfridge, J. L. "Factors of Fermat Numbers." *Math. Comput.* **7**, 274–275, 1953.

Shanks, D. *Solved and Unsolved Problems in Number Theory, 4th ed.* New York: Chelsea, pp. 13 and 78–80, 1993.

Sloane, N. J. A. Sequence A000215/M2503 in "An On-Line Version of the Encyclopedia of Integer Sequences."

Weisstein, E. W. "Fermat Numbers." `http://www.astro.virginia.edu/~eww6n/math/notebooks/Fermat.m`.

Wrathall, C. P. "New Factors of Fermat Numbers." *Math. Comput.* **18**, 324–325, 1964.

Young, J. and Buell, D. A. "The Twentieth Fermat Number is Composite." *Math. Comput.* **50**, 261–263, 1988.

Fermat Number (Lucas)

A number of the form $2^n - 1$ obtained by setting $x = 1$ in a FERMAT POLYNOMIAL is called a MERSENNE NUMBER.

see also FERMAT-LUCAS NUMBER, MERSENNE NUMBER

Fermat Point

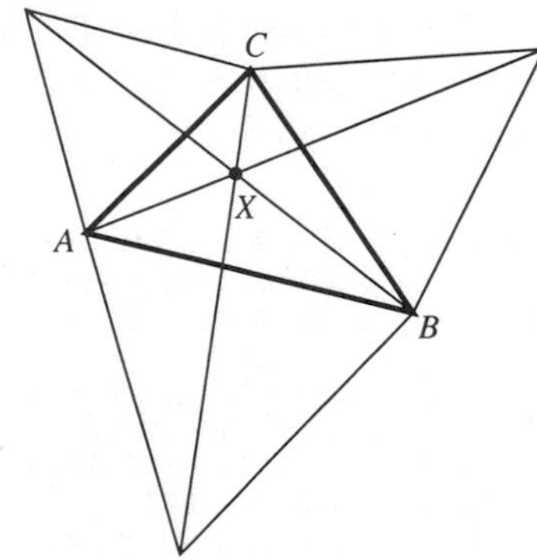

Also known as the first ISOGONIC CENTER and the TORRICELLI POINT. In a given ACUTE TRIANGLE ΔABC, the Fermat point is the point X which minimizes the sum of distances from A, B, and C,

$$|AX| + |BX| + |CX|. \tag{1}$$

This problem is called FERMAT'S PROBLEM or STEINER'S PROBLEM (Courant and Robbins 1941) and was proposed by Fermat to Torricelli. Torricelli's solution was published by his pupil Viviani in 1659 (Johnson 1929).

If all ANGLES of the TRIANGLE are less than 120° ($2\pi/3$), then the Fermat point is the interior point X from which each side subtends an ANGLE of 120°, i.e.,

$$\angle BXC = \angle CXA = \angle AXB = 120°. \tag{2}$$

The Fermat point can also be constructed by drawing EQUILATERAL TRIANGLES on the outside of the given TRIANGLE and connecting opposite VERTICES. The three diagonals in the figure then intersect in the Fermat point. The TRIANGLE CENTER FUNCTION of the Fermat point is

$$\alpha = \csc(A + \tfrac{1}{3}\pi) \quad (3)$$
$$= bc[c^2a^2 + (c^2 + a^2 - b^2)^2][a^2b^2 - (a^2 + b^2 - c^2)^2] \times [4\Delta - \sqrt{3}(b^2 + c^2 - a^2)]. \quad (4)$$

The ANTIPEDAL TRIANGLE is EQUILATERAL and has AREA

$$\Delta' = 2\Delta\left[1 + \cot\omega\cot\left(\frac{\pi}{3}\right)\right], \quad (5)$$

where ω is the BROCARD ANGLE.

Given three POSITIVE REAL NUMBERS l, m, n, the "generalized" Fermat point is the point P of a given ACUTE TRIANGLE ΔABC such that

$$l \cdot PA + m \cdot PB + n \cdot PC \quad (6)$$

is a minimum (Greenberg and Robertello 1965, van de Lindt 1966, Tong and Chua 1995)

see also ISOGONIC CENTERS

References

Courant, R. and Robbins, H. *What is Mathematics?, 2nd ed.* Oxford, England: Oxford University Press, 1941.
Gallatly, W. *The Modern Geometry of the Triangle, 2nd ed.* London: Hodgson, p. 107, 1913.
Greenberg, I. and Robertello, R. A. "The Three Factory Problem." *Math. Mag.* **38**, 67–72, 1965.
Honsberger, R. *Mathematical Gems I.* Washington, DC: Math. Assoc. Amer., pp. 24–34, 1973.
Johnson, R. A. *Modern Geometry: An Elementary Treatise on the Geometry of the Triangle and the Circle.* Boston, MA: Houghton Mifflin, pp. 221–222, 1929.
Kimberling, C. "Central Points and Central Lines in the Plane of a Triangle." *Math. Mag.* **67**, p. 174, 1994.
Kimberling, C. "Fermat Point." `http://www.evansville.edu/~ck6/tcenters/class/fermat.html`.
Mowaffaq, H. "An Advanced Calculus Approach to Finding the Fermat Point." *Math. Mag.* **67**, 29–34, 1994.
Pottage, J. *Geometrical Investigations.* Reading, MA: Addison-Wesley, 1983.
Spain, P. G. "The Fermat Point of a Triangle." *Math. Mag.* **69**, 131–133, 1996.
Tong, J. and Chua, Y. S. "The Generalized Fermat's Point." *Math. Mag.* **68**, 214–215, 1995.
van de Lindt, W. J. "A Geometrical Solution of the Three Factory Problem." *Math. Mag.* **39**, 162–165, 1966.

Fermat's Polygonal Number Theorem

In 1638, Fermat proposed that every POSITIVE INTEGER is a sum of *at most* three TRIANGULAR NUMBERS, four SQUARE NUMBERS, five PENTAGONAL NUMBERS, and n n-POLYGONAL NUMBERS. Fermat claimed to have a proof of this result, although Fermat's proof has never been found. Gauss proved the triangular case, and noted the event in his diary on July 10, 1796, with the notation

$$**\,E\Upsilon RHKA \qquad num = \Delta + \Delta + \Delta.$$

This case is equivalent to the statement that every number of the form $8m + 3$ is a sum of three ODD SQUARES (Duke 1997). More specifically, a number is a sum of three SQUARES IFF it is not of the form $4^b(8m + 7)$ for $b \geq 0$, as first proved by Legendre in 1798.

Euler was unable to prove the square case of Fermat's theorem, but he left partial results which were subsequently used by Lagrange. The square case was finally proved by Jacobi and independently by Lagrange in 1772. It is therefore sometimes known as LAGRANGE'S FOUR-SQUARE THEOREM. In 1813, Cauchy proved the proposition in its entirety.

see also FIFTEEN THEOREM, VINOGRADOV'S THEOREM, LAGRANGE'S FOUR-SQUARE THEOREM, WARING'S PROBLEM

References

Cassels, J. W. S. *Rational Quadratic Forms.* New York: Academic Press, 1978.
Conway, J. H.; Guy, R. K.; Schneeberger, W. A.; and Sloane, N. J. A. "The Primary Pretenders." *Acta Arith.* **78**, 307–313, 1997.
Duke, W. "Some Old Problems and New Results about Quadratic Forms." *Not. Amer. Math. Soc.* **44**, 190–196, 1997.
Shanks, D. *Solved and Unsolved Problems in Number Theory, 4th ed.* New York: Chelsea, pp. 143–144, 1993.
Smith, D. E. *A Source Book in Mathematics.* New York: Dover, p. 91, 1984.

Fermat Polynomial

The POLYNOMIALS obtained by setting $p(x) = 3x$ and $q(x) = -2$ in the LUCAS POLYNOMIAL SEQUENCES. The first few Fermat polynomials are

$$\mathcal{F}_1(x) = 1$$
$$\mathcal{F}_2(x) = 3x$$
$$\mathcal{F}_3(x) = 9x^2 - 2$$
$$\mathcal{F}_4(x) = 27x^3 - 12x$$
$$\mathcal{F}_5(x) = 81x^4 - 54x^2 + 4,$$

and the first few Fermat-Lucas polynomials are

$$f_1(x) = 3x$$
$$f_2(x) = 9x^2 - 4$$
$$f_3(x) = 27x^3 - 18x$$
$$f_4(x) = 81x^4 - 72x^2 + 8$$
$$f_5(x) = 243x^5 - 270x^3 + 60x.$$

Fermat and Fermat-Lucas POLYNOMIALS satisfy

$$\mathcal{F}_n(1) = \mathcal{F}_n$$

$$f_n(1) = f_n$$

where $\mathcal{F}_n$ are FERMAT NUMBERS and f_n are FERMAT-LUCAS NUMBERS.

Fermat's Primality Test

see FERMAT'S LITTLE THEOREM

Fermat Prime

A FERMAT NUMBER $F_n = 2^{2^n} + 1$ which is PRIME.

see also CONSTRUCTIBLE POLYGON, FERMAT NUMBER

Fermat's Problem

In a given ACUTE TRIANGLE ΔABC, locate a point whose distances from A, B, and C have the smallest possible sum. The solution is the point from which each side subtends an angle of 120°, known as the FERMAT POINT.

see also ACUTE TRIANGLE, FERMAT POINT

Fermat Pseudoprime

A Fermat pseudoprime to a base a, written psp(a), is a COMPOSITE NUMBER n such that $a^{n-1} \equiv 1 \pmod n$ (i.e., it satisfies FERMAT'S LITTLE THEOREM, sometimes with the requirement that n must be ODD; Pomerance *et al.* 1980). psp(2)s are called POULET NUMBERS or, less commonly, SARRUS NUMBERS or FERMATIANS (Shanks 1993). The first few EVEN psp(2)s (including the PRIME 2 as a pseudoprime) are 2, 161038, 215326, ... (Sloane's A006935).

If base 3 is used in addition to base 2 to weed out potential COMPOSITE NUMBERS, only 4709 COMPOSITE NUMBERS remain $< 25 \times 10^9$. Adding base 5 leaves 2552, and base 7 leaves only 1770 COMPOSITE NUMBERS.

see also FERMAT'S LITTLE THEOREM, POULET NUMBER, PSEUDOPRIME

References

Pomerance, C.; Selfridge, J. L.; and Wagstaff, S. S. "The Pseudoprimes to $25 \cdot 10^9$." *Math. Comput.* **35**, 1003–1026, 1980. Available electronically from ftp://sable.ox.ac.uk/pub/math/primes/ps2.Z.

Shanks, D. *Solved and Unsolved Problems in Number Theory, 4th ed.* New York: Chelsea, p. 115, 1993.

Sloane, N. J. A. Sequence A006935/M2190 in "An On-Line Version of the Encyclopedia of Integer Sequences."

Fermat Quotient

The Fermat quotient for a number a and a PRIME base p is defined as

$$q_p(a) \equiv \frac{a^{p-1} - 1}{p}. \quad (1)$$

If $p \nmid ab$, then

$$q_p(ab) = q_p(a) + q_p(b) \quad (2)$$

$$q_p(p \pm 1) = \mp 1 \quad (3)$$

$$q_p(2) = \frac{1}{p}\left(1 - \frac{1}{2} + \frac{1}{3} - \frac{1}{4} + \ldots - \frac{1}{p-1}\right), \quad (4)$$

all (mod p). The quantity $q_p(2) = (2^{p-1} - 1)/p$ is known to be SQUARE for only two PRIMES: the so-called WIEFERICH PRIMES 1093 and 3511 (Lehmer 1981, Crandall 1986).

see also WIEFERICH PRIME

References

Crandall, R. *Projects in Scientific Computation.* New York: Springer-Verlag, 1986.

Lehmer, D. H. "On Fermat's Quotient, Base Two." *Math. Comput.* **36**, 289–290, 1981.

Fermat's Right Triangle Theorem

The AREA of a RATIONAL RIGHT TRIANGLE cannot be a SQUARE NUMBER. This statement is equivalent to "a CONGRUUM cannot be a SQUARE NUMBER."

see also CONGRUUM, RATIONAL TRIANGLE, RIGHT TRIANGLE, SQUARE NUMBER

Fermat's Sigma Problem

Solve

$$\sigma(x^3) = y^2$$

and

$$\sigma(x^2) = y^3,$$

where σ is the DIVISOR FUNCTION.

see also WALLIS'S PROBLEM

Fermat's Simple Theorem

see FERMAT'S LITTLE THEOREM

Fermat's Spiral

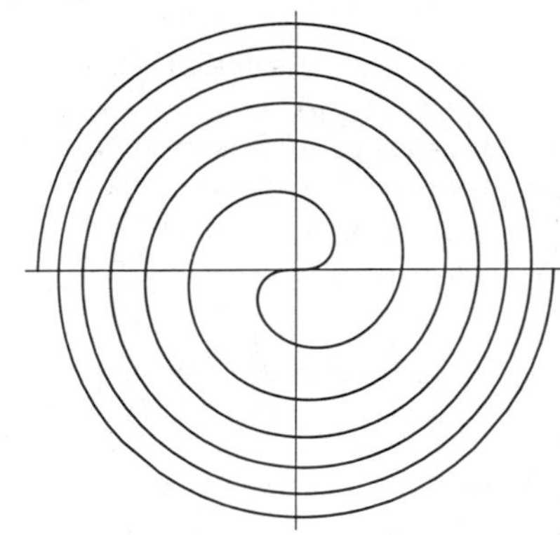

An ARCHIMEDEAN SPIRAL with $m = 2$ having polar equation

$$r = a\theta^{1/2},$$

discussed by Fermat in 1636 (MacTutor Archive). It is also known as the PARABOLIC SPIRAL. For any given POSITIVE value of θ, there are two corresponding values of r of opposite signs. The resulting spiral is therefore symmetrical about the line $y = -x$. The CURVATURE is

$$\kappa(\theta) = \frac{\frac{3a^2}{4\theta} + a^2\theta}{\left(\frac{a^2}{4\theta} + a^2\theta\right)^{3/2}}.$$

References

Dixon, R. *Mathographics.* New York: Dover, p. 121, 1991.

Gray, A. *Modern Differential Geometry of Curves and Surfaces.* Boca Raton, FL: CRC Press, pp. 69–70, 1993.
Lee, X. "Equiangular Spiral." http://www.best.com/~xah/SpecialPlaneCurves_dir/EquiangularSpiral_dir/equiangularSpiral.html.
Lockwood, E. H. *A Book of Curves.* Cambridge, England: Cambridge University Press, p. 175, 1967.
MacTutor History of Mathematics Archive. "Fermat's Spiral." http://www-groups.dcs.st-and.ac.uk/~history/Curves/Fermats.html.
Wells, D. *The Penguin Dictionary of Curious and Interesting Geometry.* Middlesex, England: Penguin Books, 1991.

Fermat Spiral Inverse Curve

The INVERSE CURVE of FERMAT'S SPIRAL with the origin taken as the INVERSION CENTER is the LITUUS.

References
Lawrence, J. D. *A Catalog of Special Plane Curves.* New York: Dover, pp. 186–187, 1972.

Fermat Sum Theorem

The only whole number solution to the DIOPHANTINE EQUATION

$$y^3 = x^2 + 2$$

is $y = 3$, $x = \pm 5$. This theorem was offered as a problem by Fermat, who suppressed his own proof.

Fermat's Theorem

A PRIME p can be represented in an essentially unique manner in the form $x^2 + y^2$ for integral x and y IFF $p \equiv 1 \pmod 4$ or $p = 2$. It can be restated by letting

$$Q(x, y) \equiv x^2 + y^2,$$

then all RELATIVELY PRIME solutions (x, y) to the problem of representing $Q(x, y) = m$ for m any INTEGER are achieved by means of successive applications of the GENUS THEOREM and COMPOSITION THEOREM. There is an analog of this theorem for EISENSTEIN INTEGERS.

see also EISENSTEIN INTEGER, SQUARE NUMBER

References
Shanks, D. *Solved and Unsolved Problems in Number Theory, 4th ed.* New York: Chelsea, pp. 142–143, 1993.

Fermat's Two-Square Theorem

see FERMAT'S THEOREM

Fermatian

see POULET NUMBER

Fermi-Dirac Distribution

A distribution which arises in the study of half-integral spin particles in physics,

$$P(k) = \frac{k^s}{e^{k-\mu} + 1}.$$

Its integral is

$$\int_0^\infty \frac{k^s\,dk}{e^{k-\mu} + 1} = e^\mu \Gamma(s+1)\Phi(-e^\mu, s+1, 1),$$

where Φ is the LERCH TRANSCENDENT.

Fern

see BARNSLEY'S FERN

Ferrari's Identity

$$(a^2 + 2ac - 2bc - b^2)^4 + (b^2 - 2ab - 2ac - c^2)^4$$
$$+(c^2 + 2ab + 2bc - a^2)^4 = 2(a^2 + b^2 + c^2 - ab + ac + bc)^4.$$

see also DIOPHANTINE EQUATION—QUARTIC

References
Berndt, B. C. *Ramanujan's Notebooks, Part IV.* New York: Springer-Verlag, pp. 96–97, 1994.

Ferrers Diagram

see YOUNG DIAGRAM

Ferrers' Function

An alternative name for an associated LEGENDRE POLYNOMIAL.

see also LEGENDRE POLYNOMIAL

References
Sansone, G. *Orthogonal Functions, rev. English ed.* New York: Dover, p. 246, 1991.

Ferrier's Prime

According to Hardy and Wright (1979), the largest PRIME found before the days of electronic computers is the 44-digit number

$$F \equiv \tfrac{1}{17}(2^{148} + 1)$$
$$= 20988936657440586486151264256610222593863921,$$

which was found using only a mechanical calculator.

References
Hardy, G. H. and Wright, E. M. *An Introduction to the Theory of Numbers, 5th ed.* Oxford, England: Clarendon Press, pp. 16–22, 1979.

Feuerbach Circle

see NINE-POINT CIRCLE

Feuerbach's Conic Theorem

The LOCUS of the centers of all CONICS through the VERTICES and ORTHOCENTER of a TRIANGLE (which are RECTANGULAR HYPERBOLAS when not degenerate), is a CIRCLE through the MIDPOINTS of the sides, the points half way from the ORTHOCENTER to the VERTICES, and the feet of the ALTITUDE.

see also ALTITUDE, CONIC SECTION, FEUERBACH'S THEOREM, KIEPERT'S HYPERBOLA, MIDPOINT, ORTHOCENTER, RECTANGULAR HYPERBOLA

References
Coolidge, J. L. *A Treatise on Algebraic Plane Curves.* New York: Dover, p. 198, 1959.

Feuerbach Point

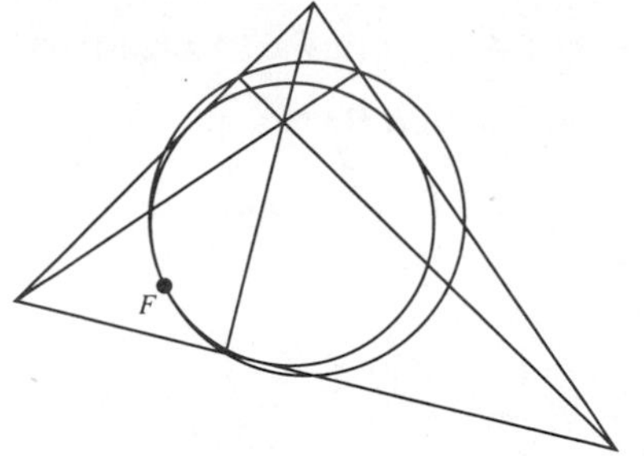

The point F at which the INCIRCLE and NINE-POINT CIRCLE are tangent. It has TRIANGLE CENTER FUNCTION

$$\alpha = 1 - \cos(B - C).$$

see also FEUERBACH'S THEOREM

References

Johnson, R. A. *Modern Geometry: An Elementary Treatise on the Geometry of the Triangle and the Circle.* Boston, MA: Houghton Mifflin, p. 200, 1929.
Kimberling, C. "Central Points and Central Lines in the Plane of a Triangle." *Math. Mag.* **67**, 163–187, 1994.
Salmon, G. *Conic Sections, 6th ed.* New York: Chelsea, p. 127, 1954.

Feuerbach's Theorem

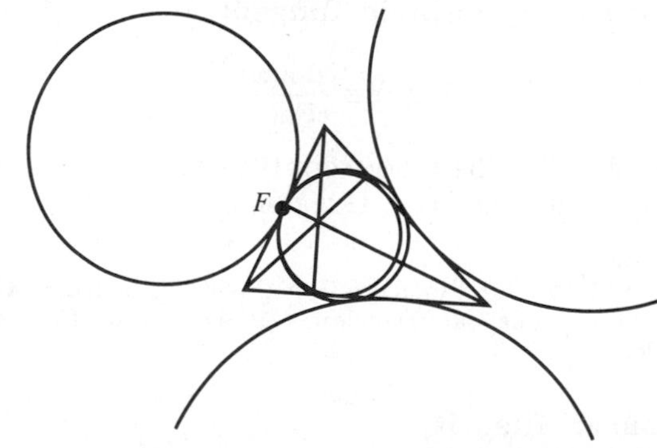

1. The CIRCLE which passes through the feet of the PERPENDICULARS dropped from the VERTICES of any TRIANGLE on the sides opposite them passes also through the MIDPOINTS of these sides as well as through the MIDPOINT of the segments which join the VERTICES to the point of intersection of the PERPENDICULAR (a NINE-POINT CIRCLE).
2. The NINE-POINT CIRCLE of any TRIANGLE is TANGENT internally to the INCIRCLE and TANGENT externally to the three EXCIRCLES.

see also EXCIRCLE, FEUERBACH POINT, INCIRCLE, MIDPOINT, NINE-POINT CIRCLE, PERPENDICULAR, TANGENT

References

Coxeter, H. S. M. and Greitzer, S. L. *Geometry Revisited.* Washington, DC: Math. Assoc. Amer., pp. 117–119, 1967.
Dixon, R. *Mathographics.* New York: Dover, p. 59, 1991.

Feynman Point

The sequence of six 9s which begins at the 762th decimal place of PI,

$$\pi = 3.14159\ldots134\underbrace{999999}837\ldots.$$

see also PI

FFT

see FAST FOURIER TRANSFORM

Fiber

A quantity F corresponding to a FIBER BUNDLE, where the FIBER BUNDLE is a MAP $f : E \to B$, with E the TOTAL SPACE of the FIBER BUNDLE and B the BASE SPACE of the FIBER BUNDLE.

see also FIBER BUNDLE, WHITNEY SUM

Fiber Bundle

A fiber bundle (also called simply a BUNDLE) with FIBER F is a MAP $f : E \to B$ where E is called the TOTAL SPACE of the fiber bundle and B the BASE SPACE of the fiber bundle. The main condition for the MAP to be a fiber bundle is that every point in the BASE SPACE $b \in B$ has a NEIGHBORHOOD U such that $f^{-1}(U)$ is HOMEOMORPHIC to $U \times F$ in a special way. Namely, if

$$h : f^{-1}(U) \to U \times F$$

is the HOMEOMORPHISM, then

$$\text{proj}_U \circ h = f_{|f^{-1}(U)|},$$

where the MAP proj_U means projection onto the U component. The homeomorphisms h which "commute with projection" are called local TRIVIALIZATIONS for the fiber bundle f. In other words, E looks like the product $B \times F$ (at least locally), except that the fibers $f^{-1}(x)$ for $x \in B$ may be a bit "twisted."

Examples of fiber bundles include any product $B \times F \to B$ (which is a bundle over B with FIBER F), the MÖBIUS STRIP (which is a fiber bundle over the CIRCLE with FIBER given by the unit interval [0,1]; i.e, the BASE SPACE is the CIRCLE), and $\mathbb{S}^3$ (which is a bundle over $\mathbb{S}^2$ with fiber $\mathbb{S}^1$). A special class of fiber bundle is the VECTOR BUNDLE, in which the FIBER is a VECTOR SPACE.

see also BUNDLE, FIBER SPACE, FIBRATION

Fiber Space

A fiber space, depending on context, means either a FIBER BUNDLE or a FIBRATION.

see also FIBER BUNDLE, FIBRATION

Fibonacci Dual Theorem

Let F_n be the nth FIBONACCI NUMBER. Then the sequence $\{F_n\}_{n=2}^{\infty} = \{1, 2, 3, 5, 8, \ldots\}$ is COMPLETE, even if one is restricted to subsequences in which no two consecutive terms are both passed over (until the desired total is reached; Brown 1965, Honsberger 1985).

see also COMPLETE SEQUENCE, FIBONACCI NUMBER.

References

Brown, J. L. Jr. "A New Characterization of the Fibonacci Numbers." *Fib. Quart.* **3**, 1–8, 1965.

Honsberger, R. *Mathematical Gems III.* Washington, DC: Math. Assoc. Amer., p. 130, 1985.

Fibonacci Hyperbolic Cosine

Let

$$\psi \equiv 1 + \phi = \tfrac{1}{2}(3 + \sqrt{5}\,) \approx 2.618034 \quad (1)$$

where ϕ is the GOLDEN RATIO, and

$$\alpha = \ln \phi \approx 0.4812118. \quad (2)$$

Then define

$$\begin{aligned} \mathrm{cFh}(x) &\equiv \frac{\psi^{x+1/2} + \psi^{-(x+1/2)}}{\sqrt{5}} && (3) \\ &= \frac{\phi^{(2x+1)} + \phi^{-(2x+1)}}{\sqrt{5}} && (4) \\ &= \frac{2}{\sqrt{5}} \cosh[(2x+1)\alpha]. && (5) \end{aligned}$$

This function satisfies

$$\mathrm{cFh}(-x) = \mathrm{cFh}(x-1). \quad (6)$$

For $n \in \mathbb{Z}$, $\mathrm{cFh}(n) = F_{2n+1}$ where F_n is a FIBONACCI NUMBER.

References

Trzaska, Z. W. "On Fibonacci Hyperbolic Trigonometry and Modified Numerical Triangles." *Fib. Quart.* **34**, 129–138, 1996.

Fibonacci Hyperbolic Cotangent

$$\mathrm{ctFh}(x) \equiv \frac{\mathrm{cFh}(x)}{\mathrm{sFh}(x)},$$

where $\mathrm{cFh}(x)$ is the FIBONACCI HYPERBOLIC COSINE and $\mathrm{sFh}(x)$ is the FIBONACCI HYPERBOLIC SINE.

References

Trzaska, Z. W. "On Fibonacci Hyperbolic Trigonometry and Modified Numerical Triangles." *Fib. Quart.* **34**, 129–138, 1996.

Fibonacci Hyperbolic Sine

Let

$$\psi \equiv 1 + \phi = \tfrac{1}{2}(3 + \sqrt{5}\,) \approx 2.618034 \quad (1)$$

where ϕ is the GOLDEN RATIO, and

$$\alpha = \ln \phi \approx 0.4812118. \quad (2)$$

Then define

$$\begin{aligned} \mathrm{sFh}(x) &\equiv \frac{\psi^{x} - \psi^{-x}}{\sqrt{5}} && (3) \\ &= \frac{\phi^{2x} - \phi^{-2x}}{\sqrt{5}} && (4) \\ &= \frac{2}{\sqrt{5}} \sinh[2x\alpha]. && (5) \end{aligned}$$

For $n \in \mathbb{Z}$, $\mathrm{sFh}(n) = F_{2n}$ where F_n is a FIBONACCI NUMBER. The function satisfies

$$\mathrm{sFh}(-x) = -\,\mathrm{sFh}(x). \quad (6)$$

References

Trzaska, Z. W. "On Fibonacci Hyperbolic Trigonometry and Modified Numerical Triangles." *Fib. Quart.* **34**, 129–138, 1996.

Fibonacci Hyperbolic Tangent

$$\mathrm{tFh}(x) \equiv \frac{\mathrm{sFh}(x)}{\mathrm{cFh}(x)},$$

where $\mathrm{sFh}(x)$ is the FIBONACCI HYPERBOLIC SINE and $\mathrm{cFh}(x)$ is the FIBONACCI HYPERBOLIC COSINE.

References

Trzaska, Z. W. "On Fibonacci Hyperbolic Trigonometry and Modified Numerical Triangles." *Fib. Quart.* **34**, 129–138, 1996.

Fibonacci Identity

Since

$$|(a+ib)(c+id)| = |a+ib|\,|c+di| \quad (1)$$

$$|(ac-bd) + i(bc+ad)| = \sqrt{a^2+b^2}\sqrt{c^2+d^2}, \quad (2)$$

it follows that

$$(a^2+b^2)(c^2+d^2) = (ac-bd)^2 + (bc+ad)^2 \equiv e^2 + f^2. \quad (3)$$

This identity implies the 2-D CAUCHY-SCHWARZ SUM INEQUALITY.

see also CAUCHY-SCHWARZ SUM INEQUALITY, EULER FOUR-SQUARE IDENTITY

References

Petkovšek, M.; Wilf, H. S.; and Zeilberger, D. *A=B.* Wellesley, MA: A. K. Peters, p. 9, 1996.

Fibonacci Matrix

A SQUARE MATRIX related to the FIBONACCI NUMBERS. The simplest is the FIBONACCI Q-MATRIX.

Fibonacci n-Step Number

An n-step Fibonacci sequence is given by defining $F_k = 0$ for $k \leq 0$, $F_1 = F_2 = 1$, $F_3 = 2$, and

$$F_k = \sum_{i=1}^{k} F_{n-i} \tag{1}$$

for $k > 3$. The case $n = 1$ corresponds to the degenerate 1, 1, 2, 2, 2, 2 ..., $n = 2$ to the usual FIBONACCI NUMBERS 1, 1, 2, 3, 5, 8, ... (Sloane's A000045), $n = 3$ to the TRIBONACCI NUMBERS 1, 1, 2, 4, 7, 13, 24, 44, 81, ... (Sloane's A000073), $n = 4$ to the TETRANACCI NUMBERS 1, 1, 2, 4, 8, 15, 29, 56, 108, ... (Sloane's A000078), etc.

The limit $\lim_{k\to\infty} F_k/F_{k-1}$ is given by solving

$$x^n(2-x) = 1 \tag{2}$$

for x and taking the REAL ROOT $x > 1$. If $n = 2$, the equation reduces to

$$x^2(2-x) = 1 \tag{3}$$

$$x^3 - 2x^2 + 1 = (x-1)(x^2 - x - 1) = 0, \tag{4}$$

giving solutions

$$x = 1, \tfrac{1}{2}(1 \pm \sqrt{5}\,). \tag{5}$$

The ratio is therefore

$$x = \tfrac{1}{2}(1 + \sqrt{5}\,) = \phi = 1.618\ldots, \tag{6}$$

which is the GOLDEN RATIO, as expected. Solutions for $n = 1, 2, \ldots$ are given numerically by 1, 1.61803, 1.83929, 1.92756, 1.96595, ..., approaching 2 as $n \to \infty$.

see also FIBONACCI NUMBER, TRIBONACCI NUMBER

References

Sloane, N. J. A. Sequences A000045/M0692, A000073/M1074, and A000078/M1108 in "An On-Line Version of the Encyclopedia of Integer Sequences."

Fibonacci Number

The sequence of numbers defined by the U_n in the LUCAS SEQUENCE. They are companions to the LUCAS NUMBERS and satisfy the same RECURRENCE RELATION,

$$F_n \equiv F_{n-2} + F_{n-1} \tag{1}$$

for $n = 3, 4, \ldots$, with $F_1 = F_2 = 1$. The first few Fibonacci numbers are 1, 1, 2, 3, 5, 8, 13, 21, ... (Sloane's A000045). The Fibonacci numbers give the number of pairs of rabbits n months after a single pair begins breeding (and newly born bunnies are assumed to begin breeding when they are two months old).

The ratios of alternate Fibonacci numbers are given by the convergents to ϕ^{-2}, where ϕ is the GOLDEN RATIO, and are said to measure the fraction of a turn between successive leaves on the stalk of a plant (PHYLLOTAXIS): 1/2 for elm and linden, 1/3 for beech and hazel, 2/5 for oak and apple, 3/8 for poplar and rose, 5/13 for willow and almond, etc. (Coxeter 1969, Ball and Coxeter 1987). The Fibonacci numbers are sometimes called PINE CONE NUMBERS (Pappas 1989, p. 224)

Another RECURRENCE RELATION for the Fibonacci numbers is

$$F_{n+1} = \left\lfloor \frac{F_n(1+\sqrt{5})+1}{2} \right\rfloor = \left\lfloor \phi F_n + \tfrac{1}{2} \right\rfloor, \tag{2}$$

where $\lfloor x \rfloor$ is the FLOOR FUNCTION and ϕ is the GOLDEN RATIO. This expression follows from the more general RECURRENCE RELATION that

$$\begin{vmatrix} F_n & F_{n+1} & \cdots & F_{n+k} \\ F_{n+k+1} & F_{n+k+2} & \cdots & F_{n+2k} \\ \vdots & \vdots & \ddots & \vdots \\ F_{n+k(k-1)+1} & F_{n+k(k-1)+2} & \cdots & F_{n+k^2} \end{vmatrix} = 0. \tag{3}$$

The GENERATING FUNCTION for the Fibonacci numbers is

$$g(x) = \sum_{n=0}^{\infty} F_n x^n = \frac{1}{1-x-x^2}. \tag{4}$$

Yuri Matijasevič (1970) proved that the equation $n = F_{2m}$ is a DIOPHANTINE EQUATION. This led to the proof of the impossibility of the tenth of HILBERT'S PROBLEMS (does there exist a general method for solving DIOPHANTINE EQUATIONS?) by Julia Robinson and Martin Davis in 1970.

The Fibonacci number F_{n+1} gives the number of ways for 2×1 DOMINOES to cover a $2 \times n$ CHECKERBOARD, as illustrated in the following diagrams (Dickau).

The number of ways of picking a SET (including the EMPTY SET) from the numbers 1, 2, ..., n without picking two consecutive numbers is F_{n+2}. The number of ways of picking a set (including the EMPTY SET) from the numbers 1, 2, ..., n without picking two consecutive numbers (where 1 and n are now consecutive) is $L_n = F_{n+1} + F_{n-1}$, where L_n is a LUCAS NUMBER. The probability of not getting two heads in a row in n tosses of a COIN is $F_{n+2}/2^n$ (Honsberger 1985, pp. 120–122). Fibonacci numbers are also related to the number of ways in which n COIN TOSSES can be made such that there are not three consecutive heads or tails. The number of ideals of an n-element FENCE POSET is the Fibonacci number F_n.

Sum identities are

$$\sum_{k=1}^{n} F_k = F_{n+2} - 1. \tag{5}$$

$$F_1 + F_3 + F_5 + \ldots + F_{2k+1} = F_{2k+2} \tag{6}$$

$$1 + F_2 + F_4 + F_6 + \ldots + F_{2k} = F_{2k+1} \tag{7}$$

$$\sum_{k=1}^{n} {F_k}^2 = F_n F_{n+1} \tag{8}$$

$$F_{2n} = {F_{n+1}}^2 - {F_{n-1}}^2 \tag{9}$$

$$F_{3n} = {F_{n+1}}^3 + {F_n}^3 + {F_{n-1}}^3. \tag{10}$$

Additional RECURRENCE RELATIONS are CASSINI'S IDENTITY

$$F_{n-1}F_{n+1} - {F_n}^2 = (-1)^n \tag{11}$$

and the relations

$$F_{2n+1} = 1 + F_2 + F_4 + \ldots + F_{2n} \tag{12}$$

$${F_{n+1}}^2 = 4F_nF_{n-1} + {F_{n-2}}^2 \tag{13}$$

(Brousseau 1972),

$$F_{n+m} = F_{n-1}F_m + F_nF_{m+1} \tag{14}$$

$$F_{(k+1)n} = F_{n-1}F_{kn} + F_nF_{kn+1} \tag{15}$$

(Honsberger 1985, p. 107),

$$F_n = F_lF_{n-l+1} + F_{l-1}F_{n-l}, \tag{16}$$

so if $l = n - l + 1$, then $2l = n + 1$ and $l = (n + 1)/2$

$$F_n = {F_{(n+1)/2}}^2 + {F_{(n-1)/2}}^2. \tag{17}$$

Letting $k \equiv (n - 1)/2$,

$$F_{2k+1} = {F_{k+1}}^2 + {F_k}^2 \tag{18}$$

$${F_{n+2}}^2 - {F_{n+1}}^2 = F_nF_{n+3} \tag{19}$$

$${F_n}^2 = {F_{n-1}}^2 + 3{F_{n-2}}^2 + 2F_{n-2}F_{n-3}. \tag{20}$$

Sum FORMULAS for F_n include

$$F_n = \frac{1}{2^{n-1}}\left[\binom{n}{1} + 5\binom{n}{3} + 5^2\binom{n}{5} + \ldots\right] \tag{21}$$

$$F_{n+1} = \binom{n}{0} + \binom{n-1}{1} + \binom{n-2}{2} + \ldots. \tag{22}$$

Cesàro derived the FORMULAS

$$\sum_{k=0}^{n} \binom{n}{k} F_k = F_{2n} \tag{23}$$

$$\sum_{k=0}^{n} \binom{n}{k} 2^k F_k = F_{3n} \tag{24}$$

(Honsberger 1985, pp. 109–110). Additional identities can be found throughout the *Fibonacci Quarterly* journal. A list of 47 generalized identities are given by Halton (1965).

In terms of the LUCAS NUMBER L_n,

$$F_{2n} = F_nL_n \tag{25}$$

$$F_{2n}({L_{2n}}^2 - 1) = F_{6n} \tag{26}$$

$$F_{m+p} + (-1)^{p+1}F_{m-p} = F_pL_m \tag{27}$$

$$\sum_{k=a+1}^{a+4n} F_k = F_{a+4n+2} - F_{a+2} = F_{2n}L_{a+2n+2} \tag{28}$$

(Honsberger 1985, pp. 111–113). A remarkable identity is

$$\exp(L_1x + \tfrac{1}{2}L_2x^2 + \tfrac{1}{3}L_3x^3 + \ldots) = F_1 + F_2x + F_3x^2 + \ldots \tag{29}$$

(Honsberger 1985, pp. 118-119). It is also true that

$$5{F_n}^2 = {L_n}^2 - 4(-1)^n \tag{30}$$

and

$$\frac{{L_n}^2 - (-1)^a{L_{n+a}}^2}{{F_n}^2 - (-1)^a{F_{n+a}}^2} = 5 \tag{31}$$

for a ODD, and

$$\frac{{L_n}^2 + {L_{n+a}}^2 - 8(-1)^n}{{F_n}^2 + {F_{n+a}}^2} = 5 \tag{32}$$

for a EVEN (Freitag 1996).

The equation (1) is a LINEAR RECURRENCE SEQUENCE

$$x_n = Ax_{x-1} + Bx_{n-2} \qquad n \geq 3, \tag{33}$$

so the closed form for F_n is given by

$$F_n = \frac{\alpha^n - \beta^n}{\alpha - \beta}, \tag{34}$$

where α and β are the roots of $x^2 = Ax + B$. Here, $A = B = 1$, so the equation becomes

$$x^2 - x - 1 = 0, \tag{35}$$

which has ROOTS

$$x = \tfrac{1}{2}(1 \pm \sqrt{5}). \tag{36}$$

The closed form is therefore given by

$$F_n = \frac{(1+\sqrt{5})^n - (1-\sqrt{5})^n}{2^n \sqrt{5}}. \tag{37}$$

This is known as BINET'S FORMULA. Another closed form is

$$F_n = \left[\frac{1}{\sqrt{5}}\left(\frac{1+\sqrt{5}}{2}\right)^n\right] = \left[\frac{\phi^n}{\sqrt{5}}\right], \tag{38}$$

where $[x]$ is the NINT function.

From (1), the RATIO of consecutive terms is

$$\begin{aligned}\frac{F_n}{F_{n-1}} &= 1 + \frac{F_{n-2}}{F_{n-1}} = 1 + \frac{1}{\frac{F_{n-1}}{F_{n-2}}} \\ &= 1 + \cfrac{1}{1 + \frac{1}{\frac{F_{n-3}}{F_{n-2}}}} = \left[1, 1, \ldots, \frac{F_2}{F_1}\right] \\ &= [\underbrace{1, 1, \ldots, 1}_{n-1}],\end{aligned} \tag{39}$$

which is just the first few terms of the CONTINUED FRACTION for the GOLDEN RATIO ϕ. Therefore,

$$\lim_{n\to\infty} \frac{F_n}{F_{n-1}} = \phi. \tag{40}$$

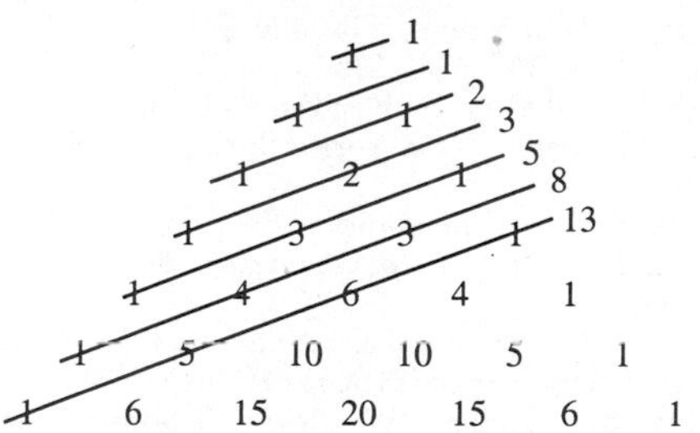

The "SHALLOW DIAGONALS" of PASCAL'S TRIANGLE sum to Fibonacci numbers (Pappas 1989),

$$\sum_{k=1}^{n} \binom{k}{n-k} = \frac{(-1)^n {}_3F_2(1, 2, 1-n; \frac{1}{2}(3-n), 2 - \frac{1}{2}n; -\frac{1}{4})}{\pi(2 - 3n + n^2)} = F_{n+1}, \tag{41}$$

where ${}_3F_2(a, b, c; d, e; z)$ is a GENERALIZED HYPERGEOMETRIC FUNCTION.

The sequence of final digits in Fibonacci numbers repeats in cycles of 60. The last two digits repeat in 300, the last three in 1500, the last four in 15,000, etc.

$$\sum_{n=1}^{\infty} \frac{(-1)^n}{F_n F_{n+2}} = 2 - \sqrt{5} \tag{42}$$

(Clark 1995). A very curious addition of the Fibonacci numbers is the following addition tree,

```
0
 1
  1
   2
    3
     5
      8
      13
       21
        34
         55
          89
            ⋱
0112359550561...
```

which is equal to the fractional digits of 1/89,

$$\sum_{n=0}^{\infty} \frac{F_n}{10^{n+1}} = \frac{1}{89}. \tag{43}$$

For $n \geq 3$, $F_n | F_m$ IFF $n | m$. $L_n | L_m$ IFF n divides into m an EVEN number of times. $(F_m, F_n) = F_{(m,n)}$ (Michael 1964; Honsberger 1985, pp. 131-132). No ODD Fibonacci number is divisible by 17 (Honsberger 1985, pp. 132 and 242). No Fibonacci number > 8 is ever of the form $p - 1$ or $p + 1$ where p is a PRIME number (Honsberger 1985, p. 133).

Consider the sum

$$s_k = \sum_{n=2}^{k} \frac{1}{F_{n-1}F_{n+1}} = \sum_{n=2}^{k} \left(\frac{1}{F_{n-1}F_n} - \frac{1}{F_n F_{n+1}}\right). \tag{44}$$

This is a TELESCOPING SUM, so

$$s_k = 1 - \frac{1}{F_{k+1}F_{k+2}}, \tag{45}$$

thus

$$S \equiv \lim_{k\to\infty} s_k = 1 \tag{46}$$

(Honsberger 1985, pp. 134–135). Using BINET'S FORMULA, it also follows that

$$\frac{F_{n+r}}{F_n} = \frac{\alpha^{n+r} - \beta^{n+r}}{\alpha^n - \beta^n} = \frac{\alpha^{n+r}}{\alpha^n} \frac{1 - \left(\frac{\beta}{\alpha}\right)^{n+r}}{1 - \left(\frac{\beta}{\alpha}\right)^n}, \tag{47}$$

where

$$\alpha = \tfrac{1}{2}(1 + \sqrt{5}) \tag{48}$$

$$\beta = \tfrac{1}{2}(1 - \sqrt{5}) \tag{49}$$

so

$$\frac{F_{n+r}}{F_n} = \alpha^r. \quad (50)$$

$$S' = \sum_{n=1}^{\infty} \frac{F_n}{F_{n+1}F_{n+2}} = 1 \quad (51)$$

(Honsberger 1985, pp. 138 and 242–243). The MILLIN SERIES has sum

$$S'' \equiv \sum_{n=0}^{\infty} \frac{1}{F_{2^n}} = \tfrac{1}{2}(7-\sqrt{5}) \quad (52)$$

(Honsberger 1985, pp. 135–137).

The Fibonacci numbers are COMPLETE. In fact, dropping one number still leaves a COMPLETE SEQUENCE, although dropping two numbers does not (Honsberger 1985, pp. 123 and 126). Dropping two terms from the Fibonacci numbers produces a sequence which is not even WEAKLY COMPLETE (Honsberger 1985, p. 128). However, the sequence

$$F'_n \equiv F_n - (-1)^n \quad (53)$$

is WEAKLY COMPLETE, even with any finite subsequence deleted (Graham 1964). $\{{F_n}^2\}$ is not COMPLETE, but $\{{F_n}^2\} + \{{F_n}^2\}$ are. 2^{N-1} copies of $\{{F_n}^N\}$ are COMPLETE.

For a discussion of SQUARE Fibonacci numbers, see Cohn (1964), who proved that the only SQUARE NUMBER Fibonacci numbers are 1 and $F_{12} = 144$ (Cohn 1964, Guy 1994). Ming (1989) proved that the only TRIANGULAR Fibonacci numbers are 1, 3, 21, and 55. The Fibonacci and LUCAS NUMBERS have no common terms except 1 and 3. The only CUBIC Fibonacci numbers are 1 and 8.

$$(F_nF_{n+3}, 2F_{n+1}F_{n+2}, F_{2n+3} = {F_{n+1}}^2 + {F_{n+2}}^2) \quad (54)$$

is a PYTHAGOREAN TRIPLE.

$${F_{4n}}^2 + 8F_{2n}(F_{2n} + F_{6n}) = (3F_{4n})^2 \quad (55)$$

is always a SQUARE NUMBER (Honsberger 1985, p. 243).

In 1975, James P. Jones showed that the Fibonacci numbers are the POSITIVE INTEGER values of the POLYNOMIAL

$$P(x,y) = -y^5 + 2y^4x + y^3x^2 - 2y^2x^3 - y(x^4 - 2) \quad (56)$$

for GAUSSIAN INTEGERS x and y (Le Lionnais 1983). If n and k are two POSITIVE INTEGERS, then between n^k and n^{k+1}, there can never occur more than n Fibonacci numbers (Honsberger 1985, pp. 104–105).

Every F_n that is PRIME has a PRIME n, but the converse is not necessarily true. The first few PRIME Fibonacci numbers are for n = 3, 4, 5, 7, 11, 13, 17, 23, 29, 43, 47, 83, 131, 137, 359, 431, 433, 449, 509, 569, 571, ... (Sloane's A001605; Dubner and Keller 1998). Gardner's statement that F_{531} is prime is incorrect, especially since 531 is not even PRIME (Gardner 1979, p. 161). It is not known if there are an INFINITE number of Fibonacci primes.

The Fibonacci numbers F_n, are SQUAREFUL for n = 6, 12, 18, 24, 25, 30, 36, 42, 48, 50, 54, 56, 60, 66, ..., 300, 306, 312, 324, 325, 330, 336, ... (Sloane's A037917) and SQUAREFREE for n = 1, 2, 3, 4, 5, 7, 8, 9, 10, 11, 13, ... (Sloane's A037918). The largest known SQUAREFUL Fibonacci number is F_{336}, and no SQUAREFUL Fibonacci numbers F_p are known with p PRIME.

see also CASSINI'S IDENTITY, FAST FIBONACCI TRANSFORM, FIBONACCI DUAL THEOREM, FIBONACCI n-STEP NUMBER, FIBONACCI Q-MATRIX, GENERALIZED FIBONACCI NUMBER, INVERSE TANGENT, LINEAR RECURRENCE SEQUENCE, LUCAS SEQUENCE, NEAR NOBLE NUMBER, PELL SEQUENCE, RABBIT CONSTANT, STOLARSKY ARRAY, TETRANACCI NUMBER, TRIBONACCI NUMBER, WYTHOFF ARRAY, ZECKENDORF REPRESENTATION, ZECKENDORF'S THEOREM

References

Ball, W. W. R. and Coxeter, H. S. M. *Mathematical Recreations and Essays, 13th ed.* New York: Dover, pp. 56–57, 1987.

Basin, S. L. and Hoggatt, V. E. Jr. "A Primer on the Fibonacci Sequence." *Fib. Quart.* **1**, 1963.

Basin, S. L. and Hoggatt, V. E. Jr. "A Primer on the Fibonacci Sequence—Part II." *Fib. Quart.* **1**, 61–68, 1963.

Borwein, J. M. and Borwein, P. B. *Pi & the AGM: A Study in Analytic Number Theory and Computational Complexity.* New York: Wiley, pp. 94–101, 1987.

Brillhart, J.; Montgomery, P. L.; and Silverman, R. D. "Tables of Fibonacci and Lucas Factorizations." *Math. Comput.* **50**, 251–260 and S1–S15, 1988.

Brook, M. "Fibonacci Formulas." *Fib. Quart.* **1**, 60, 1963.

Brousseau, A. "Fibonacci Numbers and Geometry." *Fib. Quart.* **10**, 303–318, 1972.

Clark, D. Solution to Problem 10262. *Amer. Math. Monthly* **102**, 467, 1995.

Cohn, J. H. E. "On Square Fibonacci Numbers." *J. London Math. Soc.* **39**, 537–541, 1964.

Conway, J. H. and Guy, R. K. "Fibonacci Numbers." In *The Book of Numbers.* New York: Springer-Verlag, pp. 111–113, 1996.

Coxeter, H. S. M. "The Golden Section and Phyllotaxis." Ch. 11 in *Introduction to Geometry, 2nd ed.* New York: Wiley, 1969.

Dickau, R. M. "Fibonacci Numbers." `http://www.prairienet.org/~pops/fibboard.html`.

Dubner, H. and Keller, W. "New Fibonacci and Lucas Primes." *Math. Comput.* 1998.

Freitag, H. Solution to Problem B-772. "An Integral Ratio." *Fib. Quart.* **34**, 82, 1996.

Gardner, M. *Mathematical Circus: More Puzzles, Games, Paradoxes and Other Mathematical Entertainments from Scientific American.* New York: Knopf, 1979.

Graham, R. "A Property of Fibonacci Numbers." *Fib. Quart.* **2**, 1–10, 1964.

Guy, R. K. "Fibonacci Numbers of Various Shapes." §D26 in *Unsolved Problems in Number Theory, 2nd ed.* New York: Springer-Verlag, pp. 194–195, 1994.

Halton, J. H. "On a General Fibonacci Identity." *Fib. Quart.* **3**, 31–43, 1965.

Hoggatt, V. E. Jr. *The Fibonacci and Lucas Numbers.* Boston, MA: Houghton Mifflin, 1969.

Hoggatt, V. E. Jr. and Ruggles, I. D. "A Primer on the Fibonacci Sequence—Part III." *Fib. Quart.* **1**, 61–65, 1963.

Hoggatt, V. E. Jr. and Ruggles, I. D. "A Primer on the Fibonacci Sequence—Part IV." *Fib. Quart.* **1**, 65–71, 1963.

Hoggatt, V. E. Jr. and Ruggles, I. D. "A Primer on the Fibonacci Sequence—Part V." *Fib. Quart.* **2**, 59–66, 1964.

Hoggatt, V. E. Jr.; Cox, N.; and Bicknell, M. "A Primer for the Fibonacci Numbers: Part XII." *Fib. Quart.* **11**, 317–331, 1973.

Honsberger, R. "A Second Look at the Fibonacci and Lucas Numbers." Ch. 8 in *Mathematical Gems III.* Washington, DC: Math. Assoc. Amer., 1985.

Knott, R. "Fibonacci Numbers and the Golden Section." `http://www.mcs.surrey.ac.uk/Personal/R.Knott/Fibonacci/fib.html`.

Le Lionnais, F. *Les nombres remarquables.* Paris: Hermann, p. 146, 1983.

Leyland, P. `ftp://sable.ox.ac.uk/pub/math/factors/fibonacci.Z`.

Matijasevič, Yu. V. "Solution to of the Tenth Problem of Hilbert." *Mat. Lapok* **21**, 83–87, 1970.

Matijasevich, Yu. V. *Hilbert's Tenth Problem.* Cambridge, MA: MIT Press, 1993.

Michael, G. "A New Proof for an Old Property." *Fib. Quart.* **2**, 57–58, 1964.

Ming, L. "On Triangular Fibonacci Numbers." *Fib. Quart.* **27**, 98–108, 1989.

Ogilvy, C. S. and Anderson, J. T. "Fibonacci Numbers." Ch. 11 in *Excursions in Number Theory.* New York: Dover, pp. 133–144, 1988.

Pappas, T. "Fibonacci Sequence," "Pascal's Triangle, the Fibonacci Sequence & Binomial Formula," "The Fibonacci Trick," and "The Fibonacci Sequence & Nature." *The Joy of Mathematics.* San Carlos, CA: Wide World Publ./Tetra, pp. 28–29, 40–41, 51, 106, and 222–225, 1989.

Schroeder, M. *Fractals, Chaos, Power Laws: Minutes from an Infinite Paradise.* New York: W. H. Freeman, pp. 49–57, 1991.

Sloane, N. J. A. Sequences A037917, A037918, A000045/M0692, and A001605/M2309 in "An On-Line Version of the Encyclopedia of Integer Sequences."

Vorob'ev, N. N. *Fibonacci Numbers.* New York: Blaisdell Publishing Co., 1961.

Fibonacci Polynomial

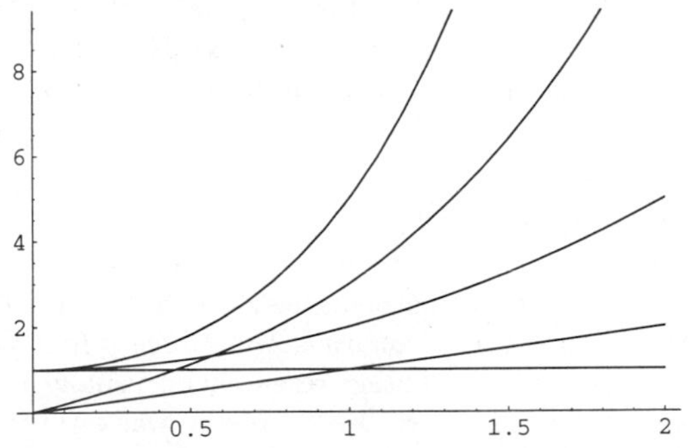

The W POLYNOMIALS obtained by setting $p(x) = x$ and $q(x) = 1$ in the LUCAS POLYNOMIAL SEQUENCE. (The corresponding w POLYNOMIALS are called LUCAS POLYNOMIALS.) The Fibonacci polynomials are defined by the RECURRENCE RELATION

$$F_{n+1}(x) = xF_n(x) + F_{n-1}(x), \tag{1}$$

with $F_1(x) = 1$ and $F_2(x) = x$. They are also given by the explicit sum formula

$$F_n(x) = \sum_{j=0}^{\lfloor (n-1)/2 \rfloor} \binom{n-j-1}{j} x^{n-2j-1}, \tag{2}$$

where $\lfloor x \rfloor$ is the FLOOR FUNCTION and $\binom{n}{m}$ is a BINOMIAL COEFFICIENT. The first few Fibonacci polynomials are

$$\begin{aligned} F_1(x) &= 1 \\ F_2(x) &= x \\ F_3(x) &= x^2 + 1 \\ F_4(x) &= x^3 + 2x \\ F_5(x) &= x^4 + 3x^2 + 1. \end{aligned}$$

The Fibonacci polynomials are normalized so that

$$F_n(1) = F_n, \tag{3}$$

where the F_ns are FIBONACCI NUMBERS.

The Fibonacci polynomials are related to the MORGAN-VOYCE POLYNOMIALS by

$$F_{2n+1}(x) = b_n(x^2) \tag{4}$$

$$F_{2n+n2}(x) = xB_n(x^2) \tag{5}$$

(Swamy 1968).

see also BRAHMAGUPTA POLYNOMIAL, FIBONACCI NUMBER, MORGAN-VOYCE POLYNOMIAL

References

Swamy, M. N. S. "Further Properties of Morgan-Voyce Polynomials." *Fib. Quart.* **6**, 167–175, 1968.

Fibonacci Pseudoprime

Consider a LUCAS SEQUENCE with $P > 0$ and $Q = \pm 1$. A Fibonacci pseudoprime is a COMPOSITE NUMBER n such that

$$V_n \equiv P \pmod{n}.$$

There exist no EVEN Fibonacci pseudoprimes with parameters $P = 1$ and $Q = -1$ (Di Porto 1993) or $P = Q = 1$ (André-Jeannin 1996). André-Jeannin (1996) also proved that if $(P,Q) \neq (1,-1)$ and $(P,Q) \neq (1,1)$, then there exists at least one EVEN Fibonacci pseudoprime with parameters P and Q.

see also PSEUDOPRIME

References

André-Jeannin, R. "On the Existence of Even Fibonacci Pseudoprimes with Parameters P and Q." *Fib. Quart.* **34**, 75–78, 1996.

Di Porto, A. "Nonexistence of Even Fibonacci Pseudoprimes of the First Kind." *Fib. Quart.* **31**, 173–177, 1993.

Ribenboim, P. "Fibonacci Pseudoprimes." §2.X.A in *The New Book of Prime Number Records, 3rd ed.* New York: Springer-Verlag, pp. 127–129, 1996.

Fibonacci Q-Matrix

A FIBONACCI MATRIX of the form

$$\mathsf{M} = \begin{bmatrix} m & 1 \\ 1 & 0 \end{bmatrix}. \quad (1)$$

If U and V are defined as BINET FORMS

$$U_n = mU_{n-1} + U_{n-2} \quad (U_0 = 0, U_1 = 1) \quad (2)$$

$$V_n = mV_{n-1} + V_{n-2} \quad (V_0 = 2, V_1 = m), \quad (3)$$

then

$$\mathsf{M} = \begin{bmatrix} U_{n+1} & U_n \\ U_n & U_{n-1} \end{bmatrix} \quad (4)$$

$$\mathsf{M}^{-1} = \mathsf{M} - m\mathsf{I} = \begin{bmatrix} 0 & 1 \\ 1 & -m \end{bmatrix}. \quad (5)$$

Defining

$$\mathsf{Q} \equiv \begin{bmatrix} F_2 & F_1 \\ F_1 & F_0 \end{bmatrix} = \begin{bmatrix} 1 & 1 \\ 1 & 0 \end{bmatrix}, \quad (6)$$

then

$$\mathsf{Q}^n = \begin{bmatrix} F_{n+1} & F_n \\ F_n & F_{n-1} \end{bmatrix} \quad (7)$$

(Honsberger 1985, pp. 106–107).

see also BINET FORMS, FIBONACCI NUMBER

References

Honsberger, R. "A Second Look at the Fibonacci and Lucas Numbers." Ch. 8 in *Mathematical Gems III.* Washington, DC: Math. Assoc. Amer., 1985.

Fibonacci Sequence

see FIBONACCI NUMBER

Fibration

If $f : E \to B$ is a FIBER BUNDLE with B a PARACOMPACT TOPOLOGICAL SPACE, then f satisfies the HOMOTOPY LIFTING PROPERTY with respect to all TOPOLOGICAL SPACES. In other words, if $g : [0,1] \times X \to B$ is a HOMOTOPY from g_0 to g_1, and if g_0' is a LIFT of the MAP g_0 with respect to f, then g has a LIFT to a MAP g' with respect to f. Therefore, if you have a HOMOTOPY of a MAP into B, and if the beginning of it has a LIFT, then that LIFT can be extended to a LIFT of the HOMOTOPY itself.

A fibration is a MAP between TOPOLOGICAL SPACES $f : E \to B$ such that it satisfies the HOMOTOPY LIFTING PROPERTY.

see also FIBER BUNDLE, FIBER SPACE

Field

A field is any set of elements which satisfies the FIELD AXIOMS for both addition and multiplication and is a commutative DIVISION ALGEBRA. An archaic word for a field is RATIONAL DOMAIN. A field with a finite number of members is known as a FINITE FIELD or GALOIS FIELD.

Because the identity condition must be different for addition and multiplication, every field must have at least two elements. Examples include the COMPLEX NUMBERS ($\mathbb{C}$), RATIONAL NUMBERS ($\mathbb{Q}$), and REAL NUMBERS ($\mathbb{R}$), but *not* the INTEGERS ($\mathbb{Z}$), which form a RING. It has been proven by Hilbert and Weierstraß that all generalizations of the field concept to triplets of elements are equivalent to the field of COMPLEX NUMBERS.

see also ADJUNCTION, ALGEBRAIC NUMBER FIELD, COEFFICIENT FIELD, CYCLOTOMIC FIELD, FIELD AXIOMS, FIELD EXTENSION, FUNCTION FIELD, GALOIS FIELD, MAC LANE'S THEOREM, MODULE, NUMBER FIELD, QUADRATIC FIELD, RING, SKEW FIELD, VECTOR FIELD

Field Axioms

The field axioms are generally written in additive and multiplicative pairs.

Name	Addition	Multiplication
Commutivity	$a + b = b + a$	$ab = ba$
Associativity	$(a + b) + c = a + (b + c)$	$(ab)c = a(bc)$
Distributivity	$a(b + c) = ab + ac$	$(a + b)c = ac + bc$
Identity	$a + 0 = a = 0 + a$	$a \cdot 1 = a = 1 \cdot a$
Inverses	$a + (-a) = 0 = (-a) + a$	$aa^{-1} = 1 = a^{-1}a$ if $a \neq 0$

see also ALGEBRA, FIELD

Field Extension

A FIELD L is said to be a field extension of field K if K is a SUBFIELD of L. This is denoted L/K (note that this NOTATION conflicts with that of a QUOTIENT GROUP). The COMPLEX NUMBERS are a field extension of the REAL NUMBERS, and the REAL NUMBERS are a field extension of the RATIONAL NUMBERS.

see also FIELD

Fields Medal

The mathematical equivalent of the Nobel Prize (there is no Nobel Prize in mathematics) which is awarded by the International Mathematical Union every four years to one or more outstanding researchers, usually under 40 years of age. The first Fields Medal was awarded in 1936.

see also BURNSIDE PROBLEM, MATHEMATICS PRIZES, POINCARÉ CONJECTURE, ROTH'S THEOREM, TAU CONJECTURE

References

MacTutor History of Mathematics Archives. "The Fields Medal." http://www-groups.dcs.st-and.ac.uk/~history/Societies/FieldsMedal.html.

Monastyrsky, M. *Modern Mathematics in the Light of the Fields Medals.* Wellesley, MA: A. K. Peters, 1997.

Fifteen Theorem

A theorem due to Conway *et al.* (1997) which states that, if a POSITIVE definite QUADRATIC FORM with integral matrix entries represents all natural numbers up to 15, then it represents all natural numbers. This theorem contains LAGRANGE'S FOUR-SQUARE THEOREM, since every number up to 15 is the sum of at most four SQUARES.

see also INTEGER-MATRIX FORM, LAGRANGE'S FOUR-SQUARE THEOREM, QUADRATIC FORM

References

Conway, J. H.; Guy, R. K.; Schneeberger, W. A.; and Sloane, N. J. A. "The Primary Pretenders." *Acta Arith.* **78**, 307–313, 1997.

Duke, W. "Some Old Problems and New Results about Quadratic Forms." *Not. Amer. Math. Soc.* **44**, 190–196, 1997.

Figurate Number

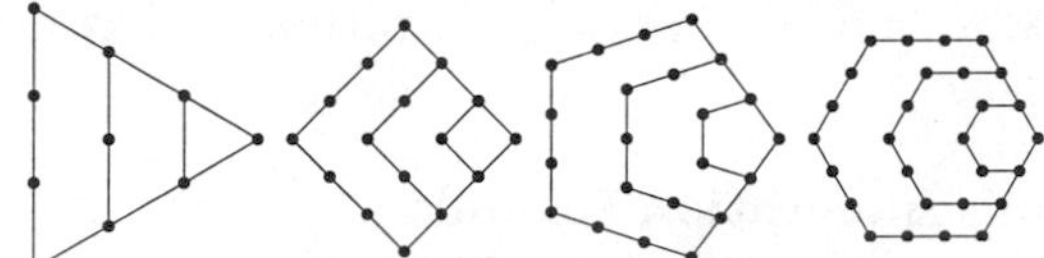

A number which can be represented by a regular geometrical arrangement of equally spaced points. If the arrangement forms a REGULAR POLYGON, the number is called a POLYGONAL NUMBER. The polygonal numbers illustrated above are called triangular, square, pentagonal, and hexagon numbers, respectively. Figurate numbers can also form other shapes such as centered polygons, L-shapes, 3-dimensional solids, etc. The following table lists the most common types of figurate numbers.

Name	Formula
biquadratic	n^4
centered cube	$(2n-1)(n^2-n+1)$
centered pentagonal	$\frac{1}{2}(5n^2-5n+2)$
centered square	$n^2+(n-1)^2$
centered triangular	$\frac{1}{2}(3n^2-3n+2)$
cubic	n^3
decagonal	$4n^2-3n$
gnomic	$2n-1$
heptagonal	$\frac{1}{2}n(5n-3)$
heptagonal pyramidal	$\frac{1}{6}n(n+1)(5n-2)$
hex	$3n^2-3n+1$
hexagonal	$n(2n-1)$
hexagonal pyramidal	$\frac{1}{6}n(n+1)(4n-1)$
octagonal	$n(3n-2)$
octahedral	$\frac{1}{3}n(2n^2+1)$
pentagonal	$\frac{1}{2}n(3n-1)$
pentagonal pyramidal	$\frac{1}{2}n^2(n+1)$
pentatope	$\frac{1}{24}n(n+1)(n+2)(n+3)$
pronic number	$n(n+1)$
rhombic dodecahedral	$(2n-1)(2n^2-2n+1)$
square	n^2
stella octangula	$n(2n^2-1)$
tetrahedral	$\frac{1}{6}n(n+1)(n+2)$
triangular	$\frac{1}{2}n(n+1)$
truncated octahedral	$16n^3-33n^2+24n-6$
truncated tetrahedral	$\frac{1}{6}n(23n^2-27n+10)$

An n-D FIGURATE NUMBER can be defined by

$$f_{m,s}^r = \frac{(rs+m-s)(r+m-2)}{m!(r-1)!}.$$

see also BIQUADRATIC NUMBER, CENTERED CUBE NUMBER, CENTERED PENTAGONAL NUMBER, CENTERED POLYGONAL NUMBER, CENTERED SQUARE NUMBER, CENTERED TRIANGULAR NUMBER, CUBIC NUMBER, DECAGONAL NUMBER, FIGURATE NUMBER TRIANGLE, GNOMIC NUMBER, HEPTAGONAL NUMBER, HEPTAGONAL PYRAMIDAL NUMBER, HEX NUMBER, HEX PYRAMIDAL NUMBER, HEXAGONAL NUMBER, HEXAGONAL PYRAMIDAL NUMBER, NEXUS NUMBER, OCTAGONAL NUMBER, OCTAHEDRAL NUMBER, PENTAGONAL NUMBER, PENTAGONAL PYRAMIDAL NUMBER, PENTATOPE NUMBER, POLYGONAL NUMBER, PRONIC NUMBER, PYRAMIDAL NUMBER, RHOMBIC DODECAHEDRAL NUMBER, SQUARE NUMBER, STELLA OCTANGULA NUMBER, TETRAHEDRAL NUMBER, TRIANGULAR NUMBER, TRUNCATED OCTAHEDRAL NUMBER, TRUNCATED TETRAHEDRAL NUMBER

References

Conway, J. H. and Guy, R. K. *The Book of Numbers.* New York: Springer-Verlag, pp. 30–62, 1996.

Dickson, L. E. "Polygonal, Pyramidal, and Figurate Numbers." Ch. 1 in *History of the Theory of Numbers, Vol. 2: Diophantine Analysis.* New York: Chelsea, pp. 1–39, 1952.

Goodwin, P. "A Polyhedral Sequence of Two." *Math. Gaz.* **69**, 191–197, 1985.
Guy, R. K. "Figurate Numbers." §D3 in *Unsolved Problems in Number Theory, 2nd ed.* New York: Springer-Verlag, pp. 147–150, 1994.
Kraitchik, M. "Figurate Numbers." §3.4 in *Mathematical Recreations.* New York: W. W. Norton, pp. 66–69, 1942.

Figurate Number Triangle

A PASCAL'S TRIANGLE written in a square grid and padded with zeroes, as written by Jakob Bernoulli (Smith 1984). The figurate number triangle therefore has entries

$$a_{ij} = \binom{i}{j},$$

where i is the row number, j the column number, and $\binom{i}{j}$ a BINOMIAL COEFFICIENT. Written out explicitly (beginning each row with $j = 0$),

$$\begin{bmatrix} 1 & 0 & 0 & 0 & 0 & 0 & 0 & \cdots \\ 1 & 1 & 0 & 0 & 0 & 0 & 0 & \cdots \\ 1 & 2 & 1 & 0 & 0 & 0 & 0 & \cdots \\ 1 & 3 & 3 & 1 & 0 & 0 & 0 & \cdots \\ 1 & 4 & 6 & 4 & 1 & 0 & 0 & \cdots \\ 1 & 5 & 10 & 10 & 5 & 1 & 0 & \cdots \\ 1 & 6 & 15 & 20 & 15 & 6 & 1 & \cdots \\ 1 & 7 & 21 & 35 & 35 & 21 & 7 & \ddots \\ \vdots & \vdots & \vdots & \vdots & \vdots & \vdots & \vdots & \ddots \end{bmatrix}$$

Then we have the sum identities

$$\sum_{j=0}^{i} a_{ij} = 2^i$$

$$\sum_{j=1}^{i} a_{ij} = 2^i - 1$$

$$\sum_{i=0}^{n} a_{ij} = a_{(n+1),(j+1)} = \frac{n+1}{j+1} a_{nj}.$$

see also BINOMIAL COEFFICIENT, FIGURATE NUMBER, PASCAL'S TRIANGLE

References
Smith, D. E. *A Source Book in Mathematics.* New York: Dover, p. 86, 1984.

Figure Eight Knot

see FIGURE-OF-EIGHT KNOT

Figure Eight Surface

see EIGHT SURFACE

Figure-of-Eight Knot

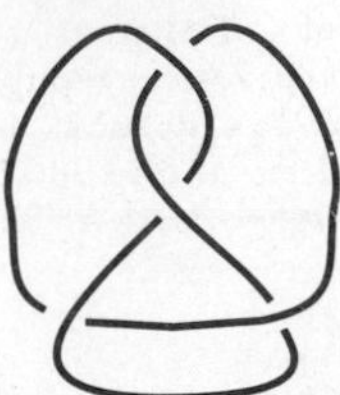

The knot 04_{001}, which is the unique PRIME KNOT of four crossings, and which is a 2-EMBEDDABLE KNOT. It is AMPHICHIRAL. It is also known as the LEMISH KNOT and SAVOY KNOT, and it has BRAID WORD $\sigma_1{\sigma_2}^{-1}\sigma_1{\sigma_2}^{-1}$.

References
Owen, P. *Knots.* Philadelphia, PA: Courage, p. 16, 1993.

Figures

A number x is said to have "n figures" if it takes n DIGITS to express it. The number of figures is therefore equal to one more than the POWER of 10 in the SCIENTIFIC NOTATION representation of the number. The word is most frequently used in reference to monetary amounts, e.g., a "six-figure salary" would fall in the range of $100,000 to $999,999.

see also DIGIT, SCIENTIFIC NOTATION, SIGNIFICANT FIGURES

Filon's Integration Formula

A formula for NUMERICAL INTEGRATION,

$$\int_{x_0}^{x_n} f(x)\cos(tx)\,dx = h\{\alpha(th)[f_{2n}\sin(tx_{2n}) - f_0\sin(tx_0)] + \beta(th)C_{2n} + \gamma(th)C_{2n-1} + \tfrac{2}{45}th^4 S'_{2n-1}\} - R_n, \quad (1)$$

where

$$C_{2n} = \sum_{i=0}^{n} f_{2i}\cos(tx_{2i}) - \tfrac{1}{2}[f_{2n}\cos(tx_{2n}) + f_0\cos(tx_0)] \quad (2)$$

$$C_{2n-1} = \sum_{i=1}^{n} f_{2i-1}\cos(tx_{2i-1}) \quad (3)$$

$$S'_{2n-1} = \sum_{i=1}^{n} f_{2i-1}^{(3)}\sin(tx_{2i-1}) \quad (4)$$

$$\alpha(\theta) = \frac{1}{\theta} + \frac{\sin(2\theta)}{2\theta^2} - \frac{2\sin^2\theta}{\theta^3} \quad (5)$$

$$\beta(\theta) = 2\left[\frac{1+\cos^2\theta}{\theta^2} - \frac{\sin(2\theta)}{\theta^3}\right] \quad (6)$$

$$\gamma(\theta) = 4\left(\frac{\sin\theta}{\theta^3} - \frac{\cos\theta}{\theta^2}\right), \quad (7)$$

and the remainder term is

$$R_n = \tfrac{1}{90}nh^5 f^{(4)}(\xi) + \mathcal{O}(th^7). \quad (8)$$

References

Abramowitz, M. and Stegun, C. A. (Eds.). *Handbook of Mathematical Functions with Formulas, Graphs, and Mathematical Tables, 9th printing.* New York: Dover, pp. 890–891, 1972.

Tukey, J. W. In *On Numerical Approximation: Proceedings of a Symposium Conducted by the Mathematics Research Center, United States Army, at the University of Wisconsin, Madison, April 21–23, 1958* (Ed. R. E. Langer). Madison, WI: University of Wisconsin Press, p. 400, 1959.

Filter

Formally, a filter is defined in terms of a SET X and a SET Φ of SUBSETS of X. Then Φ is called a filter if

1. $X \in \Phi$,
2. the EMPTY SET $\varnothing \notin \Phi$,
3. $A \subset B \subset X$ and $A \in \Phi$ IMPLIES $B \in \Phi$,
4. and $A, B \in \Phi$ IMPLIES $A \cup B \in \Phi$.

Informally, a filter is a function or procedure which removes unwanted parts of a signal. The concept of filtering and filter functions is particularly useful in engineering. One particularly elegant method of filtering FOURIER TRANSFORMS a signal into frequency space, performs the filtering operation there, then transforms back into the original space (Press *et al.* 1992).

see also SAVITZKY-GOLAY FILTER, WIENER FILTER

References

Press, W. H.; Flannery, B. P.; Teukolsky, S. A.; and Vetterling, W. T. "Digital Filtering in the Time Domain." §13.5 in *Numerical Recipes in FORTRAN: The Art of Scientific Computing, 2nd ed.* Cambridge, England: Cambridge University Press, pp. 551–556, 1992.

Fine's Equation

$$\prod_{n=1} \frac{(1-q^{2n})(1-q^{3n})(1-q^{8n})(1-q^{12n})}{(1-q^n)(1-q^{24n})} = 1 + \sum_{N=1} E_{1,5,7,11}(N;24)q^N,$$

where E is the sum of the DIVISORS of N CONGRUENT to 1, 5, 7, and 11 (mod 24) minus the sum of DIVISORS of N CONGRUENT to -1, -5, -7, and -11 (mod 24).

see also q-SERIES

Finite

A SET which contains a NONNEGATIVE integral number of elements is said to be finite. A SET which is not finite is said to be INFINITE. A finite or COUNTABLY INFINITE SET is said to be COUNTABLE. While the meaning of the term "finite" is fairly clear in common usage, precise definitions of FINITE and INFINITE are needed in technical mathematics and especially in SET THEORY.

see also COUNTABLE SET, COUNTABLY INFINITE SET, INFINITE, SET THEORY, UNCOUNTABLY INFINITE SET

Finite Difference

The finite difference is the discrete analog of the DERIVATIVE. The finite FORWARD DIFFERENCE of a function f_p is defined as

$$\Delta f_p \equiv f_{p+1} - f_p, \tag{1}$$

and the finite BACKWARD DIFFERENCE as

$$\nabla f_p \equiv f_p - f_{p-1}. \tag{2}$$

If the values are tabulated at spacings h, then the notation

$$f_p \equiv f(x_0 + ph) \equiv f(x) \tag{3}$$

is used. The kth FORWARD DIFFERENCE would then be written as $\Delta^k f_p$, and similarly, the kth BACKWARD DIFFERENCE as $\nabla^k f_p$.

However, when f_p is viewed as a discretization of the continuous function $f(x)$, then the finite difference is sometimes written

$$\Delta f(x) \equiv f(x+\tfrac{1}{2}) - f(x-\tfrac{1}{2}) = 2\,\mathrm{I_I}(x) * f(x), \tag{4}$$

where $*$ denotes CONVOLUTION and $\mathrm{I_I}(x)$ is the odd IMPULSE PAIR. The finite difference operator can therefore be written

$$\tilde{\Delta} = 2\,\mathrm{I_I} *. \tag{5}$$

An nth POWER has a constant nth finite difference. For example, take $n = 3$ and make a DIFFERENCE TABLE,

$$\begin{array}{cccccc} x & x^3 & & & & \\ & & \Delta & & & \\ 1 & 1 & & \Delta^2 & & \\ & & 7 & & \Delta^3 & \\ 2 & 8 & & 12 & & \Delta^4 \\ & & 19 & & 6 & \\ 3 & 27 & & 18 & & 0 \\ & & 37 & & 6 & \\ 4 & 64 & & 24 & & \\ & & 61 & & & \\ 5 & 125 & & & & \end{array}. \tag{6}$$

The Δ^3 column is the constant 6.

Finite difference formulas can be very useful for extrapolating a finite amount of data in an attempt to find the general term. Specifically, if a function $f(n)$ is known at only a few discrete values $n = 0, 1, 2, \ldots$ and it is desired to determine the analytical form of f, the following procedure can be used if f is assumed to be a POLYNOMIAL function. Denote the nth value in the SEQUENCE of interest by a_n. Then define b_n as the FORWARD DIFFERENCE $\Delta_n \equiv a_{n+1} - a_n$, c_n as the second FORWARD DIFFERENCE $\Delta_n^2 \equiv b_{n+1} - b_n$, etc., constructing a table as follows

$$\begin{array}{ccccc} a_0 \equiv f(0) & a_1 \equiv f(1) & a_2 \equiv f(2) & \ldots & a_p \equiv f(p) \\ b_0 \equiv a_1 - a_0 & b_1 \equiv a_2 - a_1 & \ldots & b_{p-1} \equiv a_p - a_{p-1} & \\ & c_0 \equiv b_2 - b_1 & \ldots & \ldots & \\ & & \ddots & & \end{array} \tag{7}$$

Continue computing d_0, e_0, etc., until a 0 value is obtained. Then the POLYNOMIAL function giving the values a_n is given by

$$f(n) = \sum_{k=0}^{p} \alpha_k \binom{n}{k} \quad (8)$$

$$= a_0 + b_0 n + \frac{c_0 n(n-1)}{2} + \frac{d_0 n(n-1)(n-2)}{2 \cdot 3} + \ldots \quad (9)$$

When the notation $\Delta_0 \equiv a_0$, $\Delta_0^2 \equiv b_0$, etc., is used, this beautiful equation is called NEWTON'S FORWARD DIFFERENCE FORMULA. To see a particular example, consider a SEQUENCE with first few values of 1, 19, 143, 607, 1789, 4211, and 8539. The difference table is then given by

$$\begin{array}{ccccccccccccc} 1 & & 19 & & 143 & & 607 & & 1789 & & 4211 & & 8539 \\ & 18 & & 124 & & 464 & & 1182 & & 2422 & & 4328 & \\ & & 106 & & 340 & & 718 & & 1240 & & 1906 & & \\ & & & 234 & & 378 & & 522 & & 666 & & & \\ & & & & 144 & & 144 & & 144 & & & & \\ & & & & & 0 & & 0 & & & & & \end{array}$$

Reading off the first number in each row gives $a_0 = 1$, $b_0 = 18$, $c_0 = 106$, $d_0 = 234$, $e_0 = 144$. Plugging these in gives the equation

$$f(n) = 1 + 18n + 53n(n-1) + 39n(n-1)(n-2) + 6n(n-1)(n-2)(n-3), \quad (10)$$

which simplifies to $f(n) = 6n^4 + 3n^3 + 2n^2 + 7n + 1$, and indeed fits the original data exactly!

Beyer (1987) gives formulas for the derivatives

$$h^n \frac{d^n f(x_0 + ph)}{dx^n} \equiv h^n \frac{d^n f_p}{dx^n} \equiv \frac{d^n f_p}{dp^n} \quad (11)$$

(Beyer 1987, pp. 449–451) and integrals

$$\int_{x_0}^{x_n} f(x)\,dx = h \int_0^n f_p\,dp \quad (12)$$

(Beyer 1987, pp. 455-456) of finite differences.

Finite differences lead to DIFFERENCE EQUATIONS, finite analogs of DIFFERENTIAL EQUATIONS. In fact, UMBRAL CALCULUS displays many elegant analogs of well-known identities for continuous functions. Common finite difference schemes for PARTIAL DIFFERENTIAL EQUATIONS include the so-called Crank-Nicholson, Du Fort-Frankel, and Laasonen methods.

see also BACKWARD DIFFERENCE, BESSEL'S FINITE DIFFERENCE FORMULA, DIFFERENCE EQUATION, DIFFERENCE TABLE, EVERETT'S FORMULA, FORWARD DIFFERENCE, GAUSS'S BACKWARD FORMULA, GAUSS'S FORWARD FORMULA, INTERPOLATION, JACKSON'S DIFFERENCE FAN, NEWTON'S BACKWARD DIFFERENCE FORMULA, NEWTON-COTES FORMULAS, NEWTON'S DIVIDED DIFFERENCE INTERPOLATION FORMULA, NEWTON'S FORWARD DIFFERENCE FORMULA, QUOTIENT-DIFFERENCE TABLE, STEFFENSON'S FORMULA, STIRLING'S FINITE DIFFERENCE FORMULA, UMBRAL CALCULUS

References

Abramowitz, M. and Stegun, C. A. (Eds.). "Differences." §25.1 in *Handbook of Mathematical Functions with Formulas, Graphs, and Mathematical Tables, 9th printing.* New York: Dover, pp. 877–878, 1972.

Beyer, W. H. *CRC Standard Mathematical Tables, 28th ed.* Boca Raton, FL: CRC Press, pp. 429–515, 1987.

Boole, G. and Moulton, J. F. *A Treatise on the Calculus of Finite Differences, 2nd rev. ed.* New York: Dover, 1960.

Conway, J. H. and Guy, R. K. "Newton's Useful Little Formula." In *The Book of Numbers.* New York: Springer-Verlag, pp. 81–83, 1996.

Iyanaga, S. and Kawada, Y. (Eds.). "Interpolation." Appendix A, Table 21 in *Encyclopedic Dictionary of Mathematics.* Cambridge, MA: MIT Press, pp. 1482–1483, 1980.

Jordan, K. *Calculus of Finite Differences, 2nd ed.* New York: Chelsea, 1950.

Levy, H. and Lessman, F. *Finite Difference Equations.* New York: Dover, 1992.

Milne-Thomson, L. M. *The Calculus of Finite Differences.* London: Macmillan, 1951.

Richardson, C. H. *An Introduction to the Calculus of Finite Differences.* New York: Van Nostrand, 1954.

Spiegel, M. *Calculus of Finite Differences and Differential Equations.* New York: McGraw-Hill, 1971.

Finite Field

A finite field is a FIELD with a finite ORDER (number of elements), also called a GALOIS FIELD. The order of a finite field is always a PRIME or a POWER of a PRIME (Birkhoff and Mac Lane 1965). For each PRIME POWER, there exists exactly one (up to an ISOMORPHISM) finite field $GF(p^n)$, often written as $\mathbb{F}_{p^n}$ in current usage. $GF(p)$ is called the PRIME FIELD of order p, and is the FIELD of RESIDUE CLASSES modulo p, where the p elements are denoted 0, 1, ..., $p-1$. $a = b$ in $GF(p)$ means the same as $a \equiv b \pmod{p}$. Note, however, that $2 \times 2 \equiv 0 \pmod{4}$ in the RING of residues modulo 4, so 2 has no reciprocal, and the RING of residues modulo 4 is distinct from the finite field with four elements. Finite fields are therefore denoted $GF(p^n)$, instead of $GF(p_1 \cdots p_n)$ for clarity.

The finite field $GF(2)$ consists of elements 0 and 1 which satisfy the following addition and multiplications tables.

+	0	1
0	0	1
1	1	0

×	0	1
0	0	0
1	0	1

If a subset S of the elements of a finite field F satisfies the above AXIOMS with the same operators of F, then S

is called a SUBFIELD. Finite fields are used extensively in the study of ERROR-CORRECTING CODES.

When $n > 1$, $\mathrm{GF}(p^n)$ can be represented as the FIELD of EQUIVALENCE CLASSES of POLYNOMIALS whose COEFFICIENTS belong to $\mathrm{GF}(p)$. Any IRREDUCIBLE POLYNOMIAL of degree n yields the same FIELD up to an ISOMORPHISM. For example, for $\mathrm{GF}(2^3)$, the modulus can be taken as $x^3+x^2+1=0$, x^3+x+1, or any other IRREDUCIBLE POLYNOMIAL of degree 3. Using the modulus x^3+x+1, the elements of $\mathrm{GF}(2^3)$—written 0, x^0, x^1, ...—can be represented as POLYNOMIALS with degree less than 3. For instance,

$$\begin{aligned}
x^3 &\equiv -x-1 \equiv x+1 \\
x^4 &\equiv x(x^3) \equiv x(x+1) \equiv x^2+x \\
x^5 &\equiv x(x^2+x) \equiv x^3+x^2 \equiv x^2-x-1 \equiv x^2+x+1 \\
x^6 &\equiv x(x^2+x+1) \equiv x^3+x^2+x \equiv x^2-1 \equiv x^2+1 \\
x^7 &\equiv x(x^2+1) \equiv x^3+x \equiv -1 \equiv 1 \equiv x_0.
\end{aligned}$$

Now consider the following table which contains several different representations of the elements of a finite field. The columns are the power, polynomial representation, triples of polynomial representation COEFFICIENTS (the vector representation), and the binary INTEGER corresponding to the vector representation (the regular representation).

	Representation		
Power	Polynomial	Vector	Regular
0	0	(000)	0
x^0	1	(001)	1
x^1	x	(010)	2
x^2	x^2	(100)	4
x^3	$x+1$	(011)	3
x^4	x^2+x	(110)	6
x^5	x^2+x+1	(111)	7
x^6	x^2+1	(101)	5

The set of POLYNOMIALS in the second column is closed under ADDITION and MULTIPLICATION modulo x^3+x+1, and these operations on the set satisfy the AXIOMS of finite field. This particular finite field is said to be an extension field of degree 3 of $\mathrm{GF}(2)$, written $\mathrm{GF}(2^3)$, and the field $\mathrm{GF}(2)$ is called the base field of $\mathrm{GF}(2^3)$. If an IRREDUCIBLE POLYNOMIAL generates all elements in this way, it is called a PRIMITIVE IRREDUCIBLE POLYNOMIAL. For any PRIME or PRIME POWER q and any POSITIVE INTEGER n, there exists a PRIMITIVE IRREDUCIBLE POLYNOMIAL of degree n over $\mathrm{GF}(q)$.

For any element c of $\mathrm{GF}(q)$, $c^q = c$, and for any NONZERO element d of $\mathrm{GF}(q)$, $d^{q-1} = 1$. There is a smallest POSITIVE INTEGER n satisfying the sum condition $n \cdot 1 = 0$ in $\mathrm{GF}(q)$, which is called the characteristic of the finite field $\mathrm{GF}(q)$. The characteristic is a PRIME NUMBER for every finite field, and it is true that

$$(x+y)^p = x^p + y^p$$

over a finite field with characteristic p.

see also FIELD, HADAMARD MATRIX, RING, SUBFIELD

References

Ball, W. W. R. and Coxeter, H. S. M. *Mathematical Recreations and Essays, 13th ed.* New York: Dover, pp. 73–75, 1987.

Birkhoff, G. and Mac Lane, S. *A Survey of Modern Algebra, 3rd ed.* New York: Macmillan, p. 413, 1965.

Dickson, L. E. *History of the Theory of Numbers, Vol. 1: Divisibility and Primality.* New York: Chelsea, p. viii, 1952.

Finite Game

A GAME in which each player has a finite number of moves and a finite number of of choices at each move.

see also GAME, ZERO-SUM GAME

References

Dresher, M. *The Mathematics of Games of Strategy: Theory and Applications.* New York: Dover, p. 2, 1981.

Finite Group

A GROUP of finite ORDER. Examples of finite groups are the MODULO MULTIPLICATION GROUPS and the POINT GROUPS. The CLASSIFICATION THEOREM of finite SIMPLE GROUPS states that the finite SIMPLE GROUPS can be classified completely into one of five types.

There is no known FORMULA to give the number of possible finite groups as a function of the ORDER h. It is possible, however, to determine the number of ABELIAN GROUPS using the KRONECKER DECOMPOSITION THEOREM, and there is at least one ABELIAN GROUP for every finite order h.

The following table gives the numbers and names of the first few groups of ORDER h. In the table, N_{NA} denotes the number of non-Abelian groups, N_A denotes the number of ABELIAN GROUPS, and N the total number of groups. In addition, Z_n denotes an CYCLIC GROUP of ORDER n, A_n an ALTERNATING GROUP, D_n a DIHEDRAL GROUP, Q_8 the group of the QUATERNIONS, T the cubic group, and $\otimes$ a DIRECT PRODUCT.

h	Name	N_{NA}	N_A	N
1	$\langle e \rangle$	1	0	1
2	Z_2	1	0	1
3	Z_3	1	0	1
4	$Z_2 \otimes Z_2$, Z_4	2	0	2
5	Z_5	1	0	1
6	Z_6, D_3	1	1	2
7	Z_7	1	0	1
8	$Z_2 \otimes Z_2 \otimes Z_2$, $Z_2 \otimes Z_4$, Z_8, Q_8, D_4	3	2	5
9	$Z_3 \otimes Z_3$, Z_9	2	0	2
10	Z_{10}, D_5	1	1	2
11	Z_{11}	1	0	1
12	$Z_2 \otimes Z_6$, Z_{12}, A_4, D_6, T	2	3	5
13	Z_{13}	1	0	1
14	Z_{14}, D_7	1	1	2
15	Z_{15}	1	0	1

Miller (1930) gave the number of groups for orders 1–100, including an erroneous 297 as the number of groups of ORDER 64. Senior and Lunn (1934, 1935) subsequently completed the list up to 215, but omitted 128 and 192. The number of groups of ORDER 64 was corrected in Hall and Senior (1964). James *et al.* (1990) found 2328 groups in 115 ISOCLINISM families of ORDER 128, correcting previous work, and O'Brien (1991) found the number of groups of ORDER 256. The number of groups is known for orders up to 1000, with the possible exception of 512 and 768. Besche and Eick (1998) have determined the number of finite groups of orders less than 1000 which are not powers of 2 or 3. These numbers appear in the *Magma*® database. The numbers of nonisomorphic finite groups N of each ORDER h for the first few hundred orders are given in the following table (Sloane's A000001—the very first sequence).

The number of ABELIAN GROUPS of ORDER h is denoted N_A (Sloane's A000688). The smallest order for which there exist $n = 1, 2, \ldots$ nonisomorphic groups are 1, 4, 75, 28, 8, 42, ... (Sloane's A046057). The incrementally largest number of nonisomorphic finite groups are 1, 2, 5, 14, 15, 51, 52, 267, 2328, ... (Sloane's A046058), which occur for orders 1, 4, 8, 16, 24, 32, 48, 64, 128, ... (Sloane's A046059).

h	N	N_A	h	N	N_A	h	N	N_A	h	N	N_A
1	1	1	51	1	1	101	1	1	151	1	1
2	1	1	52	5	2	102	4	1	152	12	3
3	1	1	53	1	1	103	1	1	153	2	2
4	2	2	54	15	3	104	14	3	154	4	1
5	1	1	55	2	1	105	2	1	155	2	1
6	2	1	56	13	3	106	2	1	156	18	2
7	1	1	57	2	1	107	1	1	157	1	1
8	5	3	58	2	1	108	45	6	158	2	1
9	2	2	59	1	1	109	1	1	159	1	1
10	2	1	60	13	2	110	6	1	160	238	7
11	1	1	61	1	1	111	2	1	161	1	1
12	5	2	62	2	1	112	43	5	162	55	5
13	1	1	63	4	2	113	1	1	163	1	1
14	2	1	64	267	11	114	6	1	164	5	2
15	1	1	65	1	1	115	1	1	165	2	1
16	14	5	66	4	1	116	5	2	166	2	1
17	1	1	67	1	1	117	4	2	167	1	1
18	5	2	68	5	2	118	2	1	168	57	3
19	1	1	69	1	1	119	1	1	169	2	2
20	5	2	70	4	1	120	47	3	170	4	1
21	2	1	71	1	1	121	2	2	171	5	2
22	2	1	72	50	6	122	2	1	172	4	2
23	1	1	73	1	1	123	1	1	173	1	1
24	15	3	74	2	1	124	4	2	174	4	1
25	2	2	75	3	2	125	5	3	175	2	2
26	2	1	76	4	2	126	16	2	176	42	5
27	5	3	77	1	1	127	1	1	177	1	1
28	4	2	78	6	1	128	2328	15	178	2	1
29	1	1	79	1	1	129	2	1	179	1	1
30	4	1	80	52	5	130	4	1	180	37	4
31	1	1	81	15	5	131	1	1	181	1	1
32	51	7	82	2	1	132	10	2	182	4	1
33	1	1	83	1	1	133	1	1	183	2	1
34	2	1	84	15	2	134	2	1	184	12	3
35	1	1	85	1	1	135	5	3	185	1	1
36	14	4	86	2	1	136	15	3	186	6	1
37	1	1	87	1	1	137	1	1	187	1	1
38	2	1	88	12	3	138	4	1	188	4	2
39	2	1	89	1	1	139	1	1	189	13	3
40	14	3	90	10	2	140	11	2	190	4	1
41	1	1	91	1	1	141	1	1	191	1	1
42	6	1	92	4	2	142	2	1	192	1543	11
43	1	1	93	2	1	143	1	1	193	1	1
44	4	2	94	2	1	144	197	10	194	2	1
45	2	2	95	1	1	145	1	1	195	2	1
46	2	1	96	230	7	146	2	1	196	17	4
47	1	1	97	1	1	147	6	2	197	1	1
48	52	5	98	5	2	148	5	2	198	10	2
49	2	2	99	2	2	149	1	1	199	1	1
50	2	2	100	16	4	150	13	2	200	52	6

h	N	N_A	h	N	N_A	h	N	N_A	h	N	N_A
201	2	1	251	1	1	301	2	1	351	14	3
202	2	1	252	46	4	302	2	1	352	195	7
203	2	1	253	2	1	303	1	1	353	1	1
204	12	2	254	2	1	304	42	5	354	4	1
205	2	1	255	1	1	305	2	1	355	2	1
206	2	1	256	56092	22	306	10	2	356	5	2
207	2	2	257	1	1	307	1	1	357	2	1
208	51	5	258	6	1	308	9	2	358	2	1
209	1	1	259	1	1	309	2	1	359	1	1
210	12	1	260	15	2	310	6	1	360	162	6
211	1	1	261	2	2	311	1	1	361	2	2
212	5	2	262	2	1	312	61	3	362	2	1
213	1	1	263	1	1	313	1	1	363	3	2
214	2	1	264	39	3	314	2	1	364	11	2
215	1	1	265	1	1	315	4	2	365	1	1
216	177	9	266	4	1	316	4	2	366	6	1
217	1	1	267	1	1	317	1	1	367	1	1
218	2	1	268	4	2	318	4	1	368	42	5
219	2	1	269	1	1	319	1	1	369	2	2
220	15	2	270	30	3	320	1640	11	370	4	1
221	1	1	271	1	1	321	1	1	371	1	1
222	6	1	272	54	5	322	4	1	372	15	2
223	1	1	273	5	1	323	1	1	373	1	1
224	197	7	274	2	1	324	176	10	374	4	1
225	6	4	275	4	2	325	2	2	375	7	3
226	2	1	276	10	2	326	2	1	376	12	3
227	1	1	277	1	1	327	2	1	377	1	1
228	15	2	278	2	1	328	15	3	378	60	3
229	1	1	279	4	2	329	1	1	379	1	1
230	4	1	280	40	3	330	12	1	380	11	2
231	2	1	281	1	1	331	1	1	381	2	1
232	14	3	282	4	1	332	4	2	382	2	1
233	1	1	283	1	1	333	5	2	383	1	1
234	16	2	284	4	2	334	2	1	384	20169	15
235	1	1	285	2	1	335	1	1	385	2	1
236	4	2	286	4	1	336	228	5	386	2	1
237	2	1	287	1	1	337	1	1	387	4	2
238	4	1	288	1045	14	338	5	2	388	5	2
239	1	1	289	2	2	339	1	1	389	1	1
240	208	5	290	4	1	340	15	2	390	12	1
241	1	1	291	2	1	341	1	1	391	1	1
242	5	2	292	5	2	342	18	2	392	44	6
243	67	7	293	1	1	343	5	3	393	1	1
244	5	2	294	23	2	344	12	3	394	2	1
245	2	2	295	1	1	345	1	1	395	1	1
246	4	1	296	14	3	346	2	1	396	30	4
247	1	1	297	5	3	347	1	1	397	1	1
248	12	3	298	2	1	348	12	2	398	2	1
249	1	1	299	1	1	349	1	1	399	5	1
250	15	3	300	49	4	350	10	2	400	221	10

see also ABELIAN GROUP, ABEL'S THEOREM, ABHYANKAR'S CONJECTURE, ALTERNATING GROUP, BURNSIDE'S LEMMA, BURNSIDE PROBLEM, CHEVALLEY GROUPS, CLASSIFICATION THEOREM, COMPOSITION SERIES, DIHEDRAL GROUP, GROUP, JORDAN-HÖLDER THEOREM, KRONECKER DECOMPOSITION THEOREM, LIE GROUP, LIE-TYPE GROUP, LINEAR GROUP, MODULO MULTIPLICATION GROUP, ORDER (GROUP), ORTHOGONAL GROUP, p-GROUP, POINT GROUPS, SIMPLE GROUP, SPORADIC GROUP, SYMMETRIC GROUP, SYMPLECTIC GROUP, TWISTED CHEVALLEY GROUPS, UNITARY GROUP

References

Arfken, G. "Discrete Groups." §4.9 in *Mathematical Methods for Physicists, 3rd ed.* Orlando, FL: Academic Press, pp. 243–251, 1985.

Artin, E. "The Order of the Classical Simple Groups." *Comm. Pure Appl. Math.* **8**, 455–472, 1955.

Aschbacher, M. *Finite Group Theory.* Cambridge, England: Cambridge University Press, 1994.

Ball, W. W. R. and Coxeter, H. S. M. *Mathematical Recreations and Essays, 13th ed.* New York: Dover, pp. 73–75, 1987.

Besche and Eick. "Construction of Finite Groups." To Appear in *J. Symb. Comput.*

Besche and Eick. "The Groups of Order at Most 1000." To Appear in *J. Symb. Comput.*

Conway, J. H.; Curtis, R. T.; Norton, S. P.; Parker, R. A.; and Wilson, R. A. *Atlas of Finite Groups: Maximal Subgroups and Ordinary Characters for Simple Groups.* Oxford, England: Clarendon Press, 1985.

Hall, M. Jr. and Senior, J. K. *The Groups of Order* $2^n (n \leq 6)$. New York: Macmillan, 1964.

James, R.; Newman, M. F.; and O'Brien, E. A. "The Groups of Order 128." *J. Algebra* **129**, 136–158, 1990.

Miller, G. A. "Determination of All the Groups of Order 64." *Amer. J. Math.* **52**, 617–634, 1930.

O'Brien, E. A. "The Groups of Order 256." *J. Algebra* **143**, 219–235, 1991.

O'Brien, E. A. and Short, M. W. "Bibliography on Classification of Finite Groups." Manuscript, Australian National University, 1988.

Senior, J. K. and Lunn, A. C. "Determination of the Groups of Orders 101–161, Omitting Order 128." *Amer. J. Math.* **56**, 328–338, 1934.

Senior, J. K. and Lunn, A. C. "Determination of the Groups of Orders 162–215, Omitting Order 192." *Amer. J. Math.* **57**, 254–260, 1935.

Simon, B. *Representations of Finite and Compact Groups.* Providence, RI: Amer. Math. Soc., 1996.

Sloane, N. J. A. Sequences A000001/M0098 and A000688/M0064 in "An On-Line Version of the Encyclopedia of Integer Sequences."

University of Sydney Computational Algebra Group. "The Magma Computational Algebra for Algebra, Number Theory and Geometry." `http://www.maths.usyd.edu.au:8000/u/magma/`.

Weisstein, E. W. "Groups." `http://www.astro.virginia.edu/~eww6n/math/notebooks/Groups.m`.

Wilson, R. A. "ATLAS of Finite Group Representation." `http://for.mat.bham.ac.uk/atlas`.

Finite Group—D_3

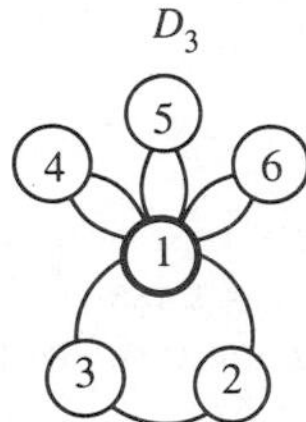

The DIHEDRAL GROUP D_3 is one of the two groups of ORDER 6. It the non-Abelian group of smallest ORDER. Examples of D_3 include the POINT GROUPS known as C_{3h}, C_{3v}, S_3, D_3, the symmetry group of the EQUILATERAL TRIANGLE, and the group of permutation of three

objects. Its elements A_i satisfy $A_i^3 = 1$, and four of its elements satisfy $A_i^2 = 1$, where 1 is the IDENTITY ELEMENT. The CYCLE GRAPH is shown above, and the MULTIPLICATION TABLE is given below.

D_3	1	A	B	C	D	E
1	1	A	B	C	D	E
A	A	1	D	E	B	C
B	B	E	1	D	C	A
C	C	D	E	1	A	B
D	D	C	A	B	E	1
E	E	B	C	A	1	D

The CONJUGACY CLASSES are $\{1\}$, $\{A, B, C\}$

$$A^{-1}AA = A \tag{1}$$

$$B^{-1}AB = C \tag{2}$$

$$C^{-1}AC = B \tag{3}$$

$$D^{-1}AD = C \tag{4}$$

$$E^{-1}AE = B, \tag{5}$$

and $\{D, E\}$,

$$DA^{-1}D = E \tag{6}$$

$$B^{-1}DB = D. \tag{7}$$

A reducible 2-D representation using REAL MATRICES can be found by performing the spatial rotations corresponding to the symmetry elements of C_{3v}. Take the z-AXIS along the C_3 axis.

$$I = R_z(0) = \begin{bmatrix} 1 & 0 \\ 0 & 1 \end{bmatrix} \tag{8}$$

$$A = R_z(\tfrac{2}{3}\pi) = \begin{bmatrix} \cos(\frac{2}{3}\pi) & \sin(\frac{2}{3}\pi) \\ -\sin(\frac{2}{3}\pi) & \cos(\frac{2}{3}\pi) \end{bmatrix}$$

$$= \begin{bmatrix} -\frac{1}{2} & -\frac{1}{2}\sqrt{3} \\ \frac{1}{2}\sqrt{3} & -\frac{1}{2} \end{bmatrix} \tag{9}$$

$$B = R_z(\tfrac{4}{3}\pi) = \begin{bmatrix} -\frac{1}{2} & \frac{1}{2}\sqrt{3} \\ -\frac{1}{2}\sqrt{3} & -\frac{1}{2} \end{bmatrix} \tag{10}$$

$$C = R_C(\pi) = \begin{bmatrix} -1 & 0 \\ 0 & 1 \end{bmatrix} \tag{11}$$

$$D = R_D(\pi) = CB = \begin{bmatrix} \frac{1}{2} & -\frac{1}{2}\sqrt{3} \\ -\frac{1}{2}\sqrt{3} & -\frac{1}{2} \end{bmatrix} \tag{12}$$

$$E = R_E(\pi) = CA = \begin{bmatrix} \frac{1}{2} & \frac{1}{2}\sqrt{3} \\ \frac{1}{2}\sqrt{3} & -\frac{1}{2} \end{bmatrix}. \tag{13}$$

To find the irreducible representation, note that there are three CONJUGACY CLASSES. Rule 5 requires that there be three irreducible representations satisfying

$$h = l_1{}^2 + l_2{}^2 + l_3{}^2 = 6, \tag{14}$$

so it must be true that

$$l_1 = l_2 = 1, l_3 = 2. \tag{15}$$

By rule 6, we can let the first representation have all 1s.

D_3	1	A	B	C	D	E
Γ_1	1	1	1	1	1	1

To find representation orthogonal to the totally symmetric representation, we must have three +1 and three −1 CHARACTERS. We can also add the constraint that the components of the IDENTITY ELEMENT 1 be positive. The three CONJUGACY CLASSES have 1, 2, and 3 elements. Since we need a total of three +1s and we have required that a +1 occur for the CONJUGACY CLASS of ORDER 1, the remaining +1s must be used for the elements of the CONJUGACY CLASS of ORDER 2, i.e., A and B.

D_3	1	A	B	C	D	E
Γ_1	1	1	1	1	1	1
Γ_2	1	1	1	−1	−1	−1

Using the rule 1, we see that

$$1^2 + 1^2 + \chi_3{}^2(1) = 6, \tag{16}$$

so the final representation for 1 has CHARACTER 2. Orthogonality with the first two representations (rule 3) then yields the following constraints:

$$1 \cdot 1 \cdot 2 + 1 \cdot 2 \cdot \chi_2 + 1 \cdot 3 \cdot \chi_3 = 2 + 2\chi_2 + 3\chi_3 = 0 \tag{17}$$

$$1 \cdot 1 \cdot 2 + 1 \cdot 2 \cdot \chi_2 + (-1) \cdot 3 \cdot \chi_3 = 2 + 2\chi_2 - 3\chi_3 = 0. \tag{18}$$

Solving these simultaneous equations by adding and subtracting (18) from (17), we obtain $\chi_2 = -1$, $\chi_3 = 0$. The full CHARACTER TABLE is then

D_3	1	A	B	C	D	E
Γ_1	1	1	1	1	1	1
Γ_2	1	1	1	−1	−1	−1
Γ_3	2	−1	−1	0	0	0

Since there are only three CONJUGACY CLASSES, this table is conventionally written simply as

D_3	1	$A = B$	$C = D = E$
Γ_1	1	1	1
Γ_2	1	1	−1
Γ_3	2	−1	0

Writing the irreducible representations in matrix form then yields

$$1 = \begin{bmatrix} 1 & 0 & 0 & 0 \\ 0 & 1 & 0 & 0 \\ 0 & 0 & 1 & 0 \\ 0 & 0 & 0 & 1 \end{bmatrix} \tag{19}$$

$$A = \begin{bmatrix} -\frac{1}{2} & -\frac{1}{2}\sqrt{3} & 0 & 0 \\ \frac{1}{2}\sqrt{3} & -\frac{1}{2} & 0 & 0 \\ 0 & 0 & 1 & 0 \\ 0 & 0 & 0 & 1 \end{bmatrix} \tag{20}$$

$$B = \begin{bmatrix} -\frac{1}{2} & \frac{1}{2}\sqrt{3} & 0 & 0 \\ -\frac{1}{2}\sqrt{3} & -\frac{1}{2} & 0 & 0 \\ 0 & 0 & 1 & 0 \\ 0 & 0 & 0 & 1 \end{bmatrix} \quad (21)$$

$$C = \begin{bmatrix} -1 & 0 & 0 & 0 \\ 0 & 1 & 0 & 0 \\ 0 & 0 & 1 & 0 \\ 0 & 0 & 0 & -1 \end{bmatrix} \quad (22)$$

$$D = \begin{bmatrix} \frac{1}{2} & -\frac{1}{2}\sqrt{3} & 0 & 0 \\ -\frac{1}{2}\sqrt{3} & -\frac{1}{2} & 0 & 0 \\ 0 & 0 & 1 & 0 \\ 0 & 0 & 0 & -1 \end{bmatrix} \quad (23)$$

$$E = \begin{bmatrix} \frac{1}{2} & \frac{1}{2}\sqrt{3} & 0 & 0 \\ \frac{1}{2}\sqrt{3} & -\frac{1}{2} & 0 & 0 \\ 0 & 0 & 1 & 0 \\ 0 & 0 & 0 & -1 \end{bmatrix}. \quad (24)$$

see also DIHEDRAL GROUP, FINITE GROUP—D_4, FINITE GROUP—Z_6

Finite Group—D_4

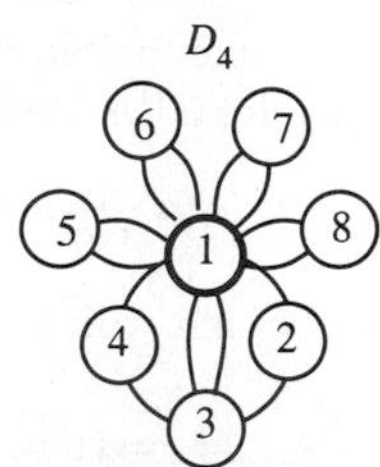

The DIHEDRAL GROUP D_4 is one of the two non-Abelian groups of the five groups total of ORDER 8. It is sometimes called the octic group. Examples of D_4 include the symmetry group of the SQUARE. The CYCLE GRAPH is shown above.

see also DIHEDRAL GROUP, FINITE GROUP—D_3, FINITE GROUP—Z_8, FINITE GROUP—$Z_2 \otimes Z_2 \otimes Z_2$, FINITE GROUP—$Z_2 \otimes Z_4$, FINITE GROUP—$Z_8$,

Finite Group—$\langle e \rangle$

The unique (and trivial) group of ORDER 1 is denoted $\langle e \rangle$. It is (trivially) ABELIAN and CYCLIC. Examples include the POINT GROUP C_1 and the integers modulo 1 under addition.

$\langle e \rangle$	1
1	1

The only class is $\{1\}$.

Finite Group—Q_8

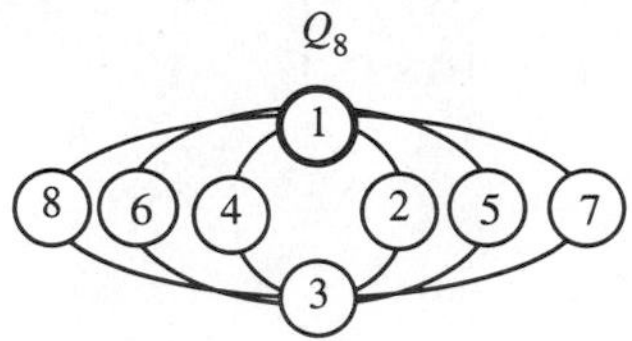

One of the three Abelian groups of the five groups total of ORDER 8. The group Q_8 has the MULTIPLICATION TABLE of $\pm 1, i, j, k$, where 1, i, j, and k are the QUATERNIONS. The CYCLE GRAPH is shown above.

see also FINITE GROUP—D_4, FINITE GROUP—$Z_2 \otimes Z_2 \otimes Z_2$, FINITE GROUP—$Z_2 \otimes Z_4$, FINITE GROUP—$Z_8$, QUATERNION

Finite Group—Z_2

The unique group of ORDER 2. Z_2 is both ABELIAN and CYCLIC. Examples include the POINT GROUPS C_s, C_i, and C_2, the integers modulo 2 under addition, and the MODULO MULTIPLICATION GROUPS M_3, M_4, and M_6. The elements A_i satisfy ${A_i}^2 = 1$, where 1 is the IDENTITY ELEMENT. The CYCLE GRAPH is shown above, and the MULTIPLICATION TABLE is given below.

Z_2	1	A
1	1	A
A	A	1

The CONJUGACY CLASSES are $\{1\}$ and $\{A\}$. The irreducible representation for the C_2 group is $\{1, -1\}$.

Finite Group—$Z_2 \otimes Z_2$

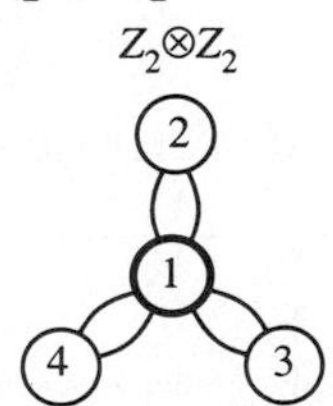

One of the two groups of ORDER 4. The name of this group derives from the fact that it is a DIRECT PRODUCT of two Z_2 SUBGROUPS. Like the group Z_4, $Z_2 \otimes Z_2$ is an ABELIAN GROUP. Unlike Z_4, however, it is not CYCLIC. In addition to satisfying ${A_i}^4 = 1$ for each element A_i, it also satisfies ${A_i}^2 = 1$, where 1 is the IDENTITY ELEMENT. Examples of the $Z_2 \otimes Z_2$ group include the VIERGRUPPE, POINT GROUPS D_2, C_{2h}, and C_{2v}, and the MODULO MULTIPLICATION GROUPS M_8 and M_{12}. That M_8, the RESIDUE CLASSES prime to 8 given by $\{1, 3, 5, 7\}$, are a group of type $Z_2 \otimes Z_2$ can be shown by verifying that

$$1^2 = 1 \quad 3^2 = 9 \equiv 1 \quad 5^2 = 25 \equiv 1 \quad 7^2 = 49 \equiv 1 \pmod{8} \quad (1)$$

and

$$3 \cdot 5 = 15 \equiv 7 \quad 3 \cdot 7 = 21 \equiv 5 \quad 5 \cdot 7 = 35 \equiv 3 \pmod{8}. \quad (2)$$

$Z_2 \otimes Z_2$ is therefore a MODULO MULTIPLICATION GROUP.

The CYCLE GRAPH is shown above, and the multiplication table for the $Z_2 \otimes Z_2$ group is given below.

$Z_2 \otimes Z_2$	1	A	B	C
1	1	A	B	C
A	A	1	C	B
B	B	C	1	A
C	C	B	A	1

The CONJUGACY CLASSES are $\{1\}$, $\{A\}$,

$$A^{-1}AA = A \tag{3}$$
$$B^{-1}AB = A \tag{4}$$
$$C^{-1}AC = A, \tag{5}$$

$\{B\}$,

$$A^{-1}BA = B \tag{6}$$
$$C^{-1}BC = B, \tag{7}$$

and $\{C\}$.

Now explicitly consider the elements of the C_{2v} POINT GROUP.

C_{2v}	E	C_2	σ_v	σ_v
E	E	C_2	σ_v	σ_v'
C_2	C_2	E	σ_v'	σ_v
σ_v	σ_v	σ_v'	E	C_2
σ_v'	σ_v'	σ_v	C_2	E

In terms of the VIERGRUPPE elements

V	I	V_1	V_2	V_3
I	V_1	V_2	V_3	V_4
V_1	V_1	I	V_3	V_2
V_2	V_2	V_3	I	V_1
V_3	V_3	V_2	V_1	I

A reducible representation using 2-D REAL MATRICES is

$$1 = \begin{bmatrix} 1 & 0 \\ 0 & 1 \end{bmatrix} \tag{8}$$

$$A = \begin{bmatrix} -1 & 0 \\ 0 & -1 \end{bmatrix} \tag{9}$$

$$B = \begin{bmatrix} 0 & 1 \\ 1 & 0 \end{bmatrix} \tag{10}$$

$$C = \begin{bmatrix} 0 & -1 \\ -1 & 0 \end{bmatrix}. \tag{11}$$

Another reducible representation using 3-D REAL MATRICES can be obtained from the symmetry elements of the D_2 group (1, $C_2(z)$, $C_2(y)$, and $C_2(x)$) or C_{2v} group (1, C_2, σ_v, and σ_v'). Place the C_2 axis along the z-axis, σ_v in the x-y plane, and σ_v' in the y-z plane.

$$1 = E = E = \begin{bmatrix} 1 & 0 & 0 \\ 0 & 1 & 0 \\ 0 & 0 & 1 \end{bmatrix} \tag{12}$$

$$A = R_x(\pi) = \sigma_v = \begin{bmatrix} 1 & 0 & 0 \\ 0 & -1 & 0 \\ 0 & 0 & 1 \end{bmatrix} \tag{13}$$

$$C = R_z(\pi) = C_2 = \begin{bmatrix} -1 & 0 & 0 \\ 0 & -1 & 0 \\ 0 & 0 & 1 \end{bmatrix} \tag{14}$$

$$B = R_y(\pi) = \sigma_v' = \begin{bmatrix} -1 & 0 & 0 \\ 0 & 1 & 0 \\ 0 & 0 & 1 \end{bmatrix}. \tag{15}$$

In order to find the irreducible representations, note that the traces are given by $\chi(1) = 3, \chi(C_2) = -1$, and $\chi(\sigma_v) = \chi(\sigma_v') = 1$. Therefore, there are at least three distinct CONJUGACY CLASSES. However, we see from the MULTIPLICATION TABLE that there are actually four CONJUGACY CLASSES, so group rule 5 requires that there must be four irreducible representations. By rule 1, we are looking for POSITIVE INTEGERS which satisfy

$$l_1{}^2 + l_2{}^2 + l_3{}^2 + l_4{}^2 = 4. \tag{16}$$

The only combination which will work is

$$l_1 = l_2 = l_3 = l_4 = 1, \tag{17}$$

so there are four one-dimensional representations. Rule 2 requires that the sum of the squares equal the ORDER $h = 4$, so each 1-D representation must have CHARACTER ± 1. Rule 6 requires that a totally symmetric representation always exists, so we are free to start off with the first representation having all 1s. We then use orthogonality (rule 3) to build up the other representations. The simplest solution is then given by

C_{2v}	1	C_2	σ_v	σ_v'
Γ_1	1	1	1	1
Γ_2	1	-1	-1	1
Γ_3	1	-1	1	-1
Γ_4	1	1	-1	-1

These can be put into a more familiar form by switching Γ_1 and Γ_3, giving the CHARACTER TABLE

C_{2v}	1	C_2	σ_v	σ_v'
Γ_3	1	-1	1	-1
Γ_2	1	-1	-1	1
Γ_1	1	1	1	1
Γ_4	1	1	-1	-1

The matrices corresponding to this representation are now

$$1 = \begin{bmatrix} 1 & 0 & 0 & 0 \\ 0 & 1 & 0 & 0 \\ 0 & 0 & 1 & 0 \\ 0 & 0 & 0 & 1 \end{bmatrix} \quad (18)$$

$$C_2 = \begin{bmatrix} -1 & 0 & 0 & 0 \\ 0 & -1 & 0 & 0 \\ 0 & 0 & 1 & 0 \\ 0 & 0 & 0 & 1 \end{bmatrix} \quad (19)$$

$$\sigma_v = \begin{bmatrix} 1 & 0 & 0 & 0 \\ 0 & -1 & 0 & 0 \\ 0 & 0 & 1 & 0 \\ 0 & 0 & 0 & -1 \end{bmatrix} \quad (20)$$

$$\sigma_v' = \begin{bmatrix} -1 & 0 & 0 & 0 \\ 0 & 1 & 0 & 0 \\ 0 & 0 & 1 & 0 \\ 0 & 0 & 0 & -1 \end{bmatrix}, \quad (21)$$

which consist of the previous representation with an additional component. These matrices are now orthogonal, and the order equals the matrix dimension. As before, $\chi(\sigma_v) = \chi(\sigma_v')$.

see also FINITE GROUP—Z_4

Finite Group—$Z_2 \otimes Z_2 \otimes Z_2$

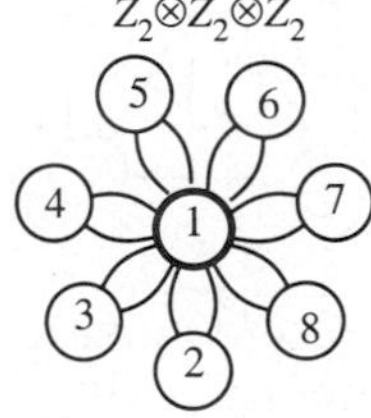

One of the three Abelian groups of the five groups total of ORDER 8. Examples include the MODULO MULTIPLICATION GROUP M_{24}. The elements A_i of this group satisfy $A_i{}^2 = 1$, where 1 is the IDENTITY ELEMENT. The CYCLE GRAPH is shown above.

see also FINITE GROUP—D_4, FINITE GROUP—Q_8, FINITE GROUP—$Z_2 \otimes Z_4$, FINITE GROUP—Z_8

Finite Group—$Z_2 \otimes Z_4$

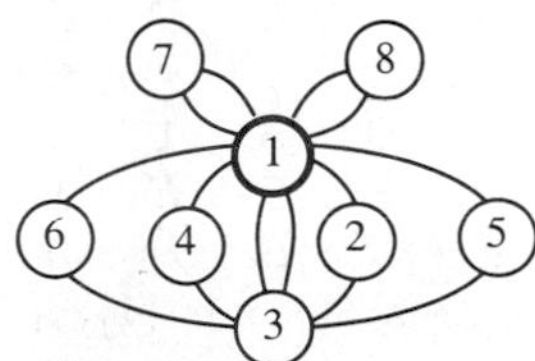

One of the three Abelian groups of the five groups total of ORDER 8. Examples include the MODULO MULTIPLICATION GROUPS M_{15}, M_{16}, M_{20}, and M_{30}. The elements A_i of this group satisfy $A_i{}^4 = 1$, where 1 is the IDENTITY ELEMENT, and four of the elements satisfy $A_i{}^2 = 1$. The CYCLE GRAPH is shown above.

see also FINITE GROUP—D_4, FINITE GROUP—Q_8, FINITE GROUP—$Z_2 \otimes Z_2 \otimes Z_2$, FINITE GROUP—$Z_8$

Finite Group—Z_3

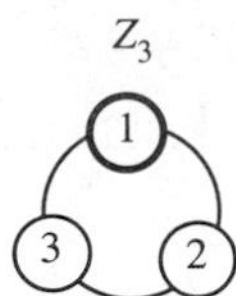

The unique group of ORDER 3. It is both ABELIAN and CYCLIC. Examples include the POINT GROUPS C_3 and D_3 and the integer modulo 3. The elements A_i of the group satisfy $A_i{}^3 = 1$ where 1 is the IDENTITY ELEMENT. The CYCLE GRAPH is shown above, and the MULTIPLICATION TABLE is given below.

Z_3	1	A	B
1	1	A	B
A	A	B	1
B	B	1	A

The CONJUGACY CLASSES are $\{1\}$, $\{A\}$,

$$A^{-1}AA = A$$
$$B^{-1}AB = A,$$

and $\{B\}$,

$$A^{-1}BA = B$$
$$B^{-1}BB = B.$$

The irreducible representation (CHARACTER TABLE) is therefore

Γ	1	A	B
Γ_1	1	1	1
Γ_2	1	1	-1
Γ_3	1	-1	1

Finite Group—Z_4

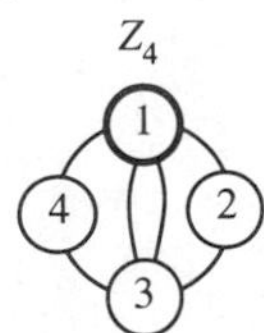

One of the two groups of ORDER 4. Like $Z_2 \otimes Z_2$, it is ABELIAN, but unlike $Z_2 \otimes Z_2$, it is a CYCLIC. Examples include the POINT GROUPS C_4 and S_4 and the MODULO MULTIPLICATION GROUPS M_5 and M_{10}. Elements A_i of the group satisfy $A_i{}^4 = 1$, where 1 is the IDENTITY ELEMENT, and two of the elements satisfy $A_i{}^2 = 1$.

The CYCLE GRAPH is shown above. The MULTIPLICATION TABLE for this group may be written in three equivalent ways—denoted here by $Z_4^{(1)}$, $Z_4^{(2)}$, and $Z_4^{(3)}$—by permuting the symbols used for the group elements.

$Z_4^{(1)}$	1	A	B	C
1	1	A	B	C
A	A	B	C	1
B	B	C	1	A
C	C	1	A	B

The MULTIPLICATION TABLE for $Z_4^{(2)}$ is obtained from $Z_4^{(1)}$ by interchanging A and B.

$Z_4^{(2)}$	1	A	B	C
1	1	A	B	C
A	A	1	C	B
B	B	C	A	1
C	C	B	1	A

The MULTIPLICATION TABLE for $Z_4^{(3)}$ is obtained from $Z_4^{(1)}$ by interchanging A and C.

$Z_4^{(3)}$	1	A	B	C
1	1	A	B	C
A	A	C	1	B
B	B	1	C	A
C	C	B	A	1

The CONJUGACY CLASSES of Z_4 are $\{1\}$, $\{A\}$,

$$A^{-1}AA = A \tag{1}$$
$$B^{-1}AB = A \tag{2}$$
$$C^{-1}AC = A, \tag{3}$$

$\{B\}$,

$$A^{-1}BA = B \tag{4}$$
$$B^{-1}BB = B \tag{5}$$
$$C^{-1}BC = B, \tag{6}$$

and $\{C\}$.

The group may be given a reducible representation using COMPLEX NUMBERS

$$1 = 1 \tag{7}$$
$$A = i \tag{8}$$
$$B = -1 \tag{9}$$
$$C = -i, \tag{10}$$

or REAL MATRICES

$$1 = \begin{bmatrix} 1 & 0 \\ 0 & 1 \end{bmatrix} \tag{11}$$
$$A = \begin{bmatrix} 0 & -1 \\ 1 & 0 \end{bmatrix} \tag{12}$$
$$B = \begin{bmatrix} -1 & 0 \\ 0 & -1 \end{bmatrix} \tag{13}$$
$$C = \begin{bmatrix} 0 & 1 \\ -1 & 0 \end{bmatrix}. \tag{14}$$

see also FINITE GROUP—$Z_2 \otimes Z_2$

Finite Group—Z_5

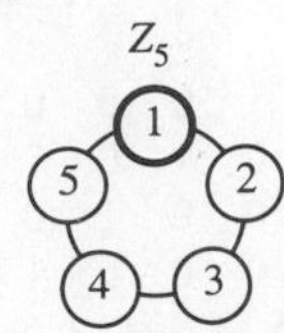

The unique GROUP of ORDER 5, which is ABELIAN. Examples include the POINT GROUP C_5 and the integers mod 5 under addition. The elements A_i satisfy $A_i^5 = 1$, where 1 is the IDENTITY ELEMENT. The CYCLE GRAPH is shown above, and the MULTIPLICATION TABLE is illustrated below.

Z_5	1	A	B	C	D
1	1	A	B	C	D
A	A	B	C	D	1
B	B	C	D	1	A
C	C	D	1	A	B
D	D	1	A	B	C

The CONJUGACY CLASSES are $\{1\}$, $\{A\}$, $\{B\}$, $\{C\}$, and $\{D\}$.

Finite Group—Z_6

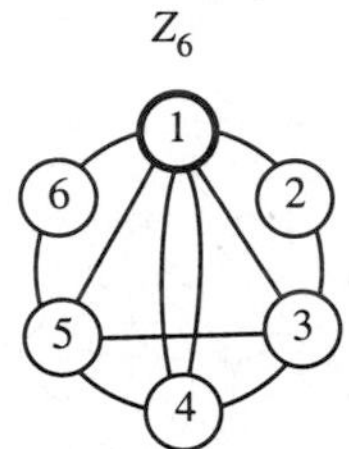

One of the two groups of ORDER 6 which, unlike D_3, is ABELIAN. It is also a CYCLIC. It is isomorphic to $Z_2 \otimes Z_3$. Examples include the POINT GROUPS C_6 and S_6, the integers modulo 6 under addition, and the MODULO MULTIPLICATION GROUPS M_7, M_9, and M_{14}. The elements A_i of the group satisfy $A_i^6 = 1$, where 1 is the IDENTITY ELEMENT, three elements satisfy $A_i^3 = 1$, and two elements satisfy $A_i^2 = 1$. The CYCLE GRAPH is shown above, and the MULTIPLICATION TABLE is given below.

Z_6	1	A	B	C	D	E
1	1	A	B	C	D	E
A	A	1	E	D	B	C
B	B	E	1	A	C	D
C	C	D	A	1	E	B
D	D	B	C	E	1	A
E	E	C	D	B	A	1

The CONJUGACY CLASSES are $\{1\}$, $\{A\}$, $\{B\}$, $\{C\}$, $\{D\}$, and $\{E\}$.

see also FINITE GROUP—D_3

Finite Group—Z_7

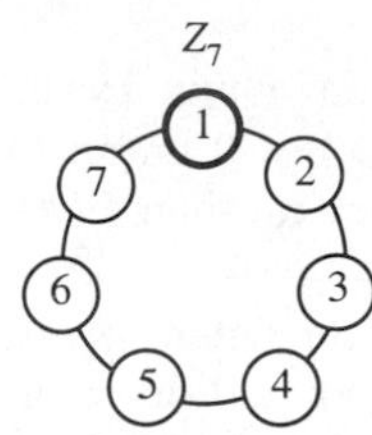

The unique GROUP of ORDER 7. It is ABELIAN and CYCLIC. Examples include the POINT GROUP C_7 and the integers modulo 7 under addition. The elements A_i of the group satisfy $A_i{}^7 = 1$, where 1 is the IDENTITY ELEMENT. The CYCLE GRAPH is shown above.

Z_7	1	A	B	C	D	E	F
1	1	A	B	C	D	E	F
A	A	B	C	D	E	F	1
B	B	C	D	E	F	1	A
C	C	D	E	F	1	A	B
D	D	E	F	1	A	B	C
E	E	F	1	A	B	C	D
F	F	1	A	B	C	D	E

The CONJUGACY CLASSES are $\{1\}$, $\{A\}$, $\{B\}$, $\{C\}$, $\{D\}$, $\{E\}$, and $\{F\}$.

Finite Group—Z_8

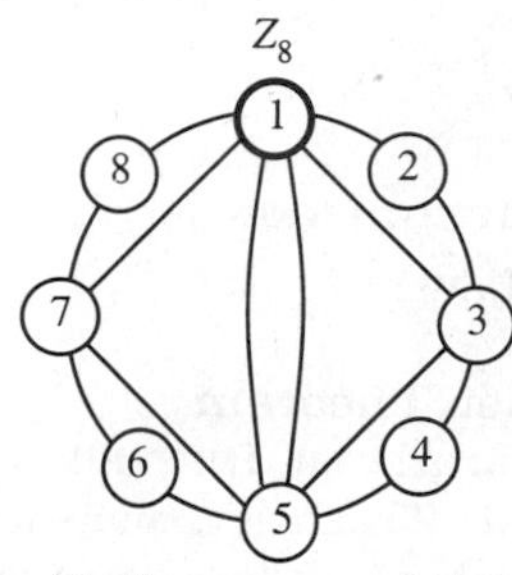

One of the three Abelian groups of the five groups total of ORDER 8. An example is the residue classes modulo 17 which QUADRATIC RESIDUES, i.e., {1, 2, 4, 8, 9, 13, 15, 16} under multiplication modulo 17. The elements A_i satisfy $A_i{}^8 = 1$, four of them satisfy $A_i{}^4 = 1$, and two satisfy $A_i{}^2 = 1$. The CYCLE GRAPH is shown above.

see also FINITE GROUP—D_4, FINITE GROUP—Q_8, FINITE GROUP—$Z_2 \otimes Z_4$, FINITE GROUP—$Z_2 \otimes Z_2 \otimes Z_2$

Finite Mathematics

The branch of mathematics which does not involve infinite sets, limits, or continuity.

see also COMBINATORICS, DISCRETE MATHEMATICS

References

Hildebrand, F. H. and Johnson, C. G. *Finite Mathematics.* Boston, MA: Prindle, Weber, and Schmidt, 1970.

Kemeny, J. G.; Snell, J. L.; and Thompson, G. L. *Introduction to Finite Mathematics, 3rd ed.* Englewood Cliffs, NJ: Prentice-Hall, 1974.

Marcus, M. *A Survey of Finite Mathematics.* New York: Dover, 1993.

Finite Simple Group

see SIMPLE GROUP

Finite Simple Group Classification Theorem

see CLASSIFICATION THEOREM

Finite-to-One Factor

A MAP $\psi : M \to M$, where M is a MANIFOLD, is a finite-to-one factor of a MAP $\Psi : X \to X$ if there exists a continuous ONTO MAP $\pi : X \to M$ such that $\psi \circ \pi = \pi \circ \Psi$ and $\pi^{-1}(x) \subset X$ is finite for each $x \in M$.

Finsler Geometry

The geometry of FINSLER SPACE.

Finsler Manifold

see FINSLER SPACE

Finsler Metric

A continuous real function $L(x, y)$ defined on the TANGENT BUNDLE $T(M)$ of an n-D DIFFERENTIABLE MANIFOLD M is said to be a Finsler metric if

1. $L(x, y)$ is DIFFERENTIABLE at $x \neq y$,
2. $L(x, \lambda y) = |\lambda| L(x, y)$ for any element $(x, y) \in T(M)$ and any REAL NUMBER λ,
3. Denoting the METRIC

$$g_{ij}(x, y) = \frac{1}{2} \frac{\partial^2 [L(x, y)]^2}{\partial y^i \partial y^j},$$

then g_{ij} is a POSITIVE DEFINITE MATRIX.

A DIFFERENTIABLE MANIFOLD M with a Finsler metric is called a FINSLER SPACE.

see also DIFFERENTIABLE MANIFOLD, FINSLER SPACE, TANGENT BUNDLE

References

Iyanaga, S. and Kawada, Y. (Eds.). "Finsler Spaces." §161 in *Encyclopedic Dictionary of Mathematics.* Cambridge, MA: MIT Press, p. 540–542, 1980.

Finsler Space

A general space based on the LINE ELEMENT

$$ds = F(x^1, \ldots, x^n; dx^1, \ldots, dx^n),$$

with $F(x, y) > 0$ for $y \neq 0$ a function on the TANGENT BUNDLE $T(M)$, and homogeneous of degree 1 in y. Formally, a Finsler space is a DIFFERENTIABLE MANIFOLD possessing a FINSLER METRIC. Finsler geometry is RIEMANNIAN GEOMETRY *without* the restriction that the LINE ELEMENT be quadratic of the form

$$F^2 = g_{ij}(x)\, dx^i\, dx^j.$$

A compact boundaryless Finsler space is locally Minkowskian IFF it has 0 "flag curvature."

see also FINSLER METRIC, HODGE'S THEOREM, RIEMANNIAN GEOMETRY, TANGENT BUNDLE

References

Akbar-Zadeh, H. "Sur les espaces de Finsler à courbures sectionnelles constantes." *Acad. Roy. Belg. Bull. Cl. Sci.* **74**, 281–322, 1988.

Bao, D.; Chern, S.-S.; and Shen, Z. (Eds.). *Finsler Geometry.* Providence, RI: Amer. Math. Soc., 1996.

Chern, S.-S. "Finsler Geometry is Just Riemannian Geometry without the Quadratic Restriction." *Not. Amer. Math. Soc.* **43**, 959–963, 1996.

Iyanaga, S. and Kawada, Y. (Eds.). "Finsler Spaces." §161 in *Encyclopedic Dictionary of Mathematics.* Cambridge, MA: MIT Press, p. 540–542, 1980.

Finsler-Hadwiger Theorem

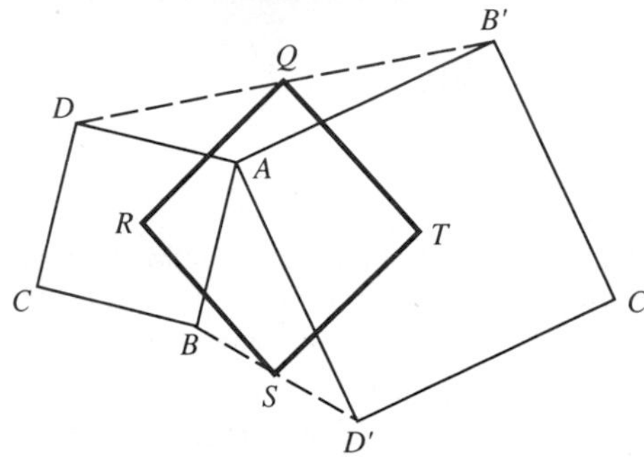

Let the SQUARES $\square ABCD$ and $\square AB'C'D'$ share a common VERTEX A. The midpoints Q and S of the segments $B'D$ and BD' together with the centers of the original squares R and T then form another square $\square QRST$. This theorem is a special case of the FUNDAMENTAL THEOREM OF DIRECTLY SIMILAR FIGURES (Detemple and Harold 1996).

see also FUNDAMENTAL THEOREM OF DIRECTLY SIMILAR FIGURES, SQUARE

References

Detemple, D. and Harold, S. "A Round-Up of Square Problems." *Math. Mag.* **69**, 15–27, 1996.

Finsler, P. and Hadwiger, H. "Einige Relationen im Dreieck." *Comment. Helv.* **10**, 316–326, 1937.

Fisher, J. C.; Ruoff, D.; and Shileto, J. "Polygons and Polynomials." In *The Geometric Vein: The Coxeter Festschrift.* New York: Springer-Verlag, 321–333, 1981.

First-Countable Space

A TOPOLOGICAL SPACE in which every point has a countable BASE for its neighborhood system.

First Curvature

see CURVATURE

First Derivative Test

$f'(x)<0$, $f''(x)>0$; $f'(x)=0$, $f''(x)=0$; $f'(x)<0$, $f''(x)<0$ — *stationary point*

$f'(x)<0$; $f'(x)>0$; $f'(x)=0$ — *minimum*

$f'(x)=0$; $f'(x)>0$; $f'(x)<0$ — *maximum*

Suppose $f(x)$ is CONTINUOUS at a STATIONARY POINT x_0.

1. If $f'(x)>0$ on an OPEN INTERVAL extending left from x_0 and $f'(x)<0$ on an OPEN INTERVAL extending right from x_0, then f has a RELATIVE MAXIMUM (possibly a GLOBAL MAXIMUM) at x_0.
2. If $f'(x)<0$ on an OPEN INTERVAL extending left from x_0 and $f'(x)>0$ on an OPEN INTERVAL extending right from x_0, then f has a RELATIVE MINIMUM (possibly a GLOBAL MINIMUM) at x_0.
3. If $f'(x)$ has the same sign on an OPEN INTERVAL extending left from x_0 and on an OPEN INTERVAL extending right from x_0, then f does not have a RELATIVE EXTREMUM at x_0.

see also EXTREMUM, GLOBAL MAXIMUM, GLOBAL MINIMUM, INFLECTION POINT, MAXIMUM, MINIMUM, RELATIVE EXTREMUM, RELATIVE MAXIMUM, RELATIVE MINIMUM, SECOND DERIVATIVE TEST, STATIONARY POINT

References

Abramowitz, M. and Stegun, C. A. (Eds.). *Handbook of Mathematical Functions with Formulas, Graphs, and Mathematical Tables, 9th printing.* New York: Dover, p. 14, 1972.

First Digit Law

see BENFORD'S LAW

First Digit Phenomenon

see BENFORD'S LAW

First Multiplier Theorem

Let D be a planar Abelian DIFFERENCE SET and t be any DIVISOR of n. Then t is a numerical multiplier of D, where a multiplier is defined as an automorphism α of G which takes D to a translation $g+D$ of itself for some $g \in G$. If α is of the form $\alpha : x \to tx$ for $t \in \mathbb{Z}$ relatively prime to the order of G, then α is called a numerical multiplier.

References

Gordon, D. M. "The Prime Power Conjecture is True for $n < 2,000,000$." *Electronic J. Combinatorics* **1**, R6, 1–7, 1994. `http://www.combinatorics.org/Volume_1/volume1.html#R6`.

Fischer's Baby Monster Group

see BABY MONSTER GROUP

Fischer Groups

The SPORADIC GROUPS Fi_{22}, Fi_{23}, and Fi'_{24}. These groups were discovered during the investigation of 3-TRANSPOSITION GROUPS.

see also SPORADIC GROUP

References

Wilson, R. A. "ATLAS of Finite Group Representation." `http://for.mat.bham.ac.uk/atlas/F22.html`, `F23.html`, and `F24.html`.

Fish Bladder

see LENS

Fisher-Behrens Problem

The determination of a test for the equality of MEANS for two NORMAL DISTRIBUTIONS with different VARIANCES given samples from each. There exists an exact test which, however, does not give a unique answer because it does not use all the data. There also exist approximate tests which do not use all the data.

see also NORMAL DISTRIBUTION

References

Fisher, R. A. "The Fiducial Argument in Statistical Inference." *Ann. Eugenics* **6**, 391–398, 1935.

Kenney, J. F. and Keeping, E. S. "The Behrens-Fisher Test." §9.8 in *Mathematics of Statistics, Pt. 2, 2nd ed.* Princeton, NJ: Van Nostrand, pp. 257–260 and 261–264, 1951.

Sukhatme, P. V. "On Fisher and Behrens' Test of Significance of the Difference in Means of Two Normal Samples." *Sankhya* **4**, 39, 1938.

Fisher's Block Design Inequality

A balanced incomplete BLOCK DESIGN (v, k, λ, r, b) exists only or $b \geq v$ (or, equivalently, $r \geq k$).

see also BRUCK-RYSER-CHOWLA THEOREM

References

Dinitz, J. H. and Stinson, D. R. "A Brief Introduction to Design Theory." Ch. 1 in *Contemporary Design Theory: A Collection of Surveys* (Ed. J. H. Dinitz and D. R. Stinson). New York: Wiley, pp. 1–12, 1992.

Fisher's Estimator Inequality

Given T an UNBIASED ESTIMATOR of θ so that $\langle T\rangle = \theta$. Then

$$\operatorname{var}(T) \geq \frac{1}{N\int_{-\infty}^{\infty}\left[\frac{\partial(\ln f)}{\partial\theta}\right]^2 f\,dx},$$

where var is the VARIANCE.

Fisher's Exact Test

A STATISTICAL TEST used to determine if there are nonrandom associations between two CATEGORICAL VARIABLES. Let there exist two such variables X and Y, with m and n observed states, respectively. Now form an $n\times m$ MATRIX in which the entries a_{ij} represent the number of observations in which $x = i$ and $y = j$. Calculate the row and column sums R_i and C_j, respectively, and the total sum

$$N = \sum_i R_i = \sum_j C_j$$

of the MATRIX. Then calculate the conditional LIKELIHOOD (P-VALUE) of getting the actual matrix given the particular row and column sums, given by

$$P_{\text{crit}} = \frac{(R_1!R_2!\cdots R_m!)(C_1!C_2!\cdots C_n!)}{N!\prod_{i,j} a_{ij}!}$$

(which is a HYPERGEOMETRIC DISTRIBUTION). Now find all possible MATRICES of NONNEGATIVE INTEGERS consistent with the row and column sums R_i and C_j. For each one, calculate the associated P-VALUE using (0) (where the sum of these probabilities must be 1). Then the P-VALUE of the test is given by the sum of all P-VALUES which are $\leq P_{\text{crit}}$.

The test is most commonly applied to a 2×2 MATRICES, and is computationally unwieldy for large m or n.

As an example application of the test, let X be a journal, say either *Mathematics Magazine* or *Science*, and let Y be the number of articles on the topics of mathematics and biology appearing in a given issue of one of these journals. If *Mathematics Magazine* has five articles on math and one on biology, and *Science* has none on math and four on biology, then the relevant matrix would be

	Math. Mag.	*Science*	
math	5	0	$R_1 = 5$
biology	1	4	$R_2 = 5$
	$C_1 = 6$	$C_2 = 4$	$N = 10.$

Computing P_{crit} gives

$$P_{\text{crit}} = \frac{5!^2 6!4!}{10!(5!0!1!4!)} = 0.0238,$$

and the other possible matrices and their Ps are

$$\begin{bmatrix} 4 & 1 \\ 2 & 3 \end{bmatrix} \quad P = 0.2381$$

$$\begin{bmatrix} 3 & 2 \\ 3 & 2 \end{bmatrix} \quad P = 0.4762$$

$$\begin{bmatrix} 2 & 3 \\ 4 & 1 \end{bmatrix} \quad P = 0.2381$$

$$\begin{bmatrix} 1 & 4 \\ 5 & 0 \end{bmatrix} \quad P = 0.0238,$$

which indeed sum to 1, as required. The sum of P-values less than or equal to $P_{\text{crit}} = 0.0238$ is then 0.0476 which, because it is less than 0.05, is SIGNIFICANT. Therefore, in this case, there would be a statistically significant association between the journal and type of article appearing.

Fisher Index

The statistical INDEX

$$P_B \equiv \sqrt{P_L P_P},$$

where P_L is LASPEYRES' INDEX and P_P is PAASCHE'S INDEX.

see also INDEX

References

Kenney, J. F. and Keeping, E. S. *Mathematics of Statistics, Pt. 1, 3rd ed.* Princeton, NJ: Van Nostrand, p. 66, 1962.

Fisher Kurtosis

$$\gamma_2 \equiv b_2 \equiv \frac{\mu_4}{{\mu_2}^2} - 3 = \frac{\mu_4}{\sigma^4} - 3,$$

where μ_i is the ith MOMENT about the MEAN and $\sigma = \sqrt{\mu_2}$ is the STANDARD DEVIATION.

see also FISHER SKEWNESS, KURTOSIS, PEARSON KURTOSIS

Fisher Sign Test

A robust nonparametric test which is an alternative to the PAIRED t-TEST. This test makes the basic assumption that there is information only in the signs of the differences between paired observations, not in their sizes. Take the paired observations, calculate the differences, and count the number of +s n_+ and −s n_-, where

$$N \equiv n_+ + n_-$$

is the sample size. Calculate the BINOMIAL COEFFICIENT

$$B \equiv \binom{N}{n_+}.$$

Then $B/2^N$ gives the probability of getting exactly this many +s and −s if POSITIVE and NEGATIVE values are equally likely. Finally, to obtain the P-VALUE for the test, sum all the COEFFICIENTS that are $\leq B$ and divide by 2^N.

see also HYPOTHESIS TESTING

Fisher Skewness

$$\gamma_1 = \frac{\mu_3}{{\mu_2}^{3/2}} = \frac{\mu_3}{\sigma^3},$$

where μ_i is the i MOMENT about the MEAN, and $\sigma = \sqrt{\mu_2}$ is the STANDARD DEVIATION.

see also FISHER KURTOSIS, MOMENT, SKEWNESS, STANDARD DEVIATION

Fisher's Theorem

Let A be a sum of squares of n independent normal standardized variates x_i, and suppose $A = B + C$ where B is a quadratic form in the x_i, distributed as CHI-SQUARED with h DEGREES OF FREEDOM. Then C is distributed as χ^2 with $n - h$ DEGREES OF FREEDOM and is independent of B. The converse of this theorem is known as COCHRAN'S THEOREM.

see also CHI-SQUARED DISTRIBUTION, COCHRAN'S THEOREM

Fisher-Tippett Distribution

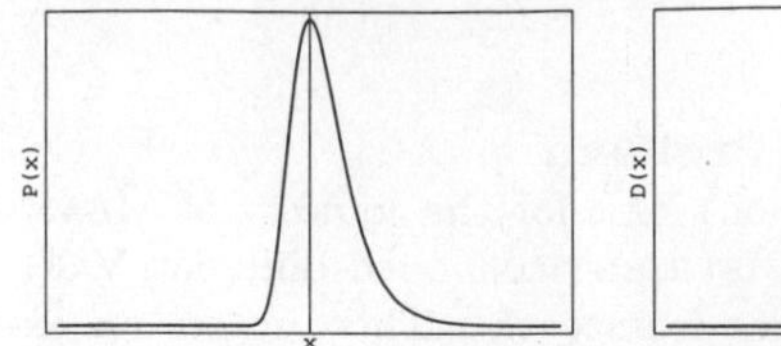

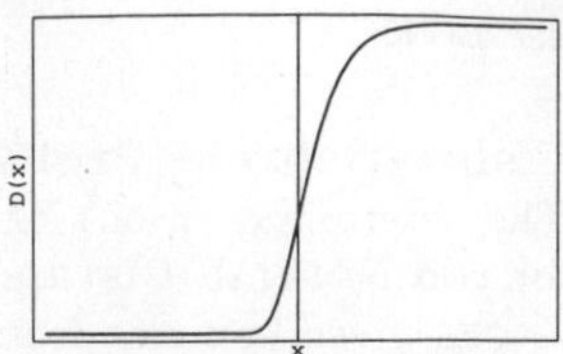

Also called the EXTREME VALUE DISTRIBUTION and LOG-WEIBULL DISTRIBUTION. It is the limiting distribution for the smallest or largest values in a large sample drawn from a variety of distributions.

$$P(x) = \frac{e^{(a-x)/b - e^{(a-x)/b}}}{b} \tag{1}$$

$$D(x) = e^{-e^{(a-x)/b}}. \tag{2}$$

These can be computed directly be defining

$$z \equiv \exp\left(\frac{a-x}{b}\right) \tag{3}$$

$$x = a - b\ln z \tag{4}$$

$$dz = -\frac{1}{b}\exp\left(\frac{a-x}{b}\right)\,dx. \tag{5}$$

Then the MOMENTS are

$$\begin{aligned}\mu_n &\equiv \int_{-\infty}^{\infty} x^n P(x)\,dx \\ &= \frac{1}{b}\int_{-\infty}^{\infty} x^n \exp\left(\frac{a}{x-b}\right)\exp[-e^{(a-x)/b}]\,dx \\ &= -\int_{\infty}^{0} (a - b\ln z)^n e^{-z}\,dz \\ &= \int_0^{\infty} (a - b\ln z)^n e^{-z}\,dz \\ &= \sum_{k=0}^{n} \binom{n}{k} (-1)^k a^{n-k} b^k \int_0^{\infty} (\ln z)^k e^{-z}\,dz \\ &= \sum_{k=0}^{n} \binom{n}{k} a^{n-k} b^k I(k),\end{aligned} \tag{6}$$

where $I(k)$ are EULER-MASCHERONI INTEGRALS. Plugging in the EULER-MASCHERONI INTEGRALS $I(k)$ gives

$$\mu_0 = 1 \tag{7}$$

$$\mu_1 = a + b\gamma \tag{8}$$

$$\mu_2 = a^2 + 2ab\gamma + b^2(\gamma^2 + \tfrac{1}{6}\pi^2) \tag{9}$$

$$\begin{aligned}\mu_3 &= a^3 + 3a^2b\gamma + 3ab^2(\gamma^2 + \tfrac{1}{6}\pi^2) \\ &\quad + b^3[\gamma^3 + \tfrac{1}{2}\gamma\pi^2 + 2\zeta(3)]\end{aligned} \tag{10}$$

$$\begin{aligned}\mu_4 &= a^4 + 4a^3b\gamma + 6a^2b^2(\gamma^2 + \tfrac{1}{6}\pi^2) \\ &\quad + 4ab^3[\gamma^3 + \tfrac{1}{2}\gamma\pi^2 + 2\zeta(3)] \\ &\quad + b^4[\gamma^4 + \gamma^2\pi^2 + \tfrac{3}{20}\pi^4 + 8\gamma\zeta(3)],\end{aligned} \tag{11}$$

where γ is the EULER-MASCHERONI CONSTANT and $\zeta(3)$ is APÉRY'S CONSTANT. The MEAN, VARIANCE, SKEWNESS, and KURTOSIS are therefore

$$\mu = a + b\gamma \tag{12}$$

$$\sigma^2 = \mu_2 - \mu_1{}^2 = \tfrac{1}{6}\pi^2 b^2 \tag{13}$$

$$\begin{aligned}\gamma_1 &= \frac{\mu_3}{\sigma^3}\\ &= \frac{6\sqrt{6}}{b^3\pi^3}\{a^3 + 3a^2 b\gamma + 3ab^2(\gamma^2 + \tfrac{1}{6}\pi^2)\\ &\quad + b^3[\gamma^3 + \tfrac{1}{2}\gamma\pi^2 + 2\zeta(3)]\}\end{aligned} \tag{14}$$

$$\begin{aligned}\gamma_2 &= \frac{\mu_4}{\sigma^4} - 3\\ &= \frac{36}{b^4\pi^4}\{a^4 + 4a^3 b\gamma + a^2 b^2(6\gamma^2 + \pi^2)\\ &\quad + 4ab^3[\gamma^3 + \tfrac{1}{2}\gamma\pi^2 + 2\zeta(3)]\\ &\quad + b^4[\gamma^4 + \gamma^2\pi^2 + \tfrac{3}{20}\pi^4 + 8\gamma\zeta(3)]\}.\end{aligned} \tag{15}$$

The CHARACTERISTIC FUNCTION is

$$\phi(t) = \Gamma(1 - i\beta t)e^{i\alpha t}, \tag{16}$$

where $\Gamma(z)$ is the GAMMA FUNCTION. The special case of the Fisher-Tippett distribution with $a = 0$, $b = 1$ is called GUMBEL'S DISTRIBUTION.

see also EULER-MASCHERONI INTEGRALS, GUMBEL'S DISTRIBUTION

Fisher's z-Distribution

$$g(z) = \frac{2{n_1}^{n_1/2}{n_2}^{n_2/2}}{B\left(\frac{n_1}{2}, \frac{n_2}{2}\right)} \frac{e^{n_1 z}}{(n_1 e^{2z} + n_2)^{(n_1+n_1)/2}} \tag{1}$$

(Kenney and Keeping 1951). This general distribution includes the CHI-SQUARED DISTRIBUTION and STUDENT'S t-DISTRIBUTION as special cases. Let u^2 and v^2 be INDEPENDENT UNBIASED ESTIMATORS of the VARIANCE of a NORMALLY DISTRIBUTED variate. Define

$$z \equiv \ln\left(\frac{u}{v}\right) = \tfrac{1}{2}\ln\left(\frac{u^2}{v^2}\right). \tag{2}$$

Then let

$$F \equiv \frac{u^2}{v^2} = \frac{\frac{N{s_1}^2}{n_1}}{\frac{N{s_2}^2}{n_2}} \tag{3}$$

so that $n_1 F/n_2$ is a ratio of CHI-SQUARED variates

$$\frac{n_1 F}{n_2} = \frac{\chi^2(n_1)}{\chi^2(n_2)}, \tag{4}$$

which makes it a ratio of GAMMA DISTRIBUTION variates, which is itself a BETA PRIME DISTRIBUTION variate,

$$\frac{\gamma\left(\frac{n_1}{2}\right)}{\gamma\left(\frac{n_2}{2}\right)} = \beta'\left(\tfrac{n_1}{2}, \tfrac{n_2}{2}\right), \tag{5}$$

giving

$$f(F) = \frac{\left(\frac{n_1 F}{n_2}\right)^{n_1/2-1}\left(1 + \frac{n_1 F}{n_2}\right)^{-(n_1+n_2)/2}\frac{n_1}{n_2}}{B\left(\frac{n_1}{2}, \frac{n_2}{2}\right)}. \tag{6}$$

The MEAN is

$$\langle F\rangle = \frac{n_2}{n_2 - 2}, \tag{7}$$

and the MODE is

$$\frac{n_2}{n_2 + 2}\,\frac{n_1 - 2}{n_1}. \tag{8}$$

see also BETA DISTRIBUTION, BETA PRIME DISTRIBUTION, CHI-SQUARED DISTRIBUTION, GAMMA DISTRIBUTION, NORMAL DISTRIBUTION, STUDENT'S t-DISTRIBUTION

References

Kenney, J. F. and Keeping, E. S. *Mathematics of Statistics, Pt. 2, 2nd ed.* Princeton, NJ: Van Nostrand, pp. 180–181, 1951.

Fisher's z'-Transformation

Let r be the CORRELATION COEFFICIENT. Then defining

$$z' \equiv \tanh^{-1} r \tag{1}$$

$$\zeta \equiv \tanh^{-1} \rho, \tag{2}$$

gives

$$\sigma_{z'} = (N-3)^{-1/2} \tag{3}$$

$$\operatorname{var}(z') = \frac{1}{n} + \frac{4-\rho^2}{2n^2} + \ldots \tag{4}$$

$$\gamma_1 = \frac{\rho\left|\rho^2 - \frac{9}{16}\right|}{n^{3/2}} \tag{5}$$

$$\gamma_2 = \frac{32 - 3\rho^4}{16N}, \tag{6}$$

where $n \equiv N - 1$.

see also CORRELATION COEFFICIENT

Fitting Subgroup

The unique smallest NORMAL NILPOTENT SUBGROUP of H, denoted $F(H)$. The generalized fitting subgroup is defined by $F^*(H) = F(H)E(H)$, where $E(H)$ is the commuting product of all components of H, and F is the fitting subgroup of H.

Five Cubes

see CUBE 5-COMPOUND

Five Disks Problem

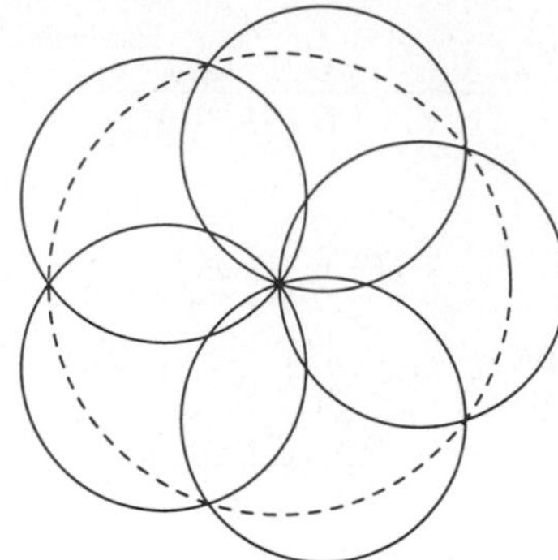

Given five *equal* DISKS placed *symmetrically* about a given center, what is the smallest RADIUS r for which the RADIUS of the circular AREA covered by the five disks is 1? The answer is $r = \phi - 1 = 1/\phi = 0.6180340\ldots$, where ϕ is the GOLDEN RATIO, and the centers c_i of the disks $i = 1, \ldots, 5$ are located at

$$c_i = \begin{bmatrix} \frac{1}{\phi}\cos\left(\frac{2\pi i}{5}\right) \\ \frac{1}{\phi}\sin\left(\frac{2\pi i}{5}\right) \end{bmatrix}.$$

The GOLDEN RATIO enters here through its connection with the regular PENTAGON. If the requirement that the disks be symmetrically placed is dropped (the general DISK COVERING PROBLEM), then the RADIUS for $n = 5$ disks can be reduced slightly to 0.609383... (Neville 1915).

see also ARC, DISK COVERING PROBLEM, FLOWER OF LIFE, SEED OF LIFE

References

Ball, W. W. R. and Coxeter, H. S. M. "The Five-Disc Problem." In *Mathematical Recreations and Essays, 13th ed.* New York: Dover, pp. 97–99, 1987.

Neville, E. H. "On the Solution of Numerical Functional Equations, Illustrated by an Account of a Popular Puzzle and of its Solution." *Proc. London Math. Soc.* **14**, 308–326, 1915.

Five Tetrahedra Compound

see TETRAHEDRON 5-COMPOUND

Fixed

When referring to a planar object, "fixed" means that the object is regarded as fixed in the plane so that it may not be picked up and flipped. As a result, MIRROR IMAGES are not necessarily equivalent for fixed objects.

see also FREE, MIRROR IMAGE

Fixed Element

see FIXED POINT (MAP)

Fixed Point

A point which does not change upon application of a MAP, system of DIFFERENTIAL EQUATIONS, etc.

see also FIXED POINT (DIFFERENTIAL EQUATIONS), FIXED POINT (MAP), FIXED POINT THEOREM

References

Shashkin, Yu. A. *Fixed Points.* Providence, RI: Amer. Math. Soc., 1991.

Fixed Point (Differential Equations)

Points of an AUTONOMOUS system of ordinary differential equations at which

$$\begin{cases} \frac{dx_1}{dt} = f_1(x_1, \ldots, x_n) = 0 \\ \vdots \\ \frac{dx_n}{dt} = f_n(x_1, \ldots, x_n) = 0. \end{cases}$$

If a variable is slightly displaced from a FIXED POINT, it may (1) move back to the fixed point ("asymptotically stable" or "superstable"), (2) move away ("unstable"), or (3) move in a neighborhood of the fixed point but not approach it ("stable" but not "asymptotically stable"). Fixed points are also called CRITICAL POINTS or EQUILIBRIUM POINTS. If a variable starts at a point that is not a CRITICAL POINT, it cannot reach a critical point in a finite amount of time. Also, a trajectory passing through at least one point that is not a CRITICAL POINT cannot cross itself unless it is a CLOSED CURVE, in which case it corresponds to a periodic solution.

A fixed point can be classified into one of several classes using LINEAR STABILITY analysis and the resulting STABILITY MATRIX.

see also ELLIPTIC FIXED POINT (DIFFERENTIAL EQUATIONS), HYPERBOLIC FIXED POINT (DIFFERENTIAL EQUATIONS), STABLE IMPROPER NODE, STABLE NODE, STABLE SPIRAL POINT, STABLE STAR, UNSTABLE IMPROPER NODE, UNSTABLE NODE, UNSTABLE SPIRAL POINT, UNSTABLE STAR

Fixed Point (Map)

A point x^* which is mapped to itself under a MAP G, so that $x^* = G(x^*)$. Such points are sometimes also called INVARIANT POINTS, or FIXED ELEMENTS (Woods 1961). Stable fixed points are called elliptical. Unstable fixed points, corresponding to an intersection of a stable and unstable invariant MANIFOLD, are called HYPERBOLIC (or SADDLE). Points may also be called asymptotically stable (a.k.a. superstable).

see also CRITICAL POINT, INVOLUNTARY

References

Shashkin, Yu. A. *Fixed Points.* Providence, RI: Amer. Math. Soc., 1991.

Woods, F. S. *Higher Geometry: An Introduction to Advanced Methods in Analytic Geometry.* New York: Dover, p. 14, 1961.

Fixed Point Theorem

If g is a continuous function $g(x) \in [a, b]$ FOR ALL $x \in [a, b]$, then g has a FIXED POINT in $[a, b]$. This can be proven by noting that

$$g(a) \geq a \qquad g(b) \leq b$$

$$g(a) - a \geq 0 \qquad g(b) - b \leq 0.$$

Since g is continuous, the INTERMEDIATE VALUE THEOREM guarantees that there exists a $c \in [a, b]$ such that

$$g(c) - c = 0,$$

so there must exist a c such that

$$g(c) = c,$$

so there must exist a FIXED POINT $\in [a, b]$.

see also BANACH FIXED POINT THEOREM, BROUWER FIXED POINT THEOREM, KAKUTANI'S FIXED POINT THEOREM, LEFSHETZ FIXED POINT FORMULA, LEFSHETZ TRACE FORMULA, POINCARÉ-BIRKHOFF FIXED POINT THEOREM, SCHAUDER FIXED POINT THEOREM

Fixed Point (Transformation)

see FIXED POINT (MAP)

Flag

A collection of FACES of an n-D POLYTOPE or simplicial COMPLEX, one of each DIMENSION $0, 1, \ldots, n-1$, which all have a common nonempty INTERSECTION. In normal 3-D, the flag consists of a half-plane, its bounding RAY, and the RAY's endpoint.

Flag Manifold

For any SEQUENCE of INTEGERS $0 < n_1 < \ldots < n_k$, there is a flag manifold of type $(n_1, \ldots, n_k)$ which is the collection of ordered pairs of vector SUBSPACES of $\mathbb{R}^{n_k}$ $(V_1, \ldots, V_k)$ with $\dim(V_i) = n_i$ and V_i a SUBSPACE of V_{i+1}. There are also COMPLEX flag manifolds with COMPLEX subspaces of $\mathbb{C}^{n_k}$ instead of REAL SUBSPACES of a REAL n_k-space. These flag manifolds admit the structure of MANIFOLDS in a natural way and are used in the theory of LIE GROUPS.

see also GRASSMANN MANIFOLD

References

Lu, J.-H. and Weinstein, A. "Poisson Lie Groups, Dressing Transformations, and the Bruhat Decomposition." *J. Diff. Geom.* **31**, 501–526, 1990.

Flat

A set in $\mathbb{R}^d$ formed by translating an affine subspace or by the intersection of a set of HYPERPLANES.

Flat Norm

The flat norm on a CURRENT is defined by

$$\mathcal{F}(S) = \int \{\text{Area } T + \text{vol } R : S - T = \partial R\},$$

where ∂R is the boundary of R.

see also COMPACTNESS THEOREM, CURRENT

References

Morgan, F. "What Is a Surface?" *Amer. Math. Monthly* **103**, 369–376, 1996.

Flat Space Theorem

If it is possible to transform a coordinate system to a form where the metric elements $g_{\mu\nu}$ are constants independent of x^μ, then the space is flat.

Flat Surface

A REGULAR SURFACE and special class of MINIMAL SURFACE for the GAUSSIAN CURVATURE vanishes everywhere. A TANGENT DEVELOPABLE, GENERALIZED CONE, and GENERALIZED CYLINDER are all flat surfaces.

see also MINIMAL SURFACE

References

Gray, A. *Modern Differential Geometry of Curves and Surfaces.* Boca Raton, FL: CRC Press, p. 280, 1993.

Flattening

The flattening of a SPHEROID (also called OBLATENESS) is denoted ε or f. It is defined as

$$\varepsilon \equiv \begin{cases} \frac{a-c}{a} = 1 - \frac{c}{a} & \text{oblate} \\ \frac{c-a}{a} = \frac{c}{a} - 1 & \text{prolate,} \end{cases}$$

where c is the polar RADIUS and a is the equatorial RADIUS.

see also ECCENTRICITY, ELLIPSOID, OBLATE SPHEROID, PROLATE SPHEROID, SPHEROID

Flemish Knot

see FIGURE-OF-EIGHT KNOT

Fletcher Point

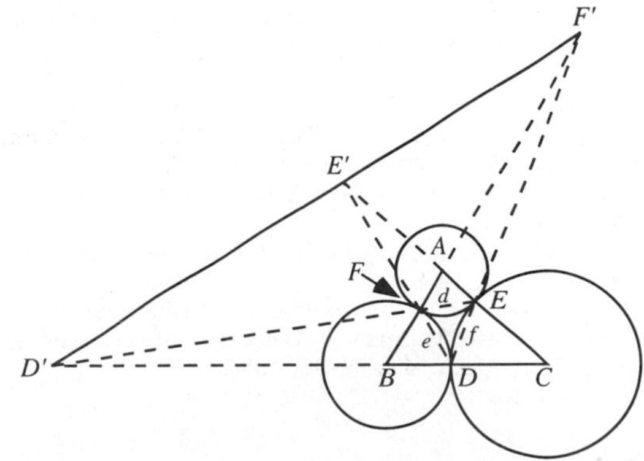

The intersection of the GERGONNE LINE and the SODDY LINE. It has TRILINEAR COORDINATES given by

$$Fl = I - \frac{1}{3}\left(\frac{1}{d} + \frac{1}{e} + \frac{1}{f}\right) Ge,$$

where I is the INCENTER, Ge the GERGONNE POINT, and d, e, and f are the lengths of the sides of the CONTACT TRIANGLE ΔDEF.

see also CONTACT TRIANGLE, GERGONNE LINE, GERGONNE POINT, SODDY LINE

References

Oldknow, A. "The Euler-Gergonne-Soddy Triangle of a Triangle." *Amer. Math. Monthly* **103**, 319–329, 1996.

Flexible Polyhedron

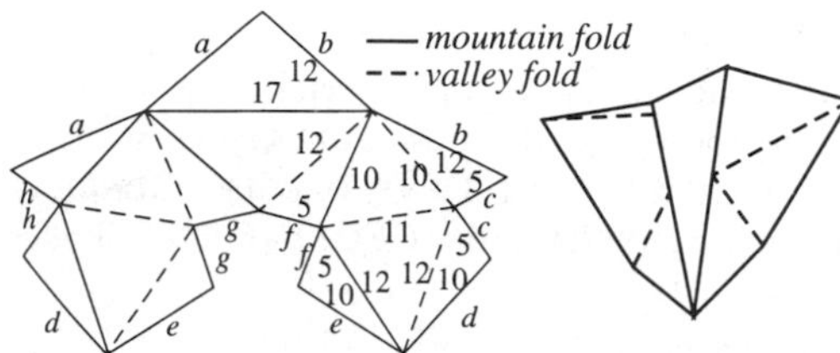

The RIGIDITY THEOREM states that if the faces of a *convex* POLYHEDRON are made of metal plates and the EDGES are replaced by hinges, the POLYHEDRON would be RIGID. The theorem was stated by Cauchy (1813), although a mistake in this paper went unnoticed for more than 50 years. Concave polyhedra need not be RIGID, and such nonrigid polyhedra are called flexible polyhedra. Connelly (1978) found the first example of a reflexible polyhedron, consisting of 18 triangular faces. A flexible polyhedron with only 14 triangular faces and 9 vertices (shown above), believed to be the simplest possible composed of only triangles, was subsequently found by Steffen (Mackenzie 1998). There also exists a six-vertex eight-face flexible polyhedron (Wunderlich and Schwabe 1986, Cromwell 1997).

Connelly *et al.* (1997) proved that a flexible polyhedron must keep its VOLUME constant (Mackenzie 1998).

see also POLYHEDRON, QUADRICORN, RIGID, RIGIDITY THEOREM

References

Cauchy, A. L. "Sur les polygons et le polyhéders." *XVIe Cahier* **IX**, 87–89, 1813.
Connelly, R. "A Flexible Sphere." *Math. Intel.* **1**, 130–131, 1978.
Connelly, R.; Sabitov, I.; and Walz, A. "The Bellows Conjecture." *Contrib. Algebra Geom.* **38**, 1–10, 1997.
Cromwell, P. R. *Polyhedra.* New York: Cambridge University Press, 1997.
Mackenzie, D. "Polyhedra Can Bend But Not Breathe." *Science* **279**, 1637, 1998.
Wunderlich, W. and Schwabe, C. "Eine Familie von geschlossen gleichflachigen Polyhedern, die fast beweglich sind." *Elem. Math.* **41**, 88–98, 1986.

Flexagon

An object created by FOLDING a piece of paper along certain lines to form loops. The number of states possible in an n-FLEXAGON is a CATALAN NUMBER. By manipulating the folds, it is possible to hide and reveal different faces.

see also FLEXATUBE, FOLDING, HEXAFLEXAGON, TETRAFLEXAGON

References

Crampin, J. "On Note 2449." *Math. Gazette* **41**, 55–56, 1957.
Cundy, H. and Rollett, A. *Mathematical Models, 3rd ed.* Stradbroke, England: Tarquin Pub., pp. 205–207, 1989.
Madachy, J. S. *Madachy's Mathematical Recreations.* New York: Dover, pp. 62–84, 1979.
Gardner, M. "Hexaflexagons." Ch. 1 in *The Scientific American Book of Mathematical Puzzles & Diversions.* New York: Simon and Schuster, 1959.
Gardner, M. Ch. 2 in *The Second Scientific American Book of Mathematical Puzzles & Diversions: A New Selection.* New York: Simon and Schuster, pp. 24–31, 1961.
Maunsell, F. G. "The Flexagon and the Hexaflexagon." *Math. Gazette* **38**, 213–214, 1954.
Oakley, C. O. and Wisner, R. J. "Flexagons." *Amer. Math. Monthly* **64**, 143–154, 1957.
Wheeler, R. F. "The Flexagon Family." *Math. Gaz.* **42**, 1–6, 1958.

Flexatube

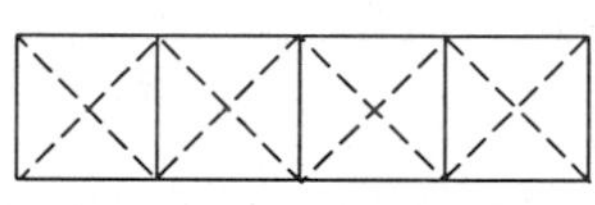
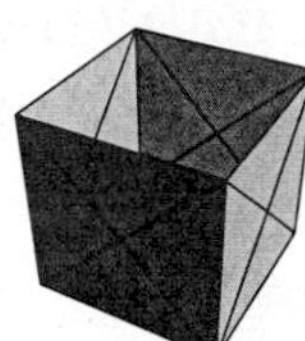

A FLEXAGON-like structure created by connecting the ends of a strip of four squares after folding along 45° diagonals. Using a number of folding movements, it is possible to flip the flexatube inside out so that the faces originally facing inward face outward. Gardner (1961) illustrated one possible solution, and Steinhaus (1983) gives a second.

see also FLEXAGON, HEXAFLEXAGON, TETRAFLEXAGON

References

Cundy, H. and Rollett, A. *Mathematical Models, 3rd ed.* Stradbroke, England: Tarquin Pub., p. 205, 1989.
Gardner, M. *The Second Scientific American Book of Mathematical Puzzles & Diversions: A New Selection.* New York: Simon and Schuster, pp. 29–31, 1961.
Steinhaus, H. *Mathematical Snapshots, 3rd American ed.* New York: Oxford University Press, pp. 177–181 and 190, 1983.

Flip Bifurcation

Let $f : \mathbb{R} \times \mathbb{R} \to \mathbb{R}$ be a one-parameter family of C^3 maps satisfying

$$f(0,0) = 0$$

$$\left[\frac{\partial f}{\partial x}\right]_{\mu=0,x=0} = -1$$

$$\left[\frac{\partial^2 f}{\partial x^2}\right]_{\mu=0,x=0} < 0$$

$$\left[\frac{\partial^3 f}{\partial x^3}\right]_{\mu=0,x=0} < 0.$$

Then there are intervals $(\mu_1, 0)$, $(0, \mu_2)$, and $\epsilon > 0$ such that

1. If $\mu \in (0, \mu_2)$, then $f_\mu(x)$ has one unstable fixed point and one stable orbit of period two for $x \in (-\epsilon, \epsilon)$, and
2. If $\mu \in (\mu_1, 0)$, then $f_\mu(x)$ has a single stable fixed point for $x \in (-\epsilon, \epsilon)$.

This type of BIFURCATION is known as a flip bifurcation. An example of an equation displaying a flip bifurcation is

$$f(x) = \mu - x - x^2.$$

see also BIFURCATION

References

Rasband, S. N. *Chaotic Dynamics of Nonlinear Systems.* New York: Wiley, pp. 27–30, 1990.

Floor Function

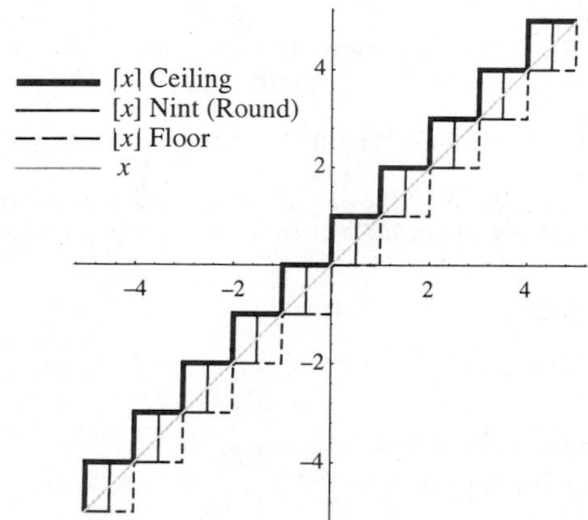

The function $\lfloor x \rfloor$ is the largest INTEGER $\leq x$, shown as the dashed curve in the above plot, and also called the GREATEST INTEGER FUNCTION. In many computer languages, the floor function is called the INTEGER PART function and is denoted `int(x)`. The name and symbol for the floor function were coined by K. E. Iverson (Graham *et al.* 1990).

Unfortunately, in many older and current works (e.g., Shanks 1993, Ribenboim 1996), the symbol $[x]$ is used instead of $\lfloor x \rfloor$. Because of the elegant symmetry of the floor function and CEILING FUNCTION symbols $\lfloor x \rfloor$ and $\lceil x \rceil$, and because $[x]$ is such a useful symbol when interpreted as an IVERSON BRACKET, the use of $[x]$ to denote the floor function should be deprecated. In this work, the symbol $[x]$ is used to denote the nearest integer NINT function since it naturally falls between the $\lfloor x \rfloor$ and $\lceil x \rceil$ symbols.

see also CEILING FUNCTION, FRACTIONAL PART, INT, IVERSON BRACKET, NINT

References

Graham, R. L.; Knuth, D. E.; and Patashnik, O. "Integer Functions." Ch. 3 in *Concrete Mathematics: A Foundation for Computer Science.* Reading, MA: Addison-Wesley, pp. 67–101, 1990.

Iverson, K. E. *A Programming Language.* New York: Wiley, p. 12, 1962.

Ribenboim, P. *The New Book of Prime Number Records.* New York: Springer-Verlag, pp. 180–182, 1996.

Shanks, D. *Solved and Unsolved Problems in Number Theory, 4th ed.* New York: Chelsea, p. 14, 1993.

Spanier, J. and Oldham, K. B. "The Integer-Value Int(x) and Fractional-Value frac(x) Functions." Ch. 9 in *An Atlas of Functions.* Washington, DC: Hemisphere, pp. 71–78, 1987.

Floquet Analysis

Given a system of periodic ORDINARY DIFFERENTIAL EQUATIONS of the form

$$\frac{d}{dt}\begin{bmatrix} x \\ y \\ v_x \\ v_y \end{bmatrix} = -\begin{bmatrix} 0 & 0 & -1 & 0 \\ 0 & 0 & 0 & -1 \\ \Phi_{xx} & \Phi_{yy} & 0 & 0 \\ \Phi_{xy} & \Phi_{yy} & 0 & 0 \end{bmatrix}\begin{bmatrix} x \\ y \\ v_x \\ v_y \end{bmatrix}, \quad (1)$$

the solution can be written as a linear combination of functions of the form

$$\begin{bmatrix} x(t) \\ y(t) \\ v_x(t) \\ v_y(t) \end{bmatrix} = \begin{bmatrix} x_0 \\ y_0 \\ v_{x0} \\ v_{y0} \end{bmatrix} e^{\mu t} P_\mu(t), \quad (2)$$

where $P_\mu(t)$ is a function periodic with the same period T as the equations themselves. Given an ORDINARY DIFFERENTIAL EQUATION of the form

$$\ddot{x} + g(t)x = 0, \quad (3)$$

where $g(t)$ is periodic with period T, the ODE has a pair of independent solutions given by the REAL and IMAGINARY PARTS of

$$x(t) = w(t)e^{i\psi(t)} \quad (4)$$

$$\dot{x} = (\dot{w} + iw\dot{\psi})e^{i\psi} \quad (5)$$

$$\ddot{x} = [\ddot{w} + i\dot{w}\dot{\psi} + i(\dot{w}\dot{\psi} + w\ddot{\psi} + iw\dot{\psi}^2)]e^{i\psi}$$
$$= [(\ddot{w} - w\dot{\psi}^2) + i(2\dot{w}\dot{\psi} + w\ddot{\psi})]e^{i\psi}. \quad (6)$$

Plugging these into (3) gives

$$\ddot{w} + 2i\dot{w}\dot{\psi} + w(g + i\ddot{\psi} - \dot{\psi}^2) = 0, \quad (7)$$

so the REAL and IMAGINARY PARTS are

$$\ddot{w} + w(g - \dot{\psi}^2) = 0 \quad (8)$$

$$2\dot{w}\dot{\psi} + w\ddot{\psi} = 0. \quad (9)$$

From (9),

$$\frac{2\dot{w}}{w} + \frac{\ddot{\psi}}{\dot{\psi}} = 2\frac{d}{dt}(\ln w) + \frac{d}{dt}[\ln(\dot{\psi})]$$
$$= \frac{d}{dt}\ln(\dot{\psi}w^2) = 0. \quad (10)$$

Integrating gives

$$\dot{\psi} = \frac{C}{w^2}, \tag{11}$$

where C is a constant which must equal 1, so ψ is given by

$$\psi = \int_{t_0}^{t} \frac{dt}{w^2}. \tag{12}$$

The REAL solution is then

$$x(t) = w(t)\cos[\psi(t)], \tag{13}$$

so

$$\begin{aligned}\dot{x} &= \dot{w}\cos\psi - w\dot{\psi}\sin\psi = \dot{w}\frac{x}{w} - w\dot{\psi}\sin\psi \\ &= \dot{w}\frac{x}{w} - w\frac{1}{w^2}\sin\psi = \dot{w}\frac{x}{w} - \frac{1}{w}\sin\psi \end{aligned} \tag{14}$$

and

$$\begin{aligned} 1 = \cos^2\psi + \sin^2\psi &= x^2 w^{-2} + \left[w\left(\dot{w}\frac{x}{w} - \dot{x}\right)\right]^2 \\ &= x^2 w^{-2} + (\dot{w}x - w\dot{x})^2 \equiv I(x, \dot{x}, t), \end{aligned} \tag{15}$$

which is an integral of motion. Therefore, although $w(t)$ is not explicitly known, an integral I always exists. Plugging (10) into (8) gives

$$\ddot{w} + g(t)w - \frac{1}{w^3} = 0, \tag{16}$$

which, however, is not any easier to solve than (3).

References

Abramowitz, M. and Stegun, C. A. (Eds.). *Handbook of Mathematical Functions with Formulas, Graphs, and Mathematical Tables, 9th printing.* New York: Dover, p. 727, 1972.

Binney, J. and Tremaine, S. *Galactic Dynamics.* Princeton, NJ: Princeton University Press, p. 175, 1987.

Lichtenberg, A. and Lieberman, M. *Regular and Stochastic Motion.* New York: Springer-Verlag, p. 32, 1983.

Margenau, H. and Murphy, G. M. *The Mathematics of Physics and Chemistry, 2 vols.* Princeton, NJ: Van Nostrand, 1956–64.

Morse, P. M. and Feshbach, H. *Methods of Theoretical Physics, Part I.* New York: McGraw-Hill, pp. 556–557, 1953.

Floquet's Theorem

see FLOQUET ANALYSIS

Flow

An ACTION with $G = \mathbb{R}$. Flows are generated by VECTOR FIELDS and vice versa.

see also ACTION, AMBROSE-KAKUTANI THEOREM, ANOSOV FLOW, AXIOM A FLOW, CASCADE, GEODESIC FLOW, SEMIFLOW

Flow Line

A flow line for a map on a VECTOR FIELD $\mathbf{F}$ is a path $\sigma(t)$ such that $\sigma'(t) = \mathbf{F}(\sigma(t))$.

Flower

see DAISY, FLOWER OF LIFE, ROSE

Flower of Life

One of the beautiful arrangements of CIRCLES found at the Temple of Osiris at Abydos, Egypt (Rawles 1997). The CIRCLES are placed with six-fold symmetry, forming a mesmerizing pattern of CIRCLES and LENSES.

see also FIVE DISKS PROBLEM, REULEAUX TRIANGLE, SEED OF LIFE, VENN DIAGRAM

References

Rawles, B. *Sacred Geometry Design Sourcebook: Universal Dimensional Patterns.* Nevada City, CA: Elysian Pub., p. 15, 1997.

Wein, J. "The Flower of Life." `http://www2.cruzio.com/~flower`.

Weisstein, E. W. "Flower of Life." `http://www.astro.virginia.edu/~eww6n/math/notebooks/FlowerOfLife.m`.

Flowsnake

see PEANO-GOSPER CURVE

Flowsnake Fractal

see GOSPER ISLAND

Floyd's Algorithm

An algorithm for finding the shortest path between two VERTICES.

see also DIJKSTRA'S ALGORITHM

Fluent

Newton's term for a variable in his method of FLUXIONS (differential calculus).

Fluxion

The term for DERIVATIVE in Newton's CALCULUS.

Flype

A 180° rotation of a TANGLE.

see also FLYPING CONJECTURE, TANGLE

Flyping Conjecture

Also called the TAIT FLYPING CONJECTURE. Given two reduced alternating projections of the same knot, they are equivalent on the SPHERE IFF they are related by a series of FLYPES. It was proved by Menasco and Thistlethwaite (1991). It allows all possible REDUCED alternating projections of a given ALTERNATING KNOT to be drawn.

References
Adams, C. C. *The Knot Book: An Elementary Introduction to the Mathematical Theory of Knots.* New York: W. H. Freeman, pp. 164–165, 1994.
Menasco, W. and Thistlethwaite, M. "The Tait Flyping Conjecture." *Bull. Amer. Math. Soc.* **25**, 403–412, 1991.
Stewart, I. *The Problems of Mathematics, 2nd ed.* Oxford, England: Oxford University Press, pp. 284–285, 1987.

Focus

A point related to the construction and properties of CONIC SECTIONS.

see also ELLIPSE, ELLIPSOID, HYPERBOLA, HYPERBOLOID, PARABOLA, PARABOLOID, REFLECTION PROPERTY

References
Coxeter, H. S. M. and Greitzer, S. L. *Geometry Revisited.* Washington, DC: Math. Assoc. Amer., pp. 141–144, 1967.

Fold Bifurcation

Let $f : \mathbb{R} \times \mathbb{R} \to \mathbb{R}$ be a one-parameter family of C^2 MAP satisfying

$$\begin{aligned} f(0,0) &= 0 \\ \left[\frac{\partial f}{\partial x}\right]_{\mu=0,x=0} &= 1 \\ \left[\frac{\partial^2 f}{\partial x^2}\right]_{\mu=0,x=0} &> 0 \\ \left[\frac{\partial f}{\partial \mu}\right]_{\mu=0,x=0} &> 0, \end{aligned}$$

then there exist intervals $(\mu_1, 0)$, $(0, \mu_2)$ and $\epsilon > 0$ such that

1. If $\mu \in (\mu_1, 0)$, then $f_\mu(x)$ has two fixed points in $(-\epsilon, \epsilon)$ with the positive one being unstable and the negative one stable, and
2. If $\mu \in (0, \mu_2)$, then $f_\mu(x)$ has no fixed points in $(-\epsilon, \epsilon)$.

This type of BIFURCATION is known as a fold bifurcation, sometimes also called a SADDLE-NODE BIFURCATION or TANGENT BIFURCATION. An example of an equation displaying a fold bifurcation is

$$\dot{x} = \mu - x^2$$

(Guckenheimer and Holmes 1997, p. 145).

see also BIFURCATION

References
Guckenheimer, J. and Holmes, P. *Nonlinear Oscillations, Dynamical Systems, and Bifurcations of Vector Fields, 3rd ed.* New York: Springer-Verlag, pp. 145–149, 1997.
Rasband, S. N. *Chaotic Dynamics of Nonlinear Systems.* New York: Wiley, pp. 27–28, 1990.

Fold Catastrophe

A CATASTROPHE which can occur for one control factor and one behavior axis.

Folding

The points accessible from c by a single fold which leaves $a_1, \ldots, a_n$ fixed are exactly those points interior to or on the boundary of the intersection of the CIRCLES through c with centers at a_i, for $i = 1, \ldots, n$. Given any three points in the plane a, b, and c, there is an EQUILATERAL TRIANGLE with VERTICES x, y, and z for which a, b, and c are the images of x, y, and z under a single fold. Given any four points in the plane a, b, c, and d, there is some SQUARE with VERTICES x, y, z, and w for which a, b, c, and d are the images of x, y, z, and w under a sequence of at most three folds. Also, any four collinear points are the images of the VERTICES of a suitable SQUARE under at most two folds. Every five (six) points are the images of the VERTICES of suitable regular PENTAGON (HEXAGON) under at most five (six) folds. The least number of folds required for $n \geq 4$ is not known, but some bounds are. In particular, every set of n points is the image of a suitable REGULAR n-gon under at most $F(n)$ folds, where

$$F(n) \leq \begin{cases} \frac{1}{2}(3n-2) & \text{for } n \text{ even} \\ \frac{1}{2}(3n-3) & \text{for } n \text{ odd.} \end{cases}$$

The first few values are 0, 2, 3, 5, 6, 8, 9, 11, 12, 14, 15, 17, 18, 20, 21, ... (Sloane's A007494).

see also FLEXAGON, MAP FOLDING, ORIGAMI

References
Sabinin, P. and Stone, M. G. "Transforming n-gons by Folding the Plane." *Amer. Math. Monthly* **102**, 620–627, 1995.
Sloane, N. J. A. Sequence A007494 in "An On-Line Version of the Encyclopedia of Integer Sequences."

Foliation

Let M^n be an n-MANIFOLD and let $\mathsf{F} = \{F_\alpha\}$ denote a PARTITION of M into DISJOINT path-connected SUBSETS. Then F is called a foliation of M of codimension c (with $0 < c < n$) if there EXISTS a COVER of M by OPEN SETS U, each equipped with a HOMEOMORPHISM $h : U \to \mathbb{R}^n$ or $h : U \to \mathbb{R}^n_+$ which throws each nonempty component of $F_\alpha \cap U$ onto a parallel translation of the standard HYPERPLANE $\mathbb{R}^{n-c}$ in $\mathbb{R}^n$. Each F_α is then called a LEAF and is not necessarily closed or compact.

see also LEAF (FOLIATION), REEB FOLIATION

References
Rolfsen, D. *Knots and Links.* Wilmington, DE: Publish or Perish Press, p. 284, 1976.

Folium

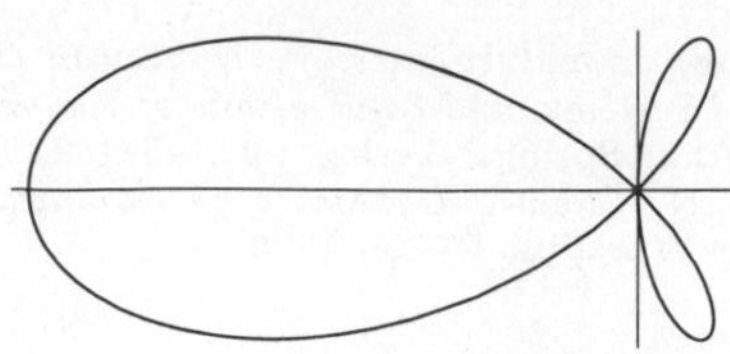

The word "folium" means leaf-shaped. The polar equation is

$$r = \cos\theta(4a\sin^2\theta - b).$$

If $b \geq 4a$, it is a single folium. If $b = 0$, it is a BIFOLIUM. If $0 < b < 4a$, it is a TRIFOLIUM. The simple folium is the PEDAL CURVE of the DELTOID where the PEDAL POINT is one of the CUSPS.

see also BIFOLIUM, FOLIUM OF DESCARTES, KEPLER'S FOLIUM, QUADRIFOLIUM, ROSE, TRIFOLIUM

References

Lawrence, J. D. *A Catalog of Special Plane Curves.* New York: Dover, pp. 152–153, 1972.

MacTutor History of Mathematics Archive. "Folium." http://www-groups.dcs.st-and.ac.uk/~history/Curves/Folium.html.

Folium of Descartes

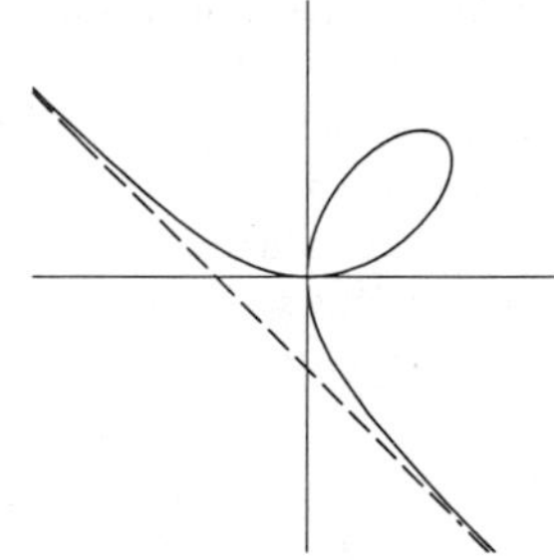

A plane curve proposed by Descartes to challenge Fermat's extremum-finding techniques. In parametric form,

$$x = \frac{3at}{1+t^3} \tag{1}$$

$$y = \frac{3at^2}{1+t^3}. \tag{2}$$

The curve has a discontinuity at $t = -1$. The left wing is generated as t runs from -1 to 0, the loop as t runs from 0 to ∞, and the right wing as t runs from $-\infty$ to -1.

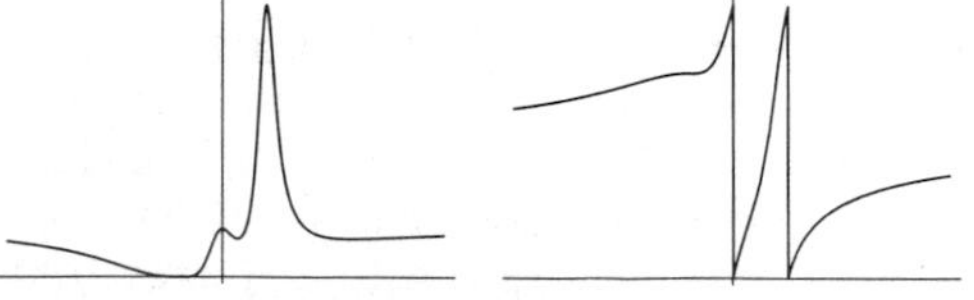

The CURVATURE and TANGENTIAL ANGLE of the folium of Descartes, illustrated above, are

$$\kappa(t) = \frac{2(1+t^3)^4}{3(1+4t^2-4t^3-4t^5+4t^6+t^8)^{3/2}} \tag{3}$$

$$\phi(t) = \frac{1}{2}\left[\pi + \tan^{-1}\left(\frac{1-2t^3}{t^4-2t}\right) - \tan^{-1}\left(\frac{2t^3-1}{t^4-2t}\right)\right]. \tag{4}$$

Converting the parametric equations to POLAR COORDINATES gives

$$r^2 = \frac{(3at)^2(1+t^2)}{(1+t^3)^2} \tag{5}$$

$$\theta = \tan^{-1}\left(\frac{y}{x}\right) = \tan^{-1} t, \tag{6}$$

so

$$d\theta = \frac{dt}{1+t^2}. \tag{7}$$

The AREA enclosed by the curve is

$$A = \tfrac{1}{2}\int r^2\,d\theta = \tfrac{1}{2}\int_0^\infty \frac{(3at)^2(1+t^2)}{(1+t^3)^2}\frac{dt}{1+t^2} = \tfrac{3}{2}a^2\int_0^\infty \frac{3t^2\,dt}{(1+t^3)^2}. \tag{8}$$

Now let $u \equiv 1+t^3$ so $du = 3t^2\,dt$

$$A = \tfrac{3}{2}a^2\int_1^\infty \frac{du}{u^2} = \tfrac{3}{2}a^2\left[-\frac{1}{u}\right]_1^\infty = \tfrac{3}{2}a^2(-0+1) = \tfrac{3}{2}a^2. \tag{9}$$

In CARTESIAN COORDINATES,

$$x^3 + y^3 = \frac{(3at)^3(1+t^3)}{(1+t^3)^3} = \frac{(3at)^3}{(1+t^3)^2} = 3axy \tag{10}$$

(MacTutor Archive). The equation of the ASYMPTOTE is

$$y = -a - x. \tag{11}$$

References

Gray, A. *Modern Differential Geometry of Curves and Surfaces.* Boca Raton, FL: CRC Press, pp. 59–62, 1993.

Lawrence, J. D. *A Catalog of Special Plane Curves.* New York: Dover, pp. 106–109, 1972.

MacTutor History of Mathematics Archive. "Folium of Descartes." http://www-groups.dcs.st-and.ac.uk/~history/Curves/Foliumd.html.

Stroeker, R. J. "Brocard Points, Circulant Matrices, and Descartes' Folium." *Math. Mag.* **61**, 172–187, 1988.

Yates, R. C. "Folium of Descartes." In *A Handbook on Curves and Their Properties.* Ann Arbor, MI: J. W. Edwards, pp. 98–99, 1952.

Follows

see SUCCEEDS

Fontené Theorems

1. If the sides of the PEDAL TRIANGLE of a point P meet the corresponding sides of a TRIANGLE $\Delta O_1O_2O_3$ at X_1, X_2, and X_3, respectively, then P_1X_1, P_2X_2, P_3X_3 meet at a point L common to the CIRCLES $O_1O_2O_3$ and $P_1P_2P_3$. In other words, L is one of the intersections of the NINE-POINT CIRCLE of $A_1A_2A_3$ and the PEDAL CIRCLE of P.
2. If a point moves on a fixed line through the CIRCUMCENTER, then its PEDAL CIRCLE passes through a fixed point on the NINE-POINT CIRCLE.

3. The PEDAL CIRCLE of a point is tangent to the NINE-POINT CIRCLE IFF the point and its ISOGONAL CONJUGATE lie on a LINE through the ORTHOCENTER. FEUERBACH'S THEOREM is a special case of this theorem.

see also CIRCUMCENTER, FEUERBACH'S THEOREM, ISOGONAL CONJUGATE, NINE-POINT CIRCLE, ORTHOCENTER, PEDAL CIRCLE

References

Johnson, R. A. *Modern Geometry: An Elementary Treatise on the Geometry of the Triangle and the Circle.* Boston, MA: Houghton Mifflin, pp. 245–247, 1929.

Foot

see PERPENDICULAR FOOT

For All

If a proposition P is true for all B, this is written $P\forall B$.

see also ALMOST ALL, EXISTS, QUANTIFIER

Forcing

A technique in SET THEORY invented by P. Cohen (1963, 1964, 1966) and used to prove that the AXIOM OF CHOICE and CONTINUUM HYPOTHESIS are independent of one another in ZERMELO-FRAENKEL SET THEORY.

see also AXIOM OF CHOICE, CONTINUUM HYPOTHESIS, SET THEORY, ZERMELO-FRAENKEL SET THEORY

References

Cohen, P. J. "The Independence of the Continuum Hypothesis." *Proc. Nat. Acad. Sci. U. S. A.* **50**, 1143–1148, 1963.

Cohen, P. J. "The Independence of the Continuum Hypothesis. II." *Proc. Nat. Acad. Sci. U. S. A.* **51**, 105–110, 1964.

Cohen, P. J. *Set Theory and the Continuum Hypothesis.* New York: W. A. Benjamin, 1966.

Ford Circle

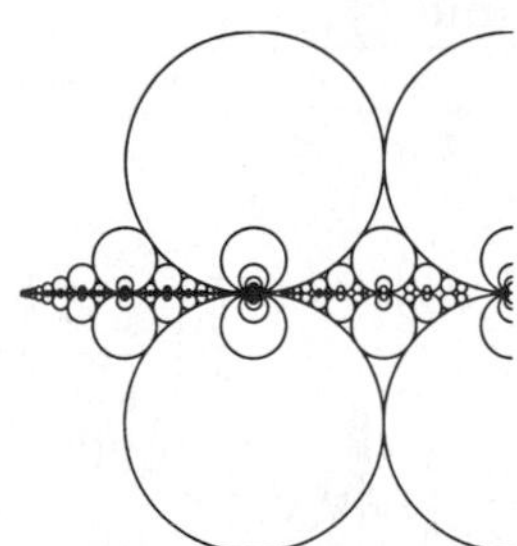

Pick any two INTEGERS h and k, then the CIRCLE of RADIUS $1/(2k^2)$ centered at $(h/k, 1/(2k^2))$ is known as a Ford circle. No matter what and how many hs and ks are picked, none of the Ford circles intersect (and all are tangent to the x-AXIS). This can be seen by examining the squared distance between the centers of the circles with (h,k) and (h',k'),

$$d^2 = \left(\frac{h'}{k'} - \frac{h}{k}\right)^2 + \left(\frac{1}{2k'^2} - \frac{1}{2k^2}\right). \quad (1)$$

Let s be the sum of the radii

$$s = r_1 + r_2 = \frac{1}{2k^2} + \frac{1}{2k'^2}, \quad (2)$$

then

$$d^2 - s^2 = \frac{(h'k - hk')^2 - 1}{k^2 k'^2}. \quad (3)$$

But $(h'k - k'h)^2 \geq 1$, so $d^2 - s^2 \geq 0$ and the distance between circle centers is $\geq$ the sum of the CIRCLE RADII, with equality (and therefore tangency) IFF $|h'k - k'h| = 1$. Ford circles are related to the FAREY SEQUENCE (Conway and Guy 1996).

see also ADJACENT FRACTION, FAREY SEQUENCE, STERN-BROCOT TREE

References

Conway, J. H. and Guy, R. K. "Farey Fractions and Ford Circles." *The Book of Numbers.* New York: Springer-Verlag, pp. 152–154, 1996.

Ford, L. R. "Fractions." *Amer. Math. Monthly* **45**, 586–601, 1938.

Pickover, C. A. "Fractal Milkshakes and Infinite Archery." Ch. 14 in *Keys to Infinity.* New York: W. H. Freeman, pp. 117–125, 1995.

Rademacher, H. *Higher Mathematics from an Elementary Point of View.* Boston, MA: Birkhäuser, 1983.

Ford's Theorem

Let a, b, and k be INTEGERS with $k \geq 1$. For $j = 0$, 1, 2, let

$$S_j \equiv \sum_{\substack{i=0 \\ i\equiv j \pmod 3}} (-1)^j \binom{k}{i} a^{k-i} b^i.$$

Then

$$2(a^2 + ab + b^2)^{2k} = (S_0 - S_1)^4 + (S_1 - S_2)^4 + (S_2 - S_0)^4.$$

see also BHARGAVA'S THEOREM, DIOPHANTINE EQUATION—QUARTIC

References

Berndt, B. C. *Ramanujan's Notebooks, Part IV.* New York: Springer-Verlag, pp. 100–101, 1994.

Forest

A GRAPH without any CIRCUITS (CYCLES), which therefore consists only of TREES. A forest with k components and n nodes has $n - k$ EDGES.

Fork

see TREE

Form

see CANONICAL FORM, CUSP FORM, DIFFERENTIAL k-FORM, FORM (GEOMETRIC), FORM (POLYNOMIAL), MODULAR FORM, NORMAL FORM, PFAFFIAN FORM, QUADRATIC FORM

Form (Geometric)

A 1-D geometric object such as a PENCIL or RANGE.

Form (Polynomial)

A HOMOGENEOUS POLYNOMIAL in two or more variables.

see also DISCONNECTED FORM, k-FORM

Formal Logic

see SYMBOLIC LOGIC

Formosa Theorem

see CHINESE REMAINDER THEOREM

Formula

A mathematical equation or a formal logical expression. The correct Latin plural form of formula is "formulae," although the less pretentious-sounding "formulas" is used more commonly.

see also ARCHIMEDES' RECURRENCE FORMULA, BAYES' FORMULA, BENSON'S FORMULA, BESSEL'S FINITE DIFFERENCE FORMULA, BESSEL'S INTERPOLATION FORMULA, BESSEL'S STATISTICAL FORMULA, BINET'S FORMULA, BINOMIAL FORMULA, BRAHMAGUPTA'S FORMULA, BRENT-SALAMIN FORMULA, BRETSCHNEIDER'S FORMULA, BRIOSCHI FORMULA, CALDERÓN'S FORMULA, CARDANO'S FORMULA, CAUCHY'S FORMULA, CAUCHY'S COSINE INTEGRAL FORMULA, CAUCHY INTEGRAL FORMULA, CHASLES-CAYLEY-BRILL FORMULA, CHEBYSHEV APPROXIMATION FORMULA, CHRISTOFFEL-DARBOUX FORMULA, CHRISTOFFEL FORMULA, CLAUSEN FORMULA, CLENSHAW RECURRENCE FORMULA, DESCARTES-EULER POLYHEDRAL FORMULA, DESCARTES' FORMULA, DIRICHLET'S FORMULA, DIXON-FERRAR FORMULA, DOBIŃSKI'S FORMULA, DUPLICATION FORMULA, ENNEPER-WEIERSTRAß PARAMETERIZATION, EULER CURVATURE FORMULA, EULER FORMULA, EULER-MACLAURIN INTEGRATION FORMULAS, EULER POLYHEDRAL FORMULA, EULER TRIANGLE FORMULA, EVERETT'S FORMULA, EXPONENTIAL SUM FORMULAS, FAULHABER'S FORMULA, FRENET FORMULAS, GAUSS'S BACKWARD FORMULA, GAUSS-BONNET FORMULA, GAUSS'S FORMULA, GAUSS'S FORWARD FORMULA, GAUSS MULTIPLICATION FORMULA, GAUSS-SALAMIN FORMULA, GIRARD'S SPHERICAL EXCESS FORMULA, GOODMAN'S FORMULA, GREGORY'S FORMULA, GRENZ-FORMEL, GRINBERG FORMULA, HALLEY'S IRRATIONAL FORMULA, HALLEY'S RATIONAL FORMULA, HANSEN-BESSEL FORMULA, HERON'S FORMULA, HOOK LENGTH FORMULA, JACOBI ELLIPTIC FUNCTIONS, JENSEN'S FORMULA, JONAH FORMULA, KAC FORMULA, KNESER-SOMMERFELD FORMULA, KUMMER'S FORMULAS, LAISANT'S RECURRENCE FORMULA, LANDEN'S FORMULA, LEFSHETZ FIXED POINT FORMULA, LEFSHETZ TRACE FORMULA, LEGENDRE DUPLICATION FORMULA, LEGENDRE'S FORMULA, LEHMER'S FORMULA, LICHNEROWICZ FORMULA, LICHNEROWICZ-WEITZENBOCK FORMULA, LOBACHEVSKY'S FORMULA, LOGARITHMIC BINOMIAL FORMULA, LUDWIG'S INVERSION FORMULA, MACHIN'S FORMULA, MACHIN-LIKE FORMULAS, MEHLER'S BESSEL FUNCTION FORMULA, MEHLER'S HERMITE POLYNOMIAL FORMULA, MEISSEL'S FORMULA, MENSURATION FORMULA, MÖBIUS INVERSION FORMULA, MORLEY'S FORMULA, NEWTON'S BACKWARD DIFFERENCE FORMULA, NEWTON-COTES FORMULAS, NEWTON'S FORWARD DIFFERENCE FORMULA, NICHOLSON'S FORMULA, PASCAL'S FORMULA, PICK'S FORMULA, POINCARÉ FORMULA, POISSON'S BESSEL FUNCTION FORMULA, POISSON'S HARMONIC FUNCTION FORMULA, POISSON SUM FORMULA, POLYHEDRAL FORMULA, PROSTHAPHAERESIS FORMULAS, QUADRATIC FORMULA, QUADRATURE FORMULAS, RAYLEIGH'S FORMULAS, RIEMANN'S FORMULA, RODRIGUES FORMULA, ROTATION FORMULA, SCHLÄFLI'S FORMULA, SCHRÖTER'S FORMULA, SCHWENK'S FORMULA, SEGNER'S RECURRENCE FORMULA, SERRET-FRENET FORMULAS, SHERMAN-MORRISON FORMULA, SOMMERFELD'S FORMULA, SONINE-SCHAFHEITLIN FORMULA, STEFFENSON'S FORMULA, STIRLING'S FINITE DIFFERENCE FORMULA, STIRLING'S FORMULA, STRASSEN FORMULAS, THIELE'S INTERPOLATION FORMULA, WALLIS FORMULA, WATSON'S FORMULA, WATSON-NICHOLSON FORMULA, WEBER'S FORMULA, WEBER-SONINE FORMULA, WEYRICH'S FORMULA, WOODBURY FORMULA

References

Carr, G. S. *Formulas and Theorems in Pure Mathematics.* New York: Chelsea, 1970.

Spiegel, M. R. *Mathematical Handbook of Formulas and Tables.* New York: McGraw-Hill, 1968.

Tallarida, R. J. *Pocket Book of Integrals and Mathematical Formulas, 3rd ed.* Boca Raton, FL: CRC Press, 1992.

Fortunate Prime

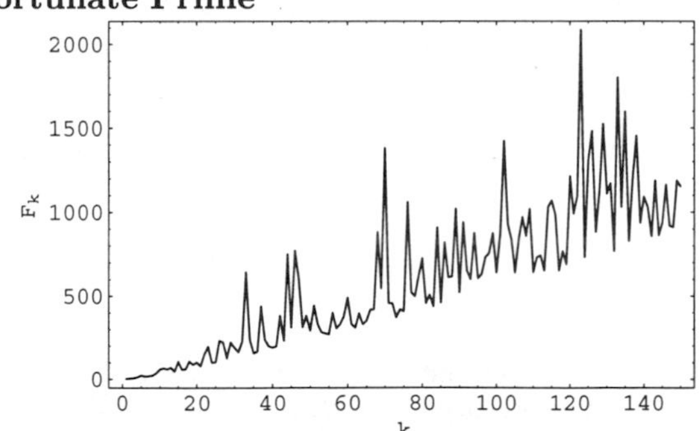

Let

$$X_k \equiv 1 + p_k\#,$$

where p_k is the kth PRIME and $p\#$ is the PRIMORIAL, and let q_k be the NEXT PRIME (i.e., the smallest PRIME greater than X_k),

$$q_k = p_{1+\pi(X_k)} = p_{1+\pi(1+p_k\#)},$$

where $\pi(n)$ is the PRIME COUNTING FUNCTION. Then R. F. Fortune conjectured that $F_k \equiv q_k - X_k + 1$ is PRIME for all k. The first values of F_k are 3, 5, 7, 13, 23, 17, 19, 23, ... (Sloane's A005235), and all known values of F_k are indeed PRIME (Guy 1994). The indices of these primes are 2, 3, 4, 6, 9, 7, 8, 9, 12, 18, ... In numerical order with duplicates removed, the Fortunate primes are 3, 5, 7, 13, 17, 19, 23, 37, 47, 59, 61, 67, 71, 79, 89, ... (Sloane's A046066).

see also ANDRICA'S CONJECTURE, PRIMORIAL

References

Guy, R. K. *Unsolved Problems in Number Theory, 2nd ed.* New York: Springer-Verlag, p. 7, 1994.

Sloane, N. J. A. Sequences A046066 and A005235/M2418 in "An On-Line Version of the Encyclopedia of Integer Sequences."

Forward Difference

The forward difference is a FINITE DIFFERENCE defined by

$$\Delta f_p \equiv f_{p+1} - f_p. \tag{1}$$

Higher order differences are obtained by repeated operations of the forward difference operator, so

$$\begin{aligned} \Delta^2 f_p &= \Delta_p{}^2 = \Delta(\Delta_p) = \Delta(f_{p+1} - f_p) \\ &= \Delta_{p+1} - \Delta_p = f_{p+2} - 2f_{p+1} + f_p. \end{aligned} \tag{2}$$

In general,

$$\Delta_p^k \equiv \Delta^k f_p \equiv \sum_{m=0}^{k} (-1)^m \binom{k}{m} f_{p+k-m}, \tag{3}$$

where $\binom{k}{m}$ is a BINOMIAL COEFFICIENT.

NEWTON'S FORWARD DIFFERENCE FORMULA expresses f_p as the sum of the nth forward differences

$$f_p = f_0 + p\Delta_0 + \tfrac{1}{2!}p(p+1)\Delta_0^2 + \tfrac{1}{3!}p(p+1)(p+2)\Delta_0^3 + \ldots \tag{4}$$

where Δ_0^n is the first nth difference computed from the difference table.

see also BACKWARD DIFFERENCE, CENTRAL DIFFERENCE, DIFFERENCE EQUATION, DIVIDED DIFFERENCE, RECIPROCAL DIFFERENCE

References

Abramowitz, M. and Stegun, C. A. (Eds.). *Handbook of Mathematical Functions with Formulas, Graphs, and Mathematical Tables, 9th printing.* New York: Dover, p. 877, 1972.

Fountain

An (n,k) fountain is an arrangement of n coins in rows such that exactly k coins are in the bottom row and each coin in the $(i+1)$st row touches exactly two in the ith row.

References

Berndt, B. C. *Ramanujan's Notebooks, Part III.* New York: Springer-Verlag, p. 79, 1985.

Four Coins Problem

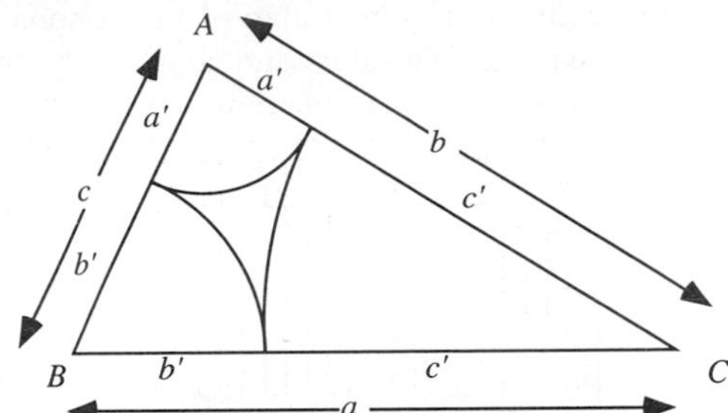

Given three coins of possibly different sizes which are arranged so that each is tangent to the other two, find the coin which is tangent to the other three coins. The solution is the inner SODDY CIRCLE.

see also APOLLONIUS CIRCLES, APOLLONIUS' PROBLEM, ARBELOS, BEND (CURVATURE), CIRCUMCIRCLE, COIN, DESCARTES CIRCLE THEOREM, HART'S THEOREM, PAPPUS CHAIN, SODDY CIRCLES, SPHERE PACKING, STEINER CHAIN

References

Oldknow, A. "The Euler-Gergonne-Soddy Triangle of a Triangle." *Amer. Math. Monthly* **103**, 319–329, 1996.

Four-Color Theorem

The four-color theorem states that any map in a PLANE can be colored using four-colors in such a way that regions sharing a common boundary (other than a single point) do not share the same color. This problem is sometimes also called GUTHRIE'S PROBLEM after F. Guthrie, who first conjectured the theorem in 1853. The CONJECTURE was then communicated to de Morgan and thence into the general community. In 1878, Cayley wrote the first paper on the conjecture.

Fallacious proofs were given independently by Kempe (1879) and Tait (1880). Kempe's proof was accepted for a decade until Heawood showed an error using a map with 18 faces (although a map with nine faces suffices to show the fallacy). The HEAWOOD CONJECTURE provided a very general result for map coloring, showing that in a GENUS 0 SPACE (i.e., either the SPHERE or PLANE), six colors suffice. This number can easily be reduced to five, but reducing the number of colors all the way to four proved very difficult.

Finally, Appel and Haken (1977) announced a computer-assisted proof that four colors were SUFFICIENT. However, because part of the proof consisted of an exhaustive analysis of many discrete cases by a computer, some mathematicians do not accept it. However, no flaws have yet been found, so the proof appears valid. A potentially independent proof has recently been constructed by N. Robertson, D. P. Sanders, P. D. Seymour, and R. Thomas.

Martin Gardner (1975) played an April Fool's joke by (incorrectly) claiming that the map of 110 regions illustrated below requires five colors and constitutes a counterexample to the four-color theorem.

see also CHROMATIC NUMBER, HEAWOOD CONJECTURE, MAP COLORING, SIX-COLOR THEOREM

References

Appel, K. and Haken, W. "Every Planar Map is Four-Colorable, I and II." *Illinois J. Math.* **21**, 429–567, 1977.

Appel, K. and Haken, W. "The Solution of the Four-Color Map Problem." *Sci. Amer.* **237**, 108–121, 1977.

Appel, K. and Haken, W. *Every Planar Map is Four-Colorable.* Providence, RI: Amer. Math. Soc., 1989.

Barnette, D. *Map Coloring, Polyhedra, and the Four-Color Problem.* Providence, RI: Math. Assoc. Amer., 1983.

Birkhoff, G. D. "The Reducibility of Maps." *Amer. Math. J.* **35**, 114–128, 1913.

Chartrand, G. "The Four Color Problem." §9.3 in *Introductory Graph Theory.* New York: Dover, pp. 209–215, 1985.

Coxeter, H. S. M. "The Four-Color Map Problem, 1840–1890." *Math. Teach.,* Apr. 1959.

Franklin, P. *The Four-Color Problem.* New York: Scripta Mathematica, Yeshiva College, 1941.

Gardner, M. "Mathematical Games: The Celebrated Four-Color Map Problem of Topology." *Sci. Amer.* **203**, 218–222, Sep. 1960.

Gardner, M. "The Four-Color Map Theorem." Ch. 10 in *Martin Gardner's New Mathematical Diversions from Scientific American.* New York: Simon and Schuster, pp. 113–123, 1966.

Gardner, M. "Mathematical Games: Six Sensational Discoveries that Somehow or Another have Escaped Public Attention." *Sci. Amer.* **232**, 127–131, Apr. 1975.

Gardner, M. "Mathematical Games: On Tessellating the Plane with Convex Polygons." *Sci. Amer.* **232**, 112–117, Jul. 1975.

Kempe, A. B. "On the Geographical Problem of Four-Colors." *Amer. J. Math.* **2**, 193–200, 1879.

Kraitchik, M. §8.4.2 in *Mathematical Recreations.* New York: W. W. Norton, p. 211, 1942.

Ore, Ø. *The Four-Color Problem.* New York: Academic Press, 1967.

Pappas, T. "The Four-Color Map Problem: Topology Turns the Tables on Map Coloring." *The Joy of Mathematics.* San Carlos, CA: Wide World Publ./Tetra, pp. 152–153, 1989.

Robertson, N.; Sanders, D. P.; and Thomas, R. "The Four-Color Theorem." `http://www.math.gatech.edu/~thomas/FC/fourcolor.html`.

Saaty, T. L. and Kainen, P. C. *The Four-Color Problem: Assaults and Conquest.* New York: Dover, 1986.

Tait, P. G. "Note on a Theorem in Geometry of Position." *Trans. Roy. Soc. Edinburgh* **29**, 657–660, 1880.

Four Travelers Problem

Let four LINES in a PLANE represent four roads in GENERAL POSITION, and let one traveler T_i be walking along each road at a constant (but not necessarily equal to any other traveler's) speed. Say that two travelers T_i and T_j have "met" if they were simultaneously at the intersection of their two roads. Then if T_1 has met all other three travelers (T_2, T_3, and T_4) and T_2, in addition to meeting T_1, has met T_3 and T_4, then T_3 and T_4 have also met!

References

Bogomolny, A. "Four Travellers Problem." `http://www.cut-the-knot.com/gproblems.html`.

Four-Vector

A four-element vector

$$a^\mu = \begin{bmatrix} a^0 \\ a^1 \\ a^2 \\ a^3 \end{bmatrix}, \quad (1)$$

which transforms under a LORENTZ TRANSFORMATION like the POSITION FOUR-VECTOR. This means it obeys

$$a'^\mu = \Lambda^\mu_\nu a^\nu \quad (2)$$

$$a_\mu \cdot b_\mu \equiv a_\mu b^\mu \quad (3)$$

$$a_\mu \cdot b^\mu = a'_\mu b'_\mu, \quad (4)$$

where Λ^μ_μ is the LORENTZ TENSOR. Multiplication of two four-vectors with the METRIC $g_{\mu\nu}$ gives products of the form

$$g_{\mu\nu} x^\mu x^\nu = (x^0)^2 - (x^1)^2 - (x^2)^2 - (x^3)^2. \quad (5)$$

In the case of the POSITION FOUR-VECTOR, $x^0 = ct$ (where c is the speed of light) and this product is an invariant known as the spacetime interval.

see also GRADIENT FOUR-VECTOR, LORENTZ TRANSFORMATION, POSITION FOUR-VECTOR, QUATERNION

References

Morse, P. M. and Feshbach, H. "The Lorentz Transformation, Four-Vectors, Spinors." §1.7 in *Methods of Theoretical Physics, Part I.* New York: McGraw-Hill, pp. 93–107, 1953.

Four-Vertex Theorem

A closed embedded smooth PLANE CURVE has at least four vertices, where a vertex is defined as an extremum of CURVATURE.

see also CURVATURE

References

Tabachnikov, S. "The Four-Vertex Theorem Revisited—Two Variations on the Old Theme." *Amer. Math. Monthly* **102**, 912–916, 1995.

Fourier-Bessel Series

see Bessel Function Fourier Expansion, Schlömilch's Series

Fourier-Bessel Transform

see Hankel Transform

Fourier Cosine Series

If $f(x)$ is an Even Function, then $b_n = 0$ and the Fourier Series collapses to

$$f(x) = \tfrac{1}{2}a_0 + \sum_{n=1}^{\infty} a_n \cos(nx), \tag{1}$$

where

$$a_0 = \frac{1}{\pi}\int_{-\pi}^{\pi} f(x)\,dx = \frac{2}{\pi}\int_0^{\pi} f(x)\,dx \tag{2}$$

$$\begin{aligned} a_n &= \frac{1}{\pi}\int_{-\pi}^{\pi} f(x)\cos(nx)\,dx \\ &= \frac{2}{\pi}\int_0^{\pi} f(x)\cos(nx)\,dx, \end{aligned} \tag{3}$$

where the last equality is true because

$$f(x)\cos(nx) = f(-x)\cos(-nx). \tag{4}$$

Letting the range go to L,

$$a_0 = \frac{1}{L}\int_0^L f(x)\,dx \tag{5}$$

$$a_n = \frac{2}{L}\int_0^L f(x)\cos\left(\frac{n\pi x}{L}\right)\,dx. \tag{6}$$

see also Even Function, Fourier Cosine Transform, Fourier Series, Fourier Sine Series

Fourier Cosine Transform

The Fourier cosine transform is the Real Part of the full complex Fourier Transform,

$$\mathcal{F}\cos[f(x)] = \Re[\mathcal{F}[f(x)]].$$

see also Fourier Sine Transform, Fourier Transform

References

Press, W. H.; Flannery, B. P.; Teukolsky, S. A.; and Vetterling, W. T. "FFT of Real Functions, Sine and Cosine Transforms." §12.3 in *Numerical Recipes in FORTRAN: The Art of Scientific Computing, 2nd ed.* Cambridge, England: Cambridge University Press, pp. 504–515, 1992.

Fourier Integral

see Fourier Transform

Fourier Matrix

The $n \times n$ Square Matrix F_n with entries given by

$$F_{jk} = e^{2\pi ijk/n} \tag{1}$$

for $j, k = 1, 2, \ldots, n$, and normalized by $1/\sqrt{n}$ to make it a Unitary. The Fourier matrix F_2 is given by

$$\mathsf{F}_2 = \frac{1}{\sqrt{2}}\begin{bmatrix} 1 & 1 \\ 1 & i^2 \end{bmatrix}, \tag{2}$$

and the F_4 matrix by

$$\begin{aligned} \mathsf{F}_4 &= \frac{1}{\sqrt{4}}\begin{bmatrix} 1 & 1 & 1 & 1 \\ 1 & i & i^2 & i^3 \\ 1 & i^2 & i^4 & i^6 \\ 1 & i^3 & i^6 & i^9 \end{bmatrix} \\ &= \frac{1}{2}\begin{bmatrix} 1 & & 1 & \\ & 1 & & i \\ 1 & & -1 & \\ & 1 & & -i \end{bmatrix}\begin{bmatrix} 1 & 1 & & \\ 1 & i^2 & & \\ & & 1 & 1 \\ & & 1 & i^2 \end{bmatrix}\begin{bmatrix} 1 & & & \\ & & 1 & \\ & 1 & & \\ & & & 1 \end{bmatrix}. \end{aligned} \tag{3}$$

In general,

$$\mathsf{F}_{2n} = \begin{bmatrix} \mathsf{I}_n & \mathsf{D}_n \\ \mathsf{I}_n & -\mathsf{D}_n \end{bmatrix}\begin{bmatrix} \mathsf{F}_n & \\ & \mathsf{F}_n \end{bmatrix}\begin{bmatrix} \text{even-odd} \\ \text{shuffle} \end{bmatrix}, \tag{4}$$

with

$$\begin{aligned} \begin{bmatrix} \mathsf{F}_n & \\ & \mathsf{F}_n \end{bmatrix} &= \begin{bmatrix} \mathsf{I}_{n/2} & \mathsf{D}_{n/2} & & \\ \mathsf{I}_{n/2} & -\mathsf{D}_{n/2} & & \\ & & \mathsf{I}_{n/2} & \mathsf{D}_{n/2} \\ & & \mathsf{I}_{n/2} & -\mathsf{D}_{n/2} \end{bmatrix} \\ &\times \begin{bmatrix} \mathsf{F}_{n/2} & & & \\ & \mathsf{F}_{n/2} & & \\ & & \mathsf{F}_{n/2} & \\ & & & \mathsf{F}_{n/2} \end{bmatrix}\begin{bmatrix} \text{even-odd} \\ 0, 2 \pmod 4 \\ \text{even-odd} \\ 1, 3 \pmod 4 \end{bmatrix}, \end{aligned} \tag{5}$$

where I_n is the $n \times n$ Identity Matrix. Note that the factorization (which is the basis of the Fast Fourier Transform) has two copies of F_2 in the center factor Matrix.

see also Fast Fourier Transform, Fourier Transform

References

Strang, G. "Wavelet Transforms Versus Fourier Transforms." *Bull. Amer. Math. Soc.* **28**, 288–305, 1993.

Fourier-Mellin Integral

The inverse of the Laplace Transform

$$F(t) = \mathcal{L}^{-1}[f(s)] = \frac{1}{2\pi i}\int_{\gamma-i\infty}^{\gamma+i\infty} e^{st} f(s)\,ds$$

$$f(s) = \mathcal{L}[F(t)] = \int_0^{\infty} F(t)e^{-st}\,dt.$$

see also Bromwich Integral, Laplace Transform

Fourier Series

Fourier series are expansions of PERIODIC FUNCTIONS $f(x)$ in terms of an infinite sum of SINES and COSINES

$$f(x) = \sum_{n=0}^{\infty} a_n \cos(nx) + \sum_{n=0}^{\infty} b_n \sin(nx). \quad (1)$$

Fourier series make use of the ORTHOGONALITY relationships of the SINE and COSINE functions, which can be used to calculate the coefficients a_n and b_n in the sum. The computation and study of Fourier series is known as HARMONIC ANALYSIS.

To compute a Fourier series, use the integral identities

$$\int_{-\pi}^{\pi} \sin(mx)\sin(nx)\,dx = \pi\delta_{mn} \quad \text{for } n, m \neq 0 \quad (2)$$

$$\int_{-\pi}^{\pi} \cos(mx)\cos(nx)\,dx = \pi\delta_{mn} \quad \text{for } n, m \neq 0 \quad (3)$$

$$\int_{-\pi}^{\pi} \sin(mx)\cos(nx)\,dx = 0 \quad (4)$$

$$\int_{-\pi}^{\pi} \sin(mx)\,dx = 0 \quad (5)$$

$$\int_{-\pi}^{\pi} \cos(mx)\,dx = 0, \quad (6)$$

where δ_{mn} is the KRONECKER DELTA. Now, expand your function $f(x)$ as an infinite series of the form

$$\begin{aligned} f(x) &= \sum_{n=0}^{\infty} a'_n \cos(nx) + \sum_{n=0}^{\infty} b_n \sin(nx) \\ &= \tfrac{1}{2}a_0 + \sum_{n=1}^{\infty} a_n \cos(nx) + \sum_{n=1}^{\infty} b_n \sin(nx), \quad (7) \end{aligned}$$

where we have relabeled the $a_0 = 2a'_0$ term for future convenience but left $a_n = a'_n$. Assume the function is periodic in the interval $[-\pi, \pi]$. Now use the orthogonality conditions to obtain

$$\begin{aligned} &\int_{-\pi}^{\pi} f(x)\,dx \\ &= \int_{-\pi}^{\pi} \left[\sum_{n=1}^{\infty} a_n \cos(nx) + \sum_{n=1}^{\infty} b_n \sin(nx) + \tfrac{1}{2}a_0\right] dx \\ &= \sum_{n=1}^{\infty} \int_{-\pi}^{\pi} [a_n \cos(nx) + b_n \sin(nx)]\,dx + \tfrac{1}{2}a_0 \int_{-\pi}^{\pi} dx \\ &= \sum_{n=1}^{\infty} (0+0) + \pi a_0 = \pi a_0 \quad (8) \end{aligned}$$

and

$$\begin{aligned} &\int_{-\pi}^{\pi} f(x)\sin(mx)\,dx \\ &= \int_{-\pi}^{\pi} \left[\sum_{n=1}^{\infty} a_n \cos(nx) + \sum_{n=1}^{\infty} b_n \sin(nx) + \tfrac{1}{2}a_0\right] \times \sin(mx)\,dx \\ &= \sum_{n=1}^{\infty} \int_{-\pi}^{\pi} [a_n \cos(nx)\sin(mx) + b_n \sin(nx)\sin(mx)]\,dx \\ &\quad + \tfrac{1}{2}a_0 \int_{-\pi}^{\pi} \sin(mx)\,dx \\ &= \sum_{n=1}^{\infty} (0 + b_n\pi\delta_{mn}) + 0 = \pi b_n, \quad (9) \end{aligned}$$

so

$$\begin{aligned} \int_{-\pi}^{\pi} f(x)\cos(mx)\,dx &= \int_{-\pi}^{\pi} \left[\sum_{n=1}^{\infty} a_n \cos(nx) + \sum_{n=1}^{\infty} b_n \sin(nx) + \tfrac{1}{2}a_0\right] \cos(mx)\,dx \\ &= \sum_{n=1}^{\infty} \int_{-\pi}^{\pi} [a_n \cos(nx)\cos(mx) \\ &\quad + b_n \sin(nx)\cos(mx)]\,dx + \tfrac{1}{2}a_0 \int_{-\pi}^{\pi} \cos(mx)\,dx \\ &= \sum_{n=1}^{\infty} (a_n\pi\delta_{mn} + 0) + 0 = \pi a_n. \quad (10) \end{aligned}$$

Plugging back into the original series then gives

$$a_0 = \frac{1}{\pi}\int_{-\pi}^{\pi} f(x)\,dx \quad (11)$$

$$a_n = \frac{1}{\pi}\int_{-\pi}^{\pi} f(x)\cos(nx)\,dx \quad (12)$$

$$b_n = \frac{1}{\pi}\int_{-\pi}^{\pi} f(x)\sin(nx)\,dx \quad (13)$$

for $n = 1, 2, 3, \ldots$. The series expansion converges to the function $\bar{f}$ (equal to the original function at points of continuity or to the average of the two limits at points of discontinuity)

$$\bar{f} \equiv \begin{cases} \tfrac{1}{2}\left[\lim_{x\to x_0^-} f(x) + \lim_{x\to x_0^+} f(x)\right] \\ \quad \text{for } -\pi < x_0 < \pi \\ \tfrac{1}{2}\left[\lim_{x\to -\pi^+} f(x) + \lim_{x\to \pi_-} f(x)\right] \\ \quad \text{for } x_0 = -\pi, \pi \end{cases} \quad (14)$$

if the function satisfies the DIRICHLET CONDITIONS.

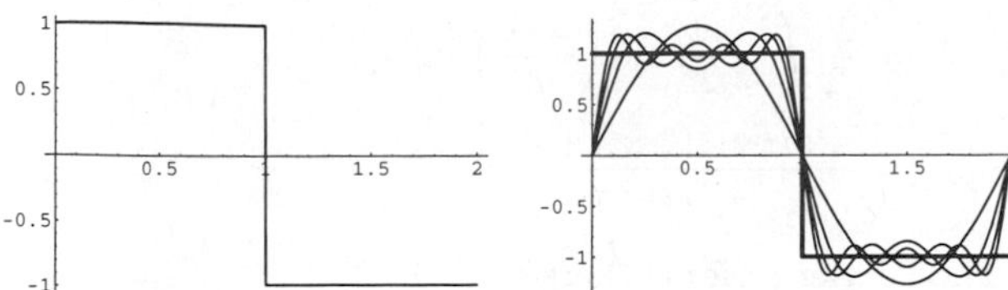

Near points of discontinuity, a "ringing" known as the GIBBS PHENOMENON, illustrated below, occurs. For a function $f(x)$ periodic on an interval $[-L, L]$, use a change of variables to transform the interval of integration to $[-1, 1]$. Let

$$x \equiv \frac{\pi x'}{L} \tag{15}$$

$$dx = \frac{\pi \, dx'}{L}. \tag{16}$$

Solving for x', $x' = Lx/\pi$. Plugging this in gives

$$f(x') = \tfrac{1}{2}a_0 + \sum_{n=1}^{\infty} a_n \cos\left(\frac{n\pi x'}{L}\right) + \sum_{n=1}^{\infty} b_n \sin\left(\frac{n\pi x'}{L}\right) \tag{17}$$

$$\begin{cases} a_0 = \frac{1}{L}\int_{-L}^{L} f(x')\,dx' \\ a_n = \frac{1}{L}\int_{-L}^{L} f(x')\cos\left(\frac{n\pi x'}{L}\right)dx' \\ b_n = \frac{1}{L}\int_{-L}^{L} f(x')\sin\left(\frac{n\pi x'}{L}\right)dx' \end{cases}. \tag{18}$$

If a function is EVEN so that $f(x) = f(-x)$, then $f(x)\sin(nx)$ is ODD. (This follows since $\sin(nx)$ is ODD and an EVEN FUNCTION times an ODD FUNCTION is an ODD FUNCTION.) Therefore, $b_n = 0$ for all n. Similarly, if a function is ODD so that $f(x) = f(-x)$, then $f(x)\cos(nx)$ is ODD. (This follows since $\cos(nx)$ is EVEN and an EVEN FUNCTION times an ODD FUNCTION is an ODD FUNCTION.) Therefore, $a_n = 0$ for all n.

Because the SINES and COSINES form a COMPLETE ORTHOGONAL BASIS, the SUPERPOSITION PRINCIPLE holds, and the Fourier series of a linear combination of two functions is the same as the linear combination of the corresponding two series. The COEFFICIENTS for Fourier series expansions for a few common functions are given in Beyer (1987, pp. 411–412) and Byerly (1959, p. 51).

The notion of a Fourier series can also be extended to COMPLEX COEFFICIENTS. Consider a real-valued function $f(x)$. Write

$$f(x) = \sum_{n=-\infty}^{\infty} A_n e^{inx}. \tag{19}$$

Now examine

$$\int_{-\pi}^{\pi} f(x)e^{-imx}\,dx = \int_{-\pi}^{\pi}\left(\sum_{n=-\infty}^{\infty} A_n e^{inx}\right)e^{-imx}\,dx$$

$$= \sum_{n=-\infty}^{\infty} A_n \int_{-\pi}^{\pi} e^{i(n-m)x}\,dx$$

$$= \sum_{n=-\infty}^{\infty} A_n \int_{-\pi}^{\pi} \{\cos[(n-m)x] + i\sin[(n-m)x]\}\,dx$$

$$= \sum_{m=-\infty}^{\infty} A_n 2\pi\delta_{mn} = 2\pi A_m, \tag{20}$$

so

$$A_n = \frac{1}{2\pi}\int_{-\pi}^{\pi} f(x)e^{-inx}\,dx. \tag{21}$$

The COEFFICIENTS can be expressed in terms of those in the FOURIER SERIES

$$A_n = \frac{1}{2\pi}\int_{-\pi}^{\pi} f(x)[\cos(nx) - i\sin(nx)]\,dx$$

$$= \begin{cases} \frac{1}{2\pi}\int_{-\pi}^{\pi} f(x)[\cos(nx) + i\sin(nx)]\,dx & n < 0 \\ \frac{1}{2\pi}\int_{-\pi}^{\pi} f(x)\,dx & n = 0 \\ \frac{1}{2\pi}\int_{-\pi}^{\pi} f(x)[\cos(nx) - i\sin(nx)]\,dx & n > 0 \end{cases}$$

$$= \begin{cases} \frac{1}{2}(a_n + ib_n) & n < 0 \\ \frac{1}{2}a_0 & n = 0 \\ \frac{1}{2}(a_n - ib_n) & n > 0. \end{cases} \tag{22}$$

For a function periodic in $[-L, L]$, these become

$$f(x) = \sum_{n=-\infty}^{\infty} A_n e^{i(2\pi nx/L)} \tag{23}$$

$$A_n = \frac{1}{L}\int_{-L/2}^{L/2} f(x)e^{-i(2\pi nx/L)}\,dx. \tag{24}$$

These equations are the basis for the extremely important FOURIER TRANSFORM, which is obtained by transforming A_n from a discrete variable to a continuous one as the length $L \to \infty$.

see also DIRICHLET FOURIER SERIES CONDITIONS, FOURIER COSINE SERIES, FOURIER SINE SERIES, FOURIER TRANSFORM, GIBBS PHENOMENON, LEBESGUE CONSTANTS (FOURIER SERIES), LEGENDRE SERIES, RIESZ-FISCHER THEOREM, SCHLÖMILCH'S SERIES

References

Arfken, G. "Fourier Series." Ch. 14 in *Mathematical Methods for Physicists, 3rd ed.* Orlando, FL: Academic Press, pp. 760–793, 1985.

Beyer, W. H. (Ed.). *CRC Standard Mathematical Tables, 28th ed.* Boca Raton, FL: CRC Press, 1987.

Brown, J. W. and Churchill, R. V. *Fourier Series and Boundary Value Problems, 5th ed.* New York: McGraw-Hill, 1993.

Byerly, W. E. *An Elementary Treatise on Fourier's Series, and Spherical, Cylindrical, and Ellipsoidal Harmonics,*

with Applications to Problems in Mathematical Physics. New York: Dover, 1959.
Carslaw, H. S. *Introduction to the Theory of Fourier's Series and Integrals, 3rd ed., rev. and enl.* New York: Dover, 1950.
Davis, H. F. *Fourier Series and Orthogonal Functions.* New York: Dover, 1963.
Dym, H. and McKean, H. P. *Fourier Series and Integrals.* New York: Academic Press, 1972.
Folland, G. B. *Fourier Analysis and Its Applications.* Pacific Grove, CA: Brooks/Cole, 1992.
Groemer, H. *Geometric Applications of Fourier Series and Spherical Harmonics.* New York: Cambridge University Press, 1996.
Körner, T. W. *Fourier Analysis.* Cambridge, England: Cambridge University Press, 1988.
Körner, T. W. *Exercises for Fourier Analysis.* New York: Cambridge University Press, 1993.
Lighthill, M. J. *Introduction to Fourier Analysis and Generalised Functions.* Cambridge, England: Cambridge University Press, 1958.
Morrison, N. *Introduction to Fourier Analysis.* New York: Wiley, 1994.
Sansone, G. "Expansions in Fourier Series." Ch. 2 in *Orthogonal Functions, rev. English ed.* New York: Dover, pp. 39–168, 1991.

Fourier Series—Power Series

For $f(x) = x^k$ on the INTERVAL $[-L, L)$ and periodic with period $2L$, the FOURIER SERIES is given by

$$\begin{aligned} a_n &= \frac{1}{L}\int_{-L}^{L} x^k \cos\left(\frac{n\pi x}{L}\right) dx \\ &= \frac{2L^k}{1+k}\,{}_1F_2\left(\begin{matrix} 1+\frac{1}{2}k \\ \frac{1}{2} \quad \frac{1}{2}(3+k) \end{matrix}; -\tfrac{1}{4}\pi^2 n^2\right) \\ b_n &= \frac{1}{L}\int_{-L}^{L} x^k \sin\left(\frac{n\pi x}{L}\right) dx \\ &= \frac{2n\pi L^k}{2+k}\,{}_1F_2\left(\begin{matrix} 1+\frac{1}{2}k \\ \frac{3}{2} \quad 2+\frac{1}{2}k \end{matrix}; -\tfrac{1}{4}\pi^2 n^2\right), \end{aligned}$$

where ${}_1F_2(a; b, c; x)$ is a generalized HYPERGEOMETRIC FUNCTION.

Fourier Series—Right Triangle

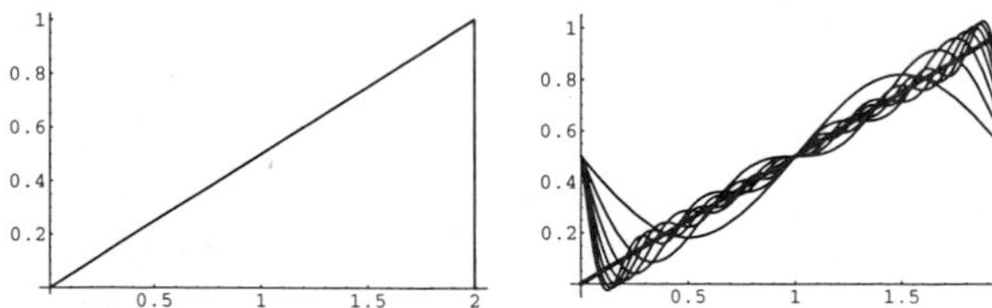

Consider a string of length $2L$ plucked at the right end, then

$$\begin{aligned} a_0 &= \frac{1}{L}\int_0^{2L} \frac{x}{2L}\,dx = \frac{1}{2L^2}[\tfrac{1}{2}x^2]_0^L = \frac{1}{4L^2}(2L)^2 = 1 \\ a_n &= \frac{1}{L}\int_0^{2L} \frac{x}{2L}\cos\left(\frac{n\pi x}{L}\right) dx \\ &= \frac{[2n\pi\cos(n\pi) - \sin(n\pi)]\sin(n\pi)}{n^2\pi^2} = 0 \end{aligned}$$

$$\begin{aligned} b_n &= \frac{1}{L}\int_0^{2L} \frac{x}{2L}\sin\left(\frac{n\pi x}{L}\right) dx \\ &= \frac{-2n\pi\cos(2n\pi) + \sin(2n\pi)}{2n^2\pi^2} = -\frac{1}{n\pi}. \end{aligned}$$

The Fourier series is therefore

$$f(x) = \tfrac{1}{2} - \frac{1}{\pi}\sum_{n=1}^{\infty} \frac{1}{n}\sin\left(\frac{n\pi x}{L}\right).$$

see also FOURIER SERIES

Fourier Series—Square Wave

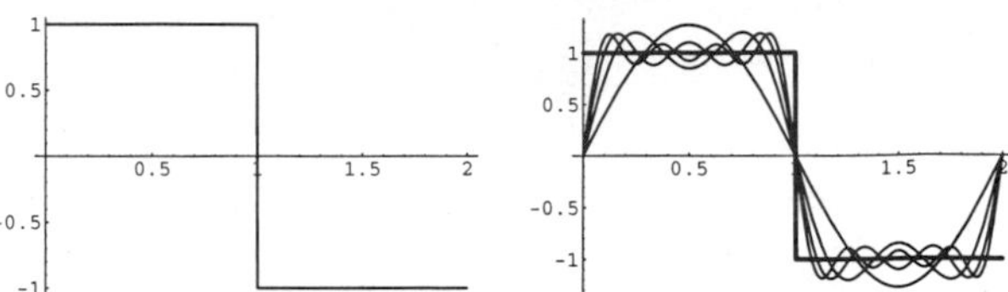

Consider a square wave of length $2L$. Since the function is ODD, $a_0 = a_n = 0$, and

$$\begin{aligned} b_n &= \frac{2}{L}\int_0^L \sin\left(\frac{n\pi x}{L}\right) dx \\ &= \frac{4}{n\pi}\sin^2(\tfrac{1}{2}n\pi) = \frac{4}{n\pi}\begin{cases} 0 & n \text{ even} \\ 1 & n \text{ odd.} \end{cases} \end{aligned}$$

The Fourier series is therefore

$$f(x) = \frac{4}{\pi}\sum_{n=1,3,5,\ldots}^{\infty} \frac{1}{n}\sin\left(\frac{n\pi x}{L}\right).$$

see also FOURIER SERIES, SQUARE WAVE

Fourier Series—Triangle

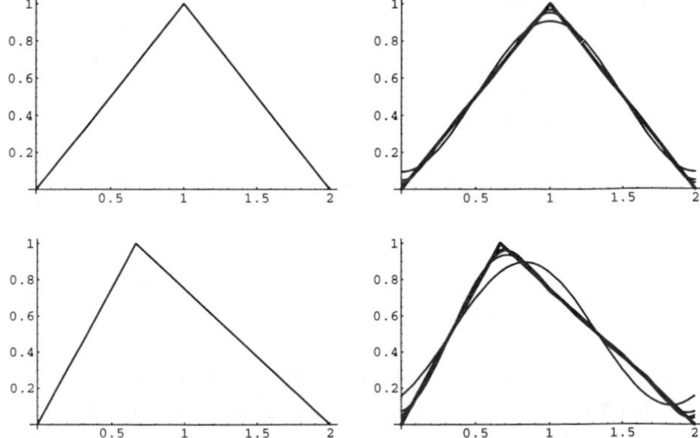

Let a string of length $2L$ have a y-displacement of unity when it is pinned an x-distance which is $(1/m)$th of the way along the string. The displacement as a function of x is then

$$f_m(x) = \begin{cases} \frac{mx}{2L} & 0 \le x \le \frac{2L}{m} \\ \frac{m}{1-m}\left(\frac{x}{2L} - 1\right) & \frac{2L}{m} \le x \le 2L. \end{cases}$$

The COEFFICIENTS are therefore

$$a_0 = \frac{1}{L}\left[\int_0^{2L/m} \frac{nx}{2L}\,dx + \int_{2L/m}^{2L} \frac{n}{1-n}\left(\frac{x}{2L}-1\right)dx\right] = 1$$

$$a_n = \frac{m\left[1-m-\cos(2\pi n)+m\cos\left(\frac{2n\pi}{m}\right)\right]}{2(m-1)n^2\pi^2} = \frac{m^2\left[\cos\left(\frac{2n\pi}{m}\right)-1\right]}{2(m-1)n^2\pi^2}$$

$$b_n = \frac{m\left[m\sin\left(\frac{2\pi n}{m}\right)-\sin(2\pi n)\right]}{2(m-1)n^2\pi^2} = \frac{m^2\sin\left(\frac{2\pi n}{m}\right)}{2(m-1)n^2\pi^2}.$$

The Fourier series is therefore

$$f_m(x) = \tfrac{1}{2} + \frac{m^2}{2(m-1)\pi^2} \times \sum_{n=1}^\infty \left\{\frac{1}{n^2}\left[\cos\left(\frac{2n\pi}{m}\right)-1\right]\cos\left(\frac{n\pi x}{L}\right) + \frac{\sin\left(\frac{2\pi n}{m}\right)}{n^2}\sin\left(\frac{n\pi x}{L}\right)\right\}.$$

If $m=2$, then a_n and b_n simplify to

$$a_n = -\frac{4}{n^2\pi^2}\sin^2(\tfrac{1}{2}n\pi) = -\frac{4}{n^2\pi^2}\begin{cases}0 & n=0,\,2,\,\ldots\\ 1 & n=1,\,3,\,\ldots\end{cases}$$

$$b_n = 0,$$

giving

$$f_2(x) = \tfrac{1}{2} - \frac{4}{\pi^2}\sum_{n=1,3,5,\ldots}^\infty \frac{1}{n^2}\cos\left(\frac{n\pi x}{L}\right).$$

see also FOURIER SERIES

Fourier Series—Triangle Wave

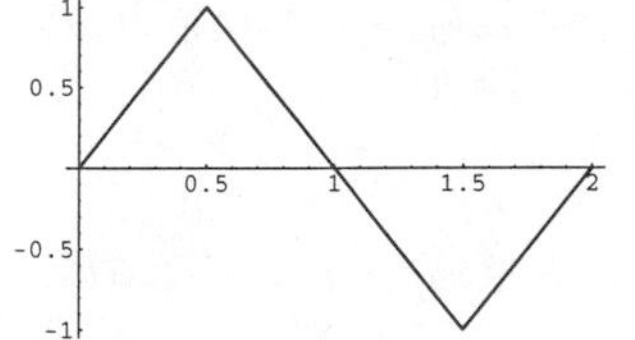

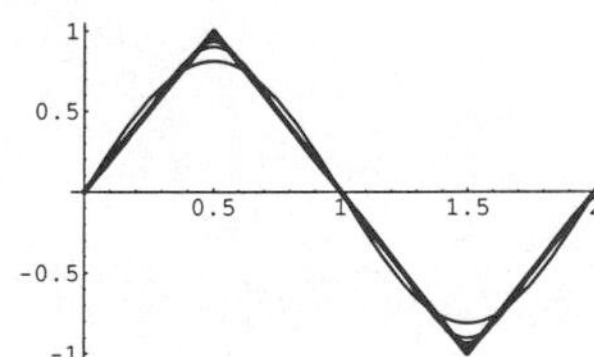

Consider a triangle wave of length $2L$. Since the function is ODD, $a_0 = a_n = 0$, and

$$b_n = \frac{2}{L}\left\{\int_0^{L/2} \frac{x}{L/2}\sin\left(\frac{n\pi x}{L}\right)dx + \int_{L/2}^{L}\left[1-\frac{2}{L}(x-\tfrac{1}{2}L)\right]\sin\left(\frac{n\pi x}{L}\right)dx\right\}dx$$

$$= \frac{32}{\pi^2 n^2}\cos(\tfrac{1}{4}n\pi)\sin^3(\tfrac{1}{4}n\pi)$$

$$= \frac{32}{\pi^2 n^2}\begin{cases}0 & n=0,\,4,\,\ldots\\ \frac{1}{4} & n=1,\,5,\,\ldots\\ 0 & n=2,\,6,\,\ldots\\ -\frac{1}{4} & n=3,\,7,\,\ldots\end{cases}$$

$$= \frac{8}{\pi^2 n^2}\begin{cases}(-1)^{(n-1)/2} & \text{for } n \text{ odd}\\ 0 & \text{for } n \text{ even.}\end{cases}$$

The Fourier series is therefore

$$f(x) = \frac{8}{\pi^2}\sum_{n=1,3,5,\ldots}^\infty \frac{(-1)^{(n-1)/2}}{n^2}\sin\left(\frac{n\pi x}{L}\right).$$

see also FOURIER SERIES

Fourier Sine Series

If $f(x)$ is an ODD FUNCTION, then $a_n = 0$ and the FOURIER SERIES collapses to

$$f(x) = \sum_{n=1}^\infty b_n \sin(nx), \tag{1}$$

where

$$b_n = \frac{1}{\pi}\int_{-\pi}^{\pi} f(x)\sin(nx)\,dx = \frac{2}{\pi}\int_0^\pi f(x)\sin(nx)\,dx \tag{2}$$

for $n = 1, 2, 3, \ldots$. The last EQUALITY is true because

$$f(x)\sin(nx) = [-f(-x)][-\sin(-nx)] = f(-x)\sin(-nx). \tag{3}$$

Letting the range go to L,

$$b_n = \frac{2}{L}\int_0^L f(x)\sin\left(\frac{n\pi x}{L}\right)dx. \tag{4}$$

see also FOURIER COSINE SERIES, FOURIER SERIES, FOURIER SINE TRANSFORM

Fourier Sine Transform

The Fourier sine transform is the IMAGINARY PART of the full complex FOURIER TRANSFORM,

$$\mathcal{F}\sin[f(x)] = \Im[\mathcal{F}[f(x)]].$$

see also FOURIER COSINE TRANSFORM, FOURIER TRANSFORM

References

Press, W. H.; Flannery, B. P.; Teukolsky, S. A.; and Vetterling, W. T. "FFT of Real Functions, Sine and Cosine Transforms." §12.3 in *Numerical Recipes in FORTRAN: The Art of Scientific Computing, 2nd ed.* Cambridge, England: Cambridge University Press, pp. 504–515, 1992.

Fourier-Stieltjes Transform

Let $f(x)$ be a positive definite, measurable function on the INTERVAL $(-\infty,\infty)$. Then there exists a monotone increasing, real-valued bounded function $\alpha(t)$ such that

$$f(x) = \int_{-\infty}^\infty e^{itx}\,d\alpha(t)$$

for "ALMOST ALL" x. If $\alpha(t)$ is nondecreasing and bounded and $f(x)$ is defined as above, then $f(x)$ is called the Fourier-Stieltjes transform of $\alpha(t)$, and is both continuous and positive definite.

see also FOURIER TRANSFORM, LAPLACE TRANSFORM

References

Iyanaga, S. and Kawada, Y. (Eds.). *Encyclopedic Dictionary of Mathematics.* Cambridge, MA: MIT Press, p. 618, 1980.

Fourier Transform

The Fourier transform is a generalization of the COMPLEX FOURIER SERIES in the limit as $L \to \infty$. Replace the discrete A_n with the continuous $F(k)\,dk$ while letting $n/L \to k$. Then change the sum to an INTEGRAL, and the equations become

$$f(x) = \int_{-\infty}^{\infty} F(k) e^{2\pi ikx}\,dk \tag{1}$$

$$F(k) = \int_{-\infty}^{\infty} f(x) e^{-2\pi ikx}\,dx. \tag{2}$$

Here,

$$F(k) = \mathcal{F}[f(x)] = \int_{-\infty}^{\infty} f(x) e^{-2\pi ikx}\,dx \tag{3}$$

is called the *forward* $(-i)$ Fourier transform, and

$$f(x) = \mathcal{F}^{-1}[F(k)] = \int_{-\infty}^{\infty} F(k) e^{2\pi ikx}\,dk \tag{4}$$

is called the *inverse* $(+i)$ Fourier transform. Some authors (especially physicists) prefer to write the transform in terms of angular frequency $\omega \equiv 2\pi\nu$ instead of the oscillation frequency ν. However, this destroys the symmetry, resulting in the transform pair

$$H(\nu) = \mathcal{F}[h(t)] = \int_{-\infty}^{\infty} h(t) e^{-i\omega t}\,dt \tag{5}$$

$$h(t) = \mathcal{F}^{-1}[H(\nu)] = \frac{1}{2\pi} \int_{-\infty}^{\infty} H(\nu) e^{i\omega t}\,d\omega. \tag{6}$$

In general, the Fourier transform pair may be defined using two arbitrary constants A and B as

$$F(\omega) = A \int_{-\infty}^{\infty} f(t) e^{Bi\omega t}\,dt \tag{7}$$

$$f(t) = \frac{B}{2\pi A} \int_{-\infty}^{\infty} F(\omega) e^{-Bi\omega t}\,d\omega. \tag{8}$$

The *Mathematica®* program (Wolfram Research, Champaign, IL) calls A the **`$FourierOverallConstant`** and B the **`$FourierFrequencyConstant`**, and defines $A = B = 1$ by default. Morse and Feshbach (1953) use $B = 1$ and $A = 1/\sqrt{2\pi}$. In this work, following Bracewell (1965, pp. 6–7), $A = 1$ and $B = -2\pi$ unless otherwise stated.

Since any function can be split up into EVEN and ODD portions $E(x)$ and $O(x)$,

$$f(x) = \tfrac{1}{2}[f(x)+f(-x)]+\tfrac{1}{2}[f(x)-f(-x)] = E(x)+O(x), \tag{9}$$

a Fourier transform can always be expressed in terms of the FOURIER COSINE TRANSFORM and FOURIER SINE TRANSFORM as

$$\mathcal{F}[f(x)] = \int_{-\infty}^{\infty} E(x)\cos(2\pi kx)\,dx - i\int_{-\infty}^{\infty} O(x)\sin(2\pi kx)\,dx. \tag{10}$$

A function $f(x)$ has a forward and inverse Fourier transform such that

$$f(x) = \begin{cases} \int_{-\infty}^{\infty} e^{2\pi ikx}\left[\int_{-\infty}^{\infty} f(x)e^{-2\pi ikx}\,dx\right] dk \\ \quad \text{for } f(x) \text{ continuous at } x \\ \frac{1}{2}[f(x_+)+f(x_-)] \\ \quad \text{for } f(x) \text{ discontinuous at } x, \end{cases} \tag{11}$$

provided that

1. $\int_{-\infty}^{\infty} |f(x)|\,dx$ exists.
2. Any discontinuities are finite.
3. The function has bounded variation. A SUFFICIENT weaker condition is fulfillment of the LIPSCHITZ CONDITION.

The smoother a function (i.e., the larger the number of continuous DERIVATIVES), the more compact its Fourier transform.

The Fourier transform is linear, since if $f(x)$ and $g(x)$ have FOURIER TRANSFORMS $F(k)$ and $G(k)$, then

$$\int [af(x)+bg(x)]e^{-2\pi ikx}\,dx = a\int_{-\infty}^{\infty} f(x)e^{-2\pi ikx}\,dx + b\int_{-\infty}^{\infty} g(x)e^{-2\pi ikx}\,dx = F(k)+G(k). \tag{12}$$

Therefore,

$$\mathcal{F}[af(x)+bg(x)] = a\mathcal{F}[f(x)]+b\mathcal{F}[g(x)] = aF(k)+bG(k). \tag{13}$$

The Fourier transform is also symmetric since $F(k) = \mathcal{F}[f(x)]$ implies $F(-k) = \mathcal{F}[f(x)]$.

Let $f * g$ denote the CONVOLUTION, then the transforms of convolutions of functions have particularly nice transforms,

$$\mathcal{F}[f*g] = \mathcal{F}[f]\mathcal{F}[g] \tag{14}$$

$$\mathcal{F}[fg] = \mathcal{F}[f]*\mathcal{F}[g] \tag{15}$$

$$\mathcal{F}[\mathcal{F}(f)+\mathcal{F}(g)] = f*g \tag{16}$$

$$\mathcal{F}[\mathcal{F}(f)*\mathcal{F}(g)] = fg. \tag{17}$$

The first of these is derived as follows:

$$\begin{aligned}\mathcal{F}[f*g] &= \int_{-\infty}^{\infty}\int_{-\infty}^{\infty} e^{-2\pi ikx} f(x')g(x-x')\,dx'\,dx \\ &= \int_{-\infty}^{\infty}\int_{-\infty}^{\infty} [e^{-2\pi ikx'} f(x')\,dx'] \\ &\quad\times [e^{-2\pi ik(x-x')} g(x-x')\,dx] \\ &= \left[\int_{-\infty}^{\infty} e^{-2\pi ikx'} f(x')\,dx'\right] \\ &\quad\times \left[\int_{-\infty}^{\infty} e^{-2\pi ikx''} g(x'')\,dx''\right] \\ &= \mathcal{F}[f]\mathcal{F}[g],\end{aligned} \tag{18}$$

where $x'' \equiv x - x'$.

There is also a somewhat surprising and extremely important relationship between the AUTOCORRELATION and the Fourier transform known as the WIENER-KHINTCHINE THEOREM. Let $\mathcal{F}[f(x)] = F(k)$, and F^* denote the COMPLEX CONJUGATE of F, then the FOURIER TRANSFORM of the ABSOLUTE SQUARE of $F(k)$ is given by

$$\mathcal{F}[|F(k)|^2] = \int_{-\infty}^{\infty} f^*(\tau)f(\tau+x)\,d\tau. \tag{19}$$

The Fourier transform of a DERIVATIVE $f'(x)$ of a function $f(x)$ is simply related to the transform of the function $f(x)$ itself. Consider

$$\mathcal{F}[f'(x)] = \int_{-\infty}^{\infty} f'(x)e^{-2\pi ikx}\,dx. \tag{20}$$

Now use INTEGRATION BY PARTS

$$\int v\,du = [uv] - \int u\,dv \tag{21}$$

with

$$du = f'(x)\,dx \qquad v = e^{-2\pi ikx} \tag{22}$$

$$u = f(x) \qquad dv = -2\pi ike^{-2\pi ikx}\,dx, \tag{23}$$

then

$$\mathcal{F}[f'(x)] = [f(x)e^{-2\pi ikx}]_{-\infty}^{\infty} - \int_{-\infty}^{\infty} f(x)(-2\pi ike^{-2\pi ikx}\,dx). \tag{24}$$

The first term consists of an oscillating function times $f(x)$. But if the function is bounded so that

$$\lim_{x\to\pm\infty} f(x) = 0 \tag{25}$$

(as any physically significant signal must be), then the term vanishes, leaving

$$\mathcal{F}[f'(x)] = 2\pi ik\int_{-\infty}^{\infty} f(x)e^{-2\pi ikx}\,dx = 2\pi ik\mathcal{F}[f(x)]. \tag{26}$$

This process can be iterated for the nth DERIVATIVE to yield

$$\mathcal{F}[f^{(n)}(x)] = (2\pi ik)^n\mathcal{F}[f(x)]. \tag{27}$$

The important MODULATION THEOREM of Fourier transforms allows $\mathcal{F}[\cos(2\pi k_0x)f(x)]$ to be expressed in terms of $\mathcal{F}[f(x)] = F(k)$ as follows,

$$\begin{aligned}\mathcal{F}[\cos(2\pi k_0x)f(x)] &\equiv \int_{-\infty}^{\infty} f(x)\cos(2\pi k_0x)e^{-2\pi ikx}\,dx \\ &= \tfrac{1}{2}\int_{-\infty}^{\infty} f(x)e^{2\pi ik_0x}e^{-2\pi ikx}\,dx \\ &\quad + \tfrac{1}{2}\int_{-\infty}^{\infty} f(x)e^{-2\pi ik_0x}e^{-2\pi ikx}\,dx \\ &= \tfrac{1}{2}\int_{-\infty}^{\infty} f(x)e^{-2\pi i(k-k_0)x}\,dx \\ &\quad + \tfrac{1}{2}\int_{-\infty}^{\infty} f(x)e^{-2\pi i(k+k_0)x}\,dx \\ &= \tfrac{1}{2}[F(k-k_0)+F(k+k_0)].\end{aligned} \tag{28}$$

Since the DERIVATIVE of the FOURIER TRANSFORM is given by

$$F'(k) \equiv \frac{d}{dk}\mathcal{F}[f(x)] = \int_{-\infty}^{\infty} (-2\pi ix)f(x)e^{-2\pi ikx}\,dx, \tag{29}$$

it follows that

$$F'(0) = -2\pi i\int_{-\infty}^{\infty} xf(x)\,dx. \tag{30}$$

Iterating gives the general FORMULA

$$\mu_n \equiv \int_{-\infty}^{\infty} x^n f(x)\,dx = \frac{F^{(n)}(0)}{(-2\pi i)^n}. \tag{31}$$

The VARIANCE of a FOURIER TRANSFORM is

$${\sigma_f}^2 = \left\langle (xf - \langle xf\rangle)^2\right\rangle, \tag{32}$$

and it is true that

$$\sigma_{f+g} = \sigma_f + \sigma_g. \tag{33}$$

If $f(x)$ has the FOURIER TRANSFORM $F(k)$, then the Fourier transform has the shift property

$$\begin{aligned}&\int_{-\infty}^{\infty} f(x-x_0)e^{-2\pi ikx}\,dx \\ &= \int_{-\infty}^{\infty} f(x-x_0)e^{-2\pi i(x-x_0)k}e^{-2\pi i(kx_0)}\,d(x-x_0) \\ &= e^{-2\pi ikx_0}F(k),\end{aligned} \tag{34}$$

so $f(x-x_0)$ has the FOURIER TRANSFORM

$$\mathcal{F}[f(x-x_0)] = e^{-2\pi ikx_0}F(k). \tag{35}$$

If $f(x)$ has a FOURIER TRANSFORM $F(k)$, then the Fourier transform obeys a similarity theorem.

$$\int_{-\infty}^{\infty} f(ax)e^{-2\pi ikx}\,dx = \frac{1}{|a|}\int_{-\infty}^{\infty} f(ax)e^{-2\pi i(ax)(k/a)}\,d(ax) = \frac{1}{|a|}F\left(\frac{k}{a}\right), \tag{36}$$

so $f(ax)$ has the FOURIER TRANSFORM $|a|^{-1}F\left(\frac{k}{a}\right)$.

The "equivalent width" of a Fourier transform is

$$w_e \equiv \frac{\int_{-\infty}^{\infty} f(x)\,dx}{f(0)} = \frac{F(0)}{\int_{-\infty}^{\infty} F(k)\,dk}. \tag{37}$$

The "autocorrelation width" is

$$w_a \equiv \frac{\int_{-\infty}^{\infty} f\star f^*\,dx}{[f\star f^*]_0} = \frac{\int_{-\infty}^{\infty} f\,dx \int_{-\infty}^{\infty} f^*\,dx}{\int_{-\infty}^{\infty} ff^*\,dx}, \tag{38}$$

where $f\star g$ denotes the CROSS-CORRELATION of f and g.

Any operation on $f(x)$ which leaves its AREA unchanged leaves $F(0)$ unchanged, since

$$\int_{-\infty}^{\infty} f(x)\,dx = \mathcal{F}[f(0)] = F(0). \tag{39}$$

In 2-D, the Fourier transform becomes

$$F(x,y) = \int_{-\infty}^{\infty}\int_{-\infty}^{\infty} f(k_x,k_y)e^{-2\pi i(k_x x+k_y y)}\,dk_x\,dk_y \tag{40}$$

$$f(k_x,k_y) = \int_{-\infty}^{\infty}\int_{-\infty}^{\infty} F(x,y)e^{2\pi i(k_x x+k_y y)}\,dx\,dy. \tag{41}$$

Similarly, the n-D Fourier transform can be defined for $\mathbf{k}, \mathbf{x} \in \mathbb{R}^n$ by

$$F(\mathbf{x}) = \underbrace{\int_{-\infty}^{\infty}\cdots\int_{-\infty}^{\infty}}_{n} f(\mathbf{k})e^{-2\pi i\mathbf{k}\cdot\mathbf{x}}\,d^n\mathbf{k} \tag{42}$$

$$f(\mathbf{k}) = \underbrace{\int_{-\infty}^{\infty}\cdots\int_{-\infty}^{\infty}}_{n} F(\mathbf{x})e^{2\pi i\mathbf{k}\cdot\mathbf{x}}\,d^n\mathbf{x}. \tag{43}$$

see also AUTOCORRELATION, CONVOLUTION, DISCRETE FOURIER TRANSFORM, FAST FOURIER TRANSFORM, FOURIER SERIES, FOURIER-STIELTJES TRANSFORM, HANKEL TRANSFORM, HARTLEY TRANSFORM, INTEGRAL TRANSFORM, LAPLACE TRANSFORM, STRUCTURE FACTOR, WINOGRAD TRANSFORM

References

Arfken, G. "Development of the Fourier Integral," "Fourier Transforms—Inversion Theorem," and "Fourier Transform of Derivatives." §15.2–15.4 in *Mathematical Methods for Physicists, 3rd ed.* Orlando, FL: Academic Press, pp. 794–810, 1985.

Blackman, R. B. and Tukey, J. W. *The Measurement of Power Spectra, From the Point of View of Communications Engineering.* New York: Dover, 1959.

Bracewell, R. *The Fourier Transform and Its Applications.* New York: McGraw-Hill, 1965.

Brigham, E. O. *The Fast Fourier Transform and Applications.* Englewood Cliffs, NJ: Prentice Hall, 1988.

James, J. F. *A Student's Guide to Fourier Transforms with Applications in Physics and Engineering.* New York: Cambridge University Press, 1995.

Körner, T. W. *Fourier Analysis.* Cambridge, England: Cambridge University Press, 1988.

Morrison, N. *Introduction to Fourier Analysis.* New York: Wiley, 1994.

Morse, P. M. and Feshbach, H. "Fourier Transforms." §4.8 in *Methods of Theoretical Physics, Part I.* New York: McGraw-Hill, pp. 453–471, 1953.

Papoulis, A. *The Fourier Integral and Its Applications.* New York: McGraw-Hill, 1962.

Press, W. H.; Flannery, B. P.; Teukolsky, S. A.; and Vetterling, W. T. *Numerical Recipes in C: The Art of Scientific Computing.* Cambridge, England: Cambridge University Press, 1989.

Sansone, G. "The Fourier Transform." §2.13 in *Orthogonal Functions, rev. English ed.* New York: Dover, pp. 158–168, 1991.

Sneddon, I. N. *Fourier Transforms.* New York: Dover, 1995.

Sogge, C. D. *Fourier Integrals in Classical Analysis.* New York: Cambridge University Press, 1993.

Spiegel, M. R. *Theory and Problems of Fourier Analysis with Applications to Boundary Value Problems.* New York: McGraw-Hill, 1974.

Strichartz, R. *Fourier Transforms and Distribution Theory.* Boca Raton, FL: CRC Press, 1993.

Titchmarsh, E. C. *Introduction to the Theory of Fourier Integrals, 3rd ed.* Oxford, England: Clarendon Press, 1948.

Tolstov, G. P. *Fourier Series.* New York: Dover, 1976.

Walker, J. S. *Fast Fourier Transforms, 2nd ed.* Boca Raton, FL: CRC Press, 1996.

Fourier Transform—1

The FOURIER TRANSFORM of the CONSTANT FUNCTION $f(x) = 1$ is given by

$$\mathcal{F}[1] = \int_{-\infty}^{\infty} e^{-2\pi ikx}\,dx = \delta(k),$$

according to the definition of the DELTA FUNCTION.

see also DELTA FUNCTION

Fourier Transform—1/x

The FOURIER TRANSFORM of the function $1/x$ is given by

$$\mathcal{F}\left(-\frac{1}{\pi x}\right) = -\frac{1}{\pi}\int_{-\infty}^{\infty} \frac{e^{-2\pi ikx}}{x}\,dx$$
$$= PV\int_{-\infty}^{\infty} \frac{\cos(2\pi kx) - i\sin(2\pi kx)}{x}\,dx$$
$$= \begin{cases} -\frac{2i}{\pi}\int_0^\infty \frac{\sin(2\pi kx)}{x}\,dx & \text{for } k<0 \\ \frac{2i}{\pi}\int_0^\infty \frac{\sin(2\pi kx)}{x}\,dx & \text{for } k>0 \end{cases}$$
$$= \begin{cases} -i & \text{for } k<0 \\ i & \text{for } k>0, \end{cases} \tag{1}$$

which can also be written as the single equation

$$\mathcal{F}\left(-\frac{1}{\pi x}\right) = i[1-2H(-k)], \tag{2}$$

where $H(x)$ is the HEAVISIDE STEP FUNCTION. The integrals follow from the identity

$$\int_0^\infty \frac{\sin(2\pi kx)}{x}\,dx = \int_0^\infty \frac{\sin(2\pi kx)}{2\pi kx}\,d(2\pi kx)$$
$$= \int_0^\infty \operatorname{sinc} z\,dz = \tfrac{1}{2}\pi. \tag{3}$$

Fourier Transform—Cosine

$$\mathcal{F}[\cos(2\pi k_0 x)] = \int_{-\infty}^{\infty} e^{-2\pi ikx}\left(\frac{e^{2\pi ik_0x}+e^{-2\pi ik_0x}}{2}\right)dx$$
$$= \tfrac{1}{2}\int_{-\infty}^{\infty} [e^{-2\pi i(k-k_0)x} + e^{-2\pi i(k+k_0)x}]\,dx$$
$$= \tfrac{1}{2}[\delta(k-k_0)+\delta(k+k_0)],$$

where $\delta(x)$ is the DELTA FUNCTION.

see also COSINE, FOURIER TRANSFORM—SINE

Fourier Transform—Delta Function

The FOURIER TRANSFORM of the DELTA FUNCTION is given by

$$\mathcal{F}[\delta(x-x_0)] = \int_{-\infty}^{\infty} \delta(x-x_0)e^{-2\pi ikx}\,dx = e^{-2\pi ikx_0}.$$

see also DELTA FUNCTION

Fourier Transform—Exponential Function

The FOURIER TRANSFORM of $e^{-k_0|x|}$ is given by

$$\mathcal{F}[e^{-k_0|x|}] = \int_{-\infty}^{\infty} e^{-k_0|x|}e^{-2\pi ikx}\,dx$$
$$= \int_{-\infty}^{0} e^{-2\pi ikx}e^{2\pi xk_0}\,dx + \int_0^\infty e^{-2\pi ikx}e^{-2\pi k_0x}\,dx$$
$$= \int_{-\infty}^{0} [\cos(2\pi kx) - i\sin(2\pi kx)]e^{2\pi k_0x}\,dx$$
$$+ \int_0^\infty [\cos(2\pi kx) - i\sin(2\pi kx)]e^{-2\pi k_0x}\,dx. \tag{1}$$

Now let $u \equiv -x$ so $du = -dx$, then

$$\mathcal{F}[e^{-k_0|x|}] = \int_0^\infty [\cos(2\pi ku) + i\sin(2\pi ku)]e^{-2\pi k_0u}\,du]$$
$$+ \int_0^\infty [\cos(2\pi ku) - i\sin(2\pi ku)]e^{-2\pi k_0u}\,du]$$
$$= 2\int_0^\infty \cos(2\pi ku)e^{-2\pi k_0u}\,du, \tag{2}$$

which, from the DAMPED EXPONENTIAL COSINE INTEGRAL, gives

$$\mathcal{F}[e^{-2\pi k_0|x|}] = \frac{1}{\pi}\frac{k_0}{k^2+{k_0}^2}, \tag{3}$$

which is a LORENTZIAN FUNCTION.

see also DAMPED EXPONENTIAL COSINE INTEGRAL, EXPONENTIAL FUNCTION, LORENTZIAN FUNCTION

Fourier Transform—Gaussian

The FOURIER TRANSFORM of a GAUSSIAN FUNCTION $f(x) \equiv e^{-ax^2}$ is given by

$$F(k) = \int_{-\infty}^{\infty} e^{-ax^2}e^{ikx}\,dx$$
$$= \int_{-\infty}^{\infty} e^{-ax^2}[\cos(kx) + i\sin(kx)]\,dx$$
$$= \int_{-\infty}^{\infty} e^{-ax^2}\cos(kx)\,dx + i\int_{-\infty}^{\infty} e^{-ax^2}\sin(kx)\,dx.$$

The second integrand is EVEN, so integration over a symmetrical range gives 0. The value of the first integral is given by Abramowitz and Stegun (1972, p. 302, equation 7.4.6)

$$F(k) = \sqrt{\frac{\pi}{a}}\,e^{-k^2/4a},$$

so a GAUSSIAN transforms to a GAUSSIAN.

see also GAUSSIAN FUNCTION

References

Abramowitz, M. and Stegun, C. A. (Eds.). *Handbook of Mathematical Functions with Formulas, Graphs, and Mathematical Tables, 9th printing.* New York: Dover, 1972.

Fourier Transform—Heaviside Step Function

$$\mathcal{F}[H(x)] = \int_{-\infty}^{\infty} e^{-2\pi ikx} H(x)\,dx = \frac{1}{2}\left[\delta(k) - \frac{i}{\pi k}\right],$$

where $H(x)$ is the HEAVISIDE STEP FUNCTION and $\delta(k)$ is the DELTA FUNCTION.

see also HEAVISIDE STEP FUNCTION

Fourier Transform—Lorentzian Function

$$\mathcal{F}\left[\frac{1}{\pi}\frac{\frac{1}{2}\Gamma}{(x-x_0)^2+(\frac{1}{2}\Gamma)^2}\right] = e^{-2\pi ikx_0-\Gamma\pi|k|}.$$

see also LORENTZIAN FUNCTION

Fourier Transform—Ramp Function

Let $R(x)$ be the RAMP FUNCTION, then the FOURIER TRANSFORM of $R(x)$ is given by

$$\mathcal{F}[R(x)] = \int_{-\infty}^{\infty} e^{-2\pi ikx} R(x)\,dx = \pi i\delta'(2\pi k) - \frac{1}{4\pi^2k^2},$$

where $\delta'(x)$ is the DERIVATIVE of the DELTA FUNCTION.

see also RAMP FUNCTION

Fourier Transform—Rectangle Function

Let $\Pi(x)$ be the RECTANGLE FUNCTION, then the FOURIER TRANSFORM is

$$\mathcal{F}[\Pi(x)] = \mathrm{sinc}(\pi k),$$

where $\mathrm{sinc}(x)$ is the SINC FUNCTION.

see also RECTANGLE FUNCTION, SINC FUNCTION

Fourier Transform—Sine

$$\begin{aligned}\mathcal{F}[\sin(2\pi k_0x)] &= \int_{-\infty}^{\infty} e^{-2\pi ik_0x}\left(\frac{e^{2\pi i\nu_0t} - e^{-2\pi ik_0x}}{2i}\right)dx \\ &= \tfrac{1}{2}i\int_{-\infty}^{\infty}[-e^{-2\pi i(k-k_0)x} + e^{-2\pi i(k+k_0)x}]\,dt \\ &= \tfrac{1}{2}i[\delta(k+k_0) - \delta(k-k_0)],\end{aligned}$$

where $\delta(x)$ is the DELTA FUNCTION.

see also FOURIER TRANSFORM—COSINE, SINE

Fox's H-Function

A very general function defined by

$$\begin{aligned}H(z) &= \mathbf{H}^{m,n}_{p,q}\left[z\left|\begin{matrix}(a_1,\alpha_1),\ldots,(a_p,\alpha_p)\\(b_1,\beta_1),\ldots,(b_p,\beta_p)\end{matrix}\right.\right] \\ &= \frac{1}{2\pi i}\int_C \frac{\prod_{j=1}^m \Gamma(b_j-\beta_i s)\prod_{j=1}^n \Gamma(1-a_j+\alpha_j s)}{\prod_{j=m+1}^q \Gamma(1-b_j+\beta_j s)\prod_{j=n+1}^{qp}\Gamma(a_j-\alpha_j s)} z^s\,ds,\end{aligned}$$

where $0 \le m \le q$, $0 \le n \le p$, $\alpha_j, \beta_j > 0$, and a_j, b_j are COMPLEX NUMBERS such that the pole of $\Gamma(b_j - \beta_j s)$ for $j = 1, 2, \ldots, m$ coincides with any POLE of $\Gamma(1 - a_j + \alpha_j s)$ for $j = 1, 2, \ldots, n$. In addition C, is a CONTOUR in the complex s-plane from $\omega - i\infty$ to $\omega + i\infty$ such that $(b_j + k)/\beta_j$ and $(a_j - 1 - k)/\alpha_j$ lie to the right and left of C, respectively.

see also MACROBERT'S E-FUNCTION, MEIJER'S G-FUNCTION

References

Carter, B. D. and Springer, M. D. "The Distribution of Products, Quotients, and Powers of Independent H-Functions." *SIAM J. Appl. Math.* **33**, 542–558, 1977.

Fox, C. "The G and H-Functions as Symmetrical Fourier Kernels." *Trans. Amer. Math. Soc.* **98**, 395–429, 1961.

Frac

see FRACTIONAL PART

Fractal

An object or quantity which displays SELF-SIMILARITY, in a somewhat technical sense, on all scales. The object need not exhibit *exactly* the same structure at all scales, but the same "type" of structures must appear on all scales. A plot of the quantity on a log-log graph versus scale then gives a straight line, whose slope is said to be the FRACTAL DIMENSION. The prototypical example for a fractal is the length of a coastline measured with different length RULERS. The shorter the RULER, the longer the length measured, a PARADOX known as the COASTLINE PARADOX.

see also BACKTRACKING, BARNSLEY'S FERN, BOX FRACTAL, BUTTERFLY FRACTAL, CACTUS FRACTAL, CANTOR SET, CANTOR SQUARE FRACTAL, CAROTID-KUNDALINI FRACTAL, CESÀRO FRACTAL, CHAOS GAME, CIRCLES-AND-SQUARES FRACTAL, COASTLINE PARADOX, DRAGON CURVE, FAT FRACTAL, FATOU SET, FLOWSNAKE FRACTAL, FRACTAL DIMENSION, H-FRACTAL, HÉNON MAP, ITERATED FUNCTION SYSTEM, JULIA FRACTAL, KAPLAN-YORKE MAP, KOCH ANTISNOWFLAKE, KOCH SNOWFLAKE, LÉVY FRACTAL, LÉVY TAPESTRY, LINDENMAYER SYSTEM, MANDELBROT SET, MANDELBROT TREE, MENGER SPONGE, MINKOWSKI SAUSAGE, MIRA FRACTAL, NEWTON'S METHOD, PENTAFLAKE, PYTHAGORAS TREE, RABINOVICH-FABRIKANT EQUATION, SAN MARCO FRACTAL, SIERPIŃSKI CARPET, SIERPIŃSKI CURVE, SIERPIŃSKI SIEVE, STAR FRACTAL, ZASLAVSKII MAP

References
Barnsley, M. F. and Rising, H. *Fractals Everywhere, 2nd ed.* Boston, MA: Academic Press, 1993.
Bogomolny, A. "Fractal Curves and Dimension." http://www.cut-the-knot.com/do_you_know/dimension.html.
Brandt, C.; Graf, S.; and Zähle, M. (Eds.). *Fractal Geometry and Stochastics.* Boston, MA: Birkhäuser, 1995.
Bunde, A. and Havlin, S. (Eds.). *Fractals and Disordered Systems, 2nd ed.* New York: Springer-Verlag, 1996.
Bunde, A. and Havlin, S. (Eds.). *Fractals in Science.* New York: Springer-Verlag, 1994.
Devaney, R. L. *Complex Dynamical Systems: The Mathematics Behind the Mandelbrot and Julia Sets.* Providence, RI: Amer. Math. Soc., 1994.
Devaney, R. L. and Keen, L. *Chaos and Fractals: The Mathematics Behind the Computer Graphics.* Providence, RI: Amer. Math. Soc., 1989.
Edgar, G. A. *Classics on Fractals.* Reading, MA: Addison-Wesley, 1994.
Eppstein, D. "Fractals." http://www.ics.uci.edu/~eppstein/junkyard/fractal.html.
Falconer, K. J. *The Geometry of Fractal Sets, 1st pbk. ed., with corr.* Cambridge, England Cambridge University Press, 1986.
Feder, J. *Fractals.* New York: Plenum Press, 1988.
Giffin, N. "The Spanky Fractal Database." http://spanky.triumf.ca/www/welcome1.html.
Hastings, H. M. and Sugihara, G. *Fractals: A User's Guide for the Natural Sciences.* New York: Oxford University Press, 1994.
Kaye, B. H. *A Random Walk Through Fractal Dimensions, 2nd ed.* New York: Wiley, 1994.
Lauwerier, H. A. *Fractals: Endlessly Repeated Geometrical Figures.* Princeton, NJ: Princeton University Press, 1991.
Mandelbrot, B. B. *Fractals: Form, Chance, & Dimension.* San Francisco, CA: W. H. Freeman, 1977.
Mandelbrot, B. B. *The Fractal Geometry of Nature.* New York: W. H. Freeman, 1983.
Massopust, P. R. *Fractal Functions, Fractal Surfaces, and Wavelets.* San Diego, CA: Academic Press, 1994.
Pappas, T. "Fractals—Real or Imaginary." *The Joy of Mathematics.* San Carlos, CA: Wide World Publ./Tetra, pp. 78–79, 1989.
Peitgen, H.-O.; Jürgens, H.; and Saupe, D. *Chaos and Fractals: New Frontiers of Science.* New York: Springer-Verlag, 1992.
Peitgen, H.-O. and Richter, D. H. *The Beauty of Fractals: Images of Complex Dynamical Systems.* New York: Springer-Verlag, 1986.
Peitgen, H.-O. and Saupe, D. (Eds.). *The Science of Fractal Images.* New York: Springer-Verlag, 1988.
Pickover, C. A. (Ed.). *The Pattern Book: Fractals, Art, and Nature.* World Scientific, 1995.
Pickover, C. A. (Ed.). *Fractal Horizons: The Future Use of Fractals.* New York: St. Martin's Press, 1996.
Rietman, E. *Exploring the Geometry of Nature: Computer Modeling of Chaos, Fractals, Cellular Automata, and Neural Networks.* New York: McGraw-Hill, 1989.
Russ, J. C. *Fractal Surfaces.* New York: Plenum, 1994.
Schroeder, M. *Fractals, Chaos, Power Law: Minutes from an Infinite Paradise.* New York: W. H. Freeman, 1991.
Sprott, J. C. "Sprott's Fractal Gallery." http://sprott.physics.wisc.edu/fractals.htm.
Stauffer, D. and Stanley, H. E. *From Newton to Mandelbrot, 2nd ed.* New York: Springer-Verlag, 1995.
Stevens, R. T. *Fractal Programming in C.* New York: Henry Holt, 1989.
Takayasu, H. *Fractals in the Physical Sciences.* Manchester, England: Manchester University Press, 1990.
Taylor, M. C. "sci.fractals FAQ." http://www.mta.ca/~mctaylor/sci.fractals-faq.
Tricot, C. *Curves and Fractal Dimension.* New York: Springer-Verlag, 1995.
Triumf Mac Fractal Programs. http://spanky.triumf.ca/pub/fractals/programs/MAC/.
Vicsek, T. *Fractal Growth Phenomena, 2nd ed.* Singapore: World Scientific, 1992.
Weisstein, E. W. "Fractals." http://www.astro.virginia.edu/~eww6n/math/notebooks/Fractal.m.
Yamaguti, M.; Hata, M.; and Kigami, J. *Mathematics of Fractals.* Providence, RI: Amer. Math. Soc., 1997.

Fractal Dimension

The term "fractal dimension" is sometimes used to refer to what is more commonly called the CAPACITY DIMENSION (which is, roughly speaking, the exponent D in the expression $n(\epsilon) = \epsilon^{-D}$, where $n(\epsilon)$ is the minimum number of OPEN SETS of diameter ϵ needed to cover the set). However, it can more generally refer to any of the dimensions commonly used to characterize fractals (e.g., CAPACITY DIMENSION, CORRELATION DIMENSION, INFORMATION DIMENSION, LYAPUNOV DIMENSION, MINKOWSKI-BOULIGAND DIMENSION).

see also BOX COUNTING DIMENSION, CAPACITY DIMENSION, CORRELATION DIMENSION, FRACTAL DIMENSION, HAUSDORFF DIMENSION, INFORMATION DIMENSION, LYAPUNOV DIMENSION, MINKOWSKI-BOULIGAND DIMENSION, POINTWISE DIMENSION, q-DIMENSION

References
Rasband, S. N. "Fractal Dimension." Ch. 4 in *Chaotic Dynamics of Nonlinear Systems.* New York: Wiley, pp. 71–83, 1990.

Fractal Land

see CAROTID-KUNDALINI FRACTAL

Fractal Process

A 1-D MAP whose increments are distributed according to a NORMAL DISTRIBUTION. Let $y(t-\Delta t)$ and $y(t+\Delta t)$ be values, then their correlation is given by the BROWN FUNCTION

$$r = 2^{2H-1} - 1.$$

When $H = 1/2$, $r = 0$ and the fractal process corresponds to 1-D Brownian motion. If $H > 1/2$, then $r > 0$ and the process is called a PERSISTENT PROCESS. If $H < 1/2$, then $r < 0$ and the process is called an ANTIPERSISTENT PROCESS.

see also ANTIPERSISTENT PROCESS, PERSISTENT PROCESS

References
von Seggern, D. *CRC Standard Curves and Surfaces.* Boca Raton, FL: CRC Press, 1993.

Fractal Sequence

Given an INFINITIVE SEQUENCE $\{x_n\}$ with associated array $a(i,j)$, then $\{x_n\}$ is said to be a fractal sequence

1. If $i+1 = x_n$, then there exists $m < n$ such that $i = x_m$,
2. If $h < i$, then, for every j, there is exactly one k such that $a(i,j) < a(h,k) < a(i,j+1)$.

(As i and j range through N, the array $A = a(i,j)$, called the associative array of x, ranges through all of N.) An example of a fractal sequence is 1, 1, 1, 1, 2, 1, 2, 1, 3, 2, 1, 3, 2, 1, 3, ...

If $\{x_n\}$ is a fractal sequence, then the associated array is an INTERSPERSION. If x is a fractal sequence, then the UPPER-TRIMMED SUBSEQUENCE is given by $\lambda(x) = x$, and the LOWER-TRIMMED SUBSEQUENCE $V(x)$ is another fractal sequence. The SIGNATURE of an IRRATIONAL NUMBER is a fractal sequence.

see also INFINITIVE SEQUENCE

References

Kimberling, C. "Fractal Sequences and Interspersions." *Ars Combin.* **45**, 157–168, 1997.

Fractal Valley

see CAROTID-KUNDALINI FUNCTION

Fraction

A RATIONAL NUMBER expressed in the form a/b, where a is called the NUMERATOR and b is called the DENOMINATOR. A PROPER FRACTION is a fraction such that $a/b < 1$, and a LOWEST TERMS FRACTION is a fraction with common terms canceled out of the NUMERATOR and DENOMINATOR.

The Egyptians expressed their fractions as sums (and differences) of UNIT FRACTIONS. Conway and Guy (1999) give a table of Roman NOTATION for fractions, in which multiples of 1/12 (the UNCIA) were given separate names.

see also ADJACENT FRACTION, ANOMALOUS CANCELLATION, CONTINUED FRACTION, DENOMINATOR, EGYPTIAN FRACTION, FAREY SEQUENCE, GOLDEN RULE, HALF, LOWEST TERMS FRACTION, MEDIANT, NUMERATOR, PROPER FRACTION, PYTHAGOREAN FRACTION, QUARTER, RATIONAL NUMBER, UNIT FRACTION

References

Conway, J. H. and Guy, R. K. *The Book of Numbers.* New York: Springer-Verlag, pp. 22–23, 1996.

Courant, R. and Robbins, H. "Decimal Fractions. Infinite Decimals." §2.2.2 in *What is Mathematics?: An Elementary Approach to Ideas and Methods, 2nd ed.* Oxford, England: Oxford University Press, pp. 61–63, 1996.

Fractional Calculus

Denote the nth DERIVATIVE D^n and the n-fold INTEGRAL D^{-n}. Then

$$D^{-1}f(t) = \int_0^t f(\xi)\,d\xi. \tag{1}$$

Now, if

$$D^{-n}f(t) = \frac{1}{(n-1)!}\int_0^t (t-\xi)^{n-1} f(\xi)\,d\xi \tag{2}$$

is true for n, then

$$\begin{aligned} D^{-(n+1)}f(t) &= D^{-1}\left[\frac{1}{(n-1)!}\int_0^t (t-\xi)^{n-1} f(\xi)\,d\xi\right] \\ &= \int_0^t \left[\frac{1}{(n-1)!}\int_0^x (x-\xi)^{n-1} f(\xi)\,d\xi\right] dx. \end{aligned} \tag{3}$$

Interchanging the order of integration gives

$$D^{-(n+1)}f(t) = \frac{1}{n!}\int_0^t (t-\xi)^n f(\xi)\,d\xi. \tag{4}$$

But (2) is true for $n = 1$, so it is also true for all n by INDUCTION. The fractional integral of $f(t)$ can then be defined by

$$D^{-\nu}f(t) = \frac{1}{\Gamma(\nu)}\int_0^t (t-\xi)^{\nu-1} f(\xi)\,d\xi, \tag{5}$$

where $\Gamma(\nu)$ is the GAMMA FUNCTION.

The fractional integral can only be given in terms of elementary functions for a small number of functions. For example,

$$D^{-\nu}t^{-\lambda} = \frac{\Gamma(\lambda+1)}{\Gamma(\lambda+\nu+1)}t^{\lambda+\nu} \quad \text{for } \lambda > -1, \nu > 0 \tag{6}$$

$$D^{-\nu}e^{at} = \frac{1}{\Gamma(\nu)}e^{at}\int_0^t x^{\nu-1}e^{-ax}\,dx \equiv E_t(\nu, a), \tag{7}$$

where $E_t(\nu, a)$ is the E_t-FUNCTION. The fractional derivative of f (if it exists) can be defined by

$$D^\mu f(t) = D^m[D^{-(m-\mu)}f(t)]. \tag{8}$$

An example is

$$\begin{aligned} D^\mu t^\lambda &= \frac{\Gamma(\lambda+1)}{\Gamma(\lambda+m-\mu+1)} \\ &= \frac{\Gamma(\lambda+1)}{\Gamma(\lambda-\mu+1)}t^{\lambda-\mu} \quad \text{for } \lambda > -1, \mu > 0 \end{aligned} \tag{9}$$

$$D^\rho E_t(\nu, a) = E_t(\nu-\rho, a) \qquad \text{for } \nu > 0, \rho \neq 0. \tag{10}$$

It is always true that, for $\mu, \nu > 0$,

$$D^{-\mu} D^{-\nu} f(t) = D^{-(\mu+\nu)}, \tag{11}$$

but *not* always true that

$$D^{\mu} D^{\nu} = D^{\mu+\nu}. \tag{12}$$

see also DERIVATIVE, INTEGRAL

References

Love, E. R. "Fractional Derivatives of Imaginary Order." *J. London Math. Soc.* **3**, 241–259, 1971.
McBride, A. C. *Fractional Calculus.* New York: Halsted Press, 1986.
Miller, K. S. "Derivatives of Noninteger Order." *Math. Mag.* **68**, 183–192, 1995.
Nishimoto, K. *Fractional Calculus.* New Haven, CT: University of New Haven Press, 1989.
Spanier, J. and Oldhan, K. B. *The Fractional Calculus.* New York: Academic Press, 1974.

Fractional Derivative

see FRACTIONAL CALCULUS

Fractional Differential Equation

The solution to the differential equation

$$[D^{2v} + aD^{v} + bD^{0}]y(t) = 0$$

is

$$y(t) = \begin{cases} e_\alpha(t) - e_\beta(t) \\ \quad \text{for } \alpha \neq \beta \\ te^{\alpha t}, \sum_{k=-(q-1)}^{q-1} \alpha^k (q - |k|) D^{1-(k+1)v} (te^{\alpha^q t}) \\ \quad \text{for } \alpha = \beta \neq 0 \\ \frac{t^{2\nu-1}}{\Gamma(2v)} \\ \quad \text{for } \alpha = \beta = 0, \end{cases}$$

where

$$q = \frac{1}{v}$$

$$e_\beta(t) = \sum_{k=0}^{q-1} \beta^{q-k-1} E_t(-kv, \beta^q),$$

$E_t(a, x)$ is the E_t-FUNCTION, and $\Gamma(n)$ is the GAMMA FUNCTION.

References

Miller, K. S. "Derivatives of Noninteger Order." *Math. Mag.* **68**, 183–192, 1995.

Fractional Fourier Transform

A z-TRANSFORM with

$$z \equiv e^{2\pi i \alpha / N}$$

for $\alpha \neq \pm 1$. This transform can be used to detect frequencies which are not INTEGER multiples of the lowest DISCRETE FOURIER TRANSFORM frequency.

see also z-TRANSFORM

References

Graham, R. L.; Knuth, D. E.; and Patashnik, O. *Concrete Mathematics, 2nd ed.* Reading, MA: Addison-Wesley, 1994.

Fractional Integral

see FRACTIONAL CALCULUS

Fractional Part

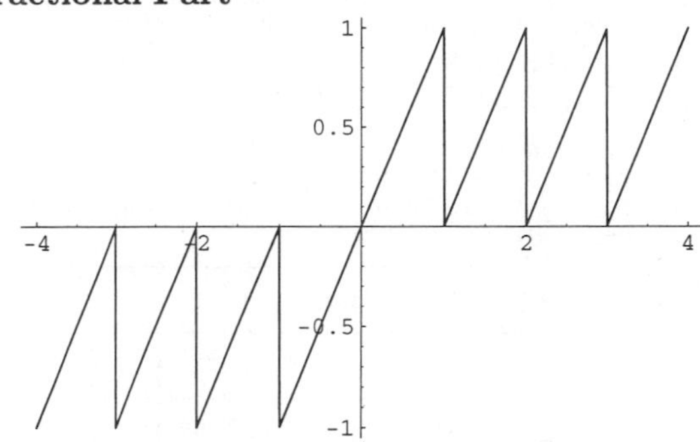

The function giving the fractional (nonintegral) part of a number and defined as

$$\operatorname{frac}(x) \equiv \begin{cases} x - \lfloor x \rfloor & x \geq 0 \\ x - \lfloor x \rfloor - 1 & x \leq 0, \end{cases}$$

where $\lfloor x \rfloor$ is the FLOOR FUNCTION.

see also CEILING FUNCTION, FLOOR FUNCTION, NINT, ROUND, TRUNCATE, WHOLE NUMBER

References

Spanier, J. and Oldham, K. B. "The Integer-Value Int(x) and Fractional-Value frac(x) Functions." Ch. 9 in *An Atlas of Functions.* Washington, DC: Hemisphere, pp. 71–78, 1987.

Fractran

Fractran is an algorithm applied to a given list f_1, f_2, ..., f_k of FRACTIONS. Given a starting INTEGER N, the Fractran algorithm proceeds by repeatedly multiplying the integer at a given stage by the first element f_i given an integer PRODUCT. The algorithm terminates when there is no such f_i.

The list

$$\frac{17}{91}, \frac{78}{85}, \frac{19}{51}, \frac{23}{38}, \frac{29}{33}, \frac{77}{29}, \frac{95}{23}, \frac{77}{19}, \frac{1}{17}, \frac{11}{13}, \frac{13}{11}, \frac{15}{2}, \frac{1}{7}, \frac{55}{1}$$

with starting integer $N = 2$ generates a sequence 2, 15, 825, 725, 1925, 2275, 425, 390, 330, 290, 770, ... Conway (1987) showed that the only other powers of 2 which occur are those with PRIME exponent: 2^2, 2^3, 2^5, 2^7, ...

References

Conway, J. H. "Unpredictable Iterations." In *Proc. Number Theory Conf., Boulder, CO,* pp. 49–52, 1972.
Conway, J. H. "Fractran: A Simple Universal Programming Language for Arithmetic." Ch. 2 in *Open Problems in Communication and Computation* (Ed. T. M. Cover and B. Gopinath). New York: Springer-Verlag, pp. 4–26, 1987.

Framework

Consider a finite collection of points $p = (p_1, \ldots, p_n)$, $p_i \in \mathbb{R}^d$ EUCLIDEAN SPACE (known as a CONFIGURATION) and a graph G whose VERTICES correspond to pairs of points that are constrained to stay the same distance apart. Then the graph G together with the configuration p, denoted $G(p)$, is called a framework.

see also BAR (EDGE), CONFIGURATION, RIGID

Franklin Magic Square

52	61	4	13	20	29	36	45
14	3	62	51	46	35	30	19
53	60	5	12	21	28	37	44
11	6	59	54	43	38	27	22
55	58	7	10	23	26	39	42
9	8	57	56	41	40	25	24
50	63	2	15	18	31	34	47
16	1	64	49	48	33	32	17

Benjamin Franklin constructed the above 8×8 PANMAGIC SQUARE having MAGIC CONSTANT 260. Any half-row or half-column in this square totals 130, and the four corners plus the middle total 260. In addition, bent diagonals (such as 52-3-5-54-10-57-63-16) also total 260 (Madachy 1979, p. 87).

see also MAGIC SQUARE, PANMAGIC SQUARE

References

Madachy, J. S. "Magic and Antimagic Squares." Ch. 4 in *Madachy's Mathematical Recreations.* New York: Dover, pp. 103–113, 1979.

Pappas, T. "The Magic Square of Benjamin Franklin." *The Joy of Mathematics.* San Carlos, CA: Wide World Publ./Tetra, p. 97, 1989.

Fransén-Robinson Constant

$$F \equiv \int_0^\infty \frac{dx}{\Gamma(x)} = 2.8077702420\ldots,$$

where $\Gamma(x)$ is the GAMMA FUNCTION. The above plots show the functions $\Gamma(x)$ and $1/\Gamma(x)$.

see also GAMMA FUNCTION

References

Finch, S. "Favorite Mathematical Constants." `http://www.mathsoft.com/asolve/constant/fran/fran.html`.

Fransén, A. "Accurate Determination of the Inverse Gamma Integral." *BIT* **19**, 137–138, 1979.

Fransén, A. "Addendum and Corrigendum to 'High-Precision Values of the Gamma Function and of Some Related Coefficients.'" *Math. Comput.* **37**, 233–235, 1981.

Fransén, A. and Wrigge, S. "High-Precision Values of the Gamma Function and of Some Related Coefficients." *Math. Comput.* **34**, 553–566, 1980.

Plouffe, S. "Fransen-Robinson Constant." `http://lacim.uqam.ca/piDATA/fransen.txt`.

Fréchet Bounds

Any bivariate distribution function with marginal distribution functions F and G satisfies

$$\max\{F(x)+G(y)-1,0\} \leq H(x,y) \leq \min\{F(x),G(y)\}.$$

Fréchet Derivative

A function f is Fréchet differentiable at a if

$$\lim_{x\to a} \frac{f(x)-f(a)}{x-a}$$

exists. This is equivalent to the statement that ϕ has a removable DISCONTINUITY at a, where

$$\phi(x) \equiv \frac{f(x)-f(a)}{x-a}.$$

Every function which is Fréchet differentiable is also Carathéodory differentiable.

see also CARATHÉODORY DERIVATIVE, DERIVATIVE

Fréchet Space

A complete metrizable SPACE, sometimes also with the restriction that the space be locally convex.

Fredholm Integral Equation of the First Kind

An INTEGRAL EQUATION of the form

$$f(x) = \int_{-\infty}^\infty K(x,t)\phi(t)\,dt$$

$$\phi(x) = \frac{1}{2\pi}\int_{-\infty}^\infty \frac{F(\omega)}{K(\omega)} e^{-i\omega x}\,d\omega.$$

see also FREDHOLM INTEGRAL EQUATION OF THE SECOND KIND, INTEGRAL EQUATION, VOLTERRA INTEGRAL EQUATION OF THE FIRST KIND, VOLTERRA INTEGRAL EQUATION OF THE SECOND KIND

References

Arfken, G. *Mathematical Methods for Physicists, 3rd ed.* Orlando, FL: Academic Press, p. 865, 1985.

Fredholm Integral Equation of the Second Kind

An INTEGRAL EQUATION of the form

$$\phi(x) = f(x) + \lambda \int_{-\infty}^\infty K(x,t)\phi(t)\,dt$$

$$\phi(x) = \frac{1}{\sqrt{2\pi}} \int_{-\infty}^\infty \frac{F(t)e^{-ixt}\,dt}{1-\sqrt{2\pi}\,\lambda K(t)}.$$

see also FREDHOLM INTEGRAL EQUATION OF THE FIRST KIND, INTEGRAL EQUATION, NEUMANN SERIES (INTEGRAL EQUATION), VOLTERRA INTEGRAL

EQUATION OF THE FIRST KIND, VOLTERRA INTEGRAL EQUATION OF THE SECOND KIND

References
Arfken, G. *Mathematical Methods for Physicists, 3rd ed.* Orlando, FL: Academic Press, p. 865, 1985.
Press, W. H.; Flannery, B. P.; Teukolsky, S. A.; and Vetterling, W. T. "Fredholm Equations of the Second Kind." §18.1 in *Numerical Recipes in FORTRAN: The Art of Scientific Computing, 2nd ed.* Cambridge, England: Cambridge University Press, pp. 782–785, 1992.

Free

When referring to a planar object, "free" means that the object is regarded as capable of being picked up out of the plane and flipped over. As a result, MIRROR IMAGES are equivalent for free objects.

A free abstract mathematical object is generated by n elements in a "free manner," i.e., such that the n elements satisfy no nontrivial relations among themselves. To make this more formal, an algebraic GADGET X is freely generated by a SUBSET G if, for any function $f : G \to Y$ where Y is any other algebraic GADGET, there exists a unique HOMOMORPHISM (which has different meanings depending on what kind of GADGETS you're dealing with) $g : X \to Y$ such that g restricted to G is f.

If the algebraic GADGETS are VECTOR SPACES, then G freely generates X IFF G is a BASIS for X. If the algebraic GADGETS are ABELIAN GROUPS, then G freely generates X IFF X is a DIRECT SUM of the INTEGERS, with G consisting of the standard BASIS.

see also FIXED, GADGET, MIRROR IMAGE, RANK

Free Group

The generators of a group G are defined to be the smallest subset of group elements such that all other elements of G can be obtained from them and their inverses. A GROUP is a free group if no relation exists between its generators (other than the relationship between an element and its inverse required as one of the defining properties of a group). For example, the additive group of whole numbers is free with a single generator, 1.

see also FREE SEMIGROUP

Free Semigroup

A SEMIGROUP with a noncommutative product in which no PRODUCT can ever be expressed more simply in terms of other ELEMENTS.

see also FREE GROUP, SEMIGROUP

Free Variable

An occurrence of a variable in a LOGIC FORMULA which is not inside the scope of a QUANTIFIER.

see also BOUND, SENTENCE

Freemish Crate

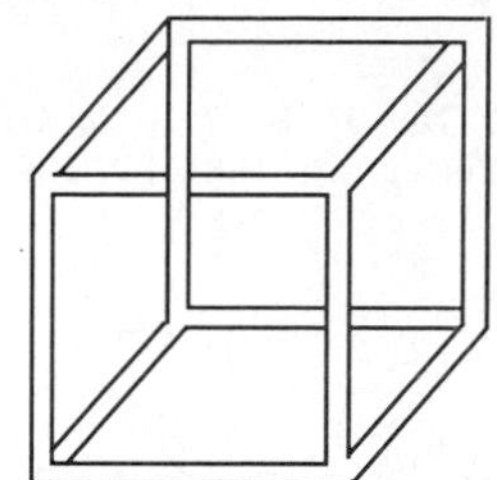

An IMPOSSIBLE FIGURE box which can be drawn but not built.

References
Fineman, M. *The Nature of Visual Illusion.* New York: Dover, p. 120–122, 1996.
Jablan, S. "Are Impossible Figures Possible?" http://members.tripod.com/~modularity/kulpa.htm.
Pappas, T. "The Impossible Tribar." *The Joy of Mathematics.* San Carlos, CA: Wide World Publ./Tetra, p. 13, 1989.

Freeth's Nephroid

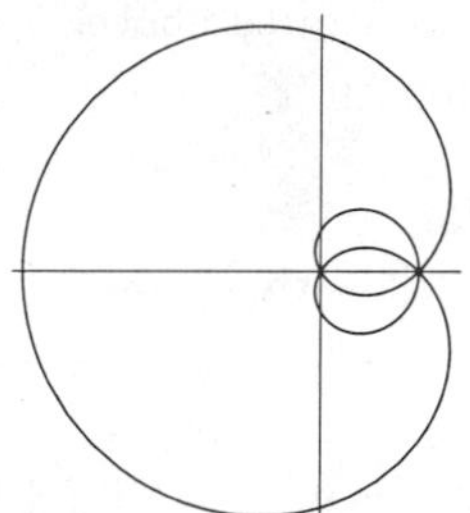

A STROPHOID of a CIRCLE with the POLE O at the CENTER of the CIRCLE and the fixed point P on the CIRCUMFERENCE of the CIRCLE. In a paper published by the London Mathematical Society in 1879, T. J. Freeth described it and various other STROPHOIDS (MacTutor Archive). If the line through P PARALLEL to the y-AXIS cuts the NEPHROID at A, then ANGLE AOP is $3\pi/7$, so this curve can be used to construct a regular HEPTAGON. The POLAR equation is

$$r = a[1 + 2\sin(\tfrac{1}{2}\theta)].$$

see also STROPHOID

References
Lawrence, J. D. *A Catalog of Special Plane Curves.* New York: Dover, pp. 175 and 177–178, 1972.
MacTutor History of Mathematics Archive. "Freeth's Nephroid." http://www-groups.dcs.st-and.ac.uk/~history/Curves/Freeths.html.

Freiman's Constant

The end of the last gap in the LAGRANGE SPECTRUM, given by

$$F \equiv \frac{2221564096 + 293748\sqrt{462}}{491993569} = 4.5278295661\ldots.$$

REAL NUMBERS greater than F are members of the MARKOV SPECTRUM.

see also LAGRANGE SPECTRUM, MARKOV SPECTRUM

References

Conway, J. H. and Guy, R. K. *The Book of Numbers.* New York: Springer-Verlag, pp. 188–189, 1996.

French Curve

French curves are plastic (or wooden) templates having an edge composed of several different curves. French curves are used in drafting (or were before computer-aided design) to draw smooth curves of almost any desired curvature in mechanical drawings. Several typical French curves are illustrated above.

see also CORNU SPIRAL

Frenet Formulas

Also known as the SERRET-FRENET FORMULAS

$$\begin{bmatrix} \dot{\mathbf{T}} \\ \dot{\mathbf{N}} \\ \dot{\mathbf{B}} \end{bmatrix} = \begin{bmatrix} 0 & \kappa & 0 \\ -\kappa & 0 & \tau \\ 0 & -\tau & 0 \end{bmatrix} \begin{bmatrix} \mathbf{T} \\ \mathbf{N} \\ \mathbf{B} \end{bmatrix},$$

where $\mathbf{T}$ is the unit TANGENT VECTOR, $\mathbf{N}$ is the unit NORMAL VECTOR, $\mathbf{B}$ is the unit BINORMAL VECTOR, τ is the TORSION, κ is the CURVATURE, and $\dot{\mathbf{x}}$ denotes $d\mathbf{x}/ds$.

see also CENTRODE, FUNDAMENTAL THEOREM OF SPACE CURVES, NATURAL EQUATION

References

Frenet, F. "Sur les courbes à double courbure." Thèse. Toulouse, 1847. Abstract in *J. de Math.* **17**, 1852.

Gray, A. *Modern Differential Geometry of Curves and Surfaces.* Boca Raton, FL: CRC Press, p. 126, 1993.

Kreyszig, E. "Formulae of Frenet." §15 in *Differential Geometry.* New York: Dover, p. 40–43, 1991.

Serret, J. A. "Sur quelques formules relatives à la théorie des courbes à double courbure." *J. de Math.* **16**, 1851.

Frequency Curve

see GAUSSIAN FUNCTION

Fresnel's Elasticity Surface

A QUARTIC SURFACE given by

$$r = \sqrt{a^2x^2 + b^2y^2 + c^2z^2},$$

where

$$r^2 \equiv x'^2 + y'^2 + z'^2,$$

also known as FRESNEL'S WAVE SURFACE. It was introduced by Fresnel in his studies of crystal optics.

References

Fischer, G. (Ed.). *Mathematical Models from the Collections of Universities and Museums.* Braunschweig, Germany: Vieweg, p. 16, 1986.

Fischer, G. (Ed.). Plates 38–39 in *Mathematische Modelle/Mathematical Models, Bildband/Photograph Volume.* Braunschweig, Germany: Vieweg, pp. 38–39, 1986.

von Seggern, D. *CRC Standard Curves and Surfaces.* Boca Raton, FL: CRC Press, p. 304, 1993.

Fresnel Integrals

In physics, the Fresnel integrals are most often defined by

$$C(u) + iS(u) \equiv \int_0^u e^{i\pi x^2/2}\,dx = \int_0^u \cos(\tfrac{1}{2}\pi x^2)\,dx + i\int_0^u \sin(\tfrac{1}{2}\pi x^2)\,dx, \quad (1)$$

so

$$C(u) \equiv \int_0^u \cos(\tfrac{1}{2}\pi x^2)\,dx \quad (2)$$

$$S(u) \equiv \int_0^u \sin(\tfrac{1}{2}\pi x^2)\,dx. \quad (3)$$

They satisfy

$$C(\pm\infty) = -\tfrac{1}{2} \quad (4)$$

$$S(\pm\infty) = \tfrac{1}{2}. \quad (5)$$

Related functions are defined as

$$C_1(z) \equiv \sqrt{\frac{2}{\pi}}\int_0^x \cos t^2\,dt \quad (6)$$

$$S_1(z) \equiv \sqrt{\frac{2}{\pi}}\int_0^x \sin t^2\,dt \quad (7)$$

$$C_2(z) \equiv \frac{1}{\sqrt{2\pi}}\int \frac{\cos t}{\sqrt{t}}\,dt \quad (8)$$

$$S_2(z) \equiv \frac{1}{\sqrt{2\pi}}\int \frac{\sin t}{\sqrt{t}}\,dt. \quad (9)$$

An asymptotic expansion for $x \gg 1$ gives

$$C(u) \approx \frac{1}{2} + \frac{1}{\pi u}\sin(\tfrac{1}{2}\pi u^2) \quad (10)$$

$$S(u) \approx \frac{1}{2} - \frac{1}{\pi u}\cos(\tfrac{1}{2}\pi u^2). \quad (11)$$

Therefore, as $u \to \infty$, $C(u) = 1/2$ and $S(u) = 1/2$. The Fresnel integrals are sometimes alternatively defined as

$$x(t) = \int_0^t \cos(v^2)\,dv \quad (12)$$

$$y(t) = \int_0^t \sin(v^2)\,dv. \quad (13)$$

Letting $x \equiv v^2$ so $dx = 2v\,dv = 2\sqrt{x}\,dv$, and $dv = x^{-1/2}\,dx/2$

$$x(t) = \tfrac{1}{2}\int_0^{\sqrt{t}} x^{-1/2}\cos x\,dx \quad (14)$$

$$y(t) = \tfrac{1}{2}\int_0^{\sqrt{t}} x^{-1/2}\sin x\,dx. \quad (15)$$

In this form, they have a particularly simple expansion in terms of SPHERICAL BESSEL FUNCTIONS OF THE FIRST KIND. Using

$$j_0(x) = \frac{\sin x}{x} \quad (16)$$

$$n_1(x) = -j_{-1}(x) = -\frac{\cos x}{x}, \quad (17)$$

where $n_1(x)$ is a SPHERICAL BESSEL FUNCTION OF THE SECOND KIND

$$x(t^2) = -\tfrac{1}{2}\int_0^t n_1(x)x^{1/2}\,dx$$

$$= \tfrac{1}{2}\int_0^t j_{-1}(x)x^{1/2}\,dx = x^{1/2}\sum_{n=0}^{\infty} j_{2n}(x) \quad (18)$$

$$y(t^2) = \tfrac{1}{2}\int_0^t j_0(x)x^{1/2}\,dx$$

$$= x^{1/2}\sum_{n=0}^{\infty} j_{2n+1}(x). \quad (19)$$

see also CORNU SPIRAL

References

Abramowitz, M. and Stegun, C. A. (Eds.). "Fresnel Integrals." §7.3 in *Handbook of Mathematical Functions with Formulas, Graphs, and Mathematical Tables, 9th printing.* New York: Dover, pp. 300–302, 1972.

Leonard, I. E. "More on Fresnel Integrals." *Amer. Math. Monthly* **95**, 431–433, 1988.

Press, W. H.; Flannery, B. P.; Teukolsky, S. A.; and Vetterling, W. T. "Fresnel Integrals, Cosine and Sine Integrals." §6.79 in *Numerical Recipes in FORTRAN: The Art of Scientific Computing, 2nd ed.* Cambridge, England: Cambridge University Press, pp. 248–252, 1992.

Spanier, J. and Oldham, K. B. "The Fresnel Integrals $S(x)$ and $C(x)$." Ch. 39 in *An Atlas of Functions.* Washington, DC: Hemisphere, pp. 373–383, 1987.

Fresnel's Wave Surface

see FRESNEL'S ELASTICITY SURFACE

Frey Curve

Let $a^p + b^p = c^p$ be a solution to FERMAT'S LAST THEOREM. Then the corresponding Frey curve is

$$y^2 = x(x - a^p)(x + b^p). \quad (1)$$

Frey showed that such curves cannot be MODULAR, so if the TANIYAMA-SHIMURA CONJECTURE were true, Frey curves couldn't exist and FERMAT'S LAST THEOREM would follow with b EVEN and $a \equiv -1 \pmod 4$. Frey curves are SEMISTABLE. Invariants include the DISCRIMINANT

$$(a^p - 0)^2(-b^p - 0)[a^p - (-b)^p]^2 = a^{2p}b^{2p}c^{2p}. \quad (2)$$

The MINIMAL DISCRIMINANT is

$$\Delta = 2^{-8}a^{2p}b^{2p}c^{2p}, \quad (3)$$

the CONDUCTOR is

$$N = \prod_{l|abc} l, \quad (4)$$

and the j-INVARIANT is

$$j = \frac{2^8(a^{2p} + b^{2p} + a^pb^p)^3}{a^{2p}b^{2p}c^{2p}} = \frac{2^8(c^{2p} - b^pc^p)^3}{(abc)^{2p}}. \quad (5)$$

see also ELLIPTIC CURVE, FERMAT'S LAST THEOREM, TANIYAMA-SHIMURA CONJECTURE

References

Cox, D. A. "Introduction to Fermat's Last Theorem." *Amer. Math. Monthly* **101**, 3–14, 1994.

Gouvêa, F. Q. "A Marvelous Proof." *Amer. Math. Monthly* **101**, 203–222, 1994.

Frey Elliptic Curve

see FREY CURVE

Friday the Thirteenth

The Gregorian calendar follows a pattern of leap years which repeats every 400 years. There are 4,800 months in 400 years, so the 13th of the month occurs 4,800 times in this interval. The number of times the 13th occurs on each weekday is given in the table below. As shown by Brown (1933), the thirteenth of the month is slightly more likely to be on a Friday than on any other day.

Day	Number of 13s	Fraction
Sunday	687	14.31%
Monday	685	14.27%
Tuesday	685	14.27%
Wednesday	687	14.31%
Thursday	684	14.25%
Friday	688	14.33%
Saturday	684	14.25%

see also 13, WEEKDAY

References

Ball, W. W. R. and Coxeter, H. S. M. *Mathematical Recreations and Essays, 13th ed.* New York: Dover, p. 27, 1987.

Brown, B. H. "Solution to Problem E36." *Amer. Math. Monthly* **40**, 607, 1933.

Press, W. H.; Flannery, B. P.; Teukolsky, S. A.; and Vetterling, W. T. *Numerical Recipes in FORTRAN: The Art of Scientific Computing, 2nd ed.* Cambridge, England: Cambridge University Press, pp. 14–15, 1992.

Friend

A friend of a number n is another number m such that (m, n) is a FRIENDLY PAIR.

see also FRIENDLY PAIR, SOLITARY NUMBER

References

Anderson, C. W. and Hickerson, D. Problem 6020. "Friendly Integers." *Amer. Math. Monthly* **84**, 65–66, 1977.

Friendly Giant Group

see MONSTER GROUP

Friendly Pair

Define

$$\Sigma(n) \equiv \frac{\sigma(n)}{n},$$

where $\sigma(n)$ is the DIVISOR FUNCTION. Then a PAIR of distinct numbers (k, m) is a friendly pair (and k is said to be a FRIEND of m) if

$$\Sigma(k) = \Sigma(m).$$

For example, 4320 and 4680 are a friendly pair, since $\sigma(4320) = 15120$, $\sigma(4680) = 16380$, and

$$\Sigma(4320) \equiv \tfrac{15120}{4320} = \tfrac{7}{2}$$
$$\Sigma(4680) \equiv \tfrac{16380}{4680} = \tfrac{7}{2}.$$

Numbers which do not have FRIENDS are called SOLITARY NUMBERS. SOLITARY NUMBERS satisfy $(\sigma(n), n) = 1$, where (a, b) is the GREATEST COMMON DIVISOR of a and b.

see also ALIQUOT SEQUENCE, FRIEND, SOLITARY NUMBER

References

Anderson, C. W. and Hickerson, D. Problem 6020. "Friendly Integers." *Amer. Math. Monthly* **84**, 65–66, 1977.

Frieze Pattern

$$\begin{matrix} & b & \\ a & & d \\ & c & \end{matrix}$$

An arrangement of numbers at the intersection of two sets of perpendicular diagonals such that $a+d = b+c+1$ (for an additive frieze pattern) or $ad = bc + 1$ (for a multiplicative frieze pattern) in each diamond.

References

Conway, J. H. and Coxeter, H. S. M. "Triangulated Polygons and Frieze Patterns." *Math. Gaz.* **57**, 87–94, 1973.

Conway, J. H. and Guy, R. K. In *The Book of Numbers.* New York: Springer-Verlag, pp. 74–76 and 96–97, 1996.

Frobenius-König Theorem

The PERMANENT of an $n \times n$ MATRIX with all entries either 0 or 1 is 0 IFF the MATRIX contains an $r \times s$ submatrix of 0s with $r + s = n + 1$. This result follows from the KÖNIG-EGEVÁRY THEOREM.

see also KÖNIG-EGEVÁRY THEOREM, PERMANENT

Frobenius Map

A map $x \mapsto x^p$ where p is a PRIME.

Frobenius Method

If x_0 is an ordinary point of the ORDINARY DIFFERENTIAL EQUATION, expand y in a TAYLOR SERIES about x_0, letting

$$y = \sum_{n=0}^{\infty} a_n x^n. \tag{1}$$

Plug y back into the ODE and group the COEFFICIENTS by POWER. Now, obtain a RECURRENCE RELATION for the nth term, and write the TAYLOR SERIES in terms of the a_ns. Expansions for the first few derivatives are

$$y = \sum_{n=0}^{\infty} a_n x^n \tag{2}$$

$$y' = \sum_{n=1}^{\infty} n a_n x^{n-1} = \sum_{n=0}^{\infty} (n+1) a_{n+1} x^n \tag{3}$$

$$y'' = \sum_{n=2}^{\infty} n(n-1) a_n x^{n-2} = \sum_{n=0}^{\infty} (n+2)(n+1) a_{n+2} x^n. \tag{4}$$

If x_0 is a regular singular point of the ORDINARY DIFFERENTIAL EQUATION,

$$P(x)y'' + Q(x)y' + R(x)y = 0, \tag{5}$$

solutions may be found by the Frobenius method or by expansion in a LAURENT SERIES. In the Frobenius method, assume a solution of the form

$$y = x^k \sum_{n=0}^{\infty} a_n x^n, \tag{6}$$

so that

$$y = x^k \sum_{n=0}^{\infty} a_n x^n = \sum_{n=0}^{\infty} a_n x^{n+k} \tag{7}$$

$$y' = \sum_{n=0}^{\infty} a_n (n+k) x^{k+n-1} \tag{8}$$

$$y'' = \sum_{n=0}^{\infty} a_n (n+k)(n+k-1) x^{k+n-2}. \tag{9}$$

Now, plug y back into the ODE and group the COEFFICIENTS by POWER to obtain a recursion FORMULA for the a_nth term, and then write the TAYLOR SERIES in terms of the a_ns. Equating the a_0 term to 0 will produce the so-called INDICIAL EQUATION, which will give the allowed values of k in the TAYLOR SERIES.

FUCHS'S THEOREM guarantees that at least one POWER series solution will be obtained when applying the Frobenius method if the expansion point is an ordinary, or regular, SINGULAR POINT. For a regular SINGULAR POINT, a LAURENT SERIES expansion can also be used. Expand y in a LAURENT SERIES, letting

$$y = c_{-n}x^{-n} + \ldots + c_{-1}x^{-1} + c_0 + c_1 x + \ldots + c_n x^n + \ldots. \tag{10}$$

Plug y back into the ODE and group the COEFFICIENTS by POWER. Now, obtain a recurrence FORMULA for the c_nth term, and write the TAYLOR EXPANSION in terms of the c_ns.

see also FUCHS'S THEOREM, ORDINARY DIFFERENTIAL EQUATION

References

Arfken, G. "Series Solutions—Frobenius' Method." §8.5 in *Mathematical Methods for Physicists, 3rd ed.* Orlando, FL: Academic Press, pp. 454–467, 1985.

Frobenius-Peron Equation

$$\rho_{n+1}(x) = \int \rho_n(y)\delta[x - M(y)]\,dy,$$

where $\delta(x)$ is a DELTA FUNCTION, $M(x)$ is a map, and ρ is the NATURAL DENSITY.

References

Ott, E. *Chaos in Dynamical Systems.* New York: Cambridge University Press, p. 51, 1993.

Frobenius Pseudoprime

Let $f(x)$ be a MONIC POLYNOMIAL of degree d with discriminant Δ. Then an ODD INTEGER n with $(n, f(0)\Delta) = 1$ is called a Frobenius pseudoprime with respect to $f(x)$ if it passes a certain algorithm given by Grantham (1996). A Frobenius pseudoprime with respect to a POLYNOMIAL $f(x) \in \mathbb{Z}[x]$ is then a composite Frobenius probably prime with respect to the POLYNOMIAL $x - a$.

While 323 is the first LUCAS PSEUDOPRIME with respect to the Fibonacci polynomial $x^2 - x - 1$, the first Frobenius pseudoprime is 5777. If $f(x) = x^3 - rx^2 + sx - 1$, then any Frobenius pseudoprime n with respect to $f(x)$ is also a PERRIN PSEUDOPRIME. Grantham (1997) gives a test based on Frobenius pseudoprimes which is passed by COMPOSITE NUMBERS with probability at most 1/7710.

see also PERRIN PSEUDOPRIME, PSEUDOPRIME, STRONG FROBENIUS PSEUDOPRIME

References

Grantham, J. "Frobenius Pseudoprimes." 1996. http://www.clark.net/pub/grantham/pseudo/pseudo.ps

Grantham, J. "A Frobenius Probable Prime Test with High Confidence." 1997. http://www.clark.net/pub/grantham/pseudo/pseudo2.ps

Grantham, J. "Pseudoprimes/Probable Primes." http://www.clark.net/pub/grantham/pseudo.

Frobenius Theorem

Let $\mathsf{A} = a_{ij}$ be a MATRIX with POSITIVE COEFFICIENTS so that $a_{ij} > 0$ for all $i, j = 1, 2, \ldots, n$, then A has a POSITIVE EIGENVALUE λ_0, and all its EIGENVALUES lie on the CLOSED DISK

$$|z| \leq \lambda_0.$$

see also CLOSED DISK, OSTROWSKI'S THEOREM

References

Gradshteyn, I. S. and Ryzhik, I. M. *Tables of Integrals, Series, and Products, 5th ed.* San Diego, CA: Academic Press, p. 1121, 1979.

Frobenius Triangle Identities

Let $C_{L,M}$ be a PADÉ APPROXIMANT. Then

$$C_{(L+1)/M}S_{(L-1)/M} - C_{L/(M+1)}S_{L/(M+1)} = C_{L/M}S_{L/M} \tag{1}$$

$$C_{L/(M+1)}S_{(L+1)/M} - C_{(L+1)/M}S_{L/(M+1)} = C_{(L+1)/(M+1)}xS_{L/M} \tag{2}$$

$$C_{(L+1)/M}S_{L/M} - C_{L/M}S_{(L+1)/M} = C_{(L+1)/(M+1)}xS_{L/(M-1)} \tag{3}$$

$$C_{L/(M+1)}S_{L/M} - C_{L/M}S_{L/(M+1)} = C_{(L+1)/(M+1)}xS_{(L-1)/M}, \tag{4}$$

where

$$S_{L/M} = G(x)P_L(x) + H(x)Q_M(x) \tag{5}$$

and C is the C-DETERMINANT.

see also C-DETERMINANT, PADÉ APPROXIMANT

References

Baker, G. A. Jr. *Essentials of Padé Approximants in Theoretical Physics.* New York: Academic Press, p. 31, 1975.

Frontier

see BOUNDARY

Frullani's Integral

If $f'(x)$ is continuous and the integral converges,

$$\int_0^\infty \frac{f(ax) - f(bx)}{x}\,dx = [f(0) - f(\infty)] \ln\left(\frac{b}{a}\right).$$

References

Spiegel, M. R. *Mathematical Handbook of Formulas and Tables.* New York: McGraw-Hill, 1968.

Frustum

The portion of a solid which lies between two PARALLEL PLANES cutting the solid. Degenerate cases are obtained for finite solids by cutting with a single PLANE only.

see also CONICAL FRUSTUM, PYRAMIDAL FRUSTUM, SPHERICAL SEGMENT

Fubini Principle

If the average number of envelopes per pigeonhole is a, then some pigeonhole will have at least a envelopes. Similarly, there must be a pigeonhole with at most a envelopes.

see also PIGEONHOLE PRINCIPLE

Fuchsian System

A system of linear differential equations

$$\frac{dy}{dz} = A(z)y,$$

with $A(z)$ an ANALYTIC $n \times n$ MATRIX, for which the MATRIX $A(z)$ is ANALYTIC in $\bar{\mathbb{C}}\backslash\{a_1, \ldots, a_N\}$ and has a POLE of order 1 at a_j for $j = 1, \ldots, N$. A system is Fuchsian IFF there exist $n \times n$ matrices $B_1, \ldots, B_N$ with entries in $\mathbb{Z}$ such that

$$A(z) = \sum_{j=1}^{N} \frac{B_j}{z - a_j}$$

$$\sum_{j=1}^{N} B_j = v.$$

Fuchs's Theorem

At least one POWER SERIES solution will be obtained when applying the FROBENIUS METHOD if the expansion point is an ordinary, or regular, SINGULAR POINT. The number of ROOTS is given by the ROOTS of the INDICIAL EQUATION.

References

Arfken, G. *Mathematical Methods for Physicists, 3rd ed.* Orlando, FL: Academic Press, pp. 462–463, 1985.

Fuhrmann Circle

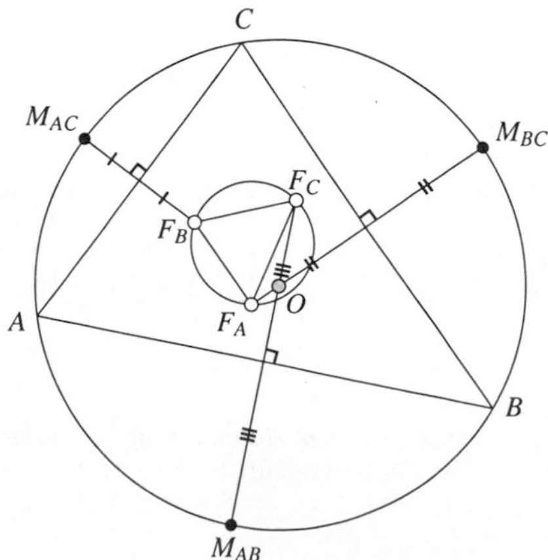

The CIRCUMCIRCLE of the FUHRMANN TRIANGLE.

see also FUHRMANN TRIANGLE, MID-ARC POINTS

References

Fuhrmann, W. *Synthetische Beweise Planimetrischer Sätze.* Berlin, p. 107, 1890.

Johnson, R. A. *Modern Geometry: An Elementary Treatise on the Geometry of the Triangle and the Circle.* Boston, MA: Houghton Mifflin, pp. 228–229, 1929.

Fuhrmann's Theorem

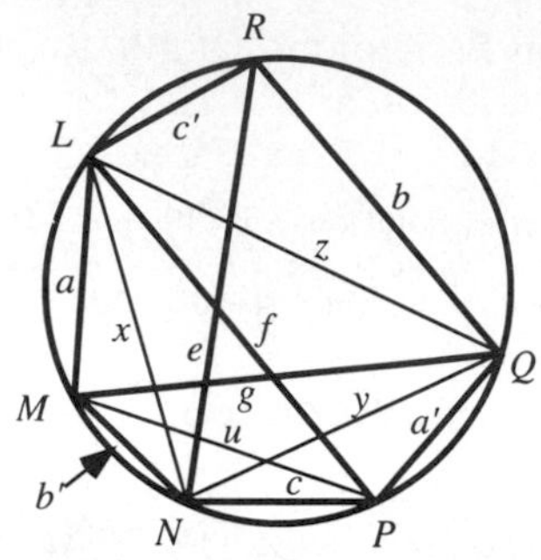

Let the opposite sides of a convex CYCLIC HEXAGON be a, a', b, b', c, and c', and let the DIAGONALS e, f, and g be so chosen that a, a', and e have no common VERTEX (and likewise for b, b', and f), then

$$efg = aa'e + bb'f + cc'g + abc + a'b'c'.$$

This is an extension of PTOLEMY'S THEOREM to the HEXAGON.

see also CYCLIC HEXAGON, HEXAGON, PTOLEMY'S THEOREM

References

Fuhrmann, W. *Synthetische Beweise Planimetrischer Sätze.* Berlin, p. 61, 1890.

Johnson, R. A. *Modern Geometry: An Elementary Treatise on the Geometry of the Triangle and the Circle.* Boston, MA: Houghton Mifflin, pp. 65–66, 1929.

Fuhrmann Triangle

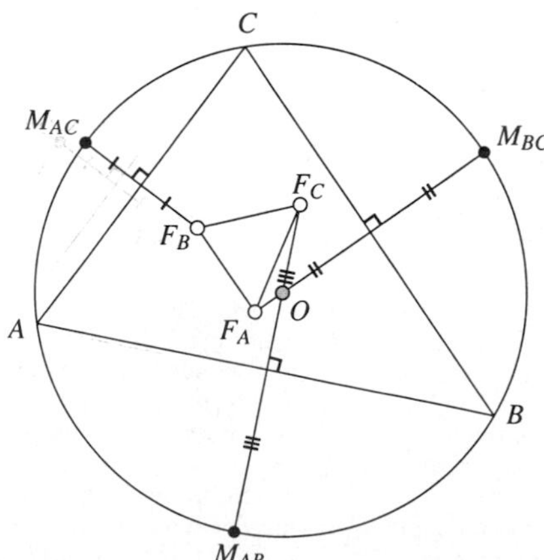

The Fuhrmann triangle of a TRIANGLE ΔABC is the TRIANGLE $\Delta F_C F_B F_A$ formed by reflecting the MID-ARC POINTS M_{AB}, M_{AC}, M_{BC} about the lines AB, AC, and BC. The CIRCUMCIRCLE of the Fuhrmann triangle is called the FUHRMANN CIRCLE, and the lines $F_A M_{BC}$, $F_B M_{AC}$, and $F_C M_{AB}$ CONCUR at the CIRCUMCENTER O.

see also FUHRMANN CIRCLE, MID-ARC POINTS

References

Fuhrmann, W. *Synthetische Beweise Planimetrischer Sätze.* Berlin, p. 107, 1890.

Johnson, R. A. *Modern Geometry: An Elementary Treatise on the Geometry of the Triangle and the Circle.* Boston, MA: Houghton Mifflin, pp. 228–229, 1929.

Full Reptend Prime

A PRIME p for which $1/p$ has a maximal period DECIMAL EXPANSION of $p-1$ DIGITS. The first few numbers with maximal decimal expansions are 7, 17, 19, 23, 29, 47, 59, 61, 97, ... (Sloane's A001913).

References

Sloane, N. J. A. Sequence A001913/M4353 in "An On-Line Version of the Encyclopedia of Integer Sequences."

Full Width at Half Maximum

The full width at half maximum (FWHM) is a parameter commonly used to describe the width of a "bump" on a curve or function. It is given by the distance between points on the curve at which the function reaches half its maximum value. The following table gives the analytic and numerical full widths for several common curves.

Function	Formula	FWHM
Bartlett	$1-\frac{\lvert x\rvert}{a}$	a
Blackman		$0.810957a$
Connes	$\left(1-\frac{x^2}{a^2}\right)$	$\sqrt{4-2\sqrt{2}}\,a$
Cosine	$\cos\left(\frac{\pi x}{2a}\right)$	$\frac{4}{3}a$
Gaussian	$e^{-x^2/(2\sigma^2)}$	$2\sqrt{2\ln 2}\,\sigma$
Hamming		$1.05543a$
Hanning		a
Lorentzian	$\frac{\frac{1}{2}\Gamma}{x^2+(\frac{1}{2}\Gamma)^2}$	Γ
Welch	$1-\frac{x^2}{a^2}$	$\sqrt{2}\,a$

see also APODIZATION FUNCTION, MAXIMUM

Fuller Dome

see GEODESIC DOME

Function

A way of associating unique objects to every point in a given SET. A function from A to B is an object f such that for every $a \in A$, there is a unique object $f(a) \in B$. Examples of functions include $\sin x$, x, x^2, etc. The term MAP is synonymous with function.

Poincaré remarked with regard to the proliferation of pathological functions, "Formerly, when one invented a new function, it was to further some practical purpose; today one invents them in order to make incorrect the reasoning of our fathers, and nothing more will ever be accomplished by these inventions."

see also ABELIAN FUNCTION, ABSOLUTE VALUE, ACKERMANN FUNCTION, AIRY FUNCTIONS, ALGEBRAIC FUNCTION, ALGEBROIDAL FUNCTION, ALPHA FUNCTION, ANDREW'S SINE, ANGER FUNCTION, APODIZATION FUNCTION, APPARATUS FUNCTION, ARGUMENT (FUNCTION), ARTIN L-FUNCTION, AUTOMORPHIC FUNCTION, BACHELIER FUNCTION, BARNES G-FUNCTION, BARTLETT FUNCTION, BASSET FUNCTION, BATEMAN FUNCTION, BEI, BER, BERNOULLI FUNCTION, BESSEL FUNCTION OF THE FIRST KIND, BESSEL FUNCTION OF THE SECOND KIND, BESSEL FUNCTION OF THE THIRD KIND, BETA FUNCTION, BETA FUNCTION (EXPONENTIAL), BINOMIAL COEFFICIENT, BLACKMAN FUNCTION, BLANCMANGE FUNCTION, BOOLEAN FUNCTION, BOURGET FUNCTION, BOXCAR FUNCTION, BROWN FUNCTION, CAL, CANTOR FUNCTION, CARMICHAEL FUNCTION, CAROTID-KUNDALINI FUNCTION, CEILING FUNCTION, CENTER FUNCTION, CENTRAL BETA FUNCTION, CHARACTERISTIC FUNCTION, CHEBYSHEV FUNCTION, CIRCULAR FUNCTIONS, CLAUSEN FUNCTION, COMB FUNCTION, COMPLETE FUNCTIONS, COMPLEX CONJUGATE, COMPUTABLE FUNCTION, CONCAVE FUNCTION, CONFLUENT HYPERGEOMETRIC FUNCTION, CONFLUENT HYPERGEOMETRIC FUNCTION OF THE FIRST KIND, CONFLUENT HYPERGEOMETRIC FUNCTION OF THE SECOND KIND, CONFLUENT HYPERGEOMETRIC LIMIT FUNCTION, CONICAL FUNCTION, CONNES FUNCTION, CONSTANT FUNCTION, CONTIGUOUS FUNCTION, CONTINUOUS FUNCTION, CONVEX FUNCTION, COPULA, COSECANT, COSINE, COSINE APODIZATION FUNCTION, COTANGENT, COULOMB WAVE FUNCTION, COVERSINE, CUBE ROOT, CUBED, CUMULANT-GENERATING FUNCTION, CUMULATIVE DISTRIBUTION FUNCTION, CUNNINGHAM FUNCTION, CYLINDER FUNCTION, CYLINDRICAL FUNCTION, DEBYE FUNCTIONS, DECREASING FUNCTION, DEDEKIND ETA FUNCTION, DEDEKIND FUNCTION, DELTA FUNCTION, DIGAMMA FUNCTION, DILOGARITHM, DIRAC DELTA FUNCTION, DIRICHLET BETA FUNCTION, DIRICHLET ETA FUNCTION, DIRICHLET FUNCTION, DIRICHLET LAMBDA FUNCTION, DISTRIBUTION FUNCTION, DIVISOR FUNCTION, DOUBLE GAMMA FUNCTION, DOUBLET FUNCTION, E_n-FUNCTION, E_t-FUNCTION, EIGENFUNCTION, EIN FUNCTION, EINSTEIN FUNCTIONS, ELEMENTARY FUNCTION, ELLIPTIC ALPHA FUNCTION, ELLIPTIC DELTA FUNCTION, ELLIPTIC EXPONENTIAL FUNCTION, ELLIPTIC FUNCTION, ELLIPTIC FUNCTIONAL, ELLIPTIC LAMBDA FUNCTION, ELLIPTIC MODULAR FUNCTION, ELLIPTIC THETA FUNCTION, ELSASSER FUNCTION, ENTIRE FUNCTION, EPSTEIN ZETA FUNCTION, ERDŐS-SELFRIDGE FUNCTION, ERF, ERROR FUNCTION, EXPONENTIAL RAMP, EULER L-FUNCTION, EVEN FUNCTION, EXPONENTIAL FUNCTION, EXPONENTIAL FUNCTION (TRUNCATED), EXPONENTIAL SUM FUNCTION, EXSECANT, FLOOR FUNCTION, FOX'S H-FUNCTION,

FUNCTION SPACE, G-FUNCTION, GAMMA FUNCTION, GATE FUNCTION, GAUSSIAN FUNCTION, GEGENBAUER FUNCTION, GENERALIZED FUNCTION, GENERALIZED HYPERBOLIC FUNCTIONS, GENERALIZED HYPERGEOMETRIC FUNCTION, GENERATING FUNCTION, GORDON FUNCTION, GREEN'S FUNCTION, GROWTH FUNCTION, GUDERMANNIAN FUNCTION, H-FUNCTION, HAAR FUNCTION, HAMMING FUNCTION, HANKEL FUNCTION, HANKEL FUNCTION OF THE FIRST KIND, HANKEL FUNCTION OF THE SECOND KIND, HANN FUNCTION, HANNING FUNCTION, HARMONIC FUNCTION, HAVERSINE, HEAVISIDE STEP FUNCTION, HECKE L-FUNCTION, HEMICYLINDRICAL FUNCTION, HEMISPHERICAL FUNCTION, HEUMAN LAMBDA FUNCTION, HH FUNCTION, HILBERT FUNCTION, HOLONOMIC FUNCTION, HOMOGENEOUS FUNCTION, HURWITZ ZETA FUNCTION, HYPERBOLIC COSECANT, HYPERBOLIC COSINE, HYPERBOLIC COTANGENT, HYPERBOLIC FUNCTIONS, HYPERBOLIC SECANT, HYPERBOLIC SINE, HYPERBOLIC TANGENT, HYPERELLIPTIC FUNCTION, HYPERGEOMETRIC FUNCTION, IDENTITY FUNCTION, IMPLICIT FUNCTION, IMPLICIT FUNCTION THEOREM, INCOMPLETE GAMMA FUNCTION, INCREASING FUNCTION, INFINITE PRODUCT, INSTRUMENT FUNCTION, INT, INVERSE COSECANT, INVERSE COSINE, INVERSE COTANGENT, INVERSE FUNCTION, INVERSE HYPERBOLIC FUNCTIONS, INVERSE SECANT, INVERSE SINE, INVERSE TANGENT, j-FUNCTION, JACOBI ELLIPTIC FUNCTIONS, JACOBI FUNCTION OF THE FIRST KIND, JACOBI FUNCTION OF THE SECOND KIND, JACOBI THETA FUNCTION, JACOBI ZETA FUNCTION, JINC FUNCTION, JOINT PROBABILITY DENSITY FUNCTION, JONQUIÈRE'S FUNCTION, K-FUNCTION, KEI, KELVIN FUNCTIONS, KER, KOEBE FUNCTION, L-FUNCTION, LAMBDA FUNCTION, LAMBDA HYPERGEOMETRIC FUNCTION, LAMBERT'S W-FUNCTION, LAMÉ FUNCTION, LEGENDRE FUNCTION OF THE FIRST KIND, LEGENDRE FUNCTION OF THE SECOND KIND, LEMNISCATE FUNCTION, LEMNISCATE FUNCTION, LENGTH DISTRIBUTION FUNCTION, LERCH TRANSCENDENT, LÉVY FUNCTION, LINEARLY DEPENDENT FUNCTIONS, LIOUVILLE FUNCTION, LIPSCHITZ FUNCTION, LOGARITHM, LOGARITHMICALLY CONVEX FUNCTION, LOGIT TRANSFORMATION, LOMMEL FUNCTION, LYAPUNOV FUNCTION, MACROBERT'S E-FUNCTION, MANGOLDT FUNCTION, MATHIEU FUNCTION, MEASURABLE FUNCTION, MEIJER'S G-FUNCTION, MEROMORPHIC, MERTENS FUNCTION, MERTZ APODIZATION FUNCTION, MITTAG-LEFFLER FUNCTION, MÖBIUS FUNCTION, MÖBIUS PERIODIC FUNCTION, MOCK THETA FUNCTION, MODIFIED BESSEL FUNCTION OF THE FIRST KIND, MODIFIED BESSEL FUNCTION OF THE SECOND KIND, MODIFIED SPHERICAL BESSEL FUNCTION, MODIFIED STRUVE FUNCTION, MODULAR FUNCTION, MODULAR GAMMA FUNCTION, MODULAR LAMBDA FUNCTION, MOMENT-GENERATING FUNCTION, MONOGENIC FUNCTION, MONOTONIC FUNCTION, MU FUNCTION, MULTIPLICATIVE FUNCTION, MULTIVALUED FUNCTION, MULTIVARIATE FUNCTION, NEUMANN FUNCTION, NINT, NU FUNCTION, NULL FUNCTION, NUMERIC FUNCTION, OBLATE SPHEROIDAL WAVE FUNCTION, ODD FUNCTION, OMEGA FUNCTION, ONE-WAY FUNCTION, PARABOLIC CYLINDER FUNCTION, PARTITION FUNCTION P, PARTITION FUNCTION Q, PARZEN APODIZATION FUNCTION, PEARSON-CUNNINGHAM FUNCTION, PEARSON'S FUNCTION, PERIODIC FUNCTION, PLANCK'S RADIATION FUNCTION, PLURISUBHARMONIC FUNCTION, POCHHAMMER SYMBOL, POINCARÉ-FUCHS-KLEIN AUTOMORPHIC FUNCTION, POISSON-CHARLIER FUNCTION, POLYGAMMA FUNCTION, POLYGENIC FUNCTION, POLYLOGARITHM, POSITIVE DEFINITE FUNCTION, POTENTIAL FUNCTION, POWER, PRIME COUNTING FUNCTION, PRIME DIFFERENCE FUNCTION, PROBABILITY DENSITY FUNCTION, PROBABILITY DISTRIBUTION FUNCTION, PROLATE SPHEROIDAL WAVE FUNCTION, PSI FUNCTION, PULSE FUNCTION, q-BETA FUNCTION, Q-FUNCTION, q-GAMMA FUNCTION, QUASIPERIODIC FUNCTION, RADEMACHER FUNCTION, RAMANUJAN FUNCTION, RAMANUJAN g- AND G- FUNCTIONS, RAMANUJAN THETA FUNCTIONS, RAMP FUNCTION, RATIONAL FUNCTION, REAL FUNCTION, RECTANGLE FUNCTION, REGULAR FUNCTION, REGULARIZED GAMMA FUNCTION, RESTRICTED DIVISOR FUNCTION, RIEMANN FUNCTION, RIEMANN-MANGOLDT FUNCTION, RIEMANN-SIEGEL FUNCTIONS, RIEMANN THETA FUNCTION, RIEMANN ZETA FUNCTION, RING FUNCTION, SAL, SAMPLING FUNCTION, SCALAR FUNCTION, SCHLOMILCH'S FUNCTION, SECANT, SEQUENCY FUNCTION, SGN, SHAH FUNCTION, SIEGEL MODULAR FUNCTION, SIGMA FUNCTION, SIGMOID FUNCTION, SIGN, SINC FUNCTION, SINE, SMARANDACHE FUNCTION, SPENCE'S FUNCTION, SPHERICAL BESSEL FUNCTION OF THE FIRST KIND, SPHERICAL BESSEL FUNCTION OF THE SECOND KIND, SPHERICAL HANKEL FUNCTION OF THE FIRST KIND, SPHERICAL HANKEL FUNCTION OF THE SECOND KIND, SPHERICAL HARMONIC, SPHEROIDAL WAVEFUNCTION, SPRAGUE-GRUNDY FUNCTION, SQUARE ROOT, SQUARED, STEP FUNCTION, STRUVE FUNCTION, STURM FUNCTION, SUMMATORY FUNCTION, SYMMETRIC FUNCTION, TAK FUNCTION, TANGENT, TAPERING FUNCTION, TAU FUNCTION, TETRACHORIC FUNCTION, THETA FUNCTION, TOROIDAL FUNCTION, TORONTO FUNCTION, TOTAL FUNCTION, TOTIENT FUNCTION, TOTIENT VALENCE FUNCTION, TRANSCENDENTAL FUNCTION, TRANSFER FUNCTION, TRAPDOOR FUNCTION, TRIANGLE CENTER FUNCTION, TRIANGLE FUNCTION, TRICOMI FUNCTION, TRIGONOMETRIC FUNCTIONS, UNIFORM APODIZATION FUNCTION, UNIVALENT FUNCTION, VECTOR FUNCTION, VERSINE, VON MANGOLDT FUNCTION, W-FUNCTION, WALSH FUNCTION, WEBER FUNCTIONS, WEIERSTRAß ELLIPTIC FUNCTION, WEIERSTRAß FUNCTION, WEIERSTRAß SIGMA FUNCTION, WEIERSTRAß ZETA FUNCTION, WEIGHTING FUNCTION, WELCH APODIZATION

FUNCTION, WHITTAKER FUNCTION, WIENER FUNCTION, WINDOW FUNCTION, XI FUNCTION, ZETA FUNCTION

References

Abramowitz, M. and Stegun, C. A. (Eds.). "Miscellaneous Functions." Ch. 27 in *Handbook of Mathematical Functions with Formulas, Graphs, and Mathematical Tables, 9th printing.* New York: Dover, pp. 997–1010, 1972.

Arfken, G. "Special Functions." Ch. 13 in *Mathematical Methods for Physicists, 3rd ed.* Orlando, FL: Academic Press, pp. 712–759, 1985.

Press, W. H.; Flannery, B. P.; Teukolsky, S. A.; and Vetterling, W. T. "Special Functions." Ch. 6 in *Numerical Recipes in FORTRAN: The Art of Scientific Computing, 2nd ed.* Cambridge, England: Cambridge University Press, pp. 205–265, 1992.

Function Field

see ALGEBRAIC FUNCTION FIELD

Function Space

$f(I)$ is the collection of all real-valued continuous functions defined on some interval I. $f^{(n)}(I)$ is the collection of all functions $\in f(I)$ with continuous nth DERIVATIVES. A function space is a TOPOLOGICAL VECTOR SPACE whose "points" are functions.

see also FUNCTIONAL, FUNCTIONAL ANALYSIS, OPERATOR

Functional

A mapping between FUNCTION SPACES if the range is on the REAL LINE or in the COMPLEX PLANE.

see also COERCIVE FUNCTIONAL, CURRENT, ELLIPTIC FUNCTIONAL, GENERALIZED FUNCTION, LAX-MILGRAM THEOREM, OPERATOR, RIESZ REPRESENTATION THEOREM

Functional Analysis

A branch of mathematics concerned with infinite dimensional spaces (mainly FUNCTION SPACES) and mappings between them. The SPACES may be of different, and possibly INFINITE, DIMENSIONS. These mappings are called OPERATORS or, if the range is on the REAL line or in the COMPLEX PLANE, FUNCTIONALS.

see also FUNCTIONAL, OPERATOR

References

Balakrishnan, A. V. *Applied Functional Analysis, 2nd ed.* New York: Springer-Verlag, 1981.

Berezansky, Y. M.; Us, G. F.; and Sheftel, Z. G. *Functional Analysis, Vol. 1.* Boston, MA: Birkhäuser, 1996.

Berezansky, Y. M.; Us, G. F.; and Sheftel, Z. G. *Functional Analysis, Vol. 2.* Boston, MA: Birkhäuser, 1996.

Birkhoff, G. and Kreyszig, E. "The Establishment of Functional Analysis." *Historia Math.* **11**, 258–321, 1984.

Hutson, V. and Pym, J. S. *Applications of Functional Analysis and Operator Theory.* New York: Academic Press, 1980.

Kreyszig, E. *Introductory Functional Analysis with Applications.* New York: Wiley, 1989.

Yoshida, K. *Functional Analysis and Its Applications.* New York: Springer-Verlag, 1971.

Zeidler, E. *Nonlinear Functional Analysis and Its Applications.* New York: Springer-Verlag, 1989.

Zeidler, E. *Applied Functional Analysis: Applications to Mathematical Physics.* New York: Springer-Verlag, 1995.

Functional Calculus

An early name for CALCULUS OF VARIATIONS.

Functional Derivative

A generalization of the concept of the DERIVATIVE to GENERALIZED FUNCTIONS.

Functor

A function between CATEGORIES which maps objects to objects and MORPHISMS to MORPHISMS. Functors exist in both covariant and contravariant types.

see also CATEGORY, EILENBERG-STEENROD AXIOMS, MORPHISM, SCHUR FUNCTOR

Fundamental Class

The canonical generator of the nonvanishing HOMOLOGY GROUP on a TOPOLOGICAL MANIFOLD.

see also CHERN NUMBER, PONTRYAGIN NUMBER, STIEFEL-WHITNEY NUMBER

Fundamental Continuity Theorem

Given two POLYNOMIALS of the same order in one variable where the first p COEFFICIENTS (but not the first $p-1$) are 0 and the COEFFICIENTS of the second approach the corresponding COEFFICIENTS of the first as limits, then the second POLYNOMIAL will have exactly p roots that increase indefinitely. Furthermore, exactly k ROOTS of the second will approach each ROOT of multiplicity k of the first as a limit.

References

Coolidge, J. L. *A Treatise on Algebraic Plane Curves.* New York: Dover, p. 4, 1959.

Fundamental Discriminant

$-D$ is a fundamental discriminant if D is a POSITIVE INTEGER which is not DIVISIBLE by any square of an ODD PRIME and which satisfies $D \equiv 3 \pmod 4$ or $D \equiv 4, 8 \pmod{16}$.

see also DISCRIMINANT

References

Atkin, A. O. L. and Morain, F. "Elliptic Curves and Primality Proving." *Math. Comput.* **61**, 29–68, 1993.

Borwein, J. M. and Borwein, P. B. *Pi & the AGM: A Study in Analytic Number Theory and Computational Complexity.* New York: Wiley, p. 294, 1987.

Cohn, H. *Advanced Number Theory.* New York: Dover, 1980.

Dickson, L. E. *History of the Theory of Numbers, Vols. 1–3.* New York: Chelsea, 1952.

Fundamental Forms

There are three types of so-called fundamental forms. The most important are the first and second (since the third can be expressed in terms of these). The fundamental forms are extremely important and useful in determining the metric properties of a surface, such as LINE ELEMENT, AREA ELEMENT, NORMAL CURVATURE, GAUSSIAN CURVATURE, and MEAN CURVATURE. Let M be a REGULAR SURFACE with $\mathbf{v_p}, \mathbf{w_p}$ points on the TANGENT SPACE $M_\mathbf{p}$ of M. Then the first fundamental form is the INNER PRODUCT of tangent vectors,

$$\mathbf{I}(\mathbf{v_p}, \mathbf{w_p}) = \mathbf{v_p} \cdot \mathbf{w_p}. \tag{1}$$

For $M \in \mathbb{R}^3$, the second fundamental form is the symmetric bilinear form on the TANGENT SPACE $M_\mathbf{p}$,

$$\mathbf{II}(\mathbf{v_p}, \mathbf{w_p}) = S(\mathbf{v_p}) \cdot \mathbf{w_p}, \tag{2}$$

where S is the SHAPE OPERATOR. The third fundamental form is given by

$$\mathbf{III}(\mathbf{v_p}, \mathbf{w_p}) = S(\mathbf{v_p}) \cdot S(\mathbf{w_p}). \tag{3}$$

The first and second fundamental forms satisfy

$$\mathbf{I}(a\mathbf{x}_u + b\mathbf{x}_v, a\mathbf{x}_u + b\mathbf{x}_v) = Ea^2 + 2Fab + Gb^2 \tag{4}$$
$$\mathbf{II}(a\mathbf{x}_u + b\mathbf{x}_v, a\mathbf{x}_u + b\mathbf{x}_v) = ea^2 + 2fab + gb^2, \tag{5}$$

and so their ratio is simply the NORMAL CURVATURE

$$\kappa(\mathbf{v_p}) = \frac{\mathbf{II}(\mathbf{v_p})}{\mathbf{I}(\mathbf{v_p})} \tag{6}$$

for any nonzero TANGENT VECTOR. The third fundamental form is given in terms of the first and second forms by

$$\mathbf{III} - 2H\mathbf{II} + K\mathbf{I} = 0, \tag{7}$$

where H is the MEAN CURVATURE and K is the GAUSSIAN CURVATURE.

The first fundamental form (or LINE ELEMENT) is given explicitly by the RIEMANNIAN METRIC

$$ds^2 = E\,du^2 + 2F\,du\,dv + G\,dv^2. \tag{8}$$

It determines the ARC LENGTH of a curve on a surface. The coefficients are given by

$$E = \mathbf{x}_{uu} = \left|\frac{\partial \mathbf{x}}{\partial u}\right|^2 \tag{9}$$
$$F = \mathbf{x}_{uv} = \frac{\partial \mathbf{x}}{\partial u} \cdot \frac{\partial \mathbf{x}}{\partial v} \tag{10}$$
$$G = \mathbf{x}_{vv} = \left|\frac{\partial \mathbf{x}}{\partial v}\right|^2. \tag{11}$$

The coefficients are also denoted $g_{uu} = E$, $g_{uv} = F$, and $g_{vv} = G$. In CURVILINEAR COORDINATES (where $F = 0$), the quantities

$$h_u \equiv \sqrt{g_{uu}} = \sqrt{E} \tag{12}$$
$$h_v \equiv \sqrt{g_{vv}} = \sqrt{G} \tag{13}$$

are called SCALE FACTORS.

The second fundamental form is given explicitly by

$$e\,du^2 + 2f\,du\,dv + g\,dv^2 \tag{14}$$

where

$$e = \sum_i X_i \frac{\partial^2 x_i}{\partial u^2} \tag{15}$$
$$f = \sum_i X_i \frac{\partial^2 x_i}{\partial u \partial v} \tag{16}$$
$$g = \sum_i X_i \frac{\partial^2 x_i}{\partial v^2}, \tag{17}$$

and X_i are the DIRECTION COSINES of the surface normal. The second fundamental form can also be written

$$e = -\mathbf{N}_u \cdot \mathbf{x}_u = \mathbf{N} \cdot \mathbf{x}_{uu} \tag{18}$$
$$f = -\mathbf{N}_v \cdot \mathbf{x}_u = \mathbf{N} \cdot \mathbf{x}_{uv} = \mathbf{N}_{vu} \cdot \mathbf{x}_{vu}$$
$$= -\mathbf{N}_u \cdot \mathbf{x}_v \tag{19}$$
$$g = -\mathbf{N}_v \cdot \mathbf{x}_v = \mathbf{N} \cdot \mathbf{x}_{vv}, \tag{20}$$

where $\mathbf{N}$ is the NORMAL VECTOR, or

$$e = \frac{\det(\mathbf{x}_{uu}\mathbf{x}_u\mathbf{x}_v)}{\sqrt{EG - F^2}} \tag{21}$$
$$f = \frac{\det(\mathbf{x}_{uv}\mathbf{x}_u\mathbf{x}_v)}{\sqrt{EG - F^2}} \tag{22}$$
$$g = \frac{\det(\mathbf{x}_{vv}\mathbf{x}_u\mathbf{x}_v)}{\sqrt{EG - F^2}}. \tag{23}$$

see also ARC LENGTH, AREA ELEMENT, GAUSSIAN CURVATURE, GEODESIC, KÄHLER MANIFOLD, LINE OF CURVATURE, LINE ELEMENT, MEAN CURVATURE, NORMAL CURVATURE, RIEMANNIAN METRIC, SCALE FACTOR, WEINGARTEN EQUATIONS

References

Gray, A. "The Three Fundamental Forms." §14.6 in *Modern Differential Geometry of Curves and Surfaces.* Boca Raton, FL: CRC Press, pp. 251–255, 259–260, 275–276, and 282–291, 1993.

Fundamental Group

The fundamental group of a CONNECTED SET S is the QUOTIENT GROUP of the GROUP of all paths with initial and final points at a given point P and the SUBGROUP of all paths HOMOTOPIC to the degenerate path consisting of the point P.

The fundamental group of the CIRCLE is the INFINITE CYCLIC GROUP. Two fundamental groups having different points P are ISOMORPHIC. If the fundamental group consists only of the identity element, then the set S is simply connected.

see also MILNOR'S THEOREM

Fundamental Homology Class

see also FUNDAMENTAL CLASS

Fundamental Lemma of Calculus of Variations

If

$$\int_a^b M(x)h(x)\,dx = 0$$

$\forall\ h(x)$ with CONTINUOUS second PARTIAL DERIVATIVES, then

$$M(x) = 0$$

on the OPEN INTERVAL (a, b).

Fundamental System

A set of ALGEBRAIC INVARIANTS for a QUANTIC such that any invariant of the QUANTIC is expressible as a POLYNOMIAL in members of the set. In 1868, Gordan proved the existence of finite fundamental systems of algebraic invariants and covariants for any binary QUANTIC. In 1890, Hilbert (1890) proved the HILBERT BASIS THEOREM, which is a finiteness theorem for the related concept of SYZYGIES.

see also HILBERT BASIS THEOREM, SYZYGY

References

Hilbert, D. "Über die Theorie der algebraischen Formen." *Math. Ann.* **36**, 473–534, 1890.

Fundamental Theorem of Algebra

Every POLYNOMIAL equation having COMPLEX COEFFICIENTS and degree ≥ 1 has at least one COMPLEX ROOT. This theorem was first proven by Gauss. It is equivalent to the statement that a POLYNOMIAL $P(z)$ of degree n has n values of z (some of them possibly degenerate) for which $P(z) = 0$. An example of a POLYNOMIAL with a single ROOT of multiplicity > 1 is $z^2 - 2z + 1 = (z-1)(z-1)$, which has $z = 1$ as a ROOT of multiplicity 2.

see also DEGENERATE, POLYNOMIAL

References

Courant, R. and Robbins, H. "The Fundamental Theorem of Algebra." §2.5.4 in *What is Mathematics?: An Elementary Approach to Ideas and Methods, 2nd ed.* Oxford, England: Oxford University Press, pp. 101–103, 1996.

Fundamental Theorem of Arithmetic

Any POSITIVE INTEGER can be represented in exactly one way as a PRODUCT of PRIMES. The theorem is also called the UNIQUE FACTORIZATION THEOREM. The fundamental theorem of algebra is a COROLLARY of the first of EUCLID'S THEOREMS (Hardy and Wright 1979).

see also EUCLID'S THEOREMS, INTEGER, PRIME NUMBER

References

Courant, R. and Robbins, H. *What is Mathematics?: An Elementary Approach to Ideas and Methods, 2nd ed.* Oxford, England: Oxford University Press, p. 23, 1996.

Hardy, G. H. and Wright, E. M. "Statement of the Fundamental Theorem of Arithmetic," "Proof of the Fundamental Theorem of Arithmetic," and "Another Proof of teh Fundamental Theorem of Arithmetic." §1.3, 2.10 and 2.11 in *An Introduction to the Theory of Numbers, 5th ed.* Oxford, England: Clarendon Press, pp. 3 and 21, 1979.

Fundamental Theorems of Calculus

The first fundamental theorem of calculus states that, if f is CONTINUOUS on the CLOSED INTERVAL $[a, b]$ and F is the ANTIDERIVATIVE (INDEFINITE INTEGRAL) of f on $[a, b]$, then

$$\int_a^b f(x)\,dx = F(b) - F(a). \tag{1}$$

The second fundamental theorem of calculus lets f be CONTINUOUS on an OPEN INTERVAL I and lets a be any point in I. If F is defined by

$$F(x) = \int_a^x f(t)\,dt, \tag{2}$$

then

$$F'(x) = f(x) \tag{3}$$

at each point in I.

The complex fundamental theorem of calculus states that if $f(z)$ has a CONTINUOUS ANTIDERIVATIVE $F(z)$ in a region R containing a parameterized curve $\gamma : z = z(t)$ for $\alpha \leq t \leq \beta$, then

$$\int_\gamma f(z)\,dz = F(z(\beta)) - F(z(\alpha)). \tag{4}$$

see also CALCULUS, DEFINITE INTEGRAL, INDEFINITE INTEGRAL, INTEGRAL

Fundamental Theorem of Curves

The CURVATURE and TORSION functions along a SPACE CURVE determine it up to an orientation-preserving ISOMETRY.

Fundamental Theorem of Directly Similar Figures

Let F_0 and F_1 denote two directly similar figures in the plane, where $P_1 \in F_1$ corresponds to $P_0 \in F_0$ under the given similarity. Let $r \in (0,1)$, and define $F_r = \{(1-r)P_0 + rP_1 : P_0 \in F_0\}$. Then F_r is also directly similar to F_0.

see also FINSLER-HADWIGER THEOREM

References

Detemple, D. and Harold, S. "A Round-Up of Square Problems." *Math. Mag.* **69**, 15–27, 1996.

Eves, H. Solution to Problem E521. *Amer. Math. Monthly* **50**, 64, 1943.

Fundamental Theorem of Gaussian Quadrature

The ABSCISSAS of the N point GAUSSIAN QUADRATURE FORMULA are precisely the ROOTS of the ORTHOGONAL POLYNOMIAL for the same INTERVAL and WEIGHTING FUNCTION.

see also GAUSSIAN QUADRATURE

Fundamental Theorem of Genera

$$2^{\omega(d)-1}|h(-d)|,$$

where $\omega(d)$ is the genus of forms and $h(-d)$ is the CLASS NUMBER of an IMAGINARY QUADRATIC FIELD.

References

Arno, S.; Robinson, M. L.; and Wheeler, F. S. "Imaginary Quadratic Fields with Small Odd Class Number." `http://www.math.uiuc.edu/Algebraic-Number-Theory/0009/`.

Cohn, H. *Advanced Number Theory.* New York: Dover, p. 224, 1980.

Gauss, C. F. *Disquisitiones Arithmeticae.* New Haven, CT: Yale University Press, 1966.

Fundamental Theorem of Plane Curves

Two unit-speed plane curves which have the same CURVATURE differ only by a EUCLIDEAN MOTION.

see also FUNDAMENTAL THEOREM OF SPACE CURVES

References

Gray, A. *Modern Differential Geometry of Curves and Surfaces.* Boca Raton, FL: CRC Press, pp. 103 and 110–111, 1993.

Fundamental Theorem of Projective Geometry

A PROJECTIVITY is determined when three points of one RANGE and the corresponding three points of the other are given.

see also PROJECTIVE GEOMETRY

Fundamental Theorem of Space Curves

If two single-valued continuous functions $\kappa(s)$ (CURVATURE) and $\tau(s)$ (TORSION) are given for $s > 0$, then there exists EXACTLY ONE SPACE CURVE, determined except for orientation and position in space (i.e., up to a EUCLIDEAN MOTION), where s is the ARC LENGTH, κ is the CURVATURE, and τ is the TORSION.

see also ARC LENGTH, CURVATURE, EUCLIDEAN MOTION, FUNDAMENTAL THEOREM OF PLANE CURVES, TORSION (DIFFERENTIAL GEOMETRY)

References

Gray, A. "The Fundamental Theorem of Space Curves." §7.7 in *Modern Differential Geometry of Curves and Surfaces.* Boca Raton, FL: CRC Press, pp. 123 and 142–145, 1993.

Struik, D. J. *Lectures on Classical Differential Geometry.* New York: Dover, p. 29, 1988.

Fundamental Theorem of Symmetric Functions

Any symmetric polynomial (respectively, symmetric rational function) can be expressed as a POLYNOMIAL (respectively, RATIONAL FUNCTION) in the ELEMENTARY SYMMETRIC FUNCTIONS on those variables.

see also ELEMENTARY SYMMETRIC FUNCTION

References

Coolidge, J. L. *A Treatise on Algebraic Plane Curves.* New York: Dover, p. 2, 1959.

Herstein, I. N. *Noncommutative Rings.* Washington, DC: Math. Assoc. Amer., 1968.

Fundamental Unit

In a real QUADRATIC FIELD, there exists a special UNIT η known as the fundamental unit such that all units ρ are given by $\rho = \pm\eta^m$, for $m = 0, \pm 1, \pm 2, \ldots$. The following table gives the fundamental units for the first few real quadratic fields.

d	$\eta(d)$	d	$\eta(d)$
2	$1+\sqrt{2}$	51	$50+7\sqrt{51}$
3	$2+\sqrt{3}$	53	$\frac{1}{2}(7+\sqrt{53})$
5	$\frac{1}{2}(1+\sqrt{5})$	55	$89+12\sqrt{55}$
6	$5+2\sqrt{6}$	57	$151+20\sqrt{57}$
7	$8+3\sqrt{7}$	58	$99+13\sqrt{58}$
10	$3+\sqrt{10}$	59	$530+69\sqrt{59}$
11	$10+3\sqrt{11}$	61	$\frac{1}{2}(39+5\sqrt{61})$
13	$\frac{1}{2}(3+\sqrt{13})$	62	$63+8\sqrt{62}$
14	$15+4\sqrt{14}$	65	$8+\sqrt{65}$
15	$4+\sqrt{15}$	66	$65+8\sqrt{66}$
17	$4+\sqrt{17}$	67	$48842+5967\sqrt{67}$
19	$170+39\sqrt{19}$	69	$\frac{1}{2}(25+3\sqrt{69})$
21	$\frac{1}{2}(5+\sqrt{21})$	70	$251+30\sqrt{70}$
22	$197+42\sqrt{22}$	71	$3480+413\sqrt{71}$
23	$24+5\sqrt{23}$	73	$1068+125\sqrt{73}$
26	$5+\sqrt{26}$	74	$43+5\sqrt{74}$
29	$\frac{1}{2}(5+\sqrt{29})$	77	$\frac{1}{2}(9+\sqrt{77})$
30	$11+2\sqrt{30}$	78	$53+6\sqrt{78}$
31	$1520+273\sqrt{31}$	79	$80+9\sqrt{79}$
33	$5+4\sqrt{33}$	82	$9+\sqrt{82}$
34	$35+6\sqrt{34}$	83	$82+9\sqrt{83}$
35	$6+\sqrt{35}$	85	$\frac{1}{2}(9+\sqrt{85})$
37	$6+\sqrt{37}$	86	$10405+1122\sqrt{86}$
38	$37+6\sqrt{38}$	87	$28+3\sqrt{87}$
39	$25+4\sqrt{39}$	89	$501+54\sqrt{89}$
41	$32+5\sqrt{41}$	91	$1574+165\sqrt{91}$
42	$13+2\sqrt{42}$	93	$\frac{1}{2}(29+3\sqrt{93})$
43	$3482+531\sqrt{43}$	95	$39+4\sqrt{95}$
46	$24335+3588\sqrt{46}$	97	$5604+569\sqrt{97}$
47	$48+7\sqrt{47}$		

see also QUADRATIC FIELD, UNIT

References
Cohn, H. "Fundamental Units" and "Construction of Fundamental Units." §6.4 and 6.5 in *Advanced Number Theory.* New York: Dover, pp. 98–102, and 261–274, 1980.
Weisstein, E. W. "Class Numbers." http://www.astro.virginia.edu/~eww6n/math/notebooks/ClassNumbers.m.

Funnel

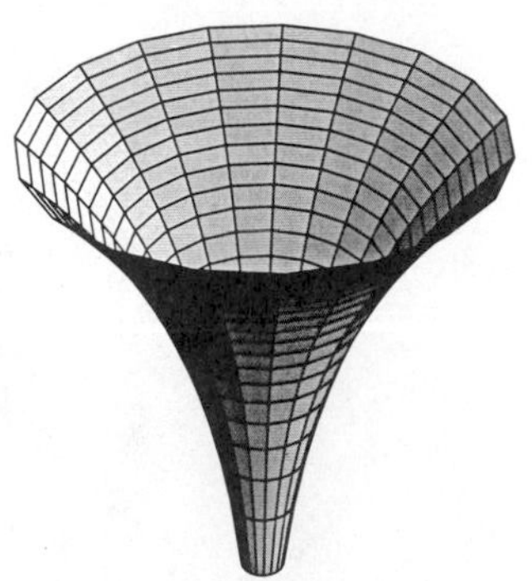

The funnel surface is a REGULAR SURFACE defined by the Cartesian equation

$$z = \frac{1}{2}\ln(x^2+y^2) \tag{1}$$

and the parametric equations

$$x(r,\theta) = r\cos\theta \tag{2}$$
$$y(r,\theta) = r\sin\theta \tag{3}$$
$$z(r,\theta) = \ln r. \tag{4}$$

see also GABRIEL'S HORN, PSEUDOSPHERE, SINCLAIR'S SOAP FILM PROBLEM

References
Gray, A. *Modern Differential Geometry of Curves and Surfaces.* Boca Raton, FL: CRC Press, pp. 325–327, 1993.

Fuss's Problem

see BICENTRIC POLYGON

Futile Game

A GAME which permits a draw ("tie") when played properly by both players.

Fuzzy Logic

An extension of two-valued LOGIC such that statements need not be TRUE or FALSE, but may have a degree of truth between 0 and 1. Such a system can be extremely useful in designing control logic for real-world systems such as elevators.

see also ALETHIC, FALSE, LOGIC, TRUE

References
McNeill, D. and Freiberger, P. *Fuzzy Logic.* New York: Simon & Schuster, 1993.
Nguyen, H. T. and Walker, E. A. *A First Course in Fuzzy Logic.* Boca Raton, FL: CRC Press, 1996.
Yager, R. R. and Zadeh, L. A. (Eds.) *An Introduction to Fuzzy Logic Applications in Intelligent Systems.* Boston, MA: Kluwer, 1992.
Zadeh, L. and Kacprzyk, J. (Eds.). *Fuzzy Logic for the Management of Uncertainty.* New York: Wiley, 1992.

FWHM

see FULL WIDTH AT HALF MAXIMUM

G

g-Function

see RAMANUJAN *g*- AND *G*-FUNCTIONS

G-Function

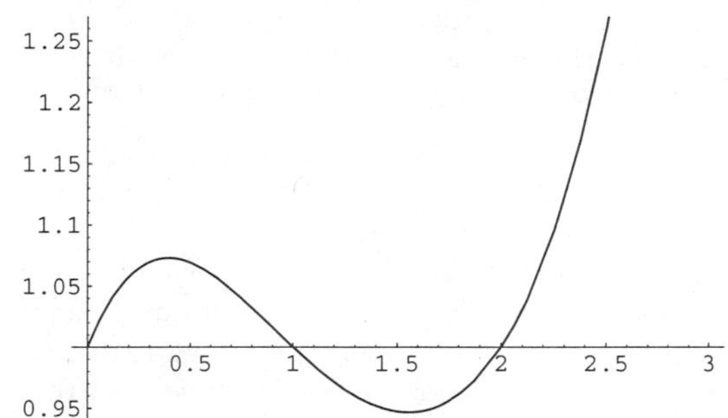

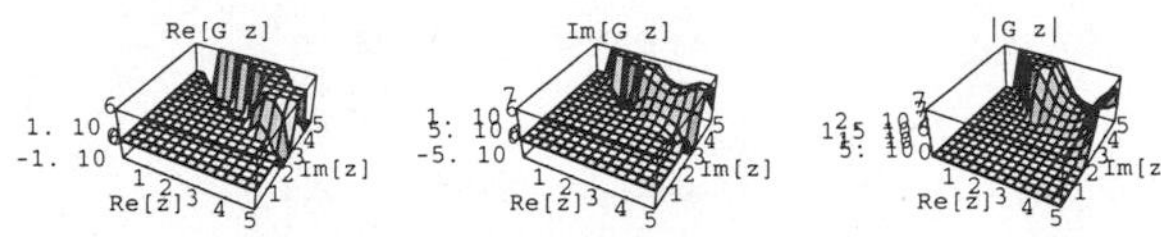

Defined in Whittaker and Watson (1990, p. 264) and also called the BARNES *G*-FUNCTION.

$$G(z+1) \equiv (2\pi)^{z/2} e^{-[z(z+1)+\gamma z^2]/2} \times \prod_{n=1}^{\infty} \left[\left(1+\frac{z}{n}\right)^n e^{-z+z^2/(2n)} \right], \quad (1)$$

where γ is the EULER-MASCHERONI CONSTANT. This is an ANALYTIC CONTINUATION of the G function defined in the construction of the GLAISHER-KINKELIN CONSTANT

$$G(n+1) \equiv \frac{(n!)^n}{K(n+1)}, \quad (2)$$

which has the special values

$$G(n) = \begin{cases} 0 & \text{if } n = 0, -1, -2, \ldots \\ 1 & \text{if } n = 1 \\ 0!1!2! \cdots (n-2)! & \text{if } n = 2, 3, 4, \ldots \end{cases} \quad (3)$$

for INTEGER n. This function is what Sloane and Plouffe (1995) call the SUPERFACTORIAL, and the first few values for $n = 1, 2, \ldots$ are 1, 1, 1, 2, 12, 288, 34560, 24883200, 125411328000, 5056584744960000, ... (Sloane's A000178).

The G-function is the reciprocal of the DOUBLE GAMMA FUNCTION. It satisfies

$$G(z+1) = \Gamma(z)G(z) \quad (4)$$

$$\frac{(n!)^n}{G(n+1)} = 1^1 \cdot 2^2 \cdot 3^3 \cdots n^n \quad (5)$$

$$\frac{G'(z+1)}{G(z+1)} = \tfrac{1}{2}\ln(2\pi) - \tfrac{1}{2} - z + z\frac{\Gamma'(z)}{\Gamma(z)} \quad (6)$$

$$\ln\left[\frac{G(1-z)}{G(1+z)}\right] = \int_0^z \pi z \cot(\pi z)\, dz - z\ln(2\pi) \quad (7)$$

and has the special values

$$G(\tfrac{1}{2}) = A^{-3/2}\pi^{-1/4}e^{1/8}2^{1/24} \quad (8)$$

$$G(1) = 1, \quad (9)$$

where

$$A = \exp\left[-\frac{\zeta'(2)}{2\pi^2} + \frac{\ln(2\pi)}{12} + \frac{\gamma}{2}\right] = 1.28242713\ldots. \quad (10)$$

The G-function can arise in spectral functions in mathematical physics (Voros 1987).

An unrelated pair of functions are denoted g_n and G_n and are known as RAMANUJAN *g*- AND *G*-FUNCTIONS.

see also EULER-MASCHERONI CONSTANT, GLAISHER-KINKELIN CONSTANT, *K*-FUNCTION, MEIJER'S *G*-FUNCTION, RAMANUJAN *g*- AND *G*-FUNCTIONS, SUPERFACTORIAL

References

Barnes, E. W. "The Theory of the *G*-Function." *Quart. J. Pure Appl. Math.* **31**, 264–314, 1900.

Glaisher, J. W. L. "On a Numerical Continued Product." *Messenger Math.* **6**, 71–76, 1877.

Glaisher, J. W. L. "On the Product $1^1 2^2 3^3 \cdots n^n$." *Messenger Math.* **7**, 43–47, 1878.

Glaisher, J. W. L. "On Certain Numerical Products." *Messenger Math.* **23**, 145–175, 1893.

Glaisher, J. W. L. "On the Constant which Occurs in the Formula for $1^1 2^2 3^3 \cdots n^n$." *Messenger Math.* **24**, 1–16, 1894.

Kinkelin. "Über eine mit der Gammafunktion verwandte Transcendente und deren Anwendung auf die Integralrechnung." *J. Reine Angew. Math.* **57**, 122–158, 1860.

Sloane, N. J. A. Sequence A000178/M2049 in "An On-Line Version of the Encyclopedia of Integer Sequences."

Voros, A. "Spectral Functions, Special Functions and the Selberg Zeta Function." *Commun. Math. Phys.* **110**, 439–465, 1987.

Whittaker, E. T. and Watson, G. N. *A Course in Modern Analysis, 4th ed.* Cambridge, England: Cambridge University Press, 1990.

G-Number

see EISENSTEIN INTEGER

G-Space

A G-space is a special type of HAUSDORFF SPACE. Consider a point x and a HOMEOMORPHISM of an open NEIGHBORHOOD V of x onto an OPEN SET of $\mathbb{R}^n$. Then a space is a G-space if, for any two such NEIGHBORHOODS V' and V'', the images of $V' \cup V''$ under the different HOMEOMORPHISMS are ISOMETRIC. If $n = 2$, the HOMEOMORPHISMS need only be conformal (but not necessarily orientation-preserving).

see also GREEN SPACE

Gabriel's Horn

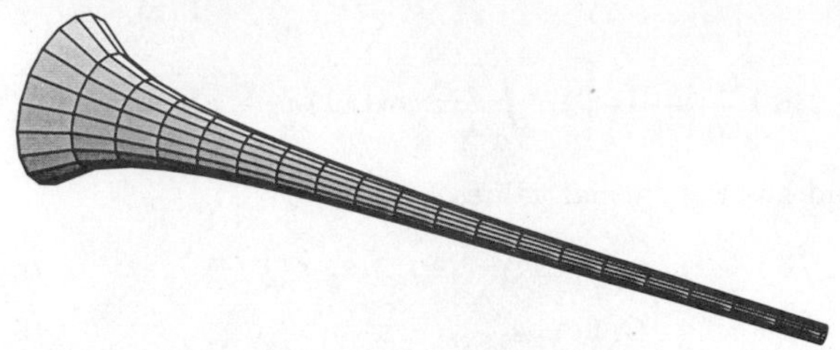

The SURFACE OF REVOLUTION of the function $y = 1/x$ about the x-axis for $x \geq 1$. It has FINITE VOLUME

$$V = \int_1^\infty \pi y^2\,dx = \pi \int_1^\infty \frac{dx}{x^2} = \pi\left[-\frac{1}{x}\right]_1^\infty = \pi[0-(-1)] = \pi,$$

but INFINITE SURFACE AREA, since

$$S = \int_1^\infty 2\pi y\sqrt{1+y'^2}\,dx > 2\pi \int_1^\infty y\,dx = 2\pi \int_1^\infty \frac{dx}{x} = 2\pi[\ln x]_1^\infty = 2\pi[\ln\infty - 0] = \infty.$$

This leads to the paradoxical consequence that while Gabriel's horn can be filled up with π cubic units of paint, an INFINITE number of square units of paint are needed to cover its surface!

see also FUNNEL, PSEUDOSPHERE

Gabriel's Staircase

The SUM

$$\sum_{k=1}^\infty kr^k = \frac{r}{(1-r)^2},$$

valid for $0 < r < 1$.

Gadget

A term of endearment used by ALGEBRAIC TOPOLOGISTS when talking about their favorite power tools such as ABELIAN GROUPS, BUNDLES, HOMOLOGY GROUPS, HOMOTOPY GROUPS, k-THEORY, MORSE THEORY, OBSTRUCTIONS, stable homotopy theory, VECTOR SPACES, etc.

see also ABELIAN GROUP, ALGEBRAIC TOPOLOGY, BUNDLE, FREE, HOMOLOGY GROUP, HOMOTOPY GROUP, k-THEORY, OBSTRUCTION, MORSE THEORY, VECTOR SPACE

Gale-Ryser Theorem

Let p and q be PARTITIONS of a POSITIVE INTEGER, then there exists a (0,1)-matrix A such that $c(A) = p$, $r(A) = q$ IFF q is dominated by p^*.

References

Brualdi, R. and Ryser, H. J. §6.2.4 in *Combinatorial Matrix Theory*. New York: Cambridge University Press, 1991.

Krause, M. "A Simple Proof of the Gale-Ryser Theorem." *Amer. Math. Monthly* **103**, 335–337, 1996.

Robinson, G. §1.4 in *The Representation Theory of the Symmetric Group*. Toronto, Canada: University of Toronto Press, 1961.

Ryser, H. J. "The Class $\mathcal{A}(\mathbf{R},\mathbf{S})$." *Combinatorial Mathematics*. Buffalo, NY: Math. Assoc. Amer., pp. 61–65, 1963.

Galilean Transformation

A transformation from one reference frame to another moving with a constant VELOCITY v with respect to the first for classical motion. However, special relativity shows that the transformation must be modified to the LORENTZ TRANSFORMATION for relativistic motion. The forward Galilean transformation is

$$\begin{bmatrix} t' \\ x' \\ y' \\ z' \end{bmatrix} = \begin{bmatrix} 1 & 0 & 0 & 0 \\ -v & 1 & 0 & 0 \\ 0 & 0 & 1 & 0 \\ 0 & 0 & 0 & 1 \end{bmatrix} \begin{bmatrix} t \\ x \\ y \\ z \end{bmatrix},$$

and the inverse transformation is

$$\begin{bmatrix} t \\ x \\ y \\ z \end{bmatrix} = \begin{bmatrix} 1 & 0 & 0 & 0 \\ v & 1 & 0 & 0 \\ 0 & 0 & 1 & 0 \\ 0 & 0 & 0 & 1 \end{bmatrix} \begin{bmatrix} t' \\ x' \\ y' \\ z' \end{bmatrix}.$$

see also LORENTZ TRANSFORMATION

Gall's Stereographic Projection

A CYLINDRICAL PROJECTION which projects the equator onto a tangent cylinder which intersects the globe at $\pm\,45°$. The transformation equations are

$$x = \lambda$$
$$y = \tan(\tfrac{1}{2}\phi),$$

where λ is the LONGITUDE and ϕ the LATITUDE.

see also STEREOGRAPHIC PROJECTION

References

Dana, P. H. "Map Projections." `http://www.utexas.edu/depts/grg/gcraft/notes/mapproj/mapproj.html`.

Gallows

Schroeder (1991) calls the CEILING FUNCTION symbols $\lceil$ and $\rceil$ the "gallows" because of their similarity in appearance to the structure used for hangings.

see also CEILING FUNCTION

References

Schroeder, M. *Fractals, Chaos, Power Laws: Minutes from an Infinite Paradise*. New York: W. H. Freeman, p. 57, 1991.

Gallucci's Theorem

If three SKEW LINES all meet three other SKEW LINES, any TRANSVERSAL to the first set of three meets any TRANSVERSAL to the second set of three.

see also SKEW LINES, TRANSVERSAL LINE

Galois Extension Field

The splitting FIELD for a separable POLYNOMIAL over a FINITE FIELD K, where L is a FIELD EXTENSION of K.

Galois Field

see FINITE FIELD

Galois Group

Let L be a FIELD EXTENSION of K, denoted L/K, and let G be the set of AUTOMORPHISMS of L/K, that is, the set of AUTOMORPHISMS σ of L such that $\sigma(x) = x$ for every $x \in K$, so that K is fixed. Then G is a GROUP of transformations of L, called the Galois group of L/K.

The Galois group of $(\mathbb{C}/\mathbb{R})$ consists of the IDENTITY ELEMENT and COMPLEX CONJUGATION. These functions both take a given REAL to the same real.

see also ABHYANKAR'S CONJECTURE, FINITE GROUP, GROUP

References

Jacobson, N. *Basic Algebra I, 2nd ed.* New York: W. H. Freeman, p. 234, 1985.

Galois Imaginary

A mathematical object invented to solve irreducible CONGRUENCES of the form

$$F(x) \equiv 0 \pmod{p},$$

where p is PRIME.

Galois's Theorem

An algebraic equation is algebraically solvable IFF its GROUP is SOLVABLE. In order that an irreducible equation of PRIME degree be solvable by radicals, it is NECESSARY and SUFFICIENT that all its ROOTS be rational functions of two ROOTS.

see also ABEL'S IMPOSSIBILITY THEOREM, SOLVABLE GROUP

Galois Theory

If there exists a ONE-TO-ONE correspondence between two SUBGROUPS and SUBFIELDS such that

$$G(E(G')) = G'$$
$$E(G(E')) = E',$$

then E is said to have a Galois theory.

Galoisian

An algebraic extension E of F for which every IRREDUCIBLE POLYNOMIAL in F which has a single ROOT in E has *all* its ROOTS in E is said to be Galoisian. Galoisian extensions are also called algebraically normal.

Gambler's Ruin

Let two players each have a finite number of pennies (say, n_1 for player one and n_2 for player two). Now, flip one of the pennies (from either player), with each player having 50% probability of winning, and give the penny to the winner. If the process is repeated indefinitely, the probability that one or the other player will *eventually* lose all his pennies is unity. However, the chances that the individual players will be rendered penniless are

$$P_1 = \frac{n_1}{n_1 + n_2}$$
$$P_2 = \frac{n_2}{n_1 + n_2}.$$

see also COIN TOSSING, MARTINGALE, SAINT PETERSBURG PARADOX

References

Cover, T. M. "Gambler's Ruin: A Random Walk on the Simplex." §5.4 in In *Open Problems in Communications and Computation.* (Ed. T. M. Cover and B. Gopinath). New York: Springer-Verlag, p. 155, 1987.

Hajek, B. "Gambler's Ruin: A Random Walk on the Simplex." §6.3 in In *Open Problems in Communications and Computation.* (Ed. T. M. Cover and B. Gopinath). New York: Springer-Verlag, pp. 204–207, 1987.

Kraitchik, M. "The Gambler's Ruin." §6.20 in *Mathematical Recreations.* New York: W. W. Norton, p. 140, 1942.

Game

A game is defined as a conflict involving gains and losses between two or more opponents who follow formal rules. The study of games belongs to a branch of mathematics known as GAME THEORY.

see also GAME THEORY

Game Expectation

Let the elements in a PAYOFF MATRIX be denoted a_{ij}, where the is are player A's STRATEGIES and the js are player B's STRATEGIES. Player A can get at least

$$\min_{j \leq n} a_{ij} \tag{1}$$

for STRATEGY i. Player B can force player A to get no more than $\max_{j \leq m} a_{ij}$ for a STRATEGY j. The best STRATEGY for player A is therefore

$$\min_{i \leq m} \min_{j \leq n} a_{ij}, \tag{2}$$

and the best STRATEGY for player B is

$$\min_{j \leq n} \max_{i \leq m} a_{ij}. \tag{3}$$

In general,

$$\min_{i \le m} \min_{j \le n} a_{ij} \le \min_{j \le n} \max_{i \le m} a_{ij}. \tag{4}$$

Equality holds only if a SADDLE POINT is present, in which case the quantity is called the VALUE of the game.

see also GAME, PAYOFF MATRIX, SADDLE POINT (GAME), STRATEGY, VALUE

Game of Life

see LIFE

Game Matrix

see PAYOFF MATRIX

Game Theory

A branch of MATHEMATICS and LOGIC which deals with the analysis of GAMES (i.e., situations in which parties are involved in situations where their interests conflict). In addition to the mathematical elegance and complete "solution" which is possible for simple games, the principles of game theory also find applications to complicated games such as cards, checkers, and chess, as well as real-world problems as diverse as economics, property division, politics, and warfare.

see also BOREL DETERMINACY THEOREM, CATEGORICAL GAME, CHECKERS, CHESS, DECISION THEORY, EQUILIBRIUM POINT, FINITE GAME, FUTILE GAME, GAME EXPECTATION, GO, HI-Q, IMPARTIAL GAME, MEX, MINIMAX THEOREM, MIXED STRATEGY, NASH EQUILIBRIUM, NASH'S THEOREM, NIM, NIM-VALUE, PARTISAN GAME, PAYOFF MATRIX, PEG SOLITAIRE, PERFECT INFORMATION, SADDLE POINT (GAME), SAFE, SPRAGUE-GRUNDY FUNCTION, STRATEGY, TACTIX, TIT-FOR-TAT, UNSAFE, VALUE, WYTHOFF'S GAME, ZERO-SUM GAME

References

Berlekamp, E. R.; Conway, J. H; and Guy, R. K. *Winning Ways, For Your Mathematical Plays, Vol. 1: Games in General.* London: Academic Press, 1982.

Berlekamp, E. R.; Conway, J. H; and Guy, R. K. *Winning Ways, For Your Mathematical Plays, Vol. 2: Games in Particular.* London: Academic Press, 1982.

Dresher, M. *The Mathematics of Games of Strategy: Theory and Applications.* New York: Dover, 1981.

Eppstein, D. "Combinatorial Game Theory." `http://www.ics.uci.edu/~eppstein/cgt/`.

Gardner, M. "Game Theory." Ch. 3 in *Mathematical Magic Show: More Puzzles, Games, Diversions, Illusions and Other Mathematical Sleight-of-Mind from Scientific American.* New York: Vintage, 1978.

Karlin, S. *Mathematical Methods and Theory in Games, Programming, and Economics, 2 Vols. Vol. 1: Matrix Games, Programming, and Mathematical Economics. Vol. 2: The Theory of Infinite Games.* New York: Dover, 1992.

Kuhn, H. W. (Ed.). *Classics in Game Theory.* Princeton, NJ: Princeton University Press, 1997.

McKinsey, J. C. C. *Introduction to the Theory of Games.* New York: McGraw-Hill, 1952.

Neumann, J. von and Morgenstern, O. *Theory of Games and Economic Behavior, 3rd ed.* New York: Wiley, 1964.

Packel, E. *The Mathematics of Games and Gambling.* Washington, DC: Math. Assoc. Amer., 1981.

Straffin, P. D. Jr. *Game Theory and Strategy.* Washington, DC: Math. Assoc. Amer., 1993.

Vajda, S. *Mathematical Games and How to Play Them.* New York: Routledge, 1992.

Walker, P. "An Outline of the History of Game Theory." `http://william-king.www.drexel.edu/top/class/histf.html`.

Williams, J. D. *The Compleat Strategyst, Being a Primer on the Theory of Games of Strategy.* New York: Dover, 1986.

Gamma Distribution

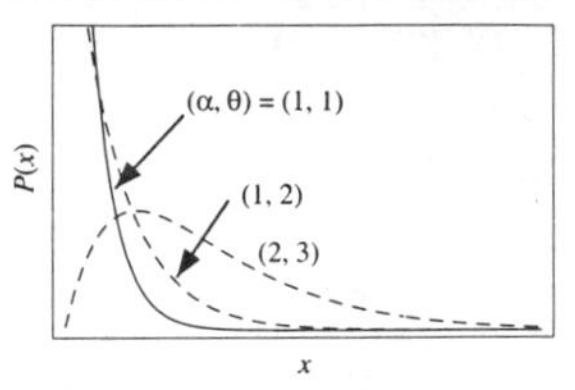

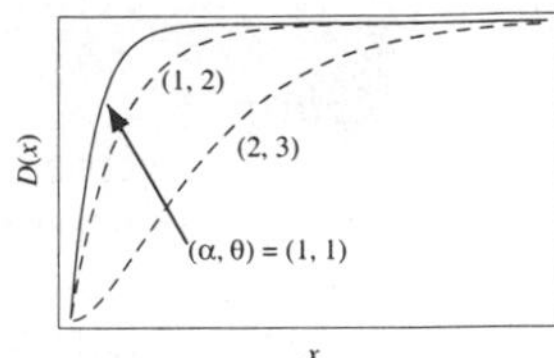

A general type of statistical DISTRIBUTION which is related to the BETA DISTRIBUTION and arises naturally in processes for which the waiting times between POISSON DISTRIBUTED events are relevant. Gamma distributions have two free parameters, labeled α and θ, a few of which are illustrated above.

Given a POISSON DISTRIBUTION with a rate of change λ, the DISTRIBUTION FUNCTION $D(x)$ giving the waiting times until the hth change is

$$\begin{aligned} D(x) &= P(X \le x) = 1 - P(X > x) \\ &= 1 - \sum_{k=0}^{h-1} \frac{(\lambda x)^k e^{-\lambda x}}{k!} \\ &= 1 - e^{-\lambda x} \sum_{k=0}^{h-1} \frac{(\lambda x)^k}{k!} \end{aligned} \tag{1}$$

for $x \ge 0$. The probability function $P(x)$ is then obtained by differentiating $D(x)$,

$$\begin{aligned} P(x) &= D'(x) \\ &= \lambda e^{-\lambda x} \sum_{k=0}^{h-1} \frac{(\lambda x)^k}{k!} - e^{-\lambda x} \sum_{k=0}^{h-1} \frac{k(\lambda x)^{k-1}\lambda}{k!} \\ &= \lambda e^{-\lambda x} + \lambda e^{-\lambda x} \sum_{k=1}^{h-1} \frac{(\lambda x)^k}{k!} - e^{-\lambda x} \sum_{k=1}^{h-1} \frac{k(\lambda x)^{k-1}\lambda}{k!} \\ &= \lambda e^{-\lambda x} - \lambda e^{-\lambda x} \sum_{k=1}^{h-1} \left[\frac{k(\lambda x)^{k-1}}{k!} - \frac{(\lambda x)^k}{k!} \right] \\ &= \lambda e^{-\lambda x} \left\{ 1 - \sum_{k=1}^{h-1} \left[\frac{(\lambda x)^{k-1}}{(k-1)!} - \frac{(\lambda x)^k}{k!} \right] \right\} \\ &= \lambda e^{-\lambda x} \left\{ 1 - \left[1 - \frac{(\lambda x)^{h-1}}{(h-1)!} \right] \right\} = \frac{\lambda(\lambda x)^{h-1}}{(h-1)!} e^{-\lambda x}. \end{aligned} \tag{2}$$

Now let $\alpha \equiv h$ and define $\theta \equiv 1/\lambda$ to be the time between changes. Then the above equation can be written

$$P(x) = \begin{cases} \frac{x^{\alpha-1}e^{-x/\theta}}{\Gamma(\alpha)\theta^\alpha} & 0 \leq x < \infty \\ 0 & x < 0. \end{cases} \tag{3}$$

The CHARACTERISTIC FUNCTION describing this distribution is

$$\phi(t) = (1 - it)^{-p}, \tag{4}$$

and the MOMENT-GENERATING FUNCTION is

$$\begin{aligned} M(t) &= \int_0^\infty \frac{e^{tx}x^{\alpha-1}e^{-x/\theta}\,dx}{\Gamma(\alpha)\theta^\alpha} \\ &= \int_0^\infty \frac{x^{\alpha-1}e^{-(1-\theta t)x/\theta}\,dx}{\Gamma(\alpha)\theta^\alpha}. \end{aligned} \tag{5}$$

In order to find the MOMENTS of the distribution, let

$$y \equiv \frac{(1-\theta t)x}{\theta} \tag{6}$$

$$dy = \frac{1-\theta t}{\theta}\,dx, \tag{7}$$

so

$$\begin{aligned} M(t) &= \int_0^\infty \left(\frac{\theta y}{1-\theta t}\right)^{\alpha-1} \frac{e^{-y}}{\Gamma(\alpha)\theta^\alpha}\frac{\theta\,dy}{1-\theta t} \\ &= \frac{1}{(1-\theta t)^\alpha\Gamma(\alpha)}\int_0^\infty y^{\alpha-1}e^{-y}\,dy \\ &= \frac{1}{(1-\theta t)^\alpha}, \end{aligned} \tag{8}$$

and the logarithmic Moment-Generating function is

$$R(t) \equiv \ln M(t) = -\alpha\ln(1-\theta t) \tag{9}$$

$$R'(t) = \frac{\alpha\theta}{1-\theta t} \tag{10}$$

$$R''(t) = \frac{\alpha\theta^2}{(1-\theta t)^2}. \tag{11}$$

The MEAN, VARIANCE, SKEWNESS, and KURTOSIS are then

$$\mu = R'(0) = \alpha\theta \tag{12}$$

$$\sigma^2 = R''(0) = \alpha\theta^2 \tag{13}$$

$$\gamma_1 = \frac{2}{\sqrt{\alpha}} \tag{14}$$

$$\gamma_2 = \frac{6}{\alpha}. \tag{15}$$

The gamma distribution is closely related to other statistical distributions. If X_1, X_2, ..., X_n are independent random variates with a gamma distribution having parameters (α_1, θ), (α_2, θ), ..., (α_n, θ), then $\sum_{i=1}^n X_i$ is distributed as gamma with parameters

$$\alpha = \sum_{i=1}^n \alpha_i \tag{16}$$

$$\theta = \theta. \tag{17}$$

Also, if X_1 and X_2 are independent random variates with a gamma distribution having parameters (α_1, θ) and (α_2, θ), then $X_1/(X_1+X_2)$ is a BETA DISTRIBUTION variate with parameters (α_1, α_2). Both can be derived as follows.

$$P(x, y) = \frac{1}{\Gamma(\alpha_1)\Gamma(\alpha_2)}e^{x_1+x_2}{x_1}^{\alpha_1-1}{x_2}^{\alpha_2-1}. \tag{18}$$

Let

$$u = x_1 + x_2 \qquad x_1 = uv \tag{19}$$

$$v = \frac{x_1}{x_1+x_2} \qquad x_2 = u(1-v), \tag{20}$$

then the JACOBIAN is

$$J\left(\frac{x_1, x_2}{u, v}\right) = \begin{vmatrix} v & u \\ 1-v & -u \end{vmatrix} = -u, \tag{21}$$

so

$$g(u, v)\,du\,dv = f(x, y)\,dx\,dy = f(x, y)u\,du\,dv. \tag{22}$$

$$\begin{aligned} g(u, v) &= \frac{u}{\Gamma(\alpha_1)\Gamma(\alpha_2)}e^{-u}(uv)^{\alpha_1-1}u^{\alpha_2-1}(1-v)^{\alpha_2-1} \\ &= \frac{1}{\Gamma(\alpha_1)\Gamma(\alpha_2)}e^{-u}u^{\alpha_1+\alpha_2-1}v^{\alpha_1-1}(1-v)^{\alpha_2-1}. \end{aligned} \tag{23}$$

The sum $X_1 + X_2$ therefore has the distribution

$$f(u) = f(x_1 + x_2) = \int_0^1 g(u, v)\,dv = \frac{e^{-u}u^{\alpha_1+\alpha_2-1}}{\Gamma(\alpha_1+\alpha_2)}, \tag{24}$$

which is a gamma distribution, and the ratio $X_1/(X_1 + X_2)$ has the distribution

$$\begin{aligned} h(v) = h\left(\frac{x_1}{x_1+x_2}\right) &= \int_0^\infty g(u, v)\,du \\ &= \frac{v^{\alpha_1-1}(1-v)^{\alpha_2-1}}{B(\alpha_1, \alpha_2)}, \end{aligned} \tag{25}$$

where B is the BETA FUNCTION, which is a BETA DISTRIBUTION.

If X and Y are gamma variates with parameters α_1 and α_2, the X/Y is a variate with a BETA PRIME DISTRIBUTION with parameters α_1 and α_2. Let

$$u = x + y \qquad v = \frac{x}{y}, \tag{26}$$

then the JACOBIAN is

$$J\left(\frac{u,v}{x,y}\right) = \begin{vmatrix} 1 & 1 \\ \frac{1}{y} & -\frac{x}{y^2} \end{vmatrix} = -\frac{x+y}{y^2} = -\frac{(1+v)^2}{u}, \quad (27)$$

so

$$dx\,dy = \frac{u}{(1+v)^2}\,du\,dv \quad (28)$$

$$g(u,v) = \frac{1}{\Gamma(\alpha_1)\Gamma(\alpha_2)} e^{-u}\left(\frac{uv}{1+v}\right)^{\alpha_1-1} \left(\frac{u}{1+v}\right)^{\alpha_2-1} \frac{u}{(1+v)^2}$$
$$= \frac{1}{\Gamma(\alpha_1)\Gamma(\alpha_2)} e^{-u} u^{\alpha_1+\alpha_2-1} v^{\alpha_2-1}(1+v)^{-\alpha_1-\alpha_2}. \quad (29)$$

The ratio X/Y therefore has the distribution

$$h(v) = \int_0^\infty g(u,v)\,du = \frac{v^{\alpha_1-1}(1+v)^{-\alpha_1-\alpha_2}}{B(\alpha_1,\alpha_2)}, \quad (30)$$

which is a BETA PRIME DISTRIBUTION with parameters (α_1, α_2).

The "standard form" of the gamma distribution is given by letting $y \equiv x/\theta$, so $dy = dx/\theta$ and

$$P(y)\,dy = \frac{x^{\alpha-1}e^{-x/\theta}}{\Gamma(\alpha)\theta^\alpha}\,dx = \frac{(\theta y)^{\alpha-1}e^{-y}}{\Gamma(\alpha)\theta^\alpha}\,(\theta\,dy)$$
$$= \frac{y^{\alpha-1}e^{-y}}{\Gamma(\alpha)}\,dy, \quad (31)$$

so the MOMENTS about 0 are

$$\nu_r = \frac{1}{\Gamma(\alpha)}\int_0^\infty e^{-x}x^{\alpha-1+r}\,dx = \frac{\Gamma(\alpha+r)}{\Gamma(\alpha)} = (\alpha)_r, \quad (32)$$

where $(\alpha)_r$ is the POCHHAMMER SYMBOL. The MOMENTS about $\mu = \mu_1$ are then

$$\mu_1 = \alpha \quad (33)$$
$$\mu_2 = \alpha \quad (34)$$
$$\mu_3 = 2\alpha \quad (35)$$
$$\mu_4 = 3\alpha^2 + 6\alpha. \quad (36)$$

The MOMENT-GENERATING FUNCTION is

$$M(t) = \frac{1}{(1-t)^\alpha}, \quad (37)$$

and the CUMULANT-GENERATING FUNCTION is

$$K(t) = \alpha\ln(1-t) = \alpha\left(t + \tfrac{1}{2}t^2 + \tfrac{1}{3}t^3 + \ldots\right), \quad (38)$$

so the CUMULANTS are

$$\kappa_r = \alpha\Gamma(r). \quad (39)$$

If x is a NORMAL variate with MEAN μ and STANDARD DEVIATION σ, then

$$y \equiv \frac{(x-\mu)^2}{2\sigma^2} \quad (40)$$

is a standard gamma variate with parameter $\alpha = 1/2$.

see also BETA DISTRIBUTION, CHI-SQUARED DISTRIBUTION

References

Beyer, W. H. *CRC Standard Mathematical Tables, 28th ed.* Boca Raton, FL: CRC Press, p. 534, 1987.

Gamma Function

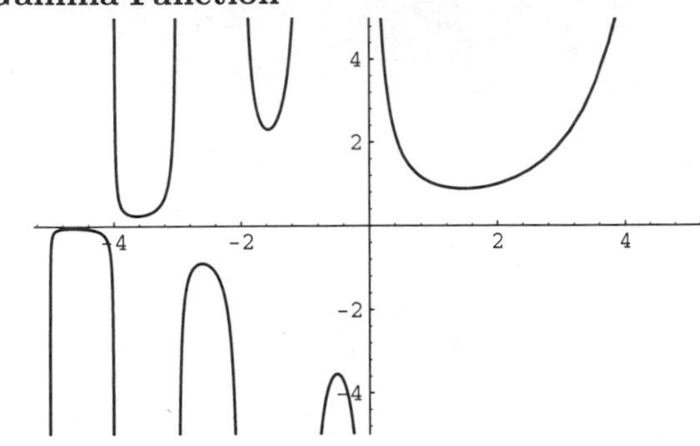

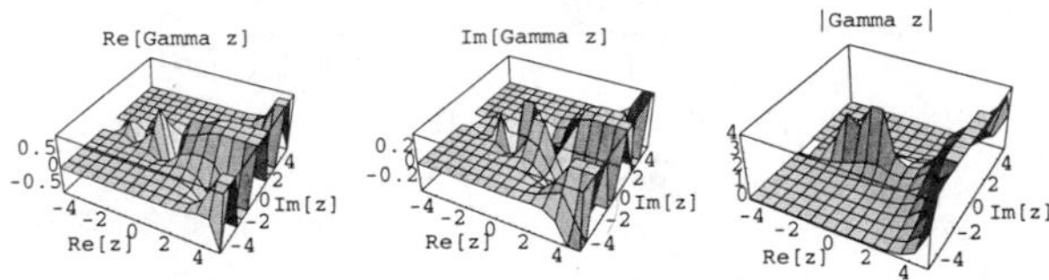

The complete gamma function is defined to be an extension of the FACTORIAL to COMPLEX and REAL NUMBER arguments. It is ANALYTIC everywhere except at $z = 0, -1, -2, \ldots$. It can be defined as a DEFINITE INTEGRAL for $\Re[z] > 0$ (Euler's integral form)

$$\Gamma(z) \equiv \int_0^\infty t^{z-1}e^{-t}\,dt \quad (1)$$
$$= 2\int_0^\infty e^{-t^2}t^{2z-1}\,dt, \quad (2)$$

or

$$\Gamma(z) \equiv \int_0^1 \left[\ln\left(\frac{1}{t}\right)\right]^{z-1} dt. \quad (3)$$

INTEGRATING (1) by parts for a REAL argument, it can be seen that

$$\Gamma(x) = \int_0^\infty t^{x-1}e^{-t}\,dt$$
$$= [-t^{x-1}e^{-t}]_0^\infty + \int_0^\infty (x-1)t^{x-2}e^{-t}\,dt$$
$$= (x-1)\int_0^\infty t^{x-2}e^{-t}\,dt = (x-1)\Gamma(x-1). \quad (4)$$

If x is an INTEGER $n = 1, 2, 3, \ldots$ then

$$\begin{aligned}\Gamma(n) &= (n-1)\Gamma(n-1) = (n-1)(n-2)\Gamma(n-2) \\ &= (n-1)(n-2)\cdots 1 = (n-1)!, \end{aligned} \tag{5}$$

so the gamma function reduces to the FACTORIAL for a POSITIVE INTEGER argument.

BINET'S FORMULA is

$$\ln\Gamma(a) = (a-\tfrac{1}{2})\ln a - a + \tfrac{1}{2}\ln(2\pi) + 2\int_0^\infty \frac{\tan\left(\frac{z}{a}\right)}{e^{2\pi z}-1}\,dz \tag{6}$$

for $\Re[a] > 0$ (Whittaker and Watson 1990, p. 251). The gamma function can also be defined by an INFINITE PRODUCT form (Weierstraß Form)

$$\Gamma(z) \equiv \left[ze^{\gamma z}\prod_{r=1}^{\infty}\left(1+\frac{z}{r}\right)e^{-z/r}\right]^{-1}, \tag{7}$$

where γ is the EULER-MASCHERONI CONSTANT. This can be written

$$\Gamma(z) = \frac{1}{z}\exp\left[\sum_{k=1}^{\infty}\frac{(-1)^k s_k}{k}x^k\right], \tag{8}$$

where

$$s_1 \equiv \gamma \tag{9}$$

$$s_k \equiv \zeta(k) \tag{10}$$

for $k \geq 2$, where ζ is the RIEMANN ZETA FUNCTION (Finch). Taking the logarithm of both sides of (7),

$$-\ln[\Gamma(z)] = \ln z + \gamma z + \sum_{n=1}^{\infty}\left[\ln\left(1+\frac{z}{n}\right) - \frac{z}{n}\right]. \tag{11}$$

Differentiating,

$$\begin{aligned}-\frac{\Gamma'(z)}{\Gamma(z)} &= \frac{1}{z} + \gamma + \sum_{n=1}^{\infty}\left(\frac{\frac{1}{n}}{1+\frac{z}{n}} - \frac{1}{n}\right) \\ &= \frac{1}{z} + \gamma + \sum_{n=1}^{\infty}\left(\frac{1}{n+z} - \frac{1}{n}\right)\end{aligned} \tag{12}$$

$$\Gamma'(z) = -\Gamma(z)\left[\frac{1}{z} + \gamma + \sum_{n=1}^{\infty}\left(\frac{1}{n+z} - \frac{1}{n}\right)\right] \tag{13}$$

$$\equiv \Gamma(z)\Psi(z) = \Gamma(z)\psi_0(z) \tag{14}$$

$$\begin{aligned}\Gamma'(1) &= -\Gamma(1) - \{1 + \gamma + [(\tfrac{1}{2} - 1) + (\tfrac{1}{3} - \tfrac{1}{2}) \\ &\quad + \ldots + \left(\frac{1}{n+1} - \frac{1}{n}\right) + \ldots]\} \\ &= -(1 + \gamma - 1) = -\gamma\end{aligned} \tag{15}$$

$$\begin{aligned}\Gamma'(n) &= -\Gamma(n)\left\{\frac{1}{n} + \gamma + \left[\left(\frac{1}{1+n} - 1\right) + \left(\frac{1}{2+n} - \frac{1}{2}\right)\right.\right. \\ &\quad \left.\left. + \left(\frac{1}{3+n} - \frac{1}{3}\right) + \ldots\right]\right\} \\ &= -(n-1)!\left(\frac{1}{n} + \gamma - \sum_{k=1}^{n}\frac{1}{k}\right),\end{aligned} \tag{16}$$

where $\Psi(z)$ is the DIGAMMA FUNCTION and $\psi_0(z)$ is the POLYGAMMA FUNCTION. nth derivatives are given in terms of the POLYGAMMA FUNCTIONS ψ_n, ψ_{n-1}, ..., ψ_0.

The minimum value x_0 of $\Gamma(x)$ for REAL POSITIVE $x = x_0$ is achieved when

$$\Gamma'(x_0) = \Gamma(x_0)\psi_0(x_0) = 0 \tag{17}$$

$$\psi_0(x_0) = 0, \tag{18}$$

This can be solved numerically to give $x_0 = 1.46163\ldots$ (Sloane's A030169), which has CONTINUED FRACTION [1, 2, 6, 63, 135, 1, 1, 1, 1, 4, 1, 38, ...] (Sloane's A030170). At x_0, $\Gamma(x_0)$ achieves the value 0.8856031944... (Sloane's A030171), which has CONTINUED FRACTION [0, 1, 7, 1, 2, 1, 6, 1, 1, ...] (Sloane's A030172).

The Euler limit form is

$$\begin{aligned}\frac{1}{\Gamma(z)} &= z\left[\lim_{m\to\infty}e^{(1+1/2+\ldots+1/m-\ln m)z}\right] \\ &\quad \times \left[\lim_{m\to\infty}\prod_{n=1}^{m}\left\{\left(1+\frac{z}{n}\right)e^{-z/n}\right\}\right] \\ &= \frac{1}{z}\prod_{n=1}^{\infty}\left[\left(1+\frac{1}{n}\right)^z\left(1+\frac{z}{n}\right)^{-1}\right],\end{aligned} \tag{19}$$

so

$$\Gamma(z) \equiv \lim_{n\to\infty}\frac{1\cdot 2\cdot 3\cdots n}{z(z+1)(z+2)\cdots(z+n)}n^z. \tag{20}$$

The LANCZOS APPROXIMATION for $z > 0$ is

$$\begin{aligned}\Gamma(z+1) &= (z + \gamma + \tfrac{1}{2})^{z+1/2}e^{z+\gamma+1/2}\sqrt{2\pi} \\ &\quad \times \left[c_0 + \frac{c_1}{z+1} + \frac{c_2}{z+2} + \ldots + \frac{c_n}{z+n} + \epsilon\right].\end{aligned} \tag{21}$$

The complete gamma function $\Gamma(x)$ can be generalized to the incomplete gamma function $\Gamma(x,a)$ such that $\Gamma(x) = \Gamma(x,0)$. The gamma function satisfies the recurrence relations

$$\Gamma(1+z) = z\Gamma(z) \tag{22}$$

$$\Gamma(1-z) = -z\Gamma(-z). \tag{23}$$

Additional identities are

$$\Gamma(x)\Gamma(-x) = -\frac{\pi}{x\sin(\pi x)} \tag{24}$$

$$\Gamma(x)\Gamma(1-x) = \frac{\pi}{\sin(\pi x)} \tag{25}$$

$$\begin{aligned}\ln[\Gamma(x+iy+1)] &= \ln(x^2+y^2) + i\tan^{-1}\left(\frac{y}{x}\right) \\ &\quad + \ln[\Gamma(x+iy)]\end{aligned} \tag{26}$$

$$|(ix)!|^2 = \frac{\pi x}{\sinh(\pi x)} \tag{27}$$

$$|(n+ix)!| = \sqrt{\frac{\pi x}{\sinh(\pi x)}}\prod_{s=1}^{n}\sqrt{s^2+x^2}. \tag{28}$$

For integral arguments, the first few values are 1, 1, 2, 6, 24, 120, 720, 5040, 40320, 362880, ... (Sloane's A000142). For half integral arguments,

$$\Gamma(\tfrac{1}{2}) = \sqrt{\pi} \tag{29}$$

$$\Gamma(\tfrac{3}{2}) = \tfrac{1}{2}\sqrt{\pi} \tag{30}$$

$$\Gamma(\tfrac{5}{2}) = \tfrac{3}{4}\sqrt{\pi}. \tag{31}$$

In general, for m a POSITIVE INTEGER $m = 1, 2, \ldots$

$$\Gamma(\tfrac{1}{2}+m) = \frac{1\cdot 3\cdot 5\cdots(2m-1)}{2^m}\sqrt{\pi} = \frac{(2m-1)!!}{2^m}\sqrt{\pi} \tag{32}$$

$$\Gamma(\tfrac{1}{2}-m) = \frac{(-1)^m 2^m}{1\cdot 3\cdot 5\cdots(2m-1)}\sqrt{\pi} = \frac{(-1)^m 2^m}{(2m-1)!!}\sqrt{\pi}. \tag{33}$$

For $\Re[x] = -\frac{1}{2}$,

$$|(-\tfrac{1}{2}+iy)!|^2 = \frac{\pi}{\cosh(\pi y)}. \tag{34}$$

Gamma functions of argument $2z$ can be expressed using the LEGENDRE DUPLICATION FORMULA

$$\Gamma(2z) = (2\pi)^{-1/2}2^{2z-1/2}\Gamma(z)\Gamma(z+\tfrac{1}{2}). \tag{35}$$

Gamma functions of argument $3z$ can be expressed using a triplication FORMULA

$$\Gamma(3z) = (2\pi)^{-1}3^{3z-1/2}\Gamma(z)\Gamma(z+\tfrac{1}{3})\Gamma(z+\tfrac{2}{3}). \tag{36}$$

The general result is the GAUSS MULTIPLICATION FORMULA

$$\Gamma(z)\Gamma(z+\tfrac{1}{n})\cdots\Gamma(z+\tfrac{n+1}{n}) = (2\pi)^{(n-1)/2}n^{1/2-nz}\Gamma(nz). \tag{37}$$

The gamma function is also related to the RIEMANN ZETA FUNCTION ζ by

$$\Gamma\left(\frac{s}{2}\right)\pi^{-2/2}\zeta(s) = \Gamma\left(\frac{1-s}{2}\right)\pi^{-(1-s)/2}\zeta(1-s). \tag{38}$$

Borwein and Zucker (1992) give a variety of identities relating gamma functions to square roots and ELLIPTIC INTEGRAL SINGULAR VALUES k_n, i.e., MODULI k_n such that

$$\frac{K'(k_n)}{K(k_n)} = \sqrt{n}, \tag{39}$$

where $K(k)$ is a complete ELLIPTIC INTEGRAL OF THE FIRST KIND and $K'(k) = K(k') = K(\sqrt{1-k^2})$ is the complementary integral.

$$\Gamma(\tfrac{1}{3}) = 2^{7/9}3^{-1/12}\pi^{1/3}[K(k_3)]^{1/3} \tag{40}$$

$$\Gamma(\tfrac{1}{4}) = 2\pi^{1/4}[K(k_1)]^{1/2} \tag{41}$$

$$\Gamma(\tfrac{1}{6}) = 2^{-1/3}3^{1/2}\pi^{-1/2}[\Gamma(\tfrac{1}{3})]^2 \tag{42}$$

$$\Gamma(\tfrac{1}{8})\Gamma(\tfrac{3}{8}) = (\sqrt{2}-1)^{1/2}2^{13/4}\pi^{1/2}K(k_2) \tag{43}$$

$$\frac{\Gamma(\frac{1}{8})}{\Gamma(\frac{3}{8})} = 2(\sqrt{2}+1)^{1/2}\pi^{-1/4}[K(k_1)]^{1/2} \tag{44}$$

$$\Gamma(\tfrac{1}{12}) = 2^{-1/4}3^{3/8}(\sqrt{3}+1)^{1/2}\pi^{-1/2}\Gamma(\tfrac{1}{4})\Gamma(\tfrac{1}{3}) \tag{45}$$

$$\Gamma(\tfrac{5}{12}) = 2^{1/4}3^{-1/8}(\sqrt{3}-1)^{1/2}\pi^{1/2}\frac{\Gamma(\frac{1}{4})}{\Gamma(\frac{1}{3})} \tag{46}$$

$$\frac{\Gamma(\frac{1}{24})\Gamma(\frac{11}{24})}{\Gamma(\frac{5}{24})\Gamma(\frac{7}{24})} = \sqrt{3}\sqrt{2+\sqrt{3}} \tag{47}$$

$$\frac{\Gamma(\frac{1}{24})\Gamma(\frac{5}{24})}{\Gamma(\frac{7}{24})\Gamma(\frac{11}{24})} = 4\cdot 3^{1/4}(\sqrt{3}+\sqrt{2})\pi^{-1/2}K(k_1) \tag{48}$$

$$\frac{\Gamma(\frac{1}{24})\Gamma(\frac{7}{24})}{\Gamma(\frac{5}{24})\Gamma(\frac{11}{24})} = 2^{25/18}3^{1/3}(\sqrt{2}+1)\pi^{-1/3}[K(k_3)]^{2/3} \tag{49}$$

$$\Gamma(\tfrac{1}{24})\Gamma(\tfrac{5}{25})\Gamma(\tfrac{7}{24})\Gamma(\tfrac{11}{24}) = 384(\sqrt{2}+1)(\sqrt{3}-\sqrt{2})(2-\sqrt{3})\pi[K(k_6)]^2 \tag{50}$$

$$\Gamma(\tfrac{1}{10}) = 2^{-7/10}5^{1/4}(\sqrt{5}+1)^{1/2}\pi^{-1/2}\Gamma(\tfrac{1}{5})\Gamma(\tfrac{2}{5}) \tag{51}$$

$$\Gamma(\tfrac{3}{10}) = 2^{-3/5}(\sqrt{5}-1)\pi^{1/2}\frac{\Gamma(\frac{1}{5})}{\Gamma(\frac{2}{5})} \tag{52}$$

$$\frac{\Gamma(\frac{1}{15})\Gamma(\frac{4}{15})\Gamma(\frac{7}{15})}{\Gamma(\frac{2}{15})} = 2\cdot 3^{1/2}5^{1/6}\sin(\tfrac{2}{15}\pi)[\Gamma(\tfrac{1}{3})]^2 \tag{53}$$

$$\frac{\Gamma(\frac{1}{15})\Gamma(\frac{2}{15})\Gamma(\frac{7}{15})}{\Gamma(\frac{4}{15})} = 2^2\cdot 3^{2/5}\sin(\tfrac{1}{5}\pi)\sin(\tfrac{4}{15}\pi)[\Gamma(\tfrac{1}{5})]^2 \tag{54}$$

$$\frac{\Gamma(\frac{2}{15})\Gamma(\frac{4}{15})\Gamma(\frac{7}{15})}{\Gamma(\frac{1}{15})} = \frac{2^{-3/2}3^{-1/5}5^{1/4}(\sqrt{5}-1)^{1/2}[\Gamma(\frac{2}{5})]^2}{\sin(\frac{4}{15}\pi)} \tag{55}$$

$$\frac{\Gamma(\frac{1}{15})\Gamma(\frac{2}{15})\Gamma(\frac{4}{15})}{\Gamma(\frac{7}{15})} = 60(\sqrt{5}-1)\sin(\tfrac{7}{15}\pi)[K(k_{15})]^2 \tag{56}$$

$$\frac{\Gamma(\frac{1}{20})\Gamma(\frac{9}{20})}{\Gamma(\frac{3}{20})\Gamma(\frac{7}{20})} = 2^{-1}5^{1/4}(\sqrt{5}+1) \tag{57}$$

$$\frac{\Gamma(\frac{1}{20})\Gamma(\frac{3}{20})}{\Gamma(\frac{7}{20})\Gamma(\frac{9}{20})} = 2^{4/5}(10-2\sqrt{5})^{1/2}\pi^{-1}\sin(\tfrac{7}{20}\pi)\sin(\tfrac{9}{20}\pi)[\Gamma(\tfrac{1}{5})]^2 \tag{58}$$

$$\frac{\Gamma(\frac{1}{20})\Gamma(\frac{7}{20})}{\Gamma(\frac{3}{20})\Gamma(\frac{9}{20})}$$

$$= 2^{3/5}(10 + 2\sqrt{5})^{1/2}\pi^{-1}\sin(\tfrac{3}{20}\pi)\sin(\tfrac{9}{20}\pi)[\Gamma(\tfrac{2}{5})]^2 \tag{59}$$

$$\Gamma(\tfrac{1}{20})\Gamma(\tfrac{3}{20})\Gamma(\tfrac{7}{20})\Gamma(\tfrac{9}{20}) = 160(\sqrt{5}-2)^{1/2}\pi[K(k_5)]^2. \tag{60}$$

A few curious identities include

$$\prod_{n=1}^{8}\Gamma(\tfrac{1}{3}n) = \frac{640}{3^6}\left(\frac{\pi}{\sqrt{3}}\right)^3 \tag{61}$$

$$\frac{[\Gamma(\frac{1}{4})]^4}{16\pi^2} = \frac{3^2}{3^2-1}\frac{5^2-1}{5^2}\frac{7^2}{7^2-1}\cdots \tag{62}$$

$$\frac{\Gamma'(1)}{\Gamma(1)} - \frac{\Gamma'(\frac{1}{2})}{\Gamma(\frac{1}{2})} = 2\ln 2 \tag{63}$$

(Magnus and Oberhettinger 1949, p. 1). Ramanujan also gave a number of fascinating identities:

$$\frac{\Gamma^2(n+1)}{\Gamma(n+xi+1)\Gamma(n-xi+1)} = \prod_{k=1}^{\infty}\left[1+\frac{x^2}{(n+k)^2}\right] \tag{64}$$

$$\phi(m,n)\phi(n,m) = \frac{\Gamma^3(m+1)\Gamma^3(n+1)}{\Gamma(2m+n+1)\Gamma(2n+m+1)} \times\frac{\cosh[\pi(m+n)\sqrt{3}\,] - \cos[\pi(m-n)]}{2\pi^2(m^2+mn+n^2)}, \tag{65}$$

where

$$\phi(m,n) \equiv \prod_{k=1}^{\infty}\left[1+\left(\frac{m+n}{k+m}\right)^3\right], \tag{66}$$

$$\prod_{k=1}^{\infty}\left[1+\left(\frac{n}{k}\right)^3\right]\prod_{k=1}^{\infty}\left[1+3\left(\frac{n}{n+2k}\right)^2\right] = \frac{\Gamma(\frac{1}{2}n)}{\Gamma[\frac{1}{2}(n+1)]}\frac{\cosh(\pi n\sqrt{3}\,)-\cos(\pi n)}{2^{n+2}\pi^{3/2}n} \tag{67}$$

(Berndt 1994).

The following ASYMPTOTIC SERIES is occasionally useful in probability theory (e.g., the 1-D RANDOM WALK):

$$\frac{\Gamma(J+\frac{1}{2})}{\Gamma(J)} = \sqrt{J}\left(1-\frac{1}{8J}+\frac{1}{128J^2} +\frac{5}{1024J^3}-\frac{21}{32768J^4}+\ldots\right) \tag{68}$$

(Graham *et al.* 1994). This series also gives a nice asymptotic generalization of STIRLING NUMBERS OF THE FIRST KIND to fractional values.

It has long been known that $\Gamma(\frac{1}{4})\pi^{-1/4}$ is TRANSCENDENTAL (Davis 1959), as is $\Gamma(\frac{1}{3})$ (Le Lionnais 1983), and Chudnovsky has apparently recently proved that $\Gamma(\frac{1}{4})$ is itself TRANSCENDENTAL.

The upper incomplete gamma function is given by

$$\Gamma(a,x) \equiv \int_x^{\infty} t^{a-1}e^{-t}\,dt = 1-\gamma(a,x), \tag{69}$$

where γ is the lower incomplete gamma function. For a an INTEGER n

$$\Gamma(n,x) = (n-1)!e^{-x}\sum_{s=0}^{n-1}\frac{x^s}{s!} = (n-1)!e^{-x}\,\mathrm{es}_{n-1}(x), \tag{70}$$

where es is the EXPONENTIAL SUM FUNCTION. The lower incomplete gamma function is given by

$$\begin{aligned}\gamma(a,x) &\equiv \Gamma(a)-\Gamma(a,x) = \int_0^x e^{-t}t^{a-1}\,dt \\ &= a^{-1}x^a e^{-x}{}_1F_1(1;1+a;x) \\ &= a^{-1}x^a{}_1F_1(a;1+a;-x),\end{aligned} \tag{71}$$

where ${}_1F_1(a;b;x)$ is the CONFLUENT HYPERGEOMETRIC FUNCTION OF THE FIRST KIND. For a an INTEGER n,

$$\begin{aligned}\gamma(n,x) &= (n-1)!\left(1-e^{-x}\sum_{s=0}^{n-1}\frac{x^s}{s!}\right) \\ &= (n-1)!\,[1-\mathrm{es}_{n-1}(x)].\end{aligned} \tag{72}$$

The function $\Gamma(a,z)$ is denoted **Gamma[a,z]** and the function $\gamma(a,z)$ is denoted **Gamma[a,0,z]** in *Mathematica*® (Wolfram Research, Champaign, IL).

see also DIGAMMA FUNCTION, DOUBLE GAMMA FUNCTION, FRANSÉN-ROBINSON CONSTANT G-FUNCTION, GAUSS MULTIPLICATION FORMULA, LAMBDA FUNCTION, LEGENDRE DUPLICATION FORMULA, MU FUNCTION, NU FUNCTION, PEARSON'S FUNCTION, POLYGAMMA FUNCTION, REGULARIZED GAMMA FUNCTION, STIRLING'S SERIES

References

Abramowitz, M. and Stegun, C. A. (Eds.). "Gamma (Factorial) Function" and "Incomplete Gamma Function." §6.1 and 6.5 in *Handbook of Mathematical Functions with Formulas, Graphs, and Mathematical Tables, 9th printing.* New York: Dover, pp. 255–258 and 260–263, 1972.

Arfken, G. "The Gamma Function (Factorial Function)." Ch. 10 in *Mathematical Methods for Physicists, 3rd ed.* Orlando, FL: Academic Press, pp. 339–341 and 539–572, 1985.

Artin, E. *The Gamma Function.* New York: Holt, Rinehart, and Winston, 1964.

Berndt, B. C. *Ramanujan's Notebooks, Part IV.* New York: Springer-Verlag, pp. 334–342, 1994.

Borwein, J. M. and Zucker, I. J. "Elliptic Integral Evaluation of the Gamma Function at Rational Values of Small Denominator." *IMA J. Numerical Analysis* **12**, 519–526, 1992.

Davis, H. T. *Tables of the Higher Mathematical Functions.* Bloomington, IN: Principia Press, 1933.

Davis, P. J. "Leonhard Euler's Integral: A Historical Profile of the Gamma Function." *Amer. Math. Monthly* **66**, 849–869, 1959.

Finch, S. "Favorite Mathematical Constants." `http://www.mathsoft.com/asolve/constant/fran/fran.html`.

Graham, R. L.; Knuth, D. E.; and Patashnik, O. Answer to problem 9.60 in *Concrete Mathematics: A Foundation for Computer Science.* Reading, MA: Addison-Wesley, 1994.

Le Lionnais, F. *Les nombres remarquables.* Paris: Hermann, p. 46, 1983.

Magnus, W. and Oberhettinger, F. *Formulas and Theorems for the Special Functions of Mathematical Physics.* New York: Chelsea, 1949.

Nielsen, H. *Die Gammafunktion.* New York: Chelsea, 1965.

Press, W. H.; Flannery, B. P.; Teukolsky, S. A.; and Vetterling, W. T. "Gamma Function, Beta Function, Factorials, Binomial Coefficients" and "Incomplete Gamma Function, Error Function, Chi-Square Probability Function, Cumulative Poisson Function." §6.1 and 6.2 in *Numerical Recipes in FORTRAN: The Art of Scientific Computing, 2nd ed.* Cambridge, England: Cambridge University Press, pp. 206–209 and 209–214, 1992.

Sloane, N. J. A. Sequences A030169, A030170, A030171, A030172, and A000142/M1675 in "An On-Line Version of the Encyclopedia of Integer Sequences."

Spanier, J. and Oldham, K. B. "The Gamma Function $\Gamma(x)$" and "The Incomplete Gamma $\gamma(\nu; x)$ and Related Functions." Chs. 43 and 45 in *An Atlas of Functions.* Washington, DC: Hemisphere, pp. 411–421 and 435–443, 1987.

Whittaker, E. T. and Watson, G. N. *A Course in Modern Analysis, 4th ed.* Cambridge, England: Cambridge University Press, 1990.

Gamma Group

The gamma group Γ is the set of all transformations w of the form

$$w(t) = \frac{at+b}{ct+d},$$

where a, b, c, and d are INTEGERS and $ad - bc = 1$.

see also KLEIN'S ABSOLUTE INVARIANT, LAMBDA GROUP, THETA FUNCTION

References

Borwein, J. M. and Borwein, P. B. *Pi & the AGM: A Study in Analytic Number Theory and Computational Complexity.* New York: Wiley, pp. 127–132, 1987.

Gamma-Modular

see MODULAR GAMMA FUNCTION

Gamma Statistic

$$\gamma_r \equiv \frac{\kappa_r}{\sigma^{r+2}},$$

where κ_r are CUMULANTS and σ is the STANDARD DEVIATION.

see also KURTOSIS, SKEWNESS

Garage Door

see ASTROID

Gårding's Inequality

Gives a lower bound for the inner product (Lu, u), where L is a linear elliptic REAL differential operator of order m, and u has compact support.

References

Knapp, A. W. "Group Representations and Harmonic Analysis, Part II." *Not. Amer. Math. Soc.* **43**, 537–549, 1996.

Garman-Kohlhagen Formula

$$V_t = e^{-y\tau} S_t N(d_1) - e^{-r\tau} K N(d_2),$$

where N is the cumulative NORMAL DISTRIBUTION and

$$d_1, d_2 = \frac{\log\left(\frac{S_t}{K}\right) + (r - y \pm \frac{1}{2}\sigma^2)\tau}{\sigma\sqrt{\tau}}.$$

If $y = 0$, this is the standard form of the Black-Scholes formula.

see also BLACK-SCHOLES THEORY

References

Garman, M. B. and Kohlhagen, S. W. "Foreign Currency Option Values." *J. International Money and Finance* **2**, 231–237, 1983.

Price, J. F. "Optional Mathematics is Not Optional." *Not. Amer. Math. Soc.* **43**, 964–971, 1996.

Gate Function

Bracewell's term for the RECTANGLE FUNCTION.

References

Bracewell, R. *The Fourier Transform and Its Applications.* New York: McGraw-Hill, 1965.

Gauche Conic

see SKEW CONIC

Gaullist Cross

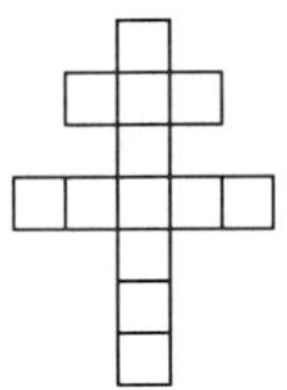

A CROSS also called the CROSS OF LORRAINE or PATRIARCHAL CROSS.

see also CROSS, DISSECTION

Gauss's Backward Formula

$$f_p = f_0 + p\delta_{-1/2} + G_2^*\delta_0^2 + G_3\delta_{-1/2}^3 + G_4^*\delta_0^4 + G_5\delta_{-1/2}^5 + \ldots,$$

for $p \in [0, 1]$, where δ is the CENTRAL DIFFERENCE and

$$G_{2n}^* = \binom{p+n}{2n}$$

$$G_{2n+1} = \binom{p+n}{2n+1},$$

where $\binom{n}{k}$ is a BINOMIAL COEFFICIENT.

see also CENTRAL DIFFERENCE, GAUSS'S FORWARD FORMULA

References

Beyer, W. H. *CRC Standard Mathematical Tables, 28th ed.* Boca Raton, FL: CRC Press, p. 433, 1987.

Gauss-Bodenmiller Theorem

The CIRCLES on the DIAGONALS of a COMPLETE QUADRILATERAL as DIAMETERS are COAXAL. Furthermore, the ORTHOCENTERS of the four TRIANGLES of a COMPLETE QUADRILATERAL are COLLINEAR on the RADICAL AXIS of the COAXAL CIRCLES.

see also COAXAL CIRCLES, COLLINEAR, COMPLETE QUADRILATERAL, DIAGONAL (POLYGON), ORTHOCENTER, RADICAL AXIS

References

Johnson, R. A. *Modern Geometry: An Elementary Treatise on the Geometry of the Triangle and the Circle.* Boston, MA: Houghton Mifflin, p. 172, 1929.

Gauss-Bolyai-Lobachevsky Space

A non-Euclidean space with constant NEGATIVE GAUSSIAN CURVATURE.

see also LOBACHEVSKY-BOLYAI-GAUSS GEOMETRY, NON-EUCLIDEAN GEOMETRY

Gauss-Bonnet Formula

The Gauss-Bonnet formula has several formulations. The simplest one expresses the total GAUSSIAN CURVATURE of an embedded triangle in terms of the total GEODESIC CURVATURE of the boundary and the JUMP ANGLES at the corners.

More specifically, if M is any 2-D RIEMANNIAN MANIFOLD (like a surface in 3-space) and if T is an embedded triangle, then the Gauss-Bonnet formula states that the integral over the whole triangle of the GAUSSIAN CURVATURE with respect to AREA is given by 2π minus the sum of the JUMP ANGLES minus the integral of the GEODESIC CURVATURE over the whole of the boundary of the triangle (with respect to ARC LENGTH),

$$\iint_T K\,dA = 2\pi - \sum \alpha_i - \int_{\partial T} \kappa_g\,ds, \tag{1}$$

where K is the GAUSSIAN CURVATURE, dA is the AREA measure, the α_is are the JUMP ANGLES of ∂T, and κ_g is the GEODESIC CURVATURE of ∂T, with ds the ARC LENGTH measure.

The next most common formulation of the Gauss-Bonnet formula is that for any compact, boundaryless 2-D RIEMANNIAN MANIFOLD, the integral of the GAUSSIAN CURVATURE over the entire MANIFOLD with respect to AREA is 2π times the EULER CHARACTERISTIC of the MANIFOLD,

$$\iint_M K\,dA = 2\pi\chi(M). \tag{2}$$

This is somewhat surprising because the total GAUSSIAN CURVATURE is differential-geometric in character, but the EULER CHARACTERISTIC is topological in character and does not depend on differential geometry at all. So if you distort the surface and change the curvature at any location, regardless of how you do it, the same total curvature is maintained.

Another way of looking at the Gauss-Bonnet theorem for surfaces in 3-space is that the GAUSS MAP of the surface has DEGREE given by half the EULER CHARACTERISTIC of the surface

$$\iint_M K\,dA = 2\pi\chi(M) - \sum \alpha_i - \int_{\partial M} \kappa_g\,ds, \tag{3}$$

which works only for ORIENTABLE SURFACES. This makes the Gauss-Bonnet theorem a simple consequence of the POINCARE-HOPF INDEX THEOREM, which is a nice way of looking at things if you're a topologist, but not so nice for a differential geometer. This proof can be found in Guillemin and Pollack (1974). Millman and Parker (1977) give a standard differential-geometric proof of the Gauss-Bonnet theorem, and Singer and Thorpe (1996) give a GAUSS'S THEOREMA EGREGIUM-inspired proof which is entirely intrinsic, without any reference to the ambient EUCLIDEAN SPACE.

A general Gauss-Bonnet formula that takes into account both formulas can also be given. For any compact 2-D RIEMANNIAN MANIFOLD with corners, the integral of the GAUSSIAN CURVATURE over the 2-MANIFOLD with respect to AREA is 2π times the EULER CHARACTERISTIC of the MANIFOLD minus the sum of the JUMP ANGLES and the total GEODESIC CURVATURE of the boundary.

References

Chavel, I. *Riemannian Geometry: A Modern Introduction.* New York: Cambridge University Press, 1994.

Guillemin, V. and Pollack, A. *Differential Topology.* Englewood Cliffs, NJ: Prentice-Hall, 1974.

Millman, R. S. and Parker, G. D. *Elements of Differential Geometry.* Prentice-Hall, 1977.

Reckziegel, H. In *Mathematical Models from the Collections of Universities and Museums* (Ed. G. Fischer). Braunschweig, Germany: Vieweg, p. 31, 1986.

Singer, I. M. and Thorpe, J. A. *Lecture Notes on Elementary Topology and Geometry.* New York: Springer-Verlag, 1996.

Gauss-Bonnet Theorem

see GAUSS-BONNET FORMULA

Gauss's Circle Problem

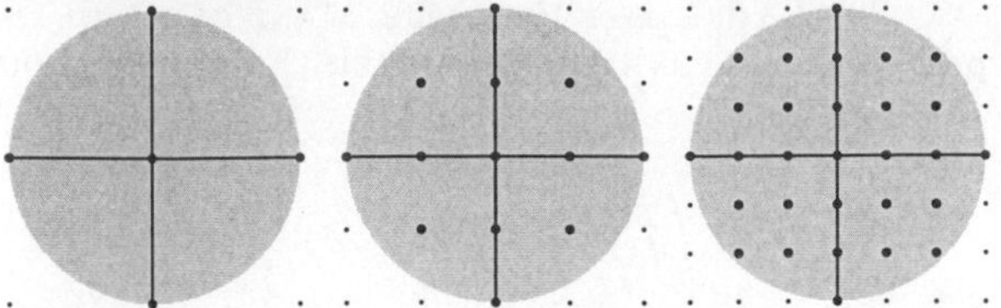

Count the number of LATTICE POINTS $N(r)$ inside the boundary of a CIRCLE of RADIUS r with center at the origin. The exact solution is given by the SUM

$$N(r) = 1 + 4\lfloor r \rfloor + 4\sum_{i=1}^{\lfloor r \rfloor} \left\lfloor \sqrt{r^2 - i^2} \right\rfloor. \qquad (1)$$

The first few values are 1, 5, 13, 29, 49, 81, 113, 149, ... (Sloane's A000328).

Gauss showed that

$$N(r) = \pi r^2 + E(r), \qquad (2)$$

where

$$|E(r)| \leq 2\sqrt{2}\,\pi r. \qquad (3)$$

Writing $|E(r)| \leq Cr^\theta$, the best bounds on θ are $1/2 < \theta \leq 46/73 \approx 0.630137$ (Huxley 1990). The problem has also been extended to CONICS and higher dimensions. The limit 1/2 was obtained by Hardy and Landau (1915), and the limit 46/73 improves previous values of $24/37 \approx 0.64864$ (Cheng 1963) and $34/53 \approx 0.64150$ (Vinogradov), and $7/11 \approx 0.63636$.

see also CIRCLE LATTICE POINTS

References

Cheng, J. R. "The Lattice Points in a Circle." *Sci. Sinica* **12**, 633–649, 1963.

Cilleruello, J. "The Distribution of Lattice Points on Circles." *J. Number Th.* **43**, 198–202, 1993.

Guy, R. K. "Gauß's Lattice Point Problem." §F1 in *Unsolved Problems in Number Theory, 2nd ed.* New York: Springer-Verlag, pp. 240–2417, 1994.

Huxley, M. N. "Exponential Sums and Lattice Points." *Proc. London Math. Soc.* **60**, 471–502, 1990.

Huxley, M. N. "Corrigenda: 'Exponential Sums and Lattice Points'." *Proc. London Math. Soc.* **66**, 70, 1993.

Le Lionnais, F. *Les nombres remarquables.* Paris: Hermann, p. 24, 1983.

Sloane, N. J. A. Sequence A000328/M3829 in "An On-Line Version of the Encyclopedia of Integer Sequences."

Weisstein, E. W. "Circle Lattice Points." `http://www.astro.virginia.edu/~eww6n/math/notebooks/CircleLatticePoints.m`.

Gauss's Class Number Conjecture

In his monumental treatise *Disquisitiones Arithmeticae,* Gauss conjectured that the CLASS NUMBER $h(-d)$ of an IMAGINARY quadratic field with DISCRIMINANT $-d$ tends to infinity with d. A proof was finally given by Heilbronn (1934), and Siegel (1936) showed that for any $\epsilon > 0$, there exists a constant $c_\epsilon > 0$ such that

$$h(-d) > c_\epsilon d^{1/2-\epsilon}$$

as $d \to \infty$. However, these results were not effective in actually determining the values for a given m of a complete list of fundamental discriminants $-d$ such that $h(-d) = m$, a problem known as GAUSS'S CLASS NUMBER PROBLEM.

Goldfeld (1976) showed that if there exists a "Weil curve" whose associated DIRICHLET L-SERIES has a zero of at least third order at $s = 1$, then for any $\epsilon > 0$, there exists an effectively computable constant c_ϵ such that

$$h(-d) > c_\epsilon (\ln d)^{1-\epsilon}.$$

Gross and Zaiger (1983) showed that certain curves must satisfy the condition of Goldfeld, and Goldfeld's proof was simplified by Oesterlé (1985).

see also CLASS NUMBER, GAUSS'S CLASS NUMBER PROBLEM, HEEGNER NUMBER

References

Arno, S.; Robinson, M. L.; and Wheeler, F. S. "Imaginary Quadratic Fields with Small Odd Class Number." `http://www.math.uiuc.edu/Algebraic-Number-Theory/0009/`.

Böcherer, S. "Das Gauß'sche Klassenzahlproblem." *Mitt.. Math. Ges. Hamburg* **11**, 565–589, 1988.

Gauss, C. F. *Disquisitiones Arithmeticae.* New Haven, CT: Yale University Press, 1966.

Goldfeld, D. M. "The Class Number of Quadratic Fields and the Conjectures of Birch and Swinnerton-Dyer." *Ann. Scuola Norm. Sup. Pisa* **3**, 623–663, 1976.

Gross, B. and Zaiger, D. "Points de Heegner et derivées de fonctions L." *C. R. Acad. Sci. Paris* **297**, 85–87, 1983.

Heilbronn, H. "On the Class Number in Imaginary Quadratic Fields." *Quart. J. Math. Oxford Ser.* **25**, 150–160, 1934.

Oesterlé, J. "Nombres de classes des corps quadratiques imaginaires." *Astérique* **121-122**, 309–323, 1985.

Siegel, C. L. "Über die Klassenzahl quadratischer Zahlkörper." *Acta. Arith.* **1**, 83–86, 1936.

Gauss's Class Number Problem

For a given m, determine a complete list of fundamental DISCRIMINANTS $-d$ such that the CLASS NUMBER is given by $h(-d) = m$. Heegner (1952) gave a solution for $m = 1$, but it was not completely accepted due to a number of apparent gaps. However, subsequent examination of Heegner's proof show it to be "essentially" correct (Conway and Guy 1996). Conway and Guy (1996) therefore call the nine values of $n(-d)$ having $h(-d) = 1$ where $-d$ is the DISCRIMINANT corresponding to a QUADRATIC FIELD $a + b\sqrt{-n}$ ($n = -1, -2, -3, -7, -11, -19, -43, -67$, and -163; Sloane's A003173) the HEEGNER NUMBERS. The HEEGNER NUMBERS have a number of fascinating properties.

Stark (1967) and Baker (1966) gave independent proofs of the fact that only nine such numbers exist; both proofs were accepted. Baker (1971) and Stark (1975) subsequently and independently solved the generalized class number problem completely for $m = 2$. Oesterlé (1985) solved the case $m = 3$, and Arno (1992) solved the case $m = 4$. Wagner (1996) solve the cases $n = 5$, 6, and 7. Arno *et al.* (1993) solved the problem for ODD m satisfying $5 \leq m \leq 23$.

see also CLASS NUMBER, GAUSS'S CLASS NUMBER CONJECTURE, HEEGNER NUMBER

References

Arno, S. "The Imaginary Quadratic Fields of Class Number 4." *Acta Arith.* **40**, 321–334, 1992.

Arno, S.; Robinson, M. L.; and Wheeler, F. S. "Imaginary Quadratic Fields with Small Odd Class Number." Dec. 1993. `http://www.math.uiuc.edu/Algebraic-Number-Theory/0009/`.

Baker, A. "Linear Forms in the Logarithms of Algebraic Numbers. I." *Mathematika* **13**, 204–216, 1966.

Baker, A. "Imaginary Quadratic Fields with Class Number 2." *Ann. Math.* **94**, 139–152, 1971.

Conway, J. H. and Guy, R. K. "The Nine Magic Discriminants." In *The Book of Numbers.* New York: Springer-Verlag, pp. 224–226, 1996.

Goldfeld, D. M. "Gauss' Class Number Problem for Imaginary Quadratic Fields." *Bull. Amer. Math. Soc.* **13**, 23–37, 1985.

Heegner, K. "Diophantische Analysis und Modulfunktionen." *Math. Z.* **56**, 227–253, 1952.

Heilbronn, H. A. and Linfoot, E. H. "On the Imaginary Quadratic Corpora of Class-Number One." *Quart. J. Math. (Oxford)* **5**, 293–301, 1934.

Lehmer, D. H. "On Imaginary Quadratic Fields whose Class Number is Unity." *Bull. Amer. Math. Soc.* **39**, 360, 1933.

Montgomery, H. and Weinberger, P. "Notes on Small Class Numbers." *Acta. Arith.* **24**, 529–542, 1974.

Oesterlé, J. "Nombres de classes des corps quadratiques imaginaires." *Astérique* **121–122**, 309–323, 1985.

Oesterlé, J. "Le problème de Gauss sur le nombre de classes." *Enseign Math.* **34**, 43–67, 1988.

Serre, J.-P. $\Delta = b^2 - 4ac$." *Math. Medley* **13**, 1–10, 1985.

Shanks, D. "On Gauss's Class Number Problems." *Math. Comput.* **23**, 151–163, 1969.

Sloane, N. J. A. Sequence A003173/M0827 in "An On-Line Version of the Encyclopedia of Integer Sequences."

Stark, H. M. "A Complete Determination of the Complex Quadratic Fields of Class Number One." *Michigan Math. J.* **14**, 1–27, 1967.

Stark, H. M. "On Complex Quadratic Fields with Class Number Two." *Math. Comput.* **29**, 289–302, 1975.

Wagner, C. "Class Number 5, 6, and 7." *Math. Comput.* **65**, 785–800, 1996.

Gauss's Constant

The RECIPROCAL of the ARITHMETIC-GEOMETRIC MEAN of 1 and $\sqrt{2}$,

$$\begin{aligned}\frac{1}{M(1,\sqrt{2})} &= \frac{2}{\pi}\int_0^1 \frac{1}{\sqrt{1-x^4}}\,dx && (1)\\ &= \frac{2}{\pi}\int_0^{\pi/2}\frac{d\theta}{\sqrt{1+\sin^2\theta}} && (2)\\ &= \frac{\sqrt{2}}{\pi}K\left(\frac{1}{\sqrt{2}}\right)\\ &= \frac{1}{(2\pi)^{3/2}}[\Gamma(\tfrac{1}{4})]^2 && (3)\\ &= 0.83462684167\ldots, && (4)\end{aligned}$$

where $K(k)$ is the complete ELLIPTIC INTEGRAL OF THE FIRST KIND and $\Gamma(z)$ is the GAMMA FUNCTION.

see also ARITHMETIC-GEOMETRIC MEAN, GAUSS-KUZMIN-WIRSING CONSTANT

References

Finch, S. "Favorite Mathematical Constants." `http://www.mathsoft.com/asolve/constant/gauss/gauss.html`.

Gauss's Criterion

Let p be an ODD PRIME and b a POSITIVE INTEGER not divisible by p. Then for each POSITIVE ODD INTEGER $2k-1<p$, let r_i be

$$r_k \equiv (2k-1)b \pmod p$$

with $0<r_k<p$, and let t be the number of EVEN r_is. Then

$$(b/p) = (-1)^t,$$

where (b/p) is the LEGENDRE SYMBOL.

References

Shanks, D. "Gauss's Criterion." §1.17 in *Solved and Unsolved Problems in Number Theory, 4th ed.* New York: Chelsea, pp. 38–40, 1993.

Gauss's Double Point Theorem

If a sequence of DOUBLE POINTS is passed as a CLOSED CURVE is traversed, each DOUBLE POINT appears once in an EVEN place and once in an ODD place.

References

Rademacher, H. and Toeplitz, O. *The Enjoyment of Mathematics: Selections from Mathematics for the Amateur.* Princeton, NJ: Princeton University Press, pp. 61–66, 1957.

Gauss Equations

If $\mathbf{x}$ is a regular patch on a REGULAR SURFACE in $\mathbb{R}^3$ with normal $\hat{\mathbf{N}}$, then

$$\begin{aligned}\mathbf{x}_{uu} &= \Gamma_{11}^1\mathbf{x}_u + \Gamma_{11}^2\mathbf{x}_v + e\hat{\mathbf{N}} && (1)\\ \mathbf{x}_{uv} &= \Gamma_{12}^1\mathbf{x}_u + \Gamma_{12}^2\mathbf{x}_v + f\hat{\mathbf{N}} && (2)\\ \mathbf{x}_{vv} &= \Gamma_{22}^1\mathbf{x}_u + \Gamma_{22}^2\mathbf{x}_v + g\hat{\mathbf{N}}, && (3)\end{aligned}$$

where e, f, and g are coefficients of the second FUNDAMENTAL FORM and Γ_{ij}^k are CHRISTOFFEL SYMBOLS OF THE SECOND KIND.

see also CHRISTOFFEL SYMBOL OF THE SECOND KIND, FUNDAMENTAL FORMS, MAINARDI-CODAZZI EQUATIONS

References

Gray, A. *Modern Differential Geometry of Curves and Surfaces.* Boca Raton, FL: CRC Press, pp. 398–400, 1993.

Gauss's Equation (Radius Derivatives)

Expresses the second derivatives of $\mathbf{r}$ in terms of the CHRISTOFFEL SYMBOL OF THE SECOND KIND.

$$\mathbf{r}_{ij} = \Gamma_{ij}^k\mathbf{r}_k + (\mathbf{r}_{ij}\cdot\mathbf{n})\mathbf{n}.$$

Gauss's Formula

$$4\frac{x^p - y^p}{x - y} = R^2(x,y) - (-1)^{(p-1)/2} p S^2(x,y),$$

where R and S are HOMOGENEOUS POLYNOMIALS in x and y with integral COEFFICIENTS.

see also AURIFEUILLEAN FACTORIZATION, GAUSS'S BACKWARD FORMULA, GAUSS'S FORWARD FORMULA

References
Shanks, D. *Solved and Unsolved Problems in Number Theory, 4th ed.* New York: Chelsea, p. 105, 1993.

Gauss's Formulas

Let a SPHERICAL TRIANGLE have sides a, b, and c with A, B, and C the corresponding opposite angles. Then

$$\frac{\sin[\frac{1}{2}(a-b)]}{\sin(\frac{1}{2}c)} = \frac{\sin[\frac{1}{2}(A-B)]}{\cos(\frac{1}{2}C)} \quad (1)$$

$$\frac{\sin[\frac{1}{2}(a+b)]}{\sin(\frac{1}{2}c)} = \frac{\cos[\frac{1}{2}(A-B)]}{\sin(\frac{1}{2}C)} \quad (2)$$

$$\frac{\cos[\frac{1}{2}(a-b)]}{\cos(\frac{1}{2}c)} = \frac{\sin[\frac{1}{2}(A+B)]}{\cos(\frac{1}{2}C)} \quad (3)$$

$$\frac{\cos[\frac{1}{2}(a+b)]}{\cos(\frac{1}{2}c)} = \frac{\cos[\frac{1}{2}(A+B)]}{\sin(\frac{1}{2}C)}. \quad (4)$$

see also SPHERICAL TRIGONOMETRY

Gauss's Forward Formula

$$f_p = f_0 + p\delta_{1/2} + G_2\delta_0^2 + G_3\delta_{1/2}^3 + G_4\delta_0^4 + G_5\delta_{1/2}^5 + \ldots,$$

for $p \in [0,1]$, where δ is the CENTRAL DIFFERENCE and

$$G_{2n} = \binom{p+n-1}{2n}$$

$$G_{2n+1} = \binom{p+n}{2n+1},$$

where $\binom{n}{k}$ is a BINOMIAL COEFFICIENT.

see also CENTRAL DIFFERENCE, GAUSS'S BACKWARD FORMULA

References
Beyer, W. H. *CRC Standard Mathematical Tables, 28th ed.* Boca Raton, FL: CRC Press, p. 433, 1987.

Gauss's Harmonic Function Theorem

If a function ϕ is HARMONIC in a SPHERE, then the value of ϕ at the center of the SPHERE is the ARITHMETIC MEAN of its value on the surface.

Gauss's Hypergeometric Theorem

$${}_2F_1(a,b;c;1) = \frac{\Gamma(c)\Gamma(c-a-b)}{\Gamma(c-a)\Gamma(c-b)}$$

for $\Re[c-a-b] > 0$, where ${}_2F_1(a,b;c;x)$ is a HYPERGEOMETRIC FUNCTION. If a is a NEGATIVE INTEGER $-n$, this becomes

$${}_2F_1(-n,b;c;1) = \frac{(c-b)_n}{(c)_n},$$

which is known as the VANDERMONDE THEOREM.

see also GENERALIZED HYPERGEOMETRIC FUNCTION, HYPERGEOMETRIC FUNCTION

References
Petkovšek, M.; Wilf, H. S.; and Zeilberger, D. *A=B.* Wellesley, MA: A. K. Peters, pp. 42 and 126, 1996.

Gauss's Inequality

If a distribution has a single MODE at μ_0, then

$$P(|x-\mu_0| \geq \lambda\tau) \leq \frac{4}{9\lambda^2},$$

where

$$\tau^2 \equiv \sigma^2 + (\mu - \mu_0)^2.$$

Gauss's Interpolation Formula

$$f(x) \approx t_n(x) = \sum_{k=0}^{2n} f_k \zeta_k(x),$$

where $t_n(x)$ is a trigonometric POLYNOMIAL of degree n such that $t_n(x_k) = f_k$ for $k = 0, \ldots, 2n$, and

$$\zeta_k(x) = \frac{\sin[\frac{1}{2}(x-x_0)]\cdots\sin[\frac{1}{2}(x-x_{k-1})]}{\sin[\frac{1}{2}(x_k-x_0)]\cdots\sin[\frac{1}{2}(x_k-x_{k-1})]} \frac{\sin[\frac{1}{2}(x-x_{k+1})]\cdots\sin[\frac{1}{2}(x-x_{2n})]}{\sin[\frac{1}{2}(x_k-x_{k+1})]\cdots\sin[\frac{1}{2}(x_k-x_{2n})]}.$$

References
Abramowitz, M. and Stegun, C. A. (Eds.). *Handbook of Mathematical Functions with Formulas, Graphs, and Mathematical Tables, 9th printing.* New York: Dover, p. 881, 1972.
Beyer, W. H. (Ed.) *CRC Standard Mathematical Tables, 28th ed.* Boca Raton, FL: CRC Press, pp. 442–443, 1987.

Gauss-Jacobi Mechanical Quadrature

If $x_1 < x_2 < \ldots < x_n$ denote the zeros of $p_n(x)$, there exist REAL NUMBERS $\lambda_1, \lambda_2, \ldots, \lambda_n$ such that

$$\int_a^b \rho(x)\,d\alpha(x) = \lambda_1\rho(x_1) + \lambda_2\rho(x_2) + \ldots + \lambda_n\rho(x_n),$$

for an arbitrary POLYNOMIAL of order $2n-1$ and the $\lambda_n' s$ are called CHRISTOFFEL NUMBERS. The distribution $d\alpha(x)$ and the INTEGER n uniquely determine these numbers λ_ν.

References
Szegő, G. *Orthogonal Polynomials, 4th ed.* Providence, RI: Amer. Math. Soc., p. 47, 1975.

Gauss-Jordan Elimination

A method for finding a MATRIX INVERSE. To apply Gauss-Jordan elimination, operate on a MATRIX

$$[\mathsf{A} \quad \mathsf{I}] \equiv \begin{bmatrix} a_{11} & \cdots & a_{1n} & 1 & 0 & \cdots & 0 \\ a_{21} & \cdots & a_{2n} & 0 & 1 & \cdots & 0 \\ \vdots & \ddots & \vdots & \vdots & \vdots & \ddots & \vdots \\ a_{n1} & \cdots & a_{nn} & 0 & 0 & \cdots & 1 \end{bmatrix},$$

where I is the IDENTITY MATRIX, to obtain a MATRIX of the form

$$\begin{bmatrix} 1 & 0 & \cdots & 0 & b_{11} & \cdots & b_{1n} \\ 0 & 1 & \cdots & 0 & b_{21} & \cdots & b_{2n} \\ \vdots & \vdots & \ddots & \vdots & \vdots & \ddots & \vdots \\ 0 & 0 & \cdots & 1 & b_{n1} & \cdots & b_{nn} \end{bmatrix}.$$

The MATRIX

$$\mathsf{B} \equiv \begin{bmatrix} b_{11} & \cdots & b_{1n} \\ b_{21} & \cdots & b_{2n} \\ \vdots & \ddots & \vdots \\ b_{n1} & \cdots & b_{nn} \end{bmatrix}$$

is then the MATRIX INVERSE of A. The procedure is numerically unstable unless PIVOTING (exchanging rows and columns as appropriate) is used. Picking the largest available element as the pivot is usually a good choice.

see also GAUSSIAN ELIMINATION, LU DECOMPOSITION, MATRIX EQUATION

References

Press, W. H.; Flannery, B. P.; Teukolsky, S. A.; and Vetterling, W. T. "Gauss-Jordan Elimination" and "Gaussian Elimination with Backsubstitution." §2.1 and 2.2 in *Numerical Recipes in FORTRAN: The Art of Scientific Computing, 2nd ed.* Cambridge, England: Cambridge University Press, pp. 27–32 and 33–34, 1992.

Gauss-Kummer Series

$$_2F_1(-\tfrac{1}{2}, -\tfrac{1}{2}; 1; h^2) = \sum_{n=0}^{\infty} \binom{\frac{1}{2}}{n}^2 h^{2n}$$
$$= 1 + \tfrac{1}{4}h^2 + \tfrac{1}{64}h^4 + \tfrac{1}{256}h^6 + \ldots,$$

where $_2F_1(a, b; c; x)$ is a HYPERGEOMETRIC FUNCTION. This can be derived using KUMMER'S QUADRATIC TRANSFORMATION.

Gauss-Kuzmin-Wirsing Constant

N.B. A detailed on-line essay by S. Finch was the starting point for this entry.

Let x_0 be a random number from $[0, 1]$ written as a simple CONTINUED FRACTION

$$x_0 = 0 + \cfrac{1}{a_1 + \cfrac{1}{a_2 + \cfrac{1}{a_3 + \ldots}}}. \tag{1}$$

Define

$$x_n = 0 + \cfrac{1}{a_{n+1} + \cfrac{1}{a_{n+2} + \cfrac{1}{a_{n+3} + \ldots}}}$$
$$= \frac{1}{x_{n-1}} - \left\lfloor \frac{1}{x_{n-1}} \right\rfloor. \tag{2}$$

Gauss (1800) showed that if $F(n, x)$ is the probability that $x_n < x$, then

$$\lim_{n\to\infty} F(n, x) = \frac{\ln(1 + x)}{\ln 2}. \tag{3}$$

Kuzmin (1928) published the first proof, which was subsequently improved by Lévy (1929). Wirsing (1974) showed, among other results, that

$$\lim_{n\to\infty} \frac{F(n, x) - \frac{\ln(1+x)}{\ln 2}}{(-\lambda)^n} = \Psi(x), \tag{4}$$

where $\lambda = 0.3036630029\ldots$ and $\Psi(x)$ is an analytic function with $\Psi(0) = \Psi(1) = 0$. This constant is connected to the efficiency of the EUCLIDEAN ALGORITHM (Knuth 1981).

References

Babenko, K. I. "On a Problem of Gauss." *Soviet Math. Dokl.* **19**, 136–140, 1978.

Daudé, H.; Flajolet, P.; and Vallée, B. "An Average-Case Analysis of the Gaussian Algorithm for Lattice Reduction." Submitted.

Durner, A. "On a Theorem of Gauss-Kuzmin-Lévy." *Arch. Math.* **58**, 251–256, 1992.

Finch, S. "Favorite Mathematical Constants." `http://www.mathsoft.com/asolve/constant/kuzmin/kuzmin.html`.

Flajolet, P. and Vallée, B. "On the Gauss-Kuzmin-Wirsing Constant." Unpublished memo. 1995. `http://pauillac.inria.fr/algo/flajolet/Publications/gauss-kuzmin.ps`.

Knuth, D. E. *The Art of Computer Programming, Vol. 2: Seminumerical Algorithms, 2nd ed.* Reading, MA: Addison-Wesley, 1981.

MacLeod, A. J. "High-Accuracy Numerical Values of the Gauss-Kuzmin Continued Fraction Problem." *Computers Math. Appl.* **26**, 37–44, 1993.

Wirsing, E. "On the Theorem of Gauss-Kuzmin-Lévy and a Frobenius-Type Theorem for Function Spaces." *Acta Arith.* **24**, 507–528, 1974.

Gauss-Laguerre Quadrature

see LAGUERRE-GAUSS QUADRATURE

Gauss's Lemma

Let the multiples $m, 2m, \ldots, [(p-1)/2]m$ of an INTEGER such that $p \nmid m$ be taken. If there are an EVEN number r of least POSITIVE RESIDUES mod p of these numbers $> p/2$, then m is a QUADRATIC RESIDUE of p. If r is ODD, m is a QUADRATIC NONRESIDUE. Gauss's lemma can therefore be stated as $(m|p) = (-1)^r$, where $(m|p)$ is the LEGENDRE SYMBOL. It was proved by Gauss as a step along the way to the QUADRATIC RECIPROCITY THEOREM.

see also QUADRATIC RECIPROCITY THEOREM

Gauss's Machin-Like Formula

The MACHIN-LIKE FORMULA

$$\tfrac{1}{4}\pi = 12\cot^{-1}18 + 8\cot^{-1}57 - 5\cot^{-1}239.$$

Gauss-Manin Connection

A connection defined on a smooth ALGEBRAIC VARIETY defined over the COMPLEX NUMBERS.

References

Iyanaga, S. and Kawada, Y. (Eds.). *Encyclopedic Dictionary of Mathematics.* Cambridge, MA: MIT Press, p. 81, 1980.

Gauss Map

The Gauss map is a function from an ORIENTABLE SURFACE in EUCLIDEAN SPACE to a SPHERE. It associates to every point on the surface its oriented NORMAL VECTOR. For surfaces in 3-space, the Gauss map of the surface has DEGREE given by half the EULER CHARACTERISTIC of the surface

$$\iint_M K\,dA = 2\pi\chi(M) - \sum \alpha_i - \int_{\partial M} \kappa_g\,ds,$$

which works only for ORIENTABLE SURFACES.

see also CURVATURE, NIRENBERG'S CONJECTURE, PATCH

References

Gray, A. "The Local Gauss Map" and "The Gauss Map via Mathematica." §10.3 and §15.3 in *Modern Differential Geometry of Curves and Surfaces.* Boca Raton, FL: CRC Press, pp. 193–194 and 310–316, 1993.

Gauss's Mean-Value Theorem

Let $f(z)$ be an ANALYTIC FUNCTION in $|z-a|<R$. Then

$$f(z) = \frac{1}{2\pi}\int_0^{2\pi} f(z+re^{i\theta})\,d\theta$$

for $0<r<R$.

Gauss Measure

The standard Gauss measure of a finite dimensional REAL HILBERT SPACE H with norm $||\cdot||_H$ has the BOREL MEASURE

$$\mu_H(dh) = (\sqrt{2\pi})^{-\dim(H)}\exp(\tfrac{1}{2}||h||_H^2)\lambda_H(dh),$$

where λ_H is the LEBESGUE MEASURE on H.

Gauss Multiplication Formula

$$\begin{aligned}(2n\pi)^{(n-1)/2}n^{1/2-nz}\Gamma(nz) &= \Gamma(z)\Gamma(z+\tfrac{1}{n})\Gamma(z+\tfrac{2}{n})\cdots\Gamma(z+\tfrac{n-1}{n})\\ &= \prod_{k=0}^{n-1}\Gamma\left(z+\frac{k}{n}\right),\end{aligned}$$

where $\Gamma(z)$ is the GAMMA FUNCTION.

see also GAMMA FUNCTION, LEGENDRE DUPLICATION FORMULA, POLYGAMMA FUNCTION

References

Abramowitz, M. and Stegun, C. A. (Eds.). *Handbook of Mathematical Functions with Formulas, Graphs, and Mathematical Tables, 9th printing.* New York: Dover, p. 256, 1972.

Gauss Plane

see COMPLEX PLANE

Gauss's Polynomial Theorem

If a POLYNOMIAL

$$f(x) = x^N + C_1x^{N-1} + C_2x^{N-2} + \ldots + C_N$$

with integral COEFFICIENTS is divisible into a product of two POLYNOMIALS $f=\psi\phi$

$$\begin{aligned}\psi &= x^m + \alpha_1x^{m-1} + \ldots + \alpha_m\\ \phi &= x^n + \beta_1x^{n-1} + \ldots + \beta_n,\end{aligned}$$

then the COEFFICIENTS of this POLYNOMIAL are INTEGERS.

see also ABEL'S IRREDUCIBILITY THEOREM, ABEL'S LEMMA, KRONECKER'S POLYNOMIAL THEOREM, POLYNOMIAL, SCHOENEMANN'S THEOREM

References

Dörrie, H. *100 Great Problems of Elementary Mathematics: Their History and Solutions.* New York: Dover, p. 119, 1965.

Gauss's Reciprocity Theorem

see QUADRATIC RECIPROCITY THEOREM

Gauss-Salamin Formula

see BRENT-SALAMIN FORMULA

Gauss's Test

If $u_n > 0$ and given $B(n)$ a bounded function of n as $n\to\infty$, express the ratio of successive terms as

$$\frac{u_n}{u_{n+1}} = 1 + \frac{h}{n} + \frac{B(n)}{n^2}.$$

The SERIES converges for $h>1$ and diverges for $h\le 1$.

see also CONVERGENCE TESTS

References

Arfken, G. *Mathematical Methods for Physicists, 3rd ed.* Orlando, FL: Academic Press, pp. 287–288, 1985.

Gauss's Theorem

see DIVERGENCE THEOREM

Gauss's Theorema Egregium

Gauss's theorema egregium states that the GAUSSIAN CURVATURE of a surface embedded in 3-space may be understood intrinsically to that surface. "Residents" of the surface may observe the GAUSSIAN CURVATURE of the surface without ever venturing into full 3-dimensional space; they can observe the curvature of the surface they live in without even knowing about the 3-dimensional space in which they are embedded.

In particular, GAUSSIAN CURVATURE can be measured by checking how closely the ARC LENGTH of small RADIUS CIRCLES correspond to what they should be in EUCLIDEAN SPACE, $2\pi r$. If the ARC LENGTH of CIRCLES tends to be smaller than what is expected in EUCLIDEAN SPACE, then the space is positively curved; if larger, negatively; if the same, 0 GAUSSIAN CURVATURE.

Gauss (effectively) expressed the theorema egregium by saying that the GAUSSIAN CURVATURE at a point is given by $-R(v,w)v,w$, where R is the RIEMANN TENSOR, and v and w are an orthonormal basis for the TANGENT SPACE.

see also CHRISTOFFEL SYMBOL OF THE SECOND KIND, GAUSS EQUATIONS, GAUSSIAN CURVATURE

References

Gray, A. "Gauss's Theorema Egregium." §20.2 in *Modern Differential Geometry of Curves and Surfaces.* Boca Raton, FL: CRC Press, pp. 395–397, 1993.

Reckziegel, H. In *Mathematical Models from the Collections of Universities and Museums* (Ed. G. Fischer). Braunschweig, Germany: Vieweg, pp. 31–32, 1986.

Gauss's Transformation

If

$$(1 + x\sin^2\alpha)\sin\beta = (1+x)\sin\alpha,$$

then

$$(1+x)\int_0^\alpha \frac{d\phi}{\sqrt{1-x^2\sin^2\phi}} = \int_0^\beta \frac{d\phi}{\sqrt{1-\frac{4x}{(1+x)^2}\sin^2\phi}}.$$

see also ELLIPTIC INTEGRAL OF THE FIRST KIND, LANDEN'S TRANSFORMATION

Gaussian Approximation Algorithm

see ARITHMETIC-GEOMETRIC MEAN

Gaussian Bivariate Distribution

The Gaussian bivariate distribution is given by

$$P(x_1,x_2) = \frac{1}{2\pi\sigma_1\sigma_2\sqrt{1-\rho^2}} \exp\left[-\frac{z}{2(1-\rho^2)}\right], \quad (1)$$

where

$$z \equiv \frac{(x_1-\mu_1)^2}{{\sigma_1}^2} - \frac{2\rho(x_1-\mu_1)(x_2-\mu_2)}{\sigma_1\sigma_2} + \frac{(x_2-\mu_2)^2}{{\sigma_2}^2}, \quad (2)$$

and

$$\rho \equiv \operatorname{cov}(x_1,x_2) = \frac{\langle x_1x_2\rangle - \langle x_1\rangle\langle x_2\rangle}{\sigma_1\sigma_2} \quad (3)$$

is the COVARIANCE. Let X_1 and X_2 be normally and independently distributed variates with MEAN 0 and VARIANCE 1. Then define

$$Y_1 \equiv \mu_1 + \sigma_{11}X_1 + \sigma_{12}X_2 \quad (4)$$

$$Y_2 \equiv \mu_2 + \sigma_{21}X_1 + \sigma_{22}X_2. \quad (5)$$

These new variates are normally distributed with MEAN μ_1 and μ_2, VARIANCE

$${\sigma_1}^2 \equiv {\sigma_{11}}^2 + {\sigma_{12}}^2 \quad (6)$$

$${\sigma_2}^2 \equiv {\sigma_{21}}^2 + {\sigma_{22}}^2, \quad (7)$$

and COVARIANCE

$$V_{12} \equiv \sigma_{11}\sigma_{21} + \sigma_{12}\sigma_{22}. \quad (8)$$

The COVARIANCE matrix is

$$V_{ij} = \begin{bmatrix} {\sigma_1}^2 & \rho\sigma_1\sigma_2 \\ \rho\sigma_1\sigma_2 & {\sigma_2}^2 \end{bmatrix}, \quad (9)$$

where

$$\rho \equiv \frac{V_{12}}{\sigma_1\sigma_2} = \frac{\sigma_{11}\sigma_{21}+\sigma_{12}\sigma_{22}}{\sigma_1\sigma_2}. \quad (10)$$

The joint probability density function for x_1 and x_2 is

$$f(x_1,x_2)\,dx_1\,dx_2 = \frac{1}{2\pi}e^{-({x_1}^2+{x_2}^2)/2}\,dx_1\,dx_2. \quad (11)$$

However, from (4) and (5) we have

$$\begin{bmatrix} y_1-\mu_1 \\ y_2-\mu_2 \end{bmatrix} = \begin{bmatrix} \sigma_{11} & \sigma_{12} \\ \sigma_{21} & \sigma_{22} \end{bmatrix} \begin{bmatrix} x_1 \\ x_2 \end{bmatrix}. \quad (12)$$

Now, if

$$\begin{vmatrix} \sigma_{11} & \sigma_{12} \\ \sigma_{21} & \sigma_{22} \end{vmatrix} \neq 0, \quad (13)$$

then this can be inverted to give

$$\begin{bmatrix} x_1 \\ x_2 \end{bmatrix} = \begin{bmatrix} \sigma_{11} & \sigma_{12} \\ \sigma_{21} & \sigma_{22} \end{bmatrix}^{-1} \begin{bmatrix} y_1-\mu_1 \\ y_2-\mu_2 \end{bmatrix} = \frac{1}{\sigma_{11}\sigma_{22}-\sigma_{12}\sigma_{21}} \begin{bmatrix} \sigma_{22} & -\sigma_{12} \\ -\sigma_{21} & \sigma_{11} \end{bmatrix} \begin{bmatrix} y_1-\mu_1 \\ y_2-\mu_2 \end{bmatrix}. \quad (14)$$

Therefore,

$${x_1}^2 + {x_2}^2 = \frac{[\sigma_{22}(y_1-\mu_1)-\sigma_{12}(y_2-\mu_2)]^2}{(\sigma_{11}\sigma_{22}-\sigma_{12}\sigma_{21})^2} + \frac{[-\sigma_{21}(y_1-\mu_1)+\sigma_{11}(y_2-\mu_2)]^2}{(\sigma_{11}\sigma_{22}-\sigma_{12}\sigma_{21})^2}. \quad (15)$$

Expanding the NUMERATOR gives

$$\begin{aligned}\sigma_{22}^2(y_1-\mu_1)^2 &- 2\sigma_{12}\sigma_{22}(y_1-\mu_1)(y_2-\mu_2)\\ &+\sigma_{12}^2(y_2-\mu_2)^2+\sigma_{21}^2(y_1-\mu_1)^2\\ &-2\sigma_{11}\sigma_{21}(y_1-\mu_1)(y_2-\mu_2)+\sigma_{11}^2(y_2-\mu_2)^2,\end{aligned} \tag{16}$$

so

$$\begin{aligned}(x_1^2+x_2^2)&(\sigma_{11}\sigma_{22}-\sigma_{12}\sigma_{21})^2\\ &=(y_1-\mu_1)^2(\sigma_{21}^2+\sigma_{22}^2)\\ &\quad -2(y_1-\mu_1)(y_2-\mu_2)(\sigma_{11}\sigma_{21}+\sigma_{12}\sigma_{22})\\ &\quad +(y_2-\mu_2)^2(\sigma_{11}^2+\sigma_{12}^2)\\ &=\sigma_2^2(y_1-\mu_1)^2-2(y_1-\mu_1)(y_2-\mu_2)(\rho\sigma_1\sigma_2)\\ &\quad +\sigma_1^2(y_2-\mu_2)^2\\ &=\sigma_1^2\sigma_2^2\left[\frac{(y_1-\mu_1)^2}{\sigma_1^2}-\frac{2\rho(y_1-\mu_1)(y_2-\mu_2)}{\sigma_1\sigma_2}+\frac{(y_2-\mu_2)^2}{\sigma_2^2}\right].\end{aligned} \tag{17}$$

But

$$\begin{aligned}\frac{1}{1-\rho^2}&=\frac{1}{1-\frac{V_{12}^2}{\sigma_1^2\sigma_2^2}}=\frac{\sigma_1^2\sigma_2^2}{\sigma_1^2\sigma_2^2-V_{12}^2}\\ &=\frac{\sigma_1^2\sigma_2^2}{(\sigma_{11}^2+\sigma_{12}^2)(\sigma_{21}^2+\sigma_{22}^2)-(\sigma_{11}\sigma_{21}+\sigma_{12}\sigma_{22})^2}.\end{aligned} \tag{18}$$

The DENOMINATOR is

$$\begin{aligned}\sigma_{11}^2\sigma_{21}^2&+\sigma_{11}^2\sigma_{22}^2+\sigma_{12}^2\sigma_{21}^2+\sigma_{12}^2\sigma_{22}^2-\sigma_{11}^2\sigma_{21}^2\\ &-2\sigma_{11}\sigma_{12}\sigma_{21}\sigma_{22}-\sigma_{12}^2\sigma_{22}^2=(\sigma_{11}\sigma_{22}-\sigma_{12}\sigma_{21})^2,\end{aligned} \tag{19}$$

so

$$\frac{1}{1-\rho^2}=\frac{\sigma_1^2\sigma_2^2}{(\sigma_{11}\sigma_{22}-\sigma_{12}\sigma_{21})^2} \tag{20}$$

and

$$\begin{aligned}x_1^2+x_2^2&=\frac{1}{1-\rho^2}\\ &\times\left[\frac{(y_1-\mu_1)^2}{\sigma_1^2}-\frac{2\rho(y_1-\mu_1)(y_2-\mu_2)}{\sigma_1\sigma_2}+\frac{(y_2-\mu_2)^2}{\sigma_2^2}\right].\end{aligned} \tag{21}$$

Solving for x_1 and x_2 and defining

$$\rho'\equiv\frac{\sigma_1\sigma_2\sqrt{1-\rho^2}}{\sigma_{11}\sigma_{22}-\sigma_{12}\sigma_{21}} \tag{22}$$

gives

$$x_1=\frac{\sigma_{22}(y_1-\mu_1)-\sigma_{12}(y_2-\mu_2)}{\rho'} \tag{23}$$

$$x_2=\frac{-\sigma_{21}(y_1-\mu_1)+\sigma_{11}(y_2-\mu_2)}{\rho'}. \tag{24}$$

The JACOBIAN is

$$\begin{aligned}J\left(\frac{x_1,x_2}{y_1,y_2}\right)&=\begin{vmatrix}\frac{\partial x_1}{\partial y_1} & \frac{\partial x_1}{\partial y_2}\\ \frac{\partial x_2}{\partial y_1} & \frac{\partial x_2}{\partial y_2}\end{vmatrix}=\begin{vmatrix}\frac{\sigma_{22}}{\rho'} & -\frac{\sigma_{12}}{\rho'}\\ -\frac{\sigma_{21}}{\rho'} & \frac{\sigma_{11}}{\rho'}\end{vmatrix}\\ &=\frac{1}{\rho'^2}(\sigma_{11}\sigma_{22}-\sigma_{12}\sigma_{21})\\ &=\frac{1}{\rho'}=\frac{1}{\sigma_1\sigma_2\sqrt{1-\rho^2}}.\end{aligned} \tag{25}$$

Therefore,

$$dx_1\,dx_2=\frac{dy_1\,dy_2}{\sigma_1\sigma_2\sqrt{1-\rho^2}} \tag{26}$$

and

$$\begin{aligned}\frac{1}{2\pi}&e^{-(x_1^2+x_2^2)/2}\,dx_1\,dx_2\\ &=\frac{1}{2\pi\sigma_1\sigma_2\sqrt{1-\rho^2}}e^{-v/2}\,dy_1\,dy_2,\end{aligned} \tag{27}$$

where

$$\begin{aligned}v&\equiv\frac{1}{1-\rho^2}\\ &\times\left[\frac{(y_1-\mu_1)^2}{\sigma_1^2}-\frac{2\rho(y_1-\mu_1)(y_2-\mu_2)}{\sigma_1\sigma_2}+\frac{(y_2-\mu_2)^2}{\sigma_2^2}\right].\end{aligned} \tag{28}$$

Now, if

$$\begin{vmatrix}\sigma_{11} & \sigma_{12}\\ \sigma_{21} & \sigma_{22}\end{vmatrix}=0, \tag{29}$$

then

$$\sigma_{11}\sigma_{12}=\sigma_{12}\sigma_{21} \tag{30}$$

$$y_1=\mu_1+\sigma_{11}x_1+\sigma_{12}x_2 \tag{31}$$

$$\begin{aligned}y_2&=\mu_2+\frac{\sigma_{12}\sigma_{21}}{\sigma_{11}}x_2=\mu_2+\frac{\sigma_{11}\sigma_{21}x_1+\sigma_{12}\sigma_{21}x_2}{\sigma_{11}}\\ &=\mu_2+\frac{\sigma_{21}}{\sigma_{11}}(\sigma_{11}x_1+\sigma_{12}x_2),\end{aligned} \tag{32}$$

so

$$y_1=\mu_1+x_3 \tag{33}$$

$$y_2=\mu_2+\frac{\sigma_{21}}{\sigma_{11}}x_3, \tag{34}$$

where

$$x_3=y_1-\mu_1=\frac{\sigma_{11}}{\sigma_{21}}(y_2-\mu_2). \tag{35}$$

The CHARACTERISTIC FUNCTION is given by

$$\begin{aligned}\phi(t_1,t_2)&\equiv\int_{-\infty}^{\infty}\int_{-\infty}^{\infty}e^{i(t_1x_1+t_2x_2)}P(x_1,x_2)\,dx_1\,dx_2\\ &=N\int_{-\infty}^{\infty}\int_{-\infty}^{\infty}e^{i(t_1x_1+t_2x_2)}\exp\left[-\frac{z}{2(1-\rho^2)}\right]dx_1\,dx_2,\end{aligned} \tag{36}$$

where

$$z \equiv \left[\frac{(x_1-\mu_1)^2}{\sigma_1{}^2} - \frac{2\rho(x_1-\mu_1)(x_2-\mu_2)}{\sigma_1\sigma_2} + \frac{(x_2-\mu_2)^2}{\sigma_2{}^2}\right] \tag{37}$$

and

$$N \equiv \frac{1}{2\pi\sigma_1\sigma_2\sqrt{1-\rho^2}}. \tag{38}$$

Now let

$$u \equiv x_1 - \mu_1 \tag{39}$$

$$w \equiv x_2 - \mu_2. \tag{40}$$

Then

$$\phi(t_1,t_2) = N' \int_{-\infty}^{\infty} \left(e^{it_2w} \exp\left[-\frac{1}{2(1-\rho^2)}\frac{w^2}{\sigma_2{}^2}\right]\right) \times \int_{-\infty}^{\infty} e^v e^{t_1u}\,du\,dw, \tag{41}$$

where

$$v \equiv -\frac{1}{2(1-\rho^2)}\frac{1}{\sigma_1{}^2}\left[u^2 - \frac{2\rho\sigma_1 w}{\sigma_2}u\right]$$

$$N' \equiv \frac{e^{i(t_1\mu_1+t_2\mu_2)}}{2\pi\sigma_1\sigma_2\sqrt{1-\rho^2}}. \tag{42}$$

COMPLETE THE SQUARE in the inner integral

$$\int_{-\infty}^{\infty} \exp\left\{-\frac{1}{2(1-\rho^2)}\frac{1}{\sigma_1{}^2}\left[u^2 - \frac{2\rho\sigma_1 w}{\sigma_2}u\right]\right\} e^{t_1u}\,du$$

$$= \int_{-\infty}^{\infty} \exp\left\{-\frac{1}{2\sigma_1{}^2(1-\rho^2)}\left[u - \frac{\rho_1\sigma_1 w}{\sigma^2}\right]^2\right\} \times \left\{\frac{1}{2\sigma_1{}^2(1-\rho^2)}\left(\frac{\rho_1\sigma_1 w}{\sigma_2}\right)^2\right\} e^{it_1u}\,du. \tag{43}$$

Rearranging to bring the exponential depending on w outside the inner integral, letting

$$v \equiv u - \rho\frac{\sigma_1 w}{\sigma_2}, \tag{44}$$

and writing

$$e^{it_1u} = \cos(t_1u) + i\sin(t_1u) \tag{45}$$

gives

$$\phi(t_1,t_2) = N' \int_{-\infty}^{\infty} e^{it_2w} \exp\left[-\frac{1}{2\sigma_2{}^2(1-\rho^2)}w^2\right] \times \exp\left[\frac{\rho^2}{2\sigma_2{}^2(1-\rho^2)}w^2\right] \int_{-\infty}^{\infty} \exp\left[-\frac{1}{2\sigma_2{}^2(1-\rho^2)}v^2\right] \times \left\{\cos\left[t_1\left(v+\frac{\rho\sigma_1 w}{\sigma_2}\right)\right] + i\sin\left[t_1\left(v+\frac{\rho\sigma_1 w}{\sigma_2}\right)\right]\right\} dv\,dw. \tag{46}$$

Expanding the term in braces gives

$$\left[\cos(t_1v)\cos\left(\frac{\rho\sigma_1 w t_1}{\sigma_2}\right) - \sin(t_1v)\sin\left(\frac{\rho\sigma_1 w}{\sigma_2 t_1}\right)\right] + i\left[\sin(t_1v)\cos\left(\frac{\rho\sigma_1 w}{\sigma_2 t_1}\right) + \cos(t_1v)\sin\left(\frac{\rho\sigma_1 w t_1}{\sigma_2}\right)\right]$$

$$= \left[\cos\left(\frac{\rho\sigma_1 w t_1}{\sigma_2}\right) + i\sin\left(\frac{\rho\sigma_1 w t_1}{\sigma_2}\right)\right] [\cos(t_1v) + i\sin(t_1v)]$$

$$= \exp\left(\frac{i\rho\sigma_1 w}{\sigma_2}t_1\right)[\cos(t_1v) + i\sin(t_1v)]. \tag{47}$$

But $e^{-ax^2}\sin(bx)$ is ODD, so the integral over the sine term vanishes, and we are left with

$$\phi(t_1,t_2) = N' \int_{-\infty}^{\infty} e^{it_2w}\exp\left[-\frac{w^2}{2\sigma_2{}^2}\right] \times \exp\left[\frac{\rho^2 w^2}{2\sigma_2{}^2(1-\rho^2)}\right] \exp\left[\frac{i\rho\sigma_1 w t_1}{\sigma_2}\right] dw \times \int_{-\infty}^{\infty} \exp\left[-\frac{v^2}{2\sigma_1{}^2(1-\rho^2)}\right]\cos(t_1v)\,dv$$

$$= N' \int_{-\infty}^{\infty} \exp\left[iw\left(t_2 + t_1\left(\rho\frac{\sigma_1}{\sigma_2}\right)\right)\right] \exp\left[-\frac{w^2}{2\sigma_2{}^2}\right] dw \times \int_{-\infty}^{\infty} \exp\left[-\frac{v^2}{2\sigma_1{}^2(1-\rho^2)}\right]\cos(t_1v)\,dv. \tag{48}$$

Now evaluate the GAUSSIAN INTEGRAL

$$\int_{-\infty}^{\infty} e^{ikx}e^{-ax^2}\,dx = \int_{-\infty}^{\infty} e^{-ax^2}\cos(kx)\,dx = \sqrt{\frac{\pi}{a}}e^{-k^2/4a} \tag{49}$$

to obtain the explicit form of the CHARACTERISTIC FUNCTION,

$$\phi(t_1,t_2) = \frac{e^{i(t_1\mu_1+t_2\mu_2)}}{2\pi\sigma_1\sigma_2\sqrt{1-\rho^2}} \times \left\{\sigma_2\sqrt{2\pi}\exp\left[-\frac{1}{4}\left(t_2+\rho\frac{\sigma_1}{\sigma_2}t_1\right)^2 2\sigma_2{}^2\right]\right\} \times \left\{\sigma_1\sqrt{2\pi(1-\rho^2)}\exp\left[-\tfrac{1}{4}t_1{}^2 2\sigma_1{}^2(1-\rho^2)\right]\right\}$$

$$= e^{i(t_1\mu_1+t_2\mu_2)}\exp\{-\tfrac{1}{2}[t_2{}^2\sigma_2{}^2 + 2\rho\sigma_1\sigma_2t_1t_2 + \rho^2\sigma_1{}^2t_1{}^2 + (1-\rho^2)\sigma_1{}^2t_1{}^2]\}$$

$$= \exp[i(t_1\mu_1 + t_2\mu_2) - \tfrac{1}{2}(\sigma_1{}^2t_1{}^2 + 2\rho\sigma_1\sigma_2t_1t_2 + \sigma_1{}^2t_1{}^2)]. \tag{50}$$

Let z_1 and z_2 be two independent Gaussian variables with MEANS $\mu_i = 0$ and $\sigma_i{}^2 = 1$ for $i = 1, 2$. Then the variables a_1 and a_2 defined below are Gaussian bivariates with unit VARIANCE and CROSS-CORRELATION COEFFICIENT ρ:

$$a_1 = \sqrt{\frac{1+\rho}{2}}z_1 + \sqrt{\frac{1-\rho}{2}}z_2 \tag{51}$$

$$a_2 = \sqrt{\frac{1+\rho}{2}}\, z_1 - \sqrt{\frac{1-\rho}{2}}\, z_2. \tag{52}$$

The conditional distribution is

$$P(x_2|x_1) = \frac{1}{\sigma_2\sqrt{2\pi(1-\rho^2)}} \exp\left[-\frac{(x^2 - \mu'^2)^2}{2{\sigma'_2}^2}\right], \tag{53}$$

where

$$\mu'_2 \equiv \mu_2 + \rho\frac{\sigma_2}{\sigma_1}(x_1 - \mu_1) \tag{54}$$

$$\sigma'_2 \equiv \sigma_2\sqrt{1-\rho_2}\,. \tag{55}$$

The marginal probability density is

$$P(x_2) = \int_{-\infty}^{\infty} P(x_1, x_2)\, dx_1 = \frac{1}{\sigma_2\sqrt{2\pi}} \exp\left[-\frac{(x_2 - \mu_2)^2}{2{\sigma_2}^2}\right]. \tag{56}$$

see also BOX-MULLER TRANSFORMATION, GAUSSIAN DISTRIBUTION, MCMOHAN'S THEOREM, NORMAL DISTRIBUTION

References

Abramowitz, M. and Stegun, C. A. (Eds.). *Handbook of Mathematical Functions with Formulas, Graphs, and Mathematical Tables, 9th printing.* New York: Dover, pp. 936–937, 1972.

Spiegel, M. R. *Theory and Problems of Probability and Statistics.* New York: McGraw-Hill, p. 118, 1992.

Gaussian Brackets

Published by Gauss in *Disquisitiones Arithmeticae.* They are defined as follows.

$$[\,] = 1 \tag{1}$$

$$[a_1] = a_1 \tag{2}$$

$$[a_1, a_2] = [a_1]a_2 + [\,] \tag{3}$$

$$[a_1, a_2, \ldots, a_n] = [a_1, a_2, \ldots, a_{n-1}]a_n + [a_1, a_2, \ldots, a_{n-2}]. \tag{4}$$

Gaussian brackets are useful for treating CONTINUED FRACTIONS because

$$\cfrac{1}{a_1 + \cfrac{1}{a_2 + \cfrac{1}{a_3 + \ldots + \cfrac{1}{a_n}}}} = \frac{[a_2, a_n]}{[a_1, a_n]}. \tag{5}$$

The NOTATION $[x]$ conflicts with that of GAUSSIAN POLYNOMIALS and the NINT function.

References

Herzberger, M. *Modern Geometrical Optics.* New York: Interscience Publishers, pp. 457–462, 1958.

Gaussian Coefficient

see q-BINOMIAL COEFFICIENT

Gaussian Coordinate System

A coordinate system which has a METRIC satisfying $g_{ii} = -1$ and $\partial g_{ij}/\partial x_j = 0$.

Gaussian Curvature

An intrinsic property of a space independent of the coordinate system used to describe it. The Gaussian curvature of a REGULAR SURFACE in $\mathbb{R}^3$ at a point $\mathbf{p}$ is formally defined as

$$K(\mathbf{p}) = \tfrac{1}{2}\det(S(\mathbf{p})), \tag{1}$$

where S is the SHAPE OPERATOR and det denotes the DETERMINANT.

If $\mathbf{x} : U \to \mathbb{R}^3$ is a REGULAR PATCH, then the Gaussian curvature is given by

$$K = \frac{eg - f^2}{EG - F^2}, \tag{2}$$

where E, F, and G are coefficients of the first FUNDAMENTAL FORM and e, f, and g are coefficients of the second FUNDAMENTAL FORM (Gray 1993, p. 282). The Gaussian curvature can be given entirely in terms of the first FUNDAMENTAL FORM

$$ds^2 = E\,du^2 + 2F\,du\,dv + G\,dv^2 \tag{3}$$

and the DISCRIMINANT

$$g \equiv EG - F^2 \tag{4}$$

by

$$K = \frac{1}{\sqrt{g}}\left[\frac{\partial}{\partial v}\left(\frac{\sqrt{g}}{E}\Gamma_{11}^2\right) - \frac{\partial}{\partial u}\left(\frac{\sqrt{g}}{E}\Gamma_{12}^2\right)\right], \tag{5}$$

where Γ_{ij}^k are the CONNECTION COEFFICIENTS. Equivalently,

$$K = \frac{1}{g^2}\begin{vmatrix} E & F & \frac{\partial F}{\partial v} - \frac{1}{2}\frac{\partial G}{\partial u} \\ F & G & \frac{1}{2}\frac{\partial G}{\partial v} \\ \frac{1}{2}\frac{\partial E}{\partial u} & k_{23} & k_{33} \end{vmatrix} - \frac{1}{g^2}\begin{vmatrix} E & F & \frac{1}{2}\frac{\partial E}{\partial v} \\ F & G & \frac{1}{2}\frac{\partial G}{\partial u} \\ \frac{1}{2}\frac{\partial E}{\partial v} & \frac{1}{2}\frac{\partial G}{\partial v} & 0 \end{vmatrix}, \tag{6}$$

where

$$k_{23} \equiv \frac{\partial F}{\partial u} - \frac{1}{2}\frac{\partial E}{\partial v} \tag{7}$$

$$k_{33} \equiv -\frac{1}{2}\frac{\partial^2 E}{\partial v^2} + \frac{\partial^2 F}{\partial u \partial v} - \frac{1}{2}\frac{\partial^2 G}{\partial u^2}. \tag{8}$$

Writing this out,

$$K = \frac{1}{2g}\left[2\frac{\partial^2 F}{\partial u \partial v} - \frac{\partial^2 E}{\partial v^2} - \frac{\partial^2 G}{\partial u^2}\right]$$
$$- \frac{G}{4g^2}\left[\frac{\partial E}{\partial u}\left(2\frac{\partial F}{\partial v} - \frac{\partial G}{\partial u}\right) - \left(\frac{\partial E}{\partial v}\right)^2\right]$$
$$+ \frac{F}{4g_2}\left[\frac{\partial E}{\partial u}\frac{\partial G}{\partial v} - 2\frac{\partial E}{\partial v}\frac{\partial G}{\partial u}\right.$$
$$\left.+ \left(2\frac{\partial F}{\partial u} - \frac{\partial E}{\partial v}\right)\left(2\frac{\partial F}{\partial v} - \frac{\partial G}{\partial u}\right)\right]$$
$$- \frac{E}{4g^2}\left[\frac{\partial G}{\partial v}\left(2\frac{\partial F}{\partial u} - \frac{\partial E}{\partial v}\right) - \left(\frac{\partial G}{\partial u}\right)^2\right]. \quad (9)$$

The Gaussian curvature is also given by

$$K = \frac{\det(\mathbf{x}_{uu}\mathbf{x}_u\mathbf{x}_v)\det(\mathbf{x}_{vv}\mathbf{x}_u\mathbf{x}_v) - [\det(\mathbf{x}_{uv}\mathbf{x}_u\mathbf{x}_v)]^2}{[|\mathbf{x}_u|^2|\mathbf{x}_v|^2 - (\mathbf{x}_u \cdot \mathbf{x}_v)^2]^2} \quad (10)$$

(Gray 1993, p. 285), as well as

$$K = \frac{[\hat{\mathbf{N}}\,\hat{\mathbf{N}}_1\,\hat{\mathbf{N}}_2]}{\sqrt{g}} = \frac{\epsilon^{ij}[\hat{\mathbf{N}}\,\hat{\mathbf{T}}\,\hat{\mathbf{T}}_i]_j}{\sqrt{g}}, \quad (11)$$

where ϵ^{ij} is the LEVI-CIVITA SYMBOL, $\hat{\mathbf{N}}$ is the unit NORMAL VECTOR and $\hat{\mathbf{T}}$ is the unit TANGENT VECTOR. The Gaussian curvature is also given by

$$K = -\frac{R}{2} = \kappa_1\kappa_2 = \frac{1}{R_1R_2}, \quad (12)$$

where R is the CURVATURE SCALAR, κ_1 and κ_2 the PRINCIPAL CURVATURES, and R_1 and R_2 the PRINCIPAL RADII OF CURVATURE. For a MONGE PATCH with $z = h(u, v)$,

$$K = \frac{h_{uu}h_{vv} - {h_{uv}}^2}{(1 + {h_u}^2 + {h_v}^2)^2}. \quad (13)$$

The Gaussian curvature K and MEAN CURVATURE H satisfy

$$H^2 \geq K, \quad (14)$$

with equality only at UMBILIC POINTS, since

$$H^2 - K^2 = \tfrac{1}{4}(\kappa_1 - \kappa_2)^2. \quad (15)$$

If $\mathbf{p}$ is a point on a REGULAR SURFACE $M \subset \mathbb{R}^3$ and $\mathbf{v_p}$ and $\mathbf{w_p}$ are tangent vectors to M at $\mathbf{p}$, then the Gaussian curvature of M at $\mathbf{p}$ is related to the SHAPE OPERATOR S by

$$S(\mathbf{v_p}) \times S(\mathbf{w_p}) = K(\mathbf{p})\mathbf{v_p} \times \mathbf{w_p}. \quad (16)$$

Let $\mathbf{Z}$ be a nonvanishing VECTOR FIELD on M which is everywhere PERPENDICULAR to M, and let V and W be VECTOR FIELDS tangent to M such that $V \times W = \mathbf{Z}$, then

$$K = \frac{\mathbf{Z} \cdot (D_V\mathbf{Z} \times D_W\mathbf{Z})}{2|\mathbf{Z}|^4} \quad (17)$$

(Gray 1993, pp. 291–292).

For a SPHERE, the Gaussian quadrature is $K = 1/a^2$. For EUCLIDEAN SPACE, the Gaussian quadrature is $K = 0$. For GAUSS-BOLYAI-LOBACHEVSKY SPACE, the Gaussian quadrature is $K = -1/a^2$. A FLAT SURFACE is a REGULAR SURFACE and special class of MINIMAL SURFACE on which Gaussian curvature vanishes everywhere.

A point $\mathbf{p}$ on a REGULAR SURFACE $M \in \mathbb{R}^3$ is classified based on the sign of $K(\mathbf{p})$ as given in the following table (Gray 1993, p. 280), where S is the SHAPE OPERATOR.

Sign	Point
$K(\mathbf{p}) > 0$	elliptic point
$K(\mathbf{p}) < 0$	hyperbolic point
$K(\mathbf{p}) = 0$ but $S(\mathbf{p}) \neq 0$	parabolic point
$K(\mathbf{p}) = 0$ and $S(\mathbf{p}) = 0$	planar point

A surface on which the Gaussian curvature K is everywhere POSITIVE is called SYNCLASTIC, while a surface on which K is everywhere NEGATIVE is called ANTICLASTIC. Surfaces with constant Gaussian curvature include the CONE, CYLINDER, KUEN SURFACE, PLANE, PSEUDOSPHERE, and SPHERE. Of these, the CONE and CYLINDER are the only FLAT SURFACES OF REVOLUTION.

see also ANTICLASTIC, BRIOSCHI FORMULA, DEVELOPABLE SURFACE, ELLIPTIC POINT, FLAT SURFACE, HYPERBOLIC POINT, INTEGRAL CURVATURE, MEAN CURVATURE, METRIC TENSOR, MINIMAL SURFACE, PARABOLIC POINT, PLANAR POINT, SYNCLASTIC, UMBILIC POINT

References

Geometry Center. "Gaussian Curvature." `http://www.geom.umn.edu/zoo/diffgeom/surfspace/concepts/curvatures/gauss-curv.html`.

Gray, A. "The Gaussian and Mean Curvatures" and "Surfaces of Constant Gaussian Curvature." §14.5 and Ch. 19 in *Modern Differential Geometry of Curves and Surfaces*. Boca Raton, FL: CRC Press, pp. 279–285 and 375–387, 1993.

Gaussian Differential Equation

see HYPERGEOMETRIC DIFFERENTIAL EQUATION

Gaussian Distribution

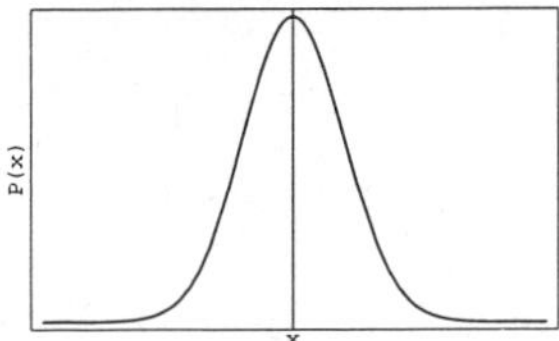

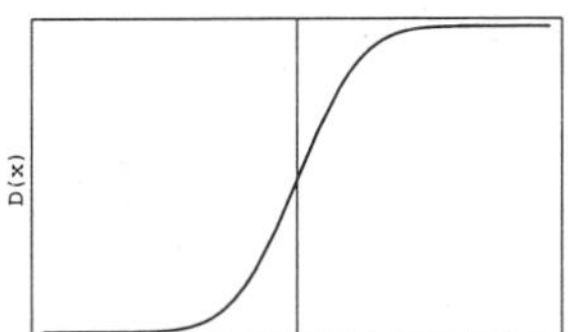

The Gaussian probability distribution with MEAN μ and STANDARD DEVIATION σ is a GAUSSIAN FUNCTION of the form

$$P(x) = \frac{1}{\sigma\sqrt{2\pi}} e^{-(x-\mu)^2/2\sigma^2}, \qquad (1)$$

where $P(x)\,dx$ gives the probability that a variate with a Gaussian distribution takes on a value in the range $[x, x+dx]$. This distribution is also called the NORMAL DISTRIBUTION or, because of its curved flaring shape, the BELL CURVE. The distribution $P(x)$ is properly normalized for $x \in (-\infty, \infty)$ since

$$\int_{-\infty}^{\infty} P(x)\,dx = 1. \qquad (2)$$

The cumulative DISTRIBUTION FUNCTION, which gives the probability that a variate will assume a value $\leq x$, is then

$$D(x) \equiv \int_{-\infty}^{x} P(x)\,dx = \frac{1}{\sigma\sqrt{2\pi}} \int_{-\infty}^{x} e^{-(x-\mu)^2/2\sigma^2}\,dx. \qquad (3)$$

Gaussian distributions have many convenient properties, so random variates with unknown distributions are often assumed to be Gaussian, especially in physics and astronomy. Although this can be a dangerous assumption, it is often a good approximation due to a surprising result known as the CENTRAL LIMIT THEOREM. This theorem proves that the MEAN of any set of variates with any distribution having a finite MEAN and VARIANCE tends to the Gaussian distribution. Many common attributes such as test scores, height, etc., follow roughly Gaussian distributions, with few members at the high and low ends and many in the middle.

Making the transformation

$$z \equiv \frac{x-\mu}{\sigma} \qquad (4)$$

so that $dz = dz/\sigma$ gives a variate with unit VARIANCE and 0 MEAN

$$P(x)\,dx = \frac{1}{\sqrt{2\pi}} e^{-z^2/2}\,dz, \qquad (5)$$

known as a standard NORMAL DISTRIBUTION. So defined, z is known as a z-SCORE).

The NORMAL DISTRIBUTION FUNCTION gives the probability that a standard normal variate assumes a value in the interval $[0, z]$,

$$\Phi(x) \equiv \frac{1}{\sqrt{2\pi}} \int_0^x e^{-z^2/2}\,dz = \tfrac{1}{2}\operatorname{erf}\left(\frac{x}{\sqrt{2}}\right). \qquad (6)$$

Here, ERF is a function sometimes called the error function. Neither $\Phi(z)$ nor ERF can be expressed in terms of finite additions, subtractions, multiplications, and root extractions, and so both must be either computed numerically or otherwise approximated. The value of a for which $P(x)$ falls within the interval $[-a, a]$ with a given probability P is called the P CONFIDENCE INTERVAL.

The Gaussian distribution is also a special case of the CHI-SQUARED DISTRIBUTION, since substituting

$$z \equiv \frac{(x-\mu)^2}{\sigma^2} \qquad (7)$$

so that

$$dz = \tfrac{1}{2}\frac{2(x-\mu)}{\sigma}\,dx = \sqrt{\frac{z}{\sigma}}\,dx \qquad (8)$$

(where an extra factor of 1/2 has been added to dz since z runs from 0 to ∞ instead of from $-\infty$ to ∞), gives

$$\begin{aligned} P(x)\,dx &= \frac{1}{\sqrt{2\pi}} e^{-(z/\sigma)/2} \left(\frac{z}{\sigma}\right)^{-1/2} d\left(\frac{z}{\sigma}\right)\,dz \\ &= \frac{1}{2^{1/2}\Gamma(\frac{1}{2})} e^{-(z/\sigma)/2} \left(\frac{z}{\sigma}\right)^{-1/2} d\left(\frac{z}{\sigma}\right)\,dz, \end{aligned} \qquad (9)$$

which is a CHI-SQUARED DISTRIBUTION in z/σ with $r = 1$ (i.e., a GAMMA DISTRIBUTION with $\alpha = 1/2$ and $\theta = 2$).

Cramer showed in 1936 that if X and Y are INDEPENDENT variates and $X+Y$ has a Gaussian distribution, then both X and Y must be Gaussian (CRAMER'S THEOREM).

The ratio X/Y of independent Gaussian-distributed variates with zero MEAN is distributed with a CAUCHY DISTRIBUTION. This can be seen as follows. Let X and Y both have MEAN 0 and standard deviations of σ_x and σ_y, respectively, then the joint probability density function is the GAUSSIAN BIVARIATE DISTRIBUTION with $\rho = 0$,

$$f(x,y) = \frac{1}{2\pi\sigma_x\sigma_y} e^{-[x^2/(2\sigma_x{}^2)+y^2/(2\sigma_y{}^2)]}. \qquad (10)$$

From RATIO DISTRIBUTION, the distribution of $U = Y/X$ is

$$\begin{aligned} P(u) &= \int_{-\infty}^{\infty} |x| f(x, ux)\,dx \\ &= \frac{1}{2\pi\sigma_x\sigma_y} \int_{-\infty}^{\infty} |x| e^{-[x^2/(2\sigma_x{}^2)+u^2x^2/(2\sigma_y{}^2)]} \\ &= \frac{1}{\pi\sigma_x\sigma_y} \int_0^{\infty} x \exp\left[-x^2\left(\frac{1}{2\sigma_x{}^2} + \frac{u^2}{2\sigma_y{}^2}\right)\right] dx. \end{aligned} \qquad (11)$$

But

$$\int_0^{\infty} x e^{-ax^2}\,dx = \left[-\frac{1}{2a} e^{-ax^2}\right]_0^{\infty} = \frac{1}{2a}[0-(-1)] = \frac{1}{2a}, \qquad (12)$$

so

$$P(u) = \frac{1}{\pi\sigma_x\sigma_y}\frac{1}{2\left(\frac{1}{2\sigma_x{}^2}+\frac{u^2}{2\sigma_y{}^2}\right)} = \frac{1}{\pi}\frac{\sigma_x\sigma_y}{u^2\sigma_x{}^2+\sigma_y{}^2}$$

$$= \frac{1}{\pi}\frac{\frac{\sigma_y}{\sigma_x}}{u^2+\left(\frac{\sigma_y}{\sigma_x}\right)^2}, \tag{13}$$

which is a CAUCHY DISTRIBUTION with MEAN $\mu = 0$ and full width

$$\Gamma = \frac{2\sigma_y}{\sigma_x}. \tag{14}$$

The CHARACTERISTIC FUNCTION for the Gaussian distribution is

$$\phi(t) = e^{imt-\sigma^2t^2/2}, \tag{15}$$

and the MOMENT-GENERATING FUNCTION is

$$M(t) = \langle e^{tx}\rangle = \int_{-\infty}^{\infty}\frac{e^{tx}}{\sigma\sqrt{2\pi}}e^{-(x-\mu)^2/2\sigma^2}\,dx$$

$$= \frac{1}{\sigma\sqrt{2\pi}}\int_{-\infty}^{\infty}\exp\left\{-\frac{1}{2\sigma^2}\left[x^2-2(\mu+\sigma^2t)x+\mu^2\right]\right\}dx. \tag{16}$$

COMPLETING THE SQUARE in the exponent,

$$\frac{1}{2\sigma^2}[x^2-2(\mu+\sigma^2t)x+\mu^2]$$

$$= \frac{1}{2\sigma^2}\left\{[x-(\mu+\sigma^2t)]^2+[\mu^2-(\mu+\sigma^2t)^2]\right\}. \tag{17}$$

Let

$$y \equiv x-(\mu+\sigma^2t) \tag{18}$$

$$dy = dx \tag{19}$$

$$a \equiv \frac{1}{2\sigma^2}. \tag{20}$$

The integral then becomes

$$M(t) = \frac{1}{\sigma\sqrt{2\pi}}\int_{-\infty}^{\infty}\exp\left[-ay^2+\frac{2\mu\sigma^2t+\sigma^4t^2}{2\sigma^2}\right]dy$$

$$= \frac{1}{\sigma\sqrt{2\pi}}\int_{-\infty}^{\infty}\exp[-ay^2+\mu t+\tfrac{1}{2}\sigma^2t^2]\,dy$$

$$= \frac{1}{\sigma\sqrt{2\pi}}e^{\mu t+\sigma^2t^2/2}\int_{-\infty}^{\infty}e^{-ay^2}\,dy$$

$$= \frac{1}{\sigma\sqrt{2\pi}}\sqrt{\frac{\pi}{a}}\,e^{\mu t+\sigma^2t^2/2}$$

$$= \frac{\sqrt{2\sigma^2\pi}}{\sigma\sqrt{2\pi}}e^{\mu t+\sigma^2t^2/2} = e^{\mu t+\sigma^2t^2/2}, \tag{21}$$

so

$$M'(t) = (\mu+\sigma^2t)e^{\mu t+\sigma^2t^2/2} \tag{22}$$

$$M''(t) = \sigma^2e^{\mu t+\sigma^2t^2/2}+e^{\mu t+\sigma^2t^2/2}(\mu+t\sigma^2)^2, \tag{23}$$

and

$$\mu = M'(0) = \mu \tag{24}$$

$$\sigma^2 = M''(0)-[M'(0)]^2$$
$$= (\sigma^2+\mu^2)-\mu^2 = \sigma^2. \tag{25}$$

These can also be computed using

$$R(t) = \ln[M(t)] = \mu t+\tfrac{1}{2}\sigma^2t^2 \tag{26}$$

$$R'(t) = \mu+\sigma^2t \tag{27}$$

$$R''(t) = \sigma^2, \tag{28}$$

yielding, as before,

$$\mu = R'(0) = \mu \tag{29}$$

$$\sigma^2 = R''(0) = \sigma^2. \tag{30}$$

The moments can also be computed directly by computing the MOMENTS about the origin $\mu_n' \equiv \langle x^n\rangle$,

$$\mu_n' = \frac{1}{\sigma\sqrt{2\pi}}\int_{-\infty}^{\infty}x^ne^{-(x-\mu)^2/2\sigma^2}\,dx. \tag{31}$$

Now let

$$u \equiv \frac{x-\mu}{\sqrt{2}\sigma} \tag{32}$$

$$du = \frac{dx}{\sqrt{2}\sigma} \tag{33}$$

$$x = \sigma u\sqrt{2}+\mu, \tag{34}$$

giving

$$\mu_n' = \frac{\sqrt{2}\,\sigma}{\sigma\sqrt{2\pi}}\int_{-\infty}^{\infty}x^ne^{-u^2}\,du = \frac{1}{\sqrt{\pi}}\int_{-\infty}^{\infty}x^ne^{-u^2}\,du, \tag{35}$$

so

$$\mu_0' = 1 \tag{36}$$

$$\mu_1' = \frac{1}{\sqrt{\pi}}\int_{-\infty}^{\infty}xe^{-u^2}\,du$$
$$= \frac{1}{\sqrt{\pi}}\int_{-\infty}^{\infty}(\sqrt{2}\,\sigma u+\mu)e^{-u^2}\,du$$
$$= [\sqrt{2}\sigma H_1(1)+\mu H_0(1)] = (0+\mu) = \mu \tag{37}$$

$$\mu_2' = \frac{1}{\sqrt{\pi}}\int_{-\infty}^{\infty}x^2e^{-u^2}\,du$$
$$= \frac{1}{\sqrt{\pi}}\int_{-\infty}^{\infty}(2\sigma^2u^2+2\sqrt{2}\,\sigma\mu u+\mu^2)e^{-u^2}\,du$$
$$= [2\sigma^2H_2(1)+2\sqrt{2}\,\sigma\mu H_1(1)+\mu^2H_0(1)]$$
$$= (2\sigma^2\tfrac{1}{2}+0+\mu^2) = \mu^2+\sigma^2 \tag{38}$$

$$\mu_3' = \frac{1}{\sqrt{\pi}}\int_{-\infty}^{\infty}x^3e^{-u^2}\,du$$

$$= \frac{1}{\sqrt{\pi}} \int_{-\infty}^{\infty} (2\sqrt{2}\,\sigma^3 u^3 + 6\mu\sigma^2 u^2 + 3\sqrt{2}\,\mu^2 \sigma u + \mu^3) e^{-u^2}\, du$$
$$= [2\sqrt{2}\,\sigma^3 H_3(1) + 6\mu\sigma^2 H_2(1) + 3\sqrt{2}\,\mu^2 \sigma H_1(1) + \mu^3 H_0(1)]$$
$$= (0 + 6\mu^2\sigma^2 \tfrac{1}{2} + 0 + \mu^3) = \mu(\mu^2 + 3\sigma^2) \qquad (39)$$

$$\mu_4' = \frac{1}{\sqrt{\pi}} \int_{-\infty}^{\infty} x^3 e^{-u^2}\, du$$
$$= \frac{1}{\sqrt{\pi}} \int_{-\infty}^{\infty} (4\sigma^4 u^4 + 8\sqrt{2}\,\mu\sigma^3 u^3 + 12\mu^2\sigma^2 u^2 + 4\sqrt{2}\,\mu^3 \sigma u + \mu^4) e^{-u^2}\, du$$
$$= [4\sigma^4 H_4(1) + 8\sqrt{2}\,\mu\sigma^3 H_3(1) + 12\mu^2\sigma^2 H_2(1) + 4\sqrt{2}\,\mu^3 \sigma H_1(1) + \mu^4 H_0(1)]$$
$$= (4\sigma^4 \tfrac{3}{4} + 0 + 12\mu^2\sigma^2 \tfrac{1}{2} + 0 + \mu^4)$$
$$= \mu^4 + 6\mu^2\sigma^2 + 3\sigma^4, \qquad (40)$$

where $H_n(a)$ are GAUSSIAN INTEGRALS.

Now find the MOMENTS about the MEAN,

$$\mu_1 \equiv 0 \qquad (41)$$
$$\mu_2 \equiv \mu_2' - (\mu_1')^2 = (\mu^2 + \sigma^2) - \mu^2 = \sigma^2 \qquad (42)$$
$$\mu_3 \equiv \mu_3' - 3\mu_2'\mu_1' + 2(\mu_1')^3 = \mu(\mu^2 + 3\sigma^2) - 3(\sigma^2 + \mu^2)\mu + 2\mu^3 = 0 \qquad (43)$$
$$\mu_4 \equiv \mu_4' - 4\mu_3'\mu_1' + 6\mu_2'(\mu_1')^2 - 3(\mu_1')^4 = (\mu^4 + 6\mu^2\sigma^2 + 3\sigma^4) - 4(\mu^3 + 3\mu\sigma^2)\mu + 6(\mu^2 + \sigma^2)\mu^2 - 3\mu^4 = 3\sigma^4, \qquad (44)$$

so the VARIANCE, STANDARD DEVIATION, SKEWNESS, and KURTOSIS are given by

$$\mathrm{var}(x) \equiv \mu_2 = \sigma^2 \qquad (45)$$
$$\mathrm{stdv}\,(x) \equiv \sqrt{\mathrm{var}(x)} = \sigma \qquad (46)$$
$$\gamma_1 = \frac{\mu_3}{\sigma^3} = 0 \qquad (47)$$
$$\gamma_2 = \frac{\mu_4}{\sigma^4} - 3 = \frac{3\sigma^4}{\sigma^4} - 3 = 0. \qquad (48)$$

The VARIANCE of the SAMPLE VARIANCE s^2 for a sample taken from a population with a Gaussian distribution is

$$\mathrm{var}(s^2) = \frac{(N-1)[(N-1)\mu_4' - (N-3)\mu_2'^2]}{N^3}$$
$$= \frac{(N-1)}{N^3}[(N-1)(\mu^4 + 6\mu^2\sigma^2 + 3\sigma^4) - (N-3)(\mu^2 + \sigma^2)^2]$$
$$= \frac{2(N-1)(\mu^4 + 2\mu^2 N\sigma^2 + N\sigma^4)}{N^3}. \qquad (49)$$

If $\mu = 0$, this expression simplifies to

$$\mathrm{var}(s^2) = \frac{2(N-1)N\sigma^4}{N^3} = \frac{2\sigma^4(N-1)}{N^2}, \qquad (50)$$

and the STANDARD ERROR is

$$[\text{standard error}] = \frac{\sqrt{2(N-1)}}{N}. \qquad (51)$$

The CUMULANT-GENERATING FUNCTION for a Gaussian distribution is

$$K(h) = \ln(e^{\nu_1 h} e^{\sigma^2 h^2/2}) = \nu_1 h + \tfrac{1}{2}\sigma^2 h^2, \qquad (52)$$

so

$$\kappa_1 = \nu_1 \qquad (53)$$
$$\kappa_2 = \sigma^2 \qquad (54)$$
$$\kappa_r = 0 \quad \text{for } r > 2. \qquad (55)$$

For Gaussian variates, $\kappa_r = 0$ for $r > 2$, so the variance of k-STATISTIC k_3 is

$$\mathrm{var}(k_3) = \frac{\kappa_6}{N} + \frac{9\kappa_2\kappa_4}{N-1} + \frac{9\kappa_3{}^2}{N-1} + \frac{6\kappa_2{}^3}{N(N-1)(N-2)} = \frac{6\kappa_2{}^3}{N(N-1)(N-2)}. \qquad (56)$$

Also,

$$\mathrm{var}(k_4) = \frac{24k_2{}^4 N(N-1)^2}{(N-3)(N-2)(N+3)(N+5)} \qquad (57)$$
$$\mathrm{var}(g_1) = \frac{6N(N-1)}{(N-2)(N+1)(N+3)} \qquad (58)$$
$$\mathrm{var}(g_2) = \frac{24N(N-1)^2}{(N-3)(N-2)(N+3)(N+5)}, \qquad (59)$$

where

$$g_1 \equiv \frac{k_3}{k_2{}^{3/2}} \qquad (60)$$
$$g_2 \equiv \frac{k_4}{k_2{}^2}. \qquad (61)$$

If $P(x)$ is a Gaussian distribution, then

$$D(x) = \frac{1}{2}\left[1 + \mathrm{erf}\left(\frac{x-\mu}{\sigma\sqrt{2}}\right)\right], \qquad (62)$$

so variates x_i with a Gaussian distribution can be generated from variates y_i having a UNIFORM DISTRIBUTION in (0,1) via

$$x_i = \sigma\sqrt{2}\,\mathrm{erf}^{-1}(2y_i - 1) + \mu. \qquad (63)$$

However, a simpler way to obtain numbers with a Gaussian distribution is to use the BOX-MULLER TRANSFORMATION.

The Gaussian distribution is an approximation to the BINOMIAL DISTRIBUTION in the limit of large numbers,

$$P(n_1) = \frac{1}{\sqrt{2\pi Npq}} \exp\left[-\frac{(n_1 - Np)^2}{2Npq}\right], \quad (64)$$

where n_1 is the number of steps in the POSITIVE direction, N is the number of trials ($N \equiv n_1 + n_2$), and p and q are the probabilities of a step in the POSITIVE direction and NEGATIVE direction ($q \equiv 1 - p$).

The differential equation having a Gaussian distribution as its solution is

$$\frac{dy}{dx} = \frac{y(\mu - x)}{\sigma^2}, \quad (65)$$

since

$$\frac{dy}{y} = \frac{\mu - x}{\sigma^2}\, dx \quad (66)$$

$$\ln\left(\frac{y}{y_0}\right) = -\frac{1}{2\sigma^2}(\mu - x)^2 \quad (67)$$

$$y = y_0 e^{-(x-\mu)^2/2\sigma^2} \quad (68)$$

This equation has been generalized to yield more complicated distributions which are named using the so-called PEARSON SYSTEM.

see also BINOMIAL DISTRIBUTION, CENTRAL LIMIT THEOREM, ERF, GAUSSIAN BIVARIATE DISTRIBUTION, LOGIT TRANSFORMATION, NORMAL DISTRIBUTION, NORMAL DISTRIBUTION FUNCTION, PEARSON SYSTEM, RATIO DISTRIBUTION, z-SCORE

References

Beyer, W. H. *CRC Standard Mathematical Tables, 28th ed.* Boca Raton, FL: CRC Press, pp. 533–534, 1987.

Kraitchik, M. "The Error Curve." §6.4 in *Mathematical Recreations.* New York: W. W. Norton, pp. 121–123, 1942.

Spiegel, M. R. *Theory and Problems of Probability and Statistics.* New York: McGraw-Hill, p. 109–111, 1992.

Gaussian Distribution—Linear Combination of Variates

If x is NORMALLY DISTRIBUTED with MEAN μ and VARIANCE σ^2, then a linear function of x,

$$y = ax + b, \quad (1)$$

is also NORMALLY DISTRIBUTED. The new distribution has MEAN $a\mu + b$ and VARIANCE $a^2\sigma^2$, as can be derived using the MOMENT-GENERATING FUNCTION

$$M(t) = \left\langle e^{t(ax+b)} \right\rangle = e^{tb}\left\langle e^{atx}\right\rangle = e^{tb} e^{\mu at + \sigma^2 (at)^2/2}$$
$$= e^{tb + \mu at + \sigma^2 a^2 t^2/2} = e^{(b+a\mu)t + a^2\sigma^2 t^2/2}, \quad (2)$$

which is of the standard form with

$$\mu' = b + a\mu \quad (3)$$

$$\sigma'^2 = a^2\sigma^2. \quad (4)$$

For a weighted sum of independent variables

$$y \equiv \sum_{i=1}^n a_i x_i, \quad (5)$$

the expectation is given by

$$M(t) = \left\langle e^{yt}\right\rangle = \left\langle \exp\left(t \sum_{i=1}^n a_i x_i\right)\right\rangle$$
$$= \left\langle e^{a_1 t x_1} e^{a_2 t x_2} \cdots e^{a_n t x_n}\right\rangle$$
$$= \prod_{i=1}^n \left\langle e^{a_i t x_i}\right\rangle = \prod_{i=1}^n \exp(a_i \mu_i t + \tfrac{1}{2} {a_i}^2 {\sigma_i}^2 t^2). \quad (6)$$

Setting this equal to

$$\exp(\mu t + \tfrac{1}{2}\sigma^2 t^2) \quad (7)$$

gives

$$\mu \equiv \sum_{i=1}^n a_i \mu_i \quad (8)$$

$$\sigma^2 \equiv \sum_{i=1}^n {a_i}^2 {\sigma_i}^2. \quad (9)$$

Therefore, the MEAN and VARIANCE of the weighted sums of n RANDOM VARIABLES are their weighted sums.

If x_i are INDEPENDENT and NORMALLY DISTRIBUTED with MEAN 0 and VARIANCE σ^2, define

$$y_i \equiv \sum_j c_{ij} x_j, \quad (10)$$

where c obeys the ORTHOGONALITY CONDITION

$$c_{ik} c_{jk} = \delta_{ij}, \quad (11)$$

with δ the KRONECKER DELTA. Then y_i are also independent and normally distributed with MEAN 0 and VARIANCE σ^2.

Gaussian Elimination

A method for solving MATRIX EQUATIONS of the form

$$\mathsf{A}\mathbf{x} = \mathbf{b}. \quad (1)$$

Starting with the system of equations

$$\begin{bmatrix} a_{11} & a_{12} & \cdots & a_{1k} \\ a_{21} & a_{22} & \cdots & a_{2k} \\ \vdots & \vdots & \ddots & \vdots \\ a_{k1} & a_{k2} & \cdots & a_{kk} \end{bmatrix} \begin{bmatrix} x_1 \\ x_2 \\ \vdots \\ x_k \end{bmatrix} = \begin{bmatrix} b_1 \\ b_2 \\ \vdots \\ b_k \end{bmatrix}, \tag{2}$$

compose the augmented MATRIX equation

$$\left[\begin{array}{cccc|c} a_{11} & a_{12} & \cdots & a_{1k} & b_1 \\ a_{21} & a_{22} & \cdots & a_{2k} & b_2 \\ \vdots & \vdots & \ddots & \vdots & \vdots \\ a_{k1} & a_{k2} & \cdots & a_{kk} & b_k \end{array}\right] \begin{bmatrix} x_1 \\ x_2 \\ \vdots \\ x_k \end{bmatrix}. \tag{3}$$

Then, perform MATRIX operations to put the augmented MATRIX into the form

$$\left[\begin{array}{cccc|c} a'_{11} & a'_{12} & \cdots & a'_{1k} & b'_1 \\ 0 & a'_{22} & \cdots & a'_{2k} & b'_2 \\ \vdots & \vdots & \ddots & \vdots & \vdots \\ 0 & 0 & \cdots & a'_{kk} & b'_k \end{array}\right] \begin{bmatrix} x'_1 \\ x'_2 \\ \vdots \\ x'_k \end{bmatrix}. \tag{4}$$

Solve for a'_{kk}, then substitute back in to obtain solutions for $n = 1, 2, \ldots, k-1$,

$$x_i = \frac{1}{a'_{ii}} \left(b'_i - \sum_{j=i+1}^{k} a'_{ij} x'_j \right). \tag{5}$$

see also GAUSS-JORDAN ELIMINATION, LU DECOMPOSITION, MATRIX EQUATION, SQUARE ROOT METHOD

Gaussian Function

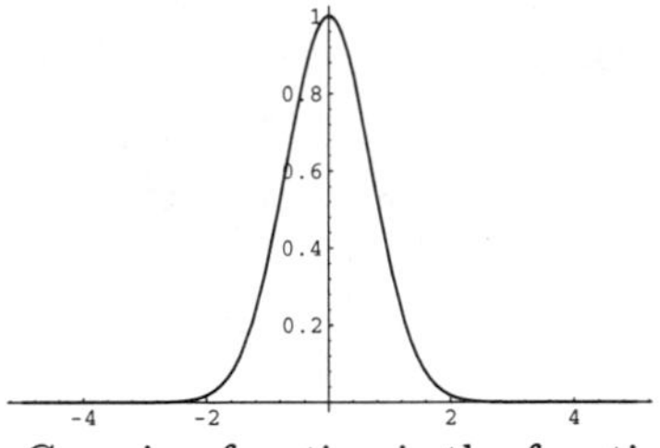

In 1-D, the Gaussian function is the function from the GAUSSIAN DISTRIBUTION,

$$f(x) = \frac{1}{\sigma\sqrt{2\pi}} e^{-(x-\mu)^2/2\sigma^2}, \tag{1}$$

sometimes also called the FREQUENCY CURVE. The FULL WIDTH AT HALF MAXIMUM (FWHM) for a Gaussian is found by finding the half-maximum points x_0. The constant scaling factor can be ignored, so we must solve

$$e^{-(x_0-\mu)^2/2\sigma^2} = \tfrac{1}{2} f(x_{\max}) \tag{2}$$

But $f(x_{\max})$ occurs at $x_{\max} = \mu$, so

$$e^{-(x_0-\mu)^2/2\sigma^2} = \tfrac{1}{2} f(\mu) = \tfrac{1}{2}. \tag{3}$$

Solving,

$$e^{-(x_0-\mu)^2/2\sigma^2} = 2^{-1} \tag{4}$$

$$-\frac{(x_0-\mu)^2}{2\sigma^2} = -\ln 2 \tag{5}$$

$$(x_0-\mu)^2 = 2\sigma^2 \ln 2 \tag{6}$$

$$x_0 = \pm\sigma\sqrt{2\ln 2} + \mu. \tag{7}$$

The FULL WIDTH AT HALF MAXIMUM is therefore given by

$$\text{FWHM} \equiv x_+ - x_- = 2\sqrt{2\ln 2}\,\sigma \approx 2.3548\sigma. \tag{8}$$

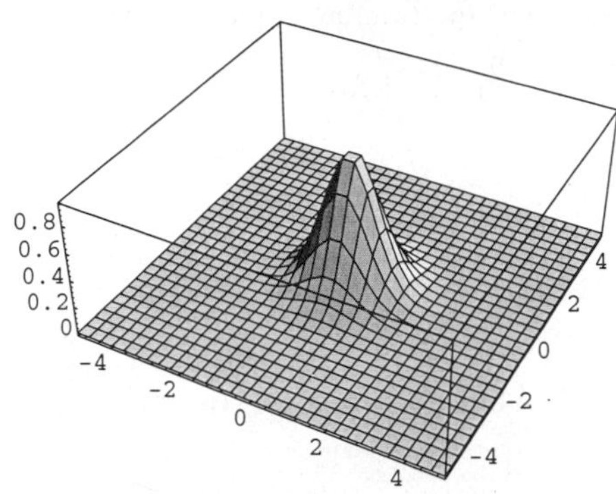

In 2-D, the circular Gaussian function is the distribution function for uncorrelated variables x and y having a GAUSSIAN BIVARIATE DISTRIBUTION and equal STANDARD DEVIATION $\sigma = \sigma_x = \sigma_y$,

$$f(x,y) = \frac{1}{2\pi\sigma^2} e^{-[(x-\mu_x)^2+(y-\mu_y)^2]/2\sigma^2}. \tag{9}$$

The corresponding elliptical Gaussian function corresponding to $\sigma_x \neq \sigma_y$ is given by

$$f(x,y) = \frac{1}{2\pi\sigma_x\sigma_y} e^{-[(x-\mu_x)^2/2\sigma_x{}^2+[(y-\mu_y)^2/2\sigma_y{}^2]}. \tag{10}$$

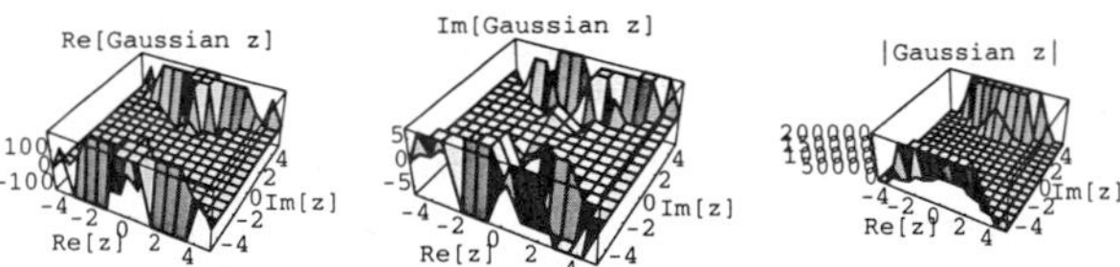

The above plots show the real and imaginary parts of $(2\pi)^{-1/2}e^{-z^2}$ together with the complex absolute value $|(2\pi)^{-1/2}e^{-z^2}|$.

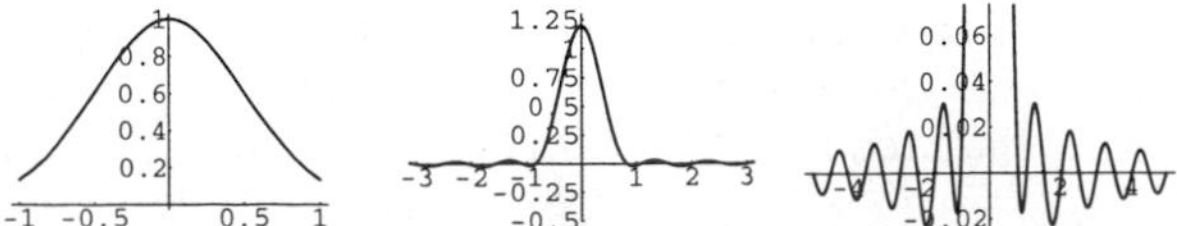

The Gaussian function can also be used as an APODIZATION FUNCTION, shown above with the corresponding INSTRUMENT FUNCTION.

The HYPERGEOMETRIC FUNCTION is also sometimes known as the Gaussian function.

see also ERF, ERFC, FOURIER TRANSFORM—GAUSSIAN, GAUSSIAN BIVARIATE DISTRIBUTION, GAUSSIAN DISTRIBUTION, NORMAL DISTRIBUTION

References

MacTutor History of Mathematics Archive. "Frequency Curve." http://www-groups.dcs.st-and.ac.uk/~history/Curves/Frequency.html.

Gaussian Hypergeometric Series

see HYPERGEOMETRIC FUNCTION

Gaussian Integer

A COMPLEX NUMBER $a+bi$ where a and b are INTEGERS. The Gaussian integers are members of the QUADRATIC FIELD $\mathbb{Q}(\sqrt{-1})$. The sum, difference, and product of two Gaussian integers are Gaussian integers, but $a+bi|c+di$ only if there is an $e+fi$ such that

$$(a+bi)(e+fi)=(ae-bf)+(af+be)i=c+di.$$

Gaussian INTEGERS can be uniquely factored in terms of other Gaussian INTEGERS up to POWERS of i and rearrangements.

The norm of a Gaussian integer is defined by

$$n(x+iy)=x^2+y^2.$$

GAUSSIAN PRIMES are Gaussian integers $a+ib$ for which $n(a+ib)=a^2+b^2$ is PRIME and a a PRIME INTEGER a such that $a\equiv 3 \pmod 4$.

1. If $2|n(x+iy)$, then $1+i$ and $1-i|x+iy$. These factors are equivalent since $-i(i-1)=i+1$. For example, $2=(1+i)(1-i)$ is not a Gaussian prime.
2. If $n(x+iy)\equiv 3 \pmod 4 |n(x+iy)$, then $n(a+ib)|x+iy$.
3. If $n(x+iy)\equiv 1 \pmod 4 |n(x+iy)$, then $a+ib$ or $b+ia|x+iy$. If both do, then $n(a+ib)|x+iy$.

The Gaussian primes with $|a|,|b|\leq 5$ are given by $-5-4i$, $-5-2i$, $-5+2i$, $-5+4i$, $-4-5i$, $-4-i$, $-4+i$, $-4+5i$, $-3-2i$, -3, $-3+2i$, $-2-5i$, $-2-3i$, $-2-i$, $-2+i$, $-2+3i$, $-2+5i$, $-1-4i$, $-1-2i$, $-1-i$, $-1+i$, $-1+2i$, $-1+4i$, $-3i$, $3i$, $1-4i$, $1-2i$, $1-i$, $1+i$, $1+2i$, $1+4i$, $2-5i$, $2-3i$, $2-i$, $2+i$, $2+3i$, $2+5i$, $3-2i$, 3, $3+2i$, $4-5i$, $4-i$, $4+i$, $4+5i$, $5-4i$, $5-2i$, $5+2i$, $5+4i$.

Every Gaussian integer is within $|n|/\sqrt{2}$ of a multiple of a Gaussian integer n.

see also COMPLEX NUMBER, EISENSTEIN INTEGER, GAUSSIAN PRIME, INTEGER, OCTONION

References

Conway, J. H. and Guy, R. K. "Gauss's Whole Numbers." In *The Book of Numbers.* New York: Springer-Verlag, pp. 217–223, 1996.

Shanks, D. "Gaussian Integers and Two Applications." §50 in *Solved and Unsolved Problems in Number Theory, 4th ed.* New York: Chelsea, pp. 149–151, 1993.

Gaussian Integral

The Gaussian integral, also called the PROBABILITY INTEGRAL, is the integral of the 1-D Gaussian over $(-\infty,\infty)$. It can be computed using the trick of combining two 1-D Gaussians

$$\int_{-\infty}^{\infty} e^{-x^2}\,dx=\sqrt{\left(\int_{-\infty}^{\infty} e^{-y^2}\,dy\right)\left(\int_{-\infty}^{\infty} e^{-x^2}\,dx\right)}=\sqrt{\int_{-\infty}^{\infty}\int_{-\infty}^{\infty} e^{-(x^2+y^2)}\,dy\,dx} \quad (1)$$

and switching to POLAR COORDINATES,

$$\int_{-\infty}^{\infty} e^{-x^2}\,dx=\sqrt{\int_0^{2\pi}\int_0^{\infty} e^{-r^2}r\,dr\,d\theta}=\sqrt{2\pi\left[-\tfrac{1}{2}e^{-r^2}r\right]_0^{\infty}}=\sqrt{\pi}. \quad (2)$$

However, a simple proof can also be given which does not require transformation to POLAR COORDINATES (Nicholas and Yates 1950).

The integral from 0 to a finite upper limit a can be given by the CONTINUED FRACTION

$$\int_0^a e^{-x^2}\,dx=\frac{\sqrt{\pi}}{2}\,\frac{1}{a+}\,\frac{2}{2a+}\,\frac{3}{a+}\,\frac{4}{2a+\ldots}. \quad (3)$$

The general class of integrals of the form

$$I_n(a)\equiv\int_0^{\infty} e^{-ax^2}x^n\,dx \quad (4)$$

can be solved analytically by setting

$$x\equiv a^{-1/2}y \quad (5)$$

$$dx=a^{-1/2}\,dy \quad (6)$$

$$y^2=ax^2. \quad (7)$$

Then

$$I_n(a)=a^{-1/2}\int_0^{\infty} e^{-y^2}(a^{-1/2})^n\,dy=a^{-(1+n)/2}\int_0^{\infty} e^{-y^2}y^n\,dy. \quad (8)$$

For $n=0$, this is just the usual Gaussian integral, so

$$I_0(a)=\frac{\sqrt{\pi}}{2}a^{-1/2}=\frac{1}{2}\sqrt{\frac{\pi}{a}}. \quad (9)$$

For $n=1$, the integrand is integrable by quadrature,

$$I_1(a)=a^{-1}\int_0^{\infty} e^{-y^2}y\,dy=a^{-1}[-\tfrac{1}{2}e^{-y^2}]_0^{\infty}=\tfrac{1}{2}a^{-1}. \quad (10)$$

To compute $I_n(a)$ for $n > 1$, use the identity

$$-\frac{\partial}{\partial a}I_{n-2}(a) = -\frac{\partial}{\partial a}\int_0^\infty e^{-ax^2}x^{n-2}\,dx$$
$$= -\int_0^\infty -x^2e^{-ax^2}x^{n-2}\,dx$$
$$= \int_0^\infty e^{-ax^2}x^n\,dx = I_n(a). \quad (11)$$

For $n = 2s$ EVEN,

$$I_n(a) = \left(-\frac{\partial}{\partial a}\right)I_{n-2}(a) = \left(-\frac{\partial}{\partial a}\right)^2 I_{n-4}$$
$$= \ldots = \left(-\frac{\partial}{\partial a}\right)^{n/2} I_0(a)$$
$$= \frac{\partial^{n/2}}{\partial a^{n/2}}I_0(a) = \frac{\sqrt{\pi}}{2}\frac{\partial^{n/2}}{\partial a^{n/2}}a^{-1/2}, \quad (12)$$

so

$$\int_0^\infty x^{2s}e^{-ax^2}\,dx = \frac{(s-\frac{1}{2})!}{2a^{s+1/2}} = \frac{(2s-1)!!}{2^{s+1}a^s}\sqrt{\frac{\pi}{a}}. \quad (13)$$

If $n = 2s+1$ is ODD, then

$$I_n(a) = \left(-\frac{\partial}{\partial a}\right)I_{n-2}(a) = \left(-\frac{\partial}{\partial a}\right)^2 I_{n-4}(a)$$
$$= \ldots = \left(-\frac{\partial}{\partial a}\right)^{(n-1)/2} I_1(a)$$
$$= \frac{\partial^{(n-1)/2}}{\partial a^{(n-1)/2}}I_1(a) = \frac{1}{2}\frac{\partial^{(n-1)/2}}{\partial a^{(n-1)/2}}a^{-1}, \quad (14)$$

so

$$\int_0^\infty x^{2s+1}e^{-ax^2}\,dx = \frac{s!}{2a^{s+1}}. \quad (15)$$

The solution is therefore

$$\int_0^\infty e^{-ax^2}x^n\,dx = \begin{cases} \frac{(n-1)!!}{2^{n/2+1}a^{n/2}}\sqrt{\frac{\pi}{a}} & \text{for } n \text{ even} \\ \frac{[(n+1)/2]!}{2a^{(n+1)/2}} & \text{for } n \text{ odd.} \end{cases} \quad (16)$$

The first few values are therefore

$$I_0(a) = \frac{1}{2}\sqrt{\frac{\pi}{a}} \quad (17)$$
$$I_1(a) = \frac{1}{2a} \quad (18)$$
$$I_2(a) = \frac{1}{4a}\sqrt{\frac{\pi}{a}} \quad (19)$$
$$I_3(a) = \frac{1}{2a^2} \quad (20)$$
$$I_4(a) = \frac{3}{8a^2}\sqrt{\frac{\pi}{a}} \quad (21)$$
$$I_5(a) = \frac{1}{a^3} \quad (22)$$
$$I_6(a) = \frac{15}{16a^3}\sqrt{\frac{\pi}{a}}. \quad (23)$$

A related, often useful integral is

$$H_n(a) \equiv \frac{1}{\sqrt{\pi}}\int_{-\infty}^\infty e^{-ax^2}x^n\,dx, \quad (24)$$

which is simply given by

$$H_n = \begin{cases} \frac{2I_n(a)}{\sqrt{\pi}} & \text{for } n \text{ even} \\ 0 & \text{for } n \text{ odd.} \end{cases} \quad (25)$$

References

Nicholas, C. B. and Yates, R. C. "The Probability Integral." *Amer. Math. Monthly* **57**, 412–413, 1950.

Gaussian Integral (Linking Number)

see LINKING NUMBER

Gaussian Joint Variable Theorem

Also called the MULTIVARIATE THEOREM. Given an EVEN number of variates from a NORMAL DISTRIBUTION with MEANS all 0,

$$\langle x_1x_2\rangle = \langle x_1\rangle\langle x_2\rangle, \quad (1)$$

$$\langle x_1x_2x_3x_4\rangle$$
$$= \langle x_1x_2\rangle\langle x_3x_4\rangle + \langle x_1x_3\rangle\langle x_2x_4\rangle + \langle x_1x_4\rangle\langle x_2x_3\rangle, \quad (2)$$

etc. Given an ODD number of variates,

$$\langle x_1\rangle = 0, \quad (3)$$

$$\langle x_1x_2x_3\rangle = 0, \quad (4)$$

etc.

Gaussian Mountain Range

see CAROTID-KUNDALINI FUNCTION

Gaussian Multivariate Distribution

see also GAUSSIAN BIVARIATE DISTRIBUTION, JOINT THEOREM, MULTIVARIATE THEOREM

Gaussian Polynomial

Defined by

$$[l] \equiv \frac{1-q^l}{1-q} \quad (1)$$

for integral l, and

$$\begin{bmatrix} n \\ k \end{bmatrix} \equiv \begin{cases} \prod_{l=1}^k \frac{[n-l+1]}{[l]} & \text{for } 0 \leq k \leq n \\ 0 & \text{otherwise.} \end{cases} \quad (2)$$

Unfortunately, the NOTATION conflicts with that of GAUSSIAN BRACKETS and the NEAREST INTEGER

FUNCTION. Gaussian POLYNOMIALS satisfy the identities

$$\frac{\begin{bmatrix} n+1 \\ k+1 \end{bmatrix}}{\begin{bmatrix} n \\ k+1 \end{bmatrix}} = \frac{1-q^{n+1}}{1-q^{n-k}} \tag{3}$$

$$\frac{\begin{bmatrix} n+1 \\ k+1 \end{bmatrix}}{\begin{bmatrix} n+1 \\ k \end{bmatrix}} = \frac{1-q^{n-k+1}}{1-q^{k+1}}. \tag{4}$$

For $q = 1$, the Gaussian polynomial turns into the BINOMIAL COEFFICIENT.

see also BINOMIAL COEFFICIENT, GAUSSIAN COEFFICIENT, q-SERIES

Gaussian Prime

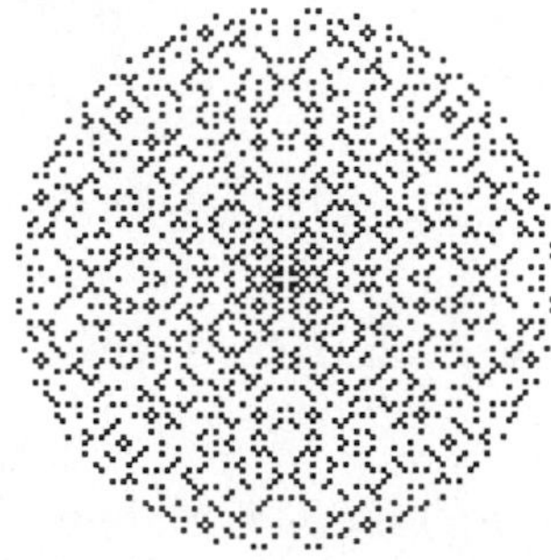

Gaussian primes are GAUSSIAN INTEGERS $a + ib$ for which $n(a + ib) = a^2 + b^2$ is PRIME and a a PRIME INTEGER a such that $a \equiv 3 \pmod 4$. The above plot of the COMPLEX PLANE shows the Gaussian primes as filled squares.

see also EISENSTEIN INTEGER, GAUSSIAN INTEGER

References

Guy, R. K. "Gaussian Primes. Eisenstein-Jacobi Primes." §A16 in *Unsolved Problems in Number Theory, 2nd ed.* New York: Springer-Verlag, pp. 33–36, 1994.

Wagon, S. "Gaussian Primes." §9.4 in *Mathematica in Action.* New York: W. H. Freeman, pp. 298–303, 1991.

Gaussian Quadrature

Seeks to obtain the best numerical estimate of an integral by picking optimal ABSCISSAS x_i at which to evaluate the function $f(x)$. The FUNDAMENTAL THEOREM OF GAUSSIAN QUADRATURE states that the optimal ABSCISSAS of the m-point GAUSSIAN QUADRATURE FORMULAS are precisely the roots of the orthogonal POLYNOMIAL for the same interval and WEIGHTING FUNCTION. Gaussian quadrature is optimal because it fits all POLYNOMIALS up to degree $2m$ exactly. Slightly less optimal fits are obtained from RADAU QUADRATURE and LAGUERRE QUADRATURE.

$W(x)$	Interval	x_i Are Roots Of
1	$(-1,1)$	$P_n(x)$
e^{-t}	$(0,\infty)$	$L_n(x)$
e^{-t^2}	$(-\infty,\infty)$	$H_n(x)$
$(1-t^2)^{-1/2}$	$(-1,1)$	$T_n(x)$
$(1-t^2)^{1/2}$	$(-1,1)$	$U_n(x)$
$x^{1/2}$	$(0,1)$	$x^{-1/2}P_{2n+1}(\sqrt{x})$
$x^{-1/2}$	$(0,1)$	$P_n(\sqrt{x})$

To determine the weights corresponding to the Gaussian ABSCISSAS, compute a LAGRANGE INTERPOLATING POLYNOMIAL for $f(x)$ by letting

$$\pi(x) \equiv \prod_{j=1}^{m}(x - x_j) \tag{1}$$

(where Chandrasekhar 1967 uses F instead of π), so

$$\pi'(x_j) = \left[\frac{d\pi}{dx}\right]_{x=x_j} = \prod_{\substack{i=1 \\ i\neq j}}^{m}(x_j - x_i). \tag{2}$$

Then fitting a LAGRANGE INTERPOLATING POLYNOMIAL through the m points gives

$$\phi(x) = \sum_{j=1}^{m} \frac{\pi(x)}{(x - x_j)\pi'(x_j)} f(x_j) \tag{3}$$

for arbitrary points x_i. We are therefore looking for a set of points x_j and weights w_j such that for a WEIGHTING FUNCTION $W(x)$,

$$\begin{aligned}\int_a^b \phi(x)W(x)\,dx &= \int_a^b \sum_{j=1}^{m} \frac{\pi(x)W(x)}{(x-x_j)\pi'(x_j)}\,dx\,f(x_j) \\ &\equiv \sum_{j=1}^{m} w_j f(x_j),\end{aligned} \tag{4}$$

with WEIGHT

$$w_j = \frac{1}{\pi'(x_j)} \int_a^b \frac{\pi(x)W(x)}{x - x_j}\,dx. \tag{5}$$

The weights w_j are sometimes also called the CHRISTOFFEL NUMBER (Chandrasekhar 1967). For orthogonal POLYNOMIALS $\phi_j(x)$ with j=1, ..., n,

$$\phi_j(x) = A_j\pi(x) \tag{6}$$

(Hildebrand 1956, p. 322), where A_n is the COEFFICIENT of x^n in $\phi_n(x)$, then

$$\begin{aligned} w_j &= \frac{1}{\phi_n'(x_j)} \int_a^b W(x)\frac{\phi(x)}{x - x_j}\,dx \\ &= -\frac{A_{n+1}\gamma_n}{A_n\phi_n'(x_j)\phi_{n+1}(x)},\end{aligned} \tag{7}$$

where

$$\gamma_m \equiv \int [\phi_m(x)]^2 W(x)\,dx. \qquad (8)$$

Using the relationship

$$\phi_{n+1}(x_i) = -\frac{A_{n+1}A_{n-1}}{{A_n}^2}\frac{\gamma_n}{\gamma_{n-1}}\phi_{n-1}(x_i) \qquad (9)$$

(Hildebrand 1956, p. 323) gives

$$w_j = \frac{A_n}{A_{n-1}}\frac{\gamma_{n-1}}{\phi_n'(x_j)\phi_{n-1}(x_j)}. \qquad (10)$$

(Note that Press *et al.* 1992 omit the factor A_n/A_{n-1}.) In Gaussian quadrature, the weights are all POSITIVE. The error is given by

$$E_n = \frac{f^{(2n)}(\xi)}{(2n)!}\int_a^b W(x)[\pi(x)]^2\,dx = \frac{\gamma_n}{{A_n}^2}\frac{f^{(2n)}(\xi)}{(2n)!}, \qquad (11)$$

where $a < \xi < b$ (Hildebrand 1956, pp. 320–321).

Other curious identities are

$$\sum_{k=0}^{m}\frac{[\phi_k(x)]^2}{\gamma_k} = \frac{A_m}{A_{m+1}\gamma_m}[\phi_{m+1}'(x)\phi_m(x) - \phi_m'(x)\phi_{m+1}(x)] \qquad (12)$$

and

$$\sum_{k=0}^{m}\frac{[\phi_k(x)]^2}{\gamma_k} = -\frac{A_m\phi_m'(x_i)\phi_{m+1}(x_i)}{A_{m+1}\gamma_m} = \frac{1}{w_i} \qquad (13)$$

(Hildebrand 1956, p. 323).

In the NOTATION of Szegő (1975), let $x_{1n} < \ldots < x_{nn}$ be an ordered set of points in $[a,b]$, and let $\lambda_{1n}, \ldots, \lambda_{nn}$ be a set of REAL NUMBERS. If $f(x)$ is an arbitrary function on the CLOSED INTERVAL $[a,b]$, write the MECHANICAL QUADRATURE as

$$Q_n(f) = \sum_{\nu=1}^{n}\lambda_{\nu n} f(x_{\nu n}). \qquad (14)$$

Here $x_{\nu n}$ are the ABSCISSAS and $\lambda_{\nu n}$ are the COTES NUMBERS.

see also CHEBYSHEV QUADRATURE, CHEBYSHEV-GAUSS QUADRATURE, CHEBYSHEV-RADAU QUADRATURE, FUNDAMENTAL THEOREM OF GAUSSIAN QUADRATURE, HERMITE-GAUSS QUADRATURE, JACOBI-GAUSS QUADRATURE, LAGUERRE-GAUSS QUADRATURE, LEGENDRE-GAUSS QUADRATURE, LOBATTO QUADRATURE, MEHLER QUADRATURE, RADAU QUADRATURE

References

Abramowitz, M. and Stegun, C. A. (Eds.). *Handbook of Mathematical Functions with Formulas, Graphs, and Mathematical Tables, 9th printing.* New York: Dover, pp. 887–888, 1972.

Acton, F. S. *Numerical Methods That Work, 2nd printing.* Washington, DC: Math. Assoc. Amer., p. 103, 1990.

Arfken, G. "Appendix 2: Gaussian Quadrature." *Mathematical Methods for Physicists, 3rd ed.* Orlando, FL: Academic Press, pp. 968–974, 1985.

Beyer, W. H. *CRC Standard Mathematical Tables, 28th ed.* Boca Raton, FL: CRC Press, p. 461, 1987.

Chandrasekhar, S. *An Introduction to the Study of Stellar Structure.* New York: Dover, 1967.

Hildebrand, F. B. *Introduction to Numerical Analysis.* New York: McGraw-Hill, pp. 319–323, 1956.

Press, W. H.; Flannery, B. P.; Teukolsky, S. A.; and Vetterling, W. T. "Gaussian Quadratures and Orthogonal Polynomials." §4.5 in *Numerical Recipes in FORTRAN: The Art of Scientific Computing, 2nd ed.* Cambridge, England: Cambridge University Press, pp. 140–155, 1992.

Szegő, G. *Orthogonal Polynomials, 4th ed.* Providence, RI: Amer. Math. Soc., pp. 37–48 and 340–349, 1975.

Whittaker, E. T. and Robinson, G. *The Calculus of Observations: A Treatise on Numerical Mathematics, 4th ed.* New York: Dover, pp. 152–163, 1967.

Gaussian Sum

$$S(p,q) \equiv \sum_{r=0}^{q-1} e^{-\pi i r^2 p/q}, \qquad (1)$$

where p and q are RELATIVELY PRIME INTEGERS. If $(n,n')=1$, then

$$S(m,nn') = S(mn',n)S(mn,n'). \qquad (2)$$

Gauss showed

$$\sum_{r=0}^{q-1} e^{2\pi i r^2/q} = \frac{1-i^q}{1-i}\sqrt{q} \qquad (3)$$

for ODD q. A more general result was obtained by Schaar. For p and q of opposite PARITY (i.e., one is EVEN and the other is ODD), SCHAAR'S IDENTITY states

$$\frac{1}{\sqrt{q}}\sum_{r=0}^{q-1} e^{-\pi i r^2 p/q} = \frac{e^{-\pi i/4}}{\sqrt{p}}\sum_{r=0}^{p-1} e^{\pi i r^2 q/p}. \qquad (4)$$

Such sums are important in the theory of QUADRATIC RESIDUES.

see also KLOOSTERMAN'S SUM, SCHAAR'S IDENTITY, SINGULAR SERIES

References

Evans, R. and Berndt, B. "The Determination of Gauss Sums." *Bull. Amer. Math. Soc.* **5**, 107–129, 1981.

Katz, N. M. *Gauss Sums, Kloosterman Sums, and Monodromy Groups.* Princeton, NJ: Princeton University Press, 1987.

Riesel, H. *Prime Numbers and Computer Methods for Factorization, 2nd ed.* Boston, MA: Birkhäuser, pp. 132–134, 1994.

Gear Graph

A WHEEL GRAPH with a VERTEX added between each pair of adjacent VERTICES.

Gegenbauer Function

see ULTRASPHERICAL FUNCTION

Gegenbauer Polynomial

see ULTRASPHERICAL POLYNOMIAL

Gelfond-Schneider Constant

The number $2^{\sqrt{2}} = 2.66514414\ldots$ which is known to be TRANSCENDENTAL by GELFOND'S THEOREM.

References

Courant, R. and Robbins, H. *What is Mathematics?: An Elementary Approach to Ideas and Methods, 2nd ed.* Oxford, England: Oxford University Press, p. 107, 1996.

Gelfond-Schneider Theorem

see GELFOND'S THEOREM

Gelfond's Theorem

Also called the GELFOND-SCHNEIDER THEOREM. a^b is TRANSCENDENTAL if

1. a is ALGEBRAIC $\neq 0$, 1 and
2. b is ALGEBRAIC and IRRATIONAL.

This provides the solution to the seventh of HILBERT'S PROBLEMS.

see also ALGEBRAIC NUMBER, HILBERT'S PROBLEMS, IRRATIONAL NUMBER, TRANSCENDENTAL NUMBER

References

Baker, A. *Transcendental Number Theory.* London: Cambridge University Press, 1990.

Courant, R. and Robbins, H. *What is Mathematics?: An Elementary Approach to Ideas and Methods, 2nd ed.* Oxford, England: Oxford University Press, p. 107, 1996.

Genaille Rods

Numbered rods which can be used to perform multiplication.

see also NAPIER'S BONES

References

Gardner, M. "Napier's Bones." Ch. 7 in *Knotted Doughnuts and Other Mathematical Entertainments.* New York: W. H. Freeman, 1986.

Genera

see FUNDAMENTAL THEOREM OF GENERA

General Linear Group

The general linear group $GL_n(q)$ is the set of $n \times n$ MATRICES with entries in the FIELD $\mathbb{F}_q$ which have NONZERO DETERMINANT.

see also LANGLANDS RECIPROCITY, PROJECTIVE GENERAL LINEAR GROUP, PROJECTIVE SPECIAL LINEAR GROUP, SPECIAL LINEAR GROUP

References

Conway, J. H.; Curtis, R. T.; Norton, S. P.; Parker, R. A.; and Wilson, R. A. "The Groups $GL_n(q)$, $SL_n(q)$, $PGL_n(q)$, and $PSL_n(q) = L_n(q)$." §2.1 in *Atlas of Finite Groups: Maximal Subgroups and Ordinary Characters for Simple Groups.* Oxford, England: Clarendon Press, p. x, 1985.

General Orthogonal Group

The general orthogonal group $GO_n(q,F)$ is the SUBGROUP of all elements of the PROJECTIVE GENERAL LINEAR GROUP that fix the particular nonsingular QUADRATIC FORM F. The determinant of such an element is ± 1.

see also PROJECTIVE GENERAL LINEAR GROUP

References

Conway, J. H.; Curtis, R. T.; Norton, S. P.; Parker, R. A.; and Wilson, R. A. "The Groups $GO_n(q)$, $SO_n(q)$, $PGO_n(q)$, and $PSO_n(q)$, and $O_n(q)$." §2.4 in *Atlas of Finite Groups: Maximal Subgroups and Ordinary Characters for Simple Groups.* Oxford, England: Clarendon Press, pp. xi–xii, 1985.

General Position

An arrangement of points with no three COLLINEAR, or of lines with no three concurrent.

see also ORDINARY LINE, NEAR-PENCIL

References

Guy, R. K. "Unsolved Problems Come of Age." *Amer. Math. Monthly* **96**, 903–909, 1989.

General Prismatoid

A solid such that the AREA A_y of any section parallel to and a distance y from a fixed PLANE can be expressed as

$$A_y = ay^3 + by^2 + cy + d.$$

The volume of such a solid is the same as for a PRISMATOID,

$$V = \tfrac{1}{6}h(A_1 + 4M + A_2).$$

Examples include the CONE, CYLINDER, PRISMATOID, SPHERE, and SPHEROID.

see also PRISMATOID, PRISMOID

References

Beyer, W. H. *CRC Standard Mathematical Tables, 28th ed.* Boca Raton, FL: CRC Press, p. 132, 1987.

General Unitary Group

The general unitary group $GU_n(q)$ is the SUBGROUP of all elements of the GENERAL LINEAR GROUP $GL(q^2)$ that fix a given nonsingular Hermitian form. This is equivalent, in the canonical case, to the definition of GU_n as the group of UNITARY MATRICES.

References

Conway, J. H.; Curtis, R. T.; Norton, S. P.; Parker, R. A.; and Wilson, R. A. "The Groups $GU_n(q)$, $SU_n(q)$, $PGU_n(q)$, and $PSU_n(q) = U_n(q)$." §2.2 in *Atlas of Finite Groups: Maximal Subgroups and Ordinary Characters for Simple Groups.* Oxford, England: Clarendon Press, p. x, 1985.

Generalized Cone

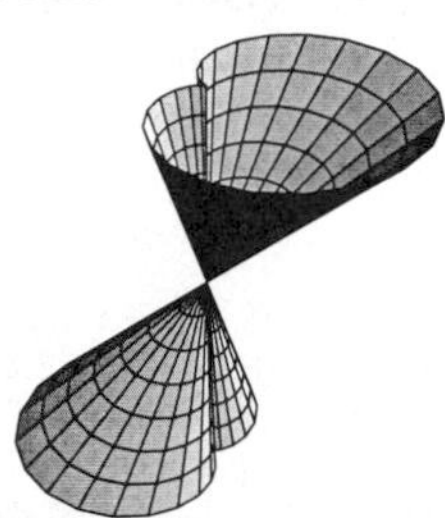

A RULED SURFACE is called a generalized cone if it can be parameterized by $\mathbf{x}(u,v) = \mathbf{p} + v\mathbf{y}(u)$, where $\mathbf{p}$ is a fixed point which can be regarded as the vertex of the cone. A generalized cone is a REGULAR SURFACE wherever $v\mathbf{y}\times\mathbf{y}' \neq \mathbf{0}$. The above surface is a generalized cylinder over a CARDIOID. A generalized cone is a FLAT SURFACE.

see also CONE

References

Gray, A. *Modern Differential Geometry of Curves and Surfaces.* Boca Raton, FL: CRC Press, pp. 341–342, 1993.

Generalized Cylinder

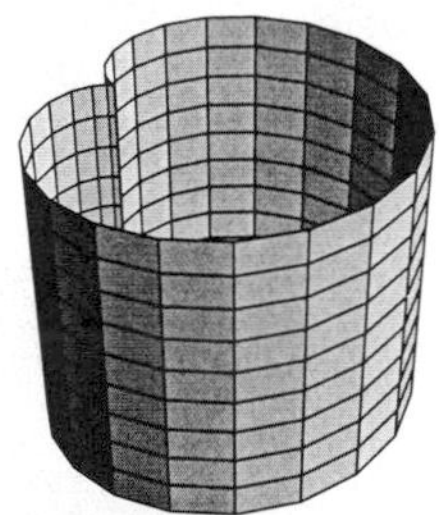

A RULED SURFACE is called a generalized cylinder if it can be parameterized by $\mathbf{x}(u,v) = v\mathbf{p} + \mathbf{y}(u)$, where $\mathbf{p}$ is a fixed point. A generalized cylinder is a REGULAR SURFACE wherever $\mathbf{y}' \times \mathbf{p} \neq \mathbf{0}$. The above surface is a generalized cylinder over a CARDIOID. A generalized cylinder is a FLAT SURFACE.

see also CYLINDER

References

Gray, A. *Modern Differential Geometry of Curves and Surfaces.* Boca Raton, FL: CRC Press, pp. 341–342, 1993.

Generalized Fibonacci Number

A generalization of the FIBONACCI NUMBERS defined by $1 = G_1 = G_2 = \ldots = G_{c-1}$ and the RECURRENCE RELATION

$$G_n = G_{n-1} + G_{n-c}. \quad (1)$$

These are the sums of elements on successive diagonals of a left-justified PASCAL'S TRIANGLE beginning in the left-most column and moving in steps of $c-1$ up and 1 right. The case $c = 2$ equals the usual FIBONACCI NUMBER. These numbers satisfy the identities

$$G_1 + G_2 + G_3 + \ldots + G_n = G_{n+3} - 1 \quad (2)$$

$$G_3 + G_6 + G_9 + \ldots + G_{3k} = G_{3k+1} - 1 \quad (3)$$

$$G_1 + G_4 + G_7 + \ldots + G_{3k+1} = G_{3k+2} \quad (4)$$

$$G_2 + G_5 + G_8 + \ldots + G_{3k+2} = G_{3k+3} \quad (5)$$

(Bicknell-Johnson and Spears 1996). For the special case $c = 3$,

$$G_{n+w} = G_{w-2}G_n + G_{w-3}G_{n+1} + G_{w-1}G_{n+2}. \quad (6)$$

Bicknell-Johnson and Spears (1996) give many further identities.

Horadam (1965) defined the generalized Fibonacci numbers $\{w_n\}$ as $w_n = w_n(a,b;p,q)$, where a, b, p, and q are INTEGERS, $w_0 = a$, $w_1 = b$, and $w_n = pw_{n-1} - qw_{n-2}$ for $n \geq 2$. They satisfy the identities

$$w_n w_{n+2r} - eq^n U_r = {w_{n+r}}^2 \quad (7)$$

$$4w_n {w_{n+1}}^2 w_{n+2} + (wq^n)^2 = (w_n w_{n+2} + {w_{n+1}}^2)^2 \quad (8)$$

$$w_n w_{n+1} w_{n+3} w_{n+4} = {w_{n+2}}^4 + eq^n(p^2+q){w_{n+2}}^2 + e^2 q^{2n+1} p^2 \quad (9)$$

$$4w_n w_{n+1} w_{n+2} w_{n+4} w_{n+5} w_{n+6} + e^2 q^{2n}(w_n U_4 U_5 - w_{n+1} U_2 U_6 - w_n U_1 U_8)^2 = (w_{n+1} w_{n+2} w_{n+6} + w_n w_{n+4} w_{n+5})^2, \quad (10)$$

where

$$e \equiv pab - qa^2 - b^2 \quad (11)$$

$$U_n \equiv w_n(0,1;p,q). \quad (12)$$

The final above result is due to Morgado (1987) and is called the MORGADO IDENTITY.

Another generalization of the Fibonacci numbers is denoted x_n. Given x_1 and x_2, define the generalized Fibonacci number by $x_n \equiv x_{n-2} + x_{n-1}$ for $n \geq 3$,

$$\sum_{i=1}^{n} x_n = x_{n+2} - x_2 \quad (13)$$

$$\sum_{i=1}^{10} x_n = 11x_7 \tag{14}$$

$$x_n{}^2 - x_{n-1}x_{n+2} = (-1)^n(x_2{}^2 - x_1{}^2 - x_1x_2), \tag{15}$$

where the plus and minus signs alternate.

see also FIBONACCI NUMBER

References

Bicknell, M. "A Primer for the Fibonacci Numbers, Part VIII: Sequences of Sums from Pascal's Triangle." *Fib. Quart.* **9**, 74–81, 1971.

Bicknell-Johnson, M. and Spears, C. P. "Classes of Identities for the Generalized Fibonacci Numbers $G_n = G_{n-1} + G_{n-c}$ for Matrices with Constant Valued Determinants." *Fib. Quart.* **34**, 121–128, 1996.

Dujella, A. "Generalized Fibonacci Numbers and the Problem of Diophantus." *Fib. Quart.* **34**, 164–175, 1996.

Horadam, A. F. "Generating Functions for Powers of a Certain Generalized Sequence of Numbers." *Duke Math. J.* **32**, 437–446, 1965.

Horadam, A. F. "Generalization of a Result of Morgado." *Portugaliae Math.* **44**, 131–136, 1987.

Horadam, A. F. and Shannon, A. G. "Generalization of Identities of Catalan and Others." *Portugaliae Math.* **44**, 137–148, 1987.

Morgado, J. "Note on Some Results of A. F. Horadam and A. G. Shannon Concerning a Catalan's Identity on Fibonacci Numbers." *Portugaliae Math.* **44**, 243–252, 1987.

Generalized Function

The class of all regular sequences of PARTICULARLY WELL-BEHAVED FUNCTIONS equivalent to a given regular sequence (sometimes also called a DISTRIBUTION or FUNCTIONAL). A generalized function $p(x)$ has the properties

$$\int_{-\infty}^{\infty} p'(x)f(x)\,dx = -\int_{-\infty}^{\infty} p(x)f'(x)\,dx$$

$$\int_{-\infty}^{\infty} p^{(n)}f(x)\,dx = (-1)^n \int_{-\infty}^{\infty} p(x)f^{(n)}(x)\,dx.$$

The DELTA FUNCTION is a generalized function.

see also DELTA FUNCTION

Generalized Helicoid

The SURFACE generated by a twisted curve C when rotated about a fixed axis A and, at the same time, displaced PARALLEL to A so that the velocity of displacement is always proportional to the ANGULAR VELOCITY of ROTATION.

see also GENERALIZED HELIX, HELICOID, HELIX

References

do Carmo, M. P.; Fischer, G.; Pinkall, U.; and Reckziegel, H. "General Helicoids." §3.4.3 in *Mathematical Models from the Collections of Universities and Museums* (Ed. G. Fischer). Braunschweig, Germany: Vieweg, pp. 36–37, 1986.

Fischer, G. (Ed.). Plate 89 in *Mathematische Modelle/Mathematical Models, Bildband/Photograph Volume.* Braunschweig, Germany: Vieweg, p. 85, 1986.

Kreyszig, E. *Differential Geometry.* New York: Dover, p. 88, 1991.

Generalized Helix

The GEODESICS on a general cylinder generated by lines PARALLEL to a line l with which the TANGENT makes a constant ANGLE.

see also HELIX

Generalized Hyperbolic Functions

In 1757, V. Riccati first recorded the generalizations of the HYPERBOLIC FUNCTIONS defined by

$$F_{n,r}^{\alpha}(x) \equiv C\sum_{k=0}^{\infty} \frac{\alpha^k}{(nk+r)!} x^{nk+r}, \tag{1}$$

for $r = 0, \ldots, n-1$, where α is COMPLEX, and where the normalization is taken so that

$$F_{n,0}^{\alpha}(0) = 1. \tag{2}$$

This is called the α-hyperbolic function of order n of the kth kind. The functions $F_{n,r}^{\alpha}$ satisfy

$$F_{n,r}^{\alpha}(x) = (\sqrt[n]{\alpha})^{-r}(\sqrt[n]{\alpha}\,x) \tag{3}$$

and

$$f^{(k)}(x) = \alpha f(x), \tag{4}$$

where

$$f^{(k)}(0) = \begin{cases} 0 & k \neq r,\ 0 \leq k \leq n-1, \\ 1 & k = r. \end{cases} \tag{5}$$

In addition,

$$\frac{d}{dx}F_{n,r}^{\alpha}(x) = \begin{cases} F_{n,r-1}^{\alpha}(x) & \text{for } 0 < r \leq n-1 \\ \alpha F_{n,n-1}^{\alpha}(x) & \text{for } r = 0. \end{cases} \tag{6}$$

The functions give a generalized EULER FORMULA

$$e^{\sqrt[n]{\alpha}} = \sum_{r=0}^{n-1} (\sqrt[n]{\alpha})^r F_{n,r}^{\alpha}(x). \tag{7}$$

Since there are n nth roots of α, this gives a system of n linear equations. Solving for $F_{n,r}^{\alpha}$ gives

$$F_{n,r}^{\alpha}(x) = \frac{1}{n}(\sqrt[n]{\alpha})^{-r} \sum_{k=0}^{n-1} \omega_n{}^{-rk} \exp(\omega_n{}^k \sqrt[n]{\alpha}\,x), \tag{8}$$

where

$$\omega_n = \exp\left(\frac{2\pi i}{n}\right) \tag{9}$$

is a PRIMITIVE ROOT OF UNITY.

The LAPLACE TRANSFORM is

$$\int_0^{\infty} e^{-st} F_{n,r}^{\alpha}(at)\,dt = \frac{s^{n-r-1}a^r}{s^n + \alpha a_n}. \tag{10}$$

The generalized hyperbolic function is also related to the MITTAG-LEFFLER FUNCTION $E_\gamma(x)$ by

$$F^1_{n,0}(x) = E_n(x^n). \quad (11)$$

The values $n = 1$ and $n = 2$ give the exponential and circular/hyperbolic functions (depending on the sign of α), respectively.

$$F^\alpha_{1,0}(x) = e^{\alpha x} \quad (12)$$

$$F^\alpha_{2,0}(x) = \cosh(\sqrt{\alpha}\,x) \quad (13)$$

$$F^\alpha_{2,1}(x) = \frac{\sinh(\sqrt{\alpha}\,x)}{\sqrt{\alpha}}. \quad (14)$$

For $\alpha = 1$, the first few functions are

$$\begin{aligned}
F^1_{1,0}(x) &= e^x \\
F^1_{2,0}(x) &= \cosh x \\
F^1_{2,1}(x) &= \sinh x \\
F^1_{3,0}(x) &= \tfrac{1}{3}[e^x + 2e^{-x/2}\cos(\tfrac{1}{2}\sqrt{3}\,x)] \\
F^1_{3,1}(x) &= \tfrac{1}{3}[e^x + 2e^{-x/2}\cos(\tfrac{1}{2}\sqrt{3}\,x + \tfrac{1}{3}\pi)] \\
F^1_{3,2}(x) &= \tfrac{1}{3}[e^x + 2e^{-x/2}\cos(\tfrac{1}{2}\sqrt{3}\,x - \tfrac{1}{3}\pi)] \\
F^1_{4,0}(x) &= \tfrac{1}{2}(\cosh x + \cos x) \\
F^1_{4,1}(x) &= \tfrac{1}{2}(\sinh x + \sin x) \\
F^1_{4,2}(x) &= \tfrac{1}{2}(\cosh x - \cos x) \\
F^1_{4,3}(x) &= \tfrac{1}{2}(\sinh x - \sin x).
\end{aligned}$$

see also HYPERBOLIC FUNCTIONS, MITTAG-LEFFLER FUNCTION

References

Kaufman, H. "A Biographical Note on the Higher Sine Functions." *Scripta Math.* **28**, 29–36, 1967.

Muldoon, M. E. and Ungar, A. A. "Beyond Sin and Cos." *Math. Mag.* **69**, 3–14, 1996.

Petkovšek, M.; Wilf, H. S.; and Zeilberger, D. *A=B*. Wellesley, MA: A. K. Peters, 1996.

Ungar, A. "Generalized Hyperbolic Functions." *Amer. Math. Monthly* **89**, 688–691, 1982.

Ungar, A. "Higher Order Alpha-Hyperbolic Functions." *Indian J. Pure. Appl. Math.* **15**, 301–304, 1984.

Generalized Hypergeometric Function

The generalized hypergeometric function is given by a HYPERGEOMETRIC SERIES, i.e., a series for which the ratio of successive terms can be written

$$\frac{a_{k+1}}{a_k} = \frac{P(k)}{Q(k)} = \frac{(k+a_1)(k+a_2)\cdots(k+a_p)}{(k+b_1)(k+b_2)\cdots(k+b_q)(k+1)}x. \quad (1)$$

(The factor of $k+1$ in the DENOMINATOR is present for historical reasons of notation.) The resulting generalized hypergeometric function is written

$$\sum_{k=0} a_k x^k = {}_pF_q\left[\begin{matrix} a_1, a_2, \cdots, a_p \\ b_1, b_2, \cdots, b_q \end{matrix}; x\right] \quad (2)$$

$$= \sum_{k=0}^{\infty} \frac{(a_1)_k (a_2)_k \cdots (a_p)_k}{(b_1)_k b(b_2)_k \cdots (b_q)_k} \frac{x^k}{k!}, \quad (3)$$

where $(a)_k$ is the POCHHAMMER SYMBOL or RISING FACTORIAL

$$(a)_k \equiv \frac{\Gamma(a+k)}{\Gamma(a)} = a(a+1)\cdots(a+k-1). \quad (4)$$

If the argument $x = 1$, then the function is abbreviated

$${}_pF_q\left[\begin{matrix} a_1, a_2 \ldots, a_p \\ b_1, b_2, \ldots, b_q \end{matrix}\right] \equiv {}_pF_q\left[\begin{matrix} a_1, a_2 \ldots, a_p \\ b_1, b_2, \ldots, b_q \end{matrix}; x\right]. \quad (5)$$

${}_2F_1(a, b; c; z)$ is "the" HYPERGEOMETRIC FUNCTION, and ${}_1F_1(a; b; z) \equiv M(z)$ is the CONFLUENT HYPERGEOMETRIC FUNCTION. A function of the form ${}_0F_1(; b; z)$ is called a CONFLUENT HYPERGEOMETRIC LIMIT FUNCTION.

The generalized hypergeometric function

$${}_{p+1}F_p\left[\begin{matrix} a_1, a_2, \ldots, a_{p+1} \\ b_1, b_2, \ldots, b_p \end{matrix}; z\right] \quad (6)$$

is a solution to the DIFFERENTIAL EQUATION

$$\begin{aligned}
&[\vartheta(\vartheta + b - 1)\cdots(\vartheta + b_p - 1) \\
&\quad -z(\vartheta + a_1)(\vartheta + a_2)\cdots(\vartheta + a_{p+1})]y = 0, \quad (7)
\end{aligned}$$

where

$$\vartheta = z\frac{d}{dz}. \quad (8)$$

The other linearly independent solution is

$$z^{1-b_1}{}_{p+1}F_p\left[\begin{matrix} 1 + a_1 - b_1, 1 - a_2 - b_2, \\ \ldots, 1 + a_{p+1} - b_1 \\ 2 - b_1, 1 - b_2 - b_1, \ldots, \\ 1 - b_p - b_1 \end{matrix}; z\right]. \quad (9)$$

A generalized hypergeometric equation is termed "well posed" if

$$1 + a_1 = b_1 + a_2 = \ldots = b_p + a_{p+1}. \quad (10)$$

Many sums can be written as generalized hypergeometric functions by inspection of the ratios of consecutive terms in the generating HYPERGEOMETRIC SERIES. For example, for

$$f(n) \equiv \sum_k (-1)^k \binom{2n}{k}^2, \quad (11)$$

the ratio of successive terms is

$$\frac{a_{k+1}}{a_k} = \frac{(-1)^{k+1}\binom{2n}{k+1}^2}{(-1)^k\binom{2n}{k}^2} = -\frac{(k-2n)^2}{(k+1)^2}, \quad (12)$$

yielding

$$f(n) = {}_2F_1\left[\begin{matrix}-2n, -2n\\ 1\end{matrix}; -1\right] = {}_2F_1(-2n, -2n; 1; -1) \tag{13}$$

(Petkovšek 1996, pp. 44–45).

Gosper (1978) discovered a slew unusual hypergeometric function identities, many of which were subsequently proven by Gessel and Stanton (1982). An important generalization of Gosper's technique, called ZEILBERGER'S ALGORITHM, in turn led to the powerful machinery of the WILF-ZEILBERGER PAIR (Zeilberger 1990).

Special hypergeometric identities include GAUSS'S HYPERGEOMETRIC THEOREM

$${}_2F_1(a, b; c; 1) = \frac{\Gamma(c)\Gamma(c-a-b)}{\Gamma(c-a)\Gamma(c-b)} \tag{14}$$

for $\Re[c-a-b] > 0$, KUMMER'S FORMULA

$${}_2F_1(a, b; c; -1) = \frac{\Gamma(\frac{1}{2}b+1)\Gamma(b-a+1)}{\Gamma(b+1)\Gamma(\frac{1}{2}b-a+1)}, \tag{15}$$

where $a - b + c = 1$ and b is a positive integer, SAALSCHÜTZ'S THEOREM

$${}_3F_2(a, b, c; d, e; 1) = \frac{(d-a)_{|c|}(d-b)_{|c|}}{d_{|c|}(d-a-b)_{|c|}} \tag{16}$$

for $d + e = a + b + c + 1$ with c a negative integer and $(a)_n$ the POCHHAMMER SYMBOL, DIXON'S THEOREM

$${}_3F_2(a, b, c; d, e; 1) = \frac{(\frac{1}{2}a)!(a-b)!(a-c)!(\frac{1}{2}a-b-c)!}{a!(\frac{1}{2}a-b)!(\frac{1}{2}a-c)!(a-b-c)!}, \tag{17}$$

where $1 + a/2 - b - c$ has a positive REAL PART, $d = a - b + 1$, and $e = a - c + 1$, the CLAUSEN FORMULA

$${}_4F_3\left[\begin{matrix}a & b & c & d\\ e & f & g\end{matrix}; 1\right] = \frac{(2a)_{|d|}(a+b)_{|d|}(2b)_{|d|}}{(2a+2b)_{|d|}a_{|d|}b_{|d|}}, \tag{18}$$

for $a+b+c-d = 1/2$, $e = a+b+1/2$, $a+f = d+1 = b+g$, d a nonpositive integer, and the DOUGALL-RAMANUJAN IDENTITY

$$\begin{aligned} &{}_7F_6\left[\begin{matrix}a_1, a_2, a_3, a_4, a_5, a_6, a_7\\ b_1, b_2, b_3, b_4, b_5, b_6\end{matrix}; 1\right] \\ &= \frac{(a_1+1)_n(a_1-a_2-a_3+1)_n}{(a_1-a_2+1)_n(a_1-a_3+1)_n} \\ &\quad \times \frac{(a_1-a_2-a_4+1)_n(a_1-a_3-a_4+1)_n}{(a_1-a_4+1)_n(a_1-a_2-a_3-a_4+1)_n}, \end{aligned} \tag{19}$$

where $n = 2a_1 + 1 = a_2 + a_3 + a_4 + a_5$, $a_6 = 1 + a_1/2$, $a_7 = -n$, and $b_i = 1 + a_1 - a_{i+1}$ for $i = 1, 2, \ldots, 6$. For all these identities, $(a)_n$ is the POCHHAMMER SYMBOL.

Gessel (1994) found a slew of new identities using WILF-ZEILBERGER PAIRS, including the following:

$${}_5F_4\left[\begin{matrix}-a-b, n+1, n+c+1, 2n-a-b+1, n+\frac{1}{2}(3-a-b)\\ n-a-b-c+1, n-a-b+1, 2n+2, n+\frac{1}{2}(1-a-b)\end{matrix}; 1\right] = 0 \tag{20}$$

$${}_3F_2\left[\begin{matrix}-3n, \frac{2}{3}-c, 3n+2\\ \frac{3}{2}, 1-3c\end{matrix}; \frac{3}{4}\right] = \frac{(c+\frac{2}{3})_n(\frac{1}{3})_n}{(1-c)_n(\frac{4}{3})_n} \tag{21}$$

$${}_3F_2\left[\begin{matrix}-3b, -\frac{3}{2}n, \frac{1}{2}(1-3n)\\ -3n, \frac{2}{3}-b-n\end{matrix}; \frac{4}{3}\right] = \frac{(\frac{1}{3}-b)_n}{(\frac{1}{3}+b)_n} \tag{22}$$

$${}_4F_3\left[\begin{matrix}\frac{3}{2}+\frac{1}{5}n, \frac{2}{3}, -n, 2n+2\\ n+\frac{11}{6}, \frac{4}{3}, \frac{1}{5}n+\frac{1}{2}\end{matrix}; \frac{2}{27}\right] = \frac{(\frac{5}{2})_n(\frac{11}{6})_n}{(\frac{3}{2})_n(\frac{7}{2})_n} \tag{23}$$

(Petkovšek *et al.* 1996, pp. 135–137).

see also CARLSON'S THEOREM, CLAUSEN FORMULA, CONFLUENT HYPERGEOMETRIC FUNCTION, CONFLUENT HYPERGEOMETRIC LIMIT FUNCTION, DIXON'S THEOREM, DOUGALL-RAMANUJAN IDENTITY, DOUGALL'S THEOREM, GOSPER'S ALGORITHM, HEINE HYPERGEOMETRIC SERIES, HYPERGEOMETRIC FUNCTION, HYPERGEOMETRIC IDENTITY, HYPERGEOMETRIC SERIES, JACKSON'S IDENTITY, KUMMER'S THEOREM, RAMANUJAN'S HYPERGEOMETRIC IDENTITY, SAALSCHÜTZ'S THEOREM, SAALSCHÜTZIAN, SISTER CELINE'S METHOD, THOMAE'S THEOREM, WATSON'S THEOREM, WHIPPLE'S TRANSFORMATION, WILF-ZEILBERGER PAIR, ZEILBERGER'S ALGORITHM

References

Bailey, W. N. *Generalised Hypergeometric Series.* Cambridge, England: Cambridge University Press, 1935.

Dwork, B. *Generalized Hypergeometric Functions.* Oxford, England: Clarendon Press, 1990.

Exton, H. *Multiple Hypergeometric Functions and Applications.* New York: Wiley, 1976.

Gessel, I. "Finding Identities with the WZ Method." *Theoret. Comput. Sci.* To appear.

Gessel, I. and Stanton, D. "Strange Evaluations of Hypergeometric Series." *SIAM J. Math. Anal.* **13**, 295–308, 1982.

Gosper, R. W. "Decision Procedures for Indefinite Hypergeometric Summation." *Proc. Nat. Acad. Sci. USA* **75**, 40–42, 1978.

Petkovšek, M.; Wilf, H. S.; and Zeilberger, D. *A=B.* Wellesley, MA: A. K. Peters, 1996.

Saxena, R. K. and Mathai, A. M. *Generalized Hypergeometric Functions with Applications in Statistics and Physical Sciences.* New York: Springer-Verlag, 1973.

Slater, L. J. *Generalized Hypergeometric Functions.* Cambridge, England: Cambridge University Press, 1966.

Zeilberger, D. "A Fast Algorithm for Proving Terminating Hypergeometric Series Identities." *Discrete Math.* **80**, 207–211, 1990.

Generalized Matrix Inverse

see MOORE-PENROSE GENERALIZED MATRIX INVERSE

Generalized Mean

A generalized version of the MEAN

$$m(t) \equiv \left(\frac{1}{n}\sum_{k=1}^{n} {a_k}^t\right)^{1/t} \tag{1}$$

with parameter t which gives the GEOMETRIC MEAN, ARITHMETIC MEAN, and HARMONIC MEAN as special cases:

$$\lim_{t\to 0} m(t) = G \tag{2}$$

$$m(1) = A \tag{3}$$

$$m(-1) = H. \tag{4}$$

see also MEAN

Generalized Remainder Method

An algorithm for computing a UNIT FRACTION.

Generating Function

A POWER SERIES

$$f(x) = \sum_{n=0}^{\infty} a_n x^n$$

whose COEFFICIENTS give the SEQUENCE $\{a_0, a_1, \ldots\}$. The *Mathematica*® (Wolfram Research, Champaign, IL) function `DiscreteMath'RSolve'PowerSum` gives the generating function of a given expression, and `ExponentialPowerSum` gives the exponential generating function.

Generating functions for the first few powers are

$$\begin{array}{lll} 1: & \frac{x}{1-x} & = x + x^2 + x^3 + \ldots \\ n: & \frac{x}{(1-x)^2} & = x + 2x^2 + 3x^3 + 4x^4 + \ldots \\ n^2: & \frac{x(x+1)}{(1-x)^3} & = x + 4x^2 + 9x^3 + 16x^4 + \ldots \\ n^3: & \frac{x(x^2+4x+1)}{(x-1)^4} & = x + 8x^2 + 27x^3 + \ldots \\ n^4: & \frac{x(x+1)(x^2+10x+1)}{(x-1)^5} & = x + 16x^2 + 81x^3 + \ldots. \end{array}$$

see also MOMENT-GENERATING FUNCTION, RECURRENCE RELATION

References

Wilf, H. S. *Generatingfunctionology, 2nd ed.* New York: Academic Press, 1990.

Generation

In population studies, the direct offspring of a reference population (roughly) constitutes a single generation. For a CELLULAR AUTOMATON, the fundamental unit of time during which the rules of reproduction are applied once is called a generation.

Generator (Digitadition)

An INTEGER used to generate a DIGITADITION. A number can have more than one generator. If a number has no generator, it is called a SELF NUMBER.

Generator (Group)

An element of a CYCLIC GROUP, the POWERS of which generate the entire GROUP.

References

Arfken, G. "Generators." §4.11 in *Mathematical Methods for Physicists, 3rd ed.* Orlando, FL: Academic Press, pp. 261–267, 1985.

Genetic Algorithm

An adaptive ALGORITHM involving search and optimization first used by John Holland. Holland created an electronic organism as a binary string ("chromosome"), and then used genetic and evolutionary principles of fitness-proportionate selection for reproduction (including random crossover and mutation) to search enormous solution spaces efficiently. So-called genetic programming languages apply the same principles, using an expression tree instead of a bit string as the "chromosome."

see also CELLULAR AUTOMATON

Genocchi Number

A number given by the GENERATING FUNCTION

$$\frac{2t}{e^t+1} = \sum_{n=1}^{\infty} G_n \frac{t^n}{n!}.$$

It satisfies $G_1 = 1$, $G_3 = G_5 = G_7 = \ldots$, and even coefficients are given by

$$\begin{aligned} G_{2n} &= 2(1-2^{2n})B_{2n} \\ &= 2nE_{2n-1}(0), \end{aligned}$$

where B_n is a BERNOULLI NUMBER and $E_n(x)$ is an EULER POLYNOMIAL. The first few Genocchi numbers for n EVEN are -1, 1, -3, 17, -155, 2073, ... (Sloane's A001469).

see also BERNOULLI NUMBER, EULER POLYNOMIAL

References

Comtet, L. *Advanced Combinatorics: The Art of Finite and Infinite Expansions, rev. enl. ed.*ordrecht, Netherlands: Reidel, p. 49, 1974.

Kreweras, G. "An Additive Generation for the Genocchi Numbers and Two of its Enumerative Meanings." *Bull. Inst. Combin. Appl.* **20**, 99–103, 1997.

Kreweras, G. "Sur les permutations comptées par les nombres de Genocchi de 1-ière et 2-ième espèce." *Europ. J. Comb.* **18**, 49–58, 1997.

Sloane, N. J. A. Sequence A001469/M3041 in "An On-Line Version of the Encyclopedia of Integer Sequences."

Gentle Diagonal

see PASCAL'S TRIANGLE

Gentle Giant Group

see MONSTER GROUP

Genus (Curve)

One of the PLÜCKER CHARACTERISTICS, defined by

$$p \equiv \tfrac{1}{2}(n-1)(n-2)-(\delta+\kappa) = \tfrac{1}{2}(m-1)(m-2)-(\tau+\iota),$$

where m is the class, n the order, δ the number of nodes, κ the number of CUSPS, ι the number of stationary tangents (INFLECTION POINTS), and τ the number of BITANGENTS.

see also RIEMANN CURVE THEOREM

References

Coolidge, J. L. *A Treatise on Algebraic Plane Curves.* New York: Dover, p. 100, 1959.

Genus (Knot)

The least genus of any SEIFERT SURFACE for a given KNOT. The UNKNOT is the only KNOT with genus 0.

Genus (Surface)

A topologically invariant property of a surface defined as the largest number of nonintersecting simple closed curves that can be drawn on the surface without separating it. Roughly speaking, it is the number of HOLES in a surface.

see also EULER CHARACTERISTIC

Genus Theorem

A DIOPHANTINE EQUATION

$$x^2 + y^2 = p$$

can be solved for p a PRIME IFF $p \equiv 1 \pmod 4$ or $p = 2$. The representation is unique except for changes of sign or rearrangements of x and y.

see also COMPOSITION THEOREM, FERMAT'S THEOREM

Geocentric Latitude

An AUXILIARY LATITUDE given by

$$\phi_g = \tan^{-1}[(1-e^2)\tan\phi].$$

The series expansion is

$$\phi_g = \phi - e_2 \sin(2\phi) + \tfrac{1}{2}{e_2}^2 \sin(4\phi) + \tfrac{1}{3}{e_2}^3 \sin(6\phi) + \ldots,$$

where

$$e_2 \equiv \frac{e^2}{2-e^2}.$$

see also LATITUDE

References

Adams, O. S. "Latitude Developments Connected with Geodesy and Cartography with Tables, Including a Table for Lambert Equal-Area Meridional Projections." Spec. Pub. No. 67. U. S. Coast and Geodetic Survey, 1921.

Snyder, J. P. *Map Projections—A Working Manual.* U. S. Geological Survey Professional Paper 1395. Washington, DC: U. S. Government Printing Office, pp. 17–18, 1987.

Geodesic

Given two points on a surface, the geodesic is defined as the shortest path on the surface connecting them. Geodesics have many interesting properties. The NORMAL VECTOR to any point of a GEODESIC arc lies along the normal to a surface at that point (Weinstock 1974, p. 65).

Furthermore, no matter how badly a SPHERE is distorted, there exist an infinite number of closed geodesics on it. This general result, demonstrated in the early 1990s, extended earlier work by Birkhoff, who proved in 1917 that there exists at least one closed geodesic on a distorted sphere, and Lyusternik and Schirelmann, who proved in 1923 that there exist at least three closed geodesics on such a sphere (Cipra 1993).

For a surface $g(x,y,z) = 0$, the geodesic can be found by minimizing the ARC LENGTH

$$L \equiv \int ds = \int \sqrt{dx^2 + dy^2 + dz^2}. \qquad (1)$$

But

$$dx = \frac{\partial x}{\partial u} du + \frac{\partial x}{\partial v} dv \qquad (2)$$

$$dx^2 = \left(\frac{\partial x}{\partial u}\right)^2 du^2 + 2\frac{\partial x}{\partial u}\frac{\partial x}{\partial v} du\,dv + \left(\frac{\partial x}{\partial v}\right)^2 dv^2, \qquad (3)$$

and similarly for dy^2 and dz^2. Plugging in,

$$L = \int \left\{ \left[\left(\frac{\partial x}{\partial u}\right)^2 + \left(\frac{\partial y}{\partial u}\right)^2 + \left(\frac{\partial z}{\partial u}\right)^2 \right] du^2 + 2\left[\frac{\partial x}{\partial u}\frac{\partial x}{\partial v} + \frac{\partial y}{\partial u}\frac{\partial y}{\partial v} + \frac{\partial z}{\partial u}\frac{\partial z}{\partial v} \right] du\,dv + \left[\left(\frac{\partial x}{\partial v}\right)^2 + \left(\frac{\partial y}{\partial v}\right)^2 + \left(\frac{\partial z}{\partial v}\right)^2 \right] dv^2 \right\}^{1/2}. \qquad (4)$$

This can be rewritten as

$$L = \int \sqrt{P + 2Qv' + Rv'^2}\, du \qquad (5)$$

$$= \int \sqrt{Pu'^2 + 2Qu' + R}\, dv, \qquad (6)$$

where

$$v' \equiv \frac{dv}{du} \qquad (7)$$

$$u' \equiv \frac{du}{dv} \qquad (8)$$

and

$$P \equiv \left(\frac{\partial x}{\partial u}\right)^2 + \left(\frac{\partial y}{\partial u}\right)^2 + \left(\frac{\partial z}{\partial u}\right)^2 \qquad (9)$$

$$Q \equiv \frac{\partial x}{\partial u}\frac{\partial x}{\partial v} + \frac{\partial y}{\partial u}\frac{\partial y}{\partial v} + \frac{\partial z}{\partial u}\frac{\partial z}{\partial v} \qquad (10)$$

$$R \equiv \left(\frac{\partial x}{\partial v}\right)^2 + \left(\frac{\partial y}{\partial v}\right)^2 + \left(\frac{\partial z}{\partial v}\right)^2. \qquad (11)$$

Taking derivatives,

$$\frac{\partial L}{\partial v} = \tfrac{1}{2}(P + 2Qv' + Rv'^2)^{-1/2} \times \left(\frac{\partial P}{\partial v} + 2\frac{\partial Q}{\partial v}v' + \frac{\partial R}{\partial v}v'^2\right) \quad (12)$$

$$\frac{\partial L}{\partial v'} = \tfrac{1}{2}(P + 2Qv' + Rv'^2)^{-1/2}(2Q + 2Rv'), \quad (13)$$

so the EULER-LAGRANGE DIFFERENTIAL EQUATION then gives

$$\frac{\frac{\partial P}{\partial v} + 2v'\frac{\partial Q}{\partial v} + v'^2\frac{\partial R}{\partial v}}{2\sqrt{P + 2Qv' + Rv'^2}} - \frac{d}{du}\left(\frac{Q + Rv'}{\sqrt{P + 2Qv' + Rv'^2}}\right) = 0. \quad (14)$$

In the special case when P, Q, and R are explicit functions of u only,

$$\frac{Q + Rv'}{\sqrt{P + 2Qv' + Rv'^2}} = c_1 \quad (15)$$

$$\frac{Q^2 + 2QRv' + R^2v'^2}{P + 2Qv' + Rv'^2} = c_1{}^2 \quad (16)$$

$$v'^2R(R - c_1{}^2) + 2v'Q(R - c_1{}^2) + (Q^2 - Pc_1{}^2) = 0 \quad (17)$$

$$v' = \frac{1}{2R(R - c_1{}^2)}[2Q(c_1{}^2 - R) \pm\sqrt{4Q^2(R - c_1{}^2)^2 - 4R(R - c_1{}^2)(Q^2 - Pc_1{}^2)}\,]. \quad (18)$$

Now, if P and R are explicit functions of u only *and* $Q = 0$,

$$v' = \frac{\sqrt{4R(R - c_1{}^2)Pc_1{}^2}}{2R(R - c_1{}^2)} = c_1\sqrt{\frac{P}{R(R - c_1{}^2)}}, \quad (19)$$

so

$$v = c_1\int\sqrt{\frac{P}{R(R - c_1{}^2)}}\,du. \quad (20)$$

In the case $Q = 0$ where P and R are explicit functions of v only, then

$$\frac{\frac{\partial P}{\partial v} + v'^2\frac{\partial R}{\partial v}}{2\sqrt{P + Rv'^2}} - \frac{d}{du}\left(\frac{Rv'}{\sqrt{P + Rv'^2}}\right) = 0, \quad (21)$$

so

$$\frac{\partial P}{\partial v} + v'^2\frac{\partial R}{\partial v} - 2\sqrt{P + Rv'^2}\,R \times\left[\frac{v''}{\sqrt{P + Rv'^2}} + (-\tfrac{1}{2})\frac{v'(2Rv'v'')}{(P + Rv'^2)^{3/2}}\right] = 0 \quad (22)$$

$$\frac{\partial P}{\partial v} + v'^2\frac{\partial R}{\partial v} - 2Rv'' + \frac{2R^2v'^2v''}{P + Rv'^2} = 0 \quad (23)$$

$$\frac{Rv'^2}{\sqrt{P + Rv'^2}} - \sqrt{P + Rv'^2} = c_1 \quad (24)$$

$$Rv'^2 - (P + Rv'^2) - c_1\sqrt{P + Rv'^2} \quad (25)$$

$$\left(-\frac{P}{c_1}\right)^2 = P + Rv'^2 \quad (26)$$

$$\frac{P^2 - c_1{}^2P}{Rc_1{}^2} = v'^2, \quad (27)$$

and

$$u = c_1\int\sqrt{\frac{R}{P^2 - c_1{}^2P}}\,dv. \quad (28)$$

For a surface of revolution in which $y = g(x)$ is rotated about the x-axis so that the equation of the surface is

$$y^2 + z^2 = g^2(x), \quad (29)$$

the surface can be parameterized by

$$x = u \quad (30)$$

$$y = g(u)\cos v \quad (31)$$

$$z = g(u)\sin v. \quad (32)$$

The equation of the geodesics is then

$$v = c_1\int\frac{\sqrt{1 + [g'(u)]^2}\,du}{g(u)\sqrt{[g(u)]^2 - c_1{}^2}}. \quad (33)$$

see also ELLIPSOID GEODESIC, GEODESIC CURVATURE, GEODESIC DOME, GEODESIC EQUATION, GEODESIC TRIANGLE, GREAT CIRCLE, HARMONIC MAP, OBLATE SPHEROID GEODESIC, PARABOLOID GEODESIC

References

Cipra, B. *What's Happening in the Mathematical Sciences, Vol. 1.* Providence, RI: Amer. Math. Soc., pp. 27, 1993.

Weinstock, R. *Calculus of Variations, with Applications to Physics and Engineering.* New York: Dover, pp. 26–28 and 45–46, 1974.

Geodesic Curvature

For a unit speed curve on a surface, the length of the surface-tangential component of acceleration is the geodesic curvature κ_g. Curves with $\kappa_g = 0$ are called GEODESICS. For a curve parameterized as $\boldsymbol{\alpha}(t) = \mathbf{x}(u(t), v(t))$, the geodesic curvature is given by

$$\kappa_g = \sqrt{EG - F^2}[-\Gamma^2_{11}u'^3 + \Gamma^1_{22}v'^3 - (2\Gamma^2_{12} - \Gamma^1_{11})u'^2v' + (2\Gamma^1_{12} - \Gamma^2_{22})u'v'^2 + u''v' - v''u'],$$

where E, F, and G are coefficients of the first FUNDAMENTAL FORM and Γ^k_{ij} are CHRISTOFFEL SYMBOLS OF THE SECOND KIND.

see also GEODESIC

References

Gray, A. "Geodesic Curvature." §20.5 in *Modern Differential Geometry of Curves and Surfaces.* Boca Raton, FL: CRC Press, pp. 402–407, 1993.

Geodesic Dome

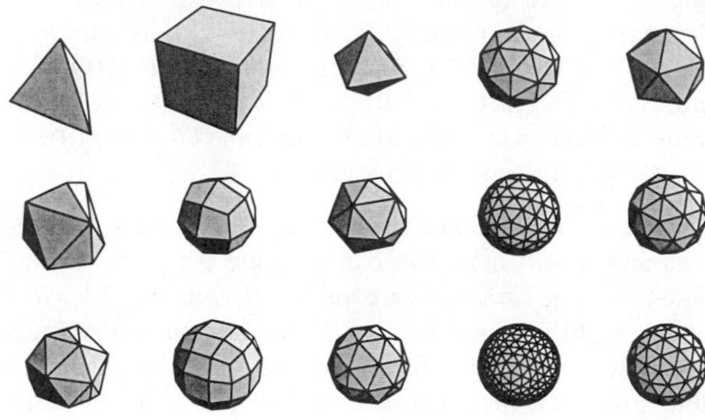

A TRIANGULATION of a PLATONIC SOLID or other POLYHEDRON to produce a close approximation to a SPHERE. The nth order geodesation operation replaces each polygon of the polyhedron by the projection onto the CIRCUMSPHERE of the order n regular tessellation of that polygon. The above figure shows geodesations of orders 1 to 3 (from top to bottom) of the TETRAHEDRON, CUBE, OCTAHEDRON, DODECAHEDRON, and ICOSAHEDRON (from left to right).

R. Buckminster Fuller designed the first geodesic dome (i.e., geodesation of a HEMISPHERE). Fuller's dome was constructed from an ICOSAHEDRON by adding ISOSCELES TRIANGLES about each VERTEX and slightly repositioning the VERTICES. In such domes, neither the VERTICES nor the centers of faces necessarily lie at exactly the same distances from the center. However, these conditions are approximately satisfied.

In the geodesic domes discussed by Kniffen (1994), the sum of VERTEX angles is chosen to be a constant. Given a PLATONIC SOLID, let $e' \equiv 2e/v$ be the number of EDGES meeting at a VERTEX and n be the number of EDGES of the constituent POLYGON. Call the angle of the old VERTEX point A and the angle of the new VERTEX point F. Then

$$A = B \tag{1}$$

$$2e'A = nF \tag{2}$$

$$2A + F = 180^\circ. \tag{3}$$

Solving for A gives

$$2A + \frac{2e'}{n}A = 2A\left(1 + \frac{e'}{n}\right) = 180^\circ \tag{4}$$

$$A = 90^\circ \frac{n}{e'+n}, \tag{5}$$

and

$$F = \frac{2e'}{n}A = 180^\circ \frac{e'}{e'+n}. \tag{6}$$

The VERTEX sum is

$$\Sigma = nF = 180^\circ \frac{e'n}{e'+n}. \tag{7}$$

Solid	f	v	e'	n	A	F	Σ
tetrahedron			3	3	45°	90°	270°
cube	24	14	3	4	$51\frac{3}{7}{}^\circ$	$81\frac{3}{7}{}^\circ$	$308\frac{4}{7}{}^\circ$
octahedron			4	3	$38\frac{4}{7}{}^\circ$	$108\frac{4}{7}{}^\circ$	$308\frac{4}{7}{}^\circ$
dodecahedron	60	32	3	5	$56\frac{1}{4}{}^\circ$	$71\frac{1}{4}{}^\circ$	$337\frac{1}{2}{}^\circ$
icosahedron			5	3	$33\frac{3}{4}{}^\circ$	$118\frac{3}{4}{}^\circ$	$337\frac{1}{2}{}^\circ$

see also TRIANGULAR SYMMETRY GROUP

References

Kenner, H. *Geodesic Math and How to Use It.* Berkeley, CA: University of California Press, 1976.

Kniffen, D. "Geodesic Domes for Amateur Astronomers." *Sky and Telescope,* pp. 90–94, Oct. 1994.

Pappas, T. "Geodesic Dome of Leonardo da Vinci." *The Joy of Mathematics.* San Carlos, CA: Wide World Publ./Tetra, p. 81, 1989.

Geodesic Equation

$$d\tau^2 = -\eta_{\alpha\beta}\, d\xi^\alpha\, d\xi^\beta,$$

or

$$\frac{d^2\xi^\alpha}{d\tau^2} = 0.$$

see also GEODESIC

Geodesic Flow

A type of FLOW technically defined in terms of the TANGENT BUNDLE of a MANIFOLD.

see also DYNAMICAL SYSTEM

Geodesic Triangle

A TRIANGLE formed by the arcs of three GEODESICS on a smooth surface.

see also INTEGRAL CURVATURE

Geodetic Latitude

see LATITUDE

Geographic Latitude

see LATITUDE

Geometric Construction

In antiquity, geometric constructions of figures and lengths were restricted to use of only a STRAIGHTEDGE and COMPASS. Although the term "RULER" is sometimes used instead of "STRAIGHTEDGE," no markings which could be used to make measurements were allowed according to the Greek prescription. Furthermore, the "COMPASS" could not even be used to mark off distances by setting it and then "walking" it along, so the COMPASS had to be considered to automatically collapse when not in the process of drawing a CIRCLE.

Because of the prominent place Greek geometric constructions held in Euclid's *Elements*, these constructions are sometimes also known as EUCLIDEAN CONSTRUCTIONS. Such constructions lay at the heart of the GEOMETRIC PROBLEMS OF ANTIQUITY of CIRCLE SQUARING, CUBE DUPLICATION, and TRISECTION of an ANGLE. The Greeks were unable to solve these problems, but it was not until hundreds of years later that the problems were proved to be actually impossible under the limitations imposed.

Simple algebraic operations such as $a+b$, $a-b$, ra (for r a RATIONAL NUMBER), a/b, ab, and $\sqrt{x}$ can be performed using geometric constructions (Courant and Robbins 1996). Other more complicated constructions, such as the solution of APOLLONIUS' PROBLEM and the construction of INVERSE POINTS can also accomplished.

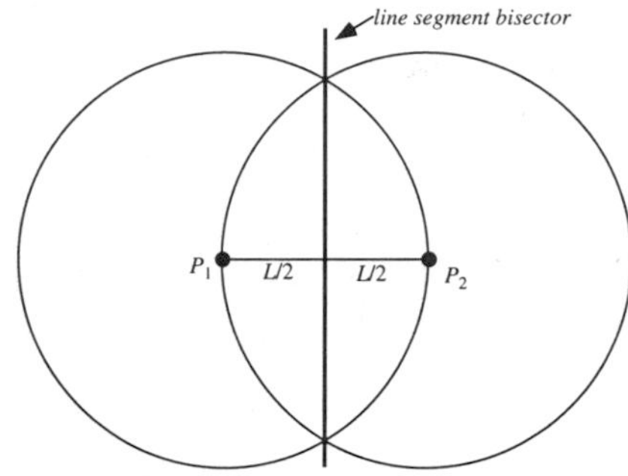

One of the simplest geometric constructions is the construction of a BISECTOR of a LINE SEGMENT, illustrated above.

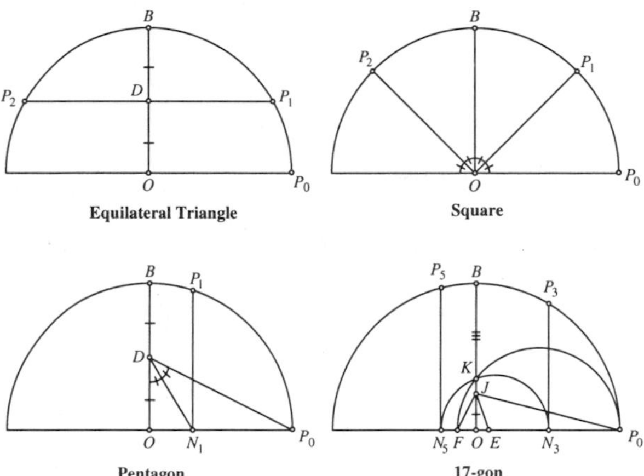

The Greeks were very adept at constructing POLYGONS, but it took the genius of Gauss to mathematically determine which constructions were possible and which were not. As a result, Gauss determined that a series of POLYGONS (the smallest of which has 17 sides; the HEPTADECAGON) had constructions unknown to the Greeks. Gauss showed that the CONSTRUCTIBLE POLYGONS (several of which are illustrated above) were closely related to numbers called the FERMAT PRIMES.

Wernick (1982) gave a list of 139 sets of three located points from which a TRIANGLE was to be constructed. Of Wernick's original list of 139 problems, 20 had not yet been solved as of 1996 (Meyers 1996).

It is possible to construct RATIONAL NUMBERS and EUCLIDEAN NUMBERS using a STRAIGHTEDGE and COMPASS construction. In general, the term for a number which can be constructed using a COMPASS and STRAIGHTEDGE is a CONSTRUCTIBLE NUMBER. Some IRRATIONAL NUMBERS, but *no* TRANSCENDENTAL NUMBERS, can be constructed.

It turns out that all constructions possible with a COMPASS and STRAIGHTEDGE can be done with a COMPASS alone, as long as a line is considered constructed when its two endpoints are located. The reverse is also true, since Jacob Steiner showed that all constructions possible with STRAIGHTEDGE and COMPASS can be done using only a straightedge, as long as a fixed CIRCLE and its center (or two intersecting CIRCLES without their centers, or three nonintersecting CIRCLES) have been drawn beforehand. Such a construction is known as a STEINER CONSTRUCTION.

GEOMETROGRAPHY is a quantitative measure of the simplicity of a geometric construction. It reduces geometric constructions to five types of operations, and seeks to reduce the total number of operations (called the "SIMPLICITY") needed to effect a geometric construction.

Dixon (1991, pp. 34–51) gives approximate constructions for some figures (the HEPTAGON and NONAGON) and lengths (PI) which cannot be rigorously constructed. Ramanujan (1913–14) and Olds (1963) give geometric constructions for $355/113 \approx \pi$. Gardner (1966, pp. 92–93) gives a geometric construction for

$$3+\tfrac{16}{113}=3.1415929\ldots\approx\pi.$$

Constructions for π are approximate (but inexact) forms of CIRCLE SQUARING.

see also CIRCLE SQUARING, COMPASS, CONSTRUCTIBLE NUMBER, CONSTRUCTIBLE POLYGON, CUBE DUPLICATION, ELEMENTS, FERMAT PRIME, GEOMETRIC PROBLEMS OF ANTIQUITY, GEOMETROGRAPHY, MASCHERONI CONSTRUCTION, NAPOLEON'S PROBLEM, NEUSIS CONSTRUCTION, PLANE GEOMETRY, POLYGON, PONCELET-STEINER THEOREM, RECTIFICATION, SIMPLICITY, STEINER CONSTRUCTION, STRAIGHTEDGE, TRISECTION

References

Ball, W. W. R. and Coxeter, H. S. M. *Mathematical Recreations and Essays, 13th ed.* New York: Dover, pp. 96–97, 1987.

Conway, J. H. and Guy, R. K. *The Book of Numbers.* New York: Springer-Verlag, pp. 191–202, 1996.

Courant, R. and Robbins, H. "Geometric Constructions. The Algebra of Number Fields." Ch. 3 in *What is Mathematics?: An Elementary Approach to Ideas and Methods, 2nd ed.* Oxford, England: Oxford University Press, pp. 117–164, 1996.

Dantzig, T. *Number, The Language of Science.* New York: Macmillan, p. 316, 1954.

Dixon, R. *Mathographics.* New York: Dover, 1991.

Eppstein, D. "Geometric Models." http://www.ics.uci.edu/~eppstein/junkyard/model.html.
Gardner, M. "The Transcendental Number Pi." Ch. 8 in *Martin Gardner's New Mathematical Diversions from Scientific American.* New York: Simon and Schuster, 1966.
Gardner, M. "Mascheroni Constructions." Ch. 17 in *Mathematical Circus: More Puzzles, Games, Paradoxes and Other Mathematical Entertainments from Scientific American.* New York: Knopf, pp. 216–231, 1979.
Herterich, K. *Die Konstruktion von Dreiecken.* Stuttgart: Ernst Klett Verlag, 1986.
Krötenheerdt, O. "Zur Theorie der Dreieckskonstruktionen." *Wissenschaftliche Zeitschrift der Martin-Luther-Univ. Halle-Wittenberg, Math. Naturw. Reihe* **15**, 677–700, 1966.
Meyers, L. F. "Update on William Wernick's 'Triangle Constructions with Three Located Points.'" *Math. Mag.* **69**, 46–49, 1996.
Olds, C. D. *Continued Fractions.* New York: Random House, pp. 59–60, 1963.
Petersen, J. "Methods and Theories for the Solution of Problems of Geometrical Constructions." Reprinted in *String Figures and Other Monographs.* New York: Chelsea, 1960.
Plouffe, S.. "The Computation of Certain Numbers Using a Ruler and Compass." Dec. 12, 1997. http://www.research.att.com/~njas/sequences/JIS/compass.html.
Posamentier, A. S. and Wernick, W. *Advanced Geometric Constructions.* Palo Alto, CA: Dale Seymour Pub., 1988.
Ramanujan, S. "Modular Equations and Approximations to π." *Quart. J. Pure. Appl. Math.* **45**, 350–372, 1913–1914.
Wernick, W. "Triangle Constructions with Three Located Points." *Math. Mag.* **55**, 227–230, 1982.

Geometric Distribution

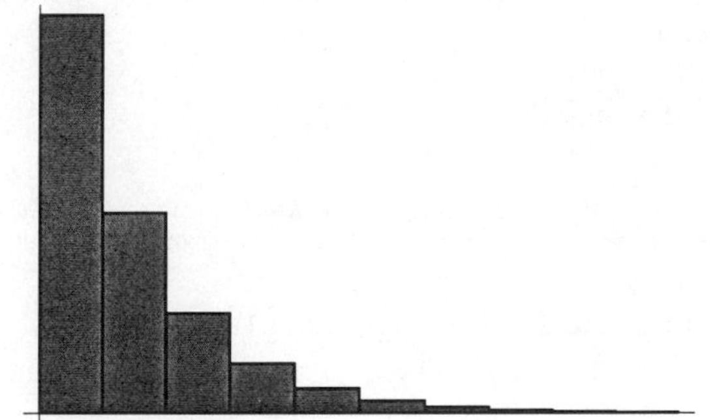

A distribution such that

$$P(n) = q^{n-1}p = p(1-p)^{n-1}, \tag{1}$$

where $q \equiv 1-p$ and for $n = 1, 2, \ldots$. The distribution is normalized since

$$\sum_{n=1}^{\infty} P(n) = \sum_{n=1}^{\infty} q^{n-1}p = p\sum_{n=0}^{\infty} q^n = \frac{p}{1-q} = \frac{p}{p} = 1. \tag{2}$$

The MOMENT-GENERATING FUNCTION is

$$\phi(t) = p[1-(1-p)e^{it}]^{-1}, \tag{3}$$

or

$$M(t) = \langle e^{tn}\rangle = \sum_{n=1}^{\infty} e^{tn}pq^{n-1} = p\sum_{n=0}^{\infty} e^{t(n+1)}q^n$$

$$= pe^t\sum_{n=0}^{\infty}(e^tq)^n = \frac{pe^t}{1-e^tq} \tag{4}$$

$$M'(t) = p\left[\frac{(1-e^tq)e^t - e^t(-e^tq)}{(1-e^tq)^2}\right]$$

$$= \frac{p(e^t - qe^{2t} + qe^{2t})}{(1-e^tq)^2} = \frac{pe^t}{(1-e^tq)^2} \tag{5}$$

$$M''(t) = p\frac{(1-e^tq)^2e^t - e^t2(1-e^tq)(-e^tq)}{(1-e^tq)^4}$$

$$= p\frac{(1-2e^tq+e^{2t}q^2)e^t + 2qe^{2t}(1-e^tq)}{(1-e^tq)^4}$$

$$= p\frac{e^t - 2e^{2t}q + e^{3t}q^2 + 2qe^{2t} - 2q^2e^{3t}}{(1-e^tq)^4}$$

$$= p\frac{e^t - q^2e^{3t}}{(1-e^tq)^4} = \frac{pe^t(1-q^2e^{2t})}{(1-e^tq)^4}$$

$$= \frac{pe^t(1+qe^t)}{(1-e^tq)^3} \tag{6}$$

$$M'''(t) = \frac{pe^t[1+4e^t(1-p)+e^{2t}(1-p)^2]}{(1-e^t+e^tp)^4}. \tag{7}$$

Therefore,

$$M'(0) = \mu_1' = \mu = \frac{p}{(1-q)^2} = \frac{p}{p^2} = \frac{1}{p} \tag{8}$$

$$M''(0) = \mu_2' = \frac{p(1+q)}{(1+q)^3} = \frac{p(2-p)}{p^3} = \frac{2-p}{p^2} \tag{9}$$

$$M'''(0) = \mu_3' = \frac{(6-6p+p^2)}{p^3} \tag{10}$$

$$M^{(4)}(0) = \mu_4' = \frac{(p-2)(-p^2+12p-12)}{p^4}, \tag{11}$$

and

$$\mu_2 \equiv \mu_2' - (\mu_1')^2 = \left(\frac{2}{p^2} - \frac{1}{p}\right) - \frac{1}{p^2} = \frac{1-p}{p^2}$$

$$= \frac{q}{p^2} \tag{12}$$

$$\mu_3 \equiv \mu_3' - 3\mu_2'\mu_1' + 2(\mu_1')^3$$

$$= \frac{6-6p+p^2}{p^3} - 3\frac{2-p}{p^2}\frac{1}{p} + 2\left(\frac{1}{p}\right)^3$$

$$= \frac{6-6p+p^2-3(2-p)+2}{p^3}$$

$$= \frac{(p-1)(p-2)}{p^3} \tag{13}$$

$$\mu_4 \equiv \mu_4' - 4\mu_3'\mu_1' + 6\mu_2'(\mu_1')^2 - 3(\mu_1')^4$$

$$= \frac{(p-2)(-p^2+12p-12)}{p^4} - 4\frac{6-6p+p^2}{p^3}\frac{1}{p}$$

$$+ 6\frac{2-p}{p^2}\left(\frac{1}{p}\right)^2 - 3\left(\frac{1}{p}\right)^4$$

$$= \frac{(p-1)(-p^2+9p-9)}{p^4}, \tag{14}$$

so the MEAN, VARIANCE, SKEWNESS, and KURTOSIS are given by

$$\mu \equiv \mu_1' = \frac{1}{p} \tag{15}$$

$$\sigma^2 = \mu_2 = \frac{q}{p^2} \tag{16}$$

$$\begin{aligned}\gamma_1 &= \frac{\mu_3}{\sigma^3} = \frac{(p-1)(p-2)}{p^3}\left(\frac{p^2}{1-p}\right)^{3/2} \\ &= \frac{(p-1)(p-2)}{(1-p)\sqrt{1-p}} = \frac{2-p}{\sqrt{1-p}} = \frac{2-p}{\sqrt{q}}\end{aligned} \tag{17}$$

$$\begin{aligned}\gamma_2 &= \frac{\mu_4}{\sigma^4} - 3 = \frac{(p-1)(-p^2+9p-9)}{p^4\frac{(1-p)^2}{p^4}} - 3 \\ &= \frac{-9+9p-p^2}{(p-1)} - 3 \\ &= \frac{p^2-6p+6}{1-p}.\end{aligned} \tag{18}$$

In fact, the moments of the distribution are given analytically in terms of the POLYLOGARITHM function,

$$\mu_k' \equiv \sum_{n=1}^{\infty} P(n)n^k = \sum_{n=1}^{\infty} p(1-p)^{n-1}n^k = \frac{p\,\mathrm{Li}_{-k}(1-p)}{1-p}. \tag{19}$$

For the case $p = 1/2$ (corresponding to the distribution of the number of COIN TOSSES needed to win in the SAINT PETERSBURG PARADOX) this formula immediately gives

$$\mu_1' = 2 \tag{20}$$

$$\mu_2' = 6 \tag{21}$$

$$\mu_3' = 26 \tag{22}$$

$$\mu_4' = 150, \tag{23}$$

so the MEAN, VARIANCE, SKEWNESS, and KURTOSIS in this case are

$$\mu = 2 \tag{24}$$

$$\sigma^2 = 2 \tag{25}$$

$$\gamma_1 = \tfrac{3}{2}\sqrt{2} \tag{26}$$

$$\gamma_2 = \tfrac{13}{2}. \tag{27}$$

The first CUMULANT of the geometric distribution is

$$\kappa_1 = \frac{1-p}{p}, \tag{28}$$

and subsequent CUMULANTS are given by the RECURRENCE RELATION

$$\kappa_{r+1} = (1-p)\frac{d\kappa_r}{dp}. \tag{29}$$

see also SAINT PETERSBURG PARADOX

References

Beyer, W. H. *CRC Standard Mathematical Tables, 28th ed.* Boca Raton, FL: CRC Press, pp. 531–532, 1987.

Spiegel, M. R. *Theory and Problems of Probability and Statistics.* New York: McGraw-Hill, p. 118, 1992.

Geometric Mean

$$G \equiv \left(\prod_{i=1}^{n} a_i\right)^{1/n}.$$

Hoehn and Niven (1985) show that

$$G(a_1+c, a_2+c, \ldots, a_n+c) > c + G(a_1, a_2, \ldots, a_n)$$

for any POSITIVE constant c.

see also ARITHMETIC MEAN, ARITHMETIC-GEOMETRIC MEAN, CARLEMAN'S INEQUALITY, HARMONIC MEAN, MEAN, ROOT-MEAN-SQUARE

References

Abramowitz, M. and Stegun, C. A. (Eds.). *Handbook of Mathematical Functions with Formulas, Graphs, and Mathematical Tables, 9th printing.* New York: Dover, p. 10, 1972.

Hoehn, L. and Niven, I. "Averages on the Move." *Math. Mag.* **58**, 151–156, 1985.

Geometric Mean Index

The statistical INDEX

$$P_G \equiv \left[\prod \left(\frac{p_n}{p_0}\right)^{v_0}\right]^{1/\Sigma v_0},$$

where p_n is the price per unit in period n, q_n is the quantity produced in period n, and $v_n \equiv p_n q_n$ the value of the n units.

see also INDEX

References

Kenney, J. F. and Keeping, E. S. *Mathematics of Statistics, Pt. 1, 3rd ed.* Princeton, NJ: Van Nostrand, p. 69, 1962.

Geometric Probability Constants

see CUBE POINT PICKING, CUBE TRIANGLE PICKING

Geometric Problems of Antiquity

The Greek problems of antiquity were a set of geometric problems whose solution was sought using only COMPASS and STRAIGHTEDGE:

1. CIRCLE SQUARING.
2. CUBE DUPLICATION.
3. TRISECTION of an ANGLE.

Only in modern times, more than 2,000 years after they were formulated, were all three ancient problems proved insoluble using only COMPASS and STRAIGHTEDGE.

Another ancient geometric problem not proved impossible until 1997 is ALHAZEN'S BILLIARD PROBLEM. As Ogilvy (1990) points out, constructing the general REGULAR POLYHEDRON was really a "fourth" unsolved problem of antiquity.

see also ALHAZEN'S BILLIARD PROBLEM, CIRCLE SQUARING, COMPASS, CONSTRUCTIBLE NUMBER, CONSTRUCTIBLE POLYGON, CUBE DUPLICATION, GEOMETRIC CONSTRUCTION, REGULAR POLYHEDRON, STRAIGHTEDGE, TRISECTION

References

Conway, J. H. and Guy, R. K. "Three Greek Problems." In *The Book of Numbers.* New York: Springer-Verlag, pp. 190–191, 1996.
Courant, R. and Robbins, H. "The Unsolvability of the Three Greek Problems." §3.3 in *What is Mathematics?: An Elementary Approach to Ideas and Methods, 2nd ed.* Oxford, England: Oxford University Press, pp. 117–118 and 134–140, 1996.
Ogilvy, C. S. *Excursions in Geometry.* New York: Dover, pp. 135–138, 1990.
Pappas, T. "The Impossible Trio." *The Joy of Mathematics.* San Carlos, CA: Wide World Publ./Tetra, pp. 130–132, 1989.
Jones, A.; Morris, S.; and Pearson, K. *Abstract Algebra and Famous Impossibilities.* New York: Springer-Verlag, 1991.

Geometric Progression

see GEOMETRIC SEQUENCE

Geometric Sequence

A geometric sequence is a SEQUENCE $\{a_k\}$, $k = 1, 2, \ldots$, such that each term is given by a multiple r of the previous one. Another equivalent definition is that a sequence is geometric IFF it has a zero BIAS. If the multiplier is r, then the kth term is given by

$$a_k = ra_{k-1} = r^2 a_{k-2} = a_0 r^k.$$

Without loss of generality, take $a_0 = 1$, giving

$$a_k = r^k.$$

Geometric Series

A geometric series $\sum_k a_k$ is a series for which the ratio of each two consecutive terms a_{k+1}/a_k is a constant function of the summation index k, say r. Then the terms a_k are of the form $a_k = a_0 r^k$, so $a_{k+1}/a_k = r$. If $\{a_k\}$, with $k = 1, 2, \ldots$, is a GEOMETRIC SEQUENCE with multiplier $-1 < r < 1$ and $a_0 = 1$, then the geometric series

$$S_n = \sum_{k=0}^{n} a_k = \sum_{k=0}^{n} r^k \tag{1}$$

is given by

$$S_n \equiv \sum_{k=0}^{n} r^k = 1 + r + r^2 + \ldots + r^n, \tag{2}$$

so

$$rS_n = r + r^2 + r^3 + \ldots + r^{n+1}. \tag{3}$$

Subtracting

$$\begin{aligned}(1-r)S_n &= (1 + r + r^2 + \ldots + r^n) \\ &\quad - (r + r^2 + r^3 + \ldots + r^{n+1}) \\ &= 1 - r^{n+1},\end{aligned} \tag{4}$$

so

$$S_n \equiv \sum_{k=0}^{n} r^k = \frac{1 - r^{n+1}}{1 - r}. \tag{5}$$

As $n \to \infty$, then

$$S \equiv S_\infty = \sum_{k=0}^{\infty} r^k = \frac{1}{1 - r}. \tag{6}$$

see also ARITHMETIC SERIES, GABRIEL'S STAIRCASE, HARMONIC SERIES, HYPERGEOMETRIC SERIES, WHEAT AND CHESSBOARD PROBLEM

References

Abramowitz, M. and Stegun, C. A. (Eds.). *Handbook of Mathematical Functions with Formulas, Graphs, and Mathematical Tables, 9th printing.* New York: Dover, p. 10, 1972.
Arfken, G. *Mathematical Methods for Physicists, 3rd ed.* Orlando, FL: Academic Press, pp. 278–279, 1985.
Beyer, W. H. *CRC Standard Mathematical Tables, 28th ed.* Boca Raton, FL: CRC Press, p. 8, 1987.
Courant, R. and Robbins, H. "The Geometric Progression." §1.2.3 in *What is Mathematics?: An Elementary Approach to Ideas and Methods, 2nd ed.* Oxford, England: Oxford University Press, pp. 13–14, 1996.
Pappas, T. "Perimeter, Area & the Infinite Series." *The Joy of Mathematics.* San Carlos, CA: Wide World Publ./ Tetra, pp. 134–135, 1989.

Geometrization Conjecture

see THURSTON'S GEOMETRIZATION CONJECTURE

Geometrography

A quantitative measure of the simplicity of a GEOMETRIC CONSTRUCTION which reduces geometric constructions to five steps. It was devised by È. Lemoine.

- S_1 Place a STRAIGHTEDGE'S EDGE through a given POINT,
- S_2 Draw a straight LINE,
- C_1 Place a POINT of a COMPASS on a given POINT,
- C_2 Place a POINT of a COMPASS on an indeterminate POINT on a LINE,
- C_3 Draw a CIRCLE.

Geometrography seeks to reduce the number of operations (called the "SIMPLICITY") needed to effect a construction. If the number of the above operations are denoted m_1, m_2, n_1, n_2, and n_3, respectively, then the SIMPLICITY is $m_1 + m_2 + n_1 + n_2 + n_3$ and the symbol is $m_1 S_1 + m_2 S_2 + n_1 C_1 + n_2 C_2 + n_3 C_3$. It is apparently an unsolved problem to determine if a given GEOMETRIC CONSTRUCTION is of the smallest possible simplicity.

see also SIMPLICITY

References

De Temple, D. W. "Carlyle Circles and the Lemoine Simplicity of Polygonal Constructions." *Amer. Math. Monthly* **98**, 97–108, 1991.

Eves, H. *An Introduction to the History of Mathematics, 6th ed.* New York: Holt, Rinehart, and Winston, 1990.

Geometry

Geometry is the study of figures in a SPACE of a given number of dimensions and of a given type. The most common types of geometry are PLANE GEOMETRY (dealing with objects like the LINE, CIRCLE, TRIANGLE, and POLYGON), SOLID GEOMETRY (dealing with objects like the LINE, SPHERE, and POLYHEDRON), and SPHERICAL GEOMETRY (dealing with objects like the SPHERICAL TRIANGLE and SPHERICAL POLYGON).

Historically, the study of geometry proceeds from a small number of accepted truths (AXIOMS or POSTULATES), then builds up true statements using a systematic and rigorous step-by-step PROOF. However, there is much more to geometry than this relatively dry textbook approach, as evidenced by some of the beautiful and unexpected results of PROJECTIVE GEOMETRY (not to mention Schubert's powerful but questionable ENUMERATIVE GEOMETRY).

Formally, a geometry is defined as a complete locally homogeneous RIEMANNIAN METRIC. In $\mathbb{R}^2$, the possible geometries are Euclidean planar, hyperbolic planar, and elliptic planar. In $\mathbb{R}^3$, the possible geometries include Euclidean, hyperbolic, and elliptic, but also include five other types.

see also ABSOLUTE GEOMETRY, AFFINE GEOMETRY, COORDINATE GEOMETRY, DIFFERENTIAL GEOMETRY, ENUMERATIVE GEOMETRY, FINSLER GEOMETRY, INVERSIVE GEOMETRY, MINKOWSKI GEOMETRY, NIL GEOMETRY, NON-EUCLIDEAN GEOMETRY, ORDERED GEOMETRY, PLANE GEOMETRY, PROJECTIVE GEOMETRY, SOL GEOMETRY, SOLID GEOMETRY, SPHERICAL GEOMETRY, THURSTON'S GEOMETRIZATION CONJECTURE

References

Altshiller-Court, N. *College Geometry: A Second Course in Plane Geometry for Colleges and Normal Schools, 2nd ed., rev. enl.* New York: Barnes and Noble, 1952.

Brown, K. S. "Geometry." `http://www.seanet.com/~ksbrown/igeometr.htm`.

Coxeter, H. S. M. *Introduction to Geometry, 2nd ed.* New York: Wiley, 1969.

Croft, H. T.; Falconer, K. J.; and Guy, R. K. *Unsolved Problems in Geometry.* New York: Springer-Verlag, 1994.

Eppstein, D. "Geometry Junkyard." `http://www.ics.uci.edu/~eppstein/junkyard/`.

Eppstein, D. "Many-Dimensional Geometry." `http://www.ics.uci.edu/~eppstein/junkyard/highdim.html`.

Eppstein, D. "Planar Geometry." `http://www.ics.uci.edu/~eppstein/junkyard/2d.html`.

Eppstein, D. "Three-Dimensional Geometry." `http://www.ics.uci.edu/~eppstein/junkyard/3d.html`.

Eves, H. W. *A Survey of Geometry, rev. ed.* Boston, MA: Allyn and Bacon, 1972.

Geometry Center. `http://www.geom.umn.edu`.

Ghyka, M. C. *The Geometry of Art and Life, 2nd ed.* New York: Dover, 1977.

Hilbert, D. *The Foundations of Geometry, 2nd ed.* Chicago, IL: The Open Court Publishing Co., 1921.

Johnson, R. A. *Advanced Euclidean Geometry: An Elementary Treatise on the Geometry of the Triangle and the Circle.* New York: Dover, 1960.

King, J. and Schattschneider, D. (Eds.). *Geometry Turned On: Dynamic Software in Learning, Teaching and Research.* Washington, DC: Math. Assoc. Amer., 1997.

Klein, F. *Famous Problems of Elementary Geometry and Other Monographs.*082840108X New York: Dover, 1956.

Melzak, Z. A. *Invitation to Geometry.* New York: Wiley, 1983.

Moise, E. E. *Elementary Geometry from an Advanced Standpoint, 3rd ed.* Reading, MA: Addison-Wesley, 1990.

Ogilvy, C. S. "Some Unsolved Problems of Modern Geometry." Ch. 11 in *Excursions in Geometry.* New York: Dover, pp. 143–153, 1990.

Simon, M. *Über die Entwicklung der Elementargeometrie im XIX Jahrhundert.* Berlin, pp. 97–105, 1906.

Woods, F. S. *Higher Geometry: An Introduction to Advanced Methods in Analytic Geometry.* New York: Dover, 1961.

Gergonne Line

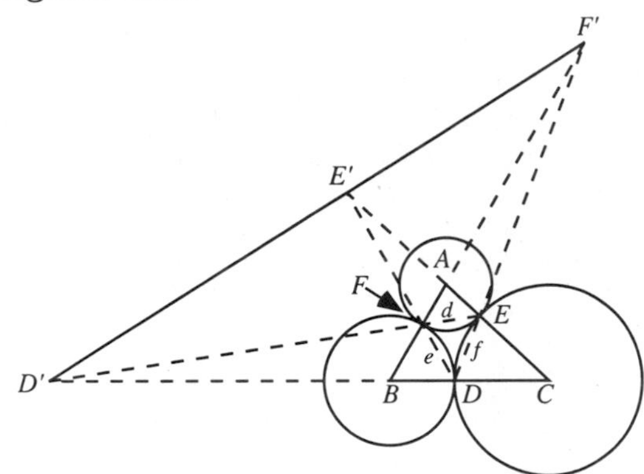

The perspective line for the CONTACT TRIANGLE ΔDEF and its TANGENTIAL TRIANGLE ΔABC. It is determined by the NOBBS POINTS D', E', and F'. In addition to the NOBBS POINTS, the FLETCHER POINT and EVANS POINT also lie on the Gergonne line where it intersects the SODDY LINE and EULER LINE, respectively. The D and D' coordinates are given by

$$D = B + \frac{f}{e}C$$
$$D' = B - \frac{f}{e}C,$$

so $BDCD'$ form a HARMONIC RANGE. The equation of the Gergonne line is

$$\frac{\alpha}{d} + \frac{\beta}{e} + \frac{\gamma}{f} = 0.$$

see also CONTACT TRIANGLE, EULER LINE, EVANS POINT, FLETCHER POINT, NOBBS POINTS, SODDY LINE, TANGENTIAL TRIANGLE

References

Oldknow, A. "The Euler-Gergonne-Soddy Triangle of a Triangle." *Amer. Math. Monthly* **103**, 319–329, 1996.

Gergonne Point

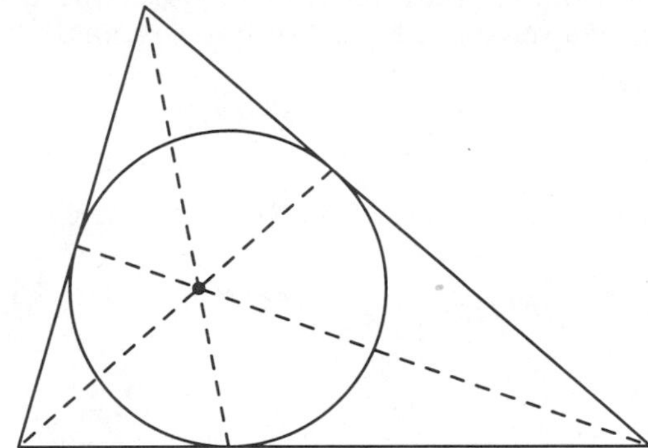

The common point of the CONCURRENT lines from the TANGENT points of a TRIANGLE'S INCIRCLE to the opposite VERTICES. It has TRIANGLE CENTER FUNCTION

$$\alpha = [a(b+c-a)]^{-1} = \tfrac{1}{2}\sec^2 A.$$

It is the ISOTOMIC CONJUGATE POINT of the NAGEL POINT. The CONTACT TRIANGLE and TANGENTIAL TRIANGLE are perspective from the Gergonne point.

see also GERGONNE LINE

References

Altshiller-Court, N. *College Geometry: A Second Course in Plane Geometry for Colleges and Normal Schools, 2nd ed.* New York: Barnes and Noble, pp. 160–164, 1952.

Coxeter, H. S. M. and Greitzer, S. L. *Geometry Revisited.* New York: Random House, pp. 11–13, 1967.

Eves, H. W. *A Survey of Geometry, rev. ed.* Boston, MA: Allyn and Bacon, p. 83, 1972.

Gallatly, W. *The Modern Geometry of the Triangle, 2nd ed.* London: Hodgson, p. 22, 1913.

Johnson, R. A. *Modern Geometry: An Elementary Treatise on the Geometry of the Triangle and the Circle.* Boston, MA: Houghton Mifflin, pp. 184 and 216, 1929.

Kimberling, C. "Gergonne Point." `http://www.evansville.edu/~ck6/tcenters/class/gergonne.html`.

Germain Primes

see SOPHIE GERMAIN PRIME

Gerono Lemniscate

see EIGHT CURVE

Gerŝgorin Circle Theorem

Gives a region in the COMPLEX PLANE containing all the EIGENVALUES of a COMPLEX SQUARE MATRIX. Let

$$|x_k| = \max\{|x_i| : 1 \leq i \leq n\} > 0 \tag{1}$$

and define

$$R_i = \sum_{\substack{i=1\\ j\neq i}}^{n} |a_{ij}|. \tag{2}$$

Then each EIGENVALUE of the MATRIX A of order n is in at least one of the disks

$$\{z : |z - a_{ii}| \leq R_i\}. \tag{3}$$

The theorem can be made stronger as follows. Let r be an INTEGER with $1 \leq r \leq n$, then each EIGENVALUE of A is either in one of the disks Γ_1

$$\{z : |z - a_{jj}| \leq S_j^{(r-1)}\}, \tag{4}$$

or in one of the regions

$$\left\{ z : \sum_{i=1}^{r} |z - a_{ii}| \leq \sum_{i=1}^{r} R_i \right\}, \tag{5}$$

where $S_j^{(r-1)}$ is the sum of magnitudes of the $r-1$ largest off-diagonal elements in column j.

References

Buraldi, R. A. and Mellendorf, S. "Regions in the Complex Plane Containing the Eigenvalues of a Matrix." *Amer. Math. Monthly* **101**, 975–985, 1994.

Gradshteyn, I. S. and Ryzhik, I. M. *Tables of Integrals, Series, and Products, 5th ed.* San Diego, CA: Academic Press, pp. 1120–1121, 1979.

Taussky-Todd, O. "A Recurring Theorem on Determinants." *Amer. Math. Monthly* **56**, 672–676, 1949.

Ghost

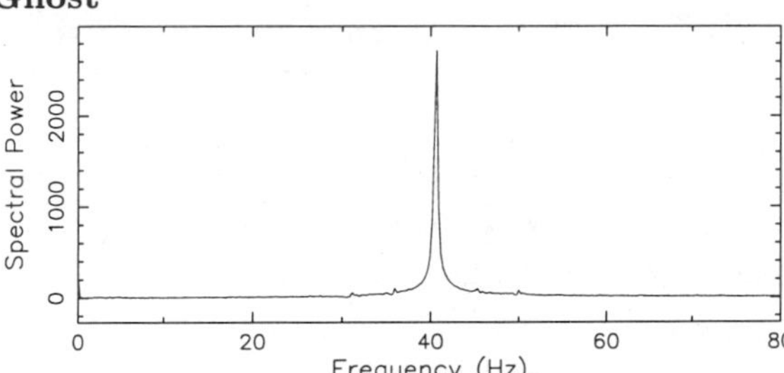

If the sampling of an interferogram is modulated at a definite frequency instead of being uniformly sampled, spurious spectral features called "ghosts" are produced (Brault 1985). Periodic ruling or sampling errors introduce a modulation superposed on top of the expected fringe pattern due to uniform stage translation. Because modulation is a multiplicative process, spurious features are generated in spectral space at the sum and difference of the true fringe and ghost fringe frequencies, thus throwing power out of its spectral band.

Ghosts are copies of the actual spectrum, but appear at reduced strength. The above shows the power spectrum for a pure sinusoidal signal sampled by translating a Fourier transform spectrometer mirror at constant speed. The small blips on either side of the main peaks are ghosts.

In order for a ghost to appear, the process producing it must exist for most of the interferogram. However, if the ruling errors are not truly sinusoidal but vary across the length of the screw, a longer travel path can reduce their effect.

see also JITTER

References
Brault, J. W. "Fourier Transform Spectroscopy." In *High Resolution in Astronomy: 15th Advanced Course of the Swiss Society of Astronomy and Astrophysics* (Ed. A. Benz, M. Huber, and M. Mayor). Geneva Observatory, Sauverny, Switzerland, 1985.

Gibbs Constant

see WILBRAHAM-GIBBS CONSTANT

Gibbs Effect

see GIBBS PHENOMENON

Gibbs Phenomenon

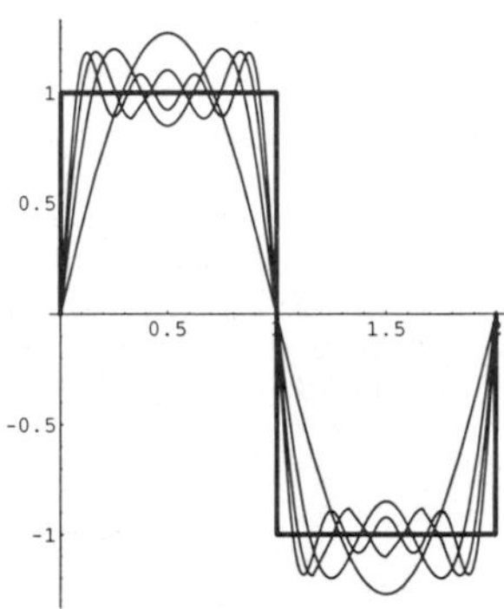

An overshoot of FOURIER SERIES and other EIGENFUNCTION series occurring at simple DISCONTINUITIES. It can be removed with the LANCZOS σ FACTOR.

see also FOURIER SERIES

References
Arfken, G. "Gibbs Phenomenon." §14.5 in *Mathematical Methods for Physicists, 3rd ed.* Orlando, FL: Academic Press, pp. 783–787, 1985.
Foster, J. and Richards, F. B. "The Gibbs Phenomenon for Piecewise-Linear Approximation." *Amer. Math. Monthly* **98**, 47–49, 1991.
Gibbs, J. W. "Fourier Series." *Nature* **59**, 200 and 606, 1899.
Hewitt, E. and Hewitt, R. "The Gibbs-Wilbraham Phenomenon: An Episode in Fourier Analysis." *Arch. Hist. Exact Sci.* **21**, 129–160, 1980.
Sansone, G. "Gibbs' Phenomenon." §2.10 in *Orthogonal Functions, rev. English ed.* New York: Dover, pp. 141–148, 1991.

Gigantic Prime

A PRIME with 10,000 or more decimal digits. As of Nov. 15, 1995, 127 were known.

see also TITANIC PRIME

References
Caldwell, C. "The Ten Largest Known Primes." `http://www.utm.edu/research/primes/largest.html#largest`.

Gilbrat's Distribution

A CONTINUOUS DISTRIBUTION in which the LOGARITHM of a variable x has a NORMAL DISTRIBUTION,

$$P(x) = \frac{1}{\sqrt{2\pi}} e^{-(\ln x)^2/2}.$$

It is a special case of the LOG NORMAL DISTRIBUTION

$$P(x) = \frac{1}{S\sqrt{2\pi}} e^{-(\ln x - M)^2/2S^2}.$$

with $S = 1$ and $M = 0$.

see also LOG NORMAL DISTRIBUTION

Gilbreath's Conjecture

Let the DIFFERENCE of successive PRIMES be defined by $d_n \equiv p_{n+1} - p_n$, and d_n^k by

$$d_n^k \equiv \begin{cases} d_n & \text{for } k = 1 \\ |d_{n+1}^{k-1} - d_n^{k-1}| & \text{for } k > 1. \end{cases}$$

N. L. Gilbreath claimed that $d_1^k = 1$ for all k (Guy 1994). It has been verified for $k < 63419$ and all PRIMES up to $\pi(10^{13})$, where π is the PRIME COUNTING FUNCTION.

References
Guy, R. K. "Gilbreath's Conjecture." §A10 in *Unsolved Problems in Number Theory, 2nd ed.* New York: Springer-Verlag, pp. 25–26, 1994.

Gill's Method

A formula for numerical solution of differential equations,

$$y_{n+1} = y_n + \tfrac{1}{6}[k_1 + (2 - \sqrt{2})k_2 + (2 + \sqrt{2})k_3 + k_4) + \mathcal{O}(h^5),$$

where

$$\begin{aligned} k_1 &= hf(x_n, y_n) \\ k_2 &= hf(x_n + \tfrac{1}{2}h, y_n + \tfrac{1}{2}k_1) \\ k_3 &= hf(x_n + \tfrac{1}{2}h, y_n + \tfrac{1}{2}(-1 + \sqrt{2})k_1 + (1 - \tfrac{1}{2}\sqrt{2})k_2) \\ k_4 &= hf(x_n + h, y_n - \tfrac{1}{2}\sqrt{2}\,k_2 + (1 + \tfrac{1}{2}\sqrt{2})k_3). \end{aligned}$$

see also ADAMS' METHOD, MILNE'S METHOD, PREDICTOR-CORRECTOR METHODS, RUNGE-KUTTA METHOD

References
Abramowitz, M. and Stegun, C. A. (Eds.). *Handbook of Mathematical Functions with Formulas, Graphs, and Mathematical Tables, 9th printing.* New York: Dover, p. 896, 1972.

Gingerbreadman Map

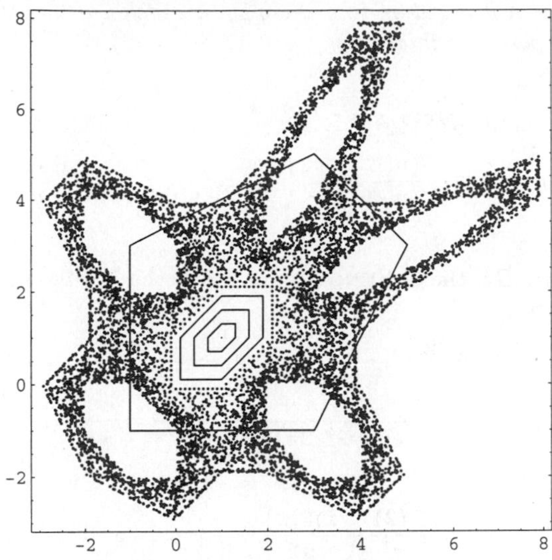

A 2-D piecewise linear MAP defined by

$$x_{n+1} = 1 - y_n + |x_n|$$
$$y_{n+1} = x_n.$$

The map is chaotic in the filled region above and stable in the six hexagonal regions. Each point in the interior hexagon defined by the vertices (0, 0), (1, 0), (2, 1), (2, 2), (1, 2), and (0, 1) has an orbit with period six (except the point (1, 1), which has period 1). Orbits in the other five hexagonal regions circulate from one to the other. There is a unique orbit of period five, with all others having period 30. The points having orbits of period five are (−1, 3), (−1, −1), (3, −1), (5, 3), and (3, 5), indicated in the above figure by the black line. However, there are infinitely many distinct periodic orbits which have an arbitrarily long period.

References

Devaney, R. L. "A Piecewise Linear Model for the Zones of Instability of an Area Preserving Map." *Physica D* **10**, 387–393, 1984.

Peitgen, H.-O. and Saupe, D. (Eds.). "A Chaotic Gingerbreadman." §3.2.3 in *The Science of Fractal Images*. New York: Springer-Verlag, pp. 149–150, 1988.

Girard's Spherical Excess Formula

Let a SPHERICAL TRIANGLE Δ have angles A, B, and C. Then the SPHERICAL EXCESS is given by

$$\Delta = A + B + C - \pi.$$

see also ANGULAR DEFECT, L'HUILIER'S THEOREM, SPHERICAL EXCESS

References

Coxeter, H. S. M. *Introduction to Geometry, 2nd ed.* New York: Wiley, pp. 94–95, 1969.

Girko's Circular Law

Let λ be EIGENVALUES of a set of RANDOM $n \times n$ MATRICES. Then $\lambda/\sqrt{n}$ is uniformly distributed on the DISK.

References

Girko, V. L. *Theory of Random Determinants*. Boston, MA: Kluwer, 1990.

Girth

The length of the shortest CYCLE in a GRAPH.

Girth	Example
3	tetrahedron
4	cube, $K_{3,3}$
5	Petersen graph

Giuga's Conjecture

If $n > 1$ and

$$n|1^{n-1} + 2^{n-1} + \ldots + (n-1)^{n-1} + 1,$$

is n necessarily a PRIME? In other words, defining

$$s_n \equiv \sum_{k=1}^{n-1} k^{n-1},$$

does there exist a COMPOSITE n such that $s_n \equiv -1 \pmod{n}$? It is known that $s_n \equiv -1 \pmod{n}$ IFF for each prime divisor p of n, $(p-1)|(n/p-1)$ and $p|(n/p-1)$ (Giuga 1950, Borwein *et al.* 1996); therefore, any counterexample must be SQUAREFREE. A composite INTEGER n satisfies $s_n \equiv -1 \pmod{n}$ IFF it is both a CARMICHAEL NUMBER and a GIUGA NUMBER. Giuga showed that there are no exceptions to the conjecture up to 10^{1000}. This was later improved to 10^{1700} (Bedocchi 1985) and 10^{13800} (Borwein *et al.* 1996).

see also ARGOH'S CONJECTURE

References

Bedocchi, E. "The $\mathbb{Z}(\sqrt{14})$ Ring and the Euclidean Algorithm." *Manuscripta Math.* **53**, 199–216, 1985.

Borwein, D.; Borwein, J. M.; Borwein, P. B.; and Girgensohn, R. "Giuga's Conjecture on Primality." *Amer. Math. Monthly* **103**, 40–50, 1996.

Giuga, G. "Su una presumibile propertietà caratteristica dei numeri primi." *Ist. Lombardo Sci. Lett. Rend. A* **83**, 511–528, 1950.

Ribenboim, P. *The Book of Prime Number Records, 2nd ed.* New York: Springer-Verlag, pp. 20–21, 1989.

Giuga Number

Any COMPOSITE NUMBER n with $p|(n/p-1)$ for all PRIME DIVISORS p of n. n is a Giuga number IFF

$$\sum_{k=1}^{n-1} k^{\phi(n)} \equiv -1 \pmod{n}$$

where ϕ is the TOTIENT FUNCTION and IFF

$$\sum_{p|n} \frac{1}{p} - \prod_{p|n} \frac{1}{p} \in \mathbb{N}.$$

n is a Giuga number IFF

$$nB_{\phi(n)} \equiv -1 \pmod{n},$$

where B_k is a BERNOULLI NUMBER and ϕ is the TOTIENT FUNCTION. Every counterexample to Giuga's conjecture is a contradiction to ARGOH'S CONJECTURE and vice versa. The smallest known Giuga numbers are 30 (3 factors), 858, 1722 (4 factors), 66198 (5 factors), 2214408306, 24423128562 (6 factors), 432749205173838, 14737133470010574, 550843391309130318 (7 factors),

$$244197000982499715087866346,$$
$$554079914617070801288578559178$$

(8 factors), ... (Sloane's A007850).

It is not known if there are an infinite number of Giuga numbers. All the above numbers have sum minus product equal to 1, and any Giuga number of higher order must have at least 59 factors. The smallest ODD Giuga number must have at least nine PRIME factors.

see also ARGOH'S CONJECTURE, BERNOULLI NUMBER, TOTIENT FUNCTION

References

Borwein, D.; Borwein, J. M.; Borwein, P. B.; and Girgensohn, R. "Giuga's Conjecture on Primality." *Amer. Math. Monthly* **103**, 40–50, 1996.

Sloane, N. J. A. Sequence A007850 in "An On-Line Version of the Encyclopedia of Integer Sequences."

Giuga Sequence

A finite, increasing sequence of INTEGERS $\{n_1, \ldots, n_m\}$ such that

$$\sum_{i=1}^{m} \frac{1}{n_i} - \prod_{i=1}^{m} \frac{1}{n_i} \in \mathbb{N}.$$

A sequence is a Giuga sequence IFF it satisfies

$$n_i | (n_1 \cdots n_{i-1} \cdot n_{i+1} \cdot n_m - 1)$$

for $i = 1, \ldots, m$. There are no Giuga sequences of length 2, one of length 3 ($\{2, 3, 5\}$), two of length 4 ($\{2, 3, 7, 41\}$ and $\{2, 3, 11, 13\}$), 3 of length 5 ($\{2, 3, 7, 43, 1805\}$, $\{2, 3, 7, 83, 85\}$, and $\{2, 3, 11, 17, 59\}$), 17 of length 6, 27 of length 7, and hundreds of length 8. There are infinitely many Giuga sequences. It is possible to generate longer Giuga sequences from shorter ones satisfying certain properties.

see also CARMICHAEL SEQUENCE

References

Borwein, D.; Borwein, J. M.; Borwein, P. B.; and Girgensohn, R. "Giuga's Conjecture on Primality." *Amer. Math. Monthly* **103**, 40–50, 1996.

Glaisher-Kinkelin Constant

N.B. A detailed on-line essay by S. Finch was the starting point for this entry.

Define

$$K(n+1) \equiv 0^0 1^1 2^2 3^3 \cdots n^n \tag{1}$$

$$G(n+1) \equiv \frac{(n!)^n}{K(n+1)} = \begin{cases} 1 & \text{if } n = 0 \\ 0!1!2! \cdots (n-1)! & \text{if } n > 0. \end{cases} \tag{2}$$

where G is the G-FUNCTION and K is the K-FUNCTION. Then

$$\lim_{n\to\infty} \frac{K(n+1)}{n^{n^2/2+n/2+1/2} e^{-n^2/4}} = A \tag{3}$$

$$\lim_{n\to\infty} \frac{G(n+1)}{n^{n^2/2-1/12} (2\pi)^{n/2} e^{-3n^2/4}} = \frac{e^{1/12}}{A}, \tag{4}$$

where

$$A = \exp\left[-\frac{\zeta'(2)}{2\pi^2} + \frac{\ln(2\pi)}{12} + \frac{\gamma}{2}\right] = 1.28242713\ldots, \tag{5}$$

where $\zeta(z)$ is the RIEMANN ZETA FUNCTION, π is PI, and γ is the EULER-MASCHERONI CONSTANT (Kinkelin 1860, Glaisher 1877, 1878, 1893, 1894). Glaisher (1877) also obtained

$$A = 2^{7/36} \pi^{-1/6} \exp\left\{\frac{1}{3} + \frac{2}{3} \int_0^{1/2} \ln[\Gamma(x+1)]\, dx\right\}. \tag{6}$$

Glaisher (1894) showed that

$$1^{1/1} 2^{1/4} 3^{1/9} 4^{1/16} 5^{1/25} \cdots = \left(\frac{A^{12}}{2\pi e^{\gamma}}\right)^{\pi^2/6} \tag{7}$$

$$1^{1/1} 3^{1/9} 5^{1/25} 7^{1/49} 9^{1/81} \cdots = \left(\frac{A^{12}}{2^{4/3} \pi e^{\gamma}}\right)^{\pi^2/8} \tag{8}$$

$$\frac{1^{1/1} 5^{1/125} 9^{1/729} \cdots}{3^{1/27} 7^{1/343} 11^{1/1331} \cdots} = \left(\frac{A}{2^{5/32} \pi^{1/32} e^{3/32+\gamma/48+s/4}}\right)^{\pi^3}, \tag{9}$$

where

$$s \equiv \frac{\zeta(3)}{3 \cdot 4 \cdot 5} \frac{1}{4^3} + \frac{\zeta(5)}{5 \cdot 6 \cdot 7} \frac{1}{4^5} + \frac{\zeta(7)}{7 \cdot 8 \cdot 9} \frac{1}{4^7} + \ldots. \tag{10}$$

see also G-FUNCTION, HYPERFACTORIAL, K-FUNCTION

References

Finch, S. "Favorite Mathematical Constants." `http://www.mathsoft.com/asolve/constant/glshkn/glshkn.html`.

Glaisher, J. W. L. "On a Numerical Continued Product." *Messenger Math.* **6**, 71–76, 1877.

Glaisher, J. W. L. "On the Product $1^1 2^2 3^3 \cdots n^n$." *Messenger Math.* **7**, 43–47, 1878.

Glaisher, J. W. L. "On Certain Numerical Products." *Messenger Math.* **23**, 145–175, 1893.

Glaisher, J. W. L. "On the Constant which Occurs in the Formula for $1^1 2^2 3^3 \cdots n^n$." *Messenger Math.* **24**, 1–16, 1894.

Kinkelin. "Über eine mit der Gammafunktion verwandte Transcendente und deren Anwendung auf die Integralrechnung." *J. Reine Angew. Math.* **57**, 122–158, 1860.

Glide

A product of a REFLECTION in a line and TRANSLATION along the same line.

see also REFLECTION, TRANSLATION

Glissette

The locus of a point P (or the envelope of a line) fixed in relation to a curve C which slides between fixed curves. For example, if C is a line segment and P a point on the line segment, then P describes an ELLIPSE when C slides so as to touch two ORTHOGONAL straight LINES. The glissette of the LINE SEGMENT C itself is, in this case, an ASTROID.

see also ROULETTE

References

Besant, W. H. *Notes on Roulettes and Glissettes, 2nd enl. ed.* Cambridge, England: Deighton, Bell & Co., 1890.

Lockwood, E. H. "Glissettes." Ch. 20 in *A Book of Curves.* Cambridge, England: Cambridge University Press, pp. 160–165, 1967.

Yates, R. C. "Glissettes." *A Handbook on Curves and Their Properties.* Ann Arbor, MI: J. W. Edwards, pp. 108–112, 1952.

Global $C(G;T)$ Theorem

If a SYLOW 2-SUBGROUP T of G lies in a unique maximal 2-local P of G, then P is a "strongly embedded" SUBGROUP of G, and G is known.

Global Extremum

A GLOBAL MINIMUM or GLOBAL MAXIMUM.

see also LOCAL EXTREMUM

Global Maximum

The largest overall value of a set, function, etc., over its entire range.

see also GLOBAL MINIMUM, LOCAL MAXIMUM, MAXIMUM

Global Minimum

The smallest overall value of a set, function, etc., over its entire range.

see also GLOBAL MAXIMUM, KUHN-TUCKER THEOREM, LOCAL MINIMUM, MINIMUM

Globe

A SPHERE which acts as a model of a spherical (or ellipsoidal) celestial body, especially the Earth, and on which the outlines of continents, oceans, etc. are drawn.

see also LATITUDE, LONGITUDE, SPHERE

Glove Problem

Let there be m doctors and $n \leq m$ patients, and let all mn possible combinations of examinations of patients by doctors take place. Then what is the minimum number of surgical gloves needed $G(m,n)$ so that no doctor must wear a glove contaminated by a patient and no patient is exposed to a glove worn by another doctor? In this problem, the gloves can be turned inside out and even placed on top of one another if necessary, but no "decontamination" of gloves is permitted. The optimal solution is

$$g(m,n) = \begin{cases} 2 & m = n = 2 \\ \frac{1}{2}(m+1) & n = 1,\ m = 2k+1 \\ \left\lceil \frac{1}{2}(m) + \frac{2}{3}n \right\rceil & \text{otherwise,} \end{cases}$$

where $\lceil x \rceil$ is the CEILING FUNCTION (Vardi 1991). The case $m = n = 2$ is straightforward since two gloves have a total of four surfaces, which is the number needed for $mn = 4$ examinations.

References

Gardner, M. *Aha! Aha! Insight.* New York: Scientific American, 1978.

Gardner, M. *Science Fiction Puzzle Tales.* New York: Crown, pp. 5, 67, and 104–150, 1981.

Hajnal, A. and Lovász, L. "An Algorithm to Prevent the Propagation of Certain Diseases at Minimum Cost." §10.1 in *Interfaces Between Computer Science and Operations Research* (Ed. J. K. Lenstra, A. H. G. Rinnooy Kan, and P. van Emde Boas). Amsterdam: Matematisch Centrum, 1978.

Orlitzky, A. and Shepp, L. "On Curbing Virus Propagation." Exercise 10.2 in Technical Memo. Bell Labs, 1989.

Vardi, I. "The Condom Problem." Ch. 10 in *Computational Recreations in Mathematica.* Redwood City, CA: Addison-Wesley, p. 203–222, 1991.

Glue Vector

A VECTOR specifying how layers are stacked in a LAMINATED LATTICE.

Gnomic Number

A FIGURATE NUMBER of the form $g_n = 2n - 1$ which are the areas of square gnomons, obtained by removing a SQUARE of side $n-1$ from a SQUARE of side n,

$$g_n = n^2 - (n-1)^2 = 2n - 1.$$

The gnomic numbers are therefore equivalent to the ODD NUMBERS, and the first few are 1, 3, 5, 7, 9, 11, ... (Sloane's A005408). The GENERATING FUNCTION for the gnomic numbers is

$$\frac{x(1+x)}{(x-1)^2} = x + 3x^2 + 5x^3 + 7x^4 + \ldots.$$

see also ODD NUMBER

References

Sloane, N. J. A. Sequence A005408/M2400 in "An On-Line Version of the Encyclopedia of Integer Sequences."

Gnomic Projection

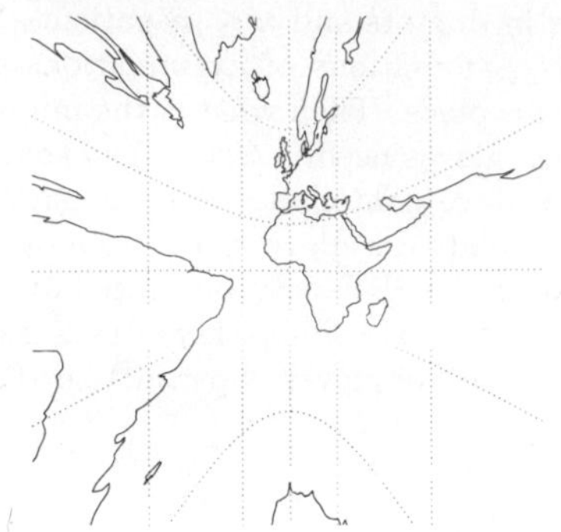

A nonconformal projection from a SPHERE's center in which ORTHODROMES are straight LINES.

$$x = \frac{\cos\phi \sin(\lambda - \lambda_0)}{\cos c} \quad (1)$$

$$y = \frac{\cos\phi_1 \sin\phi - \sin\phi_1 \cos\phi \cos(\lambda - \lambda_0)}{\cos c}, \quad (2)$$

where

$$\cos c = \sin\phi_1 \sin\phi + \cos\phi_1 \cos\phi \cos(\lambda - \lambda_0). \quad (3)$$

The inverse FORMULAS are

$$\phi = \sin^{-1}(\cos c \sin\phi_1 + y \sin c \cos c \cos\phi_1) \quad (4)$$

$$\lambda = \lambda_0 + \tan^{-1}\left(\frac{x}{\cos\phi_1 - y \sin\phi_1}\right). \quad (5)$$

References

Coxeter, H. S. M. and Greitzer, S. L. *Geometry Revisited.* Washington, DC: Math. Assoc. Amer., pp. 150–153, 1967.

Snyder, J. P. *Map Projections—A Working Manual.* U. S. Geological Survey Professional Paper 1395. Washington, DC: U. S. Government Printing Office, pp. 164–168, 1987.

Gnomon

A shape which, when added to a figure, yields another figure SIMILAR to the original.

References

Shanks, D. *Solved and Unsolved Problems in Number Theory, 4th ed.* New York: Chelsea, p. 123, 1993.

Gnomon Magic Square

A 3×3 array of numbers in which the elements in each 2×2 corner have the same sum.

see also MAGIC SQUARE

Go

There are estimated to be about 4.63×10^{170} possible positions on a 19×19 board (Flammenkamp). The number of n-move Go games are 1, 362, 130683, 47046242, ... (Sloane's A007565).

References

Beeler, M.; Gosper, R. W.; and Schroeppel, R. *HAKMEM.* Cambridge, MA: MIT Artificial Intelligence Laboratory, Memo AIM-239, Item 96, Feb. 1972.

Flammenkamp, A. "A Short, Concise Ruleset of Go." `http://www.minet.uni-jena.de/~achim/gorules.html`.

Kraitchik, M. "Go." §12.4 in *Mathematical Recreations.* New York: W. W. Norton, pp. 279–280, 1942.

Sloane, N. J. A. Sequence A007565/M5447 in "An On-Line Version of the Encyclopedia of Integer Sequences."

Göbel's Sequence

Consider the RECURRENCE RELATION

$$x_n = \frac{1 + {x_0}^2 + {x_1}^2 + \ldots + {x_{n-1}}^2}{n}, \quad (1)$$

with $x_0 = 1$. The first few iterates of x_n are 1, 2, 3, 5, 10, 28, 154, ... (Sloane's A003504). The terms grow extremely rapidly, but are given by the asymptotic formula

$$x_n \approx (n^2+2n-1+4n^{-1}-21n^{-2}+137n^{-3}-\ldots)C^{2^n}, \quad (2)$$

where

$$C = 1.04783144757641122955990946274313755459\ldots \quad (3)$$

(Zagier). It is more convenient to work with the transformed sequence

$$s_n = 2 + {x_1}^2 + {x_2}^2 + \ldots + {x_{n-1}}^2 = nx_n, \quad (4)$$

which gives the new recurrence

$$s_{n+1} = s_n + \frac{{s_n}^2}{n^2} \quad (5)$$

with initial condition $s_1 = 2$. Now, s_{n+1} will be nonintegral IFF $n\nmid s_n$. The smallest p for which $s_p \not\equiv 0 \pmod p$ therefore gives the smallest nonintegral s_{p+1}. In addition, since $p\nmid s_p$, $x_p = s_p/p$ is also the smallest nonintegral x_p.

For example, we have the sequences $\{s_n \pmod k\}_{n=1}^k$:

$$2, 6 \equiv 2, \tfrac{5}{4} \equiv 0, 0, 0 \pmod 5 \quad (6)$$

$$2, 6, 15 \equiv 1, \tfrac{5}{4} \equiv 0, 0, 0, 0 \pmod 7 \quad (7)$$

$$2, 6, 15 \equiv 4, \tfrac{52}{9} \equiv 7, \tfrac{161}{16} \equiv 8, \tfrac{264}{5} \equiv 0, 0, \ldots, 0 \pmod{11}. \quad (8)$$

Testing values of k shows that the first nonintegral x_n is x_{43}. Note that a direct verification of this fact is impossible since

$$x_{43} \approx 5.4093 \times 10^{178485291567} \quad (9)$$

(calculated using the asymptotic formula) is much too large to be computed and stored explicitly.

A sequence even more striking for remaining integral over many terms is the 3-Göbel sequence

$$x_n = \frac{1 + {x_0}^3 + {x_1}^3 + \ldots + {x_{n-1}}^3}{n}. \quad (10)$$

The first few terms of this sequence are 1, 2, 5, 45, 22815, ... (Sloane's A005166).

The Göbel sequences can be generalized to k powers by

$$x_n = \frac{1 + {x_0}^k + {x_1}^k + \ldots + {x_{n-1}}^k}{n}. \quad (11)$$

see also SOMOS SEQUENCE

References

Guy, R. K. "The Strong Law of Small Numbers." *Amer. Math. Monthly* **95**, 697–712, 1988.

Guy, R. K. "A Recursion of Göbel." §E15 in *Unsolved Problems in Number Theory, 2nd ed.* New York: Springer-Verlag, pp. 214–215, 1994.

Sloane, N. J. A. Sequences A003504/M0728 and A005166/M1551 in "An On-Line Version of the Encyclopedia of Integer Sequences."

Zaiger, D. "Solution: Day 5, Problem 3." `http://www-groups.dcs.st-and.ac.uk/~john/Zagier/Solution5.3.html`.

Goblet Illusion

An ILLUSION in which the eye alternately sees two black faces, or a white goblet.

References

Fineman, M. *The Nature of Visual Illusion.* New York: Dover, pp. 111 and 115, 1996.

Rubin, E. *Synoplevede Figurer.* Copenhagen, Denmark: Gyldendalske, 1915.

What's Up with Kids Magazine. "Reversible Goblet." `http://wuwk.spurtek.com/COI_reversible_goblet.htm`.

Gödel's Completeness Theorem

If T is a set of AXIOMS in a first-order language, and a statement p holds for any structure M satisfying T, then p can be formally deduced from T in some appropriately defined fashion.

see also GÖDEL'S INCOMPLETENESS THEOREM

Gödel's Incompleteness Theorem

Informally, Gödel's incompleteness theorem states that all consistent axiomatic formulations of NUMBER THEORY include undecidable propositions (Hofstadter 1989). This is is sometimes called Gödel's first incompleteness theorem, and answers in the negative HILBERT'S PROBLEM asking whether mathematics is "complete" (in the sense that every statement in the language of NUMBER THEORY can be either proved or disproved). Formally, Gödel's theorem states, "To every ω-consistent recursive class κ of FORMULAS, there correspond recursive class-signs r such that neither (v Gen r) nor Neg(v Gen r) belongs to Flg(κ), where v is the FREE VARIABLE of r" (Gödel 1931).

A statement sometimes known as Gödel's second incompleteness theorem states that if NUMBER THEORY is consistent, then a proof of this fact does not exist using the methods of first-order PREDICATE CALCULUS. Stated more colloquially, any formal system that is interesting enough to formulate its own consistency can prove its own consistency IFF it is inconsistent.

Gerhard Gentzen showed that the consistency and completeness of arithmetic can be proved if "transfinite" induction is used. However, this approach does not allow proof of the consistency of all mathematics.

see also GÖDEL'S COMPLETENESS THEOREM, HILBERT'S PROBLEMS, KREISEL CONJECTURE, NATURAL INDEPENDENCE PHENOMENON, NUMBER THEORY, RICHARDSON'S THEOREM, UNDECIDABLE

References

Barrow, J. D. *Pi in the Sky: Counting, Thinking, and Being.* Oxford, England: Clarendon Press, p. 121, 1992.

Gödel, K. "Über Formal Unentscheidbare Sätze der *Principia Mathematica* und Verwandter Systeme, I." *Monatshefte für Math. u. Physik* **38**, 173–198, 1931.

Gödel, K. *On Formally Undecidable Propositions of Principia Mathematica and Related Systems.* New York: Dover, 1992.

Hofstadter, D. R. *Gödel, Escher, Bach: An Eternal Golden Braid.* New York: Vintage Books, p. 17, 1989.

Kolata, G. "Does Gödel's Theorem Matter to Mathematics?" *Science* **218**, 779–780, 1982.

Smullyan, R. M. *Gödel's Incompleteness Theorems.* New York: Oxford University Press, 1992.

Whitehead, A. N. and Russell, B. *Principia Mathematica.* New York: Cambridge University Press, 1927.

Gödel Number

A Gödel number is a unique number associated to a statement about arithmetic. It is formed as the PRODUCT of successive PRIMES raised to the POWER of the number corresponding to the individual symbols that comprise the sentence. For example, the statement $(\exists x)(x = sy)$ that reads "there EXISTS an x such that x is the immediate successor of y" is coded

$$(2^8)(3^4)(5^{13})(7^9)(11^8)(13^{13})(17^5)(19^7)(23^{16})(29^9),$$

where the numbers in the set (8, 4, 13, 9, 8, 13, 5, 7, 16, 9) correspond to the symbols that make up $(\exists x)(x = sy)$.

see also GÖDEL'S INCOMPLETENESS THEOREM

References

Hofstadter, D. R. *Gödel, Escher, Bach: An Eternal Golden Braid.* New York: Vintage Books, p. 18, 1989.

Goldbach Conjecture

Goldbach's original conjecture, written in a 1742 letter to Euler, states that every INTEGER > 5 is the SUM of three PRIMES. As re-expressed by Euler, an equivalent of this CONJECTURE (called the "strong" Goldbach conjecture) asserts that all POSITIVE EVEN INTEGERS ≥ 4 can be expressed as the SUM of two PRIMES. Schnirelmann (1931) proved that every EVEN number can be written as the sum of not more than 300,000 PRIMES (Dunham 1990), which seems a rather far cry from a proof for *four* PRIMES! The strong Goldbach conjecture has been shown to be true up to 4×10^{11} by Sinisalo (1993). Pogorzelski (1977) claimed to have proven the Goldbach conjecture, but his proof is not generally accepted (Shanks 1993).

The conjecture that all ODD numbers ≥ 9 are the SUM of three ODD PRIMES is called the "weak" Goldbach conjecture. Vinogradov proved that all ODD INTEGERS starting at some sufficiently large value are the SUM of three PRIMES (Guy 1994). The original "sufficiently large" $N \geq 3^{3^{15}} = e^{e^{16.573}}$ was subsequently reduced to $e^{e^{11.503}}$ by Chen and Wang (1989). Chen (1973, 1978) also showed that all sufficiently large EVEN NUMBERS are the sum of a PRIME and the PRODUCT of at most two PRIMES (Guy 1994, Courant and Robbins 1996).

It has been shown that if the weak Goldbach conjecture is false, then there are only a FINITE number of exceptions.

Other variants of the Goldbach conjecture include the statements that every EVEN number ≥ 6 is the SUM of two ODD PRIMES, and every INTEGER > 17 the sum of exactly three distinct PRIMES. Let $R(n)$ be the number of representations of an EVEN INTEGER n as the sum of two PRIMES. Then the "extended" Goldbach conjecture states that

$$R(n) \sim 2\Pi_2 \prod_{\substack{k=2 \\ p_k | n}} \frac{p_k - 1}{p_k - 2} \int_2^x \frac{dx}{(\ln x)^2},$$

where Π_2 is the TWIN PRIMES CONSTANT (Halberstam and Richert 1974).

If the Goldbach conjecture is true, then for every number m, there are PRIMES p and q such that

$$\phi(p) + \phi(q) = 2m,$$

where $\phi(x)$ is the TOTIENT FUNCTION (Guy 1994, p. 105).

Vinogradov (1937ab, 1954) proved that every sufficiently large ODD NUMBER is the sum of three PRIMES, and Estermann (1938) proves that almost all EVEN NUMBERS are the sums of two PRIMES.

References

Ball, W. W. R. and Coxeter, H. S. M. *Mathematical Recreations and Essays, 13th ed.* New York: Dover, p. 64, 1987.

Chen, J.-R. "On the Representation of a Large Even Number as the Sum of a Prime and the Product of at Most Two Primes." *Sci. Sinica* **16**, 157–176, 1973.

Chen, J.-R. "On the Representation of a Large Even Number as the Sum of a Prime and the Product of at Most Two Primes, II." *Sci. Sinica* **21**, 421–430, 1978.

Chen, J.-R. and Wang, T.-Z. "On the Goldbach Problem." *Acta Math. Sinica* **32**, 702–718, 1989.

Courant, R. and Robbins, H. *What is Mathematics?: An Elementary Approach to Ideas and Methods, 2nd ed.* Oxford, England: Oxford University Press, pp. 30–31, 1996.

Devlin, K. *Mathematics: The New Golden Age.* London: Penguin Books, 1988.

Dunham, W. *Journey Through Genius: The Great Theorems of Mathematics.* New York: Wiley, p. 83, 1990.

Estermann, T. "On Goldbach's Problem: Proof that Almost All Even Positive Integers are Sums of Two Primes." *Proc. London Math. Soc. Ser. 2* **44**, 307–314, 1938.

Guy, R. K. "Goldbach's Conjecture." §C1 in *Unsolved Problems in Number Theory, 2nd ed.* New York: Springer-Verlag, pp. 105–107, 1994.

Hardy, G. H. and Littlewood, J. E. "Some Problems of Partitio Numerorum (V): A Further Contribution to the Study of Goldbach's Problem." *Proc. London Math. Soc. Ser. 2* **22**, 46–56, 1924.

Halberstam, H. and Richert, H.-E. *Sieve Methods.* New York: Academic Press, 1974.

Pogorzelski, H. A. "Goldbach Conjecture." *J. Reine Angew. Math.* **292**, 1–12, 1977.

Shanks, D. *Solved and Unsolved Problems in Number Theory, 4th ed.* New York: Chelsea, pp. 30–31 and 222, 1985.

Sinisalo, M. K. "Checking the Goldbach Conjecture up to $4 \cdot 10^{11}$." *Math. Comput.* **61**, 931–934, 1993.

Vinogradov, I. M. "Representation of an Odd Number as a Sum of Three Primes." *Comtes rendus (Doklady) de l'Académie des Sciences de l'U.R.S.S.* **15**, 169–172, 1937a.

Vinogradov, I. "Some Theorems Concerning the Theory of Primes." *Recueil Math.* **2**, 179–195, 1937b.

Vinogradov, I. M. *The Method of Trigonometrical Sums in the Theory of Numbers.* London: Interscience, p. 67, 1954.

Yuan, W. *Goldbach Conjecture.* Singapore: World Scientific, 1984.

Golden Mean

see GOLDEN RATIO

Golden Ratio

A number often encountered when taking the ratios of distances in simple geometric figures such as the DECAGON and DODECAGON. It is denoted ϕ, or sometimes τ (which is an abbreviation of the Greek "tome," meaning "to cut"). ϕ is also known as the DIVINE PROPORTION, GOLDEN MEAN, and GOLDEN SECTION and is a PISOT-VIJAYARAGHAVAN CONSTANT. It has surprising connections with CONTINUED FRACTIONS and the

EUCLIDEAN ALGORITHM for computing the GREATEST COMMON DIVISOR of two INTEGERS.

Given a RECTANGLE having sides in the ratio $1 : \phi$, ϕ is defined such that partitioning the original RECTANGLE into a SQUARE and new RECTANGLE results in a new RECTANGLE having sides with a ratio $1 : \phi$. Such a RECTANGLE is called a GOLDEN RECTANGLE, and successive points dividing a GOLDEN RECTANGLE into SQUARES lie on a LOGARITHMIC SPIRAL. This figure is known as a WHIRLING SQUARE.

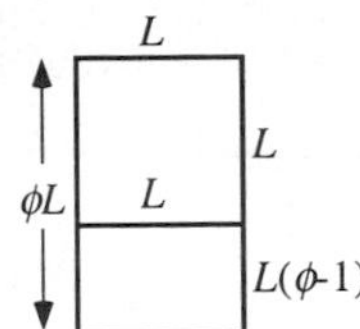

This means that

$$\frac{1}{\phi - 1} = \phi \tag{1}$$

$$\phi^2 - \phi - 1 = 0. \tag{2}$$

So, by the QUADRATIC EQUATION,

$$\phi = \tfrac{1}{2}(1 \pm \sqrt{1+4}) = \tfrac{1}{2}(1+\sqrt{5}) \tag{3}$$

$$= 1.6180339887498948482045868343656381177720\ldots \tag{4}$$

(Sloane's A001622).

x 1
A B C

A geometric definition can be given in terms of the above figure. Let the ratio $x \equiv AB/BC$. The NUMERATOR and DENOMINATOR can then be taken as $\overline{AB} = x$ and $\overline{BC} = 1$ without loss of generality. Now define the position of B by

$$\frac{BC}{AB} = \frac{AB}{AC}. \tag{5}$$

Plugging in gives

$$\frac{1}{x} = \frac{x}{1+x}, \tag{6}$$

or

$$x^2 - x - 1 = 0, \tag{7}$$

which can be solved using the QUADRATIC EQUATION to obtain

$$\phi \equiv x = \frac{1 \pm \sqrt{1^2 - (-4)}}{2} = \tfrac{1}{2}(1+\sqrt{5}). \tag{8}$$

ϕ is the "most" IRRATIONAL number because it has a CONTINUED FRACTION representation

$$\phi = [1, 1, 1, \ldots] \tag{9}$$

(Sloane's A000012). Another infinite representation in terms of a CONTINUED SQUARE ROOT is

$$\phi = \sqrt{1+\sqrt{1+\sqrt{1+\sqrt{1+\ldots}}}} \tag{10}$$

Ramanujan gave the curious CONTINUED FRACTION identities

$$\frac{1}{(\sqrt{\phi\sqrt{5}})e^{2\pi/5}} = 1 + \cfrac{e^{-2\pi}}{1 + \cfrac{e^{-4\pi}}{1 + \cfrac{e^{-6\pi}}{1 + \cfrac{e^{-8\pi}}{1 + \cfrac{e^{-10\pi}}{1+\ldots}}}}} \tag{11}$$

$$\frac{1}{\left\{\frac{\sqrt{5}}{1+\left[5^{3/4}(\phi-1)^{5/2}-1\right]-\phi}\right\} e^{2\pi/\sqrt{5}}} = 1 + \cfrac{e^{-2\pi\sqrt{5}}}{1 + \cfrac{e^{-4\pi\sqrt{5}}}{1 + \cfrac{e^{-6\pi\sqrt{5}}}{1 + \cfrac{e^{-8\pi\sqrt{5}}}{1 + \cfrac{e^{-10\pi\sqrt{5}}}{1+\ldots}}}}} \tag{12}$$

(Ramanathan 1984).

The legs of a GOLDEN TRIANGLE are in a golden ratio to its base. In fact, this was the method used by Pythagoras to construct ϕ. Euclid used the following construction.

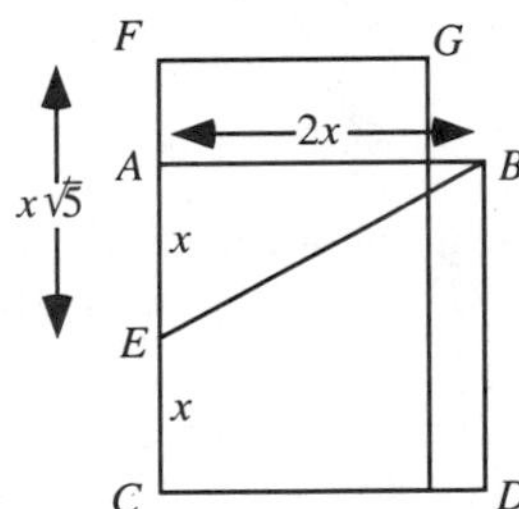

Draw the SQUARE $\square ABDC$, call E the MIDPOINT of AC, so that $AE = EC \equiv x$. Now draw the segment BE, which has length

$$x\sqrt{2^2+1^2} = x\sqrt{5}, \tag{13}$$

and construct EF with this length. Now construct $FG = EF$, then

$$\phi = \frac{FC}{CD} = \frac{EF+CE}{CD} = \frac{x(\sqrt{5}+1)}{2x} = \tfrac{1}{2}(\sqrt{5}+1). \tag{14}$$

The ratio of the CIRCUMRADIUS to the length of the side of a DECAGON is also ϕ,

$$\frac{R}{s} = \tfrac{1}{2}\csc\left(\frac{\pi}{10}\right) = \tfrac{1}{2}(1+\sqrt{5}) = \phi. \tag{15}$$

Similarly, the legs of a GOLDEN TRIANGLE (an ISOSCELES TRIANGLE with a VERTEX ANGLE of 36°) are in a GOLDEN RATIO to the base. Bisecting a GAULLIST CROSS also gives a golden ratio (Gardner 1961, p. 102).

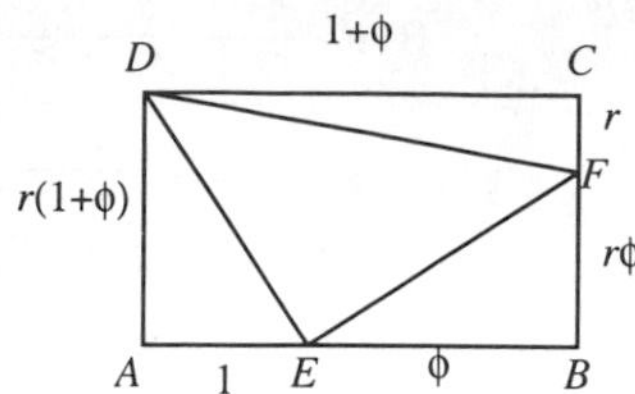

In the figure above, three TRIANGLES can be INSCRIBED in the RECTANGLE $\square ABCD$ of arbitrary aspect ratio $1:r$ such that the three RIGHT TRIANGLES have equal areas by dividing AB and BC in the golden ratio. Then

$$K_{\Delta ADE} = \tfrac{1}{2}\cdot r(1+\phi)\cdot 1 = \tfrac{1}{2}r\phi^2 \tag{16}$$
$$K_{\Delta BEF} = \tfrac{1}{2}\cdot r\phi\cdot\phi = \tfrac{1}{2}r\phi^2 \tag{17}$$
$$K_{\Delta CDF} = \tfrac{1}{2}(1+\phi)\cdot r = \tfrac{1}{2}r\phi^2, \tag{18}$$

which are all equal.

The golden ratio also satisfies the RECURRENCE RELATION

$$\phi^{n+1} = \phi^{n-1} + \phi^n. \tag{19}$$

Taking $n = 0$ gives

$$\phi = \phi^{-1} + 1 \tag{20}$$

$$\phi^2 = 1 + \phi. \tag{21}$$

But this is the definition equation for ϕ (when the root with the plus sign is used). Squaring gives

$$\phi^2 = \tfrac{1}{4}(5+2\sqrt{5}+1) = \tfrac{1}{4}(6+2\sqrt{5}) = \tfrac{1}{2}(3+\sqrt{5})$$
$$= \tfrac{1}{2}(\sqrt{5}+1)+1 = \phi^0 + \phi^1 \tag{22}$$
$$\phi^3 = (\phi^0+\phi^1)\phi^1 = \phi^0\phi^1 + (\phi^1)^2 = \phi^1 + \phi^2, \tag{23}$$

and so on.

For the difference equations

$$\begin{cases} x_0 = 1 \\ x_n = 1 + \dfrac{1}{x_{n-1}} \quad \text{for } n = 1, 2, 3, \end{cases} \tag{24}$$

ϕ is also given by

$$\phi = \lim_{n\to\infty} x_n. \tag{25}$$

In addition,

$$\phi = \lim_{n\to\infty} \frac{F_n}{F_{n-1}}, \tag{26}$$

where F_n is the nth FIBONACCI NUMBER.

The SUBSTITUTION MAP

$$0 \to 01 \tag{27}$$
$$1 \to 0 \tag{28}$$

gives

$$0 \to 01 \to 010 \to 01001 \to \ldots, \tag{29}$$

giving rise to the sequence

$$0100101001001010010100100101\ldots \tag{30}$$

(Sloane's A003849). Here, the zeros occur at positions 1, 3, 4, 6, 8, 9, 11, 12, ... (Sloane's A000201), and the ones occur at positions 2, 5, 7, 10, 13, 15, 18, ... (Sloane's A001950). These are complementary BEATTY SEQUENCES generated by $\lfloor n\phi \rfloor$ and $\lfloor n\phi^2 \rfloor$. The sequence also has many connections with the FIBONACCI NUMBERS.

Salem showed that the set of PISOT-VIJAYARAGHAVAN CONSTANTS is closed, with ϕ the smallest accumulation point of the set (Le Lionnais 1983).

see also BERAHA CONSTANTS, DECAGON, FIVE DISKS PROBLEM, GOLDEN RATIO CONJUGATE, GOLDEN TRIANGLE, ICOSIDODECAHEDRON, NOBLE NUMBER, PENTAGON, PENTAGRAM, PHI NUMBER SYSTEM, SECANT METHOD

References

Boyer, C. B. *History of Mathematics.* New York: Wiley, p. 56, 1968.

Coxeter, H. S. M. "The Golden Section, Phyllotaxis, and Wythoff's Game." *Scripta Mathematica* **19**, 135–143, 1953.

Dixon, R. *Mathographics.* New York: Dover, pp. 30–31 and 50, 1991.

Finch, S. "Favorite Mathematical Constants." `http://www.mathsoft.com/asolve/constant/cntfrc/cntfrc.html`.

Finch, S. "Favorite Mathematical Constants." `http://www.mathsoft.com/asolve/constant/gold/gold.html`.

Gardner, M. "Phi: The Golden Ratio." Ch. 8 in *The Second Scientific American Book of Mathematical Puzzles & Diversions, A New Selection.* New York: Simon and Schuster, 1961.

Gardner, M. "Notes on a Fringe-Watcher: The Cult of the Golden Ratio." *Skeptical Inquirer* **18**, 243–247, 1994.

Herz-Fischler, R. *A Mathematical History of the Golden Number.* New York: Dover, 1998.

Huntley, H. E. *The Divine Proportion.* New York: Dover, 1970.

Knott, R. "Fibonacci Numbers and the Golden Section." `http://www.mcs.surrey.ac.uk/Personal/R.Knott/Fibonacci/fib.html`.

Le Lionnais, F. *Les nombres remarquables.* Paris: Hermann, p. 40, 1983.

Markowsky, G. "Misconceptions About the Golden Ratio." *College Math. J.* **23**, 2–19, 1992.

Ogilvy, C. S. *Excursions in Geometry.* New York: Dover, pp. 122–134, 1990.

Pappas, T. "Anatomy & the Golden Section." *The Joy of Mathematics.* San Carlos, CA: Wide World Publ./Tetra, pp. 32–33, 1989.
Ramanathan, K. G. "On Ramanujan's Continued Fraction." *Acta. Arith.* **43**, 209–226, 1984.
Sloane, N. J. A. Sequences A003849, A000012/M0003, A000201/M2322, A001622/M4046, and A001950/M1332 in "An On-Line Version of the Encyclopedia of Integer Sequences."

Golden Ratio Conjugate

The quantity

$$\phi_C \equiv \frac{1}{\phi} = \phi - 1 = \frac{\sqrt{5}-1}{2} \approx 0.6180339885, \qquad (1)$$

where ϕ is the GOLDEN RATIO. The golden ratio conjugate is sometimes also called the SILVER RATIO. A quantity similar to the FEIGENBAUM CONSTANT can be found for the nth CONTINUED FRACTION representation

$$[a_0, a_1, a_2, \ldots]. \qquad (2)$$

Taking the limit of

$$\delta_n \equiv \frac{\sigma_n - \sigma_{n-1}}{\sigma_n - \sigma_{n+1}} \qquad (3)$$

gives

$$\delta \equiv \lim_{n\to\infty} = 1 + \phi = 2 + \phi_C. \qquad (4)$$

see also GOLDEN RATIO, SILVER RATIO

Golden Rectangle

Given a RECTANGLE having sides in the ratio $1:\phi$, the GOLDEN RATIO ϕ is defined such that partitioning the original RECTANGLE into a SQUARE and new RECTANGLE results in a new RECTANGLE having sides with a ratio $1:\phi$. Such a RECTANGLE is called a golden rectangle, and successive points dividing a golden rectangle into SQUARES lie on a LOGARITHMIC SPIRAL.

see also GOLDEN RATIO, LOGARITHMIC SPIRAL, RECTANGLE

References
Pappas, T. "The Golden Rectangle." *The Joy of Mathematics.* San Carlos, CA: Wide World Publ./Tetra, pp. 102–106, 1989.

Golden Rule

The mathematical golden rule states that, for any FRACTION, both NUMERATOR and DENOMINATOR may be multiplied by the same number without changing the fraction's value.

see also DENOMINATOR, FRACTION, NUMERATOR

References
Conway, J. H. and Guy, R. K. *The Book of Numbers.* New York: Springer-Verlag, p. 151, 1996.

Golden Section

see GOLDEN RATIO

Golden Theorem

see QUADRATIC RECIPROCITY THEOREM

Golden Triangle

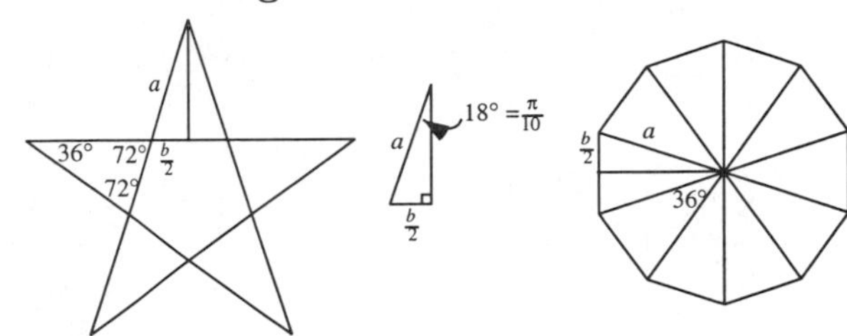

An ISOSCELES TRIANGLE with VERTEX angles 36°. Such TRIANGLES occur in the PENTAGRAM and DECAGON. The legs are in a GOLDEN RATIO to the base. For such a TRIANGLE,

$$\sin(18°) = \sin(\tfrac{1}{10}\pi) = \frac{\frac{1}{2}b}{l} \qquad (1)$$

$$b = 2a\sin(\tfrac{1}{10}\pi) = 2a\frac{\sqrt{5}-1}{4} = \tfrac{1}{2}a(\sqrt{5}-1) \qquad (2)$$

$$b + l = \tfrac{1}{2}a(\sqrt{5}+1) \qquad (3)$$

$$\frac{b+a}{a} = \frac{\sqrt{5}+1}{2} = \phi. \qquad (4)$$

see also DECAGON, GOLDEN RATIO, ISOSCELES TRIANGLE, PENTAGRAM

References
Pappas, T. "The Pentagon, the Pentagram & the Golden Triangle." *The Joy of Mathematics.* San Carlos, CA: Wide World Publ./Tetra, pp. 188–189, 1989.

Goldschmidt Solution

The discontinuous solution of the SURFACE OF REVOLUTION AREA minimization problem for surfaces connecting two CIRCLES. When the CIRCLES are sufficiently far apart, the usual CATENOID is no longer stable and the surface will break and form two surfaces with the CIRCLES as boundaries.

see also CALCULUS OF VARIATIONS, SURFACE OF REVOLUTION

Golomb Constant

see GOLOMB-DICKMAN CONSTANT

Golomb-Dickman Constant

N.B. A detailed on-line essay by S. Finch was the starting point for this entry.

Let Π be a PERMUTATION of n elements, and let α_i be the number of CYCLES of length i in this PERMUTATION. Picking Π at RANDOM gives

$$\left\langle \sum_{j=1}^{\infty} \alpha_j \right\rangle = \sum_{i=1}^{n} \frac{1}{i} = \ln n + \gamma + \mathcal{O}\left(\frac{1}{n}\right) \tag{1}$$

$$\operatorname{var}\left(\sum_{j=1}^{\infty} \alpha_j\right) = \sum_{i=1}^{n} \frac{i-1}{i^2} = \ln n + \gamma - \tfrac{1}{6}\pi^2 + \mathcal{O}\left(\frac{1}{n}\right) \tag{2}$$

$$\lim_{n\to\infty} P(\alpha_1 = 0) = \frac{1}{e} \tag{3}$$

(Shepp and Lloyd 1966, Wilf 1990). Goncharov (1942) showed that

$$\lim_{n\to\infty} P(\alpha_j = k) = \frac{1}{k!} e^{-1/j} j^{-k}, \tag{4}$$

which is a POISSON DISTRIBUTION, and

$$\lim_{n\to\infty} P\left[\left(\sum_{j=1}^{\infty} \alpha_j - \ln n\right)(\ln n)^{-1/2} \leq x\right] = \Phi(x), \tag{5}$$

which is a NORMAL DISTRIBUTION, γ is the EULER-MASCHERONI CONSTANT, and Φ is the NORMAL DISTRIBUTION FUNCTION. Let

$$M(\alpha) \equiv \max_{1\leq j<\infty} \alpha_j \tag{6}$$

$$m(\alpha) \equiv \min_{1\leq j<\infty} \alpha_j. \tag{7}$$

Golomb (1959) derived

$$\lambda \equiv \lim_{n\to\infty} \frac{\langle M(\alpha)\rangle}{n} = 0.6243299885\ldots, \tag{8}$$

which is known as the GOLOMB CONSTANT or Golomb-Dickman constant. Knuth (1981) asked for the constants b and c such that

$$\lim_{n\to\infty} n^b[\langle M(\alpha)\rangle - \lambda n - \tfrac{1}{2}\lambda] = c, \tag{9}$$

and Gourdon (1996) showed that

$$\langle M(\alpha)\rangle = \lambda(n + \tfrac{1}{2}) - \frac{e^\gamma}{24n} + \frac{\frac{1}{48}e^\gamma - \frac{1}{8}(-1)^n}{n^2} + \frac{\frac{17}{3840}e^\gamma + \frac{1}{8}(-1)^n + \frac{1}{6}j^{1+2n} + \frac{1}{6}j^{2+n}}{n^3}, \tag{10}$$

where

$$j \equiv e^{2\pi i/3}. \tag{11}$$

λ can be expressed in terms of the function $f(x)$ defined by $f(x) = 1$ for $1 \leq x \leq 2$ and

$$\frac{df}{dx} = -\frac{f(x-1)}{x-1} \tag{12}$$

for $x > 2$, by

$$\lambda = \int_1^\infty \frac{f(x)}{x^2}\,dx. \tag{13}$$

Shepp and Lloyd (1966) derived

$$\begin{aligned}\lambda &= \int_0^\infty \exp\left(-x - \int_x^\infty \frac{e^{-y}}{y}\,dy\right) \\ &= \int_0^1 \exp\left(\int_0^x \frac{dy}{\ln y}\right) dx.\end{aligned} \tag{14}$$

Mitchell (1968) computed λ to 53 decimal places.

Surprisingly enough, there is a connection between λ and PRIME FACTORIZATION (Knuth and Pardo 1976, Knuth 1981, pp. 367–368, 395, and 611). Dickman (1930) investigated the probability $P(x,n)$ that the largest PRIME FACTOR p of a random INTEGER between 1 and n satisfies $p < n^x$ for $x \in (0,1)$. He found that

$$F(x) \equiv \lim_{n\to\infty} P(x,n) = \begin{cases} 1 & \text{if } x \geq 1 \\ \int_0^x F\left(\frac{t}{1-t}\right)\frac{dt}{t} & \text{if } 0 \leq x < 1. \end{cases} \tag{15}$$

Dickman then found the average value of x such that $p = n^x$, obtaining

$$\begin{aligned}\mu \equiv \lim_{n\to\infty} \langle x\rangle &= \lim_{n\to\infty}\left\langle \frac{\ln p}{\ln n}\right\rangle = \int_0^1 x\frac{dF}{dx}\,dx \\ &= \int_0^1 F\left(\frac{1}{1-t}\right) dt = 0.62432999,\end{aligned} \tag{16}$$

which is λ.

References

Finch, S. "Favorite Mathematical Constants." `http://www.mathsoft.com/asolve/constant/golomb/golomb.html`.

Gourdon, X. 1996. `http://www.mathsoft.com/asolve/constant/golomb/gourdon.html`.

Knuth, D. E. *The Art of Computer Programming, Vol. 1: Fundamental Algorithms, 2nd ed.* Reading, MA: Addison-Wesley, 1973.

Knuth, D. E. *The Art of Computer Programming, Vol. 2: Seminumerical Algorithms, 2nd ed.* Reading, MA: Addison-Wesley, 1981.

Knuth, D. E. and Pardo, L. T. "Analysis of a Simple Factorization Algorithm." *Theor. Comput. Sci.* **3**, 321–348, 1976.

Mitchell, W. C. "An Evaluation of Golomb's Constant." *Math. Comput.* **22**, 411–415, 1968.

Purdom, P. W. and Williams, J. H. "Cycle Length in a Random Function." *Trans. Amer. Math. Soc.* **133**, 547–551, 1968.

Shepp, L. A. and Lloyd, S. P. "Ordered Cycle Lengths in Random Permutation." *Trans. Amer. Math. Soc.* **121**, 350–557, 1966.

Wilf, H. S. *Generatingfunctionology, 2nd ed.* New York: Academic Press, 1993.

Golomb Ruler

A Golomb ruler is a set of NONNEGATIVE integers such that all pairwise POSITIVE differences are distinct. The optimum Golomb ruler with n marks is the Golomb ruler having the smallest possible maximum element ("length"). The set (0, 1, 3, 7) is an order four Golomb ruler since its differences are ($1 = 1 - 0$, $2 = 3 - 1$, $3 = 3 - 0$, $4 = 7 - 3$, $6 = 7 - 1$, $7 = 7 - 0$), all of which are distinct. However, the optimum 4-mark Golomb ruler is (0, 1, 4, 6), which measures the distances (1, 2, 3, 4, 5, 6) (and is therefore also a PERFECT RULER).

The lengths of the optimal n-mark Golomb rulers for $n = 2, 3, 4, \ldots$ are 1, 3, 6, 11, 17, 25, 34, ... (Sloane's A003022, Vanderschel and Garry). The lengths of the optimal n-mark Golomb rulers are not known for $n \geq 20$.

see also PERFECT DIFFERENCE SET, PERFECT RULER, RULER, TAYLOR'S CONDITION, WEIGHINGS

References

Atkinson, M. D.; Santoro, N.; and Urrutia, J. "Integer Sets with Distinct Sums and Differences and Carrier Frequency Assignments for Nonlinear Repeaters." *IEEE Trans. Comm.* **34**, 614–617, 1986.

Colbourn, C. J. and Dinitz, J. H. (Eds.) *CRC Handbook of Combinatorial Designs.* Boca Raton, FL: CRC Press, p. 315, 1996.

Guy, R. K. "Modular Difference Sets and Error Correcting Codes." §C10 in *Unsolved Problems in Number Theory, 2nd ed.* New York: Springer-Verlag, pp. 118–121, 1994.

Lam, A. W. and D. V. Sarwate, D. V. "On Optimum Time Hopping Patterns." *IEEE Trans. Comm.* **36**, 380–382, 1988.

Robinson, J. P. and Bernstein, A. J. "A Class of Binary Recurrent Codes with Limited Error Propagation." *IEEE Trans. Inform. Th. 13*, 106–113, 1967.

Sloane, N. J. A. Sequence A003022/M2540 in "An On-Line Version of the Encyclopedia of Integer Sequences."

Vanderschel, D. and Garry, M. "In Search of the Optimal 20 & 21 Mark Golomb Rulers." `http://members.aol.com/golomb20/`.

Golygon

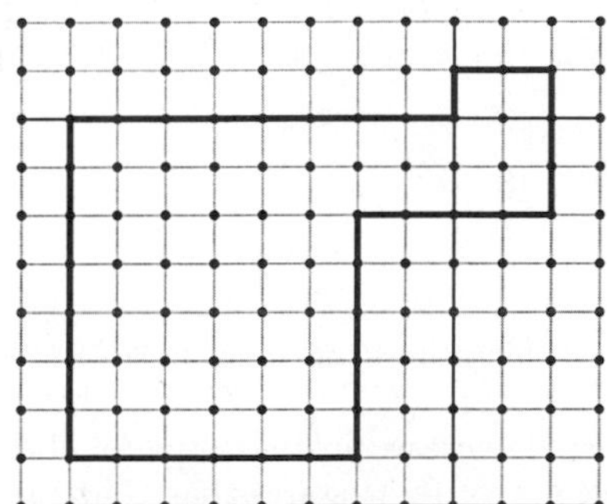

A PLANE path on a set of equally spaced LATTICE POINTS, starting at the ORIGIN, where the first step is one unit to the north or south, the second step is two units to the east or west, the third is three units to the north or south, etc., and continuing until the ORIGIN is again reached. No crossing or backtracking is allowed. The simplest golygon is (0, 0), (0, 1), (2, 1), (2, −2), (−2, −2), (−2, −7), (−8, −7), (−8, 0), (0, 0).

A golygon can be formed if there exists an EVEN INTEGER n such that

$$\pm 1 \pm 3 \pm \ldots \pm (n-1) = 0 \qquad (1)$$

$$\pm 2 \pm 4 \pm \ldots \pm n = 0 \qquad (2)$$

(Vardi 1991). Gardner proved that all golygons are of the form $n = 8k$. The number of golygons of length n (EVEN), with each initial direction counted separately, is the PRODUCT of the COEFFICIENT of $x^{n^2/8}$ in

$$(1+x)(1+x^3)\cdots(1+x^{n-1}), \qquad (3)$$

with the COEFFICIENT of $x^{n(n/2+1)/8}$ in

$$(1+x)(1+x^2)\cdots(1+x^{n/2}). \qquad (4)$$

The number of golygons $N(n)$ of length $8n$ for the first few n are 4, 112, 8432, 909288, ... (Sloane's A006718) and is asymptotic to

$$N(n) \sim \frac{3 \cdot 2^{8n-4}}{\pi n^2(4n+1)} \qquad (5)$$

(Sallows *et al.* 1991, Vardi 1991).

see also LATTICE PATH

References

Dudeney, A. K. "An Odd Journey Along Even Roads Leads to Home in Golygon City." *Sci. Amer.* **263**, 118–121, July 1990.

Sallows, L. C. F. "New Pathways in Serial Isogons." *Math. Intell.* **14**, 55–67, 1992.

Sallows, L.; Gardner, M.; Guy, R. K.; and Knuth, D. "Serial Isogons of 90 Degrees." *Math Mag.* **64**, 315–324, 1991.

Sloane, N. J. A. Sequence A006718/M3707 in "An On-Line Version of the Encyclopedia of Integer Sequences."

Vardi, I. "American Science." §5.3 in *Computational Recreations in Mathematica.* Redwood City, CA: Addison-Wesley, pp. 90–96, 1991.

Gomory's Theorem

Regardless of where one white and one black square are deleted from an ordinary 8×8 CHESSBOARD, the reduced board can always be covered exactly with 31 DOMINOES (of dimension 2×1).

see also CHESSBOARD

Gompertz Constant

$$G \equiv \int_0^\infty \frac{e^{-u}}{1+u}\,du = -e\,\mathrm{ei}(-1) = 0.596347362\ldots,$$

where $\mathrm{ei}(x)$ is the EXPONENTIAL INTEGRAL. Stieltjes showed it has the CONTINUED FRACTION representation

$$G = \frac{1}{2-}\frac{1^2}{4-}\frac{2^2}{6-}\frac{3^2}{8-}\cdots.$$

see also EXPONENTIAL INTEGRAL

References

Le Lionnais, F. *Les nombres remarquables.* Paris: Hermann, p. 29, 1983.

Gompertz Curve

The function defined by

$$y = ab^{q^x}.$$

It is used in actuarial science for specifying a simplified mortality law. Using $s(x)$ as the probability that a newborn will achieve age x, the Gompertz law (1825) is

$$s(x) = \exp[-m(c^x - 1)],$$

for $c > 1$, $x \geq 0$.

see also LIFE EXPECTANCY, LOGISTIC GROWTH CURVE, MAKEHAM CURVE, POPULATION GROWTH

References

Bowers, N. L. Jr.; Gerber, H. U.; Hickman, J. C.; Jones, D. A.; and Nesbitt, C. J. *Actuarial Mathematics.* Itasca, IL: Society of Actuaries, p. 71, 1997.

Gompertz, B. "On the Nature of the Function Expressive of the Law of Human Mortality." *Phil. Trans. Roy. Soc. London,* 1825.

Gonal Number

see POLYGONAL NUMBER

Good Path

see p-GOOD PATH

Good Prime

A PRIME p_n is called "good" if

$$p_n{}^2 > p_{n-i}p_{n+i}$$

for all $1 \leq i \leq n-1$ (there is a typo in Guy 1994 in which the is are replaced by 1s). There are infinitely many good primes, and the first few are 5, 11, 17, 29, 37, 41, 53, ... (Sloane's A028388).

see also ANDRICA'S CONJECTURE, PÓLYA CONJECTURE

References

Guy, R. K. "'Good' Primes and the Prime Number Graph." §A14 in *Unsolved Problems in Number Theory, 2nd ed.* New York: Springer-Verlag, pp. 32–33, 1994.

Sloane, N. J. A. Sequence A028388 in "An On-Line Version of the Encyclopedia of Integer Sequences."

Goodman's Formula

A two-coloring of a COMPLETE GRAPH K_n of n nodes which contains exactly the number of MONOCHROMATIC FORCED TRIANGLES and no more (i.e., a minimum of $R+B$ where R and B are the number of red and blue TRIANGLES) is called an EXTREMAL GRAPH. Goodman (1959) showed that for an extremal graph,

$$R+B = \begin{cases} \frac{1}{3}m(m-1)(m-2) & \text{for } n=2m \\ \frac{2}{3}m(m-1)(4m+1) & \text{for } n=4m+1 \\ \frac{2}{3}m(m+1)(4m-1) & \text{for } n=4m+3. \end{cases}$$

Schwenk (1972) rewrote the equation in the form

$$R+B = \binom{n}{3} - \left\lfloor \tfrac{1}{2}n \left\lfloor \tfrac{1}{4}(n-1)^2 \right\rfloor \right\rfloor,$$

where $\binom{n}{k}$ is a BINOMIAL COEFFICIENT and $\lfloor x \rfloor$ is the FLOOR FUNCTION.

see also BLUE-EMPTY GRAPH, EXTREMAL GRAPH, MONOCHROMATIC FORCED TRIANGLE

References

Goodman, A. W. "On Sets of Acquaintances and Strangers at Any Party." *Amer. Math. Monthly* **66**, 778–783, 1959.

Schwenk, A. J. "Acquaintance Party Problem." *Amer. Math. Monthly* **79**, 1113–1117, 1972.

Goodstein Sequence

Given a HEREDITARY REPRESENTATION of a number n in BASE, let $B[b](n)$ be the NONNEGATIVE INTEGER which results if we syntactically replace each b by $b+1$ (i.e., $B[b]$ is a base change operator that 'bumps the base' from b up to $b+1$). The HEREDITARY REPRESENTATION of 266 in base 2 is

$$\begin{aligned} 266 &= 2^8 + 2^3 + 2 \\ &= 2^{2^{2+1}} + 2^{2+1} + 2, \end{aligned}$$

so bumping the base from 2 to 3 yields

$$B[2](266) = 3^{3^{3+1}} + 3^{3+1} + 3.$$

Now repeatedly bump the base and subtract 1,

$$\begin{aligned} G_0(266) &= 266 = 2^{2^{2+1}} + 2^{2+1} + 2 \\ G_1(266) &= B[2](266) - 1 = 3^{3^{3+1}} + 3^{3+1} + 2 \\ G_2(266) &= B[3](G_1) - 1 = 4^{4^{4+1}} + 4^{4+1} + 1 \\ G_3(266) &= B[4](G_2) - 1 = 5^{5^{5+1}} + 5^{5+1} \\ G_4(266) &= B[5](G_3) - 1 = 6^{6^{6+1}} + 6^{6+1} - 1 \\ &= 6^{6^{6+1}} + 5\cdot 6^6 + 5\cdot 6^5 + \ldots + 5\cdot 6 + 5 \\ G_5(266) &= B[6](G_4) - 1 \\ &= 7^{7^{7+1}} + 5\cdot 7^7 + 5\cdot 7^5 + \ldots + 5\cdot 7 + 4, \end{aligned}$$

etc. Starting this procedure at an INTEGER n gives the Goodstein sequence $\{G_k(n)\}$. Amazingly, despite the apparent rapid increase in the terms of the sequence, GOODSTEIN'S THEOREM states that $G_k(n)$ is 0 for any n and any sufficiently large k.

see also GOODSTEIN'S THEOREM, HEREDITARY REPRESENTATION

Goodstein's Theorem

For all n, there exists a k such that the kth term of the GOODSTEIN SEQUENCE $G_k(n) = 0$. In other words, every GOODSTEIN SEQUENCE converges to 0.

The secret underlying Goodstein's theorem is that the HEREDITARY REPRESENTATION of n in base b mimics an ordinal notation for ordinals less than some number. For such ordinals, the base bumping operation leaves the ordinal fixed whereas the subtraction of one decreases the ordinal. But these ordinals are well-ordered, and this allows us to conclude that a Goodstein sequence eventually converges to zero.

Goodstein's theorem cannot be proved in PEANO ARITHMETIC (i.e., formal NUMBER THEORY).

see also NATURAL INDEPENDENCE PHENOMENON, PEANO ARITHMETIC

Googol

A LARGE NUMBER equal to 10^{100}, or

$$\begin{array}{c}10000000000000000000000000\\0000000000000000000000000\\0000000000000000000000000\\0000000000000000000000000.\end{array}$$

see also GOOGOLPLEX, LARGE NUMBER

References

Kasner, E. and Newman, J. R. *Mathematics and the Imagination.* Redmond, WA: Tempus Books, pp. 20–27, 1989.

Pappas, T. "Googol & Googolplex." *The Joy of Mathematics.* San Carlos, CA: Wide World Publ./Tetra, p. 76, 1989.

Googolplex

A LARGE NUMBER equal to $10^{10^{100}}$.

see also GOOGOL, LARGE NUMBER

References

Kasner, E. and Newman, J. R. *Mathematics and the Imagination.* Redmond, WA: Tempus Books, pp. 23–27, 1989.

Pappas, T. "Googol & Googolplex." *The Joy of Mathematics.* San Carlos, CA: Wide World Publ./Tetra, p. 76, 1989.

Gordon Function

Another name for the CONFLUENT HYPERGEOMETRIC FUNCTION OF THE SECOND KIND, defined by

$$G(a|c|z) = e^{i\pi a}\frac{\Gamma(c)}{\Gamma(a)}\left\{\frac{\Gamma(1-c)}{\Gamma(1-a)}\left[e^{-\pi c} + \frac{\sin[\pi(a-c)]}{\sin(\pi a)}\right]\right.$$
$$\left.\times {}_1F_1(a;c;z) - 2\frac{\Gamma(c-1)}{\Gamma(c-a)}z^{1-c}{}_1F_1(a-c+1;2-c;z)\right\},$$

where $\Gamma(x)$ is the GAMMA FUNCTION and ${}_1F_1(a;b;z)$ is the CONFLUENT HYPERGEOMETRIC FUNCTION OF THE FIRST KIND.

see also CONFLUENT HYPERGEOMETRIC FUNCTION OF THE SECOND KIND

References

Morse, P. M. and Feshbach, H. *Methods of Theoretical Physics, Part I.* New York: McGraw-Hill, pp. 671–672, 1953.

Gorenstein Ring

An algebraic RING which appears in treatments of duality in ALGEBRAIC GEOMETRY. Let A be a local ARTINIAN RING with $m \subset A$ its maximal IDEAL. Then A is a Gorenstein ring if the ANNIHILATOR of m has DIMENSION 1 as a VECTOR SPACE over $K = A/m$.

see also CAYLEY-BACHARACH THEOREM

References

Eisenbud, D.; Green, M.; and Harris, J. "Cayley-Bacharach Theorems and Conjectures." *Bull. Amer. Math. Soc.* **33**, 295–324, 1996.

Gosper's Algorithm

An ALGORITHM for finding closed form HYPERGEOMETRIC IDENTITIES The algorithm treats sums whose successive terms have ratios which are RATIONAL FUNCTIONS. Not only does it decide conclusively whether there exists a hypergeometric sequence z_n such that

$$t_n = z_{n+1} - z_n,$$

but actually produces z_n if it exists. If not, it produces $\sum_{k=0}^{n-1} t_n$. An outline of the algorithm follows (Petkovšek 1996):

1. For the ratio $r(n) = t_{n+1}/t_n$ which is a RATIONAL FUNCTION of n.
2. Write
$$r(n) = \frac{a(n)}{b(n)}\frac{c(n+1)}{c(n)},$$
where $a(n)$, $b(n)$, and $c(n)$ are polynomials satisfying
$$\mathrm{GCD}(a(n), b(n+h)) = 1$$
for all nonnegative integers h.
3. Find a nonzero polynomial solution $x(n)$ of
$$a(n)x(n+1) - b(n-1)x(n) = c(n),$$
if one exists.
4. Return $b(n-1)x(n)/c(n)t_n$ and stop.

Petkovšek *et al.* (1996) describe the algorithm as "one of the landmarks in the history of computerization of the problem of closed form summation." Gosper's algorithm is vital in the operation of ZEILBERGER'S ALGORITHM and the machinery of WILF-ZEILBERGER PAIRS.

see also HYPERGEOMETRIC IDENTITY, SISTER CELINE'S METHOD, WILF-ZEILBERGER PAIR, ZEILBERGER'S ALGORITHM

References

Gessel, I. and Stanton, D. "Strange Evaluations of Hypergeometric Series." *SIAM J. Math. Anal.* **13**, 295–308, 1982.

Gosper, R. W. "Decision Procedure for Indefinite Hypergeometric Summation." *Proc. Nat. Acad. Sci. USA* **75**, 40–42, 1978.

Graham, R. L.; Knuth, D. E.; and Patashnik, O. *Concrete Mathematics: A Foundation for Computer Science, 2nd ed.* Reading, MA: Addison-Wesley, 1994.

Lafron, J. C. "Summation in Finite Terms." In *Computer Algebra Symbolic and Algebraic Computation, 2nd ed.* (Ed. B. Buchberger, G. E. Collins, and R. Loos). New York: Springer-Verlag, 1983.
Paule, P. and Schorn, M. "A Mathematica Version of Zeilberger's Algorithm for Proving Binomial Coefficient Identities." *J. Symb. Comput.* **20**, 673–698, 1995.
Petkovšek, M.; Wilf, H. S.; and Zeilberger, D. "Gosper's Algorithm." Ch. 5 in *A=B*. Wellesley, MA: A. K. Peters, pp. 73–99, 1996.
Zeilberger, D. "The Method of Creative Telescoping." *J. Symb. Comput.* **11**, 195–204, 1991.

Gosper Island

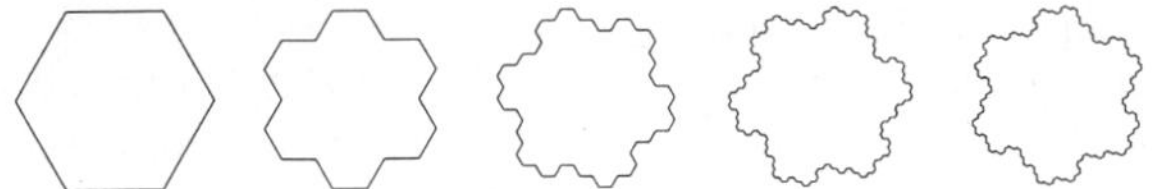

A modification of the KOCH SNOWFLAKE which has FRACTAL DIMENSION

$$D = \frac{2\ln 3}{\ln 7} = 1.12915\ldots.$$

The term "Gosper island" was used by Mandelbrot (1977) because this curve bounds the space filled by the PEANO-GOSPER CURVE; Gosper and Gardner use the term FLOWSNAKE FRACTAL instead. Gosper islands can TILE the PLANE.

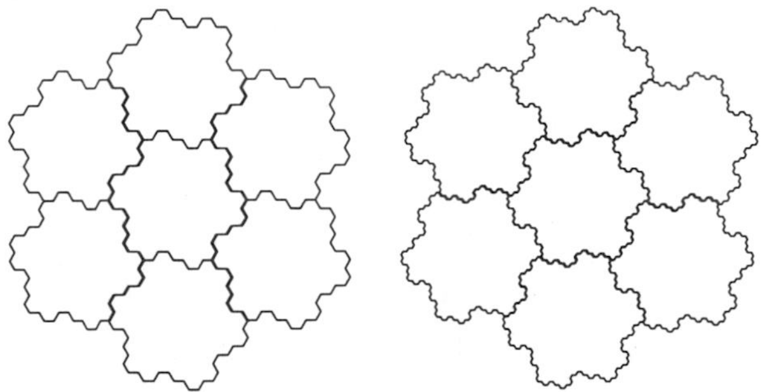

see also KOCH SNOWFLAKE, PEANO-GOSPER CURVE

References

Mandelbrot, B. B. *Fractals: Form, Chance, & Dimension.* San Francisco, CA: W. H. Freeman, Plate 46, 1977.

Gosper's Method

see GOSPER'S ALGORITHM

Graceful Graph

A LABELLED GRAPH which can be "gracefully numbered" is called a graceful graph. Label the nodes with distinct NONNEGATIVE INTEGERS. Then label the EDGES with the absolute differences between node values. If the EDGE numbers then run from 1 to e, the graph is gracefully numbered. In order for a graph to be graceful, it must be without loops or multiple EDGES.

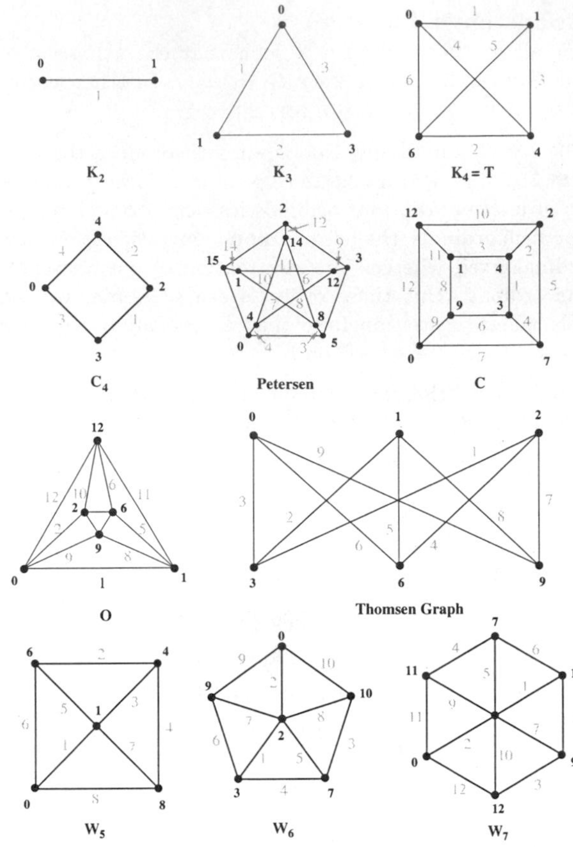

Golomb showed that the number of EDGES connecting the EVEN-numbered and ODD-numbered sets of nodes is $\lfloor (e+1)/2 \rfloor$, where e is the number of EDGES. In addition, if the nodes of a graph are all of EVEN ORDER, then the graph is graceful only if $\lfloor (e+1)/2 \rfloor$ is EVEN. The only ungraceful simple graphs with ≤ 5 nodes are shown below.

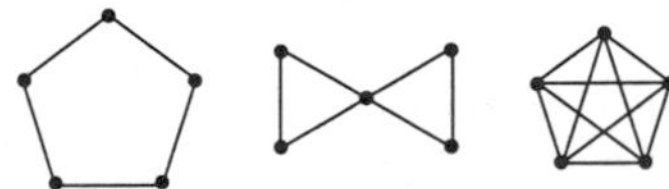

There are exactly $e!$ graceful graphs with e EDGES (Sheppard 1976), where $e!/2$ of these correspond to different labelings of the same graph. Golomb (1974) showed that all complete bipartite graphs are graceful. CATERPILLAR GRAPHS; COMPLETE GRAPHS K_2, K_3, $K_4 = W_4 = T$ (and only these; Golomb 1974); CYCLIC GRAPHS C_n when $n \equiv 0$ or 3 (mod 4), when the number of consecutive chords $k = 2$, 3, or $n - 3$ (Koh and Punim 1982), or when they contain a P_k chord (Delorme *et al.* 1980, Koh and Yap 1985, Punnim and Pabhapote 1987); GEAR GRAPHS; PATH GRAPHS; the PETERSEN GRAPH; POLYHEDRAL GRAPHS $T = K_4 = W_4$, C, O, D, and I (Gardner 1983); STAR GRAPHS; the THOMSEN GRAPH (Gardner 1983); and WHEEL GRAPHS (Frucht 1988) are all graceful.

Some graceful graphs have only one numbering, but others have more than one. It is conjectured that all trees are graceful (Bondy and Murty 1976), but this has only been proved for trees with ≤ 16 VERTICES. It has also been conjectured that all unicyclic graphs are graceful.

An excellent on-line resource is Brundage (`http://www.math.washington.edu/~brundage/oldgraceful/`).

see also HARMONIOUS GRAPH, LABELLED GRAPH

References

Abraham, J. and Kotzig, A. "All 2-Regular Graphs Consisting of 4-Cycles are Graceful." *Disc. Math.* **135**, 1–24, 1994.

Abraham, J. and Kotzig, A. "Extensions of Graceful Valuations of 2-Regular Graphs Consisting of 4-Gons." *Ars Combin.* **32**, 257–262, 1991.

Bloom, G. S. and Golomb, S. W. "Applications of Numbered Unidirected Graphs." *Proc. IEEE* **65**, 562–570, 1977.

Bolian, L. and Xiankun, Z. "On Harmonious Labellings of Graphs." *Ars Combin.* **36**, 315–326, 1993.

Brualdi, R. A. and McDougal, K. F. "Semibandwidth of Bipartite Graphs and Matrices." *Ars Combin.* **30**, 275–287, 1990.

Brundage, M. "Graceful Graphs." `http://www.math.washington.edu/~brundage/oldgraceful/`.

Cahit, I. "Are All Complete Binary Trees Graceful?" *Amer. Math. Monthly* **83**, 35–37, 1976.

Delorme, C.; Maheo, M.; Thuillier, H.; Koh, K. M.; and Teo, H. K. "Cycles with a Chord are Graceful." *J. Graph Theory* **4**, 409–415, 1980.

Frucht, R. W. and Gallian, J. A. "Labelling Prisms." *Ars Combin.* **26**, 69–82, 1988.

Gallian, J. A. "A Survey: Recent Results, Conjectures, and Open Problems in Labelling Graphs." *J. Graph Th.* **13**, 491–504, 1989.

Gallian, J. A. "Open Problems in Grid Labeling." *Amer. Math. Monthly* **97**, 133–135, 1990.

Gallian, J. A. "A Guide to the Graph Labelling Zoo." *Disc. Appl. Math.* **49**, 213–229, 1994.

Gallian, J. A.; Prout, J.; and Winters, S. "Graceful and Harmonious Labellings of Prism Related Graphs." *Ars Combin.* **34**, 213–222, 1992.

Gardner, M. "Golomb's Graceful Graphs." Ch. 15 in *Wheels, Life, and Other Mathematical Amusements.* New York: W. H. Freeman, pp. 152–165, 1983.

Golomb, S. W. "The Largest Graceful Subgraph of the Complete Graph." *Amer. Math. Monthly* **81**, 499–501, 1974.

Guy, R. "Monthly Research Problems, 1969–75." *Amer. Math. Monthly* **82**, 995–1004, 1975.

Guy, R. "Monthly Research Problems, 1969–1979." *Amer. Math. Monthly* **86**, 847–852, 1979.

Guy, R. K. "The Corresponding Modular Covering Problem. Harmonious Labelling of Graphs." §C13 in *Unsolved Problems in Number Theory, 2nd ed.* New York: Springer-Verlag, pp. 127–128, 1994.

Huang, J. H. and Skiena, S. "Gracefully Labelling Prisms." *Ars Combin.* **38**, 225–242, 1994.

Koh, K. M. and Punnim, N. "On Graceful Graphs: Cycles with 3-Consecutive Chords." *Bull. Malaysian Math. Soc.* **5**, 49–64, 1982.

Jungreis, D. S. and Reid, M. "Labelling Grids." *Ars Combin.* **34**, 167–182, 1992.

Koh, K. M. and Yap, K. Y. "Graceful Numberings of Cycles with a P_3-Chord." *Bull. Inst. Math. Acad. Sinica* **13**, 41–48, 1985.

Morris, P. A. "On Graceful Trees." `http://www.math.washington.edu/~brundage/math/graceful/source/on_graceful_trees.ps`.

Moulton, D. "Graceful Labellings of Triangular Snakes." *Ars Combin.* **28**, 3–13, 1989.

Murty, U. S. R. and Bondy, J. A. *Graph Theory with Applications.* New York: North Holland, p. 248, 1976.

Punnim, N. and Pabhapote, N. "On Graceful Graphs: Cycles with a P_k-Chord, $k \geq 4$." *Ars Combin. A* **23**, 225–228, 1987.

Rosa, A. "On Certain Valuations of the Vertices of a Graph." In *Theory of Graphs, International Symposium, Rome, July 1966.* New York: Gordon and Breach, pp. 349–355, 1967.

Sheppard, D. A. "The Factorial Representation of Balanced Labelled Graphs." *Discr. Math.* **15**, 379–388, 1976.

Sierksma, G. and Hoogeveen, H. "Seven Criteria for Integer Sequences Being Graphic." *J. Graph Th.* **15**, 223–231, 1991.

Slater, P. J. "Note on k-Graceful, Locally Finite Graphs." *J. Combin. Th. Ser. B* **35**, 319–322, 1983.

Snevily, H. S. "New Families of Graphs That Have α-Labellings." Preprint.

Snevily, H. S. "Remarks on the Graceful Tree Conjecture." Preprint.

Xie, L. T. and Liu, G. Z. "A Survey of the Problem of Graceful Trees." *Qufu Shiyuan Xuebao* **1**, 8–15, 1984.

Graded Algebra

If A is a graded module and there EXISTS a degree-preserving linear map $\phi : A \otimes A \to A$, then (A, ϕ) is called a graded algebra.

References

Jacobson, N. *Lie Algebras.* New York: Dover, p. 163, 1979.

Gradian

A unit of angular measure in which the angle of an entire CIRCLE is 400 gradians. A RIGHT ANGLE is therefore 100 gradians.

see also DEGREE, RADIAN

Gradient

The gradient is a VECTOR operator denoted ∇ and sometimes also called DEL or NABLA. It most often is applied to a real function of three variables $f(u_1, u_2, u_3)$, and may be denoted

$$\nabla f \equiv \text{grad(f)}. \tag{1}$$

For general CURVILINEAR COORDINATES, the gradient is given by

$$\nabla\phi = \frac{1}{h_1}\frac{\partial\phi}{\partial u_1}\hat{\mathbf{u}}_1 + \frac{1}{h_2}\frac{\partial\phi}{\partial u_2}\hat{\mathbf{u}}_2 + \frac{1}{h_3}\frac{\partial\phi}{\partial u_3}\hat{\mathbf{u}}_3, \tag{2}$$

which simplifies to

$$\nabla\phi(x, y, z) = \frac{\partial\phi}{\partial x}\hat{\mathbf{x}} + \frac{\partial\phi}{\partial y}\hat{\mathbf{y}} + \frac{\partial\phi}{\partial z}\hat{\mathbf{z}} \tag{3}$$

in CARTESIAN COORDINATES.

The direction of ∇f is the orientation in which the DIRECTIONAL DERIVATIVE has the largest value and $|\nabla f|$ is the value of that DIRECTIONAL DERIVATIVE. Furthermore, if $\nabla f \neq 0$, then the gradient is PERPENDICULAR to the LEVEL CURVE through (x_0, y_0) if $z = f(x, y)$ and PERPENDICULAR to the level surface through (x_0, y_0, z_0) if $F(x, y, z) = 0$.

In TENSOR notation, let

$$ds^2 = g_\mu \, dx_\mu{}^2 \tag{4}$$

be the LINE ELEMENT in principal form. Then

$$\nabla_{\vec{e}_\alpha} \vec{e}_\beta = \nabla_\alpha \vec{e}_\beta = \frac{1}{\sqrt{g_\alpha}} \frac{\partial}{\partial x_\alpha} \vec{e}_\beta. \tag{5}$$

For a MATRIX A,

$$\nabla |\mathsf{Ax}| = \frac{(\mathsf{Ax})^{\mathrm{T}} \mathsf{A}}{|\mathsf{Ax}|}. \tag{6}$$

For expressions giving the gradient in particular coordinate systems, see CURVILINEAR COORDINATES.

see also CONVECTIVE DERIVATIVE, CURL, DIVERGENCE, LAPLACIAN, VECTOR DERIVATIVE

References

Arfken, G. "Gradient, ∇" and "Successive Applications of ∇." §1.6 and 1.9 in *Mathematical Methods for Physicists, 3rd ed.* Orlando, FL: Academic Press, pp. 33–37 and 47–51, 1985.

Gradient Four-Vector

The 4-dimensional version of the GRADIENT, encountered frequently in general relativity and special relativity, is

$$\nabla_\mu = \begin{bmatrix} \frac{1}{c}\frac{\partial}{\partial t} \\ \frac{\partial}{\partial x} \\ \frac{\partial}{\partial y} \\ \frac{\partial}{\partial z} \end{bmatrix},$$

which can be written

$$(\nabla^\mu)^2 \equiv \Box^2,$$

where $\Box^2$ is the D'ALEMBERTIAN OPERATOR.

see also D'ALEMBERTIAN OPERATOR, GRADIENT, TENSOR, VECTOR

References

Morse, P. M. and Feshbach, H. "The Differential Operator ∇." §1.4 in *Methods of Theoretical Physics, Part I.* New York: McGraw-Hill, pp. 31–44, 1953.

Gradient Theorem

$$\int_b^a (\nabla f) \cdot d\mathbf{s} = f(b) - f(a),$$

where ∇ is the GRADIENT, and the integral is a LINE INTEGRAL. It is this relationship which makes the definition of a scalar potential function f so useful in gravitation and electromagnetism as a concise way to encode information about a VECTOR FIELD.

see also DIVERGENCE THEOREM, GREEN'S THEOREM, LINE INTEGRAL

Graeco-Latin Square

see EULER SQUARE

Graeco-Roman Square

see EULER SQUARE

Graeffe's Method

A ROOT-finding method which proceeds by multiplying a POLYNOMIAL $f(x)$ by $f(-x)$ and noting that

$$f(x) = (x - a_1)(x - a_2) \cdots (x - a_n) \tag{1}$$

$$f(-x) = (-1)^n (x + a_1)(x + a_2) \cdots (x + a_n) \tag{2}$$

so the result is

$$f(x) f(-x) = (-1)^n (x^2 - a_1{}^2)(x^2 - a_2{}^2) \cdots (x^2 - a_n{}^2). \tag{3}$$

Repeat ν times, then write this in the form

$$y^n + b_1 y^{n-1} + \ldots + b_n = 0 \tag{4}$$

where $y \equiv x^{2\nu}$. Since the coefficients are given by NEWTON'S RELATIONS

$$b_1 = -(y_1 + y_2 + \ldots + y_n) \tag{5}$$

$$b_2 = (y_1 y_2 + y_1 y_3 + \ldots + y_{n-1} y_n) \tag{6}$$

$$b_n = (-1)^n y_1 y_2 \cdots y_n, \tag{7}$$

and since the squaring procedure has separated the roots, the first term is larger than rest. Therefore,

$$b_1 \approx -y_1 \tag{8}$$

$$b_2 \approx y_1 y_2 \tag{9}$$

$$b_n \approx (-1)^n y_1 y_2 \cdots y_n, \tag{10}$$

giving

$$y_1 \approx -b_1 \tag{11}$$

$$y_2 \approx -\frac{b_2}{b_1} \tag{12}$$

$$y_n \approx -\frac{b_n}{b_{n-1}}. \tag{13}$$

Solving for the original roots gives

$$a_1 \approx \sqrt[2\nu]{-b_1} \tag{14}$$

$$a_2 \approx \sqrt[2\nu]{-\frac{b_2}{b_1}} \tag{15}$$

$$a_n \approx \sqrt[2\nu]{-\frac{b_n}{b_{n-1}}}. \tag{16}$$

This method works especially well if all roots are real.

References

von Kármán, T. and Biot, M. A. "Squaring the Roots (Graeffe's Method)." §5.8.c in *Mathematical Methods in Engineering: An Introduction to the Mathematical Treatment of Engineering Problems.* New York: McGraw-Hill, pp. 194–196, 1940.

Graham's Biggest Little Hexagon

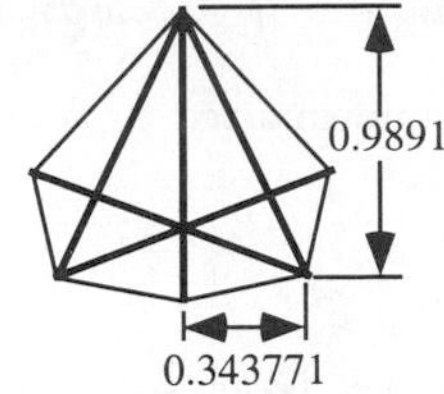

The largest possible (not necessarily regular) HEXAGON for which no two of the corners are more than unit distance apart. In the above figure, the heavy lines are all of unit length. The AREA of the hexagon is $A = 0.674981\ldots$, where A is a ROOT of

$$4096A^{10} - 8192A^9 - 3008A^8 - 30,848A^7 + 21,056A^6 + 146,496A^5 - 221,360A^4 + 1232A^3 + 144,464A^2 - 78,488A + 11,993 = 0.$$

see also CALABI'S TRIANGLE

References

Conway, J. H. and Guy, R. K. "Graham's Biggest Little Hexagon." In *The Book of Numbers.* New York: Springer-Verlag, pp. 206–207, 1996.

Graham, R. L. "The Largest Small Hexagon." *J. Combin. Th. Ser. A* **18**, 165–170, 1975.

Graham's Number

The smallest dimension of a HYPERCUBE such that if the lines joining all pairs of corners are two-colored, a PLANAR COMPLETE GRAPH K_4 of one color will be forced. That an answer exists was proved by R. L. Graham and B. L. Rothschild. The actual answer is believed to be 6, but the best bound proved is

$$64\left\{\begin{array}{c}\underbrace{3\uparrow\uparrow\uparrow\uparrow 3}\\ \underbrace{3\uparrow 3}\\ \vdots\\ \underbrace{}\\ 3\uparrow 3,\end{array}\right.$$

where $\uparrow$ is stacked ARROW NOTATION. It is less than $3 \to 3 \to 3 \to 3$, where CHAINED ARROW NOTATION has been used.

see also ARROW NOTATION, CHAINED ARROW NOTATION, SKEWES NUMBER

References

Conway, J. H. and Guy, R. K. *The Book of Numbers.* New York: Springer-Verlag, pp. 61–62, 1996.

Gardner, M. "Mathematical Games." *Sci. Amer.* **237**, 18–28, Nov. 1977.

Gram-Charlier Series

Approximates a distribution in terms of a NORMAL DISTRIBUTION. Let

$$\phi(t) \equiv \frac{1}{\sqrt{2\pi}} e^{-t^2/2},$$

then

$$f(t) = \phi(t) + \tfrac{1}{6}\gamma_1\phi^{(3)}(t) + \tfrac{1}{24}\gamma_2\phi^{(4)}(t) + \ldots.$$

see also EDGEWORTH SERIES

References

Kenney, J. F. and Keeping, E. S. *Mathematics of Statistics, Pt. 2, 2nd ed.* Princeton, NJ: Van Nostrand, pp. 107–108, 1951.

Gram Determinant

The DETERMINANT

$$G(f_1, f_2, \ldots, f_n) = \begin{vmatrix} \int f_1^2\,dt & \int f_1 f_2\,dt & \cdots & \int f_1 f_n\,dt \\ \int f_2 f_1\,dt & \int f_2^2\,dt & \cdots & \int f_2 f_n\,dt \\ \vdots & \vdots & \ddots & \vdots \\ \int f_1 f_n\,dt & \int f_1 f_n\,dt & \cdots & \int f_n^2\,dt \end{vmatrix}.$$

see also GRAM-SCHMIDT ORTHONORMALIZATION, WRONSKIAN

References

Sansone, G. *Orthogonal Functions, rev. English ed.* New York: Dover, p. 2, 1991.

Gram's Inequality

Let $f_1(x), \ldots, f_n(x)$ be REAL INTEGRABLE FUNCTIONS over the CLOSED INTERVAL $[a, b]$, then the DETERMINANT of their integrals satisfies

$$\begin{vmatrix} \int_a^b {f_1}^2(x)\,dx & \int_a^b f_1(x)f_2(x)\,dx & \cdots & \int_a^b f_1(x)f_n(x)\,dx \\ \int_a^b f_2(x)f_1(x)\,dx & \int_a^b {f_2}^2(x)\,dx & \cdots & \int_a^b f_2(x)f_n(x)\,dx \\ \vdots & \vdots & \ddots & \vdots \\ \int_a^b f_n(x)f_1(x)\,dx & \int_a^b f_n(x)f_2(x)\,dx & \cdots & \int_a^b f_n(x)f_n(x)\,dx \end{vmatrix} \geq 0.$$

see also GRAM-SCHMIDT ORTHONORMALIZATION

References

Gradshteyn, I. S. and Ryzhik, I. M. *Tables of Integrals, Series, and Products, 5th ed.* San Diego, CA: Academic Press, p. 1100, 1979.

Gram Matrix

Given m points with n-D vector coordinates $\mathbf{v}_i$, let M be the $n \times m$ matrix whose jth column consists of the coordinates of the vector $\mathbf{v}_j$, with $j = 1, \ldots, m$. Then define the $m \times m$ Gram matrix of dot products $a_{ij} = \mathbf{v}_i \cdot \mathbf{v}_j$ as

$$\mathsf{A} = \mathsf{M}^{\mathrm{T}}\mathsf{M},$$

where A^{T} denotes the TRANSPOSE. The Gram matrix determines the vectors $\mathbf{v}_i$ up to ISOMETRY.

Gram-Schmidt Orthonormalization

A procedure which takes a nonorthogonal set of LINEARLY INDEPENDENT functions and constructs an ORTHOGONAL BASIS over an arbitrary interval with respect to an arbitrary WEIGHTING FUNCTION $w(x)$. Given an original set of linearly independent functions $\{u_n\}$, let $\{\psi_n\}$ denote the orthogonalized (but not normalized) functions and $\{\phi_n\}$ the orthonormalized functions.

$$\psi_0(x) \equiv u_1(x) \tag{1}$$

$$\phi_0(x) \equiv \frac{\psi_0(x)}{\sqrt{\int \psi_0{}^2(x)w(x)\,dx}}. \tag{2}$$

Take

$$\psi_1(x) = u_1(x) + a_{10}\phi_0(x), \tag{3}$$

where we require

$$\int \psi_1\phi_0 w\,dx = \int u_1\phi_0 w\,dx + a_{10}\int \phi_0{}^2 w\,dx = 0. \tag{4}$$

By definition,

$$\int \phi_0{}^2 w\,dx = 1, \tag{5}$$

so

$$a_{10} = -\int u_1\phi_0 w\,dx. \tag{6}$$

The first orthogonalized function is therefore

$$\psi_1 = u_1(x) - \left[\int u_1\phi_0 w\,dx\right]\phi_0, \tag{7}$$

and the corresponding normalized function is

$$\phi_1 = \frac{\psi_1(x)}{\sqrt{\int \psi_1{}^2 w\,dx}}. \tag{8}$$

By mathematical induction, it follows that

$$\phi_i(x) = \frac{\psi_i(x)}{\sqrt{\int \psi_i{}^2 w\,dx}}, \tag{9}$$

where

$$\psi_i(x) = u_i + a_{i0}\phi_0 + a_{i1}\phi_1 \ldots + a_{i,i-1}\phi_{i-1} \tag{10}$$

and

$$a_{ij} \equiv -\int u_i\phi_j w\,dx. \tag{11}$$

If the functions are normalized to N_j instead of 1, then

$$\int_a^b [\phi_j(x)]^2 w\,dx = N_j{}^2 \tag{12}$$

$$\phi_i(x) = N_i\frac{\psi_i(x)}{\sqrt{\int \psi_i{}^2 w\,dx}} \tag{13}$$

$$a_{ij} = -\frac{\int u_i\phi_j w\,dx}{N_j{}^2}. \tag{14}$$

ORTHOGONAL POLYNOMIALS are especially easy to generate using GRAM-SCHMIDT ORTHONORMALIZATION. Use the notation

$$\langle x_i|x_j\rangle \equiv \langle x_i|w|x_j\rangle \equiv \int_a^b x_i(x)x_j(x)w(x)\,dx, \tag{15}$$

where $w(x)$ is a WEIGHTING FUNCTION, and define the first few POLYNOMIALS,

$$p_0(x) \equiv 1 \tag{16}$$

$$p_1(x) = \left[x - \frac{\langle xp_0|p_0\rangle}{\langle p_0|p_0\rangle}\right]p_0. \tag{17}$$

As defined, p_0 and p_1 are ORTHOGONAL POLYNOMIALS, as can be seen from

$$\begin{aligned}\langle p_0|p_1\rangle &= \left\langle\left[x - \frac{\langle xp_0|p_0\rangle}{\langle p_0|p_0\rangle}\right]p_0\right\rangle\\ &= \langle xp_0\rangle - \frac{\langle xp_0|p_0\rangle}{\langle p_0|p_0\rangle}\langle p_0\rangle\\ &= \langle xp_0\rangle - \langle xp_0\rangle = 0.\end{aligned} \tag{18}$$

Now use the RECURRENCE RELATION

$$p_{i+1}(x) = \left[x - \frac{\langle xp_i|p_i\rangle}{\langle p_i|p_i\rangle}\right]p_i - \left[\frac{\langle p_i|p_i\rangle}{\langle p_{i-1}|p_{i-1}\rangle}\right]p_{i-1} \tag{19}$$

to construct all higher order POLYNOMIALS.

To verify that this procedure does indeed produce ORTHOGONAL POLYNOMIALS, examine

$$\begin{aligned}\langle p_{i+1}|p_i\rangle &= \left\langle\left[x - \frac{\langle xp_i|p_i\rangle}{\langle p_i|p_i\rangle}\right]p_i\middle|p_i\right\rangle\\ &\quad - \left\langle\frac{\langle p_i|p_i\rangle}{\langle p_{i-1}|p_{i-1}\rangle}p_{i-1}\middle|p_i\right\rangle\\ &= \langle xp_i|p_i\rangle - \frac{\langle xp_i|p_i\rangle}{\langle p_i|p_i\rangle}\langle p_i|p_i\rangle\\ &\quad - \frac{\langle p_i|p_i\rangle}{\langle p_{i-1}|p_{i-1}\rangle}\langle p_{i-1}|p_i\rangle\\ &= -\frac{\langle p_i|p_i\rangle}{\langle p_{i-1}|p_{i-1}\rangle}\langle p_{i-1}|p_i\rangle\\ &= -\frac{\langle p_i|p_i\rangle}{\langle p_{i-1}|p_{i-1}\rangle}\left[-\frac{\langle p_{i-1}|p_{j-1}\rangle}{\langle p_{j-2}|p_{j-2}\rangle}\langle p_{j-2}|p_{j-1}\rangle\right]\\ &= \ldots = (-1)^j\frac{\langle p_j|p_j\rangle}{\langle p_0|p_0\rangle}\langle p_0|p_1\rangle = 0,\end{aligned} \tag{20}$$

since $\langle p_0|p_1\rangle = 0$. Therefore, all the POLYNOMIALS $p_i(x)$ are orthogonal. Many common ORTHOGONAL POLYNOMIALS of mathematical physics can be generated in this manner. However, the process is numerically unstable (Golub and van Loan 1989).

see also GRAM DETERMINANT, GRAM'S INEQUALITY, ORTHOGONAL POLYNOMIALS

References
Arfken, G. "Gram-Schmidt Orthogonalization." §9.3 in *Mathematical Methods for Physicists, 3rd ed.* Orlando, FL: Academic Press, pp. 516–520, 1985.
Golub, G. H. and van Loan, C. F. *Matrix Computations, 3rd ed.* Baltimore, MD: Johns Hopkins, 1989.

Gram Series

$$R(x) = 1 + \sum_{k=1}^{\infty} \frac{(\ln x)^k}{k k! \zeta(k+1)},$$

where ζ is the RIEMANN ZETA FUNCTION. This approximation to the PRIME COUNTING FUNCTION is 10 times better than $\mathrm{Li}(x)$ for $x < 10^9$ but has been proven to be worse infinitely often by Littlewood (Ingham 1990). An equivalent formulation due to Ramanujan is

$$G(x) \equiv \frac{4}{\pi} \sum_{k=1}^{\infty} \frac{(-1)^{k-1} k}{B_{2k}(2k-1)} \left(\frac{\ln x}{2\pi}\right)^{2k-1} \sim \pi(x)$$

(Berndt 1994), where B_{2k} is a BERNOULLI NUMBER. The integral analog, also found by Ramanujan, is

$$J(x) \equiv \int_0^{\infty} \frac{(\ln x)^t \, dt}{t\Gamma(t+1)\zeta(t+1)} \sim \pi(x)$$

(Berndt 1994).

References
Berndt, B. C. *Ramanujan's Notebooks, Part IV.* New York: Springer-Verlag, pp. 124–129, 1994.
Gram, J. P. "Undersøgelser angaaende Maengden af Primtal under en given Graeense." *K. Videnskab. Selsk. Skr.* **2**, 183–308, 1884.
Ingham, A. E. Ch. 5 in *The Distribution of Prime Numbers.* New York: Cambridge, 1990.
Vardi, I. *Computational Recreations in Mathematica.* Reading, MA: Addison-Wesley, p. 74, 1991.

Granny Knot

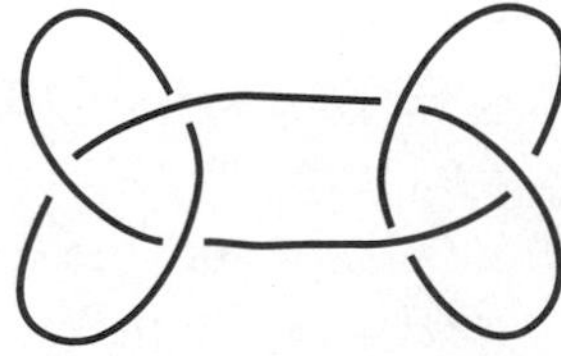

A COMPOSITE KNOT of seven crossings consisting of a KNOT SUM of TREFOILS. The granny knot has the same ALEXANDER POLYNOMIAL $(x^2 - x + 1)^2$ as the SQUARE KNOT.

Graph (Function)

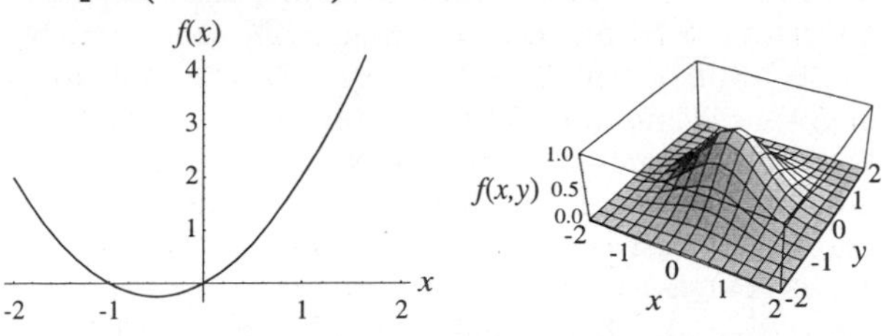

Technically, the graph of a function is its RANGE (a.k.a. image). Informally, given a FUNCTION $f(x_1, \ldots, x_n)$ defined on a DOMAIN U, the graph of f is defined as a CURVE or SURFACE showing the values taken by f over U (or some portion of U),

$$\mathrm{graph}\, f(x) \equiv \{(x, F(x)) \in \mathbb{R}^2 : x \in U\}$$

$$\mathrm{graph}\, f(x_1, \ldots, x_n) \equiv \{(x_1, \ldots, x_n, f(x_1, \ldots, x_n)) \in \mathbb{R}^{n+1} : (x_1, \ldots, x_n) \in U\}.$$

A graph is sometimes also called a PLOT.

Good routines for plotting graphs use adaptive algorithms which plot more points in regions where the function varies most rapidly (Wagon 1991, Math Works 1992, Heck 1993, Wickham-Jones 1994).

see also CURVE, EXTREMUM, GRAPH (GRAPH THEORY), HISTOGRAM, MAXIMUM, MINIMUM

References
Cleveland, W. S. *The Elements of Graphing Data, rev. ed.* Summit, NJ: Hobart, 1994.
Heck, A. *Introduction to Maple, 2nd ed.* New York: Springer-Verlag, pp. 303–304, 1993.
Math Works. *Matlab Reference Guide.* Natick, MA: The Math Works, p. 216, 1992.
Tufte, E. R. *The Visual Display of Quantitative Information.* Cheshire, CN: Graphics Press, 1983.
Tufte, E. R. *Envisioning Information.* Cheshire, CN: Graphics Press, 1990.
Wagon, S. *Mathematica in Action.* New York: W. H. Freeman, pp. 24–25, 1991.
Wickham-Jones, T. *Computer Graphics with Mathematica.* Santa Clara, CA: TELOS, pp. 579–584, 1994.
Yates, R. C. "Sketching." *A Handbook on Curves and Their Properties.* Ann Arbor, MI: J. W. Edwards, pp. 188–205, 1952.

Graph (Graph Theory)

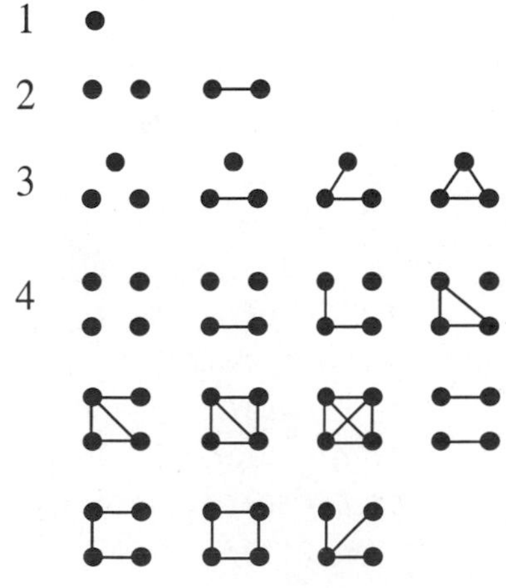

A mathematical object composed of points known as VERTICES or NODES and lines connecting some (possibly empty) SUBSET of them, known as EDGES. The study of graphs is known as GRAPH THEORY. Graphs are 1-D COMPLEXES, and there are always an EVEN number of ODD NODES in a graph. The number of nonisomorphic graphs with v NODES is given by the PÓLYA ENUMERATION THEOREM. The first few values for $n = 1, 2, \ldots$, are 1, 2, 4, 11, 34, 156, 1044, ... (Sloane's A000088; see above figure).

Graph sums, differences, powers, and products can be defined, as can graph eigenvalues.

Before applying PÓLYA ENUMERATION THEOREM, define the quantity

$$h_{\mathbf{j}} = \frac{p!}{\prod_{i=1}^{p} i^{j_i} j_i!}, \tag{1}$$

where $p!$ is the FACTORIAL of p, and the related polynomial

$$Z_p(S) = \sum_i h_{\mathbf{j}_i} \prod_{k=1}^{p} f_k{}^{(\mathbf{j}_i)_k}, \tag{2}$$

where the $\mathbf{j}_i = (j_1, \ldots, j_p)_i$ are all of the p-VECTORS satisfying

$$j_1 + 2j_2 + 3j_3 + \ldots + pj_p = p. \tag{3}$$

For example, for $p = 3$, the three possible values of $\mathbf{j}$ are

$$\mathbf{j}_1 = (3,0,0), \text{ since } (1\cdot 3) + (2\cdot 0) + (3\cdot 0) = 3,$$

$$\text{giving } h_{\mathbf{j}_1} = \frac{3!}{(1^3 3!)(2^0 0!)(3^0 0!)} = 1 \tag{4}$$

$$\mathbf{j}_2 = (1,1,0), \text{ since } (1\cdot 1) + (2\cdot 1) + (3\cdot 0) = 3,$$

$$\text{giving } h_{\mathbf{j}_2} = \frac{3!}{(1^1 1!)(2^1 1!)(3^0 0!)} = 3, \tag{5}$$

$$\mathbf{j}_3 = (0,0,1), \text{ since } (1\cdot 0) + (2\cdot 0) + (3\cdot 1) = 3$$

$$\text{giving } h_{\mathbf{j}_3} = \frac{3!}{(1^0 0!)(2^0 0!)(3^1 1!)} = 2. \tag{6}$$

Therefore,

$$Z_3(S) = f_1{}^3 + 3f_1 f_2 + 2f_3. \tag{7}$$

For small p, the first few values of $Z_p(S)$ are given by

$$Z_2(S) = f_1{}^2 + f_2 \tag{8}$$

$$Z_3(S) = f_1{}^3 + 3f_1 f_2 + 2f_3 \tag{9}$$

$$Z_4(S) = f_1{}^4 + 6f_1{}^2 f_2 + 3f_2{}^2 + 8f_1 f_3 + 6f_4 \tag{10}$$

$$Z_5(S) = f_1{}^5 + 10f_1{}^3 f_2 + 15f_1 f_2{}^2 + 20f_1{}^2 f_3 + 20f_2 f_3 + 30f_1 f_4 + 24f_5 \tag{11}$$

$$Z_6(S) = f_1{}^6 + 15f_1{}^4 f_2 + 45f_1{}^2 f_2{}^2 + 15f_2{}^3 + 40f_1{}^3 f_3 + 120f_1 f_2 f_3 + 40f_3{}^2 + 90f_1{}^2 f_4 + 90f_2 f_4 + 144f_1 f_5 + 120f_6 \tag{12}$$

$$Z_7(S) = f_1{}^7 + 21f_1{}^5 f_2 + 105f_1{}^3 f_2{}^2 + 105f_1 f_2{}^3 + 70f_1{}^4 f_3 + 420f_1{}^2 f_2 f_3 + 210f_2{}^2 f_3 + 280f_1 f_3{}^2 + 210f_1{}^3 f_4 + 630f_1 f_2 f_4 + 420f_3 f_4 + 504f_1{}^2 f_5 + 504f_2 f_5 + 840f_1 f_6 + 720f_7. \tag{13}$$

Application of the PÓLYA ENUMERATION THEOREM then gives the formula

$$Z(R) = \frac{1}{p!} \sum_{(j)} h_{\mathbf{j}} \prod_{n=0}^{\lfloor (p-1)/2 \rfloor} g_{2n+1}{}^{n j_{2n+1} + (2n+1)\binom{j_{2n+1}}{2}} \times \prod_{n=1}^{\lfloor p/2 \rfloor} [(g_n g_{2n})^{n-1}]^{j_{2n}} g_{2n}{}^{2n\binom{j_{2n}}{2}} \times \prod_{q=1}^{p} \prod_{r=q+1}^{p} g_{\mathrm{LCM}(q,r)}{}^{j_q j_r \mathrm{GCD}(q,r)}, \tag{14}$$

where $\lfloor x \rfloor$ is the FLOOR FUNCTION, $\binom{n}{m}$ is a BINOMIAL COEFFICIENT, LCM is the LEAST COMMON MULTIPLE, GCD is the GREATEST COMMON DIVISOR, and the SUM (j) is over all $\mathbf{j}_i$ satisfying the sum identity described above. The first few generating functions $Z_p(R)$ are

$$Z_2(R) = 2g_1 \tag{15}$$

$$Z_3(R) = g_1{}^3 + 3g_1 g_2 + 2g_3 \tag{16}$$

$$Z_4(R) = g_1{}^6 + 9g_1{}^2 g_2{}^2 + 8g_3{}^2 + 6g_2 g_4 \tag{17}$$

$$Z_5(R) = g_1{}^{10} + 10g_1{}^4 g_2{}^3 + 15g_1{}^2 g_2{}^4 + 20g_1 g_3{}^3 + 30g_2 g_4{}^2 + 24g_5{}^2 + 20g_1 g_3 g_6 \tag{18}$$

$$Z_6(R) = g_1{}^{15} + 15g_1{}^7 g_2{}^4 + 60g_1{}^3 g_2{}^6 + 40g_1{}^3 g_3{}^4 + 40g_3{}^5 + 180g_1 g_2 g_4{}^3 + 144g_5{}^3 + 120g_1 g_2 g_3{}^2 g_6 + 120g_3 g_6{}^2 \tag{19}$$

$$Z_7(R) = g_1{}^{21} + 21g_1{}^{11} g_2{}^5 + 105g_1{}^5 g_2{}^8 + 105g_1{}^3 g_2{}^9 + 70g_1{}^6 g_3{}^5 + 280g_3{}^7 + 210g_1{}^3 g_2 g_4{}^4 + 630g_1 g_2{}^2 g_4{}^4 + 504g_1 g_5{}^4 + 420g_1{}^2 g_2{}^2 g_3{}^3 g_6 + 210g_1{}^2 g_2{}^2 g_3 g_6{}^2 + 840g_3 g_6{}^3 + 720g_7{}^3 + 504g_1 g_5{}^2 g_{10} + 420g_2 g_3 g_4 g_{12}. \tag{20}$$

Letting $g_i = 1 + x^i$ then gives a POLYNOMIAL $S_i(x)$, which is a GENERATING FUNCTION for (i.e., the terms of x^i give) the number of graphs with i EDGES. The total number of graphs having i edges is $S_i(1)$. The first few $S_i(x)$ are

$$S_2 = 1 + x \tag{21}$$

$$S_3 = 1 + x + x^2 + x^3 \tag{22}$$

$$S_4 = 1 + x + 2x^2 + 3x^3 + 2x^4 + x^5 + x^6 \tag{23}$$

$$S_5 = 1 + x + 2x^2 + 4x^3 + 6x^4 + 6x^5 + 6x^6 + 4x^7 + 2x^8 + x^9 + x^{10} \tag{24}$$

$$S_6 = 1 + x + 2x^2 + 5x^3 + 9x^4 + 15x^5 + 21x^6 + 24x^7 + 24x^8 + 21x^9 + 15x^{10} + 9x^{11} + 5x^{12} + 2x^{13} + x^{14} + x^{15} \tag{25}$$

$$S_7 = 1 + x + 2x^2 + 5x^3 + 10x^4 + 21x^5 + 21x^6 + 24x^7 + 41x^6 + 65x^7 + 97x^8$$

$$+ 131x^9 + 148x^{10} + 148x^{11} + 131x^{12} + 97x^{13} + 65x^{14} + 41x^{15} + 21x^{16} + 10x^{17} + 5x^{18} + 2x^{19} + x^{20} + x^{21}, \quad (26)$$

giving the number of graphs with n nodes as 1, 2, 4, 11, 34, 156, 1044, ... (Sloane's A000088). King and Palmer (cited in Read 1981) have calculated S_n up to $n = 24$, for which

$$S_{24} = 195,704,906,302,078,447,922,174,862,416,\cdots \cdots 726,256,004,122,075,267,063,365,754,368. \quad (27)$$

see also BIPARTITE GRAPH, CATERPILLAR GRAPH, CAYLEY GRAPH, CIRCULANT GRAPH, COCKTAIL PARTY GRAPH, COMPARABILITY GRAPH, COMPLEMENT GRAPH, COMPLETE GRAPH, CONE GRAPH, CONNECTED GRAPH, COXETER GRAPH, CUBICAL GRAPH, DE BRUIJN GRAPH, DIGRAPH, DIRECTED GRAPH, DODECAHEDRAL GRAPH, EULER GRAPH, EXTREMAL GRAPH, GEAR GRAPH, GRACEFUL GRAPH, GRAPH THEORY, HANOI GRAPH, HARARY GRAPH, HARMONIOUS GRAPH, HOFFMAN-SINGLETON GRAPH, ICOSAHEDRAL GRAPH, INTERVAL GRAPH, ISOMORPHIC GRAPHS, LABELLED GRAPH, LADDER GRAPH, LATTICE GRAPH, MATCHSTICK GRAPH, MINOR GRAPH, MOORE GRAPH, NULL GRAPH, OCTAHEDRAL GRAPH, PATH GRAPH, PETERSEN GRAPHS, PLANAR GRAPH, RANDOM GRAPH, REGULAR GRAPH, SEQUENTIAL GRAPH, SIMPLE GRAPH, STAR GRAPH, SUBGRAPH, SUPERGRAPH, SUPERREGULAR GRAPH, SYLVESTER GRAPH, TETRAHEDRAL GRAPH, THOMASSEN GRAPH, TOURNAMENT, TRIANGULAR GRAPH, TURAN GRAPH, TUTTE'S GRAPH, UNIVERSAL GRAPH, UTILITY GRAPH, WEB GRAPH, WHEEL GRAPH

References

Bogomolny, A. "Graph Puzzles." http://www.cut-the-knot.com/do_you_know/graphs2.html.
Fujii, J. N. *Puzzles and Graphs.* Washington, DC: National Council of Teachers, 1966.
Harary, F. "The Number of Linear, Directed, Rooted, and Connected Graphs." *Trans. Amer. Math. Soc.* **78**, 445–463, 1955.
Pappas, T. "Networks." *The Joy of Mathematics.* San Carlos, CA: Wide World Publ./Tetra, pp. 126–127, 1989.
Read, R. "The Graph Theorists Who Count—and What They Count." In *The Mathematical Gardner* (Ed. D. Klarner). Boston, MA: Prindle, Weber, and Schmidt, pp. 326–345, 1981.
Sloane, N. J. A. Sequences A000088/M1253 in "An On-Line Version of the Encyclopedia of Integer Sequences."
Sloane, N. J. A. and Plouffe, S. Extended entry in *The Encyclopedia of Integer Sequences.* San Diego: Academic Press, 1995.

Graph Theory

The mathematical study of the properties of the formal mathematical structures called GRAPHS.

see also ADJACENCY MATRIX, ADJACENCY RELATION, ARTICULATION VERTEX, BLUE-EMPTY COLORING, BRIDGE (GRAPH), CHROMATIC NUMBER, CHROMATIC POLYNOMIAL, CIRCUIT RANK, CROSSING NUMBER (GRAPH), CYCLE (GRAPH), CYCLOMATIC NUMBER, DEGREE, DIAMETER (GRAPH), DIJKSTRA'S ALGORITHM, ECCENTRICITY, EDGE-COLORING, EDGE CONNECTIVITY, EULERIAN CIRCUIT, EULERIAN TRAIL, FACTOR (GRAPH), FLOYD'S ALGORITHM, GIRTH, GRAPH TWO-COLORING, GROUP THEORY, HAMILTONIAN CIRCUIT, HASSE DIAGRAM, HUB, INDEGREE, INTEGRAL DRAWING, ISTHMUS, JOIN (GRAPH), LOCAL DEGREE, MONOCHROMATIC FORCED TRIANGLE, OUTDEGREE, PARTY PROBLEM, PÓLYA ENUMERATION THEOREM, PÓLYA POLYNOMIAL, RADIUS (GRAPH), RAMSEY NUMBER, RE-ENTRANT CIRCUIT, SEPARATING EDGE, TAIT COLORING, TAIT CYCLE, TRAVELING SALESMAN PROBLEM, TREE, TUTTE'S THEOREM, UNICURSAL CIRCUIT, VALENCY, VERTEX COLORING, WALK

References

Berge, C. *The Theory of Graphs.* New York: Wiley, 1962.
Bogomolny, A. "Graphs." http://www.cut-the-knot.com/do_you_know/graphs.html.
Bollobás, B. *Graph Theory: An Introductory Course.* New York: Springer-Verlag, 1979.
Chartrand, G. *Introductory Graph Theory.* New York: Dover, 1985.
Foulds, L. R. *Graph Theory Applications.* New York: Springer-Verlag, 1992.
Chung, F. and Graham, R. *Erdős on Graphs: His Legacy of Unsolved Problems.* New York: A. K. Peters, 1998.
Grossman, I. and Magnus, W. *Groups and Their Graphs.* Washington, DC: Math. Assoc. Amer., 1965.
Harary, F. *Graph Theory.* Reading, MA: Addison-Wesley, 1994.
Hartsfield, N. and Ringel, G. *Pearls in Graph Theory: A Comprehensive Introduction, 2nd ed.* San Diego, CA: Academic Press, 1994.
Ore, Ø. *Graphs and Their Uses.* New York: Random House, 1963.
Ruskey, F. "Information on (Unlabelled) Graphs." http://sue.csc.uvic.ca/~cos/inf/grap/GraphInfo.html.
Saaty, T. L. and Kainen, P. C. *The Four-Color Problem: Assaults and Conquest.* New York: Dover, 1986.
Skiena, S. S. *Implementing Discrete Mathematics: Combinatorics and Graph Theory with Mathematica.* Redwood City, CA: Addison-Wesley, 1988.
Trudeau, R. J. *Introduction to Graph Theory.* New York: Dover, 1994.

Graph Two-Coloring

Assignment of each EDGE of a GRAPH to one of two color classes ("red" or "green").

see also BLUE-EMPTY GRAPH, MONOCHROMATIC FORCED TRIANGLE

Graphical Partition

A graphical partition of order n is the DEGREE SEQUENCE of a GRAPH with $n/2$ EDGES and no isolated VERTICES. For $n = 2, 4, 6, \ldots$, the number of graphical partitions is 1, 2, 5, 9, 17, ... (Sloane's A000569).

References

Barnes, T. M. and Savage, C. D. "A Recurrence for Counting Graphical Partitions." *Electronic J. Combinatorics* **2**, R11, 1–10, 1995. http://www.combinatorics.org/Volume_2/volume2.html#R11.

Barnes, T. M. and Savage, C. D. "Efficient Generation of Graphical Partitions." Submitted.

Ruskey, F. "Information on Graphical Partitions." http://sue.csc.uvic.ca/~cos/inf/nump/GraphicalPartition.html.

Sloane, N. J. A. Sequence A000569 in "An On-Line Version of the Encyclopedia of Integer Sequences."

Grassmann Algebra

see EXTERIOR ALGEBRA

Grassmann Coordinates

An $(m+1)$-D SUBSPACE W of an $(n+1)$-D VECTOR SPACE V can be specified by an $(m+1) \times (n+1)$ MATRIX whose rows are the coordinates of a BASIS of W. The set of all $\binom{n+1}{m+1}$ $(m+1) \times (m+1)$ MINORS of this MATRIX are then called the Grassmann coordinates of w (where $\binom{a}{b}$ is a BINOMIAL COEFFICIENT).

see also CHOW COORDINATES

References

Wilson, W. S.; Chern, S. S.; Abhyankar, S. S.; Lang, S.; and Igusa, J.-I. "Wei-Liang Chow." *Not. Amer. Math. Soc.* **43**, 1117–1124, 1996.

Grassmann Manifold

A special case of a FLAG MANIFOLD. A Grassmann manifold is a certain collection of vector SUBSPACES of a VECTOR SPACE. In particular, $G_{n,k}$ is the Grassmann manifold of k-dimensional subspaces of the VECTOR SPACE $\mathbb{R}^n$. It has a natural MANIFOLD structure as an orbit-space of the STIEFEL MANIFOLD $V_{n,k}$ of orthonormal k-frames in $\mathbb{R}^n$. One of the main things about Grassmann manifolds is that they are classifying spaces for VECTOR BUNDLES.

Gray Code

An encoding of numbers so that adjacent numbers have a single DIGIT differing by 1. A BINARY Gray code with n DIGITS corresponds to a HAMILTONIAN PATH on an n-D HYPERCUBE (including direction reversals). The term Gray code is often used to refer to a "reflected" code, or more specifically still, the binary reflected Gray code.

To convert a BINARY number $d_1 d_2 \cdots d_{n-1} d_n$ to its corresponding binary reflected Gray code, start at the right with the digit d_n (the nth, or last, DIGIT). If the d_{n-1} is 1, replace d_n by $1 - d_n$; otherwise, leave it unchanged. Then proceed to d_{n-1}. Continue up to the first DIGIT d_1, which is kept the same since d_0 is assumed to be a 0. The resulting number $g_1 g_2 \cdots g_{n-1} g_n$ is the reflected binary Gray code.

To convert a binary reflected Gray code $g_1 g_2 \cdots g_{n-1} g_n$ to a BINARY number, start again with the nth digit, and compute

$$\Sigma_n \equiv \sum_{i=1}^{n-1} g_i \pmod 2.$$

If Σ_n is 1, replace g_n by $1 - g_n$; otherwise, leave it the unchanged. Next compute

$$\Sigma_{n-1} \equiv \sum_{i=1}^{n-2} g_i \pmod 2,$$

and so on. The resulting number $d_1 d_2 \cdots d_{n-1} d_n$ is the BINARY number corresponding to the initial binary reflected Gray code.

The code is called reflected because it can be generated in the following manner. Take the Gray code 0, 1. Write it forwards, then backwards: 0, 1, 1, 0. Then append 0s to the first half and 1s to the second half: 00, 01, 11, 10. Continuing, write 00, 01, 11, 10, 10, 11, 01, 00 to obtain: 000, 001, 011, 010, 110, 111, 101, 100, ... (Sloane's A014550). Each iteration therefore doubles the number of codes. The Gray codes corresponding to the first few nonnegative integers are given in the following table.

0	0	20	11110	40	111100
1	1	21	11111	41	111101
2	11	22	11101	42	111111
3	10	23	11100	43	111110
4	110	24	10100	44	111010
5	111	25	10101	45	111011
6	101	26	10111	46	111001
7	100	27	10110	47	111000
8	1100	28	10010	48	101000
9	1101	29	10011	49	101001
10	1111	30	10001	50	101011
11	1110	31	10000	51	101010
12	1010	32	110000	52	101110
13	1011	33	110001	53	101111
14	1001	34	110011	54	101101
15	1000	35	110010	55	101100
16	11000	36	110110	56	100100
17	11001	37	110111	57	100101
18	11011	38	110101	58	100111
19	11010	39	110100	59	100110

The binary reflected Gray code is closely related to the solution of the TOWERS OF HANOI as well as the BAGUENAUDIER.

see also BAGUENAUDIER, BINARY, HILBERT CURVE, MORSE-THUE SEQUENCE, RYSER FORMULA, TOWERS OF HANOI

References

Gardner, M. "The Binary Gray Code." Ch. 2 in *Knotted Doughnuts and Other Mathematical Entertainments.* New York: W. H. Freeman, 1986.

Press, W. H.; Flannery, B. P.; Teukolsky, S. A.; and Vetterling, W. T. "Gray Codes." §20.2 in *Numerical Recipes in FORTRAN: The Art of Scientific Computing, 2nd ed.* Cambridge, England: Cambridge University Press, pp. 886–888, 1992.

Sloane, N. J. A. Sequence A014550 in "An On-Line Version of the Encyclopedia of Integer Sequences."

Vardi, I. *Computational Recreations in Mathematica.* Redwood City, CA: Addison-Wesley, pp. 111–112 and 246, 1991.

Great Circle

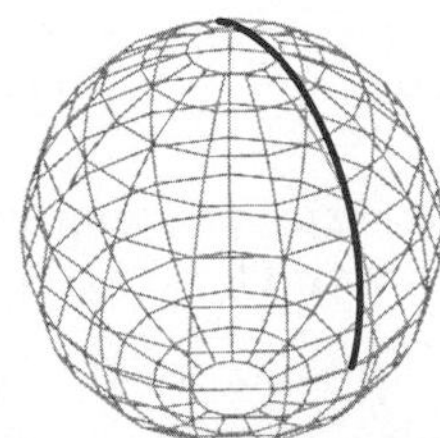

The shortest path between two points on a SPHERE, also known as an ORTHODROME. To find the great circle (GEODESIC) distance between two points located at LATITUDE δ and LONGITUDE λ of (δ_1, λ_1) and (δ_2, λ_2) on a SPHERE of RADIUS a, convert SPHERICAL COORDINATES to CARTESIAN COORDINATES using

$$\mathbf{r}_i = a \begin{bmatrix} \cos\lambda_i \cos\delta_i \\ \sin\lambda_i \cos\delta_i \\ \sin\delta_i \end{bmatrix}. \tag{1}$$

(Note that the LATITUDE δ is related to the COLATITUDE ϕ of SPHERICAL COORDINATES by $\delta = 90° - \phi$, so the conversion to CARTESIAN COORDINATES replaces $\sin\phi$ and $\cos\phi$ by $\cos\delta$ and $\sin\delta$, respectively.) Now find the ANGLE α between $\mathbf{r}_1$ and $\mathbf{r}_2$ using the DOT PRODUCT,

$$\begin{aligned} \cos\alpha &= \hat{\mathbf{r}}_1 \cdot \hat{\mathbf{r}}_2 \\ &= \cos\delta_1 \cos\delta_2 (\sin\lambda_1 \sin\lambda_2 + \cos\lambda_1 \cos\lambda_2) \\ &\quad + \sin\delta_1 \sin\delta_2 \\ &= \cos\delta_1 \cos\delta_2 \cos(\lambda_1 - \lambda_2) + \sin\delta_1 \sin\delta_2. \end{aligned} \tag{2}$$

The great circle distance is then

$$d = a\cos^{-1}[\cos\delta_1 \cos\delta_2 \cos(\lambda_1 - \lambda_2) + \sin\delta_1 \sin\delta_2]. \tag{3}$$

For the Earth, the *equatorial* RADIUS is $a \approx 6378$ km, or 3963 (statute) miles. Unfortunately, the FLATTENING of the Earth cannot be taken into account in this simple derivation, since the problem is considerable more complicated for a SPHEROID or ELLIPSOID (each of which has a RADIUS which is a function of LATITUDE).

The equation of the great circle can be explicitly computed using the GEODESIC formalism. Writing

$$u = \lambda \tag{4}$$

$$v = \delta = \tfrac{1}{2}\pi - \phi \tag{5}$$

gives the P, Q, and R parameters of the GEODESIC (which are just combinations of the PARTIAL DERIVATIVES) as

$$P \equiv \left(\frac{\partial x}{\partial u}\right)^2 + \left(\frac{\partial y}{\partial u}\right)^2 + \left(\frac{\partial z}{\partial u}\right)^2 = a^2 \sin^2 v \tag{6}$$

$$Q \equiv \frac{\partial x}{\partial u}\frac{\partial x}{\partial v} + \frac{\partial y}{\partial u}\frac{\partial y}{\partial v} + \frac{\partial z}{\partial u}\frac{\partial z}{\partial v} = 0 \tag{7}$$

$$R \equiv \left(\frac{\partial x}{\partial v}\right)^2 + \left(\frac{\partial y}{\partial v}\right)^2 + \left(\frac{\partial z}{\partial v}\right)^2 = a^2. \tag{8}$$

The GEODESIC differential equation then becomes

$$\cos v \sin^4 v + 2\cos v \sin^2 v v'^2 + \cos v v'^4 - \sin v v'' = 0. \tag{9}$$

However, because this is a special case of $Q = 0$ with P and R explicit functions of v only, the GEODESIC solution takes on the special form

$$\begin{aligned} v &= c_1 \int \sqrt{\frac{R}{P^2 - {c_1}^2 P}}\, dv = c_1 \int \frac{dv}{a^2 \sin^4 v - {c_1}^2 \sin^2 v} \\ &= \int \frac{dv}{\sin v \sqrt{\left(\frac{a}{c_1}\right)^2 \sin^2 v - 1}} \\ &= -\tan^{-1}\left[\frac{\cos v}{\sqrt{\left(\frac{a}{c_1}\right)^2 - 1}}\right] + c_2 \end{aligned} \tag{10}$$

(Gradshteyn and Ryzhik 1979, p. 174, eqn. 2.599.6), which can be rewritten as

$$v = -\sin^{-1}\left(\frac{\cot v}{\sqrt{\left(\frac{a}{c_1}\right)^2 - 1}}\right) + c_2. \tag{11}$$

It therefore follows that

$$(\sin c_2) a \sin v \cos u - (\cos c_2) a \sin v \sin u - \frac{a\cos v}{\sqrt{\left(\frac{a}{c_1}\right)^2 - 1}} = 0. \tag{12}$$

This equation can be written in terms of the CARTESIAN COORDINATES as

$$x \sin c_2 - y \cos c_2 - \frac{z}{\sqrt{\left(\frac{a}{c_1}\right)^2 - 1}} = 0, \tag{13}$$

which is simply a PLANE passing through the center of the SPHERE and the two points on the surface of the SPHERE.

see also GEODESIC, GREAT SPHERE, LOXODROME, MIKUSIŃSKI'S PROBLEM, ORTHODROME, POINT-POINT DISTANCE—2-D, SPHERE

References

Gradshteyn, I. S. and Ryzhik, I. M. *Tables of Integrals, Series, and Products, 5th ed.* San Diego, CA: Academic Press, 1979.

Weinstock, R. *Calculus of Variations, with Applications to Physics and Engineering.* New York: Dover, pp. 26–28 and 62–63, 1974.

Great Cubicuboctahedron

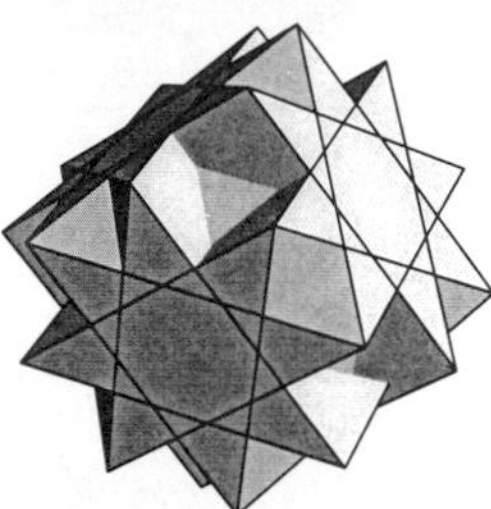

The UNIFORM POLYHEDRON U_{14} whose DUAL POLYHEDRON is the GREAT HEXACRONIC ICOSITETRAHEDRON. It has WYTHOFF SYMBOL $3\,4\,|\,\frac{4}{3}$. Its faces are $8\{3\}+6\{4\}+6\{\frac{8}{3}\}$. It is a FACETED version of the CUBE. The CIRCUMRADIUS of a great cubicuboctahedron with unit edge length is

$$R = \tfrac{1}{2}\sqrt{5-2\sqrt{2}}.$$

References

Wenninger, M. J. *Polyhedron Models.* Cambridge, England: Cambridge University Press, pp. 118–119, 1989.

Great Deltoidal Hexecontahedron

The DUAL of the GREAT RHOMBICOSIDODECAHEDRON (UNIFORM).

Great Deltoidal Icositetrahedron

The DUAL of the GREAT RHOMBICUBOCTAHEDRON (UNIFORM).

Great Dirhombicosidodecacron

The DUAL of the GREAT DIRHOMBICOSIDODECAHEDRON.

Great Dirhombicosidodecahedron

The UNIFORM POLYHEDRON U_{75} whose DUAL is the GREAT DIRHOMBICOSIDODECACRON. This POLYHEDRON is exceptional because it cannot be derived from SCHWARZ TRIANGLES and because it is the only UNIFORM POLYHEDRON with more than six POLYGONS surrounding each VERTEX (four SQUARES alternating with two TRIANGLES and two PENTAGRAMS). It has WYTHOFF SYMBOL $|\,\frac{3}{2}\,\frac{5}{3}\,3\,\frac{5}{2}$. Its faces are $40\{3\}+60\{4\}+24\{\frac{5}{2}\}$, and its CIRCUMRADIUS for unit edge length is

$$R = \tfrac{1}{2}\sqrt{2}.$$

References

Wenninger, M. J. *Polyhedron Models.* Cambridge, England: Cambridge University Press, pp. 200–203, 1989.

Great Disdyakis Dodecahedron

The DUAL of the GREAT TRUNCATED CUBOCTAHEDRON.

Great Disdyakis Triacontahedron

The DUAL of the GREAT TRUNCATED ICOSIDODECAHEDRON.

Great Ditrigonal Dodecacronic Hexecontahedron

The DUAL of the GREAT DITRIGONAL DODECICOSIDODECAHEDRON.

Great Ditrigonal Dodecicosidodecahedron

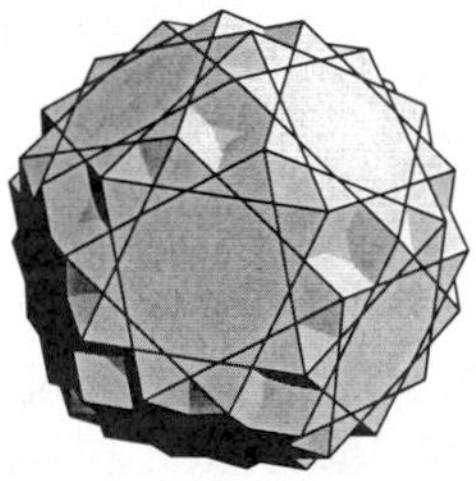

The UNIFORM POLYHEDRON U_{42} whose DUAL is the GREAT DITRIGONAL DODECACRONIC HEXECONTAHEDRON. It has WYTHOFF SYMBOL $3\,5\,|\,\frac{5}{3}$. Its faces are $20\{3\}+12\{5\}+12\{\frac{10}{3}\}$, and its CIRCUMRADIUS for unit edge length is

$$R = \tfrac{1}{4}\sqrt{34-6\sqrt{5}}.$$

References

Wenninger, M. J. *Polyhedron Models.* Cambridge, England: Cambridge University Press, p. 125, 1989.

Great Ditrigonal Icosidodecahedron

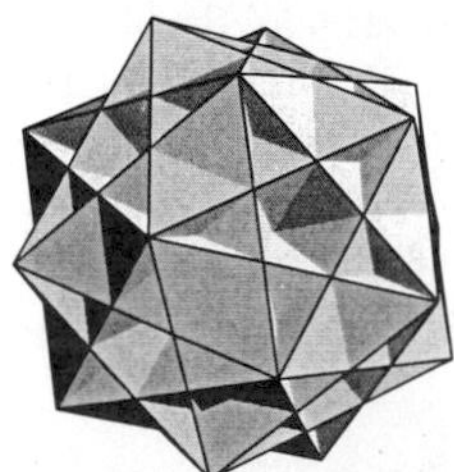

The UNIFORM POLYHEDRON U_{47} whose DUAL is the GREAT TRIAMBIC ICOSAHEDRON. It has WYTHOFF SYMBOL $\frac{3}{2}\,|\,3\,5$. Its faces are $20\{3\}+12\{5\}$, and its CIRCUMRADIUS for unit edge length is

$$R=\tfrac{1}{2}\sqrt{3}.$$

References

Wenninger, M. J. *Polyhedron Models.* Cambridge, England: Cambridge University Press, pp. 135–136, 1989.

Great Dodecacronic Hexecontahedron

The DUAL of the GREAT DODECICOSIDODECAHEDRON.

Great Dodecadodecahedron

see DODECADODECAHEDRON

Great Dodecahedron

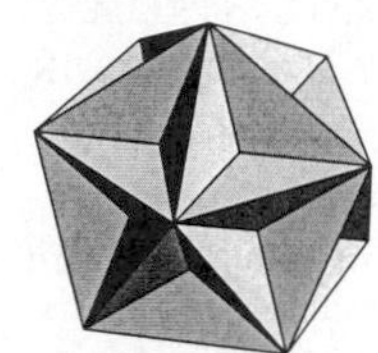

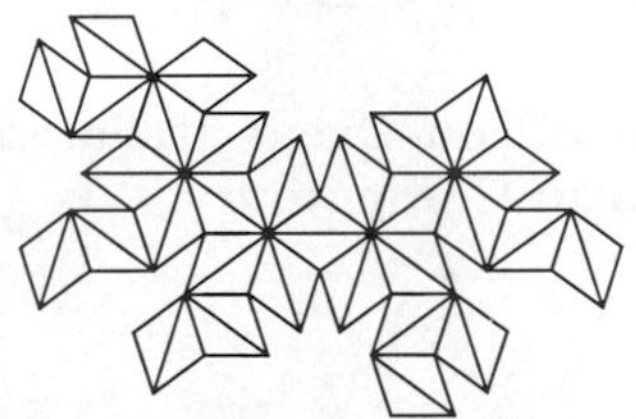

The UNIFORM POLYHEDRON U_{35} which is the DUAL of the SMALL STELLATED DODECAHEDRON and one of the KEPLER-POINSOT SOLIDS. Its faces are $12\{5\}$. Its SCHLÄFLI SYMBOL is $\{5,\frac{5}{2}\}$, and its WYTHOFF SYMBOL is $\frac{5}{2}\,|\,2\,5$. Its faces are $12\{5\}$. Its CIRCUMRADIUS for unit edge length is

$$R=\tfrac{1}{2}5^{1/4}\phi^{1/2}a=\tfrac{1}{4}5^{1/4}\sqrt{2(1+\sqrt{5})},$$

where ϕ is the GOLDEN RATIO.

see also GREAT ICOSAHEDRON, GREAT STELLATED DODECAHEDRON, KEPLER-POINSOT SOLID, SMALL STELLATED DODECAHEDRON

References

Fischer, G. (Ed.). Plate 105 in *Mathematische Modelle/Mathematical Models, Bildband/Photograph Volume.* Braunschweig, Germany: Vieweg, p. 104, 1986.

Great Dodecahedron-Small Stellated Dodecahedron Compound

A POLYHEDRON COMPOUND in which the GREAT DODECAHEDRON is interior to the SMALL STELLATED DODECAHEDRON.

see also POLYHEDRON COMPOUND

Great Dodecahemicosacron

The DUAL of the GREAT DODECAHEMICOSAHEDRON.

Great Dodecahemicosahedron

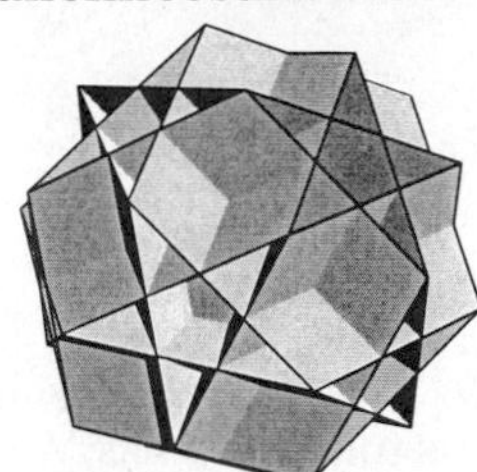

The UNIFORM POLYHEDRON U_{65} whose DUAL is the GREAT DODECAHEMICOSACRON. It has WYTHOFF SYMBOL $\frac{5}{3}\,\frac{5}{2}\,|\,\frac{5}{3}$. Its faces are $12\{\frac{5}{2}\}+6\{\frac{10}{3}\}$. It is a FACETED DODECADODECAHEDRON. The CIRCUMRADIUS for unit edge length is

$$R=\tfrac{1}{2}\sqrt{3}.$$

References

Wenninger, M. J. *Polyhedron Models.* Cambridge, England: Cambridge University Press, pp. 106–107, 1989.

Great Dodecahemidodecacron

The DUAL of the GREAT DODECAHEMIDODECAHEDRON.

Great Dodecahemidodecahedron

The UNIFORM POLYHEDRON U_{70} whose DUAL is the GREAT DODECAHEMIDODECACRON. It has WYTHOFF SYMBOL $\frac{5}{3}\,\frac{5}{2}\,|\,\frac{5}{3}$. Its faces are $12\{\frac{5}{2}\}+6\{\frac{10}{3}\}$. Its CIRCUMRADIUS for unit edge length is

$$R=\phi^{-1},$$

where ϕ is the GOLDEN RATIO.

References

Wenninger, M. J. *Polyhedron Models.* Cambridge, England: Cambridge University Press, p. 165, 1989.

Great Dodecicosacron

The DUAL of the GREAT DODECICOSAHEDRON.

Great Dodecicosahedron

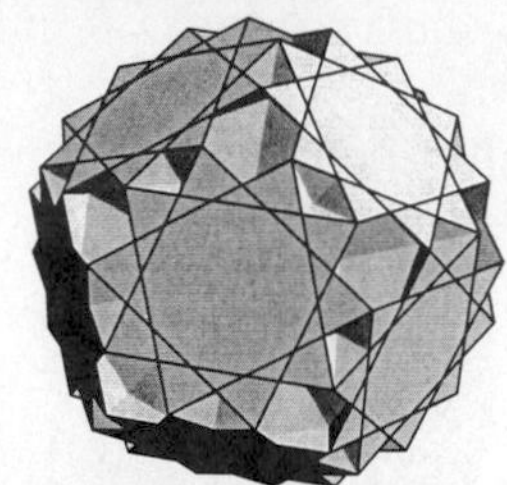

The UNIFORM POLYHEDRON U_{63} whose DUAL is the GREAT DODECICOSACRON. It has WYTHOFF SYMBOL $3\,\frac{5}{3}\mid \left.\frac{\frac{3}{2}}{\frac{5}{2}}\right|$. Its faces are $20\{6\} + 12\{\frac{10}{3}\}$. Its CIRCUMRADIUS for unit edge length is

$$R = \tfrac{1}{4}\sqrt{34 - 6\sqrt{5}}\,.$$

References
Wenninger, M. J. *Polyhedron Models.* Cambridge, England: Cambridge University Press, pp. 156–157, 1989.

Great Dodecicosidodecahedron

The UNIFORM POLYHEDRON U_{61} whose DUAL is the GREAT DODECACRONIC HEXECONTAHEDRON. Its WYTHOFF SYMBOL is $2\,\frac{5}{2}\,|\,3$. Its faces are $20\{6\}+12\{\frac{5}{2}\}$, and its CIRCUMRADIUS for unit edge length is

$$R = \tfrac{1}{4}\sqrt{58 - 18\sqrt{5}}\,.$$

References
Wenninger, M. J. *Polyhedron Models.* Cambridge, England: Cambridge University Press, p. 148, 1989.

Great Hexacronic Icositetrahedron
The DUAL of the GREAT CUBICUBOCTAHEDRON.

Great Hexagonal Hexecontahedron
The DUAL of the GREAT SNUB DODECICOSIDODECAHEDRON.

Great Icosacronic Hexecontahedron
The DUAL of the GREAT ICOSICOSIDODECAHEDRON.

Great Icosahedron

One of the KEPLER-POINSOT SOLIDS whose DUAL is the GREAT STELLATED DODECAHEDRON. Its faces are $20\{3\}$. It is also UNIFORM POLYHEDRON U_{53} and has WYTHOFF SYMBOL $3\,\frac{5}{2}\mid\frac{5}{3}$. Its faces are $20\{3\}+12\{\frac{5}{2}\}+12\{\frac{10}{3}\}$. Its CIRCUMRADIUS for unit edge length is

$$R = \tfrac{1}{2}\sqrt{11 - 4\sqrt{5}}\,.$$

see also GREAT DODECAHEDRON, GREAT ICOSAHEDRON, GREAT STELLATED DODECAHEDRON, KEPLER-POINSOT SOLID, SMALL STELLATED DODECAHEDRON, TRUNCATED GREAT ICOSAHEDRON

References
Fischer, G. (Ed.). Plate 106 in *Mathematische Modelle/Mathematical Models, Bildband/Photograph Volume.* Braunschweig, Germany: Vieweg, p. 105, 1986.
Wenninger, M. J. *Polyhedron Models.* Cambridge, England: Cambridge University Press, p. 154, 1989.

Great Icosahedron-Great Stellated Dodecahedron Compound

A POLYHEDRON COMPOUND most easily constructed by adding the VERTICES of a GREAT ICOSAHEDRON to a GREAT STELLATED DODECAHEDRON.

see also POLYHEDRON COMPOUND

References
Cundy, H. and Rollett, A. *Mathematical Models, 3rd ed.* Stradbroke, England: Tarquin Pub., pp. 132–133, 1989.

Great Icosicosidodecahedron

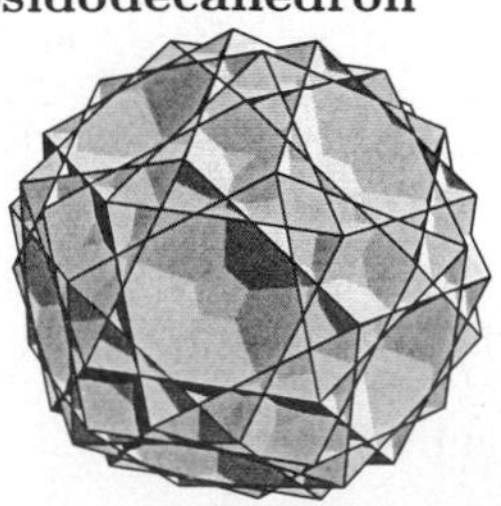

The UNIFORM POLYHEDRON U_{48} whose DUAL is the GREAT ICOSACRONIC HEXECONTAHEDRON. It has WYTHOFF SYMBOL $\frac{3}{2}\,5\,|\,3$. Its faces are $20\{3\}+20\{6\}+12\{5\}$. Its CIRCUMRADIUS for unit edge length is

$$R = \tfrac{1}{4}\sqrt{34-6\sqrt{5}}.$$

References
Wenninger, M. J. *Polyhedron Models.* Cambridge, England: Cambridge University Press, pp. 137–139, 1989.

Great Icosidodecahedron

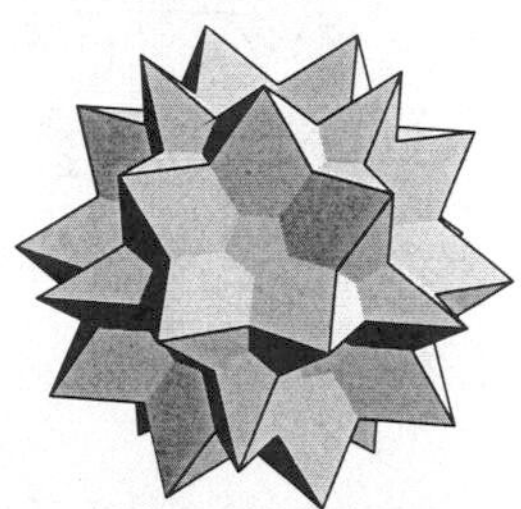

A UNIFORM POLYHEDRON U_{54} whose DUAL is the GREAT RHOMBIC TRIACONTAHEDRON (also called the GREAT STELLATED TRIACONTAHEDRON). It is a STELLATED ARCHIMEDEAN SOLID. It has SCHLÄFLI SYMBOL $\left\{\begin{smallmatrix}3\\ \frac{5}{2}\end{smallmatrix}\right\}$ and WYTHOFF SYMBOL $2\,|\,3\,\frac{5}{2}$. Its faces are $20\{3\}+12\{\frac{5}{2}\}$. Its CIRCUMRADIUS for unit edge length is

$$R = \phi^{-1},$$

where ϕ is the GOLDEN RATIO.

References
Cundy, H. and Rollett, A. *Mathematical Models, 3rd ed.* Stradbroke, England: Tarquin Pub., p. 124, 1989.
Wenninger, M. J. *Polyhedron Models.* Cambridge, England: Cambridge University Press, p. 147, 1989.

Great Icosihemidodecacron

The DUAL of the GREAT ICOSIHEMIDODECAHEDRON.

Great Icosihemidodecahedron

The UNIFORM POLYHEDRON U_{71} whose DUAL is the GREAT ICOSIHEMIDODECACRON. It has WYTHOFF SYMBOL $\frac{3}{2}\,3\,|\,\frac{5}{3}$. Its faces are $20\{3\}+6\{\frac{10}{3}\}$. For unit edge length, its CIRCUMRADIUS is

$$R = \phi^{-1},$$

where ϕ is the GOLDEN RATIO.

References
Wenninger, M. J. *Polyhedron Models.* Cambridge, England: Cambridge University Press, p. 164, 1989.

Great Inverted Pentagonal Hexecontahedron

The DUAL of the GREAT INVERTED SNUB ICOSIDODECAHEDRON.

Great Inverted Retrosnub Icosidodecahedron

see GREAT RETROSNUB ICOSIDODECAHEDRON

Great Inverted Snub Icosidodecahedron

The UNIFORM POLYHEDRON U_{69} whose DUAL is the GREAT INVERTED PENTAGONAL HEXECONTAHEDRON. It has WYTHOFF SYMBOL $|\,2\,3\,\frac{5}{2}$. Its faces are $80\{3\}+12\{\frac{5}{2}\}$. For unit edge length, it has CIRCUMRADIUS

$$R = \frac{1}{2}\sqrt{\frac{8\cdot 2^{2/3}-16x+2^{1/3}x^2}{8\cdot 2^{2/3}-10x+2^{1/3}x^2}} = 0.816080674799923,$$

where

$$x \equiv \left(49-27\sqrt{5}+3\sqrt{6}\sqrt{93-49\sqrt{5}}\right)^{1/3}.$$

References
Wenninger, M. J. *Polyhedron Models.* Cambridge, England: Cambridge University Press, p. 179, 1989.

Great Pentagonal Hexecontahedron

The DUAL of the GREAT SNUB ICOSIDODECAHEDRON.

Great Pentagrammic Hexecontahedron

The DUAL of the GREAT RETROSNUB ICOSIDODECAHEDRON.

Great Pentakis Dodecahedron

The DUAL of the SMALL STELLATED TRUNCATED DODECAHEDRON.

Great Quasitruncated Icosidodecahedron

see GREAT TRUNCATED ICOSIDODECAHEDRON

Great Retrosnub Icosidodecahedron

The UNIFORM POLYHEDRON U_{74}, also called the GREAT INVERTED RETROSNUB ICOSIDODECAHEDRON, whose DUAL is the GREAT PENTAGRAMMIC HEXECONTAHEDRON. It has WYTHOFF SYMBOL $|2\,\frac{3}{2}\,\frac{5}{3}$. Its faces are $80\{3\}+12\{\frac{5}{2}\}$. For unit edge length, it has CIRCUMRADIUS

$$R = \tfrac{1}{2}\sqrt{\frac{2-x}{1-x}} \approx 0.5800015,$$

where x is the smaller NEGATIVE root of

$$x^3 + 2x^2 - \phi^{-2} = 0,$$

with ϕ the GOLDEN MEAN.

References

Wenninger, M. J. *Polyhedron Models.* Cambridge, England: Cambridge University Press, pp. 189–193, 1989.

Great Rhombic Triacontahedron

A ZONOHEDRON which is the DUAL of the GREAT ICOSIDODECAHEDRON. It is also called the GREAT STELLATED TRIACONTAHEDRON.

References

Cundy, H. and Rollett, A. *Mathematical Models, 3rd ed.* Stradbroke, England: Tarquin Pub., p. 126, 1989.

Great Rhombicosidodecahedron (Archimedean)

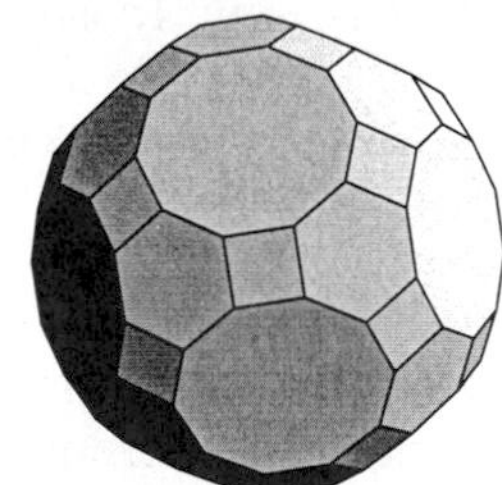

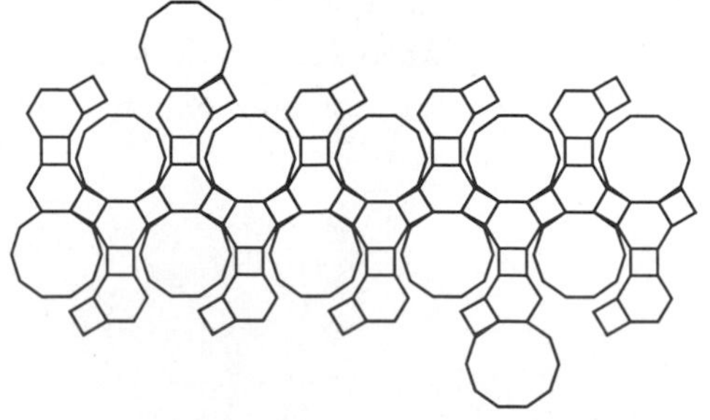

An ARCHIMEDEAN SOLID also known as the RHOMBITRUNCATED ICOSIDODECAHEDRON. It is sometimes improperly called the TRUNCATED ICOSIDODECAHEDRON, a name which is inappropriate since TRUNCATION would yield RECTANGULAR instead of SQUARE. The great rhombicosidodecahedron is also UNIFORM POLYHEDRON U_{28}. Its DUAL is the DISDYAKIS TRIACONTAHEDRON, also called the HEXAKIS ICOSAHEDRON. It has SCHLÄFLI SYMBOL $t\{\begin{smallmatrix}3\\5\end{smallmatrix}\}$ and WYTHOFF SYMBOL $2\,3\,5\,|$. The INRADIUS, MIDRADIUS, and CIRCUMRADIUS for $a=1$ are

$$r = \tfrac{1}{241}(105+6\sqrt{5})\sqrt{31+12\sqrt{5}} \approx 3.73665$$

$$\rho = \tfrac{1}{2}\sqrt{30+12\sqrt{5}} \approx 3.76938$$

$$R = \tfrac{1}{2}\sqrt{31+12\sqrt{5}} \approx 3.80239.$$

see also SMALL RHOMBICOSIDODECAHEDRON

References

Ball, W. W. R. and Coxeter, H. S. M. *Mathematical Recreations and Essays, 13th ed.* New York: Dover, p. 137, 1987.

Great Rhombicosidodecahedron (Uniform)

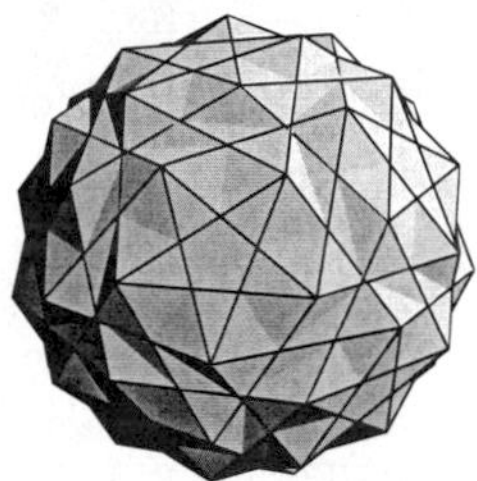

The UNIFORM POLYHEDRON U_{67}, also called the QUASIRHOMBICOSIDODECAHEDRON, whose DUAL is the GREAT DELTOIDAL HEXECONTAHEDRON. It has SCHLÄFLI SYMBOL $r'\left\{\begin{smallmatrix}3\\\frac{5}{2}\end{smallmatrix}\right\}$. It has WYTHOFF SYMBOL $3\,\frac{5}{3}\,|\,2$. Its faces are $20\{3\}+30\{4\}+12\{\frac{5}{2}\}$. For unit edge length, its CIRCUMRADIUS is

$$R = \tfrac{1}{2}\sqrt{11-4\sqrt{5}}.$$

References

Wenninger, M. J. *Polyhedron Models.* Cambridge, England: Cambridge University Press, pp. 162–163, 1989.

Great Rhombicuboctahedron (Archimedean)

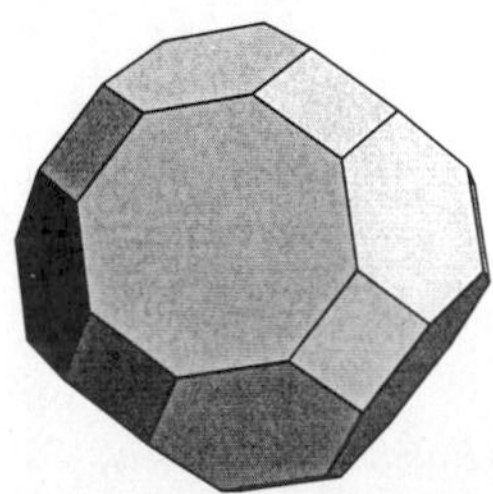

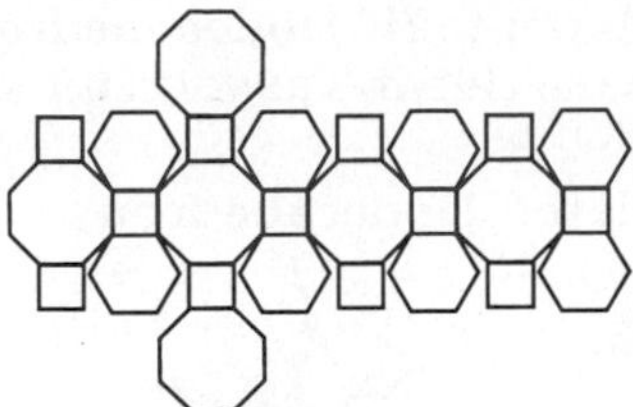

An ARCHIMEDEAN SOLID sometimes (improperly) called the TRUNCATED CUBOCTAHEDRON and also called the RHOMBITRUNCATED CUBOCTAHEDRON. Its DUAL is the DISDYAKIS DODECAHEDRON, also called the HEXAKIS OCTAHEDRON. It has SCHLÄFLI SYMBOL $t\left\{\begin{matrix}3\\4\end{matrix}\right\}$. It is also UNIFORM POLYHEDRON U_{11} and has WYTHOFF SYMBOL $2\,3\,4\,|$. Its faces are $8\{6\}+12\{4\}+6\{8\}$. The SMALL CUBICUBOCTAHEDRON is a FACETED version. The INRADIUS, MIDRADIUS, and CIRCUMRADIUS for unit edge length are

$$r = \tfrac{3}{97}(14+\sqrt{2})\sqrt{13+6\sqrt{2}} \approx 2.20974$$
$$\rho = \tfrac{1}{2}\sqrt{12+6\sqrt{2}} \approx 2.26303$$
$$R = \tfrac{1}{2}\sqrt{13+6\sqrt{2}} \approx 2.31761.$$

Additional quantities are

$$t = \tan(\tfrac{1}{8}\pi) = \sqrt{2}-1$$
$$l = 2t = 2(\sqrt{2}-1)$$
$$h = 1 + l\sin(\tfrac{1}{4}\pi) = 3-\sqrt{2}.$$

see also SMALL RHOMBICUBOCTAHEDRON, GREAT TRUNCATED CUBOCTAHEDRON

References

Ball, W. W. R. and Coxeter, H. S. M. *Mathematical Recreations and Essays, 13th ed.* New York: Dover, p. 138, 1987.

Great Rhombicuboctahedron (Uniform)

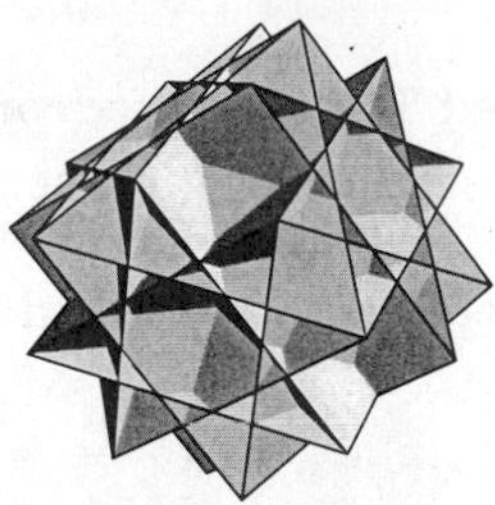

The UNIFORM POLYHEDRON U_{17}, also known as the QUASIRHOMBICUBOCTAHEDRON, whose DUAL is the GREAT DELTOIDAL ICOSITETRAHEDRON. It has SCHLÄFLI SYMBOL $r'\{\tfrac{3}{4}\}$ and WYTHOFF SYMBOL $\tfrac{3}{2}\,4\,|\,2$. Its faces are $8\{3\}+20\{4\}$. Its CIRCUMRADIUS for unit edge length is

$$R = \tfrac{1}{2}\sqrt{5-2\sqrt{2}}.$$

References

Wenninger, M. J. *Polyhedron Models.* Cambridge, England: Cambridge University Press, pp. 132–133, 1989.

Great Rhombidodecacron

The DUAL of the GREAT RHOMBIDODECAHEDRON.

Great Rhombidodecahedron

The UNIFORM POLYHEDRON U_{73} whose DUAL is the Great Rhombidodecacron. It WYTHOFF SYMBOL $2\,\tfrac{5}{3}\,|\,\begin{matrix}\frac{3}{2}\\ \frac{5}{4}\end{matrix}\Big|$. Its faces are $30\{4\}+12\{\tfrac{10}{3}\}$. Its CIRCUMRADIUS for unit edge length is

$$R = \tfrac{1}{2}\sqrt{11-4\sqrt{5}}.$$

References

Wenninger, M. J. *Polyhedron Models.* Cambridge, England: Cambridge University Press, pp. 168–170, 1989.

Great Rhombihexacron

The DUAL of the GREAT RHOMBIHEXAHEDRON.

Great Rhombihexahedron

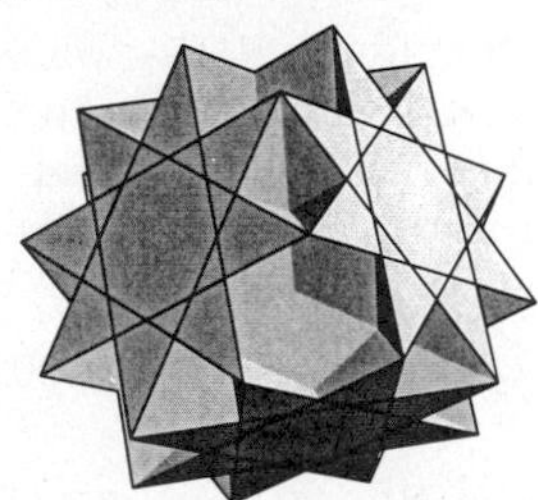

The UNIFORM POLYHEDRON U_{21} whose DUAL is the GREAT RHOMBIHEXACRON. It has WYTHOFF SYMBOL $2\,\tfrac{4}{3}\,\begin{matrix}\frac{3}{2}\\ \frac{4}{2}\end{matrix}\Big|$. Its faces are $12\{4\}+6\{\tfrac{5}{3}\}$. Its CIRCUMRADIUS for unit edge length is

$$R = \tfrac{1}{2}\sqrt{5-2\sqrt{2}}.$$

References

Wenninger, M. J. *Polyhedron Models.* Cambridge, England: Cambridge University Press, pp. 159–160, 1989.

Great Snub Dodecicosidodecahedron

The UNIFORM POLYHEDRON U_{64} whose DUAL is the GREAT HEXAGONAL HEXECONTAHEDRON. It has WYTHOFF SYMBOL $|\,3\,\frac{5}{3}\,\frac{5}{2}$. Its faces are $80\{3\}+24\{\frac{5}{2}\}$. Its CIRCUMRADIUS for unit edge length is

$$R = \tfrac{1}{2}\sqrt{2}.$$

References
Wenninger, M. J. *Polyhedron Models.* Cambridge, England: Cambridge University Press, pp. 183–185, 1989.

Great Snub Icosidodecahedron

The UNIFORM POLYHEDRON U_{57} whose DUAL is the GREAT PENTAGONAL HEXECONTAHEDRON. It has WYTHOFF SYMBOL $|\,2\,3\,\frac{5}{3}$. Its faces are $80\{3\}+12\{\frac{5}{2}\}$. For unit edge length, it has CIRCUMRADIUS

$$R = \tfrac{1}{2}\sqrt{\frac{2-x}{1-x}} \approx 0.6450202,$$

where x is the most NEGATIVE ROOT of

$$x^3 + 2x^2 - \phi^{-2} = 0,$$

with ϕ the GOLDEN RATIO.

References
Wenninger, M. J. *Polyhedron Models.* Cambridge, England: Cambridge University Press, pp. 186–188, 1989.

Great Sphere

The great sphere on the surface of a HYPERSPHERE is the 3-D analog of the GREAT CIRCLE on the surface of a SPHERE. Let $2h$ be the number of reflecting SPHERES, and let great spheres divide a HYPERSPHERE into g 4-D TETRAHEDRA. Then for the POLYTOPE with SCHLÄFLI SYMBOL $\{p,q,r\}$,

$$\frac{64h}{g} = 12 - p - 2q - r + \frac{4}{p} + \frac{4}{r}.$$

see also GREAT CIRCLE

Great Stellapentakis Dodecahedron

The DUAL of the GREAT TRUNCATED ICOSAHEDRON.

Great Stellated Dodecahedron

One of the KEPLER-POINSOT SOLIDS whose DUAL is the GREAT ICOSAHEDRON. Its SCHLÄFLI SYMBOL is $\{\frac{5}{2},3\}$. It is also UNIFORM POLYHEDRON U_{52} and has WYTHOFF SYMBOL $3\,|\,2\,\frac{5}{2}$. Its faces are $12\{\frac{5}{2}\}$. Its CIRCUMRADIUS for unit edge length is

$$R = \tfrac{1}{2}\sqrt{3}\phi^{-1} = \tfrac{1}{4}\sqrt{3}(\sqrt{5}-1).$$

The easiest way to construct it is to make 12 TRIANGULAR PYRAMIDS

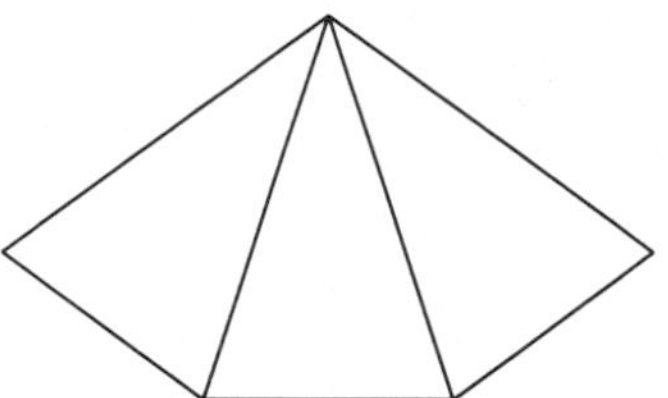

with side length $\phi = (1+\sqrt{5})/2$ (the GOLDEN RATIO) times the base and attach them to the sides of an ICOSAHEDRON.

see also GREAT DODECAHEDRON, GREAT ICOSAHEDRON, GREAT STELLATED TRUNCATED DODECAHEDRON, KEPLER-POINSOT SOLID, SMALL STELLATED DODECAHEDRON

References
Fischer, G. (Ed.). Plate 104 in *Mathematische Modelle/Mathematical Models, Bildband/Photograph Volume.* Braunschweig, Germany: Vieweg, p. 103, 1986.

Great Stellated Triacontahedron

see GREAT RHOMBIC TRIACONTAHEDRON

Great Stellated Truncated Dodecahedron

The UNIFORM POLYHEDRON U_{66}, also called the QUASITRUNCATED GREAT STELLATED DODECAHEDRON, whose DUAL is the GREAT TRIAKIS ICOSAHEDRON. It has SCHLÄFLI SYMBOL t'$\{\frac{5}{2},3\}$ and WYTHOFF SYMBOL

$2\,3\,|\,\frac{5}{3}$. Its faces are $20\{3\}+12\{\frac{10}{3}\}$. Its CIRCUMRADIUS for unit edge length is

$$R = \tfrac{1}{4}\sqrt{74 - 30\sqrt{5}}\,.$$

References
Wenninger, M. J. *Polyhedron Models.* Cambridge, England: Cambridge University Press, p. 161, 1989.

Great Triakis Icosahedron

The DUAL of the GREAT STELLATED TRUNCATED DODECAHEDRON.

Great Triakis Octahedron

The DUAL of the STELLATED TRUNCATED HEXAHEDRON.

see also SMALL TRIAKIS OCTAHEDRON

Great Triambic Icosahedron

The DUAL of the GREAT DITRIGONAL ICOSIDODECAHEDRON.

Great Truncated Cuboctahedron

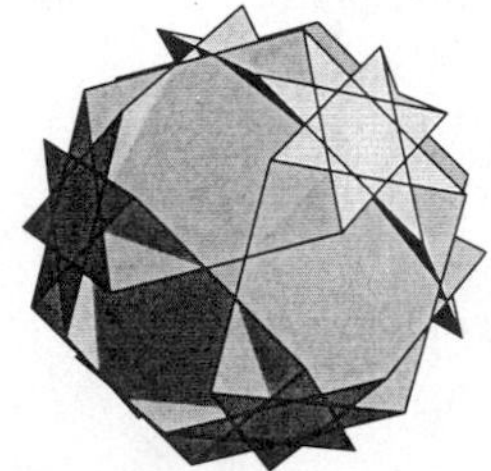

The UNIFORM POLYHEDRON U_{20}, also called the QUASITRUNCATED CUBOCTAHEDRON, whose DUAL is the GREAT DISDYAKIS DODECAHEDRON. It has SCHLÄFLI SYMBOL $t'\{\frac{3}{4}\}$ and WYTHOFF SYMBOL $2\,3\,\frac{4}{3}\,|$. Its CIRCUMRADIUS for unit edge length is

$$R = \tfrac{1}{2}\sqrt{13 - 6\sqrt{2}}\,.$$

References
Wenninger, M. J. *Polyhedron Models.* Cambridge, England: Cambridge University Press, pp. 145–146, 1989.

Great Truncated Icosahedron

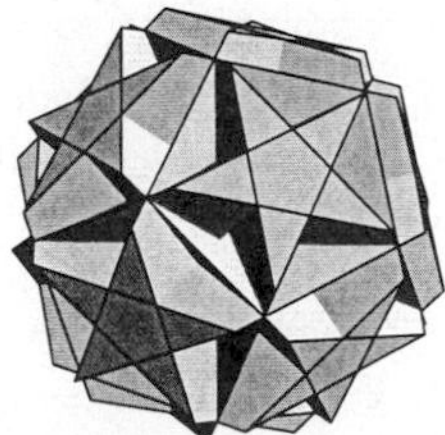

The UNIFORM POLYHEDRON U_{55}, also called the TRUNCATED GREAT ICOSAHEDRON, whose DUAL is the GREAT STELLAPENTAKIS DODECAHEDRON. It has SCHLÄFLI SYMBOL $t\{3, \frac{5}{2}\}$ and WYTHOFF SYMBOL $2\,\frac{5}{2}\,|\,3$. Its faces are $20\{6\}+12\{\frac{5}{2}\}$. Its CIRCUMRADIUS for unit edge length is

$$R = \tfrac{1}{4}\sqrt{58 - 18\sqrt{5}}\,.$$

References
Wenninger, M. J. *Polyhedron Models.* Cambridge, England: Cambridge University Press, p. 148, 1989.

Great Truncated Icosidodecahedron

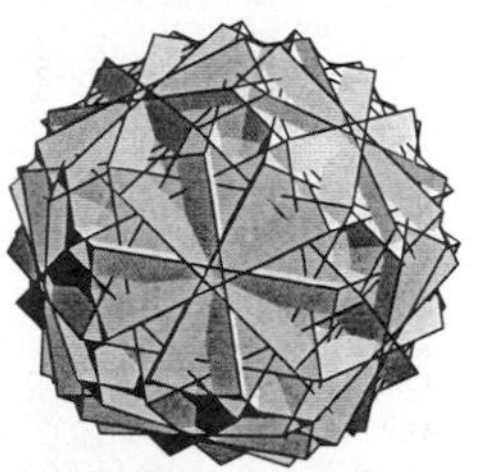

The UNIFORM POLYHEDRON U_{68}, also called the GREAT QUASITRUNCATED ICOSIDODECAHEDRON, whose DUAL is the GREAT DISDYAKIS TRIACONTAHEDRON. It has SCHLÄFLI SYMBOL $t'\left\{\begin{matrix}3\\ \frac{5}{2}\end{matrix}\right\}$ and WYTHOFF SYMBOL $2\,3\,\frac{5}{3}\,|$. Its faces are $20\{6\}+30\{4\}+12\{\frac{10}{3}\}$. Its CIRCUMRADIUS for unit edge length is

$$R = \tfrac{1}{2}\sqrt{31 - 12\sqrt{5}}\,.$$

References
Wenninger, M. J. *Polyhedron Models.* Cambridge, England: Cambridge University Press, pp. 166–167, 1989.

Greater

A quantity a is said to be greater than b if a is larger than b, written $a > b$. If a is greater than or EQUAL to b, the relationship is written $a \geq b$. If a is MUCH GREATER than b, this is written $a \gg b$. Statements involving greater than and LESS than symbols are called INEQUALITIES.

see also EQUAL, GREATER THAN/LESS THAN SYMBOL, INEQUALITY, LESS, MUCH GREATER

Greater Than/Less Than Symbol

When applied to a system possessing a length R at which solutions in a variable r change character (such as the gravitational field of a sphere as r runs from the interior to the exterior), the symbols

$$r_> \equiv \max(r, R)$$

$$r_< \equiv \min(r, R)$$

are sometimes used.

see also EQUAL, GREATER, LESS

Greatest Common Denominator

see GREATEST COMMON DIVISOR

Greatest Common Divisor

The greatest common divisor of a and b GCD(a,b), sometimes written (a,b), is the largest DIVISOR common to a and b. Symbolically, let

$$a \equiv \prod_i p_i{}^{\alpha_i} \tag{1}$$

$$b \equiv \prod_i p_i{}^{\beta_i} \tag{2}$$

Then the greatest common divisor is given by

$$(a,b) = \prod_i p_i{}^{\min(\alpha_i,\beta_i)}, \tag{3}$$

where min denotes the MINIMUM. The GCD is DISTRIBUTIVE

$$(ma, mb) = m(a,b) \tag{4}$$

$$(ma, mb, mc) = m(a,b,c), \tag{5}$$

and ASSOCIATIVE

$$(a,b,c) = ((a,b),c) = (a,(b,c)) \tag{6}$$

$$(ab, cd) = (a,c)(b,d)\left(\frac{a}{(a,c)}, \frac{d}{(b,d)}\right)\left(\frac{c}{(a,c)}, \frac{b}{(b,d)}\right). \tag{7}$$

If $a = a_1(a,b)$ and $b = b_1(a,b)$, then

$$(a,b) = (a_1(a,b), b_1(a,b)) = (a,b)(a_1,b_1), \tag{8}$$

so $(a_1,b_1) = 1$ and a_1 and b_1 are said to be RELATIVELY PRIME. The GCD is also IDEMPOTENT

$$(a,a) = a, \tag{9}$$

COMMUTATIVE

$$(a,b) = (b,a), \tag{10}$$

and satisfies the ABSORPTION LAW

$$[a,(a,b)] = a. \tag{11}$$

The probability that two INTEGERS picked at random are RELATIVELY PRIME is $[\zeta(2)]^{-1} = 6/\pi^2$, where $\zeta(z)$ is the RIEMANN ZETA FUNCTION. Polezzi (1997) observed that $(m,n) = k$, where k is the number of LATTICE POINTS in the PLANE on the straight LINE connecting the VECTORS (0, 0) and (m,n) (excluding (m,n) itself). This observation is intimately connected with the probability of obtaining RELATIVELY PRIME integers, and also with the geometric interpretation of a REDUCED FRACTION y/x as a string through a LATTICE of points with ends at (1,0) and (x,y). The pegs it presses against (x_i,y_i) give alternate CONVERGENTS y_i/x_i of the CONTINUED FRACTION for y/x, while the other CONVERGENTS are obtained from the pegs it presses against with the initial end at (0, 1).

Knuth showed that

$$(2^p - 1, q^q - 1) = 2^{(p,q)} - 1 \tag{12}$$

for p, q PRIME.

The extended greatest common divisor of two INTEGERS m and n can be defined as the greatest common divisor of m and n which also satisfies the constraint $g = rm + sn$ for r and s given INTEGERS. It is used in solving LINEAR DIOPHANTINE EQUATIONS.

see also BEZOUT NUMBERS, EUCLIDEAN ALGORITHM, LEAST PRIME FACTOR

References

Polezzi, M. "A Geometrical Method for Finding an Explicit Formula for the Greatest Common Divisor." *Amer. Math. Monthly* **104**, 445–446, 1997.

Greatest Common Divisor Theorem

Given m and n, it is possible to choose c and d such that $cm + dn$ is a common factor of m and n.

Greatest Common Factor

see GREATEST COMMON DIVISOR

Greatest Integer Function

see FLOOR FUNCTION

Greatest Lower Bound

see INFIMUM, LEAST UPPER BOUND

Greatest Prime Factor

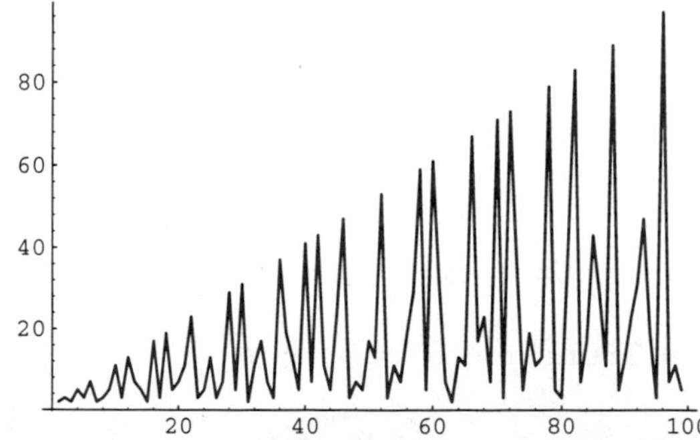

For an INTEGER $n \geq 2$, let gpf(x) denote the greatest prime factor of n, i.e., the number p_k in the factorization

$$n = p_1{}^{a_1} \cdots p_k{}^{a_k},$$

with $p_i < p_j$ for $i < j$. For $n = 2, 3, \ldots$, the first few are 2, 3, 2, 5, 3, 7, 2, 3, 5, 11, 3, 13, 7, 5, ... (Sloane's A006530). The greatest *multiple* prime factors for SQUAREFUL integers are 2, 2, 3, 2, 2, 3, 2, 2, 5, 3, 2, 2, 3, ... (Sloane's A046028).

see also DISTINCT PRIME FACTORS, FACTOR, LEAST COMMON MULTIPLE, LEAST PRIME FACTOR, MANGOLDT FUNCTION, PRIME FACTORS, TWIN PEAKS

References

Erdős, P. and Pomerance, C. "On the Largest Prime Factors of n and $n+1$." *Aequationes Math.* **17**, 211–321, 1978.
Guy, R. K. "The Largest Prime Factor of n." §B46 in *Unsolved Problems in Number Theory, 2nd ed.* New York: Springer-Verlag, pp. 101, 1994.
Heath-Brown, D. R. "The Largest Prime Factor of the Integers in an Interval." *Sci. China Ser. A* **39**, 449–476, 1996.
Mahler, K. "On the Greatest Prime Factor of $ax^m + by^n$." *Nieuw Arch. Wiskunde* **1**, 113–122, 1953.
Sloane, N. J. A. Sequence A006530/M0428 in "An On-Line Version of the Encyclopedia of Integer Sequences."

Grebe Point

see LEMOINE POINT

Greedy Algorithm

An algorithm used to recursively construct a SET of objects from the smallest possible constituent parts.

Given a SET of k INTEGERS $(a_1, a_2, \ldots, a_k)$ with $a_1 < a_2 < \ldots < a_k$, a greedy algorithm can be used to find a VECTOR of coefficients $(c_1, c_2, \ldots, c_k)$ such that

$$\sum_{i=1}^{k} c_i a_i = \mathbf{c} \cdot \mathbf{a} = n, \qquad (1)$$

where $\mathbf{c} \cdot \mathbf{a}$ is the DOT PRODUCT, for some given INTEGER n. This can be accomplished by letting $c_i = 0$ for $i = 1, \ldots, k-1$ and setting

$$c_k = \left\lfloor \frac{n}{a_k} \right\rfloor. \qquad (2)$$

Now define the difference between the representation and n as

$$\Delta \equiv n - \mathbf{c} \cdot \mathbf{a}. \qquad (3)$$

If $\Delta = 0$ at any step, a representation has been found. Otherwise, decrement the NONZERO a_i term with least i, set all $a_j = 0$ for $j < i$, and build up the remaining terms from

$$c_j = \left\lfloor \frac{\Delta_j}{a_k} \right\rfloor \qquad (4)$$

for $j = i-1, \ldots, 1$ until $\Delta = 0$ or all possibilities have been exhausted.

For example, MCNUGGET NUMBERS are numbers which are representable using only $(a_1, a_2, a_3) = (6, 9, 20)$. Taking $n = 62$ and applying the algorithm iteratively gives the sequence (0, 0, 3), (0, 2, 2), (2, 1, 2), (3, 0, 2), (1, 4, 1), at which point $\Delta = 0$. 62 is therefore a MCNUGGET NUMBER with

$$62 = (1 \cdot 6) + (4 \cdot 9) + (1 \cdot 20). \qquad (5)$$

If *any* INTEGER n can be represented with $c_i = 0$ or 1 using a sequence $(a_1, a_2, \ldots)$, then this sequence is called a COMPLETE SEQUENCE.

A greedy algorithm can also be used to break down arbitrary fractions into UNIT FRACTIONS in a finite number of steps. For a FRACTION a/b, find the least INTEGER x_1 such that $1/x_1 \leq a/b$, i.e.,

$$x_1 = \frac{\lceil b \rceil}{a}, \qquad (6)$$

where $\lceil x \rceil$ is the CEILING FUNCTION. Then find the least INTEGER x_2 such that $1/x_2 \leq a/b - 1/x_1$. Iterate until there is no remainder. The ALGORITHM gives two or fewer terms for $1/n$ and $2/n$, three or fewer terms for $3/n$, and four or fewer for $4/n$.

Paul Erdős and E. G. Strays have conjectured that the DIOPHANTINE EQUATION

$$\frac{4}{n} = \frac{1}{a} + \frac{1}{b} + \frac{1}{c} \qquad (7)$$

always can be solved, and W. Sierpiński conjectured that

$$\frac{5}{n} = \frac{1}{a} + \frac{1}{b} + \frac{1}{c} \qquad (8)$$

can be solved.

see also COMPLETE SEQUENCE, INTEGER RELATION, LEVINE-O'SULLIVAN GREEDY ALGORITHM, MCNUGGET NUMBER, REVERSE GREEDY ALGORITHM, SQUARE NUMBER, SYLVESTER'S SEQUENCE, UNIT FRACTION

References

Greek Cross

An irregular DODECAHEDRON CROSS in the shape of a PLUS SIGN.

see also CROSS, DISSECTION, DODECAHEDRON, LATIN CROSS, PLUS SIGN, SAINT ANDREW'S CROSS

Greek Problems

see GEOMETRIC PROBLEMS OF ANTIQUITY

Green's Function

Let

$$\tilde{L} = \tilde{D}^n + a_{n-1}(t)\tilde{D}^{n-1} + \ldots + a_1(t)\tilde{D} + a_0(t) \qquad (1)$$

be a differential OPERATOR in 1-D, with $a_i(t)$ CONTINUOUS for $i = 0, 1, \ldots, n-1$ on the interval I, and assume we wish to find the solution $y(t)$ to the equation

$$\tilde{L}y(t) = h(t), \qquad (2)$$

where $h(t)$ is a given CONTINUOUS on I. To solve equation (2), we look for a function $g : C^n(I) \mapsto C(I)$ such that $\tilde{L}(g(h)) = h$, where

$$y(t) = g(h(t)). \tag{3}$$

This is a CONVOLUTION equation of the form

$$y = g * h, \tag{4}$$

so the solution is

$$y(t) = \int_{t_0}^{t} g(t-x)h(x)\,dx, \tag{5}$$

where the function $g(t)$ is called the Green's function for $\tilde{L}$ on I.

Now, note that if we take $h(t) = \delta(t)$, then

$$y(t) = \int_{t_0}^{t} g(t-x)\delta(x)\,dx = g(t), \tag{6}$$

so the Green's function can be defined by

$$\tilde{L}g(t) = \delta(t). \tag{7}$$

However, the Green's function can be uniquely determined only if some initial or boundary conditions are given.

For an arbitrary linear differential operator $\tilde{L}$ in 3-D, the Green's function $G(\mathbf{r},\mathbf{r}')$ is defined by analogy with the 1-D case by

$$\tilde{L}G(\mathbf{r},\mathbf{r}') = \delta(\mathbf{r}-\mathbf{r}'). \tag{8}$$

The solution to $\tilde{L}\phi = f$ is then

$$\phi(\mathbf{r}) = \int G(\mathbf{r},\mathbf{r}')f(\mathbf{r}')\,d^3\mathbf{r}'. \tag{9}$$

Explicit expressions for $G(\mathbf{r},\mathbf{r}')$ can often be found in terms of a basis of given eigenfunctions $\phi_n(\mathbf{r}_1)$ by expanding the Green's function

$$G(\mathbf{r}_1,\mathbf{r}_2) = \sum_{n=0}^{\infty} a_n(\mathbf{r}_2)\phi_n(\mathbf{r}_1) \tag{10}$$

and DELTA FUNCTION,

$$\delta^3(\mathbf{r}_1-\mathbf{r}_2) = \sum_{n=0}^{\infty} b_n\phi_n(\mathbf{r}_1). \tag{11}$$

Multiplying both sides by $\phi_m(\mathbf{r}_2)$ and integrating over $\mathbf{r}_1$ space,

$$\int \phi_m(\mathbf{r}_2)\delta^3(\mathbf{r}_1-\mathbf{r}_2)\,d^3\mathbf{r}_1 = \sum_{n=0}^{\infty} b_n \int \phi_m(\mathbf{r}_2)\phi_n(\mathbf{r}_1)\,d^3\mathbf{r}_1 \tag{12}$$

$$\phi_m(\mathbf{r}_2) = \sum_{n=0}^{\infty} b_n\delta_{nm} = b_m, \tag{13}$$

so

$$\delta^3(\mathbf{r}_1-\mathbf{r}_2) = \sum_{n=0}^{\infty} \phi_n(\mathbf{r}_1)\phi_n(\mathbf{r}_2). \tag{14}$$

By plugging in the differential operator, solving for the a_ns, and substituting into G, the original nonhomogeneous equation then can be solved.

References

Arfken, G. "Nonhomogeneous Equation—Green's Function," "Green's Functions—One Dimension," and "Green's Functions—Two and Three Dimensions." §8.7 and §16.5–16.6 in *Mathematical Methods for Physicists, 3rd ed.* Orlando, FL: Academic Press, pp. 480–491 and 897–924, 1985.

Green's Function—Helmholtz Differential Equation

The inhomogeneous HELMHOLTZ DIFFERENTIAL EQUATION is

$$\nabla^2\psi(\mathbf{r}) + k^2\psi(\mathbf{r}) = \rho(\mathbf{r}), \tag{1}$$

where the Helmholtz operator is defined as $\tilde{L} \equiv \nabla^2 + k^2$. The Green's function is then defined by

$$(\nabla^2 + k^2)G(\mathbf{r}_1,\mathbf{r}_2) = \delta^3(\mathbf{r}_1-\mathbf{r}_2). \tag{2}$$

Define the basis functions ϕ_n as the solutions to the homogeneous HELMHOLTZ DIFFERENTIAL EQUATION

$$\nabla^2\phi_n(\mathbf{r}) + {k_n}^2\phi_n(\mathbf{r}) = 0. \tag{3}$$

The Green's function can then be expanded in terms of the ϕ_ns,

$$G(\mathbf{r}_1,\mathbf{r}_2) = \sum_{n=0}^{\infty} a_n(\mathbf{r}_2)\phi_n(\mathbf{r}_1), \tag{4}$$

and the DELTA FUNCTION as

$$\delta^3(\mathbf{r}_1-\mathbf{r}_2) = \sum_{n=0}^{\infty} \phi_n(\mathbf{r}_1)\phi_n(\mathbf{r}_2). \tag{5}$$

Plugging (4) and (5) into (2) gives

$$\nabla^2\left[\sum_{n=0}^{\infty} a_n(\mathbf{r}_2)\phi_n(\mathbf{r}_1)\right] + k^2\sum_{n=0}^{\infty} a_n(\mathbf{r}_2)\phi_n(\mathbf{r}_1) = \sum_{n=0}^{\infty} \phi_n(\mathbf{r}_1)\phi_n(\mathbf{r}_2). \tag{6}$$

Using (3) gives

$$-\sum_{n=0}^{\infty} a_n(\mathbf{r}_2){k_n}^2\phi_n(\mathbf{r}) + k^2\sum_{n=0}^{\infty} a_n(\mathbf{r}_2)\phi_n(\mathbf{r}_1) = \sum_{n=0}^{\infty} \phi_n(\mathbf{r}_1)\phi_n(\mathbf{r}_2) \tag{7}$$

$$\sum_{n=0}^{\infty} a_n(\mathbf{r}_2)\phi_n(\mathbf{r}_1)(k^2 - k_n{}^2) = \sum_{n=0}^{\infty} \phi_n(\mathbf{r}_1)\phi_n(\mathbf{r}_2). \quad (8)$$

This equation must hold true for each n, so

$$a_n(\mathbf{r}_2)\phi_n(\mathbf{r}_1)(k^2 - k_n{}^2) = \phi_n(\mathbf{r}_1)\phi_n(\mathbf{r}_2) \quad (9)$$

$$a_n(\mathbf{r}_2) = \frac{\phi_n(\mathbf{r}_2)}{k^2 - k_n{}^2}, \quad (10)$$

and (4) can be written

$$G(\mathbf{r}_1, \mathbf{r}_2) = \sum_{n=0}^{\infty} \frac{\phi_n(\mathbf{r}_1)\phi_n(\mathbf{r}_2)}{k^2 - k_n{}^2}. \quad (11)$$

The general solution to (1) is therefore

$$\psi(\mathbf{r}_1) = \int G(\mathbf{r}_1, \mathbf{r}_2)\rho(\mathbf{r}_2)\, d^3\mathbf{r}_2$$
$$= \sum_{n=0}^{\infty} \int \frac{\phi_n(\mathbf{r}_1)\phi_n(\mathbf{r}_2)\rho(\mathbf{r}_2)}{k^2 - k_n{}^2}\, d^3\mathbf{r}_2. \quad (12)$$

References

Arfken, G. *Mathematical Methods for Physicists, 3rd ed.* Orlando, FL: Academic Press, pp. 529–530, 1985.

Green's Function—Poisson's Equation

POISSON'S EQUATION equation is

$$\nabla^2\phi = 4\pi\rho, \quad (1)$$

where ϕ is often called a potential function and ρ a density function, so the differential operator in this case is $\tilde{L} = \nabla^2$. As usual, we are looking for a Green's function $G(\mathbf{r}_1, \mathbf{r}_2)$ such that

$$\nabla^2 G(\mathbf{r}_1, \mathbf{r}_2) = \delta^3(\mathbf{r}_1 - \mathbf{r}_2). \quad (2)$$

But from LAPLACIAN,

$$\nabla^2\left(\frac{1}{|\mathbf{r} - \mathbf{r}'|}\right) = -4\pi\delta^3(\mathbf{r} - \mathbf{r}'), \quad (3)$$

so

$$G(\mathbf{r}, \mathbf{r}') = -\frac{1}{4\pi|\mathbf{r} - \mathbf{r}'|}, \quad (4)$$

and the solution is

$$\phi(\mathbf{r}) = \int G(\mathbf{r}, \mathbf{r}')[4\pi\rho(\mathbf{r}')]\, d^3\mathbf{r}' = -\int \frac{\rho(\mathbf{r}')\, d^3\mathbf{r}'}{|\mathbf{r} - \mathbf{r}'|}. \quad (5)$$

Expanding $G(\mathbf{r}_1, \mathbf{r}_2)$ in the SPHERICAL HARMONICS Y_l^m gives

$$G(\mathbf{r}_1, \mathbf{r}_2)$$
$$= \sum_{l=0}^{\infty} \sum_{m=-l}^{l} \frac{1}{2l+1} \frac{r_<^l}{r_>^{l+1}} Y_l^m(\theta_1, \phi_1) Y_l^{m*}(\theta_2, \phi_2), \quad (6)$$

where $r_<$ and $r_>$ are GREATER THAN/LESS THAN SYMBOLS. This expression simplifies to

$$G(\mathbf{r}_1, \mathbf{r}_2) = \frac{1}{4\pi} \sum_{l=0}^{\infty} \frac{r_<^l}{r_>^{l+1}} P_l(\cos\gamma), \quad (7)$$

where P_l are LEGENDRE POLYNOMIALS, and $\cos\gamma \equiv \mathbf{r}_1 \cdot \mathbf{r}_2$. Equations (6) and (7) give the addition theorem for LEGENDRE POLYNOMIALS.

In CYLINDRICAL COORDINATES, the Green's function is much more complicated,

$$G(\mathbf{r}_1, \mathbf{r}_2) = \frac{1}{2\pi^2} \sum_{m=-\infty}^{\infty}$$
$$\int_0^\infty I_m(k\rho_<) K_m(k\rho_>) e^{im(\phi_1 - \phi_2)} \cos[k(z_1 - z_2)]\, dk, \quad (8)$$

where $I_m(x)$ and $K_m(x)$ are MODIFIED BESSEL FUNCTIONS OF THE FIRST and SECOND KINDS (Arfken 1985).

References

Arfken, G. *Mathematical Methods for Physicists, 3rd ed.* Orlando, FL: Academic Press, pp. 485–486, 905, and 912, 1985.

Green's Identities

Green's identities are a set of three vector derivative/integral identities which can be derived starting with the vector derivative identities

$$\nabla \cdot (\psi\nabla\phi) = \psi\nabla^2\phi + (\nabla\psi) \cdot (\nabla\phi) \quad (1)$$

and

$$\nabla \cdot (\phi\nabla\psi) = \phi\nabla^2\psi + (\nabla\phi) \cdot (\nabla\psi), \quad (2)$$

where $\nabla\cdot$ is the DIVERGENCE, ∇ is the GRADIENT, ∇^2 is the LAPLACIAN, and $\mathbf{a} \cdot \mathbf{b}$ is the DOT PRODUCT. From the DIVERGENCE THEOREM,

$$\int_V (\nabla \cdot \mathbf{F})\, dV = \int_S \mathbf{F} \cdot d\mathbf{a}. \quad (3)$$

Plugging (2) into (3),

$$\int_S \phi(\nabla\psi) \cdot d\mathbf{a} = \int_V [\phi\nabla^2\psi + (\nabla\phi) \cdot (\nabla\psi)]\, dV. \quad (4)$$

This is Green's first identity.

Subtracting (2) from (1),

$$\nabla \cdot (\phi\nabla\psi - \psi\nabla\phi) = \phi\nabla^2\psi - \psi\nabla^2\phi. \quad (5)$$

Therefore,

$$\int_V (\phi\nabla^2\psi - \psi\nabla^2\phi)\, dV = \int_S (\phi\nabla\psi - \psi\nabla\phi) \cdot d\mathbf{a}. \quad (6)$$

This is Green's second identity.

Let u have continuous first PARTIAL DERIVATIVES and be HARMONIC inside the region of integration. Then Green's third identity is

$$u(x,y) = \frac{1}{2\pi} \oint_C \left[\ln\left(\frac{1}{r}\right) \frac{\partial u}{\partial n} - u \frac{\partial}{\partial n} \ln\left(\frac{1}{r}\right) \right] ds \quad (7)$$

(Kaplan 1991, p. 361).

References

Kaplan, W. *Advanced Calculus, 4th ed.* Reading, MA: Addison-Wesley, 1991.

Greene's Method

A method for predicting the onset of widespread CHAOS. It is based on the hypothesis that the dissolution of an invariant torus can be associated with the sudden change from stability to instability of nearly closed orbits (Tabor 1989, p. 163).

see also OVERLAPPING RESONANCE METHOD

References

Tabor, M. *Chaos and Integrability in Nonlinear Dynamics: An Introduction.* New York: Wiley, 1989.

Green Space

A G-SPACE provides local notions of harmonic, hyperharmonic, and superharmonic functions. When there exists a nonconstant superharmonic function greater than 0, it is a called a Green space. Examples are $\mathbb{R}^n$ (for $n \geq 3$) and any bounded domain of $\mathbb{R}^n$.

Green's Theorem

Green's theorem is a vector identity which is equivalent to the CURL THEOREM in the PLANE. Over a region D in the plane with boundary ∂D,

$$\int_{\partial D} f(x,y)\,dx + g(x,y)\,dy = \iint_D \left(\frac{\partial g}{\partial x} - \frac{\partial f}{\partial y} \right) dx\,dy$$

$$\int_{\partial D} \mathbf{F} \cdot d\mathbf{s} = \iint_D (\nabla \times \mathbf{F}) \cdot \mathbf{k}\,dA.$$

If the region D is on the left when traveling around ∂D, then AREA of D can be computed using

$$A = \tfrac{1}{2} \int_{\partial D} x\,dy - y\,dx.$$

see also CURL THEOREM, DIVERGENCE THEOREM

References

Arfken, G. "Gauss's Theorem." §1.11 in *Mathematical Methods for Physicists, 3rd ed.* Orlando, FL: Academic Press, pp. 57–61, 1985.

Gregory's Formula

A series FORMULA for PI found by Gregory and Leibniz,

$$\frac{\pi}{4} = 1 - \frac{1}{3} + \frac{1}{5} + \ldots.$$

It converges very slowly, but its convergence can be accelerated using certain transformations, in particular

$$\pi = \sum_{k=1}^{\infty} \frac{3^k - 1}{4^k} \zeta(k+1),$$

where $\zeta(z)$ is the RIEMANN ZETA FUNCTION (Vardi 1991).

see also MACHIN'S FORMULA, MACHIN-LIKE FORMULAS, PI

References

Vardi, I. *Computational Recreations in Mathematica.* Reading, MA: Addison-Wesley, pp. 157–158, 1991.

Gregory Number

A number

$$t_x = \tan^{-1}(\tfrac{1}{x}) = \cot^{-1} x,$$

where x is an INTEGER or RATIONAL NUMBER, $\tan^{-1} x$ is the INVERSE TANGENT, and $\cot^{-1} x$ is the INVERSE COTANGENT. Gregory numbers arise in the determination of MACHIN-LIKE FORMULAS. Every Gregory number t_x can be expressed uniquely as a sum of t_ns where the ns are STØRMER NUMBERS.

References

Conway, J. H. and Guy, R. K. "Gregory's Numbers" In *The Book of Numbers.* New York: Springer-Verlag, pp. 241–242, 1996.

Grelling's Paradox

A semantic PARADOX, also called the HETEROLOGICAL PARADOX, which arises by defining "heterological" to mean "a word which does not describe itself." The word "heterological" is therefore heterological IFF it is not.

see also RUSSELL'S PARADOX

References

Hofstadter, D. R. *Gödel, Escher, Bach: An Eternal Golden Braid.* New York: Vintage Books, pp. 20–21, 1989.

Grenz-Formel

An equation derived by Kronecker:

$$\sum{}' (x^2 + y^2 + dz^2)^{-s} = 4\zeta(s)\eta(s) + \frac{2\pi}{s-1}\frac{\zeta(2s-2)}{d^{s-1}}$$

$$+ \frac{2\pi^s}{\Gamma(s)} d^{(1-s)/2} \sum_{n=1}^{\infty} n^{(s-1)/2} \sum_{u^2|n} \frac{r\left(\frac{n}{u^2}\right)}{u^{2s-2}} \int_0^{\infty} e^{\pi\sqrt{nd}\,(y+y^{-1})} y^{s-2}\,dy,$$

where

$$r(n) = 4 \sum_{d|n} \sin(\tfrac{1}{2}\pi d),$$

$\zeta(z)$, is the RIEMANN ZETA FUNCTION, $\eta(z)$ is the DIRICHLET ETA FUNCTION, $\Gamma(z)$ is the GAMMA FUNCTION, and the primed sum omits infinite terms (Selberg and Chowla 1967).

References

Borwein, J. M. and Borwein, P. B. *Pi & the AGM: A Study in Analytic Number Theory and Computational Complexity.* New York: Wiley, pp. 296–297, 1987.

Selberg, A. and Chowla, S. "On Epstein's Zeta-Function." *J. Reine. Angew. Math.* **227**, 86–110, 1967.

Griffiths Points

"The" Griffiths point is the fixed point in GRIFFITHS' THEOREM. Given four points on a CIRCLE and a line through the center of the CIRCLE, the four corresponding Griffiths points are COLLINEAR (Tabov 1995).

The points

$$Gr = I + 4Ge$$

$$Gr' = I - 4Ge,$$

are known as the first and second Griffiths points, where I is the INCENTER and Ge is the GERGONNE POINT.

see also GERGONNE POINT, GRIFFITHS' THEOREM, INCENTER, OLDKNOW POINTS, RIGBY POINTS

References

Oldknow, A. "The Euler-Gergonne-Soddy Triangle of a Triangle." *Amer. Math. Monthly* **103**, 319–329, 1996.

Tabov, J. "Four Collinear Griffiths Points." *Math. Mag.* **68**, 61–64, 1995.

Griffiths' Theorem

When a point P moves along a line through the CIRCUMCENTER of a given TRIANGLE Δ, the CIRCUMCIRCLE of the PEDAL TRIANGLE of P with respect to Δ passes through a fixed point (the GRIFFITHS POINT) on the NINE-POINT CIRCLE of Δ.

see also CIRCUMCENTER, GRIFFITHS POINTS, NINE-POINT CIRCLE, PEDAL TRIANGLE

Grimm's Conjecture

Grimm conjectures that if $n+1$, $n+2$, ..., $n+k$ are all COMPOSITE NUMBERS, then there are distinct PRIMES p_{i_j} such that $p_{i_j}|(n+j)$ for $1 \leq j \leq k$.

References

Guy, R. K. "Grimm's Conjecture." §B32 in *Unsolved Problems in Number Theory, 2nd ed.* New York: Springer-Verlag, p. 86, 1994.

Grinberg Formula

A formula satisfied by all HAMILTONIAN CIRCUITS with n nodes. Let f_j be the number of regions inside the circuit with j sides, and let g_j be the number of regions outside the circuit with j sides. If there are d interior diagonals, then there must be $d+1$ regions

$$[\#\text{ regions in interior}] = d+1 = f_2 + f_3 + \ldots + f_n. \quad (1)$$

Any region with j sides is bounded by j EDGES, so such regions contribute jf_j to the total. However, this counts each diagonal twice (and each EDGE only once). Therefore,

$$2f_2 + 3f_3 + \ldots + nf_n = 2d + n. \quad (2)$$

Take (2) $- 2 \times$ (1),

$$f_3 + 2f_4 + 3f_5 + \ldots + (n-2)f_n = n - 2. \quad (3)$$

Similarly,

$$g_3 + 2g_4 + \ldots + (n-2)g_n = n - 2, \quad (4)$$

so

$$(f_3-g_3)+2(f_4-g_4)+3(f_5-g_5)+\ldots+(n-2)(f_n-g_n) = 0. \quad (5)$$

Gröbner Basis

A Gröbner basis for a system of POLYNOMIAL equations is an equivalence system that possesses useful properties. It is very roughly analogous to computing an ORTHONORMAL BASIS from a set of BASIS VECTORS and can be described roughly as a combination of GAUSSIAN ELIMINATION (for linear systems) and the EUCLIDEAN ALGORITHM (for UNIVARIATE POLYNOMIALS over a FIELD).

Gröbner bases are useful in the construction of symbolic algebra algorithms. The algorithm for computing Gröbner bases is known as BUCHBERGER'S ALGORITHM.

see also BUCHBERGER'S ALGORITHM, COMMUTATIVE ALGEBRA

References

Adams, W. W. and Loustaunau, P. *An Introduction to Gröbner Bases.* Providence, RI: Amer. Math. Soc., 1994.

Becker, T. and Weispfennig, V. *Gröbner Bases: A Computational Approach to Commutative Algebra.* New York: Springer-Verlag, 1993.

Cox, D.; Little, J.; and O'Shea, D. *Ideals, Varieties, and Algorithms: An Introduction to Algebraic Geometry and Commutative Algebra, 2nd ed.* New York: Springer-Verlag, 1996.

Eisenbud, D. *Commutative Algebra with a View toward Algebraic Geometry.* New York: Springer-Verlag, 1995.

Mishra, B. *Algorithmic Algebra.* New York: Springer-Verlag, 1993.

Groemer Packing

A honeycomb-like packing that forms HEXAGONS.

see also GROEMER THEOREM

References

Stewart, I. "A Bundling Fool Beats the Wrap." *Sci. Amer.* **268**, 142–144, 1993.

Groemer Theorem

Given n CIRCLES and a PERIMETER p, the total AREA of the CONVEX HULL is

$$A_{\text{Convex Hull}} = 2\sqrt{3}(n-1) + p(1 - \tfrac{1}{2}\sqrt{3}) + \pi(\sqrt{3}-1).$$

Furthermore, the actual AREA equals this value IFF the packing is a GROEMER PACKING. The theorem was proved in 1960 by Helmut Groemer.

see also CONVEX HULL

Gronwall's Theorem

Let $\sigma(n)$ be the DIVISOR FUNCTION. Then

$$\overline{\lim_{n\to\infty}} \frac{\sigma(n)}{n \ln\ln n} = e^{\gamma},$$

where γ is the EULER-MASCHERONI CONSTANT. Ramanujan independently discovered a less precise version of this theorem (Berndt 1994). Robin (1984) showed that the validity of the inequality

$$\sigma(n) < e^{\gamma} n \ln\ln n$$

for $n \geq 5041$ is equivalent to the RIEMANN HYPOTHESIS.

References

Berndt, B. C. *Ramanujan's Notebooks: Part I.* New York: Springer-Verlag, p. 94, 1985.

Gronwall, T. H. "Some Asymptotic Expressions in the Theory of Numbers." *Trans. Amer. Math. Soc.* **37**, 113–122, 1913.

Nicholas, J.-L. "On Highly Composite Numbers." In *Ramanujan Revisited: Proceedings of the Centenary Conference* (Ed. G. E. Andrews, B. C. Berndt, and R. A. Rankin). Boston, MA: Academic Press, pp. 215–244, 1988.

Robin, G. "Grandes Valeurs de la foction somme des diviseurs et hypothèse de Riemann." *J. Math. Pures Appl.* **63**, 187–213, 1984.

Gross

A DOZEN DOZEN, or the SQUARE NUMBER 144.

see also 12, DOZEN

Grossencharacter

In the original formulation, a quantity associated with ideal class groups. According to Chevalley's formulation, a Grossencharacter is a MULTIPLICATIVE CHARACTER of the group of ADÉLES that is trivial on the diagonally embedded $k^{\times}$, where k is a number FIELD.

References

Knapp, A. W. "Group Representations and Harmonic Analysis, Part II." *Not. Amer. Math. Soc.* **43**, 537–549, 1996.

Grossman's Constant

Define the sequence $a_0 = 1$, $a_1 = x$, and

$$a_{n+2} = \frac{a_n}{1 + a_{n+1}}$$

for $n \geq 0$. Janssen and Tjaden (1987) showed that this sequence converges for exactly one value of x, $x = 0.73733830336929\ldots$, confirming Grossman's conjecture.

References

Finch, S. "Favorite Mathematical Constants." `http://www.mathsoft.com/asolve/constant/grssmn/grssmn.html`.

Janssen, A. J. E. M. and Tjaden, D. L. A. Solution to Problem 86-2. *Math. Intel.* **9**, 40–43, 1987.

Grothendieck's Majorant

The best known majorant of Grothendieck's constant. Let A be an $n \times n$ REAL SQUARE MATRIX such that

$$\left| \sum_{1 \leq i,j \leq n} a_{ij} x_i y_j \right| \tag{1}$$

in which x_i and y_j have REAL ABSOLUTE VALUES $<$ 1. Grothendieck has shown there exists a number K_G independent of A and n satisfying

$$\left| \sum_{1 \leq i,j \leq n} a_{ij} \langle x_i, y_j \rangle \right| \tag{2}$$

in which the vectors x_i and y_j have a norm < 1 in HILBERT SPACE. The Grothendieck constant is the smallest REAL NUMBER for which this inequality has been proven. Krivine (1977) showed that

$$1.676\ldots \leq K_G \leq 1.782\ldots, \tag{3}$$

and has postulated that

$$K_G \equiv \frac{\pi}{2\ln(1+\sqrt{2})} = 1.7822139\ldots. \tag{4}$$

It is related to KHINTCHINE'S CONSTANT.

References

Krivine, J. L. "Sur las constante de Grothendieck." *C. R. A. S.* **284**, 8, 1977.

Le Lionnais, F. *Les nombres remarquables.* Paris: Hermann, p. 42, 1983.

Grothendieck's Theorem

Let E and F be paired spaces with S a family of absolutely convex bounded sets of F such that the sets of S generate F and, if $B_1, B_2 \in S$, then there exists a $B_3 \in S$ such that $B_3 \supset B_1$ and $B_3 \supset B_2$. Then E_S is complete IFF algebraic linear functional $f(y)$ of F that is weakly continuous on every $B \in S$ is expressed as $f(y) = \langle x, y \rangle$ for some $x \in E$. When E_S is not complete, the space of all linear functionals satisfying this condition gives the completion $\hat{E}_S$ of E_S.

see also MACKEY'S THEOREM

References

Iyanaga, S. and Kawada, Y. (Eds.). "Grothendieck's Theorem." §407L in *Encyclopedic Dictionary of Mathematics.* Cambridge, MA: MIT Press, p. 1274, 1980.

Ground Set

A PARTIALLY ORDERED SET is defined as an ordered pair $P = (X, \leq)$. Here, X is called the GROUND SET of P and $\leq$ is the PARTIAL ORDER of P.

see also PARTIAL ORDER, PARTIALLY ORDERED SET

Group

A group G is defined as a finite or infinite set of OPERANDS (called "elements") A, B, C, ... that may be combined or "multiplied" via a BINARY OPERATOR to form well-defined products and which furthermore satisfy the following conditions:

1. Closure: If A and B are two elements in G, then the product AB is also in G.
2. Associativity: The defined multiplication is associative, i.e., for all $A, B, C \in G$, $(AB)C = A(BC)$.
3. Identity: There is an IDENTITY ELEMENT I (a.k.a. 1, E, or e) such that $IA = AI = A$ for every element $A \in G$.
4. Inverse: There must be an inverse or reciprocal of each element. Therefore, the set must contain an element $B = A^{-1}$ such that $AA^{-1} = A^{-1}A = I$ for each element of G.

A group is therefore a MONOID for which every element is invertible. A group must contain at least one element.

The study of groups is known as GROUP THEORY. If there are a finite number of elements, the group is called a FINITE GROUP and the number of elements is called the ORDER of the group.

Since each element A, B, C, ..., X, and Y is a member of the GROUP, GROUP property 1 requires that the product

$$D \equiv ABC \cdots XY \tag{1}$$

must also be a member. Now apply D to $Y^{-1}X^{-1} \cdots C^{-1}B^{-1}A^{-1}$,

$$D(Y^{-1}X^{-1} \cdots C^{-1}B^{-1}A^{-1}) = (ABC \cdots XY)(Y^{-1}X^{-1} \cdots C^{-1}B^{-1}A^{-1}). \tag{2}$$

But

$$ABC \cdots XYY^{-1}X^{-1} \cdots C^{-1}B^{-1}A^{-1} = ABC \cdots XIX^{-1} \cdots C^{-1}B^{-1}A^{-1} = ABC \cdots C^{-1}B^{-1}A^{-1} = \ldots = AA^{-1} = I, \tag{3}$$

so

$$I = D(Y^{-1}X^{-1} \cdots C^{-1}B^{-1}A^{-1}), \tag{4}$$

which means that

$$D^{-1} = Y^{-1}X^{-1} \cdots C^{-1}B^{-1}A^{-1} \tag{5}$$

and

$$(ABC \cdots XY)^{-1} = Y^{-1}X^{-1} \cdots C^{-1}B^{-1}A^{-1}. \tag{6}$$

An IRREDUCIBLE REPRESENTATION of a group is a representation for which there exists no UNITARY TRANSFORMATION which will transform the representation MATRIX into block diagonal form. The IRREDUCIBLE REPRESENTATION has some remarkable properties. Let the ORDER of a GROUP be h, and the dimension of the ith representation (the order of each constituent matrix) be l_i (a POSITIVE INTEGER). Let any operation be denoted R, and let the mth row and nth column of the matrix corresponding to a matrix R in the ith IRREDUCIBLE REPRESENTATION be $\Gamma_i(R)_{mn}$. The following properties can be derived from the GROUP ORTHOGONALITY THEOREM,

$$\sum_R \Gamma_i(R)_{mn}\Gamma_j(R)_{m'n'}{}^* = \frac{h}{\sqrt{l_i l_j}}\delta_{ij}\delta_{mm'}\delta_{nn'}. \tag{7}$$

1. The DIMENSIONALITY THEOREM:
$$h = \sum_i {l_i}^2 = {l_1}^2 + {l_2}^2 + {l_3}^2 + \ldots = \sum_i {\chi_i}^2(I), \tag{8}$$
where each l_i must be a POSITIVE INTEGER and χ is the CHARACTER (trace) of the representation.
2. The sum of the squares of the CHARACTERS in any IRREDUCIBLE REPRESENTATION i equals h,
$$h = \sum_R {\chi_i}^2(R). \tag{9}$$
3. ORTHOGONALITY of different representations
$$\sum_R \chi_i(R)\chi_j(R) = 0 \quad \text{for } i \neq j. \tag{10}$$
4. In a given representation, reducible or irreducible, the CHARACTERS of all MATRICES belonging to operations in the same class are identical (but differ from those in other representations).
5. The number of IRREDUCIBLE REPRESENTATIONS of a GROUP is equal to the number of CONJUGACY CLASSES in the GROUP. This number is the dimension of the Γ MATRIX (although some may have zero elements).
6. A one-dimensional representation with all 1s (totally symmetric) will always exist for any GROUP.
7. A 1-D representation for a GROUP with elements expressed as MATRICES can be found by taking the CHARACTERS of the MATRICES.
8. The number a_i of IRREDUCIBLE REPRESENTATIONS χ_i present in a reducible representation c is given by
$$a_i = \frac{1}{h}\sum_R \chi(R)\chi_i(R), \tag{11}$$

where h is the ORDER of the GROUP and the sum must be taken over all elements in each class. Written explicitly,

$$a_i = \frac{1}{h} \sum_R \chi(R) \chi_i'(R) n_R, \tag{12}$$

where χ_i' is the CHARACTER of a single entry in the CHARACTER TABLE and n_R is the number of elements in the corresponding CONJUGACY CLASS.

see also ABELIAN GROUP, ADÉLE GROUP, AFFINE GROUP, ALTERNATING GROUP, ARTINIAN GROUP, ASCHBACHER'S COMPONENT THEOREM, B_p-THEOREM, BABY MONSTER GROUP, BETTI GROUP, BIMONSTER, BORDISM GROUP, BRAID GROUP, BRAUER GROUP, BURNSIDE PROBLEM, CENTER (GROUP), CENTRALIZER, CHARACTER (GROUP), CHARACTER (MULTIPLICATIVE), CHEVALLEY GROUPS, CLASSICAL GROUPS, COBORDISM GROUP, COHOMOTOPY GROUP, COMPONENT, CONJUGACY CLASS, COSET, CONWAY GROUPS, COXETER GROUP, CYCLIC GROUP, DIHEDRAL GROUP, DIMENSIONALITY THEOREM, DYNKIN DIAGRAM, ELLIPTIC GROUP MODULO p, ENGEL'S THEOREM, EUCLIDEAN GROUP, FEIT-THOMPSON THEOREM, FINITE GROUP, FISCHER GROUPS, FISCHER'S BABY MONSTER GROUP, FUNDAMENTAL GROUP, GENERAL LINEAR GROUP, GENERAL ORTHOGONAL GROUP, GENERAL UNITARY GROUP, GLOBAL $C(G;T)$ THEOREM, GROUPOID, GROUP ORTHOGONALITY THEOREM, HALL-JANKO GROUP, HAMILTONIAN GROUP, HARADA-NORTON GROUP, HEISENBERG GROUP, HELD GROUP, HERMANN-MAUGUIN SYMBOL, HIGMAN-SIMS GROUP, HOMEOMORPHIC GROUP, HYPERGROUP, ICOSAHEDRAL GROUP, IRREDUCIBLE REPRESENTATION, ISOMORPHIC GROUPS, JANKO GROUPS, JORDAN-HÖLDER THEOREM, KLEINIAN GROUP, KUMMER GROUP, $L_{p'}$-BALANCE THEOREM, LAGRANGE'S GROUP THEOREM, LOCAL GROUP THEORY, LINEAR GROUP, LYONS GROUP, MATHIEU GROUPS, MATRIX GROUP, MCLAUGHLIN GROUP, MÖBIUS GROUP, MODULAR GROUP, MODULO MULTIPLICATION GROUP, MONODROMY GROUP, MONOID, MONSTER GROUP, MULLIKEN SYMBOLS, NÉRON-SEVERI GROUP, NILPOTENT GROUP, NONCOMMUTATIVE GROUP, NORMAL SUBGROUP, NORMALIZER, O'NAN GROUP, OCTAHEDRAL GROUP, ORDER (GROUP), ORTHOGONAL GROUP, ORTHOGONAL ROTATION GROUP, OUTER AUTOMORPHISM GROUP, p-GROUP, p'-GROUP, p-LAYER, POINT GROUPS, POSITIVE DEFINITE FUNCTION, PRIME GROUP, PROJECTIVE GENERAL LINEAR GROUP, PROJECTIVE GENERAL ORTHOGONAL GROUP, PROJECTIVE GENERAL UNITARY GROUP, PROJECTIVE SPECIAL LINEAR GROUP, PROJECTIVE SPECIAL ORTHOGONAL GROUP, PROJECTIVE SPECIAL UNITARY GROUP, PROJECTIVE SYMPLECTIC GROUP, PSEUDOGROUP, QUASIGROUP, QUASISIMPLE GROUP, QUASITHIN THEOREM, QUASI-UNIPOTENT GROUP, REPRESENTATION, RESIDUE CLASS, RUBIK'S CUBE, RUDVALIS GROUP, SCHÖNFLIES SYMBOL, SCHUR MULTIPLIER, SEMISIMPLE, SIGNALIZER FUNCTOR THEOREM, SELMER GROUP, SEMIGROUP, SIMPLE GROUP, SOLVABLE GROUP, SPACE GROUPS, SPECIAL LINEAR GROUP, SPECIAL ORTHOGONAL GROUP, SPECIAL UNITARY GROUP, SPORADIC GROUP, STOCHASTIC GROUP, STRONGLY EMBEDDED THEOREM, SUBGROUP, SUBNORMAL, SUPPORT, SUZUKI GROUP, SYMMETRIC GROUP, SYMPLECTIC GROUP, TETRAHEDRAL GROUP, THOMPSON GROUP, TIGHTLY EMBEDDED, TITS GROUP, TRIANGULAR SYMMETRY GROUP, TWISTED CHEVALLEY GROUPS, UNIMODULAR GROUP, UNIPOTENT, UNITARY GROUP, VIERGRUPPE, VON DYCK'S THEOREM

References

Arfken, G. *Mathematical Methods for Physicists, 3rd ed.* Orlando, FL: Academic Press, pp. 237–276, 1985.

Farmer, D. *Groups and Symmetry.* Providence, RI: Amer. Math. Soc., 1995.

Weisstein, E. W. "Groups." `http://www.astro.virginia.edu/~eww6n/math/notebooks/Groups.m`.

Weyl, H. *The Classical Groups: Their Invariants and Representations.* Princeton, NJ: Princeton University Press, 1997.

Wybourne, B. G. *Classical Groups for Physicists.* New York: Wiley, 1974.

Group Convolution

The convolution of two COMPLEX-valued functions on a GROUP G is defined as

$$(a * b)(g) = \sum_{k \in G} a(k) b(k^{-1} g)$$

where the SUPPORT (set which is not zero) of each function is finite.

References

Weinstein, A. "Groupoids: Unifying Internal and External Symmetry." *Not. Amer. Math. Soc.* **43**, 744–752, 1996.

Group Orthogonality Theorem

Let Γ be a representation for a GROUP of ORDER h, then

$$\sum_R \Gamma_i(R)_{mn} \Gamma_j(R)_{m'n'}{}^* = \frac{h}{\sqrt{l_i l_j}} \delta_{ij} \delta_{mm'} \delta_{nn'}.$$

The proof is nontrivial and may be found in Eyring *et al.* (1944).

References

Eyring, H.; Walker, J.; and Kimball, G. E. *Quantum Chemistry.* New York: Wiley, p. 371, 1944.

Group Ring

The set of sums $\sum_x a_x x$ ranging over a multiplicative GROUP and a_i are elements of a FIELD with all but a finite number of $a_i = 0$.

Group Theory

The study of GROUPS. Gauss developed but did not publish parts of the mathematics of group theory, but Galois is generally considered to have been the first to develop the theory. Group theory is a powerful formal method for analyzing abstract and physical systems in which SYMMETRY is present and has surprising importance in physics, especially quantum mechanics.

see also FINITE GROUP, GROUP, PLETHYSM, SYMMETRY

References

Arfken, G. "Introduction to Group Theory." §4.8 in *Mathematical Methods for Physicists, 3rd ed.* Orlando, FL: Academic Press, pp. 237–276, 1985.

Burnside, W. *Theory of Groups of Finite Order, 2nd ed.* New York: Dover, 1955.

Burrow, M. *Representation Theory of Finite Groups.* New York: Dover, 1993.

Carmichael, R. D. *Introduction to the Theory of Groups of Finite Order.* New York: Dover, 1956.

Conway, J. H.; Curtis, R. T.; Norton, S. P.; Parker, R. A.; and Wilson, R. A. *Atlas of Finite Groups: Maximal Subgroups and Ordinary Characters for Simple Groups.* Oxford, England: Clarendon Press, 1985.

Cotton, F. A. *Chemical Applications of Group Theory, 3rd ed.* New York: Wiley, 1990.

Dixon, J. D. *Problems in Group Theory.* New York: Dover, 1973.

Grossman, I. and Magnus, W. *Groups and Their Graphs.* Washington, DC: Math. Assoc. Amer., 1965.

Hamermesh, M. *Group Theory and Its Application to Physical Problems.* New York: Dover, 1989.

Lomont, J. S. *Applications of Finite Groups.* New York: Dover, 1987.

Magnus, W.; Karrass, A.; and Solitar, D. *Combinatorial Group Theory: Presentations of Groups in Terms of Generators and Relations.* New York: Dover, 1976.

Robinson, D. J. S. *A Course in the Theory of Groups, 2nd ed.* New York: Springer-Verlag, 1995.

Rose, J. S. *A Course on Group Theory.* New York: Dover, 1994.

Rotman, J. J. *An Introduction to the Theory of Groups, 4th ed.* New York: Springer-Verlag, 1995.

Groupoid

There are at least two definitions of "groupoid" currently in use.

The first type of groupoid is an algebraic structure on a SET with a BINARY OPERATOR. The only restriction on the operator is CLOSURE (i.e., applying the BINARY OPERATOR to two elements of a given set S returns a value which is itself a member of S). Associativity, commutativity, etc., are not required (Rosenfeld 1968, pp. 88–103). A groupoid can be empty. The numbers of nonisomorphic groupoids of this type having n elements are 1, 1, 10, 3330, 178981952, ... (Sloane's A001329), and the numbers of nonisomorphic and nonantiisimorphic groupoids are 1, 7, 1734, 89521056, ... (Sloane's A001424). An associative groupoid is called a SEMIGROUP.

The second type of groupoid is an algebraic structure first defined by Brandt (1926) and also known as a VIRTUAL GROUP. A groupoid with base B is a set G with mappings α and β from G onto B and a partially defined binary operation $(g,h) \mapsto gh$, satisfying the following four conditions:

1. gh is defined only when $\beta(G) = \alpha(h)$ for certain maps α and β from G onto $\mathbb{R}^2$ with $\alpha : (x,\gamma,y) \mapsto x$ and $\beta : (x,\gamma,y) \mapsto y$.
2. ASSOCIATIVITY: If either $(gh)k$ or $g(hk)$ is defined, then so is the other and $(gh)k = g(hk)$.
3. For each g in G, there are left and right IDENTITY ELEMENTS λ_g and ρ_g such that $\lambda_g g = g = g\rho_g$.
4. Each g in G has an inverse g^{-1} for which $gg^{-1} = \lambda_g$ and $g^{-1}g = \rho_g$

(Weinstein 1996). A groupoid is a small CATEGORY with every morphism invertible.

see also BINARY OPERATOR, INVERSE SEMIGROUP, LIE ALGEBROID, LIE GROUPOID, MONOID, QUASIGROUP, SEMIGROUP, TOPOLOGICAL GROUPOID

References

Brandt, W. "Über eine Verallgemeinerung des Gruppengriffes." *Math. Ann.* **96**, 360–366, 1926.

Brown, R. "From Groups to Groupoids: A Brief Survey." *Bull. London Math. Soc.* **19**, 113–134, 1987.

Brown, R. *Topology: A Geometric Account of General Topology, Homotopy Types, and the Fundamental Groupoid.* New York: Halsted Press, 1988.

Higgins, P. J. *Notes on Categories and Groupoids.* London: Van Nostrand Reinhold, 1971.

Ramsay, A.; Chiaramonte, R.; and Woo, L. "Groupoid Home Page." `http://amath-www.colorado.edu:80/math/researchgroups/groupoids/groupoids.shtml`.

Rosenfeld, A. *An Introduction to Algebraic Structures.* New York: Holden-Day, 1968.

Sloane, N. J. A. Sequences A001329/M4760 and A001424/M4465 in "An On-Line Version of the Encyclopedia of Integer Sequences."

Weinstein, A. "Groupoids: Unifying Internal and External Symmetry." *Not. Amer. Math. Soc.* **43**, 744–752, 1996.

Growth

A general term which refers to an increase (or decrease in the case of the oxymoron "NEGATIVE growth") in a given quantity.

see also GROWTH FUNCTION, GROWTH SPIRAL

Growth Function

see BLOCK GROWTH

Growth Spiral

see LOGARITHMIC SPIRAL

Grundy's Game

A special case of NIM played by the following rules. Given a heap of size n, two players alternately select a heap and divide it into two unequal heaps. A player loses when he cannot make a legal move because all heaps have size 1 or 2. Flammenkamp gives a table of the extremal SPRAGUE-GRUNDY VALUES for this game. The first few values of Grundy's game are 0, 0, 0, 1, 0, 2, 1, 0, 2, ... (Sloane's A002188).

References

Flammenkamp, A. "Sprague-Grundy Values of Grundy's Game." `http://www.minet.uni-jena.de/~achim/grundy.html`.

Sloane, N. J. A. Sequence A002188/M0044 in "An On-Line Version of the Encyclopedia of Integer Sequences."

Grundy-Sprague Number

see NIM-VALUE

Gudermannian Function

Denoted either $\gamma(x)$ or $\mathrm{gd}(x)$.

$$\mathrm{gd}(x) \equiv \tan^{-1}(\sinh x) = 2\tan^{-1}(e^x) - \tfrac{1}{2}\pi \quad (1)$$

$$\mathrm{gd}^{-1}(x) = \ln[\tan(\tfrac{1}{4}\pi + \tfrac{1}{2}x)] = \ln(\sec x + \tan x). \quad (2)$$

The derivatives are given by

$$\frac{d}{dx}\,\mathrm{gd}(x) = \operatorname{sech} x \quad (3)$$

$$\frac{d}{dx}\,\mathrm{gd}^{-1}(x) = \sec x. \quad (4)$$

Guldinus Theorem

see PAPPUS'S CENTROID THEOREM

Gumbel's Distribution

A special case of the FISHER-TIPPETT DISTRIBUTION with $a = 0$, $b = 1$. The MEAN, VARIANCE, SKEWNESS, and KURTOSIS are

$$\begin{aligned} \mu &= \gamma \\ \sigma^2 &= \tfrac{1}{6}\pi^2 \\ \gamma_1 &= \frac{12\sqrt{6}\,\zeta(3)}{\pi^3} \\ \gamma_2 &= \tfrac{12}{5}. \end{aligned}$$

where γ is the EULER-MASCHERONI CONSTANT, and $\zeta(3)$ is APÉRY'S CONSTANT.

see also FISHER-TIPPETT DISTRIBUTION

Guthrie's Problem

The problem of deciding if four-colors are sufficient to color any map on a plane or SPHERE.

see also FOUR-COLOR THEOREM

Gutschoven's Curve

see KAPPA CURVE

Guy's Conjecture

Guy's conjecture, which has not yet been proven or disproven, states that the CROSSING NUMBER for a COMPLETE GRAPH of order n is

$$\frac{1}{4}\left\lfloor\frac{n}{2}\right\rfloor\left\lfloor\frac{n-1}{2}\right\rfloor\left\lfloor\frac{n-2}{2}\right\rfloor\left\lfloor\frac{n-3}{2}\right\rfloor,$$

where $\lfloor x\rfloor$ is the FLOOR FUNCTION, which can be rewritten

$$\begin{cases} \frac{1}{64}n(n-2)^2(n-4) & \text{for } n \text{ even} \\ \frac{1}{64}(n-1)^2(n-3)^2 & \text{for } n \text{ odd.} \end{cases}$$

The first few values are 0, 0, 0, 0, 1, 3, 9, 18, 36, 60, ... (Sloane's A000241).

see also CROSSING NUMBER (GRAPH)

References

Sloane, N. J. A. Sequence A000241/M2772 in "An On-Line Version of the Encyclopedia of Integer Sequences."

Gyrate Bidiminished Rhombicosidodecahedron

see JOHNSON SOLID

Gyrate Rhombicosidodecahedron

see JOHNSON SOLID

Gyrobicupola

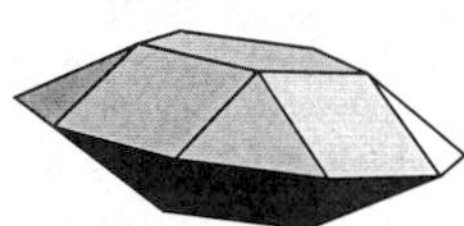
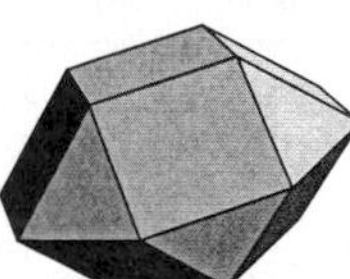

A BICUPOLA in which the bases are in opposite orientations.

see also BICUPOLA, PENTAGONAL GYROBICUPOLA, SQUARE GYROBICUPOLA

Gyrobifastigium

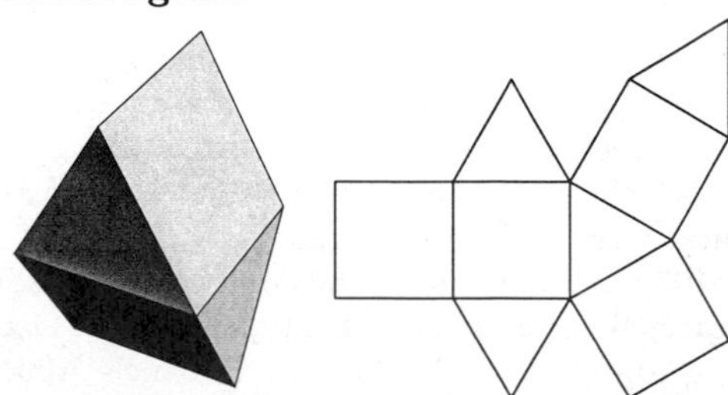

JOHNSON SOLID J_{26}, consisting of two joined triangular PRISMS.

Gyrobirotunda

A BIROTUNDA in which the bases are in opposite orientations.

Gyrocupolarotunda

A CUPOLAROTUNDA in which the bases are in opposite orientations.

see also ORTHOCUPOLAROTUNDA

Gyroelongated Cupola

A n-gonal CUPOLA adjoined to a $2n$-gonal ANTIPRISM.

see also GYROELONGATED PENTAGONAL CUPOLA, GYROELONGATED SQUARE CUPOLA, GYROELONGATED TRIANGULAR CUPOLA

Gyroelongated Dipyramid

see GYROELONGATED PYRAMID, GYROELONGATED SQUARE DIPYRAMID

Gyroelongated Pentagonal Bicupola

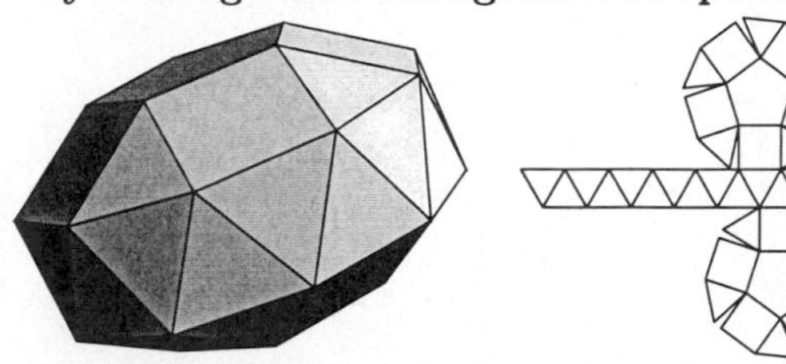

JOHNSON SOLID J_{46}, which consists of a PENTAGONAL ROTUNDA adjoined to a decagonal ANTIPRISM.

Gyroelongated Pentagonal Birotunda

see JOHNSON SOLID

Gyroelongated Pentagonal Cupola

see JOHNSON SOLID

Gyroelongated Pentagonal Cupolarotunda

see JOHNSON SOLID

Gyroelongated Pentagonal Pyramid

see JOHNSON SOLID

Gyroelongated Pentagonal Rotunda

see JOHNSON SOLID

Gyroelongated Pyramid

An n-gonal pyramid adjoined to an n-gonal ANTIPRISM.

see also ELONGATED PYRAMID, GYROELONGATED DIPYRAMID, GYROELONGATED PENTAGONAL PYRAMID, GYROELONGATED SQUARE DIPYRAMID, GYROELONGATED SQUARE PYRAMID

Gyroelongated Rotunda

see GYROELONGATED PENTAGONAL ROTUNDA

Gyroelongated Square Bicupola

see JOHNSON SOLID

Gyroelongated Square Cupola

see JOHNSON SOLID

Gyroelongated Square Dipyramid

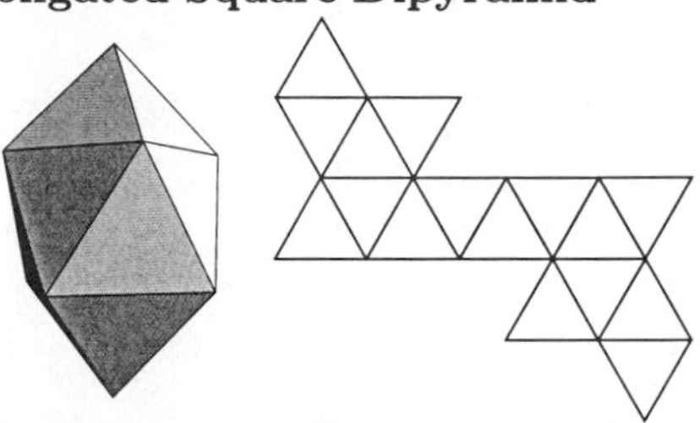

One of the eight convex DELTAHEDRA. It consists of two oppositely faced SQUARE PYRAMIDS rotated 45° to each other and separated by a ribbon of eight side-to-side TRIANGLES. It is JOHNSON SOLID J_{17}.

Call the coordinates of the upper PYRAMID bases ($\pm$ 1, $\pm$ 1, h_1) and of the lower ($\pm\sqrt{2}$, 0, $-h_1$) and (0, $\pm\sqrt{2}$, $-h_1$). Call the PYRAMID apexes (0, 0, $\pm(h_1+h_2)$). Consider the points (1, 1, 0) and (0, 0, h_1+h_2). The height of the PYRAMID is then given by

$$\sqrt{1^2+1^2+{h_2}^2}=\sqrt{2+{h_2}^2}=2 \tag{1}$$

$$h_2=\sqrt{2}. \tag{2}$$

Now consider the points (1, 1, h_1) and ($\sqrt{2}$, 0, $-h_1$). The height of the base is given by

$$\begin{aligned}(1-\sqrt{2})^2+1^2+(2h_1)^2&=1-2\sqrt{2}+2+1+4{h_1}^2\\&=4-2\sqrt{2}+4{h_1}^2=2^2=4\end{aligned} \tag{3}$$

$$4{h_1}^2=2\sqrt{2} \tag{4}$$

$${h_1}^2=\frac{\sqrt{2}}{2}=\frac{1}{\sqrt{2}}=2^{-1/2}, \tag{5}$$

so

$$h_1=2^{-1/4} \tag{6}$$

$$h_2=2^{1/2}. \tag{7}$$

Gyroelongated Square Pyramid

see JOHNSON SOLID

Gyroelongated Triangular Bicupola

see JOHNSON SOLID

Gyroelongated Triangular Cupola

see JOHNSON SOLID

H

h-Cobordism

An *h*-cobordism is a COBORDISM W between two MANIFOLDS M_1 and M_2 such that W is SIMPLY CONNECTED and the inclusion maps $M_1 \to W$ and $M_2 \to W$ are HOMOTOPY equivalences.

h-Cobordism Theorem

If W is a SIMPLY CONNECTED, COMPACT MANIFOLD with a boundary that has two components, M_1 and M_2, such that inclusion of each is a HOMOTOPY equivalence, then W is DIFFEOMORPHIC to the product $M_1 \times [0,1]$ for $\dim(M_1) \geq 5$. In other words, if M and M' are two simply connected MANIFOLDS of DIMENSION ≥ 5 and there exists an *h*-COBORDISM W between them, then W is a product $M \times I$ and M is DIFFEOMORPHIC to M'.

The proof of the *h*-cobordism theorem can be accomplished using SURGERY. A particular case of the *h*-cobordism theorem is the POINCARÉ CONJECTURE in dimension $n \geq 5$. Smale proved this theorem in 1961.

see also DIFFEOMORPHISM, POINCARÉ CONJECTURE, SURGERY

References

Smale, S. "Generalized Poincaré's Conjecture in Dimensions Greater than Four." *Ann. Math.* **74**, 391–406, 1961.

H-Fractal

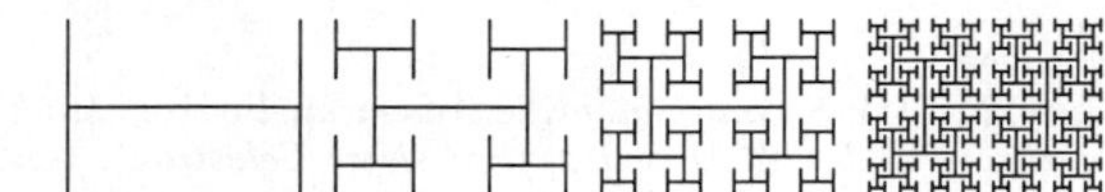

The FRACTAL illustrated above.

References

Lauwerier, H. *Fractals: Endlessly Repeated Geometric Figures.* Princeton, NJ: Princeton University Press, pp. 1–2, 1991.

Weisstein, E. W. "Fractals." `http://www.astro.virginia.edu/~eww6n/math/notebooks/Fractal.m`.

H-Function

see FOX'S *H*-FUNCTION

H-Spread

The difference $H_2 - H_1$, where H_1 and H_2 are HINGES. It is the same as the INTERQUARTILE RANGE for $N = 5$, 9, 13, ... points.

see also HINGE, INTERQUARTILE RANGE, STEP

References

Tukey, J. W. *Explanatory Data Analysis.* Reading, MA: Addison-Wesley, p. 44, 1977.

H-Transform

A 2-D generalization of the HAAR TRANSFORM which is used for the compression of astronomical images. The algorithm consists of dividing the $2^N \times 2^N$ image into blocks of 2×2 pixels, calling the pixels in the block a_{00}, a_{10}, a_{01}, and a_{11}. For each block, compute the four coefficients

$$h_0 \equiv \tfrac{1}{2}(a_{11} + a_{10} + a_{01} + a_{00})$$
$$h_x \equiv \tfrac{1}{2}(a_{11} + a_{10} - a_{01} - a_{00})$$
$$h_y \equiv \tfrac{1}{2}(a_{11} - a_{10} + a_{01} - a_{00})$$
$$h_c \equiv \tfrac{1}{2}(a_{11} - a_{10} - a_{01} + a_{00}).$$

Construct a $2^{N-1} \times 2^{N-1}$ image from the h_0 values, and repeat until only one h_0 value remains. The H-transform can be performed in place and requires about $16N^2/3$ additions for an $N \times N$ image.

see also HAAR TRANSFORM

References

Capaccioli, M.; Held, E. V.; Lorenz, H.; Richter, G. M.; and Ziener, R. "Application of an Adaptive Filtering Technique to Surface Photometry of Galaxies. I. The Method Tested on NGC 3379." *Astron. Nachr.* **309**, 69–80, 1988.

Fritze, K.; Lange, M.; Möstle, G.; Oleak, H.; and Richter, G. M. "A Scanning Microphotometer with an On-Line Data Reduction for Large Field Schmidt Plates." *Astron. Nachr.* **298**, 189–196, 1977.

Richter, G. M. "The Evaluation of Astronomical Photographs with the Automatic Area Photometer." *Astron. Nachr.* **299**, 283–303, 1978.

White, R. L.; Postman, M.; and Lattanzi, M. G. "Compression of the Guide Star Digitised Schmidt Plates." In *Digitised Optical Sky Surveys: Proceedings of the Conference on "Digitised Optical Sky Surveys" held in Edinburgh, Scotland, 18–21 June 1991* (Ed. H. T. MacGillivray and E. B. Thompson). Dordrecht, Netherlands: Kluwer, pp. 167–175, 1992.

Haar Function

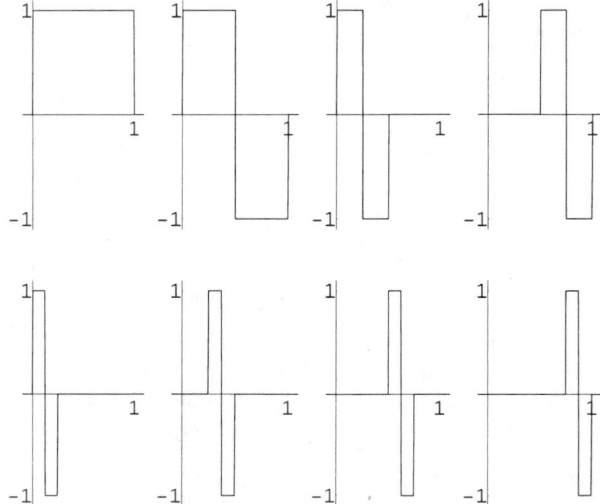

Define

$$\psi(x) \equiv \begin{cases} 1 & 0 \leq x \leq \frac{1}{2} \\ -1 & \frac{1}{2} \leq x \leq 1 \\ 0 & \text{otherwise} \end{cases} \tag{1}$$

and

$$\psi_{jk}(x) \equiv \psi(2^j x - k), \tag{2}$$

where the FUNCTIONS plotted above are

$$\begin{aligned} \psi_{00} &= \psi(x) \\ \psi_{10} &= \psi(2x) \\ \psi_{11} &= \psi(2x-1) \\ \psi_{20} &= \psi(4x) \\ \psi_{21} &= \psi(4x-1) \\ \psi_{21} &= \psi(4x-2) \\ \psi_{21} &= \psi(4x-3). \end{aligned}$$

Then a FUNCTION $f(x)$ can be written as a series expansion by

$$f(x) = c_0 + \sum_{j=0}^{\infty} \sum_{k=0}^{2^j-1} c_{jk}\psi_{jk}(x). \tag{3}$$

The FUNCTIONS ψ_{jk} and ψ are all ORTHOGONAL in $[0,1]$, with

$$\int_0^1 \phi(x)\phi_{jk}(x)\,dx = 0 \tag{4}$$

$$\int_0^1 \phi_{jk}(x)\phi_{lm}(x)\,dx = 0. \tag{5}$$

These functions can be used to define WAVELETS. Let a FUNCTION be defined on n intervals, with n a POWER of 2. Then an arbitrary function can be considered as an n-VECTOR $\mathbf{f}$, and the COEFFICIENTS in the expansion $\mathbf{b}$ can be determined by solving the MATRIX equation

$$\mathbf{f} = \mathsf{W}_n \mathbf{b} \tag{6}$$

for $\mathbf{b}$, where W is the MATRIX of ψ basis functions. For example,

$$\begin{aligned} \mathsf{W}_4 &= \begin{bmatrix} 1 & 1 & 1 & 0 \\ 1 & 1 & -1 & 0 \\ 1 & -1 & 0 & 1 \\ 1 & -1 & 0 & -1 \end{bmatrix} \\ &= \begin{bmatrix} 1 & 1 & & \\ 1 & -1 & & \\ & & 1 & 1 \\ & & 1 & -1 \end{bmatrix} \begin{bmatrix} 1 & & & \\ & & 1 & \\ & 1 & & \\ & & & 0 \end{bmatrix} \begin{bmatrix} 1 & 1 & & \\ 1 & -1 & & \\ & & 1 & \\ & & & 1 \end{bmatrix}. \end{aligned} \tag{7}$$

The WAVELET MATRIX can be computed in $\mathcal{O}(n)$ steps, compared to $\mathcal{O}(n \lg n)$ for the FOURIER MATRIX.

see also WAVELET, WAVELET TRANSFORM

References

Haar, A. "Zur Theorie der orthogonalen Funktionensysteme." *Math. Ann.* **69**, 331–371, 1910.

Strang, G. "Wavelet Transforms Versus Fourier Transforms." *Bull. Amer. Math. Soc.* **28**, 288–305, 1993.

Haar Integral

The INTEGRAL associated with the HAAR MEASURE.

see also HAAR MEASURE

Haar Measure

Any locally compact Hausdorff topological group has a unique (up to scalars) NONZERO left invariant measure which is finite on compact sets. If the group is Abelian or compact, then this measure is also right invariant and is known as the Haar measure.

Haar Transform

A 1-D transform which makes use of the HAAR FUNCTIONS.

see H-TRANSFORM, HAAR FUNCTION

References

Haar, A. "Zur Theorie der orthogonalen Funktionensysteme." *Math. Ann.* **69**, 331–371, 1910.

Haberdasher's Problem

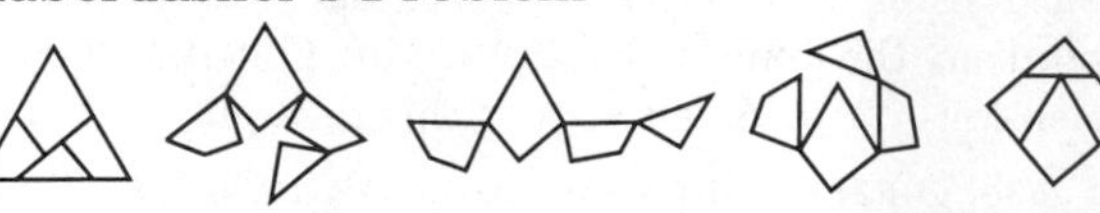

With four cuts, DISSECT an EQUILATERAL TRIANGLE into a SQUARE. First proposed by Dudeney (1907) and discussed in Gardner (1961, p. 34) and Stewart (1987, p. 169). The solution can be hinged so that the three pieces collapse into either the TRIANGLE or the SQUARE.

see also DISSECTION

References

Gardner, M. *The Second Scientific American Book of Mathematical Puzzles & Diversions: A New Selection.* New York: Simon and Schuster, 1961.

Stewart, I. *The Problems of Mathematics, 2nd ed.* Oxford, England: Oxford University Press, 1987.

Hadamard Design

A SYMMETRIC BLOCK DESIGN $(4n+3, n+1, n)$ which is equivalent to a HADAMARD MATRIX of order $4n+4$. It is conjectured that Hadamard designs exist from all integers $n > 0$, but this has not yet been proven. This elusive proof (or disproof) remains one of the most important unsolved problems in COMBINATORICS.

References

Dinitz, J. H. and Stinson, D. R. "A Brief Introduction to Design Theory." Ch. 1 in *Contemporary Design Theory: A Collection of Surveys* (Ed. J. H. Dinitz and D. R. Stinson). New York: Wiley, pp. 1–12, 1992.

Hadamard's Inequality

Let $\mathsf{A} = a_{ii}$ be an arbitrary $n \times n$ nonsingular MATRIX with REAL elements and DETERMINANT $|\mathsf{A}|$, then

$$|\mathsf{A}|^2 \leq \prod_{i=1}^{n} \left(\sum_{k=1}^{n} {a_{ik}}^2 \right).$$

see also HADAMARD'S THEOREM

References

Gradshteyn, I. S. and Ryzhik, I. M. *Tables of Integrals, Series, and Products, 5th ed.* San Diego, CA: Academic Press, p. 1110, 1979.

Hadamard Matrix

A class of SQUARE MATRIX invented by Sylvester (1867) under the name of ANALLAGMATIC PAVEMENT. A Hadamard matrix is a SQUARE MATRIX containing only 1s and -1s such that when any two columns or rows are placed side by side, HALF the adjacent cells are the same SIGN and half the other (excepting from the count an L-shaped "half-frame" bordering the matrix on two sides which is composed entirely of 1s). When viewed as pavements, cells with 1s are colored black and those with -1s are colored white. Therefore, the $n \times n$ Hadamard matrix H_n must have $n(n-1)/2$ white squares (-1s) and $n(n+1)/2$ black squares (1s).

This is equivalent to the definition

$$\mathsf{H}_n \mathsf{H}_n{}^{\mathrm{T}} = n\mathsf{I}_n, \tag{1}$$

where I_n is the $n \times n$ IDENTITY MATRIX. A Hadamard matrix of order $4n + 4$ corresponds to a HADAMARD DESIGN ($4n + 3$, $2n + 1$, n).

PALEY'S THEOREM guarantees that there always exists a Hadamard matrix H_n when n is divisible by 4 and of the form $2^e(q^m + 1)$, where p is an ODD PRIME. In such cases, the MATRICES can be constructed using a PALEY CONSTRUCTION. The PALEY CLASS k is undefined for the following values of $m < 1000$: 92, 116, 156, 172, 184, 188, 232, 236, 260, 268, 292, 324, 356, 372, 376, 404, 412, 428, 436, 452, 472, 476, 508, 520, 532, 536, 584, 596, 604, 612, 652, 668, 712, 716, 732, 756, 764, 772, 808, 836, 852, 856, 872, 876, 892, 904, 932, 940, 944, 952, 956, 964, 980, 988, 996.

Sawade (1985) constructed H_{268}. It is conjectured (and verified up to $n < 428$) that H_n exists for all n DIVISIBLE by 4 (van Lint and Wilson 1993). However, the proof of this CONJECTURE remains an important problem in CODING THEORY. The number of Hadamard matrices of order $4n$ are 1, 1, 1, 5, 3, 60, 487, ... (Sloane's A007299).

If H_n and H_m are known, then H_{nm} can be obtained by replacing all 1s in H_m by H_n and all -1s by $-\mathsf{H}_n$. For $n \leq 100$, Hadamard matrices with $n = 12$, 20, 28, 36, 44, 52, 60, 68, 76, 84, 92, and 100 cannot be built up from lower order Hadamard matrices.

$$\mathsf{H}_2 = \begin{bmatrix} 1 & 1 \\ -1 & 1 \end{bmatrix} \tag{2}$$

$$\mathsf{H}_4 = \begin{bmatrix} \mathsf{H}_2 & \mathsf{H}_2 \\ -\mathsf{H}_2 & \mathsf{H}_2 \end{bmatrix} = \begin{bmatrix} \begin{bmatrix} 1 & 1 \\ -1 & 1 \end{bmatrix} & \begin{bmatrix} 1 & 1 \\ -1 & 1 \end{bmatrix} \\ -\begin{bmatrix} 1 & 1 \\ -1 & 1 \end{bmatrix} & \begin{bmatrix} 1 & 1 \\ -1 & 1 \end{bmatrix} \end{bmatrix}$$

$$= \begin{bmatrix} 1 & 1 & 1 & 1 \\ -1 & 1 & -1 & 1 \\ -1 & -1 & 1 & 1 \\ 1 & -1 & -1 & 1 \end{bmatrix}. \tag{3}$$

H_8 can be similarly generated from H_4. Hadamard matrices can also be expressed in terms of the WALSH FUNCTIONS Cal and Sal

$$H_8 = \begin{bmatrix} \mathrm{Cal}(0,t) \\ \mathrm{Sal}(4,t) \\ \mathrm{Sal}(2,t) \\ \mathrm{Cal}(2,t) \\ \mathrm{Sal}(1,t) \\ \mathrm{Cal}(3,t) \\ \mathrm{Cal}(1,t) \\ \mathrm{Sal}(3,t) \end{bmatrix}. \tag{4}$$

Hadamard matrices can be used to make ERROR-CORRECTING CODES.

see also HADAMARD DESIGN, PALEY CONSTRUCTION, PALEY'S THEOREM, WALSH FUNCTION

References

Ball, W. W. R. and Coxeter, H. S. M. *Mathematical Recreations and Essays, 13th ed.* New York: Dover, pp. 107–109 and 274, 1987.

Beth, T.; Jungnickel, D.; and Lenz, H. *Design Theory.* New York: Cambridge University Press, 1986.

Colbourn, C. J. and Dinitz, J. H. (Eds.) "Hadamard Matrices and Designs." Ch. 24 in *CRC Handbook of Combinatorial Designs.* Boca Raton, FL: CRC Press, pp. 370–377, 1996.

Geramita, A. V. *Orthogonal Designs: Quadratic Forms and Hadamard Matrices.* New York: Marcel Dekker, 1979.

Golomb, S. W. and Baumert, L. D. "The Search for Hadamard Matrices." *Amer. Math. Monthly* **70**, 12–17, 1963.

Hall, M. Jr. *Combinatorial Theory, 2nd ed.* New York: Wiley, p. 207, 1986.

Hedayat, A. and Wallis, W. D. "Hadamard Matrices and Their Applications." *Ann. Stat.* **6**, 1184–1238, 1978.

Kimura, H. "Classification of Hadamard Matrices of Order 28." *Disc. Math.* **133**, 171–180, 1994.

Kimura, H. "Classification of Hadamard Matrices of Order 28 with Hall Sets." *Disc. Math.* **128**, 257–269, 1994.

Kitis, L. "Paley's Construction of Hadamard Matrices." `http://www.mathsource.com/cgi-bin/MathSource/Applications/Mathematics/0205-760`.

Ogilvie, G. A. "Solution to Problem 2511." *Math. Questions and Solutions* **10**, 74–76, 1868.

Paley, R. E. A. C. "On Orthogonal Matrices." *J. Math. Phys.* **12**, 311–320, 1933.

Ryser, H. J. *Combinatorial Mathematics.* Buffalo, NY: Math. Assoc. Amer., pp. 104–122, 1963.

Sawade, K. "A Hadamard Matrix of Order-268." *Graphs Combinatorics* **1**, 185–187, 1985.
Seberry, J. and Yamada, M. "Hadamard Matrices, Sequences, and Block Designs." Ch. 11 in *Contemporary Design Theory: A Collection of Surveys* (Eds. J. H. Dinitz and D. R. Stinson). New York: Wiley, pp. 431–560, 1992.
Sloane, N. J. A. Sequence A007299/M3736 in "An On-Line Version of the Encyclopedia of Integer Sequences."
Spence, E. "Classification of Hadamard Matrices of Order 24 and 28." *Disc. Math* **140**, 185–243, 1995.
Sylvester, J. J. "Thoughts on Orthogonal Matrices, Simultaneous Sign-Successions, and Tessellated Pavements in Two or More Colours, with Applications to Newton's Rule, Ornamental Tile-Work, and the Theory of Numbers." *Phil. Mag.* **34**, 461–475, 1867.
Sylvester, J. J. "Problem 2511." *Math. Questions and Solutions* **10**, 74, 1868.
van Lint, J. H. and Wilson, R. M. *A Course in Combinatorics.* New York: Cambridge University Press, 1993.

Hadamard's Theorem

Let $|A|$ be an $n \times n$ DETERMINANT with COMPLEX (or REAL) elements a_{ij}, then $|A| \neq 0$ if

$$|a_{ii}| > \sum_{\substack{j=1 \\ j \neq i}}^{n} |a_{ij}|.$$

see also HADAMARD'S INEQUALITY

References

Gradshteyn, I. S. and Ryzhik, I. M. *Tables of Integrals, Series, and Products, 5th ed.* San Diego, CA: Academic Press, p. 1110, 1979.

Hadamard Transform

A FAST FOURIER TRANSFORM-like ALGORITHM which produces a hologram of an image.

Hadamard-Vallée Poussin Constants

N.B. A detailed on-line essay by S. Finch was the starting point for this entry.

The sum of RECIPROCALS of PRIMES diverges, but

$$\lim_{n\to\infty} \left[\sum_{k=1}^{\pi(n)} \frac{1}{p_k} - \ln(\ln n) \right] = \gamma + \sum_{k=1}^{\infty} \left[\ln\left(1 - \frac{1}{p_k}\right) + \frac{1}{p_k} \right] \equiv C_1 = 0.2614972128\ldots, \quad (1)$$

where $\pi(n)$ is the PRIME COUNTING FUNCTION and γ is the EULER-MASCHERONI CONSTANT (Le Lionnais 1983). Hardy and Wright (1985) show that, if $\omega(n)$ is the number of distinct PRIME factors of n, then

$$\lim_{n\to\infty} \left[\frac{1}{n} \sum_{k=1}^{n} \omega(k) - \ln(\ln n) \right] = C_1. \quad (2)$$

Furthermore, if $\Omega(n)$ is the total number of PRIME factors of n, then

$$\lim_{n\to\infty} \left[\frac{1}{n} \sum_{k=1}^{n} \Omega(k) - \ln(\ln n) \right] = C_1 + \sum_{k=1}^{\infty} \frac{1}{p_k(p_k - 1)} = 1.0346538819\ldots. \quad (3)$$

Similarly,

$$\lim_{n\to\infty} \left(\sum_{k=1}^{\pi(n)} \frac{\ln p_k}{p_k} - \ln n \right) = -\gamma - \sum_{j=2}^{\infty} \sum_{k=1}^{\infty} \frac{\ln p_k}{{p_k}^j} \equiv -C_2 = -1.3325822757\ldots. \quad (4)$$

References

Finch, S. "Favorite Mathematical Constants." http://www.mathsoft.com/asolve/constant/hdmrd/hdmrd.html.
Hardy, G. H. and Wright, E. M. *An Introduction to the Theory of Numbers, 5th ed.* Oxford, England: Clarendon Press, 1985.
Le Lionnais, F. *Les nombres remarquables.* Paris: Hermann, p. 24, 1983.
Rosser, J. B. and Schoenfeld, L. "Approximate Formulas for Some Functions of Prime Numbers." *Ill. J. Math.* **6**, 64–94, 1962.

Hadwiger's Principal Theorem

The VECTORS $\pm\mathbf{a}_1, \ldots, \pm\mathbf{a}_n$ in a 3-space form a normalized EUTACTIC STAR IFF $T\mathbf{x} = \mathbf{x}$ for all $\mathbf{x}$ in the 3-space.

Hadwiger Problem

What is the largest number of subcubes (not necessarily different) into which a CUBE cannot be divided by plane cuts? The answer is 47.

see also CUBE DISSECTION

Hafner-Sarnak-McCurley Constant

N.B. A detailed on-line essay by S. Finch was the starting point for this entry.

Given two randomly chosen INTEGER $n \times n$ matrices, what is the probability $D(n)$ that the corresponding determinants are coprime? Hafner *et al.* (1993) showed that

$$D(n) = \prod_{p_k} \left\{ 1 - \left[1 - \prod_{j=1}^{n} (1 - {p_k}^{-j}) \right]^2 \right\}, \quad (1)$$

where the product is over PRIMES. The case $D(1)$ is just the probability that two random INTEGERS are coprime,

$$D(1) = \frac{6}{\pi^2} = 0.6079271019\ldots. \quad (2)$$

Vardi (1991) computed the limit

$$\sigma \equiv \lim_{n\to\infty} D(n) = 0.3532363719\ldots. \quad (3)$$

The speed of convergence is roughly $\sim 0.57^n$ (Flajolet and Vardi 1996).

References

Finch, S. "Favorite Mathematical Constants." http://www.mathsoft.com/asolve/constant/hafner/hafner.html.

Flajolet, P. and Vardi, I. "Zeta Function Expansions of Classical Constants." Unpublished manuscript. 1996. http://pauillac.inria.fr/algo/flajolet/Publications/landau.ps.

Hafner, J. L.; Sarnak, P.; and McCurley, K. "Relatively Prime Values of Polynomials." In *Contemporary Mathematics* Vol. 143 (Ed. M. Knopp and M. Seingorn). Providence, RI: Amer. Math. Soc., 1993.

Vardi, I. *Computational Recreations in Mathematica.* Redwood City, CA: Addison-Wesley, 1991.

Hahn-Banach Theorem

A linear FUNCTIONAL defined on a SUBSPACE of a VECTOR SPACE V and which is dominated by a sublinear function defined on V has a linear extension which is also dominated by the sublinear function.

References

Zeidler, E. *Applied Functional Analysis: Applications to Mathematical Physics.* New York: Springer-Verlag, 1995.

Hailstone Number

Sequences of INTEGERS generated in the COLLATZ PROBLEM. For example, for a starting number of 7, the sequence is 7, 22, 11, 34, 17, 52, 26, 13, 40, 20, 10, 5, 16, 8, 4, 2, 1, 4, 2, 1, ... Such sequences are called hailstone sequences because the values typically rise and fall, somewhat analogously to a hailstone inside a cloud.

While a hailstone eventually becomes so heavy that it falls to ground, every starting INTEGER ever tested has produced a hailstone sequence that eventually drops down to the number 1 and then "bounces" into the small loop 4, 2, 1, ...

see also COLLATZ PROBLEM

References

Schwartzman, S. *The Words of Mathematics: An Etymological Dictionary of Mathematical Terms Used in English.* Washington, DC: Math. Assoc. Amer., 1994.

Hairy Ball Theorem

There does not exist an everywhere NONZERO VECTOR FIELD on the 2-SPHERE $\mathbb{S}^2$. This implies that somewhere on the surface of the Earth, there is a point with zero horizontal wind velocity.

Half

The UNIT FRACTION 1/2.

see also QUARTER, SQUARE ROOT, UNIT FRACTION

Half-Closed Interval

An INTERVAL in which one endpoint is included but not the other. A half-closed interval is denoted $[a, b)$ or $(a, b]$ and is also called a HALF-OPEN INTERVAL.

see also CLOSED INTERVAL, OPEN INTERVAL

Half-Normal Distribution

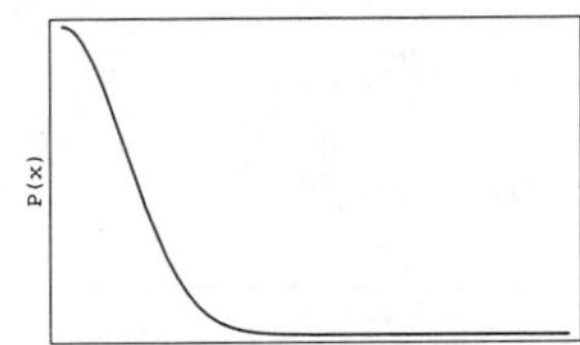

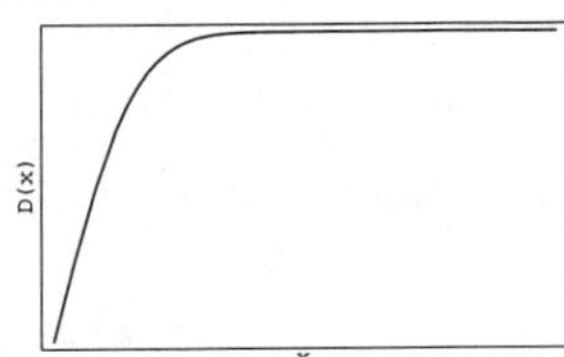

A NORMAL DISTRIBUTION with MEAN 0 and STANDARD DEVIATION $1/\theta$ limited to the domain $[0, \infty)$.

$$P(x) = \frac{2\theta}{\pi} e^{-x^2\theta^2/\pi} \quad (1)$$

$$D(x) = \operatorname{erf}\left(\frac{tx}{\sqrt{\pi}}\right). \quad (2)$$

The MOMENTS are

$$\mu_1 = \frac{1}{t} \quad (3)$$

$$\mu_2 = \frac{\pi}{2t^2} \quad (4)$$

$$\mu_3 = \frac{\pi}{t^3} \quad (5)$$

$$\mu_4 = \frac{3\pi^2}{4t^4}, \quad (6)$$

so the MEAN, VARIANCE, SKEWNESS, and KURTOSIS are

$$\mu = \frac{1}{\theta} \quad (7)$$

$$\sigma^2 \equiv \mu_2 - {\mu_1}^2 = \frac{\pi - 2}{2t^2} \quad (8)$$

$$\gamma_1 = 2\sqrt{\frac{2}{\pi}} \quad (9)$$

$$\gamma_2 = 0. \quad (10)$$

see also NORMAL DISTRIBUTION

Half-Open Interval

see HALF-CLOSED INTERVAL

Hall-Janko Group

The SPORADIC GROUP HJ, also denoted J_2.

see also JANKO GROUPS

Halley's Irrational Formula

A ROOT-finding ALGORITHM which makes use of a third-order TAYLOR SERIES

$$f(x) = f(x_n) + f'(x_n)(x - x_n) + \tfrac{1}{2}f''(x_n)(x - x_n)^2 + \ldots. \tag{1}$$

A ROOT of $f(x)$ satisfies $f(x) = 0$, so

$$0 \approx f(x_n) + f'(x_n)(x_{n+1} - x_n) + \tfrac{1}{2}f''(x_n)(x_{n+1} - x_n)^2. \tag{2}$$

Using the QUADRATIC EQUATION then gives

$$x_{n+1} = x_n + \frac{-f'(x_n) \pm \sqrt{[f'(x_n)]^2 - 2f(x_n)f''(x_n)}}{f''(x_n)}. \tag{3}$$

Picking the plus sign gives the iteration function

$$C_f(x) = x - \frac{1 - \sqrt{1 - \frac{2f(x)f''(x)}{[f'(x)]^2}}}{\frac{f''(x)}{f'(x)}}. \tag{4}$$

This equation can be used as a starting point for deriving HALLEY'S METHOD.

If the alternate form of the QUADRATIC EQUATION is used instead in solving (2), the iteration function becomes instead

$$C_f(x) = x - \frac{2f(x)}{f'(x) \pm \sqrt{[f'(x)]^2 - 2f(x)f''(x)}}. \tag{5}$$

This form can also be derived by setting $n = 2$ in LAGUERRE'S METHOD. Numerically, the SIGN in the DENOMINATOR is chosen to maximize its ABSOLUTE VALUE. Note that in the above equation, if $f''(x) = 0$, then NEWTON'S METHOD is recovered. This form of Halley's irrational formula has cubic convergence, and is usually found to be substantially more stable than NEWTON'S METHOD. However, it does run into difficulty when both $f(x)$ and $f'(x)$ or $f'(x)$ and $f''(x)$ are simultaneously near zero.

see also HALLEY'S METHOD, LAGUERRE'S METHOD, NEWTON'S METHOD

References

Qiu, H. "A Robust Examination of the Newton-Raphson Method with Strong Global Convergence Properties." Master's Thesis. University of Central Florida, 1993.

Scavo, T. R. and Thoo, J. B. "On the Geometry of Halley's Method." *Amer. Math. Monthly* **102**, 417–426, 1995.

Halley's Method

Also known as the TANGENT HYPERBOLAS METHOD or HALLEY'S RATIONAL FORMULA. As in HALLEY'S IRRATIONAL FORMULA, take the second-order TAYLOR POLYNOMIAL

$$f(x) = f(x_n) + f'(x_n)(x - x_n) + \tfrac{1}{2}f''(x_n)(x - x_n)^2 + \ldots. \tag{1}$$

A ROOT of $f(x)$ satisfies $f(x) = 0$, so

$$0 \approx f(x_n) + f'(x_n)(x_{n+1} - x_n) + \tfrac{1}{2}f''(x_n)(x_{n+1} - x_n)^2. \tag{2}$$

Now write

$$0 = f(x_n) + (x_{n+1} - x_n)[f'(x_n) + \tfrac{1}{2}f''(x_n)(x_{n+1} - x_n)], \tag{3}$$

giving

$$x_{n+1} = x_n - \frac{f(x_n)}{f'(x_n) + \frac{1}{2}f''(x_n)(x_{n+1} - x_n)}. \tag{4}$$

Using the result from NEWTON'S METHOD,

$$x_{n+1} - x_n = -\frac{f(x_n)}{f'(x_n)}, \tag{5}$$

gives

$$x_{n+1} = x_n - \frac{2f(x_n)f'(x_n)}{2[f'(x_n)]^2 - f(x_n)f''(x_n)}, \tag{6}$$

so the iteration function is

$$H_f(x) = x - \frac{2f(x)f'(x)}{2[f'(x)]^2 - f(x)f''(x)}. \tag{7}$$

This satisfies $H_f'(\alpha) = H_f''(\alpha) = 0$ where α is a ROOT, so it is third order for simple zeros. Curiously, the third derivative

$$H_f'''(\alpha) = -\left\{\frac{f'''(\alpha)}{f'(\alpha)} - \frac{3}{2}\left[\frac{f''(\alpha)}{f'(\alpha)}\right]^2\right\} \tag{8}$$

is the SCHWARZIAN DERIVATIVE. Halley's method may also be derived by applying NEWTON'S METHOD to $ff'^{-1/2}$. It may also be derived by using an OSCULATING CURVE of the form

$$y(x) = \frac{(x - x_n) + c}{a(x - x_n) + b}. \tag{9}$$

Taking derivatives,

$$f(x_n) = \frac{c}{b} \tag{10}$$

$$f'(x_n) = \frac{b - ac}{b^2} \tag{11}$$

$$f''(x_n) = \frac{2a(ac - b)}{b^3}, \tag{12}$$

which has solutions

$$a = -\frac{f''(x_n)}{2[f'(x_n)]^2 - f(x_n)f''(x_n)} \tag{13}$$

$$b = \frac{2f'(x_n)}{2[f'(x_n)]^2 - f(x_n)f''(x_n)} \tag{14}$$

$$c = \frac{2f(x_n)f'(x_n)}{2[f'(x_n)]^2 - f(x_n)f''(x_n)}, \tag{15}$$

so at a ROOT, $y(x_{n+1}) = 0$ and

$$x_{n+1} = x_n - c, \tag{16}$$

which is Halley's method.

see also HALLEY'S IRRATIONAL FORMULA, LAGUERRE'S METHOD, NEWTON'S METHOD

References
Scavo, T. R. and Thoo, J. B. "On the Geometry of Halley's Method." *Amer. Math. Monthly* **102**, 417–426, 1995.

Halley's Rational Formula

see HALLEY'S METHOD

Halphen Constant

see ONE-NINTH CONSTANT

Halphen's Transformation

A curve and its polar reciprocal with regard to the fixed CONIC have the same Halphen transformation.

References
Coolidge, J. L. *A Treatise on Algebraic Plane Curves.* New York: Dover, pp. 346–347, 1959.

Halting Problem

The determination of whether a TURING MACHINE will come to a halt given a particular input program. This problem is formally UNDECIDABLE, as first proved by Turing.

see also BUSY BEAVER, CHAITIN'S CONSTANT, TURING MACHINE, UNDECIDABLE

References
Chaitin, G. J. "Computing the Busy Beaver Function." §4.4 in *Open Problems in Communication and Computation* (Ed. T. M. Cover and B. Gopinath). New York: Springer-Verlag, pp. 108–112, 1987.
Davis, M. "What It a Computation." In *Mathematics Today: Twelve Informal Essays* (Ed. L. A. Steen). New York: Springer-Verlag, pp. 241–267, 1978.
Penrose, R. *The Emperor's New Mind: Concerning Computers, Minds, and the Laws of Physics.* Oxford, England: Oxford University Press, pp. 63–66, 1989.

Ham Sandwich Theorem

The volumes of any n n-D solids can always be simultaneously bisected by a $(n-1)$-D HYPERPLANE. Proving the theorem for $n = 2$ (where it is known as the PANCAKE THEOREM) is simple and can be found in Courant and Robbins (1978). The theorem was proved for $n > 3$ by Stone and Tukey (1942).

see also PANCAKE THEOREM

References
Chinn, W. G. and Steenrod, N. E. *First Concepts of Topology.* Washington, DC: Math. Assoc. Amer., 1966.
Courant, R. and Robbins, H. *What is Mathematics?: An Elementary Approach to Ideas and Methods.* Oxford, England: Oxford University Press, 1978.
Davis, P. J. and Hersh, R. *The Mathematical Experience.* Boston, MA: Houghton Mifflin, pp. 274–284, 1981.
Hunter, J. A. H. and Madachy, J. S. *Mathematical Diversions.* New York: Dover, pp. 67–69, 1975.
Stone, A. H. and Tukey, J. W. "Generalized 'Sandwich' Theorems." *Duke Math. J.* **9**, 356–359, 1942.

Hamilton's Equations

The equations defined by

$$\dot{q} = \frac{\partial H}{\partial p} \tag{1}$$

$$\dot{p} = -\frac{\partial H}{\partial q}, \tag{2}$$

where $\dot{x} \equiv dx/dt$ and H is the so-called Hamiltonian, are called Hamilton's equations. These equations frequently arise in problems of celestial mechanics. Another formulation related to Hamilton's equation is

$$p = \frac{\partial L}{\partial \dot{q}}, \tag{3}$$

where L is the so-called Lagrangian.

References
Morse, P. M. and Feshbach, H. "Hamilton's Principle and Classical Dynamics." §3.2 in *Methods of Theoretical Physics, Part I.* New York: McGraw-Hill, pp. 280–301, 1953.

Hamilton's Rules

The rules for the MULTIPLICATION of QUATERNIONS.

see also QUATERNION

Hamiltonian Circuit

A closed loop through a GRAPH that visits each node exactly once and ends adjacent to the initial point. The Hamiltonian circuit is named after Sir William Rowan Hamilton, who devised a puzzle in which such a path along the EDGES of an ICOSAHEDRON was sought (the ICOSIAN GAME).

All PLATONIC SOLIDS have a Hamiltonian circuit, as do planar 4-connected graphs. However, no foolproof method is known for determining whether a given general GRAPH has a Hamiltonian circuit. The number of Hamiltonian circuits on an n-HYPERCUBE is 2, 8, 96, 43008, ... (Sloane's A006069, Gardner 1986, pp. 23–24).

see also CHVÁTAL'S THEOREM, DIRAC'S THEOREM, EULER GRAPH, GRINBERG FORMULA, HAMILTONIAN GRAPH, HAMILTONIAN PATH, ICOSIAN GAME, KOZYREV-GRINBERG THEORY, ORE'S THEOREM, PÓSA'S THEOREM, SMITH'S NETWORK THEOREM

References
Chartrand, G. *Introductory Graph Theory.* New York: Dover, p. 68, 1985.
Gardner, M. "The Binary Gray Code." In *Knotted Doughnuts and Other Mathematical Entertainments.* New York: W. H. Freeman, pp. 23–24, 1986.
Sloane, N. J. A. Sequence A006069/M1903 in "An On-Line Version of the Encyclopedia of Integer Sequences."

Hamiltonian Cycle

see HAMILTONIAN CIRCUIT

Hamiltonian Graph

A GRAPH possessing a HAMILTONIAN CIRCUIT.

see also HAMILTONIAN CIRCUIT, HAMILTONIAN PATH

References

Chartrand, G. *Introductory Graph Theory.* New York: Dover, p. 68, 1985.

Chartrand, G.; Kapoor, S. F.; and Kronk, H. V. "The Many Facets of Hamiltonian Graphs." *Math. Student* **41**, 327–336, 1973.

Hamiltonian Group

A non-Abelian GROUP all of whose SUBGROUPS are self-conjugate.

References

Carmichael, R. D. "Hamiltonian Groups." §31 in *Introduction to the Theory of Groups of Finite Order.* New York: Dover, p. 113–116, 1956.

Hamiltonian Map

Consider a 1-D Hamiltonian MAP of the form

$$H(p,q)=\tfrac{1}{2}p^2+V(q), \tag{1}$$

which satisfies HAMILTON'S EQUATIONS

$$\dot q=\frac{\partial H}{\partial p} \tag{2}$$

$$\dot p=-\frac{\partial H}{\partial q}. \tag{3}$$

Now, write

$$\dot q_i=\frac{(q_{i+1}-q_i)}{\Delta t}, \tag{4}$$

where

$$q_i=q(t) \tag{5}$$

$$q_{i+1}=q(t+\Delta t). \tag{6}$$

Then the equations of motion become

$$q_{i+1}=q_i+p_i\Delta t \tag{7}$$

$$p_{i+1}=p_i-\Delta t\left(\frac{\partial V}{\partial q_i}\right)_{q=q_i}. \tag{8}$$

Note that equations (7) and (8) are not AREA-PRESERVING, since

$$\frac{\partial(q_{i+1},p_{i+1})}{\partial(q_i,p_i)}=\begin{vmatrix}1 & -\Delta t\frac{\partial^2 V}{\partial {q_i}^2}\\ \Delta t & 1\end{vmatrix}=1+(\Delta t)^2\frac{\partial^2 V}{\partial {q_i}^2}\neq 1. \tag{9}$$

However, if we take instead of (7) and (8),

$$q_{i+1}=q_i+p_i\Delta t \tag{10}$$

$$p_{i+1}=p_i-\Delta t\left(\frac{\partial V}{\partial q_i}\right)_{q=q_{i+1}} \tag{11}$$

$$\begin{aligned}\frac{\partial(q_{i+1},p_{i+1})}{\partial(q_i,p_i)}&=\begin{vmatrix}1 & -\Delta t\frac{\partial}{\partial q_i}\left(\frac{\partial V}{\partial q}\right)_{q=q_{i+1}}\\ \Delta t & 1\end{vmatrix}\\ &=1+(\Delta t)^2\frac{\partial^2 V}{\partial {q_i}^2}=1,\end{aligned} \tag{12}$$

which is AREA-PRESERVING.

Hamiltonian Path

A loop through a GRAPH that visits each node exactly once but does *not* end adjacent to the initial point. The number of Hamiltonian paths on an n-HYPERCUBE is 0, 0, 48, 48384, ... (Sloane's A006070, Gardner 1986, pp. 23–24).

see also HAMILTONIAN CIRCUIT, HAMILTONIAN GRAPH

References

Gardner, M. "The Binary Gray Code." In *Knotted Doughnuts and Other Mathematical Entertainments.* New York: W. H. Freeman, pp. 23–24, 1986.

Sloane, N. J. A. Sequence A006070/M5295 in "An On-Line Version of the Encyclopedia of Integer Sequences."

Hamiltonian System

A system of variables which can be written in the form of HAMILTON'S EQUATIONS.

Hammer-Aitoff Equal-Area Projection

A MAP PROJECTION whose inverse is defined using the intermediate variable

$$z\equiv\sqrt{1-(\tfrac{1}{4}x)^2-(\tfrac{1}{2}y)^2}.$$

Then the longitude and latitude are given by

$$\lambda=2\tan^{-1}\left(\frac{zx}{2(2z^2-1)}\right)$$

$$\phi=\sin^{-1}(yz).$$

Hamming Function

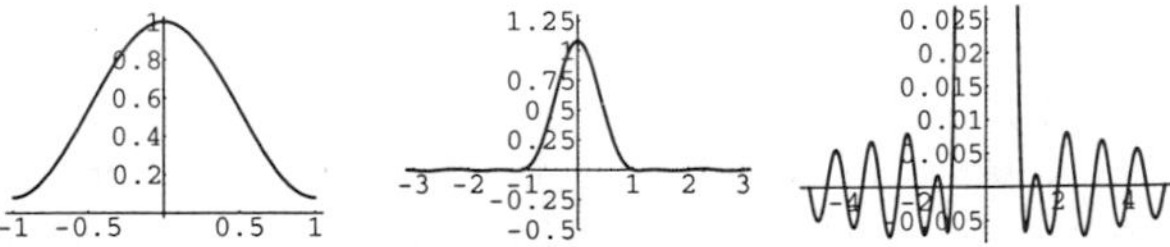

An APODIZATION FUNCTION chosen to minimize the height of the highest sidelobe. The Hamming function is given by

$$A(x)=0.54+0.46\cos\left(\frac{\pi x}{a}\right), \tag{1}$$

Its FULL WIDTH AT HALF MAXIMUM is $1.05543a$. The corresponding INSTRUMENT FUNCTION is

$$I(k)=\frac{a(1.08-0.64a^2k^2)\operatorname{sinc}(2\pi ak)}{1-4a^2k^2}. \tag{2}$$

This APODIZATION FUNCTION is close to the one produced by the requirement that the APPARATUS FUNCTION goes to 0 at $ka = 5/4$. From APODIZATION FUNCTION, a general symmetric apodization function $A(x)$ can be written as a FOURIER SERIES

$$A(x) = a_0 + 2\sum_{n=1}^{\infty} a_n \cos\left(\frac{n\pi x}{b}\right), \quad (3)$$

where the COEFFICIENTS satisfy

$$a_0 + 2\sum_{n=1}^{\infty} a_n = 1. \quad (4)$$

The corresponding apparatus function is

$$I(t) = 2b\{a_0 \operatorname{sinc}(2\pi kb) + \sum_{n=1}^{\infty}[\operatorname{sinc}(2\pi kb + n\pi) + \operatorname{sinc}(2\pi kb - n\pi)]\}. \quad (5)$$

To obtain an APODIZATION FUNCTION with zero at $ka = 3/4$, use

$$a_0 + 2a_1 = 1, \quad (6)$$

so

$$a_0 \operatorname{sinc}(\tfrac{5}{2}\pi) + a_1[\operatorname{sinc}(\tfrac{7}{2}\pi) + \operatorname{sinc}(\tfrac{3}{2}\pi) = 0 \quad (7)$$

$$(1-2a_1)\frac{2}{5\pi} - a_1\left(\frac{2}{7\pi} + \frac{2}{3\pi}\right) = (1-2a_1)\tfrac{1}{5} - a_1(\tfrac{1}{7}+\tfrac{1}{3}) = 0 \quad (8)$$

$$a_1(\tfrac{1}{7} + \tfrac{1}{3} + \tfrac{2}{5}) = \tfrac{1}{5} \quad (9)$$

$$a_1 = \frac{\frac{1}{5}}{\frac{2}{5} + \frac{1}{7} + \frac{1}{3}} = \frac{7 \cdot 3}{2 \cdot 3 \cdot 7 + 3 \cdot 5 + 5 \cdot 7} = \tfrac{21}{92} \approx 0.2283 \quad (10)$$

$$a_0 = 1 - 2a_1 = \frac{92 - 2 \cdot 21}{92} = \frac{92 - 42}{92} = \tfrac{50}{92} = \tfrac{25}{46} \approx 0.5435. \quad (11)$$

The FWHM is 1.81522, the peak is 1.08, the peak NEGATIVE and POSITIVE sidelobes (in units of the peak) are -0.00689132 and 0.00734934, respectively.

see also APODIZATION FUNCTION, HANNING FUNCTION, INSTRUMENT FUNCTION

References

Blackman, R. B. and Tukey, J. W. "Particular Pairs of Windows." In *The Measurement of Power Spectra, From the Point of View of Communications Engineering.* New York: Dover, pp. 98–99, 1959.

Handedness

Objects which are identical except for a mirror reflection are said to display handedness and to be CHIRAL.

see also AMPHICHIRAL, CHIRAL, ENANTIOMER, MIRROR IMAGE

Handkerchief Surface

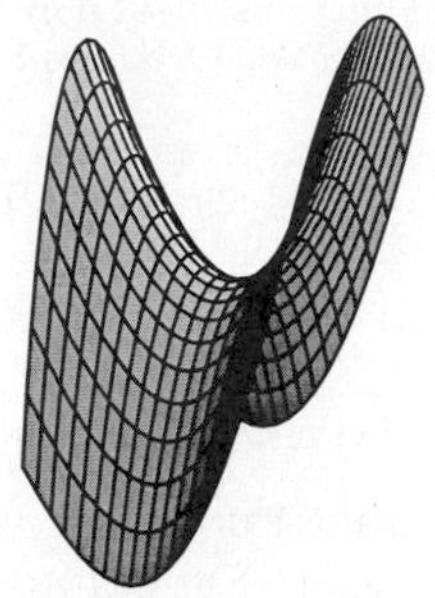

A surface given by the parametric equations

$$x(u,v) = u$$
$$y(u,v) = v$$
$$z(u,v) = \tfrac{1}{3}u^3 + uv^2 + 2(u^2 - v^2).$$

References

Gray, A. *Modern Differential Geometry of Curves and Surfaces.* Boca Raton, FL: CRC Press, p. 628, 1993.

Handle

Handles are to MANIFOLDS as CELLS are to CW-COMPLEXES. If M is a MANIFOLD together with a $(k-1)$-SPHERE $\mathbb{S}^{k-1}$ embedded in its boundary with a trivial TUBULAR NEIGHBORHOOD, we attach a k-handle to M by gluing the tubular NEIGHBORHOOD of the $(k-1)$-SPHERE $\mathbb{S}^{k-1}$ to the TUBULAR NEIGHBORHOOD of the standard $(k-1)$-SPHERE $\mathbb{S}^{k-1}$ in the $\dim(M)$-dimensional DISK.

In this way, attaching a k-handle is essentially just the process of attaching a fattened-up k-DISK to M along the $(k-1)$-SPHERE $\mathbb{S}^{k-1}$. The embedded DISK in this new MANIFOLD is called the k-handle in the UNION of M and the handle.

see also HANDLEBODY, SURGERY, TUBULAR NEIGHBORHOOD

Handlebody

A handlebody of type (n, k) is an n-D MANIFOLD that is attained from the standard n-DISK by attaching only k-D HANDLES.

see also HANDLE, HEEGAARD SPLITTING, SURGERY

References

Rolfsen, D. *Knots and Links.* Wilmington, DE: Publish or Perish Press, p. 46, 1976.

Hankel Function

A COMPLEX function which is a linear combination of BESSEL FUNCTIONS OF THE FIRST and SECOND KINDS.

see also HANKEL FUNCTION OF THE FIRST KIND, HANKEL FUNCTION OF THE SECOND KIND, SPHERICAL HANKEL FUNCTION OF THE FIRST KIND, SPHERICAL HANKEL FUNCTION OF THE SECOND KIND

References
Arfken, G. "Hankel Functions." §11.4 in *Mathematical Methods for Physicists, 3rd ed.* Orlando, FL: Academic Press, pp. 604–610, 1985.
Morse, P. M. and Feshbach, H. *Methods of Theoretical Physics, Part I.* New York: McGraw-Hill, pp. 623–624, 1953.

Hankel Function of the First Kind

$$H_n^{(1)}(z) \equiv J_n(z) + iY_n(z),$$

where $J_n(z)$ is a BESSEL FUNCTION OF THE FIRST KIND and $Y_n(z)$ is a BESSEL FUNCTION OF THE SECOND KIND. Hankel functions of the first kind can be represented as a CONTOUR INTEGRAL using

$$H_n^{(1)}(z) = \frac{1}{i\pi} \int_{0\ \text{[upper half plane]}}^{\infty} \frac{e^{(z/2)(t-1/t)}}{t^{n+1}}\, dt.$$

see also DEBYE'S ASYMPTOTIC REPRESENTATION, WATSON-NICHOLSON FORMULA, WEYRICH'S FORMULA

References
Arfken, G. "Hankel Functions." §11.4 in *Mathematical Methods for Physicists, 3rd ed.* Orlando, FL: Academic Press, pp. 604–610, 1985.
Morse, P. M. and Feshbach, H. *Methods of Theoretical Physics, Part I.* New York: McGraw-Hill, pp. 623–624, 1953.

Hankel Function of the Second Kind

$$H_n^{(2)}(z) \equiv J_n(z) - iY_n(z),$$

where $J_n(z)$ is a BESSEL FUNCTION OF THE FIRST KIND and $Y_n(z)$ is a BESSEL FUNCTION OF THE SECOND KIND. Hankel functions of the second kind can be represented as a CONTOUR INTEGRAL using

$$H_n^{(2)}(z) = \frac{1}{i\pi} \int_{-\infty\ \text{[lower half plane]}}^{0} \frac{e^{(z/2)(t-1/t)}}{t^{n+1}}\, dt.$$

see also WATSON-NICHOLSON FORMULA

References
Arfken, G. "Hankel Functions." §11.4 in *Mathematical Methods for Physicists, 3rd ed.* Orlando, FL: Academic Press, pp. 604–610, 1985.
Morse, P. M. and Feshbach, H. *Methods of Theoretical Physics, Part I.* New York: McGraw-Hill, pp. 623–624, 1953.

Hankel's Integral

$$J_m(x) = \frac{x^m}{2^{m-1}\sqrt{\pi}\,\Gamma\left(m+\frac{1}{2}\right)} \times \int_0^1 \cos(xt)(1-t^2)^{m-1/2}\, dt,$$

where $J_m(x)$ is a BESSEL FUNCTION OF THE FIRST KIND and $\Gamma(z)$ is the GAMMA FUNCTION. Hankel's integral can be derived from SONINE'S INTEGRAL.

see also POISSON INTEGRAL, SONINE'S INTEGRAL

Hankel Matrix

A MATRIX with identical values for each element in a given diagonal. Define H_n to be the Hankel matrix with leading column made up of the INTEGERS $1, \ldots, n$, then

$$\mathsf{H}_2 = \begin{bmatrix} 1 & 2 \\ 2 & 0 \end{bmatrix}$$
$$\mathsf{H}_3 = \begin{bmatrix} 1 & 2 & 3 \\ 2 & 3 & 0 \\ 3 & 0 & 0 \end{bmatrix}.$$

Hankel Transform

Equivalent to a 2-D FOURIER TRANSFORM with a radially symmetric KERNEL, and also called the FOURIER-BESSEL TRANSFORM.

$$g(u,v) = \mathcal{F}[f(r)] = \int_{-\infty}^{\infty}\int_{-\infty}^{\infty} f(r) e^{-2\pi i(ux+vy)}\, dx\, dy. \tag{1}$$

Let

$$x + iy = re^{i\theta} \tag{2}$$
$$u + iv = qe^{i\phi} \tag{3}$$

so that

$$x = r\cos\theta \tag{4}$$
$$y = r\sin\theta \tag{5}$$
$$r = \sqrt{x^2+y^2} \tag{6}$$

$$u = q\cos\phi \tag{7}$$
$$v = q\sin\phi \tag{8}$$
$$q = \sqrt{u^2+v^2}. \tag{9}$$

Then

$$\begin{aligned} g(q) &= \int_0^\infty \int_0^{2\pi} f(r) e^{-2\pi irq(\cos\phi\cos\theta + \sin\phi\sin\theta)} r\, dr\, d\theta \\ &= \int_0^\infty \int_0^{2\pi} f(r) e^{-2\pi irq\cos(\theta-\phi)} r\, dr\, d\theta \\ &= \int_0^\infty \int_{-\phi}^{2\pi-\phi} f(r) e^{-2\pi irq\cos\theta} r\, dr\, d\theta \\ &= \int_0^\infty \int_0^{2\pi} f(r) e^{-2\pi irq\cos\theta} r\, dr\, d\theta \\ &= \int_0^\infty f(r) \left[\int_0^{2\pi} e^{-2\pi irq\cos\theta}\, d\theta\right] r\, dr \\ &= 2\pi \int_0^\infty f(r) J_0(2\pi qr) r\, dr, \end{aligned} \tag{10}$$

where $J_0(z)$ is a zeroth order BESSEL FUNCTION OF THE FIRST KIND. Therefore, the Hankel transform pairs are

$$g(k) = \int_0^\infty f(x) J_0(kx) x \, dx \tag{11}$$

$$f(x) = \int_0^\infty g(k) J_0(kx) k \, dk. \tag{12}$$

see also BESSEL FUNCTION OF THE FIRST KIND, FOURIER TRANSFORM, LAPLACE TRANSFORM

References

Arfken, G. *Mathematical Methods for Physicists, 3rd ed.* Orlando, FL: Academic Press, p. 795, 1985.

Bracewell, R. *The Fourier Transform and Its Applications.* New York: McGraw-Hill, pp. 244–250, 1965.

Hann Function

see HANNING FUNCTION

Hanning Function

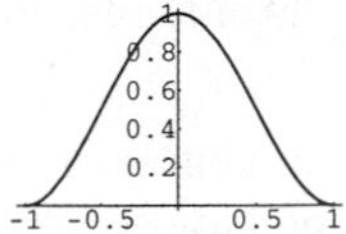
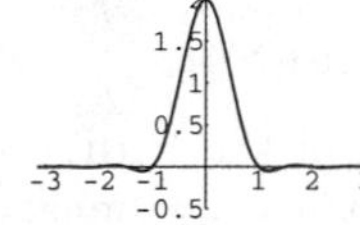
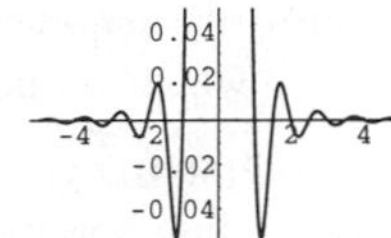

An APODIZATION FUNCTION, also called the HANN FUNCTION, frequently used to reduce ALIASING in FOURIER TRANSFORMS. The illustrations above show the Hanning function, its INSTRUMENT FUNCTION, and a blowup of the INSTRUMENT FUNCTION sidelobes. The Hanning function is given by

$$f(x) = \cos^2\left(\frac{\pi x}{2a}\right) = \tfrac{1}{2} - \tfrac{1}{2}\cos\left(\frac{\pi x}{a}\right). \tag{1}$$

The INSTRUMENT FUNCTION for Hanning apodization can also be written

$$a[\operatorname{sinc}(2\pi ka) + \tfrac{1}{2}\operatorname{sinc}(2\pi ka - \pi) + \tfrac{1}{2}\operatorname{sinc}(2\pi ka + \pi)]. \tag{2}$$

Its FULL WIDTH AT HALF MAXIMUM is a. It has APPARATUS FUNCTION

$$\begin{aligned} A(x) &= \int_{-a}^{a} \left[\tfrac{1}{2} - \tfrac{1}{2}\cos\left(\frac{\pi x}{a}\right)\right] e^{-2\pi ikx} \, dx \\ &= \tfrac{1}{2}\int_{-a}^{a} e^{-2\pi ikx} \, dx - \tfrac{1}{2}\int_{-a}^{a} e^{-2\pi ikx} \, dx \\ &\equiv \tfrac{1}{2}(A_1 + A_2). \end{aligned} \tag{3}$$

The first integral is

$$I_1 = \int_{-a}^{a} e^{-2\pi ikx} \, dx = \frac{\sin(2\pi ka)}{\pi k} = 2a \operatorname{sinc}(2\pi ka). \tag{4}$$

The second integral can be rewritten

$$\begin{aligned} I_2 &= \int_{-a}^{0} \cos\left(\frac{\pi x}{a}\right) e^{-2\pi ikx} \, dx \\ &\quad + \int_0^a \cos\left(\frac{\pi x}{a}\right) e^{-2\pi ikx} \, dx \\ &= \int_0^a \cos\left(\frac{\pi x}{a}\right) (e^{2\pi ikx} + e^{-2\pi ikx}) \, dx \\ &= 2\int_0^a \cos\left(\frac{\pi x}{a}\right) \cos(2\pi kx) \, dx \\ &= 2\left\{\frac{\sin\left(\frac{\pi}{a} - 2\pi k\right) x}{2\left(\frac{\pi}{a} - 2\pi k\right)} + \frac{\sin\left(\frac{\pi}{a} + 2\pi k\right) x}{2\left(\frac{\pi}{a} + 2\pi k\right)}\right\}_0^a \\ &= a\left[\frac{\sin(\pi - 2\pi ka)}{\pi - 2\pi ka} + \frac{\sin(\pi + 2\pi ka)}{\pi + 2\pi ka}\right] \\ &= \frac{a}{\pi}\left[\frac{\sin(2\pi ka)}{1 - 2ka} - \frac{\sin(2\pi ka)}{1 + 2ka}\right] \\ &= a[\operatorname{sinc}(\pi - 2\pi ka) + \operatorname{sinc}(\pi + 2\pi ka)]. \end{aligned} \tag{5}$$

Combining (4) and (5) gives

$$\begin{aligned} A(x) = a[\operatorname{sinc}(2\pi ka) &+ \tfrac{1}{2}\operatorname{sinc}(\pi - 2\pi ka) \\ &+ \tfrac{1}{2}\operatorname{sinc}(\pi + 2\pi ka)]. \end{aligned} \tag{6}$$

To find the extrema, define $x \equiv 2\pi ka$ and rewrite (6) as

$$A(x) = a[\sin x + \tfrac{1}{2}\operatorname{sinc}(x - \pi) + \tfrac{1}{2}\operatorname{sinc}(x + \pi)]. \tag{7}$$

Then solve

$$\frac{dA}{dx} = \frac{\pi^2(-x^3 \cos x + 3x^2 \sin x + \pi^2 x \cos x - \pi^2 \sin x)}{x^2(\pi^2 - x^2)^2} = 0 \tag{8}$$

to find the extrema. The roots are $x = 7.42023$ and 10.7061, giving a peak NEGATIVE sidelobe of -0.026708 and a peak POSITIVE sidelobe (in units of a) of 0.00843441. The peak in units of a is 1, and the full-width at half maximum is given by setting (7) equal to $1/2$ and solving for x, yielding

$$x_{1/2} = 2\pi k_{1/2} a = \pi. \tag{9}$$

Therefore, with $L \equiv 2a$, the FULL WIDTH AT HALF MAXIMUM is

$$\text{FWHM} = 2k_{1/2} = \frac{1}{a} = \frac{2}{L}. \tag{10}$$

see also APODIZATION FUNCTION, HAMMING FUNCTION

Hanoi Graph

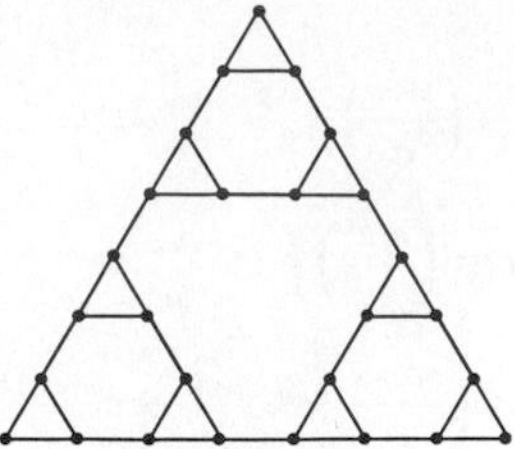

A GRAPH H_n arising in conjunction with the TOWERS OF HANOI problem. The above figure is the Hanoi graph H_3.

see also TOWERS OF HANOI

Hanoi Towers

see TOWERS OF HANOI

Hansen-Bessel Formula

$$\begin{aligned} J_n(z) &= \frac{1}{2\pi}\int_{-\pi}^{\pi} e^{iz\cos t} e^{in(t-\pi/2)}\,dt \\ &= \frac{i^{-n}}{\pi}\int_0^{\pi} e^{iz\cos t} cos(nt)\,dt \\ &= \frac{1}{\pi}\int_0^{\pi}\cos(z\sin t - nt)\,dt \end{aligned}$$

for $n = 0, 1, 2, \ldots$, where $J_n(z)$ is a BESSEL FUNCTION OF THE FIRST KIND.

References

Iyanaga, S. and Kawada, Y. (Eds.). *Encyclopedic Dictionary of Mathematics.* Cambridge, MA: MIT Press, p. 1472, 1980.

Hansen Chain

An ADDITION CHAIN for which there is a SUBSET H of members such that each member of the chain uses the largest element of H which is less than the member.

see also ADDITION CHAIN, BRAUER CHAIN, HANSEN NUMBER

References

Guy, R. K. "Addition Chains. Brauer Chains. Hansen Chains." §C6 in *Unsolved Problems in Number Theory, 2nd ed.* New York: Springer-Verlag, pp. 111–113, 1994.

Hansen Number

A number n for which a shortest chain exists which is a HANSEN CHAIN is called a Hansen number.

References

Guy, R. K. *Unsolved Problems in Number Theory, 2nd ed.* New York: Springer-Verlag, pp. 111–112, 1994.

Hansen's Problem

A SURVEYING PROBLEM: from the position of two known but inaccessible points A and B, determine the position of two unknown accessible points P and P' by bearings from A, B, P' to P and A, B, P to P'.

see also SURVEYING PROBLEMS

References

Dörrie, H. "Annex to a Survey." §40 in *100 Great Problems of Elementary Mathematics: Their History and Solutions.* New York: Dover, pp. 193–197, 1965.

Happy Number

Let the sum of the SQUARES of the DIGITS of a POSITIVE INTEGER s_0 be represented by s_1. In a similar way, let the sum of the SQUARES of the DIGITS of s_1 be represented by s_2, and so on. If some $s_i = 1$ for $i \geq 1$, then the original INTEGER s_0 is said to be happy.

Once it is known whether a number is happy (or not), then any number in the sequence s_1, s_2, s_3, ... will also be happy (or not). A number which is not happy is called UNHAPPY. Unhappy numbers have EVENTUALLY PERIODIC sequences of s_i 4, 16, 37, 58, 89, 145, 42, 20, 4, ... which do not reach 1.

Any PERMUTATION of the DIGITS of an UNHAPPY or happy number must also be unhappy or happy. This follows from the fact that ADDITION is COMMUTATIVE. The first few happy numbers are 1, 7, 10, 13, 19, 23, 28, 31, 32, 44, 49, 68, 70, 79, 82, 86, 91, 94, 97, 100, ... (Sloane's A007770). These are also the numbers whose 2-RECURRING DIGITAL INVARIANT sequences have period 1.

see also KAPREKAR NUMBER, RECURRING DIGITAL INVARIANT , UNHAPPY NUMBER

References

Dudeney, H. E. Problem 143 in *536 Puzzles & Curious Problems.* New York: Scribner, pp. 43 and 258–259, 1967.

Guy, R. K. "Happy Numbers." §E34 in *Unsolved Problems in Number Theory, 2nd ed.* New York: Springer-Verlag, pp. 234–235, 1994.

Madachy, J. S. *Madachy's Mathematical Recreations.* New York: Dover, pp. 163–165, 1979.

Schwartzman, S. *The Words of Mathematics: An Etymological Dictionary of Mathematical Terms Used in English.* Washington, DC: Math. Assoc. Amer., 1994.

Sloane, N. J. A. Sequence A007770 in "An On-Line Version of the Encyclopedia of Integer Sequences."

Harada-Norton Group

The SPORADIC GROUP *HN*.

References

Wilson, R. A. "ATLAS of Finite Group Representation." `http://for.mat.bham.ac.uk/atlas/HN.html`.

Harary Graph

The smallest k-connected GRAPH with n VERTICES.

Hard Hexagon Entropy Constant

N.B. A detailed on-line essay by S. Finch was the starting point for this entry.

A constant related to the HARD SQUARE ENTROPY CONSTANT. This constant is given by

$$\kappa_h \equiv \lim_{N\to\infty} [G(N)]^{1/N} = 1.395485972\ldots, \qquad (1)$$

where $G(N)$ is the number of configurations of nonattacking KINGS on an $n \times n$ chessboard with regular hexagonal cells, where $N \equiv n^2$. Amazingly, κ_h is algebraic and given by

$$\kappa_h \equiv \kappa_1\kappa_2\kappa_3\kappa_4, \qquad (2)$$

where

$$\kappa_1 \equiv 4^{-1}3^{5/4}11^{-5/12}c^{-2} \qquad (3)$$

$$\kappa_2 \equiv [1-\sqrt{1-c}+\sqrt{2+c+2\sqrt{1+c+c^2}}\,]^2 \qquad (4)$$

$$\kappa_3 \equiv [-1-\sqrt{1-c}+\sqrt{2+c+2\sqrt{1+c+c^2}}\,]^2 \qquad (5)$$

$$\kappa_4 \equiv [\sqrt{1-a}+\sqrt{2+a+2\sqrt{1+a+a^2}}\,]^{-1/2} \qquad (6)$$

$$a \equiv -\tfrac{124}{363}11^{1/3} \qquad (7)$$

$$b \equiv \tfrac{2501}{11979}33^{1/2} \qquad (8)$$

$$c \equiv \{\tfrac{1}{4}+\tfrac{3}{8}a[(b+1)^{1/3}-(b-1)^{1/3}]\}^{1/3}. \qquad (9)$$

(Baxter 1980, Joyce 1988).

References

Baxter, R. J. "Hard Hexagons: Exact Solution." *J. Physics A* **13**, 1023–1030, 1980.

Finch, S. "Favorite Mathematical Constants." `http://www.mathsoft.com/asolve/constant/square/square.html`.

Joyce, G. S. "On the Hard Hexagon Model and the Theory of Modular Functions." *Phil. Trans. Royal Soc. London A* **325**, 643–702, 1988.

Plouffe, S. "Hard Hexagons Constant." `http://lacim.uqam.ca/piDATA/hardhex.html`.

Hard Square Entropy Constant

N.B. A detailed on-line essay by S. Finch was the starting point for this entry.

Let $F(n^2)$ be the number of binary $n\times n$ MATRICES with no adjacent 1s (in either columns or rows). Define $N \equiv n^2$, then the hard square entropy constant is defined by

$$\kappa \equiv \lim_{N\to\infty} [F(N)]^{1/N} = 1.503048082\ldots.$$

The quantity $\ln\kappa$ arises in statistical physics (Baxter *et al.* 1980, Pearce and Seaton 1988), and is known as the entropy per site of hard squares. A related constant known as the HARD HEXAGON ENTROPY CONSTANT can also be defined.

References

Baxter, R. J.; Enting, I. G.; and Tsang, S. K. "Hard-Square Lattice Gas." *J. Statist. Phys.* **22**, 465–489, 1980.

Finch, S. "Favorite Mathematical Constants." `http://www.mathsoft.com/asolve/constant/square/square.html`.

Pearce, P. A. and Seaton, K. A. "A Classical Theory of Hard Squares." *J. Statist. Phys.* **53**, 1061–1072, 1988.

Hardy's Inequality

Let $\{a_n\}$ be a NONNEGATIVE SEQUENCE and $f(x)$ a NONNEGATIVE integrable FUNCTION. Define

$$A_n = \sum_{k=1}^{n} a_k \qquad (1)$$

$$B_n = \sum_{k=n}^{\infty} a_k \qquad (2)$$

and

$$F(x) = \int_0^x f(t)\,dt \qquad (3)$$

$$G(x) = \int_x^\infty f(t)\,dt, \qquad (4)$$

and take $p > 1$. For sums,

$$\sum_{n=1}^{\infty}\left(\frac{A_n}{n}\right)^p < \left(\frac{p}{p-1}\right)^p \sum_{n=1}^{\infty}(a_n)^p \qquad (5)$$

(unless all $a_n = 0$), and for integrals,

$$\int_0^\infty \left[\frac{F(x)}{x}\right]^p dx < \left(\frac{p}{p-1}\right)^p \int_0^\infty [f(x)]^p\,dx \qquad (6)$$

(unless f is identically 0).

References

Hardy, G. H.; Littlewood, J. E.; and Pólya, G. *Inequalities, 2nd ed.* Cambridge, England: Cambridge University Press, pp. 239–243, 1988.

Mitrinovic, D. S.; Pecaric, J. E.; and Fink, A. M. *Inequalities Involving Functions and Their Integrals and Derivatives.* New York: Kluwer, 1991.

Opic, B. and Kufner, A. *Hardy-Type Inequalities.* Essex, England: Longman, 1990.

Hardy-Littlewood Conjectures

The first Hardy-Littlewood conjecture is called the k-TUPLE CONJECTURE. It states that the asymptotic number of PRIME CONSTELLATIONS can be computed explicitly.

The second Hardy-Littlewood conjecture states that

$$\pi(x+y) - \pi(x) \leq \pi(y)$$

for all x and y, where $\pi(x)$ is the PRIME COUNTING FUNCTION. Although it is not obvious, Richards (1974) proved that this conjecture is incompatible with the first Hardy-Littlewood conjecture.

see also PRIME CONSTELLATION, PRIME COUNTING FUNCTION

References

Richards, I. "On the Incompatibility of Two Conjectures Concerning Primes." *Bull. Amer. Math. Soc.* **80**, 419–438, 1974.

Riesel, H. *Prime Numbers and Computer Methods for Factorization, 2nd ed.* Boston, MA: Birkhäuser, pp. 61–62 and 68–69, 1994.

Hardy-Littlewood Constants

see PRIME CONSTELLATION

Hardy-Littlewood Tauberian Theorem

Let $a_n \geq 0$ and suppose

$$\sum_{n=1}^{\infty} a_n e^{-an} \sim \frac{1}{a}$$

as $a \to 0^+$. Then

$$\sum_{n \leq x} a_n \sim x$$

as $x \to \infty$.

see also TAUBERIAN THEOREM

References

Berndt, B. C. *Ramanujan's Notebooks, Part IV.* New York: Springer-Verlag, pp. 118–119, 1994.

Hardy-Littlewood k-Tuple Conjecture

see PRIME PATTERNS CONJECTURE

Hardy-Ramanujan Number

The smallest nontrivial TAXICAB NUMBER, i.e., the smallest number representable in two ways as a sum of two CUBES. It is given by

$$1729 = 1^3 + 12^3 = 9^3 + 10^3.$$

The number derives its name from the following story G. H. Hardy told about Ramanujan. "Once, in the taxi from London, Hardy noticed its number, 1729. He must have thought about it a little because he entered the room where Ramanujan lay in bed and, with scarcely a hello, blurted out his disappointment with it. It was, he declared, 'rather a dull number,' adding that he hoped that wasn't a bad omen. 'No, Hardy,' said Ramanujan, 'it is a very interesting number. It is the smallest number expressible as the sum of two [POSITIVE] cubes in two different ways'" (Hofstadter 1989, Kanigel 1991, Snow 1993).

see also DIOPHANTINE EQUATION—CUBIC, TAXICAB NUMBER

References

Guy, R. K. "Sums of Like Powers. Euler's Conjecture." §D1 in *Unsolved Problems in Number Theory, 2nd ed.* New York: Springer-Verlag, pp. 139–144, 1994.

Hardy, G. H. *Ramanujan: Twelve Lectures on Subjects Suggested by His Life and Work, 3rd ed.* New York: Chelsea, p. 68, 1959.

Hofstadter, D. R. *Gödel, Escher, Bach: An Eternal Golden Braid.* New York: Vintage Books, p. 564, 1989.

Kanigel, R. *The Man Who Knew Infinity: A Life of the Genius Ramanujan.* New York: Washington Square Press, p. 312, 1991.

Snow, C. P. Foreword to Hardy, G. H. *A Mathematician's Apology, reprinted with a foreword by C. P. Snow.* New York: Cambridge University Press, p. 37, 1993.

Hardy-Ramanujan Theorem

Let $\omega(n)$ be the number of DISTINCT PRIME FACTORS of n. If $\Psi(x)$ tends steadily to infinity with x, then

$$\ln \ln x - \Psi(x)\sqrt{\ln \ln x} < \omega(n) < \ln \ln x + \Psi(x)\sqrt{\ln \ln x}$$

for ALMOST ALL numbers $n < x$. "ALMOST ALL" means here the frequency of those INTEGERS n in the interval $1 \leq n \leq x$ for which

$$|\omega(n) - \ln \ln x| > \Psi(x)\sqrt{\ln \ln x}$$

approaches 0 as $x \to \infty$.

see also DISTINCT PRIME FACTORS, ERDŐS-KAC THEOREM

Hardy's Rule

Let the values of a function $f(x)$ be tabulated at intervals equally spaced by h about x_0, so that $f_{-3} = f(x_0 - 3h)$, $f_{-2} = f(x_0 - 2h)$, etc. Then Hardy's rule gives the approximation to the integral of $f(x)$ as

$$\int_{x_0-3h}^{x_0+3h} f(x)\,dx = \tfrac{1}{100}h(28f_{-3} + 162f_{-2} + 22f_0 + 162f_2 + 28f_3) + \tfrac{9}{1400}h^7[2f^{(4)}(\xi_2) - h^2 f^{(8)}(\xi_1)],$$

where the final term gives the error, with $\xi_1, \xi_2 \in [x_0 - 3h, x_0 + 3h]$.

see also BODE'S RULE, DURAND'S RULE, NEWTON-COTES FORMULAS, SIMPSON'S 3/8 RULE, SIMPSON'S RULE, TRAPEZOIDAL RULE, WEDDLE'S RULE

Harmonic Addition Theorem

To convert an equation of the form

$$f(\theta) = a\cos\theta + b\sin\theta \tag{1}$$

to the form

$$f(\theta) = c\cos(\theta + \delta), \tag{2}$$

expand (2) using the trigonometric addition formulas to obtain

$$f(\theta) = c\cos\theta\cos\delta - c\sin\theta\sin\delta. \tag{3}$$

Now equate the COEFFICIENTS of (1) and (3)

$$a = c\cos\delta \tag{4}$$

$$b = -c\sin\delta, \tag{5}$$

so

$$\tan\delta = -\frac{b}{a} \tag{6}$$

$$a^2 + b^2 = c^2, \tag{7}$$

and we have

$$\delta = \tan^{-1}\left(-\frac{b}{a}\right) \tag{8}$$

$$c = \sqrt{a^2 + b^2}. \tag{9}$$

Given two general sinusoidal functions with frequency ω:

$$\psi_1 = A_1 \sin(\omega t + \delta_1) \tag{10}$$

$$\psi_2 = A_2 \sin(\omega t + \delta_2), \tag{11}$$

their sum ψ can be expressed as a sinusoidal function with frequency ω

$$\begin{aligned}\psi \equiv \psi_1 + \psi_2 &= A_1[\sin(\omega t)\cos\delta_1 + \sin\delta_1\cos(\omega t)] \\ &\quad + A_2[\sin(\omega t)\cos\delta_2 + \sin\delta_2\cos(\omega t)] \\ &= [A_1\cos\delta_1 + A_2\cos\delta_2]\sin(\omega t) \\ &\quad + [A_1\sin\delta_1 + A_2\sin\delta_2]\cos(\omega t).\end{aligned} \tag{12}$$

Now, define

$$A\cos\delta \equiv A_1\cos\delta_1 + A_2\cos\delta_2 \tag{13}$$

$$A\sin\delta \equiv A_1\sin\delta_1 + A_2\sin\delta_2. \tag{14}$$

Then (12) becomes

$$A\cos\delta\sin(\omega t) + A\sin\delta\cos(\omega t) = A\sin(\omega t + \delta). \tag{15}$$

Square and add (13) and (14)

$$A_2 = {A_1}^2 + {A_2}^2 + 2A_1A_2\cos(\delta_2 - \delta_1). \tag{16}$$

Also, divide (14) by (13)

$$\tan\delta = \frac{A_1\sin\delta_1 + A_2\sin\delta_2}{A_1\cos\delta_1 + A_2\cos\delta_2}, \tag{17}$$

so

$$\psi = A\sin(\omega t + \delta), \tag{18}$$

where A and δ are defined by (16) and (17).

This procedure can be generalized to a sum of n harmonic waves, giving

$$\psi = \sum_{i=1}^{n} A_i\cos(\omega t + \delta_i) = A\cos(\omega t + \delta), \tag{19}$$

where

$$A^2 \equiv \sum_{i=1}^{n} {A_i}^2 + 2\sum_{j>i}^{n}\sum_{i=1}^{n} A_iA_j\cos(\delta_i - \delta_j) \tag{20}$$

and

$$\tan\delta = -\frac{\sum_{i=1}^{n} A_i\sin\delta_i}{\sum_{i=1}^{n} A_i\cos\delta_i}. \tag{21}$$

Harmonic Analysis

see also FOURIER SERIES

Harmonic Brick

A right-angled PARALLELEPIPED with dimensions $a \times ab \times abc$, where a, b, and c are INTEGERS.

see also BRICK, DE BRUIJN'S THEOREM, EULER BRICK

Harmonic Conjugate Function

The harmonic conjugate to a given function $u(x, y)$ is a function $v(x, y)$ such that

$$f(x, y) = u(x, y) + v(x, y)$$

is COMPLEX DIFFERENTIABLE (i.e., satisfies the CAUCHY-RIEMANN EQUATIONS). It is given by

$$v(z) = \int ux\,dy - uy\,dx.$$

Harmonic Conjugate Points

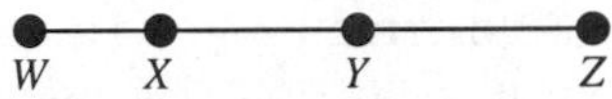

Given COLLINEAR points W, X, Y, and Z, Y and Z are harmonic conjugates with respect to W and X if

$$\frac{|WY|}{|YX|} = \frac{|WZ|}{|ZX|}.$$

The distances between such points are said to be in HARMONIC RATIO, and the LINE SEGMENT depicted above is called a HARMONIC SEGMENT.

Harmonic conjugate points are also defined for a TRIANGLE. If W and X have TRILINEAR COORDINATES $\alpha : \beta : \gamma$ and $\alpha' : \beta' : \gamma'$, then the TRILINEAR COORDINATES of the harmonic conjugates are

$$Y = \alpha + \alpha' : \beta + \beta' : \gamma + \gamma'$$
$$Z = \alpha - \alpha' : \beta - \beta' : \gamma - \gamma'$$

(Kimberling 1994).

see also HARMONIC RANGE, HARMONIC RATIO

References

Kimberling, C. "Central Points and Central Lines in the Plane of a Triangle." *Math. Mag.* **67**, 163–187, 1994.

Ogilvy, C. S. *Excursions in Geometry.* New York: Dover, pp. 13–14, 1990.

Phillips, A. W. and Fisher, I. *Elements of Geometry.* New York: American Book Co., 1896.

Wells, D. *The Penguin Dictionary of Curious and Interesting Geometry.* New York: Viking Penguin, p. 92, 1992.

Harmonic Coordinates

Satisfy the condition

$$\Gamma^\lambda \equiv g^{\mu\nu}{\Gamma_{\mu\nu}}^\lambda = 0, \tag{1}$$

or equivalently,

$$\frac{\partial}{\partial x^\kappa}\left(\sqrt{g}\,g^{\lambda\kappa}\right) = 0. \tag{2}$$

It is always possible to choose such a system. Using the D'ALEMBERTIAN OPERATOR,

$$\Box^2\phi \equiv (g^{\lambda\kappa}\phi_{;\lambda})_{;\kappa} = g^{\lambda\kappa}\frac{\partial^2\phi}{\partial x^\lambda \partial x^\kappa} - \Gamma^\lambda \frac{\partial\phi}{\partial x^\lambda}. \tag{3}$$

But since $\Gamma^\lambda \equiv 0$ for harmonic coordinates,

$$\Box^2 x^\mu = 0. \tag{4}$$

Harmonic Decomposition

A POLYNOMIAL function in the elements of $\mathbf{x}$ can be uniquely decomposed into a sum of harmonic POLYNOMIALS times POWERS of $|\mathbf{x}|$.

Harmonic Divisor Number

A number n for which the HARMONIC MEAN of the DIVISORS of n, i.e., $nd(n)/\sigma(n)$, is an INTEGER, where $d(n)$ is the number of POSITIVE integral DIVISORS of n and $\sigma(n)$ is the DIVISOR FUNCTION. For example, the divisors of $n = 140$ are 1, 2, 4, 5, 7, 10, 14, 20, 28, 35, 70, and 140, giving

$$d(140) = 12$$
$$\sigma(140) = 336$$
$$\frac{140d(140)}{\sigma(140)} = \frac{140 \cdot 12}{335} = 5,$$

so 140 is a harmonic divisor number. Harmonic divisor numbers are also called ORE NUMBERS. Garcia (1954) gives the 45 harmonic divisor numbers less than 10^7. The first few are 1, 6, 140, 270, 672, 1638, ... (Sloane's A007340).

For distinct PRIMES p and q, harmonic divisor numbers are equivalent to EVEN PERFECT NUMBERS for numbers of the form $p^r q$. Mills (1972) proved that if there exists an ODD POSITIVE harmonic divisor number n, then n has a prime-POWER factor greater than 10^7.

Another type of number called "harmonic" is the HARMONIC NUMBER.

see also DIVISOR FUNCTION, HARMONIC NUMBER

References

Edgar, H. M. W. "Harmonic Numbers." *Amer. Math. Monthly* **99**, 783–789, 1992.

Garcia, M. "On Numbers with Integral Harmonic Mean." *Amer. Math. Monthly* **61**, 89–96, 1954.

Guy, R. K. "Almost Perfect, Quasi-Perfect, Pseudoperfect, Harmonic, Weird, Multiperfect and Hyperperfect Numbers." §B2 in *Unsolved Problems in Number Theory, 2nd ed.* New York: Springer-Verlag, pp. 45–53, 1994.

Mills, W. H. "On a Conjecture of Ore." *Proceedings of the 1972 Number Theory Conference.* University of Colorado, Boulder, pp. 142–146, 1972.

Ore, Ø. "On the Averages of the Divisors of a Number." *Amer. Math. Monthly* **55**, 615–619, 1948.

Pomerance, C. "On a Problem of Ore: Harmonic Numbers." Unpublished manuscript, 1973.

Sloane, N. J. A. Sequences A007340/M4299 in "An On-Line Version of the Encyclopedia of Integer Sequences."

Sloane, N. J. A. and Plouffe, S. Extended entry in *The Encyclopedia of Integer Sequences.* San Diego: Academic Press, 1995.

Zachariou, A. and Zachariou, E. "Perfect, Semi-Perfect and Ore Numbers." *Bull. Soc. Math. Grèce (New Ser.)* **13**, 12–22, 1972.

Harmonic Equation

see LAPLACE'S EQUATION

Harmonic Function

Any real-valued function $u(x, y)$ with continuous second PARTIAL DERIVATIVES which satisfies LAPLACE'S EQUATION

$$\nabla^2 u(x, y) = 0 \tag{1}$$

is called a harmonic function. Harmonic functions are called POTENTIAL FUNCTIONS in physics and engineering. Potential functions are extremely useful, for example, in electromagnetism, where they reduce the study of a 3-component VECTOR FIELD to a 1-component SCALAR FUNCTION. A scalar harmonic function is called a SCALAR POTENTIAL, and a vector harmonic function is called a VECTOR POTENTIAL.

To find a class of such functions in the PLANE, write the LAPLACE'S EQUATION in POLAR COORDINATES

$$u_{rr} + \frac{1}{r}u_r + \frac{1}{r^2}u_{\theta\theta} = 0, \tag{2}$$

and consider only radial solutions

$$u_{rr} + \frac{1}{r}u_r = 0. \tag{3}$$

This is integrable by quadrature, so define $v \equiv du/dr$,

$$\frac{dv}{dr} + \frac{1}{r}v = 0 \tag{4}$$

$$\frac{dv}{v} = -\frac{dr}{r} \tag{5}$$

$$\ln\left(\frac{v}{A}\right) = -\ln r \tag{6}$$

$$\frac{v}{A} = \frac{1}{r} \tag{7}$$

$$v = \frac{du}{dr} = \frac{A}{r} \tag{8}$$

$$du = A\frac{dr}{r}, \tag{9}$$

so the solution is

$$u = A\ln r. \tag{10}$$

Ignoring the trivial additive and multiplicative constants, the general pure radial solution then becomes

$$u = \ln[(x-a)^2+(y-b)^2]^{1/2} = \tfrac{1}{2}\ln\left[(x-a)^2 + (y-b)^2\right]. \tag{11}$$

Other solutions may be obtained by differentiation, such as

$$u = \frac{x-a}{(x-a)^2+(y-b)^2} \tag{12}$$

$$v = \frac{y-b}{(x-a)^2+(y-b)^2}, \tag{13}$$

$$u = e^x \sin y \tag{14}$$

$$v = e^x \cos y, \tag{15}$$

and

$$\tan^{-1}\left(\frac{y-b}{x-a}\right). \tag{16}$$

Harmonic functions containing azimuthal dependence include

$$u = r^n \cos(n\theta) \tag{17}$$

$$v = r^n \sin(n\theta). \tag{18}$$

The POISSON KERNEL

$$u(r,R,\theta,\phi) = \frac{R^2-r^2}{R^2-2rR\cos(\theta-\phi)+r^2} \tag{19}$$

is another harmonic function.

see also SCALAR POTENTIAL, VECTOR POTENTIAL

References

Ash, J. M. (Ed.) *Studies in Harmonic Analysis.* Washington, DC: Math. Assoc. Amer., 1976.

Axler, S.; Pourdon, P.; and Ramey, W. *Harmonic Function Theory.* Springer-Verlag, 1992.

Benedetto, J. J. *Harmonic Analysis and Applications.* Boca Raton, FL: CRC Press, 1996.

Cohn, H. *Conformal Mapping on Riemann Surfaces.* New York: Dover, 1980.

Harmonic-Geometric Mean

Let

$$\alpha_{n+1} = \frac{2\alpha_n\beta_n}{\alpha_n+\beta_n}$$

$$\beta_{n+1} = \sqrt{\alpha_n\beta_n},$$

then

$$H(\alpha_0,\beta_0) \equiv \lim_{n\to\infty} a_n = \frac{1}{M(\alpha_0{}^{-1},\beta_0{}^{-1})},$$

where M is the ARITHMETIC-GEOMETRIC MEAN.

see also ARITHMETIC MEAN, ARITHMETIC-GEOMETRIC MEAN, GEOMETRIC MEAN, HARMONIC MEAN

Harmonic Homology

A PERSPECTIVE COLLINEATION with center O and axis o not incident is called a HOMOLOGY. A HOMOLOGY is said to be harmonic if the points A and A' on a line through O are harmonic conjugates with respect to O and $o\cdot a$. Every PERSPECTIVE COLLINEATION of period two is a harmonic homology.

see also HOMOLOGY (GEOMETRY), PERSPECTIVE COLLINEATION

References

Coxeter, H. S. M. *Introduction to Geometry, 2nd ed.* New York: Wiley, p. 248, 1969.

Harmonic Logarithm

For all INTEGERS n and NONNEGATIVE INTEGERS t, the harmonic logarithms $\lambda_n^{(t)}(x)$ of order t and degree n are defined as the unique functions satisfying

1. $\lambda_0^{(t)}(x) = (\ln x)^t$,
2. $\lambda_n^{(t)}(x)$ has no constant term except $\lambda_0^{(0)}(x) = 1$,
3. $\frac{d}{dx}\lambda_n^{(t)}(x) = \lfloor n\rceil\,\lambda_{n-1}^{(t)}(x)$,

where the "ROMAN SYMBOL" $\lfloor n\rceil$ is defined by

$$\lfloor n\rceil \equiv \begin{cases} n & \text{for } n\neq 0\\ 1 & \text{for } n=0\end{cases} \tag{1}$$

(Roman 1992). This gives the special cases

$$\lambda_n^{(0)}(x) = \begin{cases} x^n & \text{for } n\geq 0\\ 0 & \text{for } n<0\end{cases} \tag{2}$$

$$\lambda_n^{(1)}(x) = \begin{cases} x^n(\ln x - H_n) & \text{for } n\geq 0\\ x^n & \text{for } n<0,\end{cases} \tag{3}$$

where H_n is a HARMONIC NUMBER

$$H_n \equiv \sum_{k=1}^n \frac{1}{k}. \tag{4}$$

The harmonic logarithm has the INTEGRAL

$$\int \lambda_n^{(1)}(x)\,dx = \frac{1}{\lfloor n+1\rceil}\lambda_{n+1}^{(1)}(x). \tag{5}$$

The harmonic logarithm can be written

$$\lambda_n^{(t)}(x) = \lfloor n\rceil!\tilde{D}^{-n}(\ln x)^t, \tag{6}$$

where $\tilde{D}$ is the DIFFERENTIAL OPERATOR, (so $\tilde{D}^{-n}$ is the nth INTEGRAL). Rearranging gives

$$\tilde{D}^k\lambda_n^{(t)}(x) = \left\lfloor \frac{\lfloor n\rceil!}{\lfloor n-k\rceil}\right\rceil!\lambda_{n-k}^{(t)}(x). \tag{7}$$

This formulation gives an analog of the BINOMIAL THEOREM called the LOGARITHMIC BINOMIAL FORMULA. Another expression for the harmonic logarithm is

$$\lambda_n^{(t)}(x) = x^n\sum_{j=0}^t (-1)^j (t)_j c_n^{(j)}(\ln x)^{t-j}, \tag{8}$$

where $(t)_j = t(t-1)\cdots(t-j+1)$ is a POCHHAMMER SYMBOL and $c_n^{(j)}$ is a two-index HARMONIC NUMBER (Roman 1992).

see also LOGARITHM, ROMAN FACTORIAL

References

Loeb, D. and Rota, G.-C. "Formal Power Series of Logarithmic Type." *Advances Math.* **75**, 1–118, 1989.

Roman, S. "The Logarithmic Binomial Formula." *Amer. Math. Monthly* **99**, 641–648, 1992.

Harmonic Map

A harmonic map between RIEMANNIAN MANIFOLDS can be viewed as a generalization of a GEODESIC when the domain DIMENSION is one, or of a HARMONIC FUNCTION when the range is a EUCLIDEAN SPACE.

see also BOCHNER IDENTITY, EUCLIDEAN SPACE, GEODESIC, HARMONIC FUNCTION, RIEMANNIAN MANIFOLD

References

Burstal, F.; Lemaire, L.; and Rawnsley, J. "Harmonic Maps Bibliography." http://www.bath.ac.uk/~masfeb/harmonic.html.

Eels, J. and Lemaire, L. "A Report on Harmonic Maps." *Bull. London Math. Soc.* **10**, 1–68, 1978.

Eels, J. and Lemaire, L. "Another Report on Harmonic Maps." *Bull. London Math. Soc.* **20**, 385–524, 1988.

Harmonic Mean

The harmonic mean $H(x_1, \ldots, x_n)$ of n points x_i (where $i = 1, \ldots, n$) is

$$\frac{1}{H} \equiv \frac{1}{n}\sum_{i=1}^{n}\frac{1}{x_i}. \quad (1)$$

The special case of $n = 2$ is therefore

$$\frac{1}{H} = \frac{1}{2}\left(\frac{1}{x_1} + \frac{1}{x_2},\right) \quad (2)$$

or

$$\frac{1}{H} = \frac{x_1 + x_2}{2x_1x_2}. \quad (3)$$

The VOLUME-to-SURFACE AREA ratio for a cylindrical container with height h and radius r and the MEAN CURVATURE of a general surface are related to the harmonic mean.

Hoehn and Niven (1985) show that

$$H(a_1+c, a_2+c, \ldots, a_n+c) > c + H(a_1, a_2, \ldots, a_n) \quad (4)$$

for any POSITIVE constant c.

see also ARITHMETIC MEAN, ARITHMETIC-GEOMETRIC MEAN, GEOMETRIC MEAN, HARMONIC-GEOMETRIC MEAN, ROOT-MEAN-SQUARE

References

Abramowitz, M. and Stegun, C. A. (Eds.). *Handbook of Mathematical Functions with Formulas, Graphs, and Mathematical Tables, 9th printing.* New York: Dover, p. 10, 1972.

Hoehn, L. and Niven, I. "Averages on the Move." *Math. Mag.* **58**, 151–156, 1985.

Harmonic Mean Index

The statistical INDEX

$$P_H \equiv \frac{\sum p_0 q_0}{\sum \frac{{p_0}^2 q_0}{p_n}},$$

where p_n is the price per unit in period n, q_n is the quantity produced in period n, and $v_n \equiv p_n q_n$ the value of the n units.

see also INDEX

References

Kenney, J. F. and Keeping, E. S. *Mathematics of Statistics, Pt. 1, 3rd ed.* Princeton, NJ: Van Nostrand, p. 69, 1962.

Harmonic Number

A number of the form

$$H_n = \sum_{k=1}^{n}\frac{1}{k}. \quad (1)$$

This can be expressed analytically as

$$H_n = \gamma + \psi_0(n+1), \quad (2)$$

where γ is the EULER-MASCHERONI CONSTANT and $\Psi(x) = \psi_0(x)$ is the DIGAMMA FUNCTION. The number formed by taking alternate signs in the sum also has an analytic solution

$$H_n' = \sum_{k=1}^{n}\frac{(-1)^{k+1}}{k} \quad (3)$$

$$= \ln 2 + \tfrac{1}{2}(-1)^n[\psi_0(\tfrac{1}{2}n + \tfrac{1}{2}) - \psi_0(\tfrac{1}{2}n + 1)]. \quad (4)$$

The first few harmonic numbers H_n are 1, 3/2, 11/6, 25/12, 137/60, ... (Sloane's A001008 and A002805). The HARMONIC NUMBER H_n is never an INTEGER (except for H_1), which can be proved by using the strong triangle inequality to show that the 2-ADIC VALUE of H_n is greater than 1 for $n > 1$. The harmonic numbers have ODD NUMERATORS and EVEN DENOMINATORS. The nth harmonic number is given asymptotically by

$$H_n \sim \ln n + \gamma + \frac{1}{2n}, \quad (5)$$

where γ is the EULER-MASCHERONI CONSTANT (Conway and Guy 1996). Gosper gave the interesting identity

$$\sum_{i=0}^{\infty}\frac{z^i H_i}{i!} = -e^z\sum_{k=1}^{\infty}\frac{(-z)^k}{kk!} = e^z[\ln z + \Gamma(0,z) + \gamma], \quad (6)$$

where $\Gamma(0,z)$ is the incomplete GAMMA FUNCTION and γ is the EULER-MASCHERONI CONSTANT. Borwein and Borwein (1995) show that

$$\sum_{n=1}^{\infty} \frac{{H_n}^2}{(n+1)^2} = \tfrac{11}{4}\zeta(4) = \tfrac{11}{360}\pi^4 \tag{7}$$

$$\sum_{n=1}^{\infty} \frac{{H_n}^2}{n^2} = \tfrac{17}{4}\zeta(4) = \tfrac{17}{360}\pi^4 \tag{8}$$

$$\sum_{n=1}^{\infty} \frac{H_n}{n^3} = \tfrac{5}{4}\zeta(4) = \tfrac{1}{72}\pi^4, \tag{9}$$

where $\zeta(z)$ is the RIEMANN ZETA FUNCTION. The first of these had been previously derived by de Doelder (1991), and the last by Euler (1775). These identities are corollaries of the identity

$$\frac{1}{\pi}\int_0^\pi x^2\{\ln[2\cos(\tfrac{1}{2}x)]\}^2\,dx = \tfrac{11}{2}\zeta(4) = \tfrac{11}{180}\pi^4 \tag{10}$$

(Borwein and Borwein 1995). Additional identities due to Euler are

$$\sum_{n=1}^{\infty} \frac{H_n}{n^2} = 2\zeta(3) \tag{11}$$

$$2\sum_{n=1}^{\infty} \frac{H_n}{n^m} = (m+2)\zeta(m+1) - \sum_{n=1}^{m-2}\zeta(m-n)\zeta(n+1) \tag{12}$$

for $m = 2, 3, \ldots$ (Borwein and Borwein 1995), where $\zeta(3)$ is APÉRY'S CONSTANT. These sums are related to so-called EULER SUMS.

Conway and Guy (1996) define the second harmonic number by

$$H_n^{(2)} \equiv \sum_{i=1}^{n} H_i = (n+1)(H_{n+1}-1) = (n+1)(H_{n+1}-H_1), \tag{13}$$

the third harmonic number by

$$H_n^{(3)} \equiv \sum_{i=1}^{n} H_i^{(2)} = \binom{n+2}{2}(H_{n+2}-H_2), \tag{14}$$

and the nth harmonic number by

$$H_n^{(k)} = \binom{n+k-1}{k-1}(H_{n+k-1}-H_{k-1}). \tag{15}$$

A slightly different definition of a two-index harmonic number $c_n^{(j)}$ is given by Roman (1992) in connection with the HARMONIC LOGARITHM. Roman (1992) defines this by

$$c_n^{(0)} = \begin{cases} 1 & \text{for } n \geq 0 \\ 0 & \text{for } n < 0 \end{cases} \tag{16}$$

$$c_0^{(j)} = \begin{cases} 1 & \text{for } j = 0 \\ 0 & \text{for } j \neq 0 \end{cases} \tag{17}$$

plus the recurrence relation

$$cn_n^{(j)} = c_n^{(j-1)} + nc_{n-1}^{(j)}. \tag{18}$$

For general $n > 0$ and $j > 0$, this is equivalent to

$$c_n^{(j)} = \sum_{i=1}^{n} \frac{1}{i} c_i^{(j-1)}, \tag{19}$$

and for $n > 0$, it simplifies to

$$c_n^{(j)} = \sum_{i=1}^{n} \binom{n}{i}(-1)^{i-1} i^{-j}. \tag{20}$$

For $n < 0$, the harmonic number can be written

$$c_n^{(j)} = (-1)^j \lfloor n \rceil! s(-n,j), \tag{21}$$

where $\lfloor n \rceil!$ is the ROMAN FACTORIAL and s is a STIRLING NUMBER OF THE FIRST KIND.

A separate type of number sometimes also called a "harmonic number" is a HARMONIC DIVISOR NUMBER (or ORE NUMBER).

see also APÉRY'S CONSTANT, EULER SUM, HARMONIC LOGARITHM, HARMONIC SERIES, ORE NUMBER

References

Borwein, D. and Borwein, J. M. "On an Intriguing Integral and Some Series Related to $\zeta(4)$." *Proc. Amer. Math. Soc.* **123**, 1191–1198, 1995.

Conway, J. H. and Guy, R. K. *The Book of Numbers.* New York: Springer-Verlag, pp. 143 and 258–259, 1996.

de Doelder, P. J. "On Some Series Containing $\Psi(x) - \Psi(y)$ and $(\Psi(x) - \Psi(y))^2$ for Certain Values of x and y." *J. Comp. Appl. Math.* **37**, 125–141, 1991.

Roman, S. "The Logarithmic Binomial Formula." *Amer. Math. Monthly* **99**, 641–648, 1992.

Sloane, N. J. A. Sequences A001008/M2885 and A002805/M1589 in "An On-Line Version of the Encyclopedia of Integer Sequences."

Harmonic Progression

see Harmonic Series

Harmonic Range

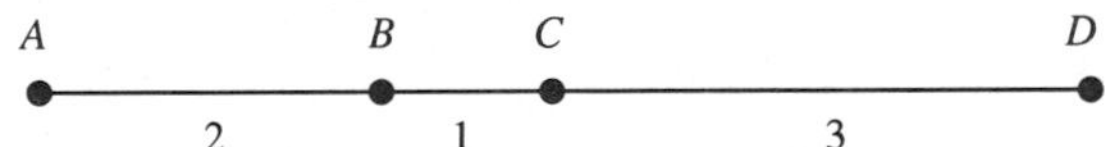

A set of four COLLINEAR points A, B, C, and D arranged such that

$$AB : BC = 2 : 1$$

$$AD : DC = 6 : 3.$$

Hardy (1967) uses the term HARMONIC SYSTEM OF POINTS to refer to a harmonic range.

see also EULER LINE, GERGONNE LINE, HARMONIC CONJUGATE POINTS, SODDY LINE

References

Hardy, G. H. *A Course of Pure Mathematics, 10th ed.* Cambridge, England: Cambridge University Press, pp. 99 and 106, 1967.

Harmonic Ratio

see HARMONIC CONJUGATE POINTS

Harmonic Segment

see HARMONIC CONJUGATE POINTS

Harmonic Series

The SUM

$$\sum_{k=1}^{\infty} \frac{1}{k} \quad (1)$$

is called the harmonic series. It can be shown to DIVERGE using the INTEGRAL TEST by comparison with the function $1/x$. The divergence, however, is very slow. In fact, the sum

$$\sum_{p} \frac{1}{p} \quad (2)$$

taken over all PRIMES also diverges. The generalization of the harmonic series

$$\zeta(n) \equiv \sum_{k=1}^{\infty} \frac{1}{k^n} \quad (3)$$

is known as the RIEMANN ZETA FUNCTION.

The sum of the first few terms of the harmonic series is given analytically by the nth HARMONIC NUMBER

$$H_n = \sum_{j=1}^{n} \frac{1}{j} = \gamma + \psi_0(n+1), \quad (4)$$

where γ is the EULER-MASCHERONI CONSTANT and $\Psi(x) = \psi_0(x)$ is the DIGAMMA FUNCTION. The number of terms needed to exceed 1, 2, 3, ... are 1, 4, 11, 31, 83, 227, 616, 1674, 4550, 12367, 33617, 91380, 248397, ... (Sloane's A004080). Using the analytic form shows that after 2.5×10^8 terms, the sum is still less than 20. Furthermore, to achieve a sum greater than 100, more than 1.509×10^{43} terms are needed!

Progressions of the form

$$\frac{1}{a_1}, \frac{1}{a_1+d}, \frac{1}{a_1+2d}, \ldots \quad (5)$$

are also sometimes called harmonic series (Beyer 1987).

The modified harmonic series, given by the sum

$$T = \sum_{k=1}^{\infty} \frac{1}{p_k}, \quad (6)$$

where p_k is the kth PRIME, diverges.

see also ARITHMETIC SERIES, BERNOULLI'S PARADOX, BOOK STACKING PROBLEM, EULER SUM, ZIPF'S LAW

References

Arfken, G. *Mathematical Methods for Physicists, 3rd ed.* Orlando, FL: Academic Press, pp. 279–280, 1985.

Beyer, W. H. (Ed.). *CRC Standard Mathematical Tables, 28th ed.* Boca Raton, FL: CRC Press, p. 8, 1987.

Boas, R. P. and Wrench, J. W. "Partial Sums of the Harmonic Series." *Amer. Math. Monthly* **78**, 864–870, 1971.

Honsberger, R. "An Intriguing Series." Ch. 10 in *Mathematical Gems II.* Washington, DC: Math. Assoc. Amer., pp. 98–103, 1976.

Sloane, N. J. A. Sequence A004080 in "An On-Line Version of the Encyclopedia of Integer Sequences."

Harmonic System of Points

see HARMONIC RANGE

Harmonious Graph

A connected LABELLED GRAPH with n EDGES in which all VERTICES can be labeled with distinct INTEGERS (mod n) so that the sums of the PAIRS of numbers at the ends of each EDGE are also distinct (mod n). The LADDER GRAPH, FAN, WHEEL GRAPH, PETERSEN GRAPH, TETRAHEDRAL GRAPH, DODECAHEDRAL GRAPH, and ICOSAHEDRAL GRAPH are all harmonious (Graham and Sloane 1980).

see also GRACEFUL GRAPH, LABELLED GRAPH, POSTAGE STAMP PROBLEM, SEQUENTIAL GRAPH

References

Gallian, J. A. "Open Problems in Grid Labeling." *Amer. Math. Monthly* **97**, 133–135, 1990.

Gardner, M. *Wheels, Life, and other Mathematical Amusements.* New York: W. H. Freeman, p. 164, 1983.

Graham, R. L. and Sloane, N. "On Additive Bases and Harmonious Graphs." *SIAM J. Algebraic Discrete Math.* **1**, 382–404, 1980.

Guy, R. K. "The Corresponding Modular Covering Problem. Harmonious Labelling of Graphs." §C13 in *Unsolved Problems in Number Theory, 2nd ed.* New York: Springer-Verlag, pp. 127–128, 1994.

Harnack's Inequality

Let $D = D(z_0, R)$ be an OPEN DISK, and let u be a HARMONIC FUNCTION on D such that $u(z) \geq 0$ for all $z \in D$. Then for all $z \in D$, we have

$$0 \leq u(z) \leq \left(\frac{R}{R - |z - z_0|}\right)^2 u(z_0).$$

see also LIOUVILLE'S CONFORMALITY THEOREM

References

Flanigan, F. J. "Harnack's Inequality." §2.5.1 in *Complex Variables: Harmonic and Analytic Functions.* New York: Dover, pp. 88–90, 1983.

Harnack's Theorems

Harnack's first theorem states that a real irreducible curve of order n cannot have more than

$$\tfrac{1}{2}(n-1)(n-2) - \sum s_i(s_i - 1) + 1$$

circuits (Coolidge 1959, p. 57).

Harnack's second theorem states that there exists a curve of every order with the maximum number of circuits compatible with that order and with a certain number of double points, provided that number is not permissible for a curve of lower order (Coolidge 1959, p. 61).

References

Coolidge, J. L. *A Treatise on Algebraic Plane Curves.* New York: Dover, 1959.

Harshad Number

A POSITIVE INTEGER which is DIVISIBLE by the sum of its DIGITS, also called a NIVEN NUMBER (Kennedy *et al.* 1980). The first few are 1, 2, 3, 4, 5, 6, 7, 8, 9, 10, 12, 18, 20, 21, 24, ... (Sloane's A005349). Grundman (1994) proved that there is no sequence of more than 20 consecutive Harshad numbers, and found the smallest sequence of 20 consecutive Harshad numbers, each member of which has 44,363,342,786 digits.

Grundman (1994) defined an n-Harshad (or n-Niven) number to be a POSITIVE INTEGER which is DIVISIBLE by the sum of its digits in base $n \geq 2$. Cai (1996) showed that for $n = 2$ or 3, there exists an infinite family of sequences of consecutive n-Harshad numbers of length $2n$.

References

Cai, T. "On 2-Niven Numbers and 3-Niven Numbers." *Fib. Quart.* **34**, 118–120, 1996.

Cooper, C. N. and Kennedy, R. E. "Chebyshev's Inequality and Natural Density." *Amer. Math. Monthly* **96**, 118–124, 1989.

Cooper, C. N. and Kennedy, R. "On Consecutive Niven Numbers." *Fib. Quart.* **21**, 146–151, 1993.

Grundman, H. G. "Sequences of Consecutive n-Niven Numbers." *Fib. Quart.* **32**, 174–175, 1994.

Kennedy, R.; Goodman, R.; and Best, C. "Mathematical Discovery and Niven Numbers." *MATYC J.* **14**, 21–25, 1980.

Sloane, N. J. A. Sequence A005349/M0481 in "An On-Line Version of the Encyclopedia of Integer Sequences."

Vardi, I. "Niven Numbers." §2.3 in *Computational Recreations in Mathematica.* Redwood City, CA: Addison-Wesley, pp. 19 and 28–31, 1991.

Hart's Inversor

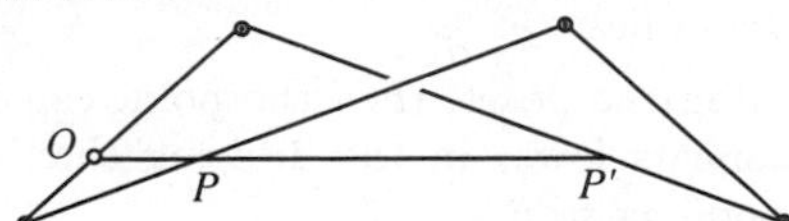

A linkage which draws the inverse of a given curve. It can also convert circular to linear motion. The rods satisfy $AB = CD$ and $BC = DA$, and O, P, and P' remain COLLINEAR. Coxeter (1969, p. 428) shows that if $AO = \mu AB$, then

$$OP \times OP' = \mu(1-\mu)(AD^2 - AB^2).$$

see also PEAUCELLIER INVERSOR

References

Courant, R. and Robbins, H. *What is Mathematics?: An Elementary Approach to Ideas and Methods.* Oxford, England: Oxford University Press, p. 157, 1978.

Coxeter, H. S. M. *Introduction to Geometry, 2nd ed.* New York: Wiley, pp. 82–83, 1969.

Rademacher, H. and Toeplitz, O. *The Enjoyment of Mathematics: Selections from Mathematics for the Amateur.* Princeton, NJ: Princeton University Press, pp. 124–129, 1957.

Hart's Theorem

Any one of the eight APOLLONIUS CIRCLES of three given CIRCLES is TANGENT to a CIRCLE C, as are the other three APOLLONIUS CIRCLES having (1) like contact with two of the given CIRCLES and (2) unlike contact with the third.

see also APOLLONIUS CIRCLES

References

Johnson, R. A. *Modern Geometry: An Elementary Treatise on the Geometry of the Triangle and the Circle.* Boston, MA: Houghton Mifflin, pp. 127–128, 1929.

Hartley Transform

An INTEGRAL TRANSFORM which shares some features with the FOURIER TRANSFORM, but which (in the discrete case), multiplies the KERNEL by

$$\cos\left(\frac{2\pi kn}{N}\right) - \sin\left(\frac{2\pi kn}{N}\right) \tag{1}$$

instead of

$$e^{-2\pi ikn/N} = \cos\left(\frac{2\pi kn}{N}\right) - i\sin\left(\frac{2\pi kn}{N}\right). \tag{2}$$

The Hartley transform produces REAL output for a REAL input, and is its own inverse. It therefore can have computational advantages over the DISCRETE FOURIER TRANSFORM, although analytic expressions are usually more complicated for the Hartley transform.

The discrete version of the Hartley transform can be written explicitly as

$$\mathcal{H}[a] \equiv \frac{1}{\sqrt{N}} \sum_{n=0}^{N-1} a_n \left[\cos\left(\frac{2\pi kn}{N}\right) - \sin\left(\frac{2\pi kn}{N}\right)\right] \tag{3}$$

$$= \Re\mathcal{F}[a] - \Im\mathcal{F}[a], \tag{4}$$

where $\mathcal{F}$ denotes the FOURIER TRANSFORM. The Hartley transform obeys the CONVOLUTION property

$$\mathcal{H}[a*b]_k = \tfrac{1}{2}(A_k B_k - \bar{A}_k \bar{B}_k + A_k \bar{B}_k + \bar{A}_k B_k), \tag{5}$$

where

$$\bar{a}_0 \equiv a_0 \tag{6}$$
$$\bar{a}_{n/2} \equiv a_{n/2} \tag{7}$$
$$\bar{a}_k \equiv a_{n-k} \tag{8}$$

(Arndt). Like the FAST FOURIER TRANSFORM, there is a "fast" version of the Hartley transform. A decimation in time algorithm makes use of

$$\mathcal{H}_n^{\text{left}}[a] = \mathcal{H}_{n/2}[a^{\text{even}}] + \mathcal{X}\mathcal{H}_{n/2}[a^{\text{odd}}] \tag{9}$$
$$\mathcal{H}_n^{\text{right}}[a] = \mathcal{H}_{n/2}[a^{\text{even}}] - \mathcal{X}\mathcal{H}_{n/2}[a^{\text{odd}}], \tag{10}$$

where $\mathcal{X}$ denotes the sequence with elements

$$a_n \cos\left(\frac{\pi n}{N}\right) - \bar{a}_n \sin\left(\frac{\pi n}{N}\right). \tag{11}$$

A decimation in frequency algorithm makes use of

$$\mathcal{H}_n^{\text{even}}[a] = \mathcal{H}_{n/2}[a^{\text{left}} + a^{\text{right}}], \tag{12}$$
$$\mathcal{H}_n^{\text{odd}}[a] = \mathcal{H}_{n/2}[\mathcal{X}(a^{\text{left}} - a^{\text{right}})]. \tag{13}$$

The DISCRETE FOURIER TRANSFORM

$$A_k \equiv \mathcal{F}[a] = \sum_{n=0}^{N-1} e^{-2\pi ikn/N} a_n \tag{14}$$

can be written

$$\begin{bmatrix} A_k \\ A_{-k} \end{bmatrix} = \sum_{n=0}^{N-1} \underbrace{\begin{bmatrix} e^{-2\pi ikn/N} & 0 \\ 0 & e^{2\pi ikn/N} \end{bmatrix}}_{\mathsf{F}} \begin{bmatrix} a_n \\ a_n \end{bmatrix} \tag{15}$$

$$= \sum_{n=0}^{N-1} \underbrace{\frac{1}{2}\begin{bmatrix} 1-i & 1+i \\ 1+i & 1-i \end{bmatrix}}_{\mathsf{T}^{-1}} \underbrace{\begin{bmatrix} \cos\left(\frac{2\pi kn}{N}\right) & \sin\left(\frac{2\pi kn}{N}\right) \\ -\sin\left(\frac{2\pi kn}{N}\right) & \cos\left(\frac{2\pi kn}{N}\right) \end{bmatrix}}_{\mathsf{H}}$$
$$\times \underbrace{\frac{1}{2}\begin{bmatrix} 1+i & 1-i \\ 1-i & 1+i \end{bmatrix}}_{\mathsf{T}} \begin{bmatrix} a_n \\ a_n \end{bmatrix}, \tag{16}$$

so

$$\mathsf{F} = \mathsf{T}^{-1}\mathsf{H}\mathsf{T}. \tag{17}$$

see also DISCRETE FOURIER TRANSFORM, FAST FOURIER TRANSFORM, FOURIER TRANSFORM

References

Arndt, J. "The Hartley Transform (HT)." Ch. 2 in "Remarks on FFT Algorithms." http://www.jjj.de/fxt/.

Bracewell, R. N. *The Fourier Transform and Its Applications.* New York: McGraw-Hill, 1965.

Bracewell, R. N. *The Hartley Transform.* New York: Oxford University Press, 1986.

HashLife

A LIFE ALGORITHM that achieves remarkable speed by storing subpatterns in a hash table, and using them to skip forward, sometimes thousands of generations at a time. HashLife takes tremendous amounts of memory and can't show patterns at every step, but can quickly calculate the outcome of a pattern that takes millions of generations to complete.

References

Hensel, A. "A Brief Illustrated Glossary of Terms in Conway's Game of Life." http://www.cs.jhu.edu/~callahan/glossary.html.

Hasse's Algorithm

see COLLATZ PROBLEM

Hasse's Conjecture

Define the ZETA FUNCTION of a VARIETY over a NUMBER FIELD by taking the product over all PRIME IDEALS of the ZETA FUNCTIONS of this VARIETY reduced modulo the PRIMES. Hasse conjectured that this product has a MEROMORPHIC continuation over the whole plane and a functional equation.

References

Lang, S. "Some History of the Shimura-Taniyama Conjecture." *Not. Amer. Math. Soc.* **42**, 1301–1307, 1995.

Hasse-Davenport Relation

Let F be a FINITE FIELD with q elements, and let F_s be a FIELD containing F such that $[F_s : F] = s$. Let χ be a nontrivial MULTIPLICATIVE CHARACTER of F and $\chi' = \chi \circ N_{F_s/F}$ a character of F_s. Then

$$(-g(\chi))^s = -g(\chi'),$$

where $g(x)$ is a GAUSSIAN SUM.

see also GAUSSIAN SUM, MULTIPLICATIVE CHARACTER

References

Ireland, K. and Rosen, M. "A Proof of the Hasse-Davenport Relation." §11.4 in *A Classical Introduction to Modern Number Theory, 2nd ed.* New York: Springer-Verlag, pp. 162–165, 1990.

Hasse Diagram

A graphical rendering of a PARTIALLY ORDERED SET displayed via the COVER relation of the PARTIALLY ORDERED SET with an implied upward orientation. A point is drawn for each element of the POSET, and line segments are drawn between these points according to the following two rules:

1. If $x < y$ in the poset, then the point corresponding to x appears lower in the drawing than the point corresponding to y.
2. The line segment between the points corresponding to any two elements x and y of the poset is included in the drawing IFF x covers y or y covers x.

Hasse diagrams are also called UPWARD DRAWINGS.

Hasse-Minkowski Theorem

Two nonsingular forms are equivalent over the rationals IFF they have the same DETERMINANT and the same p-SIGNATURES for all p.

Hasse Principle

A collection of equations satisfies the Hasse principle if, whenever one of the equations has solutions in $\mathbb{R}$ and all the $\mathbb{Q}_p$, then the equations have solutions in the RATIONALS $\mathbb{Q}$. Examples include the set of equations

$$ax^2 + bxy + cy^2 = 0$$

with a, b, and c INTEGERS, and the set of equations

$$x^2 + y^2 = a$$

for a rational. The trivial solution $x = y = 0$ is usually not taken into account when deciding if a collection of homogeneous equations satisfies the Hasse principle. The Hasse principle is sometimes called the LOCAL-GLOBAL PRINCIPLE.

see also LOCAL FIELD

Hasse's Resolution Modulus Theorem

The JACOBI SYMBOL $(a/y) = \chi(y)$ as a CHARACTER can be extended to the KRONECKER SYMBOL $(f(a)/y) = \chi^*(y)$ so that $\chi^*(y) = \chi(y)$ whenever $\chi(y) \neq 0$. When y is RELATIVELY PRIME to $f(a)$, then $\chi^*(y) \neq 0$, and for NONZERO values $\chi^*(y_1) = \chi^*(y_2)$ IFF $y_1 \equiv y_2 \bmod^+ f(a)$. In addition, $|f(a)|$ is the minimum value for which the latter congruence property holds in any extension symbol for $\chi(y)$.

see also CHARACTER (NUMBER THEORY), JACOBI SYMBOL, KRONECKER SYMBOL

References

Cohn, H. *Advanced Number Theory.* New York: Dover, pp. 35–36, 1980.

Hat

The hat is a caret-shaped symbol most commonly used to denote a UNIT VECTOR ($\hat{\mathbf{v}}$) or an ESTIMATOR ($\hat{x}$).

see also ESTIMATOR, UNIT VECTOR

Haupt-Exponent

The smallest exponent e for which $b^e \equiv 1 \pmod{p}$, where b and p are given numbers, is the haupt-exponent of $b \pmod{p}$. The number of bases having a haupt-exponent e is $\phi(e)$, where $\phi(e)$ is the TOTIENT FUNCTION. Cunningham (1922) published the haupt-exponents for primes to 25409 and bases 2, 3, 5, 6, 7, 10, 11, and 12.

see also COMPLETE RESIDUE SYSTEM, RESIDUE INDEX

References

Cunningham, A. *Haupt-Exponents, Residue Indices, Primitive Roots.* London: F. Hodgson, 1922.

Hausdorff Axioms

Describe subsets of elements x in a NEIGHBORHOOD SET E of x. The NEIGHBORHOOD is assumed to satisfy:

1. There corresponds to each point x at least one NEIGHBORHOOD $U(x)$, and each NEIGHBORHOOD $U(x)$ contains the point x.
2. If $U(x)$ and $V(x)$ are two NEIGHBORHOODS of the same point x, there must exist a NEIGHBORHOOD $W(x)$ that is a subset of both.
3. If the point y lies in $U(x)$, there must exist a NEIGHBORHOOD $U(y)$ that is a SUBSET of $U(x)$.
4. For two different points x and y, there are two corresponding NEIGHBORHOODS $U(x)$ and $U(y)$ with no points in common.

Hausdorff-Besicovitch Dimension

see CAPACITY DIMENSION

Hausdorff Dimension

Let A be a SUBSET of a METRIC SPACE X. Then the Hausdorff dimension $D(A)$ of A is the INFIMUM of $d \geq 0$ such that the d-dimensional HAUSDORFF MEASURE of A is 0. Note that this need not be an INTEGER.

In many cases, the Hausdorff dimension correctly describes the correction term for a resonator with FRACTAL PERIMETER in Lorentz's conjecture. However, in general, the proper dimension to use turns out to be the MINKOWSKI-BOULIGAND DIMENSION (Schroeder 1991).

see also CAPACITY DIMENSION, FRACTAL DIMENSION, MINKOWSKI-BOULIGAND DIMENSION

References

Federer, H. *Geometric Measure Theory.* New York: Springer-Verlag, 1969.
Hausdorff, F. "Dimension und äußeres Maß." *Math. Ann.* **79**, 157–179, 1919.
Ott, E. "Appendix: Hausdorff Dimension." *Chaos in Dynamical Systems.* New York: Cambridge University Press, pp. 100–103, 1993.
Schroeder, M. *Fractals, Chaos, Power Laws: Minutes from an Infinite Paradise.* New York: W. H. Freeman, pp. 41–45, 1991.

Hausdorff Measure

Let X be a METRIC SPACE, A be a SUBSET of X, and d a number ≥ 0. The d-dimensional Hausdorff measure of A, $H^d(A)$, is the INFIMUM of POSITIVE numbers y such that for every $r > 0$, A can be covered by a countable family of closed sets, each of diameter less than r, such that the sum of the dth POWERS of their diameters is less than y. Note that $H^d(A)$ may be infinite, and d need not be an INTEGER.

References

Federer, H. *Geometric Measure Theory.* New York: Springer-Verlag, 1969.
Ott, E. *Chaos in Dynamical Systems.* Cambridge, England: Cambridge University Press, p. 103, 1993.

Hausdorff Paradox

For $n \geq 3$, there exist no additive finite and invariant measures for the group of displacements in $\mathbb{R}^n$.

References

Le Lionnais, F. *Les nombres remarquables.* Paris: Hermann, p. 49, 1983.

Hausdorff Space

A TOPOLOGICAL SPACE in which any two points have disjoint NEIGHBORHOODS.

Haversine

$$\text{hav}(z) \equiv \tfrac{1}{2}\,\text{vers}(z) = \tfrac{1}{2}(1 - \cos z),$$

where vers(z) is the VERSINE and cos is the COSINE. Using a trigonometric identity, the haversine is equal to

$$\text{hav}(z) = \sin^2(\tfrac{1}{2}z).$$

see also COSINE, COVERSINE, EXSECANT, VERSINE

References

Abramowitz, M. and Stegun, C. A. (Eds.). *Handbook of Mathematical Functions with Formulas, Graphs, and Mathematical Tables, 9th printing.* New York: Dover, p. 78, 1972.

Heads Minus Tails Distribution

A fair COIN is tossed $2n$ times. Let $D \equiv |H - T|$ be the absolute difference in the number of heads and tails obtained. Then the probability distribution is given by

$$P(D = 2k) = \begin{cases} (\frac{1}{2})^{2n}\binom{2n}{n} & k = 0 \\ 2(\frac{1}{2})^{2n}\binom{2n}{n+k} & k = 1, 2, \ldots, \end{cases}$$

where $P(D = 2k - 1) = 0$. The most probable value of D is $D = 2$, and the expectation value is

$$\langle D \rangle = \frac{n\binom{2n}{n}}{2^{2n-1}}.$$

see also BERNOULLI DISTRIBUTION, COIN, COIN TOSSING

References

Handelsman, M. B. Solution to Problem 436, "Distributing 'Heads' Minus 'Tails.'" *College Math. J.* **22**, 444–446, 1991.

Heap

A SET of N members forms a heap if it satisfies $a_{\lfloor j/2 \rfloor} \geq a_j$ for $1 \leq \lfloor j/2 \rfloor < j \leq N$, where $\lfloor x \rfloor$ is the FLOOR FUNCTION.

see also HEAPSORT

Heapsort

An $N \lg N$ SORTING ALGORITHM which is not quite as fast as QUICKSORT. It is a "sort-in-place" algorithm and requires no auxiliary storage, which makes it particularly concise and elegant to implement.

see also QUICKSORT, SORTING

References

Press, W. H.; Flannery, B. P.; Teukolsky, S. A.; and Vetterling, W. T. "Heapsort." §8.3 in *Numerical Recipes in FORTRAN: The Art of Scientific Computing, 2nd ed.* Cambridge, England: Cambridge University Press, pp. 327–329, 1992.

Heart Surface

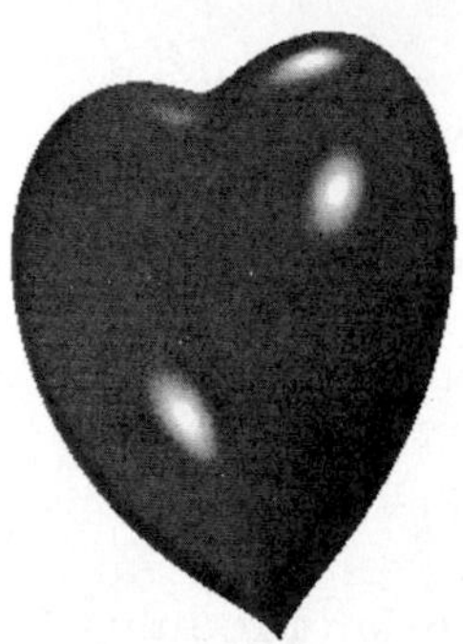

A heart-shaped surface given by the SEXTIC EQUATION

$$(2x^2 + 2y^2 + z^2 - 1)^3 - \tfrac{1}{10}x^2z^3 - y^2z^3 = 0.$$

see also BONNE PROJECTION, PIRIFORM

References

Nordstrand, T. "Heart." `http://www.uib.no/people/nfytn/hearttxt.htm`.

Heat Conduction Equation

A diffusion equation of the form

$$\frac{\partial T}{\partial t} = \kappa \nabla^2 T. \tag{1}$$

Physically, the equation commonly arises in situations where κ is the thermal diffusivity and T the temperature.

The 1-D heat conduction equation is

$$\frac{\partial T}{\partial t} = \kappa \frac{\partial^2 T}{\partial x^2}. \tag{2}$$

This can be solved by SEPARATION OF VARIABLES using

$$T(x,t) = X(x)T(t). \tag{3}$$

Then

$$X\frac{dT}{dt} = \kappa T \frac{d^2 X}{dx^2}. \tag{4}$$

Dividing both sides by κXT gives

$$\frac{1}{\kappa T}\frac{dT}{dt} = \frac{1}{X}\frac{d^2X}{dx^2} = -\frac{1}{\lambda^2}, \qquad (5)$$

where each side must be equal to a constant. Anticipating the exponential solution in T, we have picked a negative separation constant so that the solution remains finite at all times and λ has units of length. The T solution is

$$T(t) = Ae^{-\kappa t/\lambda^2}, \qquad (6)$$

and the X solution is

$$X(x) = C\cos\left(\frac{x}{\lambda}\right) + D\sin\left(\frac{x}{\lambda}\right). \qquad (7)$$

The general solution is then

$$\begin{aligned} T(x,t) &= T(t)X(x) \\ &= Ae^{-\kappa t/\lambda^2}\left[C\cos\left(\frac{x}{\lambda}\right) + D\sin\left(\frac{x}{\lambda}\right)\right] \\ &= e^{-\kappa t/\lambda^2}\left[D\cos\left(\frac{x}{\lambda}\right) + E\sin\left(\frac{x}{\lambda}\right)\right]. \qquad (8) \end{aligned}$$

If we are given the boundary conditions

$$T(0,t) = 0 \qquad (9)$$

and

$$T(L,t) = 0, \qquad (10)$$

then applying (9) to (8) gives

$$D\cos\left(\frac{x}{\lambda}\right) = 0 \Rightarrow D = 0, \qquad (11)$$

and applying (10) to (8) gives

$$E\sin\left(\frac{L}{\lambda}\right) = 0 \Rightarrow \frac{L}{\lambda} = n\pi \Rightarrow \lambda = \frac{L}{n\pi}, \qquad (12)$$

so (8) becomes

$$T_n(x,t) = E_n e^{-\kappa(n\pi/L)^2 t}\sin\left(\frac{n\pi x}{L}\right). \qquad (13)$$

Since the general solution can have any n,

$$T(x,t) = \sum_{n=1}^{\infty} c_n \sin\left(\frac{n\pi x}{L}\right) e^{-\kappa(n\pi/L)^2 t}. \qquad (14)$$

Now, if we are given an initial condition $T(x,0)$, we have

$$T(x,0) = \sum_{n=1}^{\infty} c_n \sin\left(\frac{n\pi x}{L}\right). \qquad (15)$$

Multiplying both sides by $\sin(m\pi x/L)$ and integrating from 0 to L gives

$$\begin{aligned} &\int_0^L \sin\left(\frac{m\pi x}{L}\right) T(x,0)\,dx \\ &\quad = \int_0^L \sum_{n=1}^{\infty} c_n \sin\left(\frac{m\pi x}{L}\right)\sin\left(\frac{n\pi x}{L}\right) dx. \qquad (16) \end{aligned}$$

Using the ORTHOGONALITY of $\sin(nx)$ and $\sin(mx)$,

$$\begin{aligned} \sum_{n=1}^{\infty} c_n \int_0^L \sin\left(\frac{n\pi x}{L}\right)\sin\left(\frac{m\pi x}{L}\right) dx &= \sum_{n=1}^{\infty} \tfrac{1}{2}\pi\delta_{mn}c_n \\ = \tfrac{1}{2}\pi c_m &= \int_0^L \sin\left(\frac{m\pi x}{L}\right) T(x,0)\,dx, \qquad (17) \end{aligned}$$

so

$$c_n = \frac{2}{\pi}\int_0^L \sin\left(\frac{m\pi x}{L}\right) T(x,0)\,dx. \qquad (18)$$

If the boundary conditions are replaced by the requirement that the derivative of the temperature be zero at the edges, then (9) and (10) are replaced by

$$\left.\frac{\partial T}{\partial x}\right|_{(0,t)} = 0 \qquad (19)$$

$$\left.\frac{\partial T}{\partial x}\right|_{(L,t)} = 0. \qquad (20)$$

Following the same procedure as before, a similar answer is found, but with sine replaced by cosine:

$$T(x,t) = \sum_{n=1}^{\infty} c_n \cos\left(\frac{n\pi x}{L}\right) e^{-\kappa(n\pi/L)^2 t}, \qquad (21)$$

where

$$c_n = \frac{2}{\pi}\int_0^L \cos\left(\frac{m\pi x}{L}\right) \left.\frac{\partial T(x,0)}{\partial x}\right|_{t=0} dx. \qquad (22)$$

Heat Conduction Equation—Disk

To solve the HEAT CONDUCTION EQUATION on a 2-D disk of radius $R = 1$, try to separate the equation using

$$T(r,\theta,t) = R(r)\Theta(\theta)T(t). \qquad (1)$$

Writing the θ and r terms of the LAPLACIAN in SPHERICAL COORDINATES gives

$$\nabla^2 = \frac{d^2R}{dr^2} + \frac{2}{r}\frac{dR}{dr} + \frac{1}{r^2}\frac{d^2\Theta}{d\theta^2}, \qquad (2)$$

so the HEAT CONDUCTION EQUATION becomes

$$\frac{R\Theta}{\kappa}\frac{d^2T}{dt^2} = \frac{d^2R}{dr^2}\Theta T + \frac{2}{r}\frac{dR}{dr}\Theta T + \frac{1}{r^2}\frac{d^2\Theta}{d\theta^2}RT. \qquad (3)$$

Multiplying through by $r^2/R\Theta T$ gives

$$\frac{r^2}{\kappa T}\frac{d^2T}{dt^2} = \frac{r^2}{R}\frac{d^2R}{dr^2} + \frac{2r}{R}\frac{dR}{dr} + \frac{d^2\Theta}{d\theta^2}\frac{1}{\Theta}. \quad (4)$$

The θ term can be separated.

$$\frac{d^2\Theta}{d\theta^2}\frac{1}{\Theta} = -n(n+1), \quad (5)$$

which has a solution

$$\Theta(\theta) = A\cos\left[\sqrt{n(n+1)}\,\theta\right] + B\sin\left[\sqrt{n(n+1)}\,\theta\right]. \quad (6)$$

The remaining portion becomes

$$\frac{r^2}{\kappa T}\frac{d^2T}{dt^2} = \frac{r^2}{R}\frac{d^2R}{dr^2} + \frac{2r}{R}\frac{dR}{dr} - n(n+1). \quad (7)$$

Dividing by r^2 gives

$$\frac{1}{\kappa T}\frac{d^2T}{dt^2} = \frac{1}{R}\frac{d^2R}{dr^2} + \frac{2}{rR}\frac{dR}{dr} - \frac{n(n+1)}{r^2} = -\frac{1}{\lambda^2}, \quad (8)$$

where a NEGATIVE separation constant has been chosen so that the t portion remains finite

$$T(t) = Ce^{-\kappa t/\lambda^2}. \quad (9)$$

The radial portion then becomes

$$\frac{1}{R}\frac{d^2R}{dr^2} + \frac{2}{rR}\frac{dR}{dr} - \frac{n(n+1)}{r^2} + \frac{1}{\lambda^2} = 0 \quad (10)$$

$$r^2\frac{d^2R}{dr^2} + 2r\frac{dR}{dr} + \left[\frac{r^2}{\lambda^2} - n(n+1)\right]R = 0, \quad (11)$$

which is the SPHERICAL BESSEL DIFFERENTIAL EQUATION. If the initial temperature is $T(r,0) = 0$ and the boundary condition is $T(1,t) = 1$, the solution is

$$T(r,t) = 1 - 2\sum_{n=1}^{\infty}\frac{J_0(\alpha_n r)}{\alpha_n J_1(\alpha_n)}e^{\alpha_n{}^2 t}, \quad (12)$$

where α_n is the nth POSITIVE zero of the BESSEL FUNCTION OF THE FIRST KIND J_0.

Heaviside Calculus

A method of solving differential equations using FOURIER TRANSFORMS and LAPLACE TRANSFORMS.

see also FOURIER TRANSFORM, LAPLACE TRANSFORM

Heaviside Step Function

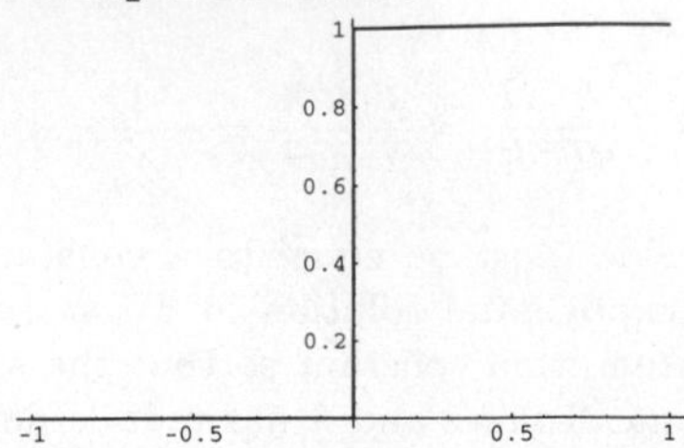

A discontinuous "step" function, also called the UNIT STEP, and defined by

$$H(x) = \begin{cases} 0 & x < 0 \\ \frac{1}{2} & x = 0 \\ 1 & x > 0. \end{cases} \quad (1)$$

It is related to the BOXCAR FUNCTION. The DERIVATIVE is given by

$$\frac{d}{dx}H(x) = \delta(x), \quad (2)$$

where $\delta(x)$ is the DELTA FUNCTION, and the step function is related to the RAMP FUNCTION $R(x)$ by

$$\frac{d}{dx}R(x) = -H(x). \quad (3)$$

Bracewell (1965) gives many identities, some of which include the following. Letting $*$ denote the CONVOLUTION,

$$H(x) * f(x) = \int_{-\infty}^{x} f(x')\,dx' \quad (4)$$

$$H(T) * H(T) = \int_{-\infty}^{\infty} H(u)H(T-u)\,du \quad (5)$$

$$= H(0)\int_0^{\infty} H(T-u)\,du$$

$$= H(0)H(T)\int_0^{T} du = TH(T). \quad (6)$$

Additional identities are

$$H(x)H(y) = \begin{cases} H(x) & x > y \\ H(y) & x < y \end{cases} \quad (7)$$

$$H(ax+b) = H\left(x + \frac{b}{a}\right)H(a) + H\left(-x - \frac{b}{a}\right)H(-a)$$

$$= \begin{cases} H\left(x + \frac{b}{a}\right) & a > 0 \\ H\left(-x - \frac{b}{a}\right) & a < 0. \end{cases} \quad (8)$$

The step function obeys the integral identities

$$\int_{-a}^{b} H(u-u_0)f(u)\,du = H(u_0)\int_{u_0}^{b} f(u)\,du \quad (9)$$

$$\int_{-a}^{b} H(u_1-u)f(u)\,du = H(u_1)\int_{-a}^{u_1} f(u)\,du \quad (10)$$

$$\int_{-a}^{b} H(u-u_0)H(u_1-u)f(u)\,du = H(u_0)H(u_1)\int_{u_0}^{u_1} f(u)\,du. \quad (11)$$

The Heaviside step function can be defined by the following limits,

$$H(x) = \lim_{t\to 0}\left[\tfrac{1}{2}+\frac{1}{\pi}\tan^{-1}\left(-\frac{x}{t}\right)\right] \quad (12)$$
$$= \tfrac{1}{2}\lim_{t\to 0}\operatorname{erfc}(-\tfrac{x}{t}) \quad (13)$$
$$= \frac{1}{\sqrt{\pi}}\lim_{t\to 0}\int_{-s}^{\infty} t^{-1}e^{-u^2/t}\,du \quad (14)$$
$$= \tfrac{1}{2}+\tfrac{1}{\pi}\lim_{t\to 0}\operatorname{si}(\tfrac{\pi x}{t}) \quad (15)$$
$$= \lim_{t\to 0}\int_{-\infty}^{x} t^{-1}\operatorname{sinc}\left(\frac{u}{t}\right)du \quad (16)$$
$$= \lim_{t\to 0}\begin{cases}\tfrac{1}{2}(1-e^{-x/t}) & x>0\\ -\tfrac{1}{2}(1-e^{-x/t}) & x<0\end{cases} \quad (17)$$
$$= \lim_{t\to 0}\int_{-\infty}^{x} t^{-1}\Lambda\left(\frac{x-\tfrac{1}{2}t}{t}\right)dx, \quad (18)$$

where Λ is the one-argument TRIANGLE FUNCTION and $\operatorname{si}(x)$ is the SINE INTEGRAL.

The FOURIER TRANSFORM of the Heaviside step function is given by

$$\mathcal{F}[H(x)] = \int_{-\infty}^{\infty} e^{-2\pi ikx}H(x)\,dx = \frac{1}{2}\left[\delta(k)-\frac{i}{\pi k}\right], \quad (19)$$

where $\delta(k)$ is the DELTA FUNCTION.

see also BOXCAR FUNCTION, DELTA FUNCTION, FOURIER TRANSFORM—HEAVISIDE STEP FUNCTION, RAMP FUNCTION, RAMP FUNCTION, RECTANGLE FUNCTION, SQUARE WAVE

References

Bracewell, R. *The Fourier Transform and Its Applications.* New York: McGraw-Hill, 1965.

Spanier, J. and Oldham, K. B. "The Unit-Step $u(x-a)$ and Related Functions." Ch. 8 in *An Atlas of Functions.* Washington, DC: Hemisphere, pp. 63–69, 1987.

Heawood Conjecture

The bound for the number of colors which are SUFFICIENT for MAP COLORING on a surface of GENUS g,

$$\chi(g) = \left\lfloor \tfrac{1}{2}(7+\sqrt{48g+1})\right\rfloor$$

is the best possible, where $\lfloor x\rfloor$ is the FLOOR FUNCTION. $\chi(g)$ is called the CHROMATIC NUMBER, and the first few values for $g=0, 1, \ldots$ are 4, 7, 8, 9, 10, 11, 12, 12, 13, 13, 14, ... (Sloane's A000934).

The fact that $\chi(g)$ is also NECESSARY was proved by Ringel and Youngs (1968) with two exceptions: the SPHERE (PLANE), and the KLEIN BOTTLE (for which the Heawood FORMULA gives seven, but the correct bound is six). When the FOUR-COLOR THEOREM was proved in 1976, the KLEIN BOTTLE was left as the only exception. The four most difficult cases to prove were $g = 59$, 83, 158, and 257.

see also CHROMATIC NUMBER, FOUR-COLOR THEOREM, MAP COLORING, SIX-COLOR THEOREM, TORUS COLORING

References

Ringel, G. *Map Color Theorem.* New York: Springer-Verlag, 1974.

Ringel, G. and Youngs, J. W. T. "Solution of the Heawood Map-Coloring Problem." *Proc. Nat. Acad. Sci. USA* **60**, 438–445, 1968.

Sloane, N. J. A. Sequence A000934/M3292 in "An On-Line Version of the Encyclopedia of Integer Sequences."

Wagon, S. "Map Coloring on a Torus." §7.5 in *Mathematica in Action.* New York: W. H. Freeman, pp. 232–237, 1991.

Hebesphenomegacorona

see JOHNSON SOLID

Hecke Algebra

An associative RING, also called a HECKE RING, which has a technical definition in terms of commensurable SUBGROUPS.

Hecke L-Function

A generalization of the EULER L-FUNCTION associated with a GROSSENCHARACTER.

References

Knapp, A. W. "Group Representations and Harmonic Analysis, Part II." *Not. Amer. Math. Soc.* **43**, 537–549, 1996.

Hecke Operator

A family of operators on each SPACE of MODULAR FORMS. Hecke operators COMMUTE with each other.

Hecke Ring

see HECKE ALGEBRA

Hectogon

A 100-sided POLYGON.

Hedgehog

An envelope parameterized by its GAUSS MAP. The parametric equations for a hedgehog are

$$x = p(\theta)\cos\theta + p'(\theta)\sin\theta$$
$$y = p(\theta)\sin\theta + p'(\theta)\cos\theta.$$

A plane convex hedgehog has at least four VERTICES where the CURVATURE has a stationary value. A plane convex hedgehog of constant width has at least six VERTICES (Martinez-Maure 1996).

References
Langevin, R.; Levitt, G.; and Rosenberg, H. "Hérissons et Multihérissons (Enveloppes paramétrées par leu application de Gauss." Warsaw: Singularities, 245–253, 1985. Banach Center Pub. 20, PWN Warsaw, 1988.
Martinez-Maure, Y. "A Note on the Tennis Ball Theorem." *Amer. Math. Monthly* **103**, 338–340, 1996.

Heegaard Diagram

A diagram expressing how the gluing operation that connects the HANDLEBODIES involved in a HEEGAARD SPLITTING proceeds, usually by showing how the meridians of the HANDLEBODY are mapped.

see also HANDLEBODY, HEEGAARD SPLITTING

References
Rolfsen, D. *Knots and Links.* Wilmington, DE: Publish or Perish Press, p. 239, 1976.

Heegaard Splitting

A Heegaard splitting of a connected orientable 3-MANIFOLD M is any way of expressing M as the UNION of two (3,1)-HANDLEBODIES along their boundaries. The boundary of such a (3,1)-HANDLEBODY is an orientable SURFACE of some GENUS, which determines the number of HANDLES in the (3,1)-HANDLEBODIES. Therefore, the HANDLEBODIES involved in a Heegaard splitting are the same, but they may be glued together in a strange way along their boundary. A diagram showing how the gluing is done is known as a HEEGAARD DIAGRAM.

References
Adams, C. C. *The Knot Book: An Elementary Introduction to the Mathematical Theory of Knots.* New York: W. H. Freeman, p. 255, 1994.

Heegner Number

The values of $-d$ for which QUADRATIC FIELDS $\mathbb{Q}(\sqrt{-d})$ are uniquely factorable into factors of the form $a + b\sqrt{-d}$. Here, a and b are half-integers, except for $d = 1$ and 2, in which case they are INTEGERS. The Heegner numbers therefore correspond to DISCRIMINANTS $-d$ which have CLASS NUMBER $h(-d)$ equal to 1, except for Heegner numbers -1 and -2, which correspond to $d = -4$ and -8, respectively.

The determination of these numbers is called GAUSS'S CLASS NUMBER PROBLEM, and it is now known that there are only nine Heegner numbers: -1, -2, -3, -7, -11, -19, -43, -67, and -163 (Sloane's A003173), corresponding to discriminants -4, -8, -3, -7, -11, -19, -43, -67, and -163, respectively.

Heilbronn and Linfoot (1934) showed that if a larger d existed, it must be $> 10^9$. Heegner (1952) published a proof that only nine such numbers exist, but his proof was not accepted as complete at the time. Subsequent examination of Heegner's proof show it to be "essentially" correct (Conway and Guy 1996).

The Heegner numbers have a number of fascinating connections with amazing results in PRIME NUMBER theory. In particular, the j-FUNCTION provides stunning connections between e, π, and the ALGEBRAIC INTEGERS. They also explain why Euler's PRIME-GENERATING POLYNOMIAL $n^2 - n + 41$ is so surprisingly good at producing PRIMES.

see also CLASS NUMBER, DISCRIMINANT (BINARY QUADRATIC FORM), GAUSS'S CLASS NUMBER PROBLEM, j-Function, PRIME-GENERATING POLYNOMIAL, QUADRATIC FIELD

References
Conway, J. H. and Guy, R. K. "The Nine Magic Discriminants." In *The Book of Numbers.* New York: Springer-Verlag, pp. 224–226, 1996.
Heegner, K. "Diophantische Analysis und Modulfunktionen." *Math. Z.* **56**, 227–253, 1952.
Heilbronn, H. A. and Linfoot, E. H. "On the Imaginary Quadratic Corpora of Class-Number One." *Quart. J. Math. (Oxford)* **5**, 293–301, 1934.
Sloane, N. J. A. Sequence A003173/M0827 in "An On-Line Version of the Encyclopedia of Integer Sequences."

Heesch Number

The Heesch number of a closed plane figure is the maximum number of times that figure can be completely surrounded by copies of itself. The determination of the maximum possible (finite) Heesch number is known as HEESCH'S PROBLEM. The Heesch number of a TRIANGLE, QUADRILATERAL, regular HEXAGON, or any other shape that can TILE or TESSELLATE the plane, is infinity. Conversely, any shape with infinite Heesch number must tile the plane (Eppstein). The largest known (finite) Heesch number is 3, and corresponds to a tile invented by R. Ammann (Senechal 1995).

References
Eppstein, D. "Heesch's Problem." `http://www.ics.uci.edu/~eppstein/junkyard/heesch/`.
Fontaine, A. "An Infinite Number of Plane Figures with Heesch Number Two." *J. Comb. Th. A* **57**, 151–156, 1991.
Senechal, M. *Quasicrystals and Geometry.* New York: Cambridge University Press, 1995.

Heesch's Problem

How many times can a shape be completely surrounded by copies of itself without being able to TILE the entire plane, i.e., what is the maximum (finite) HEESCH NUMBER?

References
Eppstein, D. "Heesch's Problem." `http://www.ics.uci.edu/~eppstein/junkyard/heesch/`.

Height

The vertical length of an object from top to bottom.

see also LENGTH (SIZE), WIDTH (SIZE)

Heilbronn Triangle Problem

N.B. A detailed on-line essay by S. Finch was the starting point for this entry.

Given any arrangement of n points within a UNIT SQUARE, let H_n be the smallest value for which there is at least one TRIANGLE formed from three of the points with AREA $\leq H_n$. The first few values are

$$\begin{aligned} H_3 &= \tfrac{1}{2} \\ H_4 &= \tfrac{1}{2} \\ H_5 &= \tfrac{1}{9}\sqrt{3} \\ H_6 &= \tfrac{1}{8} \\ H_7 &\geq \tfrac{1}{12} \\ H_8 &\geq \tfrac{1}{4}(2-\sqrt{3}) \\ H_9 &\geq \tfrac{1}{21} \\ H_{10} &\geq \tfrac{1}{32}(3\sqrt{17}-11) \\ H_{11} &\geq \tfrac{1}{27} \\ H_{12} &\geq \tfrac{1}{33} \\ H_{13} &\geq 0.030 \\ H_{14} &\geq 0.022 \\ H_{15} &\geq 0.020 \\ H_{16} &\geq 0.0175. \end{aligned}$$

Komlós *et al.* (1981, 1982) have shown that there are constants c such that

$$\frac{c \ln n}{n^2} \leq H_n \leq \frac{C}{n^{8/7} - \epsilon},$$

for any $\epsilon > 0$ and all sufficiently large n.

Using an EQUILATERAL TRIANGLE of unit AREA instead gives the constants

$$\begin{aligned} h_3 &= 1 \\ h_4 &= \tfrac{1}{3} \\ h_5 &= 3 - 2\sqrt{2} \\ h_6 &= \tfrac{1}{8}. \end{aligned}$$

References

Finch, S. "Favorite Mathematical Constants." `http://www.mathsoft.com/asolve/constant/hlb/hlb.html`.

Goldberg, M. "Maximizing the Smallest Triangle Made by N Points in a Square." *Math. Mag.* **45**, 135–144, 1972.

Guy, R. K. *Unsolved Problems in Number Theory, 2nd ed.* New York: Springer-Verlag, pp. 242–244, 1994.

Komloś, J.; Pintz, J.; and Szemerédi, E. "On Heilbronn's Triangle Problem." *J. London Math. Soc.* **24**, 385–396, 1981.

Komloś, J.; Pintz, J.; and Szemerédi, E. "A Lower Bound for Heilbronn's Triangle Problem." *J. London Math. Soc.* **25**, 13–24, 1982.

Roth, K. F. "Developments in Heilbronn's Triangle Problem." *Adv. Math.* **22**, 364–385, 1976.

Heine-Borel Theorem

If a CLOSED SET of points on a line can be covered by a set of intervals so that every point of the set is an interior point of at least one of the intervals, then there exist a finite number of intervals with the covering property.

Heine Hypergeometric Series

$${}_r\phi_s\left[\begin{matrix}\alpha_1, \alpha_2, \ldots, \alpha_r \\ \beta_1, \ldots, \beta_s\end{matrix}; z\right] \equiv \sum_{n=0}^{\infty} \frac{(\alpha_1;q)_n(\alpha_2;q)_n\cdots(\alpha_r;q)_n}{(q;q)_n(\beta_1;q)_n\cdots(\beta_s;q)_n} z^n, \quad (1)$$

where

$$(a;q)_n = (1-a)(1-aq)(1-aq^2)\cdots(1-aq^{n-1}), \quad (2)$$

$$(a;q)_0 = 1. \quad (3)$$

In particular,

$${}_2\psi_1(a,b;c;q,z) = \sum_{n=0}^{\infty} \frac{(a;q)_n(b;q)_n z^n}{(q;q)_n(c;q)_n} \quad (4)$$

(Andrews 1986, p. 10). Heine proved the transformation formula

$${}_2\phi_1(a,b;c;q,z) = \frac{(b;q)_\infty(az;q)_\infty}{(c;q)_\infty(z;q)_\infty} {}_2\phi_1(c/b,a;az;q,b), \quad (5)$$

and Rogers (1893) obtained the formulas

$${}_2\phi_1(a,b;c;q,z) = \frac{(c/b;q)_\infty(bz;q)_\infty}{(z;q)_\infty(c;q)_\infty} {}_2\phi_1(b,abz/c;bz;q,c/b) \quad (6)$$

$${}_2\phi_1(a,b,c;q,z) = (abz/c;q)_\infty(z;q)_\infty {}_2\phi_1(c/a,c/b;c;q,abz/c) \quad (7)$$

(Andrews 1986, pp. 10–11).

see also q-SERIES

References

Andrews, G. E. *q-Series: Their Development and Application in Analysis, Number Theory, Combinatorics, Physics, and Computer Algebra.* Providence, RI: Amer. Math. Soc., p. 10, 1986.

Heine, E. "Über die Reihe $1 + \frac{(q^\alpha-1)(q^\beta-1)}{(q-1)(q^\gamma-1)}x + \frac{(q^\alpha-1)(q^{\alpha+1}-1)(q^\beta-1)(q^{\beta+1}-1)}{(q-1)(q^2-1)(q^\gamma-1)(q^{\gamma+1}-1)}x^2 + \ldots$" *J. reine angew. Math.* **32**, 210–212, 1846.

Heine, E. "Untersuchungen über die Reihe $1 + \frac{(1-q^\alpha)(1-q^\beta)}{(1-q)(1-q^\gamma)} \cdot x + \frac{(1-q^\alpha)(1-q^{\alpha+1})(1-q^\beta)(1-q^{\beta+1})}{(1-q)(1-q^2)(1-q^\gamma)(1-q^{\gamma+1})} \cdot x^2 + \ldots$" *J. reine angew. Math.* **34**, 285–328, 1847.

Heine, E. *Theorie der Kugelfunctionen und der verwandten Functionen, Vol. 1.* Berlin: Reimer, 1878.

Rogers, L. J. "On a Three-Fold Symmetry in the Elements of Heine's Series." *Proc. London Math. Soc.* **24**, 171–179, 1893.

Heisenberg Group

The Heisenberg group H^n in n COMPLEX variables is the GROUP of all (z,t) with $z \in \mathbb{C}^n$ and $t \in \mathbb{R}$ having multiplication

$$(w,t)(z,t') = (w+z, t+t'+\Im[w^{\mathrm{T}}z])$$

where w^{T} is the conjugate transpose. The Heisenberg group is ISOMORPHIC to the group of MATRICES

$$\begin{bmatrix} 1 & z^{\mathrm{T}} & \frac{1}{2}|z|^2+it \\ 0 & 1 & z \\ 0 & 0 & 1 \end{bmatrix},$$

and satisfies

$$(z,t)^{-1} = (-z,-t).$$

Every finite-dimensional unitary representation is trivial on Z and therefore factors to a REPRESENTATION of the quotient $\mathbb{C}^n$.

see also NIL GEOMETRY

References

Knapp, A. W. "Group Representations and Harmonic Analysis, Part II." *Not. Amer. Math. Soc.* **43**, 537–549, 1996.

Heisenberg Space

The boundary of COMPLEX HYPERBOLIC 2-SPACE.

see also HYPERBOLIC SPACE

Held Group

The SPORADIC GROUP *He*.

References

Wilson, R. A. "ATLAS of Finite Group Representation." `http://for.mat.bham.ac.uk/atlas/He.html`.

Helen of Geometers

see CYCLOID

Helicoid

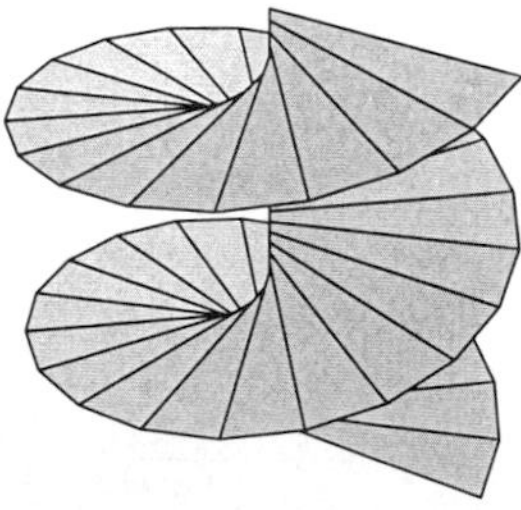

The MINIMAL SURFACE having a HELIX as its boundary. It is the only RULED MINIMAL SURFACE other than the PLANE (Catalan 1842, do Carmo 1986). For many years, the helicoid remained the only known example of a complete embedded MINIMAL SURFACE of finite topology with infinite CURVATURE. However, in 1992 a second example, known as HOFFMAN'S MINIMAL SURFACE and consisting of a helicoid with a HOLE, was discovered (*Sci. News* 1992).

The equation of a helicoid in CYLINDRICAL COORDINATES is

$$z = c\theta. \tag{1}$$

In CARTESIAN COORDINATES, it is

$$\frac{y}{x} = \tan\left(\frac{z}{c}\right). \tag{2}$$

It can be given in parametric form by

$$x = u\cos v \tag{3}$$
$$y = u\sin v \tag{4}$$
$$z = cu, \tag{5}$$

which has an obvious generalization to the ELLIPTIC HELICOID. The differentials are

$$dx = \cos v\,du - u\sin v\,dv \tag{6}$$
$$dy = \sin v\,du + u\cos v\,dv \tag{7}$$
$$dz = 2cu\,dy, \tag{8}$$

so the LINE ELEMENT on the surface is

$$\begin{aligned} ds^2 &= dx^2+dy^2+dz^2 \\ &= \cos^2 v\,du^2 - 2u\sin v\cos v\,du\,dv + u^2\sin^2 v\,dv^2 \\ &\quad + \sin^2 v\,du^2 + 2u\sin v\cos v\,du\,dv + u^2\cos^2 v\,dv^2 \\ &\quad + 4c^2u^2\,du^2 \\ &= (1+4c^2u^2)\,du^2 + u^2\,dv^2, \end{aligned} \tag{9}$$

and the METRIC components are

$$g_{uu} = 1+4c^2u^2 \tag{10}$$
$$g_{uv} = 0 \tag{11}$$
$$g_{vv} = u^2. \tag{12}$$

From GAUSS'S THEOREMA EGREGIUM, the GAUSSIAN CURVATURE is then

$$K = \frac{4c^2}{(1+4c^2u^2)^2}. \tag{13}$$

The MEAN CURVATURE is

$$H = 0, \tag{14}$$

and the equation for the LINES OF CURVATURE is

$$u = \pm c\sinh(v-k). \tag{15}$$

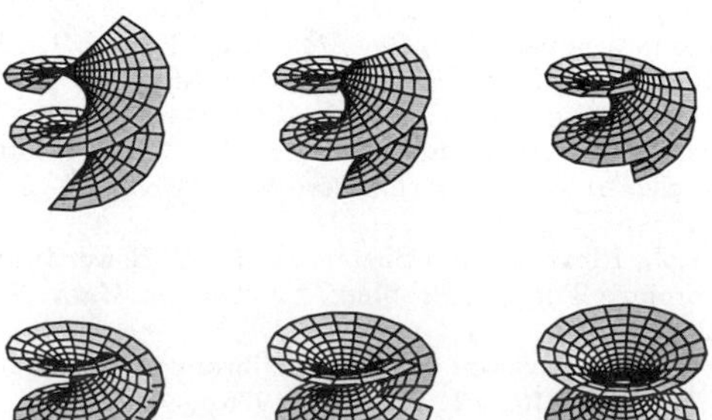

The helicoid can be continuously deformed into a CATENOID by the transformation

$$x(u,v) = \cos\alpha \sinh v \sin u + \sin\alpha \cosh v \cos u \quad (16)$$
$$y(u,v) = -\cos\alpha \sinh v \cos u + \sin\alpha \cosh v \sin u \quad (17)$$
$$z(u,v) = u\cos\alpha + v\sin\alpha, \quad (18)$$

where $\alpha = 0$ corresponds to a helicoid and $\alpha = \pi/2$ to a CATENOID.

If a twisted curve C (i.e., one with TORSION $\tau \neq 0$) rotates about a fixed axis A and, at the same time, is displaced parallel to A such that the speed of displacement is always proportional to the angular velocity of rotation, then C generates a GENERALIZED HELICOID.

see also CALCULUS OF VARIATIONS, CATENOID, ELLIPTIC HELICOID, GENERALIZED HELICOID, HELIX, HOFFMAN'S MINIMAL SURFACE, MINIMAL SURFACE

References

Catalan E. "Sur les surfaces régléess dont l'aire est un minimum." *J. Math. Pure Appl.* **7**, 203–211, 1842.

do Carmo, M. P. "The Helicoid." §3.5B in *Mathematical Models from the Collections of Universities and Museums* (Ed. G. Fischer). Braunschweig, Germany: Vieweg, pp. 44–45, 1986.

Fischer, G. (Ed.). Plate 91 in *Mathematische Modelle/Mathematical Models, Bildband/Photograph Volume.* Braunschweig, Germany: Vieweg, p. 87, 1986.

Geometry Center. "The Helicoid." `http://www.geom.umn.edu/zoo/diffgeom/surfspace/helicoid/`.

Gray, A. *Modern Differential Geometry of Curves and Surfaces.* Boca Raton, FL: CRC Press, p. 264, 1993.

Kreyszig, E. *Differential Geometry.* New York: Dover, p. 88, 1991.

Meusnier, J. B. "Mémoire sur la courbure des surfaces." *Mém. des savans étrangers* **10** (lu 1776), 477–510, 1785.

Peterson, I. "Three Bites in a Doughnut." *Sci. News* **127**, 168, Mar. 16, 1985.

"Putting a Handle on a Minimal Helicoid." *Sci. News* **142**, 276, Oct. 24, 1992.

Wolfram, S. *The Mathematica Book, 3rd ed.* Champaign, IL: Wolfram Media, p. 164, 1996.

Helix

A helix is also called a CURVE OF CONSTANT SLOPE. It can be defined as a curve for which the TANGENT makes a constant ANGLE with a fixed line. The helix is a SPACE CURVE with parametric equations

$$x = r\cos t \quad (1)$$
$$y = r\sin t \quad (2)$$
$$z = ct, \quad (3)$$

where c is a constant. The CURVATURE of the helix is given by

$$\kappa = \frac{r}{r^2+c^2}, \quad (4)$$

and the LOCUS of the centers of CURVATURE of a helix is another helix. The ARC LENGTH is given by

$$s = \int \sqrt{x'^2+y'^2+z'^2}\,dt = \sqrt{r^2+c^2}\,t. \quad (5)$$

The TORSION of a helix is given by

$$\tau = \frac{1}{r^2(r^2+c^2)} \begin{vmatrix} -r\sin t & -r\cos t & r\sin t \\ r\cos t & -r\sin t & -r\cos t \\ c & 0 & 0 \end{vmatrix} = \frac{c}{r^2+c^2}, \quad (6)$$

so

$$\frac{\kappa}{\tau} = \frac{\frac{r}{r^2+c^2}}{\frac{c}{r^2+c^2}} = \frac{r}{c}, \quad (7)$$

which is a constant. In fact, LANCRET'S THEOREM states that a NECESSARY and SUFFICIENT condition for a curve to be a helix is that the ratio of CURVATURE to TORSION be constant. The OSCULATING PLANE of the helix is given by

$$\begin{vmatrix} z_1 - r\cos t & z_2 - r\sin t & z_3 - ct \\ -r\sin t & r\cos t & c \\ -r\cos t & -r\sin t & 0 \end{vmatrix} = 0 \quad (8)$$

$$z_1 c\sin t - z_2 c\cos t + (z_3 - ct)r = 0. \quad (9)$$

The MINIMAL SURFACE of a helix is a HELICOID.

see also GENERALIZED HELIX, HELICOID, SPHERICAL HELIX

References

Geometry Center. "The Helix." `http://www.geom.umn.edu/zoo/diffgeom/surfspace/helicoid/helix.html`.

Gray, A. "The Helix and Its Generalizations." §7.5 in *Modern Differential Geometry of Curves and Surfaces.* Boca Raton, FL: CRC Press, pp. 138–140, 1993.

Isenberg, C. Plate 4.11 in *The Science of Soap Films and Soap Bubbles.* New York: Dover, 1992.

Pappas, T. "The Helix—Mathematics & Genetics." *The Joy of Mathematics.* San Carlos, CA: Wide World Publ./Tetra, pp. 166–168, 1989.

Wolfram, S. *The Mathematica Book, 3rd ed.* Champaign, IL: Wolfram Media, p. 163, 1996.

Helly Number

Given a Euclidean n-space,

$$H_n \equiv n+1.$$

see also EUCLIDEAN SPACE, HELLY'S THEOREM

Helly's Theorem

If F is a family of more than n bounded closed convex sets in Euclidean n-space $\mathbb{R}^n$, and if every H_n (where H_n is the HELLY NUMBER) members of F have at least one point in common, then all the members of F have at least one point in common.

see also CARATHÉODORY'S FUNDAMENTAL THEOREM, HELLY NUMBER

Helmholtz Differential Equation

A PARTIAL DIFFERENTIAL EQUATION which can be written in a SCALAR version

$$\nabla^2\psi + k^2\psi = 0, \tag{1}$$

or VECTOR form,

$$\nabla^2\mathbf{A} + k^2\mathbf{A} = 0, \tag{2}$$

where ∇^2 is the LAPLACIAN. When $k = 0$, the Helmholtz differential equation reduces to LAPLACE'S EQUATION. When $k^2 < 0$, the equation becomes the space part of the diffusion equation.

The Helmholtz differential equation can be solved by SEPARATION OF VARIABLES in only 11 coordinate systems, 10 of which (with the exception of CONFOCAL PARABOLOIDAL COORDINATES) are particular cases of the CONFOCAL ELLIPSOIDAL system: CARTESIAN, CONFOCAL ELLIPSOIDAL, CONFOCAL PARABOLOIDAL, CONICAL, CYLINDRICAL, ELLIPTIC CYLINDRICAL, OBLATE SPHEROIDAL, PARABOLOIDAL, PARABOLIC CYLINDRICAL, PROLATE SPHEROIDAL, and SPHERICAL COORDINATES (Eisenhart 1934). LAPLACE'S EQUATION (the Helmholtz differential equation with $k = 0$) is separable in the two additional BISPHERICAL COORDINATES and TOROIDAL COORDINATES.

If Helmholtz's equation is separable in a 3-D coordinate system, then Morse and Feshbach (1953, pp. 509–510) show that

$$\frac{h_1h_2h_3}{h_n{}^2} = f_n(u_n)g_n(u_i,u_j), \tag{3}$$

where $i \neq j \neq n$. The LAPLACIAN is therefore of the form

$$\begin{aligned}\nabla^2 = \frac{1}{h_1h_2h_3}\biggl\{ & g_1(u_2,u_3)\frac{\partial}{\partial u_1}\left[f_1(u_1)\frac{\partial}{\partial u_1}\right] \\ & + g_2(u_1,u_3)\frac{\partial}{\partial u_2}\left[f_2(u_2)\frac{\partial}{\partial u_2}\right] \\ & + g_3(u_1,u_2)\frac{\partial}{\partial u_3}\left[f_3(u_3)\frac{\partial}{\partial u_3}\right]\biggr\},\end{aligned} \tag{4}$$

which simplifies to

$$\begin{aligned}\nabla^2 = & \frac{1}{h_1{}^2 f_1}\frac{\partial}{\partial u_1}\left[f_1(u_1)\frac{\partial}{\partial u_1}\right] \\ & + \frac{1}{h_2{}^2 f_2}\frac{\partial}{\partial u_2}\left[f_2(u_2)\frac{\partial}{\partial u_2}\right] \\ & + \frac{1}{h_3{}^2 f_3}\frac{\partial}{\partial u_3}\left[f_3(u_3)\frac{\partial}{\partial u_3}\right].\end{aligned} \tag{5}$$

Such a coordinate system obeys the ROBERTSON CONDITION, which means that the STÄCKEL DETERMINANT is of the form

$$S = \frac{h_1h_2h_3}{f_1(u_1)f_2(u_2)f_3(u_3)}. \tag{6}$$

Coordinate System	Variables	Solution Functions
Cartesian	$X(x)Y(y)Z(z)$	exponential, circular, hyperbolic
circular cylindrical	$R(r)\Theta(\theta)Z(z)$	Bessel, exponential, circular
conical		ellipsoidal harmonics, power
ellipsoidal	$\Lambda(\lambda)M(\mu)N(\nu)$	ellipsoidal harmonics
elliptic cylindrical	$U(u)V(v)Z(z)$	Mathieu, circular
oblate spheroidal	$\Lambda(\lambda)M(\mu)N(\nu)$	Legendre, circular
parabolic		Bessel, circular
parabolic cylindrical		Parabolic cylinder, Bessel, circular
paraboloidal	$U(u)V(v)\Theta(\theta)$	Baer functions, circular
prolate spheroidal	$\Lambda(\lambda)M(\mu)N(\nu)$	Legendre, circular
spherical	$R(r)\Theta(\theta)\Phi(\phi)$	Legendre, power, circular

see also LAPLACE'S EQUATION, POISSON'S EQUATION, SEPARATION OF VARIABLES, SPHERICAL BESSEL DIFFERENTIAL EQUATION

References

Eisenhart, L. P. "Separable Systems in Euclidean 3-Space." *Physical Review* **45**, 427–428, 1934.

Eisenhart, L. P. "Separable Systems of Stäckel." *Ann. Math.* **35**, 284–305, 1934.

Eisenhart, L. P. "Potentials for Which Schroedinger Equations Are Separable." *Phys. Rev.* **74**, 87–89, 1948.

Morse, P. M. and Feshbach, H. *Methods of Theoretical Physics, Part I.* New York: McGraw-Hill, pp. 125–126 and 509–510, 1953.

Helmholtz Differential Equation—Bipolar Coordinates

In BIPOLAR COORDINATES, the HELMHOLTZ DIFFERENTIAL EQUATION is not separable, but LAPLACE'S EQUATION is.

see also LAPLACE'S EQUATION—BIPOLAR COORDINATES

Helmholtz Differential Equation—Cartesian Coordinates

In 2-D CARTESIAN COORDINATES, attempt SEPARATION OF VARIABLES by writing

$$F(x,y) = X(x)Y(y), \tag{1}$$

then the HELMHOLTZ DIFFERENTIAL EQUATION becomes

$$\frac{d^2X}{dx^2}Y + \frac{d^2Y}{dy^2}X + k^2XY = 0. \quad (2)$$

Dividing both sides by XY gives

$$\frac{1}{X}\frac{d^2X}{dx^2} + \frac{1}{Y}\frac{d^2Y}{dy^2} + k^2 = 0. \quad (3)$$

This leads to the two coupled ordinary differential equations with a separation constant m^2,

$$\frac{1}{X}\frac{d^2X}{dx^2} = m^2 \quad (4)$$

$$\frac{1}{Y}\frac{d^2Y}{dy^2} = -(m^2 + k^2), \quad (5)$$

where X and Y could be interchanged depending on the boundary conditions. These have solutions

$$X = A_m e^{mx} + B_m e^{-mx} \quad (6)$$

$$Y = C_m e^{i\sqrt{m^2+k^2}\,y} + D_m e^{-i\sqrt{m^2+k^2}\,y}$$
$$= E_m \sin(\sqrt{m^2+k^2}\,y) + F_m \cos(\sqrt{m^2+k^2}\,y). \quad (7)$$

The general solution is then

$$F(x,y) = \sum_{m=1}^{\infty}(A_m e^{mx} + B_m e^{-mx})$$
$$\times[E_m \sin(\sqrt{m^2+k^2}\,y) + F_m \cos(\sqrt{m^2+k^2}\,y)]. \quad (8)$$

In 3-D CARTESIAN COORDINATES, attempt SEPARATION OF VARIABLES by writing

$$F(x,y,z) = X(x)Y(y)Z(z), \quad (9)$$

then the HELMHOLTZ DIFFERENTIAL EQUATION becomes

$$\frac{d^2X}{dx^2}YZ + \frac{d^2Y}{dy^2}XZ + \frac{d^2Z}{dz^2}XY + k^2XY = 0. \quad (10)$$

Dividing both sides by XYZ gives

$$\frac{1}{X}\frac{d^2X}{dx^2} + \frac{1}{Y}\frac{d^2Y}{dy^2} + \frac{1}{Z}\frac{d^2Z}{dz^2} + k^2 = 0. \quad (11)$$

This leads to the three coupled differential equations

$$\frac{1}{X}\frac{d^2X}{dx^2} = l^2 \quad (12)$$

$$\frac{1}{Y}\frac{d^2Y}{dy^2} = m^2 \quad (13)$$

$$\frac{1}{Z}\frac{d^2Z}{dz^2} = -(k^2 + l^2 + m^2), \quad (14)$$

where X, Y, and Z could be permuted depending on boundary conditions. The general solution is therefore

$$F(x,y,z)$$
$$= \sum_{l=1}^{\infty}\sum_{m=1}^{\infty}(A_l e^{lx} + B_l e^{-lx})(C_m e^{my} + D_m e^{-my})$$
$$\times(E_{lm} e^{-i\sqrt{k^2+l^2+m^2}\,z} + F_{lm} e^{i\sqrt{k^2+l^2+m^2}\,z}). \quad (15)$$

References

Morse, P. M. and Feshbach, H. *Methods of Theoretical Physics, Part I.* New York: McGraw-Hill, pp. 501–502, 513–514 and 656, 1953.

Helmholtz Differential Equation—Circular Cylindrical Coordinates

In CYLINDRICAL COORDINATES, the SCALE FACTORS are $h_r = 1$, $h_\theta = r$, $h_z = 1$ and the separation functions are $f_1(r) = r$, $f_2(\theta) = 1$, $f_3(z) = 1$, so the STÄCKEL DETERMINANT is 1. Attempt SEPARATION OF VARIABLES by writing

$$F(r,\theta,z) = R(r)\Theta(\theta)Z(z), \quad (1)$$

then the HELMHOLTZ DIFFERENTIAL EQUATION becomes

$$\frac{d^2R}{dr^2}\Theta Z + \frac{1}{r}\frac{dR}{dr}\Theta Z + \frac{1}{r^2}\frac{d^2\Theta}{d\theta^2}RZ + \frac{d^2Z}{dz^2}R\Theta + k^2R\Theta Z = 0. \quad (2)$$

Now divide by $R\Theta Z$,

$$\left(\frac{r^2}{R}\frac{d^2R}{dr^2} + \frac{r}{R}\frac{dR}{dr}\right) + \frac{d^2\Theta}{d\theta^2}\frac{1}{\Theta} + \frac{d^2Z}{dz^2}\frac{r^2}{Z} + k^2 = 0, \quad (3)$$

so the equation has been separated. Since the solution must be periodic in Θ from the definition of the circular cylindrical coordinate system, the solution to the second part of (3) must have a NEGATIVE separation constant

$$\frac{d^2\Theta}{d\theta^2}\frac{1}{\Theta} = -(k^2 + m^2), \quad (4)$$

which has a solution

$$\Theta(\theta) = C_m e^{-i\sqrt{k^2+m^2}\,\theta} + D_m e^{i\sqrt{k^2+m^2}\,\theta}. \quad (5)$$

Plugging (5) back into (3) gives

$$\frac{r^2}{R}\frac{d^2R}{dr^2} + \frac{r}{R}\frac{dR}{dr} - m^2 + \frac{d^2Z}{dz^2}\frac{r^2}{Z} = 0 \quad (6)$$

$$\frac{1}{R}\frac{d^2R}{dr^2} + \frac{1}{rR}\frac{dR}{dr} - \frac{m^2}{r^2} + \frac{d^2Z}{dz^2}\frac{1}{Z} = 0. \quad (7)$$

The solution to the second part of (7) must not be sinusoidal at $\pm\infty$ for a physical solution, so the differential equation has a POSITIVE separation constant

$$\frac{d^2Z}{dz^2}\frac{1}{Z} = n^2, \quad (8)$$

and the solution is

$$Z(z) = E_n e^{-nz} + F_n e^{nz}. \tag{9}$$

Plugging (9) back into (7) and multiplying through by R yields

$$\frac{d^2R}{dr^2} + \frac{1}{r}\frac{dR}{dr} + \left(n^2 - \frac{m^2}{r^2}\right) R = 0 \tag{10}$$

$$\frac{1}{n^2}\frac{d^2R}{dr^2} + \frac{1}{(nr)}\frac{1}{n}\frac{dR}{dr} + \left[1 - \frac{m^2}{(nr)^2}\right] R = 0 \tag{11}$$

$$\frac{d^2R}{d(nr)^2} + \frac{1}{(nr)}\frac{dR}{d(nr)} + \left[1 - \frac{m^2}{(nr)^2}\right] R = 0. \tag{12}$$

This is the BESSEL DIFFERENTIAL EQUATION, which has a solution

$$R(r) = A_{mn} J_m(nr) + B_{mn} Y_m(nr), \tag{13}$$

where $J_n(x)$ and $Y_n(x)$ are BESSEL FUNCTIONS OF THE FIRST and SECOND KINDS, respectively. The general solution is therefore

$$F(r,\theta,z) = \sum_{m=0}^{\infty}\sum_{n=0}^{\infty}[A_{mn}J_m(nr) + B_{mn}Y_m(nr)] \times (C_m e^{-i\sqrt{k^2+m^2}\,\theta} + D_m e^{i\sqrt{k^2+m^2}\,\theta})(E_n e^{-nz} + F_n e^{nz}). \tag{14}$$

Actually, the HELMHOLTZ DIFFERENTIAL EQUATION is separable for general k of the form

$$k^2(r,\theta,z) = f(r) + \frac{g(\theta)}{r^2} + h(z) + k'^2. \tag{15}$$

see also CYLINDRICAL COORDINATES, HELMHOLTZ DIFFERENTIAL EQUATION

References

Morse, P. M. and Feshbach, H. *Methods of Theoretical Physics, Part I.* New York: McGraw-Hill, pp. 514 and 656–657, 1953.

Helmholtz Differential Equation—Confocal Ellipsoidal Coordinates

Using the NOTATION of Byerly (1959, pp. 252–253), LAPLACE'S EQUATION can be reduced to

$$\nabla^2 F = (\mu^2-\nu^2)\frac{\partial^2 F}{\partial\alpha^2} + (\lambda^2-\nu^2)\frac{\partial^2 F}{\partial\beta^2} + (\lambda^2-\mu^2)\frac{\partial^2 F}{\partial\gamma^2} = 0, \tag{1}$$

where

$$\alpha = c\int_c^\lambda \frac{d\lambda}{\sqrt{(\lambda^2-b^2)(\lambda^2-c^2)}} = F\left[\frac{b}{c},\frac{\pi}{2}\right) - F\left(\frac{b}{c},\sin^{-1}\left(\frac{c}{\lambda}\right)\right] \tag{2}$$

$$\beta = c\int_b^\mu \frac{d\mu}{\sqrt{(c^2-\mu^2)(\mu^2-b^2)}} = F\left[\sqrt{1-b^2-c^2},\sin^{-1}\left(\sqrt{\frac{1-\frac{b^2}{\mu^2}}{1-\frac{b^2}{c^2}}}\right)\right] \tag{3}$$

$$\gamma = c\int_0^\nu \frac{d\nu}{\sqrt{(b^2-\nu^2)(c^2-\nu^2)}} = F\left(\frac{b}{c},\sin^{-1}\left(\frac{\nu}{b}\right)\right). \tag{4}$$

In terms of α, β, and γ,

$$\lambda = c\,\mathrm{dc}\left(\alpha,\frac{b}{c}\right) \tag{5}$$

$$\mu = b\,\mathrm{nd}\left(\beta,\sqrt{1-\frac{b^2}{c^2}}\right) \tag{6}$$

$$\nu = b\,\mathrm{sn}\left(\gamma,\frac{b}{c}\right). \tag{7}$$

Equation (1) is not separable using a function of the form

$$F = L(\alpha)M(\beta)N(\gamma), \tag{8}$$

but it is if we let

$$\frac{1}{L}\frac{d^2L}{d\alpha^2} = \sum a_k\lambda^k \tag{9}$$

$$\frac{1}{M}\frac{d^2M}{d\beta^2} = \sum b_k\mu^k \tag{10}$$

$$\frac{1}{N}\frac{d^2N}{d\gamma^2} = \sum c_k\nu^k. \tag{11}$$

These give

$$a_0 = -b_0 = c_0 \tag{12}$$

$$a_2 = -b_2 = c_2, \tag{13}$$

and all others terms vanish. Therefore (1) can be broken up into the equations

$$\frac{d^2L}{d\alpha^2} = (a_0 + a_2\lambda^2)L \tag{14}$$

$$\frac{d^2M}{d\beta^2} = -(a_0 + a_2\mu^2)M \tag{15}$$

$$\frac{d^2N}{d\gamma^2} = (a_0 + a_2\nu^2)N. \tag{16}$$

For future convenience, now write

$$a_0 = -(b^2 + c^2)p \tag{17}$$

$$a_2 = m(m+1), \tag{18}$$

then

$$\frac{d^2L}{d\alpha^2} - [m(m+1)\lambda^2 - (b^2+c^2)p]L = 0 \tag{19}$$

$$\frac{d^2M}{d\beta^2} + [m(m+1)\mu^2 - (b^2+c^2)p]M = 0 \tag{20}$$

$$\frac{d^2N}{d\gamma^2} - [m(m+1)\nu^2 - (b^2+c^2)p]N = 0. \tag{21}$$

Now replace α, β, and γ to obtain

$$(\lambda^2-b^2)(\lambda^2-c^2)\frac{d^2L}{d\lambda^2} + \lambda(\lambda^2-b^2+\lambda^2-c^2)\frac{dL}{d\lambda} - [m(m+1)\lambda^2 - (b^2+c^2)p]L = 0 \tag{22}$$

$$(\mu^2-b^2)(\mu^2-c^2)\frac{d^2M}{d\mu^2} + \mu(\mu^2-b^2+\mu^2-c^2)\frac{dM}{d\mu} - [m(m+1)\mu^2 - (b^2+c^2)p]M = 0 \tag{23}$$

$$(\nu^2-b^2)(\nu^2-c^2)\frac{d^2N}{d\nu^2} + \nu(\nu^2-b^2+\nu^2-c^2)\frac{dN}{d\nu} - [m(m+1)\nu^2 - (b^2+c^2)p]N = 0. \tag{24}$$

Each of these is a LAMÉ'S DIFFERENTIAL EQUATION, whose solution is called an ELLIPSOIDAL HARMONIC. Writing

$$L(\lambda) = E_m^p(\lambda) \tag{25}$$

$$M(\lambda) = E_m^p(\mu) \tag{26}$$

$$N(\lambda) = E_m^p(\nu) \tag{27}$$

gives the solution to (1) as a product of ELLIPSOIDAL HARMONICS $E_m^p(x)$.

$$F = E_m^p(\lambda)E_m^p(\mu)E_m^p(\nu). \tag{28}$$

References

Arfken, G. "Confocal Ellipsoidal Coordinates (ξ_1, ξ_2, ξ_3)." §2.15 in *Mathematical Methods for Physicists, 2nd ed.* Orlando, FL: Academic Press, pp. 117–118, 1970.

Byerly, W. E. *An Elementary Treatise on Fourier's Series, and Spherical, Cylindrical, and Ellipsoidal Harmonics, with Applications to Problems in Mathematical Physics.* New York: Dover, pp. 251–258, 1959.

Morse, P. M. and Feshbach, H. *Methods of Theoretical Physics, Part I.* New York: McGraw-Hill, p. 663, 1953.

Helmholtz Differential Equation—Confocal Paraboloidal Coordinates

As shown by Morse and Feshbach (1953), the HELMHOLTZ DIFFERENTIAL EQUATION is separable in CONFOCAL PARABOLOIDAL COORDINATES.

see also CONFOCAL PARABOLOIDAL COORDINATES

References

Morse, P. M. and Feshbach, H. *Methods of Theoretical Physics, Part I.* New York: McGraw-Hill, p. 664, 1953.

Helmholtz Differential Equation—Conical Coordinates

In CONICAL COORDINATES, LAPLACE'S EQUATION can be written

$$\frac{\partial^2 V}{\partial\alpha^2} + \frac{\partial^2 V}{\partial\beta^2} + (\mu^2-\nu^2)\frac{\partial}{\partial\lambda}\left(\lambda^2\frac{\partial V}{\partial\lambda}\right) = 0, \tag{1}$$

where

$$\alpha = \int_a^\mu \frac{d\mu}{\sqrt{(\mu^2-a^2)(b^2-\mu^2)}} \tag{2}$$

$$\beta = \int_0^\nu \frac{d\nu}{\sqrt{(a^2-\nu^2)(b^2-\nu^2)}} \tag{3}$$

(Byerly 1959). Letting

$$V = U(u)R(r) \tag{4}$$

breaks (1) into the two equations,

$$\frac{d}{dr}\left(r^2\frac{dR}{dr}\right) = m(m+1)R \tag{5}$$

$$\frac{\partial^2 U}{\partial\alpha^2} + \frac{\partial^2 U}{\partial\beta^2} + m(m+1)(\mu^2-\nu^2)U = 0. \tag{6}$$

Solving these gives

$$R(r) = Ar^m + Br^{-m-1} \tag{7}$$

$$U(u) = E_m^p(\mu)E_m^p(\nu), \tag{8}$$

where E_m^p are ELLIPSOIDAL HARMONICS. The regular solution is therefore

$$V = Ar^m E_m^p(\mu)E_m^p(\nu). \tag{9}$$

However, because of the cylindrical symmetry, the solution $E_m^p(\mu)E_m^p(\nu)$ is an mth degree SPHERICAL HARMONIC.

References

Arfken, G. "Conical Coordinates (ξ_1, ξ_2, ξ_3)." §2.16 in *Mathematical Methods for Physicists, 2nd ed.* Orlando, FL: Academic Press, pp. 118–119, 1970.

Byerly, W. E. *An Elementary Treatise on Fourier's Series, and Spherical, Cylindrical, and Ellipsoidal Harmonics, with Applications to Problems in Mathematical Physics.* New York: Dover, p. 263, 1959.

Morse, P. M. and Feshbach, H. *Methods of Theoretical Physics, Part I.* New York: McGraw-Hill, pp. 514 and 659, 1953.

Helmholtz Differential Equation—Elliptic Cylindrical Coordinates

In ELLIPTIC CYLINDRICAL COORDINATES, the SCALE FACTORS are $h_u = h_v = a\sqrt{\sinh^2 u + \sin^2 v}$, $h_z = 1$, and the separation functions are $f_1(u) = f_2(v) = f_3(z) = 1$, giving a STÄCKEL DETERMINANT of $S = a^2(\sin^2 v + \sinh^2 u)$. The Helmholtz differential equation is

$$\frac{1}{a^2(\sinh^2 u + \sin^2 v)}\left(\frac{\partial^2 F}{\partial u^2} + \frac{\partial^2 F}{\partial v^2}\right) + \frac{\partial^2 F}{\partial z^2} + k^2 = 0. \tag{1}$$

Attempt SEPARATION OF VARIABLES by writing

$$F(u, v, z) = U(u)V(v)Z(z), \tag{2}$$

then the HELMHOLTZ DIFFERENTIAL EQUATION becomes

$$\frac{Z}{\sinh^2 u + \sin^2 v}\left(V\frac{d^2U}{du^2} + U\frac{d^2V}{dv^2}\right) + UV\frac{d^2Z}{dz^2} + k^2 UVZ = 0. \tag{3}$$

Now divide by UVZ to give

$$\frac{1}{\sinh^2 u + \sin^2 v}\left(\frac{1}{U}\frac{d^2U}{du^2} + \frac{1}{V}\frac{d^2V}{dv^2}\right) + \frac{1}{Z}\frac{d^2Z}{dz^2} + k^2 = 0. \tag{4}$$

Separating the Z part,

$$\frac{1}{Z}\frac{d^2Z}{dz^2} = -(k^2 + m^2) \tag{5}$$

$$\frac{1}{\sinh^2 u + \sin^2 v}\left(\frac{1}{U}\frac{d^2U}{du^2} + \frac{1}{V}\frac{d^2V}{dv^2}\right) = m^2, \tag{6}$$

so

$$\frac{d^2Z}{dz^2} = -(k^2 + m^2)Z, \tag{7}$$

which has the solution

$$Z(z) = A\cos(\sqrt{k^2 + m^2}\, z) + B\sin(\sqrt{k^2 + m^2}\, z). \tag{8}$$

Rewriting (6) gives

$$\left(\frac{1}{U}\frac{d^2U}{du^2} - m^2\sinh^2 u\right) + \left(\frac{1}{V}\frac{d^2V}{dv^2} - m^2\sin^2 v\right) = 0, \tag{9}$$

which can be separated into

$$\frac{1}{U}\frac{d^2U}{du^2} - m^2\sinh^2 u = c \tag{10}$$

$$c + \frac{1}{V}\frac{d^2V}{dv^2} - m^2\sin^2 v = 0, \tag{11}$$

so

$$\frac{d^2U}{du^2} - (c + m^2\sinh^2 u)U = 0 \tag{12}$$

$$\frac{d^2V}{dv^2} + (c - m^2\sin^2 v)V = 0. \tag{13}$$

Now use

$$\sinh^2 u = \tfrac{1}{2}[1 - \cosh(2u)] \tag{14}$$

$$\sin^2 v = \tfrac{1}{2}[1 - \cos(2v)] \tag{15}$$

to obtain

$$\frac{d^2U}{du^2} - \{c + \tfrac{1}{2}m^2[1 - \cosh(2u)]\}U = 0 \tag{16}$$

$$\frac{d^2V}{dv^2} + \{c + \tfrac{1}{2}m^2[1 - \cos(2v)]\}V = 0. \tag{17}$$

Regrouping gives

$$\frac{d^2U}{du^2} - [(c + \tfrac{1}{2}m^2) - \tfrac{1}{4}m^2 2\cosh(2u)]U = 0 \tag{18}$$

$$\frac{d^2V}{dv^2} + [(c + \tfrac{1}{2}m^2) - \tfrac{1}{4}m^2 2\cos(2v)]V = 0. \tag{19}$$

Let $b \equiv \frac{1}{2}m^2 + c$ and $q \equiv \frac{1}{4}m^2$, then these become

$$\frac{d^2U}{du^2} - [b - 2q\cosh(2u)]U = 0 \tag{20}$$

$$\frac{d^2V}{dv^2} + [b - 2q\cos(2v)]V = 0. \tag{21}$$

Here, (21) is the MATHIEU DIFFERENTIAL EQUATION and (20) is the modified MATHIEU DIFFERENTIAL EQUATION. These solutions are known as MATHIEU FUNCTIONS.

see also ELLIPTIC CYLINDRICAL COORDINATES, MATHIEU DIFFERENTIAL EQUATION, MATHIEU FUNCTION

References

Morse, P. M. and Feshbach, H. *Methods of Theoretical Physics, Part I.* New York: McGraw-Hill, pp. 514 and 657, 1953.

Helmholtz Differential Equation—Oblate Spheroidal Coordinates

As shown by Morse and Feshbach (1953) and Arfken (1970), the HELMHOLTZ DIFFERENTIAL EQUATION is separable in OBLATE SPHEROIDAL COORDINATES.

References

Arfken, G. "Oblate Spheroidal Coordinates (u, v, φ)." §2.11 in *Mathematical Methods for Physicists, 2nd ed.* Orlando, FL: Academic Press, pp. 107–109, 1970.

Morse, P. M. and Feshbach, H. *Methods of Theoretical Physics, Part I.* New York: McGraw-Hill, p. 662, 1953.

Helmholtz Differential Equation—Parabolic Coordinates

The SCALE FACTORS are $h_u = h_v = \sqrt{u^2 + v^2}$, $h_\theta = uv$ and the separation functions are $f_1(u) = u$, $f_2(v) = v$, $f_3(\theta) = 1$, given a STÄCKEL DETERMINANT of $S = u^2 + v^2$. The LAPLACIAN is

$$\frac{1}{u^2 + v^2}\left(\frac{1}{u}\frac{\partial F}{\partial u} + \frac{\partial^2 F}{\partial u^2} + \frac{1}{v}\frac{\partial F}{\partial v} + \frac{\partial^2 F}{\partial v^2}\right) + \frac{1}{u^2 v^2}\frac{\partial^2 F}{\partial \theta^2} + k^2 = 0. \tag{1}$$

Attempt SEPARATION OF VARIABLES by writing

$$F(u,v,z) \equiv U(u)V(v)\Theta(\theta), \tag{2}$$

then the HELMHOLTZ DIFFERENTIAL EQUATION becomes

$$\frac{1}{u^2+v^2}\left[V\Theta\left(\frac{1}{u}\frac{dU}{du}+\frac{d^2U}{du^2}\right) + U\Theta\left(\frac{1}{v}\frac{dV}{dv}+\frac{d^2V}{dv^2}\right)\right] + k^2UV\Theta = 0. \tag{3}$$

Now divide by $UV\Theta$,

$$\frac{u^2v^2}{u^2+v^2}\left[\frac{1}{U}\left(\frac{1}{u}\frac{dU}{du}+\frac{d^2U}{du^2}\right)+\frac{1}{V}\left(\frac{1}{v}\frac{dV}{dv}+\frac{d^2V}{dv^2}\right)\right] + \frac{1}{\Theta}\frac{d^2\Theta}{d\theta^2}+k^2 = 0. \tag{4}$$

Separating the Θ part,

$$\frac{1}{\Theta}\frac{d^2\Theta}{f\theta^2} = -(k^2+m^2) \tag{5}$$

$$\frac{u^2v^2}{u^2+v^2}\left[\frac{1}{U}\left(\frac{1}{u}\frac{dU}{du}+\frac{d^2U}{du^2}\right)+\frac{1}{V}\left(\frac{1}{v}\frac{dV}{dv}+\frac{d^2V}{dv^2}\right)\right] = k^2, \tag{6}$$

so

$$\frac{d^2\Theta}{d\theta^2} = -(k^2+m^2)\Theta, \tag{7}$$

which has solution

$$\Theta(\theta) = A\cos(\sqrt{k^2+m^2}\,\theta) + B\sin(\sqrt{k^2+m^2}\,\theta), \tag{8}$$

and

$$\left[\frac{1}{U}\left(\frac{1}{u}\frac{dU}{du}+\frac{d^2U}{du^2}\right)\right]+\left[\frac{1}{V}\left(\frac{1}{v}\frac{dV}{dv}+\frac{d^2V}{dv^2}\right)\right] - k^2\frac{u^2+v^2}{u^2v^2} = 0 \tag{9}$$

$$\left[\frac{1}{U}\left(\frac{1}{u}\frac{dU}{du}+\frac{d^2U}{du^2}\right)-\frac{k^2}{u^2}\right] + \left[\frac{1}{V}\left(\frac{1}{v}\frac{dV}{dv}+\frac{d^2V}{dv^2}\right)-\frac{k^2}{v^2}\right] = 0. \tag{10}$$

This can be separated

$$\frac{1}{U}\left(\frac{1}{u}\frac{dU}{du}+\frac{d^2U}{du^2}\right)-\frac{k^2}{u^2} = c \tag{11}$$

$$\frac{1}{V}\left(\frac{1}{v}\frac{dV}{dv}+\frac{d^2V}{dv^2}\right)-\frac{k^2}{v^2} = -c, \tag{12}$$

so

$$u^2\frac{d^2U}{du^2}+\frac{dU}{du}-(c+k^2)U = 0 \tag{13}$$

$$v^2\frac{d^2V}{dv^2}+\frac{dV}{dv}+(c-k^2)V = 0. \tag{14}$$

References

Arfken, G. "Parabolic Coordinates (ξ,η,ϕ)." §2.12 in *Mathematical Methods for Physicists, 2nd ed.* Orlando, FL: Academic Press, pp. 109–111, 1970.

Morse, P. M. and Feshbach, H. *Methods of Theoretical Physics, Part I.* New York: McGraw-Hill, pp. 514–515 and 660, 1953.

Helmholtz Differential Equation—Parabolic Cylindrical Coordinates

In PARABOLIC CYLINDRICAL COORDINATES, the SCALE FACTORS are $h_u = h_v = \sqrt{u^2+v^2}$, $h_z = 1$ and the separation functions are $f_1(u) = f_2(v) = f_3(z) = 1$, giving STÄCKEL DETERMINANT of $S = u^2+v^2$. The HELMHOLTZ DIFFERENTIAL EQUATION is

$$\frac{1}{u^2+v^2}\left(\frac{\partial^2F}{\partial u^2}+\frac{\partial^2F}{\partial v^2}\right)+\frac{\partial^2F}{\partial z^2}+k^2 = 0. \tag{1}$$

Attempt SEPARATION OF VARIABLES by writing

$$F(u,v,z) \equiv U(u)V(v)Z(z), \tag{2}$$

then the HELMHOLTZ DIFFERENTIAL EQUATION becomes

$$\frac{1}{u^2+v^2}\left(VZ\frac{d^2U}{du^2}+UZ\frac{d^2V}{dv^2}\right)+UV\frac{d^2Z}{dz^2} + k^2UVZ = 0. \tag{3}$$

Divide by UVZ,

$$\frac{1}{u^2+v^2}\left(\frac{1}{U}\frac{d^2U}{du^2}+\frac{1}{V}\frac{d^2V}{dv^2}\right)+\frac{1}{Z}\frac{d^2Z}{dz^2}+k^2 = 0. \tag{4}$$

Separating the Z part,

$$\frac{1}{Z}\frac{d^2Z}{dz^2} = -(k^2+m^2) \tag{5}$$

$$\frac{1}{u^2+v^2}\left(\frac{1}{U}\frac{d^2U}{du^2}+\frac{1}{V}\frac{d^2V}{dv^2}\right)-k^2 = 0 \tag{6}$$

$$\frac{1}{U}\frac{d^2U}{du^2}+\frac{1}{V}\frac{d^2V}{dv^2}-k^2(u^2+v^2) = 0, \tag{7}$$

so

$$\frac{d^2Z}{dz^2} = -(k^2+m^2)Z, \tag{8}$$

which has solution

$$Z(z) = A\cos(\sqrt{k^2+m^2}\,z) + B\sin(\sqrt{k^2+m^2}\,z), \tag{9}$$

and

$$\left(\frac{1}{U}\frac{d^2U}{du^2} - k^2u^2\right) + \left(\frac{1}{V}\frac{d^2V}{dv^2} - k^2v^2\right) = 0. \quad (10)$$

This can be separated

$$\frac{1}{U}\frac{d^2U}{du^2} - k^2u^2 = c \quad (11)$$

$$\frac{1}{V}\frac{d^2V}{dv^2} - k^2v^2 = -c, \quad (12)$$

so

$$\frac{d^2U}{du^2} - (c + k^2u^2)U = 0 \quad (13)$$

$$\frac{d^2V}{dv^2} + (c - k^2v^2)V = 0. \quad (14)$$

These are the WEBER DIFFERENTIAL EQUATIONS, and the solutions are known as PARABOLIC CYLINDER FUNCTIONS.

see also PARABOLIC CYLINDER FUNCTION, PARABOLIC CYLINDRICAL COORDINATES, WEBER DIFFERENTIAL EQUATIONS

References

Morse, P. M. and Feshbach, H. *Methods of Theoretical Physics, Part I.* New York: McGraw-Hill, pp. 515 and 658, 1953.

Helmholtz Differential Equation—Polar Coordinates

In 2-D POLAR COORDINATES, attempt SEPARATION OF VARIABLES by writing

$$F(r, \theta) = R(r)\Theta(\theta), \quad (1)$$

then the HELMHOLTZ DIFFERENTIAL EQUATION becomes

$$\frac{d^2R}{dr^2}\Theta + \frac{1}{r}\frac{dR}{dr}\Theta + \frac{1}{r^2}\frac{d^2\Theta}{d\theta^2}R + k^2R\Theta = 0. \quad (2)$$

Divide both sides by $R\Theta$

$$\left(\frac{r^2}{R}\frac{d^2R}{dr^2} + \frac{r}{R}\frac{dR}{dr}\right) + \left(\frac{1}{\Theta}\frac{d^2\Theta}{d\theta^2} + k^2\right) = 0. \quad (3)$$

The solution to the second part of (3) must be periodic, so the differential equation is

$$\frac{d^2\Theta}{d\theta^2}\frac{1}{\Theta} = -(k^2 + m^2), \quad (4)$$

which has solutions

$$\begin{aligned}\Theta(\theta) &= c_1 e^{i\sqrt{k^2+m^2}\,\theta} + c_2 e^{-i\sqrt{k^2+m^2}\,\theta}\\ &= c_3 \sin(\sqrt{k^2+m^2}\,\theta) + c_4 \cos(\sqrt{k^2+m^2}\,\theta).\end{aligned} \quad (5)$$

Plug (4) back into (3)

$$r^2R'' + rR' - m^2R = 0. \quad (6)$$

This is an EULER DIFFERENTIAL EQUATION with $\alpha \equiv 1$ and $\beta \equiv -m^2$. The roots are $r = \pm m$. So for $m = 0$, $r = 0$ and the solution is

$$R(r) = c_1 + c_2 \ln r. \quad (7)$$

But since $\ln r$ blows up at $r = 0$, the only possible physical solution is $R(r) = c_1$. When $m > 0$, $r = \pm m$, so

$$R(r) = c_1 r^m + c_2 r^{-m}. \quad (8)$$

But since r^{-m} blows up at $r = 0$, the only possible physical solution is $R_m(r) = c_1 r^m$. The solution for R is then

$$R_m(r) = c_m r^m \quad (9)$$

for $m = 0, 1, \ldots$ and the general solution is

$$\begin{aligned}F(r, \theta) = \sum_{m=0}^{\infty}[a_m r^m \sin(\sqrt{k^2+m^2}\,\theta)\\ + b_m r^m \cos(\sqrt{k^2+m^2}\,\theta)].\end{aligned} \quad (10)$$

References

Morse, P. M. and Feshbach, H. *Methods of Theoretical Physics, Part I.* New York: McGraw-Hill, pp. 502–504, 1953.

Helmholtz Differential Equation—Prolate Spheroidal Coordinates

As shown by Morse and Feshbach (1953) and Arfken (1970), the HELMHOLTZ DIFFERENTIAL EQUATION is separable in PROLATE SPHEROIDAL COORDINATES.

References

Arfken, G. "Prolate Spheroidal Coordinates (u, v, φ)." §2.10 in *Mathematical Methods for Physicists, 2nd ed.* Orlando, FL: Academic Press, pp. 103–107, 1970.

Morse, P. M. and Feshbach, H. *Methods of Theoretical Physics, Part I.* New York: McGraw-Hill, p. 661, 1953.

Helmholtz Differential Equation—Spherical Coordinates

In SPHERICAL COORDINATES, the SCALE FACTORS are $h_r = 1$, $h_\theta = r\sin\phi$, $h_\phi = r$, and the separation functions are $f_1(r) = r^2$, $f_2(\theta) = 1$, $f_3(\phi) = \sin\phi$, giving a STÄCKEL DETERMINANT of $S = 1$. The LAPLACIAN is

$$\begin{aligned}\nabla^2 \equiv \frac{1}{r^2}\frac{\partial}{\partial r}\left(r^2\frac{\partial}{\partial r}\right) + \frac{1}{r^2\sin^2\phi}\frac{\partial^2}{\partial\theta^2}\\ + \frac{1}{r^2\sin\phi}\frac{\partial}{\partial\phi}\left(\sin\phi\frac{\partial}{\partial\phi}\right).\end{aligned} \quad (1)$$

To solve the HELMHOLTZ DIFFERENTIAL EQUATION in SPHERICAL COORDINATES, attempt SEPARATION OF VARIABLES by writing

$$F(r, \theta, \phi) = R(r)\Theta(\theta)\Phi(\phi). \quad (2)$$

Then the HELMHOLTZ DIFFERENTIAL EQUATION becomes

$$\frac{d^2R}{dr^2}\Phi\Theta + \frac{2}{r}\frac{dR}{dr}\Phi\Theta + \frac{1}{r^2\sin^2\phi}\frac{d^2\Theta}{d\theta^2}\Phi R + \frac{\cos\phi}{r^2\sin\phi}\frac{d\Phi}{d\phi}\Theta R + \frac{1}{r^2}\frac{d^2\Phi}{d\phi^2}\Theta R = 0. \quad (3)$$

Now divide by $R\Theta\Phi$,

$$\frac{r^2\sin^2\phi}{\Phi R\Theta}\Phi\Theta\frac{d^2R}{dr^2} + \frac{2}{r}\frac{r^2\sin^2\phi}{\Phi R\Theta}\Phi\Theta\frac{dR}{dr} + \frac{1}{r^2\sin^2\phi}\frac{r^2\sin^2\phi}{\Phi R\Theta}\Phi R\frac{d^2\Theta}{d\theta^2} + \frac{\cos\phi}{r^2\sin\phi}\frac{r^2\sin^2\phi}{\Phi\Theta R}\frac{d\Phi}{d\phi}\Theta R + \frac{1}{r^2}\frac{r^2\sin^2\phi}{\Phi R\Theta}\frac{d^2\Phi}{d\phi^2}\Theta R = 0 \quad (4)$$

$$\left(\frac{r^2\sin^2\phi}{R}\frac{d^2R}{dr^2} + \frac{2r\sin^2\phi}{R}\frac{dR}{dr}\right) + \left(\frac{1}{\Theta}\frac{d^2\Theta}{d\theta^2}\right) + \left(\frac{\cos\phi\sin\phi}{\Phi}\frac{d\Phi}{d\phi} + \frac{\sin^2\phi}{\Phi}\frac{d^2\Phi}{d\phi^2}\right) = 0. \quad (5)$$

The solution to the second part of (5) must be sinusoidal, so the differential equation is

$$\frac{d^2\Theta}{d\theta^2}\frac{1}{\Theta} = -m^2, \quad (6)$$

which has solutions which may be defined either as a COMPLEX function with $m = -\infty, \ldots, \infty$

$$\Theta(\theta) = A_m e^{im\theta}, \quad (7)$$

or as a sum of REAL sine and cosine functions with $m = -\infty, \ldots, \infty$

$$\Theta(\theta) = S_m \sin(m\theta) + C_m \cos(m\theta). \quad (8)$$

Plugging (6) back into (7),

$$\frac{r^2}{R}\frac{d^2R}{dr^2} + \frac{2r}{R}\frac{dR}{dr} - \frac{1}{\sin^2\phi}\left(m^2 + \frac{\cos\phi\sin\phi}{\Phi}\right)\frac{d\Phi}{d\phi} + \frac{\sin^2\phi}{\Phi}\frac{d^2\Phi}{d\phi^2} = 0. \quad (9)$$

The radial part must be equal to a constant

$$\frac{r^2}{R}\frac{d^2R}{dr^2} + \frac{2r}{R}\frac{dR}{dr} = l(l+1) \quad (10)$$

$$r^2\frac{d^2R}{dr^2} + 2r\frac{dR}{dr} = l(l+1)R. \quad (11)$$

But this is the EULER DIFFERENTIAL EQUATION, so we try a series solution of the form

$$R = \sum_{n=0}^{\infty} a_n r^{n+c}. \quad (12)$$

Then

$$r^2\sum_{n=0}^{\infty}(n+c)(n+c-1)a_n r^{n+c-2} + 2r\sum_{n=0}^{\infty}(n+c)a_n r^{n+c-1} - l(l+1)\sum_{n=0}^{\infty}a_n r^{n+c} = 0 \quad (13)$$

$$\sum_{n=0}^{\infty}(n+c)(n+c-1)a_n r^{n+c} + 2\sum_{n=0}^{\infty}(n+c)a_n r^{n+c} - l(l+1)\sum_{n=0}^{\infty}a_n r^{n+c} = 0 \quad (14)$$

$$\sum_{n=0}^{\infty}[(n+c)(n+c+1) - l(l+1)]a_n r^{n+c} = 0. \quad (15)$$

This must hold true for all POWERS of r. For the r^c term (with $n = 0$),

$$c(c+1) = l(l+1), \quad (16)$$

which is true only if $c = l, -l-1$ and all other terms vanish. So $a_n = 0$ for $n \neq l,\ -l-1$. Therefore, the solution of the R component is given by

$$R_l(r) = A_l r^l + B_l r^{-l-1}. \quad (17)$$

Plugging (17) back into (9),

$$l(l+1) - \frac{m^2}{\sin^2\phi} + \frac{\cos\phi}{\sin\phi}\frac{1}{\Phi}\frac{d\Phi}{d\phi} + \frac{1}{\Phi}\frac{d^2\Phi}{d\phi^2} = 0 \quad (18)$$

$$\Phi'' + \frac{\cos\phi}{\sin\phi}\Phi' + \left[l(l+1) - \frac{m^2}{\sin^2\phi}\right]\Phi = 0, \quad (19)$$

which is the associated LEGENDRE DIFFERENTIAL EQUATION for $x = \cos\phi$ and $m = 0, \ldots, l$. The general COMPLEX solution is therefore

$$\sum_{l=0}^{\infty}\sum_{m=-l}^{l}(A_l r^l + B_l r^{-l-1})P_l^m(\cos\phi)e^{-im\theta} \equiv \sum_{l=0}^{\infty}\sum_{m=-1}^{l}(A_l r^l + B_l r^{-l-1})Y_l^m(\theta,\phi), \quad (20)$$

where

$$Y_l^m(\theta,\phi) \equiv P_l^m(\cos\phi)e^{-im\theta} \quad (21)$$

are the (COMPLEX) SPHERICAL HARMONICS. The general REAL solution is

$$\sum_{l=0}^{\infty}\sum_{m=0}^{l}(A_l r^l + B_l r^{-l-1})P_l^m(\cos\phi) \times [S_m\sin(m\theta) + C_m\cos(m\theta)]. \quad (22)$$

Some of the normalization constants of P_l^m can be absorbed by S_m and C_m, so this equation may appear in the form

$$\sum_{l=0}^{\infty}\sum_{m=0}^{l}(A_l r^l + B_l r^{-l-1})P_l^m(\cos\phi) \times [S_l^m \sin(m\theta) + C_l^m \cos(m\theta)]$$

$$\equiv \sum_{l=0}^{\infty}\sum_{m=0}^{l}(A_l r^l + B_l r^{-l-1}) \times [S_l^m Y_l^{m(o)}(\theta,\phi) + C_l^m Y_l^{m(e)}(\theta,\phi)], \quad (23)$$

where

$$Y_l^{m(o)}(\theta,\phi) \equiv P_l^m(\cos\theta)\sin(m\theta) \quad (24)$$

$$Y_l^{m(e)}(\theta,\phi) \equiv P_l^m(\cos\theta)\cos(m\theta) \quad (25)$$

are the EVEN and ODD (real) SPHERICAL HARMONICS. If azimuthal symmetry is present, then $\Theta(\theta)$ is constant and the solution of the Φ component is a LEGENDRE POLYNOMIAL $P_l(\cos\phi)$. The general solution is then

$$F(r,\phi) = \sum_{l=0}^{\infty}(A_l r^l + B_l r^{-l-1})P_l(\cos\phi). \quad (26)$$

Actually, the equation is separable under the more general condition that k^2 is of the form

$$k^2(r,\theta,\phi) = f(r) + \frac{g(\theta)}{r^2} + \frac{h(\phi)}{r^2\sin\theta} + k'^2. \quad (27)$$

References

Morse, P. M. and Feshbach, H. *Methods of Theoretical Physics, Part I.* New York: McGraw-Hill, p. 514 and 658, 1953.

Helmholtz Differential Equation—Spherical Surface

On the surface of a SPHERE, attempt SEPARATION OF VARIABLES in SPHERICAL COORDINATES by writing

$$F(\theta,\phi) = \Theta(\theta)\Phi(\phi), \quad (1)$$

then the HELMHOLTZ DIFFERENTIAL EQUATION becomes

$$\frac{1}{\sin^2\phi}\frac{d^2\Theta}{d\theta^2}\Phi + \frac{\cos\phi}{\sin\phi}\frac{d\Phi}{d\phi}\Theta + \frac{d^2\Phi}{d\phi^2}\Theta + k^2\Theta\Phi = 0. \quad (2)$$

Dividing both sides by $\Phi\Theta$,

$$\left(\frac{\cos\phi\sin\phi}{\Phi}\frac{d\Phi}{d\phi} + \frac{\sin^2\phi}{\Phi}\frac{d^2\Phi}{d\phi^2}\right) + \left(\frac{1}{\Theta}\frac{d^2\Theta}{d\theta^2} + k^2\right) = 0, \quad (3)$$

which can now be separated by writing

$$\frac{d^2\Theta}{d\theta^2}\frac{1}{\Theta} = -(k^2 + m^2). \quad (4)$$

The solution to this equation must be periodic, so m must be an INTEGER. The solution may then be defined either as a COMPLEX function

$$\Theta(\theta) = A_m e^{i\sqrt{k^2+m^2}\,\theta} + B_m e^{-i\sqrt{k^2+m^2}\,\theta} \quad (5)$$

for $m = -\infty, \ldots, \infty$, or as a sum of REAL sine and cosine functions

$$\Theta(\theta) = S_m \sin(\sqrt{k^2+m^2}\,\theta) + C_m \cos(\sqrt{k^2+m^2}\,\theta) \quad (6)$$

for $m = 0, \ldots, \infty$. Plugging (4) into (3) gives

$$\frac{\cos\phi\sin\phi}{\Phi}\frac{d\Phi}{d\phi} + \frac{\sin^2\phi}{\Phi}\frac{d^2\Phi}{d\phi^2} + m^2 = 0 \quad (7)$$

$$\Phi'' + \frac{\cos\phi}{\sin\phi}\Phi' + \frac{m^2}{\sin^2\phi}\Phi = 0, \quad (8)$$

which is the LEGENDRE DIFFERENTIAL EQUATION for $x = \cos\phi$ with

$$m^2 \equiv l(l+1), \quad (9)$$

giving

$$l^2 + l - m^2 = 0 \quad (10)$$

$$l = \tfrac{1}{2}(-1 \pm \sqrt{1+4m^2}\,). \quad (11)$$

Solutions are therefore LEGENDRE POLYNOMIALS with a COMPLEX index. The general COMPLEX solution is then

$$F(\theta,\phi) = \sum_{m=-\infty}^{\infty} P_l(\cos\phi)(A_m e^{im\theta} + B_m e^{-im\theta}), \quad (12)$$

and the general REAL solution is

$$F(\theta,\phi) = \sum_{m=0}^{\infty} P_l(\cos\phi)[S_m\sin(m\theta) + C_m\cos(m\theta)]. \quad (13)$$

Note that these solutions depend on only a single variable m. However, on the surface of a sphere, it is usual to express solutions in terms of the SPHERICAL HARMONICS derived for the 3-D spherical case, which depend on the two variables l and m.

Helmholtz Differential Equation—Toroidal Coordinates

The HELMHOLTZ DIFFERENTIAL EQUATION is not separable.

see LAPLACE'S EQUATION—TOROIDAL COORDINATES

Helmholtz's Theorem

Any VECTOR FIELD **v** satisfying

$$[\nabla \cdot \mathbf{v}]_\infty = 0 \quad (1)$$

$$[\nabla \times \mathbf{v}]_\infty = 0 \quad (2)$$

may be written as the sum of an IRROTATIONAL part and a SOLENOIDAL part,

$$\mathbf{v} = -\nabla\phi + \nabla \times \mathbf{A}, \quad (3)$$

where for a VECTOR FIELD F,

$$\phi = -\int_V \frac{\nabla \cdot \mathbf{F}}{4\pi|\mathbf{r}' - \mathbf{r}|} d^3\mathbf{r}' \quad (4)$$

$$\mathbf{A} = \int_V \frac{\nabla \times \mathbf{F}}{4\pi|\mathbf{r}' - \mathbf{r}|} d^3\mathbf{r}'. \quad (5)$$

see also IRROTATIONAL FIELD, SOLENOIDAL FIELD, VECTOR FIELD

References

Arfken, G. "Helmholtz's Theorem." §1.15 in *Mathematical Methods for Physicists, 3rd ed.* Orlando, FL: Academic Press, pp. 78–84, 1985.

Gradshteyn, I. S. and Ryzhik, I. M. *Tables of Integrals, Series, and Products, 5th ed.* San Diego, CA: Academic Press, pp. 1084, 1980.

Helson-Szegő Measure

An absolutely continuous measure on ∂D whose density has the form $\exp(x + \bar{y})$, where x and y are real-valued functions in L^∞, $||y||_\infty < \pi/2$, exp is the EXPONENTIAL FUNCTION, and $||y||$ is the NORM.

Hemicylindrical Function

A function $S_n(z)$ which satisfies the RECURRENCE RELATION

$$S_{n-1}(z) - S_{n+1}(z) = 2S_n'(z)$$

together with

$$S_1(z) = -S_0'(z)$$

is called a hemicylindrical function.

References

Sonine, N. "Recherches sur les fonctions cylindriques et le développement des fonctions continues en séries." *Math. Ann.* **16**, 1–9 and 71–80, 1880.

Watson, G. N. "Hemi-Cylindrical Functions." §10.8 in *A Treatise on the Theory of Bessel Functions, 2nd ed.* Cambridge, England: Cambridge University Press, p. 353, 1966.

Hemisphere

Half of a SPHERE cut by a PLANE passing through its CENTER. A hemisphere of RADIUS r can be given by the usual SPHERICAL COORDINATES

$$x = r\cos\theta\sin\phi \quad (1)$$

$$y = r\sin\theta\sin\phi \quad (2)$$

$$z = r\cos\phi, \quad (3)$$

where $\theta \in [0, 2\pi)$ and $\phi \in [0, \pi/2]$. All CROSS-SECTIONS passing through the z-axis are SEMICIRCLES.

The VOLUME of the hemisphere is

$$V = \pi \int_0^r (r^2 - z^2)\,dz = \tfrac{2}{3}\pi r^3. \quad (4)$$

The weighted mean of z over the hemisphere is

$$\langle z \rangle = \pi \int_0^r z(r^2 - z^2)\,dz = \tfrac{1}{4}\pi r^2. \quad (5)$$

The CENTROID is then given by

$$\bar{z} = \frac{\langle z \rangle}{V} = \tfrac{3}{8}r \quad (6)$$

(Beyer 1987).

see also SEMICIRCLE, SPHERE

References

Beyer, W. H. (Ed.) *CRC Standard Mathematical Tables, 28th ed.* Boca Raton, FL: CRC Press, p. 133, 1987.

Hemispherical Function

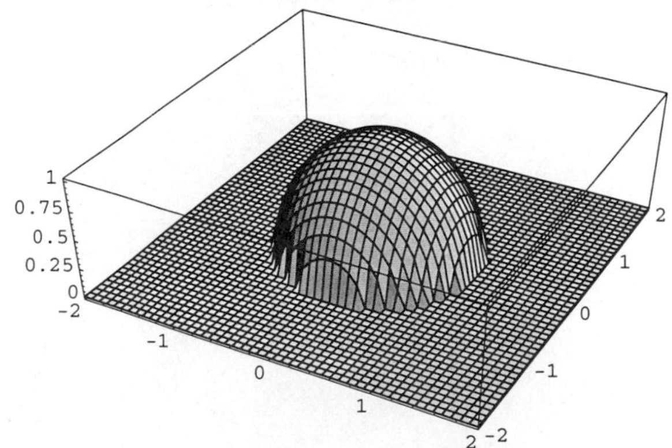

The hemisphere function is defined as

$$H(x,y) = \begin{cases} \sqrt{a - x^2 - y^2} & \text{for } \sqrt{x^2+y^2} \leq a \\ 0 & \text{for } \sqrt{x^2+y^2} > a. \end{cases}$$

Watson (1966) defines a hemispherical function as a function S which satisfies the RECURRENCE RELATIONS

$$S_{n-1}(z) - S_{n+1}(z) = 2S_n'(z)$$

with

$$S_1(z) = -S_0'(z).$$

see also CYLINDER FUNCTION, CYLINDRICAL FUNCTION

References

Watson, G. N. *A Treatise on the Theory of Bessel Functions, 2nd ed.* Cambridge, England: Cambridge University Press, p. 353, 1966.

Hempel's Paradox

A purple cow is a confirming instance of the hypothesis that all crows are black.

References

Carnap, R. *Logical Foundations of Probability.* Chicago, IL: University of Chicago Press, pp. 224 and 469, 1950.

Gardner, M. *The Scientific American Book of Mathematical Puzzles & Diversions.* New York: Simon and Schuster, pp. 52–54, 1959.

Goodman, N. Ch. 3 in *Fact, Fiction, and Forecast.* Cambridge, MA: Harvard University Press, 1955.

Hempel, C. G. "A Purely Syntactical Definition of Confirmation." *J. Symb. Logic* **8**, 122–143, 1943.

Hempel, C. G. "Studies in Logic and Confirmation." *Mind* **54**, 1–26, 1945.

Hempel, C. G. "Studies in Logic and Confirmation. II." *Mind* **54**, 97–121, 1945.

Hempel, C. G. "A Note on the Paradoxes of Confirmation." *Mind* **55**, 1946.

Hosiasson-Lindenbaum, J. "On Confirmation." *J. Symb. Logic* **5**, 133–148, 1940.

Whiteley, C. H. "Hempel's Paradoxes of Confirmation." *Mind* **55**, 156–158, 1945.

Hendecagon

see UNDECAGON

Henneberg's Minimal Surface

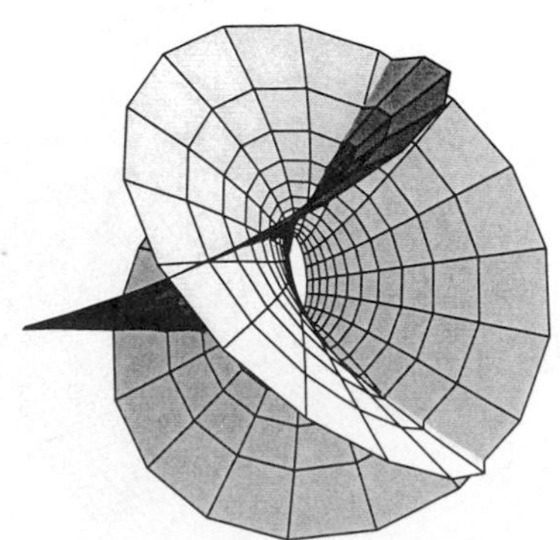

A double algebraic surface of 15th order and fifth class which can be given by parametric equations

$$x(u,v) = 2\sinh u \cos v - \tfrac{2}{3}\sinh(3u)\cos(3v) \quad (1)$$

$$y(u,v) = 2\sinh u \sin v - \tfrac{2}{3}\sinh(3u)\sin(3v) \quad (2)$$

$$z(u,v) = 2\cosh(2u)\cos(2v). \quad (3)$$

It can also be obtained from the ENNEPER-WEIERSTRAß PARAMETERIZATION with

$$f = 2 - 2z^{-4} \quad (4)$$

$$g = z. \quad (5)$$

see also MINIMAL SURFACE

References

Eisenhart, L. P. *A Treatise on the Differential Geometry of Curves and Surfaces.* New York: Dover, p. 267, 1960.

Gray, A. *Modern Differential Geometry of Curves and Surfaces.* Boca Raton, FL: CRC Press, pp. 446–448, 1993.

Nitsche, J. C. C. *Introduction to Minimal Surfaces.* Cambridge, England: Cambridge University Press, p. 144, 1989.

Hénon Attractor

see HÉNON MAP

Hénon-Heiles Equation

A nonlinear nonintegrable HAMILTONIAN SYSTEM with

$$\ddot{x} = -\frac{\partial V}{\partial x} \quad (1)$$

$$\ddot{y} = -\frac{\partial V}{\partial y}, \quad (2)$$

where

$$V(x,y) = \tfrac{1}{2}(x^2 + y^2 + 2x^2y - \tfrac{2}{3}y^3) \quad (3)$$

$$V(r,\theta) = \tfrac{1}{2}r^2 + \frac{1}{3}r^3\sin(3\theta). \quad (4)$$

The energy is

$$E = V(x,y) + \tfrac{1}{2}(\dot{x}^2 + \dot{y}^2). \quad (5)$$

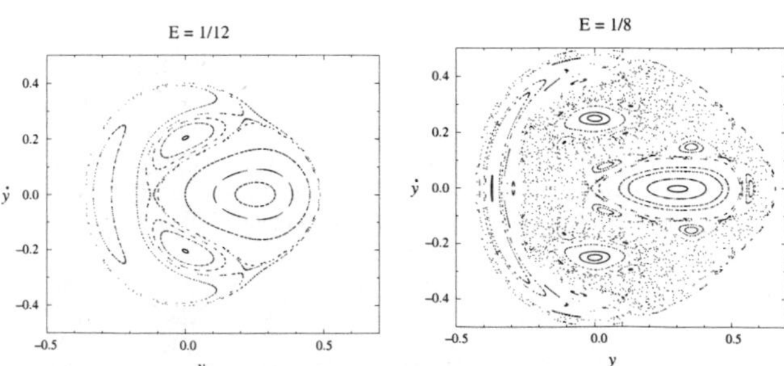

The above plots are SURFACES OF SECTION for $E = 1/12$ and $E = 1/8$. The Hamiltonian for a generalized Hénon-Heiles potential is

$$H = \tfrac{1}{2}(p_x^2 + p_y^2 + Ax^2 + By^2) + Dx^2y - \tfrac{1}{3}Cy^3. \quad (6)$$

The equations of motion are integrable only for

1. $D/C = 0$,
2. $D/C = -1, A/B = 1$,
3. $D/C = -1/6$, and
4. $D/C = -1/16, A/B = 1/6$.

References

Gleick, J. *Chaos: Making a New Science.* New York: Penguin Books, pp. 144–153, 1988.

Hénon, M. and Heiles, C. "The Applicability of the Third Integral of Motion: Some Numerical Experiments." *Astron. J.* **69**, 73–79, 1964.

Hénon Map

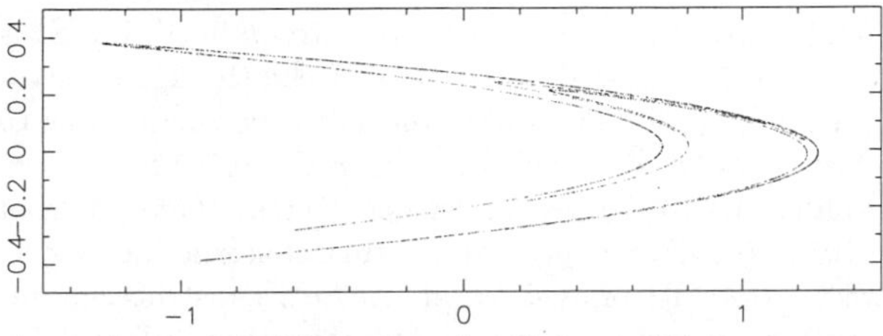

A quadratic 2-D MAP given by the equations

$$x_{n+1} = 1 - \alpha {x_n}^2 + y_n \quad (1)$$
$$y_{n+1} = \beta x_n \quad (2)$$

or

$$x_{n+1} = x_n \cos\alpha - (y_n - {x_n}^2)\sin\alpha \quad (3)$$
$$y_{n+1} = x_n \sin\alpha + (y_n - {x_n}^2)\cos\alpha. \quad (4)$$

The above map is for $\alpha = 1.4$ and $\beta = 0.3$. The Hénon map has CORRELATION EXPONENT 1.25 ± 0.02 (Grassberger and Procaccia 1983) and CAPACITY DIMENSION 1.261 ± 0.003 (Russell *et al.* 1980). Hitzl and Zele (1985) give conditions for the existence of periods 1 to 6.

see also BOGDANOV MAP, LOZI MAP, QUADRATIC MAP

References

Dickau, R. M. "The Hénon Attractor." `http://forum.swarthmore.edu/advanced/robertd/henon.html`.

Gleick, J. *Chaos: Making a New Science.* New York: Penguin Books, pp. 144–153, 1988.

Grassberger, P. and Procaccia, I. "Measuring the Strangeness of Strange Attractors." *Physica D* **9**, 189–208, 1983.

Hitzl, D. H. and Zele, F. "An Exploration of the Hénon Quadratic Map." *Physica D* **14**, 305–326, 1985.

Lauwerier, H. *Fractals: Endlessly Repeated Geometric Figures.* Princeton, NJ: Princeton University Press, pp. 128–133, 1991.

Peitgen, H.-O. and Saupe, D. (Eds.). "A Chaotic Set in the Plane." §3.2.2 in *The Science of Fractal Images.* New York: Springer-Verlag, pp. 146–148, 1988.

Russell, D. A.; Hanson, J. D.; and Ott, E. "Dimension of Strange Attractors." *Phys. Rev. Let.* **45**, 1175–1178, 1980.

Hensel's Lemma

An important result in VALUATION THEORY which gives information on finding roots of POLYNOMIALS. Hensel's lemma is formally stated as follow. Let $(K, |\cdot|)$ be a complete non-Archimedean valuated field, and let R be the corresponding VALUATION RING. Let $f(x)$ be a POLYNOMIAL whose COEFFICIENTS are in R and suppose a_0 satisfies

$$|f(a_0)| < |f'(a_0)|^2, \quad (1)$$

where f' is the (formal) DERIVATIVE of f. Then there exists a unique element $a \in R$ such that $f(a) = 0$ and

$$|a - a_0| \leq \left|\frac{f(a_0)}{f'(a_0)}\right|. \quad (2)$$

Less formally, if $f(x)$ is a POLYNOMIAL with "INTEGER" COEFFICIENTS and $f(a_0)$ is "small" compared to $f'(a_0)$, then the equation $f(x) = 0$ has a solution "near" a_0. In addition, there are no other solutions near a_0, although there may be other solutions. The proof of the LEMMA is based around the Newton-Raphson method and relies on the non-Archimedean nature of the valuation.

Consider the following example in which Hensel's lemma is used to determine that the equation $x^2 = -1$ is solvable in the 5-adic numbers $\mathbb{Q}_5$ (and so we can embed the GAUSSIAN INTEGERS inside $\mathbb{Q}_5$ in a nice way). Let K be the 5-adic numbers $\mathbb{Q}_5$, let $f(x) = x^2 + 1$, and let $a_0 = 2$. Then we have $f(2) = 5$ and $f'(2) = 4$, so

$$|f(2)|_5 = \tfrac{1}{5} < |f'(2)|_5^2 = 1, \quad (3)$$

and the condition is satisfied. Hensel's lemma then tells us that there is a 5-adic number a such that $a^2 + 1 = 0$ and

$$|a - 2|_5 <= |\tfrac{5}{4}|_5 = \tfrac{1}{5}. \quad (4)$$

Similarly, there is a 5-adic number b such that $b^2 + 1 = 0$ and

$$|b - 3|_5 <= |\tfrac{10}{7}|_5 = \tfrac{1}{5}. \quad (5)$$

Therefore, we have found both the square roots of -1 in $\mathbb{Q}_5$. It is possible to find the roots of any POLYNOMIAL using this technique.

Henstock-Kurzweil Integral

see HK INTEGRAL

Heptacontagon

A 70-sided POLYGON.

Heptadecagon

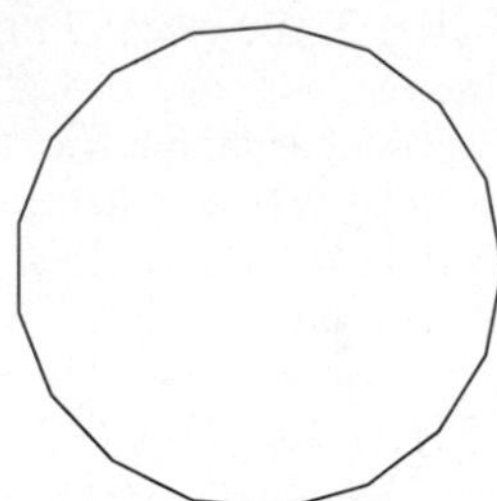

The REGULAR POLYGON of 17 sides is called the HEPTADECAGON, or sometimes the HEPTAKAIDECAGON. Gauss proved in 1796 (when he was 19 years old) that the heptadecagon is CONSTRUCTIBLE with a COMPASS and STRAIGHTEDGE. Gauss's proof appears in his monumental work *Disquisitiones Arithmeticae.* The proof relies on the property of irreducible POLYNOMIAL equations that ROOTS composed of a finite number of SQUARE ROOT extractions only exist when the order of the equation is a product of the form $2^a 3^b F_c \cdot F_d \cdots F_e$, where the F_n are distinct PRIMES of the form

$$F_n = 2^{2^n} + 1,$$

known as FERMAT PRIMES. Constructions for the regular TRIANGLE (3^1), SQUARE (2^2), PENTAGON ($2^{2^1} + 1$), HEXAGON ($2^1 3^1$), etc., had been given by Euclid, but constructions based on the FERMAT PRIMES ≥ 17 were unknown to the ancients. The first explicit construction of a heptadecagon was given by Erchinger in about 1800.

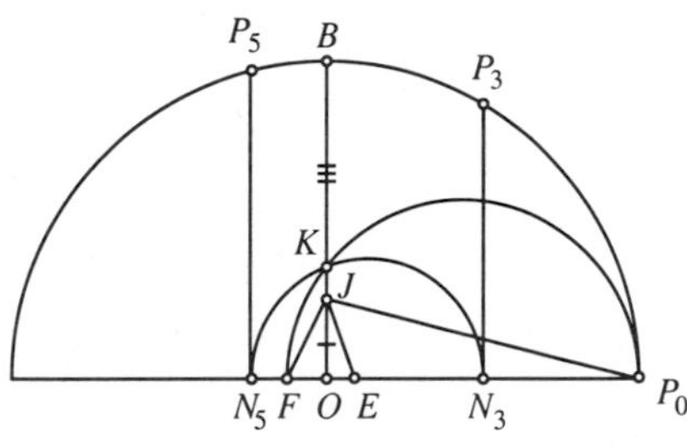

17-gon

The following elegant construction for the heptadecagon (Yates 1949, Coxeter 1969, Stewart 1977, Wells 1992) was first given by Richmond (1893).

1. Given an arbitrary point O, draw a CIRCLE centered on O and a DIAMETER drawn through O.
2. Call the right end of the DIAMETER dividing the CIRCLE into a SEMICIRCLE P_0.
3. Construct the DIAMETER PERPENDICULAR to the original DIAMETER by finding the PERPENDICULAR BISECTOR OB.
4. Find J a QUARTER the way up OB.
5. Join JP_0 and find E so that $\angle OJE$ is a QUARTER of $\angle OJP_0$.
6. Find F so that $\angle EJF$ is 45°.
7. Construct the SEMICIRCLE with DIAMETER FP_0.
8. This SEMICIRCLE cuts OB at K.
9. Draw a SEMICIRCLE with center E and RADIUS EK.
10. This cuts the extension of OP_0 at N_3.
11. Construct a line PERPENDICULAR to OP_0 through N_3.
12. This line meets the original SEMICIRCLE at P_3.
13. You now have points P_0 and P_3 of a heptadecagon.
14. Use P_0 and P_3 to get the remaining 15 points of the heptadecagon around the original CIRCLE by constructing P_0, P_3, P_6, P_9, P_{12}, P_{15}, P_1, P_4, P_7, P_{10}, P_{13}, P_{16}, P_2, P_5, P_8, P_{11}, and P_{14}.
15. Connect the adjacent points P_i.

This construction, when suitably streamlined, has SIMPLICITY 53. The construction of Smith (1920) has a greater SIMPLICITY of 58. Another construction due to Tietze (1965) and reproduced in Hall (1970) has a SIMPLICITY of 50. However, neither Tietze (1965) nor Hall (1970) provides a proof that this construction is correct. Both Richmond's and Tietze's constructions require extensive calculations to prove their validity. De Temple (1991) gives an elegant construction involving the CARLYLE CIRCLES which has GEOMETROGRAPHY symbol $8S_1 + 4S_2 + 22C_1 + 11C_3$ and SIMPLICITY 45. The construction problem has now been automated to some extent (Bishop 1978).

see also 257-GON, 65537-GON, COMPASS, CONSTRUCTIBLE POLYGON, FERMAT NUMBER, FERMAT PRIME, REGULAR POLYGON, STRAIGHTEDGE, TRIGONOMETRY VALUES—$\pi/17$

References

Archibald, R. C. "The History of the Construction of the Regular Polygon of Seventeen Sides." *Bull. Amer. Math. Soc.* **22**, 239–246, 1916.

Archibald, R. C. "Gauss and the Regular Polygon of Seventeen Sides." *Amer. Math. Monthly* **27**, 323–326, 1920.

Ball, W. W. R. and Coxeter, H. S. M. *Mathematical Recreations and Essays, 13th ed.* New York: Dover, pp. 95–96, 1987.

Bishop, W. "How to Construct a Regular Polygon." *Amer. Math. Monthly* **85**, 186–188, 1978.

Conway, J. H. and Guy, R. K. *The Book of Numbers.* New York: Springer-Verlag, pp. 201 and 229–230, 1996.

Coxeter, H. S. M. *Introduction to Geometry, 2nd ed.* New York: Wiley, pp. 26–28, 1969.

De Temple, D. W. "Carlyle Circles and the Lemoine Simplicity of Polygonal Constructions." *Amer. Math. Monthly* **98**, 97–108, 1991.

Dixon, R. "Gauss Extends Euclid." §1.4 in *Mathographics.* New York: Dover, pp. 52–54, 1991.

Gauss, C. F. §365 and 366 in *Disquisitiones Arithmeticae.* Leipzig, Germany, 1801. New Haven, CT: Yale University Press, 1965.

Hall, T. *Carl Friedrich Gauss: A Biography.* Cambridge, MA: MIT Press, 1970.

Klein, F. *Famous Problems of Elementary Geometry and Other Monographs.* New York: Chelsea, 1956.

Ore, Ø. *Number Theory and Its History.* New York: Dover, 1988.

Rademacher, H. *Lectures on Elementary Number Theory.* New York: Blaisdell, 1964.

Richmond, H. W. "A Construction for a Regular Polygon of Seventeen Sides." *Quart. J. Pure Appl. Math.* **26**, 206–207, 1893.
Smith, L. L. "A Construction of the Regular Polygon of Seventeen Sides." *Amer. Math. Monthly* **27**, 322–323, 1920.
Stewart, I. "Gauss." *Sci. Amer.* **237**, 122–131, 1977.
Tietze, H. *Famous Problems of Mathematics.* New York: Graylock Press, 1965.
Wells, D. *The Penguin Dictionary of Curious and Interesting Geometry.* New York: Viking Penguin, 1992.
Yates, R. C. *Geometrical Tools.* St. Louis, MO: Educational Publishers, 1949.

Heptagon

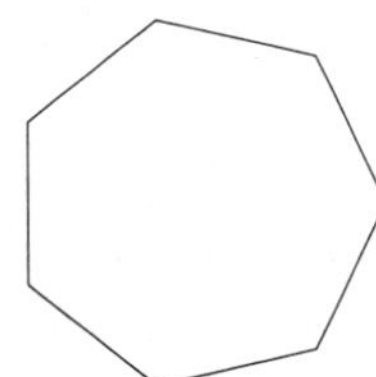

The unconstructible regular seven-sided POLYGON, illustrated above, has SCHLÄFLI SYMBOL $\{7\}$.

Although the regular heptagon is not a CONSTRUCTIBLE POLYGON, Dixon (1991) gives several close approximations. While the ANGLE subtended by a side is $360°/7 \approx 51.428571°$, Dixon gives constructions containing angles of $2\sin^{-1}(\sqrt{3}/4) \approx 51.317812°$, $\tan^{-1}(5/4) \approx 51.340191°$, and $30° + \sin^{-1}((\sqrt{3}-1)/2) \approx 51.470701°$.

Madachy (1979) illustrates how to construct a heptagon by folding and knotting a strip of paper.

see also EDMONDS' MAP, TRIGONOMETRY VALUES—$\pi/7$

References

Courant, R. and Robbins, H. "The Regular Heptagon." §3.3.4 in *What is Mathematics?: An Elementary Approach to Ideas and Methods, 2nd ed.* Oxford, England: Oxford University Press, pp. 138–139, 1996.
Dixon, R. *Mathographics.* New York: Dover, pp. 35–40, 1991.
Madachy, J. S. *Madachy's Mathematical Recreations.* New York: Dover, pp. 59–61, 1979.

Heptagonal Number

A FIGURATE NUMBER of the form $n(5n-3)/2$. The first few are 1, 7, 18, 34, 55, 81, 112, ... (Sloane's A000566). The GENERATING FUNCTION for the heptagonal numbers is

$$\frac{x(4x+1)}{(1-x)^3} = x + 7x^2 + 18x^3 + 34x^4 + \ldots.$$

References

Sloane, N. J. A. Sequence A000566/M4358 in "An On-Line Version of the Encyclopedia of Integer Sequences."

Heptagonal Pyramidal Number

A PYRAMIDAL NUMBER of the form $n(n+1)(5n-2)/6$. The first few are 1, 8, 26, 60, 115, ... (Sloane's A002413). The GENERATING FUNCTION for the heptagonal pyramidal numbers is

$$\frac{x(4x+1)}{(x-1)^4} = x + 8x^2 + 26x^3 + 60x^4 + \ldots.$$

References

Sloane, N. J. A. Sequence A002413/M4498 in "An On-Line Version of the Encyclopedia of Integer Sequences."

Heptahedron

The regular heptahedron is a one-sided surface made from four TRIANGLES and three QUADRILATERALS. It is topologically equivalent to the ROMAN SURFACE (Wells 1991). While all of the faces are regular and vertices equivalent, the heptahedron is self-intersecting and is therefore not considered an ARCHIMEDEAN SOLID. There are three semiregular heptahedra: the pentagonal and pentagrammic PRISMS, and a FACETED OCTAHEDRON (Holden 1991).

References

Holden, A. *Shapes, Space, and Symmetry.* New York: Dover, p. 95, 1991.
Wells, D. *The Penguin Dictionary of Curious and Interesting Geometry.* New York: Viking Penguin, p. 98, 1992.

Heptakaidecagon

see HEPTADECAGON

Heptaparallelohedron

see CUBOCTAHEDRON

Heptomino

The heptominoes are the 7-POLYOMINOES. There are 108 different heptominoes.

see also HERSCHEL, PI HEPTOMINO, POLYOMINO

Herbrand's Theorem

Let an ideal class be in $\mathcal{A}$ if it contains an IDEAL whose lth power is PRINCIPAL. Let i be an ODD INTEGER $1 \leq i \leq l$ and define j by $i + j = 1$. Then $\mathcal{A}_1 = \langle e \rangle$. If $i \geq 3$ and $l \nmid B_j$, then $\mathcal{A}_i = \langle e \rangle$.

References

Ireland, K. and Rosen, M. "Herbrand's Theorem." §15.3 in *A Classical Introduction to Modern Number Theory, 2nd ed.* New York: Springer-Verlag, pp. 241–248, 1990.

Hereditary Representation

The representation of a number as a sum of powers of a BASE b, followed by expression of each of the exponents as a sum of powers of b, etc., until the process stops. For example, the hereditary representation of 266 in base 2 is

$$\begin{aligned}266 &= 2^8 + 2^3 + 2 \\ &= 2^{2^{2+1}} + 2^{2+1} + 2.\end{aligned}$$

see also GOODSTEIN SEQUENCE

Heredity

A property of a SPACE which is also true of each of its SUBSPACES. Being "COUNTABLE" is hereditary, but having a given GENUS is not.

Hermann's Formula

The MACHIN-LIKE FORMULA

$$\tfrac{1}{4}\pi = 2\tan^{-1}(\tfrac{1}{2}) - \tan^{-1}(\tfrac{1}{7}).$$

The other 2-term MACHIN-LIKE FORMULAS are EULER'S MACHIN-LIKE FORMULA, HUTTON'S FORMULA, and MACHIN'S FORMULA.

Hermann Grid Illusion

A regular 2-D arrangement of squares separated by vertical and horizontal "canals." Looking at the grid produces the illusion of gray spots in the white AREA between square VERTICES. The illusion was noted by Hermann (1870) while reading a book on sound by J. Tyndall.

References
Fineman, M. *The Nature of Visual Illusion.* New York: Dover, pp. 139–140, 1996.

Hermann-Hering Illusion

The illusion in view by staring at the small black dot for a half minute or so, then switching to the white dot. The black squares appear stationary when staring at the white dot, but a fainter grid of moving squares also appears to be present.

Hermann-Mauguin Symbol

A symbol used to represent the point and space groups (e.g., $2/m\bar{3}$). Some symbols have abbreviated form. The equivalence between Hermann-Mauguin symbols ("crystallographic symbol") and SCHÖNFLIES SYMBOLS for the POINT GROUPS is given by Cotton (1990).

see also POINT GROUPS

References
Cotton, F. A. *Chemical Applications of Group Theory, 3rd ed.* New York: Wiley, p. 379, 1990.

Hermit Point

see ISOLATED POINT

Hermite Constants

N.B. A detailed on-line essay by S. Finch was the starting point for this entry.

The Hermite constant is defined for DIMENSION n as the value

$$\gamma_n = \frac{\sup_f \min_{x_i} f(x_1, x_2, \ldots, x_n)}{[\text{discriminant}(f)]^{1/n}}$$

(Le Lionnais 1983). In other words, they are given by

$$\gamma_n = 4\left(\frac{\delta_n}{V_n}\right)^{2/n},$$

where δ_n is the maximum *lattice* PACKING DENSITY for HYPERSPHERE PACKING and V_n is the CONTENT of the n-HYPERSPHERE. The first few values of $(\gamma_n)^n$ are 1, 4/3, 2, 4, 8, 64/3, 64, 256, ... Values for larger n are not known.

For sufficiently large n,

$$\frac{1}{2\pi e} \leq \frac{\gamma_n}{n} \leq \frac{1.744\ldots}{2\pi e}.$$

see also HYPERSPHERE PACKING, KISSING NUMBER, SPHERE PACKING

References
Finch, S. "Favorite Mathematical Constants." `http://www.mathsoft.com/asolve/constant/hermit/hermit.html`.
Conway, J. H. and Sloane, N. J. A. *Sphere Packings, Lattices, and Groups, 2nd ed.* New York: Springer-Verlag, p. 20, 1993.
Le Lionnais, F. *Les nombres remarquables.* Paris: Hermann, p. 38, 1983.

Hermite Differential Equation

$$\frac{d^2y}{dx^2} - 2x\frac{dy}{dx} + \lambda y = 0. \tag{1}$$

This differential equation has an irregular singularity at ∞. It can be solved using the series method

$$\sum_{n=0}^{\infty}(n+2)(n+1)a_{n+2}x^n - \sum_{n=1}^{\infty} 2na_n x^n + \sum_{n=0}^{\infty} \lambda a_n x^n = 0 \tag{2}$$

$$(2a_2+\lambda a_4)+\sum_{n=1}^{\infty}[(n+2)(n+1)a_{n+2}-2na_n+\lambda a_n]x^n=0. \tag{3}$$

Therefore,

$$a_2=-\frac{\lambda a_0}{2} \tag{4}$$

and

$$a_{n+2}=\frac{2n-\lambda}{(n+2)(n+1)}a_n \tag{5}$$

for $n = 1, 2, \ldots$. Since (4) is just a special case of (5),

$$a_{n+2}=\frac{2n-\lambda}{(n+2)(n+1)}a_n \tag{6}$$

for $n = 0, 1, \ldots$. The linearly independent solutions are then

$$y_1=a_0\left[1-\frac{\lambda}{2!}x^2-\frac{(4-\lambda)\lambda}{4!}x^4 -\frac{(8-\lambda)(4-\lambda)\lambda}{6!}x^6-\ldots\right] \tag{7}$$

$$y_2=a_1\left[x+\frac{(2-\lambda)}{3!}x^3+\frac{(6-\lambda)(2-\lambda)}{5!}x^5+\ldots\right]. \tag{8}$$

If $\lambda \equiv 4n = 0, 4, 8, \ldots$, then y_1 terminates with the POWER x^λ, and y_1 (normalized so that the COEFFICIENT of x^n is 2^n) is the regular solution to the equation, known as the HERMITE POLYNOMIAL. If $\lambda \equiv 4n+2 = 2, 6, 10, \ldots$, then y_2 terminates with the POWER x^λ, and y_2 (normalized so that the COEFFICIENT of x^n is 2^n) is the regular solution to the equation, known as the HERMITE POLYNOMIAL.

If $\lambda = 0$, then Hermite's differential equation becomes

$$y''-2xy'=0, \tag{9}$$

which is of the form $P_2(x)y''+P_1(x)y'=0$ and so has solution

$$\begin{aligned} y &= c_1\int\frac{dx}{\exp\left(\int\frac{P_1}{P_2}\,dx\right)}+c_2 \\ &= c_1\int\frac{dx}{\exp\int -2x\,dx}+c_2 \\ &= c_1\int\frac{dx}{e^{-x^2}}+c_2=c_1\int e^{x^2}\,dx+c_2. \end{aligned} \tag{10}$$

Hermite-Gauss Quadrature

Also called HERMITE QUADRATURE. A GAUSSIAN QUADRATURE over the interval $(-\infty,\infty)$ with WEIGHTING FUNCTION $W(x)=e^{-x^2}$. The ABSCISSAS for quadrature order n are given by the roots of the HERMITE POLYNOMIALS $H_n(x)$, which occur symmetrically about 0. The WEIGHTS are

$$w_i=-\frac{A_{n+1}\gamma_n}{A_nH_n'(x_i)H_{n+1}(x_i)}=\frac{A_n}{A_{n-1}}\frac{\gamma_{n-1}}{H_{n-1}(x_i)H_n'(x_i)}, \tag{1}$$

where A_n is the COEFFICIENT of x^n in $H_n(x)$. For HERMITE POLYNOMIALS,

$$A_n=2^n, \tag{2}$$

so

$$\frac{A_{n+1}}{A_n}=2. \tag{3}$$

Additionally,

$$\gamma_n=\sqrt{\pi}\,2^nn!, \tag{4}$$

so

$$\begin{aligned} w_i &= -\frac{2^{n+1}n!\sqrt{\pi}}{H_{n+1}(x_i)H_n'(x_i)} \\ &= \frac{2^n(n-1)!\sqrt{\pi}}{H_{n-1}(x_i)H_n'(x_i)}. \end{aligned} \tag{5}$$

Using the RECURRENCE RELATION

$$H_n'(x)=2nH_{n-1}(x)=2xH_n(x)-H_{n+1}(x) \tag{6}$$

yields

$$H_n'(x_i)=2nH_{n-1}(x_i)=-H_{n+1}(x_i) \tag{7}$$

and gives

$$w_i=\frac{2^{n+1}n!\sqrt{\pi}}{[H_n'(x_i)]^2}=\frac{2^{n+1}n!\sqrt{\pi}}{[H_{n+1}(x_i)]^2}. \tag{8}$$

The error term is

$$E=\frac{n!\sqrt{\pi}}{2^n(2n)!}f^{(2n)}(\xi). \tag{9}$$

Beyer (1987) gives a table of ABSCISSAS and weights up to n=12.

n	x_i	w_i
2	±0.707107	0.886227
3	0	1.18164
	±1.22474	0.295409
4	±0.524648	0.804914
	±1.65068	0.0813128
5	0	0.945309
	±0.958572	0.393619
	±2.02018	0.0199532

The ABSCISSAS and weights can be computed analytically for small n.

n	x_i	w_i
2	$\pm\frac{1}{2}\sqrt{2}$	$\frac{1}{2}\sqrt{\pi}$
3	0	$\frac{2}{3}\sqrt{\pi}$
	$\pm\frac{1}{2}\sqrt{6}$	$\frac{1}{6}\sqrt{\pi}$
4	$\pm\sqrt{\frac{3-\sqrt{6}}{2}}$	$\frac{\sqrt{\pi}}{4(3-\sqrt{6})}$
	$\pm\sqrt{\frac{3+\sqrt{6}}{2}}$	$\frac{\sqrt{\pi}}{4(3+\sqrt{6})}$

References

Beyer, W. H. *CRC Standard Mathematical Tables, 28th ed.* Boca Raton, FL: CRC Press, p. 464, 1987.

Hildebrand, F. B. *Introduction to Numerical Analysis.* New York: McGraw-Hill, pp. 327–330, 1956.

Hermite Interpolation

see HERMITE'S INTERPOLATING FUNDAMENTAL POLYNOMIAL

Hermite's Interpolating Fundamental Polynomial

Let $l(x)$ be an nth degree POLYNOMIAL with zeros at $x_1, \ldots, x_m$. Then the fundamental POLYNOMIALS are

$$h_\nu^{(1)}(x) = \left[1 - \frac{l''(x_\nu)}{l'(x_\nu)}\right][l_\nu(x)]^2 \tag{1}$$

and

$$h_\nu^{(2)}(x) = (x - x_\nu)[l_\nu(x)]^2. \tag{2}$$

They have the properties

$$h_\nu^{(1)}(x_\mu) = \delta_{\nu\mu} \tag{3}$$

$$h^{(1)\prime}{}_\nu(x_\mu) = 0 \tag{4}$$

$$h^{(2)}(x_\mu) = 0 \tag{5}$$

$$h^{(2)\prime}(x_\mu) = \delta_{\nu\mu}. \tag{6}$$

Now let $f_1, \ldots, f_n$ and $f_1', \ldots, f_\nu'$ be values. Then the expansion

$$W_n(x) = \sum_{\nu=1}^n f_\nu h_\nu^{(1)}(x) + \sum_{\nu=1}^n f_\nu' h^{(2)}(x) \tag{7}$$

gives the unique HERMITE'S INTERPOLATING FUNDAMENTAL POLYNOMIAL for which

$$W_n(x_\nu) = f_\nu \tag{8}$$

$$W_n'(x_\nu) = f_\nu'. \tag{9}$$

If $f_\nu' = 0$, these are called STEP POLYNOMIALS. The fundamental POLYNOMIALS satisfy

$$h_1(x) + \ldots + h_n(x) = 1 \tag{10}$$

and

$$\sum_{\nu=1}^n x_\nu h_\nu^{(1)}(x) + \sum_{\nu=1}^n h_\nu^{(2)}(x) = x. \tag{11}$$

Also,

$$\int_a^b h_\nu^{(1)}(x)\,d\alpha(x) = \lambda_\nu \tag{12}$$

$$\int_a^b h_\nu^{(1)}(x)\,d\alpha(x) = 0 \tag{13}$$

$$\int_a^b x h_\nu'(x)\,d\alpha(x) = 0 \tag{14}$$

$$\int_a^b h_\nu^{(2)}(x)\,d\alpha(x) = 0 \tag{15}$$

$$\int_a^b h^{(2)\prime}{}_\nu\,d\alpha(x) = \lambda_\nu \tag{16}$$

$$\int_a^b x h^{(2)\prime}{}_\nu(x)\,dx = \lambda_\nu x_\nu, \tag{17}$$

for $\nu = 1, \ldots, n$.

References

Hildebrand, F. B. *Introduction to Numerical Analysis.* New York: McGraw-Hill, pp. 314–319, 1956.

Szegő, G. *Orthogonal Polynomials, 4th ed.* Providence, RI: Amer. Math. Soc., pp. 330–332, 1975.

Hermite-Lindemann Theorem

The expression

$$A_1 e^{\alpha_1} + A_2 e^{\alpha_2} + A_3 e^{\alpha_3} + \ldots,$$

in which the COEFFICIENTS A_i differ from zero and in which the exponents α_i are ALGEBRAIC NUMBERS differing from each other, cannot equal zero.

see also ALGEBRAIC NUMBER, CONSTANT PROBLEM, INTEGER RELATION, LINDEMANN-WEIERSTRAß THEOREM

References

Dörrie, H. "The Hermite-Lindemann Transcendence Theorem." §26 in *100 Great Problems of Elementary Mathematics: Their History and Solutions.* New York: Dover, pp. 128–137, 1965.

Hermite Polynomial

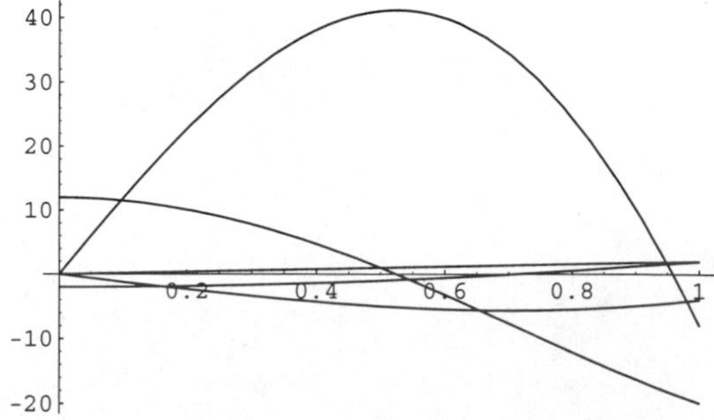

A set of ORTHOGONAL POLYNOMIALS. The Hermite polynomials $H_n(x)$ are illustrated above for $x \in [0,1]$ and $n = 1, 2, \ldots, 5$.

The GENERATING FUNCTION for Hermite polynomials is

$$\exp(2xt - t^2) \equiv \sum_{n=0}^{\infty} \frac{H_n(x)t^n}{n!}. \tag{1}$$

Using a TAYLOR SERIES shows that,

$$\begin{aligned} H_n(x) &= \left[\left(\frac{\partial}{\partial t}\right)^n \exp(2xt - t^2)\right]_{t=0} \\ &= \left[e^{x^2}\left(\frac{\partial}{\partial t}\right)^n e^{-(x-t)^2}\right]_{t=0}. \end{aligned} \tag{2}$$

Since $\partial f(x-t)/\partial t = -\partial f(x-t)/\partial x$,

$$\begin{aligned} H_n(x) &= (-1)^n e^{x^2}\left[\left(\frac{\partial}{\partial x}\right)^n e^{-(x-t)^2}\right]_{t=0} \\ &= (-1)^n e^{x^2} \frac{d^n}{dx^n} e^{-x^2}. \end{aligned} \tag{3}$$

Now define operators

$$\tilde{O}_1 \equiv -e^{x^2} \frac{d}{dx} e^{-x^2} \tag{4}$$

$$\tilde{O}_2 \equiv e^{x^2/2}\left(x - \frac{d}{dx}\right) e^{-x^2/2}. \tag{5}$$

It follows that

$$\tilde{O}_1 f = -e^{x^2} \frac{d}{dx}[fe^{-x^2}] = 2xf - \frac{df}{dx} \tag{6}$$

$$\begin{aligned} \tilde{O}_2 f &= e^{x^2/2}\left(x - \frac{d}{dx}\right)[fe^{-x^2/2}] \\ &= xf + xf - \frac{df}{dx} = 2xf - \frac{df}{dx}, \end{aligned} \tag{7}$$

so

$$\tilde{O}_1 = \tilde{O}_2, \tag{8}$$

and

$$-e^{x^2}\frac{d}{dx}e^{-x^2} = e^{x^2/2}\left(x - \frac{d}{dx}\right)e^{-x^2/2}, \tag{9}$$

which means the following definitions are equivalent:

$$\exp(2xt - t^2) \equiv \sum_{n=0}^{\infty} \frac{H_n(x)t^n}{n!} \tag{10}$$

$$H_n(x) \equiv (-1)^n e^{x^2} \frac{d^n}{dx^n} e^{-x^2} \tag{11}$$

$$H_n(x) \equiv e^{x^2/2}\left(x - \frac{d}{dx}\right) n e^{-x^2/2}. \tag{12}$$

The Hermite POLYNOMIALS are related to the derivative of the ERROR FUNCTION by

$$H_n(z) = (-1)^2 \frac{\sqrt{\pi}}{2} e^{z^2} \frac{d^{n+1}}{dz^{n+1}} \operatorname{erf}(z). \tag{13}$$

They have a contour integral representation

$$H_n(x) = \frac{n!}{2\pi i} \int e^{-t^2+2tx} t^{-n-1}\, dt. \tag{14}$$

They are orthogonal in the range $(-\infty, \infty)$ with respect to the WEIGHTING FUNCTION e^{-x^2}

$$\int_{-\infty}^{\infty} H_n(x) H_m(x) e^{-x^2}\, dx = \delta_{mn} 2^n n! \sqrt{\pi}. \tag{15}$$

Define the associated functions

$$u_n(x) \equiv \sqrt{\frac{a}{\pi^{1/2} n! 2^n}}\, H_n(ax) e^{-a^2x^2/2}. \tag{16}$$

These obey the orthogonality conditions

$$\int_{-\infty}^{\infty} u_n(x) \frac{du_m}{dx}\, dx = \begin{cases} a\sqrt{\frac{n+1}{2}} & m = n+1 \\ -a\sqrt{\frac{n}{2}} & m = n-1 \\ 0 & \text{otherwise} \end{cases} \tag{17}$$

$$\int_{-\infty}^{\infty} u_m(x) u_n(x)\, dx = \delta_{mn} \tag{18}$$

$$\int_{-\infty}^{\infty} u_m(x) x u_n(x)\, dx = \begin{cases} \frac{1}{a}\sqrt{\frac{n+!}{2}} & m = n+1 \\ \frac{1}{a}\sqrt{\frac{n}{2}} & m = n-1 \\ 0 & \text{otherwise} \end{cases} \tag{19}$$

$$\int_{-\infty}^{\infty} u_m(x) x^2 u_n(x)\, dx = \begin{cases} \frac{2n+1}{2a^2} & m = n \\ \frac{\sqrt{(n+1)(n+2)}}{2a^2} & m = n+2 \\ 0 & m \neq n \neq n \pm 2 \end{cases} \tag{20}$$

$$\int_{-\infty}^{\infty} e^{-x^2} H_\alpha H_\beta H_\gamma\, dx = \sqrt{\pi} \frac{2^s \alpha! \beta! \gamma!}{(s-\alpha)!(s-\beta)!(s-\gamma)!}, \tag{21}$$

if $\alpha + \beta + \gamma = 2s$ is EVEN and $s \geq \alpha$, $s \geq \beta$, and $s \geq \gamma$. Otherwise, the last integral is 0 (Szegő 1975, p. 390).

They also satisfy the RECURRENCE RELATIONS

$$H_{n+1} = 2xH_n(x) - 2nH_{n-1}(x) \tag{22}$$

$$H_n'(x) = 2nH_{n-1}(x). \tag{23}$$

The DISCRIMINANT is

$$D_n = 2^{3n(n-1)/2} \prod_{\nu=1}^{n} \nu^\nu \tag{24}$$

(Szegő 1975, p. 143).

An interesting identity is

$$\sum_{\nu=0}^{n} \binom{n}{\nu} H_\nu(x) H_{n-\nu}(y) = 2^{n/2} H_n[2^{-1/2}(x+y)]. \tag{25}$$

The first few POLYNOMIALS are

$$H_0(x) = 1$$
$$H_1(x) = 2x$$
$$H_2(x) = 4x^2 - 2$$
$$H_3(x) = 8x^3 - 12x$$
$$H_4(x) = 16x^4 - 48x^2 + 12$$
$$H_5(x) = 32x^5 - 160x^3 + 120x$$
$$H_6(x) = 64x^6 - 480x^4 + 720x^2 - 120$$
$$H_7(x) = 128x^7 - 1344x^5 + 3360x^3 - 1680x$$
$$H_8(x) = 256x^8 - 3594x^6 + 13440x^4 - 13440x^2 + 160$$
$$H_9(x) = 512x^9 - 9216x^7 + 48384x^5 - 80640x^3 + 30240x$$
$$H_{10}(x) = 1024x^{10} - 23040x^8 + 161280x^6 - 403200x^4 + 302400x^2 - 30240.$$

A class of generalized Hermite POLYNOMIALS $\gamma_n^m(x)$ satisfying

$$e^{mxt-t^m} = \sum_{n=0}^{\infty} \gamma_n^m(x)t^n \tag{26}$$

was studied by Subramanyan (1990). A class of related POLYNOMIALS defined by

$$h_{n,m} = \gamma_n^m\left(\frac{2x}{m}\right) \tag{27}$$

and with GENERATING FUNCTION

$$e^{2xt-t^m} = \sum_{n=0}^{\infty} h_{n,m}(x)t^n \tag{28}$$

was studied by Djordjević (1996). They satisfy

$$H_n(x) = n!h_{n,2}(x). \tag{29}$$

A modified version of the HERMITE POLYNOMIAL is sometimes defined by

$$\mathrm{He}_n(x) \equiv H_n\left(\frac{x}{\sqrt{2}}\right). \tag{30}$$

see also MEHLER'S HERMITE POLYNOMIAL FORMULA, WEBER FUNCTIONS

References

Abramowitz, M. and Stegun, C. A. (Eds.). "Orthogonal Polynomials." Ch. 22 in *Handbook of Mathematical Functions with Formulas, Graphs, and Mathematical Tables, 9th printing.* New York: Dover, pp. 771–802, 1972.

Arfken, G. "Hermite Functions." §13.1 in *Mathematical Methods for Physicists, 3rd ed.* Orlando, FL: Academic Press, pp. 712–721, 1985.

Chebyshev, P. L. "Sur le développement des fonctions à une seule variable." *Bull. ph.-math., Acad. Imp. Sc. St. Pétersbourg* **1**, 193–200, 1859.

Chebyshev, P. L. *Oeuvres, Vol. 1.* New York: Chelsea, pp. 49–508, 1987.

Djordjević, G. "On Some Properties of Generalized Hermite Polynomials." *Fib. Quart.* **34**, 2–6, 1996.

Hermite, C. "Sur un nouveau développement en série de fonctions." *Compt. Rend. Acad. Sci. Paris* **58**, 93–100 and 266–273, 1864. Reprinted in Hermite, C. *Oeuvres complètes, Vol. 2.* Paris, pp. 293–308, 1908.

Hermite, C. *Oeuvres complètes, Vol. 3.* Paris, p. 432, 1912.

Iyanaga, S. and Kawada, Y. (Eds.). "Hermite Polynomials." Appendix A, Table 20.IV in *Encyclopedic Dictionary of Mathematics.* Cambridge, MA: MIT Press, pp. 1479–1480, 1980.

Sansone, G. "Expansions in Laguerre and Hermite Series." Ch. 4 in *Orthogonal Functions, rev. English ed.* New York: Dover, pp. 295–385, 1991.

Spanier, J. and Oldham, K. B. "The Hermite Polynomials $H_n(x)$." Ch. 24 in *An Atlas of Functions.* Washington, DC: Hemisphere, pp. 217–223, 1987.

Subramanyan, P. R. "Springs of the Hermite Polynomials." *Fib. Quart.* **28**, 156–161, 1990.

Szegő, G. *Orthogonal Polynomials, 4th ed.* Providence, RI: Amer. Math. Soc., 1975.

Hermite Quadrature

see HERMITE-GAUSS QUADRATURE

Hermite's Theorem

e is TRANSCENDENTAL.

Hermitian Form

A combination of variables x and y given by

$$axx^* + bxy^* + b^*x^*y + cyy^*,$$

where x^* and y^* are COMPLEX CONJUGATES.

Hermitian Matrix

If a MATRIX is SELF-ADJOINT, it is said to be a Hermitian matrix. Therefore, a Hermitian MATRIX is defined as one for which

$$\mathsf{A} = \mathsf{A}^\dagger, \tag{1}$$

where † denotes the ADJOINT MATRIX. Hermitian MATRICES have REAL EIGENVALUES with ORTHOGONAL EIGENVECTORS. For REAL MATRICES, Hermitian is the same as symmetrical. Any MATRIX C which is not Hermitian can be expressed as the sum of two Hermitian matrices

$$\mathsf{C} = \tfrac{1}{2}(\mathsf{C} + \mathsf{C}^\dagger) + \tfrac{1}{2}(\mathsf{C} - \mathsf{C}^\dagger). \tag{2}$$

Let U be a UNITARY MATRIX and A be a Hermitian matrix. Then the ADJOINT MATRIX of a SIMILARITY TRANSFORMATION is

$$(\mathsf{UAU}^{-1})^\dagger = [(\mathsf{UA})(\mathsf{U}^{-1})]^\dagger = (\mathsf{U}^{-1})^\dagger(\mathsf{UA})^\dagger = (\mathsf{U}^\dagger)^\dagger(\mathsf{A}^\dagger\mathsf{U}^\dagger) = \mathsf{UAU}^\dagger = \mathsf{UAU}^{-1}. \tag{3}$$

The specific matrix

$$\mathsf{H}(x,y,z) = \begin{bmatrix} z & x+iy \\ x-iy & -z \end{bmatrix} = x\mathsf{P}_1 + y\mathsf{P}_2 + z\mathsf{P}_3, \quad (4)$$

where P_i are PAULI SPIN MATRICES, is sometimes called "the" Hermitian matrix.

see also ADJOINT MATRIX, HERMITIAN OPERATOR, PAULI SPIN MATRICES

References

Arfken, G. "Hermitian Matrices, Unitary Matrices." §4.5 in *Mathematical Methods for Physicists, 3rd ed.* Orlando, FL: Academic Press, pp. 209–217, 1985.

Hermitian Operator

A Hermitian OPERATOR $\tilde{L}$ is one which satisfies

$$\int_a^b v^* \tilde{L} u \, dx = \int_a^b u \tilde{L} v^* \, dx. \quad (1)$$

As shown in STURM-LIOUVILLE THEORY, if $\tilde{L}$ is SELF-ADJOINT and satisfies the boundary conditions

$$[v^* p u']_{x=a} = [v^* p u']_{x=b}, \quad (2)$$

then it is automatically Hermitian. Hermitian operators have REAL EIGENVALUES, ORTHOGONAL EIGENFUNCTIONS, and the corresponding EIGENFUNCTIONS form a COMPLETE set when $\tilde{L}$ is second-order and linear. In order to prove that EIGENVALUES must be REAL and EIGENFUNCTIONS ORTHOGONAL, consider

$$\tilde{L} u_i + \lambda_i w u_i = 0. \quad (3)$$

Assume there is a second EIGENVALUE λ_j such that

$$\tilde{L} u_j + \lambda_j w u_j = 0 \quad (4)$$

$$\tilde{L} u_j{}^* + \lambda_j{}^* w u_j{}^* = 0. \quad (5)$$

Now multiply (3) by $u_j{}^*$ and (5) by u_i

$$u_j{}^* \tilde{L} u_i + u_j{}^* \lambda_i w u_i = 0 \quad (6)$$

$$u_i \tilde{L} u_j{}^* + u_i \lambda_j{}^* w u_j{}^* = 0 \quad (7)$$

$$u_j{}^* \tilde{L} u_i - u_i \tilde{L} u_j{}^* = (\lambda_j{}^* - \lambda_i) w u_i u_j{}^*. \quad (8)$$

Now integrate

$$\int_a^b u_j{}^* \tilde{L} u_i - \int_a^b u_i \tilde{L} u_j{}^* = (\lambda_j{}^* - \lambda_i) \int_a^b w u_i u_j{}^*. \quad (9)$$

But because $\tilde{L}$ is Hermitian, the left side vanishes.

$$(\lambda_j{}^* - \lambda_i) \int_a^b w u_i u_j{}^* = 0. \quad (10)$$

If EIGENVALUES λ_i and λ_j are not degenerate, then $\int_a^b w u_i u_j{}^* = 0$, so the EIGENFUNCTIONS are ORTHOGONAL. If the EIGENVALUES are degenerate, the EIGENFUNCTIONS are not necessarily orthogonal. Now take $i = j$.

$$(\lambda_i{}^* - \lambda_i) \int_a^b w u_i u_i{}^* = 0. \quad (11)$$

The integral cannot vanish unless $u_i = 0$, so we have $\lambda_i{}^* = \lambda_i$ and the EIGENVALUES are real.

For a Hermitian operator $\tilde{O}$,

$$\langle \phi | \tilde{O} \psi \rangle = \langle \phi | \tilde{O} \psi \rangle^* = \langle \tilde{O} \phi | \psi \rangle. \quad (12)$$

In integral notation,

$$\int (\tilde{A}\phi)^* \psi \, dx = \int \phi^* \tilde{A} \psi \, dx. \quad (13)$$

Given Hermitian operators $\tilde{A}$ and $\tilde{B}$,

$$\langle \phi | \tilde{A}\tilde{B} \psi \rangle = \langle \tilde{A}\phi | \tilde{B}\psi \rangle = \langle \tilde{B}\tilde{A}\phi | \psi \rangle = \langle \phi | \tilde{B}\tilde{A}\psi \rangle^*. \quad (14)$$

Because, for a Hermitian operator $\tilde{A}$ with EIGENVALUE a,

$$\langle \psi | \tilde{A} \psi \rangle = \langle \tilde{A} \psi | \psi \rangle \quad (15)$$

$$a \langle \psi | \psi \rangle = a^* \langle \psi | \psi \rangle. \quad (16)$$

Therefore, either $\langle \psi | \psi \rangle = 0$ or $a = a^*$. But $\langle \psi | \psi \rangle = 0$ IFF $\psi = 0$, so

$$\langle \psi | \psi \rangle \neq 0, \quad (17)$$

for a nontrivial EIGENFUNCTION. This means that $a = a^*$, namely that Hermitian operators produce REAL expectation values. Every observable must therefore have a corresponding Hermitian operator. Furthermore,

$$\langle \psi_n | \tilde{A} \psi_m \rangle = \langle \tilde{A} \psi_n | \psi_m \rangle \quad (18)$$

$$a_m \langle \psi_n | \psi_m \rangle = a_n{}^* \langle \psi_n | \psi_m \rangle = a_n \langle \psi_n | \psi_m \rangle, \quad (19)$$

since $a_n = a_n{}^*$. Then

$$(a_m - a_n) \langle \psi_n | \psi_m \rangle = 0 \quad (20)$$

For $a_m \neq a_n$ (i.e., $\psi_n \neq \psi_m$),

$$\langle \psi_n | \psi_m \rangle = 0. \quad (21)$$

For $a_m = a_n$ (i.e., $\psi_n = \psi_m$),

$$\langle \psi_n | \psi_m \rangle = \langle \psi_n | \psi_n \rangle \equiv 1. \quad (22)$$

Therefore,

$$\langle \psi_n | \psi_m \rangle = \delta_{nm}, \quad (23)$$

so the basis of EIGENFUNCTIONS corresponding to a Hermitian operator are ORTHONORMAL. Given two Hermitian operators $\tilde{A}$ and $\tilde{B}$,

$$(\tilde{A}\tilde{B})^\dagger = \tilde{B}^\dagger \tilde{A}^\dagger = \tilde{B}\tilde{A} = \tilde{A}\tilde{B} + [\tilde{B}, \tilde{A}], \qquad (24)$$

the operator $\tilde{A}\tilde{B}$ equals $(\tilde{A}\tilde{B})^\dagger$, and is therefore Hermitian, only if

$$[\tilde{B}, \tilde{A}] = 0. \qquad (25)$$

Given an arbitrary operator $\tilde{A}$,

$$\begin{aligned} \langle \psi_1 | (\tilde{A} + \tilde{A}^\dagger)\psi_2 \rangle &= \langle (\tilde{A}^\dagger + \tilde{A})\psi_1 | \psi_2 \rangle \\ &= \langle (\tilde{A} + \tilde{A}^\dagger)\psi_1 | \psi_2 \rangle, \qquad (26) \end{aligned}$$

so $\tilde{A} + \tilde{A}^\dagger$ is Hermitian.

$$\begin{aligned} \langle \psi_1 | i(\tilde{A} - \tilde{A}^\dagger)\psi_2 \rangle &= \langle -i(\tilde{A}^\dagger - \tilde{A})\psi_1 | \psi_2 \rangle \\ &= \langle i(\tilde{A} - \tilde{A}^\dagger)\psi_1 | \psi_2 \rangle, \qquad (27) \end{aligned}$$

so $i(\tilde{A} - \tilde{A}^\dagger)$ is Hermitian. Similarly,

$$\langle \psi_1 | (\tilde{A}\tilde{A}^\dagger)\psi_2 \rangle = \langle \tilde{A}^\dagger \psi_1 | \tilde{A} \dagger \psi_2 \rangle = \langle (\tilde{A}\tilde{A}^\dagger)\psi_1 | \psi_2 \rangle, \qquad (28)$$

so $\tilde{A}\tilde{A}^\dagger$ is Hermitian.

Define the Hermitian conjugate operator $\tilde{A}^\dagger$ by

$$\langle \tilde{A}\psi | \psi \rangle \equiv \langle \psi | \tilde{A}^\dagger \psi \rangle. \qquad (29)$$

For a Hermitian operator, $\tilde{A} = \tilde{A}^\dagger$. Furthermore, given two Hermitian operators $\tilde{A}$ and $\tilde{B}$,

$$\begin{aligned} \langle \psi_2 | (\tilde{A}\tilde{B})^\dagger \psi_1 \rangle &= \langle (\tilde{A}\tilde{B})\psi_2 | \psi_1 \rangle = \langle \tilde{B}\psi_2 | \tilde{A}^\dagger \psi_1 \rangle \\ &= \langle \psi_2 | \tilde{B}^\dagger \tilde{A}^\dagger \psi_1 \rangle, \qquad (30) \end{aligned}$$

so

$$(\tilde{A}\tilde{B})^\dagger = \tilde{B}^\dagger \tilde{A}^\dagger. \qquad (31)$$

By further iterations, this can be generalized to

$$(\tilde{A}\tilde{B}\cdots\tilde{Z})^\dagger = \tilde{Z}^\dagger \cdots \tilde{B}^\dagger \tilde{A}^\dagger. \qquad (32)$$

see also ADJOINT OPERATOR, HERMITIAN MATRIX, SELF-ADJOINT OPERATOR, STURM-LIOUVILLE THEORY

References

Arfken, G. "Hermitian (Self-Adjoint) Operators." §9.2 in *Mathematical Methods for Physicists, 3rd ed.* Orlando, FL: Academic Press, pp. 504–506 and 510–516, 1985.

Heron's Formula

Gives the AREA of a TRIANGLE in terms of the lengths of the sides a, b, and c and the SEMIPERIMETER

$$s \equiv \tfrac{1}{2}(a + b + c). \qquad (1)$$

Heron's formula then states

$$\Delta = \sqrt{s(s-a)(s-b)(s-c)}. \qquad (2)$$

Expressing the side lengths a, b, and c in terms of the radii a', b', and c' of the mutually tangent circles centered on the TRIANGLE vertices (which define the SODDY CIRCLES),

$$\begin{aligned} a &= b' + c' \qquad (3) \\ b &= a' + c' \qquad (4) \\ c &= a' + b', \qquad (5) \end{aligned}$$

gives the particularly pretty form

$$\Delta = \sqrt{a'b'c'(a' + b' + c')} \qquad (6)$$

The proof of this fact was discovered by Heron (ca. 100 BC–100 AD), although it was already known to Archimedes prior to 212 BC (Kline 1972). Heron's proof (Dunham 1990) is ingenious but extremely convoluted, bringing together a sequence of apparently unrelated geometric identities and relying on the properties of CYCLIC QUADRILATERALS and RIGHT TRIANGLES.

Heron's proof can be found in Proposition 1.8 of his work *Metrica*. This manuscript had been lost for centuries until a fragment was discovered in 1894 and a complete copy in 1896 (Dunham 1990, p. 118). More recently, writings of the Arab scholar Abu'l Raihan Muhammed al-Biruni have credited the formula to Heron's predecessor Archimedes (Dunham 1990, p. 127).

A much more accessible algebraic proof proceeds from the LAW OF COSINES,

$$\cos A = \frac{b^2 + c^2 - a^2}{2bc}. \qquad (7)$$

Then

$$\sin A = \frac{\sqrt{-a^4 - b^4 - c^4 + 2b^2c^2 + 2c^2a^2 + 2a^2b^2}}{2bc}, \qquad (8)$$

giving

$$\begin{aligned} \Delta &= \tfrac{1}{2} bc \sin A \qquad (9) \\ &= \tfrac{1}{4}\sqrt{-a^4 - b^4 - c^4 + 2b^2c^2 + 2c^2a^2 + 2a^2b^2} \qquad (10) \\ &= \tfrac{1}{4}[(a+b+c)(-a+b+c)(a-b+c)(a+b-c)]^{1/2} \qquad (11) \\ &= \sqrt{s(s-a)(s-b)(s-c)} \qquad (12) \end{aligned}$$

(Coxeter 1969). Heron's formula contains the PYTHAGOREAN THEOREM.

see also BRAHMAGUPTA'S FORMULA, BRETSCHNEIDER'S FORMULA, HERONIAN TETRAHEDRON, HERONIAN TRIANGLE, SODDY CIRCLES, SSS THEOREM, TRIANGLE

References

Coxeter, H. S. M. *Introduction to Geometry, 2nd ed.* New York: Wiley, p. 12, 1969.

Dunham, W. "Heron's Formula for Triangular Area." Ch. 5 in *Journey Through Genius: The Great Theorems of Mathematics.* New York: Wiley, pp. 113–132, 1990.

Kline, M. *Mathematical Thought from Ancient to Modern Times.* New York: Oxford University Press, 1972.

Pappas, T. "Heron's Theorem." *The Joy of Mathematics.* San Carlos, CA: Wide World Publ./Tetra, p. 62, 1989.

Heron Triangle

see HERONIAN TRIANGLE

Heronian Tetrahedron

A TETRAHEDRON with RATIONAL sides, FACE AREAS, and VOLUME. The smallest examples have pairs of opposite sides (148, 195, 203), (533, 875, 888), (1183, 1479, 1804), (2175, 2296, 2431), (1825, 2748, 2873), (2180, 2639, 3111), (1887, 5215, 5512), (6409, 6625, 8484), and (8619, 10136, 11275).

see also HERON'S FORMULA, HERONIAN TRIANGLE

References

Guy, R. K. "Simplexes with Rational Contents." §D22 in *Unsolved Problems in Number Theory, 2nd ed.* New York: Springer-Verlag, pp. 190–192, 1994.

Heronian Triangle

A TRIANGLE with RATIONAL side lengths and RATIONAL AREA. Brahmagupta gave a parametric solution for *integer* Heronian triangles (the three side lengths and area can be multiplied by their LEAST COMMON MULTIPLE to make them all INTEGERS): side lengths $c(a^2+b^2)$, $b(a^2+c^2)$, and $(b+c)(a^2-bc)$, giving SEMIPERIMETER

$$s = a^2(b+c)$$

and AREA

$$\Delta = abc(a+b)(a^2-bc).$$

The first few integer Hernonian triangles, sorted by increasing maximal side lengths, are (3, 4, 5), (6, 8, 10), (5, 12, 13), (9, 12, 15), (4, 13, 15), (13, 14, 15), (9, 10, 17), ... (Sloane's A046128, A046129, and A046130), having areas 6, 24, 30, 54, 24, 84, 36, ... (Sloane's A046131).

Schubert (1905) claimed that Heronian triangles with two rational MEDIANS do not exist (Dickson 1952). This was shown to be incorrect by Buchholz and Rathbun (1997), who discovered six such triangles.

see also HERON'S FORMULA, MEDIAN (TRIANGLE), PYTHAGOREAN TRIPLE, TRIANGLE

References

Buchholz, R. H. *On Triangles with Rational Altitudes, Angle Bisectors or Medians.* Doctoral Dissertation. Newcastle, England: Newcastle University, 1989.

Buchholz, R. H. and Rathbun, R. L. "An Infinite Set of Heron Triangles with Two Rational Medians." *Amer. Math. Monthly* **104**, 107–115, 1997.

Dickson, L. E. *History of the Theory of Numbers, Vol. 2: Diophantine Analysis.* New York: Chelsea, pp. 199 and 208, 1952.

Guy, R. K. "Simplexes with Rational Contents." §D22 in *Unsolved Problems in Number Theory, 2nd ed.* New York: Springer-Verlag, pp. 190–192, 1994.

Kraitchik, M. "Heronian Triangles." §4.13 in *Mathematical Recreations.* New York: W. W. Norton, pp. 104–108, 1942.

Schubert, H. "Die Ganzzahligkeit in der algebraischen Geometrie." In *Festgabe 48 Versammlung d. Philologen und Schulmänner zu Hamburg.* Leipzig, Germany, pp. 1–16, 1905.

Wells, D. G. *The Penguin Dictionary of Curious and Interesting Puzzles.* London: Penguin Books, p. 34, 1992.

Herschel

A HEPTOMINO shaped like the astronomical symbol for Uranus (which was discovered by William Herschel).

Herschfeld's Convergence Theorem

For real, NONNEGATIVE terms x_n and REAL p with $0 < p < 1$, the expression

$$\lim_{k\to\infty} x_0 + (x_1 + (x_2 + (\ldots + (x_k)^p)^p)^p)^p$$

converges IFF $(x_n)^{p^n}$ is bounded.

see also CONTINUED SQUARE ROOT

References

Herschfeld, A. "On Infinite Radicals." *Amer. Math. Monthly* **42**, 419–429, 1935.

Jones, D. J. "Continued Powers and a Sufficient Condition for Their Convergence." *Math. Mag.* **68**, 387–392, 1995.

Hesse's Theorem

If two pairs of opposite VERTICES of a COMPLETE QUADRILATERAL are pairs of CONJUGATE POINTS, then the third pair of opposite VERTICES is likewise a pair of CONJUGATE POINTS.

Hessenberg Matrix

A matrix of the form

$$\begin{bmatrix} a_{11} & a_{12} & a_{13} & \cdots & a_{1(n-1)} & a_{1n} \\ a_{21} & a_{22} & a_{23} & \cdots & a_{2(n-1)} & a_{2n} \\ 0 & a_{32} & a_{33} & \cdots & a_{3(n-1)} & a_{3n} \\ 0 & 0 & a_{43} & \cdots & a_{4(n-1)} & a_{4n} \\ 0 & 0 & 0 & \cdots & a_{5(n-1)} & a_{5n} \\ \vdots & \vdots & \vdots & \ddots & \vdots & \vdots \\ 0 & 0 & 0 & 0 & a_{(n-1)(n-1)} & a_{(n-1)n} \\ 0 & 0 & 0 & 0 & a_{n(n-1)} & a_{nn} \end{bmatrix}.$$

References

Press, W. H.; Flannery, B. P.; Teukolsky, S. A.; and Vetterling, W. T. "Reduction of a General Matrix to Hessenberg Form." §11.5 in *Numerical Recipes in FORTRAN: The Art of Scientific Computing, 2nd ed.* Cambridge, England: Cambridge University Press, pp. 476–480, 1992.

Hessian Covariant

$$H \equiv |aa'a''|a_{x^{n-2}}a'_{x^{n-2}}a''_{x^{n-2}} = 0.$$

The nonsingular inflections of a curve are its nonsingular intersections with the Hessian.

References

Coolidge, J. L. *A Treatise on Algebraic Plane Curves.* New York: Dover, pp. 79, 95–98, and 151–161, 1959.

Hessian Determinant

The DETERMINANT

$$Hf(x,y) \equiv \begin{vmatrix} \frac{\partial^2 f}{\partial x^2} & \frac{\partial^2 f}{\partial x \partial y} \\ \frac{\partial^2 f}{\partial y \partial x} & \frac{\partial^2 f}{\partial y^2} \end{vmatrix}$$

appearing in the SECOND DERIVATIVE TEST as $D \equiv Hf(x,y)$.

see also SECOND DERIVATIVE TEST

References

Gradshteyn, I. S. and Ryzhik, I. M. *Tables of Integrals, Series, and Products, 5th ed.* San Diego, CA: Academic Press, pp. 1112–1113, 1979.

Heteroclinic Point

If intersecting stable and unstable MANIFOLDS (SEPARATRICES) emanate from FIXED POINTS of different families, they are called heteroclinic points.

see also HOMOCLINIC POINT

Heterogeneous Numbers

Two numbers are heterogeneous if their PRIME factors are distinct.

see also DISTINCT PRIME FACTORS, HOMOGENEOUS NUMBERS

References

Le Lionnais, F. *Les nombres remarquables.* Paris: Hermann, p. 146, 1983.

Heterological Paradox

see GRELLING'S PARADOX

Heteroscedastic

A set of STATISTICAL DISTRIBUTIONS having different VARIANCES.

see also HOMOSCEDASTIC, VARIANCE

Heterosquare

9	8	7
2	1	6
3	4	5

1	2	3	4
5	6	7	8
9	10	11	12
13	14	16	15

A heterosquare is an $n \times n$ ARRAY of the integers from 1 to n^2 such that the rows, columns, and diagonals have different sums. (By contrast, in a MAGIC SQUARE, they have the *same* sum.) There are no heterosquares of order two, but heterosquares of every ODD order exist. They can be constructed by placing consecutive INTEGERS in a SPIRAL pattern (Fults 1974, Madachy 1979).

An ANTIMAGIC SQUARE is a special case of a heterosquare for which the sums of rows, columns, and main diagonals form a SEQUENCE of consecutive integers.

see also ANTIMAGIC SQUARE, MAGIC SQUARE, TALISMAN SQUARE

References

Duncan, D. "Problem 86." *Math. Mag.* **24**, 166, 1951.

Fults, J. L. *Magic Squares.* Chicago, IL: Open Court, 1974.

Madachy, J. S. *Madachy's Mathematical Recreations.* New York: Dover, pp. 101–103, 1979.

Weisstein, E. W. "Magic Squares." http://www.astro.virginia.edu/~eww6n/math/notebooks/MagicSquares.m.

Heuman Lambda Function

$$\Lambda_0(\phi|m) \equiv \frac{F(\phi|1-m)}{K(1-m)} + \frac{2}{\pi}K(m)Z(\phi|1-m),$$

where ϕ is the AMPLITUDE, m is the PARAMETER, Z is the JACOBI ZETA FUNCTION, and $F(\phi|m')$ and $K(m)$ are incomplete and complete ELLIPTIC INTEGRALS OF THE FIRST KIND.

see also ELLIPTIC INTEGRAL OF THE FIRST KIND, JACOBI ZETA FUNCTION

References

Abramowitz, M. and Stegun, C. A. (Eds.). *Handbook of Mathematical Functions with Formulas, Graphs, and Mathematical Tables, 9th printing.* New York: Dover, p. 595, 1972.

Heun's Differential Equation

$$\frac{d^2w}{dx^2} + \left(\frac{\gamma}{x} + \frac{\delta}{x-1} + \frac{\epsilon}{x-a}\right)\frac{dw}{dx} + \frac{\alpha\beta x - q}{x(x-1)(x-a)}w = 0,$$

where

$$\alpha + \beta - \gamma - \delta - \epsilon + 1 = 0.$$

References

Erdelyi, A.; Magnus, W.; Oberhettinger, F.; and Tricomi, F. G. *Higher Transcendental Functions, Vol. 3.* Krieger, pp. 57–62, 1981.

Whittaker, E. T. and Watson, G. N. *A Course in Modern Analysis, 4th ed.* Cambridge, England: Cambridge University Press, p. 576, 1990.

Heuristic

(1) Based on or involving trial and error. (2) Convincing without being rigorous.

Hex Game

A two-player GAME. There is a winning strategy for the first player if there is an even number of cells on each side; otherwise, there is a winning strategy for the second player.

References

Gardner, M. Ch. 8 in *The Scientific American Book of Mathematical Puzzles & Diversions.* New York, NY: Simon and Schuster, 1959.

Hex Number

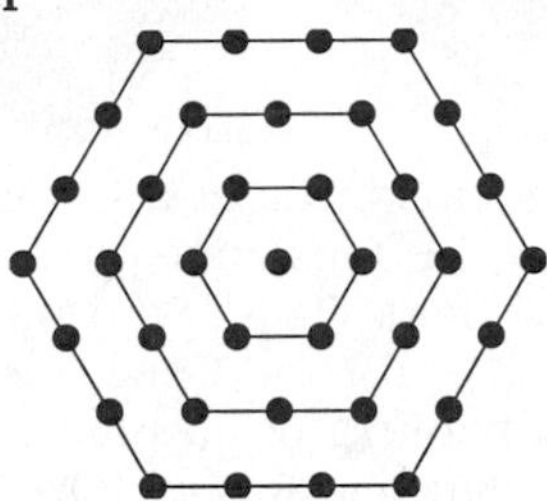

The CENTERED HEXAGONAL NUMBER given by

$$H_n = 1 + 6T_n = 2H_{n-1} - H_{n-2} + 6 = 3n^2 - 3n + 1,$$

where T_n is the nth TRIANGULAR NUMBER. The first few hex numbers are 1, 7, 19, 37, 61, 91, 127, 169, ... (Sloane's A003215). The GENERATING FUNCTION of the hex numbers is

$$\frac{x(x^2 + 4x + 1)}{(1 - x)^3} = x + 7x^2 + 19x^3 + 37x^4 + \ldots.$$

The first TRIANGULAR hex numbers are 1 and 91, and the first few SQUARE ones are 1, 169, 32761, 6355441, ... (Sloane's A006051). SQUARE hex numbers are obtained by solving the DIOPHANTINE EQUATION

$$3x^2 + 1 = y^2.$$

The only hex number which is SQUARE and TRIANGULAR is 1. There are no CUBIC hex numbers.

see also MAGIC HEXAGON, CENTERED SQUARE NUMBER, STAR NUMBER, TALISMAN HEXAGON

References

Conway, J. H. and Guy, R. K. *The Book of Numbers.* New York: Springer-Verlag, p. 41, 1996.

Gardner, M. "Hexes and Stars." Ch. 2 in *Time Travel and Other Mathematical Bewilderments.* New York: W. H. Freeman, 1988.

Hindin, H. "Stars, Hexes, Triangular Numbers, and Pythagorean Triples." *J. Recr. Math.* **16**, 191–193, 1983–1984.

Sloane, N. J. A. Sequences A003215/M4362 and A006051/M5409 in "An On-Line Version of the Encyclopedia of Integer Sequences."

Hex (Polyhex)

see POLYHEX

Hex Pyramidal Number

A FIGURATE NUMBER which is equal to the CUBIC NUMBER n^3. The first few are 1, 8, 27, 64, ... (Sloane's A000578).

References

Conway, J. H. and Guy, R. K. *The Book of Numbers.* New York: Springer-Verlag, pp. 42–44, 1996.

Sloane, N. J. A. Sequence A000578/M4499 in "An On-Line Version of the Encyclopedia of Integer Sequences."

Hexa

see POLYHEX

Hexabolo

A 6-POLYABOLO.

Hexacontagon

A 60-sided POLYGON.

Hexacronic Icositetrahedron

see GREAT HEXACRONIC ICOSITETRAHEDRON, SMALL HEXACRONIC ICOSITETRAHEDRON

Hexad

A SET of six.

see also MONAD, QUARTET, QUINTET, TETRAD, TRIAD

Hexadecagon

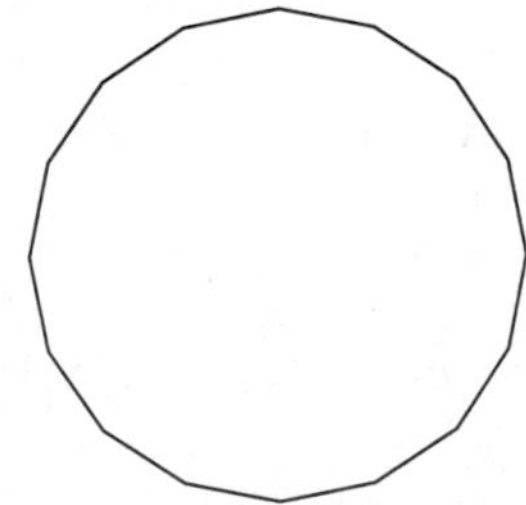

A 16-sided POLYGON, sometimes also called a HEXAKAIDECAGON.

see also POLYGON, REGULAR POLYGON, TRIGONOMETRY VALUES—$\pi/16$

Hexadecimal

The base 16 notational system for representing REAL NUMBERS. The digits used to represent numbers using hexadecimal NOTATION are 0, 1, 2, 3, 4, 5, 6, 7, 8, 9, A, B, C, D, E, and F.

see also BASE (NUMBER), BINARY, DECIMAL, METADROME, OCTAL, QUATERNARY, TERNARY, VIGESIMAL

References

Weisstein, E. W. "Bases." `http://www.astro.virginia.edu/~eww6n/math/notebooks/Bases.m`.

Hexaflexagon

A FLEXAGON made by folding a strip into adjacent EQUILATERAL TRIANGLES. The number of states possible in a hexaflexagon is the CATALAN NUMBER $C_4 = 42$.

see also FLEXAGON, FLEXATUBE, TETRAFLEXAGON

References

Cundy, H. and Rollett, A. *Mathematical Models, 3rd ed.* Stradbroke, England: Tarquin Pub., pp. 205–207, 1989.
Gardner, M. Ch. 1 in *The Scientific American Book of Mathematical Puzzles & Diversions.* New York: Simon and Schuster, 1959.
Gardner, M. Ch. 2 in *The Second Scientific American Book of Mathematical Puzzles & Diversions: A New Selection.* New York: Simon and Schuster, 1961.
Maunsell, F. G. "The Flexagon and the Hexaflexagon." *Math. Gazette* **38**, 213–214, 1954.
Wheeler, R. F. "The Flexagon Family." *Math. Gaz.* **42**, 1–6, 1958.

Hexagon

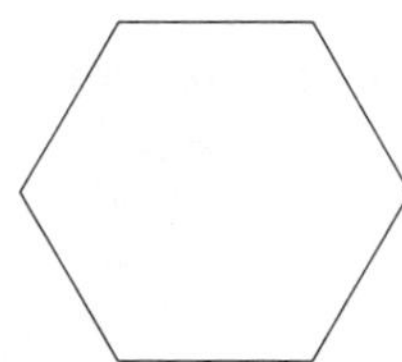

A six-sided POLYGON. In proposition IV.15, Euclid showed how to inscribe a regular hexagon in a CIRCLE. The INRADIUS r, CIRCUMRADIUS R, and AREA A can be computed directly from the formulas for a general regular POLYGON with side length s and $n = 6$ sides,

$$r = \tfrac{1}{2}s\cot\left(\frac{\pi}{6}\right) = \tfrac{1}{2}\sqrt{3}\,s \quad (1)$$

$$R = \tfrac{1}{2}s\csc\left(\frac{\pi}{6}\right) = s \quad (2)$$

$$A = \tfrac{1}{4}ns^2\cot\left(\frac{\pi}{6}\right) = \tfrac{3}{2}\sqrt{3}\,s^2. \quad (3)$$

Therefore, for a regular hexagon,

$$\frac{R}{r} = \sec\left(\frac{\pi}{6}\right) = \frac{2}{\sqrt{3}}, \quad (4)$$

so

$$\frac{A_R}{A_r} = \left(\frac{R}{r}\right)^2 = \frac{4}{3}. \quad (5)$$

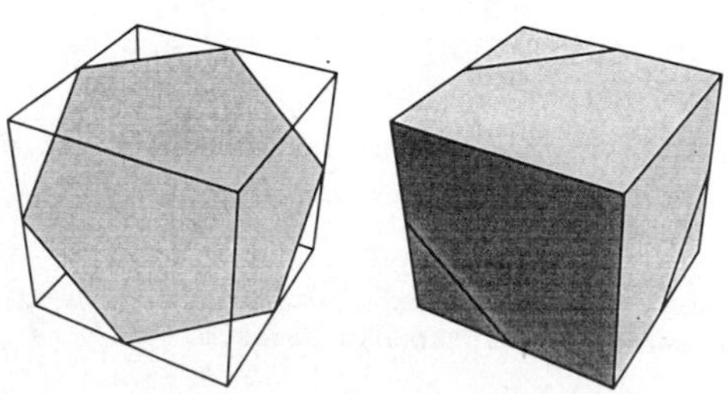

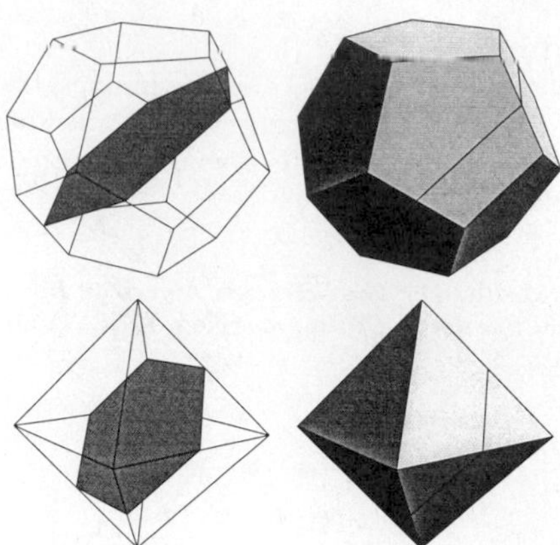

A PLANE PERPENDICULAR to a C_3 axis of a CUBE, DODECAHEDRON, or ICOSAHEDRON cuts the solid in a regular HEXAGONAL CROSS-SECTION (Holden 1991, pp. 22–23 and 27). For the CUBE, the PLANE passes through the MIDPOINTS of opposite sides (Steinhaus 1983, p. 170; Cundy and Rollett 1989, p. 157; Holden 1991, pp. 22–23). Since there are four such axes for the CUBE and OCTAHEDRON, there are four possibly hexagonal cross-sections. Since there are four such axes in each case, there are also four possibly hexagonal cross-sections.

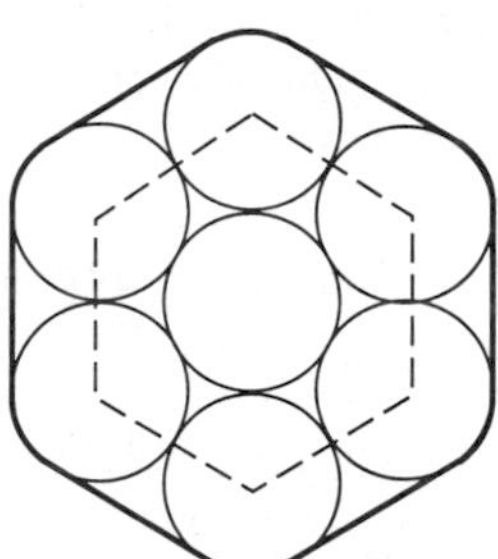

Take seven CIRCLES and close-pack them together in a hexagonal arrangement. The PERIMETER obtained by wrapping a band around the CIRCLE then consists of six straight segments of length d (where d is the DIAMETER) and 6 arcs with total length 1/6 of a CIRCLE. The PERIMETER is therefore

$$p = (12 + 2\pi)r = 2(6 + \pi)r. \quad (6)$$

see also CUBE, CYCLIC HEXAGON, DISSECTION, DODECAHEDRON, GRAHAM'S BIGGEST LITTLE HEXAGON, HEXAGON POLYIAMOND, HEXAGRAM, MAGIC HEXAGON, OCTAHEDRON, PAPPUS'S HEXAGON THEOREM, PASCAL'S THEOREM, TALISMAN HEXAGON

References

Cundy, H. and Rollett, A. "Hexagonal Section of a Cube." §3.15.1 in *Mathematical Models, 3rd ed.* Stradbroke, England: Tarquin Pub., p. 157, 1989.
Dixon, R. *Mathographics.* New York: Dover, p. 16, 1991.
Holden, A. *Shapes, Space, and Symmetry.* New York: Dover, 1991.
Pappas, T. "Hexagons in Nature." *The Joy of Mathematics.* San Carlos, CA: Wide World Publ./Tetra, pp. 74–75, 1989.
Steinhaus, H. *Mathematical Snapshots, 3rd American ed.* New York: Oxford University Press, 1983.

Hexagon Polyiamond

A 6-POLYIAMOND.

see also HEXAGON

References

Golomb, S. W. *Polyominoes: Puzzles, Patterns, Problems, and Packings, 2nd ed.* Princeton, NJ: Princeton University Press, p. 92, 1994.

Hexagonal Number

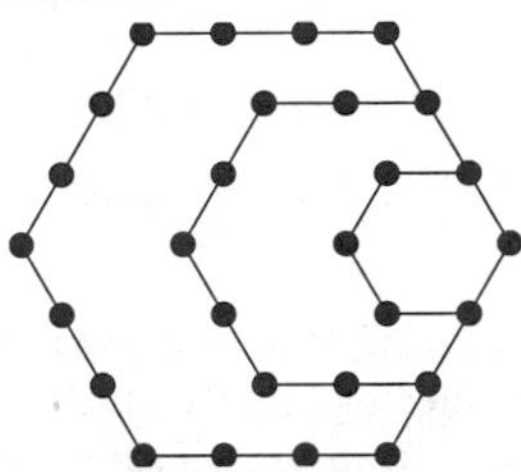

A FIGURATE NUMBER and 6-POLYGONAL NUMBER of the form $n(2n-1)$. The first few are 1, 6, 15, 28, 45, ... (Sloane's A000384). The GENERATING FUNCTION of the hexagonal numbers

$$\frac{x(3x+1)}{(1-x)^3} = x + 6x^2 + 15x^3 + 28x^4 + \ldots.$$

Every hexagonal number is a TRIANGULAR NUMBER since

$$r(2r-1) = \tfrac{1}{2}(2r-1)[(2r-1)+1].$$

In 1830, Legendre (1979) proved that every number larger than 1791 is a sum of four hexagonal numbers, and Duke and Schulze-Pillot (1990) improved this to three hexagonal numbers for every sufficiently large integer. The numbers 11 and 26 can only be represented as a sum using the maximum possible of six hexagonal numbers:

$$11 = 1+1+1+1+1+6$$
$$26 = 1+1+6+6+6+6.$$

see also FIGURATE NUMBER, HEX NUMBER, TRIANGULAR NUMBER

References

Duke, W. and Schulze-Pillot, R. "Representations of Integers by Positive Ternary Quadratic Forms and Equidistribution of Lattice Points on Ellipsoids." *Invent. Math.* **99**, 49–57, 1990.

Guy, R. K. "Sums of Squares." §C20 in *Unsolved Problems in Number Theory, 2nd ed.* New York: Springer-Verlag, pp. 136–138, 1994.

Legendre, A.-M. *Théorie des nombres, 4th ed., 2 vols.* Paris: A. Blanchard, 1979.

Sloane, N. J. A. Sequence A000384/M4108 in "An On-Line Version of the Encyclopedia of Integer Sequences."

Hexagonal Pyramidal Number

A PYRAMIDAL NUMBER of the form $n(n+1)(4n-1)/6$, The first few are 1, 7, 22, 50, 95, ... (Sloane's A002412). The GENERATING FUNCTION of the hexagonal pyramidal numbers is

$$\frac{x(3x+1)}{(x-1)^4} = x + 7x^2 + 22x^3 + 50x^4 + \ldots.$$

References

Sloane, N. J. A. Sequence A002412/M4374 in "An On-Line Version of the Encyclopedia of Integer Sequences."

Hexagonal Scalenohedron

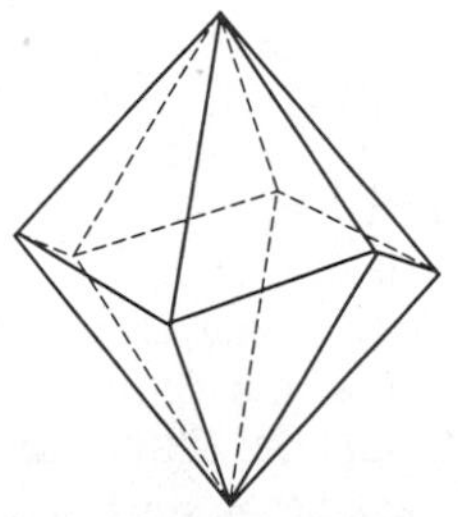

An irregular DODECAHEDRON which is also a TRAPEZOHEDRON.

see also DODECAHEDRON, TRAPEZOHEDRON

References

Cotton, F. A. *Chemical Applications of Group Theory, 3rd ed.* New York: Wiley, p. 63, 1990.

Hexagonal Tiling

see TILING

Hexagram

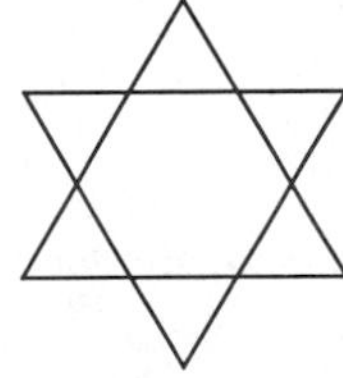

The STAR POLYGON $\{6,2\}$, also known as the STAR OF DAVID.

see also DISSECTION, PENTAGRAM, SOLOMON'S SEAL KNOT, STAR FIGURE, STAR OF LAKSHMI

Hexagrammum Mysticum Theorem

see PASCAL'S THEOREM

Hexahedron

A hexahedron is a six-sided POLYHEDRON. The regular hexahedron is the CUBE, although there are seven topologically different CONVEX hexahedra (Guy 1994).

see also CUBE

References

Guy, R. K. *Unsolved Problems in Number Theory, 2nd ed.* New York: Springer-Verlag, p. 189, 1994.

Hexahemioctahedron

The DUAL POLYHEDRON of the CUBOHEMIOCTAHEDRON.

Hexakaidecagon

see HEXADECAGON

Hexakis Icosahedron

see DISDYAKIS TRIACONTAHEDRON

Hexakis Octahedron

see DISDYAKIS DODECAHEDRON

Hexlet

Also called SODDY'S HEXLET. Consider three mutually tangent SPHERES A, B, and C. Then construct a chain of SPHERES tangent to each of A, B, and C threading and interlocking with the $A-B-C$ ring. Surprisingly, every chain closes into a "necklace" after six SPHERES regardless of where the first SPHERE is placed. This is a special case of KOLLROS' THEOREM. The centers of a Soddy hexlet always lie on an ELLIPSE (Ogilvy 1990, p. 63).

see also COXETER'S LOXODROMIC SEQUENCE OF TANGENT CIRCLES, KOLLROS' THEOREM, STEINER CHAIN

References

Coxeter, H. S. M. "Interlocking Rings of Spheres." *Scripta Math.* **18**, 113–121, 1952.

Gosset, T. "The Hexlet." *Nature* **139**, 251–252, 1937.

Honsberger, R. *Mathematical Gems II.* Washington, DC: Math. Assoc. Amer., pp. 49–50, 1976.

Morley, F. "The Hexlet." *Nature* **139**, 72–73, 1937.

Ogilvy, C. S. *Excursions in Geometry.* New York: Dover, pp. 60–72, 1990.

Soddy, F. "The Bowl of Integers and the Hexlet." *Nature* **139**, 77–79, 1937.

Soddy, F. "The Hexlet." *Nature* **139**, 154 and 252, 1937.

HexLife

An alternative LIFE game similar to Conway's, which is played on a hexagonal grid. No set of rules has yet emerged as uniquely interesting.

see also HIGHLIFE

References

Hensel, A. "A Brief Illustrated Glossary of Terms in Conway's Game of Life." `http://www.cs.jhu.edu/~callahan/glossary.html`.

Hexomino

One of the 35 6-POLYOMINOES.

References

Pappas, T. "Triangular, Square & Pentagonal Numbers." *The Joy of Mathematics.* San Carlos, CA: Wide World Publ./Tetra, p. 214, 1989.

Heyting Algebra

An ALGEBRA which is a special case of a LOGOS.

see also LOGOS, TOPOS

Hh Function

Let

$$Z(x) \equiv \frac{1}{\sqrt{2\pi}} e^{-x^2/2} \tag{1}$$

$$Q(x) \equiv \frac{1}{\sqrt{2\pi}} \int_x^\infty e^{-t^2/2}\,dt, \tag{2}$$

where Z and Q are closely related to the NORMAL DISTRIBUTION FUNCTION, then

$$\mathrm{Hh}_{-n}(x) = (-1)^{n-1}\sqrt{2\pi}\,z^{(n-1)}(x) \tag{3}$$

$$\mathrm{Hh}_n(x) = \frac{(-1)^n}{n!}\mathrm{Hh}_{-1}(x)\frac{d^n}{dx^n}\left[\frac{Q(x)}{Z(x)}\right]. \tag{4}$$

see also NORMAL DISTRIBUTION FUNCTION, TETRACHORIC FUNCTION

Hi-Q

A triangular version of PEG SOLITAIRE with 15 holes and 14 pegs. Numbering hole 1 at the apex of the triangle and thereafter from left to right on the next lower row, etc., the following table gives possible ending holes for a single peg removed (Beeler *et al.* 1972, Item 76). Because of symmetry, only the first five pegs need be considered. Also because of symmetry, removing peg 2 is equivalent to removing peg 3 and flipping the board horizontally.

remove	possible ending pegs
1	1, 7 = 10, 13
2	2, 6, 11, 14
4	3 = 12, 4, 9, 15
5	13

References

Beeler, M.; Gosper, R. W.; and Schroeppel, R. Item 75 in *HAKMEM.* Cambridge, MA: MIT Artificial Intelligence Laboratory, Memo AIM-239, Feb. 1972.

Higher Arithmetic

An archaic term for NUMBER THEORY.

Highest Weight Theorem

A theorem proved by É. Cartan (1913) which classifies the irreducible representations of COMPLEX semisimple LIE ALGEBRAS.

References

Knapp, A. W. "Group Representations and Harmonic Analysis, Part II." *Not. Amer. Math. Soc.* **43**, 537–549, 1996.

HighLife

An alternate set of LIFE rules similar to Conway's, but with the additional rule that six neighbors generate a birth. Most of the interest in this variant is due to the presence of a so-called replicator.

see also HEXLIFE, LIFE

References

Hensel, A. "A Brief Illustrated Glossary of Terms in Conway's Game of Life." `http://www.cs.jhu.edu/~callahan/glossary.html`.

Highly Abundant Number

see HIGHLY COMPOSITE NUMBER

Highly Composite Number

A COMPOSITE NUMBER (also called a SUPERABUNDANT NUMBER) is a number n which has more FACTORS than any other number less than n. In other words, $\sigma(n)/n$ exceeds $\sigma(k)/k$ for all $k < n$, where $\sigma(n)$ is the DIVISOR FUNCTION. They were called highly composite numbers by Ramanujan, who found the first 100 or so, and superabundant by Alaoglu and Erdős (1944).

There are an infinite numbers of highly composite numbers, and the first few are 2, 4, 6, 12, 24, 36, 48, 60, 120, 180, 240, 360, 720, 840, 1260, 1680, 2520, 5040, ... (Sloane's A002182). Ramanujan (1915) listed 102 up to 6746328388800 (but omitted 293, 318, 625, 600, and 29331862500). Robin (1983) gives the first 5000 highly composite numbers, and a comprehensive survey is given by Nicholas (1988).

If

$$N = 2^{a_2}3^{a_3}\cdots p^{a_p} \tag{1}$$

is the prime decomposition of a highly composite number, then

1. The PRIMES 2, 3, ..., p form a string of consecutive PRIMES,
2. The exponents are nonincreasing, so $a_2 \geq a_3 \geq \ldots \geq a_p$, and
3. The final exponent a_p is always 1, except for the two cases $N = 4 = 2^2$ and $N = 36 = 2^2 \cdot 3^2$, where it is 2.

Let $Q(x)$ be the number of highly composite numbers $\leq x$. Ramanujan (1915) showed that

$$\lim_{x\to\infty} \frac{Q(x)}{\ln x} = \infty. \tag{2}$$

Erdős (1944) showed that there exists a constant $c_1 > 0$ such that

$$Q(x) \geq (\ln x)^{1+c_1} \tag{3}$$

Nicholas proved that there exists a constant $c_2 > 0$ such that

$$Q(x) \ll (\ln x)^{c_2}. \tag{4}$$

see also ABUNDANT NUMBER

References

Alaoglu, L. and Erdős, P. "On Highly Composite and Similar Numbers." *Trans. Amer. Math. Soc.* **56**, 448–469, 1944.

Andree, R. V. "Ramanujan's Highly Composite Numbers." *Abacus* **3**, 61–62, 1986.

Berndt, B. C. *Ramanujan's Notebooks, Part IV.* New York: Springer-Verlag, p. 53, 1994.

Dickson, L. E. *History of the Theory of Numbers, Vol. 1: Divisibility and Primality.* New York: Chelsea, p. 323, 1952.

Flammenkamp, A. `http://www.minet.uni-jena.de/~achim/highly.html`.

Honsberger, R. *Mathematical Gems I.* Washington, DC: Math. Assoc. Amer., p. 112, 1973.

Honsberger, R. "An Introduction to Ramanujan's Highly Composite Numbers." Ch. 14 in *Mathematical Gems III.* Washington, DC: Math. Assoc. Amer., pp. 193–207, 1985.

Kanigel, R. *The Man Who Knew Infinity: A Life of the Genius Ramanujan.* New York: Washington Square Press, p. 232, 1991.

Nicholas, J.-L. "On Highly Composite Numbers." In *Ramanujan Revisited: Proceedings of the Centenary Conference* (Ed. G. E. Andrews, B. C. Berndt, and R. A. Rankin). Boston, MA: Academic Press, pp. 215–244, 1988.

Ramanujan, S. "Highly Composite Numbers." *Proc. London Math. Soc.* **14**, 347–409, 1915.

Ramanujan, S. *Collected Papers.* New York: Chelsea, 1962.

Robin, G. "Méthodes d'optimalisation pour un problème de théories des nombres." *RAIRO Inform. Théor.* **17**, 239–247, 1983.

Sloane, N. J. A. Sequence A002182/M1025 in "An On-Line Version of the Encyclopedia of Integer Sequences."

Wells, D. *The Penguin Dictionary of Curious and Interesting Numbers.* New York: Penguin Books, p. 128, 1986.

Higman-Sims Group

The SPORADIC GROUP *HS*.

References

Wilson, R. A. "ATLAS of Finite Group Representation." `http://for.mat.bham.ac.uk/atlas/HS.html`.

Hilbert's Axioms

The 21 assumptions which underlie the GEOMETRY published in Hilbert's classic text *Grundlagen der Geometrie.* The eight INCIDENCE AXIOMS concern collinearity and intersection and include the first of EUCLID'S POSTULATES. The four ORDERING AXIOMS concern the arrangement of points, the five CONGRUENCE AXIOMS concern geometric equivalence, and the three CONTINUITY AXIOMS concern continuity. There is also a single parallel axiom equivalent to Euclid's PARALLEL POSTULATE.

see also CONGRUENCE AXIOMS, CONTINUITY AXIOMS, INCIDENCE AXIOMS, ORDERING AXIOMS, PARALLEL POSTULATE

References

Hilbert, D. *The Foundations of Geometry, 2nd ed.* Chicago, IL: Open Court, 1980.

Iyanaga, S. and Kawada, Y. (Eds.). "Hilbert's System of Axioms." §163B in *Encyclopedic Dictionary of Mathematics.* Cambridge, MA: MIT Press, pp. 544–545, 1980.

Hilbert Basis Theorem

If R is a NOETHERIAN RING, then $S = R[X]$ is also a NOETHERIAN RING.

see also ALGEBRAIC VARIETY, FUNDAMENTAL SYSTEM, SYZYGY

References

Hilbert, D. "Über die Theorie der algebraischen Formen." *Math. Ann.* **36**, 473–534, 1890.

Hilbert's Constants

N.B. A detailed on-line essay by S. Finch was the starting point for this entry.

Extend HILBERT'S INEQUALITY by letting $p, q > 1$ and

$$\frac{1}{p} + \frac{1}{q} \geq 1, \tag{1}$$

so that

$$0 < \lambda = 2 - \frac{1}{p} - \frac{1}{q} \leq 1. \tag{2}$$

Levin (1937) and SteČkin (1949) showed that

$$\sum_{m=1}^{\infty}\sum_{n=1}^{\infty} \frac{a_m b_n}{(m+n)^\lambda} \leq \left\{\pi \csc\left[\frac{\pi(q-1)}{\lambda q}\right]\right\}^\lambda \times \left(\int_0^\infty [f(x)]^p\, dx\right)^{1/p} \left(\int_0^\infty [g(x)]^q\, dx\right)^{1/q} \tag{3}$$

and

$$\int_0^\infty \int_0^\infty \frac{f(x)g(y)}{(x+y)^\lambda}\, dx\, dy < \pi \csc\left[\frac{\pi(q-1)}{p}\right]^\lambda \times \left(\int_0^\infty [f(x)]^p\, dx\right)^{1/p} \left(\int_0^\infty [g(x)]^q\, dx\right)^{1/q}. \tag{4}$$

Mitrinovic *et al.* (1991) indicate that this constant is the best possible.

see also HILBERT'S INEQUALITY

References

Finch, S. "Favorite Mathematical Constants." `http://www.mathsoft.com/asolve/constant/hilbert/hilbert.html`.

Mitrinovic, D. S.; Pecaric, J. E.; and Fink, A. M. *Inequalities Involving Functions and Their Integrals and Derivatives.* Dordrecht, Netherlands: Kluwer, 1991.

SteČkin, S. B. "On the Degree of Best Approximation to Continuous Functions." *Dokl. Akad. Nauk SSSR* **65**, 135–137, 1949.

Hilbert Curve

A LINDENMAYER SYSTEM invented by Hilbert (1891) whose limit is a PLANE-FILLING CURVE which fills a square. Traversing the VERTICES of an n-D HYPERCUBE in GRAY CODE order produces a generator for the n-D Hilbert curve (Goetz). The Hilbert curve can be simply encoded with initial string `"L"`, STRING REWRITING rules `"L" -> "+RF-LFL-FR+"`, `"R"->"-LF+RFR+FL-"`, and angle 90° (Peitgen and Saupe 1988, p. 278).

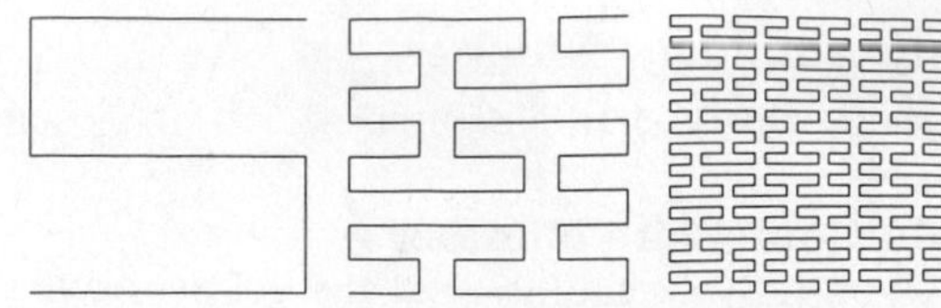

A related curve is the Hilbert II curve, shown above (Peitgen and Saupe 1988, p. 284). It is also a LINDENMAYER SYSTEM and the curve can be encoded with initial string `"X"`, STRING REWRITING rules `"X" -> "XFYFX+F+YFXFY-F-XFYFX"`, `"Y" -> "YFXFY-F-XFYFX+F+YFXFY"`, and angle 90°.

see also LINDENMAYER SYSTEM, PEANO CURVE, PLANE-FILLING CURVE, SIERPIŃSKI CURVE, SPACE-FILLING CURVE

References

Bogomolny, A. "Plane Filling Curves." `http://www.cut-the-knot.com/do_you_know/hilbert.html`.

Dickau, R. M. "Two-Dimensional L-Systems." `http://forum.swarthmore.edu/advanced/robertd/lsys2d.html`.

Dickau, R. M. "Three-Dimensional L-Systems." `http://forum.swarthmore.edu/advanced/robertd/lsys3d.html`.

Goetz, P. "Phil's Good Enough Complexity Dictionary." `http://www.cs.buffalo.edu/~goetz/dict.html`.

Hilbert, D. "Über die stetige Abbildung einer Linie auf ein Flachenstück." *Math. Ann.* **38**, 459–460, 1891.

Peitgen, H.-O. and Saupe, D. (Eds.). *The Science of Fractal Images.* New York: Springer-Verlag, pp. 278 and 284, 1988.

Wagon, S. *Mathematica in Action.* New York: W. H. Freeman, pp. 198–206, 1991.

Hilbert Function

Let $\Gamma = \{p_1, \ldots, p_m\} \subset \mathbb{P}^2$ be a collection of m distinct points. Then the number of conditions imposed by Γ on forms of degree d is called the Hilbert function h_Γ of Γ. If curves X_1 and X_2 of degrees d and e meet in a collection Γ of $d \cdot e$ points, then for any k, the number $h_\Gamma(k)$ of conditions imposed by Γ on forms of degree k is independent of X_1 and X_2 and is given by

$$h_\Gamma(k) = \binom{k+2}{2} - \binom{k-d+2}{2} - \binom{k-e+2}{2} + \binom{k-d-e+2}{2},$$

where the BINOMIAL COEFFICIENT $\binom{a}{2}$ is taken as 0 if $a < 2$ (Cayley 1843).

References

Eisenbud, D.; Green, M.; and Harris, J. "Cayley-Bacharach Theorems and Conjectures." *Bull. Amer. Math. Soc.* **33**, 295–324, 1996.

Hilbert Hotel

Let a hotel have a DENUMERABLE set of rooms numbered 1, 2, 3, ... Then any finite number n of guests can be accommodated without evicting the current guests by moving the current guests from room i to room $i + n$. Furthermore, a DENUMERABLE number of guests can be similarly accommodated by moving the existing guests from i to $2i$, freeing up a DENUMERABLE number of rooms $2i - 1$.

References

Lauwerier, H. "Hilbert Hotel." In *Fractals: Endlessly Repeated Geometric Figures.* Princeton, NJ: Princeton University Press, p. 22, 1991.

Pappas, T. "Hotel Infinity." *The Joy of Mathematics.* San Carlos, CA: Wide World Publ./Tetra, p. 37, 1989.

Hilbert's Inequality

Given a POSITIVE SEQUENCE $\{a_n\}$,

$$\sqrt{\sum_{j=-\infty}^{\infty}\left|\sum_{\substack{n=-\infty\\ n\neq j}}^{\infty}\frac{a_n}{j-n}\right|^2} \leq \pi\sqrt{\sum_{n=-\infty}^{\infty}|a_n|^2},$$

where the a_ns are REAL and "square summable."

Another INEQUALITY known as Hilbert's applies to NONNEGATIVE sequences $\{a_n\}$ and $\{b_n\}$,

$$\sum_{m=1}^{\infty}\sum_{n=1}^{\infty}\frac{a_m b_m}{m+n} < \pi\csc\left(\frac{\pi}{p}\right)\left(\sum_{m=1}^{\infty}{a_m}^p\right)^{1/p}\left(\sum_{n=1}^{\infty}{b_n}^q\right)^{1/q}$$

unless all a_n or all b_n are 0. If $f(x)$ and $g(x)$ are NONNEGATIVE integrable functions, then the integral form is

$$\int_0^\infty\int_0^\infty\frac{f(x)g(y)}{x+y}\,dx\,dy < \pi\csc\left(\frac{\pi}{p}\right) \times\left(\int_0^\infty [f(x)]^p\,dx\right)^{1/p}\left(\int_0^\infty [g(x)]^q\,dx\right)^{1/q}.$$

The constant $\pi\csc(\pi/P)$ is the best possible, in the sense that counterexamples can be constructed for any smaller value.

References

Hardy, G. H.; Littlewood, J. E.; and Pólya, G. *Inequalities, 2nd ed.* Cambridge, England: Cambridge University Press, pp. 308–309, 1988.

Hilbert Matrix

A MATRIX H with elements

$$H_{ij} \equiv (i+j-1)^{-1}$$

for $i, j = 1, 2, \ldots, n$. Although the MATRIX INVERSE is given analytically by

$$(H^{-1})_{ij} = \frac{(-1)^{i+j}}{i+j-1}\frac{(n+i-1)!(n+j-1)!}{[(i-1)!(j-1)!]^2(n-i)!(n-j)!},$$

Hilbert matrices are difficult to invert numerically. The DETERMINANTS for the first few values of H_n are given in the following table.

n	$\det(\mathsf{H}_n)$
1	1
2	8.33333×10^{-2}
3	4.62963×10^{-4}
4	1.65344×10^{-7}
5	3.74930×10^{-12}
6	5.36730×10^{-18}

Hilbert's Nullstellansatz

Let K be an algebraically closed field and let I be an IDEAL in $K(x)$, where $x = (x_1, x_2, \ldots, x_n)$ is a finite set of indeterminates. Let $p \in K(x)$ be such that for any $(c_1, \ldots, c_n)$ in K^n, if every element of I vanishes when evaluated if we set each $(x_i = c_i)$, then p also vanishes. Then p^j lies in I for some j. Colloquially, the theory of algebraically closed fields is a complete model.

Hilbert Number

see GELFOND-SCHNEIDER CONSTANT

Hilbert Polynomial

Let Γ be an ALGEBRAIC CURVE in a projective space of DIMENSION n, and let p be the PRIME IDEAL defining Γ, and let $\chi(p, m)$ be the number of linearly independent forms of degree m modulo p. For large m, $\chi(p, m)$ is a POLYNOMIAL known as the Hilbert polynomial.

References

Iyanaga, S. and Kawada, Y. (Eds.). *Encyclopedic Dictionary of Mathematics.* Cambridge, MA: MIT Press, p. 36, 1980.

Hilbert's Problems

A set of (originally) unsolved problems in mathematics proposed by Hilbert. Of the 23 total, ten were presented at the Second International Congress in Paris in 1900. These problems were designed to serve as examples for the kinds of problems whose solutions would lead to the furthering of disciplines in mathematics.

1a. Is there a transfinite number between that of a DENUMERABLE SET and the numbers of the CONTINUUM? This question was answered by Gödel and Cohen to the effect that the answer depends on the particular version of SET THEORY assumed.

1b. Can the CONTINUUM of numbers be considered a WELL-ORDERED SET? This question is related to Zermelo's AXIOM OF CHOICE. In 1963, the AXIOM OF CHOICE was demonstrated to be independent of all other AXIOMS in SET THEORY, so there appears to be no universally valid solution to this question either.

2. Can it be proven that the AXIOMS of logic are consistent? GÖDEL'S INCOMPLETENESS THEOREM indicated that the answer is "no," in the sense

that any formal system interesting enough to formulate its own consistency can prove its own consistency IFF it is inconsistent.

3. Give two TETRAHEDRA which cannot be decomposed into congruent TETRAHEDRA directly or by adjoining congruent TETRAHEDRA. Max Dehn showed this could not be done in 1902. W. F. Kagon obtained the same result independently in 1903.
4. Find GEOMETRIES whose AXIOMS are closest to those of EUCLIDEAN GEOMETRY if the ORDERING and INCIDENCE AXIOMS are retained, the CONGRUENCE AXIOMS weakened, and the equivalent of the PARALLEL POSTULATE omitted. This problem was solved by G. Hamel.
5. Can the assumption of differentiability for functions defining a continuous transformation GROUP be avoided? (This is a generalization of the CAUCHY FUNCTIONAL EQUATION.) Solved by John von Neumann in 1930 for bicompact groups. Also solved for the ABELIAN case, and for the solvable case in 1952 with complementary results by Montgomery and Zipin (subsequently combined by Yamabe in 1953). Andrew Glean showed in 1952 that the answer is also "yes" for all locally bicompact groups.
6. Can physics be axiomized?
7. Let $\alpha \neq 1 \neq 0$ be ALGEBRAIC and β IRRATIONAL. Is α^β then TRANSCENDENTAL? Proved true in 1934 by Aleksander Gelfond (GELFOND'S THEOREM; Courant and Robins 1996).
8. Prove the RIEMANN HYPOTHESIS. The CONJECTURE has still been neither proved nor disproved.
9. Construct generalizations of the RECIPROCITY THEOREM of NUMBER THEORY.
10. Does there exist a universal algorithm for solving DIOPHANTINE EQUATIONS? The impossibility of obtaining a general solution was proven by Julia Robinson and Martin Davis in 1970, following proof of the result that the equation $n = F_{2m}$ (where F_{2m} is a FIBONACCI NUMBER) is Diophantine by Yuri Matijasevich (Matijasevič 1970, Davis 1973, Davis and Hersh 1973, Matijasevič 1993).
11. Extend the results obtained for quadratic fields to arbitrary INTEGER algebraic fields.
12. Extend a theorem of Kronecker to arbitrary algebraic fields by explicitly constructing Hilbert class fields using special values. This calls for the construction of HOLOMORPHIC FUNCTIONS in several variables which have properties analogous to the exponential function and elliptic modular functions (Holtzapfel 1995).
13. Show the impossibility of solving the general seventh degree equation by functions of two variables.
14. Show the finiteness of systems of relatively integral functions.
15. Justify Schubert's ENUMERATIVE GEOMETRY (Bell 1945).
16. Develop a topology of REAL algebraic curves and surfaces. The SHIMURA-TANIYAMA CONJECTURE postulates just this connection. See Ilyashenko and Yakovenko (1995) and Gudkov and Utkin (1978).
17. Find a representation of definite form by SQUARES.
18. Build spaces with congruent POLYHEDRA.
19. Analyze the analytic character of solutions to variational problems.
20. Solve general BOUNDARY VALUE PROBLEMS.
21. Solve differential equations given a MONODROMY GROUP. More technically, prove that there always exists a FUCHSIAN SYSTEM with given singularities and a given MONODROMY GROUP. Several special cases had been solved, but a NEGATIVE solution was found in 1989 by B. Bolibruch (Anasov and Bolibruch 1994).
22. Uniformization.
23. Extend the methods of CALCULUS OF VARIATIONS.

References

Anasov, D. V. and Bolibruch, A. A. *The Riemann-Hilbert Problem.* Braunschweig, Germany: Vieweg, 1994.
Bell, E. T. *The Development of Mathematics, 2nd ed.* New York: McGraw-Hill, p. 340, 1945.
Borowski, E. J. and Borwein, J. M. (Eds.). "Hilbert Problems." Appendix 3 in *The Harper Collins Dictionary of Mathematics.* New York: Harper-Collins, p. 659, 1991.
Boyer, C. and Merzbach, U. "The Hilbert Problems." *History of Mathematics, 2nd ed.* New York: Wiley, pp. 610–614, 1991.
Browder, Felix E. (Ed.). *Mathematical Developments Arising from Hilbert Problems.* Providence, RI: Amer. Math. Soc., 1976.
Courant, R. and Robbins, H. *What is Mathematics?: An Elementary Approach to Ideas and Methods, 2nd ed.* Oxford, England: Oxford University Press, p. 107, 1996.
Davis, M. "Hilbert's Tenth Problem is Unsolvable." *Amer. Math. Monthly* **80**, 233–269, 1973.
Davis, M. and Hersh, R. "Hilbert's 10th Problem." *Sci. Amer.*, pp. 84–91, Nov. 1973.
Gudkov, D. and Utkin, G. A. *Nine Papers on Hilbert's 16th Problem.* Providence, RI: Amer. Math. Soc., 1978.
Holtzapfel, R.-P. *The Ball and Some Hilbert Problems.* Boston, MA: Birkhäuser, 1995.
Ilyashenko, Yu. and Yakovenko, S. (Eds.). *Concerning the Hilbert 16th Problem.* Providence, RI: Amer. Math. Soc., 1995.
Matijasevič, Yu. V. "Solution to of the Tenth Problem of Hilbert." *Mat. Lapok* **21**, 83–87, 1970.
Matijasevich, Yu. V. *Hilbert's Tenth Problem.* Cambridge, MA: MIT Press, 1993.

Hilbert-Schmidt Norm

The Hilbert-Schmidt norm of a MATRIX A is defined as

$$|\mathsf{A}|_2 \equiv \sqrt{\sum_{i,j} a_{ij}}.$$

Hilbert-Schmidt Theory

The study of linear integral equations of the Fredholm type with symmetric kernels

$$K(x,t) = K(t,x).$$

References

Arfken, G. "Hilbert-Schmidt Theory." §16.4 in *Mathematical Methods for Physicists, 3rd ed.* Orlando, FL: Academic Press, pp. 890–897, 1985.

Hilbert Space

A Hilbert space is VECTOR SPACE H with an INNER PRODUCT $\langle f, g\rangle$ such that the NORM defined by

$$|f| = \sqrt{\langle f, f\rangle}$$

turns H into a COMPLETE METRIC SPACE. If the INNER PRODUCT does not so define a NORM, it is instead known as an INNER PRODUCT SPACE.

Examples of FINITE-dimensional Hilbert spaces include

1. The REAL NUMBERS $\mathbb{R}^n$ with $\langle v, u\rangle$ the vector DOT PRODUCT of v and u.
2. The COMPLEX NUMBERS $\mathbb{C}^n$ with $\langle v, u\rangle$ the vector DOT PRODUCT of v and the COMPLEX CONJUGATE of u.

An example of an INFINITE-dimensional Hilbert space is L^2, the SET of all FUNCTIONS $f : \mathbb{R} \to \mathbb{R}$ such that the INTEGRAL of f^2 over the whole REAL LINE is FINITE. In this case, the INNER PRODUCT is

$$\langle f, g\rangle = \int f(x) g(x)\, dx.$$

A Hilbert space is always a BANACH SPACE, but the converse need not hold.

see also BANACH SPACE, L_2-NORM, L_2-SPACE, LIOUVILLE SPACE, PARALLELOGRAM LAW, VECTOR SPACE

References

Sansone, G. "Elementary Notions of Hilbert Space." §1.3 in *Orthogonal Functions, rev. English ed.* New York: Dover, pp. 5–10, 1991.

Stone, M. H. *Linear Transformations in Hilbert Space and Their Applications Analysis.* Providence, RI: Amer. Math. Soc., 1932.

Hilbert's Theorem

Every MODULAR SYSTEM has a MODULAR SYSTEM BASIS consisting of a finite number of POLYNOMIALS. Stated another way, for every order n there exists a non-singular curve with the maximum number of circuits and the maximum number for any one nest.

References

Coolidge, J. L. *A Treatise on Algebraic Plane Curves.* New York: Dover, p. 61, 1959.

Hilbert Transform

$$g(y) = \frac{1}{\pi} \int_{-\infty}^{\infty} \frac{f(x)\, dx}{x - y}$$

$$f(x) = \frac{1}{\pi} \int_{-\infty}^{\infty} \frac{g(y)\, dy}{y - x}.$$

see also TITCHMARSH THEOREM

References

Bracewell, R. *The Fourier Transform and Its Applications.* New York: McGraw-Hill, pp. 267–272, 1965.

Hill Determinant

A DETERMINANT which arises in the solution of the second-order ORDINARY DIFFERENTIAL EQUATION

$$x^2 \frac{d^2\psi}{dx^2} + x\frac{d\psi}{dx} + \left(\tfrac{1}{4}h^2x^2 + \tfrac{1}{2}h^2 - b + \frac{h^2}{4x^2}\right)\psi = 0. \quad (1)$$

Writing the solution as a POWER SERIES

$$\psi = \sum_{n=-\infty}^{\infty} a_n x^{s+2n} \quad (2)$$

gives a RECURRENCE RELATION

$$h^2 a_{n+1} + [2h^2 - 4b + 16(n + \tfrac{1}{2}s)^2] a_n + h^2 a_{n-1} = 0. \quad (3)$$

The value of s can be computed using the Hill determinant

$$\Delta(s) = \begin{vmatrix} \ddots & \vdots & \vdots & \vdots & \vdots & ⋰ \\ \cdots & \frac{(\sigma+2)-\alpha^2}{4-\alpha^2} & \frac{\beta^2}{4-\alpha^2} & 0 & 0 & \cdots \\ \cdots & 0 & -\frac{\beta^2}{\alpha^2} & -\frac{\sigma^2-\alpha^2}{\alpha^2} & -\frac{\beta^2}{\alpha^2} & \cdots \\ \cdots & 0 & 0 & -\frac{\beta^2}{1-\alpha^2} & \frac{(\sigma-1)^2-\alpha^2}{1-\alpha^2} & \cdots \\ ⋰ & \vdots & \vdots & \vdots & \vdots & \ddots \end{vmatrix}, \quad (4)$$

where

$$\sigma = \tfrac{1}{2}s \quad (5)$$

$$\alpha^2 = \tfrac{1}{4}b - \tfrac{1}{8}h^2 \quad (6)$$

$$\beta = \tfrac{1}{4}h, \quad (7)$$

and σ is the variable to solve for. The determinant can be given explicitly by the amazing formula

$$\Delta(s) = \Delta(0) - \frac{\sin^2(\pi s/2)}{\sin^2(\frac{1}{2}\pi\sqrt{b - \frac{1}{2}h^2})}, \tag{8}$$

where

$$\Delta(0) = \begin{vmatrix} \ddots & \vdots & \vdots & \vdots & \vdots & \cdot^{\cdot^{\cdot}} \\ \cdots & 1 & \frac{h^2}{144+2h^2-4b} & 0 & 0 & \cdots \\ \cdots & \frac{h^2}{64+2h^2-4b} & 1 & \frac{h^2}{64+2h^2-4b} & 0 & \cdots \\ \cdots & 0 & \frac{h^2}{16+2h^2-4b} & 1 & \frac{h^2}{16+2h^2-4b} & \cdots \\ \cdots & 0 & 0 & \frac{h^2}{2h^2-4b} & 1 & \cdots \\ \cdots & 0 & 0 & 0 & \frac{h^2}{16+2h^2-4b} & \cdots \\ \cdot^{\cdot^{\cdot}} & \vdots & \vdots & \vdots & \vdots & \ddots \end{vmatrix}, \tag{9}$$

leading to the implicit equation for s,

$$\sin^2(\tfrac{1}{2}\pi s) = \Delta(0)\sin^2\left(\tfrac{1}{2}\pi\sqrt{b - \tfrac{1}{2}h^2}\right). \tag{10}$$

see also HILL'S DIFFERENTIAL EQUATION

References

Morse, P. M. and Feshbach, H. *Methods of Theoretical Physics, Part I.* New York: McGraw-Hill, pp. 555–562, 1953.

Hill's Differential Equation

$$\frac{d^2x}{dt^2} = \phi(t)x,$$

where ϕ is periodic. It can be written as

$$\frac{d^2y}{dx^2} + \left[\theta_0 + 2\sum_{n=1}^{\infty}\theta_n\cos(2nz)\right] = 0,$$

where θ_n are known constants. A solution can be given by taking the "DETERMINANT" of an infinite MATRIX.

see also HILL DETERMINANT

Hillam's Theorem

If $f : [a,b] \to [a,b]$ (where $[a,b]$ denotes the CLOSED INTERVAL from a to b on the REAL LINE) satisfies a LIPSCHITZ CONDITION with constant K, i.e., if

$$|f(x) - f(y)| \leq K|x - y|$$

for all $x, y \in [a,b]$, then the iteration scheme

$$x_{n+1} = (1 - \lambda)x_n + \lambda f(x_n),$$

where $\lambda = 1/(K+1)$, converges to a FIXED POINT of f.

References

Falkowski, B.-J. "On the Convergence of Hillam's Iteration Scheme." *Math. Mag.* **69**, 299–303, 1996.

Geist, R.; Reynolds, R.; and Suggs, D. "A Markovian Framework for Digital Halftoning." *ACM Trans. Graphics* **12**, 136–159, 1993.

Hillam, B. P. "A Generalization of Krasnoselski's Theorem on the Real Line." *Math. Mag.* **48**, 167–168, 1975.

Krasnoselski, M. A. "Two Remarks on the Method of Successive Approximations." *Uspehi Math. Nauk (N. S.)* **10**, 123–127, 1955.

Hindu Check

see CASTING OUT NINES

Hinge

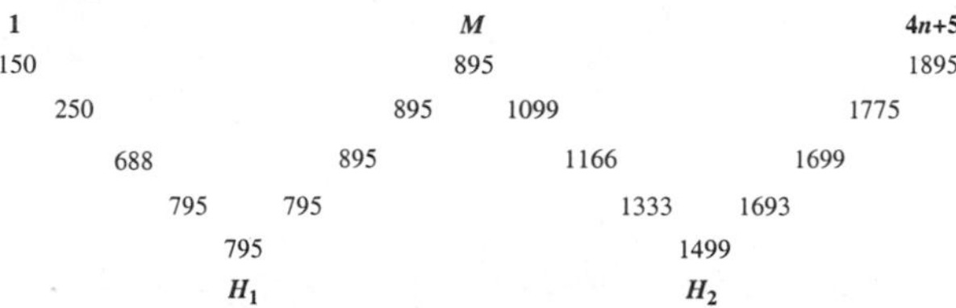

The upper and lower hinges are descriptive statistics of a set of N data values, where N is of the form $N = 4n+5$ with $n = 0, 1, 2, \ldots$. The hinges are obtained by ordering the data in increasing order $a_1, \ldots, a_N$, and writing them out in the shape of a "w" as illustrated above. The values at the bottom legs are called the hinges H_1 and H_2 (and the central peak is the MEDIAN). In this ordering,

$$H_1 = a_{n+2} = a_{(N+3)/4}$$
$$M = a_{2n+3} = a_{(N+1)/2}$$
$$H_2 = a_{3n+4} = a_{(3N+1)/4}.$$

For N of the form $4n + 5$, the hinges are identical to the QUARTILES. The difference $H_2 - H_1$ is called the H-SPREAD.

see also H-SPREAD, HABERDASHER'S PROBLEM, MEDIAN (STATISTICS), ORDER STATISTIC, QUARTILE, TRIMEAN

References

Tukey, J. W. *Explanatory Data Analysis.* Reading, MA: Addison-Wesley, pp. 32–34, 1977.

Hippias' Quadratrix

see QUADRATRIX OF HIPPIAS

Hippopede

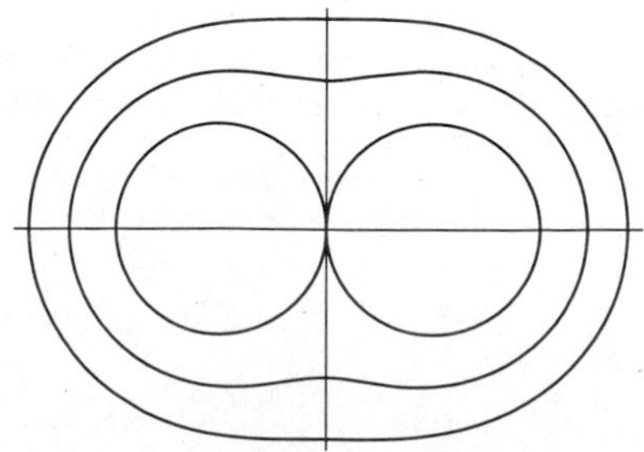

A curve also known as a HORSE FETTER and given by the polar equation

$$r^2 = 4b(a - b\sin^2\theta).$$

References
Lawrence, J. D. *A Catalog of Special Plane Curves.* New York: Dover, pp. 144–146, 1972.

Histogram

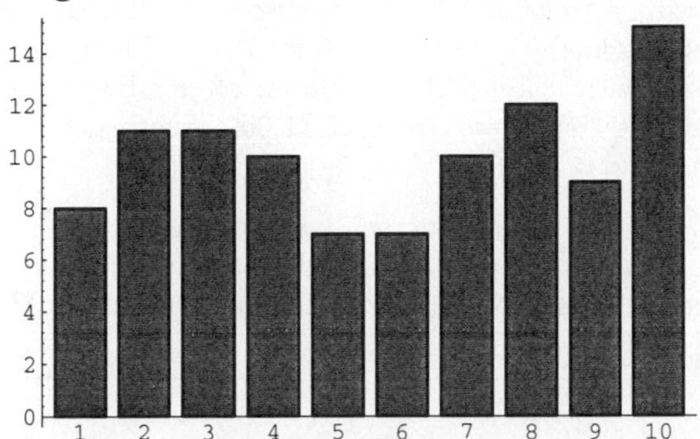

The grouping of data into bins (spaced apart by the so-called CLASS INTERVAL) plotting the number of members in each bin versus the bin number. The above histogram shows the number of variates in bins with CLASS INTERVAL 1 for a sample of 100 real variates with a UNIFORM DISTRIBUTION from 0 and 10. Therefore, bin 1 gives the number of variates in the range 0–1, bin 2 gives the number of variates in the range 1–2, etc.

see also OGIVE

Hitch

A KNOT that secures a rope to a post, ring, another rope, etc., but does not keep its shape by itself.

see also CLOVE HITCH, KNOT, LINK, LOOP (KNOT)

References
Owen, P. *Knots.* Philadelphia, PA: Courage, p. 17, 1993.

Hitting Set

Let S be a collection S of subsets of a finite set X. The smallest subset Y of X that meets every member of S is called the hitting set or VERTEX COVER. Finding the hitting set is an NP-COMPLETE PROBLEM.

Hjelmslev's Theorem

When all the points P on one line are related by an ISOMETRY to all points P' on another, the MIDPOINTS of the segments PP' are either distinct and collinear or coincident.

HJLS Algorithm

An algorithm for finding INTEGER RELATIONS whose running time is bounded by a polynomial in the number of real variables. It is much faster than other algorithms such as the FERGUSON-FORCADE ALGORITHM, LLL ALGORITHM, and PSOS ALGORITHM.

Unfortunately, it is numerically unstable and therefore requires extremely high precision. The cause of this instability is not known (Ferguson and Bailey 1992), but is believed to derive from its reliance on GRAM-SCHMIDT ORTHONORMALIZATION, which is know to be numerically unstable (Golub and van Loan 1989).

see also FERGUSON-FORCADE ALGORITHM, INTEGER RELATION, LLL ALGORITHM, PSLQ ALGORITHM, PSOS ALGORITHM

References
Ferguson, H. R. P. and Bailey, D. H. "A Polynomial Time, Numerically Stable Integer Relation Algorithm." RNR Techn. Rept. RNR-91-032, Jul. 14, 1992.
Golub, G. H. and van Loan, C. F. *Matrix Computations, 3rd ed.* Baltimore, MD: Johns Hopkins, 1996.
Hastad, J.; Just, B.; Lagarias, J. C.; and Schnorr, C. P. "Polynomial Time Algorithms for Finding Integer Relations Among Real Numbers." *SIAM J. Comput.* **18**, 859–881, 1988.

HK Integral

Named after Henstock and Kurzweil. Every LEBESGUE INTEGRABLE function is HK integrable with the same value.

References
Shenitzer, A. and Steprans, J. "The Evolution of Integration." *Amer. Math. Monthly* **101**, 66–72, 1994.

Hodge Star

On an oriented n-D RIEMANNIAN MANIFOLD, the Hodge star is a linear FUNCTION which converts alternating DIFFERENTIAL k-FORMS to alternating $(n-k)$-forms. If w is an alternating k-FORM, its Hodge star is given by

$$w(v_1, \ldots, v_k) = (^*w)(v_{k+1}, \ldots, v_n)$$

when $v_1, \ldots, v_n$ is an oriented orthonormal basis.

see also STOKES' THEOREM

Hodge's Theorem

On a compact oriented FINSLER MANIFOLD without boundary, every COHOMOLOGY class has a UNIQUE harmonic representative. The DIMENSION of the SPACE of all harmonic forms of degree p is the pth BETTI NUMBER of the MANIFOLD.

see also BETTI NUMBER, COHOMOLOGY, DIMENSION, FINSLER MANIFOLD

References
Chern, S.-S. "Finsler Geometry is Just Riemannian Geometry without the Quadratic Restriction." *Not. Amer. Math. Soc.* **43**, 959–963, 1996.

Hoehn's Theorem

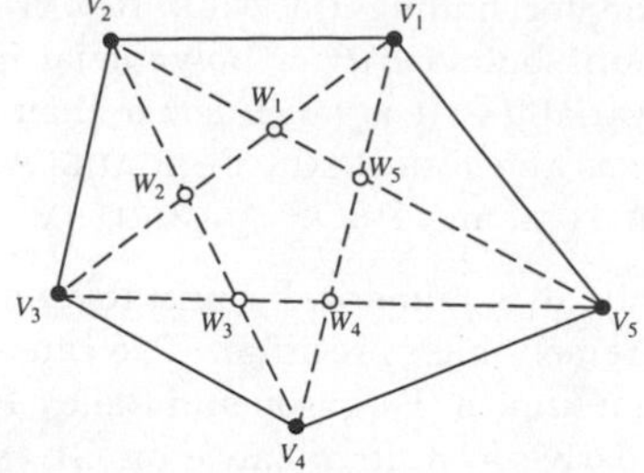

A geometric theorem related to the PENTAGRAM and also called the PRATT-KASAPI THEOREM.

$$\frac{|V_1W_1|}{|W_2V_3|}\frac{|V_2W_2|}{|W_3V_4|}\frac{|V_3W_3|}{|W_4V_5|}\frac{|V_4W_4|}{|W_5V_1|}\frac{|V_5W_5|}{|W_1V_2|}=1$$

$$\frac{|V_1W_2|}{|W_1V_3|}\frac{|V_2W_3|}{|W_2V_4|}\frac{|V_3W_4|}{|W_3V_5|}\frac{|V_4W_5|}{|W_4V_1|}\frac{|V_5W_1|}{|W_5V_2|}=1.$$

In general, it is also true that

$$\frac{|V_iW_i|}{|W_{i+1}V_{i+2}|}=\frac{|V_iV_{i+1}V_{i+4}|}{|V_iV_{i+1}V_{i+2}V_{i+4}|}\frac{|V_iV_{i+1}V_{i+2}V_{i+3}|}{|V_{i+2}V_{i+3}V_{i+1}|}.$$

This type of identity was generalized to other figures in the plane and their duals by Pinkernell (1996).

References
Chou, S. C. *Mechanical Geometry Theorem Proving.* Dordrecht, Netherlands: Reidel, 1987.
Grünbaum, B. and Shepard, G. C. "Ceva, Menelaus, and the Area Principle." *Math. Mag.* **68**, 254–268, 1995.
Hoehn, L. "A Menelaus-Type Theorem for the Pentagram." *Math. Mag.* **68**, 254–268, 1995.
Pinkernell, G. M. "Identities on Point-Line Figures in the Euclidean Plane." *Math. Mag.* **69**, 377–383, 1996.

Hoffman's Minimal Surface

A minimal embedded surface discovered in 1992 consisting of a helicoid with a HOLE and HANDLE (*Science News* 1992). It has the same topology as a punctured sphere with a handle, and is only the second complete embedded minimal surface of finite topology and infinite total curvature discovered (the HELICOID being the first).

A three-ended minimal surface GENUS 1 is sometimes also called Hoffman's minimal surface (Peterson 1988).

see also HELICOID

References
Peterson, I. *Mathematical Tourist: Snapshots of Modern Mathematics.* New York: W. H. Freeman, pp. 57–59, 1988.
"Putting a Handle on a Minimal Helicoid." *Sci. News* **142**, 276, Oct. 24, 1992.

Hoffman-Singleton Graph

The only GRAPH of DIAMETER 2, GIRTH 5, and VALENCY 7. It contains many copies of the PETERSEN GRAPH.

References
Hoffman, A. J. and Singleton, R. R. "On Moore Graphs of Diameter Two and Three." *IBM J. Res. Develop.* **4**, 497–504, 1960.

Hofstadter-Conway $10,000 Sequence

The INTEGER SEQUENCE defined by the RECURRENCE RELATION

$$a(n)=a(a(n-1))+a(n-a(n-1)),$$

with $a(1)=a(2)=1$. The first few values are 1, 1, 2, 2, 3, 4, 4, 4, 5, 6, ... (Sloane's A004001). Plotting $a(n)/n$ against n gives the BATRACHION plotted below. Conway (1988) showed that $\lim_{n\to\infty}a(n)/n=1/2$ and offered a prize of $10,000 to the discoverer of a value of n for which $|a(i)/i-1/2|<1/20$ for $i>n$. The prize was subsequently claimed by Mallows, after adjustment to Conway's "intended" prize of $1,000 (Schroeder 1991), who found $n=1489$.

$a(n)/n$ takes a value of $1/2$ for n of the form 2^k with $k=1, 2, \ldots$. Pickover (1996) gives a table of analogous values of n corresponding to different values of $|a(n)/n-1/2|<e$.

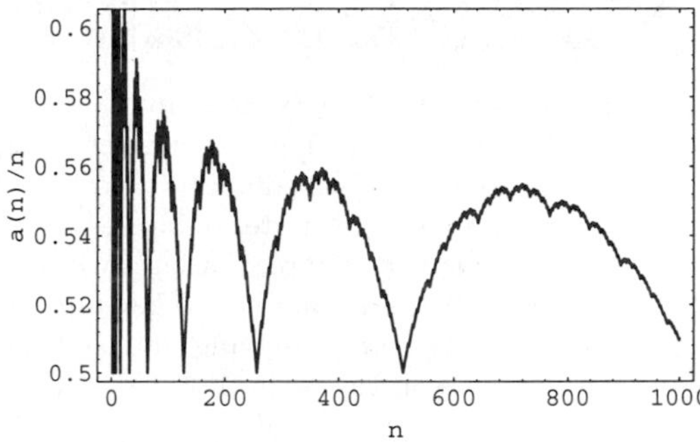

see also BLANCMANGE FUNCTION, HOFSTADTER'S Q-SEQUENCE, MALLOW'S SEQUENCE

References
Conolly, B. W. "Meta-Fibonacci Sequences." In *Fibonacci and Lucas Numbers, and the Golden Section* (Ed. S. Vajda). New York: Halstead Press, pp. 127–138, 1989.
Conway, J. "Some Crazy Sequences." Lecture at AT&T Bell Labs, July 15, 1988.
Guy, R. K. "Three Sequences of Hofstadter." §E31 in *Unsolved Problems in Number Theory, 2nd ed.* New York: Springer-Verlag, pp. 231–232, 1994.
Kubo, T. and Vakil, R. "On Conway's Recursive Sequence." *Disc. Math.* **152**, 225–252, 1996.
Mallows, C. "Conway's Challenging Sequence." *Amer. Math. Monthly* **98**, 5–20, 1991.
Pickover, C. A. "The Drums of Ulupu." In *Mazes for the Mind: Computers and the Unexpected.* New York: St. Martin's Press, 1993.
Pickover, C. A. "The Crying of Fractal Batrachion 1,489." Ch. 25 in *Keys to Infinity.* New York: W. H. Freeman, pp. 183–191, 1995.
Schroeder, M. "John Horton Conway's 'Death Bet.'" *Fractals, Chaos, Power Laws.* New York: W. H. Freeman, pp. 57–59, 1991.
Sloane, N. J. A. Sequence A004001/M0276 in "An On-Line Version of the Encyclopedia of Integer Sequences."

Hofstadter Figure-Figure Sequence

Define $F(1) = 1$ and $S(1) = 2$ and write

$$F(n) = F(n-1) + S(n-1),$$

where the sequence $\{S(n)\}$ consists of those integers not already contained in $\{F(n)\}$. For example, $F(2) = F(1) + S(1) = 3$, so the next term of $S(n)$ is $S(2) = 4$, giving $F(3) = F(2) + S(2) = 7$. The next integer is 5, so $S(3) = 5$ and $F(4) = F(3) + S(3) = 12$. Continuing in this manner gives the "figure" sequence $F(n)$ as 1, 3, 7, 12, 18, 26, 35, 45, 56, ... (Sloane's A005228) and the "space" sequence as 2, 4, 5, 6, 8, 9, 10, 11, 13, 14, ... (Sloane's A030124).

References

Hofstadter, D. R. *Gödel, Escher, Bach: An Eternal Golden Braid.* New York: Vintage Books, p. 73, 1989.

Sloane, N. J. A. Sequences A030124 and A005288/M2629 in "An On-Line Version of the Encyclopedia of Integer Sequences."

Hofstadter G-Sequence

The sequence defined by $G(0) = 0$ and

$$G(n) = n - G(G(n-1)).$$

The first few terms are 1, 1, 2, 3, 3, 4, 4, 5, 6, 6, 7, 8, 8, 9, 9, ... (Sloane's A005206).

References

Hofstadter, D. R. *Gödel, Escher, Bach: An Eternal Golden Braid.* New York: Vintage Books, p. 137, 1989.

Sloane, N. J. A. Sequence A005206/M0436 in "An On-Line Version of the Encyclopedia of Integer Sequences."

Hofstadter H-Sequence

The sequence defined by $H(0) = 0$ and

$$H(n) = n - H(H(H(n-1))).$$

The first few terms are 1, 1, 2, 3, 4, 4, 5, 5, 6, 7, 7, 8, 9, 10, 10, 11, 12, 13, 13, 14, ... (Sloane's A005374).

References

Hofstadter, D. R. *Gödel, Escher, Bach: An Eternal Golden Braid.* New York: Vintage Books, p. 137, 1989.

Sloane, N. J. A. Sequence A005374/M0449 in "An On-Line Version of the Encyclopedia of Integer Sequences."

Hofstadter Male-Female Sequences

The pair of sequences defined by $F(0) = 1$, $M(0) = 0$, and

$$F(n) = n - M(F(n-1))$$
$$M(n) = n - F(M(n-1)).$$

The first few terms of the "male" sequence $M(n)$ are 0, 1, 2, 2, 3, 4, 4, 5, 6, 6, 7, 7, 8, 9, 9, ... (Sloane's A005379), and the first few terms of the "female" sequence $F(n)$ are 1, 2, 2, 3, 3, 4, 5, 5, 6, 6, 7, 8, 8, 9, 9, ... (Sloane's A005378).

References

Hofstadter, D. R. *Gödel, Escher, Bach: An Eternal Golden Braid.* New York: Vintage Books, p. 137, 1989.

Sloane, N. J. A. Sequences A005378/M0263 and A005379/M0278 in "An On-Line Version of the Encyclopedia of Integer Sequences."

Hofstadter Point

The r-HOFSTADTER TRIANGLE of a given TRIANGLE ΔABC is perspective to ΔABC, and the PERSPECTIVE CENTER is called the Hofstadter point. The TRIANGLE CENTER FUNCTION is

$$\alpha = \frac{\sin(rA)}{\sin(r - rA)}.$$

As $r \to 0$, the TRIANGLE CENTER FUNCTION approaches

$$\alpha = \frac{A}{a},$$

and as $r \to 1$, the TRIANGLE CENTER FUNCTION approaches

$$\alpha = \frac{a}{A}.$$

see also HOFSTADTER TRIANGLE

References

Kimberling, C. "Hofstadter Points." *Nieuw Arch. Wiskunder* **12**, 109–114, 1994.

Kimberling, C. "Major Centers of Triangles." *Amer. Math. Monthly* **104**, 431–438, 1997.

Kimberling, C. "Hofstadter Points." `http://www.evansville.edu/~ck6/tcenters/recent/hofstad.html`.

Hofstadter's Q-Sequence

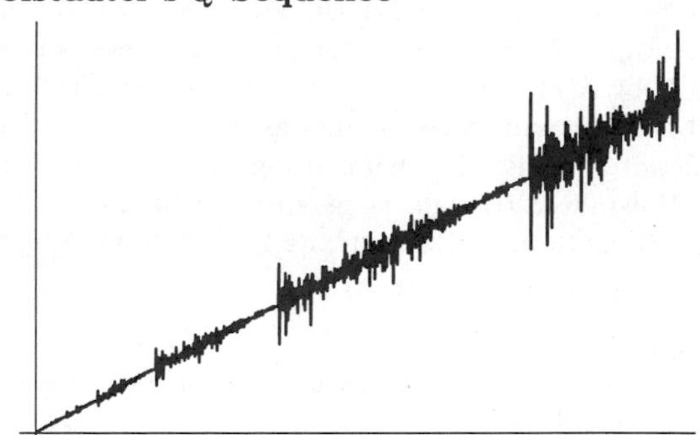

The INTEGER SEQUENCE given by

$$Q(n) = Q(n - Q(n-1)) + Q(n - Q(n-2)),$$

with $Q(1) = Q(2) = 1$. The first few values are 1, 1, 2, 3, 3, 4, 5, 5, 6, 6, ... (Sloane's A005185; illustrated above). These numbers are sometimes called Q-NUMBERS.

see also HOFSTADTER-CONWAY \$10,000 SEQUENCE, MALLOW'S SEQUENCE

References

Conolly, B. W. "Meta-Fibonacci Sequences." In *Fibonacci and Lucas Numbers, and the Golden Section* (Ed. S. Vajda). New York: Halstead Press, pp. 127–138, 1989.

Guy, R. "Some Suspiciously Simple Sequences." *Amer. Math. Monthly* **93**, 186–191, 1986.
Hofstadter, D. R. *Gödel, Escher Bach: An Eternal Golden Braid.* New York: Vintage Books, pp. 137–138, 1980.
Pickover, C. A. "The Crying of Fractal Batrachion 1,489." Ch. 25 in *Keys to Infinity.* New York: W. H. Freeman, pp. 183–191, 1995.
Sloane, N. J. A. Sequence A005185/M0438 in "An On-Line Version of the Encyclopedia of Integer Sequences."

Hofstadter Sequences

Let $b_1 = 1$ and $b_2 = 2$ and for $n \geq 3$, let b_n be the least INTEGER $> b_{n-1}$ which can be expressed as the SUM of two or more consecutive terms. The resulting sequence is 1, 2, 3, 5, 6, 8, 10, 11, 14, 16, ... (Sloane's A005243). Let $c_1 = 2$ and $c_2 = 3$, form all possible expressions of the form $c_i c_j - 1$ for $1 \leq i \leq j \leq n$, and append them. The resulting sequence is 2, 3, 5, 9, 14, 16, 17, 18, ... (Sloane's A05244).

see also HOFSTADTER-CONWAY \$10,000 SEQUENCE, HOFSTADTER'S Q-SEQUENCE

References

Guy, R. K. "Three Sequences of Hofstadter." §E31 in *Unsolved Problems in Number Theory, 2nd ed.* New York: Springer-Verlag, pp. 231–232, 1994.
Sloane, N. J. A. Sequences A005243/M0623 and A00524/M0705 in "An On-Line Version of the Encyclopedia of Integer Sequences."

Hofstadter Triangle

For a NONZERO REAL NUMBER r and a TRIANGLE ΔABC, swing LINE SEGMENT BC about the vertex B towards vertex A through an ANGLE rB. Call the line along the rotated segment L. Construct a second line L' by rotating LINE SEGMENT BC about vertex C through an ANGLE rC. Now denote the point of intersection of L and L' by $A(r)$. Similarly, construct $B(r)$ and $C(r)$. The TRIANGLE having these points as vertices is called the Hofstadter r-triangle. Kimberling (1994) showed that the Hofstadter triangle is perspective to ΔABC, and calls PERSPECTIVE CENTER the HOFSTADTER POINT.

see also HOFSTADTER POINT

References

Kimberling, C. "Hofstadter Points." *Nieuw Arch. Wiskunde* **12**, 109–114, 1994.
Kimberling, C. "Hofstadter Points." `http://www.evansville.edu/~ck6/tcenters/recent/hofstad.html`.

Hölder Condition

A function $\phi(t)$ satisfies the Hölder condition on two points t_1 and t_2 on an arc L when

$$|\phi(t_2) - \phi(t_1)| \leq A|t_2 - t_1|^\mu,$$

with A and μ POSITIVE REAL constants.

Hölder Integral Inequality

If

$$\frac{1}{p} + \frac{1}{q} = 1$$

with p, $q > 1$, then

$$\int_a^b |f(x)g(x)|\,dx \leq \left[\int_a^b |f(x)|^p\,dx\right]^{1/p} \left[\int_a^b |g(x)|^q\,dx\right]^{1/q},$$

with equality when

$$|g(x)| = c|f(x)|^{p-1}.$$

If $p = q = 2$, this inequality becomes SCHWARZ'S INEQUALITY.

References

Abramowitz, M. and Stegun, C. A. (Eds.). *Handbook of Mathematical Functions with Formulas, Graphs, and Mathematical Tables, 9th printing.* New York: Dover, p. 11, 1972.
Gradshteyn, I. S. and Ryzhik, I. M. *Tables of Integrals, Series, and Products, 5th ed.* San Diego, CA: Academic Press, p. 1099, 1993.
Hölder, O. "Über einen Mittelwertsatz." *Göttingen Nachr.*, 44, 1889.
Riesz, F. "Untersuchungen über Systeme integrierbarer Funktionen." *Math. Ann.* **69**, 456, 1910.
Riesz, F. "Su alcune disuguaglianze." *Boll. Un. Mat. It.* **7**, 77–79, 1928.
Sansone, G. *Orthogonal Functions, rev. English ed.* New York: Dover, pp. 32–33, 1991.

Hölder Sum Inequality

If

$$\frac{1}{p} + \frac{1}{q} = 1$$

with p, $q > 1$, then

$$\sum_{k=1}^n |a_k b_k| \leq \left(\sum_{k=1}^n |a_k|^p\right)^{1/p} \left(\sum_{k=1}^n |b_k|^q\right)^{1/q},$$

with equality when $|b_k| = c|a_k|^{p-1}$. If $p = q = 2$, this becomes the CAUCHY INEQUALITY.

References

Abramowitz, M. and Stegun, C. A. (Eds.). *Handbook of Mathematical Functions with Formulas, Graphs, and Mathematical Tables, 9th printing.* New York: Dover, p. 11, 1972.
Gradshteyn, I. S. and Ryzhik, I. M. *Tables of Integrals, Series, and Products, 5th ed.* San Diego, CA: Academic Press, p. 1092, 1979.
Hardy, G. H.; Littlewood, J. E.; and Pólya, G. *Inequalities, 2nd ed.* Cambridge, England: Cambridge University Press, pp. 10–15, 1988.

Hole

A hole in a mathematical object is a TOPOLOGICAL structure which prevents the object from being continuously shrunk to a point. When dealing with TOPOLOGICAL SPACES, a DISCONNECTIVITY is interpreted as a hole in the space. Examples of holes are things like the hole in the "center" of a SPHERE or a CIRCLE and the hole produced in EUCLIDEAN SPACE cutting a KNOT out from it.

Singular HOMOLOGY GROUPS form a MEASURE of the hole structure of a SPACE, but they are one particular measure and they don't always pick up everything. HOMOTOPY GROUPS of a SPACE are another measure of holes in a SPACE, as well as BORDISM GROUPS, k-THEORY, COHOMOTOPY GROUPS, and so on.

There are many ways to measure holes in a space. Some holes are picked up by HOMOTOPY GROUPS that are not picked up by HOMOLOGY GROUPS, and some holes are picked up by HOMOLOGY GROUPS that are not picked up by HOMOTOPY GROUPS. (For example, in the TORUS, HOMOTOPY GROUPS "miss" the two-dimensional hole that is given by the TORUS itself, but the second HOMOLOGY GROUP picks that hole up.) In addition, HOMOLOGY GROUPS don't pick up the varying hole structures of the complement of KNOTS in 3-space, but the first HOMOTOPY GROUP (the fundamental group) does.

see also BRANCH CUT, BRANCH POINT, CORK PLUG, CROSS-CAP, GENUS (SURFACE), SINGULAR POINT (FUNCTION), SPHERICAL RING, TORUS

Holomorphic Function

A synonym for ANALYTIC FUNCTION.

see also ANALYTIC FUNCTION, HOMEOMORPHIC

Holonomic Constant

A limiting value of a HOLONOMIC FUNCTION near a SINGULAR POINT. Holonomic constants include APÉRY'S CONSTANT, CATALAN'S CONSTANT, PÓLYA'S RANDOM WALK CONSTANTS for $d > 2$, and PI.

Holonomic Function

A solution of a linear homogeneous ORDINARY DIFFERENTIAL EQUATION with POLYNOMIAL COEFFICIENTS.

see also HOLONOMIC CONSTANT

References

Zeilberger, D. "A Holonomic Systems Approach to Special Function Identities." *J. Comput. Appl. Math.* **32**, 321–348, 1990.

Holonomy

A general concept in CATEGORY THEORY involving the globalization of topological or differential structures.

see also MONODROMY

Home Plate

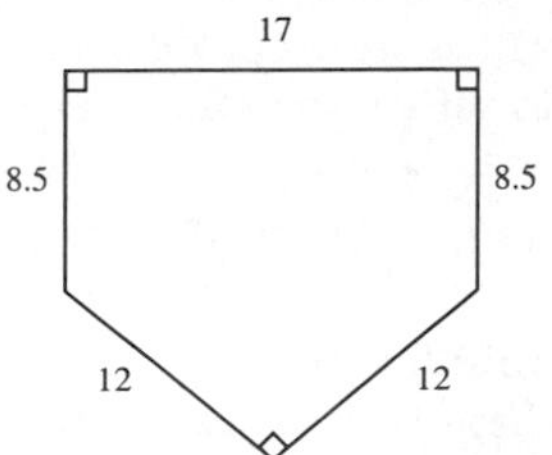

Home plate in the game of BASEBALL is an irregular PENTAGON. However, the Little League rulebook's specification of the shape of home plate (Kreutzer and Kerley 1990), illustrated above, is not physically realizable, since it requires the existence of a (12, 12, 17) RIGHT TRIANGLE, whereas

$$12^2 + 12^2 = 288 \neq 289 = 17^2$$

(Bradley 1996).

see also BASEBALL COVER

References

Bradley, M. J. "Building Home Plate: Field of Dreams or Reality?" *Math. Mag.* **69**, 44–45, 1996.

Kreutzer, P. and Kerley, T. *Little League's Official How-to-Play Baseball Book.* New York: Doubleday, 1990.

Homeoid

A shell bounded by two similar ELLIPSOIDS having a constant ratio of axes. Given a CHORD passing through a homeoid, the distance between inner and outer intersections is equal on both sides. Since a spherical shell is a symmetric case of a homeoid, this theorem is also true for spherical shells (CONCENTRIC CIRCLES in the PLANE), for which it is easily proved by symmetry arguments.

see also CHORD, ELLIPSOID

Homeomorphic

There are two possible definitions:

1. Possessing similarity of form,
2. Continuous, ONE-TO-ONE, ONTO, and having a continuous inverse.

The most common meaning is possessing intrinsic topological equivalence. Two objects are homeomorphic if they can be deformed into each other by a continuous, invertible mapping. Homeomorphism ignores the space in which surfaces are embedded, so the deformation can be completed in a higher dimensional space than the surface was originally embedded. MIRROR IMAGES are homeomorphic, as are MÖBIUS BANDS with an EVEN number of half twists, and MÖBIUS BANDS with an ODD number of twists.

In CATEGORY THEORY terms, homeomorphisms are ISOMORPHISMS in the CATEGORY of TOPOLOGICAL SPACES and continuous maps.

see also HOMOMORPHIC, POLISH SPACE

Homeomorphic Group

If the ELEMENTS of two GROUPS are n to 1 and the correspondences satisfy the same GROUP multiplication table, the GROUPS are said to be homeomorphic.

see also ISOMORPHIC GROUPS

Homeomorphic Type

The following three pieces of information completely determine the homeomorphic type of the surface (Massey 1967):

1. Orientability,
2. Number of boundary components,
3. EULER CHARACTERISTIC.

see also ALGEBRAIC TOPOLOGY, EULER CHARACTERISTIC

References

Massey, W. S. *Algebraic Topology: An Introduction.* New York: Springer-Verlag, 1996.

Homeomorphism

see HOMEOMORPHIC, HOMEOMORPHIC GROUP, HOMEOMORPHIC TYPE, TOPOLOGICALLY CONJUGATE

HOMFLY Polynomial

A 2-variable oriented KNOT POLYNOMIAL $P_L(a,z)$ motivated by the JONES POLYNOMIAL (Freyd *et al.* 1985). Its name is an acronym for the last names of its co-discoverers: Hoste, Ocneanu, Millett, Freyd, Lickorish, and Yetter (Freyd *et al.* 1985). Independent work related to the HOMFLY polynomial was also carried out by Prztycki and Traczyk (1987). HOMFLY polynomial is defined by the SKEIN RELATIONSHIP

$$a^{-1}P_{L_+}(a,z) - aP_{L_-}(a,z) = zP_{L_0}(a,z) \qquad (1)$$

(Doll and Hoste 1991), where v is sometimes written instead of a (Kanenobu and Sumi 1993) or, with a slightly different relationship, as

$$\alpha P_{L_+}(\alpha,z) - \alpha^{-1}P_{L_-}(\alpha,z) = zP_{L_0}(\alpha,z) \qquad (2)$$

(Kauffman 1991). It is also defined as $P_L(\ell,m)$ in terms of SKEIN RELATIONSHIP

$$\ell P_{L_+} + \ell^{-1}P_{L_-} + mP_{L_0} = 0 \qquad (3)$$

(Lickorish and Millett 1988). It can be regarded as a nonhomogeneous POLYNOMIAL in two variables or a homogeneous POLYNOMIAL in three variables. In three variables the SKEIN RELATIONSHIP is written

$$xP_{L_+}(x,y,z) + yP_{L_-}(x,y,z) + zP_{L_0}(x,y,z) = 0. \quad (4)$$

It is normalized so that $P_{\text{unknot}} = 1$. Also, for n unlinked unknotted components,

$$P_L(x,y,z) = \left(-\frac{x+y}{z}\right)^{n-1}. \qquad (5)$$

This POLYNOMIAL usually detects CHIRALITY but does not detect the distinct ENANTIOMERS of the KNOTS 09_{042}, 10_{048}, 10_{071}, 10_{091}, 10_{104}, and 10_{125} (Jones 1987). The HOMFLY polynomial of an oriented KNOT is the same if the orientation is reversed. It is a generalization of the JONES POLYNOMIAL $V(t)$, satisfying

$$V(t) = P(a=t, z=t^{1/2}-t^{-1/2}) \qquad (6)$$

$$V(t) = P(\ell = it^{-1}, m = i(t^{-1/2}-t^{1/2})). \qquad (7)$$

It is also a generalization of the ALEXANDER POLYNOMIAL $\nabla(z)$, satisfying

$$\Delta(z) = P(a=1, z=t^{1/2}-t^{-1/2}). \qquad (8)$$

The HOMFLY POLYNOMIAL of the MIRROR IMAGE K^* of a KNOT K is given by

$$P_{K^*}(\ell,m) = P_K(\ell^{-1},m), \qquad (9)$$

so P usually but not always detects CHIRALITY.

A split union of two links (i.e., bringing two links together without intertwining them) has HOMFLY polynomial

$$P(L_1 \cup L_2) = -(\ell+\ell^{-1})m^{-1}P(L_1)P(L_2). \qquad (10)$$

Also, the composition of two links

$$P(L_1 \# L_2) = P(L_1)P(L_2), \qquad (11)$$

so the POLYNOMIAL of a COMPOSITE KNOT factors into POLYNOMIALS of its constituent knots (Adams 1994).

MUTANTS have the same HOMFLY polynomials. In fact, there are infinitely many distinct KNOTS with the same HOMFLY POLYNOMIAL (Kanenobu 1986). Examples include (05_{001}, 10_{132}), (08_{008}, 10_{129}) (08_{016}, 10_{156}), and (10_{025}, 10_{056}) (Jones 1987). Incidentally, these also have the same JONES POLYNOMIAL.

M. B. Thistlethwaite has tabulated the HOMFLY polynomial for KNOTS up to 13 crossings.

see also ALEXANDER POLYNOMIAL, JONES POLYNOMIAL, KNOT POLYNOMIAL

References

Adams, C. C. *The Knot Book: An Elementary Introduction to the Mathematical Theory of Knots.* New York: W. H. Freeman, pp. 171–172, 1994.

Doll, H. and Hoste, J. "A Tabulation of Oriented Links." *Math. Comput.* **57**, 747–761, 1991.

Freyd, P.; Yetter, D.; Hoste, J.; Lickorish, W. B. R.; Millett, K.; and Oceanu, A. "A New Polynomial Invariant of Knots and Links." *Bull. Amer. Math. Soc.* **12**, 239–246, 1985.

Jones, V. "Hecke Algebra Representations of Braid Groups and Link Polynomials." *Ann. Math.* **126**, 335–388, 1987.
Kanenobu, T. "Infinitely Many Knots with the Same Polynomial." *Proc. Amer. Math. Soc.* **97**, 158–161, 1986.
Kanenobu, T. and Sumi, T. "Polynomial Invariants of 2-Bridge Knots through 22 Crossings." *Math. Comput.* **60**, 771–778 and S17–S28, 1993.
Kauffman, L. H. *Knots and Physics.* Singapore: World Scientific, p. 52, 1991.
Lickorish, W. B. R. and Millett, B. R. "The New Polynomial Invariants of Knots and Links." *Math. Mag.* **61**, 1–23, 1988.
Morton, H. R. and Short, H. B. "Calculating the 2-Variable Polynomial for Knots Presented as Closed Braids." *J. Algorithms* **11**, 117–131, 1990.
Przytycki, J. and Traczyk, P. "Conway Algebras and Skein Equivalence of Links." *Proc. Amer. Math. Soc.* **100**, 744–748, 1987.
Stoimenow, A. "Jones Polynomials." `http://www.informatik.hu-berlin.de/~stoimeno/ptab/j10.html`.
Weisstein, E. W. "Knots and Links." `http://www.astro.virginia.edu/~eww6n/math/notebooks/Knots.m`.

Homoclinic Point

A point where a stable and an unstable separatrix (invariant manifold) from the same fixed point or same family intersect. Therefore, the limits

$$\lim_{k\to\infty} f^k(X)$$

and

$$\lim_{k\to -\infty} f^k(X)$$

exist and are equal.

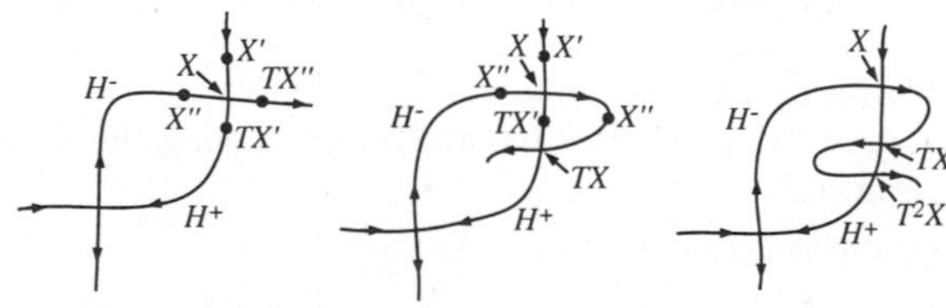

Refer to the above figure. Let X be the point of intersection, with X_1 ahead of X on one MANIFOLD and X_2 ahead of X of the other. The mapping of each of these points TX_1 and TX_2 must be ahead of the mapping of X, TX. The only way this can happen is if the MANIFOLD loops back and crosses itself at a new homoclinic point. Another loop must be formed, with T^2X another homoclinic point. Since T^2X is closer to the hyperbolic point than TX, the distance between T^2X and TX is less than that between X and TX. Area preservation requires the AREA to remain the same, so each new curve (which is closer than the previous one) must extend further. In effect, the loops become longer and thinner. The network of curves leading to a dense AREA of homoclinic points is known as a homoclinic tangle or tendril. Homoclinic points appear where CHAOTIC regions touch in a hyperbolic FIXED POINT.

A small DISK centered near a homoclinic point includes infinitely many periodic points of different periods. Poincaré showed that if there is a single homoclinic point, there are an infinite number. More specifically, there are infinitely many homoclinic points in each small disk (Nusse and Yorke 1996).

see also HETEROCLINIC POINT

References

Nusse, H. E. and Yorke, J. A. "Basins of Attraction." *Science* **271**, 1376–1380, 1996.
Tabor, M. *Chaos and Integrability in Nonlinear Dynamics: An Introduction.* New York: Wiley, p. 145, 1989.

Homogeneous Coordinates

see TRILINEAR COORDINATES

Homogeneous Function

A function which satisfies

$$f(tx, ty) = t^n f(x, y)$$

for a fixed n. MEANS, the WEIERSTRAß ELLIPTIC FUNCTION, and TRIANGLE CENTER FUNCTIONS are homogeneous functions. A transformation of the variables of a TENSOR changes the TENSOR into another whose components are linear homogeneous functions of the components of the original TENSOR.

see also EULER'S HOMOGENEOUS FUNCTION THEOREM

Homogeneous Numbers

Two numbers are homogeneous if they have identical PRIME FACTORS. An example of a homogeneous pair is (6, 36), both of which share PRIME FACTORS 2 and 3:

$$6 = 2 \cdot 3$$
$$36 = 2^2 \cdot 3^2.$$

see also HETEROGENEOUS NUMBERS, PRIME FACTORS, PRIME NUMBER

References

Le Lionnais, F. *Les nombres remarquables.* Paris: Hermann, p. 146, 1983.

Homogeneous Polynomial

A multivariate polynomial (i.e., a POLYNOMIAL in more than one variable) with all terms having the same degree. For example, $x^3 + xyz + y^2z + z^3$ is a homogeneous polynomial of degree three.

see also POLYNOMIAL

Homographic

see MÖBIUS TRANSFORMATION

Homography

A CIRCLE-preserving transformation composed of an EVEN number of inversions.

see also ANTIHOMOGRAPHY

Homological Algebra

An abstract ALGEBRA concerned with results valid for many different kinds of SPACES.

References

Hilton, P. and Stammbach, U. *A Course in Homological Algebra, 2nd ed.* New York: Springer-Verlag, 1997.

Weibel, C. A. *An Introduction to Homological Algebra.* New York: Cambridge University Press, 1994.

Homologous Points

The extremities of PARALLEL RADII of two CIRCLES are called homologous with respect to the SIMILITUDE CENTER collinear with them.

see also ANTIHOMOLOGOUS POINTS

References

Johnson, R. A. *Modern Geometry: An Elementary Treatise on the Geometry of the Triangle and the Circle.* Boston, MA: Houghton Mifflin, p. 19, 1929.

Homolographic Equal Area Projection

see MOLLWEIDE PROJECTION

Homology (Geometry)

A PERSPECTIVE COLLINEATION in which the center and axis are not incident.

see also ELATION, HARMONIC HOMOLOGY, PERSPECTIVE COLLINEATION

Homology Group

The term "homology group" usually means a singular homology group, which is an ABELIAN GROUP which partially counts the number of HOLES in a TOPOLOGICAL SPACE. In particular, singular homology groups form a MEASURE of the HOLE structure of a SPACE, but they are one particular measure and they don't always pick up everything.

In addition, there are "generalized homology groups" which are *not* singular homology groups.

Homology (Topology)

Historically, the term "homology" was first used in a topological sense by Poincaré. To him, it meant pretty much what is now called a COBORDISM, meaning that a homology was thought of as a relation between MANIFOLDS mapped into a MANIFOLD. Such MANIFOLDS form a homology when they form the boundary of a higher-dimensional MANIFOLD inside the MANIFOLD in question.

To simplify the definition of homology, Poincaré simplified the spaces he dealt with. He assumed that all the spaces he dealt with had a triangulation (i.e., they were "SIMPLICIAL COMPLEXES"). Then instead of talking about general "objects" in these spaces, he restricted himself to subcomplexes, i.e., objects in the space made up only on the simplices in the TRIANGULATION of the space. Eventually, Poincaré's version of homology was dispensed with and replaced by the more general SINGULAR HOMOLOGY. SINGULAR HOMOLOGY is the concept mathematicians mean when they say "homology."

In modern usage, however, the word homology is used to mean HOMOLOGY GROUP. For example, if someone says "X did Y by computing the homology of Z," they mean "X did Y by computing the HOMOLOGY GROUPS of Z." But sometimes homology is used more loosely in the context of a "homology in a SPACE," which corresponds to *singular* homology groups.

Singular homology groups of a SPACE measure the extent to which there are finite (compact) boundaryless GADGETS in that SPACE, such that these GADGETS are not the boundary of other finite (compact) GADGETS in that SPACE.

A generalized homology or cohomology theory must satisfy all of the EILENBERG-STEENROD AXIOMS with the exception of the DIMENSION AXIOM.

see also COHOMOLOGY, DIMENSION AXIOM, EILENBERG-STEENROD AXIOMS, GADGET, HOMOLOGICAL ALGEBRA, HOMOLOGY GROUP, SIMPLICIAL COMPLEX, SIMPLICIAL HOMOLOGY, SINGULAR HOMOLOGY

Homomorphic

Related to one another by a HOMOMORPHISM.

Homomorphism

A term used in CATEGORY THEORY to mean a general MORPHISM.

see also HOMEOMORPHISM, MORPHISM

Homoscedastic

A set of STATISTICAL DISTRIBUTIONS having the same VARIANCE.

see also HETEROSCEDASTIC

Homothecy

see DILATION

Homothetic

Two figures are homothetic if they are related by a DILATION (a dilation is also known as a HOMOTHECY). This means that they lie in the same plane and corresponding sides are PARALLEL; such figures have connectors of corresponding points which are CONCURRENT at a point known as the HOMOTHETIC CENTER. The HOMOTHETIC CENTER divides each connector in the same ratio k, known as the SIMILITUDE RATIO. For figures which are similar but do not have PARALLEL sides, a SIMILITUDE CENTER exists.

see also DILATION, HOMOTHETIC CENTER, PERSPECTIVE, SIMILITUDE RATIO

References

Johnson, R. A. *Modern Geometry: An Elementary Treatise on the Geometry of the Triangle and the Circle.* Boston, MA: Houghton Mifflin, 1929.

Homothetic Center

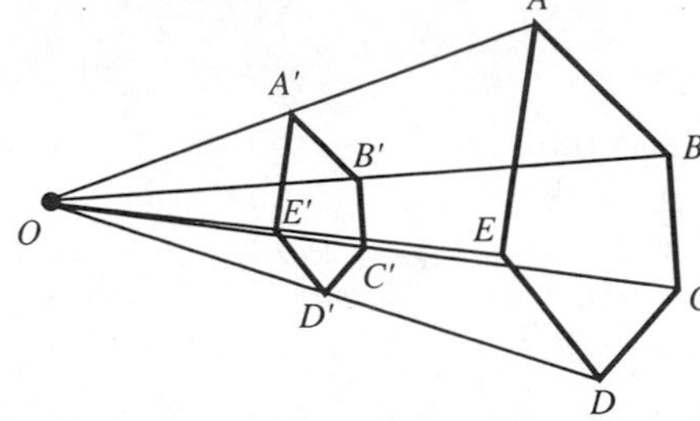

The meeting point of lines that connect corresponding points from HOMOTHETIC figures. In the above figure, O is the homothetic center of the HOMOTHETIC figures $ABCDE$ and $A'B'C'D'E'$. For figures which are similar but do not have PARALLEL sides, a SIMILITUDE CENTER exists (Johnson 1929, pp. 16–20).

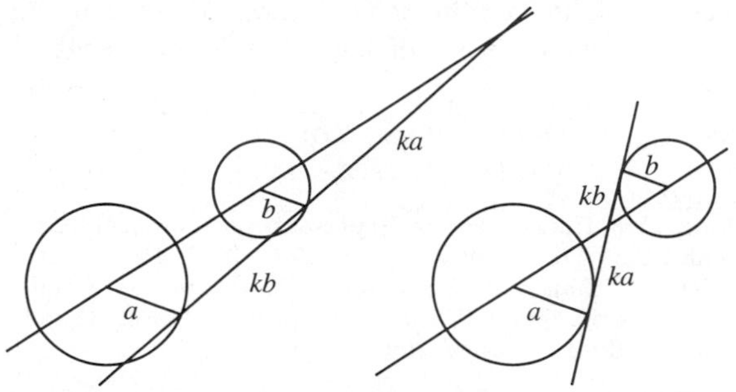

Given two nonconcentric CIRCLES, draw RADII PARALLEL and in the same direction. Then the line joining the extremities of the RADII passes through a fixed point on the line of centers which divides that line externally in the ratio of RADII. This point is called the external homothetic center, or external center of similitude (Johnson 1929, pp. 19–20 and 41).

If RADII are drawn PARALLEL but instead in opposite directions, the extremities of the RADII pass through a fixed point on the line of centers which divides that line internally in the ratio of RADII (Johnson 1929, pp. 19–20 and 41). This point is called the internal homothetic center, or internal center of similitude (Johnson 1929, pp. 19–20 and 41).

The position of the homothetic centers for two circles of radii r_i, centers (x_i, y_i), and segment angle θ are given by solving the simultaneous equations

$$y - y_2 = \frac{y_2 - y_1}{x_2 - x_1}(x - x_2)$$

$$y - y_2^\pm = \frac{y_2^\pm - y_1^\pm}{x_2^\pm - x_1^\pm}(x - x_2^\pm)$$

for (x, y), where

$$x_i^\pm \equiv x_i + (-1)^i r_i \cos\theta$$

$$y_i^\pm \equiv y_i + (-1)^i r_i \sin\theta,$$

and the plus signs give the external homothetic center, while the minus signs give the internal homothetic center.

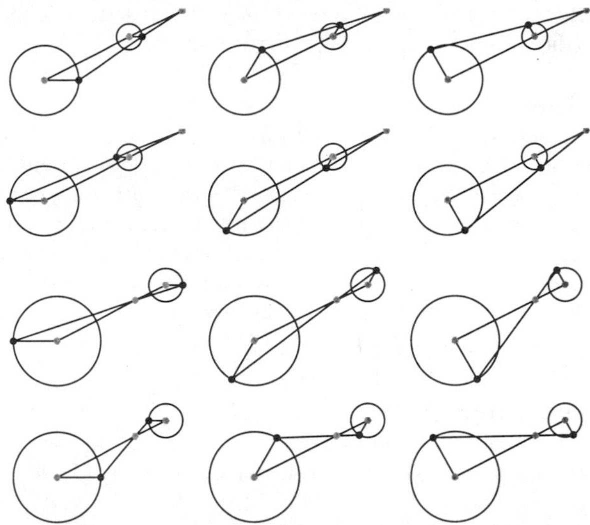

As the above diagrams show, as the angles of the parallel segments are varied, the positions of the homothetic centers remain the same. This fact provides a (slotted) LINKAGE for converting circular motion with one radius to circular motion with another.

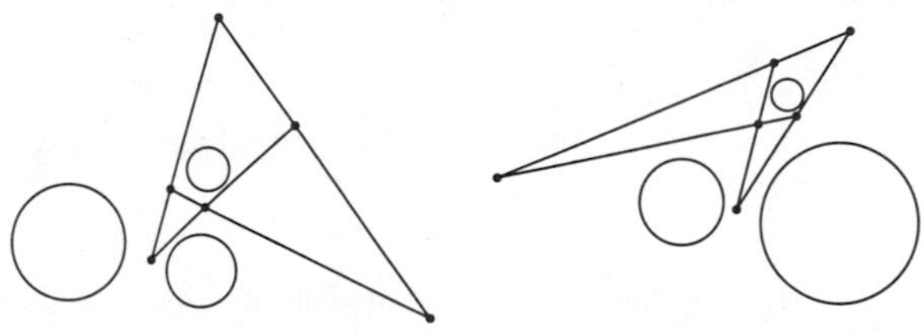

The six homothetic centers of three circles lie three by three on four lines (Johnson 1929, p. 120), which "enclose" the smallest circle.

The homothetic center of triangles is the PERSPECTIVE CENTER of HOMOTHETIC TRIANGLES. It is also called the SIMILITUDE CENTER (Johnson 1929, pp. 16–17).

see also APOLLONIUS' PROBLEM, PERSPECTIVE, SIMILITUDE CENTER

References

Johnson, R. A. *Modern Geometry: An Elementary Treatise on the Geometry of the Triangle and the Circle.* Boston, MA: Houghton Mifflin, 1929.

Weisstein, E. W. "Plane Geometry." `http://www.astro.virginia.edu/~eww6n/math/notebooks/PlaneGeometry.m`.

Homothetic Position

Two similar figures with PARALLEL homologous LINES and connectors of HOMOLOGOUS POINTS CONCURRENT at the HOMOTHETIC CENTER are said to be in homothetic position. If two SIMILAR figures are in the same plane but the corresponding sides are not PARALLEL, there exists a self-HOMOLOGOUS POINT which occupies the same homologous position with respect to the two figures.

Homothetic Triangles

Nonconcurrent TRIANGLES with PARALLEL sides are always HOMOTHETIC. Homothetic triangles are always PERSPECTIVE TRIANGLES. Their PERSPECTIVE CENTER is called their HOMOTHETIC CENTER.

Homotopy

A continuous transformation from one FUNCTION to another. A homotopy between two functions f and g from a SPACE X to a SPACE Y is a continuous MAP G from $X \in [0,1] \mapsto Y$ such that $G(x,0) = f(x)$ and $G(x,1) = g(x)$. Another way of saying this is that a homotopy is a path in the mapping SPACE $\mathrm{Map}(X,Y)$ from the first FUNCTION to the second.

see also h-COBORDISM

Homotopy Axiom

One of the EILENBERG-STEENROD AXIOMS which states that, if $f:(X,A) \to (Y,B)$ is homotopic to $g:(X,A) \to (Y,B)$, then their INDUCED MAPS $f_*: H_n(X,A) \to H_n(Y,B)$ and $g_*: H_n(X,A) \to H_n(Y,B)$ are the same.

Homotopy Group

A GROUP related to the HOMOTOPY classes of MAPS from SPHERES $\mathbb{S}^n$ into a SPACE X.

see also COHOMOTOPY GROUP

Homotopy Theory

The branch of ALGEBRAIC TOPOLOGY which deals with HOMOTOPY GROUPS.

References

Aubry, M. *Homotopy Theory and Models.* Boston, MA: Birkhäuser, 1995.

Honeycomb

A TESSELLATION in n-D, for $n \geq 3$. The only regular honeycomb in 3-D is $\{4,3,4\}$, which consists of eight cubes meeting at each VERTEX. The only quasiregular honeycomb (with regular cells and semiregular VERTEX FIGURES) has each VERTEX surrounded by eight TETRAHEDRA and six OCTAHEDRA and is denoted $\left\{\begin{matrix} 3 \\ 3,4 \end{matrix}\right\}$.

There are many semiregular honeycombs, such as $\left\{\begin{matrix} 3,3 \\ 4 \end{matrix}\right\}$, in which each VERTEX consists of two OCTAHEDRA $\{3,4\}$ and four CUBOCTAHEDRA $\left\{\begin{matrix} 3 \\ 4 \end{matrix}\right\}$.

see also SPONGE, TESSELLATION

References

Bulatov, V. "Infinite Regular Polyhedra." `http://www.physics.orst.edu/~bulatov/polyhedra/infinite/`.

Hoof

see CYLINDRICAL WEDGE

Hook

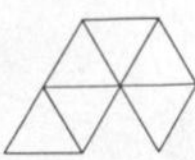

A 6-POLYIAMOND.

References

Golomb, S. W. *Polyominoes: Puzzles, Patterns, Problems, and Packings, 2nd ed.* Princeton, NJ: Princeton University Press, p. 92, 1994.

Hook Length Formula

A FORMULA for the number of YOUNG TABLEAUX associated with a given YOUNG DIAGRAM. In each box, write the sum of one plus the number of boxes horizontally to the right and vertically below the box (the "hook length"). The number of tableaux is then $n!$ divided by the product of all "hook lengths". The `Combinatorica'NumberOfTableaux` function in *Mathematica®* implements the hook length formula.

see also YOUNG DIAGRAM, YOUNG TABLEAU

References

Jones, V. "Hecke Algebra Representations of Braid Groups and Link Polynomials." *Ann. Math.* **126**, 335–388, 1987.

Skiena, S. *Implementing Discrete Mathematics: Combinatorics and Graph Theory with Mathematica.* Reading, MA: Addison-Wesley, 1990.

Hopf Algebra

Let a graded module A have a multiplication ϕ and a co-multiplication ψ. Then if ϕ and ψ have the unity of k as unity and $\psi:(A,\phi) \to (A,\phi) \otimes (A,\phi)$ is an algebra homomorphism, then (A,ϕ,ψ) is called a Hopf algebra.

Hopf Bifurcation

The BIFURCATION of a FIXED POINT to a LIMIT CYCLE (Tabor 1989).

References

Guckenheimer, J. and Holmes, P. *Nonlinear Oscillations, Dynamical Systems, and Bifurcations of Vector Fields, 3rd ed.* New York: Springer-Verlag, pp. 150–154, 1997.

Marsden, J. and McCracken, M. *Hopf Bifurcation and Its Applications.* New York: Springer-Verlag, 1976.

Tabor, M. *Chaos and Integrability in Nonlinear Dynamics: An Introduction.* New York: Wiley, p. 197, 1989.

Hopf Circle

see HOPF MAP

Hopf Link

The LINK 2_1^2 which has JONES POLYNOMIAL

$$V(t) = -t - t^{-1}$$

and HOMFLY POLYNOMIAL

$$P(z, \alpha) = z^{-1}(\alpha^{-1} - \alpha^{-3}) + z\alpha^{-1}.$$

It has BRAID WORD ${\sigma_1}^2$.

Hopf Map

The first example discovered of a MAP from a higher-dimensional SPHERE to a lower-dimensional SPHERE which is not null-HOMOTOPIC. Its discovery was a shock to the mathematical community, since it was believed at the time that all such maps were null-HOMOTOPIC, by analogy with HOMOLOGY GROUPS. The Hopf map takes points (X_1, X_2, X_3, X_4) on a 3-sphere to points on a 2-sphere (x_1, x_2, x_3)

$$x_1 = 2(X_1X_2 + X_3X_4)$$
$$x_2 = 2(X_1X_4 - X_2X_3)$$
$$x_3 = ({X_1}^2 + {X_3}^2) - ({X_2}^2 + {X_4}^2).$$

Every point on the two SPHERES corresponds to a CIRCLE called the HOPF CIRCLE on the 3-SPHERE.

Hopf's Theorem

A NECESSARY and SUFFICIENT condition for a MEASURE which is quasi-invariant under a transformation to be equivalent to an invariant PROBABILITY MEASURE is that the transformation cannot (in a measure theoretic sense) compress the SPACE.

Horizontal

Oriented in position PERPENDICULAR to up-down, and therefore PARALLEL to a flat surface.

see also VERTICAL

Horizontal-Vertical Illusion

see VERTICAL-HORIZONTAL ILLUSION

Horn Angle

The configuration formed by two curves starting at a point, called the VERTEX V, in a common direction. They are concrete illustrations of non-Archimedean geometries.

References

Kasner, E. "The Recent Theory of the Horn Angle." *Scripta Math* **11**, 263–267, 1945.

Horn Cyclide

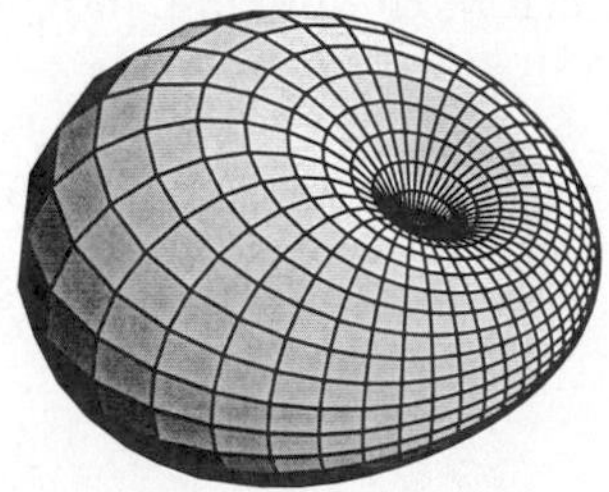

The inversion of a HORN TORUS. If the inversion center lies on the torus, then the horn cyclide degenerates to a PARABOLIC HORN CYCLIDE.

see also CYCLIDE, HORN TORUS, PARABOLIC CYCLIDE, RING CYCLIDE, SPINDLE CYCLIDE, TORUS

Horn Torus

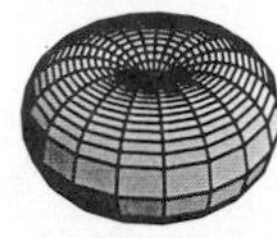

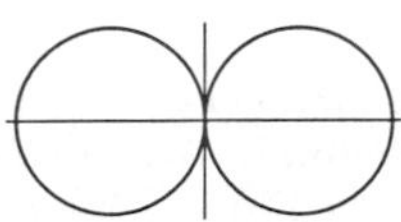

One of the three STANDARD TORI given by the parametric equations

$$x = (c + a\cos v)\cos u \quad (1)$$
$$y = (c + a\cos v)\sin u \quad (2)$$
$$z = a\sin v \quad (3)$$

with $a = c$. The inversion of a horn torus is a HORN CYCLIDE (or PARABOLIC HORN CYCLIDE). The above left figure shows a horn torus, the middle a cutaway, and the right figure shows a CROSS-SECTION of the horn torus through the xz-plane.

see also CYCLIDE, HORN CYCLIDE, RING TORUS, SPINDLE TORUS, STANDARD TORI, TORUS

References

Gray, A. "Tori." §11.4 in *Modern Differential Geometry of Curves and Surfaces.* Boca Raton, FL: CRC Press, pp. 218–220, 1993.
Pinkall, U. "Cyclides of Dupin." §3.3 in *Mathematical Models from the Collections of Universities and Museums* (Ed. G. Fischer). Braunschweig, Germany: Vieweg, pp. 28–30, 1986.

Horned Sphere

see ALEXANDER'S HORNED SPHERE, ANTOINE'S HORNED SPHERE

Horner's Method

Let

$$P(x) = a_nx^n + \ldots + a_0 \quad (1)$$

and $b_n \equiv a_n$. If we then define

$$b_k \equiv a_k + b_{k-1}x_0 \quad (2)$$

for $k = n-1$, $n-2$, ..., 0, we obtain $b_0 = P(x_0)$. It therefore follows that

$$P(x) = (x - x_0)Q(x) + b_0, \quad (3)$$

where

$$Q(x) \equiv b_nx^{n-1} + b_{n-1}x^{n-2} + \ldots + b_2x + b_1. \quad (4)$$

In addition,

$$P'(x) = Q(x) + (x - x_0)Q'(x) \quad (5)$$
$$P'(x_0) = Q(x_0). \quad (6)$$

Horner's Rule

A rule for POLYNOMIAL computation which both reduces the number of necessary multiplications and results in less numerical instability due to potential subtraction of one large number from another. The rule simply factors out POWERS of x, giving

$$a_n x^n + a_{n-1} x^{n-1} + \ldots + a_0 = ((a_n x + a_{n-1})x + \ldots)x + a_0.$$

References
Vardi, I. *Computational Recreations in Mathematica.* Reading, MA: Addison-Wesley, p. 9, 1991.

Horocycle

The LOCUS of a point which is derived from a fixed point Q by continuous parallel displacement.

References
Coxeter, H. S. M. *Introduction to Geometry, 2nd ed.* New York: Wiley, p. 300, 1969.

Horse Fetter

see HIPPOPEDE

Horseshoe Map

see SMALE HORSESHOE MAP

Hough Transform

A technique used to detect boundaries in digital images.

Householder's Method

A ROOT-finding algorithm based on the iteration formula

$$x_{n+1} = x_n - \frac{f(x_n)}{f'(x_n)} \left\{ 1 - \frac{[f(x_n)]^2 f''(x_n)}{2[f'(x_n)]^2} \right\}.$$

This method, like NEWTON'S METHOD, has poor convergence properties near any point where the DERIVATIVE $f'(x) = 0$.

see also NEWTON'S METHOD

References
Householder, A. S. *The Numerical Treatment of a Single Nonlinear Equation.* New York: McGraw-Hill, 1970.

Howell Design

Let S be a set of $n+1$ symbols, then a Howell design $H(s, 2n)$ on symbol set S is an $s \times s$ array H such that

1. Every cell of H is either empty or contains an unordered pair of symbols from S,
2. Every symbol of S occurs once in each row and column of H, and
3. Every unordered pair of symbols occurs in at most one cell of H.

References
Colbourn, C. J. and Dinitz, J. H. (Eds.) "Howell Designs." Ch. 26 in *CRC Handbook of Combinatorial Designs.* Boca Raton, FL: CRC Press, pp. 381–385, 1996.

Hub

The central point in a WHEEL GRAPH W_n. The hub has DEGREE $n-1$.

see also WHEEL GRAPH

Huffman Coding

A lossless data compression algorithm which uses a small number of bits to encode common characters. Huffman coding approximates the probability for each character as a POWER of 1/2 to avoid complications associated with using a nonintegral number of bits to encode characters using their actual probabilities.

References
Press, W. H.; Flannery, B. P.; Teukolsky, S. A.; and Vetterling, W. T. "Huffman Coding and Compression of Data." Ch. 20.4 in *Numerical Recipes in FORTRAN: The Art of Scientific Computing, 2nd ed.* Cambridge, England: Cambridge University Press, pp. 896–901, 1992.

Hull

see AFFINE HULL, CONVEX HULL

Humbert's Theorem

The NECESSARY and SUFFICIENT condition that an algebraic curve has an algebraic INVOLUTE is that the ARC LENGTH is a two-valued algebraic function of the coordinates of the extremities. Furthermore, this function is a ROOT of a QUADRATIC EQUATION whose COEFFICIENTS are rational functions of x and y.

References
Coolidge, J. L. *A Treatise on Algebraic Plane Curves.* New York: Dover, p. 195, 1959.

Hundkurve

see TRACTRIX

Hundred

$100 = 10^2$. Madachy (1979) gives a number of algebraic equations using the digits 1 to 9 which evaluate to 100, such as

$$\begin{aligned} (7-5)^2 + 96 + 8 - 4 - 3 - 1 &= 100 \\ 3^2 + 91 + 7 + 8 - 6 - 5 - 4 &= 100 \\ \sqrt{9} - 6 + 72 - (1)(3!) - 8 + 45 &= 100 \\ 123 - 45 - 67 + 89 &= 100, \end{aligned}$$

and so on.

see also 10, BILLION, HUNDRED, LARGE NUMBER, MILLION, THOUSAND

References
Madachy, J. S. *Madachy's Mathematical Recreations.* New York: Dover, pp. 156–159, 1979.

Hunt's Surface

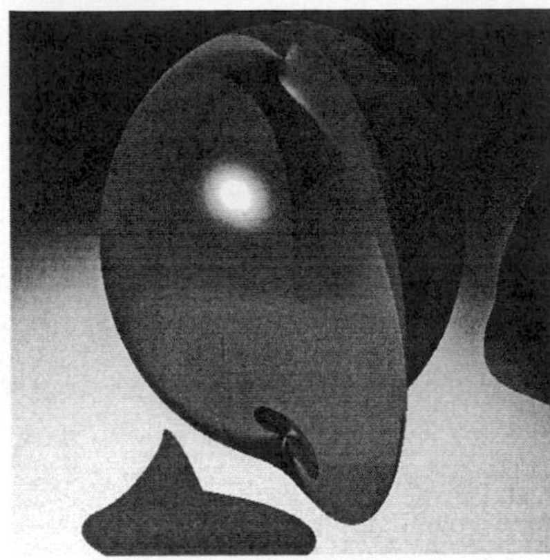

An ALGEBRAIC SURFACE given by the implicit equation

$$4(x^2+y^2+z^2-13)^3+27(3x^2+y^2-4z^2-12)^2=0.$$

References

Hunt, B. "Algebraic Surfaces." http://www.mathematik.uni-kl.de/~wwwagag/Galerie.html.

Nordstrand, T. "Hunt's Surface." http://www.uib.no/people/nfytn/hunttxt.htm.

Huntington Equation

An equation proposed by Huntington (1933) as part of his definition of a BOOLEAN ALGEBRA,

$$n(n(x)+y)+n(n(x)+n(y))=x.$$

see also ROBBINS ALGEBRA, ROBBINS EQUATION

References

Huntington, E. V. "New Sets of Independent Postulates for the Algebra of Logic, with Special Reference to Whitehead and Russell's *Principia Mathematica*." *Trans. Amer. Math. Soc.* **35**, 274–304, 1933.

Huntington, E. V. "Boolean Algebra. A Correction." *Trans. Amer. Math. Soc.* **35**, 557–558, 1933.

Hurwitz Equation

The DIOPHANTINE EQUATION

$${x_1}^2+{x_2}^2+\ldots+{x_n}^2=ax_1x_2\cdots x_n$$

which has no INTEGER solutions for $a>n$.

see also LAGRANGE NUMBER (DIOPHANTINE EQUATION)

References

Guy, R. K. "Markoff Numbers." §D12 in *Unsolved Problems in Number Theory, 2nd ed.* New York: Springer-Verlag, pp. 166–168, 1994.

Hurwitz's Irrational Number Theorem

As Lagrange showed, any IRRATIONAL NUMBER α has an infinity of rational approximations p/q which satisfy

$$\left|\alpha-\frac{p}{q}\right|<\frac{1}{\sqrt{5}\,q^2}. \qquad (1)$$

Similarly, if $\alpha\neq\frac{1}{2}(1+\sqrt{5}\,)$,

$$\left|\alpha-\frac{p}{q}\right|<\frac{1}{\sqrt{8}\,q^2}, \qquad (2)$$

and if $\alpha\neq\frac{1}{2}(1+\sqrt{5}\,)\neq\sqrt{2}$,

$$\left|\alpha-\frac{p}{q}\right|<\frac{5}{\sqrt{221}}\frac{1}{q^2}. \qquad (3)$$

In general, even tighter bounds of the form

$$\left|\alpha-\frac{p}{q}\right|<\frac{1}{L_nq^2} \qquad (4)$$

can be obtained for the best rational approximation possible for an arbitrary irrational number α, where the L_n are called LAGRANGE NUMBERS and get steadily larger for each "bad" set of irrational numbers which is excluded.

see also HURWITZ'S IRRATIONAL NUMBER THEOREM, LIOUVILLE'S RATIONAL APPROXIMATION THEOREM, LIOUVILLE-ROTH CONSTANT, MARKOV NUMBER, ROTH'S THEOREM, SEGRE'S THEOREM, THUE-SIEGEL-ROTH THEOREM

References

Ball, W. W. R. and Coxeter, H. S. M. *Mathematical Recreations and Essays, 13th ed.* New York: Dover, p. 40, 1987.

Chandrasekharan, K. *An Introduction to Analytic Number Theory.* Berlin: Springer-Verlag, p. 23, 1968.

Conway, J. H. and Guy, R. K. *The Book of Numbers.* New York: Springer-Verlag, pp. 187–189, 1996.

Hurwitz Number

A number with a continued fraction whose terms are the values of one or more POLYNOMIALS evaluated on consecutive INTEGERS and then interleaved. This property is preserved by MÖBIUS TRANSFORMATIONS (Beeler *et al.* 1972, p. 44).

References

Beeler, M.; Gosper, R. W.; and Schroeppel, R. *HAKMEM.* Cambridge, MA: MIT Artificial Intelligence Laboratory, Memo AIM-239, Feb. 1972.

Hurwitz Polynomial

A POLYNOMIAL with REAL POSITIVE COEFFICIENTS and ROOTS which are either NEGATIVE or pairwise conjugate with NEGATIVE REAL PARTS.

Hurwitz-Radon Theorem

Determined the possible values of r and n for which there is an IDENTITY of the form

$$({x_1}^2 + \ldots + {x_r}^2)({y_1}^2 + \ldots + {y_r}^2) = {z_1}^2 + \ldots + {z_n}^2.$$

Hurwitz's Root Theorem

Let $\{f(x)\}$ be a SEQUENCE of ANALYTIC FUNCTIONS REGULAR in a region G, and let this sequence be UNIFORMLY CONVERGENT in every CLOSED SUBSET of G. If the ANALYTIC FUNCTION

$$\lim_{n\to\infty} f_n(x) = f(x)$$

does not vanish identically, then if $x = a$ is a zero of $f(x)$ of order k, a NEIGHBORHOOD $|x-a| < \delta$ of $x = a$ and a number N exist such that if $n > N$, $f_n(x)$ has exactly k zeros in $|x-a| < \delta$.

References

Szegő, G. *Orthogonal Polynomials, 4th ed.* Providence, RI: Amer. Math. Soc., p. 22, 1975.

Hurwitz Zeta Function

A generalization of the RIEMANN ZETA FUNCTION with a FORMULA

$$\zeta(s,a) \equiv \sum_{k=0}^{\infty} \frac{1}{(k+a)^s}, \tag{1}$$

where any term with $k+a=0$ is excluded. The Hurwitz zeta function can also be given by the functional equation

$$\zeta\left(s, \frac{p}{q}\right) = 2\Gamma(1-s)(2\pi q)^{s-1} \sum_{n=1}^{q} \sin\left(\frac{\pi s}{2} + \frac{2\pi np}{q}\right) \zeta\left(1-s, \frac{n}{q}\right) \tag{2}$$

(Apostol 1976, Miller and Adamchik), or the integral

$$\zeta(s,a) = \tfrac{1}{2}a^{-3} + \frac{a^{1-s}}{s-1} + 2\int_0^\infty (a^2+y^2)^{-s/2} \left\{\sin\left[s\tan^{-1}\left(\frac{y}{a}\right)\right]\right\} \frac{dy}{e^{2\pi y}-1}. \tag{3}$$

If $\Re[z] < 0$, then

$$\zeta(z,a) = \frac{2\Gamma(1-z)}{(2\pi)^{1-z}} \left[\sin\left(\frac{\pi z}{2}\right) \sum_{n=1}^{\infty} \frac{\cos(2\pi an)}{n^{1-z}} + \cos\left(\frac{\pi z}{2}\right) \sum_{n=1}^{\infty} \frac{\sin(2\pi an)}{n^{1-z}}\right]. \tag{4}$$

The Hurwitz zeta function satisfies

$$\zeta(0,a) = \tfrac{1}{2} - a \tag{5}$$

$$\frac{d}{ds}\zeta(0,a) = \ln[\Gamma(a)] - \tfrac{1}{2}\ln(2\pi) \tag{6}$$

$$\frac{d}{ds}\zeta(0,0) = \tfrac{1}{2}\ln(2\pi), \tag{7}$$

where $\Gamma(z)$ is the GAMMA FUNCTION. The POLYGAMMA FUNCTION $\psi_m(z)$ can be expressed in terms of the Hurwitz zeta function by

$$\psi_m(z) = (-1)^{m+1} m! \zeta(1+m, z). \tag{8}$$

For POSITIVE integers k, p, and $q > p$,

$$\zeta'\left(-2k+1, \frac{p}{q}\right) = \frac{[\psi(2k) - \ln(2\pi q)]B_{2k}(p/q)}{2k} - \frac{[\psi(2k)-\ln(2\pi)]B_{2k}}{q^{2k}2k} + \frac{(-1)^{k+1}\pi}{(2\pi q)^{2k}} \sum_{n=1}^{q-1} \sin\left(\frac{2\pi pn}{q}\right) \psi_{(2k-1)}\left(\frac{n}{q}\right) + \frac{(-1)^{k+1}2(2k-1)!}{(2\pi q)^{2k}} \sum_{n=1}^{q-1} \cos\left(\frac{2\pi pn}{q}\right) \zeta'\left(2k, \frac{n}{q}\right) + \frac{\zeta'(-2k+1)}{q^{2k}}, \tag{9}$$

where B_n is a BERNOULLI NUMBER, $B_n(x)$ a BERNOULLI POLYNOMIAL, $\psi_n(z)$ is a POLYGAMMA FUNCTION, and $\zeta(z)$ is a RIEMANN ZETA FUNCTION (Miller and Adamchik). Miller and Adamchik also give the closed-form expressions

$$\zeta'(-2k+1, \tfrac{1}{2}) = -\frac{B_{2k}\ln 2}{4^k k} - \frac{(2^{2k-1}-1)\zeta'(-2k+1)}{2^{2k-1}} \tag{10}$$

$$\zeta'\left(-2k+1, \begin{matrix}\frac{1}{3}\\ \frac{2}{3}\end{matrix}\right) = \mp\frac{(9^k-1)B_{2k}\pi}{\sqrt{3}(3^{2k-1}-1)8k} - \frac{B_{2k}\ln 3}{(3^{2k-1})4k} \mp \frac{(-1)^k\psi_{2k-1}(\frac{1}{3})}{2\sqrt{3}(6\pi)^{2k-1}} - \frac{(3^{2k-1}-1)\zeta'(-2k+1)}{2(3^{2k-1})} \tag{11}$$

$$\zeta'\left(-2k+1, \begin{matrix}\frac{1}{4}\\ \frac{3}{4}\end{matrix}\right) = \mp\frac{(4^k+1)B_{2k}\pi}{4^{k+1}k} + \frac{(4^{k-1}-1)B_{2k}\ln 2}{2^{3k-1}k} \mp \frac{(-1)^k\psi_{2k-1}(\frac{1}{4})}{4(8\pi)^{2k-1}} - \frac{(2^{2k-1}-1)\zeta'(-2k+1)}{2^{4k-1}} \tag{12}$$

$$\zeta'\left(-2k+1, \begin{matrix}\frac{1}{6}\\ \frac{5}{6}\end{matrix}\right) = \mp\frac{(9^k-1)(2^{2k-1}+1)B_{2k}\pi}{\sqrt{3}(6^{2k-1})8k} + \frac{B_{2k}(3^{2k-1}-1)\ln 2}{(6^{2k-1})4k} + \frac{B_{2k}(2^{2k-1}-1)\ln 3}{(6^{2k-1})4k} \mp \frac{(-1)^k(2^{2k-1}+1)\psi_{2k-1}(\frac{1}{3})}{2\sqrt{3}\,(12\pi)^{2k-1}} + \frac{(2^{2k-1}-1)(3^{2k-1}-1)\zeta'(-2k+1)}{2(6^{2k-1})}. \tag{13}$$

see also KHINTCHINE'S CONSTANT, POLYGAMMA FUNCTION, PSI FUNCTION, RIEMANN ZETA FUNCTION, ZETA FUNCTION

References

Apostol, T. M. *Introduction to Analytic Number Theory.* New York: Springer-Verlag, 1995.

Elizalde, E.; Odintsov, A. D.; and Romeo, A. *Zeta Regularization Techniques with Applications.* River Edge, NJ: World Scientific, 1994.

Knopfmacher, J. "Generalised Euler Constants." *Proc. Edinburgh Math. Soc.* **21**, 25–32, 1978.

Magnus, W. and Oberhettinger, F. *Formulas and Theorems for the Special Functions of Mathematical Physics, 3rd ed.* New York: Springer-Verlag, 1966.

Miller, J. and Adamchik, V. "Derivatives of the Hurwitz Zeta Function for Rational Arguments." Submitted to *J. Symb. Comput.* http://www.wolfram.com/~victor/articles/hurwitz.html.

Spanier, J. and Oldham, K. B. "The Hurwitz Function $\zeta(\nu; u)$." Ch. 62 in *An Atlas of Functions.* Washington, DC: Hemisphere, pp. 653–664, 1987.

Whittaker, E. T. and Watson, G. N. *A Course in Modern Analysis, 4th ed.* Cambridge, England: Cambridge University Press, pp. 268–269, 1950.

Hutton's Formula

The MACHIN-LIKE FORMULA

$$\tfrac{1}{4}\pi = 2\tan^{-1}(\tfrac{1}{3}) + \tan^{-1}(\tfrac{1}{7}).$$

The other two-term MACHIN-LIKE FORMULAS are EULER'S MACHIN-LIKE FORMULA, HERMANN'S FORMULA, and MACHIN'S FORMULA.

Hutton's Method

see LAMBERT'S METHOD

Hyperbola

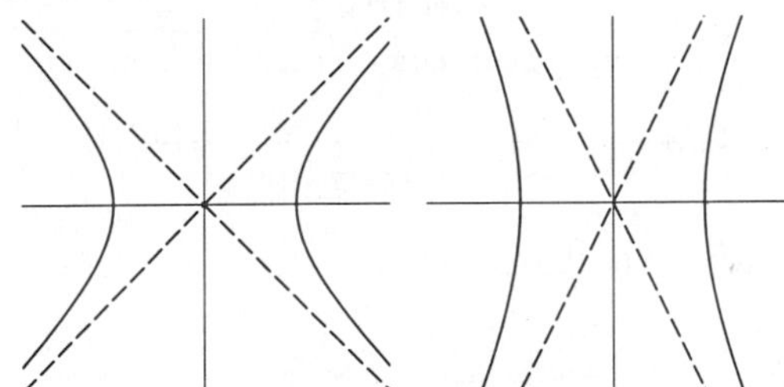

In general, a hyperbola is defined as the LOCUS of all points in the PLANE the difference of whose distance from two fixed points (the FOCI F_1 and F_2) separated by a distance $2c$, where

$$c \equiv \sqrt{a^2 + b^2}, \tag{1}$$

is a given POSITIVE constant. By analogy with the definition of the ELLIPSE, the equation for a hyperbola with SEMIMAJOR AXIS a parallel to the x-AXIS and SEMIMINOR AXIS b parallel to the y-AXIS is given by

$$\frac{(x - x_0)^2}{a^2} - \frac{(y - y_0)^2}{b^2} = 1. \tag{2}$$

Unlike the ELLIPSE, no points of the hyperbola actually lie on the SEMIMINOR AXIS, but rather the ratio b/a determined the vertical scaling of the hyperbola. The ECCENTRICITY of the hyperbola is defined as

$$e \equiv \frac{c}{a} = \sqrt{1 + \frac{b^2}{a^2}}. \tag{3}$$

In the standard equation of the hyperbola, the center is located at (x_0, y_0), the FOCI are at $(x_0 \pm c, y_0)$, and the vertices are at $(x_0 \pm a, y_0)$. The so-called ASYMPTOTES (shown as the dashed lines in the above figures) can be found by substituting 0 for the 1 on the right side of the general equation (2),

$$y = \pm\frac{b}{a}(x - x_0) + y_0, \tag{4}$$

and therefore have SLOPES $\pm b/a$.

The special case $a = b$ (the left diagram above) is known as a RIGHT HYPERBOLA because the ASYMPTOTES are PERPENDICULAR.

In POLAR COORDINATES, the equation of a hyperbola centered at the ORIGIN (i.e., with $x_0 = y_0 = 0$) is

$$r^2 = \frac{a^2 b^2}{b^2 \cos^2\theta - a^2 \sin^2\theta}. \tag{5}$$

In POLAR COORDINATES centered at a FOCUS,

$$r = \frac{a(e^2 - 1)}{1 - e\cos\theta}. \tag{6}$$

The two-center BIPOLAR COORDINATES equation with origin at a FOCUS is

$$r_1 - r_2 = \pm 2a. \tag{7}$$

The parametric equations for the hyperbola are

$$x = \pm a\cosh t \tag{8}$$

$$y = b\sinh t. \tag{9}$$

The CURVATURE and TANGENTIAL ANGLE are

$$\kappa(t) = -[\cosh(2t)]^{-3/2} \tag{10}$$

$$\phi(t) = -\tan^{-1}(\tanh t). \tag{11}$$

The special case of the RIGHT HYPERBOLA was first studied by Menaechmus. Euclid and Aristaeus wrote about the general hyperbola, but only studied one branch of it. The hyperbola was given its present name by Apollonius, who was the first to study both branches. The FOCUS and DIRECTRIX were considered by Pappus (MacTutor Archive). The hyperbola is the shape of an orbit of a body on an escape trajectory (i.e., a body with positive energy), such as some comets, about a fixed mass, such as the sun.

The LOCUS of the apex of a variable CONE containing an ELLIPSE fixed in 3-space is a hyperbola through the FOCI of the ELLIPSE. In addition, the LOCUS of the apex of a CONE containing that hyperbola is the original ELLIPSE. Furthermore, the ECCENTRICITIES of the ELLIPSE and hyperbola are reciprocals.

see also CONIC SECTION, ELLIPSE, HYPERBOLOID, JERABEK'S HYPERBOLA, KIEPERT'S HYPERBOLA, PARABOLA, QUADRATIC CURVE, RECTANGULAR HYPERBOLA, REFLECTION PROPERTY, RIGHT HYPERBOLA

References

Beyer, W. H. *CRC Standard Mathematical Tables, 28th ed.* Boca Raton, FL: CRC Press, pp. 199–200, 1987.

Casey, J. "The Hyperbola." Ch. 7 in *A Treatise on the Analytical Geometry of the Point, Line, Circle, and Conic Sections, Containing an Account of Its Most Recent Extensions, with Numerous Examples, 2nd ed., rev. enl.* Dublin: Hodges, Figgis, & Co., pp. 250–284, 1893.

Courant, R. and Robbins, H. *What is Mathematics?: An Elementary Approach to Ideas and Methods, 2nd ed.* Oxford, England: Oxford University Press, pp. 75–76, 1996.

Lawrence, J. D. *A Catalog of Special Plane Curves.* New York: Dover, pp. 79–82, 1972.

Lee, X. "Hyperbola." `http://www.best.com/~xah/SpecialPlaneCurves_dir/Hyperbola_dir/hyperbola.html`.

Lockwood, E. H. "The Hyperbola." Ch. 3 in *A Book of Curves.* Cambridge, England: Cambridge University Press, pp. 24–33, 1967.

MacTutor History of Mathematics Archive. "Hyperbola." `http://www-groups.dcs.st-and.ac.uk/~history/Curves/Hyperbola.html`.

Hyperbola Evolute

The EVOLUTE of a RECTANGULAR HYPERBOLA is the LAMÉ CURVE

$$(ax)^{2/3} - (by)^{2/3} = (a+b)^{2/3}.$$

From a point between the two branches of the EVOLUTE, two NORMALS can be drawn to the HYPERBOLA. However, from a point beyond the EVOLUTE, four NORMALS can be drawn.

Hyperbola Inverse Curve

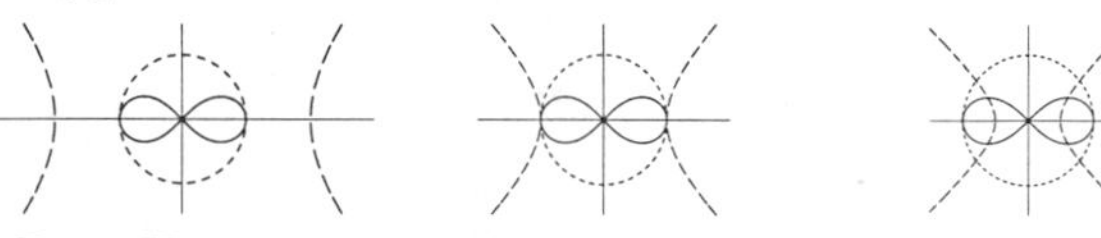

For a HYPERBOLA with $a = b$ with INVERSION CENTER at the center, the INVERSE CURVE

$$x = \frac{2k\cos t}{a[3-\cos(2t)]} \quad (1)$$

$$y = \frac{k\sin(2t)}{a[3-\cos(2t)]} \quad (2)$$

is a LEMNISCATE.

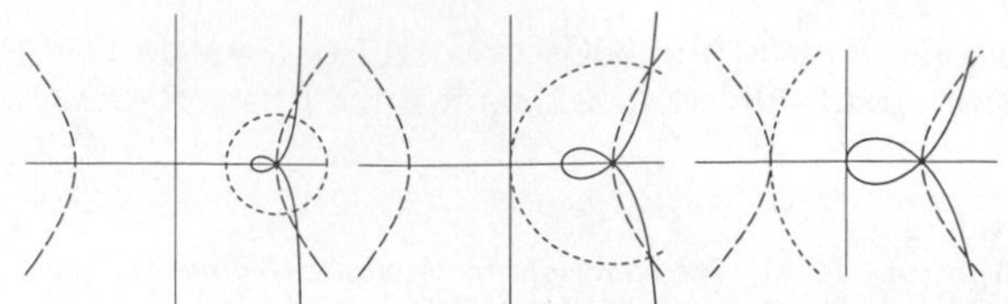

For an INVERSION CENTER at the VERTEX, the INVERSE CURVE

$$x = a + \frac{4k\cos t\sin^2(\tfrac{1}{2}t)}{a[5-4\cos t+\cos(2t)-2\sin(2t)]} \quad (3)$$

$$y = a + \frac{k(\tan t - 1)}{a[(\sec t - 1)^2 + (\tan t - 1)^2]} \quad (4)$$

is a RIGHT STROPHOID.

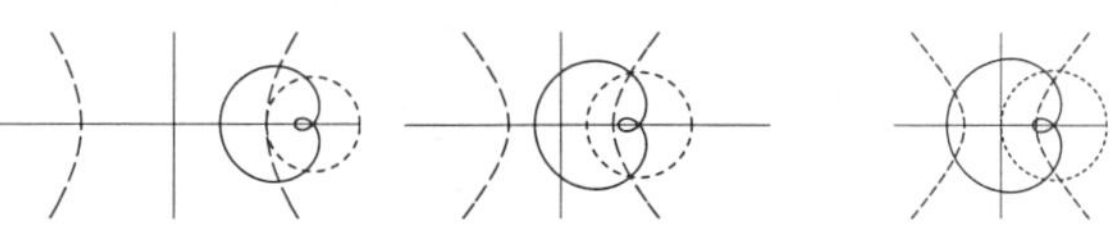

For an INVERSION CENTER at the FOCUS, the INVERSE CURVE

$$x = ae = \frac{k\cos t(1-e\cos t)}{a(\cos t - e)^2} \quad (5)$$

$$y = \frac{\sqrt{e^2-1}\,k\sin(2t)}{2a(\cos t - e)^2} \quad (6)$$

is a LIMAÇON, where e is the ECCENTRICITY.

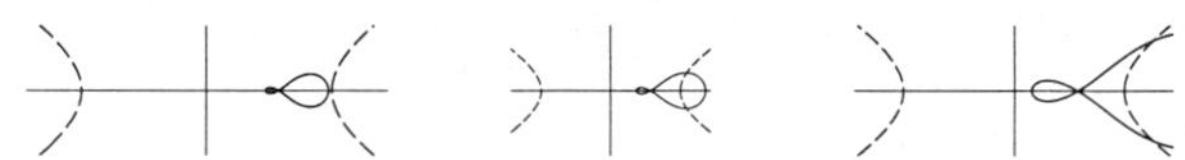

For a HYPERBOLA with $a = \sqrt{3}\,b$ and INVERSION CENTER at the VERTEX, the INVERSE CURVE

$$x = b + \frac{2k\cos t(\sqrt{3}-\cos t)}{b[9-4\sqrt{3}\cos t+\cos(2t)-2\sin(2t)]} \quad (7)$$

$$y = b + \frac{k(\tan t - 1)}{b[(\sqrt{3}\sec t - 1)^2 + (\tan t - 1)^2]} \quad (8)$$

is a MACLAURIN TRISECTRIX.

References

Lawrence, J. D. *A Catalog of Special Plane Curves.* New York: Dover, p. 203, 1972.

Hyperbola Pedal Curve

The PEDAL CURVE of a HYPERBOLA with the PEDAL POINT at the FOCUS is a CIRCLE. The PEDAL CURVE of a RECTANGULAR HYPERBOLA with PEDAL POINT at the center is a LEMNISCATE.

Hyperbolic Automorphism

see ANOSOV AUTOMORPHISM

Hyperbolic Cosecant

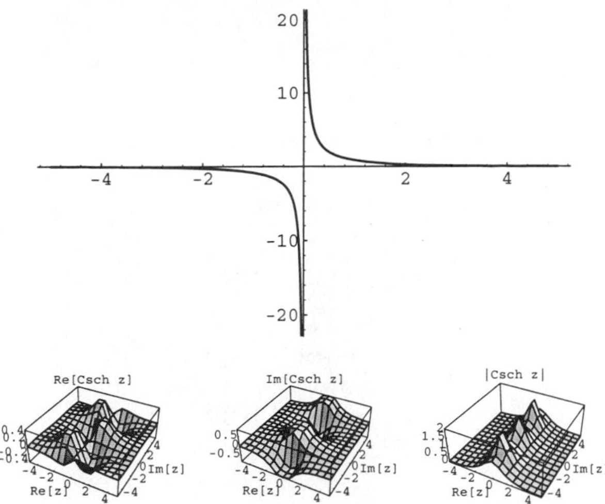

The hyperbolic cosecant is defined as

$$\operatorname{csch} x \equiv \frac{1}{\sinh x} = \frac{2}{e^x - e^{-x}}.$$

see also BERNOULLI NUMBER, BIPOLAR COORDINATES, BIPOLAR CYLINDRICAL COORDINATES, COSECANT, HELMHOLTZ DIFFERENTIAL EQUATION—TOROIDAL COORDINATES, HYPERBOLIC SINE, POINSOT'S SPIRALS, SURFACE OF REVOLUTION, TOROIDAL FUNCTION

References

Abramowitz, M. and Stegun, C. A. (Eds.). "Hyperbolic Functions." §4.5 in *Handbook of Mathematical Functions with Formulas, Graphs, and Mathematical Tables, 9th printing.* New York: Dover, pp. 83–86, 1972.

Spanier, J. and Oldham, K. B. "The Hyperbolic Secant sech(x) and Cosecant csch(x) Functions." Ch. 29 in *An Atlas of Functions.* Washington, DC: Hemisphere, pp. 273–278, 1987.

Hyperbolic Cosine

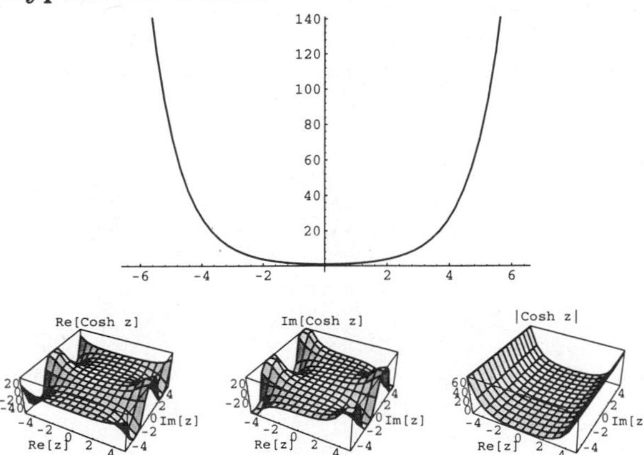

The hyperbolic cosine is defined as

$$\cosh x \equiv \tfrac{1}{2}(e^x + e^{-x}).$$

This function describes the shape of a hanging cable, known as the CATENARY.

see also BIPOLAR COORDINATES, BIPOLAR CYLINDRICAL COORDINATES, BISPHERICAL COORDINATES, CATENARY, CATENOID, CHI, CONICAL FUNCTION, CORRELATION COEFFICIENT—GAUSSIAN BIVARIATE DISTRIBUTION, COSINE, CUBIC EQUATION, DE MOIVRE'S IDENTITY, ELLIPTIC CYLINDRICAL COORDINATES, ELSASSER FUNCTION, FIBONACCI HYPERBOLIC COSINE, FIBONACCI HYPERBOLIC SINE, HYPERBOLIC GEOMETRY, HYPERBOLIC LEMNISCATE FUNCTION, HYPERBOLIC SINE, HYPERBOLIC SECANT, HYPERBOLIC TANGENT, INVERSIVE DISTANCE, LAPLACE'S EQUATION—BIPOLAR COORDINATES, LAPLACE'S EQUATION—BISPHERICAL COORDINATES, LAPLACE'S EQUATION—TOROIDAL COORDINATES, LEMNISCATE FUNCTION, LORENTZ GROUP, MATHIEU DIFFERENTIAL EQUATION, MEHLER'S BESSEL FUNCTION FORMULA, MERCATOR PROJECTION, MODIFIED BESSEL FUNCTION OF THE FIRST KIND, OBLATE SPHEROIDAL COORDINATES, PROLATE SPHEROIDAL COORDINATES, PSEUDOSPHERE, RAMANUJAN COS/COSH IDENTITY, SINE-GORDON EQUATION, SURFACE OF REVOLUTION, TOROIDAL COORDINATES

References

Abramowitz, M. and Stegun, C. A. (Eds.). "Hyperbolic Functions." §4.5 in *Handbook of Mathematical Functions with Formulas, Graphs, and Mathematical Tables, 9th printing.* New York: Dover, pp. 83–86, 1972.

Spanier, J. and Oldham, K. B. "The Hyperbolic Sine sinh(x) and Cosine cosh(x) Functions." Ch. 28 in *An Atlas of Functions.* Washington, DC: Hemisphere, pp. 263–271, 1987.

Hyperbolic Cotangent

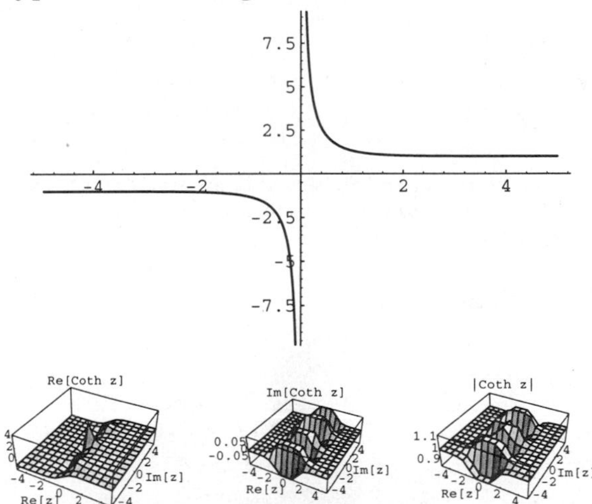

The hyperbolic cotangent is defined as

$$\coth x \equiv \frac{e^x + e^{-x}}{e^x - e^{-x}} = \frac{e^{2x} + 1}{e^{2x} - 1}.$$

Its LAURENT SERIES is

$$\coth x = \frac{1}{x} + \tfrac{1}{3}x - \tfrac{1}{45}x^3 + \ldots.$$

see also BERNOULLI NUMBER, BIPOLAR COORDINATES, BIPOLAR CYLINDRICAL COORDINATES, COTANGENT, FIBONACCI HYPERBOLIC COTANGENT, HYPERBOLIC TANGENT, LAPLACE'S EQUATION—TOROIDAL COORDINATES, LEBESGUE CONSTANTS (FOURIER SERIES), PROLATE SPHEROIDAL COORDINATES, SURFACE OF REVOLUTION, TOROIDAL COORDINATES, TOROIDAL FUNCTION

References

Abramowitz, M. and Stegun, C. A. (Eds.). "Hyperbolic Functions." §4.5 in *Handbook of Mathematical Functions with Formulas, Graphs, and Mathematical Tables, 9th printing.* New York: Dover, pp. 83–86, 1972.

Spanier, J. and Oldham, K. B. "The Hyperbolic Tangent $\tanh(x)$ and Cotangent $\coth(x)$ Functions." Ch. 30 in *An Atlas of Functions.* Washington, DC: Hemisphere, pp. 279–284, 1987.

Hyperbolic Cube

A hyperbolic version of the Euclidean CUBE.

see also HYPERBOLIC DODECAHEDRON, HYPERBOLIC OCTAHEDRON, HYPERBOLIC TETRAHEDRON

References

Rivin, I. "Hyperbolic Polyhedron Graphics." `http://www.mathsource.com/cgi-bin/MathSource/Applications/Graphics/3D/0201-788`.

Hyperbolic Cylinder

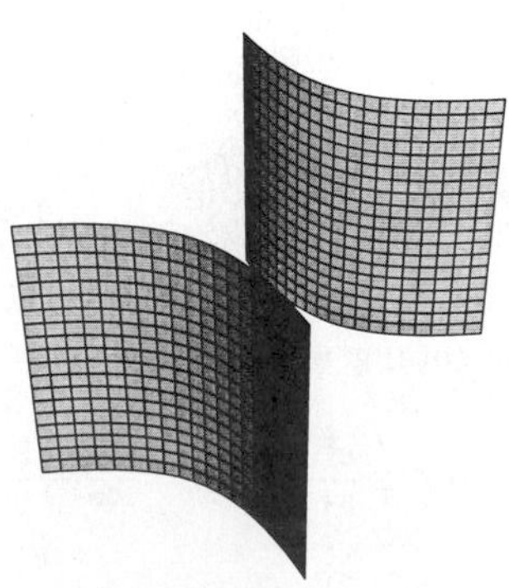

A QUADRATIC SURFACE given by the equation

$$\frac{x^2}{a^2} - \frac{y^2}{b^2} = -1.$$

see also ELLIPTIC PARABOLOID, PARABOLOID

Hyperbolic Dodecahedron

A hyperbolic version of the Euclidean DODECAHEDRON.

see also HYPERBOLIC CUBE, HYPERBOLIC OCTAHEDRON, HYPERBOLIC TETRAHEDRON

References

Rivin, I. "Hyperbolic Polyhedron Graphics." `http://www.mathsource.com/cgi-bin/MathSource/Applications/Graphics/3D/0201-788`.

Hyperbolic Fixed Point (Differential Equations)

A FIXED POINT for which the STABILITY MATRIX has EIGENVALUES $\lambda_1 < 0 < \lambda_2$, also called a SADDLE POINT.

see also ELLIPTIC FIXED POINT (DIFFERENTIAL EQUATIONS), FIXED POINT, STABLE IMPROPER NODE, STABLE SPIRAL POINT, STABLE STAR, UNSTABLE IMPROPER NODE, UNSTABLE NODE, UNSTABLE SPIRAL POINT, UNSTABLE STAR

References

Tabor, M. "Classification of Fixed Points." §1.4.b in *Chaos and Integrability in Nonlinear Dynamics: An Introduction.* New York: Wiley, pp. 22–25, 1989.

Hyperbolic Fixed Point (Map)

A FIXED POINT of a LINEAR TRANSFORMATION (MAP) for which the rescaled variables satisfy

$$(\delta - \alpha)^2 + 4\beta\gamma > 0.$$

see also ELLIPTIC FIXED POINT (MAP), LINEAR TRANSFORMATION, PARABOLIC FIXED POINT

Hyperbolic Functions

The hyperbolic functions sinh, cosh, tanh, csch, sech, coth (HYPERBOLIC SINE, HYPERBOLIC COSINE, etc.) share many properties with the corresponding CIRCULAR FUNCTIONS. The hyperbolic functions arise in many problems of mathematics and mathematical physics in which integrals involving $\sqrt{1+x^2}$ arise (whereas the CIRCULAR FUNCTIONS involve $\sqrt{1-x^2}$).

For instance, the HYPERBOLIC SINE arises in the gravitational potential of a cylinder and the calculation of the Roche limit. The HYPERBOLIC COSINE function is the shape of a hanging cable (the so-called CATENARY). The HYPERBOLIC TANGENT arises in the calculation of magnetic moment and rapidity of special relativity. All three appear in the Schwarzschild metric using external isotropic Kruskal coordinates in general relativity. The HYPERBOLIC SECANT arises in the profile of a laminar jet. The HYPERBOLIC COTANGENT arises in the Langevin function for magnetic polarization.

The hyperbolic functions are defined by

$$\sinh z \equiv \frac{e^z - e^{-z}}{2} = -\sinh(-z) \tag{1}$$

$$\cosh z \equiv \frac{e^z + e^{-z}}{2} = \cosh(-z) \tag{2}$$

$$\tanh z \equiv \frac{e^z - e^{-z}}{e^z + e^{-z}} = \frac{e^{2z}-1}{e^{2z}+1} \tag{3}$$

$$\operatorname{csch} z \equiv \frac{2}{e^z - e^{-z}} \tag{4}$$

$$\operatorname{sech} z \equiv \frac{2}{e^z + e^{-z}} \tag{5}$$

$$\coth z \equiv \frac{e^z + e^{-z}}{e^z - e^{-z}} = \frac{e^{2z}+1}{e^{2z}-1}. \tag{6}$$

For purely IMAGINARY arguments,

$$\sinh(iz) = i \sin z \tag{7}$$

$$\cosh(iz) = \cos z. \tag{8}$$

The hyperbolic functions satisfy many identities anomalous to the trigonometric identities (which can be inferred using OSBORNE'S RULE) such as

$$\cosh^2 x - \sinh^2 x = 1 \tag{9}$$

$$\cosh x + \sinh x = e^x \tag{10}$$

$$\cosh x - \sinh x = e^{-x}. \tag{11}$$

See also Beyer (1987, p. 168). Some half-angle FORMULAS are

$$\tanh\left(\frac{z}{2}\right) = \frac{\sinh x + i \sin y}{\cosh x + \cos y} \tag{12}$$

$$\coth\left(\frac{z}{2}\right) = \frac{\sinh x - i \sin y}{\cosh x - \cos y}. \tag{13}$$

Some double-angle FORMULAS are

$$\sinh(2x) = 2 \sinh x \cosh x \tag{14}$$

$$\cosh(2x) = 2\cosh^2 x - 1 = 1 + 2\sinh^2 x. \tag{15}$$

Identities for COMPLEX arguments include

$$\sinh(x+iy) = \sinh x \cos y + i \cosh x \sin y \tag{16}$$

$$\cosh(x+iy) = \cosh x \cos y + i \sinh x \sin y. \tag{17}$$

The ABSOLUTE SQUARES for COMPLEX arguments are

$$|\sinh(z)|^2 = \sinh^2 x + \sin^2 y \tag{18}$$

$$|\cosh(z)|^2 = \sinh^2 x + \cos^2 y. \tag{19}$$

Integrals involving hyperbolic functions include

$$\begin{aligned}\int \frac{dx}{x\sqrt{a+bx}} &= \ln\left|\frac{\sqrt{a+bx}-\sqrt{a}}{\sqrt{a+bx}+\sqrt{a}}\right| \\ &= \ln\left|\frac{(\sqrt{a+bx}-\sqrt{a})^2}{(a+bx)-a}\right| \\ &= \ln\left|\frac{(a+bx)-2\sqrt{a(a+bx)}+a}{bx}\right|.\end{aligned} \tag{20}$$

If $b > 0$, then

$$\begin{aligned}\int \frac{dx}{x\sqrt{a+bx}} &= \ln\left|\frac{2a+bx-2\sqrt{a(a+bx)}}{bx}\right| \\ &= \ln\left|\left(\frac{2a}{bx}+1\right) - 2\sqrt{\frac{a}{bx}\left(\frac{a}{bx}+1\right)}\right|.\end{aligned} \tag{21}$$

Let $z \equiv 2a/bx + 1$, and $a/bx = (z-1)/2$ and

$$\begin{aligned}\int \frac{dx}{x\sqrt{a+bx}} &= \ln\left[z - 2\sqrt{\tfrac{1}{2}(z-1)\tfrac{1}{2}(z+1)}\right] \\ &= \ln\left[z - \sqrt{(z-1)(z+1)}\right] \\ &= \ln\left(z - \sqrt{z^2-1}\right) = \cosh^{-1}(z) \\ &= \cosh^{-1}\left(1 + \frac{2a}{bx}\right) \\ &= 2\tanh\left(-\sqrt{\frac{a}{a+bx}}\right).\end{aligned} \tag{22}$$

see also HYPERBOLIC COSECANT, HYPERBOLIC COSINE, HYPERBOLIC COTANGENT, GENERALIZED HYPERBOLIC FUNCTIONS, HYPERBOLIC INVERSE FUNCTIONS, HYPERBOLIC SECANT, HYPERBOLIC SINE, HYPERBOLIC TANGENT, HYPERBOLIC INVERSE FUNCTIONS, OSBORNE'S RULE

References
Abramowitz, M. and Stegun, C. A. (Eds.). "Hyperbolic Functions." §4.5 in *Handbook of Mathematical Functions with Formulas, Graphs, and Mathematical Tables, 9th printing.* New York: Dover, pp. 83–86, 1972.
Beyer, W. H. "Hyperbolic Function." *CRC Standard Mathematical Tables, 28th ed.* Boca Raton, FL: CRC Press, pp. 168–186, 1987.
Coxeter, H. S. M. and Greitzer, S. L. *Geometry Revisited.* Washington, DC: Math. Assoc. Amer., pp. 126–131, 1967.
Yates, R. C. "Hyperbolic Functions." *A Handbook on Curves and Their Properties.* Ann Arbor, MI: J. W. Edwards, pp. 113–118, 1952.

Hyperbolic Geometry

A NON-EUCLIDEAN GEOMETRY, also called LOBACHEVSKY-BOLYAI-GAUSS GEOMETRY, having constant SECTIONAL CURVATURE -1. This GEOMETRY satisfies all of EUCLID'S POSTULATES *except* the PARALLEL POSTULATE, which is modified to read: For any infinite straight LINE L and any POINT P not on it, there are *many other* infinitely extending straight LINES that pass through P and which do not intersect L.

In hyperbolic geometry, the sum of ANGLES of a TRIANGLE is less than 180°, and TRIANGLES with the same angles have the same areas. Furthermore, not all TRIANGLES have the same ANGLE sum (c.f. the AAA THEOREM for TRIANGLES in Euclidean 2-space). The best-known example of a hyperbolic space are SPHERES in Lorentzian 4-space. The POINCARÉ HYPERBOLIC DISK is a hyperbolic 2-space. Hyperbolic geometry is well understood in 2-D, but not in 3-D.

Geometric models of hyperbolic geometry include the KLEIN-BELTRAMI MODEL, which consists of an OPEN DISK in the Euclidean plane whose open chords correspond to hyperbolic lines. A 2-D model is the POINCARÉ HYPERBOLIC DISK. Felix Klein constructed an analytic hyperbolic geometry in 1870 in which a POINT is represented by a pair of REAL NUMBERS (x_1, x_2) with

$$x_1{}^2 + x_2{}^2 < 1$$

(i.e., points of an OPEN DISK in the COMPLEX PLANE) and the distance between two points is given by

$$d(x, X) = a\cosh^{-1}\left[\frac{1 - x_1X_1 - x_2X_2}{\sqrt{1 - x_1{}^2 - x_2{}^2}\,\sqrt{1 - X_1{}^2 - X_2{}^2}}\right].$$

The geometry generated by this formula satisfies all of EUCLID'S POSTULATES except the fifth. The METRIC of this geometry is given by the CAYLEY-KLEIN-HILBERT METRIC,

$$g_{11} = \frac{a^2(1 - x_2{}^2)}{(1 - x_1{}^2 - x_2{}^2)^2}$$

$$g_{12} = \frac{a^2 x_1 x_2}{(1 - x_1{}^2 - x_2{}^2)^2}$$

$$g_{22} = \frac{a^2(1 - x_1{}^2)}{(1 - x_1{}^2 - x_2{}^2)^2}.$$

Hilbert extended the definition to general bounded sets in a EUCLIDEAN SPACE.

see also ELLIPTIC GEOMETRY, EUCLIDEAN GEOMETRY, HYPERBOLIC METRIC, KLEIN-BELTRAMI MODEL, NON-EUCLIDEAN GEOMETRY, SCHWARZ-PICK LEMMA

References
Dunham, W. *Journey Through Genius: The Great Theorems of Mathematics.* New York: Wiley, pp. 57–60, 1990.
Eppstein, D. "Hyperbolic Geometry." http://www.ics.uci.edu/~eppstein/junkyard/hyper.html.
Stillwell, J. *Sources of Hyperbolic Geometry.* Providence, RI: Amer. Math. Soc., 1996.

Hyperbolic Inverse Functions

$$\sinh^{-1}\left(\frac{a}{b}\right) = \ln\left(a + \sqrt{a^2 + b^2}\right) \tag{1}$$

$$\cosh^{-1} z = \ln\left(z \pm \sqrt{z^2 - 1}\right) \tag{2}$$

$$\tanh^{-1}\left(\frac{a}{b}\right) = \tfrac{1}{2}\ln\left(\frac{b + a}{b - a}\right) \tag{3}$$

$$\operatorname{csch}^{-1} z = \ln\left(1 \pm \sqrt{1 + z^2}\right) \tag{4}$$

$$\operatorname{sech}^{-1} z = \ln\left(\frac{1 \pm \sqrt{1 - z^2}}{z}\right) \tag{5}$$

$$\coth^{-1} z = \tfrac{1}{2}\ln\left(\frac{z + 1}{z - 1}\right). \tag{6}$$

References
Abramowitz, M. and Stegun, C. A. (Eds.). "Hyperbolic Functions." §4.6 in *Handbook of Mathematical Functions with Formulas, Graphs, and Mathematical Tables, 9th printing.* New York: Dover, pp. 86–89, 1972.

Hyperbolic Knot

A hyperbolic knot is a KNOT that has a complement that can be given a metric of constant curvature -1. The only KNOTS which are not hyperbolic are TORUS KNOTS and SATELLITE KNOTS (including COMPOSITE KNOTS), as proved by Thurston in 1978. Therefore, all but six of the PRIME KNOTS with 10 or fewer crossings are hyperbolic. The exceptions with nine or fewer crossings are 03_{001} (the$(3,2)$-TORUS KNOT), 05_{001}, 07_{001}, 08_{019} (the $(4,3)$-TORUS KNOT), and 09_{001}.

Almost all hyperbolic knots can be distinguished by their hyperbolic volumes (exceptions being 05_{002} and a certain 12-crossing knot; see Adams 1994, p. 124). It has been conjectured that the smallest hyperbolic volume is 2.0298..., that of the FIGURE-OF-EIGHT KNOT.

MUTANT KNOTS have the same hyperbolic knot volume.

References
Adams, C. C. *The Knot Book: An Elementary Introduction to the Mathematical Theory of Knots.* New York: W. H. Freeman, pp. 119–127, 1994.
Adams, C.; Hildebrand, M.; and Weeks, J. "Hyperbolic Invariants of Knots and Links." *Trans. Amer. Math. Soc.* **326**, 1–56, 1991.
Weisstein, E. W. "Knots and Links." http://www.astro.virginia.edu/~eww6n/math/notebooks/Knots.m.

Hyperbolic Lemniscate Function

By analogy with the LEMNISCATE FUNCTIONS, hyperbolic lemniscate functions can also be defined

$$\operatorname{arcsinhlemn} x \equiv \int_0^x (1+t^4)^{1/2}\,dt \qquad (1)$$

$$\operatorname{arccoshlemn} x \equiv \int_x^1 (1+t^4)^{1/2}\,dt. \qquad (2)$$

Let $0 \leq \theta \leq \pi/2$ and $0 \leq v \leq 1$, and write

$$\frac{\theta\mu}{2} = \int_0^v \frac{dt}{\sqrt{1+t^2}}, \qquad (3)$$

where μ is the constant obtained by setting $\theta = \pi/2$ and $v = 1$. Then

$$\mu = \frac{2}{\pi} K\left(\frac{1}{\sqrt{2}}\right), \qquad (4)$$

where $K(k)$ is a complete ELLIPTIC INTEGRAL OF THE FIRST KIND, and Ramanujan showed

$$2\tan^{-1} v = \theta + \sum_{n=1}^\infty \frac{\sin(2n\theta)}{n\cosh(n\pi)}, \qquad (5)$$

$$\tfrac{1}{8}\pi - \tfrac{1}{2}\tan^{-1}(v^2) = \sum_{n=0}^\infty \frac{(-1)^n \cos[(2n+1)\theta]}{(2n+1)\cosh[\frac{1}{2}(2n+1)\pi]} \qquad (6)$$

and

$$\ln\left(\frac{1+v}{1-v}\right) = \ln[\tan(\tfrac{1}{4}\pi + \tfrac{1}{2}\theta)] + 4\sum_{n=0}^\infty \frac{(-1)^n \sin[(2n+1)\theta]}{(2n+1)[e^{(2n+1)\pi}-1]} \qquad (7)$$

(Berndt 1994).

see also LEMNISCATE FUNCTION

References

Berndt, B. C. *Ramanujan's Notebooks, Part IV.* New York: Springer-Verlag, pp. 255–258, 1994.

Hyperbolic Map

A linear MAP $\mathbb{R}^n$ is hyperbolic if none of its EIGENVALUES have modulus 1. This means that $\mathbb{R}^n$ can be written as a direct sum of two A-invariant SUBSPACES E^s and E^u (where s stands for stable and u for unstable). This means that there exist constants $C > 0$ and $0 < \lambda < 1$ such that

$$||A^n v|| \leq C\lambda^n ||v|| \quad \text{if } v \in E^s$$

$$||A^{-n} v|| \leq C\lambda^n ||v|| \quad \text{if } v \in E^u$$

for $n = 0, 1, \ldots$.

see also PESIN THEORY

Hyperbolic Metric

The METRIC for the POINCARÉ HYPERBOLIC DISK, a model for HYPERBOLIC GEOMETRY. The hyperbolic metric is invariant under conformal maps of the disk onto itself.

see also HYPERBOLIC GEOMETRY, POINCARÉ HYPERBOLIC DISK

References

Bear, H. S. "Part Metric and Hyperbolic Metric." *Amer. Math. Monthly* **98**, 109–123, 1991.

Hyperbolic Octahedron

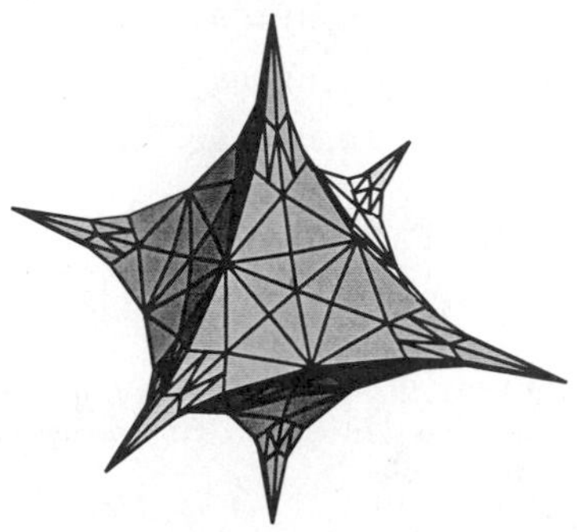

A hyperbolic version of the Euclidean OCTAHEDRON, which is a special case of the ASTROIDAL ELLIPSOID with $a = b = c = 1$. It is given by the parametric equations

$$x = (\cos u \cos v)^3$$
$$y = (\sin u \cos v)^3$$
$$z = \sin^3 v$$

for $u \in [-\pi/2, \pi/2]$ and $v \in [-\pi, \pi]$.

see also ASTROIDAL ELLIPSOID, HYPERBOLIC CUBE, HYPERBOLIC DODECAHEDRON, HYPERBOLIC TETRAHEDRON

References

Gray, A. *Modern Differential Geometry of Curves and Surfaces.* Boca Raton, FL: CRC Press, pp. 305–306, 1993.

Nordstrand, T. "Astroidal Ellipsoid." `http://www.uib.no/people/nfytn/asttxt.htm`.

Rivin, I. "Hyperbolic Polyhedron Graphics." `http://www.mathsource.com/cgi-bin/MathSource/Applications/Graphics/3D/0201-788`.

Hyperbolic Paraboloid

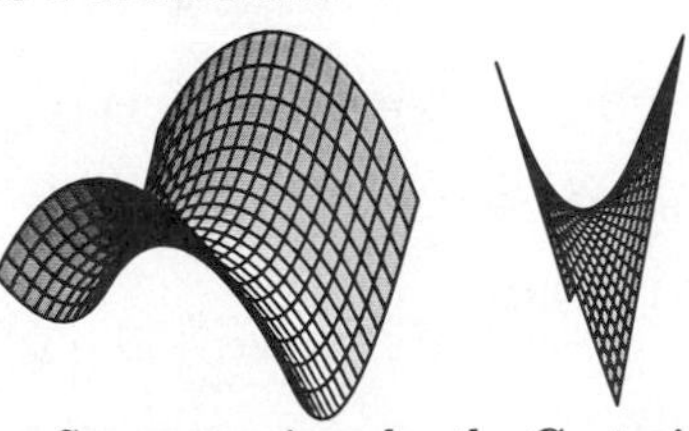

A QUADRATIC SURFACE given by the Cartesian equation

$$z = \frac{y^2}{b^2} - \frac{x^2}{a^2} \qquad (1)$$

(left figure). This form has parametric equations

$$x(u,v) = a(u+v) \quad (2)$$
$$y(u,v) = \pm bv \quad (3)$$
$$z(u,v) = u^2 + 2uv \quad (4)$$

(Gray 1993, p. 336). An alternative form is

$$z = xy \quad (5)$$

(right figure; Fischer 1986), which has parametric equations

$$x(u,v) = u \quad (6)$$
$$y(u,v) = v \quad (7)$$
$$z(u,v) = uv. \quad (8)$$

see also ELLIPTIC PARABOLOID, PARABOLOID, RULED SURFACE

References

Fischer, G. (Ed.). *Mathematical Models from the Collections of Universities and Museums.* Braunschweig, Germany: Vieweg, pp. 3–4, 1986.

Fischer, G. (Ed.). Plates 7–9 in *Mathematische Modelle/Mathematical Models, Bildband/Photograph Volume.* Braunschweig, Germany: Vieweg, pp. 8–10, 1986.

Gray, A. *Modern Differential Geometry of Curves and Surfaces.* Boca Raton, FL: CRC Press, pp. 211–212 and 336, 1993.

Meyer, W. "Spezielle algebraische Flächen." *Encylopädie der Math. Wiss. III*, **22B**, 1439–1779.

Salmon, G. *Analytic Geometry of Three Dimensions.* New York: Chelsea, 1979.

Hyperbolic Partial Differential Equation

A PARTIAL DIFFERENTIAL EQUATION of second-order, i.e., one of the form

$$Au_{xx} + 2Bu_{xy} + Cu_{yy} + Du_x + Eu_y + F = 0, \quad (1)$$

is called hyperbolic if the MATRIX

$$\mathsf{Z} \equiv \begin{bmatrix} A & B \\ B & C \end{bmatrix} \quad (2)$$

satisfies $\det(\mathsf{Z}) < 0$. The WAVE EQUATION is an example of a hyperbolic partial differential equation. Initial-boundary conditions are used to give

$$u(x,y,t) = g(x,y,t) \quad \text{for } x \in \partial\Omega, t > 0 \quad (3)$$
$$u(x,y,0) = v_0(x,y) \quad \text{in } \Omega \quad (4)$$
$$u_t(x,y,0) = v_1(x,y) \quad \text{in } \Omega, \quad (5)$$

where

$$u_{xy} = f(u_x, u_t, x, y) \quad (6)$$

holds in Ω.

see also ELLIPTIC PARTIAL DIFFERENTIAL EQUATION, PARABOLIC PARTIAL DIFFERENTIAL EQUATION, PARTIAL DIFFERENTIAL EQUATION

Hyperbolic Plane

In the hyperbolic plane $\mathbb{H}^2$, a pair of LINES can be PARALLEL (diverging from one another in one direction and intersecting at an IDEAL POINT at infinity in the other), can intersect, or can be HYPERPARALLEL (diverge from each other in both directions).

see also EUCLIDEAN PLANE, RIGID MOTION

Hyperbolic Point

A point $\mathbf{p}$ on a REGULAR SURFACE $M \in \mathbb{R}^3$ is said to be hyperbolic if the GAUSSIAN CURVATURE $K(\mathbf{p}) < 0$ or equivalently, the PRINCIPAL CURVATURES κ_1 and κ_2, have opposite signs.

see also ANTICLASTIC, ELLIPTIC POINT, GAUSSIAN CURVATURE, HYPERBOLIC FIXED POINT (DIFFERENTIAL EQUATIONS), HYPERBOLIC FIXED POINT (MAP), PARABOLIC POINT, PLANAR POINT, SYNCLASTIC

References

Gray, A. *Modern Differential Geometry of Curves and Surfaces.* Boca Raton, FL: CRC Press, p. 280, 1993.

Hyperbolic Polyhedron

A POLYHEDRON in a HYPERBOLIC GEOMETRY.

see HYPERBOLIC CUBE, HYPERBOLIC DODECAHEDRON, HYPERBOLIC OCTAHEDRON, HYPERBOLIC TETRAHEDRON

Hyperbolic Rotation

Also known as the LORENTZ TRANSFORMATION or PROCRUSTIAN STRETCH. Leaves each branch of the HYPERBOLA $x'y' = xy$ invariant and transforms CIRCLES into ELLIPSES with the same AREA.

$$x' = \mu^{-1}x$$
$$y' = \mu y.$$

Hyperbolic Rotation (Crossed)

Exchanges branches of the HYPERBOLA $x'y' = xy$.

$$x' = \mu^{-1}x$$
$$y' = -\mu y.$$

Hyperbolic Secant

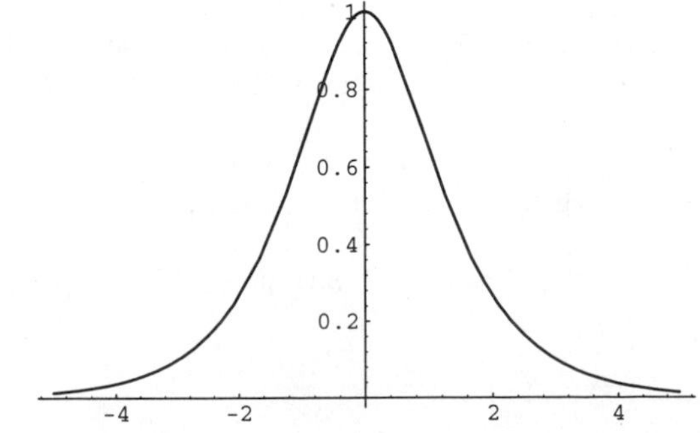

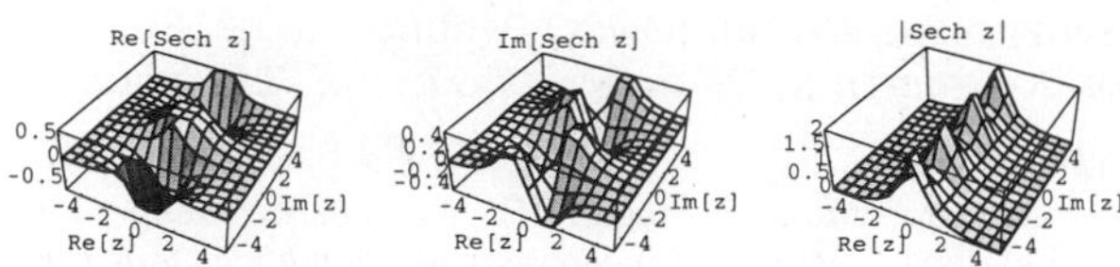

The hyperbolic secant is defined as

$$\operatorname{sech} x \equiv \frac{1}{\cosh x} = \frac{2}{e^x + e^{-x}}.$$

It has a MAXIMUM at $x = 0$ and inflection points at $x = \pm\operatorname{sech}^{-1}(1/\sqrt{2}) \approx 0.881374$.

see also BENSON'S FORMULA, CATENARY, CATENOID, EULER NUMBER, HYPERBOLIC COSINE, OBLATE SPHEROIDAL COORDINATES, PSEUDOSPHERE, SECANT, SURFACE OF REVOLUTION, TRACTRIX, TRACTROID

References
Abramowitz, M. and Stegun, C. A. (Eds.). "Hyperbolic Functions." §4.5 in *Handbook of Mathematical Functions with Formulas, Graphs, and Mathematical Tables, 9th printing.* New York: Dover, pp. 83–86, 1972.
Spanier, J. and Oldham, K. B. "The Hyperbolic Secant sech(x) and Cosecant csch(x) Functions." Ch. 29 in *An Atlas of Functions.* Washington, DC: Hemisphere, pp. 273–278, 1987.

Hyperbolic Sine

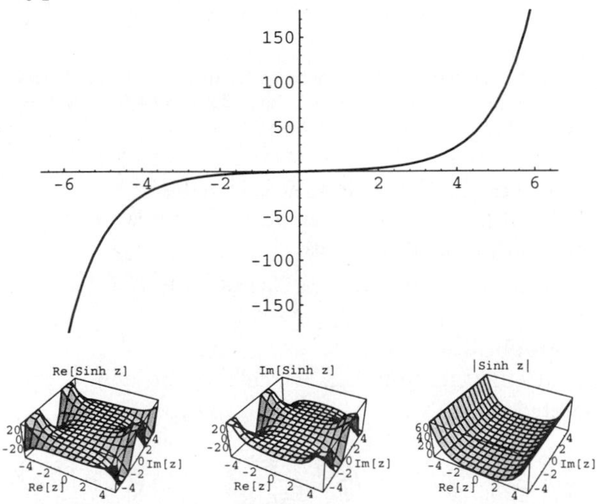

The hyperbolic sine is defined as

$$\sinh x \equiv \tfrac{1}{2}(e^x - e^{-x}).$$

see also BETA FUNCTION (EXPONENTIAL), BIPOLAR COORDINATES, BIPOLAR CYLINDRICAL COORDINATES, BISPHERICAL COORDINATES, CATENARY, CATENOID, CONICAL FUNCTION, CUBIC EQUATION, DE MOIVRE'S IDENTITY, DIXON-FERRAR FORMULA, ELLIPTIC CYLINDRICAL COORDINATES, ELSASSER FUNCTION, FIBONACCI HYPERBOLIC COSINE, FIBONACCI HYPERBOLIC SINE, GUDERMANNIAN FUNCTION, HELICOID, HELMHOLTZ DIFFERENTIAL EQUATION—ELLIPTIC CYLINDRICAL COORDINATES, HYPERBOLIC COSECANT, LAPLACE'S EQUATION—BISPHERICAL COORDINATES, LAPLACE'S EQUATION—TOROIDAL COORDINATES, LEBESGUE CONSTANTS (FOURIER SERIES), LORENTZ GROUP, MERCATOR PROJECTION, MILLER CYLINDRICAL PROJECTION, MODIFIED BESSEL FUNCTION OF THE SECOND KIND, MODIFIED SPHERICAL BESSEL FUNCTION, MODIFIED STRUVE FUNCTION, NICHOLSON'S FORMULA, OBLATE SPHEROIDAL COORDINATES, PARABOLA INVOLUTE, PARTITION FUNCTION P, POINSOT'S SPIRALS, PROLATE SPHEROIDAL COORDINATES, RAMANUJAN'S TAU FUNCTION, SCHLÄFLI'S FORMULA, SHI, SINE, SINE-GORDON EQUATION, SURFACE OF REVOLUTION, TOROIDAL COORDINATES, TOROIDAL FUNCTION, TRACTRIX, WATSON'S FORMULA

References
Abramowitz, M. and Stegun, C. A. (Eds.). "Hyperbolic Functions." §4.5 in *Handbook of Mathematical Functions with Formulas, Graphs, and Mathematical Tables, 9th printing.* New York: Dover, pp. 83–86, 1972.
Spanier, J. and Oldham, K. B. "The Hyperbolic Sine sinh(x) and Cosine cosh(x) Functions." Ch. 28 in *An Atlas of Functions.* Washington, DC: Hemisphere, pp. 263–271, 1987.

Hyperbolic Space

see HYPERBOLIC GEOMETRY

Hyperbolic Spiral

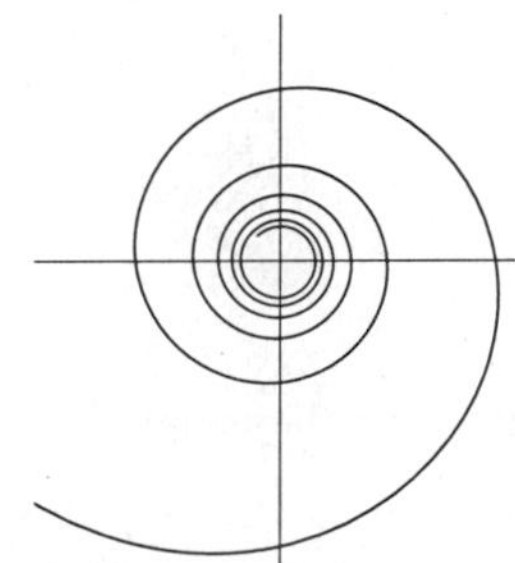

An ARCHIMEDEAN SPIRAL with POLAR equation

$$r = \frac{a}{\theta}.$$

The hyperbolic spiral originated with Pierre Varignon in 1704 and was studied by Johann Bernoulli between 1710 and 1713, as well as by Cotes in 1722 (MacTutor Archive).

see also ARCHIMEDEAN SPIRAL, SPIRAL

References
Gray, A. *Modern Differential Geometry of Curves and Surfaces.* Boca Raton, FL: CRC Press, pp. 69–70, 1993.
Lawrence, J. D. *A Catalog of Special Plane Curves.* New York: Dover, pp. 186 and 188, 1972.
Lockwood, E. H. *A Book of Curves.* Cambridge, England: Cambridge University Press, p. 175, 1967.
MacTutor History of Mathematics Archive. "Hyperbolic Spiral." `http://www-groups.dcs.st-and.ac.uk/~history/Curves/Hyperbolic.html`.

Hyperbolic Spiral Inverse Curve

Taking the pole as the INVERSION CENTER, the HYPERBOLIC SPIRAL inverts to ARCHIMEDES' SPIRAL

$$r = a\theta.$$

Hyperbolic Spiral Roulette

The ROULETTE of the pole of a HYPERBOLIC SPIRAL rolling on a straight line is a TRACTRIX.

Hyperbolic Substitution

A substitution which can be used to transform integrals involving square roots into a more tractable form.

Form	Substitution
$\sqrt{x^2+a^2}$	$x = a\sinh u$
$\sqrt{x^2-a^2}$	$x = a\cosh u$

see also TRIGONOMETRIC SUBSTITUTION

Hyperbolic Tangent

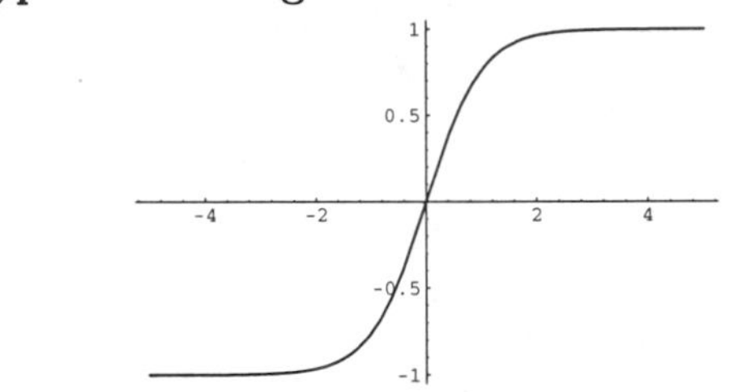

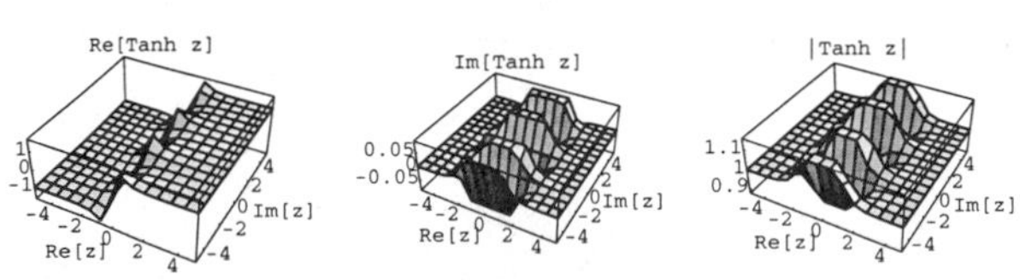

By way of analogy with the usual TANGENT

$$\tan x \equiv \frac{\sin x}{\cos x},$$

the hyperbolic tangent is defined as

$$\tanh x \equiv \frac{\sinh x}{\cosh x} = \frac{e^x - e^{-x}}{e^x + e^{-x}} = \frac{e^{2x} - 1}{e^{2x} + 1},$$

where $\sinh x$ is the HYPERBOLIC SINE and $\cosh x$ is the HYPERBOLIC COSINE. The hyperbolic tangent can be written using a CONTINUED FRACTION as

$$\tanh x = \cfrac{x}{1 + \cfrac{x^2}{3 + \cfrac{x^2}{5 + \ldots}}}.$$

see also BERNOULLI NUMBER, CATENARY, CORRELATION COEFFICIENT—GAUSSIAN BIVARIATE DISTRIBUTION, FIBONACCI HYPERBOLIC TANGENT, FISHER'S z'-TRANSFORMATION, HYPERBOLIC COTANGENT, LORENTZ GROUP, MERCATOR PROJECTION, OBLATE SPHEROIDAL COORDINATES, PSEUDOSPHERE, SURFACE OF REVOLUTION, TANGENT, TRACTRIX, TRACTROID

References

Abramowitz, M. and Stegun, C. A. (Eds.). "Hyperbolic Functions." §4.5 in *Handbook of Mathematical Functions with Formulas, Graphs, and Mathematical Tables, 9th printing.* New York: Dover, pp. 83–86, 1972.

Spanier, J. and Oldham, K. B. "The Hyperbolic Tangent tanh(x) and Cotangent coth(x) Functions." Ch. 30 in *An Atlas of Functions.* Washington, DC: Hemisphere, pp. 279–284, 1987.

Hyperbolic Tetrahedron

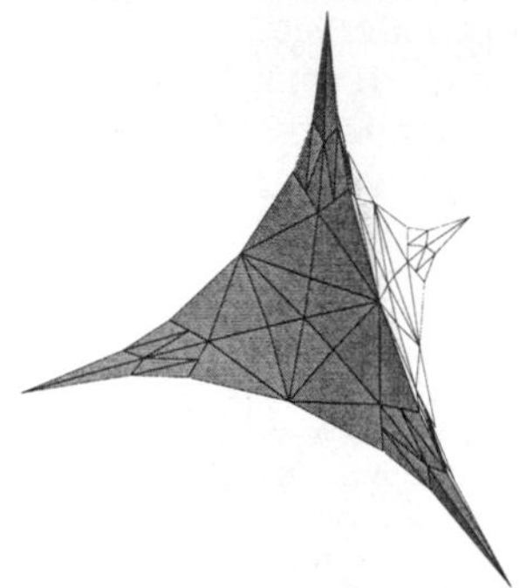

A hyperbolic version of the Euclidean TETRAHEDRON.

see also HYPERBOLIC CUBE, HYPERBOLIC DODECAHEDRON, HYPERBOLIC OCTAHEDRON

References

Rivin, I. "Hyperbolic Polyhedron Graphics." `http://www.mathsource.com/cgi-bin/MathSource/Applications/Graphics/3D/0201-788`.

Hyperbolic Umbilic Catastrophe

A CATASTROPHE which can occur for three control factors and two behavior axes.

see also ELLIPTIC UMBILIC CATASTROPHE

Hyperboloid

A QUADRATIC SURFACE which may be one- or two-sheeted.

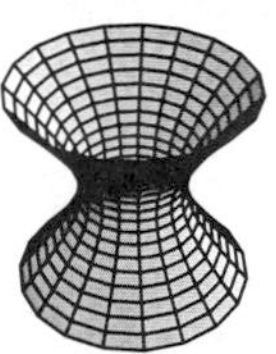

The one-sheeted circular hyperboloid is a doubly RULED SURFACE. When oriented along the z-AXIS, the one-sheeted circular hyperboloid has CARTESIAN COORDINATES equation

$$\frac{x^2}{a^2} + \frac{y^2}{a^2} - \frac{z^2}{c^2} = 1, \tag{1}$$

and parametric equation

$$x = a\sqrt{1+u^2}\cos v \tag{2}$$
$$y = a\sqrt{1+u^2}\sin v \tag{3}$$
$$z = cu \tag{4}$$

for $v \in [0, 2\pi)$ (left figure). Other parameterizations include

$$x(u,v) = a(\cos u \mp v \sin u) \tag{5}$$
$$y(u,v) = a(\sin u \pm v \cos u) \tag{6}$$
$$z(u,v) = \pm cv, \tag{7}$$

(middle figure), or

$$x(u,v) = a\cosh v \cos u \tag{8}$$
$$y(u,v) = a\cosh v \sin u \tag{9}$$
$$z(u,v) = c\sinh v \tag{10}$$

(right figure). An obvious generalization gives the one-sheeted ELLIPTIC HYPERBOLOID.

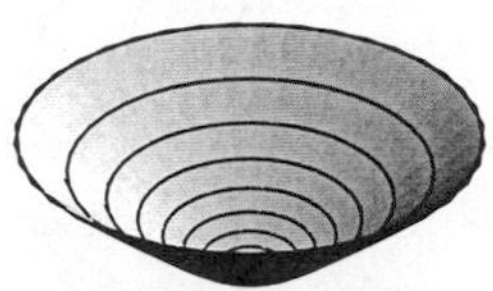

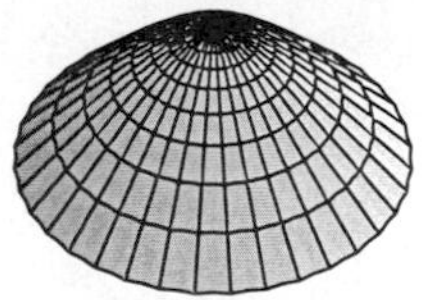

A two-sheeted circular hyperboloid oriented along the z-AXIS has CARTESIAN COORDINATES equation

$$\frac{x^2}{a^2} + \frac{y^2}{a^2} - \frac{z^2}{c^2} = -1. \tag{11}$$

The parametric equations are

$$x = a\sinh u \cos v \tag{12}$$
$$y = a\sinh u \sin v \tag{13}$$
$$z = \pm c\cosh u \tag{14}$$

for $v \in [0, 2\pi)$. Note that the plus and minus signs in z correspond to the upper and lower sheets. The two-sheeted circular hyperboloid oriented along the x-AXIS has Cartesian equation

$$\frac{x^2}{a^2} - \frac{y^2}{a^2} - \frac{z^2}{c^2} = 1 \tag{15}$$

and parametric equations

$$x = \pm a\cosh u \cosh v \tag{16}$$
$$y = a\sinh u \cosh v \tag{17}$$
$$z = c\sinh v \tag{18}$$

(Gray 1993, p. 313). Again, an obvious generalization gives the two-sheeted ELLIPTIC HYPERBOLOID.

The SUPPORT FUNCTION of the hyperboloid of one sheet

$$\frac{x^2}{a^2} + \frac{y^2}{b^2} - \frac{z^2}{c^2} = 1 \tag{19}$$

is

$$h = \left(\frac{x^2}{a^4} + \frac{y^2}{b^4} + \frac{z^2}{c^4}\right)^{-1/2}, \tag{20}$$

and the GAUSSIAN CURVATURE is

$$K = -\frac{h^4}{a^2b^2c^2}. \tag{21}$$

The SUPPORT FUNCTION of the hyperboloid of two sheets

$$\frac{x^2}{a^2} - \frac{y^2}{b^2} - \frac{z^2}{c^2} = 1 \tag{22}$$

is

$$h = \left(\frac{x^2}{a^4} - \frac{y^2}{b^4} + \frac{z^2}{c^4}\right)^{-1/2}, \tag{23}$$

and the GAUSSIAN CURVATURE is

$$K = \frac{h^4}{a^2b^2c^2} \tag{24}$$

(Gray 1993, pp. 296–297).

see also CATENOID, ELLIPSOID, ELLIPTIC HYPERBOLOID, HYPERBOLOID EMBEDDING, PARABOLOID, RULED SURFACE

References

Fischer, G. (Ed.). Plates 67 and 69 in *Mathematische Modelle/Mathematical Models, Bildband/Photograph Volume.* Braunschweig, Germany: Vieweg, pp. 62 and 64, 1986.

Gray, A. "The Hyperboloid of Revolution." §18.5 in *Modern Differential Geometry of Curves and Surfaces.* Boca Raton, FL: CRC Press, pp. 296–297, 311–314, and 369–370, 1993.

Hyperboloid Embedding

A 4-HYPERBOLOID has NEGATIVE CURVATURE, with

$$R^2 = x^2 + y^2 + z^2 - w^2 \tag{1}$$

$$2x\frac{dx}{dw} + 2y\frac{dy}{dw} + 2z\frac{dz}{dw} - 2w = 0. \tag{2}$$

Since

$$\mathbf{r} \equiv x\hat{\mathbf{x}} + y\hat{\mathbf{y}} + z\hat{\mathbf{z}}, \tag{3}$$

$$dw = \frac{x\,dx + y\,dy + z\,dz}{w} = \frac{\mathbf{r}\cdot d\mathbf{r}}{\sqrt{r^2 - R^2}}. \tag{4}$$

To stay on the surface of the HYPERBOLOID,

$$\begin{aligned} ds^2 &= dx^2 + dy^2 + dz^2 - dw^2 \\ &= dx^2 + dy^2 + dz^2 - \frac{r^2\,dr^2}{r^2 - R^2} \\ &= dr^2 + r^2 d\Omega^2 + \frac{dr^2}{1 - \frac{R^2}{r^2}}. \end{aligned} \tag{5}$$

Hypercomplex Number

A number having properties departing from those of the REAL and COMPLEX NUMBERS. The most common examples are BIQUATERNIONS, EXTERIOR ALGEBRAS, GROUP algebras, MATRICES, OCTONIONS, and QUATERNIONS.

References

van der Waerden, B. L. *A History of Algebra from al-Khwarizmi to Emmy Noether.* New York: Springer-Verlag, pp. 177–217, 1985.

Hypercube

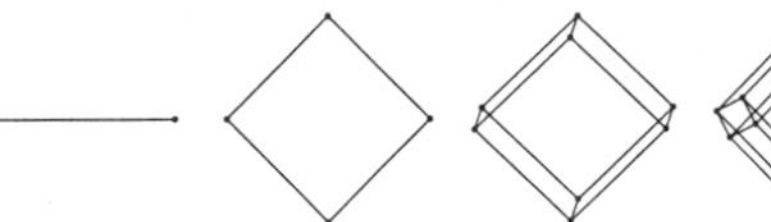

The generalization of a 3-CUBE to n-D, also called a MEASURE POLYTOPE. It is a regular POLYTOPE with mutually PERPENDICULAR sides, and is therefore an ORTHOTOPE. It is denoted γ_n and has SCHLÄFLI SYMBOL $\{4, \underbrace{3,3}_{n-2}\}$. The number of k-cubes contained in an n-cube can be found from the COEFFICIENTS of $(2k+1)^n$.

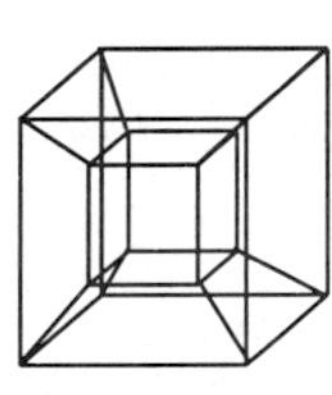

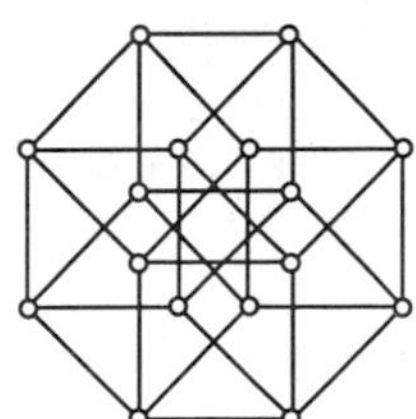

The 1-hypercube is a LINE SEGMENT, the 2-hypercube is the SQUARE, and the 3-hypercube is the CUBE. The hypercube in $\mathbb{R}^4$, called a TESSERACT, has the SCHLÄFLI SYMBOL $\{4,3,3\}$ and VERTICES $(\pm1, \pm1, \pm1, \pm1)$. The above figures show two visualizations of the TESSERACT. The figure on the left is a projection of the TESSERACT in 3-space (Gardner 1977), and the figure on the right is the GRAPH of the TESSERACT symmetrically projected into the PLANE (Coxeter 1973). A TESSERACT has 16 VERTICES, 32 EDGES, four SQUARES, and eight CUBES.

see also CROSS POLYTOPE, CUBE, HYPERSPHERE, ORTHOTOPE, PARALLELEPIPED, POLYTOPE, SIMPLEX, TESSERACT

References

Coxeter, H. S. M. *Regular Polytopes, 3rd ed.* New York: Dover, p. 123, 1973.

Gardner, M. "Hypercubes." Ch. 4 in *Mathematical Carnival: A New Round-Up of Tantalizers and Puzzles from Scientific American.* New York: Vintage Books, 1977.

Geometry Center. "The Tesseract (or Hypercube)." http://www.geom.umn.edu/docs/outreach/4-cube/.

Pappas, T. "How Many Dimensions are There?" *The Joy of Mathematics.* San Carlos, CA: Wide World Publ./Tetra, pp. 204–205, 1989.

Hyperdeterminant

A technically defined extension of the ordinary DETERMINANT to "higher dimensional" HYPERMATRICES. Cayley (1845) originally coined the term, but subsequently used it to refer to an ALGEBRAIC INVARIANT of a multilinear form. The hyperdeterminant of the $2\times2\times2$ HYPERMATRIX $A = a_{ijk}$ (for $i, j, k = 0, 1$) is given by

$$\begin{aligned} \det(A) = {} & ({a_{000}}^2 {a_{111}}^2 + {a_{001}}^2 {a_{110}}^2 + {a_{010}}^2 {a_{101}}^2 + {a_{011}}^2 {a_{100}}^2) \\ & - 2(a_{000}a_{001}a_{110}a_{111} + a_{000}a_{010}a_{101}a_{111} + a_{000}a_{011}a_{100}a_{111} \\ & + a_{001}a_{010}a_{101}a_{110} + a_{001}a_{011}a_{110}a_{100} + a_{010}a_{011}a_{101}a_{100}) \\ & + 4(a_{000}a_{011}a_{101}a_{110} + a_{001}a_{010}a_{100}a_{111}). \end{aligned}$$

The above hyperdeterminant vanishes IFF the following system of equations in six unknowns has a nontrivial solution,

$$\begin{aligned} a_{000}x_0y_0 + a_{010}x_0y_1 + a_{100}x_1y_0 + a_{110}x_1y_1 &= 0 \\ a_{001}x_0y_0 + a_{011}x_0y_1 + a_{101}x_1y_0 + a_{111}x_1y_1 &= 0 \\ a_{000}x_0z_0 + a_{001}x_0z_1 + a_{100}x_1z_0 + a_{101}x_1z_1 &= 0 \\ a_{010}x_0z_0 + a_{011}x_0z_1 + a_{110}x_1z_0 + a_{111}x_1z_1 &= 0 \\ a_{000}y_0z_0 + a_{001}y_0z_1 + a_{010}y_1z_0 + a_{011}y_1z_1 &= 0 \\ a_{100}y_0z_0 + a_{101}y_0z_1 + a_{110}y_1z_0 + a_{111}y_1z_1 &= 0. \end{aligned}$$

see also DETERMINANT, HYPERMATRIX

References

Cayley, A. "On the Theory of Linear Transformations." *Cambridge Math. J.* **4**, 193–209, 1845.

Gel'fand, I. M.; Kapranov, M. M.; and Zelevinsky, A. V. "Hyperdeterminants." *Adv. Math.* **96**, 226–263, 1992.

Schläfli, L. "Über die Resultante eine Systemes mehrerer algebraischer Gleichungen." *Denkschr. Kaiserl. Akad. Wiss., Math.-Naturwiss. Klasse* **4**, 1852.

Hyperellipse

$$y^{n/m} + c\left|\frac{x}{a}\right|^{n/m} - c = 0,$$

with $n/m > 2$. If $n/m < 2$, the curve is a HYPOELLIPSE.

see also ELLIPSE, HYPOELLIPSE, SUPERELLIPSE

References

von Seggern, D. *CRC Standard Curves and Surfaces.* Boca Raton, FL: CRC Press, p. 82, 1993.

Hyperelliptic Function

see ABELIAN FUNCTION

Hyperelliptic Integral

see ABELIAN INTEGRAL

Hyperfactorial

The function defined by

$$H(n) \equiv K(n+1) \equiv 1^1 2^2 3^3 \cdots n^n,$$

where K is the K-FUNCTION and the first few values for $n = 1, 2, \ldots$ are 1, 4, 108, 27648, 86400000, 4031078400000, 3319766398771200000, ... (Sloane's A002109), and these numbers are called hyperfactorials by Sloane and Plouffe (1995).

see also G-FUNCTION, GLAISHER-KINKELIN CONSTANT, K-FUNCTION

References

Sloane, N. J. A. Sequence A002109/M3706 in "An On-Line Version of the Encyclopedia of Integer Sequences."

Hypergeometric Differential Equation

$$x(x-1)\frac{d^2y}{dx^2} + [(1+\alpha+\beta)x - \gamma]\frac{dy}{dx} + \alpha\beta y = 0.$$

It has REGULAR SINGULAR POINTS at 0, 1, and ∞. Every ORDINARY DIFFERENTIAL EQUATION of second-order with at most three REGULAR SINGULAR POINTS can be transformed into the hypergeometric differential equation.

see also CONFLUENT HYPERGEOMETRIC DIFFERENTIAL EQUATION, CONFLUENT HYPERGEOMETRIC FUNCTION, HYPERGEOMETRIC FUNCTION

References

Morse, P. M. and Feshbach, H. *Methods of Theoretical Physics, Part I.* New York: McGraw-Hill, pp. 542–543, 1953.

Hypergeometric Distribution

Let there be n ways for a successful and m ways for an unsuccessful trial out of a total of $n+m$ possibilities. Take N samples and let x_i equal 1 if selection i is successful and 0 if it is not. Let x be the total number of successful selections,

$$x \equiv \sum_{i=1}^N x_i. \tag{1}$$

The probability of i successful selections is then

$$\begin{aligned} P(x=i) &= \\ &\frac{[\text{\# ways for } i \text{ successes}][\text{\# ways for } N-i \text{ unsuccesses}]}{[\text{total number of ways to select}]} \\ &= \frac{\binom{n}{i}\binom{m}{N-i}}{\binom{n+m}{N}} = \frac{\frac{n!}{i!(n-i!)}\frac{m!}{(m+i-N)!(N-i)!}}{\frac{(n+m)!}{N!(N-n-m)!}} \\ &= \frac{n!m!N!(N-m-n)!}{i!(n-i)!(m+i-N)!(N-i)!(n+m)!}. \end{aligned} \tag{2}$$

The ith selection has an equal likelihood of being in any trial, so the fraction of acceptable selections p is

$$p \equiv \frac{n}{n+m} \tag{3}$$

$$P(x_i = 1) = \frac{n}{n+m} \equiv p. \tag{4}$$

The expectation value of x is

$$\begin{aligned} \mu \equiv \langle x \rangle &= \left\langle \sum_{i=1}^N x_i \right\rangle = \sum_{i=1}^N \langle x_i \rangle \\ &= \sum_{i=1}^N \frac{n}{n+m} = \frac{nN}{n+m} = Np. \end{aligned} \tag{5}$$

The VARIANCE is

$$\operatorname{var}(x) \equiv \sum_{i=1}^N \operatorname{var}(x_i) + \sum_{i=1}^N \sum_{\substack{j=1 \\ j\neq i}}^N \operatorname{cov}(x_i, x_j). \tag{6}$$

Since x_i is a BERNOULLI variable,

$$\begin{aligned} \operatorname{var}(x_i) = p(1-p) &= \frac{n}{n+m}\left(1 - \frac{n}{n+m}\right) \\ &= \frac{n}{n+m}\left(1 - \frac{n}{n+m}\right) \\ &= \frac{n}{n+m}\left(\frac{n+m-n}{n+m}\right) = \frac{nm}{(n+m)^2}, \end{aligned} \tag{7}$$

so

$$\sum_{i=1}^N \operatorname{var}(x_i) = \frac{Nnm}{(n+m)^2}. \tag{8}$$

For $i < j$, the COVARIANCE is

$$\operatorname{cov}(x_i, x_j) = \langle x_i x_j \rangle - \langle x_i \rangle \langle x_j \rangle. \tag{9}$$

The probability that both i and j are successful for $i \neq j$ is

$$\begin{aligned} P(x_i = 1, x_j = 1) &= P(x_i = 1)P(x_j = 1 | x_i = 1) \\ &= \frac{n}{n+m}\frac{n-1}{n+m-1} \\ &= \frac{n(n-1)}{(n+m)(n+m-1)}. \end{aligned} \tag{10}$$

But since x_i and x_j are random BERNOULLI variables (each 0 or 1), their product is also a BERNOULLI variable. In order for $x_i x_j$ to be 1, both x_i and x_j must be 1,

$$\begin{aligned} \langle x_i x_j \rangle = P(x_i x_j = 1) &= P(x_i = 1, x_j = 1) \\ &= \frac{n}{n+m}\frac{n-1}{n+m-1} \\ &= \frac{n(n-1)}{(n+m)(n+m-1)}. \end{aligned} \tag{11}$$

Combining (11) with

$$\langle x_i \rangle \langle x_j \rangle = \frac{n}{n+m}\frac{n}{n+m} = \frac{n^2}{(n+m)^2}, \tag{12}$$

gives

$$\begin{aligned}\operatorname{cov}(x_i, x_j) &= \frac{(n+m)(n^2-n) - n^2(n+m-1)}{(n+m)^2(n+m-1)} \\ &= \frac{n^3+mn^2-n^2-mn-n^3-n^2m+n^2}{(n+m)^2(n+m-1)} \\ &= -\frac{mn}{(n+m)^2(n+m-1)}.\end{aligned} \tag{13}$$

There are a total of N^2 terms in a double summation over N. However, $i = j$ for N of these, so there are a total of $N^2 - N = N(N-1)$ terms in the COVARIANCE summation

$$\sum_{i=1}^{N}\sum_{\substack{j=1\\ j\neq i}}^{N} \operatorname{cov}(x_i, x_j) = -\frac{N(N-1)mn}{(n+m)^2(n+m-1)}. \tag{14}$$

Combining equations (6), (8), (11), and (14) gives the VARIANCE

$$\begin{aligned}\operatorname{var}(x) &= \frac{Nmn}{(n+m)^2} - \frac{N(N-1)mn}{(n+m)^2(n+m-1)} \\ &= \frac{Nmn}{(m+n)^2}\left(1 - \frac{N-1}{n+m-1}\right) \\ &= \frac{Nmn}{(n+m)^2}\left(\frac{N+m-1-N+1}{n+m-1}\right) \\ &= \frac{Nmn(n+m-N)}{(n+m)^2(n+m-1)},\end{aligned} \tag{15}$$

so the final result is

$$\langle x \rangle = Np \tag{16}$$

and, since

$$1 - p = \frac{m}{n+m} \tag{17}$$

and

$$np(1-p) = \frac{mn}{(n+m)^2}, \tag{18}$$

we have

$$\begin{aligned}\sigma^2 = \operatorname{var}(x) &= Np(1-p)\left(1 - \frac{N-1}{n+m-1}\right) \\ &= \frac{mnN(m+n-N)}{(m+n)^2(m+n-1)}.\end{aligned} \tag{19}$$

The SKEWNESS is

$$\begin{aligned}\gamma_1 &= \frac{q-p}{\sqrt{npq}}\sqrt{\frac{N-1}{N-m}}\left(\frac{N-2n}{N-2}\right) \\ &= \frac{(m-n)(m+n-2N)}{m+n-2}\sqrt{\frac{m+n-1}{mnN(m+n-N)}},\end{aligned} \tag{20}$$

and the KURTOSIS

$$\gamma_2 = \frac{F(m,n,N)}{mnN(-3+m+n)(-2+m+n)(-m-n+N)}, \tag{21}$$

where

$$\begin{aligned}F(m,n,N) = {}& m^3 - m^5 + 3m^2n - 6m^3n + m^4n + 3mn^2 \\ & - 12m^2n^2 + 8m^3n^2 + n^3 - 6mn^3 + 8m^2n^3 \\ & + mn^4 - n^5 - 6m^3N + 6m^4N + 18m^2nN \\ & - 6m^3nN + 18mn^2N - 24m^2n^2N - 6n^3N \\ & - 6mn^3N + 6n^4N + 6m^2N^2 - 6m^3N^2 \\ & - 24mnN^2 + 12m^2nN^2 + 6n^2N^2 \\ & + 12mn^2N^2 - 6n^3N^2.\end{aligned} \tag{22}$$

The GENERATING FUNCTION is

$$\phi(t) = \frac{\binom{m}{N}}{\binom{n+m}{N}}\, {}_2F_1(-N, -n; m-N+1; e^{it}), \tag{23}$$

where ${}_2F_1(a,b;c;z)$ is the HYPERGEOMETRIC FUNCTION.

If the hypergeometric distribution is written

$$h_n(x,s) = \frac{\binom{np}{x}\binom{nq}{s-x}}{\binom{n}{s}}, \tag{24}$$

then

$$\sum_{x=0}^{s} h_n(x,s)u^x = A\, {}_2F_1(-s, -np; nq-s+1; u). \tag{25}$$

References

Beyer, W. H. *CRC Standard Mathematical Tables, 28th ed.* Boca Raton, FL: CRC Press, pp. 532–533, 1987.

Spiegel, M. R. *Theory and Problems of Probability and Statistics.* New York: McGraw-Hill, pp. 113–114, 1992.

Hypergeometric Function

A GENERALIZED HYPERGEOMETRIC FUNCTION ${}_pF_q(a_1, \ldots, a_p; b_1, \ldots, b_q; x)$ is a function which can be defined in the form of a HYPERGEOMETRIC SERIES, i.e., a series for which the ratio of successive terms can be written

$$\frac{a_{k+1}}{a_k} = \frac{P(k)}{Q(k)} = \frac{(k+a_1)(k+a_2)\cdots(k+a_p)}{(k+b_1)(k+b_2)\cdots(k+b_q)(k+1)}x. \tag{1}$$

(The factor of $k+1$ in the DENOMINATOR is present for historical reasons of notation.) The function ${}_2F_1(a,b;c;x)$ corresponding to $p = 2$, $q = 1$ is the first hypergeometric function to be studied (and, in general, arises the most frequently in physical problems), and so is frequently known as "the" hypergeometric equation.

To confuse matters even more, the term "hypergeometric function" is less commonly used to mean CLOSED FORM.

The hypergeometric functions are solutions to the HYPERGEOMETRIC DIFFERENTIAL EQUATION, which has a REGULAR SINGULAR POINT at the ORIGIN. To derive the hypergeometric function based on the HYPERGEOMETRIC DIFFERENTIAL EQUATION, plug

$$y = \sum_{n=0}^{\infty} A_n z^n \tag{2}$$

$$y' = \sum_{n=0}^{\infty} n A_n z^{n-1} \tag{3}$$

$$y'' = \sum_{n=0}^{\infty} n(n-1) A_n z^{n-2} \tag{4}$$

into

$$z(1-z)y'' + [c-(a+b+1)a]y' - aby = 0 \tag{5}$$

to obtain

$$\sum_{n=0}^{\infty} n(n-1)A_n z^{n-1} - \sum_{n=0}^{\infty} n(n-1)A_n z^n + c\sum_{n=0}^{\infty} nA_n z^{n-1} + (a+b+1)\sum_{n=0}^{\infty} nA_n z^n - ab\sum_{n=0}^{\infty} A_n z^n = 0 \tag{6}$$

$$\sum_{n=2}^{\infty} n(n-1)A_n z^{n-1} - \sum_{n=0}^{\infty} n(n-1)A_n z^n + c\sum_{n=1}^{\infty} nA_n z^{n-1} - (a+b+1)\sum_{n=1}^{\infty} nA_n z^n - ab\sum_{n=0}^{\infty} A_n z^n = 0 \tag{7}$$

$$\sum_{n=0}^{\infty} (n+1)nA_{n+1} z^n - \sum_{n=0}^{\infty} n(n-1)A_n z^n + c\sum_{n=0}^{\infty} (n+1)A_{n+1} z^n - (a+b+1)\sum_{n=0}^{\infty} nA_n z^n - ab\sum_{n=0}^{\infty} A_n z^n = 0 \tag{8}$$

$$\sum_{n=0}^{\infty} [n(n+1)A_{n+1} - n(n-1)A_n + c(n+1)A_{n-1} - (a+b+1)nA_n - abA_n] z^n = 0 \tag{9}$$

$$\sum_{n=0}^{\infty} \{(n+1)(n+c)A_{n+1} - [n(n-1+a+b+1)+ab]A_n\} z^n = 0 \tag{10}$$

$$\sum_{n=0}^{\infty} \{(n+1)(n+c)A_{n+1} - [n^2+(a+b)n+ab]A_n\} z^n = 0, \tag{11}$$

so

$$A_{n+1} = \frac{(n+a)(n+b)}{(n+1)(n+c)} A_n \tag{12}$$

and

$$y = A_0 \left[1 + \frac{ab}{1!c} z + \frac{a(a+1)b(b+1)}{2!c(c+1)} z^2 + \ldots \right]. \tag{13}$$

This is the regular solution and is denoted

$$\begin{aligned} {}_2F_1(a,b;c;z) &= 1 + \frac{ab}{1!c} z + \frac{a(a+1)b(b+1)}{2!c(c+1)} z^2 + \ldots \\ &= \sum_{n=0}^{\infty} \frac{(a)_n (b)_n}{(c)_n} \frac{z^n}{n!}, \end{aligned} \tag{14}$$

where $(a)_n$ are POCHHAMMER SYMBOLS. The hypergeometric series is convergent for REAL $-1 < z < 1$, and for $z = \pm 1$ if $c > a + b$. The complete solution to the HYPERGEOMETRIC DIFFERENTIAL EQUATION is

$$y = A\,{}_2F_1(a,b;c;z) + Bz^{1-c}\,{}_2F_1(a+1-c, b+1-c; 2-c; z). \tag{15}$$

Derivatives are given by

$$\frac{d\,{}_2F_1(a,b;c;z)}{dz} = \frac{ab}{c}\,{}_2F_1(a+1,b+1;c+1;z) \tag{16}$$

$$\frac{d^2\,{}_2F_1(a,b;c;z)}{dz^2} = \frac{a(a+1)b(b+1)}{c(c+1)} \times {}_2F_1(a+2,b+2;c+2;z) \tag{17}$$

(Magnus and Oberhettinger 1949, p. 8). An integral giving the hypergeometric function is

$${}_2F_1(a,b;c;z) = \frac{\Gamma(c)}{\Gamma(b)\Gamma(c-b)} \int_0^1 \frac{t^{b-1}(1-t)^{c-b-1}}{(1-tz)^a}\, dt \tag{18}$$

as shown by Euler in 1748.

A hypergeometric function can be written using EULER'S HYPERGEOMETRIC TRANSFORMATIONS

$$t \to t \tag{19}$$

$$t \to 1-t \tag{20}$$

$$t \to (1-z-tz)^{-1} \tag{21}$$

$$t \to \frac{1-t}{1-tz} \tag{22}$$

in any one of four equivalent forms

$$ {}_2F_1(a,b;c;z) = (1-z)^{-a}\,{}_2F_1(a,c-b;c;z/(z-1)) \quad (23)$$
$$ = (1-z)^{-b}\,{}_2F_1(c-a,b;c;z/(z-1)) \quad (24)$$
$$ = (1-z)^{c-a-b}\,{}_2F_1(c-a,c-b;c;z). \quad (25)$$

It can also be written as a linear combination

$$ {}_2F_1(a,b;c;z) = \frac{\Gamma(c)\Gamma(c-a-b)}{\Gamma(c-a)\Gamma(c-b)}\,{}_2F_1(a,b;a+b+1-c;1-z) + \frac{\Gamma(c)\Gamma(a+b-c)}{\Gamma(a)\Gamma(b)}(1-z)^{c-a-b} \times {}_2F_1(c-a,c-b;1+c-a-b;1-z). \quad (26)$$

Kummer found all six solutions (not necessarily regular at the origin) to the HYPERGEOMETRIC DIFFERENTIAL EQUATION,

$$u_1(x) = {}_2F_1(a,b;c;z)$$
$$u_2(x) = {}_2F_1(a,b;a+b+1-c;1-z)$$
$$u_3(x) = z^{-a}\,{}_2F_1(a,a+1-c;a+1-b;1/z)$$
$$u_4(x) = z^{-b}\,{}_2F_1(b+1-c,b;b+1-a;1/z)$$
$$u_5(x) = z^{1-c}\,{}_2F_1(b+1-c,a+1-c;2-c;z)$$
$$u_6(x) = (1-z)^{c-a-b}\,{}_2F_1(c-a,c-b;c+1-a-b;1-z).$$

Applying EULER'S HYPERGEOMETRIC TRANSFORMATIONS to the Kummer solutions then gives all 24 possible forms which are solutions to the HYPERGEOMETRIC DIFFERENTIAL EQUATION

$$u_1^{(1)}(x) = {}_2F_1(a,b;c;z)$$
$$u_1^{(2)}(x) = (1-z)^{-a}\,{}_2F_1(a,c-b;c;z/(z-1))$$
$$u_1^{(3)}(x) = (1-z)^{-b}\,{}_2F_1(c-a,b;c;z/(z-1))$$
$$u_1^{(4)}(x) = (1-z)^{c-a-b}\,{}_2F_1(c-a,c-b;c;z)$$
$$u_2^{(1)}(x) = {}_2F_1(a,b;a+b+1-c;1-z)$$
$$u_2^{(2)}(x) = z^{-a}\,{}_2F_1(a,a+1-c;a+b+1-c;1-1/z)$$
$$u_2^{(3)}(x) = z^{-b}\,{}_2F_1(b+1-c,b;a+b+1-c;1-1/z)$$
$$u_2^{(4)}(x) = z^{1-c}\,{}_2F_1(b+1-c,a+1-c;a+b+1-c;1-z)$$
$$u_3^{(1)}(x) = z^{-a}\,{}_2F_1(a,a+1-c;a+1-b;1/z)$$
$$u_3^{(2)}(x) = z^{-a}(1-1/1z)^{-a}\,{}_2F_1(a,c-b;a+1-b;1/(1-z))$$
$$u_3^{(3)}(x) = z^{-a}(1-1/z)^{c-a-1} \times {}_2F_1(1-b,a+1-c;a+1-b;1/(1-z))$$
$$u_3^{(4)}(x) = z^{-a}(1-1/z)^{c-a-b}\,{}_2F_1(1-b,c-b;a+1-b;1/z)$$
$$u_4^{(1)}(x) = z^{-a}\,{}_2F_1(b+1-c,b;b+1-a;1/z)$$
$$u_4^{(2)}(x) = z^{-b}(1-1/z)^{c-b-1} \times {}_2F_1(b_1-c,1-a;b+1-a;1/(1-z))$$
$$u_4^{(3)}(x) = z^{-b}(1-1/z)^{-b}\,{}_2F_1(c-a,b;b+1-a;1/(1-z))$$
$$u_4^{(4)}(x) = z^{-b}(1-1/z)^{c-a-b}\,{}_2F_1(c-a,1-a;b+1-a;1/z)$$
$$u_5^{(1)}(x) = z^{1-c}\,{}_2F_1(b+1-c,a+1-c;2-c;z)$$
$$u_5^{(2)}(x) = z^{1-c}(1-z)^{c-b-1}\,{}_2F_1(b+1-c,1-a;2-c;z/(z-1))$$
$$u_5^{(2)}(x) = z^{1-c}(1-z)^{c-a-1}\,{}_2F_1(1-b,a+1-c;2-c;z/(z-1))$$
$$u_5^{(4)}(x) = z^{1-c}(1-z)^{c-a-b}\,{}_2F_1(1-b,1-a;2-c;z)$$
$$u_6^{(1)}(x) = (1-z)^{c-a-b}\,{}_2F_1(c-a,c-b;c+1-a-b;1-z)$$
$$u_6^{(2)}(x) = z^{a-c}(1-z)^{c-a-b} \times {}_2F_1(c-a,1-a;c+1-a-b;1-1/z)$$
$$u_6^{(2)}(x) = z^{b-c}(1-z)^{c-a-b} \times {}_2F_1(1-b,c-b;c+1-a-b;1-1/z)$$
$$u_6^{(4)}(x) = z^{c-a-b}(1-z)^{c-a-b} \times {}_2F_1(1-b,1-a;c+1-a-b;1-z).$$

Goursat (1881) gives many hypergeometric transformation FORMULAS, including several cubic transformation FORMULAS.

Many functions of mathematical physics can be expressed as special cases of the hypergeometric functions. For example,

$$ {}_2F_1(-l,l+1,1;(1-z)/2) = P_l(z), \quad (27)$$

where $P_l(z)$ is a LEGENDRE POLYNOMIAL.

$$(1+z)^n = {}_2F_1(-n,b;b;-z) \quad (28)$$

$$\ln(1+z) = z\,{}_2F_1(1,1;2;-z) \quad (29)$$

Complete ELLIPTIC INTEGRALS and the RIEMANN P-SERIES can also be expressed in terms of ${}_2F_1(a,b;c;z)$. Special values include

$$ {}_2F_1(a,b;a-b+1;-1) = 2^{-a}\sqrt{\pi}\frac{\Gamma(1+a+b)}{\Gamma(1+\frac{1}{2}a-b)\Gamma(\frac{1}{2}+\frac{1}{2}a)} \quad (30)$$

$$ {}_2F_1(1,-a;a;-1) = \frac{\sqrt{\pi}}{2}\frac{\Gamma(a)}{\Gamma(a+\frac{1}{2})}+1 \quad (31)$$

$$ {}_2F_1(a,b;c;\tfrac{1}{2}) = 2^a\,{}_2F_1(a,c-b;c;-1) \quad (32)$$

$$ {}_2F_1(a,b;\tfrac{1}{2}(a+b+1);\tfrac{1}{2}) = \frac{\Gamma(\frac{1}{2})\Gamma[(\frac{1}{2}(1+a+b)]}{\Gamma[\frac{1}{2}(1+a)]\Gamma[\frac{1}{2}(1+b)]} \quad (33)$$

$$ {}_2F_1(a,1-a;c;\tfrac{1}{2}) = \frac{\Gamma(\frac{1}{2}c)\Gamma[\frac{1}{2}(c+1)]}{\Gamma[\frac{1}{2}(a+c)]\Gamma[\frac{1}{2}(1+c-a)]} \quad (34)$$

$$ {}_2F_1(a,b;c;1) = \frac{\Gamma(c)\Gamma(c-a-b)}{\Gamma(c-a)\Gamma(c-b)}. \quad (35)$$

KUMMER'S FIRST FORMULA gives

$$ {}_2F_1(\tfrac{1}{2}+m-k,-n;2m+1;1) = \frac{\Gamma(2m+1)\Gamma(m+\frac{1}{2}+k+n)}{\Gamma(m+\frac{1}{2}+k)\Gamma(2m+1+n)}, \quad (36)$$

where $m \neq -1/2, -1, -3/2, \ldots$. Many additional identities are given by Abramowitz and Stegun (1972, p. 557).

Hypergeometric functions can be generalized to GENERALIZED HYPERGEOMETRIC FUNCTIONS

$$ {}_nF_m(a_1,\ldots,a_n;b_1,\ldots,b_m;z). \qquad (37)$$

A function of the form ${}_1F_1(a;b;z)$ is called a CONFLUENT HYPERGEOMETRIC FUNCTION, and a function of the form ${}_0F_1(;b;z)$ is called a CONFLUENT HYPERGEOMETRIC LIMIT FUNCTION.

see also APPELL HYPERGEOMETRIC FUNCTION, BARNES' LEMMA, BRADLEY'S THEOREM, CAYLEY'S HYPERGEOMETRIC FUNCTION THEOREM, CLAUSEN FORMULA, CLOSED FORM, CONFLUENT HYPERGEOMETRIC FUNCTION, CONFLUENT HYPERGEOMETRIC LIMIT FUNCTION, CONTIGUOUS FUNCTION, DARLING'S PRODUCTS, GENERALIZED HYPERGEOMETRIC FUNCTION, GOSPER'S ALGORITHM, HYPERGEOMETRIC IDENTITY, HYPERGEOMETRIC SERIES, JACOBI POLYNOMIAL, KUMMER'S FORMULAS, KUMMER'S QUADRATIC TRANSFORMATION, KUMMER'S RELATION, ORR'S THEOREM, RAMANUJAN'S HYPERGEOMETRIC IDENTITY, SAALSCHÜTZIAN, SISTER CELINE'S METHOD, ZEILBERGER'S ALGORITHM

References

Abramowitz, M. and Stegun, C. A. (Eds.). "Hypergeometric Functions." Ch. 15 in *Handbook of Mathematical Functions with Formulas, Graphs, and Mathematical Tables, 9th printing.* New York: Dover, pp. 555–566, 1972.

Arfken, G. "Hypergeometric Functions." §13.5 in *Mathematical Methods for Physicists, 3rd ed.* Orlando, FL: Academic Press, pp. 748–752, 1985.

Fine, N. J. *Basic Hypergeometric Series and Applications.* Providence, RI: Amer. Math. Soc., 1988.

Gasper, G. and Rahman, M. *Basic Hypergeometric Series.* Cambridge, England: Cambridge University Press, 1990.

Gauss, C. F. "Disquisitiones Generales Circa Seriem Infinitam $\left[\frac{\alpha\beta}{1\cdot\gamma}\right]x+\left[\frac{\alpha(\alpha+1)\beta(\beta+1)}{1\cdot 2\cdot\gamma(\gamma+1)}\right]x^2$ $+\left[\frac{\alpha(\alpha+1)(\alpha+2)\beta(\beta+1)(\beta+2)}{1\cdot 2\cdot 3\cdot\gamma(\gamma+1)(\gamma+2)}\right]x^3+$ etc. Pars Prior." *Commentationes Societiones Regiae Scientiarum Gottingensis Recentiores, Vol. II.* 1813.

Gessel, I. and Stanton, D. "Strange Evaluations of Hypergeometric Series." *SIAM J. Math. Anal.* **13**, 295–308, 1982.

Gosper, R. W. "Decision Procedures for Indefinite Hypergeometric Summation." *Proc. Nat. Acad. Sci. USA* **75**, 40–42, 1978.

Goursat, M. E. "Sur l'équation différentielle linéaire qui admet pour intégrale la série hypergéométrique." *Ann. Sci. École Norm. Super. Sup.* **10**, S3–S142, 1881.

Iyanaga, S. and Kawada, Y. (Eds.). "Hypergeometric Functions and Spherical Functions." Appendix A, Table 18 in *Encyclopedic Dictionary of Mathematics.* Cambridge, MA: MIT Press, pp. 1460–1468, 1980.

Kummer, E. E. "Über die Hypergeometrische Reihe." *J. für die Reine Angew. Mathematik* **15**, 39–83 and 127–172, 1837.

Magnus, W. and Oberhettinger, F. *Formulas and Theorems for the Special Functions of Mathematical Physics.* New York: Chelsea, 1949.

Morse, P. M. and Feshbach, H. *Methods of Theoretical Physics, Part I.* New York: McGraw-Hill, pp. 541–547, 1953.

Petkovšek, M.; Wilf, H. S.; and Zeilberger, D. *A=B.* Wellesley, MA: A. K. Peters, 1996.

Press, W. H.; Flannery, B. P.; Teukolsky, S. A.; and Vetterling, W. T. "Hypergeometric Functions." §6.12 in *Numerical Recipes in FORTRAN: The Art of Scientific Computing, 2nd ed.* Cambridge, England: Cambridge University Press, pp. 263–265, 1992.

Seaborn, J. B. *Hypergeometric Functions and Their Applications.* New York: Springer-Verlag, 1991.

Snow, C. *Hypergeometric and Legendre Functions with Applications to Integral Equations of Potential Theory.* Washington, DC: U. S. Government Printing Office, 1952.

Spanier, J. and Oldham, K. B. "The Gauss Function $F(a,b;c;x)$." Ch. 60 in *An Atlas of Functions.* Washington, DC: Hemisphere, pp. 599–607, 1987.

Hypergeometric Identity

A relation expressing a sum potentially involving BINOMIAL COEFFICIENTS, FACTORIALS, RATIONAL FUNCTIONS, and power functions in terms of a simple result. Thanks to results by Fasenmyer, Gosper, Zeilberger, Wilf, and Petkovšek, the problem of determining whether a given hypergeometric sum is expressible in simple closed form and, if so, finding the form, is now (subject to a mild restriction) completely solved. The algorithm which does so has been implemented in several computer algebra packages and is called ZEILBERGER'S ALGORITHM.

see also GENERALIZED HYPERGEOMETRIC FUNCTION, GOSPER'S ALGORITHM, HYPERGEOMETRIC SERIES, SISTER CELINE'S METHOD, WILF-ZEILBERGER PAIR, ZEILBERGER'S ALGORITHM

References

Petkovšek, M.; Wilf, H. S.; and Zeilberger, D. *A=B.* Wellesley, MA: A. K. Peters, p. 18, 1996.

Hypergeometric Polynomial

see JACOBI POLYNOMIAL

Hypergeometric Series

A hypergeometric series $\sum_k a_k$ is a series for which $a_0 = 1$ and the ratio of consecutive terms is a RATIONAL FUNCTION of the summation index k, i.e., one for which

$$\frac{a_{k+1}}{a_k} = \frac{P(k)}{Q(k)},$$

with $P(k)$ and $Q(k)$ POLYNOMIALS. The functions generated by hypergeometric series are called HYPERGEOMETRIC FUNCTIONS or, more generally, GENERALIZED HYPERGEOMETRIC FUNCTIONS. If the polynomials are completely factored, the ratio of successive terms can be written

$$\frac{a_{k+1}}{a_k} = \frac{P(k)}{Q(k)} = \frac{(k+a_1)(k+a_2)\cdot(k+a_p)}{(k+b_1)(k+b_2)\cdots(k+b_q)(k+1)}x,$$

where the factor of $k+1$ in the DENOMINATOR is present for historical reasons of notation, and the resulting GENERALIZED HYPERGEOMETRIC FUNCTION is written

$${}_pF_q\begin{bmatrix} a_1 & a_2 & \cdots & a_p \\ b_1 & b_2 & \cdots & b_q \end{bmatrix}; x\Bigg] = \sum_{k=0} a_k x^k.$$

If $p = 2$ and $q = 1$, the function becomes a traditional HYPERGEOMETRIC FUNCTION ${}_2F_1(a, b; c; x)$.

Many sums can be written as GENERALIZED HYPERGEOMETRIC FUNCTIONS by inspections of the ratios of consecutive terms in the generating hypergeometric series.

see also GENERALIZED HYPERGEOMETRIC FUNCTION, GEOMETRIC SERIES, HYPERGEOMETRIC FUNCTION, HYPERGEOMETRIC IDENTITY

References
Petkovšek, M.; Wilf, H. S.; and Zeilberger, D. "Hypergeometric Series," "How to Identify a Series as Hypergeometric," and "Software That Identifies Hypergeometric Series." §3.2–3.4 in *A=B*. Wellesley, MA: A. K. Peters, pp. 34–42, 1996.

Hypergroup

A MEASURE ALGEBRA which has many properties associated with the convolution MEASURE ALGEBRA of a GROUP, but no algebraic structure is assumed for the underlying SPACE.

References
Bloom, W. R.; and Heyer, H. *The Harmonic Analysis of Probability Measures on Hypergroups.* Berlin: de Gruyter, 1995.
Jewett, R. I. "Spaces with an Abstract Convolution of Measures." *Adv. Math.* **18**, 1–101, 1975.

Hypermatrix

A generalization of the MATRIX to an $n_1 \times n_2 \times \cdots$ array of numbers.

see also HYPERDETERMINANT

References
Gel'fand, I. M.; Kapranov, M. M.; and Zelevinsky, A. V. "Hyperdeterminants." *Adv. Math.* **96**, 226–263, 1992.

Hyperparallel

Two lines in HYPERBOLIC GEOMETRY which diverge from each other in both directions.

see also ANTIPARALLEL, IDEAL POINT, PARALLEL

Hyperperfect Number

A number n is called k-hyperperfect if

$$n = 1 + k \sum_i d_i,$$

where the summation is over the PROPER DIVISORS with $1 < d_i < n$, giving

$$k\sigma(n) = (k+1)n + k + 1,$$

where $\sigma(n)$ is the DIVISOR FUNCTION. The first few hyperperfect numbers are 21, 301, 325, 697, 1333, ... (Sloane's A007592). 2-hyperperfect numbers include 21, 2133, 19521, 176661, ... (Sloane's A007593), and the first 3-hyperperfect number is 325.

References
Guy, R. K. "Almost Perfect, Quasi-Perfect, Pseudoperfect, Harmonic, Weird, Multiperfect and Hyperperfect Numbers." §B2 in *Unsolved Problems in Number Theory, 2nd ed.* New York: Springer-Verlag, pp. 45–53, 1994.
Sloane, N. J. A. Sequences A007592/M5113 and A007593/M5121 in "An On-Line Version of the Encyclopedia of Integer Sequences."

Hyperplane

Let a_1, a_2, ..., a_n be SCALARS not all equal to 0. Then the SET S consisting of all VECTORS

$$\mathbf{X} = \begin{bmatrix} x_1 \\ x_2 \\ \vdots \\ x_n \end{bmatrix}$$

in $\mathbb{R}^n$ such that

$$a_1x_1 + a_2x_2 + \ldots + a_nx_n = 0$$

is a SUBSPACE of $\mathbb{R}^n$ called a hyperplane. More generally, a hyperplane is any co-dimension 1 vector SUBSPACE of a VECTOR SPACE. Equivalently, a hyperplane V in a VECTOR SPACE W is any SUBSPACE such that W/V is 1-dimensional. Equivalently, a hyperplane is the KERNEL of any NONZERO linear MAP from the VECTOR SPACE to the underlying FIELD.

Hyperreal Number

Hyperreal numbers are an extension of the REAL NUMBERS to include certain classes of infinite and infinitesimal numbers. A hyperreal number is said to be finite IFF $|x| < n$ for some INTEGER n. x is said to be infinitesimal IFF $|x| < 1/n$ for all INTEGERS n.

see also AX-KOCHEN ISOMORPHISM THEOREM, NONSTANDARD ANALYSIS

References
Apps, P. "The Hyperreal Line." http://www.math.wisc.edu/~apps/line.html.
Keisler, H. J. "The Hyperreal Line." In *Real Numbers, Generalizations of the Reals, and Theories of Continua* (Ed. P. Ehrlich). Norwell, MA: Kluwer, 1994.

Hyperspace

A SPACE having DIMENSION $n > 3$.

Hypersphere

The n-hypersphere (often simply called the n-sphere) is a generalization of the CIRCLE ($n = 2$) and SPHERE ($n = 3$) to dimensions $n \geq 4$. It is therefore defined as the set of n-tuples of points (x_1, x_2, ..., x_n) such that

$$x_1{}^2 + x_2{}^2 + \ldots + x_n{}^2 = R^2, \tag{1}$$

where R is the RADIUS of the hypersphere. The CONTENT (i.e., n-D VOLUME) of an n-hypersphere of RADIUS R is given by

$$V_n = \int_0^R S_n r^{n-1}\,dr = \frac{S_n R^n}{n}, \tag{2}$$

where S_n is the hyper-SURFACE AREA of an n-sphere of unit radius. But, for a unit hypersphere, it must be true that

$$S_n \int_0^\infty e^{-r^2} r^{n-1}\,dr = \underbrace{\int_{-\infty}^\infty \cdots \int_{-\infty}^\infty}_{n} e^{-({x_1}^2+\ldots+{x_n}^2)}\,dx_1\cdots dx_m = \left(\int_{-\infty}^\infty e^{-x^2}\,dx\right)^n. \tag{3}$$

But the GAMMA FUNCTION can be defined by

$$\Gamma(m) = 2\int_0^\infty e^{-r^2} r^{2m-1}\,dr, \tag{4}$$

so

$$\tfrac{1}{2}S_n\Gamma(\tfrac{1}{2}n) = [\Gamma(\tfrac{1}{2})]^n = (\pi^{1/2})^n \tag{5}$$

$$S_n = \frac{2\pi^{n/2}}{\Gamma(\frac{1}{2}n)}. \tag{6}$$

This gives the RECURRENCE RELATION

$$S_{n+2} = \frac{2\pi S_n}{n}. \tag{7}$$

Using $\Gamma(n+1) = n\Gamma(n)$ then gives

$$V_n = \frac{S_n R^n}{n} = \frac{\pi^{n/2} R^n}{(\frac{1}{2}n)\Gamma(\frac{1}{2}n)} = \frac{\pi^{n/2} R^n}{\Gamma(1+\frac{1}{2}n)} \tag{8}$$

(Conway and Sloane 1993).

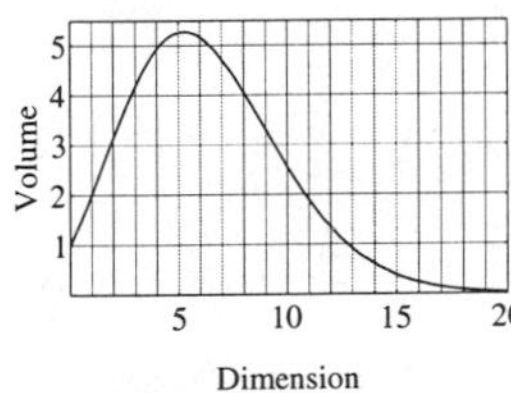

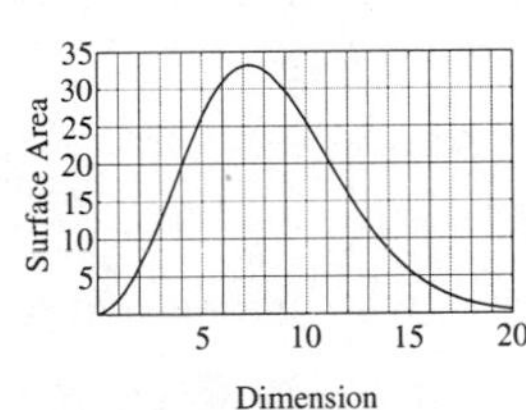

Strangely enough, the hyper-SURFACE AREA and CONTENT reach MAXIMA and then decrease towards 0 as n increases. The point of MAXIMAL hyper-SURFACE AREA satisfies

$$\frac{dS_n}{dn} = \frac{\pi^{n/2}[\ln\pi - \psi_0(\frac{1}{2}n)]}{\Gamma(\frac{1}{2}n)} = 0, \tag{9}$$

where $\psi_0(x) \equiv \Psi(x)$ is the DIGAMMA FUNCTION. The point of MAXIMAL CONTENT satisfies

$$\frac{dV_n}{dn} = \frac{\pi^{n/2}[\ln\pi - \psi_0(1+\frac{1}{2}n)]}{2\Gamma(1+\frac{1}{2}n)} = 0. \tag{10}$$

Neither can be solved analytically for n, but the numerical solutions are $n = 7.25695\ldots$ for hyper-SURFACE AREA and $n = 5.25695\ldots$ for CONTENT. As a result, the 7-D and 5-D hyperspheres have MAXIMAL hyper-SURFACE AREA and CONTENT, respectively (Le Lionnais 1983).

n	V_n	$V_n/V_{n-\text{cube}}$	S_n
0	1	1	0
1	2	1	2
2	π	$\frac{1}{4}\pi$	2π
3	$\frac{4}{3}\pi$	$\frac{1}{6}\pi$	4π
4	$\frac{1}{2}\pi^2$	$\frac{1}{32}\pi^2$	$2\pi^2$
5	$\frac{8}{15}\pi^2$	$\frac{1}{60}\pi^2$	$\frac{8}{3}\pi^2$
6	$\frac{1}{6}\pi^3$	$\frac{1}{384}\pi^3$	π^3
7	$\frac{16}{105}\pi^3$	$\frac{1}{840}\pi^3$	$\frac{16}{15}\pi^3$
8	$\frac{1}{24}\pi^4$	$\frac{1}{6144}\pi^4$	$\frac{1}{3}\pi^4$
9	$\frac{32}{945}\pi^4$	$\frac{1}{15120}\pi^4$	$\frac{32}{105}\pi^4$
10	$\frac{1}{120}\pi^5$	$\frac{1}{122880}\pi^5$	$\frac{1}{12}\pi^5$

In 4-D, the generalization of SPHERICAL COORDINATES is defined by

$$x_1 = R\sin\psi\sin\phi\cos\theta \tag{11}$$

$$x_2 = R\sin\psi\sin\phi\sin\theta \tag{12}$$

$$x_3 = R\sin\psi\cos\phi \tag{13}$$

$$x_4 = R\cos\psi. \tag{14}$$

The equation for a 4-sphere is

$${x_1}^2 + {x_2}^2 + {x_3}^2 + {x_4}^2 = R^2, \tag{15}$$

and the LINE ELEMENT is

$$ds^2 = R^2[d\psi^2 + \sin^2\psi(d\phi^2 + \sin^2\phi\,d\theta^2)]. \tag{16}$$

By defining $r \equiv R\sin\psi$, the LINE ELEMENT can be rewritten

$$ds^2 = \frac{dr^2}{(1-\frac{r^2}{R^2})} + r^2(d\phi^2 + \sin^2\phi\,d\theta^2). \tag{17}$$

The hyper-SURFACE AREA is therefore given by

$$S_4 = \int_0^\pi R\,d\psi \int_0^\pi R\sin\psi\,d\phi \int_0^{2\pi} R\sin\psi\sin\phi\,d\theta = 2\pi^2 R^3. \tag{18}$$

see also CIRCLE, HYPERCUBE, HYPERSPHERE PACKING, MAZUR'S THEOREM, SPHERE, TESSERACT

References

Conway, J. H. and Sloane, N. J. A. *Sphere Packings, Lattices, and Groups, 2nd ed.* New York: Springer-Verlag, p. 9, 1993.

Le Lionnais, F. *Les nombres remarquables.* Paris: Hermann, p. 58, 1983.

Peterson, I. *The Mathematical Tourist: Snapshots of Modern Mathematics.* New York: W. H. Freeman, pp. 96–101, 1988.

Hypersphere Packing

Draw unit n-spheres in an n-D space centered at all ± 1 coordinates. Then place an additional HYPERSPHERE at the origin tangent to the other HYPERSPHERES. Then the central HYPERSPHERE is contained with the HYPERSPHERE with VERTICES at the center of the other spheres for n between 2 and 8. However, for $n = 9$, the central HYPERSPHERE just touches the bounding HYPERSPHERE, and for $n > 9$, the HYPERSPHERE is partially outside the hypercube. This can be seen by finding the distance from the origin to the center of one of the HYPERSPHERES

$$\underbrace{\sqrt{(\pm 1)^2 + \ldots + (\pm 1)^2}}_{n} = \sqrt{n}.$$

The radius of the central sphere is therefore $\sqrt{n} - 1$. The distance from the origin to the center of the bounding hypercube is always 2 (two radii), so the center HYPERSPHERE is tangent when $\sqrt{n} - 1 = 2$, or $n = 9$, and outside for $n > 9$.

The analog of face-centered cubic packing is the densest lattice in 4- and 5-D. In 8-D, the densest lattice packing is made up of two copies of face-centered cubic. In 6- and 7-D, the densest lattice packings are cross-sections of the 8-D case. In 24-D, the densest packing appears to be the LEECH LATTICE. For high dimensions ($\sim$ 1000-D), the densest known packings are nonlattice. The densest lattice packings in n-D have been rigorously proved to have PACKING DENSITY 1, $\pi/(2\sqrt{3})$, $\pi/(3\sqrt{2})$, $\pi^2/16$, $\pi^2/(15\sqrt{2})$, $\pi^3/(48\sqrt{3})$, $\pi^3/105$, and $\pi^4/384$ (Finch).

The largest number of unit CIRCLES which can touch another is six. For SPHERES, the maximum number is 12. Newton considered this question long before a proof was published in 1874. The maximum number of hyperspheres that can touch another in n-D is the so-called KISSING NUMBER.

see also KISSING NUMBER, LEECH LATTICE, SPHERE PACKING

References

Finch, S. "Favorite Mathematical Constants." `http://www.mathsoft.com/asolve/constant/hermit/hermit.html`.

Gardner, M. *Martin Gardner's New Mathematical Diversions from Scientific American.* New York: Simon and Schuster, pp. 89–90, 1966.

Hypervolume

see CONTENT

Hypocycloid

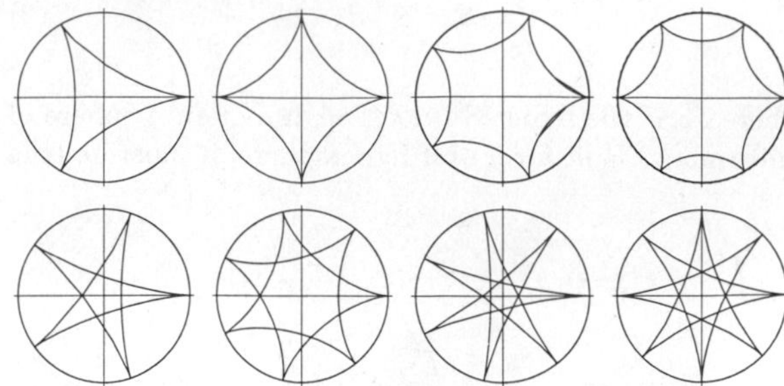

The curve produced by a small CIRCLE of RADIUS b rolling around the inside of a large CIRCLE of RADIUS $a > b$. A hypocycloid is a HYPOTROCHOID with $h = b$. To derive the equations of the hypocycloid, call the ANGLE by which a point on the small CIRCLE rotates about its center ϑ, and the ANGLE from the center of the large CIRCLE to that of the small CIRCLE ϕ. Then

$$(a - b)\phi = b\vartheta, \tag{1}$$

so

$$\vartheta = \frac{a - b}{b}\phi. \tag{2}$$

Call $\rho \equiv a - 2b$. If $x(0) = \rho$, then the first point is at minimum radius, and the Cartesian parametric equations of the hypocycloid are

$$\begin{aligned} x &= (a - b)\cos\phi - b\cos\vartheta \\ &= (a - b)\cos\phi - b\cos\left(\frac{a - b}{b}\phi\right) \qquad (3) \\ y &= (a - b)\sin\phi + b\sin\vartheta \\ &= (a - b)\sin\phi + b\sin\left(\frac{a - b}{b}\phi\right). \qquad (4) \end{aligned}$$

If $x(0) = a$ instead so the first point is at maximum radius (on the CIRCLE), then the equations of the hypocycloid are

$$x = (a - b)\cos\phi + b\cos\left(\frac{a - b}{b}\phi\right) \tag{5}$$

$$y = (a - b)\sin\phi - b\sin\left(\frac{a - b}{b}\phi\right). \tag{6}$$

An n-cusped non-self-intersecting hypocycloid has $a/b = n$. A 2-cusped hypocycloid is a LINE SEGMENT, as can be seen by setting $a = b$ in equations (3) and (4) and noting that the equations simplify to

$$x = a\sin\phi \tag{7}$$

$$y = 0. \tag{8}$$

A 3-cusped hypocycloid is called a DELTOID or TRICUSPOID, and a 4-cusped hypocycloid is called an ASTROID. If a/b is rational, the curve closes on itself and has b cusps. If a/b is IRRATIONAL, the curve never closes and fills the entire interior of the CIRCLE.

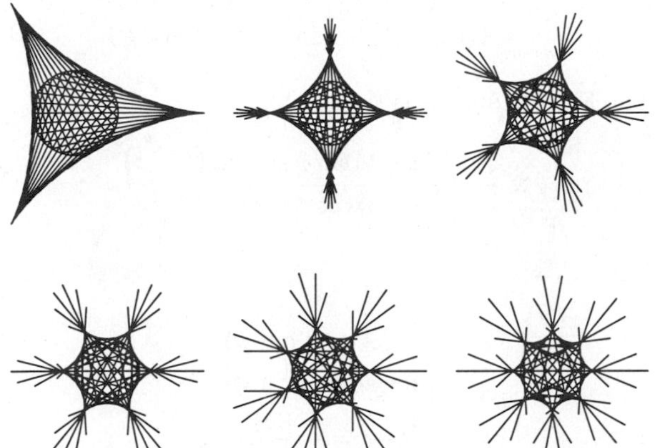

n-hypocycloids can also be constructed by beginning with the DIAMETER of a CIRCLE, offsetting one end by a series of steps while at the same time offsetting the other end by steps n times as large in the opposite direction and extending beyond the edge of the CIRCLE. After traveling around the CIRCLE once, an n-cusped hypocycloid is produced, as illustrated above (Madachy 1979).

Let r be the radial distance from a fixed point. For RADIUS OF TORSION ρ and ARC LENGTH s, a hypocycloid can given by the equation

$$s^2 + \rho^2 = 16r^2 \tag{9}$$

(Kreyszig 1991, pp. 63–64). A hypocycloid also satisfies

$$\sin^2\psi = \frac{\rho^2}{a^2-\rho^2}\frac{a^2-r^2}{r^2}, \tag{10}$$

where

$$r\frac{dr}{d\theta} = \tan\psi \tag{11}$$

and ψ is the ANGLE between the RADIUS VECTOR and the TANGENT to the curve.

The ARC LENGTH of the hypocycloid can be computed as follows

$$\begin{aligned} x' &= -(a-b)\sin\phi - (a-b)\sin\left(\frac{a-b}{b}\phi\right) \\ &= (a-b)\left[\sin\phi + \sin\left(\frac{a-b}{b}\phi\right)\right] \end{aligned} \tag{12}$$

$$\begin{aligned} y' &= (a-b)\cos\phi - (a-b)\cos\left(\frac{a-b}{a}\phi\right) \\ &= (a-b)\left[\cos\phi - \cos\left(\frac{a-b}{b}\phi\right)\right] \end{aligned} \tag{13}$$

$$\begin{aligned} x'^2 + y'^2 &= (a-b)^2\left[\sin^2\phi + 2\sin\phi\sin\left(\frac{a-b}{b}\phi\right)\right. \\ &\quad + \sin^2\left(\frac{a-b}{b}\phi\right) + \cos^2\phi - 2\cos\phi\cos\left(\frac{a-b}{b}\phi\right) \\ &\quad \left. + \cos^2\left(\frac{a-b}{b}\phi\right)\right] \\ &= (a-b)^2\left\{2 + 2\left[\sin\phi\sin\left(\frac{a-b}{a}\phi\right)\right.\right. \\ &\quad \left.\left. - \cos\phi\cos\left(\frac{a-b}{b}\phi\right)\right]\right\} \\ &= 2(a-b)^2\left[1 - \cos\left(\phi + \frac{a-b}{b}\phi\right)\right] \\ &= 4(a-b)^2\tfrac{1}{2}\left[1-\cos\left(\frac{a}{b}\phi\right)\right] = 4(a-b)^2\sin^2\left(\frac{a\phi}{2b}\right), \end{aligned} \tag{14}$$

so

$$ds = \sqrt{x'^2+y'^2}\,d\phi = 2(a-b)\sin\left(\frac{a\phi}{2b}\right)d\phi \tag{15}$$

for $\phi \leq (b/2a)\pi$. Integrating,

$$\begin{aligned} s(\phi) &= \int_0^\phi ds = 2(a-b)\left[-\frac{2b}{a}\cos\left(\frac{a\phi}{2b}\right)\right]_0^\phi \\ &= \frac{4b(a-b)}{a}\left[-\cos\left(\frac{a}{2b}\phi\right)+1\right] \\ &= \frac{8b(a-b)}{a}\sin^2\left(\frac{a}{4b}\phi\right). \end{aligned} \tag{16}$$

The length of a single cusp is then

$$s\left(2\pi\frac{b}{a}\right) = \frac{8b(a-b)}{a}\sin^2\left(\frac{\pi}{2}\right) = \frac{8b(a-b)}{a}. \tag{17}$$

If $n \equiv a/b$ is rational, then the curve closes on itself without intersecting after n cusps. For $n \equiv a/b$ and with $x(0) = a$, the equations of the hypocycloid become

$$x = \frac{1}{n}[(n-1)\cos\phi - \cos[(n-1)\phi]a, \tag{18}$$

$$y = \frac{1}{n}[(n-1)\sin\phi + \sin[(n-1)\phi]a, \tag{19}$$

and

$$s_n = n\frac{8b(bn-b)}{nb} = 8b(n-1) = \frac{8a(n-1)}{n}. \tag{20}$$

Compute

$$\begin{aligned} xy' - yx' &= \left[(a-b)\cos\phi + b\cos\left(\frac{a-b}{a}\phi\right)\right](b-a) \\ &\quad \times \left[\sin\phi + \sin\left(\frac{a-b}{b}\phi\right)\right] \\ &\quad - \left[(a-b)\sin\phi - b\sin\left(\frac{a-b}{b}\phi\right)\right](a-b) \\ &\quad \times \left[\cos\phi - \cos\left(\frac{a-b}{b}\phi\right)\right] \\ &= 2(a^2-3ab+2b^2)\sin^2\left(\frac{a\phi}{2b}\right). \end{aligned} \tag{21}$$

The AREA of one cusp is then

$$\begin{aligned} A &= \tfrac{1}{2}\int_0^{2\pi b/a} (xy' - yx')\,d\phi \\ &= (a^2 - 3ab + 2b^2)\left[\frac{at - b\sin\left(\frac{at}{b}\right)}{2a}\right]_a^{2\pi b/a} \\ &= (a^2 - 3ab + 2b^2)\left[\frac{a\left(2\pi\frac{b}{a}\right)}{2a}\right] \\ &= \frac{b(a^2 - 3ab + 2b^2)}{a}\pi. \end{aligned} \tag{22}$$

If $n = a/b$ is rational, then after n cusps,

$$\begin{aligned} A_n &= n\pi\frac{b(a^2 - 3ab + 2b^2)}{a} = n\pi\frac{\frac{a}{n}\left(a^2 - 3a\frac{a}{n} + 2\frac{a^2}{n^2}\right)}{a} \\ &= \frac{n^2 - 3n + 2}{n^2}\pi a^2 = \frac{(n-1)(n-2)}{n^2}\pi a^2. \end{aligned} \tag{23}$$

The equation of the hypocycloid can be put in a form which is useful in the solution of CALCULUS OF VARIATIONS problems with radial symmetry. Consider the case $x(0) = \rho$, then

$$\begin{aligned} r^2 &= x^2 + y^2 \\ &= \left[(a-b)^2\cos^2\phi - 2(a-b)b\cos\phi\cos\left(\frac{a-b}{b}\phi\right)\right. \\ &\quad + b^2\cos^2\left(\frac{a-b}{b}\phi\right) \\ &\quad + (a-b)^2\sin^2\phi + 2(a-b)b\sin\phi\sin\left(\frac{a-b}{b}\phi\right) \\ &\quad \left. + b^2\sin^2\left(\frac{a-b}{b}\phi\right)\right] \\ &= \left\{(a-b)^2 + b^2 - 2(a-b)b\right. \\ &\quad \times\left.\left[\cos\phi\cos\left(\frac{a-b}{b}\phi\right) - \sin\phi\sin\left(\frac{a-b}{b}\phi\right)\right]\right\} \\ &= (a-b)^2 + b^2 - 2(a-b)b\cos\left(\frac{a}{b}\phi\right). \end{aligned} \tag{24}$$

But $\rho = a - 2b$, so $b = (a - \rho)/2$, which gives

$$\begin{aligned} (a-b)^2 + b^2 &= [a - \tfrac{1}{2}(a-\rho)]^2 + [\tfrac{1}{2}(a-\rho)]^2 \\ &= [\tfrac{1}{2}(a+\rho)]^2 + [\tfrac{1}{2}(a-\rho)]^2 \\ &= \tfrac{1}{4}(a^2 + 2a\rho + \rho^2 + a^2 - 2a\rho + \rho^2) \\ &= \tfrac{1}{2}(a^2 + \rho^2) \end{aligned} \tag{25}$$

$$\begin{aligned} 2(a-b)b &= 2[a - \tfrac{1}{2}(a-\rho)]\tfrac{1}{2}(a-\rho) \\ &= \tfrac{1}{2}(a+\rho)(a-\rho) = \tfrac{1}{2}(a^2 - \rho^2). \end{aligned} \tag{26}$$

Now let

$$2\Omega t \equiv \frac{a}{b}\phi, \tag{27}$$

so

$$\phi = \frac{a-\rho}{a}\Omega t \tag{28}$$

$$\frac{\phi}{a-\rho} = \frac{\Omega t}{a}, \tag{29}$$

then

$$\begin{aligned} r^2 &= \tfrac{1}{2}(a^2 + \rho^2) - \tfrac{1}{2}(a^2 - \rho^2)\cos\left(\frac{a}{b}\phi\right) \\ &= \tfrac{1}{2}(a^2 + \rho^2) - \tfrac{1}{2}(a^2 - \rho^2)\cos(2\Omega t). \end{aligned} \tag{30}$$

The POLAR ANGLE is

$$\tan\theta \equiv \frac{y}{x} = \frac{(a-b)\sin\phi + b\sin\left(\frac{a-b}{a}\phi\right)}{(a-b)\cos\phi - b\cos\left(\frac{a-b}{a}\phi\right)}. \tag{31}$$

But

$$b = \tfrac{1}{2}(a - \rho) \tag{32}$$

$$a - b = \tfrac{1}{2}(a + \rho) \tag{33}$$

$$\frac{a-b}{b} = \frac{a+\rho}{a-\rho}, \tag{34}$$

so

$$\begin{aligned} \tan\theta &= \frac{\frac{1}{2}(a+\rho)\sin\phi + \frac{1}{2}(a-\rho)\sin\left(\frac{a+\rho}{z-\rho}\phi\right)}{\frac{1}{2}(a+\rho)\cos\phi - \frac{1}{2}(a-\rho)\cos\left(\frac{a+\rho}{a-\rho}\phi\right)} \\ &= \frac{(a+\rho)\sin\left(\frac{a-\rho}{a}\Omega t\right) + (a-\rho)\sin\left(\frac{a+\rho}{a}\Omega t\right)}{(a+\rho)\cos\left(\frac{a-\rho}{a}\Omega t\right) - (a-\rho)\cos\left(\frac{a+\rho}{a}\Omega t\right)} \\ &= \frac{a\left[\sin\left(\frac{a-\rho}{a}\Omega t\right) + \sin\left(\frac{a+\rho}{a}\Omega t\right)\right] + \rho\left[\sin\left(\frac{a-\rho}{a}\Omega t\right) - \sin\left(\frac{a+\rho}{a}\Omega t\right)\right]}{a\left[\cos\left(\frac{a-\rho}{a}\Omega t\right) - \cos\left(\frac{a+\rho}{a}\Omega t\right)\right] + \rho\left[\cos\left(\frac{a-\rho}{a}\Omega t\right) + \cos\left(\frac{a+\rho}{a}\Omega t\right)\right]} \\ &= \frac{2a\sin(\Omega t)\cos\left(\frac{\rho}{a}\Omega t\right) - 2\rho\cos(\Omega t)\sin\left(\frac{\rho}{a}\Omega t\right)}{2a\sin(\Omega t)\sin\left(\frac{\rho}{a}\Omega t\right) + 2\rho\cos(\Omega t)\cos\left(\frac{\rho}{a}\Omega t\right)} \\ &= \frac{a\tan(\Omega t) - \rho\tan\left(\frac{\rho}{a}\Omega t\right)}{a\tan(\Omega t)\tan\left(\frac{\rho}{a}\Omega t\right) + \rho}. \end{aligned} \tag{35}$$

Computing

$$\begin{aligned} \tan\left(\theta + \frac{\rho}{a}\Omega t\right) &= \frac{\left[a\tan(\Omega t) - \rho\tan\left(\frac{\rho}{a}\Omega t\right) + \tan\left(\frac{\rho}{a}\Omega t\right)\right] \times \left[a\tan(\Omega t)\tan\left(\frac{\rho}{a}\Omega t\right) + \rho\right]}{\left[a\tan(\Omega t)\tan\left(\frac{\rho}{a}\Omega t\right) + \rho\right] - \left[a\tan(\Omega t) - \rho\tan\left(\frac{\rho}{a}\Omega t\right)\right]\tan\left(\frac{\rho}{a}\Omega t\right)} \\ &= \frac{a\tan(\Omega t)\left[1 + \tan^2\left(\frac{\rho}{a}\Omega t\right)\right]}{\rho\left[1 + \tan^2\left(\frac{\rho}{a}\Omega t\right)\right]} \\ &= \frac{a}{\rho}\tan(\Omega t), \end{aligned} \tag{36}$$

then gives

$$\theta = \tan^{-1}\left[\frac{a}{\rho}\tan(\Omega t)\right] - \frac{\rho}{a}\Omega t. \tag{37}$$

Finally, plugging back in gives

$$\theta = \tan^{-1}\left[\frac{a}{\rho}\tan\left(\frac{a}{a-\rho}\phi\right)\right] - \frac{\rho}{a}\frac{a}{a-\rho}\phi$$
$$= \tan^{-1}\left[\frac{a}{\rho}\tan\left(\frac{a}{a-\rho}\phi\right)\right] - \frac{\rho}{a-\rho}\phi. \quad (38)$$

This form is useful in the solution of the SPHERE WITH TUNNEL problem, which is the generalization of the BRACHISTOCHRONE PROBLEM, to find the shape of a tunnel drilled through a SPHERE (with gravity varying according to Gauss's law for gravitation) such that the travel time between two points on the surface of the SPHERE under the force of gravity is minimized.

see also CYCLOID, EPICYCLOID

References
Bogomolny, A. "Cycloids." http://www.cut-the-knot.com/pythagoras/cycloids.html.
Kreyszig, E. *Differential Geometry.* New York: Dover, 1991.
Lawrence, J. D. *A Catalog of Special Plane Curves.* New York: Dover, pp. 171–173, 1972.
Lee, X. "Epicycloid and Hypocycloid." http://www.best.com/~xah/SpecialPlaneCurves_dir/EpiHypocycloid_dir/epiHypocycloid.html.
MacTutor History of Mathematics Archive. "Hypocycloid." http://www-groups.dcs.st-and.ac.uk/~history/Curves/Hypocycloid.html.
Madachy, J. S. *Madachy's Mathematical Recreations.* New York: Dover, pp. 225–231, 1979.
Wagon, S. *Mathematica in Action.* New York: W. H. Freeman, pp. 50–52, 1991.
Yates, R. C. "Epi- and Hypo-Cycloids." *A Handbook on Curves and Their Properties.* Ann Arbor, MI: J. W. Edwards, pp. 81–85, 1952.

Hypocycloid—3-Cusped

see DELTOID

Hypocycloid—4-Cusped

see ASTROID

Hypocycloid Evolute

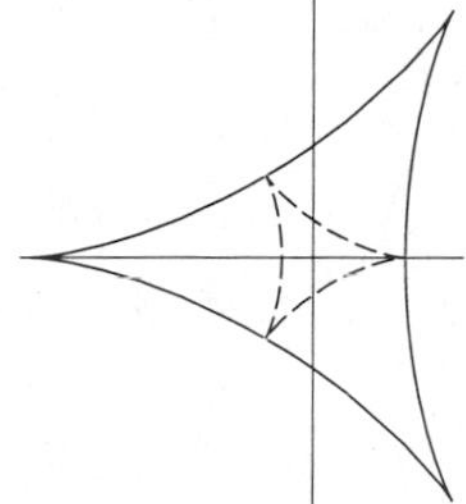

For $x(0) = a$,

$$x = \frac{a}{a-2b}\left[(a-b)\cos\phi - b\cos\left(\frac{a-b}{b}\phi\right)\right]$$
$$y = \frac{a}{a-2b}\left[(a-b)\sin\phi + b\sin\left(\frac{a-b}{b}\phi\right)\right].$$

If $a/b = n$, then

$$x = \frac{1}{n-2}[(n-1)\cos\phi - \cos[(n-1)\phi]a$$
$$y = \frac{1}{n-2}[(n-1)\sin\phi + \sin[(n-1)\phi]a.$$

This is just the original HYPOCYCLOID scaled by the factor $(n-2)/n$ and rotated by $1/(2n)$ of a turn.

Hypocycloid Involute

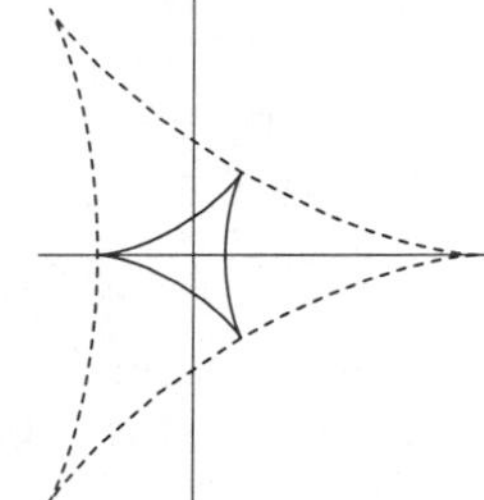

The HYPOCYCLOID

$$x = \frac{a}{a-2b}\left[(a-b)\cos\phi - b\cos\left(\frac{a-b}{b}\phi\right)\right]$$
$$y = \frac{a}{a-2b}\left[(a-b)\sin\phi + b\sin\left(\frac{a-b}{b}\phi\right)\right]$$

has INVOLUTE

$$x = \frac{a-2b}{a}\left[(a-b)\cos\phi + b\cos\left(\frac{a-b}{b}\phi\right)\right]$$
$$y = \frac{a-2b}{a}\left[(a-b)\sin\phi - b\sin\left(\frac{a-b}{b}\phi\right)\right],$$

which is another HYPOCYCLOID.

Hypocycloid Pedal Curve

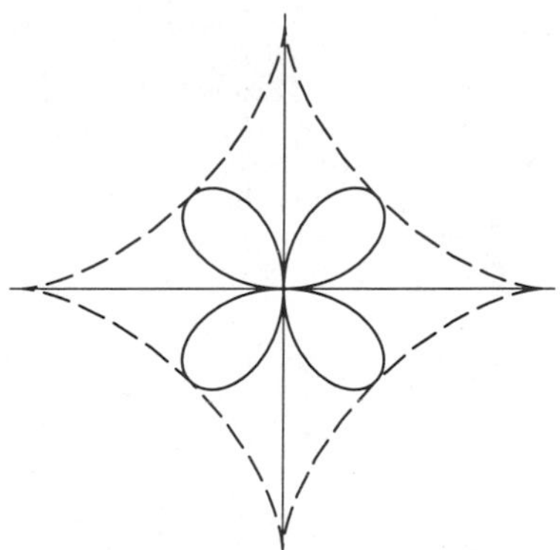

The PEDAL CURVE for a PEDAL POINT at the center is a ROSE.

Hypoellipse

$$y^{n/m} + c\left|\frac{x}{a}\right|^{n/m} - c = 0,$$

with $n/m < 2$. If $n/m > 2$, the curve is a HYPERELLIPSE.

see also ELLIPSE, HYPERELLIPSE, SUPERELLIPSE

References

von Seggern, D. *CRC Standard Curves and Surfaces.* Boca Raton, FL: CRC Press, p. 82, 1993.

Hypotenuse

The longest LEG of a RIGHT TRIANGLE (which is the side opposite the RIGHT ANGLE).

Hypothesis

A proposition that is consistent with known data, but has been neither verified nor shown to be false. It is synonymous with CONJECTURE.

see also BOURGET'S HYPOTHESIS, CHINESE HYPOTHESIS, CONTINUUM HYPOTHESIS, HYPOTHESIS TESTING, NESTED HYPOTHESIS, NULL HYPOTHESIS, POSTULATE, RAMANUJAN'S HYPOTHESIS, RIEMANN HYPOTHESIS, SCHINZEL'S HYPOTHESIS, SOUSLIN'S HYPOTHESIS

Hypothesis Testing

The use of statistics to determine the probability that a given hypothesis is true.

see also BONFERRONI CORRECTION, ESTIMATE, FISHER SIGN TEST, PAIRED t-TEST, STATISTICAL TEST, TYPE I ERROR, TYPE II ERROR, WILCOXON SIGNED RANK TEST

References

Hoel, P. G.; Port, S. C.; and Stone, C. J. "Testing Hypotheses." Ch. 3 in *Introduction to Statistical Theory.* New York: Houghton Mifflin, pp. 52–110, 1971.

Iyanaga, S. and Kawada, Y. (Eds.). "Statistical Estimation and Statistical Hypothesis Testing." Appendix A, Table 23 in *Encyclopedic Dictionary of Mathematics.* Cambridge, MA: MIT Press, pp. 1486–1489, 1980.

Shaffer, J. P. "Multiple Hypothesis Testing." *Ann. Rev. Psych.* **46**, 561–584, 1995.

Hypotrochoid

The ROULETTE traced by a point P attached to a CIRCLE of radius b rolling around the inside of a fixed CIRCLE of radius a. The parametric equations for a hypotrochoid are

$$x = n\cos t + h\cos\left(\frac{n}{b}t\right) \tag{1}$$

$$y = n\sin t - h\sin\left(\frac{n}{b}t\right), \tag{2}$$

where $n \equiv a - b$ and h is the distance from P to the center of the rolling CIRCLE. Special cases include the HYPOCYCLOID with $h = b$, the ELLIPSE with $a = 2b$, and the ROSE with

$$a = \frac{2nh}{n+1} \tag{3}$$

$$b = \frac{(n-1)h}{n+1}. \tag{4}$$

see also EPITROCHOID, HYPOCYCLOID, SPIROGRAPH

References

Lawrence, J. D. *A Catalog of Special Plane Curves.* New York: Dover, pp. 165–168, 1972.

Lee, X. "Hypotrochoid." http://www.best.com/~xah/SpecialPlaneCurves_dir/Hypotrochoid_dir/hypotrochoid.html.

Lee, X. "Epitrochoid and Hypotrochoid Movie Gallery." http://www.best.com/~xah/SpecialPlaneCurves_dir/EpiHypoTMovieGallery_dir/epiHypoTMovieGallery.html.

MacTutor History of Mathematics Archive. "Hypotrochoid." http://www-groups.dcs.st-and.ac.uk/~history/Curves/Hypotrochoid.html.

Hypotrochoid Evolute

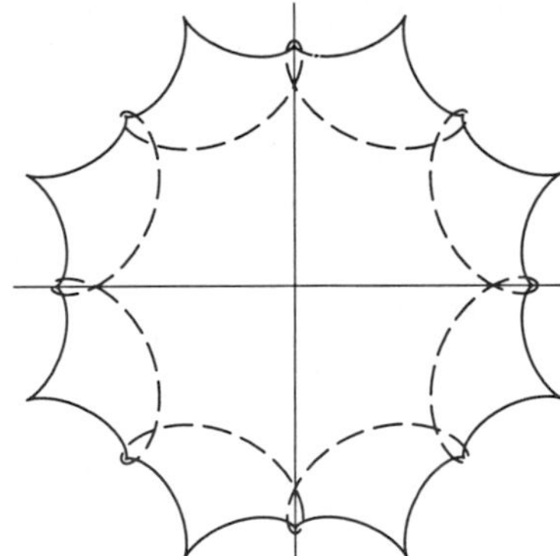

The EVOLUTE of the HYPOTROCHOID is illustrated above.

Hyzer's Illusion

see FREEMISH CRATE

I

i

The IMAGINARY NUMBER i is defined as $i \equiv \sqrt{-1}$. However, for some reason engineers and physicists prefer the symbol j to i. Numbers of the form $z = x + iy$ where x and y are REAL NUMBERS are called COMPLEX NUMBERS, and when z is used to denote a COMPLEX NUMBER, it is sometimes (in older texts) called an "AFFIX."

The SQUARE ROOT of i is

$$\sqrt{i} = \pm\frac{i+1}{\sqrt{2}}, \tag{1}$$

since

$$\left[\frac{1}{\sqrt{2}}(i+1)\right]^2 = \tfrac{1}{2}(i^2 + 2i + 1) = i. \tag{2}$$

This can be immediately derived from the EULER FORMULA with $x = \pi/2$,

$$i = e^{i\pi/2} \tag{3}$$

$$\sqrt{i} = \sqrt{e^{i\pi/2}} = e^{i\pi/4} = \cos(\tfrac{1}{4}\pi) + i\sin(\tfrac{1}{4}\pi) = \frac{1+i}{\sqrt{2}}. \tag{4}$$

The PRINCIPAL VALUE of i^i is

$$i^i = (e^{i\pi/2})^i = e^{i^2\pi/2} = e^{-\pi/2} = 0.207879\ldots. \tag{5}$$

see also COMPLEX NUMBER, IMAGINARY IDENTITY, IMAGINARY NUMBER, REAL NUMBER, SURREAL NUMBER

References

Courant, R. and Robbins, H. *What is Mathematics?: An Elementary Approach to Ideas and Methods, 2nd ed.* Oxford, England: Oxford University Press, p. 89, 1996.

$\mathbb{I}$

see $\mathbb{Z}$

I-Signature

see SIGNATURE (RECURRENCE RELATION)

Iamond

see POLYIAMOND

Ice Fractal

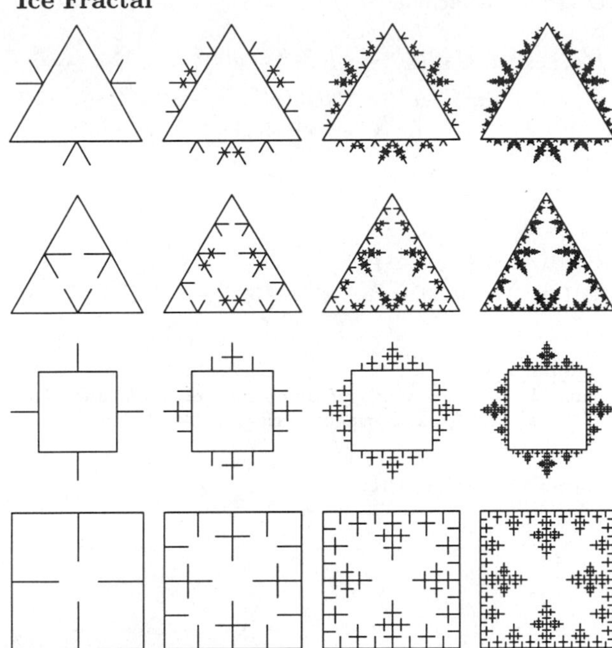

A FRACTAL (square, triangle, etc.) based on a simple generating motif. The above plots show the ice triangle, antitriangle, square, and antisquare. The base curves and motifs for the fractals illustrated above are shown below.

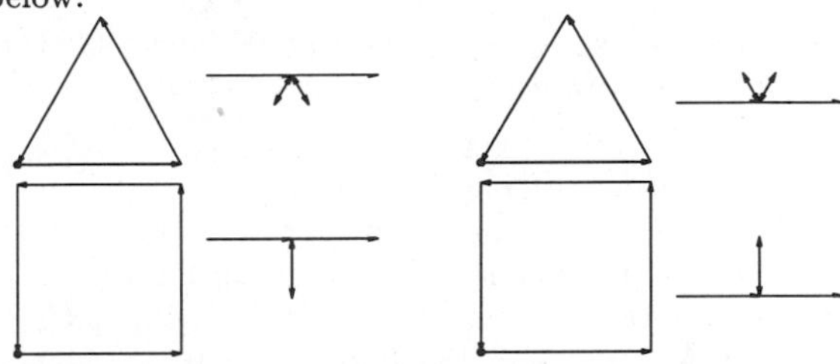

see also FRACTAL

References

Lauwerier, H. *Fractals: Endlessly Repeated Geometric Figures.* Princeton, NJ: Princeton University Press, p. 44, 1991.

Weisstein, E. W. "Fractals." http://www.astro.virginia.edu/~eww6n/math/notebooks/Fractal.m.

Icosagon

A 20-sided POLYGON. The SWASTIKA is an irregular icosagon.

see also SWASTIKA

Icosahedral Equation

Hunt (1996) gives the "dehomogenized" icosahedral equation as

$$[(z^{20}+1) - 228(z^{15} - z^5) + 494z^{10})^3 + 1728uz^5(z^{10} + 11z^5 - 1)^5 = 0.$$

Other forms include

$$I(u,v,Z) = u^5v^5(u^{10}+11u^5v^5-v^{10})^5 - [u^{30}+v^{30}-10005(u^{20}v^{10}+u^{10}v^{20}) + 522(u^25v^5-u^5v^25)]^2 Z = 0$$

and

$$I(z,1,Z) = z^5(-1+11z^5+z^{10})^5 - [1+z^{30}-10005(z^{10}+z^{20})+522(-z^5+z^{25})]^2 Z = 0.$$

References

Hunt, B. *The Geometry of Some Special Arithmetic Quotients.* New York: Springer-Verlag, p. 146, 1996.

Icosahedral Graph

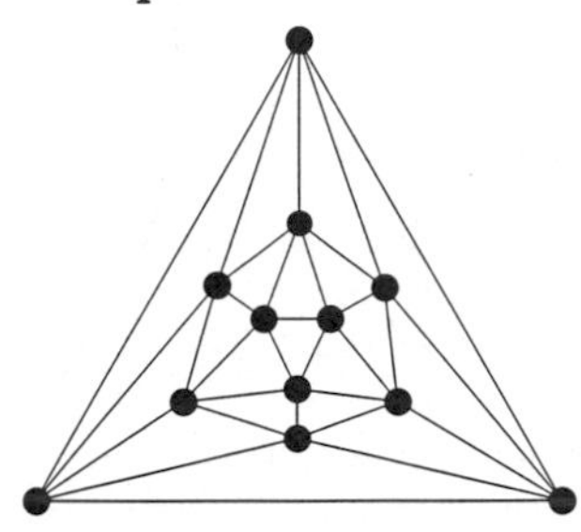

A POLYHEDRAL GRAPH.

see also CUBICAL GRAPH, DODECAHEDRAL GRAPH, OCTAHEDRAL GRAPH, TETRAHEDRAL GRAPH

Icosahedral Group

The GROUP I_h of symmetries of the ICOSAHEDRON and DODECAHEDRON. The icosahedral group consists of the symmetry operations E, $12C_5$, $12C_5^2$, $20C_3$, $15C_2$, i, $12S_{10}$, $12S_{10}^3$, $20S_6$, and 15σ (Cotton 1990).

see also DODECAHEDRON, ICOSAHEDRON, OCTAHEDRAL GROUP, TETRAHEDRAL GROUP

References

Cotton, F. A. *Chemical Applications of Group Theory, 3rd ed.* New York: Wiley, p. 48–50, 1990.

Lomont, J. S. "Icosahedral Group." §3.10.E in *Applications of Finite Groups.* New York: Dover, p. 82, 1987.

Icosahedron

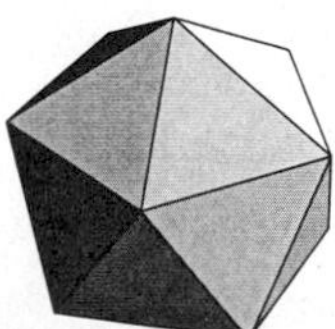

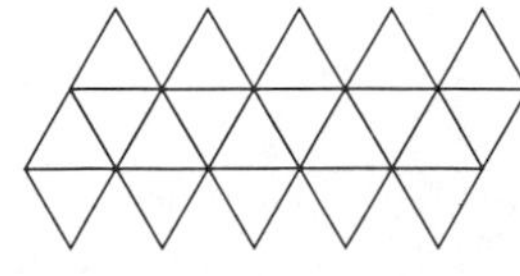

A PLATONIC SOLID (P_5) with 12 VERTICES, 30 EDGES, and 20 equivalent EQUILATERAL TRIANGLE faces $20\{3\}$. It is described by the SCHLÄFLI SYMBOL $\{3,5\}$. It is also UNIFORM POLYHEDRON U_{22} and has WYTHOFF SYMBOL $5\,|\,2\,3$. The icosahedron has the ICOSAHEDRAL GROUP I_h of symmetries.

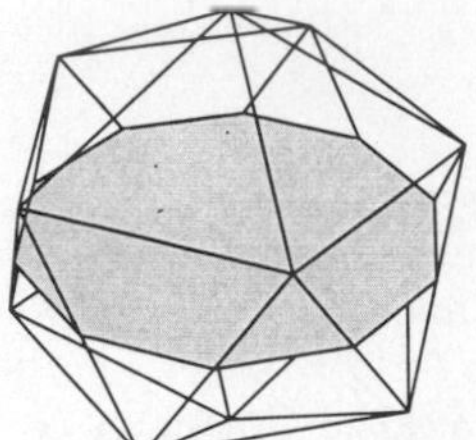

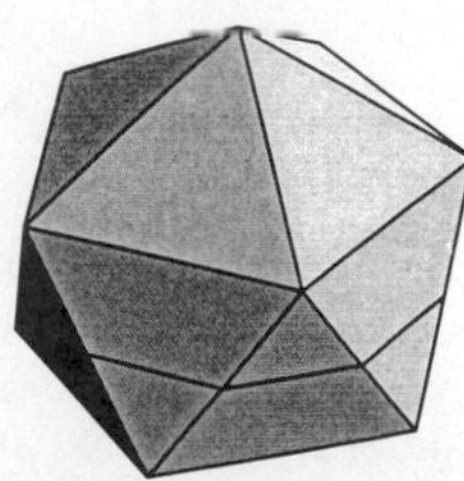

A plane PERPENDICULAR to a C_5 axis of an icosahedron cuts the solid in a regular DECAGONAL CROSS-SECTION (Holden 1991, pp. 24–25).

A construction for an icosahedron with side length $a = \sqrt{50-10\sqrt{5}}/5$ places the end vertices at $(0,0,\pm1)$ and the central vertices around two staggered CIRCLES of RADII $\frac{2}{5}\sqrt{5}$ and heights $\pm\frac{1}{5}\sqrt{5}$, giving coordinates

$$\pm\left(\tfrac{2}{5}\sqrt{5}\cos(\tfrac{2}{5}i\pi), \tfrac{2}{5}\sqrt{5}\sin(\tfrac{2}{5}i\pi), \tfrac{1}{5}\sqrt{5}\right) \quad (1)$$

for $i = 0, 1, \ldots, 4$, where all the plus signs or minus signs are taken together. Explicitly, these coordinates are

$$\mathbf{x}_0^\pm = \pm(\tfrac{2}{5}\sqrt{5}, 0, \tfrac{1}{5}\sqrt{5}) \quad (2)$$

$$\mathbf{x}_1^\pm = \pm(\tfrac{1}{10}(5-\sqrt{5}), \tfrac{1}{10}\sqrt{50+10\sqrt{5}}, \tfrac{1}{5}\sqrt{5}) \quad (3)$$

$$\mathbf{x}_2^\pm = \pm(-\tfrac{1}{10}(\sqrt{5}+5), \tfrac{1}{10}\sqrt{50-10\sqrt{5}}, \tfrac{1}{5}\sqrt{5}) \quad (4)$$

$$\mathbf{x}_3^\pm = \pm(-\tfrac{1}{10}(\sqrt{5}-5), -\tfrac{1}{10}\sqrt{50-10\sqrt{5}}, \tfrac{1}{5}\sqrt{5}) \quad (5)$$

$$\mathbf{x}_4^\pm = \pm(\tfrac{1}{10}(5-\sqrt{5}), -\tfrac{1}{10}\sqrt{50+10\sqrt{5}}, \tfrac{1}{5}\sqrt{5}). \quad (6)$$

By a suitable rotation, the VERTICES of an icosahedron of side length 2 can also be placed at $(0,\pm\phi,\pm1)$, $(\pm1,0,\pm\phi)$, and $(\pm\phi,\pm1,0)$, where ϕ is the GOLDEN RATIO. These points divide the EDGES of an OCTAHEDRON into segments with lengths in the ratio $\phi:1$.

The DUAL POLYHEDRON of the icosahedron is the DODECAHEDRON. There are 59 distinct icosahedra when each TRIANGLE is colored differently (Coxeter 1969).

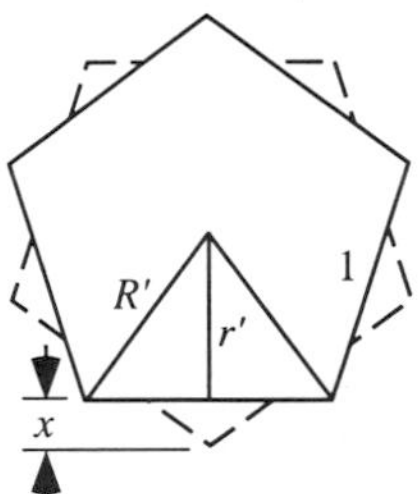

To derive the VOLUME of an icosahedron having edge length a, consider the orientation so that two VERTICES are oriented on top and bottom. The vertical distance between the top and bottom PENTAGONAL DIPYRAMIDS is then given by

$$z = \sqrt{\ell^2 - x^2}, \quad (7)$$

where

$$\ell = \tfrac{1}{2}\sqrt{3}\,a \quad (8)$$

is the height of an ISOSCELES TRIANGLE, and the SAGITTA $x \equiv R' - r'$ of the pentagon is

$$x = \tfrac{1}{2}a\tfrac{1}{10}\sqrt{25 - 10\sqrt{5}}\,a, \tag{9}$$

giving

$$x^2 = \tfrac{1}{20}\sqrt{5 - 2\sqrt{5}}\,a^2. \tag{10}$$

Plugging (8) and (10) into (7) gives

$$\begin{aligned} z &= a\sqrt{\tfrac{3}{4} - \tfrac{1}{20}(5 - 2\sqrt{5})} = a\sqrt{\frac{15 - (5 - 2\sqrt{5})}{20}} \\ &= a\sqrt{\frac{10 + 2\sqrt{5}}{20}} = \tfrac{1}{2}a\sqrt{\frac{10 + 2\sqrt{5}}{5}} \\ &= \tfrac{1}{10}\sqrt{50 + 10\sqrt{5}}\,a, \end{aligned} \tag{11}$$

which is identical to the radius of a PENTAGON of side a. The CIRCUMRADIUS is then

$$R = h + \tfrac{1}{2}z, \tag{12}$$

where

$$h = \tfrac{1}{10}\sqrt{50 - 10\sqrt{5}}\,a \tag{13}$$

is the height of a PENTAGONAL DIPYRAMID. Therefore,

$$\begin{aligned} R^2 &= (h + \tfrac{1}{2}z)^2 \\ &= (\tfrac{1}{10}\sqrt{50 - 10\sqrt{5}} + \tfrac{1}{20}\sqrt{50 + 10\sqrt{5}}\,)^2 a^2 \\ &= \left(\frac{5}{8} - \frac{3}{8\sqrt{5}} + \frac{\sqrt{20}}{10}\right) a^2 = \tfrac{1}{8}(5 + \sqrt{5}\,)a. \end{aligned} \tag{14}$$

Taking the square root gives the CIRCUMRADIUS

$$R = \sqrt{\tfrac{1}{8}(5 + \sqrt{5})}\,a = \tfrac{1}{4}\sqrt{10 + 2\sqrt{5}}\,a \approx 0.95105a. \tag{15}$$

The INRADIUS is

$$r = \tfrac{1}{12}(3\sqrt{3} + \sqrt{15}\,)a \approx 0.75576a. \tag{16}$$

The square of the INTERRADIUS is

$$\begin{aligned} \rho^2 &= (\tfrac{1}{2}z)^2 + {x_l}^2 \\ &= [(\tfrac{1}{4})(\tfrac{1}{100})(50 + 10\sqrt{5}\,) + \tfrac{1}{100}(25 + 10\sqrt{5}\,)]a^2 \\ &= \tfrac{1}{8}(3 + \sqrt{5})a^2, \end{aligned} \tag{17}$$

so

$$\rho = \sqrt{\tfrac{1}{8}(3 + \sqrt{5}\,)}\,a = \tfrac{1}{4}(1 + \sqrt{5}\,)a \approx 0.80901a. \tag{18}$$

The AREA of one face is the AREA of an EQUILATERAL TRIANGLE

$$A = \tfrac{1}{4}a^2\sqrt{3}. \tag{19}$$

The volume can be computed by taking 20 pyramids of height r

$$\begin{aligned} V &= 20[(\tfrac{1}{3}A)r] = 20\tfrac{1}{3}\tfrac{1}{4}\sqrt{3}\,a^2\tfrac{1}{12}(3\sqrt{3} + \sqrt{15}\,)a \\ &= \tfrac{5}{12}(3 + \sqrt{5}\,)a^3. \end{aligned} \tag{20}$$

Apollonius showed that

$$\frac{V_{\text{icosahedron}}}{V_{\text{dodecahedron}}} = \frac{A_{\text{icosahedron}}}{A_{\text{dodecahedron}}}, \tag{21}$$

where V is the volume and A the surface area.

see also AUGMENTED TRIDIMINISHED ICOSAHEDRON, DECAGON, DODECAHEDRON, GREAT ICOSAHEDRON, ICOSAHEDRON STELLATIONS, METABIDIMINISHED ICOSAHEDRON, TRIDIMINISHED ICOSAHEDRON, TRIGONOMETRY VALUES—$\pi/5$

References

Coxeter, H. S. M. *Introduction to Geometry, 2nd ed.* New York: Wiley, 1969.

Davie, T. "The Icosahedron." `http://www.dcs.st-and.ac.uk/~ad/mathrecs/polyhedra/icosahedron.html`.

Holden, A. *Shapes, Space, and Symmetry.* New York: Dover, 1991.

Klein, F. *Lectures on the Icosahedron.* New York: Dover, 1956.

Pappas, T. "The Icosahedron & the Golden Rectangle." *The Joy of Mathematics.* San Carlos, CA: Wide World Publ./Tetra, p. 115, 1989.

Icosahedron Stellations

Applying the STELLATION process to the ICOSAHEDRON gives

$$20 + 30 + 60 + 20 + 60 + 120 + 12 + 30 + 60 + 60$$

cells of ten different shapes and sizes in addition to the ICOSAHEDRON itself. After application of five restrictions due to J. C. P. Miller to define which forms should be considered distinct, 59 stellations are found to be possible. Miller's restrictions are

1. The faces must lie in the twenty bounding planes of the icosahedron.
2. The parts of the faces in the twenty planes must be congruent, but those parts lying in one place may be disconnected.
3. The parts lying in one plane must have threefold rotational symmetry with or without reflections.
4. All parts must be accessible, i.e., lie on the outside of the solid.
5. Compounds are excluded that can be divided into two sets, each of which has the full symmetry of the whole.

Of these, 32 have full icosahedral symmetry and 27 are ENANTIOMERIC forms. Four are POLYHEDRON COMPOUNDS, one is a KEPLER-POINSOT SOLID, and one is the DUAL POLYHEDRON of an ARCHIMEDEAN SOLID.

The only STELLATIONS of PLATONIC SOLIDS which are UNIFORM POLYHEDRA are the three DODECAHEDRON STELLATIONS the GREAT ICOSAHEDRON (stellation # 11).

n	name
1	icosahedron
2	triakisicosahedron
3	octahedron 5-compound
4	echidnahedron
11	great icosahedron
18	tetrahedron 10-compound
20	deltahedron-60
36	tetrahedron 5-compound

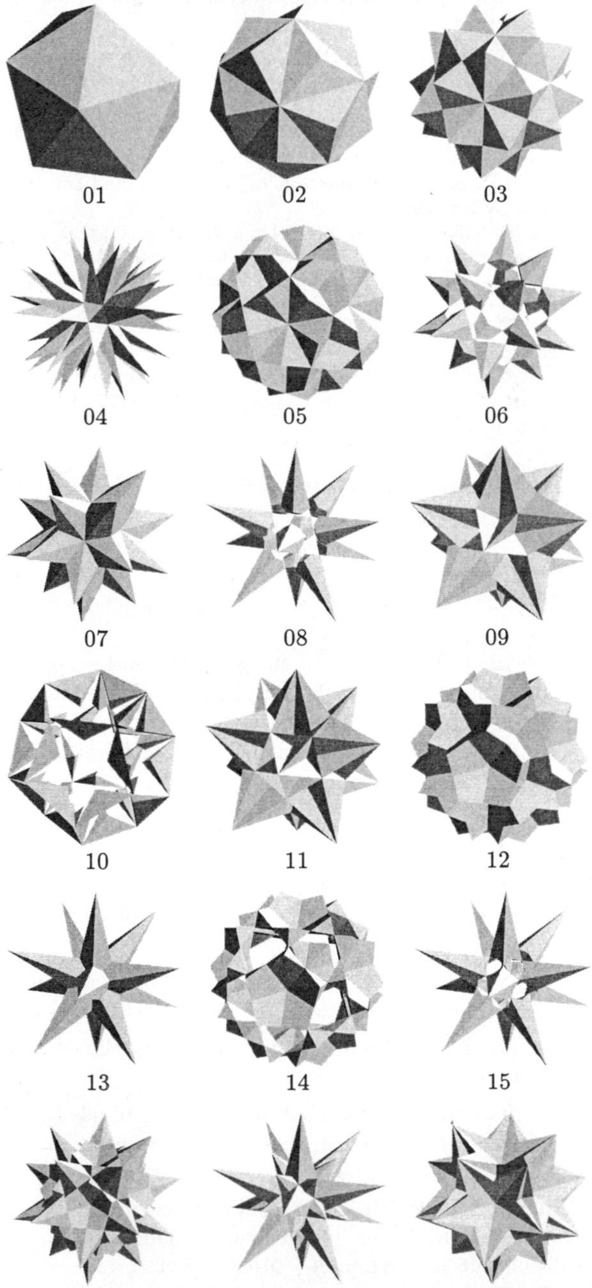

01 02 03

04 05 06

07 08 09

10 11 12

13 14 15

16 17 18

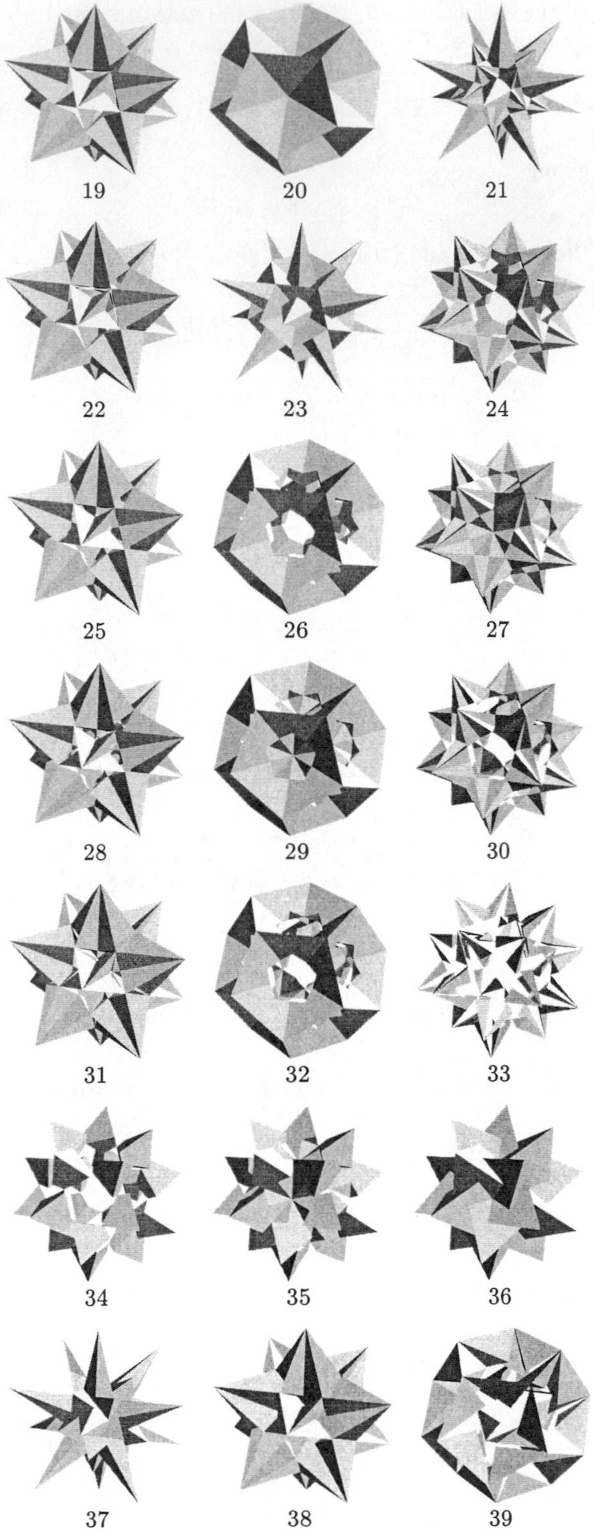

19 20 21

22 23 24

25 26 27

28 29 30

31 32 33

34 35 36

37 38 39

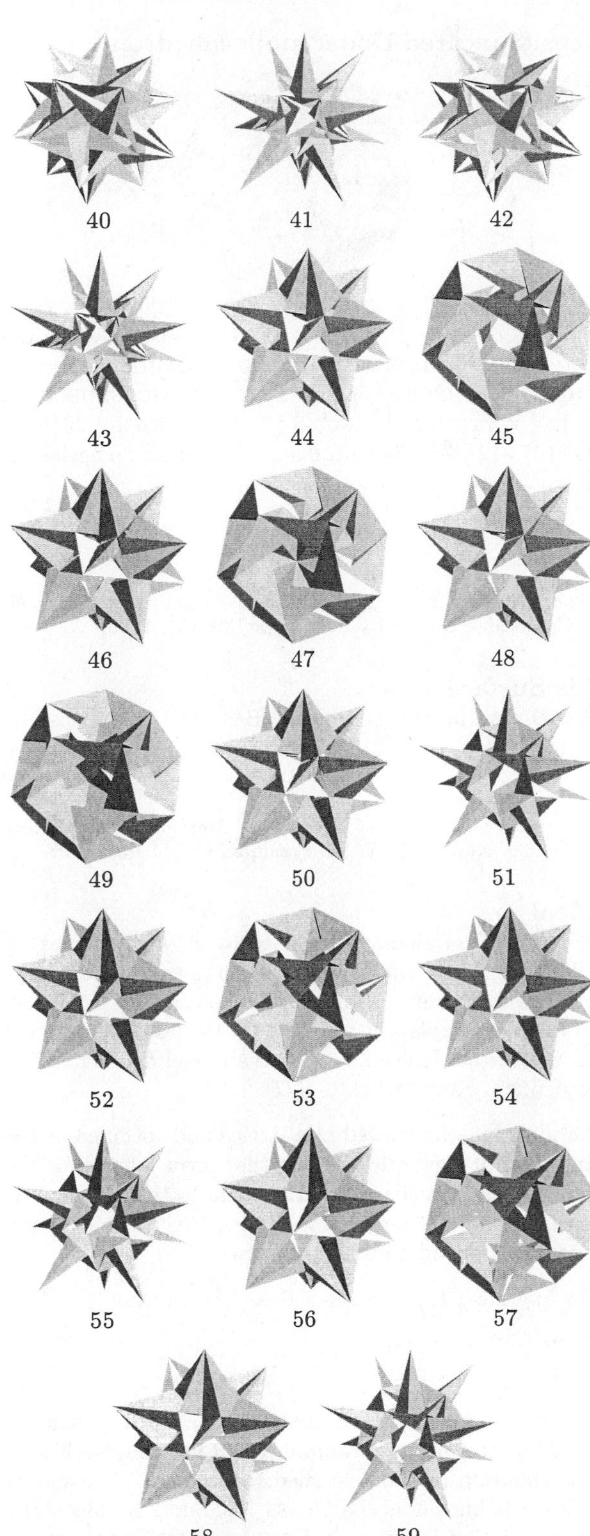

see also ARCHIMEDEAN SOLID STELLATION, DODECAHEDRON STELLATIONS, STELLATION

References
Ball, W. W. R. and Coxeter, H. S. M. *Mathematical Recreations and Essays, 13th ed.* New York: Dover, pp. 146–147, 1987.
Bulatov, V. "Stellations of Icosahedron." `http://www.physics.orst.edu/~bulatov/polyhedra/icosahedron/`.
Coxeter, H. S. M. *The Fifty-Nine Icosahedra.* New York: Springer-Verlag, 1982.
Hart, G. W. "59 Stellations of the Icosahedron." `http://www.li.net/~george/virtual-polyhedra/stellations-icosahedron-index.html`.
Maeder, R. E. `Icosahedra.m` notebook. `http://www.inf.ethz.ch/department/TI/rm/programs.html`.
Maeder, R. E. "The Stellated Icosahedra." *Mathematica in Education* **3**, 1994. `ftp://ftp.inf.ethz.ch/doc/papers/ti/scs/icosahedra94.ps.gz`.
Maeder, R. E. "Stellated Icosahedra." `http://www.mathconsult.ch/showroom/icosahedra/`.
Wang, P. "Polyhedra." `http://www.ugcs.caltech.edu/~peterw/portfolio/polyhedra/`.
Wenninger, M. J. *Polyhedron Models.* New York: Cambridge University Press, pp. 41–65, 1989.
Wheeler, A. H. "Certain Forms of the Icosahedron and a Method for Deriving and Designating Higher Polyhedra." *Proc. Internat. Math. Congress* **1**, 701–708, 1924.

Icosian Game

The problem of finding a HAMILTONIAN CIRCUIT along the edges of an ICOSAHEDRON, i.e., a path such that every vertex is visited a single time, no edge is visited twice, and the ending point is the same as the starting point.

see also HAMILTONIAN CIRCUIT, ICOSAHEDRON

References
Herschel, A. S. "Sir Wm. Hamilton's Icosian Game." *Quart. J. Pure Applied Math.* **5**, 305, 1862.

Icosidodecadodecahedron

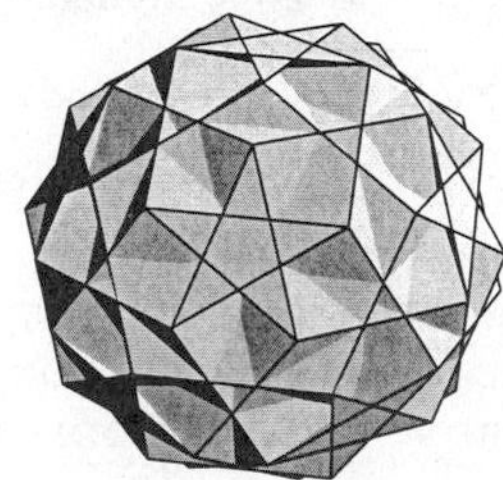

The UNIFORM POLYHEDRON U_{44} whose DUAL POLYHEDRON is the MEDIAL ICOSACRONIC HEXECONTAHEDRON. It has WYTHOFF SYMBOL $\frac{5}{3}\,5\,|\,3$. Its faces are $20\{6\}+12\{\frac{5}{2}\}+12\{5\}$. Its CIRCUMRADIUS for unit edge length is

$$R = \tfrac{1}{2}\sqrt{7}\,.$$

References
Wenninger, M. J. *Polyhedron Models.* Cambridge, England: Cambridge University Press, pp. 128–129, 1989.

Icosidodecahedron

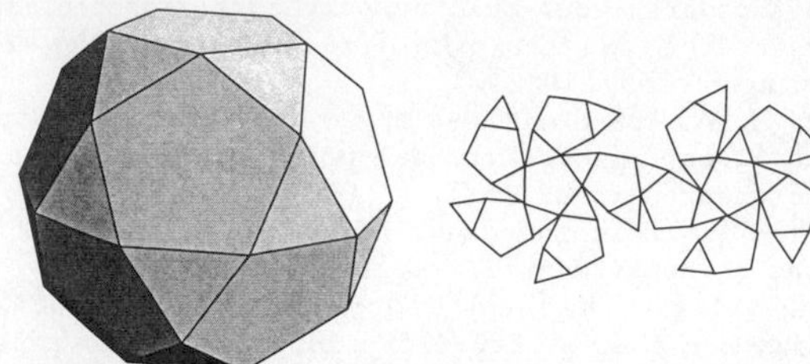

An ARCHIMEDEAN SOLID whose DUAL POLYHEDRON is the RHOMBIC TRIACONTAHEDRON. It is one of the two convex QUASIREGULAR POLYHEDRA and has SCHLÄFLI SYMBOL $\left\{ {3 \atop 5} \right\}$. It is also UNIFORM POLYHEDRON U_{24} and has WYTHOFF SYMBOL $2\,|\,3\,5$. Its faces are $20\{3\}+12\{5\}$. The VERTICES of an icosidodecahedron of EDGE length $2\phi^{-1}$ are $(\pm 2,0,0)$, $(0,\pm 2,0)$, $(0,0,\pm 2)$, $(\pm 1,\pm\phi^{-1},\pm 1)$, $(\pm 1,\pm\phi,\pm\phi^{-1})$, $(\pm\phi^{-1},\pm 1,\pm\phi)$. The 30 VERTICES of an OCTAHEDRON 5-COMPOUND form an icosidodecahedron (Ball and Coxeter 1987). FACETED versions include the SMALL ICOSIHEMIDODECAHEDRON and SMALL DODECAHEMIDODECAHEDRON.

The faces of the icosidodecahedron consist of 20 triangles and 12 pentagons. Furthermore, its 60 edges are bisected perpendicularly by those of the reciprocal RHOMBIC TRIACONTAHEDRON (Ball and Coxeter 1987).

The INRADIUS, MIDRADIUS, and CIRCUMRADIUS for unit edge length are

$$r = \tfrac{1}{8}(5+3\sqrt{5}\,) \approx 1.46353$$
$$\rho = \tfrac{1}{2}\sqrt{5+2\sqrt{5}} \approx 1.53884$$
$$R = \tfrac{1}{2}(1+\sqrt{5}\,) = \phi \approx 1.61803.$$

see also ARCHIMEDEAN SOLID, GREAT ICOSIDODECAHEDRON, QUASIREGULAR POLYHEDRON, SMALL ICOSIHEMIDODECAHEDRON, SMALL DODECAHEMIDODECAHEDRON

References

Ball, W. W. R. and Coxeter, H. S. M. *Mathematical Recreations and Essays, 13th ed.* New York: Dover, p. 137, 1987.

Wenninger, M. J. *Polyhedron Models.* Cambridge, England: Cambridge University Press, p. 73, 1989.

Icosidodecahedron Stellation

The first stellation is a DODECAHEDRON-ICOSAHEDRON COMPOUND.

References

Wenninger, M. J. *Polyhedron Models.* Cambridge, England: Cambridge University Press, pp. 73–96, 1989.

Icosidodecatruncated Icosidodecahedron

see ICOSITRUNCATED DODECADODECAHEDRON

Icositruncated Dodecadodecahedron

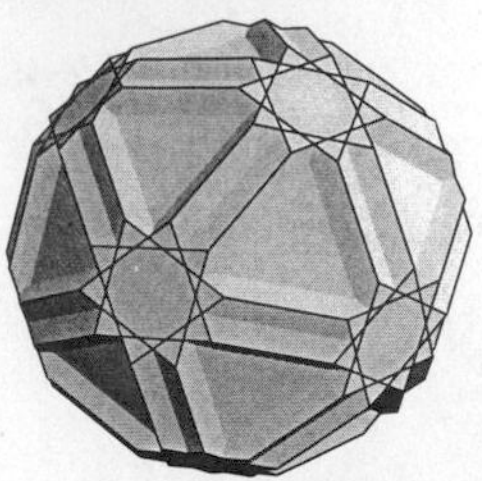

The UNIFORM POLYHEDRON U_{45} also called the ICOSIDODECATRUNCATED ICOSIDODECAHEDRON whose DUAL POLYHEDRON is the TRIDYAKIS ICOSAHEDRON. It has WYTHOFF SYMBOL $3\,\frac{5}{3}\,5\,|$. Its faces are $20\{6\}+12\{10\}+12\{\frac{10}{3}\}$. Its CIRCUMRADIUS for unit edge length is

$$R = 2.$$

References

Wenninger, M. J. *Polyhedron Models.* Cambridge, England: Cambridge University Press, pp. 130–131, 1989.

Ida Surface

A 3-D shadow of a 4-D KLEIN BOTTLE.

see also KLEIN BOTTLE

References

Peterson, I. *Islands of Truth: A Mathematical Mystery Cruise.* New York: W. H. Freeman, pp. 44–45, 1990.

Ideal

A subset I of elements in a RING R which forms an additive GROUP and has the property that, whenever x belongs to R and y belongs to I, then xy and yx belong to I. For example, the set of EVEN INTEGERS is an ideal in the RING of INTEGERS. Given an ideal I, it is possible to define a FACTOR RING R/I.

An ideal may be viewed as a lattice and specified as the finite list of algebraic integers that form a basis for the lattice. Any two bases for the same lattice are equivalent. Ideals have multiplication, and this is basically the Kronecker product of the two bases.

For any ideal I, there is an ideal I_i such that

$$II_i = z,$$

where z is a PRINCIPAL IDEAL, (i.e., an ideal of rank 1). Moreover there is a finite list of ideals I_i such that this equation may be satisfied for every I. The size of this list is known as the CLASS NUMBER. In effect, the above relation imposes an EQUIVALENCE RELATION on ideals, and the number of ideals modulo this relation is the class number. When the CLASS NUMBER is 1, the corresponding number RING has unique factorization and, in a sense, the class number is a measure of the failure of unique factorization in the original number ring.

Dedekind (1871) showed that every NONZERO ideal in the domain of INTEGERS of a FIELD is a unique product of PRIME IDEALS.

see also CLASS NUMBER, DIVISOR THEORY, IDEAL NUMBER, MAXIMAL IDEAL, PRIME IDEAL, PRINCIPAL IDEAL

References

Malgrange, B. *Ideals of Differentiable Functions.* London: Oxford University Press, 1966.

Ideal Number

A type of number involving the ROOTS OF UNITY which was developed by Kummer while trying to solve FERMAT'S LAST THEOREM. Although factorization over the INTEGERS is unique (the FUNDAMENTAL THEOREM OF ALGEBRA), factorization is *not* unique over the COMPLEX NUMBERS. Over the ideal numbers, however, factorization in terms of the COMPLEX NUMBERS becomes unique. Ideal numbers were so powerful that they were generalized by Dedekind into the more abstract IDEALS in general RINGS which are a key part of modern abstract ALGEBRA.

see also DIVISOR THEORY, FERMAT'S LAST THEOREM, IDEAL

Ideal (Partial Order)

An ideal I of a PARTIAL ORDER P is a subset of the elements of P which satisfy the property that if $y \in I$ and $x < y$, then $x \in I$. For k disjoint chains in which the ith chain contains n_i elements, there are $(1+n_1)(1+n_2)\cdots(1+n_k)$ ideals. The number of ideals of a n-element FENCE POSET is the FIBONACCI NUMBER F_n.

References

Ruskey, F. "Information on Ideals of Partially Ordered Sets." http://sue.csc.uvic.ca/~cos/inf/pose/Ideals.html.

Steiner, G. "An Algorithm to Generate the Ideals of a Partial Order." *Operat. Res. Let.* **5**, 317–320, 1986.

Ideal Point

A type of POINT AT INFINITY in which parallel lines in the HYPERBOLIC PLANE intersect at infinity in one direction, while diverging from one another in the other.

see also HYPERPARALLEL

Idele

The multiplicative subgroup of all elements in the product of the multiplicative groups $k_\nu^\times$ whose absolute value is 1 at all but finitely many ν, where k is a number FIELD and ν a PLACE.

see also ADÉLE

References

Knapp, A. W. "Group Representations and Harmonic Analysis, Part II." *Not. Amer. Math. Soc.* **43**, 537–549, 1996.

Idemfactor

see DYADIC

Idempotent

An OPERATOR $\tilde{A}$ such that $\tilde{A}^2 = \tilde{A}$ or an element of an ALGEBRA x such that $x^2 = x$.

see also AUTOMORPHIC NUMBER, BOOLEAN ALGEBRA, GROUP, SEMIGROUP

Identity

An identity is a mathematical relationship equating one quantity to another (which may initially appear to be different).

see also ABEL'S IDENTITY, ANDREWS-SCHUR IDENTITY, BAC-CAB IDENTITY, BEAUZAMY AND DÉGOT'S IDENTITY, BELTRAMI IDENTITY, BIANCHI IDENTITIES, BOCHNER IDENTITY, BRAHMAGUPTA IDENTITY, CASSINI'S IDENTITY, CAUCHY-LAGRANGE IDENTITY, CHRISTOFFEL-DARBOUX IDENTITY, CHU-VANDERMONDE IDENTITY, DE MOIVRE'S IDENTITY, DOUGALL-RAMANUJAN IDENTITY, EULER FOUR-SQUARE IDENTITY, EULER IDENTITY, EULER POLYNOMIAL IDENTITY, FERRARI'S IDENTITY, FIBONACCI IDENTITY, FROBENIUS TRIANGLE IDENTITIES, GREEN'S IDENTITIES, HYPERGEOMETRIC IDENTITY, IMAGINARY IDENTITY, JACKSON'S IDENTITY, JACOBI IDENTITIES, JACOBI'S DETERMINANT IDENTITY, LAGRANGE'S IDENTITY, LE CAM'S IDENTITY, LEIBNIZ IDENTITY, LIOUVILLE POLYNOMIAL IDENTITY, MATRIX POLYNOMIAL IDENTITY, MORGADO IDENTITY, NEWTON'S IDENTITIES, QUINTUPLE PRODUCT IDENTITY, RAMANUJAN 6-10-8 IDENTITY, RAMANUJAN COS/COSH IDENTITY, RAMANUJAN'S IDENTITY, RAMANUJAN'S SUM IDENTITY, REZNIK'S IDENTITY, ROGERS-RAMANUJAN IDENTITIES, SCHAAR'S IDENTITY, STREHL IDENTITY, SYLVESTER'S DETERMINANT IDENTITY, TRINOMIAL IDENTITY, VISIBLE POINT VECTOR IDENTITY, WATSON QUINTUPLE PRODUCT IDENTITY, WORPITZKY'S IDENTITY

Identity Element

The identity element I (also denoted E, e, or 1) of a GROUP or related mathematical structure S is the unique elements such that $IA = AI = I$ for every element $A \in S$. The symbol "E" derives from the German word for unity, "Einheit."

see also BINARY OPERATOR, GROUP, INVOLUTION (GROUP), MONOID

Identity Function

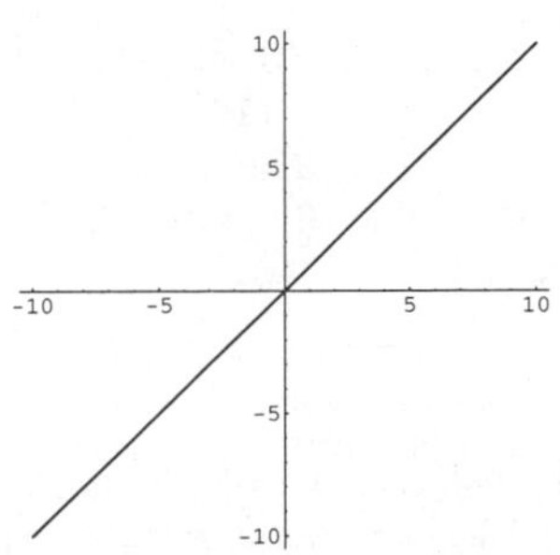

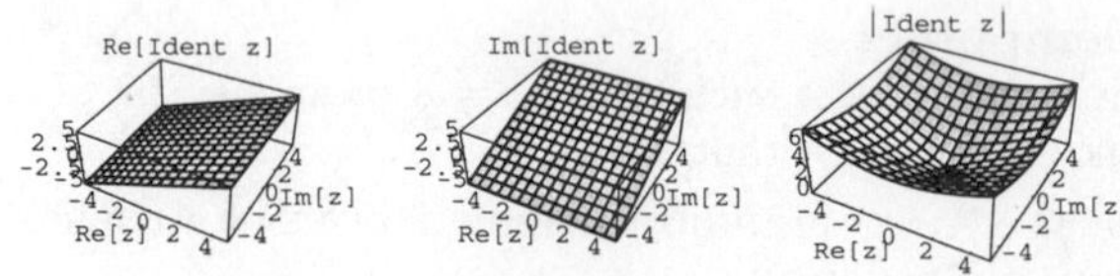

The function $f(x) = x$ which assigns every REAL NUMBER x to the same REAL NUMBER x. It is identical to the IDENTITY MAP.

Identity Map

The MAP which assigns every REAL NUMBER to the same REAL NUMBER $\mathrm{id}_{\mathbb{R}}$. It is identical to the IDENTITY FUNCTION.

Identity Matrix

The identity matrix is defined as the MATRIX **1** (or I) such that

$$\mathsf{I}(\mathbf{X}) \equiv \mathbf{X}$$

for all VECTORS **X**. The identity matrix is

$$I_{ij} = \delta_{ij}$$

for $i, j = 1, 2, \ldots, n$, where δ_{ij} is the KRONECKER DELTA. Written explicitly,

$$\mathsf{I} = \begin{bmatrix} 1 & 0 & \cdots & 0 \\ 0 & 1 & \cdots & 0 \\ \vdots & \vdots & \ddots & \vdots \\ 0 & 0 & \cdots & 1 \end{bmatrix}.$$

Identity Operator

The OPERATOR $\tilde{I}$ which takes a REAL NUMBER to the same REAL NUMBER $\tilde{I}r = r$.

see also IDENTITY FUNCTION, IDENTITY MAP

Idoneal Number

A POSITIVE value of D for which the fact that a number is a MONOMORPH (i.e., the number is expressible in only one way as x^2+Dy^2 or x^2-Dy^2 where x^2 is RELATIVELY PRIME to Dy^2) guarantees it to be a PRIME, POWER of a PRIME, or twice one of these. The numbers are also called EULER'S IDONEAL NUMBERS, or SUITABLE NUMBERS.

The 65 idoneal numbers found by Gauss and Euler and conjectured to be the *only* such numbers (Shanks 1969) are 1, 2, 3, 4, 5, 6, 7, 8, 9, 10, 12, 13, 15, 16, 18, 21, 22, 24, 25, 28, 30, 33, 37, 40, 42, 45, 48, 57, 58, 60, 70, 72, 78, 85, 88, 93, 102, 105, 112, 120, 130, 133, 165, 168, 177, 190, 210, 232, 240, 253, 273, 280, 312, 330, 345, 357, 385, 408, 462, 520, 760, 840, 1320, 1365, and 1848 (Sloane's A000926).

References

Shanks, D. "On Gauss's Class Number Problems." *Math. Comput.* **23**, 151–163, 1969.

Sloane, N. J. A. Sequence A000926/M0476 in "An On-Line Version of the Encyclopedia of Integer Sequences."

Iff

If and only if (i.e., NECESSARY and SUFFICIENT). The terms "JUST IF" or "EXACTLY WHEN" are sometimes used instead. A iff B is written symbolically as $A \leftrightarrow B$. A iff B is also equivalent to $A \Rightarrow B$, together with $B \Rightarrow A$, where the symbol $\Rightarrow$ denotes "IMPLIES."

J. H. Conway believes that the word originated with P. Halmos and was transmitted through Kelley (1975). Halmos has stated, "To the best of my knowledge, I DID invent the silly thing, but I wouldn't swear to it in a court of law. So there—give me credit for it anyway" (Asimov 1997).

see also EQUIVALENT, EXACTLY ONE, IMPLIES, NECESSARY, SUFFICIENT

References

Asimov, D. "Iff." math-fun@cs.arizona.edu posting, Sept. 19, 1997.

Kelley, J. L. *General Topology.* New York: Springer-Verlag, 1975.

Ill-Conditioned

A system is ill-conditioned if the CONDITION NUMBER is too large (and singular if it is INFINITE).

see also CONDITION NUMBER

Illumination Problem

In the early 1950s, Ernst Straus asked

1. Is every POLYGONAL region illuminable from every point in the region?
2. Is every POLYGONAL region illuminable from at least one point in the region?

Here, illuminable means that there is a path from every point to every other by repeated reflections. Tokarsky (1995) showed that unilluminable rooms exist in the plane and 3-D, but question (2) remains open. The smallest known counterexample to (1) in the PLANE has 26 sides.

see also ART GALLERY THEOREM

References

Klee, V. "Is Every Polygonal Region Illuminable from Some Point?" *Math. Mag.* **52**, 180, 1969.

Tokarsky, G. W. "Polygonal Rooms Not Illuminable from Every Point." *Amer. Math. Monthly* **102**, 867–879, 1995.

Illusion

An object or drawing which appears to have properties which are physically impossible, deceptive, or counterintuitive.

see also BENHAM'S WHEEL, FREEMISH CRATE, GOBLET ILLUSION, HERMANN GRID ILLUSION, HERMANN-HERING ILLUSION, HYZER'S ILLUSION, IMPOSSIBLE FIGURE, IRRADIATION ILLUSION, KANIZSA TRIANGLE, MÜLLER-LYER ILLUSION, NECKER CUBE, ORBISON'S ILLUSION, PARALLELOGRAM ILLUSION, PENROSE

STAIRWAY, POGGENDORFF ILLUSION, PONZO'S ILLUSION, RABBIT-DUCK ILLUSION, TRIBAR, VERTICAL-HORIZONTAL ILLUSION, YOUNG GIRL-OLD WOMAN ILLUSION, ZOLLNER'S ILLUSION

References

Ausbourne, B. "A Sensory Adventure." http://www.lainet.com/illusions/.

Ausbourne, B. "Optical Illusions: A Collection." http://www.lainet.com/~ausbourn/.

Ernst, B. *Optical Illusions.* New York: Taschen, 1996.

Fineman, M. *The Nature of Visual Illusion.* New York: Dover, 1996.

Gardner, M. "Optical Illusions." Ch. 1 in *Mathematical Circus: More Puzzles, Games, Paradoxes and Other Mathematical Entertainments from Scientific American.* New York: Knopf, 1979.

Gregory, R. L. *Eye and Brain, 5th ed.* Princeton, NJ: Princeton University Press, 1997.

"Illusions: Central Station." http://www.heureka.fi/i/Illusions_ctrl_station.html.en.

Landrigad, D. "Gallery of Illusions." http://valley.uml.edu/psychology/illusion.html.

Luckiesh, M. *Visual Illusions: Their Causes, Characteristics, and Applications.* New York: Dover, 1965.

Pappas, T. "History of Optical Illusions." *The Joy of Mathematics.* San Carlos, CA: Wide World Publ./Tetra, pp. 172–173, 1989.

Tolansky, S. *Optical Illusions.* New York: Pergamon Press, 1964.

Image

see RANGE (IMAGE)

Imaginary Identity

see i

Imaginary Number

A COMPLEX NUMBER which has zero REAL PART, so that it can be written as a REAL NUMBER multiplied by the "imaginary unit" i (equal to $\sqrt{-1}$).

see also COMPLEX NUMBER, GALOIS IMAGINARY, GAUSSIAN INTEGER, i, REAL NUMBER

References

Conway, J. H. and Guy, R. K. *The Book of Numbers.* New York: Springer-Verlag, pp. 211–216, 1996.

Imaginary Part

The imaginary part $\Im$ of a COMPLEX NUMBER $z = x+iy$ is the REAL NUMBER multiplying i, so $\Im[x+iy] = y$. In terms of z itself,

$$\Im[z] = \frac{z - z^*}{2i},$$

where z^* is the COMPLEX CONJUGATE of z.

see also ABSOLUTE SQUARE, COMPLEX CONJUGATE, REAL PART

References

Abramowitz, M. and Stegun, C. A. (Eds.). *Handbook of Mathematical Functions with Formulas, Graphs, and Mathematical Tables, 9th printing.* New York: Dover, p. 16, 1972.

Imaginary Point

A pair of values x and y one or both of which is COMPLEX.

References

Woods, F. S. *Higher Geometry: An Introduction to Advanced Methods in Analytic Geometry.* New York: Dover, p. 2, 1961.

Imaginary Quadratic Field

A QUADRATIC FIELD $\mathbb{Q}(\sqrt{D})$ with $D < 0$.

see also QUADRATIC FIELD

Immanant

For an $n \times n$ matrix, let S denote any permutation e_1, e_2, ..., e_n of the set of numbers 1, 2, ..., n, and let $\chi^{(\lambda)}(S)$ be the character of the symmetric group corresponding to the partition (λ). Then the immanant $|a_{mn}|^{(\lambda)}$ is defined as

$$|a_{mn}|^{(\lambda)} = \sum \chi^{(\lambda)}(S) P_S$$

where the summation is over the $n!$ permutations of the SYMMETRIC GROUP and

$$P_s = a_{1e_1} a_{2e_2} \cdots a_{ne_n}.$$

see also DETERMINANT, PERMANENT

References

Littlewood, D. E. and Richardson, A. R. "Group Characters and Algebra." *Philos. Trans. Roy. Soc. London A* **233**, 99–141, 1934.

Littlewood, D. E. and Richardson, A. R. "Immanants of Some Special Matrices." *Quart. J. Math. (Oxford)* **5**, 269–282, 1934.

Wybourne, B. G. "Immanants of Matrices." §2.19 in *Symmetry Principles and Atomic Spectroscopy.* New York: Wiley, pp. 12–13, 1970.

Immersed Minimal Surface

see ENNEPER'S SURFACES

Immersion

A special nonsingular MAP from one MANIFOLD to another such that at every point in the domain of the map, the DERIVATIVE is an injective linear map. This is equivalent to saying that every point in the DOMAIN has a NEIGHBORHOOD such that, up to DIFFEOMORPHISMS of the TANGENT SPACE, the map looks like the inclusion map from a lower-dimensional EUCLIDEAN SPACE to a higher-dimensional EUCLIDEAN SPACE.

see also BOY SURFACE, EVERSION, SMALE-HIRSCH THEOREM

References

Boy, W. "Über die Curvatura integra und die Topologie geschlossener Flächen." *Math. Ann* **57**, 151–184, 1903.

Pinkall, U. "Models of the Real Projective Plane." Ch. 6 in *Mathematical Models from the Collections of Universities and Museums* (Ed. G. Fischer). Braunschweig, Germany: Vieweg, pp. 63–67, 1986.

Impartial Game

A GAME in which the possible moves are the same for each player in any position. All positions in all impartial GAMES form an additive ABELIAN GROUP. For impartial games in which the last player wins (normal form games), the nim-value of the sum of two GAMES is the nim-sum of their nim-values. If the last player loses, the GAME is said to be in misère form and the analysis is much more difficult.

see also FAIR GAME, GAME, PARTISAN GAME

Implicit Function

A function which is not defined explicitly, but rather is defined in terms of an algebraic relationship (which can not, in general, be "solved" for the function in question). For example, the ECCENTRIC ANOMALY E of a body orbiting on an ELLIPSE with ECCENTRICITY e is defined implicitly in terms of the mean anomaly M by KEPLER'S EQUATION

$$M = E - e \sin E.$$

Implicit Function Theorem

Given

$$F_1(x, y, z, u, v, w) = 0$$

$$F_2(x, y, z, u, v, w) = 0$$

$$F_3(x, y, z, u, v, w) = 0,$$

if the JACOBIAN

$$JF(u, v, w) = \frac{\partial(F_1, F_2, F_3)}{\partial(u, v, w)} \neq 0,$$

then u, v, and w can be solved for in terms of x, y, and z and PARTIAL DERIVATIVES of u, v, w with respect to x, y, and z can be found by differentiating implicitly.

More generally, let A be an OPEN SET in $\mathbb{R}^{n+k}$ and let $f : A \to \mathbb{R}^n$ be a C^r FUNCTION. Write f in the form $f(x, y)$, where x and y are elements of $\mathbb{R}^k$ and $\mathbb{R}^n$. Suppose that (a, b) is a point in A such that $f(a, b) = 0$ and the DETERMINANT of the $n \times n$ MATRIX whose elements are the DERIVATIVES of the n component FUNCTIONS of f with respect to the n variables, written as y, evaluated at (a, b), is not equal to zero. The latter may be rewritten as

$$\operatorname{rank}(Df(a, b)) = n.$$

Then there exists a NEIGHBORHOOD B of a in $\mathbb{R}^k$ and a unique C^r FUNCTION $g : B \to \mathbb{R}^n$ such that $g(a) = b$ and $f(x, g(x)) = 0$ for all $x \in B$.

see also CHANGE OF VARIABLES THEOREM, JACOBIAN

References

Munkres, J. R. *Analysis on Manifolds.* Reading, MA: Addison-Wesley, 1991.

Implies

The symbol $\Rightarrow$ means "implies" in the mathematical sense. Let A be true. If this implies that B is also true, then the statement is written symbolically as $A \Rightarrow B$, or sometimes $A \subset B$. If $A \Rightarrow B$ and $B \Rightarrow A$ (i.e, $A \Rightarrow B \wedge B \Rightarrow A$), then A and B are said to be EQUIVALENT, a relationship which is written symbolically as $A \Leftrightarrow B$ or $A \rightleftharpoons B$.

see also EQUIVALENT

Impossible Figure

A class of ILLUSION in which an object which is physically unrealizable is apparently depicted.

see also FREEMISH CRATE, HOME PLATE, ILLUSION, NECKER CUBE, PENROSE STAIRWAY, TRIBAR

References

Cowan, T. M. "The Theory of Braids and the Analysis of Impossible Figures." *J. Math. Psych.* **11**, 190–212, 1974.

Cowan, T. M. "Supplementary Report: Braids, Side Segments, and Impossible Figures." *J. Math. Psych.* **16**, 254–260, 1977.

Cowan, T. M. "Organizing the Properties of Impossible Figures." *Perception* **6**, 41–56, 1977.

Cowan, T. M. and Pringle, R. "An Investigation of the Cues Responsible for Figure Impossibility." *J. Exper. Psych. Human Perception Performance* **4**, 112–120, 1978.

Ernst, B. *Adventures with Impossible Figures.* Stradbroke, England: Tarquin, 1987.

Harris, W. F. "Perceptual Singularities in Impossible Pictures Represent Screw Dislocations." *South African J. Sci.* **69**, 10–13, 1973.

Fineman, M. *The Nature of Visual Illusion.* New York: Dover, pp. 119–122, 1996.

Jablan, S. "Impossible Figures." `http://members.tripod.com/~modularity/impos.htm` and "Are Impossible Figures Possible?" `http://members.tripod.com/~modularity/kulpa.htm`.

Kulpa, Z. "Are Impossible Figures Possible?" *Signal Processing* **5**, 201–220, 1983.

Kulpa, Z. "Putting Order in the Impossible." *Perception* **16**, 201–214, 1987.

Sugihara, K. "Classification of Impossible Objects." *Perception* **11**, 65–74, 1982.

Terouanne, E. "Impossible Figures and Interpretations of Polyhedral Figures." *J. Math. Psych.* **27**, 370–405, 1983.

Terouanne, E. "On a Class of 'Impossible' Figures: A New Language for a New Analysis." *J. Math. Psych.* **22**, 24–47, 1983.

Thro, E. B. "Distinguishing Two Classes of Impossible Objects." *Perception* **12**, 733–751, 1983.

Wilson, R. "Stamp Corner: Impossible Figures." *Math. Intell.* **13**, 80, 1991.

Impredicative

Definitions about a SET which depend on the entire SET.

Improper Integral

An INTEGRAL which has either or both limits INFINITE or which has an INTEGRAND which approaches INFINITY at one or more points in the range of integration.

see also DEFINITE INTEGRAL, INDEFINITE INTEGRAL, INTEGRAL, PROPER INTEGRAL

References

Press, W. H.; Flannery, B. P.; Teukolsky, S. A.; and Vetterling, W. T. "Improper Integrals." §4.4 in *Numerical Recipes in FORTRAN: The Art of Scientific Computing, 2nd ed.* Cambridge, England: Cambridge University Press, pp. 135–140, 1992.

Improper Node

A FIXED POINT for which the STABILITY MATRIX has equal nonzero EIGENVECTORS.

see also STABLE IMPROPER NODE, UNSTABLE IMPROPER NODE

Improper Rotation

The SYMMETRY OPERATION corresponding to a a ROTATION followed by an INVERSION OPERATION, also called a ROTOINVERSION. This operation is denoted $\bar{n}$ for an improper rotation by $360°/n$, so the CRYSTALLOGRAPHY RESTRICTION gives only $\bar{1}$, $\bar{2}$, $\bar{3}$, $\bar{4}$, $\bar{6}$ for crystals. The MIRROR PLANE symmetry operation is $(x, y, z) \to (x, y, -z)$, etc., which is equivalent to $\bar{2}$.

Impulse Pair

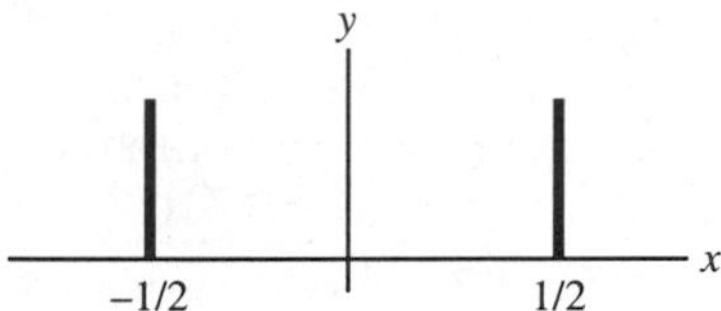

The even impulse pair is the FOURIER TRANSFORM of $\cos(\pi k)$,

$$\text{II}(x) \equiv \tfrac{1}{2}\delta(x+\tfrac{1}{2}) + \tfrac{1}{2}\delta(x-\tfrac{1}{2}). \tag{1}$$

It satisfies

$$\text{II}(x) * f(x) = \tfrac{1}{2}f(x+\tfrac{1}{2}) + \tfrac{1}{2}f(x-\tfrac{1}{2}), \tag{2}$$

where $*$ denotes CONVOLUTION, and

$$\int_{-\infty}^{\infty} \text{II}(x)\,dx = 1. \tag{3}$$

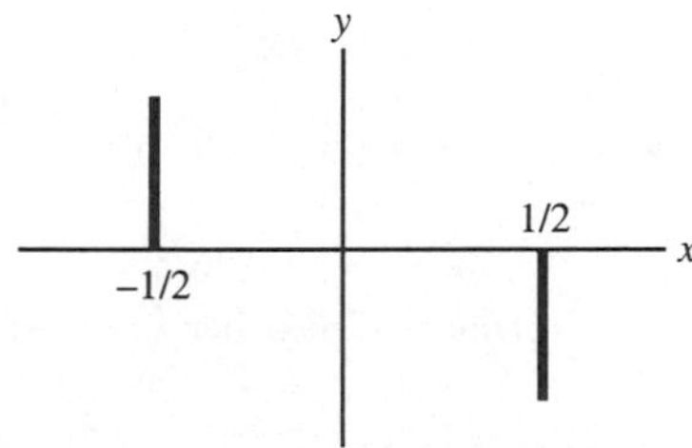

The odd impulse pair is the FOURIER TRANSFORM of $i\sin(\pi s)$,

$$\text{I}_\text{I}(x) \equiv \tfrac{1}{2}\delta(x+\tfrac{1}{2}) - \tfrac{1}{2}\delta(x-\tfrac{1}{2}). \tag{4}$$

Impulse Symbol

Bracewell's term for the DELTA FUNCTION.

see also IMPULSE PAIR

References

Bracewell, R. *The Fourier Transform and Its Applications.* New York: McGraw-Hill, 1965.

In-and-Out Curve

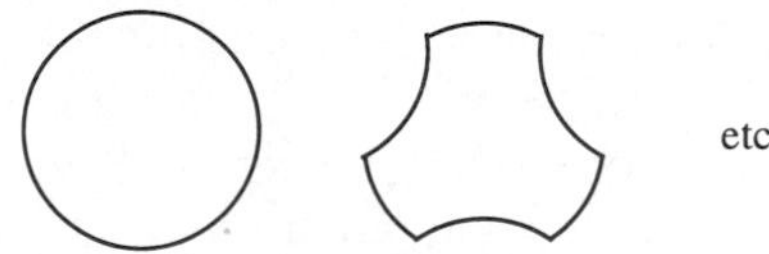

A curve created by starting with a circle, dividing it into six arcs, and flipping three alternating arcs. The process is then repeated an infinite number of times.

Inaccessible Cardinal

An inaccessible cardinal is a CARDINAL NUMBER which cannot be expressed in terms of a smaller number of smaller cardinals.

Inaccessible Cardinals Axiom

see also LEBESGUE MEASURABILITY PROBLEM

Inadmissible

A word or string which is not ADMISSIBLE.

Incenter

The center I of a TRIANGLE'S INCIRCLE. It can be found as the intersection of ANGLE BISECTORS, and it is the interior point for which distances to the sidelines are equal. Its TRILINEAR COORDINATES are 1:1:1. The distance between the incenter and CIRCUMCENTER is $\sqrt{R(R-2r)}$.

The incenter lies on the EULER LINE only for an ISOSCELES TRIANGLE. It does, however, lie on the SODDY LINE. For an EQUILATERAL TRIANGLE, the CIRCUMCENTER O, CENTROID G, NINE-POINT CENTER F, ORTHOCENTER H, and DE LONGCHAMPS POINT Z all coincide with I.

The incenter and EXCENTERS of a TRIANGLE are an ORTHOCENTRIC SYSTEM. The POWER of the incenter with respect to the CIRCUMCIRCLE is

$$p = \frac{a_1 a_2 a_3}{a_1 + a_2 + a_3}$$

(Johnson 1929, p. 190). If the incenters of the TRIANGLES $\Delta A_1 H_2 H_3$, $\Delta A_2 H_3 A_1$, and $\Delta A_3 H_1 H_2$ are X_1, X_2, and X_3, then $X_2 X_3$ is equal and parallel to $I_2 I_3$, where H_i are the FEET of the ALTITUDES and I_i are the incenters of the TRIANGLES. Furthermore, X_1, X_2, X_3, are the reflections of I with respect to the sides of the TRIANGLE $\Delta I_1 I_2 I_3$ (Johnson 1929, p. 193).

If four points are on a CIRCLE (i.e., they are CONCYCLIC), the incenters of the four TRIANGLES form a RECTANGLE whose sides are parallel to the lines connecting the middle points of opposite arcs. Furthermore, the connectors pass through the center of the RECTANGLE (Fuhrmann 1890, p. 50; Johnson 1929, pp. 254–255). More generally, the 16 incenters and excenters of the TRIANGLES whose VERTICES are four points on a CIRCLE, are the intersections of two sets of four PARALLEL lines which are mutually PERPENDICULAR (Johnson 1929, p. 255).

see also CENTROID (ORTHOCENTRIC SYSTEM), CIRCUMCENTER, EXCENTER, GERGONNE POINT, INCIRCLE, INRADIUS, ORTHOCENTER

References
Carr, G. S. *Formulas and Theorems in Pure Mathematics, 2nd ed.* New York: Chelsea, p. 622, 1970.
Dixon, R. *Mathographics.* New York: Dover, p. 58, 1991.
Fuhrmann, W. *Synthetische Beweise Planimetrischer Sätze.* Berlin, 1890.
Johnson, R. A. *Modern Geometry: An Elementary Treatise on the Geometry of the Triangle and the Circle.* Boston, MA: Houghton Mifflin, pp. 182–194, 1929.
Kimberling, C. "Central Points and Central Lines in the Plane of a Triangle." *Math. Mag.* **67**, 163–187, 1994.
Kimberling, C. "Incenter." `http://www.evansville.edu/~ck6/tcenters/class/incenter.html`.

Incenter-Excenter Circle

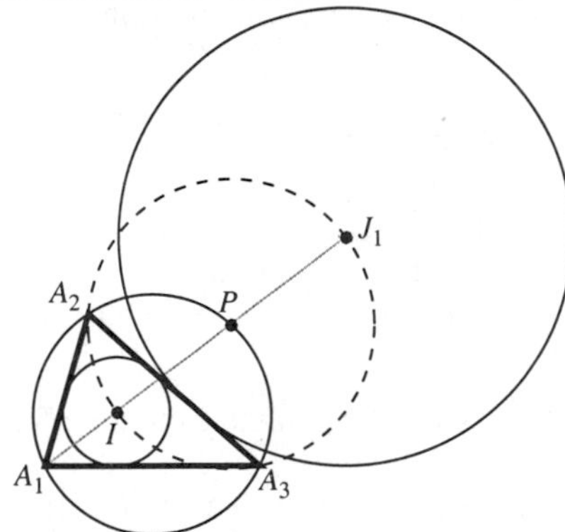

Given a triangle $\Delta A_1A_2A_3$, the points A_1, I, and J_1 lie on a line, where I is the INCENTER and J_1 is the EXCENTER corresponding to A_1. Furthermore, the CIRCLE with IJ_1 as the DIAMETER has P as its center, where P is the intersection of A_1J_1 with the CIRCUMCIRCLE of $\Delta A_1A_2A_3$, and passes through A_2 and A_3. This CIRCLE has RADIUS

$$r = \tfrac{1}{2}a_1 \sec(\tfrac{1}{2}\alpha_1) = 2R\sin(\tfrac{1}{2}\alpha_1).$$

It arises because $IJ_1J_2J_3$ forms an ORTHOCENTRIC SYSTEM.

see also CIRCUMCIRCLE, EXCENTER, EXCENTER-EXCENTER CIRCLE, INCENTER, ORTHOCENTRIC SYSTEM

References
Johnson, R. A. *Modern Geometry: An Elementary Treatise on the Geometry of the Triangle and the Circle.* Boston, MA: Houghton Mifflin, p. 185, 1929.

Incidence Axioms

The eight of HILBERT'S AXIOMS which concern collinearity and intersection; they include the first four of EUCLID'S POSTULATES.

see also ABSOLUTE GEOMETRY, CONGRUENCE AXIOMS, CONTINUITY AXIOMS, EUCLID'S POSTULATES, HILBERT'S AXIOMS, ORDERING AXIOMS, PARALLEL POSTULATE

References
Hilbert, D. *The Foundations of Geometry, 2nd ed.* Chicago, IL: Open Court, 1980.
Iyanaga, S. and Kawada, Y. (Eds.). "Hilbert's System of Axioms." §163B in *Encyclopedic Dictionary of Mathematics.* Cambridge, MA: MIT Press, pp. 544–545, 1980.

Incidence Matrix

For a k-D POLYTOPE Π_k, the incidence matrix is defined by

$$\eta_{ij}^k = \begin{cases} 1 & \text{if } \Pi_{k-1}^i \text{ belongs to } \Pi_k^j \\ 0 & \text{if } \Pi_{k-1}^i \text{ does not belong to } \Pi_k^j. \end{cases}$$

The ith row shows which Π_ks surround Π_{k-1}^i, and the jth column shows which Π_{k-1}s bound Π_k^j. Incidence matrices are also used to specify PROJECTIVE PLANES. The incidence matrices for a TETRAHEDRON $ABCD$ are

η^0	1	A	B	C
1	1	1	1	1

η^1	AD	BD	CD	BC	AC	AB
A	1	0	0	0	1	1
B	0	1	0	1	0	1
C	0	0	1	1	1	0
D	1	1	1	0	0	0

η^2	BCD	ACD	ABD	ABC
AD	0	1	1	0
BD	1	0	1	0
CD	1	1	0	0
BC	1	0	0	1
AC	0	1	0	1
AB	0	0	1	1

η^3	$ABCD$
BCD	1
ACD	1
ABD	1
ABC	1

see also ADJACENCY MATRIX, k-CHAIN, k-CIRCUIT

Incident

Two objects which touch each other are said to be incident.

see also INCIDENCE MATRIX

Incircle

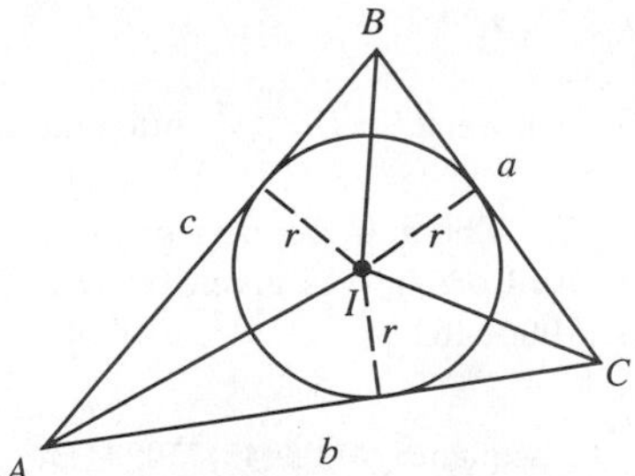

The INSCRIBED CIRCLE of a TRIANGLE ΔABC. The center I is called the INCENTER and the RADIUS r the INRADIUS. The points of intersection of the incircle with T are the VERTICES of the PEDAL TRIANGLE of T with the INCENTER as the PEDAL POINT (c.f. TANGENTIAL TRIANGLE). This TRIANGLE is called the CONTACT TRIANGLE.

The AREA K of the TRIANGLE ΔABC is given by

$$\begin{aligned} K &= \Delta AIC + \Delta CIB + \Delta AIB \\ &= \tfrac{1}{2}br + \tfrac{1}{2}ar + \tfrac{1}{2}cr = \tfrac{1}{2}(a+b+c)r = sr, \end{aligned}$$

where s is the SEMIPERIMETER.

Using the incircle of a TRIANGLE as the INVERSION CENTER, the sides of the TRIANGLE and its CIRCUMCIRCLE are carried into four equal CIRCLES (Honsberger 1976, p. 21). Pedoe (1995, p. xiv) gives a GEOMETRIC CONSTRUCTION for the incircle.

see also CIRCUMCIRCLE, CONGRUENT INCIRCLES POINT, CONTACT TRIANGLE, INRADIUS, TRIANGLE TRANSFORMATION PRINCIPLE

References

Coxeter, H. S. M. and Greitzer, S. L. *Geometry Revisited.* Washington, DC: Math. Assoc. Amer., pp. 11–13, 1967.

Honsberger, R. *Mathematical Gems II.* Washington, DC: Math. Assoc. Amer., 1976.

Johnson, R. A. *Modern Geometry: An Elementary Treatise on the Geometry of the Triangle and the Circle.* Boston, MA: Houghton Mifflin, pp. 182–194, 1929.

Pedoe, D. *Circles: A Mathematical View, rev. ed.* Washington, DC: Math. Assoc. Amer., 1995.

Inclusion-Exclusion Principle

If $A_1, \ldots, A_k$ are finite sets, then

$$\left| \bigcup_{i=1}^{k} A_i \right| = \sum_{i=1}^{k} (-1)^{i+1} \xi_i,$$

where ξ_i is the sum of the CARDINALITIES of the intersections of the sets taken i at a time.

Inclusion Map

Given a SUBSET B of a SET A, the INJECTION $f: B \to A$ defined by $f(b) = b$ for all $b \in B$ is called the inclusion map.

see also LONG EXACT SEQUENCE OF A PAIR AXIOM

Incommensurate

Two lengths are called incommensurate or incommensurable if their ratio cannot be expressed as a ratio of whole numbers. IRRATIONAL NUMBERS and TRANSCENDENTAL NUMBERS are incommensurate with the integers.

see also IRRATIONAL NUMBER, PYTHAGORAS'S CONSTANT, TRANSCENDENTAL NUMBER

Incomplete Gamma Function

see GAMMA FUNCTION

Incompleteness

A formal theory is said to be incomplete if it contains fewer theorems than would be possible while still retaining CONSISTENCY.

see also CONSISTENCY, GÖDEL'S INCOMPLETENESS THEOREM

References

Chaitin, G. J. "G. J. Chaitin's Home Page." `http://www.cs.auckland.ac.nz/CDMTCS/chaitin.`

Increasing Function

A function $f(x)$ increases on an INTERVAL I if $f(b) > f(a)$ for all $b > a$, where $a, b \in I$. Conversely, a function $f(x)$ decreases on an INTERVAL I if $f(b) < f(a)$ for all $b > a$ with $a, b \in I$.

If the DERIVATIVE $f'(x)$ of a CONTINUOUS FUNCTION $f(x)$ satisfies $f'(x) > 0$ on an OPEN INTERVAL (a, b), then $f(x)$ is increasing on (a, b). However, a function may increase on an interval without having a derivative defined at all points. For example, the function $x^{1/3}$ is increasing everywhere, including the origin $x = 0$, despite the fact that the DERIVATIVE is not defined at that point.

see also DECREASING FUNCTION, DERIVATIVE, NONDECREASING FUNCTION, NONINCREASING FUNCTION

Increasing Sequence

For a SEQUENCE $\{a_n\}$, if $a_{n+1} - a_n > 0$ for $n \geq x$, then a is increasing for $n \geq x$. Conversely, if $a_{n+1} - a_n < 0$ for $n \geq x$, then a is DECREASING for $n \geq x$. If $a_{n+1}/a_n > 1$ for all $n \geq x$, then a is increasing for $n \geq x$. Conversely, if $a_{n+1}/a_n < 1$ for all $n \geq x$, then a is decreasing for $n \geq x$.

Indefinite Integral

An INTEGRAL

$$\int f(x)\,dx$$

without upper and lower limits, also called an ANTIDERIVATIVE. The first FUNDAMENTAL THEOREM OF CALCULUS allows DEFINITE INTEGRALS to be computed in terms of indefinite integrals. If F is the indefinite integral for $f(x)$, then

$$\int_a^b f(x)\,dx = F(b) - F(a).$$

see also ANTIDERIVATIVE, CALCULUS, DEFINITE INTEGRAL, FUNDAMENTAL THEOREMS OF CALCULUS, INTEGRAL

Indefinite Quadratic Form

A QUADRATIC FORM $Q(\mathbf{x})$ is indefinite if it is less than 0 for some values and greater than 0 for others. The QUADRATIC FORM, written in the form $(\mathbf{x}, \mathsf{A}\mathbf{x})$, is indefinite if EIGENVALUES of the MATRIX A are of both signs.

see also POSITIVE DEFINITE QUADRATIC FORM, POSITIVE SEMIDEFINITE QUADRATIC FORM

References

Gradshteyn, I. S. and Ryzhik, I. M. *Tables of Integrals, Series, and Products, 5th ed.* San Diego, CA: Academic Press, p. 1106, 1979.

Indegree

The number of inward directed EDGES from a given VERTEX in a DIRECTED GRAPH.

see also LOCAL DEGREE, OUTDEGREE

Independence Axiom

A rational choice between two alternatives should depend only on how they differ.

Independence Complement Theorem

If sets E and F are INDEPENDENT, then so are E and F', where F' is the complement of F (i.e., the set of all possible outcomes not contained in F). Let $\cup$ denote "or" and $\cap$ denote "and." Then

$$P(E) = P(EF \cup EF') \tag{1}$$
$$= P(EF) + P(EF') - P(EF \cap EF'), \tag{2}$$

where AB is an abbreviation for $A \cap B$. But E and F are independent, so

$$P(EF) = P(E)P(F). \tag{3}$$

Also, since F and F' are complements, they contain no common elements, which means that

$$P(EF \cap EF') = 0 \tag{4}$$

for any E. Plugging (4) and (3) into (2) then gives

$$P(E) = P(E)P(F) + P(EF'). \tag{5}$$

Rearranging,

$$P(EF') = P(E)[1 - P(F)] = P(E)P(F'), \tag{6}$$

Q.E.D.

see also INDEPENDENT STATISTICS

Independence Number

The number

$$\alpha(G) = \max(|U| : U \subset V \text{ independent})$$

for a GRAPH G. The independence number of the DE BRUIJN GRAPH of order n is given by 1, 2, 3, 7, 13, 28, ... (Sloane's A006946).

References

Sloane, N. J. A. Sequence A006946/M0834 in "An On-Line Version of the Encyclopedia of Integer Sequences."

Independent Equations

see LINEARLY INDEPENDENT

Independent Sequence

see STRONGLY INDEPENDENT, WEAKLY INDEPENDENT

Independent Statistics

Two variates A and B are statistically independent IFF the CONDITIONAL PROBABILITY $P(A|B)$ of A given B satisfies

$$P(A|B) = P(A), \tag{1}$$

in which case the probability of A and B is just

$$P(AB) = P(A \cap B) = P(A)P(B). \tag{2}$$

Similarly, n events $A_1, A_2, \ldots, A_n$ are independent IFF

$$P\left(\bigcap_{i=1}^n A_i\right) = \prod_{i=1}^n P(A_i). \tag{3}$$

Statistically independent variables are always UNCORRELATED, but the converse is not necessarily true.

see also BAYES' FORMULA, CONDITIONAL PROBABILITY, INDEPENDENCE COMPLEMENT THEOREM, UNCORRELATED

Independent Vertices

A set of VERTICES A of a GRAPH with EDGES V is independent if it contains no EDGES.

see also INDEPENDENCE NUMBER

Indeterminate Problems

see DIOPHANTINE EQUATION—LINEAR

Index

A statistic which assigns a single number to several individual statistics in order to quantify trends. The best-known index in the United States is the consumer price index, which gives a sort of "average" value for inflation based on the price changes for a group of selected products.

Let p_n be the price per unit in period n, q_n be the quantity produced in period n, and $v_n \equiv p_n q_n$ be the value of the n units. Let q_a be the estimated relative importance of a product. There are several types of indices defined, among them those listed in the following table.

Index	Abbr.	Formula
Bowley index	P_B	$\frac{1}{2}(P_L + P_P)$
Fisher index	P_F	$\sqrt{P_L P_P}$
Geometric mean index	P_G	$\left[\prod\left(\frac{p_n}{p_0}\right)^{v_0}\right]^{1/\sum v_0}$
Harmonic mean index	P_H	$\frac{\sum p_0 q_0}{\sum \frac{p_0{}^2 q_0}{p_n}}$
Laspeyres's index	P_L	$\frac{\sum p_n q_0}{\sum p_0 q_0}$
Marshall-Edgeworth index	P_{ME}	$\frac{\sum p_n(q_0+q_n)}{\sum(v_0+v_n)}$
Mitchell index	P_M	$\frac{\sum p_n q_a}{\sum p_0 q_a}$
Paasche's index	P_P	$\frac{\sum p_n q_n}{\sum p_0 q_n}$
Walsh index	P_W	$\frac{\sum \sqrt{q_0 q_n}\, p_n}{\sum \sqrt{q_0 q_n}\, p_0}$

see also BOWLEY INDEX, FISHER INDEX, GEOMETRIC MEAN INDEX, HARMONIC MEAN INDEX, LASPEYRES' INDEX, MARSHALL-EDGEWORTH INDEX, MITCHELL INDEX, PAASCHE'S INDEX, RESIDUE INDEX, WALSH INDEX

References

Fisher, I. *The Making of Index Numbers: A Study of Their Varieties, Tests and Reliability, 3rd ed.* New York: Augustus M. Kelly, 1967.

Kenney, J. F. and Keeping, E. S. "Index Numbers." Ch. 5 in *Mathematics of Statistics, Pt. 1, 3rd ed.* Princeton, NJ: Van Nostrand, pp. 64–74, 1962.

Mudgett, B. D. *Index Numbers.* New York: Wiley, 1951.

Index Set

A STOCHASTIC PROCESS is a family of RANDOM VARIABLES $\{x(t,\bullet), t \in \mathcal{J}\}$ from some PROBABILITY SPACE $(S, \mathbb{S}, P)$ into a STATE SPACE $(S', \mathbb{S}')$, where $\mathcal{J}$ is the index set of the process.

References

Doob, J. L. "The Development of Rigor in Mathematical Probability (1900–1950)." *Amer. Math. Monthly* **103**, 586–595, 1996.

Index Theory

A branch of TOPOLOGY dealing with topological invariants of MANIFOLDS.

References

Roe, J. *Index Theory, Coarse Geometry, and Topology of Manifolds.* Providence, RI: Amer. Math. Soc., 1996.

Upmeier, H. *Toeplitz Operators and Index Theory in Several Complex Variables.* Boston, MA: Birkhäuser, 1996.

Indicatrix

A spherical image of a curve. The most common indicatrix is DUPIN'S INDICATRIX.

see also DUPIN'S INDICATRIX

Indicial Equation

The RECURRENCE RELATION obtained during application of the FROBENIUS METHOD of solving a second-order ordinary differential equation. The indicial equation (also called the CHARACTERISTIC EQUATION) is obtained by noting that, by definition, the lowest order term x^k (that corresponding to $n = 0$) must have a COEFFICIENT of zero. For an example of the construction of an indicial equation, see BESSEL DIFFERENTIAL EQUATION.

1. If the two ROOTS are equal, only one solution can be obtained.
2. If the two ROOTS differ by a noninteger, two solutions can be obtained.
3. If the two ROOTS differ by an INTEGER, the larger will yield a solution. The smaller may or may not.

References

Morse, P. M. and Feshbach, H. *Methods of Theoretical Physics, Part I.* New York: McGraw-Hill, pp. 532–534, 1953.

Indifference Principle

see INSUFFICIENT REASON PRINCIPLE

Induced Map

If $f : (X, A) \to (Y, B)$ is homotopic to $g : (X, A) \to (Y, B)$, then $f_* : H_n(X, A) \to H_n(Y, B)$ and $g_* : H_n(X, A) \to H_n(Y, B)$ are said to be the induced maps.

see also EILENBERG-STEENROD AXIOMS

Induced Norm

see NATURAL NORM

Induction

The use of the INDUCTION PRINCIPLE in a PROOF. Induction used in mathematics is often called MATHEMATICAL INDUCTION.

References

Buck, R. C. "Mathematical Induction and Recursive Definitions." *Amer. Math. Monthly* **70**, 128–135, 1963.

Induction Axiom

The fifth of PEANO'S AXIOMS, which states: If a SET S of numbers contains zero and also the successor of every number in S, then every number is in S.

see also PEANO'S AXIOMS

Induction Principle

The truth of an INFINITE sequence of propositions P_i for $i = 1, \ldots, \infty$ is established if (1) P_1 is true, and (2) P_k IMPLIES P_{k+1} for all k.

References

Courant, R. and Robbins, H. "The Principle of Mathematical Induction" and "Further Remarks on Mathematical Induction." §1.2.1 and 1.7 in *What is Mathematics?: An Elementary Approach to Ideas and Methods, 2nd ed.* Oxford, England: Oxford University Press, pp. 9–11 and 18–20, 1996.

Inequality

A mathematical statement that one quantity is greater than or less than another. "a is less than b" is denoted $a < b$, and "a is greater than b" is denoted $a > b$. "a is less than or equal to b" is denoted $a \leq b$, and "a is greater than or equal to b" is denoted $a \geq b$. The symbols $a \ll b$ and $a \gg b$ are used to denote "a is much less than b" and "a is much greater than b," respectively.

Solutions to the inequality $|x - a| < b$ consist of the set $\{x : -b < x - a < b\}$, or equivalently $\{x : a - b < x < a + b\}$. Solutions to the inequality $|x - a| > b$ consist of the set $\{x : x - a > b\} \cup \{x : x - a < -b\}$. If a and b are both POSITIVE or both NEGATIVE and $a < b$, then $1/a > 1/b$.

see also ABC CONJECTURE, ARITHMETIC-LOGARITHMIC-GEOMETRIC MEAN INEQUALITY, BERNOULLI INEQUALITY, BERNSTEIN'S INEQUALITY, BERRY-OSSEEN INEQUALITY, BIENAYMÉ-CHEBYSHEV INEQUALITY, BISHOP'S INEQUALITY, BOGOMOLOV-MIYAOKA-YAU INEQUALITY, BOMBIERI'S INEQUALITY, BONFERRONI'S INEQUALITY, BOOLE'S INEQUALITY, CARLEMAN'S INEQUALITY, CAUCHY INEQUALITY, CHEBYSHEV INEQUALITY, CHI INEQUALITY, COPSON'S INEQUALITY, ERDŐS-MORDELL THEOREM, EXPONENTIAL INEQUALITY, FISHER'S BLOCK DESIGN INEQUALITY, FISHER'S ESTIMATOR INEQUALITY, GÅRDING'S INEQUALITY, GAUSS'S INEQUALITY, GRAM'S INEQUALITY, HADAMARD'S INEQUALITY, HARDY'S INEQUALITY, HARNACK'S INEQUALITY, HÖLDER INTEGRAL INEQUALITY, HÖLDER'S SUM INEQUALITY, ISOPERIMETRIC INEQUALITY, JARNICK'S INEQUALITY, JENSEN'S INEQUALITY, JORDAN'S INEQUALITY, KANTROVICH INEQUALITY, MARKOV'S INEQUALITY, MINKOWSKI INTEGRAL INEQUALITY, MINKOWSKI SUM INEQUALITY, MORSE INEQUALITIES, NAPIER'S INEQUALITY, NOSARZEWSKA'S INEQUALITY, OSTROWSKI'S INEQUALITY, PTOLEMY INEQUALITY, ROBBIN'S INEQUALITY, SCHRÖDER-BERNSTEIN THEOREM, SCHUR'S INEQUALITIES, SCHWARZ'S INEQUALITY, SQUARE ROOT INEQUALITY, STEFFENSEN'S INEQUALITY, STOLARSKY'S INEQUALITY, STRONG SUBADDITIVITY INEQUALITY, TRIANGLE INEQUALITY, TURÁN'S INEQUALITIES, WEIERSTRAß PRODUCT INEQUALITY, WIRTINGER'S INEQUALITY, YOUNG INEQUALITY

References

Abramowitz, M. and Stegun, C. A. (Eds.). *Handbook of Mathematical Functions with Formulas, Graphs, and Mathematical Tables, 9th printing.* New York: Dover, p. 16, 1972.
Beckenbach, E. F. and Bellman, Richard E. *An Introduction to Inequalities.* New York: Random House, 1961.
Beckenbach, E. F. and Bellman, Richard E. *Inequalities, 2nd rev. print.* Berlin: Springer-Verlag, 1965.
Hardy, G. H.; Littlewood, J. E.; and Pólya, G. *Inequalities, 2nd ed.* Cambridge, England: Cambridge University Press, 1952.
Kazarinoff, N. D. *Geometric Inequalities.* New York: Random House, 1961.
Mitrinovic, D. S. *Analytic Inequalities.* New York: Springer-Verlag, 1970.
Mitrinovic, D. S.; Pecaric, J. E.; and Fink, A. M. *Classical & New Inequalities in Analysis.* Dordrecht, Netherlands: Kluwer, 1993.
Mitrinovic, D. S.; Pecaric, J. E.; Fink, A. M. *Inequalities Involving Functions & Their Integrals & Derivatives.* Dordrecht, Netherlands: Kluwer, 1991.
Mitrinovic, D. S.; Pecaric, J. E.; and Volenec, V. *Recent Advances in Geometric Inequalities.* Dordrecht, Netherlands: Kluwer, 1989.

Inexact Differential

An infinitesimal which is not the differential of an actual function and which cannot be expressed as

$$dz = \left(\frac{\partial z}{\partial x}\right)_y dx + \left(\frac{\partial z}{\partial y}\right)_z dy,$$

the way an EXACT DIFFERENTIAL can. Inexact differentials are denoted with a bar through the d. The most common example of an inexact differential is the change in heat $đQ$ encountered in thermodynamics.

see also EXACT DIFFERENTIAL, PFAFFIAN FORM

References

Zemansky, M. W. *Heat and Thermodynamics, 5th ed.* New York: McGraw-Hill, p. 38, 1968.

Inf

see INFIMUM, INFIMUM LIMIT

Infimum

The greatest lower bound of a set. It is denoted

$$\inf_S .$$

see also INFIMUM LIMIT, SUPREMUM

Infimum Limit

The limit infimum of a set is the greatest lower bound of the CLOSURE of a set. It is denoted

$$\liminf_S .$$

see also INFIMUM, SUPREMUM

Infinary Divisor

p^x is an infinary divisor of p^y (with $y > 0$) if $p^x|_{y-1}p^y$. This generalizes the concept of the k-ARY DIVISOR.

see also INFINARY PERFECT NUMBER, k-ARY DIVISOR

References

Cohen, G. L. "On an Integer's Infinary Divisors." *Math. Comput.* **54**, 395–411, 1990.

Cohen, G. and Hagis, P. "Arithmetic Functions Associated with the Infinary Divisors of an Integer." *Internat. J. Math. Math. Sci.* **16**, 373–383, 1993.

Guy, R. K. *Unsolved Problems in Number Theory, 2nd ed.* New York: Springer-Verlag, p. 54, 1994.

Infinary Multiperfect Number

Let $\sigma_\infty(n)$ be the SUM of the INFINARY DIVISORS of a number n. An infinary k-multiperfect number is a number n such that $\sigma_\infty(n) = kn$. Cohen (1990) found 13 infinary 3-multiperfects, seven 4-multiperfects, and two 5-multiperfects.

see also INFINARY PERFECT NUMBER

References

Cohen, G. L. "On an Integer's Infinary Divisors." *Math. Comput.* **54**, 395–411, 1990.

Guy, R. K. *Unsolved Problems in Number Theory, 2nd ed.* New York: Springer-Verlag, p. 54, 1994.

Infinary Perfect Number

Let $\sigma_\infty(n)$ be the SUM of the INFINARY DIVISORS of a number n. An infinary perfect number is a number n such that $\sigma_\infty(n) = 2n$. Cohen (1990) found 14 such numbers. The first few are 6, 60, 90, 36720, ... (Sloane's A007257).

see also INFINARY MULTIPERFECT NUMBER

References

Cohen, G. L. "On an Integer's Infinary Divisors." *Math. Comput.* **54**, 395–411, 1990.

Guy, R. K. *Unsolved Problems in Number Theory, 2nd ed.* New York: Springer-Verlag, p. 54, 1994.

Sloane, N. J. A. Sequence A007257/M4267 in "An On-Line Version of the Encyclopedia of Integer Sequences."

Infinite

Greater than any assignable quantity of the sort in question. In mathematics, the concept of the infinite is made more precise through the notion of an INFINITE SET.

see also COUNTABLE SET, COUNTABLY INFINITE SET, FINITE, INFINITE SET, INFINITESIMAL, INFINITY

Infinite Product

N.B. A detailed on-line essay by S. Finch was the starting point for this entry.

A PRODUCT involving an INFINITE number of terms. Such products can converge. In fact, for POSITIVE a_n, the PRODUCT $\prod_{n=1}^\infty a_n$ converges to a NONZERO number IFF $\sum_{n=1}^\infty \ln a_n$ converges.

Infinite products can be used to define the COSINE

$$\cos x = \prod_{n=1}^\infty \left[1 - \frac{4x^2}{\pi^2(2n-1)^2}\right], \tag{1}$$

GAMMA FUNCTION

$$\Gamma(z) = \left[ze^{\gamma z}\prod_{r=1}^\infty \left(1+\frac{z}{r}\right)e^{-z/r}\right]^{-1}, \tag{2}$$

SINE, and SINC FUNCTION. They also appear in the POLYGON CIRCUMSCRIBING CONSTANT

$$K = \prod_{n=3}^\infty \frac{1}{\cos\left(\frac{\pi}{n}\right)}. \tag{3}$$

An interesting infinite product formula due to Euler which relates π and the nth PRIME p_n is

$$\pi = \frac{2}{\prod_{i=n}^\infty \left[1 + \frac{\sin(\frac{1}{2}\pi p_n)}{p_n}\right]} \tag{4}$$

$$= \frac{2}{\prod_{i=n}^\infty \left[1 + \frac{(-1)^{(p_n-1)/2}}{p_n}\right]} \tag{5}$$

(Blatner 1997).

The product

$$\prod_{n=1}^\infty \left(1 + \frac{1}{n^p}\right) \tag{6}$$

has closed form expressions for small POSITIVE integral $p \geq 2$,

$$\prod_{n=1}^\infty \left(1+\frac{1}{n^2}\right) = \frac{\sinh\pi}{\pi} \tag{7}$$

$$\prod_{n=1}^\infty \left(1+\frac{1}{n^3}\right) = \frac{1}{\pi}\cosh(\tfrac{1}{2}\pi\sqrt{3}) \tag{8}$$

$$\prod_{n=1}^\infty \left(1+\frac{1}{n^4}\right) = \frac{\cosh(\pi\sqrt{2}) - \cos(\pi\sqrt{2})}{2\pi^2} \tag{9}$$

$$\prod_{n=1}^\infty \left(1+\frac{1}{n^5}\right) = \left|\Gamma[\exp(\tfrac{2}{5}\pi i)]\Gamma[\exp(\tfrac{6}{5}\pi i)]\right|^{-2}. \tag{10}$$

The d-ANALOG expression

$$[\infty!]_d = \prod_{n=3}^\infty \left(1 - \frac{2^d}{n^d}\right) \tag{11}$$

also has closed form expressions,

$$\prod_{n=3}^{\infty}\left(1-\frac{4}{n^2}\right)=\frac{1}{6} \tag{12}$$

$$\prod_{n=3}^{\infty}\left(1-\frac{8}{n^3}\right)=\frac{\sinh(\pi\sqrt{3})}{42\pi\sqrt{3}} \tag{13}$$

$$\prod_{n=3}^{\infty}\left(1-\frac{16}{n^4}\right)=\frac{\sinh(2\pi)}{120\pi} \tag{14}$$

$$\prod_{n=3}^{\infty}\left(1-\frac{32}{n^5}\right)=\left|\Gamma[\exp(\tfrac{1}{5}\pi i)]\Gamma[2\exp(\tfrac{7}{5}\pi i)]\right|^{-2}. \tag{15}$$

see also COSINE, DIRICHLET ETA FUNCTION, EULER IDENTITY, GAMMA FUNCTION, ITERATED EXPONENTIAL CONSTANTS, POLYGON CIRCUMSCRIBING CONSTANT, POLYGON INSCRIBING CONSTANT, Q-FUNCTION, SINE

References

Abramowitz, M. and Stegun, C. A. (Eds.). *Handbook of Mathematical Functions with Formulas, Graphs, and Mathematical Tables, 9th printing.* New York: Dover, p. 75, 1972.

Arfken, G. "Infinite Products." §5.11 in *Mathematical Methods for Physicists, 3rd ed.* Orlando, FL: Academic Press, pp. 346–351, 1985.

Blatner, D. *The Joy of Pi.* New York: Walker, p. 119, 1997.

Finch, S. "Favorite Mathematical Constants." `http://www.mathsoft.com/asolve/constant/infprd/infprd.html`.

Hansen, E. R. *A Table of Series and Products.* Englewood Cliffs, NJ: Prentice-Hall, 1975.

Whittaker, E. T. and Watson, G. N. §7.5 and 7.6 in *A Course in Modern Analysis, 4th ed.* Cambridge, England: Cambridge University Press, 1990.

Infinite Series

A SERIES with an INFINITE number of terms.

see also SERIES

Infinite Set

A SET of S elements is said to be infinite if the elements of a PROPER SUBSET S' can be put into ONE-TO-ONE correspondence with the elements of S. An infinite set whose elements can be put into a ONE-TO-ONE correspondence with the set of INTEGERS is said to be COUNTABLY INFINITE; otherwise, it is called UNCOUNTABLY INFINITE.

see also ALEPH-0, ALEPH-1, CARDINAL NUMBER, COUNTABLY INFINITE SET, CONTINUUM, FINITE, INFINITE, INFINITY, ORDINAL NUMBER, TRANSFINITE NUMBER, UNCOUNTABLY INFINITE SET

References

Courant, R. and Robbins, H. *What is Mathematics?: An Elementary Approach to Ideas and Methods, 2nd ed.* Oxford, England: Oxford University Press, p. 77, 1996.

Infinitesimal

A quantity which yields 0 after the application of some LIMITING process. The understanding of infinitesimals was a major roadblock to the acceptance of CALCULUS and its placement on a firm mathematical foundation.

see also INFINITE, INFINITY, NONSTANDARD ANALYSIS

Infinitesimal Analysis

An archaic term for CALCULUS.

Infinitesimal Matrix Change

Let B, A, and e be square matrices with e small, and define

$$\mathsf{B}\equiv\mathsf{A}(\mathsf{I}+\mathsf{e}), \tag{1}$$

where I is the IDENTITY MATRIX. Then the inverse of B is approximately

$$\mathsf{B}^{-1}=(\mathsf{I}-\mathsf{e})\mathsf{A}^{-1}. \tag{2}$$

This can be seen by multiplying

$$\begin{aligned}\mathsf{B}\mathsf{B}^{-1}&=(\mathsf{A}+\mathsf{A}\mathsf{e})(\mathsf{A}^{-1}-\mathsf{e}\mathsf{A}^{-1})\\&=\mathsf{A}\mathsf{A}^{-1}-\mathsf{A}\mathsf{e}\mathsf{A}^{-1}+\mathsf{A}\mathsf{e}\mathsf{A}^{-1}-\mathsf{A}\mathsf{e}^2\mathsf{A}^{-1}\\&=\mathsf{I}-\mathsf{A}\mathsf{e}^2\mathsf{A}^{-1}\approx 1.\end{aligned} \tag{3}$$

Note that if we instead let $\mathsf{B}'\equiv\mathsf{A}+\mathsf{e}$, and look for an inverse of the form $\mathsf{B}'^{-1}=\mathsf{A}^{-1}+\mathsf{C}$, we obtain

$$\begin{aligned}\mathsf{B}'\mathsf{B}'^{-1}&=(\mathsf{A}+\mathsf{e})(\mathsf{A}^{-1}+\mathsf{C})=\mathsf{A}\mathsf{A}^{-1}+\mathsf{A}\mathsf{C}+\mathsf{e}\mathsf{A}^{-1}+\mathsf{e}\mathsf{C}\\&=\mathsf{I}+\mathsf{A}\mathsf{C}+\mathsf{e}(\mathsf{C}+\mathsf{A}^{-1})\equiv\mathsf{I}.\end{aligned} \tag{4}$$

In order to eliminate the e term, we require $\mathsf{C}=-\mathsf{A}^{-1}$. However, then $\mathsf{A}\mathsf{C}=-\mathsf{I}$, so $\mathsf{B}\mathsf{B}^{-1}=0$ so there can be no inverse of this form.

The exact inverse of B can be found as follows.

$$\mathsf{B}=\mathsf{A}(\mathsf{I}+\mathsf{e})=\mathsf{A}(\mathsf{I}+\mathsf{A}^{-1}\mathsf{e}), \tag{5}$$

so

$$\mathsf{B}^{-1}=[\mathsf{A}(\mathsf{I}+\mathsf{A}^{-1}\mathsf{e})]^{-1}. \tag{6}$$

Using a general MATRIX INVERSE identity then gives

$$\mathsf{B}^{-1}=(\mathsf{I}+\mathsf{A}^{-1}\mathsf{e})^{-1}\mathsf{A}^{-1}. \tag{7}$$

Infinitesimal Rotation

An infinitesimal transformation of a VECTOR $\mathbf{r}$ is given by

$$\mathbf{r}'=(\mathsf{I}+\mathsf{e})\mathbf{r}, \tag{1}$$

where the MATRIX e is infinitesimal and I is the IDENTITY MATRIX. (Note that the infinitesimal transformation may not correspond to an inversion, since inversion is a discontinuous process.) The COMMUTATIVITY of infinitesimal transformations e_1 and e_2 is established by the equivalence of

$$(\mathsf{I}+\mathsf{e}_1)(\mathsf{I}+\mathsf{e}_2) = \mathsf{I}^2 + \mathsf{e}_1\mathsf{I} + \mathsf{I}\mathsf{e}_2 + \mathsf{e}_1\mathsf{e}_2 \approx \mathsf{I} + \mathsf{e}_1 + \mathsf{e}_2 \quad (2)$$

$$(\mathsf{I}+\mathsf{e}_2)(\mathsf{I}+\mathsf{e}_1) = \mathsf{I}^2 + \mathsf{e}_2\mathsf{I} + \mathsf{I}\mathsf{e}_1 + \mathsf{e}_2\mathsf{e}_1 \approx \mathsf{I} + \mathsf{e}_2 + \mathsf{e}_1. \quad (3)$$

Now let

$$\mathsf{A} \equiv \mathsf{I} + \mathsf{e}. \quad (4)$$

The inverse A^{-1} is then $\mathsf{I} - \mathsf{e}$, since

$$\mathsf{A}\mathsf{A}^{-1} = (\mathsf{I}+\mathsf{e})(\mathsf{I}-\mathsf{e}) = \mathsf{I}^2 - \mathsf{e}^2 \approx \mathsf{I}. \quad (5)$$

Since we are defining our infinitesimal transformation to be a rotation, ORTHOGONALITY of ROTATION MATRICES requires that

$$\mathsf{A}^{\mathrm{T}} = \mathsf{A}^{-1}, \quad (6)$$

but

$$\mathsf{A}^{-1} = \mathsf{I} - \mathsf{e} \quad (7)$$

$$(\mathsf{I}+\mathsf{e})^{\mathrm{T}} = \mathsf{I}^{\mathrm{T}} + \mathsf{e}^{\mathrm{T}} = \mathsf{I} + \mathsf{e}^{\mathrm{T}}, \quad (8)$$

so $\mathsf{e} = -\mathsf{e}^{\mathrm{T}}$ and the infinitesimal rotation is ANTISYMMETRIC. It must therefore have a MATRIX of the form

$$\mathsf{e} = \begin{bmatrix} 0 & d\Omega_3 & -d\Omega_2 \\ -d\Omega_3 & 0 & d\Omega_1 \\ d\Omega_2 & -d\Omega_1 & 0 \end{bmatrix}. \quad (9)$$

The differential change in a vector $\mathbf{r}$ upon application of the ROTATION MATRIX is then

$$d\mathbf{r} \equiv \mathbf{r}' - \mathbf{r} = (\mathsf{I}+\mathsf{e})\mathbf{r} - \mathbf{r} = \mathsf{e}\mathbf{r}. \quad (10)$$

Writing in MATRIX form,

$$\begin{aligned} d\mathbf{r} &= \begin{bmatrix} x \\ y \\ z \end{bmatrix} \begin{bmatrix} 0 & d\Omega_3 & -d\Omega_2 \\ -d\Omega_3 & 0 & d\Omega_1 \\ d\Omega_2 & -d\Omega_1 & 0 \end{bmatrix} \\ &= \begin{bmatrix} y\,d\Omega_3 - z\,d\Omega_2 \\ z\,d\Omega_1 - x\,d\Omega_3 \\ x\,d\Omega_2 - y\,d\Omega_1 \end{bmatrix} \quad (11) \\ &= (y\,d\Omega_3 - z\,d\Omega_2)\hat{\mathbf{x}} + (z\,d\Omega_1 - x\,d\Omega_3)\hat{\mathbf{y}} \\ &\quad + (x\,d\Omega_2 - y\,d\Omega_1)\hat{\mathbf{z}} \\ &= \mathbf{r} \times d\mathbf{\Omega}. \quad (12) \end{aligned}$$

Therefore,

$$\left(\frac{d\mathbf{r}}{dt}\right)_{\text{rotation, body}} = \mathbf{r} \times \frac{d\mathbf{\Omega}}{dt} = \mathbf{r} \times \boldsymbol{\omega}, \quad (13)$$

where

$$\boldsymbol{\omega} \equiv \frac{d\mathbf{\Omega}}{dt} = \hat{\mathbf{n}}\frac{d\phi}{dt}. \quad (14)$$

The total rotation observed in the stationary frame will be a sum of the rotational velocity and the velocity in the rotating frame. However, note that an observer in the stationary frame will see a velocity opposite in direction to that of the observer in the frame of the rotating body, so

$$\left(\frac{d\mathbf{r}}{dt}\right)_{\text{space}} = \left(\frac{d\mathbf{r}}{dt}\right)_{\text{body}} + \boldsymbol{\omega} \times \mathbf{r}. \quad (15)$$

This can be written as an operator equation, known as the ROTATION OPERATOR, defined as

$$\left(\frac{d}{dt}\right)_{\text{space}} = \left(\frac{d}{dt}\right)_{\text{body}} + \boldsymbol{\omega} \times . \quad (16)$$

see also ACCELERATION, EULER ANGLES, ROTATION, ROTATION MATRIX, ROTATION OPERATOR

Infinitive Sequence

A sequence $\{x_n\}$ is called an infinitive sequence if, for every i, $x_n = i$ for infinitely many n. Write $a(i,j)$ for the jth index n for which $x_n = i$. Then as i and j range through N, the array $A = a(i,j)$, called the associative array of x, ranges through all of N.

see also FRACTAL SEQUENCE

References

Kimberling, C. "Fractal Sequences and Interspersions." *Ars Combin.* **45**, 157–168, 1997.

Infinitude of Primes

see EUCLID'S THEOREMS

Infinity

An unbounded number greater than every REAL NUMBER, most often denoted as ∞. The symbol ∞ had been used as an alternative to M (1,000) in ROMAN NUMERALS until 1655, when John Wallis suggested it be used instead for infinity.

Infinity is a very tricky concept to work with, as evidenced by some of the counterintuitive results which follow from Georg Cantor's treatment of INFINITE SETS. Informally, $1/\infty = 0$, a statement which can be made rigorous using the LIMIT concept,

$$\lim_{x\to\infty} \frac{1}{x} = 0.$$

Similarly,

$$\lim_{x\to 0^+} \frac{1}{x} = \infty,$$

where the notation 0^+ indicates that the LIMIT is taken from the POSITIVE side of the REAL LINE.

see also ALEPH, ALEPH-0, ALEPH-1, CARDINAL NUMBER, CONTINUUM, CONTINUUM HYPOTHESIS, HILBERT HOTEL, INFINITE, INFINITE SET, INFINITESIMAL, LINE AT INFINITY, L'HOSPITAL'S RULE, POINT AT INFINITY, TRANSFINITE NUMBER, UNCOUNTABLY INFINITE SET, ZERO

References

Conway, J. H. and Guy, R. K. *The Book of Numbers.* New York: Springer-Verlag, p. 19, 1996.

Courant, R. and Robbins, H. "The Mathematical Analysis of Infinity." §2.4 in *What is Mathematics?: An Elementary Approach to Ideas and Methods, 2nd ed.* Oxford, England: Oxford University Press, pp. 77–88, 1996.
Hardy, G. H. *Orders of Infinity, the 'infinitarcalcul' of Paul Du Bois-Reymond, 2nd ed.* Cambridge, England: Cambridge University Press, 1924.
Lavine, S. *Understanding the Infinite.* Cambridge, MA: Harvard University Press, 1994.
Maor, E. *To Infinity and Beyond: A Cultural History of the Infinite.* Boston, MA: Birkhäuser, 1987.
Moore, A. W. *The Infinite.* New York: Routledge, 1991.
Morris, R. *Achilles in the Quantum Universe: The Definitive History of Infinity.* New York: Henry Holt, 1997.
Péter, R. *Playing with Infinity.* New York: Dover, 1976.
Smail, L. L. *Elements of the Theory of Infinite Processes.* New York: McGraw-Hill, 1923.
Vilenskin, N. Ya. *In Search of Infinity.* Boston, MA: Birkhäuser, 1995.
Wilson, A. M. *The Infinite in the Finite.* New York: Oxford University Press, 1996.
Zippin, L. *Uses of Infinity.* New York: Random House, 1962.

Inflection Point

A point on a curve at which the SIGN of the CURVATURE (i.e., the concavity) changes. The FIRST DERIVATIVE TEST can sometimes distinguish inflection points from EXTREMA for DIFFERENTIABLE functions $f(x)$.

see also CURVATURE, DIFFERENTIABLE, EXTREMUM, FIRST DERIVATIVE TEST, STATIONARY POINT

Information Dimension

Define the "information function" to be

$$I \equiv -\sum_{i=1}^{N} P_i(\epsilon) \ln[P_i(\epsilon)], \tag{1}$$

where $P_i(\epsilon)$ is the NATURAL MEASURE, or probability that element i is populated, normalized such that

$$\sum_{i=1}^{N} P_i(\epsilon) = 1. \tag{2}$$

The information dimension is then defined by

$$\begin{aligned} d_{\text{inf}} &\equiv -\lim_{\epsilon\to 0^+} \frac{I}{\ln(\epsilon)} \\ &= \lim_{\epsilon\to 0^+} \sum_{i=1}^{N} \frac{P_i(\epsilon)\ln[P_i(\epsilon)]}{\ln(\epsilon)}. \end{aligned} \tag{3}$$

If every element is equally likely to be visited, then $P_i(\epsilon)$ is independent of i, and

$$\sum_{i=1}^{N} P_i(\epsilon) = N P_i(\epsilon) = 1, \tag{4}$$

so

$$P_i(\epsilon) = \frac{1}{N}, \tag{5}$$

and

$$\begin{aligned} d_{\text{inf}} &= \lim_{\epsilon\to 0^+} \frac{\sum_{i=1}^{N} \frac{1}{N}\ln\left(\frac{1}{N}\right)}{\ln\epsilon} \\ &= \lim_{\epsilon\to 0^+} \frac{\ln(N^{-1})}{\ln\epsilon} = -\lim_{\epsilon\to 0^+} \frac{\ln N}{\ln(\epsilon)} = d_{\text{cap}}, \end{aligned} \tag{6}$$

where d_{cap} is the CAPACITY DIMENSION.

see also CORRELATION EXPONENT

References

Farmer, J. D. "Chaotic Attractors of an Infinite-dimensional Dynamical System." *Physica D* **4**, 366–393, 1982.
Nayfeh, A. H. and Balachandran, B. *Applied Nonlinear Dynamics: Analytical, Computational, and Experimental Methods.* New York: Wiley, pp. 545–547, 1995.

Information Entropy

see ENTROPY

Information Theory

The branch of mathematics dealing with the efficient and accurate storage, transmission, and representation of information.

see also CODING THEORY, ENTROPY

References

Goldman, S. *Information Theory.* New York: Dover, 1953.
Lee, Y. W. *Statistical Theory of Communication.* New York: Wiley, 1960.
Pierce, J. R. *An Introduction to Information Theory.* New York: Dover, 1980.
Reza, F. M. *An Introduction to Information Theory.* New York: Dover, 1994.
Singh, J. *Great Ideas in Information Theory, Language and Cybernetics.* New York: Dover, 1966.
Zayed, A. I. *Advances in Shannon's Sampling Theory.* Boca Raton, FL: CRC Press, 1993.

Initial Value Problem

An initial value problem is a problem that has its conditions specified at some time $t = t_0$. Usually, the problem is an ORDINARY DIFFERENTIAL EQUATION or a PARTIAL DIFFERENTIAL EQUATION. For example,

$$\begin{cases} \frac{\partial^2 u}{\partial t^2} - \nabla^2 u = f & \text{in } \Omega \\ u = u_0 & t = t_0 \\ u = u_1 & \text{on } \partial\Omega, \end{cases}$$

where $\partial\Omega$ denotes the boundary of Ω, is an initial value problem.

see also BOUNDARY CONDITIONS, BOUNDARY VALUE PROBLEM, PARTIAL DIFFERENTIAL EQUATION

References

Eriksson, K.; Estep, D.; Hansbo, P.; and Johnson, C. *Computational Differential Equations.* Lund, Sweden: Studentlitteratur, 1996.

Injection

see ONE-TO-ONE

Injective

A MAP is injective when it is ONE-TO-ONE, i.e., f is injective when $x \neq y$ IMPLIES $f(x) \neq f(y)$.

see also ONE-TO-ONE, SURJECTIVE

Injective Patch

An injective patch is a PATCH such that $\mathbf{x}(u_1, v_1) = \mathbf{x}(u_2, v_2)$ implies that $u_1 = u_2$ and $v_1 = v_2$. An example of a PATCH which is injective but not REGULAR is the function defined by (u^3, v^3, uv) for $u, v \in (-1, 1)$. However, if $\mathbf{x} : U \to \mathbb{R}^n$ is an injective regular patch, then $\mathbf{x}$ maps U diffeomorphically onto $\mathbf{x}(U)$.

see also PATCH, REGULAR PATCH

References

Gray, A. *Modern Differential Geometry of Curves and Surfaces.* Boca Raton, FL: CRC Press, p. 187, 1993.

Inner Automorphism Group

A particular type of AUTOMORPHISM GROUP which exists only for GROUPS. For a GROUP G, the inner automorphism group is defined by

$$\mathrm{Inn}(G) = \{\sigma_a : a \in G\} \subset \mathrm{Aut}(G)$$

where σ_a is an AUTOMORPHISM of G defined by

$$\sigma_a(x) = axa^{-1}.$$

see also AUTOMORPHISM, AUTOMORPHISM GROUP

Inner Product

see DOT PRODUCT

Inner Product Space

An inner product space is a VECTOR SPACE which has an INNER PRODUCT. If the INNER PRODUCT defines a NORM, then the inner product space is called a HILBERT SPACE.

see also HILBERT SPACE, INNER PRODUCT, NORM

Inradius

The radius of a TRIANGLE'S INCIRCLE or of a POLYHEDRON'S INSPHERE, denoted r. For a TRIANGLE,

$$r = \frac{1}{2}\sqrt{\frac{(b+c-a)(c+a-b)(a+b-c)}{a+b+c}} = \frac{\Delta}{s} \tag{1}$$

$$= 4R\sin(\tfrac{1}{2}\alpha_1)\sin(\tfrac{1}{2}\alpha_2)\sin(\tfrac{1}{2}\alpha_3), \tag{2}$$

where Δ is the AREA of the TRIANGLE, a, b, and c are the side lengths, s is the SEMIPERIMETER, and R is the CIRCUMRADIUS (Johnson 1929, p. 189).

Equation (1) can be derived easily using TRILINEAR COORDINATES. Since the INCENTER is equally spaced from all three sides, its trilinear coordinates are 1:1:1, and its exact trilinear coordinates are $r : r : r$. The ratio k of the exact trilinears to the homogeneous coordinates is given by

$$k = \frac{2\Delta}{a+b+c} = \frac{\Delta}{s}. \tag{3}$$

But since $k = r$ in this case,

$$r = k = \frac{\Delta}{s}, \tag{4}$$

Q. E. D.

Other equations involving the inradius include

$$Rr = \frac{abc}{4s} \tag{5}$$

$$\Delta^2 = rr_1r_2r_3 \tag{6}$$

$$\cos A + \cos B + \cos C = 1 + \frac{r}{R} \tag{7}$$

$$r = 2R\cos A\cos B\cos C \tag{8}$$

$$a^2 + b^2 + c^2 = 4rR + 8R^2, \tag{9}$$

where r_i are the EXRADII (Johnson 1929, pp. 189–191).

As shown in RIGHT TRIANGLE, the inradius of a RIGHT TRIANGLE of integral side lengths x, y, and z is also integral, and is given by

$$r = \frac{xy}{x+y+z}, \tag{10}$$

where z is the HYPOTENUSE. Let d be the distance between inradius r and CIRCUMRADIUS R, $d = \overline{rR}$. Then

$$R^2 - d^2 = 2Rr \tag{11}$$

$$\frac{1}{R-d} + \frac{1}{R+d} = \frac{1}{r} \tag{12}$$

(Mackay 1886–87). These and many other identities are given in Johnson (1929, pp. 186–190).

Expressing the MIDRADIUS ρ and CIRCUMRADIUS R in terms of the midradius gives

$$r = \frac{\rho^2}{\sqrt{\rho^2 + \frac{1}{4}a^2}} \tag{13}$$

$$r = \frac{R^2 - \frac{1}{4}a^2}{R} \tag{14}$$

for an ARCHIMEDEAN SOLID.

see also CARNOT'S THEOREM, CIRCUMRADIUS, MIDRADIUS

References

Johnson, R. A. *Modern Geometry: An Elementary Treatise on the Geometry of the Triangle and the Circle.* Boston, MA: Houghton Mifflin, 1929.

Mackay, J. S. "Historical Notes on a Geometrical Theorem and its Developments [18th Century]." *Proc. Edinburgh Math. Soc.* **5**, 62–78, 1886–1887.

Mackay, J. S. "Formulas Connected with the Radii of the Incircle and Excircles of a Triangle." *Proc. Edinburgh Math. Soc.* **12**, 86–105.

Mackay, J. S. "Formulas Connected with the Radii of the Incircle and Excircles of a Triangle." *Proc. Edinburgh Math. Soc.* **13**, 103–104.

Inscribed
A geometric figure which touches only the sides (or interior) of another figure.

see also CIRCUMSCRIBED, INCENTER, INCIRCLE, INRADIUS

Inscribed Angle

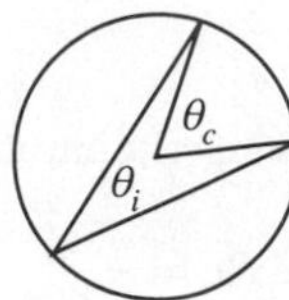

The ANGLE with VERTEX on a CIRCLE's CIRCUMFERENCE formed by two points on a CIRCLE's CIRCUMFERENCE. For ANGLES with the same endpoints,

$$\theta_c = 2\theta_i,$$

where θ_c is the CENTRAL ANGLE.

see also CENTRAL ANGLE

References

Pedoe, D. *Circles: A Mathematical View, rev. ed.* Washington, DC: Math. Assoc. Amer., pp. xxi–xxii, 1995.

Inside-Outside Theorem
Let $P(z)$ and $Q(z)$ be POLYNOMIALS with COMPLEX arguments and $\deg(Q) \geq \deg(P+2)$. Then

$$\int_\gamma \frac{P(z)}{Q(z)}\,dz = \begin{cases} 2\pi i \sum_{\text{inside }\gamma} \operatorname{Res} f(z) & \text{inside } \gamma \\ -2\pi i \sum_{\text{outside }\gamma} \operatorname{Res} f(z) & \text{outside } \gamma, \end{cases}$$

where Res are the RESIDUES.

Insphere
A SPHERE INSCRIBED in a given solid.

see also CIRCUMSPHERE, MIDSPHERE

Instrument Function
The finite FOURIER COSINE TRANSFORM of an APODIZATION FUNCTION, also known as an APPARATUS FUNCTION. The instrument function $I(k)$ corresponding to a given APODIZATION FUNCTION $A(x)$ is then given by

$$I(k) = \int_{-a}^{a} \cos(2\pi kx) A(x)\,dx.$$

see also APODIZATION FUNCTION, FOURIER COSINE TRANSFORM

Insufficient Reason Principle
A principle also called the INDIFFERENCE PRINCIPLE which was first enunciated by Johann Bernoulli. The insufficient reason principle states that, if we are ignorant of the ways an event can occur and therefore have no reason to believe that one way will occur preferentially to another, it will occur equally likely in any way.

Int
see INTEGER PART

Integer
One of the numbers ..., -2, -1, 0, 1, 2, ... The SET of INTEGERS forms a RING which is denoted $\mathbb{Z}$. A given INTEGER n may be NEGATIVE ($n \in \mathbb{Z}^-$), NONNEGATIVE ($n \in \mathbb{Z}^*$), ZERO ($n = 0$), or POSITIVE ($n \in \mathbb{Z}^+ = \mathbb{N}$). The RING $\mathbb{Z}$ has CARDINALITY of $\aleph_0$. The GENERATING FUNCTION for the POSITIVE INTEGERS is

$$f(x) = \frac{1}{(1-x)^2} = x + 2x^2 + 3x^3 + 4x^4 + \ldots.$$

There are several symbols used to perform operations having to do with conversion between REAL NUMBERS and integers. The symbol $\lfloor x \rfloor$ ("FLOOR x") means "the largest integer not greater than x," i.e., `int(x)` in computer parlance. The symbol $[x]$ means "the nearest integer to x" (NINT), i.e., `nint(x)` in computer parlance. The symbol $\lceil x \rceil$ ("CEILING x") means "the smallest integer not smaller x," or `-int(-x)`, where `int(x)` is the INTEGER PART of x.

see also ALGEBRAIC INTEGER, ALMOST INTEGER, COMPLEX NUMBER, COUNTING NUMBER, CYCLOTOMIC INTEGER, EISENSTEIN INTEGER, GAUSSIAN INTEGER, $\mathbb{N}$, NATURAL NUMBER, NEGATIVE, POSITIVE, RADICAL INTEGER, REAL NUMBER, WHOLE NUMBER, $\mathbb{Z}$, $\mathbb{Z}^-$, $\mathbb{Z}^+$, $\mathbb{Z}^*$, ZERO

Integer Division
DIVISION in which the fractional part (remainder) is discarded is called integer division and is sometimes denoted $\backslash$. Integer division can be defined as $a\backslash b \equiv \lfloor a/b \rfloor$, where "/" denotes normal division and $\lfloor x \rfloor$ is the FLOOR FUNCTION. For example,

$$10/3 = 3 + 1/3$$
$$10\backslash 3 = 3.$$

Integer Factorization
see PRIME FACTORIZATION

Integer-Matrix Form
Let $Q(x) \equiv Q(\mathbf{x}) = Q(x_1, x_2, \ldots, x_n)$ be an integer-valued n-ary QUADRATIC FORM, i.e., a POLYNOMIAL with integer COEFFICIENTS which satisfies $Q(x) > 0$ for REAL $x \neq 0$. Then $Q(x)$ can be represented by

$$Q(x) = \mathbf{x}^{\mathrm{T}} \mathsf{A} \mathbf{x},$$

where

$$\mathsf{A} = \frac{1}{2} \frac{\partial^2 Q(x)}{\partial x_i \partial x_j}$$

is a POSITIVE symmetric matrix (Duke 1997). If A has POSITIVE entries, then $Q(x)$ is called an integer matrix form. Conway *et al.* (1997) have proven that, if a POSITIVE integer matrix quadratic form represents each of 1, 2, 3, 5, 6, 7, 10, 14, and 15, then it represents all POSITIVE INTEGERS.

see also FIFTEEN THEOREM

References
Conway, J. H.; Guy, R. K.; Schneeberger, W. A.; and Sloane, N. J. A. "The Primary Pretenders." *Acta Arith.* **78**, 307–313, 1997.
Duke, W. "Some Old Problems and New Results about Quadratic Forms." *Not. Amer. Math. Soc.* **44**, 190–196, 1997.

Integer Module

see ABELIAN GROUP

Integer Part

The function $\text{int}(x)$ gives the INTEGER PART of x. In many computer languages, the function is denoted `int(x)`, but in mathematics, it is usually called the FLOOR FUNCTION and denoted $\lfloor x \rfloor$.

see also CEILING FUNCTION, FLOOR FUNCTION, NINT

Integer Relation

A set of REAL NUMBERS $x_1, \ldots, x_n$ is said to possess an integer relation if there exist integers a_i such that

$$a_1x_1 + a_2x_2 + \ldots + a_nx_n = 0,$$

with not all $a_i = 0$. An interesting example of such a relation is the 17-VECTOR $(1, x, x^2, \ldots, x^{16})$ with $x = 3^{1/4} - 2^{1/4}$, which has an integer relation (1, 0, 0, 0, −3860, 0, 0, 0, −666, 0, 0, 0, −20, 0, 0, 0, 1), i.e.,

$$1 - 3860x^4 - 666x^8 - 20x^{12} + x^{16} = 0.$$

This is a special case of finding the polynomial of degree $n = rs$ satisfied by $x = 3^{1/r} - 2^{1/s}$.

Algorithms for finding integer relations include the FERGUSON-FORCADE ALGORITHM, HJLS ALGORITHM, LLL ALGORITHM, PSLQ ALGORITHM, PSOS ALGORITHM, and the algorithm of Lagarias and Odlyzko (1985). Perhaps the simplest (and unfortunately most inefficient) such algorithm is the GREEDY ALGORITHM. Plouffe's "Inverse Symbolic Calculator" site includes a huge 54 million database of REAL NUMBERS which are algebraically related to fundamental mathematical constants.

see also CONSTANT PROBLEM, FERGUSON-FORCADE ALGORITHM, GREEDY ALGORITHM, HERMITE-LINDEMANN THEOREM, HJLS ALGORITHM, LATTICE REDUCTION, LLL ALGORITHM, PSLQ ALGORITHM, PSOS ALGORITHM, REAL NUMBER, LINDEMANN-WEIERSTRAß THEOREM

References
Bailey, D. and Plouffe, S. "Recognizing Numerical Constants." `http://www.cecm.sfu.ca/organics/papers/bailey`.
Lagarias, J. C. and Odlyzko, A. M. "Solving Low-Density Subset Sum Problems." *J. ACM* **32**, 229–246, 1985.
Plouffe, S. "Inverse Symbolic Calculator." `http://www.cecm.sfu.ca/projects/ISC/`.

Integer Sequence

A SEQUENCE whose terms are INTEGERS. The most complete printed references for such sequences are Sloane (1973) and its update, Sloane and Plouffe (1995). Sloane also maintains the sequences from both works together with many additional sequences in an on-line listing. In this listing, sequences are identified by a unique 6-DIGIT A-number. Sequences appearing in Sloane and Plouffe (1995) are ordered lexicographically and identified with a 4-DIGIT M-number, and those appearing in Sloane (1973) are identified with a 4-DIGIT N-number.

Sloane's huge (and enjoyable) database is accessible by either e-mail or web browser. To look up sequences by e-mail, send a message to either `sequences@research.att.com` or `superseeker@research.att.com` containing lines of the form `lookup 5 14 42 132 ....` To use the browser version, point to `http://www.research.att.com/~njas/sequences/eisonline.html`.

see also ARONSON'S SEQUENCE, COMBINATORICS, CONSECUTIVE NUMBER SEQUENCES, CONWAY SEQUENCE, EBAN NUMBER, HOFSTADTER-CONWAY $10,000 SEQUENCE, HOFSTADTER'S Q-SEQUENCE, LEVINE-O'SULLIVAN SEQUENCE, LOOK AND SAY SEQUENCE, MALLOW'S SEQUENCE, MIAN-CHOWLA SEQUENCE, MORSE-THUE SEQUENCE, NEWMAN-CONWAY SEQUENCE, NUMBER, PADOVAN SEQUENCE, PERRIN SEQUENCE, RATS SEQUENCE, SEQUENCE, SMARANDACHE SEQUENCES

References
Aho, A. V. and Sloane, N. J. A. "Some Doubly Exponential Sequences." *Fib. Quart.* **11**, 429–437, 1973.
Bernstein, M. and Sloane, N. J. A. "Some Canonical Sequences of Integers." *Linear Algebra and Its Applications* **226–228**, 57–72, 1995.
Erdős, P.; Sárközy, E.; and Szemerédi, E. "On Divisibility Properties of Sequences of Integers." In *Number Theory, Colloq. Math. Soc. János Bolyai, Vol. 2.* Amsterdam, Netherlands: North-Holland, pp. 35–49, 1970.
Guy, R. K. "Sequences of Integers." Ch. E in *Unsolved Problems in Number Theory, 2nd ed.* New York: Springer-Verlag, pp. 199–239, 1994.
Krattenthaler, C. "RATE: A Mathematica Guessing Machine." `http://radon.mat.univie.ac.at/People/kratt/rate/rate.html`.
Ostman, H. *Additive Zahlentheorie I, II.* Heidelberg, Germany: Springer-Verlag, 1956.
Pomerance, C. and Sárközy, A. "Combinatorial Number Theory." In *Handbook of Combinatorics* (Ed. R. Graham, M. Grötschel, and L. Lovász). Amsterdam, Netherlands: North-Holland, 1994.
Ruskey, F. "The (Combinatorial) Object Server." `http://sue.csc.uvic.ca/~cos`.
Sloane, N. J. A. *A Handbook of Integer Sequences.* Boston, MA: Academic Press, 1973.

Sloane, N. J. A. "Find the Next Term." *J. Recr. Math.* **7**, 146, 1974.
Sloane, N. J. A. "An On-Line Version of the Encyclopedia of Integer Sequences." *Elec. J. Combin.* **1**, F1 1–5, 1994. http://www.combinatorics.org/Volume_1/volume1.html#F1.
Sloane, N. J. A. "Some Important Integer Sequences." In *CRC Standard Mathematical Tables and Formulae* (Ed. D. Zwillinger). Boca Raton, FL: CRC Press, 1995.
Sloane, N. J. A. "An On-Line Version of the Encyclopedia of Integer Sequences." http://www.research.att.com/~njas/sequences/eisonline.html.
Sloane, N. J. A. and Plouffe, S. *The Encyclopedia of Integer Sequences.* San Diego, CA: Academic Press, 1995.
Stöhr, A. "Gelöste und ungelöste Fragen über Basen der natürlichen Zahlenreihe I, II." *J. reine angew. Math.* **194**, 40–65 and 111–140, 1955.
Turán, P. (Ed.). *Number Theory and Analysis: A Collection of Papers in Honor of Edmund Landau (1877–1938).* New York: Plenum Press, 1969.
Weisstein, E. W. "Integer Sequences." http://www.astro.virginia.edu/~eww6n/math/notebooks/IntegerSequences.m.

Integrable

A function for which the INTEGRAL can be computed is said to be integrable.

see also DIFFERENTIABLE, INTEGRAL, INTEGRATION

Integral

An integral is a mathematical object which can be interpreted as an AREA or a generalization of AREA. Integrals, together with DERIVATIVES, are the fundamental objects of CALCULUS. Other words for integral include ANTIDERIVATIVE and PRIMITIVE. The RIEMANN INTEGRAL is the simplest integral definition and the only one usually encountered in elementary CALCULUS. The RIEMANN INTEGRAL of the function $f(x)$ over x from a to b is written

$$\int_a^b f(x)\,dx. \tag{1}$$

Every definition of an integral is based on a particular MEASURE. For instance, the RIEMANN INTEGRAL is based on JORDAN MEASURE, and the LEBESGUE INTEGRAL is based on LEBESGUE MEASURE. The process of computing an integral is called INTEGRATION (a more archaic term for INTEGRATION is QUADRATURE), and the approximate computation of an integral is termed NUMERICAL INTEGRATION.

There are two classes of (Riemann) integrals: DEFINITE INTEGRALS

$$\int_a^b f(x)\,dx, \tag{2}$$

which have upper and lower limits, and INDEFINITE INTEGRALS, which are written without limits. The first FUNDAMENTAL THEOREM OF CALCULUS allows DEFINITE INTEGRALS to be computed in terms of INDEFINITE INTEGRALS, since if F is the INDEFINITE INTEGRAL for $f(x)$, then

$$\int_a^b f(x)\,dx = F(b) - F(a). \tag{3}$$

Wolfram Research (http://www.integrals.com) maintains a web site which will integrate many common (and not so common) functions. However, it cannot solve some simple integrals such as

$$\int \left[\frac{d}{dx}(x\sqrt{\sin x})\right] dx = \int \left(\frac{x\cos x}{2\sqrt{\sin x}} + \sqrt{\sin x}\right) dx \tag{4}$$

$$\int \left[\frac{d}{dx}L_2(x\ln x)\right] dx = -\int \left[\frac{(\ln x + 1)\ln(1 - x\ln x)}{x\ln x}\right] dx, \tag{5}$$

where L_2 is the DILOGARITHM. Furthermore, it gives an incorrect answer of $\pi^{1-2\sqrt{3}}/(\sqrt{3}\cdot 4^{\sqrt{3}})$ to

$$I(\sqrt{3}) = \int_0^{\pi/2} \frac{dx}{1 + (\tan x)^{\sqrt{3}}} = \tfrac{1}{2}\pi. \tag{6}$$

This integral and, in fact, the generalized integral for arbitrary a

$$I(a) = \int_0^{\pi/2} \frac{dx}{1 + (\tan x)^a}, \tag{7}$$

have a "trick" solution which takes advantage of the trigonometric identity

$$\tan(\tfrac{1}{2}\pi - x) = \cot x. \tag{8}$$

Letting $z \equiv (\tan x)^a$,

$$\begin{aligned} I(a) &= \int_0^{\pi/4} \frac{dx}{1+z} + \int_{\pi/4}^{\pi/2} \frac{dx}{1+1} \\ &= \int_0^{\pi/4} \frac{dx}{1+z} + \int_0^{\pi/4} \frac{dx}{1+\frac{1}{z}} \\ &= \int_0^{\pi/4} \left(\frac{1}{1+z} + \frac{1}{1+\frac{1}{z}}\right) = \int_0^{\pi/4} dx \\ &= \tfrac{1}{4}\pi. \end{aligned} \tag{9}$$

Here is a list of common INDEFINITE INTEGRALS:

$$\int x^r\,dx = \frac{x^{r+1}}{r+1} + C \tag{10}$$

$$\int \frac{dx}{x} = \ln|x| + C \tag{11}$$

$$\int a^x\,dx = \frac{a^x}{\ln a} + C \tag{12}$$

$$\int \sin x\,dx = -\cos x + C \tag{13}$$

$$\int \cos x\,dx = \sin x + C \tag{14}$$

$$\int \tan x\,dx = \ln|\sec x| + C \tag{15}$$

$$\int \csc x\,dx = \ln|\csc x - \cot x| + C \tag{16}$$

$$= \ln\left[\tan(\tfrac{1}{2}x)\right] + C \tag{17}$$

$$= \frac{1}{2}\ln\left(\frac{1-\cos x}{1+\cos x}\right) + C \tag{18}$$

$$\int \sec x\,dx = \ln|\sec x + \tan x| + C$$

$$= \mathrm{gd}^{-1}(x) + C \tag{19}$$

$$\int \cot x\,dx = \ln|\sin x| + C \tag{20}$$

$$\int \sec^2 x\,dx = \tan x + C \tag{21}$$

$$\int \csc^2 x\,dx = -\cot x + C \tag{22}$$

$$\int \sec x \tan x\,dx = \sec x + C \tag{23}$$

$$\int \cos^{-1} x\,dx = x\cos^{-1} x - \sqrt{1-x^2} + C \tag{24}$$

$$\int \sin^{-1} x\,dx = x\sin^{-1} x + \sqrt{1-x^2} + C \tag{25}$$

$$\int \tan^{-1} x\,dx = x\tan^{-1} x - \tfrac{1}{2}\ln(1+x^2) + C \tag{26}$$

$$\int \frac{dx}{\sqrt{a^2-x^2}} = \sin^{-1}\left(\frac{x}{a}\right) + C \tag{27}$$

$$\int \frac{dx}{\sqrt{a^2-x^2}} = \cos^{-1}\left(\frac{x}{a}\right) + C \tag{28}$$

$$\int \frac{dx}{a^2+x^2} = \frac{1}{a}\tan^{-1}\left(\frac{x}{a}\right) + C \tag{29}$$

$$\int \frac{dx}{a^2+x^2} = -\frac{1}{a}\cot^{-1}\left(\frac{x}{a}\right) + C \tag{30}$$

$$\int \frac{dx}{x\sqrt{x^2-a^2}} = \frac{1}{a}\sec^{-1}\left(\frac{x}{a}\right) + C \tag{31}$$

$$\int \frac{dx}{x\sqrt{x^2-a^2}} = -\frac{1}{a}\csc^{-1}\left(\frac{x}{a}\right) + C \tag{32}$$

$$\int \sin^2(ax)\,dx = \frac{x}{2} - \frac{1}{4a}\sin(2ax) + C \tag{33}$$

$$\int \mathrm{sn}\,u\,du = k^{-1}\ln(\mathrm{dn}\,u - k\,\mathrm{cn}\,u) + C \tag{34}$$

$$\int \mathrm{sn}^2\,u\,du = \frac{u-E(u)}{k^2} + C \tag{35}$$

$$\int \mathrm{cn}\,u\,du = k^{-1}\sin^{-1}(k\,\mathrm{sn}\,u) + C \tag{36}$$

$$\int \mathrm{dn}\,u\,du = \sin^{-1}(\mathrm{sn}\,u) + C, \tag{37}$$

where $\sin x$ is the SINE; $\cos x$ is the COSINE; $\tan x$ is the TANGENT; $\csc x$ is the COSECANT; $\sec x$ is the SECANT; $\cot x$ is the COTANGENT; $\cos^{-1} x$ is the INVERSE COSINE; $\sin^{-1} x$ is the INVERSE SINE; $\tan^{-1}$ is the INVERSE TANGENT; $\mathrm{sn}\,u$, $\mathrm{cn}\,u$, and $\mathrm{dn}\,u$ are JACOBI ELLIPTIC FUNCTIONS; $E(u)$ is a complete ELLIPTIC INTEGRAL OF THE SECOND KIND; and $\mathrm{gd}(x)$ is the GUDERMANNIAN FUNCTION.

To derive (15), let $u \equiv \cos x$, so $du = -\sin x\,dx$ and

$$\int \tan x = \int \frac{\sin x}{\cos x}\,dx = -\int \frac{du}{u}$$
$$= -\ln|u| + C = -\ln|\cos x| + C$$
$$= \ln|\cos x|^{-1} + C = \ln|\sec x| + C. \tag{38}$$

To derive (18), let $u \equiv \csc x - \cot x$, so $du = (-\csc x \cot x + \csc^2 x)\,dx$ and

$$\int \csc x\,dx = \int \csc x \frac{\csc x - \cot x}{\csc x - \cot x}\,dx$$
$$= \int \frac{\csc^2 x + \cot x \csc x}{\csc x + \cot x}\,dx$$
$$= \int \frac{du}{u} = \ln|u| + C$$
$$= \ln|\csc x - \cot x| + C. \tag{39}$$

To derive (19), let

$$u \equiv \sec x + \tan x, \tag{40}$$

so

$$du = (\sec x \tan x + \sec^2 x)\,dx \tag{41}$$

and

$$\int \sec x\,dx = \int \sec x \frac{\sec x + \tan x}{\sec x + \tan x}\,dx$$
$$= \int \frac{\sec^2 x + \sec x + \tan x}{\sec x + \tan x}\,dx$$
$$= \int \frac{du}{u} = \ln|u| + C$$
$$= \ln|\sec x + \tan x| + C. \tag{42}$$

To derive (20), let $u \equiv \sin x$, so $du = \cos x\, dx$ and

$$\int \cot x\, dx = \int \frac{\cos x}{\sin x}\, dx = \int \frac{du}{u} = \ln|u| + C = \ln|\sin x| + C. \quad (43)$$

Differentiating integrals leads to some useful and powerful identities, for instance

$$\frac{d}{dx}\int_a^x f(x)\, dx = f(x), \quad (44)$$

which is the first FUNDAMENTAL THEOREM OF CALCULUS.

$$\frac{d}{dx}\int_x^b f(x)\, dx = -f(x) \quad (45)$$

$$\frac{d}{dx}\int_a^b f(x,t)\, dt = \int_a^b \frac{\partial}{\partial x} f(x,t)\, dt \quad (46)$$

$$\frac{d}{dx}\int_a^x f(x,t)\, dt = f(x,t) + \int_a^x \frac{\partial}{\partial x} f(x,t)\, dt. \quad (47)$$

If $f(x,t)$ is singular or INFINITE, then

$$\frac{d}{dx}\int_a^x f(x,t)\, dx = \frac{1}{x-a}\int_a^x \left[(x-a)\frac{\partial f}{\partial x} + (t-a)\frac{\partial f}{\partial t} + f\right] dt. \quad (48)$$

The LEIBNIZ IDENTITY is

$$\frac{d}{dx}\int_{u(x)}^{v(x)} f(x,t)\, dt = v'(x) f(x, v(x)) - u' f(x, u(x)) + \int_{u(x)}^{v(x)} \frac{\partial}{\partial x} f(x,t)\, dt. \quad (49)$$

Other integral identities include

$$\int_a^x \int_a^x f(t)\, dt\, dx = \int_a^x (x-t) f(t)\, dt \quad (50)$$

$$\int_0^x dt_n \int_0^{t_n} dt_{n-1} \cdots \int_0^{t_3} dt_2 \int_0^{t_2} f(t_1)\, dt_1 = \frac{1}{(n-1)!}\int_0^x (x-t)^{n-1} f(t)\, dt \quad (51)$$

$$\frac{\partial}{\partial x_k}(x_j J_k) = \delta_{jk} J_k + x_j \frac{\partial}{\partial x_k} J_k = \mathbf{J} + \mathbf{r}\nabla\cdot\mathbf{J} \quad (52)$$

$$\int_V \mathbf{J}\, d^3\mathbf{r} = \int_V \frac{\partial}{\partial x_k}(x_i J_k) - \int_V \mathbf{r}\nabla\cdot\mathbf{J}\, d^3\mathbf{r} = -\int_V \mathbf{r}\nabla\cdot\mathbf{J}\, d^3\mathbf{r}. \quad (53)$$

Integrals of the form

$$\int_a^b f(x)\, dx \quad (54)$$

with one INFINITE LIMIT and the other NONZERO may be expressed as finite integrals over transformed functions. If $f(x)$ decreases at least as fast as $1/x^2$, then let

$$t \equiv \frac{1}{x} \quad (55)$$

$$dt = -\frac{dx}{x^2} \quad (56)$$

$$dx = -x^2\, dt = -\frac{dt}{t^2}, \quad (57)$$

and

$$\int_a^b f(x)\, dx = -\int_{1/a}^{1/b} \frac{1}{t^2} f\left(\frac{1}{t}\right) dt = \int_{1/b}^{1/a} \frac{1}{t^2} f\left(\frac{1}{t}\right) dt. \quad (58)$$

If $f(x)$ diverges as $(x-a)^\gamma$ for $\gamma \in [0, 1]$, let

$$x \equiv t^{1/(1-\gamma)} + a \quad (59)$$

$$dx = \frac{1}{1-\gamma} t^{(1/1-\gamma)-1}\, dt = \frac{1}{1-\gamma} t^{[1-(1-\gamma)]/(1-\gamma)}\, dt = \frac{1}{\gamma-1} t^{\gamma/(1-\gamma)}\, dt \quad (60)$$

$$t = (x-a)^{1-\gamma}, \quad (61)$$

and

$$\int_a^b f(x)\, dx = \frac{1}{1-\gamma} = \int_0^{(b-a)^{1-\gamma}} t^{\gamma/(1-\gamma)} f(t^{1/(1-\gamma)} + a)\, dt. \quad (62)$$

If $f(x)$ diverges as $(x+b)^\gamma$ for $\gamma \in [0, 1]$, let

$$x \equiv b - t^{1/(1-\gamma)} \quad (63)$$

$$dx = -\frac{1}{\gamma-1} t^{\gamma/(1-\gamma)}\, dt \quad (64)$$

$$t = (b-x)^{1-\gamma}, \quad (65)$$

and

$$\int_a^b f(x)\, dx = \frac{1}{1-\gamma} = \int_0^{(b-a)^{1-\gamma}} t^{\gamma/(1-\gamma)} f(b - t^{1/(1-\gamma)})\, dt. \quad (66)$$

If the integral diverges exponentially, then let

$$t \equiv e^{-x} \quad (67)$$

$$dt = -e^{-x}\, dx \quad (68)$$

$$x = -\ln t, \quad (69)$$

and

$$\int_a^\infty f(x)\,dx = \int_0^{e^{-a}} f(-\ln t)\frac{dt}{t}. \qquad (70)$$

Integrals with rational exponents can often be solved by making the substitution $u = x^{1/n}$, where n is the LEAST COMMON MULTIPLE of the DENOMINATOR of the exponents.

Integration rules include

$$\int_a^a f(x)\,dx = 0 \qquad (71)$$

$$\int_a^b f(x)\,dx = -\int_b^a f(x)\,dx. \qquad (72)$$

For $c \in (a,b)$,

$$\int_a^b f(x)\,dx = \int_a^c f(x)\,dx + \int_c^b f(x)\,dx. \qquad (73)$$

If g' is continuous on $[a,b]$ and f is continuous and has an antiderivative on an INTERVAL containing the values of $g(x)$ for $a \leq x \leq b$, then

$$\int_a^b f(g(x))g'(x)\,dx = \int_{g(a)}^{g(b)} f(u)\,du. \qquad (74)$$

Liouville showed that the integrals

$$\int e^{-x^2}\,dx \quad \int \frac{e^x}{x}\,dx \quad \int \frac{\sin x}{x}\,dx \quad \int \frac{dx}{\ln x} \qquad (75)$$

cannot be expressed as terms of a finite number of elementary functions. Other irreducibles include

$$\int x^x\,dx \quad \int x^{-x}\,dx \quad \int \sqrt{\sin x}\,dx. \qquad (76)$$

Chebyshev proved that if U, V, and W are RATIONAL NUMBERS, then

$$\int x^U (A + Bx^V)^W\,dx \qquad (77)$$

is integrable in terms of elementary functions IFF $(U+1)/V$, W, or $W + (U+1)/V$ is an INTEGER (Ritt 1948, Shanks 1993).

There are a wide range of methods available for NUMERICAL INTEGRATION. A good source for such techniques is Press *et al.* (1992). The most straightforward numerical integration technique uses the NEWTON-COTES FORMULAS (also called QUADRATURE FORMULAS), which approximate a function tabulated at a sequence of regularly spaced INTERVALS by various degree POLYNOMIALS. If the endpoints are tabulated, then the 2- and 3-point formulas are called the TRAPEZOIDAL RULE and SIMPSON'S RULE, respectively. The 5-point formula is called BODE'S RULE. A generalization of the TRAPEZOIDAL RULE is ROMBERG INTEGRATION, which can yield accurate results for many fewer function evaluations.

If the analytic form of a function is known (instead of its values merely being tabulated at a fixed number of points), the best numerical method of integration is called GAUSSIAN QUADRATURE. By picking the optimal ABSCISSAS at which to compute the function, Gaussian quadrature produces the most accurate approximations possible. However, given the speed of modern computers, the additional complication of the GAUSSIAN QUADRATURE formalism often makes it less desirable than the brute-force method of simply repeatedly calculating twice as many points on a regular grid until convergence is obtained. An excellent reference for GAUSSIAN QUADRATURE is Hildebrand (1956).

see also A-INTEGRABLE, ABELIAN INTEGRAL, CALCULUS, CHEBYSHEV-GAUSS QUADRATURE, CHEBYSHEV QUADRATURE, DARBOUX INTEGRAL, DEFINITE INTEGRAL, DENJOY INTEGRAL, DERIVATIVE, DOUBLE EXPONENTIAL INTEGRATION, EULER INTEGRAL, FUNDAMENTAL THEOREM OF GAUSSIAN QUADRATURE, GAUSS-JACOBI MECHANICAL QUADRATURE, GAUSSIAN QUADRATURE, HAAR INTEGRAL, HERMITE-GAUSS QUADRATURE, HERMITE QUADRATURE, HK INTEGRAL, INDEFINITE INTEGRAL, INTEGRATION, JACOBI-GAUSS QUADRATURE, JACOBI QUADRATURE, LAGUERRE-GAUSS QUADRATURE, LAGUERRE QUADRATURE, LEBESGUE INTEGRAL, LEBESGUE-STIELTJES INTEGRAL, LEGENDRE-GAUSS QUADRATURE, LEGENDRE QUADRATURE, LOBATTO QUADRATURE, MECHANICAL QUADRATURE, MEHLER QUADRATURE, NEWTON-COTES FORMULAS, NUMERICAL INTEGRATION, PERON INTEGRAL, QUADRATURE, RADAU QUADRATURE, RECURSIVE MONOTONE STABLE QUADRATURE, RIEMANN-STIELTJES INTEGRAL, ROMBERG INTEGRATION, RIEMANN INTEGRAL, STIELTJES INTEGRAL

References

Beyer, W. H. "Integrals." *CRC Standard Mathematical Tables, 28th ed.* Boca Raton, FL: CRC Press, pp. 233–296, 1987.

Bronstein, M. *Symbolic Integration I: Transcendental Functions.* New York: Springer-Verlag, 1996.

Gordon, R. A. *The Integrals of Lebesgue, Denjoy, Perron, and Henstock.* Providence, RI: Amer. Math. Soc., 1994.

Gradshteyn, I. S. and Ryzhik, I. M. *Tables of Integrals, Series, and Products, 5th ed.* San Diego, CA: Academic Press, 1993.

Hildebrand, F. B. *Introduction to Numerical Analysis.* New York: McGraw-Hill, pp. 319–323, 1956.

Piessens, R.; de Doncker, E.; Uberhuber, C. W.; and Kahaner, D. K. *QUADPACK: A Subroutine Package for Automatic Integration.* New York: Springer-Verlag, 1983.

Press, W. H.; Flannery, B. P.; Teukolsky, S. A.; and Vetterling, W. T. "Integration of Functions." Ch. 4 in *Numerical Recipes in FORTRAN: The Art of Scientific Computing,*

2nd ed. Cambridge, England: Cambridge University Press, pp. 123–158, 1992.
Ritt, J. F. *Integration in Finite Terms.* New York: Columbia University Press, p. 37, 1948.
Shanks, D. *Solved and Unsolved Problems in Number Theory, 4th ed.* New York: Chelsea, p. 145, 1993.
Wolfram Research. "The Integrator." `http://www.integrals.com`

Integral Brick

see EULER BRICK

Integral Calculus

That portion of "the" CALCULUS dealing with INTEGRALS.

see also CALCULUS, DIFFERENTIAL CALCULUS, INTEGRAL

Integral Cuboid

see EULER BRICK

Integral Current

A RECTIFIABLE CURRENT whose boundary is also a RECTIFIABLE CURRENT.

Integral Curvature

Given a GEODESIC TRIANGLE (a triangle formed by the arcs of three GEODESICS on a smooth surface),

$$\int_{ABC} K\,da = A + B + C - \pi.$$

Given the EULER CHARACTERISTIC χ,

$$\iint K\,da = 2\pi\chi,$$

so the integral curvature of a closed surface is not altered by a topological transformation.

see also GAUSS-BONNET FORMULA, GEODESIC TRIANGLE

Integral Domain

A RING that is COMMUTATIVE under multiplication, has a unit element, and has no divisors of 0. The INTEGERS form an integral domain.

see also FIELD, RING

Integral Drawing

A GRAPH drawn such that the EDGES have only INTEGER lengths. It is conjectured that every PLANAR GRAPH has an integral drawing.

References

Harborth, H. and Möller, M. "Minimum Integral Drawings of the Platonic Graphs." *Math. Mag.* **67**, 355–358, 1994.

Integral Equation

If the limits are fixed, an integral equation is called a Fredholm integral equation. If one limit is variable, it is called a Volterra integral equation. If the unknown function is only under the integral sign, the equation is said to be of the "first kind." If the function is both inside and outside, the equation is called of the "second kind." A Fredholm equation of the first kind is of the form

$$f(x) = \int_a^b K(x,t)\phi(t)\,dt. \tag{1}$$

A Fredholm equation of the second kind is of the form

$$\phi(x) = f(x) + \lambda \int_a^b K(x,t)\phi(t)\,dt. \tag{2}$$

A Volterra equation of the first kind is of the form

$$f(x) = \int_a^x K(x,t)\phi(t)\,dt. \tag{3}$$

A Volterra equation of the second kind is of the form

$$\phi(x) = f(x) + \int_a^x K(x,t)\phi(t)\,dt, \tag{4}$$

where the functions $K(x,t)$ are known as KERNELS. Integral equations may be solved directly if they are SEPARABLE. Otherwise, a NEUMANN SERIES must be used.

A KERNEL is separable if

$$K(x,t) = \lambda \sum_{j=1}^n M_j(x) N_j(t). \tag{5}$$

This condition is satisfied by all POLYNOMIALS and many TRANSCENDENTAL FUNCTIONS. A FREDHOLM INTEGRAL EQUATION OF THE SECOND KIND with separable KERNEL may be solved as follows:

$$\begin{aligned}\phi(x) &= f(x) + \int_a^b K(x,t)\phi(t)\,dt\\ &= f(x) + \lambda \sum_{j=1}^n M_j(x) \int_a^b N_j(t)\phi(t)\,dt\\ &= f(x) + \lambda \sum_{j=1}^n c_j M_j(x),\end{aligned} \tag{6}$$

where

$$c_j \equiv \int_a^b N_j(t)\phi(t)\,dt. \tag{7}$$

Now multiply both sides of (7) by $N_i(x)$ and integrate over dx.

$$\int_a^b \phi(x) N_i(x)\,dx = \int_a^b f(x) N_i(x)\,dx + \lambda \sum_{j=1}^n c_j \int_a^b M_j(x) N_i(x)\,dx. \tag{8}$$

By (7), the first term is just c_i. Now define

$$b_i \equiv \int_a^b N_i(x) f(x)\, dx \tag{9}$$

$$a_{ij} = \int_a^b N_i(x) M_j(x)\, dx, \tag{10}$$

so (8) becomes

$$c_i = b_i + \lambda \sum_{j=1}^{n} a_{ij} c_j. \tag{11}$$

Writing this in matrix form,

$$\mathbf{C} = \mathbf{B} + \lambda \mathsf{A} \mathbf{C}, \tag{12}$$

so

$$(\mathsf{I} - \lambda \mathsf{A})\mathbf{C} = \mathbf{B} \tag{13}$$

$$\mathbf{C} = (\mathsf{I} - \lambda \mathsf{A})^{-1} \mathbf{B}. \tag{14}$$

see also FREDHOLM INTEGRAL EQUATION OF THE FIRST KIND, FREDHOLM INTEGRAL EQUATION OF THE SECOND KIND, VOLTERRA INTEGRAL EQUATION OF THE FIRST KIND, VOLTERRA INTEGRAL EQUATION OF THE SECOND KIND

References

Corduneanu, C. *Integral Equations and Applications.* Cambridge, England: Cambridge University Press, 1991.

Davis, H. T. *Introduction to Nonlinear Differential and Integral Equations.* New York: Dover, 1962.

Kondo, J. *Integral Equations.* Oxford, England: Clarendon Press, 1992.

Lovitt, W. V. *Linear Integral Equations.* New York: Dover, 1950.

Mikhlin, S. G. *Integral Equations and Their Applications to Certain Problems in Mechanics, Mathematical Physics and Technology, 2nd rev. ed.* New York: Macmillan, 1964.

Mikhlin, S. G. *Linear Integral Equations.* New York: Gordon & Breach, 1961.

Pipkin, A. C. *A Course on Integral Equations.* New York: Springer-Verlag, 1991.

Porter, D. and Stirling, D. S. G. *Integral Equations: A Practical Treatment, from Spectral Theory to Applications.* Cambridge, England: Cambridge University Press, 1990.

Press, W. H.; Flannery, B. P.; Teukolsky, S. A.; and Vetterling, W. T. "Integral Equations and Inverse Theory." Ch. 18 in *Numerical Recipes in FORTRAN: The Art of Scientific Computing, 2nd ed.* Cambridge, England: Cambridge University Press, pp. 779–817, 1992.

Tricomi, F. G. *Integral Equations.* New York: Dover, 1957.

Integral of Motion

A function of the coordinates which is constant along a trajectory in PHASE SPACE. The number of DEGREES OF FREEDOM of a DYNAMICAL SYSTEM such as the DUFFING DIFFERENTIAL EQUATION can be decreased by one if an integral of motion can be found. In general, it is very difficult to discover integrals of motion.

Integral Sign

The symbol $\int$ used to denote an INTEGRAL $\int f(x)\,dx$. The symbol was chosen to be a stylized script "S" to stand for "summation."

see also INTEGRAL

Integral Test

Let $\sum u_k$ be a series with POSITIVE terms and let $f(x)$ be the function that results when k is replaced by x in the FORMULA for u_k. If f is decreasing and continuous for $x \geq 1$ and

$$\lim_{x\to\infty} f(x) = 0,$$

then

$$\sum_{k=1}^{\infty} u_k$$

and

$$\int_t^{\infty} f(x)\, dx$$

both converge or diverge, where $1 \leq t < \infty$. The test is also called the CAUCHY INTEGRAL TEST or MACLAURIN INTEGRAL TEST.

see also CONVERGENCE TESTS

References

Arfken, G. *Mathematical Methods for Physicists, 3rd ed.* Orlando, FL: Academic Press, pp. 283–284, 1985.

Integral Transform

A general integral transform is defined by

$$g(\alpha) = \int_a^b f(t) K(\alpha, t)\, dt,$$

where $K(\alpha, t)$ is called the KERNEL of the transform.

see also FOURIER TRANSFORM, FOURIER-STIELTJES TRANSFORM, H-TRANSFORM, HADAMARD TRANSFORM, HANKEL TRANSFORM, HARTLEY TRANSFORM, HOUGH TRANSFORM, OPERATIONAL MATHEMATICS, RADON TRANSFORM, WAVELET TRANSFORM, Z-TRANSFORM

References

Arfken, G. "Integral Transforms." Ch. 16 in *Mathematical Methods for Physicists, 3rd ed.* Orlando, FL: Academic Press, pp. 794–864, 1985.

Carslaw, H. S. and Jaeger, J. C. *Operational Methods in Applied Mathematics.*

Davies, B. *Integral Transforms and Their Applications, 2nd ed.* New York: Springer-Verlag, 1985.

Poularikas, A. D. (Ed.). *The Transforms and Applications Handbook.* Boca Raton, FL: CRC Press, 1995.

Zayed, A. I. *Handbook of Function and Generalized Function Transformations.* Boca Raton, FL: CRC Press, 1996.

Integrand

The quantity being INTEGRATED, also called the KERNEL. For example, in $\int f(x)\,dx$, $f(x)$ is the integrand.

see also INTEGRAL, INTEGRATION

Integrating Factor

A FUNCTION by which an ORDINARY DIFFERENTIAL EQUATION is multiplied in order to make it integrable.

see also ORDINARY DIFFERENTIAL EQUATION

References

Morse, P. M. and Feshbach, H. *Methods of Theoretical Physics, Part I.* New York: McGraw-Hill, pp. 526–529, 1953.

Integration

The process of computing or obtaining an INTEGRAL. A more archaic term for integration is QUADRATURE.

see also CONTOUR INTEGRATION, INTEGRAL, INTEGRATION BY PARTS, MEASURE THEORY, NUMERICAL INTEGRATION

Integration Lattice

A discrete subset of $\mathbb{R}^s$ which is closed under addition and subtraction and which contains $\mathbb{Z}^s$ as a SUBSET.

see also LATTICE

References

Sloan, I. H. and Joe, S. *Lattice Methods for Multiple Integration.* New York: Oxford University Press, 1994.

Integration by Parts

A first-order (single) integration by parts uses

$$d(uv) = u\,dv + v\,du \tag{1}$$

$$\int d(uv) = uv = \int u\,dv + \int v\,du, \tag{2}$$

so

$$\int u\,dv = uv - \int v\,du \tag{3}$$

and

$$\int_a^b u\,dv = [uv]_a^b - \int_{f(a)}^{f(b)} v\,du. \tag{4}$$

Now apply this procedure n times to $\int f^{(n)}(x)g(x)\,dx$.

$$u = g(x) \qquad dv = f^{(n)}(x)\,dx \tag{5}$$

$$du = g'(x)\,dx \qquad v = f^{(n-1)}(x). \tag{6}$$

Therefore,

$$\int f^{(n)}g(x)\,dx = g(x)f^{(n-1)}(x) - \int f^{(n-1)}(x)g'(x)\,dx. \tag{7}$$

But

$$\int f^{(n-1)}(x)g'(x)\,dx = g'(x)f^{(n-2)}(x) - \int f^{(n-2)}(x)g''(x)\,dx \tag{8}$$

$$\int f^{(n-2)}(x)g''(x)\,dx = g''(x)f^{(n-3)}(x) - \int f^{(n-3)}(x)g^{(3)}(x)\,dx, \tag{9}$$

so

$$\int f^{(n)}(x)g(x)\,dx = g(x)f^{(n-1)}(x) - g'(x)f^{(n-2)}(x) + g''(x)f^{(n-3)}(x) - \ldots + (-1)^n \int f(x)g^{(n)}(x)\,dx. \tag{10}$$

Now consider this in the slightly different form $\int f(x)g(x)\,dx$. Integrate by parts a first time

$$u = f(x) \qquad dv = g(x)\,dx \tag{11}$$

$$du = f'(x)\,dx \qquad v = \int g(x)\,dx, \tag{12}$$

so

$$\int f(x)g(x)\,dx = f(x)\int g(x)\,dx - \int \left[\int g(x)\,dx\right] f'(x)\,dx. \tag{13}$$

Now integrate by parts a second time,

$$u = f'(x) \qquad dv = \int g(x)(dx)^2 \tag{14}$$

$$du = f''(x)\,dx \qquad v = \iint g(x)(dx)^2, \tag{15}$$

so

$$\int f(x)g(x)\,dx = f(x)\int g(x)\,dx - f'(x)\iint g(x)(dx)^2 + \int \left[\iint g(x)(dx)^2\right] f''(x)\,dx. \tag{16}$$

Repeating a third time,

$$\int f(x)g(x)\,dx = f(x)\int g(x)\,dx - f'(x)\iint g(x)(dx)^2 + f''(x)\iiint g(x)(dx)^3 - \int \left[\iiint g(x)(dx)^3\right] f'''(x)\,dx. \tag{17}$$

Therefore, after n applications,

$$\int f(x)g(x)\,dx = f(x)\int g(x)\,dx - f'(x)\iint g(x)(dx)^2 + f''(x)\iiint g(x)(dx)^3 - \ldots + (-1)^{n+1} f^{(n)}(x) \underbrace{\int\cdots\int}_{n+1} g(x)(dx)^{n+1} + (-1)^n \int \left[\underbrace{\int\cdots\int}_{n+1} g(x)(dx)^{n+1}\right] f^{(n+1)}(x)\,dx. \quad (18)$$

If $f^{(n+1)}(x) = 0$ (e.g., for an nth degree POLYNOMIAL), the last term is 0, so the sum terminates after n terms and

$$\int f(x)g(x)\,dx = f(x)\int g(x)\,dx - f'(x)\iint g(x)(dx)^2 + f''(x)\iiint g(x)(dx)^3 - \ldots + (-1)^{n+1} f^{(n)}(x) \underbrace{\int\cdots\int}_{n+1} g(x)(dx)^{n+1}. \quad (19)$$

References

Abramowitz, M. and Stegun, C. A. (Eds.). *Handbook of Mathematical Functions with Formulas, Graphs, and Mathematical Tables, 9th printing.* New York: Dover, p. 12, 1972.

Intension

A definition of a SET by mentioning a defining property.

see also EXTENSION

References

Russell, B. "Definition of Number." *Introduction to Mathematical Philosophy.* New York: Simon and Schuster, 1971.

Interchange Graph

see LINE GRAPH

Interest

Interest is a fee (or payment) made for the borrowing (or lending) of money. The two most common types of interest are SIMPLE INTEREST, for which interest is paid only on the initial PRINCIPAL, and COMPOUND INTEREST, for which interest earned can be re-invested to generate further interest.

see also COMPOUND INTEREST, CONVERSION PERIOD, RULE OF 72, SIMPLE INTEREST

References

Kellison, S. G. *Theory of Interest, 2nd ed.* Burr Ridge, IL: Richard D. Irwin, 1991.

Interior

That portion of a region lying "inside" a specified boundary. For example, the interior of the SPHERE is a BALL.

see also EXTERIOR

Interior Angle Bisector

see ANGLE BISECTOR

Intermediate Value Theorem

If f is continuous on a CLOSED INTERVAL $[a, b]$ and c is any number between $f(a)$ and $f(b)$ inclusive, there is at least one number x in the CLOSED INTERVAL such that $f(x) = c$.

see also WEIERSTRAß INTERMEDIATE VALUE THEOREM

Internal Bisectors Problem

see STEINER-LEHMUS THEOREM

Internal Knot

One of the knots $t_{p+1}, \ldots, t_{m-p-1}$ of a B-SPLINE with control points $\mathbf{P}_0, \ldots, \mathbf{P}_n$ and KNOT VECTOR

$$\mathbf{T} = \{t_0, t_1, \ldots, t_m\},$$

where

$$p \equiv m - n - 1.$$

see also B-SPLINE, KNOT VECTOR

Interpolation

The computation of points or values between ones that are known or tabulated using the surrounding points or values.

see also AITKEN INTERPOLATION, BESSEL'S INTERPOLATION FORMULA, EVERETT INTERPOLATION, EXTRAPOLATION, FINITE DIFFERENCE, GAUSS'S INTERPOLATION FORMULA, HERMITE INTERPOLATION, LAGRANGE INTERPOLATING POLYNOMIAL, NEWTON-COTES FORMULAS, NEWTON'S DIVIDED DIFFERENCE INTERPOLATION FORMULA, OSCULATING INTERPOLATION, THIELE'S INTERPOLATION FORMULA

References

Abramowitz, M. and Stegun, C. A. (Eds.). "Interpolation." §25.2 in *Handbook of Mathematical Functions with Formulas, Graphs, and Mathematical Tables, 9th printing.* New York: Dover, pp. 878–882, 1972.

Iyanaga, S. and Kawada, Y. (Eds.). "Interpolation." Appendix A, Table 21 in *Encyclopedic Dictionary of Mathematics.* Cambridge, MA: MIT Press, pp. 1482–1483, 1980.

Press, W. H.; Flannery, B. P.; Teukolsky, S. A.; and Vetterling, W. T. "Interpolation and Extrapolation." Ch. 3 in *Numerical Recipes in FORTRAN: The Art of Scientific Computing, 2nd ed.* Cambridge, England: Cambridge University Press, pp. 99–122, 1992.

Interquartile Range

Divide a set of data into two groups (high and low) of equal size at the MEDIAN if there is an EVEN number of data points, or two groups consisting of points on either side of the MEDIAN itself plus the MEDIAN if there is an ODD number of data points. Find the MEDIANS of the low and high groups, denoting these first and third quartiles by Q_1 and Q_3. The interquartile range is then defined by

$$\mathrm{IQR} \equiv Q_3 - Q_1.$$

see also H-SPREAD, HINGE, MEDIAN (STATISTICS)

Interradius

see MIDRADIUS

Intersection

The intersection of two sets A and B is the set of elements common to A and B. This is written $A \cap B$, and is pronounced "A intersection B" or "A cap B." The intersection of sets A_1 through A_n is written $\bigcap_{i=1}^n A_i$. The intersection of lines AB and CD is written $AB \cap CD$.

see also AND, UNION

Interspersion

An ARRAY $A = a_{ij}$, $i, j \geq 1$ of POSITIVE INTEGERS is called an interspersion if

1. The rows of A comprise a PARTITION of the POSITIVE INTEGERS,
2. Every row of A is an increasing sequence,
3. Every column of A is a (possibly FINITE) increasing sequence,
4. If (u_j) and (v_j) are distinct rows of A and if p and q are any indices for which $u_p < v_q < u_{p+1}$, then $u_{p+1} < v_{q+1} < u_{p+2}$.

If an array $A = a_{ij}$ is an interspersion, then it is a DISPERSION. If an array $A = a(i,j)$ is an interspersion, then the sequence $\{x_n\}$ given by $\{x_n = i : n = (i,j)\}$ for some j is a FRACTAL SEQUENCE. Examples of interspersion are the STOLARSKY ARRAY and WYTHOFF ARRAY.

see also DISPERSION (SEQUENCE), FRACTAL SEQUENCE, STOLARSKY ARRAY

References

Kimberling, C. "Interspersions and Dispersions." *Proc. Amer. Math. Soc.* **117**, 313–321, 1993.

Kimberling, C. "Fractal Sequences and Interspersions." *Ars Combin.* **45**, 157–168, 1997.

Intersphere

see MIDSPHERE

Interval

A collection of points on a LINE SEGMENT. If the endpoints a and b are FINITE and are included, the interval is called CLOSED and is denoted $[a, b]$. If one of the endpoints is $\pm\infty$, then the interval still contains all of its LIMIT POINTS, so $[a, \infty)$ and $(-\infty, b]$ are also closed intervals. If the endpoints are not included, the interval is called OPEN and denoted (a, b). If one endpoint is included but not the other, the interval is denoted $[a, b)$ or $(a, b]$ and is called a HALF-CLOSED (or HALF-OPEN) interval.

see also CLOSED INTERVAL, HALF-CLOSED INTERVAL, LIMIT POINT, OPEN INTERVAL, PENCIL

Interval Graph

A GRAPH $G = (V, E)$ is an interval graph if it captures the INTERSECTION RELATION for some set of INTERVALS on the REAL LINE. Formally, P is an interval graph provided that one can assign to each $v \in V$ an interval I_v such that $I_u \cap I_v$ is nonempty precisely when $uv \in E$.

see also COMPARABILITY GRAPH

References

Booth, K. S. and Lueker, G. S. "Testing for the Consecutive Ones Property, Interval Graphs, and Graph Planarity using PQ-Tree Algorithms." *J. Comput. System Sci.* **13**, 335–379, 1976.

Fishburn, P. C. *Interval Orders and Interval Graphs: A Study of Partially Ordered Sets.* New York: Wiley, 1985.

Gilmore, P. C. and Hoffman, A. J. "A Characterization of Comparability Graphs and of Interval Graphs." *Canad. J. Math.* **16**, 539–548, 1964.

Lekkerkerker, C. G. and Boland, J. C. "Representation of a Finite Graph by a Set of Intervals on the Real Line." *Fund. Math.* **51**, 45–64, 1962.

Interval Order

A POSET $P = (X, \leq)$ is an interval order if it is ISOMORPHIC to some set of INTERVALS on the REAL LINE ordered by left-to-right precedence. Formally, P is an interval order provided that one can assign to each $x \in X$ an INTERVAL $[x_L, x_R]$ such that $x_R < y_L$ in the REAL NUMBERS IFF $x < y$ in P.

see also PARTIALLY ORDERED SET

References

Fishburn, P. C. *Interval Orders and Interval Graphs: A Study of Partially Ordered Sets.* New York: Wiley, 1985.

Wiener, N. "A Contribution to the Theory of Relative Position." *Proc. Cambridge Philos. Soc.* **17**, 441–449, 1914.

Intrinsic Curvature

A CURVATURE such as GAUSSIAN CURVATURE which is detectable to the "inhabitants" of a surface and not just outside observers. An EXTRINSIC CURVATURE, on the other hand, is not detectable to someone who can't study the 3-dimensional space surrounding the surface on which he resides.

see also CURVATURE, EXTRINSIC CURVATURE, GAUSSIAN CURVATURE

Intrinsic Equation

An equation which specifies a CURVE in terms of intrinsic properties such as ARC LENGTH, RADIUS OF CURVATURE, and TANGENTIAL ANGLE instead of with reference to artificial coordinate axes. Intrinsic equations are also called NATURAL EQUATIONS.

see also CESÀRO EQUATION, NATURAL EQUATION, WHEWELL EQUATION

References

Yates, R. C. "Intrinsic Equations." *A Handbook on Curves and Their Properties.* Ann Arbor, MI: J. W. Edwards, pp. 123–126, 1952.

Intrinsically Linked

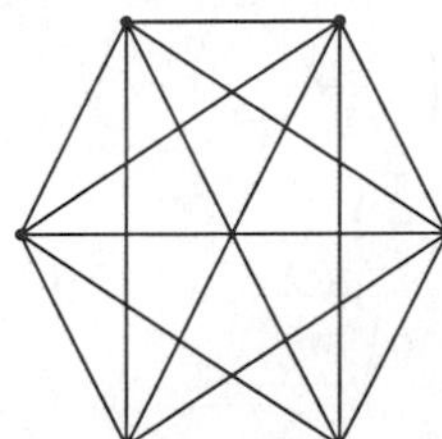

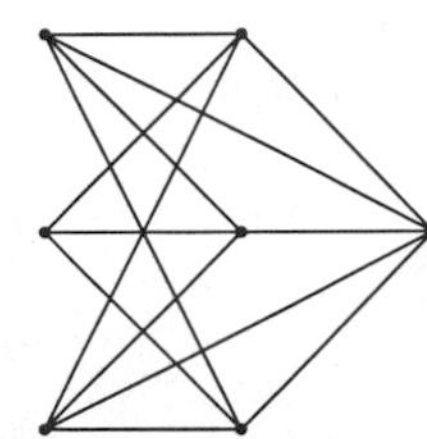

A GRAPH is intrinsically linked if any embedding of it in 3-D contains a nontrivial LINK. A GRAPH is intrinsically linked IFF it contains one of the seven PETERSEN GRAPHS (Robertson *et al.* 1993).

The COMPLETE GRAPH K_6 (left) is intrinsically linked because it contains at least two linked TRIANGLES. The COMPLETE k-PARTITE GRAPH $K_{3,3,1}$ (right) is also intrinsically linked.

see also COMPLETE GRAPH, COMPLETE k-PARTITE GRAPH, PETERSEN GRAPHS

References

Adams, C. C. *The Knot Book: An Elementary Introduction to the Mathematical Theory of Knots.* New York: W. H. Freeman, pp. 217–221, 1994.

Robertson, N.; Seymour, P. D.; and Thomas, R. "Linkless Embeddings of Graphs in 3-Space." *Bull. Amer. Math. Soc.* **28**, 84–89, 1993.

Invariant

A quantity which remains unchanged under certain classes of transformations. Invariants are extremely useful for classifying mathematical objects because they usually reflect intrinsic properties of the object of study.

see ADIABATIC INVARIANT, ALEXANDER INVARIANT, ALGEBRAIC INVARIANT, ARF INVARIANT, INTEGRAL OF MOTION

References

Hunt, B. "Invariants." Appendix B.1 in *The Geometry of Some Special Arithmetic Quotients.* New York: Springer-Verlag, pp. 282–290, 1996.

Invariant Density

see NATURAL INVARIANT

Invariant (Elliptic Function)

The invariants of a WEIERSTRAß ELLIPTIC FUNCTION are defined by

$$g_2 \equiv 60\Sigma'\Omega_{mn}{}^{-4}$$
$$g_3 \equiv 140\Sigma'\Omega_{mn}{}^{-6}.$$

Here,

$$\Omega_{mn} \equiv 2m\omega_1 - 2n\omega_2,$$

where ω_1 and ω_2 are the periods of the ELLIPTIC FUNCTION.

Invariant Manifold

When stable and unstable invariant MANIFOLDS intersect, they do so in a HYPERBOLIC FIXED POINT (SADDLE POINT). The invariant MANIFOLDS are then called SEPARATRICES. A HYPERBOLIC FIXED POINT is characterized by two ingoing stable MANIFOLDS and two outgoing unstable MANIFOLDS. In integrable systems, incoming W^s and outgoing W^u MANIFOLDS all join up smoothly.

A stable invariant MANIFOLD W^s of a FIXED POINT Y^* is the set of all points Y_0 such that the trajectory passing through Y_0 tends to Y^* as $j \to \infty$.

An unstable invariant MANIFOLD W^u of a FIXED POINT Y^* is the set of all points Y_0 such that the trajectory passing through Y_0 tends to Y^* as $j \to -\infty$.

see also HOMOCLINIC POINT

Invariant Point

see FIXED POINT (TRANSFORMATION)

Invariant Subgroup

see NORMAL SUBGROUP

Inverse Cosecant

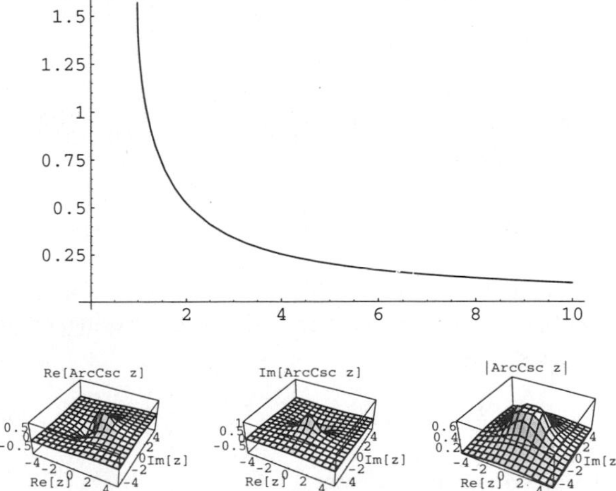

The function $\csc^{-1} x$, also denoted $\operatorname{arccsc}(x)$, where $\csc x$ is the COSECANT and the SUPERSCRIPT -1 denotes an

INVERSE FUNCTION, *not* the multiplicative inverse. The inverse cosecant satisfies

$$\csc^{-1} x = \sec^{-1}\left(\frac{x}{\sqrt{x^2-1}}\right) \quad (1)$$

for POSITIVE or NEGATIVE x, and

$$\csc^{-1} x = \pi + \csc^{-1}(-x) \quad (2)$$

for $x \geq 0$. The inverse cosecant is given in terms of other inverse trigonometric functions by

$$\csc^{-1} = \cos^{-1}\left(\frac{\sqrt{x^2-1}}{x}\right) \quad (3)$$

$$= \cot^{-1}(\sqrt{x^2-1}) \quad (4)$$

$$= \tfrac{1}{2}\pi - \sec^{-1} x = -\tfrac{1}{2}\pi - \sec^{-1}(-x) \quad (5)$$

$$= \sin^{-1}\left(\frac{1}{x}\right) \quad (6)$$

for $x \geq 0$.

see also COSECANT INVERSE SINE, SINE

References

Beyer, W. H. *CRC Standard Mathematical Tables, 28th ed.* Boca Raton, FL: CRC Press, pp. 142–143, 1987.

Inverse Cosine

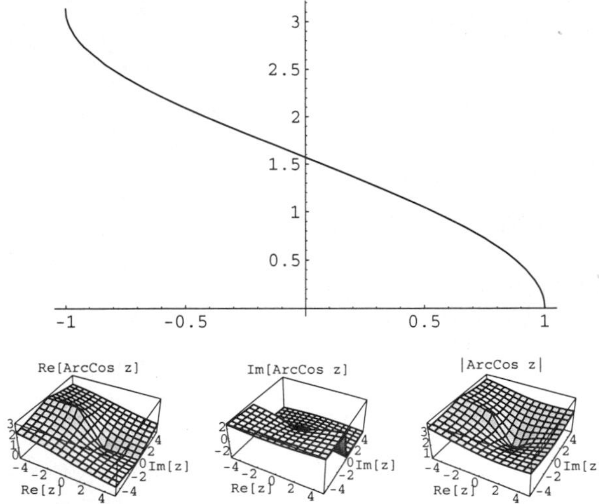

The function $\cos^{-1} x$, also denoted $\arccos(x)$, where $\cos x$ is the COSINE and the superscript -1 denotes an INVERSE FUNCTION, *not* the multiplicative inverse. The MACLAURIN SERIES for the inverse cosine range $-1 < x < 1$ is

$$\cos^{-1} x = \tfrac{1}{2}\pi - x - \tfrac{1}{6}x^3 - \tfrac{3}{40}x^5 - \tfrac{5x}{112}x^7 - \tfrac{35}{1152}x^9 - \ldots. \quad (1)$$

The inverse cosine satisfies

$$\cos^{-1} x = \pi - \cos^{-1}(-x) \quad (2)$$

for POSITIVE and NEGATIVE x, and

$$\cos^{-1} = \tfrac{1}{2}\pi - \cos^{-1}(\sqrt{1-x^2}) \quad (3)$$

for $x \geq 0$. The inverse cosine is given in terms of other inverse trigonometric functions by

$$\cos^{-1} x = \cot^{-1}\left(\frac{x}{\sqrt{1-x^2}}\right) \quad (4)$$

$$= \tfrac{1}{2}\pi + \sin^{-1}(-x) = \tfrac{1}{2}\pi - \sin^{-1} x \quad (5)$$

$$= \tfrac{1}{2}\pi - \tan^{-1}\left(\frac{x}{\sqrt{1-x^2}}\right) \quad (6)$$

for POSITIVE or NEGATIVE x, and

$$\cos^{-1} x = \csc^{-1}\left(\frac{1}{\sqrt{1-x^2}}\right) \quad (7)$$

$$= \sec^{-1}\left(\frac{1}{x}\right) \quad (8)$$

$$= \sin^{-1}(\sqrt{1-x^2}) \quad (9)$$

$$= \tan^{-1}\left(\frac{\sqrt{1-x^2}}{x}\right) \quad (10)$$

for $x \geq 0$.

see also COSINE, INVERSE SECANT

References

Abramowitz, M. and Stegun, C. A. (Eds.). "Inverse Circular Functions." §4.4 in *Handbook of Mathematical Functions with Formulas, Graphs, and Mathematical Tables, 9th printing.* New York: Dover, pp. 79–83, 1972.

Beyer, W. H. *CRC Standard Mathematical Tables, 28th ed.* Boca Raton, FL: CRC Press, pp. 142–143, 1987.

Inverse Cotangent

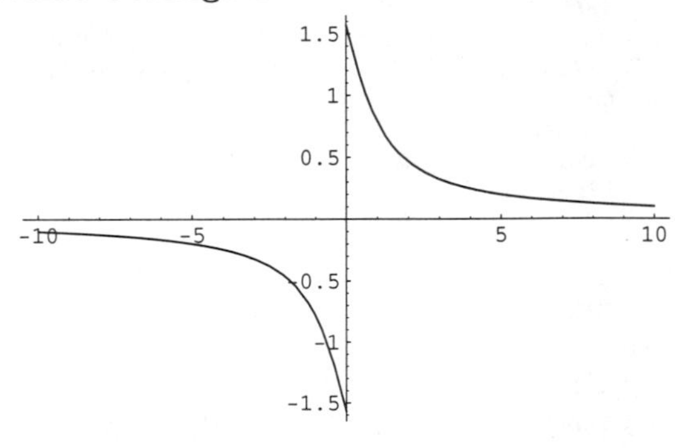

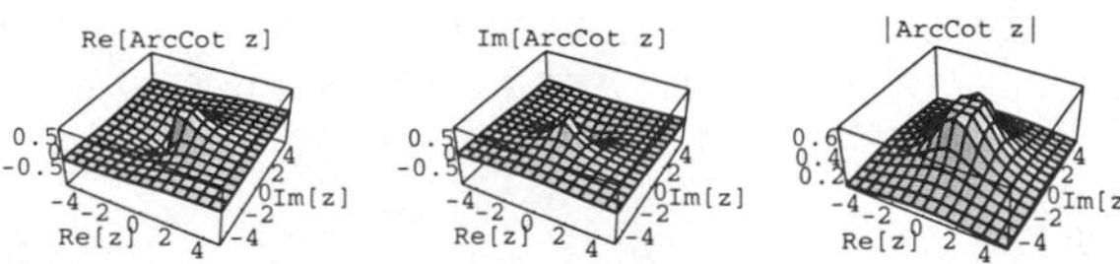

The function $\cot^{-1} x$, also denoted $\text{arccot}(x)$, where $\cot x$ is the COTANGENT and the superscript -1 denotes an INVERSE FUNCTION and *not* the multiplicative inverse. The MACLAURIN SERIES is given by

$$\cot^{-1} x = \tfrac{1}{2}\pi - x + \tfrac{1}{3}x^3 - \tfrac{1}{5}x^5 + \tfrac{1}{7}x^7 - \tfrac{1}{9}x^9 + \ldots, \quad (1)$$

and POWER SERIES by

$$\cot^{-1} x = x^{-1} - \tfrac{1}{3}x^{-3} + \tfrac{1}{5}x^{-5} - \tfrac{1}{7}x^{-7} + \tfrac{1}{9}x^{-9} + \ldots. \quad (2)$$

Euler derived the INFINITE series

$$\cot^{-1} x = x\left[\frac{1}{x^2+1} + \frac{2}{3(x^2+1)^2} + \frac{2\cdot 4}{3\cdot 5(x^2+1)^3} + \ldots\right] \quad (3)$$

(Wetherfield 1996).

The inverse cotangent satisfies

$$\cot^{-1} x = \pi - \cot^{-1}(-x) \quad (4)$$

for POSITIVE and NEGATIVE x, and

$$\cot^{-1} = \tfrac{1}{2}\pi - \cot^{-1}\left(\frac{1}{x}\right) \quad (5)$$

for $x \geq 0$. The inverse cotangent is given in terms of other inverse trigonometric functions by

$$\cot^{-1} x = \cos^{-1}\left(\frac{x}{\sqrt{x^2+1}}\right) \quad (6)$$

$$= \tfrac{1}{2}\pi - \sin^{-1}\left(\frac{x}{\sqrt{x^2+1}}\right) \quad (7)$$

$$= \tfrac{1}{2}\pi + \tan^{-1}(-x) = \tfrac{1}{2}\pi - \tan^{-1} x \quad (8)$$

for POSITIVE or NEGATIVE x, and

$$\cot^{-1} x = \csc^{-1}(\sqrt{x^2+1}) \quad (9)$$

$$= \sec^{-1}\left(\frac{\sqrt{x^2+1}}{x}\right) \quad (10)$$

$$= \sin^{-1}\left(\frac{1}{\sqrt{x^2+1}}\right) \quad (11)$$

$$= \tan^{-1}\left(\frac{1}{x}\right) \quad (12)$$

for $x \geq 0$.

A number

$$t_x = \cot^{-1} x, \quad (13)$$

where x is an INTEGER or RATIONAL NUMBER, is sometimes called a GREGORY NUMBER. Lehmer (1938a) showed that $\cot^{-1}(a/b)$ can be expressed as a finite sum of inverse cotangents of INTEGER arguments

$$\cot^{-1}\left(\frac{a}{b}\right) = \sum_{i=1}^{k} (-1)^{i-1} \cot^{-1} n_i, \quad (14)$$

where

$$n_i = \left\lfloor \frac{a_i}{b_i} \right\rfloor, \quad (15)$$

with $\lfloor x \rfloor$ the FLOOR FUNCTION, and

$$a_{i+1} = a_i n + i + b_i \quad (16)$$

$$b_{i+1} = a_i - n_i b_i, \quad (17)$$

with $a_0 = a$ and $b_0 = b$, and where the recurrence is continued until $b_{k+1} = 0$. If an INVERSE TANGENT sum is written as

$$\tan^{-1} n = \sum_{k=1} f_k \tan^{-1} n_k + f \tan^{-1}, \quad (18)$$

then equation (14) becomes

$$\cot^{-1} n = \sum_{k=1} f_k \cot^{-1} n_k + c \cot^{-1} 1, \quad (19)$$

where

$$c = 2 - f - 2\sum_{k=1} f_r. \quad (20)$$

Inverse cotangent sums can be used to generate MACHIN-LIKE FORMULAS.

An interesting inverse cotangent identity attributed to Charles Dodgson (Lewis Carroll) by Lehmer (1938b; Bromwich 1965, Castellanos 1988ab) is

$$\cot^{-1}(p+r) + \tan^{-1}(p+q) = \tan^{-1} p, \quad (21)$$

where

$$1 + p^2 = qr. \quad (22)$$

Other inverse cotangent identities include

$$2\cot^{-1}(2x) - \cot^{-1} x = \cot^{-1}(4x^3 + 3x) \quad (23)$$

$$3\cot^{-1}(3x) - \cot^{-1} x = \cot^{-1}\left(\frac{27x^4 + 18x^2 - 1}{8x}\right), \quad (24)$$

as well as many others (Bennett 1926, Lehmer 1938b).

see also COTANGENT, INVERSE TANGENT, MACHIN'S FORMULA, MACHIN-LIKE FORMULAS, TANGENT

References

Abramowitz, M. and Stegun, C. A. (Eds.). "Inverse Circular Functions." §4.4 in *Handbook of Mathematical Functions with Formulas, Graphs, and Mathematical Tables, 9th printing.* New York: Dover, pp. 79–83, 1972.

Bennett, A. A. "The Four Term Diophantine Arccotangent Relation." *Ann. Math.* **27**, 21–24, 1926.

Beyer, W. H. *CRC Standard Mathematical Tables, 28th ed.* Boca Raton, FL: CRC Press, pp. 142–143, 1987.

Bromwich, T. J. I. and MacRobert, T. M. *An Introduction to the Theory of Infinite Series, 3rd ed.* New York: Chelsea, 1991.

Castellanos, D. "The Ubiquitous Pi. Part I." *Math. Mag.* **61**, 67–98, 1988a.

Castellanos, D. "The Ubiquitous Pi. Part II." *Math. Mag.* **61**, 148–163, 1988b.

Lehmer, D. H. "A Cotangent Analogue of Continued Fractions." *Duke Math. J.* **4**, 323–340, 1938a.

Lehmer, D. H. "On Arccotangent Relations for π." *Amer. Math. Monthly* **45**, 657–664, 1938b.

Weisstein, E. W. "Arccotangent Series." `http://www.astro.virginia.edu/~eww6n/math/notebooks/CotSeries.m`.

Wetherfield, M. "The Enhancement of Machin's Formula by Todd's Process." *Math. Gaz.*, 333–344, July 1996.

Inverse Curve

Given a CIRCLE C with CENTER O and RADIUS k, then two points P and Q are inverse with respect to C if $OP \cdot OQ = k^2$. If P describes a curve C_1, then Q describes a curve C_2 called the inverse of C_1 with respect to the circle C (with INVERSION CENTER O). If the POLAR equation of C is $r(\theta)$, then the inverse curve has polar equation

$$r = \frac{k^2}{r(\theta)}.$$

If $O = (x_0, y_0)$ and $P = (f(t), g(t))$, then the inverse has equations

$$x = x_0 + \frac{k^2(f - x_0])}{(f - x_0)^2 + (g - y_0)^2}$$

$$y = y_0 + \frac{k^2(g - y_0)}{(f - x_0)^2 + (g - y_0)^2}.$$

Curve	Inversion Center	Inverse Curve
Archimedean spiral	origin	Archimedean spiral
cardioid	cusp	parabola
circle	any pt.	another circle
cissoid of Diocles	cusp	parabola
cochleoid	origin	quadratrix of Hippias
epispiral	origin	Rose
Fermat's spiral	origin	lituus
hyperbola	center	lemniscate
hyperbola	vertex	right strophoid
hyperbola with $a = \sqrt{3}$	vertex	Maclaurin trisectrix
lemniscate	center	hyperbola
lituus	origin	Fermat spiral
logarithmic spiral	origin	logarithmic spiral
Maclaurin trisectrix	focus	Tschirnhausen's cubic
parabola	focus	cardioid
parabola	vertex	cissoid of Diocles
quadratrix of Hippias		cochleoid
right strophoid	origin	the same right strophoid
sinusoidal spiral	origin	sinusoidal spiral inverse curve
Tschirnhausen cubic		sinusoidal spiral

see also INVERSION, INVERSION CENTER, INVERSION CIRCLE

References

Lee, X. "Inversion." `http://www.best.com/~xah/SpecialPlaneCurves_dir/Inversion_dir/inversion.html`.

Lee, X. "Inversion Gallery." `http://www.best.com/~xah/SpecialPlaneCurves_dir/InversionGallery_dir/inversionGallery.html`.

Yates, R. C. "Inversion." *A Handbook on Curves and Their Properties.* Ann Arbor, MI: J. W. Edwards, pp. 127–134, 1952.

Inverse Filter

A linear DECONVOLUTION ALGORITHM.

Inverse Function

Given a FUNCTION $f(x)$, its inverse $f^{-1}(x)$ is defined by $f(f^{-1}(x)) \equiv x$. Therefore, $f(x)$ and $f^{-1}(x)$ are reflections about the line $y = x$.

Inverse Hyperbolic Cosecant

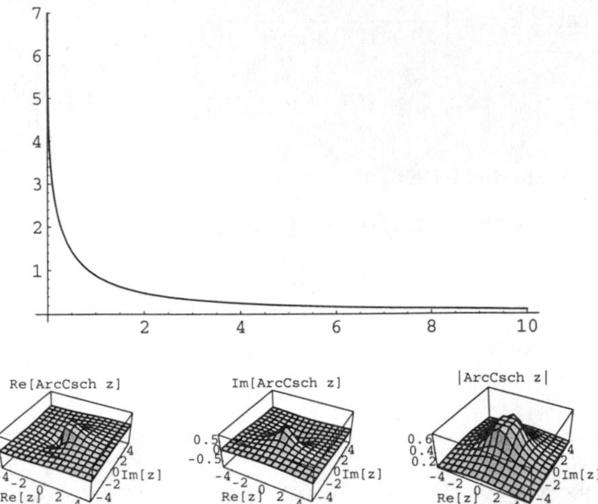

The INVERSE FUNCTION of the HYPERBOLIC COSECANT, denoted $\operatorname{csch}^{-1} x$.

see also HYPERBOLIC COSECANT

Inverse Hyperbolic Cosine

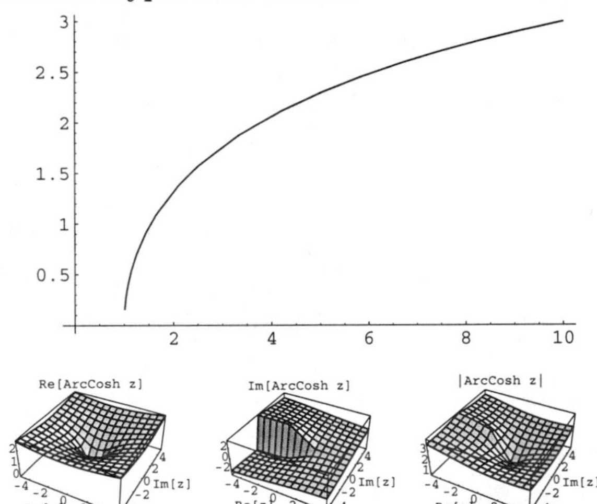

The INVERSE FUNCTION of the HYPERBOLIC COSINE, denoted $\cosh^{-1} x$.

see also HYPERBOLIC COSINE

Inverse Hyperbolic Cotangent

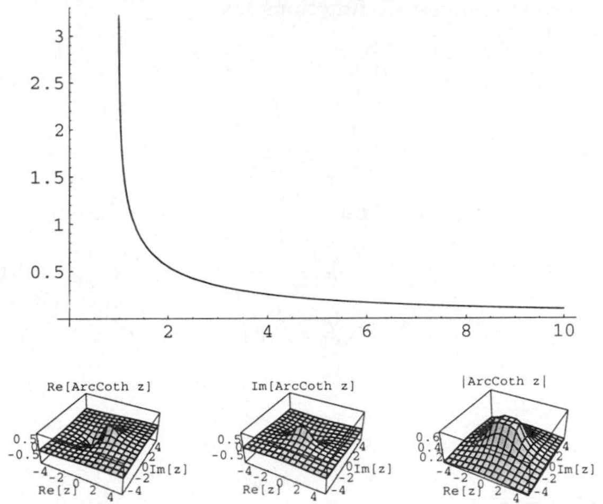

The INVERSE FUNCTION of the HYPERBOLIC COTANGENT, denoted $\coth^{-1} x$.

see also HYPERBOLIC COTANGENT

Inverse Hyperbolic Functions

The INVERSE of the HYPERBOLIC FUNCTIONS, denoted $\cosh^{-1} x$, $\coth^{-1} x$, $\operatorname{csch}^{-1} x$, $\operatorname{sech}^{-1} x$, $\sinh^{-1} x$, and $\tanh^{-1} x$.

see also HYPERBOLIC FUNCTIONS

References

Spanier, J. and Oldham, K. B. "The Inverse Hyperbolic Functions." Ch. 31 in *An Atlas of Functions.* Washington, DC: Hemisphere, pp. 285–293, 1987.

Inverse Hyperbolic Secant

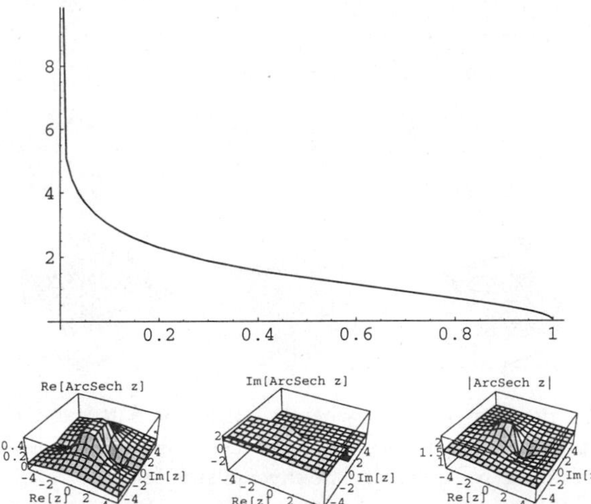

The INVERSE FUNCTION of the HYPERBOLIC SECANT, denoted $\operatorname{sech}^{-1} x$.

see also HYPERBOLIC SECANT

Inverse Hyperbolic Sine

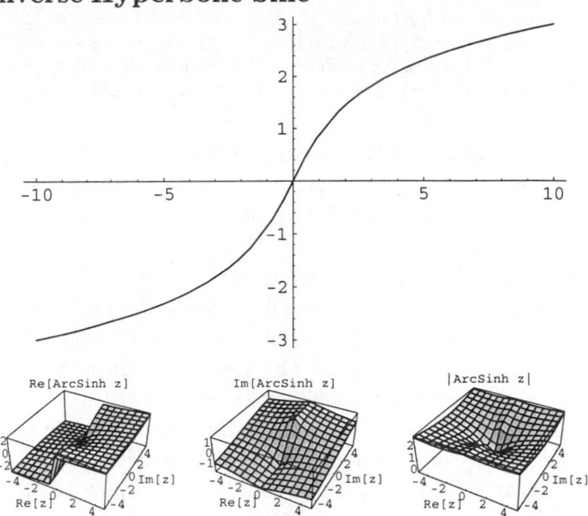

The INVERSE FUNCTION of the HYPERBOLIC SINE, denoted $\sinh^{-1} x$.

see also HYPERBOLIC SINE

Inverse Hyperbolic Tangent

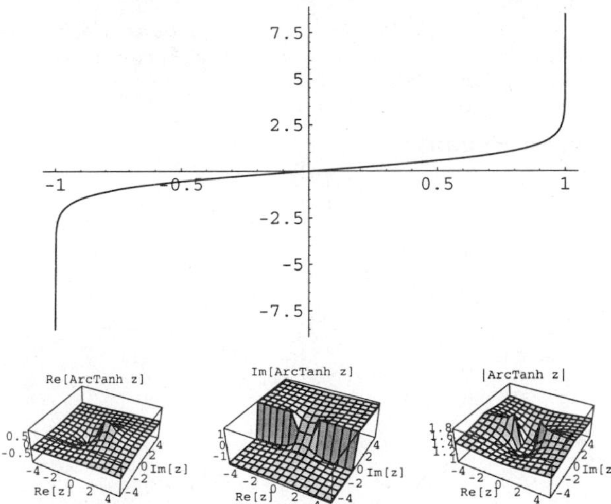

The INVERSE FUNCTION of the HYPERBOLIC TANGENT, denoted $\tanh^{-1} x$.

see also HYPERBOLIC TANGENT

Inverse Matrix

see also MATRIX INVERSE

Inverse Points

Points which are transformed into each other through INVERSION about a given INVERSION CIRCLE. The point P' which is the inverse point of a given point P with respect to an INVERSION CIRCLE C may be constructed geometrically using a COMPASS only (Courant and Robbins 1996).

see also GEOMETRIC CONSTRUCTION, INVERSION, POLAR, POLE (GEOMETRY)

References
Courant, R. and Robbins, H. "Geometrical Construction of Inverse Points." §3.4.3 in *What is Mathematics?: An Elementary Approach to Ideas and Methods, 2nd ed.* Oxford, England: Oxford University Press, pp. 144–145, 1996.

Inverse Quadratic Interpolation

The use of three prior points in a ROOT-finding ALGORITHM to estimate the zero crossing.

Inverse Scattering Method

A method which can be used to solve the initial value problem for certain classes of nonlinear PARTIAL DIFFERENTIAL EQUATIONS. The method reduces the initial value problem to a linear INTEGRAL EQUATION in which time appears only implicitly. However, the solutions $u(x,t)$ and various of their derivatives must approach zero as $x \to \pm\infty$ (Infeld and Rowlands 1990).

see also ABLOWITZ-RAMANI-SEGUR CONJECTURE, BÄCKLUND TRANSFORMATION

References
Infeld, E. and Rowlands, G. "Inverse Scattering Method." §7.4 in *Nonlinear Waves, Solitons, and Chaos.* Cambridge, England: Cambridge University Press, pp. 192–196, 1990.
Miura, R. M. (Ed.) *Bäcklund Transformations, the Inverse Scattering Method, Solitons, and Their Applications.* New York: Springer-Verlag, 1974.

Inverse Secant

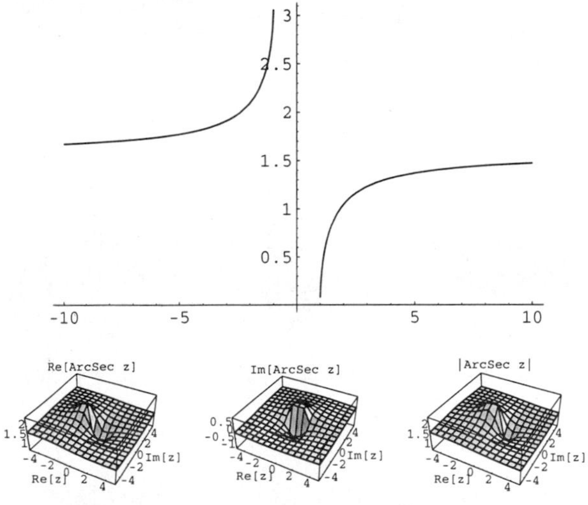

The function $\sec^{-1} x$, where $\sec x$ is the SECANT and the superscript -1 denotes the INVERSE FUNCTION, *not* the multiplicative inverse. The inverse secant satisfies

$$\sec^{-1} x = \csc^{-1}\left(\frac{x}{\sqrt{x^2-1}}\right) \qquad (1)$$

for POSITIVE or NEGATIVE x, and

$$\sec^{-1} x = \pi + \sec^{-1}(-x) \qquad (2)$$

for $x \geq 0$. The inverse secant is given in terms of other inverse trigonometric functions by

$$\sec^{-1} x = \cos^{-1}\left(\frac{1}{x}\right) \qquad (3)$$

$$= \cot^{-1}\left(\frac{1}{\sqrt{x^2-1}}\right) \qquad (4)$$

$$= \tfrac{1}{2}\pi - \csc^{-1} x = -\tfrac{1}{2}\pi - \csc^{-1}(-x) \qquad (5)$$

$$= \sin^{-1}\left(\frac{\sqrt{x^2-1}}{x}\right) \qquad (6)$$

$$= \tan^{-1}(\sqrt{x^2-1}) \qquad (7)$$

for $x \geq 0$.

see also INVERSE COSECANT, SECANT

References
Beyer, W. H. *CRC Standard Mathematical Tables, 28th ed.* Boca Raton, FL: CRC Press, pp. 141–143, 1987.

Inverse Semigroup

The abstract counterpart of a PSEUDOGROUP formed by certain subsets of a GROUPOID which admit a MULTIPLICATION.

References
Weinstein, A. "Groupoids: Unifying Internal and External Symmetry." *Not. Amer. Math. Soc.* **43**, 744–752, 1996.

Inverse Sine

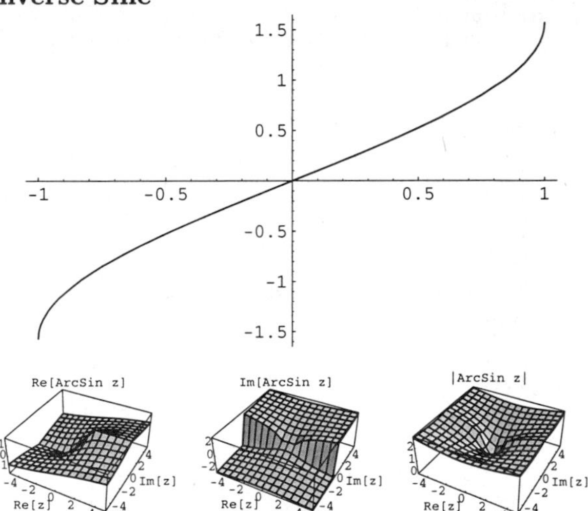

The function $\sin^{-1} x$, where $\sin x$ is the SINE and the superscript -1 denotes the INVERSE FUNCTION, *not* the multiplicative inverse. The inverse sine satisfies

$$\sin^{-1} x = -\sin^{-1}(-x) \qquad (1)$$

for POSITIVE and NEGATIVE x, and

$$\sin^{-1} = \tfrac{1}{2}\pi - \sin^{-1}(\sqrt{1-x^2}) \qquad (2)$$

for $x \geq 0$. The inverse sine is given in terms of other inverse trigonometric functions by

$$\sin^{-1} x = \cos^{-1}(-x) - \tfrac{1}{2}\pi = \tfrac{1}{2}\pi - \cos^{-1} x \qquad (3)$$

$$= \tfrac{1}{2}\pi - \cot^{-1}\left(\frac{x}{\sqrt{1-x^2}}\right) \qquad (4)$$

$$= \tan^{-1}\left(\frac{x}{\sqrt{1-x^2}}\right) \qquad (5)$$

for POSITIVE or NEGATIVE x, and

$$\sin^{-1} x = \cos^{-1}(\sqrt{1-x^2}\,) \qquad (6)$$

$$= \cot^{-1}\left(\frac{\sqrt{1-x^2}}{x}\right) \qquad (7)$$

$$= \csc^{-1}\left(\frac{1}{x}\right) \qquad (8)$$

$$= \sec^{-1}\left(\frac{1}{\sqrt{1-x^2}}\right) \qquad (9)$$

for $x \geq 0$.

see also INVERSE COSINE, SINE

References

Abramowitz, M. and Stegun, C. A. (Eds.). "Inverse Circular Functions." §4.4 in *Handbook of Mathematical Functions with Formulas, Graphs, and Mathematical Tables, 9th printing.* New York: Dover, pp. 79–83, 1972.

Beyer, W. H. *CRC Standard Mathematical Tables, 28th ed.* Boca Raton, FL: CRC Press, pp. 142–143, 1987.

Inverse Tangent

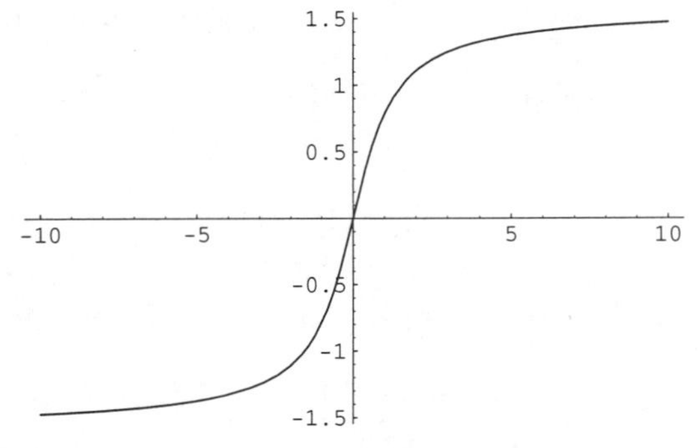

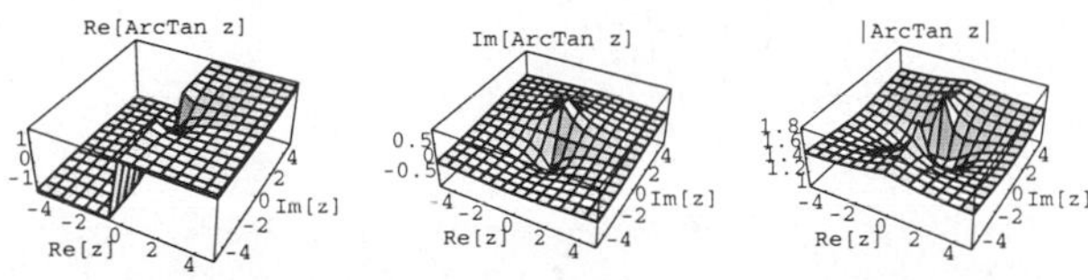

The inverse tangent is also called the arctangent and is denoted either $\tan^{-1} x$ or $\arctan x$. It has the MACLAURIN SERIES

$$\tan^{-1} x = \sum_{n=0}^{\infty} \frac{(-1)^n x^{2n+1}}{2n+1} = x - \tfrac{1}{3}x^3 + \tfrac{1}{5}x^5 - \tfrac{1}{7}x^7 + \ldots. \qquad (1)$$

A more rapidly converging form due to Euler is given by

$$\tan^{-1} x = \sum_{n=0}^{\infty} \frac{2^{2n}(n!)^2}{(2n+1)!} \frac{x^{2n+1}}{(+x^2)^{n+1}} \qquad (2)$$

(Castellanos 1988). The inverse tangent satisfies

$$\tan^{-1} x = -\tan^{-1}(-x) \qquad (3)$$

for POSITIVE and NEGATIVE x, and

$$\tan^{-1} = \tfrac{1}{2}\pi - \tan^{-1}\left(\frac{1}{x}\right) \qquad (4)$$

for $x \geq 0$. The inverse tangent is given in terms of other inverse trigonometric functions by

$$\tan^{-1} x = \tfrac{1}{2}\pi - \cos^{-1}\left(\frac{x}{\sqrt{x^2+1}}\right) \qquad (5)$$

$$= \cot^{-1}(-x) - \tfrac{1}{2}\pi = \tfrac{1}{2}\pi - \cot^{-1} x \qquad (6)$$

$$= \sin^{-1}\left(\frac{x}{\sqrt{x^2+1}}\right) \qquad (7)$$

for POSITIVE or NEGATIVE x, and

$$\tan^{-1} x = \cos^{-1}\left(\frac{1}{\sqrt{x^2+1}}\right) \qquad (8)$$

$$= \cot^{-1}\left(\frac{1}{x}\right) \qquad (9)$$

$$= \csc^{-1}\left(\frac{\sqrt{x^2+1}}{x}\right) \qquad (10)$$

$$= \sec^{-1}(\sqrt{x^2+1}\,) \qquad (11)$$

for $x \geq 0$.

In terms of the HYPERGEOMETRIC FUNCTION,

$$\tan^{-1} x = x\,{}_2F_1(1, \tfrac{1}{2}; \tfrac{3}{2}; -x^2) \qquad (12)$$

$$= \frac{x}{1+x^2}\,{}_2F_1\left(1, 1; \tfrac{3}{2}; \frac{x^2}{1+x^2}\right) \qquad (13)$$

(Castellanos 1988). Castellanos (1986, 1988) also gives some curious formulas in terms of the FIBONACCI NUMBERS,

$$\tan^{-1} x = \sum_{n=0}^{\infty} \frac{(-1)^n F_{2n+1} t^{2n+1}}{5^n(2n+1)} \qquad (14)$$

$$= 5\sum_{n=0}^{\infty} \frac{(-1)^n {F_{2n+1}}^2}{(2n+1)(u+\sqrt{u^2+1}\,)^{2n+1}} \qquad (15)$$

$$= \sum_{n=0}^{\infty} \frac{(-1)^n 5^{n+2} {F_{2n+1}}^3}{(2n+1)(v+\sqrt{v^2+5}\,)^{2n+1}}, \qquad (16)$$

where

$$t \equiv \frac{2x}{1+\sqrt{\frac{4x^2}{5}}} \qquad (17)$$

$$u \equiv \frac{5}{4x}\left(1+\sqrt{1+\tfrac{24}{25}x^2}\right), \qquad (18)$$

and v is the largest POSITIVE ROOT of

$$8xv^4 - 100v^3 - 450xv^2 + 875v + 625x = 0. \qquad (19)$$

The inverse tangent satisfies the addition FORMULA

$$\tan^{-1} x + \tan^{-1} y = \tan^{-1}\left(\frac{x+y}{1-xy}\right) \qquad (20)$$

as well as the more complicated FORMULAS

$$\tan^{-1}\left(\frac{1}{a-b}\right) = \tan^{-1}\left(\frac{1}{a}\right) + \tan^{-1}\left(\frac{b}{a^2-ab+1}\right) \qquad (21)$$

$$\tan^{-1}\left(\frac{1}{a}\right) = 2\tan^{-1}\left(\frac{1}{2a}\right) - \tan^{-1}\left(\frac{1}{4a^3+3a}\right) \qquad (22)$$

$$\tan^{-1}\left(\frac{1}{p}\right) = \tan^{-1}\left(\frac{1}{p+q}\right) + \tan^{-1}\left(\frac{q}{p^2+pq+1}\right), \qquad (23)$$

the latter of which was known to Euler. The inverse tangent FORMULAS are connected with many interesting approximations to PI

$$\tan^{-1}(1+x) = \tfrac{1}{4}\pi + \tfrac{1}{2}x - \tfrac{1}{4}x^2 + \tfrac{1}{12}x^3 + \tfrac{1}{40}x^5 + \tfrac{1}{48}x^6 + \tfrac{1}{112}x^7 + \ldots. \qquad (24)$$

Euler gave

$$\tan^{-1} x = \frac{y}{x}\left(\frac{2}{3}y + \frac{2\cdot 4}{3\cdot 5}y^2 + \frac{2\cdot 4\cdot 6}{3\cdot 5\cdot 7}y^3 + \ldots\right), \qquad (25)$$

where

$$y \equiv \frac{x^2}{1+x^2}. \qquad (26)$$

The inverse tangent has CONTINUED FRACTION representations

$$\tan^{-1} x = \cfrac{x}{1+\cfrac{x^2}{3+\cfrac{4x^2}{5+\cfrac{9x^2}{7+\cfrac{16x^2}{9+\ldots}}}}} \qquad (27)$$

$$= \cfrac{x}{1+\cfrac{x^2}{3-x^2+\cfrac{9x^2}{5-3x^2+\cfrac{25x^2}{7-5x^2+\ldots}}}}. \qquad (28)$$

To find $\tan^{-1} x$ numerically, the following ARITHMETIC-GEOMETRIC MEAN-like ALGORITHM can be used. Let

$$a_0 = (1+x^2)^{-1/2} \qquad (29)$$

$$b_0 = 1. \qquad (30)$$

Then compute

$$a_{i+1} = \tfrac{1}{2}(a_i + b_i) \qquad (31)$$

$$b_{i+1} = \sqrt{a_{i+1}b_i}, \qquad (32)$$

and the inverse tangent is given by

$$\tan^{-1} x = \lim_{n\to\infty} \frac{x}{\sqrt{1+x^2}\, a_n} \qquad (33)$$

(Acton 1990).

An inverse tangent $\tan^{-1} n$ with integral n is called reducible if it is expressible as a finite sum of the form

$$\tan^{-1} n = \sum_{k=1} f_k \tan^{-1} n_k, \qquad (34)$$

where f_k are POSITIVE or NEGATIVE INTEGERS and n_i are iINTEGERS $< n$. $\tan^{-1} m$ is reducible IFF all the PRIME factors of $1+m^2$ occur among the PRIME factors of $1+n^2$ for $n = 1, \ldots, m-1$. A second NECESSARY and SUFFICIENT condition is that the largest PRIME factor of $1+m^2$ is less than $2m$. Equivalent to the second condition is the statement that every GREGORY NUMBER $t_x = \cot^{-1} x$ can be uniquely expressed as a sum in terms of t_ms for which m is a STØRMER NUMBER (Conway and Guy 1996). To find this decomposition, write

$$\arg(1+in) = \arg\prod_{k=1}(1+n_k i)^{f_k}, \qquad (35)$$

so the ratio

$$r = \frac{\prod_{k=1}(1+n_k i)^{f_k}}{1+in} \qquad (36)$$

is a RATIONAL NUMBER. Equation (36) can also be written

$$r^2(1+n^2) = \prod_{k=1}(1+{n_k}^2)^{f_k}. \qquad (37)$$

Writing (34) in the form

$$\tan^{-1} n = \sum_{k=1} f_k \tan^{-1} n_k + f\tan^{-1} 1 \qquad (38)$$

allows a direct conversion to a corresponding INVERSE COTANGENT FORMULA

$$\cot^{-1} n = \sum_{k=1} f_k \cot^{-1} n_k + c\cot^{-1} 1, \qquad (39)$$

where

$$c = 2 - f - 2\sum_{k=1} f_r. \qquad (40)$$

Todd (1949) gives a table of decompositions of $\tan^{-1} n$ for $n \leq 342$. Conway and Guy (1996) give a similar table in terms of STØRMER NUMBERS.

Arndt and Gosper give the remarkable inverse tangent identity

$$\sin\left(\sum_{k=1}^{2n+1} \tan^{-1} a_k\right) = \frac{(-1)^n}{2n+1} \frac{\sum_{k=1}^{2n+1} \prod_{j=1}^{2n+1} \left[a_j - \tan\left(\frac{\pi(j-k)}{2n+1}\right)\right]}{\sqrt{\prod_{j=1}^{2n+1} (a_j{}^2 + 1)}}. \quad (41)$$

see also INVERSE COTANGENT, TANGENT

References

Abramowitz, M. and Stegun, C. A. (Eds.). "Inverse Circular Functions." §4.4 in *Handbook of Mathematical Functions with Formulas, Graphs, and Mathematical Tables, 9th printing.* New York: Dover, pp. 79–83, 1972.

Acton, F. S. "The Arctangent." In *Numerical Methods that Work, upd. and rev.* Washington, DC: Math. Assoc. Amer., pp. 6–10, 1990.

Arndt, J. "Completely Useless Formulas." `http://www.jjj.de/hfloat/hfloatpage.html#formulas`.

Beeler, M.; Gosper, R. W.; and Schroeppel, R. *HAKMEM.* Cambridge, MA: MIT Artificial Intelligence Laboratory, Memo AIM-239, Item 137, Feb. 1972.

Beyer, W. H. *CRC Standard Mathematical Tables, 28th ed.* Boca Raton, FL: CRC Press, pp. 142–143, 1987.

Castellanos, D. "Rapidly Converging Expansions with Fibonacci Coefficients." *Fib. Quart.* **24**, 70–82, 1986.

Castellanos, D. "The Ubiquitous Pi. Part I." *Math. Mag.* **61**, 67–98, 1988.

Conway, J. H. and Guy, R. K. "Størmer's Numbers." *The Book of Numbers.* New York: Springer-Verlag, pp. 245–248, 1996.

Todd, J. "A Problem on Arc Tangent Relations." *Amer. Math. Monthly* **56**, 517–528, 1949.

Inverse Trigonometric Functions

INVERSE FUNCTIONS of the TRIGONOMETRIC FUNCTIONS written $\cos^{-1} x$, $\cot^{-1} x$, $\csc^{-1} x$, $\sec^{-1} x$, $\sin^{-1} x$, and $\tan^{-1} x$.

see also INVERSE COSECANT, INVERSE COSINE, INVERSE COTANGENT, INVERSE SECANT, INVERSE SINE, INVERSE TANGENT

References

Abramowitz, M. and Stegun, C. A. (Eds.). "Inverse Circular Functions." §4.4 in *Handbook of Mathematical Functions with Formulas, Graphs, and Mathematical Tables, 9th printing.* New York: Dover, pp. 79–83, 1972.

Spanier, J. and Oldham, K. B. "Inverse Trigonometric Functions." Ch. 35 in *An Atlas of Functions.* Washington, DC: Hemisphere, pp. 331–341, 1987.

Inversely Similar

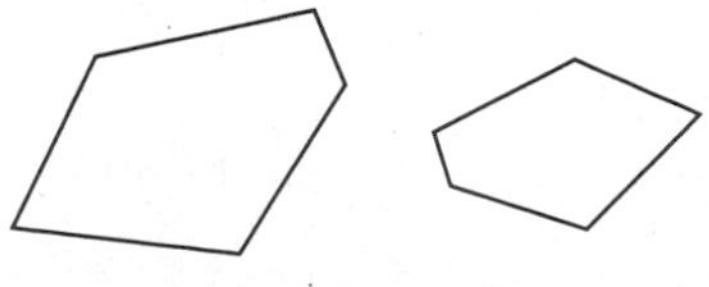

inversely similar

Two figures are said to be SIMILAR when all corresponding ANGLES are equal, and are inversely similar when all corresponding ANGLES are equal and described in the opposite rotational sense.

see also DIRECTLY SIMILAR, SIMILAR

Inversion

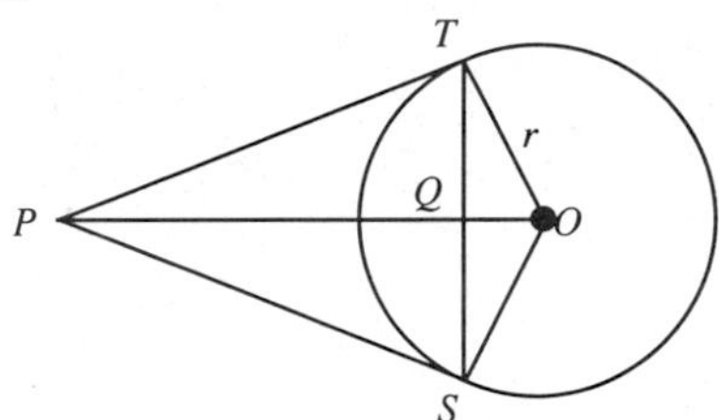

Inversion is the process of transforming points to their INVERSE POINTS. This sort of inversion was first systematically investigated by Jakob Steiner. Two points are said to be inverses with respect to an INVERSION CIRCLE with INVERSION CENTER $O = (x_0, y_0)$ and INVERSION RADIUS k if PT and PS are line segments symmetric about OP and tangent to the CIRCLE, and Q is the intersection of OP and ST. The curve to which a given curve is transformed under inversion is called its INVERSE CURVE.

Note that a point on the CIRCUMFERENCE of the INVERSION CIRCLE is its own inverse point. The inverse points obey

$$\frac{OP}{k} = \frac{k}{OQ}, \quad (1)$$

or

$$k^2 = OP \times OQ, \quad (2)$$

where k^2 is called the POWER. The equation for the inverse of the point (x, y) relative to the INVERSION CIRCLE with INVERSION CENTER (x_0, y_0) and inversion radius k is therefore

$$x' = x_0 + \frac{k^2(x - x_0)}{(x - x_0)^2 + (y - y_0)^2} \quad (3)$$

$$y' = y_0 + \frac{k^2(y - y_0)}{(x - x_0)^2 + (y - y_0)^2}. \quad (4)$$

In vector form,

$$\mathbf{x}' = \mathbf{x}_0 + \frac{k^2(\mathbf{x} - \mathbf{x}_0)}{|\mathbf{x} - \mathbf{x}_0|^2}. \quad (5)$$

Any ANGLE inverts to an opposite ANGLE.

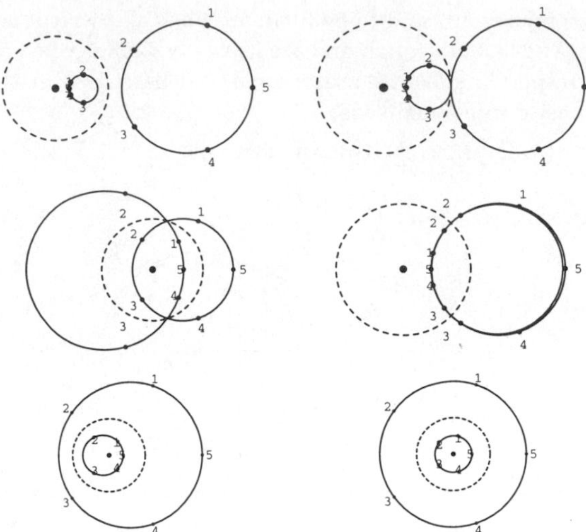

Treating LINES as CIRCLES of INFINITE RADIUS, all CIRCLES invert to CIRCLES. Furthermore, any two nonintersecting circles can be inverted into concentric circles by taking the INVERSION CENTER at one of the two limiting points (Coxeter 1969), and ORTHOGONAL CIRCLES invert to ORTHOGONAL CIRCLES (Coxeter 1969).

The inverse of a CIRCLE of RADIUS a with CENTER (x, y) with respect to an inversion circle with INVERSION CENTER $(0,0)$ and INVERSION RADIUS k is another CIRCLE with CENTER $(x', y') = (sx, sy)$ and RADIUS $r' = |s|a$, where

$$s \equiv \frac{k^2}{x^2 + y^2 - a^2}. \tag{6}$$

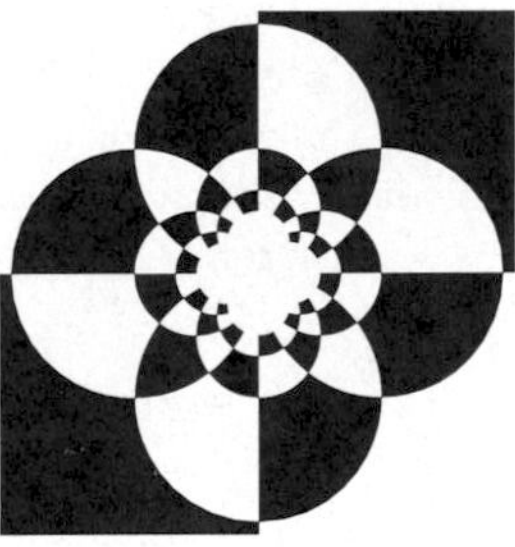

The above plot shows a checkerboard centered at (0, 0) and its inverse about a small circle also centered at (0, 0) (Dixon 1991).

see also ARBELOS, HEXLET, INVERSE CURVE, INVERSION CIRCLE, INVERSION OPERATION, INVERSION RADIUS, INVERSIVE DISTANCE, INVERSIVE GEOMETRY, MIDCIRCLE, PAPPUS CHAIN, PEAUCELLIER INVERSOR, POLAR, POLE (GEOMETRY), POWER (CIRCLE), RADICAL LINE, STEINER CHAIN, STEINER'S PORISM

References

Courant, R. and Robbins, H. "Geometrical Transformations. Inversion." §3.4 in *What is Mathematics?: An Elementary Approach to Ideas and Methods, 2nd ed.* Oxford, England: Oxford University Press, pp. 140–146, 1996.

Coxeter, H. S. M. "Inversion in a Circle" and "Inversion of Lines and Circles." §6.1 and 6.3 in *Introduction to Geometry, 2nd ed.* New York: Wiley, p. 77–83, 1969.

Coxeter, H. S. M. and Greitzer, S. L. *Geometry Revisited.* Washington, DC: Math. Assoc. Amer., pp. 108–114, 1967.

Dixon, R. "Inverse Points and Mid-Circles." §1.6 in *Mathographics.* New York: Dover, pp. 62–73, 1991.

Johnson, R. A. *Modern Geometry: An Elementary Treatise on the Geometry of the Triangle and the Circle.* Boston, MA: Houghton Mifflin, pp. 43–57, 1929.

Lockwood, E. H. "Inversion." Ch. 23 in *A Book of Curves.* Cambridge, England: Cambridge University Press, pp. 176–181, 1967.

Ogilvy, C. S. *Excursions in Geometry.* New York: Dover, pp. 25–31, 1990.

Weisstein, E. W. "Plane Geometry." `http://www.astro.virginia.edu/~eww6n/math/notebooks/PlaneGeometry.m`.

Inversion Center

The point that INVERSION OF A CURVE is performed with respect to.

see also INVERSE POINTS, INVERSION CIRCLE, INVERSION RADIUS, INVERSIVE DISTANCE, POLAR, POLE (GEOMETRY), POWER (CIRCLE)

Inversion Circle

The CIRCLE with respect to which a INVERSE CURVE is computed or relative to which INVERSE POINTS are computed.

see also INVERSE POINTS, INVERSION CENTER, INVERSION RADIUS, INVERSIVE DISTANCE, MIDCIRCLE, POLAR, POLE (GEOMETRY), POWER (CIRCLE)

Inversion Operation

The SYMMETRY OPERATION $(x, y, z) \to (-x, -y, -z)$. When used in conjunction with a ROTATION, it becomes an IMPROPER ROTATION.

Inversion Radius

The RADIUS used in performing an INVERSION with respect to an INVERSION CIRCLE.

see also INVERSE POINTS, INVERSION CENTER, INVERSION CIRCLE, INVERSIVE DISTANCE, POLAR, POLE (GEOMETRY), POWER (CIRCLE)

Inversive Distance

The inversive distance is the NATURAL LOGARITHM of the ratio of two concentric circles into which the given circles can be inverted. Let c be the distance between the centers of two nonintersecting CIRCLES of RADII a and $b < a$. Then the inversive distance is

$$\delta = \cosh^{-1}\left|\frac{a^2 + b^2 - c^2}{2ab}\right|$$

(Coxeter and Greitzer 1967).

The inversive distance between the SODDY CIRCLES is given by

$$\delta = 2\cosh^{-1} 2,$$

and the CIRCUMCIRCLE and INCIRCLE of a TRIANGLE with CIRCUMRADIUS R and INRADIUS r are at inversive distance

$$\delta = 2\sinh^{-1}\left(\frac{1}{2}\sqrt{\frac{r}{R}}\right)$$

(Coxeter and Greitzer 1967, pp. 130–131).

References

Coxeter, H. S. M. and Greitzer, S. L. *Geometry Revisited.* Washington, DC: Math. Assoc. Amer., pp. 123–124 and 127–131, 1967.

Inversive Geometry

The GEOMETRY resulting from the application of the INVERSION operation. It can be especially powerful for solving apparently difficult problems such as STEINER'S PORISM and APOLLONIUS' PROBLEM.

see also HEXLET, INVERSE CURVE, INVERSION, PEAUCELLIER INVERSOR, POLAR, POLE (GEOMETRY), POWER (CIRCLE), RADICAL LINE

References

Ogilvy, C. S. "Inversive Geometry" and "Applications of Inversive Geometry." Chs. 3—4 in *Excursions in Geometry.* New York: Dover, pp. 24–55, 1990.

Inverted Funnel

see also FUNNEL, SINCLAIR'S SOAP FILM PROBLEM

Inverted Snub Dodecadodecahedron

The UNIFORM POLYHEDRON U_{60} whose DUAL POLYHEDRON is the MEDIAL INVERTED PENTAGONAL HEXECONTAHEDRON. It has WYTHOFF SYMBOL $|2\,\frac{5}{3}\,5$. Its faces are $12\{\frac{5}{3}\}+60\{3\}+12\{5\}$. It has CIRCUMRADIUS for unit edge length of

$$R \approx 0.8516302.$$

References

Wenninger, M. J. *Polyhedron Models.* Cambridge, England: Cambridge University Press, pp. 180–182, 1989.

Invertible Knot

A knot which can be deformed into itself but with the orientation reversed. The simplest noninvertible knot is 08_{017}. No general technique is known for determining if a KNOT is invertible. Burde and Zieschang (1985) give a tabulation from which it is possible to extract the invertible knots up to 10 crossings.

see also AMPHICHIRAL KNOT

References

Burde, G. and Zieschang, H. *Knots.* Berlin: de Gruyter, 1985.

Involuntary

A LINEAR TRANSFORMATION of period two. Since a LINEAR TRANSFORMATION has the form,

$$\lambda' = \frac{\alpha\lambda+\beta}{\gamma\lambda+\delta}, \tag{1}$$

applying the transformation a second time gives

$$\lambda'' = \frac{\alpha\lambda'+\beta}{\gamma\lambda'+\delta} = \frac{(\alpha^2+\beta\gamma)\lambda+\beta(\alpha+\delta)}{(\alpha+\delta)\gamma\lambda+\beta\gamma+\delta^2}. \tag{2}$$

For an involuntary, $\lambda''=\lambda$, so

$$\gamma(\alpha+\delta)\lambda^2+(\delta^2-\alpha^2)\lambda-(\alpha+\delta)\beta=0. \tag{3}$$

Since each COEFFICIENT must vanish separately,

$$\alpha\gamma+\gamma\delta=0 \tag{4}$$

$$\delta^2-\alpha^2=0 \tag{5}$$

$$\alpha\beta+\beta\delta=0. \tag{6}$$

The first equation gives $\delta=\pm\alpha$. Taking $\delta=\alpha$ would require $\gamma=\beta=0$, giving $\lambda=\lambda'$, the identity transformation. Taking $\delta=-\alpha$ gives $\delta=-\alpha$, so

$$\lambda' = \frac{\alpha\lambda+\beta}{\gamma\lambda-\alpha}, \tag{7}$$

the general form of an INVOLUTION.

see also CROSS-RATIO, INVOLUTION (LINE)

References

Woods, F. S. *Higher Geometry: An Introduction to Advanced Methods in Analytic Geometry.* New York: Dover, pp. 14–15, 1961.

Involute

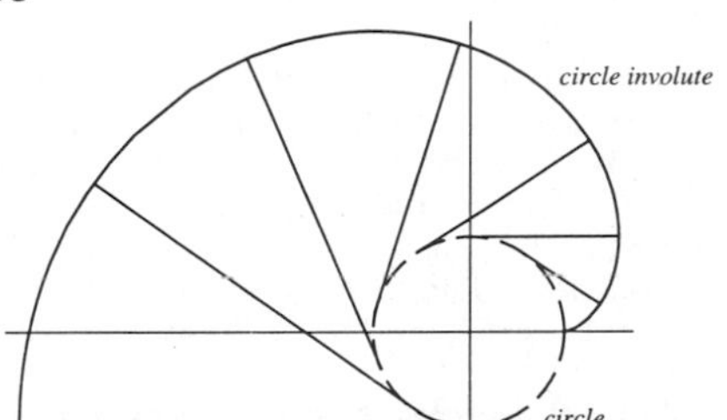

Attach a string to a point on a curve. Extend the string so that it is tangent to the curve at the point of attachment. Then wind the string up, keeping it always taut. The LOCUS of points traced out by the end of the string is the involute of the original curve, and the original curve is called the EVOLUTE of its involute. Although a curve has a unique EVOLUTE, it has infinitely many involutes corresponding to different choices of initial point. An involute can also be thought of as any curve ORTHOGONAL to all the TANGENTS to a given curve.

The equation of the involute is

$$\mathbf{r}_i = \mathbf{r} - s\hat{\mathbf{T}}, \tag{1}$$

where $\hat{\mathbf{T}}$ is the TANGENT VECTOR

$$\hat{\mathbf{T}} = \frac{\frac{d\mathbf{r}}{dt}}{\left|\frac{d\mathbf{r}}{dt}\right|} \tag{2}$$

and s is the ARC LENGTH

$$s = \int ds = \int \frac{ds}{dt}\, dt = \int \frac{\sqrt{ds^2}}{dt}\, dt = \int \sqrt{f'^2 + g'^2}\, dt. \tag{3}$$

This can be written for a parametrically represented function $(f(t), g(t))$ as

$$x(t) = f - \frac{sf'}{\sqrt{f'^2 + g'^2}} \tag{4}$$

$$y(t) = g - \frac{sg'}{\sqrt{f'^2 + g'^2}}. \tag{5}$$

Curve	Involute
astroid	astroid 1/2 as large
cardioid	cardioid 3 times as large
catenary	tractrix
circle catacaustic for a point source	limaçon
circle	circle involute (a spiral)
cycloid	equal cycloid
deltoid	deltoid 1/3 as large
ellipse	ellipse involute
epicycloid	reduced epicycloid
hypocycloid	similar hypocycloid
logarithmic spiral	equal logarithmic spiral
Neile's parabola	parabola
nephroid	Cayley's sextic
nephroid	nephroid 2 times as large

see also EVOLUTE, HUMBERT'S THEOREM

References

Cundy, H. and Rollett, A. "Roulettes and Involutes." §2.6 in *Mathematical Models, 3rd ed.* Stradbroke, England: Tarquin Pub., pp. 46–55, 1989.

Dixon, R. "String Drawings." Ch. 2 in *Mathographics.* New York: Dover, pp. 75–78, 1991.

Gray, A. "Involutes." §5.4 in *Modern Differential Geometry of Curves and Surfaces.* Boca Raton, FL: CRC Press, pp. 81–85, 1993.

Lawrence, J. D. *A Catalog of Special Plane Curves.* New York: Dover, pp. 40–42 and 202, 1972.

Lee, X. "Involute." `http://www.best.com/~xah/SpecialPlaneCurves_dir/Involute_dir/involute.html`.

Lockwood, E. H. "Evolutes and Involutes." Ch. 21 in *A Book of Curves.* Cambridge, England: Cambridge University Press, pp. 166–171, 1967.

Pappas, T. "The Involute." *The Joy of Mathematics.* San Carlos, CA: Wide World Publ./Tetra, p. 187, 1989.

Yates, R. C. "Involutes." *A Handbook on Curves and Their Properties.* Ann Arbor, MI: J. W. Edwards, pp. 135–137, 1952.

Involution (Group)

An element of order 2 in a GROUP (i.e., an element A of a GROUP such that $A^2 = I$, where I is the IDENTITY ELEMENT).

see also GROUP, IDENTITY ELEMENT

Involution (Line)

Pairs of points of a line, the product of whose distances from a FIXED POINT is a given constant. This is more concisely defined as a PROJECTIVITY of period two.

see also INVOLUNTARY

Involution (Operator)

An OPERATOR of period 2, i.e., an OPERATOR $*$ which satisfies $((a)^*)^* = a$.

Involution (Set)

An involution of a SET S is a PERMUTATION of S which does not contain any cycles of length > 2. The PERMUTATION MATRICES of an involution are SYMMETRIC. The number of involutions $I(n)$ of a SET containing the first n integers is given by the RECURRENCE RELATION

$$I(n) = I(n-1) + (n-1)I(n-2).$$

For $n = 1, 2, \ldots$, the first few values of $I(n)$ are 1, 2, 4, 10, 26, 76, ... (Sloane's A000085). The number of involutions on n symbols cannot be expressed as a fixed number of hypergeometric terms (Petkovšek *et al.* 1996, p. 160).

see also PERMUTATION

References

Petkovšek, M.; Wilf, H. S.; and Zeilberger, D. *A=B.* Wellesley, MA: A. K. Peters, 1996.

Ruskey, F. "Information on Involutions." `http://sue.csc.uvic.ca/~cos/inf/perm/Involutions.html`.

Sloane, N. J. A. Sequence A00085/M1221 in "An On-Line Version of the Encyclopedia of Integer Sequences."

Involution (Transformation)

A TRANSFORMATION of period 2.

Irradiation Illusion

The ILLUSION shown above which was discovered by Helmholtz in the 19th century. Despite the fact that the two above figures are identical in size, the white hole looks bigger than the black one in this ILLUSION.

References

Pappas, T. "Irradiation Optical Illusion." *The Joy of Mathematics.* San Carlos, CA: Wide World Publ./Tetra, p. 199, 1989.

Irrational Number

A number which cannot be expressed as a FRACTION p/q for any INTEGERS p and q. Every TRANSCENDENTAL NUMBER is irrational. Numbers of the form $n^{1/m}$ are irrational unless n is the mth POWER of an INTEGER.

Numbers of the form $\log_n m$, where log is the LOGARITHM, are irrational if m and n are INTEGERS, one of which has a PRIME factor which the other lacks. e^r is irrational for rational $r \neq 0$. The irrationality of e was proven by Lambert in 1761; for the general case, see Hardy and Wright (1979, p. 46). π^n is irrational for POSITIVE integral n. The irrationality of π was proven by Lambert in 1760; for the general case, see Hardy and Wright (1979, p. 47). APÉRY'S CONSTANT $\zeta(3)$ (where $\zeta(z)$ is the RIEMANN ZETA FUNCTION) was proved irrational by Apéry (Apéry 1979, van der Poorten 1979).

From GELFOND'S THEOREM, a number of the form a^b is TRANSCENDENTAL (and therefore irrational) if a is ALGEBRAIC $\neq 0$, 1 and b is irrational and ALGEBRAIC. This establishes the irrationality of e^π (since $(-1)^{-i} = (e^{i\pi})^{-i} = e^\pi$), $2^{\sqrt{2}}$, and $e\pi$. Nesterenko (1996) proved that $\pi + e^\pi$ is irrational. In fact, he proved that π, e^π and $\Gamma(1/4)$ are algebraically independent, but it was not previously known that $\pi + e^\pi$ was irrational.

Given a POLYNOMIAL equation

$$x^m + c_{m-1}x^{m-1} + \ldots + c_0, \qquad (1)$$

where c_i are INTEGERS, the roots x_i are either integral or irrational. If $\cos(2\theta)$ is irrational, then so are $\cos\theta$, $\sin\theta$, and $\tan\theta$.

Irrationality has not yet been established for 2^e, π^e, $\pi^{\sqrt{2}}$, or γ (where γ is the EULER-MASCHERONI CONSTANT).

QUADRATIC SURDS are irrational numbers which have periodic CONTINUED FRACTIONS.

HURWITZ'S IRRATIONAL NUMBER THEOREM gives bounds of the form

$$\left|\alpha - \frac{p}{q}\right| < \frac{1}{L_n q^2} \qquad (2)$$

for the best rational approximation possible for an arbitrary irrational number α, where the L_n are called LAGRANGE NUMBERS and get steadily larger for each "bad" set of irrational numbers which is excluded.

The SERIES

$$\sum_{n=1}^\infty \frac{\sigma_k(n)}{n!}, \qquad (3)$$

where $\sigma_k(n)$ is the DIVISOR FUNCTION, is irrational for $k = 1$ and 2. The series

$$\sum_{n=1}^\infty \frac{1}{2^n - 1} = \sum_{n=1}^\infty \frac{d(n)}{2^n}, \qquad (4)$$

where $d(n)$ is the number of divisors of n, is also irrational, as are

$$\sum_{n=1}^\infty \frac{1}{q^n + r} \quad \text{for} \quad \sum_{n=1}^\infty \frac{(-1)^n}{q^n + r} \qquad (5)$$

for q an INTEGER other than p, ± 1, and r a RATIONAL NUMBER other than 0 or $-q^n$ (Guy 1994).

see also ALGEBRAIC INTEGER, ALGEBRAIC NUMBER, ALMOST INTEGER, DIRICHLET FUNCTION, FERGUSON-FORCADE ALGORITHM, GELFOND'S THEOREM, HURWITZ'S IRRATIONAL NUMBER THEOREM, NEAR NOBLE NUMBER, NOBLE NUMBER, PYTHAGORAS'S THEOREM, QUADRATIC IRRATIONAL NUMBER, RATIONAL NUMBER, SEGRE'S THEOREM, TRANSCENDENTAL NUMBER

References

Apéry, R. "Irrationalité de $\zeta(2)$ et $\zeta(3)$." *Astérisque* **61**, 11–13, 1979.

Courant, R. and Robbins, H. "Incommensurable Segments, Irrational Numbers, and the Concept of Limit." §2.2 in *What is Mathematics?: An Elementary Approach to Ideas and Methods, 2nd ed.* Oxford, England: Oxford University Press, pp. 58-61, 1996.

Guy, R. K. "Some Irrational Series." §B14 in *Unsolved Problems in Number Theory, 2nd ed.* New York: Springer-Verlag, p. 69, 1994.

Hardy, G. H. and Wright, E. M. *An Introduction to the Theory of Numbers, 5th ed.* Oxford, England: Clarendon Press, 1979.

Manning, H. P. *Irrational Numbers and Their Representation by Sequences and Series.* New York: Wiley, 1906.

Nesterenko, Yu. "Modular Functions and Transcendence Problems." *C. R. Acad. Sci. Paris Sér. I Math.* **322**, 909–914, 1996.

Nesterenko, Yu. V. "Modular Functions and Transcendence Questions." *Mat. Sb.* **187**, 65–96, 1996.

Niven, I. M. *Irrational Numbers.* New York: Wiley, 1956.

Niven, I. M. *Numbers: Rational and Irrational.* New York: Random House, 1961.

Pappas, T. "Irrational Numbers & the Pythagoras Theorem." *The Joy of Mathematics.* San Carlos, CA: Wide World Publ./Tetra, pp. 98–99, 1989.

van der Poorten, A. "A Proof that Euler Missed... Apéry's Proof of the Irrationality of $\zeta(3)$." *Math. Intel.* **1**, 196–203, 1979.

Irrationality Measure

see LIOUVILLE-ROTH CONSTANT

Irrationality Sequence

A sequence of POSITIVE INTEGERS $\{a_n\}$ such that $\sum 1/(a_n b_n)$ is IRRATIONAL for all integer sequences $\{b_n\}$. Erdős showed that $\{2^{2^n}\}$ is an irrationality sequence.

References

Guy, R. K. "Irrationality Sequence." §E24 in *Unsolved Problems in Number Theory, 2nd ed.* New York: Springer-Verlag, p. 225, 1994.

Irreducible Matrix

A SQUARE MATRIX which is not REDUCIBLE is said to be irreducible.

Irreducible Polynomial

A POLYNOMIAL or polynomial equation is said to be irreducible if it cannot be factored into polynomials of lower degree over the same FIELD.

The number of binary irreducible polynomials of degree n is equal to the number of n-bead fixed NECKLACES of two colors: 1, 2, 3, 4, 6, 8, 14, 20, 36, ... (Sloane's A000031), the first few of which are given in the following table.

n	Polynomials
1	x
2	$x,\ x+1$
3	$x,\ x^2+x+1,\ x+1$
4	$x,\ x^3+x+1,\ x^3+x^2+1,\ x+1$

see also FIELD, GALOIS FIELD, NECKLACE, POLYNOMIAL, PRIMITIVE IRREDUCIBLE POLYNOMIAL

References

Sloane, N. J. A. Sequences A000031/M0564 in "An On-Line Version of the Encyclopedia of Integer Sequences."

Sloane, N. J. A. and Plouffe, S. Extended entry in *The Encyclopedia of Integer Sequences*. San Diego: Academic Press, 1995.

Irreducible Representation

An irreducible representation of a GROUP is a representation for which there exists no UNITARY TRANSFORMATION which will transform the representation MATRIX into block diagonal form. The irreducible representation has a number of remarkable properties.

see also GROUP, ITÔ'S THEOREM, UNITARY TRANSFORMATION

Irreducible Semiperfect Number

see PRIMITIVE PSEUDOPERFECT NUMBER

Irreducible Tensor

Given a general second RANK TENSOR A_{ij} and a METRIC g_{ij}, define

$$\theta \equiv A_{ij}g^{ij} = A_i^i \tag{1}$$

$$\omega^i \equiv \epsilon^{ijk}A_{jk} \tag{2}$$

$$\sigma_{ij} \equiv \tfrac{1}{2}(A_{ij}+A_{ji}) - \tfrac{1}{3}g_{ij}A_k^k, \tag{3}$$

where δ_{ij} is the KRONECKER DELTA and ϵ^{ijk} is the LEVI-CIVITA SYMBOL. Then

$$\begin{aligned}\sigma_{ij} &+ \tfrac{1}{3}\theta g_{ij} + \tfrac{1}{2}\epsilon_{ijk}\omega^k \\ &= [\tfrac{1}{2}(A_{ij}+A_{ji}) - \tfrac{1}{3}g_{ij}A_k^k] + \tfrac{1}{3}A_k^k g_{ij} + \tfrac{1}{2}\epsilon_{ijk}[\epsilon^{\lambda\mu k}A_{\lambda\mu}] \\ &= \tfrac{1}{2}(A_{ij}+A_{ji}) + \tfrac{1}{2}(\delta_i^\lambda\delta_j^\mu - \delta_i^\mu\delta_j^\lambda)A_{\lambda\mu} \\ &= \tfrac{1}{2}(A_{ij}+A_{ji}) + \tfrac{1}{2}(A_{ij}-A_{ji}) = A_{ij},\end{aligned} \tag{4}$$

where θ, ω^i, and σ_{ij} are TENSORS of RANK 0, 1, and 2.

see also TENSOR

Irredundant Ramsey Number

Let G_1, G_2, ..., G_t be a t-EDGE coloring of the COMPLETE GRAPH K_n, where for each $i = 1, 2, \ldots, t$, G_i is the spanning SUBGRAPH of K_n consisting of all EDGES colored with the ith color. The irredundant Ramsey number $s(q_1, \ldots, q_t)$ is the smallest INTEGER n such that for any t-EDGE coloring of K_n, the COMPLEMENT GRAPH $\overline{G_i}$ has an irredundant set of size q_i for at least one $i = 1, \ldots, t$. Irredundant Ramsey numbers were introduced by Brewster *et al.* (1989) and satisfy

$$s(q_1, \ldots, q_t) \leq R(q_1, \ldots, q_t).$$

For a summary, see Mynhardt (1992).

s	Bounds	Reference
$s(3,3)$	6	Brewster *et al.* 1989
$s(3,4)$	8	Brewster *et al.* 1989
$s(3,5)$	12	Brewster *et al.* 1989
$s(3,6)$	15	Brewster *et al.* 1990
$s(3,7)$	18	Chen and Rousseau 1995, Cockayne *et al.* 1991
$s(4,4)$	13	Cockayne *et al.* 1992
$s(3,3,3)$	13	Cockayne and Mynhardt 1994

References

Brewster, R. C.; Cockayne, E. J.; and Mynhardt, C. M. "Irredundant Ramsey Numbers for Graphs." *J. Graph Theory* **13**, 283–290, 1989.

Brewster, R. C.; Cockayne, E. J.; and Mynhardt, C. M. "The Irredundant Ramsey Number $s(3,6)$." *Quaest. Math.* **13**, 141–157, 1990.

Chen, G. and Rousseau, C. C. "The Irredundant Ramsey Number $s(3,7)$." *J. Graph. Th.* **19**, 263–270, 1995.

Cockayne, E. J.; Exoo, G.; Hattingh, J. H.; and Mynhardt, C. M. "The Irredundant Ramsey Number $s(4,4)$." *Util. Math.* **41**, 119–128, 1992.

Cockayne, E. J.; Hattingh, J. H.; and Mynhardt, C. M. "The Irredundant Ramsey Number $s(3,7)$." *Util. Math.* **39**, 145–160, 1991.

Cockayne, E. J. and Mynhardt, C. M. "The Irredundant Ramsey Number $s(3,3,3) = 13$." *J. Graph. Th.* **18**, 595–604, 1994.

Hattingh, J. H. "On Irredundant Ramsey Numbers for Graphs." *J. Graph Th.* **14**, 437–441, 1990.

Mynhardt, C. M. "Irredundant Ramsey Numbers for Graphs: A Survey." *Congres. Numer.* **86**, 65–79, 1992.

Irreflexive

A RELATION R on a SET S is irreflexive provided that no element is related to itself; in other words, xRx for no x in S.

see also RELATION

Irregular Pair

If p divides the NUMERATOR of the BERNOULLI NUMBER B_{2k} for $0 < 2k < p-1$, then $(p, 2k)$ is called an irregular pair. For $p < 30000$, the irregular pairs of various forms are $p = 16843$ for $(p, p-3)$, $p = 37$ for $(p, p-5)$, none for $(p, p-7)$, and $p = 67,877$ for $(p, p-9)$.

see also BERNOULLI NUMBER, IRREGULAR PRIME

References

Johnson, W. "Irregular Primes and Cyclotomic Invariants." *Math. Comput.* **29**, 113–120, 1975.

Irregular Prime

PRIMES for which Kummer's theorem on the unsolvability of FERMAT'S LAST THEOREM does not apply. An irregular prime p divides the NUMERATOR of one of the BERNOULLI NUMBERS B_{10}, B_{12}, ..., B_{2p-2}, as shown by Kummer in 1850. The FERMAT EQUATION has no solutions for REGULAR PRIMES.

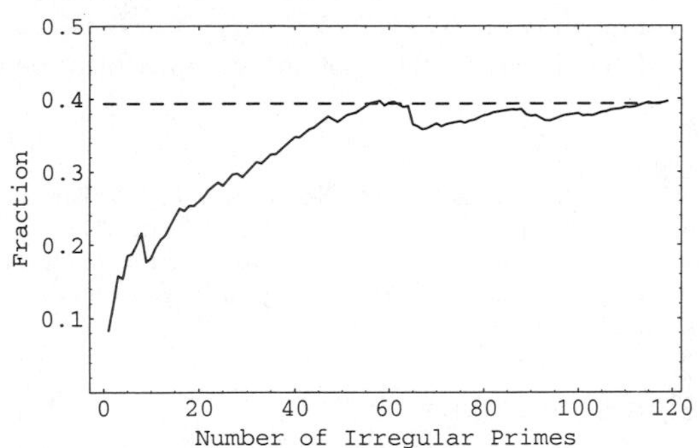

An INFINITE number of irregular PRIMES exist, as proven in 1915 by Jensen. The first few irregular primes are 37, 59, 67, 101, 103, 131, 149, 157, ... (Sloane's A000928). Of the 283,145 PRIMES less than 4×10^6, 111,597 (or 39.41%) are regular. The conjectured FRACTION is $1 - e^{-1/2} \approx 39.35\%$ (Ribenboim 1996, p. 415).

see also BERNOULLI NUMBER, FERMAT'S LAST THEOREM, IRREGULAR PAIR, REGULAR PRIME

References

Buhler, J.; Crandall, R.; Ernvall, R.; and Metsänkylä, T. "Irregular Primes and Cyclotomic Invariants to Four Million." *Math. Comput.* **60**, 151–153, 1993.

Hardy, G. H. and Wright, E. M. *An Introduction to the Theory of Numbers, 5th ed.* Oxford, England: Clarendon Press, p. 202, 1979.

Johnson, W. "Irregular Primes and Cyclotomic Invariants." *Math. Comput.* **29**, 113–120, 1975.

Ribenboim, P. *The New Book of Prime Number Records.* New York: Springer-Verlag, pp. 325–329 and 414–425, 1996.

Sloane, N. J. A. Sequence A000928/M5260 in "An On-Line Version of the Encyclopedia of Integer Sequences."

Stewart, C. L. "A Note on the Fermat Equation." *Mathematika* **24**, 130–132, 1977.

Irregular Singularity

Consider a second-order ORDINARY DIFFERENTIAL EQUATION

$$y'' + P(x)y' + Q(x)y = 0.$$

If $P(x)$ and $Q(x)$ remain FINITE at $x = x_0$, then x_0 is called an ORDINARY POINT. If either $P(x)$ or $Q(x)$ diverges as $x \to x_0$, then x_0 is called a singular point. If $P(x)$ diverges more quickly than $1/(x - x_0)$, so $(x - x_0)P(x)$ approaches INFINITY as $x \to x_0$, or $Q(x)$ diverges more quickly than $1/(x - x_0)^2 Q$ so that $(x - x_0)^2 Q(x)$ goes to INFINITY as $x \to x_0$, then x_0 is called an IRREGULAR SINGULARITY (or ESSENTIAL SINGULARITY).

see also ORDINARY POINT, REGULAR SINGULAR POINT, SINGULAR POINT (DIFFERENTIAL EQUATION)

References

Arfken, G. "Singular Points." §8.4 in *Mathematical Methods for Physicists, 3rd ed.* Orlando, FL: Academic Press, pp. 451–453 and 461–463, 1985.

Irrotational Field

A VECTOR FIELD $\mathbf{v}$ for which the CURL vanishes,

$$\nabla \times \mathbf{v} = \mathbf{0}.$$

see also BELTRAMI FIELD, CONSERVATIVE FIELD, SOLENOIDAL FIELD, VECTOR FIELD

Isarithm

see EQUIPOTENTIAL CURVE

ISBN

Publisher	Digits
Addison-Wesley	0201
Amer. Math. Soc.	0821
Cambridge University Press	0521
CRC Press	0849
Dover	0486
McGraw-Hill	0070
Oxford University Press	0198
Springer-Verlag	0387
Wiley	0471

The International Standard Book Number (ISBN) is a 10-digit CODE which is used to identify a book uniquely. The first four digits specify the publisher, the next five digits the book, and the last digit d_{10} is a check digit which may be in the range 0–9 or X (where X equals 10). The check digit is computed from the equation

$$10d_1 + 9d_2 + 8d_3 + \ldots + 2d_9 + d_{10} \equiv 0 \pmod{11}.$$

For example, the number for this book is 0-8493-9640-9, and

$$\begin{aligned} &10 \cdot 0 + 9 \cdot 8 + 8 \cdot 4 + 7 \cdot 9 + 6 \cdot 3 + 5 \cdot 9 \\ &+4 \cdot 6 + 3 \cdot 4 + 2 \cdot 0 + 1 \cdot 9 = 275 = 25 \cdot 11 \equiv 0 \pmod{11}, \end{aligned}$$

as required.

see also CODE

References

Press, W. H.; Flannery, B. P.; Teukolsky, S. A.; and Vetterling, W. T. *Numerical Recipes in FORTRAN: The Art of Scientific Computing, 2nd ed.* Cambridge, England: Cambridge University Press, p. 894, 1992.

Island

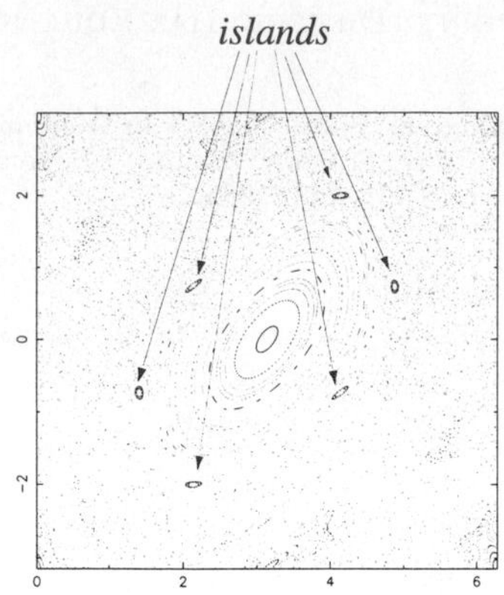

If an integrable QUASIPERIODIC system is slightly perturbed so that it becomes nonintegrable, only a finite number of n-CYCLES remain as a result of MODE LOCKING. One will be elliptical and one will be hyperbolic.

Surrounding the ELLIPTIC FIXED POINT is a region of stable ORBITS which circle it, as illustrated above in the STANDARD MAP with $K = 1.5$. As the map is iteratively applied, the island is mapped to a similar structure surrounding the next point of the elliptic cycle. The map thus has a chain of islands, with the FIXED POINT alternating between ELLIPTIC (at the center of the islands) and HYPERBOLIC (between islands). Because the unperturbed system goes through an INFINITY of rational values, the perturbed system must have an INFINITE number of island chains.

see also MODE LOCKING, ORBIT (MAP), QUASIPERIODIC FUNCTION

Isobaric Polynomial

A POLYNOMIAL in which the sum of SUBSCRIPTS is the same in each term.

see also HOMOGENEOUS POLYNOMIAL

Isochronous Curve

see SEMICUBICAL PARABOLA, TAUTOCHRONE PROBLEM

Isoclinal

see ISOCLINE

Isocline

A graphical method of solving an ORDINARY DIFFERENTIAL EQUATION of the form

$$\frac{dy}{dx} = f(x, y)$$

by plotting a series of curves $f(x, y) = [\text{const}]$, then drawing a curve PERPENDICULAR to each curve such that it satisfies the initial condition. This curve is the solution to the ORDINARY DIFFERENTIAL EQUATION.

References

Kármán, T. von and Biot, M. A. *Mathematical Methods in Engineering: An Introduction to the Mathematical Treatment of Engineering Problems.* New York: McGraw-Hill, pp. 3 and 7, 1940.

Isoclinic Groups

Two GROUPS G and H are said to be isoclinic if there are isomorphisms $G/Z(G) \to H/Z(H)$ and $G' \to H'$, where $Z(G)$ is the CENTER of the group, which identify the two commutator maps.

References

Conway, J. H.; Curtis, R. T.; Norton, S. P.; Parker, R. A.; and Wilson, R. A. "Isoclinism." §6.7 in *Atlas of Finite Groups: Maximal Subgroups and Ordinary Characters for Simple Groups.* Oxford, England: Clarendon Press, pp. xxiii–xxiv, 1985.

Isodynamic Points

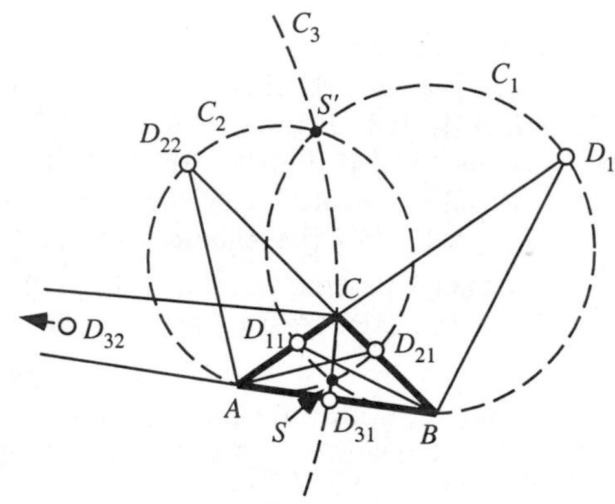

The first and second isodynamic points of a TRIANGLE ΔABC can be constructed by drawing the triangle's ANGLE BISECTORS and EXTERIOR ANGLE BISECTORS. Each pair of bisectors intersects a side of the triangle (or its extension) in two points D_{i1} and D_{i2}, for $i = 1$, 2, 3. The three CIRCLES having $D_{11}D_{12}$, $D_{21}D_{22}$, and $D_{31}D_{32}$ as DIAMETERS are the APOLLONIUS CIRCLES C_1, C_2, and C_3. The points S and S' in which the three APOLLONIUS CIRCLES intersect are the first and second isodynamic points, respectively.

S and S' have TRIANGLE CENTER FUNCTIONS

$$\alpha = \sin(A \pm \tfrac{1}{3}\pi),$$

respectively. The ANTIPEDAL TRIANGLES of both points are EQUILATERAL and have AREAS

$$\Delta' = 2\Delta[\cot\omega \cot(\tfrac{1}{3}\pi)],$$

where ω is the BROCARD ANGLE.

The isodynamic points are ISOGONAL CONJUGATES of the ISOGONIC CENTERS. They lie on the BROCARD AXIS. The distances from either isodynamic point to the VERTICES are inversely proportional to the sides. The PEDAL TRIANGLE of either isodynamic point is an EQUILATERAL TRIANGLE. An INVERSION with either isodynamic point as the INVERSION CENTER transforms the triangle into an EQUILATERAL TRIANGLE.

The CIRCLE which passes through both the isodynamic points and the CENTROID of a TRIANGLE is known as the PARRY CIRCLE.

see also APOLLONIUS CIRCLES, BROCARD AXIS, CENTROID (TRIANGLE), ISOGONIC CENTERS, PARRY CIRCLE

References

Gallatly, W. *The Modern Geometry of the Triangle, 2nd ed.* London: Hodgson, p. 106, 1913.

Johnson, R. A. *Modern Geometry: An Elementary Treatise on the Geometry of the Triangle and the Circle.* Boston, MA: Houghton Mifflin, pp. 295–297, 1929.

Kimberling, C. "Central Points and Central Lines in the Plane of a Triangle." *Math. Mag.* **67**, 163–187, 1994.

Isoenergetic Nondegeneracy

The condition for isoenergetic nondegeneracy for a Hamiltonian

$$H = H_0(\mathbf{I}) + \epsilon H_1(\mathbf{I}, \boldsymbol{\theta})$$

is

$$\begin{vmatrix} \frac{\partial^2 H_0}{\partial I_i \partial I_j} & \frac{\partial H_0}{\partial I_i} \\ \frac{\partial H_0}{\partial I_j} & 0 \end{vmatrix} \neq 0,$$

which guarantees the EXISTENCE on every energy level surface of a set of invariant tori whose complement has a small MEASURE.

References

Tabor, M. *Chaos and Integrability in Nonlinear Dynamics: An Introduction.* New York: Wiley, pp. 113–114, 1989.

Isogonal Conjugate

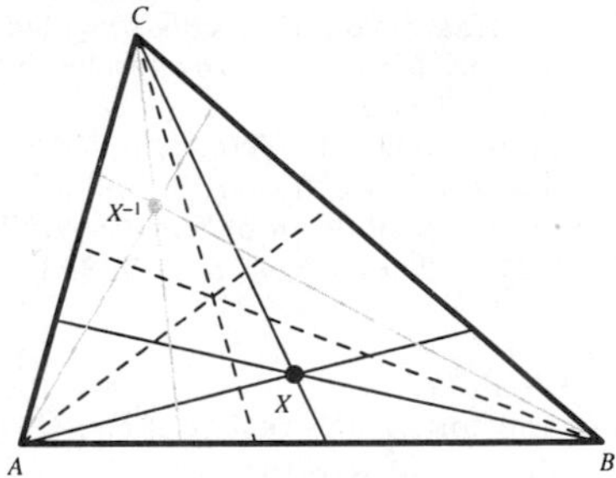

The isogonal conjugate X^{-1} of a point X in the plane of the TRIANGLE ΔABC is constructed by reflecting the lines AX, BX, and CX about the ANGLE BISECTORS at A, B, and C. The three reflected lines CONCUR at the isogonal conjugate. The TRILINEAR COORDINATES of the isogonal conjugate of the point with coordinates

$$\alpha : \beta : \gamma$$

are

$$\alpha^{-1} : \beta^{-1} : \gamma^{-1}.$$

Isogonal conjugation maps the interior of a TRIANGLE onto itself. This mapping transforms lines onto CONIC SECTIONS that CIRCUMSCRIBE the TRIANGLE. The type of CONIC SECTION is determined by whether the line d meets the CIRCUMCIRCLE C',

1. If d does not intersect C', the isogonal transform is an ELLIPSE;
2. If d is tangent to C', the transform is a PARABOLA;
3. If d cuts C', the transform is a HYPERBOLA, which is a RECTANGULAR HYPERBOLA if the line passes through the CIRCUMCENTER

(Casey 1893, Vandeghen 1965).

The isogonal conjugate of a point on the CIRCUMCIRCLE is a POINT AT INFINITY (and conversely). The sides of the PEDAL TRIANGLE of a point are PERPENDICULAR to the connectors of the corresponding VERTICES with the isogonal conjugate. The isogonal conjugate of a set of points is the LOCUS of their isogonal conjugate points.

The product of ISOTOMIC and isogonal conjugation is a COLLINEATION which transforms the sides of a TRIANGLE to themselves (Vandeghen 1965).

see also ANTIPEDAL TRIANGLE, COLLINEATION, ISOGONAL LINE, ISOTOMIC CONJUGATE POINT, LINE AT INFINITY, SYMMEDIAN LINE

References

Casey, J. *A Treatise on the Analytical Geometry of the Point, Line, Circle, and Conic Sections, Containing an Account of Its Most Recent Extensions with Numerous Examples, 2nd rev. enl. ed.* Dublin: Hodges, Figgis, & Co., 1893.

Johnson, R. A. *Modern Geometry: An Elementary Treatise on the Geometry of the Triangle and the Circle.* Boston, MA: Houghton Mifflin, pp. 153–158, 1929.

Vandeghen, A. "Some Remarks on the Isogonal and Cevian Transforms. Alignments of Remarkable Points of a Triangle." *Amer. Math. Monthly* **72**, 1091–1094, 1965.

Isogonal Line

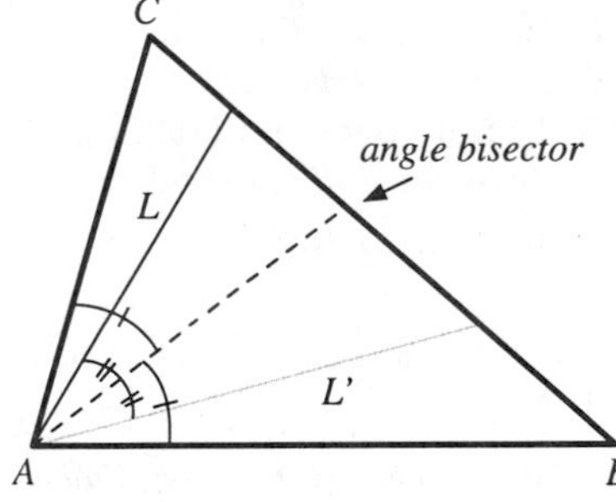

The line L' through a TRIANGLE VERTEX obtained by reflecting an initial line L (also through a VERTEX) about the ANGLE BISECTOR. If three lines from the VERTICES of a TRIANGLE ΔABC are CONCURRENT at $X = L_1L_2L_3$, then their isogonal lines are also CONCURRENT, and the point of concurrence $X' = L_1'L_2'L_3'$ is called the ISOGONAL CONJUGATE point.

see also ISOGONAL CONJUGATE

References

Johnson, R. A. *Modern Geometry: An Elementary Treatise on the Geometry of the Triangle and the Circle.* Boston, MA: Houghton Mifflin, pp. 153–157, 1929.

Isogonic Centers

The first isogonic center F_1 of a TRIANGLE is the FERMAT POINT. The second isogonic center F_2 is constructed analogously with the first isogonic center except that for F_2, the EQUILATERAL TRIANGLES are constructed on the same side of the opposite VERTEX. The second isogonic center has TRIANGLE CENTER FUNCTION

$$\alpha = \csc(A - \tfrac{1}{3}\pi).$$

Its ANTIPEDAL TRIANGLE is EQUILATERAL and has AREA

$$2\Delta = [-1 + \cot\omega\cot(\tfrac{1}{3}\pi)],$$

where ω is the BROCARD ANGLE.

The first and second isogonic centers are ISOGONAL CONJUGATES of the ISODYNAMIC POINTS.

see also BROCARD ANGLE, EQUILATERAL TRIANGLE, FERMAT POINT, ISODYNAMIC POINTS, ISOGONAL CONJUGATE

References

Gallatly, W. *The Modern Geometry of the Triangle, 2nd ed.* London: Hodgson, p. 107, 1913.

Kimberling, C. "Central Points and Central Lines in the Plane of a Triangle." *Math. Mag.* **67**, 163–187, 1994.

Isograph

The substitution of $re^{i\theta}$ for z in a POLYNOMIAL $p(z)$. $p(z)$ is then plotted as a function of θ for a given r in the COMPLEX PLANE. By varying r so that the curve passes through the ORIGIN, it is possible to determine a value for one ROOT of the POLYNOMIAL.

Isohedral Tiling

Let $S(T)$ be the group of symmetries which map a MONOHEDRAL TILING T onto itself. The TRANSITIVITY CLASS of a given tile T is then the collection of all tiles to which T can be mapped by one of the symmetries of $S(T)$. If T has k TRANSITIVITY CLASSES, then T is said to be k-isohedral. Berglund (1993) gives examples of k-isohedral tilings for $k = 1$, 2, and 4.

see also ANISOHEDRAL TILING

References

Berglund, J. "Is There a k-Anisohedral Tile for $k \geq 5$?" *Amer. Math. Monthly* **100**, 585-588, 1993.

Grünbaum, B. and Shephard, G. C. "The 81 Types of Isohedral Tilings of the Plane." *Math. Proc. Cambridge Philos. Soc.* **82**, 177-196, 1977.

Isohedron

A convex POLYHEDRON with symmetries acting transitively on its faces. Every isohedron has an EVEN number of faces (Grünbaum 1960).

References

Grünbaum, B. "On Polyhedra in E^3 Having All Faces Congruent." *Bull. Research Council Israel* **8F**, 215–218, 1960.

Grünbaum, B. and Shepard, G. C. "Spherical Tilings with Transitivity Properties." In *The Geometric Vein: The Coxeter Festschrift* (Ed. C. Davis, B. Grünbaum, and F. Shenk). New York: Springer-Verlag, 1982.

Isolated Point

A point on a curve, also known as an ACNODE or HERMIT POINT, which has no other points in its NEIGHBORHOOD.

Isolated Singularity

An isolated singularity is a SINGULARITY for which there exists a (small) REAL NUMBER ϵ such that there are no other SINGULARITIES within a NEIGHBORHOOD of radius ϵ centered about the SINGULARITY.

The types of isolated singularities possible for CUBIC SURFACES have been classified (Schläfli 1864, Cayley 1869, Bruce and Wall 1979) and are summarized in the following table from Fischer (1986).

Double Pt. Name	Symbol	Normal Form	Coxeter Diagram
conic	C_2	$x^2+y^2+z^2$	A_1
biplanar	B_3	$x^2+y^2+z^3$	A_2
biplanar	B_4	$x^2+y^2+z^4$	A_3
biplanar	B_5	$x^2+y^2+z^5$	A_4
biplanar	B_6	$x^2+y^2+z^6$	A_5
uniplanar	U_6	$x^2+z(y^2+z^2)$	D_4
uniplanar	U_7	$x^2+z(y^2+z^3)$	D_5
uniplanar	U_8	$x^2+y^3+z^4$	E_6
elliptic cone pt.	—	xy^2-4z^3 $-g_2x^2y+g_3x^3$	$\tilde{E}_6$

see also CUBIC SURFACE, RATIONAL DOUBLE POINT, SINGULARITY

References

Bruce, J. and Wall, C. T. C. "On the Classification of Cubic Surfaces." *J. London Math. Soc.* **19**, 245–256, 1979.

Cayley, A. "A Memoir on Cubic Surfaces." *Phil. Trans. Roy. Soc.* **159**, 231–326, 1869.

Fischer, G. (Ed.). *Mathematical Models from the Collections of Universities and Museums.* Braunschweig, Germany: Vieweg, pp. 12–13, 1986.

Morse, P. M. and Feshbach, H. *Methods of Theoretical Physics, Part I.* New York: McGraw-Hill, pp. 380–381, 1953.

Schläfli, L. "On the Distribution of Surfaces of Third Order into Species." *Phil. Trans. Roy. Soc.* **153**, 193–247, 1864.

Isolating Integral

An integral of motion which restricts the PHASE SPACE available to a DYNAMICAL SYSTEM.

Isometry

A BIJECTIVE MAP between two METRIC SPACES that preserves distances, i.e.,

$$d(f(x), f(y)) = d(x, y),$$

where f is the MAP and $d(a, b)$ is the DISTANCE function.

An isometry of the PLANE is a linear transformation which preserves length. Isometries include ROTATION, TRANSLATION, REFLECTION, GLIDES, and the IDENTITY MAP. If an isometry has more than one FIXED

POINT, it must be either the identity transformation or a reflection. Every isometry of period two (two applications of the transformation preserving lengths in the original configuration) is either a reflection or a half turn rotation. Every isometry in the plane is the product of at most three reflections (at most two if there is a FIXED POINT). Every finite group of isometries has at least one FIXED POINT.

see also DISTANCE, EUCLIDEAN MOTION, HJELMSLEV'S THEOREM, LENGTH (CURVE), REFLECTION, ROTATION, TRANSLATION

References

Gray, A. "Isometries of Surfaces." §13.2 in *Modern Differential Geometry of Curves and Surfaces.* Boca Raton, FL: CRC Press, pp. 255–258, 1993.

Isometric Latitude

An AUXILIARY LATITUDE which is directly proportional to the spacing of parallels of LATITUDE from the equator on an ellipsoidal MERCATOR PROJECTION. It is defined by

$$\psi = \ln\left|\tan(\tfrac{1}{4}\pi + \tfrac{1}{2}\phi)\left(\frac{1-e\sin\phi}{1+e\sin\phi}\right)^{e/2}\right|, \qquad (1)$$

where the symbol τ is sometimes used instead of ψ. The isometric latitude is related to the CONFORMAL LATITUDE by

$$\psi = \ln\tan(\tfrac{1}{4}\pi + \tfrac{1}{2}\chi). \qquad (2)$$

The inverse is found by iterating

$$\phi = 2\tan^{-1}\left[\exp(\psi)\left(\frac{1+e\sin\phi}{1-e\sin\phi}\right)^{e/2}\right] - \tfrac{1}{2}\pi, \qquad (3)$$

with the first trial as

$$\phi_0 = 2\tan^{-1}(e^\psi) - \tfrac{1}{2}\pi. \qquad (4)$$

see also LATITUDE

References

Adams, O. S. "Latitude Developments Connected with Geodesy and Cartography with Tables, Including a Table for Lambert Equal-Area Meridional Projections." Spec. Pub. No. 67. U. S. Coast and Geodetic Survey, 1921.

Snyder, J. P. *Map Projections—A Working Manual.* U. S. Geological Survey Professional Paper 1395. Washington, DC: U. S. Government Printing Office, p. 15, 1987.

Isomorphic Graphs

Two GRAPHS which contain the same number of VERTICES connected in the same way are said to be isomorphic. Formally, two graphs G and H with VERTICES $V_n = \{1, 2, \ldots, n\}$ are said to be isomorphic if there is a PERMUTATION p of V_n such that $\{u, v\}$ is in the set of EDGES $E(G)$ IFF $\{p(u), p(v)\}$ is in the set of EDGES $E(H)$.

References

Chartrand, G. "Isomorphic Graphs." §2.2 in *Introductory Graph Theory.* New York: Dover, pp. 32–40, 1985.

Isomorphic Groups

Two GROUPS are isomorphic if the correspondence between them is ONE-TO-ONE and the "multiplication" table is preserved. For example, the POINT GROUPS C_2 and D_1 are isomorphic GROUPS, written $C_2 \cong D_1$ or $C_2 \rightleftharpoons D_1$ (Shanks 1993). Note that the symbol $\cong$ is also used to denote geometric CONGRUENCE.

References

Shanks, D. *Solved and Unsolved Problems in Number Theory, 4th ed.* New York: Chelsea, 1993.

Isomorphic Posets

Two POSETS are said to be isomorphic if their "structures" are entirely analogous. Formally, POSETS $P = (X, \leq)$ and $Q = (X', \leq')$ are isomorphic if there is a BIJECTION f from X to X' such that $x \leq x'$ precisely when $f(x) \leq' f(x')$.

Isomorphism

Isomorphism is a very general concept which appears in several areas of mathematics. Formally, an isomorphism is BIJECTIVE MORPHISM. Informally, an isomorphism is a map which preserves sets and relations among elements.

A space isomorphism is a VECTOR SPACE in which addition and scalar multiplication are preserved. An isomorphism of a TOPOLOGICAL SPACE is called a HOMEOMORPHISM.

Two groups G_1 and G_2 with binary operators $+$ and $\times$ are isomorphic if there exists a map $f: G_1 \mapsto G_2$ which satisfies

$$f(x+y) = f(x) \times f(y).$$

An isomorphism preserves the identities and inverses of a GROUP. A GROUP which is isomorphic to itself is called an AUTOMORPHISM.

see also AUTOMORPHISM, AX-KOCHEN ISOMORPHISM THEOREM, HOMEOMORPHISM, MORPHISM

Isoperimetric Inequality

Let a PLANE figure have AREA A and PERIMETER p. Let the CIRCLE of PERIMETER p have RADIUS r. Then

$$\frac{4\pi A}{p^2} \leq 1,$$

where the quantity on the left is known as the ISOPERIMETRIC QUOTIENT.

Isoperimetric Point

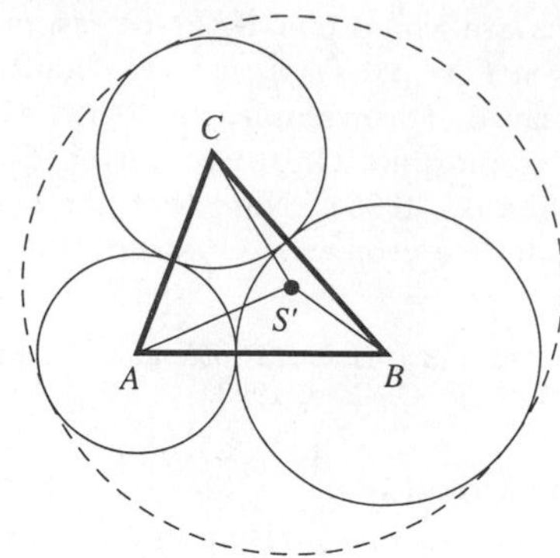

The point S' which makes the PERIMETERS of the TRIANGLES $\Delta BS'C$, $\Delta CS'A$, and $\Delta AS'B$ equal. The isoperimetric point exists IFF the largest ANGLE of the triangle satisfies

$$\max(A, B, C) < 2\sin^{-1}(\tfrac{4}{5}) \approx 1.85459 \text{ rad} \approx 106.26°,$$

or equivalently

$$a + b + c > 4R + r,$$

where a, b, and c are the side lengths of ΔABC, r is the INRADIUS, and R is the CIRCUMRADIUS. The isoperimetric point is also the center of the outer SODDY CIRCLE of ΔABC and has TRIANGLE CENTER FUNCTION

$$\alpha = 1 - \frac{2\Delta}{a(b + c - a)} = \sec(\tfrac{1}{2}A)\cos(\tfrac{1}{2}B)\cos(\tfrac{1}{2}C) - 1.$$

see also EQUAL DETOUR POINT, PERIMETER, SODDY CIRCLES

References

Kimberling, C. "Central Points and Central Lines in the Plane of a Triangle." *Math. Mag.* **67**, 163–187, 1994.

Kimberling, C. "Isoperimetric Point and Equal Detour Point." `http://www.evansville.edu/~ck6/tcenters/recent/isoper.html`.

Kimberling, C. and Wagner, R. W. "Problem E 3020 and Solution." *Amer. Math. Monthly* **93**, 650–652, 1986.

Veldkamp, G. R. "The Isoperimetric Point and the Point(s) of Equal Detour." *Amer. Math. Monthly* **92**, 546–558, 1985.

Isoperimetric Problem

Find a closed plane curve of a given length which encloses the greatest AREA. The solution is a CIRCLE. If the class of curves to be considered is limited to smooth curves, the isoperimetric problem can be stated symbolically as follows: find an arc with parametric equations $x = x(t)$, $y = y(t)$ for $t \in [t_1, t_2]$ such that $x(t_1) = x(t_2)$, $y(t_1) = y(t_2)$ (where no further intersections occur) constrained by

$$l = \int_{t_1}^{t_2} \sqrt{x'^2 + y'^2}\, dt$$

such that

$$A = \tfrac{1}{2}\int_{t_1}^{t_2} (xy' - x'y)\, dt$$

is a MAXIMUM.

see also DIDO'S PROBLEM, ISOVOLUME PROBLEM

References

Bogomolny, A. "Isoperimetric Theorem and Inequality." `http://www.cut-the-knot.com/do_you_know/isoperimetric.html`.

Isenberg, C. Appendix V in *The Science of Soap Films and Soap Bubbles.* New York: Dover, 1992.

Isoperimetric Quotient

A quantity defined in the ISOPERIMETRIC INEQUALITY

$$Q \equiv \frac{4\pi A}{p^2}.$$

see also ISOPERIMETRIC INEQUALITY

Isoperimetric Theorem

Of all convex n-gons of a given PERIMETER, the one which maximizes AREA is the regular n-gon.

see also ISOPERIMETRIC INEQUALITY, ISOPERIMETRIC PROBLEM

Isopleth

see EQUIPOTENTIAL CURVE

Isoptic Curve

For a given curve C, consider the locus of the point P from where the TANGENTS from P to C meet at a fixed given ANGLE. This is called an isoptic curve of the given curve.

Curve	Isoptic
cycloid	curtate or prolate cycloid
epicycloid	epitrochoid
hypocycloid	hypotrochoid
parabola	hyperbola
sinusoidal spiral	sinusoidal spiral

see also ORTHOPTIC CURVE

References

Lawrence, J. D. *A Catalog of Special Plane Curves.* New York: Dover, pp. 58–59 and 206, 1972.

Yates, R. C. "Isoptic Curves." *A Handbook on Curves and Their Properties.* Ann Arbor, MI: J. W. Edwards, pp. 138–140, 1952.

Isosceles Tetrahedron

A nonregular TETRAHEDRON in which each pair of opposite EDGES are equal such that all triangular faces are congruent. A TETRAHEDRON is isosceles IFF the sum of the face angles at each VERTEX is 180°, and IFF its INSPHERE and CIRCUMSPHERE are concentric.

The only way for all the faces of a TETRAHEDRON to have the same PERIMETER or to have the same AREA is for them to be fully congruent, in which case the tetrahedron is isosceles.

see also CIRCUMSPHERE, INSPHERE, ISOSCELES TRIANGLE, TETRAHEDRON

References

Brown, B. H. "Theorem of Bang. Isosceles Tetrahedra." *Amer. Math. Monthly* **33**, 224–226, 1926.

Honsberger, R. "A Theorem of Bang and the Isosceles Tetrahedron." Ch. 9 in *Mathematical Gems II*. Washington, DC: Math. Assoc. Amer., pp. 90–97, 1976.

Isosceles Triangle

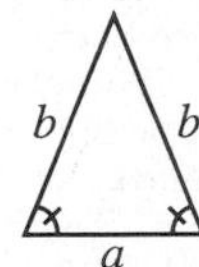

A TRIANGLE with two equal sides (and two equal ANGLES). The name derives from the Greek *iso* (same) and *skelos* (LEG). The height of the above isosceles triangle can be found from the PYTHAGOREAN THEOREM as

$$h = \sqrt{b^2 - \tfrac{1}{4}a^2}. \qquad (1)$$

The AREA is therefore given by

$$A = \tfrac{1}{2}ah = \tfrac{1}{2}a\sqrt{b^2 - \tfrac{1}{4}a^2}. \qquad (2)$$

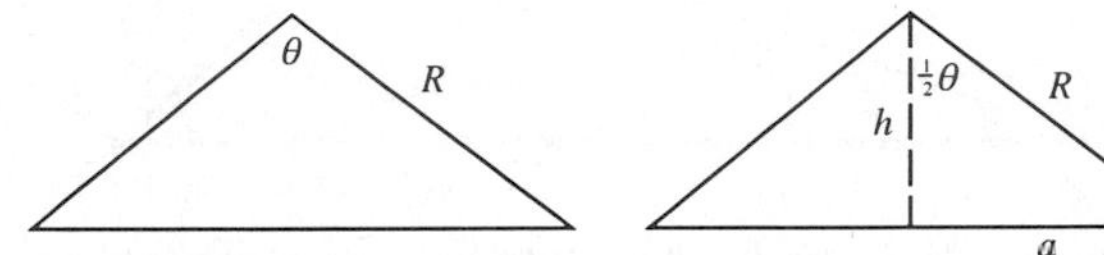

There is a surprisingly simple relationship between the AREA and VERTEX ANGLE θ. As shown in the above diagram, simple TRIGONOMETRY gives

$$h = R\cos(\tfrac{1}{2}\theta) \qquad (3)$$

$$a = R\sin(\tfrac{1}{2}\theta), \qquad (4)$$

so the AREA is

$$A = \tfrac{1}{2}(2a)h = ah = R^2\cos(\tfrac{1}{2}\theta)\sin(\tfrac{1}{2}\theta) = \tfrac{1}{2}R^2\sin\theta. \qquad (5)$$

No set of $n > 6$ points in the PLANE can determine only ISOSCELES TRIANGLES.

see also ACUTE TRIANGLE, EQUILATERAL TRIANGLE, INTERNAL BISECTORS PROBLEM, ISOSCELES TETRAHEDRON, ISOSCELIZER, OBTUSE TRIANGLE, POINT PICKING, PONS ASINORUM, RIGHT TRIANGLE, SCALENE TRIANGLE, STEINER-LEHMUS THEOREM

Isoscelizer

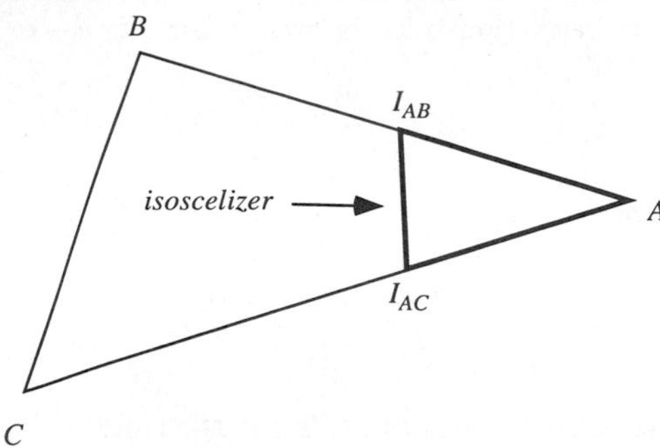

An isoscelizer of an ANGLE A in a TRIANGLE ΔABC is a LINE SEGMENT $I_{AB}I_{AC}$ where I_{AB} lies on AB and I_{AC} on AC such that $\Delta AI_{AB}I_{AC}$ is an ISOSCELES TRIANGLE.

see also CONGRUENT ISOSCELIZERS POINT, ISOSCELES TRIANGLE, YFF CENTER OF CONGRUENCE

Isospectral Manifolds

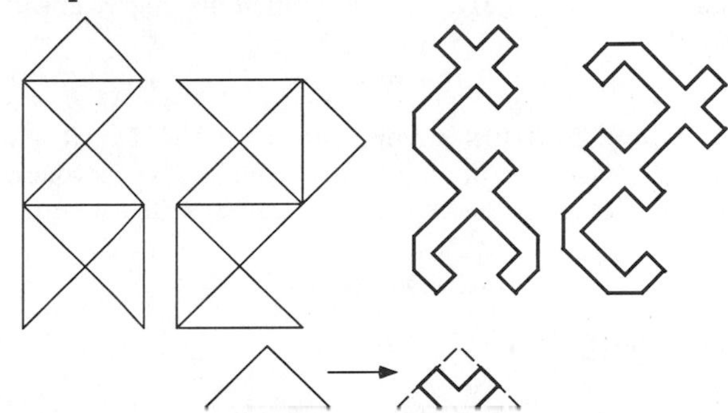

DRUMS that sound the same, i.e., have the same eigenfrequency spectrum. Two drums with differing AREA, PERIMETER, or GENUS can always be distinguished. However, Kac (1966) asked if it was possible to construct differently shaped drums which have the same eigenfrequency spectrum. This question was answered in the affirmative by Gordon *et al.* (1992). Two such isospectral manifolds are shown in the right figure above (Cipra 1992).

References

Chapman, S. J. "Drums That Sound the Same." *Amer. Math. Monthly* **102**, 124–138, 1995.

Cipra, B. "You Can't Hear the Shape of a Drum." *Science* **255**, 1642–1643, 1992.

Gordon, C.; Webb, D.; and Wolpert, S. "Isospectral Plane Domains and Surfaces via Riemannian Orbifolds." *Invent. Math.* **110**, 1–22, 1992.

Gordon, C.; Webb, D.; and Wolpert, S. "You Cannot Hear the Shape of a Drum." *Bull. Amer. Math. Soc.* **27**, 134–138, 1992.

Kac, M. "Can One Hear the Shape of a Drum?" *Amer. Math. Monthly* **73**, 1–23, 1966.

Isothermal Parameterization

A parameterization is isothermal if, for $\zeta \equiv u + iv$ and

$$\phi_k(\zeta) = \frac{\partial x_k}{\partial u} - i\frac{\partial x_k}{\partial v},$$

the identity

$$\phi_1{}^2(\zeta) + \phi_2{}^2(\zeta) + \phi_3{}^2(\zeta) = 0$$

holds.

see also MINIMAL SURFACE, TEMPERATURE

Isotomic Conjugate Point

The point of concurrence Q of the ISOTOMIC LINES relative to a point P. The isotomic conjugate $\alpha' : \beta' : \gamma'$ of a point with TRILINEAR COORDINATES $\alpha : \beta : \gamma$ is

$$(a^2\alpha)^{-1} : (b^2\beta)^{-1} : (c^2\gamma)^{-1}. \qquad (1)$$

The isotomic conjugate of a LINE d having trilinear equation

$$l\alpha + m\beta + n\gamma \qquad (2)$$

is a CONIC SECTION circumscribed on the TRIANGLE ΔABC (Casey 1893, Vandeghen 1965). The isotomic conjugate of the LINE AT INFINITY having trilinear equation

$$a\alpha + b\beta + c\gamma = 0 \qquad (3)$$

is STEINER'S ELLIPSE

$$\frac{\beta'\gamma'}{a} + \frac{\gamma'\alpha'}{b} + \frac{\alpha'\beta'}{c} = 0 \qquad (4)$$

(Vandeghen 1965). The type of CONIC SECTION to which d is transformed is determined by whether the line d meets STEINER'S ELLIPSE E.

1. If d does not intersect E, the isotomic transform is an ELLIPSE.
2. If d is tangent to E, the transform is a PARABOLA.
3. If d cuts E, the transform is a HYPERBOLA, which is a RECTANGULAR HYPERBOLA if the line passes through the isotomic conjugate of the ORTHOCENTER

(Casey 1893, Vandeghen 1965).

There are four points which are isotomically self-conjugate: the CENTROID M and each of the points of intersection of lines through the VERTICES PARALLEL to the opposite sides. The isotomic conjugate of the EULER LINE is called JERABEK'S HYPERBOLA (Casey 1893, Vandeghen 1965).

Vandeghen (1965) calls the transformation taking points to their isotomic conjugate points the CEVIAN TRANSFORM. The product of isotomic and ISOGONAL is a COLLINEATION which transforms the sides of a TRIANGLE to themselves (Vandeghen 1965).

see also CEVIAN TRANSFORM, GERGONNE POINT, ISOGONAL CONJUGATE, JERABEK'S HYPERBOLA, NAGEL POINT, STEINER'S ELLIPSE

References

Casey, J. *A Treatise on the Analytical Geometry of the Point, Line, Circle, and Conic Sections, Containing an Account of Its Most Recent Extensions with Numerous Examples, 2nd rev. enl. ed.* Dublin: Hodges, Figgis, & Co., 1893.

Eddy, R. H. and Fritsch, R. "The Conics of Ludwig Kiepert: A Comprehensive Lesson in the Geometry of the Triangle." *Math. Mag.* **67**, 188–205, 1994.

Johnson, R. A. *Modern Geometry: An Elementary Treatise on the Geometry of the Triangle and the Circle.* Boston, MA: Houghton Mifflin, pp. 157–159, 1929.

Vandeghen, A. "Some Remarks on the Isogonal and Cevian Transforms. Alignments of Remarkable Points of a Triangle." *Amer. Math. Monthly* **72**, 1091–1094, 1965.

Isotomic Lines

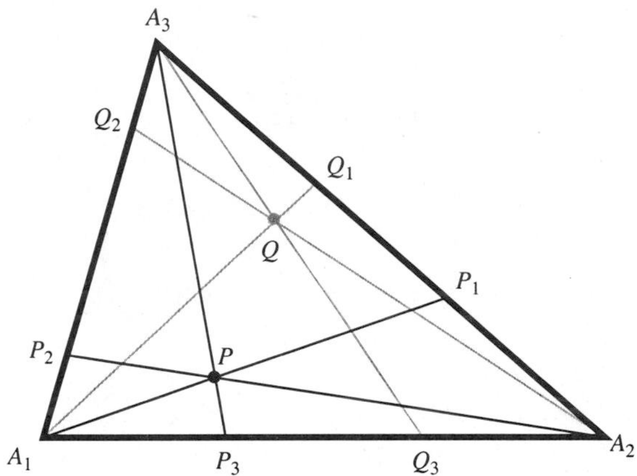

Given a point P in the interior of a TRIANGLE $\Delta A_1A_2A_3$, draw the CEVIANS through P from each VERTEX which meet the opposite sides at P_1, P_2, and P_3. Now, mark off point Q_1 along side A_2A_3 such that $A_3P_1 = A_2Q_1$, etc., i.e., so that Q_i and P_i are equidistance from the MIDPOINT of A_jA_k. The lines A_1Q_1, A_2Q_2, and A_3Q_3 then coincide in a point Q known as the ISOTOMIC CONJUGATE POINT.

see also CEVIAN, ISOTOMIC CONJUGATE POINT, MIDPOINT

Isotone Map

A MAP which is monotone increasing and therefore order-preserving.

Isotope

To rearrange without cutting or pasting.

Isotopy

A HOMOTOPY from one embedding of a MANIFOLD M in N to another such that at every time, it is an embedding. The notion of isotopy is category independent, so notions of topological, piecewise-linear, smooth, isotopy (and so on) exist. When no explicit mention is made, "isotopy" usually means "smooth isotopy."

see also AMBIENT ISOTOPY, REGULAR ISOTOPY

Isotropic Tensor

A TENSOR which has the same components in all rotated coordinate systems.

rank	isotropic tensors
0	all
1	none
2	Kronecker delta
3	1
4	3

Isovolume Problem

Find the surface enclosing the maximum volume per unit surface AREA $I \equiv V/S$. The solution is a SPHERE, which has

$$I_{\rm sphere} = \frac{\frac{4}{3}\pi r^3}{4\pi r^2} = \tfrac{1}{3}r.$$

see also DIDO'S PROBLEM, ISOPERIMETRIC PROBLEM

References

Bogomolny, A. "Isoperimetric Theorem and Inequality." `http://www.cut-the-knot.com/do_you_know/isoperimetric.html`.

Isenberg, C. Appendix VI in *The Science of Soap Films and Soap Bubbles.* New York: Dover, 1992.

Isthmus

see BRIDGE (GRAPH)

Iterated Exponential Constants

N.B. A detailed on-line essay by S. Finch was the starting point for this entry.

Euler (Le Lionnais 1983) and Eisenstein (1844) showed that the function $h(x) = x^{x^{x^{\cdot^{\cdot^{\cdot}}}}}$, where x^{x^x} is an abbreviation for $x^{(x^x)}$, converges only for $e^{-e} \leq x \leq e^{1/e}$, that is, $0.0659\ldots \leq x \leq 1.44466\ldots$. The value it converges to is the inverse of $x^{1/x}$, which has a closed form expression in terms of LAMBERT'S W-FUNCTION,

$$h(z) = \frac{W(-\ln z)}{-\ln z} \tag{1}$$

(Corless *et al.*). Knoebel (1981) gives

$$h(z) = 1 + \ln x + \frac{3^2(\ln z)^2}{3!} + \frac{4^3(\ln z)^3}{4!} + \ldots \tag{2}$$

(Vardi 1991). A CONTINUED FRACTION due to Khovanskii (1963) is

$$x^{1/x} = 1 + \cfrac{2(x-1)}{x^2+1-\cfrac{(x^2-1)(x-1)^2}{3x(x+1)-\cfrac{(4x^2-1)(x-1)^2}{5x(x+1)-\cfrac{(9x^2-1)(x-1)^2}{7x(x+1)-\ldots}}}}. \tag{3}$$

The function $g(x) = x^{(1/x)^{(1/x)^{\cdot^{\cdot^{\cdot}}}}}$ converges only for $e^{-1/e} \leq x \leq e^e$, that is, $0.692\ldots \leq x \leq 15.154\ldots$. The value it converges to is the inverse of x^x.

Some interesting related integrals are

$$\int_0^1 x^x\,dx = \sum_{n=1}^\infty \frac{(-1)^{n+1}}{n^n} = 0.7834305107\ldots \tag{4}$$

$$\int_0^1 x^{-x}\,dx = \sum_{n=1}^\infty \frac{1}{n^n} = 1.2912859971\ldots \tag{5}$$

(Spiegel 1968, Abramowitz and Stegun 1972).

see also LAMBERT'S W-FUNCTION

References

Abramowitz, M. and Stegun, C. A. (Eds.). *Handbook of Mathematical Functions with Formulas, Graphs, and Mathematical Tables, 9th printing.* New York: Dover, 1972.

Baker, I. N. and Rippon, P. J. "A Note on Complex Iteration." *Amer. Math. Monthly* **92**, 501–504, 1985.

Barrows, D. F. "Infinite Exponentials." *Amer. Math. Monthly* **43**, 150–160, 1936.

Corless, R. M.; Gonnet, G. H.; Hare, D. E. G.; and Jeffrey, D. J. "On Lambert's W Function." `ftp://watdragon.uwaterloo.ca/cs-archive/CS-93-03/W.ps.Z`.

Creutz, M. and Sternheimer, R. M. "On the Convergence of Iterated Exponentiation, Part I." *Fib. Quart.* **18**, 341–347, 1980.

Creutz, M. and Sternheimer, R. M. "On the Convergence of Iterated Exponentiation, Part II." *Fib. Quart.* **19**, 326–335, 1981.

de Villiers, J. M. and Robinson, P. N. "The Interval of Convergence and Limiting Functions of a Hyperpower Sequence." *Amer. Math. Monthly* **93**, 13–23, 1986.

Eisenstein, G. "Entwicklung von $\alpha^{\alpha^{\alpha^{\cdot^{\cdot^{\cdot}}}}}$." *J. Reine angew. Math.* **28**, 49–52, 1844.

Finch, S. "Favorite Mathematical Constants." `http://www.mathsoft.com/asolve/constant/itrexp/itrexp.html`.

Khovanskii, A. N. *The Application of Continued Fractions and Their Generalizations to Problems in Approximation Theory.* Groningen, Netherlands: P. Noordhoff, 1963.

Knoebel, R. A. "Exponentials Reiterated." *Amer. Math. Monthly* **88**, 235–252, 1981.

Le Lionnais, F. *Les nombres remarquables.* Paris: Hermann, pp. 22 and 39, 1983.

Mauerer, H. "Über die Funktion $x^{x^{\cdot^{\cdot^{\cdot}}}}$ für ganzzahliges Argument (Abundanzen)." *Mitt. Math. Gesell. Hamburg* **4**, 33–50, 1901.

Spiegel, M. R. *Mathematical Handbook of Formulas and Tables.* New York: McGraw-Hill, 1968.

Vardi, I. *Computational Recreations in Mathematica.* Reading, MA: Addison-Wesley, p. 12, 1991.

Iterated Function System

A finite set of contraction maps w_i for $i = 1, 2, \ldots, N$, each with a contractivity factor $s < 1$, which map a compact METRIC SPACE onto itself. It is the basis for FRACTAL image compression techniques.

see also BARNSLEY'S FERN, SELF-SIMILARITY

References
Barnsley, M. F. "Fractal Image Compression." *Not. Amer. Math. Soc.* **43**, 657–662, 1996.
Barnsley, M. *Fractals Everywhere, 2nd ed.* Boston, MA: Academic Press, 1993.
Barnsley, M. F. and Demko, S. G. "Iterated Function Systems and the Global Construction of Fractals." *Proc. Roy. Soc. London, Ser. A* **399**, 243–275, 1985.
Barnsley, M. F. and Hurd, L. P. *Fractal Image Compression.* Wellesley, MA: A. K. Peters, 1993.
Diaconis, P. M. and Shashahani, M. "Products of Random Matrices and Computer Image Generation." *Contemp. Math.* **50**, 173–182, 1986.
Fisher, Y. *Fractal Image Compression.* New York: Springer-Verlag, 1995.
Hutchinson, J. "Fractals and Self-Similarity." *Indiana Univ. J. Math.* **30**, 713–747, 1981.
Wagon, S. "Iterated Function Systems." §5.2 in *Mathematica in Action.* New York: W. H. Freeman, pp. 149–156, 1991.

Iterated Radical

see NESTED RADICAL

Iteration Sequence

A SEQUENCE $\{a_j\}$ of POSITIVE INTEGERS is called an iteration sequence if there EXISTS a strictly increasing sequence $\{s_k\}$ of POSITIVE INTEGERS such that $a_1 = s_1 \geq 2$ and $a_j = s_{a_{j-1}}$ for $j = 2, 3, \ldots$. A NECESSARY and SUFFICIENT condition for $\{a_j\}$ to be an iteration sequence is

$$a_j \geq 2a_{j-1} - a_{j-2}$$

for all $j \geq 3$.

References
Kimberling, C. "Interspersions and Dispersions." *Proc. Amer. Math. Soc.* **117**, 313–321, 1993.

Itô's Lemma

$$V_t - V_0 = \int_0^t f_x(S_u, T-u)\,dS_u - \int_0^t f_\tau(S_u, T-u)\,du + \tfrac{1}{2}\sigma^2 \int_0^t {S_u}^2 f_{xx}(S_u, T-u)\,du,$$

where $V_t = f(S_t, \tau)$ for $0 \leq \tau \equiv T - t \leq T$, and $f \in C^{2,1}((0,\infty) \times [0,T])$.

References
Price, J. F. "Optional Mathematics is Not Optional." *Not. Amer. Math. Soc.* **43**, 964–971, 1996.

Itô's Theorem

The dimension d of any IRREDUCIBLE REPRESENTATION of a GROUP G must be a DIVISOR of the index of each maximal normal Abelian SUBGROUP of G.

see also ABELIAN GROUP, IRREDUCIBLE REPRESENTATION, SUBGROUP

References
Lomont, J. S. *Applications of Finite Groups.* New York: Dover, p. 55, 1993.

Iverson Bracket

Let S be a mathematical statement, then the Iverson bracket is defined by

$$[S] \equiv \begin{cases} 0 & \text{if } S \text{ is true} \\ 1 & \text{if } S \text{ is false.} \end{cases}$$

This notation conflicts with the brackets sometimes used to denote the FLOOR FUNCTION. For this reason, and because of the elegant symmetry of the FLOOR FUNCTION and CEILING FUNCTION symbols $\lfloor x \rfloor$ and $\lceil x \rceil$, the use of $[x]$ to denote the FLOOR FUNCTION should be deprecated.

see also CEILING FUNCTION, FLOOR FUNCTION

References
Graham, R. L.; Knuth, D. E.; and Patashnik, O. *Concrete Mathematics: A Foundation for Computer Science.* Reading, MA: Addison-Wesley, p. 24, 1990.
Iverson, K. E. *A Programming Language.* New York: Wiley, p. 11, 1962.

Iwasawa's Theorem

Every finite-dimensional LIE ALGEBRA of characteristic $p \neq 0$ has a faithful finite-dimensional representation.

References
Jacobson, N. *Lie Algebras.* New York: Dover, pp. 204–205, 1979.

J

j

The symbol used by engineers and some physicists to denote i, the IMAGINARY NUMBER $\sqrt{-1}$.

j-Conductor

see FREY CURVE

j-Function

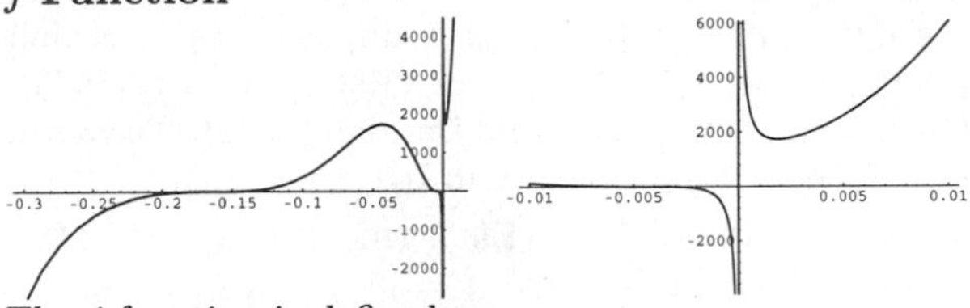

The j-function is defined as

$$j(q) \equiv 1728 J(\sqrt{q}), \tag{1}$$

where

$$J(q) \equiv \frac{4}{27} \frac{[1-\lambda(q)+\lambda^2(q)]^3}{\lambda^2(q)[1-\lambda(q)]^2} \tag{2}$$

is KLEIN'S ABSOLUTE INVARIANT, $\lambda(q)$ the ELLIPTIC LAMBDA FUNCTION

$$\lambda(q) \equiv k^2(q) = \left[\frac{\vartheta_2(q)}{\vartheta_3(q)}\right]^4, \tag{3}$$

and ϑ_i a THETA FUNCTION. This function can also be specified in terms of the WEBER FUNCTIONS f, f_1, f_2, γ_2, and γ_3 as

$$j(z) = \frac{[f^{24}(z)-16]^3}{f^{24}(z)} \tag{4}$$

$$= \frac{[{f_1}^{24}(z)+16]^3}{{f_1}^{24}(z)} \tag{5}$$

$$= \frac{[{f_2}^{24}(z)+16]^3}{{f_2}^{24}(z)} \tag{6}$$

$$= {\gamma_2}^3(z) \tag{7}$$

$$= {\gamma_3}^2(z) + 1728 \tag{8}$$

(Weber 1902, p. 179; Atkin and Morain 1993).

The j-function is MEROMORPHIC function on the upper half of the COMPLEX PLANE which is invariant with respect to the SPECIAL LINEAR GROUP $SL(2,Z)$. It has a FOURIER SERIES

$$j(q) = \sum_{n=-\infty}^{\infty} a_n q^n, \tag{9}$$

for the NOME

$$q \equiv e^{2\pi i t} \tag{10}$$

with $\Im[t] > 0$. The coefficients in the expansion of the j-function satisfy:

1. $a_n = 0$ for $n < -1$ and $a_{-1} = 1$,
2. all a_ns are INTEGERS with fairly limited growth with respect to n, and
3. $j(q)$ is an ALGEBRAIC NUMBER, sometimes a RATIONAL NUMBER, and sometimes even an INTEGER at certain very special values of q (or t).

The latter result is the end result of the massive and beautiful theory of COMPLEX multiplication and the first step of Kronecker's so-called "JUGENDTRAUM."

Then all of the COEFFICIENTS in LAURENT SERIES

$$j(q) = \frac{1}{q} + 744 + 196884q + 21494760q^2$$
$$+864299970q^3 + 20245856256q^4 + 333202640600q^5 + \ldots \tag{11}$$

(Sloane's A000521) are POSITIVE INTEGERS (Rankin 1977). Let d be a POSITIVE SQUAREFREE INTEGER, and define

$$t \equiv \begin{cases} i\sqrt{d} & \text{for } d \equiv 1 \text{ or } 2 \pmod 4 \\ \frac{1}{2}(1+i\sqrt{d}) & \text{for } d \equiv 3 \pmod 4. \end{cases} \tag{12}$$

Then the NOME is

$$q \equiv e^{i\pi\tau} = \begin{cases} e^{2\pi i(i\sqrt{d})} \\ e^{2\pi i(1+i\sqrt{d})/2} \end{cases}$$
$$= \begin{cases} e^{-2\pi\sqrt{d}} & \text{for } d \equiv 1 \text{ or } 2 \pmod 4 \\ -e^{-\pi\sqrt{d}} & \text{for } d \equiv 3 \pmod 4. \end{cases} \tag{13}$$

It then turns out that $j(q)$ is an ALGEBRAIC INTEGER of degree $h(-d)$, where $h(-d)$ is the CLASS NUMBER of the DISCRIMINANT $-d$ of the QUADRATIC FIELD $\mathbb{Q}(\sqrt{n})$ (Silverman 1986). The first term in the LAURENT SERIES is then $q^{-1} = e^{-2\pi\sqrt{n}}$ or $-e^{-\pi\sqrt{n}}$, and all the later terms are POWERS of q^{-1}, which are small numbers. The larger n, the faster the series converges. If $h(-d) = 1$, then $j(q)$ is a ALGEBRAIC INTEGER of degree 1, i.e., just a plain INTEGER. Furthermore, the INTEGER is a perfect CUBE.

The numbers whose LAURENT SERIES give INTEGERS are those with CLASS NUMBER 1. But these are precisely the HEEGNER NUMBERS -1, -2, -3, -7, -11, -19, -43, -67, -163. The greater (in ABSOLUTE VALUE) the HEEGNER NUMBER d, the closer to an INTEGER is the expression $e^{\pi\sqrt{-n}}$, since the initial term in $j(q)$ is the largest and subsequent terms are the smallest. The best approximations with $h(-d) = 1$ are therefore

$$e^{\pi\sqrt{43}} \approx 960^3 + 744 - 2.2 \times 10^{-4} \tag{14}$$

$$e^{\pi\sqrt{67}} \approx 5280^3 + 744 - 1.3 \times 10^{-6} \tag{15}$$

$$e^{\pi\sqrt{163}} \approx 640320^3 + 744 - 7.5 \times 10^{-13}. \tag{16}$$

The exact values of $j(q)$ corresponding to the HEEGNER NUMBERS are

$$j(-e^{-\pi}) = 12^3 \quad (17)$$
$$j(e^{-2\pi\sqrt{2}}) = 20^3 \quad (18)$$
$$j(-e^{-\pi\sqrt{3}}) = 0^3 \quad (19)$$
$$j(-e^{-\pi\sqrt{7}}) = -15^3 \quad (20)$$
$$j(-e^{-\pi\sqrt{11}}) = -32^3 \quad (21)$$
$$j(-e^{-\pi\sqrt{19}}) = -96^3 \quad (22)$$
$$j(-e^{-\pi\sqrt{43}}) = -960^3 \quad (23)$$
$$j(-e^{-\pi\sqrt{67}}) = -5280^3 \quad (24)$$
$$j(-e^{-\pi\sqrt{163}}) = -640320^3. \quad (25)$$

(The number 5280 is particularly interesting since it is also the number of feet in a mile.) The ALMOST INTEGER generated by the last of these, $e^{\pi\sqrt{163}}$ (corresponding to the field $\mathbb{Q}(\sqrt{-163})$ and the IMAGINARY quadratic field of maximal discriminant), is known as the RAMANUJAN CONSTANT.

$e^{\pi\sqrt{22}}$, $e^{\pi\sqrt{37}}$, and $e^{\pi\sqrt{58}}$ are also ALMOST INTEGERS. These correspond to binary quadratic forms with discriminants -88, -148, and -232, all of which have CLASS NUMBER two and were noted by Ramanujan (Berndt 1994).

It turns out that the j-function also is important in the CLASSIFICATION THEOREM for finite simple groups, and that the factors of the orders of the SPORADIC GROUPS, including the celebrated MONSTER GROUP, are also related.

see also ALMOST INTEGER, KLEIN'S ABSOLUTE INVARIANT, WEBER FUNCTIONS

References

Atkin, A. O. L. and Morain, F. "Elliptic Curves and Primality Proving." *Math. Comput.* **61**, 29–68, 1993.

Berndt, B. C. *Ramanujan's Notebooks, Part IV.* New York: Springer-Verlag, pp. 90–91, 1994.

Borwein, J. M. and Borwein, P. B. *Pi & the AGM: A Study in Analytic Number Theory and Computational Complexity.* New York: Wiley, pp. 117–118, 1987.

Cohn, H. *Introduction to the Construction of Class Fields.* New York: Dover, p. 73, 1994.

Conway, J. H. and Guy, R. K. "The Nine Magic Discriminants." In *The Book of Numbers.* New York: Springer-Verlag, pp. 224–226, 1996.

Morain, F. "Implementation of the Atkin-Goldwasser-Kilian Primality Testing Algorithm." Rapport de Récherche 911, INRIA, Oct. 1988.

Rankin, R. A. *Modular Forms.* New York: Wiley, 1985.

Rankin, R. A. *Modular Forms and Functions.* Cambridge, England: Cambridge University Press, p. 199, 1977.

Serre, J. P. *Cours d'arithmétique.* Paris: Presses Universitaires de France, 1970.

Silverman, J. H. *The Arithmetic of Elliptic Curves.* New York: Springer-Verlag, p. 339, 1986.

Sloane, N. J. A. Sequence A000521/M5477 in "An On-Line Version of the Encyclopedia of Integer Sequences."

Weber, H. *Lehrbuch der Algebra, Vols. I–II.* New York: Chelsea, 1979.

Weisstein, E. W. "j-Function." http://www.astro.virginia.edu/~eww6n/math/notebooks/jFunction.m.

j-Invariant

An invariant of an ELLIPTIC CURVE closely related to the DISCRIMINANT and defined by

$$j(E) \equiv \frac{2^8 3^3 a^3}{4a^3 + 27b^2}.$$

The determination of j as an ALGEBRAIC INTEGER in the QUADRATIC FIELD $\mathbb{Q}(j)$ is discussed by Greenhill (1891), Weber (1902), Berwick (1928), Watson (1938), Gross and Zaiger (1985), and Dorman (1988). The norm of j in $\mathbb{Q}(j)$ is the CUBE of an INTEGER in $\mathbb{Z}$.

see also DISCRIMINANT (ELLIPTIC CURVE), ELLIPTIC CURVE, FREY CURVE

References

Berwick, W. E. H. "Modular Invariants Expressible in Terms of Quadratic and Cubic Irrationalities." *Proc. London Math. Soc.* **28**, 53–69, 1928.

Dorman, D. R. "Special Values of the Elliptic Modular Function and Factorization Formulae." *J. reine angew. Math.* **383**, 207–220, 1988.

Greenhill, A. G. "Table of Complex Multiplication Moduli." *Proc. London Math. Soc.* **21**, 403–422, 1891.

Gross, B. H. and Zaiger, D. B. "On Singular Moduli." *J. reine angew. Math.* **355**, 191–220, 1985.

Watson, G. N. "Ramanujans Vermutung über Zerfällungsanzahlen." *J. reine angew. Math.* **179**, 97–128, 1938.

Weber, H. *Lehrbuch der Algebra, Vols. I–II.* New York: Chelsea, 1979.

Jackson's Difference Fan

If, after constructing a DIFFERENCE TABLE, no clear pattern emerges, turn the paper through an ANGLE of 60° and compute a new table. If necessary, repeat the process. Each ROTATION reduces POWERS by 1, so the sequence $\{k^n\}$ multiplied by any POLYNOMIAL in n is reduced to 0s by a k-fold difference fan.

References

Conway, J. H. and Guy, R. K. "Jackson's Difference Fans." In *The Book of Numbers.* New York: Springer-Verlag, pp. 84–85, 1996.

Jackson's Identity

A q-SERIES identity involving

$$\frac{(aq)_q^m (aqde)_q^m (adec)_q^m (aqcd)_q^m}{(aqc)_q^m (aqd)_q^m (aqe)_q^m (aqcde)_q^m},$$

where

$$(a)_q^n = (1-a)(1-aq)\cdots(1-aq^{n-1}).$$

see also q-SERIES

References

Hardy, G. H. *Ramanujan: Twelve Lectures on Subjects Suggested by His Life and Work, 3rd ed.* New York: Chelsea, pp. 109–110, 1959.

Jackson, F. H. "Summation of q-Hypergeometric Series." *Messenger Math.* **47**, 101–112, 1917.

Jackson's Theorem

Jackson's theorem is a statement about the error $E_n(f)$ of the best uniform approximation to a REAL FUNCTION $f(x)$ on $[-1,1]$ by REAL POLYNOMIALS of degree at most n. Let $f(x)$ be of bounded variation in $[-1,1]$ and let M' and V' denote the least upper bound of $|f(x)|$ and the total variation of $f(x)$ in $[-1,1]$, respectively. Given the function

$$F(x) = F(-1) + \int_{-1}^{x} f(x)\,dx, \tag{1}$$

then the coefficients

$$a_n = \tfrac{1}{2}(2n+1)\int_{-1}^{1} F(x)P_n(x)\,dx \tag{2}$$

of its LEGENDRE SERIES, where $P_n(x)$ is a LEGENDRE POLYNOMIAL, satisfy the inequalities

$$|a_n| < \begin{cases} \frac{6}{\sqrt{\pi}}(M'+V')n^{-3/2} & \text{for } n \geq 1 \\ \frac{4}{\sqrt{\pi}}(M'+V')n^{-3/2} & \text{for } n \geq 2. \end{cases} \tag{3}$$

Moreover, the LEGENDRE SERIES of $F(x)$ converges uniformly and absolutely to $F(x)$ in $[-1,1]$.

Bernstein strengthened Jackson's theorem to

$$2nE_{2n}(\alpha) \leq \frac{4n}{\pi(2n+1)} < \frac{2}{\pi} = 0.6366. \tag{4}$$

A specific application of Jackson's theorem shows that if

$$\alpha(x) = |x|, \tag{5}$$

then

$$E_n(\alpha) \leq \frac{6}{n}. \tag{6}$$

see also LEGENDRE SERIES, PICONE'S THEOREM

References

Cheney, E. W. *Introduction to Approximation Theory.* New York: McGraw-Hill, 1966.

Jackson, D. *The Theory of Approximation.* New York: Amer. Math. Soc., p. 76, 1930.

Rivlin, T. J. *An Introduction to the Approximation of Functions.* New York: Dover, 1981.

Sansone, G. *Orthogonal Functions, rev. English ed.* New York: Dover, pp. 205–208, 1991.

Jaco-Shalen-Johannson Torus Decomposition

Irreducible orientable COMPACT 3-MANIFOLDS have a canonical (up to ISOTOPY) minimal collection of disjointly EMBEDDED incompressible TORI such that each component of the 3-MANIFOLD removed by the TORI is either "atoroidal" or "Seifert-fibered."

Jacobi Algorithm

A method which can be used to solve a TRIDIAGONAL MATRIX equation with largest absolute values in each row and column dominated by the diagonal element. Each diagonal element is solved for, and an approximate value plugged in. The process is then iterated until it converges. This algorithm is a stripped-down version of the JACOBI METHOD of matrix diagonalization.

see also JACOBI METHOD, TRIDIAGONAL MATRIX

References

Acton, F. S. *Numerical Methods That Work, 2nd printing.* Washington, DC: Math. Assoc. Amer., pp. 161–163, 1990.

Jacobi-Anger Expansion

$$e^{iz\cos\theta} = \sum_{n=-\infty}^{\infty} i^n J_n(z)e^{in\theta},$$

where $J_n(z)$ is a BESSEL FUNCTION OF THE FIRST KIND. The identity can also be written

$$e^{iz\cos\theta} = J_0(z) + 2\sum_{n=1}^{\infty} i^n J_n(z)\cos(n\theta).$$

This expansion represents an expansion of plane waves into a series of cylindrical waves.

see also BESSEL FUNCTION OF THE FIRST KIND

Jacobi's Curvature Theorem

The principal normal indicatrix of a closed space curve with nonvanishing curvature bisects the AREA of the unit sphere if it is embedded.

Jacobi's Determinant Identity

Let

$$\mathsf{A} = \begin{bmatrix} \mathsf{B} & \mathsf{D} \\ \mathsf{E} & \mathsf{C} \end{bmatrix} \tag{1}$$

$$\mathsf{A}^{-1} = \begin{bmatrix} \mathsf{W} & \mathsf{X} \\ \mathsf{Y} & \mathsf{Z} \end{bmatrix}, \tag{2}$$

where B and W are $k \times k$ MATRICES. Then

$$(\det \mathsf{Z})(\det \mathsf{A}) = \det \mathsf{B}. \tag{3}$$

The proof follows from equating determinants on the two sides of the block matrices

$$\begin{bmatrix} \mathsf{B} & \mathsf{D} \\ \mathsf{E} & \mathsf{C} \end{bmatrix}\begin{bmatrix} \mathsf{I} & \mathsf{X} \\ \mathsf{O} & \mathsf{Z} \end{bmatrix} = \begin{bmatrix} \mathsf{B} & \mathsf{O} \\ \mathsf{E} & \mathsf{I} \end{bmatrix}, \tag{4}$$

where I is the IDENTITY MATRIX and O is the zero matrix.

References

Gantmacher, F. R. *The Theory of Matrices, Vol. 1.* New York: Chelsea, p. 21, 1960.

Horn, R. A. and Johnson, C. R. *Matrix Analysis.* Cambridge, England: Cambridge University Press, p. 21, 1985.

Jacobi Differential Equation

$$(1-x^2)y''+[\beta-\alpha-(\alpha+\beta+2)x]y'+n(n+\alpha+\beta+1)y=0 \quad (1)$$

or

$$\frac{d}{dx}[(1-x)^{\alpha+1}(1+x)^{\beta+1}y'] +n(n+\alpha+\beta+1)(1-x)^{\alpha}(1+x)^{\beta}y=0. \quad (2)$$

The solutions are JACOBI POLYNOMIALS. They can be transformed to

$$\frac{d^2u}{dx^2}+\left[\frac{1}{4}\frac{1-\alpha^2}{(1-x)^2}+\frac{1}{4}\frac{1-\beta^2}{(1+x)^2} +\frac{n(n+\alpha+\beta+1)+\frac{1}{2}(\alpha+1)(\beta+1)}{1-x^2}\right]u=0, \quad (3)$$

where

$$u=u(x)=(1-x)^{(\alpha+1)/2}(1+x)^{(\beta+1)/2}P_n^{(\alpha,\beta)}(x), \quad (4)$$

and

$$\frac{d^2u}{d\theta^2}+\left[\frac{\frac{1}{4}-\alpha^2}{4\sin^2(\frac{1}{2}\theta)}+\frac{\frac{1}{4}-\beta^2}{4\cos^2(\frac{1}{2}\theta)} +\left(n+\frac{\alpha+\beta+1}{2}\right)^2\right]u=0, \quad (5)$$

where

$$u=u(\theta)=\sin^{\alpha+1/2}(\tfrac{1}{2}\theta)\cos^{\beta+1/2}(\tfrac{1}{2}\theta)P_n^{(\alpha,\beta)}(\cos\theta). \quad (6)$$

Jacobi Differential Equation (Calculus of Variations)

$$\frac{d}{dx}\Omega_{\eta'}-\Omega_\eta=\frac{d}{dx}(f_{y'y}\eta+f_{y'y}\eta')-(f_{yy}\eta+f_{yy'}\eta')=0,$$

where

$$\Omega(x,\eta,\eta')\equiv\tfrac{1}{2}(f_{yy}\eta^2+2f_{yy'}\eta\eta'+f_{y'y}\eta'^2).$$

This equations arises in the CALCULUS OF VARIATIONS.

References

Bliss, G. A. *Calculus of Variations.* Chicago, IL: Open Court, pp. 162–163, 1925.

Jacobi Elliptic Functions

The Jacobi elliptic functions are standard forms of ELLIPTIC FUNCTIONS. The three basic functions are denoted $\mathrm{cn}(u,k)$, $\mathrm{dn}(u,k)$, and $\mathrm{sn}(u,k)$, where k is known as the MODULUS. In terms of THETA FUNCTIONS,

$$\mathrm{sn}(u,k)=\frac{\vartheta_3}{\vartheta_4}\frac{\vartheta_1(u\vartheta_3{}^{-2})}{\vartheta_4(u\vartheta_3{}^{-2})} \quad (1)$$

$$\mathrm{cn}(u,k)=\frac{\vartheta_4}{\vartheta_2}\frac{\vartheta_2(u\vartheta_3{}^{-2})}{\vartheta_4(u\vartheta_3{}^{-2})} \quad (2)$$

$$\mathrm{dn}(u,k)=\frac{\vartheta_4}{\vartheta_3}\frac{\vartheta_3(u\vartheta_3{}^{-2})}{\vartheta_4(u\vartheta_3{}^{-2})} \quad (3)$$

(Whittaker and Watson 1990, p. 492), where $\vartheta_i\equiv\vartheta_i(0)$ (Whittaker and Watson 1990, p. 464). Ratios of Jacobi elliptic functions are denoted by combining the first letter of the NUMERATOR elliptic function with the first of the DENOMINATOR elliptic function. The multiplicative inverses of the elliptic functions are denoted by reversing the order of the two letters. These combinations give a total of 12 functions: cd, cn, cs, dc, dn, ds, nc, nd, ns, sc, sd, and sn. The AMPLITUDE ϕ is defined in terms of $\mathrm{sn}\,u$ by

$$y=\sin\phi=\mathrm{sn}(u,k). \quad (4)$$

The k argument is often suppressed for brevity so, for example, $\mathrm{sn}(u,k)$ can be written $\mathrm{sn}\,u$.

The Jacobi elliptic functions are periodic in $K(k)$ and $K'(k)$ as

$$\mathrm{sn}(u+2mK+2niK',k)=(-1)^m\,\mathrm{sn}(u,k) \quad (5)$$

$$\mathrm{cn}(u+2mK+2niK',k)=(-1)^{m+n}\,\mathrm{cn}(u,k) \quad (6)$$

$$\mathrm{dn}(u+2mK+2niK',k)=(-1)^n\,\mathrm{dn}(u,k), \quad (7)$$

where $K(k)$ is the complete ELLIPTIC INTEGRAL OF THE FIRST KIND, $K'(k)\equiv K(k')$, and $k'\equiv\sqrt{1-k^2}$ (Whittaker and Watson 1990, p. 503).

The $\mathrm{cn}\,x$, $\mathrm{dn}\,x$, and $\mathrm{sn}\,x$ functions may also be defined as solutions to the differential equations

$$\frac{d^2y}{dx^2}=-(1+k^2)y+2k^2y^3 \quad (8)$$

$$\frac{d^2y}{dx^2}=-(1-2k^2)y-2k^2y^3 \quad (9)$$

$$\frac{d^2y}{dx^2}=(2-k^2)y-2y^3. \quad (10)$$

The standard Jacobi elliptic functions satisfy the identities

$$\mathrm{sn}^2 u+\mathrm{cn}^2 u=1 \quad (11)$$

$$k^2\,\mathrm{sn}^2 u+\mathrm{dn}^2 u=1 \quad (12)$$

$$k^2\,\mathrm{cn}^2 u+k'^2=\mathrm{dn}^2 u \quad (13)$$

$$\mathrm{cn}^2 u+k'^2\,\mathrm{sn}^2 u=\mathrm{dn}^2 u. \quad (14)$$

Special values are

$$\operatorname{cn}(0) = 1 \quad (15)$$
$$\operatorname{dn}(0) = 1 \quad (16)$$
$$\operatorname{cn}(K) = 0 \quad (17)$$
$$\operatorname{dn}(K) = k' \equiv \sqrt{1-k^2}, \quad (18)$$
$$\operatorname{sn}(K) = 1, \quad (19)$$

where $K = K(k)$ is a complete ELLIPTIC INTEGRAL OF THE FIRST KIND and k' is the complementary MODULUS (Whittaker and Watson 1990, pp. 498–499).

In terms of integrals,

$$u = \int_0^{\operatorname{sn} u} (1-t^2)^{1-/2}(1-k^2t^2)^{-1/2}\,dt \quad (20)$$
$$= \int_{\operatorname{ns} u}^{\infty} (t^2-1)^{-1/2}(t^2-l^2)^{-1/2}\,dt \quad (21)$$
$$= \int_{\operatorname{cn} u}^{1} (1-t^2)^{-1/2}(k'^2+k^2t^2)^{-1/2}\,dt \quad (22)$$
$$= \int_1^{\operatorname{nc} u} (t^2-1)^{-1/2}(k'^2t^2+k^2)^{-1/2}\,dt \quad (23)$$
$$= \int_{\operatorname{dn} u}^{1} (1-t^2)^{-1/2}(t^2-k'^2)^{-1/2}\,dt \quad (24)$$
$$= \int_1^{\operatorname{nd} u} (t^2-1)^{-1/2}(1-k'^2t^2)^{-1/2}\,dt \quad (25)$$
$$= \int_0^{U} (1+t^2)^{-1/2}(1+k'^2t^2)^{-1/2}\,dt \quad (26)$$
$$= \int_{\operatorname{cs} u}^{\infty} (t^2+1)^{-1/2}(t^2+k'^2)^{-1/2}\,dt \quad (27)$$
$$= \int_0^{\operatorname{sd} u} (1-k'^2t^2)^{-1/2}(1+k^2t^2)^{-1/2}\,dt \quad (28)$$
$$= \int_{\operatorname{ds} u}^{\infty} (t^2-k'^2)^{-1/2}(t^2+k^2)^{-1/2}\,dt \quad (29)$$
$$= \int_1^{\operatorname{cd} u} (1-t^2)^{-1/2}(1-k^2t^2)^{-1/2}\,dt \quad (30)$$
$$= \int_{\operatorname{dc} u}^{1} (t^2-1)^{-1/2}(t^2-k^2)^{-1/2}\,dt \quad (31)$$

(Whittaker and Watson 1990, p. 494).

Jacobi elliptic functions addition formulas include

$$\operatorname{sn}(u+v) = \frac{\operatorname{sn} u \operatorname{cn} v \operatorname{dn} v + \operatorname{sn} v \operatorname{cn} u \operatorname{dn} u}{1-k^2\operatorname{sn}^2 u \operatorname{sn}^2 v} \quad (32)$$
$$\operatorname{cn}(u+v) = \frac{\operatorname{cn} u \operatorname{cn} v - \operatorname{sn} u \operatorname{sn} v \operatorname{dn} u \operatorname{dn} v}{1-k^2\operatorname{sn}^2 u \operatorname{sn}^2 v} \quad (33)$$
$$\operatorname{dn}(u+v) = \frac{\operatorname{dn} u \operatorname{dn} v - k^2 \operatorname{sn} u \operatorname{sn} v \operatorname{cn} u \operatorname{cn} v}{1-k^2\operatorname{sn}^2 u \operatorname{sn}^2 v}. \quad (34)$$

Extended to integral periods,

$$\operatorname{sn}(u+K) = \frac{\operatorname{cn} u}{\operatorname{dn} u} \quad (35)$$
$$\operatorname{cn}(u+K) = -\frac{k' \operatorname{sn} u}{\operatorname{dn} u} \quad (36)$$
$$\operatorname{dn}(u+K) = \frac{k'}{\operatorname{dn} u} \quad (37)$$

$$\operatorname{sn}(u+2K) = -\operatorname{sn} u \quad (38)$$
$$\operatorname{cn}(u+2K) = -\operatorname{cn} u \quad (39)$$
$$\operatorname{dn}(u+2K) = \operatorname{dn} u \quad (40)$$

For COMPLEX arguments,

$$\operatorname{sn}(u+iv) = \frac{\operatorname{sn}(u,k)\operatorname{dn}(v,k')}{1-\operatorname{dn}^2(u,k)\operatorname{sn}^2(v,k')} + \frac{i\operatorname{cn}(u,k)\operatorname{dn}(u,k)\operatorname{sn}(v,k')\operatorname{cn}(v,k')}{1-\operatorname{dn}^2(u,k)\operatorname{sn}^2(v,k')} \quad (41)$$
$$\operatorname{cn}(u+iv) = \frac{\operatorname{cn}(u,k)\operatorname{cn}(v,k')}{1-\operatorname{dn}^2(u,k)\operatorname{sn}^2(v,k')} - \frac{i\operatorname{sn}(u,k)\operatorname{dn}(u,k)\operatorname{sn}(v,k')\operatorname{dn}(v,k')}{1-\operatorname{dn}^2(u,k)\operatorname{sn}^2(v,k')} \quad (42)$$
$$\operatorname{dn}(u+iv) = \frac{\operatorname{dn}(u,k)\operatorname{cn}(v,k')\operatorname{dn}(v,k')}{1-\operatorname{dn}^2(u,k)\operatorname{sn}^2(v,k')} - \frac{ik^2\operatorname{sn}(u,k)\operatorname{cn}(u,k)\operatorname{sn}(v,k')}{1-\operatorname{dn}^2(u,k)\operatorname{sn}^2(v,k')}. \quad (43)$$

DERIVATIVES of the Jacobi elliptic functions include

$$\frac{d\operatorname{sn} u}{du} = \operatorname{cn} u \operatorname{dn} u \quad (44)$$
$$\frac{d\operatorname{cn} u}{du} = \operatorname{sn} u \operatorname{dn} u \quad (45)$$
$$\frac{d\operatorname{dn} u}{du} = -k^2 \operatorname{sn} u \operatorname{cn} u. \quad (46)$$

Double-period formulas involving the Jacobi elliptic functions include

$$\operatorname{sn}(2u) = \frac{2\operatorname{sn} u \operatorname{cn} u \operatorname{dn} u}{1-k^2\operatorname{sn}^4 u} \quad (47)$$
$$\operatorname{cn}(2u) = \frac{1-2\operatorname{sn}^2 u + k^2\operatorname{sn}^4 u}{1-k^2\operatorname{sn}^4 u} \quad (48)$$
$$\operatorname{dn}(2u) = \frac{1-2k^2\operatorname{sn}^2 u + k^2\operatorname{sn}^4 u}{1-k^2\operatorname{sn}^4 u}. \quad (49)$$

Half-period formulas involving the Jacobi elliptic functions include

$$\operatorname{sn}(\tfrac{1}{2}K) = \frac{1}{\sqrt{1+k'}} \quad (50)$$
$$\operatorname{cn}(\tfrac{1}{2}K) = \sqrt{\frac{k'}{1+k'}} \quad (51)$$
$$\operatorname{dn}(\tfrac{1}{2}K) = \sqrt{k'}. \quad (52)$$

Squared formulas include

$$\operatorname{sn}^2 u = \frac{1-\operatorname{cn}(2u)}{1+\operatorname{dn}(2u)} \quad (53)$$

$$\operatorname{cn}^2 u = \frac{\operatorname{dn}(2u)+\operatorname{cn}(2u)}{1+\operatorname{dn}(2u)} \quad (54)$$

$$\operatorname{dn}^2 u = \frac{\operatorname{dn}(2u)+\operatorname{cn}(2u)}{1+\operatorname{cn}(2u)}. \quad (55)$$

see also AMPLITUDE, ELLIPTIC FUNCTION, JACOBI'S IMAGINARY TRANSFORMATION, THETA FUNCTION, WEIERSTRASS ELLIPTIC FUNCTION

References

Abramowitz, M. and Stegun, C. A. (Eds.). "Jacobian Elliptic Functions and Theta Functions." Ch. 16 in *Handbook of Mathematical Functions with Formulas, Graphs, and Mathematical Tables, 9th printing.* New York: Dover, pp. 567–581, 1972.

Bellman, R. E. *A Brief Introduction to Theta Functions.* New York: Holt, Rinehart and Winston, 1961.

Morse, P. M. and Feshbach, H. *Methods of Theoretical Physics, Part I.* New York: McGraw-Hill, p. 433, 1953.

Press, W. H.; Flannery, B. P.; Teukolsky, S. A.; and Vetterling, W. T. "Elliptic Integrals and Jacobi Elliptic Functions." §6.11 in *Numerical Recipes in FORTRAN: The Art of Scientific Computing, 2nd ed.* Cambridge, England: Cambridge University Press, pp. 254–263, 1992.

Spanier, J. and Oldham, K. B. "The Jacobian Elliptic Functions." Ch. 63 in *An Atlas of Functions.* Washington, DC: Hemisphere, pp. 635–652, 1987.

Whittaker, E. T. and Watson, G. N. *A Course in Modern Analysis, 4th ed.* Cambridge, England: Cambridge University Press, 1990.

Jacobi Function of the First Kind

see JACOBI POLYNOMIAL

Jacobi Function of the Second Kind

$$Q_n^{(\alpha,\beta)}(x) = 2^{-n-1}(x-1)^{-\alpha}(x+1)^{-\beta} \times \int_{-1}^{1} (1-t)^{n+\alpha}(1+t)^{n+\beta}(x-t)^{-n-1}\,dt.$$

In the exceptional case $n = 0$, $\alpha+\beta+1 = 0$, a nonconstant solution is given by

$$Q^{(\alpha)}(x) = \ln(x+1) + \pi^{-1}\sin(\pi\alpha)(x-1)^{-\alpha}(x+1)^{-\beta} \times \int_{-1}^{1} \frac{(1-t)^{\alpha}(1+t)^{\beta}}{x-t}\ln(1+t)\,dt.$$

References

Szegő, G. "Jacobi Polynomials." Ch. 4 in *Orthogonal Polynomials, 4th ed.* Providence, RI: Amer. Math. Soc., pp. 73–79, 1975.

Jacobi-Gauss Quadrature

Also called JACOBI QUADRATURE or MEHLER QUADRATURE. A GAUSSIAN QUADRATURE over the interval $[-1,1]$ with WEIGHTING FUNCTION $W(x) = (1-x)^\alpha(1+x)^\beta$. The ABSCISSAS for quadrature order n are given by the roots of the JACOBI POLYNOMIALS $P_n^{(\alpha,\beta)}(x)$. The weights are

$$w_i = -\frac{A_{n+1}\gamma_n}{A_n P_n^{(\alpha,\beta)\prime}(x_i)P_{n+1}^{(\alpha,\beta)}(x_i)} = \frac{A_n}{A_{n-1}}\frac{\gamma_{n-1}}{P_{n-1}^{(\alpha,\beta)}(x_i)P_n^{(\alpha,\beta)\prime}(x_i)}, \quad (1)$$

where A_n is the COEFFICIENT of x^n in $P_n^{(\alpha,\beta)}(x)$. For JACOBI POLYNOMIALS,

$$A_n = \frac{\Gamma(2n+\alpha+\beta+1)}{2^n n!\,\Gamma(n+\alpha+\beta+1)}, \quad (2)$$

where $\Gamma(z)$ is a GAMMA FUNCTION. Additionally,

$$\gamma_n = \frac{1}{2^{2n}(n!)^2}\frac{2^{2n+\alpha+\beta+1}n!}{2n+\alpha+\beta+1} \times \frac{\Gamma(n+\alpha+1)\Gamma(n+\beta+1)}{\Gamma(n+\alpha+\beta+1)}, \quad (3)$$

so

$$w_i = \frac{2n+\alpha+\beta+2}{n+\alpha+\beta+1}\frac{\Gamma(n+\alpha+1)\Gamma(n+\beta+1)}{\Gamma(n+\alpha+\beta+1)} \times \frac{2^{2n+\alpha+\beta+1}n!}{V_n'(x_i)V_{n+1}(x_i)} \quad (4)$$

$$= \frac{\Gamma(n+\alpha+1)\Gamma(n+\beta+1)}{\Gamma(n+\alpha+\beta+1)}\frac{2^{2n+\alpha+\beta+1}n!}{(1-x_i{}^2)[V_n'(x_i)]^2}, \quad (5)$$

where

$$V_m \equiv P_n^{(\alpha,\beta)}(x)\frac{2^n n!}{(-1)^n}. \quad (6)$$

The error term is

$$E_n = \frac{\Gamma(n+\alpha+1)\Gamma(n+\beta+1)\Gamma(n+\alpha+\beta+1)}{(2n+\alpha+\beta+1)[\Gamma(2n+\alpha+\beta+1)]^2} \times \frac{2^{2n+\alpha+\beta+1}n!}{(2n)!}f^{(2n)}(\xi) \quad (7)$$

(Hildebrand 1959).

References

Hildebrand, F. B. *Introduction to Numerical Analysis.* New York: McGraw-Hill, pp. 331–334, 1956.

Jacobi Identities

"The" Jacobi identity is a relationship

$$[A,[B,C]]+[B,[C,A]]+[C,[A,B]]=0, \quad (1),$$

between three elements A, B, and C, where $[A,B]$ is the COMMUTATOR. The elements of a LIE GROUP satisfy this identity.

Relationships between the Q-FUNCTIONS Q_i are also known as Jacobi identities:

$$Q_1Q_2Q_3=1, \quad (2)$$

equivalent to the JACOBI TRIPLE PRODUCT (Borwein and Borwein 1987, p. 65) and

$$Q_2{}^8=16qQ_1{}^8+Q_3{}^8, \quad (3)$$

where

$$q\equiv e^{-\pi K'(k)/K(k)}, \quad (4)$$

$K=K(k)$ is the complete ELLIPTIC INTEGRAL OF THE FIRST KIND, and $K'(k)=K(k')=K(\sqrt{1-k^2})$. Using WEBER FUNCTIONS

$$f_1=q^{-1/24}Q_3 \quad (5)$$

$$f_2=2^{1/2}q^{1/12}Q_1 \quad (6)$$

$$f=q^{-1/24}Q_2, \quad (7)$$

(5) and (6) become

$$f_1f_2f=\sqrt{2} \quad (8)$$

$$f^8=f_1{}^8+f_2{}^8 \quad (9)$$

(Borwein and Borwein 1987, p. 69).

see also COMMUTATOR, JACOBI TRIPLE PRODUCT, Q-FUNCTION, WEBER FUNCTIONS

References

Borwein, J. M. and Borwein, P. B. *Pi & the AGM: A Study in Analytic Number Theory and Computational Complexity.* New York: Wiley, 1987.

Jacobi's Imaginary Transformation

For JACOBI ELLIPTIC FUNCTIONS $\operatorname{sn} u$, $\operatorname{cn} u$, and $\operatorname{dn} u$,

$$\operatorname{sn}(iu,k)=i\frac{\operatorname{sn}(u,k')}{\operatorname{cn}(u,k')}$$

$$\operatorname{cn}(iu,k)=\frac{1}{\operatorname{cn}(u,k')}$$

$$\operatorname{dn}(iu,k)=\frac{\operatorname{dn}(u,k')}{\operatorname{cn}(u,k')}$$

(Abramowitz and Stegun 1972, Whittaker and Watson 1990).

see also JACOBI ELLIPTIC FUNCTIONS

References

Abramowitz, M. and Stegun, C. A. (Eds.). *Handbook of Mathematical Functions with Formulas, Graphs, and Mathematical Tables, 9th printing.* New York: Dover, pp. 592 and 595, 1972.

Whittaker, E. T. and Watson, G. N. *A Course in Modern Analysis, 4th ed.* Cambridge, England: Cambridge University Press, p. 505, 1990.

Jacobi Matrix

see JACOBI ROTATION MATRIX, JACOBIAN

Jacobi Method

A method of diagonalizing MATRICES using JACOBI ROTATION MATRICES. It consists of a sequence of ORTHOGONAL SIMILARITY TRANSFORMATIONS, each of which eliminates one off-diagonal element.

see also JACOBI ALGORITHM, JACOBI ROTATION MATRIX

References

Press, W. H.; Flannery, B. P.; Teukolsky, S. A.; and Vetterling, W. T. "Jacobi Transformation of a Symmetric Matrix." §11.1 in *Numerical Recipes in FORTRAN: The Art of Scientific Computing, 2nd ed.* Cambridge, England: Cambridge University Press, pp. 456–462, 1992.

Jacobi Polynomial

Also known as the HYPERGEOMETRIC POLYNOMIALS, they occur in the study of ROTATION GROUPS and in the solution to the equations of motion of the symmetric top. They are solutions to the JACOBI DIFFERENTIAL EQUATION. Plugging

$$y=\sum_{\nu=0}^{\infty}a_\nu(x-1)^\nu \quad (1)$$

into the differential equation gives the RECURRENCE RELATION

$$[\gamma-\nu(\nu+\alpha+\beta+1)]a_\nu-2(\nu+1)(\nu+\alpha+1)a_{\nu+1}=0 \quad (2)$$

for $\nu=0, 1, \ldots$, where

$$\gamma\equiv n(n+\alpha+\beta+1). \quad (3)$$

Solving the RECURRENCE RELATION gives

$$P_n^{(\alpha,\beta)}(x)=\frac{(-1)^n}{2^n n!}(1-x)^{-\alpha}(1+x)^{-\beta}\times\frac{d^n}{dx^n}[(1-x)^{\alpha+n}(1+x)^{\beta+n}] \quad (4)$$

for $\alpha,\beta>-1$. They form a complete orthogonal system in the interval $[-1,1]$ with respect to the weighting function

$$w_n(x)=(1-x)^\alpha(1+x)^\beta, \quad (5)$$

and are normalized according to

$$P_n^{(\alpha,\beta)}(1)=\binom{n+\alpha}{n}, \quad (6)$$

where $\binom{n}{k}$ is a BINOMIAL COEFFICIENT. Jacobi polynomials can also be written

$$P_n^{(\alpha,\beta)}=\frac{\Gamma(2n+\alpha+\beta+1)}{n!\Gamma(n+\alpha+\beta+1)}\times G_n(\alpha+\beta+1,\beta+1,\tfrac{1}{2}(x+1)), \quad (7)$$

where $\Gamma(z)$ is the GAMMA FUNCTION and

$$G_n(p,q,x) \equiv \frac{n!\Gamma(n+p)}{\Gamma(2n+p)} P_n^{(p-q,q-1)}(2x-1). \quad (8)$$

Jacobi polynomials are ORTHOGONAL satisfying

$$\int_{-1}^{1} P_m^{(\alpha,\beta)} P_n^{(\alpha,\beta)} (1-x)^\alpha (1+x)^\beta \, dx$$
$$= \frac{2^{\alpha+\beta+1}}{2n+\alpha+\beta+1} \frac{\Gamma(n+\alpha+1)\Gamma(n+\beta+1)}{n!\Gamma(n+\alpha+\beta+1)} \delta_{mn}. \quad (9)$$

The COEFFICIENT of the term x^n in $P_n^{(\alpha,beta)}(x)$ is given by

$$A_n = \frac{\Gamma(2n+\alpha+\beta+1)}{2^n n!\Gamma(n+\alpha+\beta+1)}. \quad (10)$$

They satisfy the RECURRENCE RELATION

$$2(n+1)(n+\alpha+\beta+1)(2n+\alpha+\beta)P_{n+1}^{(\alpha,\beta)}(x)$$
$$= [(2n+\alpha+\beta+1)(\alpha^2-\beta^2) + (2n+\alpha+\beta)_3 x] P_n^{(\alpha,\beta)}(x)$$
$$-2(n+\alpha)(n+\beta)(2n+\alpha+\beta+2)P_{n-1}^{(\alpha,\beta)}(x), \quad (11)$$

where $(m)_n$ is the RISING FACTORIAL

$$(m)_n \equiv m(m+1)\cdots(m+n-1) = \frac{(m+n-1)!}{(m-1)!}. \quad (12)$$

The DERIVATIVE is given by

$$\frac{d}{dx}[P_n^{(\alpha,\beta)}(x)] = \tfrac{1}{2}(n+\alpha+\beta+1)P_{n-1}^{(\alpha+1,\beta+1)}(x). \quad (13)$$

The ORTHOGONAL POLYNOMIALS with WEIGHTING FUNCTION $(b-x)^\alpha(x-a)^\beta$ on the CLOSED INTERVAL $[a,b]$ can be expressed in the form

$$[\text{const.}] P_n^{(\alpha,\beta)}\left(2\frac{x-a}{b-a} - 1\right) \quad (14)$$

(Szegő 1975, p. 58).

Special cases with $\alpha = \beta$ are

$$P_{2\nu}^{(\alpha,\alpha)}(x) = \frac{\Gamma(2\nu+\alpha+1)\Gamma(\nu+1)}{\Gamma(\nu+\alpha+1)\Gamma(2\nu+1)} P_\nu^{(\alpha,-1/2)}(2x^2-1)$$
$$= (-1)^\nu \frac{\Gamma(2\nu+\alpha+1)\Gamma(\nu+1)}{\Gamma(\nu+\alpha+1)\Gamma(2\nu+1)} P_\nu^{(-1/2,\alpha)}(1-2x^2) \quad (15)$$

$$P_{2\nu+1}^{(\alpha,\alpha)}(x) = \frac{\Gamma(2\nu+\alpha+2)\Gamma(\nu+1)}{\Gamma(\nu+\alpha+1)\Gamma(2\nu+2)} x P_\nu^{(\alpha,1/2)}(2x^2-1)$$
$$= (-1)^\nu \frac{\Gamma(2\nu+\alpha+2)\Gamma(\nu+1)}{\Gamma(\nu+\alpha+1)\Gamma(2\nu+2)} x P_\nu^{(1/2,\alpha)}(1-2x^2). \quad (16)$$

Further identities are

$$P_n^{(\alpha+1,\beta)}(x) = \frac{2}{2n+\alpha+\beta+2} \times \frac{(n+\alpha+1)P_n^{(\alpha,\beta)} - (n+1)P_{n+1}^{(\alpha,\beta)}(x)}{1-x} \quad (17)$$

$$P_n^{(\alpha,\beta+1)}(x) = \frac{2}{2n+\alpha+\beta+2} \frac{(n+\beta+1)P_n^{(\alpha,\beta)}(x) + (n+1)P_{n+1}^{(\alpha,\beta)}(x)}{1+x} \quad (18)$$

$$\sum_{\nu=0}^{n} \frac{2\nu+\alpha+\beta+1}{2^{\alpha+\beta+1}} \frac{\Gamma(\nu+1)\Gamma(\nu+\alpha+\beta+1)}{\Gamma(\nu+\alpha+1)\Gamma(\nu+\beta+1)} \times P_\nu^{(\alpha,\beta)}(x) Q_\nu^{(\alpha,\beta)}(y)$$
$$= \frac{1}{2} \frac{(y-1)^{-\alpha}(y+1)^{-\beta}}{y-x} + \frac{2^{-\alpha-\beta}}{2n+\alpha+\beta+2} \times \frac{\Gamma(n+2)\Gamma(n+\alpha+\beta+2)}{\Gamma(n+\alpha+1)\Gamma(n+\beta+1)}$$
$$\frac{P_{n+1}^{(\alpha,\beta)}(x)Q_n^{(\alpha,\beta)}(y) - P_n^{(\alpha,\beta)}(x)Q_{n+1}^{\alpha,\beta}(y)}{x-y} \quad (19)$$

(Szegő 1975, p. 79).

The KERNEL POLYNOMIAL is

$$K_n^{(\alpha,\beta)}(x,y) = \frac{2^{-\alpha-\beta}}{2n+\alpha+\beta+2} \times \frac{\Gamma(n+2)\Gamma(n+\alpha+\beta+2)}{\Gamma(n+\alpha+1)\Gamma(n+\beta+1)} \times \frac{P_{n+1}^{(\alpha,\beta)}(x)P_n^{(\alpha,\beta)}(y) - P_n^{(\alpha,\beta)}(x)P_{n+1}^{(\alpha,\beta)}(y)}{x-y} \quad (20)$$

(Szegő 1975, p. 71).

The DISCRIMINANT is

$$D_n^{(\alpha,\beta)} = 2^{-n(n-1)} \prod_{\nu=1}^{n} \nu^{\nu-2n+2}(\nu+\alpha)^{\nu-1}(\nu+\beta)^{\nu-1} \times (n+\nu+\alpha+\beta)^{n-\nu} \quad (21)$$

(Szegő 1975, p. 143).

For $\alpha = \beta = 0$, $P_n^{(0,0)}(x)$ reduces to a LEGENDRE POLYNOMIAL. The GEGENBAUER POLYNOMIAL

$$G_n(p,q,x) = \frac{n!\Gamma(n+p)}{\Gamma(2n+p)} P_n^{(p-q,q-1)}(2x-1) \quad (22)$$

and CHEBYSHEV POLYNOMIAL OF THE FIRST KIND can also be viewed as special cases of the Jacobi POLYNOMIALS. In terms of the HYPERGEOMETRIC FUNCTION,

$$P_n^{(\alpha,\beta)}(x) = \binom{n+\alpha}{n} \times {}_2F_1(-n, n+\alpha+\beta; \alpha+1; \tfrac{1}{2}(1-x)) \quad (23)$$

$$P_n^{(\alpha,\beta)}(x) = \binom{n+\alpha}{n} \left(\frac{x+1}{2}\right)^2 \times {}_2F_1\left(-n, -n-\beta; \alpha+1; \frac{x-1}{x+1}\right). \quad (24)$$

Let N_1 be the number of zeros in $x \in (-1,1)$, N_2 the number of zeros in $x \in (-\infty,-1)$, and N_3 the number of zeros in $x \in (1,\infty)$. Define Klein's symbol

$$E(u) = \begin{cases} 0 & \text{if } u \leq 0 \\ \lfloor u \rfloor & \text{if } u \text{ positive and nonintegral} \\ u-1 & \text{if } u = 1, 2, \ldots, \end{cases} \quad (25)$$

where $\lfloor x \rfloor$ is the FLOOR FUNCTION, and

$$X(\alpha,\beta) = E[\tfrac{1}{2}(|2n+\alpha+\beta+1| - |\alpha| - |\beta| + 1)] \quad (26)$$

$$Y(\alpha,\beta) = E[\tfrac{1}{2}(-|2n+\alpha+\beta+1| + |\alpha| - |\beta| + 1)] \quad (27)$$

$$Z(\alpha,\beta) = E[\tfrac{1}{2}(-|2n+\alpha+\beta+1| - |\alpha| + |\beta| + 1)]. \quad (28)$$

If the cases $\alpha = -1, -2, \ldots, -n$, $\beta = -1, -2, \ldots, -n$, and $n+\alpha+\beta = -1, -2, \ldots, -n$ are excluded, then the number of zeros of $P_n^{(\alpha,\beta)}$ in the respective intervals are

$$N_1(\alpha,\beta) = \begin{cases} 2\left\lfloor \frac{1}{2}(X+1) \right\rfloor & \text{for } (-1)^n \binom{n+\alpha}{n}\binom{n+\beta}{n} > 0 \\ 2\left\lfloor \frac{1}{2}X \right\rfloor + 1 & \text{for } (-1)^n \binom{n+\alpha}{n}\binom{n+\beta}{n} < 0 \end{cases} \quad (29)$$

$$N_2(\alpha,\beta) = \begin{cases} 2\left\lfloor \frac{1}{2}(Y+1) \right\rfloor & \text{for } \binom{2n+\alpha+\beta}{n}\binom{n+\beta}{n} > 0 \\ 2\left\lfloor \frac{1}{2}Y \right\rfloor + 1 & \text{for } \binom{2n+\alpha+\beta}{n}\binom{n+\beta}{n} < 0 \end{cases} \quad (30)$$

$$N_3(\alpha,\beta) = \begin{cases} 2\left\lfloor \frac{1}{2}(Z+1) \right\rfloor & \text{for } \binom{2n+\alpha+\beta}{n}\binom{n+\alpha}{n} > 0 \\ 2\left\lfloor \frac{1}{2}Z \right\rfloor + 1 & \text{for } \binom{2n+\alpha+\beta}{n}\binom{n+\alpha}{n} < 0 \end{cases} \quad (31)$$

(Szegő 1975, pp. 144–146).

The first few POLYNOMIALS are

$$P_0^{(\alpha,\beta)}(x) = 1$$

$$P_1^{(\alpha,\beta)}(x) = \tfrac{1}{2}[2(\alpha+1) + (\alpha+\beta+2)(x-1)]$$

$$P_2^{(\alpha,\beta)}(x) = \tfrac{1}{8}[4(\alpha+1)_2 + 4(\alpha+\beta+3)(\alpha+2)(x-1) + (\alpha+\beta+2)_2(x-1)^2],$$

where $(m)_n$ is a RISING FACTORIAL. See Abramowitz and Stegun (1972, pp. 782–793) and Szegő (1975, Ch. 4) for additional identities.

see also CHEBYSHEV POLYNOMIAL OF THE FIRST KIND, GEGENBAUER POLYNOMIAL, JACOBI FUNCTION OF THE SECOND KIND, RISING FACTORIAL, ZERNIKE POLYNOMIAL

References

Abramowitz, M. and Stegun, C. A. (Eds.). "Orthogonal Polynomials." Ch. 22 in *Handbook of Mathematical Functions with Formulas, Graphs, and Mathematical Tables, 9th printing.* New York: Dover, pp. 771–802, 1972.

Iyanaga, S. and Kawada, Y. (Eds.). "Jacobi Polynomials." Appendix A, Table 20.V in *Encyclopedic Dictionary of Mathematics.* Cambridge, MA: MIT Press, p. 1480, 1980.

Szegő, G. "Jacobi Polynomials." Ch. 4 in *Orthogonal Polynomials, 4th ed.* Providence, RI: Amer. Math. Soc., 1975.

Jacobi Quadrature

see JACOBI-GAUSS QUADRATURE

Jacobi Rotation Matrix

A MATRIX used in the JACOBI TRANSFORMATION method of diagonalizing MATRICES. It contains $\cos\phi$ in p rows and columns and $\sin\phi$ in q rows and columns,

$$P_{pq} \equiv \begin{bmatrix} 1 & & & & & & & & 0 \\ & \ddots & & & \vdots & & & \iddots & \\ & & \cos\phi & \cdots & 0 & \cdots & \sin\phi & & \\ & \cdots & 0 & \cdots & 1 & \cdots & 0 & \cdots & \\ & & \sin\phi & \cdots & 0 & \cdots & \cos\phi & & \\ & \iddots & & & \vdots & & & \ddots & \\ 0 & & & & & & & & 1 \end{bmatrix}.$$

see also JACOBI TRANSFORMATION

References

Press, W. H.; Flannery, B. P.; Teukolsky, S. A.; and Vetterling, W. T. "Jacobi Transformation of a Symmetric Matrix." §11.1 in *Numerical Recipes in FORTRAN: The Art of Scientific Computing, 2nd ed.* Cambridge, England: Cambridge University Press, pp. 456–462, 1992.

Jacobi Symbol

The product of LEGENDRE SYMBOLS (n/p_i) for each of the PRIME factors p_i such that $m = \prod_i p_i$, denoted (n/m). When m is a PRIME, the Jacobi symbol reduces to the LEGENDRE SYMBOL. The Jacobi symbol satisfies the same rules as the LEGENDRE SYMBOL

$$(n/m)(n/m') = (n/(mm')) \quad (1)$$

$$(n/m)(n'/m) = ((nn')/m) \quad (2)$$

$$(n^2/m) = (n/m^2) = 1 \quad \text{if } (m,n) = 1 \quad (3)$$

$$(n/m) = (n'/m) \quad \text{if } n \equiv n' \pmod{m} \quad (4)$$

$$(-1/m) = (-1)^{(m-1)/2} = \begin{cases} 1 & \text{for } m \equiv 1 \pmod 4 \\ -1 & \text{for } m \equiv -1 \pmod 4 \end{cases} \quad (5)$$

$$(2/m) = (-1)^{(m^2-1)/8} = \begin{cases} 1 & \text{for } m \equiv \pm 1 \pmod 8 \\ -1 & \text{for } m \equiv \pm 3 \pmod 8 \end{cases} \quad (6)$$

$$(n/m) = \begin{cases} (m/n) & \text{for } m \text{ or } n \equiv 1 \pmod 4 \\ -(m/n) & \text{for } m, n \equiv 3 \pmod 4. \end{cases} \quad (7)$$

Written another way, for m and n RELATIVELY PRIME ODD INTEGERS with $n \geq 3$,

$$(m/n) = (-1)^{(m-1)(n-1)/4}(n/m). \quad (8)$$

The Jacobi symbol is not defined if $m \leq 0$ or m is EVEN.

Bach and Shallit (1996) show how to compute the Jacobi symbol in terms of the SIMPLE CONTINUED FRACTION of a RATIONAL NUMBER a/b.

see also KRONECKER SYMBOL

References

Bach, E. and Shallit, J. *Algorithmic Number Theory, Vol. 1: Efficient Algorithms.* Cambridge, MA: MIT Press, pp. 343–344, 1996.

Guy, R. K. "Quadratic Residues. Schur's Conjecture." §F5 in *Unsolved Problems in Number Theory, 2nd ed.* New York: Springer-Verlag, pp. 244–245, 1994.

Riesel, H. "Jacobi's Symbol." *Prime Numbers and Computer Methods for Factorization, 2nd ed.* Boston, MA: Birkhäuser, pp. 281–284, 1994.

Jacobi Tensor

$$J^{\mu}_{\nu\alpha\beta} = J^{\mu}_{\nu\beta\alpha} \equiv \tfrac{1}{2}(R^{\mu}_{\alpha\nu\beta} + R^{\mu}_{\beta\nu\alpha}),$$

where R is the RIEMANN TENSOR.

see also RIEMANN TENSOR

Jacobi's Theorem

Let M_r be an r-rowed MINOR of the nth order DETERMINANT $|\mathsf{A}|$ associated with an $n \times n$ MATRIX $\mathsf{A} = a_{ij}$ in which the rows i_1, i_2, ..., i_r are represented with columns k_1, k_2, ..., k_r. Define the complementary minor to M_r as the $(n-k)$-rowed MINOR obtained from $|\mathsf{A}|$ by deleting all the rows and columns associated with M_r and the signed complementary minor $M^{(r)}$ to M_r to be

$$\begin{aligned} M^{(r)} &= (-1)^{i_1+i_2+\ldots+i_r+k_1+k_2+\ldots+k_r} \\ &\quad \times [\text{complementary minor to} M_r]. \end{aligned}$$

Let the MATRIX of cofactors be given by

$$\Delta = \begin{vmatrix} A_{11} & A_{12} & \cdots & A_{1n} \\ A_{21} & A_{22} & \cdots & A_{2n} \\ \vdots & \vdots & \ddots & \vdots \\ A_{n1} & A_{n2} & \cdots & A_{nn} \end{vmatrix},$$

with M_r and M_r' the corresponding r-rowed minors of $|\mathsf{A}|$ and Δ, then it is true that

$$M_r' = |\mathsf{A}|^{r-1} M^{(r)}.$$

References

Gradshteyn, I. S. and Ryzhik, I. M. *Tables of Integrals, Series, and Products, 5th ed.* San Diego, CA: Academic Press, pp. 1109–1100, 1979.

Jacobi Theta Function

see THETA FUNCTION

Jacobi Transformation

see JACOBI METHOD

Jacobi Triple Product

The Jacobi triple product is the beautiful identity

$$\prod_{n=1}^{\infty}(1-x^{2n})(1+x^{2n-1}z^2)\left(1+\frac{x^{2n-1}}{z^2}\right) = \sum_{m=-\infty}^{\infty} x^{m^2} z^{2m}. \quad (1)$$

In terms of the Q-FUNCTIONS, (1) is written

$$Q_1 Q_2 Q_3 = 1, \quad (2)$$

which is one of the two JACOBI IDENTITIES. For the special case of $z = 1$, (1) becomes

$$\begin{aligned} \varphi(x) \equiv G(1) &= \prod_{n=1}^{\infty}(1+x^{2n-1})^2(1-x^{2n}) \\ &= \sum_{m=-\infty}^{\infty} x^{m^2} = 1 + 2\sum_{m=0}^{\infty} x^{m^2}, \end{aligned} \quad (3)$$

where $\varphi(x)$ is the one-variable RAMANUJAN THETA FUNCTION.

To prove the identity, define the function

$$\begin{aligned} F(z) &\equiv \prod_{n=1}^{\infty}(1+x^{2n-1}z^2)\left(1+\frac{x^{2n-1}}{z^2}\right) \\ &= (1+xz^2)\left(1+\frac{x}{z^2}\right)(1+x^3z^2)\left(1+\frac{x^3}{z^2}\right) \\ &\quad \times (1+x^5z^2)\left(1+\frac{x^5}{z^2}\right)\cdots. \end{aligned} \quad (4)$$

Then

$$\begin{aligned} F(xz) &= (1+x^3z^2)\left(1+\frac{1}{xz^2}\right)(1+x^5z^2)\left(1+\frac{x}{z^2}\right) \\ &\quad (1+x^7z^2)\left(1+\frac{x^3}{z^2}\right)\cdots. \end{aligned} \quad (5)$$

Taking (5) ÷ (4),

$$\begin{aligned} \frac{F(xz)}{F(z)} &= \left(1+\frac{1}{xz^2}\right)\left(\frac{1}{1+xz^2}\right) \\ &= \frac{xz^2+1}{xz^2}\frac{1}{1+xz^2} = \frac{1}{xz^2}, \end{aligned} \quad (6)$$

which yields the fundamental relation

$$xz^2 F(xz) = F(z). \quad (7)$$

Now define

$$G(z) \equiv F(z)\prod_{n=1}^{\infty}(1-x^{2n}) \quad (8)$$

$$G(xz) = F(xz)\prod_{n=1}^{\infty}(1-x^{2n}). \tag{9}$$

Using (7), (9) becomes

$$G(xz) = \frac{F(z)}{xz^2}\prod_{n=1}^{\infty}(1-x^{2n}) = \frac{G(z)}{xz^2}, \tag{10}$$

so

$$G(z) = xz^2 G(xz). \tag{11}$$

Expand G in a LAURENT SERIES. Since G is an EVEN FUNCTION, the LAURENT SERIES contains only even terms.

$$G(z) = \sum_{m=-\infty}^{\infty} a_m z^{2m}. \tag{12}$$

Equation (11) then requires that

$$\sum_{m=-\infty}^{\infty} a_m z^{2m} = xz^2 \sum_{m=-\infty}^{\infty} a_m (xz)^{2m}$$
$$= \sum_{m=-\infty}^{\infty} a_m x^{2m+1} z^{2m+2}. \tag{13}$$

This can be re-indexed with $m' \equiv m-1$ on the left side of (13)

$$\sum_{m=-\infty}^{\infty} a_m z^{2m} = \sum_{m=-\infty}^{\infty} a_m x^{2m-1} z^{2m}, \tag{14}$$

which provides a RECURRENCE RELATION

$$a_m = a_{m-1}x^{2m-1}, \tag{15}$$

so

$$a_1 = a_0 x \tag{16}$$
$$a_2 = a_1 x^3 = a_0 x^{3+1} = a_0 x^4 = a_0 x^{2^2} \tag{17}$$
$$a_3 = a_2 x^5 = a_0 x^{5+4} = a_0 x^9 = a_0 x^{3^2}. \tag{18}$$

The exponent grows greater by $(2m-1)$ for each increase in m of 1. It is given by

$$\sum_{n=1}^{m}(2m-1) = 2\frac{m(m+1)}{2} - m = m^2. \tag{19}$$

Therefore,

$$a_m = a_0 x^{m^2}. \tag{20}$$

This means that

$$G(z) = a_0 \sum_{m=-\infty}^{\infty} x^{m^2} z^{2m}. \tag{21}$$

The COEFFICIENT a_0 must be determined by going back to (4) and (8) and letting $z = 1$. Then

$$F(1) = \prod_{n=1}^{\infty}(1+x^{2n-1})(1+x^{2n-1})$$
$$= \prod_{n=1}^{\infty}(1+x^{2n-1})^2 \tag{22}$$

$$G(1) = F(1)\prod_{n=1}^{\infty}(1-x^{2n})$$
$$= \prod_{n=1}^{\infty}(1+x^{2n-1})^2 \prod_{n=1}^{\infty}(1-x^{2n})$$
$$= \prod_{n=1}^{\infty}(1+x^{2n-1})^2(1-x^{2n}), \tag{23}$$

since multiplication is ASSOCIATIVE. It is clear from this expression that the a_0 term must be 1, because all other terms will contain higher POWERS of x. Therefore,

$$a_0 = 1, \tag{24}$$

so we have the Jacobi triple product,

$$G(z) = \prod_{n=1}^{\infty}(1-x^{2n})(1+x^{2n-1}z^2)\left(1+\frac{x^{2n-1}}{z^2}\right)$$
$$= \sum_{m=-\infty}^{\infty} x^{m^2} z^{2m}. \tag{25}$$

see also EULER IDENTITY, JACOBI IDENTITIES, Q-FUNCTION, QUINTUPLE PRODUCT IDENTITY, RAMANUJAN PSI SUM, RAMANUJAN THETA FUNCTIONS, SCHRÖTER'S FORMULA, THETA FUNCTION

References

Andrews, G. E. *q-Series: Their Development and Application in Analysis, Number Theory, Combinatorics, Physics, and Computer Algebra.* Providence, RI: Amer. Math. Soc., pp. 63–64, 1986.

Borwein, J. M. and Borwein, P. B. "Jacobi's Triple Product and Some Number Theoretic Applications." Ch. 3 in *Pi & the AGM: A Study in Analytic Number Theory and Computational Complexity.* New York: Wiley, pp. 62–101, 1987.

Jacobi, C. G. J. *Fundamentia Nova Theoriae Functionum Ellipticarum.* Regiomonti, Sumtibus fratrum Borntraeger, p. 90, 1829.

Whittaker, E. T. and Watson, G. N. *A Course in Modern Analysis, 4th ed.* Cambridge, England: Cambridge University Press, p. 470, 1990.

Jacobi Zeta Function

Denoted $\mathrm{zn}(u,k)$ or $Z(u)$.

$$Z(\phi|m) \equiv E(\phi|m) - \frac{E(m)F(\phi|m)}{K(m)},$$

where ϕ is the AMPLITUDE, m is the PARAMETER, and F and K are ELLIPTIC INTEGRALS OF THE FIRST KIND, and E is an ELLIPTIC INTEGRAL OF THE SECOND KIND. See Gradshteyn and Ryzhik (1980, p. xxxi) for expressions in terms of THETA FUNCTIONS.

see also ZETA FUNCTION

References

Abramowitz, M. and Stegun, C. A. (Eds.). *Handbook of Mathematical Functions with Formulas, Graphs, and Mathematical Tables, 9th printing.* New York: Dover, p. 595, 1972.

Gradshteyn, I. S. and Ryzhik, I. M. *Tables of Integrals, Series, and Products, 5th ed.* San Diego, CA: Academic Press, 1979.

Jacobian

Given a set $\mathbf{y} = \mathbf{f}(\mathbf{x})$ of n equations in n variables x_1, ..., x_n, written explicitly as

$$\mathbf{y} \equiv \begin{bmatrix} f_1 \\ f_2 \\ \vdots \\ f_n \end{bmatrix}, \tag{1}$$

or more explicitly as

$$\begin{cases} y_1 = f_1(x_1, \ldots, x_n) \\ \vdots \\ y_n = f_n(x_1, \ldots, x_n), \end{cases} \tag{2}$$

the Jacobian matrix, sometimes simply called "the Jacobian" (Simon and Blume 1994) is defined by

$$\mathsf{J}(x_1, x_2, x_3) = \begin{bmatrix} \frac{\partial y_1}{\partial x_1} & \cdots & \frac{\partial y_1}{\partial x_n} \\ \vdots & \ddots & \vdots \\ \frac{\partial y_n}{\partial x_1} & \cdots & \frac{\partial y_n}{\partial x_n} \end{bmatrix}. \tag{3}$$

The DETERMINANT of J is the JACOBIAN DETERMINANT (confusingly, often called "the Jacobian" as well) and is denoted

$$J = \left| \tfrac{\partial(y_1, \ldots, y_n)}{\partial(x_1, \ldots, x_n)} \right|. \tag{4}$$

Taking the differential

$$d\mathbf{y} = \mathbf{y_x}\, d\mathbf{x} \tag{5}$$

shows that J is the DETERMINANT of the MATRIX $\mathbf{y_x}$, and therefore gives the ratios of n-D volumes (CONTENTS) in y and x,

$$dy_1 \cdots dy_n = \left| \frac{\partial(y_1, \ldots, y_n)}{\partial(x_1, \ldots, x_n)} \right| dx_1 \cdots dx_n. \tag{6}$$

The concept of the Jacobian can also be applied to n functions in more than n variables. For example, considering $f(u, v, w)$ and $g(u, v, w)$, the Jacobians

$$\frac{\partial(f, g)}{\partial(u, v)} = \begin{vmatrix} f_u & f_v \\ g_u & g_v \end{vmatrix} \tag{7}$$

$$\frac{\partial(f, g)}{\partial(u, w)} = \begin{vmatrix} f_u & f_w \\ g_u & g_w \end{vmatrix} \tag{8}$$

can be defined (Kaplan 1984, p. 99).

For the case of $n = 3$ variables, the Jacobian takes the special form

$$Jf(x_1, x_2, x_3) \equiv \left| \frac{\partial \mathbf{y}}{\partial x_1} \cdot \frac{\partial \mathbf{y}}{\partial x_2} \times \frac{\partial \mathbf{y}}{\partial x_3} \right|, \tag{9}$$

where $\mathbf{a} \cdot \mathbf{b}$ is the DOT PRODUCT and $\mathbf{b} \times \mathbf{c}$ is the CROSS PRODUCT, which can be expanded to give

$$\left| \frac{\partial(y_1, y_2, y_3)}{\partial(x_1, x_2, x_3)} \right| = \begin{vmatrix} \frac{\partial y_1}{\partial x_1} & \frac{\partial y_1}{\partial x_2} & \frac{\partial y_1}{\partial x_3} \\ \frac{\partial y_2}{\partial x_1} & \frac{\partial y_2}{\partial x_2} & \frac{\partial y_2}{\partial x_3} \\ \frac{\partial y_3}{\partial x_1} & \frac{\partial y_3}{\partial x_2} & \frac{\partial y_3}{\partial x_3} \end{vmatrix}. \tag{10}$$

see also CHANGE OF VARIABLES THEOREM, CURVILINEAR COORDINATES, IMPLICIT FUNCTION THEOREM

References

Kaplan, W. *Advanced Calculus, 3rd ed.* Reading, MA: Addison-Wesley, pp. 98–99, 123, and 238–245, 1984.

Simon, C. P. and Blume, L. E. *Mathematics for Economists.* New York: W. W. Norton, 1994.

Jacobian Conjecture

If $\det[F'(x)] = 1$ for a POLYNOMIAL mapping F (where det is the DETERMINANT), then F is BIJECTIVE with POLYNOMIAL inverse.

Jacobian Curve

The Jacobian of a linear net of curves of order n is a curve of order $3(n-1)$. It passes through all points common to all curves of the net. It is the LOCUS of points where the curves of the net touch one another and of singular points of the curve.

see also CAYLEYIAN CURVE, HESSIAN COVARIANT, STEINERIAN CURVE

References

Coolidge, J. L. *A Treatise on Algebraic Plane Curves.* New York: Dover, p. 149, 1959.

Jacobian Determinant

see JACOBIAN

Jacobian Group

The Jacobian group of a 1-D linear series is given by intersections of the base curve with the JACOBIAN CURVE of itself and two curves cutting the series.

References

Coolidge, J. L. *A Treatise on Algebraic Plane Curves.* New York: Dover, p. 283, 1959.

Jacobsthal-Lucas Number

see JACOBSTHAL NUMBER

Jacobsthal-Lucas Polynomial

see JACOBSTHAL POLYNOMIAL

Jacobsthal Number

The Jacobsthal numbers are the numbers obtained by the U_ns in the LUCAS SEQUENCE with $P = 1$ and $Q = -2$, corresponding to $a = 2$ and $b = -1$. They and the Jacobsthal-Lucas numbers (the V_ns) satisfy the RECURRENCE RELATION

$$J_n = J_{n-1} + 2J_{n-2}. \quad (1)$$

The Jacobsthal numbers satisfy $J_0 = 0$ and $J_1 = 1$ and are 0, 1, 1, 3, 5, 11, 21, 43, 85, 171, 341, ... (Sloane's A001045). The Jacobsthal-Lucas numbers satisfy $j_0 = 2$ and $j_1 = 1$ and are 2, 1, 5, 7, 17, 31, 65, 127, 257, 511, 1025, ... (Sloane's A014551). The properties of these numbers are summarized in Horadam (1996). They are given by the closed form expressions

$$J_n = \sum_{r=0}^{\lfloor (n-1)/2 \rfloor} \binom{n-1-r}{r} 2^r \quad (2)$$

$$j_n = \sum_{r=0}^{\lfloor n/2 \rfloor} \frac{n}{n-r} \binom{n-r}{r} 2^r, \quad (3)$$

where $\lfloor x \rfloor$ is the FLOOR FUNCTION and $\binom{n}{k}$ is a BINOMIAL COEFFICIENT. The Binet forms are

$$J_n = \tfrac{1}{3}(a^n - b^n) = \tfrac{1}{3}[2^n - (-1)^n] \quad (4)$$

$$j_n = a^n + b^n = 2^n + (-1)^n. \quad (5)$$

The GENERATING FUNCTIONS are

$$\sum_{i=1}^{\infty} J_i x^{i-1} = (1 - x - 2x^2)^{-1} \quad (6)$$

$$\sum_{i=1}^{\infty} j_i x^{i-1} = (1 + 4x)(1 - x - 2x^2)^{-1}. \quad (7)$$

The Simson FORMULAS are

$$J_{n+1}J_{n-1} - J_n{}^2 = (-1)^n 2^{n-1} \quad (8)$$

$$j_{n+1}j_{n-1} - j_n{}^2 = 9(-1)^{n-1}2^{n-1} = -9(J_{n+1}J_{n-1} - J_n{}^2). \quad (9)$$

Summation FORMULAS include

$$\sum_{i=2}^{n} J_i = \tfrac{1}{2}(J_{n+2} - 3) \quad (10)$$

$$\sum_{i=1}^{n} j_i = \tfrac{1}{2}(j_{n+2} - 5). \quad (11)$$

Interrelationships are

$$j_n J_n = J_{2n} \quad (12)$$

$$j_n = J_{n+1} + 2J_{n-1} \quad (13)$$

$$9J_n = j_{n+1} + 2j_{n-1} \quad (14)$$

$$j_{n+1} + j_n = 3(J_{n+1} + J_n) = 3 \cdot 2^n \quad (15)$$

$$j_{n+1} - j_n = 3(J_{n+1} - J_n) + 4(-1)^{n+1} = 2^n + 2(-1)^{n+1} \quad (16)$$

$$j_{n+1} - 2j_n = 3(2J_n - J_{n+1}) = 3(-1)^{n+1} \quad (17)$$

$$2j_{n+1} + j_{n-1} = 3(2J_{n+1} + J_{n-1}) + 6(-1)^{n+1} \quad (18)$$

$$j_{n+r} + j_{n-r} = 3(J_{n+r} + J_{n-r}) + 4(-1)^{n-r} \quad (19)$$

$$= 2^{n-r}(2^{2r} + 1) + 2(-1)^{n-r} \quad (20)$$

$$j_{n+r} - j_{n-r} = 3(J_{n+r} - J_{n-r}) = 2^{n-r}(2^{2r} - 1) \quad (21)$$

$$j_n = 3J_n + 2(-1)^n \quad (22)$$

$$3J_n + j_n = 2^{n+1} \quad (23)$$

$$J_n + j_n = 2J_{n+1} \quad (24)$$

$$j_{n+2}j_{n-2} - j_n{}^2 = -9(J_{n+2}J_{n-2} - J_n)^2 = 9(-1)^n 2^{n-2} \quad (25)$$

$$J_m j_n + J_n j_m = 2J_{m+n} \quad (26)$$

$$j_m j_n + 9J_m J_n = 2j_{m+n} \quad (27)$$

$$j_n{}^2 + 9J_n{}^2 = 2j_{2n} \quad (28)$$

$$J_m j_n - J_n j_m = (-1)^n 2^{n+1} J_{m-n} \quad (29)$$

$$j_m j_n - 9J_m J_n = (-1)^n 2^{n+1} j_{m-n} \quad (30)$$

$$j_n{}^2 - 9J_n{}^2 = (-1)^n 2^{n+2} \quad (31)$$

(Horadam 1996).

References

Horadam, A. F. "Jacobsthal and Pell Curves." *Fib. Quart.* **26**, 79–83, 1988.

Horadam, A. F. "Jacobsthal Representation Numbers." *Fib. Quart.* **34**, 40–54, 1996.

Sloane, N. J. A. Sequences A014551 and A001045/M2482 in "An On-Line Version of the Encyclopedia of Integer Sequences."

Jacobsthal Polynomial

The Jacobsthal polynomials are the POLYNOMIALS obtained by setting $p(x) = 1$ and $q(x) = 2x$ in the LUCAS POLYNOMIAL SEQUENCE. The first few Jacobsthal polynomials are

$$J_1(x) = 1$$
$$J_2(x) = 1$$
$$J_3(x) = 1 + 2x$$
$$J_4(x) = 1 + 4x$$
$$J_5(x) = 4x^2 + 6x + 1,$$

and the first few Jacobsthal-Lucas polynomials are

$$\begin{aligned} j_1(x) &= 1 \\ j_2(x) &= 4x+1 \\ j_3(x) &= 6x+1 \\ j_4(x) &= 8x^2+8x+1 \\ j_5(x) &= 20x^2+10x+1. \end{aligned}$$

Jacobsthal and Jacobsthal-Lucas polynomials satisfy

$$J_n(1) = J_n$$

$$j_n(1) = j_n$$

where J_n is a JACOBSTHAL NUMBER and j_n is a JACOBSTHAL-LUCAS NUMBER.

Janko Groups

The SPORADIC GROUPS J_1, J_2, J_3 and J_4. The Janko group J_2 is also known as the HALL-JANKO GROUP.

see also SPORADIC GROUP

References

Wilson, R. A. "ATLAS of Finite Group Representation." `http://for.mat.bham.ac.uk/atlas/J1.html`, `J2.html`, `J3.html`, and `J4.html`.

Japanese Triangulation Theorem

Let a convex CYCLIC POLYGON be TRIANGULATED in any manner, and draw the INCIRCLE to each TRIANGLE so constructed. Then the sum of the INRADII is a constant independent of the TRIANGULATION chosen. This theorem can be proved using CARNOT'S THEOREM. It is also true that if the sum of INRADII does not depend on the TRIANGULATION of a POLYGON, then the POLYGON is CYCLIC.

see also CARNOT'S THEOREM, CYCLIC POLYGON, INCIRCLE, INRADIUS, TRIANGULATION

References

Honsberger, R. *Mathematical Gems III.* Washington, DC: Math. Assoc. Amer., pp. 24–26, 1985.

Lambert, T. "The Delaunay Triangulation Maximizes the Mean Inradius." *Proc. Sixth Canadian Conf. Comput. Geometry.* Saskatoon, Saskatchewan, Canada, pp. 201–206, August 1994.

Jarnick's Inequality

Given a CONVEX plane region with AREA A and PERIMETER p, then

$$|N - A| < p,$$

where N is the number of enclosed LATTICE POINTS.

see also LATTICE POINT, NOSARZEWSKA'S INEQUALITY

Jeep Problem

Maximize the distance a jeep can penetrate into the desert using a given quantity of fuel. The jeep is allowed to go forward, unload some fuel, and then return to its base using the fuel remaining in its tank. At its base, it may refuel and set out again. When it reaches fuel it has previously stored, it may then use it to partially fill its tank. This problem is also called the EXPLORATION PROBLEM (Ball and Coxeter 1987).

Given $n+f$ (with $0 \leq f < 1$) drums of fuel at the edge of the desert and a jeep capable of holding one drum (and storing fuel in containers along the way), the maximum one-way distance which can be traveled (assuming the jeep travels one unit of distance per drum of fuel expended) is

$$\begin{aligned} d &= \frac{f}{2n+1} + \sum_{i=1}^{n} \frac{1}{2i-1} \\ &= \frac{f}{2n+1} + \tfrac{1}{2}[\gamma + 2\ln 2 + \psi_0(\tfrac{1}{2}+n)], \end{aligned}$$

where γ is the EULER-MASCHERONI CONSTANT and $\psi_n(z)$ the POLYGAMMA FUNCTION.

For example, the farthest a jeep with $n=1$ drum can travel is obviously 1 unit. However, with $n=2$ drums of gas, the maximum distance is achieved by filling up the jeep's tank with the first drum, traveling 1/3 of a unit, storing 1/3 of a drum of fuel there, and then returning to base with the remaining 1/3 of a tank. At the base, the tank is filled with the second drum. The jeep then travels 1/3 of a unit (expending 1/3 of a drum of fuel), refills the tank using the 1/3 of a drum of fuel stored there, and continues an additional 1 unit of distance on a full tank, giving a total distance of 4/3. The solutions for $n = 1, 2, \ldots$ drums are 1, 4/3, 23/15, 176/105, 563/315, ..., which can also be written as $a(n)/b(n)$, where

$$\begin{aligned} a(n) &= \left(\frac{1}{1} + \frac{1}{3} + \ldots + \frac{1}{2n-1}\right) \operatorname{LCM}(1,3,5,\ldots,2n-1) \\ b(n) &= \operatorname{LCM}(1,3,5,\ldots,2n-1) \end{aligned}$$

(Sloane's A025550 and A025547).

see also HARMONIC NUMBER

References

Alway, G. C. "Crossing the Desert." *Math. Gaz.* **41**, 209, 1957.

Ball, W. W. R. and Coxeter, H. S. M. *Mathematical Recreations and Essays, 13th ed.* New York: Dover, p. 32, 1987.

Bellman, R. Exercises 54–55 *Dynamic Programming.* Princeton, NJ: Princeton University Press, p. 103, 1955.

Fine, N. J. "The Jeep Problem." *Amer. Math. Monthly* **54**, 24–31, 1947.

Gale, D. "The Jeep Once More or Jeeper by the Dozen." *Amer. Math. Monthly* **77**, 493–501, 1970.

Gardner, M. *The Second Scientific American Book of Mathematical Puzzles & Diversions: A New Selection.* New York: Simon and Schuster, pp. 152 and 157–159, 1961.

Haurath, A.; Jackson, B.; Mitchem, J.; and Schmeichel, E. "Gale's Round-Trip Jeep Problem." *Amer. Math. Monthly* **102**, 299–309, 1995.
Helmer, O. "A Problem in Logistics: The Jeep Problem." Project Rand Report No. Ra 15015, Dec. 1947.
Phipps, C. G. "The Jeep Problem, A More General Solution." *Amer. Math. Monthly* **54**, 458–462, 1947.

Jenkins-Traub Method

A complicated POLYNOMIAL ROOT-finding algorithm which is used in the *IMSL*® (IMSL, Houston, TX) library and which Press *et al.* (1992) describe as "practically a standard in black-box POLYNOMIAL ROOT-finders."

References

IMSL, Inc. *IMSL Math/Library User's Manual.* Houston, TX: IMSL, Inc.
Press, W. H.; Flannery, B. P.; Teukolsky, S. A.; and Vetterling, W. T. *Numerical Recipes in FORTRAN: The Art of Scientific Computing, 2nd ed.* Cambridge, England: Cambridge University Press, p. 369, 1992.
Ralston, A. and Rabinowitz, P. §8.9–8.13 in *A First Course in Numerical Analysis, 2nd ed.* New York: McGraw-Hill, 1978.

Jensen's Formula

$$\int_0^{2\pi} \ln|z+e^{i\theta}|\,d\theta = 2\pi \ln^+|z|,$$

where

$$\ln^+ \equiv \max(0, \ln x)$$

and $\ln x$ is the NATURAL LOGARITHM.

Jensen's Inequality

For a REAL CONTINUOUS CONCAVE FUNCTION

$$\frac{\sum f(x_i)}{n} \leq f\left(\frac{\sum x_i}{n}\right)$$

if f is concave down,

$$\frac{\sum f(x_i)}{n} \geq f\left(\frac{\sum x_i}{n}\right)$$

if f is concave up, and

$$\frac{\sum f(x_i)}{n} = f\left(\frac{\sum x_i}{n}\right)$$

IFF $x_1 = x_2 = \ldots = x_n$. A special case is

$$\sqrt[n]{x_1 x_2 \cdots x_n} \leq \frac{x_1 + x_2 + \ldots + x_n}{n},$$

with equality IFF $x_1 = x_2 = \ldots = x_n$.

see also CONCAVE FUNCTION, MAHLER'S MEASURE

References

Gradshteyn, I. S. and Ryzhik, I. M. *Tables of Integrals, Series, and Products, 5th ed.* San Diego, CA: Academic Press, p. 1101, 1979.

Jensen Polynomial

Let $f(x)$ be a real ENTIRE FUNCTION of the form

$$f(x) = \sum_{k=0}^{\infty} \gamma_k \frac{x^k}{k!},$$

where the γ_ks are POSITIVE and satisfy TURÁN'S INEQUALITIES

$$\gamma_k{}^2 - \gamma_{k-1}\gamma_{k+1} \geq 0$$

for $k = 1, 2, \ldots$. The Jensen polynomial $g(t)$ associated with $f(x)$ is then given by

$$g_n(t) = \sum_{k=0}^{n} \binom{n}{k} \gamma_k t^k,$$

where $\binom{a}{b}$ is a BINOMIAL COEFFICIENT.

References

Csordas, G.; Varga, R. S.; and Vincze, I. "Jensen Polynomials with Applications to the Riemann ζ-Function." *J. Math. Anal. Appl.* **153**, 112–135, 1990.

Jerabek's Hyperbola

The ISOGONAL CONJUGATE of the EULER LINE. It passes through the the vertices of a TRIANGLE, the ORTHOCENTER, CIRCUMCENTER, the LEMOINE POINT, and the ISOGONAL CONJUGATE points of the NINE-POINT CENTER and DE LONGCHAMPS POINT.

see also CIRCUMCENTER, DE LONGCHAMPS POINT, EULER LINE, ISOGONAL CONJUGATE, LEMOINE POINT, NINE-POINT CENTER, ORTHOCENTER

References

Casey, J. *A Treatise on the Analytical Geometry of the Point, Line, Circle, and Conic Sections, Containing an Account of Its Most Recent Extensions with Numerous Examples, 2nd rev. enl. ed.* Dublin: Hodges, Figgis, & Co., 1893.
Pinkernell, G. M. "Cubic Curves in the Triangle Plane." *J. Geom.* **55**, 141–161, 1996.
Vandeghen, A. "Some Remarks on the Isogonal and Cevian Transforms. Alignments of Remarkable Points of a Triangle." *Amer. Math. Monthly* **72**, 1091–1094, 1965.

Jerk

The jerk $\mathbf{j}$ is defined as the time DERIVATIVE of the VECTOR ACCELERATION $\mathbf{a}$,

$$\mathbf{j} \equiv \frac{d\mathbf{a}}{dt}.$$

see also ACCELERATION, VELOCITY

Jinc Function

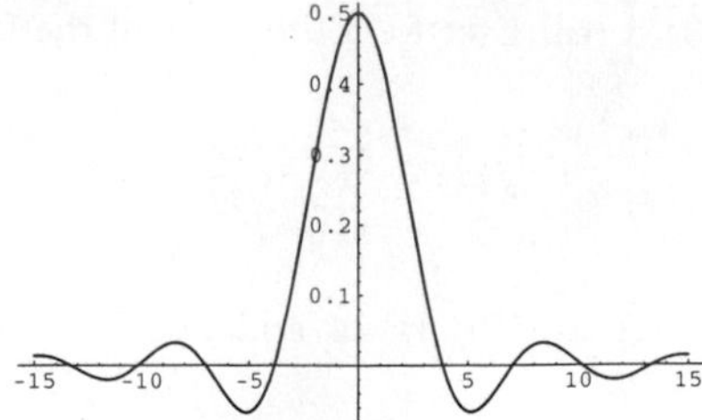

The jinc function is defined as

$$\text{jinc}(x) \equiv \frac{J_1(x)}{x},$$

where $J_1(x)$ is a BESSEL FUNCTION OF THE FIRST KIND, and satisfies $\lim_{x\to 0} \text{jinc}(x) = 1/2$. The DERIVATIVE of the jinc function is given by

$$\text{jinc}'(x) = -\frac{J_2(x)}{x}.$$

The function is sometimes normalized by multiplying by a factor of 2 so that $\text{jinc}(0) = 1$ (Siegman 1986, p. 729).

see also BESSEL FUNCTION OF THE FIRST KIND, SINC FUNCTION

References
Siegman, A. E. *Lasers.* Sausalito, CA: University Science Books, 1986.

Jitter

A SAMPLING phenomenon produced when a waveform is not sampled uniformly at an interval t each time, but rather at a series of slightly shifted intervals $t+\Delta t_i$ such that the average $\langle \Delta t_i \rangle = 0$.

see also GHOST, SAMPLING

Joachimsthal's Equation

Using CLEBSCH-ARONHOLD NOTATION,

$$\xi_1^n a_y^n + \xi_1^{n-1}\xi_2 a_y^{n-1} a_x + \tfrac{1}{2}n(n-1)\xi_1^{n-2}\xi_2^2 a_y^{n-2} a_x^2 + \ldots + n\xi_1\xi_2^{n-1} a_y a_x^{n-1} + \xi_2^n a_x^n = 0.$$

References
Coolidge, J. L. *A Treatise on Algebraic Plane Curves.* New York: Dover, p. 89, 1959.

Johnson Circle

The CIRCUMCIRCLE in JOHNSON'S THEOREM.

see also JOHNSON'S THEOREM

Johnson's Equation

The PARTIAL DIFFERENTIAL EQUATION

$$\frac{\partial}{\partial x}\left(u_t + uu_x + \tfrac{1}{2}u_{xxx} + \frac{u}{2t}\right) + \frac{3\alpha^2}{2t^2}u_{yy} = 0$$

which arises in the study of water waves.

References
Infeld, E. and Rowlands, G. *Nonlinear Waves, Solitons, and Chaos.* Cambridge, England: Cambridge University Press, p. 223, 1990.

Johnson Solid

The Johnson solids are the CONVEX POLYHEDRA having regular faces (with the exception of the completely regular PLATONIC SOLIDS, the "SEMIREGULAR" ARCHIMEDEAN SOLIDS, and the two infinite families of PRISMS and ANTIPRISMS). There are 28 simple (i.e., cannot be dissected into two other regular-faced polyhedra by a plane) regular-faced polyhedra in addition to the PRISMS and ANTIPRISMS (Zalgaller 1969), and Johnson (1966) proposed and Zalgaller (1969) proved that there exist exactly 92 Johnson solids in all.

A database of solids and VERTEX NETS of these solids is maintained on the Bell Laboratories Netlib server, but a few errors exist in several entries. A concatenated and corrected version of the files is given by Weisstein, together with *Mathematica*® (Wolfram Research, Champaign, IL) code to display the solids and nets. The following table summarizes the names of the Johnson solids and gives their images and nets.

1. Square pyramid

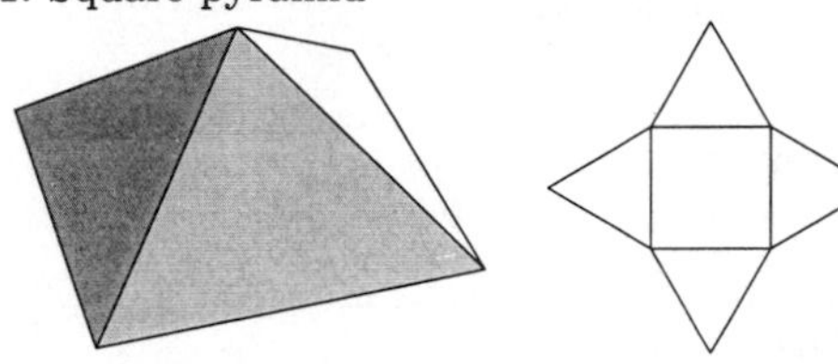

2. Pentagonal pyramid

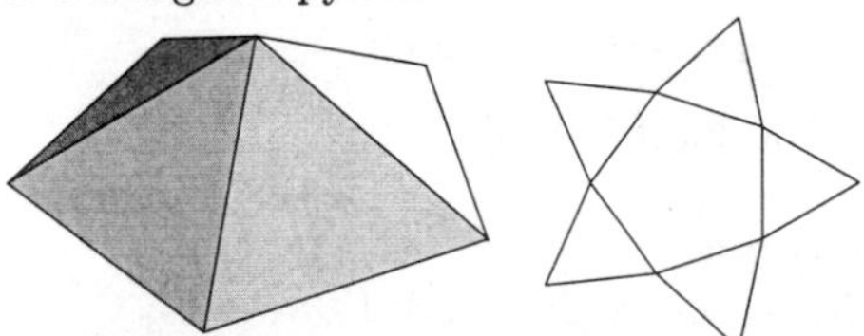

3. Triangular cupola

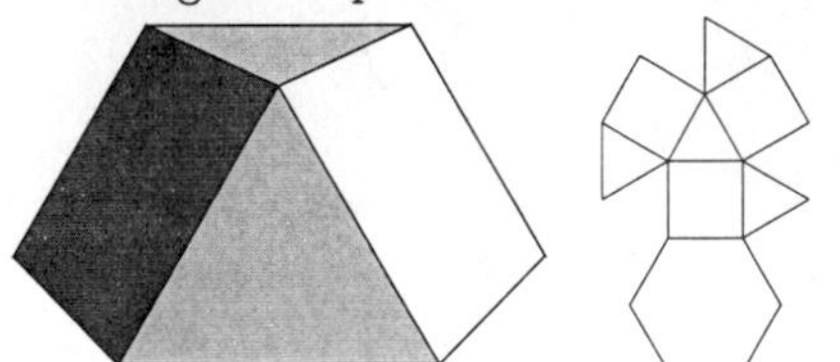

4. Square cupola

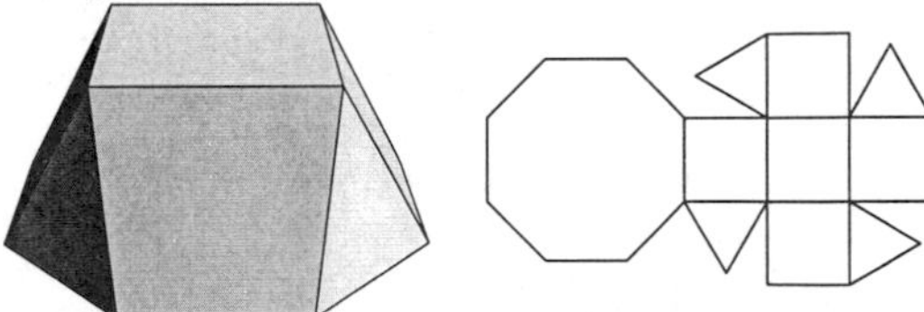

5. Pentagonal cupola

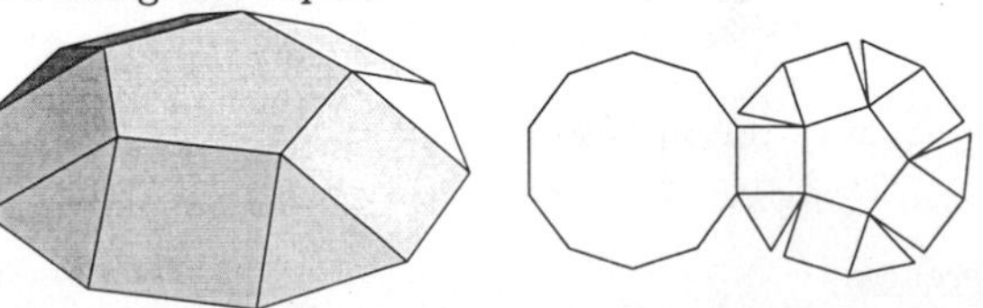

6. Pentagonal rotunda

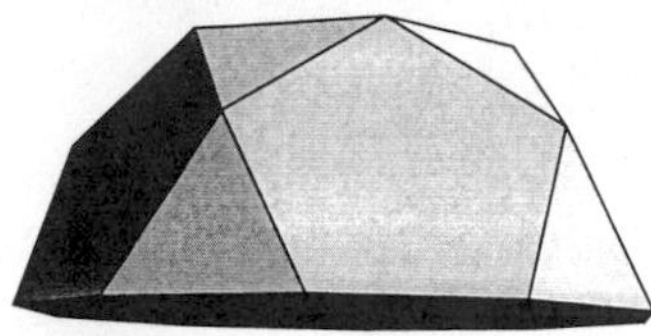

7. Elongated triangular pyramid

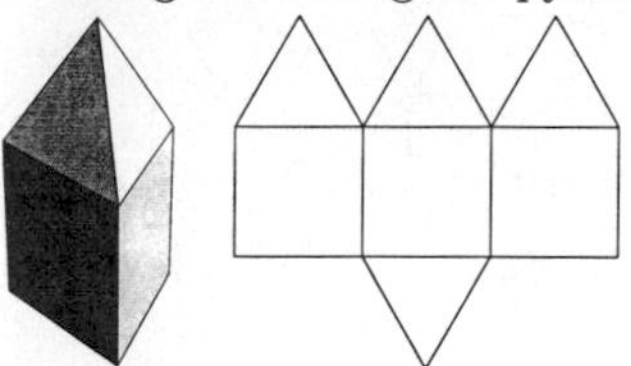

8. Elongated square pyramid

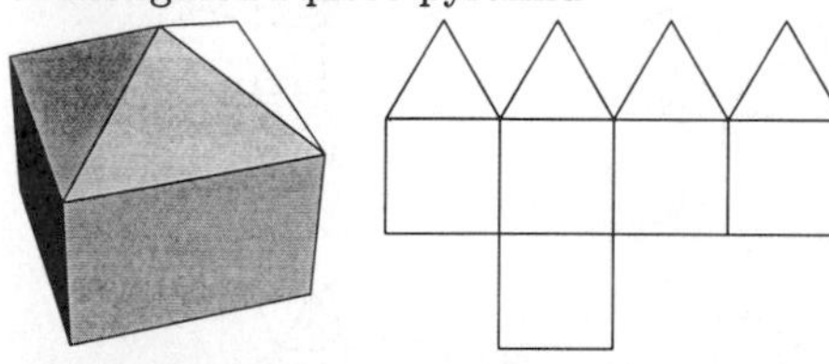

9. Elongated pentagonal pyramid

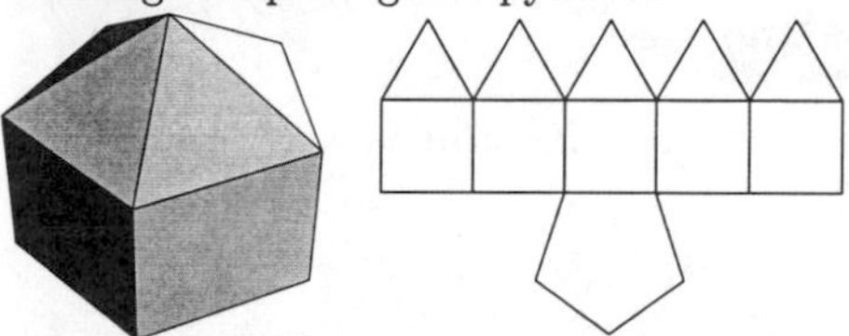

10. Gyroelongated square pyramid

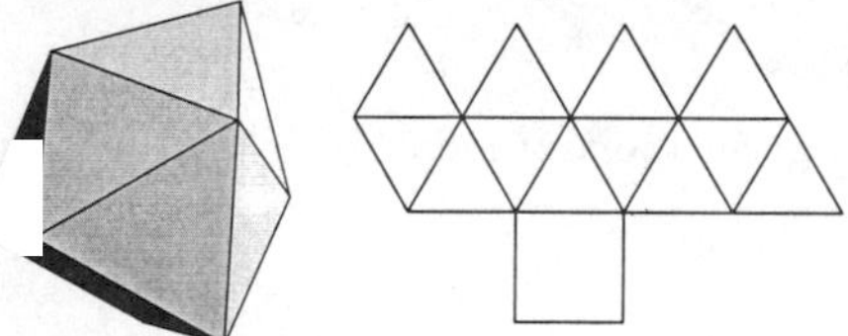

11. Gyroelongated pentagonal pyramid

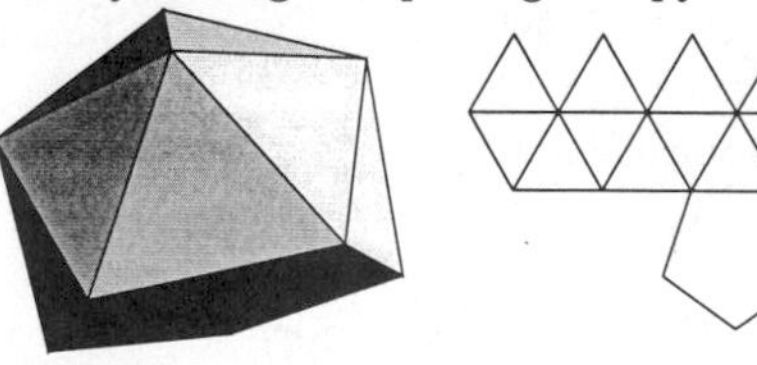

12. Triangular dipyramid

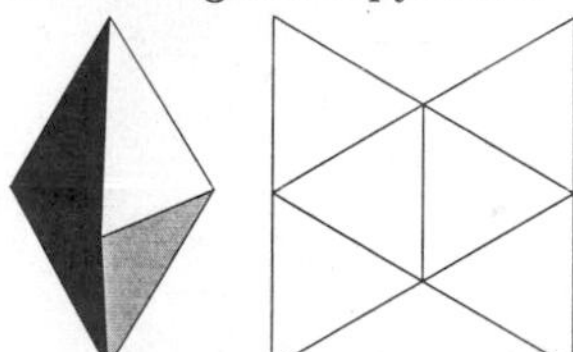

13. Pentagonal dipyramid

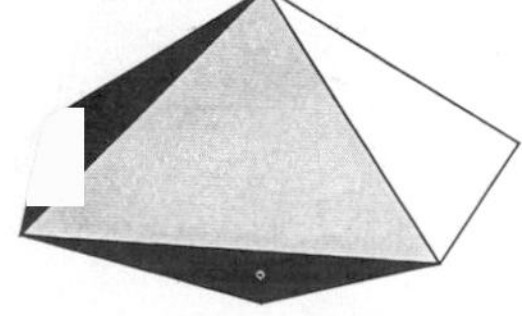
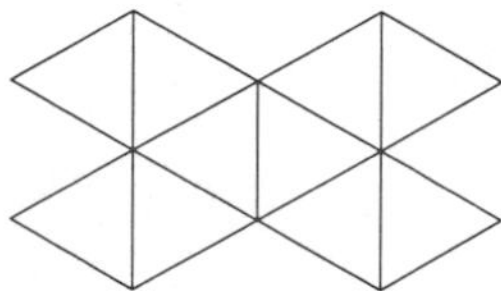

14. Elongated triangular dipyramid

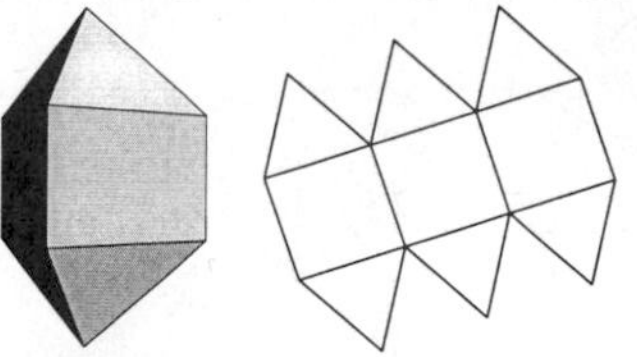

15. Elongated square dipyramid

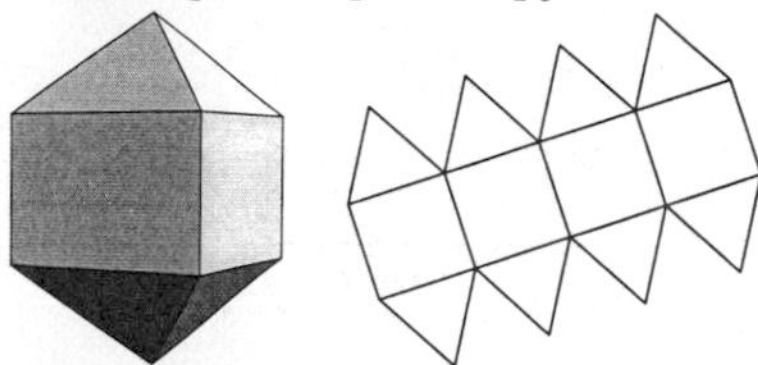

16. Elongated pentagonal dipyramid

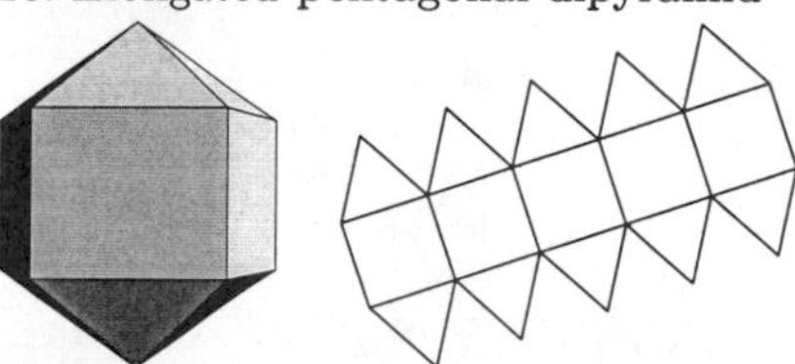

17. Gyroelongated square dipyramid

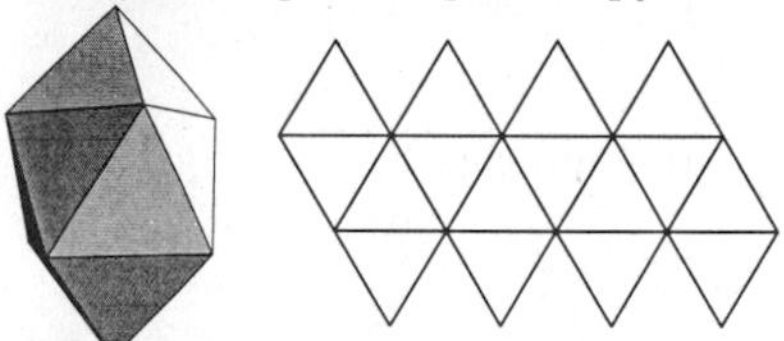

18. Elongated triangular cupola

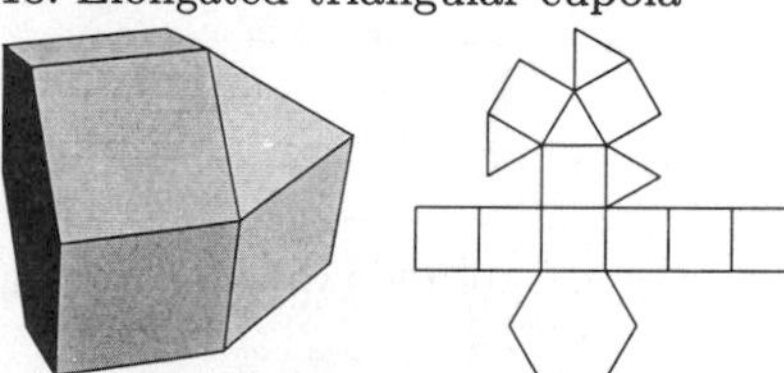

19. Elongated square cupola

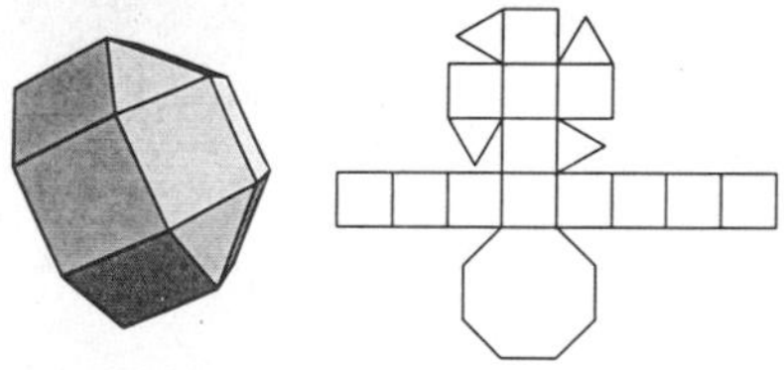

20. Elongated pentagonal cupola

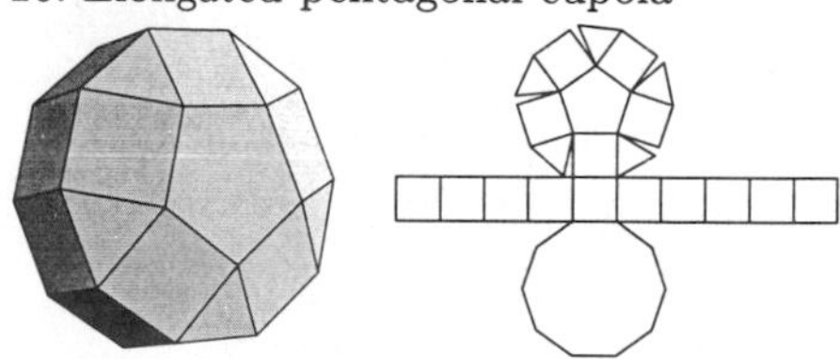

21. Elongated pentagonal rotunda

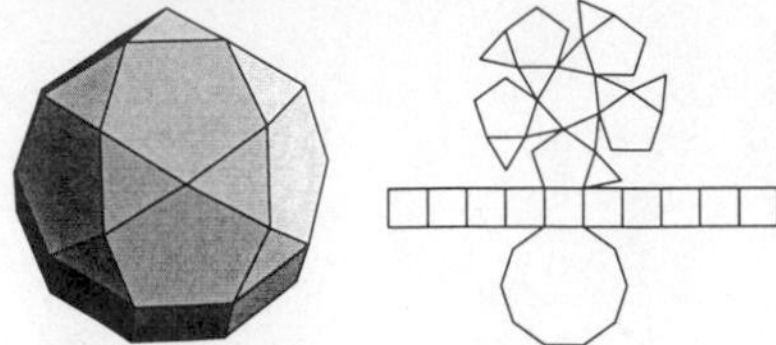

22. Gyroelongated triangular cupola

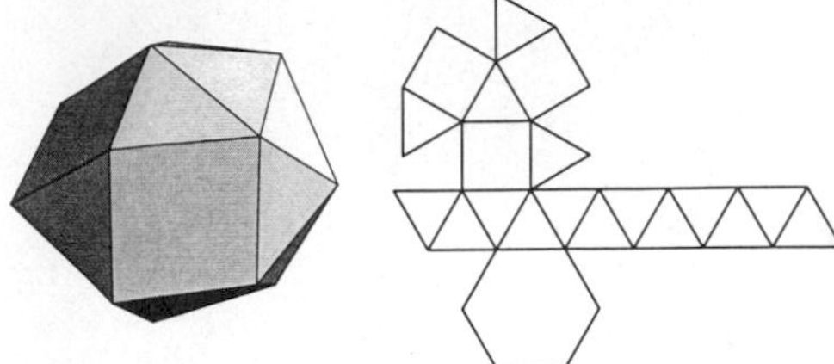

23. Gyroelongated square cupola

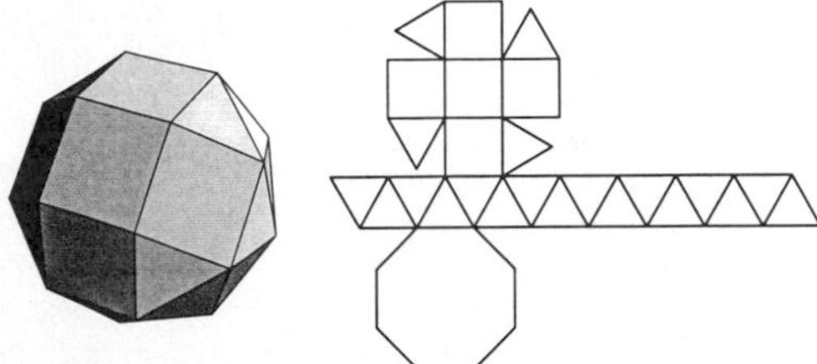

24. Gyroelongated pentagonal cupola

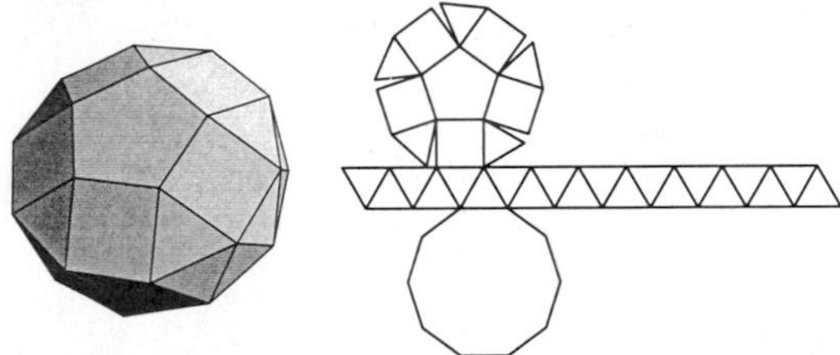

25. Gyroelongated pentagonal rotunda

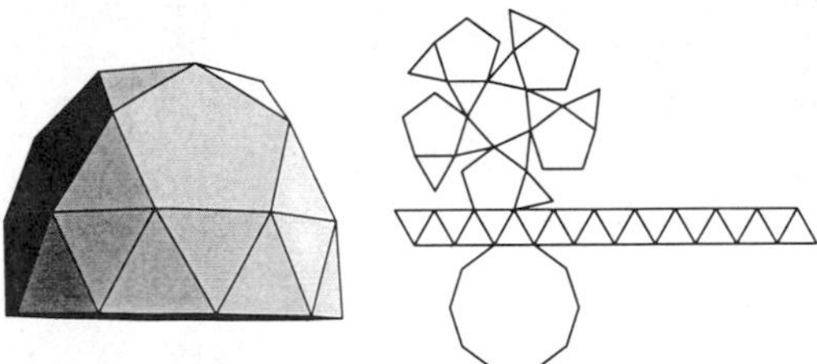

26. Gyrobifastigium

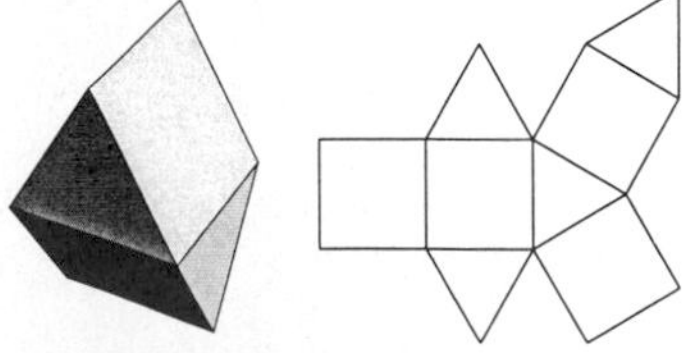

27. Triangular orthobicupola

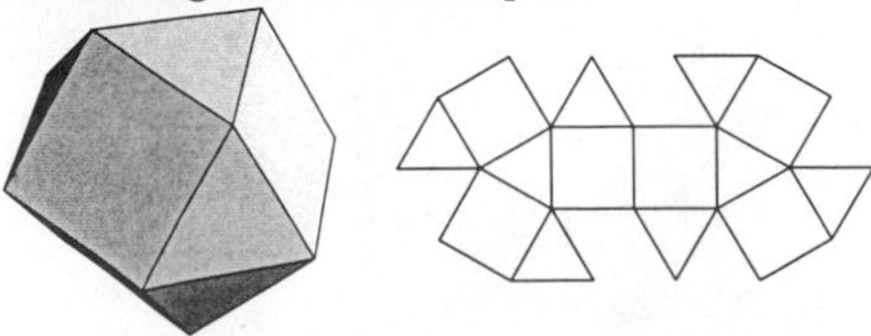

28. Square orthobicupola

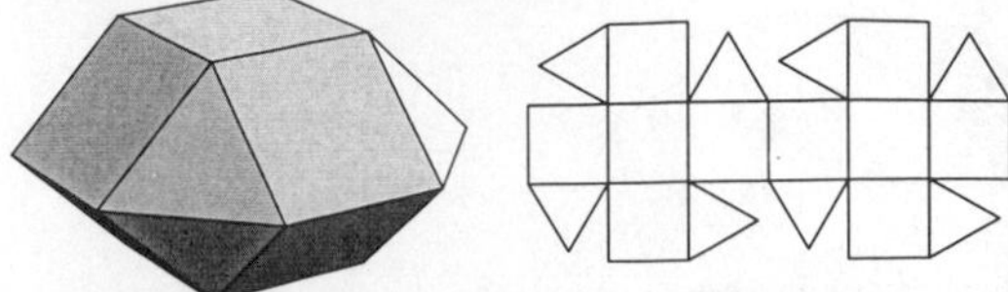

29. Square gyrobicupola

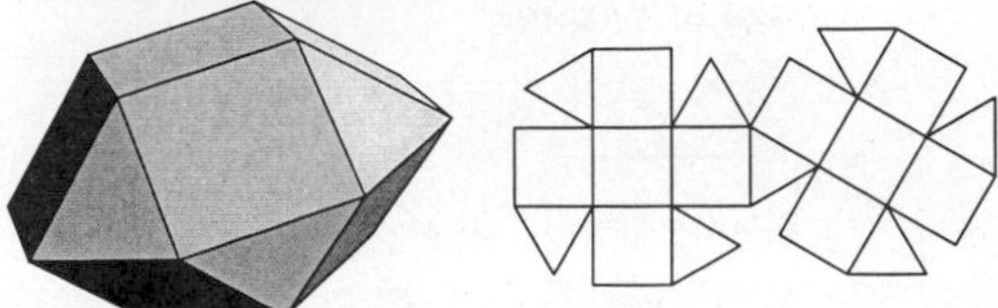

30. Pentagonal orthobicupola

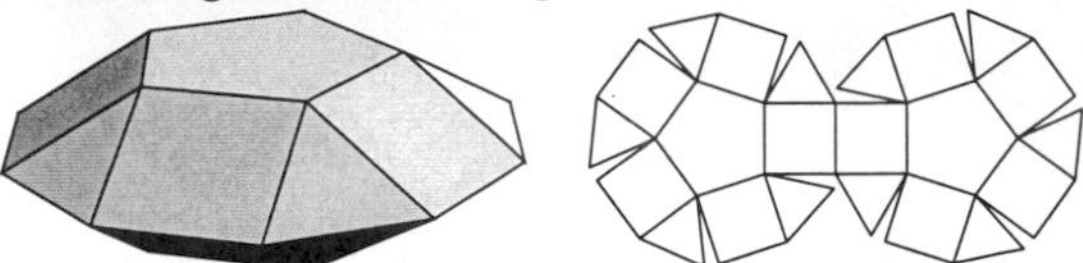

31. Pentagonal gyrobicupola

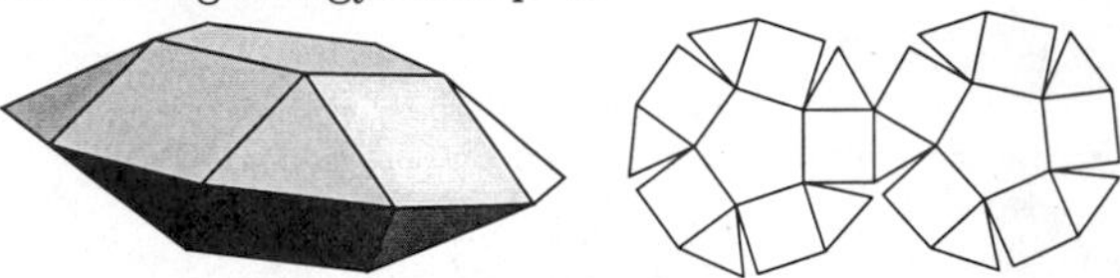

32. Pentagonal orthocupolarontunda

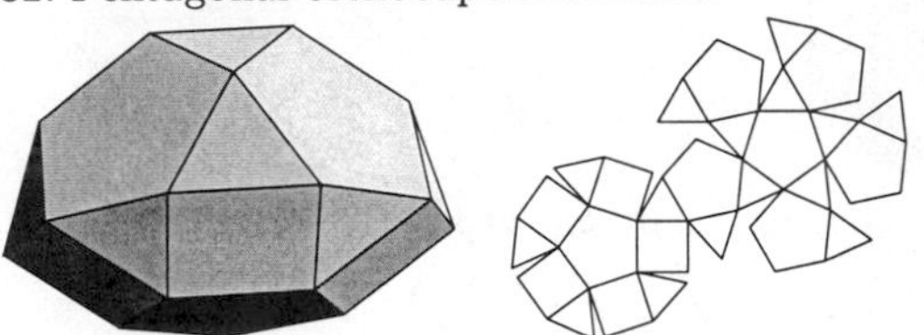

33. Pentagonal gyrocupolarotunda

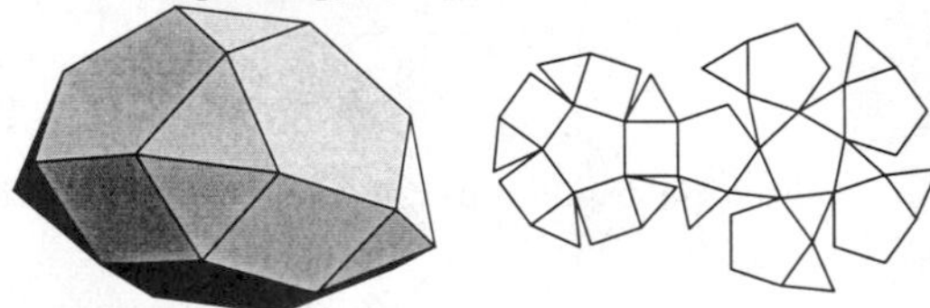

34. Pentagonal orthobirotunda

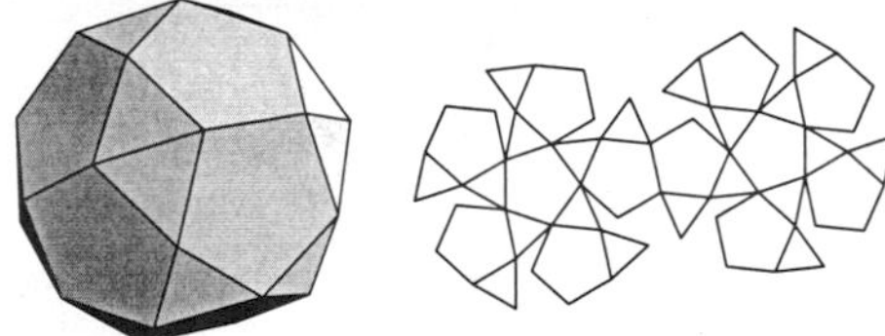

35. Elongated triangular orthobicupola

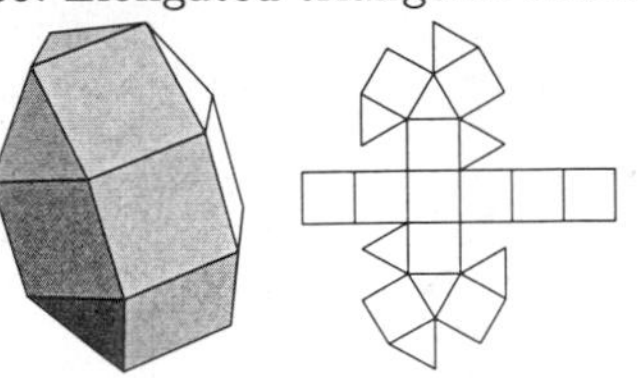

36. Elongated triangular gyrobicupola

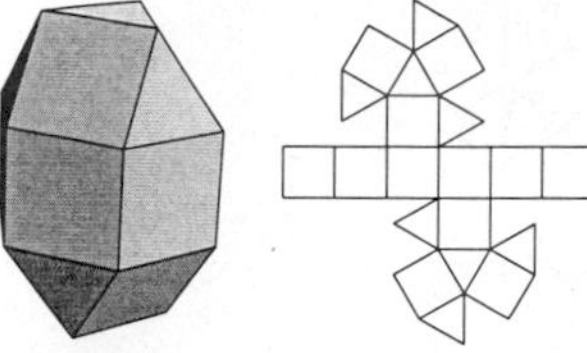

37. Elongated square gyrobicupola

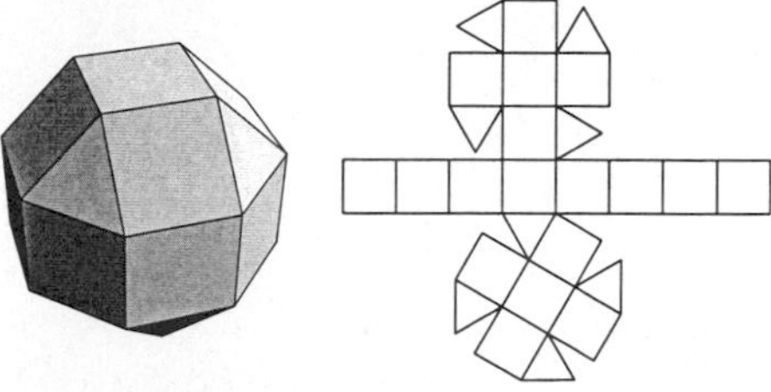

38. Elongated pentagonal orthobicupola

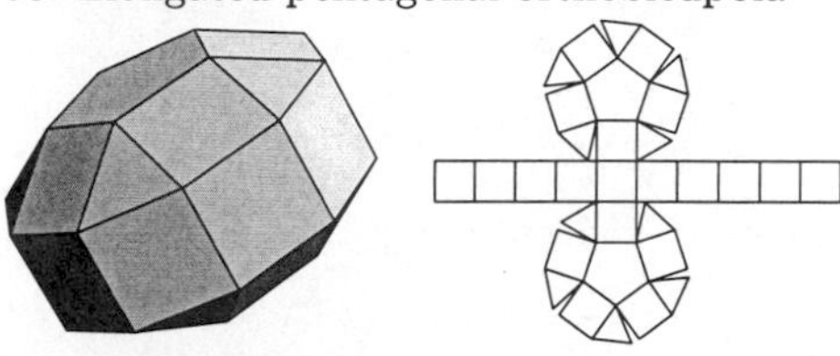

39. Elongated pentagonal gyrobicupola

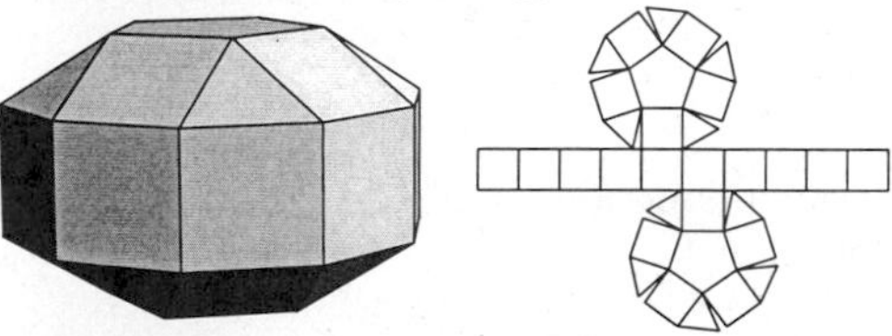

40. Elongated pentagonal orthocupolarotunda

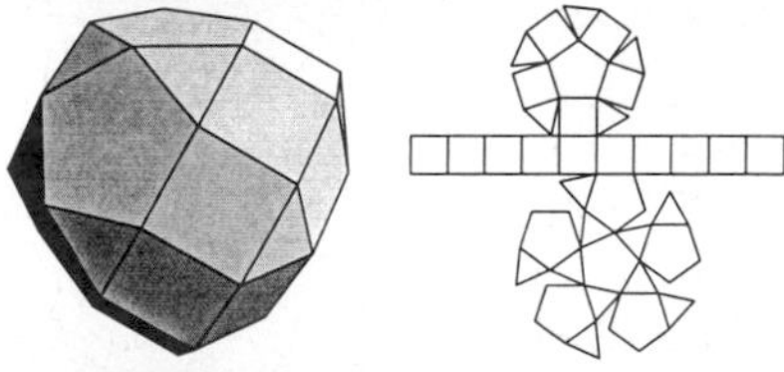

41. Elongated pentagonal gyrocupolarotunda

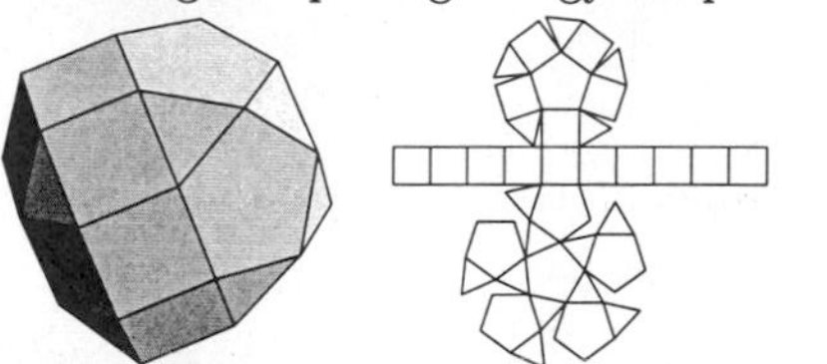

42. Elongated pentagonal orthobirotunda

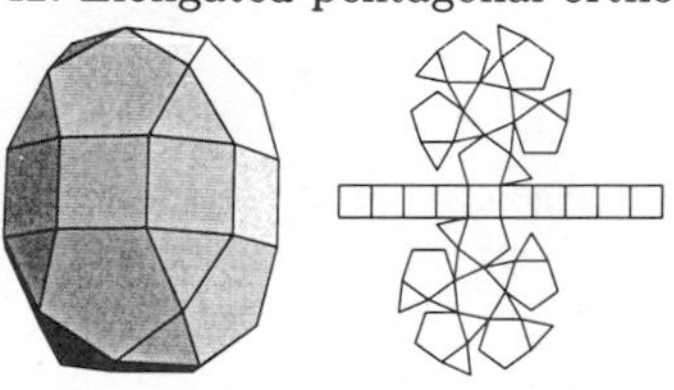

43. Elongated pentagonal gyrobirotunda

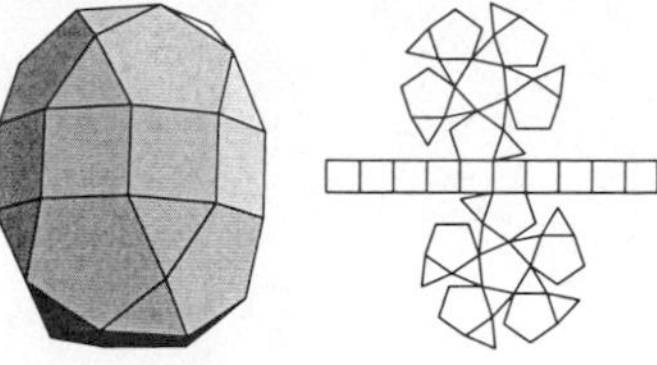

44. Gyroelongated triangular bicupola

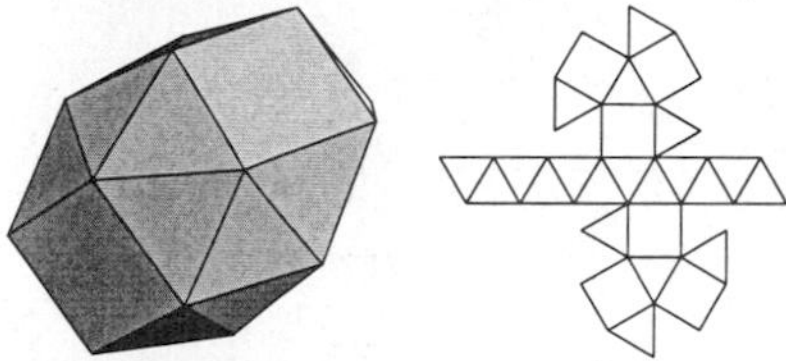

45. Gyroelongated square bicupola

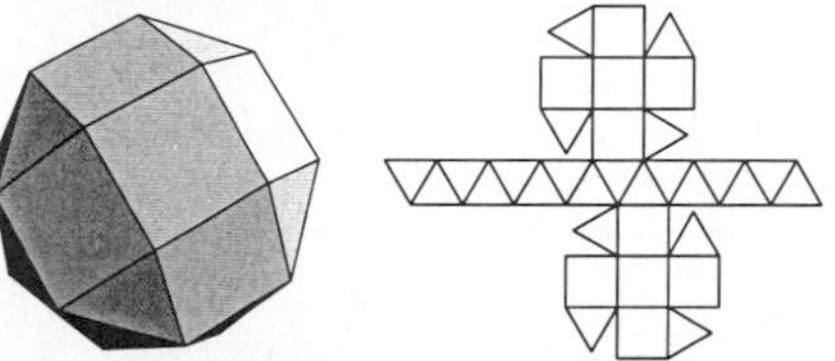

46. Gyroelongated pentagonal bicupola

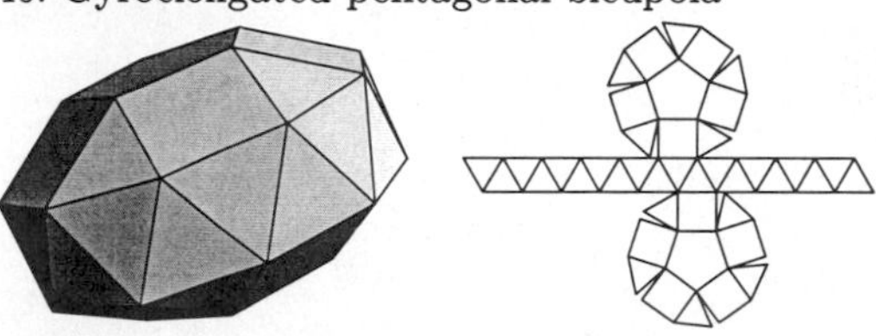

47. Gyroelongated pentagonal cupolarotunda

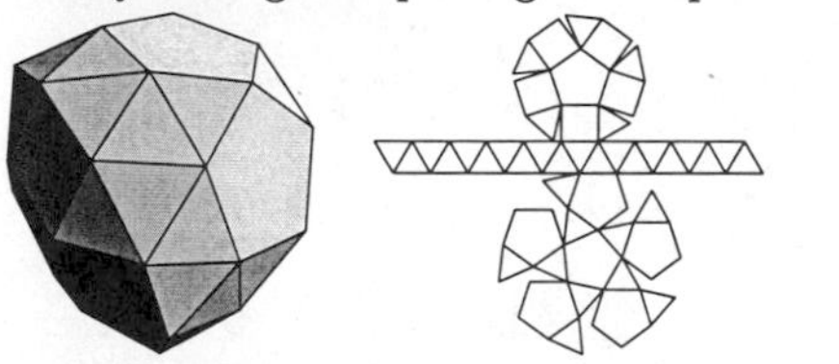

48. Gyroelongated pentagonal birotunda

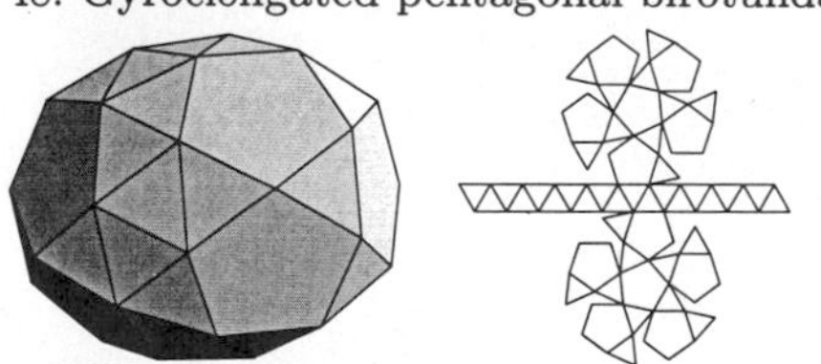

49. Augmented triangular prism

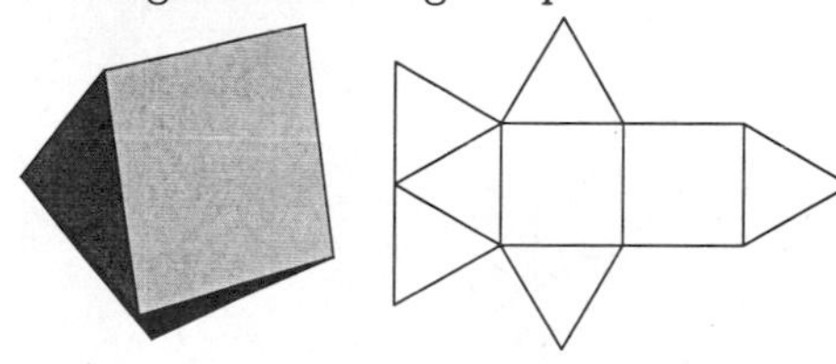

50. Biaugmented triangular prism

51. Triaugmented triangular prism

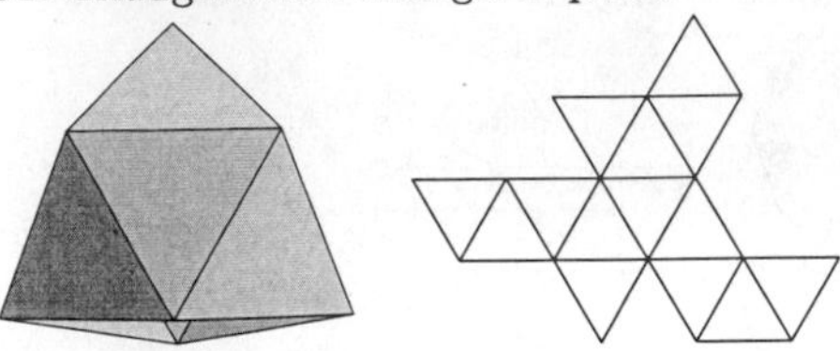

52. Augmented pentagonal prism

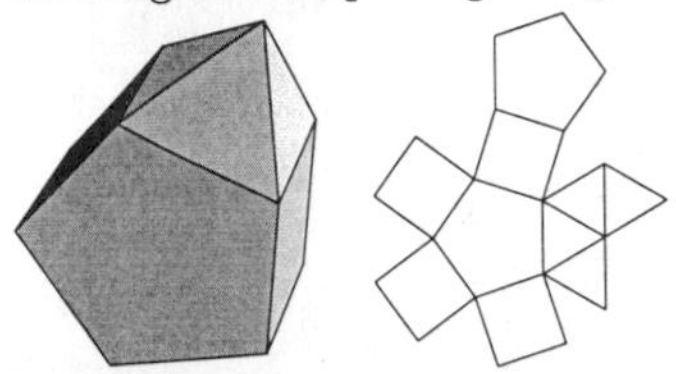

53. Biaugmented pentagonal prism

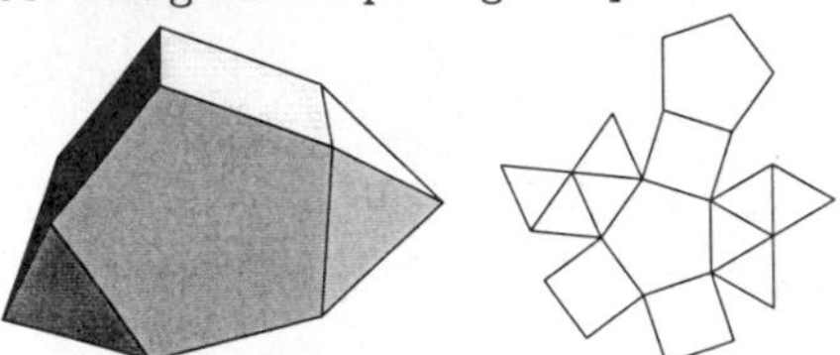

54. Augmented hexagonal prism

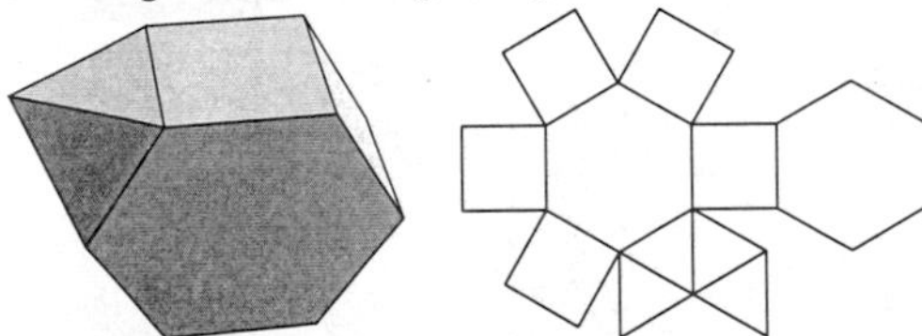

55. Parabiaugmented hexagonal prism

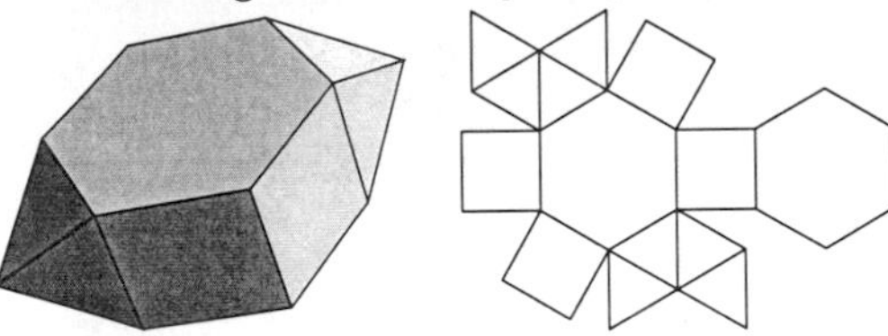

56. Metabiaugmented hexagonal prism

57. Triaugmented hexagonal prism

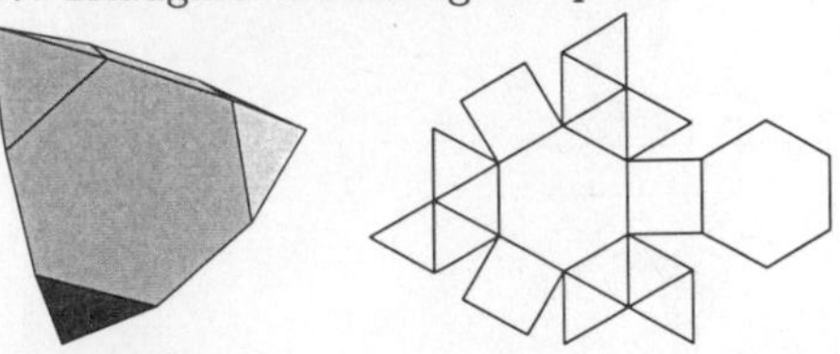

58. Augmented dodecahedron

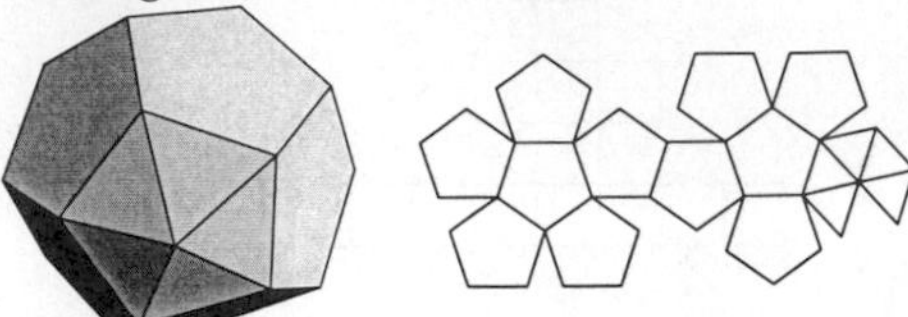

59. Parabiaugmented dodecahedron

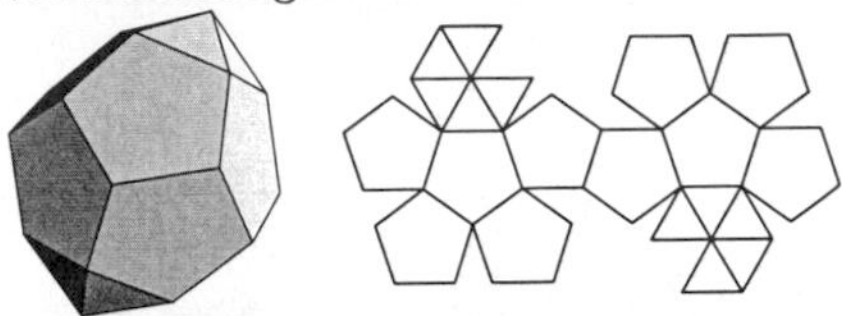

60. Metabiaugmented dodecahedron

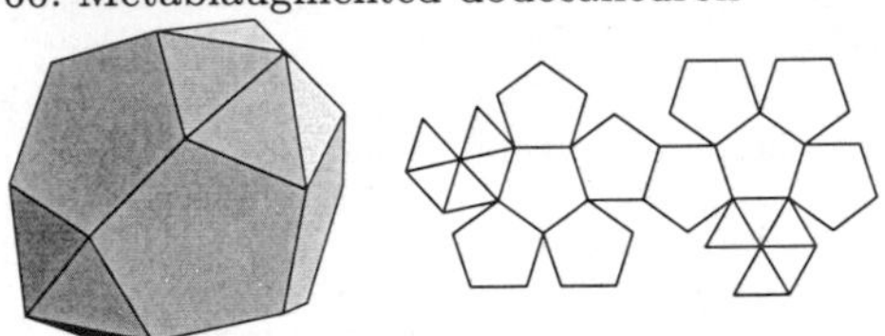

61. Triaugmented dodecahedron

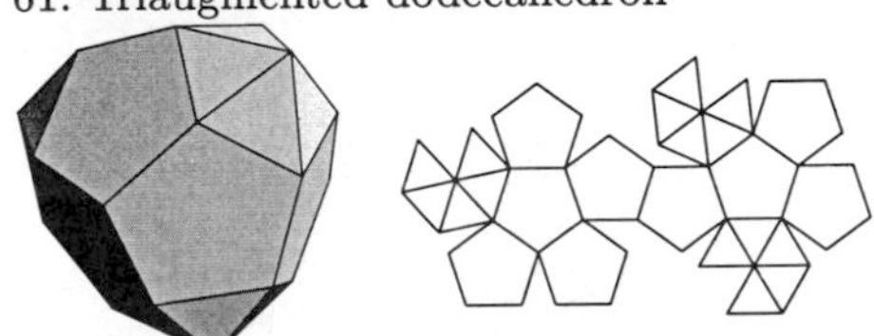

62. Metabidiminished icosahedron

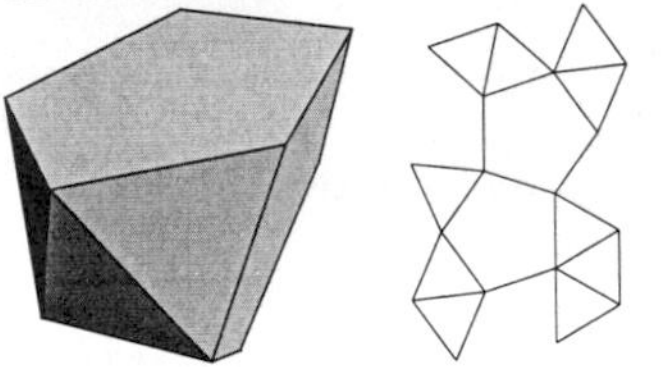

63. Tridiminished icosahedron

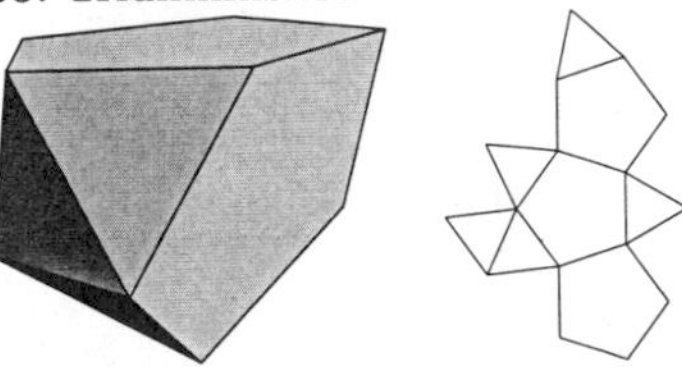

64. Augmented tridiminished icosahedron

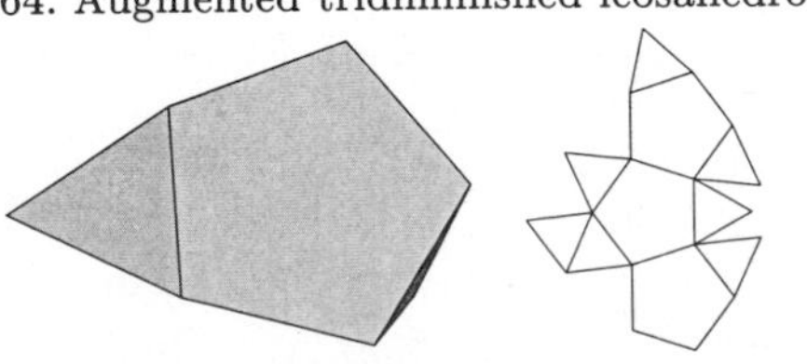

65. Augmented truncated tetrahedron

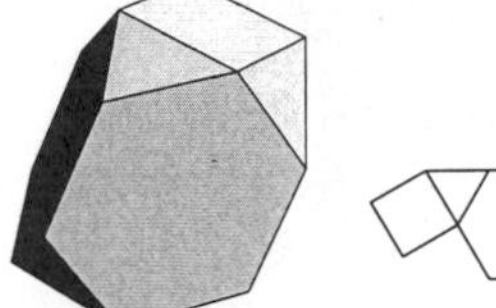
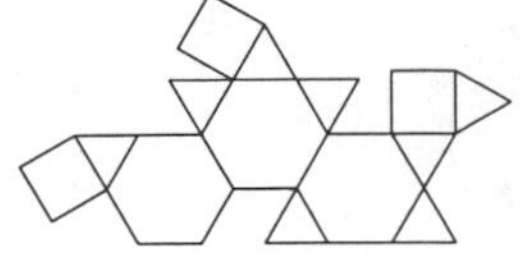

66. Augmented truncated cube

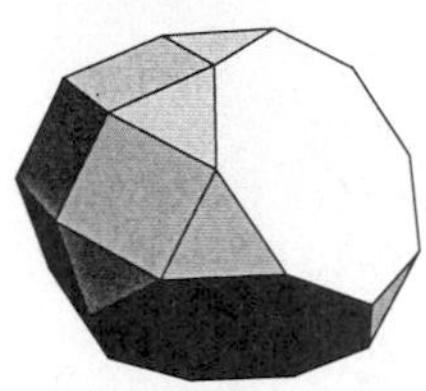
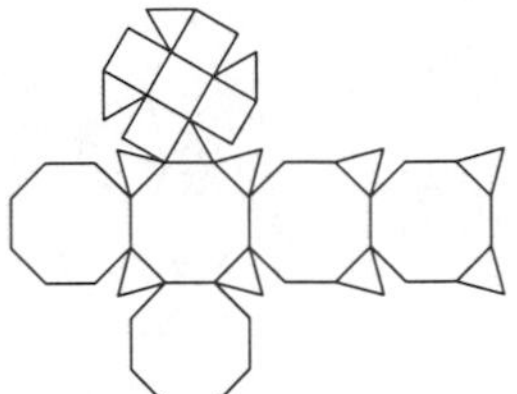

67. Biaugmented truncated cube

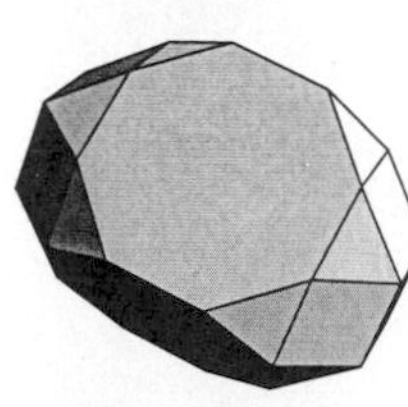
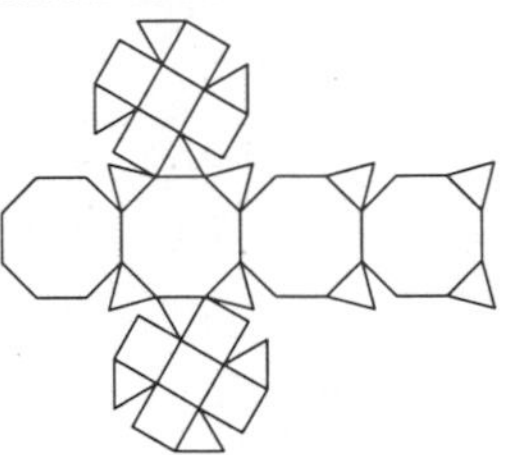

68. Augmented truncated dodecahedron

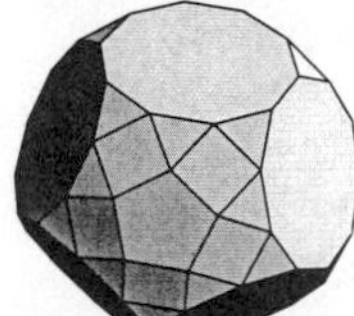
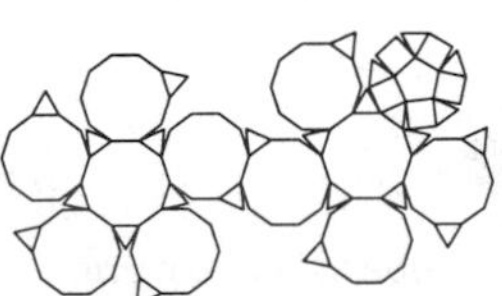

69. Parabiaugmented truncated dodecahedron

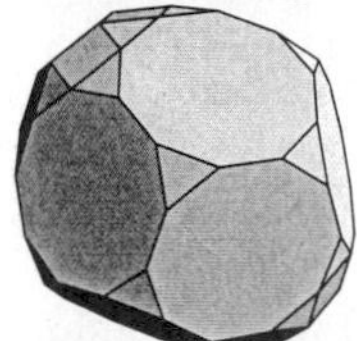
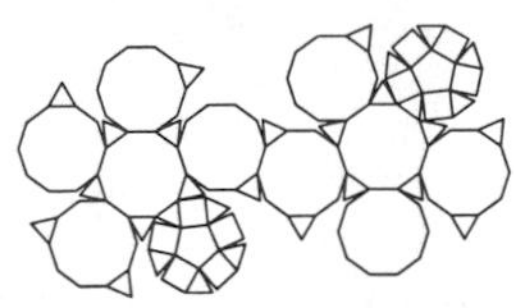

70. Metabiaugmented truncated dodecahedron

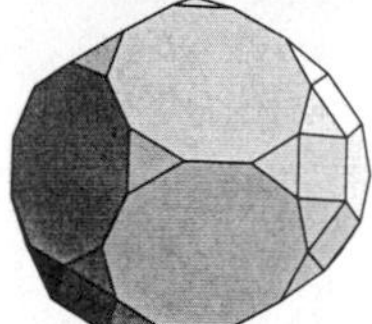
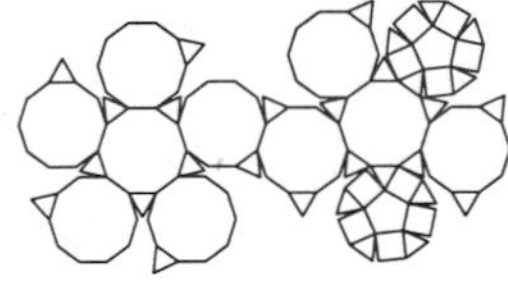

71. Triaugmented truncated dodecahedron

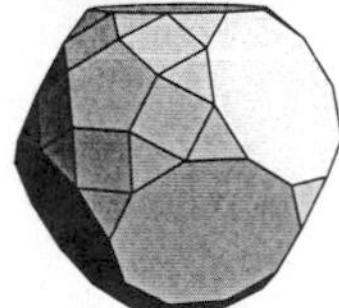
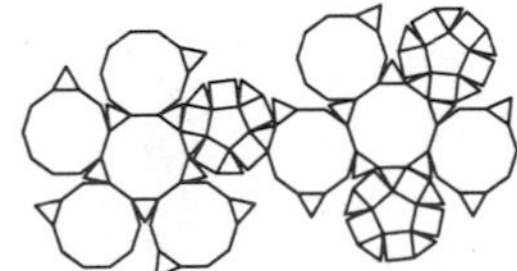

72. Gyrate rhombicosidodecahedron

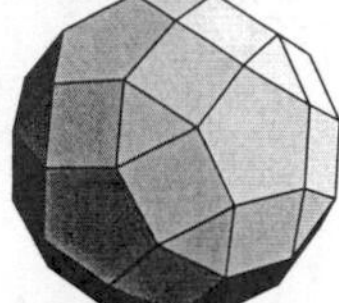
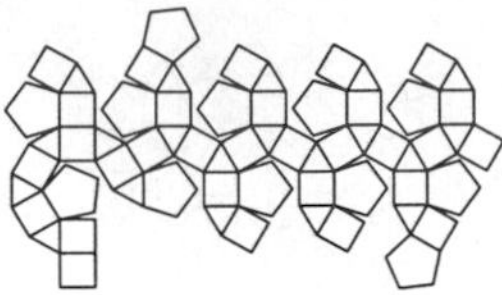

73. Parabigyrate rhombicosidodecahedron

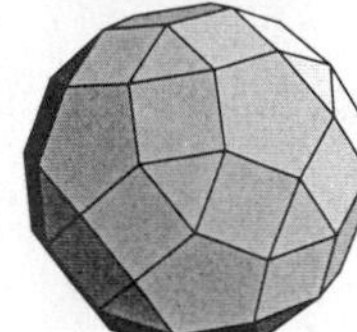
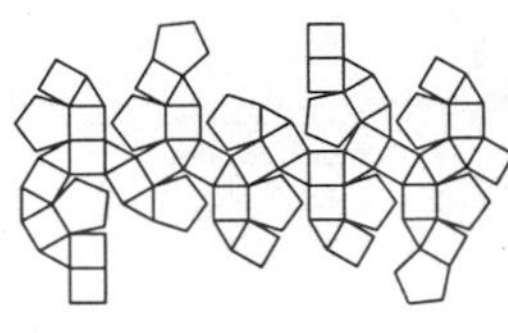

74. Metabigyrate rhombicosidodecahedron

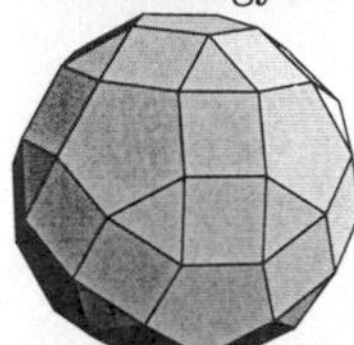
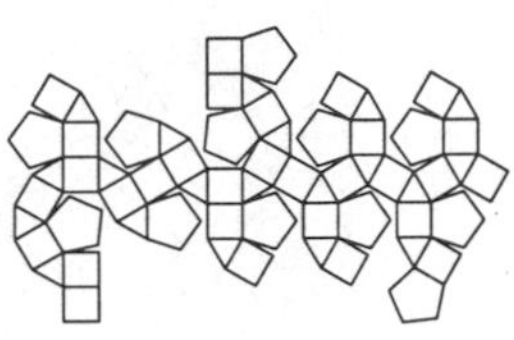

75. Trigyrate rhombicosidodecahedron

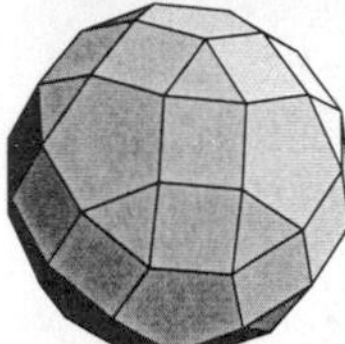
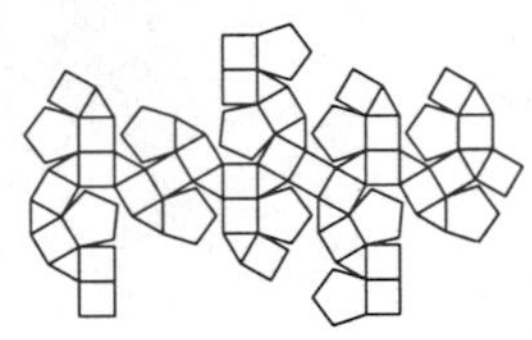

76. Diminished rhombicosidodecahedron

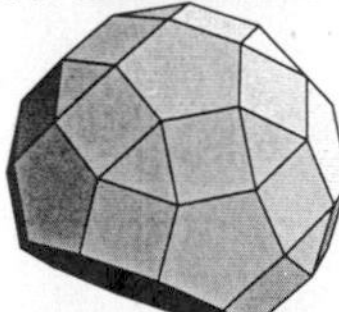
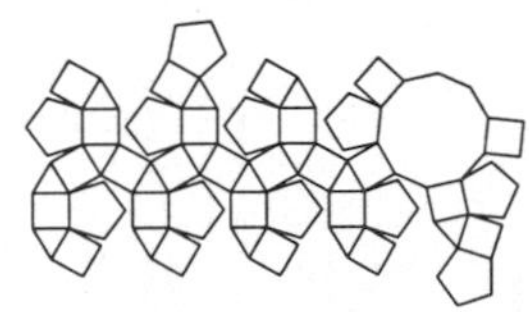

77. Paragyrate diminished rhombicosidodecahedron

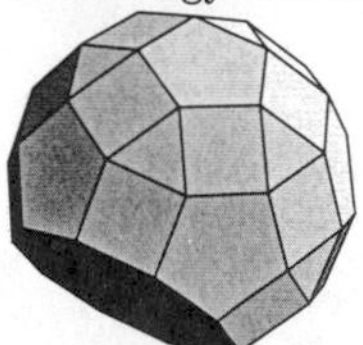
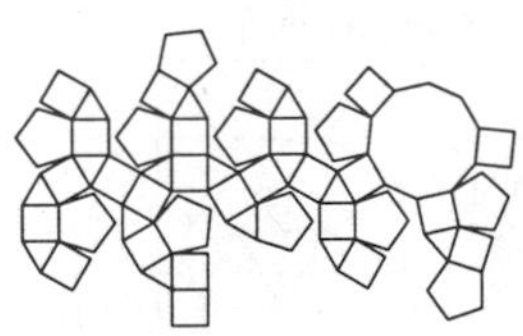

78. Metagyrate diminished rhombicosidodecahedron

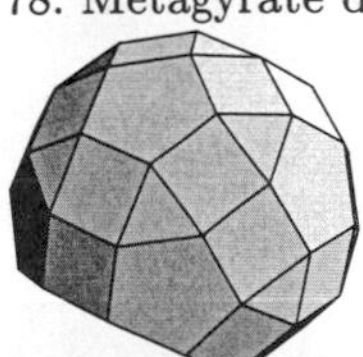
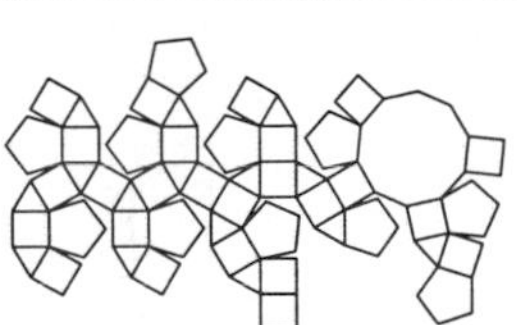

79. Bigyrate diminished rhombicosidodecahedron

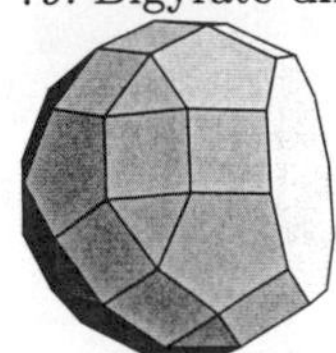
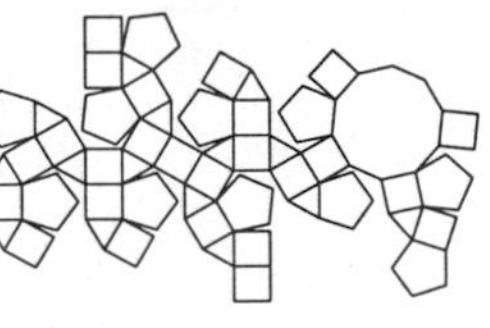

80. Parabidiminished rhombicosidodecahedron

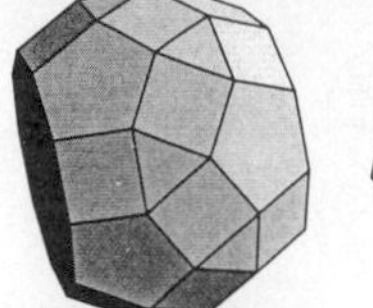
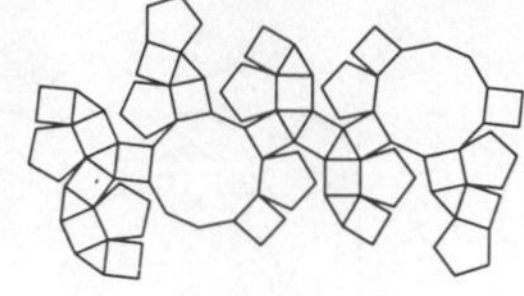

81. Metabidiminished rhombicosidodecahedron

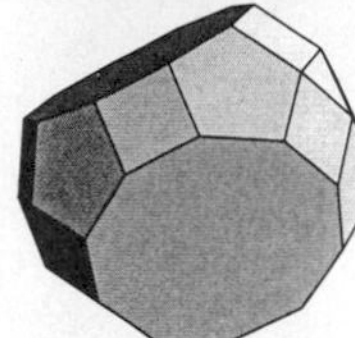
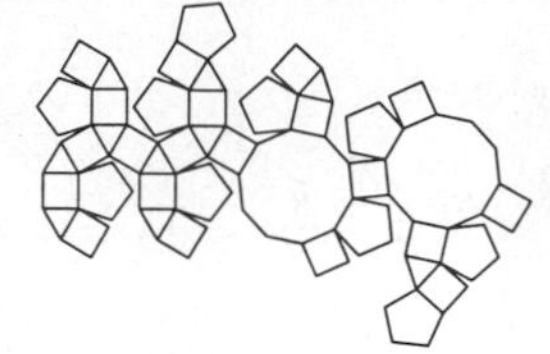

82. Gyrate bidiminished rhombicosidodecahedron

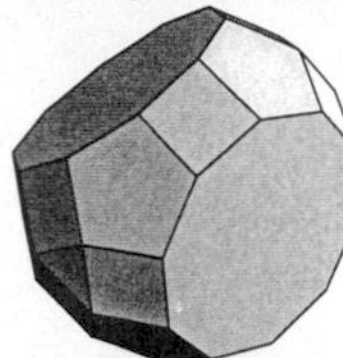
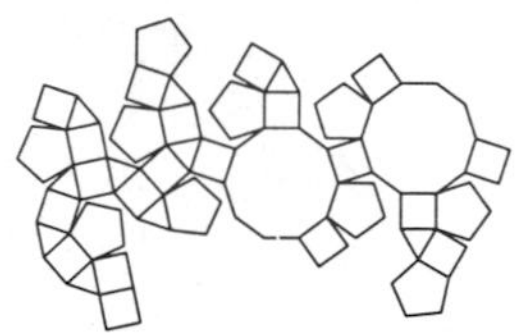

83. Tridiminished rhombicosidodecahedron

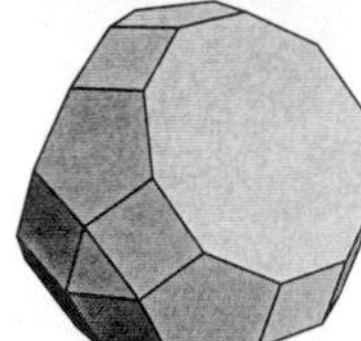
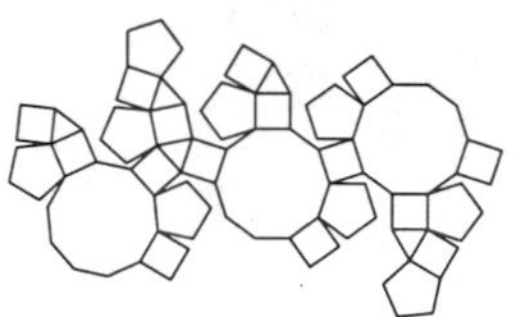

84. Snub disphenoid

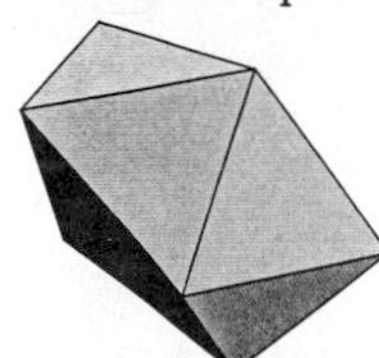
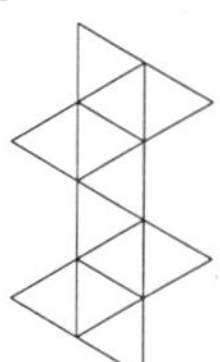

85. Snub square antiprism

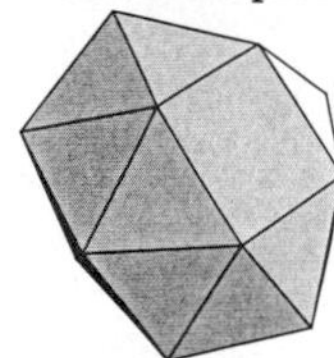
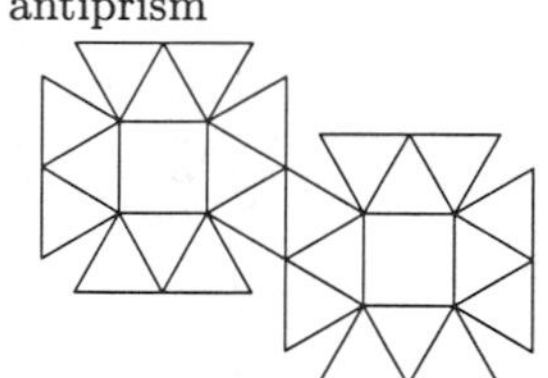

86. Sphenocorona

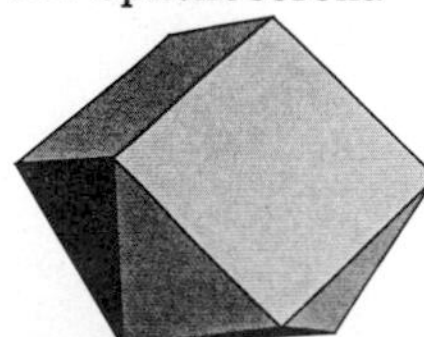
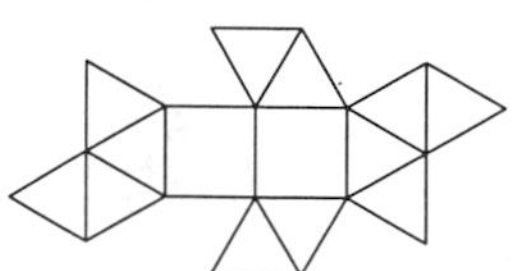

87. Augmented sphenocorona

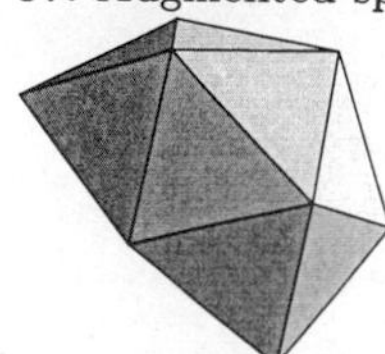
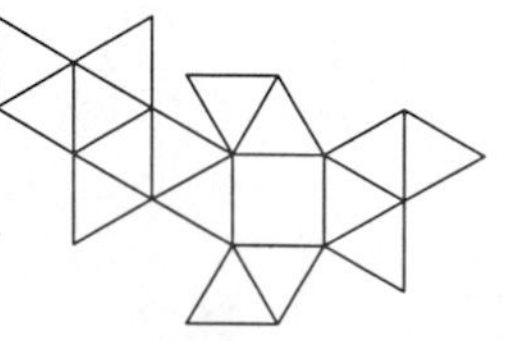

88. Sphenomegacorona

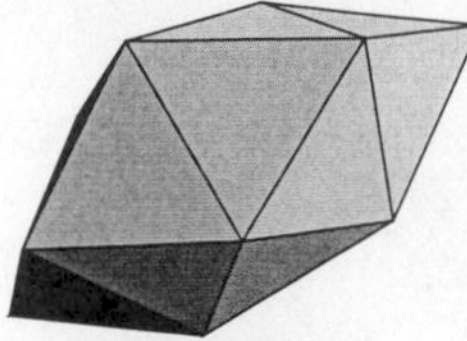

89. Hebesphenomegacorona

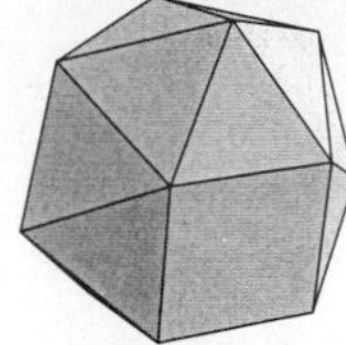
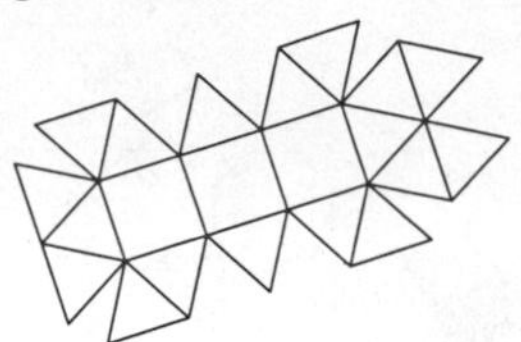

90. Disphenocingulum

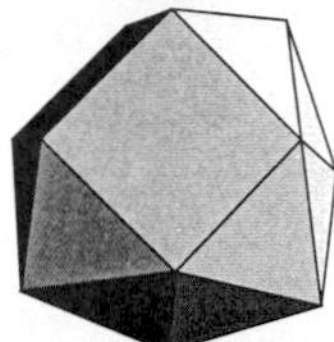
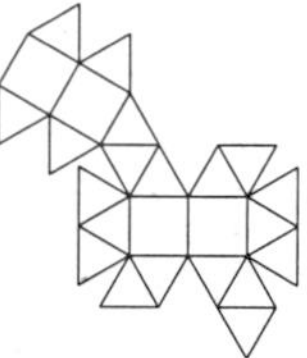

91. Bilunabirotunda

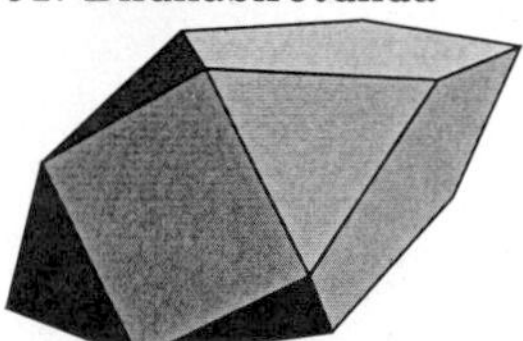

92. Triangular hebesphenorotunda

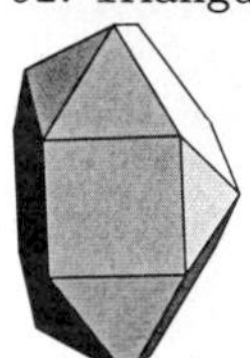

The number of constituent n-gons ($\{n\}$) for each Johnson solid are given in the following table.

J_n	{3}	{4}	{5}	{6}	{8}	{10}
1	4	1				
2	5		1			
3	4	3		1		
4	4	5			1	
5	5	5	1			1
6	10		6			1
7	4	3				
8	4	5				
9	5	5	1			
10	12	1				
11	15		1			
12	6					
13	10					
14	6	3				
15	8	4				
16	10	5				
17	16					
18	4	9		1		
19	4	13			1	
20	5	15	1			1
21	10	10	6			1
22	16	3		1		
23	20	5			1	
24	25	5	1			1
25	30		6			1
26	4	4				
27	8	6				
28	8	10				
29	8	10				
30	10	10	2			
31	10	10	2			
32	15	5	7			
33	15	5	7			
34	20		12			
35	8	12				
36	8	12				
37	8	18				
38	10	20	2			
39	10	20	2			
40	15	15	7			
41	15	15	7			
42	20	10	12			
43	20	10	12			
44	20	6				
45	24	10				
46	30	10	2			

J_n	{3}	{4}	{5}	{6}	{8}	{10}
47	35	5	7			
48	40		12			
49	6	2				
50	10	1				
51	14					
52	4	4	2			
53	8	3	2			
54	4	5		2		
55	8	4		2		
56	8	4		2		
57	12	3		2		
58	5		11			
59	10		10			
60	10		10			
61	15		9			
62	10		2			
63	5		3			
64	7		3			
65	8	3		3		
66	12	5			5	
67	16	10			4	
68	25	5	1			11
69	30	10	2			10
70	30	10	2			10
71	35	15	3			9
72	20	30	12			
73	20	30	12			
74	20	30	12			
75	20	30	12			
76	15	25	11			1
77	15	25	11			1
78	15	25	11			1
79	15	25	11			1
80	10	20	10			2
81	10	20	10			2
82	10	20	10			2
83	5	15	9			3
84	12					
85	24	2				
86	12	2				
87	16	1				
88	16	2				
89	18	3				
90	20	4				
91	8	2	4			
92	13	3	3	1		

see also ANTIPRISM, ARCHIMEDEAN SOLID, CONVEX POLYHEDRON, KEPLER-POINSOT SOLID, POLYHEDRON, PLATONIC SOLID, PRISM, UNIFORM POLYHEDRON

References

Bell Laboratories. `http://netlib.bell-labs.com/netlib/polyhedra/`.

Bulatov, V. "Johnson Solids." `http://www.physics.orst.edu/~bulatov/polyhedra/johnson/`.

Cromwell, P. R. *Polyhedra.* New York: Cambridge University Press, pp. 86–92, 1997.

Hart, G. W. "NetLib Polyhedra DataBase." http://www.li.net/~george/virtual-polyhedra/netlib-info.html.
Holden, A. *Shapes, Space, and Symmetry.* New York: Dover, 1991.
Hume, A. *Exact Descriptions of Regular and Semi-Regular Polyhedra and Their Duals.* Computer Science Technical Report #130. Murray Hill, NJ: AT&T Bell Laboratories, 1986.
Johnson, N. W. "Convex Polyhedra with Regular Faces." *Canad. J. Math.* **18**, 169–200, 1966.
Pugh, A. "Further Convex Polyhedra with Regular Faces." Ch. 3 in *Polyhedra: A Visual Approach.* Berkeley, CA: University of California Press, pp. 28–35, 1976.
Weisstein, E. W. "Johnson Solids." http://www.astro.virginia.edu/~eww6n/math/notebooks/JohnsonSolids.m.
Weisstein, E. W. "Johnson Solid Netlib Database." http://www.astro.virginia.edu/~eww6n/math/notebooks/JohnsonSolids.dat.
Zalgaller, V. *Convex Polyhedra with Regular Faces.* New York: Consultants Bureau, 1969.

Johnson's Theorem

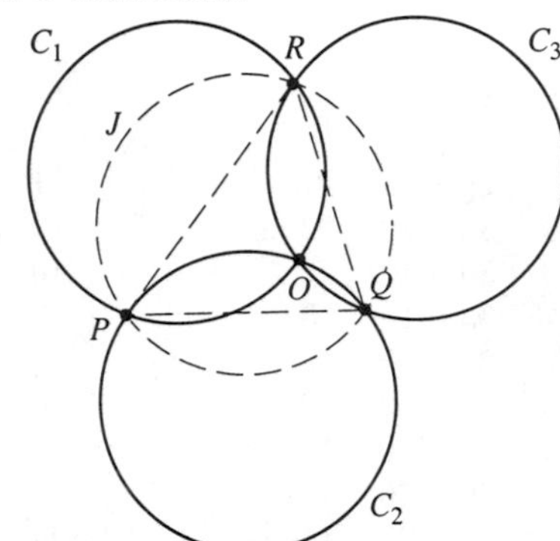

Let three equal CIRCLES with centers C_1, C_2, and C_3 intersect in a single point O and intersect pairwise in the points P, Q, and R. Then the CIRCUMCIRCLE J of ΔPQR (the so-called JOHNSON CIRCLE) is congruent to the original three.

see also CIRCUMCIRCLE, JOHNSON CIRCLE

References

Emch, A. "Remarks on the Foregoing Circle Theorem." *Amer. Math. Monthly* **23**, 162–164, 1916.
Honsberger, R. *Mathematical Gems II.* Washington, DC: Math. Assoc. Amer., pp. 18–21, 1976.
Johnson, R. "A Circle Theorem." *Amer. Math. Monthly* **23**, 161–162, 1916.

Join (Graph)

Let x and y be distinct nodes of G which are not joined by an EDGE. Then the graph G/xy which is formed by adding the EDGE (x, y) to G is called a join of G.

Join (Spaces)

Let X and Y be TOPOLOGICAL SPACES. Then their join is the factor space

$$X * Y = (X \times Y \times I)/\sim,$$

where $\sim$ is the EQUIVALENCE RELATION

$$(x, y, t) \sim (x', y', t') \Leftrightarrow \begin{cases} t = t' = 0 \text{ and } x = x' \\ \text{or} \\ t = t' = 1 \text{ and } y = y. \end{cases}$$

see also CONE (SPACE), SUSPENSION

References

Rolfsen, D. *Knots and Links.* Wilmington, DE: Publish or Perish Press, p. 6, 1976.

Joint Distribution Function

A joint distribution function is a DISTRIBUTION FUNCTION in two variables defined by

$$D(x, y) \equiv P(X \leq x, Y \leq y) \quad (1)$$

$$D_x(x) \equiv D(x, \infty) \quad (2)$$

$$D_y(y) \equiv D(\infty, y) \quad (3)$$

so that the joint probability function

$$P[(x, y) \in C)] = \iint_{(x,y) \in C} P(x, y)\, dx\, dy \quad (4)$$

$$P(x \in A, y \in B) = \int_B \int_A P(x, y)\, dx\, dy \quad (5)$$

$$P(x, y) = P\{x \in (-\infty, x], y \in (-\infty, y]\} = \int_{-\infty}^{b} \int_{-\infty}^{a} P(x, y)\, dx\, dy \quad (6)$$

$$P(a \leq x \leq a + da, b \leq y \leq b + db) = \int_b^{b+db} \int_a^{a+da} P(x, y)\, dx\, dy \approx P(a, b)\, da\, db. \quad (7)$$

A multiple distribution function is of the form

$$D(a_1, \ldots, a_n) \equiv P(x_1 \leq a_1, \ldots, x_n \leq a_n). \quad (8)$$

see also DISTRIBUTION FUNCTION

Joint Probability Density Function

see JOINT DISTRIBUTION FUNCTION

Joint Theorem

see GAUSSIAN JOINT VARIABLE THEOREM

Jonah Formula

A formula for the generalized CATALAN NUMBER ${}_pd_{qi}$. The general formula is

$$\binom{n-q}{k-1} = \sum_{i=1}^{k} {}_pd_{qi}\binom{n-pi}{k-i},$$

where $\binom{n}{k}$ is a BINOMIAL COEFFICIENT, although Jonah's original formula corresponded to $p=2$, $q=0$ (Hilton and Pederson 1991).

References

Hilton, P. and Pederson, J. "Catalan Numbers, Their Generalization, and Their Uses." *Math. Intel.* **13**, 64–75, 1991.

Jones Polynomial

The second KNOT POLYNOMIAL discovered. Unlike the first-discovered ALEXANDER POLYNOMIAL, the Jones polynomial can sometimes distinguish handedness (as can its more powerful generalization, the HOMFLY POLYNOMIAL). Jones polynomials are LAURENT POLYNOMIALS in t assigned to an $\mathbb{R}^3$ KNOT. The Jones polynomials are denoted $V_L(t)$ for LINKS, $V_K(t)$ for KNOTS, and normalized so that

$$V_{\text{unknot}}(t) = 1. \qquad (1)$$

For example, the Jones polynomial of the TREFOIL KNOT is given by

$$V_{\text{trefoil}}(t) = t + t^3 - t^4. \qquad (2)$$

If a LINK has an ODD number of components, then V_L is a LAURENT POLYNOMIAL over the INTEGERS; if the number of components is EVEN, $V_L(t)$ is $t^{1/2}$ times a LAURENT POLYNOMIAL. The Jones polynomial of a KNOT SUM $L_1\#L_2$ satisfies

$$V_{L_1\#L_2} = (V_{L_1})(V_{L_2}). \qquad (3)$$

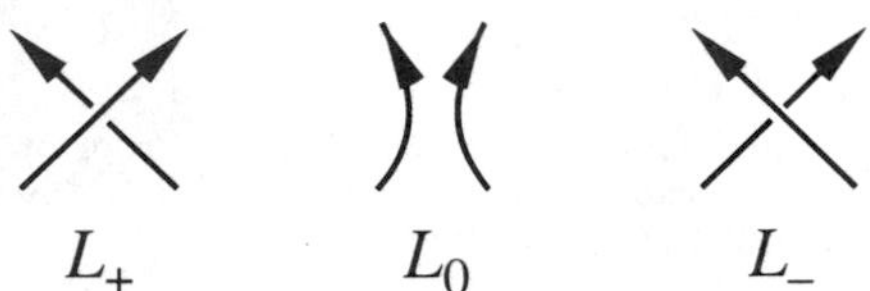

L_+ $\qquad$ L_0 $\qquad$ L_-

The SKEIN RELATIONSHIP for under- and overcrossings is

$$t^{-1}V_{L_+} - tV_{L_-} = (t^{1/2} - t^{-1/2})V_{L_0}. \qquad (4)$$

Combined with the link sum relationship, this allows Jones polynomials to be built up from simple knots and links to more complicated ones.

Some interesting identities from Jones (1985) follow. For any LINK L,

$$V_L(-1) = \Delta_L(-1), \qquad (5)$$

where Δ_L is the ALEXANDER POLYNOMIAL, and

$$V_L(1) = (-2)^{p-1}, \qquad (6)$$

where p is the number of components of L. For any KNOT K,

$$V_K(e^{2\pi i/3}) = 1 \qquad (7)$$

and

$$\frac{d}{dt}V_K(1) = 0. \qquad (8)$$

Let K^* denote the MIRROR IMAGE of a KNOT K. Then

$$V_{K^*}(t) = V_K(t^{-1}). \qquad (9)$$

For example, the right-hand and left-hand TREFOIL KNOTS have polynomials

$$V_{\text{trefoil}}(t) = t + t^3 - t^4 \qquad (10)$$

$$V_{\text{trefoil}^*}(t) = t^{-1} + t^{-3} - t^{-4}. \qquad (11)$$

Jones defined a simplified trace invariant for knots by

$$W_K(t) = \frac{1 - V_K(t)}{(1-t^3)(1-t)}. \qquad (12)$$

The ARF INVARIANT of W_K is given by

$$\operatorname{Arf}(K) = W_K(i) \qquad (13)$$

(Jones 1985), where i is $\sqrt{-1}$. A table of the W polynomials is given by Jones (1985) for knots of up to eight crossings, and by Jones (1987) for knots of up to 10 crossings. (Note that in these papers, an additional polynomial which Jones calls V is also tabulated, but it is not the conventionally defined Jones polynomial.)

Jones polynomials were subsequently generalized to the two-variable HOMFLY POLYNOMIALS, the relationship being

$$V(t) = P(a = t, x = t^{1/2} - t^{-1/2}) \qquad (14)$$

$$V(t) = P(\ell = it, m = i(t^{-1/2} - t^{1/2})). \qquad (15)$$

They are related to the KAUFFMAN POLYNOMIAL F by

$$V(t) = F(-t^{-3/4}, t^{-1/4} + t^{1/4}). \qquad (16)$$

Jones (1987) gives a table of BRAID WORDS and W polynomials for knots up to 10 crossings. Jones polynomials for KNOTS up to nine crossings are given in Adams (1994) and for oriented links up to nine crossings by Doll and Hoste (1991). All PRIME KNOTS with 10 or fewer crossings have distinct Jones polynomials. It is not known if there is a nontrivial knot with Jones polynomial 1. The Jones polynomial of an (m,n)-TORUS KNOT is

$$\frac{t^{(m-1)(n-1)/2}(1 - t^{m+1} - t^{n+1} + t^{m+n})}{1-t^2}. \qquad (17)$$

Let k be one component of an oriented LINK L. Now form a new oriented LINK L^* by reversing the orientation of k. Then

$$V_{L^*} = t^{-3\lambda} V(L),$$

where V is the Jones polynomial and λ is the LINKING NUMBER of k and $L - k$. No such result is known for HOMFLY POLYNOMIALS (Lickorish and Millett 1988).

Birman and Lin (1993) showed that substituting the POWER SERIES for e^x as the variable in the Jones polynomial yields a POWER SERIES whose COEFFICIENTS are VASSILIEV POLYNOMIALS.

Let L be an oriented connected LINK projection of n crossings, then

$$n \geq \text{span } V(L), \tag{18}$$

with equality if L is ALTERNATING and has no REMOVABLE CROSSING (Lickorish and Millett 1988).

There exist distinct KNOTS with the same Jones polynomial. Examples include (05_{001}, 10_{132}), (08_{008}, 10_{129}), (08_{016}, 10_{156}), (10_{025}, 10_{056}), (10_{022}, 10_{035}), (10_{041}, 10_{094}), (10_{043}, 10_{091}), (10_{059}, 10_{106}), (10_{060}, 10_{083}), (10_{071}, 10_{104}), (10_{073}, 10_{086}), (10_{081}, 10_{109}), and (10_{137}, 10_{155}) (Jones 1987). Incidentally, the first four of these also have the same HOMFLY POLYNOMIAL.

Witten (1989) gave a heuristic definition in terms of a topological quantum field theory, and Sawin (1996) showed that the "quantum group" $U_q(sl_2)$ gives rise to the Jones polynomial.

see also ALEXANDER POLYNOMIAL, HOMFLY POLYNOMIAL, KAUFFMAN POLYNOMIAL F, KNOT, LINK, VASSILIEV POLYNOMIAL

References

Adams, C. C. *The Knot Book: An Elementary Introduction to the Mathematical Theory of Knots.* New York: W. H. Freeman, 1994.

Birman, J. S. and Lin, X.-S. "Knot Polynomials and Vassiliev's Invariants." *Invent. Math.* **111**, 225–270, 1993.

Doll, H. and Hoste, J. "A Tabulation of Oriented Links." *Math. Comput.* **57**, 747–761, 1991.

Jones, V. "A Polynomial Invariant for Knots via von Neumann Algebras." *Bull. Am. Math. Soc.* **12**, 103–111, 1985.

Jones, V. "Hecke Algebra Representations of Braid Groups and Link Polynomials." *Ann. Math.* **126**, 335–388, 1987.

Lickorish, W. B. R. and Millett, B. R. "The New Polynomial Invariants of Knots and Links." *Math. Mag.* **61**, 1–23, 1988.

Murasugi, K. "Jones Polynomials and Classical Conjectures in Knot Theory." *Topology* **26**, 297–307, 1987.

Praslov, V. V. and Sossinsky, A. B. *Knots, Links, Braids and 3-Manifolds: An Introduction to the New Invariants in Low-Dimensional Topology.* Providence, RI: Amer. Math. Soc., 1996.

Sawin, S. "Links, Quantum Groups, and TQFTS." *Bull. Amer. Math. Soc.* **33**, 413–445, 1996.

Stoimenow, A. "Jones Polynomials." `http://www.informatik.hu-berlin.de/~stoimeno/ptab/j10.html`.

Thistlethwaite, M. "A Spanning Tree Expansion for the Jones Polynomial." *Topology* **26**, 297–309, 1987.

Weisstein, E. W. "Knots and Links." `http://www.astro.virginia.edu/~eww6n/math/notebooks/Knots.m`.

Witten, E. "Quantum Field Theory and the Jones Polynomial." *Comm. Math. Phys.* **121**, 351–399, 1989.

Jonquière's Function

see POLYGAMMA FUNCTION

Jordan Algebra

A nonassociative algebra with the product of elements A and B defined by the ANTICOMMUTATOR $\{A, B\} = AB + BA$.

see also ANTICOMMUTATOR

Jordan Curve

A Jordan curve is a plane curve which is topologically equivalent to (a HOMEOMORPHIC image of) the UNIT CIRCLE.

It is not known if every Jordan curve contains all four VERTICES of some SQUARE, but it has been proven true for "sufficiently smooth" curves and closed convex curves (Schnirelmann). For every TRIANGLE T and Jordan curve J, J has an INSCRIBED TRIANGLE similar to T.

see also JORDAN CURVE THEOREM, UNIT CIRCLE

Jordan Curve Theorem

If J is a simple closed curve in $\mathbb{R}^2$, then $\mathbb{R}^2 - J$ has two components (an "inside" and "outside"), with J the BOUNDARY of each.

see also JORDAN CURVE, SCHÖNFLIES THEOREM

References

Rolfsen, D. *Knots and Links.* Wilmington, DE: Publish or Perish Press, p. 9, 1976.

Jordan Decomposition Theorem

Let $V \neq (0)$ be a finite dimensional VECTOR SPACE over the COMPLEX NUMBERS, and let A be a linear operator on V. Then V can be expressed as a DIRECT SUM of cyclic subspaces.

References

Gohberg, I. and Goldberg, S. "A Simple Proof of the Jordan Decomposition Theorem for Matrices." *Amer. Math. Monthly* **103**, 157–159, 1996.

Jordan-Hölder Theorem

The composition quotient groups belonging to two COMPOSITION SERIES of a FINITE GROUP G are, apart from their sequence, ISOMORPHIC in pairs. In other words, if

$$I \subset H_s \subset \ldots \subset H_2 \subset H_1 \subset G$$

is one COMPOSITION SERIES and

$$I \subset K_t \subset \ldots \subset K_2 \subset K_1 \subset G$$

is another, then $t = s$, and corresponding to any composition quotient group K_j/K_{j+1}, there is a composition quotient group H_i/H_{i+1} such that

$$\frac{K_j}{K_{j+1}} = \frac{H_i}{H_{i+1}}.$$

This theorem was proven in 1869–1889.

see also COMPOSITION SERIES, FINITE GROUP, ISOMORPHIC GROUPS

References

Lomont, J. S. *Applications of Finite Groups.* New York: Dover, p. 26, 1993.

Jordan's Inequality

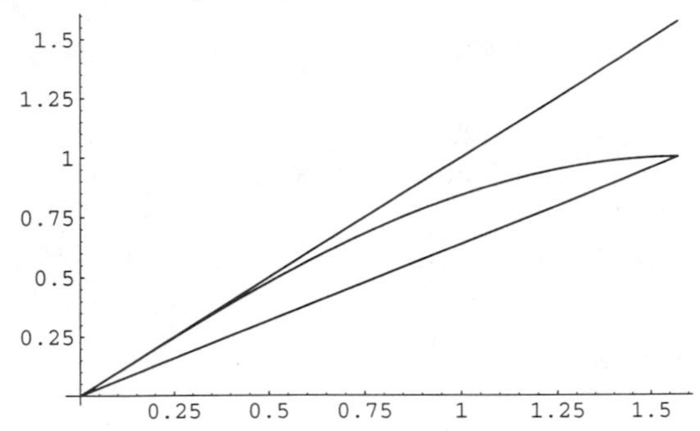

For $0 \leq x \leq \pi/2$,

$$\frac{2}{\pi}x \leq \sin x \leq x.$$

References

Yuefeng, F. "Jordan's Inequality." *Math. Mag.* **69**, 126, 1996.

Jordan's Lemma

Jordan's lemma shows the value of the INTEGRAL

$$I \equiv \int_{-\infty}^{\infty} f(x)e^{iax}\,dx \tag{1}$$

along the REAL AXIS is 0 for "nice" functions which satisfy $\lim_{R\to\infty} |f(Re^{i\theta})| = 0$. This is established using a CONTOUR INTEGRAL I_R which satisfies

$$\lim_{R\to\infty} |I_R| \leq \frac{\pi}{a} \lim_{R\to\infty} \epsilon = 0. \tag{2}$$

To derive the lemma, write

$$x \equiv Re^{i\theta} = R(\cos\theta + i\sin\theta) \tag{3}$$
$$dx = iRe^{i\theta}\,d\theta \tag{4}$$

and define the CONTOUR INTEGRAL

$$I_R = \int_0^{\pi} f(Re^{i\theta})e^{iaR\cos\theta - aR\sin\theta}iRe^{i\theta}\,d\theta \tag{5}$$

Then

$$\begin{aligned} |I_R| &= R\int_0^{\pi} |f(Re^{i\theta})|\,|e^{iaR\cos\theta}|\,|e^{-aR\sin\theta}|\,|i|\,|e^{i\theta}|\,d\theta \\ &= R\int_0^{\pi} |f(Re^{i\theta})|e^{-aR\sin\theta}\,d\theta \\ &= 2R\int_0^{\pi/2} |f(Re^{i\theta})|e^{-aR\sin\theta}\,d\theta. \end{aligned} \tag{6}$$

Now, if $\lim_{R\to\infty} |f(Re^{i\theta})| = 0$, choose an ϵ such that $|f(Re^{i\theta})| \leq \epsilon$, so

$$|I_R| \leq 2R\epsilon \int_0^{\pi/2} e^{-aR\sin\theta}\,d\theta. \tag{7}$$

But, for $\theta \in [0, \pi/2]$,

$$\frac{2}{\pi}\theta \leq \sin\theta, \tag{8}$$

so

$$\begin{aligned} |I_R| &\leq 2R\epsilon \int_0^{\pi/2} e^{-2aR\theta/\pi}\,d\theta \\ &= 2\epsilon R\frac{1-e^{-aR}}{\frac{2aR}{\pi}} = \frac{\pi\epsilon}{a}(1-e^{-aR}). \end{aligned} \tag{9}$$

As long as $\lim_{R\to\infty} |f(z)| = 0$, Jordan's lemma

$$\lim_{R\to\infty} |I_R| \leq \frac{\pi}{a} \lim_{R\to\infty} \epsilon = 0 \tag{10}$$

then follows.

see also CONTOUR INTEGRATION

References

Arfken, G. *Mathematical Methods for Physicists, 3rd ed.* Orlando, FL: Academic Press, pp. 406–408, 1985.

Jordan Measure

Let the set M correspond to a bounded, NONNEGATIVE function f on an interval $0 \leq f(x) \leq c$ for $x \in [a, b]$. The Jordan measure, when it exists, is the common value of the outer and inner Jordan measures of M.

The outer Jordan measure is the greatest lower bound of the areas of the covering of M, consisting of finite unions of RECTANGLES. The inner Jordan measure of M is the difference between the AREA $c(a-b)$ of the RECTANGLE S with base $[a, b]$ and height c, and the outer measure of the complement of M in S.

References

Shenitzer, A. and Steprans, J. "The Evolution of Integration." *Amer. Math. Monthly* **101**, 66–72, 1994.

Jordan Polygon

see SIMPLE POLYGON

Josephus Problem

Given a group of n men arranged in a CIRCLE under the edict that every mth man will be executed going around the CIRCLE until only one remains, find the position $L(n,m)$ in which you should stand in order to be the last survivor (Ball and Coxeter 1987). The original problem consisted of a CIRCLE of 41 men with every third man killed ($n = 41$, $m = 3$). In order for the lives of the last two men to be spared, they must be placed at positions 31 (last) and 16 (second-to-last).

The following array gives the original position of the last survivor out of a group of $n = 1, 2, \ldots$, if every mth man is killed:

1									
2	1								
3	3	2							
4	1	1	2						
5	3	4	1	2					
6	5	1	5	1	4				
7	7	4	2	6	3	5			
8	1	7	6	3	1	4	4		
9	3	1	1	8	7	2	3	8	
10	5	4	5	3	3	9	1	7	8

(Sloane's A032434). The survivor for $m = 2$ can be given analytically by

$$L(n,2) = 1 + 2n - 2^{1+\lfloor \lg n \rfloor},$$

where $\lfloor n \rfloor$ is the FLOOR FUNCTION and LG is the LOGARITHM to base 2. The first few solutions are therefore 1, 1, 3, 1, 3, 5, 7, 1, 3, 5, 7, 9, 11, 13, 15, 1, ... (Sloane's A006257).

Mott-Smith (1954) discusses a card game called "Out and Under" in which cards at the top of a deck are alternately discarded and placed at the bottom. This is a Josephus problem with parameter $m = 2$, and Mott-Smith hints at the above closed-form solution.

The original position of the second-to-last survivor is given in the following table for $n = 2, 3, \ldots$:

1	1								
2	1	1							
3	1	1	2						
4	3	2	1	2					
5	1	1	5	1	4				
6	3	1	2	1	3	4			
7	1	4	6	3	1	3	4		
8	3	1	1	2	7	1	3	7	
9	5	4	5	3	3	8	1	6	4

(Sloane's A032435).

Another version of the problem considers a CIRCLE of two groups (say, "A" and "B") of 15 men each, with every ninth man cast overboard. To save all the members of the "A" group, the men must be placed at positions 1, 2, 3, 4, 10, 11, 13, 14, 15, 17, 20, 21, 25, 28, 29, giving the ordering

$$AAAABBBBBAABAAABABBAABBBABBAAB$$

which can be remembered with the aid of the MNEMONIC "From numbers' aid and art, never will fame depart." Consider the vowels only, assign $a = 1$, $e = 2$, $i = 3$, $o = 4$, $u = 5$, and alternately add a number of letters corresponding to a vowel value, so 4A (o), 5B (u), 2A (e), etc. (Ball and Coxeter 1987).

If every tenth man is instead thrown overboard, the men from the "A" group must be placed in positions 1, 2, 4, 5, 6, 12, 13, 16, 17, 18, 19, 21, 25, 28, 29, giving the sequence

$$AABAAABBBBBAABBAAAABABBBABBAAB$$

which can be constructed using the MNEMONIC "Rex paphi cum gente bona dat signa serena" (Ball and Coxeter 1987).

see also KIRKMAN'S SCHOOLGIRL PROBLEM, NECKLACE

References

Bachet, C. G. Problem 23 in *Problèmes plaisans et délectables, 2nd ed.* p. 174, 1624.

Ball, W. W. R. and Coxeter, H. S. M. *Mathematical Recreations and Essays, 13th ed.* New York: Dover, pp. 32–36, 1987.

Kraitchik, M. "Josephus' Problem." §3.13 in *Mathematical Recreations.* New York: W. W. Norton, pp. 93–94, 1942.

Mott-Smith, G. *Mathematical Puzzles for Beginners and Enthusiasts.* New York: Dover, 1954.

Sloane, N. J. A. Sequence A006257/M2216 in "An On-Line Version of the Encyclopedia of Integer Sequences."

Jug

see THREE JUG PROBLEM

Jugendtraum

Kronecker proved that all the Galois extensions of the RATIONALS $\mathbb{Q}$ with ABELIAN Galois groups are SUBFIELDS of cyclotomic fields $Q(\mu_n)$, where μ_n is the group of nth ROOTS OF UNITY. He then sought to find a similar function whose division values would generate the Abelian extensions of an arbitrary NUMBER FIELD. He discovered that the j-FUNCTION works for IMAGINARY quadratic number fields K, but the completion of this problem, known as Kronecker's Jugendtraum ("dream of youth"), for other fields remains one of the great unsolved problems in NUMBER THEORY.

see also j-FUNCTION

References

Shimura, G. *Introduction to the Arithmetic Theory of Automorphic Functions.* Princeton, NJ: Princeton University Press, 1981.

Juggling

The throwing and catching of multiple objects such that at least one is always in the air. Some aspects of juggling turn out to be quite mathematical. The best examples are the two-handed asynchronous juggling sequences known as "SITESWAPS."

see also SITESWAP

References

Buhler, J.; Eisenbud, D.; Graham, R.; and Wright, C. "Juggling Drops and Descents." *Amer. Math. Monthly* **101**, 507–519, 1994.

Donahue, B. "Jugglers Now Juggle Numbers to Compute New Tricks for Ancient Art." *New York Times,* pp. B5 and B10, Apr. 16, 1996.

Juggling Information Service. "Siteswaps." `http://www.juggling.org/help/siteswap`.

Julia Fractal

see JULIA SET

Julia Set

Let $R(z)$ be a rational function

$$R(z) \equiv \frac{P(z)}{Q(z)}, \tag{1}$$

where $z \in \mathbb{C}^*$, $\mathbb{C}^*$ is the RIEMANN SPHERE $\mathbb{C}\cup\{\infty\}$, and P and Q are POLYNOMIALS without common divisors. The "filled-in" Julia set J_R is the set of points z which do not approach infinity after $R(z)$ is repeatedly applied. The true Julia set is the boundary of the filled-in set (the set of "exceptional points"). There are two types of Julia sets: connected sets and CANTOR SETS.

For a Julia set J_c with $c \ll 1$, the CAPACITY DIMENSION is

$$d_{\text{capacity}} = 1 + \frac{|c|^2}{4\ln 2} + \mathcal{O}(|c|^3). \tag{2}$$

For small c, J_c is also a JORDAN CURVE, although its points are not COMPUTABLE.

Quadratic Julia sets are generated by the quadratic mapping

$$z_{n+1} = {z_n}^2 + c \tag{3}$$

for fixed c. The special case $c = -0.123 + 0.745i$ is called DOUADY'S RABBIT FRACTAL, $c = -0.75$ is called the SAN MARCO FRACTAL, and $c = -0.391 - 0.587i$ is the SIEGEL DISK FRACTAL. For every c, this transformation generates a FRACTAL. It is a CONFORMAL TRANSFORMATION, so angles are preserved. Let J be the JULIA SET, then $x' \mapsto x$ leaves J invariant. If a point P is on J, then all its iterations are on J. The transformation has a two-valued inverse. If $b = 0$ and y is started at 0, then the map is equivalent to the LOGISTIC MAP. The set of all points for which J is connected is known as the MANDELBROT SET.

see also DENDRITE FRACTAL, DOUADY'S RABBIT FRACTAL, FATOU SET, MANDELBROT SET, NEWTON'S METHOD, SAN MARCO FRACTAL, SIEGEL DISK FRACTAL

References

Dickau, R. M. "Julia Sets." `http://forum.swarthmore.edu/advanced/robertd/julias.html`.

Dickau, R. M. "Another Method for Calculating Julia Sets." `http://forum.swarthmore.edu/advanced/robertd/inversejulia.html`.

Douady, A. "Julia Sets and the Mandelbrot Set." In *The Beauty of Fractals: Images of Complex Dynamical Systems* (Ed. H.-O. Peitgen and D. H. Richter). Berlin: Springer-Verlag, p. 161, 1986.

Lauwerier, H. *Fractals: Endlessly Repeated Geometric Figures.* Princeton, NJ: Princeton University Press, pp. 124–126, 138–148, and 177–179, 1991.

Peitgen, H.-O. and Saupe, D. (Eds.). "The Julia Set," "Julia Sets as Basin Boundaries," "Other Julia Sets," and "Exploring Julia Sets." §3.3.2 to 3.3.5 in *The Science of Fractal Images.* New York: Springer-Verlag, pp. 152–163, 1988.

Schroeder, M. *Fractals, Chaos, Power Laws.* New York: W. H. Freeman, p. 39, 1991.

Wagon, S. "Julia Sets." §5.4 in *Mathematica in Action.* New York: W. H. Freeman, pp. 163–178, 1991.

Jump

A point of DISCONTINUITY.

see also DISCONTINUITY, JUMP ANGLE, JUMPING CHAMPION

Jump Angle

A GEODESIC TRIANGLE with oriented boundary yields a curve which is piecewise DIFFERENTIABLE. Furthermore, the TANGENT VECTOR varies continuously at all but the three corner points, where it changes suddenly. The angular difference of the tangent vectors at these corner points are called the jump angles.

see also ANGULAR DEFECT, GAUSS-BONNET FORMULA

Jumping Champion

An integer n is called a JUMPING CHAMPION if n is the most frequently occurring difference between consecutive primes $n \leq N$ for some N (Odlyzko *et al.*). This term was coined by J. H. Conway in 1993. There are occasionally several jumping champions in a range. Odlyzko *et al.* give a table of jumping champions for $n \leq 1000$, consisting mainly of 2, 4, and 6. 6 is the jumping champion up to about $n \approx 1.74 \times 10^{35}$, at which point 30 dominates. At $n \approx 10^{425}$, 210 becomes champion, and subsequent PRIMORIALS are conjectured to take over at larger and larger n. Erdős and Straus (1980) proved that the jumping champions tend to infinity under the assumption of a quantitative form of the k-tuples conjecture.

see also PRIME DIFFERENCE FUNCTION, PRIME GAPS, PRIME NUMBER, PRIMORIAL

References

Erdős, P.; and Straus, E. G. "Remarks on the Differences Between Consecutive Primes." *Elem. Math.* **35**, 115–118, 1980.

Guy, R. K. *Unsolved Problems in Number Theory, 2nd ed.* New York: Springer-Verlag, 1994.
Nelson, H. "Problem 654." *J. Recr. Math.* **11**, 231, 1978–1979.
Odlyzko, A.; Rubinstein, M.; and Wolf, M. "Jumping Champions." `http://www.research.att.com/~amo/doc/recent.html`.

Jung's Theorem

Every finite set of points with SPAN d has an enclosing CIRCLE with RADIUS no greater than $\sqrt{3}\,d/3$.

References

Le Lionnais, F. *Les nombres remarquables.* Paris: Hermann, p. 28, 1983.
Rademacher, H. and Toeplitz, O. *The Enjoyment of Mathematics: Selections from Mathematics for the Amateur.* Princeton, NJ: Princeton University Press, pp. 103–110, 1957.

Just If

see IFF

Just One

see EXACTLY ONE

K

k-ary Divisor

Let a DIVISOR d of n be called a 1-ary divisor if $d \perp n/d$. Then d is called a k-ary divisor of n, written $d|_k n$, if the GREATEST COMMON $(k-1)$-ary divisor of d and (n/d) is 1.

In this notation, $d|n$ is written $d|_0 n$, and $d||n$ is written $d|_1 n$. p^x is an INFINARY DIVISOR of p^y (with $y > 0$) if $p^x|_{y-1}p^y$.

see also DIVISOR, GREATEST COMMON DIVISOR, INFINARY DIVISOR

References

Guy, R. K. *Unsolved Problems in Number Theory, 2nd ed.* New York: Springer-Verlag, p. 54, 1994.

k-Chain

Any sum of a selection of Π_ks, where Π_k denotes a k-D POLYTOPE.

see also k-CIRCUIT

k-Circuit

A k-CHAIN whose bounding $(k-1)$-CHAIN vanishes.

k-Coloring

A k-coloring of a GRAPH G is an assignment of one of k possible colors to each vertex of G such that no two adjacent vertices receive the same color.

see also COLORING, EDGE-COLORING

References

Saaty, T. L. and Kainen, P. C. *The Four-Color Problem: Assaults and Conquest.* New York: Dover, p. 13, 1986.

k-Form

see DIFFERENTIAL k-FORM

K-Function

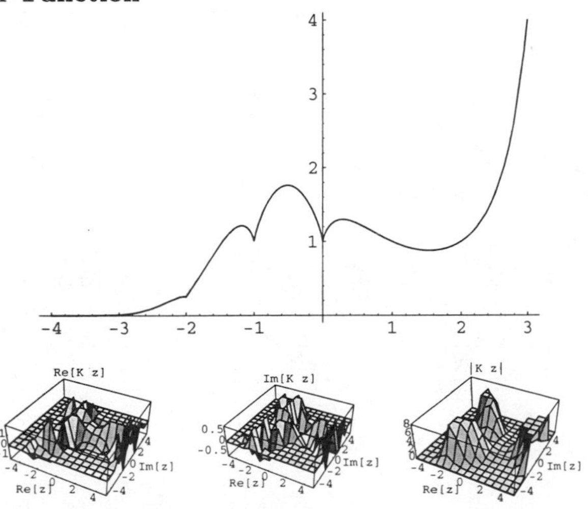

An extension of the K-function

$$K(n+1) \equiv 0^0 1^1 2^2 3^3 \cdots n^n \tag{1}$$

defined by

$$K(z) = \frac{[\Gamma(z)]^z}{G(z)}. \tag{2}$$

Here, $G(z)$ is the G-FUNCTION defined by

$$G(n+1) \equiv \frac{(n!)^n}{K(n+1)} = \begin{cases} 1 & \text{if } n = 0 \\ 0!1!2!\cdots(n-1)! & \text{if } n > 0. \end{cases} \tag{3}$$

The K-function is given by the integral

$$K(z) = (2\pi)^{-(z-1)/2} \exp\left[\binom{z}{2} + \int_0^{z-1} \ln(t!)\,dt\right] \tag{4}$$

and the closed-form expression

$$K(z) = \exp[\zeta'(-1,z) - \zeta'(-1)], \tag{5}$$

where $\zeta(z)$ is the RIEMANN ZETA FUNCTION, $\zeta'(z)$ its DERIVATIVE, $\zeta(a,z)$ is the HURWITZ ZETA FUNCTION, and

$$\zeta'(a,z) \equiv \left[\frac{d\zeta(s,z)}{ds}\right]_{s=a}. \tag{6}$$

$K(z)$ also has a STIRLING-like series

$$K(z+1) = (2^{1/3}\pi_1 z)^{1/12} z^{\binom{z+1}{2}} \times \exp\left(\tfrac{1}{4}z^2 + \tfrac{1}{12} - \frac{B_4}{2\cdot 3\cdot 4z^2} - \frac{B_6}{4\cdot 5\cdot 6z^4} - \ldots\right), \tag{7}$$

where

$$\pi_1 \equiv [K(\tfrac{1}{2})]^8 \tag{8}$$

$$= e^{-(\ln 2)/3 - 12\zeta'(-1)} \tag{9}$$

$$= 2^{2/3}\pi e^{\gamma - 1 - \zeta'(2)/\zeta(2)}, \tag{10}$$

and γ is the EULER-MASCHERONI CONSTANT (Gosper).

The first few values of $K(n)$ for $n = 2, 3, \ldots$ are 1, 4, 108, 27648, 86400000, 4031078400000, ... (Sloane's A002109). These numbers are called HYPERFACTORIALS by Sloane and Plouffe (1995).

see also G-FUNCTION, GLAISHER-KINKELIN CONSTANT, HYPERFACTORIAL, STIRLING'S SERIES

References

Sloane, N. J. A. Sequence A002109/M3706 in "An On-Line Version of the Encyclopedia of Integer Sequences."

K-Graph

The GRAPH obtained by dividing a set of VERTICES $\{1, \ldots, n\}$ into $k-1$ pairwise disjoint subsets with VERTICES of degree $n_1, \ldots, n_{k-1}$, satisfying

$$n = n_1 + \ldots + n_{k-1},$$

and with two VERTICES joined IFF they lie in distinct VERTEX sets. Such GRAPHS are denoted $K_{n_1,\ldots,n_k}$.

see also BIPARTITE GRAPH, COMPLETE GRAPH, COMPLETE k-PARTITE GRAPH, k-PARTITE GRAPH

k-Matrix

A k-matrix is a kind of CUBE ROOT of the IDENTITY MATRIX defined by

$$\mathsf{k} = \begin{bmatrix} 0 & 0 & -i \\ i & 0 & 0 \\ 0 & 1 & 0 \end{bmatrix}.$$

It satisfies

$$\mathsf{k}^3 = \mathsf{I},$$

where I is the IDENTITY MATRIX.

see also CUBE ROOT, QUATERNION

k-Partite Graph

A k-partite graph is a GRAPH whose VERTICES can be partitioned into k disjoint sets so that no two vertices within the same set are adjacent.

see also COMPLETE k-PARTITE GRAPH, K-GRAPH

References

Saaty, T. L. and Kainen, P. C. *The Four-Color Problem: Assaults and Conquest.* New York: Dover, p. 12, 1986.

k-Statistic

An UNBIASED ESTIMATOR of the CUMULANTS κ_i of a DISTRIBUTION. The expectation values of the k-statistics are therefore given by the corresponding CUMULANTS

$$\langle k_1 \rangle = \kappa_1 \tag{1}$$

$$\langle k_2 \rangle = \kappa_2 \tag{2}$$

$$\langle k_3 \rangle = \kappa_3 \tag{3}$$

$$\langle k_4 \rangle = \kappa_4 \tag{4}$$

(Kenney and Keeping 1951, p. 189). For a sample of size, N, the first few k-statistics are given by

$$k_1 = m_1 \tag{5}$$

$$k_2 = \frac{N}{N-1} m_2 \tag{6}$$

$$k_3 = \frac{N^2}{(N-1)(N-2)} m_3 \tag{7}$$

$$k_4 = \frac{N^2[(N+1)m_4 - 3(N-1){m_2}^2]}{(N-1)(N-2)(N-3)}, \tag{8}$$

where m_1 is the sample MEAN, m_2 is the sample VARIANCE, and m_i is the sample ith MOMENT about the MEAN (Kenney and Keeping 1951, pp. 109–110, 163–165, and 189; Kenney and Keeping 1962). These statistics are obtained from inverting the relationships

$$\langle m_1 \rangle = \mu \tag{9}$$

$$\langle m_2 \rangle = \frac{N-1}{N} \mu_2 \tag{10}$$

$$\langle {m_2}^2 \rangle = \frac{(N-1)[(N-1)\mu_4 + (N^2-2N+3){\mu_2}^2]}{N^3} \tag{11}$$

$$\langle m_3 \rangle = \frac{(N-1)(N-2)}{N^2} \mu_3 \tag{12}$$

$$\langle m_4 \rangle = \frac{(N-1)[(N^2-3N+3)\mu_4 + 3(2N-3){\mu_2}^2]}{N^3}. \tag{13}$$

The first moment (sample MEAN) is

$$m_1 \equiv \langle x \rangle = \frac{1}{N} \sum_{i=1}^{N} x_i, \tag{14}$$

and the expectation value is

$$\langle m_1 \rangle = \left\langle \frac{1}{N} \sum_{i=1}^{N} x_i \right\rangle = \mu. \tag{15}$$

The second MOMENT (sample STANDARD DEVIATION) is

$$\begin{aligned} m_2 &\equiv \langle (x-\mu)^2 \rangle = \langle x^2 \rangle - 2\mu \langle x \rangle + \mu^2 = \langle x^2 \rangle - \mu^2 \\ &= \frac{1}{N} \sum_{i=1}^{N} {x_i}^2 - \left(\frac{1}{N} \sum_{i=1}^{N} x_i \right)^2 \\ &= \frac{1}{N} \sum_{i=1}^{N} {x_i}^2 - \frac{1}{N^2} \left(\sum_{i=1}^{N} {x_i}^2 + \sum_{\substack{i,j=1 \\ i \neq j}}^{N} x_i x_j \right) \\ &= \frac{N-1}{N^2} \sum_{i=1}^{N} {x_i}^2 - \frac{1}{N^2} \sum_{\substack{i,j=1 \\ i \neq j}}^{N} x_i x_j, \end{aligned} \tag{16}$$

and the expectation value is

$$\begin{aligned} \langle m_2 \rangle &= \frac{N-1}{N} \left\langle \frac{1}{N} \sum_{i=1}^{N} {x_i}^2 \right\rangle - \frac{1}{N^2} \left\langle \sum_{\substack{i,j=1 \\ i \neq j}}^{N} x_i x_j \right\rangle \\ &= \frac{N-1}{N} \mu_2' - \frac{N(N-1)}{N^2} \mu^2, \end{aligned} \tag{17}$$

since there are $N(N-1)$ terms $x_i x_j$, using

$$\langle x_i x_j \rangle = \langle x_i \rangle \langle x_j \rangle = \langle x_i \rangle^2, \tag{18}$$

and where μ_2' is the MOMENT about 0. Using the identity

$$\mu_2' = \mu_2 + \mu^2 \tag{19}$$

to convert to the MOMENT μ_2 about the MEAN and simplifying then gives

$$\langle m_2 \rangle = \frac{N-1}{N}\mu_2 \tag{20}$$

The factor $(N-1)/N$ is known as BESSEL'S CORRECTION.

The third MOMENT is

$$\begin{aligned} m_3 &\equiv \langle (x-\mu)^3 \rangle = \langle x^3 - 3\mu x^2 + 3\mu^2 x - \mu^3 \rangle \\ &= \langle x^3 \rangle - 3\mu \langle x^2 \rangle + 3\mu^2 \langle x \rangle - \mu^3 \\ &= \langle x^3 \rangle - 3\mu \langle x^2 \rangle + 3\mu^3 - \mu^3 \\ &= \langle x^3 \rangle - 3\mu \langle x^2 \rangle + 2\mu^3 \\ &= \frac{1}{N}\sum x_i^3 - 3\left(\frac{1}{N}\sum x_i\right)\left(\frac{1}{N}\sum x_j^2\right) \\ &\quad + 2\left(\frac{1}{N}\sum x_i\right)^3 \\ &= \frac{1}{N}\sum x_i^3 - \frac{3}{N^2}\left(\sum x_i\right)\left(\sum x_j^2\right) \\ &\quad + \frac{2}{N^3}\left(\sum x_i\right)^3. \end{aligned} \tag{21}$$

Now use the identities

$$\left(\sum x_i^2\right)\left(\sum x_j\right) = \sum x_i^3 + \sum x_i^2 x_j \tag{22}$$

$$\left(\sum x_i\right)^3 = \sum x_i^3 + 3\sum x_i^2 x_j + 6\sum x_i x_j x_k, \tag{23}$$

where it is understood that sums over products of variables exclude equal indices. Plugging in

$$\begin{aligned} m_3 &= \left(\frac{1}{N} - \frac{3}{N^2} + \frac{2}{N^3}\right)\sum x_i^3 \\ &+ \left(-\frac{3}{N^2} + 3\cdot\frac{2}{N^3}\right)\sum x_i^2 x_j + 6\cdot\frac{2}{N^3}\sum x_i x_j x_k. \end{aligned} \tag{24}$$

The expectation value is then given by

$$\begin{aligned} \langle m_3 \rangle &= \left(\frac{1}{N} - \frac{3}{N^2} + \frac{2}{N^3}\right) N\mu_3' \\ &+ \left(-\frac{3}{N^2} + \frac{6}{N^3}\right) N(N-1)\mu_2'\mu + \frac{12}{N^3}\tfrac{1}{6}N(N-1)(N-2)\mu^3 \end{aligned} \tag{25}$$

where μ_2' is the MOMENT about 0. Plugging in the identities

$$\mu_2' = \mu_2 + \mu^2 \tag{26}$$

$$\mu_3' = \mu_3 + 3\mu_2\mu + \mu^3 \tag{27}$$

and simplifying then gives

$$\langle m_3 \rangle = \frac{(N-1)(N-2)}{N^2}\mu_3 \tag{28}$$

(Kenney and Keeping 1951, p. 189).

The fourth MOMENT is

$$\begin{aligned} m_4 &= \langle (x-\mu)^4 \rangle = \langle x^4 - 4x^3\mu + 6x^2\mu^2 - 4x\mu^3 + \mu^4 \rangle \\ &= \langle x^4 \rangle - 4\mu \langle x^3 \rangle + 6\mu^2 \langle x^2 \rangle - 3\mu^4 \\ &= \frac{1}{N}\sum x_i^4 - \frac{4}{N^2}\left(\sum x_i\right)\left(\sum x_j^3\right) \\ &\quad + \frac{6}{N^3}\left(\sum x_i\right)^2\left(\sum x_j^2\right) - \frac{3}{N^4}\left(\sum x_i\right)^4 \end{aligned} \tag{29}$$

Now use the identities

$$\left(\sum x_i\right)\left(\sum x_j^3\right) = \sum x_i^4 + \sum x_i^3 x_j \tag{30}$$

$$\begin{aligned} \left(\sum x_i\right)^2\left(\sum x_j^2\right) &= \sum x_i^4 + 2\sum x_i^3 x_j \\ &\quad + 2\sum x_i^2 x_j^2 + 2\sum x_i^2 x_j x_k \end{aligned} \tag{31}$$

$$\begin{aligned} \left(\sum x_i\right)^4 &= \sum x_i^4 + 4\sum x_i^3 x_j + 6\sum x_i^2 x_j^2 \\ &\quad + 12\sum_{x_i}^{2} x_j x_k + 24\sum x_i x_j x_k x_l. \end{aligned} \tag{32}$$

Plugging in,

$$\begin{aligned} m_4 &= \left(\frac{1}{N} - \frac{4}{N^2} + \frac{6}{N^3} - \frac{3}{N^4}\right)\sum x_i^4 \\ &+ \left(-\frac{4}{N^2} + 2\cdot\frac{6}{N^3} - 4\cdot\frac{3}{N^4}\right)\sum x_i^3 x_j \\ &+ \left(2\cdot\frac{6}{N^3} - 6\cdot\frac{3}{N^4}\right)\sum x_i^2 x_j^2 \\ &+ \left(2\cdot\frac{6}{N^3} - 12\cdot\frac{3}{N^4}\right)\sum x_i^2 x_j x_k \\ &- 24\cdot\frac{3}{N^4}\sum x_i x_j x_k x_l. \end{aligned} \tag{33}$$

The expectation value is then given by

$$\begin{aligned} \langle m_4 \rangle &= \left(\frac{1}{N} - \frac{4}{N^2} + \frac{6}{N^3} - \frac{3}{N^4}\right) N\mu_4' \\ &+ \left(-\frac{4}{N^2} + \frac{12}{N^3} - \frac{12}{N^4}\right) N(N-1)\mu_3'\mu \\ &+ \left(\frac{12}{N^3} - \frac{18}{N^4}\right)\tfrac{1}{2}N(N-1)\mu_2'^2 \\ &+ \left(\frac{18}{N^3} - \frac{36}{N^4}\right)\tfrac{1}{2}N(N-1)(N-2)\mu_2'\mu^2 \\ &- \frac{72}{N^4}\tfrac{1}{24}N(N-1)(N-2)(N-3)\mu^4, \end{aligned} \tag{34}$$

where μ_i' are MOMENTS about 0. Using the identities

$$\mu_2' = \mu_2 + \mu^2 \tag{35}$$

$$\mu_3' = \mu_3 + 3\mu_2\mu + \mu^3 \tag{36}$$

$$\mu_4' = \mu_4 + 4\mu_3\mu + 6\mu_2\mu^2 + \mu^4 \tag{37}$$

and simplifying gives

$$\langle m_4 \rangle = \frac{(N-1)[(N^2-3N+3)\mu_4 + 3(2N-3){\mu_2}^2]}{N^3} \tag{38}$$

(Kenney and Keeping 1951, p. 189).

The square of the second moment is

$$\begin{aligned} {m_2}^2 &= (\langle x^2 \rangle - \mu^2)^2 = \langle x^2 \rangle^2 - 2\mu^2 \langle x^2 \rangle + \mu^4 \\ &= \left(\frac{1}{N}\sum {x_i}^2\right)^2 - 2\left(\frac{1}{N}\sum x_i\right)^2 \left(\frac{1}{N}\sum {x_i}^2\right) \\ &\quad + \left(\frac{1}{N}\sum x_i\right)^4 \\ &= \frac{1}{N^2}\left(\sum {x_i}^2\right)^2 - \frac{2}{N^3}\left(\sum x_i\right)^2 \left(\sum {x_j}^2\right) \\ &\quad + \frac{1}{N^4}\left(\sum x_i\right)^4. \end{aligned} \tag{39}$$

Now use the identities

$$\left(\sum {x_i}^2\right)^2 = \sum {x_i}^4 + 2\sum {x_i}^2 {x_j}^2 \tag{40}$$

$$\begin{aligned} \left(\sum x_i\right)^2 \left(\sum {x_j}^2\right) &= \sum {x_i}^4 + 2\sum {x_i}^2 {x_j}^2 \\ &\quad + 2\sum {x_i}^3 x_j + 2\sum {x_i}^2 x_j x_k \end{aligned} \tag{41}$$

$$\begin{aligned} \left(\sum x_i\right)^4 &= \sum {x_i}^4 + 6\sum {x_i}^2 {x_j}^2 \\ &\quad + 4\sum {x_i}^3 x_j + 12\sum_{x_i}^{2} x_j x_k + 24\sum x_i x_j x_k x_l. \end{aligned} \tag{42}$$

Plugging in,

$$\begin{aligned} {m_2}^2 &= \left(\frac{1}{N^2} - \frac{2}{N^3} + \frac{1}{N^4}\right)\sum {x_i}^4 \\ &\quad + \left(2\cdot\frac{1}{N^2} - 2\cdot\frac{2}{N^3} + 6\cdot\frac{1}{N^4}\right)\sum {x_i}^2 {x_j}^2 \\ &\quad + \left(-2\cdot\frac{2}{N^3} + 4\cdot\frac{1}{N^4}\right)\sum {x_i}^3 x_j \\ &\quad + \left(-2\cdot\frac{2}{N^3} + 12\cdot\frac{1}{N^4}\right)\sum {x_i}^2 x_j x_k \\ &\quad + \frac{24}{N^4}\sum x_i x_j x_k x_l \end{aligned} \tag{43}$$

The expectation value is then given by

$$\begin{aligned} \langle {m_2}^2 \rangle &= \left(\frac{1}{N^2} - \frac{2}{N^3} + \frac{1}{N^4}\right) N\mu'_4 \\ &\quad + \left(\frac{2}{N^2} - \frac{4}{N^3} + \frac{6}{N^4}\right)\tfrac{1}{2}N(N-1)\mu'^2_2 \\ &\quad + \left(-\frac{4}{N^3} + \frac{4}{N^4}\right)N(N-1)\mu'_3\mu \\ &\quad + \left(-\frac{4}{N^3} + \frac{12}{N^4}\right)\tfrac{1}{2}N(N-1)(N-2)\mu'_2\mu^2 \\ &\quad + \frac{24}{N^4}\tfrac{1}{24}N(N-1)(N-2)(N-3)\mu^4 \end{aligned} \tag{44}$$

where μ'_i are MOMENTS about 0. Using the identities

$$\mu'_2 = \mu_2 + \mu^2 \tag{45}$$

$$\mu'_3 = \mu_3 + 3\mu_2\mu + \mu^3 \tag{46}$$

$$\mu'_4 = \mu_4 + 4\mu_3\mu + 6\mu_2\mu^2 + \mu^4 \tag{47}$$

and simplifying gives

$$\langle {m_2}^2 \rangle = \frac{(N-1)[(N-1)\mu_4 + (N^2-2N+3){\mu_2}^2]}{N^3} \tag{48}$$

(Kenney and Keeping 1951, p. 189).

The VARIANCE of k_2 is given by

$$\operatorname{var}(k_2) = \frac{\kappa_4}{N} + \frac{2}{(N-1){\kappa_2}^2}, \tag{49}$$

so an unbiased estimator of $\operatorname{var}(k_2)$ is given by

$$\hat{\operatorname{var}}(k_2) = \frac{2{k_2}^2 N + (N-1)k_4}{N(N+1)} \tag{50}$$

(Kenney and Keeping 1951, p. 189). The VARIANCE of k_3 can be expressed in terms of CUMULANTS by

$$\operatorname{var}(k_3) = \frac{\kappa_6}{N} + \frac{9\kappa_2\kappa_4}{N-1} + \frac{9{\kappa_3}^2}{N-1} + \frac{6{\kappa_2}^3}{N(N-1)(N-2)}. \tag{51}$$

An UNBIASED ESTIMATOR for $\operatorname{var}(k_3)$ is

$$\hat{\operatorname{var}}(k_3) = \frac{6{k_2}^2 N(N-1)}{(N-2)(N+1)(N+3)} \tag{52}$$

(Kenney and Keeping 1951, p. 190).

Now consider a finite population. Let a sample of N be taken from a population of M. Then UNBIASED ESTIMATORS M_2 for the population MEAN μ, M_2 for the population VARIANCE μ_2, G_1 for the population SKEWNESS γ_1, and G_2 for the population KURTOSIS γ_2 are

$$M_1 = \mu \tag{53}$$

$$M_2 = \frac{M-N}{N(M-1)}\mu_2 \tag{54}$$

$$G_1 = \frac{M-2N}{M-2}\sqrt{\frac{M-1}{N(M-N)}}\gamma_1 \tag{55}$$

$$\begin{aligned} G_2 &= \frac{(M-1)(M^2-6MN+M+6N^2)\gamma_2}{N(M-2)(M-3)(M-N)} \\ &\quad - \frac{6M(MN+M-N^2-1)}{N(M-2)(M-3)(M-N)} \end{aligned} \tag{56}$$

(Church 1926, p. 357; Carver 1930; Irwin and Kendall 1944; Kenney and Keeping 1951, p. 143), where γ_1 is the sample SKEWNESS and γ_2 is the sample KURTOSIS.

see also GAUSSIAN DISTRIBUTION, KURTOSIS, MEAN, MOMENT, SKEWNESS, VARIANCE

References

Carver, H. C. (Ed.). "Fundamentals of the Theory of Sampling." *Ann. Math. Stat.* **1**, 101–121, 1930.

Church, A. E. R. "On the Means and Squared Standard-Deviations of Small Samples from Any Population." *Biometrika* **18**, 321–394, 1926.

Irwin, J. O. and Kendall, M. G. "Sampling Moments of Moments for a Finite Population." *Ann. Eugenics* **12**, 138–142, 1944.

Kenney, J. F. and Keeping, E. S. *Mathematics of Statistics, Pt. 2, 2nd ed.* Princeton, NJ: Van Nostrand, 1951.

Kenney, J. F. and Keeping, E. S. "The k-Statistics." §7.9 in *Mathematics of Statistics, Pt. 1, 3rd ed.* Princeton, NJ: Van Nostrand, pp. 99–100, 1962.

k-Subset

A k-subset is a SUBSET containing exactly k elements.

see also SUBSET

k-Theory

A branch of mathematics which brings together ideas from algebraic geometry, LINEAR ALGEBRA, and NUMBER THEORY. In general, there are two main types of k-theory: topological and algebraic.

Topological k-theory is the "true" k-theory in the sense that it came first. Topological k-theory has to do with VECTOR BUNDLES over TOPOLOGICAL SPACES. Elements of a k-theory are STABLE EQUIVALENCE classes of VECTOR BUNDLES over a TOPOLOGICAL SPACE. You can put a RING structure on the collection of STABLY EQUIVALENT bundles by defining ADDITION through the WHITNEY SUM, and MULTIPLICATION through the TENSOR PRODUCT of VECTOR BUNDLES. This defines "the reduced real topological k-theory of a space."

"The reduced k-theory of a space" refers to the same construction, but instead of REAL VECTOR BUNDLES, COMPLEX VECTOR BUNDLES are used. Topological k-theory is significant because it forms a generalized COHOMOLOGY theory, and it leads to a solution to the vector fields on spheres problem, as well as to an understanding of the J-homeomorphism of HOMOTOPY THEORY.

Algebraic k-theory is somewhat more involved. Swan (1962) noticed that there is a correspondence between the CATEGORY of suitably nice TOPOLOGICAL SPACES (something like regular HAUSDORFF SPACES) and C^*-ALGEBRAS. The idea is to associate to every SPACE the C^*-ALGEBRA of CONTINUOUS MAPS from that SPACE to the REALS.

A VECTOR BUNDLE over a SPACE has sections, and these sections can be multiplied by CONTINUOUS FUNCTIONS to the REALS. Under Swan's correspondence, VECTOR BUNDLES correspond to modules over the C^*-ALGEBRA of CONTINUOUS FUNCTIONS, the MODULES being the modules of sections of the VECTOR BUNDLE. This study of MODULES over C^*-ALGEBRA is the starting point of algebraic k-theory.

The QUILLEN-LICHTENBAUM CONJECTURE connects algebraic k-theory to Étale cohomology.

see also C^*-ALGEBRA

References

Srinivas, V. *Algebraic k-Theory, 2nd ed.* Boston, MA: Birkhäuser, 1995.

Swan, R. G. "Vector Bundles and Projective Modules." *Trans. Amer. Math. Soc.* **105**, 264–277, 1962.

k-Tuple Conjecture

The first of the HARDY-LITTLEWOOD CONJECTURES. The k-tuple conjecture states that the asymptotic number of PRIME CONSTELLATIONS can be computed explicitly. In particular, unless there is a trivial divisibility condition that stops p, $p+a_1$, ..., $p+a_k$ from consisting of PRIMES infinitely often, then such PRIME CONSTELLATIONS will occur with an asymptotic density which is computable in terms of a_1, ..., a_k. Let $0 < m_1 < m_2 < \ldots < m_k$, then the k-tuple conjecture predicts that the number of PRIMES $p \leq x$ such that $p+2m_1$, $p+2m_2$, ..., $p+2m_k$ are all PRIME is

$$P(x; m_1, m_2, \ldots, m_k) \sim C(m_1, m_2, \ldots, m_k) \int_2^x \frac{dt}{\ln^{k+1} t}, \tag{1}$$

where

$$C(m_1, m_2, \ldots, m_k) = 2^k \prod_q \frac{1 - \frac{w(q; m_1, m_2, \ldots, m_k)}{q}}{\left(1 - \frac{1}{q}\right)^{k+1}}, \tag{2}$$

the product is over ODD PRIMES q, and

$$w(q; m_1, m_2, \ldots, m_k) \tag{3}$$

denotes the number of distinct residues of 0, m_1, ..., m_k (mod q) (Halberstam and Richert 1974, Odlyzko). If $k = 1$, then this becomes

$$C(m) = 2 \prod_q \frac{q(q-2)}{(q-1)^2} \prod_{q|m} \frac{q-1}{q-2}. \tag{4}$$

This conjecture is generally believed to be true, but has not been proven (Odlyzko *et al.*). The following special case of the conjecture is sometimes known as the PRIME PATTERNS CONJECTURE. Let S be a FINITE set of INTEGERS. Then it is conjectured that there exist infinitely many k for which $\{k+s : s \in S\}$ are all PRIME IFF S does not include all the RESIDUES of any PRIME. The TWIN PRIME CONJECTURE is a special case of the prime patterns conjecture with $S = \{0, 2\}$. This conjecture also implies that there are arbitrarily long ARITHMETIC PROGRESSIONS of PRIMES.

see also ARITHMETIC PROGRESSION, DIRICHLET'S THEOREM, HARDY-LITTLEWOOD CONJECTURES, k-TUPLE CONJECTURE, PRIME ARITHMETIC PROGRESSION, PRIME CONSTELLATION, PRIME QUADRUPLET,

PRIME PATTERNS CONJECTURE, TWIN PRIME CONJECTURE, TWIN PRIMES

References

Brent, R. P. "The Distribution of Small Gaps Between Successive Primes." *Math. Comput.* **28**, 315–324, 1974.

Brent, R. P. "Irregularities in the Distribution of Primes and Twin Primes." *Math. Comput.* **29**, 43–56, 1975.

Halberstam, E. and Richert, H.-E. *Sieve Methods.* New York: Academic Press, 1974.

Hardy, G. H. and Littlewood, J. E. "Some Problems of 'Partitio Numerorum.' III. On the Expression of a Number as a Sum of Primes." *Acta Math.* **44**, 1–70, 1922.

Odlyzko, A.; Rubinstein, M.; and Wolf, M. "Jumping Champions."

Riesel, H. *Prime Numbers and Computer Methods for Factorization, 2nd ed.* Boston, MA: Birkhäuser, pp. 66–68, 1994.

Kabon Triangles

The largest number $N(n)$ of nonoverlapping TRIANGLES which can be produced by n straight LINE SEGMENTS. The first few terms are 1, 2, 5, 7, 11, 15, 21, ... (Sloane's A006066).

References

Sloane, N. J. A. Sequence A006066/M1334 in "An On-Line Version of the Encyclopedia of Integer Sequences."

Kac Formula

The expected number of REAL zeros E_n of a RANDOM POLYNOMIAL of degree n is

$$E_n = \frac{1}{\pi}\int_{-\infty}^{\infty}\sqrt{\frac{1}{(t^2-1)^2}-\frac{(n+1)^2t^{2n}}{(t^{2n+2}-1)^2}}\,dt \quad (1)$$

$$= \frac{4}{\pi}\int_0^1\sqrt{\frac{1}{(1-t^2)^2}-\frac{(n+1)^2t^{2n}}{(1-t^{2n+2})^2}}\,dt. \quad (2)$$

As $n \to \infty$,

$$E_n = \frac{2}{\pi}\ln n + C_1 + \frac{2}{\pi n} + \mathcal{O}(n^{-2}), \quad (3)$$

where

$$C_1 = \frac{2}{\pi}\left[\ln 2 + \int_0^\infty\left(\sqrt{\frac{1}{x^2}-\frac{4e^{-2x}}{(1-e^{-2x})^2}}-\frac{1}{x+1}\right)dx\right] = 0.6257358072\ldots. \quad (4)$$

The initial term was derived by Kac (1943).

References

Edelman, A. and Kostlan, E. "How Many Zeros of a Random Polynomial are Real?" *Bull. Amer. Math. Soc.* **32**, 1–37, 1995.

Kac, M. "On the Average Number of Real Roots of a Random Algebraic Equation." *Bull. Amer. Math. Soc.* **49**, 314–320, 1943.

Kac, M. "A Correction to 'On the Average Number of Real Roots of a Random Algebraic Equation'." *Bull. Amer. Math. Soc.* **49**, 938, 1943.

Kac Matrix

The $(n+1)\times(n+1)$ TRIDIAGONAL MATRIX (also called the CLEMENT MATRIX) defined by

$$S_n = \begin{bmatrix} 0 & n & 0 & 0 & \cdots & 0 \\ 1 & 0 & n-1 & 0 & \cdots & 0 \\ 0 & 2 & 0 & n-2 & \cdots & 0 \\ \vdots & \vdots & \ddots & \ddots & \ddots & \vdots \\ 0 & 0 & 0 & n-1 & 0 & 1 \\ 0 & 0 & 0 & 0 & n & 0 \end{bmatrix}.$$

The EIGENVALUES are $2k-n$ for $k = 0, 1, \ldots, n$.

Kähler Manifold

A manifold for which the EXTERIOR DERIVATIVE of the FUNDAMENTAL FORM Ω associated with the given Hermitian metric vanishes, so $d\Omega = 0$.

References

Amorós, J. *Fundamental Groups of Compact Kähler Manifolds.* Providence, RI: Amer. Math. Soc., 1996.

Iyanaga, S. and Kawada, Y. (Eds.). "Kähler Manifolds." §232 in *Encyclopedic Dictionary of Mathematics.* Cambridge, MA: MIT Press, pp. 732–734, 1980.

Kakeya Needle Problem

What is the plane figure of least AREA in which a line segment of width 1 can be freely rotated (where translation of the segment is also allowed)? Besicovitch (1928) proved that there is *no* MINIMUM AREA. This can be seen by rotating a line segment inside a DELTOID, star-shaped 5-oid, star-shaped 7-oid, etc. When the figure is restricted to be convex, Cunningham and Schoenberg (1965) found there is still *no* minimum AREA. However, the smallest *simple convex* domain in which one can put a segment of length 1 which will coincide with itself when rotated by 180° is

$$\tfrac{1}{24}(5-2\sqrt{2})\pi = 0.284258\ldots$$

(Le Lionnais 1983).

see also CURVE OF CONSTANT WIDTH, LEBESGUE MINIMAL PROBLEM, REULEAUX POLYGON, REULEAUX TRIANGLE

References

Ball, W. W. R. and Coxeter, H. S. M. *Mathematical Recreations and Essays, 13th ed.* New York: Dover, pp. 99–101, 1987.

Besicovitch, A. S. "On Kakeya's Problem and a Similar One." *Math. Z.* **27**, 312–320, 1928.

Besicovitch, A. S. "The Kakeya Problem." *Amer. Math. Monthly* **70**, 697–706, 1963.

Cunningham, F. Jr. and Schoenberg, I. J. "On the Kakeya Constant." *Canad. J. Math.* **17**, 946–956, 1965.

Le Lionnais, F. *Les nombres remarquables.* Paris: Hermann, p. 24, 1983.

Ogilvy, C. S. *A Calculus Notebook.* Boston: Prindle, Weber, & Schmidt, 1968.

Ogilvy, C. S. *Excursions in Geometry.* New York: Dover, pp. 147–153, 1990.

Pál, J. "Ein Minimumproblem für Ovale." *Math. Ann.* **88**, 311–319, 1921.
Plouffe, S. "Kakeya Constant." `http://lacim.uqam.ca/piDATA/kakeya.txt`.
Wagon, S. *Mathematica in Action.* New York: W. H. Freeman, pp. 50–52, 1991.

Kakutani's Fixed Point Theorem

Every correspondence that maps a compact convex subset of a locally convex space into itself with a closed graph and convex nonempty images has a fixed point.

see also FIXED POINT THEOREM

Kakutani's Problem

see COLLATZ PROBLEM

Kalman Filter

An ALGORITHM in CONTROL THEORY introduced by R. Kalman in 1960 and refined by Kalman and R. Bucy. It is an ALGORITHM which makes optimal use of imprecise data on a linear (or nearly linear) system with Gaussian errors to continuously update the best estimate of the system's current state.

see also WIENER FILTER

References

Chui, C. K. and Chen, G. *Kalman Filtering: With Real-Time Applications, 2nd ed.* Berlin: Springer-Verlag, 1991.
Grewal, M. S. *Kalman FIltering: Theory & Practice.* Englewood Cliffs, NJ: Prentice-Hall, 1993.

KAM Theorem

see KOLMOGOROV-ARNOLD-MOSER THEOREM

Kampyle of Eudoxus

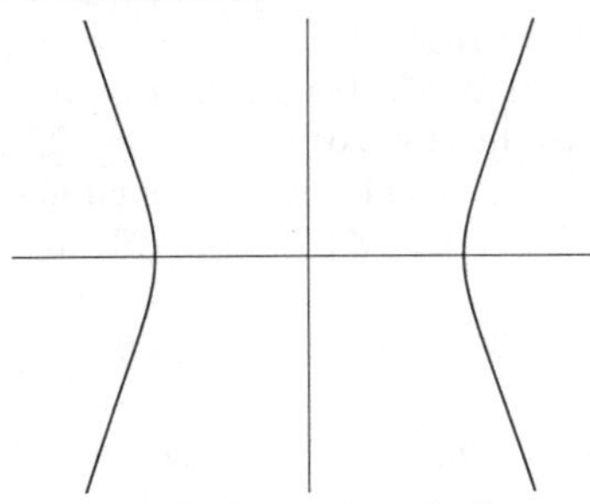

A curve studied by Eudoxus in relation to the classical problem of CUBE DUPLICATION. It is given by the polar equation

$$r\cos^2\theta = a,$$

and the parametric equations

$$x = a\sec t$$
$$y = a\tan t\sec t$$

with $t \in [-\pi/2, \pi/2]$.

References

Lawrence, J. D. *A Catalog of Special Plane Curves.* New York: Dover, pp. 141–143, 1972.
MacTutor History of Mathematics Archive. "Kampyle of Eudoxus." `http://www-groups.dcs.st-and.ac.uk/~history/Curves/Kampyle.html`.

Kanizsa Triangle

An optical ILLUSION, illustrated above, in which the eye perceives a white upright EQUILATERAL TRIANGLE where none is actually drawn.

see also ILLUSION

References

Bradley, D. R. and Petry, H. M. "Organizational Determinants of Subjective Contour." *Amer. J. Psychology* **90**, 253–262, 1977.
Fineman, M. *The Nature of Visual Illusion.* New York: Dover, pp. 26, 137, and 156, 1996.

Kantrovich Inequality

Suppose $x_1 < x_2 < \ldots < x_n$ are given POSITIVE numbers. Let $\lambda_1, \ldots, \lambda_n \geq 0$ and $\sum \lambda_j = 1$. Then

$$\left(\sum \lambda_j x_j\right)\left(\sum \lambda_j {x_j}^{-1}\right) \leq A^2 G^{-2},$$

where

$$A = \tfrac{1}{2}(x_1 + x_n)$$
$$G = \sqrt{x_1 x_n}$$

are the ARITHMETIC and GEOMETRIC MEAN, respectively, of the first and last numbers.

References

Pták, V. "The Kantrovich Inequality." *Amer. Math. Monthly* **102**, 820–821, 1995.

Kaplan-Yorke Conjecture

There are several versions of the Kaplan-Yorke conjecture, with many of the higher dimensional ones remaining unsettled. The original Kaplan-Yorke conjecture (Kaplan and Yorke 1979) proposed that, for a two-dimensional mapping, the CAPACITY DIMENSION D equals the KAPLAN-YORKE DIMENSION D_{KY},

$$D = D_{KY} = d_{\text{Lya}} = 1 + \frac{\sigma_1}{\sigma_2},$$

where σ_1 and σ_2 are the LYAPUNOV CHARACTERISTIC EXPONENTS. This was subsequently proven to be true in 1982. A later conjecture held that the KAPLAN-YORKE DIMENSION is generically equal to a probabilistic dimension which appears to be identical to the INFORMATION DIMENSION (Frederickson *et al.* 1983). This conjecture is partially verified by Ledrappier (1981). For invertible 2-D maps, $\nu = \sigma = D$, where ν is the CORRELATION EXPONENT, σ is the INFORMATION DIMENSION, and D is the CAPACITY DIMENSION (Young 1984).

see also CAPACITY DIMENSION, KAPLAN-YORKE DIMENSION, LYAPUNOV CHARACTERISTIC EXPONENT, LYAPUNOV DIMENSION

References

Chen, Z. M. "A Note on Kaplan-Yorke-Type Estimates on the Fractal Dimension of Chaotic Attractors." *Chaos, Solitons, and Fractals* **3**, 575–582, 1994.

Frederickson, P.; Kaplan, J. L.; Yorke, E. D.; and Yorke, J. A. "The Liapunov Dimension of Strange Attractors." *J. Diff. Eq.* **49**, 185–207, 1983.

Kaplan, J. L. and Yorke, J. A. In *Functional Differential Equations and Approximations of Fixed Points* (Ed. H.-O. Peitgen and H.-O. Walther). Berlin: Springer-Verlag, p. 204, 1979.

Ledrappier, F. "Some Relations Between Dimension and Lyapunov Exponents." *Commun. Math. Phys.* **81**, 229–238, 1981.

Worzbusekros, A. "Remark on a Conjecture of Kaplan and Yorke." *Proc. Amer. Math. Soc.* **85**, 381–382, 1982.

Young, L. S. "Dimension, Entropy, and Lyapunov Exponents in Differentiable Dynamical Systems." *Phys. A* **124**, 639–645, 1984

Kaplan-Yorke Dimension

$$D_{\text{KY}} \equiv j + \frac{\sigma_1 + \ldots + \sigma_j}{|\sigma_{j+1}|},$$

where $\sigma_1 \leq \sigma_n$ are LYAPUNOV CHARACTERISTIC EXPONENTS and j is the largest INTEGER for which

$$\lambda_1 + \ldots + \lambda_j \geq 0.$$

If $\nu = \sigma = D$, where ν is the CORRELATION EXPONENT, σ the INFORMATION DIMENSION, and D the HAUSDORFF DIMENSION, then

$$D \leq D_{\text{KY}}$$

(Grassberger and Procaccia 1983).

References

Grassberger, P. and Procaccia, I. "Measuring the Strangeness of Strange Attractors." *Physica D* **9**, 189–208, 1983.

Kaplan-Yorke Map

$$x_{n+1} = 2x_n$$
$$y_{n+1} = \alpha y_n + \cos(4\pi x_n),$$

where x_n, y_n are computed mod 1. (Kaplan and Yorke 1979). The Kaplan-Yorke map with $\alpha = 0.2$ has CORRELATION EXPONENT 1.42 ± 0.02 (Grassberger Procaccia 1983) and CAPACITY DIMENSION 1.43 (Russell *et al.* 1980).

References

Grassberger, P. and Procaccia, I. "Measuring the Strangeness of Strange Attractors." *Physica D* **9**, 189–208, 1983.

Kaplan, J. L. and Yorke, J. A. In *Functional Differential Equations and Approximations of Fixed Points* (Ed. H.-O. Peitgen and H.-O. Walther). Berlin: Springer-Verlag, p. 204, 1979.

Russell, D. A.; Hanson, J. D.; and Ott, E. "Dimension of Strange Attractors." *Phys. Rev. Let.* **45**, 1175–1178, 1980.

Kappa Curve

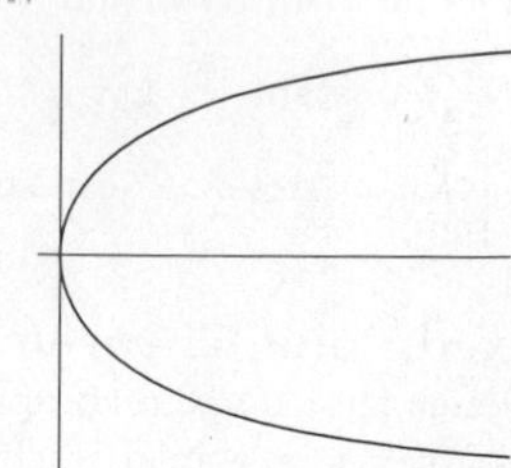

A curve also known as GUTSCHOVEN'S CURVE which was first studied by G. van Gutschoven around 1662 (MacTutor Archive). It was also studied by Newton and, some years later, by Johann Bernoulli. It is given by the Cartesian equation

$$(x^2 + y^2)y^2 = a^2x^2, \tag{1}$$

by the polar equation

$$r = a \cot\theta, \tag{2}$$

and the parametric equations

$$x = a \cos t \cot t \tag{3}$$
$$y = a \cos t. \tag{4}$$

References

Lawrence, J. D. *A Catalog of Special Plane Curves.* New York: Dover, pp. 136 and 139–141, 1972.

MacTutor History of Mathematics Archive. "Kappa Curve." http://www-groups.dcs.st-and.ac.uk/~history/Curves/Kappa.html.

Kaprekar Number

Consider an n-digit number k. Square it and add the right n digits to the left n or $n-1$ digits. If the resultant sum is k, then k is called a Kaprekar number. The first few are 1, 9, 45, 55, 99, 297, 703, ... (Sloane's A006886).

$$9^2 = 81 \qquad 8 + 1 = 9$$

$$297^2 = 88{,}209 \qquad 88 + 209 = 297.$$

see also DIGITAL ROOT, DIGITADITION, HAPPY NUMBER, KAPREKAR ROUTINE, NARCISSISTIC NUMBER, RECURRING DIGITAL INVARIANT

References

Sloane, N. J. A. Sequence A006886/M4625 in "An On-Line Version of the Encyclopedia of Integer Sequences."

Kaprekar Routine

A routine discovered in 1949 by D. R. Kaprekar for 4-digit numbers, but which can be generalized to k-digit numbers. To apply the Kaprekar routine to a number n, arrange the digits in descending (n') and ascending (n'') order. Now compute $K(n) \equiv n' - n''$ and iterate. The algorithm reaches 0 (a degenerate case), a constant, or a cycle, depending on the number of digits in k and the value of n.

For a 3-digit number n in base 10, the Kaprekar routine reaches the number 495 in at most six iterations. In base r, there is a unique number $((r-2)/2, r-1, r/2)_r$ to which n converges in at most $(r+2)/2$ iterations IFF r is EVEN. For any 4-digit number n in base-10, the routine terminates on the number 6174 after seven or fewer steps (where it enters the 1-cycle $K(6174) = 6174$).

2. 0, 0, 9, 21, {(45), (49)}, ...,
3. 0, 0, (32, 52), 184, (320, 580, 484), ...,
4. 0, 30, {201, (126, 138)}, (570, 765), {(2550), (3369), (3873)}, ...,
5. 8, (48, 72), 392, (1992, 2616, 2856, 2232), (7488, 10712, 9992, 13736, 11432), ...,
6. 0, 105, (430, 890, 920, 675, 860, 705), {5600, (4305, 5180)}, {(27195), (33860), (42925), (16840, 42745, 35510)}, ...,
7. 0, (144, 192), (1068, 1752, 1836), (9936, 15072, 13680, 13008, 10608), (55500, 89112, 91800, 72012, 91212, 77388), ...,
8. 21, 252, {(1589, 3178, 2723), (1022, 3122, 3290, 2044, 2212)}, {(17892, 20475), (21483, 25578, 26586, 21987)}, ...,
9. (16, 48), (320, 400), {(2256, 5312, 3856), (3712, 5168, 5456)}, {41520, (34960, 40080, 55360, 49520, 42240)}, ...,
10. 0, 495, 6174, {(53955, 59994), (61974, 82962, 75933, 63954), (62964, 71973, 83952, 74943)}, ...,

see also 196-ALGORITHM, KAPREKAR NUMBER, RATS SEQUENCE

References

Eldridge, K. E. and Sagong, S. "The Determination of Kaprekar Convergence and Loop Convergence of All 3-Digit Numbers." *Amer. Math. Monthly* **95**, 105–112, 1988.

Kaprekar, D. R. "An Interesting Property of the Number 6174." *Scripta Math.* **15**, 244-245, 1955.

Trigg, C. W. "All Three-Digit Integers Lead to..." *The Math. Teacher*, **67**, 41–45, 1974.

Young, A. L. "A Variation on the 2-digit Kaprekar Routine." *Fibonacci Quart.* **31**, 138–145, 1993.

Kaps-Rentrop Methods

A generalization of the RUNGE-KUTTA METHOD for solution of ORDINARY DIFFERENTIAL EQUATIONS, also called ROSENBROCK METHODS.

see also RUNGE-KUTTA METHOD

References

Press, W. H.; Flannery, B. P.; Teukolsky, S. A.; and Vetterling, W. T. *Numerical Recipes in FORTRAN: The Art of Scientific Computing, 2nd ed.* Cambridge, England: Cambridge University Press, pp. 730–735, 1992.

Kapteyn Series

A series of the form

$$\sum_{n=0}^{\infty} \alpha_n J_{\nu+n}[(\nu+n)z],$$

where $J_n(z)$ is a BESSEL FUNCTION OF THE FIRST KIND. Examples include Kapteyn's original series

$$\frac{1}{1-z} = 1 + 2\sum_{n=1}^{\infty} J_n(nz)$$

and

$$\frac{z^2}{2(1-z^2)} = \sum_{n=1}^{\infty} J_{2n}(2nz).$$

see also BESSEL FUNCTION OF THE FIRST KIND, NEUMANN SERIES (BESSEL FUNCTION)

References

Iyanaga, S. and Kawada, Y. (Eds.). *Encyclopedic Dictionary of Mathematics.* Cambridge, MA: MIT Press, p. 1473, 1980.

Karatsuba Multiplication

It is possible to perform MULTIPLICATION of LARGE NUMBERS in (many) fewer operations than the usual brute-force technique of "long multiplication." As discovered by Karatsuba and Ofman (1962), MULTIPLICATION of two n-DIGIT numbers can be done with a BIT COMPLEXITY of less than n^2 using identities of the form

$$\begin{aligned}(a + b\cdot 10^n)(c + d\cdot 10^n)\\ = ac + [(a+b)(c+d) - ac - bd]10^n + bd\cdot 10^{2n}.\end{aligned} \quad (1)$$

Proceeding recursively then gives BIT COMPLEXITY $\mathcal{O}(n^{\lg 3})$, where $\lg 3 = 1.58\ldots < 2$ (Borwein *et al.* 1989). The best known bound is $\mathcal{O}(n \lg n \lg\lg n)$ steps for $n \gg 1$ (Schönhage and Strassen 1971, Knuth 1981). However, this ALGORITHM is difficult to implement, but a procedure based on the FAST FOURIER TRANSFORM is straightforward to implement and gives BIT COMPLEXITY $\mathcal{O}((\lg n)^{2+\epsilon} n)$ (Brigham 1974, Borodin and Munro 1975, Knuth 1981, Borwein *et al.* 1989).

As a concrete example, consider MULTIPLICATION of two numbers each just two "digits" long in base w,

$$N_1 = a_0 + a_1 w \quad (2)$$

$$N_2 = b_0 + b_1 w, \quad (3)$$

then their PRODUCT is

$$\begin{aligned}P &\equiv N_1 N_2\\ &= a_0 b_0 + (a_0 b_1 + a_1 b_0)w + a_1 b_1 w^2\\ &= p_0 + p_1 w + p_2 w^2.\end{aligned} \quad (4)$$

Instead of evaluating products of individual digits, now write

$$q_0 = a_0 b_0 \tag{5}$$
$$q_1 = (a_0 + a_1)(b_0 + b_1) \tag{6}$$
$$q_2 = a_1 b_1. \tag{7}$$

The key term is q_1, which can be expanded, regrouped, and written in terms of the p_j as

$$q_1 = p_1 + p_0 + p_2. \tag{8}$$

However, since $p_0 = q_0$, and $p_2 = q_2$, it immediately follows that

$$p_0 = q_0 \tag{9}$$
$$p_1 = q_1 - q_0 - q_2 \tag{10}$$
$$p_2 = q_2, \tag{11}$$

so the three "digits" of p have been evaluated using three multiplications rather than four. The technique can be generalized to multidigit numbers, with the trade-off being that more additions and subtractions are required.

Now consider four-"digit" numbers

$$N_1 = a_0 + a_1 w + a_2 w^2 + a_3 w^3, \tag{12}$$

which can be written as a two-"digit" number represented in the base w^2,

$$N_1 = (a_0 + a_1 w) + (a_2 + a_3 w) * w^2. \tag{13}$$

The "digits" in the new base are now

$$a_0' = a_0 + a_1 w \tag{14}$$
$$a_1' = a_2 + a_3 w, \tag{15}$$

and the Karatsuba algorithm can be applied to N_1 and N_2 in this form. Therefore, the Karatsuba algorithm is not restricted to multiplying two-digit numbers, but more generally expresses the multiplication of two numbers in terms of multiplications of numbers of half the size. The asymptotic speed the algorithm obtains by recursive application to the smaller required subproducts is $\mathcal{O}(n^{\lg 3})$ (Knuth 1981).

When this technique is recursively applied to multidigit numbers, a point is reached in the recursion when the overhead of additions and subtractions makes it more efficient to use the usual $\mathcal{O}(n^2)$ MULTIPLICATION algorithm to evaluate the partial products. The most efficient overall method therefore relies on a combination of Karatsuba and conventional multiplication.

see also COMPLEX MULTIPLICATION, MULTIPLICATION, STRASSEN FORMULAS

References

Borodin, A. and Munro, I. *The Computational Complexity of Algebraic and Numeric Problems.* New York: American Elsevier, 1975.

Borwein, J. M.; Borwein, P. B.; and Bailey, D. H. "Ramanujan, Modular Equations, and Approximations to Pi, or How to Compute One Billion Digits of Pi." *Amer. Math. Monthly* **96**, 201–219, 1989.

Brigham, E. O. *The Fast Fourier Transform.* Englewood Cliffs, NJ: Prentice-Hall, 1974.

Brigham, E. O. *Fast Fourier Transform and Applications.* Englewood Cliffs, NJ: Prentice-Hall, 1988.

Cook, S. A. *On the Minimum Computation Time of Functions.* Ph.D. Thesis. Cambridge, MA: Harvard University, pp. 51–77, 1966.

Hollerbach, U. "Fast Multiplication & Division of Very Large Numbers." `sci.math.research` posting, Jan. 23, 1996.

Karatsuba, A. and Ofman, Yu. "Multiplication of Many-Digital Numbers by Automatic Computers." *Doklady Akad. Nauk SSSR* **145**, 293–294, 1962. Translation in *Physics-Doklady* **7**, 595–596, 1963.

Knuth, D. E. *The Art of Computing, Vol. 2: Seminumerical Algorithms, 2nd ed.* Reading, MA: Addison-Wesley, pp. 278–286, 1981.

Schönhage, A. and Strassen, V. "Schnelle Multiplikation Grosser Zahlen." *Computing* **7**, 281–292, 1971.

Toom, A. L. "The Complexity of a Scheme of Functional Elements Simulating the Multiplication of Integers." *Dokl. Akad. Nauk SSSR* **150**, 496–498, 1963. English translation in *Soviet Mathematics* **3**, 714–716, 1963.

Zuras, D. "More on Squaring and Multiplying Large Integers." *IEEE Trans. Comput.* **43**, 899–908, 1994.

Katona's Problem

Find the minimum number $f(n)$ of subsets in a SEPARATING FAMILY for a SET of n elements, where a SEPARATING FAMILY is a SET of SUBSETS in which each pair of adjacent elements is found separated, each in one of two disjoint subsets. For example, the 26 letters of the alphabet can be separated by a family of nine:

$$\begin{matrix} (abcdefghi) & (jklmnopqr) & (stuvwxyz) \\ (abcjklstu) & (defmnovwx) & (ghipqryz) \\ (adgjmpsvy) & (behknqtwz) & (cfilorux) \end{matrix}.$$

The problem was posed by Katona (1973) and solved by C. Mao-Cheng in 1982,

$$f(n) = \min\left\{2p + 3\left\lceil \log_3\left(\frac{n}{2^p}\right)\right\rceil : p = 0, 1, 2\right\},$$

where $\lceil x \rceil$ is the CEILING FUNCTION. $f(n)$ is nondecreasing, and the values for $n = 1, 2, \ldots$ are 0, 2, 3, 4, 5, 5, 6, 6, 6, 7, ... (Sloane's A07600). The values at which $f(n)$ increases are 1, 2, 3, 4, 5, 7, 10, 13, 19, 28, 37, ... (Sloane's A007601), so $f(26) = 9$, as illustrated in the preceding example.

see also SEPARATING FAMILY

References

Honsberger, R. "Cai Mao-Cheng's Solution to Katona's Problem on Families of Separating Subsets." Ch. 18 in *Mathematical Gems III.* Washington, DC: Math. Assoc. Amer., pp. 224–239, 1985.

Katona, G. O. H. "Combinatorial Search Problem." In *A Survey of Combinatorial Theory* (Ed. J. N. Srivasta *et al.*). Amsterdam, Netherlands: North-Holland, pp. 285–308, 1973.

Sloane, N. J. A. Sequences A007600/M0456 and A007601/M0525 in "An On-Line Version of the Encyclopedia of Integer Sequences."

Kauffman Polynomial F

A semi-oriented 2-variable KNOT POLYNOMIAL defined by

$$F_L(a,z) = a^{-w(L)}\langle|L|\rangle, \tag{1}$$

where L is an oriented LINK DIAGRAM, $w(L)$ is the WRITHE of L, $|L|$ is the unoriented diagram corresponding to L, and $\langle L\rangle$ is the BRACKET POLYNOMIAL. It was developed by Kauffman by extending the BLM/HO POLYNOMIAL Q to two variables, and satisfies

$$F(1,x) = Q(x). \tag{2}$$

The Kauffman POLYNOMIAL is a generalization of the JONES POLYNOMIAL $V(t)$ since it satisfies

$$V(t) = F(-t^{-3/4}, t^{-1/4} + t^{1/4}), \tag{3}$$

but its relationship to the HOMFLY POLYNOMIAL is not well understood. In general, it has more terms than the HOMFLY POLYNOMIAL, and is therefore more powerful for discriminating KNOTS. It is a semi-oriented POLYNOMIAL because changing the orientation only changes F by a POWER of a. In particular, suppose L^* is obtained from L by reversing the orientation of component k, then

$$F_{L^*} = a^{4\lambda}F_L, \tag{4}$$

where λ is the LINKING NUMBER of k with $L-k$ (Lickorish and Millett 1988). F is unchanged by MUTATION.

$$F_{L_1+F_{L_2}} = F(L_1)F(L_2) \tag{5}$$

$$F_{L_1\cup L_2} = [(a^{-1}+a)x^{-1} - 1]F_{L_1}F_{L_2}. \tag{6}$$

M. B. Thistlethwaite has tabulated the Kauffman 2-variable POLYNOMIAL for KNOTS up to 13 crossings.

References

Lickorish, W. B. R. and Millett, B. R. "The New Polynomial Invariants of Knots and Links." *Math. Mag.* **61**, 1–23, 1988.

Stoimenow, A. "Kauffman Polynomials." http://www.informatik.hu-berlin.de/~stoimeno/ptab/k10.html.

Weisstein, E. W. "Knots and Links." http://www.astro.virginia.edu/~eww6n/math/notebooks/Knots.m.

Kauffman Polynomial X

A 1-variable KNOT POLYNOMIAL denoted X or $\mathcal{L}$.

$$\mathcal{L}_L(A) \equiv (-A^3)^{-w(L)}\langle L\rangle, \tag{1}$$

where $\langle L\rangle$ is the BRACKET POLYNOMIAL and $w(L)$ is the WRITHE of L. This POLYNOMIAL is invariant under AMBIENT ISOTOPY, and relates MIRROR IMAGES by

$$\mathcal{L}_{L^*} = \mathcal{L}_L(A^{-1}). \tag{2}$$

It is identical to the JONES POLYNOMIAL with the change of variable

$$\mathcal{L}(t^{-1/4}) = V(t). \tag{3}$$

The X POLYNOMIAL of the MIRROR IMAGE K^* is the same as for K but with A replaced by A^{-1}.

References

Kauffman, L. H. *Knots and Physics.* Singapore: World Scientific, p. 33, 1991.

Kei

The IMAGINARY PART of

$$e^{-\nu\pi i/2}K_\nu(xe^{\pi i/4}) = \operatorname{ker}_\nu(x) + i\operatorname{kei}_\nu(x).$$

see also BEI, BER, KER, KELVIN FUNCTIONS

References

Abramowitz, M. and Stegun, C. A. (Eds.). "Kelvin Functions." §9.9 in *Handbook of Mathematical Functions with Formulas, Graphs, and Mathematical Tables, 9th printing.* New York: Dover, pp. 379–381, 1972.

Keith Number

A Keith number is an n-digit INTEGER N such that if a Fibonacci-like sequence (in which each term in the sequence is the sum of the n previous terms) is formed with the first n terms taken as the decimal digits of the number N, then N itself occurs as a term in the sequence. For example, 197 is a Keith number since it generates the sequence 1, 9, 7, 17, 33, 57, 107, 197, ... (Keith). Keith numbers are also called REPFIGIT NUMBERS.

There is no known general technique for finding Keith numbers except by exhaustive search. Keith numbers are much rarer than the PRIMES, with only 52 Keith numbers with < 15 digits: 14, 19, 28, 47, 61, 75, 197, 742, 1104, 1537, 2208, 2580, 3684, 4788, 7385, 7647, 7909, ... (Sloane's A007629). In addition, three 15-digit Keith numbers are known (Keith 1994). It is not known if there are an INFINITE number of Keith numbers.

References

Esche, H. A. "Non-Decimal Replicating Fibonacci Digits." *J. Recr. Math.* **26**, 193–194, 1994.

Heleen, B. "Finding Repfigits—A New Approach." *J. Recr. Math.* **26**, 184–187, 1994.

Keith, M. "Repfigit Numbers." *J. Recr. Math.* **19**, 41–42, 1987.
Keith, M. "All Repfigit Numbers Less than 100 Billion (10^{11})." *J. Recr. Math.* **26**, 181–184, 1994.
Keith, M. "Keith Numbers." http://users.aol.com/s6sj7gt/mikekeit.htm.
Robinson, N. M. "All Known Replicating Fibonacci Digits Less than One Thousand Billion (10^{12})." *J. Recr. Math.* **26**, 188–191, 1994.
Shirriff, K. "Computing Replicating Fibonacci Digits." *J. Recr. Math.* **26**, 191–193, 1994.
Sloane, N. J. A. Sequence A007629/M4922 in "An On-Line Version of the Encyclopedia of Integer Sequences."
"Table: Repfigit Numbers (Base 10^*) Less than 10^{15}." *J. Recr. Math.* **26**, 195, 1994.

Keller's Conjecture

Keller conjectured that tiling an n-D space with n-D HYPERCUBES of equal size yields an arrangement in which at least two hypercubes have an entire $(n-1)$-D "side" in common. The CONJECTURE has been proven true for $n=1$ to 6, but disproven for $n \geq 10$.

References

Cipra, B. "If You Can't See It, Don't Believe It." *Science* **259**, 26–27, 1993.
Cipra, B. *What's Happening in the Mathematical Sciences, Vol. 1.* Providence, RI: Amer. Math. Soc., pp. 24, 1993.

Kelvin's Conjecture

What space-filling arrangement of similar polyhedral cells of equal volume has minimal surface AREA? Kelvin proposed the 14-sided TRUNCATED OCTAHEDRON. Wearie and Phelan (1994) discovered another 14-sided POLYHEDRON that has 3% less SURFACE AREA.

References

Gray, J. "Parsimonious Polyhedra." *Nature* **367**, 598–599, 1994.
Wearie, D. and Phelan, R. "A Counter-Example to Kelvin's Conjecture on Minimal Surfaces." *Philos. Mag. Let.* **69**, 107–110, 1994.

Kelvin Functions

Kelvin defined the Kelvin functions BEI and BER according to

$$J_\nu(xe^{3\pi i/4}) = \operatorname{ber}_\nu(x) + i\operatorname{bei}_\nu(x), \qquad (1)$$

where $J_\nu(s)$ is a BESSEL FUNCTION OF THE FIRST KIND, and the functions KEI and KER by

$$e^{-\nu\pi i/2}K_\nu(xe^{\pi i/4}) = \operatorname{ker}_\nu(x) + i\operatorname{kei}_\nu(x), \qquad (2)$$

where $K_\nu(x)$ is a MODIFIED BESSEL FUNCTION OF THE SECOND KIND. For the special case $\nu = 0$,

$$J_0(i\sqrt{i}\,x) = J_0(\tfrac{1}{2}\sqrt{2}(i-1)x) \equiv \operatorname{ber}(x) + i\operatorname{bei}(x). \qquad (3)$$

see also BEI, BER, KEI, KER

References

Abramowitz, M. and Stegun, C. A. (Eds.). "Kelvin Functions." §9.9 in *Handbook of Mathematical Functions with Formulas, Graphs, and Mathematical Tables, 9th printing.* New York: Dover, pp. 379–381, 1972.
Spanier, J. and Oldham, K. B. "The Kelvin Functions." Ch. 55 in *An Atlas of Functions.* Washington, DC: Hemisphere, pp. 543–554, 1987.

Kelvin Transformation

The transformation

$$v(x_1', \ldots, x_n') = \left(\frac{a}{r'}\right)^{n-2} u\left(\frac{a^2 x_1'}{r'^2}, \ldots, \frac{a^2 x_n'}{r'^2}\right),$$

where

$$r'^2 = x_1'^2 + \ldots + x_n'^2.$$

If $u(x_1, \ldots, x_n)$ is a HARMONIC FUNCTION on a DOMAIN D of $\mathbb{R}^n$ (with $n \geq 3$), then $v(x_1', \ldots, x_n')$ is HARMONIC on D'.

References

Iyanaga, S. and Kawada, Y. (Eds.). *Encyclopedic Dictionary of Mathematics.* Cambridge, MA: MIT Press, p. 623, 1980.

Kempe Linkage

A double rhomboid LINKAGE which gives rectilinear motion from circular without an inversion.

References

Rademacher, H. and Toeplitz, O. *The Enjoyment of Mathematics: Selections from Mathematics for the Amateur.* Princeton, NJ: Princeton University Press, pp. 126–127, 1957.

Kepler Conjecture

In 1611, Kepler proposed that close packing (cubic or hexagonal) is the densest possible SPHERE PACKING (has the greatest η), and this assertion is known as the Kepler conjecture. Finding the densest (not necessarily periodic) packing of spheres is known as the KEPLER PROBLEM.

A putative proof of the Kepler conjecture was put forward by W.-Y. Hsiang (Hsiang 1992, Cipra 1993), but was subsequently determined to be flawed (Conway *et al.* 1994, Hales 1994). According to J. H. Conway, nobody who has read Hsiang's proof has any doubts about its validity: it is nonsense.

see also DODECAHEDRAL CONJECTURE, KEPLER PROBLEM

References

Cipra, B. "Gaps in a Sphere Packing Proof?" *Science* **259**, 895, 1993.
Conway, J. H.; Hales, T. C.; Muder, D. J.; and Sloane, N. J. A. "On the Kepler Conjecture." *Math. Intel.* **16**, 5, Spring 1994.
Eppstein, D. "Sphere Packing and Kissing Numbers." http://www.ics.uci.edu/~eppstein/junkyard/spherepack.html.
Hales, T. C. "The Sphere Packing Problem." *J. Comput. Appl. Math.* **44**, 41–76, 1992.
Hales, T. C. "Remarks on the Density of Sphere Packings in 3 Dimensions." *Combinatori* **13**, 181–197, 1993.
Hales, T. C. "The Status of the Kepler Conjecture." *Math. Intel.* **16**, 47–58, Summer 1994.
Hales, T. C. 'The Kepler Conjecture." http://www.math.lsa.umich.edu/~hales/kepler.html.
Hsiang, W.-Y. "On Soap Bubbles and Isoperimetric Regions in Noncompact Symmetrical Spaces. 1." *Tôhoku Math. J.* **44**, 151–175, 1992.
Hsiang, W.-Y. "A Rejoinder to Hales's Article." *Math. Intel.* **17**, 35–42, Winter 1995.

Kepler's Equation

Let M be the mean anomaly and E the ECCENTRIC ANOMALY of a body orbiting on an ELLIPSE with ECCENTRICITY e, then

$$M = E - e \sin E. \tag{1}$$

For M not a multiple of π, Kepler's equation has a unique solution, but is a TRANSCENDENTAL EQUATION and so cannot be inverted and solved directly for E given an arbitrary M. However, many algorithms have been derived for solving the equation as a result of its importance in celestial mechanics.

Writing a E as a POWER SERIES in e gives

$$E = M + \sum_{n=1}^{\infty} a_n e^n, \tag{2}$$

where the coefficients are given by the LAGRANGE INVERSION THEOREM as

$$a_n = \frac{1}{2^{n-1} n!} \sum_{k=0}^{\lfloor n/2 \rfloor} (-1)^k \binom{n}{k} (n-2k)^{n-1} \sin[(n-2k)M] \tag{3}$$

(Wintner 1941, Moulton 1970, Henrici 1974, Finch). Surprisingly, this series diverges for

$$e > 0.6627434193\ldots, \tag{4}$$

a value known as the LAPLACE LIMIT. In fact, E converges as a GEOMETRIC SERIES with ratio

$$r = \frac{e}{1+\sqrt{1+e^2}} \exp(\sqrt{1+e^2}\,) \tag{5}$$

(Finch).

There is also a series solution in BESSEL FUNCTIONS OF THE FIRST KIND,

$$E = M + \sum_{n=1}^{\infty} \frac{2}{n} J_n(ne) \sin(nM). \tag{6}$$

This series converges for all $e < 1$ like a GEOMETRIC SERIES with ratio

$$r = \frac{e}{1+\sqrt{1-e^2}} \exp(\sqrt{1-e^2}\,). \tag{7}$$

The equation can also be solved by letting ψ be the ANGLE between the planet's motion and the direction PERPENDICULAR to the RADIUS VECTOR. Then

$$\tan\psi = \frac{e \sin E}{\sqrt{1-e^2}}. \tag{8}$$

Alternatively, we can define e in terms of an intermediate variable ϕ

$$e \equiv \sin\phi, \tag{9}$$

then

$$\sin[\tfrac{1}{2}(v-E)] = \sqrt{\frac{r}{p}} \sin(\tfrac{1}{2}\phi) \sin v \tag{10}$$

$$\sin[\tfrac{1}{2}(v+E)] = \sqrt{\frac{r}{p}} \cos(\tfrac{1}{2}\phi) \sin v. \tag{11}$$

Iterative methods such as the simple

$$E_{i+1} = M + e \sin E_i \tag{12}$$

with $E_0 = 0$ work well, as does NEWTON'S METHOD,

$$E_{i+1} = E_i + \frac{M + e \sin E_i - E_i}{1 - e \cos E_i}. \tag{13}$$

In solving Kepler's equation, Stieltjes required the solution to

$$e^x (x-1) = e^{-x}(x+1), \tag{14}$$

which is 1.1996678640257734... (Goursat 1959, Le Lionnais 1983).

see also ECCENTRIC ANOMALY

References

Danby, J. M. *Fundamentals of Celestial Mechanics, 2nd ed., rev. ed.* Richmond, VA: Willmann-Bell, 1988.

Dörrie, H. "The Kepler Equation." §81 in *100 Great Problems of Elementary Mathematics: Their History and Solutions.* New York: Dover, pp. 330–334, 1965.

Finch, S. "Favorite Mathematical Constants." `http://www.mathsoft.com/asolve/constant/lpc/lpc.html`.

Goldstein, H. *Classical Mechanics, 2nd ed.* Reading, MA: Addison-Wesley, pp. 101–102 and 123–124, 1980.

Goursat, E. *A Course in Mathematical Analysis, Vol. 2.* New York: Dover, p. 120, 1959.

Henrici, P. *Applied and Computational Complex Analysis, Vol. 1: Power Series–Integration–Conformal Mapping–Location of Zeros.* New York: Wiley, 1974.

Ioakimids, N. I. and Papadakis, K. E. "A New Simple Method for the Analytical Solution of Kepler's Equation." *Celest. Mech.* **35**, 305–316, 1985.

Ioakimids, N. I. and Papadakis, K. E. "A New Class of Quite Elementary Closed-Form Integrals Formulae for Roots of Nonlinear Systems." *Appl. Math. Comput.* **29**, 185–196, 1989.

Le Lionnais, F. *Les nombres remarquables.* Paris: Hermann, p. 36, 1983.

Marion, J. B. and Thornton, S. T. "Kepler's Equations." §7.8 in *Classical Dynamics of Particles & Systems, 3rd ed.* San Diego, CA: Harcourt Brace Jovanovich, pp. 261–266, 1988.

Moulton, F. R. *An Introduction to Celestial Mechanics, 2nd rev. ed.* New York: Dover, pp. 159–169, 1970.

Siewert, C. E. and Burniston, E. E. "An Exact Analytical Solution of Kepler's Equation." *Celest. Mech.* **6**, 294–304, 1972.

Wintner, A. *The Analytic Foundations of Celestial Mechanics.* Princeton, NJ: Princeton University Press, 1941.

Kepler's Folium

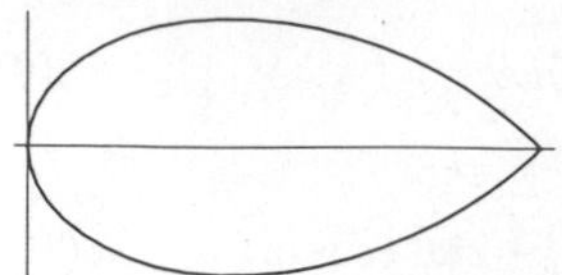

The curve with implicit equation

$$[(x-b)^2+y^2][x(x-b)+y^2]-4a(x-b)y^2.$$

References

Gray, A. *Modern Differential Geometry of Curves and Surfaces.* Boca Raton, FL: CRC Press, pp. 71–72, 1993.

Kepler-Poinsot Solid

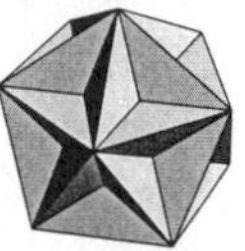

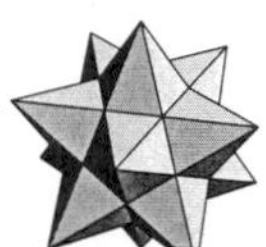

The Kepler-Poinsot solids are the four regular CONCAVE POLYHEDRA with intersecting facial planes. They are composed of regular CONCAVE POLYGONS and were unknown to the ancients. Kepler discovered two of them about 1619. These two were subsequently rediscovered by Poinsot, who also discovered the other two, in 1809. As shown by Cauchy, they are stellated forms of the DODECAHEDRON and ICOSAHEDRON.

The Kepler-Poinsot solids, illustrated above, are known as the GREAT DODECAHEDRON, GREAT ICOSAHEDRON, GREAT STELLATED DODECAHEDRON, and SMALL STELLATED DODECAHEDRON. Cauchy (1813) proved that these four exhaust all possibilities for regular star polyhedra (Ball and Coxeter 1987).

A table listing these solids, their DUALS, and COMPOUNDS is given below.

Polyhedron	Dual
great dodecahedron	small stellated dodec.
great Icosahedron	great stellated dodec.
great stellated dodec.	great icosahedron
small stellated dodec.	great dodecahedron

Polyhedron	Compound
great dodecahedron	great dodecahedron-small stellated dodec.
great icosahedron	great icosahedron-great stellated dodec.
great stellated dodec.	great icosahedron-great stellated dodec.
small stellated dodec.	great dodecahedron-small stellated dodec.

The polyhedra $\{\frac{5}{2},5\}$ and $\{5,\frac{5}{2}\}$ fail to satisfy the POLYHEDRAL FORMULA

$$V-E+F=2,$$

where V is the number of faces, E the number of edges, and F the number of faces, despite the fact that formula holds for all ordinary polyhedra (Ball and Coxeter 1987). This unexpected result led none less than Schläfli (1860) to conclude that they could not exist.

In 4-D, there are 10 Kepler-Poinsot solids, and in n-D with $n \geq 5$, there are none. In 4-D, nine of the solids have the same VERTICES as $\{3,3,5\}$, and the tenth has the same as $\{5,3,3\}$. Their SCHLÄFLI SYMBOLS are $\{\frac{5}{2},5,3\}$, $\{3,5,\frac{5}{2}\}$, $\{5,\frac{5}{2},5\}$, $\{\frac{5}{2},3,5\}$, $\{5,3,\frac{5}{2}\}$, $\{\frac{5}{2},5,\frac{5}{2}\}$, $\{5,\frac{5}{2},3\}$, $\{3,\frac{5}{2},5\}$, $\{\frac{5}{2},3,3\}$, and $\{3,3,\frac{5}{2}\}$.

Coxeter *et al.* (1954) have investigated star "Archimedean" polyhedra.

see also ARCHIMEDEAN SOLID, DELTAHEDRON, JOHNSON SOLID, PLATONIC SOLID, POLYHEDRON COMPOUND, UNIFORM POLYHEDRON

References

Ball, W. W. R. and Coxeter, H. S. M. *Mathematical Recreations and Essays, 13th ed.* New York: Dover, pp. 144–146, 1987.

Cauchy, A. L. "Recherches sur les polyèdres." *J. de l'École Polytechnique* **9**, 68–86, 1813.

Coxeter, H. S. M.; Longuet-Higgins, M. S.; and Miller, J. C. P. "Uniform Polyhedra." *Phil. Trans. Roy. Soc. London Ser. A* **246**, 401–450, 1954.

Pappas, T. "The Kepler-Poinsot Solids." *The Joy of Mathematics.* San Carlos, CA: Wide World Publ./Tetra, p. 113, 1989.

Quaisser, E. "Regular Star-Polyhedra." Ch. 5 in *Mathematical Models from the Collections of Universities and Museums* (Ed. G. Fischer). Braunschweig, Germany: Vieweg, pp. 56–62, 1986.

Schläfli. *Quart. J. Math.* **3**, 66–67, 1860.

Kepler Problem

Finding the densest not necessarily periodic SPHERE PACKING.

see also KEPLER CONJECTURE, SPHERE PACKING

Kepler Solid

see KEPLER-POINSOT SOLID

Ker

The REAL PART of

$$e^{-\nu\pi i/2}K_\nu(xe^{\pi i/4}) = \operatorname{ker}_\nu(x) + i\operatorname{kei}_\nu(x),$$

where $K_\nu(x)$ is a MODIFIED BESSEL FUNCTION OF THE SECOND KIND.

see also BEI, BER, KEI, KELVIN FUNCTIONS

References

Abramowitz, M. and Stegun, C. A. (Eds.). "Kelvin Functions." §9.9 in *Handbook of Mathematical Functions with Formulas, Graphs, and Mathematical Tables, 9th printing.* New York: Dover, pp. 379–381, 1972.

Keratoid Cusp

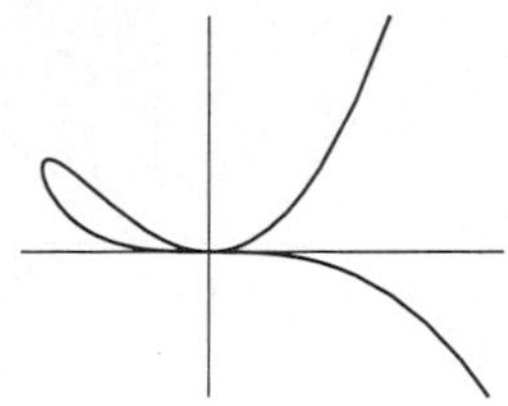

The PLANE CURVE given by the Cartesian equation

$$y^2 = x^2 y + x^5.$$

References
Cundy, H. and Rollett, A. *Mathematical Models, 3rd ed.* Stradbroke, England: Tarquin Pub., p. 72, 1989.

Kernel (Integral)

The function $K(\alpha, t)$ in an INTEGRAL or INTEGRAL TRANSFORM

$$g(\alpha) = \int_a^b f(t)K(\alpha, t)\,dt.$$

see also BERGMAN KERNEL, POISSON KERNEL

Kernel (Linear Algebra)

see NULLSPACE

Kernel Polynomial

The function

$$K_n(x_0, x) = \overline{K_n(x, x_0)} = K_n(\bar{x}, \bar{x}_0)$$

which is useful in the study of many POLYNOMIALS.

References
Szegő, G. *Orthogonal Polynomials, 4th ed.* Providence, RI: Amer. Math. Soc., 1975.

Kervaire's Characterization Theorem

Let G be a GROUP, then there exists a piecewise linear KNOT K^{n-2} in $\mathbb{S}^n$ for $n \geq 5$ with $G = \pi_1(\mathbb{S}^n - K)$ IFF G satisfies

1. G is finitely presentable,
2. The Abelianization of G is infinite cyclic,
3. The normal closure of some single element is all of G,
4. $H_2(G) = 0$; the second homology of the group is trivial.

References
Rolfsen, D. *Knots and Links.* Wilmington, DE: Publish or Perish Press, pp. 350–351, 1976.

Ket

A CONTRAVARIANT VECTOR, denoted $|\psi\rangle$. The ket is DUAL to the COVARIANT BRA 1-VECTOR $\langle\psi|$. Taken together, the BRA and ket form an ANGLE BRACKET (bra+ket = bracket) $\langle\psi|\psi\rangle$. The ket is commonly encountered in quantum mechanics.

see also ANGLE BRACKET, BRA, BRACKET PRODUCT, CONTRAVARIANT VECTOR, COVARIANT VECTOR, DIFFERENTIAL k-FORM, ONE-FORM

Khinchin Constant

see KHINTCHINE'S CONSTANT

Khintchine's Constant

N.B. A detailed on-line essay by S. Finch was the starting point for this entry.

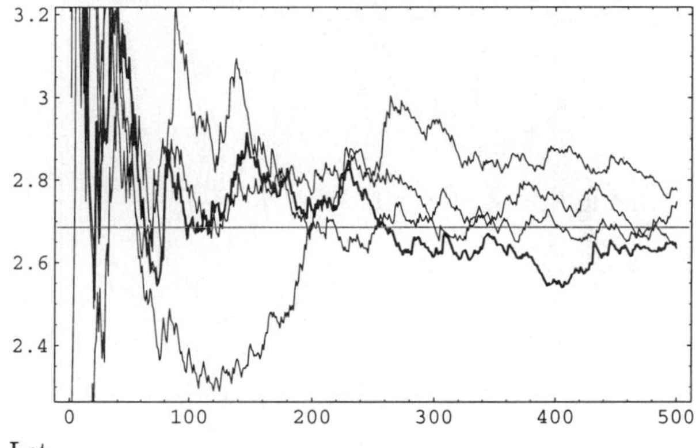

Let

$$x = [q_0, q_1, \ldots] = q_0 + \cfrac{1}{q_1 + \cfrac{1}{q_2 + \cfrac{1}{q_3 + \ldots}}} \tag{1}$$

be the SIMPLE CONTINUED FRACTION of a REAL NUMBER x, where the numbers q_i are called PARTIAL QUOTIENTS. Khintchine (1934) considered the limit of the GEOMETRIC MEAN

$$G_n(x) = (q_1 q_2 \cdots q_n)^{1/n} \tag{2}$$

as $n \to \infty$. Amazingly enough, this limit is a constant *independent* of x—except if x belongs to a set of MEASURE 0–given by

$$K = 2.685452001\ldots \tag{3}$$

(Sloane's A002210), as proved in Kac (1959). The values $G_n(x)$ are plotted above for $n = 1$ to 500 and $x = \pi$, $1/\pi$, $\sin 1$, the EULER-MASCHERONI CONSTANT γ, and the COPELAND-ERDŐS CONSTANT. REAL NUMBERS x for which $\lim_{n\to\infty} G_n(x) \neq K$ include $x = e$, $\sqrt{2}$, $\sqrt{3}$, and the GOLDEN RATIO ϕ, all of which have periodic PARTIAL QUOTIENTS, plotted below.

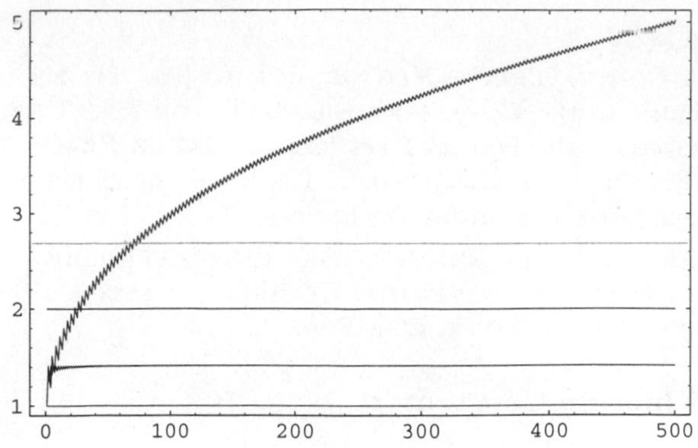

The CONTINUED FRACTION for K is [2, 1, 2, 5, 1, 1, 2, 1, 1, ...] (Sloane's A002211). It is not known if K is IRRATIONAL, let alone TRANSCENDENTAL. Bailey *et al.* (1995) have computed K to 7350 DIGITS.

Explicit expressions for K include

$$K = \prod_{n=1}^{\infty} \left[1 + \frac{1}{n(n+2)}\right]^{\ln n/\ln 2} \quad (4)$$

$$\ln 2 \ln K = \tfrac{1}{12}\pi^2 + \tfrac{1}{2}(\ln 2)^2 + \int_0^\pi \frac{\ln(\theta|\cot\theta|)\,d\theta}{\theta} \quad (5)$$

$$\ln K = \frac{1}{\ln 2} \sum_{m=1}^{\infty} \frac{h_{m-1}}{m}[\zeta(2m) - 1], \quad (6)$$

where $\zeta(z)$ is the RIEMANN ZETA FUNCTION and

$$h_m = \sum_{j=1}^{m} \frac{(-1)^{j-1}}{j} \quad (7)$$

(Shanks and Wrench 1959). Gosper gave

$$\ln K = \frac{1}{\ln 2} \sum_{j=2}^{\infty} \frac{(-1)^j (2 - 2^j)\zeta'(j)}{j}, \quad (8)$$

where $\zeta'(z)$ is the DERIVATIVE of the RIEMANN ZETA FUNCTION. An extremely rapidly converging sum also due to Gosper is

$$\begin{aligned} \ln K = \frac{1}{\ln 2} \sum_{k=0}^{\infty} \Bigg\{ & -\ln(k+1)[\ln(k+3) \\ & -2\ln(k+2) + \ln(k+1)] \\ & - \frac{(-1)^k(2-2^{k+2})}{k+2}\left[\frac{\ln(k+1)}{(k+1)^{k+2}} - \zeta'(k+2, k+2)\right] \\ & + \ln(k+1)\left[\sum_{s=1}^{k+2} \frac{(-1)^s(2-2^s)}{(k+1)^s s}\right]\Bigg\}, \end{aligned} \quad (9)$$

where $\zeta(s, a)$ is the HURWITZ ZETA FUNCTION.

Khintchine's constant is also given by the integral

$$\ln 2 \ln(\tfrac{1}{2}K) = \int_0^1 \frac{1}{x(1+x)} \ln\left[\frac{\pi x(1-x^2)}{\sin(\pi x)}\right] dx. \quad (10)$$

If P_n/Q_n is the nth CONVERGENT of the CONTINUED FRACTION of x, then

$$\lim_{n\to\infty} (Q_n)^{1/n} = \lim_{n\to\infty} \left(\frac{P_n}{x}\right)^{1/n} = e^{\pi^2/(12\ln 2)} \approx 3.27582 \quad (11)$$

for almost all REAL x (Lévy 1936, Finch). This number is sometimes called the LÉVY CONSTANT, and the argument of the exponential is sometimes called the KHINTCHINE-LÉVY CONSTANT.

Define the following quantity in terms of the kth partial quotient q_k,

$$M(s, n, x) = \left(\frac{1}{n}\sum_{k=1}^{n} {q_k}^s\right)^{1/s}. \quad (12)$$

Then

$$\lim_{n\to\infty} M(1, n, x) = \infty \quad (13)$$

for almost all real x (Khintchine, Knuth 1981, Finch), and

$$M(1, n, x) \sim \mathcal{O}(\ln n). \quad (14)$$

Furthermore, for $s < 1$, the limiting value

$$\lim_{n\to\infty} M(s, n, x) = K(s) \quad (15)$$

exists and is a constant $K(s)$ with probability 1 (Rockett and Szüsz 1992, Khintchine 1997).

see also CONTINUED FRACTION, CONVERGENT, KHINTCHINE-LÉVY CONSTANT, LÉVY CONSTANT, PARTIAL QUOTIENT, SIMPLE CONTINUED FRACTION

References

Bailey, D. H.; Borwein, J. M.; and Crandall, R. E. "On the Khintchine Constant." *Math. Comput.* **66**, 417–431, 1997.

Finch, S. "Favorite Mathematical Constants." `http://www.mathsoft.com/asolve/constant/khntchn/khntchn.html`.

Kac, M. *Statistical Independence and Probability, Analysts and Number Theory.* Providence, RI: Math. Assoc. Amer., 1959.

Khinchin, A. Ya. *Continued Fractions.* New York: Dover, 1997.

Knuth, D. E. Exercise 24 in *The Art of Computer Programming, Vol. 2: Seminumerical Algorithms, 2nd ed.* Reading, MA: Addison-Wesley, p. 604, 1981.

Le Lionnais, F. *Les nombres remarquables.* Paris: Hermann, p. 46, 1983.

Lehmer, D. H. "Note on an Absolute Constant of Khintchine." *Amer. Math. Monthly* **46**, 148–152, 1939.

Phillipp, W. "Some Metrical Theorems in Number Theory." *Pacific J. Math.* **20**, 109–127, 1967.

Plouffe, S. "Plouffe's Inverter: Table of Current Records for the Computation of Constants." `http://lacim.uqam.ca/pi/records.html`.

Rockett, A. M. and Szüsz, P. *Continued Fractions.* Singapore: World Scientific, 1992.
Shanks, D. and Wrench, J. W. "Khintchine's Constant." *Amer. Math. Monthly* **66**, 148–152, 1959.
Sloane, N. J. A. Sequences A002210/M1564 and A002211/M0118 in "An On-Line Version of the Encyclopedia of Integer Sequences."
Vardi, I. "Khinchin's Constant." §8.4 in *Computational Recreations in Mathematica.* Reading, MA: Addison-Wesley, pp. 163–171, 1991.
Wrench, J. W. "Further Evaluation of Khintchine's Constant." *Math. Comput.* **14**, 370–371, 1960.

Khintchine-Lévy Constant

A constant related to KHINTCHINE'S CONSTANT defined by

$$KL \equiv \frac{\pi^2}{12\ln 2} = 1.1865691104\ldots.$$

see also KHINTCHINE'S CONSTANT, LÉVY CONSTANT

References

Plouffe, S. "Khintchine-Levy Constant." `http://lacim.uqam.ca/piDATA/klevy.txt`.

Khovanski's Theorem

If $f_1, \ldots, f_m : \mathbb{R}^n \to \mathbb{R}$ are exponential polynomials, then $\{x \in \mathbb{R}^n : f_1(x) = \cdots f_n(x) = 0\}$ has finitely many connected components.

References

Marker, D. "Model Theory and Exponentiation." *Not. Amer. Math. Soc.* **43**, 753–759, 1996.

Kiepert's Conics

see KIEPERT'S HYPERBOLA, KIEPERT'S PARABOLA

Kiepert's Hyperbola

A curve which is related to the solution of LEMOINE'S PROBLEM and its generalization to ISOSCELES TRIANGLES constructed on the sides of a given TRIANGLE. The VERTICES of the constructed TRIANGLES are

$$A' = -\sin\phi : \sin(C+\phi) : \sin(B+\phi) \quad (1)$$
$$B' = \sin(\mathbf{C}+\phi) : -\sin\phi : \sin(A+\phi) \quad (2)$$
$$C' = \sin(B+\phi) : \sin(A+\phi) : -\sin\phi, \quad (3)$$

where ϕ is the base ANGLE of the ISOSCELES TRIANGLE. Kiepert showed that the lines connecting the VERTICES of the given TRIANGLE and the corresponding peaks of the ISOSCELES TRIANGLES CONCUR. The TRILINEAR COORDINATES of the point of concurrence are

$$\sin(B+\phi)\sin(C+\phi) : \sin(C+\phi)\sin(A+\phi) : \sin(A+\phi)\sin(B+\phi). \quad (4)$$

The locus of this point as the base ANGLE varies is given by the curve

$$\frac{\sin(B-C)}{\alpha} + \frac{\sin(C-A)}{\beta} + \frac{\sin(A-B)}{\gamma} = \frac{bc(c^2-c^2)}{\alpha} + \frac{ca(c^2-a^2)}{\beta} + \frac{ab(a^2-b^2)}{\gamma} = 0. \quad (5)$$

Writing the TRILINEAR COORDINATES as

$$\alpha_i = d_i s_i, \quad (6)$$

where d_i is the distance to the side opposite α_i of length s_i and using the POINT-LINE DISTANCE FORMULA with (x_0, y_0) written as (x, y),

$$d_i = \frac{|(y_{i+2}-y_{i+1})(x-x_{i+1})}{s_i} - \frac{(x_{i+2}-x_{i+1})(y-y_{i+1})|}{s_i}, \quad (7)$$

where $y_4 \equiv y_1$ and $y_5 \equiv y_2$ gives the FORMULA

$$\sum_{i=1}^{3} s_{i+1}s_{i+2}(s_{i+1}^2 - s_{i+2}^2) \times \frac{s_i}{(y_{i+2}-y_{i+1})(x-x_{i+1}) - (x_{i+2}-x_{i+1})(y-y_{i+1})} = 0 \quad (8)$$

$$\sum_{i=1}^{3} \frac{(s_{i+1}^2 - s_{i+2}^2)}{(y_{i+2}-y_{i+1})(x-x_{i+1}) - (x_{i+2}-x_{i+1})(y-y_{i+1})} = 0. \quad (9)$$

Bringing this equation over a common DENOMINATOR then gives a quadratic in x and y, which is a CONIC SECTION (in fact, a HYPERBOLA). The curve can also be written as $\csc(A+t) : \csc(B+t) : \csc(C+t)$, as t varies over $[-\pi/4, \pi/4]$.

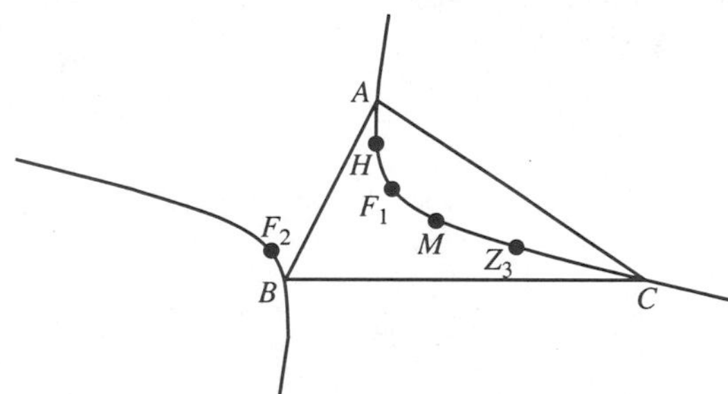

Kiepert's hyperbola passes through the triangle's CENTROID M ($\phi = 0$), ORTHOCENTER H ($\phi = \pi/2$), VERTICES A ($\phi = -\alpha$ if $\alpha \leq \pi/2$ and $\phi = \pi - \alpha$ if $\alpha > \pi/2$), B ($\phi = -\beta$), C ($\phi = -\gamma$), FERMAT POINT F_1 ($\phi = \pi/3$), second ISOGONIC CENTER F_2 ($\phi = -\pi/3$), ISOGONAL CONJUGATE of the BROCARD MIDPOINT ($\phi = \omega$), and BROCARD'S THIRD POINT Z_3 ($\phi = \omega$), where ω is the BROCARD ANGLE (Eddy and Fritsch 1994, p. 193).

The ASYMPTOTES of Kiepert's hyperbola are the SIMSON LINES of the intersections of the BROCARD AXIS with the CIRCUMCIRCLE. Kiepert's hyperbola is a RECTANGULAR HYPERBOLA. In fact, all nondegenerate conics through the VERTICES and ORTHOCENTER of a TRIANGLE are RECTANGULAR HYPERBOLAS the centers of which lie halfway between the ISOGONIC CENTERS and on the NINE-POINT CIRCLE. The LOCUS of centers of these HYPERBOLAS is the NINE-POINT CIRCLE.

The ISOGONAL CONJUGATE curve of Kiepert's hyperbola is the BROCARD AXIS. The center of the INCIRCLE of the TRIANGLE constructed from the MIDPOINTS of the sides of a given TRIANGLE lies on Kiepert's hyperbola of the original TRIANGLE.

see also BROCARD ANGLE, BROCARD AXIS, BROCARD POINTS, CENTROID (TRIANGLE), CIRCUMCIRCLE, ISOGONAL CONJUGATE, ISOGONIC CENTERS, ISOSCELES TRIANGLE, LEMOINE'S PROBLEM, NINE-POINT CIRCLE, ORTHOCENTER, SIMSON LINE

References

Casey, J. *A Treatise on the Analytical Geometry of the Point, Line, Circle, and Conic Sections, Containing an Account of Its Most Recent Extensions with Numerous Examples, 2nd rev. enl. ed.* Dublin: Hodges, Figgis, & Co., 1893.

Eddy, R. H. and Fritsch, R. "The Conics of Ludwig Kiepert: A Comprehensive Lesson in the Geometry of the Triangle." *Math. Mag.* **67**, 188–205, 1994.

Kelly, P. J. and Merriell, D. "Concentric Polygons." *Amer. Math. Monthly* **71**, 37–41, 1964.

Mineuer, A. "Sur les asymptotes de l'hyperbole de Kiepert." *Mathesis* **49**, 30–33, 1935.

Rigby, J. F. "A Concentrated Dose of Old-Fashioned Geometry." *Math. Gaz.* **57**, 296–298, 1953.

Vandeghen, A. "Some Remarks on the Isogonal and Cevian Transforms. Alignments of Remarkable Points of a Triangle." *Amer. Math. Monthly* **72**, 1091–1094, 1965.

Kiepert's Parabola

Let three similar ISOSCELES TRIANGLES $\Delta A'BC$, $\Delta AB'C$, and $\Delta ABC'$ be constructed on the sides of a TRIANGLE ΔABC. Then the ENVELOPE of the axis of the TRIANGLES ΔABC and $\Delta A'B'C'$ is Kiepert's parabola, given by

$$\frac{\sin A(\sin^2 B-\sin^2 C)}{u}+\frac{\sin B(\sin^2 C-\sin^2 A)}{v}+\frac{\sin C(\sin^2 A-\sin^2 B)}{w}=0 \quad (1)$$

$$\frac{a(b^2-c^2)}{u}+\frac{b(c^2-a^2)}{v}+\frac{c(a^2-b^2)}{w}=0, \quad (2)$$

where $[u,v,w]$ are the TRILINEAR COORDINATES for a line tangent to the parabola. It is tangent to the sides of the TRIANGLE, the line at infinity, and the LEMOINE LINE. The FOCUS has TRIANGLE CENTER FUNCTION

$$\alpha=\csc(B-C). \quad (3)$$

The EULER LINE of a triangle is the DIRECTRIX of Kiepert's parabola. In fact, the DIRECTRICES of all parabolas inscribed in a TRIANGLE pass through the ORTHOCENTER. The BRIANCHON POINT for Kiepert's parabola is the STEINER POINT.

see also BRIANCHON POINT, ENVELOPE, EULER LINE, ISOSCELES TRIANGLE, LEMOINE LINE, STEINER POINTS

Kieroid

Let the center B of a CIRCLE of RADIUS a move along a line BA. Let O be a fixed point located a distance c away from AB. Draw a SECANT LINE through O and D, the MIDPOINT of the chord cut from the line DE (which is parallel to AB) and a distance b away. Then the LOCUS of the points of intersection of OD and the CIRCLE P_1 and P_2 is called a kieroid.

Special Case	Curve
$b=0$	conchoid of Nicomedes
$b=a$	cissoid plus asymptote
$b=a=-c$	strophoid plus asymptote

References

Yates, R. C. "Kieroid." *A Handbook on Curves and Their Properties.* Ann Arbor, MI: J. W. Edwards, pp. 141–142, 1952.

Killing's Equation

The equation defining KILLING VECTORS.

$$\mathcal{L}_X g_{ab}=X_{a;b}+X_{b;a}=2X_{(a;b)}=0,$$

where $\mathcal{L}$ is the LIE DERIVATIVE.

see also KILLING VECTORS

Killing Vectors

If any set of points is displaced by $X^i dx_i$ where all distance relationships are unchanged (i.e., there is an ISOMETRY), then the VECTOR field is called a Killing vector.

$$g_{ab}=\frac{\partial x'^c}{\partial x^a}\frac{\partial x'^d}{\partial x^b}g_{cd}(x'), \quad (1)$$

so let

$$x'^a=x^a+\epsilon x^a$$

$$\frac{\partial x'^a}{\partial x^b}=\delta^a_b+\epsilon x^a{}_{,b} \quad (2)$$

$$\begin{aligned}g_{ab}(x)&=(\delta^c_a+\epsilon x^c{}_{,a})\left(\delta^d_b+\epsilon x^d{}_{,b}\right)g_{cd}(x^e+\epsilon X^e)\\&=(\delta^c_a+\epsilon x^c{}_{,a})\left(\delta^d_b+\epsilon x^d{}_{,b}\right)[g_{cd}(x)+\epsilon X^e g_{cd}(x)_{,e}+\ldots]\\&=g_{ab}(x)+\epsilon[g_{ad}X^d{}_{,b}+g_{bd}X^d{}_{,a}+X^e g_{ab,e}]+\mathcal{O}(\epsilon^2)\\&=\mathcal{L}_X g_{ab},\end{aligned} \quad (3)$$

where $\mathcal{L}$ is the LIE DERIVATIVE. An ordinary derivative can be replaced with a covariant derivative in a LIE DERIVATIVE, so we can take as the definition

$$g_{ab;c}=0 \quad (4)$$

$$g_{ab}g^{bc}=\delta^c_a, \quad (5)$$

which gives KILLING'S EQUATION

$$\mathcal{L}_X g_{ab}=X_{a;b}+X_{b;a}=2X_{(a;b)}=0. \quad (6)$$

A Killing vector X^b satisfies

$$g^{bc}X_{c;ab} - R_{ab}X^b = 0 \tag{7}$$

$$X_{a;bc} = R_{abcd}X^d \tag{8}$$

$$X^{a;b}{}_{;b} + R^a_c X^c = 0, \tag{9}$$

where R_{ab} is the RICCI TENSOR and R_{abcd} is the RIEMANN TENSOR.

A 2-sphere with METRIC

$$ds^2 = d\theta^2 + \sin^2\theta\, d\phi^2 \tag{10}$$

has three Killing vectors, given by the angular momentum operators

$$\tilde{L}_x = -\cos\phi\frac{\partial}{\partial\theta} + \cot\theta\sin\phi\frac{\partial}{\partial\phi} \tag{11}$$

$$\tilde{L}_y = \sin\phi\frac{\partial}{\partial\theta} + \cot\theta\cos\phi\frac{\partial}{\partial\phi} \tag{12}$$

$$\tilde{L}_z = \frac{\partial}{\partial\phi}. \tag{13}$$

The Killing vectors in Euclidean 3-space are

$$x^1 = \frac{\partial}{\partial x} \tag{14}$$

$$x^2 = \frac{\partial}{\partial y} \tag{15}$$

$$x^3 = \frac{\partial}{\partial z} \tag{16}$$

$$x^4 = y\frac{\partial}{\partial z} - z\frac{\partial}{\partial y} \tag{17}$$

$$x^5 = z\frac{\partial}{\partial x} - x\frac{\partial}{\partial z} \tag{18}$$

$$x^6 = x\frac{\partial}{\partial y} - y\frac{\partial}{\partial x}. \tag{19}$$

In MINKOWSKI SPACE, there are 10 Killing vectors

$$X_i^\mu = a_i{}^\mu \quad \text{for } i = 1,2,3,4 \tag{20}$$

$$X_k^0 = 0 \tag{21}$$

$$X_k^l = \epsilon^{lkm}x_m \quad \text{for } k = 1,2,3 \tag{22}$$

$$X_\mu^k = \delta_\mu{}^{[0_x k]} \quad \text{for } k = 1,2,3. \tag{23}$$

The first group is TRANSLATION, the second ROTATION, and the final corresponds to a "boost."

Kimberling Sequence

A sequence generated by beginning with the POSITIVE integers, then iteratively applying the following algorithm:

1. In iteration i, discard the ith element,
2. Alternately write the $i+k$ and $i-k$th elements until $k = i$,
3. Write the remaining elements in order.

The first few iterations are therefore

$$\begin{array}{ccccccccccc}
\boxed{1} & 2 & 3 & 4 & 5 & 6 & 7 & 8 & 9 & 10 & 11 \\
2 & \boxed{3} & 4 & 5 & 6 & 7 & 8 & 9 & 10 & 11 & 12 \\
4 & 2 & \boxed{5} & 6 & 7 & 8 & 9 & 10 & 11 & 12 & 13. \\
6 & 2 & 7 & \boxed{4} & 8 & 9 & 10 & 11 & 12 & 13 & 14 \\
8 & 7 & 9 & 2 & \boxed{10} & 6 & 11 & 12 & 13 & 14 & 15
\end{array}$$

The diagonal elements form the sequence 1, 3, 5, 4, 10, 7, 15, ... (Sloane's A007063).

References

Guy, R. K. "The Kimberling Shuffle." §E35 in *Unsolved Problems in Number Theory, 2nd ed.* New York: Springer-Verlag, pp. 235–236, 1994.

Kimberling, C. "Problem 1615." *Crux Math.* **17**, 44, 1991.

Sloane, N. J. A. Sequence A007063/M2387 in "An On-Line Version of the Encyclopedia of Integer Sequences."

Kimberling Shuffle

see also KIMBERLING SEQUENCE

Kings Problem

Kg		Kg		Kg		Kg	
Kg		Kg		Kg		Kg	
Kg		Kg		Kg		Kg	
Kg		Kg		Kg		Kg	

The problem of determining how many nonattacking kings can be placed on an $n \times n$ CHESSBOARD. For $n = 8$, the solution is 16, as illustrated above (Madachy 1979). In general, the solutions are

$$K(n) = \begin{cases} \frac{1}{4}n^2 & n \text{ even} \\ \frac{1}{4}(n+1)^2 & n \text{ odd} \end{cases} \tag{1}$$

(Madachy 1979), giving the sequence of doubled squares 1, 1, 4, 4, 9, 9, 16, 16, ... (Sloane's A008794). This sequence has GENERATING FUNCTION

$$\frac{1+x^2}{(1-x^2)^2(1-x)} = 1 + x + 4x^2 + 4x^3 + 9x^4 + 9x^5 + \ldots. \tag{2}$$

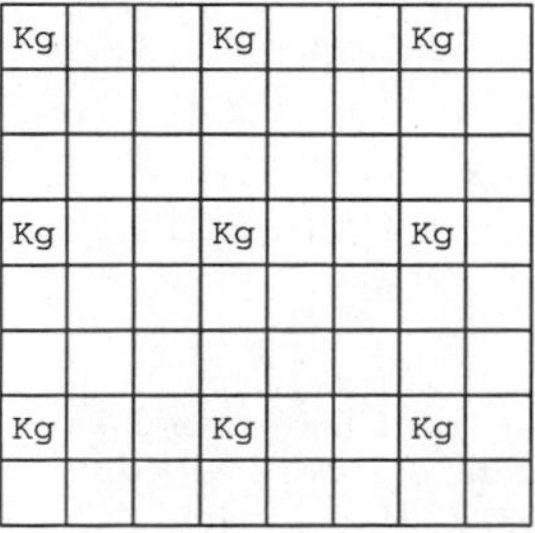

The minimum number of kings needed to attack or occupy all squares on an 8×8 CHESSBOARD is nine, illustrated above (Madachy 1979).

see also BISHOPS PROBLEM, CHESS, HARD HEXAGON ENTROPY CONSTANT, KNIGHTS PROBLEM, QUEENS PROBLEM, ROOKS PROBLEM

References
Madachy, J. S. *Madachy's Mathematical Recreations.* New York: Dover, p. 39, 1979.

King Walk

see DELANNOY NUMBER

Kinney's Set

A set of plane MEASURE 0 that contains a CIRCLE of every RADIUS.

References
Falconer, K. J. *The Geometry of Fractal Sets.* New York: Cambridge University Press, 1985.
Fejzić, H. "On Thin Sets of Circles." *Amer. Math. Monthly* **103**, 582–585, 1996.
Kinney, J. R. "A Thin Set of Circles." *Amer. Math. Monthly* **75**, 1077–1081, 1968.

Kinoshita-Terasaka Knot

The KNOT with BRAID WORD

$$\sigma_1{}^3\sigma_3{}^2\sigma_2\sigma_3{}^{-1}\sigma_1{}^{-2}\sigma_2\sigma_1{}^{-1}\sigma_3{}^{-1}\sigma_2{}^{-1}.$$

Its JONES POLYNOMIAL is

$$t^{-4}(-1+2t-2t^2+2t^3+t^6-2t^7+2t^8-2t^9+t^{10}),$$

the *same* as for CONWAY'S KNOT. It has the same ALEXANDER POLYNOMIAL as the UNKNOT.

References
Kinoshita, S. and Terasaka, H. "On Unions of Knots." *Osaka Math. J.* **9**, 131–153, 1959.

Kinoshita-Terasaka Mutants

References
Adams, C. C. *The Knot Book: An Elementary Introduction to the Mathematical Theory of Knots.* New York: W. H. Freeman, pp. 49–50, 1994.

Kirby Calculus

The manipulation of DEHN SURGERY descriptions by a certain set of operations.

References
Adams, C. C. *The Knot Book: An Elementary Introduction to the Mathematical Theory of Knots.* New York: W. H. Freeman, p. 263, 1994.

Kirby's List

A list of problems in low-dimensional TOPOLOGY maintained by R. C. Kirby. The list currently runs about 380 pages.

References
Kirby, R. "Problems in Low-Dimensional Topology." `http://math.berkeley.edu/~kirby/`.

Kirkman's Schoolgirl Problem

In a boarding school there are fifteen schoolgirls who always take their daily walks in rows of threes. How can it be arranged so that each schoolgirl walks in the same row with every other schoolgirl exactly once a week? Solution of this problem is equivalent to constructing a KIRKMAN TRIPLE SYSTEM of order $n = 2$. The following table gives one of the 7 distinct (up to permutations of letters) solutions to the problem.

Sun	Mon	Tue	Wed	Thu	Fri	Sat
ABC	ADE	AFG	AHI	AJK	ALM	ANO
DHL	BIK	BHJ	BEG	CDF	BEF	BDG
EJN	CMO	CLN	BMN	CLO	CIJ	CHK
FIO	FHN	DIM	DJO	EHM	DKN	EIL
GKM	GJL	EKO	FKL	GIN	GHO	FJM

(The table of Dörrie 1965 contains a misprint in which the $a_1 = B$ and $a_2 = C$ entries for Wednesday and Thursday are written simply as a.)

see also JOSEPHUS PROBLEM, KIRKMAN TRIPLE SYSTEM, STEINER TRIPLE SYSTEM

References
Abel, R. J. R. and Furino, S. C. "Kirkman Triple Systems." §I.6.3 in *The CRC Handbook of Combinatorial Designs* (Ed. C. J. Colbourn and J. H. Dinitz). Boca Raton, FL: CRC Press, pp. 88–89, 1996.
Ball, W. W. R. and Coxeter, H. S. M. *Mathematical Recreations and Essays, 13th ed.* New York: Dover, pp. 287–289, 1987.
Dörrie, H. §5 in *100 Great Problems of Elementary Mathematics: Their History and Solutions.* New York: Dover, pp. 14–18, 1965.
Frost, A. "General Solution and Extension of the Problem of the 15 Schoolgirls." *Quart. J. Pure Applied Math.* **11**, 1871.
Kirkman, T. P. "On a Problem in COmbinatorics." *Cambridge and Dublin Math. J.* **2**, 191–204, 1847.
Kirkman, T. P. *Lady's and Gentleman's Diary.* 1850.
Kraitchik, M. §9.3.1 in *Mathematical Recreations.* New York: W. W. Norton, pp. 226–227, 1942.
Peirce, B. "Cyclic Solutions of the School-Girl Puzzle." *Astron. J.* **6**, 169–174, 1859–1861.
Ryser, H. J. *Combinatorial Mathematics.* Buffalo, NY: Math. Assoc. Amer., pp. 101–102, 1963.

Kirkman Triple System

A Kirkman triple system of order $v = 6n + 3$ is a STEINER TRIPLE SYSTEM with parallelism (Ball and Coxeter 1987), i.e., one with the following additional stipulation: the set of $b = (2n + 1)(3n + 1)$ triples is partitioned into $3n+1$ components such that each component is a $(2n + 1)$-subset of triples and each of the v elements appears exactly once in each component. The

STEINER TRIPLE SYSTEMS of order 3 and 9 are Kirkman triple systems with $n = 0$ and 1. Solution to KIRKMAN'S SCHOOLGIRL PROBLEM requires construction of a Kirkman triple system of order $n = 2$.

Ray-Chaudhuri and Wilson (1971) showed that there exists at least one Kirkman triple system for every NONNEGATIVE order n. Earlier editions of Ball and Coxeter (1987) gave constructions of Kirkman triple systems with $9 \leq v \leq 99$. For $n = 1$, there is a single unique (up to an isomorphism) solution, while there are 7 different systems for $n = 2$ (Mulder 1917, Cole 1922, Ball and Coxeter 1987).

see also STEINER TRIPLE SYSTEM

References

Abel, R. J. R. and Furino, S. C. "Kirkman Triple Systems." §I.6.3 in *The CRC Handbook of Combinatorial Designs* (Ed. C. J. Colbourn and J. H. Dinitz). Boca Raton, FL: CRC Press, pp. 88–89, 1996.

Ball, W. W. R. and Coxeter, H. S. M. *Mathematical Recreations and Essays, 13th ed.* New York: Dover, pp. 287–289, 1987.

Cole, F. N. *Bull. Amer. Math. Soc.* **28**, 435–437, 1922.

Kirkman, T. P. *Cambridge and Dublin Math. J.* **2**, 191–204, 1947.

Lindner, C. C. and Rodger, C. A. *Design Theory.* Boca Raton, FL: CRC Press, 1997.

Mulder, P. *Kirkman-Systemen.* Groningen Dissertation. Leiden, Netherlands, 1917.

Ray-Chaudhuri, D. K. and Wilson, R. M. "Solution of Kirkman's Schoolgirl Problem." *Combinatorics, Proc. Sympos. Pure Math., Univ. California, Los Angeles, Calif., 1968* **19**, 187–203, 1971.

Ryser, H. J. *Combinatorial Mathematics.* Buffalo, NY: Math. Assoc. Amer., pp. 101–102, 1963.

Kiss Surface

The QUINTIC SURFACE given by the equation

$$\tfrac{1}{2}x^5 + \tfrac{1}{2}x^4 - (y^2 + z^2) = 0.$$

References

Nordstrand, T. "Surfaces." http://www.uib.no/people/nfytn/surfaces.htm.

Kissing Circles Problem

see DESCARTES CIRCLE THEOREM, SODDY CIRCLES

Kissing Number

The number of equivalent HYPERSPHERES in n-D which can touch an equivalent HYPERSPHERE without any intersections, also sometimes called the NEWTON NUMBER, CONTACT NUMBER, COORDINATION NUMBER, or LIGANCY. Newton correctly believed that the kissing number in 3-D was 12, but the first proofs were not produced until the 19th century (Conway and Sloane 1993, p. 21) by Bender (1874), Hoppe (1874), and Günther (1875). More concise proofs were published by Schütte and van der Waerden (1953) and Leech (1956). Exact values for *lattice packings* are known for $n = 1$ to 9 and $n = 24$ (Conway and Sloane 1992, Sloane and Nebe). Odlyzko and Sloane (1979) found the exact value for 24-D.

The following table gives the largest known kissing numbers in DIMENSION D for lattice (L) and nonlattice (NL) packings (if a nonlattice packing with higher number exists). In nonlattice packings, the kissing number may vary from sphere to sphere, so the largest value is given below (Conway and Sloane 1993, p. 15). An more extensive and up-to-date tabulation is maintained by Sloane and Nebe.

D	L	NL	D	L	NL
1	2		13	≥ 918	$\geq 1,130$
2	6		14	$\geq 1,422$	$\geq 1,582$
3	12		15	$\geq 2,340$	
4	24		16	$\geq 4,320$	
5	40		17	$\geq 5,346$	
6	72		18	$\geq 7,398$	
7	126		19	$\geq 10,668$	
8	240		20	$\geq 17,400$	
9	272	≥ 306	21	$\geq 27,720$	
10	≥ 336	≥ 500	22	$\geq 49,896$	
11	≥ 438	≥ 582	23	$\geq 93,150$	
12	≥ 756	≥ 840	24	196,560	

The lattices having maximal packing numbers in 12- and 24-D have special names: the COXETER-TODD LATTICE and LEECH LATTICE, respectively. The general form of the lower bound of n-D lattice densities given by

$$\eta \geq \frac{\zeta(n)}{2^{n-1}},$$

where $\zeta(n)$ is the RIEMANN ZETA FUNCTION, is known as the MINKOWSKI-HLAWKA THEOREM.

see also COXETER-TODD LATTICE, HERMITE CONSTANTS, HYPERSPHERE PACKING, LEECH LATTICE, MINKOWSKI-HLAWKA THEOREM

References

Bender, C. "Bestimmung der grössten Anzahl gleich Kugeln, welche sich auf eine Kugel von demselben Radius, wie die übrigen, auflegen lassen." *Archiv Math. Physik (Grunert)* **56**, 302–306, 1874.

Conway, J. H. and Sloane, N. J. A. "The Kissing Number Problem" and "Bounds on Kissing Numbers." §1.2 and Ch. 13 in *Sphere Packings, Lattices, and Groups, 2nd ed.* New York: Springer-Verlag, pp. 21–24 and 337–339, 1993.

Edel, Y.; Rains, E. M.; Sloane, N. J. A. "On Kissing Numbers in Dimensions 32 to 128." *Electronic J. Combinatorics* **5**, No. 1, R22, 1–5, 1998. `http://www.combinatorics.org/Volume_5/v5i1toc.html`.
Günther, S. "Ein stereometrisches Problem." *Archiv Math. Physik* **57**, 209–215, 1875.
Hoppe, R. "Bemerkung der Redaction." *Archiv Math. Physik. (Grunert)* **56**, 307–312, 1874.
Kuperberg, G. "Average Kissing Numbers for Sphere Packings." Preprint.
Kuperberg, G. and Schramm, O. "Average Kissing Numbers for Non-Congruent Sphere Packings." *Math. Res. Let.* **1**, 339–344, 1994.
Leech, J. "The Problem of Thirteen Spheres." *Math. Gaz.* **40**, 22–23, 1956.
Odlyzko, A. M. and Sloane, N. J. A. "New Bounds on the Number of Unit Spheres that Can Touch a Unit Sphere in n Dimensions." *J. Combin. Th. A* **26**, 210–214, 1979.
Schütte, K. and van der Waerden, B. L. "Das Problem der dreizehn Kugeln." *Math. Ann.* **125**, 325–334, 1953.
Sloane, N. J. A. Sequence A001116/M1585 in "An On-Line Version of the Encyclopedia of Integer Sequences."
Sloane, N. J. A. and Nebe, G. "Table of Highest Kissing Numbers Presently Known." `http://www.research.att.com/~njas/lattices/kiss.html`.
Stewart, I. *The Problems of Mathematics, 2nd ed.* Oxford, England: Oxford University Press, pp. 82–84, 1987.

Kite

see DIAMOND, LOZENGE, PARALLELOGRAM, PENROSE TILES, QUADRILATERAL, RHOMBUS

Klarner-Rado Sequence

The thinnest sequence which contains 1, and whenever it contains x, also contains $2x$, $3x+2$, and $6x+3$: 1, 2, 4, 5, 8, 9, 10, 14, 15, 16, 17, ... (Sloane's A005658).

References

Guy, R. K. "Klarner-Rado Sequences." §E36 in *Unsolved Problems in Number Theory, 2nd ed.* New York: Springer-Verlag, p. 237, 1994.
Klarner, D. A. and Rado, R. "Linear Combinations of Sets of Consecutive Integers." *Amer. Math. Monthly* **80**, 985–989, 1973.
Sloane, N. J. A. Sequence A005658/M0969 in "An On-Line Version of the Encyclopedia of Integer Sequences."

Klarner's Theorem

An $a \times b$ RECTANGLE can be packed with $1 \times n$ strips IFF $n|a$ or $n|b$.

see also BOX-PACKING THEOREM, CONWAY PUZZLE, DE BRUIJN'S THEOREM, SLOTHOUBER-GRAATSMA PUZZLE

References

Honsberger, R. *Mathematical Gems II.* Washington, DC: Math. Assoc. Amer., p. 88, 1976.

Klein's Absolute Invariant

$$J(q) \equiv \frac{4}{27}\frac{[1-\lambda(q)+\lambda^2(q)]^3}{\lambda^2(q)[1-\lambda(q)]^2} = \frac{[E_4(q)]^3}{[E_4(q)]^3 - [E_6(q)^2]}$$

(Cohn 1994), where $q \equiv e^{i\pi t}$ is the NOME, $\lambda(q)$ is the ELLIPTIC LAMBDA FUNCTION

$$\lambda(q) \equiv k^2(q) = \left[\frac{\vartheta_2(q)}{\vartheta_3(q)}\right]^4,$$

$\vartheta_i(q)$ is a THETA FUNCTION, and the $E_i(q)$ are RAMANUJAN-EISENSTEIN SERIES. $J(t)$ is GAMMA-MODULAR.

see also ELLIPTIC LAMBDA FUNCTION, j-FUNCTION, PI, RAMANUJAN-EISENSTEIN SERIES, THETA FUNCTION

References

Borwein, J. M. and Borwein, P. B. *Pi & the AGM: A Study in Analytic Number Theory and Computational Complexity.* New York: Wiley, pp. 115 and 179, 1987.
Cohn, H. *Introduction to the Construction of Class Fields.* New York: Dover, p. 73, 1994.
Weisstein, E. W. "j-Function." `http://www.astro.virginia.edu/~eww6n/math/notebooks/jFunction.m`.

Klein-Beltrami Model

The Klein-Beltrami model of HYPERBOLIC GEOMETRY consists of an OPEN DISK in the Euclidean plane whose open chords correspond to hyperbolic lines. Two lines l and m are then considered parallel if their chords fail to intersect and are PERPENDICULAR under the following conditions,

1. If at least one of l and m is a diameter of the DISK, they are hyperbolically perpendicular IFF they are perpendicular in the Euclidean sense.
2. If neither is a diameter, l is perpendicular to m IFF the Euclidean line extending l passes through the pole of m (defined as the point of intersection of the tangents to the disk at the "endpoints" of m).

There is an isomorphism between the POINCARÉ HYPERBOLIC DISK model and the Klein-Beltrami model. Consider a Klein disk in Euclidean 3-space with a SPHERE of the same radius seated atop it, tangent at the ORIGIN. If we now project chords on the disk orthogonally upward onto the SPHERE's lower HEMISPHERE, they become arcs of CIRCLES orthogonal to the equator. If we then stereographically project the SPHERE's lower HEMISPHERE back onto the plane of the Klein disk from the north pole, the equator will map onto a disk somewhat larger than the Klein disk, and the chords of the original Klein disk will now be arcs of CIRCLES orthogonal to this larger disk. That is, they will be Poincaré lines. Now we can say that two Klein lines or angles are congruent iff their corresponding Poincaré lines and angles under this isomorphism are congruent in the sense of the Poincaré model.

see also HYPERBOLIC GEOMETRY, POINCARÉ HYPERBOLIC DISK

Klein Bottle

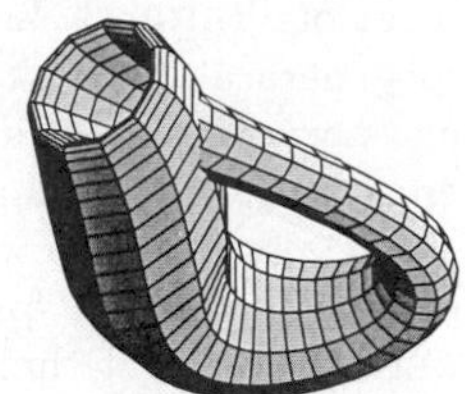

A closed NONORIENTABLE SURFACE of GENUS one having no inside or outside. It can be physically realized only in 4-D (since it must pass through itself without the presence of a HOLE). Its TOPOLOGY is equivalent to a pair of CROSS-CAPS with coinciding boundaries. It can be cut in half along its length to make two MÖBIUS STRIPS.

The above picture is an IMMERSION of the Klein bottle in $\mathbb{R}^3$ (3-space). There is also another possible IMMERSION called the "figure-8" IMMERSION (Geometry Center).

The equation for the usual IMMERSION is given by the implicit equation

$$(x^2+y^2+z^2+2y-1)[(x^2+y^2+z^2-2y-1)^2-8z^2] +16xz(x^2+y^2+z^2-2y-1)=0 \quad (1)$$

(Stewart 1991). Nordstrand gives the parametric form

$$x=\cos u[\cos(\tfrac{1}{2}u)(\sqrt{2}+\cos v)+\sin(\tfrac{1}{2}u)\sin v\cos v] \quad (2)$$

$$y=\sin u[\cos(\tfrac{1}{2}u)(\sqrt{2}+\cos v)+\sin(\tfrac{1}{2}u)\sin v\cos v] \quad (3)$$

$$z=-\sin(\tfrac{1}{2}u)(\sqrt{2}+\cos v)+\cos(\tfrac{1}{2}u)\sin v\cos v. \quad (4)$$

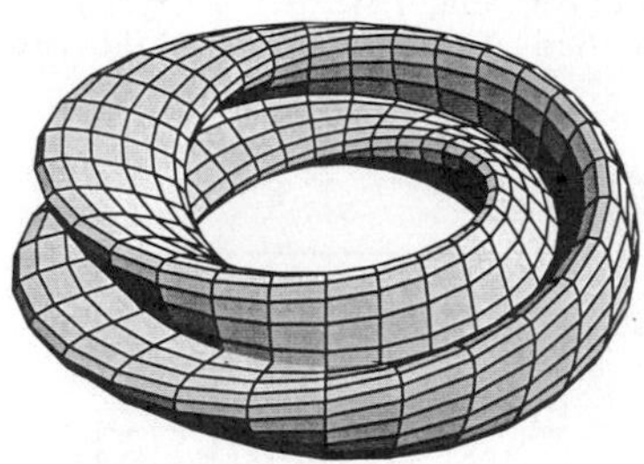

The "figure-8" form of the Klein bottle is obtained by rotating a figure eight about an axis while placing a twist in it, and is given by parametric equations

$$x(u,v)=[a+\cos(\tfrac{1}{2}u)\sin(v)-\sin(\tfrac{1}{2}u)\sin(2v)]\cos(u) \quad (5)$$

$$y(u,v)=[a+\cos(\tfrac{1}{2}u)\sin(v)-\sin(\tfrac{1}{2}u)\sin(2v)]\sin(u) \quad (6)$$

$$z(u,v)=\sin(\tfrac{1}{2}u)\sin(v)+\cos(\tfrac{1}{2}u)\sin(2v) \quad (7)$$

for $u\in[0,2\pi)$, $v\in[0,2\pi)$, and $a>2$ (Gray 1993).

The image of the CROSS-CAP map of a TORUS centered at the ORIGIN is a Klein bottle (Gray 1993, p. 249).

Any set of regions on the Klein bottle can be colored using ss colors only (Franklin 1934, Saaty 1986).

see also CROSS-CAP, ETRUSCAN VENUS SURFACE, IDA SURFACE, MAP COLORING MÖBIUS STRIP

References

Dickson, S. "Klein Bottle Graphic." `http://www.mathsource.com/cgi-bin/MathSource/Applications/Graphics/3D/0201-801`.

Franklin, P. "A Six Colour Problem." *J. Math. Phys.* **13**, 363–369, 1934.

Geometry Center. "The Klein Bottle." `http://www.geom.umn.edu/zoo/toptype/klein/`.

Geometry Center. "The Klein Bottle in Four-Space." `http://www.geom.umn.edu/~banchoff/Klein4D/Klein4D.html`.

Gray, A. "The Klein Bottle." §12.4 in *Modern Differential Geometry of Curves and Surfaces.* Boca Raton, FL: CRC Press, pp. 239–240, 1993.

Nordstrand, T. "The Famed Klein Bottle." `http://www.uib.no/people/nfytn/kleintxt.htm`.

Pappas, T. "The Moebius Strip & the Klein Bottle." *The Joy of Mathematics.* San Carlos, CA: Wide World Publ./Tetra, pp. 44–46, 1989.

Saaty, T. L. and Kainen, P. C. *The Four-Color Problem: Assaults and Conquest.* New York: Dover, p. 45, 1986.

Stewart, I. *Game, Set and Math.* New York: Viking Penguin, 1991.

Wang, P. "Renderings." `http://www.ugcs.caltech.edu/~peterw/portfolio/renderings/`.

Klein's Equation

If a REAL curve has no singularities except nodes and CUSPS, BITANGENTS, and INFLECTION POINTS, then

$$n+2\tau_2'+\iota'=m+2\delta_2'+\kappa',$$

where n is the order, τ' is the number of conjugate tangents, ι' is the number of REAL inflections, m is the class, δ' is the number of REAL conjugate points, and κ' is the number of REAL CUSPS. This is also called KLEIN'S THEOREM.

see also PLÜCKER'S EQUATION

References

Coolidge, J. L. *A Treatise on Algebraic Plane Curves.* New York: Dover, p. 114, 1959.

Klein Four-Group

see VIERGRUPPE

Klein-Gordon Equation

$$\frac{1}{c^2}\frac{\partial^2\psi}{\partial t^2}=\frac{\partial^2\psi}{\partial x^2}-\mu^2\psi.$$

see also SINE-GORDON EQUATION, WAVE EQUATION

Klein Quartic

The 3-holed TORUS.

Klein's Theorem

see KLEIN'S EQUATION

Kleinian Group

A finitely generated discontinuous group of linear fractional transformation acting on a domain in the COMPLEX PLANE.

References

Iyanaga, S. and Kawada, Y. (Eds.). *Encyclopedic Dictionary of Mathematics.* Cambridge, MA: MIT Press, p. 425, 1980.

Kra, I. *Automorphic Forms and Kleinian Groups.* Reading, MA: W. A. Benjamin, 1972.

Kloosterman's Sum

$$S(u,v,n) \equiv \sum_n \exp\left[\frac{2\pi i(uh+v\bar{h})}{n}\right], \qquad (1)$$

where h runs through a complete set of residues RELATIVELY PRIME to n, and $\bar{h}$ is defined by

$$h\bar{h} \equiv 1 \pmod{n}. \qquad (2)$$

If $(n,n')=1$ (if n and n' are RELATIVELY PRIME), then

$$S(u,v,n)S(u,v',n') = S(u, vn'^2 + v'n^2, nn'). \qquad (3)$$

Kloosterman's sum essentially solves the problem introduced by Ramanujan of representing sufficiently large numbers by QUADRATIC FORMS $ax_1{}^2 + bx_2{}^2 + cx_3{}^2 + dx_4{}^2$. Weil improved on Kloosterman's estimate for Ramanujan's problem with the best possible estimate

$$|S(u,u,n)| \leq 2\sqrt{n} \qquad (4)$$

(Duke 1997).

see also GAUSSIAN SUM

References

Duke, W. "Some Old Problems and New Results about Quadratic Forms." *Not. Amer. Math. Soc.* **44**, 190–196, 1997.

Hardy, G. H. and Wright, E. M. *An Introduction to the Theory of Numbers, 5th ed.* Oxford, England: Clarendon Press, p. 56, 1979.

Katz, N. M. *Gauss Sums, Kloosterman Sums, and Monodromy Groups.* Princeton, NJ: Princeton University Press, 1987.

Kloosterman, H. D. "On the Representation of Numbers in the Form $ax^2+by^2+cz^2+dt^2$." *Acta Math.* **49**, 407–464, 1926.

Ramanujan, S. "On the Expression of a Number in the Form $ax^2+by^2+cz^2+du^2$." *Collected Papers.* New York: Chelsea, 1962.

Knapsack Problem

Given a SUM and a set of WEIGHTS, find the WEIGHTS which were used to generate the SUM. The values of the weights are then encrypted in the sum. The system relies on the existence of a class of knapsack problems which can be solved trivially (those in which the weights are separated such that they can be "peeled off" one at a time using a GREEDY-like algorithm), and transformations which convert the trivial problem to a difficult one and vice versa. Modular multiplication is used as the TRAPDOOR FUNCTION. The simple knapsack system was broken by Shamir in 1982, the Graham-Shamir system by Adleman, and the iterated knapsack by Ernie Brickell in 1984.

References

Coppersmith, D. "Knapsack Used in Factoring." §4.6 in *Open Problems in Communication and Computation* (Ed. T. M. Cover and B. Gopinath). New York: Springer-Verlag, pp. 117–119, 1987.

Honsberger, R. *Mathematical Gems III.* Washington, DC: Math. Assoc. Amer., pp. 163–166, 1985.

Kneser-Sommerfeld Formula

Let J_ν be a BESSEL FUNCTION OF THE FIRST KIND, N_ν a NEUMANN FUNCTION, and $j_{\nu,n}$ the zeros of $z^{-\nu}J_\nu(z)$ in order of ascending REAL PART. Then for $0<x<X<1$ and $\Re[z]>0$,

$$\frac{\pi J_\nu(xz)}{4J_\nu(z)}[J_\nu(z)N_\nu(Xz) - N_\nu(z)J_\nu(Xz)] = \sum_{n=1}^\infty \frac{J_\nu(j_{\nu,n}x)J_\nu(j_{\nu,n}X)}{(z^2 - j_{\nu,n}{}^2)J'_{\nu,n}{}^2(j_{\nu,n})}.$$

References

Iyanaga, S. and Kawada, Y. (Eds.). *Encyclopedic Dictionary of Mathematics.* Cambridge, MA: MIT Press, p. 1474, 1980.

Knights Problem

Kt		Kt		Kt		Kt	
	Kt		Kt		Kt		Kt
Kt		Kt		Kt		Kt	
	Kt		Kt		Kt		Kt
Kt		Kt		Kt		Kt	
	Kt		Kt		Kt		Kt
Kt		Kt		Kt		Kt	
	Kt		Kt		Kt		Kt

The problem of determining how many nonattacking knights $K(n)$ can be placed on an $n\times n$ CHESSBOARD. For $n=8$, the solution is 32 (illustrated above). In general, the solutions are

$$K(n) = \begin{cases} \frac{1}{2}n^2 & n>2 \text{ even} \\ \frac{1}{2}(n^2+1) & n>1 \text{ odd,} \end{cases}$$

giving the sequence 1, 4, 5, 8, 13, 18, 25, ... (Sloane's A030978, Dudeney 1970, p. 96; Madachy 1979).

					Kt		
	Kt	Kt		Kt	Kt		
		Kt					
					Kt		
		Kt	Kt		Kt	Kt	
		Kt					

The minimal number of knights needed to occupy or attack every square on an $n \times n$ CHESSBOARD is given by 1, 4, 4, 4, 5, 8, 10, ... (Sloane's A006075). The number of such solutions are given by 1, 1, 2, 3, 8, 22, 3, ... (Sloane's A006076).

see also BISHOPS PROBLEM, CHESS, KINGS PROBLEM, KNIGHT'S TOUR, QUEENS PROBLEM, ROOKS PROBLEM

References

Dudeney, H. E. "The Knight-Guards." §319 in *Amusements in Mathematics.* New York: Dover, p. 95, 1970.

Madachy, J. S. *Madachy's Mathematical Recreations.* New York: Dover, pp. 38–39, 1979.

Moser, L. "King Paths on a Chessboard." *Math. Gaz.* **39**, 54, 1955.

Sloane, N. J. A. Sequences A030978, A006076/M0884, and A006075/M3224 in "An On-Line Version of the Encyclopedia of Integer Sequences."

Sloane, N. J. A. and Plouffe, S. Extended entry for M3224 in *The Encyclopedia of Integer Sequences.* San Diego: Academic Press, 1995.

Vardi, I. *Computational Recreations in Mathematica.* Redwood City, CA: Addison-Wesley, pp. 196–197, 1991.

Wilf, H. S. "The Problem of Kings." *Electronic J. Combinatorics* **2**, 3, 1–7, 1995. `http://www.combinatorics.org/Volume_2/volume2.html#3`.

Knights of the Round Table

see NECKLACE

Knight's Tour

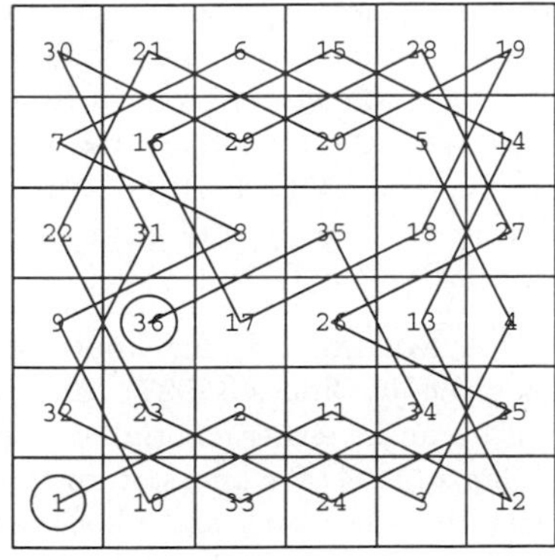

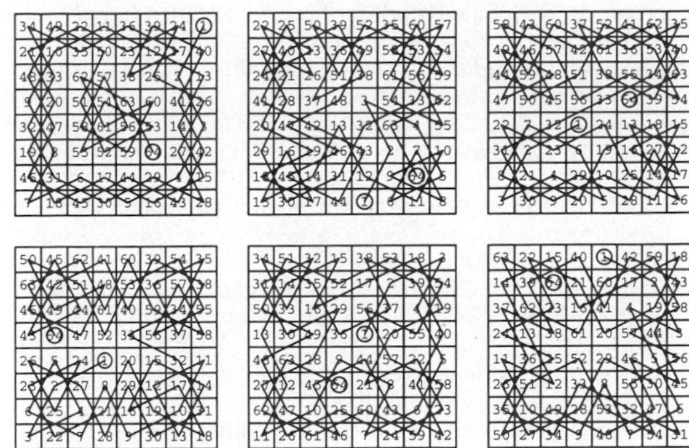

A knight's tour of a CHESSBOARD (or any other grid) is a sequence of moves by a knight CHESS piece (which may only make moves which simultaneously shift one square along one axis and two along the other) such that each square of the board is visited exactly once (i.e., a HAMILTONIAN CIRCUIT). If the final position is a knight's move away from the first position, the tour is called re-entrant. The first figure above shows a knight's tour on a 6×6 CHESSBOARD. The second set of figures shows six knight's tours on an 8×8 CHESSBOARD, all but the first of which are re-entrant. The final tour has the additional property that it is a SEMIMAGIC SQUARE with row and column sums of 260 and main diagonal sums of 348 and 168.

Löbbing and Wegener (1996) computed the number of cycles covering the directed knight's graph for an 8×8 CHESSBOARD. They obtained α^2, where $\alpha = 2{,}849{,}759{,}680$, i.e., 8,121,130,233,753,702,400. They also computed the number of undirected tours, obtaining an incorrect answer 33,439,123,484,294 (which is not divisible by 4 as it must be), and so are currently redoing the calculation.

The following results are given by Kraitchik (1942). The number of possible tours on a $4k \times 4k$ board for $k = 3$, 4, ... are 8, 0, 82, 744, 6378, 31088, 189688, 1213112, ... (Kraitchik 1942, p. 263). There are 14 tours on the 3×7 rectangle, two of which are symmetrical. There are 376 tours on the 3×8 rectangle, none of which is closed. There are 16 symmetric tours on the 3×9 rectangle and 8 closed tours on the 3×10 rectangle. There are 58 symmetric tours on the 3×11 rectangle and 28 closed tours on the 3×12 rectangle. There are five doubly symmetric tours on the 6×6 square. There are 1728 tours on the 5×5 square, 8 of which are symmetric. The longest "uncrossed" knight's tours on an $n \times n$ board for $n = 3$, 4, ... are 2, 5, 10, 17, 24, 35, ... (Sloane's A003192).

see also CHESS, KINGS PROBLEM, KNIGHTS PROBLEM, MAGIC TOUR, QUEENS PROBLEM, TOUR

References

Ball, W. W. R. and Coxeter, H. S. M. *Mathematical Recreations and Essays, 13th ed.* New York: Dover, pp. 175–186, 1987.

Chartrand, G. "The Knight's Tour." §6.2 in *Introductory Graph Theory.* New York: Dover, pp. 133–135, 1985.
Gardner, M. "Knights of the Square Table." Ch. 14 in *Mathematical Magic Show: More Puzzles, Games, Diversions, Illusions and Other Mathematical Sleight-of-Mind from Scientific American.* New York: Vintage, pp. 188–202, 1978.
Guy, R. K. "The *n* Queens Problem." §C18 in *Unsolved Problems in Number Theory, 2nd ed.* New York: Springer-Verlag, pp. 133–135, 1994.
Kraitchik, M. "The Problem of the Knights." Ch. 11 in *Mathematical Recreations.* New York: W. W. Norton, pp. 257–266, 1942.
Madachy, J. S. *Madachy's Mathematical Recreations.* New York: Dover, pp. 87–89, 1979.
Ruskey, F. "Information on the *n* Knight's Tour Problem." http://sue.csc.uvic.ca/~cos/inf/misc/Knight.html.
Sloane, N. J. A. Sequences A003192/M1369 and A006075/M3224 in "An On-Line Version of the Encyclopedia of Integer Sequences."
van der Linde, A. *Geschichte und Literatur des Schachspiels, Vol. 2.* Berlin, pp. 101–111, 1874.
Volpicelli, P. "Soluzione completa e generale, mediante la geometria di situazione, del problema relativo alle corse del cavallo sopra qualunque scacchiere." *Atti della Reale Accad. dei Lincei* **25**, 87–162, 1872.
Wegener, I. and Löbbing, M. "The Number of Knight's Tours Equals 33,439,123,484,294—Counting with Binary Decision Diagrams." *Electronic J. Combinatorics* **3**, R5, 1–4, 1996. http://www.combinatorics.org/Volume_3/volume3.html#R5.

Knödel Numbers

For every $k \geq 1$, let C_k be the set of COMPOSITE numbers $n > k$ such that if $1 < a < n$, $\text{GCD}(a, n) = 1$ (where GCD is the GREATEST COMMON DIVISOR), then $a^{n-k} \equiv 1 \pmod{n}$. C_1 is the set of CARMICHAEL NUMBERS. Makowski (1962/1963) proved that there are infinitely many members of C_k for $k \geq 2$.

see also CARMICHAEL NUMBER, D-NUMBER, GREATEST COMMON DIVISOR

References

Makowski, A. "Generalization of Morrow's D-Numbers." *Simon Stevin* **36**, 71, 1962/1963.
Ribenboim, P. *The Book of Prime Number Records, 2nd ed.* New York: Springer-Verlag, p. 101, 1989.

Knot

A knot is defined as a closed, non-self-intersecting curve embedded in 3-D. A knot is a single component LINK. Klein proved that knots cannot exist in an EVEN-numbered dimensional space ≥ 4. It has since been shown that a knot cannot exist in *any* dimension ≥ 4. Two distinct knots cannot have the same KNOT COMPLEMENT (Gordon and Luecke 1989), but two LINKS can! (Adams 1994, p. 261). The KNOT SUM of any number of knots cannot be the UNKNOT unless each knot in the sum is the UNKNOT.

Knots can be cataloged based on the minimum number of crossings present. Knots are usually further broken down into PRIME KNOTS. Knot theory was given its first impetus when Lord Kelvin proposed a theory that atoms were vortex loops, with different chemical elements consisting of different knotted configurations (Thompson 1867). P. G. Tait then cataloged possible knots by trial and error.

Thistlethwaite has used DOWKER NOTATION to enumerate the number of PRIME KNOTS of up to 13 crossings, and ALTERNATING KNOTS up to 14 crossings. In this compilation, MIRROR IMAGES are counted as a single knot type. The number of distinct PRIME KNOTS $N(n)$ for knots from $n = 3$ to 13 crossings are 1, 1, 2, 3, 7, 21, 49, 165, 552, 2176, 9988 (Sloane's A002863). Combining PRIME KNOTS gives one additional type of knot each for knots six and seven crossings.

Let $C(n)$ be the number of distinct PRIME KNOTS of n crossings, *counting* CHIRAL *versions of the same knot separately.* Then

$$\tfrac{1}{3}(2^{n-2} - 1) \leq N(n) \lesssim e^n$$

(Ernst and Summers 1987). Welsh has shown that the number of knots is bounded by an exponential in n.

A pictorial enumeration of PRIME KNOTS of up to 10 crossings appears in Rolfsen (1976, Appendix C). Note, however, that in this table, the PERKO PAIR 10_{161} and 10_{162} are actually identical, and the uppermost crossing in 10_{144} should be changed (Jones 1987). The kth knot having n crossings in this (arbitrary) ordering of knots is given the symbol n_k. Another possible representation for knots uses the BRAID GROUP. A knot with $n + 1$ crossings is a member of the BRAID GROUP n. There is no general method known for deciding whether two given knots are equivalent or interlocked. There is no general ALGORITHM to determine if a tangled curve is a knot. Haken (1961) has given an ALGORITHM, but it is too complex to apply to even simple cases.

If a knot is AMPHICHIRAL, the "amphichirality" is $A = 1$, otherwise $A = 0$ (Jones 1987). ARF INVARIANTS are designated a. BRAID WORDS are denoted b (Jones 1987). CONWAY'S KNOT NOTATION C for knots up to 10 crossings is given by Rolfsen (1976). Hyperbolic volumes are given (Adams, Hildebrand, and Weeks 1991; Adams 1994). The BRAID INDEX i is given by Jones (1987). ALEXANDER POLYNOMIALS Δ are given in Rolfsen (1976), but with the POLYNOMIALS for 10_{083} and 10_{086} reversed (Jones 1987). The ALEXANDER POLYNOMIALS are normalized according to Conway, and given in abbreviated form $[a_1, a_2, \ldots$ for $a_1 + a_2(x^{-1} + x) + \ldots$.

The JONES POLYNOMIALS W for knots of up to 10 crossings are given by Jones (1987), and the JONES POLYNOMIALS V can be either computed from these, or taken from Adams (1994) for knots of up to 9 crossings (although most POLYNOMIALS are associated with the wrong knot in the first printing). The JONES POLYNOMIALS are listed in the abbreviated form $\{n\}\, a_0\, a_1 \ldots$ for $t^{-n}(a_0 + a_1 t + \ldots)$, and correspond either to the knot depicted by Rolfsen or its MIRROR IMAGE, whichever has the lower Power of t^{-1}. The HOMFLY Polynomial $P(\ell, m)$ and Kauffman Polynomial $F(a, x)$ are given in Lickorish and Millett (1988) for knots of up to 7 crossings.

M. B. Thistlethwaite has tabulated the HOMFLY Polynomial and Kauffman Polynomial F for Knots of up to 13 crossings.

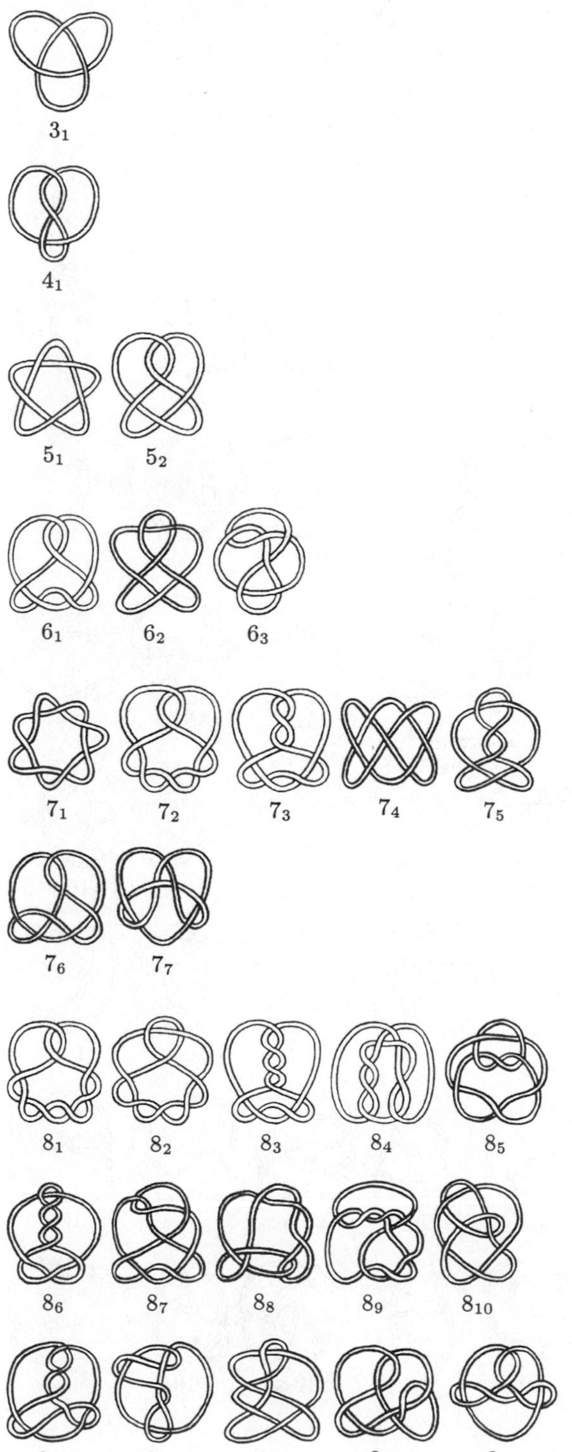

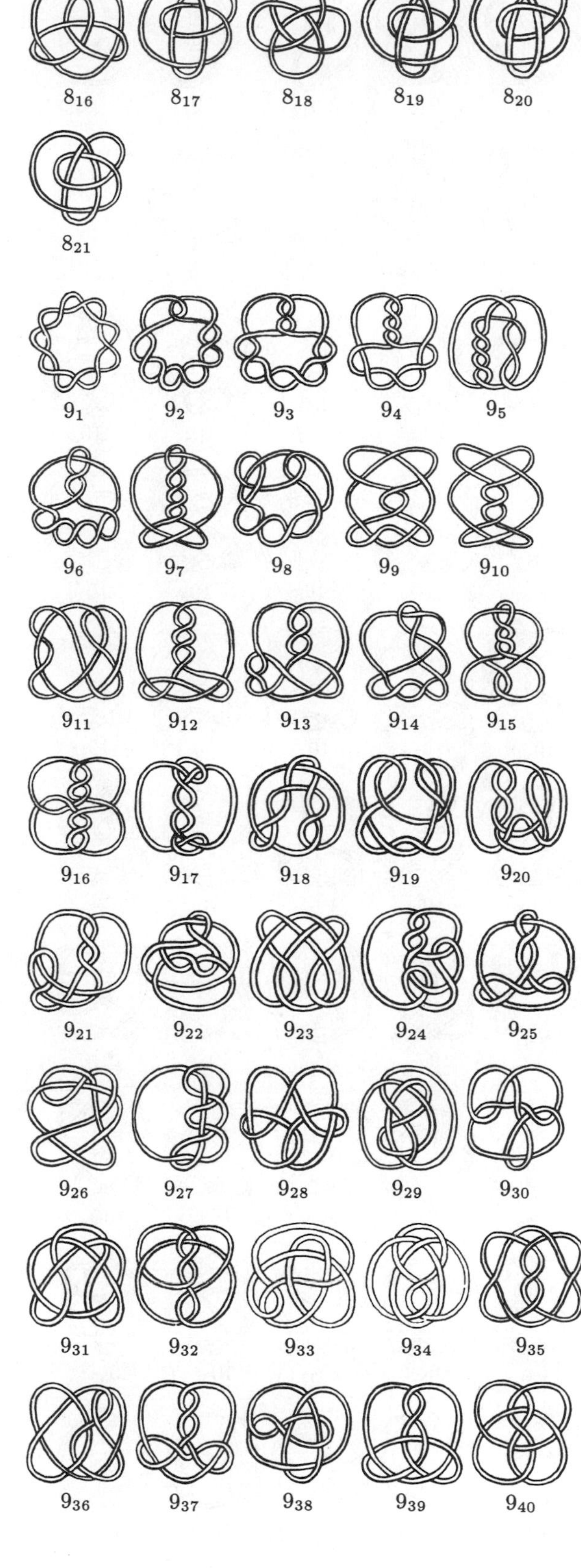

9_{41} 9_{42} 9_{43} 9_{44} 9_{45}
9_{46} 9_{47} 9_{48} 9_{49}
10_{1} 10_{2} 10_{3} 10_{4} 10_{5}
10_{6} 10_{7} 10_{8} 10_{9} 10_{10}
10_{11} 10_{12} 10_{13} 10_{14} 10_{15}
10_{16} 10_{17} 10_{18} 10_{19} 10_{20}
10_{21} 10_{22} 10_{23} 10_{24} 10_{25}
10_{26} 10_{27} 10_{28} 10_{29} 10_{30}
10_{31} 10_{32} 10_{33} 10_{34} 10_{35}
10_{36} 10_{37} 10_{38} 10_{39} 10_{40}
10_{41} 10_{42} 10_{43} 10_{44} 10_{45}
10_{46} 10_{47} 10_{48} 10_{49} 10_{50}
10_{51} 10_{52} 10_{53} 10_{54} 10_{55}
10_{56} 10_{57} 10_{58} 10_{59} 10_{60}
10_{61} 10_{62} 10_{63} 10_{64} 10_{65}
10_{66} 10_{67} 10_{68} 10_{69} 10_{70}
10_{71} 10_{72} 10_{73} 10_{74} 10_{75}
10_{76} 10_{77} 10_{78} 10_{79} 10_{80}
10_{81} 10_{82} 10_{83} 10_{84} 10_{85}
10_{86} 10_{87} 10_{88} 10_{89} 10_{90}

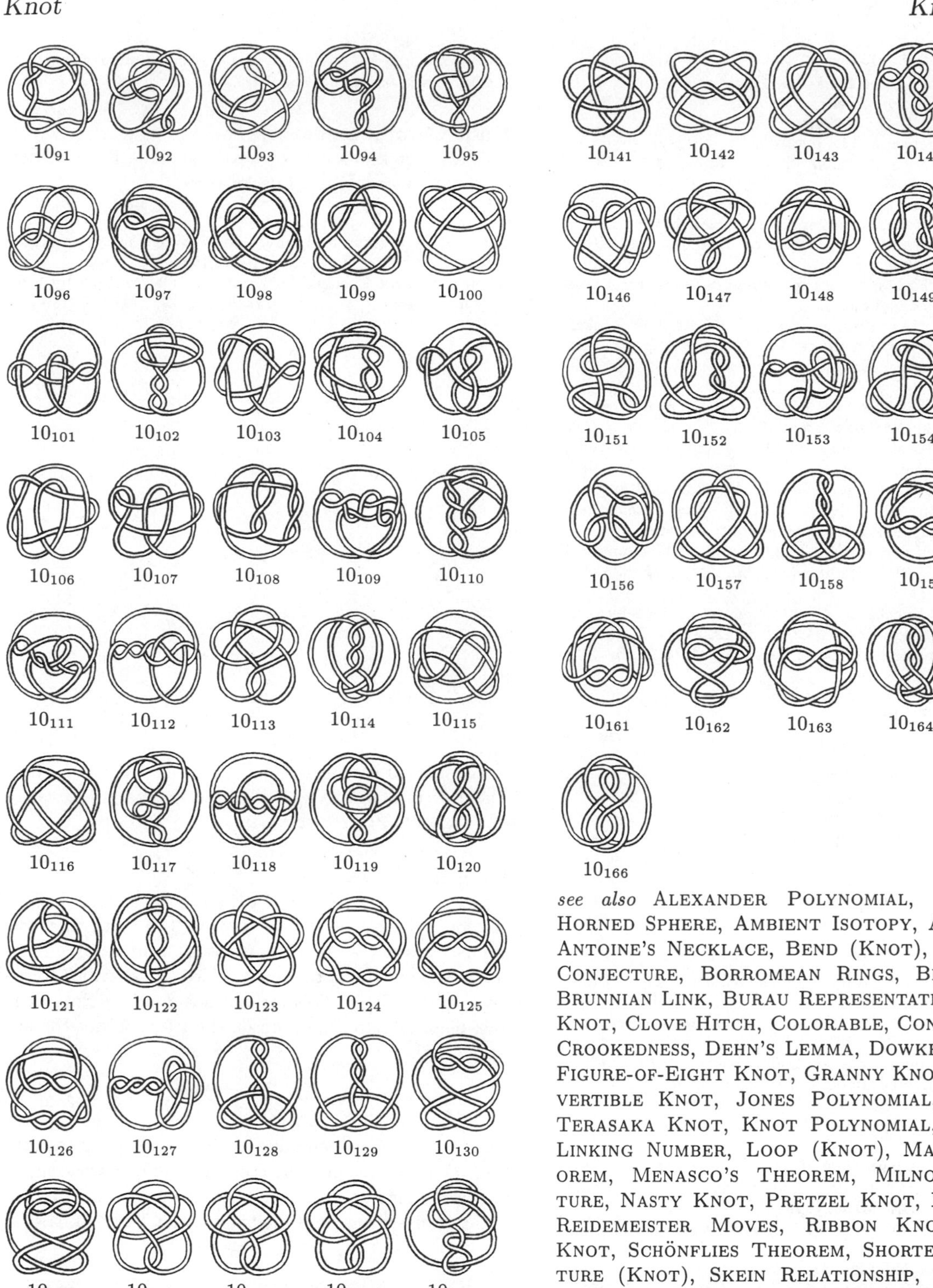

see also ALEXANDER POLYNOMIAL, ALEXANDER'S HORNED SPHERE, AMBIENT ISOTOPY, AMPHICHIRAL, ANTOINE'S NECKLACE, BEND (KNOT), BENNEQUIN'S CONJECTURE, BORROMEAN RINGS, BRAID GROUP, BRUNNIAN LINK, BURAU REPRESENTATION, CHEFALO KNOT, CLOVE HITCH, COLORABLE, CONWAY'S KNOT, CROOKEDNESS, DEHN'S LEMMA, DOWKER NOTATION, FIGURE-OF-EIGHT KNOT, GRANNY KNOT, HITCH, INVERTIBLE KNOT, JONES POLYNOMIAL, KINOSHITA-TERASAKA KNOT, KNOT POLYNOMIAL, KNOT SUM, LINKING NUMBER, LOOP (KNOT), MARKOV'S THEOREM, MENASCO'S THEOREM, MILNOR'S CONJECTURE, NASTY KNOT, PRETZEL KNOT, PRIME KNOT, REIDEMEISTER MOVES, RIBBON KNOT, RUNNING KNOT, SCHÖNFLIES THEOREM, SHORTENING, SIGNATURE (KNOT), SKEIN RELATIONSHIP, SLICE KNOT, SLIP KNOT, SMITH CONJECTURE, SOLOMON'S SEAL KNOT, SPAN (LINK), SPLITTING, SQUARE KNOT, STEVEDORE'S KNOT, STICK NUMBER, STOPPER KNOT, TAIT'S KNOT CONJECTURES, TAME KNOT, TANGLE, TORSION NUMBER, TREFOIL KNOT, UNKNOT, UNKNOTTING NUMBER, VASSILIEV POLYNOMIAL, WHITEHEAD LINK

References
Adams, C. C. *The Knot Book: An Elementary Introduction to the Mathematical Theory of Knots.* New York: W. H. Freeman, pp. 280–286, 1994.
Adams, C.; Hildebrand, M.; and Weeks, J. "Hyperbolic Invariants of Knots and Links." *Trans. Amer. Math. Soc.* **1**, 1–56, 1991.
Anderson, J. "The Knotting Dictionary of Kännet." http://www.netg.se/~jan/knopar/english/index.htm.
Ashley, C. W. *The Ashley Book of Knots.* New York: McGraw-Hill, 1996.
Bogomolny, A. "Knots...." http://www.cut-the-knot.com/do_you_know/knots.html.
Conway, J. H. "An Enumeration of Knots and Links." In *Computational Problems in Abstract Algebra* (Ed. J. Leech). Oxford, England: Pergamon Press, pp. 329–358, 1970.
Eppstein, D. "Knot Theory." http://www.ics.uci.edu/~eppstein/junkyard/knot.html.
Eppstein, D. "Knot Theory." http://www.ics.uci.edu/~eppstein/junkyard/knot/.
Erdener, K.; Candy, C.; and Wu, D. "Verification and Extension of Topological Knot Tables." ftp://chs.cusd.claremont.edu/pub/knot/FinalReport.sit.hqx.
Ernst, C. and Sumner, D. W. "The Growth of the Number of Prime Knots." *Proc. Cambridge Phil. Soc.* **102**, 303–315, 1987.
Gordon, C. and Luecke, J. "Knots are Determined by their Complements." *J. Amer. Math. Soc.* **2**, 371–415, 1989.
Haken, W. "Theorie der Normalflachen." *Acta Math.* **105**, 245–375, 1961.
Kauffman, L. *Knots and Applications.* River Edge, NJ: World Scientific, 1995.
Kauffman, L. *Knots and Physics.* Teaneck, NJ: World Scientific, 1991.
Lickorish, W. B. R. and Millett, B. R. "The New Polynomial Invariants of Knots and Links." *Math. Mag.* **61**, 1–23, 1988.
Livingston, C. *Knot Theory.* Washington, DC: Math. Assoc. Amer., 1993.
Praslov, V. V. and Sossinsky, A. B. *Knots, Links, Braids and 3-Manifolds: An Introduction to the New Invariants in Low-Dimensional Topology.* Providence, RI: Amer. Math. Soc., 1996.
Rolfsen, D. "Table of Knots and Links." Appendix C in *Knots and Links.* Wilmington, DE: Publish or Perish Press, pp. 280–287, 1976.
"Ropers Knots Page." http://huizen.dds.nl/~erpprs/kne/kroot.htm.
Sloane, N. J. A. Sequences A002863/M0851 in "An On-Line Version of the Encyclopedia of Integer Sequences."
Sloane, N. J. A. and Plouffe, S. Extended entry in *The Encyclopedia of Integer Sequences.* San Diego: Academic Press, 1995.
Stoimenow, A. "Polynomials of Knots with Up to 10 Crossings." Rev. March 16, 1998. http://www.informatik.hu-berlin.de/~stoimeno/poly.ps.
Suber, O. "Knots on the Web." http://www.earlham.edu/suber/knotlink.htm.
Tait, P. G. "On Knots I, II, and III." *Scientific Papers, Vol. 1.* Cambridge: University Press, pp. 273–347, 1898.
Thistlethwaite, M. B. "Knot Tabulations and Related Topics." In *Aspects of Topology in Memory of Hugh Dowker 1912–1982* (Ed. I. M. James and E. H. Kronheimer). Cambridge, England: Cambridge University Press, pp. 2–76, 1985.
Thistlethwaite, M. B. ftp://chs.cusd.claremont.edu/pub/knot/Thistlethwaite_Tables/.
Thompson, W. T. "On Vortex Atoms." *Philos. Mag.* **34**, 15–24, 1867.
Weisstein, E. W. "Knots." http://www.astro.virginia.edu/~eww6n/math/notebooks/Knots.m.

Knot Complement

Two distinct knots cannot have the same KNOT COMPLEMENT (Gordon and Luecke 1989).

References
Cipra, B. "To Have and Have Knot: When are Two Knots Alike?" *Science* **241**, 1291–1292, 1988.
Gordon, C. and Luecke, J. "Knots are Determined by their Complements." *J. Amer. Math. Soc.* **2**, 371–415, 1989.

Knot Curve

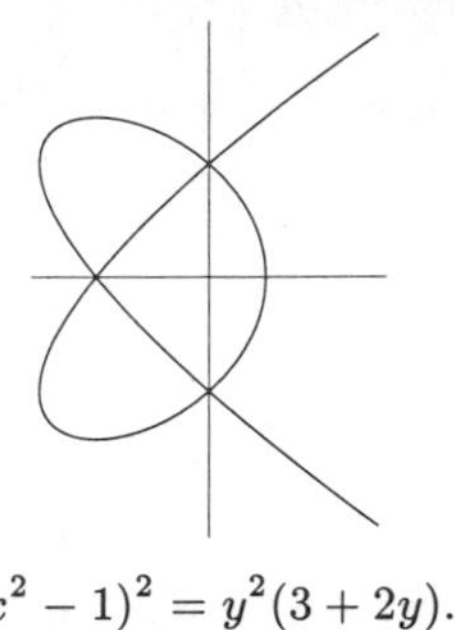

$$(x^2-1)^2 = y^2(3+2y).$$

References
Cundy, H. and Rollett, A. *Mathematical Models, 3rd ed.* Stradbroke, England: Tarquin Pub., p. 72, 1989.

Knot Determinant

The determinant of a knot is $|\Delta(-1)|$, where $\Delta(z)$ is the ALEXANDER POLYNOMIAL.

Knot Diagram

A picture of a projection of a KNOT onto a PLANE. Usually, only double points are allowed (no more than two points are allowed to be superposed), and the double or crossing points must be "genuine crossings" which transverse in the plane. This means that double points must look like the below diagram on the left, and not the one on the right.

Also, it is usually demanded that a knot diagram contain the information if the crossings are overcrossings or undercrossings so that the original knot can be reconstructed. Here is a knot diagram of the TREFOIL KNOT,

KNOT POLYNOMIALS can be computed from knot diagrams. Such POLYNOMIALS often (but not always) allow the knots corresponding to given diagrams to be uniquely identified.

Knot Exterior

The COMPLEMENT of an open solid TORUS knotted at the KNOT. The removed open solid TORUS is called a tubular NEIGHBORHOOD.

Knot Linking

In general, it is possible to link two n-D HYPERSPHERES in $(n+2)$-D space in an infinite number of inequivalent ways. In dimensions greater than $n+2$ in the piecewise linear category, it is true that these spheres are themselves unknotted. However, they may still form nontrivial links. In this way, they are something like higher dimensional analogs of two 1-spheres in 3-D. The following table gives the number of nontrivial ways that two n-D HYPERSPHERES can be linked in k-D.

D of spheres	D of space	Distinct Linkings
23	40	239
31	48	959
102	181	3
102	182	10438319
102	183	3

Two 10-D HYPERSPHERES link up in 12, 13, 14, 15, and 16-D, then unlink in 17-D, link up again in 18, 19, 20, and 21-D. The proof of these results consists of an "easy part" (Zeeman 1962) and a "hard part" (Ravenel 1986). The hard part is related to the calculation of the (stable and unstable) HOMOTOPY GROUPS of SPHERES.

References

Bing, R. H. *The Geometric Topology of 3-Manifolds.* Providence, RI: Amer. Math. Soc., 1983.

Ravenel, D. *Complex Cobordism and Stable Homotopy Groups of Spheres.* New York: Academic Press, 1986.

Rolfsen, D. *Knots and Links.* Wilmington, DE: Publish or Perish Press, p. 7, 1976.

Zeeman. "Isotopies and Knots in Manifolds." In *Topology of 3-Manifolds and Related Topics* (Ed. M. K. Fort). Englewood Cliffs, NJ: Prentice-Hall, 1962.

Knot Polynomial

A knot invariant in the form of a POLYNOMIAL such as the ALEXANDER POLYNOMIAL, BLM/HO POLYNOMIAL, BRACKET POLYNOMIAL, CONWAY POLYNOMIAL, JONES POLYNOMIAL, KAUFFMAN POLYNOMIAL F, KAUFFMAN POLYNOMIAL X, and VASSILIEV POLYNOMIAL.

References

Lickorish, W. B. R. and Millett, K. C. "The New Polynomial Invariants of Knots and Links." *Math. Mag.* **61**, 3–23, 1988.

Knot Problem

The problem of deciding if two KNOTS in 3-space are equivalent such that one can be continuously deformed into another.

Knot Shadow

A LINK DIAGRAM which does not specify whether crossings are under- or overcrossings.

Knot Sum

Two oriented knots (or links) can be summed by placing them side by side and joining them by straight bars so that orientation is preserved in the sum. This operation is denoted #, so the knot sum of knots K_1 and K_2 is written

$$K_1 \# K_2 = K_2 \# K_1.$$

see also CONNECTED SUM

Knot Theory

The mathematical study of KNOTS. Knot theory considers questions such as the following:

1. Given a tangled loop of string, is it really knotted or can it, with enough ingenuity and/or luck, be untangled without having to cut it?
2. More generally, given two tangled loops of string, when are they deformable into each other?
3. Is there an effective algorithm (or any algorithm to speak of) to make these determinations?

Although there has been almost explosive growth in the number of important results proved since the discovery of the JONES POLYNOMIAL, there are still many "knotty" problems and conjectures whose answers remain unknown.

see also KNOT, LINK

Knot Vector

see B-SPLINE

Koch Antisnowflake

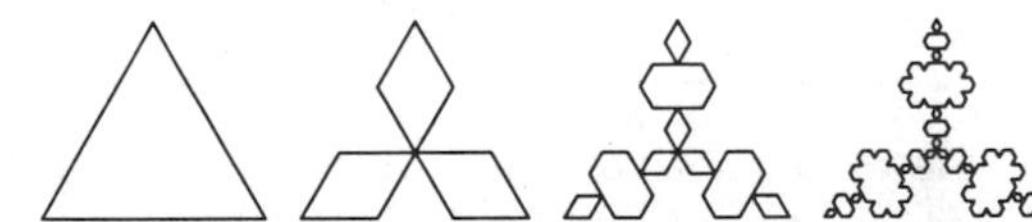

A FRACTAL derived from the KOCH SNOWFLAKE. The base curve and motif for the fractal are illustrated below.

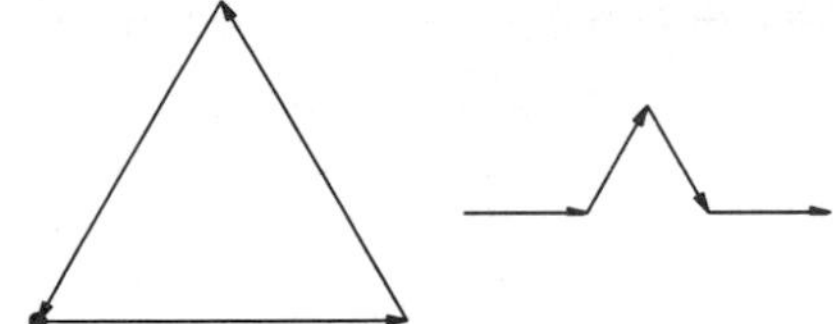

The AREA after the nth iteration is

$$A_n = A_{n-1} - \frac{1}{3}\frac{\ell_{n-1}}{a}\frac{\Delta}{3^n},$$

where Δ is the area of the original EQUILATERAL TRIANGLE, so from the derivation for the KOCH SNOWFLAKE,

$$A \equiv \lim_{n\to\infty} A_n = (1 - \tfrac{3}{5})\Delta = \tfrac{2}{5}\Delta.$$

see also EXTERIOR SNOWFLAKE, FLOWSNAKE FRACTAL, KOCH SNOWFLAKE, PENTAFLAKE, SIERPIŃSKI CURVE

References
Cundy, H. and Rollett, A. *Mathematical Models, 3rd ed.* Stradbroke, England: Tarquin Pub., pp. 66–67, 1989.
Lauwerier, H. *Fractals: Endlessly Repeated Geometric Figures.* Princeton, NJ: Princeton University Press, pp. 36–37, 1991.
Weisstein, E. W. "Fractals." http://www.astro.virginia.edu/~eww6n/math/notebooks/Fractal.m.

Koch Island

see KOCH SNOWFLAKE

Koch Snowflake

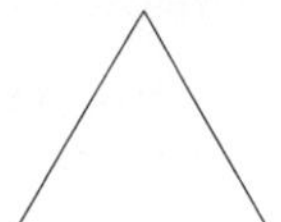  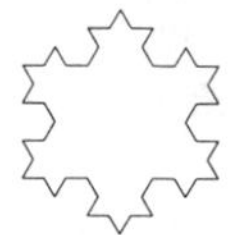 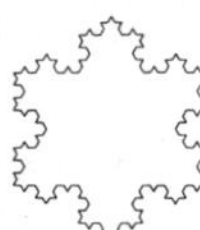

A FRACTAL, also known as the KOCH ISLAND, which was first described by Helge von Koch in 1904. It is built by starting with an EQUILATERAL TRIANGLE, removing the inner third of each side, building another EQUILATERAL TRIANGLE at the location where the side was removed, and then repeating the process indefinitely. The Koch snowflake can be simply encoded as a LINDENMAYER SYSTEM with initial string "F--F--F", STRING REWRITING rule "F" -> "F+F--F+F", and angle 60°. The zeroth through third iterations of the construction are shown above. The fractal can also be constructed using a base curve and motif, illustrated below.

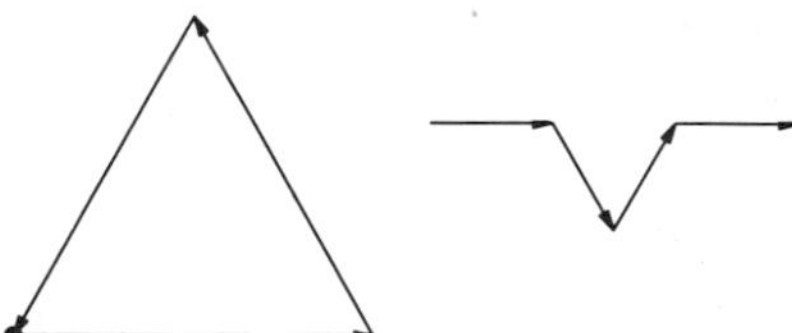

Let N_n be the number of sides, L_n be the length of a single side, ℓ_n be the length of the PERIMETER, and A_n the snowflake's AREA after the nth iteration. Further, denote the AREA of the initial $n = 0$ TRIANGLE Δ, and the length of an initial $n = 0$ side 1. Then

$$N_n = 3 \cdot 4^n \tag{1}$$

$$L_n = \left(\tfrac{1}{3}\right)^n = 3^{-n} \tag{2}$$

$$\ell_n \equiv N_n L_n = 3\left(\tfrac{4}{3}\right)^n \tag{3}$$

$$\begin{aligned} A_n &= A_{n-1} + \tfrac{1}{4} N_n {L_n}^2 \Delta = A_{n-1} + \frac{3 \cdot 4^n}{4}\left(\frac{1}{3}\right)^{2n} \Delta \\ &= A_{n-1} + \frac{3 \cdot 4^{n-1}}{9^n}\Delta = A_{n-1} + \frac{3 \cdot 4^{4-1}}{9 \cdot 9^{n-1}}\Delta \\ &= A_{n-1} + \tfrac{1}{3}\left(\tfrac{4}{9}\right)^{n-1}\Delta. \end{aligned} \tag{4}$$

The CAPACITY DIMENSION is then

$$\begin{aligned} d_{\text{cap}} &= -\lim_{n\to\infty} \frac{\ln N_n}{\ln L_n} = -\lim_{n\to\infty} \frac{\ln(3 \cdot 4^n)}{\ln(3^{-n})} \\ &= \lim_{n\to\infty} \frac{\ln 3 + n \ln 4}{n \ln 3} \\ &= \frac{\ln 4}{\ln 3} = \frac{2 \ln 2}{\ln 3} = 1.261859507\ldots. \end{aligned} \tag{5}$$

Now compute the AREA explicitly,

$$A_0 = \Delta \tag{6}$$

$$A_1 = A_0 + \frac{1}{3}\left(\frac{4}{9}\right)^0 \Delta = \Delta\left\{1 + \frac{1}{3}\left(\frac{4}{9}\right)^0\right\} \tag{7}$$

$$A_2 = A_1 + \frac{1}{3}\left(\frac{4}{9}\right)^1 \Delta = \Delta\left\{1 + \frac{1}{3}\left[\left(\frac{4}{9}\right)^0 + \left(\frac{4}{9}\right)^1\right]\right\} \tag{8}$$

$$A_n = \left[1 + \frac{1}{3}\sum_{k=0}^{n}\left(\frac{4}{9}\right)^k\right]\Delta, \tag{9}$$

so as $n \to \infty$,

$$\begin{aligned} A \equiv A_\infty &= \left[1 + \frac{1}{3}\sum_{k=1}^{\infty}\left(\frac{4}{9}\right)^k\right] = \left(1 + \tfrac{1}{3}\frac{1}{1-\frac{4}{9}}\right)\Delta \\ &= \tfrac{8}{5}\Delta. \end{aligned} \tag{10}$$

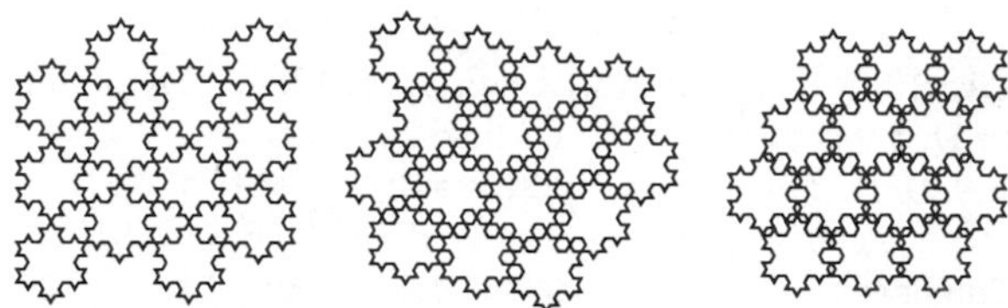

Some beautiful TILINGS, a few examples of which are illustrated above, can be made with iterations toward Koch snowflakes.

In addition, two sizes of Koch snowflakes in AREA ratio 1:3 TILE the PLANE, as shown above (Gosper).

Another beautiful modification of the Koch snowflake involves inscribing the constituent triangles with filled-in triangles, possibly rotated at some angle. Some sample results are illustrated above for 3 and 4 iterations.

see also CESÀRO FRACTAL, EXTERIOR SNOWFLAKE, GOSPER ISLAND, KOCH ANTISNOWFLAKE, PEANO-GOSPER CURVE, PENTAFLAKE, SIERPIŃSKI SIEVE

References

Cundy, H. and Rollett, A. *Mathematical Models, 3rd ed.* Stradbroke, England: Tarquin Pub., pp. 65–66, 1989.

Dickau, R. M. "Two-Dimensional L-Systems." `http://forum.swarthmore.edu/advanced/robertd/lsys2d.html`.

Dixon, R. *Mathographics.* New York: Dover, pp. 175–177 and 179, 1991.

Lauwerier, H. *Fractals: Endlessly Repeated Geometric Figures.* Princeton, NJ: Princeton University Press, pp. 28–29 and 32–36, 1991.

Pappas, T. "The Snowflake Curve." *The Joy of Mathematics.* San Carlos, CA: Wide World Publ./Tetra, pp. 78 and 160–161, 1989.

Peitgen, H.-O.; Jürgens, H.; and Saupe, D. *Chaos and Fractals: New Frontiers of Science.* New York: Springer-Verlag, 1992.

Peitgen, H.-O. and Saupe, D. (Eds.). "The von Koch Snowflake Curve Revisited." §C.2 in *The Science of Fractal Images.* New York: Springer-Verlag, pp. 275–279, 1988.

Wagon, S. *Mathematica in Action.* New York: W. H. Freeman, pp. 185–195, 1991.

Weisstein, E. W. "Fractals." `http://www.astro.virginia.edu/~eww6n/math/notebooks/Fractal.m`.

Kochansky's Approximation

The approximation for PI,

$$\pi \approx \sqrt{\frac{40}{3} - \sqrt{12}} = 3.141533\ldots.$$

Koebe's Constant

A CONSTANT equal to one QUARTER, 1/4.

see also QUARTER

References

Le Lionnais, F. *Les nombres remarquables.* Paris: Hermann, p. 24, 1983.

Koebe Function

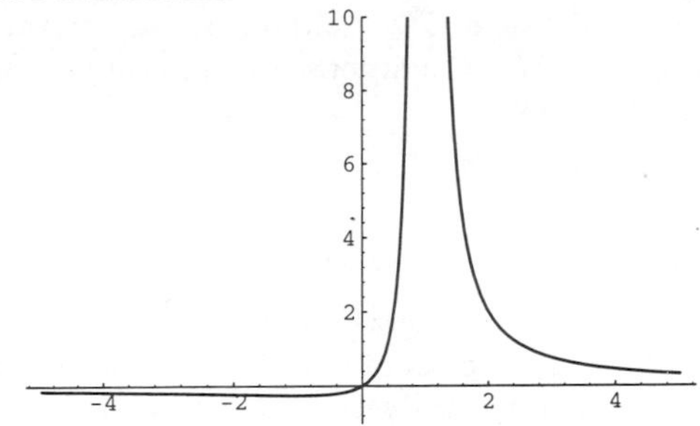

The function

$$f(z) \equiv \frac{z}{(1-z)^2}.$$

It has a MINIMUM at $z = -1$, where

$$f'(z) = -\frac{1+z}{(z-1)^3} = 0,$$

and an INFLECTION POINT at $z = -2$, where

$$f''(z) = \frac{2(2+z)}{(z-1)^4} = 0.$$

References

Stewart, I. *From Here to Infinity: A Guide to Today's Mathematics.* Oxford, England: Oxford University Press, pp. 164–165, 1996.

Kollros' Theorem

For every ring containing p SPHERES, there exists a ring of q SPHERES, each touching each of the p SPHERES, where

$$\frac{1}{p} + \frac{1}{q} = \frac{1}{3}.$$

The HEXLET is a special case with $p = 3$.

see also HEXLET, SPHERE

References

Honsberger, R. *Mathematical Gems II.* Washington, DC: Math. Assoc. Amer., p. 50, 1976.

Kolmogorov-Arnold-Moser Theorem

A theorem outlined in 1954 by Kolmogorov which was subsequently proved in the 1960s by Arnold and Moser (Tabor 1989, p. 105). It gives conditions under which CHAOS is restricted in extent. Moser's 1962 proof was valid for TWIST MAPS

$$\theta' = \theta + 2\pi f(I) + g(\theta, I) \tag{1}$$

$$I' = I + f(\theta, I). \tag{2}$$

In 1963, Arnold produced a proof for Hamiltonian systems

$$H = H_0(\mathbf{I}) + \epsilon H_1(\mathbf{I}). \tag{3}$$

The original theorem required perturbations $\epsilon \sim 10^{-48}$, although this has since been significantly increased. Arnold's proof required C^∞, and Moser's original proof required C^{333}. Subsequently, Moser's version has been reduced to C^6, then $C^{2+\epsilon}$, although counterexamples are known for C^2. Conditions for applicability of the KAM theorem are:

1. small perturbations,
2. smooth perturbations, and
3. sufficiently irrational WINDING NUMBER.

Moser considered an integrable Hamiltonian function H_0 with a TORUS T_0 and set of frequencies ω having an incommensurate frequency vector $\boldsymbol{\omega}^*$ (i.e., $\boldsymbol{\omega}\cdot\mathbf{k} \neq 0$ for all INTEGERS k_i). Let H_0 be perturbed by some periodic function H_1. The KAM theorem states that, if H_1 is small enough, then for almost every ω^* there exists an invariant TORUS $T(\omega^*)$ of the perturbed system such that $T(\omega^*)$ is "close to" $T_0(\omega^*)$. Moreover, the TORI $T(\omega^*)$ form a set of POSITIVE measures whose complement has a measure which tends to zero as $|H_1| \to 0$. A useful paraphrase of the KAM theorem is, "For sufficiently small perturbation, almost all TORI (excluding those with rational frequency vectors) are preserved." The theorem thus explicitly excludes TORI with rationally related frequencies, that is, $n-1$ conditions of the form

$$\boldsymbol{\omega}\cdot\mathbf{k} = 0. \tag{4}$$

These TORI are destroyed by the perturbation. For a system with two DEGREES OF FREEDOM, the condition of closed orbits is

$$\sigma = \frac{\omega_1}{\omega_2} = \frac{r}{s}. \tag{5}$$

For a QUASIPERIODIC ORBIT, σ is IRRATIONAL. KAM shows that the preserved TORI satisfy the irrationality condition

$$\left|\frac{\omega_1}{\omega_2} - \frac{r}{s}\right| > \frac{K(\epsilon)}{s^{2.5}} \tag{6}$$

for all r and s, although not much is known about $K(\epsilon)$.

The KAM theorem broke the deadlock of the small divisor problem in classical perturbation theory, and provides the starting point for an understanding of the appearance of CHAOS. For a HAMILTONIAN SYSTEM, the ISOENERGETIC NONDEGENERACY condition

$$\left|\frac{\partial^2 H_0}{\partial I_j \partial I_j}\right| \neq 0 \tag{7}$$

guarantees preservation of most invariant TORI under small perturbations $\epsilon \ll 1$. The Arnold version states that

$$\left|\sum_{k=1}^{n} m_k\omega_k\right| > K(\epsilon)\left(\sum_{k=1}^{n}|m_k|\right)^{-n-1} \tag{8}$$

for all $m_k \in \mathbb{Z}$. This condition is less restrictive than Moser's, so fewer points are excluded.

see also CHAOS, HAMILTONIAN SYSTEM, QUASIPERIODIC FUNCTION, TORUS

References

Tabor, M. *Chaos and Integrability in Nonlinear Dynamics: An Introduction.* New York: Wiley, 1989.

Kolmogorov Complexity

The complexity of a pattern parameterized as the shortest ALGORITHM required to reproduce it. Also known as ALGORITHMIC COMPLEXITY.

References

Goetz, P. "Phil's Good Enough Complexity Dictionary." `http://www.cs.buffalo.edu/~goetz/dict.html`.

Kolmogorov Constant

The exponent 5/3 in the spectrum of homogeneous turbulence, $k^{-5/3}$.

References

Le Lionnais, F. *Les nombres remarquables.* Paris: Hermann, p. 41, 1983.

Kolmogorov Entropy

Also known as METRIC ENTROPY. Divide PHASE SPACE into D-dimensional HYPERCUBES of CONTENT ϵ^D. Let $P_{i_0,\ldots,i_n}$ be the probability that a trajectory is in HYPERCUBE i_0 at $t=0$, i_1 at $t=T$, i_2 at $t=2T$, etc. Then define

$$K_n = h_K = -\sum_{i_0,\ldots,i_n} P_{i_0,\ldots,i_n} \ln P_{i_0,\ldots,i_n}, \tag{1}$$

where $K_{N+1} - K_N$ is the information needed to predict which HYPERCUBE the trajectory will be in at $(n+1)T$ given trajectories up to nT. The Kolmogorov entropy is then defined by

$$K \equiv \lim_{T\to 0}\lim_{\epsilon\to 0^+}\lim_{N\to\infty}\frac{1}{NT}\sum_{n=0}^{N-1}(K_{n+1} - K_n). \tag{2}$$

The Kolmogorov entropy is related to LYAPUNOV CHARACTERISTIC EXPONENTS by

$$h_K = \int_P \sum_{\sigma_i>0} \sigma_i \, d\mu. \tag{3}$$

see also HYPERCUBE, LYAPUNOV CHARACTERISTIC EXPONENT

References

Ott, E. *Chaos in Dynamical Systems.* New York: Cambridge University Press, p. 138, 1993.
Schuster, H. G. *Deterministic Chaos: An Introduction, 3rd ed.* New York: Wiley, p. 112, 1995.

Kolmogorov-Sinai Entropy

see KOLMOGOROV ENTROPY, METRIC ENTROPY

Kolmogorov-Smirnov Test

A goodness-of-fit test for any DISTRIBUTION. The test relies on the fact that the value of the sample cumulative density function is asymptotically normally distributed.

To apply the Kolmogorov-Smirnov test, calculate the cumulative frequency (normalized by the sample size) of the observations as a function of class. Then calculate the cumulative frequency for a true distribution (most commonly, the NORMAL DISTRIBUTION). Find the greatest discrepancy between the observed and expected cumulative frequencies, which is called the "D-STATISTIC." Compare this against the critical D-STATISTIC for that sample size. If the calculated D-STATISTIC is greater than the critical one, then reject the NULL HYPOTHESIS that the distribution is of the expected form. The test is an R-ESTIMATE.

see also ANDERSON-DARLING STATISTIC, D-STATISTIC, KUIPER STATISTIC, NORMAL DISTRIBUTION, R-ESTIMATE

References

Boes, D. C.; Graybill, F. A.; and Mood, A. M. *Introduction to the Theory of Statistics, 3rd ed.* New York: McGraw-Hill, 1974.

Knuth, D. E. §3.3.1B in *The Art of Computer Programming, Vol. 2: Seminumerical Algorithms, 2nd ed.* Reading, MA: Addison-Wesley, pp. 45–52, 1981.

Press, W. H.; Flannery, B. P.; Teukolsky, S. A.; and Vetterling, W. T. "Kolmogorov-Smirnov Test." In *Numerical Recipes in FORTRAN: The Art of Scientific Computing, 2nd ed.* Cambridge, England: Cambridge University Press, pp. 617–620, 1992.

König-Egeváry Theorem

A theorem on BIPARTITE GRAPHS.

see also BIPARTITE GRAPH, FROBENIUS-KÖNIG THEOREM

König's Theorem

If an ANALYTIC FUNCTION has a single simple POLE at the RADIUS OF CONVERGENCE of its POWER SERIES, then the ratio of the coefficients of its POWER SERIES converges to that POLE.

see also POLE

References

König, J. "Über eine Eigenschaft der Potenzreihen." *Math. Ann.* **23**, 447–449, 1884.

Königsberg Bridge Problem

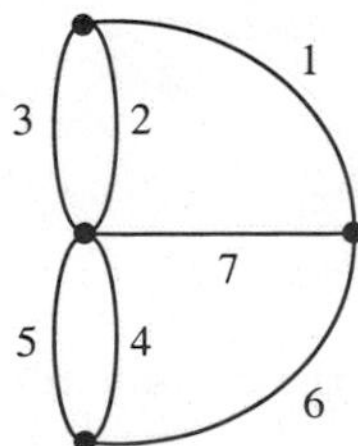

The Königsberg bridges cannot all be traversed in a single trip without doubling back. This problem was solved by Euler, and represented the beginning of GRAPH THEORY.

see also EULERIAN CIRCUIT, GRAPH THEORY

References

Bogomolny, A. "Graphs." `http://www.cut-the-knot.com/do_you_know/graphs.html`.

Chartrand, G. "The Königsberg Bridge Problem: An Introduction to Eulerian Graphs." §3.1 in *Introductory Graph Theory.* New York: Dover, pp. 51–66, 1985.

Kraitchik, M. §8.4.1 in *Mathematical Recreations.* New York: W. W. Norton, pp. 209–211, 1942.

Newman, J. "Leonhard Euler and the Königsberg Bridges." *Sci. Amer.* **189**, 66–70, 1953.

Pappas, T. "Königsberg Bridge Problem & Topology." *The Joy of Mathematics.* San Carlos, CA: Wide World Publ./Tetra, pp. 124–125, 1989.

Korselt's Criterion

n DIVIDES $a^n - a$ for all INTEGERS a IFF n is SQUAREFREE and $(p-1)|n/p-1$ for all PRIME DIVISORS p of n. CARMICHAEL NUMBERS satisfy this CRITERION.

References

Borwein, D.; Borwein, J. M.; Borwein, P. B.; and Girgensohn, R. "Giuga's Conjecture on Primality." *Amer. Math. Monthly* **103**, 40–50, 1996.

Kovalevskaya Exponent

see LEADING ORDER ANALYSIS

Kozyrev-Grinberg Theory

A theory of HAMILTONIAN CIRCUITS.

see also GRINBERG FORMULA, HAMILTONIAN CIRCUIT

Kramers Rate

The characteristic escape rate from a stable state of a potential in the absence of signal.

see also STOCHASTIC RESONANCE

References

Bulsara, A. R. and Gammaitoni, L. "Tuning in to Noise." *Phys. Today* **49**, 39–45, March 1996.

Krawtchouk Polynomial

Let $\alpha(x)$ be a STEP FUNCTION with the JUMP

$$j(x) = \binom{N}{x} p^x q^{N-x} \tag{1}$$

at $x = 0, 1, \ldots, N$, where $p > 0, q > 0$, and $p + q = 1$. Then

$$k_n^{(p)}(x) = \left[\binom{N}{n}\right]^{-1/2} (pq)^{-n/2} \times \sum_{\nu=0}^{n} (-1)^{n-\nu} \binom{N-x}{n-\nu} \binom{x}{\nu} p^{n-\nu} q^{\nu}, \tag{2}$$

for $n = 0, 1, \ldots, N$. It has WEIGHT FUNCTION

$$w = \frac{N!p^x q^{N-x}}{\Gamma(1+x)\Gamma(N+1-x)}, \tag{3}$$

where $\Gamma(x)$ is the GAMMA FUNCTION, RECURRENCE RELATION

$$(n+1)k_{n+1}^{(p)}(x) + pq(N-n+1)k_{n-1}^{(p)}(x) = [x-n-(N-2)]k_n^{(p)}(x), \tag{4}$$

and squared norm

$$\frac{N!}{n!(N-n)!}(pq)^n. \tag{5}$$

It has the limit

$$\lim_{n\to\infty}\left(\frac{2}{Npq}\right)^{n/2} n!k_n^{(p)}(Np+\sqrt{2Npqs}) = H_n(s), \tag{6}$$

where $H_n(x)$ is a HERMITE POLYNOMIAL, and is related to the HYPERGEOMETRIC FUNCTION by

$$\begin{aligned} k_n^{(p)}(x,N) &= k_n^{(p)}(x,N) \\ &= (-1)^n \binom{N}{n} p^n\,{}_2F_1(-n,-x;-N;1/p) \\ &\quad \frac{(-1)^n p^n}{n!}\frac{\Gamma(N-x+1)}{\Gamma(N-x-n+1)} \\ &\quad \times {}_2F_1(-n,-x;N-x-n+1;-q/p). \end{aligned} \tag{7}$$

see also ORTHOGONAL POLYNOMIALS

References

Nikiforov, A. F.; Uvarov, V. B.; and Suslov, S. S. *Classical Orthogonal Polynomials of a Discrete Variable.* New York: Springer-Verlag, 1992.

Szegő, G. *Orthogonal Polynomials, 4th ed.* Providence, RI: Amer. Math. Soc., pp. 35–37, 1975.

Zelenkov, V. "Krawtchouk Polynomial Home Page." `http://www.isir.minsk.by/~zelenkov/physmath/kr_polyn/`.

Kreisel Conjecture

A CONJECTURE in DECIDABILITY theory which postulates that, if there is a uniform bound to the lengths of shortest proofs of instances of $S(n)$, then the universal generalization is necessarily provable in PEANO ARITHMETIC. The CONJECTURE was proven true by M. Baaz in 1988 (Baaz and Pudlák 1993).

see also DECIDABLE

References

Baaz, M. and Pudlák P. "Kreisel's Conjecture for $L\exists_1$. In *Arithmetic, Proof Theory, and Computational Complexity, Papers from the Conference Held in Prague, July 2–5, 1991* (Ed. P. Clote and J. Krajiček). New York: Oxford University Press, pp. 30–60, 1993.

Dawson, J. "The Gödel Incompleteness Theorem from a Length of Proof Perspective." *Amer. Math. Monthly* **86**, 740–747, 1979.

Kreisel, G. "On the Interpretation of Nonfinitistic Proofs, II." *J. Symbolic Logic* **17**, 43–58, 1952.

Kronecker Decomposition Theorem

Every FINITE ABELIAN GROUP can be written as a DIRECT PRODUCT of CYCLIC GROUPS of PRIME POWER ORDERS. In fact, the number of nonisomorphic ABELIAN FINITE GROUPS $a(n)$ of any given ORDER n is given by writing n as

$$n = \prod_i {p_i}^{\alpha_i},$$

where the p_i are distinct PRIME FACTORS, then

$$a(n) = \prod_i P(\alpha_i),$$

where P is the PARTITION FUNCTION. This gives 1, 1, 1, 2, 1, 1, 1, 3, 2, ... (Sloane's A000688).

see also ABELIAN GROUP, FINITE GROUP, ORDER (GROUP), PARTITION FUNCTION P

References

Sloane, N. J. A. Sequence A000688/M0064 in "An On-Line Version of the Encyclopedia of Integer Sequences."

Kronecker Delta

The simplest interpretation of the Kronecker delta is as the discrete version of the DELTA FUNCTION defined by

$$\delta_{ij} \equiv \begin{cases} 0 & \text{for } i \neq j \\ 1 & \text{for } i = j. \end{cases} \tag{1}$$

It has the COMPLEX GENERATING FUNCTION

$$\delta_{mn} = \frac{1}{2\pi i}\int z^{m-n-1}\,dz, \tag{2}$$

where m and n are INTEGERS. In 3-space, the Kronecker delta satisfies the identities

$$\delta_{ii} = 3 \tag{3}$$

$$\delta_{ij}\epsilon_{ijk} = 0 \tag{4}$$

$$\epsilon_{ipq}\epsilon_{jpq} = 2\delta_{ij} \tag{5}$$

$$\epsilon_{ijk}\epsilon_{pqk} = \delta_{ip}\delta_{jq} - \delta_{iq}\delta_{jp}, \tag{6}$$

where EINSTEIN SUMMATION is implicitly assumed, $i, j = 1, 2, 3$, and ϵ is the PERMUTATION SYMBOL.

Technically, the Kronecker delta is a TENSOR defined by the relationship

$$\delta_l^k \frac{\partial x'_i}{\partial x_k}\frac{\partial x_l}{\partial x'_j} = \frac{\partial x'_i}{\partial x_k}\frac{\partial x_k}{\partial x'_j} = \frac{\partial x'_i}{\partial x'_j}. \tag{7}$$

Since, by definition, the coordinates x_i and x_j are independent for $i \neq j$,

$$\frac{\partial x'_i}{\partial x'_j} = \delta'^i{}_j, \tag{8}$$

so

$$\delta'^{i}_{j} = \frac{\partial x'_i}{\partial x_k}\frac{\partial x_l}{\partial x'_j}\delta^k_l, \tag{9}$$

and δ^i_j is really a mixed second RANK TENSOR. It satisfies

$$\delta_{ab}{}^{jk} = \epsilon_{abi}\epsilon^{jki} = \delta^j_a\delta^k_b - \delta^k_a\delta^j_b \tag{10}$$

$$\delta_{abjk} = g_{aj}g_{bk} - g_{ak}g_{bj} \tag{11}$$

$$\epsilon_{aij}\epsilon^{bij} = \delta_{ai}{}^{bi} = 2\delta^b_a. \tag{12}$$

see also DELTA FUNCTION, PERMUTATION SYMBOL

Kronecker's Polynomial Theorem

An algebraically soluble equation of ODD PRIME degree which is irreducible in the natural FIELD possesses either

1. Only a single REAL ROOT, or
2. All REAL ROOTS.

see also ABEL'S IRREDUCIBILITY THEOREM, ABEL'S LEMMA, SCHOENEMANN'S THEOREM

References
Dörrie, H. *100 Great Problems of Elementary Mathematics: Their History and Solutions.* New York: Dover p. 127, 1965.

Kronecker Product

see DIRECT PRODUCT (MATRIX)

Kronecker Symbol

An extension of the JACOBI SYMBOL (n/m) to all INTEGERS. It can be computed using the normal rules for the JACOBI SYMBOL

$$\begin{aligned}\left(\frac{ab}{cd}\right) = \left(\frac{a}{cd}\right)\left(\frac{b}{cd}\right) &= \left(\frac{ab}{c}\right)\left(\frac{ab}{d}\right)\\ &= \left(\frac{a}{c}\right)\left(\frac{b}{c}\right)\left(\frac{a}{d}\right)\left(\frac{b}{d}\right)\end{aligned}$$

plus additional rules for $m = -1$,

$$(n/-1) = \begin{cases} -1 & \text{for } n < 0 \\ 1 & \text{for } n > 0, \end{cases}$$

and $m = 2$. The definition for $(n/2)$ is variously written as

$$(n/2) \equiv \begin{cases} 0 & \text{for } n \text{ even} \\ 1 & \text{for } n \text{ odd, } n \equiv \pm 1 \pmod 8 \\ -1 & \text{for } n \text{ odd, } n \equiv \pm 3 \pmod 8 \end{cases}$$

or

$$(n/2) \equiv \begin{cases} 0 & \text{for } 4|n \\ 1 & \text{for } n \equiv 1 \pmod 8 \\ -1 & \text{for } n \equiv 5 \pmod 8 \\ \text{undefined} & \text{otherwise} \end{cases}$$

(Cohn 1980). Cohn's form "undefines" $(n/2)$ for SINGLY EVEN NUMBERS $n \equiv 4 \pmod 2$ and $n \equiv -1, 3 \pmod 8$, probably because no other values are needed in applications of the symbol involving the DISCRIMINANTS d of QUADRATIC FIELDS, where $m > 0$ and d always satisfies $d \equiv 0, 1 \pmod 4$.

The KRONECKER SYMBOL is a REAL CHARACTER modulo n, and is, in fact, essentially the only type of REAL primitive character (Ayoub 1963).

see also CHARACTER (NUMBER THEORY), CLASS NUMBER, DIRICHLET L-SERIES, JACOBI SYMBOL, LEGENDRE SYMBOL

References
Ayoub, R. G. *An Introduction to the Analytic Theory of Numbers.* Providence, RI: Amer. Math. Soc., 1963.
Cohn, H. *Advanced Number Theory.* New York: Dover, p. 35, 1980.

Krull Dimension

If R is a RING (commutative with 1), the height of a PRIME IDEAL p is defined as the SUPREMUM of all n so that there is a chain $p_0 \subset \cdots p_{n-1} \subset p_n = p$ where all p_i are distinct PRIME IDEALS. Then, the Krull dimension of R is defined as the SUPREMUM of all the heights of all its PRIME IDEALS.

see also PRIME IDEAL

References
Eisenbud, D. *Commutative Algebra with a View Toward Algebraic Geometry.* New York: Springer-Verlag, 1995.
Macdonald, I. G. and Atiyah, M. F. *Introduction to Commutative Algebra.* Reading, MA: Addison-Wesley, 1969.

Kruskal's Algorithm

An ALGORITHM for finding a GRAPH's spanning TREE of minimum length.

see also KRUSKAL'S TREE THEOREM

References
Gardner, M. *Mathematical Magic Show: More Puzzles, Games, Diversions, Illusions and Other Mathematical Sleight-of-Mind from Scientific American.* New York: Vintage, pp. 248–249, 1978.

Kruskal's Tree Theorem

A theorem which plays a fundamental role in computer science because it is one of the main tools for showing that certain orderings on TREES are well-founded. These orderings play a crucial role in proving the termination of rewriting rules and the correctness of the Knuth-Bendix equational completion procedures.

see also KRUSKAL'S ALGORITHM, NATURAL INDEPENDENCE PHENOMENON, TREE

References
Gallier, J. "What's so Special about Kruskal's Theorem and the Ordinal Gamma[0]? A Survey of Some Results in Proof Theory." *Ann. Pure and Appl. Logic* **53**, 199–260, 1991.

KS Entropy

see METRIC ENTROPY

Kuen Surface

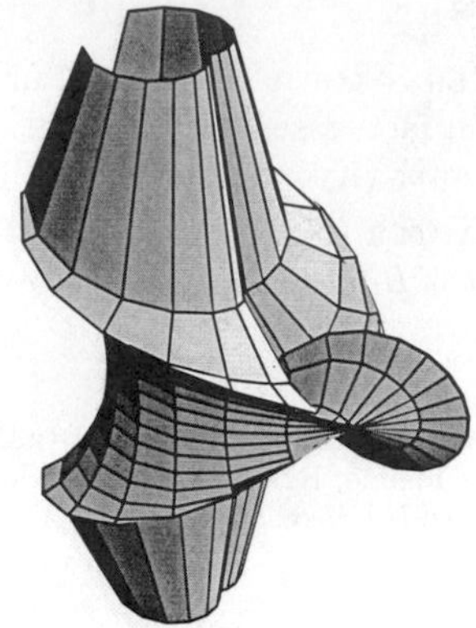

A special case of ENNEPER'S SURFACES which can be given parametrically by

$$x = \frac{2(\cos u + u \sin u)\sin v}{1+u^2\sin^2 v} \quad (1)$$

$$= \frac{2\sqrt{1+u^2}\cos(u-\tan^{-1}u)\sin v}{1+u^2\sin^2 v} \quad (2)$$

$$y = \frac{2(\sin u - u\cos u)\sin v}{1+u^2\sin^2 v} \quad (3)$$

$$= \frac{2\sqrt{1+u^2}\sin(u-\tan^{-1}u)\sin v}{1+u^2\sin^2 v} \quad (4)$$

$$z = \ln[\tan(\tfrac{1}{2}v)] + \frac{2\cos v}{1+u^2\sin^2 v} \quad (5)$$

for $v \in [0,\pi)$, $u \in [0, 2\pi)$ (Reckziegel *et al.* 1986). The Kuen surface has constant NEGATIVE GAUSSIAN CURVATURE of $K = -1$. The PRINCIPAL CURVATURES are given by

$$\kappa_1 = -\frac{u\cos(\tfrac{1}{2}v)[-2-u^2+u^2\cos(2v)]^4\sin(\tfrac{1}{2}v)}{2[2-u^2+u^2\cos(2v)](1+u^2\sin^2 v)^4} \quad (6)$$

$$\kappa_2 = \frac{[-2-u^2+u^2\cos(2v)]^4[2-u^2+u^2\cos(2v)]\csc(v)}{64u(1+u^2\sin^2 v)^4}. \quad (7)$$

see also ENNEPER'S SURFACES, REMBS' SURFACES, SIEVERT'S SURFACE

References

Fischer, G. (Ed.). Plate 86 in *Mathematische Modelle/Mathematical Models, Bildband/Photograph Volume.* Braunschweig, Germany: Vieweg, p. 82, 1986.

Gray, A. "Kuen's Surface." §19.4 in *Modern Differential Geometry of Curves and Surfaces.* Boca Raton, FL: CRC Press, pp. 384–386, 1993.

Kuen, T. "Ueber Flächen von constantem Krümmungsmaass." *Sitzungsber. d. königl. Bayer. Akad. Wiss. Math.-phys. Classe,* Heft II, 193–206, 1884.

Nordstrand, T. "Kuen's Surface." `http://www.uib.no/people/nfytn/kuentxt.htm`.

Reckziegel, H. "Kuen's Surface." §3.4.4.2 in *Mathematical Models from the Collections of Universities and Museums* (Ed. G. Fischer). Braunschweig, Germany: Vieweg, p. 38, 1986.

Kuhn-Tucker Theorem

A theorem in nonlinear programming which states that if a regularity condition holds and f and the functions h_j are convex, then a solution x^0 which satisfies the conditions h_j for a VECTOR of multipliers λ is a GLOBAL MINIMUM. The Kuhn-Tucker theorem is a generalization of LAGRANGE MULTIPLIERS. FARKAS'S LEMMA is key in proving this theorem.

see also FARKAS'S LEMMA, LAGRANGE MULTIPLIER

Kuiper Statistic

A statistic defined to improve the KOLMOGOROV-SMIRNOV TEST in the TAILS.

see also ANDERSON-DARLING STATISTIC

References

Press, W. H.; Flannery, B. P.; Teukolsky, S. A.; and Vetterling, W. T. *Numerical Recipes in FORTRAN: The Art of Scientific Computing, 2nd ed.* Cambridge, England: Cambridge University Press, p. 621, 1992.

Kulikowski's Theorem

For every POSITIVE INTEGER n, there exists a SPHERE which has exactly n LATTICE POINTS on its surface. The SPHERE is given by the equation

$$(x-a)^2 + (y-b)^2 + (z-\sqrt{2})^2 = c^2 + 2,$$

where a and b are the coordinates of the center of the so-called SCHINZEL CIRCLE

$$\begin{cases} (x-\tfrac{1}{2})^2 + y^2 = \tfrac{1}{4}5^{k-1} & \text{for } n = 2k \text{ even} \\ (x-\tfrac{1}{3})^2 + y^2 = \tfrac{1}{9}5^{2k} & \text{for } n = 2k+1 \text{ odd} \end{cases}$$

and c is its RADIUS.

see also CIRCLE LATTICE POINTS, LATTICE POINT, SCHINZEL'S THEOREM

References

Honsberger, R. "Circles, Squares, and Lattice Points." Ch. 11 in *Mathematical Gems I.* Washington, DC: Math. Assoc. Amer., pp. 117–127, 1973.

Kulikowski, T. "Sur l'existence d'une sphère passant par un nombre donné aux coordonnées entières." *L'Enseignement Math. Ser. 2* **5**, 89–90, 1959.

Schinzel, A. "Sur l'existence d'un cercle passant par un nombre donné de points aux coordonnées entières." *L'Enseignement Math. Ser. 2* **4**, 71–72, 1958.

Sierpiński, W. "Sur quelques problèmes concernant les points aux coordonnées entières." *L'Enseignement Math. Ser. 2* **4**, 25–31, 1958.

Sierpiński, W. "Sur un problème de H. Steinhaus concernant les ensembles de points sur le plan." *Fund. Math.* **46**, 191–194, 1959.

Sierpiński, W. *A Selection of Problems in the Theory of Numbers.* New York: Pergamon Press, 1964.

Kummer's Conjecture

A conjecture concerning PRIMES.

Kummer's Differential Equation

see CONFLUENT HYPERGEOMETRIC DIFFERENTIAL EQUATION

Kummer's Formulas

Kummer's first formula is

$$ {}_2F_1(\tfrac{1}{2}+m-k,-n;2m+1;1) = \frac{\Gamma(2m+1)\Gamma(m+\frac{1}{2}+k+n)}{\Gamma(m+\frac{1}{2}+k)\Gamma(2m+1+n)}, \quad (1) $$

where ${}_2F_1(a,b;c;z)$ is the HYPERGEOMETRIC FUNCTION with $m \neq -1/2, -1, -3/2, \ldots$, and $\Gamma(z)$ is the GAMMA FUNCTION. The identity can be written in the more symmetrical form as

$$ {}_2F_1(a,b;c;-1) = \frac{\Gamma(\frac{1}{2}b+1)\Gamma(b-a+1)}{\Gamma(b+1)\Gamma(\frac{1}{2}b-a+1)}, \quad (2) $$

where $a-b+c=1$ and b is a positive integer. If b is a negative integer, the identity takes the form

$$ {}_2F_1(a,b;c;-1) = 2\cos(\tfrac{1}{2}\pi b)\frac{\Gamma(|b|)\Gamma(b-a+1)}{\Gamma(\frac{1}{2}b-a+1)} \quad (3) $$

(Petkovšek *et al.* 1996).

Kummer's second formula is

$$ {}_1F_1(\tfrac{1}{2}+m;2m+1;z) = M_{0,m}(z) = z^{m+1/2}\left[1+\sum_{p=1}^{\infty}\frac{z^{2p}}{2^{4p}p!(m+1)(m+2)\cdots(m+p)}\right], \quad (4) $$

where ${}_1F_1(a;b;z)$ is the CONFLUENT HYPERGEOMETRIC FUNCTION and $m \neq -1/2, -1, -3/2, \ldots$.

References
Petkovšek, M.; Wilf, H. S.; and Zeilberger, D. *A=B*. Wellesley, MA: A. K. Peters, pp. 42–43 and 126, 1996.

Kummer's Function

see CONFLUENT HYPERGEOMETRIC FUNCTION

Kummer Group

A GROUP of LINEAR FRACTIONAL TRANSFORMATIONS which transform the arguments of Kummer solutions to the HYPERGEOMETRIC DIFFERENTIAL EQUATION into each other. Define

$$ A(z) = 1-z $$
$$ B(z) = 1/z, $$

then the elements of the group are $\{I, A, B, AB, BA, ABA = BAB\}$.

Kummer's Quadratic Transformation

A transformation of a HYPERGEOMETRIC FUNCTION,

$$ {}_2F_1\left(\alpha,\beta;2\beta;\frac{4z}{(1+z)^2}\right) = (1+z)^{2\alpha}\,{}_2F_1(\alpha,\alpha+\tfrac{1}{2}-\beta;\beta+\tfrac{1}{2};z^2). $$

Kummer's Relation

An identity which relates HYPERGEOMETRIC FUNCTIONS,

$$ {}_2F_1(2a,2b;a+b+\tfrac{1}{2};x) = {}_2F_1(a,b;a+b+\tfrac{1}{2},4x(1-x)). $$

Kummer's Series

see HYPERGEOMETRIC FUNCTION

Kummer's Series Transformation

Let $\sum_{k=0}^{\infty} a_k = a$ and $\sum_{k=0}^{\infty} c_k = c$ be convergent series such that

$$ \lim_{k\to\infty}\frac{a_k}{c_k} = \lambda \neq 0. $$

Then

$$ a = \lambda c + \sum_{k=0}^{\infty}\left(1-\lambda\frac{c_k}{a_k}\right)a_k. $$

References
Abramowitz, M. and Stegun, C. A. (Eds.). *Handbook of Mathematical Functions with Formulas, Graphs, and Mathematical Tables, 9th printing.* New York: Dover, p. 16, 1972.

Kummer Surface

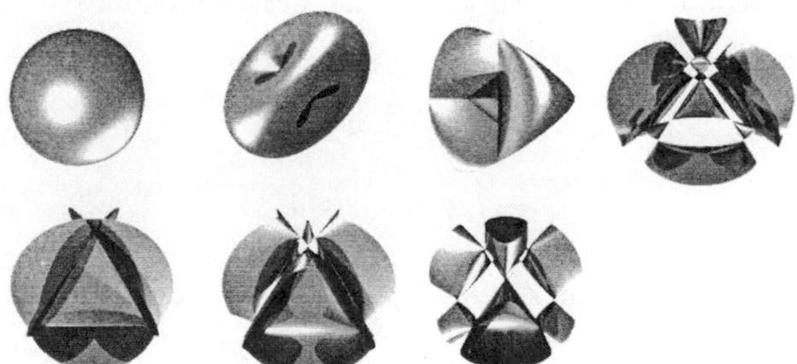

The Kummer surfaces are a family of QUARTIC SURFACES given by the algebraic equation

$$ (x^2+y^2+z^2-\mu^2w^2)^2 - \lambda pqrs = 0, \quad (1) $$

where

$$ \lambda \equiv \frac{3\mu^2-1}{3-\mu^2}, \quad (2) $$

p, q, r, and s are the TETRAHEDRAL COORDINATES

$$ p = w - z - \sqrt{2}\,x \quad (3) $$
$$ q = w - z + \sqrt{2}\,x \quad (4) $$
$$ r = w + z + \sqrt{2}\,y \quad (5) $$
$$ s = w + z - \sqrt{2}\,y, \quad (6) $$

and w is a parameter which, in the above plots, is set to $w = 1$. The above plots correspond to $\mu^2 = 1/3$

$$(3x^2 + 3y^2 + 3z^2 + 1)^2 = 0,$$

(double sphere), 2/3, 1

$$x^4 - 2x^2y^2 + y^4 + 4x^2z + 4y^2z + 4x^2z^2 + 4y^2z^2 = 0 \quad (7)$$

(ROMAN SURFACE), $\sqrt{2}$, $\sqrt{3}$

$$[(z-1)^2 - 2x^2][y^2 - (z+1)^2] = 0 \quad (8)$$

(four planes), 2, and 5. The case $0 \leq \mu^2 \leq 1/3$ corresponds to four real points.

The following table gives the number of ORDINARY DOUBLE POINTS for various ranges of μ^2, corresponding to the preceding illustrations.

Range	Real Nodes	Complex Nodes
$0 \leq \mu^2 \leq \frac{1}{3}$	4	12
$\mu^2 = \frac{1}{3}$		
$\frac{1}{3} \leq \mu^2 < 1$	4	12
$\mu^2 = 1$		
$1 < \mu^2 < 3$	16	0
$\mu^2 = 3$		
$\mu^2 > 3$	16	0

The Kummer surfaces can be represented parametrically by hyperelliptic THETA FUNCTIONS. Most of the Kummer surfaces admit 16 ORDINARY DOUBLE POINTS, the maximum possible for a QUARTIC SURFACE. A special case of a Kummer surface is the TETRAHEDROID.

Nordstrand gives the implicit equations as

$$x^4+y^4+z^4-x^2-y^2-z^2-x^2y^2-x^2z^2-y^2z^2+1 = 0 \quad (9)$$

or

$$x^4 + y^4 + z^4 + a(x^2 + y^2 + z^2) + b(x^2y^2 + x^2z^2 + y^2z^2) + cxyz - 1 = 0. \quad (10)$$

see also QUARTIC SURFACE, ROMAN SURFACE, TETRAHEDROID

References

Endraß, S. "Flächen mit vielen Doppelpunkten." *DMV-Mitteilungen* **4**, 17–20, Apr. 1995.

Endraß, S. "Kummer Surfaces." `http://www.mathematik.uni-mainz.de/Algebraische Geometrie/docs/Ekummer.shtml`.

Fischer, G. (Ed.). *Mathematical Models from the Collections of Universities and Museums.* Braunschweig, Germany: Vieweg, pp. 14–19, 1986.

Fischer, G. (Ed.). Plates 34–37 in *Mathematische Modelle/Mathematical Models, Bildband/Photograph Volume.* Braunschweig, Germany: Vieweg, pp. 33–37, 1986.

Guy, R. K. *Unsolved Problems in Number Theory, 2nd ed.* New York: Springer-Verlag, p. 183, 1994.

Hudson, R. *Kummer's Quartic Surface.* Cambridge, England: Cambridge University Press, 1990.

Kummer, E. "Über die Flächen vierten Grades mit sechszehn singulären Punkten." *Ges. Werke* **2**, 418–432.

Kummer, E. "Über Strahlensysteme, deren Brennflächen Flächen vierten Grades mit sechszehn singulären Punkten sind." *Ges. Werke* **2**, 418–432.

Nordstrand, T. "Kummer's Surface." `http://www.uib.no/people/nfytn/kummtxt.htm`.

Kummer's Test

Given a SERIES of POSITIVE terms u_i and a sequence of finite POSITIVE constants a_i, let

$$\rho \equiv \lim_{n\to\infty} \left(a_n \frac{u_n}{u_{n+1}} - a_{n+1} \right).$$

1. If $\rho > 0$, the series converges.
2. If $\rho < 0$, the series diverges.
3. If $\rho = 0$, the series may converge or diverge.

The test is a general case of BERTRAND'S TEST, the ROOT TEST, GAUSS'S TEST, and RAABE'S TEST. With $a_n = n$ and $a_{n+1} = n + 1$, the test becomes RAABE'S TEST.

see also CONVERGENCE TESTS, RAABE'S TEST

References

Arfken, G. *Mathematical Methods for Physicists, 3rd ed.* Orlando, FL: Academic Press, pp. 285–286, 1985.

Jingcheng, T. "Kummer's Test Gives Characterizations for Convergence or Divergence of All Series." *Amer. Math. Monthly* **101**, 450–452, 1994.

Samelson, H. "More on Kummer's Test." *Amer. Math. Monthly* **102**, 817–818, 1995.

Kummer's Theorem

$${}_2F_1(x, -x; x + n + 1; -1) = \frac{\Gamma(x+n+1)\Gamma(\frac{1}{2}n+1)}{\Gamma(x+\frac{1}{2}n+1)\Gamma(n+1)}$$

$${}_2F_1(\alpha, \beta; 1 + \alpha - \beta; -1) = \frac{\Gamma(1+\alpha-\beta)\Gamma(1+\frac{1}{2}\alpha)}{\Gamma(1+\alpha)\Gamma(1+\frac{1}{2}\alpha-\beta)},$$

where ${}_2F_1$ is a HYPERGEOMETRIC FUNCTION and $\Gamma(z)$ is the GAMMA FUNCTION.

Kuratowski's Closure-Component Problem

Let X be an arbitrary TOPOLOGICAL SPACE. Denote the CLOSURE of a SUBSET A of X by A^- and the complement of A by A'. Then at most 14 different SETS can be derived from A by repeated application of closure and complementation (Berman and Jordan 1975, Fife 1991). The problem was first proved by Kuratowski (1922) and popularized by Kelley (1955).

see also KURATOWSKI REDUCTION THEOREM

References

Anusiak, J. and Shum, K. P. "Remarks on Finite Topological Spaces." *Colloq. Math.* **23**, 217–223, 1971.

Aull, C. E. "Classification of Topological Spaces." *Bull. de l'Acad. Pol. Sci. Math. Astron. Phys.* **15**, 773–778, 1967.
Baron, S. Advanced Problem 5569. *Amer. Math. Monthly* **75**, 199, 1968.
Berman, J. and Jordan, S. L. "The Kuratowski Closure-Complement Problem." *Amer. Math. Monthly* **82**, 841–842, 1975.
Buchman, E. "Problem E 3144." *Amer. Math. Monthly* **93**, 299, 1986.
Chagrov, A. V. "Kuratowski Numbers, Application of Functional Analysis in Approximation Theory." Kalinin: Kalinin Gos. Univ., pp. 186–190, 1982.
Chapman, T. A. "A Further Note on Closure and Interior Operators." *Amer. Math. Monthly* **69**, 524–529, 1962.
Fife, J. H. "The Kuratowski Closure-Complement Problem." *Math. Mag.* **64**, 180–182, 1991.
Fishburn, P. C. "Operations on Binary Relations." *Discrete Math.* **21**, 7–22, 1978.
Graham, R. L.; Knuth, D. E.; and Motzkin, T. S. "Complements and Transitive Closures." *Discrete Math.* **2**, 17–29, 1972.
Hammer, P. C. "Kuratowski's Closure Theorem." *Nieuw Arch. Wisk.* **8**, 74–80, 1960.
Herda, H. H. and Metzler, R. C. "Closure and Interior in Finite Topological Spaces." *Colloq. Math.* **15**, 211–216, 1966.
Kelley, J. L. *General Topology.* Princeton: Van Nostrand, p. 57, 1955.
Koenen, W. "The Kuratowski Closure Problem in the Topology of Convexity." *Amer. Math. Monthly* **73**, 704–708, 1966.
Kuratowski, C. "Sur l'operation A de l'analysis situs." *Fund. Math.* **3**, 182–199, 1922.
Langford, E. "Characterization of Kuratowski 14-Sets." *Amer. Math. Monthly* **78**, 362–367, 1971.
Levine, N. "On the Commutativity of the Closure and Interior Operators in Topological Spaces." *Amer. Math. Monthly* **68**, 474–477, 1961.
Moser, L. E. "Closure, Interior, and Union in Finite Topological Spaces." *Colloq. Math.* **38**, 41–51, 1977.
Munkres, J. R. *Topology: A First Course.* Englewood Cliffs, NJ: Prentice-Hall, 1975.
Peleg, D. "A Generalized Closure and Complement Phenomenon." *Discrete Math.* **50**, 285–293, 1984.
Shum, K. P. "On the Boundary of Kuratowski 14-Sets in Connected Spaces." *Glas. Mat. Ser. III* **19**, 293–296, 1984.
Shum, K. P. "The Amalgamation of Closure and Boundary Functions on Semigroups and Partially Ordered Sets." In *Proceedings of the Conference on Ordered Structures and Algebra of Computer Languages.* Singapore: World Scientific, pp. 232–243, 1993.
Smith, A. Advanced Problem 5996. *Amer. Math. Monthly* **81**, 1034, 1974.
Soltan, V. P. "On Kuratowski's Problem." *Bull. Acad. Polon. Sci. Ser. Sci. Math.* **28**, 369–375, 1981.
Soltan, V. P. "Problems of Kuratowski Type." *Mat. Issled.* **65**, 121–131 and 155, 1982.

Kuratowski Reduction Theorem

Every nonplanar graph is a SUPERGRAPH of an expansion of the UTILITY GRAPH $UG = K_{3,3}$ or the COMPLETE GRAPH K_5. This theorem was also proven earlier by Pontryagin (1927–1928), and later by Frink and Smith (1930). Kennedy *et al.* (1985) give a detailed history of the theorem, and there exists a generalization known as the ROBERTSON-SEYMOUR THEOREM.

see also COMPLETE GRAPH, PLANAR GRAPH, ROBERTSON-SEYMOUR THEOREM, UTILITY GRAPH

References

Kennedy, J. W.; Quintas, L. V.; and Syslo, M. M. "The Theorem on Planar Graphs." *Historia Math.* **12**, 356–368, 1985.
Kuratowski, C. "Sur l'operation A de l'analysis situs." *Fund. Math.* **3**, 182–199, 1922.
Thomassen, C. "Kuratowski's Theorem." *J. Graph Th.* **5**, 225–241, 1981.
Thomassen, C. "A Link Between the Jordan Curve Theorem and the Kuratowski Planarity Criterion." *Amer. Math. Monthly* **97**, 216–218, 1990.

Kuratowski's Theorem

see KURATOWSKI REDUCTION THEOREM

Kürschák's Tile

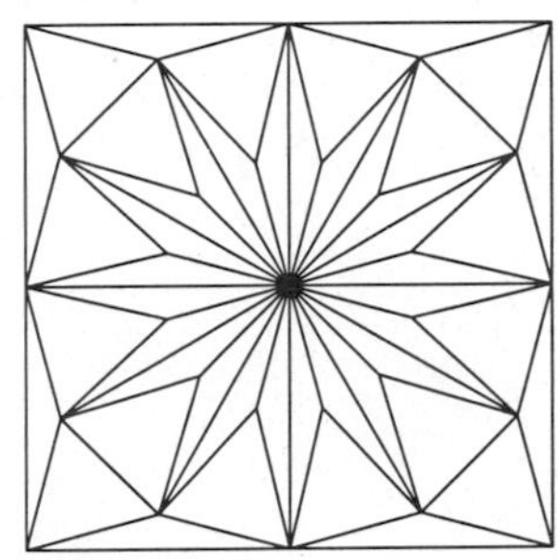

An attractive tiling of the SQUARE composed of two types of triangular tiles.

References

Alexanderson, G. L. and Seydel, K. "Kürschák's Tile." *Math. Gaz.* **62**, 192–196, 1978.
Honsberger, R. *Mathematical Gems III.* Washington, DC: Math. Assoc. Amer., pp. 30–32, 1985.
Schoenberg, I. *Mathematical Time Exposures.* Washington, DC: Math. Assoc. Amer., p. 7, 1982.
Weisstein, E. W. "Kürschák's Tile." `http://www.astro.virginia.edu/~eww6n/math/notebooks/KurschaksTile.m`.

Kurtosis

The degree of peakedness of a distribution, also called the EXCESS or EXCESS COEFFICIENT. Kurtosis is denoted γ_2 (or b_2) or β_2 and computed by taking the fourth MOMENT of a distribution. A distribution with a high peak ($\gamma_2 > 0$) is called LEPTOKURTIC, a flat-topped curve ($\gamma_2 < 0$) is called PLATYKURTIC, and the normal distribution ($\gamma_2 = 0$) is called MESOKURTIC. Let μ_i denote the ith MOMENT $\langle x^i \rangle$. The FISHER KURTOSIS is defined by

$$\gamma_2 \equiv b_2 \equiv \frac{\mu_4}{{\mu_2}^2} - 3 = \frac{\mu_4}{\sigma^4} - 3, \tag{1}$$

and the PEARSON KURTOSIS is defined by

$$\beta_2 \equiv \alpha_4 \equiv \frac{\mu_4}{\sigma^4}. \tag{2}$$

An ESTIMATOR for the γ_2 FISHER KURTOSIS is given by

$$g_2 = \frac{k_4}{{k_2}^2}, \tag{3}$$

where the ks are k-STATISTICS. The STANDARD DEVIATION of the estimator is

$$\sigma_{g_2}{}^2 \approx \frac{24}{N}. \tag{4}$$

see also FISHER KURTOSIS, MEAN, PEARSON KURTOSIS, SKEWNESS, STANDARD DEVIATION

References

Abramowitz, M. and Stegun, C. A. (Eds.). *Handbook of Mathematical Functions with Formulas, Graphs, and Mathematical Tables, 9th printing.* New York: Dover, p. 928, 1972.

Press, W. H.; Flannery, B. P.; Teukolsky, S. A.; and Vetterling, W. T. "Moments of a Distribution: Mean, Variance, Skewness, and So Forth." §14.1 in *Numerical Recipes in FORTRAN: The Art of Scientific Computing, 2nd ed.* Cambridge, England: Cambridge University Press, pp. 604–609, 1992.

L

L_1-Norm

A VECTOR NORM defined for a VECTOR

$$\mathbf{x} = \begin{bmatrix} x_1 \\ x_2 \\ \vdots \\ x_n \end{bmatrix},$$

with COMPLEX entries by

$$||\mathbf{x}||_1 = \sum_{r=1}^{n} |x_r|.$$

see also L_2-NORM, L_∞-NORM, VECTOR NORM

References

Gradshteyn, I. S. and Ryzhik, I. M. *Tables of Integrals, Series, and Products, 5th ed.* San Diego, CA: Academic Press, pp. 1114–1125, 1979.

L_2-Norm

A VECTOR NORM defined for a VECTOR

$$\mathbf{x} = \begin{bmatrix} x_1 \\ x_2 \\ \vdots \\ x_n \end{bmatrix},$$

with COMPLEX entries by

$$||\mathbf{x}||_2 = \sqrt{\sum_{r=1}^{n} |x_r|^2}.$$

The L_2-norm is also called the EUCLIDEAN NORM. The L_2-norm is defined for a function $\phi(x)$ by

$$||\phi(x)|| \equiv \phi(x) \cdot \phi(x) \equiv \left\langle [\phi(x)]^2 \right\rangle \equiv \int_a^b [\phi(x)]^2 \, dx.$$

see also L_1-NORM, L_2-SPACE, L_∞-NORM, PARALLELOGRAM LAW, VECTOR NORM

References

Gradshteyn, I. S. and Ryzhik, I. M. *Tables of Integrals, Series, and Products, 5th ed.* San Diego, CA: Academic Press, pp. 1114–1125, 1979.

L_2-Space

A HILBERT SPACE in which a BRACKET PRODUCT is defined by

$$\langle \phi | \psi \rangle \equiv \int \psi^* \phi \, dx \tag{1}$$

and which satisfies the following conditions

$$\langle \phi | \psi \rangle^* = \langle \psi | \phi \rangle \, e \tag{2}$$

$$\langle \phi | \lambda_1 \psi_1 + \lambda_2 \psi_2 \rangle = \lambda_1 \langle \phi | \psi_1 \rangle + \lambda_2 \langle \phi | \psi_2 \rangle \tag{3}$$

$$\langle \lambda_1 \phi_1 + \lambda_2 \phi_2 | \psi \rangle = \lambda_1{}^* \langle \phi_1 | \psi \rangle + \lambda_2{}^* \langle \phi_2 | \psi \rangle \tag{4}$$

$$\langle \psi | \psi \rangle \in \mathbb{R} \geq 0 \tag{5}$$

$$| \langle \psi_1 | \psi_2 \rangle |^2 \leq \langle \psi_1 | \psi_1 \rangle \langle \psi_2 | \psi_2 \rangle. \tag{6}$$

The last of these is SCHWARZ'S INEQUALITY.

see also BRACKET PRODUCT, HILBERT SPACE, L_2-NORM, RIESZ-FISCHER THEOREM, SCHWARZ'S INEQUALITY

L_∞-Norm

A VECTOR NORM defined for a VECTOR

$$\mathbf{x} = \begin{bmatrix} x_1 \\ x_2 \\ \vdots \\ x_n \end{bmatrix},$$

with COMPLEX entries by

$$||\mathbf{x}||_\infty = \max_i |x_i|.$$

see also L_1-NORM, L_2-NORM, VECTOR NORM

References

Gradshteyn, I. S. and Ryzhik, I. M. *Tables of Integrals, Series, and Products, 5th ed.* San Diego, CA: Academic Press, pp. 1114–1125, 1979.

$L_{p'}$-Balance Theorem

If every component L of $X/O_{p'}(X)$ satisfies the "Schreler property," then

$$L_{p'}(Y) \leq L_{p'}(X)$$

for every p-local SUBGROUP Y of X, where $L_{p'}$ is the p-LAYER.

see also p-LAYER, SUBGROUP

L-Estimate

A ROBUST ESTIMATION based on linear combinations of ORDER STATISTICS. Examples include the MEDIAN and TUKEY'S TRIMEAN.

see also M-ESTIMATE, R-ESTIMATE

References

Press, W. H.; Flannery, B. P.; Teukolsky, S. A.; and Vetterling, W. T. "Robust Estimation." §15.7 in *Numerical Recipes in FORTRAN: The Art of Scientific Computing, 2nd ed.* Cambridge, England: Cambridge University Press, pp. 694–700, 1992.

L-Function

see ARTIN *L*-FUNCTION, DIRICHLET *L*-SERIES, EULER *L*-FUNCTION, HECKE *L*-FUNCTION

L-Polyomino

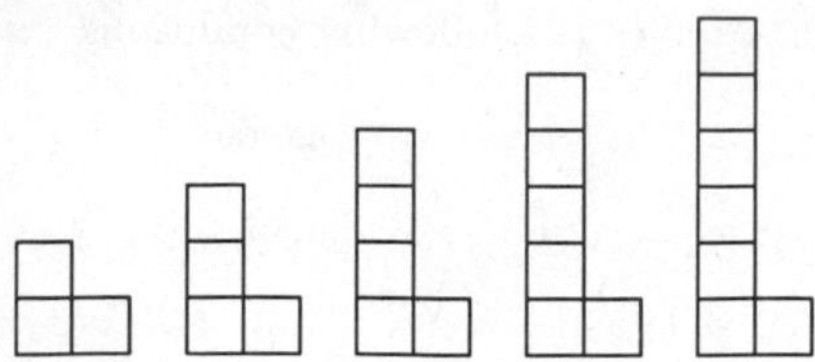

The order $n \geq 2$ L-polyomino consists of a vertical line of n SQUARES with a single additional SQUARE attached at the bottom.

see also L-POLYOMINO, SKEW POLYOMINO, SQUARE, SQUARE POLYOMINO, STRAIGHT POLYOMINO

L-Series

see DIRICHLET *L*-SERIES

L-System

see LINDENMAYER SYSTEM

L'Hospital's Cubic

see TSCHIRNHAUSEN CUBIC

L'Hospital's Rule

Let lim stand for the LIMIT $\lim_{x\to c}$, $\lim_{x\to c^-}$, $\lim_{x\to c^+}$, $\lim_{x\to\infty}$, or $\lim_{x\to-\infty}$, and suppose that $\lim f(x)$ and $\lim g(x)$ are both ZERO or are both $\pm\infty$. If

$$\lim \frac{f'(x)}{g'(x)}$$

has a finite value or if the LIMIT is $\pm\infty$, then

$$\lim \frac{f(x)}{g(x)} = \lim \frac{f'(x)}{g'(x)}.$$

L'Hospital's rule occasionally fails to yield useful results, as in the case of the function $\lim_{u\to\infty} u(u^2+1)^{-1/2}$. Repeatedly applying the rule in this case gives expressions which oscillate and never converge,

$$\begin{aligned}\lim_{u\to\infty} \frac{u}{(u^2+1)^{1/2}} &= \lim_{u\to\infty} \frac{1}{u(u^2+1)^{-1/2}} \\ &= \lim_{u\to\infty} \frac{(u^2+1)^{1/2}}{u} = \lim_{u\to\infty} \frac{u(u^2+1)^{-1/2}}{1} \\ &= \lim_{u\to\infty} \frac{u}{(u^2+1)^{1/2}}.\end{aligned}$$

(The actual LIMIT is 1.)

References

Abramowitz, M. and Stegun, C. A. (Eds.). *Handbook of Mathematical Functions with Formulas, Graphs, and Mathematical Tables, 9th printing.* New York: Dover, p. 13, 1972.

L'Hospital, G. de *L'analyse des infiniment petits pour l'intelligence des lignes courbes.* 1696.

L'Huilier's Theorem

Let a SPHERICAL TRIANGLE have sides of length a, b, and c, and SEMIPERIMETER s. Then the SPHERICAL EXCESS Δ is given by

$$\begin{aligned}&\tan(\tfrac{1}{4}\Delta) \\ &\quad = \sqrt{\tan(\tfrac{1}{2}s)\tan[\tfrac{1}{2}(s-a)]\tan[\tfrac{1}{2}(s-b)]\tan[\tfrac{1}{2}(s-c)]}.\end{aligned}$$

see also GIRARD'S SPHERICAL EXCESS FORMULA, SPHERICAL EXCESS,SPHERICAL TRIANGLE

References

Beyer, W. H. *CRC Standard Mathematical Tables, 28th ed.* Boca Raton, FL: CRC Press, p. 148, 1987.

Labelled Graph

A labelled graph $G = (V, E)$ is a finite series of VERTICES V with a set of EDGES E of 2-SUBSETS of V. Given a VERTEX set $V_n = \{1, 2, \ldots, n\}$, the number of labelled graphs is given by $2^{n(n-1)/2}$. Two graphs G and H with VERTICES $V_n = \{1, 2, \ldots, n\}$ are said to be ISOMORPHIC if there is a PERMUTATION p of V_n such that $\{u, v\}$ is in the set of EDGES $E(G)$ IFF $\{p(u), p(v)\}$ is in the set of EDGES $E(H)$.

see also CONNECTED GRAPH, GRACEFUL GRAPH, GRAPH (GRAPH THEORY), HARMONIOUS GRAPH, MAGIC GRAPH, TAYLOR'S CONDITION, WEIGHTED TREE

References

Cahit, I. "Homepage for the Graph Labelling Problems and New Results." `http://193.140.42.134/~cahit/CORDIAL.html`.

Gallian, J. A. "Graph Labelling." *Elec. J. Combin.* DS6, 1–43, Mar. 5, 1998. `http://www.combinatorics.org/Surveys/`.

Lacunarity

Quantifies deviation from translational invariance by describing the distribution of gaps within a set at multiple scales. The more lacunar a set, the more heterogeneous the spatial arrangement of gaps.

Ladder

see ASTROID, CROSSED LADDERS PROBLEM, LADDER GRAPH

Ladder Graph

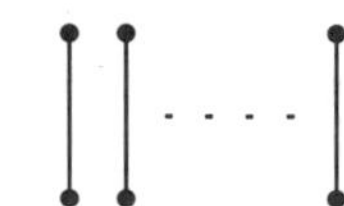

A GRAPH consisting of two rows of paired nodes each connected by an EDGE. Its complement is the COCKTAIL PARTY GRAPH.

see also COCKTAIL PARTY GRAPH

Lagrange Bracket

Let F and G be infinitely differentiable functions of x, u, and p. Then the Lagrange bracket is defined by

$$[F,G] = \sum_{\nu=1}^{n}\left[\frac{\partial F}{\partial p_\nu}\left(\frac{\partial G}{\partial x_p}+p_\nu\frac{\partial G}{\partial u}\right) - \frac{\partial G}{\partial p_\nu}\left(\frac{\partial F}{\partial x_\nu}+p_\nu\frac{\partial F}{\partial u}\right)\right]. \quad (1)$$

The Lagrange bracket satisfies

$$[F,G] = -[G,F] \quad (2)$$

$$[[F,G],H]+[[G,H],F]+[[H,F],G] = \frac{\partial F}{\partial u}[G,H]+\frac{\partial G}{\partial u}[H,F]+\frac{\partial H}{\partial u}[F,G]. \quad (3)$$

If F and G are functions of x and p only, then the Lagrange bracket $[F,G]$ collapses the POISSON BRACKET (F,G).

see also LIE BRACKET, POISSON BRACKET

References

Iyanaga, S. and Kawada, Y. (Eds.). *Encyclopedic Dictionary of Mathematics.* Cambridge, MA: MIT Press, p. 1004, 1980.

Lagrange-Bürmann Theorem

see LAGRANGE INVERSION THEOREM

Lagrangian Coefficient

COEFFICIENTS which appear in LAGRANGE INTERPOLATING POLYNOMIALS where the points are equally spaced along the ABSCISSA.

Lagrange's Continued Fraction Theorem

The REAL ROOTS of quadratic expressions with integral COEFFICIENTS have periodic CONTINUED FRACTIONS, as first proved by Lagrange.

Lagrangian Derivative

see CONVECTIVE DERIVATIVE

Lagrange's Equation

The PARTIAL DIFFERENTIAL EQUATION

$$(1+{f_y}^2)f_{xx}+2f_xf_yf_{xy}+(1+{f_x}^2)f_{yy}=0,$$

whose solutions are called MINIMAL SURFACES.

see also MINIMAL SURFACE

References

do Carmo, M. P. "Minimal Surfaces." §3.5 in *Mathematical Models from the Collections of Universities and Museums* (Ed. G. Fischer). Braunschweig, Germany: Vieweg, pp. 41–43, 1986.

Lagrange Expansion

Let $y=f(x)$ and $y_0=f(x_0)$ where $f'(x_0)\neq 0$, then

$$x = x_0+\sum_{k=1}^{\infty}\frac{(y-y_0)^k}{k!}\left\{\frac{d^{k-1}}{dx^{k-1}}\left[\frac{x-x_0}{f(x)-y_0}\right]^k\right\}_{x=x_0}$$

$$g(x) = g(x_0) + \sum_{k=1}^{\infty}\frac{(y-y_0)^k}{k!}\left\{\frac{d^{k-1}}{dx^{k-1}}\left[g'(x)\left(\frac{x-x_0}{f(x)-y_0}\right)^k\right]\right\}_{x=x_0}$$

see also MACLAURIN SERIES, TAYLOR SERIES

References

Abramowitz, M. and Stegun, C. A. (Eds.). *Handbook of Mathematical Functions with Formulas, Graphs, and Mathematical Tables, 9th printing.* New York: Dover, p. 14, 1972.

Lagrange's Four-Square Theorem

A theorem also known as BACHET'S CONJECTURE which was stated but not proven by Diophantus. It states that every POSITIVE INTEGER can be written as the SUM of at most four SQUARES. Although the theorem was proved by Fermat using infinite descent, the proof was suppressed. Euler was unable to prove the theorem. The first published proof was given by Lagrange in 1770 and made use of the EULER FOUR-SQUARE IDENTITY.

see also EULER FOUR-SQUARE IDENTITY, FERMAT'S POLYGONAL NUMBER THEOREM, FIFTEEN THEOREM, VINOGRADOV'S THEOREM, WARING'S PROBLEM

Lagrange's Group Theorem

Also known as LAGRANGE'S LEMMA. If A is an ELEMENT of a FINITE GROUP of order n, then $A^n=1$. This implies that $e|n$ where e is the smallest exponent such that $A^e=1$. Stated another way, the ORDER of a SUBGROUP divides the ORDER of the GROUP. The converse of Lagrange's theorem is not, in general, true (Gallian 1993, 1994).

References

Birkhoff, G. and Mac Lane, S. *A Brief Survey of Modern Algebra, 2nd ed.* New York: Macmillan, p. 111, 1965.

Gallian, J. A. "On the Converse of Lagrange's Theorem." *Math. Mag.* **63**, 23, 1993.

Gallian, J. A. *Contemporary Abstract Algebra, 3rd ed.* Lexington, MA: D. C. Heath, 1994.

Herstein, I. N. *Abstract Algebra, 2nd ed.* New York: Macmillan, p. 66, 1990.

Hogan, G. T. "More on the Converse of Lagrange's Theorem." *Math. Mag.* **69**, 375–376, 1996.

Shanks, D. *Solved and Unsolved Problems in Number Theory, 4th ed.* New York: Chelsea, p. 86, 1993.

Lagrange's Identity

The vector identity

$$(\mathbf{A}\times\mathbf{B})\cdot(\mathbf{C}\times\mathbf{D}) = (\mathbf{A}\cdot\mathbf{C})(\mathbf{B}\cdot\mathbf{D})-(\mathbf{A}\cdot\mathbf{D})(\mathbf{B}\cdot\mathbf{C}). \quad (1)$$

This identity can be generalized to n-D,

$$(\mathbf{a}_1 \times \cdots \times \mathbf{a}_{n-1}) \cdot (\mathbf{b}_1 \times \cdots \times \mathbf{b}_{n-1}) = \begin{vmatrix} \mathbf{a}_1 \cdot \mathbf{b}_1 & \cdots & \mathbf{a}_1 \cdot \mathbf{b}_{n-1} \\ \vdots & \ddots & \vdots \\ \mathbf{a}_{n-1} \cdot \mathbf{b}_1 & \cdots & \mathbf{a}_{n-1} \cdot \mathbf{b}_{n-1} \end{vmatrix}, \quad (2)$$

where $|\mathsf{A}|$ is the DETERMINANT of A, or

$$\left(\sum_{k=1}^n a_k b_k\right)^2 = \left(\sum_{k=1}^n {a_k}^2\right)\left(\sum_{k=1}^n {b_k}^2\right) - \sum_{1\leq k\leq j\leq n} (a_k b_j - a_j b_k)^2. \quad (3)$$

see also VECTOR TRIPLE PRODUCT, VECTOR QUADRUPLE PRODUCT

References

Gradshteyn, I. S. and Ryzhik, I. M. *Tables of Integrals, Series, and Products, 5th ed.* San Diego, CA: Academic Press, p. 1093, 1979.

Lagrange's Interpolating Fundamental Polynomial

Let $l(x)$ be an nth degree POLYNOMIAL with zeros at $x_1, \ldots, x_m$. Then the fundamental POLYNOMIALS are

$$l_\nu(x) = \frac{l(x)}{l'(x_\nu)(x - x_\nu)}. \quad (1)$$

They have the property

$$l_\nu(x) = \delta_{\nu\mu}, \quad (2)$$

where $\delta_{\nu\mu}$ is the KRONECKER DELTA. Now let $f_1, \ldots, f_n$ be values. Then the expansion

$$L_n(x) = \sum_{\nu=1}^n f_\nu l_\nu(x) \quad (3)$$

gives the unique LAGRANGE INTERPOLATING POLYNOMIAL assuming the values f_ν at x_ν. Let $d\alpha(x)$ be an arbitrary distribution on the interval $[a, b]$, $\{p_n(x)\}$ the associated ORTHOGONAL POLYNOMIALS, and $l_1(x), \ldots, l_n(x)$ the fundamental POLYNOMIALS corresponding to the set of zeros of $p_n(x)$. Then

$$\int_a^b l_\nu(x) l_\mu(x)\, d\alpha(x) = \lambda_\mu \delta_{\nu\mu} \quad (4)$$

for $\nu, \mu = 1, 2, \ldots, n$, where λ_ν are CHRISTOFFEL NUMBERS.

References

Szegő, G. *Orthogonal Polynomials, 4th ed.* Providence, RI: Amer. Math. Soc., pp. 329 and 332, 1975.

Lagrange Interpolating Polynomial

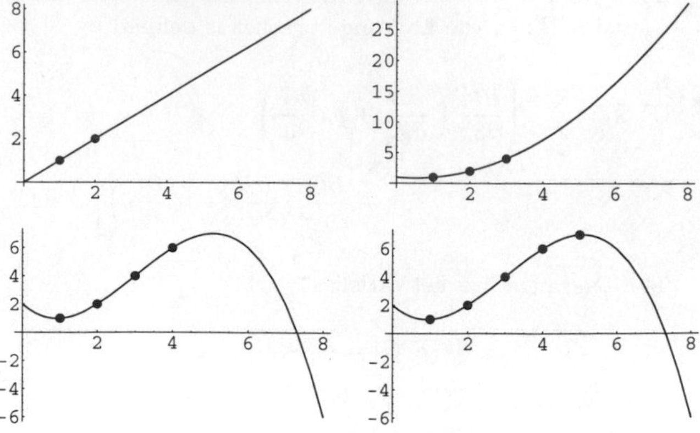

The Lagrange interpolating polynomial is the POLYNOMIAL of degree $n-1$ which passes through the n points $y_1 = f(x_1)$, $y_2 = f(x_2)$, ..., $y_n = f(x_n)$. It is given by

$$P(x) = \sum_{j=1}^n P_j(x), \quad (1)$$

where

$$P_j(x) = \prod_{\substack{k=1\\k\neq j}}^n \frac{x - x_k}{x_j - x_k} y_j. \quad (2)$$

Written explicitly,

$$\begin{aligned} P(x) = {} & \frac{(x-x_2)(x-x_3)\cdots(x-x_n)}{(x_1-x_2)(x_1-x_3)\cdots(x_1-x_n)} y_1 \\ & + \frac{(x-x_1)(x-x_3)\cdots(x-x_n)}{(x_2-x_1)(x_2-x_3)\cdots(x_2-x_n)} y_2 + \cdots \\ & + \frac{(x-x_1)(x-x_2)\cdots(x-x_{n-1})}{(x_n-x_1)(x_n-x_2)\cdots(x_n-x_{n-1})} y_n. \quad (3) \end{aligned}$$

For $n = 3$ points,

$$\begin{aligned} P(x) = {} & \frac{(x-x_2)(x-x_3)}{(x_1-x_2)(x_1-x_3)} y_1 + \frac{(x-x_1)(x-x_3)}{(x_2-x_1)(x_2-x_3)} y_2 \\ & + \frac{(x-x_1)(x-x_2)}{(x_3-x_1)(x_3-x_2)} y_3 \quad (4) \end{aligned}$$

$$\begin{aligned} P'(x) = {} & \frac{2x-x_2-x_3}{(x_1-x_2)(x_1-x_3)} y_1 + \frac{2x-x_1-x_3}{(x_2-x_1)(x_2-x_3)} y_2 \\ & + \frac{2x-x_1-x_2}{(x_3-x_1)(x_3-x_2)} y_3. \quad (5) \end{aligned}$$

Note that the function $P(x)$ passes through the points (x_i, y_i), as can be seen for the case $n = 3$,

$$\begin{aligned} P(x_1) = {} & \frac{(x_1-x_2)(x_1-x_3)}{(x_1-x_2)(x_1-x_3)} y_1 + \frac{(x_1-x_1)(x_1-x_3)}{(x_2-x_1)(x_2-x_3)} y_2 \\ & + \frac{(x_1-x_1)(x_1-x_2)}{(x_3-x_1)(x_3-x_2)} y_3 = y_1 \quad (6) \end{aligned}$$

$$P(x_2) = \frac{(x_2-x_2)(x_2-x_3)}{(x_1-x_2)(x_1-x_3)} y_1 + \frac{(x_2-x_1)(x_2-x_3)}{(x_2-x_1)(x_2-x_3)} y_2$$

$$+\frac{(x_2-x_1)(x_2-x_2)}{(x_3-x_1)(x_3-x_2)}y_3 = y_2 \quad (7)$$

$$P(x_3) = \frac{(x_3-x_2)(x_3-x_3)}{(x_1-x_2)(x_1-x_3)}y_1 + \frac{(x_3-x_1)(x_3-x_3)}{(x_2-x_1)(x_2-x_3)}y_2 + \frac{(x_3-x_1)(x_3-x_2)}{(x_3-x_1)(x_3-x_2)}y_3 = y_3. \quad (8)$$

Generalizing to arbitrary n,

$$P(x_j) = \sum_{k=1}^{n} P_k(x_j) = \sum_{k=1}^{n} \delta_{jk} y_k = y_j. \quad (9)$$

The Lagrange interpolating polynomials can also be written using

$$\pi(x) \equiv \prod_{k=1}^{n}(x-x_k), \quad (10)$$

$$\pi(x_j) = \prod_{k=1}^{n}(x_j-x_k), \quad (11)$$

$$\pi'(x_j) = \left[\frac{d\pi}{dx}\right]_{x=x_j} = \prod_{\substack{k=1\\k\neq j}}^{n}(x_j-x_k), \quad (12)$$

so

$$P(x) = \sum_{k=1}^{n} \frac{\pi(x)}{(x-x_k)\pi'(x_k)} y_k. \quad (13)$$

Lagrange interpolating polynomials give no error estimate. A more conceptually straightforward method for calculating them is NEVILLE'S ALGORITHM.

see also AITKEN INTERPOLATION, LEBESGUE CONSTANTS (LAGRANGE INTERPOLATION), NEVILLE'S ALGORITHM, NEWTON'S DIVIDED DIFFERENCE INTERPOLATION FORMULA

References

Abramowitz, M. and Stegun, C. A. (Eds.). *Handbook of Mathematical Functions with Formulas, Graphs, and Mathematical Tables, 9th printing.* New York: Dover, pp. 878–879 and 883, 1972.

Beyer, W. H. (Ed.) *CRC Standard Mathematical Tables, 28th ed.* Boca Raton, FL: CRC Press, p. 439, 1987.

Press, W. H.; Flannery, B. P.; Teukolsky, S. A.; and Vetterling, W. T. "Polynomial Interpolation and Extrapolation" and "Coefficients of the Interpolating Polynomial." §3.1 and 3.5 in *Numerical Recipes in FORTRAN: The Art of Scientific Computing, 2nd ed.* Cambridge, England: Cambridge University Press, pp. 102–104 and 113–116, 1992.

Lagrange Inversion Theorem

Let z be defined as a function of w in terms of a parameter α by

$$z = w + \alpha\phi(z).$$

Then any function of z can be expressed as a POWER SERIES in α which converges for sufficiently small α and has the form

$$F(z) = F(w) + \frac{\alpha}{1}\phi(w)F'(w) + \frac{\alpha^2}{1\cdot 2}\frac{\partial}{\partial w}\{[\phi(w)]^2 F'(w)\} + \ldots + \frac{\alpha^{n+1}}{(n+1)!}\frac{\partial^n}{\partial w^n}\{[\phi(w)]^{n+1}F'(w)\} + \ldots.$$

References

Goursat, E. *Functions of a Complex Variable, Vol. 2, Pt. 1.* New York: Dover, 1959.

Moulton, F. R. *An Introduction to Celestial Mechanics, 2nd rev. ed.* New York: Dover, p. 161, 1970.

Williamson, B. "Remainder in Lagrange's Series." §119 in *An Elementary Treatise on the Differential Calculus, 9th ed.* London: Longmans, pp. 158–159, 1895.

Lagrange's Lemma

see LAGRANGE'S FOUR-SQUARE THEOREM

Lagrange Multiplier

Used to find the EXTREMUM of $f(x_1, x_2, \ldots, x_n)$ subject to the constraint $g(x_1, x_2, \ldots, x_n) = C$, where f and g are functions with continuous first PARTIAL DERIVATIVES on the OPEN SET containing the curve $g(x_1, x_2, \ldots, x_n) = 0$, and $\nabla g \neq \mathbf{0}$ at any point on the curve (where ∇ is the GRADIENT). For an EXTREMUM to exist,

$$df = \frac{\partial f}{\partial x_1}dx_1 + \frac{\partial f}{\partial x_2}dx_2 + \ldots + \frac{\partial f}{\partial x_n}dx_n = 0. \quad (1)$$

But we also have

$$dg = \frac{\partial g}{\partial x_1}dx_1 + \frac{\partial g}{\partial x_2}dx_2 + \ldots + \frac{\partial g}{\partial x_n}dx_n = 0. \quad (2)$$

Now multiply (2) by the as yet undetermined parameter λ and add to (1),

$$\left(\frac{\partial f}{\partial x_1} + \lambda\frac{\partial g}{\partial x_1}\right)dx_1 + \left(\frac{\partial f}{\partial x_2} + \lambda\frac{\partial g}{\partial x_2}\right)dx_2 + \ldots + \left(\frac{\partial f}{\partial x_n} + \lambda\frac{\partial g}{\partial x_n}\right)dx_n = 0. \quad (3)$$

Note that the differentials are all independent, so we can set any combination equal to 0, and the remainder must still give zero. This requires that

$$\frac{\partial f}{\partial x_k} + \lambda\frac{\partial g}{\partial x_k} = 0 \quad (4)$$

for all $k = 1, \ldots, n$. The constant λ is called the Lagrange multiplier. For multiple constraints, $g_1 = 0$, $g_2 = 0, \ldots,$

$$\nabla f = \lambda_1 \nabla g_1 + \lambda_2 \nabla g_2 + \ldots. \quad (5)$$

see also KUHN-TUCKER THEOREM

References

Arfken, G. "Lagrange Multipliers." §17.6 in *Mathematical Methods for Physicists, 3rd ed.* Orlando, FL: Academic Press, pp. 945–950, 1985.

Lagrange Number (Diophantine Equation)

Given a FERMAT DIFFERENCE EQUATION (a quadratic DIOPHANTINE EQUATION)

$$x^2 - r^2 y^2 = 4$$

with r a QUADRATIC SURD, assign to each solution $x|y$ the Lagrange number

$$z \equiv \tfrac{1}{2}(x + yr).$$

The product and quotient of two Lagrange numbers are also Lagrange numbers. Furthermore, every Lagrange number is a POWER of the smallest Lagrange number with an integral exponent.

see also PELL EQUATION

References

Dörrie, H. *100 Great Problems of Elementary Mathematics: Their History and Solutions.* New York: Dover, pp. 94–95, 1965.

Lagrange Number (Rational Approximation)

HURWITZ'S IRRATIONAL NUMBER THEOREM gives the best rational approximation possible for an arbitrary irrational number α as

$$\left|\alpha - \frac{p}{q}\right| < \frac{1}{L_n q^2}.$$

The L_n are called Lagrange numbers and get steadily larger for each "bad" set of irrational numbers which is excluded.

n	Exclude	L_n
1	none	$\sqrt{5}$
2	ϕ	$\sqrt{8}$
3	$\sqrt{2}$	$\frac{\sqrt{221}}{5}$

Lagrange numbers are of the form

$$\sqrt{9 - \frac{4}{m^2}},$$

where m is a MARKOV NUMBER. The Lagrange numbers form a SPECTRUM called the LAGRANGE SPECTRUM.

see also HURWITZ'S IRRATIONAL NUMBER THEOREM, LIOUVILLE'S RATIONAL APPROXIMATION THEOREM, LIOUVILLE-ROTH CONSTANT, MARKOV NUMBER, ROTH'S THEOREM, SPECTRUM SEQUENCE, THUE-SIEGEL-ROTH THEOREM

References

Conway, J. H. and Guy, R. K. *The Book of Numbers.* New York: Springer-Verlag, pp. 187–189, 1996.

Lagrange Polynomial

see LAGRANGE INTERPOLATING POLYNOMIAL

Lagrange Remainder

Given a TAYLOR SERIES, the error after n terms is bounded by

$$R_n = \frac{f^{(n)}(\xi)}{n!}(x - a)^n$$

for some $\xi \in (a, x)$.

see also CAUCHY REMAINDER FORM, TAYLOR SERIES

Lagrange Resolvent

A quantity involving primitive cube roots of unity which can be used to solve the CUBIC EQUATION.

References

Faucette, W. M. "A Geometric Interpretation of the Solution of the General Quartic Polynomial." *Amer. Math. Monthly* **103**, 51–57, 1996.

Lagrange Spectrum

A SPECTRUM formed by the LAGRANGE NUMBERS. The only ones less than three are the LAGRANGE NUMBERS, but the last gaps end at FREIMAN'S CONSTANT. REAL NUMBERS larger than FREIMAN'S CONSTANT are in the MARKOV SPECTRUM.

see also FREIMAN'S CONSTANT, LAGRANGE NUMBER (RATIONAL APPROXIMATION), MARKOV SPECTRUM, SPECTRUM SEQUENCE

References

Conway, J. H. and Guy, R. K. *The Book of Numbers.* New York: Springer-Verlag, pp. 187–189, 1996.

Laguerre Differential Equation

$$xy'' + (1 - x)y' + \lambda y = 0. \tag{1}$$

The Laguerre differential equation is a special case of the more general "associated Laguerre differential equation"

$$xy'' + (\nu + 1 - x)y' + \lambda y = 0 \tag{2}$$

with $\nu = 0$. Note that if $\lambda = 0$, then the solution to the associated Laguerre differential equation is of the form

$$y''(x) + P(x)y'(x) = 0, \tag{3}$$

and the solution can be found using an INTEGRATING FACTOR

$$\begin{aligned}\mu &= \exp\left(\int P(x)\,dx\right) = \exp\left(\int \frac{\nu + 1 - x}{x}\,dx\right)\\ &= \exp[(\nu + 1)\ln x - x] = x^{\nu+1}e^{-x},\end{aligned} \tag{4}$$

so

$$y = C_1 \int \frac{dx}{\mu} + C_2 = C_1 \int \frac{e^x}{x^{\nu+1}}\,dx + C_2. \tag{5}$$

The associated Laguerre differential equation has a REGULAR SINGULAR POINT at 0 and an IRREGULAR

SINGULARITY at ∞. It can be solved using a series expansion,

$$x\sum_{n=2}^{\infty} n(n-1)a_n x^{n-2} + (\nu+1)\sum_{n=1}^{\infty} na_n x^{n-1} - x\sum_{n=1}^{\infty} na_n x^{n-1} + \lambda\sum_{n=0}^{\infty} a_n x^n = 0 \quad (6)$$

$$\sum_{n=2}^{\infty} n(n-1)a_n x^{n-1} + (\nu+1)\sum_{n=1}^{\infty} na_n x^{n-1} - \sum_{n=1}^{\infty} na_n x^n + \lambda\sum_{n=0}^{\infty} a_n x^n = 0 \quad (7)$$

$$\sum_{n=1}^{\infty}(n+1)na_{n+1}x^n + (\nu+1)\sum_{n=0}^{\infty}(n+1)a_{n+1}x^n - \sum_{n=1}^{\infty} na_n x^n + \lambda\sum_{n=0}^{\infty} a_n x^n = 0 \quad (8)$$

$$[(n+1)a_1 + \lambda a_0] + \sum_{n=1}^{\infty}\{[(n+1)n + (\nu+1)(n+1)]a_{n+1} - na_n + \lambda a_n\}x^n = 0 \quad (9)$$

$$[(n+1)a_1 + \lambda a_0] + \sum_{n=1}^{\infty}[(n+1)(n+\nu+1)a_{n+1} + (\lambda - n)a_n]x^n = 0. \quad (10)$$

This requires

$$a_1 = -\frac{\lambda}{\nu+1}a_0 \quad (11)$$

$$a_{n+1} = \frac{n-\lambda}{(n+1)(n+\nu+1)}a_n \quad (12)$$

for $n > 1$. Therefore,

$$a_{n+1} = \frac{n-\lambda}{(n+1)(n+\nu+1)}a_n \quad (13)$$

for $n = 1, 2, \ldots$, so

$$y = a_0\left[1 - \frac{\lambda}{\nu+1}x - \frac{\lambda(1-\lambda)}{2(\nu+1)(\nu+2)}x^2 - \frac{\lambda(1-\lambda)(2-\lambda)}{2\cdot 3(\nu+1)(\nu+2)(\nu+3)} + \cdots\right]. \quad (14)$$

If λ is a POSITIVE INTEGER, then the series terminates and the solution is a POLYNOMIAL, known as an associated LAGUERRE POLYNOMIAL (or, if $\nu = 0$, simply a LAGUERRE POLYNOMIAL).

see also LAGUERRE POLYNOMIAL

Laguerre-Gauss Quadrature

Also called GAUSS-LAGUERRE QUADRATURE or LAGUERRE QUADRATURE. A GAUSSIAN QUADRATURE over the interval $[0, \infty)$ with WEIGHTING FUNCTION $W(x) = e^{-x}$. The ABSCISSAS for quadrature order n are given by the ROOTS of the LAGUERRE POLYNOMIALS $L_n(x)$. The weights are

$$w_i = -\frac{A_{n+1}\gamma_n}{A_n L_n'(x_i)L_{n+1}(x_i)} = \frac{A_n}{A_{n-1}}\frac{\gamma_{n-1}}{L_{n-1}(x_i)L_n'(x_i)}, \quad (1)$$

where A_n is the COEFFICIENT of x^n in $L_n(x)$. For LAGUERRE POLYNOMIALS,

$$A_n = (-1)^n n!, \quad (2)$$

where $n!$ is a FACTORIAL, so

$$\frac{A_{n+1}}{A_n} = -(n+1). \quad (3)$$

Additionally,

$$\gamma_n = 1, \quad (4)$$

so

$$w_i = \frac{n+1}{L_{n+1}(x_i)L_n'(x_i)} = -\frac{n}{L_{n-1}(x_i)L_n'(x_i)}. \quad (5)$$

(Note that the normalization used here is different than that in Hildebrand 1956.) Using the recurrence relation

$$xL_n'(x) = nL_n(x) - nL_{n-1}(x) = (x-n-1)L_n(x) + (n+1)L_{n+1}(x) \quad (6)$$

which implies

$$x_i L_n'(x_i) = -nL_{n-1}(x_i) = (n+1)L_{n+1}(x_i) \quad (7)$$

gives

$$w_i = \frac{1}{x_i[L_n'(x_i)]^2} = \frac{x_i}{(n+1)^2[L_{n+1}(x_i)]^2}. \quad (8)$$

The error term is

$$E = \frac{(n!)^2}{(2n)!}f^{(2n)}(\xi). \quad (9)$$

Beyer (1987) gives a table of ABSCISSAS and weights up to $n = 6$.

n	x_i	w_i
2	0.585786	0.853553
	3.41421	0.146447
3	0.415775	0.711093
	2.29428	0.278518
	6.28995	0.0103893
4	0.322548	0.603154
	1.74576	0.357419
	4.53662	0.0388879
	9.39507	0.000539295
5	0.26356	0.521756
	1.4134	0.398667
	3.59643	0.0759424
	7.08581	0.00361176
	12.6408	0.00002337

The ABSCISSAS and weights can be computed analytically for small n.

n	x_i	w_i
2	$2-\sqrt{2}$	$\frac{1}{4}(2+\sqrt{2})$
	$2+\sqrt{2}$	$\frac{1}{4}(2-\sqrt{2})$

For the associated Laguerre polynomial $L_n^\beta(x)$ with WEIGHTING FUNCTION $w(x)=x^\beta e^{-x}$,

$$A_n=(-1)^n \tag{10}$$

and

$$\gamma_n=n!\int_0^\infty x^{\beta+n}e^{-x}\,dx=n!\Gamma(n+\beta+1). \tag{11}$$

The weights are

$$w_i=\frac{n!\Gamma(n+\beta+1)}{x_i[{L_m^\beta}'(x_i)]^2}=\frac{n!\Gamma(n+\beta+1)x_i}{[L_{n+1}^\beta(x_i)]^2}, \tag{12}$$

where $\Gamma(z)$ is the GAMMA FUNCTION, and the error term is

$$E_n=\frac{n!\Gamma(n+\beta+1)}{(2n)!}f^{(2n)}(\xi). \tag{13}$$

References

Beyer, W. H. *CRC Standard Mathematical Tables, 28th ed.* Boca Raton, FL: CRC Press, p. 463, 1987.

Chandrasekhar, S. *Radiative Transfer.* New York: Dover, pp. 64–65, 1960.

Hildebrand, F. B. *Introduction to Numerical Analysis.* New York: McGraw-Hill, pp. 325–327, 1956.

Laguerre's Method

A ROOT-finding algorithm which converges to a COMPLEX ROOT from any starting position.

$$P_n(x)=(x-x_1)(x-x_2)\cdots(x-x_n) \tag{1}$$

$$\ln|P_n(x)|=\ln|x-x_1|+\ln|x-x_2|+\ldots+\ln|x-x_n| \tag{2}$$

$$\begin{aligned}P_n'(x)&=(x-x_2)\cdots(x-x_n)+(x-x_1)\cdots(x-x_n)+\ldots\\&=P_n(x)\left(\frac{1}{x-x_1}+\ldots+\frac{1}{x-x_n}\right)\end{aligned} \tag{3}$$

$$\begin{aligned}\frac{d\ln|P_n(x)|}{dx}&=\frac{1}{x-x_1}+\frac{1}{x-x_2}+\ldots+\frac{1}{x-x_n}\\&=\frac{P_n'(x)}{P_n(x)}\equiv G(x)\end{aligned} \tag{4}$$

$$\begin{aligned}-&\frac{d^2\ln|P_n(x)|}{dx^2}\\&=\frac{1}{(x-x_1)^2}+\frac{1}{(x-x_2)^2}+\ldots+\frac{1}{(x-x_n)^2}\\&=\left[\frac{P_n'(x)}{P_n(x)}\right]^2-\frac{P_n''(x)}{P_n(x)}\equiv H(x).\end{aligned} \tag{5}$$

Now let $a\equiv x-x_1$ and $b\equiv x-x_1$. Then

$$G\equiv\frac{1}{a}+\frac{n-1}{b} \tag{6}$$

$$H\equiv\frac{1}{a^2}+\frac{n-1}{b^2}, \tag{7}$$

so

$$a=\frac{n}{\max\left[G\pm\sqrt{(n-1)(nH-G^2)}\right]}. \tag{8}$$

Setting $n=2$ gives HALLEY'S IRRATIONAL FORMULA.

see also HALLEY'S IRRATIONAL FORMULA, HALLEY'S METHOD, NEWTON'S METHOD, ROOT

References

Press, W. H.; Flannery, B. P.; Teukolsky, S. A.; and Vetterling, W. T. *Numerical Recipes in FORTRAN: The Art of Scientific Computing, 2nd ed.* Cambridge, England: Cambridge University Press, pp. 365–366, 1992.

Ralston, A. and Rabinowitz, P. §8.9–8.13 in *A First Course in Numerical Analysis, 2nd ed.* New York: McGraw-Hill, 1978.

Laguerre Polynomial

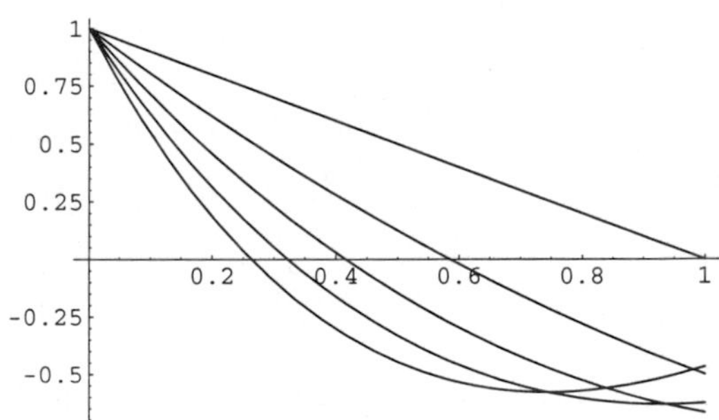

Solutions to the LAGUERRE DIFFERENTIAL EQUATION with $\nu=0$ are called Laguerre polynomials. The Laguerre polynomials $L_n(x)$ are illustrated above for $x\in[0,1]$ and $n=1, 2, \ldots, 5$.

The Rodrigues formula for the Laguerre polynomials is

$$L_n(x)=\frac{e^x}{n!}\frac{d^n}{dx^n}(x^ne^{-x}) \tag{1}$$

and the GENERATING FUNCTION for Laguerre polynomials is

$$\begin{aligned}g(x,z)&=\frac{\exp\left(-\frac{xz}{1-z}\right)}{1-z}=1+(-x+1)z\\&+(\tfrac{1}{2}x^2-2x+1)z^2+(-\tfrac{1}{6}x^3+\tfrac{3}{2}x^2-3x+1)z^3+\ldots.\end{aligned} \tag{2}$$

A CONTOUR INTEGRAL is given by

$$L_n(x)=\frac{1}{2\pi i}\int\frac{e^{-xz/(1-z)}}{(1-z)z^{n+1}}\,dz. \tag{3}$$

The Laguerre polynomials satisfy the RECURRENCE RELATIONS

$$(n+1)L_{n+1}(x) = (2n+1-x)L_n(x) - nL_{n-1}(x) \quad (4)$$

(Petkovšek *et al.* 1996) and

$$xL_n'(x) = nL_n(x) - nL_{n-1}(x). \quad (5)$$

The first few Laguerre polynomials are

$$\begin{aligned} L_0(x) &= 1 \\ L_1(x) &= -x+1 \\ L_2(x) &= \tfrac{1}{2}(x^2-4x+2) \\ L_3(x) &= \tfrac{1}{6}(-x^3+9x^2-18x+6). \end{aligned}$$

Solutions to the associated LAGUERRE DIFFERENTIAL EQUATION with $\nu \neq 0$ are called associated Laguerre polynomials $L_n^k(x)$. In terms of the normal Laguerre polynomials,

$$L_n(x) = L_n^0(x). \quad (6)$$

The Rodrigues formula for the associated Laguerre polynomials is

$$\begin{aligned} L_n^k(x) &= \frac{e^x x^{-k}}{n!}\frac{d^n}{dx^n}(e^{-x}x^{n+k}) \\ &= (-1)^n \frac{d^n}{dx^n}[L_{n+k}(x)] \quad (7) \\ &= \sum_{m=0}^{\infty}(-1)^m \frac{(n+k)!}{(n-m)!(k+m)!m!}x^m \quad (8) \end{aligned}$$

and the GENERATING FUNCTION is

$$g(x,z) = \frac{\exp\left(-\frac{xz}{1-z}\right)}{(1-z)^{k+1}}$$
$$1+(k+1-x)z+\tfrac{1}{2}[x^2-2(k+2)x+(k+1)(k+2)]z^2+\ldots. \quad (9)$$

The associated Laguerre polynomials are orthogonal over $[0,\infty)$ with respect to the WEIGHTING FUNCTION $x^n e^{-x}$.

$$\int_0^\infty e^{-x}x^k L_n^k(x)L_m^k(x)\,dx = \frac{(n+k)!}{n!}\delta_{mn}, \quad (10)$$

where δ_{mn} is the KRONECKER DELTA. They also satisfy

$$\int_0^\infty e^{-x}x^{k+1}[L_n^k(x)]^2\,dx = \frac{(n+k)!}{n!}(2n+k+1). \quad (11)$$

RECURRENCE RELATIONS include

$$\sum_{\nu=0}^{n} L_\nu^{(\alpha)}(x) = L_n^{(\alpha+1)}(x) \quad (12)$$

and

$$L_n^{(\alpha)}(x) = L_n^{(\alpha+1)}(x) - L_{n-1}^{(\alpha+1)}(x). \quad (13)$$

The DERIVATIVE is given by

$$\begin{aligned} \frac{d}{dx}L_n^{(\alpha)}(x) &= -L_{n-1}^{(\alpha+1)}(x) \\ &= x^{-1}[nL_n^{(\alpha)}(x) - (n+\alpha)L_{n-1}^{(\alpha)}(x). \quad (14) \end{aligned}$$

In terms of the CONFLUENT HYPERGEOMETRIC FUNCTION,

$$L_n^k(x) = \frac{(k+1)_n}{n!}\,{}_1F_1(-b;k+1;x). \quad (15)$$

An interesting identity is

$$\sum_{n=0}^{\infty}\frac{L_n^{(\alpha)}(x)}{\Gamma(n+\alpha+1)}w^n = e^w(xw)^{-\alpha/2}J_\alpha(2\sqrt{xw}), \quad (16)$$

where $\Gamma(z)$ is the GAMMA FUNCTION and $J_\alpha(z)$ is the BESSEL FUNCTION OF THE FIRST KIND (Szegő 1975, p. 102). An integral representation is

$$e^{-x}x^{\alpha/2}L_n^{(\alpha)}(x) = \frac{1}{n!}\int_0^\infty e^{-t}t^{n+\alpha/2}J_\alpha(2\sqrt{tx})\,dt \quad (17)$$

for $n = 0, 1, \ldots$ and $\alpha > -1$. The DISCRIMINANT is

$$D_n^{(\alpha)} = \prod_{\nu=1}^{n}\nu^{\nu-2n+2}(\nu+\alpha)^{\nu-1} \quad (18)$$

(Szegő 1975, p. 143). The KERNEL POLYNOMIAL is

$$K_n^{(\alpha)}(x,y) = \frac{n+1}{\Gamma(\alpha+1)}\binom{n+\alpha}{n}^{-1} \frac{L_n^{(\alpha)}(x)L_{n+1}^{(\alpha)}(y) - L_{n+1}^{(\alpha)}(x)L_n(\alpha)(y)}{x-y}, \quad (19)$$

where $\binom{n}{k}$ is a BINOMIAL COEFFICIENT (Szegő 1975, p. 101).

The first few associated Laguerre polynomials are

$$\begin{aligned} L_0^k(x) &= 1 \\ L_1^k(x) &= -x+k+1 \\ L_2^k(x) &= \tfrac{1}{2}[x^2-2(k+2)x+(k+1)(k+2)] \\ L_3^k(x) &= \tfrac{1}{6}[-x^3+3(k+3)x^2-3(k+2)(k+3)x \\ &\quad +(k+1)(k+2)(k+3)]. \end{aligned}$$

see also SONINE POLYNOMIAL

References

Abramowitz, M. and Stegun, C. A. (Eds.). "Orthogonal Polynomials." Ch. 22 in *Handbook of Mathematical Functions with Formulas, Graphs, and Mathematical Tables, 9th printing.* New York: Dover, pp. 771–802, 1972.

Arfken, G. "Laguerre Functions." §13.2 in *Mathematical Methods for Physicists, 3rd ed.* Orlando, FL: Academic Press, pp. 721–731, 1985.
Chebyshev, P. L. "Sur le développement des fonctions à une seule variable." *Bull. Ph.-Math., Acad. Imp. Sc. St. Pétersbourg* **1**, 193–200, 1859.
Chebyshev, P. L. *Oeuvres, Vol. 1.* New York: Chelsea, pp. 499–508, 1987.
Iyanaga, S. and Kawada, Y. (Eds.). "Laguerre Functions." Appendix A, Table 20.VI in *Encyclopedic Dictionary of Mathematics.* Cambridge, MA: MIT Press, p. 1481, 1980.
Laguerre, E. de. "Sur l'intégrale $\int_x^{+\infty} x^{-1}e^{-x}\,dx$." *Bull. Soc. math. France* **7**, 72–81, 1879. Reprinted in *Oeuvres, Vol. 1.* New York: Chelsea, pp. 428–437, 1971.
Petkovšek, M.; Wilf, H. S.; and Zeilberger, D. *A=B.* Wellesley, MA: A. K. Peters, pp. 61–62, 1996.
Sansone, G. "Expansions in Laguerre and Hermite Series." Ch. 4 in *Orthogonal Functions, rev. English ed.* New York: Dover, pp. 295–385, 1991.
Spanier, J. and Oldham, K. B. "The Laguerre Polynomials $L_n(x)$." Ch. 23 in *An Atlas of Functions.* Washington, DC: Hemisphere, pp. 209–216, 1987.
Szegő, G. *Orthogonal Polynomials, 4th ed.* Providence, RI: Amer. Math. Soc., 1975.

Laguerre Quadrature

A GAUSSIAN QUADRATURE-like FORMULA for numerical estimation of integrals. It fits exactly all POLYNOMIALS of degree $2m-1$.

References

Chandrasekhar, S. *Radiative Transfer.* New York: Dover, p. 61, 1960.

Laguerre's Repeated Fraction

The CONTINUED FRACTION

$$\frac{(x+1)^n-(x-1)^n}{(x+1)^n+(x-1)^n}=\frac{n}{x+}\,\frac{n^2-1}{3x+}\,\frac{n^2-2^2}{5x+\ldots}.$$

References

Hardy, G. H. *Ramanujan: Twelve Lectures on Subjects Suggested by His Life and Work, 3rd ed.* New York: Chelsea, p. 13, 1959.

Laisant's Recurrence Formula

The RECURRENCE RELATION

$$(n-1)A_{n+1}=(n^2-1)A_n+(n+1)A_{n-1}+4(-1)^n$$

with $A(1)=A(2)=1$ which solves the MARRIED COUPLES PROBLEM.

see also MARRIED COUPLES PROBLEM

Lakshmi Star

see STAR OF LAKSHMI

Lal's Constant

Let $P(N)$ denote the number of PRIMES of the form n^2+1 for $1\leq n\leq N$, then

$$P(N)\sim 0.68641\,\mathrm{li}(N), \qquad (1)$$

where $\mathrm{li}(N)$ is the LOGARITHMIC INTEGRAL (Shanks 1960, pp. 321–332). Let $Q(N)$ denote the number of PRIMES of the form n^4+1 for $1\leq n\leq N$, then

$$Q(N)\sim \tfrac{1}{4}s_1\,\mathrm{li}(N)=0.66974\,\mathrm{li}(N) \qquad (2)$$

(Shanks 1961, 1962). Let $R(N)$ denote the number of pairs of PRIMES $(n-1)^2+1$ and $(n+1)^2+1$ for $n\leq N-1$, then

$$R(N)\sim 0.48762\,\mathrm{li}_2(N), \qquad (3)$$

where

$$\mathrm{li}_2(N)\equiv\int_2^N\frac{dn}{(\ln n)^2} \qquad (4)$$

(Shanks 1960, pp. 201–203). Finally, let $S(N)$ denote the number of pairs of PRIMES $(n-1)^4+1$ and $(n+1)^4+1$ for $n\leq N-1$, then

$$S(N)\sim\lambda\,\mathrm{li}_2(N) \qquad (5)$$

(Lal 1967), where λ is called Lal's constant. Shanks (1967) showed that $\lambda\approx 0.79220$.

References

Lal, M. "Primes of the Form n^4+1." *Math. Comput.* **21**, 245–247, 1967.
Shanks, D. "On the Conjecture of Hardy and Littlewood Concerning the Number of Primes of the Form n^2+a." *Math. Comput.* **14**, 321–332, 1960.
Shanks, D. "On Numbers of the Form n^4+1." *Math. Comput.* **15**, 186–189, 1961.
Shanks, D. Corrigendum to "On the Conjecture of Hardy and Littlewood Concerning the Number of Primes of the Form n^2+a." *Math. Comput.* **16**, 513, 1962.
Shanks, D. "Lal's Constant and Generalization." *Math. Comput.* **21**, 705–707, 1967.

Lam's Problem

Given an 111×111 MATRIX, fill 11 spaces in each row in such a way that all columns also have 11 spaces filled. Furthermore, each pair of rows must have exactly one filled space in the same column. This problem is equivalent to finding a PROJECTIVE PLANE of order 10. Using a computer program, Lam showed that no such arrangement exists.

see also PROJECTIVE PLANE

Laman's Theorem

Let a GRAPH G have exactly $2n-3$ EDGES, where n is the number of VERTICES in G. Then G is "generically" RIGID in $\mathbb{R}^2$ IFF $e'\leq 2n'-3$ for every SUBGRAPH of G having n' VERTICES and r' EDGES.

see also RIGID

References

Laman, G. "On Graphs and Rigidity of Plane Skeletal Structures." *J. Engineering Math.* **4**, 331–340, 1970.

Lambda Calculus

Developed by Alonzo Church and Stephen Kleene to address the COMPUTABLE NUMBER problem. In the lambda calculus, λ is defined as the ABSTRACTION OPERATOR. Three theorems of lambda calculus are λ-conversion, α-conversion, and η-conversion.

see also ABSTRACTION OPERATOR, COMPUTABLE NUMBER

References

Penrose, R. *The Emperor's New Mind: Concerning Computers, Minds, and the Laws of Physics.* Oxford, England: Oxford University Press, pp. 66–70, 1989.

Lambda Function

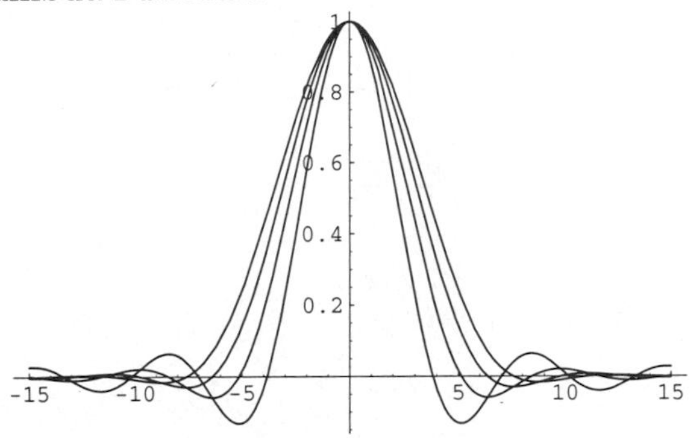

The lambda function defined by Jahnke and Emden (1945) is

$$\Lambda_\nu(z) \equiv \Gamma(\nu+1)\frac{J_\nu(z)}{(\frac{1}{2}z)^\nu} \tag{1}$$

$$\Lambda_1(z) \equiv \frac{J_1(z)}{\frac{z}{2}} = 2\,\mathrm{jinc}(z), \tag{2}$$

where $J_1(z)$ is a BESSEL FUNCTION OF THE FIRST KIND and $\mathrm{jinc}(z)$ is the JINC FUNCTION.

A two-variable lambda function defined by Gradshteyn and Ryzhik (1979) is

$$\lambda(x,y) \equiv \int_0^y \frac{\Gamma(u+1)\,du}{x^u}, \tag{3}$$

where $\Gamma(z)$ is the GAMMA FUNCTION.

see also AIRY FUNCTIONS, DIRICHLET LAMBDA FUNCTION, ELLIPTIC LAMBDA FUNCTION, JINC FUNCTION, LAMBDA HYPERGEOMETRIC FUNCTION, MANGOLDT FUNCTION, MU FUNCTION, NU FUNCTION

References

Gradshteyn, I. S. and Ryzhik, I. M. *Tables of Integrals, Series, and Products, 5th ed.* San Diego, CA: Academic Press, p. 1079, 1979.
Jahnke, E. and Emde, F. *Tables of Functions with Formulae and Curves, 4th ed.* New York: Dover, 1945.

Lambda Group

The set of linear fractional transformations w which satisfy

$$w(t) = \frac{at+b}{ct+d},$$

where a and d are ODD and b and c are EVEN. Also called the THETA SUBGROUP. It is a SUBGROUP of the GAMMA GROUP.

see also GAMMA GROUP

Lambda Hypergeometric Function

$$\lambda(t) = 16q\prod_{n=1}^{\infty}\left(\frac{1+q^{2n}}{1+q^{2n-1}}\right)^8, \tag{1}$$

where q is the NOME. The lambda hypergeometric functions satisfy the recurrence relationships

$$\lambda(t+2) = \lambda(t) \tag{2}$$

$$\lambda\left(\frac{t}{2t+1}\right) = \lambda(t). \tag{3}$$

Lambert Azimuthal Equal-Area Projection

$$x = k'\cos\phi\sin(\lambda-\lambda_0) \tag{1}$$

$$y = k'[\cos\phi_1\sin\phi - \sin\phi_1\cos\phi\cos(\lambda-\lambda_0)], \tag{2}$$

where

$$k' = \sqrt{\frac{2}{1+\sin\phi_1\sin\phi+\cos\phi_1\cos\phi\cos(\lambda-\lambda_0)}}. \tag{3}$$

The inverse FORMULAS are

$$\phi = \sin^{-1}\left(\cos c\sin\phi_1 + \frac{y\sin c\cos\phi_1}{\rho}\right) \tag{4}$$

$$\lambda = \lambda_0 + \tan^{-1}\left(\frac{x\sin c}{\rho\cos\phi_1\cos c - y\sin\phi_1\sin c}\right), \tag{5}$$

where

$$\rho = \sqrt{x^2+y^2} \tag{6}$$

$$c = 2\sin^{-1}(\tfrac{1}{2}\rho). \tag{7}$$

References

Snyder, J. P. *Map Projections—A Working Manual.* U. S. Geological Survey Professional Paper 1395. Washington, DC: U. S. Government Printing Office, pp. 182–190, 1987.

Lambert Conformal Conic Projection

$$x = \rho \sin[n(\lambda - \lambda_0)] \tag{1}$$
$$y = \rho_0 - \rho \cos[n(\lambda - \lambda_0)], \tag{2}$$

where

$$\rho = F \cot^n(\tfrac{1}{4}\pi + \tfrac{1}{2}\phi) \tag{3}$$
$$\rho_0 = F \cot^n(\tfrac{1}{4}\pi + \tfrac{1}{2}\phi_0) \tag{4}$$
$$F = \frac{\cos\phi_1 \tan^n(\tfrac{1}{4}\pi + \tfrac{1}{2}\phi_1)}{n} \tag{5}$$
$$n = \frac{\ln(\cos\phi_1 \sec\phi_2)}{\ln[\tan(\tfrac{1}{4}\pi + \tfrac{1}{2}\phi_2)\cot(\tfrac{1}{4}\pi + \tfrac{1}{2}\phi_1)]}. \tag{6}$$

The inverse FORMULAS are

$$\phi = 2\tan^{-1}\left[\left(\frac{F}{\rho}\right)^{1/n}\right] - \tfrac{1}{2}\pi \tag{7}$$
$$\lambda = \lambda_0 + \frac{\theta}{n}, \tag{8}$$

where

$$\rho = \operatorname{sgn}(n)\sqrt{x^2 + (\rho_0 - y)^2} \tag{9}$$
$$\theta = \tan^{-1}\left(\frac{x}{\rho_0 - y}\right). \tag{10}$$

References

Snyder, J. P. *Map Projections—A Working Manual.* U. S. Geological Survey Professional Paper 1395. Washington, DC: U. S. Government Printing Office, pp. 104–110, 1987.

Lambert's Method

A ROOT-finding method also called BAILEY'S METHOD and HUTTON'S METHOD. If $g(x) = x^d - r$, then

$$H_g(x) = \frac{(d-1)x^d + (d+1)r}{(d+1)x^d + (d-1)r}x.$$

References

Scavo, T. R. and Thoo, J. B. "On the Geometry of Halley's Method." *Amer. Math. Monthly* **102**, 417–426, 1995.

Lambert Series

A series of the form

$$F(x) \equiv \sum_{n=1}^{\infty} a_n \frac{x^n}{1 - x^n} \tag{1}$$

for $|x| < 1$. Then

$$F(x) = \sum_{n=1}^{\infty} a_n \sum_{m=1}^{\infty} x^{mn} = \sum_{N=1}^{\infty} b_N x^N, \tag{2}$$

where

$$b_N \equiv \sum_{n|N} a_n. \tag{3}$$

Some beautiful series of this type include

$$\sum_{n=1}^{\infty} \frac{\mu(n)x^n}{1 - x^n} = x \tag{4}$$
$$\sum_{n=1}^{\infty} \frac{\phi(n)x^n}{1 - x^n} = \frac{x}{(1-x)^2} \tag{5}$$
$$\sum_{n=1}^{\infty} \frac{x^n}{1 - x^n} = \sum_{n=1}^{\infty} d(n)x^n \tag{6}$$
$$\sum_{n=1}^{\infty} \frac{n^k x^n}{1 - x^n} = \sum_{n=1}^{\infty} \sigma_k(n)x^n \tag{7}$$
$$\sum_{n=1}^{\infty} \frac{4(-1)^{n+1}x^n}{1 - x^n} = \sum_{n=1}^{\infty} r(n)x^n, \tag{8}$$

where $\mu(n)$ is the MÖBIUS FUNCTION, $\phi(n)$ is the TOTIENT FUNCTION, $d(n) = \sigma_0(n)$ is the number of divisors of n, $\sigma_k(n)$ is the DIVISOR FUNCTION, and $r(n)$ is the number of representations of n in the form $n = A^2 + B^2$ where A and B are rational integers (Hardy and Wright 1979).

References

Abramowitz, M. and Stegun, C. A. (Eds.). "Number Theoretic Functions." §24.3.1 in *Handbook of Mathematical Functions with Formulas, Graphs, and Mathematical Tables, 9th printing.* New York: Dover, pp. 826–827, 1972.

Hardy, G. H. and Wright, E. M. *An Introduction to the Theory of Numbers, 5th ed.* Oxford, England: Clarendon Press, pp. 257–258, 1979.

Lambert's Transcendental Equation

An equation proposed by Lambert (1758) and studied by Euler in 1779 (Euler 1921).

$$x^\alpha - x^\beta = (\alpha - \beta)vx^{\alpha+\beta}.$$

When $\alpha \to \beta$, the equation becomes

$$\ln x = vx^\beta,$$

which has the solution

$$x = \exp\left[-\frac{W(-\beta v)}{\beta}\right],$$

where W is LAMBERT'S W-FUNCTION.

References

Corless, R. M.; Gonnet, G. H.; Hare, D. E. G.; and Jeffrey, D. J. "On Lambert's W Function." ftp://watdragon.uwaterloo.ca/cs-archive/CS-93-03/W.ps.Z.

de Bruijn, N. G. *Asymptotic Methods in Analysis.* Amsterdam, Netherlands: North-Holland, pp. 27–28, 1961.

Euler, L. "De Serie Lambertina Plurismique Eius Insignibus Proprietatibus." *Leonhardi Euleri Opera Omnia, Ser. 1. Opera Mathematica,* Bd. 6, 1921.

Lambert, J. H. "Observations variae in Mathesin Puram." *Acta Helvitica, physico-mathematico-anatomico-botanico-medica* **3**, 128–168, 1758.

Lambert's W-Function

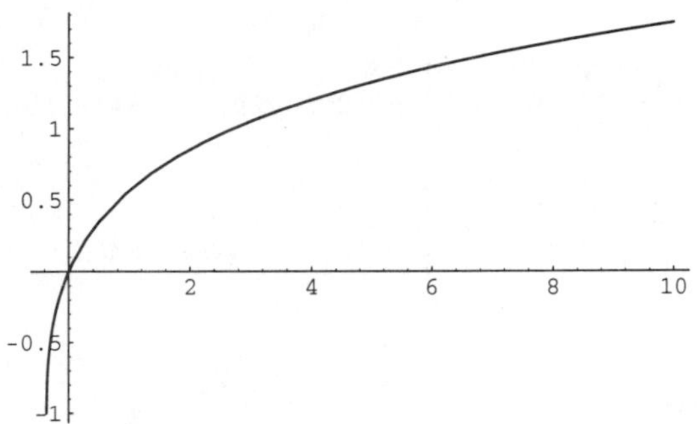

The inverse of the function

$$f(W) = We^W, \tag{1}$$

also called the OMEGA FUNCTION. The function is implemented as the *Mathematica*® (Wolfram Research, Champaign, IL) function `ProductLog[z]`. $W(1)$ is called the OMEGA CONSTANT and can be considered a sort of "GOLDEN RATIO" of exponentials since

$$\exp[-W(1)] = W(1), \tag{2}$$

giving

$$\ln\left[\frac{1}{W(1)}\right] = W(1). \tag{3}$$

Lambert's W-Function has the series expansion

$$W(x) = \sum_{n=1}^{\infty} \frac{(-1)^{n-1} n^{n-2}}{(n-1)!} x^n = x - x^2 + \tfrac{3}{2}x^3 - \tfrac{8}{3}x^4 + \tfrac{125}{24}x^5 - \tfrac{54}{5}x^6 + \tfrac{16807}{720}x^7 + \ldots. \tag{4}$$

The LAGRANGE INVERSION THEOREM gives the equivalent series expansion

$$W_0(z) = \sum_{n=1}^{\infty} \frac{(-n)^{n-1}}{n!} z^n, \tag{5}$$

where $n!$ is a FACTORIAL. However, this series oscillates between ever larger POSITIVE and NEGATIVE values for REAL $z \gtrsim 0.4$, and so cannot be used for practical numerical computation. An asymptotic FORMULA which yields reasonably accurate results for $z \gtrsim 3$ is

$$\begin{aligned} W(z) &= \operatorname{Ln} z - \ln \operatorname{Ln} z \\ &\quad + \sum_{k=0}^{\infty} \sum_{m=0}^{\infty} c_{km} (\ln \operatorname{Ln} z)^{m+1} (\operatorname{Ln} z)^{-k-m-1} \\ &= L_1 - L_2 + \frac{L_2}{L_1} + \frac{L_2(-2+L_2)}{2{L_1}^2} \\ &\quad + \frac{L_2(6 - 9L_2 + 2{L_2}^2)}{6{L_1}^2} \\ &\quad + \frac{L_2(-12 + 36L_2 - 22{L_2}^2 + 3{L_2}^3)}{12{L_1}^4} \\ &\quad + \frac{L_2(60 - 300L_2 + 350{L_2}^2 - 125{L_2}^3 + 12{L_2}^4)}{60{L_1}^5} \\ &\quad + \mathcal{O}\left[\left(\frac{L_2}{L_1}\right)^6\right], \end{aligned} \tag{6}$$

where

$$L_1 = \operatorname{Ln} z \tag{7}$$

$$L_2 = \ln \operatorname{Ln} z \tag{8}$$

(Corless *et al.*), correcting a typographical error in de Bruijn (1961). Another expansion due to Gosper is the DOUBLE SUM

$$W(x) = a + \sum_{n=0}^{\infty} \left\{ \sum_{k=0}^{n} \frac{S_1(n,k)}{\left[\ln\left(\frac{x}{a}\right) - a\right]^{k-1} (n-k+1)!} \right\} \times \left[1 - \frac{\ln\left(\frac{x}{a}\right)}{a}\right]^n, \tag{9}$$

where S_1 is a *nonnegative* STIRLING NUMBER OF THE FIRST KIND and a is a first approximation which can be used to select between branches. Lambert's W-function is two-valued for $-1/e \leq x < 0$. For $W(x) \geq -1$, the function is denoted $W_0(x)$ or simply $W(x)$, and this is called the principal branch. For $W(x) \leq -1$, the function is denoted $W_{-1}(x)$. The DERIVATIVE of W is

$$W'(x) = \frac{1}{[1 + W(x)] \exp[W(x)]} = \frac{W(x)}{x[1 + W(x)]} \tag{10}$$

for $x \neq 0$. For the principal branch when $z > 0$,

$$\ln W(z) = \ln z - W(z). \tag{11}$$

see also ITERATED EXPONENTIAL CONSTANTS, OMEGA CONSTANT

References

Corless, R. M.; Gonnet, G. H.; Hare, D. E. G.; and Jeffrey, D. J. "On Lambert's W Function." ftp://watdragon.uwaterloo.ca/cs-archive/CS-93-03/W.ps.Z.

de Bruijn, N. G. *Asymptotic Methods in Analysis.* Amsterdam, Netherlands: North-Holland, pp. 27–28, 1961.

Lamé Curve

A curve with Cartesian equation

$$\left(\frac{x}{a}\right)^n + \left(\frac{y}{b}\right)^n = c$$

first discussed in 1818 by Lamé. If n is a rational, then the curve is algebraic. However, for irrational n, the curve is transcendental. For EVEN INTEGERS n, the curve becomes closer to a rectangle as n increases. For ODD INTEGER values of n, the curve looks like the EVEN case in the POSITIVE quadrant but goes to infinity in both the second and fourth quadrants (MacTutor Archive). The EVOLUTE of an ELLIPSE,

$$(ax)^{2/3} + (by)^{2/3} = (a^2 - b^2)^{2/3}.$$

n	Curve
$\frac{2}{3}$	astroid
$\frac{5}{2}$	superellipse
3	witch of Agnesi

see also ASTROID, SUPERELLIPSE, WITCH OF AGNESI

References

MacTutor History of Mathematics Archive. "Lamé Curves." http://www-groups.dcs.st-and.ac.uk/~history/Curves/Lame.html.

Lamé's Differential Equation

$$(x^2 - b^2)(x^2 - c^2)\frac{d^2z}{dx^2} + x(x^2 - b^2 + x^2 - c^2)\frac{dz}{dx} - [m(m+1)x^2 - (b^2 + c^2)p]z = 0. \quad (1)$$

(Byerly 1959, p. 255). The solution is denoted $E_m^p(x)$ and is known as a LAMÉ FUNCTION or an ELLIPSOIDAL HARMONIC. Whittaker and Watson (1990, pp. 554–555) give the alternative forms

$$4\Delta_\lambda \frac{d}{d\lambda}\left[\Delta_\lambda \frac{d\Lambda}{d\lambda}\right] = [n(n+1)\lambda + C]\Lambda \quad (2)$$

$$\frac{d^2\Lambda}{d\lambda^2} + \left[\frac{\frac{1}{2}}{a^2+\lambda} + \frac{\frac{1}{2}}{b^2+\lambda} + \frac{\frac{1}{2}}{c^2}\right]\frac{d\Lambda}{d\lambda} = \frac{[n(n+1)\lambda + C]\Lambda}{4\Delta_\lambda} \quad (3)$$

$$\frac{d^2\Lambda}{du^2} = [n(n+1)\wp(u) + C - \tfrac{1}{3}n(n+1)(a^2+b^2+c^2)]\Lambda \quad (4)$$

$$\frac{d^2\Lambda}{d{z_1}^2} = [n(n+1)k^2 \operatorname{sn}^2\alpha + A]\Lambda, \quad (5)$$

where $\wp$ is a WEIERSTRAß ELLIPTIC FUNCTION and

$$\Lambda(\theta) \equiv \prod_{q=1}^m (\theta - \theta_q) \quad (6)$$

$$\Delta_\lambda \equiv \sqrt{(a^2+\lambda)(b^2+\lambda)(c^2+\lambda)} \quad (7)$$

$$A \equiv \frac{C - \frac{1}{3}n(n+1)(a^2+b^2+c^2) + e_3 n(n+1)}{e_1 - e_3}. \quad (8)$$

References

Byerly, W. E. *An Elementary Treatise on Fourier's Series, and Spherical, Cylindrical, and Ellipsoidal Harmonics, with Applications to Problems in Mathematical Physics.* New York: Dover, 1959.

Whittaker, E. T. and Watson, G. N. *A Course in Modern Analysis, 4th ed.* Cambridge, England: Cambridge University Press, 1990.

Lamé's Differential Equation (Types)

Whittaker and Watson (1990, pp. 539–540) write Lamé's differential equation for ELLIPSOIDAL HARMONICS of the four types as

$$4\Delta(\theta)\frac{d}{d\theta}\left[F(\theta)\frac{d\Lambda(\theta)}{d\theta}\right] = [2m(2m+1)\theta + C]\Lambda(\theta) \quad (1)$$

$$4\Delta(\theta)\frac{d}{d\theta}\left[F(\theta)\frac{d\Lambda(\theta)}{d\theta}\right] = [(2m+1)(2m+2)\theta + C]\Lambda(\theta) \quad (2)$$

$$4\Delta(\theta)\frac{d}{d\theta}\left[F(\theta)\frac{d\Lambda(\theta)}{d\theta}\right] = [(2m+2)(2m+3)\theta + C]\Lambda(\theta) \quad (3)$$

$$4\Delta(\theta)\frac{d}{d\theta}\left[F(\theta)\frac{d\Lambda(\theta)}{d\theta}\right] = [(2m+3)(2m+4)\theta + C]\Lambda(\theta), \quad (4)$$

where

$$\Delta(\theta) \equiv \sqrt{(a^2+\theta)(b^2+\theta)(c^2+\theta)} \quad (5)$$

$$\Lambda(\theta) \equiv \prod_{q=1}^m (\theta - \theta_q). \quad (6)$$

References

Whittaker, E. T. and Watson, G. N. *A Course in Modern Analysis, 4th ed.* Cambridge, England: Cambridge University Press, 1990.

Lamé Function

see ELLIPSOIDAL HARMONIC

Lamé's Theorem

If a is the smallest INTEGER for which there is a smaller INTEGER b such that a and b generate a EUCLIDEAN ALGORITHM remainder sequence with n steps, then a is the FIBONACCI NUMBER F_{n+2}. Furthermore, the number of steps in the EUCLIDEAN ALGORITHM never exceeds 5 times the number of digits in the smaller number.

see also EUCLIDEAN ALGORITHM

References

Honsberger, R. "A Theorem of Gabriel Lamé." Ch. 7 in *Mathematical Gems II.* Washington, DC: Math. Assoc. Amer., pp. 54–57, 1976.

Lamina

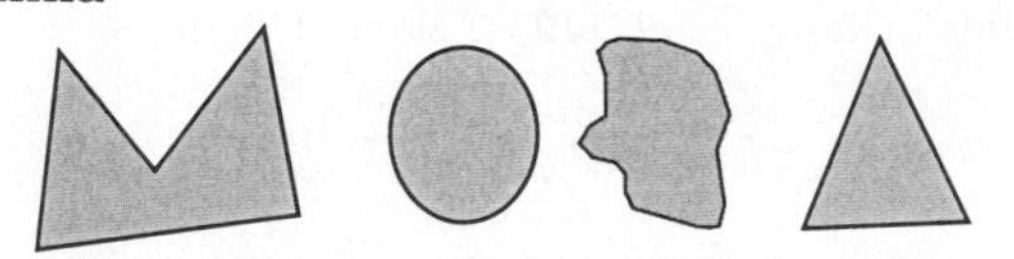

A 2-D planar closed surface L which has a mass M and a surface density $\sigma(x, y)$ (in units of mass per areas squared) such that

$$M = \int_L \sigma(x, y)\, dx\, dy.$$

The CENTER OF MASS of a lamina is called its CENTROID.

see also CENTROID (GEOMETRIC), CROSS-SECTION, SOLID

Laminated Lattice

A LATTICE which is built up of layers of n-D lattices in $(n+1)$-D space. The VECTORS specifying how layers are stacked are called GLUE VECTORS.

see also GLUE VECTOR, LATTICE

References

Conway, J. H. and Sloane, N. J. A. "Laminated Lattices." Ch. 6 in *Sphere Packings, Lattices, and Groups, 2nd ed.* New York: Springer-Verlag, pp. 157–180, 1993.

Lancret Equation

$$ds_N{}^2 = ds_T{}^2 + ds_B{}^2,$$

where N is the NORMAL VECTOR, T is the TANGENT, and B is the BINORMAL VECTOR.

Lancret's Theorem

A NECESSARY and SUFFICIENT condition for a curve to be a HELIX is that the ratio of CURVATURE to TORSION be constant.

Lanczos Approximation

see GAMMA FUNCTION

Lanczos σ Factor

Writing a FOURIER SERIES as

$$f(\theta) = \tfrac{1}{2}a_0 + \sum_{n=1}^{m} \operatorname{sinc}\left(\frac{n\pi}{2m}\right)[a_n \cos(n\theta) + b_n \sin(n\theta)],$$

where m is the last term and the $\operatorname{sinc} x$ terms are the Lanczos σ factor, removes the GIBBS PHENOMENON (Acton 1990).

see also FOURIER SERIES, GIBBS PHENOMENON, SINC FUNCTION

References

Acton, F. S. *Numerical Methods That Work, 2nd printing.* Washington, DC: Math. Assoc. Amer., p. 228, 1990.

Landau Constant

N.B. A detailed on-line essay by S. Finch was the starting point for this entry.

Let F be the set of COMPLEX analytic functions f defined on an open region containing the closure of the unit disk $D = \{z : |z| < 1\}$ satisfying $f(0) = 0$ and $df/dz(0) = 1$. For each f in F, let $l(f)$ be the SUPREMUM of all numbers r such that $f(D)$ contains a disk of radius r. Then

$$L \equiv \inf\{l(f) : f \in F\}.$$

This constant is called the Landau constant, or the BLOCH-LANDAU CONSTANT. Robinson (1938, unpublished) and Rademacher (1943) derived the bounds

$$\tfrac{1}{2} < L \le \frac{\Gamma(\frac{1}{3})\Gamma(\frac{5}{6})}{\Gamma(\frac{1}{6})} = 0.5432588\ldots,$$

where $\Gamma(z)$ is the GAMMA FUNCTION, and conjectured that the second inequality is actually an equality,

$$L = \frac{\Gamma(\frac{1}{3})\Gamma(\frac{5}{6})}{\Gamma(\frac{1}{6})} = 0.5432588\ldots.$$

see also BLOCH CONSTANT

References

Finch, S. "Favorite Mathematical Constants." http://www.mathsoft.com/asolve/constant/bloch/bloch.html.
Rademacher, H. "On the Bloch-Landau Constant." *Amer. J. Math.* **65**, 387–390, 1943.

Landau-Kolmogorov Constants

N.B. A detailed on-line essay by S. Finch was the starting point for this entry.

Let $||f||$ be the SUPREMUM of $|f(x)|$, a real-valued function f defined on $(0, \infty)$. If f is twice differentiable and both f and f'' are bounded, Landau (1913) showed that

$$||f'|| \le 2||f||^{1/2}||f''||^{1/2}, \tag{1}$$

where the constant 2 is the best possible. Schoenberg (1973) extended the result to the nth derivative of f defined on $(0, \infty)$ if both f and $f^{(n)}$ are bounded,

$$||f^{(k)}|| \le C(n, k)||f||^{1-k/n}||f^{(n)}||^{k/n}. \tag{2}$$

An explicit FORMULA for $C(n, k)$ is not known, but particular cases are

$$C(3,1) = \left(\frac{243}{8}\right)^{1/3} \tag{3}$$

$$C(3,2) = 24^{1/3} \tag{4}$$

$$C(4,1) = 4.288\ldots \tag{5}$$

$$C(4,2) = 5.750\ldots \tag{6}$$

$$C(4,3) = 3.708\ldots. \tag{7}$$

Let $||f||$ be the SUPREMUM of $|f(x)|$, a real-valued function f defined on $(-\infty, \infty)$. If f is twice differentiable and both f and f'' are bounded, Hadamard (1914) showed that

$$||f'|| \leq \sqrt{2}||f||^{1/2}||f''||^{1/2}, \quad (8)$$

where the constant $\sqrt{2}$ is the best possible. Kolmogorov (1962) determined the best constants $C(n,k)$ for

$$||f^{(k)}|| \leq C(n,k)||f||^{1-k/n}||f^{(n)}||^{k/n} \quad (9)$$

in terms of the FAVARD CONSTANTS

$$a_n = \frac{4}{\pi}\sum_{j=0}^{\infty}\left[\frac{(-1)^j}{2j+1}\right]^{n+1} \quad (10)$$

by

$$C(n,k) = a_{n-k}a_n{}^{-1+k/n}. \quad (11)$$

Special cases derived by Shilov (1937) are

$$C(3,1) = \left(\frac{9}{8}\right)^{1/3} \quad (12)$$

$$C(3,2) = 3^{1/3} \quad (13)$$

$$C(4,1) = \left(\frac{512}{375}\right)^{1/4} \quad (14)$$

$$C(4,2) = \sqrt{\frac{6}{5}} \quad (15)$$

$$C(4,3) = \left(\frac{24}{5}\right)^{1/4} \quad (16)$$

$$C(5,1) = \left(\frac{1953125}{1572864}\right)^{1/5} \quad (17)$$

$$C(5,2) = \left(\frac{125}{72}\right)^{1/5}. \quad (18)$$

For a real-valued function f defined on $(-\infty, \infty)$, define

$$||f|| = \sqrt{\int_{-\infty}^{\infty}[f(x)]^2\,dx}. \quad (19)$$

If f is n differentiable and both f and $f^{(n)}$ are bounded, Hardy *et al.* (1934) showed that

$$||f^{(k)}|| \leq ||f||^{1-k/n}\,||f^{(n)}||^{k/n}, \quad (20)$$

where the constant 1 is the best possible for all n and $0 < k < n$.

For a real-valued function f defined on $(0, \infty)$, define

$$||f|| = \sqrt{\int_0^{\infty}[f(x)]^2\,dx}. \quad (21)$$

If f is twice differentiable and both f and f'' are bounded, Hardy *et al.* (1934) showed that

$$||f'|| \leq \sqrt{2}\,||f||^{1/2}||f^{(n)}||^{1/2}, \quad (22)$$

where the constant $\sqrt{2}$ is the best possible. This inequality was extended by Ljubic (1964) and Kupcov (1975) to

$$||f^{(k)}|| \leq C(n,k)\,||f||^{1-k/n}||f^{(n)}||^{k/n} \quad (23)$$

where $C(n,k)$ are given in terms of zeros of POLYNOMIALS. Special cases are

$$C(3,1) = C(3,2) = 3^{1/2}[2(2^{1/2}-1)]^{-1/3}$$
$$= 1.84420\ldots \quad (24)$$

$$C(4,1) = C(4,3) = \sqrt{\frac{3^{1/4}+3^{-3/4}}{a}}$$
$$= 2.27432\ldots \quad (25)$$

$$C(4,2) = \sqrt{\frac{2}{b}} = 2.97963\ldots \quad (26)$$

$$C(4,3) = \left(\frac{24}{5}\right)^{1/4} \quad (27)$$

$$C(5,1) = C(5,4) = 2.70247\ldots \quad (28)$$

$$C(5,2) = C(5,3) = 4.37800\ldots, \quad (29)$$

where a is the least POSITIVE ROOT of

$$x^8 - 6x^4 - 8x^2 + 1 = 0 \quad (30)$$

and b is the least POSITIVE ROOT of

$$x^4 - 2x^2 - 4x + 1 = 0 \quad (31)$$

(Franco *et al.* 1985, Neta 1980). The constants $C(n,1)$ are given by

$$C(n,1) = \sqrt{\frac{(n-1)^{1/n}+(n+1)^{-1+1/n}}{c}}, \quad (32)$$

where c is the least POSITIVE ROOT of

$$\int_0^c\int_0^{\infty}\frac{dx\,dy}{(x^{2n}-yx^2+1)\sqrt{y}} = \frac{\pi^2}{2n}. \quad (33)$$

An explicit FORMULA of this type is not known for $k > 1$.

The cases $p = 1, 2, \infty$ are the only ones for which the best constants have exact expressions (Kwong and Zettl 1992, Franco *et al.* 1983).

References

Finch, S. "Favorite Mathematical Constants." `http://www.mathsoft.com/asolve/constant/lk/lk.html`.

Franco, Z. M.; Kaper, H. G.; Kwong, M. N.; and Zettl, A. "Bounds for the Best Constants in Landau's Inequality on the Line." *Proc. Roy. Soc. Edinburgh* **95A**, 257–262, 1983.

Franco, Z. M.; Kaper, H. G.; Kwong, M. N.; and Zettl, A. "Best Constants in Norm Inequalities for Derivatives on a Half Line." *Proc. Roy. Soc. Edinburgh* **100A**, 67–84, 1985.
Hardy, G. H.; Littlewood, J. E.; and Pólya, G. *Inequalities.* Cambridge, England: Cambridge University Press, 1934.
Kolmogorov, A. "On Inequalities Between the Upper Bounds of the Successive Derivatives of an Arbitrary Function on an Infinite Integral." *Amer. Math. Soc. Translations, Ser. 1* **2**, 233–243, 1962.
Kupcov, N. P. "Kolmogorov Estimates for Derivatives in $L_2(0,\infty)$." *Proc. Steklov Inst. Math.* **138**, 101–125, 1975.
Kwong, M. K. and Zettl, A. *Norm Inequalities for Derivatives and Differences.* New York: Springer-Verlag, 1992.
Landau, E. "Einige Ungleichungen für zweimal differentzierbare Funktionen." *Proc. London Math. Soc. Ser. 2* **13**, 43–49, 1913.
Landau, E. "Die Ungleichungen für zweimal differentzierbare Funktionen." *Danske Vid. Selsk. Math. Fys. Medd.* **6**, 1–49, 1925.
Ljubic, J. I. "On Inequalities Between the Powers of a Linear Operator." *Amer. Math. Soc. Trans. Ser. 2* **40**, 39–84, 1964.
Neta, B. "On Determinations of Best Possible Constants in Integral Inequalities Involving Derivatives." *Math. Comput.* **35**, 1191–1193, 1980.
Schoenberg, I. J. "The Elementary Case of Landau's Problem of Inequalities Between Derivatives." *Amer. Math. Monthly* **80**, 121–158, 1973.

Landau-Ramanujan Constant

N.B. A detailed on-line essay by S. Finch was the starting point for this entry.

Let $S(x)$ denote the number of POSITIVE INTEGERS not exceeding x which can be expressed as a sum of two squares, then

$$\lim_{x\to\infty} \frac{\sqrt{\ln x}}{x} S(x) = K, \tag{1}$$

as proved by Landau (1908) and stated by Ramanujan. The value of K (also sometimes called λ) is

$$K = \sqrt{\tfrac{1}{2} \prod_{\substack{p \text{ a prime} \\ \equiv 3 \pmod 4}} \frac{1}{1-p^{-2}}} = 0.764223653\ldots \tag{2}$$

(Hardy 1940, Berndt 1994). Ramanujan found the approximate value $K = 0.764$. Flajolet and Vardi (1996) give a beautiful FORMULA with fast convergence

$$K = \frac{1}{\sqrt{2}} \prod_{n=1}^{\infty} \left[\left(1 - \frac{1}{2^{2^n}}\right) \frac{\zeta(2^n)}{\beta(2^n)}\right]^{1/(2^n+1)}, \tag{3}$$

where

$$\beta(s) \equiv \frac{1}{4^s}[\zeta(s, \tfrac{1}{4}) - \zeta(s, \tfrac{3}{4})] \tag{4}$$

is the DIRICHLET BETA FUNCTION, and $\zeta(z,a)$ is the HURWITZ ZETA FUNCTION. Landau proved the even stronger fact

$$\lim_{x\to\infty} \frac{(\ln x)^{3/2}}{Kx} \left[S(x) - \frac{Kx}{\sqrt{\ln x}}\right] = C, \tag{5}$$

where

$$C \equiv \frac{1}{2}\left[1 - \ln\left(\frac{\pi e^\gamma}{L}\right)\right] - \frac{1}{4}\frac{d}{ds}\left[\ln\left(\prod_{\substack{p \text{ prime} \\ p=4k+3}} \frac{1}{p^{-2s}}\right)\right]_{s=1}$$
$$= 0.581948659\ldots. \tag{6}$$

Here,

$$L = 5.2441151086\ldots \tag{7}$$

is the ARC LENGTH of a LEMNISCATE with $a = 1$ (the LEMNISCATE CONSTANT to within a factor of 2 or 4), and γ is the EULER-MASCHERONI CONSTANT.

References

Berndt, B. C. *Ramanujan's Notebooks, Part IV.* New York: Springer-Verlag, pp. 60–66, 1994.
Finch, S. "Favorite Mathematical Constants." `http://www.mathsoft.com/asolve/constant/lr/lr.html`.
Flajolet, P. and Vardi, I. "Zeta Function Expansions of Classical Constants." Unpublished manuscript. 1996. `http://pauillac.inria.fr/algo/flajolet/Publications/landau.ps`.
Hardy, G. H. *Ramanujan: Twelve Lectures on Subjects Suggested by His Life and Work, 3rd ed.* New York: Chelsea, pp. 61–63, 1940.
Landau, E. "Über die Einteilung der positiven ganzen Zahlen in vier Klassen nach der Mindeszahl der zu ihrer additiven Zusammensetzung erforderlichen Quadrate." *Arch. Math. Phys.* **13**, 305–312, 1908.
Shanks, D. "The Second-Order Term in the Asymptotic Expansion of $B(x)$." *Math. Comput.* **18**, 75–86, 1964.
Shanks, D. "Non-Hypotenuse Numbers." *Fibonacci Quart.* **13**, 319–321, 1975.
Shanks, D. and Schmid, L. P. "Variations on a Theorem of Landau. I." *Math. Comput.* **20**, 551–569, 1966.
Shiu, P. "Counting Sums of Two Squares: The Meissel-Lehmer Method." *Math. Comput.* **47**, 351–360, 1986.

Landau Symbol

Let $f(z)$ be a function $\neq 0$ in an interval containing $z = 0$. Let $g(z)$ be another function also defined in this interval such that $g(z)/f(z) \to 0$ as $z \to 0$. Then $g(z)$ is said to be $\mathcal{O}(f(z))$.

Landen's Formula

$$\frac{\vartheta_3(z,t)\vartheta_4(z,t)}{\vartheta_4(2z,2t)} = \frac{\vartheta_3(0,t)\vartheta_4(0,t)}{\vartheta_4(0,2t)} = \frac{\vartheta_2(z,t)\vartheta_1(z,t)}{\vartheta_1(2z,2t)},$$

where ϑ_i are THETA FUNCTIONS. This transformation was used by Gauss to show that ELLIPTIC INTEGRALS could be computed using the ARITHMETIC-GEOMETRIC MEAN.

Landen's Transformation

If $x\sin\alpha = \sin(2\beta-\alpha)$, then

$$(1+x)\int_0^\alpha \frac{d\phi}{\sqrt{1-x^2\sin^2\phi}} = 2\int_0^\beta \frac{d\phi}{\sqrt{1-\frac{4x}{(1+x)^2}\sin^2\phi}}.$$

see also ELLIPTIC INTEGRAL OF THE FIRST KIND, GAUSS'S TRANSFORMATION

References

Abramowitz, M. and Stegun, C. A. (Eds.). "Ascending Landen Transformation" and "Landen's Transformation." §16.14 and 17.5 in *Handbook of Mathematical Functions with Formulas, Graphs, and Mathematical Tables, 9th printing.* New York: Dover, pp. 573–574 and 597–598, 1972.

Lane-Emden Differential Equation

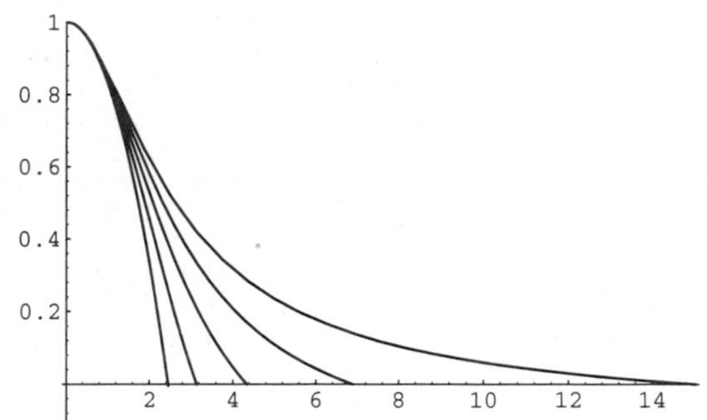

A second-order ORDINARY DIFFERENTIAL EQUATION arising in the study of stellar interiors. It is given by

$$\frac{1}{\xi^2}\frac{d}{d\xi}\left(\xi^2\frac{d\theta}{d\xi}\right)+\theta^n=0 \tag{1}$$

$$\frac{1}{\xi^2}\left(2\xi\frac{d\theta}{d\xi}+\xi^2\frac{d^2\theta}{d\xi^2}\right)+\theta^n=\frac{d^2\theta}{d\xi^2}+\frac{2}{\xi}\frac{d\theta}{d\xi}+\theta^n=0. \tag{2}$$

It has the BOUNDARY CONDITIONS

$$\theta(0)=1 \tag{3}$$

$$\left[\frac{d\theta}{d\xi}\right]_{\xi=0}=0. \tag{4}$$

Solutions $\theta(\xi)$ for $n=0$, 1, 2, 3, and 4 are shown above. The cases $n=0$, 1, and 5 can be solved analytically (Chandrasekhar 1967, p. 91); the others must be obtained numerically.

For $n=0$ ($\gamma=\infty$), the LANE-EMDEN DIFFERENTIAL EQUATION is

$$\frac{1}{\xi^2}\frac{d}{d\xi}\left(\xi^2\frac{d\theta}{d\xi}\right)+1=0 \tag{5}$$

(Chandrasekhar 1967, pp. 91–92). Directly solving gives

$$\frac{d}{d\xi}\left(\xi^2\frac{d\theta}{d\xi}\right)=-\xi^2 \tag{6}$$

$$\int d\left(\xi^2\frac{d\theta}{d\xi^2}\right)=\int\xi^2\,d\xi \tag{7}$$

$$\xi^2\frac{d\theta}{d\xi}=c_1-\tfrac{1}{3}\xi^3 \tag{8}$$

$$\frac{d\theta}{d\xi}=\frac{c_1-\frac{1}{3}\xi^3}{\xi^2} \tag{9}$$

$$\theta(\xi)=\int d\theta=\int\frac{c_1-\frac{1}{3}\xi^3}{\xi^2}\,d\xi \tag{10}$$

$$\theta(\xi)=\theta_0-c_1\xi^{-1}-\tfrac{1}{6}\xi^2. \tag{11}$$

The BOUNDARY CONDITION $\theta(0)=1$ then gives $\theta_0=1$ and $c_1=0$, so

$$\theta_1(\xi)=1-\tfrac{1}{6}\xi^2, \tag{12}$$

and $\theta_1(\xi)$ is PARABOLIC.

For $n=1$ ($\gamma=2$), the differential equation becomes

$$\frac{1}{\xi^2}\frac{d}{d\xi}\left(\xi^2\frac{d\theta}{d\xi}\right)+\theta=0 \tag{13}$$

$$\frac{d}{d\xi}\left(\xi^2\frac{d\theta}{d\xi}\right)+\theta\xi^2=0, \tag{14}$$

which is the SPHERICAL BESSEL DIFFERENTIAL EQUATION

$$\frac{d}{dr}\left(r^2\frac{dR}{dr}\right)+[k^2r^2-n(n+1)]R=0 \tag{15}$$

with $k=1$ and $n=0$, so the solution is

$$\theta(\xi)=Aj_0(\xi)+Bn_0(\xi). \tag{16}$$

Applying the BOUNDARY CONDITION $\theta(0)=1$ gives

$$\theta_2(\xi)=j_0(\xi)=\frac{\sin\xi}{\xi}, \tag{17}$$

where $j_0(x)$ is a SPHERICAL BESSEL FUNCTION OF THE FIRST KIND (Chandrasekhar 1967, pp. 92).

For $n=5$, make Emden's transformation

$$\theta=Ax^\omega z \tag{18}$$

$$\omega=\frac{2}{n-1}, \tag{19}$$

which reduces the Lane-Emden equation to

$$\frac{d^2z}{dt^2}+(2\omega-1)\frac{dz}{dt}+\omega(\omega-1)z+A^{n-1}z^n=0 \tag{20}$$

(Chandrasekhar 1967, p. 90). After further manipulation (not reproduced here), the equation becomes

$$\frac{d^2z}{dt^2}=\tfrac{1}{4}z(1-z^4) \tag{21}$$

and then, finally,

$$\theta_3(\xi)(1+\tfrac{1}{3}\xi^2)^{-1/2}. \tag{22}$$

References

Chandrasekhar, S. *An Introduction to the Study of Stellar Structure.* New York: Dover, pp. 84–182, 1967.

Langford's Problem

Arrange copies of the n digits 1, ..., n such that there is one digit between the 1s, two digits between the 2s, etc. For example, the $n = 3$ solution is 312132 and the $n = 4$ solution is 41312432. Solutions exist only if $n \equiv 0, 3 \pmod 4$. The number of solutions for $n = 3$, 4, 5, ... are 1, 1, 0, 0, 26, 150, 0, 0, 17792, 108144, ... (Sloane's A014552).

References

Gardner, M. *Mathematical Magic Show: More Puzzles, Games, Diversions, Illusions and Other Mathematical Sleight-of-Mind from Scientific American.* New York: Vintage, pp. 70 and 77–78, 1978.

Sloane, N. J. A. Sequence A014551 in "An On-Line Version of the Encyclopedia of Integer Sequences."

Langlands Program

A grand unified theory of mathematics which includes the search for a generalization of ARTIN RECIPROCITY (known as LANGLANDS RECIPROCITY) to non-Abelian Galois extensions of NUMBER FIELDS. Langlands proposed in 1970 that the mathematics of algebra and analysis are intimately related. He was a co-recipient of the 1996 Wolf Prize for this formulation.

see also ARTIN RECIPROCITY, LANGLANDS RECIPROCITY

References

American Mathematical Society. "Langlands and Wiles Share Wolf Prize." *Not. Amer. Math. Soc.* **43**, 221–222, 1996.

Knapp, A. W. "Group Representations and Harmonic Analysis from Euler to Langlands." *Not. Amer. Math. Soc.* **43**, 410–415, 1996.

Langlands Reciprocity

The conjecture that the ARTIN L-FUNCTION of any n-D GALOIS GROUP representation is an L-FUNCTION obtained from the GENERAL LINEAR GROUP $GL_1(\mathbb{A})$.

References

Knapp, A. W. "Group Representations and Harmonic Analysis, Part II." *Not. Amer. Math. Soc.* **43**, 537–549, 1996.

Langton's Ant

A CELLULAR AUTOMATON. The COHEN-KUNG THEOREM guarantees that the ant's trajectory is unbounded.

see also CELLULAR AUTOMATON, COHEN-KUNG THEOREM

References

Stewart, I. "The Ultimate in Anty-Particles." *Sci. Amer.* **271**, 104–107, 1994.

Laplace-Beltrami Operator

A self-adjoint elliptic differential operator defined somewhat technically as

$$\Delta = d\delta + \delta d,$$

where d is the EXTERIOR DERIVATIVE and d and δ are adjoint to each other with respect to the INNER PRODUCT.

References

Iyanaga, S. and Kawada, Y. (Eds.). *Encyclopedic Dictionary of Mathematics.* Cambridge, MA: MIT Press, p. 628, 1980.

Laplace Distribution

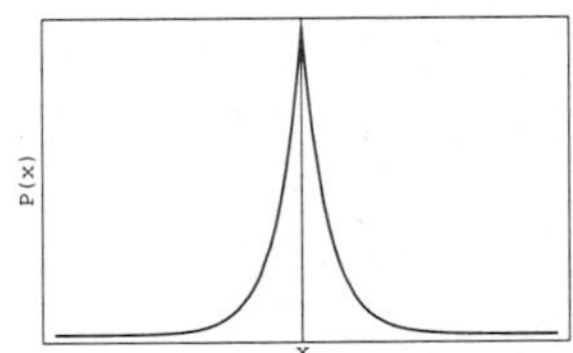

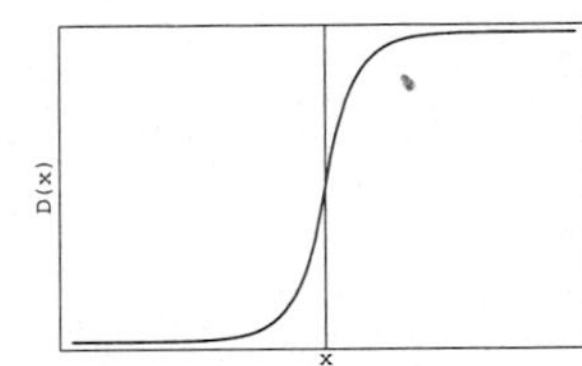

Also called the DOUBLE EXPONENTIAL DISTRIBUTION. It is the distribution of differences between two independent variates with identical EXPONENTIAL DISTRIBUTIONS (Abramowitz and Stegun 1972, p. 930).

$$P(x) = \frac{1}{2b} e^{-|x-\mu|/b} \tag{1}$$

$$D(x) = \tfrac{1}{2}[1 + \operatorname{sgn}(x-\mu)(1 - e^{-|x-\mu|/b})]. \tag{2}$$

The MOMENTS about the MEAN μ_n are related to the MOMENTS about 0 by

$$\mu_n = \sum_{j=0}^{n} \binom{n}{j} (-1)^{n-j} \mu_j' \mu^{n-j}, \tag{3}$$

where $\binom{n}{k}$ is a BINOMIAL COEFFICIENT, so

$$\begin{aligned}\mu_n &= \sum_{j=0}^{n} \sum_{k=0}^{\lfloor j/2 \rfloor} (-1)^{n-j} \binom{n}{j} \binom{j}{2k} b^{2k} \mu^{n-2k} \Gamma(2k+1) \\ &= \begin{cases} n! b^n & \text{for } n \text{ even} \\ 0 & \text{for } n \text{ odd,} \end{cases}\end{aligned} \tag{4}$$

where $\lfloor x \rfloor$ is the FLOOR FUNCTION and $\Gamma(2k+1)$ is the GAMMA FUNCTION.

The MOMENTS can also be computed using the CHARACTERISTIC FUNCTION,

$$\phi(t) \equiv \int_{-\infty}^{\infty} e^{itx} P(x)\,dx = \frac{1}{2b} \int_{-\infty}^{\infty} e^{itx} e^{-|x-\mu|/b}\,dx. \tag{5}$$

Using the FOURIER TRANSFORM OF THE EXPONENTIAL FUNCTION

$$\mathcal{F}[e^{-2\pi k_0 |x|}] = \frac{1}{\pi} \frac{k_0}{k^2 + {k_0}^2} \tag{6}$$

gives

$$\phi(t) = \frac{e^{i\mu t}}{2b} \frac{\frac{2}{b}}{t^2 + \left(\frac{1}{b}\right)^2} = \frac{e^{i\mu t}}{1 + b^2 t^2}. \tag{7}$$

The MOMENTS are therefore

$$\mu_n = (-i)^n \phi(0) = (-i)^n \left[\frac{d^n \phi}{dt^n}\right]_{t=0}. \tag{8}$$

The MEAN, VARIANCE, SKEWNESS, and KURTOSIS are

$$\mu = \mu \tag{9}$$
$$\sigma^2 = 2b^2 \tag{10}$$
$$\gamma_1 = 0 \tag{11}$$
$$\gamma_2 = 3. \tag{12}$$

References

Abramowitz, M. and Stegun, C. A. (Eds.). *Handbook of Mathematical Functions with Formulas, Graphs, and Mathematical Tables, 9th printing.* New York: Dover, 1972.

Laplace's Equation

The scalar form of Laplace's equation is the PARTIAL DIFFERENTIAL EQUATION

$$\nabla^2 \psi = 0. \tag{1}$$

It is a special case of the HELMHOLTZ DIFFERENTIAL EQUATION

$$\nabla^2 \psi + k^2 \psi = 0 \tag{2}$$

with $k = 0$, or POISSON'S EQUATION

$$\nabla^2 \psi = -4\pi\rho \tag{3}$$

with $\rho = 0$. The vector Laplace's equation is given by

$$\nabla^2 \mathbf{F} = \mathbf{0}. \tag{4}$$

A FUNCTION ψ which satisfies Laplace's equation is said to be HARMONIC. A solution to Laplace's equation has the property that the average value over a spherical surface is equal to the value at the center of the SPHERE (GAUSS'S HARMONIC FUNCTION THEOREM). Solutions have no local maxima or minima. Because Laplace's equation is linear, the superposition of any two solutions is also a solution.

A solution to Laplace's equation is uniquely determined if (1) the value of the function is specified on all boundaries (DIRICHLET BOUNDARY CONDITIONS) or (2) the normal derivative of the function is specified on all boundaries (NEUMANN BOUNDARY CONDITIONS).

Laplace's equation can be solved by SEPARATION OF VARIABLES in all 11 coordinate systems that the HELMHOLTZ DIFFERENTIAL EQUATION can. In addition, separation can be achieved by introducing a multiplicative factor in two additional coordinate systems. The separated form is

$$\psi = \frac{X_1(u_1)X_2(u_2)X_3(u_3)}{R(u_1, u_2, u_3)}, \tag{5}$$

and setting

$$\frac{h_1 h_2 h_3}{{h_i}^2} = g_i(u_{i+1}, u_{i+2}) f_i(u_i) R^2, \tag{6}$$

where h_i are SCALE FACTORS, gives the Laplace's equation

$$\sum_{i=1}^{3} \frac{1}{{h_i}^2 X_i}\left[\frac{1}{f_i}\frac{d}{du_i}\left(f_i \frac{dX_i}{du_i}\right)\right] = \sum_{i=1}^{3} \frac{1}{{h_i}^2 R}\left[\frac{1}{f_i}\frac{\partial}{\partial u_i}\left(f_i \frac{\partial R}{\partial u_i}\right)\right]. \tag{7}$$

If the right side is equal to $-{k_1}^2/F(u_1, u_2, u_3)$, where k_1 is a constant and F is any function, and if

$$h_1 h_2 h_3 = S f_1 f_2 f_3 R^2 F, \tag{8}$$

where S is the STÄCKEL DETERMINANT, then the equation can be solved using the methods of the HELMHOLTZ DIFFERENTIAL EQUATION. The two systems where this is the case are BISPHERICAL and TOROIDAL, bringing the total number of separable systems for Laplace's equation to 13 (Morse and Feshbach 1953, pp. 665–666).

In 2-D BIPOLAR COORDINATES, Laplace's equation is separable, although the HELMHOLTZ DIFFERENTIAL EQUATION is not.

see also BOUNDARY CONDITIONS, HARMONIC EQUATION, HELMHOLTZ DIFFERENTIAL EQUATION, PARTIAL DIFFERENTIAL EQUATION, POISSON'S EQUATION, SEPARATION OF VARIABLES, STÄCKEL DETERMINANT

References

Abramowitz, M. and Stegun, C. A. (Eds.). *Handbook of Mathematical Functions with Formulas, Graphs, and Mathematical Tables, 9th printing.* New York: Dover, p. 17, 1972.

Morse, P. M. and Feshbach, H. *Methods of Theoretical Physics, Part I.* New York: McGraw-Hill, pp. 125–126, 1953.

Laplace's Equation—Bipolar Coordinates

In 2-D BIPOLAR COORDINATES, LAPLACE'S EQUATION is

$$\frac{(\cosh v - \cos u)^2}{a^2}\left(\frac{\partial F^2}{\partial u^2} + \frac{\partial F^2}{\partial v^2}\right) = 0, \tag{1}$$

which simplifies to

$$\frac{\partial F^2}{\partial u^2} + \frac{\partial F^2}{\partial v^2} = 0, \tag{2}$$

so LAPLACE'S EQUATION is separable, although the HELMHOLTZ DIFFERENTIAL EQUATION is not.

Laplace's Equation—Bispherical Coordinates

$$\left[\frac{-\cos u\cot^2 u + 3\cosh v\cot^2 u}{\cosh v - \cos u} - \frac{3\cosh^2 v\cot u\csc u + \cosh^3 v\csc^2 u}{\cosh v - \cos u}\right]\frac{\partial}{\partial\phi^2}$$
$$+(\cos u - \cosh v)\sinh v\frac{\partial}{\partial v} + (\cosh^2 v - \cos u)^2\frac{\partial^2}{\partial v^2}$$
$$+(\cosh v - \cos u)(\cosh v\cot u - \sin u - \cos u\cot u)\frac{\partial}{\partial u}$$
$$+(\cosh^2 v - \cos u)^2\frac{\partial^2}{\partial u^2} = 0. \quad (1)$$

Let

$$F(u, v, \phi) = \sqrt{\cosh u - \cos v}\, U(u)V(v)\Phi(\phi), \quad (2)$$

then LAPLACE'S EQUATION is partially separable, although the HELMHOLTZ DIFFERENTIAL EQUATION is not.

References

Morse, P. M. and Feshbach, H. *Methods of Theoretical Physics, Part I.* New York: McGraw-Hill, pp. 665–666, 1953.

Laplace's Equation—Toroidal Coordinates

$$\nabla^2 f = \frac{(\cosh v - \cos u)^3}{a^2}\frac{\partial}{\partial u}\left(\frac{1}{\cosh v - \cos u}\frac{\partial f}{\partial u}\right)$$
$$+\frac{(\cosh v - \cos u)^3}{a^2\sinh v}\frac{\partial}{\partial v}\left(\frac{\sinh v}{\cosh v - \cos u}\frac{\partial f}{\partial v}\right)$$
$$+\frac{(\cosh v - \cos u)^2}{a^2\sinh v}\frac{\partial^2 f}{\partial\phi^2} \quad (1)$$
$$= \left[\frac{-3\cos\coth^2 v + \cosh v\coth^2 v}{\cosh v - \cos u} + \frac{3\cos^2 u\coth v\operatorname{csch} v - \cos^3 u\operatorname{csch}^2 v}{\cosh v - \cos u}\right]\frac{\partial^2}{\partial\phi^2}$$
$$+(\cos u - \cosh v)\sin u\frac{\partial}{\partial u} + (\cosh v - \cos u)^2\frac{\partial^2}{\partial u^2}$$
$$+(\cosh v - \cos u)(\cosh v\coth v - \sinh v - \cos u\coth v)\frac{\partial}{\partial v}$$
$$+(\cosh^2 v - \cos u)^2\frac{\partial^2}{\partial v^2}. \quad (2)$$

Let

$$f(\xi, \eta, \phi) = \sqrt{\cosh\eta - \cos\xi}\, X(\xi)H(\eta)\Psi(\psi), \quad (3)$$

then

$$X(\xi) = \frac{\sin}{\cos}(n\xi) \quad (4)$$

$$\Psi(\psi) = \frac{\sin}{\cos}(m\phi), \quad (5)$$

and the equation in η becomes

$$\frac{1}{\sinh\eta}\frac{d}{d\eta}\left(\sinh\eta\frac{dH}{d\eta}\right) - \frac{m^2}{\sinh^2\eta}H - (n^2 - \tfrac{1}{4})H = 0. \quad (6)$$

LAPLACE'S EQUATION is partially separable, although the HELMHOLTZ DIFFERENTIAL EQUATION is not.

References

Arfken, G. "Toroidal Coordinates (ξ, η, ϕ)." §2.13 in *Mathematical Methods for Physicists, 2nd ed.* Orlando, FL: Academic Press, pp. 112–114, 1970.

Byerly, W. E. *An Elementary Treatise on Fourier's Series, and Spherical, Cylindrical, and Ellipsoidal Harmonics, with Applications to Problems in Mathematical Physics.* New York: Dover, p. 264, 1959.

Morse, P. M. and Feshbach, H. *Methods of Theoretical Physics, Part I.* New York: McGraw-Hill, p. 666, 1953.

Laplace's Integral

$$P_n(x) = \frac{1}{\pi}\int_0^p i\frac{du}{\left(x + \sqrt{x^2 - 1}\cos u\right)^{n+1}}$$
$$= \frac{1}{\pi}\int_0^\pi (x + \sqrt{x^2 - 1}\cos u)^n\, du.$$

Laplace Limit

The value $e = 0.6627434193\ldots$ (Sloane's A033259) for which Laplace's formula for solving KEPLER'S EQUATION begins diverging. The constant is defined as the value e at which the function

$$f(x) = \frac{x\exp(\sqrt{1 + x^2})}{1 + \sqrt{1 + x^2}}$$

equals $f(\lambda) = 1$. The CONTINUED FRACTION of e is given by [0, 1, 1, 1, 27, 1, 1, 1, 8, 2, 154, ...] (Sloane's A033260). The positions of the first occurrences of n in the CONTINUED FRACTION of e are 2, 10, 35, 13, 15, 32, 101, 9, ... (Sloane's A033261). The incrementally largest terms in the CONTINUED FRACTION are 1, 27, 154, 1601, 2135, ... (Sloane's A033262), which occur at positions 2, 5, 11, 19, 1801, ... (Sloane's A033263).

see also ECCENTRIC ANOMALY, KEPLER'S EQUATION

References

Finch, S. "Favorite Mathematical Constants." `http://www.mathsoft.com/asolve/constant/lpc/lpc.html`.

Plouffe, S. "Laplace Limit Constant." `http://lacim.uqam.ca/piDATA/laplace.txt`.

Sloane, N. J. A. Sequences A033259, A033260, A033261, A033262, and A033263 in "An On-Line Version of the Encyclopedia of Integer Sequences."

Laplace-Mehler Integral

$$\begin{aligned} P_n(\cos\theta) &= \frac{1}{\pi}\int_0^{2\pi}(\cos\theta + i\sin\theta\cos\phi)^n\,d\phi \\ &= \frac{\sqrt{2}}{\pi}\int_0^{\theta}\frac{\cos[(n+\frac{1}{2})\phi]}{\sqrt{\cos\phi-\cos\theta}}\,d\phi \\ &= \frac{\sqrt{2}}{\pi}\int_\theta^{\pi}\frac{\sin[(n+\frac{1}{2})\phi]}{\sqrt{\cos\theta-\cos\phi}}\,d\phi. \end{aligned}$$

References

Iyanaga, S. and Kawada, Y. (Eds.). *Encyclopedic Dictionary of Mathematics.* Cambridge, MA: MIT Press, p. 1463, 1980.

Laplace Series

A function $f(\theta,\phi)$ expressed as a double sum of SPHERICAL HARMONICS is called a Laplace series. Taking f as a COMPLEX FUNCTION,

$$f(\theta,\phi) = \sum_{l=0}^{\infty}\sum_{m=-1}^{l} a_{lm}Y_l^m(\theta,\phi). \tag{1}$$

Now multiply both sides by $Y_{l'}^{m'*}\sin\theta$ and integrate over $d\theta$ and $d\phi$.

$$\begin{aligned} &\int_0^{2\pi}\int_0^{\pi} f(\theta,\phi)Y_{l'}^{m'*}\sin\theta\,d\theta\,d\phi \\ &= \sum_{l=0}^{\infty}\sum_{m=-l}^{l} a_{lm}\int_0^{2\pi}\int_0^{\pi} Y_{l'}^{m'*}(\theta,\phi)Y_l^m(\theta,\phi)\sin\theta\,d\theta\,d\phi. \end{aligned} \tag{2}$$

Now use the ORTHOGONALITY of the SPHERICAL HARMONICS

$$\int_0^{2\pi}\int_0^{\pi} Y_l^m(\theta,\phi)Y_{l'}^{m'*}\sin\theta\,d\theta\,d\phi = \delta_{mm'}\delta_{ll'}, \tag{3}$$

so (2) becomes

$$\begin{aligned} &\int_0^{2\pi}\int_0^{\pi} f(\theta,\phi)Y_{l'}^{m'*}\sin\theta\,d\theta\,d\phi \\ &\qquad = \sum_{l=0}^{\infty}\sum_{m=-1}^{l} a_{lm}\delta_{mm'}\delta_{ll'} = a_{lm}, \end{aligned} \tag{4}$$

where δ_{mn} is the KRONECKER DELTA.

For a REAL series, consider

$$\begin{aligned} &f(\theta,\phi) \\ &= \sum_{l=0}^{\infty}\sum_{m=-1}^{l}[C_l^m\cos(m\phi) + S_l^m\sin(m\phi)]P_l^m(\cos\theta). \end{aligned} \tag{5}$$

Proceed as before, using the orthogonality relationships

$$\begin{aligned} &\int_0^{2\pi}\int_0^{\pi} P_l^m(\cos\theta)\cos(m\phi)P_{l'}^{m'}(\cos\theta) \\ &\quad\times\cos(m'\phi)\sin(\theta)\,d\theta\,d\phi = -\frac{2\pi(l+m)!}{(2l+1)(l-m)!}\delta_{mm'}\delta_{ll'} \end{aligned} \tag{6}$$

$$\begin{aligned} &\int_0^{2\pi}\int_0^{\pi} P_l^m(\cos\theta)\sin(m\phi)P_{l'}^{m'}(\cos\theta) \\ &\quad\times\sin(m'\phi)\sin\theta\,d\theta\,d\phi = -\frac{2\pi(l+m)!}{(2l+1)(l-m)!}\delta_{mm'}\delta_{ll'}. \end{aligned} \tag{7}$$

So C_l^m and S_l^m are given by

$$\begin{aligned} C_l^m &= -\frac{(2l+1)(l-m)!}{2\pi(l+m)!} \\ &\quad\times\int_0^{2\pi}\int_0^{\pi} f(\theta,\phi)P_l^m\cos\theta\cos(m\phi)\sin\theta\,d\theta\,d\phi \end{aligned} \tag{8}$$

$$\begin{aligned} S_l^m &= -\frac{(2l+1)(l-m)!}{2\pi(l+m)!} \\ &\quad\times\int_0^{2\pi}\int_0^{\pi} f(\theta,\phi)P_l^m\cos\theta\sin(m\phi)\sin\theta\,d\theta\,d\phi. \end{aligned} \tag{9}$$

Laplace-Stieltjes Transform

An integral transform which is often written as an ordinary LAPLACE TRANSFORM involving the DELTA FUNCTION.

see also LAPLACE TRANSFORM

References

Abramowitz, M. and Stegun, C. A. (Eds.). *Handbook of Mathematical Functions with Formulas, Graphs, and Mathematical Tables, 9th printing.* New York: Dover, p. 1029, 1972.

Morse, P. M. and Feshbach, H. *Methods of Theoretical Physics, Part I.* New York: McGraw-Hill, 1953.

Widder, D. V. *The Laplace Transform.* Princeton, NJ: Princeton University Press, 1941.

Laplace Transform

The Laplace transform is an INTEGRAL TRANSFORM perhaps second only to the FOURIER TRANSFORM in its utility in solving physical problems. Due to its useful properties, the Laplace transform is particularly useful in solving linear ORDINARY DIFFERENTIAL EQUATIONS such as those arising in the analysis of electronic circuits.

The (one-sided) Laplace transform $\mathcal{L}$ (not to be confused with the LIE DERIVATIVE) is defined by

$$\mathcal{L}(s) = \mathcal{L}(f(t)) \equiv \int_0^{\infty} f(t)e^{-st}\,dt, \tag{1}$$

where $f(t)$ is defined for $t \geq 0$. A two-sided Laplace transform is sometimes also defined by

$$\mathcal{L}(s) = \mathcal{L}(f(t)) = \int_{-\infty}^{\infty} f(t)e^{-st}\,dt. \quad (2)$$

The Laplace transform existence theorem states that, if $f(t)$ is piecewise CONTINUOUS on every finite interval in $[0, \infty)$ satisfying

$$|f(t)| \leq Me^{at} \quad (3)$$

for all $t \in [0, \infty)$, then $\mathcal{L}(f(t))$ exists for all $s > a$. The Laplace transform is also UNIQUE, in the sense that, given two functions $F_1(t)$ and $F_2(t)$ with the same transform so that

$$\mathcal{L}[F_1(t)] = \mathcal{L}[F_2(t)] \equiv f(s), \quad (4)$$

then LERCH'S THEOREM guarantees that the integral

$$\int_0^a N(t)\,dt = 0 \quad (5)$$

vanishes for all $a > 0$ for a NULL FUNCTION defined by

$$N(t) \equiv F_1(t) - F_2(t). \quad (6)$$

The Laplace transform is LINEAR since

$$\begin{aligned}\mathcal{L}[af(t) + bg(t)] &= \int_0^{\infty} [af(t) + bg(t)]e^{-st}\,dt \\ &= a\int_0^{\infty} f(t)e^{-st}\,dt + b\int_0^{\infty} g(t)e^{-st}\,dt \\ &= a\mathcal{L}[f(t)] + b\mathcal{L}[g(t)]. \end{aligned} \quad (7)$$

The inverse Laplace transform is given by the BROMWICH INTEGRAL (see also DUHAMEL'S CONVOLUTION PRINCIPLE). A table of several important Laplace transforms follows.

$f(t)$	$\mathcal{L}[f(t)]$	Range
1	$\frac{1}{s}$	$s > 0$
t	$\frac{1}{s^2}$	$s > 0$
t^n	$\frac{n!}{s^{n+1}}$	$n \in \mathbb{Z} > 0$
t^a	$\frac{\Gamma(a+1)}{s^{a+1}}$	$a > 0$
e^{at}	$\frac{1}{s-a}$	$s > a$
$\cos(\omega t)$	$\frac{s}{s^2+\omega^2}$	$s > 0$
$\sin(\omega t)$	$\frac{\omega}{s^2+\omega^2}$	$s > 0$
$\cosh(\omega t)$	$\frac{s}{s^2-\omega^2}$	$s > \|a\|$
$\sinh(\omega t)$	$\frac{\omega}{s^2-\omega^2}$	$s > \|a\|$
$e^{at}\sin(bt)$	$\frac{b}{(s-a)^2+b^2}$	$s > a$
$e^{at}\cos(bt)$	$\frac{s-a}{(s-a)^2+b^2}$	$s > a$
$\delta(t-c)$	e^{-cs}	
$H_c(t)$	$\frac{e^{-cs}}{s}$	$s > 0$
$J_0(t)$	$\frac{1}{\sqrt{s^2+1}}$	

In the above table, $J_0(t)$ is the zeroth order BESSEL FUNCTION OF THE FIRST KIND, $\delta(t)$ is the DELTA FUNCTION, and $H_c(t)$ is the HEAVISIDE STEP FUNCTION. The Laplace transform has many important properties.

The Laplace transform of a CONVOLUTION is given by

$$\mathcal{L}(f(t) * g(t)) = \mathcal{L}(f(t))\mathcal{L}(g(t)) \quad (8)$$

$$\mathcal{L}^{-1}(F(s)G(s)) = \mathcal{L}^{-1}(F(s)) * \mathcal{L}^{-1}(G(s)). \quad (9)$$

Now consider DIFFERENTIATION. Let $f(t)$ be continuously differentiable $n-1$ times in $[0, \infty)$. If $|f(t)| \leq Me^{at}$, then

$$\begin{aligned}\mathcal{L}[f^{(n)}(t)] &= s^n\mathcal{L}(f(t)) - s^{n-1}f(0) \\ &\quad - s^{n-2}f'(0) - \ldots - f^{(n-1)}(0). \end{aligned} \quad (10)$$

This can be proved by INTEGRATION BY PARTS,

$$\begin{aligned}\mathcal{L}[f'(t)] &= \lim_{a\to 0}\int_0^a e^{-st}f'(t)\,dt \\ &= \lim_{a\to 0}[e^{-st}f(t)]_0^a + s\int_0^a e^{-st}f(t)\,dt] \\ &= \lim_{a\to 0}[e^{-sa}f(a) - f(0) + s\int_0^a e^{-st}f(t)\,dt] \\ &= s\mathcal{L}[f(t)] - f(0). \end{aligned} \quad (11)$$

Continuing for higher order derivatives then gives

$$\mathcal{L}[f''(t)] = s^2\mathcal{L}[f(t)] - sf(0) - f'(0). \quad (12)$$

This property can be used to transform differential equations into algebraic equations, a procedure known as the HEAVISIDE CALCULUS, which can then be inverse transformed to obtain the solution. For example, applying the Laplace transform to the equation

$$f''(t) + a_1f'(t) + a_0f(t) = 0 \quad (13)$$

gives

$$\begin{aligned}\{s^2\mathcal{L}[f(t)] - sf(0) - f'(0)\} + a_1\{s\mathcal{L}[f(t)] - f(0)\} \\ + a_0\mathcal{L}[f(t)] = 0 \end{aligned} \quad (14)$$

$$\mathcal{L}[f(t)](s^2 + a_1s + a_0) - sf(0) - f'(0) - a_1f(0) = 0, \quad (15)$$

which can be rearranged to

$$\mathcal{L}[f(t)] = \frac{sf(0) + [f'(0) + a_1f(0)]}{s^2 + a_1s + a_0}. \quad (16)$$

If this equation can be inverse Laplace transformed, then the original differential equation is solved.

Consider EXPONENTIATION. If $\mathcal{L}(f(t)) = F(s)$ for $s > \alpha$, then $\mathcal{L}(e^{at} f(t)) = F(s-a)$ for $s > a + \alpha$.

$$F(s-a) = \int_0^\infty f(t) e^{-(s-a)t}\,dt = \int_0^\infty [f(t)e^{at}]e^{-st}\,dt$$
$$= \mathcal{L}(e^{at} f(t)). \tag{17}$$

Consider INTEGRATION. If $f(t)$ is piecewise continuous and $|f(t)| \leq Me^{at}$, then

$$\mathcal{L}\left[\int_0^t f(t)\,dt\right] = \frac{1}{s}\mathcal{L}[f(t)]. \tag{18}$$

The inverse transform is known as the BROMWICH INTEGRAL, or sometimes the FOURIER-MELLIN INTEGRAL.

see also BROMWICH INTEGRAL, FOURIER-MELLIN INTEGRAL, FOURIER TRANSFORM, INTEGRAL TRANSFORM, LAPLACE-STIELTJES TRANSFORM, OPERATIONAL MATHEMATICS

References

Abramowitz, M. and Stegun, C. A. (Eds.). "Laplace Transforms." Ch. 29 in *Handbook of Mathematical Functions with Formulas, Graphs, and Mathematical Tables, 9th printing.* New York: Dover, pp. 1019–1030, 1972.

Arfken, G. *Mathematical Methods for Physicists, 3rd ed.* Orlando, FL: Academic Press, pp. 824–863, 1985.

Churchill, R. V. *Operational Mathematics.* New York: McGraw-Hill, 1958.

Doetsch, G. *Introduction to the Theory and Application of the Laplace Transformation.* Berlin: Springer-Verlag, 1974.

Franklin, P. *An Introduction to Fourier Methods and the Laplace Transformation.* New York: Dover, 1958.

Jaeger, J. C. and Newstead, G. H. *An Introduction to the Laplace Transformation with Engineering Applications.* London: Methuen, 1949.

Morse, P. M. and Feshbach, H. *Methods of Theoretical Physics, Part I.* New York: McGraw-Hill, pp. 467–469, 1953.

Spiegel, M. R. *Theory and Problems of Laplace Transforms.* New York: McGraw-Hill, 1965.

Widder, D. V. *The Laplace Transform.* Princeton, NJ: Princeton University Press, 1941.

Laplacian

The Laplacian operator for a SCALAR function ϕ is defined by

$$\nabla^2\phi = \frac{1}{h_1h_2h_3}\left[\frac{\partial}{\partial u_1}\left(\frac{h_2h_3}{h_1}\frac{\partial}{\partial u_1}\right) + \frac{\partial}{\partial u_2}\left(\frac{h_1h_3}{h_2}\frac{\partial}{\partial u_2}\right) + \frac{\partial}{\partial u_3}\left(\frac{h_1h_2}{h_3}\frac{\partial}{\partial u_3}\right)\right]\phi \tag{1}$$

in VECTOR notation, where the h_i are the SCALE FACTORS of the coordinate system. In TENSOR notation, the Laplacian is written

$$\nabla^2\phi = (g^{\lambda\kappa}\phi_{;\lambda})_{;\kappa} = g^{\lambda\kappa}\frac{\partial^2\phi}{\partial x^\lambda \partial x^\kappa} - \Gamma^\lambda \frac{\partial\phi}{\partial x^\lambda}$$
$$= \frac{1}{\sqrt{g}}\frac{\partial}{\partial x^j}\left(\sqrt{g}g^{ij}\frac{\partial\phi}{\partial x^i}\right), \tag{2}$$

where $g_{;\kappa}$ is a COVARIANT DERIVATIVE and

$$\Gamma^\lambda \equiv \tfrac{1}{2}g^{\mu\nu}g^{\lambda\kappa}\left(\frac{\partial g_{\kappa\mu}}{\partial x^\nu} + \frac{\partial g_{\kappa\nu}}{\partial x^\mu} - \frac{\partial g_{\mu\nu}}{\partial x^\kappa}\right). \tag{3}$$

The finite difference form is

$$\nabla^2\psi(x,y,z) = \frac{1}{h^2}[\psi(x+h,y,z) + \psi(x-h,y,z) + \psi(x,y+h,z) + \psi(x,y-h,z) + \psi(x,y,z+h) + \psi(x,y,z-h) - 6\psi(x,y,z)]. \tag{4}$$

For a pure radial function $g(r)$,

$$\nabla^2 g(r) \equiv \nabla\cdot[\nabla g(r)]$$
$$= \nabla\cdot\left[\frac{\partial g(r)}{\partial r}\hat{\mathbf{r}} + \frac{1}{r}\frac{\partial g(r)}{\partial\theta}\hat{\boldsymbol{\theta}} + \frac{1}{r\sin\theta}\frac{\partial g(r)}{\partial\phi}\hat{\boldsymbol{\phi}}\right]$$
$$= \nabla\cdot\left(\hat{\mathbf{r}}\frac{dg}{dr}\right). \tag{5}$$

Using the VECTOR DERIVATIVE identity

$$\nabla\cdot(f\mathbf{A}) = f(\nabla\cdot\mathbf{A}) + (\nabla f)\cdot(\mathbf{A}), \tag{6}$$

so

$$\nabla^2 g(r) \equiv \nabla\cdot[\nabla g(r)] = \frac{dg}{dr}\nabla\cdot\hat{\mathbf{r}} + \nabla\left(\frac{dg}{dr}\right)\cdot\hat{\mathbf{r}}$$
$$= \frac{2}{r}\frac{dg}{dr} + \frac{d^2g}{dr^2}. \tag{7}$$

Therefore, for a radial POWER law,

$$\nabla^2 r^n = \frac{2}{r}nr^{n-1} + n(n-1)r^{n-2} = [2n + n(n-1)]r^{n-2}$$
$$= n(n+1)r^{n-2}. \tag{8}$$

A VECTOR Laplacian can also be defined for a VECTOR $\mathbf{A}$ by

$$\nabla^2\mathbf{A} = \nabla(\nabla\cdot\mathbf{A}) - \nabla\times(\nabla\times\mathbf{A}) \tag{9}$$

in vector notation. In tensor notation, $\mathbf{A}$ is written A_μ, and the identity becomes

$$\nabla^2 A_\mu = A_{\mu;\lambda}{}^{;\lambda} = (g^{\lambda\kappa}A_{\mu;\lambda})_{;\kappa}$$
$$= g^{\lambda}{}_{\kappa;\kappa}A_{\mu;\lambda} + g^{\lambda\kappa}A_{\mu;\lambda\kappa}. \tag{10}$$

Similarly, a TENSOR Laplacian can be given by

$$\nabla^2 A_{\alpha\beta} = A_{\alpha\beta;\lambda}{}^{;\lambda}. \tag{11}$$

An identity satisfied by the Laplacian is

$$\nabla^2|\mathsf{xA}| = \frac{|\mathsf{A}|_2{}^2 - |(\mathsf{xA})\mathsf{A}^\mathrm{T}|^2}{|\mathsf{xA}|^3}, \tag{12}$$

where $|\mathsf{A}|_2$ is the HILBERT-SCHMIDT NORM, $\mathbf{x}$ is a row VECTOR, and A^T is the MATRIX TRANSPOSE of A.

To compute the LAPLACIAN of the inverse distance function $1/r$, where $r \equiv |\mathbf{r}-\mathbf{r}'|$, and integrate the LAPLACIAN over a volume,

$$\int_V \nabla^2 \left(\frac{1}{|\mathbf{r}-\mathbf{r}'|}\right) d^3\mathbf{r}. \tag{13}$$

This is equal to

$$\begin{aligned}\int_V \nabla^2 \frac{1}{r} d^3\mathbf{r} &= \int_V \nabla \cdot \left(\nabla \frac{1}{r}\right) d^3\mathbf{r} = \int_S \left(\nabla \frac{1}{r}\right) \cdot d\mathbf{a} \\ &= \int_S \frac{\partial}{\partial r}\left(\frac{1}{r}\right) \hat{\mathbf{r}} \cdot d\mathbf{a} = \int_S -\frac{1}{r^2} \hat{\mathbf{r}} \cdot d\mathbf{a} \\ &= -4\pi \frac{R^2}{r^2},\end{aligned} \tag{14}$$

where the integration is over a small SPHERE of RADIUS R. Now, for $r > 0$ and $R \to 0$, the integral becomes 0. Similarly, for $r = R$ and $R \to 0$, the integral becomes -4π. Therefore,

$$\nabla^2 \left(\frac{1}{|\mathbf{r}-\mathbf{r}'|}\right) = -4\pi\delta^3(\mathbf{r}-\mathbf{r}'), \tag{15}$$

where δ is the DELTA FUNCTION.

see also ANTILAPLACIAN

Laplacian Determinant Expansion by Minors

see DETERMINANT EXPANSION BY MINORS

Large Number

There are a wide variety of large numbers which crop up in mathematics. Some are contrived, but some actually arise in proofs. Often, it is possible to prove existence theorems by deriving some potentially huge upper limit which is frequently greatly reduced in subsequent versions (e.g., GRAHAM'S NUMBER, KOLMOGOROV-ARNOLD-MOSER THEOREM, MERTENS CONJECTURE, SKEWES NUMBER, WANG'S CONJECTURE).

Large decimal numbers beginning with 10^9 are named according to two mutually conflicting nomenclatures: the American system (in which the prefix stands for n in 10^{3+3n}) and the British system (in which the prefix stands for n in 10^{6n}). The following table gives the names assigned to various POWERS of 10 (Woolf 1982).

American	British	Power of 10
million	million	10^6
billion	milliard	10^9
trillion	billion	10^{12}
quadrillion		10^{15}
quintillion	trillion	10^{18}
sextillion		10^{21}
septillion	quadrillion	10^{24}
octillion		10^{27}
nonillion	quintillion	10^{30}
decillion		10^{33}
undecillion	sexillion	10^{36}
duodecillion		10^{39}
tredecillion	septillion	10^{42}
quattuordecillion		10^{45}
quindecillion	octillion	10^{48}
sexdecillion		10^{51}
septendecillion	nonillion	10^{54}
octodecillion		10^{57}
novemdecillion	decillion	10^{60}
vigintillion		10^{63}
	undecillion	10^{66}
	duodecillion	10^{72}
	tredecillion	10^{78}
	quattuordecillion	10^{84}
	quindecillion	10^{90}
	sexdecillion	10^{96}
	septendecillion	10^{102}
	octodecillion	10^{108}
	novemdecillion	10^{114}
	vigintillion	10^{120}
centillion		10^{303}
	centillion	10^{600}

see also 10, ACKERMANN NUMBER, ARROW NOTATION, BILLION, CIRCLE NOTATION, EDDINGTON NUMBER, G-FUNCTION, GÖBEL'S SEQUENCE, GOOGOL, GOOGOLPLEX, GRAHAM'S NUMBER, HUNDRED, HYPERFACTORIAL, JUMPING CHAMPION, LAW OF TRULY LARGE NUMBERS, MEGA, MEGISTRON, MILLION, MONSTER GROUP, MOSER, n-PLEX, POWER TOWER, SKEWES NUMBER, SMALL NUMBER, STEINHAUS-MOSER NOTATION, STRONG LAW OF LARGE NUMBERS, SUPERFACTORIAL, THOUSAND, WEAK LAW OF LARGE NUMBERS, ZILLION

References

Conway, J. H. and Guy, R. K. *The Book of Numbers.* New York: Springer-Verlag, pp. 59–62, 1996.

Crandall, R. E. "The Challenge of Large Numbers." *Sci. Amer.* **276**, 74–79, Feb. 1997.

Davis, P. J. *The Lore of Large Numbers.* New York: Random House, 1961.

Knuth, D. E. "Mathematics and Computer Science: Coping with Finiteness. Advances in Our Ability to Compute Are Bringing Us Substantially Closer to Ultimate Limitations." *Science* **194**, 1235–1242, 1976.

Munafo, R. "Large Numbers." `http://home.earthlink.net/~mrob/largenum.`

Spencer, J. "Large Numbers and Unprovable Theorems." *Amer. Math. Monthly* **90**, 669–675, 1983.

Woolf, H. B. (Ed. in Chief). *Webster's New Collegiate Dictionary.* Springfield, MA: Merriam, p. 782, 1980.

Large Prime

see GIGANTIC PRIME, LARGE NUMBER, TITANIC PRIME

Laspeyres' Index

The statistical INDEX

$$P_L \equiv \frac{\sum p_n q_0}{\sum p_0 q_0},$$

where p_n is the price per unit in period n and q_n is the quantity produced in period n.

see also INDEX

References

Kenney, J. F. and Keeping, E. S. *Mathematics of Statistics, Pt. 1, 3rd ed.* Princeton, NJ: Van Nostrand, pp. 65–67, 1962.

Latin Cross

An irregular DODECAHEDRON CROSS in the shape of a dagger †. The six faces of a CUBE can be cut along seven EDGES and unfolded into a Latin cross (i.e., the Latin cross is the NET of the CUBE). Similarly, eight hypersurfaces of a HYPERCUBE can be cut along 17 SQUARES and unfolded to form a 3-D Latin cross.

Another cross also called the Latin cross is illustrated above. It is a GREEK CROSS with flared ends, and is also known as the crux immissa or cross patée.

see also CROSS, DISSECTION, DODECAHEDRON, GREEK CROSS

Latin Rectangle

A $k \times n$ Latin rectangle is a $k \times n$ MATRIX with elements $a_{ij} \in \{1, 2, \ldots, n\}$ such that entries in each row and column are distinct. If $k = n$, the special case of a LATIN SQUARE results. A normalized Latin rectangle has first row $\{1, 2, \ldots, n\}$ and first column $\{1, 2, \ldots, k\}$. Let $L(k, n)$ be the number of normalized $k \times n$ Latin rectangles, then the total number of $k \times n$ Latin rectangles is

$$N(k,n) = \frac{n!(n-1)!L(k,n)}{(n-k)!}$$

(McKay and Rogoyski 1995), where $n!$ is a FACTORIAL. Kerewala (1941) found a RECURRENCE RELATION for $L(3, n)$, and Athreya, Pranesachar, and Singhi (1980) found a summation FORMULA for $L(4, n)$.

The asymptotic value of $L(o(n^{6/7}), n)$ was found by Godsil and McKay (1990). The numbers of $k \times n$ Latin rectangles are given in the following table from McKay and Rogoyski (1995). The entries $L(1, n)$ and $L(n, n)$ are omitted, since

$$L(1,n) = 1$$
$$L(n,n) = L(n-1,n),$$

but $L(1, 1)$ and $L(2, 1)$ are included for clarity. The values of $L(k, n)$ are given as a "wrap-around" series by Sloane's A001009.

n	k	$L(k,n)$
1	1	1
2	1	1
3	2	1
4	2	3
4	3	4
5	2	11
5	3	46
5	4	56
6	2	53
6	3	1064
6	4	6552
6	5	9408
7	2	309
7	3	36792
7	4	1293216
7	5	11270400
7	6	16942080
8	2	2119
8	3	1673792
8	4	420906504
8	5	27206658048
8	6	335390189568
8	7	535281401856
9	2	16687
9	3	103443808
9	4	207624560256
9	5	112681643983776
9	6	12962605404381184
9	7	224382967916691456
9	8	377597570964258816
10	2	148329
10	3	8154999232
10	4	147174521059584
10	5	746988383076286464
10	6	870735405591003709440
10	7	177144296983054185922560
10	8	4292039421591854273003520
10	9	7580721482160132811489280

References

Athreya, K. B.; Pranesachar, C. R.; and Singhi, N. M. "On the Number of Latin Rectangles and Chromatic Polynomial of $L(K_{r,s})$." *Europ. J. Combin.* **1**, 9–17, 1980.

Colbourn, C. J. and Dinitz, J. H. (Eds.) *CRC Handbook of Combinatorial Designs.* Boca Raton, FL: CRC Press, 1996.
Godsil, C. D. and McKay, B. D. "Asymptotic Enumeration of Latin Rectangles." *J. Combin. Th. Ser. B* **48**, 19–44, 1990.
Kerawla, S. M. "The Enumeration of Latin Rectangle of Depth Three by Means of Difference Equation" [sic]. *Bull. Calcutta Math. Soc.* **33**, 119–127, 1941.
McKay, B. D. and Rogoyski, E. "Latin Squares of Order 10." *Electronic J. Combinatorics* **2**, N3, 1–4, 1995. `http://www.combinatorics.org/Volume_2/volume2.html#N3`.
Ryser, H. J. "Latin Rectangles." §3.3 in *Combinatorial Mathematics.* Buffalo, NY: Math. Assoc. of Amer., pp. 35–37, 1963.
Sloane, N. J. A. Sequence A001009 in "An On-Line Version of the Encyclopedia of Integer Sequences."

Latin Square

An $n \times n$ Latin square is a LATIN RECTANGLE with $k = n$. Specifically, a Latin square consists of n sets of the numbers 1 to n arranged in such a way that no orthogonal (row or column) contains the same two numbers. The numbers of Latin squares of order $n = 1, 2, \ldots$ are 1, 2, 12, 576, ... (Sloane's A002860). A pair of Latin squares is said to be orthogonal if the n^2 pairs formed by juxtaposing the two arrays are all distinct.

Two of the Latin squares of order 3 are

$$\begin{bmatrix} 3 & 2 & 1 \\ 2 & 1 & 3 \\ 1 & 3 & 2 \end{bmatrix} \qquad \begin{bmatrix} 2 & 3 & 1 \\ 1 & 2 & 3 \\ 3 & 1 & 2 \end{bmatrix},$$

which are orthogonal. Two of the 576 Latin squares of order 4 are

$$\begin{bmatrix} 1 & 2 & 3 & 4 \\ 2 & 1 & 4 & 3 \\ 3 & 4 & 1 & 2 \\ 4 & 3 & 2 & 1 \end{bmatrix} \qquad \begin{bmatrix} 1 & 2 & 3 & 4 \\ 3 & 4 & 1 & 2 \\ 4 & 3 & 2 & 1 \\ 2 & 1 & 4 & 3 \end{bmatrix}.$$

A normalized, or reduced, Latin square is a Latin square with the first row and column given by $\{1, 2, \ldots, n\}$. General FORMULAS for the number of *normalized* Latin squares $L(n, n)$ are given by Nechvatal (1981), Gessel (1987), and Shao and Wei (1992). The total number of Latin squares of order n can then be computed from

$$N(n, n) = n!(n-1)!L(n, n) = n!(n-1)!L(n-1, n).$$

The numbers of normalized Latin square of order $n = 1, 2, \ldots$, are 1, 1, 1, 4, 56, 9408, ... (Sloane's A000315). McKay and Rogoyski (1995) give the number of normalized LATIN RECTANGLES $L(k, n)$ for $n = 1, \ldots, 10$, as well as estimates for $L(n, n)$ with $n = 11, 12, \ldots, 15$.

n	$L(n, n)$
11	5.36×10^{33}
12	1.62×10^{44}
13	2.51×10^{56}
14	2.33×10^{70}
15	1.5×10^{86}

see also EULER SQUARE, KIRKMAN TRIPLE SYSTEM, PARTIAL LATIN SQUARE, QUASIGROUP

References
Colbourn, C. J. and Dinitz, J. H. *CRC Handbook of Combinatorial Designs.* Boca Raton, FL: CRC Press, 1996.
Gessel, I. "Counting Latin Rectangles." *Bull. Amer. Math. Soc.* **16**, 79–83, 1987.
Hunter, J. A. H. and Madachy, J. S. *Mathematical Diversions.* New York: Dover, pp. 33–34, 1975.
Kraitchik, M. "Latin Squares." §7.11 in *Mathematical Recreations.* New York: W. W. Norton, p. 178, 1942.
Lindner, C. C. and Rodger, C. A. *Design Theory.* Boca Raton, FL: CRC Press, 1997.
McKay, B. D. and Rogoyski, E. "Latin Squares of Order 10." *Electronic J. Combinatorics* **2**, N3, 1–4, 1995. `http://www.combinatorics.org/Volume_2/volume2.html#N3`.
Nechvatal, J. R. "Asymptotic Enumeration of Generalised Latin Rectangles." *Util. Math.* **20**, 273–292, 1981.
Ryser, H. J. "Latin Rectangles." §3.3 in *Combinatorial Mathematics.* Buffalo, NY: Math. Assoc. Amer., pp. 35–37, 1963.
Shao, J.-Y. and Wei, W.-D. "A Formula for the Number of Latin Squares." *Disc. Math.* **110**, 293–296, 1992.
Sloane, N. J. A. Sequences A002860/M2051 and A000315/M3690 in "An On-Line Version of the Encyclopedia of Integer Sequences."

Latin-Graeco Square

see EULER SQUARE

Latitude

The latitude of a point on a SPHERE is the elevation of the point from the PLANE of the equator. The latitude δ is related to the COLATITUDE (the polar angle in SPHERICAL COORDINATES) by $\delta = \phi - 90°$. More generally, the latitude of a point on an ELLIPSOID is the ANGLE between a LINE PERPENDICULAR to the surface of the ELLIPSOID at the given point and the PLANE of the equator (Snyder 1987).

The equator therefore has latitude 0°, and the north and south poles have latitude ±90°, respectively. Latitude is also called GEOGRAPHIC LATITUDE or GEODETIC LATITUDE in order to distinguish it from several subtly different varieties of AUXILIARY LATITUDES.

The shortest distance between any two points on a SPHERE is the so-called GREAT CIRCLE distance, which can be directly computed from the latitudes and LONGITUDES of the two points.

see also AUXILIARY LATITUDE, COLATITUDE, CONFORMAL LATITUDE, GREAT CIRCLE, ISOMETRIC LATITUDE, LATITUDE, LONGITUDE, SPHERICAL COORDINATES

References
Snyder, J. P. *Map Projections—A Working Manual.* U. S. Geological Survey Professional Paper 1395. Washington, DC: U. S. Government Printing Office, p. 13, 1987.

Lattice

A lattice is a system K such that $\forall A \in K$, $A \subset A$, and if $A \subset B$ and $B \subset A$, then $A = B$, where $=$ here means "is included in." Lattices offer a natural way to formalize and study the ordering of objects using a general concept known as the POSET (partially ordered set). The study of lattices is called LATTICE THEORY. Note that this type of lattice is an abstraction of the regular array of points known as LATTICE POINTS.

The following inequalities hold for any lattice:

$$(x \wedge y) \vee (x \wedge z) \leq x \wedge (y \vee z)$$

$$x \vee (y \wedge z) \leq (x \vee y) \wedge (x \vee z)$$

$$(x \wedge y) \vee (y \wedge z) \vee (z \wedge x) \leq (x \vee y) \wedge (y \vee z) \wedge (z \vee x)$$

$$(x \wedge y) \vee (x \wedge z) \leq x \wedge (y \vee (x \wedge z))$$

(Grätzer 1971, p. 35). The first three are the distributive inequalities, and the last is the modular identity.

see also DISTRIBUTIVE LATTICE, INTEGRATION LATTICE, LATTICE THEORY, MODULAR LATTICE, TORIC VARIETY

References

Grätzer, G. *Lattice Theory: First Concepts and Distributive Lattices.* San Francisco, CA: W. H. Freeman, 1971.

Lattice Algebraic System

A generalization of the concept of set unions and intersections.

Lattice Animal

A distinct (including reflections and rotations) arrangement of adjacent squares on a grid, also called fixed POLYOMINOES.

see also PERCOLATION THEORY, POLYOMINO

Lattice Distribution

A DISCRETE DISTRIBUTION of a random variable such that every possible value can be represented in the form $a + bn$, where $a, b \neq 0$ and n is an INTEGER.

References

Abramowitz, M. and Stegun, C. A. (Eds.). *Handbook of Mathematical Functions with Formulas, Graphs, and Mathematical Tables, 9th printing.* New York: Dover, p. 927, 1972.

Lattice Graph

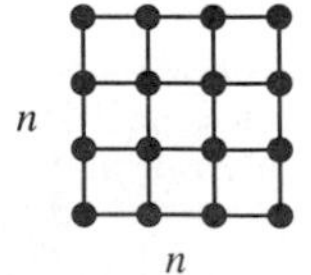

The lattice graph with n nodes on a side is denoted $L(n)$.

see also TRIANGULAR GRAPH

Lattice Groups

In the plane, there are 17 lattice groups, eight of which are pure translation. In $\mathbb{R}^3$, there are 32 POINT GROUPS and 230 SPACE GROUPS. In $\mathbb{R}^4$, there are 4783 space lattice groups.

see also POINT GROUPS, SPACE GROUPS, WALLPAPER GROUPS

Lattice Path

A path composed of connected horizontal and vertical line segments, each passing between adjacent LATTICE POINTS. A lattice path is therefore a SEQUENCE of points P_0, P_1, ..., P_n with $n \geq 0$ such that each P_i is a LATTICE POINT and P_{i+1} is obtained by offsetting one unit east (or west) or one unit north (or south).

The number of paths of length $a + b$ from the ORIGIN (0,0) to a point (a, b) which are restricted to east and north steps is given by the BINOMIAL COEFFICIENT $\binom{a+b}{a}$.

see also BALLOT PROBLEM, GOLYGON, KINGS PROBLEM, LATTICE POINT, p-GOOD PATH, RANDOM WALK

References

Dickau, R. M. "Shortest-Path Diagrams." `http://forum.swarthmore.edu/advanced/robertd/manhattan.html`.

Hilton, P. and Pederson, J. "Catalan Numbers, Their Generalization, and Their Uses." *Math. Intel.* **13**, 64–75, 1991.

Lattice Point

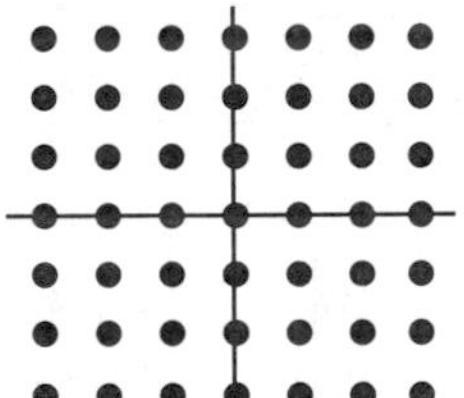

A POINT at the intersection of two or more grid lines in a ruled array. (The array of grid lines could be oriented to form unit cells in the shape of a square, rectangle, hexagon, etc.) However, unless otherwise specified, lattice points are generally taken to refer to points in a square array, i.e., points with coordinates $(m, n, \ldots)$, where m, n, ... are INTEGERS.

An n-D $\mathbb{Z}[\omega]$-lattice L_n lattice can be formally defined as a free $\mathbb{Z}[\omega]$-MODULE in complex n-D space $\mathbb{C}^n$.

The FRACTION of lattice points VISIBLE from the ORIGIN, as derived in Castellanos (1988, pp. 155–156), is

$$\frac{N'(r)}{N(r)} = \frac{\frac{24}{\pi^2}r^2 + \mathcal{O}(r \ln r)}{4r^2 + \mathcal{O}(r)} = \frac{\frac{6}{\pi^2} + \mathcal{O}\left(\frac{\ln r}{r}\right)}{1 + \mathcal{O}\left(\frac{1}{r}\right)} = \frac{6}{\pi^2}.$$

Therefore, this is also the probability that two randomly picked integers will be RELATIVELY PRIME to one another.

For $2 \leq n \leq 32$, it is possible to select $2n$ lattice points with $x, y \in [1, n]$ such that no three are in a straight

LINE. The number of distinct solutions (not counting reflections and rotations) for $n = 1, 2, \ldots$, are 1, 1, 4, 5, 11, 22, 57, 51, 156 ... (Sloane's A000769). For large n, it is conjectured that it is only possible to select at most $(c+\epsilon)n$ lattice points with no three COLLINEAR, where

$$c = (2\pi^2/3)^{1/3} \approx 1.85$$

(Guy and Kelly 1968; Guy 1994, p. 242). The number of the n^2 lattice points $x, y \in [1, n]$ which can be picked with no four CONCYCLIC is $\mathcal{O}(n^{2/3} - \epsilon)$ (Guy 1994, p. 241).

A special set of POLYGONS defined on the regular lattice are the GOLYGONS. A NECESSARY and SUFFICIENT condition that a linear transformation transforms a lattice to itself is that it be UNIMODULAR. M. Ajtai has shown that there is no efficient ALGORITHM for finding any fraction of a set of spanning vectors in a lattice having the shortest lengths unless there is an efficient algorithm for all of them (of which none is known). This result has potential applications to cryptography and authentication (Cipra 1996).

see also BARNES-WALL LATTICE, BLICHFELDT'S THEOREM, BROWKIN'S THEOREM, CIRCLE LATTICE POINTS, COXETER-TODD LATTICE, EHRHART POLYNOMIAL, GAUSS'S CIRCLE PROBLEM, GOLYGON, INTEGRATION LATTICE, JARNICK'S INEQUALITY, LATTICE PATH, LATTICE SUM, LEECH LATTICE, MINKOWSKI CONVEX BODY THEOREM, MODULAR LATTICE, N-CLUSTER, NOSARZEWSKA'S INEQUALITY, PICK'S THEOREM, POSET, RANDOM WALK, SCHINZEL'S THEOREM, SCHRÖDER NUMBER, VISIBLE POINT, VORONOI POLYGON

References

Apostol, T. *Introduction to Analytic Number Theory.* New York: Springer-Verlag, 1995.

Castellanos, D. "The Ubiquitous Pi." *Math. Mag.* **61**, 67–98, 1988.

Cipra, B. "Lattices May Put Security Codes on a Firmer Footing." *Science* **273**, 1047–1048, 1996.

Eppstein, D. "Lattice Theory and Geometry of Numbers." `http://www.ics.uci.edu/~eppstein/junkyard/lattice.html`.

Guy, R. K. "Gauß's Lattice Point Problem," "Lattice Points with Distinct Distances," "Lattice Points, No Four on a Circle," and "The No-Three-in-a-Line Problem." §F1, F2, F3, and F4 in *Unsolved Problems in Number Theory, 2nd ed.* New York: Springer-Verlag, pp. 240–244, 1994.

Guy, R. K. and Kelly, P. A. "The No-Three-in-Line-Problem." *Canad. Math. Bull.* **11**, 527–531, 1968.

Hammer, J. *Unsolved Problems Concerning Lattice Points.* London: Pitman, 1977.

Sloane, N. J. A. Sequence A000769/M3252 in "An On-Line Version of the Encyclopedia of Integer Sequences."

Lattice Reduction

The process finding a reduced set of basis vectors for a given LATTICE having certain special properties. Lattice reduction is implemented in Mathematica® (Wolfram Research, Champaign, IL) using the function `LatticeReduce`. Lattice reduction algorithms are used in a number of modern number theoretical applications, including in the discovery of a SPIGOT ALGORITHM for PI.

see also INTEGER RELATION, PSLQ ALGORITHM

Lattice Sum

Cubic lattice sums include the following:

$$b_2(2s) \equiv \sum_{i,j=-\infty}^{\infty}{}' \frac{(-1)^{i+j}}{(i^2+j^2)^s} \tag{1}$$

$$b_3(2s) \equiv \sum_{i,j,k=-\infty}^{\infty}{}' \frac{(-1)^{i+j+k}}{(i^2+j^2+k^2)^s} \tag{2}$$

$$b_n(2s) \equiv \sum_{k_1,\ldots,k_n=-\infty}^{\infty}{}' \frac{(-1)^{k_1+\ldots+k_n}}{({k_1}^2+\ldots+{k_n}^2)^s}, \tag{3}$$

where the prime indicates that summation over $(0,0,0)$ is excluded. As shown in Borwein and Borwein (1987, pp. 288–301), these have closed forms for even n

$$b_2(2s) = -4\beta(s)\eta(s) \tag{4}$$

$$b_4(2s) = -8\eta(s)\eta(s-1) \tag{5}$$

$$b_8(2s) = -16\zeta(s)\eta(s-3), \quad \text{for } \Re[s] > 1 \tag{6}$$

where $\beta(z)$ is the DIRICHLET BETA FUNCTION, $\eta(z)$ is the DIRICHLET ETA FUNCTION, and $\zeta(z)$ is the RIEMANN ZETA FUNCTION. The lattice sums evaluated at $s = 1$ are called the MADELUNG CONSTANTS. Borwein and Borwein (1986) prove that $b_8(2)$ converges (the closed form for $b_8(2s)$ above does not apply for $s = 1$), but its value has not been computed.

For hexagonal sums, Borwein and Borwein (1987, p. 292) give

$$h_2(2s) \equiv \frac{4}{3} \sum_{m,n=-\infty}^{\infty} \frac{\sin[(n+1)\theta]\sin[(m+1)\theta] - \sin(n\theta)\sin[(m-1)\theta]}{\left[(n+\frac{1}{2}m)^2 + 3(\frac{1}{2}m)^2\right]^s}, \tag{7}$$

where $\theta \equiv 2\pi/3$. This MADELUNG CONSTANT is expressible in closed form for $s = 1$ as

$$h_2(2) = \pi \ln 3\sqrt{3}. \tag{8}$$

see also BENSON'S FORMULA, MADELUNG CONSTANTS

References

Borwein, D. and Borwein, J. M. "On Some Trigonometric and Exponential Lattice Sums." *J. Math. Anal.* **188**, 209–218, 1994.

Borwein, D.; Borwein, J. M.; and Shail, R. "Analysis of Certain Lattice Sums." *J. Math. Anal.* **143**, 126–137, 1989.

Borwein, D. and Borwein, J. M. "A Note on Alternating Series in Several Dimensions." *Amer. Math. Monthly* **93**, 531–539, 1986.
Borwein, J. M. and Borwein, P. B. *Pi & the AGM: A Study in Analytic Number Theory and Computational Complexity.* New York: Wiley, 1987.
Finch, S. "Favorite Mathematical Constants." `http://www.mathsoft.com/asolve/constant/mdlung/mdlung.html`.
Glasser, M. L. and Zucker, I. J. "Lattice Sums." In *Perspectives in Theoretical Chemistry: Advances and Perspectives* **5**, 67–139, 1980.

Lattice Theory

Lattice theory is the study of sets of objects known as LATTICES. It is an outgrowth of the study of BOOLEAN ALGEBRAS, and provides a framework for unifying the study of classes or ordered sets in mathematics. Its study was given a great boost by a series of papers and subsequent textbook written by Birkhoff (1967).

see also LATTICE

References

Birkhoff, G. *Lattice Theory, 3rd ed.* Providence, RI: Amer. Math. Soc., 1967.
Grätzer, G. *Lattice Theory: First Concepts and Distributive Lattices.* San Francisco, CA: W. H. Freeman, 1971.

Latus Rectum

Twice the SEMILATUS RECTUM.

see also PARABOLA

Laurent Polynomial

A Laurent polynomial with COEFFICIENTS in the FIELD $\mathbb{F}$ is an algebraic object that is typically expressed in the form

$$\ldots + a_{-n}t^{-n} + a_{-(n-1)}t^{-(n-1)} + \ldots$$
$$+ a_{-1}t^{-1} + a_0 + a_1 t + \ldots + a_n t^n + \ldots,$$

where the a_i are elements of $\mathbb{F}$, and only finitely many of the a_i are NONZERO. A Laurent polynomial is an algebraic object in the sense that it is treated as a POLYNOMIAL except that the indeterminant "t" can also have NEGATIVE POWERS.

Expressed more precisely, the collection of Laurent polynomials with COEFFICIENTS in a FIELD $\mathbb{F}$ form a RING, denoted $\mathbb{F}[t, t^{-1}]$, with RING operations given by componentwise addition and multiplication according to the relation

$$at^n \cdot bt^m = abt^{n+m}$$

for all n and m in the INTEGERS. Formally, this is equivalent to saying that $\mathbb{F}[t, t^{-1}]$ is the GROUP RING of the INTEGERS and the FIELD $\mathbb{F}$. This corresponds to $\mathbb{F}[t]$ (the POLYNOMIAL ring in one variable for $\mathbb{F}$) being the GROUP RING or MONOID ring for the MONOID of natural numbers and the FIELD $\mathbb{F}$.

see also POLYNOMIAL

References

Lang, S. *Undergraduate Algebra, 2nd ed.* New York: Springer-Verlag, 1990.

Laurent Series

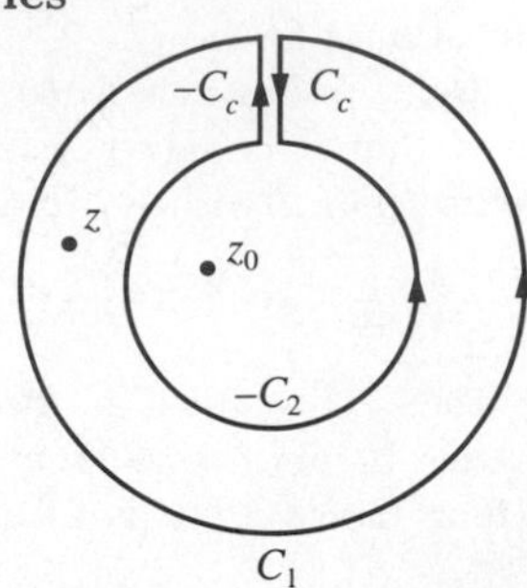

Let there be two circular contours C_2 and C_1, with the radius of C_1 larger than that of C_2. Let z_0 be interior to C_1 and C_2, and z be between C_1 and C_2. Now create a cut line C_c between C_1 and C_2, and integrate around the path $C \equiv C_1 + C_c - C_2 - C_c$, so that the plus and minus contributions of C_c cancel one another, as illustrated above. From the CAUCHY INTEGRAL FORMULA,

$$\begin{aligned} f(z) &= \frac{1}{2\pi i}\int_C \frac{f(z')}{z'-z}\,dz' \\ &= \frac{1}{2\pi i}\int_{C_1} \frac{f(z')}{z'-z}\,dz' + \frac{1}{2\pi i}\int_{C_c} \frac{f(z')}{z'-z}\,dz' \\ &\quad - \frac{1}{2\pi i}\int_{C_2} \frac{f(z')}{z'-z}\,dz' - \frac{1}{2\pi i}\int_{C_c} \frac{f(z')}{z'-z}\,dz' \\ &= \frac{1}{2\pi i}\int_{C_1} \frac{f(z')}{z'-z}\,dz' - \frac{1}{2\pi i}\int_{C_2} \frac{f(z')}{z'-z}\,dz'. \end{aligned} \quad (1)$$

Now, since contributions from the cut line in opposite directions cancel out,

$$\begin{aligned} f(z) &= \frac{1}{2\pi i}\int_{C_1} \frac{f(z')}{(z'-z_0)-(z-z_0)}dz' \\ &\quad - \frac{1}{2\pi i}\int_{C_2} \frac{f(z')}{(z'-z_0)-(z-z_0)}dz' \\ &= \frac{1}{2\pi i}\int_{C_1} \frac{f(z')}{(z'-z_0)\left(1-\frac{z-z_0}{z'-z_0}\right)}\,dz' \\ &\quad - \frac{1}{2\pi i}\int_{C_2} \frac{f(z')}{(z-z_0)\left(\frac{z'-z_0}{z-z_0}-1\right)}\,dz' \\ &= \frac{1}{2\pi i}\int_{C_1} \frac{f(z')}{(z'-z_0)\left(1-\frac{z-z_0}{z'-z_0}\right)}\,dz' \\ &\quad + \frac{1}{2\pi i}\int_{C_2} \frac{f(z')}{(z-z_0)\left(1-\frac{z'-z_0}{z-z_0}\right)}\,dz'. \end{aligned} \quad (2)$$

For the first integral, $|z'-z_0| > |z-z_0|$. For the second, $|z'-z_0| < |z-z_0|$. Now use the TAYLOR EXPANSION (valid for $|t| < 1$)

$$\frac{1}{1-t} = \sum_{n=0}^{\infty} t^n \quad (3)$$

to obtain

$$\begin{aligned} f(z) &= \frac{1}{2\pi i}\left[\int_{C_1}\frac{f(z')}{z'-z_0}\sum_{n=0}^{\infty}\left(\frac{z-z_0}{z'-z_0}\right)^n dz'\right.\\ &\quad\left.+\int_{C_2}\frac{f(z')}{z-z_0}\sum_{n=0}^{\infty}\left(\frac{z'-z_0}{z-z_0}\right)^n dz'\right]\\ &= \frac{1}{2\pi i}\sum_{n=0}^{\infty}(z-z_0)^n\int_{C_1}\frac{f(z')}{(z'-z_0)^{n+1}}\,dz'\\ &\quad+\frac{1}{2\pi i}\sum_{n=0}^{\infty}(z-z_0)^{-n-1}\int_{C_2}(z'-z_0)^n f(z')\,dz'\\ &= \frac{1}{2\pi i}\sum_{n=0}^{\infty}(z-z_0)^n\int_{C_1}\frac{f(z')}{(z'-z_0)^{n+1}}\,dz'\\ &\quad+\frac{1}{2\pi i}\sum_{n=1}^{\infty}(z-z_0)^{-n}\int_{C_2}(z'-z_0)^{n+1}f(z')\,dz', \end{aligned} \tag{4}$$

where the second term has been re-indexed. Re-indexing again,

$$\begin{aligned} f(z) &= \frac{1}{2\pi i}\sum_{n=0}^{\infty}(z-z_0)^n\int_{C_1}\frac{f(z')}{(z'-z_0)^{n+1}}\,dz'\\ &\quad+\frac{1}{2\pi i}\sum_{n=-\infty}^{-1}(z-z_0)^n\int_{C_2}\frac{f(z')}{(z'-z_0)^{n+1}}\,dz'. \end{aligned} \tag{5}$$

Now, use the CAUCHY INTEGRAL THEOREM, which requires that any CONTOUR INTEGRAL of a function which encloses no POLES has value 0. But $1/(z'-z_0)^{n+1}$ is never singular inside C_2 for $n \geq 0$, and $1/(z'-z_0)^{n+1}$ is never singular inside C_1 for $n \leq -1$. Similarly, there are no POLES in the closed cut $C_c - C_c$. We can therefore replace C_1 and C_2 in the above integrals by C without altering their values, so

$$\begin{aligned} f(z) &= \frac{1}{2\pi i}\sum_{n=0}^{\infty}(z-z_0)^n\int_{C}\frac{f(z')}{(z'-z_0)^{n+1}}\,dz'\\ &\quad+\frac{1}{2\pi i}\sum_{n=-\infty}^{-1}(z-z_0)^n\int_{C}\frac{f(z')}{(z'-z_0)^{n+1}}\,dz'\\ &= \frac{1}{2\pi i}\sum_{n=-\infty}^{\infty}(z-z_0)^n\int_{C}\frac{f(z')}{(z'-z_0)^{n+1}}\,dz'\\ &\equiv \sum_{n=-\infty}^{\infty}a_n(z-z_0)^n. \end{aligned} \tag{6}$$

The only requirement on C is that it encloses z, so we are free to choose any contour γ that does so. The RESIDUES a_n are therefore defined by

$$a_n \equiv \frac{1}{2\pi i}\int_{\gamma}\frac{f(z')}{(z'-z_0)^{n+1}}\,dz'. \tag{7}$$

see also MACLAURIN SERIES, RESIDUE (COMPLEX ANALYSIS), TAYLOR SERIES

References

Arfken, G. "Laurent Expansion." §6.5 in *Mathematical Methods for Physicists, 3rd ed.* Orlando, FL: Academic Press, pp. 376–384, 1985.

Morse, P. M. and Feshbach, H. "Derivatives of Analytic Functions, Taylor and Laurent Series." §4.3 in *Methods of Theoretical Physics, Part I.* New York: McGraw-Hill, pp. 374–398, 1953.

Law

A law is a mathematical statement which always holds true. Whereas "laws" in physics are generally experimental observations backed up by theoretical underpinning, laws in mathematics are generally THEOREMS which can formally be proven true under the stated conditions. However, the term is also sometimes used in the sense of an empirical observation, e.g., BENFORD'S LAW.

see also ABSORPTION LAW, BENFORD'S LAW, CONTRADICTION LAW, DE MORGAN'S DUALITY LAW, DE MORGAN'S LAWS, ELLIPTIC CURVE GROUP LAW, EXCLUDED MIDDLE LAW, EXPONENT LAWS, GIRKO'S CIRCULAR LAW, LAW OF COSINES, LAW OF SINES, LAW OF TANGENTS, LAW OF TRULY LARGE NUMBERS, MORRIE'S LAW, PARALLELOGRAM LAW, PLATEAU'S LAWS, QUADRATIC RECIPROCITY LAW, STRONG LAW OF LARGE NUMBERS, STRONG LAW OF SMALL NUMBERS, SYLVESTER'S INERTIA LAW, TRICHOTOMY LAW, VECTOR TRANSFORMATION LAW, WEAK LAW OF LARGE NUMBERS, ZIPF'S LAW

Law of Anomalous Numbers

see BENFORD'S LAW

Law of Cancellation

see CANCELLATION LAW

Law of Cosines

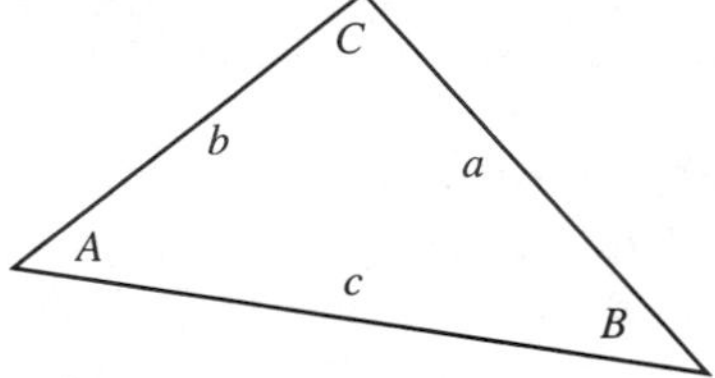

Let a, b, and c be the lengths of the legs of a TRIANGLE opposite ANGLES A, B, and C. Then the law of cosines states

$$c^2 = a^2 + b^2 - 2ab\cos C. \tag{1}$$

This law can be derived in a number of ways. The definition of the DOT PRODUCT incorporates the law of cosines, so that the length of the VECTOR from $\mathbf{X}$ to $\mathbf{Y}$ is given by

$$|\mathbf{X}-\mathbf{Y}|^2 = (\mathbf{X}-\mathbf{Y})\cdot(\mathbf{X}-\mathbf{Y}) \tag{2}$$

$$= \mathbf{X}\cdot\mathbf{X} - 2\mathbf{X}\cdot\mathbf{Y} + \mathbf{Y}\cdot\mathbf{Y} \tag{3}$$

$$= |\mathbf{X}|^2 + |\mathbf{Y}|^2 - 2|\mathbf{X}|\,|\mathbf{Y}|\cos\theta, \tag{4}$$

where θ is the ANGLE between **X** and **Y**.

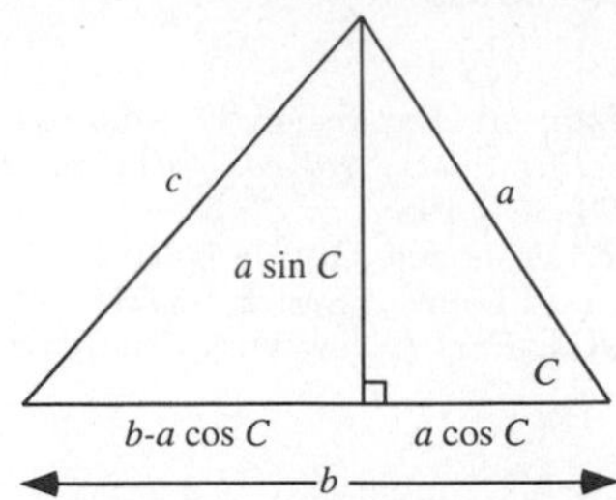

The formula can also be derived using a little geometry and simple algebra. From the above diagram,

$$\begin{aligned} c^2 &= (a\sin C)^2 + (b - a\cos C)^2 \\ &= a^2\sin^2 c + b^2 - 2ab\cos C + a^2\cos^2 C \\ &= a^2 + b^2 - 2ab\cos C. \end{aligned} \tag{5}$$

The law of cosines for the sides of a SPHERICAL TRIANGLE states that

$$\cos a = \cos b\cos c + \sin b\sin c\cos A \tag{6}$$

$$\cos b = \cos c\cos a + \sin c\sin a\cos B \tag{7}$$

$$\cos c = \cos a\cos b + \sin a\sin b\cos C \tag{8}$$

(Beyer 1987). The law of cosines for the angles of a SPHERICAL TRIANGLE states that

$$\cos A = -\cos B\cos C + \sin B\sin C\cos a \tag{9}$$

$$\cos B = -\cos C\cos A + \sin C\sin A\cos b \tag{10}$$

$$\cos C = -\cos A\cos B + \sin A\sin B\cos c \tag{11}$$

(Beyer 1987).

see also LAW OF SINES, LAW OF TANGENTS

References

Abramowitz, M. and Stegun, C. A. (Eds.). *Handbook of Mathematical Functions with Formulas, Graphs, and Mathematical Tables, 9th printing.* New York: Dover, p. 79, 1972.

Beyer, W. H. *CRC Standard Mathematical Tables, 28th ed.* Boca Raton, FL: CRC Press, pp. 148–149, 1987.

Law of Large Numbers

see LAW OF TRULY LARGE NUMBERS, STRONG LAW OF LARGE NUMBERS, WEAK LAW OF LARGE NUMBERS

Law of Sines

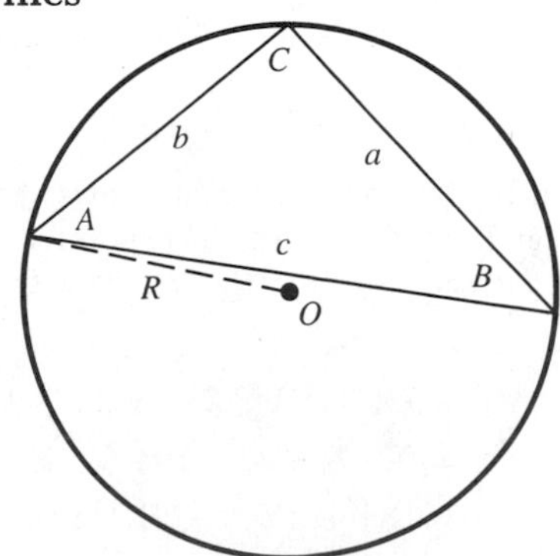

Let a, b, and c be the lengths of the LEGS of a TRIANGLE opposite ANGLES A, B, and C. Then the law of sines states that

$$\frac{a}{\sin A} = \frac{b}{\sin B} = \frac{c}{\sin C} = 2R, \tag{1}$$

where R is the radius of the CIRCUMCIRCLE. Other related results include the identities

$$a(\sin B - \sin C) + b(\sin C - \sin A) + c(\sin A - \sin B) = 0 \tag{2}$$

$$a = b\cos C + c\cos B, \tag{3}$$

the LAW OF COSINES

$$\cos A = \frac{c^2 + b^2 - a^2}{2bc}, \tag{4}$$

and the LAW OF TANGENTS

$$\frac{a+b}{a-b} = \frac{\tan[\frac{1}{2}(A+B)]}{\tan[\frac{1}{2}(A-B)]}. \tag{5}$$

The law of sines for oblique SPHERICAL TRIANGLES states that

$$\frac{\sin a}{\sin A} = \frac{\sin b}{\sin B} = \frac{\sin c}{\sin C}. \tag{6}$$

see also LAW OF COSINES, LAW OF TANGENTS

References

Abramowitz, M. and Stegun, C. A. (Eds.). *Handbook of Mathematical Functions with Formulas, Graphs, and Mathematical Tables, 9th printing.* New York: Dover, p. 79, 1972.

Beyer, W. H. *CRC Standard Mathematical Tables, 28th ed.* Boca Raton, FL: CRC Press, p. 148, 1987.

Coxeter, H. S. M. and Greitzer, S. L. *Geometry Revisited.* Washington, DC: Math. Assoc. Amer., pp. 1–3, 1967.

Law of Small Numbers

see STRONG LAW OF SMALL NUMBERS

Law of Tangents

Let a TRIANGLE have sides of lengths a, b, and c and let the ANGLES opposite these sides by A, B, and C. The law of tangents states

$$\frac{a-b}{a+b} = \frac{\tan[\frac{1}{2}(A-B)]}{\tan[\frac{1}{2}(A+B)]}.$$

An analogous result for oblique SPHERICAL TRIANGLES states that

$$\frac{\tan[\frac{1}{2}(a-b)]}{\tan[\frac{1}{2}(a+b)]} = \frac{\tan[\frac{1}{2}(A-B)]}{\tan[\frac{1}{2}(A+B)]}.$$

see also LAW OF COSINES, LAW OF SINES

References

Abramowitz, M. and Stegun, C. A. (Eds.). *Handbook of Mathematical Functions with Formulas, Graphs, and Mathematical Tables, 9th printing.* New York: Dover, p. 79, 1972.

Beyer, W. H. *CRC Standard Mathematical Tables, 28th ed.* Boca Raton, FL: CRC Press, pp. 145 and 149, 1987.

Law of Truly Large Numbers

With a large enough sample, any outrageous thing is likely to happen (Diaconis and Mosteller 1989). Littlewood (1953) considered an event which occurs one in a million times to be "surprising." Taking this definition, close to 100,000 surprising events are "expected" each year in the United States alone and, in the world at large, "we can be absolutely sure that we will see incredibly remarkable events" (Diaconis and Mosteller 1989).

see also COINCIDENCE, STRONG LAW OF LARGE NUMBERS, STRONG LAW OF SMALL NUMBERS, WEAK LAW OF LARGE NUMBERS

References

Diaconis, P. and Mosteller, F. "Methods of Studying Coincidences." *J. Amer. Statist. Assoc.* **84**, 853–861, 1989.
Littlewood, J. E. *Littlewood's Miscellany.* Cambridge, England: Cambridge University Press, 1986.

Lax-Milgram Theorem

Let ϕ be a bounded COERCIVE bilinear FUNCTIONAL on a HILBERT SPACE H. Then for every bounded linear FUNCTIONAL f on H, there exists a unique $x_f \in H$ such that

$$f(x) = \phi(x, x_f)$$

for all $x \in H$.

References

Debnath, L. and Mikusiński, P. *Introduction to Hilbert Spaces with Applications.* San Diego, CA: Academic Press, 1990.
Zeidler, E. *Applied Functional Analysis: Applications to Mathematical Physics.* New York: Springer-Verlag, 1995.

Lax Pair

A pair of linear OPERATORS L and A associated with a given PARTIAL DIFFERENTIAL EQUATION which can be used to solve the equation. However, it turns out to be very difficult to find the L and A corresponding to a given equation, so it is actually simpler to postulate a given L and A and determine to which PARTIAL DIFFERENTIAL EQUATION they correspond (Infeld and Rowlands 1990).

see also PARTIAL DIFFERENTIAL EQUATION

References

Infeld, E. and Rowlands, G. "Integrable Equations in Two Space Dimensions as Treated by the Zakharov Shabat Methods." §7.10 in *Nonlinear Waves, Solitons, and Chaos.* Cambridge, England: Cambridge University Press, pp. 216–223, 1990.

Layer

see p-LAYER

Le Cam's Identity

Let S_n be the sum of n random variates X_i with a BERNOULLI DISTRIBUTION with $P(X_i = 1) = p_i$. Then

$$\sum_{k=0}^{\infty} \left| P(S_n = k) - \frac{e^{-\lambda}\lambda^k}{k!} \right| < 2\sum_{i=1}^{n} {p_i}^2,$$

where

$$\lambda \equiv \sum_{i=1}^{n} p_i.$$

see also BERNOULLI DISTRIBUTION

References

Cox, D. A. "Introduction to Fermat's Last Theorem." *Amer. Math. Monthly* **101**, 3–14, 1994.

Leading Digit Phenomenon

see BENFORD'S LAW

Leading Order Analysis

A procedure for determining the behavior of an nth order ORDINARY DIFFERENTIAL EQUATION at a REMOVABLE SINGULARITY without actually solving the equation. Consider

$$\frac{d^n y}{dz^n} = F\left(\frac{d^{n-1}y}{dz^{n-1}}, \ldots, \frac{dy}{dx}, y, z\right), \qquad (1)$$

where F is ANALYTIC in z and rational in its other arguments. Proceed by making the substitution

$$y(z) \equiv a(z - z_0)^\alpha \qquad (2)$$

with $\alpha < 1$. For example, in the equation

$$\frac{d^2 y}{dz^2} = 6y^2 + Ay, \qquad (3)$$

making the substitution gives

$$a\alpha(\alpha-1)(z-z_0)^{\alpha-2} = 6a^2(z-z_0)^{2\alpha} + Aa(az-z_0)^\alpha. \qquad (4)$$

The most singular terms (those with the most NEGATIVE exponents) are called the "dominant balance terms," and must balance exponents and COEFFICIENTS at the SINGULARITY. Here, the first two terms are dominant, so

$$\alpha - 2 = 2\alpha \Rightarrow \alpha = -2 \qquad (5)$$

$$6a = 6a^2 \Rightarrow a = 1, \qquad (6)$$

and the solution behaves as $y(z) = (z - z_0)^{-2}$. The behavior in the NEIGHBORHOOD of the SINGULARITY is given by expansion in a LAURENT SERIES, in this case,

$$y(z) = \sum_{j=0}^{\infty} a_j (z - z_0)^{j-2}. \qquad (7)$$

Plugging this series in yields

$$\sum_{j=0}^{\infty} a_j(j-2)(j-3)(z-z_0)^{j-4}$$

$$= 6\sum_{j=0}^{\infty}\sum_{k=0}^{\infty} a_j a_k (z-z_0)^{j+k-4} + A\sum_{j=0}^{\infty} a_j (z-z_0)^{j-2}. \quad (8)$$

This gives RECURRENCE RELATIONS, in this case with a_6 arbitrary, so the $(z-z_0)^6$ term is called the resonance or KOVALEVSKAYA EXPONENT. At the resonances, the COEFFICIENT will always be arbitrary. If no resonance term is present, the POLE present is not ordinary, and the solution must be investigated using a PSI FUNCTION.

see also PSI FUNCTION

References

Tabor, M. *Chaos and Integrability in Nonlinear Dynamics: An Introduction.* New York: Wiley, p. 330, 1989.

Leaf (Foliation)

Let M^n be an n-MANIFOLD and let $\mathsf{F} = \{F_\alpha\}$ denote a PARTITION of M into DISJOINT path-connected SUBSETS. Then if F is a FOLIATION of M, each F_α is called a leaf and is not necessarily closed or compact.

see also FOLIATION

References

Rolfsen, D. *Knots and Links.* Wilmington, DE: Publish or Perish Press, p. 284, 1976.

Leaf (Tree)

An unconnected end of a TREE.

see also BRANCH, CHILD, FORK, ROOT (TREE), TREE

Leakage

see ALIASING

Least Bound

see SUPREMUM

Least Common Multiple

The least common multiple of two numbers n_1 and n_2 is denoted $\mathrm{LCM}(n_1, n_2)$ or $[n_1, n_2]$ and can be obtained by finding the PRIME factorization of each

$$n_1 = {p_1}^{a_1} \cdots {p_n}^{a_n} \quad (1)$$

$$n_2 = {p_1}^{b_1} \cdots {p_n}^{b_n}, \quad (2)$$

where the p_is are all PRIME FACTORS of n_1 and n_2, and if p_i does not occur in one factorization, then the corresponding exponent is 0. The least common multiple is then

$$\mathrm{LCM}(n_1, n_2) = [n_1, n_2] = \prod_{i=1}^{n} {p_i}^{\max(a_i, b_i)}. \quad (3)$$

Let m be a common multiple of a and b so that

$$m = ha = kb. \quad (4)$$

Write $a = a_1(a,b)$ and $b = b_1(a,b)$, where a_1 and b_1 are RELATIVELY PRIME by definition of the GREATEST COMMON DIVISOR $(a_1, b_1) = 1$. Then $ha_1 = kb_1$, and from the DIVISION LEMMA (given that ha_1 is DIVISIBLE by b and $(b_1, a_1) = 0$), we have h is DIVISIBLE by b_1, so

$$h = nb_1 \quad (5)$$

$$m = ha = nb_1 a = n\frac{ab}{(a,b)}. \quad (6)$$

The smallest m is given by $n = 1$,

$$\mathrm{LCM}(a,b) = \frac{ab}{\mathrm{GCD}(a,b)}, \quad (7)$$

so

$$\mathrm{GCD}(a,b)\ \mathrm{LCM}(a,b) = ab \quad (8)$$

$$(a,b)[a,b] = ab. \quad (9)$$

The LCM is IDEMPOTENT

$$[a,a] = a, \quad (10)$$

COMMUTATIVE

$$[a,b] = [b,a], \quad (11)$$

ASSOCIATIVE

$$[a,b,c] = [[a,b],c] = [a,[b,c]], \quad (12)$$

DISTRIBUTIVE

$$[ma, mb, mc] = m[a,b,c], \quad (13)$$

and satisfies the ABSORPTION LAW

$$(a,[a,b]) = a. \quad (14)$$

It is also true that

$$[ma, mb] = \frac{(ma)(mb)}{(ma, mb)} = m\frac{ab}{(a,b)} = m[a,b]. \quad (15)$$

see also GREATEST COMMON DIVISOR, MANGOLDT FUNCTION, RELATIVELY PRIME

References

Guy, R. K. "Density of a Sequence with L.C.M. of Each Pair Less than x." §E2 in *Unsolved Problems in Number Theory, 2nd ed.* New York: Springer-Verlag, pp. 200–201, 1994.

Least Deficient Number

A number for which

$$\sigma(n) = 2n - 1.$$

All POWERS of 2 are least deficient numbers.

see also DEFICIENT NUMBER, QUASIPERFECT NUMBER

Least Period

The smallest n for which a point x_0 is a PERIODIC POINT of a function f so that $f^n(x_0) = x_0$. For example, for the FUNCTION $f(x) = -x$, all points x have period 2 (including $x = 0$). However, $x = 0$ has a *least* period of 1. The analogous concept exists for a PERIODIC SEQUENCE, but not for a PERIODIC FUNCTION. The least period is also called the EXACT PERIOD.

Least Prime Factor

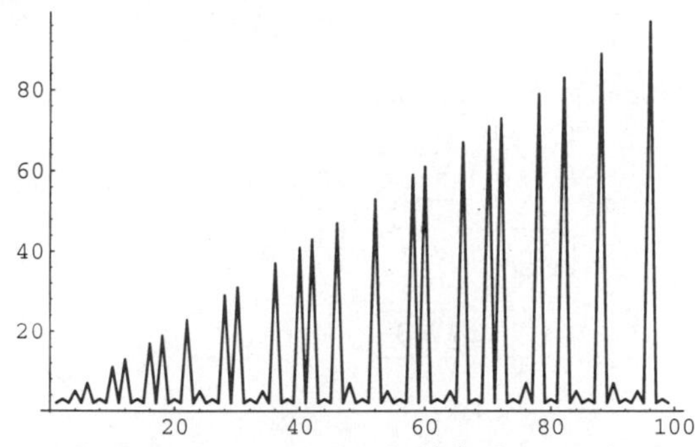

For an INTEGER $n \geq 2$, let $\text{lpf}(x)$ denote the LEAST PRIME FACTOR of n, i.e., the number p_1 in the factorization

$$n = {p_1}^{a_1} \cdots {p_k}^{a_k},$$

with $p_i < p_j$ for $i < j$. For $n = 2, 3, \ldots$, the first few are 2, 3, 2, 5, 2, 7, 2, 3, 2, 11, 2, 13, 2, 3, ... (Sloane's A020639). The above plot of the least prime factor function can be seen to resemble a jagged terrain of mountains, which leads to the appellation of "TWIN PEAKS" to a PAIR of INTEGERS (x, y) such that

1. $x < y$,
2. $\text{lpf}(x) = \text{lpf}(y)$,
3. For all z, $x < z < y$ IMPLIES $\text{lpf}(z) < \text{lpf}(x)$.

The least *multiple* prime factors for SQUAREFUL integers are 2, 2, 3, 2, 2, 3, 2, 2, 5, 3, 2, 2, 2, ... (Sloane's A046027).

see also ALLADI-GRINSTEAD CONSTANT, DISTINCT PRIME FACTORS, ERDŐS-SELFRIDGE FUNCTION, FACTOR, GREATEST PRIME FACTOR, LEAST COMMON MULTIPLE, MANGOLDT FUNCTION, PRIME FACTORS, TWIN PEAKS

References

Sloane, N. J. A. Sequence A020639 in "An On-Line Version of the Encyclopedia of Integer Sequences."

Least Squares Fitting

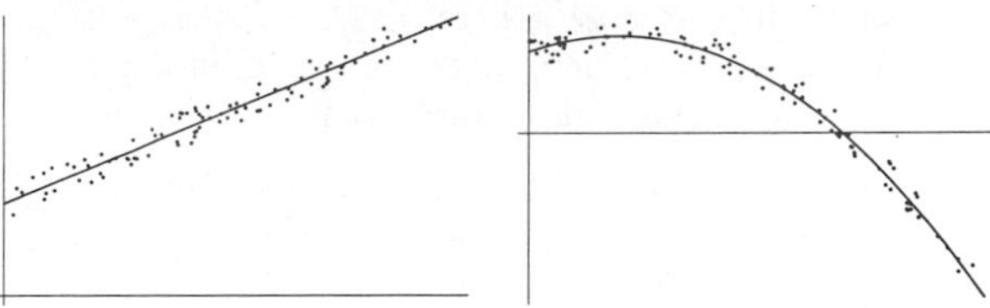

A mathematical procedure for finding the best fitting curve to a given set of points by minimizing the sum of the squares of the offsets ("the residuals") of the points from the curve. The sum of the *squares* of the offsets is used instead of the offset absolute values because this allows the residuals to be treated as a continuous differentiable quantity. However, because squares of the offsets are used, outlying points can have a disproportionate effect on the fit, a property which may or may not be desirable depending on the problem at hand.

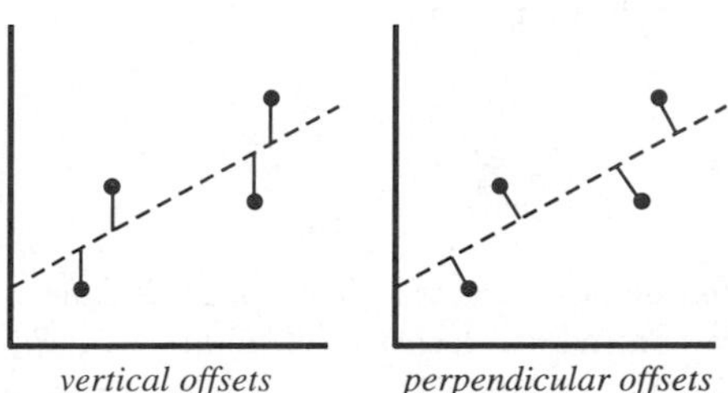

In practice, the *vertical* offsets from a line are almost always minimized instead of the *perpendicular* offsets. This allows uncertainties of the data points along the x- and y-axes to be incorporated simply, and also provides a much simpler analytic form for the fitting parameters than would be obtained using a fit based on perpendicular distances. In addition, the fitting technique can be easily generalized from a best-fit *line* to a best-fit *polynomial* when sums of vertical distances are used (which is not the case using perpendicular distances). For a reasonable number of noisy data points, the difference between vertical and perpendicular fits is quite small.

The *linear* least squares fitting technique is the simplest and most commonly applied form of LINEAR REGRESSION and provides a solution to the problem of finding the best fitting *straight* line through a set of points. In fact, if the functional relationship between the two quantities being graphed is known to within additive or multiplicative constants, it is common practice to transform the data in such a way that the resulting line *is* a straight line, say by plotting T vs. $\sqrt{\ell}$ instead of t vs. ℓ. For this reason, standard forms for EXPONENTIAL, LOGARITHMIC, and POWER laws are often explicitly computed. The formulas for linear least squares fitting were independently derived by Gauss and Legendre.

For NONLINEAR LEAST SQUARES FITTING to a number of unknown parameters, linear least squares fitting may be applied iteratively to a linearized form of the function until convergence is achieved. Depending on the type of fit and initial parameters chosen, the nonlinear fit may have good or poor convergence properties. If uncertainties (in the most general case, error ellipses) are given for the points, points can be weighted differently in order to give the high-quality points more weight.

The residuals of the best-fit line for a set of n points using *unsquared perpendicular* distances d_i of points (x_i, y_i) are given by

$$R_\perp \equiv \sum_{i=1}^n d_i. \tag{1}$$

Since the perpendicular distance from a line $y = a + bx$ to point i is given by

$$d_i = \frac{|y_i - (a + bx_i)|}{\sqrt{1+b^2}}, \tag{2}$$

the function to be minimized is

$$R_\perp \equiv \sum_{i=1}^n \frac{|y_i - (a + bx_i)|}{\sqrt{1+b^2}}. \tag{3}$$

Unfortunately, because the absolute value function does not have continuous derivatives, minimizing $R_\perp$ is not amenable to analytic solution. However, if the *square* of the perpendicular distances

$$R_\perp^2 \equiv \sum_{i=1}^n \frac{[y_i - (a + bx_i)]^2}{1+b^2} \tag{4}$$

is minimized instead, the problem can be solved in closed form. $R_\perp^2$ is a minimum when (suppressing the indices)

$$\frac{\partial R_\perp^2}{\partial a} = \frac{2}{1+b^2} \sum [y - (a+bx)](-1) = 0 \tag{5}$$

and

$$\frac{\partial R_\perp^2}{\partial b} = \frac{2}{1+b^2} \sum [y - (a+bx)](-x) + \sum \frac{[y-(a+bx)]^2(-1)(2b)}{(1+b^2)^2} = 0. \tag{6}$$

The former gives

$$a = \frac{\sum y - b\sum x}{n} = \bar{y} - b\bar{x}, \tag{7}$$

and the latter

$$(1+b^2) \sum [y - (a+bx)]x + b \sum [y - (a+bx)]^2 = 0. \tag{8}$$

But

$$\begin{aligned} [y - (a+bx)]^2 &= y^2 - 2(a+bx)y + (a+bx)^2 \\ &= y^2 - 2ay - 2bxy + a^2 + 2abx + b^2x^2, \end{aligned} \tag{9}$$

so (8) becomes

$$\begin{aligned} &(1+b^2)\left(\sum xy - a\sum x - b\sum x^2\right) \\ &+b\left(\sum y^2 - 2a\sum y - 2b\sum xy + a^2\sum 1\right. \\ &\left.+2ab\sum x + b^2\sum x^2\right) = 0 \end{aligned} \tag{10}$$

$$\begin{aligned} &[(1+b^2)(-b) + b(b^2)]\sum x^2 + [(1+b^2) - 2b^2]\sum xy \\ &+b\sum y^2 + [-a(1+b^2) + 2ab^2]\sum x - 2ab\sum y \\ &+ba^2\sum 1 = 0 \end{aligned} \tag{11}$$

$$\begin{aligned} -b\sum x^2 + (1-b^2)\sum xy + b\sum y^2 + a(b^2-1)\sum x \\ -2ab\sum y + ba^2 n = 0. \end{aligned} \tag{12}$$

Plugging (7) into (12) then gives

$$\begin{aligned} &-b\sum x^2 + (1-b^2)\sum xy + b\sum y^2 \\ &+\frac{1}{n}(b^2-1)\left(\sum y - b\sum x\right)\sum x \\ &-\frac{2}{n}\left(\sum y - b\sum x\right)b\sum y + \frac{1}{n}b\left(\sum y - b\sum x\right)^2 \\ &= 0 \end{aligned} \tag{13}$$

After a fair bit of algebra, the result is

$$b^2 + \frac{\sum y^2 - \sum x^2 + \frac{1}{n}\left[\left(\sum x\right)^2 - \left(\sum y\right)^2\right]}{\frac{1}{n}\sum x \sum y - \sum xy} b - 1 = 0. \tag{14}$$

So define

$$\begin{aligned} B &\equiv \frac{1}{2}\frac{\left[\sum y^2 - \frac{1}{n}\left(\sum y\right)^2\right] - \left[\sum x^2 - \frac{1}{n}\left(\sum x\right)^2\right]}{\frac{1}{n}\sum x\sum y - \sum xy} \\ &= \frac{1}{2}\frac{\left(\sum y^2 - n\bar{y}^2\right) - \left(\sum x^2 - n\bar{x}^2\right)}{n\sum x \sum y - \sum xy}, \end{aligned} \tag{15}$$

and the QUADRATIC FORMULA gives

$$b = -B \pm \sqrt{B^2+1}, \tag{16}$$

with a found using (7). Note the rather unwieldy form of the best-fit parameters in the formulation. In addition, minimizing $R_\perp^2$ for a second- or higher-order POLYNOMIAL leads to polynomial equations having *higher* order, so this formulation cannot be extended.

Vertical least squares fitting proceeds by finding the sum of the *squares* of the *vertical* deviations R^2 of a set of n data points

$$R^2 \equiv \sum [y_i - f(x_i, a_1, a_2, \ldots, a_n)]^2 \tag{17}$$

from a function f. Note that this procedure does *not* minimize the actual deviations from the line (which would be measured perpendicular to the given function). In addition, although the *unsquared* sum of distances might seem a more appropriate quantity to minimize, use of the absolute value results in discontinuous derivatives which cannot be treated analytically. The square deviations from each point are therefore summed, and the resulting residual is then minimized to find the best fit line. This procedure results in outlying points being given disproportionately large weighting.

The condition for R^2 to be a minimum is that

$$\frac{\partial(R^2)}{\partial a_i} = 0 \tag{18}$$

for $i = 1, \ldots, n$. For a linear fit,

$$f(a, b) = a + bx, \tag{19}$$

so

$$R^2(a, b) \equiv \sum_{i=1}^{n} [y_i - (a + bx_i)]^2 \tag{20}$$

$$\frac{\partial(R^2)}{\partial a} = -2\sum_{i=1}^{n} [y_i - (a + bx_i)] = 0 \tag{21}$$

$$\frac{\partial(R^2)}{\partial b} = -2\sum_{i=1}^{n} [y_i - (a + bx_i)]x_i = 0. \tag{22}$$

These lead to the equations

$$na + b\sum x = \sum y \tag{23}$$

$$a\sum x + b\sum x^2 = \sum xy, \tag{24}$$

where the subscripts have been dropped for conciseness. In MATRIX form,

$$\begin{bmatrix} n & \sum x \\ \sum x & \sum x^2 \end{bmatrix} \begin{bmatrix} a \\ b \end{bmatrix} = \begin{bmatrix} \sum y \\ \sum xy \end{bmatrix}, \tag{25}$$

so

$$\begin{bmatrix} a \\ b \end{bmatrix} = \begin{bmatrix} n & \sum x \\ \sum x & \sum x^2 \end{bmatrix}^{-1} \begin{bmatrix} \sum y \\ \sum xy \end{bmatrix}. \tag{26}$$

The 2×2 MATRIX INVERSE is

$$\begin{bmatrix} a \\ b \end{bmatrix} = \frac{1}{n\sum x^2 - \left(\sum x\right)^2} \begin{bmatrix} \sum y \sum x^2 - \sum x \sum xy \\ n\sum xy - \sum x \sum y \end{bmatrix}, \tag{27}$$

so

$$a = \frac{\sum y \sum x^2 - \sum x \sum xy}{n\sum x^2 - \left(\sum x\right)^2} \tag{28}$$

$$= \frac{\bar{y}\sum x^2 - \bar{x}\sum xy}{\sum x^2 - n\bar{x}^2} \tag{29}$$

$$b = \frac{n\sum xy - \sum x \sum y}{n\sum x^2 - \left(\sum x\right)^2} \tag{30}$$

$$= \frac{\sum xy - n\bar{x}\bar{y}}{\sum x^2 - n\bar{x}^2} \tag{31}$$

(Kenney and Keeping 1962). These can be rewritten in a simpler form by defining the sums of squares

$$\mathrm{ss}_{xx} = \sum_{i=1}^{n} (x_i - \bar{x})^2 = \left(\sum x^2 - n\bar{x}^2\right) \tag{32}$$

$$\mathrm{ss}_{yy} = \sum_{i=1}^{n} (y_i - \bar{y})^2 = \left(\sum y^2 - n\bar{y}^2\right) \tag{33}$$

$$\mathrm{ss}_{xy} = \sum_{i=1}^{n} (x_i - \bar{x})(y_i - \bar{y}) = \left(\sum xy - n\bar{x}\bar{y}\right), \tag{34}$$

which are also written as

$$\sigma_x^2 = \mathrm{ss}_{xx} \tag{35}$$

$$\sigma_y^2 = \mathrm{ss}_{yy} \tag{36}$$

$$\mathrm{cov}(x, y) = \mathrm{ss}_{xy}. \tag{37}$$

Here, $\mathrm{cov}(x, y)$ is the COVARIANCE and σ_x^2 and σ_y^2 are variances. Note that the quantities $\sum xy$ and $\sum x^2$ can also be interpreted as the DOT PRODUCTS

$$\sum x^2 = \mathbf{x} \cdot \mathbf{x} \tag{38}$$

$$\sum xy = \mathbf{x} \cdot \mathbf{y}. \tag{39}$$

In terms of the sums of squares, the REGRESSION COEFFICIENT b is given by

$$b = \frac{\mathrm{cov}(x, y)}{\sigma_x{}^2} = \frac{\mathrm{ss}_{xy}}{\mathrm{ss}_{xx}}, \tag{40}$$

and a is given in terms of b using (24) as

$$a = \bar{y} - b\bar{x}. \tag{41}$$

The overall quality of the fit is then parameterized in terms of a quantity known as the CORRELATION COEFFICIENT, defined by

$$r^2 = \frac{\mathrm{ss}_{xy}{}^2}{\mathrm{ss}_{xx}\mathrm{ss}_{yy}}, \tag{42}$$

which gives the proportion of ss_{yy} which is accounted for by the regression.

The STANDARD ERRORS for a and b are

$$\mathrm{SE}(a) = s\sqrt{\frac{1}{n} + \frac{\bar{x}^2}{\mathrm{ss}_{xx}}} \tag{43}$$

$$\mathrm{SE}(b) = \frac{s}{\sqrt{\mathrm{ss}_{xx}}}. \tag{44}$$

Let $\hat{y}_i$ be the vertical coordinate of the best-fit line with x-coordinate x_i, so

$$\hat{y}_i \equiv a + bx_i, \tag{45}$$

then the error between the actual vertical point y_i and the fitted point is given by

$$e_i \equiv y_i - \hat{y}_i. \tag{46}$$

Now define s^2 as an estimator for the variance in e_i,

$$s^2 = \sum_{i=1}^{n} \frac{{e_i}^2}{n-2}. \tag{47}$$

Then s can be given by

$$s = \sqrt{\frac{\mathrm{ss}_{yy} - b\,\mathrm{ss}_{xy}}{n-2}} = \sqrt{\frac{\mathrm{ss}_{yy} - \frac{s_{xy}^2}{s_{xx}}}{n-2}} \tag{48}$$

(Acton 1966, pp. 32–35; Gonick and Smith 1993, pp. 202–204).

Generalizing from a straight line (i.e., first degree polynomial) to a kth degree POLYNOMIAL

$$y = a_0 + a_1 x + \ldots + a_k x^k, \tag{49}$$

the residual is given by

$$R^2 \equiv \sum_{i=1}^{n} [y_i - (a_0 + a_1 x_i + \ldots + a_k {x_i}^k)]^2. \tag{50}$$

The PARTIAL DERIVATIVES (again dropping superscripts) are

$$\frac{\partial(R^2)}{\partial a_0} = -2\sum [y - (a_0 + a_1 x + \ldots + a_k x^k)] = 0 \tag{51}$$

$$\frac{\partial(R^2)}{\partial a_1} = -2\sum [y - (a_0 + a_1 x + \ldots + a_k x^k)]x = 0 \tag{52}$$

$$\frac{\partial(R^2)}{\partial a_k} = -2\sum [y - (a_0 + a_1 x + \ldots + a_k x^k)]x^k = 0. \tag{53}$$

These lead to the equations

$$a_0 n + a_1 \sum x + \ldots + a_k \sum x^k = \sum y \tag{54}$$

$$a_0 \sum x + a_1 \sum x^2 + \ldots + a_k \sum x^{k+1} = \sum xy \tag{55}$$

$$a_0 \sum x^k + a_1 \sum x^{k+1} + \ldots + a_k \sum x^{2k} = \sum x^k y \tag{56}$$

or, in MATRIX form

$$\begin{bmatrix} n & \sum x & \cdots & \sum x^k \\ \sum x & \sum x^2 & \cdots & \sum x^{k+1} \\ \vdots & \vdots & \ddots & \vdots \\ \sum x^k & \sum x^{k+1} & \cdots & \sum x^{2k} \end{bmatrix} \begin{bmatrix} a_0 \\ a_1 \\ \vdots \\ a_k \end{bmatrix} = \begin{bmatrix} \sum y \\ \sum xy \\ \vdots \\ \sum x^k y \end{bmatrix}. \tag{57}$$

This is a VANDERMONDE MATRIX. We can also obtain the MATRIX for a least squares fit by writing

$$\begin{bmatrix} 1 & x_1 & \cdots & {x_1}^k \\ 1 & x_2 & \cdots & {x_2}^k \\ \vdots & \vdots & \ddots & \vdots \\ 1 & x_n & \cdots & {x_n}^k \end{bmatrix} \begin{bmatrix} a_0 \\ a_1 \\ \vdots \\ a_k \end{bmatrix} = \begin{bmatrix} y_1 \\ y_2 \\ \vdots \\ y_n \end{bmatrix}. \tag{58}$$

Premultiplying both sides by the TRANSPOSE of the first MATRIX then gives

$$\begin{bmatrix} 1 & 1 & \cdots & 1 \\ x_1 & x_2 & \cdots & x_n \\ \vdots & \vdots & \ddots & \vdots \\ {x_1}^k & {x_2}^k & \cdots & {x_n}^k \end{bmatrix} \begin{bmatrix} 1 & x_1 & \cdots & {x_1}^k \\ 1 & x_2 & \cdots & {x_2}^k \\ \vdots & \vdots & \ddots & \vdots \\ 1 & x_n & \cdots & {x_n}^k \end{bmatrix} \begin{bmatrix} a_0 \\ a_1 \\ \vdots \\ a_k \end{bmatrix} = \begin{bmatrix} 1 & 1 & \cdots & 1 \\ x_1 & x_2 & \cdots & x_n \\ \vdots & \vdots & \ddots & \vdots \\ {x_1}^k & {x_2}^k & \cdots & {x_n}^k \end{bmatrix} \begin{bmatrix} y_1 \\ y_2 \\ \vdots \\ y_n \end{bmatrix}, \tag{59}$$

so

$$\begin{bmatrix} n & \sum x & \cdots & \sum x^n \\ \sum x & \sum x^2 & \cdots & \sum x^{n+1} \\ \vdots & \vdots & \ddots & \vdots \\ \sum x^n & \sum x^{n+1} & \cdots & \sum x^{2n} \end{bmatrix} \begin{bmatrix} a_0 \\ a_1 \\ \vdots \\ a_n \end{bmatrix} = \begin{bmatrix} \sum y \\ \sum xy \\ \vdots \\ \sum x^k y \end{bmatrix}. \tag{60}$$

As before, given m points (x_i, y_i) and fitting with POLYNOMIAL COEFFICIENTS $a_0, \ldots, a_n$ gives

$$\begin{bmatrix} y_1 \\ y_2 \\ \vdots \\ y_m \end{bmatrix} = \begin{bmatrix} 1 & x_1 & {x_1}^2 & \cdots & {x_1}^n \\ 1 & x_2 & {x_2}^2 & \cdots & {x_2}^n \\ \vdots & \vdots & \vdots & \ddots & \vdots \\ 1 & x_m & {x_m}^2 & \cdots & {x_m}^n \end{bmatrix} \begin{bmatrix} a_0 \\ a_0 \\ \vdots \\ a_n \end{bmatrix}, \tag{61}$$

In MATRIX notation, the equation for a polynomial fit is given by

$$\mathbf{y} = \mathsf{X}\mathbf{a}. \tag{62}$$

This can be solved by premultiplying by the MATRIX TRANSPOSE X^{T},

$$\mathsf{X}^{\mathrm{T}}\mathbf{y} = \mathsf{X}^{\mathrm{T}}\mathsf{X}\mathbf{a}. \tag{63}$$

This MATRIX EQUATION can be solved numerically, or can be inverted directly if it is well formed, to yield the solution vector

$$\mathbf{a} = (\mathsf{X}^{\mathrm{T}}\mathsf{X})^{-1}\mathsf{X}^{\mathrm{T}}\mathbf{y}. \tag{64}$$

Setting $m = 1$ in the above equations reproduces the linear solution.

see also CORRELATION COEFFICIENT, INTERPOLATION, LEAST SQUARES FITTING—EXPONENTIAL, LEAST SQUARES FITTING—LOGARITHMIC, LEAST SQUARES FITTING—POWER LAW, MOORE-PENROSE GENERALIZED MATRIX INVERSE, NONLINEAR LEAST SQUARES FITTING, REGRESSION COEFFICIENT, SPLINE

References

Acton, F. S. *Analysis of Straight-Line Data.* New York: Dover, 1966.

Bevington, P. R. *Data Reduction and Error Analysis for the Physical Sciences.* New York: McGraw-Hill, 1969.

Gonick, L. and Smith, W. *The Cartoon Guide to Statistics.* New York: Harper Perennial, 1993.

Kenney, J. F. and Keeping, E. S. "Linear Regression, Simple Correlation, and Contingency." Ch. 8 in *Mathematics of Statistics, Pt. 2, 2nd ed.* Princeton, NJ: Van Nostrand, pp. 199–237, 1951.

Kenney, J. F. and Keeping, E. S. "Linear Regression and Correlation." Ch. 15 in *Mathematics of Statistics, Pt. 1, 3rd ed.* Princeton, NJ: Van Nostrand, pp. 252–285, 1962.

Lancaster, P. and Šalkauskas, K. *Curve and Surface Fitting: An Introduction.* London: Academic Press, 1986.

Lawson, C. and Hanson, R. *Solving Least Squares Problems.* Englewood Cliffs, NJ: Prentice-Hall, 1974.

Nash, J. C. *Compact Numerical Methods for Computers: Linear Algebra and Function Minimisation, 2nd ed.* Bristol, England: Adam Hilger, pp. 21–24, 1990.

Press, W. H.; Flannery, B. P.; Teukolsky, S. A.; and Vetterling, W. T. "Fitting Data to a Straight Line" "Straight-Line Data with Errors in Both Coordinates," and "General Linear Least Squares." §15.2, 15.3, and 15.4 in *Numerical Recipes in FORTRAN: The Art of Scientific Computing, 2nd ed.* Cambridge, England: Cambridge University Press, pp. 655–675, 1992.

York, D. "Least-Square Fitting of a Straight Line." *Canad. J. Phys.* **44**, 1079–1086, 1966.

Least Squares Fitting—Exponential

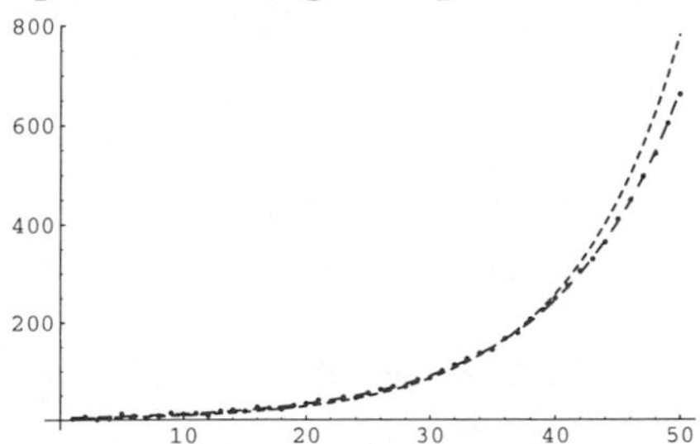

To fit a functional form

$$y = Ae^{Bx}, \tag{1}$$

take the LOGARITHM of both sides

$$\ln y = \ln A + B \ln x. \tag{2}$$

The best-fit values are then

$$a = \frac{\sum \ln y \sum x^2 - \sum x \sum x \ln y}{n \sum x^2 - \left(\sum x\right)^2} \tag{3}$$

$$b = \frac{n \sum x \ln y - \sum x \sum \ln y}{n \sum x^2 - \left(\sum x\right)^2}, \tag{4}$$

where $B \equiv b$ and $A \equiv \exp(a)$.

This fit gives greater weights to small y values so, in order to weight the points equally, it is often better to minimize the function

$$\sum y(\ln y - a - bx)^2. \tag{5}$$

Applying LEAST SQUARES FITTING gives

$$a \sum y + b \sum xy = \sum y \ln y \tag{6}$$

$$a \sum xy + b \sum x^2 y = \sum xy \ln y \tag{7}$$

$$\begin{bmatrix} \sum y & \sum xy \\ \sum xy & \sum x^2 y \end{bmatrix} \begin{bmatrix} a \\ b \end{bmatrix} = \begin{bmatrix} \sum y \ln y \\ \sum xy \ln y \end{bmatrix}. \tag{8}$$

Solving for a and b,

$$a = \frac{\sum(x^2 y) \sum(y \ln y) - \sum(xy) \sum(xy \ln y)}{\sum y \sum(x^2 y) - \left(\sum xy\right)^2} \tag{9}$$

$$b = \frac{\sum y \sum(xy \ln y) - \sum(xy) \sum(y \ln y)}{\sum y \sum(x^2 y) - \left(\sum xy\right)^2}. \tag{10}$$

In the plot above, the short-dashed curve is the fit computed from (3) and (4) and the long-dashed curve is the fit computed from (9) and (10).

see also LEAST SQUARES FITTING, LEAST SQUARES FITTING—LOGARITHMIC, LEAST SQUARES FITTING—POWER LAW

Least Squares Fitting—Logarithmic

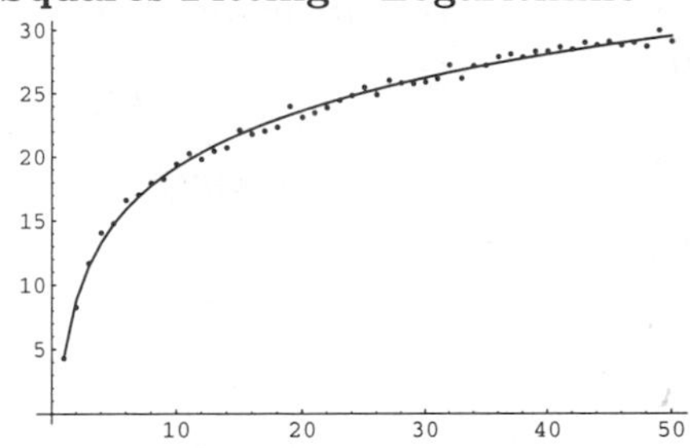

Given a function of the form

$$y = a + b \ln x, \tag{1}$$

the COEFFICIENTS can be found from LEAST SQUARES FITTING as

$$b = \frac{n \sum(y \ln x) - \sum y \sum(\ln x)}{n \sum[(\ln x)^2] - \left[\sum(\ln x)\right)^2} \tag{2}$$

$$a = \frac{\sum y - b \sum(\ln x)}{n}. \tag{3}$$

see also LEAST SQUARES FITTING, LEAST SQUARES FITTING—EXPONENTIAL, LEAST SQUARES FITTING—POWER LAW

Least Squares Fitting—Power Law

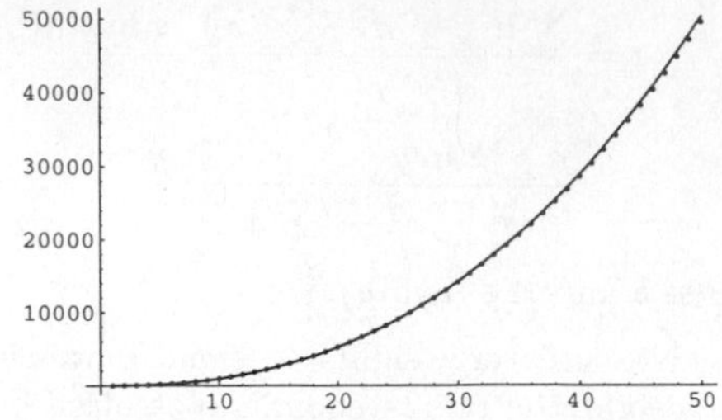

Given a function of the form

$$y = Ax^B, \tag{1}$$

LEAST SQUARES FITTING gives the COEFFICIENTS as

$$b = \frac{n\sum(\ln x \ln y) - \sum(\ln x)\sum(\ln y)}{n\sum[(\ln x)^2] - \left(\sum \ln x\right)^2} \tag{2}$$

$$a = \frac{\sum(\ln y) - b\sum(\ln x)}{n}, \tag{3}$$

where $B \equiv b$ and $A \equiv \exp(a)$.

see also LEAST SQUARES FITTING, LEAST SQUARES FITTING—EXPONENTIAL, LEAST SQUARES FITTING—LOGARITHMIC

Least Upper Bound

see SUPREMUM

Lebesgue Constants (Fourier Series)

N.B. A detailed on-line essay by S. Finch was the starting point for this entry.

Assume a function f is integrable over the interval $[-\pi, \pi]$ and $S_n(f, x)$ is the nth partial sum of the FOURIER SERIES of f, so that

$$a_k = \frac{1}{\pi}\int_{-\pi}^{\pi} f(t)\cos(kt)\,dt \tag{1}$$

$$b_k = \frac{1}{\pi}\int_{-\pi}^{\pi} f(t)\sin(kt)\,dt \tag{2}$$

and

$$S_n(f,x) = \tfrac{1}{2}a_0 + \left\{\sum_{k=1}^{n}[a_k\cos(kx) + b_k\sin(kx)]\right\}. \tag{3}$$

If

$$|f(x)| \leq 1 \tag{4}$$

for all x, then

$$S_n(f,x) \leq \frac{1}{\pi}\int_0^{\pi} \frac{|\sin[\frac{1}{2}(2n+1)\theta]|}{\sin(\frac{1}{2}\theta)}\,d\theta = L_n, \tag{5}$$

and L_n is the smallest possible constant for which this holds for all continuous f. The first few values of L_n are

$$L_0 = 1 \tag{6}$$

$$L_1 = \frac{1}{3} + \frac{2\sqrt{3}}{\pi} = 1.435991124\ldots \tag{7}$$

$$L_2 = 1.642188435\ldots \tag{8}$$

$$L_3 = 1.778322862. \tag{9}$$

Some FORMULAS for L_n include

$$L_n = \frac{1}{2n+1} + \frac{2}{\pi}\sum_{k=1}^{n}\frac{1}{k}\tan\left(\frac{\pi k}{2n+1}\right)$$

$$= \frac{16}{\pi^2}\sum_{k=1}^{\infty}\sum_{j=1}^{(2n+1)k}\frac{1}{4k^2-1}\frac{1}{2j-1} \tag{10}$$

(Zygmund 1959) and integral FORMULAS include

$$L_n = 4\int_0^{\infty}\frac{\tanh[(2n+1)x]}{\tanh x}\frac{dx}{\pi^2+4x^2}$$

$$= \frac{4}{\pi^2}\int_0^{\infty}\frac{\sinh[(2n+1)x]}{\sinh x}\ln\{\coth[\tfrac{1}{2}(2n+1)x]\}\,dx \tag{11}$$

(Hardy 1942). For large n,

$$\frac{4}{\pi^2}\ln n < L_n < 3 + \frac{4}{\pi^2}\ln n. \tag{12}$$

This result can be generalized for an r-differentiable function satisfying

$$\left|\frac{d^r f}{dx^r}\right| \leq 1 \tag{13}$$

for all x. In this case,

$$|f(x) - S_n(f,x)| \leq L_{n,r} = \frac{4}{\pi^2}\frac{\ln n}{n^r} + \mathcal{O}\left(\frac{1}{n^r}\right), \tag{14}$$

where

$$L_{n,r} = \begin{cases} \frac{1}{\pi}\int_{-\pi}^{\pi}\left|\sum_{k=n+1}^{\infty}\frac{\sin(kx)}{k^r}\right|dx & \text{for } r \geq 1 \text{ odd} \\ \frac{1}{\pi}\int_{-\pi}^{\pi}\left|\sum_{k=n+1}^{\infty}\frac{\cos(kx)}{k^r}\right|dx & \text{for } r \geq 1 \text{ even} \end{cases} \tag{15}$$

(Kolmogorov 1935, Zygmund 1959).

Watson (1930) showed that

$$\lim_{n\to\infty}\left[L_n - \frac{4}{\pi^2}\ln(2n+1)\right] = c, \tag{16}$$

where

$$c = \frac{8}{\pi^2}\left(\sum_{k=1}^{\infty}\frac{\ln k}{4k^2-1}\right) - \frac{4}{\pi^2}\frac{\Gamma'(\frac{1}{2})}{\Gamma(\frac{1}{2})} \quad (17)$$

$$= \frac{8}{\pi^2}\left[\sum_{j=0}^{\infty}\frac{\lambda(2j+2)-1}{2j+1}\right] + \frac{4}{\pi^2}(2\ln 2+\gamma) \quad (18)$$

$$= 0.9894312738\ldots, \quad (19)$$

where $\Gamma(z)$ is the GAMMA FUNCTION, $\lambda(z)$ is the DIRICHLET LAMBDA FUNCTION, and γ is the EULER-MASCHERONI CONSTANT.

References
Finch, S. "Favorite Mathematical Constants." http://www.mathsoft.com/asolve/constant/lbsg/lbsg.html.
Hardy, G. H. "Note on Lebesgue's Constants in the Theory of Fourier Series." *J. London Math. Soc.* **17**, 4–13, 1942.
Kolmogorov, A. N. "Zur Grössenordnung des Restgliedes Fourierscher reihen differenzierbarer Funktionen." *Ann. Math.* **36**, 521–526, 1935.
Watson, G. N. "The Constants of Landau and Lebesgue." *Quart. J. Math. Oxford* **1**, 310–318, 1930.
Zygmund, A. G. *Trigonometric Series, 2nd ed., Vols. 1–2.* Cambridge, England: Cambridge University Press, 1959.

Lebesgue Constants (Lagrange Interpolation)

N.B. A detailed on-line essay by S. Finch was the starting point for this entry.

Define the nth Lebesgue constant for the LAGRANGE INTERPOLATING POLYNOMIAL by

$$\Lambda_n(X) \equiv \max_{-1\le x\le 1}\sum_{k=1}^{n}\left|\prod_{j\ne k}\frac{x-x_j}{x_k-x_j}\right|. \quad (1)$$

It is true that

$$\Lambda_n > \frac{4}{\pi^2}\ln n - 1. \quad (2)$$

The efficiency of a Lagrange interpolation is related to the rate at which Λ_n increases. Erdős (1961) proved that there exists a POSITIVE constant such that

$$\Lambda_n > \frac{2}{\pi}\ln n - C \quad (3)$$

for all n. Erdős (1961) further showed that

$$\Lambda_n < \frac{2}{\pi}\ln n + 4, \quad (4)$$

so (3) cannot be improved upon.

References
Erdős, P. "Problems and Results on the Theory of Interpolation, II." *Acta Math. Acad. Sci. Hungary* **12**, 235–244, 1961.
Finch, S. "Favorite Mathematical Constants." http://www.mathsoft.com/asolve/constant/lbsg/lbsg.html.

Lebesgue Covering Dimension

An important DIMENSION and one of the first dimensions investigated. It is defined in terms of covering sets, and is therefore also called the COVERING DIMENSION. Another name for the Lebesgue covering dimension is the TOPOLOGICAL DIMENSION.

A SPACE has Lebesgue covering dimension m if for every open COVER of that space, there is an open COVER that refines it such that the refinement has order at most $m+1$. Consider how many elements of the cover contain a given point in a base space. If this has a maximum over all the points in the base space, then this maximum is called the order of the cover. If a SPACE does not have Lebesgue covering dimension m for any m, it is said to be infinite dimensional.

Results of this definition are:

1. Two homeomorphic spaces have the same dimension,
2. $\mathbb{R}^n$ has dimension n,
3. A TOPOLOGICAL SPACE can be embedded as a closed subspace of a EUCLIDEAN SPACE IFF it is locally compact, Hausdorff, second countable, and is finite dimensional (in the sense of the LEBESGUE DIMENSION), and
4. Every compact metrizable m-dimensional TOPOLOGICAL SPACE can be embedded in $\mathbb{R}^{2m+1}$.

see also LEBESGUE MINIMAL PROBLEM

References
Dieudonne, J. A. *A History of Algebraic and Differential Topology.* Boston, MA: Birkhäuser, 1994.
Iyanaga, S. and Kawada, Y. (Eds.). *Encyclopedic Dictionary of Mathematics.* Cambridge, MA: MIT Press, p. 414, 1980.
Munkres, J. R. *Topology: A First Course.* Englewood Cliffs, NJ: Prentice-Hall, 1975.

Lebesgue Dimension

see LEBESGUE COVERING DIMENSION

Lebesgue Integrable

A real-valued function f defined on the reals $\mathbb{R}$ is called Lebesgue integrable if there exists a SEQUENCE of STEP FUNCTIONS $\{f_n\}$ such that the following two conditions are satisfied:

1. $\sum_{n=1}^{\infty}\int|f_n| < \infty$,
2. $f(x) = \sum_{n=1}^{\infty}$ for every $x \in \mathbb{R}$ such that $\sum_{n=1}^{\infty}\int|f_n| < \infty$.

Here, the above integral denotes the ordinary RIEMANN INTEGRAL. Note that this definition avoids explicit use of the LEBESGUE MEASURE.

see also INTEGRAL, LEBESGUE INTEGRAL, RIEMANN INTEGRAL, STEP FUNCTION

Lebesgue Integral

The LEBESGUE INTEGRAL is defined in terms of upper and lower bounds using the LEBESGUE MEASURE of a SET. It uses a LEBESGUE SUM $S_n = \eta_i \mu(E_i)$ where η_i is the value of the function in subinterval i, and $\mu(E_i)$ is the LEBESGUE MEASURE of the SET E_i of points for which values are approximately η_i. This type of integral covers a wider class of functions than does the RIEMANN INTEGRAL.

see also A-INTEGRABLE, COMPLETE FUNCTIONS, INTEGRAL

References

Kestelman, H. "Lebesgue Integral of a Non-Negative Function" and "Lebesgue Integrals of Functions Which Are Sometimes Negative." Chs. 5–6 in *Modern Theories of Integration, 2nd rev. ed.* New York: Dover, pp. 113–160, 1960.

Lebesgue Measurability Problem

A problem related to the CONTINUUM HYPOTHESIS which was solved by Solovay (1970) using the INACCESSIBLE CARDINALS AXIOM. It has been proven by Shelah and Woodin (1990) that use of this AXIOM is essential to the proof.

see also CONTINUUM HYPOTHESIS, INACCESSIBLE CARDINALS AXIOM, LEBESGUE MEASURE

References

Shelah, S. and Woodin, H. "Large Cardinals Imply that Every Reasonable Definable Set of Reals is Lebesgue Measurable." *Israel J. Math.* **70**, 381–394, 1990.

Solovay, R. M. "A Model of Set-Theory in which Every Set of Reals is Lebesgue Measurable." *Ann. Math.* **92**, 1–56, 1970.

Lebesgue Measure

An extension of the classical notions of length and AREA to more complicated sets. Given an open set $S \equiv \sum_k (a_k, b_k)$ containing DISJOINT intervals,

$$\mu_L(S) \equiv \sum_k (b_k - a_k).$$

Given a CLOSED SET $S' \equiv [a, b] - \sum_k (a_k, b_k)$,

$$\mu_L(S') \equiv (b - a) - \sum_k (b_k - a_k).$$

A LINE SEGMENT has Lebesgue measure 1; the CANTOR SET has Lebesgue measure 0. The MINKOWSKI MEASURE of a bounded, CLOSED SET is the same as its Lebesgue measure (Ko 1995).

see also CANTOR SET, MEASURE, RIESZ-FISCHER THEOREM

References

Kestelman, H. "Lebesgue Measure." Ch. 3 in *Modern Theories of Integration, 2nd rev. ed.* New York: Dover, pp. 67–91, 1960.

Ko, K.-I. "A Polynomial-Time Computable Curve whose Interior has a Nonrecursive Measure." *Theoret. Comput. Sci.* **145**, 241–270, 1995.

Lebesgue Minimal Problem

Find the plane LAMINA of least AREA A which is capable of covering any plane figure of unit GENERAL DIAMETER. A UNIT CIRCLE is too small, but a HEXAGON circumscribed on the UNIT CIRCLE is too large. More specifically, the AREA is bounded by

$$0.8257\ldots = \tfrac{1}{8}\pi + \tfrac{1}{4}\sqrt{3} < A < \tfrac{2}{3}(3 - \sqrt{3}) = 0.8454\ldots$$

(Pal 1920).

see also AREA, BORSUK'S CONJECTURE, DIAMETER (GENERAL), KAKEYA NEEDLE PROBLEM

References

Ball, W. W. R. and Coxeter, H. S. M. *Mathematical Recreations and Essays, 13th ed.* New York: Dover, p. 99, 1987.

Coxeter, H. S. M. "Lebesgue's Minimal Problem." *Eureka* **21**, 13, 1958.

Grünbaum, B. "Borsuk's Problem and Related Questions." *Proc. Sympos. Pure Math, Vol. 7.* Providence, RI: Amer. Math. Soc., pp. 271–284, 1963.

Kakeya, S. "Some Problems on Maxima and Minima Regarding Ovals." *Sci. Reports Tôhoku Imperial Univ., Ser. 1 (Math., Phys., Chem.)* **6**, 71–88, 1917.

Ogilvy, C. S. *Excursions in Geometry.* New York: Dover, pp. 142–144, 1990.

Pál, J. *Danske videnkabernes selskab, Copenhagen Math.-fys. maddelelser* **3**, 1–35, 1920.

Yaglom, I. M. and Boltyanskii, V. G. *Convex Figures.* New York: Holt, Rinehart, & Winston, pp. 18 and 100, 1961.

Lebesgue-Radon Integral

see LEBESGUE-STIELTJES INTEGRAL

Lebesgue Singular Integrals

$$\mathcal{U}_n(f) = \int_a^b f(x) K_n(x)\,dx,$$

where $\{K_n(x)\}$ is a SEQUENCE of CONTINUOUS FUNCTIONS.

Lebesgue-Stieltjes Integral

Let $\alpha(x)$ be a monotone increasing function and define an INTERVAL $I = (x_1, x_2)$. Then define the NONNEGATIVE function

$$U(I) = \alpha(x_2 + 0) - \alpha(x_1 + 0).$$

The LEBESGUE INTEGRAL with respect to a MEASURE constructed using $U(I)$ is called the Lebesgue-Stieltjes integral, or sometimes the LEBESGUE-RADON INTEGRAL.

References

Iyanaga, S. and Kawada, Y. (Eds.). *Encyclopedic Dictionary of Mathematics.* Cambridge, MA: MIT Press, p. 326, 1980.

Lebesgue Sum

$$S_n = \eta_i \mu(E_i),$$

where $\mu(E_i)$ is the MEASURE of the SET E_i of points on the x-axis for which $f(x) \approx \eta_i$.

Leech Lattice

A 24-D Euclidean lattice. An AUTOMORPHISM of the Leech lattice modulo a center of two leads to the CONWAY GROUP Co_1. Stabilization of the 1- and 2-D sublattices leads to the CONWAY GROUPS Co_2 and Co_3, the HIGMAN-SIMS GROUP *HS* and the MCLAUGHLIN GROUP *McL*.

The Leech lattice appears to be the densest HYPERSPHERE PACKING in 24-D, and results in each HYPERSPHERE touching 195,560 others.

see also BARNES-WALL LATTICE, CONWAY GROUPS, COXETER-TODD LATTICE, HIGMAN-SIMS GROUP, HYPERSPHERE, HYPERSPHERE PACKING, KISSING NUMBER, MCLAUGHLIN GROUP

References

Conway, J. H. and Sloane, N. J. A. "The 24-Dimensional Leech Lattice Λ_{24}," "A Characterization of the Leech Lattice," "The Covering Radius of the Leech Lattice," "Twenty-Three Constructions for the Leech Lattice," "The Cellular of the Leech Lattice," "Lorentzian Forms for the Leech Lattice." §4.11, Ch. 12, and Chs. 23–26 in *Sphere Packings, Lattices, and Groups, 2nd ed.* New York: Springer-Verlag, pp. 131–135, 331–336, and 478–526, 1993.

Leech, J. "Notes on Sphere Packings." *Canad. J. Math.* **19**, 251–267, 1967.

Wilson, R. A. "Vector Stabilizers and Subgroups of Leech Lattice Groups." *J. Algebra* **127**, 387–408, 1989.

Lefshetz Fixed Point Formula

see LEFSHETZ TRACE FORMULA

Lefshetz's Theorem

Each DOUBLE POINT assigned to an irreducible curve whose GENUS is NONNEGATIVE imposes exactly one condition.

References

Coolidge, J. L. *A Treatise on Algebraic Plane Curves.* New York: Dover, p. 104, 1959.

Lefshetz Trace Formula

A formula which counts the number of FIXED POINTS for a topological transformation.

Leg

The leg of a TRIANGLE is one of its sides.

see also HYPOTENUSE, TRIANGLE

Legendre Addition Theorem

see SPHERICAL HARMONIC ADDITION THEOREM

Legendre's Chi-Function

The function defined by

$$\chi_\nu(z)=\sum_{k=0}^\infty \frac{z^{2k+1}}{(2k+1)^\nu}$$

for integral $\nu = 2, 3, \ldots$. It is related to the POLYLOGARITHM by

$$\begin{aligned}\chi_\nu(z) &= \tfrac{1}{2}[\mathrm{Li}_\nu(z)-\mathrm{Li}_\nu(-z)]\\ &= \mathrm{Li}_\nu(z)-2^{-\nu}\,\mathrm{Li}_\nu(z^2).\end{aligned}$$

see also POLYLOGARITHM

References

Cvijović, D. and Klinowski, J. "Closed-Form Summation of Some Trigonometric Series." *Math. Comput.* **64**, 205–210, 1995.

Lewin, L. *Polylogarithms and Associated Functions.* Amsterdam, Netherlands: North-Holland, pp. 282–283, 1981.

Legendre's Constant

The number 1.08366 in Legendre's guess at the PRIME NUMBER THEOREM

$$\pi(n)\sim\frac{n}{\ln n-1.08366}.$$

This expression is correct to leading term only.

References

Le Lionnais, F. *Les nombres remarquables.* Paris: Hermann, p. 147, 1983.

Wagon, S. *Mathematica in Action.* New York: W. H. Freeman, pp. 28–29, 1991.

Legendre Differential Equation

The second-order ORDINARY DIFFERENTIAL EQUATION

$$(1-x^2)\frac{d^2y}{dx^2}-2x\frac{dy}{dx}+l(l+1)y=0, \tag{1}$$

which can be rewritten

$$\frac{d}{dx}\left[(1-x^2)\frac{dy}{dx}\right]+l(l+1)y=0. \tag{2}$$

The above form is a special case of the associated Legendre differential equation with $m = 0$. The Legendre differential equation has REGULAR SINGULAR POINTS at -1, 1, and ∞. It can be solved using a series expansion,

$$y=\sum_{n=0}^\infty a_nx^n \tag{3}$$

$$y'=\sum_{n=0}^\infty na_nx^{n-1} \tag{4}$$

$$y''=\sum_{n=0}^\infty n(n-1)a_nx^{n-2}. \tag{5}$$

Plugging in,

$$(1-x^2)\sum_{n=0}^\infty n(n-1)a_nx^{n-2}-2x\sum_{n=0}^\infty na_nx^{n-1} +l(l+1)\sum_{n=0}^\infty a_nx^n=0 \tag{6}$$

$$\sum_{n=0}^{\infty} n(n-1)a_n x^{n-2} - \sum_{n=0}^{\infty} n(n-1)a_n x^n - 2x\sum_{n=0}^{\infty} na_n x^{n-1} + l(l+1)\sum_{n=0}^{\infty} a_n x^n = 0 \quad (7)$$

$$\sum_{n=2}^{\infty} n(n-1)a_n x^{n-2} - \sum_{n=0}^{\infty} n(n-1)a_n x^n - 2\sum_{n=0}^{\infty} na_n x^n + l(l+1)\sum_{n=0}^{\infty} a_n x^n = 0 \quad (8)$$

$$\sum_{n=0}^{\infty} (n+2)(n+1)a_{n+2} x^n - \sum_{n=0}^{\infty} n(n-1)a_n x^n - 2\sum_{n=0}^{\infty} na_n x^n + l(l+1)\sum_{n=0}^{\infty} a_n x^n = 0 \quad (9)$$

$$\sum_{n=0}^{\infty} \{(n+1)(n+2)a_{n+2} + [-n(n-1) - 2n + l(l+1)]a_n\} = 0, \quad (10)$$

so each term must vanish and

$$(n+1)(n+2)a_{n+2} - n(n+1) + l(l+1)]a_n = 0 \quad (11)$$

$$a_{n+2} = \frac{n(n+1) - l(l+1)}{(n+1)(n+2)} a_n = -\frac{[l+(n+1)](l-n)}{(n+1)(n+2)} a_n. \quad (12)$$

Therefore,

$$a_2 = -\frac{l(l+1)}{1 \cdot 2} a_0 \quad (13)$$

$$a_4 = -\frac{(l-2)(l+3)}{3 \cdot 4} a_2 = (-1)^2 \frac{[(l-2)l][(l+1)(l+3)]}{1 \cdot 2 \cdot 3 \cdot 4} a_0 \quad (14)$$

$$a_6 = -\frac{(l-4)(l+5)}{5 \cdot 6} a_4 = (-1)^3 \frac{[(l-4)(l-2)l][(l+1)(l+3)(l+5)]}{1 \cdot 2 \cdot 3 \cdot 4 \cdot 5 \cdot 6} a_0, \quad (15)$$

so the EVEN solution is

$$y_1(x) = 1 + \sum_{n=1}^{\infty} (-1)^n \frac{[(l-2n+2)\cdots(l-2)l][(l+1)(l+3)\cdots(l+2n-1)]}{(2n)!} x^{2n}. \quad (16)$$

Similarly, the ODD solution is

$$y_2(x) = x + \sum_{n=1}^{\infty} (-1)^n \times \frac{[(l-2n+1)\cdots(l-3)(l-1)][(l+2)(l+4)\cdots(l+2n)]}{(2n+1)!} x^{2m+1}. \quad (17)$$

If l is an EVEN INTEGER, the series y_1 reduces to a POLYNOMIAL of degree l with only EVEN POWERS of x and the series y_2 diverges. If l is an ODD INTEGER, the series y_2 reduces to a POLYNOMIAL of degree l with only ODD POWERS of x and the series y_1 diverges. The general solution for an INTEGER l is given by the LEGENDRE POLYNOMIALS

$$P_n(x) = c_n \begin{cases} y_1(x) & \text{for } l \text{ even} \\ y_2(x) & \text{for } l \text{ odd}, \end{cases} \quad (18)$$

where c_n is chosen so that $P_n(1) = 1$. If the variable x is replaced by $\cos\theta$, then the Legendre differential equation becomes

$$\frac{d^2y}{d\theta^2} + \frac{\cos\theta}{\sin\theta}\frac{dy}{dx} + l(l+1)y = 0, \quad (19)$$

as is derived for the associated Legendre differential equation with $m = 0$.

The *associated* Legendre differential equation is

$$\frac{d}{dx}\left[(1-x^2)\frac{dy}{dx}\right] + \left[l(l+1) - \frac{m^2}{1-x^2}\right]y = 0 \quad (20)$$

$$(1-x^2)\frac{d^2y}{dx^2} - 2x\frac{dy}{dx} + \left[l(l+1) - \frac{m^2}{1-x^2}\right]y = 0. \quad (21)$$

The solutions to this equation are called the associated Legendre polynomials. Writing $x \equiv \cos\theta$, first establish the identities

$$\frac{dy}{dx} = \frac{dy}{d(\cos\theta)} = -\frac{1}{\sin\theta}\frac{dy}{d\theta} \quad (22)$$

$$x\frac{dy}{dx} = -\frac{\cos\theta}{\sin\theta}\frac{dy}{d\theta}, \quad (23)$$

$$\frac{d^2y}{dx^2} = \frac{1}{\sin\theta}\frac{d}{d\theta}\left(\frac{1}{\sin\theta}\frac{dy}{d\theta}\right) = \frac{1}{\sin\theta}\left(\frac{-\cos\theta}{\sin^2\theta}\right)\frac{dy}{d\theta} + \frac{1}{\sin^2\theta}\frac{d^2y}{d\theta^2}, \quad (24)$$

and

$$1 - x^2 = 1 - \cos^2\theta = \sin^2\theta. \quad (25)$$

Therefore,

$$(1-x^2)\frac{d^2y}{dx^2} = \sin^2\theta\frac{1}{\sin\theta}\left(\frac{-\cos\theta}{\sin^2\theta}\right)\frac{dy}{d\theta} + \frac{1}{\sin^2\theta}\frac{d^2y}{d\theta^2}$$
$$= \frac{d^2y}{d\theta^2} - \frac{\cos\theta}{\sin\theta}\frac{dy}{d\theta}. \tag{26}$$

Plugging (22) into (26) and the result back into (21) gives

$$\left(\frac{d^2y}{d\theta^2} - \frac{\cos\theta}{\sin\theta}\frac{dy}{d\theta}\right)$$
$$+2\frac{\cos\theta}{\sin\theta}\frac{dy}{d\theta} + \left[l(l+1) - \frac{m^2}{\sin^2\theta}\right]y = 0 \tag{27}$$

$$\frac{d^2y}{d\theta^2} + \frac{\cos\theta}{\sin\theta}\frac{dy}{dx} + \left[l(l+1) - \frac{m^2}{\sin^2\theta}\right]y = 0. \tag{28}$$

References

Abramowitz, M. and Stegun, C. A. (Eds.). *Handbook of Mathematical Functions with Formulas, Graphs, and Mathematical Tables, 9th printing.* New York: Dover, p. 332, 1972.

Legendre Duplication Formula

GAMMA FUNCTIONS of argument $2z$ can be expressed in terms of GAMMA FUNCTIONS of smaller arguments. From the definition of the BETA FUNCTION,

$$B(m,n) = \frac{\Gamma(m)\Gamma(n)}{\Gamma(m+n)} = \int_0^1 u^{m-1}(1-u)^{n-1}\,du. \tag{1}$$

Now, let $m = n \equiv z$, then

$$\frac{\Gamma(z)\Gamma(z)}{\Gamma(2z)} = \int_0^1 u^{z-1}(1-u)^{z-1}\,du \tag{2}$$

and $u \equiv (1+x)/2$, so $du = dx/2$ and

$$\frac{\Gamma(z)\Gamma(z)}{\Gamma(2z)} = \int_0^1 \left(\frac{1+x}{2}\right)^{z-1}\left(1-\frac{1+x}{2}\right)^{z-1}(\tfrac{1}{2}\,dx)$$
$$= \frac{1}{2}\int_0^1 \left(\frac{1+x}{2}\right)^{z-1}\left(\frac{1-x}{2}\right)^{z-1}dx$$
$$= \frac{1}{2^{1+2(z-1)}}\int_0^1 (1-x^2)^{z-1}\,dx$$
$$= 2^{1-2z}\int_0^1 (1-x^2)^{z-1}\,dx. \tag{3}$$

Now, use the BETA FUNCTION identity

$$B(m,n) = 2\int_0^1 x^{2z-1}(1-x^2)^{z-1}\,dx \tag{4}$$

to write the above as

$$\frac{\Gamma(z)\Gamma(z)}{\Gamma(2z)} = 2^{1-2z}B(\tfrac{1}{2},z) = 2^{1-2z}\frac{\Gamma(\frac{1}{2})\Gamma(z)}{\Gamma(z+\frac{1}{2})}. \tag{5}$$

Solving for $\Gamma(2z)$,

$$\Gamma(2z) = \frac{\Gamma(z)\Gamma(z+\frac{1}{2})2^{2z-1}}{\Gamma(\frac{1}{2})} = \frac{\Gamma(z)\Gamma(z+\frac{1}{2})2^{2z-1}}{\sqrt{\pi}}$$
$$= (2\pi)^{-1/2}2^{2z-1/2}\Gamma(z)\Gamma(z+\tfrac{1}{2}), \tag{6}$$

since $\Gamma(\frac{1}{2}) = \sqrt{\pi}$.

see also GAMMA FUNCTION, GAUSS MULTIPLICATION FORMULA

References

Abramowitz, M. and Stegun, C. A. (Eds.). *Handbook of Mathematical Functions with Formulas, Graphs, and Mathematical Tables, 9th printing.* New York: Dover, p. 256, 1972.

Arfken, G. *Mathematical Methods for Physicists, 3rd ed.* Orlando, FL: Academic Press, pp. 561–562, 1985.

Morse, P. M. and Feshbach, H. *Methods of Theoretical Physics, Part I.* New York: McGraw-Hill, pp. 424–425, 1953.

Legendre's Factorization Method

A PRIME FACTORIZATION ALGORITHM in which a sequence of TRIAL DIVISORS is chosen using a QUADRATIC SIEVE. By using QUADRATIC RESIDUES of N, the QUADRATIC RESIDUES of the factors can also be found.

see also PRIME FACTORIZATION ALGORITHMS, QUADRATIC RESIDUE, QUADRATIC SIEVE FACTORIZATION METHOD, TRIAL DIVISOR

Legendre's Formula

Counts the number of POSITIVE INTEGERS less than or equal to a number x which are not divisible by any of the first a PRIMES,

$$\phi(x,a) = \lfloor x\rfloor - \sum\left\lfloor\frac{x}{p_i}\right\rfloor + \sum\left\lfloor\frac{x}{p_ip_j}\right\rfloor - \sum\left\lfloor\frac{x}{p_ip_jp_k}\right\rfloor + \ldots, \tag{1}$$

where $\lfloor x\rfloor$ is the FLOOR FUNCTION. Taking $a = x$ gives

$$\phi(x,x) = \pi(x) - \pi(\sqrt{x}) + 1 = \lfloor x\rfloor - \sum_{p_i\leq\sqrt{x}}\left\lfloor\frac{x}{p_i}\right\rfloor + \sum_{p_i<p_j\leq\sqrt{x}}\left\lfloor\frac{x}{p_ip_j}\right\rfloor - \sum_{p_i<p_j<p_k\leq\sqrt{x}}\left\lfloor\frac{x}{p_ip_jp_k}\right\rfloor + \ldots, \tag{2}$$

where $\pi(n)$ is the PRIME COUNTING FUNCTION. Legendre's formula holds since one more than the number of PRIMES in a range equals the number of INTEGERS minus the number of composites in the interval.

Legendre's formula satisfies the RECURRENCE RELATION

$$\phi(x,a) = \phi(x,a-1) - \phi\left(\frac{x}{p_a}, a-1\right). \tag{3}$$

Let $m_k = p_1 p_2 \cdots p_k$, then

$$\begin{aligned}\phi(m_k,k) &= \lfloor m_k \rfloor - \sum \left\lfloor \frac{m_k}{p_i} \right\rfloor + \sum \left\lfloor \frac{m_k}{p_i p_j} \right\rfloor - \ldots \\ &= m_k - \sum \frac{m_k}{p_i} + \sum \frac{m_k}{p_i p_j} - \ldots \\ &= m_k \left(1 - \frac{1}{p-1}\right)\left(1 - \frac{1}{p_2}\right)\cdots\left(1 - \frac{1}{p_k}\right) \\ &= \prod_{i=1}^{k}(p_i - 1) = \phi(m_k), \end{aligned} \tag{4}$$

where $\phi(n)$ is the TOTIENT FUNCTION, and

$$\phi(sm_k + t, k) = s\phi(m_k) + \phi(t,k), \tag{5}$$

where $0 \leq t \leq m_k$. If $t > m_k/2$, then

$$\phi(t,k) = \phi(m_k) - \phi(m_k - t - 1, k). \tag{6}$$

Note that $\phi(n,n)$ is not practical for computing $\pi(n)$ for large arguments. A more efficient modification is MEISSEL'S FORMULA.

see also LEHMER'S FORMULA, MAPES' METHOD, MEISSEL'S FORMULA, PRIME COUNTING FUNCTION

Legendre Function of the First Kind

see LEGENDRE POLYNOMIAL

Legendre Function of the Second Kind

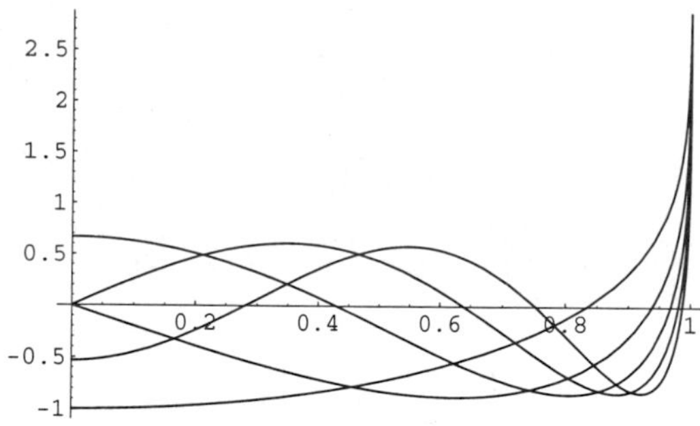

A solution to the LEGENDRE DIFFERENTIAL EQUATION which is singular at the origin. The Legendre functions of the second kind satisfy the same RECURRENCE RELATION as the LEGENDRE FUNCTIONS OF THE FIRST KIND. The first few are

$$\begin{aligned} Q_0 &= \tfrac{1}{2}\ln\left(\frac{1+x}{1-x}\right) \\ Q_1 &= \frac{x}{2}\ln\left(\frac{1+x}{1-x}\right) - 1 \\ Q_2 &= \frac{3x^2-1}{4}\ln\left(\frac{1+x}{1-x}\right) - \frac{3x}{2} \\ Q_3 &= \frac{5x^3-3x}{4}\ln\left(\frac{1+x}{1-x}\right) - \frac{5x^2}{2} + \frac{2}{3}. \end{aligned}$$

The associated Legendre functions of the second kind have DERIVATIVE about 0 of

$$\left[\frac{dQ_\nu^\mu(x)}{dx}\right]_{x=0} = \frac{2^\mu\sqrt{\pi}\,\cos[\frac{1}{2}\pi(\nu+\mu)]\Gamma(\frac{1}{2}\nu+\frac{1}{2}\mu+1)}{\Gamma(\frac{1}{2}\nu-\frac{1}{2}\mu+\frac{1}{2})}$$

(Abramowitz and Stegun 1972, p. 334). The logarithmic derivative is

$$\left[\frac{d\ln Q_\lambda^\mu(z)}{dz}\right]_{z=0} = 2\exp\{\tfrac{1}{2}\pi i\,\mathrm{sgn}(\Im[z])\}\frac{[\frac{1}{2}(\lambda+\mu)]![\frac{1}{2}(\lambda-\mu)]!}{[\frac{1}{2}(\lambda+\mu-1)]![\frac{1}{2}(\lambda-\mu-1)]!}$$

(Binney and Tremaine 1987, p. 654).

References

Abramowitz, M. and Stegun, C. A. (Eds.). "Legendre Functions." Ch. 8 in *Handbook of Mathematical Functions with Formulas, Graphs, and Mathematical Tables, 9th printing.* New York: Dover, pp. 331–339, 1972.

Arfken, G. "Legendre Functions of the Second Kind, $Q_n(x)$." *Mathematical Methods for Physicists, 3rd ed.* Orlando, FL: Academic Press, pp. 701–707, 1985.

Binney, J. and Tremaine, S. "Associated Legendre Functions." Appendix 5 in *Galactic Dynamics.* Princeton, NJ: Princeton University Press, pp. 654–655, 1987.

Morse, P. M. and Feshbach, H. *Methods of Theoretical Physics, Part I.* New York: McGraw-Hill, pp. 597–600, 1953.

Snow, C. *Hypergeometric and Legendre Functions with Applications to Integral Equations of Potential Theory.* Washington, DC: U. S. Government Printing Office, 1952.

Spanier, J. and Oldham, K. B. "The Legendre Functions $P_\nu(x)$ and $Q_\nu(x)$." Ch. 59 in *An Atlas of Functions.* Washington, DC: Hemisphere, pp. 581–597, 1987.

Legendre-Gauss Quadrature

Also called "the" GAUSSIAN QUADRATURE or LEGENDRE QUADRATURE. A GAUSSIAN QUADRATURE over the interval $[-1,1]$ with WEIGHTING FUNCTION $W(x) = 1$. The ABSCISSAS for quadrature order n are given by the roots of the LEGENDRE POLYNOMIALS $P_n(x)$, which occur symmetrically about 0. The weights are

$$w_i = -\frac{A_{n+1}\gamma_n}{A_n P_n'(x_i)P_{n+1}(x_i)} = \frac{A_n}{A_{n-1}}\frac{\gamma_{n-1}}{P_{n-1}(x_i)P_n'(x_i)}, \tag{1}$$

where A_n is the COEFFICIENT of x^n in $P_n(x)$. For LEGENDRE POLYNOMIALS,

$$A_n = \frac{(2n)!}{2^n(n!)^2}, \tag{2}$$

so

$$\begin{aligned}\frac{A_{n+1}}{A_n} &= \frac{[2(n+1)]!}{2^{n+1}[(n+1)!]^2}\frac{2^n(n!)^2}{(2n)!} \\ &= \frac{(2n+1)(2n+2)}{2(n+1)^2} = \frac{2n+1}{n+1}. \end{aligned} \tag{3}$$

Additionally,

$$\gamma_n = \frac{2}{2n+1}, \tag{4}$$

so

$$w_i = -\frac{2}{(n+1)P_{n+1}(x_i)P_n'(x_i)} = \frac{2}{nP_{n-1}(x_i)P_n'(x_i)}. \quad (5)$$

Using the RECURRENCE RELATION

$$(1-x^2)P_n'(x) = nxP_n(x) + nP_{n-1}(x) \quad (6)$$
$$= (n+1)xP_n(x) - (n+1)P_{n+1}(x) \quad (7)$$

gives

$$w_i = \frac{2}{(1-x_i{}^2)[P_n'(x_i)]^2} = \frac{2(1-x_i{}^2)}{(n+1)^2[P_{n+1}(x_i)]^2}. \quad (8)$$

The error term is

$$E = \frac{2^{2n+1}(n!)^4}{(2n+1)[(2n)!]^3} f^{(2n)}(\xi). \quad (9)$$

Beyer (1987) gives a table of ABSCISSAS and weights up to $n = 16$, and Chandrasekhar (1960) up to $n = 8$ for n EVEN.

n	x_i	w_i
2	± 0.57735	1.000000
3	0	0.888889
	± 0.774597	0.555556
4	± 0.339981	0.652145
	± 0.861136	0.347855
5	0	0.568889
	± 0.538469	0.478629
	± 0.90618	0.236927

The ABSCISSAS and weights can be computed analytically for small n.

n	x_i	w_i
2	$\pm\frac{1}{3}\sqrt{3}$	1
3	0	$\frac{8}{9}$
	$\pm\frac{1}{5}\sqrt{15}$	$\frac{5}{9}$
4	$\pm\sqrt{\frac{3-2\sqrt{\frac{6}{5}}}{7}}$	
	$\pm\sqrt{\frac{3+2\sqrt{\frac{6}{5}}}{7}}$	

References

Beyer, W. H. *CRC Standard Mathematical Tables, 28th ed.* Boca Raton, FL: CRC Press, pp. 462–463, 1987.

Chandrasekhar, S. *Radiative Transfer.* New York: Dover, pp. 56–62, 1960.

Hildebrand, F. B. *Introduction to Numerical Analysis.* New York: McGraw-Hill, pp. 323–325, 1956.

Legendre-Jacobi Elliptic Integral

Any of the three standard forms in which an ELLIPTIC INTEGRAL can be expressed.

see also ELLIPTIC INTEGRAL OF THE FIRST KIND, ELLIPTIC INTEGRAL OF THE SECOND KIND, ELLIPTIC INTEGRAL OF THE THIRD KIND

Legendre Polynomial

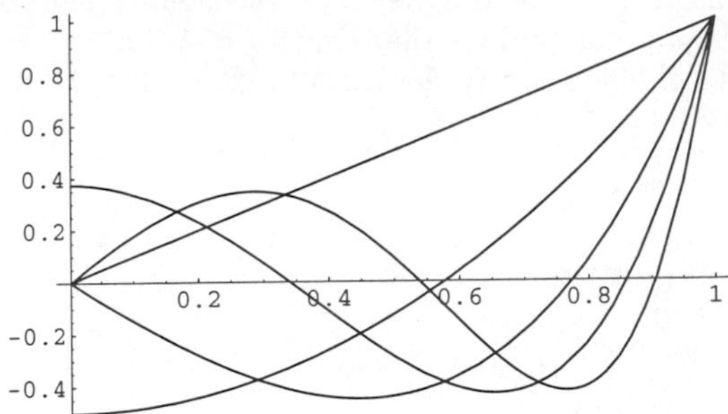

The LEGENDRE FUNCTIONS OF THE FIRST KIND are solutions to the LEGENDRE DIFFERENTIAL EQUATION. If l is an INTEGER, they are POLYNOMIALS. They are a special case of the ULTRASPHERICAL FUNCTIONS with $\alpha = 1/2$. The Legendre polynomials $P_n(x)$ are illustrated above for $x \in [0,1]$ and $n = 1, 2, \ldots, 5$.

The Rodrigues FORMULA provides the GENERATING FUNCTION

$$P_l(x) = \frac{1}{2^l l!}\frac{dl}{dx^l}(x^2-1)^l, \quad (1)$$

which yields upon expansion

$$P_l(x) = \frac{1}{2^l}\sum_{k=0}^{\lfloor n/2 \rfloor} \frac{(-1)^k(2l-2k)!}{k!(l-k)!(l-2k)!} x^{l-2k}, \quad (2)$$

where $\lfloor r \rfloor$ is the FLOOR FUNCTION. The GENERATING FUNCTION is

$$g(t,x) = (1-2xt+t^2)^{-1/2} = \sum_{n=0}^{\infty} P_n(x)t^n. \quad (3)$$

Take $\partial g/\partial t$,

$$-\tfrac{1}{2}(1-2xt+t^2)^{-3/2}(-2x+2t) = \sum_{n=0}^{\infty} nP_n(x)t^{n-1}. \quad (4)$$

Multiply (4) by $2t$,

$$-t(1-2xt+t^2)^{-3/2}(-2x+2t) = \sum_{n=0}^{\infty} 2nP_n(x)t^n \quad (5)$$

and add (3) and (5),

$$(1-2xt+t^2)^{-3/2}[(2xt-2t^2)+(1-2xt+t^2)] = \sum_{n=0}^{\infty}(2n+1)P_n(x)t^n \quad (6)$$

$$(1-2xt+t^2)^{-3/2}(1-t^2) = \sum_{n=0}^{\infty}(2n+1)P_n(x)t^n. \quad (7)$$

This expansion is useful in some physical problems, including expanding the Heyney-Greenstein phase function and computing the charge distribution on a SPHERE. They satisfy the RECURRENCE RELATION

$$(l+1)P_{l+1}(x) - (2l+1)xP_l(x) + lP_{l-1}(x) = 0. \quad (8)$$

The Legendre polynomials are orthogonal over $(-1,1)$ with WEIGHTING FUNCTION 1 and satisfy

$$\int_{-1}^{1} P_n(x)P_m(x)\,dx = \frac{2}{2n+1}\delta_{mn}, \quad (9)$$

where δ_{mn} is the KRONECKER DELTA.

A COMPLEX GENERATING FUNCTION is

$$P_l(x) = \frac{1}{2\pi i}\int (1-2zx+z^2)^{-1/2}z^{-l-1}\,dz, \quad (10)$$

and the Schläfli integral is

$$P_l(x) = \frac{(-1)^l}{2^l}\frac{1}{2\pi i}\int \frac{(1-z^2)^l}{(z-x)^{l+1}}\,dz. \quad (11)$$

Additional integrals (Byerly 1959, p. 172) include

$$\int_0^1 P_m(x)\,dx = \begin{cases} 0 & m \text{ even} \neq 0 \\ (-1)^{(m-1)/2}\frac{m!!}{m(m+1)(m-1)!!} & m \text{ odd} \end{cases} \quad (12)$$

$$\int_0^1 P_m(x)P_n(x)\,dx = \begin{cases} 0 \\ \quad m,n \text{ both even or odd } m \neq n \\ (-1)^{(m+n+1)/2}\frac{m!n!}{2^{m+n+1}(m-n)(m+n+1)(\frac{1}{2}m)!\{[\frac{1}{2}(n-1)]!\}^2} \\ \quad m \text{ even, } n \text{ odd} \\ \frac{1}{2n+1}, \\ \quad m = n. \end{cases} \quad (13)$$

An additional identity is

$$1 - [P_n(x)]^2 = \sum_{\nu=1}^{n} \frac{1-x^2}{1-x_\nu{}^2}\left[\frac{P_n(x)}{P_n'(x_\nu)(x-x_\nu)}\right]^2 \quad (14)$$

(Szegő 1975, p. 348).

The first few Legendre polynomials are

$$\begin{aligned} P_0(x) &= 1 \\ P_1(x) &= x \\ P_2(x) &= \tfrac{1}{2}(3x^2-1) \\ P_3(x) &= \tfrac{1}{2}(5x^3-3x) \\ P_4(x) &= \tfrac{1}{8}(35x^4-30x^2+3) \\ P_5(x) &= \tfrac{1}{8}(63x^5-70x^3+15x) \\ P_6(x) &= \tfrac{1}{16}(231x^6-315x^4+105x^2-5). \end{aligned}$$

The first few POWERS in terms of Legendre polynomials are

$$\begin{aligned} x &= P_1 \\ x^2 &= \tfrac{1}{3}(P_0+2P_2) \\ x^3 &= \tfrac{1}{5}(3P_1+2P_3) \\ x^4 &= \tfrac{1}{35}(7P_0+20P_2+8P_4) \\ x^5 &= \tfrac{1}{63}(27P_1+28P_3+8P_5) \\ x^6 &= \tfrac{1}{231}(33P_0+110P_2+72P_4+16P_5). \end{aligned}$$

For Legendre polynomials and POWERS up to exponent 12, see Abramowitz and Stegun (1972, p. 798).

The Legendre POLYNOMIALS can also be generated using GRAM-SCHMIDT ORTHONORMALIZATION in the OPEN INTERVAL $(-1,1)$ with the WEIGHTING FUNCTION 1.

$$P_0(x) = 1 \quad (15)$$

$$\begin{aligned} P_1(x) &= \left[x - \frac{\int_{-1}^{1} x\,dx}{\int_{-1}^{1} dx}\right]\cdot 1 \\ &= x - \frac{\frac{1}{2}[x^2]_{-1}^{1}}{[x]_{-1}^{1}} = x - \frac{\frac{1}{2}(1-1)}{1-(-1)} = x \end{aligned} \quad (16)$$

$$\begin{aligned} P_2(x) &= \left[x - \frac{\int_{-1}^{1} x^3\,dx}{\int_{-1}^{1} x^2\,dx}\right] - \left[\frac{\int_{-1}^{1} x^2\,dx}{\int_{-1}^{1} dx}\right]\cdot 1 \\ &= \left[x - \frac{\frac{1}{4}[x^4]_{-1}^{1}}{\frac{1}{3}[x^3]_{-1}^{1}}\right] x - \frac{\frac{1}{3}[x^3]_{-1}^{1}}{[x]_{-1}^{1}} = x^2 - \tfrac{1}{3} \end{aligned} \quad (17)$$

$$\begin{aligned} P_3(x) &= \left[x - \frac{\int_{-1}^{1} x(x^2-\frac{1}{3})^2\,dx}{\int_{-1}^{1}(x^2-\frac{1}{3})^2\,dx}\right](x^2-\tfrac{1}{3}) \\ &\quad - \left[\frac{\int_{-1}^{1}(x^2-\frac{1}{3})^2\,dx}{\int_{-1}^{1} x^2\,dx}\right] x \\ &= x\left[x^2 - \tfrac{1}{3} - \frac{(\frac{1}{5}-\frac{2}{9}+\frac{1}{9})x}{\frac{1}{3}}\right] \\ &= x^3 - \tfrac{1}{3}x - 3(\tfrac{1}{5}-\tfrac{1}{9}) \\ &= x^3 - x\left(\tfrac{1}{3}+\tfrac{3}{5}-\tfrac{1}{3}\right) = x^3 - \tfrac{3}{5}x. \end{aligned} \quad (18)$$

Normalizing so that $P_n(1) = 1$ gives the expected Legendre polynomials.

The "shifted" Legendre polynomials are a set of functions analogous to the Legendre polynomials, but defined on the interval (0, 1). They obey the ORTHOGONALITY relationship

$$\int_0^1 \bar{P}_m(x)\bar{P}_n(x)\,dx = \frac{1}{2n+1}\delta_{mn}. \quad (19)$$

The first few are

$$\begin{aligned} \bar{P}_0(x) &= 1 \\ \bar{P}_1(x) &= 2x-1 \\ \bar{P}_2(x) &= 6x^2-6x+1 \\ \bar{P}_3(x) &= 20x^3-30x^2+12x-1. \end{aligned}$$

The associated Legendre polynomials $P_l^m(x)$ are solutions to the associated LEGENDRE DIFFERENTIAL EQUATION, where l is a POSITIVE INTEGER and $m = 0, \ldots, l$. They can be given in terms of the unassociated polynomials by

$$P_l^m(x) = (-1)^m (1-x^2)^{m/2} \frac{d^m}{dx^m} P_l(x) = \frac{(-1)^m}{2^l l!}(1-x^2)^{m/2} \frac{d^{l+m}}{dx^{l+m}}(x^2-1)^l, \quad (20)$$

where $P_l(x)$ are the unassociated LEGENDRE POLYNOMIALS. Note that some authors (e.g., Arfken 1985, p. 668) omit the CONDON-SHORTLEY PHASE $(-1)^m$, while others include it (e.g., Abramowitz and Stegun 1972, Press *et al.* 1992, and the `LegendreP[l,m,z]` command of *Mathematica*®). Abramowitz and Stegun (1972, p. 332) use the notation

$$P_{lm}(x) \equiv (-1)^m P_m^l(x) \quad (21)$$

to distinguish these two cases.

Associated polynomials are sometimes called FERRERS' FUNCTIONS (Sansone 1991, p. 246). If $m = 0$, they reduce to the unassociated POLYNOMIALS. The associated Legendre functions are part of the SPHERICAL HARMONICS, which are the solution of LAPLACE'S EQUATION in SPHERICAL COORDINATES. They are ORTHOGONAL over $[-1,1]$ with the WEIGHTING FUNCTION 1

$$\int_{-1}^{1} P_l^m(x) P_{l'}^m(x)\,dx = \frac{2}{2l+1}\frac{(l+m)!}{(l-m)!}\delta_{ll'}, \quad (22)$$

ORTHOGONAL over $[-1,1]$ with respect to m with the WEIGHTING FUNCTION $(1-x^2)^{-2}$

$$\int_{-1}^{1} P_l^m(x) P_l^{m'}(x)\frac{dx}{1-x^2} = \frac{(l+m)!}{m(l-m)!}\delta_{mm'}. \quad (23)$$

They obey the RECURRENCE RELATIONS

$$(l-m)P_l^m(x) = x(2l-1)P_{l-1}^m(x) - (l+m-1)P_{l-2}^m(x) \quad (24)$$

$$\frac{dP_l^m}{d\theta} = -\sqrt{1-\mu^2}\,\frac{dP_l^m}{d\mu} = \tfrac{1}{2}(l-m+1)(l+m+P_l^{m-1} - P_l^{m+1}) \quad (25)$$

$$(2l+1)\mu P_l^m = (l+m)P_{l-1}^m + (l-m+1)P_{l+1}^m \quad (26)$$

$$(2l+1)\sqrt{1-\mu^2}\,P_l^m = P_{l+1}^{m+1} - P_{l-1}^{m+1}. \quad (27)$$

An identity relating associated POLYNOMIALS with NEGATIVE m to the corresponding functions with POSITIVE m is

$$P_l^{-m} = (-1)^m \frac{(l-m)!}{(l+m)!} P_l^m. \quad (28)$$

Additional identities are

$$P_l^l(x) = (-1)^l (2l-1)!!(1-x^2)^{l/2} \quad (29)$$

$$P_{l+1}^l(x) = x(2l+1)P_l^l(x). \quad (30)$$

Written in terms of x and using the convention without a leading factor of $(-1)^m$ (Arfken 1985, p. 669), the first few associated Legendre polynomials are

$$\begin{aligned}
P_0^0(x) &= 1\\
P_1^0(x) &= x\\
P_1^1(x) &= -(1-x^2)^{1/2}\\
P_2^0(x) &= \tfrac{1}{2}(3x^2-1)\\
P_2^1(x) &= -3x(1-x^2)^{1/2}\\
P_2^2(x) &= 3(1-x^2)\\
P_3^0(x) &= \tfrac{1}{2}x(5x^2-3)\\
P_3^1(x) &= \tfrac{3}{2}(1-5x^2)(1-x^2)^{1/2}\\
P_3^2(x) &= 15x(1-x^2)\\
P_3^3(x) &= -15(1-x^2)^{3/2}\\
P_4^0(x) &= \tfrac{1}{8}(35x^4-30x^2+3)\\
P_4^1(x) &= \tfrac{5}{2}x(3-7x^2)(1-x^2)^{1/2}\\
P_4^2(x) &= \tfrac{15}{2}(7x^2-1)(1-x^2)\\
P_4^3(x) &= -105x(1-x^2)^{3/2}\\
P_4^4(x) &= 105(1-x^2)^2\\
P_5^0(x) &= \tfrac{1}{8}x(63x^4-70x^2+15).
\end{aligned}$$

Written in terms $x \equiv \cos\theta$, the first few become

$$\begin{aligned}
P_0^0(\cos\theta) &= 1\\
P_1^{-1}(\cos\theta) &= \tfrac{1}{2}\sin\theta\\
P_1^0(\cos\theta) &= \cos\theta = \mu\\
P_1^1(\cos\theta) &= \sin\theta\\
P_2^{-2}(\cos\theta) &= \tfrac{1}{8}\sin^2\theta\\
P_2^{-1}(\cos\theta) &= \tfrac{1}{2}\sin\theta\cos\theta\\
P_2^0(\cos\theta) &= \tfrac{1}{2}(3\cos^2\theta-1)\\
P_2^1(\cos\theta) &= 3\sin\theta\cos\theta\\
&= \tfrac{3}{2}\sin^2\theta\\
P_2^2(\cos\theta) &= 3\sin^2\theta\\
&= \tfrac{3}{2}(1-\cos^2\theta)\\
P_3^0(\cos\theta) &= \tfrac{1}{2}\cos\theta(5\cos^2\theta-3)\\
&= \tfrac{1}{2}\cos\theta(2-5\sin^2\theta)\\
P_3^1(\cos\theta) &= \tfrac{3}{2}(5\cos^2\theta-1)\sin\theta\\
&= \tfrac{3}{8}(\sin\theta+5\sin^3\theta).
\end{aligned}$$

The derivative about the origin is

$$\left[\frac{dP_\nu^\mu(x)}{dx}\right]_{x=0} = \frac{2^{\mu+1}\sin[\frac{1}{2}\pi(\nu+\mu)]\Gamma(\frac{1}{2}\nu+\frac{1}{2}\mu+1)}{\pi^{1/2}\Gamma(\frac{1}{2}\nu-\frac{1}{2}\mu+\frac{1}{2})} \quad (31)$$

(Abramowitz and Stegun 1972, p. 334), and the logarithmic derivative is

$$\left[\frac{d\ln P_\lambda^\mu(z)}{dz}\right]_{z=0} = 2\tan[\tfrac{1}{2}\pi(\lambda+\mu)]\frac{[\frac{1}{2}(\lambda+\mu)]![\frac{1}{2}(\lambda-\mu)]!}{[\frac{1}{2}(\lambda+\mu-1)]![\frac{1}{2}(\lambda-\mu-1)]!}. \tag{32}$$

(Binney and Tremaine 1987, p. 654).

see also CONDON-SHORTLEY PHASE, CONICAL FUNCTION, GEGENBAUER POLYNOMIAL, KINGS PROBLEM, LAPLACE'S INTEGRAL, LAPLACE-MEHLER INTEGRAL, SUPER CATALAN NUMBER, TOROIDAL FUNCTION, TURÁN'S INEQUALITIES

References

Abramowitz, M. and Stegun, C. A. (Eds.). "Legendre Functions" and "Orthogonal Polynomials." Ch. 22 in Chs. 8 and 22 in *Handbook of Mathematical Functions with Formulas, Graphs, and Mathematical Tables, 9th printing.* New York: Dover, pp. 331–339 and 771–802, 1972.

Arfken, G. "Legendre Functions." Ch. 12 in *Mathematical Methods for Physicists, 3rd ed.* Orlando, FL: Academic Press, pp. 637–711, 1985.

Binney, J. and Tremaine, S. "Associated Legendre Functions." Appendix 5 in *Galactic Dynamics.* Princeton, NJ: Princeton University Press, pp. 654–655, 1987.

Byerly, W. E. *An Elementary Treatise on Fourier's Series, and Spherical, Cylindrical, and Ellipsoidal Harmonics, with Applications to Problems in Mathematical Physics.* New York: Dover, 1959.

Iyanaga, S. and Kawada, Y. (Eds.). "Legendre Function" and "Associated Legendre Function." Appendix A, Tables 18.II and 18.III in *Encyclopedic Dictionary of Mathematics.* Cambridge, MA: MIT Press, pp. 1462–1468, 1980.

Legendre, A. M. "Sur l'attraction des Sphéroides." *Mém. Math. et Phys. présentés à l'Ac. r. des. sc. par divers savants* **10**, 1785.

Morse, P. M. and Feshbach, H. *Methods of Theoretical Physics, Part I.* New York: McGraw-Hill, pp. 593–597, 1953.

Press, W. H.; Flannery, B. P.; Teukolsky, S. A.; and Vetterling, W. T. *Numerical Recipes in FORTRAN: The Art of Scientific Computing, 2nd ed.* Cambridge, England: Cambridge University Press, p. 252, 1992.

Sansone, G. "Expansions in Series of Legendre Polynomials and Spherical Harmonics." Ch. 3 in *Orthogonal Functions, rev. English ed.* New York: Dover, pp. 169–294, 1991.

Snow, C. *Hypergeometric and Legendre Functions with Applications to Integral Equations of Potential Theory.* Washington, DC: U. S. Government Printing Office, 1952.

Spanier, J. and Oldham, K. B. "The Legendre Polynomials $P_n(x)$" and "The Legendre Functions $P_\nu(x)$ and $Q_\nu(x)$." Chs. 21 and 59 in *An Atlas of Functions.* Washington, DC: Hemisphere, pp. 183–192 and 581–597, 1987.

Szegő, G. *Orthogonal Polynomials, 4th ed.* Providence, RI: Amer. Math. Soc., 1975.

Legendre Polynomial of the Second Kind

see LEGENDRE FUNCTION OF THE SECOND KIND

Legendre's Quadratic Reciprocity Law

see QUADRATIC RECIPROCITY LAW

Legendre Quadrature

see LEGENDRE-GAUSS QUADRATURE

Legendre Relation

Let $E(k)$ and $K(k)$ be complete ELLIPTIC INTEGRALS OF THE FIRST and SECOND KINDS, with $E'(k)$ and $K'(k)$ the complementary integrals. Then

$$E(k)K'(k) + E'(k)K(k) - K(k)K'(k) = \tfrac{1}{2}\pi.$$

References

Abramowitz, M. and Stegun, C. A. (Eds.). *Handbook of Mathematical Functions with Formulas, Graphs, and Mathematical Tables, 9th printing.* New York: Dover, p. 591, 1972.

Legendre Series

Because the LEGENDRE FUNCTIONS OF THE FIRST KIND form a COMPLETE ORTHOGONAL BASIS, any FUNCTION may be expanded in terms of them

$$f(x) = \sum_{n=0}^\infty a_n P_n(x). \tag{1}$$

Now, multiply both sides by $P_m(x)$ and integrate

$$\int_{-1}^1 P_m(x)f(x)\,dx = \sum_{n=0}^\infty a_n \int_{-1}^1 P_n(x)P_m(x)\,dx. \tag{2}$$

But

$$\int_{-1}^1 P_n(x)P_m(x)\,dx = \frac{2}{2m+1}\delta_{mn}, \tag{3}$$

where δ_{mn} is the KRONECKER DELTA, so

$$\int_{-1}^1 P_m(x)f(x)\,dx = \sum_{n=0}^\infty a_n \frac{2}{2m+1}\delta_{mn} = \frac{2}{2m+1}a_m \tag{4}$$

and

$$a_m = \frac{2m+1}{2}\int_{-1}^1 P_m(x)f(x)\,dx. \tag{5}$$

see also FOURIER SERIES, JACKSON'S THEOREM, LEGENDRE POLYNOMIAL, MACLAURIN SERIES, PICONE'S THEOREM, TAYLOR SERIES

Legendre Sum

see LEGENDRE'S FORMULA

Legendre Symbol

$$\left(\frac{m}{n}\right) = (m|n)$$
$$\equiv \begin{cases} 0 & \text{if } m|n \\ 1 & \text{if } n \text{ is a quadratic residue modulo } m \\ -1 & \text{if } n \text{ is a quadratic nonresidue modulo } m. \end{cases}$$

If m is an ODD PRIME, then the JACOBI SYMBOL reduces to the Legendre symbol. The Legendre symbol obeys $(ab|p) = (a|p)(b|p)$.

$$\left(\frac{3}{p}\right) = \begin{cases} 1 & \text{if } p \equiv \pm 1 \pmod{12} \\ -1 & \text{if } p \equiv \pm 5 \pmod{12}. \end{cases}$$

see also JACOBI SYMBOL, KRONECKER SYMBOL, QUADRATIC RECIPROCITY THEOREM

References

Guy, R. K. "Quadratic Residues. Schur's Conjecture." §F5 in *Unsolved Problems in Number Theory, 2nd ed.* New York: Springer-Verlag, pp. 244–245, 1994.

Shanks, D. *Solved and Unsolved Problems in Number Theory, 4th ed.* New York: Chelsea, pp. 33–34 and 40–42, 1993.

Legendre Transformation

Given a function of two variables

$$df = \frac{\partial f}{\partial x}\,dx + \frac{\partial f}{\partial y}\,dy \equiv u\,dx + v\,dy, \tag{1}$$

change the differentials from dx and dy to du and dy with the transformation

$$g \equiv f - ux \tag{2}$$

$$\begin{aligned} dg &= df - u\,dx - x\,du = u\,dx + v\,dy - u\,dx - x\,du \\ &= v\,dy - x\,du. \end{aligned} \tag{3}$$

Then

$$x \equiv -\frac{\partial g}{\partial u} \tag{4}$$

$$v \equiv \frac{\partial g}{\partial y}. \tag{5}$$

Lehmer's Constant

N.B. A detailed on-line essay by S. Finch was the starting point for this entry.

Lehmer (1938) showed that every POSITIVE IRRATIONAL NUMBER x has a unique infinite continued cotangent representation of the form

$$x = \cot\left[\sum_{k=0}^{\infty}(-1)^k \cot^{-1} b_k\right],$$

where the b_ks are NONNEGATIVE and

$$b_k \geq (b_{k-1})^2 + b_{k-1} + 1.$$

The case for which the convergence is slowest occurs when the inequality is replaced by equality, giving $c_0 = 0$ and

$$c_k = (c_{k-1})^2 + c_{k-1} + 1$$

for $k \geq 1$. The first few values are c_k are 0, 1, 3, 13, 183, 33673, ... (Sloane's A024556), resulting in the constant

$$\begin{aligned} \xi &= \cot(\cot^{-1} 0 - \cot^{-1} 1 + \cot^{-1} 3 - \cot^{-1} 13 \\ &\quad + \cot^{-1} 183 - \cot^{-1} 33673 + \cot^{-1} 1133904603 \\ &\quad - \cot^{-1} 1285739649838492213 + \ldots + (-1)^k c_k + \ldots) \\ &= \cot(\tfrac{1}{4}\pi + \cot^{-1} 3 - \cot^{-1} 13 \\ &\quad + \cot^{-1} 183 - \cot^{-1} 33673 + \cot^{-1} 1133904603 \\ &\quad - \cot^{-1} 1285739649838492213 + \ldots + (-1)^k c_k + \ldots) \\ &= 0.59263271\ldots \end{aligned}$$

(Sloane's A030125). ξ is not an ALGEBRAIC NUMBER of degree less than 4, but Lehmer's approach cannot show whether or not ξ is TRANSCENDENTAL.

see also ALGEBRAIC NUMBER, TRANSCENDENTAL NUMBER

References

Finch, S. "Favorite Mathematical Constants." http://www.mathsoft.com/asolve/constant/lehmer/lehmer.html.

Le Lionnais, F. *Les nombres remarquables.* Paris: Hermann, p. 29, 1983.

Lehmer, D. H. "A Cotangent Analogue of Continued Fractions." *Duke Math. J.* **4**, 323–340, 1938.

Plouffe, S. "The Lehmer Constant." http://lacim.uqam.ca/piDATA/lehmer.txt.

Sloane, N. J. A. Sequences A024556 and A030125 in "An On-Line Version of the Encyclopedia of Integer Sequences."

Lehmer's Formula

A FORMULA related to MEISSEL'S FORMULA.

$$\begin{aligned} \pi(x) &= \lfloor x \rfloor - \sum_{i=1}^{a} \left\lfloor \frac{x}{p_i} \right\rfloor + \sum_{1 \leq i \leq j \leq a} \left\lfloor \frac{x}{p_i p_j} \right\rfloor - \ldots \\ &\quad + \tfrac{1}{2}(b + a - 2)(b - a + 1) - \sum_{a < i \leq b} \pi\left(\frac{x}{p_i}\right) \\ &\quad - \sum_{i=a+1}^{c} \sum_{j=i}^{b_i} \left[\pi\left(\frac{x}{p_i p_j}\right) - (j-1)\right], \end{aligned}$$

where

$$\begin{aligned} a &\equiv \pi(x^{1/4}) \\ b &\equiv \pi(x^{1/2}) \\ b_i &\equiv \pi(\sqrt{x/p_i}) \\ c &\equiv \pi(x^{1/3}), \end{aligned}$$

and $\pi(n)$ is the PRIME COUNTING FUNCTION.

References

Riesel, H. "Lehmer's Formula." *Prime Numbers and Computer Methods for Factorization, 2nd ed.* Boston, MA: Birkhäuser, pp. 13–14, 1994.

Lehmer Method

see LEHMER-SCHUR METHOD

Lehmer Number

A number generated by a generalization of a LUCAS SEQUENCE. Let α and β be COMPLEX NUMBERS with

$$\alpha + \beta = \sqrt{R} \tag{1}$$

$$\alpha\beta = Q, \tag{2}$$

where Q and R are RELATIVELY PRIME NONZERO INTEGERS and α/β is a ROOT OF UNITY. Then the Lehmer numbers are

$$U_n(\sqrt{R}, Q) = \frac{\alpha^n - \beta^n}{\alpha - \beta}, \tag{3}$$

and the companion numbers

$$V_n(\sqrt{R}, Q) = \begin{cases} \frac{\alpha^n+\beta^n}{\alpha+\beta} & \text{for } n \text{ odd} \\ \alpha^n + \beta^n & \text{for } n \text{ even.} \end{cases} \tag{4}$$

References

Lehmer, D. H. "An Extended Theory of Lucas' Functions." *Ann. Math.* **31**, 419–448, 1930.

Ribenboim, P. *The Book of Prime Number Records, 2nd ed.* New York: Springer-Verlag, pp. 61 and 70, 1989.

Williams, H. C. "The Primality of $N = 2A3^n - 1$." *Canad. Math. Bull.* **15**, 585–589, 1972.

Lehmer's Phenomenon

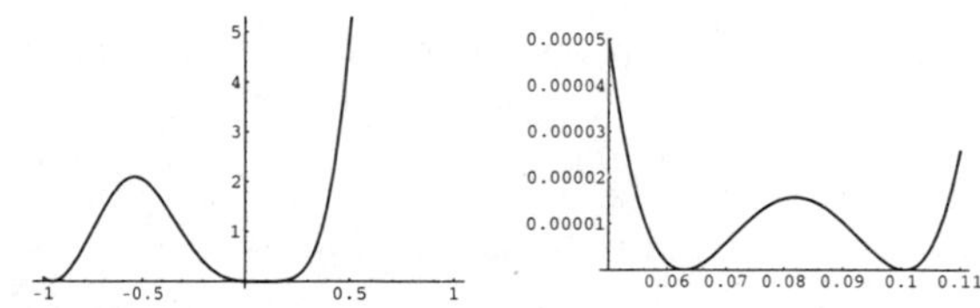

The appearance of nontrivial zeros (i.e., those along the CRITICAL STRIP with $\Re[z] = 1/2$) of the RIEMANN ZETA FUNCTION $\zeta(z)$ very close together. An example is the pair of zeros $\zeta(\frac{1}{2} + (7005 + t)i)$ given by $t_1 \approx 0.0606918$ and $t_2 \approx 0.100055$, illustrated above in the plot of $|\zeta(\frac{1}{2} + (7005 + t)i)|^2$.

see also CRITICAL STRIP, RIEMANN ZETA FUNCTION

References

Csordas, G.; Odlyzko, A. M.; Smith, W.; and Varga, R. S. "A New Lehmer Pair of Zeros and a New Lower Bound for the de Bruijn-Newman Constant." *Elec. Trans. Numer. Analysis* **1**, 104–111, 1993.

Csordas, G.; Smith, W.; and Varga, R. S. "Lehmer Pairs of Zeros, the de Bruijn-Newman Constant and the Riemann Hypothesis." *Constr. Approx.* **10**, 107–129, 1994.

Csordas, G.; Smith, W.; and Varga, R. S. "Lehmer Pairs of Zeros and the Riemann ξ-Function." In *Mathematics of Computation 1943–1993: A Half-Century of Computational Mathematics* (Vancouver, BC, 1993). *Proc. Sympos. Appl. Math.* **48**, 553–556, 1994.

Wagon, S. *Mathematica in Action.* New York: W. H. Freeman, pp. 357–358, 1991.

Lehmer's Problem

Do there exist any COMPOSITE NUMBERS n such that $\phi(n)|(n-1)$? No such numbers are known. In 1932, Lehmer showed that such an n must be ODD and SQUAREFREE, and that the number of distinct PRIME factors $d(7) \geq 7$. This was subsequently extended to $d(n) \geq 11$. The best current results are $n > 10^{20}$ and $d(n) \geq 14$ (Cohen and Hagis 1980), if $30\nmid n$, then $d(n) \geq 26$ (Wall 1980), and if $3|n$ then $d(n) \geq 213$ and 5.5×10^{570} (Lieuwens 1970).

References

Cohen, G. L. and Hagis, P. Jr. "On the Number of Prime Factors of n is $\phi(n)|(n-1)$." *Nieuw Arch. Wisk.* **28**, 177–185, 1980.

Lieuwens, E. "Do There Exist Composite Numbers for which $k\phi(M) = M - 1$ Holds?" *Nieuw. Arch. Wisk.* **18**, 165–169, 1970.

Ribenboim, P. *The Book of Prime Number Records, 2nd ed.* New York: Springer-Verlag, pp. 27–28, 1989.

Wall, D. W. "Conditions for $\phi(N)$ to Properly Divide $N-1$." In *A Collection of of Manuscripts Related to the Fibonacci Sequence* (Ed. V. E. Hoggatt and M. V. E. Bicknell-Johnson). San Jose, CA: Fibonacci Assoc., pp. 205–208, 1980.

Lehmer-Schur Method

An ALGORITHM which isolates ROOTS in the COMPLEX PLANE by generalizing 1-D bracketing.

References

Acton, F. S. *Numerical Methods That Work, 2nd printing.* Washington, DC: Math. Assoc. Amer., pp. 196–198, 1990.

Lehmer's Theorem

see FERMAT'S LITTLE THEOREM CONVERSE

Lehmus' Theorem

see STEINER-LEHMUS THEOREM

Leibniz Criterion

Also known as the ALTERNATING SERIES TEST. Given a SERIES

$$\sum_{n=1}^{\infty} (-1)^{n+1} a_n$$

with $a_n > 0$, if a_n is monotonic decreasing as $n \to \infty$ and

$$\lim_{n\to\infty} a_n = 0,$$

then the series CONVERGES.

Leibniz Harmonic Triangle

$$\begin{array}{ccccccccc} &&&& \frac{1}{1} &&&& \\ &&& \frac{1}{2} && \frac{1}{2} &&& \\ && \frac{1}{3} && \frac{1}{6} && \frac{1}{3} && \\ & \frac{1}{4} && \frac{1}{12} && \frac{1}{12} && \frac{1}{4} & \\ \frac{1}{5} && \frac{1}{20} && \frac{1}{30} && \frac{1}{20} && \frac{1}{5} \end{array}$$

In the Leibniz harmonic triangle, each FRACTION is the sum of numbers below it, with the initial and final entry on each row one over the corresponding entry in

PASCAL'S TRIANGLE. The DENOMINATORS in the second diagonals are 6, 12, 20, 30, 42, 56, ... (Sloane's A007622).

see also CATALAN'S TRIANGLE, CLARK'S TRIANGLE, EULER'S TRIANGLE, NUMBER TRIANGLE, PASCAL'S TRIANGLE, SEIDEL-ENTRINGER-ARNOLD TRIANGLE

References

Sloane, N. J. A. Sequence A007622/M4096 in "An On-Line Version of the Encyclopedia of Integer Sequences."

Leibniz Identity

$$\frac{d^n}{dx^n}(uv) = \frac{d^n u}{dx^n}v + \binom{n}{1}\frac{d^{n-1}u}{dx^{n-1}}\frac{dv}{dx} + \ldots + \binom{n}{r}\frac{d^{n-r}u}{dx^{n-r}}\frac{d^r v}{dx^r} + ud^n v + dx^n. \quad (1)$$

Therefore,

$$\frac{dx}{dy} = \frac{1}{\frac{dy}{dx}} \quad (2)$$

$$\frac{d^2x}{dy^2} = -\frac{d^2y}{dx^2}\left(\frac{dy}{dx}\right)^{-3} \quad (3)$$

$$\frac{d^3x}{dy^3} = \left[3\left(\frac{d^2y}{dx^2}\right)^2 - \frac{d^3y}{dx^3}\frac{dy}{dx}\right]\left(\frac{dy}{dx}\right)^{-5}. \quad (4)$$

References

Abramowitz, M. and Stegun, C. A. (Eds.). *Handbook of Mathematical Functions with Formulas, Graphs, and Mathematical Tables, 9th printing.* New York: Dover, p. 12, 1972.

Leibniz Integral Rule

$$\frac{\partial}{\partial z}\int_{a(z)}^{b(z)} f(x,z)\,dx = \int_{a(z)}^{b(z)} \frac{\partial f}{\partial z}\,dx + f(b(z),z)\frac{\partial b}{\partial z} - f(a(z),z)\frac{\partial a}{\partial z}.$$

References

Abramowitz, M. and Stegun, C. A. (Eds.). *Handbook of Mathematical Functions with Formulas, Graphs, and Mathematical Tables, 9th printing.* New York: Dover, p. 11, 1972.

Leibniz Series

The SERIES for the INVERSE TANGENT,

$$\tan^{-1} x = x - \tfrac{1}{3}x^3 + \tfrac{1}{5}x^5 + \ldots.$$

Lemarié's Wavelet

A wavelet used in multiresolution representation to analyze the information content of images. The WAVELET is defined by

$$H(\omega) = \left[2(1-u)^4 \frac{315 - 420u + 126u^2 - 4u^3}{315 - 420v + 126v^2 - 4v^3}\right]^{1/2},$$

where

$$u \equiv \sin^2(\tfrac{1}{2}\omega)$$

$$v \equiv \sin^2 \omega$$

(Mallat 1989).

see also WAVELET

References

Mallat, S. G. "A Theory for Multiresolution Signal Decomposition: The Wavelet Representation." *IEEE Trans. Pattern Analysis Machine Intel.* **11**, 674–693, 1989.

Mallat, S. G. "Multiresolution Approximation and Wavelet Orthonormal Bases of $L^2(\mathbb{R})$." *Trans. Amer. Math. Soc.* **315**, 69–87, 1989.

Lemma

A short THEOREM used in proving a larger THEOREM. Related concepts are the AXIOM, PORISM, POSTULATE, PRINCIPLE, and THEOREM.

see also ABEL'S LEMMA, ARCHIMEDES' LEMMA, BARNES' LEMMA, BLICHFELDT'S LEMMA, BOREL-CANTELLI LEMMA, BURNSIDE'S LEMMA, DANIELSON-LANCZOS LEMMA, DEHN'S LEMMA, DILWORTH'S LEMMA, DIRICHLET'S LEMMA, DIVISION LEMMA, FARKAS'S LEMMA, FATOU'S LEMMA, FUNDAMENTAL LEMMA OF CALCULUS OF VARIATIONS, GAUSS'S LEMMA, HENSEL'S LEMMA, ITÔ'S LEMMA, JORDAN'S LEMMA, LAGRANGE'S LEMMA, NEYMAN-PEARSON LEMMA, POINCARÉ'S HOLOMORPHIC LEMMA, POINCARÉ'S LEMMA, PÓLYA-BURNSIDE LEMMA, RIEMANN-LEBESGUE LEMMA, SCHUR'S LEMMA, SCHUR'S REPRESENTATION LEMMA, SCHWARZ-PICK LEMMA, SPIJKER'S LEMMA, ZORN'S LEMMA

Lemniscate

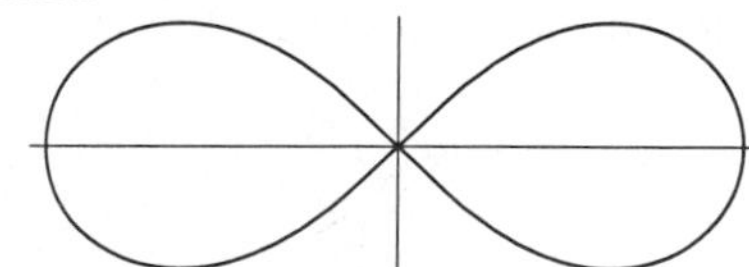

A polar curve also called LEMNISCATE OF BERNOULLI which is the LOCUS of points the product of whose distances from two points (called the FOCI) is a constant. Letting the FOCI be located at $(\pm a, 0)$, the Cartesian equation is

$$[(x-a)^2 + y^2][(x+a)^2 + y^2] = a^4, \quad (1)$$

which can be rewritten

$$x^4 + y^4 + 2x^2y^2 = 2a^2(x^2 - y^2). \quad (2)$$

Letting $a' = \sqrt{2}a$, the POLAR COORDINATES are given by

$$r^2 = a^2 \cos(2\theta). \tag{3}$$

An alternate form is

$$r^2 = a^2 \sin(2\theta). \tag{4}$$

The parametric equations for the lemniscate are

$$x = \frac{a\cos t}{1+\sin^2 t} \tag{5}$$

$$y = \frac{a \sin t \cos t}{1+\sin^2 t}. \tag{6}$$

The bipolar equation of the lemniscate is

$$rr' = \tfrac{1}{2}a^2, \tag{7}$$

and in PEDAL COORDINATES with the PEDAL POINT at the center, the equation is

$$pa^2 = r^3. \tag{8}$$

The two-center BIPOLAR COORDINATES equation with origin at a FOCUS is

$$r_1 r_2 = c^2. \tag{9}$$

Jakob Bernoulli published an article in *Acta Eruditorum* in 1694 in which he called this curve the lemniscus ("a pendant ribbon"). Jakob Bernoulli was not aware that the curve he was describing was a special case of CASSINI OVALS which had been described by Cassini in 1680. The general properties of the lemniscate were discovered by G. Fagnano in 1750 (MacTutor Archive). Gauss's and Euler's investigations of the ARC LENGTH of the curve led to later work on ELLIPTIC FUNCTIONS.

The CURVATURE of the lemniscate is

$$\kappa = \frac{3\sqrt{2}\cos t}{\sqrt{3-\cos(2t)}}. \tag{10}$$

The ARC LENGTH is more problematic. Using the polar form,

$$ds^2 = dr^2 + r^2\, d\theta^2 \tag{11}$$

so

$$ds = \sqrt{1+\left(r\frac{d\theta}{dr}\right)^2}\, dr. \tag{12}$$

But we have

$$2r\,dr = 2a^2 \sin(2\theta)\, d\theta \tag{13}$$

$$r\frac{dr}{d\theta} = \frac{r^2}{a^2\sin(2\theta)} \tag{14}$$

$$\left(r\frac{d\theta}{dr}\right)^2 = \frac{r^4}{a^4\sin^2(2\theta)} = \frac{r^4}{a^4[1-\cos^2(2\theta)]} = \frac{r^4}{a^4-r^4}, \tag{15}$$

so

$$ds = \sqrt{1+\frac{r^4}{a^4-r^4}}\,dr = \sqrt{\frac{a^4}{a^4-r^4}}\,dr = \frac{a^2}{\sqrt{a^4-r^4}}\,dr$$
$$= \frac{dr}{\sqrt{1-\left(\frac{r}{a}\right)^4}}, \tag{16}$$

and

$$L = \int_0^a ds = 2\int_0^a \frac{ds}{dr}\,dr = 2\int_0^a \frac{dr}{\sqrt{1-\left(\frac{r}{a}\right)^4}}. \tag{17}$$

Let $t \equiv r/a$, so $dt = dr/a$, and

$$L = 2a\int_0^1 (1-t^4)^{-1/2}\,dt, \tag{18}$$

which, as shown in LEMNISCATE FUNCTION, is given analytically by

$$L = \sqrt{2}\,aK\left(\frac{1}{\sqrt{2}}\right) = \frac{\Gamma^2(\frac{1}{4})}{2^{3/2}\sqrt{\pi}}\,a. \tag{19}$$

If $a = 1$, then

$$L = 5.2441151086\ldots, \tag{20}$$

which is related to GAUSS'S CONSTANT M by

$$L = \frac{2\pi}{M}. \tag{21}$$

The quantity $L/2$ or $L/4$ is called the LEMNISCATE CONSTANT and plays a role for the lemniscate analogous to that of π for the CIRCLE.

The AREA of one loop of the lemniscate is

$$A = \tfrac{1}{2}\int r^2\,d\theta = \tfrac{1}{2}a^2\int_{-\pi/4}^{\pi/4}\cos(2\theta)\,d\theta = \tfrac{1}{4}a^2[\sin(2\theta)]_{-\pi/4}^{\pi/4}$$
$$= \tfrac{1}{2}a^2[\sin(2\theta)]_0^{\pi/4} = \tfrac{1}{2}a^2[\sin(\tfrac{\pi}{2})-\sin 0] = \tfrac{1}{2}a^2. \tag{22}$$

see also LEMNISCATE FUNCTION

References

Ayoub, R. "The Lemniscate and Fagnano's Contributions to Elliptic Integrals." *Arch. Hist. Exact Sci.* **29**, 131–149, 1984.

Borwein, J. M. and Borwein, P. B. *Pi & the AGM: A Study in Analytic Number Theory and Computational Complexity.* New York: Wiley, 1987.

Gray, A. "Lemniscates of Bernoulli." §3.2 in *Modern Differential Geometry of Curves and Surfaces.* Boca Raton, FL: CRC Press, pp. 39–41, 1993.

Lawrence, J. D. *A Catalog of Special Plane Curves.* New York: Dover, pp. 120–124, 1972.

Le Lionnais, F. *Les nombres remarquables.* Paris: Hermann, p. 37, 1983.

Lee, X. "Lemniscate of Bernoulli." http://www.best.com/~xah/SpecialPlaneCurves_dir/LemniscateOfBernoulli_dir/lemniscateOfBernoulli.html.

Lockwood, E. H. *A Book of Curves.* Cambridge, England: Cambridge University Press, 1967.

MacTutor History of Mathematics Archive. "Lemniscate of Bernoulli." http://www-groups.dcs.st-and.ac.uk/~history/Curves/Lemniscate.html.

Yates, R. C. "Lemniscate." *A Handbook on Curves and Their Properties.* Ann Arbor, MI: J. W. Edwards, pp. 143–147, 1952.

Lemniscate of Bernoulli

see LEMNISCATE

Lemniscate Case

The case of the WEIERSTRAß ELLIPTIC FUNCTION with invariants $g_2 = 1$ and $g_3 = 0$.

see also EQUIANHARMONIC CASE, WEIERSTRAß ELLIPTIC FUNCTION, PSEUDOLEMNISCATE CASE

References

Abramowitz, M. and Stegun, C. A. (Eds.). "Lemniscate Case ($g_2 = 1$, $g_3 = 0$)." §18.14 in *Handbook of Mathematical Functions with Formulas, Graphs, and Mathematical Tables, 9th printing.* New York: Dover, pp. 658–662, 1972.

Lemniscate Constant

Let

$$L = \frac{1}{\sqrt{2\pi}}[\Gamma(\tfrac{1}{4})]^2 = 5.2441151086\ldots$$

be the ARC LENGTH of a LEMNISCATE with $a = 1$. Then the lemniscate constant is the quantity $L/2$ (Abramowitz and Stegun 1972), or $L/4 = 1.311028777\ldots$ (Todd 1975, Le Lionnais 1983). Todd (1975) cites T. Schneider (1937) as proving L to be a TRANSCENDENTAL NUMBER.

see also LEMNISCATE

References

Abramowitz, M. and Stegun, C. A. (Eds.). *Handbook of Mathematical Functions with Formulas, Graphs, and Mathematical Tables, 9th printing.* New York: Dover, 1972.

Borwein, J. M. and Borwein, P. B. *Pi & the AGM: A Study in Analytic Number Theory and Computational Complexity.* New York: Wiley, 1987.

Finch, S. "Favorite Mathematical Constants." `http://www.mathsoft.com/asolve/constant/gauss/gauss.html`.

Le Lionnais, F. *Les nombres remarquables.* Paris: Hermann, p. 37, 1983.

Todd, J. "The Lemniscate Constant." *Comm. ACM* **18**, 14–19 and 462, 1975.

Lemniscate Function

The lemniscate functions arise in rectifying the ARC LENGTH of the LEMNISCATE. The lemniscate functions were first studied by Jakob Bernoulli and G. Fagnano. A historical account is given by Ayoub (1984), and an extensive discussion by Siegel (1969). The lemniscate functions were the first functions defined by inversion of an integral, which was first done by Gauss.

$$L = 2a\int_0^1 (1-t^4)^{-1/2}\,dt. \tag{1}$$

Define the functions

$$\phi(x) \equiv \text{arcsinlemn}x = \int_0^x (1-t^4)^{-1/2}\,dt \tag{2}$$

$$\phi'(x) \equiv \text{arccoslemn}x = \int_x^1 (1-t^4)^{-1/2}\,dt, \tag{3}$$

where

$$\varpi \equiv \frac{L}{a}, \tag{4}$$

and write

$$x = \text{sinlemn}\,\phi \tag{5}$$

$$x = \text{coslemn}\,\phi'. \tag{6}$$

There is an identity connecting ϕ and ϕ' since

$$\phi(x) + \phi'(x) = \frac{L}{2a} = \tfrac{1}{2}\varpi, \tag{7}$$

so

$$\text{sinlemn}\,\phi = \text{coslemn}(\tfrac{1}{2}\varpi - \phi). \tag{8}$$

These functions can be written in terms of JACOBI ELLIPTIC FUNCTIONS,

$$u = \int_0^{\text{sd}(u,k)} [(1-k'^2y^2)(1+k^2y^2)]^{-1/2}\,dy. \tag{9}$$

Now, if $k = k' = 1/\sqrt{2}$, then

$$\begin{aligned} u &= \int_0^{\text{sd}(u,1/\sqrt{2})} [(1-\tfrac{1}{2}y^2)(1+\tfrac{1}{2}y^2)]^{-1/2}\,dy \\ &= \int_0^{\text{sd}(u,1/\sqrt{2})} (1-\tfrac{1}{4}y^4)^{-1/2}\,dy. \end{aligned} \tag{10}$$

Let $t \equiv y/\sqrt{2}$ so $dy = \sqrt{2}\,dt$,

$$u = \sqrt{2}\int_0^{\text{sd}(u,1/\sqrt{2})/\sqrt{2}} (1-t^4)^{-1/2}\,dt \tag{11}$$

$$\frac{u}{\sqrt{2}} = \int_0^{\text{sd}(u,1/\sqrt{2})/\sqrt{2}} (1-t^4)^{-1/2}\,dt \tag{12}$$

$$u = \int_0^{\text{sd}(u\sqrt{2},1/\sqrt{2})/\sqrt{2}} (1-t^4)^{-1/2}\,dt, \tag{13}$$

and

$$\text{sinlemn}\,\phi = \frac{1}{\sqrt{2}}\text{sd}\left(\phi\sqrt{2}, \frac{1}{\sqrt{2}}\right). \tag{14}$$

Similarly,

$$\begin{aligned} u &= \int_{\text{cn}(u,k)}^1 (1-t^2)^{-1/2}(k'^2+k^2t^2)^{-1/2}\,dt \\ &= \int_{\text{cn}(u,1/\sqrt{2})}^1 (1-t^2)^{-1/2}\left(\tfrac{1}{2}+\tfrac{1}{2}t^2\right)^{-1/2}\,dt \\ &= \sqrt{2}\int_{\text{cn}(u,1/\sqrt{2})}^1 (1-t^4)^{-1/2}\,dt \end{aligned} \tag{15}$$

$$\frac{u}{\sqrt{2}} = \int_{\text{cn}(u,1/\sqrt{2})}^1 (1-t^4)^{-1/2}\,dt \tag{16}$$

$$u = \int_{\mathrm{cn}(u\sqrt{2},1/\sqrt{2})}^{1} (1 - t^4)^{-1/2}\, dt, \quad (17)$$

and

$$\mathrm{coslemn}\,\phi = \mathrm{cn}\left(\phi\sqrt{2}, \frac{1}{\sqrt{2}}\right). \quad (18)$$

We know

$$\mathrm{coslemn}(\tfrac{1}{2}\varpi) = \mathrm{cn}\left(\tfrac{1}{2}\varpi\sqrt{2}, \frac{1}{\sqrt{2}}\right) = 0. \quad (19)$$

But it is true that

$$\mathrm{cn}(K, k) = 0, \quad (20)$$

so

$$K\left(\frac{1}{\sqrt{2}}\right) = \tfrac{1}{2}\sqrt{2}\,\varpi = \frac{1}{\sqrt{2}}\varpi \quad (21)$$

$$\frac{\Gamma^2(\frac{1}{4})}{4\sqrt{\pi}} = \frac{1}{\sqrt{2}}\varpi \quad (22)$$

$$L = a\varpi = a\sqrt{2}\,\frac{\Gamma^2(\frac{1}{4})}{4\sqrt{\pi}} = \frac{\Gamma^2(\frac{1}{4})}{2^{3/2}\sqrt{\pi}}\,a. \quad (23)$$

By expanding $(1 - t^4)^{-1/2}$ in a BINOMIAL SERIES and integrating term by term, the arcsinlemn function can be written

$$\phi(x) = \int_0^v \frac{dt}{\sqrt{1 - t^4}} = \sum_{n=0}^{\infty} \frac{(\frac{1}{2})_n x^{4n+1}}{n!(4n+1)}, \quad (24)$$

where $(a)_n$ is the RISING FACTORIAL (Berndt 1994). Ramanujan gave the following inversion FORMULA for $\phi(x)$. If

$$\frac{\theta\mu}{\sqrt{2}} = \sum_{n=0}^{\infty} \frac{(\frac{1}{2})_n x^{4n+1}}{n!(4n+1)}, \quad (25)$$

where

$$\mu = \frac{\Gamma^2(\frac{1}{4})}{2\pi^{3/2}} \quad (26)$$

is the constant obtained by letting $x = 1$ and $\theta = \pi/2$, and

$$v = 2^{-1/2}\,\mathrm{sd}(\mu\theta), \quad (27)$$

then

$$\frac{\mu^2}{2x^2} = \csc^2\theta - \frac{1}{\pi} - 8\sum_{n=1}^{\infty} \frac{n\cos(2n\theta)}{e^{2\pi n} - 1} \quad (28)$$

(Berndt 1994). Ramanujan also showed that if $0 < \theta < \pi/2$, then

$$-\frac{\mu}{\sqrt{2}} \sum_{n=0}^{\infty} \frac{(\frac{1}{2})_n v^{4n-1}}{n!(4n-1)} = \cot\theta + \frac{\theta}{\pi} + 4\sum_{n=1}^{\infty} \frac{\sin(2n\theta)}{2^{2\pi n} - 1}, \quad (29)$$

$$\ln v + \tfrac{1}{6}\pi - \tfrac{1}{2}\ln 2 + \sum_{n=1}^{\infty} \frac{(\frac{1}{4})_n v^{4n}}{(\frac{3}{4})_n 4n} = \ln(\sin\theta) + \frac{\theta^2}{2\pi} - 2\sum_{n=1}^{\infty} \frac{\cos(2n\theta)}{n(e^{2\pi n} - 1)}, \quad (30)$$

$$\tfrac{1}{2}\tan^{-1} v = \sum_{n=0}^{\infty} \frac{\sin[(2n+1)\theta]}{(2n+1)\cosh[\frac{1}{2}(2n+1)\pi]}, \quad (31)$$

$$\tfrac{1}{4}\cos^{-1}(v^2) = \sum_{n=0}^{\infty} \frac{(-1)^n \cos[(2n+1)\theta]}{(2n+1)\cosh[\frac{1}{2}(2n+1)\pi]}, \quad (32)$$

and

$$\frac{\sqrt{2}}{4\mu} \sum_{n=0}^{\infty} \frac{2^{2n}(n!)^2}{(2n+1)!(4n+3)} v^{4n+3} = \frac{\pi\theta}{8} - \sum_{n=0}^{\infty} \frac{(-1)^n \sin[(2n+1)\theta]}{(2n+1)^2\cosh[\frac{1}{2}(2n+1)\pi]} \quad (33)$$

(Berndt 1994).

A generalized version of the lemniscate function can be defined by letting $0 \le \theta \le \pi/2$ and $0 \le v \le 1$. Write

$$\tfrac{2}{3}\theta\mu = \int_0^v \frac{dt}{\sqrt{1 - t^6}}, \quad (34)$$

where μ is the constant obtained by setting $\theta = \pi/2$ and $v = 1$. Then

$$\mu = \frac{\sqrt{\pi}}{\Gamma(\frac{2}{3})\Gamma(\frac{5}{6})}, \quad (35)$$

and Ramanujan showed

$$\frac{4\mu^2}{9v^2} = \csc^2\theta - \frac{2}{\pi\sqrt{3}} + 8\sum_{n=1}^{\infty} \frac{(-1)^{n-1} n\cos(2n\theta)}{e^{\pi n\sqrt{3}} - (-1)^n} \quad (36)$$

(Berndt 1994).

see also HYPERBOLIC LEMNISCATE FUNCTION

References

Ayoub, R. "The Lemniscate and Fagnano's Contributions to Elliptic Integrals." *Arch. Hist. Exact Sci.* **29**, 131–149, 1984.

Berndt, B. C. *Ramanujan's Notebooks, Part IV.* New York: Springer-Verlag, pp. 245, and 247–255, 258–260, 1994.

Siegel, C. L. *Topics in Complex Function Theory, Vol. 1.* New York: Wiley, 1969.

Lemniscate of Gerono

see EIGHT CURVE

Lemniscate Inverse Curve

The INVERSE CURVE of a LEMNISCATE in a CIRCLE centered at the origin and touching the LEMNISCATE where it crosses the x-AXIS produces a RECTANGULAR HYPERBOLA.

Lemniscate (Mandelbrot Set)

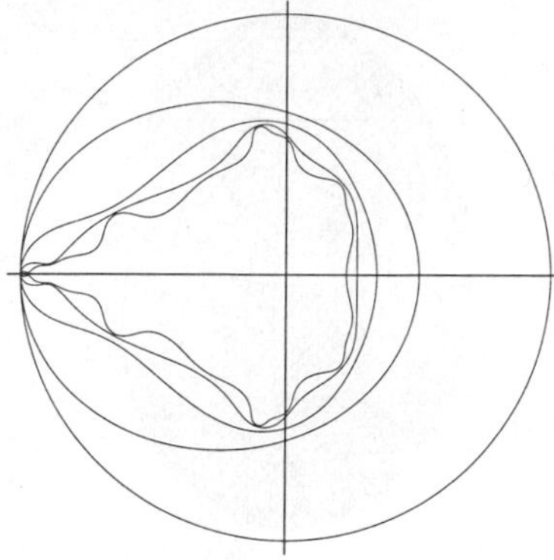

A curve on which points of a MAP z_n (such as the MANDELBROT SET) diverge to a given value $r_{\max}$ at the same rate. A common method of obtaining lemniscates is to define an INTEGER called the COUNT which is the largest n such that $|z_n| < r$ where r is usually taken as $r = 2$. Successive COUNTS then define a series of lemniscates, which are called EQUIPOTENTIAL CURVES by Peitgen and Saupe (1988).

see also COUNT, MANDELBROT SET

References

Peitgen, H.-O. and Saupe, D. (Eds.). *The Science of Fractal Images.* New York: Springer-Verlag, pp. 178–179, 1988.

Lemoine Axis

see LEMOINE LINE

Lemoine Circle

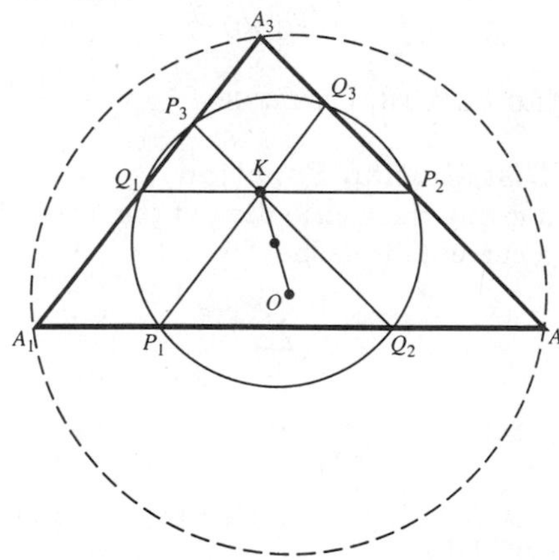

Also called the TRIPLICATE-RATIO CIRCLE. Draw lines through the LEMOINE POINT K and parallel to the sides of the triangle. The points where the parallel lines intersect the sides then lie on a CIRCLE known as the Lemoine circle. This circle has center at the MIDPOINT of OK, where O is the CIRCUMCENTER. The circle has radius

$$\tfrac{1}{2}\sqrt{R^2 + r^2} = \tfrac{1}{2} R \sec \omega,$$

where R is the CIRCUMRADIUS, r is the INRADIUS, and ω is the BROCARD ANGLE. The Lemoine circle divides any side into segments proportional to the squares of the sides

$$\overline{A_2P_2} : \overline{P_2Q_3} : \overline{Q_3A_3} = {a_3}^2 : {a_1}^2 : {a_2}^2.$$

Furthermore, the chords cut from the sides by the Lemoine circle are proportional to the squares of the sides.

The COSINE CIRCLE is sometimes called the second Lemoine circle.

see also COSINE CIRCLE, LEMOINE LINE, LEMOINE POINT, TUCKER CIRCLES

References

Johnson, R. A. *Modern Geometry: An Elementary Treatise on the Geometry of the Triangle and the Circle.* Boston, MA: Houghton Mifflin, pp. 273–275, 1929.

Lemoine Line

The Lemoine line, also called the LEMOINE AXIS, is the perspectivity axis of a TRIANGLE and its TANGENTIAL TRIANGLE, and also the TRILINEAR POLAR of the CENTROID of the triangle vertices. It is also the POLAR of K with regard to its CIRCUMCIRCLE, and is PERPENDICULAR to the BROCARD AXIS.

The centers of the APOLLONIUS CIRCLES L_1, L_2, and L_3 are COLLINEAR on the LEMOINE LINE. This line is PERPENDICULAR to the BROCARD AXIS OK and is the RADICAL AXIS of the CIRCUMCIRCLE and the BROCARD CIRCLE. It has equation

$$\frac{\alpha}{a} + \frac{\beta}{b} + \frac{\gamma}{c}$$

in terms of TRILINEAR COORDINATES (Oldknow 1996).

see also APOLLONIUS CIRCLES, BROCARD AXIS, CENTROID (TRIANGLE), CIRCUMCIRCLE, COLLINEAR, LEMOINE CIRCLE, LEMOINE POINT, POLAR, RADICAL AXIS, TANGENTIAL TRIANGLE, TRILINEAR POLAR

References

Johnson, R. A. *Modern Geometry: An Elementary Treatise on the Geometry of the Triangle and the Circle.* Boston, MA: Houghton Mifflin, p. 295, 1929.

Oldknow, A. "The Euler-Gergonne-Soddy Triangle of a Triangle." *Amer. Math. Monthly* **103**, 319–329, 1996.

Lemoine Point

The point of concurrence K of the SYMMEDIAN LINES, sometimes also called the SYMMEDIAN POINT and GREBE POINT.

Let G be the CENTROID of a TRIANGLE ΔABC, L_A, L_B, and L_C the ANGLE BISECTORS of ANGLES A, B, C, and G_A, G_B, and G_C the reflections of AG, BG, and CG about L_A, L_B, and L_C. Then K is the point of concurrence of the lines G_A, G_B, and G_C. It is the perspectivity center of a TRIANGLE and its TANGENTIAL TRIANGLE.

In AREAL COORDINATES (actual TRILINEAR COORDINATES), the Lemoine point is the point for which $\alpha^2+\beta^2+\gamma^2$ is a minimum. A center X is the CENTROID of its own PEDAL TRIANGLE IFF it is the Lemoine point.

The Lemoine point lies on the BROCARD AXIS, and its distances from the Lemoine point K to the sides of the TRIANGLE are

$$\overline{KK_i} = \tfrac{1}{2}a_i \tan\omega,$$

where ω is the BROCARD ANGLE. A BROCARD LINE, MEDIAN, and Lemoine point are concurrent, with $A_1\Omega_1$, A_2K, and A_3M meeting at a point. Similarly, $A_1\Omega'$, A_2M, and A_3K meet at a point which is the ISOGONAL CONJUGATE of the first (Johnson 1929, pp. 268–269). The line joining the MIDPOINT of any side to the midpoint of the ALTITUDE on that side passes through the Lemoine point K. The Lemoine point K is the STEINER POINT of the first BROCARD TRIANGLE.

see also ANGLE BISECTOR, BROCARD ANGLE, BROCARD AXIS, BROCARD DIAMETER, CENTROID (TRIANGLE), COSYMMEDIAN TRIANGLES, GREBE POINT, ISOGONAL CONJUGATE, LEMOINE CIRCLE, LEMOINE LINE, LINE AT INFINITY, MITTENPUNKT, PEDAL TRIANGLE, STEINER POINTS, SYMMEDIAN LINE, TANGENTIAL TRIANGLE

References
Gallatly, W. *The Modern Geometry of the Triangle, 2nd ed.* London: Hodgson, p. 86, 1913.
Honsberger, R. *Episodes in Nineteenth and Twentieth Century Euclidean Geometry.* Washington, DC: Math. Assoc. Amer., 1995.
Johnson, R. A. *Modern Geometry: An Elementary Treatise on the Geometry of the Triangle and the Circle.* Boston, MA: Houghton Mifflin, pp. 217, 268–269, and 271–272, 1929.
Kimberling, C. "Central Points and Central Lines in the Plane of a Triangle." *Math. Mag.* **67**, 163–187, 1994.
Kimberling, C. "Symmedian Point." http://www.evansville.edu/~ck6/tcenters/class/sympt.html.
Mackay, J. S. "Early History of the Symmedian Point." *Proc. Edinburgh Math. Soc.* **11**, 92–103, 1892–1893.

Lemoine's Problem

Given the vertices of the three EQUILATERAL TRIANGLES placed on the sides of a TRIANGLE T, construct T. The solution can be given using KIEPERT'S HYPERBOLA.

see also KIEPERT'S HYPERBOLA

Lemon

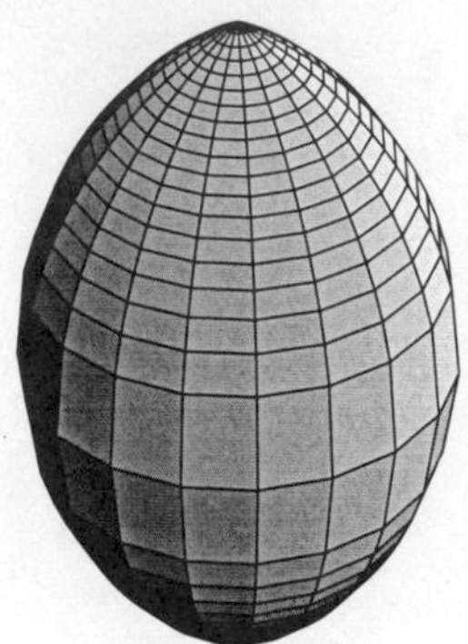

A SURFACE OF REVOLUTION defined by Kepler. It consists of less than half of a circular ARC rotated about an axis passing through the endpoints of the ARC. The equations of the upper and lower boundaries in the xz plane are

$$z_\pm = \pm\sqrt{R^2-(x+r)^2}$$

for $R>r$ and $x\in[-(R-r),R-r]$. The CROSS-SECTION of a lemon is a LENS. The lemon is the inside surface of a SPINDLE TORUS.

see also APPLE, LENS, SPINDLE TORUS

Length (Curve)

Let $\gamma(t)$ be a smooth curve in a MANIFOLD M from x to y with $\gamma(0)=x$ and $\gamma(1)=y$. Then $\gamma'(t)\in T_{\gamma(t)}$, where T_x is the TANGENT SPACE of M at x. The length of γ with respect to the Riemannian structure is given by

$$\int_0^1 ||\gamma'(t)||_{\gamma(t)}\,dt.$$

see also ARC LENGTH, DISTANCE

Length Distribution Function

A function giving the distribution of the interpoint distances of a curve. It is defined by

$$p(r)=\frac{1}{N}\sum_{ij}\delta_{r_{ij}=r}.$$

see also RADIUS OF GYRATION

References
Pickover, C. A. *Keys to Infinity.* New York: W. H. Freeman, pp. 204–206, 1995.

Length (Number)

The length of a number n in base b is the number of DIGITS in the base-b numeral for n, given by the formula

$$L(n,b)=\lfloor\log_b(n)\rfloor+1,$$

where $\lfloor x\rfloor$ is the FLOOR FUNCTION.

The MULTIPLICATIVE PERSISTENCE of an n-DIGIT is sometimes also called its length.

see also CONCATENATION, DIGIT, FIGURES, MULTIPLICATIVE PERSISTENCE

Length (Partial Order)

For a PARTIAL ORDER, the size of the longest CHAIN is called the length.

see also WIDTH (PARTIAL ORDER)

Length (Size)

The longest dimension of a 3-D object.

see also HEIGHT, WIDTH (SIZE)

Lengyel's Constant

N.B. A detailed on-line essay by S. Finch was the starting point for this entry.

Let L denote the partition lattice of the SET $\{1, 2, \ldots, n\}$. The MAXIMUM element of L is

$$M = \{\{1, 2, \ldots, n\}\} \quad (1)$$

and the MINIMUM element is

$$m = \{\{1\}, \{2\}, \ldots, \{n\}\}. \quad (2)$$

Let Z_n denote that number of chains of any length in L containing both M and m. Then Z_n satisfies the RECURRENCE RELATION

$$Z_n = \sum_{k=1}^{n-1} s(n,k) Z_k, \quad (3)$$

where $s(n,k)$ is a STIRLING NUMBER OF THE SECOND KIND. Lengyel (1984) proved that the QUOTIENT

$$r(n) = \frac{Z_n}{(n!)^2 (2\ln 2)^{-n} n^{1-(\ln 2)/3}} \quad (4)$$

is bounded between two constants as $n \to \infty$, and Flajolet and Salvy (1990) improved the result of Babai and Lengyel (1992) to show that

$$\Lambda \equiv \lim_{n\to\infty} r(n) = 1.0986858055\ldots. \quad (5)$$

References

Babai, L. and Lengyel, T. "A Convergence Criterion for Recurrent Sequences with Application to the Partition Lattice." *Analysis* **12**, 109–119, 1992.

Finch, S. "Favorite Mathematical Constants." `http://www.mathsoft.com/asolve/constant/lngy/lngy.html`.

Flajolet, P. and Salvy, B. "Hierarchal Set Partitions and Analytic Iterates of the Exponential Function." Unpublished manuscript, 1990.

Lengyel, T. "On a Recurrence Involving Stirling Numbers." *Europ. J. Comb.* **5**, 313–321, 1984.

Plouffe, S. "The Lengyel Constant." `http://lacim.uqam.ca/piDATA/lengyel.txt`.

Lens

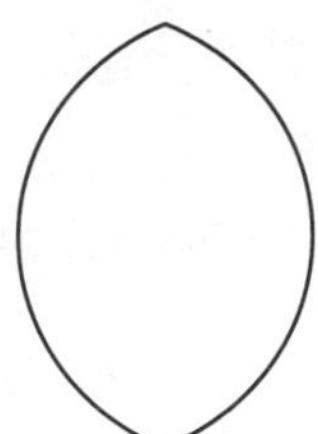

A figure composed of two equal and symmetrically placed circular ARCS. It is also known as the FISH BLADDER (Pedoe 1995, p. xii) or VESICA PISCIS. The latter term is often used for the particular lens formed by the intersection of two unit CIRCLES whose centers are offset by a unit distance (Rawles 1997). In this case, the height of the lens is given by letting $d = r = R = 1$ in the equation for a CIRCLE-CIRCLE INTERSECTION

$$a = \frac{1}{d}\sqrt{4d^2R^2 - (d^2 - r^2 + R^2)^2}, \quad (1)$$

giving $a = \sqrt{3}$. The AREA of the VESICA PISCIS is given by plugging $d = R$ into the CIRCLE-CIRCLE INTERSECTION area equation with $r = R$,

$$A = 2R^2 \cos^{-1}\left(\frac{d}{2R}\right) - \tfrac{1}{2}d\sqrt{4R^2 - d^2}, \quad (2)$$

giving

$$A = \tfrac{1}{6}(4\pi - 3\sqrt{3}) \approx 1.22837. \quad (3)$$

Renaissance artists frequently surrounded images of Jesus with the vesica piscis (Rawles 1997). An asymmetrical lens is produced by a CIRCLE-CIRCLE INTERSECTION for unequal CIRCLES.

see also CIRCLE, CIRCLE-CIRCLE INTERSECTION, FLOWER OF LIFE, LEMON, LUNE (PLANE), REULEAUX TRIANGLE, SECTOR, SEED OF LIFE, SEGMENT, VENN DIAGRAM

References

Pedoe, D. *Circles: A Mathematical View, rev. ed.* Washington, DC: Math. Assoc. Amer., 1995.

Rawles, B. *Sacred Geometry Design Sourcebook: Universal Dimensional Patterns.* Nevada City, CA: Elysian Pub., p. 11, 1997.

Lens Space

A lens space $L(p,q)$ is the 3-MANIFOLD obtained by gluing the boundaries of two solid TORI together such that the meridian of the first goes to a (p,q)-curve on the second, where a (p,q)-curve has p meridians and q longitudes.

References

Rolfsen, D. *Knots and Links.* Wilmington, DE: Publish or Perish Press, 1976.

Lenstra Elliptic Curve Method

A method of factoring INTEGERS using ELLIPTIC CURVES.

References

Montgomery, P. L. "Speeding up the Pollard and Elliptic Curve Methods of Factorization." *Math. Comput.* **48**, 243–264, 1987.

Léon Anne's Theorem

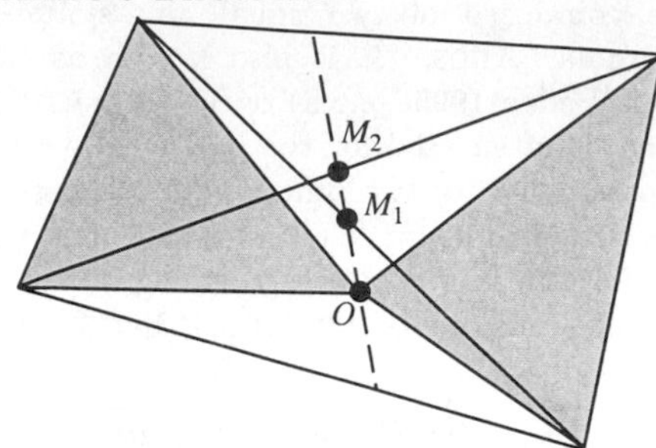

Pick a point O in the interior of a QUADRILATERAL which is not a PARALLELOGRAM. Join this point to each of the four VERTICES, then the LOCUS of points O for which the sum of opposite TRIANGLE areas is half the QUADRILATERAL AREA is the line joining the MIDPOINTS M_1 and M_2 of the DIAGONALS.

see also DIAGONAL (POLYGON), MIDPOINT, QUADRILATERAL

References

Honsberger, R. *More Mathematical Morsels.* Washington, DC: Math. Assoc. Amer., pp. 174–175, 1991.

Leonardo's Paradox

In the depiction of a row of identical columns parallel to the plane of a PERSPECTIVE drawing, the outer columns should appear wider even though they are farther away.

see also PERSPECTIVE, VANISHING POINT, ZEEMAN'S PARADOX

References

Dixon, R. *Mathographics.* New York: Dover, p. 82, 1991.

Leptokurtic

A distribution with a high peak so that the KURTOSIS satisfies $\gamma_2 > 0$.

see also KURTOSIS

Lerch's Theorem

If there are two functions $F_1(t)$ and $F_2(t)$ with the same integral transform

$$\mathcal{T}[F_1(t)] = \mathcal{T}[F_2(t)] \equiv f(s), \qquad (1)$$

then a NULL FUNCTION can be defined by

$$\delta_0(t) \equiv F_1(t) - F_2(t) \qquad (2)$$

so that the integral

$$\int_0^a \delta_0(t)\,dt = 0 \qquad (3)$$

vanishes for all $a > 0$.

see also NULL FUNCTION

Lerch Transcendent

A generalization of the HURWITZ ZETA FUNCTION and POLYLOGARITHM function. Many sums of reciprocal POWERS can be expressed in terms of it. It is defined by

$$\Phi(z,s,a) \equiv \sum_{k=0}^{\infty} \frac{z^k}{(a+k)^s}, \qquad (1)$$

where any term with $a+k=0$ is excluded.

The Lerch transcendent can be used to express the DIRICHLET BETA FUNCTION

$$\beta(s) \equiv \sum_{k=0}^{\infty} (-1)^k (2k+1)^{-s} 2^{-s} \Phi(-1, s, \tfrac{1}{2}), \qquad (2)$$

the integral of the FERMI-DIRAC DISTRIBUTION

$$\int_0^\infty \frac{k^s}{e^{k-\mu}+1}\,dk = e^\mu \Gamma(s+1)\Phi(-e^\mu, s+1, 1), \qquad (3)$$

where $\Gamma(z)$ is the GAMMA FUNCTION, and to evaluate the DIRICHLET L-SERIES.

see also DIRICHLET BETA FUNCTION, DIRICHLET L-SERIES, FERMI-DIRAC DISTRIBUTION, HURWITZ ZETA FUNCTION, POLYLOGARITHM

Less

A quantity a is said to be less than b if a is smaller than b, written $a < b$. If a is less than or EQUAL to b, the relationship is written $a \leq b$. If a is MUCH LESS than b, this is written $a \ll b$. Statements involving GREATER than and less than symbols are called INEQUALITIES.

see also EQUAL, GREATER, INEQUALITY, MUCH GREATER, MUCH LESS

Letter-Value Display

A method of displaying simple statistical parameters including HINGES, MEDIAN, and upper and lower values.

References

Tukey, J. W. *Explanatory Data Analysis.* Reading, MA: Addison-Wesley, p. 33, 1977.

Leudesdorf Theorem

Let $t(m)$ denote the set of the $\phi(m)$ numbers less than and RELATIVELY PRIME to m, where $\phi(n)$ is the TOTIENT FUNCTION. Then if

$$S_m \equiv \sum_{t(m)} \frac{1}{t},$$

then

$$\begin{cases} S_m \equiv 0 \pmod{m^2} & \text{if } 2\nmid m,\ 3\nmid m \\ S_m \equiv 0 \pmod{\frac{1}{3}m^2} & \text{if } 2\nmid m,\ 3|m \\ S_m \equiv 0 \pmod{\frac{1}{2}m^2} & 2|m,\ \nmid m,\ m \text{ not a power of 2} \\ S_m \equiv 0 \pmod{\frac{1}{6}m^2} & \text{if } 2|m,\ 3|m \\ S_m \equiv 0 \pmod{\frac{1}{4}m^2} & \text{if } m = 2^a. \end{cases}$$

see also BAUER'S IDENTICAL CONGRUENCE, TOTIENT FUNCTION

References

Hardy, G. H. and Wright, E. M. "A Theorem of Leudesdorf." §8.7 in *An Introduction to the Theory of Numbers, 5th ed.* Oxford, England: Clarendon Press, pp. 100–102, 1979.

Level Curve

A LEVEL SET in 2-D.

Level Set

The level set of c is the SET of points

$$\{(x_1, \ldots, x_n) \in U : f(x_1, \ldots, x_n) = c\} \in \mathbb{R}^n,$$

and is in the DOMAIN of the function. If $n = 2$, the level set is a plane curve (a level curve). If $n = 3$, the level set is a surface (a level surface).

References

Gray, A. "Level Surfaces in $\mathbb{R}^3$." §10.7 in *Modern Differential Geometry of Curves and Surfaces.* Boca Raton, FL: CRC Press, pp. 204–207, 1993.

Level Surface

A LEVEL SET in 3-D.

Levi-Civita Density

see PERMUTATION SYMBOL

Levi-Civita Symbol

see PERMUTATION SYMBOL

Levi-Civita Tensor

see PERMUTATION TENSOR

Leviathan Number

The number $(10^{666})!$, where 666 is the BEAST NUMBER and $n!$ denotes a FACTORIAL. The number of trailing zeros in the Leviathan number is $25 \times 10^{664} - 143$ (Pickover 1995).

see also 666, APOCALYPSE NUMBER, APOCALYPTIC NUMBER, BEAST NUMBER

References

Pickover, C. A. *Keys to Infinity.* New York: Wiley, pp. 97–102, 1995.

Levine-O'Sullivan Greedy Algorithm

For a sequence $\{\chi_i\}$, the Levine-O'Sullivan greedy algorithm is given by

$$\chi_1 = 1$$
$$\chi_i = \max_{1\le j\le i-1} (j+1)(i-\chi_j)$$

for $i > 1$.

see also GREEDY ALGORITHM, LEVINE-O'SULLIVAN SEQUENCE

References

Levine, E. and O'Sullivan, J. "An Upper Estimate for the Reciprocal Sum of a Sum-Free Sequence." *Acta Arith.* **34**, 9–24, 1977.

Levine-O'Sullivan Sequence

The sequence generated by the LEVINE-O'SULLIVAN GREEDY ALGORITHM: 1, 2, 4, 6, 9, 12, 15, 18, 21, 24, 28, 32, 36, 40, 45, 50, 55, 60, 65, ... (Sloane's A014011). The reciprocal sum of this sequence is conjectured to bound the reciprocal sum of all A-SEQUENCES.

References

Finch, S. "Favorite Mathematical Constants." `http://www.mathsoft.com/asolve/constant/erdos/erdos.html`.

Levine, E. and O'Sullivan, J. "An Upper Estimate for the Reciprocal Sum of a Sum-Free Sequence." *Acta Arith.* **34**, 9–24, 1977.

Sloane, N. J. A. Sequence A014011 in "An On-Line Version of the Encyclopedia of Integer Sequences."

Lévy Constant

Let p_n/q_n be the nth CONVERGENT of a REAL NUMBER x. Then almost all REAL NUMBERS satisfy

$$L \equiv \lim_{n\to\infty} (q_n)^{1/n} = e^{\pi^2/(12\ln 2)} = 3.27582291872\ldots.$$

see also KHINTCHINE'S CONSTANT, KHINTCHINE-LÉVY CONSTANT

References

Le Lionnais, F. *Les nombres remarquables.* Paris: Hermann, p. 51, 1983.

Lévy Distribution

$$\mathcal{F}[P_N(k)] = \exp(-N|k|^\beta),$$

where $\mathcal{F}$ is the FOURIER TRANSFORM of the probability $P_N(k)$ for N-step addition of random variables. Lévy showed that $\beta \in (0,2)$ for $P(x)$ to be NONNEGATIVE. The Lévy distribution has infinite variance and sometimes infinite mean. The case $\beta = 1$ gives a CAUCHY DISTRIBUTION, while $\beta = 2$ gives a GAUSSIAN DISTRIBUTION.

see also CAUCHY DISTRIBUTION, GAUSSIAN DISTRIBUTION

Lévy Flight

RANDOM WALK trajectories which are composed of self-similar jumps. They are described by the LÉVY DISTRIBUTION.

see also LÉVY DISTRIBUTION

References
Shlesinger, M.; Zaslavsky, G. M.; and Frisch, U. (Eds.). *Lévy Flights and Related Topics in Physics.* New York: Springer-Verlag, 1995.

Lévy Fractal

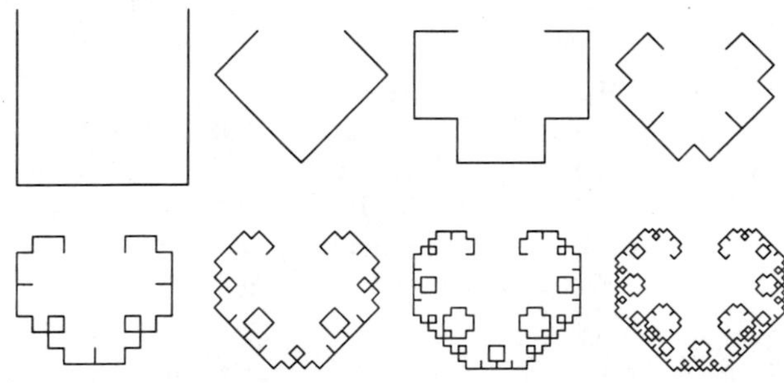

A FRACTAL curve, also called the C-CURVE (Beeler *et al.* 1972, Item 135). The base curve and motif are illustrated below.

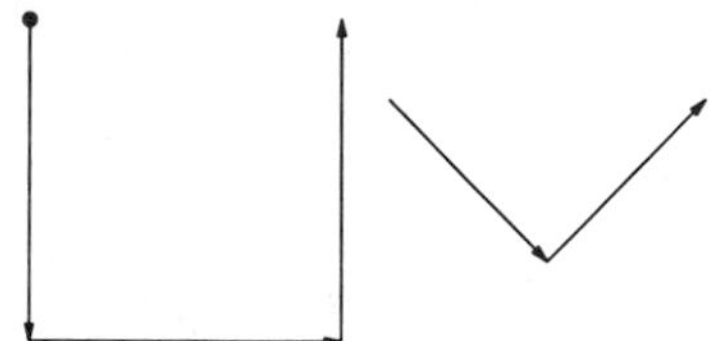

see also LÉVY TAPESTRY

References
Beeler, M.; Gosper, R. W.; and Schroeppel, R. *HAKMEM.* Cambridge, MA: MIT Artificial Intelligence Laboratory, Memo AIM-239, Feb. 1972.
Dixon, R. *Mathographics.* New York: Dover, pp. 182–183, 1991.
Lauwerier, H. *Fractals: Endlessly Repeated Geometric Figures.* Princeton, NJ: Princeton University Press, pp. 45–48, 1991.
Weisstein, E. W. "Fractals." `http://www.astro.virginia.edu/~eww6n/math/notebooks/Fractal.m`.

Lévy Function

see BROWN FUNCTION

Lévy Tapestry

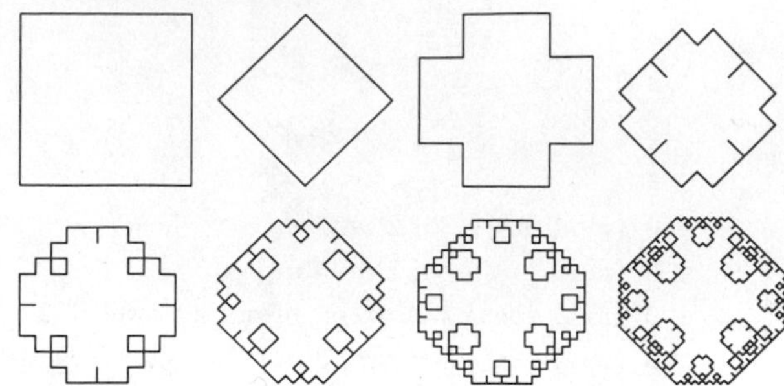

The FRACTAL curve illustrated above, with base curve and motif illustrated below.

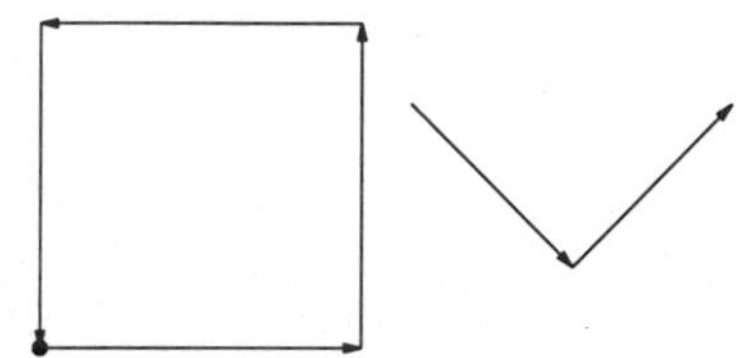

see also LÉVY FRACTAL

References
Lauwerier, H. *Fractals: Endlessly Repeated Geometric Figures.* Princeton, NJ: Princeton University Press, pp. 45–48, 1991.
Weisstein, E. W. "Fractals." `http://www.astro.virginia.edu/~eww6n/math/notebooks/Fractal.m`.

Lew k-gram

Diagrams invented by Lewis Carroll which can be used to determine the number of minimal MINIMAL COVERS of n numbers with k members.

References
Macula, A. J. "Lewis Carroll and the Enumeration of Minimal Covers." *Math. Mag.* **68**, 269–274, 1995.

Lexicographic Order

An ordering of PERMUTATIONS in which they are listed in increasing numerical order. For example, the PERMUTATIONS of $\{1,2,3\}$ in lexicographic order are 123, 132, 213, 231, 312, and 321.

see also TRANSPOSITION ORDER

References
Ruskey, F. "Information on Combinations of a Set." `http://sue.csc.uvic.ca/~cos/inf/comb/CombinationsInfo.html`.

Lexis Ratio

$$L \equiv \frac{\sigma}{\sigma_B},$$

where σ is the VARIANCE in a set of s LEXIS TRIALS and σ_B is the VARIANCE assuming BERNOULLI TRIALS.

If $L < 1$, the trials are said to be SUBNORMAL, and if $L > 1$, the trials are said to be SUPERNORMAL.

see also BERNOULLI TRIAL, LEXIS TRIALS, SUBNORMAL, SUPERNORMAL

Lexis Trials

n sets of s trials each, with the probability of success p constant in each set.

$$\operatorname{var}\left(\frac{x}{n}\right) = spq + s(s-1){\sigma_p}^2,$$

where ${\sigma_p}^2$ is the VARIANCE of p_i.

see also BERNOULLI TRIAL, LEXIS RATIO

Lg

The LOGARITHM to BASE 2 is denoted lg, i.e.,

$$\lg x \equiv \log_2 x.$$

see also BASE (LOGARITHM), E, LN, LOGARITHM, NAPIERIAN LOGARITHM, NATURAL LOGARITHM

Liar's Paradox

see EPIMENIDES PARADOX

Lichnerowicz Conditions

Second and higher derivatives of the METRIC TENSOR g_{ab} need not be continuous across a surface of discontinuity, but g_{ab} and $g_{ab,c}$ must be continuous across it.

Lichnerowicz Formula

$$D^*D\psi = \nabla^*\nabla\psi + \tfrac{1}{4}R\psi - \tfrac{1}{2}F_L^+(\psi),$$

where D is the Dirac operator $D : \Gamma(W^+) \to \Gamma(W^-)$, ∇ is the COVARIANT DERIVATIVE on SPINORS, R is the CURVATURE SCALAR, and F_L^+ is the self-dual part of the curvature of L.

see also LICHNEROWICZ-WEITZENBOCK FORMULA

References

Donaldson, S. K. "The Seiberg-Witten Equations and 4-Manifold Topology." *Bull. Amer. Math. Soc.* **33**, 45–70, 1996.

Lichnerowicz-Weitzenbock Formula

$$D^*D\psi = \nabla^*\nabla\psi + \tfrac{1}{4}R\psi,$$

where D is the Dirac operator $D : \Gamma(S^+) \to \Gamma(S^-)$, ∇ is the COVARIANT DERIVATIVE on SPINORS, and R is the CURVATURE SCALAR.

see also LICHNEROWICZ FORMULA

References

Donaldson, S. K. "The Seiberg-Witten Equations and 4-Manifold Topology." *Bull. Amer. Math. Soc.* **33**, 45–70, 1996.

Lichtenfels Surface

A MINIMAL SURFACE given by the parametric equation

$$x = \Re\left[\sqrt{2}\cos(\tfrac{1}{3}\zeta)\sqrt{\cos(\tfrac{2}{3}\zeta)}\right]$$

$$y = \Re\left[-\sqrt{2}\cos(\tfrac{1}{3}\zeta)\sqrt{\cos(\tfrac{2}{3}\zeta)}\right]$$

$$z = \Re\left[-\tfrac{1}{3}\sqrt{2}i\int_0^t \frac{d\zeta}{\sqrt{\cos(\tfrac{2}{3}\zeta)}}\right].$$

References

do Carmo, M. P. "The Helicoid." §3.5F in *Mathematical Models from the Collections of Universities and Museums* (Ed. G. Fischer). Braunschweig, Germany: Vieweg, p. 47, 1986.

Lichtenfels, O. von. "Notiz über eine transcendente Minimalfläche." *Sitzungsber. Kaiserl. Akad. Wiss. Wien* **94**, 41–54, 1889.

Lie Algebra

A NONASSOCIATIVE ALGEBRA obeyed by objects such as the LIE BRACKET and POISSON BRACKET. Elements f, g, and h of a Lie algebra satisfy

$$[f,g] = -[g,f], \quad (1)$$

$$[f+g,h] = [f,h] + [g,h], \quad (2)$$

and

$$[f,[g,h]] + [g,[h,f]] + [h,[f,g]] = 0 \quad (3)$$

(the JACOBI IDENTITY), and are *not* ASSOCIATIVE. The binary operation of a Lie algebra is the bracket

$$[fg,h] = f[g,h] + g[f,h]. \quad (4)$$

see also JACOBI IDENTITIES, LIE ALGEBROID, LIE BRACKET, IWASAWA'S THEOREM, POISSON BRACKET

References

Jacobson, N. *Lie Algebras.* New York: Dover, 1979.

Lie Algebroid

The infinitesimal algebraic object associated with a LIE GROUPOID. A Lie algebroid over a MANIFOLD B is a VECTOR BUNDLE A over B with a LIE ALGEBRA structure [,] (LIE BRACKET) on its SPACE of smooth sections together with its ANCHOR ρ.

see also LIE ALGEBRA

References

Weinstein, A. "Groupoids: Unifying Internal and External Symmetry." *Not. Amer. Math. Soc.* **43**, 744–752, 1996.

Lie Bracket

The commutation operation

$$[a,b] = ab - ba$$

corresponding to the LIE PRODUCT.

see also LAGRANGE BRACKET, POISSON BRACKET

Lie Commutator

see LIE PRODUCT

Lie Derivative

$$\mathcal{L}_x T^{ab} \equiv \lim_{\delta u \to 0} \frac{T^{ab}(x') - T'^{ab}(x)}{\delta u}.$$

Lie Group

A continuous GROUP with an infinite number of elements such that the parameters of a product element are ANALYTIC FUNCTIONS. Lie groups are also C^∞ MANIFOLDS with the restriction that the group operation maps a C^∞ map of the MANIFOLD into itself. Examples include O_3, $SU(n)$, and the LORENTZ GROUP.

see also COMPACT GROUP, LIE ALGEBRA, LIE GROUPOID, LIE-TYPE GROUP, NIL GEOMETRY, SOL GEOMETRY

References

Arfken, G. "Infinite Groups, Lie Groups." *Mathematical Methods for Physicists, 3rd ed.* Orlando, FL: Academic Press, p. 251–252, 1985.

Chevalley, C. *Theory of Lie Groups.* Princeton, NJ: Princeton University Press, 1946.

Knapp, A. W. *Lie Groups Beyond an Introduction.* Boston, MA: Birkhäuser, 1996.

Lipkin, H. J. *Lie Groups for Pedestrians, 2nd ed.* Amsterdam, Netherlands: North-Holland, 1966.

Lie Groupoid

A GROUPOID G over B for which G and B are differentiable manifolds and α, β, and multiplication are differentiable maps. Furthermore, the derivatives of α and β are required to have maximal RANK everywhere. Here, α and β are maps from G onto $\mathbb{R}^2$ with $\alpha : (x, \gamma, y) \mapsto x$ and $\beta : (x, \gamma, y) \mapsto y$.

see also LIE ALGEBROID, NILPOTENT LIE GROUP, SEMISIMPLE LIE GROUP, SOLVABLE LIE GROUP

References

Weinstein, A. "Groupoids: Unifying Internal and External Symmetry." *Not. Amer. Math. Soc.* **43**, 744–752, 1996.

Lie Product

The multiplication operation corresponding to the LIE BRACKET.

Lie-Type Group

A finite analog of LIE GROUPS. The Lie-type groups include the CHEVALLEY GROUPS [$PSL(n,q)$, $PSU(n,q)$, $PSp(2n,q)$, $P\Omega^\epsilon(n,q)$], TWISTED CHEVALLEY GROUPS, and the TITS GROUP.

see also CHEVALLEY GROUPS, FINITE GROUP, LIE GROUP, LINEAR GROUP, ORTHOGONAL GROUP, SIMPLE GROUP, SYMPLECTIC GROUP, TITS GROUP, TWISTED CHEVALLEY GROUPS, UNITARY GROUP

References

Wilson, R. A. "ATLAS of Finite Group Representation." `http://for.mat.bham.ac.uk/atlas#lie`.

Liebmann's Theorem

A SPHERE is RIGID.

see also RIGID

References

Gray, A. *Modern Differential Geometry of Curves and Surfaces.* Boca Raton, FL: CRC Press, p. 377, 1993.

O'Neill, B. *Elementary Differential Geometry, 2nd ed.* New York: Academic Press, p. 262, 1997.

Life

The most well-known CELLULAR AUTOMATON, invented by John Conway and popularized in Martin Gardner's *Scientific American* column starting in October 1970. The game was originally played (i.e., successive generations were produced) by hand with counters, but implementation on a computer greatly increased the ease of exploring patterns.

The Life AUTOMATON is run by placing a number of filled cells on a 2-D grid. Each generation then switches cells on or off depending on the state of the cells that surround it. The rules are defined as follows. All eight of the cells surrounding the current one are checked to see if they are on or not. Any cells that are on are counted, and this count is then used to determine what will happen to the current cell.

1. Death: if the count is less than 2 or greater than 3, the current cell is switched off.
2. Survival: if (a) the count is exactly 2, or (b) the count is exactly 3 and the current cell is on, the current cell is left unchanged.
3. Birth: if the current cell is off and the count is exactly 3, the current cell is switched on.

Hensel gives a Java applet (`http://www.mindspring.com/~alanh/life/`) implementing the Game of Life on his web page.

A pattern which does not change from one generation to the next is known as a Still Life, and is said to have period 1. Conway originally believed that no pattern could produce an infinite number of cells, and offered a $50 prize to anyone who could find a counterexample before the end of 1970 (Gardner 1983, p. 216). Many counterexamples were subsequently found, including Guns and Puffer Trains.

A Life pattern which has no Father Pattern is known as a Garden of Eden (for obvious biblical reasons). The first such pattern was not found until 1971, and at least 3 are now known. It is not, however, known if a pattern exists which has a Father Pattern, but no Grandfather Pattern (Gardner 1983, p. 249).

Rather surprisingly, Gosper and J. H. Conway independently showed that Life can be used to generate a UNIVERSAL TURING MACHINE (Berlekamp *et al.* 1982, Gardner 1983, pp. 250–253).

Similar CELLULAR AUTOMATON games with different rules are HASHLIFE, HEXLIFE, and HIGHLIFE.

see also CELLULAR AUTOMATON, HASHLIFE, HEXLIFE, HIGHLIFE

References

"Alife online." `http://alife.santafe.edu/alife/topics/cas/ca-faq/lifefaq/lifefaq.html`.

Berlekamp, E. R.; Conway, J. H.; and Guy, R. K. "What Is Life." Ch. 25 in *Winning Ways, For Your Mathematical Plays, Vol. 2: Games in Particular.* London: Academic Press, 1982.

Callahan, P. "Patterns, Programs, and Links for Conway's Game of Life." `http://www.cs.jhu.edu/~callahan/lifepage.html`.

Flammenkamp, A. "Game of Life." `http://www.minet.uni-jena.de/~achim/gol.html`.

"The Game of Life." *Math Horizons.* p. 9, Spring 1994.

Gardner, M. "The Game of Life, Parts I–III." Chs. 20–22 in *Wheels, Life, and other Mathematical Amusements.* New York: W. H. Freeman, 1983.

Hensel, A. "A Brief Illustrated Glossary of Terms in Conway's Game of Life." `http://www.cs.jhu.edu/~callahan/glossary.html`.

Hensel, A. "PC Life Distribution." `http://www.mindspring.com/~alanh/lifep.zip`.

Hensel, A. "Conway's Game of Life." Includes a Java applet for the Game of Life. `http://www.mindspring.com/~alanh/life/`.

Koenig, H. "Game of Life Information." `http://www.halcyon.com/hkoenig/LifeInfo/LifeInfo.html`.

McIntosh, H. V. "A Zoo of Life Forms." `http://www.cs.cinvestav.mx/mcintosh/life.html`.

Poundstone, W. *The Recursive Universe: Cosmic Complexity and the Limits of Scientific Knowledge.* New York: Morrow, 1985.

Toffoli, T. and Margolus, N. *Cellular Automata Machines: A New Environment for Modeling.* Cambridge, MA: MIT Press, 1987.

Wainwright, R. T. "LifeLine." `http://members.aol.com/lifeline/life/lifepage.htm`.

Wainwright, R. T. *LifeLine: A Quarterly Newsletter for Enthusiasts of John Conway's Game of Life.* Nos. 1–11, 1971–1973.

Life Expectancy

An l_x table is a tabulation of numbers which is used to calculate life expectancies.

x	n_x	d_x	l_x	q_x	L_x	T_x	e_x
0	1000	200	1.00	0.20	0.90	2.70	2.70
1	800	100	0.80	0.12	0.75	1.80	2.25
2	700	200	0.70	0.29	0.60	1.05	1.50
3	500	300	0.50	0.60	0.35	0.45	0.90
4	200	200	0.20	1.00	0.10	0.10	0.50
5	0	0	0.00	—	0.00	0.00	—
$\sum$		1000	2.70				

x : Age category ($x = 0, 1, \ldots, k$). These values can be in any convenient units, but must be chosen so that no observed lifespan extends past category $k-1$.

n_x : Census size, defined as the number of individuals in the study population who survive to the beginning of age category x. Therefore, $n_0 = N$ (the total population size) and $n_k = 0$.

d_x : $= n_x - n_{x+1}$; $\sum_{i=0}^k d_i = n_0$. Crude death rate, which measures the number of individuals who die within age category x.

l_x : $= n_x/n_0$. Survivorship, which measures the *proportion* of individuals who survive to the beginning of age category x.

q_x : $= d_x/n_x$; $q_{k-1} = 1$. Proportional death rate, or "risk," which measures the proportion of individuals surviving to the beginning of age category x who die within that category.

L_x : $= (l_x + l_{x+1})/2$. Midpoint survivorship, which measures the proportion of individuals surviving to the *midpoint* of age category x. Note that the simple averaging formula must be replaced by a more complicated expression if survivorship is nonlinear within age categories. The sum $\sum_{i=0}^k L_x$ gives the total number of age categories lived by the entire study population.

T_x : $= T_{x-1} - L_{x-1}$; $T_0 = \sum_{i=0}^k L_x$. Measures the total number of age categories left to be lived by all individuals who survive to the beginning of age category x.

e_x : $= T_x/l_x$; $e_{k-1} = 1/2$. Life expectancy, which is the mean number of age categories remaining until death for individuals surviving to the beginning of age category x.

For all x, $e_{x+1} + 1 > e_x$. This means that the total expected lifespan increases monotonically. For instance, in the table above, the one-year-olds have an average age at death of $2.25 + 1 = 3.25$, compared to 2.70 for newborns. In effect, the age of death of older individuals is a distribution conditioned on the fact that they have survived to their present age.

It is common to study survivorship as a semilog plot of l_x vs. x, known as a SURVIVORSHIP CURVE. A so-called $l_x m_x$ table can be used to calculate the mean generation time of a population. Two $l_x m_x$ tables are illustrated below.

Population 1

x	l_x	m_x	$l_x m_x$	$x l_x m_x$
0	1.00	0.00	0.00	0.00
1	0.70	0.50	0.35	0.35
2	0.50	1.50	0.75	1.50
3	0.20	0.00	0.00	0.00
4	0.00	0.00	0.00	0.00
			$R_0 = 1.10$	$\sum = 1.85$

$$T = \frac{\sum x l_x m_x}{\sum l_x m_x} = \frac{1.85}{1.10} = 1.68$$

$$r = \frac{\ln R_0}{T} = \frac{\ln 1.10}{1.68} = 0.057.$$

Population 2

x	l_x	m_x	$l_x m_x$	$x l_x m_x$
0	1.00	0.00	0.00	0.00
1	0.70	0.00	0.00	0.00
2	0.50	2.00	1.00	2.00
3	0.20	0.50	0.10	0.30
4	0.00	0.00	0.00	0.00
			$R_0 = 1.10$	$\sum = 2.30$

$$T = \frac{\sum x l_x m_x}{\sum l_x m_x} = \frac{2.30}{1.10} = 2.09$$

$$r = \frac{\ln R_0}{T} = \frac{\ln 1.10}{2.09} = 0.046.$$

x : Age category ($x = 0, 1, \ldots, k$). These values can be in any convenient units, but must be chosen so that no observed lifespan extends past category $k-1$ (as in an l_x table).

l_x : $= n_x/n_0$. Survivorship, which measures the *proportion* of individuals who survive to the beginning of age category x (as in an l_x table).

m_x : The average number of offspring produced by an individual in age category x *while in that age category.* $\sum_{i=0}^{k} m_x$ therefore represents the average lifetime number of offspring produced by an individual of maximum lifespan.

$l_x m_x$: The average number of offspring produced by an individual within age category x weighted by the probability of surviving to the beginning of that age category. $\sum_{i=0}^{k} l_x m_x$ therefore represents the average lifetime number of offspring produced by a member of the study population. It is called the net reproductive rate per generation and is often denoted R_0.

$x l_x m_x$: A column weighting the offspring counted in the previous column by their parents' age when they were born. Therefore, the ratio $T = \sum(x l_x m_x)/\sum(l_x m_x)$ is the mean generation time of the population.

The MALTHUSIAN PARAMETER r measures the reproductive rate per unit time and can be calculated as $r = (\ln R_0)/T$. For an exponentially increasing population, the population size $N(t)$ at time t is then given by

$$N(t) = N_0 e^{rt}.$$

In the above two tables, the populations have identical reproductive rates of $R_0 = 1.10$. However, the shift toward later reproduction in population 2 increases the generation time, thus slowing the rate of POPULATION GROWTH. Often, a slight delay of reproduction decreases POPULATION GROWTH more strongly than does even a fairly large reduction in reproductive rate.

see also GOMPERTZ CURVE, LOGISTIC GROWTH CURVE, MAKEHAM CURVE, MALTHUSIAN PARAMETER, POPULATION GROWTH, SURVIVORSHIP CURVE

Lift

Given a MAP f from a SPACE X to a SPACE Y and another MAP g from a SPACE Z to a SPACE Y, a lift is a MAP h from X to Z such that $gh = f$. In other words, a lift of f is a MAP h such that the diagram (shown below) commutes.

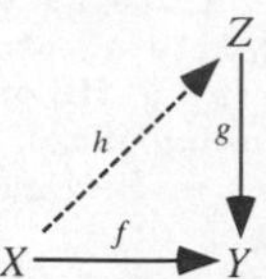

If f is the identity from Y to Y, a MANIFOLD, and if g is the bundle projection from the TANGENT BUNDLE to Y, the lifts are precisely VECTOR FIELDS. If g is a bundle projection from any FIBER BUNDLE to Y, then lifts are precisely sections. If f is the identity from Y to Y, a MANIFOLD, and g a projection from the orientation double cover of Y, then lifts exist IFF Y is an orientable MANIFOLD.

If f is a MAP from a CIRCLE to Y, an n-MANIFOLD, and g the bundle projection from the FIBER BUNDLE of alternating n-FORMS on Y, then lifts always exist IFF Y is orientable. If f is a MAP from a region in the COMPLEX PLANE to the COMPLEX PLANE (complex analytic), and if g is the exponential MAP, lifts of f are precisely LOGARITHMS of f.

see also LIFTING PROBLEM

Lifting Problem

Given a MAP f from a SPACE X to a SPACE Y and another MAP g from a SPACE Z to a SPACE Y, does there exist a MAP h from X to Z such that $gh = f$? If such a map h exists, then h is called a LIFT of f.

see also EXTENSION PROBLEM, LIFT

Ligancy

see KISSING NUMBER

Likelihood

The hypothetical PROBABILITY that an event which has already occurred would yield a specific outcome. The concept differs from that of a probability in that a probability refers to the occurrence of future events, while a likelihood refers to past events with known outcomes.

see also LIKELIHOOD RATIO, MAXIMUM LIKELIHOOD, NEGATIVE LIKELIHOOD RATIO, PROBABILITY

Likelihood Ratio

A quantity used to test NESTED HYPOTHESES. Let H' be a NESTED HYPOTHESIS with n' DEGREES OF FREEDOM within H (which has n DEGREES OF FREEDOM), then calculate the MAXIMUM LIKELIHOOD of a given outcome, first given H', then given H. Then

$$\mathrm{LR} = \frac{[\text{likelihood H'}]}{[\text{likelihood H}]}.$$

Comparison of this ratio to the critical value of the CHI-SQUARED DISTRIBUTION with $n - n'$ DEGREES OF FREEDOM then gives the SIGNIFICANCE of the increase in LIKELIHOOD.

The term likelihood ratio is also used (especially in medicine) to test nonnested complementary hypotheses as follows,

$$\mathrm{LR} = \frac{[\text{true positive rate}]}{[\text{false positive rate}]} = \frac{[\text{sensitivity}]}{1 - [\text{specificity}]}.$$

see also NEGATIVE LIKELIHOOD RATIO, SENSITIVITY, SPECIFICITY

Limaçon

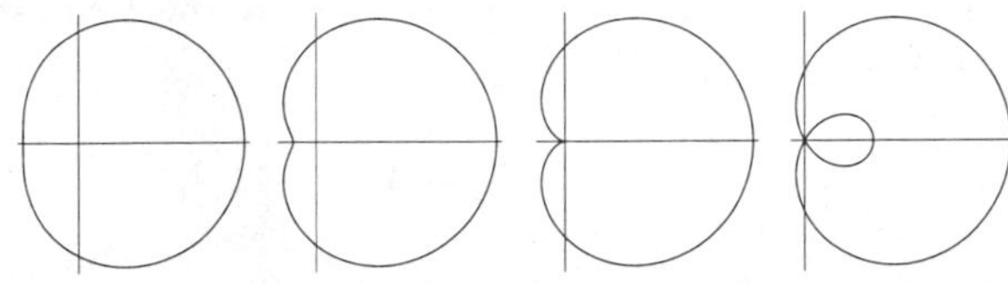

The limaçon is a polar curve of the form

$$r = b + a\cos\theta$$

also called the LIMAÇON OF PASCAL. It was first investigated by Dürer, who gave a method for drawing it in *Underweysung der Messung* (1525). It was rediscovered by Étienne Pascal, father of Blaise Pascal, and named by Gilles-Personne Roberval in 1650 (MacTutor Archive). The word "limaçon" comes from the Latin *limax*, meaning "snail."

If $b \geq 2a$, we have a convex limaçon. If $2a > b > a$, we have a dimpled limaçon. If $b = a$, the limaçon degenerates to a CARDIOID. If $b < a$, we have limaçon with an inner loop. If $b = a/2$, it is a TRISECTRIX (but *not* the MACLAURIN TRISECTRIX) with inner loop of AREA

$$A_{\text{inner loop}} = \tfrac{1}{4}a^2\left(\pi - 3\sqrt{\frac{3}{2}}\right),$$

and AREA between the loops of

$$A_{\text{between loops}} = \tfrac{1}{4}a^2(\pi + 3\sqrt{3})$$

(MacTutor Archive). The limaçon is an ANALLAGMATIC CURVE, and is also the CATACAUSTIC of a CIRCLE when the RADIANT POINT is a finite (NONZERO) distance from the CIRCUMFERENCE, as shown by Thomas de St. Laurent in 1826 (MacTutor Archive).

see also CARDIOID

References

Lawrence, J. D. *A Catalog of Special Plane Curves.* New York: Dover, pp. 113–117, 1972.

Lee, X. "Limacon of Pascal." http://www.best.com/~xah/SpecialPlaneCurves_dir/LimaconOfPascal_dir/limaconOfPascal.html.

Lee, X. "Limacon Graphics Gallery." http://www.best.com/~xah/SpecialPlaneCurves_dir/LimaconGGallery_dir/limaconGGallery.html.

Lockwood, E. H. "The Limaçon." Ch. 5 in *A Book of Curves.* Cambridge, England: Cambridge University Press, pp. 44–51, 1967.

MacTutor History of Mathematics Archive. "Limacon of Pascal." http://www-groups.dcs.st-and.ac.uk/~history/Curves/Limacon.html.

Yates, R. C. "Limacon of Pascal." *A Handbook on Curves and Their Properties.* Ann Arbor, MI: J. W. Edwards, pp. 148–151, 1952.

Limaçon Evolute

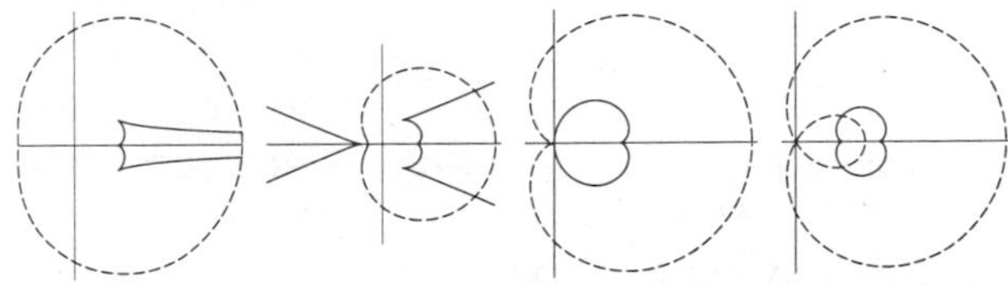

The CATACAUSTIC of a CIRCLE for a RADIANT POINT is the limaçon evolute. It has parametric equations

$$x = \frac{a[4a^2 + 4b^2 + 9ab\cos t - ab\cos(3t)]}{4(2a^2 + b^2 + 3ab\cos t)}$$

$$y = \frac{a^2 b\sin^3 t}{2a^2 + b^2 + 3ab\cos t}.$$

Limaçon of Pascal

see LIMAÇON

Limit

A function $f(z)$ is said to have a limit $\lim_{z\to a} f(z) = c$ if, for all $\epsilon > 0$, there exists a $\delta > 0$ such that $|f(z) - c| < \epsilon$ whenever $0 < |z - a| < \delta$.

A LOWER LIMIT

$$\text{lower} \lim_{n\to\infty} S_n = \underline{\lim}_{n\to\infty} S_n = h$$

is said to exist if, for every $\epsilon > 0$, $|S_n - h| < \epsilon$ for infinitely many values of n and if no number less than h has this property.

An UPPER LIMIT

$$\text{upper} \lim_{n\to\infty} S_n = \overline{\lim}_{n\to\infty} S_n = k$$

is said to exist if, for every $\epsilon > 0$, $|S_n - k| < \epsilon$ for infinitely many values of n and if no number larger than k has this property.

Indeterminate limit forms of types ∞/∞ and $0/0$ can be computed with L'HOSPITAL'S RULE. Types $0\cdot\infty$ can be converted to the form $0i/0$ by writing

$$f(x)g(x) = \frac{f(x)}{1/g(x)}.$$

Types 0^0, ∞^0, and 1^∞ are treated by introducing a dependent variable $y = f(x)g(x)$, then calculating $\lim \ln y$. The original limit then equals $e^{\lim \ln y}$.

see also CENTRAL LIMIT THEOREM, CONTINUOUS, DISCONTINUITY, L'HOSPITAL'S RULE, LOWER LIMIT, UPPER LIMIT

References
Courant, R. and Robbins, H. "Limits. Infinite Geometrical Series." §2.2.3 in *What is Mathematics?: An Elementary Approach to Ideas and Methods, 2nd ed.* Oxford, England: Oxford University Press, pp. 63–66, 1996.

Limit Comparison Test

Let $\sum a_k$ and $\sum b_k$ be two SERIES with POSITIVE terms and suppose

$$\lim_{k\to\infty} \frac{a_k}{b_k} = \rho.$$

If ρ is finite and $\rho > 0$, then the two SERIES both CONVERGE or DIVERGE.

see also CONVERGENCE TESTS

Limit Cycle

An attracting set to which orbits or trajectories converge and upon which trajectories are periodic.

see also HOPF BIFURCATION

Limit Point

A number x such that for all $\epsilon > 0$, there exists a member of the SET y different from x such that $|y - x| < \epsilon$. The topological definition of limit point P of A is that P is a point such that every OPEN SET around it intersects A.

see also CLOSED SET, OPEN SET

References
Lauwerier, H. *Fractals: Endlessly Repeated Geometric Figures.* Princeton, NJ: Princeton University Press, pp. 25–26, 1991.

Lin's Method

An ALGORITHM for finding ROOTS for QUARTIC EQUATIONS with COMPLEX ROOTS.

References
Acton, F. S. *Numerical Methods That Work, 2nd printing.* Washington, DC: Math. Assoc. Amer., pp. 198–199, 1990.

Lindeberg Condition

A SUFFICIENT condition on the LINDEBERG-FELLER CENTRAL LIMIT THEOREM. Given random variates X_1, X_2, ..., let $\langle X_i\rangle = 0$, the VARIANCE ${\sigma_i}^2$ of X_i be finite, and VARIANCE of the distribution consisting of a sum of X_is

$$S_n \equiv X_1 + X_2 + \ldots + X_n \tag{1}$$

be

$${s_n}^2 \equiv \sum_{i=1}^n {\sigma_i}^2. \tag{2}$$

Let

$$\Lambda_n(\epsilon) \equiv \sum_{k=1}^n \left\langle \left(\frac{X_k}{s_n}\right)^2 \middle| \frac{|X_k|}{s_n} \geq \epsilon \right\rangle, \tag{3}$$

then the Lindeberg condition is

$$\lim_{n\to\infty} \Lambda_n(\epsilon) = 0 \tag{4}$$

for all $\epsilon > 0$.

see also FELLER-LÉVY CONDITION

References
Zabell, S. L. "Alan Turing and the Central Limit Theorem." *Amer. Math. Monthly* **102**, 483–494, 1995.

Lindeberg-Feller Central Limit Theorem

If the random variates X_1, X_2, ... satisfy the LINDEBERG CONDITION, then for all $a < b$,

$$\lim_{n\to\infty} P\left(a < \frac{S_n}{s_n} < b\right) = \Phi(b) - \Phi(a),$$

where Φ is the NORMAL DISTRIBUTION FUNCTION.

see also CENTRAL LIMIT THEOREM, FELLER-LÉVY CONDITION, NORMAL DISTRIBUTION FUNCTION

References
Zabell, S. L. "Alan Turing and the Central Limit Theorem." *Amer. Math. Monthly* **102**, 483–494, 1995.

Lindelof's Theorem

The SURFACE OF REVOLUTION generated by the external CATENARY between a fixed point a and its conjugate on the ENVELOPE of the CATENARY through the fixed point is equal in AREA to the surface of revolution generated by its two Lindelof TANGENTS, which cross the axis of rotation at the point a and are calculable from the position of the points and CATENARY.

see also CATENARY, ENVELOPE, SURFACE OF REVOLUTION

Lindemann-Weierstraß Theorem

If α_1, ..., α_n are linearly independent over $\mathbb{Q}$, then e^{α_1}, ..., e^{α_n} are algebraically independent over $\mathbb{Q}$.

see also HERMITE-LINDEMANN THEOREM

Lindenmayer System

A STRING REWRITING system which can be used to generate FRACTALS with DIMENSION between 1 and 2. The term L-SYSTEM is often used as an abbreviation.

see also ARROWHEAD CURVE, DRAGON CURVE EXTERIOR SNOWFLAKE, FRACTAL, HILBERT CURVE, KOCH SNOWFLAKE, PEANO CURVE, PEANO-GOSPER CURVE, SIERPIŃSKI CURVE, STRING REWRITING

References
Dickau, R. M. "Two-dimensional L-systems." `http://forum.swarthmore.edu/advanced/robertd/lsys2d.html`.
Prusinkiewicz, P. and Hanan, J. *Lindenmayer Systems, Fractal, and Plants.* New York: Springer-Verlag, 1989.

Prusinkiewicz, P. and Lindenmayer, A. *The Algorithmic Beauty of Plants.* New York: Springer-Verlag, 1990.
Stevens, R. T. *Fractal Programming in C.* New York: Holt, 1989.
Wagon, S. "Recursion via String Rewriting." §6.2 in *Mathematica in Action.* New York: W. H. Freeman, pp. 190–196, 1991.

Line

Euclid defined a line as a "breadthless length," and a straight line as a line which "lies evenly with the points on itself" (Kline 1956, Dunham 1990). Lines are intrinsically 1-dimensional objects, but may be embedded in higher dimensional SPACES. An infinite line passing through points A and B is denoted $\overleftrightarrow{AB}$. A LINE SEGMENT terminating at these points is denoted $\overline{AB}$. A line is sometimes called a STRAIGHT LINE or, more archaically, a RIGHT LINE (Casey 1893), to emphasize that it has no curves anywhere along its length.

Consider first lines in a 2-D PLANE. The line with x-INTERCEPT a and y-INTERCEPT b is given by the *intercept form*

$$\frac{x}{a}+\frac{y}{b}=1. \tag{1}$$

The line through (x_1, y_1) with SLOPE m is given by the *point-slope form*

$$y-y_1=m(x-x_1). \tag{2}$$

The line with y-intercept b and slope m is given by the *slope-intercept form*

$$y=mx+b. \tag{3}$$

The line through (x_1, y_1) and (x_2, y_2) is given by the *two point form*

$$y-y_1=\frac{y_2-y_1}{x_2-x_1}(x-x_1). \tag{4}$$

Other forms are

$$a(x-x_1)+b(y-y_1)=0 \tag{5}$$

$$ax+by+c=0 \tag{6}$$

$$\begin{vmatrix} x & y & 1 \\ x_1 & y_1 & 1 \\ x_2 & y_2 & 1 \end{vmatrix}=0. \tag{7}$$

A line in 2-D can also be represented as a VECTOR. The VECTOR along the line

$$ax+by=0 \tag{8}$$

is given by

$$t\begin{bmatrix} -b \\ a \end{bmatrix}, \tag{9}$$

where $t \in \mathbb{R}$. Similarly, VECTORS of the form

$$t\begin{bmatrix} a \\ b \end{bmatrix} \tag{10}$$

are PERPENDICULAR to the line. Three points lie on a line if

$$\begin{vmatrix} x_1 & y_1 & 1 \\ x_2 & y_2 & 1 \\ x_3 & y_3 & 1 \end{vmatrix}=0. \tag{11}$$

The ANGLE between lines

$$A_1x+B_1y+C_1=0 \tag{12}$$

$$A_2x+B_2y+C_2=0 \tag{13}$$

is

$$\tan\theta=\frac{A_1B_2-A_2B_1}{A_1A_2+B_1B_2}. \tag{14}$$

The line joining points with TRILINEAR COORDINATES $\alpha_1 : \beta_1 : \gamma_1$ and $\alpha_2 : \beta_2 : \gamma_2$ is the set of point $\alpha : \beta : \gamma$ satisfying

$$\begin{vmatrix} \alpha & \beta & \gamma \\ \alpha_1 & \beta_1 & \gamma_1 \\ \alpha_2 & \beta_2 & \gamma_2 \end{vmatrix}=0 \tag{15}$$

$$(\beta_1\gamma_2-\gamma_1\beta_2)\alpha+(\gamma_1\alpha_2-\alpha_1\gamma_2)\beta+(\alpha_1\beta_2-\beta_1\alpha_2)\gamma=0. \tag{16}$$

Three lines CONCUR if their TRILINEAR COORDINATES satisfy

$$l_1\alpha+m_1\beta+n_1\gamma=0 \tag{17}$$

$$l_2\alpha+m_2\beta+n_2\gamma=0 \tag{18}$$

$$l_3\alpha+m_3\beta+n_3\gamma=0, \tag{19}$$

in which case the point is

$$m_2n_3-n_2m_3 : n_2l_3-l_2n_3 : l_2m_3-m_2l_3, \tag{20}$$

or if the COEFFICIENTS of the lines

$$A_1x+B_1y+C_1=0 \tag{21}$$

$$A_2x+B_2y+C_2=0 \tag{22}$$

$$A_3x+B_3y+C_3=0 \tag{23}$$

satisfy

$$\begin{vmatrix} A_1 & B_1 & C_1 \\ A_2 & B_2 & C_2 \\ A_3 & B_3 & C_3 \end{vmatrix}=0. \tag{24}$$

Two lines CONCUR if their TRILINEAR COORDINATES satisfy

$$\begin{vmatrix} l_1 & m_1 & n_1 \\ l_2 & m_2 & n_2 \\ l_3 & m_3 & n_3 \end{vmatrix}=0. \tag{25}$$

The line through P_1 is the direction (a_1, b_1, c_1) and the line through P_2 in direction (a_2, b_2, c_2) intersect IFF

$$\begin{vmatrix} x_2 - x_1 & y_2 - y_1 & z_2 - z_1 \\ a_1 & b_1 & c_1 \\ a_2 & b_2 & c_2 \end{vmatrix} = 0. \qquad (26)$$

The line through a point $\alpha' : \beta' : \gamma'$ PARALLEL to

$$l\alpha + m\beta + n\gamma = 0 \qquad (27)$$

is

$$\begin{vmatrix} \alpha & \beta & \gamma \\ \alpha' & \beta' & \gamma' \\ bn - cm & cl - an & am - bl \end{vmatrix} = 0. \qquad (28)$$

The lines

$$l\alpha + m\beta + n\gamma = 0 \qquad (29)$$

$$l'\alpha + m'\beta + n'\gamma = 0 \qquad (30)$$

are PARALLEL if

$$a(mn' - nm') + b(nl' - ln') + c(lm' - ml') = 0 \qquad (31)$$

for all (a, b, c), and PERPENDICULAR if

$$2abc(ll' + mm' + nn') - (mn' + m'm)\cos A - (nl' + n'l)\cos B - (lm' + l'm)\cos C = 0 \qquad (32)$$

for all (a, b, c) (Sommerville 1924). The line through a point $\alpha' : \beta' : \gamma'$ PERPENDICULAR to (32) is given by

$$\begin{vmatrix} \alpha & \beta & \gamma \\ \alpha' & \beta' & \gamma' \\ l - m\cos C & m - n\cos A & n - l\cos B \\ -n\cos B & -l\cos C & -m\cos A \end{vmatrix} = 0. \qquad (33)$$

In 3-D SPACE, the line passing through the point (x_0, y_0, z_0) and PARALLEL to the NONZERO VECTOR

$$\mathbf{v} = \begin{bmatrix} a \\ b \\ c \end{bmatrix} \qquad (34)$$

has parametric equations

$$\begin{cases} x = x_0 + at \\ y = y_0 + bt \\ z = z_0 + ct. \end{cases} \qquad (35)$$

see also ASYMPTOTE, BROCARD LINE, COLLINEAR, CONCUR, CRITICAL LINE, DESARGUES' THEOREM, ERDŐS-ANNING THEOREM, LINE SEGMENT, ORDINARY LINE, PENCIL, POINT, POINT-LINE DISTANCE—2-D, POINT-LINE DISTANCE—3-D, PLANE, RANGE (LINE SEGMENT), RAY, SOLOMON'S SEAL LINES, STEINER SET, STEINER'S THEOREM, SYLVESTER'S LINE PROBLEM

References

Casey, J. "The Right Line." Ch. 2 in *A Treatise on the Analytical Geometry of the Point, Line, Circle, and Conic Sections, Containing an Account of Its Most Recent Extensions, with Numerous Examples, 2nd ed., rev. enl.* Dublin: Hodges, Figgis, & Co., pp. 30–95, 1893.

Dunham, W. *Journey Through Genius: The Great Theorems of Mathematics.* New York: Wiley, p. 32, 1990.

Kline, M. "The Straight Line." *Sci. Amer.* **156**, 105–114, Mar. 1956.

MacTutor History of Mathematics Archive. "Straight Line." `http://www-groups.dcs.st-and.ac.uk/~history/Curves/Straight.html`.

Sommerville, D. M. Y. *Analytical Conics.* London: G. Bell, p. 186, 1924.

Spanier, J. and Oldham, K. B. "The Linear Function $bx + c$ and Its Reciprocal." Ch. 7 in *An Atlas of Functions.* Washington, DC: Hemisphere, pp. 53–62, 1987.

Line Bisector

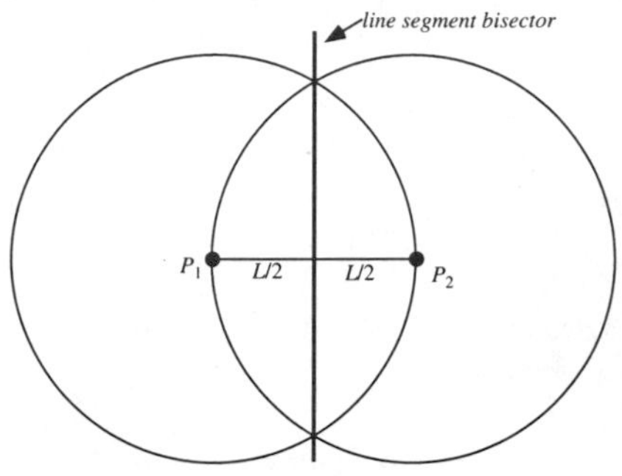

The line bisecting a given LINE SEGMENT P_1P_2 can be constructed geometrically, as illustrated above.

References

Courant, R. and Robbins, H. "How to Bisect a Segment and Find the Center of a Circle with the Compass Alone." §3.4.4 in *What is Mathematics?: An Elementary Approach to Ideas and Methods, 2nd ed.* Oxford, England: Oxford University Press, pp. 145–146, 1996.

Dixon, R. *Mathographics.* New York: Dover, p. 22, 1991.

Line of Curvature

A curve on a surface whose tangents are always in the direction of PRINCIPAL CURVATURE. The equation of the lines of curvature can be written

$$\begin{vmatrix} g_{11} & g_{12} & g_{22} \\ b_{11} & b_{12} & b_{22} \\ du^2 & -du\,dv & dv^2 \end{vmatrix} = 0,$$

where g and b are the COEFFICIENTS of the first and second FUNDAMENTAL FORMS.

see also DUPIN'S THEOREM, FUNDAMENTAL FORMS, PRINCIPAL CURVATURES

Line Element

Also known as the first FUNDAMENTAL FORM

$$ds^2 = g_{ab}\,dx^a\,dx^b.$$

In the principal axis frame for 3-D,

$$ds^2 = g_{aa}(dx^a)^2 + g_{bb}(dx^b)^2 + g_{cc}(dx^c)^2.$$

At ORDINARY POINTS on a surface, the line element is positive definite.

see also AREA ELEMENT, FUNDAMENTAL FORMS, VOLUME ELEMENT

Line Graph

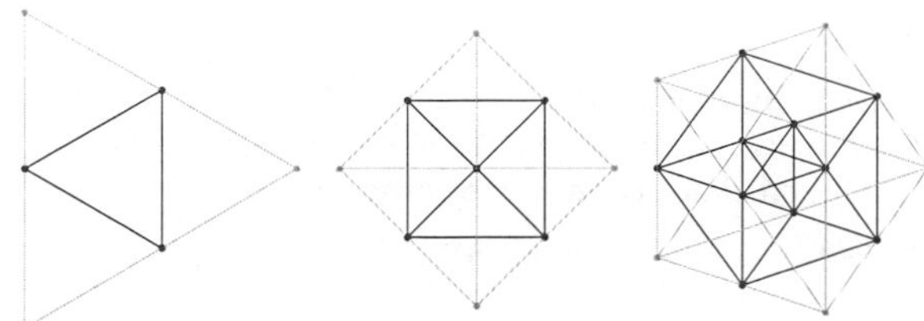

A LINE GRAPH $L(G)$ (also called an INTERCHANGE GRAPH) of a graph G is obtained by associating a vertex with each edge of the graph and connecting two vertices with an edge IFF the corresponding edges of G meet at one or both endpoints. In the three examples above, the original graphs are the COMPLETE GRAPHS K_3, K_4, and K_5 shown in gray, and their line graphs are shown in black.

References

Saaty, T. L. and Kainen, P. C. "Line Graphs." §4-3 in *The Four-Color Problem: Assaults and Conquest.* New York: Dover, pp. 108–112, 1986.

Line at Infinity

The straight line on which all POINTS AT INFINITY lie. The line at infinity is given in terms of TRILINEAR COORDINATES by

$$a\alpha + b\beta + c\gamma = 0,$$

which follows from the fact that a REAL TRIANGLE will have POSITIVE AREA, and therefore that

$$2\Delta = a\alpha + b\beta + c\gamma > 0.$$

Instead of the three reflected segments concurring for the ISOGONAL CONJUGATE of a point X on the CIRCUMCIRCLE of a TRIANGLE, they become parallel (and can be considered to meet at infinity). As X varies around the CIRCUMCIRCLE, X^{-1} varies through a line called the line at infinity. Every line is PERPENDICULAR to the line at infinity.

see also POINT AT INFINITY

Line Integral

The line integral on a curve $\boldsymbol{\sigma}$ is defined by

$$\int_{\boldsymbol{\sigma}} \mathbf{F}\cdot d\mathbf{s} = \int_a^b \mathbf{F}(\boldsymbol{\sigma}(t))\cdot\boldsymbol{\sigma}'(t)\,dt \tag{1}$$

$$= \int_C F_1\,dx + F_2\,dy + F_3\,dz, \tag{2}$$

where

$$\mathbf{F} \equiv \begin{bmatrix} F_1 \\ F_2 \\ F_3 \end{bmatrix}. \tag{3}$$

If $\nabla\cdot\mathbf{F} = 0$ (i.e., it is a DIVERGENCELESS FIELD), then the line integral is path independent and

$$\int_{(a,b,c)}^{(x,y,z)} F_1\,dx + F_2\,dy + F_3\,dz$$

$$= \int_{(a,b,c)}^{(x,b,c)} F_1\,dx + \int_{(x,b,c)}^{(x,y,c)} F_2\,dy + \int_{(x,y,c)}^{(x,y,z)} F_3\,dz. \tag{4}$$

For z COMPLEX, $\gamma : z = z(t)$, and $t \in [a,b]$,

$$\int_\gamma f\,dz = \int_a^b f(z(t))z'(t)\,dt. \tag{5}$$

see also CONTOUR INTEGRAL, PATH INTEGRAL

Line Segment

A closed interval corresponding to a FINITE portion of an infinite LINE. Line segments are generally labelled with two letters corresponding to their endpoints, say A and B, and then written AB. The length of the line segment is indicated with an overbar, so the length of the line segment AB would be written $\overline{AB}$.

Curiously, the number of points in a line segment (ALEPH-1; $\aleph_1$) is equal to that in an entire 1-D SPACE (a LINE), and also to the number of points in an n-D SPACE, as first recognized by Georg Cantor.

see also ALEPH-1 ($\aleph_1$), COLLINEAR, CONTINUUM, LINE, RAY

Line Space

see LIOUVILLE SPACE

Linear Algebra

The study of linear sets of equations and their transformation properties. Linear algebra allows the analysis of ROTATIONS in space, LEAST SQUARES FITTING, solution of coupled differential equations, determination of a circle passing through three given points, as well as many other other problems in mathematics, physics, and engineering.

The MATRIX and DETERMINANT are extremely useful tools of linear algebra. One central problem of linear algebra is the solution of the matrix equation

$$\mathsf{A}\mathbf{x} = \mathbf{b}$$

for $\mathbf{x}$. While this can, in theory, be solved using a MATRIX INVERSE

$$\mathbf{x} = \mathsf{A}^{-1}\mathbf{b},$$

other techniques such as GAUSSIAN ELIMINATION are numerically more robust.

see also CONTROL THEORY, CRAMER'S RULE, DETERMINANT, GAUSSIAN ELIMINATION, LINEAR TRANSFORMATION, MATRIX, VECTOR

References

Ayres, F. Jr. *Theory and Problems of Matrices.* New York: Schaum, 1962.
Banchoff, T. and Wermer, J. *Linear Algebra Through Geometry, 2nd ed.* New York: Springer-Verlag, 1992.
Bellman, R. E. *Introduction to Matrix Analysis, 2nd ed.* New York: McGraw-Hill, 1970.
Faddeeva, V. N. *Computational Methods of Linear Algebra.* New York: Dover, 1958.
Golub, G. and van Loan, C. *Matrix Computations, 3rd ed.* Baltimore, MD: Johns Hopkins University Press, 1996.
Halmos, P. R. *Linear Algebra Problem Book.* Providence, RI: Math. Assoc. Amer., 1995.
Lang, S. *Introduction to Linear Algebra, 2nd ed.* New York: Springer-Verlag, 1997.
Marcus, M. and Minc, H. *Introduction to Linear Algebra.* New York: Dover, 1988.
Marcus, M. and Minc, H. *A Survey of Matrix Theory and Matrix Inequalities.* New York: Dover, 1992.
Marcus, M. *Matrices and Matlab: A Tutorial.* Englewood Cliffs, NJ: Prentice-Hall, 1993.
Mirsky, L. *An Introduction to Linear Algebra.* New York: Dover, 1990.
Muir, T. *A Treatise on the Theory of Determinants.* New York: Dover, 1960.
Nash, J. C. *Compact Numerical Methods for Computers: Linear Algebra and Function Minimisation, 2nd ed.* Bristol, England: Adam Hilger, 1990.
Strang, G. *Linear Algebra and its Applications, 3rd ed.* Philadelphia, PA: Saunders, 1988.
Strang, G. *Introduction to Linear Algebra.* Wellesley, MA: Wellesley-Cambridge Press, 1993.
Strang, G. and Borre, K. *Linear Algebra, Geodesy, & GPS.* Wellesley, MA: Wellesley-Cambridge Press, 1997.

Linear Approximation

A linear approximation to a function $f(x)$ at a point x_0 can be computed by taking the first term in the TAYLOR SERIES

$$f(x_0 + \Delta x) = f(x_0) + f'(x_0)\Delta x + \ldots.$$

see also MACLAURIN SERIES, TAYLOR SERIES

Linear Code

A linear code over a FINITE FIELD with q elements F_q is a linear SUBSPACE $C \subset F_q{}^n$. The vectors forming the SUBSPACE are called code words. When code words are chosen such that the distance between them is maximized, the code is called error-correcting since slightly garbled vectors can be recovered by choosing the nearest code word.

see also CODE, CODING THEORY, ERROR-CORRECTING CODE, GRAY CODE, HUFFMAN CODING, ISBN

Linear Congruence

A linear congruence

$$ax \equiv b \pmod{m}$$

is solvable IFF the CONGRUENCE

$$b \equiv 0 \pmod{(a, m)}$$

is solvable, where $d \equiv (a, m)$ is the GREATEST COMMON DIVISOR, in which case the solutions are x_0, $x_0 + m/d$, $x_0 + 2m/d$, ..., $x_0 + (d-1)m/d$, where $x_0 < m/d$. If $d = 1$, then there is only one solution.

see also CONGRUENCE, QUADRATIC CONGRUENCE

Linear Congruence Method

A METHOD for generating RANDOM (PSEUDORANDOM) numbers using the linear RECURRENCE RELATION

$$X_{n+1} = aX_n + c \pmod{m},$$

where a and c must assume certain fixed values and X_0 is an initial number known as the SEED.

see also PSEUDORANDOM NUMBER, RANDOM NUMBER, SEED

References

Pickover, C. A. "Computers, Randomness, Mind, and Infinity." Ch. 31 in *Keys to Infinity.* New York: W. H. Freeman, pp. 233–247, 1995.

Linear Equation

An algebraic equation of the form

$$y = ax + b$$

involving only a constant and a first-order (linear) term.

see also LINE, POLYNOMIAL, QUADRATIC EQUATION

Linear Equation System

When solving a system of n linear equations with $k > n$ unknowns, use MATRIX operations to solve the system as far as possible. Then solve for the first $(k-n)$ components in terms of the last n components to find the solution space.

Linear Extension

A linear extension of a PARTIALLY ORDERED SET P is a PERMUTATION of the elements p_1, p_2, ... of P such that $i < j$ IMPLIES $p_i < p_j$. For example, the linear extensions of the PARTIALLY ORDERED SET $((1, 2), (3, 4))$ are 1234, 1324, 1342, 3124, 3142, and 3412, all of which have 1 before 2 and 3 before 4.

References

Brightwell, G. and Winkler, P. "Counting Linear Extensions." *Order* **8**, 225–242, 1991.
Preusse, G. and Ruskey, F. "Generating Linear Extensions Fast." *SIAM J. Comput.* **23**, 373–386, 1994.
Ruskey, F. "Information on Linear Extension." `http://sue.csc.uvic.ca/~cos/inf/pose/LinearExt.html`.
Varol, Y. and Rotem, D. "An Algorithm to Generate All Topological Sorting Arrangements." *Comput. J.* **24**, 83–84, 1981.

Linear Fractional Transformation

see MÖBIUS TRANSFORMATION

Linear Group

see GENERAL LINEAR GROUP, LIE-TYPE GROUP, PROJECTIVE GENERAL LINEAR GROUP, PROJECTIVE SPECIAL LINEAR GROUP, SPECIAL LINEAR GROUP

References

Wilson, R. A. "ATLAS of Finite Group Representation." `http://for.mat.bham.ac.uk/atlas#lin`.

Linear Group Theorem

Any linear system of point-groups on a curve with only ordinary singularities may be cut by ADJOINT CURVES.

References

Coolidge, J. L. *A Treatise on Algebraic Plane Curves.* New York: Dover, pp. 122 and 251, 1959.

Linear Operator

An operator $\tilde{L}$ is said to be linear if, for every pair of functions f and g and SCALAR t,

$$\tilde{L}(f+g) = \tilde{L}f + \tilde{L}g$$

and

$$\tilde{L}(tf) = t\tilde{L}f.$$

see also LINEAR TRANSFORMATION, OPERATOR

Linear Ordinary Differential Equation

see ORDINARY DIFFERENTIAL EQUATION—FIRST-ORDER, ORDINARY DIFFERENTIAL EQUATION—SECOND-ORDER

Linear Programming

The problem of maximizing a linear function over a convex polyhedron, also known as OPERATIONS RESEARCH, OPTIMIZATION THEORY, or CONVEX OPTIMIZATION THEORY. It can be solved using the SIMPLEX METHOD (Wood and Dantzig 1949, Dantzig 1949) which runs along EDGES of the visualization solid to find the best answer.

In 1979, L. G. Khachian found a $\mathcal{O}(x^5)$ POLYNOMIAL-time ALGORITHM. A much more efficient POLYNOMIAL-time ALGORITHM was found by Karmarkar (1984). This method goes through the middle of the solid and then transforms and warps, and offers many advantages over the simplex method.

see also CRISS-CROSS METHOD, ELLIPSOIDAL CALCULUS, KUHN-TUCKER THEOREM, LAGRANGE MULTIPLIER, VERTEX ENUMERATION

References

Bellman, R. and Kalaba, R. *Dynamic Programming and Modern Control Theory.* New York: Academic Press, 1965.

Dantzig, G. B. "Programming of Interdependent Activities. II. Mathematical Model." *Econometrica* **17**, 200–211, 1949.

Dantzig, G. B. *Linear Programming and Extensions.* Princeton, NJ: Princeton University Press, 1963.

Greenberg, H. J. "Mathematical Programming Glossary." `http://www-math.cudenver.edu/~hgreenbe/glossary/glossary.html`.

Karloff, H. *Linear Programming.* Boston, MA: Birkhäuser, 1991.

Karmarkar, N. "A New Polynomial-Time Algorithm for Linear Programming." *Combinatorica* **4**, 373–395, 1984.

Pappas, T. "Projective Geometry & Linear Programming." *The Joy of Mathematics.* San Carlos, CA: Wide World Publ./Tetra, pp. 216–217, 1989.

Press, W. H.; Flannery, B. P.; Teukolsky, S. A.; and Vetterling, W. T. "Linear Programming and the Simplex Method." §10.8 in *Numerical Recipes in FORTRAN: The Art of Scientific Computing, 2nd ed.* Cambridge, England: Cambridge University Press, pp. 423–436, 1992.

Sultan, A. *Linear Programming: An Introduction with Applications.* San Diego, CA: Academic Press, 1993.

Tokhomirov, V. M. "The Evolution of Methods of Convex Optimization." *Amer. Math. Monthly* **103**, 65–71, 1996.

Wood, M. K. and Dantzig, G. B. "Programming of Interdependent Activities. I. General Discussion." *Econometrica* **17**, 193–199, 1949.

Yudin, D. B. and Nemirovsky, A. S. *Problem Complexity and Method Efficiency in Optimization.* New York: Wiley, 1983.

Linear Recurrence Sequence

see RECURRENCE SEQUENCE

Linear Regression

The fitting of a straight LINE through a given set of points according to some specified goodness-of-fit criterion. The most common form of linear regression is LEAST SQUARES FITTING.

see also LEAST SQUARES FITTING, MULTIPLE REGRESSION, NONLINEAR LEAST SQUARES FITTING

References

Edwards, A. L. *An Introduction to Linear Regression and Correlation.* San Francisco, CA: W. H. Freeman, 1976.

Edwards, A. L. *Multiple Regression and the Analysis of Variance and Covariance.* San Francisco, CA: W. H. Freeman, 1979.

Linear Space

see VECTOR SPACE

Linear Stability

Consider the general system of two first-order ORDINARY DIFFERENTIAL EQUATIONS

$$\dot{x} = f(x,y) \tag{1}$$

$$\dot{y} = g(x,y). \tag{2}$$

Let x_0 and y_0 denote FIXED POINTS with $\dot{x} = \dot{y} = 0$, so

$$f(x_0,y_0) = 0 \tag{3}$$

$$g(x_0,y_0) = 0. \tag{4}$$

Then expand about (x_0, y_0) so

$$\delta\dot{x} = f_x(x_0, y_0)\delta x + f_y(x_0, y_0)\delta y + f_{xy}(x_0, y_0)\delta x\delta y + \ldots \quad (5)$$

$$\delta\dot{y} = g_x(x_0, y_0)\delta x + g_y(x_0, y_0)\delta y + g_{xy}(x_0, y_0)\delta x\delta y + \ldots. \quad (6)$$

To first-order, this gives

$$\frac{d}{dt}\begin{bmatrix}\delta x\\ \delta y\end{bmatrix} = \begin{bmatrix} f_x(x_0, y_0) & f_y(x_0, y_0)\\ g_x(x_0, y_0) & g_y(x_0, y_0)\end{bmatrix}\begin{bmatrix}\delta x\\ \delta y\end{bmatrix}, \quad (7)$$

where the 2×2 MATRIX is called the STABILITY MATRIX.

In general, given an n-D MAP $\mathbf{x}' = T(\mathbf{x})$, let $\mathbf{x}_0$ be a FIXED POINT, so that

$$T(\mathbf{x}_0) = \mathbf{x}_0. \quad (8)$$

Expand about the fixed point,

$$\begin{aligned} T(\mathbf{x}_0 + \delta\mathbf{x}) &= T(\mathbf{x}_0) + \frac{\partial T}{\partial \mathbf{x}}\delta\mathbf{x} + \mathcal{O}(\delta\mathbf{x})^2 \\ &\equiv T(\mathbf{x}_0) + \delta T, \end{aligned} \quad (9)$$

so

$$\delta T = \frac{\partial T}{\partial \mathbf{x}}\delta\mathbf{x} \equiv \mathsf{A}\,\delta\mathbf{x}. \quad (10)$$

The map can be transformed into the principal axis frame by finding the EIGENVECTORS and EIGENVALUES of the MATRIX A

$$(\mathsf{A} - \lambda\mathsf{I})\,\delta\mathbf{x} = \mathbf{0}, \quad (11)$$

so the DETERMINANT

$$|\mathsf{A} - \lambda\mathsf{I}| = 0. \quad (12)$$

The mapping is

$$\delta\mathbf{x}_{\text{princ}}' = \begin{bmatrix}\lambda_1 & \cdots & 0\\ \vdots & \ddots & \vdots\\ 0 & \cdots & \lambda_n\end{bmatrix}. \quad (13)$$

When iterated a large number of times,

$$\delta T'_{\text{princ}} \to 0 \quad (14)$$

only if $|\Re(\lambda_i)| < 1$ for $i = 1, \ldots, n$ but $\to\infty$ if any $|\lambda_i| > 1$. Analysis of the EIGENVALUES (and EIGENVECTORS) of A therefore characterizes the type of FIXED POINT. The condition for stability is $|\Re(\lambda_i)| < 1$ for $i = 1, \ldots, n$.

see also FIXED POINT, STABILITY MATRIX

References

Tabor, M. "Linear Stability Analysis." §1.4 in *Chaos and Integrability in Nonlinear Dynamics: An Introduction.* New York: Wiley, pp. 20–31, 1989.

Linear Transformation

An $n \times n$ MATRIX A is a linear transformation (linear MAP) IFF, for every pair of n-VECTORS $\mathbf{X}$ and $\mathbf{Y}$ and every SCALAR t,

$$\mathsf{A}(\mathbf{X} + \mathbf{Y}) = \mathsf{A}(\mathbf{X}) + \mathsf{A}(\mathbf{Y}) \quad (1)$$

and

$$\mathsf{A}(t\mathbf{X}) = t\mathsf{A}(\mathbf{X}). \quad (2)$$

Consider the 2-D transformation

$$\rho x_1' = a_{11}x_1 + a_{12}x_2 \quad (3)$$

$$\rho x_2' = a_{21}x_2 + a_{22}x_2. \quad (4)$$

Rescale by defining $\lambda \equiv x_1/x_2$ and $\lambda' \equiv x_1'/x_2'$, then the above equations become

$$\lambda' = \frac{\alpha\lambda + \beta}{\gamma\lambda + \delta}, \quad (5)$$

where $\alpha\delta - \beta\gamma \neq 0$ and α, β, γ and δ are defined in terms of the old constants. Solving for λ gives

$$\lambda = \frac{\delta\lambda' - \beta}{-\gamma\lambda' + \alpha}, \quad (6)$$

so the transformation is ONE-TO-ONE. To find the FIXED POINTS of the transformation, set $\lambda = \lambda'$ to obtain

$$\gamma\lambda^2 + (\delta - \alpha)\lambda - \beta = 0. \quad (7)$$

This gives two fixed points which may be distinct or coincident. The fixed points are classified as follows.

variables	type
$(\delta-\alpha)^2 + 4\beta\gamma > 0$	hyperbolic fixed point
$(\delta-\alpha)^2 + 4\beta\gamma < 0$	elliptic fixed point
$(\delta-\alpha)^2 + 4\beta\gamma = 0$	parabolic fixed point

see also ELLIPTIC FIXED POINT (MAP), HYPERBOLIC FIXED POINT (MAP), INVOLUNTARY, LINEAR OPERATOR, PARABOLIC FIXED POINT

References

Woods, F. S. *Higher Geometry: An Introduction to Advanced Methods in Analytic Geometry.* New York: Dover, pp. 13–15, 1961.

Linearly Dependent Curves

Two curves ϕ and ψ satisfying

$$\phi + \psi = 0$$

are said to be linearly dependent. Similarly, n curves ϕ_i, $i = 1, \ldots, n$ are said to be linearly dependent if

$$\sum_{i=1}^n \phi_i = 0.$$

see also BERTINI'S THEOREM, STUDY'S THEOREM

References

Coolidge, J. L. *A Treatise on Algebraic Plane Curves.* New York: Dover, pp. 32–34, 1959.

Linearly Dependent Functions

The n functions $f_1(x)$, $f_2(x)$, ..., $f_n(x)$ are linearly dependent if, for some $c_1, c_2, \ldots, c_n \in \mathbb{R}$ not all zero,

$$c_i f_i(x) = 0 \tag{1}$$

(where EINSTEIN SUMMATION is used) for all x in some interval I. If the functions are not linearly dependent, they are said to be linearly independent. Now, if the functions $\in \mathbb{R}^{n-1}$, we can differentiate (1) up to $n-1$ times. Therefore, linear dependence also requires

$$c_i f_i' = 0 \tag{2}$$

$$c_i f_i'' = 0 \tag{3}$$

$$c_i f_i^{(n-1)} = 0, \tag{4}$$

where the sums are over $i = 1, \ldots, n$. These equations have a nontrivial solution IFF the DETERMINANT

$$\begin{vmatrix} f_1 & f_2 & \cdots & f_n \\ f_1' & f_2' & \cdots & f_n' \\ \vdots & \vdots & \ddots & \vdots \\ f_1^{(n-1)} & f_2^{(n-1)} & \cdots & f_n^{(n-1)} \end{vmatrix} = 0, \tag{5}$$

where the DETERMINANT is conventionally called the WRONSKIAN and is denoted $W(f_1, f_2, \ldots, f_n)$. If the WRONSKIAN $\neq 0$ for any value c in the interval I, then the only solution possible for (2) is $c_i = 0$ ($i = 1, \ldots, n$), and the functions are linearly independent. If, on the other hand, $W = 0$ for a range, the functions are linearly dependent in the range. This is equivalent to stating that if the vectors $\mathbf{V}[f_1(c)], \ldots, \mathbf{V}[f_n(c)]$ defined by

$$\mathbf{V}[f_i(x)] = \begin{bmatrix} f_i(x) \\ f_i'(x) \\ f_i''(n) \\ \vdots \\ f_i^{(n-1)}(x) \end{bmatrix} \tag{6}$$

are linearly independent for at least one $c \in I$, then the functions f_i are linearly independent in I.

References

Sansone, G. "Linearly Independent Functions." §1.2 in *Orthogonal Functions, rev. English ed.* New York: Dover, pp. 2–3, 1991.

Linearly Dependent Vectors

n VECTORS $\mathbf{X}_1, \mathbf{X}_2, \ldots, \mathbf{X}_n$ are linearly dependent IFF there exist SCALARS $c_1, c_2, \ldots, c_n$ such that

$$c_i \mathbf{X}_i = 0, \tag{1}$$

where EINSTEIN SUMMATION is used and $i = 1, \ldots, n$. If no such SCALARS exist, then the vectors are said to be linearly independent. In order to satisfy the CRITERION for linear dependence,

$$c_1 \begin{bmatrix} x_{11} \\ x_{21} \\ \vdots \\ x_{n1} \end{bmatrix} + c_2 \begin{bmatrix} x_{12} \\ x_{22} \\ \vdots \\ x_{n2} \end{bmatrix} + \cdots + c_n \begin{bmatrix} x_{1n} \\ x_{2n} \\ \vdots \\ x_{nn} \end{bmatrix} = \begin{bmatrix} 0 \\ 0 \\ \vdots \\ 0 \end{bmatrix} \tag{2}$$

$$\begin{bmatrix} x_{11} & x_{12} & \cdots & x_{1n} \\ x_{21} & x_{22} & \cdots & x_{2n} \\ \vdots & \vdots & \ddots & \vdots \\ x_{n1} & x_{n2} & \cdots & x_{nn} \end{bmatrix} \begin{bmatrix} c_1 \\ c_2 \\ \vdots \\ c_n \end{bmatrix} = \begin{bmatrix} 0 \\ 0 \\ \vdots \\ 0 \end{bmatrix}. \tag{3}$$

In order for this MATRIX equation to have a nontrivial solution, the DETERMINANT must be 0, so the VECTORS are linearly dependent if

$$\begin{vmatrix} x_{11} & x_{12} & \cdots & x_{1n} \\ x_{21} & x_{22} & \cdots & x_{2n} \\ \vdots & \vdots & \ddots & \vdots \\ x_{n1} & x_{n2} & \cdots & x_{nn} \end{vmatrix} = 0, \tag{4}$$

and linearly independent otherwise.

Let $\mathbf{p}$ and $\mathbf{q}$ be n-D VECTORS. Then the following three conditions are equivalent (Gray 1993).

1. $\mathbf{p}$ and $\mathbf{q}$ are linearly dependent.
2. $\begin{vmatrix} \mathbf{p}\cdot\mathbf{p} & \mathbf{p}\cdot\mathbf{q} \\ \mathbf{q}\cdot\mathbf{p} & \mathbf{q}\cdot\mathbf{q} \end{vmatrix} = 0.$
3. The $2 \times n$ MATRIX $\begin{bmatrix} \mathbf{p} \\ \mathbf{q} \end{bmatrix}$ has rank less than two.

References

Gray, A. *Modern Differential Geometry of Curves and Surfaces.* Boca Raton, FL: CRC Press, pp. 186–187, 1993.

Linearly Independent

Two or more functions, equations, or vectors which are not linearly dependent are said to be linearly independent.

see also LINEARLY DEPENDENT CURVES, LINEARLY DEPENDENT FUNCTIONS, LINEARLY DEPENDENT VECTORS, MAXIMALLY LINEAR INDEPENDENT

Link

Formally, a link is one or more disjointly embedded CIRCLES in 3-space. More informally, a link is an assembly of KNOTS with mutual entanglements. Kuperberg (1994) has shown that a nontrivial KNOT or link in $\mathbb{R}^3$ has four COLLINEAR points (Eppstein). Doll and Hoste (1991) list POLYNOMIALS for oriented links of nine or fewer crossings. A listing of the first few simple links follows, arranged by CROSSING NUMBER.

0_1^2

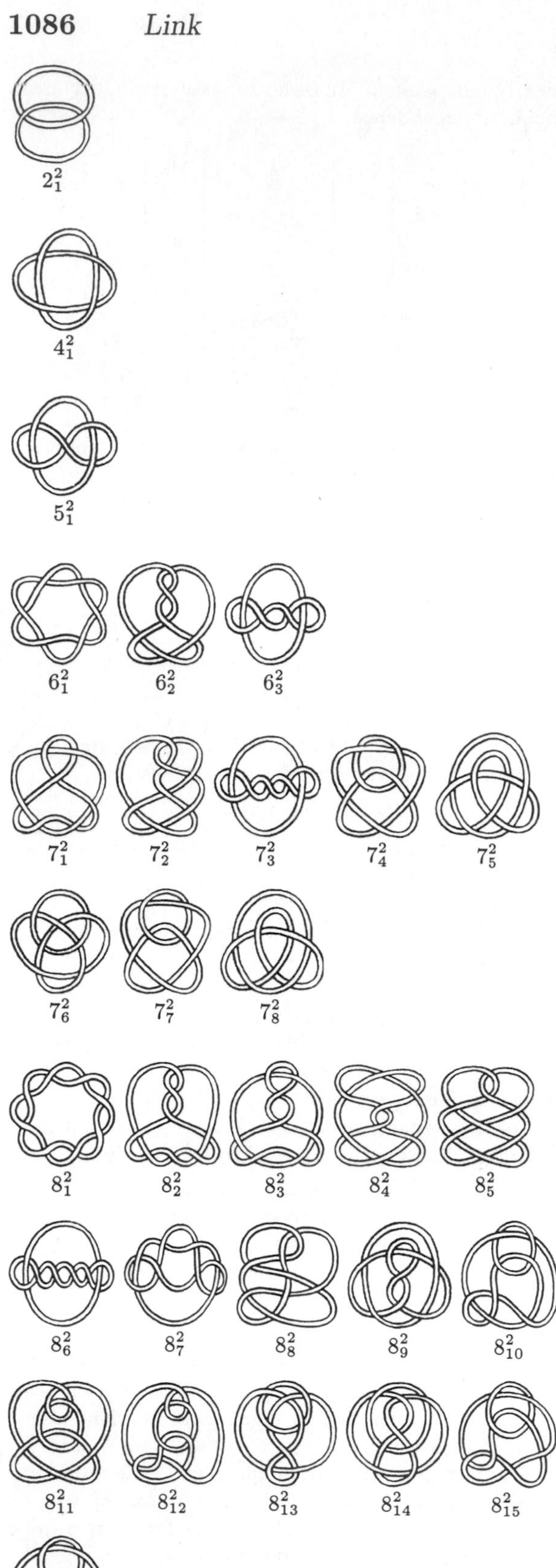
2_1^2
4_1^2
5_1^2
6_1^2 6_2^2 6_3^2
7_1^2 7_2^2 7_3^2 7_4^2 7_5^2
7_6^2 7_7^2 7_8^2
8_1^2 8_2^2 8_3^2 8_4^2 8_5^2
8_6^2 8_7^2 8_8^2 8_9^2 8_{10}^2
8_{11}^2 8_{12}^2 8_{13}^2 8_{14}^2 8_{15}^2
8_{16}^2

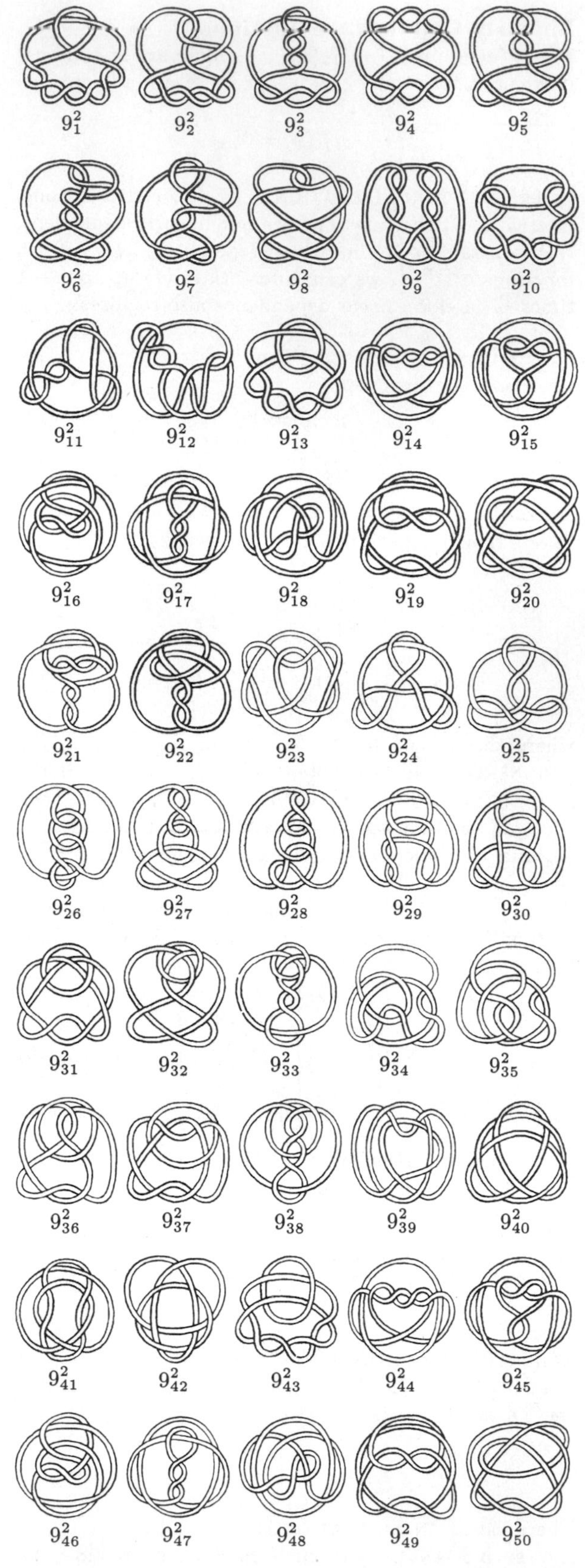
9_1^2 9_2^2 9_3^2 9_4^2 9_5^2
9_6^2 9_7^2 9_8^2 9_9^2 9_{10}^2
9_{11}^2 9_{12}^2 9_{13}^2 9_{14}^2 9_{15}^2
9_{16}^2 9_{17}^2 9_{18}^2 9_{19}^2 9_{20}^2
9_{21}^2 9_{22}^2 9_{23}^2 9_{24}^2 9_{25}^2
9_{26}^2 9_{27}^2 9_{28}^2 9_{29}^2 9_{30}^2
9_{31}^2 9_{32}^2 9_{33}^2 9_{34}^2 9_{35}^2
9_{36}^2 9_{37}^2 9_{38}^2 9_{39}^2 9_{40}^2
9_{41}^2 9_{42}^2 9_{43}^2 9_{44}^2 9_{45}^2
9_{46}^2 9_{47}^2 9_{48}^2 9_{49}^2 9_{50}^2

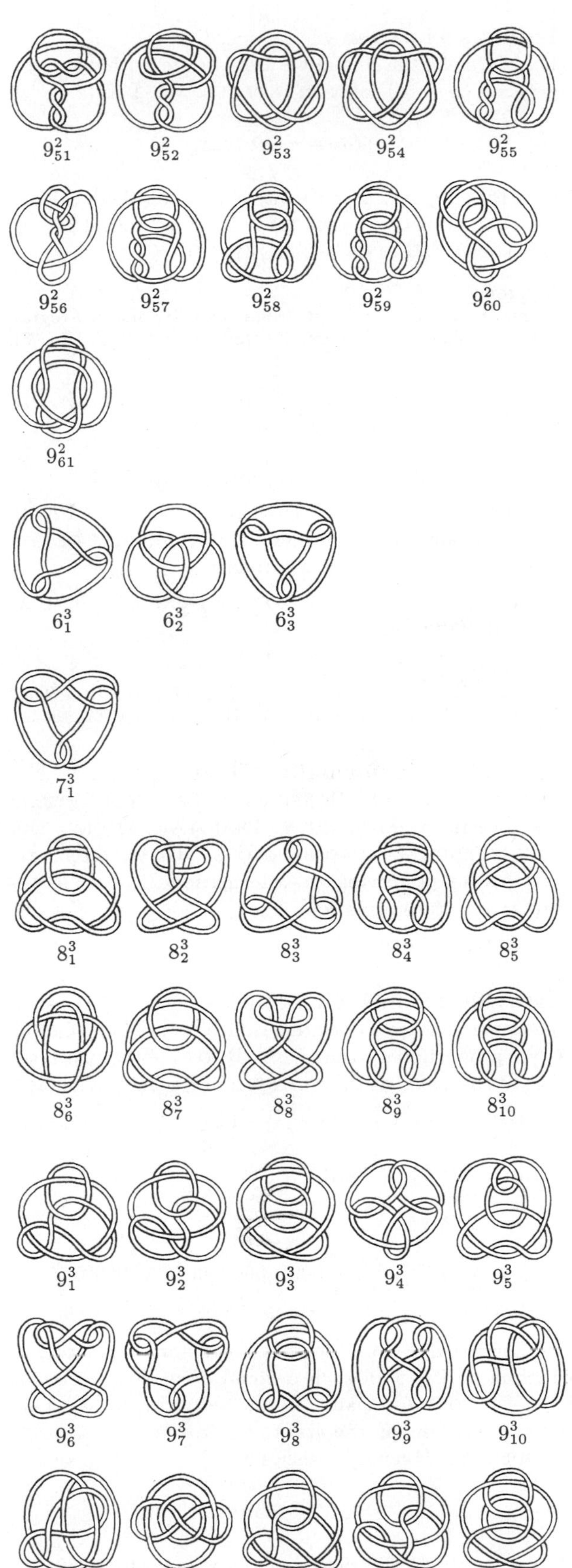

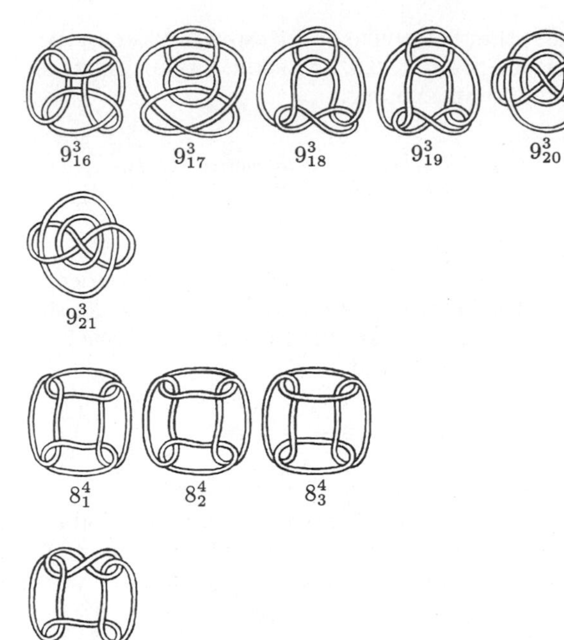

see also ANDREWS-CURTIS LINK, BORROMEAN RINGS, BRUNNIAN LINK, HOPF LINK, KNOT, WHITEHEAD LINK

References

Doll, H. and Hoste, J. "A Tabulation of Oriented Links." *Math. Comput.* **57**, 747–761, 1991.

Eppstein, D. "Colinear Points on Knots." http://www.ics.uci.edu/~eppstein/junkyard/knot-colinear.html.

Kuperberg, G. "Quadrisecants of Knots and Links." *J. Knot Theory Ramifications* **3**, 41–50, 1994.

Weisstein, E. W. "Knots." http://www.astro.virginia.edu/~eww6n/math/notebooks/Knots.m.

Link Diagram

A planar diagram depicting a LINK (or KNOT) as a sequence of segments with gaps representing undercrossings and solid lines overcrossings. In such a diagram, only two segments should ever cross at a single point. Link diagrams for the TREFOIL KNOT and FIGURE-OF-EIGHT KNOT are illustrated above.

Linkage

Sylvester, Kempe and Cayley developed the geometry associated with the theory of linkages in the 1870s. Kempe proved that every finite segment of an algebraic curve can be generated by a linkage in the manner of WATT'S CURVE.

see also HART'S INVERSOR, KEMPE LINKAGE, PANTOGRAPH, PEAUCELLIER INVERSOR, SARRUS LINKAGE, WATT'S PARALLELOGRAM

References
Cundy, H. and Rollett, A. *Mathematical Models, 3rd ed.* Stradbroke, England: Tarquin Pub., 1989.

Linking Number

A LINK invariant. Given a two-component oriented LINK, take the sum of +1 crossings and −1 crossing over all crossings between the two links and divide by 2. For components α and β,

$$L(\alpha,\beta) \equiv \tfrac{1}{2} \sum_{p\in\alpha\sqcap\beta} \epsilon(p),$$

where $\alpha \sqcap \beta$ is the set of crossings of α with β and $\epsilon(p)$ is the sign of the crossing. The linking number of a splittable two-component link is always 0.

see also JONES POLYNOMIAL, LINK

References
Rolfsen, D. *Knots and Links.* Wilmington, DE: Publish or Perish Press, pp. 132–133, 1976.

Links Curve

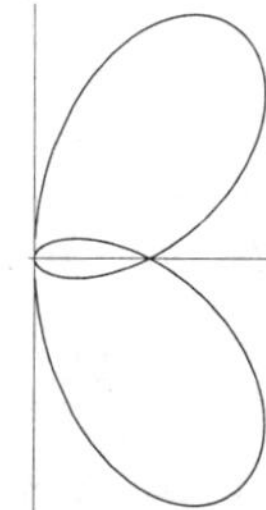

The curve given by the Cartesian equation

$$(x^2 + y^2 - 3x)^2 = 4x^2(2 - x).$$

The origin of the curve is a TACNODE.

References
Cundy, H. and Rollett, A. *Mathematical Models, 3rd ed.* Stradbroke, England: Tarquin Pub., p. 72, 1989.

Linnik's Constant

The constant L in LINNIK'S THEOREM. Heath-Brown (1992) has shown that $L \leq 5.5$, and Schinzel, Sierpiński, and Kanold (Ribenboim 1989) have conjectured that $L = 2$.

References
Finch, S. "Favorite Mathematical Constants." `http://www.mathsoft.com/asolve/constant/linnik/linnik.html`.
Guy, R. K. *Unsolved Problems in Number Theory, 2nd ed.* New York: Springer-Verlag, p. 13, 1994.
Heath-Brown, D. R. "Zero-Free Regions for Dirichlet *L*-Functions and the Least Prime in an Arithmetic Progression." *Proc. London Math. Soc.* **64**, 265–338, 1992.
Ribenboim, P. *The Book of Prime Number Records, 2nd ed.* New York: Springer-Verlag, 1989.

Linnik's Theorem

Let $p(d,a)$ be the smallest PRIME in the arithmetic progression $\{a + kd\}$ for k an INTEGER > 0. Let

$$p(d) \equiv \max p(d,a)$$

such that $1 \leq a < d$ and $(a,d) = 1$. Then there exists a $d_0 \geq 2$ and an $L > 1$ such that $p(d) < d^L$ for all $d > d_0$. L is known as LINNIK'S CONSTANT.

References
Linnik, U. V. "On the Least Prime in an Arithmetic Progression. I. The Basic Theorem." *Mat. Sbornik N. S.* **15 (57)**, 139–178, 1944.
Linnik, U. V. "On the Least Prime in an Arithmetic Progression. II. The Deuring-Heilbronn Phenomenon" *Mat. Sbornik N. S.* **15 (57)**, 347–368, 1944.

Liouville's Boundedness Theorem

A bounded ENTIRE FUNCTION in the COMPLEX PLANE $\mathbb{C}$ is constant. The FUNDAMENTAL THEOREM OF ALGEBRA follows as a simple corollary.

see also COMPLEX PLANE, ENTIRE FUNCTION, FUNDAMENTAL THEOREM OF ALGEBRA

References
Morse, P. M. and Feshbach, H. *Methods of Theoretical Physics, Part I.* New York: McGraw-Hill, pp. 381–382, 1953.

Liouville's Conformality Theorem

In SPACE, the only CONFORMAL TRANSFORMATIONS are inversions, SIMILARITY TRANSFORMATIONS, and CONGRUENCE TRANSFORMATIONS. Or, restated, every ANGLE-preserving transformation is a SPHERE-preserving transformation.

see also CONFORMAL MAP

Liouville's Conic Theorem

The lengths of the TANGENTS from a point P to a CONIC C are proportional to the CUBE ROOTS of the RADII OF CURVATURE of C at the corresponding points of contact.

see also CONIC SECTION

Liouville's Constant

$$L \equiv \sum_{n=1}^{\infty} 10^{-n!} = 0.110001000000000000000001\ldots$$

(Sloane's A012245). Liouville's constant is a decimal fraction with a 1 in each decimal place corresponding to a FACTORIAL $n!$, and ZEROS everywhere else. This number was among the first to be proven to be TRANSCENDENTAL. It nearly satisfies

$$10x^6 - 75x^3 - 190x + 21 = 0,$$

but with $x = L$, this equation gives $-0.0000000059\ldots$.

see also LIOUVILLE NUMBER

References
Conway, J. H. and Guy, R. K. "Liouville's Number." In *The Book of Numbers.* New York: Springer-Verlag, pp. 239–241, 1996.
Courant, R. and Robbins, H. "Liouville's Theorem and the Construction of Transcendental Numbers." §2.6.2 in *What is Mathematics?: An Elementary Approach to Ideas and Methods, 2nd ed.* Oxford, England: Oxford University Press, pp. 104–107, 1996.
Liouville, J. "Sur des classes très étendues de quantités dont la valeur n'est ni algébrique, ni même reductible à des irrationelles algébriques." *C. R. Acad. Sci. Paris* **18**, 883–885 and 993–995, 1844.
Liouville, J. "Sur des classes très-étendues de quantités dont la valeur n'est ni algébrique, ni même réductible à des irrationelles algébriques." *J. Math. pures appl.* **15**, 133–142, 1850.
Sloane, N. J. A. Sequence A012245 in "An On-Line Version of the Encyclopedia of Integer Sequences."

Liouville's Elliptic Function Theorem

An ELLIPTIC FUNCTION with no POLES in a fundamental cell is a constant.

Liouville Function

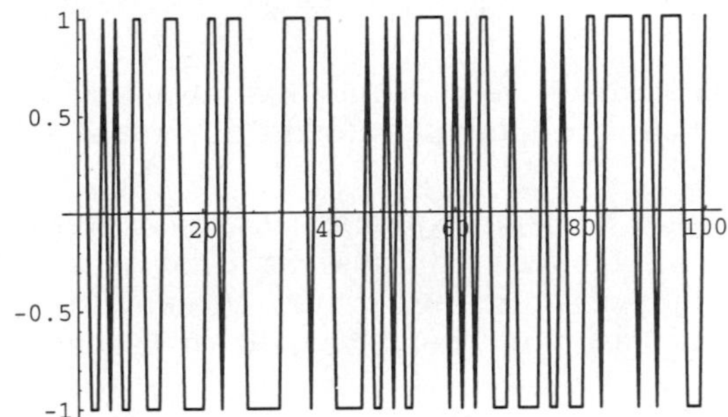

The function

$$\lambda(n) = (-1)^{r(n)}, \quad (1)$$

where $r(n)$ is the number of not necessarily distinct PRIME FACTORS of n, with $r(1) = 0$. The first few values of $\lambda(n)$ are 1, -1, -1, 1, -1, 1, -1, -1, 1, 1, -1, -1, ... The Liouville function is connected with the RIEMANN ZETA FUNCTION by the equation

$$\frac{\zeta(2s)}{\zeta(s)} = \sum_{n=1}^{\infty} \frac{\lambda(n)}{n^s} \quad (2)$$

(Lehman 1960).

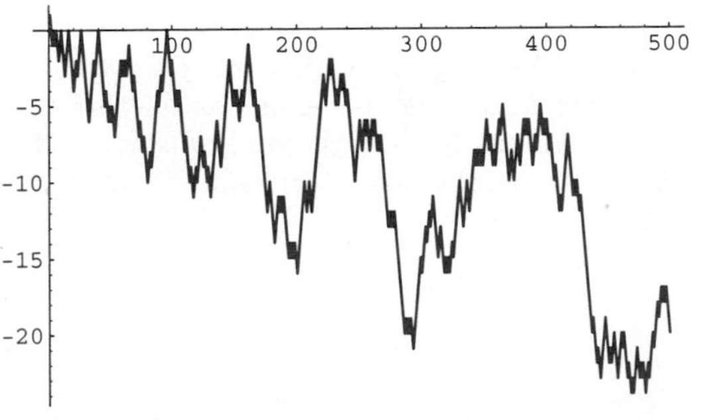

The CONJECTURE that the SUMMATORY FUNCTION

$$L(n) \equiv \sum_{k=1}^{n} \lambda(n) \quad (3)$$

satisfies $L(n) \leq 0$ for $n \geq 2$ is called the PÓLYA CONJECTURE and has been proved to be false. The first n for which $L(n) = 0$ are for $n = 2$, 4, 6, 10, 16, 26, 40, 96, 586, 906150256, ... (Sloane's A028488), and $n = 906150257$ is, in fact, the first counterexample to the PÓLYA CONJECTURE (Tanaka 1980). However, it is unknown if $L(x)$ changes sign infinitely often (Tanaka 1980). The first few values of $L(n)$ are 1, 0, -1, 0, -1, 0, -1, -2, -1, 0, -1, -2, -3, -2, -1, 0, -1, -2, -3, -4, ... (Sloane's A002819). $L(n)$ also satisfies

$$\sum_{n=1}^{x} L\left(\frac{x}{n}\right) = \lfloor \sqrt{x} \rfloor, \quad (4)$$

where $\lfloor x \rfloor$ is the FLOOR FUNCTION (Lehman 1960). Lehman (1960) also gives the formulas

$$L(x) = \sum_{m=1}^{x/w} \mu(m) \left\{ \left\lfloor \sqrt{\frac{x}{m}} \right\rfloor - \sum_{k=1}^{v-1} \lambda(k) \left(\left\lfloor \frac{x}{km} \right\rfloor - \left\lfloor \frac{x}{mv} \right\rfloor \right) \right\} - \sum_{l=x/w-1}^{x/v} L\left(\frac{x}{l}\right) \sum_{\substack{m|l \\ m=1}}^{x/w} \mu(m) \quad (5)$$

and

$$L(x) = \sum_{k=1}^{g} M\left(\frac{x}{k^2}\right) + \sum_{l=1}^{x/g^2} \mu(l) \left\lfloor \sqrt{\frac{x}{l}} \right\rfloor - M\left(\frac{x}{g^2}\right) \left\lfloor \sqrt{\frac{x}{g^2}} \right\rfloor, \quad (6)$$

where k, l, and m are variables ranging over the POSITIVE integers, $\mu(n)$ is the MÖBIUS FUNCTION, $M(x)$ is MERTENS FUNCTION, and v, w, and x are POSITIVE real numbers with $v < w < x$.

see also PÓLYA CONJECTURE, PRIME FACTORS, RIEMANN ZETA FUNCTION

References
Fawaz, A. Y. "The Explicit Formula for $L_0(x)$." *Proc. London Math. Soc.* **1**, 86–103, 1951.
Lehman, R. S. "On Liouville's Function." *Math. Comput.* **14**, 311–320, 1960.
Sloane, N. J. A. Sequences A028488 and A002819/M0042 in "An On-Line Version of the Encyclopedia of Integer Sequences."
Tanaka, M. "A Numerical Investigation on Cumulative Sum of the Liouville Function." *Tokyo J. Math.* **3**, 187–189, 1980.

Liouville Measure

$$\prod_i dp_i\, dq_i,$$

where p_i and q_i are momenta and positions of particles.

see also LIOUVILLE'S PHASE SPACE THEOREM, PHASE SPACE

Liouville Number

A Liouville number is a TRANSCENDENTAL NUMBER which is very close to a RATIONAL NUMBER. An IRRATIONAL NUMBER β is a Liouville number if, for any n, there exist an infinite number of pairs of INTEGERS p and q such that

$$0 < \left|\beta - \frac{p}{q}\right| < \frac{1}{q^n}.$$

Mahler (1953) proved that π is not a Liouville number.

see also LIOUVILLE'S CONSTANT, LIOUVILLE'S RATIONAL APPROXIMATION THEOREM, ROTH'S THEOREM, TRANSCENDENTAL NUMBER

References

Mahler, K. "On the Approximation of π." *Nederl. Akad. Wetensch. Proc. Ser. A.* **56**/*Indagationes Math.* **15**, 30–42, 1953.

Liouville's Phase Space Theorem

States that for a nondissipative HAMILTONIAN SYSTEM, phase space density (the AREA between phase space contours) is constant. This requires that, given a small time increment dt,

$$q_1 \equiv q(t_0 + dt) = q_0 + \frac{\partial H(q_0, p_0, t)}{\partial p_0} dt + \mathcal{O}(dt^2) \quad (1)$$

$$p_1 \equiv p(t_0 + dt) = p_0 - \frac{\partial H(q_0, p_0, t)}{\partial q_0} dt + \mathcal{O}(dt^2), \quad (2)$$

the JACOBIAN be equal to one:

$$\begin{aligned}\frac{\partial(q_1, p_1)}{\partial(q_0, p_0)} &= \begin{vmatrix} \frac{\partial q_1}{\partial q_0} & \frac{\partial p_1}{\partial q_0} \\ \frac{\partial q_1}{\partial p_0} & \frac{\partial p_1}{\partial p_0} \end{vmatrix} \\ &= \begin{vmatrix} 1 + \frac{\partial^2 H}{\partial q_0 \partial p_0} dt & -\frac{\partial^2 H}{\partial {q_0}^2} dt \\ \frac{\partial^2 H}{\partial {p_0}^2} dt & 1 - \frac{\partial^2 H}{\partial q_0 \partial p_0} dt \end{vmatrix} + \mathcal{O}(dt^2) \\ &= 1 + \mathcal{O}(dt^2). \end{aligned} \quad (3)$$

Expressed in another form, the integral of the LIOUVILLE MEASURE,

$$\prod_{i=1}^{N} \int dp_i\, dq_i, \quad (4)$$

is a constant of motion. SYMPLECTIC MAPS of HAMILTONIAN SYSTEMS must therefore be AREA preserving (and have DETERMINANTS equal to 1).

see also LIOUVILLE MEASURE, PHASE SPACE

References

Chavel, I. *Riemannian Geometry: A Modern Introduction.* New York: Cambridge University Press, 1994.

Liouville Polynomial Identity

$$\begin{aligned} 6({x_1}^2 + {x_2}^2 + {x_3}^2 + {x_4}^2) &= (x_1 + x_2)^4 + (x_1 + x_3)^4 \\ &+ (x_2+x_3)^4 + (x_1+x_4)^4 + (x_2+x_4)^4 + (x_3+x_4)^4 + (x_1-x_2)^4 \\ &+ (x_1 - x_3)^4 + (x_2 - x_3)^4 + (x_1 - x_4)^4 + (x_2 - x_4)^4 \\ &+ (x_3 - x_4)^4. \end{aligned}$$

This is proven in Rademacher and Toeplitz (1957).

see also WARING'S PROBLEM

References

Rademacher, H. and Toeplitz, O. *The Enjoyment of Mathematics: Selections from Mathematics for the Amateur.* Princeton, NJ: Princeton University Press, pp. 55–56, 1957.

Liouville's Rational Approximation Theorem

For any ALGEBRAIC NUMBER x of degree $n > 1$, a RATIONAL approximation $x = p/q$ must satisfy

$$\left|x - \frac{p}{q}\right| > \frac{1}{q^{n+1}}$$

for sufficiently large q. Writing $r \equiv n + 1$ leads to the definition of the LIOUVILLE-ROTH CONSTANT of a given number.

see also LAGRANGE NUMBER (RATIONAL APPROXIMATION), LIOUVILLE'S CONSTANT, LIOUVILLE NUMBER, LIOUVILLE-ROTH CONSTANT, MARKOV NUMBER, ROTH'S THEOREM, THUE-SIEGEL-ROTH THEOREM

References

Courant, R. and Robbins, H. "Liouville's Theorem and the Construction of Transcendental Numbers." §2.6.2 in *What is Mathematics?: An Elementary Approach to Ideas and Methods, 2nd ed.* Oxford, England: Oxford University Press, pp. 104–107, 1996.

Liouville-Roth Constant

N.B. A detailed on-line essay by S. Finch was the starting point for this entry.

Let x be a REAL NUMBER, and let R be the SET of POSITIVE REAL NUMBERS for which

$$\left|x - \frac{p}{q}\right| < \frac{1}{q^r} \quad (1)$$

has (at most) finitely many solutions p/q for p and q INTEGERS. Then the Liouville-Roth constant (or IRRATIONALITY MEASURE) is defined as the threshold at which LIOUVILLE'S RATIONAL APPROXIMATION THEOREM kicks in and x is no longer approximable by RATIONAL NUMBERS,

$$r(x) \equiv \inf_{r \in R} r. \quad (2)$$

There are three regimes:

$$\begin{cases} r(x) = 1 & x \text{ is rational} \\ r(x) = 2 & x \text{ is algebraic irrational} \\ r(x) \geq 2 & x \text{ is transcendental.} \end{cases} \quad (3)$$

The best known upper bounds for common constants are

$$r(L) = \infty \quad (4)$$
$$r(e) = 2 \quad (5)$$
$$r(\pi) < 8.0161 \quad (6)$$
$$r(\ln 2) < 4.13 \quad (7)$$
$$r(\pi^2) < 6.3489 \quad (8)$$
$$r(\zeta(3)) < 13.42, \quad (9)$$

where L is LIOUVILLE'S CONSTANT, $\zeta(3)$ is APÉRY'S CONSTANT, and the lower bounds are 2 for the inequalities.

see also LIOUVILLE'S RATIONAL APPROXIMATION THEOREM, ROTH'S THEOREM, THUE-SIEGEL-ROTH THEOREM

References

Borwein, J. M. and Borwein, P. B. *Pi & the AGM: A Study in Analytic Number Theory and Computational Complexity.* New York: Wiley, 1987.

Finch, S. "Favorite Mathematical Constants." http://www.mathsoft.com/asolve/constant/lvlrth/lvlrth.html.

Hardy, G. H. and Wright, E. M. *An Introduction to the Theory of Numbers, 5th ed.* Oxford: Clarendon Press, 1979.

Hata, M. "Improvement in the Irrationality Measures of π and π^2." *Proc. Japan. Acad. Ser. A Math. Sci.* **68**, 283–286, 1992.

Hata, M. "Rational Approximations to π and Some Other Numbers." *Acta Arith.* **63** 335–349, 1993.

Hata, M. "A Note on Beuker's Integral." *J. Austral. Math. Soc.* **58**, 143–153, 1995.

Stark, H. M. *An Introduction to Number Theory.* Cambridge, MA: MIT Press, 1978.

Liouville Space

Also known as LINE SPACE or "extended" HILBERT SPACE, it is the DIRECT PRODUCT of two HILBERT SPACES.

see also DIRECT PRODUCT (SET), HILBERT SPACE

Liouville's Sphere-Preserving Theorem

see LIOUVILLE'S CONFORMALITY THEOREM

Lipschitz Condition

A function $f(x)$ satisfies the Lipschitz condition of order α at $x = 0$ if

$$|f(h) - f(0)| \leq B|h|^\beta$$

for all $|h| < \epsilon$, where B and β are independent of h, $\beta > 0$, and α is an UPPER BOUND for all β for which a finite B exists.

see also HILLAM'S THEOREM, LIPSCHITZ FUNCTION

Lipschitz Function

A function f such that

$$|f(x) - f(y)| \leq C|x - y|$$

is called a Lipschitz function.

see also LIPSCHITZ CONDITION

References

Morgan, F. "What Is a Surface?" *Amer. Math. Monthly* **103**, 369–376, 1996.

Lipschitz's Integral

$$\int_0^\infty e^{-ax} J_0(bx)\,dx = \frac{1}{\sqrt{a^2 + b^2}},$$

where $J_0(z)$ is the zeroth order BESSEL FUNCTION OF THE FIRST KIND.

References

Bowman, F. *Introduction to Bessel Functions.* New York: Dover, p. 58, 1958.

Lissajous Curve

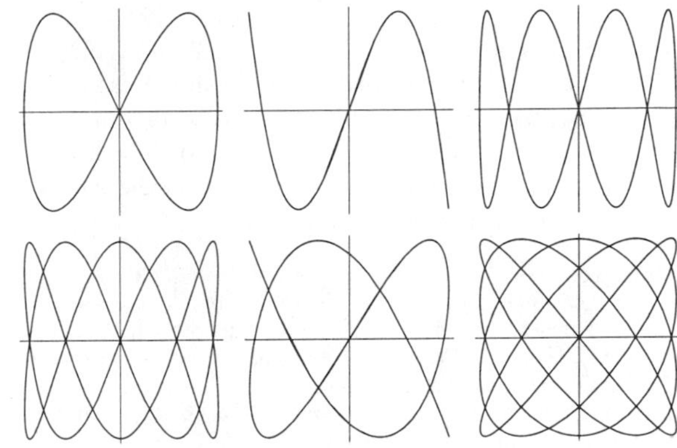

Lissajous curves are the family of curves described by the parametric equations

$$x(t) = A\cos(\omega_x t - \delta_x) \quad (1)$$
$$y(t) = B\cos(\omega_y t - \delta_y), \quad (2)$$

sometimes also written in the form

$$x(t) = a\sin(nt + c) \quad (3)$$
$$y(t) = b\sin t. \quad (4)$$

They are sometimes known as BOWDITCH CURVES after Nathaniel Bowditch, who studied them in 1815. They were studied in more detail (independently) by Jules-Antoine Lissajous in 1857 (MacTutor Archive). Lissajous curves have applications in physics, astronomy, and other sciences. The curves close IFF ω_x/ω_y is RATIONAL.

References

Gray, A. *Modern Differential Geometry of Curves and Surfaces.* Boca Raton, FL: CRC Press, pp. 53–54, 1993.

Lawrence, J. D. *A Catalog of Special Plane Curves.* New York: Dover, pp. 178–179 and 181–183, 1972.

MacTutor History of Mathematics Archive. "Lissajous Curves." http://www-groups.dcs.st-and.ac.uk/~history/Curves/Lissajous.html.

Lissajous Figure

see LISSAJOUS CURVE

List

A DATA STRUCTURE consisting of an order SET of elements, each of which may be a number, another list, etc. A list is usually denoted $(a_1, a_2, \ldots, a_n)$ or $\{a_1, a_2, \ldots, a_n\}$.

see also QUEUE, STACK

Lituus

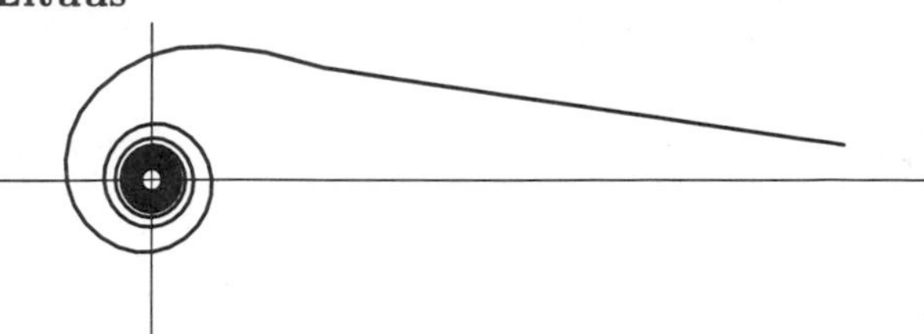

An ARCHIMEDEAN SPIRAL with $m = -2$, having polar equation

$$r^2\theta = a^2.$$

Lituus means a "crook," in the sense of a bishop's crosier. The lituus curve originated with Cotes in 1722. Maclaurin used the term lituus in his book *Harmonia Mensurarum* in 1722 (MacTutor Archive). The lituus is the locus of the point P moving such that the AREA of a circular SECTOR remains constant.

References

Gray, A. *Modern Differential Geometry of Curves and Surfaces.* Boca Raton, FL: CRC Press, pp. 69–70, 1993.

Lawrence, J. D. *A Catalog of Special Plane Curves.* New York: Dover, pp. 186 and 188, 1972.

Lockwood, E. H. *A Book of Curves.* Cambridge, England: Cambridge University Press, p. 175, 1967.

MacTutor History of Mathematics Archive. "Lituus." http://www-groups.dcs.st-and.ac.uk/~history/Curves/Lituus.html.

Lituus Inverse Curve

The INVERSE CURVE of the LITUUS is an ARCHIMEDEAN SPIRAL with $m = 2$, which is FERMAT'S SPIRAL.

see also ARCHIMEDEAN SPIRAL, FERMAT'S SPIRAL, LITUUS

LLL Algorithm

An INTEGER RELATION algorithm.

see also FERGUSON-FORCADE ALGORITHM, HJLS ALGORITHM, INTEGER RELATION, PSLQ ALGORITHM, PSOS ALGORITHM

References

Lenstra, A. K.; Lenstra, H. W.; and Lovasz, L. "Factoring Polynomials with Rational Coefficients." *Math. Ann.* **261**, 515–534, 1982.

Ln

The LOGARITHM to BASE e, also called the NATURAL LOGARITHM, is denoted ln, i.e.,

$$\ln x \equiv \log_e x.$$

see also BASE (LOGARITHM), E, LG, LOGARITHM, NAPIERIAN LOGARITHM, NATURAL LOGARITHM

Lo Shu

8	1	6
3	5	7
4	9	2

The unique MAGIC SQUARE of order three. The Lo Shu is an ASSOCIATIVE MAGIC SQUARE, but not a PANMAGIC SQUARE.

see also ASSOCIATIVE MAGIC SQUARE, MAGIC SQUARE, PANMAGIC SQUARE

References

Hunter, J. A. H. and Madachy, J. S. *Mathematical Diversions.* New York: Dover, pp. 23–24, 1975.

Lobachevsky-Bolyai-Gauss Geometry

see HYPERBOLIC GEOMETRY

Lobachevsky's Formula

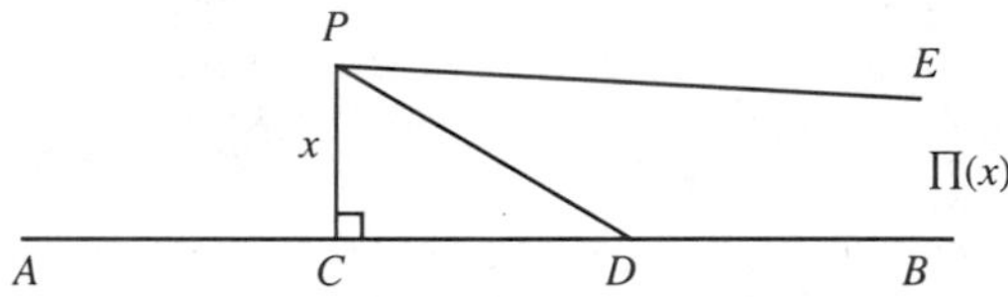

Given a point P and a LINE AB, draw the PERPENDICULAR through P and call it PC. Let PD be any other line from P which meets CB in D. In a HYPERBOLIC GEOMETRY, as D moves off to infinity along CB, then the line PD approaches the limiting line PE, which is said to be parallel to CB at P. The angle $\angle CPE$ which PE makes with PC is then called the ANGLE OF PARALLELISM for perpendicular distance x, and is given by

$$\Pi(x) = 2\tan^{-1}(e^{-x}),$$

which is called Lobachevsky's formula.

see also ANGLE OF PARALLELISM, HYPERBOLIC GEOMETRY

References

Manning, H. P. *Introductory Non-Euclidean Geometry.* New York: Dover, p. 58, 1963.

Lobatto Quadrature

Also called RADAU QUADRATURE (Chandrasekhar 1960). A GAUSSIAN QUADRATURE with WEIGHTING FUNCTION $W(x) = 1$ in which the endpoints of the interval $[-1, 1]$ are included in a total of n ABSCISSAS, giving $r = n - 2$ free abscissas. ABSCISSAS are symmetrical about the origin, and the general FORMULA is

$$\int_{-1}^{1} f(x)\,dx = w_1 f(-1) + w_n f(1) + \sum_{i=2}^{n-1} w_i f(x_i). \quad (1)$$

The free ABSCISSAS x_i for $i = 2, \ldots, n-1$ are the roots of the POLYNOMIAL $P'_{n-1}(x)$, where $P(x)$ is a LEGENDRE POLYNOMIAL. The weights of the free abscissas are

$$w_i = -\frac{2n}{(1 - x_i{}^2)P''_{n-1}(x_i)P'_m(x_i)} \quad (2)$$

$$= \frac{2}{n(n-1)[P_{n-1}(x_i)]^2}, \quad (3)$$

and of the endpoints are

$$w_{1,n} = \frac{2}{n(n-1)}. \quad (4)$$

The error term is given by

$$E = -\frac{n(n-1)^3 2^{2n-1}[(n-2)!]^4}{(2n-1)[(2n-2)!]^3} f^{(2n-2)}(\xi), \quad (5)$$

for $\xi \in (-1, 1)$. Beyer (1987) gives a table of parameters up to n=11 and Chandrasekhar (1960) up to n=9 (although Chandrasekhar's $\mu_{3,4}$ for $m = 5$ is incorrect).

n	x_i	w_i
3	0	1.33333
	±1	0.333333
4	±0.447214	0.833333
	±1	0.166667
5	0	0.711111
	±0.654654	0.544444
	±1	0.100000
6	±0.285232	0.554858
	±0.765055	0.378475
	±1	0.0666667

The ABSCISSAS and weights can be computed analytically for small n.

n	x_i	w_i
3	0	$\frac{4}{3}$
	± 1	$\frac{1}{3}$
4	$\pm\frac{1}{5}\sqrt{5}$	$\frac{1}{6}$
	± 1	$\frac{5}{6}$
5	0	$\frac{32}{45}$
	$\pm\frac{1}{7}\sqrt{21}$	$\frac{49}{90}$
	± 1	$\frac{1}{10}$

see also CHEBYSHEV QUADRATURE, RADAU QUADRATURE

References

Abramowitz, M. and Stegun, C. A. (Eds.). *Handbook of Mathematical Functions with Formulas, Graphs, and Mathematical Tables, 9th printing.* New York: Dover, pp. 888–890, 1972.

Beyer, W. H. *CRC Standard Mathematical Tables, 28th ed.* Boca Raton, FL: CRC Press, p. 465, 1987.

Chandrasekhar, S. *Radiative Transfer.* New York: Dover, pp. 63–64, 1960.

Hildebrand, F. B. *Introduction to Numerical Analysis.* New York: McGraw-Hill, pp. 343–345, 1956.

Lobster

A 6-POLYIAMOND.

References

Golomb, S. W. *Polyominoes: Puzzles, Patterns, Problems, and Packings, 2nd ed.* Princeton, NJ: Princeton University Press, p. 92, 1994.

Local Cell

The POLYHEDRON resulting from letting each SPHERE in a SPHERE PACKING expand uniformly until it touches its neighbors on flat faces.

see also LOCAL DENSITY

Local Degree

The degree of a VERTEX of a GRAPH is the number of EDGES which touch the VERTEX, also called the LOCAL DEGREE. The VERTEX degree of a point A in a GRAPH, denoted $\rho(A)$, satisfies

$$\sum_{i=1}^{n} \rho(A_i) = 2E,$$

where E is the total number of EDGES. Directed graphs have two types of degrees, known as the INDEGREE and OUTDEGREE.

see also INDEGREE, OUTDEGREE

Local Density

Let each SPHERE in a SPHERE PACKING expand uniformly until it touches its neighbors on flat faces. Call the resulting POLYHEDRON the LOCAL CELL. Then the local density is given by

$$\rho \equiv \frac{V_{\text{sphere}}}{V_{\text{local cell}}}.$$

When the LOCAL CELL is a regular DODECAHEDRON, then

$$\rho_{\text{dodecahedron}} = \frac{\pi\sqrt{5+\sqrt{5}}}{5\sqrt{10}\,(\sqrt{5}-2)} = 0.7547\ldots.$$

see also LOCAL DENSITY CONJECTURE

Local Density Conjecture

The CONJECTURE that the maximum LOCAL DENSITY is given by $\rho_{\text{dodecahedron}}$.

see also LOCAL DENSITY

Local Extremum

A LOCAL MINIMUM or LOCAL MAXIMUM.

see also EXTREMUM, GLOBAL EXTREMUM

Local Field

A FIELD which is complete with respect to a discrete VALUATION is called a local field if its FIELD of RESIDUE CLASSES is FINITE. The HASSE PRINCIPLE is one of the chief applications of local field theory.

see also HASSE PRINCIPLE, VALUATION

References

Iyanaga, S. and Kawada, Y. (Eds.). "Local Fields." §257 in *Encyclopedic Dictionary of Mathematics.* Cambridge, MA: MIT Press, pp. 811–815, 1980.

Local-Global Principle

see HASSE PRINCIPLE

Local Group Theory

The study of a FINITE GROUP G using the LOCAL SUBGROUPS of G. Local group theory plays a critical role in the CLASSIFICATION THEOREM.

see also SYLOW THEOREMS

Local Maximum

The largest value of a set, function, etc., within some local neighborhood.

see also GLOBAL MAXIMUM, LOCAL MINIMUM, MAXIMUM, PEANO SURFACE

Local Minimum

The smallest value of a set, function, etc., within some local neighborhood.

see also GLOBAL MINIMUM, LOCAL MAXIMUM, MINIMUM

Local Ring

A NOETHERIAN RING R with a Jacobson radical which has only a single maximal ideal.

References

Iyanaga, S. and Kawada, Y. (Eds.). "Local Rings." §281D in *Encyclopedic Dictionary of Mathematics.* Cambridge, MA: MIT Press, pp. 890–891, 1980.

Local Subgroup

A normalizer of a nontrivial SYLOW p-SUBGROUP of a GROUP G.

see also LOCAL GROUP THEORY

Local Surface

see PATCH

Locally Convex Space

see LOCALLY PATHWISE-CONNECTED SPACE

Locally Finite Space

A locally finite SPACE is one for which every point of a given space has a NEIGHBORHOOD that meets only finitely many elements of the COVER.

Locally Pathwise-Connected Space

A SPACE X is locally pathwise-connected if for every NEIGHBORHOOD around every point in X, there is a smaller, PATHWISE-CONNECTED NEIGHBORHOOD.

Loculus of Archimedes

see STOMACHION

Locus

The set of all points (usually forming a curve or surface) satisfying some condition. For example, the locus of points in the plane equidistant from a given point is a CIRCLE, and the set of points in 3-space equidistant from a given point is a SPHERE.

Log

The symbol $\log x$ is used by physicists, engineers, and calculator keypads to denote the BASE 10 LOGARITHM. However, mathematicians generally use the same symbol to mean the NATURAL LOGARITHM LN, $\ln x$. In this work, $\log x = \log_{10} x$, and $\ln x = \log_e x$ is used for the NATURAL LOGARITHM.

see also LG, LN, LOGARITHM, NATURAL LOGARITHM

Log Likelihood Procedure

A method for testing NESTED HYPOTHESES. To apply the procedure, given a specific model, calculate the LIKELIHOOD of observing the actual data. Then compare this likelihood to a nested model (i.e., one in which fewer parameters are allowed to vary independently).

Log Normal Distribution

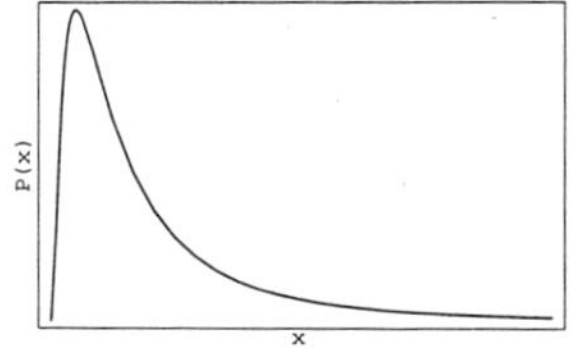

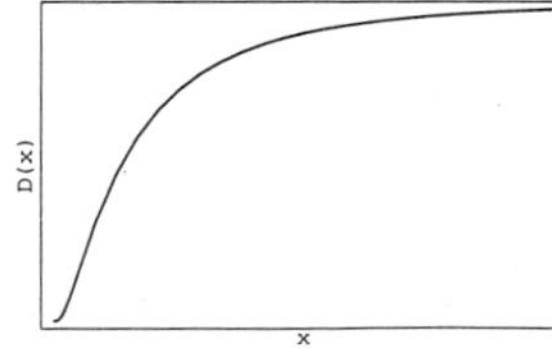

A CONTINUOUS DISTRIBUTION in which the LOGARITHM of a variable has a NORMAL DISTRIBUTION. It is a general case of GILBRAT'S DISTRIBUTION, to which the log normal distribution reduces with $S = 1$ and

$M = 0$. The probability density and cumulative distribution functions are log normal distribution

$$P(x) = \frac{1}{Sx\sqrt{2\pi}} e^{-(\ln x - M)^2/(2S^2)} \tag{1}$$

$$D(x) = \frac{1}{2}\left[1 + \operatorname{erf}\left(\frac{\ln x - M}{S\sqrt{2}}\right)\right], \tag{2}$$

where $\operatorname{erf}(x)$ is the ERF function. This distribution is normalized, since letting $y \equiv \ln x$ gives $dy = dx/x$ and $x = e^y$, so

$$\int_0^\infty P(x)\,dx = \frac{1}{S\sqrt{2\pi}} \int_{-\infty}^\infty e^{-(y-M)^2/2S^2}\,dy = 1. \tag{3}$$

The MEAN, VARIANCE, SKEWNESS, and KURTOSIS are given by

$$\mu = e^{M+S^2/2} \tag{4}$$

$$\sigma^2 = e^{S^2+2M}(e^{S^2} - 1) \tag{5}$$

$$\gamma_1 = \sqrt{e^S - 1}\,(2 + e^{S^2}) \tag{6}$$

$$\gamma_2 = e^{2S^2}(3 + 2e^{S^2} + e^{2s^2}) - 3. \tag{7}$$

These can be found by direct integration

$$\begin{aligned}
\mu &= \frac{1}{S\sqrt{2\pi}} \int_0^\infty e^{-(\ln x - M)^2/2S^2}\,dx \\
&= \frac{1}{S\sqrt{2\pi}} \int_{-\infty}^\infty e^{(y-M)^2/2S^2} e^y\,dy \\
&= \frac{1}{S\sqrt{2\pi}} \int_{-\infty}^\infty e^{-[-y+(y-M)^2/2S^2]}\,dy \\
&= \frac{1}{S\sqrt{2\pi}} \int_{-\infty}^\infty e^{-(-2S^2y+y^2-2yM+M^2)/2S^2}\,dy \\
&= \frac{1}{S\sqrt{2\pi}} \int_{-\infty}^\infty e^{-\{[y-(S^2+M)]^2+S^2(S^2+2M)\}/2S^2}\,dy \\
&= \frac{1}{S\sqrt{2\pi}} e^{M+S^2/2} \int_{-\infty}^\infty e^{-[y-(S^2+M)^2]/2S^2}\,dy \\
&= e^{M+S^2/2},
\end{aligned} \tag{8}$$

and similarly for σ^2. Examples of variates which have approximately log normal distributions include the size of silver particles in a photographic emulsion, the survival time of bacteria in disinfectants, the weight and blood pressure of humans, and the number of words written in sentences by George Bernard Shaw.

see also GILBRAT'S DISTRIBUTION, WEIBULL DISTRIBUTION

References

Aitchison, J. and Brown, J. A. C. *The Lognormal Distribution, with Special Reference to Its Use in Economics.* New York: Cambridge University Press, 1957.

Kenney, J. F. and Keeping, E. S. *Mathematics of Statistics, Pt. 2, 2nd ed.* Princeton, NJ: Van Nostrand, p. 123, 1951.

Log-Series Distribution

The terms in the series expansion of $\ln(1-\theta)$ about $\theta = 0$ are proportional to this distribution.

$$P(n) = -\frac{\theta^n}{n\ln(1-\theta)} \tag{1}$$

$$D(n) \equiv \sum_{i=1}^n P(i) = \frac{\theta^{1+n}\Phi(\theta,1,1+n) + \ln(1-\theta)}{\ln(1-\theta)}, \tag{2}$$

where Φ is the LERCH TRANSCENDENT. The MEAN, VARIANCE, SKEWNESS, and KURTOSIS

$$\mu = \frac{\theta}{(\theta-1)\ln(1-\theta)} \tag{3}$$

$$\sigma^2 = -\frac{\theta[\theta + \ln(1-\theta)]}{(\theta-1)^2[\ln(1-\theta)]^2} \tag{4}$$

$$\gamma_1 = \frac{2\theta^2 + 3\theta\ln(1-\theta) + (1+\theta)\ln^2(1-\theta)}{\ln(1-\theta)[\theta+\ln(1-\theta)]\sqrt{-\theta[\theta+\ln(1-\theta)]}}\ln(1-\theta) \tag{5}$$

$$\gamma_2 = \frac{6\theta^3 + 12\theta^2\ln(1-\theta) + \theta(7+4\theta)\ln^2(1-\theta)}{\theta[\theta+\ln(1-\theta)]^2} + \frac{(1+4\theta+\theta^2)\ln^3(1-\theta)}{\theta[\theta+\ln(1-\theta)]^2}. \tag{6}$$

Log-Weibull Distribution

see FISHER-TIPPETT DISTRIBUTION

Logarithm

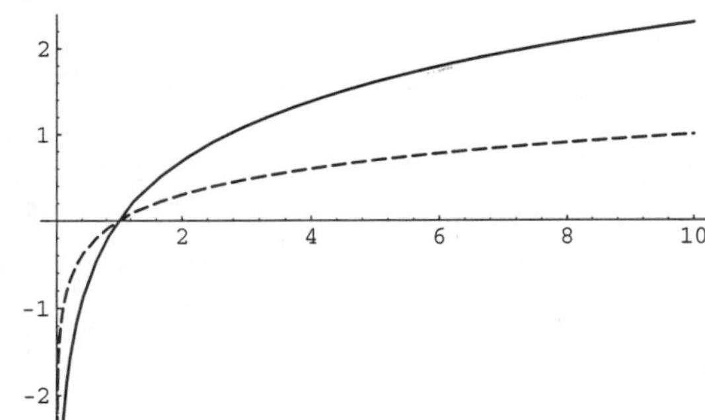

The logarithm is defined to be the INVERSE FUNCTION of taking a number to a given POWER. Therefore, for any x and b,

$$x = b^{\log_b x}, \tag{1}$$

or equivalently,

$$x = \log_b(b^x). \tag{2}$$

Here, the POWER b is known as the BASE of the logarithm. For any BASE, the logarithm function has a SINGULARITY at $x = 0$. In the above plot, the solid curve is the logarithm to BASE e (the NATURAL LOGARITHM), and the dotted curve is the logarithm to BASE 10 (LOG).

Logarithms are used in many areas of science and engineering in which quantities vary over a large range. For example, the decibel scale for the loudness of sound, the Richter scale of earthquake magnitudes, and the astronomical scale of stellar brightnesses are all logarithmic scales.

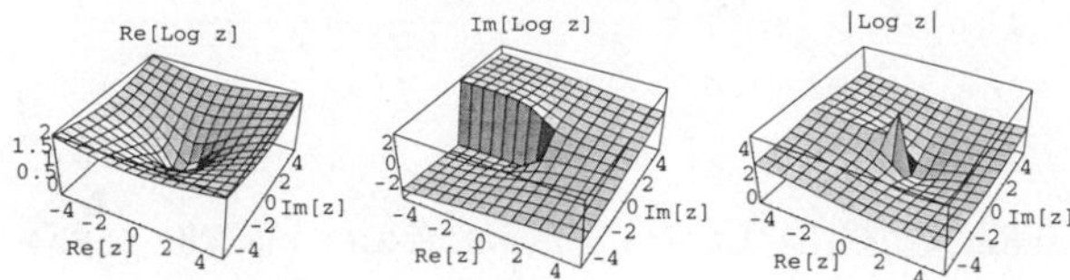

The logarithm can also be defined for COMPLEX arguments, as shown above. If the logarithm is taken as the forward function, the function taking the BASE to a given POWER is then called the ANTILOGARITHM.

For $x = \log N$, $\lfloor x \rfloor$ is called the CHARACTERISTIC and $x - \lfloor x \rfloor$ is called the MANTISSA. Division and multiplication identities follow from these

$$xy = b^{\log_b x} b^{\log_b y} = b^{\log_b x + \log_b y}, \tag{3}$$

from which it follows that

$$\log_b(xy) = \log_b x + \log_b y \tag{4}$$

$$\log_b \left(\frac{x}{y}\right) = \log_b x - \log_b y \tag{5}$$

$$\log_b x^n = n \log_b x. \tag{6}$$

There are a number of properties which can be used to change from one BASE to another

$$a = a^{\log_a b / \log_a b} = (a^{\log_a b})^{1/\log_a b} = b^{1/\log_a b} \tag{7}$$

$$\log_b a = \frac{1}{\log_a b} \tag{8}$$

$$\log_x z = \log_x (y^{\log_y z}) = \log_y z \log_x y \tag{9}$$

$$\log_y z = \frac{\log_x z}{\log_x y} \tag{10}$$

$$a^x = b^{x/\log_a b} = b^{x \log_b a}. \tag{11}$$

The logarithm BASE e is called the NATURAL LOGARITHM and is denoted $\ln x$ (LN). The logarithm BASE 10 is denoted $\log x$ (LOG), (although mathematics texts often use $\log x$ to mean $\ln x$). The logarithm BASE 2 is denoted $\lg x$ (LG).

An interesting property of logarithms follows from looking for a number y such that

$$\log_b(x+y) = -\log_b(x-y) \tag{12}$$

$$x + y = \frac{1}{x-y} \tag{13}$$

$$x^2 - y^2 = 1 \tag{14}$$

$$y = \sqrt{x^2 - 1}, \tag{15}$$

so

$$\log_b(x + \sqrt{x^2-1}) = -\log_b(x - \sqrt{x^2-1}). \tag{16}$$

Numbers of the form $\log_a b$ are IRRATIONAL if a and b are INTEGERS, one of which has a PRIME factor which the other lacks. A. Baker made a major step forward in TRANSCENDENTAL NUMBER theory by proving the transcendence of sums of numbers of the form $\alpha \ln \beta$ for α and β ALGEBRAIC NUMBERS.

see also ANTILOGARITHM, COLOGARITHM, e, EXPONENTIAL FUNCTION, HARMONIC LOGARITHM, LG, LN, LOG, LOGARITHMIC NUMBER, NAPIERIAN LOGARITHM, NATURAL LOGARITHM, POWER

References

Abramowitz, M. and Stegun, C. A. (Eds.). "Logarithmic Function." §4.1 in *Handbook of Mathematical Functions with Formulas, Graphs, and Mathematical Tables, 9th printing.* New York: Dover, pp. 67–69, 1972.

Conway, J. H. and Guy, R. K. "Logarithms." *The Book of Numbers.* New York: Springer-Verlag, pp. 248–252, 1996.

Beyer, W. H. "Logarithms." *CRC Standard Mathematical Tables, 28th ed.* Boca Raton, FL: CRC Press, pp. 159–160, 1987.

Pappas, T. "Earthquakes and Logarithms." *The Joy of Mathematics.* San Carlos, CA: Wide World Publ./Tetra, pp. 20–21, 1989.

Spanier, J. and Oldham, K. B. "The Logarithmic Function ln(x)." Ch. 25 in *An Atlas of Functions.* Washington, DC: Hemisphere, pp. 225–232, 1987.

Logarithmic Binomial Formula

see LOGARITHMIC BINOMIAL THEOREM

Logarithmic Binomial Theorem

For all integers n and $|x| < a$,

$$\lambda_n^{(t)}(x+a) = \sum_{k=0}^{\infty} \begin{bmatrix} n \\ k \end{bmatrix} \lambda_{n-k}^{(t)}(a) x^k,$$

where $\lambda_n^{(t)}$ is the HARMONIC LOGARITHM and $\begin{bmatrix} n \\ k \end{bmatrix}$ is a ROMAN COEFFICIENT. For $t = 0$, the logarithmic binomial theorem reduces to the classical BINOMIAL THEOREM for POSITIVE n, since $\lambda_{n-k}^{(0)}(a) = a^{n-k}$ for $n \geq k$, $\lambda_{n-k}^{(0)}(a) = 0$ for $n < k$, and $\begin{bmatrix} n \\ k \end{bmatrix} = \binom{n}{k}$ when $n \geq k \geq 0$.

Similarly, taking $t = 1$ and $n < 0$ gives the NEGATIVE BINOMIAL SERIES. Roman (1992) gives expressions obtained for the case $t = 1$ and $n \geq 0$ which are not obtainable from the BINOMIAL THEOREM.

see also HARMONIC LOGARITHM, ROMAN COEFFICIENT

References

Roman, S. "The Logarithmic Binomial Formula." *Amer. Math. Monthly* **99**, 641–648, 1992.

Logarithmic Distribution

A CONTINUOUS DISTRIBUTION for a variate with probability function

$$P(x) = \frac{\log x}{b(\log b - 1) - a(\log a - 1)}$$

and distribution function

$$D(x) = \frac{a(1 - \log a) - x(1 - \log x)}{a(1 - \log a) - b(1 - \log b)}.$$

The MEAN is

$$\mu = \frac{a^2(1 - 2\log a) - b^2(1 - 2\log b)}{4[a(1 - \log a) - b(1 - \log b)]},$$

but higher order moments are rather messy.

Logarithmic Integral

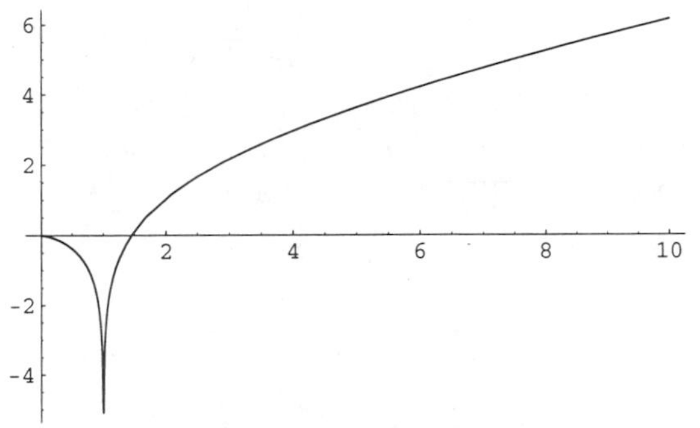

The logarithmic integral is defined by

$$\operatorname{li}(x) \equiv \int_0^x \frac{du}{\ln u}. \tag{1}$$

The offset form appearing in the PRIME NUMBER THEOREM is defined so that $\operatorname{Li}(2) = 0$:

$$\operatorname{Li}(x) \equiv \int_2^x \frac{du}{\ln u} \tag{2}$$

$$= \operatorname{li}(x) - \operatorname{li}(2) \approx \operatorname{li}(x) - 1.04516 \tag{3}$$

$$= \operatorname{ei}(\ln x), \tag{4}$$

where $\operatorname{ei}(x)$ is the EXPONENTIAL INTEGRAL. (Note that the NOTATION $\operatorname{Li}_n(z)$ is also used for the POLYLOGARITHM.) Nielsen (1965, pp. 3 and 11) showed and Ramanujan independently discovered (Berndt 1994) that

$$\int_\mu^x \frac{dt}{\ln t} = \gamma + \ln\ln x + \sum_{k=1}^\infty \frac{(\ln x)^k}{k!k}, \tag{5}$$

where γ is the EULER-MASCHERONI CONSTANT and μ is SOLDNER'S CONSTANT. Another FORMULA due to Ramanujan which converges more rapidly is

$$\int_\mu^x \frac{dt}{\ln t} = \gamma + \ln\ln x + \sqrt{x}\sum_{n=0}^\infty \frac{(-1)^{n-1}(\ln x)^n}{n!2^{n-1}} \sum_{k=0}^{[(n-1)/2]} \frac{1}{2k+1} \tag{6}$$

(Berndt 1994).

see also POLYLOGARITHM, PRIME CONSTELLATION, PRIME NUMBER THEOREM, SKEWES NUMBER

References

Berndt, B. C. *Ramanujan's Notebooks, Part IV.* New York: Springer-Verlag, pp. 126–131, 1994.

Nielsen, N. *Theorie des Integrallogarithms.* New York: Chelsea, 1965.

Vardi, I. *Computational Recreations in Mathematica.* Reading, MA: Addison-Wesley, p. 151, 1991.

Logarithmic Number

A COEFFICIENT of the MACLAURIN SERIES of

$$\frac{1}{\ln(1+x)} = \frac{1}{x} + \tfrac{1}{2} + \tfrac{1}{12}x^2 - \tfrac{19}{720}x^3 + \tfrac{3}{160}x^4 + \ldots$$

(Sloane's A002206 and A002207), the multiplicative inverse of the MERCATOR SERIES function $\ln(1+x)$.

see also MERCATOR SERIES

References

Sloane, N. J. A. Sequences A002206/M5066 and A002207/M2017 in "An On-Line Version of the Encyclopedia of Integer Sequences."

Logarithmic Spiral

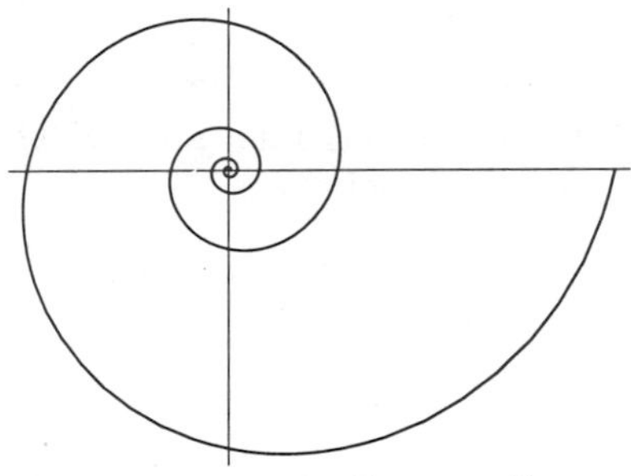

A curve whose equation in POLAR COORDINATES is given by

$$r = ae^{b\theta}, \tag{1}$$

where r is the distance from the ORIGIN, θ is the angle from the x-axis, and a and b are arbitrary constants. The logarithmic spiral is also known as the GROWTH SPIRAL, EQUIANGULAR SPIRAL, and SPIRA MIRABILIS. It can be expressed parametrically using

$$\cos\theta = \frac{1}{\sqrt{1 - \tan^2\theta}} = \frac{1}{\sqrt{1 + \frac{y^2}{x^2}}} = \frac{x}{\sqrt{x^2 + y^2}} = \frac{x}{r}, \tag{2}$$

which gives

$$x = r\cos\theta = a\cos\theta e^{b\theta} \tag{3}$$

$$y = x\tan\theta = r\sin\theta = a\sin\theta e^{b\theta}. \tag{4}$$

The logarithmic spiral was first studied by Descartes in 1638 and Jakob Bernoulli. Bernoulli was so fascinated by the spiral that he had one engraved on his tomb stone (although the engraver did not draw it true to form). Torricelli worked on it independently and found the length of the curve (MacTutor Archive).

The rate of change of RADIUS is

$$\frac{dr}{d\theta} = abe^{b\theta} = br, \tag{5}$$

and the ANGLE between the tangent and radial line at the point (r, θ) is

$$\psi = \tan^{-1}\left(\frac{r}{\frac{dr}{d\theta}}\right) = \tan^{-1}\left(\frac{1}{b}\right) = \cot^{-1} b. \tag{6}$$

So, as $b \to 0$, $\psi \to \pi/2$ and the spiral approaches a CIRCLE.

If P is any point on the spiral, then the length of the spiral from P to the origin is finite. In fact, from the point P which is at distance r from the origin measured along a RADIUS vector, the distance from P to the POLE along the spiral is just the ARC LENGTH. In addition, any RADIUS from the origin meets the spiral at distances which are in GEOMETRIC PROGRESSION (MacTutor Archive).

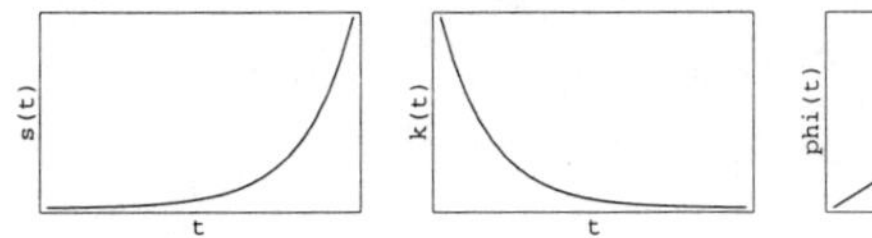

The ARC LENGTH, CURVATURE, and TANGENTIAL ANGLE of the logarithmic spiral are

$$s = \int ds = \int \sqrt{x'^2 + y'^2}\, dt = \frac{a\sqrt{1+b^2}}{b} e^{b\theta}$$

$$= \frac{r\sqrt{1+b^2}}{b} \tag{7}$$

$$\kappa = \frac{x'y'' - y'x''}{(x'^2 + y'^2)^{3/2}} = (a\sqrt{1+b^2} e^{b\theta})^{-1} \tag{8}$$

$$\phi = \int \kappa(s)\, ds = \theta. \tag{9}$$

The CESÀRO EQUATION is

$$\kappa = \frac{1}{bs}. \tag{10}$$

On the surface of a SPHERE, the analog is a LOXODROME. This SPIRAL is related to FIBONACCI NUMBERS and the GOLDEN MEAN.

References

Lawrence, J. D. *A Catalog of Special Plane Curves.* New York: Dover, pp. 184–186, 1972.

Lee, X. "EquiangularSpiral." http://www.best.com/~xah/Special Plane Curves _ dir / Equiangular Spiral _ dir / equiangularSpiral.html.

Lockwood, E. H. "The Equiangular Spiral." Ch. 11 in *A Book of Curves.* Cambridge, England: Cambridge University Press, pp. 98–109, 1967.

MacTutor History of Mathematics Archive. "Equiangular Spiral." http://www-groups.dcs.st-and.ac.uk/~history/Curves/Equiangular.html.

Logarithmic Spiral Caustic Curve

The CAUSTIC of a LOGARITHMIC SPIRAL, where the pole is taken as the RADIANT POINT, is an equal LOGARITHMIC SPIRAL.

Logarithmic Spiral Evolute

$$R = \frac{(r^2 + r_\theta{}^2)^{3/2}}{r^2 + 2r^2 r_\theta{}^2 - r r_{\theta\theta}}. \tag{1}$$

Using

$$r = ae^{b\theta} \quad r_\theta = abe^{b\theta} \quad r_{\theta\theta} = ab^2 e^{b\theta} \tag{2}$$

gives

$$\begin{aligned} R &= \frac{(a^2 e^{2b\theta} + a^2 b^2 e^{2b\theta})^{3/2}}{(ae^{b\theta})^2 + 2(abe^{b\theta})^2 - (ab^{b\theta})(ab^2 e^{b\theta})} \\ &= \frac{(1+b^2)^{3/2} a^3 e^{3b\theta}}{2a^2 b^2 e^{2b\theta} + a^2 e^{2b\theta} - a^2 b^2 e^{2b\theta}} \\ &= \frac{(1+b^2)^{3/2} a^3 e^{3b\theta}}{a^2 b^2 e^{2b\theta} + a^2 e^{2b\theta}} = \frac{(1+b^2)^{3/2} a^3 e^{3b\theta}}{a^2(1+b^2) e^{2b\theta}} \\ &= a\sqrt{1+b^2}\, e^{b\theta} \end{aligned} \tag{3}$$

and

$$\begin{aligned} \begin{bmatrix} x \\ y \end{bmatrix} &= \begin{bmatrix} ae^{b\theta}\cos\theta \\ ae^{b\theta}\sin\theta \end{bmatrix} \\ \begin{bmatrix} x' \\ y' \end{bmatrix} &= \begin{bmatrix} abe^{b\theta}\cos\theta - ae^{b\theta}\sin\theta \\ abe^{b\theta}\sin\theta + ae^{b\theta}\cos\theta \end{bmatrix} \\ &= ae^{b\theta}\begin{bmatrix} b\cos\theta - \sin\theta \\ b\sin\theta + \cos\theta \end{bmatrix}, \end{aligned} \tag{4}$$

so

$$\begin{aligned} |\mathbf{r}'| &= ae^{b\theta}\sqrt{(b\cos\theta - \sin\theta)^2 + (b\sin\theta + \cos\theta)^2} \\ &= ae^{b\theta}\sqrt{1+b^2}, \end{aligned} \tag{5}$$

and the TANGENT VECTOR is given by

$$\begin{aligned} \hat{\mathbf{T}} &= \frac{\mathbf{r}'}{|\mathbf{r}'|} = \frac{1}{ae^{b\theta}\sqrt{1+b^2}}\begin{bmatrix} ae^{b\theta}\cos\theta \\ ae^{b\theta}\sin\theta \end{bmatrix} \\ &= \frac{1}{\sqrt{1+b^2}}\begin{bmatrix} \cos\theta \\ \sin\theta \end{bmatrix}. \end{aligned} \tag{6}$$

The coordinates of the EVOLUTE are therefore

$$\xi = -abe^{b\theta}\sin\theta \tag{7}$$

$$\eta = abe^{b\theta}\cos\theta. \tag{8}$$

So the EVOLUTE is another logarithmic spiral with $a' \equiv ab$, as first shown by Johann Bernoulli. However, in some cases, the EVOLUTE is identical to the original, as can be demonstrated by making the substitution to the new variable

$$\theta \equiv \phi - \tfrac{1}{2}\pi \pm 2n\pi. \tag{9}$$

Then the above equations become

$$\xi = -abe^{b(\phi-\pi/2\pm 2n\pi)}\sin(\phi - \pi/2 \pm 2n\pi)$$
$$= abe^{b\phi}e^{b(-\pi/2\pm 2n\pi)}\cos\phi \quad (10)$$
$$\eta = abe^{b(\phi-\pi/2\pm 2n\pi)}\cos(\phi - \pi/2 \pm 2n\pi)$$
$$= abe^{b\phi}e^{b(-\pi/2\pm 2n\pi)}\sin\phi, \quad (11)$$

which are equivalent to the form of the original equation if

$$be^{b(-\frac{1}{2}\pi\pm 2n\pi)} = 1 \quad (12)$$

$$\ln b + b(-\tfrac{1}{2}\pi \pm 2n\pi) = 0 \quad (13)$$

$$\frac{\ln b}{b} = \tfrac{1}{2}\pi \mp 2n\pi = -(2n - \tfrac{1}{2})\pi, \quad (14)$$

where only solutions with the minus sign in $\mp$ exist. Solving gives the values summarized in the following table.

n	b_n	$\psi = \cot^{-1} b_n$
1	0.2744106319...	74°39′18.53"
2	0.1642700512...	80°40′16.80"
3	0.1218322508...	83°03′13.53"
4	0.0984064967...	84°22′47.53"
5	0.0832810611...	85°14′21.60"
6	0.0725974881...	85°50′51.92"
7	0.0645958183...	86°18′14.64"
8	0.0583494073...	86°39′38.20"
9	0.0533203211...	86°56′52.30"
10	0.0491732529...	87°11′05.45"

References

Lauwerier, H. *Fractals: Endlessly Repeated Geometric Figures.* Princeton, NJ: Princeton University Press, pp. 60–64, 1991.

Logarithmic Spiral Inverse Curve

The INVERSE CURVE of the LOGARITHMIC SPIRAL

$$r = e^{a\theta}$$

with INVERSION CENTER at the origin and inversion radius k is the LOGARITHMIC SPIRAL

$$r = ke^{-a\theta}.$$

Logarithmic Spiral Pedal Curve

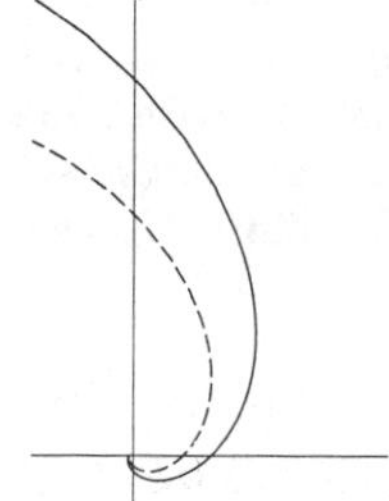

The PEDAL CURVE of a LOGARITHMIC SPIRAL with parametric equation

$$f = e^{at}\cos t \quad (1)$$
$$g = e^{at}\sin t \quad (2)$$

for a PEDAL POINT at the pole is an identical LOGARITHMIC SPIRAL

$$x = \frac{(a\sin t + \cos t)e^{at}}{1+a^2} \quad (3)$$
$$y = \frac{(\sin t - a\cos t)e^{at}}{1+a^2}, \quad (4)$$

so

$$r = \sqrt{x^2+y^2} = \frac{e^{at}}{\sqrt{1+a^2}}. \quad (5)$$

Logarithmic Spiral Radial Curve

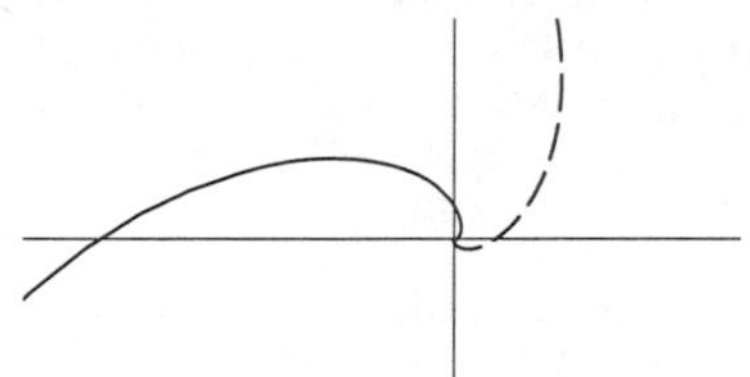

The RADIAL CURVE of the LOGARITHMIC SPIRAL is another LOGARITHMIC SPIRAL.

Logarithmically Convex Function

A function $f(x)$ is logarithmically convex on the interval $[a, b]$ if $f > 0$ and $\ln f(x)$ is CONCAVE on $[a, b]$. If $f(x)$ and $g(x)$ are logarithmically convex on the interval $[a, b]$, then the functions $f(x) + g(x)$ and $f(x)g(x)$ are also logarithmically convex on $[a, b]$.

see also CONVEX FUNCTION

References

Gradshteyn, I. S. and Ryzhik, I. M. *Tables of Integrals, Series, and Products, 5th ed.* San Diego, CA: Academic Press, p. 1100, 1980.

Logic

The formal mathematical study of the methods, structure, and validity of mathematical deduction and proof. Formal logic seeks to devise a complete, consistent formulation of mathematics such that propositions can be formally stated and proved using a small number of symbols with well-defined meanings. While this sounds like an admirable pursuit in principle, in practice the study of mathematical logic can rapidly become bogged down in pages of dense and unilluminating mathematical symbols, of which Whitehead and Russell's *Principia Mathematica* (1925) is perhaps the best (or worst) example.

A very simple form of logic is the study of "TRUTH TABLES" and digital logic circuits in which one or more outputs depend on a combination of circuit elements (AND, NAND, OR, XOR, etc.; "gates") and the input values. In such a circuit, values at each point can take on values of only TRUE (1) or FALSE (0). DE MORGAN'S DUALITY LAW is a useful principle for the analysis and simplification of such circuits.

A generalization of this simple type of logic in which possible values are TRUE, FALSE, and "undecided" is called THREE-VALUED LOGIC. A further generalization called FUZZY LOGIC treats "truth" as a continuous quantity ranging from 0 to 1.

see also ABSORPTION LAW, ALETHIC, BOOLEAN ALGEBRA, BOOLEAN CONNECTIVE, BOUND, CALIBAN PUZZLE, CONTRADICTION LAW, DE MORGAN'S DUALITY LAW, DE MORGAN'S LAWS, DEDUCIBLE, EXCLUDED MIDDLE LAW, FREE, FUZZY LOGIC, GÖDEL'S INCOMPLETENESS THEOREM, KHOVANSKI'S THEOREM, LOGICAL PARADOX, LOGOS, LÖWENHEIMER-SKOLEM THEOREM, METAMATHEMATICS, MODEL THEORY, QUANTIFIER, SENTENCE, TARSKI'S THEOREM, TAUTOLOGY, THREE-VALUED LOGIC, TOPOS, TRUTH TABLE, TURING MACHINE, UNIVERSAL STATEMENT, UNIVERSAL TURING MACHINE, VENN DIAGRAM, WILKIE'S THEOREM

References

Adamowicz, Z. and Zbierski, P. *Logic of Mathematics: A Modern Course of Classical Logic.* New York: Wiley, 1997.

Bogomolny, A. "Falsity Implies Anything." `http://www.cut-the-knot.com/do_you_know/falsity.html`.

Carnap, R. *Introduction to Symbolic Logic and Its Applications.* New York: Dover, 1958.

Church, A. *Introduction to Mathematical Logic, Vol. 1.* Princeton, NJ: Princeton University Press, 1996.

Gödel, K. *On Formally Undecidable Propositions of Principia Mathematica and Related Systems.* New York: Dover, 1992.

Jeffrey, R. C. *Formal Logic: Its Scope and Limits.* New York: McGraw-Hill, 1967.

Kac, M. and Ulam, S. M. *Mathematics and Logic: Retrospect and Prospects.* New York: Dover, 1992.

Kleene, S. C. *Introduction to Metamathematics.* Princeton, NJ: Van Nostrand, 1971.

Whitehead, A. N. and Russell, B. *Principia Mathematica, 2nd ed.* Cambridge, England: Cambridge University Press, 1962.

Logical Paradox

see PARADOX

Logistic Distribution

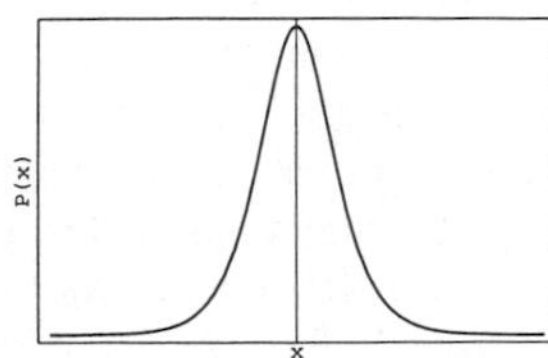

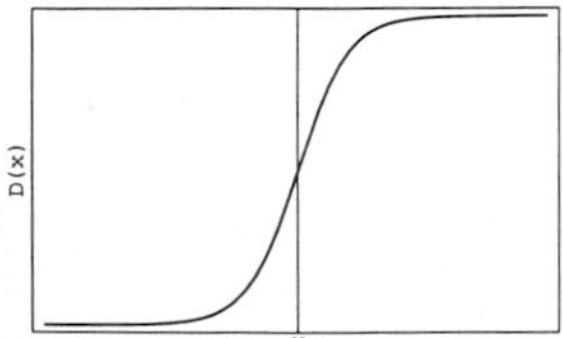

$$P(x) = \frac{e^{(x-m)/b}}{|b|[1+e^{(x-m)/b}]^2} \quad (1)$$

$$D(x) = \frac{1}{1+e^{(m-x)/|b|}}, \quad (2)$$

and the MEAN, VARIANCE, SKEWNESS, and KURTOSIS are

$$\mu = m \quad (3)$$

$$\sigma^2 = \tfrac{1}{3}\pi^2\beta^2 \quad (4)$$

$$\gamma_1 = 0 \quad (5)$$

$$\gamma_2 = \tfrac{6}{5}. \quad (6)$$

see also LOGISTIC EQUATION, LOGISTIC GROWTH CURVE

References

von Seggern, D. *CRC Standard Curves and Surfaces.* Boca Raton, FL: CRC Press, p. 250, 1993.

Logistic Equation

The logistic equation (sometimes called the VERHULST MODEL since it was first published in 1845 by the Belgian P.-F. Verhulst) is defined by

$$x_{n+1} = rx_n(1-x_n), \quad (1)$$

where r (sometimes also denoted μ) is a POSITIVE constant (the "biotic potential"). We will start x_0 in the interval $[0,1]$. In order to keep points in the interval, we must find appropriate conditions on r. The maximum value x_{n+1} can take is found from

$$\frac{dx_{n+1}}{dx_n} = r(1-2x_n) = 0, \quad (2)$$

so the largest value of x_{n+1} occurs for $x_n = 1/2$. Plugging this in, $\max(x_{n+1}) = r/4$. Therefore, to keep the MAP in the desired region, we must have $r \in (0,4]$. The JACOBIAN is

$$J = \left|\frac{dx_{n+1}}{dx_n}\right| = |r(1-2x_n)|, \quad (3)$$

and the MAP is stable at a point x_0 if $J(x_0) < 1$. Now we wish to find the FIXED POINTS of the MAP, which occur when $x_{n+1} = x_n$. Drop the n subscript on x_n

$$f(x) = rx(1-x) = x \quad (4)$$

$$x[1-r(1-x)] = x(1-r+rx) = rx\left[x-\left(1-r^{-1}\right)\right] = 0, \quad (5)$$

so the FIXED POINTS are $x_1^{(1)} = 0$ and $x_2^{(1)} = 1-r^{-1}$. An interesting thing happens if a value of r greater than 3 is chosen. The map becomes unstable and we get a PITCHFORK BIFURCATION with two stable orbits of period two corresponding to the two stable FIXED POINTS of $f^2(x)$. The fixed points of order two must satisfy $x_{n+2} = x_n$, so

$$\begin{aligned} x_{n+2} &= rx_{n+1}(1-x_{n+1}) \\ &= r[rx_n(1-x_n)][1-rx_n(1-x_n)] \\ &= r^2x_n(1-x_n)(1-rx_n+rx_n{}^2) = x_n. \end{aligned} \quad (6)$$

Now, drop the n subscripts and rewrite

$$x\{r^2[1-x(1+r)+2rx^2-rx^3]-1\}=0 \quad (7)$$

$$x[-r^3x^3+2r^3x^2-r^2(1+r)x+(r^2-1)]=0 \quad (8)$$

$$-r^3x[x-(1-r^{-1})][x^2-(1+r^{-1})x+r^{-1}(1+r^{-1})] = 0. \quad (9)$$

Notice that we have found the first-order FIXED POINTS as well, since two iterations of a first-order FIXED POINT produce a trivial second-order FIXED POINT. The true 2-CYCLES are given by solutions to the quadratic part

$$\begin{aligned} x_\pm^{(2)} &= \tfrac{1}{2}[(1+r^{-1}) \pm \sqrt{(1+r^{-1})^2-4r^{-1}(1+r^{-1})}\,] \\ &= \tfrac{1}{2}[(1+r^{-1}) \pm \sqrt{1+2r^{-1}+r^{-2}-4r^{-1}-4r^{-2}}\,] \\ &= \tfrac{1}{2}[(1+r^{-1}) \pm \sqrt{1-2r^{-1}-3r^{-2}}\,] \\ &= \tfrac{1}{2}[(1+r^{-1}) \pm r^{-1}\sqrt{(r-3)(r+1)}\,]. \quad (10) \end{aligned}$$

These solutions are only REAL for $r \geq 3$, so this is where the 2-CYCLE begins. Now look for the onset of the 4-CYCLE. To eliminate the 2- and 1-CYCLES, consider

$$\frac{f^4(x)-x}{f^2(x)-x}=0. \quad (11)$$

This gives

$$\begin{aligned} 1&+r^2+(-r^2-r^3-r^4-r^5)x \\ &+(2r^3+r^4+4r^5+r^6+2r^7)x^2 \\ &+(-r^3-5r^5-4r^6-5r^7-4r^8-r^9)x^3 \\ &+(2r^5+6r^6+4r^7+14r^8+5r^9+3r^{10})x^4 \\ &+(-4r^6-r^7-18r^8-12r^9-12r^{10}-3r^{11})x^5 \\ &+(r^6+10r^8+17r^9+18r^{10}+15r^{11}+r^{12})x^6 \\ &+(-2r^8-14r^9-12r^{10}-30r^{11}-6r^{12})x^7 \\ &+(6r^9+3r^{10}+30r^{11}+15r^{12})x^8 \\ &+(-r^9-15r^{11}-20r^{12})x^9+(3r^{11}+15r^{12})x^{10} \\ &-6r^{12}x^{11}+r^{12}x^{12}. \quad (12) \end{aligned}$$

The ROOTS of this equation are all IMAGINARY for $r < 1+\sqrt{6}$, but two of them convert to REAL roots at this value (although this is difficult to show except by plugging in). The 4-CYCLE therefore starts at $1+\sqrt{6} = 3.449490\ldots$. The BIFURCATIONS come faster and faster (8, 16, 32, ...), then suddenly break off. Beyond a certain point known as the ACCUMULATION POINT, periodicity gives way to CHAOS.

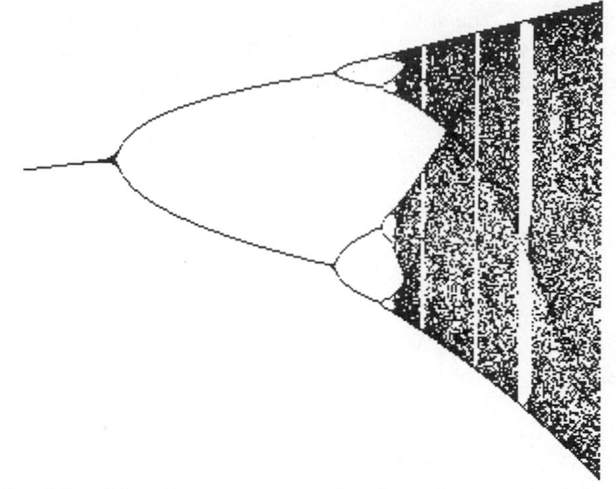

A table of the CYCLE type and value of r_n at which the cycle 2^n appears is given below.

n	cycle (2^n)	r_n
1	2	3
2	4	3.449490
3	8	3.544090
4	16	3.564407
5	32	3.568750
6	64	3.56969
7	128	3.56989
8	256	3.569934
9	512	3.569943
10	1024	3.5699451
11	2048	3.569945557
∞	acc. pt.	3.569945672

For additional values, see Rasband (1990, p. 23). Note that the table in Tabor (1989, p. 222) is incorrect, as is the $n = 2$ entry in Lauweirer 1991. In the middle of the complexity, a window suddenly appears with a regular period like 3 or 7 as a result of MODE LOCKING. The period 3 BIFURCATION occurs at $r = 1+2\sqrt{2} = 3.828427\ldots$, as is derived below. Following the 3-CYCLE, the PERIOD DOUBLINGS then begin again with CYCLES of 6, 12, ... and 7, 14, 28, ..., and then once again break off to CHAOS.

A set of $n+1$ equations which can be solved to give the onset of an arbitrary n-cycle (Saha and Strogatz 1995) is

$$\begin{cases} x_2 = rx_1(1-x_1) \\ x_3 = rx_2(1-x_2) \\ \vdots \\ x_n = rx_{n-1}(1-x_{n-1}) \\ x_1 = rx_n(1-x_n) \\ r^n \prod_{k=1}^n (1-2x_k) = 1. \end{cases} \quad (13)$$

The first n of these give $f(x)$, $f^2(x)$, ..., $f^n(x)$, and the last uses the fact that the onset of period n occurs by a TANGENT BIFURCATION, so the nth DERIVATIVE is 1.

For $n = 2$, the solutions $(x_1, \ldots, x_n, r)$ are $(0, 0, \pm 1)$ and $(2/3, 2/3, 3)$, so the desired BIFURCATION occurs at $r_2 = 3$. Taking $n = 3$ gives

$$\begin{aligned}\frac{d[f^3(x)]}{dx} &= \frac{d[f^3(x)]}{d[f^2(x)]}\frac{d[f^2(x)]}{d[f(x)]}\frac{d[f(x)]}{dx} \\ &= \frac{d[f(z)]}{dz}\frac{d[f(y)]}{dy}\frac{d[f(x)]}{dx} \\ &= r^3(1-2z)(1-2y)(1-2x). \qquad (14)\end{aligned}$$

Solving the resulting CUBIC EQUATION using computer algebra gives

$$\begin{aligned}x_1 = &-\left(\frac{2^{5^6}}{63\cdot 7^{1/3}}+\frac{1}{63\cdot 28^{1/3}}\right)c^2 \\ &-\frac{1}{9\cdot 98^{1/3}}c+\frac{10+\sqrt{2}}{21}+\left(\frac{4\cdot 2^{5/6}}{9\cdot 7^{1/3}}-\frac{2^{1/3}}{7^{1/3}}\right)c^{-1} \\ &+\frac{25\cdot 28^{1/3}-44\cdot 2^{1/6}7^{1/3}}{9}c^{-2} \qquad (15)\end{aligned}$$

$$\begin{aligned}x_2 = &\left(\frac{1}{63\cdot 28^{1/3}}+\frac{2^{5/6}}{63\cdot 7^{1/3}}\right)c^2-\frac{2^{2/3}}{9\cdot 7^{2/3}}c+\frac{10+\sqrt{2}}{21} \\ &+\left(\frac{8\cdot 2^{5/6}}{9\cdot 7^{1/3}}-\frac{2\cdot 2^{1/3}}{7^{1/3}}\right)c^{-1} \\ &+\frac{44\cdot 2^{1/6}7^{1/3}-25\cdot 28^{1/3}}{9}c^{-2} \qquad (16)\end{aligned}$$

$$x_3 = \frac{1}{3\cdot 98^{1/3}}c+\frac{10+\sqrt{2}}{21}+\frac{2^{1/3}(9-4\sqrt{2})}{3\cdot 7^{1/3}}c^{-1} \qquad (17)$$

$$r = 1+2\sqrt{2}, \qquad (18)$$

where

$$c \equiv (-25+22\sqrt{2}+3\sqrt{3}\sqrt{1100\sqrt{2}-1593})^{1/3}. \qquad (19)$$

Numerically,

$$x_1 = 0.514355\ldots \qquad (20)$$

$$x_2 = 0.956318\ldots \qquad (21)$$

$$x_3 = 0.159929\ldots \qquad (22)$$

$$r = 3.828427\ldots. \qquad (23)$$

Saha and Strogatz (1995) give a simplified algebraic treatment which involves solving

$$r^3(1-2\alpha+4\beta-8\gamma) = 1, \qquad (24)$$

together with three other simultaneous equations, where

$$\alpha \equiv x_1+x_2+x_3 \qquad (25)$$

$$\beta \equiv x_1x_2+x_1x_3+x_2x_3 \qquad (26)$$

$$\gamma \equiv x_1x_2x_3. \qquad (27)$$

Further simplifications still are provided in Bechhoeffer (1996) and Gordon (1996), but neither of these techniques generalizes easily to higher CYCLES. Bechhoeffer (1996) expresses the three additional equations as

$$2\alpha = 3+r^{-1} \qquad (28)$$

$$4\beta = \tfrac{3}{2}+5r^{-1}+\tfrac{3}{2}r^{-2} \qquad (29)$$

$$8\gamma = -\tfrac{1}{2}+\tfrac{7}{2}r^{-1}+\tfrac{5}{2}r^{-2}+\tfrac{5}{2}r^{-3}, \qquad (30)$$

giving

$$r^2-2r-7 = 0. \qquad (31)$$

Gordon (1996) derives not only the value for the onset of the 3-CYCLE, but also an upper bound for the r-values supporting stable period 3 orbits. This value is obtained by solving the CUBIC EQUATION

$$s^3-11s^2+37s-108 = 0 \qquad (32)$$

for s, then

$$r' = 1+\sqrt{s} \qquad (33)$$

$$= 1+\sqrt{\tfrac{11}{3}+(\tfrac{1915}{54}+\tfrac{5}{2}\sqrt{201})^{1/3}+(\tfrac{1915}{54}-\tfrac{5}{2}\sqrt{201})^{1/3}}$$

$$= 3.841499007543\ldots. \qquad (34)$$

The logistic equation has CORRELATION EXPONENT 0.500 ± 0.005 (Grassberger and Procaccia 1983), CAPACITY DIMENSION 0.538 (Grassberger 1981), and INFORMATION DIMENSION 0.5170976 (Grassberger and Procaccia 1983).

see also BIFURCATION, FEIGENBAUM CONSTANT, LOGISTIC DISTRIBUTION, LOGISTIC EQUATION—$r = 4$, LOGISTIC GROWTH CURVE, PERIOD THREE THEOREM, QUADRATIC MAP

References

Bechhoeffer, J. "The Birth of Period 3, Revisited." *Math. Mag.* **69**, 115–118, 1996.

Bogomolny, A. "Chaos Creation (There is Order in Chaos)." http://www.cut-the-knot.com/blue/chaos.html.

Dickau, R. M. "Bifurcation Diagram." http://forum.swarthmore.edu/advanced/robertd/bifurcation.html.

Gleick, J. *Chaos: Making a New Science.* New York: Penguin Books, pp. 69–80, 1988.

Gordon, W. B. "Period Three Trajectories of the Logistic Map." *Math. Mag.* **69**, 118–120, 1996.

Grassberger, P. "On the Hausdorff Dimension of Fractal Attractors." *J. Stat. Phys.* **26**, 173–179, 1981.

Grassberger, P. and Procaccia, I. "Measuring the Strangeness of Strange Attractors." *Physica D* **9**, 189–208, 1983.

Lauwerier, H. *Fractals: Endlessly Repeated Geometrical Figures.* Princeton, NJ: Princeton University Press, pp. 119–122, 1991.

May, R. M. "Simple Mathematical Models with Very Complicated Dynamics." *Nature* **261**, 459–467, 1976.

Peitgen, H.-O.; Jürgens, H.; and Saupe, D. *Chaos and Fractals: New Frontiers of Science.* New York: Springer-Verlag, pp. 585–653, 1992.

Rasband, S. N. *Chaotic Dynamics of Nonlinear Systems.* New York: Wiley, p. 23, 1990.

Russell, D. A.; Hanson, J. D.; and Ott, E. "Dimension of Strange Attractors." *Phys. Rev. Let.* **45**, 1175–1178, 1980.
Saha, P. and Strogatz, S. H. "The Birth of Period Three." *Math. Mag.* **68**, 42–47, 1995.
Strogatz, S. H. *Nonlinear Dynamics and Chaos.* Reading, MA: Addison-Wesley, 1994.
Tabor, M. *Chaos and Integrability in Nonlinear Dynamics: An Introduction.* New York: Wiley, 1989.
Wagon, S. "The Dynamics of the Quadratic Map." §4.4 in *Mathematica in Action.* New York: W. H. Freeman, pp. 117–140, 1991.

Logistic Equation—$r = 4$

With $r = 4$, the LOGISTIC EQUATION becomes

$$x_{n+1} = 4x_n(1 - x_n). \tag{1}$$

Now let

$$x \equiv \sin^2(\tfrac{1}{2}\pi y) = \tfrac{1}{2}[1 - \cos(\pi y)] \tag{2}$$

$$\sqrt{x} = \sin(\tfrac{1}{2}\pi y) \tag{3}$$

$$y = \frac{2}{\pi}\sin^{-1}(\sqrt{x}) \tag{4}$$

$$\frac{dy}{dx} = \frac{2}{\pi}\frac{1}{\sqrt{1-x}}\tfrac{1}{2}x^{-1/2} = \frac{1}{\pi\sqrt{x(1-x)}}. \tag{5}$$

Manipulating (2) gives

$$\begin{aligned}\sin^2(\tfrac{1}{2}\pi y_{n+1})\\ &= 4\tfrac{1}{2}[1-\cos(\pi y_n)]\{1 - \tfrac{1}{2}[1 - \tfrac{1}{2}(1-\cos(\pi y_n)]\}\\ &= 2[1 - \cos(\pi y = 1 - \cos^2(\pi y_n)\sin^2(\pi y_n),\end{aligned} \tag{6}$$

so

$$\tfrac{1}{2}\pi y_{n+1} = \pm y_n + s\pi \tag{7}$$

$$y_{n+1} = \pm 2y_n + \tfrac{1}{2}s. \tag{8}$$

But $y \in [0,1]$. Taking $y_n \in [0, 1/2]$, then $s = 0$ and

$$y_{n+1} = 2y_n. \tag{9}$$

For $y \in [1/2, 1]$, $s = 1$ and

$$y_{n+1} = 2 - 2y_n. \tag{10}$$

Combining

$$y_n = \begin{cases} 2y_n & \text{for } y_n \in [0, \tfrac{1}{2}] \\ 2 - 2y_n & \text{for } y_n \in [\tfrac{1}{2}, 1], \end{cases} \tag{11}$$

which can be written

$$y_n = 1 - 2|x_n - h|, \tag{12}$$

the TENT MAP with $\mu = 1$, so the NATURAL INVARIANT in y is

$$\rho(y) = 1. \tag{13}$$

Transforming back to x gives

$$\begin{aligned}\rho(x) &= \left|\frac{dy}{dx}\right|\rho(y(x)) = \frac{2}{\pi}\frac{1}{\sqrt{1-x}}\tfrac{1}{2}x^{-1/2}\\ &= \frac{1}{\pi\sqrt{x(1-x)}}.\end{aligned} \tag{14}$$

This can also be derived from

$$\rho(x) \equiv \lim_{N\to\infty}\frac{1}{N}\sum_{i=1}^{N}\delta(x_i - x) = \frac{1}{\pi\sqrt{x(1-x)}}, \tag{15}$$

where $\delta(x)$ is the DELTA FUNCTION.

see also LOGISTIC EQUATION

Logistic Growth Curve

The POPULATION GROWTH law which arises frequently in biology and is given by the differential equation

$$\frac{dN}{dt} = \frac{r(K-N)}{K}, \tag{1}$$

where r is the MALTHUSIAN PARAMETER and K is the so-called CARRYING CAPACITY (i.e., the maximum sustainable population). Rearranging and integrating both sides gives

$$\int_{N_0}^{N}\frac{dN}{K-N} = \frac{r}{K}\int_0^t dt \tag{2}$$

$$\ln\left(\frac{N_0 - K}{N - K}\right) = \frac{r}{K}t \tag{3}$$

$$N(t) = K + (N_0 - K)e^{-rt/K}. \tag{4}$$

The curve

$$y = \frac{a}{1 + bq^x} \tag{5}$$

is sometimes also known as the logical curve.

see also GOMPERTZ CURVE, LIFE EXPECTANCY, LOGISTIC EQUATION, MAKEHAM CURVE, MALTHUSIAN PARAMETER, POPULATION GROWTH

Logistic Map

see LOGISTIC EQUATION

Logit Transformation

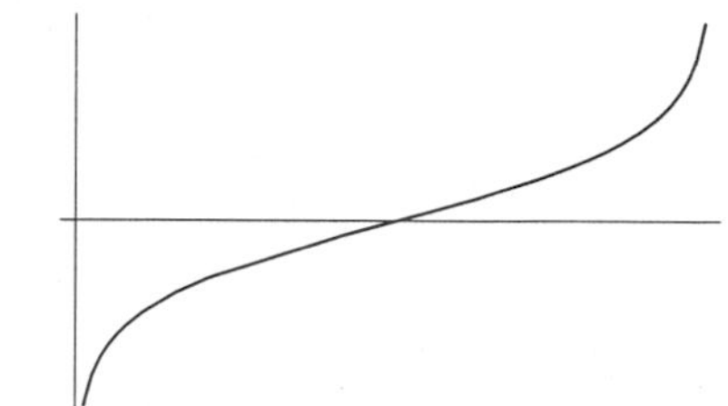

The function

$$z = f(x) = \ln\left(\frac{x}{1-x}\right).$$

This function has an inflection point at $x = 1/2$, where

$$f''(x) = \frac{2x-1}{x^2(x-1)^2} = 0.$$

Applying the logit transformation to values obtained by iterating the LOGISTIC EQUATION generates a sequence of RANDOM NUMBERS having distribution

$$P_z = \frac{1}{\pi(e^{x/2} + e^{-x/2})},$$

which is very close to a GAUSSIAN DISTRIBUTION.

References

Collins, J.; Mancilulli, M.; Hohlfeld, R.; Finch, D.; Sandri, G.; and Shtatland, E. "A Random Number Generator Based on the Logit Transform of the Logistic Variable." *Computers in Physics* **6**, 630–632, 1992.

Pickover, C. A. *Keys to Infinity.* New York: W. H. Freeman, pp. 244–245, 1995.

Logos

A generalization of a HEYTING ALGEBRA which replaces BOOLEAN ALGEBRA in "intuitionistic" LOGIC.

see also TOPOS

Lommel Differential Equation

A generalization of the BESSEL DIFFERENTIAL EQUATION (Watson 1966, p. 345),

$$z^2\frac{d^2y}{dz^2} + z\frac{dy}{dz} - (z^2+\nu^2)y = kz^{\mu+1}.$$

A further generalization gives

$$z^2\frac{d^2y}{dz^2} + z\frac{dy}{dz} - (z^2+\nu^2)y = \pm kz^{\mu+1}.$$

The solutions are LOMMEL FUNCTIONS.

see also LOMMEL FUNCTION

References

Watson, G. N. *A Treatise on the Theory of Bessel Functions, 2nd ed.* Cambridge, England: Cambridge University Press, 1966.

Lommel Function

There are several functions called "Lommel functions." One type of Lommel function is the solution to the LOMMEL DIFFERENTIAL EQUATION with a PLUS SIGN,

$$y = ks_{\mu,\nu}(z), \tag{1}$$

where

$$s_{\mu,\nu}^{(+)}(z) \equiv \tfrac{1}{2}\pi\left[Y_\nu(z)\int_0^z z^\mu J_\nu(z)\,dz - J_\nu(z)\int_0^z z^\mu Y_\nu(z)\,dz\right]. \tag{2}$$

Here, $J_\nu(z)$ and $Y_\nu(z)$ are BESSEL FUNCTIONS OF THE FIRST and SECOND KINDS (Watson 1966, p. 346). If a minus sign precedes k, then the solution is

$$s_{\mu,\nu}^{(-)} \equiv I_\nu(z)\int_z^{c_1} z^\mu K_\nu(z)\,dz - J_\nu(z)\int_{c_2}^z z^\mu I_\nu(z)\,dz, \tag{3}$$

where $K_\nu(z)$ and $I_\nu(z)$ are MODIFIED BESSEL FUNCTIONS OF THE FIRST and SECOND KINDS.

Lommel functions of two variables are related to the BESSEL FUNCTION OF THE FIRST KIND and arise in the theory of diffraction and, in particular, Mie scattering (Watson 1966, p. 537),

$$U_n(w,z) = \sum_{m=0}^\infty (-1)^m \left(\frac{w}{z}\right)^{n+2m} J_{n+2m}(z) \tag{4}$$

$$V_n(w,z) = \sum_{m=0}^\infty (-1)^m \left(\frac{w}{z}\right)^{-n-2m} J_{-n-2m}(z). \tag{5}$$

see also LOMMEL DIFFERENTIAL EQUATION, LOMMEL POLYNOMIAL

References

Chandrasekhar, S. *Radiative Transfer.* New York: Dover, p. 369, 1960.

Watson, G. N. *A Treatise on the Theory of Bessel Functions, 2nd ed.* Cambridge, England: Cambridge University Press, 1966.

Lommel's Integrals

$$(\beta^2-\alpha^2)\int xJ_n(\alpha x)J_n(\beta x)\,dx = x[\alpha J_n'(\alpha x)J_n(\beta x) - \beta J_n'(\beta x)J_n(\alpha x)]$$

$$\int x{J_n}^2(\alpha x)\,dx = \tfrac{1}{2}x^2[{J_n}^2(\alpha x) + J_{n-1}(\alpha x)J_{n+1}(\alpha x)],$$

where $J_n(x)$ is a BESSEL FUNCTION OF THE FIRST KIND.

References

Bowman, F. *Introduction to Bessel Functions.* New York: Dover, p. 101, 1958.

Lommel Polynomial

$$R_{m,\nu}(z) = \frac{\Gamma(\nu+m)}{\Gamma(\nu)(z/2)^m}\,{}_2F_3(\tfrac{1}{2}(1-m), -\tfrac{1}{2}m; \nu, -m, 1-\nu-m; z^2) \times \frac{\pi z}{2\sin(\nu\pi)}[J_{\nu+m}(z)J_{-\nu+1}(z) + (-1)^m J_{-\nu-m}(z)J_{\nu-1}(z)],$$

where $\Gamma(z)$ is a GAMMA FUNCTION, $J_n(x)$ is a BESSEL FUNCTION OF THE FIRST KIND, and ${}_2F_3(a, b; c, d, e; z)$ is a GENERALIZED HYPERGEOMETRIC FUNCTION.

see also LOMMEL FUNCTION

References

Iyanaga, S. and Kawada, Y. (Eds.). *Encyclopedic Dictionary of Mathematics.* Cambridge, MA: MIT Press, p. 1477, 1980.

Long Division

```
      7              72                 726
17|123456.      17|123456.       17|123456.
  -119            -119              -119
    44              44                44
                   -34               -34
                    105               105
                                     -102
                                       36

      7262.1          7262.11...
17|123456.0     17|123456.00
  -119            -119
    44              44
   -34             -34
    105             105
   -102            -102
     36              36
    -34             -34
      20              20
                     -17
                      30
```

Long division is an algorithm for dividing two numbers, obtaining the QUOTIENT one DIGIT at a time. The above example shows how the division of $123456/17$ is performed to obtain the result $7262.11\ldots$.

see also DIVISION

Long Exact Sequence of a Pair Axiom

One of the EILENBERG-STEENROD AXIOMS. It states that, for every pair (X, A), there is a natural long exact sequence

$$\ldots \to H_n(A) \to H_n(X) \to H_n(X, A) \to H_{n-1}(A) \to \ldots,$$

where the MAP $H_n(A) \to H_n(X)$ is induced by the INCLUSION MAP $A \to X$, $H_n(X) \to H_n(X, A)$ is induced by the INCLUSION MAP $(X, \phi) \to (X, A)$. The MAP $H_n(X, A) \to H_{n-1}(A)$ is called the BOUNDARY MAP.

see also EILENBERG-STEENROD AXIOMS

Long Prime

see DECIMAL EXPANSION

Longitude

The azimuthal coordinate on the surface of a SPHERE (θ in SPHERICAL COORDINATES) or on a SPHEROID (in PROLATE or OBLATE SPHEROIDAL COORDINATES). Longitude is defined such that $0° = 360°$. Lines of constant longitude are generally called MERIDIANS. The other angular coordinate on the surface of a SPHERE is called the LATITUDE.

The shortest distance between any two points on a SPHERE is the so-called GREAT CIRCLE distance, which can be directly computed from the LATITUDE and longitudes of two points.

see also GREAT CIRCLE, LATITUDE, MERIDIAN, OBLATE SPHEROIDAL COORDINATES, PROLATE SPHEROIDAL COORDINATES

Look and Say Sequence

The INTEGER SEQUENCE beginning with a single digit in which the next term is obtained by describing the previous term. Starting with 1, the sequence would be defined by "one 1, two 1s, one 2 two 1s," etc., and the result is 1, 11, 21, 1211, 111221, 312211, 13112221, 1113213211, ... (Sloane's A005150).

Starting the sequence instead with the digit d for $2 \leq d \leq 9$ gives d, $1d$, $111d$, $311d$, $13211d$, $111312211d$, $31131122211d$, $1321132132211d$, ... The sequences for $d = 2$ and 3 are Sloane's A006715 and A006751. The number of DIGITS in the nth term of both the sequences for $1 \leq n \leq 9$ is asymptotic to $C\lambda^n$, where C is a constant and

$$\lambda = 1.303577269034296\ldots$$

(Sloane's A014715) is CONWAY'S CONSTANT. λ is given by the largest ROOT of the POLYNOMIAL

$$\begin{aligned}0 = x^{71}&\\ &- x^{69} - 2x^{68} + 2x^{66} + 2x^{65} + x^{64} - x^{63} - x^{62} - x^{61}\\ &- x^{60} - x^{59} + 2x^{58} + 5x^{57} + 3x^{56} - 2x^{55} - 10x^{54}\\ &- 3x^{53} - 2x^{52} + 6x^{51} + 6x^{50} + x^{49} + 9x^{48} - 3x^{47}\\ &- 7x^{46} - 8x^{45} - 8x^{44} + 10x^{43} + 6x^{42} + 8x^{41} - 4x^{40}\\ &- 12x^{39} + 7x^{38} - 7x^{37} + 7x^{36} - 3x^{34} + x^{35} + 10x^{33}\\ &+ x^{32} - 6x^{31} - 2x^{30} - 10x^{29} - 3x^{28} + 2x^{27} + 9x^{26}\\ &- 3x^{25} + 14x^{24} - 8x^{23} - 7x^{21} + 9x^{20} - 3x^{19} - 4x^{18}\\ &- 10x^{17} - 7x^{16} + 12x^{15} + 7x^{14} + 2x^{13} - 12x^{12}\\ &- 4x^{11} - 2x^{10} - 5x^{9} + x^{7} - 7x^{6}\\ &+ 7x^{5} - 4x^{4} + 12x^{3} - 6x^{2} + 3x - 6.\end{aligned}$$

In fact, the constant is even more general than this, applying to *all* starting sequences (i.e., even those starting with arbitrary starting digits), with the exception of 22, a result which follows from the COSMOLOGICAL THEOREM. Conway discovered that strings sometimes factor as a concatenation of two strings whose descendants never interfere with one another. A string with no nontrivial splittings is called an "element," and other strings are called "compounds." Every string of 1s, 2s, and 3s eventually "decays" into a compound of 92 special elements, named after the chemical elements.

see also CONWAY'S CONSTANT, COSMOLOGICAL THEOREM

References

Conway, J. H. "The Weird and Wonderful Chemistry of Audioactive Decay." *Eureka,* 5–18, 1985.
Conway, J. H. "The Weird and Wonderful Chemistry of Audioactive Decay." §5.11 in *Open Problems in Communications and Computation.* (Ed. T. M. Cover and B. Gopinath). New York: Springer-Verlag, pp. 173–188, 1987.
Conway, J. H. and Guy, R. K. "The Look and Say Sequence." In *The Book of Numbers.* New York: Springer-Verlag, pp. 208–209, 1996.
Sloane, N. J. A. Sequences A005150/M4780, A006715/M2965, and A6751/M2052 in "An On-Line Version of the Encyclopedia of Integer Sequences."
Vardi, I. *Computational Recreations in Mathematica.* Reading, MA: Addison-Wesley, pp. 13–14, 1991.

Loop (Algebra)

A nonassociative ALGEBRA (and QUASIGROUP) which has a single binary operation.

Loop Gain

The loop gain is usually assigned a value between 0.1 and 0.5. The CLEAN ALGORITHM performs better for extended structures if μ is set to the lower part of this range. However, the time required for the CLEAN ALGORITHM increases rapidly for small μ. From Thompson *et al.* (1986), the number of cycles needed for one point source is

$$[\text{cycles}] = -\frac{\ln(\text{SNR})}{\ln(1-\gamma)}.$$

see also CLEAN ALGORITHM

References

Thompson, A. R.; Moran, J. M.; and Swenson, G. W. Jr. *Interferometry and Synthesis in Radio Astronomy.* New York: Wiley, p. 348, 1986.

Loop (Graph)

A degenerate edge of a graph which joins a vertex to itself.

Loop (Knot)

A KNOT or HITCH which holds its form rigidly.

References

Owen, P. *Knots.* Philadelphia, PA: Courage, p. 35, 1993.

Loop Space

Let Y^X be the set of continuous mappings $f : X \to Y$. Then the TOPOLOGICAL SPACE for Y^X supplied with a compact-open topology is called a MAPPING SPACE, and if $Y = I$ is taken as the interval $(0, 1)$, then $Y^I = \Omega(Y)$ is called a loop space (or SPACE OF CLOSED PATHS).

see also MACHINE, MAPPING SPACE, MAY-THOMASON UNIQUENESS THEOREM

References

Brylinski, J.-L. *Loop Spaces, Characteristic Classes and Geometric Quantization.* Boston, MA: Birkhäuser, 1993.
Iyanaga, S. and Kawada, Y. (Eds.). *Encyclopedic Dictionary of Mathematics.* Cambridge, MA: MIT Press, p. 658, 1980.

Lorentz Group

The Lorentz group is the GROUP L of time-preserving linear ISOMETRIES of MINKOWSKI SPACE $\mathbb{R}^4$ with the pseudo-Riemannian metric

$$d\tau^2 = -dt^2 + dx^2 + dy^2 + dz^2.$$

It is also the GROUP of ISOMETRIES of 3-D HYPERBOLIC SPACE. It is time-preserving in the sense that the unit time VECTOR $(1, 0, 0, 0)$ is sent to another VECTOR (t, x, y, z) such that $t > 0$.

A consequence of the definition of the Lorentz group is that the full GROUP of time-preserving isometries of MINKOWSKI $\mathbb{R}^4$ is the DIRECT PRODUCT of the group of translations of $\mathbb{R}^4$ (i.e., $\mathbb{R}^4$ itself, with addition as the group operation), with the Lorentz group, and that the full isometry group of the MINKOWSKI $\mathbb{R}^4$ is a group extension of $\mathbb{Z}_2$ by the product $L \otimes \mathbb{R}^4$.

The Lorentz group is invariant under space rotations and LORENTZ TRANSFORMATIONS.

see also LORENTZ TENSOR, LORENTZ TRANSFORMATION

References

Arfken, G. "Homogeneous Lorentz Group." §4.13 in *Mathematical Methods for Physicists, 3rd ed.* Orlando, FL: Academic Press, pp. 271–275, 1985.

Lorentz Tensor

The TENSOR in the LORENTZ TRANSFORMATION given by

$$\mathsf{L} \equiv \begin{bmatrix} \gamma & -\gamma\beta & 0 & 0 \\ -\gamma\beta & \gamma & 0 & 0 \\ 0 & 0 & 1 & 0 \\ 0 & 0 & 0 & 1 \end{bmatrix}, \tag{1}$$

where beta and gamma are defined by

$$\beta \equiv \frac{v}{c} \tag{2}$$

$$\gamma \equiv \frac{1}{\sqrt{1-\beta^2}}. \tag{3}$$

see also LORENTZ GROUP, LORENTZ TRANSFORMATION

Lorentz Transformation

A 4-D transformation satisfied by all FOUR-VECTORS a^ν,

$$a'^\mu = \Lambda^\mu_\nu a^\nu. \tag{1}$$

In the theory of special relativity, the Lorentz transformation replaces the GALILEAN TRANSFORMATION as the valid transformation law between reference frames moving with respect to one another at constant VELOCITY. Let x^ν be the POSITION FOUR-VECTOR with $x^0 = ct$, and let the relative motion be along the x^1 axis with VELOCITY v. Then (1) becomes

$$x'^\mu = \Lambda^\mu_\nu x^\nu, \tag{2}$$

where the LORENTZ TENSOR is given by

$$\mathsf{L} = \begin{bmatrix} \Lambda^0_0 & \Lambda^0_1 & \Lambda^0_2 & \Lambda^0_3 \\ \Lambda^1_0 & \Lambda^1_1 & \Lambda^1_2 & \Lambda^1_3 \\ \Lambda^2_0 & \Lambda^2_1 & \Lambda^2_2 & \Lambda^2_3 \\ \Lambda^3_0 & \Lambda^3_1 & \Lambda^3_2 & \Lambda^3_3 \end{bmatrix} \equiv \begin{bmatrix} \gamma & -\gamma\beta & 0 & 0 \\ -\gamma\beta & \gamma & 0 & 0 \\ 0 & 0 & 1 & 0 \\ 0 & 0 & 0 & 1 \end{bmatrix}. \tag{3}$$

Here,

$$\beta \equiv \frac{v}{c} \tag{4}$$

$$\gamma \equiv \frac{1}{\sqrt{1-\beta^2}}. \tag{5}$$

Written explicitly, the transformation between x^ν and $x^{\nu\prime}$ coordinate is

$$x^{0\prime} = \gamma(x^0 - \beta x^1) \tag{6}$$

$$x^{1\prime} = \gamma(x^1 - \beta x^0) \tag{7}$$

$$x^{2\prime} = x^2 \tag{8}$$

$$x^{3\prime} = x^3. \tag{9}$$

The DETERMINANT of the upper left 2×2 MATRIX in (3) is

$$D = (\gamma)^2 - (-\gamma\beta)^2 = \gamma^2(1-\beta^2) = \frac{\gamma^2}{\gamma^2} = 1, \tag{10}$$

so

$$\mathsf{L}^{-1} = \begin{bmatrix} (\Lambda^{-1})^0_0 & (\Lambda^{-1})^0_1 & (\Lambda^{-1})^0_2 & (\Lambda^{-1})^0_3 \\ (\Lambda^{-1})^1_0 & (\Lambda^{-1})^1_1 & (\Lambda^{-1})^1_2 & (\Lambda^{-1})^1_3 \\ (\Lambda^{-1})^2_0 & (\Lambda^{-1})^2_1 & (\Lambda^{-1})^2_2 & (\Lambda^{-1})^2_3 \\ (\Lambda^{-1})^3_0 & (\Lambda^{-1})^3_1 & (\Lambda^{-1})^3_2 & (\Lambda^{-1})^3_3 \end{bmatrix} \equiv \begin{bmatrix} \gamma & \gamma\beta & 0 & 0 \\ \gamma\beta & \gamma & 0 & 0 \\ 0 & 0 & 1 & 0 \\ 0 & 0 & 0 & 1 \end{bmatrix}. \tag{11}$$

A Lorentz transformation along the x_1-axis can also be written

$$\begin{bmatrix} x_1' \\ x_2' \\ x_3' \\ x_4' \end{bmatrix} = \begin{bmatrix} \cosh\theta & i\sinh\theta & 0 & 0 \\ 0 & 1 & 0 & 0 \\ 0 & 0 & 1 & 0 \\ -i\sinh\theta & \cosh\theta & 0 & 0 \end{bmatrix} \begin{bmatrix} x_1 \\ x_2 \\ x_3 \\ x_4 \end{bmatrix}, \tag{12}$$

where θ is called the rapidity,

$$x_4 \equiv ict, \tag{13}$$

and

$$\tanh\theta \equiv \beta \equiv \frac{v}{c} \tag{14}$$

$$\cosh\theta \equiv \gamma \equiv \frac{1}{\sqrt{1-\frac{v^2}{c^2}}} \tag{15}$$

$$\sinh\theta = \beta\gamma. \tag{16}$$

see also HYPERBOLIC ROTATION, LORENTZ GROUP, LORENTZ TENSOR

References

Fraundorf, P. "Accel-1D: Frame-Dependent Relativity at UM-StL." `http://www.umsl.edu/~fraundor/a1toc.html`.

Griffiths, D. J. *Introduction to Electrodynamics.* Englewood Cliffs, NJ: Prentice-Hall, pp. 412–414, 1981.

Morse, P. M. and Feshbach, H. "The Lorentz Transformation, Four-Vectors, Spinors." §1.7 in *Methods of Theoretical Physics, Part I.* New York: McGraw-Hill, pp. 93–107, 1953.

Lorentzian Distribution

see CAUCHY DISTRIBUTION

Lorentzian Function

The Lorentzian function is given by

$$L(x) = \frac{1}{\pi}\frac{\frac{1}{2}\Gamma}{(x-x_0)^2 + (\frac{1}{2}\Gamma)^2}.$$

Its FULL WIDTH AT HALF MAXIMUM is Γ. This function gives the shape of certain types of spectral lines and is the distribution function in the CAUCHY DISTRIBUTION. The Lorentzian function has FOURIER TRANSFORM

$$\mathcal{F}\left[\frac{1}{\pi}\frac{\frac{1}{2}\Gamma}{(x-x_0)^2 + (\frac{1}{2}\Gamma)^2}\right] = e^{-2\pi i k x_0 - \Gamma\pi|k|}.$$

see also DAMPED EXPONENTIAL COSINE INTEGRAL, FOURIER TRANSFORM—LORENTZIAN FUNCTION

Lorenz System

A simplified system of equations describing the 2-D flow of fluid of uniform depth H, with an imposed temperature difference ΔT, under gravity g, with buoyancy α, thermal diffusivity κ, and kinematic viscosity ν. The full equations are

$$\frac{\partial}{\partial t}(\nabla^2\phi) = \frac{\partial\psi}{\partial z}\frac{\partial}{\partial x}(\nabla^2\psi) - \frac{\partial\psi}{\partial x}\frac{\partial}{\partial z}(\nabla^2\psi) + \nu\nabla^2(\nabla^2\psi) + g\alpha\frac{dT}{dx} \tag{1}$$

$$\frac{\partial T}{\partial t} = \frac{\partial T}{\partial z}\frac{\partial\psi}{\partial x} - \frac{\partial\theta}{\partial x}\frac{\partial\psi}{\partial z} + \kappa\nabla^2 T + \frac{\Delta T}{H}\frac{\partial\psi}{\partial x}. \tag{2}$$

Here, ψ is the "stream function," as usual defined such that

$$u = \frac{\partial \psi}{\partial x}, \qquad v = \frac{\partial \psi}{\partial x}. \tag{3}$$

In the early 1960s, Lorenz accidentally discovered the chaotic behavior of this system when he found that, for a simplified system, periodic solutions of the form

$$\psi = \psi_0 \sin\left(\frac{\pi a x}{H}\right) \sin\left(\frac{\pi z}{H}\right) \tag{4}$$

$$\theta = \theta_0 \cos\left(\frac{\pi a x}{H}\right) \sin\left(\frac{\pi z}{H}\right) \tag{5}$$

grew for Rayleigh numbers larger than the critical value, $Ra > Ra_c$. Furthermore, vastly different results were obtained for very small changes in the initial values, representing one of the earliest discoveries of the so-called BUTTERFLY EFFECT.

Lorenz included the following terms in his system of equations,

$$X \equiv \psi_{11} \propto \text{convective intensity} \tag{6}$$

$$Y \equiv T_{11} \propto \Delta T \text{ between descending and ascending currents} \tag{7}$$

$$Z \equiv T_{02} \propto \Delta \text{ vertical temperature profile from linearity,} \tag{8}$$

and obtained the simplified equations

$$\dot{X} = \sigma(Y - X) \tag{9}$$

$$\dot{Y} = -XZ + rX - Y \tag{10}$$

$$\dot{Z} = XY - bZ, \tag{11}$$

where

$$\sigma \equiv \frac{\nu}{\kappa} = \text{Prandtl number} \tag{12}$$

$$r \equiv \frac{Ra}{Ra_c} = \text{normalized Rayleigh number} \tag{13}$$

$$b \equiv \frac{4}{1+a^2} = \text{geometric factor.} \tag{14}$$

Lorenz took $b \equiv 8/3$ and $\sigma \equiv 10$.

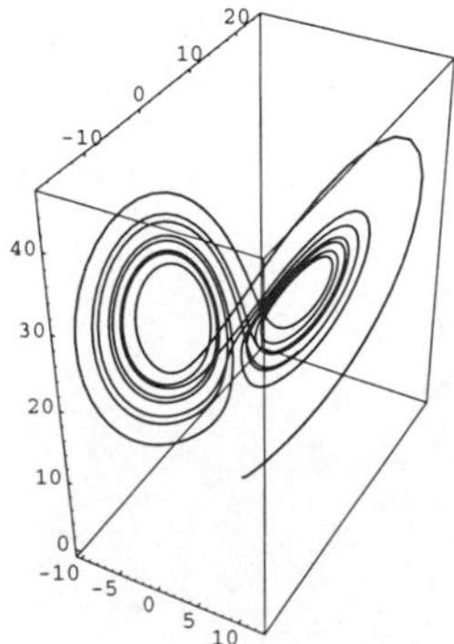

The CRITICAL POINTS at (0, 0, 0) correspond to no convection, and the CRITICAL POINTS at

$$(\sqrt{b(r-1)}, \sqrt{b(r-1)}, r-1) \tag{15}$$

and

$$(-\sqrt{b(r-1)}, -\sqrt{b(r-1)}, r-1) \tag{16}$$

correspond to steady convection. This pair is stable only if

$$r = \frac{\sigma(\sigma + b + 3)}{\sigma - b - 1}, \tag{17}$$

which can hold only for POSITIVE r if $\sigma > b + 1$. The Lorenz attractor has a CORRELATION EXPONENT of 2.05 ± 0.01 and CAPACITY DIMENSION 2.06 ± 0.01 (Grassberger and Procaccia 1983). For more details, see Lichtenberg and Lieberman (1983, p. 65) and Tabor (1989, p. 204).

see also BUTTERFLY EFFECT, RÖSSLER MODEL

References

Gleick, J. *Chaos: Making a New Science.* New York: Penguin Books, pp. 27–31, 1988.

Grassberger, P. and Procaccia, I. "Measuring the Strangeness of Strange Attractors." *Physica D* **9**, 189–208, 1983.

Lichtenberg, A. and Lieberman, M. *Regular and Stochastic Motion.* New York: Springer-Verlag, 1983.

Lorenz, E. N. "Deterministic Nonperiodic Flow." *J. Atmos. Sci.* **20**, 130–141, 1963.

Peitgen, H.-O.; Jürgens, H.; and Saupe, D. *Chaos and Fractals: New Frontiers of Science.* New York: Springer-Verlag, pp. 697–708, 1992.

Tabor, M. *Chaos and Integrability in Nonlinear Dynamics: An Introduction.* New York: Wiley, 1989.

Lorraine Cross

see GAULLIST CROSS

Lotka-Volterra Equations

An ecological model which assumes that a population x increases at a rate $dx = Ax\,dt$, but is destroyed at a rate $dx = -Bxy\,dt$. Population y decreases at a rate $dy = -Cy\,dt$, but increases at $dy = Dxy\,dt$, giving the coupled differential equations

$$\frac{dx}{dt} = Ax - Bxy$$

$$\frac{dy}{dt} = -Cy + Dxy.$$

Critical points occur when $dx/dt = dy/dt = 0$, so

$$A - By = 0$$

$$-C + Dx = 0.$$

The sole STATIONARY POINT is therefore located at $(x, y) = (C/D, A/B)$.

Low-Dimensional Topology

Low-dimensional topology usually deals with objects that are 2-, 3-, or 4-dimensional in nature. Properly speaking, low-dimensional topology should be part of DIFFERENTIAL TOPOLOGY, but the general machinery of ALGEBRAIC and DIFFERENTIAL TOPOLOGY gives only limited information. This fact is particularly noticeable in dimensions three and four, and so alternative specialized methods have evolved.

see also ALGEBRAIC TOPOLOGY, DIFFERENTIAL TOPOLOGY, TOPOLOGY

Löwenheimer-Skolem Theorem

A fundamental result in MODEL THEORY which states that if a countable theory has a model, then it has a countable model. Furthermore, it has a model of every CARDINALITY greater than or equal to $\aleph_0$ (ALEPH-0). This theorem established the existence of "nonstandard" models of arithmetic.

see also ALEPH-0 ($\aleph_0$), CARDINALITY, MODEL THEORY

References

Chang, C. C. and Keisler, H. J. *Model Theory, 3rd enl. ed.* New York: Elsevier, 1990.

Lower Bound

see GREATEST LOWER BOUND

Lower Denjoy Sum

see LOWER SUM

Lower Integral

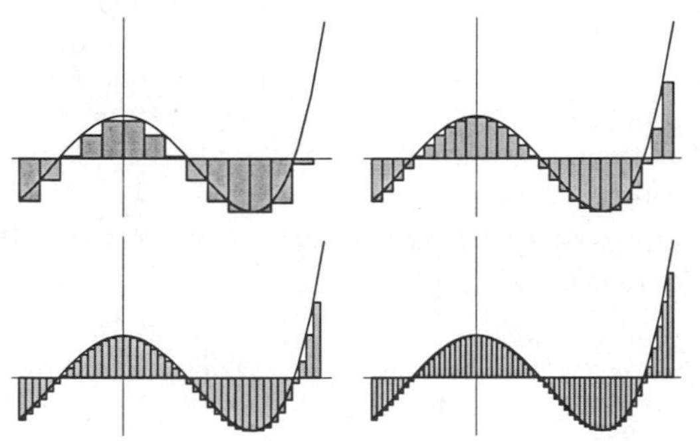

The limit of a LOWER SUM, when it exists, as the MESH SIZE approaches 0.

see also LOWER SUM, RIEMANN INTEGRAL, UPPER INTEGRAL

Lower Limit

Let the least term h of a SEQUENCE be a term which is smaller than all but a finite number of the terms which are equal to h. Then h is called the lower limit of the SEQUENCE.

A lower limit of a SERIES

$$\text{lower} \lim_{n\to\infty} S_n = \underline{\lim_{n\to\infty}} S_n = h$$

is said to exist if, for every $\epsilon > 0$, $|S_n - h| < \epsilon$ for infinitely many values of n and if no number less than h has this property.

see also LIMIT, UPPER LIMIT

References

Bromwich, T. J. I'a and MacRobert, T. M. "Upper and Lower Limits of a Sequence." §5.1 in *An Introduction to the Theory of Infinite Series, 3rd ed.* New York: Chelsea, p. 40 1991.

Lower Sum

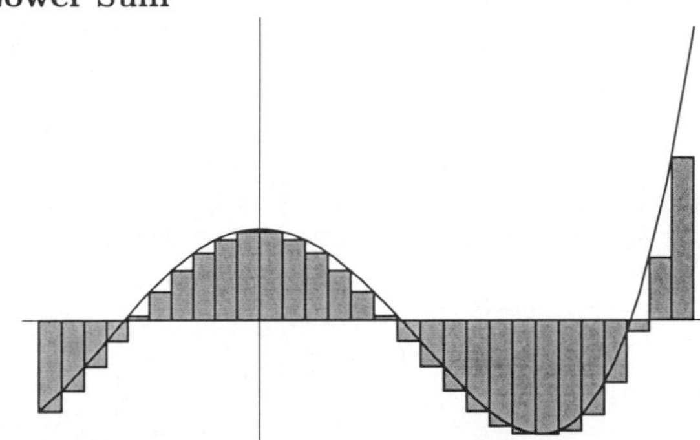

For a given function $f(x)$ over a partition of a given interval, the lower sum is the sum of box areas $f(x_k^*)\Delta x_k$ using the smallest value of the function $f(x_k^*)$ in each subinterval Δx_k.

see also LOWER INTEGRAL, RIEMANN INTEGRAL, UPPER SUM

Lower-Trimmed Subsequence

The lower-trimmed subsequence of $x = \{x_n\}$ is the sequence $V(x)$ obtained by subtracting 1 from each x_n and then removing all 0s. If x is a FRACTAL SEQUENCE, then $V(x)$ is a FRACTAL SEQUENCE. If x is a SIGNATURE SEQUENCE, then $V(x) = x$.

see also SIGNATURE SEQUENCE, UPPER-TRIMMED SUBSEQUENCE

References

Kimberling, C. "Fractal Sequences and Interspersions." *Ars Combin.* **45**, 157–168, 1997.

Lowest Terms Fraction

A FRACTION p/q for which $(p,q) = 1$, where (p,q) denotes the GREATEST COMMON DIVISOR.

Loxodrome

A path, also known as a RHUMB LINE, which cuts a MERIDIAN on a given surface (usually a SPHERE, in which case the loxodrome is also called a SPHERICAL HELIX) at any constant ANGLE but a RIGHT ANGLE. The loxodrome is the path taken when a compass is kept pointing in a constant direction. It is *not* the shortest distance between two points.

see also GREAT CIRCLE

Lozenge

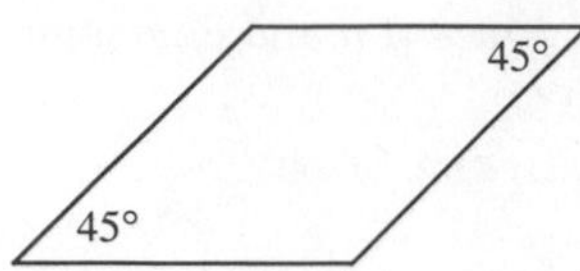

A PARALLELOGRAM whose ACUTE ANGLES are 45°.

see also DIAMOND, PARALLELOGRAM, QUADRILATERAL, RHOMBUS

Lozenge Method

A method for constructing MAGIC SQUARES of ODD order.

see also MAGIC SQUARE

Lozi Map

A 2-D map similar to the HÉNON MAP which is given by the equations

$$x_{n+1} = 1 - \alpha|x_n| + y_n$$
$$y_{n+1} = \beta x_n.$$

see also HÉNON MAP

References

Dickau, R. M. "Lozi Attractor." `http://www.prairienet.org/-pops/lozi.html`.

Peitgen, H.-O.; Jürgens, H.; and Saupe, D. §12.1 in *Chaos and Fractals: New Frontiers of Science.* New York: Springer-Verlag, p. 672, 1992.

LU Decomposition

A procedure for decomposing an $N \times N$ matrix A into a product of a lower TRIANGULAR MATRIX L and an upper TRIANGULAR MATRIX U,

$$\mathsf{LU} = \mathsf{A}. \quad (1)$$

Written explicitly for a 3×3 MATRIX, the decomposition is

$$\begin{bmatrix} l_{11} & 0 & 0 \\ l_{21} & l_{22} & 0 \\ l_{31} & l_{32} & l_{33} \end{bmatrix} \begin{bmatrix} u_{11} & u_{12} & u_{13} \\ 0 & u_{22} & u_{23} \\ 0 & 0 & u_{33} \end{bmatrix} = \begin{bmatrix} a_{11} & a_{12} & a_{13} \\ a_{21} & a_{22} & a_{23} \\ a_{31} & a_{32} & a_{33} \end{bmatrix} \quad (2)$$

$$\begin{bmatrix} l_{11}u_{11} & l_{11}u_{12} & l_{11}u_{13} \\ l_{21}u_{11} & l_{21}u_{22} + l_{22}u_{22} & l_{21}u_{13} + l_{22}u_{23} \\ l_{31}u_{11} & l_{31}u_{12} + l_{32}u_{22} & l_{31}u_{13} + l_{32}u_{23} + l_{33}u_{23} \end{bmatrix} = \begin{bmatrix} a_{11} & a_{12} & a_{13} \\ a_{21} & a_{22} & a_{23} \\ a_{31} & a_{32} & a_{33} \end{bmatrix}. \quad (3)$$

This gives three types of equations

$$i < j \qquad l_{i1}u_{ij} + l_{i2}u_{2j} + \ldots + l_{ii}u_{ij} = a_{ij} \quad (4)$$
$$i = j \qquad l_{i1}u_{ij} + l_{i2}u_{2j} + \ldots + l_{ii}u_{jj} = a_{ij} \quad (5)$$
$$i > j \qquad l_{i1}u_{1j} + l_{i2}u_{2j} + \ldots + l_{ij}u_{jj} = a_{ij}. \quad (6)$$

This gives N^2 equations for $N^2 + N$ unknowns (the decomposition is not unique), and can be solved using CROUT'S METHOD. To solve the MATRIX equation

$$\mathsf{A}\mathbf{x} = (\mathsf{LU})\mathbf{x} = \mathsf{L}(\mathsf{U}\mathbf{x}) = \mathbf{b}, \quad (7)$$

first solve $\mathsf{L}\mathbf{y} = \mathbf{b}$ for $\mathbf{y}$. This can be done by forward substitution

$$y_1 = \frac{b_1}{l_{11}} \quad (8)$$
$$y_i = \frac{1}{l_{ii}}\left(b_i - \sum_{j=1}^{i-1} l_{ij}y_j\right) \quad (9)$$

for $i = 2, \ldots, N$. Then solve $\mathsf{U}\mathbf{x} = \mathbf{y}$ for $\mathbf{x}$. This can be done by back substitution

$$x_N = \frac{y_N}{u_{NN}} \quad (10)$$
$$x_i = \frac{1}{u_{ii}}\left(y_i - \sum_{j=i+1}^{N} u_{ij}x_j\right) \quad (11)$$

for $i = N - 1, \ldots, 1$.

see also CHOLESKY DECOMPOSITION, QR DECOMPOSITION, TRIANGULAR MATRIX

References

Press, W. H.; Flannery, B. P.; Teukolsky, S. A.; and Vetterling, W. T. "LU Decomposition and Its Applications." §2.3 in *Numerical Recipes in FORTRAN: The Art of Scientific Computing, 2nd ed.* Cambridge, England: Cambridge University Press, pp. 34–42, 1992.

Lucas Correspondence

The correspondence which relates the HANOI GRAPH to the ISOMORPHIC GRAPH of the ODD BINOMIAL COEFFICIENTS in PASCAL'S TRIANGLE, where the adjacencies are determined by adjacency (either horizontal or diagonal) in PASCAL'S TRIANGLE. The proof that the correspondence is given by the LUCAS CORRESPONDENCE THEOREM.

see also BINOMIAL COEFFICIENT, HANOI GRAPH, PASCAL'S TRIANGLE

References

Poole, David G. "The Towers and Triangles of Professor Claus (or, Pascal Knows Hanoi)." *Math. Mag.* **67**, 323–344, 1994.

Lucas Correspondence Theorem

Let p be PRIME and

$$r = r_m p^m + \ldots + r_1 p + r_0 \qquad (0 \leq r_i < p) \quad (1)$$
$$k = k_m p^m + \ldots + k_1 p + k_0 \qquad (0 \leq k_i < p), \quad (2)$$

then

$$\binom{r}{k} = \prod_{i=0}^{m} \binom{r_i}{k_i} \pmod{p}. \quad (3)$$

This is proved in Fine (1947).

References

Fine, N. J. "Binomial Coefficients Modulo a Prime." *Amer. Math. Monthly* **54**, 589–592, 1947.

Lucas-Lehmer Residue

see LUCAS-LEHMER TEST

Lucas-Lehmer Test

A MERSENNE NUMBER M_p is prime IFF M_p divides s_{p-2}, where $s_0 \equiv 4$ and

$$s_i \equiv {s_{i-1}}^2 - 2 (\text{mod } 2^p - 1) \tag{1}$$

for $i \geq 1$. The first few terms of this series are 4, 14, 194, 37634, 1416317954, ... (Sloane's A003010). The remainder when s_{p-2} is divided by M_p is called the LUCAS-LEHMER RESIDUE for p. The LUCAS-LEHMER RESIDUE is 0 IFF M_p is PRIME. This test can also be extended to arbitrary INTEGERS.

A generalized version of the Lucas-Lehmer test lets

$$N+1 = \prod_{j=1}^{n} {q_j}^{\beta_j}, \tag{2}$$

with q_j the distinct PRIME factors, and β_j their respective POWERS. If there exists a LUCAS SEQUENCE U_ν such that

$$\text{GCD}(U_{(N+1)/q_j}, N) = 1 \tag{3}$$

for $j = 1, \ldots, n$ and

$$U_{N+1} \equiv 0 \pmod{N}, \tag{4}$$

then N is a PRIME. The test is particularly simple for MERSENNE NUMBERS, yielding the conventional Lucas-Lehmer test.

see also LUCAS SEQUENCE, MERSENNE NUMBER, RABIN-MILLER STRONG PSEUDOPRIME TEST

References

Sloane, N. J. A. Sequence A003010/M3494 in "An On-Line Version of the Encyclopedia of Integer Sequences."

Lucas' Married Couples Problem

see MARRIED COUPLES PROBLEM

Lucas Number

The numbers produced by the V recurrence in the LUCAS SEQUENCE with $(P, Q) = (1, -1)$ are called Lucas numbers. They are the companions to the FIBONACCI NUMBERS F_n and satisfy the same recurrence

$$L_n = L_{n-1} + L_{n-2}, \tag{1}$$

where $L_1 = 1$, $L_2 = 3$. The first few are 1, 3, 4, 7, 11, 18, 29, 47, 76, 123, ... (Sloane's A000204).

In terms of the FIBONACCI NUMBERS,

$$L_n = F_{n-1} + F_{n+1}. \tag{2}$$

The analog of BINET'S FORMULA for Lucas numbers is

$$L_n = \left(\frac{1+\sqrt{5}}{2}\right)^n + \left(\frac{1-\sqrt{5}}{2}\right)^n. \tag{3}$$

Another formula is

$$L_n = [\phi^n], \tag{4}$$

where ϕ is the GOLDEN RATIO and $[x]$ denotes the NINT function. Given L_n,

$$L_{n+1} = \left\lfloor \frac{L_n(1+\sqrt{5})+1}{2} \right\rfloor, \tag{5}$$

where $\lfloor x \rfloor$ is the FLOOR FUNCTION,

$${L_n}^2 - L_{n-1}L_{n+1} = 5(-1)^n, \tag{6}$$

and

$$\sum_{k=0}^{n} {L_k}^2 = L_n L_{n+1} - 2. \tag{7}$$

Let p be a PRIME > 3 and k be a POSITIVE INTEGER. Then L_{2p^k} ends in a 3 (Honsberger 1985, p. 113). Analogs of the Cesàro identities for FIBONACCI NUMBERS are

$$\sum_{k=0}^{n} \binom{n}{k} L_k = L_{2n} \tag{8}$$

$$\sum_{k=0}^{n} \binom{n}{k} 2^k L_k = L_{3n}, \tag{9}$$

where $\binom{n}{k}$ is a BINOMIAL COEFFICIENT.

$L_n | F_m$ (L_n DIVIDES F_m) IFF n DIVIDES into m an EVEN number of times. $L_n | L_m$ IFF n divides into m an ODD number of times. $2^n L_n$ always ends in 2 (Honsberger 1985, p. 137).

Defining

$$D_n \equiv \begin{vmatrix} 3 & i & 0 & 0 & \cdots & 0 & 0 \\ i & 1 & i & 0 & \cdots & 0 & 0 \\ 0 & i & 1 & i & \cdots & 0 & 0 \\ 0 & 0 & i & 1 & \cdots & 0 & 0 \\ \vdots & \vdots & \vdots & \vdots & \ddots & \vdots & \vdots \\ 0 & 0 & 0 & 0 & \cdots & 1 & i \\ 0 & 0 & 0 & 0 & \cdots & i & 1 \end{vmatrix} = L_{n+1} \tag{10}$$

gives

$$D_n = D_{n-1} + D_{n-2} \tag{11}$$

(Honsberger 1985, pp. 113–114).

The number of ways of picking a set (including the EMPTY SET) from the numbers 1, 2, ..., n without picking two consecutive numbers (where 1 and n are now consecutive) is L_n (Honsberger 1985, p. 122).

The only SQUARE NUMBERS in the Lucas sequence are 1 and 4, as proved by John H. E. Cohn (Alfred 1964). The only TRIANGULAR Lucas numbers are 1, 3, and 5778 (Ming 1991). The only Lucas CUBIC NUMBER is 1. The first few Lucas PRIMES L_n occur for $n = 2$, 4, 5, 7, 8, 11, 13, 16, 17, 19, 31, 37, 41, 47, 53, 61, 71, 79, 113, 313, 353, ... (Dubner and Keller 1998, Sloane's A001606).

see also FIBONACCI NUMBER

References

Alfred, Brother U. "On Square Lucas Numbers." *Fib. Quart.* **2**, 11–12, 1964.

Borwein, J. M. and Borwein, P. B. *Pi & the AGM: A Study in Analytic Number Theory and Computational Complexity.* New York: Wiley, pp. 94–101, 1987.

Brillhart, J.; Montgomery, P. L.; and Solverman, R. D. "Tables of Fibonacci and Lucas Factorizations." *Math. Comput.* **50**, 251–260 and S1–S15, 1988.

Brown, J. L. Jr. "Unique Representation of Integers as Sums of Distinct Lucas Numbers." *Fib. Quart.* **7**, 243–252, 1969.

Dubner, H. and Keller, W. "New Fibonacci and Lucas Primes." *Math. Comput.* 1998.

Guy, R. K. "Fibonacci Numbers of Various Shapes." §D26 in *Unsolved Problems in Number Theory, 2nd ed.* New York: Springer-Verlag, pp. 194–195, 1994.

Hoggatt, V. E. Jr. *The Fibonacci and Lucas Numbers.* Boston, MA: Houghton Mifflin, 1969.

Honsberger, R. "A Second Look at the Fibonacci and Lucas Numbers." Ch. 8 in *Mathematical Gems III.* Washington, DC: Math. Assoc. Amer., 1985.

Leyland, P. ftp://sable.ox.ac.uk/pub/math/factors/lucas.Z.

Ming, L. "On Triangular Lucas Numbers." *Applications of Fibonacci Numbers, Vol. 4* (Ed. G. E. Bergum, A. N. Philippou, and A. F. Horadam). Dordrecht, Netherlands: Kluwer, pp. 231–240, 1991.

Sloane, N. J. A. Sequences A000692/M2341 and A001606/M0961 in "An On-Line Version of the Encyclopedia of Integer Sequences."

Lucas Polynomial

The w POLYNOMIALS obtained by setting $p(x) = x$ and $q(x) = 1$ in the LUCAS POLYNOMIAL SEQUENCE. The first few are

$$\begin{aligned} F_1(x) &= x \\ F_2(x) &= x^2 + 2 \\ F_3(x) &= 3x^3 + 3x \\ F_4(x) &= x^4 + 4x^2 + 2 \\ F_5(x) &= x^5 + 5x^3 + 5x. \end{aligned}$$

The corresponding W POLYNOMIALS are called FIBONACCI POLYNOMIALS. The Lucas polynomials satisfy

$$L_n(1) = L_n,$$

where the L_ns are LUCAS NUMBERS.

see also FIBONACCI POLYNOMIAL, LUCAS NUMBER, LUCAS POLYNOMIAL SEQUENCE

Lucas Polynomial Sequence

A pair of generalized POLYNOMIALS which generalize the LUCAS SEQUENCE to POLYNOMIALS is given by

$$W_n^k(x) = \frac{\Delta^k(x)[a^n(x) - (-1)^k b^n(x)]}{\Delta(x)} \tag{1}$$

$$w_n^k(x) = \Delta^k(x)[a^n(x) + (-1)^k b^n(x)], \tag{2}$$

where

$$a(x) + b(x) = p(x) \tag{3}$$

$$a(x)b(x) = -q(x) \tag{4}$$

$$a(x) - b(x) = \sqrt{p^2(x) + 4q(x)} \equiv \Delta(x) \tag{5}$$

(Horadam 1996). Setting $n = 0$ gives

$$W_0^k(x) = \Delta^k(x)\frac{1 - (-1)^k}{\Delta(x)} \tag{6}$$

$$w_0^k(x) = \Delta^k(x)[1 + (-1)^k], \tag{7}$$

giving

$$W_0^0(x) = 0 \tag{8}$$

$$w_0^0(x) = 2. \tag{9}$$

When $k = 1$,

$$W_n^1(x) = w_n^0(x) = w_n(x) \tag{10}$$

$$W_n^1(x) = \Delta^2(x)W_n^0(x) = \Delta^2(x)W_n(x). \tag{11}$$

Special cases are given in the following table.

$p(x)$	$q(x)$	Polynomial 1	Polynomial 2
x	1	Fibonacci $F_n(x)$	Lucas $L_n(x)$
$2x$	1	Pell $P_n(x)$	Pell-Lucas $Q_n(x)$
1	$2x$	Jacobsthal $J_n(x)$	Jacobsthal-Lucas $j_n(x)$
$3x$	-2	Fermat $\mathcal{F}_n(x)$	Fermat-Lucas $f_n(x)$
$2x$	-1	Chebyshev $U_{n-1}(x)$	Chebyshev $2T_n(x)$

see also LUCAS SEQUENCE

References

Horadam, A. F. "Extension of a Synthesis for a Class of Polynomial Sequences." *Fib. Quart.* **34**, 68–74, 1996.

Lucas Pseudoprime

When P and Q are INTEGERS such that $D = P^2 - 4Q \neq 0$, define the LUCAS SEQUENCE $\{U_k\}$ by

$$U_k = \frac{a^k - b^k}{a - b}$$

for $k \geq 0$, with a and b the two ROOTS of $x^2 - Px + Q = 0$. Then define a Lucas pseudoprime as an ODD COMPOSITE number n such that $n \nmid Q$, the JACOBI SYMBOL $(D/n) = -1$, and $n | U_{n+1}$.

There are no EVEN Lucas pseudoprimes (Bruckman 1994). The first few Lucas pseudoprimes are 705, 2465, 2737, 3745, ... (Sloane's A005845).

see also EXTRA STRONG LUCAS PSEUDOPRIME, LUCAS SEQUENCE, PSEUDOPRIME, STRONG LUCAS PSEUDOPRIME

References

Bruckman, P. S. "Lucas Pseudoprimes are Odd." *Fib. Quart.* **32**, 155–157, 1994.

Ribenboim, P. "Lucas Pseudoprimes (lpsp(P,Q))." §2.X.B in *The New Book of Prime Number Records, 3rd ed.* New York: Springer-Verlag, p. 129, 1996.

Sloane, N. J. A. Sequence A005845/M5469 in "An On-Line Version of the Encyclopedia of Integer Sequences."

Lucas Sequence

Let P, Q be POSITIVE INTEGERS. The ROOTS of

$$x^2 - Px + Q = 0 \tag{1}$$

are

$$a \equiv \tfrac{1}{2}(P + \sqrt{D}) \tag{2}$$
$$b \equiv \tfrac{1}{2}(P - \sqrt{D}), \tag{3}$$

where

$$D \equiv P^2 - 4Q, \tag{4}$$

so

$$a + b = P \tag{5}$$
$$ab = \tfrac{1}{4}(P^2 - D) = Q \tag{6}$$
$$a - b = \sqrt{D}. \tag{7}$$

Then define

$$U_n(P,Q) \equiv \frac{a^n - b^n}{a - b} \tag{8}$$
$$V_n(P,Q) \equiv a^n + b^n. \tag{9}$$

The first few values are therefore

$$U_0(P,Q) = 0 \tag{10}$$
$$U_1(P,Q) = 1 \tag{11}$$
$$V_0(P,Q) = 2 \tag{12}$$
$$V_1(P,Q) = P. \tag{13}$$

The sequences

$$U(P,Q) = \{U_n(P,Q) : n \geq 1\} \tag{14}$$
$$V(P,Q) = \{V_n(P,Q) : n \geq 1\} \tag{15}$$

are called Lucas sequences, where the definition is usually extended to include

$$U_{-1} = \frac{a^{-1} - b^{-1}}{a - b} = \frac{-1}{ab} = -\frac{1}{Q}. \tag{16}$$

For $(P,Q) = (1,-1)$, the U_n are the FIBONACCI NUMBERS and V_n are the LUCAS NUMBERS. For $(P,Q) = (2,-1)$, the PELL NUMBERS and Pell-Lucas numbers are obtained. $(P,Q) = (1,-2)$ produces the JACOBSTHAL NUMBERS and Pell-Jacobsthal Numbers.

The Lucas sequences satisfy the general RECURRENCE RELATIONS

$$\begin{aligned} U_{m+n} &= \frac{a^{m+n} - b^{m+n}}{a - b} \\ &= \frac{(a^m - b^m)(a^n + b^n)}{a - b} - \frac{a^n b^n (a^{m-n} - b^{m-n})}{a - b} \\ &= U_m V_n - a^n b^n U_{m-n} \end{aligned} \tag{17}$$

$$\begin{aligned} V_{m+n} &= a^{m+n} + b^{m+n} \\ &= (a^m + b^m)(a^n + b^n) - a^n b^n (a^{m-n} + b^{m-n}) \\ &= V_m V_n - a^n b^n V_{m-n}. \end{aligned} \tag{18}$$

Taking $n = 1$ then gives

$$U_m(P,Q) = PU_{m-1}(P,Q) - QU_{m-2}(P,Q) \tag{19}$$
$$V_m(P,Q) = PV_{m-1}(P,Q) - QV_{m-2}(P,Q). \tag{20}$$

Other identities include

$$U_{2n} = U_n V_n \tag{21}$$
$$U_{2n+1} = U_{n+1} V_n - Q^n \tag{22}$$
$$V_{2n} = {V_n}^2 - 2(ab)^n = {V_n}^2 - 2Q^n \tag{23}$$
$$V_{2n+1} = V_{n+1} V_n - PQ^n. \tag{24}$$

These formulas allow calculations for large n to be decomposed into a chain in which only four quantities must be kept track of at a time, and the number of steps needed is $\sim \lg n$. The chain is particularly simple if n has many 2s in its factorization.

The Us in a Lucas sequence satisfy the CONGRUENCE

$$U_{p^{n-1}[p-(D/p)]} \equiv 0 \pmod{p^n} \tag{25}$$

if

$$\text{GCD}(2QcD, p) = 1, \tag{26}$$

where

$$P^2 - 4Q^2 = c^2 D. \tag{27}$$

This fact is used in the proof of the general LUCAS-LEHMER TEST.

see also FIBONACCI NUMBER, JACOBSTHAL NUMBER, LUCAS-LEHMER TEST, LUCAS NUMBER, LUCAS POLYNOMIAL SEQUENCE, PELL NUMBER, RECURRENCE SEQUENCE, SYLVESTER CYCLOTOMIC NUMBER

References

Dickson, L. E. "Recurring Series; Lucas' u_n, v_n." Ch. 17 in *History of the Theory of Numbers, Vol. 1: Divisibility and Primality.* New York: Chelsea, pp. 393–411, 1952.

Ribenboim, P. *The Little Book of Big Primes.* New York: Springer-Verlag, pp. 35–53, 1991.

Lucas's Theorem

The primitive factors $Q_n(x,y)$ of $x^n + y^n$ can be written in the form

$$Q_n(x,y) = U^2(x,y) \pm nxyV^2(x,y)$$

for SQUAREFREE n where U and V are HOMOGENEOUS POLYNOMIALS with the sign chosen according to

$$\begin{cases} + & \text{for } n = 4l+1 \\ - & \text{for } n = 4l+3 \\ \text{either} & \text{for } n = 4l+2. \end{cases}$$

Lucky Number

Write out all the ODD numbers: 1, 3, 5, 7, 9, 11, 13, 15, 17, 19, ... The first ODD number > 1 is 3, so strike out every third number from the list: 1, 3, 7, 9, 13, 15, 19, ... The first ODD number greater than 3 in the list is 7, so strike out every seventh number: 1, 3, 7, 9, 13, 15, 21, 25, 31, ...

Numbers remaining after this procedure has been carried out completely are called lucky numbers. The first few are 1, 3, 7, 9, 13, 15, 21, 25, 31, 33, 37, ... (Sloane's A000959). Many asymptotic properties of the PRIME NUMBERS are shared by the lucky numbers. The asymptotic density is $1/\ln N$, just as the PRIME NUMBER THEOREM, and the frequency of TWIN PRIMES and twin lucky numbers are similar. A version of the GOLDBACH CONJECTURE also seems to hold.

It therefore appears that the SIEVING process accounts for many properties of the PRIMES.

see also GOLDBACH CONJECTURE, LUCKY NUMBER OF EULER, PRIME NUMBER, PRIME NUMBER THEOREM, SIEVE

References

Gardner, M. "Mathematical Games: Tests Show whether a Large Number can be Divided by a Number from 2 to 12." *Sci. Amer.* **207**, 232, Sep. 1962.

Gardner, M. "Lucky Numbers and 2187." *Math. Intell.* **19**, 26, 1997.

Guy, R. K. "Lucky Numbers." §C3 in *Unsolved Problems in Number Theory, 2nd ed.* New York: Springer-Verlag, pp. 108–109, 1994.

Ogilvy, C. S. and Anderson, J. T. *Excursions in Number Theory.* New York: Dover, pp. 100–102, 1988.

Peterson, I. "MathTrek: Martin Gardner's Luck Number." `http://www.sciencenews.org/sn_arc97/9_6_97/mathland.htm`.

Sloane, N. J. A. Sequence A000959/M2616 in "An On-Line Version of the Encyclopedia of Integer Sequences."

Ulam, S. M. *A Collection of Mathematical Problems.* New York: Interscience Publishers, p. 120, 1960.

Wells, D. G. *The Penguin Dictionary of Curious and Interesting Numbers.* London: Penguin, p. 32, 1986.

Lucky Number of Euler

A number p such that the PRIME-GENERATING POLYNOMIAL

$$n^2 - n + p$$

is PRIME for $n = 0, 1, \ldots, p-2$. Such numbers are related to the COMPLEX QUADRATIC FIELD in which the RING of INTEGERS is factorable. Specifically, the Lucky numbers of Euler (excluding the trivial case $p = 3$) are those numbers p such that the QUADRATIC FIELD $\mathbb{Q}(\sqrt{1-4p})$ has CLASS NUMBER 1 (Rabinowitz 1913, Le Lionnais 1983, Conway and Guy 1996).

As established by Stark (1967), there are only nine numbers $-d$ such that $h(-d) = 1$ (the HEEGNER NUMBERS -2, -3, -7, -11, -19, -43, -67, and -163), and of these, only 7, 11, 19, 43, 67, and 163 are of the required form. Therefore, the only Lucky numbers of Euler are 2, 3, 5, 11, 17, and 41 (Le Lionnais 1983, Sloane's A014556), and there does not exist a better PRIME-GENERATING POLYNOMIAL of Euler's form.

see also CLASS NUMBER, HEEGNER NUMBER, PRIME-GENERATING POLYNOMIAL

References

Conway, J. H. and Guy, R. K. "The Nine Magic Discriminants." In *The Book of Numbers.* New York: Springer-Verlag, pp. 224–226, 1996.

Le Lionnais, F. *Les nombres remarquables.* Paris: Hermann, pp. 88 and 144, 1983.

Rabinowitz, G. "Eindeutigkeit der Zerlegung in Primzahlfaktoren in quadratischen Zahlkörpern." *Proc. Fifth Internat. Congress Math. (Cambridge)* **1**, 418–421, 1913.

Sloane, N. J. A. Sequence A014556 in "An On-Line Version of the Encyclopedia of Integer Sequences."

Stark, H. M. "A Complete Determination of the Complex Quadratic Fields of Class Number One." *Michigan Math. J.* **14**, 1–27, 1967.

LUCY

A nonlinear DECONVOLUTION technique used in deconvolving images from the Hubble Space Telescope before corrective optics were installed.

see also CLEAN ALGORITHM, DECONVOLUTION, MAXIMUM ENTROPY METHOD

Ludolph's Constant

see PI

Ludwig's Inversion Formula

Expresses a function in terms of its RADON TRANSFORM,

$$\begin{aligned} f(x,y) &= \mathcal{R}^{-1}(\mathcal{R}f)(x,y) \\ &= \frac{1}{\pi}\frac{1}{2\pi}\int_{-\infty}^{\infty} \frac{\frac{\partial}{\partial p}(\mathcal{R}f)(p,\alpha)}{x\cos\alpha + y\sin\alpha - p}\,dp\,d\alpha. \end{aligned}$$

see also RADON TRANSFORM

Lukács Theorem

Let $\rho(x)$ be an mth degree POLYNOMIAL which is NONNEGATIVE in $[-1,1]$. Then $\rho(x)$ can be represented in the form

$$\begin{cases} [A(x)]^2 + (1-x^2)[B(x)]^2 & \text{for } m \text{ even} \\ (1+x)[C(x)]^2 + (1-x)[D(x)]^2 & \text{for } m \text{ odd,} \end{cases}$$

where $A(x)$, $B(x)$, $C(x)$, and $D(x)$ are REAL POLYNOMIALS whose degrees do not exceed m.

References

Szegő, G. *Orthogonal Polynomials, 4th ed.* Providence, RI: Amer. Math. Soc., p. 4, 1975.

Lune (Plane)

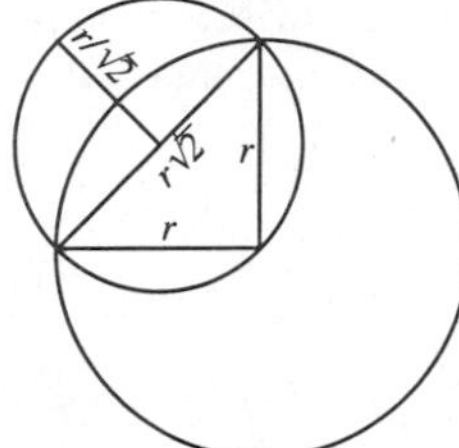

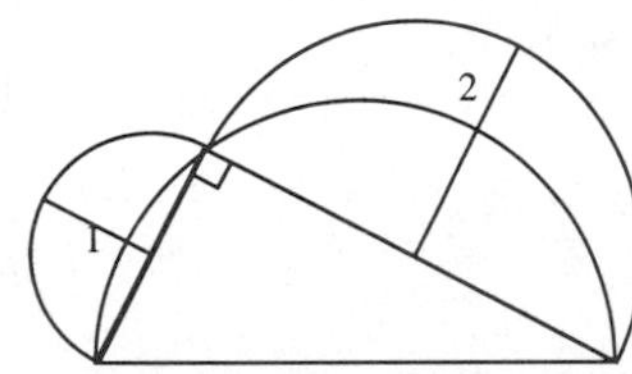

A figure bounded by two circular ARCS of unequal RADII. Hippocrates of Chios SQUARED the above left lune, as well as two others, in the fifth century BC. Two more SQUARABLE lunes were found by T. Clausen in the 19th century (Dunham 1990 attributes these discoveries to Euler in 1771). In the 20th century, N. G. Tschebatorew and A. W. Dorodnow proved that these are the only five squarable lunes (Shenitzer and Steprans 1994). The left lune above is squared as follows,

$$A_{\text{half small circle}} = \tfrac{1}{2}\pi\left(\frac{r}{\sqrt{2}}\right)^2 = \tfrac{1}{4}\pi r^2$$

$$\begin{aligned} A_{\text{lens}} &= A_{\text{quarter big circle}} - A_{\text{triangle}} \\ &= \tfrac{1}{4}\pi r^2 - \tfrac{1}{2}r^2 \\ A_{\text{lune}} &= A_{\text{half small circle}} - A_{\text{lens}} = \tfrac{1}{2}r^2 \\ &= A_{\text{triangle}}, \end{aligned}$$

so the lune and TRIANGLE have the same AREA. In the right figure, $A_1 + A_2 = A_\Delta$.

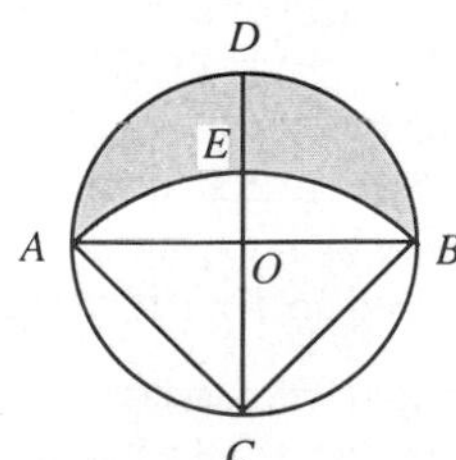

For the above lune,

$$A_{\text{lune}} = 2A_{\Delta OBC}.$$

see also ANNULUS, ARC, CIRCLE, LUNE (SURFACE)

References

Dunham, W. "Hippocrates' Quadrature of the Lune." Ch. 1 in *Journey Through Genius: The Great Theorems of Mathematics.* New York: Wiley, pp. 1–20, 1990.

Heath, T. L. *A History of Greek Mathematics.* New York: Dover, p. 185, 1981.

Pappas, T. "Lunes." *The Joy of Mathematics.* San Carlos, CA: Wide World Publ./Tetra, pp. 72–73, 1989.

Shenitzer, A. and Steprans, J. "The Evolution of Integration." *Amer. Math. Monthly* **101**, 66–72, 1994.

Lune (Solid)

A geometric figure consisting of two TRIANGLES attached to opposite sides of a SQUARE.

see also SQUARE, TRIANGLE

Lune (Surface)

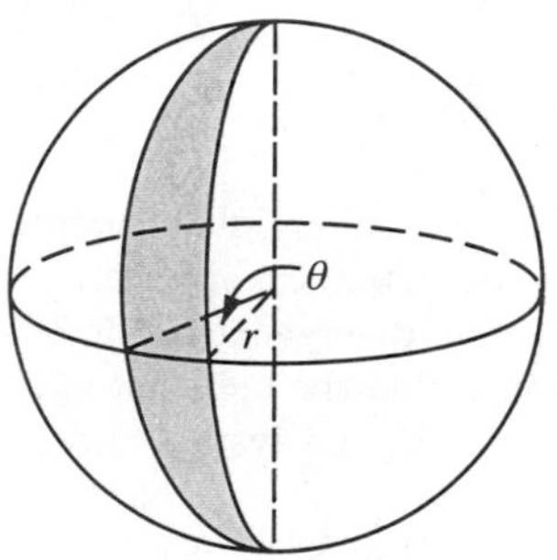

A sliver of the surface of a SPHERE of RADIUS r cut out by two planes through the azimuthal axis with DIHEDRAL ANGLE θ. The SURFACE AREA of the lune is

$$S = 2r^2\theta,$$

which is just the area of the SPHERE times $\theta/(2\pi)$.

see also LUNE (PLANE), SPHERE

References

Beyer, W. H. *CRC Standard Mathematical Tables, 28th ed.* Boca Raton, FL: CRC Press, p. 130, 1987.

Lunule

see LUNE (PLANE)

Lüroth's Theorem

If x and y are nonconstant rational functions of a parameter, the curve so defined has GENUS 0. Furthermore, x and y may be expressed rationally in terms of a parameter which is rational in them.

References

Coolidge, J. L. *A Treatise on Algebraic Plane Curves.* New York: Dover, p. 246, 1959.

Lusin's Theorem

Let $f(x)$ be a finite and MEASURABLE FUNCTION in $(-\infty, \infty)$, and let ϵ be freely chosen. Then there is a function $g(x)$ such that

1. $g(x)$ is continuous in $(-\infty, \infty)$,
2. The MEASURE of $\{x : f(x) \neq g(x)\}$ is $< \epsilon$,
3. $M(|g|; R_1) \leq M(|f|; R_1)$,

where $M(f; S)$ denotes the upper bound of the aggregate of the values of $f(P)$ as P runs through all values of S.

References

Kestelman, H. §4.4 in *Modern Theories of Integration, 2nd rev. ed.* New York: Dover, pp. 30 and 109–112, 1960.

LUX Method

A method for constructing MAGIC SQUARES of SINGLY EVEN order $n \geq 6$.

see also MAGIC SQUARE

Lyapunov Characteristic Exponent

The Lyapunov characteristic exponent [LCE] gives the rate of exponential divergence from perturbed initial conditions. To examine the behavior of an orbit around a point $\mathbf{X}^*(t)$, perturb the system and write

$$\mathbf{X}(t) = \mathbf{X}^*(t) + U(t), \tag{1}$$

where $U(t)$ is the average deviation from the unperturbed trajectory at time t. In a CHAOTIC region, the LCE σ is independent of $\mathbf{X}^*(0)$. It is given by the OSEDELEC THEOREM, which states that

$$\sigma_i = \lim_{t\to\infty} \ln |\mathbf{U}(t)|. \tag{2}$$

For an n-dimensional mapping, the Lyapunov characteristic exponents are given by

$$\sigma_i = \lim_{N\to\infty} \ln |\lambda_i(N)| \tag{3}$$

for $i = 1, \ldots, n$, where λ_i is the LYAPUNOV CHARACTERISTIC NUMBER.

One Lyapunov characteristic exponent is always 0, since there is never any divergence for a perturbed trajectory in the direction of the unperturbed trajectory. The larger the LCE, the greater the rate of exponential divergence and the wider the corresponding SEPARATRIX of the CHAOTIC region. For the STANDARD MAP, an analytic estimate of the width of the CHAOTIC zone by Chirikov (1979) finds

$$\delta I = Be^{-AK^{-1/2}}. \tag{4}$$

Since the Lyapunov characteristic exponent increases with increasing K, some relationship likely exists connecting the two. Let a trajectory (expressed as a MAP) have initial conditions (x_0, y_0) and a nearby trajectory have initial conditions $(x', y') = (x_0 + dx, y_0 + dy)$. The distance between trajectories at iteration k is then

$$dk = \|(x' - x_0, y' - y_0)\|, \tag{5}$$

and the mean exponential rate of divergence of the trajectories is defined by

$$\sigma_1 = \lim_{k\to\infty} \frac{1}{k} \ln\left(\frac{d_k}{d_0}\right). \tag{6}$$

For an n-dimensional phase space (MAP), there are n Lyapunov characteristic exponents $\sigma_1 \geq \sigma_2 \geq \ldots > \sigma_n$. However, because the largest exponent σ_1 will dominate, this limit is practically useful only for finding the largest exponent. Numerically, since d_k increases exponentially with k, after a few steps the perturbed trajectory is no longer nearby. It is therefore necessary to renormalize frequently every t steps. Defining

$$r_{k\tau} \equiv \frac{d_{k\tau}}{d_0}, \tag{7}$$

one can then compute

$$\sigma_1 = \lim_{n\to\infty} \frac{1}{n\tau} \sum_{k=1}^{n} \ln r_{k\tau}. \tag{8}$$

Numerical computation of the second (smaller) Lyapunov exponent may be carried by considering the evolution of a 2-D surface. It will behave as

$$e^{(\sigma_1+\sigma_2)t}, \tag{9}$$

so σ_2 can be extracted if σ_1 is known. The process may be repeated to find smaller exponents.

For HAMILTONIAN SYSTEMS, the LCEs exist in additive inverse pairs, so if σ is an LCE, then so is $-\sigma$. One LCE is always 0. For a 1-D oscillator (with a 2-D phase space), the two LCEs therefore must be $\sigma_1 = \sigma_2 = 0$, so the motion is QUASIPERIODIC and cannot be CHAOTIC. For higher order HAMILTONIAN SYSTEMS, there are always at least two 0 LCEs, but other LCEs may enter in plus-and-minus pairs l and $-l$. If they, too, are both zero, the motion is integrable and not CHAOTIC. If they are NONZERO, the POSITIVE LCE l results in an exponential separation of trajectories, which corresponds to a CHAOTIC region. Notice that it is not possible to have all LCEs NEGATIVE, which explains why convergence of orbits is never observed in HAMILTONIAN SYSTEMS.

Now consider a dissipative system. For an arbitrary n-D phase space, there must always be one LCE equal to 0, since a perturbation along the path results in no divergence. The LCEs satisfy $\sum_i \sigma_i < 0$. Therefore, for a 2-D phase space of a dissipative system, $\sigma_1 = 0, \sigma_2 < 0$. For a 3-D phase space, there are three possibilities:

1. (Integrable): $\sigma_1 = 0, \sigma_2 = 0, \sigma_3 < 0$,

2. (Integrable): $\sigma_1 = 0, \sigma_2, \sigma_3 < 0$,
3. (CHAOTIC): $\sigma_1 = 0, \sigma_2 > 0, \sigma_3 < -\sigma_2 < 0$.

see also CHAOS, HAMILTONIAN SYSTEM, LYAPUNOV CHARACTERISTIC NUMBER, OSEDELEC THEOREM

References

Chirikov, B. V. "A Universal Instability of Many-Dimensional Oscillator Systems." *Phys. Rep.* **52**, 264–379, 1979.

Lyapunov Characteristic Number

Given a LYAPUNOV CHARACTERISTIC EXPONENT σ_i, the corresponding Lyapunov characteristic number λ_i is defined as

$$\lambda_i \equiv e^{\sigma_i}. \quad (1)$$

For an n-dimensional linear MAP,

$$\mathbf{X}_{n+1} = \mathsf{M}\mathbf{X}_n. \quad (2)$$

The Lyapunov characteristic numbers $\lambda_1, \ldots, \lambda_n$ are the EIGENVALUES of the MAP MATRIX. For an arbitrary MAP

$$x_{n+1} = f_1(x_n, y_n) \quad (3)$$

$$y_{n+1} = f_2(x_n, y_n), \quad (4)$$

the Lyapunov numbers are the EIGENVALUES of the limit

$$\lim_{n\to\infty} [J(x_n, y_n)J(x_{n-1}, y_{n-1}) \cdots J(x_1, y_1)]^{1/n}, \quad (5)$$

where $J(x, y)$ is the JACOBIAN

$$J(x,y) \equiv \begin{vmatrix} \frac{\partial f_1(x,y)}{\partial x} & \frac{\partial f_1(x,y)}{\partial y} \\ \frac{\partial f_2(x,y)}{\partial x} & \frac{\partial f_2(x,y)}{\partial y} \end{vmatrix}. \quad (6)$$

If $\lambda_i = 0$ for all i, the system is not CHAOTIC. If $\lambda \neq 0$ and the MAP is AREA-PRESERVING (HAMILTONIAN), the product of EIGENVALUES is 1.

see also ADIABATIC INVARIANT, CHAOS, LYAPUNOV CHARACTERISTIC EXPONENT

Lyapunov Condition

If the third MOMENT exists for a DISTRIBUTION of x_i and the LEBESGUE INTEGRAL is given by

$${r_n}^3 = \sum_{i=1}^n \int_{-\infty}^\infty |x|^3 \, dF_i(x),$$

then if

$$\lim_{n\to\infty} \frac{r_n}{s_n} = 0,$$

the CENTRAL LIMIT THEOREM holds.

see also CENTRAL LIMIT THEOREM

Lyapunov Dimension

For a 2-D MAP with $\sigma_2 > \sigma_1$,

$$d_{\text{Lya}} = 1 - \frac{\sigma_1}{\sigma_2},$$

where σ_n are the LYAPUNOV CHARACTERISTIC EXPONENTS.

see also CAPACITY DIMENSION, KAPLAN-YORKE CONJECTURE

References

Frederickson, P.; Kaplan, J. L.; Yorke, E. D.; and Yorke, J. A. "The Liapunov Dimension of Strange Attractors." *J. Diff. Eq.* **49**, 185–207, 1983.

Nayfeh, A. H. and Balachandran, B. *Applied Nonlinear Dynamics: Analytical, Computational, and Experimental Methods.* New York: Wiley, p. 549, 1995.

Lyapunov's First Theorem

A NECESSARY and SUFFICIENT condition for all the EIGENVALUES of a REAL $n \times n$ matrix A to have NEGATIVE REAL PARTS is that the equation

$$\mathsf{A}^{\mathrm{T}}\mathsf{V} + \mathsf{V}\mathsf{A} = -\mathsf{I}$$

has as a solution where V is an $n \times n$ matrix and $(\mathbf{x}, \mathsf{V}\mathbf{x})$ is a positive definite quadratic form.

References

Gradshteyn, I. S. and Ryzhik, I. M. *Tables of Integrals, Series, and Products, 5th ed.* San Diego, CA: Academic Press, p. 1122, 1979.

Lyapunov Function

A function which is continuous, nonnegative, and has continuous PARTIAL DERIVATIVES. The existence of a Lyapunov function guarantees the NONLINEAR STABILITY of a FIXED POINT.

References

Jordan, D. W. and Smith, P. *Nonlinear Ordinary Differential Equations.* Oxford, England: Clarendon Press, p. 283, 1977.

Lyapunov's Second Theorem

If all the EIGENVALUES of a REAL MATRIX A have REAL PARTS, then to an arbitrary negative definite quadratic form $(\mathbf{x}, \mathsf{W}\mathbf{x})$ with $\mathbf{x} = \mathbf{x}(t)$ there corresponds a positive definite quadratic form $(\mathbf{x}, \mathsf{V}\mathbf{x})$ such that if one takes

$$\frac{d\mathbf{x}}{dt} = \mathsf{A}\mathbf{x},$$

then $(\mathbf{x}, \mathsf{W}\mathbf{x})$ and $(\mathbf{x}, \mathsf{W}\mathbf{x})$ satisfy

$$\frac{d}{dt}(\mathbf{x}, \mathsf{V}\mathbf{x}) = (\mathbf{x}, \mathsf{W}\mathbf{x}).$$

References

Gradshteyn, I. S. and Ryzhik, I. M. *Tables of Integrals, Series, and Products, 5th ed.* San Diego, CA: Academic Press, p. 1122, 1979.

Lyndon Word

A Lyndon word is an aperiodic notation for representing a NECKLACE.

see also DE BRUIJN SEQUENCE, NECKLACE

References

Ruskey, F. "Information on Necklaces, Lyndon Words, de Bruijn Sequences." `http://sue.csc.uvic.ca/~cos/inf/neck/NecklaceInfo.html`.

Sloane, N. J. A. Sequence A001037/M0116 in "An On-Line Version of the Encyclopedia of Integer Sequences."

Lyons Group

The SPORADIC GROUP *Ly*.

see also SPORADIC GROUP

References

Wilson, R. A. "ATLAS of Finite Group Representation." `http://for.mat.bham.ac.uk/atlas/Ly.html`.

M

M-Estimate

A ROBUST ESTIMATION based on maximum likelihood argument.

see also *L*-ESTIMATE, *R*-ESTIMATE

References

Press, W. H.; Flannery, B. P.; Teukolsky, S. A.; and Vetterling, W. T. "Robust Estimation." §15.7 in *Numerical Recipes in FORTRAN: The Art of Scientific Computing, 2nd ed.* Cambridge, England: Cambridge University Press, pp. 694–700, 1992.

Mac Lane's Theorem

A theorem which treats constructions of FIELDS of CHARACTERISTIC *p*.

see also CHARACTERISTIC (FIELD), FIELD

Machin's Formula

$$\tfrac{1}{4}\pi = 4\tan^{-1}(\tfrac{1}{5}) - \tan^{-1}(\tfrac{1}{239}).$$

There are a whole class of MACHIN-LIKE FORMULAS with various numbers of terms (although only four such formulas with only two terms). The properties of these formulas are intimately connected with COTANGENT identities.

see also 196-ALGORITHM, GREGORY NUMBER, MACHIN-LIKE FORMULAS, PI

Machin-Like Formulas

Machin-like formulas have the form

$$m\cot^{-1}u + n\cot^{-1}v = \tfrac{1}{4}k\pi, \tag{1}$$

where u, v, and k are POSITIVE INTEGERS and m and n are NONNEGATIVE INTEGERS. Some such FORMULAS can be found by converting the INVERSE TANGENT decompositions for which $c_n \neq 0$ in the table of Todd (1949) to INVERSE COTANGENTS. However, this gives only Machin-like formulas in which the smallest term is ± 1.

Maclaurin-like formulas can be derived by writing

$$\cot^{-1}z = \frac{1}{2i}\ln\left(\frac{z+i}{z-i}\right) \tag{2}$$

and looking for a_k and u_k such that

$$\sum_k a_k \cot^{-1}u_k = \tfrac{1}{4}\pi, \tag{3}$$

so

$$\prod_k \left(\frac{u_k+i}{u_k-i}\right)^{a_k} = e^{2\pi i/4} = i. \tag{4}$$

Machin-like formulas exist IFF (4) has a solution in INTEGERS. This is equivalent to finding INTEGER values such that

$$(1-i)^k(u+i)^m(v+i)^n \tag{5}$$

is REAL (Borwein and Borwein 1987, p. 345). An equivalent formulation is to find all integral solutions to one of

$$1 + x^2 = 2y^n \tag{6}$$

$$1 + x^2 = y^n \tag{7}$$

for $n = 3, 5, \ldots$.

There are only four such FORMULAS,

$$\tfrac{1}{4}\pi = 4\tan^{-1}(\tfrac{1}{5}) - \tan^{-1}(\tfrac{1}{239}) \tag{8}$$

$$\tfrac{1}{4}\pi = \tan^{-1}(\tfrac{1}{2}) + \tan^{-1}(\tfrac{1}{3}) \tag{9}$$

$$\tfrac{1}{4}\pi = 2\tan^{-1}(\tfrac{1}{2}) - \tan^{-1}(\tfrac{1}{7}) \tag{10}$$

$$\tfrac{1}{4}\pi = 2\tan^{-1}(\tfrac{1}{3}) + \tan^{-1}(\tfrac{1}{7}), \tag{11}$$

known as MACHIN'S FORMULA, EULER'S MACHIN-LIKE FORMULA, HERMANN'S FORMULA, and HUTTON'S FORMULA. These follow from the identities

$$\left(\frac{5+i}{5-i}\right)^4 \left(\frac{239+i}{239-i}\right)^{-1} = i \tag{12}$$

$$\left(\frac{2+i}{2-i}\right)\left(\frac{3+i}{3-i}\right) = i \tag{13}$$

$$\left(\frac{2+i}{2-i}\right)^2 \left(\frac{7+i}{7-i}\right)^{-1} = i \tag{14}$$

$$\left(\frac{3+i}{3-i}\right)^2 \left(\frac{7+i}{7-i}\right) = i. \tag{15}$$

Machin-like formulas with two terms can also be generated which do not have integral arc cotangent arguments such as Euler's

$$\tfrac{1}{4}\pi = 5\tan^{-1}(\tfrac{1}{7}) + 2\tan^{-1}(\tfrac{3}{79}) \tag{16}$$

(Wetherfield 1996), and which involve inverse SQUARE ROOTS, such as

$$\frac{\pi}{2} = 2\tan^{-1}\left(\frac{1}{\sqrt{2}}\right) + \tan^{-1}\left(\frac{1}{\sqrt{8}}\right). \tag{17}$$

Three-term Machin-like formulas include GAUSS'S MACHIN-LIKE FORMULA

$$\tfrac{1}{4}\pi = 12\cot^{-1}18 + 8\cot^{-1}57 - 5\cot^{-1}239, \tag{18}$$

STRASSNITZKY'S FORMULA

$$\tfrac{1}{4}\pi = \cot^{-1}2 + \cot^{-1}5 + \cot^{-1}8, \tag{19}$$

and tho following,

$$\tfrac{1}{4}\pi = 6\cot^{-1} 8 + 2\cot^{-1} 57 + \cot^{-1} 239 \quad (20)$$

$$\tfrac{1}{4}\pi = 4\cot^{-1} 5 - 1\cot^{-1} 70 + \cot^{-1} 99 \quad (21)$$

$$\tfrac{1}{4}\pi = 1\cot^{-1} 2 + 1\cot^{-1} 5 + \cot^{-1} 8 \quad (22)$$

$$\tfrac{1}{4}\pi = 8\cot^{-1} 10 - 1\cot^{-1} 239 - 4\cot^{-1} 515 \quad (23)$$

$$\tfrac{1}{4}\pi = 5\cot^{-1} 7 + 4\cot^{-1} 53 + 2\cot^{-1} 4443. \quad (24)$$

The first is due to Størmer, the second due to Rutherford, and the third due to Dase.

Using trigonometric identities such as

$$\cot^{-1} x = 2\cot^{-1}(2x) - \cot^{-1}(4x^3 + 3x), \quad (25)$$

it is possible to generate an infinite sequence of Machin-like formulas. Systematic searches therefore most often concentrate on formulas with particularly "nice" properties (such as "efficiency").

The efficiency of a FORMULA is the time it takes to calculate π with the POWER series for arctangent

$$\pi = a_1 \cot(b_1) + a_2 \cot(b_2) + \ldots, \quad (26)$$

and can be roughly characterized using Lehmer's "measure" formula

$$e \equiv \sum \frac{1}{\log_{10} b_i}. \quad (27)$$

The number of terms required to achieve a given precision is roughly proportional to e, so lower e-values correspond to better sums. The best currently known efficiency is 1.51244, which is achieved by the 6-term series

$$\begin{aligned}\tfrac{1}{4}\pi &= 183\cot^{-1} 239 + 32\cot^{-1} 1023 - 68\cot^{-1} 5832 \\ &\quad + 12\cot^{-1} 110443 - 12\cot^{-1} 4841182 \\ &\quad - 100\cot^{-1} 6826318 \quad (28)\end{aligned}$$

discovered by C.-L. Hwang (1997). Hwang (1997) also discovered the remarkable identities

$$\begin{aligned}\tfrac{1}{4}\pi &= P\cot^{-1} 2 - M\cot^{-1} 3 + L\cot^{-1} 5 + K\cot^{-1} 7 \\ &\quad + (N + K + L - 2M + 3P - 5)\cot^{-1} 8 \\ &\quad + (2N + M - P + 2 - L)\cot^{-1} 18 \\ &\quad - (2P - 3 - M + L + K - N)\cot^{-1} 57 - N\cot^{-1} 239, \quad (29)\end{aligned}$$

where K, L, M, N, and P are POSITIVE INTEGERS, and

$$\tfrac{1}{4}\pi = (N+2)\cot^{-1} 2 - N\cot^{-1} 3 - (N+1)\cot^{-1} N. \quad (30)$$

The following table gives the number $N(n)$ of Machin-like formulas of n terms in the compilation by Wetherfield and Hwang. Except for previously known identities (which are included), the criteria for inclusion are the following:

1. first term < 8 digits: measure < 1.8.
2. first term = 8 digits: measure < 1.9.
3. first term = 9 digits: measure < 2.0.
4. first term =10 digits: measure < 2.0.

n	$N(n)$	min e
1	1	0
2	4	1.85113
3	106	1.78661
4	39	1.58604
5	90	1.63485
6	120	1.51244
7	113	1.54408
8	18	1.65089
9	4	1.72801
10	78	1.63086
11	34	1.6305
12	188	1.67458
13	37	1.71934
14	5	1.75161
15	24	1.77957
16	51	1.81522
17	5	1.90938
18	570	1.87698
19	1	1.94899
20	11	1.95716
21	1	1.98938
Total	1500	1.51244

see also EULER'S MACHIN-LIKE FORMULA, GAUSS'S MACHIN-LIKE FORMULA, GREGORY NUMBER, HERMANN'S FORMULA, HUTTON'S FORMULA, INVERSE COTANGENT, MACHIN'S FORMULA, PI, STØRMER NUMBER, STRASSNITZKY'S FORMULA

References

Arndt, J. "Arctan Formulas." `http://jjj.spektracom.de/jjf.dvi`.

Arndt, J. "Big ArcTan Formula Bucket." `http://jjj.spektracom.de/fox.dvi`.

Ball, W. W. R. and Coxeter, H. S. M. *Mathematical Recreations and Essays, 13th ed.* New York: Dover, pp. 347–359, 1987.

Borwein, J. M. and Borwein, P. B. *Pi & the AGM: A Study in Analytic Number Theory and Computational Complexity.* New York: Wiley, 1987.

Castellanos, D. "The Ubiquitous Pi. Part I." *Math. Mag.* **61**, 67–98, 1988.

Conway, J. H. and Guy, R. K. *The Book of Numbers.* New York: Springer-Verlag, pp. 241–248, 1996.

Hwang, C.-L. "More Machin-Type Identities." *Math. Gaz.*, 120–121, March 1997.

Lehmer, D. H. "On Arccotangent Relations for π." *Amer. Math. Monthly* **45**, 657–664, 1938.

Lewin, L. *Polylogarithms and Associated Functions.* New York: North-Holland, 1981.

Lewin, L. *Structural Properties of Polylogarithms.* Providence, RI: Amer. Math. Soc., 1991.

Nielsen, N. *Der Euler'sche Dilogarithms.* Leipzig, Germany: Halle, 1909.

Størmer, C. "Sur l'Application de la Théorie des Nombres Entiers Complexes à la Solution en Nombres Rationels x_1,

$x_2, \ldots, c_1, c_2, \ldots, k$ de l'Equation...." *Archiv for Mathematik og Naturvidenskab* **B 19**, 75–85, 1896.
Todd, J. "A Problem on Arc Tangent Relations." *Amer. Math. Monthly* **56**, 517–528, 1949.
Weisstein, E. W. "Machin-Like Formulas." `http://www.astro.virginia.edu/~eww6n/math/notebooks/MachinFormulas.m`.
Wetherfield, M. "The Enhancement of Machin's Formula by Todd's Process." *Math. Gaz.* **80**, 333–344, 1996.
Wetherfield, M. "Machin Revisited." *Math. Gaz.*, 121–123, March 1997.
Williams, R. "Arctangent Formulas for Pi." `http://www.cacr.caltech.edu/~roy/pi.formulas.html`. [Contains errors].

Machine

A method for producing infinite LOOP SPACES and spectra.

see also GADGET, LOOP SPACE, MAY-THOMASON UNIQUENESS THEOREM, TURING MACHINE

Mackey's Theorem

Let E and F be paired spaces with S a family of absolutely convex bounded sets of F such that the sets of S generate F and, if $B_1, B_2 \in S$, there exists a $B_3 \in S$ such that $B_3 \supset B_1$ and $B_3 \supset B_2$. Then the dual space of E_S is equal to the union of the weak completions of λB, where $\lambda > 0$ and $B \in S$.

see also GROTHENDIECK'S THEOREM

References

Iyanaga, S. and Kawada, Y. (Eds.). "Mackey's Theorem." §407M in *Encyclopedic Dictionary of Mathematics*. Cambridge, MA: MIT Press, p. 1274, 1980.

Maclaurin-Bezout Theorem

The Maclaurin-Bèzout theorem says that two curves of degree n intersect in n^2 points, so two CUBICS intersect in nine points. This means that $n(n+3)/2$ points do not always uniquely determine a single curve of order n.

see also CRAMÉR-EULER PARADOX

Maclaurin-Cauchy Theorem

If $f(x)$ is POSITIVE and decreases to 0, then an EULER CONSTANT

$$\gamma_f \equiv \lim_{n\to\infty}\left[\sum_{k=1}^{n} f(k) - \int_a^n f(x)\,dx\right]$$

can be defined. If $f(x) = 1/x$, then

$$\gamma = \lim_{n\to\infty}\left(\sum_{k=1}^{n}\frac{1}{k} - \int_1^n \frac{dx}{x}\right) = \lim_{n\to\infty}\left(\sum_{k=1}^{n}\frac{1}{k} - \ln n\right),$$

where γ is the EULER-MASCHERONI CONSTANT.

Maclaurin Integral Test

see INTEGRAL TEST

Maclaurin Polynomial

see MACLAURIN SERIES

Maclaurin Series

A series expansion of a function about 0,

$$f(x) = f(0) + f'(0)x + \frac{f''(0)}{2!}x^2 + \frac{f^{(3)}(0)}{3!}x^3 + \ldots + \frac{f^{(n)}(0)}{n!}x^n + \ldots, \quad (1)$$

named after the Scottish mathematician Maclaurin. Maclaurin series for common functions include

$$\frac{1}{1-x} = 1 + x + x^2 + x^3 + x^4 + x^5 + \ldots \qquad -1 < x < 1 \quad (2)$$

$$\operatorname{cn}(x,k^2) = 1 - \tfrac{1}{2!}x^2 + \tfrac{1}{4!}(1+4k^2)x^4 + \ldots \quad (3)$$

$$\cos x = 1 - \tfrac{1}{2!}x^2 + \tfrac{1}{4!}x^4 - \tfrac{1}{6!}x^6 - \ldots \qquad -\infty < x < \infty \quad (4)$$

$$\cos^{-1} x = \tfrac{1}{2}\pi - x - \tfrac{1}{6}x^3 - \tfrac{3}{40}x^5 - \tfrac{5}{112}x^7 - \ldots \qquad -1 < x < 1 \quad (5)$$

$$\cosh x = 1 + \tfrac{1}{2}x^2 + \tfrac{1}{24}x^4 + \tfrac{1}{720}x^6 + \tfrac{1}{40{,}320}x^8 + \ldots \quad (6)$$

$$\cosh^{-1}(1+x) = \sqrt{2x}\,(1 - \tfrac{1}{2}x + \tfrac{3}{160}x^2 - \tfrac{5}{896}x^3 + \ldots) \quad (7)$$

$$\cot x = x^{-1} - \tfrac{1}{3}x - \tfrac{1}{45}x^3 - \tfrac{2}{945}x^5 - \tfrac{1}{4725}x^7 - \ldots \quad (8)$$

$$\cot^{-1} x = \tfrac{1}{2}\pi - x + \tfrac{1}{3}x^3 - \tfrac{1}{5}x^5 + \tfrac{1}{7}x^7 - \tfrac{1}{9}x^9 + \ldots \quad (9)$$

$$= x^{-1} - \tfrac{1}{3}x^{-3} + \tfrac{1}{5}x^{-5} - \tfrac{1}{7}x^{-7} + \tfrac{1}{9}x^{-9} + \ldots \quad (10)$$

$$\coth x = x^{-1} + \tfrac{1}{3}x - \tfrac{1}{45}x^4 + \tfrac{2}{945}x^5 - \tfrac{1}{4725}x^7 + \ldots \quad (11)$$

$$\coth^{-1}(1+x) = \tfrac{1}{2}\ln 2 - \tfrac{1}{2}\ln x + \tfrac{1}{4}x - \tfrac{1}{16}x^2 + \ldots \quad (12)$$

$$\csc x = x^{-1} + \tfrac{1}{6}x + \tfrac{7}{360}x^3 + \tfrac{31}{15120}x^5 + \ldots \quad (13)$$

$$\operatorname{csch} x = x^{-1} - \tfrac{1}{6}x + \tfrac{7}{360}x^3 + \tfrac{31}{15120}x^5 + \ldots \quad (14)$$

$$\operatorname{csch}^{-1} x = \ln 2 - \ln x + \tfrac{1}{4}x^2 - \tfrac{3}{32}x^4 + \tfrac{5}{96}x^6 - \ldots \quad (15)$$

$$\operatorname{dn}(x,k^2)x = 1 - \tfrac{1}{2!}k^2x^2 + \tfrac{1}{4!}k^2(4+k^2)x^4 + \ldots \quad (16)$$

$$\operatorname{erf} x = \frac{1}{\sqrt{\pi}}\left(2x - \tfrac{2}{3}x^3 + \tfrac{1}{5}x^5 - \tfrac{1}{21}x^7 + \ldots\right) \quad (17)$$

$$e^x = 1 + x + \tfrac{1}{2!}x^2 + \tfrac{1}{3!}x^3 + \tfrac{1}{4!}x^4 + \ldots \qquad -\infty < x < \infty \quad (18)$$

$${}_2F_1(\alpha,\beta,\gamma;x) = 1 + \frac{\alpha\beta}{1!\gamma}x + \frac{\alpha(\alpha+1)\beta(\beta+1)}{2!\gamma(\gamma+1)}x^2 + \ldots \quad (19)$$

$$\ln(1+x) = x - \tfrac{1}{2}x^2 + \tfrac{1}{3}x^3 - \tfrac{1}{4}x^4 + \ldots \qquad -1 < x < 1 \quad (20)$$

$$\ln\left(\frac{1+x}{1-x}\right) = 2x + \tfrac{2}{3}x^3 + \tfrac{2}{5}x^5 + \tfrac{2}{7}x^7 + \ldots \qquad -1 < x < 1 \quad (21)$$

$$\sec x = 1 + \tfrac{1}{2}x^2 + \tfrac{5}{24}x^4 + \tfrac{61}{720}x^6 + \tfrac{277}{8064}x^8 + \ldots \quad (22)$$

$$\operatorname{sech} x = 1 - \tfrac{1}{2}x^2 + \tfrac{5}{24}x^4 - \tfrac{61}{720}x^6 + \tfrac{277}{8064}x^8 + \ldots \quad (23)$$

$$\operatorname{sech}^{-1} x = \ln 2 - \ln x - \tfrac{1}{4}x^2 - \tfrac{3}{32}x^4 - \ldots \quad (24)$$

$$\sin x = x - \tfrac{1}{3!}x^3 + \tfrac{1}{5!}x^5 - \tfrac{1}{7!}x^7 + \ldots$$

$$-\infty < x < \infty \quad (25)$$

$$\sin^{-1} x = x + \tfrac{1}{6}x^3 + \tfrac{3}{40}x^5 + \tfrac{5}{112}x^7 + \tfrac{35}{1152}x^9 + \ldots \quad (26)$$

$$\sinh x = x + \tfrac{1}{6}x^3 + \tfrac{1}{120}x^5 + \tfrac{1}{5040}x^7 + \tfrac{1}{362,880}x^9 + \ldots \quad (27)$$

$$\sinh^{-1} x = x - \tfrac{1}{6}x^3 + \tfrac{3}{40}x^5 - \tfrac{5}{112}x^7 + \tfrac{35}{1152}x^9 - \ldots \quad (28)$$

$$\operatorname{sn}(x, k^2) = \tfrac{1}{3!}(1+k^2)x^3 + \tfrac{1}{5!}(1+14k^2+k^4)x^5 + \ldots \quad (29)$$

$$\tan x = x + \tfrac{1}{3}x^3 + \tfrac{2}{15}x^5 + \tfrac{17}{315}x^7 + \tfrac{62}{2835}x^9 + \ldots \quad (30)$$

$$\tan^{-1} x = x - \tfrac{1}{3}x^3 + \tfrac{1}{5}x^5 - \tfrac{1}{7}x^7 + \ldots$$

$$-1 < x < 1 \quad (31)$$

$$\tan^{-1}(1+x) = \tfrac{1}{4}\pi + \tfrac{1}{2}x - \tfrac{1}{4}x^2 + \tfrac{1}{12}x^3 + \tfrac{1}{40}x^5 + \ldots \quad (32)$$

$$\tanh x = x - \tfrac{1}{3}x^3 + \tfrac{2}{15}x^5 - \tfrac{17}{315}x^7 + \tfrac{62}{2835}x^9 + \ldots \quad (33)$$

$$\tanh^{-1} x = x + \tfrac{1}{3}x^3 + \tfrac{1}{5}x^5 + \tfrac{1}{7}x^7 + \tfrac{1}{9}x^9 + \ldots. \quad (34)$$

The explicit forms for some of these are

$$\frac{1}{1-x} = \sum_{n=0}^{\infty} x^n \quad (35)$$

$$\cos x = \sum_{n=0}^{\infty} \frac{(-1)^n}{(2n)!} x^{2n} \quad (36)$$

$$\csc x = \sum_{n=0}^{\infty} \frac{(-1)^{n+1} 2(2^{2n-1}-1)B_{2n}}{(2n)!} x^{2n-1} \quad (37)$$

$$e^x = \sum_{n=0}^{\infty} \frac{1}{n!} x^n \quad (38)$$

$$\ln(1+x) = \sum_{n=1}^{\infty} \frac{(-1)^{n+1}}{n} x^n \quad (39)$$

$$\ln\left(\frac{1+x}{1-x}\right) = \sum_{n=1}^{\infty} \frac{2}{(2n-1)} x^{2n-1} \quad (40)$$

$$\sec x = \sum_{n=0}^{\infty} \frac{(-1)^n E_{2n}}{(2n)!} x^{2n} \quad (41)$$

$$\sin x = \sum_{n=1}^{\infty} \frac{(-1)^{n+1}}{(2n-1)!} x^{2n-1} \quad (42)$$

$$\tan x = \sum_{n=1}^{\infty} \frac{(-1)^{n+1} 2^{2n}(2^{2n}-1)B_{2n}}{(2n)!} x^{2n-1} \quad (43)$$

$$\tan^{-1} x = \sum_{n=1}^{\infty} \frac{(-1)^{n+1}}{2n-1} x^{2n-1} \quad (44)$$

$$\tanh^{-1} x = \sum_{n=1}^{\infty} \frac{1}{2n-1} x^{2n-1}, \quad (45)$$

where B_n are BERNOULLI NUMBERS and E_n are EULER NUMBERS.

see also ALCUIN'S SEQUENCE, LAGRANGE EXPANSION, LEGENDRE SERIES, TAYLOR SERIES

References

Beyer, W. H. (Ed.) *CRC Standard Mathematical Tables, 28th ed.* Boca Raton, FL: CRC Press, pp. 299–300, 1987.

Maclaurin Trisectrix

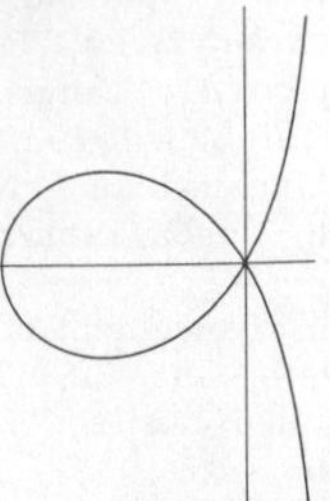

A curve first studied by Colin Maclaurin in 1742. It was studied to provide a solution to one of the GEOMETRIC PROBLEMS OF ANTIQUITY, in particular TRISECTION of an ANGLE, whence the name trisectrix. The Maclaurin trisectrix is an ANALLAGMATIC CURVE, and the origin is a CRUNODE.

The Maclaurin trisectrix has CARTESIAN equation

$$y^2 = \frac{x^2(x+3a)}{a-x}, \quad (1)$$

or the parametric equations

$$x = a\frac{t^2-3}{t^2+1} \quad (2)$$

$$y = a\frac{t(t^2-3)}{t^2+1}. \quad (3)$$

The ASYMPTOTE has equation $x = a$, and the center of the loop is as $(2a, 0)$. Draw a line L with ANGLE 3α through the loop center. Then the angle made by the line through the center and point of intersection of L with the trisectrix is α. The trisectrix is sometimes defined instead as

$$x(x^2+y^2) = a(y^2-3x^2) \quad (4)$$

$$y^2 = \frac{x^2(3a+x)}{a-x} \quad (5)$$

$$r = \frac{2a\sin(3\theta)}{\sin(2\theta)}. \quad (6)$$

Another form of the equation is the POLAR EQUATION

$$r = a\sec(\tfrac{1}{3}\theta), \quad (7)$$

where the origin is inside the loop and the crossing point is on the NEGATIVE x-AXIS.

The tangents to the curve at the origin make angles of $\pm 60°$ with the x-AXIS. The AREA of the loop is

$$A_{\text{loop}} = 3\sqrt{3}\,a^2, \quad (8)$$

and the NEGATIVE x-intercept is $(-3a, 0)$ (MacTutor Archive).

The Maclaurin trisectrix is the PEDAL CURVE of the PARABOLA where the PEDAL POINT is taken as the reflection of the FOCUS in the DIRECTRIX.

see also CATALAN'S TRISECTRIX, STROPHOID

References

Lawrence, J. D. *A Catalog of Special Plane Curves.* New York: Dover, pp. 103–106, 1972.

Lee, X. "Trisectrix." `http://www.best.com/~xah/SpecialPlaneCurves_dir/Trisectrix_dir/trisectrix.html`.

Lee, X. "Trisectrix of Maclaurin." `http://www.best.com/~xah/SpecialPlaneCurves_dir/TriOfMaclaurin_dir/triOfMaclaurin.html`.

MacTutor History of Mathematics Archive. "Trisectrix of Maclaurin." `http://www-groups.dcs.st-and.ac.uk/~history/Curves/Trisectrix.html`.

Maclaurin Trisectrix Inverse Curve

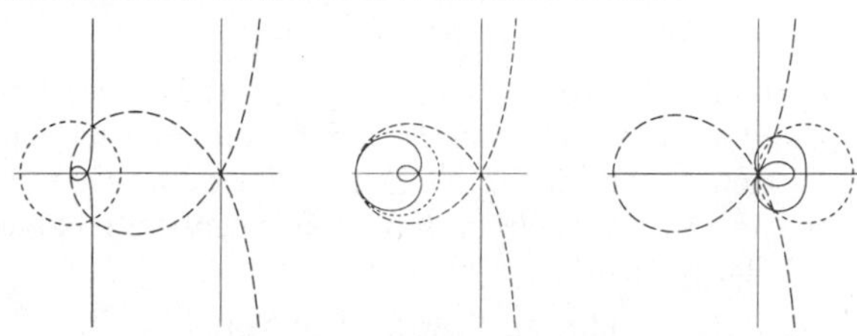

The INVERSE CURVE of the MACLAURIN TRISECTRIX with INVERSION CENTER at the NEGATIVE x-intercept is a TSCHIRNHAUSEN CUBIC.

MacMahon's Prime Number of Measurement

see PRIME NUMBER OF MEASUREMENT

MacRobert's E-Function

$$E(p;\alpha_r:\rho_s:x) \equiv \frac{\Gamma(\alpha_{q+1})}{\Gamma(\rho_1-\alpha_1)\Gamma(\rho_2-\alpha_2)\cdots\Gamma(\rho_q-\alpha_q)} \times \prod_{\mu=1}^{q}\int_0^\infty {\lambda_\mu}^{\rho_\mu-\alpha_\mu-1}(1+\lambda_\mu)^{-\rho_\mu}\,d\lambda_\mu \times \prod_{\nu=2}^{p-q-1}\int_0^\infty e^{-\lambda_{q+\nu}}{\lambda_{q+\nu}}^{\alpha_{q+\nu}-1}\,d\lambda_{q+\nu}\times \int_0^\infty e^{-\lambda_p}{\lambda_p}^{\alpha_p-1}\left[1+\frac{\lambda_{q+2}\lambda_{q+3}\cdots\lambda_p}{(1+\lambda_1)\cdots(1+\lambda_q)x}\right]^{-\alpha_q+1} d\lambda_p,$$

where $\Gamma(z)$ is the GAMMA FUNCTION and other details are discussed by Gradshteyn and Ryzhik (1980).

see also FOX'S H-FUNCTION, MEIJER'S G-FUNCTION

References

Gradshteyn, I. S. and Ryzhik, I. M. *Tables of Integrals, Series, and Products, 5th ed.* San Diego, CA: Academic Press, pp. 896–903 and 1071–1072, 1979.

Madelung Constants

The quantities obtained from cubic, hexagonal, etc., LATTICE SUMS, evaluated at $s=1$, are called Madelung constants. For cubic LATTICE SUMS, they are expressible in closed form for EVEN indices,

$$b_2(2) = -4\beta(1)\eta(1) = -4\tfrac{\pi}{4}\ln 2 = -\pi\ln 2 \quad (1)$$

$$b_4(2) = -8\eta(1)\eta(0) = -8\ln 2\cdot\tfrac{1}{2} = -4\ln 2. \quad (2)$$

$b_3(1)$ is given by BENSON'S FORMULA,

$$-b_3(1) = \sum_{i,\,j,\,k=-\infty}^{\infty}{}' \frac{(-1)^{i+j+k+1}}{\sqrt{i^2+j^2+k^2}} = 12\pi \sum_{m,\,n=1,3,\ldots}^{\infty} \operatorname{sech}^2(\tfrac{1}{2}\pi\sqrt{m^2+n^2}), \quad (3)$$

where the prime indicates that summation over (0, 0, 0) is excluded. $b_3(1)$ is sometimes called "the" Madelung constant, corresponds to the Madelung constant for a 3-D NaCl crystal, and is numerically equal to $-1.74756\ldots$.

For hexagonal LATTICE SUM, $h_2(2)$ is expressible in closed form as

$$h_2(2) = \pi\ln 3\sqrt{3}. \quad (4)$$

see also BENSON'S FORMULA, LATTICE SUM

References

Borwein, J. M. and Borwein, P. B. *Pi & the AGM: A Study in Analytic Number Theory and Computational Complexity.* New York: Wiley, 1987.

Buhler, J. and Wagon, S. "Secrets of the Madelung Constant." *Mathematica in Education and Research* **5**, 49–55, Spring 1996.

Crandall, R. E. and Buhler, J. P. "Elementary Function Expansions for Madelung Constants." *J. Phys. Ser. A: Math. and Gen.* **20**, 5497–5510, 1987.

Finch, S. "Favorite Mathematical Constants." `http://www.mathsoft.com/asolve/constant/mdlung/mdlung.html`.

Maeder's Owl Minimal Surface

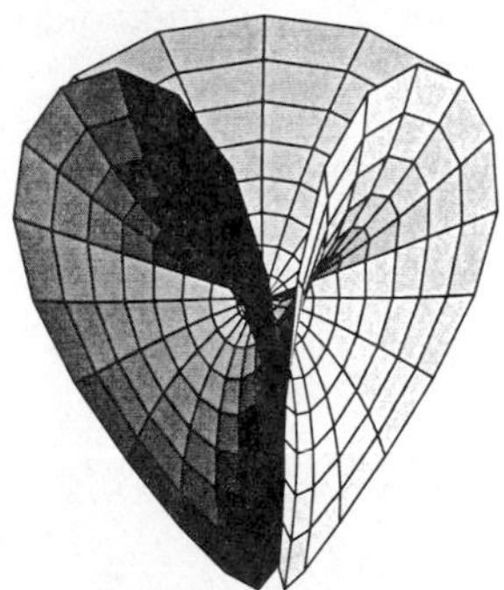

A MINIMAL SURFACE which resembles a CROSS-CAP. It is given by the polar equations

$$x = 1 \quad (1)$$

$$y = \sqrt{z} \quad (2)$$

$$z = z, \quad (3)$$

or the parametric equations

$$x = r\cos\theta - \tfrac{1}{2}r^2\cos(2\theta) \quad (4)$$
$$y = -r\sin\theta - \tfrac{1}{2}r^2\sin(2\theta), \quad (5)$$
$$z = \tfrac{4}{3}r^{3/2}\cos(\tfrac{3}{2}\theta). \quad (6)$$

see also CROSS-CAP, MINIMAL SURFACE

References
Maeder, R. *Programming in Mathematica, 3rd ed.* Reading, MA: Addison-Wesley, pp. 29–30, 1997.

Maehly's Procedure

A method for finding ROOTS which defines

$$P_j(x) = \frac{P(x)}{(x-x_1)\cdots(x-x_j)}, \quad (1)$$

so the derivative is

$$P_j'(x) = \frac{P'(x)}{(x-x_1)\cdots(x-x_j)} - \frac{P(x)}{(x-x_1)\cdots(x-x_j)}\sum_{i=1}^{j}(x-x_i)^{-1}. \quad (2)$$

One step of NEWTON'S METHOD can then be written as

$$x_{k+1} = x_k - \frac{P(x_k)}{P'(x_k) - P(x_k)\sum_{i=1}^{j}(x_k-x_i)^{-1}}. \quad (3)$$

Mainardi-Codazzi Equations

$$\frac{\partial e}{\partial v} - \frac{\partial f}{\partial u} = e\Gamma^1_{12} + f(\Gamma^2_{12} - \Gamma^1_{11}) - g\Gamma^2_{11} \quad (1)$$
$$\frac{\partial f}{\partial v} - \frac{\partial g}{\partial u} = e\Gamma^1_{22} + f(\Gamma^2_{22} - \Gamma^1_{12}) - g\Gamma^2_{12}, \quad (2)$$

where e, f, and g are coefficients of the second FUNDAMENTAL FORM and Γ^k_{ij} are CHRISTOFFEL SYMBOLS OF THE SECOND KIND. Therefore,

$$\frac{\partial e}{\partial v} = \tfrac{1}{2}E_v\left(\frac{e}{E} + \frac{g}{G}\right) \quad (3)$$
$$\frac{\partial g}{\partial u} = \tfrac{1}{2}G_u\left(\frac{e}{E} + \frac{g}{G}\right) \quad (4)$$

$$\frac{\partial(\ln f)}{\partial u} = \Gamma^1_{11} - \Gamma^2_{12} \quad (5)$$
$$\frac{\partial(\ln f)}{\partial v} = \Gamma^2_{22} - \Gamma^1_{12} \quad (6)$$

$$\frac{\partial}{\partial u}\left(\frac{\ln f}{\sqrt{EG-F^2}}\right) = -2\Gamma^2_{12} \quad (7)$$
$$\frac{\partial}{\partial v}\left(\frac{\ln f}{\sqrt{EG-F^2}}\right) = -2\Gamma^1_{12}, \quad (8)$$

where E, F, and G are coefficients of the first FUNDAMENTAL FORM.

References
Gray, A. "The Mainardi-Codazzi Equations." §20.4 in *Modern Differential Geometry of Curves and Surfaces.* Boca Raton, FL: CRC Press, pp. 401–402, 1993.
Green, A. E. and Zerna, W. *Theoretical Elasticity, 2nd ed.* New York: Dover, p. 37, 1992.

Magic Circles

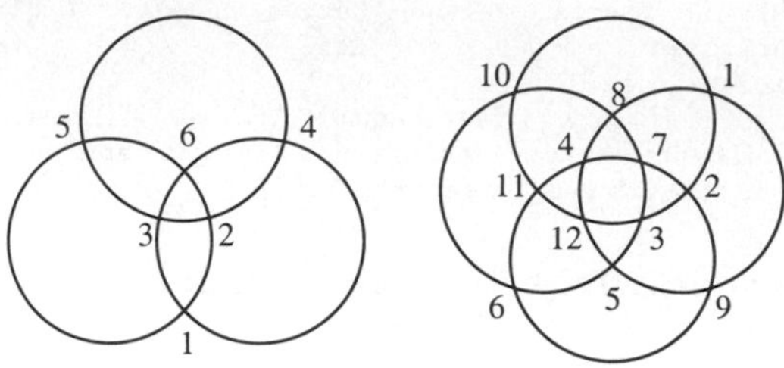

A set of n magic circles is a numbering of the intersection of the n CIRCLES such that the sum over all intersections is the same constant for all circles. The above sets of three and four magic circles have magic constants 14 and 39 (Madachy 1979).

see also MAGIC GRAPH, MAGIC SQUARE

References
Madachy, J. S. *Madachy's Mathematical Recreations.* New York: Dover, p. 86, 1979.

Magic Constant

The number

$$M_2(n) = \frac{1}{n}\sum_{k=1}^{n^2} k = \tfrac{1}{2}n(n^2+1)$$

to which the n numbers in any horizontal, vertical, or *main* diagonal line must sum in a MAGIC SQUARE. The first few values are 1, 5 (no such magic square), 15, 34, 65, 111, 175, 260, ... (Sloane's A006003). The magic constant for an nth order magic square starting with an INTEGER A and with entries in an increasing ARITHMETIC SERIES with difference D between terms is

$$M_2(n; A, D) = \tfrac{1}{2}n[2a + D(n^2-1)]$$

(Hunter and Madachy 1975, Madachy 1979). In a PANMAGIC SQUARE, in addition to the main diagonals, the broken diagonals also sum to $M_2(n)$.

For a MAGIC CUBE, the magic constant is

$$M_3(n) = \frac{1}{n^2}\sum_{k=1}^{n^3} k = \tfrac{1}{2}n(n^3+1) = \tfrac{1}{2}n(1+n)(n^2-n+1).$$

The first few values are 1, 9, 42, 130, 315, 651, 1204, ... (Sloane's A027441).

There is a corresponding multiplicative magic constant for MULTIPLICATION MAGIC SQUARES.

see also MAGIC CUBE, MAGIC GEOMETRIC CONSTANTS, MAGIC HEXAGON, MAGIC SQUARE, MULTIPLICATION MAGIC SQUARE, PANMAGIC SQUARE

References

Hunter, J. A. H. and Madachy, J. S. "Mystic Arrays." Ch. 3 in *Mathematical Diversions*. New York: Dover, pp. 23–34, 1975.

Madachy, J. S. *Madachy's Mathematical Recreations*. New York: Dover, p. 86, 1979.

Sloane, N. J. A. Sequences A027441 and A006003/M3849 in "An On-Line Version of the Encyclopedia of Integer Sequences."

Magic Cube

An $n \times n \times n$ 3-D version of the MAGIC SQUARE in which the n^2 rows, n^2 columns, n^2 pillars (or "files"), and four space diagonals each sum to a single number $M_3(n)$ known as the MAGIC CONSTANT. If the cross-section diagonals also sum to $M_3(n)$, the magic cube is called a PERFECT MAGIC CUBE; if they do not, the cube is called a SEMIPERFECT MAGIC CUBE, or sometimes an ANDREWS CUBE (Gardner 1988). A pandiagonal cube is a perfect or semiperfect magic cube which is magic not only along the main space diagonals, but also on the broken space diagonals.

A magic cube using the numbers 1, 2, ..., n^3, if it exists, has MAGIC CONSTANT

$$M_3(n) = \frac{1}{n^2}\sum_{k=1}^{n^3} k = \tfrac{1}{2}n(n^3+1) = \tfrac{1}{2}n(n+1)(n^2-n+1).$$

For $n = 1$, 2, ..., the magic constants are 1, 9, 42, 130, 315, 651, ... (Sloane's A027441).

4	12	26
11	25	6
27	5	10

20	7	15
9	14	19
13	21	8

18	23	1
22	3	17
2	16	24

60	37	12	21
13	20	61	36
56	41	8	25
1	32	49	48

7	26	55	42
50	47	2	31
11	22	59	38
62	35	14	19

57	40	9	24
16	17	64	33
53	44	5	28
4	29	52	45

6	27	54	43
51	46	3	30
10	23	58	39
63	34	15	18

The above semiperfect magic cubes of orders three (Hunter and Madachy 1975, p. 31; Ball and Coxeter 1987, p. 218) and four (Ball and Coxeter 1987, p. 220) have magic constants 42 and 130, respectively. There is a trivial semiperfect magic cube of order one, but no semiperfect cubes of orders two or three exist. Semiperfect cubes of ODD order with $n \geq 5$ and DOUBLY EVEN order can be constructed by extending the methods used for MAGIC SQUARES.

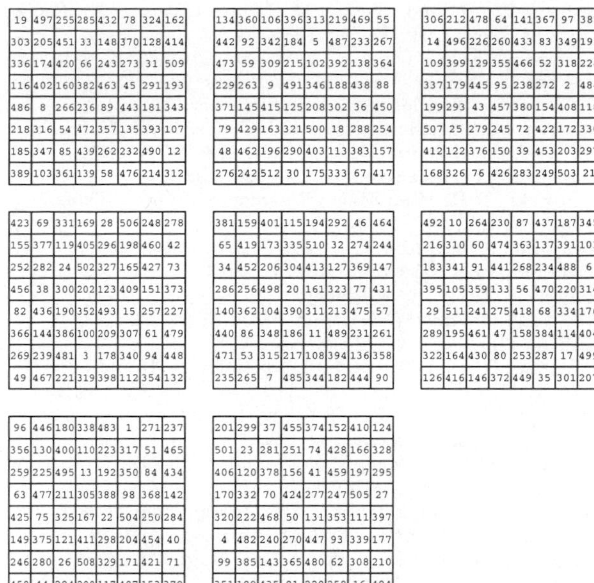

19	497	255	285	432	78	324	162
303	205	451	33	148	370	128	414
336	174	420	66	243	273	31	509
116	402	160	382	463	45	291	193
486	8	266	236	89	443	181	343
218	316	54	472	357	135	393	107
185	347	85	439	262	232	490	12
389	103	361	139	58	476	214	312

134	360	106	396	313	219	469	55
442	92	342	184	5	487	233	267
473	59	309	215	102	392	138	364
229	263	9	491	346	188	438	88
371	145	415	125	208	302	36	450
79	429	163	321	500	18	288	254
48	462	196	290	403	113	383	157
276	242	512	30	175	333	67	417

306	212	478	64	141	367	97	387
14	496	226	260	433	83	349	191
109	399	129	355	466	52	318	224
337	179	445	95	238	272	2	484
199	293	43	457	380	154	408	118
507	25	279	245	72	422	172	330
412	122	376	150	39	453	203	297
168	326	76	426	283	249	503	21

423	69	331	169	28	506	248	278
155	377	119	405	296	198	460	42
252	282	24	502	327	165	427	73
456	38	300	202	123	409	151	373
82	436	190	352	493	15	257	227
366	144	386	100	209	307	61	479
269	239	481	3	178	340	94	448
49	467	221	319	398	112	354	132

381	159	401	115	194	292	46	464
65	419	173	335	510	32	274	244
34	452	206	304	413	127	369	147
286	256	498	20	161	323	77	431
140	362	104	390	311	213	475	57
440	86	348	186	11	489	231	261
471	53	315	217	108	394	136	358
235	265	7	485	344	182	444	90

492	10	264	230	87	437	187	345
216	310	60	474	363	137	391	101
183	341	91	441	268	234	488	6
395	105	359	133	56	470	220	314
29	511	241	275	418	68	334	176
289	195	461	47	158	384	114	404
322	164	430	80	253	287	17	499
126	416	146	372	449	35	301	207

96	446	180	338	483	1	271	237
356	130	400	110	223	317	51	465
259	225	495	13	192	350	84	434
63	477	211	305	388	98	368	142
425	75	325	167	22	504	250	284
149	375	121	411	298	204	454	40
246	280	26	508	329	171	421	71
458	44	294	200	117	407	153	379

201	299	37	455	374	152	410	124
501	23	281	251	74	428	166	328
406	120	378	156	41	459	197	295
170	332	70	424	277	247	505	27
320	222	468	50	131	353	111	397
4	482	240	270	447	93	339	177
99	385	143	365	480	62	308	210
351	189	435	81	228	258	16	494

There are no perfect magic cubes of order four (Beeler *et al.* 1972, Item 50; Gardner 1988). No perfect magic cubes of order five are known, although it is known that such a cube must have a central value of 63 (Beeler *et al.* 1972, Item 51; Gardner 1988). No order-six perfect magic cubes are known, but Langman (1962) constructed a perfect magic cube of order seven. An order-eight perfect magic cube was published anonymously in 1875 (Barnard 1888, Benson and Jacoby 1981, Gardner 1988). The construction of such a cube is discussed in Ball and Coxeter (1987). Rosser and Walker rediscovered the order-eight cube in the late 1930s (but did not publish it), and Myers independently discovered the cube illustrated above in 1970 (Gardner 1988). Order 9 and 11 magic cubes have also been discovered, but none of order 10 (Gardner 1988).

Semiperfect pandiagonal cubes exist for all orders $8n$ and all ODD $n > 8$ (Ball and Coxeter 1987). A perfect pandiagonal magic cube has been constructed by Planck (1950), cited in Gardner (1988).

Berlekamp *et al.* (1982, p. 783) give a magic TESSERACT.

see also MAGIC CONSTANT, MAGIC GRAPH, MAGIC HEXAGON, MAGIC SQUARE

References

Adler, A. and Li, S.-Y. R. "Magic Cubes and Prouhet Sequences." *Amer. Math. Monthly* **84**, 618–627, 1977.

Andrews, W. S. *Magic Squares and Cubes, 2nd rev. ed.* New York: Dover, 1960.

Ball, W. W. R. and Coxeter, H. S. M. *Mathematical Recreations and Essays, 13th ed.* New York: Dover, pp. 216–224, 1987.

Barnard, F. A. P. "Theory of Magic Squares and Cubes." *Mem. Nat. Acad. Sci.* **4**, 209–270, 1888.

Beeler, M.; Gosper, R. W.; and Schroeppel, R. *HAKMEM.* Cambridge, MA: MIT Artificial Intelligence Laboratory, Memo AIM-239, Feb. 1972.

Benson, W. H. and Jacoby, O. *Magic Cubes: New Recreations*. New York: Dover, 1981.
Berlekamp, E. R.; Conway, J. H; and Guy, R. K. *Winning Ways, For Your Mathematical Plays, Vol. 2: Games in Particular*. London: Academic Press, 1982.
Gardner, M. "Magic Squares and Cubes." Ch. 17 in *Time Travel and Other Mathematical Bewilderments*. New York: W. H. Freeman, pp. 213–225, 1988.
Hendricks, J. R. "Ten Magic Tesseracts of Order Three." *J. Recr. Math.* **18**, 125–134, 1985–1986.
Hirayama, A. and Abe, G. *Researches in Magic Squares*. Osaka, Japan: Osaka Kyoikutosho, 1983.
Hunter, J. A. H. and Madachy, J. S. "Mystic Arrays." Ch. 3 in *Mathematical Diversions*. New York: Dover, p. 31, 1975.
Langman, H. *Play Mathematics*. New York: Hafner, pp. 75–76, 1962.
Lei, A. "Magic Cube and Hypercube." `http://www.cs.ust.hk/~philipl/magic/mcube2.html`.
Madachy, J. S. *Madachy's Mathematical Recreations*. New York: Dover, pp. 99–100, 1979.
Pappas, T. "A Magic Cube." *The Joy of Mathematics*. San Carlos, CA: Wide World Publ./Tetra, p. 77, 1989.
Planck, C. *Theory of Path Nasiks*. Rugby, England: Privately Published, 1905.
Rosser, J. B. and Walker, R. J. "The Algebraic Theory of Diabolical Squares." *Duke Math. J.* **5**, 705–728, 1939.
Sloane, N. J. A. Sequence A027441 in "An On-Line Version of the Encyclopedia of Integer Sequences."
Wynne, B. E. "Perfect Magic Cubes of Order 7." *J. Recr. Math.* **8**, 285–293, 1975–1976.

Magic Geometric Constants

N.B. A detailed on-line essay by S. Finch was the starting point for this entry.

Let E be a compact connected subset of d-dimensional EUCLIDEAN SPACE. Gross (1964) and Stadje (1981) proved that there is a unique REAL NUMBER $a(E)$ such that for all $x_1, x_2, \ldots, x_n \in E$, there exists $y \in E$ with

$$\frac{1}{n}\sum_{j=1}^{n}\sqrt{\sum_{k=1}^{d}(x_{j,k}-y_k)^2} = a(E). \tag{1}$$

The magic constant $m(E)$ of E is defined by

$$m(E) = \frac{a(E)}{\operatorname{diam}(E)}, \tag{2}$$

where

$$\operatorname{diam}(E) \equiv \max_{u,v\in E}\sqrt{\sum_{k=1}^{d}(u_k-v_k)^2}. \tag{3}$$

These numbers are also called DISPERSION NUMBERS and RENDEZVOUS VALUES. For any E, Gross (1964) and Stadje (1981) proved that

$$\tfrac{1}{2} \le m(E) < 1. \tag{4}$$

If I is a subinterval of the LINE and D is a circular DISK in the PLANE, then

$$m(I) = m(D) = \tfrac{1}{2}. \tag{5}$$

If C is a CIRCLE, then

$$m(C) = \frac{2}{\pi} = 0.6366\ldots. \tag{6}$$

An expression for the magic constant of an ELLIPSE in terms of its SEMIMAJOR and SEMIMINOR AXES lengths is not known. Nikolas and Yost (1988) showed that for a REULEAUX TRIANGLE T

$$0.6675276 \le m(T) \le 0.6675284. \tag{7}$$

Denote the MAXIMUM value of $m(E)$ in n-D space by $M(n)$. Then

$$M(1) = \tfrac{1}{2} \tag{8}$$

$$M(2): m(T) \le M(2) \le \frac{2+\sqrt{3}}{3\sqrt{3}} < 0.7182336 \tag{9}$$

$$M(d): \frac{d}{d+1} \le M(d) \le \frac{[\Gamma(\frac{1}{2}d)]^2 2^{d-2}\sqrt{2d}}{\Gamma(d-\frac{1}{2})\sqrt{(d+1)\pi}} < \sqrt{\frac{d}{d+1}}, \tag{10}$$

where $\Gamma(z)$ is the GAMMA FUNCTION (Nikolas and Yost 1988).

An unrelated quantity characteristic of a given MAGIC SQUARE is also known as a MAGIC CONSTANT.

References

Finch, S. "Favorite Mathematical Constants." `http://www.mathsoft.com/asolve/constant/magic/magic.html`.
Cleary, J.; Morris, S. A.; and Yost, D. "Numerical Geometry—Numbers for Shapes." *Amer. Math. Monthly* **95**, 260–275, 1986.
Croft, H. T.; Falconer, K. J.; and Guy, R. K. *Unsolved Problems in Geometry*. New York: Springer-Verlag, 1994.
Gross, O. *The Rendezvous Value of Metric Space*. Princeton, NJ: Princeton University Press, pp. 49–53, 1964.
Nikolas, P. and Yost, D. "The Average Distance Property for Subsets of Euclidean Space." *Arch. Math. (Basel)* **50**, 380–384, 1988.
Stadje, W. "A Property of Compact Connected Spaces." *Arch. Math. (Basel)* **36**, 275–280, 1981.

Magic Graph

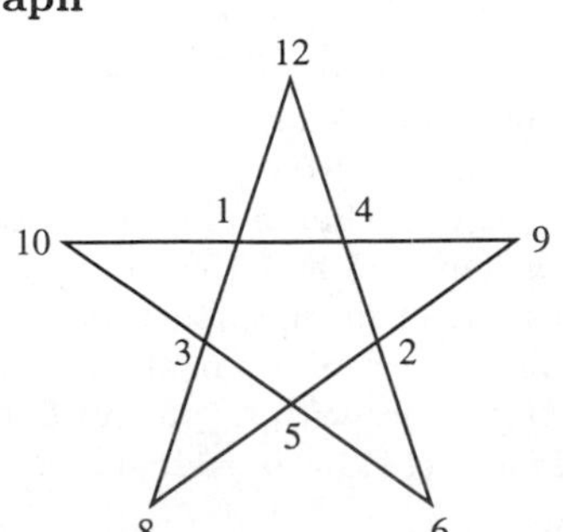

A LABELLED GRAPH with e EDGES labeled with distinct elements $\{1, 2, \ldots, e\}$ so that the sum of the EDGE labels at each VERTEX is the same. Another type of magic graph, such as the PENTAGRAM shown above, has labelled VERTICES which give the same sum along every straight line segment (Madachy 1979).

see also ANTIMAGIC GRAPH, LABELLED GRAPH, MAGIC CIRCLES, MAGIC CONSTANT, MAGIC CUBE, MAGIC HEXAGON, MAGIC SQUARE

References

Hartsfield, N. and Ringel, G. *Pearls in Graph Theory: A Comprehensive Introduction.* San Diego, CA: Academic Press, 1990.

Heinz, H. "Magic Stars." http://www.geocities.com/CapeCanaveral/Launchpad/4057/magicstar.htm.

Madachy, J. S. *Madachy's Mathematical Recreations.* New York: Dover, pp. 98–99, 1979.

Magic Hexagon

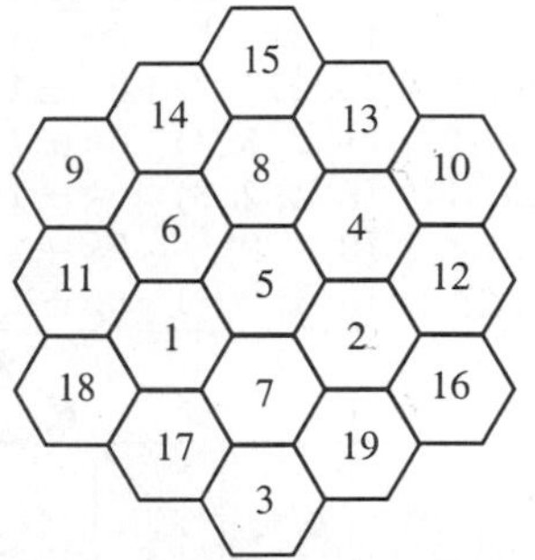

An arrangement of close-packed HEXAGONS containing the numbers 1, 2, ..., $H_n = 3n(n-1)+1$, where H_n is the nth HEX NUMBER, such that the numbers along each straight line add up to the same sum. In the above magic hexagon, each line (those of lengths both 3 and 4) adds up to 38. This is the only magic hexagon of the counting numbers for any size hexagon. It was discovered by C. W. Adam, who worked on the problem from 1910 to 1957.

see also HEX NUMBER, HEXAGON, MAGIC GRAPH, MAGIC SQUARE, TALISMAN HEXAGON

References

Beeler, M.; Gosper, R. W.; and Schroeppel, R. *HAKMEM.* Cambridge, MA: MIT Artificial Intelligence Laboratory, Memo AIM-239, Item 49, Feb. 1972.

Gardner, M. "Permutations and Paradoxes in Combinatorial Mathematics." *Sci. Amer.* **209**, 112–119, Aug. 1963.

Honsberger, R. *Mathematical Gems I.* Washington, DC: Math. Assoc. Amer., pp. 69–76, 1973.

Madachy, J. S. *Madachy's Mathematical Recreations.* New York: Dover, pp. 100–101, 1979.

Magic Labelling

It is conjectured that every TREE with e edges whose nodes are all trivalent or monovalent can be given a "magic" labelling such that the INTEGERS 1, 2, ..., e can be assigned to the edges so that the SUM of the three meeting at a node is constant.

see also MAGIC CONSTANT, MAGIC CUBE, MAGIC GRAPH, MAGIC HEXAGON, MAGIC SQUARE

References

Guy, R. K. "Unsolved Problems Come of Age." *Amer. Math. Monthly* **96**, 903–909, 1989.

Magic Number

see MAGIC CONSTANT

Magic Series

n numbers form a magic series of degree p if the sum of their kth POWERS is the MAGIC CONSTANT of degree k for all $k \in [1, p]$.

see also MAGIC CONSTANT, MAGIC SQUARE

References

Kraitchik, M. "Magic Series." §7.13.3 in *Mathematical Recreations.* New York: W. W. Norton, pp. 183–186, 1942.

Magic Square

8	1	6
3	5	7
4	9	2

16	2	3	13
5	11	10	8
9	7	6	12
4	14	15	1

17	24	1	8	15
23	5	7	14	16
4	6	13	20	22
10	12	19	21	3
11	18	25	2	9

32	29	4	1	24	21
30	31	2	3	22	23
12	9	17	20	28	25
10	11	18	19	26	27
13	16	36	33	5	8
14	15	34	35	6	7

30	39	48	1	10	19	28
38	47	7	9	18	27	29
46	6	8	17	26	35	37
5	14	16	25	34	36	45
13	15	24	33	42	44	4
21	23	32	41	43	3	12
22	31	40	49	2	11	20

64	2	3	61	60	6	7	57
9	55	54	12	13	51	50	16
17	47	46	20	21	43	42	24
40	26	27	37	36	30	31	33
32	34	35	29	28	38	39	25
41	23	22	44	45	19	18	48
49	15	14	52	53	11	10	56
8	58	59	5	4	62	63	1

A (normal) magic square consists of the distinct POSITIVE INTEGERS 1, 2, ..., n^2 such that the sum of the n numbers in any horizontal, vertical, or *main* diagonal line is always the same MAGIC CONSTANT

$$M_2(n) = \frac{1}{n}\sum_{k=1}^{n^2} k = \tfrac{1}{2}n(n^2+1).$$

The unique normal square of order three was known to the ancient Chinese, who called it the LO SHU. A version of the order 4 magic square with the numbers 15 and 14 in adjacent middle columns in the bottom row is called DÜRER'S MAGIC SQUARE. Magic squares of order 3 through 8 are shown above.

The MAGIC CONSTANT for an nth order magic square starting with an INTEGER A and with entries in an increasing ARITHMETIC SERIES with difference D between terms is

$$M_2(n; A, D) = \tfrac{1}{2}n[2a + D(n^2-1)]$$

(Hunter and Madachy 1975). If every number in a magic square is subtracted from n^2+1, another magic square is obtained called the complementary magic square. Squares which are magic under multiplication instead of addition can be constructed and are known as MULTIPLICATION MAGIC SQUARES. In addition, squares which are magic under both addition *and* multiplication can be constructed and are known as ADDITION-MULTIPLICATION MAGIC SQUARES (Hunter and Madachy 1975).

A square that fails to be magic only because one or both of the main diagonal sums do not equal the MAGIC CONSTANT is called a SEMIMAGIC SQUARE. If *all* diagonals (including those obtained by wrapping around) of a magic square sum to the MAGIC CONSTANT, the square is said to be a PANMAGIC SQUARE (also called a DIABOLICAL SQUARE or PANDIAGONAL SQUARE). If replacing each number n_i by its square n_i^2 produces another magic square, the square is said to be a BIMAGIC SQUARE (or DOUBLY MAGIC SQUARE). If a square is magic for n_i, n_i^2, and n_i^3, it is called a TREBLY MAGIC SQUARE. If all pairs of numbers symmetrically opposite the center sum to $n^2 + 1$, the square is said to be an ASSOCIATIVE MAGIC SQUARE.

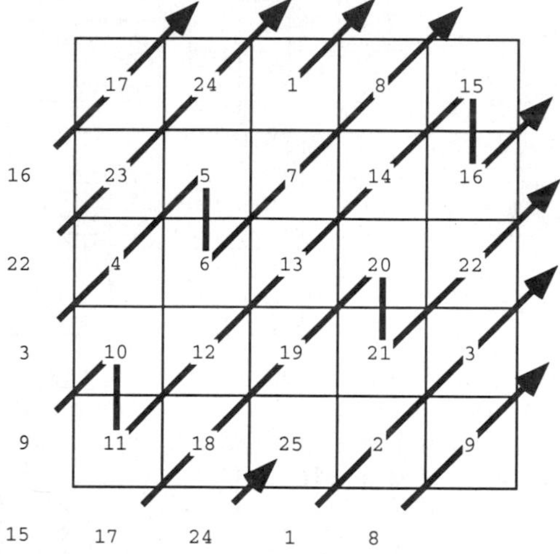

Kraitchik (1942) gives general techniques of constructing EVEN and ODD squares of order n. For n ODD, a very straightforward technique known as the Siamese method can be used, as illustrated above (Kraitchik 1942, pp. 148–149). It begins by placing a 1 in any location (in the center square of the top row in the above example), then incrementally placing subsequent numbers in the square one unit above and to the right. The counting is wrapped around, so that falling off the top returns on the bottom and falling off the right returns on the left. When a square is encountered which is already filled, the next number is instead placed *below* the previous one and the method continues as before. The method, also called de la Loubere's method, is purported to have been first reported in the West when de la Loubere returned to France after serving as ambassador to Siam.

A generalization of this method uses an "ordinary vector" (x, y) which gives the offset for each noncolliding move and a "break vector" (u, v) which gives the offset to introduce upon a collision. The standard Siamese method therefore has ordinary vector $(1, -1)$ and break vector $(0, 1)$. In order for this to produce a magic square, each break move must end up on an unfilled cell. Special classes of magic squares can be constructed by considering the absolute sums $|u+v|$, $|(u-x)+(v-y)|$, $|u-v|$, and $|(u-x)-(v-y)| = |u+y-x-v|$. Call the set of these numbers the sumdiffs (sums and differences). If all sumdiffs are RELATIVELY PRIME to n and the square is a magic square, then the square is also a PANMAGIC SQUARE. This theory originated with de la Hire. The following table gives the sumdiffs for particular choices of ordinary and break vectors.

Ordinary Vector	Break Vector	Sumdiffs	Magic Squares	Panmagic Squares
(1, −1)	(0, 1)	(1, 3)	$2k+1$	none
(1, −1)	(0, 2)	(0, 2)	$6k \pm 1$	none
(2, 1)	(1, −2)	(1, 2, 3, 4)	$6k \pm 1$	none
(2, 1)	(1, −1)	(0, 1, 2, 3)	$6k \pm 1$	$6k \pm 1$
(2, 1)	(1, 0)	(0, 1, 2)	$2k+1$	none
(2, 1)	(1, 2)	(0, 1, 2, 3)	$6k \pm 1$	none

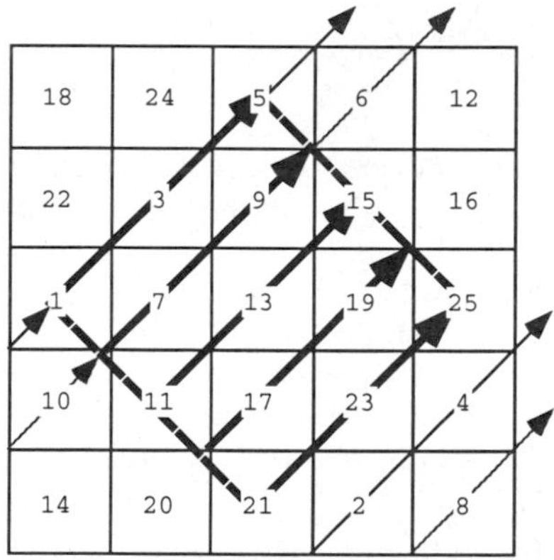

A second method for generating magic squares of ODD order has been discussed by J. H. Conway under the name of the "lozenge" method. As illustrated above, in this method, the ODD numbers are built up along diagonal lines in the shape of a DIAMOND in the central part of the square. The EVEN numbers which were missed are then added sequentially along the continuation of the diagonal obtained by wrapping around the square until the wrapped diagonal reaches its initial point. In the above square, the first diagonal therefore fills in 1, 3, 5, 2, 4, the second diagonal fills in 7, 9, 6, 8, 10, and so on.

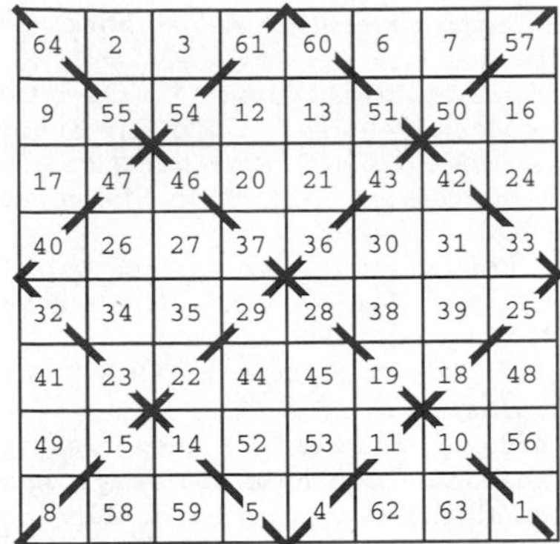

An elegant method for constructing magic squares of DOUBLY EVEN order $n = 4m$ is to draw xs through each 4×4 subsquare and fill all squares in sequence. Then replace each entry a_{ij} on a crossed-off diagonal by $(n^2 + 1) - a_{ij}$ or, equivalently, reverse the order of the crossed-out entries. Thus in the above example for $n = 8$, the crossed-out numbers are originally 1, 4, ..., 61, 64, so entry 1 is replaced with 64, 4 with 61, etc.

L		U		X	
4	1	1	4	1	4
2	3	2	3	3	2

68	65	96	93	4	1	32	29	60	57
L		L		L		L		L	
66	67	94	95	2	3	30	31	58	59
92	89	20	17	28	25	56	53	64	61
L		L		L		L		L	
90	91	18	19	26	27	54	55	62	63
16	13	24	21	49	52	80	77	88	85
L		L		U		L		L	
14	15	22	23	50	51	78	79	86	87
37	40	45	48	76	73	81	84	9	12
U		U		L		U		U	
38	39	46	47	74	75	82	83	10	11
41	44	69	72	97	100	5	8	33	36
X		X		X		X		X	
43	42	71	70	99	98	7	6	35	34

A very elegant method for constructing magic squares of SINGLY EVEN order $n = 4m + 2$ with $m \geq 1$ (there is no magic square of order 2) is due to J. H. Conway, who calls it the "LUX" method. Create an array consisting of $m + 1$ rows of Ls, 1 row of Us, and $m - 1$ rows of Xs, all of length $n/2 = 2m + 1$. Interchange the middle U with the L above it. Now generate the magic square of order $2m + 1$ using the Siamese method centered on the array of letters (starting in the center square of the top row), but fill each set of four squares surrounding a letter sequentially according to the order prescribed by the the letter. That order is illustrated on the left side of the above figure, and the completed square is illustrated to the right. The "shapes" of the letters L, U, and X naturally suggest the filling order, hence the name of the algorithm.

It is an unsolved problem to determine the number of magic squares of an arbitrary order, but the number of distinct magic squares (excluding those obtained by rotation and reflection) of order $n = 1, 2, \ldots$ are 1, 0, 1, 880, 275305224, ... (Sloane's A006052; Madachy 1979, p. 87). The 880 squares of order four were enumerated by Frenicle de Bessy in the seventeenth century, and are illustrated in Berlekamp *et al.* (1982, pp. 778–783). The number of 6×6 squares is not known.

67	1	43
13	37	61
31	73	7

3	61	19	37
43	31	5	41
7	11	73	29
67	17	23	13

The above magic squares consist only of PRIMES and were discovered by E. Dudeney (1970) and A. W. Johnson, Jr. (Dewdney 1988). Madachy (1979, pp. 93–96) and Rivera discuss other magic squares composed of PRIMES.

52	61	4	13	20	29	36	45
14	3	62	51	46	35	30	19
53	60	5	12	21	28	37	44
11	6	59	54	43	38	27	22
55	58	7	10	23	26	39	42
9	8	57	56	41	40	25	24
50	63	2	15	18	31	34	47
16	1	64	49	48	33	32	17

Benjamin Franklin constructed the above 8×8 PANMAGIC SQUARE having MAGIC CONSTANT 260. Any half-row or half-column in this square totals 130, and the four corners plus the middle total 260. In addition, bent diagonals (such as 52-3-5-54-10-57-63-16) also total 260 (Madachy 1979, p. 87).

1480028159	1480028153	1480028201
1480028213	1480028171	1480028129
1480028141	1480028189	1480028183

In addition to other special types of magic squares, a 3×3 square whose entries are consecutive PRIMES, illustrated above, has been discovered by H. Nelson (Rivera). Variations on magic squares can also be constructed using letters (either in defining the square or as entries in it), such as the ALPHAMAGIC SQUARE and TEMPLAR MAGIC SQUARE.

4	9	2
3	5	7
8	1	6

4	14	15	1
9	7	6	12
5	11	10	8
16	2	3	13

11	24	7	20	3
4	12	25	8	16
17	5	13	21	9
10	18	1	14	22
23	6	19	2	15

6	32	3	34	35	1
7	11	27	28	8	30
19	14	16	15	23	24
18	20	22	21	17	13
25	29	10	9	26	12
36	5	33	4	2	31

22	47	16	41	10	35	4
5	23	48	17	42	11	29
30	6	24	49	18	36	12
13	31	7	25	43	19	37
38	14	32	1	26	44	20
21	39	8	33	2	27	45
46	15	40	9	34	3	28

8	58	59	5	4	62	63	1
49	15	14	52	53	11	10	56
41	23	22	44	45	19	18	48
32	34	35	29	28	38	39	25
40	26	27	37	36	30	31	33
17	47	46	20	21	43	42	24
9	55	54	12	13	51	50	16
64	2	3	61	60	6	7	57

37	78	29	70	21	62	13	54	5
6	38	79	30	71	22	63	14	46
47	7	39	80	31	72	23	55	15
16	48	8	40	81	32	64	24	56
57	17	49	9	41	73	33	65	25
26	58	18	50	1	42	74	34	66
67	27	59	10	51	2	43	75	35
36	68	19	60	11	52	3	44	76
77	28	69	20	61	12	53	4	45

Various numerological properties have also been associated with magic squares. Pivari associates the squares illustrated above with Saturn, Jupiter, Mars, the Sun, Venus, Mercury, and the Moon, respectively. Attractive patterns are obtained by connecting consecutive numbers in each of the squares (with the exception of the Sun magic square).

see also ADDITION-MULTIPLICATION MAGIC SQUARE ALPHAMAGIC SQUARE, ANTIMAGIC SQUARE, ASSOCIATIVE MAGIC SQUARE, BIMAGIC SQUARE, BORDER SQUARE, DÜRER'S MAGIC SQUARE, EULER SQUARE, FRANKLIN MAGIC SQUARE, GNOMON MAGIC SQUARE, HETEROSQUARE, LATIN SQUARE, MAGIC CIRCLES, MAGIC CONSTANT, MAGIC CUBE, MAGIC HEXAGON, MAGIC LABELLING, MAGIC SERIES, MAGIC TOUR, MULTIMAGIC SQUARE, MULTIPLICATION MAGIC SQUARE, PANMAGIC SQUARE, SEMIMAGIC SQUARE, TALISMAN SQUARE, TEMPLAR MAGIC SQUARE, TRIMAGIC SQUARE

References

Abe, G. "Unsolved Problems on Magic Squares." *Disc. Math.* **127**, 3–13, 1994.

Alejandre, S. "Suzanne Alejandre's Magic Squares." `http://forum.swarthmore.edu/alejandre/magic.square.html`.

Andrews, W. S. *Magic Squares and Cubes, 2nd rev. ed.* New York: Dover, 1960.

Ball, W. W. R. and Coxeter, H. S. M. "Magic Squares." Ch. 7 in *Mathematical Recreations and Essays, 13th ed.* New York: Dover, 1987.

Barnard, F. A. P. "Theory of Magic Squares and Cubes." *Memoirs Natl. Acad. Sci.* **4**, 209–270, 1888.

Benson, W. H. and Jacoby, O. *New Recreations with Magic Squares.* New York: Dover, 1976.

Berlekamp, E. R.; Conway, J. H; and Guy, R. K. *Winning Ways, For Your Mathematical Plays, Vol. 2: Games in Particular.* London: Academic Press, 1982.

Dewdney, A. K. "Computer Recreations: How to Pan for Primes in Numerical Gravel." *Sci. Amer.* **259**, pp. 120–123, July 1988.

Dudeney, E. *Amusements in Mathematics.* New York: Dover, 1970.

Fults, J. L. *Magic Squares.* Chicago, IL: Open Court, 1974.

Gardner, M. "Magic Squares." Ch. 12 in *The Second Scientific American Book of Mathematical Puzzles & Diversions: A New Selection.* New York: Simon and Schuster, 1961.

Gardner, M. "Magic Squares and Cubes." Ch. 17 in *Time Travel and Other Mathematical Bewilderments.* New York: W. H. Freeman, 1988.

Grogono, A. W. "Magic Squares by Grog." `http://www.grogono.com/magic/`.

Heinz, H. "Magic Squares." `http://www.geocities.com/CapeCanaveral/Launchpad/4057/magicsquare.htm`.

Hirayama, A. and Abe, G. *Researches in Magic Squares.* Osaka, Japan: Osaka Kyoikutosho, 1983.

Horner, J. "On the Algebra of Magic Squares, I., II., and III." *Quart. J. Pure Appl. Math.* **11**, 57–65, 123–131, and 213–224, 1871.

Hunter, J. A. H. and Madachy, J. S. "Mystic Arrays." Ch. 3 in *Mathematical Diversions.* New York: Dover, pp. 23–34, 1975.

Kanada, Y. "Magic Square Page." `http://www.st.rim.or.jp/~kanada/puzzles/magic-square.html`.

Kraitchik, M. "Magic Squares." Ch. 7 in *Mathematical Recreations.* New York: Norton, pp. 142–192, 1942.

Lei, A. "Magic Square, Cube, Hypercube." `http://www.cs.ust.hk/~philipl/magic/magic.html`.

Madachy, J. S. "Magic and Antimagic Squares." Ch. 4 in *Madachy's Mathematical Recreations.* New York: Dover, pp. 85–113, 1979.

Moran, J. *The Wonders of Magic Squares.* New York: Vintage, 1982.

Pappas, T. "Magic Squares," "The 'Special' Magic Square," "The Pyramid Method for Making Magic Squares," "Ancient Tibetan Magic Square," "Magic 'Line.'," and "A Chinese Magic Square." *The Joy of Mathematics.* San Carlos, CA: Wide World Publ./Tetra, pp. 82–87, 112, 133, 169, and 179, 1989.

Pivari, F. "Nice Examples." `http://www.geocities.com/CapeCanaveral/Lab/3469/examples.html`.

Pivari, F. "Simple Magic Square Checker and GIF Maker." `http://www.geocities.com/CapeCanaveral/Lab/3469/squaremaker.html`.

Rivera, C. "Problems & Puzzles (Puzzles): Magic Squares with Consecutive Primes." `http://www.sci.net.mx/~crivera/ppp/puzz_003.htm`.

Rivera, C. "Problems & Puzzles (Puzzles): Prime-Magical Squares." `http://www.sci.net.mx/~crivera/ppp/puzz_004.htm`.

Sloane, N. J. A. Sequence A006052/M5482 in "An On-Line Version of the Encyclopedia of Integer Sequences."

Suzuki, M. "Magic Squares." `http://www.pse.che.tohoku.ac.jp/~msuzuki/MagicSquare.html`.

Weisstein, E. W. "Magic Squares." `http://www.astro.virginia.edu/~eww6n/math/notebooks/MagicSquares.m`.

Magic Star

see MAGIC GRAPH

Magic Tour

Let a chess piece make a TOUR on an $n \times n$ CHESSBOARD whose squares are numbered from 1 to n^2 along the path of the chess piece. Then the TOUR is called a magic tour if the resulting arrangement of numbers is a MAGIC SQUARE. If the first and last squares traversed are connected by a move, the tour is said to be closed (or "re-entrant"); otherwise it is open. The MAGIC CONSTANT for the 8×8 CHESSBOARD is 260.

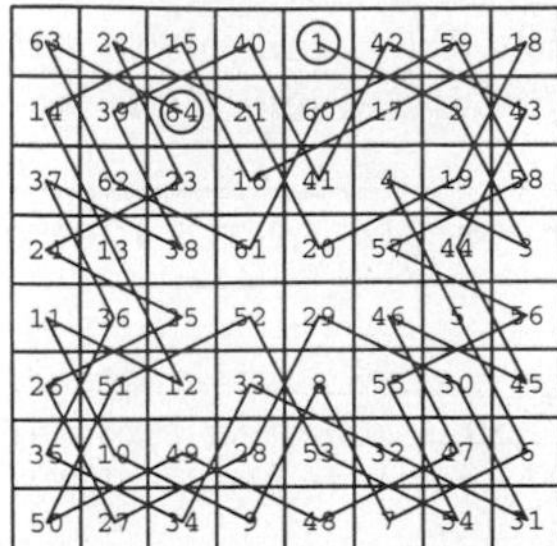

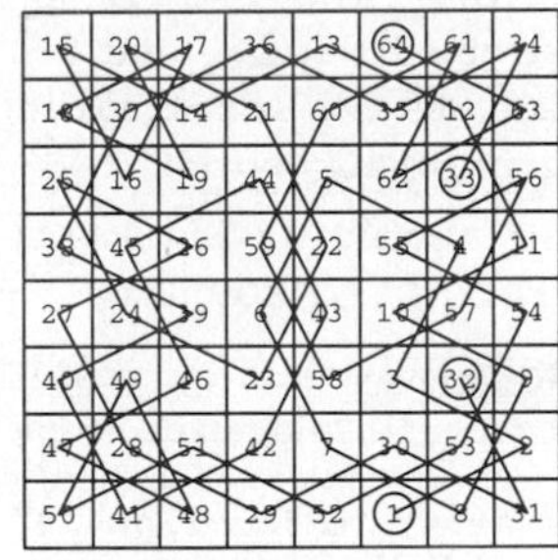

Magic KNIGHT'S TOURS are not possible on $n \times n$ boards for n ODD, and are believed to be impossible for $n = 8$. The "most magic" knight tour known on the 8×8 board is the SEMIMAGIC SQUARE illustrated in the above left figure (Ball and Coxeter 1987, p. 185) having main diagonal sums of 348 and 168. Combining two half-knights' tours one above the other as in the above right figure does, however, give a MAGIC SQUARE (Ball and Coxeter 1987, p. 185).

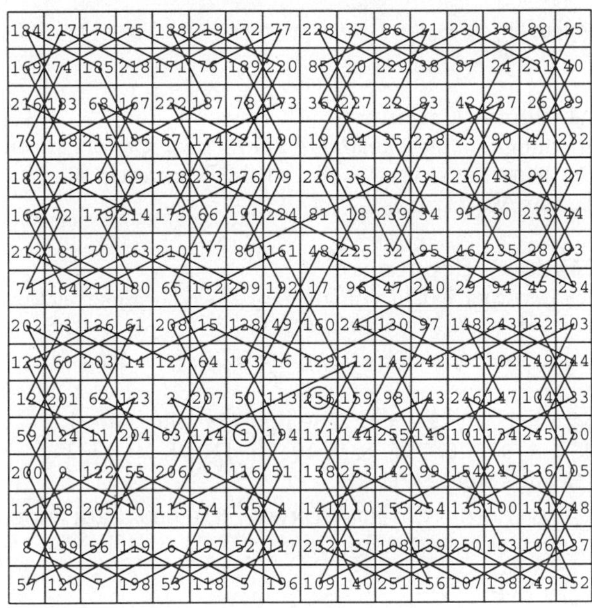

The above illustration shows a 16×16 closed magic KNIGHT'S TOUR (Madachy 1979).

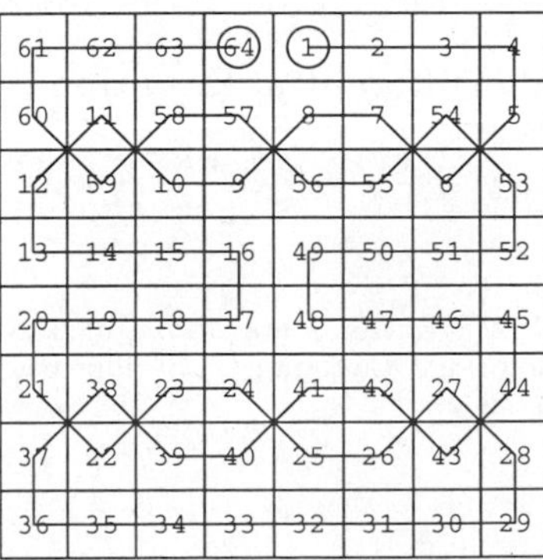

A magic tour for king moves is illustrated above (Coxeter 1987, p. 186).

see also CHESSBOARD, KNIGHT'S TOUR, MAGIC SQUARE, SEMIMAGIC SQUARE, TOUR

References

Ball, W. W. R. and Coxeter, H. S. M. *Mathematical Recreations and Essays, 13th ed.* New York: Dover, pp. 185–187, 1987.

Madachy, J. S. *Madachy's Mathematical Recreations.* New York: Dover, pp. 87–89, 1979.

Mahler-Lech Theorem

Let K be a FIELD of CHARACTERISTIC 0 (e.g., the rationals $\mathbb{Q}$) and let $\{u_n\}$ be a SEQUENCE of elements of K which satisfies a difference equation of the form

$$u_n = c_0 u_n + c_1 u_{n+1} + \ldots + c_k u_{n+k},$$

where the COEFFICIENTS c_i are fixed elements of K. Then, for any $c \in K$, we have *either* $u_n = c$ for only finitely many values of n, *or* $u_n = c$ for the values of n in some ARITHMETIC PROGRESSION.

The proof involves embedding certain fields inside the p-ADIC NUMBERS $\mathbb{Q}_p$ for some PRIME p, and using properties of zeros of POWER series over $\mathbb{Q}_p$ (STRASSMAN'S THEOREM).

see also ARITHMETIC PROGRESSION, p-ADIC NUMBER, STRASSMAN'S THEOREM

Mahler's Measure

For a POLYNOMIAL P,

$$M(P) = \exp \int_0^{2\pi} \ln |P(e^{i\theta})| \frac{d\theta}{2\pi}.$$

It is related to JENSEN'S INEQUALITY.

see also JENSEN'S INEQUALITY

Major Axis

see SEMIMAJOR AXIS

Major Triangle Center

A TRIANGLE CENTER $\alpha : \beta : \gamma$ is called a major center if the TRIANGLE CENTER FUNCTION $\alpha = f(a, b, c, A, B, C)$ is a function of ANGLE A alone, and therefore β and γ of B and C alone, respectively.

see also REGULAR TRIANGLE CENTER, TRIANGLE CENTER

References

Kimberling, C. "Major Centers of Triangles." *Amer. Math. Monthly* **104**, 431–438, 1997.

Majorant

A function used to study ORDINARY DIFFERENTIAL EQUATIONS.

Makeham Curve

The function defined by

$$y = ks^x b^{q^x}$$

which is used in actuarial science for specifying a simplified mortality law. Using $s(x)$ as the probability that a newborn will achieve age x, the Makeham law (1860) uses

$$s(x) = \exp(-Ax - m(c^x - 1))$$

for $B > 0$, $A \geq -B$, $c > 1$, $x \geq 0$.

see also GOMPERTZ CURVE, LIFE EXPECTANCY, LOGISTIC GROWTH CURVE, POPULATION GROWTH

References

Bowers, N. L. Jr.; Gerber, H. U.; Hickman, J. C.; Jones, D. A.; and Nesbitt, C. J. *Actuarial Mathematics.* Itasca, IL: Society of Actuaries, p. 71, 1997.

Makeham, W. M. "On the Law of Mortality, and the Construction of Annuity Tables." *JIA* **8**, 1860.

Malfatti Circles

Three circles packed inside a RIGHT TRIANGLE which are tangent to each other and to two sides of the TRIANGLE.

see also MALFATTI'S RIGHT TRIANGLE PROBLEM

Malfatti Points

see AJIMA-MALFATTI POINTS

Malfatti's Right Triangle Problem

Find the maximum total AREA of three CIRCLES (of possibly different sizes) which can be packed inside a RIGHT TRIANGLE of any shape without overlapping. In 1803, Malfatti gave the solution as three CIRCLES (the MALFATTI CIRCLES) tangent to each other and to two sides of the TRIANGLE. In 1929, it was shown that the MALFATTI CIRCLES were not always the best solution. Then Goldberg (1967) showed that, even worse, they are *never* the best solution.

see also MALFATTI'S TANGENT TRIANGLE PROBLEM

References

Eves, H. *A Survey of Geometry, Vol. 2.* Boston: Allyn & Bacon, p. 245, 1965.

Goldberg, M. "On the Original Malfatti Problem." *Math. Mag.* **40**, 241–247, 1967.

Ogilvy, C. S. *Excursions in Geometry.* New York: Dover, pp. 145–147, 1990.

Malfatti's Tangent Triangle Problem

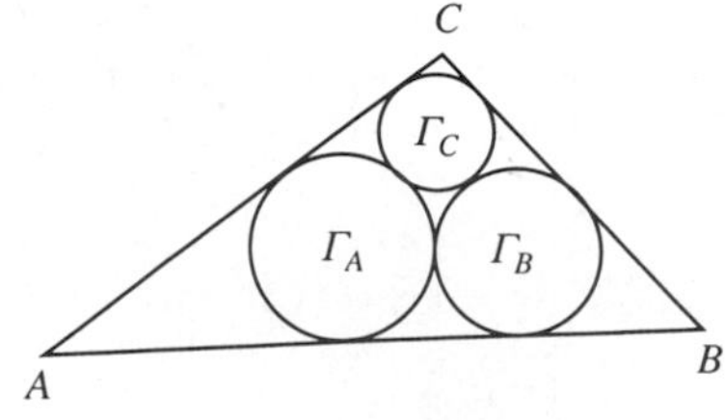

Draw within a given TRIANGLE three CIRCLES, each of which is TANGENT to the other two and to two sides of the TRIANGLE. Denote the three CIRCLES so constructed Γ_A, Γ_B, and Γ_C. Then Γ_A is tangent to AB and AC, Γ_B is tangent to BC and BA, and Γ_C is tangent to AC and BC.

see also AJIMA-MALFATTI POINTS, MALFATTI'S RIGHT TRIANGLE PROBLEM

References

Dörrie, H. "Malfatti's Problem." §30 in *100 Great Problems of Elementary Mathematics: Their History and Solutions.* New York: Dover, pp. 147–151, 1965.

Forder, H. G. *Higher Course Geometry.* Cambridge, England: Cambridge University Press, pp. 244–245, 1931.

Fukagawa, H. and Pedoe, D. *Japanese Temple Geometry Problems (San Gaku).* Winnipeg: The Charles Babbage Research Centre, pp. 106–120, 1989.

Gardner, M. *Fractal Music, Hypercards, and More Mathematical Recreations from Scientific American Magazine.* New York: W. H. Freeman, pp. 163–165, 1992.

Goldberg, M. "On the Original Malfatti Problem." *Math. Mag.* **40**, 241–247, 1967.

Lob, H. and Richmond, H. W. "On the Solution of Malfatti's Problem for a Triangle." *Proc. London Math. Soc.* **2**, 287–304, 1930.

Woods, F. S. *Higher Geometry.* New York: Dover, pp. 206–209, 1961.

Malliavin Calculus

An infinite-dimensional DIFFERENTIAL CALCULUS on the WIENER SPACE. Also called STOCHASTIC CALCULUS OF VARIATIONS.

Mallow's Sequence

An INTEGER SEQUENCE given by the recurrence relation

$$a(n) = a(a(n-2)) + a(n - a(n-2))$$

with $a(1) = a(2) = 1$. The first few values are 1, 1, 2, 3, 3, 4, 5, 6, 6, 7, 7, 8, 9, 10, 10, 11, 12, 12, 13, 14, ... (Sloane's A005229).

see also HOFSTADTER-CONWAY $10,000 SEQUENCE, HOFSTADTER'S Q-SEQUENCE

References

Mallows, C. "Conway's Challenging Sequence." *Amer. Math. Monthly* **98**, 5–20, 1991.

Sloane, N. J. A. Sequence A005229/M0441 in "An On-Line Version of the Encyclopedia of Integer Sequences."

Malmstén's Differential Equation

$$y'' + \frac{r}{z}y' = \left(Az^m + \frac{s}{z^2}\right)y.$$

References

Watson, G. N. *A Treatise on the Theory of Bessel Functions, 2nd ed.* Cambridge, England: Cambridge University Press, pp. 99–100, 1966.

Maltese Cross

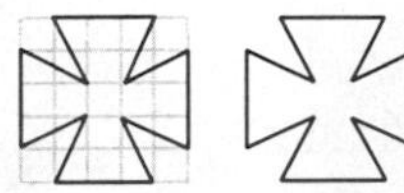

An irregular DODECAHEDRON CROSS shaped like a + sign but whose points flange out at the end: ✠. The conventional proportions as computed on a 5×5 grid as illustrated above.

see also CROSS, DISSECTION, DODECAHEDRON

References

Frederickson, G. "Maltese Crosses." Ch. 14 in *Dissections: Plane and Fancy.* New York: Cambridge University Press, pp. 157–162, 1997.

Maltese Cross Curve

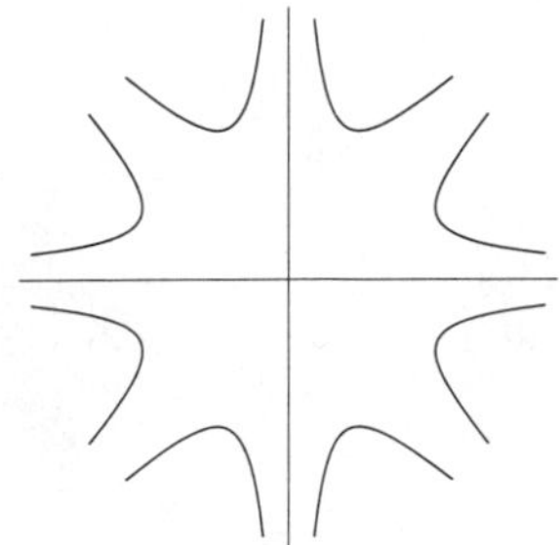

The plane curve with Cartesian equation

$$xy(x^2 - y^2) = x^2 + y^2$$

and polar equation

$$r^2 = \frac{1}{\cos\theta \sin\theta(\cos^2\theta - \sin^2\theta)}.$$

References

Cundy, H. and Rollett, A. *Mathematical Models, 3rd ed.* Stradbroke, England: Tarquin Pub., p. 71, 1989.

Malthusian Parameter

The parameter α in the exponential POPULATION GROWTH equation

$$N_1(t) = N_0 e^{\alpha t}.$$

see also LIFE EXPECTANCY, POPULATION GROWTH

Mandelbar Set

A FRACTAL set analogous to the MANDELBROT SET or its generalization to a higher power with the variable z replaced by its COMPLEX CONJUGATE z^*.

see also MANDELBROT SET

Mandelbrot Set

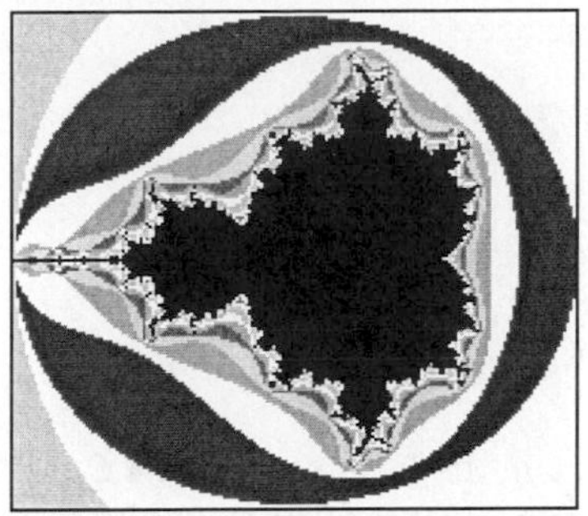

The set obtained by the QUADRATIC RECURRENCE

$$z_{n+1} = z_n{}^2 + C, \tag{1}$$

where points C for which the orbit $z = 0$ does not tend to infinity are in the SET. It marks the set of points in the COMPLEX PLANE such that the corresponding JULIA SET is CONNECTED and not COMPUTABLE. The Mandelbrot set was originally called a μ MOLECULE by Mandelbrot.

J. Hubbard and A. Douady proved that the Mandelbrot set is CONNECTED. Shishikura (1994) proved that the boundary of the Mandelbrot set is a FRACTAL with HAUSDORFF DIMENSION 2. However, it is not yet known if the Mandelbrot set is pathwise-connected. If it is pathwise-connected, then Hubbard and Douady's proof implies that the Mandelbrot set is the image of a CIRCLE and can be constructed from a DISK by collapsing certain arcs in the interior (Douady 1986).

The AREA of the set is known to lie between 1.5031 and 1.5702; it is estimated as 1.50659....

Decomposing the COMPLEX coordinate $z = x + iy$ and $z_0 = a + ib$ gives

$$x' = x^2 - y^2 + a \tag{2}$$

$$y' = 2xy + b. \tag{3}$$

In practice, the limit is approximated by

$$\lim_{n\to\infty} |z_n| \approx \lim_{n\to n_{\max}} |z_n| < r_{\max}. \tag{4}$$

Beautiful computer-generated plots can be created by coloring nonmember points depending on how quickly they diverge to $r_{\max}$. A common choice is to define an INTEGER called the COUNT to be the largest n such that $|z_n| < r$, where r is usually taken as $r = 2$, and to color points of different COUNT different colors. The boundary between successive COUNTS defines a series of "LEMNISCATES," called EQUIPOTENTIAL CURVES by Peitgen and Saupe (1988), $|L_n(C)| = r$ which have distinctive shapes. The first few LEMNISCATES are

$$L_1(C) = C \tag{5}$$

$$L_2(C) = C(C + 1) \tag{6}$$

$$L_3(C) = C + (C + C^2)^2 \tag{7}$$

$$L_4(C) = C + [C + (C + C^2)^2]^2. \tag{8}$$

When written in CARTESIAN COORDINATES, the first three of these are

$$r^2 = x^2 + y^2 \tag{9}$$

$$r^2 = (x^2 + y^2)[(x+1)^2 + y^2] \tag{10}$$

$$\begin{aligned} r^2 = (x^2 + y^2)(1 + 2x + 5x^2 + 6x^3 + 6x^4 + 4x^5 + x^6 \\ - 3y^2 - 2xy^2 + 8x^2y^2 + 8x^3y^2 \\ + 3x^4y^2 + 2y^4 + 4xy^4 + 3x^2y^4 + y^6), \end{aligned} \tag{11}$$

which are a CIRCLE, an OVAL, and a PEAR CURVE. In fact, the second LEMNISCATE L_2 can be written in terms of a new coordinate system with $x' \equiv x - 1/2$ as

$$[(x' - \tfrac{1}{2})^2 + y^2][(x' + \tfrac{1}{2})^2 + y^2] = r^2, \tag{12}$$

which is just a CASSINI OVAL with $a = 1/2$ and $b^2 = r$. The LEMNISCATES grow increasingly convoluted with higher COUNT and approach the Mandelbrot set as the COUNT tends to infinity.

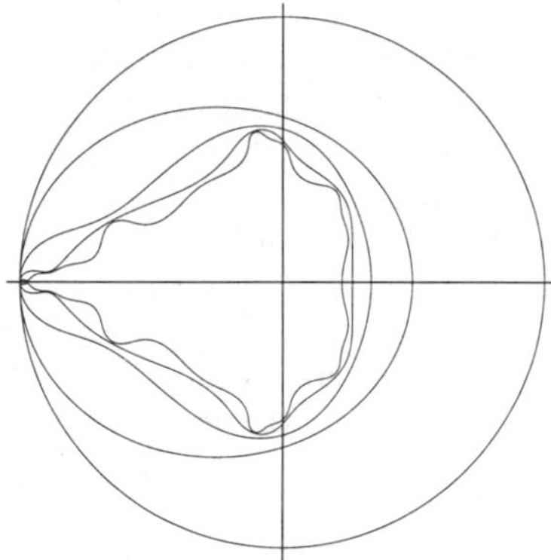

The kidney bean-shaped portion of the Mandelbrot set is bordered by a CARDIOID with equations

$$4x = 2\cos t - \cos(2t) \tag{13}$$

$$4y = 2\sin t - \sin(2t). \tag{14}$$

The adjoining portion is a CIRCLE with center at $(-1, 0)$ and RADIUS 1/4. One region of the Mandelbrot set containing spiral shapes is known as SEA HORSE VALLEY because the shape resembles the tail of a sea horse.

Generalizations of the Mandelbrot set can be constructed by replacing ${z_n}^2$ with ${z_n}^k$ or ${z_n^*}^k$, where k is a POSITIVE INTEGER and z^* denotes the COMPLEX CONJUGATE of z. The following figures show the FRACTALS obtained for $k = 2$, 3, and 4 (Dickau). The plots on the right have z replaced with z^* and are sometimes called "MANDELBAR SETS."

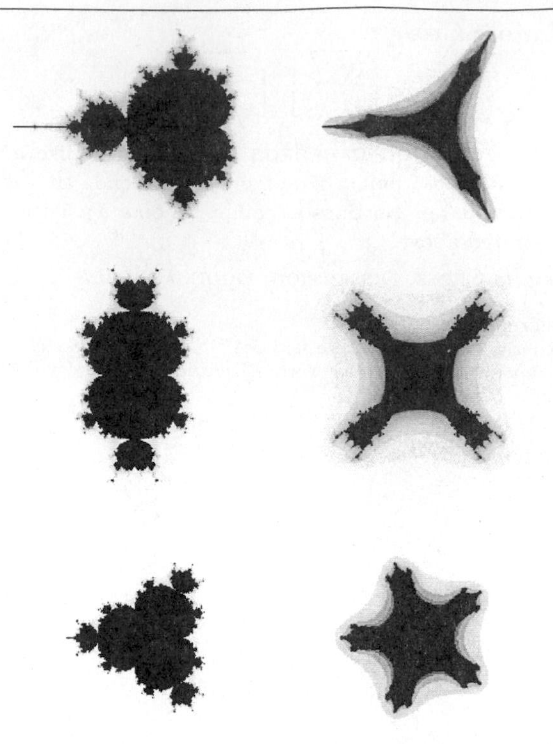

see also CACTUS FRACTAL, FRACTAL, JULIA SET, LEMNISCATE (MANDELBROT SET), MANDELBAR SET, QUADRATIC MAP, RANDELBROT SET, SEA HORSE VALLEY

References

Alfeld, P. "The Mandelbrot Set." http://www.math.utah.edu/~alfeld/math/mandelbrot/mandelbrot1.html.

Branner, B. "The Mandelbrot Set." In *Chaos and Fractals: The Mathematics Behind the Computer Graphics, Proc. Sympos. Appl. Math., Vol. 39* (Ed. R. L. Devaney and L. Keen). Providence, RI: Amer. Math. Soc., 75–105, 1989.

Dickau, R. M. "Mandelbrot (and Similar) Sets." http://forum.swarthmore.edu/advanced/robertd/mandelbrot.html.

Douady, A. "Julia Sets and the Mandelbrot Set." In *The Beauty of Fractals: Images of Complex Dynamical Systems* (Ed. H.-O. Peitgen and D. H. Richter). Berlin: Springer-Verlag, p. 161, 1986.

Eppstein, D. "Area of the Mandelbrot Set." http://www.ics.uci.edu/~eppstein/junkyard/mand-area.html.

Fisher, Y. and Hill, J. "Bounding the Area of the Mandelbrot Set." Submitted.

Hill, J. R. "Fractals and the Grand Internet Parallel Processing Project." Ch. 15 in *Fractal Horizons: The Future Use of Fractals.* New York: St. Martin's Press, pp. 299–323, 1996.

Lauwerier, H. *Fractals: Endlessly Repeated Geometric Figures.* Princeton, NJ: Princeton University Press, pp. 148–151 and 179–180, 1991.

Munafo, R. "Mu-Ency—The Encyclopedia of the Mandelbrot Set." http://home.earthlink.net/~mrob/muency.html.

Peitgen, H.-O. and Saupe, D. (Eds.). *The Science of Fractal Images.* New York: Springer-Verlag, pp. 178–179, 1988.

Shishikura, M. "The Boundary of the Mandelbrot Set has Hausdorff Dimension Two." *Astérisque*, No. 222, **7**, 389–405, 1994.

Mandelbrot Tree

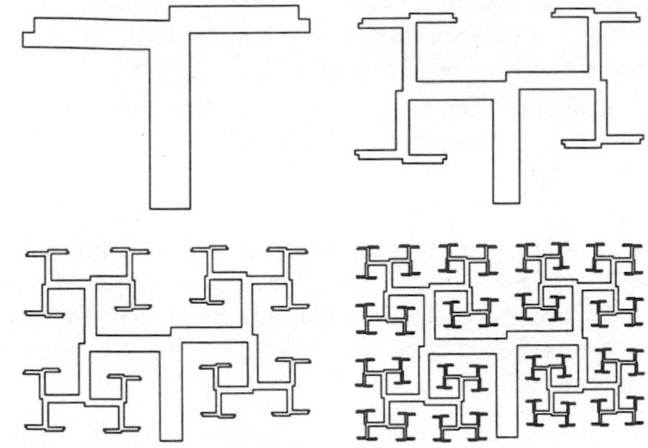

The FRACTAL illustrated above.

References

Lauwerier, H. *Fractals: Endlessly Repeated Geometric Figures.* Princeton, NJ: Princeton University Press, pp. 71–73, 1991.

Weisstein, E. W. "Fractals." http://www.astro.virginia.edu/~eww6n/math/notebooks/Fractal.m.

Mangoldt Function

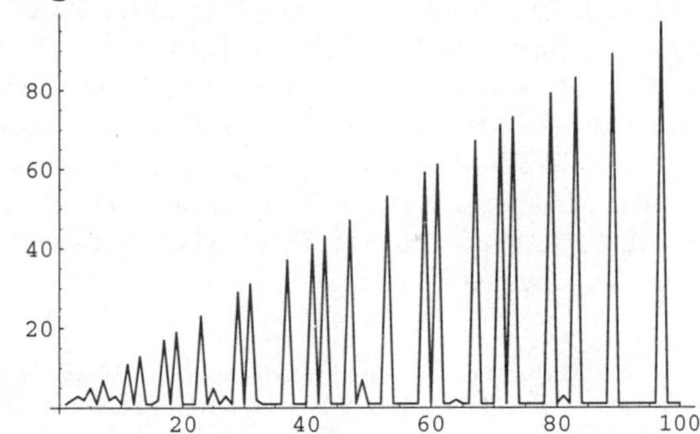

The function defined by

$$\Lambda(n) \equiv \begin{cases} \ln p & \text{if } n = p^k \text{ for } p \text{ a prime} \\ 0 & \text{otherwise.} \end{cases} \tag{1}$$

$\exp(\Lambda(n))$ is also given by $[1, 2, \ldots, n]/[1, 2, \ldots, n-1]$, where $[a, b, c, \ldots]$ denotes the LEAST COMMON MULTIPLE. The first few values of $\exp(\Lambda(n))$ for $n = 1, 2, \ldots$, plotted above, are 1, 2, 3, 2, 5, 1, 7, 2, ... (Sloane's A014963). The Mangoldt function is related to the RIEMANN ZETA FUNCTION $\zeta(z)$ by

$$-\frac{\zeta'(s)}{\zeta(s)} = \sum_{n=1}^{\infty} \frac{\Lambda(n)}{n^s}, \tag{2}$$

where $\Re[s] > 1$.

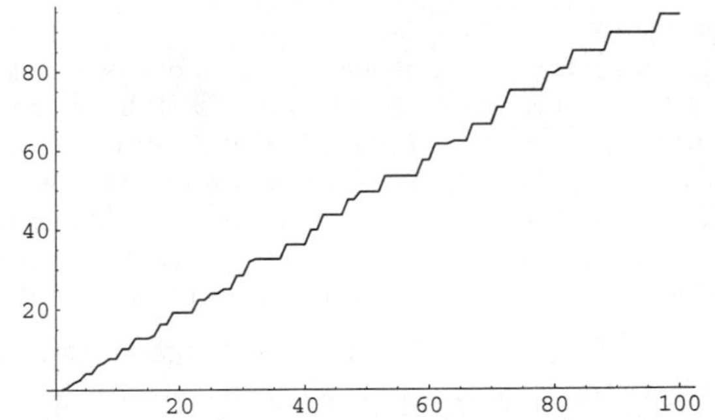

The SUMMATORY Mangoldt function, illustrated above, is defined by

$$\psi(x) \equiv \sum_{n \leq x} \Lambda(n), \tag{3}$$

where $\Lambda(n)$ is the MANGOLDT FUNCTION. This has the explicit formula

$$\psi(x) = x - \sum_{\rho} \frac{x^\rho}{\rho} - \ln(2\pi) - \tfrac{1}{2}\ln(1 - x^2), \tag{4}$$

where the second SUM is over all complex zeros ρ of the RIEMANN ZETA FUNCTION $\zeta(s)$ and interpreted as

$$\lim_{t \to \infty} \sum_{|\Im(\rho)| < t} \frac{x^\rho}{\rho}. \tag{5}$$

Vardi (1991, p. 155) also gives the interesting formula

$$\ln([x]!) = \psi(x) + \psi(\tfrac{1}{2}x) + \psi(\tfrac{1}{3}x) + \ldots, \tag{6}$$

where $[x]$ is the NINT function and $n!$ is a FACTORIAL.

Vallée Poussin's version of the PRIME NUMBER THEOREM states that

$$\psi(x) = x + \mathcal{O}(xe^{-a\sqrt{\ln x}}) \tag{7}$$

for some a (Davenport 1980, Vardi 1991). The RIEMANN HYPOTHESIS is equivalent to

$$\psi(x) = x + \mathcal{O}(\sqrt{x}\,(\ln x)^2) \tag{8}$$

(Davenport 1980, p. 114; Vardi 1991).

see also BOMBIERI'S THEOREM, GREATEST PRIME FACTOR, LAMBDA FUNCTION, LEAST COMMON MULTIPLE, LEAST PRIME FACTOR, RIEMANN FUNCTION

References

Davenport, H. *Multiplicative Number Theory, 2nd ed.* New York: Springer-Verlag, p. 110, 1980.

Sloane, N. J. A. Sequence A014963 in "An On-Line Version of the Encyclopedia of Integer Sequences."

Vardi, I. *Computational Recreations in Mathematica.* Reading, MA: Addison-Wesley, pp. 146–147, 152–153, and 249, 1991.

Manifold

Rigorously, an n-D (topological) manifold is a TOPOLOGICAL SPACE M such that any point in M has a NEIGHBORHOOD $U \subset M$ which is HOMEOMORPHIC to n-D EUCLIDEAN SPACE. The HOMEOMORPHISM is called a chart, since it lays that part of the manifold out flat, like charts of regions of the Earth. So a preferable statement is that any object which can be "charted" is a manifold.

The most important manifolds are DIFFERENTIABLE MANIFOLDS. These are manifolds where overlapping charts "relate smoothly" to each other, meaning that the inverse of one followed by the other is an infinitely differentiable map from EUCLIDEAN SPACE to itself.

Manifolds arise naturally in a variety of mathematical and physical applications as "global objects." For example, in order to precisely describe all the configurations of a robot arm or all the possible positions and momenta of a rocket, an object is needed to store all of these parameters. The objects that crop up are manifolds. From the geometric perspective, manifolds represent the profound idea having to do with global versus local properties.

Consider the ancient belief that the Earth was flat compared to the modern evidence that it is round. The discrepancy arises essentially from the fact that on the small scales that we see, the Earth does look flat. We cannot see it curve because we are too small (although the Greeks did notice that the last part of a ship to disappear over the horizon was the mast). We can detect curvature only indirectly from our vantage point on the Earth. The basic idea for this "problem" was codified by Poincaré. The problem is that on a small scale, the Earth is nearly flat. In general, any object which is nearly "flat" on small scales is a manifold, and so manifolds constitute a generalization of objects we could live on in which we would encounter the round/flat Earth problem.

see also COBORDANT MANIFOLD, COMPACT MANIFOLD, CONNECTED SUM DECOMPOSITION, DIFFERENTIABLE MANIFOLD, FLAG MANIFOLD, GRASSMANN MANIFOLD, HEEGAARD SPLITTING, ISOSPECTRAL MANIFOLDS, JACO-SHALEN-JOHANNSON TORUS DECOMPOSITION, KÄHLER MANIFOLD, POINCARÉ CONJECTURE, POISSON MANIFOLD, PRIME MANIFOLD, RIEMANNIAN MANIFOLD, SET, SMOOTH MANIFOLD, SPACE, STIEFEL MANIFOLD, STRATIFIED MANIFOLD, SUBMANIFOLD, SURGERY, SYMPLECTIC MANIFOLD, THURSTON'S GEOMETRIZATION CONJECTURE, TOPOLOGICAL MANIFOLD, TOPOLOGICAL SPACE, WHITEHEAD MANIFOLD, WIEDERSEHEN MANIFOLD

References

Conlon, L. *Differentiable Manifolds: A First Course.* Boston, MA: Birkhäuser, 1993.

Mantissa

For a REAL NUMBER x, the mantissa is defined as the POSITIVE fractional part $x - \lfloor x \rfloor =$ `frac(x)`, where $\lfloor x \rfloor$ denotes the FLOOR FUNCTION.

see also CHARACTERISTIC (REAL NUMBER), FLOOR FUNCTION, SCIENTIFIC NOTATION

Map

A way of associating unique objects to every point in a given SET. So a map from $A \mapsto B$ is an object f such that for every $a \in A$, there is a unique object $f(a) \in B$. The terms FUNCTION and MAPPING are synonymous with map.

The following table gives several common types of complex maps.

Mapping	Formula	Domain
inversion	$f(z) = \frac{1}{z}$	
magnification	$f(z) = az$	$a \in \mathbb{R} \neq 0$
magnification+rotation	$f(z) = az$	$a \in \mathbb{C} \neq 0$
Möbius	$f(z) = \frac{az+b}{cz+d}$	$a, b, c, d \in \mathbb{C}$
rotation	$f(z) = e^{i\theta z}$	
translation	$f(z) = z + a$	$a \in \mathbb{C}$

see also $2x$ MOD 1 MAP, ARNOLD'S CAT MAP, BAKER'S MAP, BOUNDARY MAP, CONFORMAL MAP, FUNCTION, GAUSS MAP, GINGERBREADMAN MAP, HARMONIC MAP, HÉNON MAP, IDENTITY MAP, INCLUSION MAP, KAPLAN-YORKE MAP, LOGISTIC MAP, MANDELBROT SET, MAP PROJECTION, PULLBACK MAP, QUADRATIC MAP, TANGENT MAP, TENT MAP, TRANSFORMATION, ZASLAVSKII MAP

References

Arfken, G. "Mapping." §6.6 in *Mathematical Methods for Physicists, 3rd ed.* Orlando, FL: Academic Press, pp. 384–392, 1985.

Lee, X. "Transformation of the Plane." http://www.best.com/~xah/MathGraphicsGallery_dir/Transform2DPlot_dir/transform2DPlot.html.

Map Coloring

Given a map with GENUS $g > 0$, Heawood showed in 1890 that the maximum number N_u of colors necessary to color a map (the CHROMATIC NUMBER) on an unbounded surface is

$$N_u \equiv \left\lfloor \tfrac{1}{2}(7 + \sqrt{48g+1}) \right\rfloor = \left\lfloor \tfrac{1}{2}(7 + \sqrt{49 - 24\chi}) \right\rfloor,$$

where $\lfloor x \rfloor$ is the FLOOR FUNCTION, g is the GENUS, and χ is the EULER CHARACTERISTIC. This is the HEAWOOD CONJECTURE. In 1968, for any orientable surface other than the SPHERE (or equivalently, the PLANE) and any nonorientable surface other than the KLEIN BOTTLE, N_u was shown to be not merely a maximum, but the actual number needed (Ringel and Youngs 1968).

When the FOUR-COLOR THEOREM was proven, the Heawood FORMULA was shown to hold also for all orientable and nonorientable surfaces with the exception of the

KLEIN BOTTLE. For this case (which has EULER CHARACTERISTIC 1, and therefore can be considered to have $g = 1/2$), the actual number of colors N needed is six—*one less than* $N_u = 7$ (Franklin 1934; Saaty 1986, p. 45).

surface	g	N_u	N
Klein bottle	1	7	6
Möbius strip	$\frac{1}{2}$	6	6
plane	0	4	4
projective plane	$\frac{1}{2}$	6	6
sphere	0	4	4
torus	1	7	7

see also CHROMATIC NUMBER, FOUR-COLOR THEOREM, HEAWOOD CONJECTURE, SIX-COLOR THEOREM, TORUS COLORING

References

Ball, W. W. R. and Coxeter, H. S. M. *Mathematical Recreations and Essays, 13th ed.* New York: Dover, pp. 237–238, 1987.

Barnette, D. *Map Coloring, Polyhedra, and the Four-Color Problem.* Washington, DC: Math. Assoc. Amer., 1983.

Franklin, P. "A Six Colour Problem." *J. Math. Phys.* **13**, 363–369, 1934.

Franklin, P. *The Four-Color Problem.* New York: Scripta Mathematica, Yeshiva College, 1941.

Ore, Ø. *The Four-Color Problem.* New York: Academic Press, 1967.

Ringel, G. and Youngs, J. W. T. "Solution of the Heawood Map-Coloring Problem." *Proc. Nat. Acad. Sci. USA* **60**, 438–445, 1968.

Saaty, T. L. and Kainen, P. C. *The Four-Color Problem: Assaults and Conquest.* New York: Dover, 1986.

Map Folding

A general FORMULA giving the number of distinct ways of folding an $N = m \times n$ rectangular map is not known. A distinct folding is defined as a permutation of N numbered cells reading from the top down. Lunnon (1971) gives values up to $n = 28$.

n	$1 \times n$	$2 \times n$	$3 \times n$	$4 \times n$	$5 \times n$
1	1	1			
2	2	8			
3	6	60	1368		
4	16	1980		300608	
5	59	19512			18698669
6	144	15552			

The limiting ratio of the number of $1 \times (n+1)$ strips to the number of $1 \times n$ strips is given by

$$\lim_{n\to\infty} \frac{[1 \times (n+1)]}{[1 \times n]} \in [3.3868, 3.9821].$$

see also STAMP FOLDING

References

Gardner, M. "The Combinatorics of Paper Folding." Ch. 7 in *Wheels, Life, and Other Mathematical Amusements.* New York: W. H. Freeman, 1983.

Koehler, J. E. "Folding a Strip of Stamps." *J. Combin. Th.* **5**, 135–152, 1968.

Lunnon, W. F. "A Map-Folding Problem." *Math. Comput.* **22**, 193–199, 1968.

Lunnon, W. F. "Multi-Dimensional Strip Folding." *Computer J.* **14**, 75–79, 1971.

Map Projection

A projection which maps a SPHERE (or SPHEROID) onto a PLANE. No projection can be simultaneously CONFORMAL and AREA-PRESERVING.

see also AIRY PROJECTION, ALBERS EQUAL-AREA CONIC PROJECTION, AXONOMETRY, AZIMUTHAL EQUIDISTANT PROJECTION, AZIMUTHAL PROJECTION, BEHRMANN CYLINDRICAL EQUAL-AREA PROJECTION, BONNE PROJECTION, CASSINI PROJECTION, CHROMATIC NUMBER, CONIC EQUIDISTANT PROJECTION, CONIC PROJECTION, CYLINDRICAL EQUAL-AREA PROJECTION, CYLINDRICAL EQUIDISTANT PROJECTION, CYLINDRICAL PROJECTION, ECKERT IV PROJECTION, ECKERT VI PROJECTION, FOUR-COLOR THEOREM, GNOMIC PROJECTION, GUTHRIE'S PROBLEM, HAMMER-AITOFF EQUAL-AREA PROJECTION, LAMBERT AZIMUTHAL EQUAL-AREA PROJECTION, LAMBERT CONFORMAL CONIC PROJECTION, MAP COLORING, MERCATOR PROJECTION, MILLER CYLINDRICAL PROJECTION, MOLLWEIDE PROJECTION, ORTHOGRAPHIC PROJECTION, POLYCONIC PROJECTION, PSEUDOCYLINDRICAL PROJECTION, RECTANGULAR PROJECTION, SINUSOIDAL PROJECTION, SIX-COLOR THEOREM, STEREOGRAPHIC PROJECTION, VAN DER GRINTEN PROJECTION, VERTICAL PERSPECTIVE PROJECTION

References

Dana, P. H. "Map Projections." `http://www.utexas.edu/depts/grg/gcraft/notes/mapproj/mapproj.html`.

Hunter College Geography. "The Map Projection Home Page." `http://everest.hunter.cuny.edu/mp/`.

Snyder, J. P. *Map Projections—A Working Manual.* U. S. Geological Survey Professional Paper 1395. Washington, DC: U. S. Government Printing Office, 1987.

Mapes' Method

A method for computing the PRIME COUNTING FUNCTION. Define the function

$$T_k(x, a) = (-1)^{\beta_0+\beta_1+\ldots+\beta_{a-1}} \left\lfloor \frac{x}{{p_1}^{\beta_0}{p_2}^{\beta_1}\cdots{p_a}^{\beta_{a-1}}} \right\rfloor, \tag{1}$$

where $\lfloor x \rfloor$ is the FLOOR FUNCTION and the β_i are the binary digits (0 or 1) in

$$k = 2^{a-1}\beta_{a-1} + 2^{a-2}\beta_{a-2} + \ldots + 2^1\beta_1 + 2^0\beta_0. \tag{2}$$

The LEGENDRE SUM can then be written

$$\phi(x, a) = \sum_{k=0}^{2^a-1} T_k(x, a). \tag{3}$$

The first few values of $T_k(x,3)$ are

$$T_0(x,3) = \lfloor x \rfloor \quad (4)$$

$$T_1(x,3) = -\left\lfloor \frac{x}{p_1} \right\rfloor \quad (5)$$

$$T_2(x,3) = -\left\lfloor \frac{x}{p_2} \right\rfloor \quad (6)$$

$$T_3(x,3) = \left\lfloor \frac{x}{p_1 p_2} \right\rfloor \quad (7)$$

$$T_4(x,3) = -\left\lfloor \frac{x}{p_3} \right\rfloor \quad (8)$$

$$T_5(x,3) = \left\lfloor \frac{x}{p_1 p_3} \right\rfloor \quad (9)$$

$$T_6(x,3) = \left\lfloor \frac{x}{p_2 p_3} \right\rfloor \quad (10)$$

$$T_7(x,3) = -\left\lfloor \frac{x}{p_1 p_2 p_3} \right\rfloor. \quad (11)$$

Mapes' method takes time $\sim x^{0.7}$, which is slightly faster than the LEHMER-SCHUR METHOD.

see also LEHMER-SCHUR METHOD, PRIME COUNTING FUNCTION

References

Mapes, D. C. "Fast Method for Computing the Number of Primes Less than a Given Limit." *Math. Comput.* **17**, 179–185, 1963.

Riesel, H. "Mapes' Method." *Prime Numbers and Computer Methods for Factorization, 2nd ed.* Boston, MA: Birkhäuser, p. 23, 1994.

Mapping (Function)

see MAP

Mapping Space

Let Y^X be the set of continuous mappings $f: X \to Y$. Then the TOPOLOGICAL SPACE for Y^X supplied with a compact-open topology is called a mapping space.

see also LOOP SPACE

References

Iyanaga, S. and Kawada, Y. (Eds.). "Mapping Spaces." §204B in *Encyclopedic Dictionary of Mathematics.* Cambridge, MA: MIT Press, p. 658, 1980.

Marginal Analysis

Let $R(x)$ be the revenue for a production x, $C(x)$ the cost, and $P(x)$ the profit. Then

$$P(x) = R(x) - C(x),$$

and the marginal profit for the x_0th unit is defined by

$$P'(x_0) = R'(x_0) - C'(x_0),$$

where $P'(x)$, $R'(x)$, and $C'(x)$ are the DERIVATIVES of $P(x)$, $R(x)$, and $C(x)$, respectively.

see also DERIVATIVE

Marginal Probability

Let S be partitioned into $r \times s$ disjoint sets E_i and F_j where the general subset is denoted $E_i \cap F_j$. Then the marginal probability of E_i is

$$P(E_i) = \sum_{j=1}^{s} P(E_i \cap F_j).$$

Markoff's Formulas

Formulas obtained from differentiating NEWTON'S FORWARD DIFFERENCE FORMULA,

$$f'(a_0 + ph) = \frac{1}{h}[\Delta_0 + \tfrac{1}{2}(2p-1)\Delta_0^2 + \tfrac{1}{6}(3p^2 - 6p + 2)\Delta_0^3 + \ldots + \frac{d}{dp}\binom{p}{n}\Delta_0^n\Big] + R_n',$$

where

$$R_n' = h^n f^{(n+1)}(\xi)\frac{d}{dp}\binom{p}{n+1} + h^{n+1}\binom{p}{n+1}\frac{d}{dx}f^{(n+1)}(\xi),$$

$\binom{n}{k}$ is a BINOMIAL COEFFICIENT, and $a_0 < \xi < a_n$. Abramowitz and Stegun (1972) and Beyer (1987) give derivatives $h^n f_0^{(n)}$ in terms of Δ^k and derivatives in terms of δ^k and ∇^k.

see also FINITE DIFFERENCE

References

Abramowitz, M. and Stegun, C. A. (Eds.). *Handbook of Mathematical Functions with Formulas, Graphs, and Mathematical Tables, 9th printing.* New York: Dover, p. 883, 1972.

Beyer, W. H. *CRC Standard Mathematical Tables, 28th ed.* Boca Raton, FL: CRC Press, pp. 449–450, 1987.

Markoff Number

see MARKOV NUMBER

Markov Algorithm

An ALGORITHM which constructs allowed mathematical statements from simple ingredients.

Markov Chain

A collection of random variables $\{X_t\}$, where the index t runs through 0, 1, ...

References

Kemeny, J. G. and Snell, J. L. *Finite Markov Chains.* New York: Springer-Verlag, 1976.

Stewart, W. J. *Introduction to the Numerical Solution of Markov Chains.* Princeton, NJ: Princeton University Press, 1995.

Markov's Inequality

If x takes only NONNEGATIVE values, then

$$P(x \geq a) \leq \frac{\langle x \rangle}{a}.$$

To prove the theorem, write

$$\langle x \rangle = \int_0^\infty x f(x)\,dx = \int_0^a x f(x)\,dx + \int_a^\infty x f(x)\,dx.$$

Since $P(x)$ is a probability density, it must be ≥ 0. We have stipulated that $x \geq 0$, so

$$\begin{aligned}\langle x \rangle &= \int_0^a x f(x)\,dx + \int_a^\infty x f(x)\,dx \\ &\geq \int_a^\infty x f(x)\,dx \geq \int_a^\infty a f(x)\,dx \\ &= a\int_a^\infty f(x)\,dx = aP(x \geq a),\end{aligned}$$

Q. E. D.

Markov Matrix

see STOCHASTIC MATRIX

Markov Moves

A type I move (CONJUGATION) takes $AB \to BA$ for A, $B \in B_n$ where B_n is a BRAID GROUP.

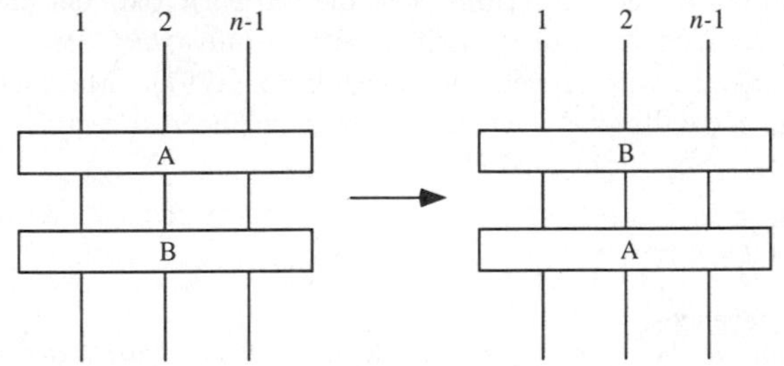

A type II move (STABILIZATION) takes $A \to Ab_n$ or $A \to Ab_n{}^{-1}$ for $A \in B_n$, and b_n, Ab_n, and $Ab_n{}^{-1} \in B_{n+1}$.

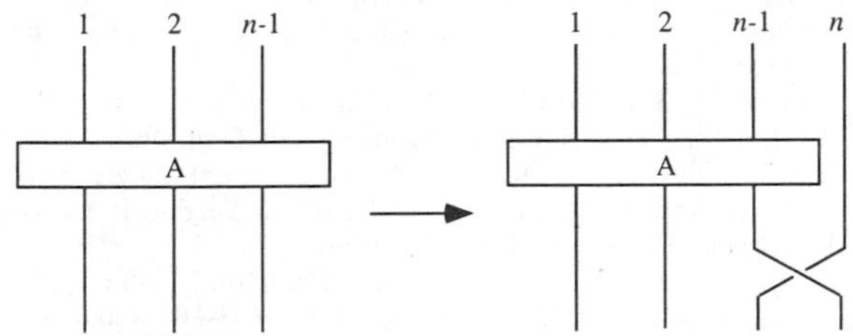

see also BRAID GROUP, CONJUGATION, REIDEMEISTER MOVES, STABILIZATION

Markov Number

The Markov numbers m occur in solutions to the DIOPHANTINE EQUATION

$$x^2 + y^2 + z^2 = 3xyz,$$

and are related to LAGRANGE NUMBERS L_n by

$$L_n = \sqrt{9 - \frac{4}{m^2}}.$$

The first few solutions are $(x, y, z) = (1, 1, 1)$, (1, 1, 2), (1, 2, 5), (1, 5, 13), (2, 5, 29), ... The solutions can be arranged in an infinite tree with two smaller branches on each trunk. It is not known if two different regions can have the same label. Strangely, the regions adjacent to 1 have alternate FIBONACCI NUMBERS 1, 2, 5, 13, 34, ..., and the regions adjacent to 2 have alternate PELL NUMBERS 1, 5, 29, 169, 985, ...

Let $M(N)$ be the number of TRIPLES with $x \leq y \leq z \leq N$, then

$$M(n) = C(\ln N) + \mathcal{O}((\ln N)^{1+\epsilon}),$$

where $C \approx 0.180717105$ (Guy 1994, p. 166).

see also HURWITZ EQUATION, HURWITZ'S IRRATIONAL NUMBER THEOREM, LAGRANGE NUMBER (RATIONAL APPROXIMATION) LIOUVILLE'S RATIONAL APPROXIMATION THEOREM, LIOUVILLE-ROTH CONSTANT, ROTH'S THEOREM, SEGRE'S THEOREM, THUE-SIEGEL-ROTH THEOREM

References

Conway, J. H. and Guy, R. K. *The Book of Numbers.* New York: Springer-Verlag, pp. 187–189, 1996.

Guy, R. K. "Markoff Numbers." §D12 in *Unsolved Problems in Number Theory, 2nd ed.* New York: Springer-Verlag, pp. 166–168, 1994.

Markov Process

A random process whose future probabilities are determined by its most recent values.

see also DOOB'S THEOREM

Markov Spectrum

A SPECTRUM containing the REAL NUMBERS larger than FREIMAN'S CONSTANT.

see also FREIMAN'S CONSTANT, SPECTRUM SEQUENCE

References

Conway, J. H. and Guy, R. K. *The Book of Numbers.* New York: Springer-Verlag, pp. 188–189, 1996.

Markov's Theorem

Published by A. A. Markov in 1935, Markov's theorem states that equivalent BRAIDS expressing the same LINK are mutually related by successive applications of two types of MARKOV MOVES. Markov's theorem is difficult to apply in practice, so it is difficult to establish the equivalence or nonequivalence of LINKS having different BRAID representations.

see also BRAID, LINK, MARKOV MOVES

Marriage Theorem

If a group of men and women may date only if they have previously been introduced, then a complete set of dates is possible IFF every subset of men has collectively been introduced to at least as many women, and vice versa.

References
Chartrand, G. *Introductory Graph Theory.* New York: Dover, p. 121, 1985.

Married Couples Problem

Also called the MÉNAGE PROBLEM. In how many ways can n married couples be seated around a circular table in such a manner than there is always one man between two women and none of the men is next to his own wife? The solution (Ball and Coxeter 1987, p. 50) uses DISCORDANT PERMUTATIONS and can be given in terms of LAISANT'S RECURRENCE FORMULA

$$(n-1)A_{n+1} = (n^2-1)A_n + (n+1)A_{n-1} + 4(-1)^n,$$

with $A_1 = A_2 = 1$. A closed form expression due to Touchard (1934) is

$$A_n = \sum_{k=0}^{n} \frac{2n}{2n-k}\binom{2n-k}{k}(n-k)!(-1)^k,$$

where $\binom{n}{k}$ is a BINOMIAL COEFFICIENT (Vardi 1991).

The first few values of A_n are -1, 1, 0, 2, 13, 80, 579, ... (Sloane's A000179), which are sometimes called MÉNAGE NUMBERS. The desired solution is then $2n!A_n$ The numbers A_n can be considered a special case of a restricted ROOKS PROBLEM.

see also DISCORDANT PERMUTATION, LAISANT'S RECURRENCE FORMULA, ROOKS PROBLEM

References
Ball, W. W. R. and Coxeter, H. S. M. *Mathematical Recreations and Essays, 13th ed.* New York: Dover, p. 50, 1987.
Dörrie, H. §8 in *100 Great Problems of Elementary Mathematics: Their History and Solutions.* New York: Dover, pp. 27–33, 1965.
Halmos, P. R.; Vaughan, H. E. "The Marriage Problem." *Amer. J. Math.* **72**, 214–215, 1950.
Lucas, E. *Théorie des Nombres.* Paris, pp. 215 and 491–495, 1891.
MacMahon, P. A. *Combinatory Analysis, Vol. 1.* London: Cambridge University Press, pp. 253–256, 1915.
Newman, D. J. "A Problem in Graph Theory." *Amer. Math. Monthly* **65**, 611, 1958.
Sloane, N. J. A. Sequence A000179/M2062 in "An On-Line Version of the Encyclopedia of Integer Sequences."
Touchard, J. "Sur un probléme de permutations." *C. R. Acad. Sci. Paris* **198**, 631–633, 1934.
Vardi, I. *Computational Recreations in Mathematica.* Reading, MA: Addison-Wesley, p. 123, 1991.

Marshall-Edgeworth Index

The statistical INDEX

$$P_{ME} \equiv \frac{\sum p_n(q_0+q_n)}{\sum(v_0+v_n)},$$

where p_n is the price per unit in period n, q_n is the quantity produced in period n, and $v_n \equiv p_n q_n$ is the value of the n units.

see also INDEX

References
Kenney, J. F. and Keeping, E. S. *Mathematics of Statistics, Pt. 1, 3rd ed.* Princeton, NJ: Van Nostrand, pp. 66–67, 1962.

Martingale

A sequence of random variates such that the CONDITIONAL PROBABILITY of x_{n+1} given x_1, x_2, ..., x_n is x_n. The term was first used to describe a type of wagering in which the bet is doubled or halved after a loss or win, respectively.

see also GAMBLER'S RUIN, SAINT PETERSBURG PARADOX

Mascheroni Constant

see EULER-MASCHERONI CONSTANT

Mascheroni Construction

A geometric construction done with a movable COMPASS alone. All constructions possible with a COMPASS and STRAIGHTEDGE are possible with a *movable* COMPASS alone, as was proved by Mascheroni (1797). Mascheroni's results are now known to have been anticipated largely by Mohr (1672).

see also COMPASS, GEOMETRIC CONSTRUCTION, NEUSIS CONSTRUCTION, STRAIGHTEDGE

References
Ball, W. W. R. and Coxeter, H. S. M. *Mathematical Recreations and Essays, 13th ed.* New York: Dover, pp. 96–97, 1987.
Bogomolny, A. "Geometric Constructions with the Compass Alone." `http://www.cut-the-knot.com/do_you_know/compass.html`.
Courant, R. and Robbins, H. "Constructions with Other Tools. Mascheroni Constructions with Compass Alone." §3.5 in *What is Mathematics?: An Elementary Approach to Ideas and Methods, 2nd ed.* Oxford, England: Oxford University Press, pp. 146–158, 1996.
Dörrie, H. "Mascheroni's Compass Problem." §33 in *100 Great Problems of Elementary Mathematics: Their History and Solutions.* New York: Dover, pp. 160–164, 1965.
Gardner, M. "Mascheroni Constructions." Ch. 17 in *Mathematical Circus: More Puzzles, Games, Paradoxes and Other Mathematical Entertainments from Scientific American.* New York: Knopf, pp. 216–231, 1979.
Mascheroni, L. *Geometry of Compass.* Pavia, Italy, 1797.
Mohr, G. *Euclides Danicus.* Amsterdam, Netherlands, 1672.

Maschke's Theorem

If a MATRIX GROUP is reducible, then it is completely reducible, i.e., if the MATRIX GROUP is equivalent to the MATRIX GROUP in which every MATRIX has the reduced form

$$\begin{bmatrix} D_i^{(1)} & X_i \\ 0 & D_i^{(2)} \end{bmatrix},$$

then it is equivalent to the MATRIX GROUP obtained by putting $X_i = 0$.

see also MATRIX GROUP

References

Lomont, J. S. *Applications of Finite Groups.* New York: Dover, p. 49, 1987.

Mason's abc Theorem

see MASON'S THEOREM

Mason's Theorem

Let there be three POLYNOMIALS $a(x)$, $b(x)$, and $c(x)$ with no common factors such that

$$a(x) + b(x) = c(x).$$

Then the number of distinct ROOTS of the three POLYNOMIALS is one or more greater than their largest degree. The theorem was first proved by Stothers (1981).

Mason's theorem may be viewed as a very special case of a Wronskian estimate (Chudnovsky and Chudnovsky 1984). The corresponding Wronskian identity in the proof by Lang (1993) is

$$c^3 * W(a, b, c) = W(W(a, c), W(b, c)),$$

so if a, b, and c are linearly dependent, then so are $W(a, c)$ and $W(b, c)$. More powerful Wronskian estimates with applications toward diophantine approximation of solutions of linear differential equations may be found in Chudnovsky and Chudnovsky (1984) and Osgood (1985).

The rational function case of FERMAT'S LAST THEOREM follows trivially from Mason's theorem (Lang 1993, p. 195).

see also ABC CONJECTURE

References

Chudnovsky, D. V. and Chudnovsky, G. V. "The Wronskian Formalism for Linear Differential Equations and Padé Approximations." *Adv. Math.* **53**, 28–54, 1984.

Lang, S. "Old and New Conjectured Diophantine Inequalities." *Bull. Amer. Math. Soc.* **23**, 37–75, 1990.

Lang, S. *Algebra, 3rd ed.* Reading, MA: Addison-Wesley, 1993.

Osgood, C. F. "Sometimes Effective Thue-Siegel-Roth-Schmidt-Nevanlinna Bounds, or Better." *J. Number Th.* **21**, 347–389, 1985.

Stothers, W. W. "Polynomial Identities and Hauptmodulen." *Quart. J. Math. Oxford Ser. II* **32**, 349–370, 1981.

Masser-Gramain Constant

N.B. A detailed on-line essay by S. Finch was the starting point for this entry.

If $f(z)$ is an ENTIRE FUNCTION such that $f(n)$ is an INTEGER for each POSITIVE INTEGER n. Pólya (1915) showed that if

$$\limsup_{r\to\infty} \frac{\ln M_r}{r} < \ln 2 = 0.693\ldots, \qquad (1)$$

where

$$M_r = \sup_{|z|\leq r} |f(x)| \qquad (2)$$

is the SUPREMUM, then f is a POLYNOMIAL. Furthermore, $\ln 2$ is the best constant (i.e., counterexamples exist for every smaller value).

If $f(z)$ is an ENTIRE FUNCTION with $f(n)$ a GAUSSIAN INTEGER for each GAUSSIAN INTEGER n, then Gelfond (1929) proved that there exists a constant α such that

$$\limsup_{r\to\infty} \frac{\ln M_r}{r^2} < \alpha \qquad (3)$$

implies that f is a POLYNOMIAL. Gramain (1981, 1982) showed that the best such constant is

$$\alpha = \frac{\pi}{2e} = 0.578\ldots. \qquad (4)$$

Maser (1980) proved the weaker result that f must be a POLYNOMIAL if

$$\limsup_{r\to\infty} \frac{\ln M_r}{r^2} < \alpha_0 = \tfrac{1}{2}\exp\left(-\delta + \frac{4c}{\pi}\right), \qquad (5)$$

where

$$c = \gamma\beta(1) + \beta'(1) = 0.6462454398948114\ldots, \qquad (6)$$

γ is the EULER-MASCHERONI CONSTANT, $\beta(z)$ is the DIRICHLET BETA FUNCTION,

$$\delta \equiv \lim_{n\to\infty} \left(\sum_{k=2}^{\infty} \frac{1}{\pi {r_k}^2} - \ln n\right), \qquad (7)$$

and r_k is the minimum NONNEGATIVE r for which there exists a COMPLEX NUMBER z for which the CLOSED DISK with center z and radius r contains at least k distinct GAUSSIAN INTEGER. Gosper gave

$$c = \pi\{-\ln[\Gamma(\tfrac{1}{4})] + \tfrac{3}{4}\pi + \tfrac{1}{2}\ln 2 + \tfrac{1}{2}\gamma\}. \qquad (8)$$

Gramain and Weber (1985, 1987) have obtained

$$1.811447299 < \delta < 1.897327117, \qquad (9)$$

which implies

$$0.1707339 < \alpha_0 < 0.1860446. \qquad (10)$$

Gramain (1981, 1982) conjectured that

$$\alpha_0 = \frac{1}{2e}, \qquad (11)$$

which would imply

$$\delta = 1 + \frac{4c}{\pi} = 1.822825249\ldots. \qquad (12)$$

References

Finch, S. "Favorite Mathematical Constants." `http://www.mathsoft.com/asolve/constant/masser/masser.html`.

Gramain, F. "Sur le théorème de Fukagawa-Gel'fond." *Invent. Math.* **63**, 495–506, 1981.

Gramain, F. "Sur le théorème de Fukagawa-Gel'fond-Gruman-Masser." *Séminaire Delange-Pisot-Poitou (Théorie des Nombres), 1980–1981.* Boston, MA: Birkhäuser, 1982.

Gramain, F. and Weber, M. "Computing and Arithmetic Constant Related to the Ring of Gaussian Integers." *Math. Comput.* **44**, 241–245, 1985.

Gramain, F. and Weber, M. "Computing and Arithmetic Constant Related to the Ring of Gaussian Integers." *Math. Comput.* **48**, 854, 1987.

Masser, D. W. "Sur les fonctions entières à valeurs entières." *C. R. Acad. Sci. Paris Sér. A-B* **291**, A1–A4, 1980.

Match Problem

Given n matches, find the number of topologically distinct planar arrangements $T(n)$ which can be made. The first few values are 1, 1, 3, 5, 10, 19, 39, ... (Sloane's A003055).

see also CIGARETTES, MATCHSTICK GRAPH

References

Gardner, M. "The Problem of the Six Matches." In *The Unexpected Hanging and Other Mathematical Diversions.* Chicago, IL: Chicago University Press, pp. 79–81, 1991.

Sloane, N. J. A. Sequence A003055/M2464 in "An On-Line Version of the Encyclopedia of Integer Sequences."

Matchstick Graph

A PLANAR GRAPH whose EDGES are all unit line segments. The minimal number of EDGES for matchstick graphs of various degrees are given in the table below. The minimal degree 1 matchstick graph is a single EDGE, and the minimal degree 2 graph is an EQUILATERAL TRIANGLE.

n	e	v
1	1	2
2	3	3
3	12	8
4		≤ 42

Mathematical Induction

see INDUCTION

Mathematics

Mathematics is a broad-ranging field of study in which the properties and interactions of idealized objects are examined. Whereas mathematics began merely as a calculational tool for computation and tabulation of quantities, it has blossomed into an extremely rich and diverse set of tools, terminologies, and approaches which range from the purely abstract to the utilitarian.

Bertrand Russell once whimsically defined mathematics as, "The subject in which we never know what we are talking about nor whether what we are saying is true" (Bergamini 1969).

The term "mathematics" is often shortened to "math" in informal American speech and, consistent with the British penchant for adding superfluous letters, "maths" in British English.

see also METAMATHEMATICS

References

Bergamini, D. *Mathematics.* New York: Time-Life Books, p. 9, 1969.

Mathematics Prizes

Several prizes are awarded periodically for outstanding mathematical achievement. There is no Nobel Prize in mathematics, and the most prestigious mathematical award is known as the FIELDS MEDAL. In rough order of importance, other awards are the $100,000 Wolf Prize of the Wolf Foundation of Israel, the Leroy P. Steele Prize of the American Mathematical Society, followed by the Bôcher Memorial Prize, Frank Nelson Cole Prizes in Algebra and Number Theory, and the Delbert Ray Fulkerson Prize, all presented by the American Mathematical Society.

see also FIELDS MEDAL

References

"AMS Funds and Prizes." `http://www.ams.org/ams/prizes.html`.

MacTutor History of Mathematics Archives. "The Fields Medal." `http://www-groups.dcs.st-and.ac.uk/~history/Societies/FieldsMedal.html`. "Winners of the Bôcher Prize of the AMS." `http://www-groups.dcs.st-and.ac.uk/~history/Societies/AMSBocherPrize.html`. "Winners of the Frank Nelson Cole Prize of the AMS." `http://www-groups.dcs.st-and.ac.uk/~history/Societies/AMSColePrize.html`.

MacTutor History of Mathematics Archives. "Mathematical Societies, Medals, Prizes, and Other Honours." `http://www-groups.dcs.st-and.ac.uk/~history/Societies/`.

Monastyrsky, M. *Modern Mathematics in the Light of the Fields Medals.* Wellesley, MA: A. K. Peters, 1997.

"Wolf Prize Recipients in Mathematics." `http://www.aquanet.co.il/wolf/wolf5.html`.

Mathieu Differential Equation

$$\frac{d^2V}{dv^2} + [b - 2q\cos(2v)]V = 0.$$

It arises in separation of variables of LAPLACE'S EQUATION in ELLIPTIC CYLINDRICAL COORDINATES. Whittaker and Watson (1990) use a slightly different form to define the MATHIEU FUNCTIONS.

The modified Mathieu differential equation

$$\frac{d^2U}{du^2} - [b - 2q\cosh(2u)]U = 0$$

arises in SEPARATION OF VARIABLES of the HELMHOLTZ DIFFERENTIAL EQUATION in ELLIPTIC CYLINDRICAL COORDINATES.

see also MATHIEU FUNCTION

References

Abramowitz, M. and Stegun, C. A. (Eds.). *Handbook of Mathematical Functions with Formulas, Graphs, and Mathematical Tables, 9th printing.* New York: Dover, p. 722, 1972.

Morse, P. M. and Feshbach, H. *Methods of Theoretical Physics, Part I.* New York: McGraw-Hill, pp. 556–557, 1953.

Whittaker, E. T. and Watson, G. N. *A Course in Modern Analysis, 4th ed.* Cambridge, England: Cambridge University Press, 1990.

Mathieu Function

The form given by Whittaker and Watson (1990, p. 405) defines the Mathieu function based on the equation

$$\frac{d^2u}{dz^2} + [a + 16q\cos(2z)]u = 0. \tag{1}$$

This equation is closely related to HILL'S DIFFERENTIAL EQUATION. For an EVEN Mathieu function,

$$G(\eta) = \lambda \int_{-\pi}^{\pi} e^{k\cos\eta\cos\theta} G(\theta)\,d\theta, \tag{2}$$

where $k \equiv \sqrt{32q}$. For an ODD Mathieu function,

$$G(\eta) = \lambda \int_{-\pi}^{\pi} \sin(k\sin\eta\sin\theta)G(\theta)\,d\theta. \tag{3}$$

Both EVEN and ODD functions satisfy

$$G(\eta) = \lambda \int_{-\pi}^{\pi} e^{ik\sin\eta\sin\theta} G(\theta)\,d\theta. \tag{4}$$

Letting $\zeta \equiv \cos^2 z$ transforms the MATHIEU DIFFERENTIAL EQUATION to

$$4\zeta(1-\zeta)\frac{d^2u}{d\zeta^2} + 2(1-2\zeta)\frac{du}{d\zeta} + (a - 16q + 32q\zeta)u = 0. \tag{5}$$

see also MATHIEU DIFFERENTIAL EQUATION

References

Abramowitz, M. and Stegun, C. A. (Eds.). "Mathieu Functions." Ch. 20 in *Handbook of Mathematical Functions with Formulas, Graphs, and Mathematical Tables, 9th printing.* New York: Dover, pp. 721–746, 1972.

Morse, P. M. and Feshbach, H. *Methods of Theoretical Physics, Part I.* New York: McGraw-Hill, pp. 562–568 and 633–642, 1953.

Whittaker, E. T. and Watson, G. N. *A Course in Modern Analysis, 4th ed.* Cambridge, England: Cambridge University Press, 1990.

Mathieu Groups

The first SIMPLE SPORADIC GROUPS discovered. M_{11}, M_{12}, M_{22}, M_{23}, M_{24} were discovered in 1861 and 1873 by Mathieu. Frobenius showed that all the Mathieu groups are SUBGROUPS of M_{24}.

The Mathieu groups are most simply defined as AUTOMORPHISM groups of STEINER SYSTEMS. M_{11} corresponds to $S(4,5,11)$ and M_{23} corresponds to $S(4,7,23)$. M_{11} and M_{23} are TRANSITIVE PERMUTATION GROUPS of 11 and 23 elements.

The ORDERS of the Mathieu groups are

$$\begin{aligned} |M_{11}| &= 2^4 \cdot 3^2 \cdot 5 \cdot 11 \\ |M_{12}| &= 2^6 \cdot 3^3 \cdot 5 \cdot 11 \\ |M_{22}| &= 2^7 \cdot 3^2 \cdot 5 \cdot 7 \cdot 11 \\ |M_{23}| &= 2^7 \cdot 3^2 \cdot 5 \cdot 7 \cdot 11 \cdot 23. \end{aligned}$$

see also SPORADIC GROUP

References

Conway, J. H. and Sloane, N. J. A. "The Golay Codes and the Mathieu Groups." Ch. 11 in *Sphere Packings, Lattices, and Groups, 2nd ed.* New York: Springer-Verlag, pp. 299–330, 1993.

Rotman, J. J. Ch. 9 in *An Introduction to the Theory of Groups, 4th ed.* New York: Springer-Verlag, 1995.

Wilson, R. A. "ATLAS of Finite Group Representation." `http://for.mat.bham.ac.uk/atlas/M11.html`, `M12.html`, `M22.html`, `M23.html`, and `M24.html`.

Matrix

The TRANSFORMATION given by the system of equations

$$\begin{aligned} x_1' &= a_{11}x_1 + a_{12}x_2 + \ldots + a_{1n}x_n \\ x_2' &= a_{21}x_1 + a_{22}x_2 + \ldots + a_{2n}x_n \\ &\vdots \\ x_m' &= a_{m1}x_1 + a_{m2}x_2 + \ldots + a_{mn}x_n \end{aligned}$$

is denoted by the MATRIX EQUATION

$$\begin{bmatrix} x_1' \\ x_2' \\ \vdots \\ x_n' \end{bmatrix} = \begin{bmatrix} a_{11} & a_{12} & \cdots & a_{1n} \\ a_{21} & a_{22} & \cdots & a_{2n} \\ \vdots & \vdots & \ddots & \vdots \\ a_{m1} & a_{m2} & \cdots & a_{mn} \end{bmatrix} \begin{bmatrix} x_1 \\ x_2 \\ \vdots \\ x_n \end{bmatrix}.$$

In concise notation, this could be written

$$\mathbf{x}' = \mathsf{A}\mathbf{x},$$

where $\mathbf{x}'$ and $\mathbf{x}$ are VECTORS and A is called an $n \times m$ matrix. A matrix is said to be SQUARE if $m = n$. Special types of SQUARE MATRICES include the IDENTITY MATRIX I, with $a_{ij} = \delta_{ij}$ (where δ_{ij} is the KRONECKER DELTA) and the DIAGONAL MATRIX $a_{ij} = c_i\delta_{ij}$ (where c_i are a set of constants).

For every linear transformation there exists one and only one corresponding matrix. Conversely, every matrix corresponds to a unique linear transformation. The matrix is an important concept in mathematics, and was first formulated by Sylvester and Cayley.

Two matrices may be added (MATRIX ADDITION) or multiplied (MATRIX MULTIPLICATION) together to yield a new matrix. Other common operations on a single matrix are diagonalization, inversion (MATRIX INVERSE), and transposition (MATRIX TRANSPOSE). The DETERMINANT $\det(\mathsf{A})$ or $|\mathsf{A}|$ of a matrix A is an very important quantity which appears in many diverse applications. Matrices provide a concise notation which is extremely useful in a wide range of problems involving linear equations (e.g., LEAST SQUARES FITTING).

see also ADJACENCY MATRIX, ADJUGATE MATRIX, ANTISYMMETRIC MATRIX, BLOCK MATRIX, CARTAN MATRIX, CIRCULANT MATRIX, CONDITION NUMBER, CRAMER'S RULE, DETERMINANT, DIAGONAL MATRIX, DIRAC MATRICES, EIGENVECTOR, ELEMENTARY MATRIX, EQUIVALENT MATRIX, FOURIER MATRIX, GRAM MATRIX, HILBERT MATRIX, HYPERMATRIX, IDENTITY MATRIX, INCIDENCE MATRIX, IRREDUCIBLE MATRIX, KAC MATRIX, LU DECOMPOSITION, MARKOV MATRIX, MATRIX ADDITION, MATRIX DECOMPOSITION THEOREM, MATRIX INVERSE, MATRIX MULTIPLICATION, MCCOY'S THEOREM, MINIMAL MATRIX, NORMAL MATRIX, PAULI MATRICES, PERMUTATION MATRIX, POSITIVE DEFINITE MATRIX, RANDOM MATRIX, RATIONAL CANONICAL FORM, REDUCIBLE MATRIX, ROTH'S REMOVAL RULE, SHEAR MATRIX, SKEW SYMMETRIC MATRIX, SMITH NORMAL FORM, SPARSE MATRIX, SPECIAL MATRIX, SQUARE MATRIX, STOCHASTIC MATRIX, SUBMATRIX, SYMMETRIC MATRIX, TOURNAMENT MATRIX

References

Arfken, G. "Matrices." §4.2 in *Mathematical Methods for Physicists, 3rd ed.* Orlando, FL: Academic Press, pp. 176–191, 1985.

Matrix Addition

Denote the sum of two MATRICES A and B (of the same dimensions) by $\mathsf{C} = \mathsf{A}+\mathsf{B}$. The sum is defined by adding entries with the same indices

$$c_{ij} \equiv a_{ij} + b_{ij}$$

over all i and j. For example,

$$\begin{bmatrix} a_{11} & a_{12} \\ a_{21} & a_{22} \end{bmatrix} + \begin{bmatrix} b_{11} & b_{12} \\ b_{21} & b_{22} \end{bmatrix} = \begin{bmatrix} a_{11}+b_{11} & a_{12}+b_{12} \\ a_{21}+b_{21} & a_{22}+b_{22} \end{bmatrix}.$$

Matrix addition is therefore both COMMUTATIVE and ASSOCIATIVE.

see also MATRIX, MATRIX MULTIPLICATION

Matrix Decomposition Theorem

Let P be a MATRIX of EIGENVECTORS of a given MATRIX A and D a MATRIX of the corresponding EIGENVALUES. Then A can be written

$$\mathsf{A} = \mathsf{PDP}^{-1}, \tag{1}$$

where D is a DIAGONAL MATRIX and the columns of P are ORTHOGONAL VECTORS. If P is not a SQUARE MATRIX, then it cannot have a MATRIX INVERSE. However, if P is $m \times n$ (with $m > n$), then A can be written

$$\mathsf{A} = \mathsf{UDV}^{\mathrm{T}}, \tag{2}$$

where U and V are $n \times n$ SQUARE MATRICES with ORTHOGONAL columns,

$$\mathsf{U}^{\mathrm{T}}\mathsf{U} = \mathsf{V}^{\mathrm{T}} = \mathsf{I}. \tag{3}$$

Matrix Diagonalization

Diagonalizing a MATRIX is equivalent to finding the EIGENVECTORS and EIGENVALUES. The EIGENVALUES make up the entries of the diagonalized MATRIX, and the EIGENVECTORS make up the new set of axes corresponding to the DIAGONAL MATRIX.

see also DIAGONAL MATRIX, EIGENVALUE, EIGENVECTOR

References

Arfken, G. "Diagonalization of Matrices." §4.6 in *Mathematical Methods for Physicists, 3rd ed.* Orlando, FL: Academic Press, pp. 217–229, 1985.

Matrix Direct Product

see DIRECT PRODUCT (MATRIX)

Matrix Equality

Two MATRICES A and B are said to be equal IFF

$$a_{ij} \equiv b_{ij}$$

for all i, j. Therefore,

$$\begin{bmatrix} 1 & 2 \\ 3 & 4 \end{bmatrix} = \begin{bmatrix} 1 & 2 \\ 3 & 4 \end{bmatrix},$$

while

$$\begin{bmatrix} 1 & 2 \\ 3 & 4 \end{bmatrix} \neq \begin{bmatrix} 0 & 2 \\ 3 & 4 \end{bmatrix}.$$

Matrix Equation

Nonhomogeneous matrix equations of the form

$$\mathsf{A}\mathbf{x} = \mathbf{b} \tag{1}$$

can be solved by taking the MATRIX INVERSE to obtain

$$\mathbf{x} = \mathsf{A}^{-1}\mathbf{b}. \tag{2}$$

This equation will have a nontrivial solution IFF the DETERMINANT $\det(\mathsf{A}) \neq 0$. In general, more numerically stable techniques of solving the equation include GAUSSIAN ELIMINATION, LU DECOMPOSITION, or the SQUARE ROOT METHOD.

For a homogeneous $n \times n$ MATRIX equation

$$\begin{bmatrix} a_{11} & a_{12} & \cdots & a_{1n} \\ a_{21} & a_{22} & \cdots & a_{2n} \\ \vdots & \vdots & \ddots & \vdots \\ a_{n1} & a_{n2} & \cdots & a_{nn} \end{bmatrix} \begin{bmatrix} x_1 \\ x_2 \\ \vdots \\ x_n \end{bmatrix} = \begin{bmatrix} 0 \\ 0 \\ \vdots \\ 0 \end{bmatrix} \tag{3}$$

to be solved for the x_is, consider the DETERMINANT

$$\begin{vmatrix} a_{11} & a_{12} & \cdots & a_{1n} \\ a_{21} & a_{22} & \cdots & a_{2n} \\ \vdots & \vdots & \ddots & \vdots \\ a_{n1} & a_{n2} & \cdots & a_{nn} \end{vmatrix}. \tag{4}$$

Now multiply by x_1, which is equivalent to multiplying the first row (or any row) by x_1,

$$x_1 \begin{vmatrix} a_{11} & a_{12} & \cdots & a_{1n} \\ a_{21} & a_{22} & \cdots & a_{2n} \\ \vdots & \vdots & \ddots & \vdots \\ a_{n1} & a_{n2} & \cdots & a_{nn} \end{vmatrix} = \begin{vmatrix} a_{11}x_1 & a_{12} & \cdots & a_{1n} \\ a_{21}x_1 & a_{22} & \cdots & a_{2n} \\ \vdots & \vdots & \ddots & \vdots \\ a_{n1}x_1 & a_{n2} & \cdots & a_{nn} \end{vmatrix}. \tag{5}$$

The value of the DETERMINANT is unchanged if multiples of columns are added to other columns. So add x_2 times column 2, ..., and x_n times column n to the first column to obtain

$$x_1 \begin{vmatrix} a_{11} & a_{12} & \cdots & a_{1n} \\ a_{21} & a_{22} & \cdots & a_{2n} \\ \vdots & \vdots & \ddots & \vdots \\ a_{n1} & a_{n2} & \cdots & a_{nn} \end{vmatrix} = \begin{vmatrix} a_{11}x_1 + a_{12}x_2 + \ldots + a_{1n}x_n & a_{12} & \cdots & a_{1n} \\ a_{21}x_1 + a_{22}x_2 + \ldots + a_{2n}x_n & a_{22} & \cdots & a_{2n} \\ \vdots & \vdots & \ddots & \vdots \\ a_{n1}x_1 + a_{n2}x_2 + \ldots + a_{nn}x_n & a_{n2} & \cdots & a_{nn} \end{vmatrix}. \tag{6}$$

But from the original MATRIX, each of the entries in the first columns is zero since

$$a_{i1}x_1 + a_{i2}x_2 + \ldots + a_{in}x_n = 0, \tag{7}$$

so

$$\begin{vmatrix} 0 & a_{12} & \cdots & a_{1n} \\ 0 & a_{22} & \cdots & a_{2n} \\ \vdots & \vdots & \ddots & \vdots \\ 0 & a_{n2} & \cdots & a_{nn} \end{vmatrix} = 0. \tag{8}$$

Therefore, if there is an $x_1 \neq 0$ which is a solution, the DETERMINANT is zero. This is also true for x_2, ..., x_n, so the original homogeneous system has a nontrivial solution for all x_is only if the DETERMINANT is 0. This approach is the basis for CRAMER'S RULE.

Given a numerical solution to a matrix equation, the solution can be iteratively improved using the following technique. Assume that the numerically obtained solution to

$$\mathsf{A}\mathbf{x} = \mathbf{b} \tag{9}$$

is $\mathbf{x}_1 = \mathbf{x} + \delta\mathbf{x}_1$, where $\delta\mathbf{x}_1$ is an error term. The first solution therefore gives

$$\mathsf{A}(\mathbf{x} + \delta\mathbf{x}_1) = \mathbf{b} + \delta\mathbf{b} \tag{10}$$

$$\mathsf{A}\delta\mathbf{x}_1 = \delta\mathbf{b}, \tag{11}$$

where $\delta\mathbf{b}$ is found by solving (10)

$$\delta\mathbf{b} = \mathsf{A}\mathbf{x}_1 - \mathbf{b}. \tag{12}$$

Combining (11) and (12) then gives

$$\delta\mathbf{x}_1 = \mathsf{A}^{-1}\delta\mathbf{b} = \mathsf{A}^{-1}(\mathsf{A}\mathbf{x}_1 - \mathbf{b}) = \mathbf{x}_1 - \mathsf{A}^{-1}\mathbf{b}, \tag{13}$$

so the next iteration to obtain $\mathbf{x}$ accurately should be

$$\mathbf{x}_2 = \mathbf{x}_1 - \delta\mathbf{x}_1. \tag{14}$$

see also CRAMER'S RULE, GAUSSIAN ELIMINATION, LU DECOMPOSITION, MATRIX, MATRIX ADDITION, MATRIX INVERSE, MATRIX MULTIPLICATION, NORMAL EQUATION, SQUARE ROOT METHOD

Matrix Exponential

Given a SQUARE MATRIX A, the matrix exponential is defined by

$$\exp(\mathsf{A}) \equiv e^{\mathsf{A}} = \sum_{n=0}^{\infty} \frac{\mathsf{A}^n}{n!} = \mathsf{I} + \mathsf{A} + \frac{\mathsf{AA}}{2!} + \frac{\mathsf{AAA}}{3!} + \ldots,$$

where I is the IDENTITY MATRIX.

see also EXPONENTIAL FUNCTION, MATRIX

Matrix Group

A GROUP in which the elements are SQUARE MATRICES, the group multiplication law is MATRIX MULTIPLICATION, and the group inverse is simply the MATRIX INVERSE. Every matrix group is equivalent to a unitary matrix group (Lomont 1987, pp. 47–48).

see also MASCHKE'S THEOREM

References

Lomont, J. S. "Matrix Groups." §3.1 in *Applications of Finite Groups.* New York: Dover, pp. 46–52, 1987.

Matrix Inverse

A MATRIX A has an inverse IFF the DETERMINANT $|\mathsf{A}| \neq 0$. For a 2×2 MATRIX

$$\mathsf{A} \equiv \begin{bmatrix} a & b \\ c & d \end{bmatrix}, \tag{1}$$

the inverse is

$$\mathsf{A}^{-1} = \frac{1}{|\mathsf{A}|} \begin{bmatrix} d & -b \\ -c & a \end{bmatrix} = \frac{1}{ad-bc} \begin{bmatrix} d & -b \\ -c & a \end{bmatrix}. \tag{2}$$

For a 3×3 MATRIX,

$$\mathsf{A}^{-1} = \frac{1}{|\mathsf{A}|} \begin{bmatrix} \begin{vmatrix} a_{22} & a_{23} \\ a_{32} & a_{33} \end{vmatrix} & \begin{vmatrix} a_{13} & a_{12} \\ a_{33} & a_{32} \end{vmatrix} & \begin{vmatrix} a_{12} & a_{13} \\ a_{22} & a_{23} \end{vmatrix} \\ \begin{vmatrix} a_{23} & a_{21} \\ a_{33} & a_{31} \end{vmatrix} & \begin{vmatrix} a_{11} & a_{13} \\ a_{31} & a_{33} \end{vmatrix} & \begin{vmatrix} a_{13} & a_{11} \\ a_{23} & a_{21} \end{vmatrix} \\ \begin{vmatrix} a_{21} & a_{22} \\ a_{31} & a_{32} \end{vmatrix} & \begin{vmatrix} a_{12} & a_{11} \\ a_{32} & a_{31} \end{vmatrix} & \begin{vmatrix} a_{11} & a_{12} \\ a_{21} & a_{22} \end{vmatrix} \end{bmatrix}. \tag{3}$$

A general $n \times n$ matrix can be inverted using methods such as the GAUSS-JORDAN ELIMINATION, GAUSSIAN ELIMINATION, or LU DECOMPOSITION.

The inverse of a PRODUCT AB of MATRICES A and B can be expressed in terms of A^{-1} and B^{-1}. Let

$$\mathsf{C} \equiv \mathsf{AB}. \tag{4}$$

Then

$$\mathsf{B} = \mathsf{A}^{-1}\mathsf{AB} = \mathsf{A}^{-1}\mathsf{C} \tag{5}$$

and

$$\mathsf{A} = \mathsf{ABB}^{-1} = \mathsf{CB}^{-1}. \tag{6}$$

Therefore,

$$\mathsf{C} = \mathsf{AB} = (\mathsf{CB}^{-1})(\mathsf{A}^{-1}\mathsf{C}) = \mathsf{CB}^{-1}\mathsf{A}^{-1}\mathsf{C}, \tag{7}$$

so

$$\mathsf{CB}^{-1}\mathsf{A}^{-1} = \mathsf{I}, \tag{8}$$

where I is the IDENTITY MATRIX, and

$$\mathsf{B}^{-1}\mathsf{A}^{-1} = \mathsf{C}^{-1} = (\mathsf{AB})^{-1}. \tag{9}$$

see also MATRIX, MATRIX ADDITION, MATRIX MULTIPLICATION, MOORE-PENROSE GENERALIZED MATRIX INVERSE, STRASSEN FORMULAS

References

Ben-Israel, A. and Greville, T. N. E. *Generalized Inverses: Theory and Applications.* New York: Wiley, 1977.

Nash, J. C. *Compact Numerical Methods for Computers: Linear Algebra and Function Minimisation, 2nd ed.* Bristol, England: Adam Hilger, pp. 24–26, 1990.

Press, W. H.; Flannery, B. P.; Teukolsky, S. A.; and Vetterling, W. T. "Is Matrix Inversion an N^3 Process?" §2.11 in *Numerical Recipes in FORTRAN: The Art of Scientific Computing, 2nd ed.* Cambridge, England: Cambridge University Press, pp. 95–98, 1992.

Matrix Multiplication

The product C of two MATRICES A and B is defined by

$$c_{ik} = a_{ij}b_{jk}, \tag{1}$$

where j is summed over for all possible values of i and k. Therefore, in order for multiplication to be defined, the dimensions of the MATRICES must satisfy

$$(n \times m)(m \times p) = (n \times p), \tag{2}$$

where $(a \times b)$ denotes a MATRIX with a rows and b columns. Writing out the product explicitly,

$$\begin{bmatrix} c_{11} & c_{12} & \cdots & c_{1p} \\ c_{21} & c_{22} & \cdots & c_{2p} \\ \vdots & \vdots & \ddots & \vdots \\ c_{n1} & c_{n2} & \cdots & c_{np} \end{bmatrix} = \begin{bmatrix} a_{11} & a_{12} & \cdots & a_{1m} \\ a_{21} & a_{22} & \cdots & a_{2m} \\ \vdots & \vdots & \ddots & \vdots \\ a_{n1} & a_{n2} & \cdots & a_{nm} \end{bmatrix} \begin{bmatrix} b_{11} & b_{12} & \cdots & b_{1p} \\ b_{21} & b_{22} & \cdots & b_{2p} \\ \vdots & \vdots & \ddots & \vdots \\ b_{m1} & b_{m2} & \cdots & b_{mp} \end{bmatrix}, \tag{3}$$

where

$$\begin{aligned} c_{11} &= a_{11}b_{11} + a_{12}b_{21} + \ldots + a_{1m}b_{m1} \\ c_{12} &= a_{11}b_{12} + a_{12}b_{22} + \ldots + a_{1m}b_{m2} \\ c_{1p} &= a_{11}b_{1p} + a_{12}b_{2p} + \ldots + a_{1m}b_{mp} \\ c_{21} &= a_{21}b_{11} + a_{22}b_{21} + \ldots + a_{2m}b_{m1} \\ c_{22} &= a_{21}b_{12} + a_{22}b_{22} + \ldots + a_{2m}b_{m2} \\ c_{2p} &= a_{21}b_{1p} + a_{22}b_{2p} + \ldots + a_{2m}b_{mp} \\ c_{n1} &= a_{n1}b_{11} + a_{n2}b_{21} + \ldots + a_{nm}b_{m1} \\ c_{n2} &= a_{n1}b_{12} + a_{n2}b_{22} + \ldots + a_{nm}b_{m2} \\ c_{np} &= a_{n1}b_{1p} + a_{n2}b_{2p} + \ldots + a_{nm}b_{mp}. \end{aligned}$$

MATRIX MULTIPLICATION is ASSOCIATIVE, as can be seen by taking

$$[(ab)c]_{ij} = (ab)_{ik}c_{kj} = (a_{il}b_{lk})c_{kj}. \tag{4}$$

Now, since a_{il}, b_{lk}, and c_{kj} are SCALARS, use the ASSOCIATIVITY of SCALAR MULTIPLICATION to write

$$(a_{il}b_{lk})c_{kj} = a_{il}(b_{lk}c_{kj}) = a_{il}(bc)_{lj} = [a(bc)]_{ij}. \tag{5}$$

Since this is true for all i and j, it must be true that

$$[(ab)c]_{ij} = [a(bc)]_{ij}. \tag{6}$$

That is, MATRIX multiplication is ASSOCIATIVE. However, MATRIX MULTIPLICATION is *not* COMMUTATIVE unless A and B are DIAGONAL (and have the same dimensions).

The product of two BLOCK MATRICES is given by multiplying each block

$$\begin{bmatrix} o & o & & & & \\ o & o & & & & \\ & & o & & & \\ & & & o & o & o \\ & & & o & o & o \\ & & & o & o & o \end{bmatrix} \begin{bmatrix} x & x & & & & \\ x & x & & & & \\ & & x & & & \\ & & & x & x & x \\ & & & x & x & x \\ & & & x & x & x \end{bmatrix}$$

$$= \begin{bmatrix} \begin{bmatrix} o & o \\ o & o \end{bmatrix} \begin{bmatrix} x & x \\ x & x \end{bmatrix} & & \\ & [o][x] & \\ & & \begin{bmatrix} o & o & o \\ o & o & o \\ o & o & o \end{bmatrix} \begin{bmatrix} x & x & x \\ x & x & x \\ x & x & x \end{bmatrix} \end{bmatrix}. \qquad (7)$$

see also MATRIX, MATRIX ADDITION, MATRIX INVERSE, STRASSEN FORMULAS

References

Arfken, G. *Mathematical Methods for Physicists, 3rd ed.* Orlando, FL: Academic Press, pp. 178–179, 1985.

Matrix Norm

Given a SQUARE MATRIX A with COMPLEX (or REAL) entries, a MATRIX NORM $||\mathsf{A}||$ is a NONNEGATIVE number associated with A having the properties

1. $||\mathsf{A}|| > 0$ when $\mathsf{A} \neq 0$ and $||\mathsf{A}|| = 0$ IFF $\mathsf{A} = 0$,
2. $||k\mathsf{A}|| = |k|\,||\mathsf{A}||$ for any SCALAR k,
3. $||\mathsf{A}+\mathsf{B}|| \leq ||\mathsf{A}|| + ||\mathsf{B}||$,
4. $||\mathsf{AB}|| \leq ||\mathsf{A}||\,||\mathsf{B}||$.

For an $n \times n$ MATRIX A and an $n \times n$ UNITARY MATRIX U,

$$||\mathsf{AU}|| = ||\mathsf{UA}|| = ||\mathsf{A}||.$$

Let $\lambda_1, \ldots, \lambda_n$ be the EIGENVALUES of A, then

$$\frac{1}{||\mathsf{A}^{-1}||} \leq |\lambda| \leq ||\mathsf{A}||.$$

The MAXIMUM ABSOLUTE COLUMN SUM NORM $||\mathsf{A}||_1$, SPECTRAL NORM $||\mathsf{A}||_2$, and MAXIMUM ABSOLUTE ROW SUM NORM $||\mathsf{A}||_\infty$ satisfy

$$(||\mathsf{A}||_2)^2 \leq ||\mathsf{A}||_1\,||\mathsf{A}||_\infty.$$

For a SQUARE MATRIX, the SPECTRAL NORM, which is the SQUARE ROOT of the maximum EIGENVALUE of $\mathsf{A}^\dagger\mathsf{A}$ (where $\mathsf{A}^\dagger$ is the ADJOINT MATRIX), is often referred to as "the" matrix norm.

see also COMPATIBLE, HILBERT-SCHMIDT NORM, MAXIMUM ABSOLUTE COLUMN SUM NORM, MAXIMUM ABSOLUTE ROW SUM NORM, NATURAL NORM, NORM, POLYNOMIAL NORM, SPECTRAL NORM, SPECTRAL RADIUS, VECTOR NORM

References

Gradshteyn, I. S. and Ryzhik, I. M. *Tables of Integrals, Series, and Products, 5th ed.* San Diego, CA: Academic Press, pp. 1114–1125, 1979.

Matrix Polynomial Identity

see CAYLEY-HAMILTON THEOREM

Matrix Product

The result of a MATRIX MULTIPLICATION.

see also PRODUCT

Matrix Transpose

see TRANSPOSE

Matroid

Roughly speaking, a finite set together with a generalization of a concept from linear algebra that satisfies a natural set of properties for that concept. For example, the finite set could be the rows of a MATRIX, and the generalizing concept could be linear dependence and independence of any subset of rows of the MATRIX. The number of matroids with n points are 1, 1, 2, 4, 9, 26, 101, 950, ... (Sloane's A002773).

References

Sloane, N. J. A. Sequences A002773/M1197 in "An On-Line Version of the Encyclopedia of Integer Sequences."

Sloane, N. J. A. and Plouffe, S. Extended entry in *The Encyclopedia of Integer Sequences.* San Diego: Academic Press, 1995.

Whitely, W. "Matroids and Rigid Structures." In Matroid Applications, *Encyclopedia of Mathematics and Its Applications* (Ed. N. White), Vol. 40. New York: Cambridge University Press, pp. 1–53, 1992.

Maurer Rose

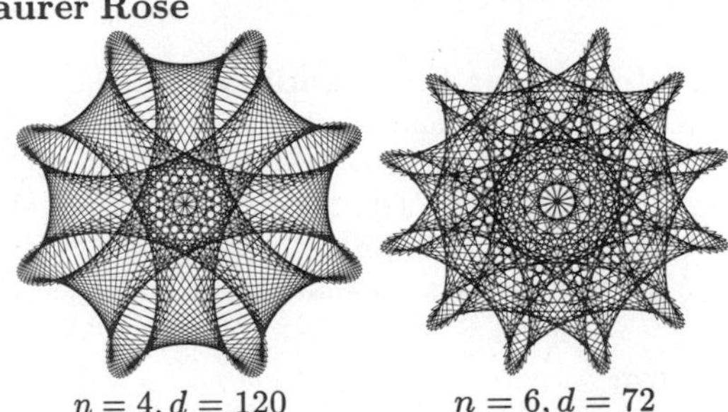

$n = 4, d = 120$ $\qquad$ $n = 6, d = 72$

A Maurer rose is a plot of a "walk" along an n- (or $2n$-) leafed ROSE in steps of a fixed number d degrees, including all cosets.

see also STARR ROSE

References

Maurer, P. "A Rose is a Rose..." *Amer. Math. Monthly* **94**, 631–645, 1987.

Wagon, S. *Mathematica in Action.* New York: W. H. Freeman, pp. 96–102, 1991.

Max Sequence

A sequence defined from a FINITE sequence $a_0, a_1, \ldots, a_n$ by defining $a_{n+1} = \max_i(a_i + a_{n-i})$.

see also MEX SEQUENCE

References

Guy, R. K. "Max and Mex Sequences." §E27 in *Unsolved Problems in Number Theory, 2nd ed.* New York: Springer-Verlag, pp. 227–228, 1994.

Maximal Ideal

A maximal ideal of a RING R is an IDEAL I, not equal to R, such that there are no IDEALS "in between" I and R. In other words, if J is an IDEAL which contains I as a SUBSET, then either $J = I$ or $J = R$. For example, $n\mathbb{Z}$ is a maximal ideal of $\mathbb{Z}$ IFF n is PRIME, where $\mathbb{Z}$ is the RING of INTEGERS.

see also IDEAL, PRIME IDEAL, REGULAR LOCAL RING, RING

Maximal Sum-Free Set

A maximal sum-free set is a set $\{a_1, a_2, \ldots, a_n\}$ of distinct NATURAL NUMBERS such that a maximum l of them satisfy $a_{i_j} + a_{i_k} \neq a_m$, for $1 \leq j < k \leq l$, $1 \leq m \leq n$.

see also MAXIMAL ZERO-SUM-FREE SET

References

Guy, R. K. "Maximal Sum-Free Sets." §C14 in *Unsolved Problems in Number Theory, 2nd ed.* New York: Springer-Verlag, pp. 128–129, 1994.

Maximal Zero-Sum-Free Set

A set having the largest number k of distinct residue classes modulo m so that no SUBSET has zero sum.

see also MAXIMAL SUM-FREE SET

References

Guy, R. K. "Maximal Zero-Sum-Free Sets." §C15 in *Unsolved Problems in Number Theory, 2nd ed.* New York: Springer-Verlag, pp. 129–131, 1994.

Maximally Linear Independent

A set of VECTORS is maximally linearly independent if including any other VECTOR in the VECTOR SPACE would make it LINEARLY DEPENDENT (i.e., if any other VECTOR in the SPACE can be expressed as a linear combination of elements of a maximal set—the BASIS).

Maximum

The largest value of a set, function, etc. The maximum value of a set of elements $A = \{a_i\}_{i=1}^N$ is denoted $\max A$ or $\max_i a_i$, and is equal to the last element of a sorted (i.e., ordered) version of A. For example, given the set {3, 5, 4, 1}, the sorted version is {1, 3, 4, 5}, so the maximum is 5. The maximum and MINIMUM are the simplest ORDER STATISTICS.

A continuous FUNCTION may assume a maximum at a single point or may have maxima at a number of points. A GLOBAL MAXIMUM of a FUNCTION is the largest value in the entire RANGE of the FUNCTION, and a LOCAL MAXIMUM is the largest value in some local neighborhood.

For a function $f(x)$ which is CONTINUOUS at a point x_0, a NECESSARY but not SUFFICIENT condition for $f(x)$ to have a RELATIVE MAXIMUM at $x = x_0$ is that x_0 be a CRITICAL POINT (i.e., $f(x)$ is either not DIFFERENTIABLE at x_0 or x_0 is a STATIONARY POINT, in which case $f'(x_0) = 0$).

The FIRST DERIVATIVE TEST can be applied to CONTINUOUS FUNCTIONS to distinguish maxima from MINIMA. For twice differentiable functions of one variable, $f(x)$, or of two variables, $f(x, y)$, the SECOND DERIVATIVE TEST can sometimes also identify the nature of an EXTREMUM. For a function $f(x)$, the EXTREMUM TEST succeeds under more general conditions than the SECOND DERIVATIVE TEST.

see also CRITICAL POINT, EXTREMUM, EXTREMUM TEST, FIRST DERIVATIVE TEST, GLOBAL MAXIMUM, INFLECTION POINT, LOCAL MAXIMUM, MIDRANGE, MINIMUM, ORDER STATISTIC, SADDLE POINT (FUNCTION), SECOND DERIVATIVE TEST, STATIONARY POINT

References

Abramowitz, M. and Stegun, C. A. (Eds.). *Handbook of Mathematical Functions with Formulas, Graphs, and Mathematical Tables, 9th printing.* New York: Dover, p. 14, 1972.

Press, W. H.; Flannery, B. P.; Teukolsky, S. A.; and Vetterling, W. T. "Minimization or Maximization of Functions." Ch. 10 in *Numerical Recipes in FORTRAN: The Art of Scientific Computing, 2nd ed.* Cambridge, England: Cambridge University Press, pp. 387–448, 1992.

Tikhomirov, V. M. *Stories About Maxima and Minima.* Providence, RI: Amer. Math. Soc., 1991.

Maximum Absolute Column Sum Norm

The NATURAL NORM induced by the L_1-NORM is called the maximum absolute column sum norm and is defined by

$$||\mathsf{A}||_1 = \max_j \sum_{i=1}^n |a_{ij}|$$

for a MATRIX A.

see also L_1-NORM, MAXIMUM ABSOLUTE ROW SUM NORM

Maximum Absolute Row Sum Norm

The NATURAL NORM induced by the L_∞-NORM is called the maximum absolute row sum norm and is defined by

$$||\mathsf{A}||_\infty = \max_i \sum_{j=1}^n |a_{ij}|$$

for a MATRIX A.

see also L_∞-NORM, MAXIMUM ABSOLUTE COLUMN SUM NORM

Maximum Clique Problem

see PARTY PROBLEM

Maximum Entropy Method

A DECONVOLUTION ALGORITHM (sometimes abbreviated MEM) which functions by minimizing a smoothness function ("ENTROPY") in an image. Maximum entropy is also called the ALL-POLES MODEL or AUTOREGRESSIVE MODEL. For images with more than a million pixels, maximum entropy is faster than the CLEAN ALGORITHM.

MEM is commonly employed in astronomical synthesis imaging. In this application, the resolution depends on the signal to NOISE ratio, which must be specified. Therefore, resolution is image dependent and varies across the map. MEM is also biased, since the ensemble average of the estimated noise is NONZERO. However, this bias is much smaller than the NOISE for pixels with a SNR $\gg 1$. It can yield super-resolution, which can usually be trusted to an order of magnitude in SOLID ANGLE.

Several definitions of "ENTROPY" normalized to the flux in the image are

$$H_1 \equiv \sum_k \ln\left(\frac{I_k}{M_k}\right) \tag{1}$$

$$H_2 \equiv -\sum_k I_k \ln\left(\frac{I_k}{M_k e}\right), \tag{2}$$

where M_k is a "default image" and I_k is the smoothed image. Some unnormalized entropy measures (Cornwell 1982, p. 3) are given by

$$H_1 \equiv -\sum f_i \ln(f_i) \tag{3}$$

$$H_2 \equiv \sum \ln(f_i) \tag{4}$$

$$H_3 \equiv -\sum \frac{1}{\ln(f_i)} \tag{5}$$

$$H_4 \equiv -\sum \frac{1}{[\ln(f_i)]^2} \tag{6}$$

$$H_5 \equiv \sum \sqrt{\ln(f_i)}. \tag{7}$$

see also CLEAN ALGORITHM, DECONVOLUTION, LUCY

References

Cornwell, T. J. "Can CLEAN be Improved?" VLA Scientific Memorandum No. 141, March 1982.

Cornwell, T. and Braun, R. "Deconvolution." Ch. 8 in *Synthesis Imaging in Radio Astronomy: Third NRAO Summer School, 1988* (Ed. R. A. Perley, F. R. Schwab, and A. H. Bridle). San Francisco, CA: Astronomical Society of the Pacific, pp. 167–183, 1989.

Christiansen, W. N. and Högbom, J. A. *Radiotelescopes, 2nd ed.* Cambridge, England: Cambridge University Press, pp. 217–218, 1985.

Narayan, R. and Nityananda, R. "Maximum Entropy Restoration in Astronomy." *Ann. Rev. Astron. Astrophys.* **24**, 127–170, 1986.

Press, W. H.; Flannery, B. P.; Teukolsky, S. A.; and Vetterling, W. T. "Power Spectrum Estimation by the Maximum Entropy (All Poles) Method" and "Maximum Entropy Image Restoration." §13.7 and 18.7 in *Numerical Recipes in FORTRAN: The Art of Scientific Computing, 2nd ed.* Cambridge, England: Cambridge University Press, pp. 565–569 and 809–817, 1992.

Thompson, A. R.; Moran, J. M.; and Swenson, G. W. Jr. §3.2 in *Interferometry and Synthesis in Radio Astronomy.* New York: Wiley, pp. 349–352, 1986.

Maximum Likelihood

The procedure of finding the value of one or more parameters for a given statistic which makes the *known* LIKELIHOOD distribution a MAXIMUM. The maximum likelihood estimate for a parameter μ is denoted $\hat{\mu}$.

For a BERNOULLI DISTRIBUTION,

$$\frac{d}{d\theta}\left[\binom{N}{Np}\theta^{Np}(1-\theta)^{Nq}\right] = Np(1-\theta) - \theta Nq = 0, \tag{1}$$

so maximum likelihood occurs for $\theta = p$. If p is not known ahead of time, the likelihood function is

$$\begin{aligned} f(x_1,\ldots,x_n|p) &= P(X_1 = x_1,\ldots,X_n = x_n|p) \\ &= p^{x_1}(1-p)^{1-x_1}\cdots p^{x_n}(1-p)^{1-x_1n} \\ &= p^{\Sigma x_i}(1-p)^{\Sigma(1-x_i)} = p^{\Sigma x_i}(1-p)^{n-\Sigma x_i}, \end{aligned} \tag{2}$$

where $x = 0$ or 1, and $i = 1, \ldots, n$.

$$\ln f = \sum x_i \ln p + \left(n - \sum x_i\right)\ln(1-p) \tag{3}$$

$$\frac{d(\ln f)}{dp} = \frac{\sum x_i}{p} - \frac{n - \sum x_i}{1-p} = 0 \tag{4}$$

$$\sum x_i - p\sum x_i = np - p\sum x_i \tag{5}$$

$$\hat{p} = \frac{\sum x_i}{n}. \tag{6}$$

For a GAUSSIAN DISTRIBUTION,

$$\begin{aligned} f(x_1,\ldots,x_n|\mu,\sigma) &= \prod \frac{1}{\sigma\sqrt{2\pi}} e^{-(x_i-\mu)^2/2\sigma^2} \\ &= \frac{(2\pi)^{-n/2}}{\sigma^n}\exp\left[-\frac{\sum(x_i-\mu)^2}{2\sigma^2}\right] \end{aligned} \tag{7}$$

$$\ln f = -\tfrac{1}{2}n\ln(2\pi) - n\ln\sigma - \frac{\sum(x_i-\mu)^2}{2\sigma^2} \tag{8}$$

$$\frac{\partial(\ln f)}{\partial\mu} = \frac{\sum(x_i-\mu)}{\sigma^2} = 0 \tag{9}$$

gives

$$\hat{\mu} = \frac{\sum x_i}{n}. \tag{10}$$

$$\frac{\partial(\ln f)}{\partial\sigma} = -\frac{n}{\sigma} + \frac{\sum(x_i-\mu)^2}{\sigma^3} \tag{11}$$

gives

$$\hat{\sigma} = \sqrt{\frac{\sum (x_i - \hat{\mu})^2}{n}}. \quad (12)$$

Note that in this case, the maximum likelihood STANDARD DEVIATION is the sample STANDARD DEVIATION, which is a BIASED ESTIMATOR for the population STANDARD DEVIATION.

For a weighted GAUSSIAN DISTRIBUTION,

$$f(x_1, \ldots, x_n | \mu, \sigma) = \prod \frac{1}{\sigma_i \sqrt{2\pi}} e^{-(x_i - \mu)^2 / 2\sigma_i^2}$$
$$= \frac{(2\pi)^{-n/2}}{\sigma^n} \exp\left[-\frac{\sum (x_i - \mu)^2}{2\sigma^2}\right] \quad (13)$$

$$\ln f = -\tfrac{1}{2} n \ln(2\pi) - n \sum \ln \sigma_i - \sum \frac{(x_i - \mu)^2}{2\sigma_i^2} \quad (14)$$

$$\frac{\partial(\ln f)}{\partial \mu} = \sum \frac{(x_i - \mu)}{\sigma_i^2} = \sum \frac{x_i}{\sigma_i^2} - \mu \sum \frac{1}{\sigma_i^2} = 0 \quad (15)$$

gives

$$\hat{\mu} = \frac{\sum \frac{x_i}{\sigma_i^2}}{\sum \frac{1}{\sigma_i^2}}. \quad (16)$$

The VARIANCE of the MEAN is then

$$\sigma_\mu^2 = \sum \sigma_i^2 \left(\frac{\partial \mu}{\partial x_i}\right)^2. \quad (17)$$

But

$$\frac{\partial \mu}{\partial x_i} = \frac{\partial}{\partial x_i} \frac{\sum (x_i / \sigma_i^2)}{\sum (1/\sigma_i^2)} = \frac{1/\sigma_i^2}{\sum (1/\sigma_i^2)}, \quad (18)$$

so

$$\sigma_\mu^2 = \sum \sigma_i^2 \left(\frac{1/\sigma_i^2}{\sum (1/\sigma_i^2)}\right)^2$$
$$= \sum \frac{1/\sigma_i^2}{\left[\sum (1/\sigma_i^2)\right]^2} = \frac{1}{\sum (1/\sigma_i^2)}. \quad (19)$$

For a POISSON DISTRIBUTION,

$$f(x_1, \ldots, x_n | \lambda) = \frac{e^{-\lambda} \lambda^{x_1}}{x_1!} \cdots \frac{e^{-\lambda} \lambda^{x_n}}{x_n!} = \frac{e^{-n\lambda} \lambda^{\sum x_i}}{x_1! \cdots x_n!} \quad (20)$$

$$\ln f = -n\lambda + (\ln \lambda) \sum x_i - \ln\left(\prod x_i!\right) \quad (21)$$

$$\frac{d(\ln f)}{\lambda} = -n + \frac{\sum x_i}{\lambda} = 0 \quad (22)$$

$$\hat{\lambda} = \frac{\sum x_i}{n}. \quad (23)$$

see also BAYESIAN ANALYSIS

References

Press, W. H.; Flannery, B. P.; Teukolsky, S. A.; and Vetterling, W. T. "Least Squares as a Maximum Likelihood Estimator." §15.1 in *Numerical Recipes in FORTRAN: The Art of Scientific Computing, 2nd ed.* Cambridge, England: Cambridge University Press, pp. 651–655, 1992.

Maxwell Distribution

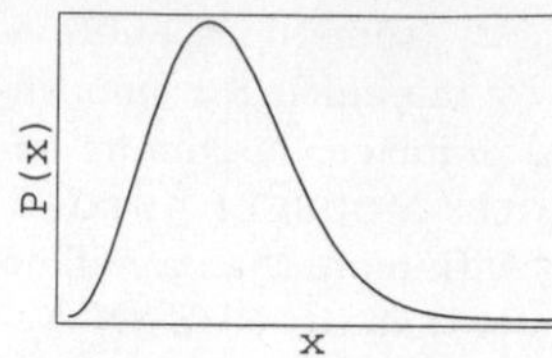

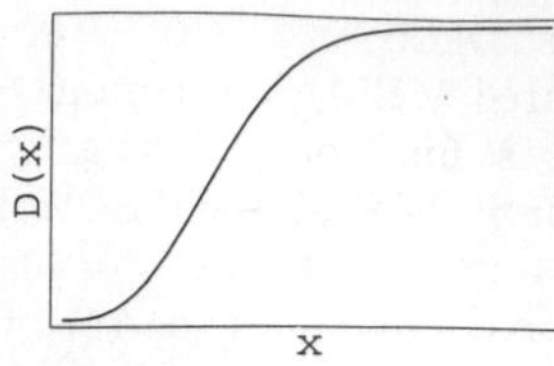

The distribution of speeds of molecules in thermal equilibrium as given by statistical mechanics. The probability and cumulative distributions are

$$P(x) = \sqrt{\frac{2}{\pi}}\, a^{3/2} x^2 e^{-ax^2/2} \quad (1)$$

$$D(x) = \frac{2\gamma(\frac{3}{2}, \frac{1}{2}ax^2)}{\sqrt{\pi}}, \quad (2)$$

where $\gamma(a, x)$ is an incomplete GAMMA FUNCTION and $x \in [0, \infty)$. The moments are

$$\mu = 2\sqrt{\frac{2}{\pi a}} \quad (3)$$

$$\mu_2 = \frac{3}{a} \quad (4)$$

$$\mu_3 = 8\sqrt{\frac{2}{a^3 \pi}} \quad (5)$$

$$\mu_4 = \tfrac{15}{2}, \quad (6)$$

and the MEAN, VARIANCE, SKEWNESS, and KURTOSIS are

$$\mu = 2\sqrt{\frac{2}{\pi a}} \quad (7)$$

$$\sigma^2 = \frac{3\pi - 8}{\pi a} \quad (8)$$

$$\gamma_1 = \frac{8}{3}\sqrt{\frac{2}{3\pi}} \quad (9)$$

$$\gamma_2 = -\tfrac{4}{3}. \quad (10)$$

see also EXPONENTIAL DISTRIBUTION, GAUSSIAN DISTRIBUTION, RAYLEIGH DISTRIBUTION

References

Spiegel, M. R. *Theory and Problems of Probability and Statistics.* New York: McGraw-Hill, p. 119, 1992.

von Seggern, D. *CRC Standard Curves and Surfaces.* Boca Raton, FL: CRC Press, p. 252, 1993.

May's Theorem

Simple majority vote is the only procedure which is ANONYMOUS, DUAL, and MONOTONIC.

References

May, K. "A Set of Independent Necessary and Sufficient Conditions for Simple Majority Decision." *Econometrica* **20**, 680–684, 1952.

May-Thomason Uniqueness Theorem

For every infinite LOOP SPACE MACHINE E, there is a natural equivalence of spectra between EX and Segal's spectrum $\mathbf{B}X$.

References

May, J. P. and Thomason, R. W. "The Uniqueness of Infinite Loop Space Machines." *Topology* **17**, 205–224, 1978.

Weibel, C. A. "The Mathematical Enterprises of Robert Thomason." *Bull. Amer. Math. Soc.* **34**, 1–13, 1996.

Maze

A maze is a drawing of impenetrable line segments (or curves) with "paths" between them. The goal of the maze is to start at one given point and find a path which reaches a second given point.

References

Gardner, M. "Mazes." Ch. 10 in *The Second Scientific American Book of Mathematical Puzzles & Diversions: A New Selection.* New York: Simon and Schuster, pp. 112–118, 1961.

Jablan, S. "Roman Mazes." `http://members.tripod.com/~modularity/mazes.htm`.

Matthews, W. H. *Mazes and Labyrinths: Their History and Development.* New York: Dover, 1970.

Pappas, T. "Mazes." *The Joy of Mathematics.* San Carlos, CA: Wide World Publ./Tetra, pp. 192–194, 1989.

Phillips, A. "The Topology of Roman Mazes." *Leonardo* **25**, 321–329, 1992.

Shepard, W. *Mazes and Labyrinths: A Book of Puzzles.* New York: Dover, 1961.

Mazur's Theorem

The generalization of the SCHÖNFLIES THEOREM to n-D. A smoothly embedded n-HYPERSPHERE in an $(n+1)$-HYPERSPHERE separates the $(n+1)$-HYPERSPHERE into two components, each HOMEOMORPHIC to $(n+1)$-BALLS. It can be proved using MORSE THEORY.

see also BALL, HYPERSPHERE

McCay Circle

If the VERTEX A_1 of a TRIANGLE describes the NEUBERG CIRCLE n_1, its MEDIAN POINT describes a circle whose radius is $1/3$ that of the NEUBERG CIRCLE. Such a CIRCLE is known as a McKay circle, and the three McCay circles are CONCURRENT at the MEDIAN POINT M. Three homologous collinear points lie on the McCay circles.

see also CIRCLE, CONCURRENT, MEDIAN POINT, NEUBERG CIRCLES

References

Johnson, R. A. *Modern Geometry: An Elementary Treatise on the Geometry of the Triangle and the Circle.* Boston, MA: Houghton Mifflin, pp. 290 and 306, 1929.

McCoy's Theorem

If two SQUARE $n \times n$ MATRICES A and B are simultaneously upper triangularizable by similarity transforms, then there is an ordering $a_1, \ldots, a_n$ of the EIGENVALUES of A and $b_1, \ldots, b_n$ of the EIGENVALUES of B so that, given any POLYNOMIAL $p(x,y)$ in noncommuting variables, the EIGENVALUES of $p(A,B)$ are the numbers $p(a_i, b_i)$ with $i = 1, \ldots, n$. McCoy's theorem states the converse: If every POLYNOMIAL exhibits the correct EIGENVALUES in a consistent ordering, then A and B are simultaneously triangularizable.

References

Luchins, E. H. and McLoughlin, M. A. "In Memoriam: Olga Taussky-Todd." *Not. Amer. Math. Soc.* **43**, 838–847, 1996.

McLaughlin Group

The SPORADIC GROUP McL.

References

Wilson, R. A. "ATLAS of Finite Group Representation." `http://for.mat.bham.ac.uk/atlas/McL.html`.

McMohan's Theorem

Consider a GAUSSIAN BIVARIATE DISTRIBUTION. Let $f(x_1, x_2)$ be an arbitrary FUNCTION. Then

$$\frac{\partial^2 \langle f\rangle}{\partial \rho^n} = \left\langle \frac{\partial^{2n} f}{\partial {x_1}^n \partial {x_2}^n} \right\rangle.$$

see also GAUSSIAN BIVARIATE DISTRIBUTION

McNugget Number

A number which can be obtained from an order of McDonald's® Chicken McNuggets™ (prior to consuming any), which originally came in boxes of 6, 9, and 20. All integers are McNugget numbers except 1, 2, 3, 4, 5, 7, 8, 10, 11, 13, 14, 16, 17, 19, 22, 23, 25, 28, 31, 34, 37, and 43. Since the Happy Meal™-sized nugget box (4 to a box) can now be purchased separately, the modern McNugget numbers are a linear combination of 4, 6, 9, and 20. These new-fangled numbers are much less interesting than before, with only 1, 2, 3, 5, 7, and 11 remaining as non-McNugget numbers.

The GREEDY ALGORITHM can be used to find a McNugget expansion of a given INTEGER.

see also COMPLETE SEQUENCE, GREEDY ALGORITHM

References

Vardi, I. *Computational Recreations in Mathematica.* Reading, MA: Addison-Wesley, pp. 19–20 and 233–234, 1991.

Wilson, D. `rec.puzzles` newsgroup posting, March 20, 1990.

Mean

A mean is HOMOGENEOUS and has the property that a mean μ of a set of numbers x_i satisfies

$$\min(x_1,\ldots,x_n) \le \mu \le \max(x_1,\ldots,x_n).$$

There are several statistical quantities called means, e.g., ARITHMETIC-GEOMETRIC MEAN, GEOMETRIC MEAN, HARMONIC MEAN, QUADRATIC MEAN, ROOT-MEAN-SQUARE. However, the quantity referred to as "the" mean is the ARITHMETIC MEAN, also called the AVERAGE.

see also ARITHMETIC-GEOMETRIC MEAN, AVERAGE, GENERALIZED MEAN, GEOMETRIC MEAN, HARMONIC MEAN, QUADRATIC MEAN, ROOT-MEAN-SQUARE

Mean Cluster Count Per Site

see s-CLUSTER

Mean Cluster Density

see s-CLUSTER

Mean Curvature

Let κ_1 and κ_2 be the PRINCIPAL CURVATURES, then their MEAN

$$H = \tfrac{1}{2}(\kappa_1 + \kappa_2) \qquad (1)$$

is called the mean curvature. Let R_1 and R_2 be the radii corresponding to the PRINCIPAL CURVATURES, then the multiplicative inverse of the mean curvature H is given by the multiplicative inverse of the HARMONIC MEAN,

$$H \equiv \frac{1}{2}\left(\frac{1}{R_1} + \frac{1}{R_2}\right) = \frac{R_1 + R_2}{2R_1R_2}. \qquad (2)$$

In terms of the GAUSSIAN CURVATURE K,

$$H = \tfrac{1}{2}(R_1 + R_2)K. \qquad (3)$$

The mean curvature of a REGULAR SURFACE in $\mathbb{R}^3$ at a point $\mathbf{p}$ is formally defined as

$$H(\mathbf{p}) = \tfrac{1}{2}\operatorname{tr}(S(\mathrm{p})), \qquad (4)$$

where S is the SHAPE OPERATOR and $\operatorname{tr}(S)$ denotes the TRACE. For a MONGE PATCH with $z = h(x,y)$,

$$H = \frac{(1+{h_v}^2)h_{uu} - 2h_uh_vh_{uv} + (1+{h_u}^2)h_{vv}}{(1+{h_u}^2+{h_v}^2)^{3/2}} \qquad (5)$$

(Gray 1993, p. 307).

If $\mathbf{x}: U \to \mathbb{R}^3$ is a REGULAR PATCH, then the mean curvature is given by

$$H = \frac{eG - 2fF + gE}{2(EG - F^2)}, \qquad (6)$$

where E, F, and G are coefficients of the first FUNDAMENTAL FORM and e, f, and g are coefficients of the second FUNDAMENTAL FORM (Gray 1993, p. 282). It can also be written

$$H = \frac{\det(\mathbf{x}_{uu}\mathbf{x}_u\mathbf{x}_v)|\mathbf{x}_v|^2 - 2\det(\mathbf{x}_{uv}\mathbf{x}_u\mathbf{x}_v)(\mathbf{x}_u\cdot\mathbf{x}_v)}{2[|\mathbf{x}_u|^2|\mathbf{x}_v|^2 - (\mathbf{x}_u\cdot\mathbf{x}_v)^2]^{3/2}} + \frac{\det(\mathbf{x}_{vv}\mathbf{x}_u\mathbf{x}_v)|\mathbf{x}_u|^2}{2[|\mathbf{x}_u|^2|\mathbf{x}_v|^2 - (\mathbf{x}_u\cdot\mathbf{x}_v)^2]^{3/2}} \qquad (7)$$

Gray (1993, p. 285).

The GAUSSIAN and mean curvature satisfy

$$H^2 \ge K, \qquad (8)$$

with equality only at UMBILIC POINTS, since

$$H^2 - K^2 = \tfrac{1}{4}(\kappa_1 - \kappa_2)^2. \qquad (9)$$

If $\mathbf{p}$ is a point on a REGULAR SURFACE $M \subset \mathbb{R}^3$ and $\mathbf{v_p}$ and $\mathbf{w_p}$ are tangent vectors to M at $\mathbf{p}$, then the mean curvature of M at $\mathbf{p}$ is related to the SHAPE OPERATOR S by

$$S(\mathbf{v_p}) \times \mathbf{w_p} + \mathbf{v_p} \times S(\mathbf{w_p}) = 2H(\mathbf{p})\mathbf{v_p} \times \mathbf{w_p}. \qquad (10)$$

Let $\mathbf{Z}$ be a nonvanishing VECTOR FIELD on M which is everywhere PERPENDICULAR to M, and let V and W be VECTOR FIELDS tangent to M such that $V \times W = \mathbf{Z}$, then

$$H = -\frac{\mathbf{Z}\cdot(D_V\mathbf{Z} \times W + V \times D_W\mathbf{Z})}{2|\mathbf{Z}|^3} \qquad (11)$$

(Gray 1993, pp. 291–292).

Wente (1985, 1986, 1987) found a nonspherical finite surface with constant mean curvature, consisting of a self-intersecting three-lobed toroidal surface. A family of such surfaces exists.

see also GAUSSIAN CURVATURE, PRINCIPAL CURVATURES, SHAPE OPERATOR

References

Gray, A. "The Gaussian and Mean Curvatures." §14.5 in *Modern Differential Geometry of Curves and Surfaces.* Boca Raton, FL: CRC Press, pp. 279–285, 1993.

Isenberg, C. *The Science of Soap Films and Soap Bubbles.* New York: Dover, p. 108, 1992.

Peterson, I. *The Mathematical Tourist: Snapshots of Modern Mathematics.* New York: W. H. Freeman, pp. 69–70, 1988.

Wente, H. C. "A Counterexample in 3-Space to a Conjecture of H. Hopf." In *Workshop Bonn 1984, Proceedings of the 25th Mathematical Workshop Held at the Max-Planck Institut für Mathematik, Bonn, June 15–22, 1984* (Ed. F. Hirzebruch, J. Schwermer, and S. Suter). New York: Springer-Verlag, pp. 421–429, 1985.

Wente, H. C. "Counterexample to a Conjecture of H. Hopf." *Pac. J. Math.* **121**, 193–243, 1986.

Wente, H. C. "Immersed Tori of Constant Mean Curvature in $\mathbb{R}^3$." In *Variational Methods for Free Surface Interfaces, Proceedings of a Conference Held in Menlo Park, CA, Sept. 7–12, 1985* (Ed. P. Concus and R. Finn). New York: Springer-Verlag, pp. 13–24, 1987.

Mean Deviation

The MEAN of the ABSOLUTE DEVIATIONS,

$$\mathrm{MD} \equiv \frac{1}{N} \sum_{i=1}^{N} |x_i - \bar{x}|,$$

where $\bar{x}$ is the MEAN of the distribution.

see also ABSOLUTE DEVIATION

Mean Distribution

For an infinite population with MEAN μ, STANDARD DEVIATION σ^2, SKEWNESS γ_1, and KURTOSIS γ_2, the corresponding quantities for the *distribution of means* are

$$\mu_{\bar{x}} = \mu \tag{1}$$

$$\sigma_{\bar{x}}{}^2 = \frac{\sigma^2}{N} \tag{2}$$

$$\gamma_{1,\bar{x}} = \frac{\gamma_1}{\sqrt{N}} \tag{3}$$

$$\gamma_{2,\bar{x}} = \frac{\gamma_2}{N}. \tag{4}$$

For a population of M (Kenney and Keeping 1962, p. 181),

$$\mu_{\bar{x}}^{(M)} = \mu \tag{5}$$

$$\sigma^{2(M)} = \frac{\sigma^2}{N} \frac{M-N}{M-1}. \tag{6}$$

References

Kenney, J. F. and Keeping, E. S. *Mathematics of Statistics, Pt. 1, 3rd ed.* Princeton, NJ: Van Nostrand, 1962.

Mean Run Count Per Site

see s-RUN

Mean Run Density

see s-RUN

Mean Square Error

see ROOT-MEAN-SQUARE

Mean-Value Theorem

Let $f(x)$ be DIFFERENTIABLE on the OPEN INTERVAL (a,b) and CONTINUOUS on the CLOSED INTERVAL $[a,b]$. Then there is at least one point c in (a,b) such that

$$f'(c) = \frac{f(b) - f(a)}{b - a}.$$

see also EXTENDED MEAN-VALUE THEOREM, GAUSS'S MEAN-VALUE THEOREM

References

Gradshteyn, I. S. and Ryzhik, I. M. *Tables of Integrals, Series, and Products, 5th ed.* San Diego, CA: Academic Press, pp. 1097–1098, 1993.

Measurable Function

A function $f : X \to Y$ for which the pre-image of every measurable set in Y is measurable in X. For a BOREL MEASURE, all continuous functions are measurable.

Measurable Set

If F is a SIGMA ALGEBRA and A is a SUBSET of X, then A is called measurable if A is a member of F. X need not have, a priori, a topological structure. Even if it does, there may be no connection between the open sets in the topology and the given SIGMA ALGEBRA.

see also MEASURABLE SPACE, SIGMA ALGEBRA

Measurable Space

A SET considered together with the SIGMA ALGEBRA on the SET.

see also MEASURABLE SET, MEASURE SPACE, SIGMA ALGEBRA

Measure

The terms "measure," "measurable," etc., have very precise technical definitions (usually involving SIGMA ALGEBRAS) which makes them a little difficult to understand. However, the technical nature of the definitions is extremely important, since it gives a firm footing to concepts which are the basis for much of ANALYSIS (including some of the slippery underpinnings of CALCULUS).

For example, every definition of an INTEGRAL is based on a particular measure: the RIEMANN INTEGRAL is based on JORDAN MEASURE, and the LEBESGUE INTEGRAL is based on LEBESGUE MEASURE. The study of measures and their application to INTEGRATION is known as MEASURE THEORY.

A measure is formally defined as a MAP $m : F \to \mathbb{R}$ (the reals) such that $m(\varnothing) = 0$ and, if A_n is a COUNTABLE SEQUENCE in F and the A_n are pairwise DISJOINT, then

$$m\left(\bigcup_n A_n\right) = \sum_n m(A_n).$$

If, in addition, $m(X) = 1$, then m is said to be a PROBABILITY MEASURE.

A measure m may also be defined on SETS other than those in the SIGMA ALGEBRA F. By adding to F all sets to which m assigns measure zero, we again obtain a SIGMA ALGEBRA and call this the "completion" of F with respect to m. Thus, the completion of a SIGMA ALGEBRA is the smallest SIGMA ALGEBRA containing F and all sets of measure zero.

see also ALMOST EVERYWHERE, BOREL MEASURE, ERGODIC MEASURE, EULER MEASURE, GAUSS MEASURE, HAAR MEASURE, HAUSDORFF MEASURE, HELSON-SZEGŐ MEASURE, INTEGRAL, JORDAN MEASURE, LEBESGUE MEASURE, LIOUVILLE MEASURE, MAHLER'S

MEASURE, MEASURABLE SPACE, MEASURE ALGEBRA, MEASURE SPACE, MINKOWSKI MEASURE, NATURAL MEASURE, PROBABILITY MEASURE, WIENER MEASURE

Measure Algebra
A Boolean SIGMA ALGEBRA which possesses a MEASURE.

Measure Polytope
see HYPERCUBE

Measure-Preserving Transformation
see ENDOMORPHISM

Measure Space
A measure space is a MEASURABLE SPACE possessing a NONNEGATIVE MEASURE. Examples of measure spaces include n-D EUCLIDEAN SPACE with LEBESGUE MEASURE and the unit interval with LEBESGUE MEASURE (i.e., probability).

see also LEBESGUE MEASURE, MEASURABLE SPACE

Measure Theory
The mathematical theory of how to perform INTEGRATION in arbitrary MEASURE SPACES.

see also CANTOR SET, FRACTAL, INTEGRAL, MEASURABLE FUNCTION, MEASURABLE SET, MEASURABLE SPACE, MEASURE, MEASURE SPACE

References
Doob, J. L. *Measure Theory.* New York: Springer-Verlag, 1994.
Evans, L. C. and Gariepy, R. F. *Measure Theory and Finite Properties of Functions.* Boca Raton, FL: CRC Press, 1992.
Gordon, R. A. *The Integrals of Lebesgue, Denjoy, Perron, and Henstock.* Providence, RI: Amer. Math. Soc., 1994.
Halmos, P. R. *Measure Theory.* New York: Springer-Verlag, 1974.
Henstock, R. *The General Theory of Integration.* Oxford, England: Clarendon Press, 1991.
Kestelman, H. *Modern Theories of Integration, 2nd rev. ed.* New York: Dover, 1960.
Rao, M. M. *Measure Theory And Integration.* New York: Wiley, 1987.
Strook, D. W. *A Concise Introduction to the Theory of Integration, 2nd ed.* Boston, MA: Birkhäuser, 1994.

Mechanical Quadrature
see GAUSSIAN QUADRATURE

Mecon
Buckminster Fuller's term for the TRUNCATED OCTAHEDRON.

see also DYMAXION

Medial Axis
The boundaries of the cells of a VORONOI DIAGRAM.

Medial Deltoidal Hexecontahedron
The DUAL of the RHOMBIDODECADODECAHEDRON.

Medial Disdyakis Triacontahedron
The DUAL of the TRUNCATED DODECADODECAHEDRON.

Medial Hexagonal Hexecontahedron
The DUAL of the SNUB ICOSIDODECADODECAHEDRON.

Medial Icosacronic Hexecontahedron
The DUAL of the ICOSIDODECADODECAHEDRON.

Medial Inverted Pentagonal Hexecontahedron
The DUAL of the INVERTED SNUB DODECADODECAHEDRON.

Medial Pentagonal Hexecontahedron
The DUAL of the SNUB DODECADODECAHEDRON.

Medial Rhombic Triacontahedron
A ZONOHEDRON which is the DUAL of the DODECADODECAHEDRON. It is also called the SMALL STELLATED TRIACONTAHEDRON.

References
Cundy, H. and Rollett, A. *Mathematical Models, 3rd ed.* Stradbroke, England: Tarquin Pub., p. 125, 1989.

Medial Triambic Icosahedron
The DUAL of the DITRIGONAL DODECADODECAHEDRON.

Medial Triangle

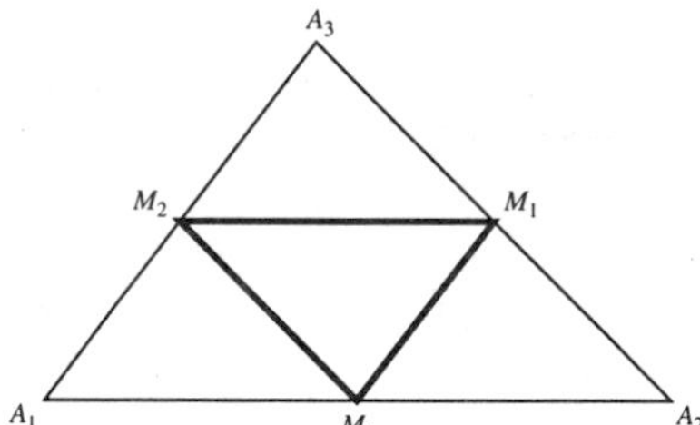

The TRIANGLE $\Delta M_1M_2M_3$ formed by joining the MIDPOINTS of the sides of a TRIANGLE $\Delta A_1A_2A_3$. The medial triangle is sometimes also called the AUXILIARY TRIANGLE (Dixon 1991). The medial triangle has TRILINEAR COORDINATES

$$A' = 0 : b^{-1} : c^{-1}$$
$$B' = a^{-1} : 0 : c^{-1}$$
$$C' = a^{-1} : b^{-1} : 0.$$

The medial triangle $\Delta M_1'M_2'M_3'$ of the medial triangle $\Delta M_1M_2M_3$ of a TRIANGLE $\Delta A_1A_2A_3$ is similar to $\Delta A_1A_2A_3$.

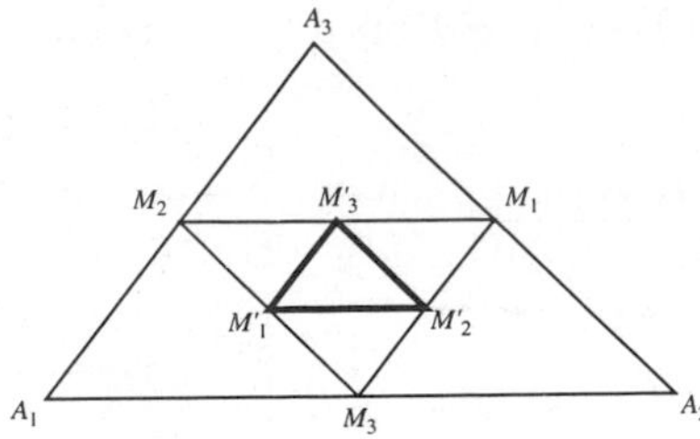

see also ANTICOMPLEMENTARY TRIANGLE

References
Coxeter, H. S. M. and Greitzer, S. L. *Geometry Revisited.* Washington, DC: Math. Assoc. Amer., pp. 18–20, 1967.
Dixon, R. *Mathographics.* New York: Dover, p. 56, 1991.

Medial Triangle Locus Theorem

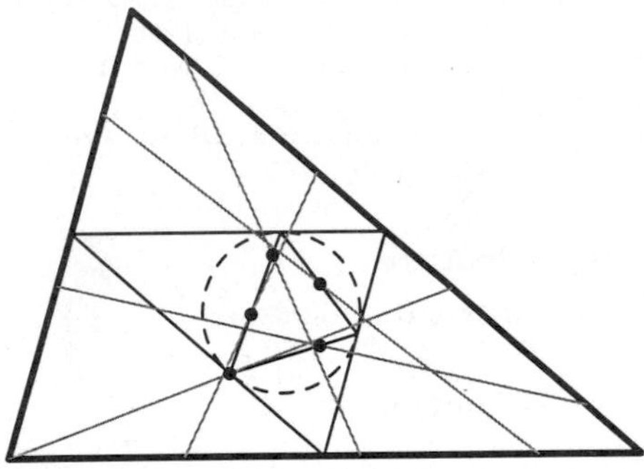

Given an original triangle (thick line), find the MEDIAL TRIANGLE (outer thin line) and its INCIRCLE. Take the PEDAL TRIANGLE (inner thin line) of the MEDIAL TRIANGLE with the INCENTER as the PEDAL POINT. Now pick any point on the original triangle, and connect it to the point located a half-PERIMETER away (gray lines). Then the locus of the MIDPOINTS of these lines (the •s in the above diagram) is the PEDAL TRIANGLE.

References
Honsberger, R. *More Mathematical Morsels.* Washington, DC: Math. Assoc. Amer., pp. 261–267, 1991.
Tsintsifas, G. "Problem 674." *Crux Math.*, p. 256, 1982.

Median Point

see CENTROID (GEOMETRIC)

Median (Statistics)

The middle value of a distribution or average of the two middle items, denoted $\mu_{1/2}$ or $\tilde{x}$. For small samples, the MEAN is more efficient than the median and approximately $\pi/2$ less. It is less sensitive to outliers than the MEAN (Kenney and Keeping 1962, p. 211).

For large N samples with population median $\tilde{x}_0$,

$$\mu_{\tilde{x}} = \tilde{x}_0$$
$$\sigma_{\tilde{x}}{}^2 = \frac{1}{8Nf^2(\tilde{x}_0)}.$$

The median is an L-ESTIMATE (Press *et al.* 1992).

see also MEAN, MIDRANGE, MODE

References
Kenney, J. F. and Keeping, E. S. *Mathematics of Statistics, Pt. 1, 3rd ed.* Princeton, NJ: Van Nostrand, 1962.
Press, W. H.; Flannery, B. P.; Teukolsky, S. A.; and Vetterling, W. T. *Numerical Recipes in FORTRAN: The Art of Scientific Computing, 2nd ed.* Cambridge, England: Cambridge University Press, p. 694, 1992.

Median (Triangle)

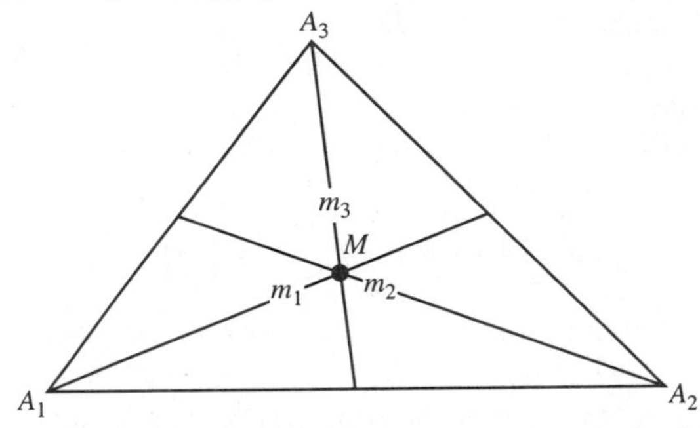

The CEVIAN from a TRIANGLE'S VERTEX to the MIDPOINT of the opposite side is called a median of the TRIANGLE. The three medians of any TRIANGLE are CONCURRENT, meeting in the TRIANGLE'S CENTROID (which has TRILINEAR COORDINATES $1/a : 1/b : 1/c$). In addition, the medians of a TRIANGLE divide one another in the ratio 2:1. A median also bisects the AREA of a TRIANGLE.

Let m_i denote the length of the median of the ith side a_i. Then

$$m_1{}^2 = \tfrac{1}{4}(2a_2{}^2 + 2a_3{}^2 - a_1{}^2) \quad (1)$$
$$m_1{}^2 + m_2{}^2 + m_3{}^2 = \tfrac{3}{4}(a_1{}^2 + a_2{}^2 + a_3{}^2) \quad (2)$$

(Johnson 1929, p. 68). The AREA of a TRIANGLE can be expressed in terms of the medians by

$$A = \tfrac{4}{3}\sqrt{s_m(s_m - m_1)(s_m - m_2)(s_m - m_3)}, \quad (3)$$

where

$$s_m \equiv \tfrac{1}{2}(m_1 + m_2 + m_3). \quad (4)$$

see also BIMEDIAN, EXMEDIAN, EXMEDIAN POINT, HERONIAN TRIANGLE, MEDIAL TRIANGLE

References
Coxeter, H. S. M. and Greitzer, S. L. *Geometry Revisited.* Washington, DC: Math. Assoc. Amer., pp. 7–8, 1967.
Johnson, R. A. *Modern Geometry: An Elementary Treatise on the Geometry of the Triangle and the Circle.* Boston, MA: Houghton Mifflin, pp. 68 and 173–175, 1929.

Median Triangle

A TRIANGLE whose sides are equal and PARALLEL to the MEDIANS of a given TRIANGLE. The median triangle of the median triangle is similar to the given TRIANGLE in the ratio 3/4.

References
Johnson, R. A. *Modern Geometry: An Elementary Treatise on the Geometry of the Triangle and the Circle.* Boston, MA: Houghton Mifflin, pp. 282–283, 1929.

Mediant

Given a FAREY SEQUENCE with consecutive terms h/k and h'/k', then the mediant is defined as the reduced form of the fraction $(h+h')/(k+k')$.

see also FAREY SEQUENCE

References

Conway, J. H. and Guy, R. K. "Farey Fractions and Ford Circles." *The Book of Numbers.* New York: Springer-Verlag, pp. 152–154, 1996.

Mega

Defined in terms of CIRCLE NOTATION by Steinhaus (1983, pp. 28–29) as

⓶ = [4] = [2 in triangle in triangle] = [2^2 in triangle] = [4^4] = [256]

where STEINHAUS-MOSER NOTATION has also been used.

see also MEGISTRON, MOSER, STEINHAUS-MOSER NOTATION

References

Steinhaus, H. *Mathematical Snapshots, 3rd American ed.* New York: Oxford University Press, 1983.

Megistron

A very LARGE NUMBER defined in terms of CIRCLE NOTATION by Steinhaus (1983) as ⑩.

see also MEGA, MOSER

References

Steinhaus, H. *Mathematical Snapshots, 3rd American ed.* New York: Oxford University Press, pp. 28–29, 1983.

Mehler's Bessel Function Formula

$$J_0(x) = \frac{2}{\pi}\int_0^\infty \sin(x\cosh t)\,dt,$$

where $J_0(x)$ is a zeroth order BESSEL FUNCTION OF THE FIRST KIND.

References

Iyanaga, S. and Kawada, Y. (Eds.). *Encyclopedic Dictionary of Mathematics.* Cambridge, MA: MIT Press, p. 1472, 1980.

Mehler-Dirichlet Integral

$$P_n(\cos\alpha) = \frac{\sqrt{2}}{\pi}\int_0^\alpha \frac{\cos[(n+\frac{1}{2})\phi]}{\sqrt{\cos\phi - \cos\alpha}}\,d\phi,$$

where $P_n(x)$ is a LEGENDRE POLYNOMIAL.

Mehler's Hermite Polynomial Formula

$$\sum_{n=0}^\infty \frac{H_n(x)H_n(y)}{n!}(\tfrac{1}{2}w)^n = (1+4w^2)^{-1/2}\exp\left[\frac{2xyw-(x^2+y^2)w^2}{1-w^2}\right],$$

where $H_n(x)$ is a HERMITE POLYNOMIAL.

References

Almqvist, G. and Zeilberger, D. "The Method of Differentiating Under the Integral Sign." *J. Symb. Comput.* **10**, 571–591, 1990.

Foata, D. "A Combinatorial Proof of the Mehler Formula." *J. Comb. Th. Ser. A* **24**, 250–259, 1978.

Petkovšek, M.; Wilf, H. S.; and Zeilberger, D. *A=B.* Wellesley, MA: A. K. Peters, pp. 194–195, 1996.

Rainville, E. D. *Special Functions.* New York: Chelsea, p. 198, 1971.

Szegő, G. *Orthogonal Polynomials, 4th ed.* Providence, RI: Amer. Math. Soc., p. 380, 1975.

Mehler Quadrature

see JACOBI-GAUSS QUADRATURE

Meijer's G-Function

$$G_{p,q}^{m,n}\left(x\middle|{}^{a_1,\ldots,a_p}_{b_1,\ldots,b_p}\right) \equiv \frac{1}{2\pi i} \times \int_{\gamma_L} \frac{\prod_{j=1}^m \Gamma(b_j - z)\prod_{j=1}^n(1-a_j+s)}{\prod_{j=m+1}^q \Gamma(1-b_j+z)\prod_{j=n+1}^q \Gamma(q_j - z)} x^z\,dz,$$

where $\Gamma(z)$ is the GAMMA FUNCTION. The CONTOUR γ_L and other details are discussed by Gradshteyn and Ryzhik (1980, pp. 896–903 and 1068–1071). Prudnikov *et al.* (1990) contains an extensive nearly 200-page listing of formulas for the Meijer G-function.

see also FOX'S H-FUNCTION, G-FUNCTION, MACROBERT'S E-FUNCTION, RAMANUJAN g- AND G-FUNCTIONS

References

Gradshteyn, I. S. and Ryzhik, I. M. *Tables of Integrals, Series, and Products, 5th ed.* San Diego, CA: Academic Press, 1979.

Luke, Y. L. *The Special Functions and Their Approximations, 2 vols.* New York: Academic Press, 1969.

Mathai, A. M. *A Handbook of Generalized Special Functions for Statistical and Physical Sciences.* New York: Oxford University Press, 1993.

Prudnikov, A. P.; Marichev, O. I.; and Brychkov, Yu. A.; *Integrals and Series, Vol. 3: More Special Functions.* Newark, NJ: Gordon and Breach, 1990.

Meissel's Formula

A modification of LEGENDRE'S FORMULA for the PRIME COUNTING FUNCTION $\pi(x)$. It starts with

$$\begin{aligned}\lfloor x\rfloor = 1 + \sum_{1\le i\le a}\left\lfloor\frac{x}{p_i}\right\rfloor - \sum_{1\le i<j\le a}\left\lfloor\frac{x}{p_ip_j}\right\rfloor \\ + \sum_{1\le i<j<k\le a}\left\lfloor\frac{x}{p_ip_jp_k}\right\rfloor - \ldots \\ + \pi(x) - a + P_2(x,a) + P_3(x,a) + \ldots,\end{aligned} \quad (1)$$

where $\lfloor x\rfloor$ is the FLOOR FUNCTION, $P_2(x,a)$ is the number of INTEGERS $p_ip_j \le x$ with $a+1 \le j \le j$, and $P_3(x,a)$ is the number of INTEGERS $p_ip_jp_k \le x$ with $a+1 \le i \le j \le k$. Identities satisfied by the Ps include

$$P_2(x,a) = \sum\left[\pi\left(\frac{x}{p_i}\right) - (i-1)\right] \quad (2)$$

for $p_a < p_i \le \sqrt{x}$ and

$$\begin{aligned}P_3(x,a) &= \sum_{i>a} P_2\left(\frac{x}{p_i}, a\right) \\ &= \sum_{i=a+1}^{c}\sum_{j=i}^{\pi(\sqrt{x/p_i})}\left[\pi\left(\frac{x}{p_ip_j}\right) - (j-1)\right]. \end{aligned}\quad (3)$$

Meissel's formula is

$$\begin{aligned}\pi(x) = \lfloor x\rfloor - \sum_{i=1}^{c}\left\lfloor\frac{x}{p_i}\right\rfloor + \sum_{1\le i\le j\le c}\left\lfloor\frac{x}{p_ip_j}\right\rfloor - \ldots \\ + \tfrac{1}{2}(b+c-2)(b-c+1) - \sum_{c\le i\le b}\pi\left(\frac{x}{p_i}\right),\end{aligned} \quad (4)$$

where

$$b \equiv \pi(x^{1/2}) \quad (5)$$

$$c \equiv \pi(x^{1/3}). \quad (6)$$

Taking the derivation one step further yields LEHMER'S FORMULA.

see also LEGENDRE'S FORMULA, LEHMER'S FORMULA, PRIME COUNTING FUNCTION

References

Riesel, H. "Meissel's Formula." *Prime Numbers and Computer Methods for Factorization, 2nd ed.* Boston, MA: Birkhäuser, p. 12, 1994.

Mellin Transform

$$\phi(z) = \int_0^\infty t^{z-1} f(t)\, dt$$

$$f(t) = \frac{1}{2\pi i}\int_{-\infty}^{\infty} t^{-z}\phi(z)\, dz.$$

see also STRASSEN FORMULAS

References

Arfken, G. *Mathematical Methods for Physicists, 3rd ed.* Orlando, FL: Academic Press, p. 795, 1985.

Bracewell, R. *The Fourier Transform and Its Applications.* New York: McGraw-Hill, pp. 254–257, 1965.

Morse, P. M. and Feshbach, H. *Methods of Theoretical Physics, Part I.* New York: McGraw-Hill, pp. 469–471, 1953.

Melnikov-Arnold Integral

$$A_m(\lambda) \equiv \int_{-\infty}^{\infty}\cos\left[\tfrac{1}{2}m\phi(t) - \lambda t\right] dt,$$

where the function

$$\phi(t) \equiv 4\tan^{-1}(e^t) - \pi$$

describes the motion along the pendulum SEPARATRIX. Chirikov (1979) has shown that this integral has the approximate value

$$A_m(\lambda) \approx \begin{cases} \frac{4\pi(2\lambda)^{m-1}}{\Gamma(m)}e^{-\pi\lambda/2} & \text{for } \lambda > 0 \\ -\frac{4e^{-\pi|\lambda|/2}}{(2|l|)^{m+1}}\Gamma(m+1)\sin(\pi m) & \text{for } \lambda < 0.\end{cases}$$

References

Chirikov, B. V. "A Universal Instability of Many-Dimensional Oscillator Systems." *Phys. Rep.* **52**, 264–379, 1979.

Melodic Series

If $a_1, a_2, a_3, \ldots$ is an ARTISTIC SERIES, then $1/a_1, 1/a_2, 1/a_3, \ldots$ is a MELODIC SERIES. The RECURRENCE RELATION obeyed by melodic series is

$$b_{i+3} = \frac{b_i {b_{i+2}}^2}{{b_{i+1}}^2} + \frac{{b_{i+2}}^2}{b_{i+1}} - b_{i+2}.$$

see also ARTISTIC SERIES

References

Duffin, R. J. "On Seeing Progressions of Constant Cross Ratio." *Amer. Math. Monthly* **100**, 38–47, 1993.

MEM

see MAXIMUM ENTROPY METHOD

Memoryless

A variable x is memoryless with respect to t if, for all s with $t \neq 0$,

$$P(x > s + t | x > t) = P(x > s). \quad (1)$$

Equivalently,

$$\frac{P(x > s + t, x > t)}{P(x > t)} = P(x > s) \quad (2)$$

$$P(x > s + t) = P(x > s)P(x > t). \quad (3)$$

The EXPONENTIAL DISTRIBUTION, which satisfies

$$P(x > t) = e^{-\lambda t} \quad (4)$$

$$P(x > s + t) = e^{-\lambda(s+t)}, \quad (5)$$

and therefore

$$\begin{aligned} P(x > s + t) = P(x > s)P(x > t) &= e^{-\lambda s}e^{-\lambda t} \\ &= e^{-\lambda(s+t)}, \end{aligned} \quad (6)$$

is the only memoryless random distribution.

see also EXPONENTIAL DISTRIBUTION

Ménage Number

see MARRIED COUPLES PROBLEM

Ménage Problem

see MARRIED COUPLES PROBLEM

Menasco's Theorem

For a BRAID with M strands, R components, P positive crossings, and N negative crossings,

$$\begin{cases} P - N \leq U_+ + M - R & \text{if } P \geq N \\ P - N \leq U_- + M - R & \text{if } P \leq N, \end{cases}$$

where $U_\pm$ are the smallest number of positive and negative crossings which must be changed to crossings of the opposite sign. These inequalities imply BENNEQUIN'S CONJECTURE. Menasco's theorem can be extended to arbitrary knot diagrams.

see also BENNEQUIN'S CONJECTURE, BRAID, UNKNOTTING NUMBER

References

Cipra, B. "From Knot to Unknot." *What's Happening in the Mathematical Sciences, Vol. 2.* Providence, RI: Amer. Math. Soc., pp. 8–13, 1994.

Menasco, W. W. "The Bennequin-Milnor Unknotting Conjectures." *C. R. Acad. Sci. Paris Sér. I Math.* **318**, 831–836, 1994.

Menelaus' Theorem

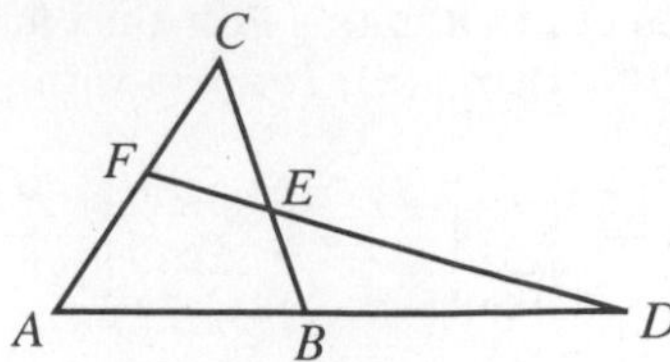

For TRIANGLES in the PLANE,

$$AD \cdot BE \cdot CF = BD \cdot CE \cdot AF. \quad (1)$$

For SPHERICAL TRIANGLES,

$$\sin AD \cdot \sin BE \cdot \sin CF = \sin BD \cdot \sin CE \cdot \sin AF. \quad (2)$$

This can be generalized to n-gons $P = [V_1, \ldots, V_n]$, where a transversal cuts the side V_iV_{i+1} in W_i for $i = 1, \ldots, n$, by

$$\prod_{i=1}^{n} \left[\frac{V_iW_i}{W_iV_{i+1}}\right] = (-1)^n. \quad (3)$$

Here, $AB||CD$ and

$$\left[\frac{AB}{CD}\right] \quad (4)$$

is the ratio of the lengths $[A, B]$ and $[C, D]$ with a PLUS or MINUS SIGN depending if these segments have the same or opposite directions (Grünbaum and Shepard 1995). The case $n = 3$ is PASCH'S AXIOM.

see also CEVA'S THEOREM, HOEHN'S THEOREM, PASCH'S AXIOM

References

Beyer, W. H. (Ed.) *CRC Standard Mathematical Tables, 28th ed.* Boca Raton, FL: CRC Press, p. 122, 1987.

Coxeter, H. S. M. and Greitzer, S. L. *Geometry Revisited.* Washington, DC: Math. Assoc. Amer., pp. 66–67, 1967.

Grünbaum, B. and Shepard, G. C. "Ceva, Menelaus, and the Area Principle." *Math. Mag.* **68**, 254–268, 1995.

Pedoe, D. *Circles: A Mathematical View, rev. ed.* Washington, DC: Math. Assoc. Amer., p. xxi, 1995.

Menger's n-Arc Theorem

Let G be a graph with A and B two disjoint n-tuples of VERTICES. Then either G contains n pairwise disjoint AB-paths, each connecting a point of A and a point of B, or there exists a set of fewer than n VERTICES that separate A and B.

References

Menger, K. *Kurventheorie.* Leipzig, Germany: Teubner, 1932.

Menger Sponge

A FRACTAL which is the 3-D analog of the SIERPIŃSKI CARPET. Let N_n be the number of filled boxes, L_n the length of a side of a hole, and V_n the fractional VOLUME after the nth iteration.

$$N_n = 20^n \tag{1}$$
$$L_n = (\tfrac{1}{3})^n = 3^{-n} \tag{2}$$
$$V_n = {L_n}^3 N_n = (\tfrac{20}{27})^n. \tag{3}$$

The CAPACITY DIMENSION is therefore

$$\begin{aligned} d_{\text{cap}} &= -\lim_{n\to\infty} \frac{\ln N_n}{\ln L_n} = -\lim_{n\to\infty} \frac{\ln(20^n)}{\ln(3^{-n})} \\ &= \frac{\ln 20}{\ln 3} = \frac{\ln(2^2 \cdot 5)}{\ln 3} = \frac{2\ln 2 + \ln 5}{\ln 3} \\ &= 2.726833028\ldots. \end{aligned} \tag{4}$$

J. Mosely is leading an effort to construct a large Menger sponge out of old business cards.

see also SIERPIŃSKI CARPET, TETRIX

References

Dickau, R. M. "Menger (Sierpinski) Sponge." `http://forum.swarthmore.edu/advanced/robertd/sponge.html`.

Mosely, J. "Menger's Sponge (Depth 3)." `http://world.std.com/~j9/sponge/`.

Menn's Surface

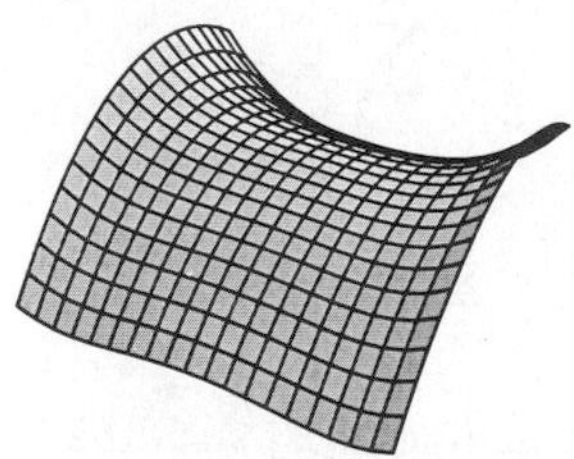

A surface given by the parametric equations

$$x(u,v) = u$$
$$y(u,v) = v$$
$$z(u,v) = au^4 + u^2v - v^2.$$

References

Gray, A. *Modern Differential Geometry of Curves and Surfaces.* Boca Raton, FL: CRC Press, p. 631, 1993.

Mensuration Formula

A mensuration formula is simply a formula for computing the length-related properties of an object (such as AREA, CIRCUMRADIUS, etc., of a POLYGON) based on other known lengths, areas, etc. Beyer (1987) gives a collection of such formulas for various plane and solid geometric figures.

References

Beyer, W. H. *CRC Standard Mathematical Tables, 28th ed.* Boca Raton, FL: CRC Press, pp. 121–133, 1987.

Mercator Projection

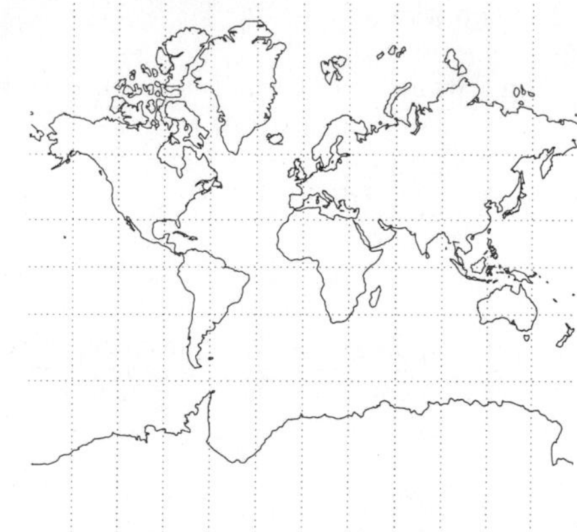

The following equations place the x-AXIS of the projection on the equator and the y-AXIS at LONGITUDE λ_0, where λ is the LONGITUDE and ϕ is the LATITUDE.

$$x = \lambda - \lambda_0 \tag{1}$$
$$y = \ln[\tan(\tfrac{1}{4}\pi + \tfrac{1}{2}\phi)] \tag{2}$$
$$= \tfrac{1}{2}\ln\left(\frac{1+\sin\phi}{1-\sin\phi}\right) \tag{3}$$
$$= \sinh^{-1}(\tan\phi) \tag{4}$$
$$= \tanh^{-1}(\sin\phi) \tag{5}$$
$$= \ln(\tan\phi + \sec\phi). \tag{6}$$

The inverse FORMULAS are

$$\phi = \tfrac{1}{2}\pi - 2\tan^{-1}(e^{-y}) = \tan^{-1}(\sinh y) \tag{7}$$
$$\lambda = x + \lambda_0. \tag{8}$$

LOXODROMES are straight lines and GREAT CIRCLES are curved.

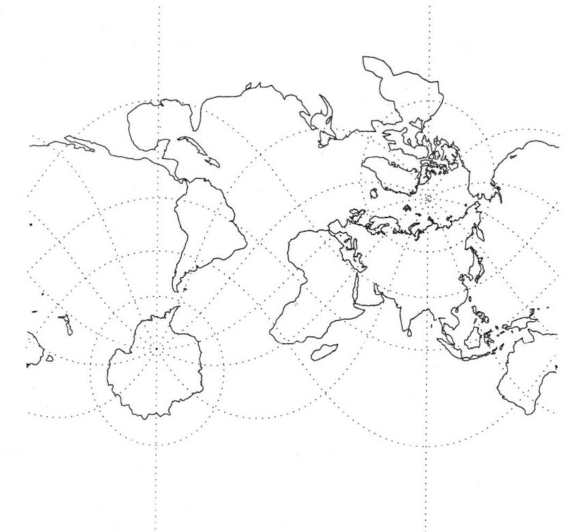

An oblique form of the Mercator projection is illustrated above. It has equations

$$x = \frac{\tan^{-1}[\tan\phi\cos\phi_p + \sin\phi_p\sin(\lambda-\lambda_0)]}{\cos(\lambda-\lambda_0)} \tag{9}$$

$$y = \tfrac{1}{2}\ln\left(\frac{1+A}{1-A}\right) = \tanh^{-1}A, \tag{10}$$

where

$$\lambda_p = \tan^{-1}\left(\frac{\cos\phi_1\sin\phi_2\cos\lambda_1 - \sin\phi_1\cos\phi_2\cos\lambda_2}{\sin\phi_1\cos\phi_2\sin\lambda_2 - \cos\phi_1\sin\phi_2\sin\lambda_1}\right) \tag{11}$$

$$\phi_p = \tan^{-1}\left(-\frac{\cos(\lambda_p-\lambda_1)}{\tan\phi_1}\right) \tag{12}$$

$$A = \sin\phi_p\sin\phi - \cos\phi_p\cos\phi\sin(\lambda-\lambda_0). \tag{13}$$

The inverse FORMULAS are

$$\phi = \sin^{-1}\left(\sin\phi_p\tanh y + \frac{\cos\phi_p\sin x}{\cosh y}\right) \tag{14}$$

$$\lambda = \lambda_0 + \tan^{-1}\left(\frac{\sin\phi_p\sin x - \cos\phi_p\sinh y}{\cos x}\right). \tag{15}$$

There is also a transverse form of the Mercator projection, illustrated above. It is given by the equations

$$x = \tfrac{1}{2}\ln\left(\frac{1+B}{1-B}\right) = \tanh^{-1}B \tag{16}$$

$$y = \tan^{-1}\left[\frac{\tan\phi}{\cos(\lambda-\lambda_0)}\right] - \phi_0 \tag{17}$$

$$\phi = \sin^{-1}\left(\frac{\sin D}{\cosh x}\right) \tag{18}$$

$$\lambda = \lambda_0 + \tan^{-1}\left(\frac{\sinh x}{\cos D}\right), \tag{19}$$

where

$$B \equiv \cos\phi\sin(\lambda-\lambda_0) \tag{20}$$

$$D \equiv y + \phi_0. \tag{21}$$

Finally, the "universal transverse Mercator projection" is a MAP PROJECTION which maps the SPHERE into 60 zones of 6° each, with each zone mapped by a transverse Mercator projection with central MERIDIAN in the center of the zone. The zones extend from 80° S to 84° N (Dana).

see also SPHERICAL SPIRAL

References

Dana, P. H. "Map Projections." http://www.utexas.edu/depts/grg/gcraft/notes/mapporoj/mapproj.html.

Snyder, J. P. *Map Projections—A Working Manual.* U. S. Geological Survey Professional Paper 1395. Washington, DC: U. S. Government Printing Office, pp. 38–75, 1987.

Mercator Series

The TAYLOR SERIES for the NATURAL LOGARITHM

$$\ln(1+x) = x - \tfrac{1}{2}x^2 + \tfrac{1}{3}x^3 - \ldots$$

which was found by Newton, but independently discovered and first published by Mercator in 1668.

see also LOGARITHMIC NUMBER, NATURAL LOGARITHM

Mercer's Theorem

see RIEMANN-LEBESGUE LEMMA

Mergelyan-Wesler Theorem

Let $P = \{D_1, D_2, \ldots\}$ be an infinite set of disjoint open DISKS D_n of radius r_n such that the union is almost the unit DISK. Then

$$\sum_{n=1}^{\infty} r_n = \infty. \tag{1}$$

Define

$$M_x(P) \equiv \sum_{n=1}^{\infty} {r_n}^x. \tag{2}$$

Then there is a number $e(P)$ such that $M_x(P)$ diverges for $x < e(P)$ and converges for $x > e(P)$. The above theorem gives

$$1 < e(P) < 2. \tag{3}$$

There exists a constant which improves the inequality, and the best value known is

$$S = 1.306951\ldots. \tag{4}$$

References

Le Lionnais, F. *Les nombres remarquables.* Paris: Hermann, pp. 36–37, 1983.

Mandelbrot, B. B. *Fractals.* San Francisco, CA: W. H. Freeman, p. 187, 1977.

Melzack, Z. A. "On the Solid Packing Constant for Circles." *Math. Comput.* **23**, 1969.

Meridian

A line of constant LONGITUDE on a SPHEROID (or SPHERE). More generally, a meridian of a SURFACE OF REVOLUTION is the intersection of the surface with a PLANE containing the axis of revolution.

see also LATITUDE, LONGITUDE, PARALLEL (SURFACE OF REVOLUTION), SURFACE OF REVOLUTION

References

Gray, A. *Modern Differential Geometry of Curves and Surfaces.* Boca Raton, FL: CRC Press, p. 358, 1993.

Meromorphic

A meromorphic FUNCTION is complex analytic in all but a discrete subset of its domain, and at those singularities it must go to infinity like a POLYNOMIAL (i.e., have no ESSENTIAL SINGULARITIES). An equivalent definition of a meromorphic function is a complex analytic MAP to the RIEMANN SPHERE.

see also ESSENTIAL SINGULARITY, RIEMANN SPHERE

References

Morse, P. M. and Feshbach, H. *Methods of Theoretical Physics, Part I.* New York: McGraw-Hill, pp. 382–383, 1953.

Mersenne Number

A number of the form

$$M_n \equiv 2^n - 1 \tag{1}$$

for n an INTEGER is known as a Mersenne number. The Mersenne numbers are therefore 2-REPDIGITS, and also the numbers obtained by setting $x = 1$ in a FERMAT POLYNOMIAL. The first few are 1, 3, 7, 15, 31, 63, 127, 255, ... (Sloane's A000225).

The number of digits D in the Mersenne number M_n is

$$D = \lfloor \log(2^n - 1) + 1 \rfloor, \tag{2}$$

where $\lfloor x \rfloor$ is the FLOOR FUNCTION, which, for large n, gives

$$D \approx \lfloor n \log 2 + 1 \rfloor \approx \lfloor 0.301029n + 1 \rfloor = \lfloor 0.301029n \rfloor + 1. \tag{3}$$

In order for the Mersenne number M_n to be PRIME, n must be PRIME. This is true since for COMPOSITE n with factors r and s, $n = rs$. Therefore, $2^n - 1$ can be written as $2^{rs} - 1$, which is a BINOMIAL NUMBER and can be factored. Since the most interest in Mersenne numbers arises from attempts to factor them, many authors prefer to define a Mersenne number as a number of the above form

$$M_p = 2^p - 1, \tag{4}$$

but with p restricted to PRIME values.

The search for MERSENNE PRIMES is one of the most computationally intensive and actively pursued areas of advanced and distributed computing.

see also CUNNINGHAM NUMBER, EBERHART'S CONJECTURE, FERMAT NUMBER, LUCAS-LEHMER TEST, MERSENNE PRIME, PERFECT NUMBER, REPUNIT, RIESEL NUMBER, SIERPIŃSKI NUMBER OF THE SECOND KIND, SOPHIE GERMAIN PRIME, SUPERPERFECT NUMBER, WIEFERICH PRIME

References

Pappas, T. "Mersenne's Number." *The Joy of Mathematics.* San Carlos, CA: Wide World Publ./Tetra, p. 211, 1989.

Shanks, D. *Solved and Unsolved Problems in Number Theory, 4th ed.* New York: Chelsea, pp. 14, 18–19, 22, and 29–30, 1993.

Sloane, N. J. A. Sequence A000225/M2655 in "An On-Line Version of the Encyclopedia of Integer Sequences."

Mersenne Prime

A MERSENNE NUMBER which is PRIME is called a Mersenne prime. In order for the Mersenne number M_n defined by

$$M_n \equiv 2^n - 1$$

for n an INTEGER to be PRIME, n must be PRIME. This is true since for COMPOSITE n with factors r and s, $n = rs$. Therefore, $2^n - 1$ can be written as $2^{rs} - 1$, which is a BINOMIAL NUMBER and can be factored. Every MERSENNE PRIME gives rise to a PERFECT NUMBER.

If $n \equiv 3 \pmod 4$ is a PRIME, then $2n + 1$ DIVIDES M_n IFF $2n+1$ is PRIME. It is also true that PRIME divisors of $2^p - 1$ must have the form $2kp + 1$ where k is a POSITIVE INTEGER and simultaneously of either the form $8n+1$ or $8n - 1$ (Uspensky and Heaslet). A PRIME factor p of a Mersenne number $M_q = 2^q - 1$ is a WIEFERICH PRIME IFF $p^2 | 2^q - 1$, Therefore, MERSENNE PRIMES are *not* WIEFERICH PRIMES. All known Mersenne numbers M_p with p PRIME are SQUAREFREE. However, Guy (1994) believes that there are M_p which are not SQUAREFREE.

TRIAL DIVISION is often used to establish the COMPOSITENESS of a potential Mersenne prime. This test immediately shows M_p to be COMPOSITE for $p = 11$, 23, 83, 131, 179, 191, 239, and 251 (with small factors 23, 47, 167, 263, 359, 383, 479, and 503, respectively). A much more powerful primality test for M_p is the LUCAS-LEHMER TEST.

It has been conjectured that there exist an infinite number of Mersenne primes, although finding them is computationally very challenging. The table below gives the index p of known Mersenne primes (Sloane's A000043) M_p, together with the number of digits, discovery years, and discoverer. A similar table has been compiled by C. Caldwell. Note that the region after the 35th known Mersenne prime has not been completely searched, so identification of "the" 36th Mersenne prime is tentative. L. Welsh maintains an extensive bibliography and history of Mersenne numbers. G. Woltman has organized a distributed search program via the Internet in which hundreds of volunteers use their personal computers to perform pieces of the search.

#	p	Digits	Year	Published Reference
1	2	1	Anc.	
2	3	1	Anc.	
3	5	2	Anc.	
4	7	3	Anc.	
5	13	4	1461	Reguis 1536, Cataldi 1603
6	17	6	1588	Cataldi 1603
7	19	6	1588	Cataldi 1603
8	31	10	1750	Euler 1772
9	61	19	1883	Pervouchine 1883, Seelhoff 1886
10	89	27	1911	Powers 1911
11	107	33	1913	Powers 1914
12	127	39	1876	Lucas 1876
13	521	157	1952	Lehmer 1952–3
14	607	183	1952	Lehmer 1952–3
15	1279	386	1952	Lehmer 1952–3
16	2203	664	1952	Lehmer 1952–3
17	2281	687	1952	Lehmer 1952–3
18	3217	969	1957	Riesel 1957
19	4253	1281	1961	Hurwitz 1961
20	4423	1332	1961	Hurwitz 1961
21	9689	2917	1963	Gillies 1964
22	9941	2993	1963	Gillies 1964
23	11213	3376	1963	Gillies 1964
24	19937	6002	1971	Tuckerman 1971
25	21701	6533	1978	Noll and Nickel 1980
26	23209	6987	1979	Noll 1980
27	44497	13395	1979	Nelson and Slowinski 1979
28	86243	25962	1982	Slowinski 1982
29	110503	33265	1988	Colquitt and Welsh 1991
30	132049	39751	1983	Slowinski 1988
31	216091	65050	1985	Slowinski 1989
32	756839	227832	1992	Gage and Slowinski 1992
33	859433	258716	1994	Gage and Slowinski 1994
34	1257787	378632	1996	Slowinski and Gage
35	1398269	420921	1996	Armengaud, Woltman, *et al.*
36?	2976221	895832	1997	Spence
37?	3021377	909526	1998	Clarkson, Woltman, *et al.*

see also CUNNINGHAM NUMBER, FERMAT-LUCAS NUMBER, FERMAT NUMBER, FERMAT NUMBER (LUCAS), FERMAT POLYNOMIAL, LUCAS-LEHMER TEST, MERSENNE NUMBER, PERFECT NUMBER, REPUNIT, SUPERPERFECT NUMBER

References

Bateman, P. T.; Selfridge, J. L.; and Wagstaff, S. S. "The New Mersenne Conjecture." *Amer. Math. Monthly* **96**, 125–128, 1989.

Ball, W. W. R. and Coxeter, H. S. M. *Mathematical Recreations and Essays, 13th ed.* New York: Dover, p. 66, 1987.

Beiler, A. H. Ch. 3 in *Recreations in the Theory of Numbers: The Queen of Mathematics Entertains.* New York: Dover, 1966.

Caldwell, C. "Mersenne Primes: History, Theorems and Lists." `http://www.utm.edu/research/primes/mersenne.shtml`.

Caldwell, C. "GIMPS Finds a Prime! $2^{1398269}-1$ is Prime." `http://www.utm.edu/research/primes/notes/1398269/`.

Colquitt, W. N. and Welsh, L. Jr. "A New Mersenne Prime." *Math. Comput.* **56**, 867–870, 1991.

Conway, J. H. and Guy, R. K. "Mersenne's Numbers." In *The Book of Numbers.* New York: Springer-Verlag, pp. 135–137, 1996.

Gillies, D. B. "Three New Mersenne Primes and a Statistical Theory." *Math Comput.* **18**, 93–97, 1964.

Guy, R. K. "Mersenne Primes. Repunits. Fermat Numbers. Primes of Shape $k \cdot 2^n + 2$ [sic]." §A3 in *Unsolved Problems in Number Theory, 2nd ed.* New York: Springer-Verlag, pp. 8–13, 1994.

Haghighi, M. "Computation of Mersenne Primes Using a Cray X-MP." *Intl. J. Comput. Math.* **41**, 251–259, 1992.

Hardy, G. H. and Wright, E. M. *An Introduction to the Theory of Numbers, 5th ed.* Oxford, England: Clarendon Press, pp. 14–16, 1979.

Kraitchik, M. "Mersenne Numbers and Perfect Numbers." §3.5 in *Mathematical Recreations.* New York: W. W. Norton, pp. 70–73, 1942.

Kravitz, S. and Berg, M. "Lucas' Test for Mersenne Numbers $6000 < p < 7000$." *Math. Comput.* **18**, 148–149, 1964.

Lehmer, D. H. "On Lucas's Test for the Primality of Mersenne's Numbers." *J. London Math. Soc.* **10**, 162–165, 1935.

Leyland, P. `ftp://sable.ox.ac.uk/pub/math/factors/mersenne`.

Mersenne, M. *Cogitata Physico-Mathematica.* 1644.

Mersenne Organization. "GIMPS Discovers 36th Known Mersenne Prime, $2^{2976221}-1$ is Now the Largest Known Prime." `http://www.mersenne.org/2976221.htm`.

Mersenne Organization. "GIMPS Discovers 37th Known Mersenne Prime, $2^{3021377}-1$ is Now the Largest Known Prime." `http://www.mersenne.org/3021377.htm`.

Noll, C. and Nickel, L. "The 25th and 26th Mersenne Primes." *Math. Comput.* **35**, 1387–1390, 1980.

Powers, R. E. "The Tenth Perfect Number." *Amer. Math. Monthly* **18**, 195–196, 1911.

Powers, R. E. "Note on a Mersenne Number." *Bull. Amer. Math. Soc.* **40**, 883, 1934.

Sloane, N. J. A. Sequence A000043/M0672 in "An On-Line Version of the Encyclopedia of Integer Sequences."

Slowinski, D. "Searching for the 27th Mersenne Prime." *J. Recreat. Math.* **11**, 258–261, 1978–1979.

Slowinski, D. *Sci. News* **139**, 191, 9/16/1989.

Tuckerman, B. "The 24th Mersenne Prime." *Proc. Nat. Acad. Sci. USA* **68**, 2319–2320, 1971.

Uhler, H. S. "A Brief History of the Investigations on Mersenne Numbers and the Latest Immense Primes." *Scripta Math.* **18**, 122–131, 1952.

Uspensky, J. V. and Heaslet, M. A. *Elementary Number Theory.* New York: McGraw-Hill, 1939.

Weisstein, E. W. "Mersenne Numbers." `http://www.astro.virginia.edu/~eww6n/math/notebooks/Mersenne.m`.

Welsh, L. "Marin Mersenne." `http://www.scruznet.com/~luke/mersenne.htm`.

Welsh, L. "Mersenne Numbers & Mersenne Primes Bibliography." `http://www.scruznet.com/~luke/biblio.htm`.

Woltman, G. "The GREAT Internet Mersenne Prime Search." `http://www.mersenne.org/prime.htm`.

Mertens Conjecture

Given MERTENS FUNCTION defined by

$$M(n) \equiv \sum_{k=1}^{n} \mu(k), \tag{1}$$

where $\mu(n)$ is the MÖBIUS FUNCTION, Mertens (1897) conjecture states that

$$|M(x)| < x^{1/2} \tag{2}$$

for $x > 1$. The conjecture has important implications, since the truth of any equality of the form

$$|M(x)| \leq cx^{1/2} \tag{3}$$

for any fixed c (the form of Mertens conjecture with $c = 1$) would imply the RIEMANN HYPOTHESIS. In 1885, Stieltjes claimed that he had a proof that $M(x)x^{-1/2}$ always stayed between two fixed bounds. However, it seems likely that Stieltjes was mistaken.

Mertens conjecture was proved false by Odlyzko and te Riele (1985). Their proof is indirect and does not produce a specific counterexample, but it does show that

$$\limsup_{x\to\infty} M(x)x^{-1/2} > 1.06 \tag{4}$$

$$\liminf_{x\to\infty} M(x)x^{-1/2} < -1.009. \tag{5}$$

Odlyzko and te Riele (1985) believe that there are no counterexamples to Mertens conjecture for $x \leq 10^{20}$, or even 10^{30}. Pintz (1987) subsequently showed that at least one counterexample to the conjecture occurs for $x \leq 10^{65}$, using a weighted integral average of $M(x)/x$ and a discrete sum involving nontrivial zeros of the RIEMANN ZETA FUNCTION.

It is still not known if

$$\limsup_{x\to\infty} |M(x)|x^{-1/2} = \infty, \tag{6}$$

although it seems very probable (Odlyzko and te Riele 1985).

see also MERTENS FUNCTION, MÖBIUS FUNCTION, RIEMANN HYPOTHESIS

References

Anderson, R. J. "On the Mertens Conjecture for Cusp Forms." *Mathematika* **26**, 236–249, 1979.

Anderson, R. J. "Corrigendum: 'On the Mertens Conjecture for Cusp Forms.'" *Mathematika* **27**, 261, 1980.

Devlin, K. "The Mertens Conjecture." *Irish Math. Soc. Bull.* **17**, 29–43, 1986.

Grupp, F. "On the Mertens Conjecture for Cusp Forms." *Mathematika* **29**, 213–226, 1982.

Jurkat, W. and Peyerimhoff, A. "A Constructive Approach to Kronecker Approximation and Its Application to the Mertens Conjecture." *J. reine angew. Math.* **286/287**, 322–340, 1976.

Mertens, F. "Über eine zahlentheoretische Funktion." *Sitzungsber. Akad. Wiss. Wien IIa* **106**, 761–830, 1897.

Odlyzko, A. M. and te Riele, H. J. J. "Disproof of the Mertens Conjecture." *J. reine angew. Math.* **357**, 138–160, 1985.

Pintz, J. "An Effective Disproof of the Mertens Conjecture." *Astérique* **147–148**, 325–333 and 346, 1987.

te Riele, H. J. J. "Some Historical and Other Notes About the Mertens Conjecture and Its Recent Disproof." *Nieuw Arch. Wisk.* **3**, 237–243, 1985.

Mertens Constant

A constant related to the TWIN PRIMES CONSTANT which appears in the FORMULA for the sum of inverse PRIMES

$$\sum_{p \text{ prime}}^{x} \frac{1}{p} = \ln\ln x + B_1 + o(1) \tag{1}$$

which is given by

$$B_1 = \gamma + \sum_{p \text{ prime}} \left[\ln(1-p^{-1}) + \frac{1}{p}\right] \approx 0.261497. \tag{2}$$

Flajolet and Vardi (1996) show that

$$e^{B_1} = e^{\gamma} \prod_{m=2}^{\infty} \zeta(m)^{\mu(m)/m}, \tag{3}$$

where γ is the EULER-MASCHERONI CONSTANT, $\zeta(n)$ is the RIEMANN ZETA FUNCTION, and $\mu(n)$ is the MÖBIUS FUNCTION. The constant B_1 also occurs in the SUMMATORY FUNCTION of the number of DISTINCT PRIME FACTORS,

$$\sum_{k=2}^{n} \omega(k) = n\ln\ln n + B_1 n + o(n) \tag{4}$$

(Hardy and Wright 1979, p. 355).

The related constant

$$B_2 = \gamma + \sum_{p \text{ prime}} \left[\ln(1-p^{-1}) + \frac{1}{p-1}\right] \approx 1.034653 \tag{5}$$

appears in the SUMMATORY FUNCTION of the DIVISOR FUNCTION $\sigma_0(n) = \Omega(n)$,

$$\sum_{k=2}^{n} \Omega(k) = n\ln\ln n + B_2 + o(n) \tag{6}$$

(Hardy and Wright 1979, p. 355).

see also BRUN'S CONSTANT, PRIME NUMBER, TWIN PRIMES CONSTANT

References

Flajolet, P. and Vardi, I. "Zeta Function Expansions of Classical Constants." Unpublished manuscript. 1996. `http://pauillac.inria.fr/algo/flajolet/Publications/landau.ps`.

Hardy, G. H. and Weight, E. M. *An Introduction to the Theory of Numbers, 5th ed.* Oxford, England: Oxford University Press, pp. 351 and 355, 1979.

Mertens Function

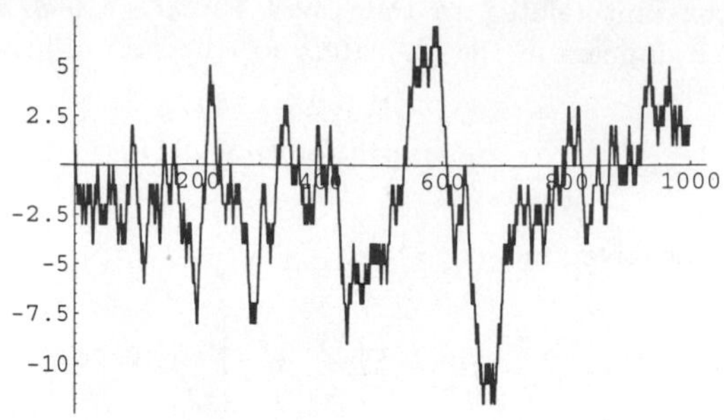

The summary function

$$M(n) \equiv \sum_{k=1}^{n} \mu(k) = \frac{6}{\pi^2}n + \mathcal{O}(\sqrt{n}),$$

where $\mu(n)$ is the MÖBIUS FUNCTION. The first few values are 1, 0, −1, −1, −2, −1, −2, −2, −2, −1, −2, −2, ... (Sloane's A002321). The first few values of n at which $M(n) = 0$ are 2, 39, 40, 58, 65, 93, 101, 145, 149, 150, ... (Sloane's A028442).

Mertens function obeys

$$\sum_{n=1}^{x} M\left(\frac{x}{n}\right) = 1$$

(Lehman 1960). The analytic form is unsolved, although MERTENS CONJECTURE that

$$|M(x)| < x^{1/2}$$

has been disproved.

Lehman (1960) gives an algorithm for computing $M(x)$ with $\mathcal{O}(x^{2/3+\epsilon})$ operations, while the Lagarias-Odlyzko (1987) algorithm for computing the PRIME COUNTING FUNCTION $\pi(x)$ can be modified to give $M(x)$ in $\mathcal{O}(x^{3/5+\epsilon})$ operations.

see also MERTENS CONJECTURE, MÖBIUS FUNCTION

References

Lagarias, J. and Odlyzko, A. "Computing $\pi(x)$: An Analytic Method." J. Algorithms **8**, 173–191, 1987.

Lehman, R. S. "On Liouville's Function." *Math. Comput.* **14**, 311–320, 1960.

Odlyzko, A. M. and te Riele, H. J. J. "Disproof of the Mertens Conjecture." *J. reine angew. Math.* **357**, 138–160, 1985.

Sloane, N. J. A. Sequence A028442/M002321 in "An On-Line Version of the Encyclopedia of Integer Sequences."0102

Mertens Theorem

$$\lim_{x\to\infty} \frac{\prod_{\substack{2\leq p\leq x\\ p \text{ prime}}} \left(1-\frac{1}{p}\right)}{\frac{e^{-\gamma}}{\ln x}} = 1,$$

where γ is the EULER-MASCHERONI CONSTANT and $e^{-\gamma} = 0.56145\ldots$.

References

Hardy, G. H. and Wright, E. M. *An Introduction to the Theory of Numbers, 5th ed.* Oxford, England: Oxford University Press, p. 351, 1979.

Riesel, H. *Prime Numbers and Computer Methods for Factorization, 2nd ed.* Boston, MA: Birkhäuser, pp. 66–67, 1994.

Mertz Apodization Function

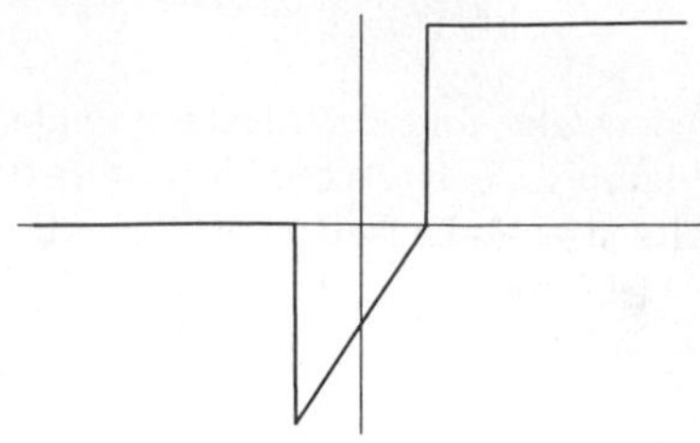

An asymmetrical APODIZATION FUNCTION defined by

$$M(x,b,d) = \begin{cases} 0 & \text{for } x < -b \\ (x-b)/(2b) & \text{for } -b < x < b \\ 1 & \text{for } b < x < b+2d \\ 0 & \text{for } x < b+2d, \end{cases}$$

where the two-sided portion is $2b$ long (total) and the one-sided portion is $b+2d$ long (Schnopper and Thompson 1974, p. 508). The APPARATUS FUNCTION is

$$M_A(k,b,d) = \frac{\sin[2\pi k(b+2d)}{2\pi k} + i\left\{\frac{\cos[2\pi k(b+2d)]}{2\pi k} - \frac{\sin(2b)}{4\pi^2k^2b}\right\}.$$

References

Schnopper, H. W. and Thompson, R. I. "Fourier Spectrometers." In *Methods of Experimental Physics* **12A**. New York: Academic Press, pp. 491–529, 1974.

Mesh Size

When a CLOSED INTERVAL $[a,b]$ is partitioned by points $a < x_1 < x_2 < \ldots < x_{n-1} < b$, the lengths of the resulting intervals between the points are denoted Δx_1, Δx_2, ..., Δx_n, and the value $\max \Delta x_k$ is called the mesh size of the partition.

see also INTEGRAL, LOWER SUM, RIEMANN INTEGRAL, UPPER SUM

Mesokurtic

A distribution with zero KURTOSIS ($\gamma_2 = 0$).

see also KURTOSIS, LEPTOKURTIC

Metabiaugmented Dodecahedron

see JOHNSON SOLID

Metabiaugmented Hexagonal Prism

see JOHNSON SOLID

Metabiaugmented Truncated Dodecahedron

see JOHNSON SOLID

Metabidiminished Icosahedron

see JOHNSON SOLID

Metabidiminished Rhombicosidodecahedron

see JOHNSON SOLID

Metabigyrate Rhombicosidodecahedron

see JOHNSON SOLID

Metadrome

A metadrome is a number whose HEXADECIMAL digits are in strict ascending order. The first few are 0, 1, 2, 3, 4, 5, 6, 7, 8, 9, 10, 11, 12, 13, 14, 15, 18, 19, 20, 21, 22, 23, 24, 25, 26, 27, ... (Sloane's A023784).

see also HEXADECIMAL

References

Sloane, N. J. A. Sequence A023784 in "An On-Line Version of the Encyclopedia of Integer Sequences."

Metagyrate Diminished Rhombicosidodecahedron

see JOHNSON SOLID

Metalogic

see METAMATHEMATICS

Metamathematics

The branch of LOGIC dealing with the study of the combination and application of mathematical symbols, sometimes called METALOGIC. Metamathematics is the study of MATHEMATICS itself, and one of its primary goals is to determine the nature of mathematical reasoning (Hofstadter 1989).

see also LOGIC, MATHEMATICS

References

Birkhoff, G. and Mac Lane, S. *A Survey of Modern Algebra, 3rd ed.* New York: Macmillan, p. 326, 1965.

Hofstadter, D. R. *Gödel, Escher, Bach: An Eternal Golden Braid.* New York: Vintage Books, p. 23, 1989.

Method

A particular way of doing something, sometimes also called an ALGORITHM or PROCEDURE. (According to Petkovšek *et al.* (1996), "a method is a trick that has worked at least twice.")

see also ADAMS-BASHFORTH-MOULTON METHOD, ADAMS' METHOD, BACKUS-GILBERT METHOD, BADER-DEUFLHARD METHOD, BAILEY'S METHOD, BAIRSTOW'S METHOD, BRENT'S FACTORIZATION METHOD, BRENT'S METHOD, CIRCLE METHOD, CONJUGATE GRADIENT METHOD, CRISS-CROSS METHOD, CROUT'S METHOD, DE LA LOUBERE'S METHOD, DIXON'S FACTORIZATION METHOD, DIXON'S RANDOM SQUARES FACTORIZATION METHOD, ELLIPTIC CURVE FACTORIZATION METHOD, EULER'S FACTORIZATION METHOD, EXCLUDENT FACTORIZATION METHOD, EXHAUSTION METHOD, FALSE POSITION METHOD, FERMAT'S FACTORIZATION METHOD, FROBENIUS METHOD, GILL'S METHOD, GOSPER'S METHOD, GRAEFFE'S METHOD, GREENE'S METHOD, HALLEY'S METHOD, HORNER'S METHOD, HUTTON'S METHOD, JACOBI METHOD, KAPS-RENTROP METHODS, LAGUERRE'S METHOD, LAMBERT'S METHOD, LEGENDRE'S FACTORIZATION METHOD, LEHMER METHOD, LEHMER-SCHUR METHOD, LENSTRA ELLIPTIC CURVE METHOD, LIN'S METHOD, LOZENGE METHOD, LUX METHOD, MAPES' METHOD, MAXIMUM ENTROPY METHOD, MILNE'S METHOD, MULLER'S METHOD, NEWTON'S METHOD, NEWTON-RAPHSON METHOD, NUMBER FIELD SIEVE FACTORIZATION METHOD, OVERLAPPING RESONANCE METHOD, POLLARD MONTE CARLO FACTORIZATION METHOD, POLLARD ρ FACTORIZATION METHOD, POLLARD $p-1$ FACTORIZATION METHOD, PREDICTOR-CORRECTOR METHODS, QUADRATIC SIEVE FACTORIZATION METHOD, RESONANCE OVERLAP METHOD, ROSENBROCK METHODS, RUNGE-KUTTA METHOD, SCHRÖDER'S METHOD, SECANT METHOD, SIAMESE METHOD, SIMPLEX METHOD, SNAKE OIL METHOD, SQUARE ROOT METHOD, STEEPEST DESCENT METHOD, TANGENT HYPERBOLAS METHOD, UNDETERMINED COEFFICIENTS METHOD, WILLIAMS $p+1$ FACTORIZATION METHOD, WYNN'S EPSILON METHOD

References

Petkovšek, M.; Wilf, H. S.; and Zeilberger, D. *A=B.* Wellesley, MA: A. K. Peters, p. 117, 1996.

Metric

A NONNEGATIVE function $g(x,y)$ describing the "DISTANCE" between neighboring points for a given SET. A metric satisfies the TRIANGLE INEQUALITY

$$g(x,y)+g(y,z)\geq g(x,z),$$

with equality IFF $x=y$, and is symmetric, so

$$g(x,y)=g(y,x).$$

A SET possessing a metric is called a METRIC SPACE. When viewed as a TENSOR, the metric is called a METRIC TENSOR.

see also CAYLEY-KLEIN-HILBERT METRIC, DISTANCE, FUNDAMENTAL FORMS, HYPERBOLIC METRIC, METRIC ENTROPY, METRIC EQUIVALENCE PROBLEM, METRIC SPACE, METRIC TENSOR, PART METRIC, RIEMANNIAN METRIC, ULTRAMETRIC

References

Gray, A. "Metrics on Surfaces." Ch. 13 in *Modern Differential Geometry of Curves and Surfaces.* Boca Raton, FL: CRC Press, pp. 251–265, 1993.

Metric Entropy

Also known as KOLMOGOROV ENTROPY, KOLMOGOROV-SINAI ENTROPY, or KS Entropy. The metric entropy is 0 for nonchaotic motion and > 0 for CHAOTIC motion.

References
Ott, E. *Chaos in Dynamical Systems.* New York: Cambridge University Press, p. 138, 1993.

Metric Equivalence Problem

1. Find a complete system of invariants, or
2. decide when two METRICS differ only by a coordinate transformation.

The most common statement of the problem is, "Given METRICS g and g', does there exist a coordinate transformation from one to the other?" Christoffel and Lipschitz (1870) showed how to decide this question for two RIEMANNIAN METRICS.

The solution by E. Cartan requires computation of the 10th order COVARIANT DERIVATIVES. The demonstration was simplified by A. Karlhede using the TETRAD formalism so that only seventh order COVARIANT DERIVATIVES need be computed. However, in many common cases, the first or second-order DERIVATIVES are SUFFICIENT to answer the question.

References
Karlhede, A. and Lindström, U. "Finding Space-Time Geometries without Using a Metric." *Gen. Relativity Gravitation* **15**, 597–610, 1983.

Metric Space

A SET S with a global distance FUNCTION (the METRIC g) which, for every two points x, y in S, gives the DISTANCE between them as a NONNEGATIVE REAL NUMBER $g(x, y)$. A metric space must also satisfy

1. $g(x, x) = 0$ IFF $x = y$,
2. $g(x, y) = g(y, x)$,
3. The TRIANGLE INEQUALITY $g(x, y) + g(y, z) \geq g(x, z)$.

References
Munkres, J. R. *Topology: A First Course.* Englewood Cliffs, NJ: Prentice-Hall, 1975.
Rudin, W. *Principles of Mathematical Analysis.* New York: McGraw-Hill, 1976.

Metric Tensor

A TENSOR, also called a RIEMANNIAN METRIC, which is symmetric and POSITIVE DEFINITE. Very roughly, the metric tensor g_{ij} is a function which tells how to compute the distance between any two points in a given SPACE. Its components can be viewed as multiplication factors which must be placed in front of the differential displacements dx_i in a generalized PYTHAGOREAN THEOREM

$$ds^2 = g_{11} dx_1{}^2 + g_{12}\, dx_1\, dx_2 + g_{22}\, dx_2{}^2 + \ldots. \quad (1)$$

In EUCLIDEAN SPACE, $g_{ij} = \delta_{ij}$ where δ is the KRONECKER DELTA (which is 0 for $i \neq j$ and 1 for $i = j$), reproducing the usual form of the PYTHAGOREAN THEOREM

$$ds^2 = dx_1{}^2 + dx_2{}^2 + \ldots. \quad (2)$$

The metric tensor is defined abstractly as an INNER PRODUCT of every TANGENT SPACE of a MANIFOLD such that the INNER PRODUCT is a symmetric, non-degenerate, bilinear form on a VECTOR SPACE. This means that it takes two VECTORS $\mathbf{v}, \mathbf{w}$ as arguments and produces a REAL NUMBER $\langle \mathbf{v}, \mathbf{w} \rangle$ such that

$$\langle k\mathbf{v}, w \rangle = k \langle \mathbf{v}, \mathbf{w} \rangle = \langle \mathbf{v}, k\mathbf{w} \rangle \quad (3)$$

$$\langle \mathbf{v} + \mathbf{w}, \mathbf{x} \rangle = \langle \mathbf{v}, \mathbf{x} \rangle + \langle \mathbf{w}, \mathbf{x} \rangle \quad (4)$$

$$\langle \mathbf{v}, \mathbf{w} + \mathbf{x} \rangle = \langle \mathbf{v}, \mathbf{w} \rangle + \langle \mathbf{v}, \mathbf{x} \rangle \quad (5)$$

$$\langle \mathbf{v}, \mathbf{w} \rangle = \langle \mathbf{w}, \mathbf{v} \rangle \quad (6)$$

$$\langle \mathbf{v}, \mathbf{v} \rangle \geq 0, \quad (7)$$

with equality IFF $\mathbf{v} = 0$.

In coordinate NOTATION (with respect to the basis),

$$g^{\alpha\beta} = \vec{e}^{\alpha} \cdot \vec{e}^{\beta} \quad (8)$$

$$g_{\alpha\beta} = \vec{e}_{\alpha} \cdot \vec{e}_{\beta}. \quad (9)$$

$$g_{\mu\nu} \equiv \frac{\partial \xi^\alpha}{\partial x^\mu} \frac{\partial \xi^\beta}{\partial x^\nu} \eta_{\alpha\beta}, \quad (10)$$

where $\eta_{\alpha\beta}$ is the MINKOWSKI METRIC. This can also be written

$$g = D^{\mathrm{T}} \eta D, \quad (11)$$

where

$$D_{\alpha\mu} \equiv \frac{\partial \xi^\alpha}{\partial x^\mu} \quad (12)$$

$$D_{\alpha\mu}{}^{\mathrm{T}} \equiv D_{\mu\alpha}. \quad (13)$$

$$\frac{\partial}{\partial x^m} g_{il} g^{lk} = \frac{\partial}{\partial x^m} \delta_i^k \quad (14)$$

gives

$$g_{il} \frac{\partial g^{lk}}{\partial x^m} = -g^{lk} \frac{\partial g_{il}}{\partial x^m}. \quad (15)$$

The metric is POSITIVE DEFINITE, so a metric's DISCRIMINANT is POSITIVE. For a metric in 2-space,

$$g \equiv g_{11} g_{22} - g_{12}{}^2 > 0. \quad (16)$$

The ORTHOGONALITY of CONTRAVARIANT and COVARIANT metrics stipulated by

$$g_{ik} g^{ij} = \delta_k^j \quad (17)$$

for $i = 1, \ldots, n$ gives n linear equations relating the $2n$ quantities g_{ij} and g^{ij}. Therefore, if n metrics are known, the others can be determined.

In 2-space,

$$g^{11} = \frac{g_{22}}{g} \qquad (18)$$

$$g^{12} = g^{21} = -\frac{g_{12}}{g} \qquad (19)$$

$$g^{22} = \frac{g_{11}}{g}. \qquad (20)$$

If g is symmetric, then

$$g_{\alpha\beta} = g_{\beta\alpha} \qquad (21)$$

$$g^{\alpha\beta} = g^{\beta\alpha}. \qquad (22)$$

In EUCLIDEAN SPACE (and all other symmetric SPACES),

$$g_\alpha^\beta = g_\alpha^\beta = \delta_\alpha^\beta, \qquad (23)$$

so

$$g_{\alpha\alpha} = \frac{1}{g^{\alpha\alpha}}. \qquad (24)$$

The ANGLE ϕ between two parametric curves is given by

$$\cos\phi = \hat{\mathbf{r}}_1 \cdot \hat{\mathbf{r}}_2 = \frac{\mathbf{r}_1}{g_1} \cdot \frac{\mathbf{r}_2}{g_2} = \frac{g_{12}}{g_1 g_2}, \qquad (25)$$

so

$$\sin\phi = \frac{\sqrt{g}}{g_1 g_2} \qquad (26)$$

and

$$|\mathbf{r}_1 \times \mathbf{r}_2| = g_1 g_2 \sin\phi = \sqrt{g}. \qquad (27)$$

The LINE ELEMENT can be written

$$ds^2 = dx_i\, dx_i = g_{ij}\, dq_i\, dq_j \qquad (28)$$

where EINSTEIN SUMMATION has been used. But

$$dx_i = \frac{\partial x_i}{\partial q_1}\, dq_1 + \frac{\partial x_i}{\partial q_2}\, dq_2 + \frac{\partial x_i}{\partial q_3}\, dq_3 = \frac{\partial x_i}{\partial q_j}\, dq_j, \qquad (29)$$

so

$$g_{ij} = \sum_k \frac{\partial^2 x_k}{\partial q_i \partial q_j}. \qquad (30)$$

For ORTHOGONAL coordinate systems, $g_{ij} = 0$ for $i \neq j$, and the LINE ELEMENT becomes (for 3-space)

$$\begin{aligned} ds^2 &= g_{11}\, dq_1{}^2 + g_{22}\, dq_2{}^2 + g_{33}\, dq_3{}^2 \\ &= (h_1\, dq_1)^2 + (h_2\, dq_2)^2 + (h_3\, dq_3)^2, \end{aligned} \qquad (31)$$

where $h_i \equiv \sqrt{g_{ii}}$ are called the SCALE FACTORS.

see also CURVILINEAR COORDINATES, DISCRIMINANT (METRIC), LICHNEROWICZ CONDITIONS, LINE ELEMENT, METRIC, METRIC EQUIVALENCE PROBLEM, MINKOWSKI SPACE, SCALE FACTOR, SPACE

Mex

The MINIMUM excluded value. The mex of a SET S of NONNEGATIVE INTEGERS is the least NONNEGATIVE INTEGER *not* in the set.

see also MEX SEQUENCE

References

Guy, R. K. "Max and Mex Sequences." §E27 in *Unsolved Problems in Number Theory, 2nd ed.* New York: Springer-Verlag, pp. 227–228, 1994.

Mex Sequence

A sequence defined from a FINITE sequence a_0, a_1, ..., a_n by defining $a_{n+1} = \text{mex}_i(a_i + a_{n-i})$, where mex is the MEX (minimum excluded value).

see also MAX SEQUENCE, MEX

References

Guy, R. K. "Max and Mex Sequences." §E27 in *Unsolved Problems in Number Theory, 2nd ed.* New York: Springer-Verlag, pp. 227–228, 1994.

Mian-Chowla Sequence

The sequence produced by starting with $a_1 = 1$ and applying the GREEDY ALGORITHM in the following way: for each $k \geq 2$, let a_k be the least INTEGER exceeding a_{k-1} for which $a_j + a_k$ are all distinct, with $1 \leq j \leq k$. This procedure generates the sequence 1, 2, 4, 8, 13, 21, 31, 45, 66, 81, 97, 123, 148, 182, 204, 252, 290, ... (Sloane's A005282). The RECIPROCAL sum of the sequence,

$$S \equiv \sum_{i=1}^\infty \frac{1}{a_i},$$

satisfies

$$2.1568 \leq S \leq 2.1596.$$

see also A-SEQUENCE, B_2-SEQUENCE

References

Guy, R. K. "B_2-Sequences." §E28 in *Unsolved Problems in Number Theory, 2nd ed.* New York: Springer-Verlag, pp. 228–229, 1994.

Sloane, N. J. A. Sequence A005282/M1094 in "An On-Line Version of the Encyclopedia of Integer Sequences."

Mice Problem

n mice start at the corners of a regular n-gon of unit side length, each heading towards its closest neighboring mouse in a counterclockwise direction at constant speed. The mice each trace out a SPIRAL, meet in the center of the POLYGON, and travel a distance

$$d_n = \frac{1}{1 - \cos\left(\frac{2\pi}{m}\right)}.$$

The first few values for $n = 2, 3, \ldots$, are

$$\tfrac{1}{2}, \tfrac{2}{3}, 1, \tfrac{1}{5}(5+\sqrt{5}), 2, \frac{1}{1-\cos\left(\frac{2\pi}{7}\right)},$$

$$2+\sqrt{2}, \frac{1}{1-\cos\left(\frac{2\pi}{9}\right)}, 3+\sqrt{5}, \ldots,$$

giving the numerical values 0.5, 0.666667, 1, 1.44721, 2, 2.65597, 3.41421, 4.27432, 5.23607, ...

see also APOLLONIUS PURSUIT PROBLEM, PURSUIT CURVE, SPIRAL, TRACTRIX

References

Bernhart, A. "Polygons of Pursuit." *Scripta Math.* **24**, 23–50, 1959.

Madachy, J. S. *Madachy's Mathematical Recreations.* New York: Dover, pp. 201–204, 1979.

Mid-Arc Points

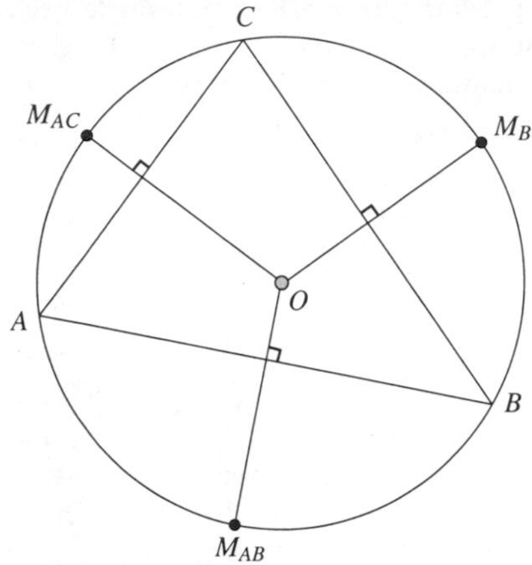

The mid-arc points M_{AB}, M_{AC}, and M_{BC} of a TRIANGLE ΔABC are the points on the CIRCUMCIRCLE of the triangle which lie half-way along each of the three ARCS determined by the vertices (Johnson 1929). These points arise in the definition of the FUHRMANN CIRCLE and FUHRMANN TRIANGLE, and lie on the extensions of the PERPENDICULAR BISECTORS of the triangle sides drawn from the CIRCUMCENTER O.

Kimberling (1988, 1994) and Kimberling and Veldkamp (1987) define the mid-arc points as the POINTS which have TRIANGLE CENTER FUNCTIONS

$$\alpha_1 = [\cos(\tfrac{1}{2}B) + \cos(\tfrac{1}{2}C)] \sec(\tfrac{1}{2}A)$$
$$\alpha_2 = [\cos(\tfrac{1}{2}B) + \cos(\tfrac{1}{2}C)] \csc(\tfrac{1}{2}A).$$

see also FUHRMANN CIRCLE, FUHRMANN TRIANGLE

References

Johnson, R. A. *Modern Geometry: An Elementary Treatise on the Geometry of the Triangle and the Circle.* Boston, MA: Houghton Mifflin, pp. 228–229, 1929.

Kimberling, C. "Problem 804." *Nieuw Archief voor Wiskunde* **6**, 170, 1988.

Kimberling, C. "Central Points and Central Lines in the Plane of a Triangle." *Math. Mag.* **67**, 163–187, 1994.

Kimberling, C. and Veldkamp, G. R. "Problem 1160 and Solution." *Crux Math.* **13**, 298–299, 1987.

Midcircle

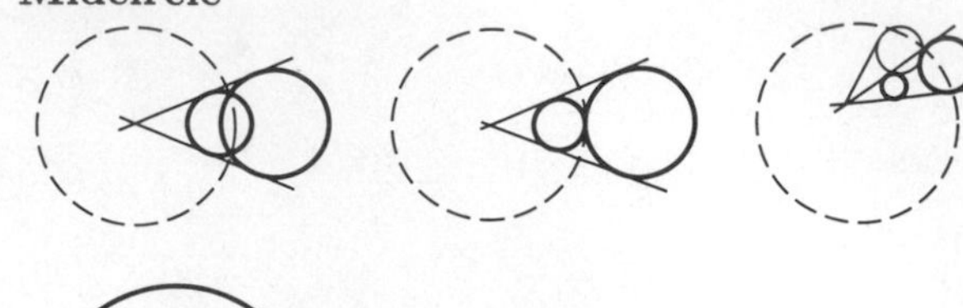

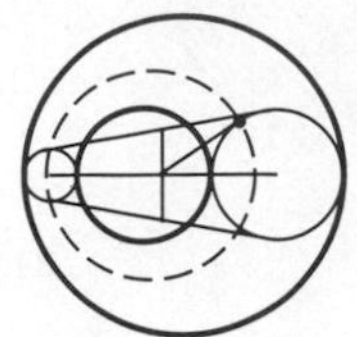

The midcircle of two given CIRCLES is the CIRCLE which would INVERT the circles into each other. Dixon (1991) gives constructions for the midcircle for four of the five possible configurations. In the case of the two given CIRCLES tangent to each other, there are two midcircles.

see also INVERSION, INVERSION CIRCLE

References

Dixon, R. *Mathographics.* New York: Dover, pp. 66–68, 1991.

Middlespoint

see MITTENPUNKT

Midpoint

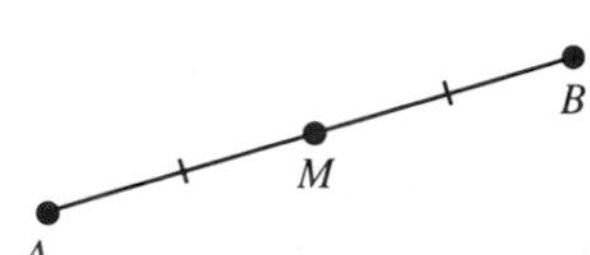

The point on a LINE SEGMENT dividing it into two segments of equal length. The midpoint of a line segment is easy to locate by first constructing a LENS using circular arcs, then connecting the cusps of the LENS. The point where the cusp-connecting line intersects the segment is then the midpoint (Pedoe 1995, p. xii). It is more challenging to locate the midpoint using only a COMPASS, but Pedoe (1995, pp. xviii–xix) gives one solution.

In a RIGHT TRIANGLE, the midpoint of the HYPOTENUSE is equidistant from the three VERTICES (Dunham 1990).

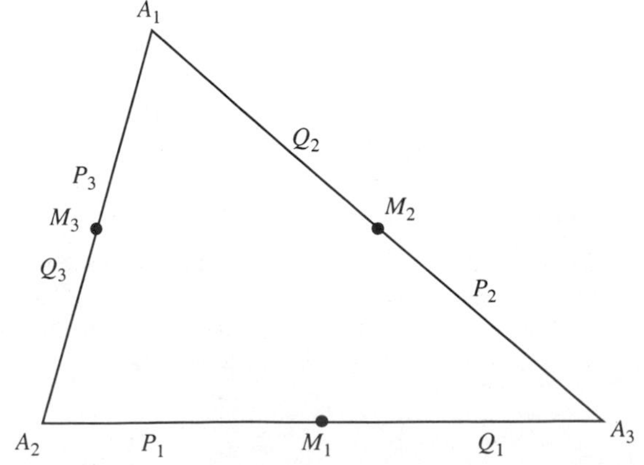

Given a TRIANGLE $\Delta A_1A_2A_3$ with AREA Δ, locate the midpoints M_i. Now inscribe two triangles $\Delta P_1P_2P_3$ and $\Delta Q_1Q_2Q_3$ with VERTICES P_i and Q_i placed so that $\overline{P_iM_i} = \overline{Q_iM_i}$. Then $\Delta P_1P_2P_3$ and $\Delta Q_1Q_2Q_3$ have equal areas

$$\Delta_P = \Delta_Q = \Delta\left[1 - \left(\frac{m_1}{a_1} + \frac{m_2}{a_2} + \frac{m_3}{a_3}\right) + \frac{m_2m_2}{a_2a_3} + \frac{m_3m_1}{a_3a_1} + \frac{m_1m_2}{a_1a_2}\right],$$

where a_i are the sides of the original triangle and m_i are the lengths of the MEDIANS (Johnson 1929).

see also ARCHIMEDES' MIDPOINT THEOREM, BROCARD MIDPOINT, CIRCLE-POINT MIDPOINT THEOREM, LINE SEGMENT, MEDIAN (TRIANGLE), MIDPOINT ELLIPSE

References

Dunham, W. *Journey Through Genius: The Great Theorems of Mathematics.* New York: Wiley, pp. 120–121, 1990.

Johnson, R. A. *Modern Geometry: An Elementary Treatise on the Geometry of the Triangle and the Circle.* Boston, MA: Houghton Mifflin, p. 80, 1929.

Pedoe, D. *Circles: A Mathematical View, rev. ed.* Washington, DC: Math. Assoc. Amer., 1995.

Midpoint Ellipse

The unique ELLIPSE tangent to the MIDPOINTS of a TRIANGLE'S LEGS. The midpoint ellipse has the maximum AREA of any INSCRIBED ELLIPSE (Chakerian 1979). Under an AFFINE TRANSFORMATION, the midpoint ellipse can be transformed into the INCIRCLE of an EQUILATERAL TRIANGLE.

see also AFFINE TRANSFORMATION, ELLIPSE, INCIRCLE, MIDPOINT, TRIANGLE

References

Central Similarities. University of Minnesota College Geometry Project. Distributed by International Film Bureau, Inc.

Chakerian, G. D. "A Distorted View of Geometry." Ch. 7 in *Mathematical Plums* (Ed. R. Honsberger). Washington, DC: Math. Assoc. Amer., pp. 135–136 and 145–146, 1979.

Pedoe, D. "Thinking Geometrically." *Amer. Math. Monthly* **77**, 711–721, 1970.

Midradius

The RADIUS of the MIDSPHERE of a POLYHEDRON, also called the INTERRADIUS. For a REGULAR POLYHEDRON with SCHLÄFLI SYMBOL $\{q,p\}$, the DUAL POLYHEDRON is $\{p,q\}$. Denote the INRADIUS r, midradius ρ, and CIRCUMRADIUS R, and let the side length be a. Then

$$r^2 = \left[a\csc\left(\frac{\pi}{p}\right)\right]^2 + R^2 = a^2 + \rho^2 \quad (1)$$

$$\rho^2 = \left[a\cot\left(\frac{\pi}{p}\right)\right]^2 + R^2. \quad (2)$$

For REGULAR POLYHEDRA and UNIFORM POLYHEDRA, the DUAL POLYHEDRON has CIRCUMRADIUS ρ^2/r and INRADIUS ρ^2/R. Let θ be the ANGLE subtended by the EDGE of an ARCHIMEDEAN SOLID. Then

$$r = \tfrac{1}{2}a\cos(\tfrac{1}{2}\theta)\cot(\tfrac{1}{2}\theta) \quad (3)$$

$$\rho = \tfrac{1}{2}a\cot(\tfrac{1}{2}\theta) \quad (4)$$

$$R = \tfrac{1}{2}a\csc(\tfrac{1}{2}\theta), \quad (5)$$

so

$$r:\rho:R = \cos(\tfrac{1}{2}\theta):1:\sec(\tfrac{1}{2}\theta) \quad (6)$$

(Cundy and Rollett 1989). Expressing the midradius in terms of the INRADIUS r and CIRCUMRADIUS R gives

$$\rho = \tfrac{1}{2}\sqrt{2}\sqrt{r^2 + r\sqrt{r^2+a^2}} = \sqrt{R^2 - \tfrac{1}{4}a^2} \quad (7)$$

for an ARCHIMEDEAN SOLID.

References

Cundy, H. and Rollett, A. *Mathematical Models, 3rd ed.* Stradbroke, England: Tarquin Pub., pp. 126–127, 1989.

Midrange

$$\operatorname{midrange}[f(x)] \equiv \tfrac{1}{2}\{\max[f(x)] + \min[f(x)]\}.$$

see also MAXIMUM, MEAN, MEDIAN (STATISTICS), MINIMUM

Midsphere

The SPHERE with respect to which the VERTICES of a POLYHEDRON are the poles of the planes of the faces of the DUAL POLYHEDRON (and vice versa). It touches all EDGES of a SEMIREGULAR POLYHEDRON or REGULAR POLYHEDRON. It is also called the INTERSPHERE or RECIPROCATING SPHERE.

see also CIRCUMSPHERE, DUAL POLYHEDRON, INSPHERE

Midy's Theorem

If the period of a REPEATING DECIMAL for a/p has an EVEN number of digits, the sum of the two halves is a string of 9s, where p is PRIME and a/p is a REDUCED FRACTION.

see also DECIMAL EXPANSION, REPEATING DECIMAL

References

Rademacher, H. and Toeplitz, O. *The Enjoyment of Mathematics: Selections from Mathematics for the Amateur.* Princeton, NJ: Princeton University Press, pp. 158–160, 1957.

Mikusiński's Problem

Is it possible to cover completely the surface of a SPHERE with congruent, nonoverlapping arcs of GREAT CIRCLES? Conway and Croft (1964) proved that it can be covered with half-open arcs, but not with open arcs. They also showed that the PLANE can be covered with congruent closed and half-open segments, but not with open ones.

References

Conway, J. H. and Croft, H. T. "Covering a Sphere with Great-Circle Arcs." *Proc. Cambridge Phil. Soc.* **60**, 787–900, 1964.

Gardner, M. "Point Sets on the Sphere." Ch. 12 in *Knotted Doughnuts and Other Mathematical Entertainments.* New York: W. H. Freeman, pp. 145–154, 1986.

Milin Conjecture

An INEQUALITY which IMPLIES the correctness of the ROBERTSON CONJECTURE (Milin 1971). de Branges (1985) proved this conjecture, which led to the proof of the full BIEBERBACH CONJECTURE.

see also BIEBERBACH CONJECTURE, ROBERTSON CONJECTURE

References

de Branges, L. "A Proof of the Bieberbach Conjecture." *Acta Math.* **154**, 137–152, 1985.

Milin, I. M. *Univalent Functions and Orthonormal Systems.* Providence, RI: Amer. Math. Soc., 1977.

Stewart, I. *From Here to Infinity: A Guide to Today's Mathematics.* Oxford, England: Oxford University Press, p. 165, 1996.

Mill

The n-roll mill curve is given by the equation

$$x^n - \binom{n}{2} x^{n-2} y^2 + \binom{n}{4} x^{n-4} y^4 - \ldots = a^n,$$

where $\binom{n}{k}$ is a BINOMIAL COEFFICIENT.

References

von Seggern, D. *CRC Standard Curves and Surfaces.* Boca Raton, FL: CRC Press, p. 86, 1993.

Miller's Algorithm

For a catastrophically unstable recurrence in one direction, any seed values for consecutive x_j and x_{j+1} will converge to the desired sequence of functions in the opposite direction times an unknown normalization factor.

Miller-Aškinuze Solid

see ELONGATED SQUARE GYROBICUPOLA

Miller Cylindrical Projection

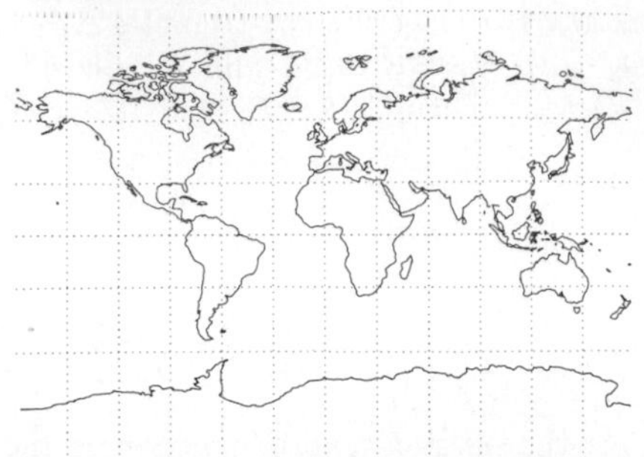

A MAP PROJECTION given by the following transformation,

$$x = \lambda - \lambda_0 \quad (1)$$
$$y = \tfrac{5}{4} \ln[\tan(\tfrac{1}{4}\pi + \tfrac{2}{5}\phi)] \quad (2)$$
$$= \tfrac{5}{4} \sinh^{-1}[\tan(\tfrac{4}{5}\phi)]. \quad (3)$$

Here x and y are the plane coordinates of a projected point, λ is the longitude of a point on the globe, λ_0 is central longitude used for the projection, and ϕ is the latitude of the point on the globe. The inverse FORMULAS are

$$\phi = \tfrac{5}{2} \tan^{-1}(e^{4y/5}) - \tfrac{5}{8}\pi = \tfrac{5}{4} \tan^{-1}[\sinh(\tfrac{4}{5}y)] \quad (4)$$
$$\lambda = \lambda_0 + x. \quad (5)$$

References

Snyder, J. P. *Map Projections—A Working Manual.* U. S. Geological Survey Professional Paper 1395. Washington, DC: U. S. Government Printing Office, pp. 86–89, 1987.

Miller's Primality Test

If a number fails this test, it is not a PRIME. If the number passes, it *may* be a PRIME. A number passing Miller's test is called a STRONG PSEUDOPRIME to base a. If a number n does not pass the test, then it is called a WITNESS for the COMPOSITENESS of n. If n is an ODD, POSITIVE COMPOSITE NUMBER, then n passes Miller's test for at most $(n-1)/4$ bases with $1 \leq a \leq -1$ (Long 1995). There is no analog of CARMICHAEL NUMBERS for STRONG PSEUDOPRIMES.

The only COMPOSITE NUMBER less than 2.5×10^{13} which does not have 2, 3, 5, or 7 as a WITNESS is 3215031751. Miller showed that any composite n has a WITNESS less than $70(\ln n)^2$ if the RIEMANN HYPOTHESIS is true.

see also ADLEMAN-POMERANCE-RUMELY PRIMALITY TEST, STRONG PSEUDOPRIME

References

Long, C. T. Th. 4.21 in *Elementary Introduction to Number Theory, 3rd ed.* Prospect Heights, IL: Waveland Press, 1995.

Miller's Solid

see ELONGATED SQUARE GYROBICUPOLA

Milliard

In British, French, and German usage, one milliard equals 10^9. American usage does not have a number called the milliard, instead using the term BILLION to denote 10^9.

see also BILLION, LARGE NUMBER, MILLION, TRILLION

Millin Series

The series with sum

$$S' \equiv \sum_{n=0}^{\infty} \frac{1}{F_{2^n}} = \tfrac{1}{2}(7 - \sqrt{5}),$$

where F_k is a FIBONACCI NUMBER (Honsberger 1985).

see also FIBONACCI NUMBER

References

Honsberger, R. *Mathematical Gems III.* Washington, DC: Math. Assoc. Amer., pp. 135–137, 1985.

Million

The number $1,000,000 = 10^6$. While one million in America means the same thing as one million in Britain, the words BILLION, TRILLION, etc., refer to *different numbers* in the two naming systems. While Americans may say "Thanks a million" to express gratitude, Norwegians offer "Thanks a thousand" ("tusen takk").

see also BILLION, LARGE NUMBER, MILLIARD, THOUSAND, TRILLION

Mills' Constant

N.B. A detailed on-line essay by S. Finch was the starting point for this entry.

Mills (1947) proved the existence of a constant $\theta = 1.3064\ldots$ such that

$$\left\lfloor \theta^{3^n} \right\rfloor \tag{1}$$

is PRIME for all $n \geq 1$, where $\lfloor x \rfloor$ is the FLOOR FUNCTION. It is not, however, known if θ is IRRATIONAL. Mills' proof was based on the following theorem by Hoheisel (1930) and Ingham (1937). Let p_n be the nth PRIME, then there exists a constant K such that

$$p_{n+1} - p_n < K{p_n}^{5/8} \tag{2}$$

for all n. This has more recently been strengthened to

$$p_{n+1} - p_n < K{p_n}^{1051/1920} \tag{3}$$

(Mozzochi 1986). If the RIEMANN HYPOTHESIS is true, then Cramér (1937) showed that

$$p_{n+1} - p_n = \mathcal{O}(\ln p_n \sqrt{p_n}\,) \tag{4}$$

(Finch).

Hardy and Wright (1979) point out that, despite the beauty of such FORMULAS, they do not have any practical consequences. In fact, unless the exact value of θ is known, the PRIMES themselves must be known in advance to determine θ. A generalization of Mills' theorem to an arbitrary sequence of POSITIVE INTEGERS is given as an exercise by Ellison and Ellison (1985). Consequently, infinitely many values for θ other than the number $1.3064\ldots$ are possible.

References

Caldwell, C. "Mills' Theorem—A Generalization." http://www.utm.edu/research/primes/notes/proofs/A3n.html.

Ellison, W. and Ellison, F. *Prime Numbers.* New York: Wiley, pp. 31–32, 1985.

Finch, S. "Favorite Mathematical Constants." http://www.mathsoft.com/asolve/constant/mills/mills.html.

Hardy, G. H. and Wright, E. M. *An Introduction to the Theory of Numbers, 5th ed.* Oxford, England: Clarendon Press, 1979.

Mills, W. H. "A Prime-Representing Function." *Bull. Amer. Math. Soc.* **53**, 604, 1947.

Mozzochi, C. J. "On the Difference Between Consecutive Primes." *J. Number Th.* **24**, 181–187, 1986.

Ribenboim, P. *The Book of Prime Number Records, 2nd ed.* New York: Springer-Verlag, pp. 135 and 191–193, 1989.

Ribenboim, P. *The Little Book of Big Primes.* New York: Springer-Verlag, pp. 109–110, 1991.

Milne's Method

A PREDICTOR-CORRECTOR METHOD for solution of ORDINARY DIFFERENTIAL EQUATIONS. The third-order equations for predictor and corrector are

$$y_{n+1} = y_{n-3} + \tfrac{4}{3}h(2y'_n - y'_{n-1} + 2y'_{n-2}) + \mathcal{O}(h^5)$$
$$y_{n+1} = y_{n-1} + \tfrac{1}{3}h(y'_{n-1} + 4y'_n + y'_{n+1}) + \mathcal{O}(h^5).$$

Abramowitz and Stegun (1972) also give the fifth order equations and formulas involving higher derivatives.

see also ADAMS' METHOD, GILL'S METHOD, PREDICTOR-CORRECTOR METHODS, RUNGE-KUTTA METHOD

References

Abramowitz, M. and Stegun, C. A. (Eds.). *Handbook of Mathematical Functions with Formulas, Graphs, and Mathematical Tables, 9th printing.* New York: Dover, pp. 896–897, 1972.

Milnor's Conjecture

The UNKNOTTING NUMBER for a TORUS KNOT (p, q) is $(p-1)(q-1)/2$. This 40-year-old CONJECTURE was proved (Adams 1994) in Kronheimer and Mrowka (1993, 1995).

see also TORUS KNOT, UNKNOTTING NUMBER

References

Adams, C. C. *The Knot Book: An Elementary Introduction to the Mathematical Theory of Knots.* New York: W. H. Freeman, p. 113, 1994.

Kronheimer, P. B. and Mrowka, T. S. "Gauge Theory for Embedded Surfaces. I." *Topology* **32**, 773–826, 1993.

Kronheimer, P. B. and Mrowka, T. S. "Gauge Theory for Embedded Surfaces. II." *Topology* **34**, 37–97, 1995.

Milnor's Theorem

If a COMPACT MANIFOLD M has NONNEGATIVE RICCI CURVATURE, then its FUNDAMENTAL GROUP has at most POLYNOMIAL growth. On the other hand, if M has NEGATIVE curvature, then its FUNDAMENTAL GROUP has exponential growth in the sense that $n(\lambda)$ grows exponentially, where $n(\lambda)$ is (essentially) the number of different "words" of length λ which can be made in the FUNDAMENTAL GROUP.

References

Chavel, I. *Riemannian Geometry: A Modern Introduction.* New York: Cambridge University Press, 1994.

Minimal Cover

A minimal cover is a COVER for which removal of one member destroys the covering property. Let $\mu(n,k)$ be the number of minimal covers of $\{1,\ldots,n\}$ with k members. Then

$$\mu(n,k)=\frac{1}{k!}\sum_{m=k}^{\alpha_k}\binom{2^k-k-1}{m-k}m!s(n,m),$$

where $\binom{n}{k}$ is a BINOMIAL COEFFICIENT, $s(n,m)$ is a STIRLING NUMBER OF THE SECOND KIND, and

$$\alpha_k=\min(n,2^k-1).$$

Special cases include $\mu(n,1)=1$ and $\mu(n,2)=s(n+1,3)$.

k	1	2	3	4	5	6	7
Sloane		000392	003468	016111			
n							
1	1						
2	1	1					
3	1	6	1				
4	1	25	22	1			
5	1	90	305	65	1		
6	1	301	3410	2540	171	1	
7	1	966	33621	77350	17066	420	1

see also COVER, LEW k-GRAM, STIRLING NUMBER OF THE SECOND KIND

References

Hearne, T. and Wagner, C. "Minimal Covers of Finite Sets." *Disc. Math.* **5**, 247–251, 1973.

Macula, A. J. "Lewis Carroll and the Enumeration of Minimal Covers." *Math. Mag.* **68**, 269–274, 1995.

Minimal Discriminant

see FREY CURVE

Minimal Matrix

A MATRIX with 0 DETERMINANT whose DETERMINANT becomes NONZERO when any element on or below the diagonal is changed from 0 to 1. An example is

$$\mathsf{M}=\begin{bmatrix}1 & -1 & 0 & 0\\ 0 & 0 & -1 & 0\\ 1 & 1 & 1 & -1\\ 0 & 0 & 1 & 0\end{bmatrix}.$$

There are 2^{n-1} minimal SPECIAL MATRICES of size $n\times n$.

see also SPECIAL MATRIX

References

Knuth, D. E. "Problem 10470." *Amer. Math. Monthly* **102**, 655, 1995.

Minimal Residue

The value b or $b-m$, whichever is smaller in ABSOLUTE VALUE, where $a\equiv b \pmod m$.

see also RESIDUE (CONGRUENCE)

Minimal Set

A SET for which the dynamics can be generated by the dynamics on any subset.

Minimal Surface

Minimal surfaces are defined as surfaces with zero MEAN CURVATURE, and therefore satisfy LAGRANGE'S EQUATION

$$(1+{f_y}^2)f_{xx}+2f_xf_yf_{xy}+(1+{f_x}^2)f_{yy}=0.$$

Minimal surfaces may also be characterized as surfaces of minimal AREA for given boundary conditions. A PLANE is a trivial MINIMAL SURFACE, and the first nontrivial examples (the CATENOID and HELICOID) were found by Meusnier in 1776 (Meusnier 1785).

Euler proved that a minimal surface is planar IFF its GAUSSIAN CURVATURE is zero at every point so that it is locally SADDLE-shaped. The EXISTENCE of a solution to the general case was independently proven by Douglas (1931) and Radó (1933), although their analysis could not exclude the possibility of singularities. Osserman (1970) and Gulliver (1973) showed that a minimizing solution *cannot* have singularities.

The only known complete (boundaryless), embedded (no self-intersections) minimal surfaces of finite topology known for 200 years were the CATENOID, HELICOID, and PLANE. Hoffman discovered a three-ended GENUS 1 minimal embedded surface, and demonstrated the existence of an infinite number of such surfaces. A four-ended embedded minimal surface has also been found. L. Bers proved that any finite isolated SINGULARITY of a single-valued parameterized minimal surface is removable.

A surface can be parameterized using a ISOTHERMAL PARAMETERIZATION. Such a parameterization is minimal if the coordinate functions x_k are HARMONIC, i.e., $\phi_k(\zeta)$ are ANALYTIC. A minimal surface can therefore be defined by a triple of ANALYTIC FUNCTIONS such that $\phi_k\phi_k=0$. The REAL parameterization is then obtained as

$$x_k=\Re\int\phi_k(\zeta)\,d\zeta. \tag{1}$$

But, for an ANALYTIC FUNCTION f and a MEROMORPHIC function g, the triple of functions

$$\phi_1(\zeta) = f(1 - g^2) \tag{2}$$
$$\phi_2(\zeta) = if(1 + g^2) \tag{3}$$
$$\phi_3(\zeta) = 2fg \tag{4}$$

are ANALYTIC as long as f has a zero of order $\geq m$ at every POLE of g of order m. This gives a minimal surface in terms of the ENNEPER-WEIERSTRAß PARAMETERIZATION

$$\Re \int \begin{bmatrix} f(1-g^2) \\ if(1+g^2) \\ 2fg \end{bmatrix} d\zeta. \tag{5}$$

see also BERNSTEIN MINIMAL SURFACE THEOREM, CALCULUS OF VARIATIONS, CATALAN'S SURFACE, CATENOID, COSTA MINIMAL SURFACE, ENNEPER-WEIERSTRAß PARAMETERIZATION, FLAT SURFACE, HENNEBERG'S MINIMAL SURFACE, HOFFMAN'S MINIMAL SURFACE, IMMERSED MINIMAL SURFACE, LICHTENFELS SURFACE, MAEDER'S OWL MINIMAL SURFACE, NIRENBERG'S CONJECTURE, PARAMETERIZATION, PLATEAU'S PROBLEM, SCHERK'S MINIMAL SURFACES, TRINOID, UNDULOID

References

Dickson, S. "Minimal Surfaces." *Mathematica J.* **1**, 38–40, 1990.

Dierkes, U.; Hildebrandt, S.; Küster, A.; and Wohlraub, O. *Minimal Surfaces, 2 vols. Vol. 1: Boundary Value Problems. Vol. 2: Boundary Regularity.* Springer-Verlag, 1992.

do Carmo, M. P. "Minimal Surfaces." §3.5 in *Mathematical Models from the Collections of Universities and Museums* (Ed. G. Fischer). Braunschweig, Germany: Vieweg, pp. 41–43, 1986.

Douglas, J. "Solution of the Problem of Plateau." *Trans. Amer. Math. Soc.* **33**, 263–321, 1931.

Fischer, G. (Ed.). Plates 93 and 96 in *Mathematische Modelle/Mathematical Models, Bildband/Photograph Volume.* Braunschweig, Germany: Vieweg, pp. 89 and 96, 1986.

Gray, A. *Modern Differential Geometry of Curves and Surfaces.* Boca Raton, FL: CRC Press, p. 280, 1993.

Gulliver, R. "Regularity of Minimizing Surfaces of Prescribed Mean Curvature." *Ann. Math.* **97**, 275–305, 1973.

Hoffman, D. "The Computer-Aided Discovery of New Embedded Minimal Surfaces." *Math. Intell.* **9**, 8–21, 1987.

Hoffman, D. and Meeks, W. H. III. *The Global Theory of Properly Embedded Minimal Surfaces.* Amherst, MA: University of Massachusetts, 1987.

Lagrange. "Essai d'une nouvelle méthode pour déterminer les maxima et les minima des formules intégrales indéfinies." 1776.

Meusnier, J. B. "Mémoire sur la courbure des surfaces." *Mém. des savans étrangers* **10** (lu 1776), 477–510, 1785.

Nitsche, J. C. C. *Introduction to Minimal Surfaces.* Cambridge, England: Cambridge University Press, 1989.

Osserman, R. *A Survey of Minimal Surfaces.* New York: Van Nostrand Reinhold, 1969.

Osserman, R. "A Proof of the Regularity Everywhere of the Classical Solution to Plateau's Problem." *Ann. Math.* **91**, 550–569, 1970.

Radó, T. "On the Problem of Plateau." *Ergeben. d. Math. u. ihrer Grenzgebiete.* Berlin: Springer-Verlag, 1933.

Minimax Approximation

A minimization of the MAXIMUM error for a fixed number of terms.

Minimax Polynomial

The approximating POLYNOMIAL which has the smallest maximum deviation from the true function. It is closely approximated by the CHEBYSHEV POLYNOMIALS OF THE FIRST KIND.

Minimax Theorem

The fundamental theorem of GAME THEORY which states that every FINITE, ZERO-SUM, two-person GAME has optimal MIXED STRATEGIES. It was proved by John von Neumann in 1928.

Formally, let $\mathbf{X}$ and $\mathbf{Y}$ be MIXED STRATEGIES for players A and B. Let A be the PAYOFF MATRIX. Then

$$\max_X \min_Y \mathbf{X}^{\mathrm{T}}\mathsf{A}\mathbf{Y} = \min_Y \max_X \mathbf{X}^{\mathrm{T}}\mathsf{A}\mathbf{Y} = v,$$

where v is called the VALUE of the GAME and $\mathbf{X}$ and $\mathbf{Y}$ are called the solutions. It also turns out that if there is more than one optimal MIXED STRATEGY, there are infinitely many.

see also MIXED STRATEGY

References

Willem, M. *Minimax Theorem.* Boston, MA: Birkhäuser, 1996.

Minimum

The smallest value of a set, function, etc. The minimum value of a set of elements $A = \{a_i\}_{i=1}^N$ is denoted $\min A$ or $\min_i a_i$, and is equal to the first element of a sorted (i.e., ordered) version of A. For example, given the set $\{3, 5, 4, 1\}$, the sorted version is $\{1, 3, 4, 5\}$, so the minimum is 1. The MAXIMUM and minimum are the simplest ORDER STATISTICS.

$f'(x) < 0$ $f'(x) > 0$ $f'(x) = 0$ — *minimum*

$f'(x) = 0$ $f'(x) > 0$ $f'(x) < 0$ — *maximum*

$f'(x) < 0$, $f''(x) > 0$; $f'(x) = 0$, $f''(x) = 0$; $f'(x) < 0$, $f''(x) < 0$ — *stationary point*

A continuous FUNCTION may assume a minimum at a single point or may have minima at a number of points. A GLOBAL MINIMUM of a FUNCTION is the smallest value in the entire RANGE of the FUNCTION, while a LOCAL MINIMUM is the smallest value in some local neighborhood.

For a function $f(x)$ which is CONTINUOUS at a point x_0, a NECESSARY but not SUFFICIENT condition for $f(x)$ to have a RELATIVE MINIMUM at $x = x_0$ is that x_0 be a CRITICAL POINT (i.e., $f(x)$ is either not DIFFERENTIABLE at x_0 or x_0 is a STATIONARY POINT, in which case $f'(x_0) = 0$).

The FIRST DERIVATIVE TEST can be applied to CONTINUOUS FUNCTIONS to distinguish minima from MAXIMA. For twice differentiable functions of one variable, $f(x)$, or of two variables, $f(x,y)$, the SECOND DERIVATIVE TEST can sometimes also identify the nature of an EXTREMUM. For a function $f(x)$, the EXTREMUM TEST succeeds under more general conditions than the SECOND DERIVATIVE TEST.

see also CRITICAL POINT, EXTREMUM, FIRST DERIVATIVE TEST, GLOBAL MAXIMUM, INFLECTION POINT, LOCAL MAXIMUM, MAXIMUM, MIDRANGE, ORDER STATISTIC, SADDLE POINT (FUNCTION), SECOND DERIVATIVE TEST, STATIONARY POINT

References

Abramowitz, M. and Stegun, C. A. (Eds.). *Handbook of Mathematical Functions with Formulas, Graphs, and Mathematical Tables, 9th printing.* New York: Dover, p. 14, 1972.

Brent, R. P. *Algorithms for Minimization Without Derivatives.* Englewood Cliffs, NJ: Prentice-Hall, 1973.

Nash, J. C. "Descent to a Minimum I–II: Variable Metric Algorithms." Chs. 15–16 in *Compact Numerical Methods for Computers: Linear Algebra and Function Minimisation, 2nd ed.* Bristol, England: Adam Hilger, pp. 186–206, 1990.

Press, W. H.; Flannery, B. P.; Teukolsky, S. A.; and Vetterling, W. T. "Minimization or Maximization of Functions." Ch. 10 in *Numerical Recipes in FORTRAN: The Art of Scientific Computing, 2nd ed.* Cambridge, England: Cambridge University Press, pp. 387–448, 1992.

Tikhomirov, V. M. *Stories About Maxima and Minima.* Providence, RI: Amer. Math. Soc., 1991.

Minkowski-Bouligand Dimension

In many cases, the HAUSDORFF DIMENSION correctly describes the correction term for a resonator with FRACTAL PERIMETER in Lorentz's conjecture. However, in general, the proper dimension to use turns out to be the Minkowski-Bouligand dimension (Schroeder 1991).

Let $F(r)$ be the AREA traced out by a small CIRCLE with RADIUS r following a fractal curve. Then, providing the LIMIT exists,

$$D_M \equiv \lim_{r\to 0} \frac{\ln F(r)}{-\ln r} + 2$$

(Schroeder 1991). It is conjectured that for all strictly self-similar fractals, the Minkowski-Bouligand dimension is equal to the HAUSDORFF DIMENSION D; otherwise $D_M > D$.

see also HAUSDORFF DIMENSION

References

Berry, M. V. "Diffractals." *J. Phys.* **A12**, 781–797, 1979.

Hunt, F. V.; Beranek, L. L.; and Maa, D. Y. "Analysis of Sound Decay in Rectangular Rooms." *J. Acoust. Soc. Amer.* **11**, 80–94, 1939.

Lapidus, M. L. and Fleckinger-Pellé, J. "Tambour fractal: vers une résolution de la conjecture de Weyl-Berry put les valeurs propres du laplacien." *Compt. Rend. Acad. Sci. Paris Math. Sér 1* **306**, 171–175, 1988.

Schroeder, M. *Fractals, Chaos, Power Laws: Minutes from an Infinite Paradise.* New York: W. H. Freeman, pp. 41–45, 1991.

Minkowski Convex Body Theorem

A bounded plane convex region symmetric about a LATTICE POINT and with AREA > 4 must contain at least three LATTICE POINTS in the interior. In n-D, the theorem can be generalized to a region with AREA $> 2^n$, which must contain at least three LATTICE POINTS. The theorem can be derived from BLICHFELDT'S THEOREM.

see also BLICHFELDT'S THEOREM

Minkowski Geometry

see MINKOWSKI SPACE

Minkowski-Hlawka Theorem

There exist lattices in n-D having HYPERSPHERE PACKING densities satisfying

$$\eta \geq \frac{\zeta(n)}{2^{n-1}},$$

where $\zeta(n)$ is the RIEMANN ZETA FUNCTION. However, the proof of this theorem is nonconstructive and it is still not known how to actually construct packings that are this dense.

see also HERMITE CONSTANTS, HYPERSPHERE PACKING

References

Conway, J. H. and Sloane, N. J. A. *Sphere Packings, Lattices, and Groups, 2nd ed.* New York: Springer-Verlag, pp. 14–16, 1993.

Minkowski Integral Inequality

If $p > 1$, then

$$\left[\int_a^b |f(x)+g(x)|^p\,dx\right]^{1/p} \leq \left[\int_a^b |f(x)|^p\,dx\right]^{1/p} + \left[\int_a^b |g(x)|^p\,dx\right]^{1/p}.$$

see also MINKOWSKI SUM INEQUALITY

References

Abramowitz, M. and Stegun, C. A. (Eds.). *Handbook of Mathematical Functions with Formulas, Graphs, and Mathematical Tables, 9th printing.* New York: Dover, p. 11, 1972.

Gradshteyn, I. S. and Ryzhik, I. M. *Tables of Integrals, Series, and Products, 5th ed.* San Diego, CA: Academic Press, p. 1099, 1993.

Hardy, G. H.; Littlewood, J. E.; and Pólya, G. *Inequalities, 2nd ed.* Cambridge, England: Cambridge University Press, pp. 146–150, 1988.

Minkowski, H. *Geometrie der Zahlen, Vol. 1.* Leipzig, Germany: pp. 115–117, 1896.

Sansone, G. *Orthogonal Functions, rev. English ed.* New York: Dover, p. 33, 1991.

Minkowski Measure

The Minkowski measure of a bounded, CLOSED SET is the same as its LEBESGUE MEASURE.

References

Ko, K.-I. "A Polynomial-Time Computable Curve whose Interior has a Nonrecursive Measure." *Theoret. Comput. Sci.* **145**, 241–270, 1995.

Minkowski Metric

In CARTESIAN COORDINATES,

$$ds^2 = dx^2 + dy^2 + dz^2 \tag{1}$$

$$d\tau^2 = -c^2\,dt^2 + dx^2 + dy^2 + dz^2, \tag{2}$$

and

$$g_{\alpha\beta} \equiv \eta_{\alpha\beta} = \begin{bmatrix} -1 & 0 & 0 & 0 \\ 0 & 1 & 0 & 0 \\ 0 & 0 & 1 & 0 \\ 0 & 0 & 0 & 1 \end{bmatrix}. \tag{3}$$

In SPHERICAL COORDINATES,

$$ds^2 = dr^2 + r^2\,d\theta^2 + r^2\sin^2\theta\,d\phi^2 \tag{4}$$

$$d\tau^2 = -c^2\,dt^2 + dr^2 + r^2\,d\theta^2 + r^2\sin^2\theta\,d\phi^2, \tag{5}$$

and

$$g = \begin{bmatrix} -1 & 0 & 0 & 0 \\ 0 & 1 & 0 & 0 \\ 0 & 0 & r^2 & 0 \\ 0 & 0 & 0 & r^2\sin^2\theta \end{bmatrix}. \tag{6}$$

see also LORENTZ TRANSFORMATION, MINKOWSKI SPACE

Minkowski Sausage

A FRACTAL created from the base curve and motif illustrated below.

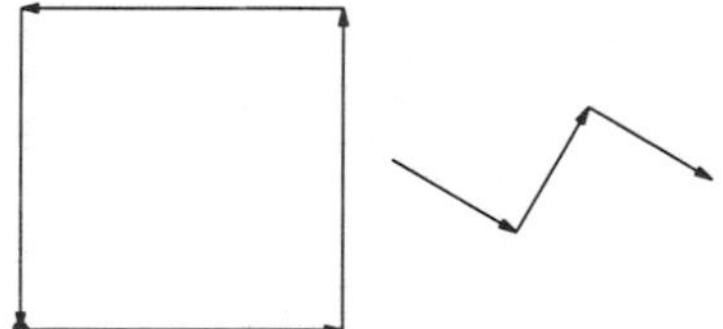

The number of segments after the nth iteration is

$$N_n = 8^n,$$

and

$$\epsilon_n = \left(\frac{1}{4}\right)^n,$$

so the CAPACITY DIMENSION is

$$D \equiv -\lim_{n\to\infty}\frac{\ln N_n}{\ln \epsilon_n} = -\lim_{n\to\infty}\frac{\ln 8^n}{\ln 4^n} = \frac{\ln 8}{\ln 4} = \frac{3\ln 2}{2\ln 2} = \frac{3}{2}.$$

References

Lauwerier, H. *Fractals: Endlessly Repeated Geometric Figures.* Princeton, NJ: Princeton University Press, pp. 37–38 and 42, 1991.
Peitgen, H.-O. and Saupe, D. (Eds.). *The Science of Fractal Images.* New York: Springer-Verlag, p. 283, 1988.
Weisstein, E. W. "Fractals." http://www.astro.virginia.edu/~eww6n/math/notebooks/Fractal.m.

Minkowski Space

A 4-D space with the MINKOWSKI METRIC. Alternatively, it can be considered to have a EUCLIDEAN METRIC, but with its VECTORS defined by

$$\begin{bmatrix} x_0 \\ x_1 \\ x_2 \\ x_3 \end{bmatrix} = \begin{bmatrix} ict \\ x \\ y \\ z \end{bmatrix}, \tag{1}$$

where c is the speed of light. The METRIC is DIAGONAL with

$$g_{\alpha\alpha} = \frac{1}{g_{\alpha\alpha}}, \tag{2}$$

so

$$\eta^{\beta\delta} = \eta_{\beta\delta}. \tag{3}$$

Let Λ be the TENSOR for a LORENTZ TRANSFORMATION. Then

$$\eta^{\beta\delta}\Lambda^\gamma{}_\delta = \Lambda^{\beta\gamma} \tag{4}$$

$$\eta_{\alpha\gamma}\Lambda^{\beta\gamma} = \Lambda_\alpha{}^\beta \tag{5}$$

$$\Lambda_\alpha{}^\beta = \eta_{\alpha\gamma}\Lambda^{\beta\gamma} = \eta_{\alpha\gamma}\eta^{\beta\delta}\Lambda^\gamma{}_\delta. \tag{6}$$

The NECESSARY and SUFFICIENT conditions for a metric $g_{\mu\nu}$ to be equivalent to the Minkowski metric $\eta_{\alpha\beta}$ are that the RIEMANN TENSOR vanishes everywhere ($R^\lambda{}_{\mu\nu\kappa} = 0$) and that at some point $g^{\mu\nu}$ has three POSITIVE and one NEGATIVE EIGENVALUES.

see also LORENTZ TRANSFORMATION, MINKOWSKI METRIC

References

Thompson, A. C. *Minkowski Geometry.* New York: Cambridge University Press, 1996.

Minkowski Sum

The sum of sets A and B in a VECTOR SPACE, equal to $\{a + b : a \in A, b \in B\}$.

Minkowski Sum Inequality

If $p > 1$ and a_k, $b_k > 0$, then

$$\left[\sum_{k=1}^n (a_k + b_k)^p\right]^{1/p} \leq \left(\sum_{k=1}^n {a_k}^p\right)^{1/p} + \left(\sum_{k=1}^n {b_k}^p\right)^{1/p}.$$

Equality holds IFF the sequences a_1, a_2, ... and b_1, b_2, ... are proportional.

see also MINKOWSKI INTEGRAL INEQUALITY

References
Abramowitz, M. and Stegun, C. A. (Eds.). *Handbook of Mathematical Functions with Formulas, Graphs, and Mathematical Tables, 9th printing.* New York: Dover, p. 11, 1972.
Gradshteyn, I. S. and Ryzhik, I. M. *Tables of Integrals, Series, and Products, 5th ed.* San Diego, CA: Academic Press, p. 1092, 1979.
Hardy, G. H.; Littlewood, J. E.; and Pólya, G. *Inequalities, 2nd ed.* Cambridge, England: Cambridge University Press, pp. 24–26, 1988.

Minor

The reduced DETERMINANT of a DETERMINANT EXPANSION, denoted M_{ij}, which is formed by omitting the ith row and jth column.

see also COFACTOR, DETERMINANT, DETERMINANT EXPANSION BY MINORS

References
Arfken, G. *Mathematical Methods for Physicists, 3rd ed.* Orlando, FL: Academic Press, pp. 169–170, 1985.

Minor Axis

see SEMIMINOR AXIS

Minor Graph

A "minor" is a sort of SUBGRAPH and is what Kuratowski means when he says "contain." It is roughly a small graph which can be mapped into the big one without merging VERTICES.

Minus

The operation of SUBTRACTION, i.e., a minus b. The operation is denoted $a - b$. The MINUS SIGN "$-$" is also used to denote a NEGATIVE number, i.e., $-x$.

see also MINUS SIGN, NEGATIVE, PLUS, PLUS OR MINUS, TIMES

Minus or Plus

see PLUS OR MINUS

Minus Sign

The symbol "$-$" which is used to denote a NEGATIVE number or SUBTRACTION.

see also MINUS, PLUS SIGN, SIGN, SUBTRACTION

Minute

see ARC MINUTE

Miquel Circles

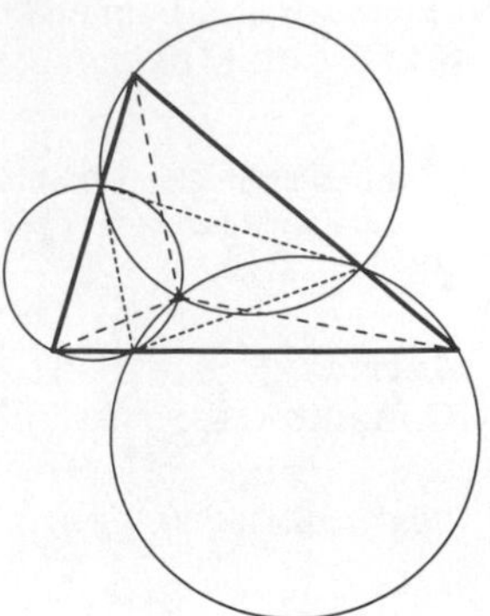

For a TRIANGLE ΔABC and three points A', B', and C', one on each of its sides, the three Miquel circles are the circles passing through each VERTEX and its neighboring side points (i.e., $AC'B'$, $BA'C'$, and $CB'A'$). According to MIQUEL'S THEOREM, the Miquel circles are CONCURRENT in a point M known as the MIQUEL POINT. Similarly, there are n Miquel circles for n lines taken $(n-1)$ at a time.

see also MIQUEL POINT, MIQUEL'S THEOREM, MIQUEL TRIANGLE

Miquel Equation

$$\measuredangle A_2MA_3 = \measuredangle A_2A_1A_3 + \measuredangle P_2P_1P_3,$$

where $\measuredangle$ is a DIRECTED ANGLE.

see also DIRECTED ANGLE, MIQUEL'S THEOREM

References
Johnson, R. A. *Modern Geometry: An Elementary Treatise on the Geometry of the Triangle and the Circle.* Boston, MA: Houghton Mifflin, pp. 131–144, 1929.

Miquel Point

The point of CONCURRENCE of the MIQUEL CIRCLES.

see also MIQUEL CIRCLES, MIQUEL'S THEOREM, MIQUEL TRIANGLE

Miquel's Theorem

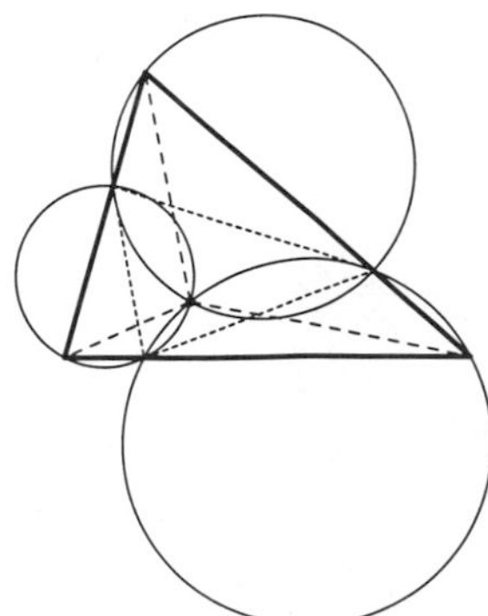

If a point is marked on each side of a TRIANGLE ΔABC, then the three MIQUEL CIRCLES (each through a VERTEX and the two marked points on the adjacent sides)

are CONCURRENT at a point M called the MIQUEL POINT. This result is a slight generalization of the so-called PIVOT THEOREM.

If M lies in the interior of the triangle, then it satisfies

$$\angle P_2MP_3 = 180^\circ - \alpha_1$$

$$\angle P_3MP_1 = 180^\circ - \alpha_2$$

$$\angle P_1MP_2 = 180^\circ - \alpha_3.$$

The lines from the MIQUEL POINT to the marked points make equal angles with the respective sides. (This is a by-product of the MIQUEL EQUATION.)

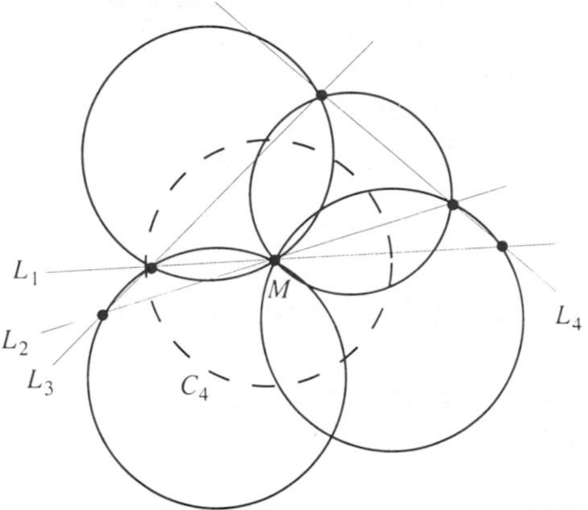

Given four lines $L_1, \ldots, L_4$ each intersecting the other three, the four MIQUEL CIRCLES passing through each subset of three intersection points of the lines meet in a point known as the 4-Miquel point M. Furthermore, the centers of these four MIQUEL CIRCLES lie on a CIRCLE C_4 (Johnson 1929, p. 139). The lines from M to given points on the sides make equal ANGLES with respect to the sides.

Similarly, given n lines taken by $(n-1)$s yield n MIQUEL CIRCLES like C_4 passing through a point P_n, and their centers lie on a CIRCLE C_{n+1}.

see also MIQUEL CIRCLES, MIQUEL EQUATION, MIQUEL TRIANGLE, NINE-POINT CIRCLE, PEDAL CIRCLE, PIVOT THEOREM

References
Johnson, R. A. *Modern Geometry: An Elementary Treatise on the Geometry of the Triangle and the Circle.* Boston, MA: Houghton Mifflin, pp. 131–144, 1929.

Miquel Triangle

Given a point P and a triangle $\Delta A_1A_2A_3$, the Miquel triangle is the triangle connecting the side points P_1, P_2, and P_3 of $\Delta A_1A_2A_3$ with respect to which P is the MIQUEL POINT. All Miquel triangles of a given point M are directly similar, and M is the SIMILITUDE CENTER in every case.

Mira Fractal

A FRACTAL based on the map

$$F(x) = ax + \frac{2(1-a)x^2}{1+x^2}.$$

References
Lauwerier, H. *Fractals: Endlessly Repeated Geometric Figures.* Princeton, NJ: Princeton University Press, p. 136, 1991.

Mirimanoff's Congruence

If the first case of FERMAT'S LAST THEOREM is false for the PRIME exponent p, then $3^{p-1} \equiv 1 \pmod{p^2}$.

see also FERMAT'S LAST THEOREM

Mirror Image

An image of an object obtained by reflecting it in a mirror so that the signs of one of its coordinates are reversed.

see AMPHICHIRAL, CHIRAL, ENANTIOMER, HANDEDNESS

Mirror Plane

The SYMMETRY OPERATION $(x, y, z) \to (x, y, -z)$, etc., which is equivalent to $\bar{2}$, where the bar denotes an IMPROPER ROTATION.

Misère Form

A version of NIM-like GAMES in which the player taking the last piece is the loser. For most IMPARTIAL GAMES, this form is much harder to analyze, but it requires only a trivial modification for the game of NIM.

Mitchell Index

The statistical INDEX

$$P_M \equiv \frac{\sum p_n q_a}{\sum p_0 q_a},$$

where p_n is the price per unit in period n and q_n is the quantity produced in period n.

see also INDEX

References
Kenney, J. F. and Keeping, E. S. *Mathematics of Statistics, Pt. 1, 3rd ed.* Princeton, NJ: Van Nostrand, pp. 66–67, 1962.

Miter Surface

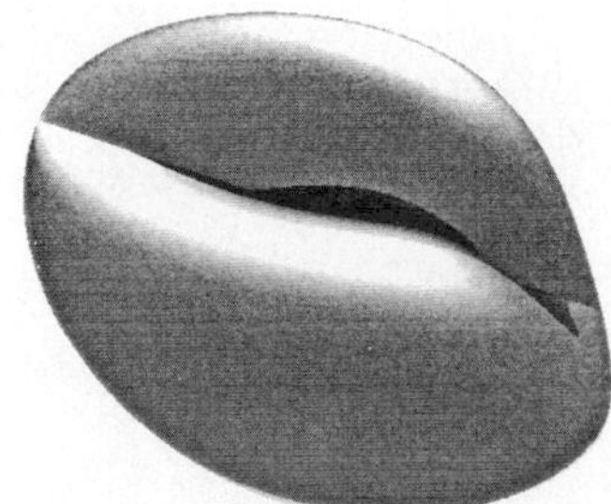

A QUARTIC SURFACE named after its resemblance to the liturgical headdress worn by bishops and given by the equation

$$4x^2(x^2+y^2+z^2)-y^2(1-y^2-z^2)=0.$$

see also QUARTIC SURFACE

References
Nordstrand, T. "Surfaces." http://www.uib.no/people/nfytn/surfaces.htm.

Mittag-Leffler Function

$$E_\gamma(x)\equiv\sum_{k=0}^\infty \frac{x^k}{\Gamma(\gamma k+1)}.$$

It is related to the GENERALIZED HYPERBOLIC FUNCTIONS by

$$F_{n,0}^1(x)=E_n(x^n).$$

References
Muldoon, M. E. and Ungar, A. A. "Beyond Sin and Cos." *Math. Mag.* **69**, 3–14, 1996.

Mittenpunkt

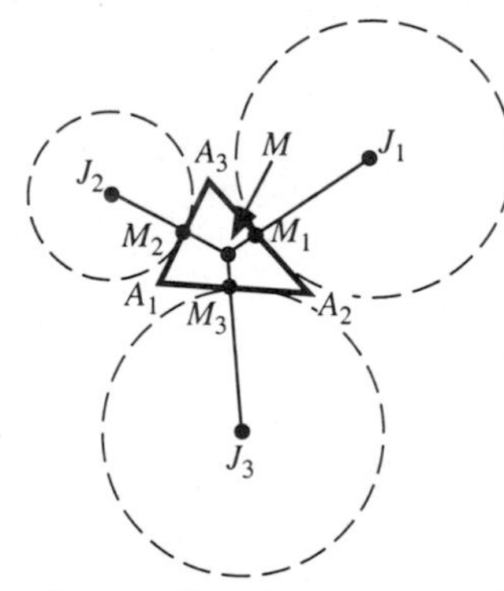

The LEMOINE POINT of the EXCENTRAL TRIANGLE, i.e., the point of concurrence M of the lines from the EXCENTERS J_i through the corresponding TRIANGLE side MIDPOINT M_i. It is also called the MIDDLESPOINT and has TRIANGLE CENTER FUNCTION

$$\alpha=b+c-a=\tfrac{1}{2}\cot A.$$

see also EXCENTER, EXCENTRAL TRIANGLE, NAGEL POINT

References
Baptist, P. *Die Entwicklung der Neueren Dreiecksgeometrie.* Mannheim: Wissenschaftsverlag, p. 72, 1992.
Eddy, R. H. "A Generalization of Nagel's Middlespoint." *Elem. Math.* **45**, 14–18, 1990.
Kimberling, C. "Central Points and Central Lines in the Plane of a Triangle." *Math. Mag.* **67**, 163–187, 1994.
Kimberling, C. "Mittenpunkt." http://www.evansville.edu/~ck6/tcenters/class/mitten.html.

Mixed Partial Derivative

A PARTIAL DERIVATIVE of second or greater order with respect to two or more different variables, for example

$$f_{xy}=\frac{\partial^2 f}{\partial x\partial y}.$$

If the mixed partial derivatives exist and are continuous at a point $\mathbf{x}_0$, then they are equal at $\mathbf{x}_0$ regardless of the order in which they are taken.

see also PARTIAL DERIVATIVE

Mixed Strategy

A collection of moves together with a corresponding set of weights which are followed probabilistically in the playing of a GAME. The MINIMAX THEOREM of GAME THEORY states that every finite, zero-sum, two-person game has optimal mixed strategies.

see also GAME THEORY, MINIMAX THEOREM, STRATEGY

Mixed Tensor

A TENSOR having CONTRAVARIANT and COVARIANT indices.

see also CONTRAVARIANT TENSOR, COVARIANT TENSOR, TENSOR

Mnemonic

A mental device used to aid memorization. Common mnemonics for mathematical constants such as e and PI consist of sentences in which the number of letters in each word give successive digits.

see also e, JOSEPHUS PROBLEM, PI

References
Luria, A. R. *The Mind of a Mnemonist: A Little Book about a Vast Memory.* Cambridge, MA: Harvard University Press, 1987.

Möbius Band

see MÖBIUS STRIP

Möbius Function

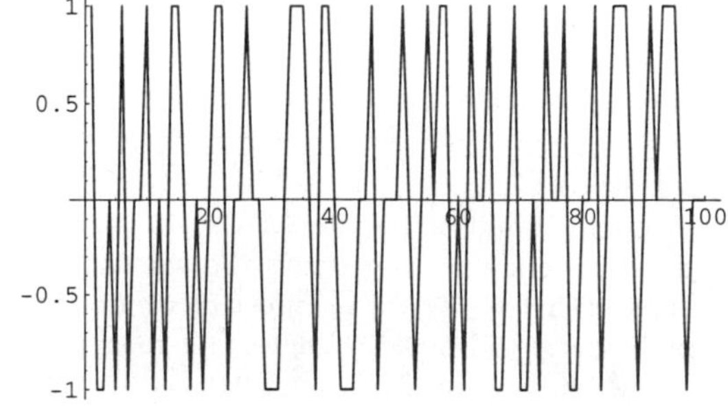

$$\mu(n) \equiv \begin{cases} 0 & \text{if } n \text{ has one or more repeated prime factors} \\ 1 & \text{if } n = 1 \\ (-1)^k & \text{if } n \text{ is a product of } k \text{ distinct primes,} \end{cases}$$

so $mu(n) \neq 0$ indicates that n is SQUAREFREE. The first few values are 1, −1, −1, 0, −1, 1, −1, 0, 0, 1, −1, 0, ... (Sloane's A008683).

The SUMMATORY FUNCTION of the Möbius function is called MERTENS FUNCTION.

see also BRAUN'S CONJECTURE, MERTENS FUNCTION, MÖBIUS INVERSION FORMULA, MÖBIUS PERIODIC FUNCTION, PRIME ZETA FUNCTION, RIEMANN FUNCTION, SQUAREFREE

References

Abramowitz, M. and Stegun, C. A. (Eds.). "The Möbius Function." §24.3.1 in *Handbook of Mathematical Functions with Formulas, Graphs, and Mathematical Tables, 9th printing.* New York: Dover, p. 826, 1972.

Deléglise, M. and Rivat, J. "Computing the Summation of the Möbius Function." *Experiment. Math.* **5**, 291–295, 1996.

Hardy, G. H. and Wright, E. M. *An Introduction to the Theory of Numbers, 5th ed.* Oxford: Clarendon Press, p. 236, 1979.

Sloane, N. J. A. Sequence A008683 in "An On-Line Version of the Encyclopedia of Integer Sequences."

Vardi, I. *Computational Recreations in Mathematica.* Redwood City, CA: Addison-Wesley, pp. 7–8 and 223–225, 1991.

Möbius Group

The equation

$$x_1{}^2 + x_2{}^2 + \ldots + x_n{}^2 - 2x_0x_\infty = 0$$

represents an n-D HYPERSPHERE $\mathbb{S}^n$ as a quadratic hypersurface in an $(n+1)$-D real projective space $\mathbb{P}^{n+1}$, where x_a are homogeneous coordinates in $\mathbb{P}^{n+1}$. Then the GROUP $M(n)$ of projective transformations which leave $\mathbb{S}^n$ invariant is called the Möbius group.

References

Iyanaga, S. and Kawada, Y. (Eds.). "Möbius Geometry." §78A in *Encyclopedic Dictionary of Mathematics.* Cambridge, MA: MIT Press, pp. 265–266, 1980.

Möbius Inversion Formula

If $g(n) \equiv \sum_{d|n} f(d)$, then

$$f(n) = \sum_{d|n} \mu(d) g\left(\frac{n}{d}\right),$$

where the sums are over all possible INTEGERS d that DIVIDE n and $\mu(d)$ is the MÖBIUS FUNCTION. The LOGARITHM of the CYCLOTOMIC POLYNOMIAL

$$\Phi_n(x) = \prod_{d|n} (1 - x^{n/d})^{\mu(d)}$$

is the Möbius inversion formula.

see also CYCLOTOMIC POLYNOMIAL, MÖBIUS FUNCTION

References

Hardy, G. H. and Wright, W. M. *An Introduction to the Theory of Numbers, 5th ed.* Oxford, England: Oxford University Press, pp. 91–93, 1979.

Schroeder, M. R. *Number Theory in Science and Communication, 3rd ed.* New York: Springer-Verlag, 1997.

Vardi, I. *Computational Recreations in Mathematica.* Redwood City, CA: Addison-Wesley, pp. 7–8 and 223–225, 1991.

Möbius Periodic Function

A function periodic with period 2π such that

$$p(\theta + \pi) = -p(\theta)$$

for all θ is said to be Möbius periodic.

Möbius Problem

Let $A = \{a_1, a_2, \ldots\}$ be a free Abelian SEMIGROUP, where a_1 is the unit element. Then do the following properties,

1. $a < b$ IMPLIES $ac < bc$ for $a, b, c \in A$, where A has the linear order $a_1 < a_2 < \ldots$,
2. $\mu(a_n) = \mu(n)$ for all n,

imply that

$$a_{mn} = a_m a_n$$

for all $m, n \geq 1$? The problem is known to be true for $mn \leq 74$ for all $n \leq 240$.

see also BRAUN'S CONJECTURE, MÖBIUS FUNCTION

References

Flath, A. and Zulauf, A. "Does the Möbius Function Determine Multiplicative Arithmetic?" *Amer. Math. Monthly* **102**, 354–256, 1995.

Möbius Shorts

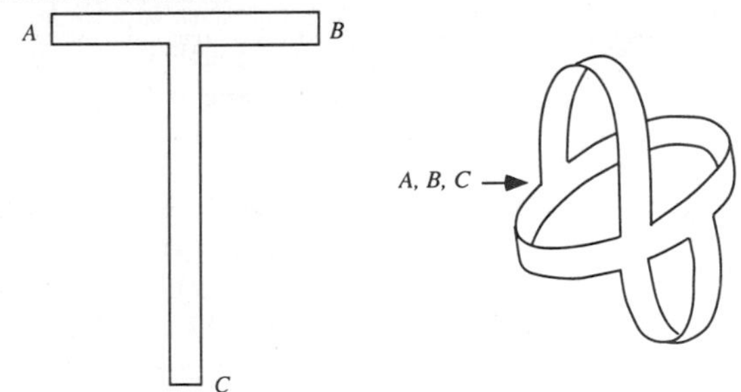

A one-sided surface reminiscent of the MÖBIUS STRIP.

see also MÖBIUS STRIP

References

Boas, R. P. Jr. "Möbius Shorts." *Math. Mag.* **68**, 127, 1995.

Möbius Strip

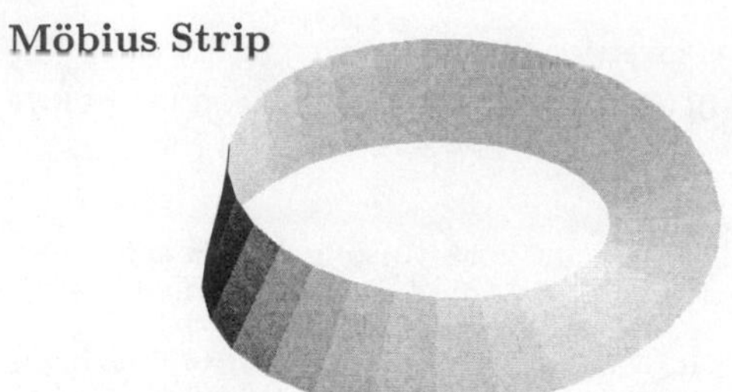

A one-sided surface obtained by cutting a band widthwise, giving it a half twist, and re-attaching the two ends. According to Madachy (1979), the B. F.Goodrich Company patented a conveyor belt in the form of a Möbius strip which lasts twice as long as conventional belts.

A Möbius strip can be represented parametrically by

$$x = [R + s\cos(\tfrac{1}{2}\theta)]\cos\theta$$
$$y = [R + s\cos(\tfrac{1}{2}\theta)]\sin\theta$$
$$z = s\sin(\tfrac{1}{2}\theta),$$

for $s \in [-1,1]$ and $\theta \in [0,2\pi)$. Cutting a Möbius strip, giving it extra twists, and reconnecting the ends produces unexpected figures called PARADROMIC RINGS (Listing and Tait 1847, Ball and Coxeter 1987) which are summarized in the table below.

half-twists	cuts	divs.	result
1	1	2	1 band, length 2
1	1	3	1 band, length 2 1 Möbius strip, length 1
1	2	4	2 bands, length 2
1	2	5	2 bands, length 2 1 Möbius strip, length 1
1	3	6	3 bands, length 2
1	3	7	3 bands, length 2 1 Möbius strip, length 1
2	1	2	2 bands, length 1
2	2	3	3 bands, length 1
2	3	4	4 bands, length 1

A TORUS can be cut into a Möbius strip with an EVEN number of half-twists, and a KLEIN BOTTLE can be cut in half along its length to make two Möbius strips. In addition, two strips on top of each other, each with a half-twist, give a single strip with four twists when disentangled.

There are three possible SURFACES which can be obtained by sewing a Möbius strip to the edge of a DISK: the BOY SURFACE, CROSS-CAP, and ROMAN SURFACE.

The Möbius strip has EULER CHARACTERISTIC 1, and the HEAWOOD CONJECTURE therefore shows that any set of regions on it can be colored using six-colors only.

see also BOY SURFACE, CROSS-CAP, MAP COLORING, PARADROMIC RINGS, PRISMATIC RING, ROMAN SURFACE

References

Ball, W. W. R. and Coxeter, H. S. M. *Mathematical Recreations and Essays, 13th ed.* New York: Dover, pp. 127–128, 1987.

Bogomolny, A. "Möbius Strip." http://www.cut-the-knot.com/do_you_know/moebius.html.

Gardner, M. "Möbius Bands." Ch. 9 in *Mathematical Magic Show: More Puzzles, Games, Diversions, Illusions and Other Mathematical Sleight-of-Mind from Scientific American.* New York: Vintage, pp. 123–136, 1978.

Geometry Center. "The Klein Bottle." http://www.geom.umn.edu/zoo/features/mobius/.

Gray, A. "The Möbius Strip." §12.3 in *Modern Differential Geometry of Curves and Surfaces.* Boca Raton, FL: CRC Press, pp. 236–238, 1993.

Hunter, J. A. H. and Madachy, J. S. *Mathematical Diversions.* New York: Dover, pp. 41–45, 1975.

Kraitchik, M. §8.4.3 in *Mathematical Recreations.* New York: W. W. Norton, pp. 212–213, 1942.

Listing and Tait. *Vorstudien zur Topologie, Göttinger Studien,* Pt. 10, 1847.

Madachy, J. S. *Madachy's Mathematical Recreations.* New York: Dover, p. 7, 1979.

Nordstrand, T. "Mobiusband." http://www.uib.no/people/nfytn/moebtxt.htm.

Pappas, T. "The Moebius Strip & the Klein Bottle," "A Twist to the Moebius Strip," "The 'Double' Moebius Strip." *The Joy of Mathematics.* San Carlos, CA: Wide World Publ./Tetra, p. 207, 1989.

Steinhaus, H. *Mathematical Snapshots, 3rd American ed.* New York: Oxford University Press, pp. 269–274, 1983.

Wagon, S. "Rotating Circles to Produce a Torus or Möbius Strip." §7.4 in *Mathematica in Action.* New York: W. H. Freeman, pp. 229–232, 1991.

Wang, P. "Renderings." http://www.ugcs.caltech.edu/~peterw/portfolio/renderings/.

Möbius Transformation

A transformation of the form

$$w = f(z) = \frac{az+b}{cz+d},$$

where a, b, c, $d \in \mathbb{C}$ and

$$ad - bc \neq 0,$$

is a CONFORMAL TRANSFORMATION and is called a Möbius transformation. It is linear in both w and z.

Every Möbius transformation except $f(z) = z$ has one or two FIXED POINTS. The Möbius transformation sends CIRCLES and lines to CIRCLES or lines. Möbius transformations preserve symmetry. The CROSS-RATIO is invariant under a Möbius transformation. A Möbius transformation is a composition of translations, rotations, magnifications, and inversions.

To determine a particular Möbius transformation, specify the map of three points which preserve orientation. A particular Möbius transformation is then uniquely determined. To determine a general Möbius transformation, pick two symmetric points α and α_S. Define $\beta \equiv f(\alpha)$, restricting β as required. Compute β_S. $f(\alpha_S)$ then equals β_S since the Möbius transformation preserves symmetry (the SYMMETRY PRINCIPLE). Plug in α and α_S into the general Möbius transformation and set equal to β and β_S. Without loss of generality, let $c = 1$ and solve for a and b in terms of β. Plug back into the general expression to obtain a Möbius transformation.

see also SYMMETRY PRINCIPLE

Möbius Triangles

SPHERICAL TRIANGLES into which a SPHERE is divided by the planes of symmetry of a UNIFORM POLYHEDRON.

see also SPHERICAL TRIANGLE, UNIFORM POLYHEDRON

Mock Theta Function

Ramanujan was the first to extensively study these THETA FUNCTION-like functions

$$f(q) = \sum_{n=0}^{\infty} \frac{q^{n^2}}{(1+q)^2(1+q^2)^2\cdots(1+q^n)^2}$$

$$\phi(q) = \sum_{n=0}^{\infty} \frac{q^{n^2}}{(1+q^2)(1+q^4)\cdots(1+q^{2n})}.$$

see also q-SERIES, THETA FUNCTION

References

Bellman, R. E. *A Brief Introduction to Theta Functions.* New York: Holt, Rinehart, and Winston, 1961.

Mod

see CONGRUENCE

Mode

The most common value obtained in a set of observations.

see also MEAN, MEDIAN (STATISTICS), ORDER STATISTIC

Mode Locking

A phenomenon in which a system being forced at an IRRATIONAL period undergoes rational, periodic motion which persists for a finite range of forcing values. It may occur for strong couplings between natural and forcing oscillation frequencies.

The phenomenon can be exemplified in the CIRCLE MAP when, after q iterations of the map, the new angle differs from the initial value by a RATIONAL NUMBER

$$\theta_{n+q} = \theta_n + \frac{p}{q}.$$

This is the form of the unperturbed CIRCLE MAP with the WINDING NUMBER

$$\Omega = \frac{p}{q}.$$

For Ω not a RATIONAL NUMBER, the trajectory is QUASIPERIODIC.

see also CHAOS, QUASIPERIODIC FUNCTION

Model Completion

Model completion is a term employed when EXISTENTIAL CLOSURE is successful. The formation of the COMPLEX NUMBERS, and the move from affine to projective geometry, are successes of this kind. The theory of existential closure gives a theoretical basis of Hilbert's "method of ideal elements."

References

Manders, K. L. "Interpretations and the Model Theory of the Classical Geometries." In *Models and Sets.* Berlin: Springer-Verlag, pp. 297–330, 1984.

Manders, K. L. "Domain Extension and the Philosophy of Mathematics." *J. Philos.* **86**, 553–562, 1989.

Model Theory

Model theory is a general theory of interpretations of an AXIOMATIC SET THEORY. It is the branch of LOGIC studying mathematical structures by considering first-order sentences which are true of those structures and the sets which are definable in those structures by first-order FORMULAS (Marker 1996).

Mathematical structures obeying axioms in a system are called "models" of the system. The usual axioms of ANALYSIS are second order and are known to have the REAL NUMBERS as their unique model. Weakening the axioms to include only the first-order ones leads to a new type of model in what is called NONSTANDARD ANALYSIS.

see also KHOVANSKI'S THEOREM, NONSTANDARD ANALYSIS, WILKIE'S THEOREM

References

Doets, K. *Basic Model Theory.* New York: Cambridge University Press, 1996.

Marker, D. "Model Theory and Exponentiation." *Not. Amer. Math. Soc.* **43**, 753–759, 1996.

Stewart, I. "Non-Standard Analysis." In *From Here to Infinity: A Guide to Today's Mathematics.* Oxford, England: Oxford University Press, pp. 80–81, 1996.

Modified Bessel Differential Equation

The second-order ordinary differential equation

$$x^2 \frac{d^2y}{dx^2} + x\frac{dy}{dx} - (x^2 + n^2)y = 0.$$

The solutions are the MODIFIED BESSEL FUNCTIONS OF THE FIRST and SECOND KINDS. If $n = 0$, the modified Bessel differential equation becomes

$$x^2 \frac{d^2y}{dx^2} + x\frac{dy}{dx} - x^2y = 0,$$

which can also be written

$$\frac{d}{dx}\left(x\frac{dy}{dx}\right) = xy.$$

Modified Bessel Function of the First Kind

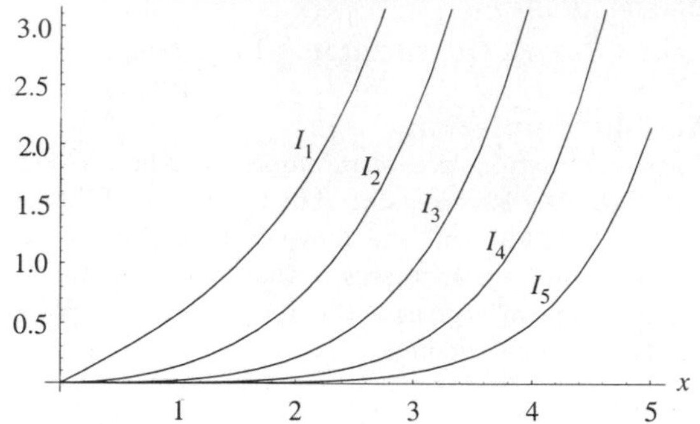

A function $I_n(x)$ which is one of the solutions to the MODIFIED BESSEL DIFFERENTIAL EQUATION and is closely related to the BESSEL FUNCTION OF THE FIRST KIND $J_n(x)$. The above plot shows $I_n(x)$ for $n = 1, 2, \ldots, 5$. In terms of $J_n(x)$,

$$I_n(x) \equiv i^{-n} J_n(ix) = e^{-n\pi i/2} J_n(xe^{i\pi/2}). \qquad (1)$$

For a REAL NUMBER ν, the function can be computed using

$$I_\nu(z) = (\tfrac{1}{2}z)^\nu \sum_{k=0}^\infty \frac{(\frac{1}{4}z^2)^k}{k!\Gamma(\nu+k+1)}, \qquad (2)$$

where $\Gamma(z)$ is the GAMMA FUNCTION. An integral formula is

$$I_\nu(z) = \frac{1}{\pi}\int_0^\pi e^{z\cos\theta}\cos(\nu\theta)\,d\theta - \frac{\sin(\nu\pi)}{\pi}\int_0^\infty e^{-z\cosh t-\nu t}\,dt, \qquad (3)$$

which simplifies for ν an INTEGER n to

$$I_n(z) = \frac{1}{\pi}\int_0^\pi e^{z\cos\theta}\cos(n\theta)\,d\theta \qquad (4)$$

(Abramowitz and Stegun 1972, p. 376).

A derivative identity for expressing higher order modified Bessel functions in terms of $I_0(x)$ is

$$I_n(x) = T_n\left(\frac{d}{dx}\right) I_0(x), \qquad (5)$$

where $T_n(x)$ is a CHEBYSHEV POLYNOMIAL OF THE FIRST KIND.

see also BESSEL FUNCTION OF THE FIRST KIND, MODIFIED BESSEL FUNCTION OF THE FIRST KIND, WEBER'S FORMULA

References

Abramowitz, M. and Stegun, C. A. (Eds.). "Modified Bessel Functions I and K." §9.6 in *Handbook of Mathematical Functions with Formulas, Graphs, and Mathematical Tables, 9th printing.* New York: Dover, pp. 374–377, 1972.

Arfken, G. "Modified Bessel Functions, $I_\nu(x)$ and $K_\nu(x)$." §11.5 in *Mathematical Methods for Physicists, 3rd ed.* Orlando, FL: Academic Press, pp. 610–616, 1985.

Finch, S. "Favorite Mathematical Constants." http://www.mathsoft.com/asolve/constant/cntfrc/cntfrc.html.

Press, W. H.; Flannery, B. P.; Teukolsky, S. A.; and Vetterling, W. T. "Bessel Functions of Fractional Order, Airy Functions, Spherical Bessel Functions." §6.7 in *Numerical Recipes in FORTRAN: The Art of Scientific Computing, 2nd ed.* Cambridge, England: Cambridge University Press, pp. 234–245, 1992.

Spanier, J. and Oldham, K. B. "The Hyperbolic Bessel Functions $I_0(x)$ and $I_1(x)$" and "The General Hyperbolic Bessel Function $I_\nu(x)$." Chs. 49–50 in *An Atlas of Functions.* Washington, DC: Hemisphere, pp. 479–487 and 489–497, 1987.

Modified Bessel Function of the Second Kind

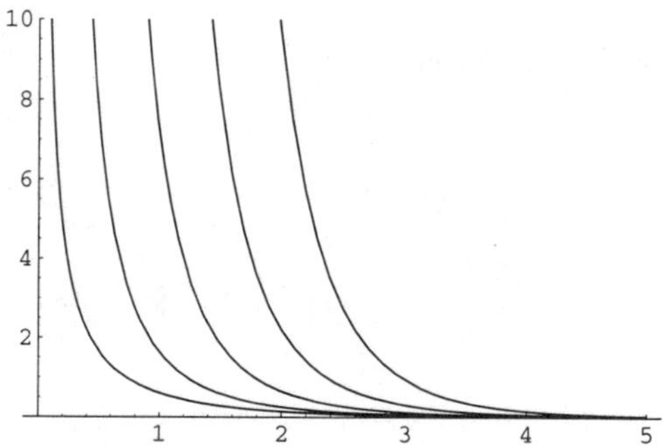

The function $K_n(x)$ which is one of the solutions to the MODIFIED BESSEL DIFFERENTIAL EQUATION. The above plot shows $K_n(x)$ for $n = 1, 2, \ldots, 5$. $K_n(x)$ is closely related to the MODIFIED BESSEL FUNCTION OF THE FIRST KIND $I_n(x)$ and HANKEL FUNCTION $H_n(x)$,

$$\begin{aligned} K_n(x) &\equiv \tfrac{1}{2}\pi i^{n+1} H_n^{(1)}(ix) & (1)\\ &= \tfrac{1}{2}\pi i^{n+1}[J_n(ix) + iN_n(ix)] & (2)\\ &= \frac{\pi}{2}\frac{I_{-n}(x) - I_n(x)}{\sin(n\pi)} & (3)\end{aligned}$$

(Watson 1966, p. 185). A sum formula for K_n is

$$K_n(z) = \tfrac{1}{2}(\tfrac{1}{2}z)^{-n}\sum_{k=0}^{n-1}\frac{(n-k-1)!}{k!}(-\tfrac{1}{4}z^2)^k + (-1)^{n+1}\ln(\tfrac{1}{2}z)I_n(z) + (-1)^n\tfrac{1}{2}(\tfrac{1}{2}z)^n\sum_{k=0}^\infty[\psi(k+1)+\psi(n+k+1)]\frac{(\frac{1}{4}z^2)^k}{k!(n+k)!}, \qquad (4)$$

where ψ is the DIGAMMA FUNCTION (Abramowitz and Stegun 1972). An integral formula is

$$K_\nu(z) = \frac{\Gamma(\nu+\frac{1}{2})(2z)^\nu}{\sqrt{\pi}}\int_0^\infty \frac{\cos t\,dt}{(t^2+z^2)^{\nu+1/2}} \qquad (5)$$

which, for $\nu = 0$, simplifies to

$$K_0(x) = \int_0^\infty \cos(x \sinh t)\,dt = \int_0^\infty \frac{\cos(xt)\,dt}{\sqrt{t^2+1}}. \quad (6)$$

Other identities are

$$K_n(z) = \frac{\sqrt{\pi}}{(n-\frac{1}{2})!}(\tfrac{1}{2}z)^n \int_1^\infty e^{-zx}(x^2-1)^{n-1/2}\,dx \quad (7)$$

for $n > -1/2$ and

$$K_n(z) = \sqrt{\frac{\pi}{2z}}\frac{e^{-z}}{(n-\frac{1}{2})!}\int_0^\infty e^{-t}t^{n-1/2}\left(1-\frac{t}{2z}\right)^{n-1/2} dt \quad (8)$$

$$= \sqrt{\frac{\pi}{2z}}\frac{e^{-z}}{(n-\frac{1}{2})!}\sum_{r=0}^{\infty}\frac{(n-\frac{1}{2})!}{r!(n-r-\frac{1}{2})!}(2z)^{-r} \times \int_0^\infty e^{-t}t^{n+r-1/2}\,dt. \quad (9)$$

The modified Bessel function of the second kind is sometimes called the BASSET FUNCTION.

References

Abramowitz, M. and Stegun, C. A. (Eds.). "Modified Bessel Functions I and K." §9.6 in *Handbook of Mathematical Functions with Formulas, Graphs, and Mathematical Tables, 9th printing.* New York: Dover, pp. 374–377, 1972.

Arfken, G. "Modified Bessel Functions, $I_\nu(x)$ and $K_\nu(x)$." §11.5 in *Mathematical Methods for Physicists, 3rd ed.* Orlando, FL: Academic Press, pp. 610–616, 1985.

Press, W. H.; Flannery, B. P.; Teukolsky, S. A.; and Vetterling, W. T. "Modified Bessel Functions of Integral Order" and "Bessel Functions of Fractional Order, Airy Functions, Spherical Bessel Functions." §6.6 and 6.7 in *Numerical Recipes in FORTRAN: The Art of Scientific Computing, 2nd ed.* Cambridge, England: Cambridge University Press, pp. 229–245, 1992.

Spanier, J. and Oldham, K. B. "The Basset $K_\nu(x)$." Ch. 51 in *An Atlas of Functions.* Washington, DC: Hemisphere, pp. 499–507, 1987.

Watson, G. N. *A Treatise on the Theory of Bessel Functions, 2nd ed.* Cambridge, England: Cambridge University Press, 1966.

Modified Spherical Bessel Differential Equation

The SPHERICAL BESSEL DIFFERENTIAL EQUATION with a NEGATIVE separation constant, given by

$$r^2\frac{d^2R}{dr^2} + 2r\frac{dR}{dr} - [k^2r^2 + n(n+1)]R = 0.$$

The solutions are called MODIFIED SPHERICAL BESSEL FUNCTIONS.

Modified Spherical Bessel Function

Solutions to the MODIFIED SPHERICAL BESSEL DIFFERENTIAL EQUATION, given by

$$i_n(x) \equiv \sqrt{\frac{\pi}{2x}}\,I_{n+1/2}(x) \quad (1)$$

$$i_0(x) = \frac{\sinh(x)}{x} \quad (2)$$

$$k_n(x) \equiv \sqrt{\frac{2\pi}{x}}\,K_{n+1/2}(x) \quad (3)$$

$$k_0(x) = \frac{e^{-x}}{x}, \quad (4)$$

where $I_n(x)$ is a MODIFIED BESSEL FUNCTION OF THE FIRST KIND and $K_n(x)$ is a MODIFIED BESSEL FUNCTION OF THE SECOND KIND.

References

Abramowitz, M. and Stegun, C. A. (Eds.). "Modified Spherical Bessel Functions." §10.2 in *Handbook of Mathematical Functions with Formulas, Graphs, and Mathematical Tables, 9th printing.* New York: Dover, pp. 443–445, 1972.

Modified Struve Function

$$\mathcal{L}_\nu(z) = \left(\tfrac{1}{2}z\right)^{\nu+1}\sum_{k=0}^{\infty}\frac{\left(\frac{z}{2}\right)^{2k}}{\Gamma\left(k+\frac{3}{2}\right)\Gamma\left(k+\nu+\frac{3}{2}\right)}$$

$$= \frac{2\left(\frac{z}{2}\right)^\nu}{\sqrt{\pi}\,\Gamma(\nu+\frac{1}{2})}\int_0^{\pi/2}\sinh(z\cos\theta)\sin^{2\nu}\theta\,d\theta,$$

where $\Gamma(z)$ is the GAMMA FUNCTION.

see also ANGER FUNCTION, STRUVE FUNCTION, WEBER FUNCTIONS

References

Abramowitz, M. and Stegun, C. A. (Eds.). "Modified Struve Function $\mathbf{L}_\nu(x)$." §12.2 in *Handbook of Mathematical Functions with Formulas, Graphs, and Mathematical Tables, 9th printing.* New York: Dover, p. 498, 1972.

Modular Angle

Given a MODULUS k in an ELLIPTIC INTEGRAL, the modular angle is defined by $k \equiv \sin\alpha$. An ELLIPTIC INTEGRAL is written $I(\phi|m)$ when the PARAMETER is used, $I(\phi, k)$ when the MODULUS is used, and $I(\phi\backslash\alpha)$ when the modular angle is used.

see also AMPLITUDE, CHARACTERISTIC (ELLIPTIC INTEGRAL), ELLIPTIC INTEGRAL, MODULUS (ELLIPTIC INTEGRAL), NOME, PARAMETER

References

Abramowitz, M. and Stegun, C. A. (Eds.). *Handbook of Mathematical Functions with Formulas, Graphs, and Mathematical Tables, 9th printing.* New York: Dover, p. 590, 1972.

Modular Equation

The modular equation of degree n gives an algebraic connection of the form

$$\frac{K'(l)}{K(l)} = n\frac{K'(k)}{K(k)} \tag{1}$$

between the TRANSCENDENTAL COMPLETE ELLIPTIC INTEGRALS OF THE FIRST KIND with moduli k and l. When k and l satisfy a modular equation, a relationship of the form

$$\frac{M(l,k)\,dy}{\sqrt{(1-y^2)(1-l^2y^2)}} = \frac{dx}{\sqrt{(1-x^2)(1-k^2x^2)}} \tag{2}$$

exists, and M is called the MODULAR FUNCTION MULTIPLIER. In general, if p is an ODD PRIME, then the modular equation is given by

$$\Omega_p(u,v) = (v-u_0)(v-u_1)\cdots(v-u_p), \tag{3}$$

where

$$u_p \equiv (-1)^{(p^2-1)/8}[\lambda(q^p)]^{1/8} \equiv (-1)^{(p^2-1)/8}u(q^p), \tag{4}$$

λ is a ELLIPTIC LAMBDA FUNCTION, and

$$q \equiv e^{i\pi t} \tag{5}$$

(Borwein and Borwein 1987, p. 126). An ELLIPTIC INTEGRAL identity gives

$$\frac{K'(k)}{K(k)} = 2\frac{K'\left(\frac{2\sqrt{k}}{1+k}\right)}{K\left(\frac{2\sqrt{k}}{1+k}\right)}, \tag{6}$$

so the modular equation of degree 2 is

$$l = \frac{2\sqrt{k}}{1+k} \tag{7}$$

which can be written as

$$l^2(1+k)^2 = 4k. \tag{8}$$

A few low order modular equations written in terms of k and l are

$$\Omega_2 = l^2(1+k)^2 - 4k = 0 \tag{9}$$

$$\Omega_7 = (kl)^{1/4} + (k'l')^{1/4} - 1 = 0 \tag{10}$$

$$\Omega_{23} = (kl)^{1/4} + (k'l')^{1/4} + 2^{2/3}(klk'l')^{1/12} - 1 = 0. \tag{11}$$

In terms of u and v,

$$\Omega_3(u,v) = u^4 - v^4 + 2uv(1-u^2v^2) = 0 \tag{12}$$

$$\begin{aligned}\Omega_5(u,v) &= v^6 - u^6 + 5u^2v^2(v^2-u^2) + 4uv(u^4v^4-1)\\ &= \left(\frac{u}{v}\right)^3 + \left(\frac{v}{u}\right)^3 = 2\left(u^2v^2 - \frac{1}{u^2v^2}\right) = 0\end{aligned} \tag{13}$$

$$\Omega_7(u,v) = (1-u^8)(1-v^8) - (1-uv)^8 = 0, \tag{14}$$

where

$$u^2 \equiv \sqrt{k} = \frac{\vartheta_2(q)}{\vartheta_3(q)} \tag{15}$$

and

$$v^2 \equiv \sqrt{l} = \frac{\vartheta_2(q^p)}{\vartheta_3(q^p)}. \tag{16}$$

Here, ϑ_i are THETA FUNCTIONS.

A modular equation of degree 2^r for $r \geq 2$ can be obtained by iterating the equation for 2^{r-1}. Modular equations for PRIME p from 3 to 23 are given in Borwein and Borwein (1987).

Quadratic modular identities include

$$\frac{\vartheta_3(q)}{\vartheta_3(q^4)} - 1 = \left[\frac{{\vartheta_3}^2(q^2)}{{\vartheta_3}^2(q^4)} - 1\right]^{1/2}. \tag{17}$$

Cubic identities include

$$\left[3\frac{\vartheta_2(q^9)}{\vartheta_2(q)} - 1\right]^3 = 9\frac{{\vartheta_2}^4(q^3)}{{\vartheta_2}^4(q)} - 1 \tag{18}$$

$$\left[3\frac{\vartheta_3(q^9)}{\vartheta_3(q)} - 1\right]^3 = 9\frac{{\vartheta_3}^4(q^3)}{{\vartheta_3}^4(q)} - 1 \tag{19}$$

$$\left[3\frac{\vartheta_4(q^9)}{\vartheta_4(q)} - 1\right]^3 = 9\frac{{\vartheta_4}^4(q^3)}{{\vartheta_4}^4(q)} - 1. \tag{20}$$

A seventh-order identity is

$$\sqrt{\vartheta_3(q)\vartheta_3(q^7)} - \sqrt{\vartheta_4(q)\vartheta_4(q^7)} = \sqrt{\vartheta_2(q)\vartheta_2(q^7)}. \tag{21}$$

From Ramanujan (1913–1914),

$$(1+q)(1+q^3)(1+q^5)\cdots = 2^{1/6}q^{1/24}(kk')^{-1/12} \tag{22}$$

$$(1-q)(1-q^3)(1-q^5)\cdots = 2^{1/6}q^{1/24}k^{-1/12}k'^{1/6}. \tag{23}$$

see also SCHLÄFLI'S MODULAR FORM

References

Borwein, J. M. and Borwein, P. B. *Pi & the AGM: A Study in Analytic Number Theory and Computational Complexity.* New York: Wiley, pp. 127–132, 1987.

Hanna, M. "The Modular Equations." *Proc. London Math. Soc.* **28**, 46–52, 1928.

Ramanujan, S. "Modular Equations and Approximations to π." *Quart. J. Pure. Appl. Math.* **45**, 350–372, 1913–1914.

Modular Form

A modular form is a function in the COMPLEX PLANE with rather spectacular and special properties resulting from a surprising array of internal symmetries. If

$$F\left(\frac{az+b}{cz+d}\right) = (cz+d)^2F(z),$$

then $F(z)$ is said to be a modular form of weight 2 and level N. If it is correctly parameterized, a modular form is ANALYTIC and vanishes at the cusps, so it is called a CUSP FORM. It is also an eigenform under a certain HECKE ALGEBRA.

A remarkable connection between rational ELLIPTIC CURVES and modular forms is given by the TANIYAMA-SHIMURA CONJECTURE, which states that any rational ELLIPTIC CURVE is a modular form in disguise. This result was the one proved by Andrew Wiles in his celebrated proof of FERMAT'S LAST THEOREM.

see also CUSP FORM, ELLIPTIC CURVE, ELLIPTIC FUNCTION, FERMAT'S LAST THEOREM, HECKE ALGEBRA, MODULAR FUNCTION, MODULAR FUNCTION MULTIPLIER, SCHLÄFLI'S MODULAR FORM, TANIYAMA-SHIMURA CONJECTURE

References

Knopp, M. I. *Modular Functions, 2nd ed.* New York: Chelsea, 1993.

Koblitz, N. *Introduction to Elliptic Curves and Modular Forms.* New York: Springer-Verlag, 1993.

Rankin, R. A. *Modular Forms and Functions.* Cambridge, England: Cambridge University Press, 1977.

Sarnack, P. *Some Applications of Modular Forms.* Cambridge, England: Cambridge University Press, 1993.

Modular Function

f is a modular function of level N on the upper half H of the COMPLEX PLANE if it is MEROMORPHIC (even at the CUSPS), $ad - bc = 1$ for all a, b, c, d, and $N|c$.

see also ELLIPTIC FUNCTION, ELLIPTIC MODULAR FUNCTION, MODULAR FORM

References

Apostol, T. M. *Modular Functions and Dirichlet Series in Number Theory.* New York: Springer-Verlag, 1976.

Askey, R. In *Ramanujan International Symposium* (Ed. N. K Thakare). pp. 1–83.

Borwein, J. M. and Borwein, P. B. *Pi and the AGM: A Study in Analytic Number Theory and Computational Complexity.* New York: Wiley, 1987.

Rankin, R. A. *Modular Forms and Functions.* Cambridge, England: Cambridge University Press, 1977.

Schoeneberg, B. *Elliptic Modular Functions: An Introduction.* Berlin: New York: Springer-Verlag, 1974.

Modular Function Multiplier

When k and l satisfy a MODULAR EQUATION, a relationship of the form

$$\frac{M(l,k)\,dy}{\sqrt{(1-y^2)(1-l^2y^2)}} = \frac{dx}{\sqrt{(1-x^2)(1-k^2x^2)}} \tag{1}$$

exists, and M is called the multiplier. The multiplier of degree n can be given by

$$M_n(l,k) \equiv \frac{{\vartheta_3}^2(q)}{{\vartheta_3}^2(q^{1/p})} = \frac{K(k)}{K(l)}, \tag{2}$$

where ϑ_i is a THETA FUNCTION and $K(k)$ is a complete ELLIPTIC INTEGRAL OF THE FIRST KIND.

The first few multipliers in terms of l and k are

$$M_2(l,k) = \frac{1}{1+k} = \frac{1+l'}{2} \tag{3}$$

$$M_3(l,k) = \frac{1-\sqrt{\frac{l^3}{k}}}{1-\sqrt{\frac{k^3}{l}}}. \tag{4}$$

In terms of the u and v defined for MODULAR EQUATIONS,

$$M_3 = \frac{v}{v+2u^3} = \frac{2v^3-u}{3u} \tag{5}$$

$$M_5 = \frac{v(1-uv^3)}{v-u^5} = \frac{u+v^5}{5u(1+u^3v)} \tag{6}$$

$$M_7 = \frac{v(1-uv)[1-uv+(uv)^2]}{v-u^7} = \frac{v^7-u}{7u(1-uv)[1-uv+(uv)^2]}. \tag{7}$$

Modular Gamma Function

The GAMMA GROUP Γ is the set of all transformations w of the form

$$w(t) = \frac{at+b}{ct+d},$$

where a, b, c, and d are INTEGERS and $ad - bc = 1$. Γ-modular functions are then defined as in Borwein and Borwein (1987, p. 114).

see also KLEIN'S ABSOLUTE INVARIANT, LAMBDA GROUP, THETA FUNCTION

References

Borwein, J. M. and Borwein, P. B. *Pi & the AGM: A Study in Analytic Number Theory and Computational Complexity.* New York: Wiley, pp. 127–132, 1987.

Modular Group

The GROUP of all MÖBIUS TRANSFORMATIONS having INTEGER coefficients and DETERMINANT equal to 1.

Modular Lambda Function

see ELLIPTIC LAMBDA FUNCTION

Modular Lattice

A LATTICE which satisfies the identity

$$(x \wedge y) \vee (x \wedge z) = x \wedge (y \vee (x \wedge z))$$

is said to be modular.

see also DISTRIBUTIVE LATTICE

References

Grätzer, G. *Lattice Theory: First Concepts and Distributive Lattices.* San Francisco, CA: W. H. Freeman, pp. 35–36, 1971.

Modular System

A set M of all POLYNOMIALS in s variables, $x_1, \ldots, x_s$ such that if P, P_1, and P_2 are members, then so are $P_1 + P_2$ and QP, where Q is any POLYNOMIAL in $x_1, \ldots, x_s$.

see also HILBERT'S THEOREM, MODULAR SYSTEM BASIS

Modular System Basis

A basis of a MODULAR SYSTEM M is any set of POLYNOMIALS $B_1, B_2, \ldots$ of M such that every POLYNOMIAL of M is expressible in the form

$$R_1B_1 + R_2B_2 + \ldots,$$

where $R_1, R_2, \ldots$ are POLYNOMIALS.

Modular Transformation

see MODULAR EQUATION

Modulation Theorem

The important property of FOURIER TRANSFORMS that $\mathcal{F}[\cos(2\pi k_0 x)f(x)]$ can be expressed in terms of $\mathcal{F}[f(x)] = F(k)$ as follows,

$$\mathcal{F}[\cos(2\pi k_0 x)f(x)] = \tfrac{1}{2}[F(k-k_0) + F(k+k_0)].$$

see also FOURIER TRANSFORM

References

Bracewell, R. "Modulation Theorem." *The Fourier Transform and Its Applications.* New York: McGraw-Hill, p. 108, 1965.

Module

A mathematical object in which things can be added together COMMUTATIVELY by multiplying COEFFICIENTS and in which most of the rules of manipulating VECTORS hold. A module is abstractly very similar to a VECTOR SPACE, although modules have COEFFICIENTS in much more general algebraic objects and use RINGS as the COEFFICIENTS instead of FIELDS.

The additive submodule of the INTEGERS is a set of quantities closed under ADDITION and SUBTRACTION (although it is SUFFICIENT to require closure under SUBTRACTION). Numbers of the form $n\alpha \pm m\alpha$ for $n, m \in \mathbb{Z}$ form a module since,

$$n\alpha \pm m\alpha = (n \pm m)\alpha.$$

Given two INTEGERS a and b, the smallest module containing a and b is $\text{GCD}(a, b)$.

References

Foote, D. and Dummit, D. *Abstract Algebra.* Englewood Cliffs, NJ: Prentice-Hall, 1990.

Modulo

see CONGRUENCE

Modulo Multiplication Group

A FINITE GROUP M_m of RESIDUE CLASSES prime to m under multiplication mod m. M_m is ABELIAN of ORDER $\phi(m)$, where $\phi(m)$ is the TOTIENT FUNCTION. The following table gives the modulo multiplication groups of small orders.

M_m	Group	$\phi(m)$	Elements
M_2	$\langle e \rangle$	1	1
M_3	Z_2	2	1, 2
M_4	Z_2	2	1, 3
M_5	Z_4	4	1, 2, 3, 4
M_6	Z_2	2	1, 5
M_7	Z_6	6	1, 2, 3, 4, 5, 6
M_8	$Z_2 \otimes Z_2$	4	1, 3, 5, 7
M_9	Z_6	6	1, 2, 4, 5, 7, 8
M_{10}	Z_4	4	1, 3, 7, 9
M_{11}	Z_{10}	10	1, 2, 3, 4, 5, 6, 7, 8, 9, 10
M_{12}	$Z_2 \otimes Z_2$	4	1, 5, 7, 11
M_{13}	Z_{12}	12	1, 2, 3, 4, 5, 6, 7, 8, 9, 10, 11
M_{14}	Z_6	6	1, 3, 5, 9, 11, 13
M_{15}	$Z_2 \otimes Z_4$	8	1, 2, 4, 7, 8, 11, 13, 14
M_{16}	$Z_2 \otimes Z_4$	8	1, 3, 5, 7, 9, 11, 13, 15
M_{17}	Z_{16}	16	1, 2, 3, ..., 16
M_{18}	Z_6	6	1, 5, 7, 11, 13, 17
M_{19}	Z_{18}	18	1, 2, 3, ..., 18
M_{20}	$Z_2 \otimes Z_4$	8	1, 3, 7, 9, 11, 13, 17, 19
M_{21}	$Z_2 \otimes Z_6$	12	1, 2, 4, 5, 7, 8, 10, 11, 13, 16, 17, 19
M_{22}	Z_{10}	10	1, 3, 5, 7, 9, 13, 15, 17, 19, 21
M_{23}	Z_{22}	22	1, 2, 3, ..., 22
M_{24}	$Z_2 \otimes Z_2 \otimes Z_2$	8	1, 5, 7, 11, 13, 17, 19, 23

M_m is a CYCLIC GROUP (which occurs exactly when m has a PRIMITIVE ROOT) IFF m is of one of the forms $m = 2$, 4, p^n, or $2p^n$, where p is an ODD PRIME and $n \geq 1$ (Shanks 1993, p. 92).

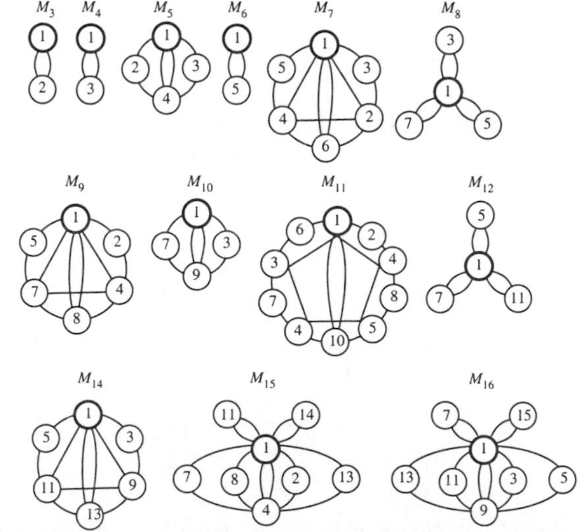

ISOMORPHIC modulo multiplication groups can be determined using a particular type of factorization of $\phi(m)$ as described by Shanks (1993, pp. 92–93). To perform this factorization (denoted ϕ_m), factor m in the standard form

$$m = p_1{}^{a_1} p_2{}^{a_2} \cdots p_n{}^{a_n}. \tag{1}$$

Now write the factorization of the TOTIENT FUNCTION involving each power of an ODD PRIME

$$\phi(p_i{}^{a_i}) = (p_i - 1)p_i{}^{a_i - 1} \tag{2}$$

as

$$\phi(p_i{}^{a_i}) = \langle q_1{}^{b_1} \rangle \langle q_2{}^{b_2} \rangle \cdots \langle q_s{}^{b_s} \rangle \langle p_i{}^{a_i-1} \rangle, \tag{3}$$

where

$$p_i - 1 = q_1{}^{b_1} q_2{}^{b_2} \cdots q_s{}^{b_s}, \tag{4}$$

$\langle q^b \rangle$ denotes the explicit expansion of q^b (i.e., $5^2 = 25$), and the last term is omitted if $a_i = 1$. If $p_1 = 2$, write

$$\phi(2^{a_1}) = \begin{cases} 2 & \text{for } a_1 = 2 \\ 2\langle 2^{a_1-2} \rangle & \text{for } a_1 > 2. \end{cases} \tag{5}$$

Now combine terms from the odd and even primes. For example, consider $m = 104 = 2^3 \cdot 13$. The only odd prime factor is 13, so factoring gives $13 - 1 = 12 = \langle 2^2 \rangle \langle 3 \rangle = 3 \cdot 4$. The rule for the powers of 2 gives $2^3 = 2\langle 2^{3-2} \rangle = 2\langle 2 \rangle = 2 \cdot 2$. Combining these two gives $\phi_{104} = 2 \cdot 2 \cdot 3 \cdot 4$. Other explicit values of ϕ_m are given below.

$$\begin{aligned} \phi_3 &= 2 \\ \phi_4 &= 2 \\ \phi_5 &= 4 \\ \phi_6 &= 2 \\ \phi_{15} &= 2 \cdot 4 \\ \phi_{16} &= 2 \cdot 4 \\ \phi_{17} &= 16 \\ \phi_{104} &= 2 \cdot 2 \cdot 3 \cdot 4 \\ \phi_{105} &= 2 \cdot 2 \cdot 3 \cdot 4. \end{aligned}$$

M_m and M_n are isomorphic IFF ϕ_m and ϕ_n are identical. More specifically, the abstract GROUP corresponding to a given M_m can be determined explicitly in terms of a DIRECT PRODUCT of CYCLIC GROUPS of the so-called CHARACTERISTIC FACTORS, whose product is denoted Φ_n. This representation is obtained from ϕ_m as the set of products of largest powers of each factor of ϕ_m. For example, for ϕ_{104}, the largest power of 2 is $4 = 2^2$ and the largest power of 3 is $3 = 3^1$, so the first characteristic factor is $4 \times 3 = 12$, leaving $2 \cdot 2$ (i.e., only powers of two). The largest power remaining is $2 = 2^1$, so the second CHARACTERISTIC FACTOR is 2, leaving 2, which is the third and last CHARACTERISTIC FACTOR. Therefore, $\Phi_{104} = 2 \cdot 2 \cdot 4$, and the group M_m is isomorphic to $Z_2 \otimes Z_2 \otimes Z_4$.

The following table summarizes the isomorphic modulo multiplication groups M_n for the first few n and identifies the corresponding abstract GROUP. No M_m is ISOMORPHIC to Z_8, Q_8, or D_4. However, every finite ABELIAN GROUP is isomorphic to a SUBGROUP of M_m for infinitely many different values of m (Shanks 1993, p. 96). CYCLE GRAPHS corresponding to M_n for small n are illustrated above, and more complicated CYCLE GRAPHS are illustrated by Shanks (1993, pp. 87–92).

Group	Isomorphic M_m
$\langle e \rangle$	M_2
Z_2	M_3, M_4, M_6
Z_4	M_5, M_{10}
$Z_2 \otimes Z_2$	M_8, M_{12}
Z_6	M_7, M_9, M_{14}, M_{18}
$Z_2 \otimes Z_4$	M_{15}, M_{16}, M_{20}, M_{30}
$Z_2 \otimes Z_2 \otimes Z_2$	M_{24}
Z_{10}	M_{11}, M_{22}
Z_{12}	M_{13}, M_{26}
$Z_2 \otimes Z_6$	M_{21}, M_{28}, M_{36}, M_{42}
Z_{16}	M_{17}, M_{34}
$Z_2 \otimes Z_8$	M_{32}
$Z_2 \otimes Z_2 \otimes Z_4$	M_{40}, M_{48}, M_{60}
Z_{18}	M_{19}, M_{27}, M_{38}, M_{54}
Z_{20}	M_{25}, M_{50}
$Z_2 \otimes Z_{10}$	M_{33}, M_{44}, M_{66}
Z_{22}	M_{23}, M_{46}
$Z_2 \otimes Z_{12}$	M_{35}, M_{39}, M_{45}, M_{52}, M_{70}, M_{78}, M_{90}
Z_{28}	M_{29}, M_{58}
Z_{30}	M_{31}, M_{62}
Z_{36}	M_{37}, M_{74}

The number of CHARACTERISTIC FACTORS r of M_m for $m = 1, 2, \ldots$ are 1, 1, 1, 1, 1, 1, 1, 2, 1, 1, 1, 2, ... (Sloane's A046072). The number of QUADRATIC RESIDUES in M_m for $m > 2$ are given by $\phi(m)/2^r$ (Shanks 1993, p. 95). The first few for $m = 1, 2, \ldots$ are 0, 1, 1, 1, 2, 1, 3, 1, 3, 2, 5, 1, 6, ... (Sloane's A046073).

In the table below, $\phi(n)$ is the TOTIENT FUNCTION (Sloane's A000010) factored into CHARACTERISTIC FACTORS, $\lambda(n)$ is the CARMICHAEL FUNCTION (Sloane's A011773), and g_i are the smallest generators of the group M_n (of which there is a number equal to the number of CHARACTERISTIC FACTORS).

n	$\phi(n)$	$\lambda(n)$	g_i	n	$\phi(n)$	$\lambda(n)$	g_i
3	2	2	2	27	18	18	2
4	2	2	3	28	$2\cdot 6$	6	13, 3
5	4	2	2	29	28	28	2
6	2	2	5	30	$2\cdot 4$	4	11, 7
7	6	6	3	31	30	30	3
8	$2\cdot 2$	2	7, 3	32	$2\cdot 8$	8	31, 3
9	6	6	2	33	$2\cdot 10$	10	10, 2
10	4	4	3	34	16	16	3
11	10	10	2	35	$2\cdot 12$	12	6, 2
12	$2\cdot 2$	2	5, 7	36	$2\cdot 6$	6	19,5
13	12	12	2	37	36	36	2
14	6	6	3	38	18	18	3
15	$2\cdot 4$	4	14, 2	39	$2\cdot 12$	12	38, 2
16	$2\cdot 4$	4	15, 3	40	$2\cdot 2\cdot 4$	4	39, 11, 3
17	16	16	3	41	40	40	6
18	6	6	5	42	$2\cdot 6$	6	13, 5
19	18	18	2	43	42	42	3
20	$2\cdot 4$	4	19, 3	44	$2\cdot 10$	10	43, 3
21	$2\cdot 6$	6	20, 2	45	$2\cdot 12$	12	44, 2
22	10	10	7	46	22	22	5
23	22	22	5	47	46	46	5
24	$2\cdot 2\cdot 2$	2	5, 7, 13	48	$2\cdot 2\cdot 4$	4	47, 7, 5
25	20	20	2	49	42	42	3
26	12	12	7	50	20	20	3

see also Characteristic Factor, Cycle Graph, Finite Group, Residue Class

References

Riesel, H. "The Structure of the Group M_n." *Prime Numbers and Computer Methods for Factorization, 2nd ed.* Boston, MA: Birkhäuser, pp. 270–272, 1994.

Shanks, D. *Solved and Unsolved Problems in Number Theory, 4th ed.* New York: Chelsea, pp. 61–62 and 92, 1993.

Sloane, N. J. A. Sequences A011773, A046072, A046073, and A000010/M0299 in "An On-Line Version of the Encyclopedia of Integer Sequences."

Weisstein, E. W. "Groups." http://www.astro.virginia.edu/~eww6n/math/notebooks/Groups.m.

Modulus (Complex Number)

The modulus of a Complex Number z is denoted $|z|$.

$$|x+iy| \equiv \sqrt{x^2+y^2} \tag{1}$$

$$|re^{i\phi}| = |r|. \tag{2}$$

Let $c_1 \equiv Ae^{i\phi_1}$ and $c_2 \equiv Be^{i\phi_2}$ be two Complex Numbers. Then

$$\left|\frac{c_1}{c_2}\right| = \left|\frac{Ae^{i\phi_1}}{Be^{i\phi_2}}\right| = \frac{A}{B}|e^{i(\phi_1-\phi_2)}| = \frac{A}{B} \tag{3}$$

$$\frac{|c_1|}{|c_2|} = \frac{|Ae^{i\phi_1}|}{|Be^{i\phi_2}|} = \frac{A}{B}\frac{|e^{i\phi_1}|}{|e^{i\phi_2}|} = \frac{A}{B}, \tag{4}$$

so

$$\left|\frac{c_1}{c_2}\right| = \frac{|c_1|}{|c_2|}. \tag{5}$$

Also,

$$|c_1c_2| = |(Ae^{i\phi_1})(Be^{i\phi_2})| = AB|e^{i(\phi_1+\phi_2)}| = AB \tag{6}$$

$$|c_1||c_2| = |Ae^{i\phi_1}||Be^{i\phi_2}| = AB|e^{i\phi_1}||e^{i\phi_2}| = AB, \tag{7}$$

so

$$|c_1c_2| = |c_1||c_2| \tag{8}$$

and, by extension,

$$|z^n| = |z|^n. \tag{9}$$

The only functions satisfying identities of the form

$$|f(x+iy)| = |f(x)+f(iy)| \tag{10}$$

are $f(z) = Az$, $f(z) = A\sin(bz)$, and $f(z) = A\sinh(bz)$ (Robinson 1957).

see also Absolute Square

References

Abramowitz, M. and Stegun, C. A. (Eds.). *Handbook of Mathematical Functions with Formulas, Graphs, and Mathematical Tables, 9th printing.* New York: Dover, p. 16, 1972.

Robinson, R. M. "A Curious Mathematical Identity." *Amer. Math. Monthly* **64**, 83–85, 1957.

Modulus (Congruence)

see Congruence

Modulus (Elliptic Integral)

A parameter k used in Elliptic Integrals and Elliptic Functions defined to be $k \equiv \sqrt{m}$, where m is the Parameter. An Elliptic Integral is written $I(\phi, k)$ when the modulus is used. It can be computed explicitly in terms of Theta Functions of zero argument:

$$k = \frac{\vartheta_2{}^2(0|\tau)}{\vartheta_3{}^2(0|\tau)}. \tag{1}$$

The Real period $K(k)$ and Imaginary period $K'(k) = K(k') = K(\sqrt{1-k^2})$ are given by

$$4K(k) = 2\pi\vartheta_3{}^2(0|\tau) \tag{2}$$

$$2iK'(k) = \pi\tau\vartheta_3{}^2(0|\tau), \tag{3}$$

where $K(k)$ is a complete Elliptic Integral of the First Kind and the complementary modulus is defined by

$$k'^2 \equiv 1-k^2, \tag{4}$$

with k the modulus.

see also Amplitude, Characteristic (Elliptic Integral), Elliptic Function, Elliptic Integral, Elliptic Integral Singular Value, Modular Angle, Nome, Parameter, Theta Function

References

Abramowitz, M. and Stegun, C. A. (Eds.). *Handbook of Mathematical Functions with Formulas, Graphs, and Mathematical Tables, 9th printing.* New York: Dover, p. 590, 1972.

Modulus (Quadratic Invariants)

The quantity $ps - rq$ obtained by letting

$$x = pX + qY \quad (1)$$
$$y = rX + sY \quad (2)$$

in

$$ax^2 + 2bxy + cy^2 \quad (3)$$

so that

$$A = ap^2 + 2bpr + cr^2 \quad (4)$$
$$B = apq + b(ps + qr) + crs \quad (5)$$
$$C = aq^2 + 2bqs + cs^2 \quad (6)$$

and

$$B^2 - AC = (ps - rq)^2(b^2 - ac), \quad (7)$$

is called the modulus.

Modulus (Set)

The name for the SET of INTEGERS modulo m, denoted $\mathbb{Z}\backslash m\mathbb{Z}$. If m is a PRIME p, then the modulus is a FINITE FIELD $\mathbb{F}_p = \mathbb{Z}\backslash p\mathbb{Z}$.

Moessner's Theorem

Write down the POSITIVE INTEGERS in row one, cross out every k_1th number, and write the partial sums of the remaining numbers in the row below. Now cross off every k_2th number and write the partial sums of the remaining numbers in the row below. Continue. For every POSITIVE INTEGER $k > 1$, if every kth number is ignored in row 1, every $(k-1)$th number in row 2, and every $(k+1-i)$th number in row i, then the kth row of partial sums will be the kth POWERS 1^k, 2^k, 3^k, ...

References

Conway, J. H. and Guy, R. K. "Moessner's Magic." In *The Book of Numbers.* New York: Springer-Verlag, pp. 63–65, 1996.

Honsberger, R. *More Mathematical Morsels.* Washington, DC: Math. Assoc. Amer., pp. 268–277, 1991.

Long, C. T. "On the Moessner Theorem on Integral Powers." *Amer. Math. Monthly* **73**, 846–851, 1966.

Long, C. T. "Strike it Out—Add it Up." *Math. Mag.* **66**, 273–277, 1982.

Moessner, A. "Eine Bemerkung über die Potenzen der natürlichen Zahlen." *S.-B. Math.-Nat. Kl. Bayer. Akad. Wiss.* **29**, 1952.

Paasche, I. "Ein neuer Beweis des moessnerischen Satzes." *S.-B. Math.-Nat. Kl. Bayer. Akad. Wiss.* **1952**, 1–5, 1953.

Paasche, I. "Ein zahlentheoretische-logarithmischer 'Rechenstab'." *Math. Naturwiss. Unterr.* **6**, 26–28, 1953–54.

Paasche, I. "Eine Verallgemeinerung des moessnerschen Satzes." *Compositio Math.* **12**, 263–270, 1956.

Mohammed Sign

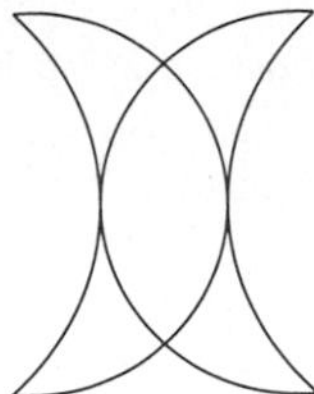

A curve consisting of two mirror-reversed intersecting crescents. This curve can be traced UNICURSALLY.

see also UNICURSAL CIRCUIT

Møiré Pattern

An interference pattern produced by overlaying similar but slightly offset templates. Møiré patterns can also be created be plotting series of curves on a computer screen. Here, the interference is provided by the discretization of the finite-sized pixels.

see also CIRCLES-AND-SQUARES FRACTAL

References

Cassin, C. *Visual Illusions in Motion with Møiré Screens: 60 Designs and 3 Plastic Screens.* New York: Dover, 1997.

Grafton, C. B. *Optical Designs in Motion with Møiré Overlays.* New York: Dover, 1976.

Mollweide's Formulas

Let a TRIANGLE have side lengths a, b, and c with opposite angles A, B, and C. Then

$$\frac{b-c}{a} = \frac{\sin[\frac{1}{2}(B-C)]}{\cos(\frac{1}{2}A)}$$
$$\frac{c-a}{b} = \frac{\sin[\frac{1}{2}(C-A)]}{\cos(\frac{1}{2}B)}$$
$$\frac{a-b}{c} = \frac{\sin[\frac{1}{2}(A-B)]}{\cos(\frac{1}{2}C)}.$$

see also NEWTON'S FORMULAS, TRIANGLE

References

Beyer, W. H. *CRC Standard Mathematical Tables, 28th ed.* Boca Raton, FL: CRC Press, p. 146, 1987.

Mollweide Projection

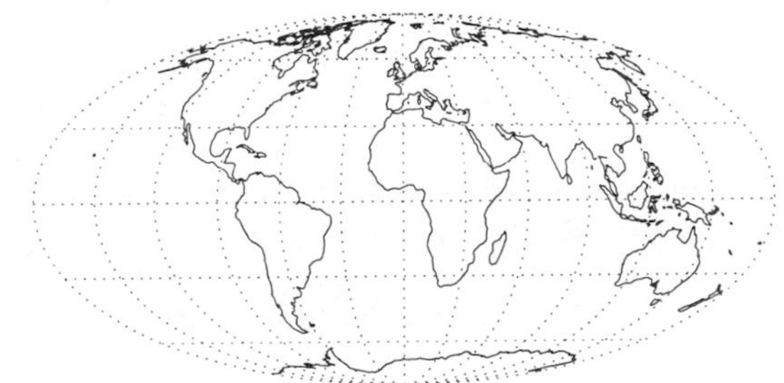

A MAP PROJECTION also called the ELLIPTICAL PROJECTION or HOMOLOGRAPHIC EQUAL AREA PROJECTION. The forward transformation is

$$x = \frac{2\sqrt{2}\,(\lambda - \lambda_0)\cos\theta}{\pi} \quad (1)$$

$$y = 2^{1/2}\sin\theta, \quad (2)$$

where θ is given by

$$2\theta + \sin(2\theta) = \pi\sin\phi. \quad (3)$$

NEWTON'S METHOD can then be used to compute θ' iteratively from

$$\Delta\theta' = -\frac{\theta' + \sin\theta' - \pi\sin\phi}{1 + \cos\theta'}, \quad (4)$$

where

$$\theta' = \tfrac{1}{2}\theta' \quad (5)$$

or, better yet,

$$\theta' = 2\sin^{-1}\left(\frac{2\phi}{\pi}\right) \quad (6)$$

can be used as a first guess.

The inverse FORMULAS are

$$\phi = \sin^{-1}\left[\frac{2\theta + \sin(2\theta)}{\pi}\right] \quad (7)$$

$$\lambda = \lambda_0 + \frac{\pi x}{2\sqrt{2}\cos\theta}, \quad (8)$$

where

$$\theta = \sin^{-1}\left(\frac{y}{\sqrt{2}}\right). \quad (9)$$

References

Snyder, J. P. *Map Projections—A Working Manual.* U. S. Geological Survey Professional Paper 1395. Washington, DC: U. S. Government Printing Office, pp. 249–252, 1987.

Moment

The nth moment of a distribution about zero μ'_n is defined by

$$\mu'_n = \langle x^n \rangle, \quad (1)$$

where

$$\langle f(x) \rangle = \begin{cases} \sum f(x)P(x) & \text{discrete distribution} \\ \int f(x)P(x)\,dx & \text{continuous distribution.} \end{cases} \quad (2)$$

μ'_1, the MEAN, is usually simply denoted $\mu = \mu_1$. If the moment is instead taken about a point a,

$$\mu_n(a) = \langle (x-a)^n \rangle = \sum (x-a)^n P(x). \quad (3)$$

The moments are most commonly taken about the MEAN. These moments are denoted μ_n and are defined by

$$\mu_n \equiv \langle (x-\mu)^n \rangle, \quad (4)$$

with $\mu_1 = 0$. The moments about zero and about the MEAN are related by

$$\mu_2 = \mu'_2 - (\mu'_1)^2 \quad (5)$$

$$\mu_3 = \mu'_3 - 3\mu'_2\mu'_1 + 2(\mu'_1)^3 \quad (6)$$

$$\mu_4 = \mu'_4 - 4\mu'_3\mu'_1 + 6\mu'_2(\mu'_1)^2 - 3(\mu'_1)^4. \quad (7)$$

The second moment about the MEAN is equal to the VARIANCE

$$\mu_2 = \sigma^2, \quad (8)$$

where $\sigma = \sqrt{\mu_2}$ is called the STANDARD DEVIATION.

The related CHARACTERISTIC FUNCTION is defined by

$$\phi^{(n)}(0) \equiv \left[\frac{d^n\phi}{dt^n}\right]_{t=0} = i^n \mu_n(0). \quad (9)$$

The moments may be simply computed using the MOMENT-GENERATING FUNCTION,

$$\mu'_n = M^{(n)}(0). \quad (10)$$

A DISTRIBUTION is not uniquely specified by its moments, although it is by its CHARACTERISTIC FUNCTION.

see also CHARACTERISTIC FUNCTION, CHARLIER'S CHECK, CUMULANT-GENERATING FUNCTION, FACTORIAL MOMENT, KURTOSIS, MEAN, MOMENT-GENERATING FUNCTION, SKEWNESS, STANDARD DEVIATION, STANDARDIZED MOMENT, VARIANCE

References

Press, W. H.; Flannery, B. P.; Teukolsky, S. A.; and Vetterling, W. T. "Moments of a Distribution: Mean, Variance, Skewness, and So Forth." §14.1 in *Numerical Recipes in FORTRAN: The Art of Scientific Computing, 2nd ed.* Cambridge, England: Cambridge University Press, pp. 604–609, 1992.

Moment-Generating Function

Given a RANDOM VARIABLE $x \in R$, if there exists an $h > 0$ such that

$$M(t) \equiv \langle e^{tx} \rangle = \begin{cases} \sum_R e^{tx}P(x) & \text{for a discrete distribution} \\ \int_{-\infty}^{\infty} e^{tx}P(x)\,dx & \text{for a continuous distribution} \end{cases} \quad (1)$$

for $|t| < h$, then

$$M(t) \equiv \langle e^{tx} \rangle \quad (2)$$

is the moment-generating function.

$$M(t) = \int_{-\infty}^{\infty} (1 + tx + \tfrac{1}{2!}t^2x^2 + \ldots)P(x)\,dx = 1 + tm_1 + \tfrac{1}{2!}t^2 m_2 + \ldots, \quad (3)$$

where m_r is the rth MOMENT about zero. The moment-generating function satisfies

$$\begin{aligned} M_{x+y}(t) &= \langle e^{t(x+y)} \rangle = \langle e^{tx} e^{ty} \rangle \\ &= \langle e^{tx} \rangle \langle e^{ty} \rangle = M_x(t) M_y(t). \end{aligned} \tag{4}$$

If $M(t)$ is differentiable at zero, then the nth MOMENTS about the ORIGIN are given by $M^n(0)$

$$M(t) = \langle e^{tx} \rangle \qquad M(0) = 1 \tag{5}$$

$$M'(t) = \langle x e^{tx} \rangle \qquad M'(0) = \langle x \rangle \tag{6}$$

$$M''(t) = \langle x^2 e^{tx} \rangle \qquad M''(0) = \langle x^2 \rangle \tag{7}$$

$$M^{(n)}(t) = \langle x^n e^{tx} \rangle \qquad M^{(n)}(0) = \langle x^n \rangle. \tag{8}$$

The MEAN and VARIANCE are therefore

$$\mu \equiv \langle x \rangle = M'(0) \tag{9}$$

$$\sigma^2 \equiv \langle x^2 \rangle - \langle x \rangle^2 = M''(0) - [M'(0)]^2. \tag{10}$$

It is also true that

$$\mu_n = \sum_{j=0}^{n} \binom{n}{j} (-1)^{n-j} \mu'_j (\mu'_1)^{n-j}, \tag{11}$$

where $\mu'_0 = 1$ and μ'_j is the jth moment about the origin.

It is sometimes simpler to work with the LOGARITHM of the moment-generating function, which is also called the CUMULANT-GENERATING FUNCTION, and is defined by

$$R(t) \equiv \ln[M(t)] \tag{12}$$

$$R'(t) = \frac{M'(t)}{M(t)} \tag{13}$$

$$R''(t) = \frac{M(t)M''(t) - [M'(t)]^2}{[M(t)]^2}. \tag{14}$$

But $M(0) = \langle 1 \rangle = 1$, so

$$\mu = M'(0) = R'(0) \tag{15}$$

$$\sigma^2 = M''(0) - [M'(0)]^2 = R''(0). \tag{16}$$

see also CHARACTERISTIC FUNCTION, CUMULANT, CUMULANT-GENERATING FUNCTION, MOMENT

References

Kenney, J. F. and Keeping, E. S. "Moment-Generating and Characteristic Functions," "Some Examples of Moment-Generating Functions," and "Uniqueness Theorem for Characteristic Functions." §4.6–4.8 in *Mathematics of Statistics, Pt. 2, 2nd ed.* Princeton, NJ: Van Nostrand, pp. 72–77, 1951.

Momental Skewness

$$\alpha^{(m)} \equiv \tfrac{1}{2}\gamma_1 = \frac{\mu_3}{2\sigma^3},$$

where γ_1 is the FISHER SKEWNESS.

see also FISHER SKEWNESS, SKEWNESS

Monad

A mathematical object which consists of a set of a single element. The YIN-YANG is also known as the monad.

see also HEXAD, QUARTET, QUINTET, TETRAD, TRIAD, YIN-YANG

Money-Changing Problem

see COIN PROBLEM

Monge-Ampère Differential Equation

A second-order PARTIAL DIFFERENTIAL EQUATION of the form

$$Hr + 2Ks + Lt + M + N(rt - s^2) = 0,$$

where H, K, L, M, and N are functions of x, y, z, p, and q, and r, s, t, p, and q are defined by

$$\begin{aligned} r &= \frac{\partial^2 z}{\partial x^2} \\ s &= \frac{\partial^2 z}{\partial x \partial y} \\ t &= \frac{\partial^2 z}{\partial y^2} \\ p &= \frac{\partial z}{\partial x} \\ q &= \frac{\partial z}{\partial y}. \end{aligned}$$

The solutions are given by a system of differential equations given by Iyanaga and Kawada (1980).

References

Iyanaga, S. and Kawada, Y. (Eds.). "Monge-Ampère Equations." §276 in *Encyclopedic Dictionary of Mathematics.* Cambridge, MA: MIT Press, pp. 879–880, 1980.

Monge's Chordal Theorem

see RADICAL CENTER

Monge's Form

A surface given by the form $z = F(x, y)$.

see also MONGE PATCH

Monge Patch

A Monge patch is a PATCH $\mathbf{x} : U \to \mathbb{R}^3$ of the form

$$\mathbf{x}(u, v) = (u, v, h(u, v)), \tag{1}$$

where U is an OPEN SET in $\mathbb{R}^2$ and $h : U \to \mathbb{R}$ is a differentiable function. The coefficients of the first FUNDAMENTAL FORM are given by

$$E = 1 + {h_u}^2 \tag{2}$$

$$F = h_u h_v \tag{3}$$

$$G = 1 + {h_v}^2 \tag{4}$$

and the second FUNDAMENTAL FORM by

$$e = \frac{h_{uu}}{\sqrt{1 + h_u{}^2 + h_v{}^2}} \quad (5)$$

$$f = \frac{h_{uv}}{\sqrt{1 + h_u{}^2 + h_v{}^2}} \quad (6)$$

$$g = \frac{g_{vv}}{\sqrt{1 + h_u{}^2 + h_v{}^2}}. \quad (7)$$

For a Monge patch, the GAUSSIAN CURVATURE and MEAN CURVATURE are

$$K = \frac{h_{uu}h_{vv} - h_{uv}{}^2}{(1 + h_u{}^2 + h_v{}^2)^2} \quad (8)$$

$$H = \frac{(1 + h_v{}^2)h_{uu} - 2h_u h_v h_{uv} + (1 + h_u{}^2)h_{vv}}{(1 + h_u{}^2 + h_v{}^2)^{3/2}}. \quad (9)$$

see also MONGE'S FORM, PATCH

References
Gray, A. *Modern Differential Geometry of Curves and Surfaces.* Boca Raton, FL: CRC Press, pp. 305–306, 1993.

Monge's Problem

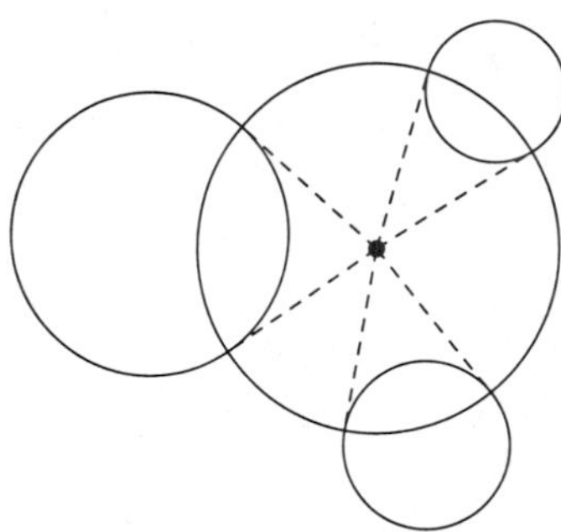

Draw a CIRCLE that cuts three given CIRCLES PERPENDICULARLY. The solution is obtained by drawing the RADICAL CENTER R of the given three CIRCLES. If it lies outside the three CIRCLES, then the CIRCLE with center R and RADIUS formed by the tangent from R to one of the given CIRCLES intersects the given CIRCLES perpendicularly. Otherwise, if R lies inside one of the circles, the problem is unsolvable.

see also CIRCLE TANGENTS, RADICAL CENTER

References
Dörrie, H. "Monge's Problem." §31 in *100 Great Problems of Elementary Mathematics: Their History and Solutions.* New York: Dover, pp. 151–154, 1965.

Monge's Shuffle

A SHUFFLE in which CARDS from the top of the deck in the left hand are alternatively moved to the bottom and top of the deck in the right hand. If the deck is shuffled m times, the final position x_m and initial position x_0 of a card are related by

$$2^{m+1}x_m = (4p+1)[2^{m-1} + (-1)^{m-1}(2^{m-2} + \ldots + 2 + 1)] + (-1)^{m-1}2x_0 + 2^m + (-1)^{m-1}$$

for a deck of $2p$ cards (Kraitchik 1942).

see also CARDS, SHUFFLE

References
Conway, J. H. and Guy, R. K. "Fractions Cycle into Decimals." In *The Book of Numbers.* New York: Springer-Verlag, pp. 157–163, 1996.
Kraitchik, M. "Monge's Shuffle." §12.2.14 in *Mathematical Recreations.* New York: W. W. Norton, pp. 321–323, 1942.

Monge's Theorem

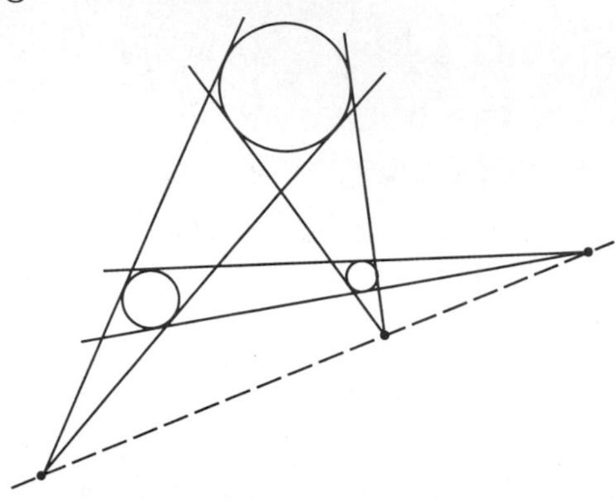

Draw three nonintersecting CIRCLES in the plane, and the common tangent line for each pair of two. The points of intersection of the three pairs of tangent lines lie on a straight line.

References
Coxeter, H. S. M. "The Problem of Apollonius." *Amer. Math. Monthly* **75**, 5–15, 1968.
Graham, L. A. Problem 62 in *Ingenious Mathematical Problems and Methods.* New York: Dover, 1959. Ogilvy, C. S. *Excursions in Geometry.* New York: Dover, pp. 115–117, 1990.
Walker, W. "Monge's Theorem in Many Dimensions." *Math. Gaz.* **60**, 185–188, 1976.

Monic Polynomial

A POLYNOMIAL in which the COEFFICIENT of the highest ORDER term is 1.

see also MONOMIAL

Monica Set

The nth Monica set M_n is defined as the set of COMPOSITE NUMBERS x for which $n|S(x) - S_p(x)$, where

$$x = a_0 + a_1(10^1) + \ldots + a_d(10^d) = p_1 p_2 \cdots p_n, \quad (1)$$

and

$$S(x) = \sum_{j=0}^{d} a_j \quad (2)$$

$$S_p(x) = \sum_{i=1}^{m} S(p_i). \quad (3)$$

Every Monica set has an infinite number of elements. The Monica set M_n is a subset of the SUZANNE SET S_n.

If x is a SMITH NUMBER, then it is a member of the Monica set M_n for all $n \in \mathbb{N}$. For any INTEGER $k > 1$, if x is a k-SMITH NUMBER, then $x \in M_{k-1}$.

see also SUZANNE SET

References

Smith, M. "Cousins of Smith Numbers: Monica and Suzanne Sets." *Fib. Quart.* **34**, 102–104, 1996.

Monkey and Coconut Problem

A DIOPHANTINE problem (i.e., one whose solution must be given in terms of INTEGERS) which seeks a solution to the following problem. Given n men and a pile of coconuts, each man in sequence takes $(1/n)$th of the coconuts and gives the m coconuts which do not divide equally to a monkey. When all n men have so divided, they divide the remaining coconuts five ways, and give the m coconuts which are left-over to the monkey. How many coconuts N were there originally? The solution is equivalent to solving the $n+1$ DIOPHANTINE EQUATIONS

$$\begin{aligned} N &= nA + m \\ (n-1)A &= nB + m \\ (n-1)B &= nC + m \\ &\vdots \\ (n-1)X &= nY + m \\ (n-1)Y &= nZ + m, \end{aligned}$$

and is given by

$$N = kn^{n+1} - m(n-1),$$

where k is an an arbitrary INTEGER (Gardner 1961).

For the particular case of $n = 5$ men and $m = 1$ left over coconuts, the 6 equations can be combined into the single DIOPHANTINE EQUATION

$$1,024N = 15,625F + 11,529,$$

where F is the number given to each man in the last division. The smallest POSITIVE solution in this case is $N = 15,621$ coconuts, corresponding to $k = 1$ and $F = 1,023$ (Gardner 1961). The following table shows how this rather large number of coconuts is divided under the scheme described above.

Removed	Given to Monkey	Left
		15,621
3,124	1	12,496
2,499	1	9,996
1,999	1	7,996
1,599	1	6,396
1,279	1	5,116
5 × 1023	1	0

If no coconuts are left for the monkey after the final n-way division (Williams 1926), then the original number of coconuts is

$$\begin{cases} (1+nk)n^n - (n-1) & n \text{ odd} \\ (n-1+nk)n^n - (n-1) & n \text{ even}. \end{cases}$$

The smallest POSITIVE solution for case $n = 5$ and $m = 1$ is $N = 3,121$ coconuts, corresponding to $k = 1$ and 1,020 coconuts in the final division (Gardner 1961). The following table shows how these coconuts are divided.

Removed	Given to Monkey	Left
		3,121
624	1	2,496
499	1	1,996
399	1	1,596
319	1	1,276
255	1	1,020
5 × 204	0	0

A different version of the problem having a solution of 79 coconuts is considered by Pappas (1989).

see also DIOPHANTINE EQUATION—LINEAR, PELL EQUATION

References

Anning, N. "Monkeys and Coconuts." *Math. Teacher* **54**, 560–562, 1951.

Bowden, J. "The Problem of the Dishonest Men, the Monkeys, and the Coconuts." In *Special Topics in Theoretical Arithmetic*. Lancaster, PA: Lancaster Press, pp. 203–212, 1936.

Gardner, M. "The Monkey and the Coconuts." Ch. 9 in *The Second Scientific American Book of Puzzles & Diversions: A New Selection.* New York: Simon and Schuster, 1961.

Kirchner, R. B. "The Generalized Coconut Problem." *Amer. Math. Monthly* **67**, 516–519, 1960.

Moritz, R. E. "Solution to Problem 3,242." *Amer. Math. Monthly* **35**, 47–48, 1928.

Ogilvy, C. S. and Anderson, J. T. *Excursions in Number Theory.* New York: Dover, pp. 52–54, 1988.

Olds, C. D. *Continued Fractions.* New York: Random House, pp. 48–50, 1963.

Pappas, T. "The Monkey and the Coconuts." *The Joy of Mathematics.* San Carlos, CA: Wide World Publ./Tetra, pp. 226–227 and 234, 1989.

Williams, B. A. "Coconuts." *The Saturday Evening Post,* Oct. 9, 1926.

Monkey Saddle

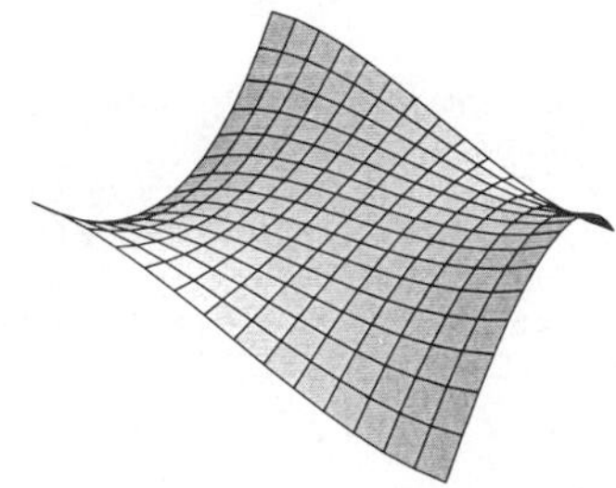

A SURFACE which a monkey can straddle with both his two legs and his tail. A simply Cartesian equation for such a surface is

$$z = x(x^2 - 3y^2), \tag{1}$$

which can also be given by the parametric equations

$$x(u,v) = u \tag{2}$$
$$y(u,v) = v \tag{3}$$
$$z(u,v) = u^3 - 3uv^2. \tag{4}$$

The coefficients of the first and second FUNDAMENTAL FORMS of the monkey saddle are given by

$$e = \frac{6u}{\sqrt{1+9u^4+18u^2v^2+9v^4}} \tag{5}$$
$$f = -\frac{6v}{\sqrt{1+9u^4+18u^2v^2+9v^4}} \tag{6}$$
$$g = -\frac{6u}{\sqrt{1+9u^4+18u^2v^2+9v^4}} \tag{7}$$
$$E = 1 + 9(u^2 - v^2)^2 \tag{8}$$
$$F = -18uv(u^2 - v^2) \tag{9}$$
$$G = 1 + 36u^2v^2, \tag{10}$$

giving RIEMANNIAN METRIC

$$ds^2 = [1 + (3u^2 - 3v^2)^2]\,du^2 - 2[18uv(u^2 - v^2)]\,du\,dv + (1 + 36u^2v^2)\,dv^2, \tag{11}$$

AREA ELEMENT

$$dA = \sqrt{1+9u^4+18u^2v^2+9v^4}\,du \wedge dv, \tag{12}$$

and GAUSSIAN and MEAN CURVATURES

$$K = -\frac{36(u^2+v^2)}{(1+9u^4+18u^2v^2+9v^4)^2} \tag{13}$$
$$H = \frac{27u(-u^4+2u^2v^2+3v^4)}{(1+9u^4+18u^2v^2+9v^4)^{3/2}} \tag{14}$$

(Gray 1993). Every point of the monkey saddle except the origin has NEGATIVE GAUSSIAN CURVATURE.

see also CROSSED TROUGH, PARTIAL DERIVATIVE

References

Coxeter, H. S. M. *Introduction to Geometry, 2nd ed.* New York: Wiley, p. 365, 1969.

Gray, A. *Modern Differential Geometry of Curves and Surfaces.* Boca Raton, FL: CRC Press, pp. 213–215, 262–263, and 288–289, 1993.

Hilbert, D. and Cohn-Vossen, S. *Geometry and the Imagination.* New York: Chelsea, p. 202, 1952.

Monochromatic Forced Triangle

Given a COMPLETE GRAPH K_n which is two-colored, the number of forced monochromatic TRIANGLES is at least

$$\begin{cases} \frac{1}{3}u(u-1)(u-2) & \text{for } n = 2u \\ \frac{2}{3}u(u-1)(4u+1) & \text{for } n = 4u+1 \\ \frac{2}{3}u(u+1)(4u-1) & \text{for } n = 4u+3. \end{cases}$$

The first few numbers of monochromatic forced triangles are 0, 0, 0, 0, 0, 2, 4, 8, 12, 20, 28, 40, ... (Sloane's A014557).

see also COMPLETE GRAPH, EXTREMAL GRAPH

References

Goodman, A. W. "On Sets of Acquaintances and Strangers at Any Party." *Amer. Math. Monthly* **66**, 778–783, 1959.

Sloane, N. J. A. Sequence A014553 in "An On-Line Version of the Encyclopedia of Integer Sequences."

Monodromy

A general concept in CATEGORY THEORY involving the globalization of local MORPHISMS.

see also HOLONOMY

Monodromy Group

A technically defined GROUP characterizing a system of linear differential equations

$$y_j' = \sum_{k=1}^{n} a_{jk}(x)y_k$$

for $j = 1, \ldots, n$, where a_{jk} are COMPLEX ANALYTIC FUNCTIONS of x in a given COMPLEX DOMAIN.

see also HILBERT'S 21ST PROBLEM, RIEMANN P-SERIES

References

Iyanaga, S. and Kawada, Y. (Eds.). "Monodromy Groups." §253B in *Encyclopedic Dictionary of Mathematics.* Cambridge, MA: MIT Press, p. 793, 1980.

Monodromy Theorem

If a COMPLEX function f is ANALYTIC in a DISK contained in a simply connected DOMAIN D and f can be ANALYTICALLY CONTINUED along every polygonal arc in D, then f can be ANALYTICALLY CONTINUED to a single-valued ANALYTIC FUNCTION on all of D!

see also ANALYTIC CONTINUATION

Monogenic Function

If

$$\lim_{z \to z_0} \frac{f(z) - f(z_0)}{z - z_0}$$

is the same for all paths in the COMPLEX PLANE, then $f(z)$ is said to be monogenic at z_0. Monogenic therefore essentially means having a single DERIVATIVE at a point. Functions are either monogenic or have infinitely many DERIVATIVES (in which case they are called POLYGENIC); intermediate cases are not possible.

see also POLYGENIC FUNCTION

References
Newman, J. R. *The World of Mathematics, Vol. 3.* New York: Simon & Schuster, p. 2003, 1956.

Monohedral Tiling

A TILING is which all tiles are congruent.

see also ANISOHEDRAL TILING, ISOHEDRAL TILING

References
Berglund, J. "Is There a k-Anisohedral Tile for $k \geq 5$?" *Amer. Math. Monthly* **100**, 585-588, 1993.
Grünbaum, B. and Shephard, G. C. "The 81 Types of Isohedral Tilings of the Plane." *Math. Proc. Cambridge Philos. Soc.* **82**, 177-196, 1977.

Monoid

A GROUP-like object which fails to be a GROUP because elements need not have an inverse within the object. A monoid S must also be ASSOCIATIVE and an IDENTITY ELEMENT $I \in S$ such that for all $a \in S$, $1a = a1 = a$. A monoid is therefore a SEMIGROUP with an identity element. A monoid must contain at least one element.

The numbers of free idempotent monoids on n letters are 1, 2, 7, 160, 332381, ... (Sloane's A005345).

see also BINARY OPERATOR, GROUP, SEMIGROUP

References
Rosenfeld, A. *An Introduction to Algebraic Structures.* New York: Holden-Day, 1968.
Sloane, N. J. A. Sequence A005345/M1820 in "An On-Line Version of the Encyclopedia of Integer Sequences."

Monomial

A POLYNOMIAL consisting of a single term.

see also BINOMIAL, MONIC POLYNOMIAL, POLYNOMIAL, TRINOMIAL

Monomino

The unique 1-POLYOMINO, consisting of a single SQUARE.

see also DOMINO, TRIOMINO

References
Gardner, M. "Polyominoes." Ch. 13 in *The Scientific American Book of Mathematical Puzzles & Diversions.* New York: Simon and Schuster, pp. 124–140, 1959.

Monomorph

An INTEGER which is expressible in only one way in the form $x^2 + Dy^2$ or $x^2 - Dy^2$ where x^2 is RELATIVELY PRIME to Dy^2. If the INTEGER is expressible in more than one way, it is called a POLYMORPH.

see also ANTIMORPH, IDONEAL NUMBER, POLYMORPH

Monomorphism

An INJECTIVE MORPHISM.

Monotone

Another word for monotonic.

see also MONOTONIC FUNCTION, MONOTONIC SEQUENCE, MONOTONIC VOTING

Monotone Decreasing

Always decreasing; never remaining constant or increasing.

Monotone Increasing

Always increasing; never remaining constant or decreasing.

Monotonic Function

A function which is either entirely nonincreasing or nondecreasing. A function is monotonic if its first DERIVATIVE (which need not be continuous) does not change sign.

Monotonic Sequence

A SEQUENCE $\{a_n\}$ such that either (1) $a_{i+1} \geq a_i$ for every $i \geq 1$, or (2) $a_{i+1} \leq a_i$ for every $i \geq 1$.

Monotonic Voting

A term in SOCIAL CHOICE THEORY meaning a change favorable for X does not hurt X.

see also ANONYMOUS, DUAL VOTING

Monster Group

The highest order SPORADIC GROUP M. It has ORDER

$$2^{46} \cdot 3^{20} \cdot 5^9 \cdot 7^6 \cdot 11^2 \cdot 13^3 \cdot 17 \cdot 19 \cdot 23 \cdot 29 \cdot 31 \cdot 41 \cdot 47 \cdot 59 \cdot 71,$$

and is also called the FRIENDLY GIANT GROUP. It was constructed in 1982 by Robert Griess as a GROUP of ROTATIONS in 196,883-D space.

see also BABY MONSTER GROUP, BIMONSTER, LEECH LATTICE

References
Conway, J. H.; Curtis, R. T.; Norton, S. P.; Parker, R. A.; and Wilson, R. A. *Atlas of Finite Groups: Maximal Subgroups and Ordinary Characters for Simple Groups.* Oxford, England: Clarendon Press, p. viii, 1985.
Conway, J. H. and Norton, S. P. "Monstrous Moonshine." *Bull. London Math. Soc.* **11**, 308–339, 1979.
Conway, J. H. and Sloane, N. J. A. "The Monster Group and its 196884-Dimensional Space" and "A Monster Lie Algebra?" Chs. 29–30 in *Sphere Packings, Lattices, and Groups, 2nd ed.* New York: Springer-Verlag, pp. 554–571, 1993.
Wilson, R. A. "ATLAS of Finite Group Representation." `http://for.mat.bham.ac.uk/atlas/M.html`.

Monte Carlo Integration

In order to integrate a function over a complicated DOMAIN D, Monte Carlo integration picks random points over some simple DOMAIN D' which is a superset of D, checks whether each point is within D, and estimates the AREA of D (VOLUME, n-D CONTENT, etc.) as the AREA of D' multiplied by the fraction of points falling within D'.

An estimate of the uncertainty produced by this technique is given by

$$\int f\,dV \approx V\langle f\rangle \pm \sqrt{\frac{\langle f^2\rangle - \langle f\rangle^2}{N}}.$$

see also MONTE CARLO METHOD

References

Press, W. H.; Flannery, B. P.; Teukolsky, S. A.; and Vetterling, W. T. "Simple Monte Carlo Integration" and "Adaptive and Recursive Monte Carlo Methods." §7.6 and 7.8 in *Numerical Recipes in FORTRAN: The Art of Scientific Computing, 2nd ed.* Cambridge, England: Cambridge University Press, pp. 295–299 and 306–319, 1992.

Monte Carlo Method

Any method which solves a problem by generating suitable random numbers and observing that fraction of the numbers obeying some property or properties. The method is useful for obtaining numerical solutions to problems which are too complicated to solve analytically. The most common application of the Monte Carlo method is MONTE CARLO INTEGRATION.

see also MONTE CARLO INTEGRATION

References

Sobol, I. M. *A Primer for the Monte Carlo Method.* Boca Raton, FL: CRC Press, 1994.

Montel's Theorem

Let $f(z)$ be an analytic function of z, regular in the half-strip S defined by $a < x < b$ and $y > 0$. If $f(z)$ is bounded in S and tends to a limit l as $y \to \infty$ for a certain fixed value ξ of x between a and b, then $f(z)$ tends to this limit l on every line $x = x_0$ in S, and $f(z) \to l$ uniformly for $a + \delta \leq x_0 \leq b - \delta$.

see also VITALI'S CONVERGENCE THEOREM

References

Titchmarsh, E. C. *The Theory of Functions, 2nd ed.* Oxford, England: Oxford University Press, p. 170, 1960.

Monty Hall Dilemma

see MONTY HALL PROBLEM

Monty Hall Problem

The Monty Hall problem is named for its similarity to the *Let's Make a Deal* television game show hosted by Monty Hall. The problem is stated as follows. Assume that a room is equipped with three doors. Behind two are goats, and behind the third is a shiny new car. You are asked to pick a door, and will win whatever is behind it. Let's say you pick door 1. Before the door is opened, however, someone who knows what's behind the doors (Monty Hall) opens *one of the other* two doors, revealing a goat, and asks you if you wish to change your selection to the third door (i.e., the door which neither you picked nor he opened). The Monty Hall problem is deciding whether you do.

The correct answer is that you *do* want to switch. If you do not switch, you have the expected 1/3 chance of winning the car, since no matter whether you initially picked the correct door, Monty will show you a door with a goat. But after Monty has eliminated one of the doors for you, you obviously do not improve your chances of winning to better than 1/3 by sticking with your original choice. If you now switch doors, however, there is a 2/3 chance you will win the car (counterintuitive though it seems).

d_1	d_2	Winning Probability
pick	stick	1/3
pick	switch	2/3

The problem can be generalized to four doors as follows. Let one door conceal the car, with goats behind the other three. Pick a door d_1. Then the host will open one of the nonwinners and give you the option of switching. Call your new choice (which could be the same as d_1 if you don't switch) d_2. The host will then open a second nonwinner, and you must decide for choice d_3 if you want to stick to d_2 or switch to the remaining door. The probabilities of winning are shown below for the four possible strategies.

d_1	d_2	d_3	Winning Probability
pick	stick	stick	4/8
pick	switch	stick	3/8
pick	stick	switch	6/8
pick	switch	switch	5/8

The above results are characteristic of the best strategy for the n-stage Monty Hall problem: stick until the last choice, then switch.

see also ALIAS' PARADOX

References

Barbeau, E. "The Problem of the Car and Goats." *CMJ* **24**, 149, 1993.
Bogomolny, A. "Monty Hall Dilemma." `http://www.cut-the-knot.com/hall.html`.
Dewdney, A. K. *200% of Nothing.* New York: Wiley, 1993.
Donovan, D. "The WWW Tackles the Monty Hall Problem." `http://math.rice.edu/~ddonovan/montyurl.html`.
Ellis, K. M. "The Monty Hall Problem." `http://www.io.com/~kmellis/monty.html`.

Gardner, M. *Aha! Gotcha: Paradoxes to Puzzle and Delight.* New York: W. H. Freeman, 1982.
Gillman, L. "The Car and the Goats." *Amer. Math. Monthly* **99**, 3, 1992.
Selvin, S. "A Problem in Probability." *Amer. Stat.* **29**, 67, 1975.
vos Savant, M. *The Power of Logical Thinking.* New York: St. Martin's Press, 1996.

Moore Graph

A GRAPH with DIAMETER d and GIRTH $2d+1$. Moore graphs have DIAMETER of at most 2. Every Moore graph is both REGULAR and distance regular. Hoffman and Singleton (1960) show that k-regular Moore graphs with DIAMETER 2 have $k \in \{2,3,7,57\}$.

References
Godsil, C. D. "Problems in Algebraic Combinatorics." *Electronic J. Combinatorics* **2**, F1, 1–20, 1995. http://www.combinatorics.org/Volume_2/volume2.html#F1.
Hoffman, A. J. and Singleton, R. R. "On Moore Graphs of Diameter Two and Three." *IBM J. Res. Develop.* **4**, 497–504, 1960.

Moore-Penrose Generalized Matrix Inverse

Given an $m \times n$ MATRIX B, the Moore-Penrose generalized MATRIX INVERSE is a unique $n \times m$ MATRIX B^+ which satisfies

$$\mathsf{B}\mathsf{B}^+\mathsf{B} = \mathsf{B} \quad (1)$$

$$\mathsf{B}^+\mathsf{B}\mathsf{B}^+ = \mathsf{B}^+ \quad (2)$$

$$(\mathsf{B}\mathsf{B}^+)^{\mathrm{T}} = \mathsf{B}\mathsf{B}^+ \quad (3)$$

$$(\mathsf{B}^+\mathsf{B})^{\mathrm{T}} = \mathsf{B}^+\mathsf{B}. \quad (4)$$

It is also true that

$$\mathbf{z} = \mathsf{B}^+\mathbf{c} \quad (5)$$

is the shortest length LEAST SQUARES solution to the problem

$$\mathsf{B}\mathbf{z} = \mathbf{c}. \quad (6)$$

If the inverse of $(\mathsf{B}^{\mathrm{T}}\mathsf{B})$ exists, then

$$\mathsf{B}^+ = (\mathsf{B}^{\mathrm{T}}\mathsf{B})^{-1}\mathsf{B}^{\mathrm{T}}, \quad (7)$$

where B^{T} is the MATRIX TRANSPOSE, as can be seen by premultiplying both sides of (7) by B^{T} to create a SQUARE MATRIX which can then be inverted,

$$\mathsf{B}^{\mathrm{T}}\mathsf{B}\mathbf{z} = \mathsf{B}^{\mathrm{T}}\mathbf{c}, \quad (8)$$

giving

$$\begin{aligned}\mathbf{z} &= (\mathsf{B}^{\mathrm{T}}\mathsf{B})^{-1}\mathsf{B}^{\mathrm{T}}\mathbf{c} \\ &\equiv \mathsf{B}^+\mathbf{c}. \quad (9)\end{aligned}$$

see also LEAST SQUARES FITTING, MATRIX INVERSE

References
Ben-Israel, A. and Greville, T. N. E. *Generalized Inverses: Theory and Applications.* New York: Wiley, 1977.
Lawson, C. and Hanson, R. *Solving Least Squares Problems.* Englewood Cliffs, NJ: Prentice-Hall, 1974.
Penrose, R. "A Generalized Inverse for Matrices." *Proc. Cambridge Phil. Soc.* **51**, 406–413, 1955.

Mordell Conjecture

DIOPHANTINE EQUATIONS that give rise to surfaces with two or more holes have only finite many solutions in GAUSSIAN INTEGERS with no common factors. Fermat's equation has $(n-1)(n-2)/2$ HOLES, so the Mordell conjecture implies that for each INTEGER $n \geq 3$, the FERMAT EQUATION has at most a finite number of solutions. This conjecture was proved by Faltings (1984).

see also FERMAT EQUATION, FERMAT'S LAST THEOREM, SAFAREVICH CONJECTURE, SHIMURA-TANIYAMA CONJECTURE

References
Faltings, G. "Die Vermutungen von Tate und Mordell." *Jahresber. Deutsch. Math.-Verein* **86**, 1–13, 1984.
Ireland, K. and Rosen, M. "The Mordell Conjecture." §20.3 in *A Classical Introduction to Modern Number Theory, 2nd ed.* New York: Springer-Verlag, pp. 340–342, 1990.

Mordell Integral

The integral

$$\phi(t,u) = \int \frac{e^{\pi i t x^2 + 2\pi i u x}}{e^{2\pi i x} - 1}\,dx$$

which is related to the THETA FUNCTIONS, MOCK THETA FUNCTIONS, and RIEMANN ZETA FUNCTION.

Mordell-Weil Theorem

For ELLIPTIC CURVES over the RATIONALS, $\mathbb{Q}$, the number of generators of the set of RATIONAL POINTS is always finite. This theorem was proved by Mordell in 1921 and extended by Weil in 1928 to ABELIAN VARIETIES over NUMBER FIELDS.

References
Ireland, K. and Rosen, M. "The Mordell-Weil Theorem." Ch. 19 in *A Classical Introduction to Modern Number Theory, 2nd ed.* New York: Springer-Verlag, pp. 319–338, 1990.

Morera's Theorem

If $f(z)$ is continuous in a simply connected region D and satisfies

$$\int_\gamma f\,dz = 0$$

for all closed CONTOURS γ in D, then $f(z)$ is ANALYTIC in D.

see also CAUCHY INTEGRAL THEOREM

References
Arfken, G. *Mathematical Methods for Physicists, 3rd ed.* Orlando, FL: Academic Press, pp. 373–374, 1985.

Morgado Identity

An identity satisfied by w GENERALIZED FIBONACCI NUMBERS:

$$\begin{aligned}4w_nw_{n+1}w_{n+2}w_{n+4}w_{n+5}w_{n+6}\\ +e^2q^{2n}(w_nU_4U_5-w_{n+1}U_2U_6-w_nU_1U_8)^2\\ =(w_{n+1}w_{n+2}w_{n+6}+w_nw_{n+4}w_{n+5})^2,\end{aligned} \quad (1)$$

where

$$e\equiv pab-qa^2-b^2 \quad (2)$$

$$U_n\equiv w_n(0,1;p,q). \quad (3)$$

see also GENERALIZED FIBONACCI NUMBER

References

Morgado, J. "Note on Some Results of A. F. Horadam and A. G. Shannon Concerning a Catalan's Identity on Fibonacci Numbers." *Portugaliae Math.* **44**, 243–252, 1987.

Morgan-Voyce Polynomial

Polynomials related to the BRAHMAGUPTA POLYNOMIALS. They are defined by the RECURRENCE RELATIONS

$$b_n(x)=xB_{n-1}(x)+b_{n-1}(x) \quad (1)$$

$$B_n(x)=(x+1)B_{n-1}(x)+b_{n-1}(x) \quad (2)$$

for $n\geq 1$, with

$$b_0(x)=B_0(x)=1. \quad (3)$$

Alternative recurrences are

$$B_{n+1}B_{n-1}-{B_n}^2=-1 \quad (4)$$

$$b_{n+1}b_{n-1}-{b_n}^2=x. \quad (5)$$

The polynomials can be given explicitly by the sums

$$B_n(x)=\sum_{k=0}^{n}\binom{n+k-1}{n-k} \quad (6)$$

$$b_n(x)=\sum_{k=0}^{n}\binom{n+k}{n-k}. \quad (7)$$

Defining the MATRIX

$$\mathsf{Q}=\begin{bmatrix}x+2 & -1\\ 1 & 0\end{bmatrix} \quad (8)$$

gives the identities

$$\mathsf{Q}^n=\begin{bmatrix}B_n & -B_{n-1}\\ B_{n-1} & -B_{n-2}\end{bmatrix} \quad (9)$$

$$\mathsf{Q}^n-\mathsf{Q}^{n-1}=\begin{bmatrix}b_n & -b_{n-1}\\ b_{n-1} & -b_{n-2}\end{bmatrix}. \quad (10)$$

Defining

$$\cos\theta=\tfrac{1}{2}(x+2) \quad (11)$$

$$\cosh\phi=\tfrac{1}{2}(x+2) \quad (12)$$

gives

$$B_n(x)=\frac{\sin[(n+1)\theta]}{\sin\theta} \quad (13)$$

$$B_n(x)=\frac{\sinh[(n+1)\phi]}{\sinh\phi} \quad (14)$$

and

$$b_n(x)=\frac{\cos[\frac{1}{2}(2n+1)\theta]}{\cos(\frac{1}{2}\theta)} \quad (15)$$

$$b_n(x)=\frac{\cosh[\frac{1}{2}(2n+1)\phi]}{\cosh(\frac{1}{2}\theta)}. \quad (16)$$

The Morgan-Voyce polynomials are related to the FIBONACCI POLYNOMIALS $F_n(x)$ by

$$b_n(x^2)=F_{2n+1}(x) \quad (17)$$

$$B_n(x^2)=\frac{1}{x}F_{2n+2}(x) \quad (18)$$

(Swamy 1968).

$B_n(x)$ satisfies the ORDINARY DIFFERENTIAL EQUATION

$$x(x+4)y''+3(x+2)y'-n(n+2)y=0, \quad (19)$$

and $b_n(x)$ the equation

$$x(x+4)y''+2(x+1)y'-n(n+1)y=0. \quad (20)$$

These and several other identities involving derivatives and integrals of the polynomials are given by Swamy (1968).

see also BRAHMAGUPTA POLYNOMIAL, FIBONACCI POLYNOMIAL

References

Lahr, J. "Fibonacci and Lucas Numbers and the Morgan-Voyce Polynomials in Ladder Networks and in Electric Line Theory." In *Fibonacci Numbers and Their Applications* (Ed. G. E. Bergum, A. N. Philippou, and A. F. Horadam). Dordrecht, Netherlands: Reidel, 1986.

Morgan-Voyce, A. M. "Ladder Network Analysis Using Fibonacci Numbers." *IRE Trans. Circuit Th.* **CT-6**, 321–322, Sep. 1959.

Swamy, M. N. S. "Properties of the Polynomials Defined by Morgan-Voyce." *Fib. Quart.* **4**, 73–81, 1966.

Swamy, M. N. S. "More Fibonacci Identities." *Fib. Quart.* **4**, 369–372, 1966.

Swamy, M. N. S. "Further Properties of Morgan-Voyce Polynomials." *Fib. Quart.* **6**, 167–175, 1968.

Morley Centers

The CENTROID of MORLEY'S TRIANGLE is called Morley's first center. It has TRIANGLE CENTER FUNCTION

$$\alpha = \cos(\tfrac{1}{3}A) + 2\cos(\tfrac{1}{3}B)\cos(\tfrac{1}{3}C).$$

The PERSPECTIVE CENTER of MORLEY'S TRIANGLE with reference TRIANGLE ABC is called Morley's second center. The TRIANGLE CENTER FUNCTION is

$$\alpha = \sec(\tfrac{1}{3}A).$$

see also CENTROID (GEOMETRIC), MORLEY'S THEOREM, PERSPECTIVE CENTER

References

Kimberling, C. "Central Points and Central Lines in the Plane of a Triangle." *Math. Mag.* **67**, 163–187, 1994.

Kimberling, C. "1st and 2nd Morley Centers." `http://www.evansville.edu/~ck6/tcenters/recent/morley.html`.

Oakley, C. O. and Baker, J. C. "The Morley Trisector Theorem." *Amer. Math. Monthly* **85**, 737–745, 1978.

Morley's Formula

$$\sum_{k=0}^{\infty} \binom{m}{k}^3 = 1 + \left(\frac{m}{1}\right)^3 + \left[\frac{m(m+1)}{1\cdot 2}\right]^3 + \ldots$$
$$= \frac{\Gamma(1-\frac{3}{2}m)}{[\Gamma(1-\frac{1}{2}m)]^3}\cos(\tfrac{1}{2}m\pi),$$

where $\binom{n}{k}$ is a BINOMIAL COEFFICIENT and $\Gamma(z)$ is the GAMMA FUNCTION.

Morley's Theorem

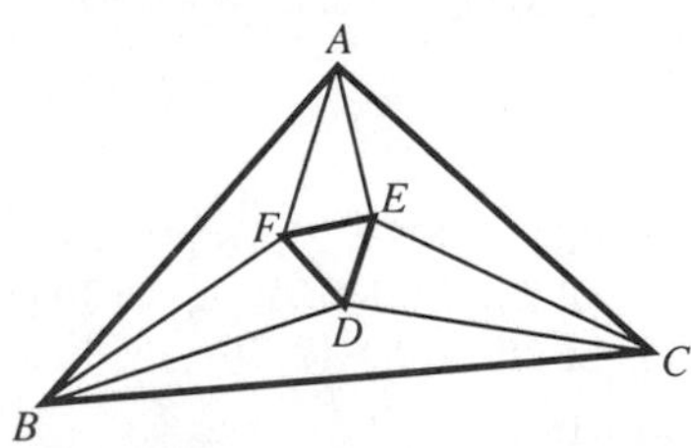

The points of intersection of the adjacent TRISECTORS of the ANGLES of any TRIANGLE ΔABC are the VERTICES of an EQUILATERAL TRIANGLE ΔDEF known as MORLEY'S TRIANGLE. Taylor and Marr (1914) give two geometric proofs and one trigonometric proof.

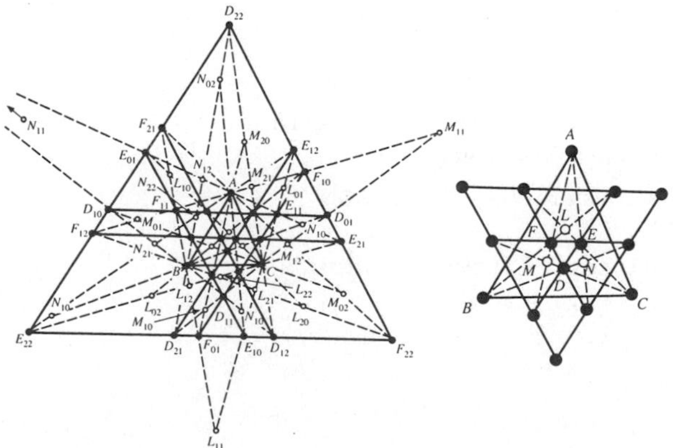

A generalization of MORLEY'S THEOREM was discovered by Morley in 1900 but first published by Taylor and Marr (1914). Each ANGLE of a TRIANGLE ΔABC has six trisectors, since each interior angle trisector has two associated lines making angles of 120° with it. The generalization of Morley's theorem states that these trisectors intersect in 27 points (denoted D_{ij}, E_{ij}, F_{ij}, for $i, j = 0, 1, 2$) which lie six by six on nine lines. Furthermore, these lines are in three triples of PARALLEL lines, ($D_{22}E_{22}$, $E_{12}D_{21}$, $F_{10}F_{01}$), ($D_{22}F_{22}$, $F_{21}D_{12}$, $E_{01}E_{10}$), and ($E_{22}F_{22}$, $F_{12}E_{21}$, $D_{10}D_{01}$), making ANGLES of 60° with one another (Taylor and Marr 1914, Johnson 1929, p. 254).

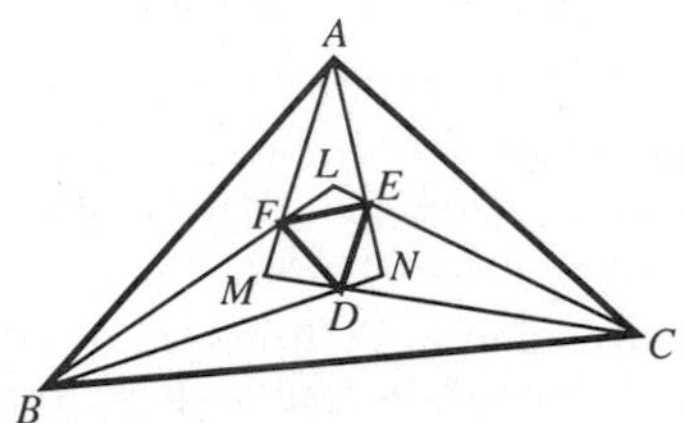

Let L, M, and N be the other trisector-trisector intersections, and let the 27 points L_{ij}, M_{ij}, N_{ij} for $i, j = 0, 1, 2$ be the ISOGONAL CONJUGATES of D, E, and F. Then these points lie 6 by 6 on 9 CONICS through ΔABC. In addition, these CONICS meet 3 by 3 on the CIRCUMCIRCLE, and the three meeting points form an EQUILATERAL TRIANGLE whose sides are PARALLEL to those of ΔDEF.

see also CONIC SECTION, MORLEY CENTERS, TRISECTION

References

Coxeter, H. S. M. and Greitzer, S. L. *Geometry Revisited.* Washington, DC: Math. Assoc. Amer., pp. 47–50, 1967.

Gardner, M. *Martin Gardner's New Mathematical Diversions from Scientific American.* New York: Simon and Schuster, pp. 198 and 206, 1966.

Honsberger, R. "Morley's Theorem." Ch. 8 in *Mathematical Gems I.* Washington, DC: Math. Assoc. Amer., pp. 92–98, 1973.

Johnson, R. A. *Modern Geometry: An Elementary Treatise on the Geometry of the Triangle and the Circle.* Boston, MA: Houghton Mifflin, pp. 253–256, 1929.

Kimberling, C. "Hofstadter Points." *Nieuw Arch. Wiskunder* **12**, 109–114, 1994.

Marr, W. L. "Morley's Trisection Theorem: An Extension and Its Relation to the Circles of Apollonius." *Proc. Edinburgh Math. Soc.* **32**, 136–150, 1914.

Oakley, C. O. and Baker, J. C. "The Morley Trisector Theorem." *Amer. Math. Monthly* **85**, 737–745, 1978.

Pappas, T. "Trisecting & the Equilateral Triangle." *The Joy of Mathematics.* San Carlos, CA: Wide World Publ./Tetra, p. 174, 1989.

Taylor, F. G. "The Relation of Morley's Theorem to the Hessian Axis and Circumcentre." *Proc. Edinburgh Math. Soc.* **32**, 132–135, 1914.

Taylor, F. G. and Marr, W. L. "The Six Trisectors of Each of the Angles of a Triangle." *Proc. Edinburgh Math. Soc.* **32**, 119–131, 1914.

Morley's Triangle

An EQUILATERAL TRIANGLE considered by MORLEY'S THEOREM with side lengths

$$8R\sin(\tfrac{1}{3}A)\sin(\tfrac{1}{3}B)\sin(\tfrac{1}{3}C),$$

where R is the CIRCUMRADIUS of the original TRIANGLE.

Morphism

A map between two objects in an abstract CATEGORY.

1. A general morphism is called a HOMOMORPHISM,
2. An injective morphism is called a MONOMORPHISM,
3. A surjective morphism is an EPIMORPHISM,
4. A bijective morphism is called an ISOMORPHISM (if there is an isomorphism between two objects, then we say they are isomorphic),
5. A surjective morphism from an object to itself is called an ENDOMORPHISM, and
6. An ISOMORPHISM between an object and itself is called an AUTOMORPHISM.

see also AUTOMORPHISM, EPIMORPHISM, HOMEOMORPHISM, HOMOMORPHISM, ISOMORPHISM, MONOMORPHISM, OBJECT

Morrie's Law

$$\cos(20°)\cos(40°)\cos(80°) = \tfrac{1}{8}.$$

This identity was referred to by Feynman (Gleick 1992). It is a special case of the general identity

$$2^k\prod_{j=0}^{k-1}\cos(2^j a) = \frac{\sin(2^k a)}{\sin a},$$

with $k = 3$ and $a = 20°$ (Beyer *et al.* 1996).

References

Anderson, E. C. "Morrie's Law and Experimental Mathematics." To appear in *J. Recr. Math.*

Beyer, W. A.; Louck, J. D.; Zeilberger, D. "A Generalization of a Curiosity that Feynman Remembered All His Life." *Math. Mag.* **69**, 43–44, 1996.

Gleick, J. *Genius: The Life and Science of Richard Feynman.* New York: Pantheon Books, p. 47, 1992.

Morse Inequalities

Topological lower bounds in terms of BETTI NUMBERS for the number of critical points form a smooth function on a smooth MANIFOLD.

Morse Theory

"CALCULUS OF VARIATIONS in the large" which uses nonlinear techniques to address problems in the CALCULUS OF VARIATIONS. Morse theory applied to a FUNCTION g on a MANIFOLD W with $g(M) = 0$ and $g(M') = 1$ shows that every COBORDISM can be realized as a finite sequence of SURGERIES. Conversely, a sequence of SURGERIES gives a COBORDISM.

see also CALCULUS OF VARIATIONS, COBORDISM, SURGERY

Morse-Thue Sequence

see THUE-MORSE SEQUENCE

Mortal

A nonempty finite set of $n \times n$ MATRICES with INTEGER entries for which there exists some product of the MATRICES in the set which is equal to the zero MATRIX.

Mortality Problem

For a given n, is the problem of determining if a set is MORTAL solvable? $n = 1$ is solvable, $n = 2$ is unknown, and $n \geq 3$ is unsolvable.

see also LIFE EXPECTANCY

Morton-Franks-Williams Inequality

Let E be the largest and e the smallest POWER of ℓ in the HOMFLY POLYNOMIAL of an oriented LINK, and i be the BRAID INDEX. Then the MORTON-FRANKS-WILLIAMS INEQUALITY holds,

$$i \geq \tfrac{1}{2}(E - e) + 1$$

(Franks and Williams 1985, Morton 1985). The inequality is sharp for all PRIME KNOTS up to 10 crossings with the exceptions of 09_{042}, 09_{049}, 10_{132}, 10_{150}, and 10_{156}.

see also BRAID INDEX

References

Franks, J. and Williams, R. F. "Braids and the Jones Polynomial." *Trans. Amer. Math. Soc.* **303**, 97–108, 1987.

Mosaic

see TESSELLATION

Moser

The very LARGE NUMBER consisting of the number 2 inside a MEGA-gon.

see also MEGA, MEGISTRON

Moser's Circle Problem

see CIRCLE CUTTING

Moss's Egg

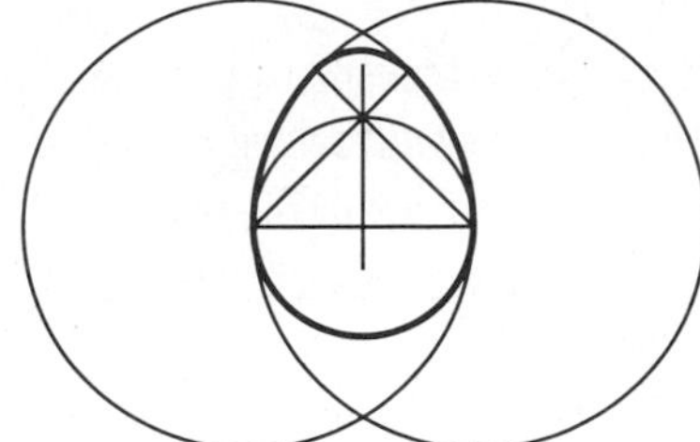

An OVAL whose construction is illustrated in the above diagram.

see also EGG, OVAL

References

Dixon, R. *Mathographics.* New York: Dover, p. 5, 1991.

Motzkin Number

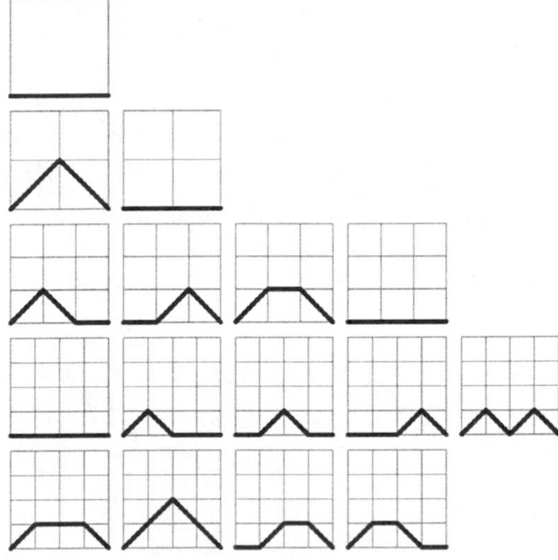

The Motzkin numbers enumerate various combinatorial objects. Donaghey and Shapiro (1977) give 14 different manifestations of these numbers. In particular, they give the number of paths from (0, 0) to $(n, 0)$ which never dip below $y = 0$ and are made up only of the steps (1, 0), (1, 1), and (1, -1), i.e., $\rightarrow$, $\nearrow$, and $\searrow$. The first are 1, 2, 4, 9, 21, 51, ... (Sloane's A001006). The Motzkin number GENERATING FUNCTION $M(z)$ satisfies

$$M = 1 + xM + x^2M^2 \tag{1}$$

and is given by

$$\begin{aligned} M(x) &= \frac{1 - x - \sqrt{1 - 2x - 3x^2}}{2x^2} \\ &= 1 + x + 2x^2 + 4x^3 + 9x^4 + 21x^5 + \ldots, \end{aligned} \tag{2}$$

or by the RECURRENCE RELATION

$$M_n = a_{n-1} + \sum_{k=0}^{n-2} M_k M_{n-2-k} \tag{3}$$

with $M_0 = 1$. The Motzkin number M_n is also given by

$$M_n = -\frac{1}{2} \sum_{\substack{a+b=n+2 \\ a\geq 0, b\geq 0}} (-3)^a \binom{\frac{1}{2}}{a} \binom{\frac{1}{2}}{b} \tag{4}$$

$$= \frac{(-1)^{n+1}}{2^{2n+5}} \sum_{\substack{a+b=n+2 \\ a\geq 0, b\geq 0}} \frac{(-3)^a}{(2a-1)(2b-1)} \binom{2a}{a} \binom{2b}{b}, \tag{5}$$

where $\binom{n}{k}$ is a BINOMIAL COEFFICIENT.

see also CATALAN NUMBER, KING WALK, SCHRÖDER NUMBER

References

Barcucci, E.; Pinzani, R.; and Sprugnoli, R. "The Motzkin Family." *Pure Math. Appl. Ser. A* **2**, 249–279, 1991.

Donaghey, R. "Restricted Plane Tree Representations of Four Motzkin-Catalan Equations." *J. Combin. Th. Ser. B* **22**, 114–121, 1977.

Donaghey, R. and Shapiro, L. W. "Motzkin Numbers." *J. Combin. Th. Ser. A* **23**, 291–301, 1977.

Kuznetsov, A.; Pak, I.; and Postnikov, A. "Trees Associated with the Motzkin Numbers." *J. Combin. Th. Ser. A* **76**, 145–147, 1996.

Motzkin, T. "Relations Between Hypersurface Cross Ratios, and a Combinatorial Formula for Partitions of a Polygon, for Permanent Preponderance, and for Nonassociative Products." *Bull. Amer. Math. Soc.* **54**, 352–360, 1948.

Sloane, N. J. A. Sequence A001006/M1184 in "An On-Line Version of the Encyclopedia of Integer Sequences."

Moufang Plane

A PROJECTIVE PLANE in which every line is a translation line is called a Moufang plane.

References

Colbourn, C. J. and Dinitz, J. H. (Eds.) *CRC Handbook of Combinatorial Designs.* Boca Raton, FL: CRC Press, p. 710, 1996.

Mousetrap

A PERMUTATION problem invented by Cayley.

References

Guy, R. K. "Mousetrap." §E37 in *Unsolved Problems in Number Theory, 2nd ed.* New York: Springer-Verlag, pp. 237–238, 1994.

Mouth

A PRINCIPAL VERTEX x_i of a SIMPLE POLYGON P is called a mouth if the diagonal $[x_{i-1}, x_{i+1}]$ is an extremal diagonal (i.e., the interior of $[x_{i-1}, x_{i+1}]$ lies in the exterior of P).

see also ANTHROPOMORPHIC POLYGON, EAR, ONE-MOUTH THEOREM

References

Toussaint, G. "Anthropomorphic Polygons." *Amer. Math. Monthly* **122**, 31–35, 1991.

Moving Average

Given a SEQUENCE $\{a_i\}_{i=1}^N$, an n-moving average is a new sequence $\{s_i\}_{i=1}^{N-n+1}$ defined from the a_i by taking the AVERAGE of subsequences of n terms,

$$s_i = \frac{1}{n}\sum_{j=i}^{i+n-1} a_j.$$

see also AVERAGE, SPENCER'S 15-POINT MOVING AVERAGE

References

Kenney, J. F. and Keeping, E. S. "Moving Averages." §14.2 in *Mathematics of Statistics, Pt. 1, 3rd ed.* Princeton, NJ: Van Nostrand, pp. 221–223, 1962.

Moving Ladder Constant

N.B. A detailed on-line essay by S. Finch was the starting point for this entry.

What is the longest ladder which can be moved around a right-angled hallway of unit width? For a straight, rigid ladder, the answer is $2\sqrt{2}$. For a smoothly-shaped ladder, the largest diameter is $\geq 2(1+\sqrt{2})$ (Finch).

see also MOVING SOFA CONSTANT, PIANO MOVER'S PROBLEM

References

Finch, S. "Favorite Mathematical Constants." `http://www.mathsoft.com/asolve/constant/sofa/sofa.html`.

Moving Sofa Constant

N.B. A detailed on-line essay by S. Finch was the starting point for this entry.

What is the sofa of greatest AREA S which can be moved around a right-angled hallway of unit width? Hammersley (Croft *et al.* 1994) showed that

$$S \geq \frac{\pi}{2} + \frac{2}{\pi} = 2.2074\ldots. \tag{1}$$

Gerver (1992) found a sofa with larger AREA and provided arguments indicating that it is either optimal or close to it. The boundary of Gerver's sofa is a complicated shape composed of 18 ARCS. Its AREA can be given by defining the constants A, B, ϕ, and θ by solving

$$A(\cos\theta - \cos\phi) - 2B\sin\phi + (\theta-\phi-1)\cos\theta - \sin\theta + \cos\phi + \sin\phi = 0 \tag{2}$$

$$A(3\sin\theta + \sin\phi) - 2B\cos\phi + 3(\theta-\phi-1)\sin\theta + 3\cos\theta - \sin\phi + \cos\phi = 0 \tag{3}$$

$$A\cos\phi - (\sin\phi + \tfrac{1}{2} - \tfrac{1}{2}\cos\phi + B\sin\phi) = 0 \tag{4}$$

$$(A+\tfrac{1}{2}\pi-\phi-\theta) - [B - \tfrac{1}{2}(\theta-\phi)(1+A) - \tfrac{1}{4}(\theta-\phi)^2] = 0. \tag{5}$$

This gives

$$A = 0.094426560843653\ldots \tag{6}$$

$$B = 1.399203727333547\ldots \tag{7}$$

$$\phi = 0.039177364790084\ldots \tag{8}$$

$$\theta = 0.681301509382725\ldots. \tag{9}$$

Now define

$$r(\alpha) \equiv \begin{cases} \tfrac{1}{2} & \text{for } 0 \leq \alpha < \phi \\ \tfrac{1}{2}(1+A+\alpha-\phi) & \text{for } \phi \leq \alpha < \theta \\ A+\alpha-\phi & \text{for } \theta \leq \alpha < \tfrac{1}{2}\pi - \theta \\ B - \tfrac{1}{2}(\tfrac{1}{2}\pi-\alpha-\phi)(1+A) - \tfrac{1}{4}(\tfrac{1}{2}\pi-\alpha-\phi)^2, & \text{for } \tfrac{1}{2}\pi-\theta \leq \alpha < \tfrac{1}{2}\pi - \phi, \end{cases} \tag{10}$$

where

$$s(\alpha) \equiv 1 - r(\alpha) \tag{11}$$

$$u(\alpha) \equiv \begin{cases} B - \tfrac{1}{2}(\alpha-\phi)(1+A) - \tfrac{1}{4}(\alpha-\phi)^2 & \text{for } \phi \leq \alpha < \theta \\ A + \tfrac{1}{2}\pi - \phi - \alpha & \text{for } \theta \leq \alpha < \tfrac{1}{4}\pi \end{cases} \tag{12}$$

$$D_u(\alpha) = \frac{du}{d\alpha} = \begin{cases} -\tfrac{1}{2}(1+A) - \tfrac{1}{2}(\alpha-\phi) & \text{for } \phi \leq \alpha \leq \theta \\ -1 & \text{if } \theta \leq \alpha < \tfrac{1}{4}\pi. \end{cases} \tag{13}$$

Finally, define the functions

$$y_1(\alpha) \equiv 1 - \int_0^\alpha r(t)\sin t\,dt \tag{14}$$

$$y_2(\alpha) \equiv 1 - \int_0^\alpha s(t)\sin t\,dt \tag{15}$$

$$y_3(\alpha) \equiv 1 - \int_0^\alpha s(t)\sin t\,dt - u(\alpha)\sin\alpha. \tag{16}$$

The AREA of the optimal sofa is given by

$$A = 2\int_0^{\pi/2-\phi} y_1(\alpha)r(\alpha)\cos\alpha\,d\alpha + 2\int_0^\theta y_2(\alpha)s(\alpha)\cos\alpha\,d\alpha + 2\int_\phi^{\pi/4} y_3(\alpha)[u(\alpha)\sin\alpha - D_u(\alpha)\cos\alpha - s(\alpha)\cos\alpha]\,d\alpha = 2.21953166887197\ldots \tag{17}$$

(Finch).

see also PIANO MOVER'S PROBLEM

References

Croft, H. T.; Falconer, K. J.; and Guy, R. K. *Unsolved Problems in Geometry.* New York: Springer-Verlag, 1994.

Finch, S. "Favorite Mathematical Constants." http://www.mathsoft.com/asolve/constant/sofa/sofa.html.

Gerver, J. L. "On Moving a Sofa Around a Corner." *Geometriae Dedicata* **42**, 267–283, 1992.

Stewart, I. *Another Fine Math You've Got Me Into....* New York: W. H. Freeman, 1992.

Mrs. Perkins' Quilt

The DISSECTION of a SQUARE of side n into a number S_n of smaller squares. Unlike a PERFECT SQUARE DISSECTION, however, the smaller SQUARES need not be all different sizes. In addition, only prime dissections are considered so that patterns which can be dissected on lower order SQUARES are not permitted. The following table gives the smallest number of coprime dissections of an $n \times n$ quilt (Sloane's A005670).

n	S_n
1	1
2	4
3	6
4	7
5	8
6–7	9
8–9	10
10–13	11
14–17	12
18–23	13
24–29	14
30-39	15
40	16
41	15
42–100	$[17, 19]$

see also PERFECT SQUARE DISSECTION

References

Conway, J. H. "Mrs. Perkins's Quilt." *Proc. Cambridge Phil. Soc.* **60**, 363–368, 1964.

Dudeney, H. E. Problem 173 in *Amusements in Mathematics.* New York: Dover, 1917.

Dudeney, H. E. Problem 177 in *536 Puzzles & Curious Problems.* New York: Scribner, 1967.

Gardner, M. "Mrs. Perkins' Quilt and Other Square-Packing Problems." Ch. 11 in *Mathematical Carnival: A New Round-Up of Tantalizers and Puzzles from Scientific American.* New York: Vintage, 1977.

Sloane, N. J. A. Sequence A005670/M3267 in "An On-Line Version of the Encyclopedia of Integer Sequences."

Trustrum, G. B. "Mrs. Perkins's Quilt." *Proc. Cambridge Phil. Soc.* **61**, 7–11, 1965.

Mu Function

$$\mu(x,\beta) \equiv \int_0^\infty \frac{x^t t^\beta \, dt}{\Gamma(\beta+1)\Gamma(t+1)}$$

$$\mu(x,\beta,\alpha) \equiv \int_0^\infty \frac{x^{\alpha+t} t^\beta \, dt}{\Gamma(\beta+1)\Gamma(\alpha+t+1)},$$

where $\Gamma(z)$ is the GAMMA FUNCTION (Gradshteyn and Ryzhik 1980, p. 1079).

see also LAMBDA FUNCTION, NU FUNCTION

References

Gradshteyn, I. S. and Ryzhik, I. M. *Tables of Integrals, Series, and Products, 5th ed.* San Diego, CA: Academic Press, 1979.

μ Molecule

see MANDELBROT SET

Much Greater

A strong INEQUALITY in which a is not only GREATER than b, but *much* greater (by some convention), is denoted $a \gg b$. For an astronomer, "much" may mean by a factor of 100 (or even 10), while for a mathematician, it might mean by a factor of 10^4 (or even much more).

see also GREATER, MUCH LESS

Much Less

A strong INEQUALITY in which a is not only LESS than b, but *much* less (by some convention) is denoted $a \ll b$.

see also LESS, MUCH GREATER

Muirhead's Theorem

A NECESSARY and SUFFICIENT condition that $[\alpha']$ should be comparable with $[\alpha]$ for all POSITIVE values of the a is that one of (α') and (α) should be majorized by the other. If $(\alpha') \prec (\alpha)$, then

$$[\alpha'] \leq [\alpha],$$

with equality only when (α') and (α) are identical or when all the a are equal. See Hardy *et al.* (1988) for a definition of notation.

References

Hardy, G. H.; Littlewood, J. E.; and Pólya, G. *Inequalities, 2nd ed.* Cambridge, England: Cambridge University Press, pp. 44–48, 1988.

Müller-Lyer Illusion

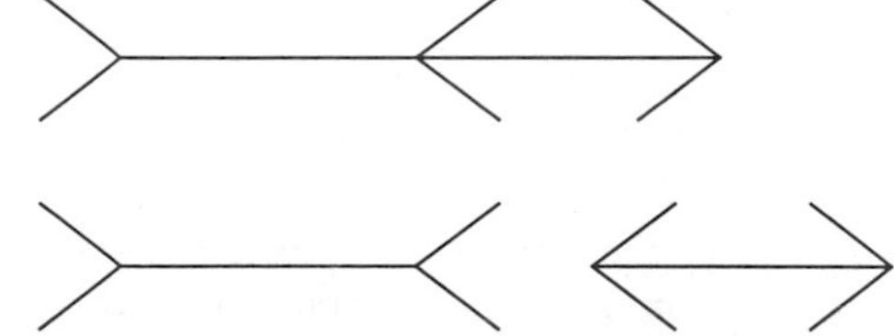

An optical ILLUSION in which the orientation of arrowheads makes one LINE SEGMENT look longer than another. In the above figure, the LINE SEGMENTS on the left and right are of equal length in both cases.

see also ILLUSION, POGGENDORFF ILLUSION, PONZO'S ILLUSION, VERTICAL-HORIZONTAL ILLUSION

References

Fineman, M. *The Nature of Visual Illusion.* New York: Dover, p. 153, 1996.

Luckiesh, M. *Visual Illusions: Their Causes, Characteristics & Applications.* New York: Dover, p. 93, 1965.

Muller's Method

Generalizes the SECANT METHOD of root finding by using quadratic 3-point interpolation

$$q \equiv \frac{x_n - x_{n-1}}{x_{n-1} - x_{n-2}}. \quad (1)$$

Then define

$$A \equiv qP(x_n) - q(1+q)P(x_{n-1}) + q^2 P(x_{n-2}) \quad (2)$$
$$B \equiv (2q+1)P(x_n) - (1+q)^2 P(x_{n-1}) + q^2 P(x_{n-2}) \quad (3)$$
$$C \equiv (1+q)P(x_n), \quad (4)$$

and the next iteration is

$$x_{n+1} = x_n - (x_n - x_{n-1}) \frac{2C}{\max(B \pm \sqrt{B^2 - 4AC})}. \quad (5)$$

This method can also be used to find COMPLEX zeros of ANALYTIC FUNCTIONS.

References

Press, W. H.; Flannery, B. P.; Teukolsky, S. A.; and Vetterling, W. T. *Numerical Recipes in FORTRAN: The Art of Scientific Computing, 2nd ed.* Cambridge, England: Cambridge University Press, p. 364, 1992.

Mulliken Symbols

Symbols used to identify irreducible representations of GROUPS:

A = singly degenerate state which is symmetric with respect to ROTATION about the principal C_n axis,

B = singly DEGENERATE state which is antisymmetric with respect to ROTATION about the principal C_n axis,

E = doubly DEGENERATE,

T = triply DEGENERATE,

X_g = (gerade, symmetric) the sign of the wavefunction does not change on INVERSION through the center of the atom,

X_u = (ungerade, antisymmetric) the sign of the wavefunction changes on INVERSION through the center of the atom,

X_1 = (on a or b) the sign of the wavefunction does not change upon ROTATION about the center of the atom,

X_2 = (on a or b) the sign of the wavefunction changes upon ROTATION about the center of the atom,

' = symmetric with respect to a horizontal symmetry plane σ_h,

" = antisymmetric with respect to a horizontal symmetry plane σ_h.

see also GROUP THEORY

Multiamicable Numbers

Two integers n and $m < n$ are (α, β)-multiamicable if

$$\sigma(m) - m = \alpha n$$

and

$$\sigma(n) - n = \beta m,$$

where $\sigma(n)$ is the DIVISOR FUNCTION and α, β are POSITIVE integers. If $\alpha = \beta = 1$, (m, n) is an AMICABLE PAIR.

m cannot have just one distinct prime factor, and if it has precisely two prime factors, then $\alpha = 1$ and m is EVEN. Small multiamicable numbers for small α, β are given by Cohen *et al.* (1995). Several of these numbers are reproduced in the below table.

α	β	m	n
1	6	76455288	183102192
1	7	52920	152280
1	7	16225560	40580280
1	7	90863136	227249568
1	7	16225560	40580280
1	7	70821324288	177124806144
1	7	199615613902848	499240550375424

see also AMICABLE PAIR, DIVISOR FUNCTION

References

Cohen, G. L; Gretton, S.; and Hagis, P. Jr. "Multiamicable Numbers." *Math. Comput.* **64**, 1743–1753, 1995.

Multifactorial

A generalization of the FACTORIAL and DOUBLE FACTORIAL,

$$n! = n(n-1)(n-2)\cdots 2 \cdot 1$$
$$n!! = n(n-2)(n-4)\cdots$$
$$n!!! = n(n-3)(n-6)\cdots,$$

etc., where the product runs through positive integers. The FACTORIALS $n!$ for $n = 1, 2, \ldots$, are 1, 2, 6, 24, 120, 720, ... (Sloane's A000142); the DOUBLE FACTORIALS $n!!$ are 1, 2, 3, 8, 15, 48, 105, ... (Sloane's A006882); the triple factorials $n!!!$ are 1, 2, 3, 4, 10, 18, 28, 80, 162, 280, ... (Sloane's A007661); and the quadruple factorials $n!!!!$ are 1, 2, 3, 4, 5, 12, 21, 32, 45, 120, ... (Sloane's A007662).

see also FACTORIAL, GAMMA FUNCTION

References

Sloane, N. J. A. Sequences A000142/M1675, A006882/M0876, A007661/M0596, and A007662/M0534 in "An On-Line Version of the Encyclopedia of Integer Sequences."

Multifractal Measure

A MEASURE for which the q-DIMENSION D_q varies with q.

References

Ott, E. *Chaos in Dynamical Systems.* New York: Cambridge University Press, 1993.

Multigrade Equation

A (k, l)-multigrade equation is a DIOPHANTINE EQUATION of the form

$$\sum_{i=1}^{l} n_i^j = \sum_{i=1}^{l} m_i^j$$

for $j = 1, \ldots, k$, where $\mathbf{m}$ and $\mathbf{n}$ are l-VECTORS. Multigrade identities remain valid if a constant is added to each element of $\mathbf{m}$ and $\mathbf{n}$ (Madachy 1979), so multigrades can always be put in a form where the minimum component of one of the vectors is 1.

Small-order examples are the (2, 3)-multigrade with $\mathbf{m} = \{1, 6, 8\}$ and $\mathbf{n} = \{2, 4, 9\}$:

$$\sum_{i=1}^{3} m_i^1 = \sum_{i=1}^{3} n_i^1 = 15$$

$$\sum_{i=1}^{3} m_i^2 = \sum_{i=1}^{3} n_i^2 = 101,$$

the (3, 4)-multigrade with $\mathbf{m} = \{1, 5, 8, 12\}$ and $\mathbf{n} = \{2, 3, 10, 11\}$:

$$\sum_{i=1}^{4} m_i^1 = \sum_{i=1}^{4} n_i^1 = 26$$

$$\sum_{i=1}^{4} m_i^2 = \sum_{i=1}^{4} n_i^2 = 234$$

$$\sum_{i=1}^{4} m_i^3 = \sum_{i=1}^{4} n_i^3 = 2366,$$

and the (4, 6)-multigrade with $\mathbf{m} = \{1, 5, 8, 12, 18, 19\}$ and $\mathbf{n} = \{2, 3, 9, 13, 16, 20\}$:

$$\sum_{i=1}^{6} m_i^1 = \sum_{i=1}^{6} n_i^1 = 63$$

$$\sum_{i=1}^{6} m_i^2 = \sum_{i=1}^{6} n_i^2 = 919$$

$$\sum_{i=1}^{6} m_i^3 = \sum_{i=1}^{6} n_i^3 = 15057$$

$$\sum_{i=1}^{6} m_i^4 = \sum_{i=1}^{6} n_i^4 = 260755$$

(Madachy 1979).

A spectacular example with $k = 9$ and $l = 10$ is given by $\mathbf{n} = \{\pm 12, \pm 11881, \pm 20231, \pm 20885, \pm 23738\}$ and $\mathbf{m} = \{\pm 436, \pm 11857, \pm 20449, \pm 20667, \pm 23750\}$ (Guy 1994), which has sums

$$\sum_{i=1}^{9} m_i^1 = \sum_{i=1}^{9} n_i^1 = 0$$

$$\sum_{i=1}^{9} m_i^2 = \sum_{i=1}^{9} n_i^2 = 3100255070$$

$$\sum_{i=1}^{9} m_i^3 = \sum_{i=1}^{9} n_i^3 = 0$$

$$\sum_{i=1}^{9} m_i^4 = \sum_{i=1}^{9} n_i^4 = 1390452894778220678$$

$$\sum_{i=1}^{9} m_i^5 = \sum_{i=1}^{9} n_i^5 = 0$$

$$\sum_{i=1}^{9} m_i^6 = \sum_{i=1}^{9} n_i^6 = 666573454337853049941719510$$

$$\sum_{i=1}^{9} m_i^7 = \sum_{i=1}^{9} n_i^7 = 0$$

$$\sum_{i=1}^{9} m_i^8 = \sum_{i=1}^{9} n_i^8$$

$$= 330958142560259813821203262692838598$$

$$\sum_{i=1}^{9} m_i^9 = \sum_{i=1}^{9} n_i^9 = 0.$$

see also DIOPHANTINE EQUATION

References

Chen, S. "Equal Sums of Like Powers: On the Integer Solution of the Diophantine System." `http://www.nease.net/~chin/eslp/`

Gloden, A. *Mehrgeradige Gleichungen.* Groningen, Netherlands: Noordhoff, 1944.

Gloden, A. "Sur la multigrade A_1, A_2, A_3, A_4, $A_5 =^k B_1$, B_2, B_3, B_4, B_5 (k = 1, 3, 5, 7)." *Revista Euclides* **8**, 383–384, 1948.

Guy, R. K. *Unsolved Problems in Number Theory, 2nd ed.* New York: Springer-Verlag, p. 143, 1994.

Kraitchik, M. "Multigrade." §3.10 in *Mathematical Recreations.* New York: W. W. Norton, p. 79, 1942.

Madachy, J. S. *Madachy's Mathematical Recreations.* New York: Dover, pp. 171–173, 1979.

Multilinear

A function, form, etc., in two or more variables is said to be multilinear if it is linear in each variable separately.

see also BILINEAR, LINEAR OPERATOR

Multimagic Series

n numbers form a p-multimagic series if the sum of their kth powers is the MAGIC CONSTANT of degree k for every $k = 1, \ldots, p$. The following table gives the number of p-multimagic series N_p of given orders n (Kraitchik 1942).

n	N_1	N_2	N_3
2	2		
3	8		
4	86	2	2
5	1,394	8	2
6	0	98	0
7	0	1,844	0
8	0	38,039	115
9	0	0	41
10	0	0	0
11	0	0	961

References

Kraitchik, M. "Multimagic Squares." §7.10 in *Mathematical Recreations.* New York: W. W. Norton, pp. 176–178, 1942.

Multimagic Square

A MAGIC SQUARE is p-multimagic if the square formed by replacing each element by its kth power for $k = 1, 2, \ldots, p$ is also magic. A 2-multimagic square is called a BIMAGIC SQUARE, and a 3-multimagic square is called a TRIMAGIC SQUARE.

see also BIMAGIC SQUARE, MAGIC SQUARE, TRIMAGIC SQUARE

References

Kraitchik, M. "Multimagic Squares." §7.10 in *Mathematical Recreations.* New York: W. W. Norton, pp. 176–178, 1942.

Multinomial Coefficient

The multinomial COEFFICIENTS

$$(x_1, x_2, \ldots) = \frac{x_1 + x_2 + \ldots}{x_1! x_2! \cdots}$$

are the terms in the MULTINOMIAL SERIES expansion. They satisfy

$$\begin{aligned}(x_1, x_2, x_3, \ldots) &= (x_1 + x_2, x_3, \ldots)(x_1, x_2) \\ &= (x_1 + x_2 + x_3, \ldots)(x_1, x_2, x_3) = \ldots\end{aligned}$$

(Beeler *et al.* 1972, Item 44).

see also BINOMIAL COEFFICIENT, MULTINOMIAL SERIES

References

Abramowitz, M. and Stegun, C. A. (Eds.). "Multinomial Coefficients." §24.1.2 in *Handbook of Mathematical Functions with Formulas, Graphs, and Mathematical Tables, 9th printing.* New York: Dover, pp. 823–824, 1972.

Beeler, M.; Gosper, R. W.; and Schroeppel, R. *HAKMEM.* Cambridge, MA: MIT Artificial Intelligence Laboratory, Memo AIM-239, Feb. 1972.

Spiegel, M. R. *Theory and Problems of Probability and Statistics.* New York: McGraw-Hill, p. 113, 1992.

Multinomial Distribution

Let a set of random variates X_1, X_2, ..., X_n have a probability function

$$P(X_1 = x_1, \ldots, X_n = x_n) = \frac{N!}{\prod_{i=1}^n x_i!} \prod_{i=1}^n \theta_i^{x_i} \qquad (1)$$

where x_i are POSITIVE INTEGERS, $\theta_i > 0$, and

$$\sum_{i=1}^n \theta_i = 1 \qquad (2)$$

$$\sum_{i=1}^n x_i = N. \qquad (3)$$

Then the joint distribution of X_1, ..., X_n is a multinomial distribution and $P(X_1 = x_1, \ldots, X_n = x_n)$ is given by the corresponding coefficient of the MULTINOMIAL SERIES

$$(\theta_1 + \theta_2 + \ldots + \theta_n)^N. \qquad (4)$$

The MEAN and VARIANCE of X_i are

$$\mu_i = N\theta_i \qquad (5)$$

$${\sigma_i}^2 = N\theta_i(1 - \theta_i). \qquad (6)$$

The COVARIANCE of X_i and X_j is

$${\sigma_{ij}}^2 = -N\theta_i\theta_j. \qquad (7)$$

see also BINOMIAL DISTRIBUTION

References

Beyer, W. H. *CRC Standard Mathematical Tables, 28th ed.* Boca Raton, FL: CRC Press, p. 532, 1987.

Multinomial Series

A generalization of the BINOMIAL SERIES discovered by Johann Bernoulli and Leibniz.

$$(a_1 + a_2 + \ldots + a_k)^n = \sum_{n_1, n_2, \ldots, n_k}^n \frac{n!}{n_1! n_2! \cdots n_k!} {a_1}^{n_1} {a_2}^{n_2} \cdots {a_k}^{n_k},$$

where $n \equiv n_1 + n_2 + \ldots + n_k$. The multinomial series arises in a generalization of the BINOMIAL DISTRIBUTION called the MULTINOMIAL DISTRIBUTION.

see also BINOMIAL SERIES, MULTINOMIAL DISTRIBUTION

Multinomial Theorem

see MULTINOMIAL SERIES

Multiperfect Number

A number n is k-multiperfect (also called a k-MULTIPLY PERFECT NUMBER or k-PLUPERFECT NUMBER) if

$$\sigma(n) = kn$$

for some INTEGER $k > 2$, where $\sigma(n)$ is the DIVISOR FUNCTION. The value of k is called the CLASS. The special case $k = 2$ corresponds to PERFECT NUMBERS P_2, which are intimately connected with MERSENNE PRIMES (Sloane's A000396). The number 120 was long known to be 3-multiply perfect (P_3) since

$$\sigma(120) = 3 \cdot 120.$$

The following table gives the first few P_n for $n = 2, 3, \ldots, 6$.

n	Sloane	P_n
2	000396	6, 28, 496, 8128, ...,
3	005820	120, 672, 523776, 459818240, ...
4	027687	30240, 32760, 2178540, 23569920, ...
5	046060	14182439040, 31998395520, ...
6	046061	154345556085770649600, ...

In 1900–1901, Lehmer proved that P_3 has at least three distinct PRIME factors, P_4 has at least four, P_5 at least six, P_6 at least nine, and P_7 at least 14.

As of of 1911, 251 pluperfect numbers were known (Carmichael and Mason 1911). As of 1929, 334 pluperfect numbers were known, many of them found by Poulet. Franqui and García (1953) found 63 additional ones (five P_5s, 29 P_6s, and 29 P_7s), several of which were known to Poulet but had not been published, bringing the total to 397. Brown (1954) discovered 110 pluperfects, including 31 discovered but not published by Poulet and 25 previously published by Franqui and García (1953), for a total of 482. Franqui and García (1954) subsequently discovered 57 additional pluperfects (3 P_6s, 52 P_7s, and 2 P_8s), increasing the total known to 539.

An outdated database is maintained by R. Schroeppel, who lists 2,094 multiperfects, and an up-to-date list by J. L. Moxham (1998). It is believed that all multiperfect numbers of index 3, 4, 5, 6, and 7 are known. The number of known n-multiperfect numbers are 1, 37, 6, 36, 65, 245, 516, 1101, 1129, 46, 0, 0, ...

If n is a P_5 number such that $3 \nmid n$, then $3n$ is a P_4 number. If $3n$ is a P_{4k} number such that $3 \nmid n$, then n is a P_{3k} number. If n is a P_3 number such that 3 (but not 5 and 9) DIVIDES n, then $45n$ is a P_4 number.

see also e-MULTIPERFECT NUMBER, FRIENDLY PAIR, HYPERPERFECT NUMBER, INFINARY MULTIPERFECT NUMBER, MERSENNE PRIME, PERFECT NUMBER, UNITARY MULTIPERFECT NUMBER

References

Brown, A. L. "Multiperfect Numbers." *Scripta Math.* **20**, 103–106, 1954.

Dickson, L. E. *History of the Theory of Numbers, Vol. 1: Divisibility and Primality.* New York: Chelsea, pp. 33–38, 1952.

Flammenkamp, A. "Multiply Perfect Numbers." `http://www.uni-bielefeld.de/~achim/mpn.html`.

Franqui, B. and García, M. "Some New Multiply Perfect Numbers." *Amer. Math. Monthly* **60**, 459–462, 1953.

Franqui, B. and García, M. "57 New Multiply Perfect Numbers." *Scripta Math.* **20**, 169–171, 1954.

Guy, R. K. "Almost Perfect, Quasi-Perfect, Pseudoperfect, Harmonic, Weird, Multiperfect and Hyperperfect Numbers." §B2 in *Unsolved Problems in Number Theory, 2nd ed.* New York: Springer-Verlag, pp. 45–53, 1994.

Helenius, F. W. "Multiperfect Numbers (MPFNs)." `http://www.netcom.com/~fredh/mpfn`.

Madachy, J. S. *Madachy's Mathematical Recreations.* New York: Dover, pp. 149–151, 1979.

Moxham, J. L. "13 New MPFN's." math-fun@cs.arizona.edu posting, Aug 13, 1998.

Poulet, P. *La Chasse aux nombres,* Vol. 1. Brussels, pp. 9–27, 1929.

Schroeppel, R. "Multiperfect Numbers–Multiply Perfect Numbers–Pluperfect Numbers–MPFNs." Rev. Dec. 13, 1995. `ftp://ftp.cs.arizona.edu/xkernel/rcs/mpfn.html`.

Schroeppel, R. (moderator). mpfn mailing list. e-mail `rcs@cs.arizona.edu` to subscribe.

Sloane, N. J. A. Sequences A000396/M4186 and A005820/M5376 in "An On-Line Version of the Encyclopedia of Integer Sequences."

Multiple Integral

A repeated integral over $n > 1$ variables

$$\underbrace{\int \cdots \int}_{n} f(x_1, \ldots, x_n)\, dx_1 \cdots dx_n$$

is called a multiple integral. An nth order integral corresponds, in general, to an n-D VOLUME (CONTENT), with $n = 2$ corresponding to an AREA. In an indefinite multiple integral, the order in which the integrals are carried out can be varied at will; for definite multiple integrals, care must be taken to correctly transform the limits if the order is changed.

see also INTEGRAL, MONTE CARLO INTEGRATION

References

Press, W. H.; Flannery, B. P.; Teukolsky, S. A.; and Vetterling, W. T. "Multidimensional Integrals." §4.6 in *Numerical Recipes in FORTRAN: The Art of Scientific Computing, 2nd ed.* Cambridge, England: Cambridge University Press, pp. 155–158, 1992.

Multiple Regression

A REGRESSION giving conditional expectation values of a given variable in terms of two or more other variables.

see also LEAST SQUARES FITTING, MULTIVARIATE ANALYSIS, NONLINEAR LEAST SQUARES FITTING

References

Edwards, A. L. *Multiple Regression and the Analysis of Variance and Covariance.* San Francisco, CA: W. H. Freeman, 1979.

Multiplication

In simple algebra, multiplication is the process of calculating the result when a number a is taken b times. The result of a multiplication is called the PRODUCT of a and b. It is denoted $a \times b$, $a \cdot b$, $(a)(b)$, or simply ab. The symbol $\times$ is known as the MULTIPLICATION SIGN. Normal multiplication is ASSOCIATIVE, COMMUTATIVE, and DISTRIBUTIVE.

More generally, multiplication can also be defined for other mathematical objects such as GROUPS, MATRICES, SETS, and TENSORS.

Karatsuba and Ofman (1962) discovered that multiplication of two n digit numbers can be done with a BIT COMPLEXITY of less than n^2 using an algorithm now known as KARATSUBA MULTIPLICATION.

see also ADDITION, BIT COMPLEXITY, COMPLEX MULTIPLICATION, DIVISION, KARATSUBA MULTIPLICATION, MATRIX MULTIPLICATION, PRODUCT, RUSSIAN MULTIPLICATION, SUBTRACTION, TIMES

References

Karatsuba, A. and Ofman, Yu. "Multiplication of Many-Digital Numbers by Automatic Computers." *Doklady Akad. Nauk SSSR* **145**, 293–294, 1962. Translation in *Physics-Doklady* **7**, 595–596, 1963.

Multiplication Magic Square

128	1	32
4	16	64
8	256	2

A square which is magic under multiplication instead of addition (the operation used to define a conventional MAGIC SQUARE) is called a multiplication magic square. Unlike (normal) MAGIC SQUARES, the n^2 entries for an nth order multiplicative magic square are not required to be consecutive. The above multiplication magic square has a multiplicative magic constant of 4,096.

see also ADDITION-MULTIPLICATION MAGIC SQUARE, MAGIC SQUARE

References

Hunter, J. A. H. and Madachy, J. S. "Mystic Arrays." Ch. 3 in *Mathematical Diversions.* New York: Dover, pp. 30–31, 1975.

Madachy, J. S. *Madachy's Mathematical Recreations.* New York: Dover, pp. 89–91, 1979.

Multiplication Principle

If one event can occur in m ways and a second can occur independently of the first in n ways, then the two events can occur in mn ways.

Multiplication Sign

The symbol $\times$ used to denote MULTIPLICATION, i.e., $a \times b$ denotes a times b.

Multiplication Table

A multiplication table is an array showing the result of applying a BINARY OPERATOR to elements of a given set S.

×	1	2	3	4	5	6	7	8	9	10
1	1	2	3	4	5	6	7	8	9	10
2	2	4	6	8	10	12	14	16	18	20
3	3	6	9	12	15	18	21	24	27	30
4	4	8	12	16	20	24	28	32	36	40
5	5	10	15	20	25	30	35	40	45	50
6	6	12	18	24	30	36	42	48	54	60
7	7	14	21	28	35	42	49	56	63	70
8	8	16	24	32	40	48	56	64	72	80
9	9	18	27	36	45	54	63	72	81	90
10	10	20	30	40	50	60	70	80	90	100

see also BINARY OPERATOR, TRUTH TABLE

Multiplicative Character

see CHARACTER (MULTIPLICATIVE)

Multiplicative Digital Root

Consider the process of taking a number, multiplying its DIGITS, then multiplying the DIGITS of numbers derived from it, etc., until the remaining number has only one DIGIT. The number of multiplications required to obtain a single DIGIT from a number n is called the MULTIPLICATIVE PERSISTENCE of n, and the DIGIT obtained is called the multiplicative digital root of n.

For example, the sequence obtained from the starting number 9876 is (9876, 3024, 0), so 9876 has a MULTIPLICATIVE PERSISTENCE of two and a multiplicative digital root of 0. The multiplicative digital roots of the first few positive integers are 1, 2, 3, 4, 5, 6, 7, 8, 9, 0, 1, 2, 3, 4, 5, 6, 7, 8, 9, 0, 2, 4, 6, 8, 0, 2, 4, 6, 8, 0, 3, 6, 9, 2, 5, 8, 2, ... (Sloane's A031347).

see also ADDITIVE PERSISTENCE, DIGITADITION, DIGITAL ROOT, MULTIPLICATIVE PERSISTENCE

References

Sloane, N. J. A. Sequence A031347 in "An On-Line Version of the Encyclopedia of Integer Sequences."

Multiplicative Function

A function $f(m)$ is called multiplicative if $(m, m') = 1$ (i.e., the statement that m and m' are RELATIVELY PRIME) implies

$$f(mm') = f(m)f(m').$$

see also QUADRATIC RESIDUE, TOTIENT FUNCTION

Multiplicative Inverse

The multiplicative of a REAL or COMPLEX NUMBER z is its RECIPROCAL $1/z$. For complex $z = x + iy$,

$$\frac{1}{z} = \frac{1}{x+iy} = \frac{x}{x^2 - y^2} - i\frac{y}{x^2 - y^2}.$$

Multiplicative Perfect Number

A number n for which the PRODUCT of DIVISORS is equal to n^2. The first few are 1, 6, 8, 10, 14, 15, 21, 22, ... (Sloane's A007422).

see also PERFECT NUMBER

References

Sloane, N. J. A. Sequence A007422/M4068 in "An On-Line Version of the Encyclopedia of Integer Sequences."

Multiplicative Persistence

Multiply all the digits of a number n by each other, repeating with the product until a single DIGIT is obtained. The number of steps required is known as the multiplicative persistence, and the final DIGIT obtained is called the MULTIPLICATIVE DIGITAL ROOT of n.

For example, the sequence obtained from the starting number 9876 is (9876, 3024, 0), so 9876 has an multiplicative persistence of two and a MULTIPLICATIVE DIGITAL ROOT of 0. The multiplicative persistences of the first few positive integers are 0, 0, 0, 0, 0, 0, 0, 0, 0, 1, 1, 1, 1, 1, 1, 1, 1, 1, 1, 1, 1, 1, 1, 2, 2, 2, 2, 2, 1, 1, 1, 1, 2, 2, 2, 2, 2, 3, 1, 1, ... (Sloane's A031346). The smallest numbers having multiplicative persistences of 1, 2, ... are 10, 25, 39, 77 679 6788 68889 267889 26888999 3778888999 277777788888899 (Sloane's A003001). There is no number $< 10^{50}$ with multiplicative persistence > 11.

The multiplicative persistence of an n-DIGIT number is also called its LENGTH. The maximum lengths for $n =$ 2-, 3-, ..., digit numbers are 4, 5, 6, 7, 7, 8, 9, 9, 10, 10, 10, ... (Sloane's A014553; (Beeler *et al.* 1972, Item 56; Gottlieb 1969–1970).

The concept of multiplicative persistence can be generalized to multiplying the kth powers of the digits of a number and iterating until the result remains constant. All numbers other than REPUNITS, which converge to 1, converge to 0. The number of iterations required for the kth powers of a number's digits to converge to 0 is called its k-multiplicative persistence. The following table gives the n-multiplicative persistences for the first few positive integers.

n	Sloane	n-Persistences
2	031348	0, 7, 6, 6, 3, 5, 5, 4, 5, 1, ...
3	031349	0, 4, 5, 4, 3, 4, 4, 3, 3, 1, ...
4	031350	0, 4, 3, 3, 3, 3, 2, 2, 3, 1, ...
5	031351	0, 4, 4, 2, 3, 3, 2, 3, 2, 1, ...
6	031352	0, 3, 3, 2, 3, 3, 3, 3, 3, 1, ...
7	031353	0, 4, 3, 3, 3, 3, 3, 2, 3, 1, ...
8	031354	0, 3, 3, 3, 2, 4, 2, 3, 2, 1, ...
9	031355	0, 3, 3, 3, 3, 2, 2, 3, 2, 1, ...
10	031356	0, 2, 2, 2, 3, 2, 3, 2, 2, 1, ...

see also 196-ALGORITHM, ADDITIVE PERSISTENCE, DIGITADITION, DIGITAL ROOT, KAPREKAR NUMBER, LENGTH (NUMBER), MULTIPLICATIVE DIGITAL ROOT, NARCISSISTIC NUMBER, RECURRING DIGITAL INVARIANT

References

Beeler, M.; Gosper, R. W.; and Schroeppel, R. *HAKMEM.* Cambridge, MA: MIT Artificial Intelligence Laboratory, Memo AIM-239, Feb. 1972.

Gottlieb, A. J. Problems 28–29 in "Bridge, Group Theory, and a Jigsaw Puzzle." *Techn. Rev.* **72**, unpaginated, Dec. 1969.

Gottlieb, A. J. Problem 29 in "Integral Solutions, Ladders, and Pentagons." *Techn. Rev.* **72**, unpaginated, Apr. 1970.

Sloane, N. J. A. "The Persistence of a Number." *J. Recr. Math.* **6**, 97–98, 1973.

Sloane, N. J. A. Sequences A014553 and A003001/M4687 in "An On-Line Version of the Encyclopedia of Integer Sequences."

Multiplicative Primitive Residue Class Group

see MODULO MULTIPLICATION GROUP

Multiplicity

The word multiplicity is a general term meaning "the number of values for which a given condition holds." The most common use of the word is as the value of the TOTIENT VALENCE FUNCTION.

see also DEGENERATE, NOETHER'S FUNDAMENTAL THEOREM, TOTIENT VALENCE FUNCTION

Multiplier

see MODULAR FUNCTION MULTIPLIER

Multiply Connected

A set which is CONNECTED but not SIMPLY CONNECTED is called multiply connected. A SPACE is n-MULTIPLY CONNECTED if it is $(n-1)$-connected and if every MAP from the n-SPHERE into it extends continuously over the $(n+1)$-DISK

A theorem of Whitehead says that a SPACE is infinitely connected IFF it is contractible.

see also CONNECTIVITY, LOCALLY PATHWISE-CONNECTED SPACE, PATHWISE-CONNECTED, SIMPLY CONNECTED

Multiply Perfect Number

see MULTIPERFECT NUMBER

Multisection

see SERIES MULTISECTION

Multivalued Function

A FUNCTION which assumes two or more distinct values at one or more points in its DOMAIN.

see also BRANCH CUT, BRANCH POINT

References

Morse, P. M. and Feshbach, H. "Multivalued Functions." §4.4 in *Methods of Theoretical Physics, Part I.* New York: McGraw-Hill, pp. 398–408, 1953.

Multivariate Analysis

The study of random distributions involving more than one variable.

see also GAUSSIAN JOINT VARIABLE THEOREM, MULTIPLE REGRESSION, MULTIVARIATE FUNCTION

References

Abramowitz, M. and Stegun, C. A. (Eds.). *Handbook of Mathematical Functions with Formulas, Graphs, and Mathematical Tables, 9th printing.* New York: Dover, pp. 927–928, 1972.

Feinstein, A. R. *Multivariable Analysis.* New Haven, CT: Yale University Press, 1996.

Hair, J. F. Jr. *Multivariate Data Analysis with Readings, 4th ed.* Englewood Cliffs, NJ: Prentice-Hall, 1995.

Sharma, S. *Applied Multivariate Techniques.* New York: Wiley, 1996.

Multivariate Function

A FUNCTION of more than one variable.

see also MULTIVARIATE ANALYSIS, UNIVARIATE FUNCTION

Multivariate Theorem

see GAUSSIAN JOINT VARIABLE THEOREM

Müntz Space

A Müntz space is a technically defined SPACE

$$M(\Lambda) \equiv \operatorname{span}\{x^{\lambda_0}, x^{\lambda_1}, \ldots\}$$

which arises in the study of function approximations.

Müntz's Theorem

Müntz's theorem is a generalization of the WEIERSTRAß APPROXIMATION THEOREM, which states that any continuous function on a closed and bounded interval can be uniformly approximated by POLYNOMIALS involving constants and any INFINITE SEQUENCE of POWERS whose RECIPROCALS diverge.

In technical language, Müntz's theorem states that the MÜNTZ SPACE $M(\Lambda)$ is dense in $C[0,1]$ IFF

$$\sum_{i=1}^{\infty} \frac{1}{\lambda_i} = \infty.$$

see also WEIERSTRAß APPROXIMATION THEOREM

Mutant Knot

Given an original KNOT K, the three knots produced by MUTATION together with K itself are called mutant knots. Mutant knots are often difficult to distinguish. For instant, mutants have the same HOMFLY POLYNOMIALS and HYPERBOLIC KNOT volume. Many but not all mutants also have the same GENUS (KNOT).

Mutation

Consider a KNOT as being formed from two TANGLES. The following three operations are called mutations.

1. Cut the knot open along four points on each of the four strings coming out of T_2, flipping T_2 over, and gluing the strings back together.
2. Cut the knot open along four points on each of the four strings coming out of T_2, flipping T_2 to the right, and gluing the strings back together.
3. Cut the knot, rotate it by 180°, and reglue. This is equivalent to performing (1), then (2).

Mutations applied to an alternating KNOT projection always yield an ALTERNATING KNOT. The mutation of a KNOT is always another KNOT (a opposed to a LINK).

References

Adams, C. C. *The Knot Book: An Elementary Introduction to the Mathematical Theory of Knots.* New York: W. H. Freeman, p. 49, 1994.

Mutual Energy

Let Ω be a SPACE with MEASURE $\mu \geq 0$, and let $\Phi(P,Q)$ be a real function on the PRODUCT SPACE $\Omega\times\Omega$. When

$$\begin{aligned}(\mu, nu) &= \iint \Phi(P,Q)\,d\mu(Q)\,d\nu(P)\\ &= \int \Phi(P,\mu)\,d\nu(P)\end{aligned}$$

exists for measures $\mu,\nu \geq 0$, (μ,ν) is called the mutual energy. (μ,μ) is then called the ENERGY.

see also ENERGY

References

Iyanaga, S. and Kawada, Y. (Eds.). "General Potential." §335.B in *Encyclopedic Dictionary of Mathematics.* Cambridge, MA: MIT Press, p. 1038, 1980.

Mutually Exclusive

Two events E_1 and E_2 are mutually exclusive if $E_1 \cap E_2 \equiv \varnothing$. n events E_1, E_2, ..., E_n are mutually exclusive if $E_i \cap E_j \equiv \varnothing$ for $i \neq j$.

Mutually Singular

Let M be a SIGMA ALGEBRA M, and let λ_1 and λ_2 be MEASURES on M. If there EXISTS a pair of disjoint SETS A and B such that λ_1 is CONCENTRATED on A and λ_2 is CONCENTRATED on B, then λ_1 and λ_2 are said to be mutually singular, written $\lambda_1 \perp \lambda_2$.

see also ABSOLUTELY CONTINUOUS, CONCENTRATED, SIGMA ALGEBRA

References

Rudin, W. *Functional Analysis.* New York: McGraw-Hill, p. 121, 1991.

Myriad

The Greek word for 10,000.

Myriagon
A 10,000-sided POLYGON.

Mystic Pentagram
see PENTAGRAM

N

ℕ

The SET of NATURAL NUMBERS (the POSITIVE INTEGERS $\mathbb{Z}^+$ 1, 2, 3, ...; Sloane's A000027), denoted $\mathbb{N}$, also called the WHOLE NUMBERS. Like whole numbers, there is no general agreement on whether 0 should be included in the list of natural numbers.

Due to lack of standard terminology, the following terms are recommended in preference to "COUNTING NUMBER," "natural number," and "WHOLE NUMBER."

Set	Name	Symbol
$\ldots, -2, -1, 0, 1, 2, \ldots$	integers	$\mathbb{Z}$
$1, 2, 3, 4, \ldots$	positive integers	$\mathbb{Z}^+$
$0, 1, 2, 3, 4 \ldots$	nonnegative integers	$\mathbb{Z}^*$
$-1, -2, -3, -4, \ldots$	negative integers	$\mathbb{Z}^-$

see also $\mathbb{C}$, CARDINAL NUMBER, COUNTING NUMBER, $\mathbb{I}$, INTEGER, $\mathbb{Q}$, $\mathbb{R}$, WHOLE NUMBER, $\mathbb{Z}$, $\mathbb{Z}^+$

References
Sloane, N. J. A. Sequence A000027/M0472 in "An On-Line Version of the Encyclopedia of Integer Sequences."

N-Cluster

A LATTICE POINT configuration with no three points COLLINEAR and no four CONCYCLIC. An example is the 6-cluster (0, 0), (132, −720), (546, −272), (960, −720), (1155, 540), (546, 1120). Call the RADIUS of the smallest CIRCLE centered at one of the points of an N-cluster which contains all the points in the N-cluster the EXTENT. Noll and Bell (1989) found 91 nonequivalent prime 6-clusters of EXTENT less than 20937, but found no 7-clusters.

References
Guy, R. K. *Unsolved Problems in Number Theory, 2nd ed.* New York: Springer-Verlag, p. 187, 1994.
Noll, L. C. and Bell, D. I. "n-clusters for $1 < n < 7$." *Math. Comput.* **53**, 439–444, 1989.

n-Cube

see HYPERCUBE, POLYCUBE

n-in-a-Row

see TIC-TAC-TOE

n-minex

n-minex is defined as 10^{-n}.

see also n-PLEX

References
Conway, J. H. and Guy, R. K. *The Book of Numbers.* New York: Springer-Verlag, p. 16, 1996.

n-Omino

see POLYOMINO

n-plex

n-plex is defined as 10^n.

see also GOOGOLPLEX, n-MINEX

References
Conway, J. H. and Guy, R. K. *The Book of Numbers.* New York: Springer-Verlag, p. 16, 1996.

n-Sphere

see HYPERSPHERE

Nabla

see DEL, LAPLACIAN

Nagel Point

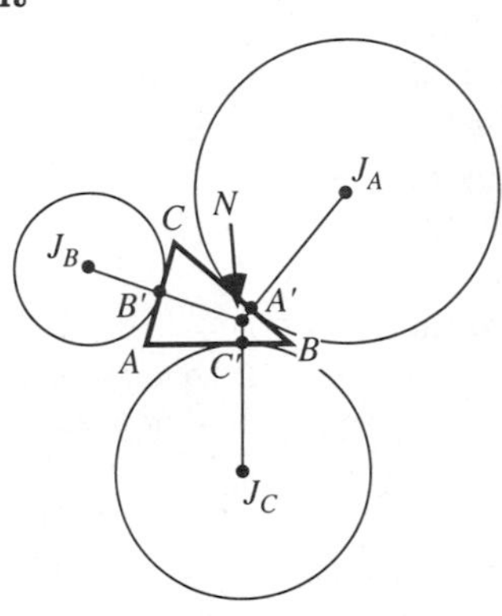

Let A' be the point at which the A-EXCIRCLE meets the side BC of a TRIANGLE ΔABC, and define B' and C' similarly. Then the lines AA', BB', and CC' CONCUR in the NAGEL POINT.

The Nagel point can also be constructed by letting A'' be the point half way around the PERIMETER of ΔABC starting at A, and B'' and C'' similarly defined. Then the lines AA'', BB'', and CC'' concur in the Nagel point. It is therefore sometimes known as the BISECTED PERIMETER POINT (Bennett *et al.* 1988, Chen *et al.* 1992, Kimberling 1994).

The Nagel point has TRIANGLE CENTER FUNCTION

$$\alpha = \frac{b+c-a}{a}.$$

It is the ISOTOMIC CONJUGATE POINT of the GERGONNE POINT.

see also EXCENTER, EXCENTRAL TRIANGLE, EXCIRCLE, MITTENPUNKT, TRISECTED PERIMETER POINT

References
Altshiller-Court, N. *College Geometry: A Second Course in Plane Geometry for Colleges and Normal Schools, 2nd ed.* New York: Barnes and Noble, pp. 160–164, 1952.
Bennett, G.; Glenn, J.; Kimberling, C.; and Cohen, J. M. "Problem E 3155 and Solution." *Amer. Math. Monthly* **95**, 874, 1988.
Chen, J.; Lo, C.-H.; and Lossers, O. P. "Problem E 3397 and Solution." *Amer. Math. Monthly* **99**, 70–71, 1992.

Eves, H. W. *A Survey of Geometry, rev. ed.* Boston, MA: Allyn and Bacon, p. 83, 1972.
Gallatly, W. *The Modern Geometry of the Triangle, 2nd ed.* London: Hodgson, p. 20, 1913.
Johnson, R. A. *Modern Geometry: An Elementary Treatise on the Geometry of the Triangle and the Circle.* Boston, MA: Houghton Mifflin, pp. 184 and 225–226, 1929.
Kimberling, C. "Central Points and Central Lines in the Plane of a Triangle." *Math. Mag.* **67**, 163–187, 1994.
Kimberling, C. "Nagel Point." `http://www.evansville.edu/~ck6/tcenters/class/nagel.html`.

Naive Set Theory

A branch of mathematics which attempts to formalize the nature of the SET using a minimal collection of independent axioms. Unfortunately, as discovered by its earliest proponents, naive set theory quickly runs into a number of PARADOXES (such as RUSSELL'S PARADOX), so a less sweeping and more formal theory known as AXIOMATIC SET THEORY must be used.

see also AXIOMATIC SET THEORY, RUSSELL'S PARADOX, SET THEORY

Napier's Analogies

Let a SPHERICAL TRIANGLE have sides a, b, and c with A, B, and C the corresponding opposite angles. Then

$$\frac{\sin[\frac{1}{2}(A-B)]}{\sin[\frac{1}{2}(A+B)]} = \frac{\tan[\frac{1}{2}(a-b)]}{\tan(\frac{1}{2}c)} \tag{1}$$

$$\frac{\cos[\frac{1}{2}(A-B)]}{\cos[\frac{1}{2}(A+B)]} = \frac{\tan[\frac{1}{2}(a+b)]}{\tan(\frac{1}{2}c)} \tag{2}$$

$$\frac{\sin[\frac{1}{2}(a-b)]}{\sin[\frac{1}{2}(a+b)]} = \frac{\tan[\frac{1}{2}(A-B)]}{\cot(\frac{1}{2}C)} \tag{3}$$

$$\frac{\cos[\frac{1}{2}(a-b)]}{\cos[\frac{1}{2}(a+b)]} = \frac{\tan[\frac{1}{2}(A+B)]}{\cot(\frac{1}{2}C)}. \tag{4}$$

see also SPHERICAL TRIGONOMETRY

Napier's Bones

Numbered rods which can be used to perform MULTIPLICATION. This process is also called RABDOLOGY.

see also GENAILLE RODS

References

Gardner, M. "Napier's Bones." Ch. 7 in *Knotted Doughnuts and Other Mathematical Entertainments.* New York: W. H. Freeman, 1986.
Pappas, T. "Napier's Bones." *The Joy of Mathematics.* San Carlos, CA: Wide World Publ./Tetra, pp. 64–65, 1989.

Napier's Constant

see e

Napier's Inequality

For $b > a > 0$,

$$\frac{1}{b} < \frac{\ln b - \ln a}{b-a} < \frac{1}{a}.$$

References

Nelsen, R. B. "Napier's Inequality (Two Proofs)." *College Math. J.* **24**, 165, 1993.

Napierian Logarithm

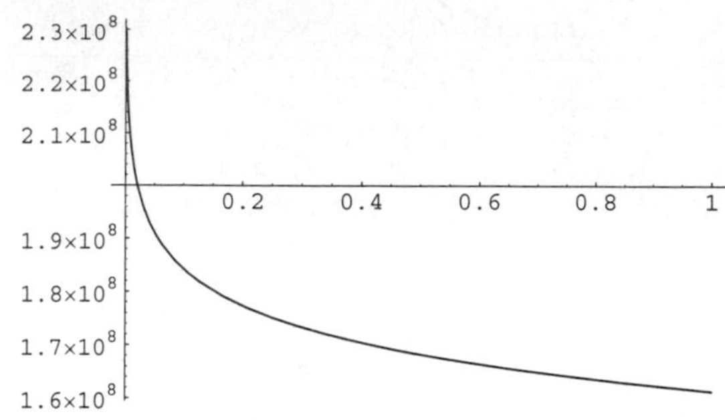

Write a number N as

$$N = 10^7(1-10^{-7})^L,$$

then L is the Napierian logarithm of N. This was the original definition of a LOGARITHM, and can be given in terms of the modern LOGARITHM as

$$L(N) = -\frac{\log\left(\frac{n}{10^7}\right)}{\log\left(\frac{10^7}{10^7-1}\right)}.$$

The Napierian logarithm decreases with increasing numbers and does not satisfy many of the fundamental properties of the modern LOGARITHM, e.g.,

$$\text{Nlog}(xy) \neq \text{Nlog}\,x + \text{Nlog}\,y.$$

Napkin Ring

see SPHERICAL RING

Napoleon Points

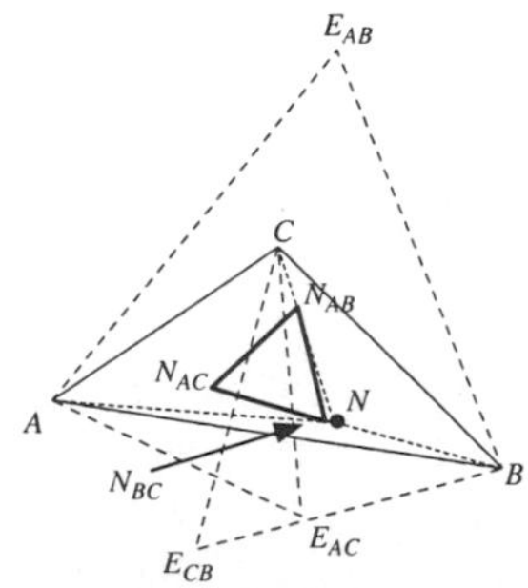

The inner Napoleon point N is the CONCURRENCE of lines drawn between VERTICES of a given TRIANGLE

ΔABC and the opposite VERTICES of the corresponding inner NAPOLEON TRIANGLE $\Delta N_{AB}N_{AC}N_{BC}$. The TRIANGLE CENTER FUNCTION of the inner Napoleon point is

$$\alpha = \csc(A - \tfrac{1}{6}\pi).$$

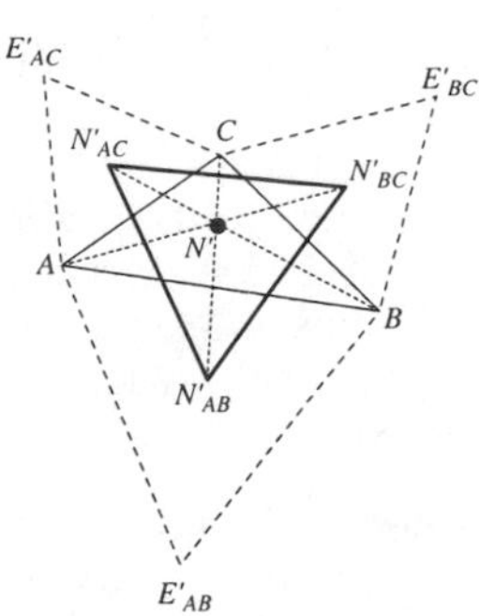

The outer Napoleon point N' is the CONCURRENCE of lines drawn between VERTICES of a given TRIANGLE ΔABC and the opposite VERTICES of the corresponding outer NAPOLEON TRIANGLE $\Delta N'_{AB}N'_{AC}N'_{BC}$. The TRIANGLE CENTER FUNCTION of the point is

$$\alpha = \csc(A + \tfrac{1}{6}\pi).$$

see also NAPOLEON'S THEOREM, NAPOLEON TRIANGLES

References

Casey, J. *Analytic Geometry, 2nd ed.* Dublin: Hodges, Figgis, & Co., pp. 442–444, 1893.

Kimberling, C. "Central Points and Central Lines in the Plane of a Triangle." *Math. Mag.* **67**, 163–187, 1994.

Napoleon's Problem

Given the center of a CIRCLE, divide the CIRCLE into four equal arcs using a COMPASS alone (a MASCHERONI CONSTRUCTION).

see also CIRCLE, COMPASS, MASCHERONI CONSTRUCTION

Napoleon's Theorem

If EQUILATERAL TRIANGLES are erected externally on the sides of any TRIANGLE, then the centers form an EQUILATERAL TRIANGLE (the outer NAPOLEON TRIANGLE). Furthermore, the inner NAPOLEON TRIANGLE is also EQUILATERAL and the difference between the areas of the outer and inner Napoleon triangles equals the AREA of the original TRIANGLE.

see also NAPOLEON POINTS, NAPOLEON TRIANGLES

References

Coxeter, H. S. M. and Greitzer, S. L. *Geometry Revisited.* Washington, DC: Math. Assoc. Amer., pp. 60–65, 1967.

Pappas, T. "Napoleon's Theorem." *The Joy of Mathematics.* San Carlos, CA: Wide World Publ./Tetra, p. 57, 1989.

Schmidt, F. "200 Jahre französische Revolution—Problem und Satz von Napoleon." *Didaktik der Mathematik* **19**, 15–29, 1990.

Wentzel, J. E. "Converses of Napoleon's Theorem." *Amer. Math. Monthly* **99**, 339–351, 1992.

Napoleon Triangles

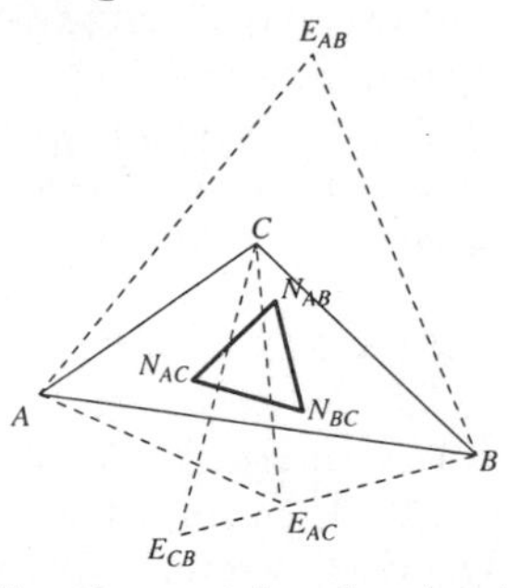

The inner Napoleon triangle is the TRIANGLE $\Delta N_{AB}N_{AC}N_{BC}$ formed by the centers of internally erected EQUILATERAL TRIANGLES ΔABE_{AB}, ΔACE_{AC}, and ΔBCE_{BC} on the sides of a given TRIANGLE ΔABC. It is an EQUILATERAL TRIANGLE.

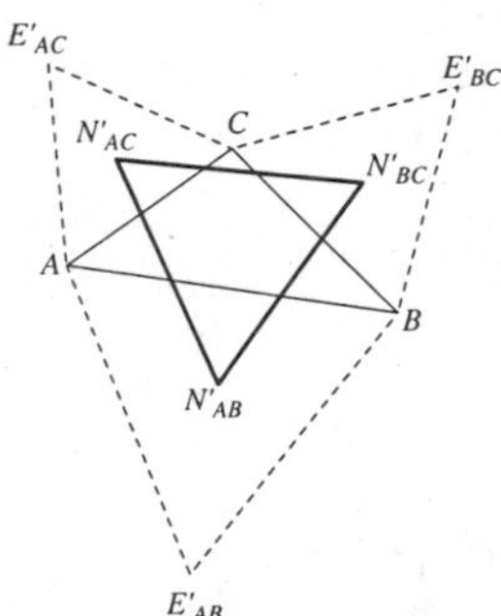

The outer Napoleon triangle is the TRIANGLE $\Delta N'_{AB}N'_{AC}N'_{BC}$ formed by the centers of externally erected EQUILATERAL TRIANGLES $\Delta ABE'_{AB}$, $\Delta ACE'_{AC}$, and $\Delta BCE'_{BC}$ on the sides of a given TRIANGLE ΔABC. It is also an EQUILATERAL TRIANGLE.

see also EQUILATERAL TRIANGLE, NAPOLEON POINTS, NAPOLEON'S THEOREM

References

Coxeter, H. S. M. and Greitzer, S. L. *Geometry Revisited.* Washington, DC: Math. Assoc. Amer., pp. 60–65, 1967.

Nappe

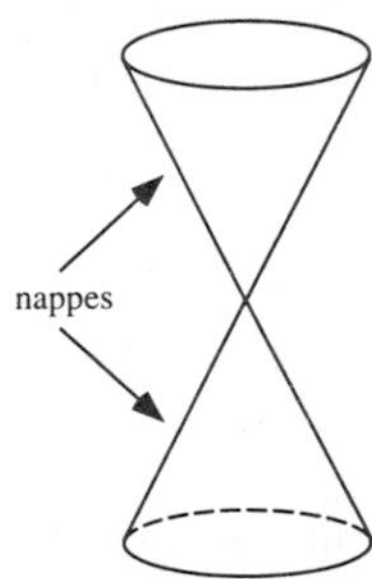

One of the two pieces of a double CONE (i.e., two CONES placed apex to apex).

see also CONE

Narcissistic Number

An n-DIGIT number which is the SUM of the nth POWERS of its DIGITS is called an n-narcissistic number, or sometimes an ARMSTRONG NUMBER or PERFECT DIGITAL INVARIANT (Madachy 1979). The smallest example other than the trivial 1-DIGIT numbers is

$$153 = 1^3 + 5^3 + 3^3.$$

The series of smallest narcissistic numbers of n digits are 0, (none), 153, 1634, 54748, 548834, ... (Sloane's A014576). Hardy (1993) wrote, "There are just four numbers, after unity, which are the sums of the cubes of their digits: $153 = 1^3+5^3+3^3$, $370 = 3^3+7^3+0^3$, $371 = 3^3+7^3+1^3$, and $407 = 4^3+0^3+7^3$. These are odd facts, very suitable for puzzle columns and likely to amuse amateurs, but there is nothing in them which appeals to the mathematician." The following table gives the generalization of these "unappealing" numbers to other POWERS (Madachy 1979, p. 164).

n	n-Narcissistic Numbers
1	0, 1, 2, 3, 4, 5, 6, 7, 8, 9
2	none
3	153, 370, 371, 407
4	1634, 8208, 9474
5	54748, 92727, 93084
6	548834
7	1741725, 4210818, 9800817, 9926315
8	24678050, 24678051, 88593477
9	146511208, 472335975, 534494836, 912985153
10	4679307774

A total of 88 narcissistic numbers exist in base-10, as proved by D. Winter in 1985 and verified by D. Hoey. These numbers exist for only 1, 3, 4, 5, 6, 7, 8, 9, 10, 11, 14, 16, 17, 19, 20, 21, 23, 24, 25, 27, 29, 31, 32, 33, 34, 35, 37, 38, and 39 digits. It can easily be shown that base-10 n-narcissistic numbers can exist only for $n \leq 60$, since

$$n \cdot 9^n < 10^{n-1}$$

for $n > 60$. The largest base-10 narcissistic number is the 39-narcissistic

$$115132219018763992565095597973971522401$$

A table of the largest known narcissistic numbers in various BASES is given by Pickover (1995). A tabulation of narcissistic numbers in various bases is given by (Corning).

A closely related set of numbers generalize the narcissistic number to n-DIGIT numbers which are the sums of *any* single POWER of their DIGITS. For example, 4150 is a 4-DIGIT number which is the sum of fifth POWERS of its DIGITS. Since the number of digits is not equal to the power to which they are taken for such numbers, it is *not* a narcissistic number. The smallest numbers which are sums of *any* single positive power of their digits are 1, 2, 3, 4, 5, 6, 7, 8, 9, 153, 370, 371, 407, 1634, 4150, 4151, 8208, 9474, ... (Sloane's A023052), with powers 1, 1, 1, 1, 1, 1, 1, 1, 1, 3, 3, 3, 3, 4, 5, 5, 4, 4, ... (Sloane's A046074).

The smallest numbers which are equal to the nth powers of their digits for n = 3, 4, ..., are 153, 1634, 4150, 548834, 1741725, ... (Sloane's A003321). The n-digit numbers equal to the sum of nth powers of their digits (a finite sequence) are called ARMSTRONG NUMBERS or PLUS PERFECT NUMBERS and are given by 1, 2, 3, 4, 5, 6, 7, 8, 9, 153, 370, 371, 407, 1634, 8208, 9474, 54748, ... (Sloane's A005188).

If the sum-of-kth-powers-of-digits operation applied iteratively to a number n eventually returns to n, the smallest number in the sequence is called a k-RECURRING DIGITAL INVARIANT.

see also ADDITIVE PERSISTENCE, DIGITAL ROOT, DIGITADITION, KAPREKAR NUMBER, MULTIPLICATIVE DIGITAL ROOT, MULTIPLICATIVE PERSISTENCE, RECURRING DIGITAL INVARIANT, VAMPIRE NUMBER

References

Corning, T. "Exponential Digital Invariants." http://members.aol.com/tec153/Edi4web/Edi.html.

Hardy, G. H. *A Mathematician's Apology.* New York: Cambridge University Press, p. 105, 1993.

Madachy, J. S. "Narcissistic Numbers." *Madachy's Mathematical Recreations.* New York: Dover, pp. 163–173, 1979.

Pickover, C. A. *Keys to Infinity.* New York: W. H. Freeman, pp. 169–170, 1995.

Rumney, M. "Digital Invariants." *Recr. Math. Mag.* No. 12, 6–8, Dec. 1962.

Sloane, N. J. A. Sequences A014576, A023052, A005188/M0488, and A003321/M5403 in "An On-Line Version of the Encyclopedia of Integer Sequences."

Weisstein, E. W. "Narcissistic Numbers." http://www.astro.virginia.edu/~eww6n/math/notebooks/Narcissistic.dat.

Nash Equilibrium

A set of MIXED STRATEGIES for finite, noncooperative GAMES of two or more players in which no player can improve his payoff by unilaterally changing strategy.

see also FIXED POINT, GAME, MIXED STRATEGY, NASH'S THEOREM

Nash's Theorem

A theorem in GAME THEORY which guarantees the existence of a NASH EQUILIBRIUM for MIXED STRATEGIES in finite, noncooperative GAMES of two or more players.

see also MIXED STRATEGY, NASH EQUILIBRIUM

Nasik Square

see PANMAGIC SQUARE

Nasty Knot

An UNKNOT which can only be unknotted by first increasing the number of crossings.

Natural Density

see NATURAL INVARIANT

Natural Equation

A natural equation is an equation which specifies a curve independent of any choice of coordinates or parameterization. The study of natural equations began with the following problem: given two functions of one parameter, find the SPACE CURVE for which the functions are the CURVATURE and TORSION.

Euler gave an integral solution for plane curves (which always have TORSION $\tau = 0$). Call the ANGLE between the TANGENT line to the curve and the x-AXIS ϕ the TANGENTIAL ANGLE, then

$$\phi = \int \kappa(s)\,ds, \tag{1}$$

where κ is the CURVATURE. Then the equations

$$\kappa = \kappa(s) \tag{2}$$
$$\tau = 0, \tag{3}$$

where τ is the TORSION, are solved by the curve with parametric equations

$$x = \int \cos\phi\,ds \tag{4}$$
$$y = \int \sin\phi\,ds. \tag{5}$$

The equations $\kappa = \kappa(s)$ and $\tau = \tau(s)$ are called the natural (or INTRINSIC) equations of the space curve. An equation expressing a plane curve in terms of s and RADIUS OF CURVATURE R (or κ) is called a CESÀRO EQUATION, and an equation expressing a plane curve in terms of s and ϕ is called a WHEWELL EQUATION.

Among the special planar cases which can be solved in terms of elementary functions are the CIRCLE, LOGARITHMIC SPIRAL, CIRCLE INVOLUTE, and EPICYCLOID. Enneper showed that each of these is the projection of a HELIX on a CONIC surface of revolution along the axis of symmetry. The above cases correspond to the CYLINDER, CONE, PARABOLOID, and SPHERE.

see also CESÀRO EQUATION, INTRINSIC EQUATION, WHEWELL EQUATION

References

Cesàro, E. *Lezioni di Geometria Intrinseca.* Napoli, Italy, 1896.

Euler, L. *Comment. Acad. Petropolit.* **8**, 66–85, 1736.

Gray, A. *Modern Differential Geometry of Curves and Surfaces.* Boca Raton, FL: CRC Press, pp. 111–112, 1993.

Melzak, Z. A. *Companion to Concrete Mathematics, Vol. 2.* New York: Wiley, 1976.

Struik, D. J. *Lectures on Classical Differential Geometry.* New York: Dover, pp. 26–28, 1988.

Natural Independence Phenomenon

A type of mathematical result which is considered by most logicians as more natural than the METAMATHEMATICAL incompleteness results first discovered by Gödel. Finite combinatorial examples include GOODSTEIN'S THEOREM, a finite form of RAMSEY'S THEOREM, and a finite form of KRUSKAL'S TREE THEOREM (Kirby and Paris 1982; Smorynski 1980, 1982, 1983; Gallier 1991).

see also GÖDEL'S INCOMPLETENESS THEOREM, GOODSTEIN'S THEOREM, KRUSKAL'S TREE THEOREM, RAMSEY'S THEOREM

References

Gallier, J. "What's so Special about Kruskal's Theorem and the Ordinal Gamma[0]? A Survey of Some Results in Proof Theory." *Ann. Pure and Appl. Logic* **53**, 199–260, 1991.

Kirby, L. and Paris, J. "Accessible Independence Results for Peano Arithmetic." *Bull. London Math. Soc.* **14**, 285–293, 1982.

Smorynski, C. "Some Rapidly Growing Functions." *Math. Intell.* **2**, 149–154, 1980.

Smorynski, C. "The Varieties of Arboreal Experience." *Math. Intell.* **4**, 182–188, 1982.

Smorynski, C. "'Big' News from Archimedes to Friedman." *Not. Amer. Math. Soc.* **30**, 251–256, 1983.

Natural Invariant

Let $\rho(x)\,dx$ be the fraction of time a typical dynamical ORBIT spends in the interval $[x, x+dx]$, and let $\rho(x)$ be normalized such that

$$\int \rho(x)\,dx = 1$$

over the entire interval of the map. Then the fraction the time an ORBIT spends in a finite interval $[a, b]$, is given by

$$\int_a^b \rho(x)\,dx.$$

The natural invariant is also called the INVARIANT DENSITY or NATURAL DENSITY.

Natural Logarithm

The LOGARITHM having base e, where

$$e = 2.718281828\ldots, \tag{1}$$

which can be defined

$$\ln x \equiv \int_1^x \frac{dt}{t} \tag{2}$$

for $x > 0$. The natural logarithm can also be defined for COMPLEX NUMBERS as

$$\ln z \equiv \ln|z| + i\arg(z), \tag{3}$$

where $|z|$ is the MODULUS and $\arg(z)$ is the ARGUMENT. The natural logarithm is especially useful in CALCULUS because its DERIVATIVE is given by the simple equation

$$\frac{d}{dx}\ln x = \frac{1}{x}, \tag{4}$$

whereas logarithms in other bases have the more complicated DERIVATIVE

$$\frac{d}{dx}\log_b x = \frac{1}{x\ln b}. \tag{5}$$

The MERCATOR SERIES

$$\ln(1+x) = x - \tfrac{1}{2}x^2 + \tfrac{1}{3}x^3 - \ldots \tag{6}$$

gives a TAYLOR SERIES for the natural logarithm.

CONTINUED FRACTION representations of logarithmic functions include

$$\ln(1+x) = \cfrac{x}{1+\cfrac{1^2x}{2+\cfrac{1^2x}{3+\cfrac{2^2x}{4+\cfrac{2^2x}{5+\cfrac{3^2x}{6+\cfrac{3^2x}{7+\ldots}}}}}}} \tag{7}$$

$$\ln\left(\frac{1+x}{1-x}\right) = \cfrac{2x}{1-\cfrac{x^2}{3-\cfrac{4x^2}{5-\cfrac{9x^2}{7-\cfrac{16x^2}{9-\ldots}}}}}. \tag{8}$$

For a COMPLEX NUMBER z, the natural logarithm satisfies

$$\ln z = \ln[re^{i(\theta+2n\pi)}] = \ln r + i(\theta + 2n\pi) \tag{9}$$

$$PV(\ln z) = \ln r + i\theta, \tag{10}$$

where PV is the PRINCIPAL VALUE.

Some special values of the natural logarithm are

$$\ln 1 = 0 \tag{11}$$

$$\ln 0 = -\infty \tag{12}$$

$$\ln(-1) = \pi i \tag{13}$$

$$\ln(\pm i) = \pm\tfrac{1}{2}\pi i. \tag{14}$$

An identity for the natural logarithm of 2 discovered using the PSLQ ALGORITHM is

$$(\ln 2)^2 = -\frac{1}{6}\sum_{k=0}^{\infty}\frac{1}{16^k}\left[-\frac{3}{(8k)^2} - \frac{16}{(8k+1)^2} - \frac{40}{(8k+2)^2} - \frac{8}{(8k+3)^2} - \frac{28}{(8k+4)^2} + \frac{4}{(8k+5)^2} - \frac{28}{(8k+4)^2} - \frac{4}{(8k+5)^2} + \frac{10}{(8k+5)^2} - \frac{2}{(8k+7)^2}\right] \tag{15}$$

(Bailey *et al.* 1995, Bailey and Plouffe).

see also e, JENSEN'S FORMULA, LG, LOGARITHM

References

Bailey, D.; Borwein, P.; and Plouffe, S. "On the Rapid Computation of Various Polylogarithmic Constants." `http://www.cecm.sfu.ca/~pborwein/PAPERS/P123.ps`.

Bailey, D. and Plouffe, S. "Recognizing Numerical Constants." `http://www.cecm.sfu.ca/organics/papers/bailey`.

Natural Measure

$\mu_i(\epsilon)$, sometimes denoted $P_i(\epsilon)$, is the probability that element i is populated, normalized such that

$$\sum_{i=1}^{N}\mu_i(\epsilon) = 1.$$

see also INFORMATION DIMENSION, q-DIMENSION

Natural Norm

Let $||\mathbf{z}||$ be a VECTOR NORM of $\mathbf{z}$ such that

$$||\mathsf{A}|| = \max_{||\mathbf{z}||=1} ||\mathsf{A}\mathbf{z}||.$$

Then $||\mathsf{A}||$ is a MATRIX NORM which is said to be the natural norm INDUCED (or SUBORDINATE) to the VECTOR NORM $||\mathbf{z}||$. For any natural norm,

$$||\mathsf{I}|| = 1,$$

where I is the IDENTITY MATRIX. The natural matrix norms induced by the L_1-NORM, L_2-NORM, and L_∞-NORM are called the MAXIMUM ABSOLUTE COLUMN SUM NORM, SPECTRAL NORM, and MAXIMUM ABSOLUTE ROW SUM NORM, respectively.

References

Gradshteyn, I. S. and Ryzhik, I. M. *Tables of Integrals, Series, and Products, 5th ed.* San Diego, CA: Academic Press, p. 1115, 1979.

Natural Number

A POSITIVE INTEGER 1, 2, 3, ... (Sloane's A000027). The set of natural numbers is denoted $\mathbb{N}$ or $\mathbb{Z}^+$. Unfortunately, 0 is sometimes also included in the list of "natural" numbers (Bourbaki 1968, Halmos 1974), and there seems to be no general agreement about whether to include it.

Due to lack of standard terminology, the following terms are recommended in preference to "COUNTING NUMBER," "natural number," and "WHOLE NUMBER."

Set	Name	Symbol
$\ldots, -2, -1, 0, 1, 2, \ldots$	integers	$\mathbb{Z}$
$1, 2, 3, 4, \ldots$	positive integers	$\mathbb{Z}^+$
$0, 1, 2, 3, 4 \ldots$	nonnegative integers	$\mathbb{Z}^*$
$-1, -2, -3, -4, \ldots$	negative integers	$\mathbb{Z}^-$

see also COUNTING NUMBER, INTEGER, $\mathbb{N}$, POSITIVE, $\mathbb{Z}$, $\mathbb{Z}^-$, $\mathbb{Z}^+$, $\mathbb{Z}^*$

References

Bourbaki, N. *Elements of Mathematics: Theory of Sets.* Paris, France: Hermann, 1968.

Courant, R. and Robbins, H. "The Natural Numbers." Ch. 1 in *What is Mathematics?: An Elementary Approach to Ideas and Methods, 2nd ed.* Oxford, England: Oxford University Press, pp. 1–20, 1996.

Halmos, P. R. *Naive Set Theory.* New York: Springer-Verlag, 1974.

Sloane, N. J. A. Sequence A000027/M0472 in "An On-Line Version of the Encyclopedia of Integer Sequences."

Naught

The British word for "ZERO." It is often used to indicate 0 subscripts, so a_0 would be spoken as "a naught."

see also ZERO

Navigation Problem

A problem in the CALCULUS OF VARIATIONS. Let a vessel traveling at constant speed c navigate on a body of water having surface velocity

$$\begin{aligned} u &= u(x,y) \\ v &= v(x,y). \end{aligned}$$

The navigation problem asks for the course which travels between two points in minimal time.

References

Sagan, H. *Introduction to the Calculus of Variations.* New York: Dover, pp. 226–228, 1992.

Near-Integer

see ALMOST INTEGER

Near Noble Number

A REAL NUMBER $0 < \nu < 1$ whose CONTINUED FRACTION is periodic, and the periodic sequence of terms is composed of a string of 1s followed by an INTEGER $n > 1$,

$$\nu = [\overline{\underbrace{1, 1, \ldots, 1}_{P}, n}]. \tag{1}$$

This can be written in the form

$$\nu = [\underbrace{1, 1, \ldots, 1}_{P}, n, \nu^{-1}], \tag{2}$$

which can be solved to give

$$\nu = \tfrac{1}{2}n\left(\sqrt{1 + 4\frac{nF_{P-1} + F_{P-2}}{n^2 F_P}} - 1\right), \tag{3}$$

where F_n is a FIBONACCI NUMBER. The special case $n = 2$ gives

$$\nu = \sqrt{\frac{F_{P+2}}{F_P}} - 1. \tag{4}$$

see also NOBLE NUMBER

References

Schroeder, M. R. *Number Theory in Science and Communication: With Applications in Cryptography, Physics, Digital Information, Computing, and Self-Similarity, 2nd enl. ed., corr. printing.* Berlin: Springer-Verlag, 1990.

Schroeder, M. "Noble and Near Noble Numbers." In *Fractals, Chaos, Power Laws: Minutes from an Infinite Paradise.* New York: W. H. Freeman, pp. 392–394, 1991.

Near-Pencil

An arrangement of $n \geq 3$ points such that $n - 1$ of them are COLLINEAR.

see also GENERAL POSITION, ORDINARY LINE, PENCIL

References

Guy, R. K. "Unsolved Problems Come of Age." *Amer. Math. Monthly* **96**, 903–909, 1989.

Nearest Integer Function

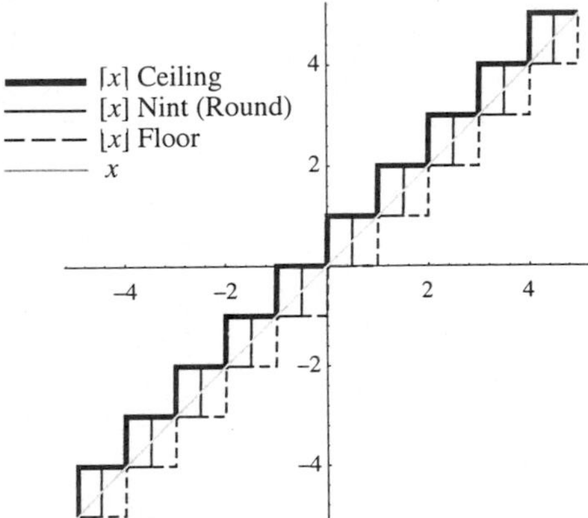

The nearest integer function `nint(x)` of x, also called NINT or the ROUND function, is defined such that $[x]$ is the INTEGER closest to x. It is shown as the thin solid curve in the above plot. Note that while $[x]$ is used to denote the nearest integer function in this work, $[x]$ is more commonly used to denote the FLOOR FUNCTION $\lfloor x \rfloor$.

see also CEILING FUNCTION, FLOOR FUNCTION

Nearest Neighbor Problem

The problem of identifying the point from a set of points which is nearest to a given point according to some measure of distance. The nearest neighborhood problem involves identifying the locus of points lying nearer to the query point than to any other point in the set.

References

Martin, E. C. "Computational Geometry." http://www.mathsource.com/cgi-bin/MathSource/Enhancements/DiscreteMath/0200-181.

Necessary

A CONDITION which must hold for a result to be true, but which does not guarantee it to be true. If a CONDITION is both NECESSARY and SUFFICIENT, then the result is said to be true IFF the CONDITION holds.

see also SUFFICIENT

Necker Cube

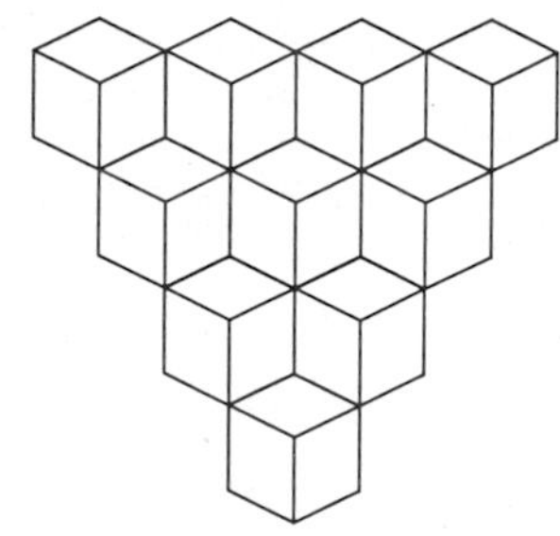

An ILLUSION in which a 2-D drawing of an array of CUBES appear to simultaneously protrude and intrude into the page.

References

Fineman, M. *The Nature of Visual Illusion.* New York: Dover, pp. 25 and 118, 1996.

Jablan, S. "Impossible Figures." http://members.tripod.com/-modularity/impos.htm.

Newbold, M. "Animated Necker Cube." http://www.sover.net/-manx/necker.html.

Necklace

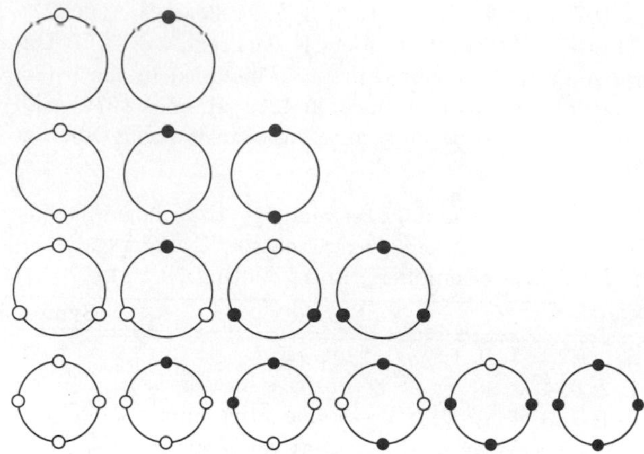

In the technical COMBINATORIAL sense, an a-ary necklace $N(n,a)$ of length n is a string of n characters, each of a possible types. Rotation is ignored, in the sense that $b_1 b_2 \ldots b_n$ is equivalent to $b_k b_{k+1} \cdots b_1 b_2 \cdots b_{k-1}$ for any k, but reversal of strings is respected. Necklaces therefore correspond to circular collections of beads in which the FIXED necklace may not be picked up out of the PLANE (so that opposite orientations are not considered equivalent).

The number of distinct FREE necklaces $N'(n,a)$ of n beads, each of a possible colors, in which opposite orientations (MIRROR IMAGES) are regarded as equivalent (so the necklace *can* be picked up out of the PLANE and flipped over) can be found as follows. Find the DIVISORS of n and label them $d_1 \equiv 1$, d_2, ..., $d_\nu(n) \equiv n$ where $\nu(n)$ is the number of DIVISORS of n. Then

$$N'(n,a) = \frac{1}{2n}\begin{cases} \sum_{i=1}^{\nu(n)} \phi(d_i) a^{n/d_i} + n a^{(n+1)/2} & \text{for } n \text{ odd} \\ \sum_{i=1}^{\nu(n)} \phi(d_i) a^{n/d_i} + \frac{1}{2} n(1+a) a^{n/2} & \text{for } n \text{ even,} \end{cases}$$

where $\phi(x)$ is the TOTIENT FUNCTION. For $a = 2$ and $n = p$ an ODD PRIME, this simplifies to

$$N'(p,2) = \frac{2^{p-1}-1}{p} + 2^{(p-1)/2} + 1.$$

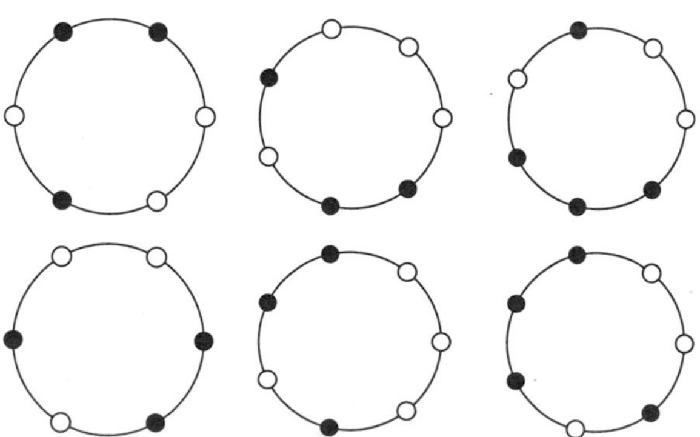

A table of the first few numbers of necklaces for $a = 2$ and $a = 3$ follows. Note that $N(n,2)$ is larger than $N'(n,2)$ for $n \geq 6$. For $n = 6$, the necklace 110100 is inequivalent to its MIRROR IMAGE 0110100, accounting for the difference of 1 between $N(6,2)$ and $N'(6,2)$. Similarly, the two necklaces 0010110 and 0101110 are inequivalent to their reversals, accounting for the difference of 2 between $N(7,2)$ and $N'(7,2)$.

n	$N(n,2)$	$N'(n,2)$	$N'(n,3)$
Sloane	000031	000029	027671
1	2	2	3
2	3	3	6
3	4	4	10
4	6	6	21
5	8	8	39
6	14	13	92
7	20	18	198
8	36	30	498
9	60	46	1219
10	108	78	3210
11	188	126	8418
12	352	224	22913
13	632	380	62415
14	1182	687	173088
15	2192	1224	481598

Ball and Coxeter (1987) consider the problem of finding the number of distinct arrangements of n people in a ring such that no person has the same two neighbors two or more times. For 8 people, there are 21 such arrangements.

see also ANTOINE'S NECKLACE, DE BRUIJN SEQUENCE, FIXED, FREE, IRREDUCIBLE POLYNOMIAL, JOSEPHUS PROBLEM, LYNDON WORD

References

Ball, W. W. R. and Coxeter, H. S. M. *Mathematical Recreations and Essays, 13th ed.* New York: Dover, pp. 49–50, 1987.

Dudeney, H. E. Problem 275 in *536 Puzzles & Curious Problems.* New York: Scribner, 1967.

Gardner, M. *Martin Gardner's New Mathematical Diversions from Scientific American.* New York: Simon and Schuster, pp. 240–246, 1966.

Gilbert, E. N. and Riordan, J. "Symmetry Types of Periodic Sequences." *Illinois J. Math.* **5**, 657–665, 1961.

Riordan, J. "The Combinatorial Significance of a Theorem of Pólya." *J. SIAM* **4**, 232–234, 1957.

Riordan, J. *An Introduction to Combinatorial Analysis.* New York: Wiley, p. 162, 1980.

Ruskey, F. "Information on Necklaces, Lyndon Words, de Bruijn Sequences." `http://sue.csc.uvic.ca/~cos/inf/neck/NecklaceInfo.html`.

Sloane, N. J. A. Sequences A000029/M0563, A000031/M0564, and A001869/M3860 in "An On-Line Version of the Encyclopedia of Integer Sequences." `http://www.research.att.com/~njas/sequences/eisonline.html`.

Sloane, N. J. A. and Plouffe, S. Extended entry for M3860 in *The Encyclopedia of Integer Sequences.* San Diego: Academic Press, 1995.

Needle

see BUFFON-LAPLACE NEEDLE PROBLEM, BUFFON'S NEEDLE PROBLEM, KAKEYA NEEDLE PROBLEM

Negation

see NOT

Negative

A quantity less than ZERO (< 0), denoted with a MINUS SIGN, i.e., $-x$.

see also NONNEGATIVE, NONPOSITIVE, NONZERO, POSITIVE, ZERO

Negative Binomial Distribution

Also known as the PASCAL DISTRIBUTION and PÓLYA DISTRIBUTION. The probability of $r-1$ successes and x failures in $x+r-1$ trials, and success on the $(x+r)$th trial is

$$\begin{aligned} p\left[\binom{x+r-1}{r-1}p^{r-1}(1-p)^{[(x+r-1)-(r-1)]}\right] &= \left[\binom{x+r-1}{r-1}p^{r-1}(1-p)^x\right]p \\ &= \binom{x+r-1}{r-1}p^r(1-p)^x, \end{aligned} \quad (1)$$

where $\binom{n}{k}$ is a BINOMIAL COEFFICIENT. Let

$$P = \frac{1-p}{p} \quad (2)$$

$$Q = \frac{1}{p}. \quad (3)$$

The CHARACTERISTIC FUNCTION is given by

$$\phi(t) = (Q - Pe^{it})^{-r}, \quad (4)$$

and the MOMENT-GENERATING FUNCTION by

$$M(t) = \langle e^{tx} \rangle = \sum_{x=0}^{\infty} e^{tx} \binom{x+r-1}{r-1} p^r (1-p)^x, \quad (5)$$

but, since $\binom{N}{n} = \binom{N}{N-m}$,

$$\begin{aligned} M(t) &= p^r \sum_{x=0}^{\infty} \binom{x+r-1}{x} [(1-p)e^t]^x \\ &= p^r[1-(1-p)e^t]^{-r} \end{aligned} \quad (6)$$

$$\begin{aligned} M'(t) &= p^r(-r)[1-(1-p)e^t]^{-r-1}(p-1)e^t \\ &= p^r(1-p)r[1-(1-p)e^t]^{-r-1}e^t \end{aligned} \quad (7)$$

$$\begin{aligned} M''(t) &= (1-p)rp^r(1-e^t+pe^t)^{-r-2} \\ &\quad \times (-1-e^t r+e^t pr)e^t \end{aligned} \quad (8)$$

$$\begin{aligned} M'''(t) &= (1-p)rp^r(1-e^t+e^t p)^{-r-3} \\ &\quad \times [1+e^t(1-p+3r-3pr) \\ &\quad + r^2 e^{2t}(1-p)^2]e^t. \end{aligned} \quad (9)$$

The MOMENTS about zero $\mu_n' = M^n(0)$ are therefore

$$\mu_1' = \mu = \frac{r(1-p)}{p} = \frac{rq}{p} \quad (10)$$

$$\mu_2' = \frac{r(1-p)[1-r(p-1)]}{p^2} = \frac{rq(1-rq)}{p^2} \quad (11)$$

$$\mu_3' = \frac{(1-p)r(2-p+3r-3pr+r^2-2pr^2+p^2r^2}{p^3} \quad (12)$$

$$\mu_4' = \frac{(-1+p)r(-6+6p-p^2-11r+15pr-4p^2r-6r^2}{p^4} + \frac{12pr^2-6p^2r^2-r^3+3pr^3-3p^2r^3+p^3r^3)}{p^4}. \quad (13)$$

(Beyer 1987, p. 487, apparently gives the MEAN incorrectly.) The MOMENTS about the mean are

$$\mu_2 = \sigma^2 = \frac{r(1-p)}{p^2} \quad (14)$$

$$\mu_3 = \frac{r(2-3p+p^2)}{p^3} = \frac{r(p-1)(p-2)}{p^3} \quad (15)$$

$$\mu_4 = \frac{r(1-p)(6-6p+p^2+3r-3pr)}{p^4}. \quad (16)$$

The MEAN, VARIANCE, SKEWNESS and KURTOSIS are then

$$\mu = \frac{r(1-p)}{p} \quad (17)$$

$$\gamma_1 = \frac{\mu_3}{\sigma^3} = \frac{r(p-1)(p-2)}{p^3}\left[\frac{p^2}{r(1-p)}\right]^{3/2} = \frac{r(2-p)(1-p)}{p^3}\frac{p^3}{r(1-p)\sqrt{1-p}} = \frac{2-p}{\sqrt{r(1-p)}} \quad (18)$$

$$\gamma_2 = \frac{\mu_4}{\sigma^4} - 3 = \frac{-6+6p-p^2-3r+3pr}{(p-1)r}, \quad (19)$$

which can also be written

$$\mu = nP \quad (20)$$

$$\mu_2 = nPQ \quad (21)$$

$$\gamma_1 = \frac{Q+P}{\sqrt{rPQ}} \quad (22)$$

$$\gamma_2 = \frac{1+6PQ}{rPQ} - 3. \quad (23)$$

The first CUMULANT is

$$\kappa_1 = nP, \quad (24)$$

and subsequent CUMULANTS are given by the recurrence relation

$$\kappa_{r+1} = PQ\frac{d\kappa_r}{dQ}. \quad (25)$$

References

Beyer, W. H. *CRC Standard Mathematical Tables, 28th ed.* Boca Raton, FL: CRC Press, p. 533, 1987.

Spiegel, M. R. *Theory and Problems of Probability and Statistics.* New York: McGraw-Hill, p. 118, 1992.

Negative Binomial Series

The SERIES which arises in the BINOMIAL THEOREM for NEGATIVE integral n,

$$(x+a)^{-n} = \sum_{k=0}^{\infty}\binom{-n}{k}x^k a^{-n-k} = \sum_{k=0}^{\infty}(-1)^k\binom{n+k-1}{k}x^k a^{-n-k}.$$

For $a=1$, the negative binomial series simplifies to

$$(x+1)^{-n} = 1 - nx + \tfrac{1}{2}n(n+1)x^2 - \tfrac{1}{6}n(n+1)(n+2) + \ldots.$$

see also BINOMIAL SERIES, BINOMIAL THEOREM

Negative Likelihood Ratio

The term NEGATIVE likelihood ratio is also used (especially in medicine) to test nonnested complementary hypotheses as follows,

$$\text{NLR} = \frac{[\text{true negative rate}]}{[\text{false negative rate}]} = \frac{[\text{specificity}]}{1-[\text{sensitivity}]}.$$

see also LIKELIHOOD RATIO, SENSITIVITY, SPECIFICITY

Negative Integer

see $\mathbb{Z}^-$

Negative Pedal Curve

Given a curve C and O a fixed point called the PEDAL POINT, then for a point P on C, draw a LINE PERPENDICULAR to OP. The ENVELOPE of these LINES as P describes the curve C is the negative pedal of C.

see also PEDAL CURVE

References

Lawrence, J. D. *A Catalog of Special Plane Curves.* New York: Dover, pp. 46–49, 1972.

Lockwood, E. H. "Negative Pedals." Ch. 19 in *A Book of Curves.* Cambridge, England: Cambridge University Press, pp. 156–159, 1967.

Neighborhood

The word neighborhood is a word with many different levels of meaning in mathematics. One of the most general concepts of a neighborhood of a point $x \in \mathbb{R}^n$ (also called an EPSILON-NEIGHBORHOOD or infinitesimal OPEN SET) is the set of points inside an n-BALL with center x and RADIUS $\epsilon > 0$.

Neile's Parabola

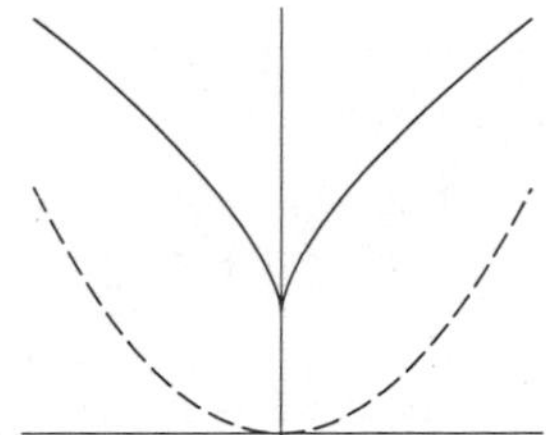

The solid curve in the above figure which is the EVOLUTE of the PARABOLA (dashed curve). In CARTESIAN COORDINATES,

$$y = \tfrac{3}{4}(2x)^{2/3} + \tfrac{1}{2}.$$

Neile's parabola is also called the SEMICUBICAL PARABOLA, and was discovered by William Neile in 1657. It was the first nontrivial ALGEBRAIC CURVE to have its ARC LENGTH computed. Wallis published the method in 1659, giving Neile the credit (MacTutor Archive).

see also PARABOLA EVOLUTE

References

Lee, X. "Semicubic Parabola." `http://www.best.com/~xah/Special Plane Curves _ dir / Semicubic Parabola _ dir / semicubicParabola.html.`

MacTutor History of Mathematics Archive. "Neile's Semi-Cubical Parabola." `http://www-groups.dcs.st-and.ac.uk/~history/Curves/Neiles.html.`

Nephroid

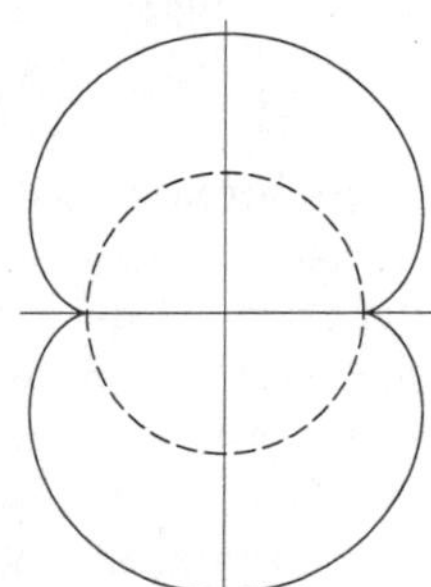

The 2-CUSPED EPICYCLOID is called a nephroid. Since $n = 2$, $a = b/2$, and the equation for r^2 in terms of the parameter ϕ is given by EPICYCLOID equation

$$r^2 = \frac{a^2}{n^2}\left[(n^2+2n+2) - 2(n+1)\cos(n\phi)\right] \quad (1)$$

with $n = 2$,

$$\begin{aligned} r^2 &= \frac{a^2}{2^2}[(2^2+2\cdot2+2) - 2(2+1)\cos(2\phi)] \\ &= \tfrac{1}{4}a^2[10 - 6\cos(2\phi)] = \tfrac{1}{2}a^2[5 - 3\cos(2\phi)], \quad (2) \end{aligned}$$

where

$$\tan\theta = \frac{3\sin\phi - \sin(3\phi)}{3\cos\phi - \cos(3\phi)}. \quad (3)$$

This can be written

$$\left(\frac{r}{2a}\right)^{2/3} = [\sin(\tfrac{1}{2}\theta)]^{2/3} + [\cos(\tfrac{1}{2}\theta)]^{2/3}. \quad (4)$$

The parametric equations are

$$x = a[3\cos t - \cos(3t)] \quad (5)$$
$$y = a[3\sin t - \sin(3t)]. \quad (6)$$

The Cartesian equation is

$$(x^2+y^2-4a^2)^3 = 108a^4y^2. \quad (7)$$

The name nephroid means "kidney shaped" and was first used for the two-cusped EPICYCLOID by Proctor in 1878 (MacTutor Archive). The nephroid has ARC LENGTH $24a$ and AREA $12\pi^2a^2$. The CATACAUSTIC for rays originating at the CUSP of a CARDIOID and reflected by it is a nephroid. Huygens showed in 1678 that the nephroid is the CATACAUSTIC of a CIRCLE when the light source is at infinity. He published this fact in *Traité de la lumière* in 1690 (MacTutor Archive).

see also ASTROID, DELTOID, FREETH'S NEPHROID

References

Lawrence, J. D. *A Catalog of Special Plane Curves.* New York: Dover, pp. 169–173, 1972.

Lee, X. "Nephroid." `http://www.best.com/~xah/SpecialPlaneCurves_dir/Nephroid_dir/nephroid.html.`

Lockwood, E. H. "The Nephroid." Ch. 7 in *A Book of Curves.* Cambridge, England: Cambridge University Press, pp. 62–71, 1967.

MacTutor History of Mathematics Archive. "Nephroid." `http://www-groups.dcs.st-and.ac.uk/~history/Curves/Nephroid.html.`

Yates, R. C. "Nephroid." *A Handbook on Curves and Their Properties.* Ann Arbor, MI: J. W. Edwards, pp. 152–154, 1952.

Nephroid Evolute

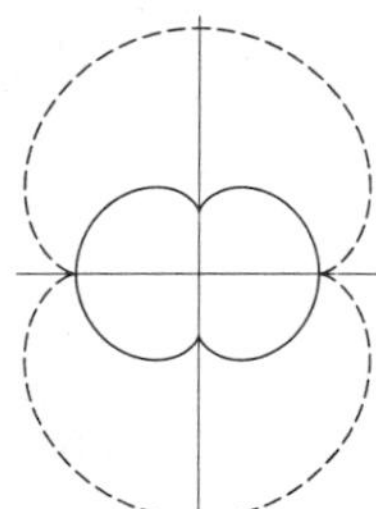

The EVOLUTE of the NEPHROID given by

$$x = \tfrac{1}{2}[3\cos t - \cos(3t)]$$
$$y = \tfrac{1}{2}[3\sin t - \sin(3t)]$$

is given by

$$x = \cos^3 t$$
$$y = \tfrac{1}{4}[3\sin t + \sin(3t)],$$

which is another NEPHROID.

Nephroid Involute

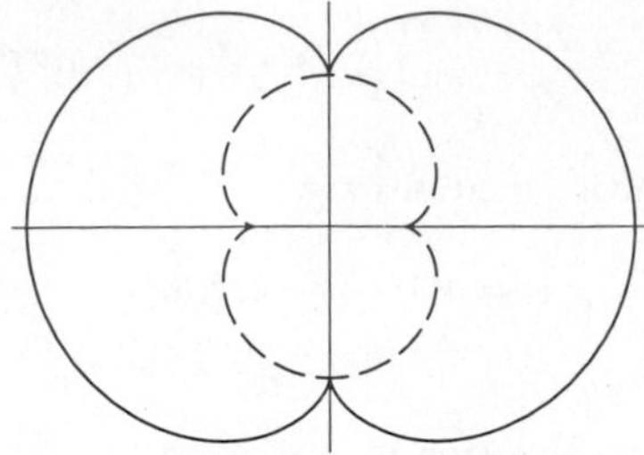

The INVOLUTE of the NEPHROID given by

$$x = \tfrac{1}{2}[3\cos t - \cos(3t)]$$
$$y = \tfrac{1}{2}[3\sin t - \sin(3t)]$$

beginning at the point where the nephroid cuts the y-AXIS is given by

$$x = 4\cos^3 t$$
$$y = 3\sin t + \sin(3t),$$

another NEPHROID. If the INVOLUTE is begun instead at the CUSP, the result is CAYLEY'S SEXTIC.

Néron-Severi Group

Let V be a complete normal VARIETY, and write $G(V)$ for the group of divisors, $G_n(V)$ for the group of divisors numerically equal to 0, and $G_a(V)$ the group of divisors algebraically equal to 0. Then the finitely generated QUOTIENT GROUP $NS(V) = G(V)/G_a(V)$ is called the Néron-Severi group.

References

Iyanaga, S. and Kawada, Y. (Eds.). *Encyclopedic Dictionary of Mathematics.* Cambridge, MA: MIT Press, p. 75, 1980.

Nerve

The SIMPLICIAL COMPLEX formed from a family of objects by taking sets that have nonempty intersections.

see also DELAUNAY TRIANGULATION, SIMPLICIAL COMPLEX

Nested Hypothesis

Let S be the set of all possibilities that satisfy HYPOTHESIS H, and let S' be the set of all possibilities that satisfy HYPOTHESIS H'. Then H' is a nested hypothesis within H IFF $S' \subset S$, where $\subset$ denotes the PROPER SUBSET.

see also LOG LIKELIHOOD PROCEDURE

Nested Radical

A RADICAL of the form

$$\sqrt{n+\sqrt{n+\sqrt{n+\ldots}}}. \tag{1}$$

For this to equal a given INTEGER x, it must be true that

$$x = \sqrt{n+\sqrt{n+\sqrt{n+\ldots}}} = \sqrt{n+x}, \tag{2}$$

so

$$x^2 = n + x \tag{3}$$

and

$$n = x(x-1). \tag{4}$$

Nested radicals in the computation of PI,

$$\frac{2}{\pi} = \sqrt{\tfrac{1}{2}}\sqrt{\tfrac{1}{2}+\sqrt{\tfrac{1}{2}}}\sqrt{\tfrac{1}{2}+\sqrt{\tfrac{1}{2}+\sqrt{\tfrac{1}{2}}}}\cdots \tag{5}$$

and in TRIGONOMETRICAL values of COSINE and SINE for arguments of the form $\pi/2^n$, e.g.,

$$\sin\left(\frac{\pi}{8}\right) = \tfrac{1}{2}\sqrt{2-\sqrt{2}} \tag{6}$$

$$\cos\left(\frac{\pi}{8}\right) = \tfrac{1}{2}\sqrt{2+\sqrt{2}} \tag{7}$$

$$\sin\left(\frac{\pi}{16}\right) = \tfrac{1}{2}\sqrt{2-\sqrt{2+\sqrt{2}}} \tag{8}$$

$$\cos\left(\frac{\pi}{16}\right) = \tfrac{1}{2}\sqrt{2+\sqrt{2+\sqrt{2}}}. \tag{9}$$

see also SQUARE ROOT

References

Berndt, B. C. *Ramanujan's Notebooks, Part IV.* New York: Springer-Verlag, pp. 14–20, 1994.

Net

A generalization of a SEQUENCE used in general topology and ANALYSIS when the spaces being dealt with are not FIRST-COUNTABLE. (Sequences provide an adequate way of dealing with CONTINUITY for FIRST-COUNTABLE SPACES.) Nets are used in the study of the RIEMANN INTEGRAL.

see also FIBER BUNDLE, FIBER SPACE, FIBRATION

Net (Polyhedron)

A plane diagram in which the EDGES of a POLYHEDRON are shown. All convex POLYHEDRA have nets, but not all concave polyhedra do (the constituent POLYGONS can overlap one another when a concave POLYHEDRON is flattened out). The GREAT DODECAHEDRON and STELLA OCTANGULA are examples of a concave polyhedron which have nets.

Netto's Conjecture

The probability that two elements P_1 and P_2 of a SYMMETRIC GROUP generate the entire GROUP tends to 3/4 as $n \to \infty$. This was proven by Dixon in 1967.

References

Le Lionnais, F. *Les nombres remarquables.* Paris: Hermann, p. 31, 1983.

Network

A DIRECTED GRAPH having a SOURCE, SINK, and a bound on each edge.

see also GRAPH (GRAPH THEORY), SINK (DIRECTED GRAPH), SMITH'S NETWORK THEOREM, SOURCE

Neuberg Circles

The LOCUS of the VERTEX A_1 of a TRIANGLE on a given base A_2A_3 and with a given BROCARD ANGLE ω is a CIRCLE on either side of A_2A_3. From the center N_1, the base A_2A_3 subtends the ANGLE 2ω. The RADIUS of the CIRCLE is

$$r = \tfrac{1}{2}a_1\sqrt{\cot^2\omega - 3}.$$

see also BROCARD ANGLE

References

Johnson, R. A. *Modern Geometry: An Elementary Treatise on the Geometry of the Triangle and the Circle.* Boston, MA: Houghton Mifflin, pp. 287–290, 1929.

Neumann Algebra

see VON NEUMANN ALGEBRA

Neumann Boundary Conditions

PARTIAL DIFFERENTIAL EQUATION BOUNDARY CONDITIONS which give the normal derivative on a surface.

see also BOUNDARY CONDITIONS, CAUCHY BOUNDARY CONDITIONS

References

Morse, P. M. and Feshbach, H. *Methods of Theoretical Physics, Part I.* New York: McGraw-Hill, p. 679, 1953.

Neumann Function

see BESSEL FUNCTION OF THE SECOND KIND

Neumann Polynomial

Polynomials which obey the RECURRENCE RELATION

$$O_{n+1}(x) = (n+1)\frac{2}{x}O_n(x) - \frac{n+1}{n-1}O_{n-1}(x) + \frac{2n}{x}\sin^2(\tfrac{1}{2}n\pi).$$

The first few are

$$O_0(x) = \frac{1}{x}$$
$$O_1(x) = \frac{1}{x^2}$$
$$O_2(x) = \frac{1}{x} + \frac{4}{x^3}.$$

see also SCHLÄFLI POLYNOMIAL

References

von Seggern, D. *CRC Standard Curves and Surfaces.* Boca Raton, FL: CRC Press, p. 196, 1993.

Neumann Series (Bessel Function)

A series of the form

$$\sum_{n=0}^\infty a_n J_{\nu+n}(z), \qquad (1)$$

where ν is a REAL and $J_{\nu+n}(z)$ is a BESSEL FUNCTION OF THE FIRST KIND. Special cases are

$$z^\nu = 2^\nu\Gamma(\tfrac{1}{2}\nu+1)\sum_{n=0}^\infty \frac{(\frac{1}{2}z)^{\nu/2+n}}{n!}J_{\nu/2+n}(z), \qquad (2)$$

where $\Gamma(z)$ is the GAMMA FUNCTION, and

$$\sum_{n=0}^\infty b_n z^{\nu+n} = \sum_{n=0}^\infty a_n\left(\tfrac{1}{2}z\right)^{(\nu+n)/2} J_{(\nu+n)/2}(z), \qquad (3)$$

where

$$a_n \equiv \sum_{m=0}^{\lfloor n/2\rfloor} \frac{2^{\nu+n-2m}\Gamma(\frac{1}{2}\nu+\frac{1}{2}n-m+1)}{m!}b_{n-2m}, \qquad (4)$$

and $\lfloor x\rfloor$ is the FLOOR FUNCTION.

see also KAPTEYN SERIES

References

Watson, G. N. *A Treatise on the Theory of Bessel Functions, 2nd ed.* Cambridge, England: Cambridge University Press, 1966.

Neumann Series (Integral Equation)

A FREDHOLM INTEGRAL EQUATION OF THE SECOND KIND

$$\phi(x) = f(x) + \int_a^b K(x,t)\phi(t)\,dt \qquad (1)$$

may be solved as follows. Take

$$\phi_0(x) \equiv f(x) \qquad (2)$$

$$\phi_1(x) = f(x) + \lambda\int_a^b K(x,t)f(t)\,dt \qquad (3)$$

$$\phi_2(x) = f(x) + \lambda\int_a^b K(x,t_1)f(t_1)\,dt_1 + \lambda^2\int_a^b\int_a^b K(x,t_1)K(t_1,t_2)f(t_2)\,dt_2\,dt_1 \qquad (4)$$

$$\phi_n(x) = \sum_{i=0}^n \lambda^i u_i(x), \qquad (5)$$

where

$$u_0(x) = f(x) \qquad (6)$$

$$u_1(x) = \int_a^b K(x,t)f(t_1)\,dt_1 \qquad (7)$$

$$u_2(x) = \int_a^b\int_a^b K(x,t_1)K(t_1,t_2)f(t_2)\,dt_2\,dt_1 \qquad (8)$$

$$u_n(x) = \int_a^b\int_a^b\int_a^b K(x,t_1)K(t_1,t_2)\cdots \times K(t_{n-1},t_n)f(t_n)\,dt_n\cdots dt_1. \qquad (9)$$

The Neumann series solution is then

$$\phi(x) = \lim_{n\to\infty} \phi_n(x) = \lim_{n\to\infty} \sum_{i=0}^{n} \lambda^i u_i(x). \tag{10}$$

References

Arfken, G. "Neumann Series, Separable (Degenerate) Kernels." §16.3 in *Mathematical Methods for Physicists, 3rd ed.* Orlando, FL: Academic Press, pp. 879–890, 1985.

Neusis Construction

A geometric construction, also called a VERGING CONSTRUCTION, which allows the classical GEOMETRIC CONSTRUCTION rules to be bent in order to permit sliding of a marked RULER. Using a Neusis construction, CUBE DUPLICATION and angle TRISECTION are soluble. Conway and Guy (1996) give Neusis constructions for the 7-, 9-, and 13-gons which are based on angle TRISECTION.

see also CUBE DUPLICATION, GEOMETRIC CONSTRUCTION, MASCHERONI CONSTRUCTION, RULER, TRISECTION

References

Conway, J. H. and Guy, R. K. *The Book of Numbers.* New York: Springer-Verlag, pp. 194–200, 1996.

Neville's Algorithm

An interpolation ALGORITHM which proceeds by first fitting a POLYNOMIAL P_k of degree 0 through the points (x_k, y_k) for $k = 0, \ldots, n$, i.e., $P_k = y_k$. A second iteration is then performed in which P_{12} is fit through pairs of points, yielding $P_{12}, P_{23}, \ldots$. The procedure is repeated, generating a "pyramid" of approximations until the final result is reached

$$\begin{matrix} P_1 & & & \\ & P_{12} & & \\ P_2 & & P_{123} & \\ & P_{23} & & P_{1234}. \\ P_3 & & P_{234} & \\ & P_{34} & & \\ P_4 & & & \end{matrix}$$

The final result is

$$P_{i(i+1)\cdots(i+m)} = \frac{(x - x_{i+m})P_{i(i+1)\cdots(i+m-1)}}{x_i - x_{i+m}} + \frac{(x_i - x)P_{(i+1)(i+2)\cdots(i+m)}}{x_i - x_{i+m}}.$$

see also BULIRSCH-STOER ALGORITHM

Neville Theta Function

The functions

$$\vartheta_s(u) = \frac{H(u)}{H'(0)} \tag{1}$$

$$\vartheta_d(u) = \frac{\Theta(u+K)}{\Theta(k)} \tag{2}$$

$$\vartheta_c(u) = \frac{H(u)}{H(K)} \tag{3}$$

$$\vartheta_n(u) = \frac{\Theta(u)}{\Theta(0)}, \tag{4}$$

where H and Θ are the JACOBI THETA FUNCTIONS and $K(u)$ is the complete ELLIPTIC INTEGRAL OF THE FIRST KIND.

see also JACOBI THETA FUNCTION, THETA FUNCTION

Newcomb's Paradox

A paradox in DECISION THEORY. Given two boxes, B1 which contains $1000 and B2 which contains either nothing or a million dollars, you may pick either B2 or both. However, at some time before the choice is made, an omniscient Being has predicted what your decision will be and filled B2 with a million dollars if he expects you to take it, or with nothing if he expects you to take both.

see also ALIAS' PARADOX

References

Gardner, M. *The Unexpected Hanging and Other Mathematical Diversions.* Chicago, IL: Chicago University Press, 1991.

Gardner, M. "Newcomb's Paradox." Ch. 13 in *Knotted Doughnuts and Other Mathematical Entertainments.* New York: W. H. Freeman, 1986.

Nozick, R. "Reflections on Newcomb's Paradox." Ch. 14 in Gardner, M. *Knotted Doughnuts and Other Mathematical Entertainments.* New York: W. H. Freeman, 1986.

Newman-Conway Sequence

The sequence 1, 1, 2, 2, 3, 4, 4, 4, 5, 6, 7, 7, ... (Sloane's A004001) defined by the recurrence $P(1) = P(2) = 1$,

$$P(n) = P(P(n-1)) + P(n - P(n-1)).$$

It satisfies

$$P(2^k) = 2^{k-1}$$

and

$$P(2n) \le 2P(n).$$

References

Bloom, D. M. "Newman-Conway Sequence." Solution to Problem 1459. *Math. Mag.* **68**, 400–401, 1995.

Sloane, N. J. A. Sequence A004001/M0276 in "An On-Line Version of the Encyclopedia of Integer Sequences."

Newton's Backward Difference Formula

$$f_p = f_0 + p\nabla_0 + \tfrac{1}{2!}p(p+1)\nabla_0^2 + \tfrac{1}{3!}p(p+1)(p+2)\nabla_0^3 + \ldots,$$

for $p \in [0, 1]$, where ∇ is the BACKWARD DIFFERENCE.

see also NEWTON'S FORWARD DIFFERENCE FORMULA

References

Beyer, W. H. *CRC Standard Mathematical Tables, 28th ed.* Boca Raton, FL: CRC Press, p. 433, 1987.

Newton-Cotes Formulas

The Newton-Cotes formulas are an extremely useful and straightforward family of NUMERICAL INTEGRATION techniques.

To integrate a function $f(x)$ over some interval $[a, b]$, divide it into n equal parts such that $f_n = f(x_n)$ and $h \equiv (b-a)/n$. Then find POLYNOMIALS which approximate the tabulated function, and integrate them to approximate the AREA under the curve. To find the fitting POLYNOMIALS, use LAGRANGE INTERPOLATING POLYNOMIALS. The resulting formulas are called Newton-Cotes formulas, or QUADRATURE FORMULAS.

Newton-Cotes formulas may be "closed" if the interval $[x_1, x_n]$ is included in the fit, "open" if the points $[x_2, x_{n-1}]$ are used, or a variation of these two. If the formula uses n points (closed or open), the COEFFICIENTS of terms sum to $n-1$.

If the function $f(x)$ is given explicitly instead of simply being tabulated at the values x_i, the best numerical method of integration is called GAUSSIAN QUADRATURE. By picking the intervals at which to sample the function, this procedure produces more accurate approximations (but is significantly more complicated to implement).

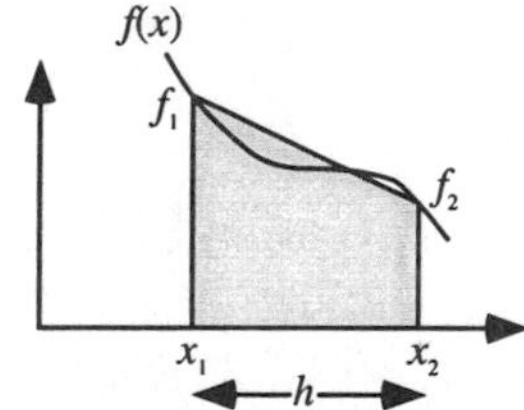

The 2-point closed Newton-Cotes formula is called the TRAPEZOIDAL RULE because it approximates the area under a curve by a TRAPEZOID with horizontal base and sloped top (connecting the endpoints x_1 and x_2). If the first point is x_1, then the other endpoint will be located at

$$x_2 = x_1 + h, \tag{1}$$

and the LAGRANGE INTERPOLATING POLYNOMIAL through the points (x_1, f_1) and (x_2, f_2) is

$$\begin{aligned} P_2(x) &= \frac{x-x_2}{x_1-x_2} f_1 + \frac{x-x_1}{x_2-x_1} f_2 \\ &= \frac{x-x_1-h}{-h} f_1 + \frac{x-x_1}{h} f_2 \\ &= \frac{x}{h}(f_2-f_1) + \left(f_1 + \frac{x_1}{h} f_1 - \frac{x_1}{h} f_2\right). \end{aligned} \tag{2}$$

Integrating over the interval (i.e., finding the area of the trapezoid) then gives

$$\begin{aligned} \int_{x_1}^{x_2} f(x)\,dx &= \int_{x_1}^{x_1+h} P_2(x)\,dx \\ &= \frac{1}{2h}(f_2-f_1)[x^2]_{x_1}^{x_2} \\ &\quad + \left(f_1 + \frac{x_1}{h} f_1 - \frac{x_1}{h} f_2\right)[x]_{x_1}^{x_2} \\ &= \frac{1}{2h}(f_2-f_1)(x_2+x_1)(x_2-x_1) \\ &\quad + (x_2-x_1)\left(f_1 + \frac{x_1}{h} f_1 - \frac{x_1}{h} f_2\right) \\ &= \tfrac{1}{2}(f_2-f_1)(2x_1+h) + f_1 h + x_1(f_1-f_2) \\ &= x_1(f_2-f_1) + \tfrac{1}{2}h(f_2-f_1) + hf_1 - x_1(f_2-f_1) \\ &= \tfrac{1}{2}h(f_1+f_2) - \tfrac{1}{2}h^3 f''(\xi). \end{aligned} \tag{3}$$

This is the trapezoidal rule, with the final term giving the amount of error (which, since $x_1 \leq \xi \leq x_2$, is no worse than the maximum value of $f''(\xi)$ in this range).

The 3-point rule is known as SIMPSON'S RULE. The ABSCISSAS are

$$x_2 = x_1 + h \tag{4}$$

$$x_3 = x_1 + 2h \tag{5}$$

and the LAGRANGE INTERPOLATING POLYNOMIAL is

$$\begin{aligned} P_3(x) &= \frac{(x-x_2)(x-x_3)}{(x_1-x_2)(x_1-x_3)} f_1 \\ &\quad + \frac{(x-x_1)(x-x_3)}{(x_2-x_1)(x_2-x_3)} f_2 + \frac{(x-x_1)(x-x_2)}{(x_3-x_1)(x_3-x_2)} f_3 \\ &= \frac{x^2 - x(x_2+x_3) + x_2x_3}{h(2h)} f_1 \\ &\quad + \frac{x^2 - x(x_1+x_3) + x_1x_3}{h(-h)} f_2 + \frac{x^2 - x(x_1+x_2) + x_1x_2}{2h(h)} f_3 \\ &= \frac{1}{h^2}\{x^2(\tfrac{1}{2}f_1 - f_2 - \tfrac{1}{2}f_3) \\ &\quad + x[-\tfrac{1}{2}(2x_1+3h)f_1 + (2x_1+2h)f_2 - \tfrac{1}{2}(2x_1+h)] \\ &\quad + [\tfrac{1}{2}(x_1+h)(x_1+2h)f_1 - x_1(x_1+2h)f_2 + \tfrac{1}{2}x_1(x_1+h)f_3]\}. \end{aligned} \tag{6}$$

Integrating and simplifying gives

$$\begin{aligned} \int_{x_1}^{x_3} f(x)\,dx &= \int_{x_1}^{x_1+2h} P_3(x)\,dx \\ &= \tfrac{1}{3}h(f_1+4f_2+f_3) - \tfrac{1}{90}h^5 f^{(4)}(\xi). \end{aligned} \tag{7}$$

The 4-point closed rule is SIMPSON'S 3/8 RULE,

$$\int_{x_1}^{x_4} f(x)\,dx = \tfrac{3}{8}h(f_1+3f_2+3f_3+f_4) - \tfrac{3}{80}h^5 f^{(4)}(\xi). \tag{8}$$

The 5-point closed rule is BODE'S RULE,

$$\int_{x_1}^{x_5} f(x)\,dx = \tfrac{2}{45}h(7f_1+32f_2+12f_3+32f_4+7f_5) - \tfrac{8}{945}h^7 f^{(6)}(\xi) \quad (9)$$

(Abramowitz and Stegun 1972, p. 886). Higher order rules include the 6-point

$$\int_{x_1}^{x_6} f(x)\,dx = \tfrac{5}{288}h(19f_1+75f_2+50f_3+50f_4+75f_5 +19f_6) - \tfrac{275}{12096}h^7 f^{(6)}(\xi), \quad (10)$$

7-point

$$\int_{x_1}^{x_7} f(x)\,dx = \tfrac{1}{140}h(41f_1+216f_2+27f_3+272f_4 +27f_5+216f_6+41f_7) - \tfrac{9}{1400}h^9 f^{(8)}(\xi), \quad (11)$$

8-point

$$\int_{x_1}^{x_8} f(x)\,dx = \tfrac{7}{17280}h(751f_1+3577f_2+1323f_2+2989f_3 +2989f_5+1323f_6+3577f_7+751f_8) - \tfrac{8183}{518400}h^9 f^{(8)}(\xi), \quad (12)$$

9-point

$$\int_{x_1}^{x_9} f(x)\,dx = \tfrac{4}{14175}h(989f_1+5888f_2-928f_3 +10496f_4-4540f_5+10496f_6-928f_7+5888f_8+989f_9) - \tfrac{2368}{467775}h^{11} f^{(10)}(\xi), \quad (13)$$

10-point

$$\int_{x_1}^{x_{10}} f(x)\,dx = \tfrac{9}{89600}h[2857(f_1+f_{10}) +15741(f_2+f_9)+1080(f_3+f_8+19344(f_4+f_7) +5788(f_5+f_6)] - \tfrac{173}{14620}h^{11} f^{(10)}(\xi), \quad (14)$$

and 11-point

$$\int_{x_1}^{x_{11}} f(x)\,dx = \tfrac{5}{299376}h[16067(f_1+f_{11}) +106300(f_2+f_{10})-48525(f_3+f_9)+272400(f_4+f_8) -260550(f_5+f_7)+427368f_6] - \tfrac{1346350}{326918592}h^{13} f^{(12)}(\xi) \quad (15)$$

rules.

Closed "extended" rules use multiple copies of lower order closed rules to build up higher order rules. By appropriately tailoring this process, rules with particularly nice properties can be constructed. For n tabulated points, using the TRAPEZOIDAL RULE $(n-1)$ times and adding the results gives

$$\int_{x_1}^{x_n} f(x)\,dx = \left(\int_{x_1}^{x_2} + \int_{x_2}^{x_3} + \ldots + \int_{x_{n-1}}^{x_n}\right) f(x)\,dx$$
$$= \tfrac{1}{2}h[(f_1+f_2)+(f_2+f_3)+\ldots+(f_{n-2}+f_{n-1}) +(f_{n-1}+f_n)] = h(\tfrac{1}{2}f_1+f_2+f_3+\ldots+f_{n-2}+f_{n-1}+\tfrac{1}{2}f_n) - \tfrac{1}{12}nh^3 f''(\xi). \quad (16)$$

Using a series of refinements on the extended TRAPEZOIDAL RULE gives the method known as ROMBERG INTEGRATION. A 3-point extended rule for ODD n is

$$\int_{x_1}^{x_n} f(x)\,dx = h[(\tfrac{1}{3}f_1+\tfrac{4}{3}f_2+\tfrac{1}{3}f_3)+(\tfrac{1}{3}f_3+\tfrac{4}{3}f_4+\tfrac{1}{3}f_5) +\ldots+(\tfrac{1}{3}f_{n-4}+\tfrac{4}{3}f_{n-3}+\tfrac{1}{3}f_{n-2}) +(\tfrac{1}{3}f_{n-2}+\tfrac{4}{3}f_{n-1}+\tfrac{1}{3}f_n)]$$
$$= \tfrac{1}{3}h(f_1+4f_2+2f_3+4f_4+2f_5+\ldots+4f_{n-1}+f_n) - \frac{n-1}{2}\tfrac{1}{90}h^5 f^{(4)}(\xi). \quad (17)$$

Applying SIMPSON'S 3/8 RULE, then SIMPSON'S RULE (3-point) twice, and adding gives

$$\left[\int_{x_1}^{x_4} + \int_{x_4}^{x_6} + \int_{x_6}^{x_8}\right] f(x)\,dx$$
$$= h[(\tfrac{3}{8}f_1+\tfrac{9}{8}f_2+\tfrac{9}{8}f_3+\tfrac{3}{8}f_4) +(\tfrac{1}{3}f_4+\tfrac{4}{3}f_5+\tfrac{1}{3}f_6)+(\tfrac{1}{3}f_6+\tfrac{4}{3}f_7+\tfrac{1}{3}f_8)]$$
$$= h[\tfrac{3}{8}f_1+\tfrac{9}{8}f_2+\tfrac{9}{8}f_3+(\tfrac{3}{8}+\tfrac{1}{3})f_4+\tfrac{4}{3}f_5 +(\tfrac{1}{3}+\tfrac{1}{3})f_6+\tfrac{4}{3}f_7+\tfrac{1}{3}f_8]$$
$$= h(\tfrac{3}{8}f_1+\tfrac{9}{8}f_2+\tfrac{9}{8}f_3+\tfrac{17}{24}f_4 +\tfrac{4}{3}f_5+\tfrac{2}{3}f_6+\tfrac{4}{3}f_7+\tfrac{1}{3}f_8). \quad (18)$$

Taking the next Simpson's 3/8 step then gives

$$\int_{x_8}^{x_{11}} f(x)\,dx = h(\tfrac{3}{8}f_8+\tfrac{9}{8}f_9+\tfrac{9}{8}f_{10}+\tfrac{3}{8}f_{11}). \quad (19)$$

Combining with the previous result gives

$$\int_{x_1}^{x_{11}} f(x)\,dx = h[\tfrac{3}{8}f_1+\tfrac{9}{8}f_2+\tfrac{9}{8}f_3+\tfrac{17}{24}f_4+\tfrac{4}{3}f_5 +\tfrac{2}{3}f_6+\tfrac{4}{3}f_7+(\tfrac{1}{3}+\tfrac{3}{8})f_8+\tfrac{9}{8}f_9+\tfrac{9}{8}f_{10}+\tfrac{3}{8}f_{11}]$$
$$= h(\tfrac{3}{8}f_1+\tfrac{9}{8}f_2+\tfrac{9}{8}f_3+\tfrac{17}{24}f_4+\tfrac{4}{3}f_5+\tfrac{2}{3}f_6+\tfrac{4}{3}f_7 +\tfrac{17}{24}f_8+\tfrac{9}{8}f_9+\tfrac{9}{8}f_{10}+\tfrac{3}{8}f_{11}), \quad (20)$$

where terms up to f_{10} have now been completely determined. Continuing gives

$$h(\tfrac{3}{8}f_1+\tfrac{9}{8}f_2+\tfrac{9}{8}f_3+\tfrac{17}{24}f_4+\tfrac{4}{3}f_5+\tfrac{2}{3}f_6+\ldots +\tfrac{2}{3}f_{n-5}+\tfrac{4}{3}f_{n-4}+\tfrac{17}{24}f_{n-3}+\tfrac{9}{8}f_{n-2}+\tfrac{9}{8}f_{n-1}+\tfrac{3}{8}f_n). \quad (21)$$

Now average with the 3-point result

$$h(\tfrac{1}{3}f_1 + \tfrac{4}{3}f_2 + \tfrac{2}{3}f_3 + \tfrac{4}{3}f_4 + \tfrac{2}{3}f_5 + \tfrac{4}{3}f_{n-1} + \tfrac{1}{3}f_n) \quad (22)$$

to obtain

$$h[\tfrac{17}{48}f_1+\tfrac{59}{48}f_2+\tfrac{43}{48}f_4+\tfrac{49}{48}f_4+(f_5+f_6+\ldots+f_{n-5}+f_{n-4}) + \tfrac{49}{48}f_{n-3} + \tfrac{43}{38}f_{n-2} + \tfrac{59}{48}f_{n-1} + \tfrac{17}{48}f_n] + \mathcal{O}(n^{-4}). \quad (23)$$

Note that all the middle terms now have unity COEFFICIENTS. Similarly, combining a 4-point with the (2+4)-point rule gives

$$h(\tfrac{5}{12}f_1+\tfrac{13}{12}f_2+f_3+f_4+\ldots+f_{n-3}+f_{n-2}+\tfrac{13}{12}f_{n-1}+\tfrac{5}{12}) + \mathcal{O}(n^{-3}). \quad (24)$$

Other Newton-Cotes rules occasionally encountered include DURAND'S RULE

$$\int_{x_1}^{x_n} f(x)\,dx = h(\tfrac{2}{5}f_1 + \tfrac{11}{10}f_2 + f_3 + \ldots + f_{n-2} + \tfrac{11}{10}f_{n-1} + \tfrac{2}{5}f_n) \quad (25)$$

(Beyer 1987), HARDY'S RULE

$$\int_{x_0-3h}^{x_0+3h} f(x)\,dx = \tfrac{1}{100}h(28f_{-3} + 162f_{-2} + 22f_0 + 162f_2 + 28f_3) + \tfrac{9}{1400}h^7[2f^{(4)}(\xi_2) - h^2 f^{(8)}(\xi_1)], \quad (26)$$

and WEDDLE'S RULE

$$\int_{x_1}^{x_{6n}} f(x)\,dx = \tfrac{3}{10}h(f_1 + 5f_2 + f_3 + 6f_4 + 5f_5 + f_6 + \ldots + 5f_{6n-1} + f_{6n}) \quad (27)$$

(Beyer 1987).

The open Newton-Cotes rules use points outside the integration interval, yielding the 1-point

$$\int_{x_0}^{x_2} f(x)\,dx = 2hf_1, \quad (28)$$

2-point

$$\int_{x_0}^{x_3} f(x)\,dx = \int_{x_1-h}^{x_1+2h} P_2(x)\,dx = \frac{1}{2h}(f_2 - f_1)[x^2]_{x_1-h}^{x_1+2h} + \left(f_1 + \frac{x_1}{h}f_1 - \frac{x_1}{h}f_2\right)[x]_{x_1-h}^{x_1+2h} = \tfrac{3}{2}h(f_1+f_2) + \tfrac{1}{4}h^3 f''(\xi), \quad (29)$$

3-point

$$\int_{x_0}^{x_4} f(x)\,dx = \tfrac{4}{3}h(2f_1 - f_2 + 2f_3) + \tfrac{28}{90}h^5 f^{(4)}(\xi), \quad (30)$$

4-point

$$\int_{x_0}^{x_5} f(x)\,dx = \tfrac{5}{24}h(11f_1+f_2+f_3+11f_4)+\tfrac{95}{144}h^5 f^{(4)}(\xi), \quad (31)$$

5-point

$$\int_{x_0}^{x_6} f(x)\,dx = \tfrac{6}{20}h(11f_1 - 14f_2 + 26f_3 - 14f_4 + 11f_5) - \tfrac{41}{140}h^7 f^{(6)}(\xi), \quad (32)$$

6-point

$$\int_{x_0}^{x_7} f(x)\,dx = \tfrac{7}{1440}h(611f_1 - 453f_2 + 562f_3 + 562f_4 - 453f_5 + 611f_6) - \tfrac{5257}{8640}h^7 f^{(6)}(\xi), \quad (33)$$

and 7-point

$$\int_{x_0}^{x_8} f(x)\,dx = \tfrac{8}{945}h(460f_1 - 954f_2 + 2196f_3 - 2459f_4 + 2196f_5 - 954f_6 + 460f_7) - \tfrac{3956}{14175}h^9 f^{(8)}(\xi) \quad (34)$$

rules.

A 2-point open extended formula is

$$\int_{x_1}^{x_n} f(x)\,dx = h[(\tfrac{1}{2}f_1 + f_2 + \ldots + f_{n-1} + \tfrac{1}{2}f_n) + \tfrac{1}{24}(-f_0 + f_2 + f_{n-1} + f_{n+1})] + \frac{11(n+1)}{720}h^5 f^{(4)}(\xi). \quad (35)$$

Single interval extrapolative rules estimate the integral in an interval based on the points around it. An example of such a rule is

$$hf_1 + \mathcal{O}(h^2 f') \quad (36)$$

$$\tfrac{1}{2}h(3f_1 - f_2) + \mathcal{O}(h^3 f'') \quad (37)$$

$$\tfrac{1}{12}h(23f_1 - 16f_2 + 5f_3) + \mathcal{O}(h^4 f^{(3)}) \quad (38)$$

$$\tfrac{1}{24}h(55f_1 - 59f_2 + 37f_3 - 9f_4) + \mathcal{O}(h^5 f^{(4)}). \quad (39)$$

see also BODE'S RULE, DIFFERENCE EQUATION, DURAND'S RULE, FINITE DIFFERENCE, GAUSSIAN QUADRATURE, HARDY'S RULE, LAGRANGE INTERPOLATING POLYNOMIAL, NUMERICAL INTEGRATION, SIMPSON'S RULE, SIMPSON'S 3/8 RULE, TRAPEZOIDAL RULE, WEDDLE'S RULE

References

Abramowitz, M. and Stegun, C. A. (Eds.). "Integration." §25.4 in *Handbook of Mathematical Functions with Formulas, Graphs, and Mathematical Tables, 9th printing.* New York: Dover, pp. 885–887, 1972.

Beyer, W. H. (Ed.) *CRC Standard Mathematical Tables, 28th ed.* Boca Raton, FL: CRC Press, p. 127, 1987.

Hildebrand, F. B. *Introduction to Numerical Analysis.* New York: McGraw-Hill, pp. 160–161, 1956.

Press, W. H.; Flannery, B. P.; Teukolsky, S. A.; and Vetterling, W. T. "Classical Formulas for Equally Spaced Abscissas." §4.1 in *Numerical Recipes in FORTRAN: The Art of Scientific Computing, 2nd ed.* Cambridge, England: Cambridge University Press, pp. 124–130, 1992.

Newton's Diverging Parabolas

Curves with CARTESIAN equation

$$ay^2 = x(x^2 - 2bx + c)$$

with $a > 0$. The above equation represents the third class of Newton's classification of CUBIC CURVES, which Newton divided into five species depending on the ROOTS of the cubic in x on the right-hand side of the equation. Newton described these cases as having the following characteristics:

1. "All the ROOTS are REAL and unequal. Then the Figure is a diverging Parabola of the Form of a Bell, with an Oval at its Vertex.
2. Two of the ROOTS are equal. A PARABOLA will be formed, either Nodated by touching an Oval, or Punctate, by having the Oval infinitely small.
3. The three ROOTS are equal. This is the NEILIAN PARABOLA, commonly called SEMI-CUBICAL.
4. Only one REAL ROOT. If two of the ROOTS are impossible, there will be a Pure PARABOLA of a Bell-like Form"

(MacTutor Archive).

References

MacTutor History of Mathematics Archive. "Newton's Diverging Parabolas." http://www-groups.dcs.st-and.ac.uk/~history/Curves/Newtons.html.

Newton's Divided Difference Interpolation Formula

Let

$$\pi_n(x) \equiv \prod_{i=1}^{n}(x - x_n), \tag{1}$$

then

$$f(x) = f_0 + \sum_{k=1}^{n} x_{k-1}(x)[x_0, x_1, \ldots, x_k] + R_n, \tag{2}$$

where $[x_1, \ldots]$ is a DIVIDED DIFFERENCE, and the remainder is

$$R_n(x) = \pi_n(x)[x_0, \ldots, x_n, x] = \pi_n(x)\frac{f^{(n+1)}(\xi)}{(n+1)} \tag{3}$$

for $x_0 < \xi < x_n$.

see also DIVIDED DIFFERENCE, FINITE DIFFERENCE

References

Abramowitz, M. and Stegun, C. A. (Eds.). *Handbook of Mathematical Functions with Formulas, Graphs, and Mathematical Tables, 9th printing.* New York: Dover, p. 880, 1972.

Hildebrand, F. B. *Introduction to Numerical Analysis.* New York: McGraw-Hill, pp.43–44 and 62–63, 1956.

Newton's Forward Difference Formula

A FINITE DIFFERENCE identity giving an interpolated value between tabulated points $\{f_p\}$ in terms of the first value f_0 and the POWERS of the FORWARD DIFFERENCE Δ. For $a \in [0,1]$, the formula states

$$f_a = f_0 + a\Delta + \tfrac{1}{2!}a(a-1)\Delta^2 + \tfrac{1}{3!}a(a-1)(a-2)\Delta^3 + \ldots.$$

When written in the form

$$f(x+a) = \sum_{n=0}^{\infty} \frac{(a)_n \Delta^n f(x)}{n!}$$

with $(a)_n$ the POCHHAMMER SYMBOL, the formula looks suspiciously like a finite analog of a TAYLOR SERIES expansion. This correspondence was one of the motivating forces for the development of UMBRAL CALCULUS.

The DERIVATIVE of Newton's forward difference formula gives MARKOFF'S FORMULAS.

see also FINITE DIFFERENCE, MARKOFF'S FORMULAS, NEWTON'S BACKWARD DIFFERENCE FORMULA, NEWTON'S DIVIDED DIFFERENCE INTERPOLATION FORMULA

References

Abramowitz, M. and Stegun, C. A. (Eds.). *Handbook of Mathematical Functions with Formulas, Graphs, and Mathematical Tables, 9th printing.* New York: Dover, p. 880, 1972.

Beyer, W. H. *CRC Standard Mathematical Tables, 28th ed.* Boca Raton, FL: CRC Press, p. 432, 1987.

Newton's Formulas

Let a TRIANGLE have side lengths a, b, and c with opposite angles A, B, and C. Then

$$\frac{b+c}{a} = \frac{\cos[\frac{1}{2}(B-C)]}{\sin(\frac{1}{2}A)}$$

$$\frac{c+a}{b} = \frac{\cos[\frac{1}{2}(C-A)]}{\sin(\frac{1}{2}B)}$$

$$\frac{a+b}{c} = \frac{\cos[\frac{1}{2}(A-B)]}{\sin(\frac{1}{2}C)}.$$

see also MOLLWEIDE'S FORMULAS, TRIANGLE

References

Beyer, W. H. *CRC Standard Mathematical Tables, 28th ed.* Boca Raton, FL: CRC Press, p. 146, 1987.

Newton's Identities

see also NEWTON'S RELATIONS

Newton's Iteration

An algorithm for the SQUARE ROOT of a number r quadratically as $\lim_{n\to\infty} x_n$,

$$x_{n+1} = \frac{1}{2}\left(x_n + \frac{r}{x_n}\right),$$

where $x_0 = 1$. The first few approximants to $\sqrt{n}$ are given by

$$1, \tfrac{1}{2}(1+n), \frac{1+6n+n^2}{4(n+1)}, \frac{1+26n+70n^2+28n^3+n^4}{8(1+n)(1+6n+n^2)}, \ldots.$$

For $\sqrt{2}$, this gives the convergents as 1, 3/2, 17/12, 577/408, 665857/470832, ...

see also SQUARE ROOT

Newton's Method

A ROOT-finding ALGORITHM which uses the first few terms of the TAYLOR SERIES in the vicinity of a suspected ROOT to zero in on the root. The TAYLOR SERIES of a function $f(x)$ about the point $x+\epsilon$ is given by

$$f(x+\epsilon) = f(x) + f'(x)\epsilon + \tfrac{1}{2}f''(x)\epsilon^2 + \ldots. \quad (1)$$

Keeping terms only to first order,

$$f(x+\epsilon) \approx f(x) + f'(x)\epsilon. \quad (2)$$

This expression can be used to estimate the amount of offset ϵ needed to land closer to the root starting from an initial guess x_0. Setting $f(x_0+\epsilon) = 0$ and solving (2) for ϵ gives

$$\epsilon_0 = -\frac{f(x_0)}{f'(x_0)}, \quad (3)$$

which is the first-order adjustment to the ROOT's position. By letting $x_1 = x_0 + \epsilon_0$, calculating a new ϵ_1, and so on, the process can be repeated until it converges to a root.

Unfortunately, this procedure can be unstable near a horizontal ASYMPTOTE or a LOCAL MINIMUM. However, with a good initial choice of the ROOT's position, the algorithm can by applied iteratively to obtain

$$x_{n+1} = x_n - \frac{f(x_n)}{f'(x_n)} \quad (4)$$

for $n = 1, 2, 3, \ldots$.

The error ϵ_{n+1} after the $(n+1)$st iteration is given by

$$\begin{aligned}\epsilon_{n+1} &= \epsilon_n + (x_{n+1} - x_n)\\ &= \epsilon_n - \frac{f(x_n)}{f'(x_n)}. \quad (5)\end{aligned}$$

But

$$\begin{aligned}f(x_n) &= f(x) + f'(x)\epsilon_n + \tfrac{1}{2}f''(x){\epsilon_n}^2 + \ldots\\ &= f'(x)\epsilon_n + \tfrac{1}{2}f''(x){\epsilon_n}^2 + \ldots \quad (6)\\ f'(x_n) &= f'(x) + f''(x)\epsilon_n + \ldots, \quad (7)\end{aligned}$$

so

$$\begin{aligned}\frac{f(x_n)}{f'(x_x)} &= \frac{f'(x)\epsilon_n + \frac{1}{2}f''(x){\epsilon_n}^2 + \ldots}{f'(x) + f''(x)\epsilon_n + \ldots}\\ &\approx \frac{f'(x)\epsilon + \frac{1}{2}f''(x){\epsilon_n}^2}{f'(x) + f''(x)\epsilon_n} = \epsilon_n + \frac{f''(x)}{2f'(x)}{\epsilon_n}^2, \quad (8)\end{aligned}$$

and (5) becomes

$$\epsilon_{n+1} = \epsilon_n - \left[\epsilon_n + \frac{f''(x)}{2f'(x)}{\epsilon_n}^2\right] = -\frac{f''(x)}{2f'(x)}{\epsilon_n}^2. \quad (9)$$

Therefore, when the method converges, it does so quadratically.

A FRACTAL is obtained by applying Newton's method to finding a ROOT of $z^n - 1 = 0$ (Mandelbrot 1983, Gleick 1988, Peitgen and Saupe 1988, Press *et al.* 1992, Dickau 1997). Iterating for a starting point z_0 gives

$$z_{i+1} = z_i - \frac{{z_i}^n - 1}{n{z_i}^{n-1}}. \quad (10)$$

Since this is an nth order POLYNOMIAL, there are n ROOTS to which the algorithm can converge.

Coloring the BASIN OF ATTRACTION (the set of initial points z_0 which converge to the same ROOT) for each ROOT a different color then gives the above plots, corresponding to $n = 2$, 3, 4, and 5.

see also HALLEY'S IRRATIONAL FORMULA, HALLEY'S METHOD, HOUSEHOLDER'S METHOD, LAGUERRE'S METHOD

References
Abramowitz, M. and Stegun, C. A. (Eds.). *Handbook of Mathematical Functions with Formulas, Graphs, and Mathematical Tables, 9th printing.* New York: Dover, p. 18, 1972.
Acton, F. S. Ch. 2 in *Numerical Methods That Work.* Washington, DC: Math. Assoc. Amer., 1990.
Arfken, G. *Mathematical Methods for Physicists, 3rd ed.* Orlando, FL: Academic Press, pp. 963–964, 1985.
Dickau, R. M. "Basins of Attraction for $z^5 = 1$ Using Newton's Method in the Complex Plane." http://forum.swarthmore.edu/advanced/robertd/newtons.html.
Dickau, R. M. "Variations on Newton's Method." http://forum . swarthmore . edu / advanced / robertd / newnewton.html.
Dickau, R. M. "Compilation of Iterative and List Operations." *Mathematica J.* **7**, 14–15, 1997.
Gleick, J. *Chaos: Making a New Science.* New York: Penguin Books, plate 6 (following pp. 114) and p. 220, 1988.
Householder, A. S. *Principles of Numerical Analysis.*ew York: McGraw-Hill, pp. 135–138, 1953.
Mandelbrot, B. B. *The Fractal Geometry of Nature.* San Francisco, CA: W. H. Freeman, 1983.
Ortega, J. M. and Rheinboldt, W. C. *Iterative Solution of Nonlinear Equations in Several Variables.* New York: Academic Press, 1970.
Peitgen, H.-O. and Saupe, D. *The Science of Fractal Images.* New York: Springer-Verlag, 1988.
Press, W. H.; Flannery, B. P.; Teukolsky, S. A.; and Vetterling, W. T. "Newton-Raphson Method Using Derivatives" and "Newton-Raphson Methods for Nonlinear Systems of Equations." §9.4 and 9.6 in *Numerical Recipes in FORTRAN: The Art of Scientific Computing, 2nd ed.* Cambridge, England: Cambridge University Press, pp. 355–362 and 372–375, 1992.
Ralston, A. and Rabinowitz, P. §8.4 in *A First Course in Numerical Analysis, 2nd ed.* New York: McGraw-Hill, 1978.

Newton Number

see KISSING NUMBER

Newton's Parallelogram

Approximates the possible values of y in terms of x if

$$\sum_{i,j=0}^{n} a_{ij} x^i y^i = 0.$$

Newton-Raphson Fractal

see NEWTON'S METHOD

Newton-Raphson Method

see NEWTON'S METHOD

Newton's Relations

Let s_i be the sum of the products of distinct ROOTS r_j of the POLYNOMIAL equation of degree n

$$a_n x^n + a_{n-1} x^{n-1} + \ldots + a_1 x + a_0 = 0, \quad (1)$$

where the roots are taken i at a time (i.e., s_i is defined as the ELEMENTARY SYMMETRIC FUNCTION $\Pi_i(r_1, \ldots, r_n)$) s_i is defined for $i = 1, \ldots, n$. For example, the first few values of s_i are

$$s_1 = r_1 + r_2 + r_3 + r_4 + \ldots \quad (2)$$

$$s_2 = r_1 r_2 + r_1 r_3 + r_1 r_4 + r_2 r_3 + \ldots \quad (3)$$

$$s_3 = r_1 r_2 r_3 + r_1 r_2 r_4 + r_2 r_3 r_4 + \ldots, \quad (4)$$

and so on. Then

$$s_i = (-1)^i \frac{a_{n-i}}{a_n}. \quad (5)$$

This can be seen for a second DEGREE POLYNOMIAL by multiplying out,

$$a_2 x^2 + a_1 x + a_0 = a_2(x - r_1)(x - r_2) = a_2[x^2 - (r_1 + r_2)x + r_1 r_2], \quad (6)$$

so

$$s_1 = \sum_{i=1}^{2} r_i = r_1 + r_2 = -\frac{a_1}{a_2} \quad (7)$$

$$s_2 = \sum_{\substack{i,j=1 \\ i \neq j}}^{2} r_i r_j = r_1 r_2 = \frac{a_0}{a_2}, \quad (8)$$

and for a third DEGREE POLYNOMIAL,

$$a_3 x^3 + a_2 x^2 + a_1 x + a_0 = a_3(x - r_1)(x - r_2)(x - r_3)$$
$$= a_3[x^3 - (r_1 + r_2 + r_3)x^2 + (r_1 r_2 + r_1 r_3 + r_2 r_3)x - r_1 r_2 r_3], \quad (9)$$

so

$$s_1 = \sum_{i=1}^{3} r_i = -\frac{a_2}{a_3} \quad (10)$$

$$s_2 = \sum_{\substack{i,j \\ i \neq j}}^{3} r_i r_j = r_1 r_2 + r_1 r_3 + r_2 r_3 = \frac{a_1}{a_3} \quad (11)$$

$$s_3 = \sum_{\substack{i,j,k \\ i \neq j \neq k}}^{3} r_i r_j r_k = r_1 r_2 r_3 = -\frac{a_0}{a_3}. \quad (12)$$

see also ELEMENTARY SYMMETRIC FUNCTION

References
Coolidge, J. L. *A Treatise on Algebraic Plane Curves.* New York: Dover, pp. 1–2, 1959.

Newton's Theorem

If each of two nonparallel transversals with nonminimal directions meets a given curve in finite points only, then the ratio of products of the distances from the two sets of intersections to the intersection of the lines is independent of the position of the latter point.

References
Coolidge, J. L. *A Treatise on Algebraic Plane Curves.* New York: Dover, p. 189, 1959.

Newtonian Form

see NEWTON'S DIVIDED DIFFERENCE INTERPOLATION FORMULA

Next Prime

The next prime function $NP(n)$ gives the smallest PRIME larger than n. The function can be given explicitly as

$$NP(n) = p_{1+\pi(n)},$$

where p_i is the ith PRIME and $\pi(n)$ is the PRIME COUNTING FUNCTION. For $n = 1, 2, \ldots$ the values are 2, 3, 5, 5, 7, 7, 11, 11, 11, 11, 13, 13, 17, 17, 17, 17, 19, ... (Sloane's A007918).

see also FORTUNATE PRIME, PRIME COUNTING FUNCTION, PRIME NUMBER

References

Sloane, N. J. A. Sequence A007918 in "An On-Line Version of the Encyclopedia of Integer Sequences."

Nexus Number

A FIGURATE NUMBER built up of the nexus of cells less than n steps away from a given cell. In k-D, the $(n+1)$th nexus number is given by

$$N_{n+1}(k) = \sum_{i=0}^{k} \binom{k}{i} n^i,$$

where $\binom{n}{n}$ is a BINOMIAL COEFFICIENT. The first few k-dimensional nexus numbers are given in the table below.

k	N_{n+1}	name
0	1	unit
1	$1+2n$	odd number
2	$1+3n+3n^2$	hex number
3	$1+4n+6n^2+4n^3$	rhombic dodecahedral number

see also HEX NUMBER, ODD NUMBER, RHOMBIC DODECAHEDRAL NUMBER

References

Conway, J. H. and Guy, R. K. *The Book of Numbers.* New York: Springer-Verlag, pp. 53–54, 1996.

Neyman-Pearson Lemma

If there exists a critical region C of size α and a NONNEGATIVE constant k such that

$$\frac{\prod_{i=1}^{n} f(x_i|\theta_1)}{\prod_{i=1}^{n} f(x_i|\theta_0)} \geq k$$

for points in C and

$$\frac{\prod_{i=1}^{n} f(x_i|\theta_1)}{\prod_{i=1}^{n} f(x_i|\theta_0)} \leq k$$

for points not in C, then C is a best critical region of size α.

References

Hoel, P. G.; Port, S. C.; and Stone, C. J. "Testing Hypotheses." Ch. 3 in *Introduction to Statistical Theory.* New York: Houghton Mifflin, pp. 56–67, 1971.

Nicholson's Formula

Let $J_\nu(z)$ be a BESSEL FUNCTION OF THE FIRST KIND, $Y_\nu(z)$ a BESSEL FUNCTION OF THE SECOND KIND, and $K_\nu(z)$ a MODIFIED BESSEL FUNCTION OF THE FIRST KIND. Also let $\Re[z] > 0$. Then

$$J_\nu^2(z) + Y_\nu^2(z) = \frac{8}{\pi^2} \int_0^\infty K_0(2z \sinh t) \cos(2\nu t)\, dt.$$

see also DIXON-FERRAR FORMULA, WATSON'S FORMULA

References

Gradshteyn, I. S. and Ryzhik, I. M. Eqn. 6.664.4 in *Tables of Integrals, Series, and Products, 5th ed.* San Diego, CA: Academic Press, p. 727, 1979.

Iyanaga, S. and Kawada, Y. (Eds.). *Encyclopedic Dictionary of Mathematics.* Cambridge, MA: MIT Press, p. 1476, 1980.

Nicomachus's Theorem

The nth CUBIC NUMBER n^3 is a sum of n consecutive ODD NUMBERS, for example

$$\begin{aligned} 1^3 &= 1 \\ 2^3 &= 3+5 \\ 3^3 &= 7+9+11 \\ 4^3 &= 13+15+17+19, \end{aligned}$$

etc. This identity follows from

$$\sum_{i=1}^{n} [n(n-1) - 1 + 2i] = n^3.$$

It also follows from this fact that

$$\sum_{k=1}^{n} k^3 = \left(\sum_{k=1}^{n} k\right)^2.$$

see also ODD NUMBER THEOREM

Nicomedes' Conchoid

see CONCHOID OF NICOMEDES

Nielsen-Ramanujan Constants

N.B. A detailed on-line essay by S. Finch was the starting point for this entry.

N. Nielsen (1909) and Ramanujan (Berndt 1985) considered the integrals

$$a_k = \int_1^2 \frac{(\ln x)^k}{x-1}\, dx. \qquad (1)$$

They found the values for $k = 1$ and 2. The general constants for $k > 3$ were found by V. Adamchik (Finch)

$$a_p = p!\zeta(p+1) - \frac{p(\ln 2)^{p+1}}{p+1} - p!\sum_{k=0}^{p-1} \frac{\mathrm{Li}_{p+1-k}(\frac{1}{2})(\ln 2)^k}{k!}, \quad (2)$$

where $\zeta(z)$ is the RIEMANN ZETA FUNCTION and $\mathrm{Li}_n(x)$ is the POLYLOGARITHM. The first few values are

$$a_1 = \tfrac{1}{2}\zeta(2) = \tfrac{1}{12}\pi^2 \quad (3)$$
$$a_2 = \tfrac{1}{4}\zeta(3) \quad (4)$$
$$a_3 = \tfrac{1}{15}\pi^4 + \tfrac{1}{4}\pi^2(\ln 2)^2 - \tfrac{1}{4}(\ln 2)^4 - 6\,\mathrm{Li}_4(\tfrac{1}{2}) - \tfrac{21}{4}\ln 2\zeta(3) \quad (5)$$
$$a_4 = \tfrac{2}{3}\pi^2(\ln 2)^3 - \tfrac{4}{5}(\ln 2)^5 - 24\ln 2\,\mathrm{Li}_4(\tfrac{1}{2}) - 24\,\mathrm{Li}_5(\tfrac{1}{2}) - \tfrac{21}{2}(\ln 2)^2\zeta(3) + 24\zeta(5). \quad (6)$$

see also POLYLOGARITHM, RIEMANN ZETA FUNCTION

References

Berndt, B. C. *Ramanujan's Notebooks, Part I.* New York: Springer-Verlag, 1985.

Finch, S. "Favorite Mathematical Constants." http://www.mathsoft.com/asolve/constant/nielram/nielram.html.

Flajolet, P. and Salvy, B. "Euler Sums and Contour Integral Representation." Submitted to *Experim. Math* 1997. http://pauillac.inria.fr/algo/flajolet/Publications/publist.html.

Nielsen's Spiral

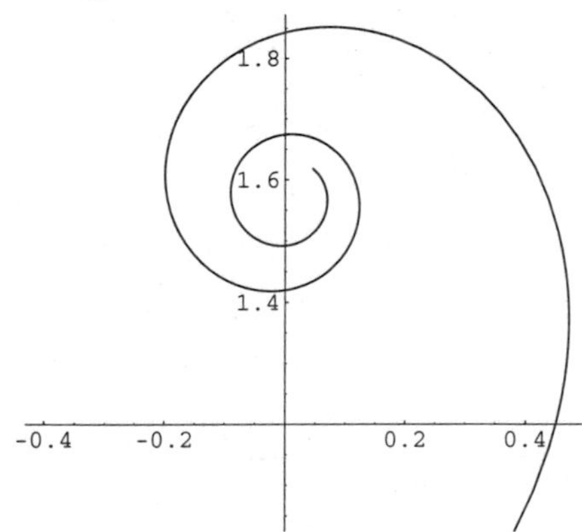

The SPIRAL with parametric equations

$$x(t) = a\,\mathrm{ci}(t) \quad (1)$$
$$y(t) = a\,\mathrm{si}(t), \quad (2)$$

where $\mathrm{ci}(t)$ is the COSINE INTEGRAL and $\mathrm{si}(t)$ is the SINE INTEGRAL. The CESÀRO EQUATION is

$$\kappa = \frac{e^{s/a}}{a}. \quad (3)$$

see also CORNU SPIRAL, COSINE INTEGRAL, SINE INTEGRAL

References

Gray, A. *Modern Differential Geometry of Curves and Surfaces.* Boca Raton, FL: CRC Press, p. 119, 1993.

Nil Geometry

The GEOMETRY of the LIE GROUP consisting of REAL MATRICES of the form

$$\begin{bmatrix} 1 & x & y \\ 0 & 1 & z \\ 0 & 0 & 1 \end{bmatrix},$$

i.e., the HEISENBERG GROUP.

see also HEISENBERG GROUP, LIE GROUP, THURSTON'S GEOMETRIZATION CONJECTURE

Nilmanifold

Let N be a NILPOTENT, connected, SIMPLY CONNECTED LIE GROUP, and let D be a discrete SUBGROUP of N with compact right QUOTIENT SPACE. Then N/D is called a nilmanifold.

Nilpotent Element

An element B of a RING is nilpotent if there exists a POSITIVE INTEGER k for which $B^k = 0$.

see also ENGEL'S THEOREM

Nilpotent Group

A GROUP G for which the chain of groups

$$I = Z_0 \subseteq Z_1 \subseteq \ldots \subseteq Z_n$$

with Z_{k+1}/Z_k (equal to the CENTER of G/Z_k) terminates finitely with $G = Z_n$ is called a nilpotent group.

see also CENTER (GROUP), NILPOTENT LIE GROUP

Nilpotent Lie Group

A LIE GROUP which has a simply connected covering group HOMEOMORPHIC to $\mathbb{R}^n$. The prototype is any connected closed subgroup of upper triangular COMPLEX matrices with 1s on the diagonal. The HEISENBERG GROUP is such a group.

References

Knapp, A. W. "Group Representations and Harmonic Analysis, Part II." *Not. Amer. Math. Soc.* **43**, 537–549, 1996.

Nilpotent Matrix

A SQUARE MATRIX whose EIGENVALUES are all 0. A related definition is a SQUARE MATRIX M such that M^n is $\mathsf{0}$ for some POSITIVE integral POWER.

see also EIGENVALUE, SQUARE MATRIX

Nim

A game, also called TACTIX, which is played by the following rules. Given one or more piles (NIM-HEAPS), players alternate by taking all or some of the counters in a single heap. The player taking the last counter or stack of counters is the winner. Nim-like games are also called TAKE-AWAY GAMES and DISJUNCTIVE GAMES.

If optimal strategies are used, the winner can be determined from any intermediate position by its associated NIM-VALUE.

see also MISÈRE FORM, NIM-VALUE, WYTHOFF'S GAME

References
Ball, W. W. R. and Coxeter, H. S. M. *Mathematical Recreations and Essays, 13th ed.* New York: Dover, pp. 36–38, 1987.
Bogomolny, A. "The Game of Nim." http://www.cut-the-knot.com/bottom_nim.html.
Bouton, C. L. "Nim, A Game with a Complete Mathematical Theory." *Ann. Math. Princeton* **3**, 35–39, 1901–1902.
Gardner, M. "Nim and Hackenbush." Ch. 14 in *Wheels, Life, and other Mathematical Amusements.* New York: W. H. Freeman, 1983.
Hardy, G. H. and Wright, E. M. *An Introduction to the Theory of Numbers, 5th ed.* Oxford, England: Oxford University Press, pp. 117–120, 1990.
Kraitchik, M. "Nim." §3.12.2 in *Mathematical Recreations.* New York: W. W. Norton, pp. 86–88, 1942.

Nim-Heap

A pile of counters in a game of NIM.

Nim-Sum

see NIM-VALUE

Nim-Value

Every position of every IMPARTIAL GAME has a nim-value, making it equivalent to a NIM-HEAP. To find the nim-value (also called the SPRAGUE-GRUNDY NUMBER), take the MEX of the nim-values of the possible moves. The nim-value can also be found by writing the number of counters in each heap in binary, adding without carrying, and replacing the digits with their values mod 2. If the nim-value is 0, the position is SAFE; otherwise, it is UNSAFE. With two heaps, safe positions are (x, x) where $x \in [1, 7]$. With three heaps, (1, 2, 3), (1, 4, 5), (1, 6, 7), (2, 4, 6), (2, 5, 7), and (3, 4, 7).

see also GRUNDY'S GAME, IMPARTIAL GAME, MEX, NIM, SAFE, UNSAFE

References
Ball, W. W. R. and Coxeter, H. S. M. *Mathematical Recreations and Essays, 13th ed.* New York: Dover, pp. 36–38, 1987.
Grundy, P. M. "Mathematics and Games." *Eureka* **2**, 6–8, 1939.
Sprague, R. "Über mathematische Kampfspiele." *Tôhoku J. Math.* **41**, 438–444, 1936.

Nine-Point Center

The center F (or N) of the NINE-POINT CIRCLE. It has TRIANGLE CENTER FUNCTION

$$\begin{aligned}\alpha &= \cos(B - C) = \cos A + 2\cos B\cos C \\ &= bc[a^2b^2 + a^2c^2 + (b^2 - c^2)^2],\end{aligned}$$

and is the MIDPOINT of the line between the CIRCUMCENTER C and ORTHOCENTER H. It lies on the EULER LINE.

see also EULER LINE, NINE-POINT CIRCLE, NINE-POINT CONIC

References
Carr, G. S. *Formulas and Theorems in Pure Mathematics, 2nd ed.* New York: Chelsea, p. 624, 1970.
Dixon, R. *Mathographics.* New York: Dover, pp. 57–58, 1991.
Kimberling, C. "Central Points and Central Lines in the Plane of a Triangle." *Math. Mag.* **67**, 163–187, 1994.
Kimberling, C. "Nine-Point Center." http://www.evansville.edu/~ck6/tcenters/class/npcenter.html.

Nine-Point Circle

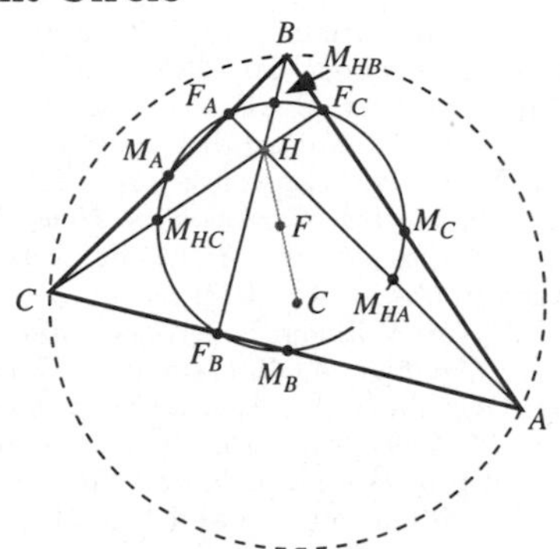

The CIRCLE, also called EULER'S CIRCLE and the FEUERBACH CIRCLE, which passes through the feet of the PERPENDICULAR F_A, F_B, and F_C dropped from the VERTICES of any TRIANGLE ΔABC on the sides opposite them. Euler showed in 1765 that it also passes through the MIDPOINTS M_A, M_B, M_C of the sides of ΔABC.

By FEUERBACH'S THEOREM, the nine-point circle also passes through the MIDPOINTS M_{HA}, M_{HB}, M_{HC} of the segments which join the VERTICES and the ORTHOCENTER H. These three triples of points make nine in all, giving the circle its name. The center F of the nine-point circle is called the NINE-POINT CENTER.

The RADIUS of the nine-point circle is $R/2$, where R is the CIRCUMRADIUS. The center of KIEPERT'S HYPERBOLA lies on the nine-point circle. The nine-point circle bisects any line from the ORTHOCENTER to a point on the CIRCUMCIRCLE. The nine-point circle of the INCENTER and EXCENTERS of a TRIANGLE is the CIRCUMCIRCLE.

The sum of the powers of the VERTICES with regard to the nine-point circle is

$$\tfrac{1}{4}({a_1}^2 + {a_2}^2 + {a_3}^2).$$

Also,

$$\overline{FA_1}^2 + \overline{FA_2}^2 + \overline{FA_3}^2 + \overline{FH}^2 = 3R^2,$$

where F is the NINE-POINT CENTER, A_i are the VERTICES, H is the ORTHOCENTER, and R is the CIRCUMRADIUS. All triangles inscribed in a given CIRCLE and having the same ORTHOCENTER have the same nine-point circle.

see also COMPLETE QUADRILATERAL, EIGHT-POINT CIRCLE THEOREM, FEUERBACH'S THEOREM, FONTENÉ THEOREMS, GRIFFITHS' THEOREM, NINE-POINT CENTER, NINE-POINT CONIC, ORTHOCENTRIC SYSTEM

References

Altshiller-Court, N. *College Geometry: A Second Course in Plane Geometry for Colleges and Normal Schools, 2nd ed., rev. enl.* New York: Barnes and Noble, pp. 93–97, 1952.

Brand, L. "The Eight-Point Circle and the Nine-Point Circle." *Amer. Math. Monthly* **51**, 84–85, 1944.

Coxeter, H. S. M. and Greitzer, S. L. *Geometry Revisited.* New York: Random House, pp. 20–22, 1967.

Dörrie, H. "The Feuerbach Circle." §28 in *100 Great Problems of Elementary Mathematics: Their History and Solutions.* New York: Dover, pp. 142–144, 1965.

Gardner, M. *Mathematical Carnival: A New Round-Up of Tantalizers and Puzzles from Scientific American.* New York: Vintage Books, p. 59, 1977.

Guggenbuhl, L. "Karl Wilhelm Feuerbach, Mathematician." Appendix to *Circles: A Mathematical View, rev. ed.* Washington, DC: Math. Assoc. Amer., pp. 89–100, 1995.

Johnson, R. A. *Modern Geometry: An Elementary Treatise on the Geometry of the Triangle and the Circle.* Boston, MA: Houghton Mifflin, pp. 165 and 195–212, 1929.

Lange, J. *Geschichte des Feuerbach'schen Kreises.* Berlin, 1894.

Mackay, J. S. "History of the Nine-Point Circle." *Proc. Edinburgh Math. Soc.* **11**, 19–61, 1892.

Ogilvy, C. S. *Excursions in Geometry.* New York: Dover, pp. 119–120, 1990.

Pedoe, D. *Circles: A Mathematical View, rev. ed.* Washington, DC: Math. Assoc. Amer., pp. 1–4, 1995.

Nine-Point Conic

A CONIC SECTION on which the MIDPOINTS of the sides of any COMPLETE QUADRANGLE lie. The three diagonal points also lie on this conic.

see also COMPLETE QUADRANGLE, CONIC SECTION, NINE-POINT CIRCLE

Nint

see NEAREST INTEGER FUNCTION

Nint Zeta Function

Let

$$S_N(s) = \sum_{n=1}^{\infty}[(n^{1/N})]^{-s}, \tag{1}$$

where $[x]$ denotes NINT, the INTEGER closest to x. For $s > 3$,

$$S_2(s) = 2\zeta(s-1) \tag{2}$$
$$S_3(s) = 3\zeta(s-2) + 4^{-s}\zeta(s) \tag{3}$$
$$S_4(s) = 4\zeta(s-3) + \zeta(s-1). \tag{4}$$

$S_N(n)$ is a POLYNOMIAL in π whose COEFFICIENTS are ALGEBRAIC NUMBERS whenever $n - N$ is ODD. The first few values are given explicitly by

$$S_3(4) = \frac{\pi^2}{2} + \frac{\pi^4}{23046} \tag{5}$$
$$S_5(6) = \frac{5\pi^2}{6} + \frac{\pi^4}{36} + \frac{\pi^6}{4^{12}} \times \left(\frac{1}{945} - \frac{170912 + 49928\sqrt{2}}{25}\sqrt{1 - \sqrt{\frac{1}{2}}}\right) \tag{6}$$
$$S_6(7) = \pi^2 + \frac{\pi^4}{18} + \frac{\pi^6}{2520} + \frac{246013 + 353664\sqrt{2}}{45}\frac{\pi^7}{2^{27}}. \tag{7}$$

References

Borwein, J. M.; Hsu, L. C.; Mabry, R.; Neu, K.; Roppert, J.; Tyler, D. B.; and de Weger, B. M. M. "Nearest Integer Zeta-Functions." *Amer. Math. Monthly* **101**, 579–580, 1994.

Nirenberg's Conjecture

If the GAUSS MAP of a complete minimal surface omits a NEIGHBORHOOD of the SPHERE, then the surface is a PLANE. This was proven by Osserman (1959). Xavier (1981) subsequently generalized the result as follows. If the GAUSS MAP of a complete MINIMAL SURFACE omits ≥ 7 points, then the surface is a PLANE.

see also GAUSS MAP, MINIMAL SURFACE, NEIGHBORHOOD

References

do Carmo, M. P. *Mathematical Models from the Collections of Universities and Museums* (Ed. G. Fischer). Braunschweig, Germany: Vieweg, p. 42, 1986.

Osserman, R. "Proof of a Conjecture of Nirenberg." *Comm. Pure Appl. Math.* **12**, 229–232, 1959.

Xavier, F. "The Gauss Map of a Complete Nonflat Minimal Surface Cannot Omit 7 Points on the Sphere." *Ann. Math.* **113**, 211–214, 1981.

Niven's Constant

N.B. A detailed on-line essay by S. Finch was the starting point for this entry.

Given a POSITIVE INTEGER $m > 1$, let its PRIME FACTORIZATION be written

$$m = p_1^{a_1} p_2^{a_2} p_3^{a_3} \cdots p_k^{a_k}. \tag{1}$$

Define the functions h and H by $h(1) = 1$, $H(1) = 1$, and

$$h(m) = \min(a_1, a_2, \ldots, a_k) \tag{2}$$
$$H(m) = \max(a_1, a_2, \ldots, a_k). \tag{3}$$

Then

$$\lim_{n\to\infty} \frac{1}{n}\sum_{m=1}^{n} h(m) = 1 \tag{4}$$

$$\lim_{n\to\infty} \frac{\sum_{m=1}^n h(m) - n}{\sqrt{n}} = \frac{\zeta(\frac{3}{2})}{\zeta(3)}, \tag{5}$$

where $\zeta(z)$ is the RIEMANN ZETA FUNCTION (Niven 1969). Niven (1969) also proved that

$$\lim_{n\to\infty} \frac{1}{n} \sum_{m=1}^{n} H(m) = C, \tag{6}$$

where

$$C = 1 + \left\{ \sum_{j=2}^{\infty} \left[1 - \frac{1}{\zeta(j)} \right] \right\} = 1.705221\ldots \tag{7}$$

(Sloane's A033150).

The CONTINUED FRACTION of Niven's constant is 1, 1, 2, 2, 1, 1, 4, 1, 1, 3, 4, 4, 8, 4, 1, ... (Sloane's A033151). The positions at which the digits 1, 2, ... first occur in the CONTINUED FRACTION are 1, 3, 10, 7, 47, 41, 34, 13, 140, 252, 20, ... (Sloane's A033152). The sequence of largest terms in the CONTINUED FRACTION is 1, 2, 4, 8, 11, 14, 29, 372, 559, ... (Sloane's A0033153), which occur at positions 1, 3, 7, 13, 20, 35, 51, 68, 96, ... (Sloane's A033154).

References
Finch, S. "Favorite Mathematical Constants." http://www.mathsoft.com/asolve/constant/niven/niven.html.
Le Lionnais, F. *Les nombres remarquables.* Paris: Hermann, p. 41, 1983.
Niven, I. "Averages of Exponents in Factoring Integers." *Proc. Amer. Math. Soc.* **22**, 356–360, 1969.
Plouffe, S. "The Niven Constant." http://www.lacim.uqam.ca/piDATA/niven.txt.

Niven Number

see HARSHAD NUMBER

Nobbs Points

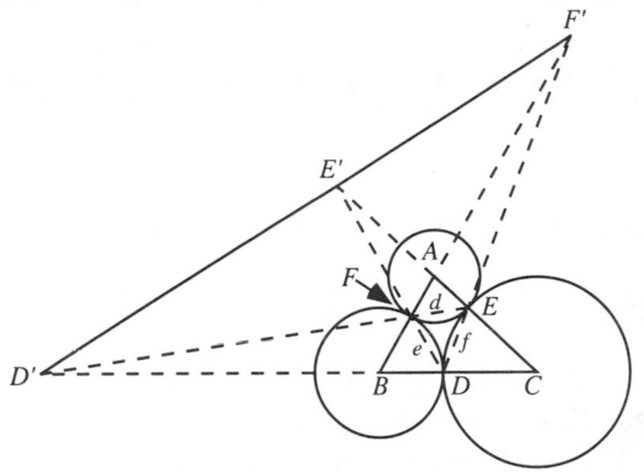

Given a TRIANGLE ΔABC, construct the CONTACT TRIANGLE ΔDEF. Then the Nobbs points are the three points D', E', and F' from which ΔABC and ΔDEF are PERSPECTIVE, as illustrated above. The Nobbs points are COLLINEAR and fall along the GERGONNE LINE.

see also COLLINEAR, CONTACT TRIANGLE, EVANS POINT, FLETCHER POINT, GERGONNE LINE, PERSPECTIVE TRIANGLES

References
Oldknow, A. "The Euler-Gergonne-Soddy Triangle of a Triangle." *Amer. Math. Monthly* **103**, 319–329, 1996.

Noble Number

A noble number is defined as an IRRATIONAL NUMBER which has a CONTINUED FRACTION which becomes an infinite sequence of 1s at some point,

$$\nu \equiv [a_1, a_2, \ldots, a_n, \bar{1}].$$

The prototype is the GOLDEN RATIO ϕ whose CONTINUED FRACTION is composed *entirely* of 1s, $[\bar{1}]$. Any noble number can written as

$$\nu = \frac{A_n + \phi A_{n-1}}{B_n + \phi B_{n+1}},$$

where A_k and B_k are the NUMERATOR and DENOMINATOR of the kth CONVERGENT of $[a_1, a_2, \ldots, a_n]$. The noble numbers are a SUBFIELD of $\mathbb{Q}(\sqrt{5})$.

see also NEAR NOBLE NUMBER

References
Hardy, G. H. and Wright, E. M. *An Introduction to the Theory of Numbers, 5th ed.* Oxford, England: Clarendon Press, p. 236, 1979.
Schroeder, M. "Noble and Near Noble Numbers." In *Fractals, Chaos, Power Laws: Minutes from an Infinite Paradise.* New York: W. H. Freeman, pp. 392–394, 1991.

Node (Algebraic Curve)

see ORDINARY DOUBLE POINT

Node (Fixed Point)

A FIXED POINT for which the STABILITY MATRIX has both EIGENVALUES of the same sign (i.e., both are POSITIVE or both are NEGATIVE). If $\lambda_1 < \lambda_2 < 0$, then the node is called STABLE; if $\lambda_1 > \lambda_2 > 0$, then the node is called an UNSTABLE NODE.

see also STABLE NODE, UNSTABLE NODE

Node (Graph)

Synonym for the VERTICES of a GRAPH, i.e., the points connected by EDGES.

see also ACNODE, CRUNODE, TACNODE

Noether's Fundamental Theorem

If two curves ϕ and ψ of MULTIPLICITIES $r_i \neq 0$ and $s_i \neq 0$ have only ordinary points or ordinary singular points and CUSPS in common, then every curve which has at least MULTIPLICITY

$$r_i + s_i - 1$$

at every point (distinct or infinitely near) can be written

$$f \equiv \phi\psi' + \psi\phi' = 0,$$

where the curves ϕ' and ψ' have MULTIPLICITIES at least $r_i - 1$ and $s_i - 1$.

References

Coolidge, J. L. *A Treatise on Algebraic Plane Curves.* New York: Dover, pp. 29–30, 1959.

Noether-Lasker Theorem

Let M be a finitely generated MODULE over a commutative NOETHERIAN RING R. Then there exists a finite set $\{N_i | 1 \leq i \leq l\}$ of submodules of M such that

1. $\cap_{i=1}^{l} N_i = 0$ and $\cap_{i \neq i_0} N_i$ is not contained in N_{i_0} for all $1 \leq i_0 \leq l$.
2. Each quotient M/N_i is primary for some prime P_i.
3. The P_i are all distinct for $1 \leq i \leq l$.
4. Uniqueness of the primary component N_i is equivalent to the statement that P_i does not contain P_j for any $j \neq i$.

Noether's Transformation Theorem

Any irreducible curve may be carried by a factorable CREMONA TRANSFORMATION into one with none but ordinary singular points.

References

Coolidge, J. L. *A Treatise on Algebraic Plane Curves.* New York: Dover, p. 207, 1959.

Noetherian Module

A MODULE M is Noetherian if every submodule is finitely generated.

see also NOETHERIAN RING

Noetherian Ring

An abstract commutative RING satisfying the abstract chain condition.

see also LOCAL RING, NOETHER-LASKER THEOREM

Noise

An error which is superimposed on top of a true signal. Noise may be random or systematic. Noise can be greatly reduced by transmitting signals digitally instead of in analog form because each piece of information is allowed only discrete values which are spaced farther apart than the contribution due to noise.

CODING THEORY studies how to encode information efficiently, and ERROR-CORRECTING CODES devise methods for transmitting and reconstructing information in the presence of noise.

see also ERROR

References

Davenport, W. B. and Root, W. L. *An Introduction to the Theory of Random Signals and Noise.* New York: IEEE Press, 1987.
McDonough, R. N. and Whalen, A. D. *Detection of Signals in Noise, 2nd ed.* Orlando, FL: Academic Press, 1995.
Pierce, J. R. *Symbols, Signals and Noise: The Nature and Process of Communication.* New York: Harper & Row, 1961.
Vainshtein, L. A. and Zubakov, V. D. *Extraction of Signals from Noise.* New York: Dover, 1970.
van der Ziel, A. *Noise: Sources, Characterization, Measurement.* New York: Prentice-Hall, 1954.
van der Ziel, A. *Noise in Measurement.* New York: Wiley, 1976.
Wax, N. *Selected Papers on Noise and Stochastic Processes.* New York: Dover, 1954.

Noise Sphere

A mapping of RANDOM NUMBER TRIPLES to points in SPHERICAL COORDINATES,

$$\theta = 2\pi X_n$$
$$\phi = \pi X_{n+1}$$
$$r = \sqrt{X_{n+2}}.$$

The graphical result can yield unexpected structure which indicates correlations between triples and therefore that the numbers are not truly RANDOM.

References

Pickover, C. A. *Computers and the Imagination.* New York: St. Martin's Press, 1991.
Pickover, C. A. "Computers, Randomness, Mind, and Infinity." Ch. 31 in *Keys to Infinity.* New York: W. H. Freeman, pp. 233–247, 1995.
Richards, T. "Graphical Representation of Pseudorandom Sequences." *Computers and Graphics* **13**, 261–262, 1989.

Nolid

An assemblage of faces forming a POLYHEDRON of zero VOLUME (Holden 1991, p. 124).

see also ACOPTIC POLYHEDRON

References

Holden, A. *Shapes, Space, and Symmetry.* New York: Dover, 1991.

Nome

Given a THETA FUNCTION, the nome is defined as

$$q(m) \equiv e^{\pi\tau i} = e^{-\pi K(1-m)/K(m)} \equiv e^{-\pi K'(m)/K(m)}, \quad (1)$$

where $K(k)$ is the complete ELLIPTIC INTEGRAL OF THE FIRST KIND, and m is the PARAMETER.

$$\vartheta_i(z,q) \equiv \vartheta(z|\tau) \quad (2)$$

$$\vartheta_i \equiv \vartheta(0,q). \quad (3)$$

Solving the nome for the PARAMETER m gives

$$m(q) = \frac{\vartheta_2{}^4(0,q)}{\vartheta_3{}^4(0,q)}, \tag{4}$$

where $\vartheta_i(z,q)$ is a THETA FUNCTION.

see also AMPLITUDE, CHARACTERISTIC (ELLIPTIC INTEGRAL), ELLIPTIC INTEGRAL, MODULAR ANGLE, MODULUS (ELLIPTIC INTEGRAL), PARAMETER

References

Abramowitz, M. and Stegun, C. A. (Eds.). *Handbook of Mathematical Functions with Formulas, Graphs, and Mathematical Tables, 9th printing.* New York: Dover, p. 591, 1972.

Nomogram

A graphical plot which can be used for solving certain types of equations.

References

Iyanaga, S. and Kawada, Y. (Eds.). "Nomograms." §282 in *Encyclopedic Dictionary of Mathematics.* Cambridge, MA: MIT Press, pp. 891–893, 1980.

Menzel, D. (Ed.). *Fundamental Formulas of Physics, Vol. 1.* New York: Dover, p. 141, 1960.

Nonagon

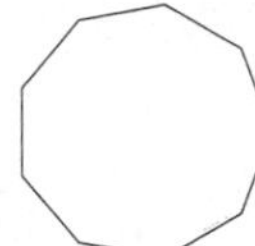

The unconstructible regular POLYGON with nine sides and SCHLÄFLI SYMBOL $\{9\}$. It is sometimes called an ENNEAGON.

Although the regular nonagon is not a CONSTRUCTIBLE POLYGON, Dixon (1991) gives several close approximations. While the ANGLE subtended by a side is $360°/9 = 40°$, Dixon gives constructions containing angles of $\tan^{-1}(5/6) \approx 39.8805571°$ and $2\tan^{-1}((\sqrt{3}-1)/2) \approx 40.207818°$.

Madachy (1979) illustrates how to construct a nonagon by folding and knotting a strip of paper.

see also NONAGRAM, TRIGONOMETRY VALUES—$\pi/9$

References

Dixon, R. *Mathographics.* New York: Dover, pp. 40–44, 1991.

Madachy, J. S. *Madachy's Mathematical Recreations.* New York: Dover, pp. 60–61, 1979.

Nonagonal Number

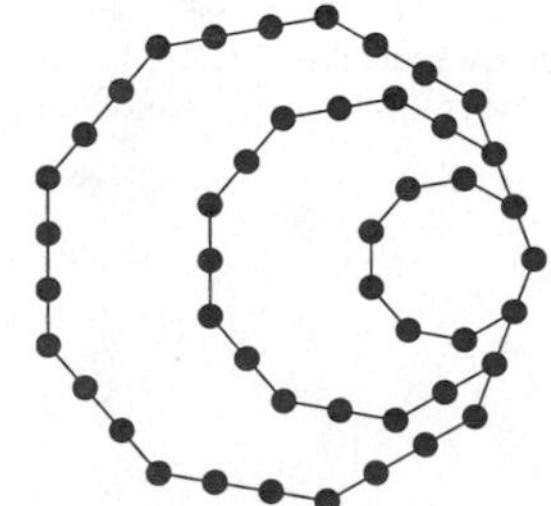

A FIGURATE NUMBER of the form $n(7n-5)/2$, also called an ENNEAGONAL NUMBER. The first few are 1, 9, 24, 46, 75, 111, 154, 204, ... (Sloane's A001106).

References

Sloane, N. J. A. Sequence A001106/M4604 in "An On-Line Version of the Encyclopedia of Integer Sequences."

Nonagram

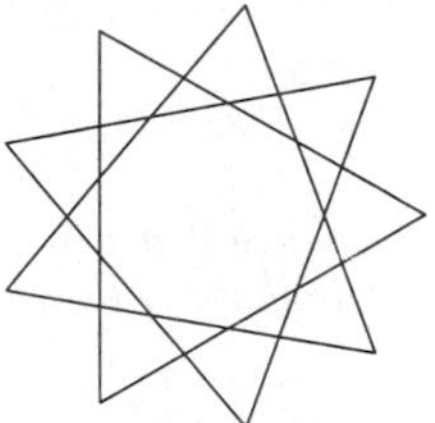

A STAR POLYGON composed of three EQUILATERAL TRIANGLES rotated at angles 0°, 40°, and 80°. It has been called the STAR OF GOLIATH by analogy with the STAR OF DAVID (HEXAGRAM).

see also HEXAGRAM, NONAGON, TRIGONOMETRY VALUES—$\pi/9$

Nonassociative Algebra

An ALGEBRA which does not satisfy

$$a(bc) = (ab)c$$

is called a nonassociative algebra. Bott and Milnor (1958) proved that the only nonassociative DIVISION ALGEBRAS are for $n = 1$, 2, 4, and 8. Each gives rise to an ALGEBRA with particularly useful physical applications (which, however, is not itself necessarily nonassociative), and these four cases correspond to REAL NUMBERS, COMPLEX NUMBERS, QUATERNIONS, and CAYLEY NUMBERS, respectively.

see also ALGEBRA, CAYLEY NUMBER, COMPLEX NUMBER, DIVISION ALGEBRA, QUATERNION, REAL NUMBER

References

Bott, R. and Milnor, J. "On the Parallelizability of the Spheres." *Bull. Amer. Math. Soc.* **64**, 87–89, 1958.

Nonassociative Product

The number of nonassociative n-products with k elements preceding the rightmost left parameter is

$$\begin{aligned} F(n,k) &= F(n-1,k) + F(n-1,k-1) \\ &= \binom{n+k-2}{k} - \binom{n+k-1}{k-1}, \end{aligned}$$

where $\binom{n}{k}$ is a BINOMIAL COEFFICIENT. The number of n-products in a nonassociative algebra is

$$F(n) = \sum_{j=0}^{n-2} F(n,j) = \frac{(2n-2)!}{n!(n-1)!}.$$

References
Niven, I. M. *Mathematics of Choice: Or, How to Count Without Counting.* Washington, DC: Math. Assoc. Amer., pp. 140–152, 1965.

Nonaveraging Sequence

N.B. A detailed on-line essay by S. Finch was the starting point for this entry.

An infinite sequence of POSITIVE INTEGERS

$$1 \leq a_1 < a_2 < a_3 < \ldots$$

is a nonaveraging sequence if it contains no three terms which are in an ARITHMETIC PROGRESSION, so that

$$a_i + a_j \neq 2a_k$$

for all distinct a_i, a_j, a_k. Wróblewski (1984) showed that

$$S(A) \equiv \sup_{\substack{\text{all nonaveraging}\\ \text{sequences}}} \sum_{k=1}^{\infty} \frac{1}{a_k} > 3.00849.$$

References
Behrend, F. "On Sets of Integers which Contain no Three Terms in an Arithmetic Progression." *Proc. Nat. Acad. Sci. USA* **32**, 331–332, 1946.
Finch, S. "Favorite Mathematical Constants." `http://www.mathsoft.com/asolve/constant/erdos/erdos.html`.
Gerver, J. L. "The Sum of the Reciprocals of a Set of Integers with No Arithmetic Progression of k Terms." *Proc. Amer. Math. Soc.* **62**, 211–214, 1977.
Gerver, J. L. and Ramsey, L. "Sets of Integers with no Long Arithmetic Progressions Generated by the Greedy Algorithm." *Math. Comput.* **33**, 1353–1360, 1979.
Guy, R. K. "Nonaveraging Sets. Nondividing Sets." §C16 in *Unsolved Problems in Number Theory, 2nd ed.* New York: Springer-Verlag, pp. 131–132, 1994.
Wróblewski, J. "A Nonaveraging Set of Integers with a Large Sum of Reciprocals." *Math. Comput.* **43**, 261–262, 1984.

Noncentral Distribution

see CHI-SQUARED DISTRIBUTION, F-DISTRIBUTION, STUDENT'S t-DISTRIBUTION

Noncommutative Group

A group whose elements do not commute. The simplest noncommutative GROUP is the DIHEDRAL GROUP D_3 of ORDER six.

see also COMMUTATIVE, FINITE GROUP—D_3

Nonconformal Mapping

Let γ be a path in $\mathbb{C}$, $w = f(z)$, and θ and ϕ be the tangents to the curves γ and $f(\gamma)$ at z_0 and w_0. If there is an N such that

$$f^{(N)}(z_0) \neq 0 \tag{1}$$

$$f^{(n)}(z_0) = 0 \tag{2}$$

for all $n < N$ (or, equivalently, if $f'(z)$ has a zero of order $N-1$), then

$$f(z) = f(z_0) + \frac{f^{(N)}(z_0)}{N!}(z-z_0)^N + \frac{f^{(N+1)}(z_0)}{(N+1)!}(z-z_0)^{N+1} + \ldots \tag{3}$$

$$f(z) - f(z_0) = (z-z_0)^N \left[\frac{f(N)(z_0)}{N!} + \frac{f^{(N+1)}(z_0)}{(N+1)!}(z-z_0) + \ldots\right], \tag{4}$$

so the ARGUMENT is

$$\arg[f(z) - f(z_0)] = N\arg(z-z_0) + \arg\left[\frac{f(N)(z_0)}{N!} + \frac{f^{(N+1)}(z_0)}{(N+1)!}(z-z_0) + \ldots\right]. \tag{5}$$

As $z \to z_0$, $\arg(z - z_0) \to \theta$ and $|\arg[f(z) - f(z_0)]| \to \phi$,

$$\phi = N\theta + \arg\left[\frac{f(N)(z_0)}{N!}\right] = N\theta + \arg[f(N)(z_0)]. \tag{6}$$

see also CONFORMAL TRANSFORMATION

Nonconstructive Proof

A PROOF which indirectly shows a mathematical object exists without providing a specific example or algorithm for producing an example.

see also PROOF

References
Courant, R. and Robbins, H. "The Indirect Method of Proof." §2.4.4 in *What is Mathematics?: An Elementary Approach to Ideas and Methods, 2nd ed.* Oxford, England: Oxford University Press, pp. 86–87, 1996.

Noncototient

A POSITIVE value of n for which $x - \phi(x) = n$ has no solution, where $\phi(x)$ is the TOTIENT FUNCTION. The first few are 10, 26, 34, 50, 52, ... (Sloane's A005278).

see also NONTOTIENT, TOTIENT FUNCTION

References
Guy, R. K. *Unsolved Problems in Number Theory, 2nd ed.* New York: Springer-Verlag, p. 91, 1994.
Sloane, N. J. A. Sequence A005278/M4688 in "An On-Line Version of the Encyclopedia of Integer Sequences."

Noncylindrical Ruled Surface

A RULED SURFACE parameterization $\mathbf{x}(u,v) = \mathbf{b}(u) + v\mathbf{g}(u)$ is called noncylindrical if $\mathbf{g} \times \mathbf{g}'$ is nowhere $\mathbf{0}$. A noncylindrical ruled surface always has a parameterization of the form

$$\mathbf{x}(u,v) = \boldsymbol{\sigma}(u) + v\boldsymbol{\delta}(u),$$

where $|\boldsymbol{\delta}| = 1$ and $\boldsymbol{\sigma}' \cdot \boldsymbol{\delta}' = 0$, where $\boldsymbol{\sigma}$ is called the STRICTION CURVE of $\mathbf{x}$ and $\boldsymbol{\delta}$ the DIRECTOR CURVE.

see also DISTRIBUTION PARAMETER, RULED SURFACE, STRICTION CURVE

References

Gray, A. "Noncylindrical Ruled Surfaces." §17.3 in *Modern Differential Geometry of Curves and Surfaces.* Boca Raton, FL: CRC Press, pp. 345–349, 1993.

Nondecreasing Function

A function $f(x)$ is said to be nondecreasing on an INTERVAL I if $f(b) \geq f(a)$ for all $b > a$, where $a, b \in I$. Conversely, a function $f(x)$ is said to be nonincreasing on an INTERVAL I if $f(b) \leq f(a)$ for all $b > a$ with $a, b \in I$.

see also DECREASING FUNCTION, NONINCREASING FUNCTION

Nondividing Set

A SET in which no element divides the SUM of any other.

References

Guy, R. K. "Nonaveraging Sets. Nondividing Sets." §C16 in *Unsolved Problems in Number Theory, 2nd ed.* New York: Springer-Verlag, pp. 131–132, 1994.

Nonessential Singularity

see REGULAR SINGULAR POINT

Non-Euclidean Geometry

In 3 dimensions, there are three classes of constant curvature GEOMETRIES. All are based on the first four of EUCLID'S POSTULATES, but each uses its own version of the PARALLEL POSTULATE. The "flat" geometry of everyday intuition is called EUCLIDEAN GEOMETRY (or PARABOLIC GEOMETRY), and the non-Euclidean geometries are called HYPERBOLIC GEOMETRY (or LOBACHEVSKY-BOLYAI-GAUSS GEOMETRY) and ELLIPTIC GEOMETRY (or RIEMANNIAN GEOMETRY). It was not until 1868 that Beltrami proved that non-Euclidean geometries were as logically consistent as EUCLIDEAN GEOMETRY.

see also ABSOLUTE GEOMETRY, ELLIPTIC GEOMETRY, EUCLID'S POSTULATES, EUCLIDEAN GEOMETRY, HYPERBOLIC GEOMETRY, PARALLEL POSTULATE

References

Borsuk, K. *Foundations of Geometry: Euclidean and Bolyai-Lobachevskian Geometry. Projective Geometry.* Amsterdam, Netherlands: North-Holland, 1960.

Carslaw, H. S. *The Elements of Non-Euclidean Plane Geometry and Trigonometry.* London: Longmans, 1916.

Coxeter, H. S. M. *Non-Euclidean Geometry, 5th ed.* Toronto: University of Toronto Press, 1965.

Dunham, W. *Journey Through Genius: The Great Theorems of Mathematics.* New York: Wiley, pp. 53–60, 1990.

Iversen, B. *An Invitation to Hyperbolic Geometry.* Cambridge, England: Cambridge University Press, 1993.

Iyanaga, S. and Kawada, Y. (Eds.). "Non-Euclidean Geometry." §283 in *Encyclopedic Dictionary of Mathematics.* Cambridge, MA: MIT Press, pp. 893–896, 1980.

Martin, G. E. *The Foundations of Geometry and the Non-Euclidean Plane.* New York: Springer-Verlag, 1975.

Pappas, T. "A Non-Euclidean World." *The Joy of Mathematics.* San Carlos, CA: Wide World Publ./Tetra, pp. 90–92, 1989.

Ramsay, A. and Richtmeyer, R. D. *Introduction to Hyperbolic Geometry.* New York: Springer-Verlag, 1995.

Sommerville, D. Y. *The Elements of Non-Euclidean Geometry.* London: Bell, 1914.

Sommerville, D. Y. *Bibliography of Non-Euclidean Geometry, 2nd ed.* New York: Chelsea, 1960.

Sved, M. *Journey into Geometries.* Washington, DC: Math. Assoc. Amer., 1991.

Trudeau, R. J. *The Non-Euclidean Revolution.* Boston, MA: Birkhäuser, 1987.

Nonillion

In the American system, 10^{30}.

see also LARGE NUMBER

Nonincreasing Function

A function $f(x)$ is said to be nonincreasing on an INTERVAL I if $f(b) \leq f(a)$ for all $b > a$, where $a, b \in I$. Conversely, a function $f(x)$ is said to be nondecreasing on an INTERVAL I if $f(b) \geq f(a)$ for all $b > a$ with $a, b \in I$.

see also INCREASING FUNCTION, NONDECREASING FUNCTION

Nonlinear Least Squares Fitting

Given a function $f(x)$ of a variable x tabulated at m values $y_1 = f(x_1)$, ..., $y_m = f(x_m)$, assume the function is of known analytic form depending on n parameters $f(x; \lambda_1, \ldots, \lambda_n)$, and consider the overdetermined set of m equations

$$y_1 = f(x_1; \lambda_1, \lambda_2, \ldots, \lambda_n) \tag{1}$$

$$y_m = f(x_m; \lambda_1, \lambda_2, \ldots, \lambda_n). \tag{2}$$

We desire to solve these equations to obtain the values λ_1, ..., λ_n which best satisfy this system of equations. Pick an initial guess for the λ_i and then define

$$d\beta_i = y_i - f(x_i; \lambda_1, \ldots, \lambda_n). \tag{3}$$

Now obtain a linearized estimate for the changes $d\lambda_i$ needed to reduce $d\beta_i$ to 0,

$$d\beta_i = \sum_{j=1}^n \frac{\partial f}{\partial \lambda_j}\, d\lambda_j \bigg|_{x_j, \lambda} \tag{4}$$

for $i = 1, \ldots, n$. This can be written in component form as

$$d\beta_i = A_{ij}\, d\lambda_i, \tag{5}$$

where A is the $m \times n$ MATRIX

$$A_{ij} = \begin{bmatrix} \left.\frac{\partial f}{\partial \lambda_1}\right|_{x_1,\lambda} & \left.\frac{\partial f}{\partial \lambda_n}\right|_{x_1,\lambda} & \cdots \\ \left.\frac{\partial f}{\partial \lambda_2}\right|_{x_2,\lambda} & \left.\frac{\partial f}{\partial \lambda_2}\right|_{x_2,\lambda} & \cdots \\ \vdots & \vdots & \ddots \\ \left.\frac{\partial f}{\partial \lambda_1}\right|_{x_m,\lambda} & \left.\frac{\partial f}{\partial \lambda_n}\right|_{x_m,\lambda} & \cdots \end{bmatrix}. \tag{6}$$

In more concise MATRIX form,

$$d\boldsymbol{\beta} = \mathsf{A}\, d\lambda, \tag{7}$$

where $d\boldsymbol{\beta}$ and $d\lambda$ are m-VECTORS. Applying the MATRIX TRANSPOSE of A to both sides gives

$$\mathsf{A}^{\mathrm{T}}\, d\boldsymbol{\beta} = (\mathsf{A}^{\mathrm{T}}\mathsf{A})\, d\lambda. \tag{8}$$

Defining

$$\mathsf{a} \equiv \mathsf{A}^{\mathrm{T}}\mathsf{A} \tag{9}$$

$$\mathbf{b} \equiv \mathsf{A}^{\mathrm{T}}\, d\boldsymbol{\beta} \tag{10}$$

in terms of the known quantities A and $d\boldsymbol{\beta}$ then gives the MATRIX EQUATION

$$\mathsf{a} d\lambda = \mathbf{b}, \tag{11}$$

which can be solved for $d\lambda$ using standard matrix techniques such as GAUSSIAN ELIMINATION. This offset is then applied to λ and a new $d\boldsymbol{\beta}$ is calculated. By iteratively applying this procedure until the elements of $d\lambda$ become smaller than some prescribed limit, a solution is obtained. Note that the procedure may not converge very well for some functions and also that convergence is often greatly improved by picking initial values close to the best-fit value. The sum of square residuals is given by $R^2 = d\boldsymbol{\beta} \cdot d\boldsymbol{\beta}$ after the final iteration.

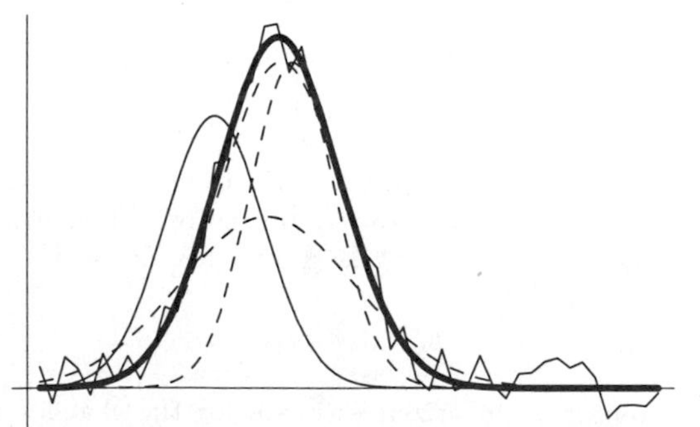

An example of a nonlinear least squares fit to a noisy GAUSSIAN FUNCTION

$$f(A, x_0, \sigma; x) = Ae^{-(x-x_0)^2/(2\sigma^2)} \tag{12}$$

is shown above, where the thin solid curve is the initial guess, the dotted curves are intermediate iterations, and the heavy solid curve is the fit to which the solution converges. The actual parameters are $(A, x_0, \sigma) = (1, 20, 5)$, the initial guess was (0.8, 15, 4), and the converged values are (1.03105, 20.1369, 4.86022), with $R^2 = 0.148461$. The PARTIAL DERIVATIVES used to construct the matrix A are

$$\frac{\partial f}{\partial A} = e^{-(x-x_0)^2/(2\sigma^2)} \tag{13}$$

$$\frac{\partial f}{\partial x_0} = \frac{A(x-x_0)}{\sigma^2} e^{-(x-x_0)^2/(2\sigma^2)} \tag{14}$$

$$\frac{\partial f}{\partial \sigma} = \frac{A(x-x_0)^2}{\sigma^3} e^{-(x-x_0)^2/(2\sigma^2)}. \tag{15}$$

The technique could obviously be generalized to multiple Gaussians, to include slopes, etc., although the convergence properties generally worsen as the number of free parameters is increased.

An analogous technique can be used to solve an overdetermined set of equations. This problem might, for example, arise when solving for the best-fit EULER ANGLES corresponding to a noisy ROTATION MATRIX, in which case there are three unknown angles, but nine correlated matrix elements. In such a case, write the n *different* functions as $f_i(\lambda_1, \ldots, \lambda_n)$ for $i = 1, \ldots, n$, call their actual values y_i, and define

$$\mathsf{A} = \begin{bmatrix} \left.\frac{\partial f_1}{\partial \lambda_1}\right|_{\lambda_i} & \left.\frac{\partial f_1}{\partial \lambda_2}\right|_{\lambda_i} & \cdots & \left.\frac{\partial f_1}{\partial \lambda_n}\right|_{\lambda_i} \\ \vdots & \vdots & \ddots & \vdots \\ \left.\frac{\partial f_m}{\partial \lambda_1}\right|_{\lambda_i} & \left.\frac{\partial f_m}{\partial \lambda_2}\right|_{\lambda_i} & \cdots & \left.\frac{\partial f_m}{\partial \lambda_n}\right|_{\lambda_i} \end{bmatrix}, \tag{16}$$

and

$$d\boldsymbol{\beta} = \mathbf{y} - f_i(\lambda_1, \ldots, \lambda_n), \tag{17}$$

where λ_i are the numerical values obtained after the ith iteration. Again, set up the equations as

$$\mathsf{A}\, d\lambda = d\boldsymbol{\beta}, \tag{18}$$

and proceed exactly as before.

see also LEAST SQUARES FITTING, LINEAR REGRESSION, MOORE-PENROSE GENERALIZED MATRIX INVERSE

Nonnegative

A quantity which is either 0 (ZERO) or POSITIVE, i.e., ≥ 0.

see also NEGATIVE, NONNEGATIVE INTEGER, NONPOSITIVE, NONZERO, POSITIVE, ZERO

Nonnegative Integer

see $\mathbb{Z}^*$

Nonnegative Partial Sum

The number of sequences with NONNEGATIVE partial sums which can be formed from n 1s and n -1s (Bailey 1996, Buraldi 1992) is given by the CATALAN NUMBERS. Bailey (1996) gives the number of NONNEGATIVE partial sums of n 1s and k -1s a_1, a_2, ..., a_{n+k}, so that

$$a_1 + a_2 + \ldots + a_i \geq 0 \tag{1}$$

for all $1 \leq i \leq n+k$. The closed form expression is

$$\left\{ \begin{matrix} n \\ 0 \end{matrix} \right\} = 1 \tag{2}$$

for $n \geq 0$,

$$\left\{ \begin{matrix} n \\ 1 \end{matrix} \right\} = n \tag{3}$$

for $n \geq 1$, and

$$\left\{ \begin{matrix} n \\ k \end{matrix} \right\} = \frac{(n+1-k)(n+2)(n+3)\cdots(n+k)}{k!}, \tag{4}$$

for $n \geq k \geq 2$. Setting $k = n$ then recovers the CATALAN NUMBERS

$$C_n = \left\{ \begin{matrix} n \\ n \end{matrix} \right\} = \frac{1}{n+1}\binom{2n}{n}. \tag{5}$$

see also CATALAN NUMBER

References

Bailey, D. F. "Counting Arrangements of 1's and -1's." *Math. Mag.* **69**, 128–131, 1996.

Buraldi, R. A. *Introductory Combinatorics, 2nd ed.* New York: Elsevier, 1992.

Nonorientable Surface

A surface such as the MÖBIUS STRIP on which there exists a closed path such that the directrix is reversed when moved around this path. The EULER CHARACTERISTIC of a nonorientable surface is ≤ 0. The real PROJECTIVE PLANE is also a nonorientable surface, as are the BOY SURFACE, CROSS-CAP, and ROMAN SURFACE, all of which are homeomorphic to the REAL PROJECTIVE PLANE (Pinkall 1986). There is a general method for constructing nonorientable surfaces which proceeds as follows (Banchoff 1984, Pinkall 1986). Choose three HOMOGENEOUS POLYNOMIALS of POSITIVE EVEN degree and consider the MAP

$$\mathbf{f} = (f_1(x,y,z), f_2(x,y,z), f_3(x,y,z)) : \mathbb{R}^3 \to \mathbb{R}^3. \tag{1}$$

Then restricting x, y, and z to the surface of a sphere by writing

$$x = \cos\theta \sin\phi \tag{2}$$

$$y = \sin\theta \sin\phi \tag{3}$$

$$z = \cos\phi \tag{4}$$

and restricting θ to $[0, 2\pi)$ and ϕ to $[0, \pi/2]$ defines a map of the REAL PROJECTIVE PLANE to $\mathbb{R}^3$.

In 3-D, there is no unbounded nonorientable surface which does not intersect itself (Kuiper 1961, Pinkall 1986).

see also BOY SURFACE, CROSS-CAP, MÖBIUS STRIP, ORIENTABLE SURFACE, PROJECTIVE PLANE, ROMAN SURFACE

References

Banchoff, T. "Differential Geometry and Computer Graphics." In *Perspectives of Mathematics: Anniversary of Oberwolfach* (Ed. W. Jager, R. Remmert, and J. Moser). Basel, Switzerland: Birkhäuser, 1984.

Gray, A. "Nonorientable Surfaces." Ch. 12 in *Modern Differential Geometry of Curves and Surfaces.* Boca Raton, FL: CRC Press, pp. 229–249, 1993.

Kuiper, N. H. "Convex Immersion of Closed Surfaces in E^3." *Comment. Math. Helv.* **35**, 85–92, 1961.

Pinkall, U. "Models of the Real Projective Plane." Ch. 6 in *Mathematical Models from the Collections of Universities and Museums* (Ed. G. Fischer). Braunschweig, Germany: Vieweg, pp. 63–67, 1986.

Nonpositive

A quantity which is either 0 (ZERO) or NEGATIVE, i.e., ≤ 0.

see also NEGATIVE, NONNEGATIVE, NONZERO, POSITIVE, ZERO

Nonsquarefree

see SQUAREFUL

Nonstandard Analysis

Nonstandard analysis is a branch of mathematical LOGIC which weakens the axioms of usual ANALYSIS to include only the first-order ones. It also introduces HYPERREAL NUMBERS to allow for the existence of "genuine INFINITESIMALS," numbers which are less than 1/2, 1/3, 1/4, 1/5, ..., but greater than 0. Abraham Robinson developed nonstandard analysis in the 1960s. The theory has since been investigated for its own sake and has been applied in areas such as BANACH SPACES, differential equations, probability theory, microeconomic theory, and mathematical physics (Apps).

see also AX-KOCHEN ISOMORPHISM THEOREM, LOGIC, MODEL THEORY

References

Albeverio, S.; Fenstad, J.; Hoegh-Krohn, R.; and Lindstrøom, T. *Nonstandard Methods in Stochastic Analysis and Mathematical Physics.* New York: Academic Press, 1986.

Anderson, R. "Nonstandard Analysis with Applications to Economics." In *Handbook of Mathematical Economics, Vol. 4.* New York: Elsevier, 1991.

Apps, P. "What is Nonstandard Analysis?" `http://www.math.wisc.edu/~apps/nonstandard.html`.

Dauben, J. W. *Abraham Robinson: The Creation of Nonstandard Analysis, A Personal and Mathematical Odyssey.* Princeton, NJ: Princeton University Press, 1998.

Davis, P. J. and Hersch, R. *The Mathematical Experience.* Boston: Birkhäuser, 1981.
Keisler, H. J. *Elementary Calculus: An Infinitesimal Approach.* Boston: PWS, 1986.
Lindstrøom, T. "An Invitation to Nonstandard Analysis." In *Nonstandard Analysis and Its Applications* (Ed. N. Cutland). New York: Cambridge University Press, 1988.
Robinson, A. *Non-Standard Analysis.* Princeton, NJ: Princeton University Press, 1996.
Stewart, I. "Non-Standard Analysis." In *From Here to Infinity: A Guide to Today's Mathematics.* Oxford, England: Oxford University Press, pp. 80–81, 1996.

Nontotient

A POSITIVE EVEN value of n for which $\phi(x) = n$, where $\phi(x)$ is the TOTIENT FUNCTION, has no solution. The first few are 14, 26, 34, 38, 50, ... (Sloane's A005277).

see also NONCOTOTIENT, TOTIENT FUNCTION

References

Guy, R. K. *Unsolved Problems in Number Theory, 2nd ed.* New York: Springer-Verlag, p. 91, 1994.
Sloane, N. J. A. Sequence A005277/M4927 in "An On-Line Version of the Encyclopedia of Integer Sequences."

Nonwandering

A point x in a MANIFOLD M is said to be nonwandering if, for every open NEIGHBORHOOD U of x, it is true that $\phi^{-n}U \cup U \neq \varnothing$ for a MAP ϕ for some $n > 0$. In other words, every point close to x has some iterate under ϕ which is also close to x. The set of all nonwandering points is denoted $\Omega(\phi)$, which is known as the nonwandering set of ϕ.

see also ANOSOV DIFFEOMORPHISM, AXIOM A DIFFEOMORPHISM, SMALE HORSESHOE MAP

Nonzero

A quantity which does not equal ZERO is said to be nonzero. A REAL nonzero number must be either POSITIVE or NEGATIVE, and a COMPLEX nonzero number can have either REAL or IMAGINARY PART nonzero.

see also NEGATIVE, NONNEGATIVE, NONPOSITIVE, POSITIVE, ZERO

Nordstrand's Weird Surface

An attractive CUBIC SURFACE defined by Nordstrand. It is given by the implicit equation

$$\begin{aligned}25[x^3(y+z)+y^3(x+z)+z^3(x+y)]+50(x^2y^2+x^2z^2\\+y^2z^2)-125(x^2yz+y^2xz+z^2xy)+60xyz\\-4(xy+xz+yz)=0.\end{aligned}$$

References

Nordstrand, T. "Weird Cube." `http://www.uib.no/people/nfytn/weirdtxt.htm`.

Norm

Given a n-D VECTOR

$$\mathbf{x} = \begin{bmatrix} x_1 \\ x_2 \\ \vdots \\ x_n \end{bmatrix},$$

a VECTOR NORM $||\mathbf{x}||$ is a NONNEGATIVE number satisfying

1. $||x|| > 0$ when $\mathbf{x} \neq \mathbf{0}$ and $||\mathbf{x}|| = 0$ IFF $\mathbf{x} = \mathbf{0}$,
2. $||k\mathbf{x}|| = |k|\,||\mathbf{x}||$ for any SCALAR k,
3. $||\mathbf{x}+\mathbf{y}|| \leq ||\mathbf{x}|| + ||\mathbf{y}||$.

The most common norm is the vector L_2-NORM, defined by

$$||\mathbf{x}||_2 = |\mathbf{x}| = \sqrt{x_1{}^2 + x_2{}^2 + \ldots + x_n{}^2}.$$

Given a SQUARE MATRIX A, a MATRIX NORM $||\mathsf{A}||$ is a NONNEGATIVE number associated with A having the properties

1. $||\mathsf{A}|| > 0$ when $\mathsf{A} \neq 0$ and $||\mathsf{A}|| = 0$ IFF $\mathsf{A} = 0$,
2. $||k\mathsf{A}|| = |k|\,||\mathsf{A}||$ for any SCALAR k,
3. $||\mathsf{A}+\mathsf{B}|| \leq ||\mathsf{A}|| + ||\mathsf{B}||$,
4. $||\mathsf{AB}|| \leq ||\mathsf{A}||\,||\mathsf{B}||$.

see also BOMBIERI NORM, COMPATIBLE, EUCLIDEAN NORM, HILBERT-SCHMIDT NORM, INDUCED NORM, L_1-NORM, L_2-NORM, L_∞-NORM, MATRIX NORM, MAXIMUM ABSOLUTE COLUMN SUM NORM, MAXIMUM ABSOLUTE ROW SUM NORM, NATURAL NORM, NORMALIZED VECTOR, NORMED SPACE, PARALLELOGRAM LAW, POLYNOMIAL NORM, SPECTRAL NORM, SUBORDINATE NORM, VECTOR NORM

References

Gradshteyn, I. S. and Ryzhik, I. M. *Tables of Integrals, Series, and Products, 5th ed.* San Diego, CA: Academic Press, pp. 1114–1125, 1979.

Norm Theorem

If a PRIME number divides a norm but not the bases of the norm, it is itself a norm.

Normal

see NORMAL CURVE, NORMAL DISTRIBUTION, NORMAL DISTRIBUTION FUNCTION, NORMAL EQUATION, NORMAL FORM, NORMAL GROUP, NORMAL MAGIC SQUARE, NORMAL MATRIX, NORMAL NUMBER, NORMAL PLANE, NORMAL SUBGROUP, NORMAL VECTOR

Normal (Algebraically)

see GALOISIAN

Normal Curvature

Let $\mathbf{u_p}$ be a unit TANGENT VECTOR of a REGULAR SURFACE $M \subset \mathbb{R}^3$. Then the normal curvature of M in the direction $\mathbf{u_p}$ is

$$\kappa(\mathbf{u_p}) = S(\mathbf{u_p}) \cdot \mathbf{u_p}, \tag{1}$$

where S is the SHAPE OPERATOR. Let $M \subset \mathbb{R}^3$ be a REGULAR SURFACE, $\mathbf{p} \in M$, $\mathbf{x}$ be an injective REGULAR PATCH of M with $\mathbf{p} = \mathbf{x}(u_0, v_0)$, and

$$\mathbf{v_p} = a\mathbf{x}_u(u_0, v_0) + b\mathbf{x}_v(u_0, v_0), \tag{2}$$

where $\mathbf{v_p} \in M_\mathbf{p}$. Then the normal curvature in the direction $\mathbf{v_p}$ is

$$\kappa(v\mathbf{p}) = \frac{ea^2 + 2fab + gb^2}{Ea^2 + 2Fab + Gb^2}, \tag{3}$$

where E, F, and G are first FUNDAMENTAL FORMS and e, f, and g second FUNDAMENTAL FORMS.

The MAXIMUM and MINIMUM values of the normal curvature on a REGULAR SURFACE at a point on the surface are called the PRINCIPAL CURVATURES κ_1 and κ_2.

see also CURVATURE, FUNDAMENTAL FORMS, GAUSSIAN CURVATURE, MEAN CURVATURE, PRINCIPAL CURVATURES, SHAPE OPERATOR, TANGENT VECTOR

References

Euler, L. "Récherches sur la coubure des surfaces." *Mém. de l'Acad. des Sciences, Berlin* **16**, 119–143, 1760.

Gray, A. "Normal Curvature." §14.2 in *Modern Differential Geometry of Curves and Surfaces*. Boca Raton, FL: CRC Press, pp. 270–273 and 277, 1993.

Meusnier, J. B. "Mémoire sur la courbure des surfaces." *Mém. des savans étrangers* **10** (lu 1776), 477–510, 1785.

Normal Curve

see GAUSSIAN DISTRIBUTION

Normal Developable

A RULED SURFACE M is a normal developable of a curve $\mathbf{y}$ if M can be parameterized by $\mathbf{x}(u,v) = \mathbf{y}(u) + v\hat{\mathbf{N}}(u)$, where $\mathbf{N}$ is the NORMAL VECTOR.

see also BINORMAL DEVELOPABLE, TANGENT DEVELOPABLE

References

Gray, A. "Developables." §17.6 in *Modern Differential Geometry of Curves and Surfaces*. Boca Raton, FL: CRC Press, pp. 352–354, 1993.

Normal Distribution

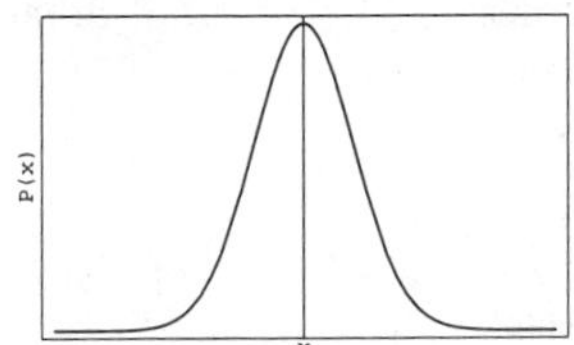

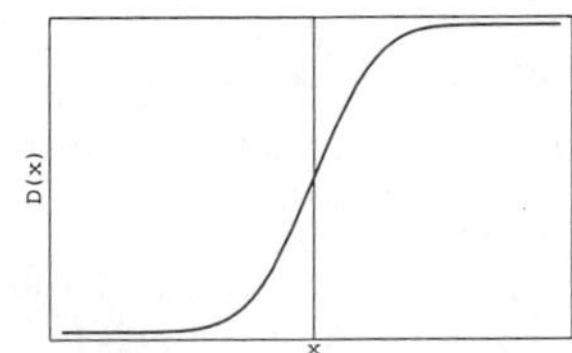

Another name for a GAUSSIAN DISTRIBUTION. Given a normal distribution in a VARIATE x with MEAN μ and VARIANCE σ^2,

$$P(x)\,dx = \frac{1}{\sigma\sqrt{2\pi}} e^{-(x-\mu)^2/2\sigma^2}\,dx,$$

the so-called "STANDARD NORMAL DISTRIBUTION" is given by taking $\mu = 0$ and $\sigma^2 = 1$. An arbitrary normal distribution can be converted to a STANDARD NORMAL DISTRIBUTION by changing variables to $z \equiv (x - \mu)/\sigma$, so $dz = dx/\sigma$, yielding

$$P(x)\,dx = \frac{1}{\sqrt{2\pi}} e^{-z^2/2}\,dz.$$

The FISHER-BEHRENS PROBLEM is the determination of a test for the equality of MEANS for two normal distributions with different VARIANCES.

see also FISHER-BEHRENS PROBLEM, GAUSSIAN DISTRIBUTION, HALF-NORMAL DISTRIBUTION, KOLMOGOROV-SMIRNOV TEST, NORMAL DISTRIBUTION FUNCTION, STANDARD NORMAL DISTRIBUTION, TETRACHORIC FUNCTION

Normal Distribution Function

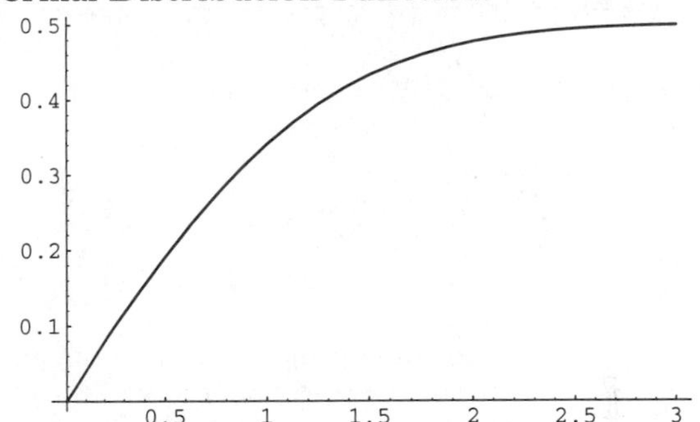

A normalized form of the cumulative GAUSSIAN DISTRIBUTION function giving the probability that a variate assumes a value in the range $[0, x]$,

$$\Phi(x) \equiv Q(x) \equiv \frac{1}{\sqrt{2\pi}} \int_0^x e^{-t^2/2}\,dt. \tag{1}$$

It is related to the PROBABILITY INTEGRAL

$$\alpha(x) \equiv \frac{1}{\sqrt{2\pi}} \int_{-x}^x e^{-t^2/2}\,dt \tag{2}$$

by

$$\Phi(x) = \tfrac{1}{2}\alpha(x). \tag{3}$$

Let $u \equiv t/\sqrt{2}$ so $du = dt/\sqrt{2}$. Then

$$\Phi(x) = \frac{1}{\sqrt{\pi}} \int_0^{x/\sqrt{2}} e^{-u^2}\,du = \tfrac{1}{2}\operatorname{erf}\left(\frac{x}{\sqrt{2}}\right). \tag{4}$$

Here, ERF is a function sometimes called the error function. The probability that a normal variate assumes a value in the range $[x_1, x_2]$ is therefore given by

$$\Phi(x_1, x_2) = \frac{1}{2}\left[\operatorname{erf}\left(\frac{x_2}{\sqrt{2}}\right) - \operatorname{erf}\left(\frac{x_1}{\sqrt{2}}\right)\right]. \quad (5)$$

Neither $\Phi(z)$ nor ERF can be expressed in terms of finite additions, subtractions, multiplications, and root extractions, and so must be either computed numerically or otherwise approximated.

Note that a function different from $\Phi(x)$ is sometimes defined as "the" normal distribution function

$$\Phi'(x) \equiv \frac{1}{2}\left[1 + \operatorname{erf}\left(\frac{x}{\sqrt{2}}\right)\right] = \tfrac{1}{2} + \Phi(x) \quad (6)$$

(Beyer 1987, p. 551), although this function is less widely encountered than the usual $\Phi(x)$.

The value of a for which $P(x)$ falls within the interval $[-a, a]$ with a given probability P is a related quantity called the CONFIDENCE INTERVAL.

For small values $x \ll 1$, a good approximation to $\Phi(x)$ is obtained from the MACLAURIN SERIES for ERF,

$$\Phi(x) = \frac{1}{\sqrt{2\pi}}(2x - \tfrac{1}{3}x^3 + \tfrac{1}{20}x^5 - \tfrac{1}{168}x^7 + \ldots). \quad (7)$$

For large values $x \gg 1$, a good approximation is obtained from the asymptotic series for ERF,

$$\Phi(x) = \frac{1}{2} + \frac{e^{-x^2/2}}{\sqrt{2\pi}}(x^{-1} - x^{-3} + 3x^{-5} - 15x^{-7} + 105x^{-9} + \ldots). \quad (8)$$

The value of $\Phi(x)$ for intermediate x can be computed using the CONTINUED FRACTION identity

$$\int_0^x e^{-u^2}\,du = \frac{\sqrt{\pi}}{2} - \cfrac{\tfrac{1}{2}e^{-x^2}}{x + \cfrac{1}{2x + \cfrac{2}{x + \cfrac{3}{2x + \cfrac{4}{x + \ldots}}}}}. \quad (9)$$

A simple approximation of $\Phi(x)$ which is good to two decimal places is given by

$$\Phi_1(x) \approx \begin{cases} 0.1x(4.4 - x) & \text{for } 0 \le x \le 2.2 \\ 0.49 & \text{for } 2.2 < x < 2.6 \\ 0.50 & \text{for } x \ge 2.6. \end{cases} \quad (10)$$

Abramowitz and Stegun (1972) and Johnson and Kotz (1970) give other functional approximations. An approximation due to Bagby (1995) is

$$\Phi_2(x) = \tfrac{1}{2}\{1 - \tfrac{1}{30}[7e^{-x^2/2} + 16e^{-x^2(2-\sqrt{2})} + (7 + \tfrac{1}{4}\pi x^2)e^{-x^2}]\}^{1/2}. \quad (11)$$

The plots below show the differences between Φ and the two approximations.

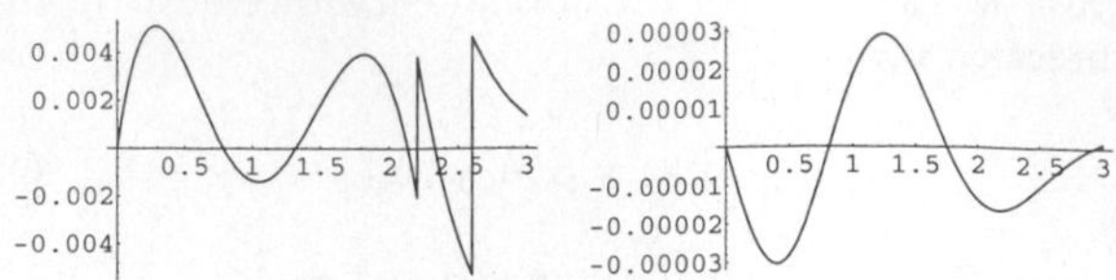

The first QUARTILE of a standard NORMAL DISTRIBUTION occurs when

$$\int_0^t \Phi(z)\,dz = \tfrac{1}{4}. \quad (12)$$

The solution is $t = 0.6745\ldots$. The value of t giving $\frac{1}{4}$ is known as the PROBABLE ERROR of a normally distributed variate.

see also CONFIDENCE INTERVAL, ERF, ERFC, FISHER-BEHRENS PROBLEM, GAUSSIAN DISTRIBUTION, GAUSSIAN INTEGRAL, HH FUNCTION, NORMAL DISTRIBUTION, PROBABILITY INTEGRAL, TETRACHORIC FUNCTION

References

Abramowitz, M. and Stegun, C. A. (Eds.). *Handbook of Mathematical Functions with Formulas, Graphs, and Mathematical Tables, 9th printing.* New York: Dover, pp. 931–933, 1972.

Bagby, R. J. "Calculating Normal Probabilities." *Amer. Math. Monthly* **102**, 46–49, 1995.

Beyer, W. H. (Ed.). *CRC Standard Mathematical Tables, 28th ed.* Boca Raton, FL: CRC Press, 1987.

Johnson, N.; Kotz, S.; and Balakrishnan, N. *Continuous Univariate Distributions, Vol. 1, 2nd ed.* Boston, MA: Houghton Mifflin, 1994.

Normal Equation

Given an overdetermined MATRIX EQUATION

$$\mathsf{A}\mathbf{x} = \mathbf{b},$$

the normal equation is that which minimizes the sum of the square differences between left and right sides

$$\mathsf{A}^{\mathrm{T}}\mathsf{A}\mathbf{x} = \mathsf{A}^{\mathrm{T}}\mathbf{b}.$$

see also LEAST SQUARES FITTING, MOORE-PENROSE GENERALIZED MATRIX INVERSE, NONLINEAR LEAST SQUARES FITTING

Normal Form

A way of representing objects so that, although each may have many different names, every possible name corresponds to exactly one object.

see also CANONICAL FORM

References

Petkovšek, M.; Wilf, H. S.; and Zeilberger, D. *A=B.* Wellesley, MA: A. K. Peters, p. 7, 1996.

Normal Function

A SQUARE INTEGRABLE function ϕ is said to be normal if

$$\int \phi^2 \, dt = 1$$

However, the NORMAL DISTRIBUTION FUNCTION is also sometimes called "the normal function."

see also NORMAL DISTRIBUTION FUNCTION, SQUARE INTEGRABLE

References

Sansone, G. *Orthogonal Functions, rev. English ed.* New York: Dover, p. 6, 1991.

Normal Group

see NORMAL SUBGROUP

Normal Magic Square

see MAGIC SQUARE

Normal Matrix

A normal matrix A is a MATRIX for which

$$[\mathsf{A}, \mathsf{A}^\dagger] = 0,$$

where $[a, b]$ is the COMMUTATOR and † denotes the ADJOINT OPERATOR.

Normal Number

An IRRATIONAL NUMBER for which any FINITE pattern of numbers occurs with the expected limiting frequency in the expansion in any base. It is not known if π or e are normal. Tests of $\sqrt{n}$ for $n = 2$, 3, 5, 6, 7, 8, 10, 11, 12, 13, 14, 15 indicate that these SQUARE ROOTS may be normal. The only numbers known to be normal are artificially constructed ones such as the CHAMPERNOWNE CONSTANT and the COPELAND-ERDŐS CONSTANT.

see also CHAMPERNOWNE CONSTANT, COPELAND-ERDŐS CONSTANT, e, PI

Normal Order

$f(n)$ has the normal order $F(n)$ if $f(n)$ is approximately $F(n)$ for ALMOST ALL values of n. More precisely, if

$$(1-\epsilon)F(n) < f(n) < (1+\epsilon)F(n)$$

for every positive ϵ and ALMOST ALL values of n, then the normal order of $f(n)$ is $F(n)$.

see also ALMOST ALL

References

Hardy, G. H. and Weight, E. M. *An Introduction to the Theory of Numbers, 5th ed.* Oxford, England: Oxford University Press, p. 356, 1979.

Normal Plane

The PLANE spanned by $\mathbf{N}$ and $\mathbf{B}$ (the NORMAL VECTOR and BINORMAL VECTOR).

see also BINORMAL VECTOR, NORMAL VECTOR, PLANE

Normal to a Plane

see NORMAL VECTOR

Normal Section

Let $M \subset \mathbb{R}^3$ be a REGULAR SURFACE and $\mathbf{u_p}$ a unit TANGENT VECTOR to M, and let $\Pi(\mathbf{u_p}, \mathbf{N}(\mathbf{p}))$ be the PLANE determined by $\mathbf{u_p}$ and the normal to the surface $\mathbf{N}(\mathbf{p})$. Then the normal section of M is defined as the intersection of $\Pi(\mathbf{u_p}, \mathbf{N}(\mathbf{p}))$ and M.

References

Gray, A. *Modern Differential Geometry of Curves and Surfaces.* Boca Raton, FL: CRC Press, p. 271, 1993.

Normal Subgroup

Let H be a SUBGROUP of a GROUP G. Then H is a normal subgroup of G, written $H \triangleleft G$, if

$$xHx^{-1} = H$$

for every element x in H. Normal subgroups are also known as INVARIANT SUBGROUPS.

see also GROUP, SUBGROUP

Normal Vector

The normal to a PLANE specified by

$$f(x, y, z) = ax + by + cz + d = 0 \tag{1}$$

is given by

$$\mathbf{N} = \nabla f = \begin{bmatrix} a \\ b \\ c \end{bmatrix}. \tag{2}$$

The normal vector at a point (x_0, y_0) on a surface $z = f(x, y)$ is

$$\mathbf{N} = \begin{bmatrix} f_x(x_0, y_0) \\ f_y(x_0, y_0) \\ -1 \end{bmatrix}. \tag{3}$$

In the PLANE, the unit normal vector is defined by

$$\hat{\mathbf{N}} \equiv \frac{d\hat{\mathbf{T}}}{d\phi}, \tag{4}$$

where $\hat{\mathbf{T}}$ is the unit TANGENT VECTOR and ϕ is the polar angle. Given a unit TANGENT VECTOR

$$\hat{\mathbf{T}} \equiv u_1\hat{\mathbf{x}} + u_2\hat{\mathbf{y}} \tag{5}$$

with $\sqrt{{u_1}^2 + {u_2}^2} = 1$, the normal is

$$\hat{\mathbf{N}} = -u_2\hat{\mathbf{x}} + u_1\hat{\mathbf{y}}. \tag{6}$$

For a function given parametrically by $(f(t), g(t))$, the normal vector relative to the point $(f(t), g(t))$ is therefore given by

$$x(t) = -\frac{g'}{\sqrt{f'^2 + g'^2}} \tag{7}$$

$$y(t) = \frac{f'}{\sqrt{f'^2 + g'^2}}. \tag{8}$$

To actually place the vector normal to the curve, it must be displaced by $(f(t), g(t))$.

In 3-D SPACE, the unit normal is

$$\hat{\mathbf{N}} \equiv \frac{\frac{d\hat{\mathbf{T}}}{ds}}{\left|\frac{d\hat{\mathbf{T}}}{ds}\right|} = \frac{\frac{d\hat{\mathbf{T}}}{dt}}{\left|\frac{d\hat{\mathbf{T}}}{dt}\right|} = \frac{1}{\kappa}\frac{d\hat{\mathbf{T}}}{ds}, \tag{9}$$

where κ is the CURVATURE. Given a 3-D surface $F(x, y, z) = 0$,

$$\hat{\mathbf{n}} = \frac{F_x + F_y + F_z}{\sqrt{{F_x}^2 + {F_y}^2 + {F_z}^2}}. \tag{10}$$

If the surface is defined parametrically in the form

$$x = x(\phi, \psi) \tag{11}$$

$$y = y(\phi, \psi) \tag{12}$$

$$z = z(\phi, \psi), \tag{13}$$

define the VECTORS

$$\mathbf{a} \equiv \begin{bmatrix} x_\phi \\ y_\phi \\ z_\phi \end{bmatrix} \tag{14}$$

$$\mathbf{b} \equiv \begin{bmatrix} x_\psi \\ y_\psi \\ z_\psi \end{bmatrix}. \tag{15}$$

Then the unit normal vector is

$$\hat{\mathbf{N}} = \frac{\mathbf{a} \times \mathbf{b}}{\sqrt{|\mathbf{a}|^2|\mathbf{b}|^2 - |\mathbf{a} \cdot \mathbf{b}|^2}}. \tag{16}$$

Let g be the discriminant of the METRIC TENSOR. Then

$$\mathbf{N} = \frac{\mathbf{r}_1 \times \mathbf{r}_2}{\sqrt{g}} = \epsilon_{ij}\mathbf{r}^j. \tag{17}$$

see also BINORMAL VECTOR, CURVATURE, FRENET FORMULAS, TANGENT VECTOR

References

Gray, A. "Tangent and Normal Lines to Plane Curves." §5.5 in *Modern Differential Geometry of Curves and Surfaces.* Boca Raton, FL: CRC Press, pp. 85–90, 1993.

Normalized Vector

The normalized vector of $\mathbf{X}$ is a VECTOR in the same direction but with NORM (length) 1. It is denoted $\hat{\mathbf{X}}$ and given by

$$\hat{\mathbf{X}} \equiv \frac{\mathbf{X}}{|\mathbf{X}|},$$

where $|\mathbf{X}|$ is the NORM of $\mathbf{X}$. It is also called a UNIT VECTOR.

see also UNIT VECTOR

Normalizer

A set of elements g of a GROUP such that

$$g^{-1}Hg = H,$$

is said to be the normalizer $N_G(H)$ with respect to a subset of group elements H.

see also CENTRALIZER, TIGHTLY EMBEDDED

Normed Space

A VECTOR SPACE possessing a NORM.

Nosarzewska's Inequality

Given a convex PLANE region with AREA A and PERIMETER p,

$$A - \tfrac{1}{2}p < N \leq A + \tfrac{1}{2}p + 1,$$

where N is the number of enclosed LATTICE POINTS (Nosarzewska 1948). This improves on JARNICK'S INEQUALITY

$$|N - A| < p.$$

see also JARNICK'S INEQUALITY, LATTICE POINT

References

Nosarzewska, M. "Évaluation de la différence entre l'aire d'une région plane convexe et le nombre des points aux coordonnées entières couverts par elle." *Colloq. Math.* **1**, 305–311, 1948.

Not

An operation in LOGIC which converts TRUE to FALSE and FALSE to TRUE. NOT A is denoted $!A$ or $\neg A$.

A	$\neg A$
F	T
T	F

see also AND, OR, TRUTH TABLE, XOR

Notation

A NOTATION is a set of well-defined rules for representing quantities and operations with symbols.

see also ARROW NOTATION, CHAINED ARROW NOTATION, CIRCLE NOTATION, CLEBSCH-ARONHOLD NOTATION, CONWAY'S KNOT NOTATION, DOWKER NOTATION, DOWN ARROW NOTATION, PETROV NOTATION, SCIENTIFIC NOTATION, STEINHAUS-MOSER NOTATION

References

Cajori, F. *A History of Mathematical Notations, Vols. 1–2.* New York: Dover, 1993.
Miller, J. "Earliest Uses of Various Mathematical Symbols." `http://members.aol.com/jeff570/mathsym.html`.
Miller, J. "Earliest Uses of Some of the Words of Mathematics." `http://members.aol.com/jeff570/mathword.html`.

Nöther

see NOETHER'S FUNDAMENTAL THEOREM, NOETHER-LASKER THEOREM, NOETHER'S TRANSFORMATION THEOREM, NOETHERIAN MODULE, NOETHERIAN RING

Novemdecillion

In the American system, 10^{60}.

see also LARGE NUMBER

NP-Complete Problem

A problem which is both NP (solvable in nondeterministic POLYNOMIAL time) and NP-HARD (can be translated into any other NP-PROBLEM). Examples of NP-hard problems include the HAMILTONIAN CYCLE and TRAVELING SALESMAN PROBLEMS.

In a landmark paper, Karp (1972) showed that 21 intractable combinatorial computational problems are all NP-complete.

see also HAMILTONIAN CYCLE, NP-HARD PROBLEM, NP-PROBLEM, P-PROBLEM, TRAVELING SALESMAN PROBLEM

References

Karp, R. M. "Reducibility Among Combinatorial Problems." In *Complexity of Computer Computations,* (Proc. Sympos. IBM Thomas J. Watson Res. Center, Yorktown Heights, N.Y., 1972). New York: Plenum, pp. 85–103, 1972.

NP-Hard Problem

A problem is NP-hard if an ALGORITHM for solving it can be translated into one for solving any other NP-PROBLEM (nondeterministic POLYNOMIAL time) problem. NP-hard therefore means "at least as hard as any NP-PROBLEM," although it might, in fact, be harder.

see also COMPLEXITY THEORY, HITTING SET, NP-COMPLETE PROBLEM, NP-PROBLEM, P-PROBLEM, SATISFIABILITY PROBLEM

NP-Problem

A problem is assigned to the NP (nondeterministic POLYNOMIAL time) class if it is solvable in polynomial time by a nondeterministic TURING MACHINE. (A nondeterministic TURING MACHINE is a "parallel" TURING MACHINE which can take many computational paths simultaneously, with the restriction that the parallel Turing machines cannot communicate.) A P-PROBLEM (whose solution time is bounded by a polynomial) is always also NP. If a solution to an NP problem is known, it can be reduced to a single P (POLYNOMIAL time) verification.

LINEAR PROGRAMMING, long known to be NP and thought *not* to be P, was shown to be P by L. Khachian in 1979. It is not known if all apparently NP problems are actually P.

A problem is said to be NP-HARD if an ALGORITHM for solving it can be translated into one for solving any other NP-problem problem. It is much easier to show that a problem is NP than to show that it is NP-HARD. A problem which is both NP and NP-HARD is called an NP-COMPLETE PROBLEM.

see also COMPLEXITY THEORY, NP-COMPLETE PROBLEM, NP-HARD PROBLEM, P-PROBLEM, TURING MACHINE

References

Borwein, J. M. and Borwein, P. B. *Pi and the AGM: A Study in Analytic Number Theory and Computational Complexity.* New York: Wiley, 1987.

Greenlaw, R.; Hoover, H. J.; and Ruzzo, W. L. *Limits to Parallel Computation: P-Completeness Theory.* Oxford, England: Oxford University Press, 1995.

NSW Number

The numbers

$$S_{2m+1} = \frac{(1+\sqrt{2})^{2m+1} + (1-\sqrt{2})^{2m+1}}{2}$$

for positive integer m. The first few terms are 1, 7, 41, 239, 1393, ... (Sloane's A002315). The indices giving PRIME NSW numbers are 3, 5, 7, 19, 29, 47, 59, 163, 257, 421, 937, 947, 1493, 1901, ... (Sloane's A005850).

References

Ribenboim, P. "The NSW Primes." §5.9 in *The New Book of Prime Number Records.* New York: Springer-Verlag, pp. 367–369, 1996.

Sloane, N. J. A. Sequences A002315/M4423 and A005850/M2426 in "An On-Line Version of the Encyclopedia of Integer Sequences."

Nu Function

$$\nu(x) \equiv \int_0^\infty \frac{x^t\,dt}{\Gamma(t+1)}$$

$$\nu(x,\alpha) \equiv \int_0^\infty \frac{x^{\alpha+t}\,dt}{\Gamma(\alpha+t+1)},$$

where $\Gamma(z)$ is the GAMMA FUNCTION. See Gradshteyn and Ryzhik (1980, p. 1079).

see also LAMBDA FUNCTION, MU FUNCTION

References

Gradshteyn, I. S. and Ryzhik, I. M. *Tables of Integrals, Series, and Products, 5th ed.* San Diego, CA: Academic Press, 1979.

Null Function

A null function $\delta^0(x)$ satisfies

$$\int_a^b \delta^0(x)\,dx = 0 \tag{1}$$

for all a, b, so

$$\int_{-\infty}^\infty |\delta^0(x)|\,dx = 0. \tag{2}$$

Like a DELTA FUNCTION, they satisfy

$$\delta_0(x) = \begin{cases} 0 & x \neq 0 \\ 1 & x = 0. \end{cases} \tag{3}$$

see also DELTA FUNCTION, LERCH'S THEOREM

Null Graph

A GRAPH containing only VERTICES and no EDGES.

Null Hypothesis

A hypothesis which is tested for possible rejection under the assumption that it is true (usually that observations are the result of chance). The concept was introduced by R. A. Fisher.

Null Tetrad

$$g_{ij} = \begin{bmatrix} 0 & 1 & 0 & 0 \\ 1 & 0 & 0 & 0 \\ 0 & 0 & 0 & -1 \\ 0 & 0 & -1 & 0 \end{bmatrix}.$$

It can be expressed as

$$g_{ab} = l_a n_b + l_b n_a - m_a \bar{m}_b - m_b \bar{m}_a.$$

see also TETRAD

References

d'Inverno, R. *Introducing Einstein's Relativity.* Oxford, England: Oxford University Press, pp. 248–249, 1992.

Nullspace

Also called the KERNEL. If T is a linear transformation of $\mathbb{R}^n$, then Null(T) is the set of all VECTORS $\mathbf{X}$ such that $T(X) = 0$, i.e.,

$$\text{Null}(T) \equiv \{\mathbf{X} : T(\mathbf{X}) = 0\}.$$

Nullstellansatz

see HILBERT'S NULLSTELLANSATZ

Number

The word "number" is a general term which refers to a member of a given (possibly ordered) SET. The meaning of "number" is often clear from context (i.e., does it refer to a COMPLEX NUMBER, INTEGER, REAL NUMBER, etc.?). Wherever possible in this work, the word "number" is used to refer to quantities which are INTEGERS, and "CONSTANT" is reserved for nonintegral numbers which have a fixed value. Because terms such as REAL NUMBER, BERNOULLI NUMBER, and IRRATIONAL NUMBER are commonly used to refer to nonintegral quantities, however, it is not possible to be entirely consistent in nomenclature.

see also ABUNDANT NUMBER, ACKERMANN NUMBER, ALGEBRAIC NUMBER, ALMOST PERFECT NUMBER, AMENABLE NUMBER, AMICABLE NUMBERS, ANTIMORPHIC NUMBER, APOCALYPSE NUMBER, APOCALYPTIC NUMBER, ARMSTRONG NUMBER, ARRANGEMENT NUMBER, BELL NUMBER, BERNOULLI NUMBER, BERTELSEN'S NUMBER, BETROTHED NUMBERS, BETTI NUMBER, BEZOUT NUMBERS, BINOMIAL NUMBER, BRAUER NUMBER, BROWN NUMBERS, CARDINAL NUMBER, CARMICHAEL NUMBER, CATALAN NUMBER, CAYLEY NUMBER, CENTERED CUBE NUMBER, CENTERED SQUARE NUMBER, CHAITIN'S NUMBER, CHERN NUMBER, CHOICE NUMBER, CHRISTOFFEL NUMBER, CLIQUE NUMBER, COLUMBIAN NUMBER, COMPLEX NUMBER, COMPUTABLE NUMBER, CONDITION NUMBER, CONGRUENT NUMBERS, CONSTRUCTIBLE NUMBER, COTES NUMBER, CROSSING NUMBER (GRAPH), CROSSING NUMBER (LINK), CUBIC NUMBER, CULLEN NUMBER, CUNNINGHAM NUMBER, CYCLIC NUMBER, CYCLOMATIC NUMBER, D-NUMBER, DE MOIVRE NUMBER, DEFICIENT NUMBER, DELANNOY NUMBER, DEMLO NUMBER, DIAGONAL RAMSEY NUMBER, e-PERFECT NUMBER, EBAN NUMBER, EDDINGTON NUMBER, EDGE NUMBER, ENNEAGONAL NUMBER, ENTRINGER NUMBER, ERDŐS NUMBER, EUCLID NUMBER, EULER'S IDONEAL NUMBER, EULER NUMBER, EULERIAN NUMBER, EULER ZIGZAG NUMBER, EVEN NUMBER, FACTORIAL NUMBER, FERMAT NUMBER, FIBONACCI NUMBER, FIGURATE NUMBER, G-NUMBER, GENOCCHI NUMBER, GIUGA NUMBER, GNOMIC NUMBER, GONAL NUMBER, GRAHAM'S NUMBER, GREGORY NUMBER, HAILSTONE NUMBER, HANSEN NUMBER, HAPPY NUMBER, HARMONIC DIVISOR NUMBER, HARMONIC NUMBER, HARSHAD NUMBER, HEEGNER NUMBER, HEESCH NUMBER, HELLY NUMBER, HEPTAGONAL NUMBER, HETEROGENEOUS NUMBERS, HEX NUMBER, HEX PYRAMIDAL NUMBER, HEXAGONAL NUMBER, HOMOGENEOUS NUMBERS, HURWITZ NUMBER, HYPERCOMPLEX NUMBER, HYPERPERFECT NUMBER, i, IDONEAL NUMBER, IMAGINARY NUMBER, INDEPENDENCE NUMBER, INFINARY MULTIPERFECT NUMBER, INFINARY PERFECT NUMBER, IRRATIONAL NUMBER, IRREDUCIBLE SEMIPERFECT NUMBER, IRREDUNDANT RAMSEY NUMBER, j, KAPREKAR NUMBER, KEITH NUMBER, KISSING NUMBER, KNÖDEL NUMBERS, LAGRANGE NUMBER (DIOPHANTINE EQUATION), LAGRANGE NUMBER (RATIONAL APPROXIMATION), LARGE NUMBER, LEAST DEFICIENT NUMBER, LEHMER NUMBER, LEVIATHAN NUMBER, LIOUVILLE NUMBER, LOGARITHMIC NUMBER, LUCAS NUMBER, LUCKY NUMBER, MACMAHON'S PRIME NUMBER OF MEASUREMENT, MARKOV NUMBER, MCNUGGET NUMBER, MÉNAGE NUMBER, MERSENNE NUMBER, MOTZKIN NUMBER, MULTIPLICATIVE PERFECT NUMBER, MULTIPLY PERFECT NUMBER, NARCISSISTIC NUMBER, NATURAL NUMBER, NEAR NOBLE NUMBER, NEXUS NUMBER, NIVEN NUMBER, NOBLE NUMBER, NONAGONAL NUMBER, NORMAL NUMBER, NSW NUMBER, NUMBER GUESSING, OBLONG NUMBER, OCTAGONAL NUMBER, OCTAHEDRAL NUMBER, ODD NUMBER, ORE NUMBER, ORDINAL NUMBER, PENTAGONAL NUMBER, PENTATOPE NUMBER, PERFECT DIGITAL INVARIANT, PERFECT NUMBER, PERSISTENT NUMBER, PLUPERFECT NUMBER, PLUS PERFECT NUMBER, PLUTARCH NUMBERS, POLYGONAL NUMBER,

PONTRYAGIN NUMBER, POULET NUMBER, POWERFUL NUMBER, PRACTICAL NUMBER, PRIMARY, PRIMITIVE ABUNDANT NUMBER, PRIMITIVE PSEUDOPERFECT NUMBER, PRIMITIVE SEMIPERFECT NUMBER, PSEUDOPERFECT NUMBER, PSEUDORANDOM NUMBER, PSEUDOSQUARE, PYRAMIDAL NUMBER, Q-NUMBER, QUASIPERFECT NUMBER, RAMSEY NUMBER, RATIONAL NUMBER, REAL NUMBER, RENCONTRES NUMBER, RECURRING DIGITAL INVARIANT, REPFIGIT NUMBER, RHOMBIC DODECAHEDRAL NUMBER, RIESEL NUMBER, ROTATION NUMBER, RSA NUMBER, SARRUS NUMBER, SCHRÖDER NUMBER, SCHUR NUMBER, SECANT NUMBER, SEGMENTED NUMBER, SELF-DESCRIPTIVE NUMBER, SELF NUMBER, SEMIPERFECT NUMBER, SIERPIŃSKI NUMBER OF THE FIRST KIND, SIERPIŃSKI NUMBER OF THE SECOND KIND, SINGLY EVEN NUMBER, SKEWES NUMBER, SMALL NUMBER, SMITH NUMBER, SMOOTH NUMBER, SOCIABLE NUMBERS, SPRAGUE-GRUNDY NUMBER, SQUARE NUMBER, SQUARE PYRAMIDAL NUMBER, STAR NUMBER, STELLA OCTANGULA NUMBER, STIEFEL-WHITNEY NUMBER, STIRLING CYCLE NUMBER, STIRLING SET NUMBER, STØRMER NUMBER, SUBLIME NUMBER, SUITABLE NUMBER, SUM-PRODUCT NUMBER, SUPER-3 NUMBER, SUPER CATALAN NUMBER, SUPERABUNDANT NUMBER, SUPERPERFECT NUMBER, SUPER-POULET NUMBER, TANGENT NUMBER, TAXICAB NUMBER, TETRAHEDRAL NUMBER, TRANSCENDENTAL NUMBER, TRANSFINITE NUMBER, TRIANGULAR NUMBER, TRIBONACCI NUMBER, TRIMORPHIC NUMBER, TRUNCATED OCTAHEDRAL NUMBER, TRUNCATED TETRAHEDRAL NUMBER, TWIST NUMBER, U-NUMBER, ULAM NUMBER, UNDULATING NUMBER, UNHAPPY NUMBER, UNITARY MULTIPERFECT NUMBER, UNITARY PERFECT NUMBER, UNTOUCHABLE NUMBER, VAMPIRE NUMBER, VAN DER WAERDEN NUMBER, VR NUMBER, WEIRD NUMBER, WHOLE NUMBER, WOODALL NUMBER, Z-NUMBER, ZAG NUMBER, ZEISEL NUMBER, ZIG NUMBER

References

Barbeau, E. J. *Power Play: A Country Walk through the Magical World of Numbers.* Providence, RI: Amer. Math. Soc., 1997.

Bogomolny, A. "What is a Number." `http://www.cut-the-knot.com/do_you_know/numbers.html`.

Borwein, J. and Borwein, P. *A Dictionary of Real Numbers.* London: Chapman & Hall, 1990.

Conway, J. H. and Guy, R. K. *The Book of Numbers.* New York: Springer-Verlag, 1996.

Dantzig, T. *Number: The Language of Science, 4th rev. ed.* New York: Free Press, 1985.

Davis, P. J. *The Lore of Large Numbers.* New York: Random House, 1961.

Frege, G. *Grundlagen der Arithmetik: Eine logisch mathematische Untersuchung über den Begriff der Zahl.* New York: Georg Olms, 1997.

Frege, G. *Foundations of Arithmetic: A Logico-Mathematical Enquiry into the Concept of Number.* Evanston, IL: Northwestern University Press, 1968.

Ifrah, G. *From One to Zero: A Universal History of Numbers.* New York: Viking, 1987.

Le Lionnais, F. *Les nombres remarquables.* Paris: Hermann, 1983.

Phillips, R. *Numbers: Facts, Figures & Fiction.* Cambridge, England: Cambridge University Press, 1994.

Russell, B. "Definition of Number." *Introduction to Mathematical Philosophy.* New York: Simon and Schuster, 1971.

Smeltzer, D. *Man and Number.* Buchanan, NY: Emerson Books, 1974.

Wells, D. W. *The Penguin Dictionary of Curious and Interesting Numbers.* Harmondsworth, England: Penguin Books, 1986.

Number Axis

see REAL LINE

Number Field

If r is an ALGEBRAIC NUMBER of degree n, then the totality of all expressions that can be constructed from r by repeated additions, subtractions, multiplications, and divisions is called a number field (or an ALGEBRAIC NUMBER FIELD) generated by r, and is denoted $F[r]$. Formally, a number field is a finite extension $\mathbb{Q}(\alpha)$ of the FIELD $\mathbb{Q}$ of RATIONAL NUMBERS.

The numbers of a number field which are ROOTS of a POLYNOMIAL

$$z^n + a_{n-1}z^{n-1} + \ldots + a_0 = 0$$

with integral coefficients and leading coefficient 1 are called the ALGEBRAIC INTEGERS of that field.

see also ALGEBRAIC FUNCTION FIELD, ALGEBRAIC INTEGER, ALGEBRAIC NUMBER, FIELD, FINITE FIELD, $\mathbb{Q}$, QUADRATIC FIELD

References

Courant, R. and Robbins, H. *What is Mathematics?: An Elementary Approach to Ideas and Methods, 2nd ed.* Oxford, England: Oxford University Press, p. 127, 1996.

Shanks, D. *Solved and Unsolved Problems in Number Theory, 4th ed.* New York: Chelsea, pp. 151–152, 1993.

Number Field Sieve Factorization Method

An extremely fast factorization method developed by Pollard which was used to factor the RSA-130 NUMBER. This method is the most powerful known for factoring general numbers, and has complexity

$$\mathcal{O}\{\exp[c(\log n)^{1/3}(\log\log n)^{2/3}]\},$$

reducing the *exponent* over the CONTINUED FRACTION FACTORIZATION ALGORITHM and QUADRATIC SIEVE FACTORIZATION METHOD. There are three values of c relevant to different flavors of the method (Pomerance 1996). For the "special" case of the algorithm applied to numbers near a large POWER,

$$c = (\tfrac{32}{9})^{1/3} = 1.523\ldots,$$

for the "general" case applicable to any ODD POSITIVE number which is not a POWER,

$$c = (\tfrac{64}{9})^{1/3} = 1.923\ldots,$$

and for a version using many POLYNOMIALS (Coppersmith 1993),

$$c = \tfrac{1}{3}(92 + 26\sqrt{13})^{1/3} = 1.902\ldots.$$

References

Coppersmith, D. "Modifications to the Number Field Sieve." *J. Cryptology* **6**, 169–180, 1993.

Coppersmith, D.; Odlyzko, A. M.; and Schroeppel, R. "Discrete Logarithms in GF(p)." *Algorithmics* **1**, 1–15, 1986.

Cowie, J.; Dodson, B.; Elkenbracht-Huizing, R. M.; Lenstra, A. K.; Montgomery, P. L.; Zayer, J. A. "World Wide Number Field Sieve Factoring Record: On to 512 Bits." In *Advances in Cryptology—ASIACRYPT '96 (Kyongju)* (Ed. K. Kim and T. Matsumoto.) New York: Springer-Verlag, pp. 382–394, 1996.

Elkenbracht-Huizing, M. "A Multiple Polynomial General Number Field Sieve." *Algorithmic Number Theory (Talence, 1996)*. New York: Springer-Verlag, pp. 99–114, 1996.

Elkenbracht-Huizing, M. "An Implementation of the Number Field Sieve." *Experiment. Math.* **5**, 231–253, 1996.

Elkenbracht-Huizing, R.-M. "Historical Background of the Number Field Sieve Factoring Method." *Nieuw Arch. Wisk.* **14**, 375–389, 1996.

Lenstra, A. K. and Lenstra, H. W. Jr. "Algorithms in Number Theory." In *Handbook of Theoretical Computer Science, Volume A: Algorithms and Complexity* (Ed. J. van Leeuwen). New York: Elsevier, pp. 673–715, 1990.

Pomerance, C. "A Tale of Two Sieves." *Not. Amer. Math. Soc.* **43**, 1473–1485, 1996.

Number Group

see FIELD

Number Guessing

By asking a small number of innocent-sounding questions about an unknown number, it is possible to reconstruct the number with absolute certainty (assuming that the questions are answered correctly). Ball and Coxeter (1987) give a number of sets of questions which can be used.

One of the simplest algorithms uses only three questions to determine an unknown number n:

1. Triple n and announce if the result $n' = 3n$ is EVEN or ODD.
2. If you were told that n' is EVEN, ask the person to reveal the number n'' which is half of n'. If you were told that n' is ODD, ask the person to reveal the number n'' which is half of $n' + 1$.
3. Ask the person to reveal the number of times k which 9 divides evenly into $n''' = 3n''$.

The original number n is then given by $2k$ if n' was EVEN, or $2k + 1$ if n' was ODD. For $n = 2m$ even, $n' = 6m$, $n'' = 3m$, $n''' = 9m$, $k = m$, so $2k = 2m = n$. For $n = 2m + 1$ odd, $n' = 6m + 3$, $n'' = 3m + 2$, $n''' = 9m + 6$, $k = m$, so $2k + 1 = 2m + 1 = n$.

Another method asks:

1. Multiply the number n by 5.
2. Add 6 to the product.
3. Multiply the sum by 4.
4. Add 9 to the product.
5. Multiply the sum by 5 and reveal the result n'.

The original number is then given by $n = (n' - 165)/100$, since the above steps give $n' = 5(4(5n+6)+9) = 100n + 165$.

References

Bachet, C. G. *Problèmes plaisans et délectables, 2nd ed.* 1624.

Ball, W. W. R. and Coxeter, H. S. M. *Mathematical Recreations and Essays, 13th ed.* New York: Dover, pp. 5–20, 1987.

Kraitchik, M. "To Guess a Selected Number." §3.3 in *Mathematical Recreations.* New York: W. W. Norton, pp. 58–66, 1942.

Number Pyramid

A set of numbers obeying a pattern like the following.

$$\begin{aligned} 91 \cdot 37 &= 3367 \\ 9901 \cdot 3367 &= 33336667 \\ 999001 \cdot 333667 &= 333333666667 \\ 99990001 \cdot 33336667 &= 3333333366666667 \end{aligned}$$

$$\begin{aligned} 4^2 &= 16 \\ 34^2 &= 1156 \\ 334^2 &= 111556 \end{aligned}$$

$$\begin{aligned} 7^2 &= 49 \\ 67^2 &= 4489 \\ 667^2 &= 444889. \end{aligned}$$

see also AUTOMORPHIC NUMBER

References

Heinz, H. "Miscellaneous Number Patters." `http://www.geocities.com/CapeCanaveral/Launchpad/4057/miscnum.htm`.

Number System

see BASE (NUMBER)

Number Theoretic Transform

Simplemindedly, a number theoretic transform is a generalization of a FAST FOURIER TRANSFORM obtained by replacing $e^{-2\pi ik/N}$ with an nth PRIMITIVE ROOT OF UNITY. This effectively means doing a transform over the QUOTIENT RING $\mathbb{Z}/p\mathbb{Z}$ instead of the COMPLEX NUMBERS $\mathbb{C}$. The theory is rather elegant and uses the language of FINITE FIELDS and NUMBER THEORY.

see also FAST FOURIER TRANSFORM, FINITE FIELD

References

Arndt, J. "Numbertheoretic Transforms (NTTs)." Ch. 4 in "Remarks on FFT Algorithms." `http://www.jjj.de/fxt/`.

Cohen, H. *A Course in Computational Algebraic Number Theory.* New York: Springer-Verlag, 1993.

Number Theory

A vast and fascinating field of mathematics consisting of the study of the properties of whole numbers. PRIMES and PRIME FACTORIZATION are especially important in number theory, as are a number of functions such as the DIVISOR FUNCTION, RIEMANN ZETA FUNCTION, and TOTIENT FUNCTION. Excellent introductions to number theory may be found in Ore (1988) and Beiler (1966). The classic history on the subject (now slightly dated) is that of Dickson (1952).

see also ARITHMETIC, CONGRUENCE, DIOPHANTINE EQUATION, DIVISOR FUNCTION, GÖDEL'S INCOMPLETENESS THEOREM, PEANO'S AXIOMS, PRIME COUNTING FUNCTION, PRIME FACTORIZATION, PRIME NUMBER, QUADRATIC RECIPROCITY THEOREM, RIEMANN ZETA FUNCTION, TOTIENT FUNCTION

References

Andrews, G. E. *Number Theory.* New York: Dover, 1994.

Andrews, G. E.; Berndt, B. C.; and Rankin, R. A. (Ed.). *Ramanujan Revisited: Proceedings of the Centenary Conference, University of Illinois at Urbana-Champaign, June 1–5, 1987.* Boston, MA: Academic Press, 1988.

Apostol, T. M. *Introduction to Analytic Number Theory.* New York: Springer-Verlag, 1976.

Ayoub, R. G. *An Introduction to the Analytic Theory of Numbers.* Providence, RI: Amer. Math. Soc., 1963.

Beiler, A. H. *Recreations in the Theory of Numbers: The Queen of Mathematics Entertains, 2nd ed.* New York: Dover, 1966.

Bellman, R. E. *Analytic Number Theory: An Introduction.* Reading, MA: Benjamin/Cummings, 1980.

Berndt, B. C. *Ramanujan's Notebooks, Part I.* New York: Springer-Verlag, 1985.

Berndt, B. C. *Ramanujan's Notebooks, Part II.* New York: Springer-Verlag, 1988.

Berndt, B. C. *Ramanujan's Notebooks, Part III.* New York: Springer-Verlag, 1997.

Berndt, B. C. *Ramanujan's Notebooks, Part IV.* New York: Springer-Verlag, 1993.

Berndt, B. C. *Ramanujan's Notebooks, Part V.* New York: Springer-Verlag, 1997.

Berndt, B. C. and Rankin, R. A. *Ramanujan: Letters and Commentary.* Providence, RI: Amer. Math. Soc, 1995.

Borwein, J. M. and Borwein, P. B. *Pi and the AGM: A Study in Analytic Number Theory and Computational Complexity.* New York: Wiley, 1987.

Brown, K. S. "Number Theory." `http://www.seanet.com/~ksbrown/inumber.htm`.

Burr, S. A. *The Unreasonable Effectiveness of Number Theory.* Providence, RI: Amer. Math. Soc., 1992.

Burton, D. M. *Elementary Number Theory, 4th ed.* Boston, MA: Allyn and Bacon, 1989.

Carmichael, R. D. *The Theory of Numbers, and Diophantine Analysis.* New York: Dover, 1959.

Cohn, H. *Advanced Number Theory.* New York: Dover, 1980.

Courant, R. and Robbins, H. "The Theory of Numbers." Supplement to Ch. 1 in *What is Mathematics?: An Elementary Approach to Ideas and Methods, 2nd ed.* Oxford, England: Oxford University Press, pp. 21–51, 1996.

Davenport, H. *The Higher Arithmetic: An Introduction to the Theory of Numbers, 6th ed.* Cambridge, England: Cambridge University Press, 1992.

Davenport, H. and Montgomery, H. L. *Multiplicative Number Theory, 2nd ed.* New York: Springer-Verlag, 1980.

Dickson, L. E. *History of the Theory of Numbers, 3 vols.* New York: Chelsea, 1952.

Dudley, U. *Elementary Number Theory.* San Francisco, CA: W. H. Freeman, 1978.

Friedberg, R. *An Adventurer's Guide to Number Theory.* New York: Dover, 1994.

Gauss, C. F. *Disquisitiones Arithmeticae.* New Haven, CT: Yale University Press, 1966.

Guy, R. K. *Unsolved Problems in Number Theory, 2nd ed.* New York: Springer-Verlag, 1994.

Hardy, G. H. and Wright, E. M. *An Introduction to the Theory of Numbers, 5th ed.* Oxford, England: Clarendon Press, 1979.

Hardy, G. H. *Ramanujan: Twelve Lectures on Subjects Suggested by His Life and Work, 3rd ed.* New York: Chelsea, 1959.

Hasse, H. *Number Theory.* Berlin: Springer-Verlag, 1980.

Ireland, K. F. and Rosen, M. I. *A Classical Introduction to Modern Number Theory, 2nd ed.* New York: Springer-Verlag, 1995.

Klee, V. and Wagon, S. *Old and New Unsolved Problems in Plane Geometry and Number Theory.* Washington, DC: Math. Assoc. Amer., 1991.

Koblitz, N. *A Course in Number Theory and Cryptography.* New York: Springer-Verlag, 1987.

Landau, E. *Elementary Number Theory, 2nd ed.* New York: Chelsea, 1988.

Lang, S. *Algebraic Number Theory, 2nd ed.* New York: Springer-Verlag, 1994.

Lenstra, H. W. and Tijdeman, R. (Eds.). *Computational Methods in Number Theory, 2 vols.* Amsterdam: Mathematisch Centrum, 1982.

LeVeque, W. J. *Fundamentals of Number Theory.* New York: Dover, 1996.

Mitrinovic, D. S. and Sandor, J. *Handbook of Number Theory.* Dordrecht, Netherlands: Kluwer, 1995.

Niven, I. M.; Zuckerman, H. S.; and Montgomery, H. L. *An Introduction to the Theory of Numbers, 5th ed.* New York: Wiley, 1991.

Ogilvy, C. S. and Anderson, J. T. *Excursions in Number Theory.* New York: Dover, 1988.

Ore, Ø. *Invitation to Number Theory.* Washington, DC: Math. Assoc. Amer., 1967.

Ore, Ø. *Number Theory and Its History.* New York: Dover, 1988.

Rose, H. E. *A Course in Number Theory, 2nd ed.* Oxford, England: Clarendon Press, 1995.

Rosen, K. H. *Elementary Number Theory and Its Applications, 3rd ed.* Reading, MA: Addison-Wesley, 1993.

Schroeder, M. R. *Number Theory in Science and Communication: With Applications in Cryptography, Physics, Digital Information, Computing, and Self-Similarity, 3rd ed.* New York: Springer-Verlag, 1997.

Shanks, D. *Solved and Unsolved Problems in Number Theory, 4th ed.* New York: Chelsea, 1993.

Sierpinski, W. *250 Problems in Elementary Number Theory.* New York: American Elsevier, 1970.

Uspensky, J. V. and Heaslet, M. A. *Elementary Number Theory.* New York: McGraw-Hill, 1939.

Vinogradov, I. M. *Elements of Number Theory, 5th rev. ed.* New York: Dover, 1954.

Weil, A. *Basic Number Theory, 3rd ed.* Berlin: Springer-Verlag, 1995.

Weil, A. *Number Theory: An Approach Through History From Hammurapi to Legendre.* Boston, MA: Birkhäuser, 1984.

Weyl, H. *Algebraic Theory of Numbers.* Princeton, NJ: Princeton University Press, 1998.

Number Triangle

see BELL TRIANGLE, CLARK'S TRIANGLE, EULER'S TRIANGLE, LEIBNIZ HARMONIC TRIANGLE, PASCAL'S TRIANGLE, SEIDEL-ENTRINGER-ARNOLD TRIANGLE, TRINOMIAL TRIANGLE

Number Wall

see QUOTIENT-DIFFERENCE TABLE

Numerator

The number p in a FRACTION p/q.

see also DENOMINATOR, FRACTION, RATIONAL NUMBER

Numeric Function

A FUNCTION $f : A \to B$ such that B is a SET of numbers.

Numerical Derivative

While it is usually much easier to compute a DERIVATIVE instead of an INTEGRAL (which is a little strange, considering that "more" functions have integrals than derivatives), there are still many applications where derivatives need to be computed numerically. The simplest approach simply uses the definition of the DERIVATIVE

$$f'(x) \equiv \lim_{h\to 0} \frac{f(x+h)-f(x)}{h}$$

for some small numerical value of $h \ll 1$.

see also NUMERICAL INTEGRATION

References

Press, W. H.; Flannery, B. P.; Teukolsky, S. A.; and Vetterling, W. T. "Numerical Derivatives." §5.7 in *Numerical Recipes in FORTRAN: The Art of Scientific Computing, 2nd ed.* Cambridge, England: Cambridge University Press, pp. 180–184, 1992.

Numerical Integration

The approximate computation of an INTEGRAL. The numerical computation of an INTEGRAL is sometimes called QUADRATURE. There are a wide range of methods available for numerical integration. A good source for such techniques is Press *et al.* (1992).

The most straightforward numerical integration technique uses the NEWTON-COTES FORMULAS (also called QUADRATURE FORMULAS), which approximate a function tabulated at a sequent of regularly spaced INTERVALS by various degree POLYNOMIALS. If the endpoints are tabulated, then the 2- and 3-point formulas are called the TRAPEZOIDAL RULE and SIMPSON'S RULE, respectively. The 5-point formula is called BODE'S RULE. A generalization of the TRAPEZOIDAL RULE is ROMBERG INTEGRATION, which can yield accurate results for many fewer function evaluations.

If the functions are known analytically instead of being tabulated at equally spaced intervals, the best numerical method of integration is called GAUSSIAN QUADRATURE. By picking the abscissas at which to evaluate the function, Gaussian quadrature produces the most accurate approximations possible. However, given the speed of modern computers, the additional complication of the GAUSSIAN QUADRATURE formalism often makes it less desirable than simply brute-force calculating twice as many points on a regular grid (which also permits the already computed values of the function to be re-used). An excellent reference for GAUSSIAN QUADRATURE is Hildebrand (1956).

see also DOUBLE EXPONENTIAL INTEGRATION, FILON'S INTEGRATION FORMULA, INTEGRAL, INTEGRATION, NUMERICAL DERIVATIVE, QUADRATURE

References

Davis, P. J. and Rabinowitz, P. *Methods of Numerical Integration, 2nd ed.* New York: Academic Press, 1984.

Hildebrand, F. B. *Introduction to Numerical Analysis.* New York: McGraw-Hill, pp. 319–323, 1956.

Milne, W. E. *Numerical Calculus: Approximations, Interpolation, Finite Differences, Numerical Integration and Curve Fitting.* Princeton, NJ: Princeton University Press, 1949.

Press, W. H.; Flannery, B. P.; Teukolsky, S. A.; and Vetterling, W. T. *Numerical Recipes in FORTRAN: The Art of Scientific Computing, 2nd ed.* Cambridge, England: Cambridge University Press, 1992.

Numerology

The study of numbers for the supposed purpose of predicting future events or seeking connections with the occult.

see also BEAST NUMBER, NUMBER THEORY

NURBS Curve

A nonuniform rational B-SPLINE curve defined by

$$\mathbf{C}(t) = \frac{\sum_{i=0}^{n} N_{i,p}(t) w_i \mathbf{P}_i}{\sum_{i=0}^{n} N_{i,p}(t) w_i},$$

where p is the order, $N_{i,p}$ are the B-SPLINE basis functions, $\mathbf{P}_i$ are control points, and the weight w_i of $\mathbf{P}_i$ is the last ordinate of the homogeneous point $\mathbf{P}_i^w$. These curves are closed under perspective transformations and can represent CONIC SECTIONS exactly.

see also B-SPLINE, BÉZIER CURVE, NURBS SURFACE

References

Piegl, L. and Tiller, W. *The NURBS Book, 2nd ed* New York: Springer-Verlag, 1997.

NURBS Surface

A nonuniform rational B-SPLINE surface of degree (p, q) is defined by

$$\mathbf{S}(u,v) = \frac{\sum_{i=0}^{m}\sum_{j=0}^{n} N_{i,p}(u) N_{j,q}(v) w_{i,j} \mathbf{P}_{i,j}}{\sum_{i=0}^{m}\sum_{j=0}^{n} N_{i,p}(u) N_{j,q}(v) w_{i,j}},$$

where $N_{i,p}$ and $N_{j,q}$ are the B-SPLINE basis functions, $\mathbf{P}_{i,j}$ are control points, and the weight $w_{i,j}$ of $\mathbf{P}_{i,j}$ is the last ordinate of the homogeneous point $\mathbf{P}_{i,j}^w$.

see also B-SPLINE, BÉZIER CURVE, NURBS CURVE

Nyquist Frequency

In order to recover all FOURIER components of a periodic waveform, it is necessary to sample twice as fast as the *highest* waveform frequency ν,

$$f_{\text{Nyquist}} = 2\nu.$$

The minimum sampling frequency is called the Nyquist frequency.

see also FOURIER SERIES, FOURIER TRANSFORM, NYQUIST SAMPLING, OVERSAMPLING, SAMPLING THEOREM

Nyquist Sampling

Sampling at the NYQUIST FREQUENCY.

O

Obelus

The symbol ÷ used to indicate DIVISION. In typography, an obelus has a more general definition as any symbol, such as the dagger (†), used to indicate a footnote.

see also DIVISION, SOLIDUS

Object

A mathematical structure (e.g., a GROUP, VECTOR SPACE, or DIFFERENTIABLE MANIFOLD) in a CATEGORY.

see also MORPHISM

Oblate Spheroid

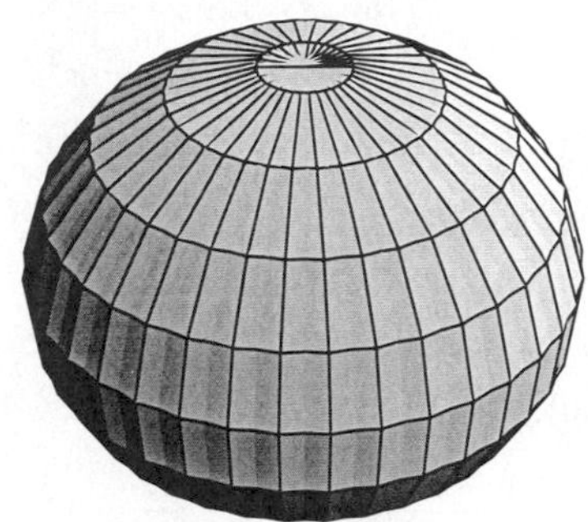

A "squashed" SPHEROID for which the equatorial radius a is greater than the polar radius c, so $a > c$. To first approximation, the shape assumed by a rotating fluid (including the Earth, which is "fluid" over astronomical time scales) is an oblate spheroid. The oblate spheroid can be specified parametrically by the usual SPHEROID equations (for a SPHEROID with z-AXIS as the symmetry axis),

$$x = a \sin v \cos u \qquad (1)$$

$$y = a \sin v \sin u \qquad (2)$$

$$z = c \cos v, \qquad (3)$$

with $a > c$, $u \in [0, 2\pi)$, and $v \in [0, \pi]$. Its Cartesian equation is

$$\frac{x^2 + y^2}{a^2} + \frac{z^2}{c^2} = 1. \qquad (4)$$

The ELLIPTICITY of an oblate spheroid is defined by

$$e \equiv \sqrt{\frac{a^2 - c^2}{a^2}}, \qquad (5)$$

so that

$$1 - e^2 = \frac{c^2}{a^2}. \qquad (6)$$

Then the radial distance from the rotation axis is given by

$$r(\delta) = a\left(1 + \frac{e^2}{1 - e^2}\sin^2\delta\right)^{-1/2} \qquad (7)$$

as a function of the LATITUDE δ.

The SURFACE AREA and VOLUME of an oblate spheroid are

$$S = 2\pi a^2 + \pi \frac{c^2}{e} \ln\left(\frac{1+e}{1-e}\right) \qquad (8)$$

$$V = \tfrac{4}{3}\pi a^2 c. \qquad (9)$$

An oblate spheroid with its origin at a FOCUS has equation

$$r = \frac{a(1 - e^2)}{1 + e\cos\phi}. \qquad (10)$$

Define k and expand up to POWERS of e^6,

$$\begin{aligned} k &\equiv e^2(1 - e^2)^{-1} = e^2(1 + e^2 - 2e^4 + 6e^6 + \ldots) \\ &= e^2 + e^4 - 2e^6 + \ldots \qquad (11) \\ k^2 &= e^4 + e^6 + \ldots \qquad (12) \\ k^3 &= e^6 + \ldots. \qquad (13) \end{aligned}$$

Expanding r in POWERS of ELLIPTICITY to e^6 therefore yields

$$\begin{aligned} \frac{r}{a} = 1 &- \tfrac{1}{2}(e^2 + e^4 - 2e^4 + 6e^6)\sin^2\delta \\ &+ \tfrac{3}{4}(e^4 + e^6)\sin^4\delta - \tfrac{15}{8}e^6\sin^6\delta + \ldots. \qquad (14) \end{aligned}$$

In terms of LEGENDRE POLYNOMIALS,

$$\begin{aligned} \frac{r}{a} &= (1 - \tfrac{1}{6}e^2 - \tfrac{11}{20}e^4 - \tfrac{103}{1680}e^6) \\ &+ (-\tfrac{1}{3}e^2 - \tfrac{5}{42}e^4 - \tfrac{3}{56}e^6)P_2 \\ &+ (\tfrac{3}{35}e^4 + \tfrac{57}{770}e^6)P_4 - \tfrac{5}{231}e^6 P_6 + \ldots. \qquad (15) \end{aligned}$$

The ELLIPTICITY may also be expressed in terms of the OBLATENESS (also called FLATTENING), denoted ϵ or f.

$$\epsilon \equiv \frac{a - c}{a} \qquad (16)$$

$$c = a(1 - \epsilon) \qquad (17)$$

$$c^2 = a^2(1 - \epsilon)^2 \qquad (18)$$

$$(1 - \epsilon)^2 = 1 - e^2, \qquad (19)$$

so

$$\epsilon = 1 - \sqrt{1 - e^2} \qquad (20)$$

and

$$e^2 = 1 - (1 - \epsilon)^2 = 1 - (1 - 2\epsilon + \epsilon^2) = 2\epsilon - \epsilon^2 \qquad (21)$$

$$r = a\left[1 + \frac{2\epsilon - \epsilon^2}{(1 - \epsilon)^2}\sin^2\delta\right]^{-1/2}. \qquad (22)$$

Define k and expand up to POWERS of ϵ^6

$$k \equiv (2\epsilon-\epsilon)(1-\epsilon)^{-2} = (2\epsilon-\epsilon^2)(1+2\epsilon-6\epsilon^2+\ldots)$$
$$= 2\epsilon+4\epsilon^4-12\epsilon^3-\epsilon^2-2\epsilon^3+\ldots$$
$$= 2\epsilon+3\epsilon^2-14\epsilon^3+\ldots \quad (23)$$
$$k^2 = 4\epsilon^2+6\epsilon^3+\ldots \quad (24)$$
$$k^3 = 8\epsilon^3+\ldots. \quad (25)$$

Expanding r in POWERS of the OBLATENESS to ϵ^3 yields

$$\frac{r}{a} = 1-\tfrac{1}{2}(2\epsilon+3\epsilon^2-14\epsilon^3)\sin^2\delta$$
$$+\tfrac{3}{4}(4\epsilon^2+6\epsilon^3)\sin^4\delta+8\epsilon^3\sin^6\delta+\ldots. \quad (26)$$

In terms of LEGENDRE POLYNOMIALS,

$$\frac{r}{a} = (1-\tfrac{1}{3}\epsilon-\tfrac{2}{5}\epsilon^2-\tfrac{13}{105}\epsilon^3)+(-\tfrac{2}{3}\epsilon-\tfrac{1}{7}\epsilon^2-\tfrac{1}{21}\epsilon^3)P_2$$
$$+(\tfrac{12}{35}\epsilon^2-\tfrac{96}{385}\epsilon^3)P_4-\tfrac{40}{231}\epsilon^3P_6+\ldots. \quad (27)$$

To find the projection of an oblate spheroid onto a PLANE, set up a coordinate system such that the z-AXIS is towards the observer, and the x-axis is in the PLANE of the page. The equation for an oblate spheroid is

$$r(\theta) = a\left[1+\frac{2\epsilon-\epsilon^2}{(1-\epsilon)^2}\cos^2\theta\right]^{-1/2}. \quad (28)$$

Define

$$k \equiv \frac{2\epsilon-\epsilon^2}{(1-\epsilon)^2}, \quad (29)$$

and $x \equiv \sin\theta$. Then

$$r(\theta) = a[1+k(1-x^2)]^{-1/2} = a(1+k-kx^2)^{-1/2}. \quad (30)$$

Now rotate that spheroid about the x-axis by an ANGLE B so that the new symmetry axes for the spheroid are $x' \equiv x$, y', and z'. The projected height of a point in the $x=0$ PLANE on the y-axis is

$$y = r(\theta)\cos(\theta-B) = r(\theta)(\cos\theta\cos B-\sin\theta\sin B)$$
$$= r(\theta)(\sqrt{1-x^2}\cos B+x\sin B). \quad (31)$$

To find the highest projected point,

$$\frac{dy}{d\theta} = \frac{a\sin(B-\theta)}{(1+k\cos^2\theta)^{1/2}}+ak\frac{\cos(B-\theta)\cos\theta\sin\theta}{(1+k\cos^2\theta)^{3/2}} = 0. \quad (32)$$

Simplifying,

$$\tan(B-\theta)(1+k\cos^2\theta)+k\cos\theta\sin\theta = 0. \quad (33)$$

But

$$\tan(B-\theta) = \frac{\tan B-\tan\theta}{1+\tan B\tan\theta} = \frac{\tan B-\frac{\sin\theta}{\sqrt{1-\sin^2\theta}}}{1+\tan B\frac{\sin\theta}{\sqrt{1-\sin^2\theta}}}$$
$$= \frac{\sqrt{1-\sin^2\theta}\tan B-\sin\theta}{\sqrt{1-\sin^2\theta}+\tan B\sin\theta}. \quad (34)$$

Plugging (34) into (33),

$$\frac{\sqrt{1-x^2}\tan B-x}{\sqrt{1-x^2}+x\tan B}[1+k(1-x^2)]+kx\sqrt{1-x^2} = 0 \quad (35)$$

and performing a number of algebraic simplifications

$$(\sqrt{1-x^2}\tan B-x)(1+k-kx^2)$$
$$+kx\sqrt{1-x^2}(\sqrt{1-x^2}+x\tan B) = 0 \quad (36)$$

$$[(1+k)\sqrt{1-x^2}\tan B-kx^2\sqrt{1-x^2}\tan B$$
$$-x-kx+kx^3]+[kx(1-x^2)+kx^2\sqrt{1-x^2}\tan B] \quad (37)$$

$$(1+k)\tan B\sqrt{1-x^2}-kx(1-x^2)-x+kx(1-x^2) = 0 \quad (38)$$
$$(1+k)\tan B\sqrt{1-x^2} = x \quad (39)$$
$$(1+k)^2\tan^2 B(1-x^2) = x^2 \quad (40)$$
$$x^2[1+(1+k)^2\tan^2 B] = (1+k)^2\tan^2 B \quad (41)$$

finally gives the expression for x in terms of B and k,

$$x^2 = \frac{\tan^2 B(1+k)^2}{1+(1+k)^2\tan^2 B}. \quad (42)$$

Combine (30) and (31) and plug in for x,

$$y = a\frac{\sqrt{1-x^2}\cos B+x\sin B}{\sqrt{1+k-kx^2}}$$
$$= a\frac{\cos B+(1+k)\frac{\sin^2 B}{\cos B}}{\sqrt{(1+k)[1+(1+k)\tan^2 B]}}$$
$$= a\frac{\cos^2 B+(1+k)\sin^2 B}{\cos B\sqrt{(1+k)[1+(1+k)\tan^2 B]}}. \quad (43)$$

Now re-express k in terms of a and c, using $\epsilon \equiv 1-c/a$,

$$k \equiv \frac{(2-\epsilon)\epsilon}{(1-\epsilon)^2} = \frac{\left(1+\frac{c}{a}\right)\left(1-\frac{c}{a}\right)}{\left(\frac{c}{a}\right)^2}$$
$$= \frac{1-\left(\frac{c}{a}\right)^2}{\left(\frac{c}{a}\right)^2} = \left(\frac{a}{c}\right)^2-1, \quad (44)$$

so

$$1+k = \left(\frac{a}{c}\right)^2. \quad (45)$$

Plug (44) and (45) into (43) to obtain the SEMIMINOR AXIS of the projected oblate spheroid,

$$c' = a\frac{\cos^2 B+\left(\frac{a}{c}\right)^2\sin^2 B}{\cos B\sqrt{\left(\frac{a}{c}\right)^2\left[1+\left(\frac{a}{c}\right)^2\tan^2 B\right]}}$$
$$= a\frac{\cos^2 B+\left(\frac{a}{c}\right)^2\sin^2 B}{\frac{a}{c}\sqrt{\cos^2 B+\left(\frac{a}{c}\right)^2\sin^2 B}}$$
$$= c\sqrt{\cos^2 B+\left(\frac{a}{c}\right)^2\sin^2 B} = \sqrt{c^2\cos^2 B+a^2\sin^2 B}$$
$$= a\sqrt{(1-\epsilon)^2\cos^2 B+\sin^2 B}. \quad (46)$$

We wish to find the equation for a spheroid which has been rotated about the $x \equiv x'$-axis by ANGLE B, then the z-axis by ANGLE P

$$\begin{bmatrix} x' \\ y' \\ z' \end{bmatrix} = \begin{bmatrix} 1 & 0 & 0 \\ 0 & \cos B & \sin B \\ 0 & -\sin B & \cos B \end{bmatrix} \begin{bmatrix} \cos P & 0 & \sin P \\ 0 & 1 & 0 \\ -\sin P & 0 & \cos P \end{bmatrix} \begin{bmatrix} x \\ y \\ z \end{bmatrix}$$
$$= \begin{bmatrix} \cos P & 0 & \sin P \\ -\sin B \sin P & \cos B & \sin B \cos P \\ -\cos B \sin P & -\sin B & \cos B \cos P \end{bmatrix} \begin{bmatrix} x \\ y \\ z \end{bmatrix}. \tag{47}$$

Now, in the original coordinates (x', y', z'), the spheroid is given by the equation

$$\frac{x'^2}{a^2} + \frac{y'^2}{c^2} + \frac{z'^2}{a^2} = 1, \tag{48}$$

which becomes in the new coordinates,

$$\frac{(x\cos P + y\sin P)^2}{a^2} + \frac{(-x\sin B\sin P + z\cos B + y\sin B\cos P)^2}{a^2} + \frac{(-x\cos B\sin P - z\sin B + y\cos B\cos P)^2}{c^2} = 1. \tag{49}$$

Collecting COEFFICIENTS,

$$Ax^2 + By^2 + Cz^2 + Dxy + Exz + Fyz = 1, \tag{50}$$

where

$$A \equiv \frac{\cos^2 P + \sin^2 B \sin^2 P}{a^2} + \frac{\cos^2 B \sin^2 P}{c^2} \tag{51}$$

$$B \equiv \frac{\sin^2 P + \sin^2 B \cos^2 P}{a^2} + \frac{\cos^2 B \cos^2 P}{c^2} \tag{52}$$

$$C \equiv \frac{\cos^2 B}{a^2} + \frac{\sin^2 B}{c^2} \tag{53}$$

$$D \equiv 2\cos P \sin P \left(\frac{1 - \sin^2 B}{a^2} - \frac{\cos^2 B}{c^2}\right)$$
$$= 2\cos P \sin P \cos^2 B \left(\frac{1}{a^2} - \frac{1}{c^2}\right) \tag{54}$$

$$E \equiv 2\sin B \cos B \sin P \left(\frac{1}{b^2} - \frac{1}{a^2}\right) \tag{55}$$

$$F \equiv 2\sin B \cos B \cos P \left(\frac{1}{a^2} - \frac{1}{b^2}\right). \tag{56}$$

If we are interested in computing z, the radial distance from the symmetry axis of the spheroid (y) corresponding to a point

$$Cz^2 + (Ex + Fy)z + (Ax^2 + By^2 + Dxy - 1)$$
$$= Cz^2 + G(x,y)z + H(x,y) = 0, \tag{57}$$

where

$$G(x,y) \equiv Ex + Fy \tag{58}$$

$$H(x,y) \equiv Ax^2 + By^2 + Dxy - 1. \tag{59}$$

z can now be computed using the quadratic equation when (x, y) is given,

$$z = \frac{-G(x,y) \pm \sqrt{G^2(x,y) - 4CG(x,y)}}{2C}. \tag{60}$$

If $P = 0$, then we have $\sin P = 0$ and $\cos P = 1$, so (51) to (56) and (58) to (59) become

$$A \equiv \frac{1}{a^2} \tag{61}$$

$$B \equiv \frac{\sin^2 B}{a^2} + \frac{\cos^2 B}{b^2} \tag{62}$$

$$C \equiv \frac{\cos^2 B}{a^2} + \frac{\sin^2 B}{b^2} \tag{63}$$

$$D \equiv 0 \tag{64}$$

$$E \equiv 0 \tag{65}$$

$$F \equiv 2\sin B \cos B \left(\frac{1}{a^2} - \frac{1}{b^2}\right) \tag{66}$$

$$G(x,y) \equiv Fy = 2y \sin B \cos B \left(\frac{1}{a^2} - \frac{1}{b^2}\right) \tag{67}$$

$$H(x,y) \equiv Ax^2 + By^2 - 1$$
$$= \frac{x^2}{a^2} + y^2\left(\frac{\sin^2 B}{a^2} + \frac{\cos^2 B}{b^2}\right) - 1. \tag{68}$$

see also DARWIN-DE SITTER SPHEROID, ELLIPSOID, OBLATE SPHEROIDAL COORDINATES, PROLATE SPHEROID, SPHERE, SPHEROID

References

Beyer, W. H. *CRC Standard Mathematical Tables, 28th ed.* Boca Raton, FL: CRC Press, p. 131, 1987.

Oblate Spheroid Geodesic

The GEODESIC on an OBLATE SPHEROID can be computed analytically for a spheroid specified parametrically by

$$x = a \sin v \cos u \tag{1}$$

$$y = a \sin v \sin u \tag{2}$$

$$z = c \cos v, \tag{3}$$

with $a > c$, although it is much more unwieldy than for a simple SPHERE. Using the first PARTIAL DERIVATIVES

$$\frac{\partial x}{\partial u} = -a \sin v \sin u \qquad \frac{\partial x}{\partial v} = a \cos v \cos u \tag{4}$$

$$\frac{\partial y}{\partial u} = a \sin v \cos u \qquad \frac{\partial y}{\partial v} = a \cos v \sin u \tag{5}$$

$$\frac{\partial z}{\partial u} = 0 \qquad \frac{\partial z}{\partial v} = -c \sin v, \tag{6}$$

and second PARTIAL DERIVATIVES

$$\frac{\partial^2 x}{\partial u^2} = -a \sin v \cos u \qquad \frac{\partial^2 x}{\partial v^2} = -a \sin v \cos u \tag{7}$$

$$\frac{\partial^2 y}{\partial u^2} = -a \sin v \sin u \qquad \frac{\partial^2 y}{\partial v^2} = -a \sin v \sin u \tag{8}$$

$$\frac{\partial^2 z}{\partial u^2} = 0 \qquad \frac{\partial^2 z}{\partial v^2} = -z \cos v, \tag{9}$$

gives the GEODESICS functions as

$$\begin{aligned} P &\equiv \left(\frac{\partial x}{\partial u}\right)^2 + \left(\frac{\partial y}{\partial u}\right)^2 + \left(\frac{\partial z}{\partial u}\right)^2 \\ &= a^2(\sin^2 v \cos^2 u + \sin^2 v \sin^2 u) \\ &= a^2 \sin^2 v \end{aligned} \tag{10}$$

$$Q \equiv \frac{\partial x}{\partial u}\frac{\partial x}{\partial v} + \frac{\partial y}{\partial u}\frac{\partial y}{\partial v} + \frac{\partial z}{\partial u}\frac{\partial z}{\partial v} = 0 \tag{11}$$

$$\begin{aligned} R &\equiv \left(\frac{\partial x}{\partial v}\right)^2 + \left(\frac{\partial y}{\partial v}\right)^2 + \left(\frac{\partial z}{\partial v}\right)^2 \\ &= a^2 + (c^2 - a^2)\sin^2 v = a^2(1 - e^2 \sin^2 v). \end{aligned} \tag{12}$$

Since $Q = 0$ and P and R are explicit functions of v only, we can use the special form of the GEODESIC equation.

$$\begin{aligned} u &= \int \sqrt{\frac{R}{P^2 - c_1{}^2 P}}\, dv \\ &= \int \sqrt{\frac{a^2(1 - e^2 \sin^2 v)}{a^4 \sin^4 v - c_1{}^2 a^2 \sin^2 v}}\, dv \\ &= c_1 \int \sqrt{\frac{1 - e^2 \sin^2 v}{\left(\frac{a}{c_1}\right)^2 \sin^2 v - 1}}\, \frac{dv}{\sin v}. \end{aligned} \tag{13}$$

Integrating gives

$$u = -c_1 \frac{e^2 F\left(\phi | \frac{(d^2-1)e^2}{d^2-e^2}\right) - b^2 \Pi\left(d^2 - 1, \phi | \frac{(d^2-1)e^2}{d^2-e^2}\right)}{\sqrt{d^2 - e^2}}, \tag{14}$$

where

$$d \equiv \frac{a}{c_1} \tag{15}$$

$$\cos\phi \equiv \frac{d \cos v}{\sqrt{d^2 - 1}}, \tag{16}$$

$F(\phi|m)$ is an ELLIPTIC INTEGRAL OF THE FIRST KIND with PARAMETER m, and $\Pi(\phi|m,k)$ is an ELLIPTIC INTEGRAL OF THE THIRD KIND.

GEODESICS other than MERIDIANS of an OBLATE SPHEROID undulate between two parallels with latitudes equidistant from the equator. Using the WEIERSTRAß SIGMA FUNCTION and WEIERSTRAß ZETA FUNCTION, the GEODESIC on the OBLATE SPHEROID can be written as

$$x + iy = \kappa \frac{\sigma(a+u)}{\sigma(u)\sigma(a)} e^{u[\eta - \zeta(\omega + a)]} \tag{17}$$

$$x - iy = \kappa \frac{\sigma(a-u)}{\sigma(u)\sigma(a)} e^{-u[\eta - \zeta(\omega + a)]} \tag{18}$$

$$z^2 = \lambda^2 \frac{\sigma(\omega'' + u)\sigma(\omega'' - u)}{\sigma^2(u)\sigma^2(a)} \tag{19}$$

(Forsyth 1960, pp. 108–109; Halphen 1886–1891).

The equation of the GEODESIC can be put in the form

$$d\phi = \frac{\sqrt{1 - e^2 \sin^2 v}\, \sin a}{\sqrt{\sin^2 v - \sin^2 a}\, \sin v}\, dv, \tag{20}$$

where a is the smallest value of v on the curve. Furthermore, the difference in longitude between points of highest and next lowest latitude on the curve is

$$\pi - 2\frac{\sqrt{1 - e^2 \sin^2 a}}{\sin a} \int_0^K \frac{\operatorname{dn} u - \operatorname{dn}^2 u}{1 + \cot^2 a \operatorname{sn}^2 u}\, du, \tag{21}$$

where the MODULUS of the ELLIPTIC FUNCTION is

$$k = \frac{e \cos a}{\sqrt{1 - e^2 \sin^2 a}} \tag{22}$$

(Forsyth 1960, p. 446).

see also ELLIPSOID GEODESIC, OBLATE SPHEROID, SPHERE GEODESIC

References

Forsyth, A. R. *Calculus of Variations.* New York: Dover, 1960.

Halphen, G. H. *Traité des fonctions elliptiques et de leurs applications fonctions elliptiques, Vol. 2.* Paris: Gauthier-Villars, pp. 238–243, 1886–1891.

Oblate Spheroidal Coordinates

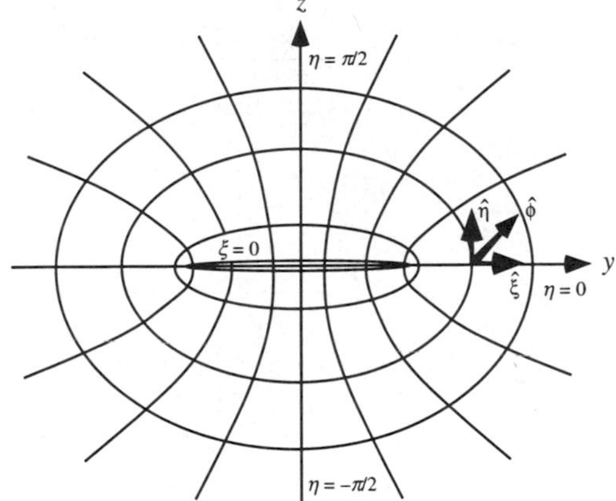

A system of CURVILINEAR COORDINATES in which two sets of coordinate surfaces are obtained by revolving the curves of the ELLIPTIC CYLINDRICAL COORDINATES about the y-AXIS which is relabeled the z-AXIS. The third set of coordinates consists of planes passing through this axis.

$$x = a \cosh\xi \cos\eta \cos\phi \tag{1}$$

$$y = a \cosh\xi \cos\eta \sin\phi \tag{2}$$

$$z = a \sinh\xi \sin\eta, \tag{3}$$

where $\xi \in [0,\infty)$, $\eta \in [-\pi/2, \pi/2]$, and $\phi \in [0, 2\pi)$. Arfken (1970) uses (u, v, φ) instead of (ξ, η, ϕ). The SCALE FACTORS are

$$h_\xi = a\sqrt{\sinh^2\xi + \sin^2\eta} \qquad (4)$$
$$h_\eta = a\sqrt{\sinh^2\xi + \sin^2\eta} \qquad (5)$$
$$h_\phi = a\cosh\xi\cos\eta. \qquad (6)$$

The LAPLACIAN is

$$\nabla^2 f = \frac{1}{a^3(\sinh^2\xi + \sin^2\eta)\cosh\xi\cos\eta} \times \left[\frac{\partial f}{\partial\xi}\left(a\cosh\xi\cos\eta\frac{\partial f}{\partial\xi}\right) + \frac{\partial f}{\partial\eta}\left(a\cosh\xi\cos\eta\frac{\partial f}{\partial\eta}\right) + \frac{a^2(\sinh^2\xi + \sin^2\eta)}{a\cosh\xi\cos\eta}\frac{\partial^2 f}{\partial\phi^2}\right]$$
$$= \frac{1}{a^3(\sinh^2\xi + \sin^2\eta)\cosh\xi\cos\eta}\left[a\sinh\xi\cos\eta\frac{\partial f}{\partial\xi} \; a\cosh\xi\cos\eta\frac{\partial^2 f}{\partial\xi^2} + a\sinh\xi\cos\eta\frac{\partial f}{\partial\eta} + a\cosh\xi\cos\eta\frac{\partial^2 f}{\partial\eta^2}\right] + \frac{1}{a^2(\sinh^2\xi + \sin^2\eta)}\frac{\partial^2 f}{\partial\phi^2}$$
$$= \frac{1}{a^2(\sinh^2\xi + \sin^2\eta)}\left[\frac{1}{\cosh\xi}\frac{\partial}{\partial\xi}\left(\cosh\xi\frac{\partial f}{\partial\xi}\right) + \frac{1}{\cos\eta}\frac{\partial}{\partial\eta}\left(\cos\eta\frac{\partial f}{\partial\eta}\right)\right] + \frac{1}{a^2(\cosh^2\xi + \cos^2\eta)}\frac{\partial^2 f}{\partial\phi^2} \qquad (7)$$
$$= \frac{1}{\sin^2\eta + \sinh^2\xi}\left[(\operatorname{sech}^2\xi\tan^2\eta + \sec^2\tanh^2\xi)\frac{\partial^2}{\partial\phi^2} + \tanh\xi\frac{\partial}{\partial\xi} + \frac{\partial^2}{\partial\xi^2} - \tan\eta\frac{\partial}{\eta} + \frac{\partial^2}{\eta^2}\right]. \qquad (8)$$

An alternate form useful for "two-center" problems is defined by

$$\xi_1 = \sinh\xi \qquad (9)$$
$$\xi_1' = \cosh\xi \qquad (10)$$
$$\xi_2 = \cos\eta \qquad (11)$$
$$\xi_3 = \phi, \qquad (12)$$

where $\xi_1 \in [1, \infty]$, $\xi_2 \in [-1, 1]$, and $\xi_3 \in [0, 2\pi)$. In these coordinates,

$$y = a\xi_1'\xi_2\sin\xi_3 \qquad (13)$$
$$z = a\sqrt{({\xi_1'}^2 - 1)(1 - {\xi_2}^2)} \qquad (14)$$
$$x = a\xi_1'\xi_2\cos\xi_3 \qquad (15)$$

(Abramowitz and Stegun 1972). The SCALE FACTORS are

$$h_{\xi_1} = a\sqrt{\frac{{\xi_1}^2 - {\xi_2}^2}{{\xi_1}^2 - 1}} \qquad (16)$$
$$h_{\xi_2} = a\sqrt{\frac{{\xi_1}^2 - {\xi_2}^2}{1 - {\xi_2}^2}} \qquad (17)$$
$$h_{\xi_3} = a\xi\eta, \qquad (18)$$

and the LAPLACIAN is

$$\nabla^2 f = \frac{1}{a^2}\left\{\frac{1}{{\xi_1}^2 + {\xi_2}^2}\frac{\partial}{\partial\xi_1}\left[({\xi_1}^2 + 1)\frac{\partial f}{\partial\xi_1}\right] + \frac{1}{{\xi_1}^2 + {\xi_2}^2}\frac{\partial}{\partial\xi_2}\left[(1 - {\xi_2}^2)\frac{\partial f}{\partial\xi_2}\right] + \frac{1}{({\xi_1}^2 + 1)(1 - {\xi_2}^2)}\frac{\partial^2 f}{\partial{\xi_3}^2}\right\}. \qquad (19)$$

The HELMHOLTZ DIFFERENTIAL EQUATION is separable.

see also HELMHOLTZ DIFFERENTIAL EQUATION—OBLATE SPHEROIDAL COORDINATES, LATITUDE, LONGITUDE, PROLATE SPHEROIDAL COORDINATES, SPHERICAL COORDINATES

References

Abramowitz, M. and Stegun, C. A. (Eds.). "Definition of Oblate Spheroidal Coordinates." §21.2 in *Handbook of Mathematical Functions with Formulas, Graphs, and Mathematical Tables, 9th printing.* New York: Dover, p. 752, 1972.

Arfken, G. "Prolate Spheroidal Coordinates (u, v, ϕ)." §2.11 in *Mathematical Methods for Physicists, 2nd ed.* Orlando, FL: Academic Press, pp. 107–109, 1970.

Morse, P. M. and Feshbach, H. *Methods of Theoretical Physics, Part I.* New York: McGraw-Hill, p. 663, 1953.

Oblate Spheroidal Wave Function

The wave equation in OBLATE SPHEROIDAL COORDINATES is

$$\nabla^2\Phi + k^2\Phi = \frac{\partial}{\partial\xi_1}\left[({\xi_1}^2 + 1)\frac{\partial\Phi}{\partial\xi_1}\right] + \frac{\partial}{\partial\xi_2}\left[(1 - {\xi_2}^2)\frac{\partial\Phi}{\partial\xi_2}\right] + \frac{{\xi_1}^2 + {\xi_2}^2}{({\xi_1}^2 + 1)(1 - {x_2}^2)}\frac{\partial^2\Phi}{\partial\phi^2} + c^2({\xi_1}^2 + {\xi_2}^2)\Phi = 0, \qquad (1)$$

where

$$c \equiv \tfrac{1}{2}ak. \qquad (2)$$

Substitute in a trial solution

$$\Phi = R_{mn}(c, \xi_1)S_{mn}(c, \xi_2)\begin{matrix}\cos\\ \sin\end{matrix}(m\phi). \qquad (3)$$

The radial differential equation is

$$\frac{d}{d\xi_2}\left[(1 + {\xi_2}^2)\frac{d}{d\xi_2}S_{mn}(c, \xi_2)\right] - \left(\lambda_{mn} - c^2{\xi_2}^2 + \frac{m^2}{1 + {\xi_2}^2}\right)R_{mn}(c, \xi_2) = 0, \qquad (4)$$

and the angular differential equation is

$$\frac{d}{d\xi_2}\left[(1-\xi_2{}^2)\frac{d}{d\xi_2}S_{mn}(c,\xi_2)\right] - \left(\lambda_{mn} - c^2\xi_2{}^2 + \frac{m^2}{1-\xi_2{}^2}\right)R_{mn}(c,\xi_2) = 0 \quad (5)$$

(Abramowitz and Stegun 1972, pp. 753–755).

see also PROLATE SPHEROIDAL WAVE FUNCTION

References

Abramowitz, M. and Stegun, C. A. (Eds.). "Spheroidal Wave Functions." Ch. 21 in *Handbook of Mathematical Functions with Formulas, Graphs, and Mathematical Tables, 9th printing.* New York: Dover, pp. 751–759, 1972.

Oblateness

see FLATTENING

Oblique Angle

An ANGLE which is not a RIGHT ANGLE.

Oblong Number

see PRONIC NUMBER

Obstruction

Obstruction theory studies the extentability of MAPS using algebraic GADGETS. While the terminology rapidly becomes technical and convoluted (as Iyanaga and Kawada note, "It is extremely difficult to discuss higher obstructions in general since they involve many complexities"), the ideas associated with obstructions are very important in modern ALGEBRAIC TOPOLOGY.

see also ALGEBRAIC TOPOLOGY, CHERN CLASS, EILENBERG-MAC LANE SPACE, STIEFEL-WHITNEY CLASS

References

Iyanaga, S. and Kawada, Y. (Eds.). "Obstructions." §300 in *Encyclopedic Dictionary of Mathematics.* Cambridge, MA: MIT Press, pp. 948–950, 1980.

Obtuse Angle

An ANGLE greater than $\pi/2$ RADIANS (90°).

see also ACUTE ANGLE, OBTUSE TRIANGLE, RIGHT ANGLE, STRAIGHT ANGLE

Obtuse Triangle

An obtuse triangle is a TRIANGLE in which one of the ANGLES is an OBTUSE ANGLE. (Obviously, only a single ANGLE in a TRIANGLE can be OBTUSE or it wouldn't be a TRIANGLE.) A triangle must be either obtuse, ACUTE, or RIGHT.

A famous problem is to find the chance that three points picked randomly in a PLANE are the VERTICES of an obtuse triangle (Eisenberg and Sullivan 1996). Unfortunately, the solution of the problem depends on the procedure used to pick the "random" points (Portnoy 1994). In fact, it is impossible to pick random variables which are uniformly distributed in the plane (Eisenberg and Sullivan 1996). Guy (1993) gives a variety of solutions to the problem. Woolhouse (1886) solved the problem by picking uniformly distributed points in the unit DISK, and obtained

$$P_2 = 1 - \left(\frac{4}{\pi^2} - \frac{1}{8}\right) = \frac{9}{8} - \frac{4}{\pi^2} = 0.719715\ldots. \quad (1)$$

The problem was generalized by Hall (1982) to n-D BALL TRIANGLE PICKING, and Buchta (1986) gave closed form evaluations for Hall's integrals.

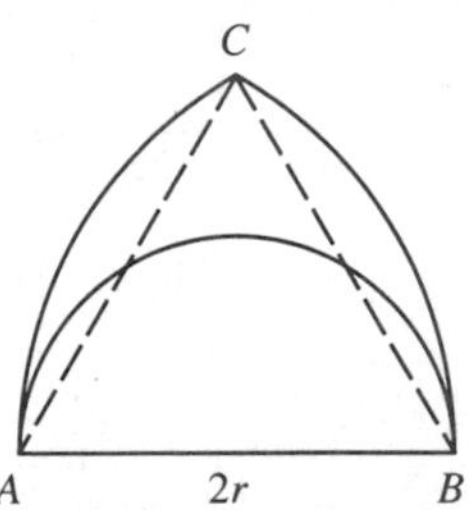

Lewis Carroll (1893) posed and gave another solution to the problem as follows. Call the longest side of a TRIANGLE AB, and call the DIAMETER $2r$. Draw arcs from A and B of RADIUS $2r$. Because the longest side of the TRIANGLE is defined to be AB, the third VERTEX of the TRIANGLE must lie within the region $ABCA$. If the third VERTEX lies within the SEMICIRCLE, the TRIANGLE is an obtuse triangle. If the VERTEX lies *on* the SEMICIRCLE (which will happen with probability 0), the TRIANGLE is a RIGHT TRIANGLE. Otherwise, it is an ACUTE TRIANGLE. The chance of obtaining an obtuse triangle is then the ratio of the AREA of the SEMICIRCLE to that of $ABCA$. The AREA of $ABCA$ is then twice the AREA of a SECTOR minus the AREA of the TRIANGLE.

$$A_{\text{whole figure}} = 2\left(\frac{4\pi r^2}{6}\right) - \sqrt{3}\,r^2 = r^2(\tfrac{4}{3}\pi - \sqrt{3}). \quad (2)$$

Therefore,

$$P = \frac{\frac{1}{2}\pi r^2}{r^2(\frac{4}{3}\pi - \sqrt{3})} = \frac{3\pi}{8\pi - 6\sqrt{3}} = 0.63938\ldots. \quad (3)$$

Let the VERTICES of a triangle in n-D be NORMAL (GAUSSIAN) variates. The probability that a Gaussian triangle in n-D is obtuse is

$$\begin{aligned} P_n &= \frac{3\Gamma(n)}{\Gamma^2(\frac{1}{2}n)}\int_0^{1/3}\frac{x^{(n-2)/2}}{(1+x)^n}\,dx \\ &= \frac{3\Gamma(n)}{\Gamma^2(\frac{1}{2}n)2^{n-1}}\int_0^{\pi/3}\sin^{n-1}\theta\,d\theta \\ &= \frac{6\Gamma(n)\,{}_2F_1(\frac{1}{2}n, n, 1+\frac{1}{2}n; -\frac{1}{3})}{3^{n/2}n\Gamma^2(\frac{1}{2}n)}, \end{aligned} \quad (4)$$

where $\Gamma(n)$ is the GAMMA FUNCTION and ${}_2F_1(a,b;c;x)$ is the HYPERGEOMETRIC FUNCTION. For EVEN $n \equiv 2k$,

$$P_{2k} = 3\sum_{j=k}^{2k-1} \binom{2k-1}{j} \left(\tfrac{1}{4}\right)^j \left(\tfrac{3}{4}\right)^{2k-1-j} \qquad (5)$$

(Eisenberg and Sullivan 1996). The first few cases are explicitly

$$P_2 = \tfrac{3}{4} = 0.75 \qquad (6)$$

$$P_3 = 1 - \frac{3\sqrt{3}}{4\pi} = 0.586503\ldots \qquad (7)$$

$$P_4 = \tfrac{15}{32} = 0.46875\ldots \qquad (8)$$

$$P_5 = 1 - \frac{9\sqrt{3}}{8\pi} = 0.379755\ldots. \qquad (9)$$

see also ACUTE ANGLE, ACUTE TRIANGLE, BALL TRIANGLE PICKING, OBTUSE ANGLE, RIGHT TRIANGLE, TRIANGLE

References
Buchta, C. "A Note on the Volume of a Random Polytope in a Tetrahedron." *Ill. J. Math.* **30**, 653–659, 1986.
Carroll, L. *Pillow Problems & A Tangled Tale.* New York: Dover, 1976.
Eisenberg, B. and Sullivan, R. "Random Triangles n Dimensions." *Amer. Math. Monthly* **103**, 308–318, 1996.
Guy, R. K. "There are Three Times as Many Obtuse-Angled Triangles as There are Acute-Angled Ones." *Math. Mag.* **66**, 175–178, 1993.
Hall, G. R. "Acute Triangles in the n-Ball." *J. Appl. Prob.* **19**, 712–715, 1982.
Portnoy, S. "A Lewis Carroll Pillow Problem: Probability on at Obtuse Triangle." *Statist. Sci.* **9**, 279–284, 1994.
Wells, D. G. *The Penguin Book of Interesting Puzzles.* London: Penguin Books, pp. 67 and 248–249, 1992.
Woolhouse, W. S. B. Solution to Problem 1350. *Mathematical Questions, with Their Solutions, from the Educational Times, 1.* London: F. Hodgson and Son, 49–51, 1886.

Ochoa Curve

The ELLIPTIC CURVE

$$3Y^2 = 2X^3 + 386X^2 + 256X - 58195,$$

given in Weierstraß form as

$$y^2 = x^3 - 440067x + 106074110.$$

The complete set of solutions to this equation consists of $(x,y) = (-761, 504)$, $(-745, 4520)$, $(-557, 13356)$, $(-446, 14616)$, $(-17, 10656)$, $(91, 8172)$, $(227, 4228)$, $(247, 3528)$, $(271, 2592)$, $(455, 200)$, $(499, 3276)$, $(523, 4356)$, $(530, 4660)$, $(599, 7576)$, $(751, 14112)$, $(1003, 25956)$, $(1862, 75778)$, $(3511, 204552)$, $(5287, 381528)$, $(23527, 3607272)$, $(64507, 16382772)$, $(100102, 31670478)$, and $(1657891, 2134685628)$ (Stroeker and de Weger 1994).

References
Guy, R. K. "The Ochoa Curve." *Crux Math.* **16**, 65–69, 1990.
Ochoa Melida, J. "La ecuacion diofántica $b_0y^3 - b_1y^2 + b_2y - b_3 = z^2$." *Gaceta Math.* 139–141, 1978.
Stroeker, R. J. and de Weger, B. M. M. "On Elliptic Diophantine Equations that Defy Thue's Method: The Case of the Ochoa Curve." *Experiment. Math.* **3**, 209–220, 1994.

Octacontagon

An 80-sided POLYGON.

Octadecagon

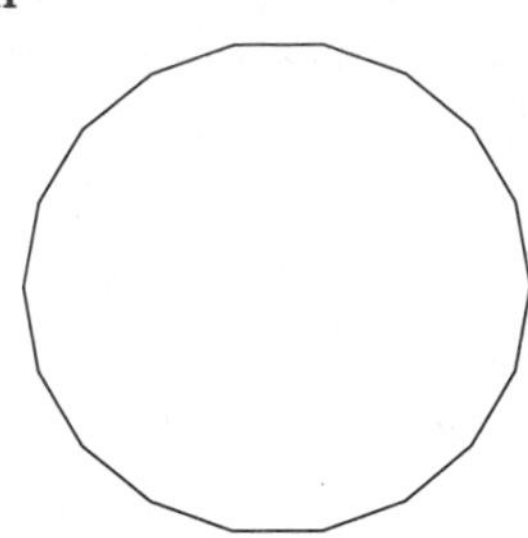

An 18-sided POLYGON, sometimes also called an OCTAKAIDECAGON.

see also POLYGON, REGULAR POLYGON, TRIGONOMETRY VALUES—$\pi/18$

Octagon

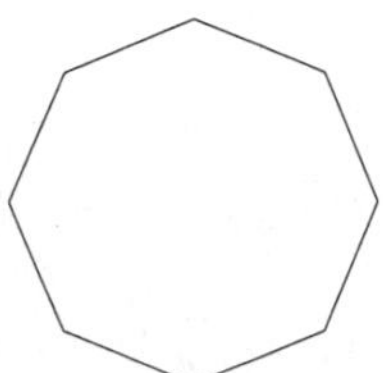

The regular 8-sided POLYGON. The INRADIUS r, CIRCUMRADIUS R, and AREA A can be computed directly from the formulas for a general regular POLYGON with side length s and $n = 8$ sides,

$$r = \tfrac{1}{2}s\cot\left(\frac{\pi}{8}\right) = \tfrac{1}{2}(1+\sqrt{2})s \qquad (1)$$

$$R = \tfrac{1}{2}s\csc\left(\frac{\pi}{8}\right) = \tfrac{1}{2}\sqrt{4+2\sqrt{2}}\,s \qquad (2)$$

$$A = \tfrac{1}{4}ns^2\cot\left(\frac{\pi}{8}\right) = 2(1+\sqrt{2})s^2. \qquad (3)$$

see also OCTAHEDRON, POLYGON, REGULAR POLYGON, TRIGONOMETRY VALUES—$\pi/8$

Octagonal Number

A POLYGONAL NUMBER of the form $n(3n-2)$. The first few are 1, 8, 21, 40, 65, 96, 133, 176, ... (Sloane's A000567). The GENERATING FUNCTION for the octagonal numbers is

$$\frac{x(5x+1)}{(1-x)^3} = x + 8x^2 + 21x^3 + 40x^4 + \ldots.$$

References
Sloane, N. J. A. Sequence A000567/M4493 in "An On-Line Version of the Encyclopedia of Integer Sequences."

Octagram

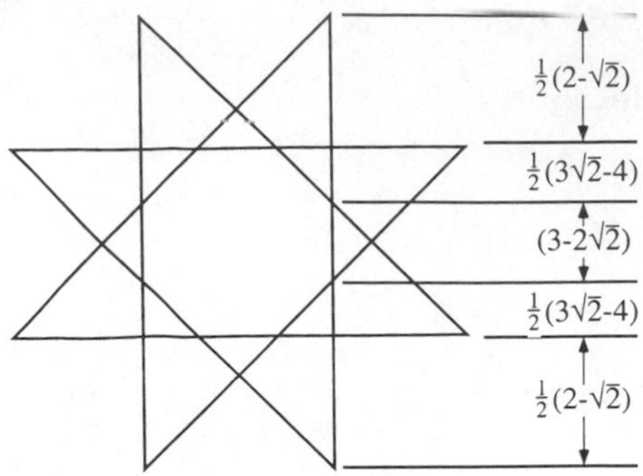

The STAR POLYGON $\{8,3\}$.

Octahedral Graph

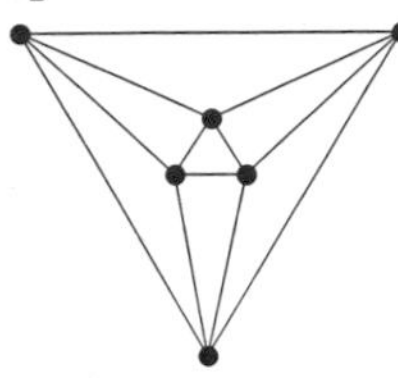

The POLYHEDRAL GRAPH having the topology of the OCTAHEDRON.

see also CUBICAL GRAPH, DODECAHEDRAL GRAPH, ICOSAHEDRAL GRAPH, OCTAHEDRON, TETRAHEDRAL GRAPH

Octahedral Group

The POINT GROUP of symmetries of the OCTAHEDRON, denoted O_h. It is also the symmetry group of the CUBE, CUBOCTAHEDRON, and TRUNCATED OCTAHEDRON. It has symmetry operations E, $8C_3$, $6C_4$, $6C_2$, $3C_2 = C_4^2$, i, $6S_4$, $8S_6$, $3\sigma_h$, and $6\sigma_4$ (Cotton 1990).

see also CUBE, CUBOCTAHEDRON, ICOSAHEDRAL GROUP, OCTAHEDRON, POINT GROUPS, TETRAHEDRAL GROUP, TRUNCATED OCTAHEDRON

References
Cotton, F. A. *Chemical Applications of Group Theory, 3rd ed.* New York: Wiley, p. 47–49, 1990.
Lomont, J. S. "Octahedral Group." §3.10.D in *Applications of Finite Groups.* New York: Dover, p. 81, 1987.

Octahedral Number

A FIGURATE NUMBER which is the sum of two consecutive PYRAMIDAL NUMBERS,

$$O_n = P_{n-1} + P_n = \tfrac{1}{3}n(2n^2+1).$$

The first few are 1, 6, 19, 44, 85, 146, 231, 344, 489, 670, 891, 1156, ... (Sloane's A005900). The GENERATING FUNCTION for the octahedral numbers is

$$\frac{x(x+1)^2}{(x-1)^4} = x + 6x^2 + 19x^3 + 44x^4 + \ldots.$$

see also TRUNCATED OCTAHEDRAL NUMBER

References
Conway, J. H. and Guy, R. K. *The Book of Numbers.* New York: Springer-Verlag, p. 50, 1996.
Sloane, N. J. A. Sequence A005900/M4128 in "An On-Line Version of the Encyclopedia of Integer Sequences."

Octahedron

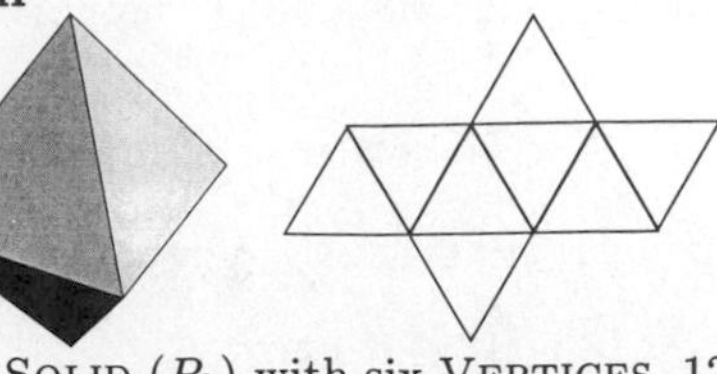

A PLATONIC SOLID (P_3) with six VERTICES, 12 EDGES, and eight equivalent EQUILATERAL TRIANGULAR faces ($8\{3\}$), given by the SCHLÄFLI SYMBOL $\{3,4\}$. It is also UNIFORM POLYHEDRON U_5 with the WYTHOFF SYMBOL $4\,|\,2\,3$. Its DUAL POLYHEDRON is the CUBE. Like the CUBE, it has the O_h OCTAHEDRAL GROUP of symmetries. The octahedron can be STELLATED to give the STELLA OCTANGULA.

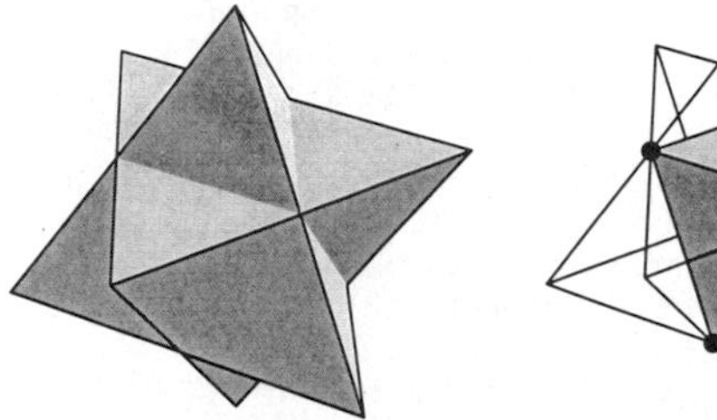

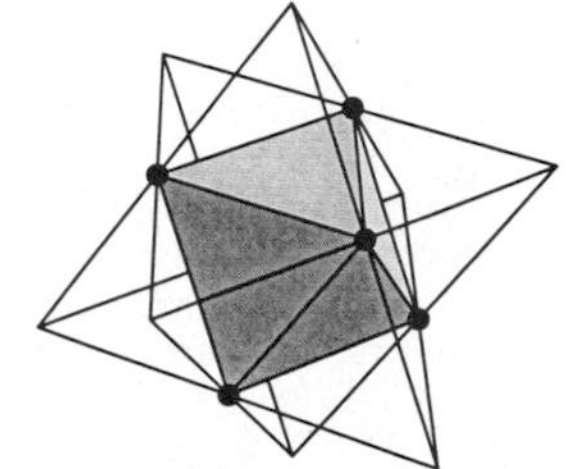

The solid bounded by the two TETRAHEDRA of the STELLA OCTANGULA (left figure) is an octahedron (right figure; Ball and Coxeter 1987).

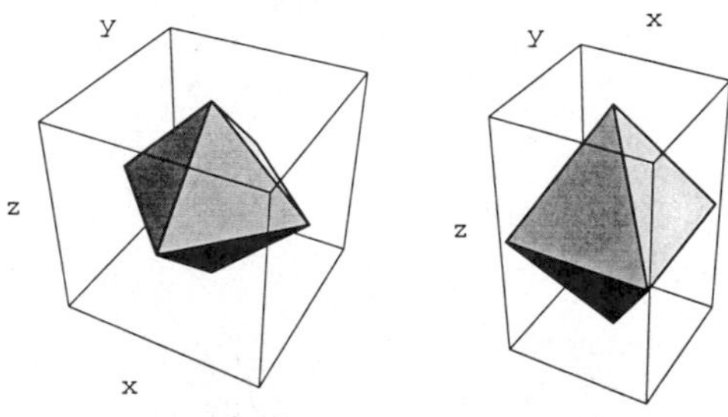

In one orientation (left figure), the VERTICES are given by $(\pm1,0,0)$, $(0,\pm1,0)$, $(0,0,\pm1)$. In another orientation (right figure), the vertices are $(\pm1,\pm1,0)$ and $(0,0,\pm\sqrt{3})$. In the latter, the constituent TRIANGLES are specified by

$$\begin{aligned}
T_1 &= \{(-1,-1,0),(1,-1,0),(0,0,\sqrt{3})\}\\
T_2 &= \{(-1,-1,0),(1,-1,0),(0,0,-\sqrt{3})\}\\
T_3 &= \{(-1,1,0),(1,1,0),(0,0,\sqrt{3})\}\\
T_4 &= \{(-1,1,0),(1,1,0),(0,0,-\sqrt{3})\}\\
T_5 &= \{(1,-1,0),(1,1,0),(0,0,\sqrt{3})\}\\
T_6 &= \{(-1,-1,0),(-1,1,0),(0,0,\sqrt{3})\}\\
T_7 &= \{(1,-1,0),(1,1,0),(0,0,-\sqrt{3})\}\\
T_8 &= \{(-1,-1,0),(-1,1,0),(0,0,-\sqrt{3})\}.
\end{aligned}$$

The face planes are $\pm x \pm y \pm z = 1$, so a solid octahedron is given by the equation

$$|x| + |y| + |z| \leq 1. \tag{1}$$

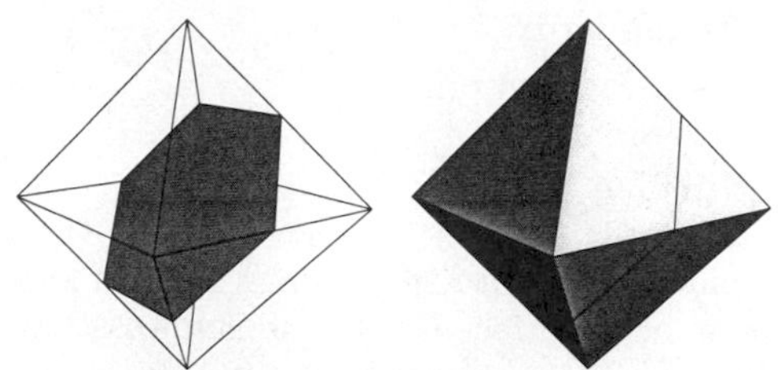

A plane PERPENDICULAR to a C_3 axis of an octahedron cuts the solid in a regular HEXAGONAL CROSS-SECTION (Holden 1991, pp. 22–23). Since there are four such axes, there are four possibly HEXAGONAL CROSS-SECTIONS. Faceted forms are the CUBOCTATRUNCATED CUBOCTAHEDRON and TETRAHEMIHEXAHEDRON.

Let an octahedron be length a on a side. The height of the top VERTEX from the square plane is also the CIRCUMRADIUS

$$R = \sqrt{a^2 - d^2}, \tag{2}$$

where

$$d = \tfrac{1}{2}\sqrt{2}\,a \tag{3}$$

is the diagonal length, so

$$R = \sqrt{a^2 - \tfrac{1}{2}a^2} = \tfrac{1}{2}\sqrt{2}\,a \approx 0.70710a. \tag{4}$$

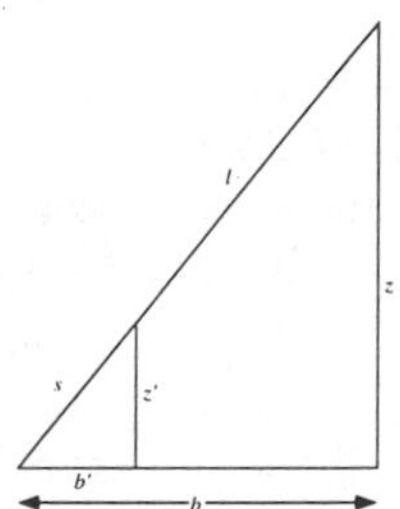

Now compute the INRADIUS.

$$\ell = \tfrac{1}{2}\sqrt{3}\,a \tag{5}$$

$$b = \tfrac{1}{2}a \tag{6}$$

$$s = \tfrac{1}{2}a\tan 30^\circ = \frac{a}{2\sqrt{3}}, \tag{7}$$

so

$$\frac{s}{\ell} = \frac{1}{2\sqrt{3}}\frac{2}{\sqrt{3}} = \tfrac{1}{3}. \tag{8}$$

Now use similar TRIANGLES to obtain

$$b' = \frac{s}{\ell}b = \tfrac{1}{6}a \tag{9}$$

$$z' = \frac{s}{\ell}z = \frac{a}{3\sqrt{2}} \tag{10}$$

$$x = b - b' = \tfrac{1}{2}a - \tfrac{1}{6}a = \tfrac{1}{3}a, \tag{11}$$

so the INRADIUS is

$$r = \sqrt{x^2 + z'^2} = a\sqrt{\tfrac{1}{9} + \tfrac{1}{18}} = \tfrac{1}{6}\sqrt{6}\,a \approx 0.40824a. \tag{12}$$

The INTERRADIUS is

$$\rho = \tfrac{1}{2}a = 0.5a. \tag{13}$$

The AREA of one face is the AREA of an EQUILATERAL TRIANGLE

$$A = \tfrac{1}{4}\sqrt{3}\,a^2. \tag{14}$$

The volume is two times the volume of a square-base pyramid,

$$V = 2(\tfrac{1}{3}a^2R) = 2(\tfrac{1}{3})(a^2)(\tfrac{1}{2}\sqrt{2}\,a) = \tfrac{1}{3}\sqrt{2}\,a^3. \tag{15}$$

The DIHEDRAL ANGLE is

$$\alpha = \cos^{-1}(-\tfrac{1}{3}) \approx 70.528779^\circ. \tag{16}$$

see also OCTAHEDRAL GRAPH, OCTAHEDRAL GROUP, OCTAHEDRON 5-COMPOUND, STELLA OCTANGULA, TRUNCATED OCTAHEDRON

References

Davie, T. "The Octahedron." http://www.dcs.st-and.ac.uk/~ad/mathrecs/polyhedra/octahedron.html.

Holden, A. *Shapes, Space, and Symmetry.* New York: Dover, 1991.

Octahedron 5-Compound

A POLYHEDRON COMPOUND composed of five OCTAHEDRA occupying the VERTICES of an ICOSAHEDRON. The 30 VERTICES of the compound form an ICOSIDODECAHEDRON (Ball and Coxeter 1987).

see also ICOSIDODECAHEDRON, OCTAHEDRON, POLYHEDRON COMPOUND

References

Ball, W. W. R. and Coxeter, H. S. M. *Mathematical Recreations and Essays, 13th ed.* New York: Dover, pp. 135 and 137, 1987.

Cundy, H. and Rollett, A. *Mathematical Models, 3rd ed.* Stradbroke, England: Tarquin Pub., pp. 137–138, 1989.

Wenninger, M. J. *Polyhedron Models.* New York: Cambridge University Press, p. 43, 1989.

Octahemioctacron

The DUAL POLYHEDRON of the OCTAHEMIOCTAHEDRON.

Octahemioctahedron

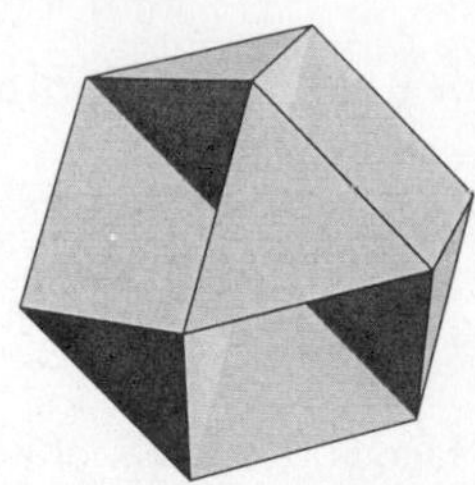

The UNIFORM POLYHEDRON U_3, also called the OCTATETRAHEDRON, whose DUAL POLYHEDRON is the OCTAHEMIOCTACRON. It has WYTHOFF SYMBOL $\frac{3}{2}\,3\,|\,3$. Its faces are $8\{3\}+4\{6\}$. It is a FACETED CUBOCTAHEDRON. For unit edge length, its CIRCUMRADIUS is

$$R = 1.$$

References

Wenninger, M. J. *Polyhedron Models.* Cambridge, England: Cambridge University Press, p. 103, 1989.

Octakaidecagon

see OCTADECAGON

Octal

The base 8 notational system for representing REAL NUMBERS. The digits used are 0, 1, 2, 3, 4, 5, 6, and 7, so that 8_{10} (8 in base 10) is represented as 10_8 $(10 = 1\cdot 8^1 + 0\cdot 8^0)$ in base 8.

see also BASE (NUMBER), BINARY, DECIMAL, HEXADECIMAL, QUATERNARY, TERNARY

References

Lauwerier, H. *Fractals: Endlessly Repeated Geometric Figures.* Princeton, NJ: Princeton University Press, pp. 9–10, 1991.

Weisstein, E. W. "Bases." `http://www.astro.virginia.edu/~eww6n/math/notebooks/Bases.m`.

Octant

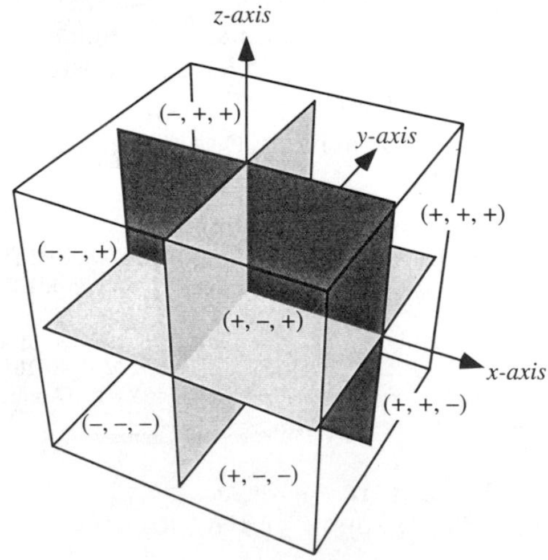

One of the eight regions of SPACE defined by the eight possible combinations of SIGNS $(\pm,\pm,\pm)$ for x, y, and z.

see also QUADRANT

Octatetrahedron

see OCTAHEMIOCTAHEDRON

Octic Surface

An ALGEBRAIC SURFACE of degree eight. The maximum number of ORDINARY DOUBLE POINTS known to exist on an octic surface is 168 (the ENDRASS OCTICS), although the rigorous upper bound is 174.

see also ALGEBRAIC SURFACE, ENDRASS OCTIC

Octillion

In the American system, 10^{27}.

see also LARGE NUMBER

Octodecillion

In the American system, 10^{57}.

see also LARGE NUMBER

Octonion

see CAYLEY NUMBER

Odd Function

An odd function is a function for which $f(x) = -f(-x)$. An EVEN FUNCTION times an odd function is odd.

Odd Number

An INTEGER of the form $N = 2n+1$, where n is an INTEGER. The odd numbers are therefore ..., −3, −1, 1, 3, 5, 7, ... (Sloane's A005408), which are also the GNOMIC NUMBERS. The GENERATING FUNCTION for the odd numbers is

$$\frac{x(1+x)}{(x-1)^2} = x + 3x^2 + 5x^3 + 7x^4 + \ldots.$$

Since the odd numbers leave a remainder of 1 when divided by two, $N \equiv 1 \pmod 2$ for odd N. Integers which are not odd are called EVEN.

see also EVEN NUMBER, GNOMIC NUMBER, NICOMACHUS'S THEOREM, ODD NUMBER THEOREM, ODD PRIME

References

Sloane, N. J. A. Sequence A005408/M2400 in "An On-Line Version of the Encyclopedia of Integer Sequences."

Odd Number Theorem

The sum of the first n ODD NUMBERS is a SQUARE NUMBER:

$$\sum_{k=1}^{n}(2k-1) = 2\sum_{k=1}^{n} k - \sum_{k=1}^{n} 1 = 2\left[\frac{n(n+1)}{2}\right] - n$$
$$= n(n+1) - n = n^2.$$

see also NICOMACHUS'S THEOREM

Odd Order Theorem

see FEIT-THOMPSON THEOREM

Odd Prime

Any PRIME NUMBER other than 2 (which is the only EVEN PRIME).

see also PRIME NUMBER

Odd Sequence

A SEQUENCE of n 0s and 1s is called an odd sequence if each of the n SUMS $\sum_{i=1}^{n-k} a_i a_{i+k}$ for $k = 0, 1, \ldots, n-1$.

References

Guy, R. K. "Odd Sequences." §E38 in *Unsolved Problems in Number Theory, 2nd ed.* New York: Springer-Verlag, pp. 238–239, 1994.

Odds

Betting odds are written in the form $r : s$ ("r to s") and correspond to the probability of winning $P = s/(r+s)$. Therefore, given a probability P, the odds of winning are $(1/P) - 1 : 1$.

see also FRACTION, RATIO, RATIONAL NUMBER

References

Kraitchik, M. "The Horses." §6.17 in *Mathematical Recreations.* New York: W. W. Norton, pp. 134–135, 1942.

ODE

see ORDINARY DIFFERENTIAL EQUATION

Offset Rings

see SURFACE OF REVOLUTION

Ogive

Any cumulative frequency curve.

see also HISTOGRAM

References

Kenney, J. F. and Keeping, E. S. "Ogive Curves." §2.7 in *Mathematics of Statistics, Pt. 1, 3rd ed.* Princeton, NJ: Van Nostrand, pp. 29–31, 1962.

Oldknow Points

The PERSPECTIVE CENTERS of a triangle and the TANGENTIAL TRIANGLES of its inner and outer SODDY CIRCLES, given by

$$Ol = I + 2Ge$$
$$Ol' = I - 2Ge,$$

where I is the INCENTER and Ge is the GERGONNE POINT.

see also GERGONNE POINT, INCENTER, PERSPECTIVE CENTER, SODDY CIRCLES, TANGENTIAL TRIANGLE

References

Oldknow, A. "The Euler-Gergonne-Soddy Triangle of a Triangle." *Amer. Math. Monthly* **103**, 319–329, 1996.

Omega Constant

$$W(1) \equiv 0.5671432904\ldots, \tag{1}$$

where $W(x)$ is LAMBERT'S W-FUNCTION. It is available in *Mathematica*® (Wolfram Research, Champaign, IL) using the function `ProductLog[1]`. $W(1)$ can be considered a sort of "GOLDEN RATIO" for exponentials since

$$\exp[-W(1)] = W(1), \tag{2}$$

giving

$$\ln\left[\frac{1}{W(1)}\right] = W(1). \tag{3}$$

see also GOLDEN RATIO, LAMBERT'S W-FUNCTION

References

Corless, R. M.; Gonnet, G. H.; Hare, D. E. G.; and Jeffrey, D. J. "On Lambert's W Function." `ftp://watdragon.uwaterloo.ca/cs-archive/CS-93-03/W.ps.Z.`

Plouffe, S. "The Omega Constant or $W(1)$." `http://lacim.uqam.ca/piDATA/omega.txt.`

Omega Function

see LAMBERT'S W-FUNCTION

Omino

see POLYOMINO

Omnific Integer

The appropriate notion of INTEGER for SURREAL NUMBERS.

O'Nan Group

The SPORADIC GROUP *O'N*.

References

Wilson, R. A. "ATLAS of Finite Group Representation." `http://for.mat.bham.ac.uk/atlas/ON.html.`

Onduloid

see UNDULOID

One

see 1

One-Form

A linear, real-valued FUNCTION of VECTORS such that $\omega^1(\mathbf{v}) \mapsto \mathbb{R}$. VECTORS and one-forms are DUAL to each other because VECTORS are CONTRAVARIANT ("KETS": $|\psi\rangle$) and one-forms are COVARIANT VECTORS ("BRAS": $\langle\phi|$), so

$$\omega^1(\mathbf{v}) \equiv \mathbf{v}(\omega^1) \equiv \langle\omega^1, \mathbf{v}\rangle = \langle\phi|\psi\rangle.$$

The operation of applying the one-form to a VECTOR $\omega^1(\mathbf{v})$ is called CONTRACTION.

see also ANGLE BRACKET, BRA, DIFFERENTIAL k-FORM, KET

One-Mouth Theorem

Except for convex polygons, every SIMPLE POLYGON has at least one MOUTH.

see also MOUTH, PRINCIPAL VERTEX, TWO-EARS THEOREM

References

Toussaint, G. "Anthropomorphic Polygons." *Amer. Math. Monthly* **122**, 31–35, 1991.

One-Ninth Constant

N.B. A detailed on-line essay by S. Finch was the starting point for this entry.

Let $\lambda_{m,n}$ be CHEBYSHEV CONSTANTS. Schönhage (1973) proved that

$$\lim_{n\to\infty} (\lambda_{0,n})^{1/n} = \tfrac{1}{3}. \qquad (1)$$

It was conjectured that

$$\Lambda \equiv \lim_{n\to\infty} (\lambda_{n,n})^{1/n} = \tfrac{1}{9}. \qquad (2)$$

Carpenter *et al.* (1984) obtained

$$\Lambda = 0.1076539192\ldots \qquad (3)$$

numerically. Gonchar and Rakhmanov (1980) showed that the limit exists and disproved the 1/9 conjecture, showing that Λ is given by

$$\Lambda = \exp\left[-\frac{\pi K(\sqrt{1-c^2})}{K(c)}\right], \qquad (4)$$

where K is the complete ELLIPTIC INTEGRAL OF THE FIRST KIND, and $c = 0.9089085575485414\ldots$ is the PARAMETER which solves

$$K(k) = 2E(k), \qquad (5)$$

and E is the complete ELLIPTIC INTEGRAL OF THE SECOND KIND. This gives the value for Λ computed by Carpenter *et al.* (1984) Λ is also given by the unique POSITIVE ROOT of

$$f(z) = \tfrac{1}{8}, \qquad (6)$$

where

$$f(z) \equiv \sum_{j=1}^{\infty} a_j z^j \qquad (7)$$

and

$$a_j = \left|\sum_{d|j} (-1)^d d\right| \qquad (8)$$

(Gonchar and Rakhmanov 1980). a_j may also be computed by writing j as

$$j = 2^m {p_1}^{m_1} {p_2}^{m_2} \cdots {p_k}^{m_k}, \qquad (9)$$

where $m \geq 0$ and $m_i \geq 1$, then

$$a_j = |2^{m+1} - 3| \frac{{p_1}^{m_1+1} - 1}{p_1 - 1} \frac{{p_2}^{m_2+1} - 1}{p_2 - 1} \cdots \frac{{p_k}^{m_k+1} - 1}{p_k - 1} \qquad (10)$$

(Gonchar 1990). Yet another equation for Λ is due to Magnus (1986). Λ is the unique solution with $x \in (0,1)$ of

$$\sum_{k=0}^{\infty} (2k+1)^2 (-x)^{k(k+1)/2} = 0, \qquad (11)$$

an equation which had been studied and whose root had been computed by Halphen (1886). It has therefore been suggested (Varga 1990) that the constant be called the HALPHEN CONSTANT. $1/\Lambda$ is sometimes called VARGA'S CONSTANT.

see also CHEBYSHEV CONSTANTS, HALPHEN CONSTANT, VARGA'S CONSTANT

References

Finch, S. "Favorite Mathematical Constants." `http://www.mathsoft.com/asolve/constant/onenin/onenin.html`.

Carpenter, A. J.; Ruttan, A.; and Varga, R. S. "Extended Numerical Computations on the '1/9' Conjecture in Rational Approximation Theory." In *Rational approximation and interpolation (Tampa, Fla., 1983)* (Ed. P. R. Graves-Morris, E. B. Saff, and R. S. Varga). New York: Springer-Verlag, pp. 383–411, 1984.

Cody, W. J.; Meinardus, G.; and Varga, R. S. "Chebyshev Rational Approximations to e^{-x} in $[0, +\infty)$ and Applications to Heat-Conduction Problems." *J. Approx. Th.* **2**, 50–65, 1969.

Dunham, C. B. and Taylor, G. D. "Continuity of Best Reciprocal Polynomial Approximation on $[0, \infty)$." *J. Approx. Th.* **30**, 71–79, 1980.

Gonchar, A. A. "Rational Approximations of Analytic Functions." *Amer. Math. Soc. Transl. Ser. 2* **147**, 25–34, 1990.

Gonchar, A. A. and Rakhmanov, E. A. "Equilibrium Distributions and Degree of Rational Approximation of Analytic Functions." *Math. USSR Sbornik* **62**, 305–348, 1980.

Magnus, A. P. "On Freud's Equations for Exponential Weights, Papers Dedicated to the Memory of Géza Freud." *J. Approx. Th.* **46**, 65–99, 1986.
Rahman, Q. I. and Schmeisser, G. "Rational Approximation to the Exponential Function." In *Padé and Rational Approximation, (Proc. Internat. Sympos., Univ. South Florida, Tampa, Fla., 1976)* (Ed. E. B. Saff and R. S. Varga). New York: Academic Press, pp. 189–194, 1977.
Schönhage, A. "Zur rationalen Approximierbarkeit von e^{-x} über $[0,\infty)$." *J. Approx. Th.* **7**, 395–398, 1973.
Varga, R. S. *Scientific Computations on Mathematical Problems and Conjectures.* Philadelphia, PA: SIAM, 1990.

One-to-One

Let f be a FUNCTION defined on a SET S and taking values in a set T. Then f is said to be one-to-one (a.k.a. an INJECTION or EMBEDDING) if, whenever $f(x) = f(y)$, it must be the case that $x = y$. In other words, f is one-to-one if it MAPS distinct objects to distinct objects.

If the function is a linear OPERATOR which assigns a unique MAP to each value in a VECTOR SPACE, it is called one-to-one. Specifically, given a VECTOR SPACE $\mathbb{V}$ with $\mathbf{X}, \mathbf{Y} \in \mathbb{V}$, then a TRANSFORMATION T defined on $\mathbb{V}$ is one-to-one if $T(\mathbf{X}) \neq T(\mathbf{Y})$ for all $\mathbf{X} \neq \mathbf{Y}$.

see also BIJECTION, ONTO

One-Way Function

Consider straight-line algorithms over a FINITE FIELD with q elements. Then the ϵ-straight line complexity $C_\epsilon(\phi)$ of a function ϕ is defined as the length of the shortest straight-line algorithm which computes a function f such that $f(x) = x$ is satisfied for at least $(1-\epsilon)q$ elements of F. A function ϕ is straight-line "one way" of range $0 \leq \delta \leq 1$ if ϕ satisfies the properties:

1. There exists an infinite set S of finite fields such that ϕ is defined in every $F \in S$ and ϵ is ONE-TO-ONE in every $F \in S$.
2. For every ϵ such that $0 \leq \epsilon \leq \delta$, $C_\epsilon(\phi^{-1})$ tends to infinity as the cardinality q of F approaches infinity.
3. For every ϵ such that $0 \leq \epsilon \leq \delta$, the "work function" η satisfies

$$\eta \equiv \liminf_{q\to\infty} \eta \equiv \liminf_{q\to\infty} \frac{\ln C\epsilon(\phi^{-1}) - \ln C\epsilon(\phi)}{\ln C\epsilon(\phi)} > 1.$$

It is not known if there is a one-way function with work factor $\eta > (\ln q)^3$.

References
Ziv, J. "In Search of a One-Way Function" §4.1 in *Open Problems in Communication and Computation* (Ed. T. M. Cover and B. Gopinath). New York: Springer-Verlag, pp. 104–105, 1987.

Only Critical Point in Town Test

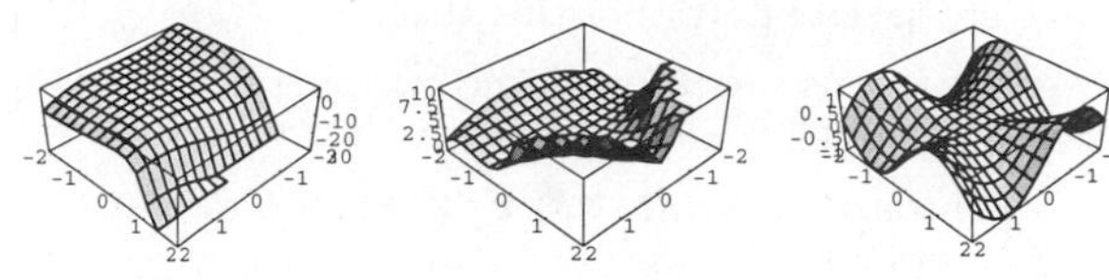

If there is only one CRITICAL POINT at an EXTREMUM, the CRITICAL POINT must be the EXTREMUM for functions of one variable. There are exceptions for two variables, but none of degree ≤ 4. Such exceptions include

$$z = 3xe^y - x^3 - e^{3y}$$

$$z = x^2(1+y)^3 + y^2$$

$$z = \begin{cases} \frac{xy(x^2-y^2)}{x^2+y^2} & \text{for } (x,y) \neq (0,0) \\ 0 & \text{for } (x,y) = (0,0) \end{cases}$$

(Wagon 1991). This latter function has discontinuous z_{xy} and z_{yx}, and $z_{yx}(0,0) = 1$ and $z_{xy}(0,0) = 1$.

References
Ash, A. M. and Sexton, H. "A Surface with One Local Minimum." *Math. Mag.* **58**, 147–149, 1985.
Calvert, B. and Vamanamurthy, M. K. "Local and Global Extrema for Functions of Several Variables." *J. Austral. Math. Soc.* **29**, 362–368, 1980.
Davies, R. Solution to Problem 1235. *Math. Mag.* **61**, 59, 1988.
Wagon, S. "Failure of the Only-Critical-Point-in-Town Test." §3.4 in *Mathematica in Action.* New York: W. H. Freeman, pp. 87–91 and 228, 1991.

Onto

Let f be a FUNCTION defined on a SET S and taking values in a set T. Then f is said to be onto (a.k.a. a SURJECTION) if, for any $t \in T$, there exists an $s \in S$ for which $t = f(s)$.

Let the function be an OPERATOR which MAPS points in the DOMAIN to every point in the RANGE and let $\mathbb{V}$ be a VECTOR SPACE with $\mathbf{X}, \mathbf{Y} \in \mathbb{V}$. Then a TRANSFORMATION T defined on $\mathbb{V}$ is onto if there is an $\mathbf{X} \in \mathbb{V}$ such that $T(\mathbf{X}) = \mathbf{Y}$ for all $\mathbf{Y}$.

see also BIJECTION, ONE-TO-ONE

Open Disk

An n-D open disk of RADIUS r is the collection of points of distance less than r from a fixed point in EUCLIDEAN n-space.

see also CLOSED DISK, DISK

Open Interval

An INTERVAL which does not include its LIMIT POINTS, denoted (a, b).

see also CLOSED INTERVAL, HALF-CLOSED INTERVAL

Open Map

A MAP which sends OPEN SETS to OPEN SETS.

see also OPEN MAPPING THEOREM

Open Mapping Theorem

There are several flavors of this theorem.

1. A continuous surjective linear mapping between BANACH SPACES is an OPEN MAP.
2. A nonconstant ANALYTIC FUNCTION on a DOMAIN D is an OPEN MAP.

References

Zeidler, E. *Applied Functional Analysis: Applications to Mathematical Physics.* New York: Springer-Verlag, 1995.

Open Set

A SET is open if every point in the set has a NEIGHBORHOOD lying in the set. An open set of RADIUS r and center $\mathbf{x}_0$ is the set of all points $\mathbf{x}$ such that $|\mathbf{x}-\mathbf{x}_0| < r$, and is denoted $D_r(\mathbf{x}_0)$. In 1-space, the open set is an OPEN INTERVAL. In 2-space, the open set is a DISK. In 3-space, the open set is a BALL.

More generally, given a TOPOLOGY (consisting of a SET X and a collection of SUBSETS T), a SET is said to be open if it is in T. Therefore, while it is not possible for a set to be both finite and open in the TOPOLOGY of the REAL LINE (a single point is a CLOSED SET), it is possible for a more general topological SET to be both finite and open.

The complement of an open set is a CLOSED SET. It is possible for a set to be neither open nor CLOSED, e.g., the interval $(0, 1]$.

see also BALL, CLOSED SET, EMPTY SET, OPEN INTERVAL

Operad

A system of parameter chain complexes used for MULTIPLICATION on differential GRADED ALGEBRAS up to HOMOTOPY.

Operand

A mathematical object upon which an OPERATOR acts. For example, in the expression 1×2, the MULTIPLICATION OPERATOR acts upon the operands 1 and 2.

see also OPERAD, OPERATOR

Operational Mathematics

The theory and applications of LAPLACE TRANSFORMS and other INTEGRAL TRANSFORMS.

References

Churchill, R. V. *Operational Mathematics, 3rd ed.* New York: McGraw-Hill, 1958.

Operations Research

A branch of mathematics which encompasses many diverse areas of minimization and optimization. Bronson (1982) describes operations research as being "concerned with the efficient allocation of scarce resources." It includes the CALCULUS OF VARIATIONS, CONTROL THEORY, CONVEX OPTIMIZATION THEORY, DECISION THEORY, GAME THEORY, LINEAR PROGRAMMING, MARKOV CHAINS, network analysis, OPTIMIZATION THEORY, queuing systems, etc. The more modern term for operations research is OPTIMIZATION THEORY.

see also CALCULUS OF VARIATIONS, CONTROL THEORY, CONVEX OPTIMIZATION THEORY, DECISION THEORY, GAME THEORY, LINEAR PROGRAMMING, MARKOV CHAIN, OPTIMIZATION THEORY, QUEUE

References

Bronson, R. *Schaum's Outline of Theory and Problems of Operations Research.* New York: McGraw-Hill, 1982.

Hiller, F. S. and Lieberman, G. J. *Introduction to Operations Research, 5th ed.* New York: McGraw-Hill, 1990.

Trick, M. "Michael Trick's Operations Research Page." `http://mat.gsia.cmu.edu`

Operator

An operator $A : f^{(n)}(I) \mapsto f(I)$ assigns to every function $f \in f^{(n)}(I)$ a function $A(f) \in f(I)$. It is therefore a mapping between two FUNCTION SPACES. If the range is on the REAL LINE or in the COMPLEX PLANE, the mapping is usually called a FUNCTIONAL instead.

see also ABSTRACTION OPERATOR, ADJOINT OPERATOR, ANTILINEAR OPERATOR, BIHARMONIC OPERATOR, BINARY OPERATOR, CASIMIR OPERATOR, CONVECTIVE OPERATOR, D'ALEMBERTIAN OPERATOR, DIFFERENCE OPERATOR, FUNCTIONAL ANALYSIS, HECKE OPERATOR, HERMITIAN OPERATOR, IDENTITY OPERATOR, LAPLACE-BELTRAMI OPERATOR, LINEAR OPERATOR, OPERAND, PERRON-FROBENIUS OPERATOR, PROJECTION OPERATOR, ROTATION OPERATOR, SCATTERING OPERATOR, SELF-ADJOINT OPERATOR, SPECTRUM (OPERATOR), THETA OPERATOR, WAVE OPERATOR

References

Gohberg, I.; Lancaster, P.; and Shivakuar, P. N. (Eds.). *Recent Developments in Operator Theory and Its Applications.* Boston, MA: Birkhäuser, 1996.

Hutson, V. and Pym, J. S. *Applications of Functional Analysis and Operator Theory.* New York: Academic Press, 1980.

Optimization Theory

see OPERATIONS RESEARCH

Or

A term in LOGIC which yields TRUE if any one of a sequence conditions is TRUE, and FALSE if *all* conditions are FALSE. A OR B is denoted $A|B$, $A + B$, or $A \vee B$. The symbol $\vee$ derives from the first letter of the Latin word "vel" meaning "or." The BINARY OR operator has the following TRUTH TABLE.

A	B	$A \vee B$
F	F	F
F	T	T
T	F	T
T	T	T

A product of ORs is called a DISJUNCTION and is denoted

$$\bigvee_{k=1}^{n} A_k.$$

Two BINARY numbers can have the operation OR performed bitwise. This operation is sometimes denoted $A||B$.

see also AND, BINARY OPERATOR, LOGIC, NOT, PREDICATE, TRUTH TABLE, UNION, XOR

Orbifold

The object obtained by identifying any two points of a MAP which are equivalent under some symmetry of the MAP'S GROUP.

Orbison's Illusion

The illusion illustrated above in which the bounding RECTANGLE and inner SQUARE both appear distorted.

see also ILLUSION, MÜLLER-LYER ILLUSION, PONZO'S ILLUSION, VERTICAL-HORIZONTAL ILLUSION

References

Fineman, M. *The Nature of Visual Illusion.* New York: Dover, p. 153, 1996.

Orbit (Group)

Given a PERMUTATION GROUP G on a set S, the orbit of an element $s \in S$ is the subset of S consisting of elements to which some element G can send s.

Orbit (Map)

The SEQUENCE generated by repeated application of a MAP. The MAP is said to have a closed orbit if it has a finite number of elements.

see also DYNAMICAL SYSTEM, SINK (MAP)

Orchard-Planting Problem

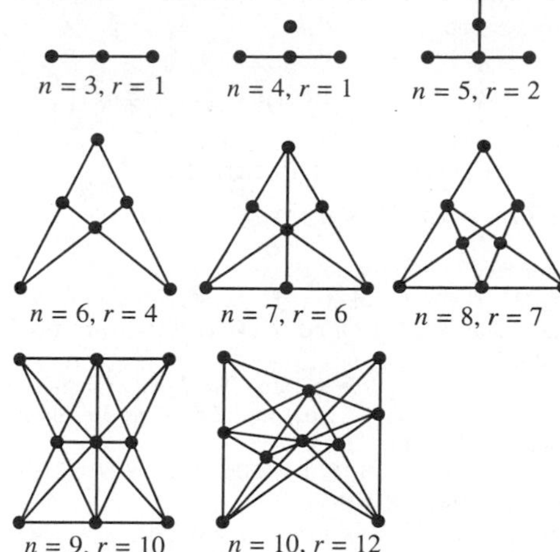

Also known as the TREE-PLANTING PROBLEM. Plant n trees so that there will be r straight rows with k trees in each row. The following table gives max(r) for various k. $k = 3$ is Sloane's A003035 and $k = 4$ is Sloane's A006065.

n	$k=3$	$k=4$	$k=5$
3	1	—	—
4	1	1	—
5	2	1	1
6	4	1	1
7	6	2	1
8	7	2	1
9	10	3	2
10	12	5	2
11	16	6	2
12	19	7	3
13	[22, 24]	≥ 9	3
14	[26, 27]	≥ 10	4
15	[31, 32]	≥ 12	≥ 6
16	37	≥ 15	≥ 6
17	[40, 42]	≥ 15	≥ 7
18	[46, 48]	≥ 18	≥ 9
19	[52, 54]	≥ 19	≥ 10
20	[57, 60]	≥ 21	≥ 11
21	[64, 67]		
22	[70, 73]		
23	[77, 81]		
24	[85, 88]		
25	[92, 96]		

Sylvester showed that

$$r(k=3) \geq \left\lfloor \tfrac{1}{6}(n-1)(n-2) \right\rfloor,$$

where $\lfloor x \rfloor$ is the FLOOR FUNCTION (Ball and Coxeter 1987). Burr, Grünbaum and Sloane (1974) have shown using cubic curves that

$$r(k=3) \leq 1 + \left\lfloor \tfrac{1}{6}n(n-3) \right\rfloor,$$

except for $n = 7$, 11, 16, and 19, and conjecture that the inequality is an equality with the exception of the preceding cases. For $n \geq 4$,

$$r(k=3) \geq \left\lfloor \tfrac{1}{3}\left[\tfrac{1}{2}n(n-1) - \left\lceil \tfrac{3}{7}n \right\rceil\right]\right\rfloor,$$

where $\lceil x \rceil$ is the CEILING FUNCTION.

see also ORCHARD VISIBILITY PROBLEM

References

Ball, W. W. R. and Coxeter, H. S. M. *Mathematical Recreations and Essays, 13th ed.* New York: Dover, pp. 104–105 and 129, 1987.

Burr, S. A. "Planting Trees." In *The Mathematical Gardner* (Ed. David Klarner). Boston, MA: Prindle, Weber, and Schmidt, pp. 90–99, 1981.

Dudeney, H. E. Problem 435 in *536 Puzzles & Curious Problems.* New York: Scribner, 1967.

Dudeney, H. E. *The Canterbury Puzzles and Other Curious Problems, 7th ed.* London: Thomas Nelson and Sons, p. 175, 1949.

Dudeney, H. E. §213 in *Amusements in Mathematics.* New York: Dover, 1970.

Gardner, M. Ch. 2 in *Mathematical Carnival: A New Round-Up of Tantalizers and Puzzles from Scientific American.* New York: Vintage Books, 1977.

Gardner, M. "Tree-Plant Problems." Ch. 22 in *Time Travel and Other Mathematical Bewilderments.* New York: W. H. Freeman, pp. 277–290, 1988.

Grünbaum, B. "New Views on Some Old Questions of Combinatorial Geometry." *Teorie Combin.* **1**, 451–468, 1976.

Grünbaum, B. and Sloane, N. J. A. "The Orchard Problem." *Geom. Dedic.* **2**, 397–424, 1974.

Jackson, J. *Rational Amusements for Winter Evenings.* London, 1821.

Macmillan, R. H. "An Old Problem." *Math. Gaz.* **30**, 109, 1946.

Sloane, N. J. A. Sequences A006065/M0290 and A003035/M0982 in "An On-Line Version of the Encyclopedia of Integer Sequences." `http://www.research.att.com/~njas/sequences/eisonline.html`.

Sloane, N. J. A. and Plouffe, S. Extended entry for M0982 in *The Encyclopedia of Integer Sequences.* San Diego: Academic Press, 1995.

Orchard Visibility Problem

A tree is planted at each LATTICE POINT in a circular orchard which has CENTER at the ORIGIN and RADIUS r. If the radius of trees exceeds $1/r$ units, one is unable to see out of the orchard in any direction. However, if the RADII of the trees are $< 1/\sqrt{r^2+1}$, one can see out at certain ANGLES.

see also LATTICE POINT, ORCHARD-PLANTING PROBLEM, VISIBILITY

References

Honsberger, R. "The Orchard Problem." Ch. 4 in *Mathematical Gems I.* Washington, DC: Math. Assoc. Amer., pp. 43–52, 1973.

Order (Algebraic Curve)

The order of the POLYNOMIAL defining the curve.

Order (Algebraic Surface)

The order n of an ALGEBRAIC SURFACE is the order of the POLYNOMIAL defining a surface, which can be geometrically interpreted as the maximum number of points in which a line meets the surface.

Order	Surface
3	cubic surface
4	quartic surface
5	quintic surface
6	sextic surface
7	heptic surface
8	octic surface
9	nonic surface
10	decic surface

see also ALGEBRAIC SURFACE

References

Fischer, G. (Ed.). *Mathematical Models from the Collections of Universities and Museums.* Braunschweig, Germany: Vieweg, p. 8, 1986.

Order (Conjugacy Class)

The number of elements of a GROUP in a given CONJUGACY CLASS.

Order (Difference Set)

Let G be GROUP of ORDER h and D be a set of k elements of G. If the set of differences $d_i - d_j$ contains every NONZERO element of G exactly λ times, then D is a (h, k, λ)-difference set in G of order $n = k - \lambda$.

Order (Field)

The number of elements in a FINITE FIELD.

Order (Group)

The number of elements in a GROUP G, denoted $|G|$. The order of an element g of a finite group G is the smallest POWER of n such that $g^n = I$, where I is the IDENTITY ELEMENT. In general, finding the order of the element of a group is at least as hard as factoring (Meijer 1996). However, the problem becomes significantly easier if $|G|$ and the factorization of $|G|$ are known. Under these circumstances, efficient ALGORITHMS are known (Cohen 1993).

see also ABELIAN GROUP, FINITE GROUP

References

Cohen, H. *A Course in Computational Algebraic Number Theory.* New York: Springer-Verlag, 1993.

Meijer, A. R. "Groups, Factoring, and Cryptography." *Math. Mag.* **69**, 103–109, 1996.

Order (Modulo)

For any INTEGER a which is not a multiple of a PRIME p, there exists a smallest exponent $h \geq 1$ such that $a^h \equiv 1 \pmod{p}$ IFF $h|k$. In that case, h is called the order of a modulo p.

see also CARMICHAEL FUNCTION

Order (Ordinary Differential Equation)

An ORDINARY DIFFERENTIAL EQUATION of order n is an equation of the form

$$F(x, y, y', \ldots, y^{(n)}) = 0.$$

Order (Permutation)

see PERMUTATION

Order (Polynomial)

The highest order POWER in a one-variable POLYNOMIAL is known as its order (or sometimes its DEGREE). For example, the POLYNOMIAL

$$a_n x^n + \ldots + a_2 x^2 + a_1 x + a_0$$

is of order n.

Order Statistic

Given a sample of n variates $X_1, \ldots, X_n$, reorder them so that $X_1' < X_2' < \ldots < X_n'$. Then the ith order statistic $X^{\langle i \rangle}$ is defined as X_i', with the special cases

$$m_n = X^{\langle 1 \rangle} = \min_j (X_j)$$
$$M_n = X^{\langle n \rangle} = \max_j (X_j).$$

A ROBUST ESTIMATION technique based on linear combinations of order statistics is called an L-ESTIMATE.

see also EXTREME VALUE DISTRIBUTION, HINGE, MAXIMUM, MINIMUM, MODE, ORDINAL NUMBER

References

Balakrishnan, N. and Cohen, A. C. *Order Statistics and Inference.* New York: Academic Press, 1991.

David, H. A. *Order Statistics, 2nd ed.* New York: Wiley, 1981.

Gibbons, J. D. and Chakraborti, S. (Eds.). *Nonparametric Statistic Inference, 3rd ed. exp. rev.* New York: Marcel Dekker, 1992.

Order (Vertex)

The number of EDGES meeting at a given node in a GRAPH is called the order of that VERTEX.

Ordered Geometry

A GEOMETRY constructed without reference to measurement. The only primitive concepts are those of points and intermediacy. There are 10 AXIOMS underlying ordered GEOMETRY.

see also ABSOLUTE GEOMETRY, AFFINE GEOMETRY, GEOMETRY

Ordered Pair

A PAIR of quantities (a, b) where ordering is significant, so (a, b) is considered distinct from (b, a) for $a \neq b$.

see also PAIR

Ordered Tree

A ROOTED TREE in which the order of the subtrees is significant. There is a ONE-TO-ONE correspondence between ordered FORESTS with n nodes and BINARY TREES with n nodes.

see also BINARY TREE, FOREST, ROOTED TREE

Ordering

The number of "ARRANGEMENTS" in an ordering of n items is given by either a COMBINATION (order is ignored) or a PERMUTATION (order is significant).

see also ARRANGEMENT, COMBINATION, CUTTING, DERANGEMENT, PARTIAL ORDER, PERMUTATION, SORTING, TOTAL ORDER

Ordering Axioms

The four of HILBERT'S AXIOMS which concern the arrangement of points.

see also CONGRUENCE AXIOMS, CONTINUITY AXIOMS, HILBERT'S AXIOMS, INCIDENCE AXIOMS, PARALLEL POSTULATE

References

Hilbert, D. *The Foundations of Geometry, 2nd ed.* Chicago, IL: Open Court, 1980.

Iyanaga, S. and Kawada, Y. (Eds.). "Hilbert's System of Axioms." §163B in *Encyclopedic Dictionary of Mathematics.* Cambridge, MA: MIT Press, pp. 544–545, 1980.

Ordinal Number

In informal usage, an ordinal number is an adjective which describes the numerical position of an object, e.g., first, second, third, etc.

In technical mathematics, an ordinal number is one of the numbers in Georg Cantor's extension of the WHOLE NUMBERS. The ordinal numbers are 0, 1, 2, ..., ω, $\omega+1$, $\omega+2$, ..., $\omega+\omega$, $\omega+\omega+1$, ... Cantor's "smallest" TRANSFINITE NUMBER ω is defined to be the earliest number greater than all WHOLE NUMBERS, and is denoted by Conway and Guy (1996) as $\omega = \{0, 1, \ldots \mid\}$. The notation of ordinal numbers can be a bit counterintuitive, e.g., even though $1 + \omega = \omega$, $\omega + 1 > \omega$.

Ordinal numbers have some other rather peculiar properties. The sum of two ordinal numbers can take on two different values, the sum of three can take on five values. The first few terms of this sequence are 2, 5, 13, 33, 81, 193, 449, 33^2, $33 \cdot 81$, 81^2, $81 \cdot 193$, 192^2, ... (Conway and Guy 1996, Sloane's A005348). The sum of n ordinals has either $193^a 81^b$ or $33 \cdot 81^a$ possible answers for $n \geq 15$ (Conway and Guy 1996).

$r \times \omega$ is the same as ω, but $\omega \times r$ is equal to $\underbrace{\omega + \ldots + \omega}_{r}$. ω^2 is larger than any number of the form $\omega \times r$, ω^3 is larger than ω^2, and so on.

There exist ordinal numbers which cannot be constructed from smaller ones by finite additions, multiplications, and exponentiations. These ordinals obey CANTOR'S EQUATION. The first such ordinal is

$$\epsilon_0 = \underbrace{\omega^{\omega^{\cdot^{\cdot^{\cdot^{\omega}}}}}}_{\omega} = 1 + \omega + \omega^\omega + \omega^{\omega^\omega} + \ldots.$$

The next is

$$\epsilon_1 = (\epsilon_0 + 1) + \omega^{\epsilon_0 + 1} + \omega^{\omega^{\epsilon_0 + 1}} + \ldots,$$

then follow $\epsilon_2, \epsilon_3, \ldots, \epsilon_\omega, \epsilon_{\omega+1}, \ldots, \epsilon_{\omega\times 2}, \ldots, \epsilon_{\omega^2}, \epsilon_{\omega^\omega}, \ldots, \epsilon_{\epsilon_0}, \epsilon_{\epsilon_0+1}, \ldots, \epsilon_{\epsilon_0+\omega}, \ldots, \epsilon_{\epsilon_0+\omega^\omega}, \ldots, \epsilon_{\epsilon_0\times 2}, \ldots, \epsilon_{\epsilon_1}, \ldots, \epsilon_{\epsilon_2}, \ldots, \epsilon_{\epsilon_\omega}, \ldots, \epsilon_{\epsilon_{\epsilon_0}}, \ldots, \epsilon_{\epsilon_{\epsilon_1}}, \ldots, \epsilon_{\epsilon_{\epsilon_\omega}}, \ldots, \epsilon_{\epsilon_{\epsilon_{\epsilon_0}}}, \ldots$ (Conway and Guy 1996).

see also AXIOM OF CHOICE, CANTOR'S EQUATION, CARDINAL NUMBER, ORDER STATISTIC, POWER SET, SURREAL NUMBER

References

Cantor, G. *Über unendliche, lineare Punktmannigfältigkeiten, Arbeiten zur Mengenlehre aus dem Jahren 1872–1884.* Leipzig, Germany: Teubner-Archiv zur Mathematik, 1884.

Conway, J. H. and Guy, R. K. "Cantor's Ordinal Numbers." In *The Book of Numbers.* New York: Springer-Verlag, pp. 266–267 and 274, 1996.

Sloane, N. J. A. Sequence A005348/M1435 in "An On-Line Version of the Encyclopedia of Integer Sequences."

Ordinary Differential Equation

An ordinary differential equation (frequently abbreviated ODE) is an equality involving a function and its DERIVATIVES. An ODE of order n is an equation of the form

$$F(x, y, y', \ldots, y^{(n)}) = 0, \tag{1}$$

where $y' = dy/dx$ is a first DERIVATIVE with respect to x and $y^{(n)} = d^n y/dx^n$ is an nth DERIVATIVE with respect to x. An ODE of order n is said to be linear if it is of the form

$$a_n(x)y^{(n)} + a_{n-1}(x)y^{(n-1)} + \ldots + a_1(x)y' + a_0(x)y = Q(x). \tag{2}$$

A linear ODE where $Q(x) = 0$ is said to be homogeneous. Confusingly, an ODE of the form

$$\frac{dy}{dx} = f\left(\frac{y}{x}\right) \tag{3}$$

is also sometimes called "homogeneous."

Simple theories exist for first-order (INTEGRATING FACTOR) and second-order (STURM-LIOUVILLE THEORY) ordinary differential equations, and arbitrary ODEs with linear constant COEFFICIENTS can be solved when they are of certain factorable forms. Integral transforms such as the LAPLACE TRANSFORM can also be used to solve classes of linear ODEs. Morse and Feshbach (1953, pp. 667–674) give canonical forms and solutions for second order ODEs.

While there are many general techniques for analytically solving classes of ODEs, the only practical solution technique for complicated equations is to use numerical methods (Milne 1970). The most popular of these is the RUNGE-KUTTA METHOD, but many others have been developed. A vast amount of research and huge numbers of publications have been devoted to the numerical solution of differential equations, both ordinary and PARTIAL (PDEs) as a result of their importance in fields as diverse as physics, engineering, economics, and electronics.

The solutions to an ODE satisfy EXISTENCE and UNIQUENESS properties. These can be formally established by PICARD'S EXISTENCE THEOREM for certain classes of ODEs. Let a system of first-order ODE be given by

$$\frac{dx_i}{dt} = f_i(x_1, \ldots, x_n, t), \tag{4}$$

for $i = 1, \ldots, n$ and let the functions $f_i(x_1, \ldots, x_n, t)$, where $i = 1, \ldots, n$, all be defined in a DOMAIN D of the $(n+1)$-D space of the variables $x_1, \ldots, x_n, t$. Let these functions be continuous in D and have continuous first PARTIAL DERIVATIVES $\partial f_i/\partial x_j$ for $i = 1, \ldots, n$ and $j = 1, \ldots, n$ in D. Let $(x_1^0, \ldots, x_n^0)$ be in D. Then there exists a solution of (4) given by

$$x_1 = x_1(t), \ldots, x_n = x_n(t) \tag{5}$$

for $t_0 - \delta < t < t_0 + \delta$ (where $\delta > 0$) satisfying the initial conditions

$$x_1(t_0) = x_1^0, \ldots, x_n(t_0) = x_n^0. \tag{6}$$

Furthermore, the solution is unique, so that if

$$x_1 = x_1^*(t), \ldots, x_n = x_n^*(t) \tag{7}$$

is a second solution of (4) for $t_0 - \delta < t < t_0 + \delta$ satisfying (6), then $x_i(t) \equiv x_i^*(t)$ for $t_0 - \delta < t < t_0 + \delta$. Because every nth-order ODE can be expressed as a system of n first-order differential equations, this theorem also applies to the single nth-order ODE.

In general, an nth-order ODE has n linearly independent solutions. Furthermore, any linear combination of LINEARLY INDEPENDENT FUNCTIONS solutions is also a solution.

An exact FIRST-ORDER ODEs is one of the form

$$p(x, y)\,dx + q(x, y)\,dy = 0, \tag{8}$$

where

$$\frac{\partial p}{\partial y} = \frac{\partial q}{\partial x}. \tag{9}$$

An equation of the form (8) with

$$\frac{\partial p}{\partial y} \neq \frac{\partial q}{\partial x} \tag{10}$$

is said to be nonexact. If

$$\frac{\frac{\partial p}{\partial y} - \frac{\partial q}{\partial x}}{q} = f(x) \tag{11}$$

in (8), it has an x-dependent integrating factor. If

$$\frac{\frac{\partial q}{\partial x} - \frac{\partial p}{\partial y}}{xp - yq} = f(xy) \tag{12}$$

in (8), it has an xy-dependent integrating factor. If

$$\frac{\frac{\partial q}{\partial x} - \frac{\partial p}{\partial y}}{p} = f(y) \tag{13}$$

in (8), it has a y-dependent integrating factor.

Other special first-order types include cross multiple equations

$$yf(xy)\,dx + xg(xy)\,dy = 0, \tag{14}$$

homogeneous equations

$$\frac{dy}{dx} = f\left(\frac{y}{x}\right), \tag{15}$$

linear equations

$$\frac{dy}{dx} + p(x)y = q(x), \tag{16}$$

and separable equations

$$\frac{dy}{dx} = X(x)Y(y). \tag{17}$$

Special classes of SECOND-ORDER ODEs include

$$\frac{d^2y}{dx^2} = f(y, y') \tag{18}$$

(x missing) and

$$\frac{d^2y}{dx^2} = f(x, y') \tag{19}$$

(y missing). A second-order linear homogeneous ODE

$$\frac{d^2y}{dx^2} + P(x)\frac{dy}{dx} + Q(x)y = 0 \tag{20}$$

for which

$$\frac{Q'(x) + 2P(x)Q(x)}{2[Q(x)]^{3/2}} = [\text{constant}] \tag{21}$$

can be transformed to one with constant coefficients.

The undamped equation of SIMPLE HARMONIC MOTION is

$$\frac{d^2y}{dx^2} + {\omega_0}^2 y = 0, \tag{22}$$

which becomes

$$\frac{d^2y}{dx^2} + \beta\frac{dy}{dx} + {\omega_0}^2 y = 0 \tag{23}$$

when damped, and

$$\frac{d^2y}{dx^2} + \beta\frac{dy}{dx} + {\omega_0}^2 y = A\cos(\omega t) \tag{24}$$

when both forced and damped.

SYSTEMS WITH CONSTANT COEFFICIENTS are of the form

$$\frac{d\mathbf{x}}{dt} = \mathsf{A}\mathbf{x}(t) + \mathbf{p}(t). \tag{25}$$

The following are examples of important ordinary differential equations which commonly arise in problems of mathematical physics.

AIRY DIFFERENTIAL EQUATION

$$\frac{d^2y}{dx^2} - xy = 0. \tag{26}$$

BERNOULLI DIFFERENTIAL EQUATION

$$\frac{dy}{dx} + p(x)y = q(x)y^n. \tag{27}$$

BESSEL DIFFERENTIAL EQUATION

$$x^2\frac{d^2y}{dx^2} + x\frac{dy}{dx} + (\lambda^2x^2 - n^2)y = 0. \tag{28}$$

CHEBYSHEV DIFFERENTIAL EQUATION

$$(1 - x^2)\frac{d^2y}{dx^2} - x\frac{dy}{dx} + \alpha^2 y = 0. \tag{29}$$

CONFLUENT HYPERGEOMETRIC DIFFERENTIAL EQUATION

$$x\frac{d^2y}{dx^2} + (\gamma - x)\frac{dy}{dx} + \alpha y = 0. \tag{30}$$

EULER DIFFERENTIAL EQUATION

$$x^2\frac{d^2y}{dx^2} + ax\frac{dy}{dx} + by = S(x). \tag{31}$$

HERMITE DIFFERENTIAL EQUATION

$$\frac{d^2y}{dx^2} - 2x\frac{dy}{dx} + \lambda y = 0. \tag{32}$$

HILL'S DIFFERENTIAL EQUATION

$$\frac{d^2y}{dx^2} + \left[\theta_0 + 2\sum_{n=1}^{\infty}\theta_n\cos(2nz)\right] = 0. \tag{33}$$

HYPERGEOMETRIC DIFFERENTIAL EQUATION

$$x(x-1)\frac{d^2y}{dx^2}+[(1+\alpha+\beta)x-\gamma]\frac{dy}{dx}+\alpha\beta y=0. \quad (34)$$

JACOBI DIFFERENTIAL EQUATION

$$(1-x^2)y''+[\beta-\alpha-(\alpha+\beta+2)x]y'+n(n+\alpha+\beta+1)y=0. \quad (35)$$

LAGUERRE DIFFERENTIAL EQUATION

$$x\frac{d^2y}{dx^2}+(1-x)\frac{dy}{dx}+\lambda y=0. \quad (36)$$

LANE-EMDEN DIFFERENTIAL EQUATION

$$\frac{1}{\xi^2}\frac{d}{d\xi}\left(\xi^2\frac{d\theta}{d\xi}\right)+\theta^n=0. \quad (37)$$

LEGENDRE DIFFERENTIAL EQUATION

$$(1-x^2)\frac{d^2y}{dx^2}-2x\frac{dy}{dx}+\alpha(\alpha+1)y=0. \quad (38)$$

LINEAR CONSTANT COEFFICIENTS

$$a_0\frac{d^ny}{dx^n}+\ldots+a_{n-1}\frac{dy}{dx}+a_ny=p(x). \quad (39)$$

MALMSTÉN'S DIFFERENTIAL EQUATION

$$y''+\frac{r}{z}y'=\left(Az^m+\frac{s}{z^2}\right)y. \quad (40)$$

RICCATI DIFFERENTIAL EQUATION

$$\frac{dw}{dx}=q_0(x)+q_1(x)w+q_2(x)w^2. \quad (41)$$

RIEMANN P-DIFFERENTIAL EQUATION

$$\frac{d^2u}{dz^2}+\left[\frac{1-\alpha-\alpha'}{z-a}+\frac{1-\beta-\beta'}{z-b}+\frac{1-\gamma-\gamma'}{z-c}\right]\frac{du}{dz}+\left[\frac{\alpha\alpha'(a-b)(a-c)}{z-a}+\frac{\beta\beta'(b-c)(b-a)}{z-b}+\frac{\gamma\gamma'(c-a)(c-b)}{z-c}\right]\frac{u}{(z-a)(z-b)(z-c)}=0. \quad (42)$$

see also ADAMS' METHOD, GREEN'S FUNCTION, ISOCLINE, LAPLACE TRANSFORM, LEADING ORDER ANALYSIS, MAJORANT, ORDINARY DIFFERENTIAL EQUATION—FIRST-ORDER, ORDINARY DIFFERENTIAL EQUATION—SECOND-ORDER, PARTIAL DIFFERENTIAL EQUATION, RELAXATION METHODS, RUNGE-KUTTA METHOD, SIMPLE HARMONIC MOTION

References

Boyce, W. E. and DiPrima, R. C. *Elementary Differential Equations and Boundary Value Problems, 5th ed.* New York: Wiley, 1992.

Braun, M. *Differential Equations and Their Applications, 3rd ed.* New York: Springer-Verlag, 1991.

Forsyth, A. R. *Theory of Differential Equations, 6 vols.* New York: Dover, 1959.

Forsyth, A. R. *A Treatise on Differential Equations.* New York: Dover, 1997.

Guterman, M. M. and Nitecki, Z. H. *Differential Equations: A First Course, 3rd ed.* Philadelphia, PA: Saunders, 1992.

Ince, E. L. *Ordinary Differential Equations.* New York: Dover, 1956.

Milne, W. E. *Numerical Solution of Differential Equations.* New York: Dover, 1970.

Morse, P. M. and Feshbach, H. "Ordinary Differential Equations." Ch. 5 in *Methods of Theoretical Physics, Part I.* New York: McGraw-Hill, pp. 492–675, 1953.

Moulton, F. R. *Differential Equations.* New York: Dover, 1958.

Press, W. H.; Flannery, B. P.; Teukolsky, S. A.; and Vetterling, W. T. "Integration of Ordinary Differential Equations." Ch. 16 in *Numerical Recipes in FORTRAN: The Art of Scientific Computing, 2nd ed.* Cambridge, England: Cambridge University Press, pp. 701–744, 1992.

Simmons, G. F. *Differential Equations, with Applications and Historical Notes, 2nd ed.* New York: McGraw-Hill, 1991.

Zwillinger, D. *Handbook of Differential Equations, 3rd ed.* Boston, MA: Academic Press, 1997.

Ordinary Differential Equation—First-Order

Given a first-order ORDINARY DIFFERENTIAL EQUATION

$$\frac{dy}{dx}=F(x,y), \quad (1)$$

if $F(x,y)$ can be expressed using SEPARATION OF VARIABLES as

$$F(x,y)=X(x)Y(y), \quad (2)$$

then the equation can be expressed as

$$\frac{dy}{Y(y)}=X(x)\,dx \quad (3)$$

and the equation can be solved by integrating both sides to obtain

$$\int\frac{dy}{Y(y)}=\int X(x)\,dx. \quad (4)$$

Any first-order ODE of the form

$$\frac{dy}{dx}+p(x)y=q(x) \quad (5)$$

can be solved by finding an INTEGRATING FACTOR $\mu=\mu(x)$ such that

$$\frac{d}{dx}(\mu y)=\mu\frac{dy}{dx}+y\frac{d\mu}{dx}=\mu q(x). \quad (6)$$

Dividing through by μy yields

$$\frac{1}{y}\frac{dy}{dx}+\frac{1}{\mu}\frac{d\mu}{dx}=\frac{q(x)}{y}. \quad (7)$$

However, this condition enables us to explicitly determine the appropriate μ for arbitrary p and q. To accomplish this, take

$$p(x) = \frac{1}{\mu}\frac{d\mu}{dx} \tag{8}$$

in the above equation, from which we recover the original equation (5), as required, in the form

$$\frac{1}{y}\frac{dy}{dx} + p(x) = \frac{q(x)}{y}. \tag{9}$$

But we can integrate both sides of (8) to obtain

$$\int p(x)\,dx = \int \frac{d\mu}{\mu} = \ln \mu + c \tag{10}$$

$$\mu = e^{\int p(x)\,dx}. \tag{11}$$

Now integrating both sides of (6) gives

$$\mu y = \int \mu q(x)\,dx + c \tag{12}$$

(with μ now a known function), which can be solved for y to obtain

$$y = \frac{\int \mu q(x)\,dx + c}{\mu} = \frac{\int e^{\int^x p(x')\,dx'} q(x)\,dx + c}{e^{\int^x p(x')\,dx'}}, \tag{13}$$

where c is an arbitrary constant of integration.

Given an nth-order linear ODE with constant COEFFICIENTS

$$\frac{d^n y}{dx^n} + a_{n-1}\frac{d^{n-1}y}{dx^{n-1}} + \ldots + a_1\frac{dy}{dx} + a_0 y = Q(x), \tag{14}$$

first solve the characteristic equation obtained by writing

$$y \equiv e^{rx} \tag{15}$$

and setting $Q(x) = 0$ to obtain the n COMPLEX ROOTS.

$$r^n e^{rx} + a_{n-1}r^{n-1}e^{rx} + \ldots + a_1 r e^{rx} + a_0 e^{rx} = 0 \tag{16}$$

$$r^n + a_{n-1}r^{n-1} + \ldots + a_1 r + a_0 = 0. \tag{17}$$

Factoring gives the ROOTS r_i,

$$(r - r_1)(r - r_2)\cdots(r - r_n) = 0. \tag{18}$$

For a nonrepeated REAL ROOT r, the corresponding solution is

$$y = e^{rx}. \tag{19}$$

If a REAL ROOT r is repeated k times, the solutions are degenerate and the linearly independent solutions are

$$y = e^{rx}, y = xe^{rx}, \ldots, y = x^{k-1}e^{rx}. \tag{20}$$

Complex ROOTS always come in COMPLEX CONJUGATE pairs, $r_\pm = a \pm ib$. For nonrepeated COMPLEX ROOTS, the solutions are

$$y = e^{ax}\cos(bx), y = e^{ax}\sin(bx). \tag{21}$$

If the COMPLEX ROOTS are repeated k times, the linearly independent solutions are

$$y = e^{ax}\cos(bx), y = e^{ax}\sin(bx), \ldots,$$
$$y = x^{k-1}e^{ax}\cos(bx), y = x^{k-1}e^{ax}\sin(bx). \tag{22}$$

Linearly combining solutions of the appropriate types with arbitrary multiplicative constants then gives the complete solution. If initial conditions are specified, the constants can be explicitly determined. For example, consider the sixth-order linear ODE

$$(\tilde{D} - 1)(\tilde{D} - 2)^3(\tilde{D}^2 + \tilde{D} + 1)y = 0, \tag{23}$$

which has the characteristic equation

$$(r - 1)(r - 2)^3(r^2 + r + 1) = 0. \tag{24}$$

The roots are 1, 2 (three times), and $(-1 \pm \sqrt{3}\,i)/2$, so the solution is

$$y = Ae^x + Be^{2x} + Cxe^{2x} + Dx^2e^{3x} + Ee^{-x/2}\cos(\tfrac{1}{2}\sqrt{3}\,x)$$
$$+Fe^{-x}\sin(\tfrac{1}{2}\sqrt{3}\,x). \tag{25}$$

If the original equation is nonhomogeneous ($Q(x) \neq 0$), now find the particular solution y^* by the method of VARIATION OF PARAMETERS. The general solution is then

$$y(x) = \sum_{i=1}^{n} c_i y_i(x) + y^*(x), \tag{26}$$

where the solutions to the linear equations are $y_1(x)$, $y_2(x)$, ..., $y_n(x)$, and $y^*(x)$ is the particular solution.

see also INTEGRATING FACTOR, ORDINARY DIFFERENTIAL EQUATION—FIRST-ORDER EXACT, SEPARATION OF VARIABLES, VARIATION OF PARAMETERS

References

Arfken, G. *Mathematical Methods for Physicists, 3rd ed.* Orlando, FL: Academic Press, pp. 440–445, 1985.

Ordinary Differential Equation—First-Order Exact

Consider a first-order ODE in the slightly different form

$$p(x, y)\,dx + q(x, y)\,dy = 0. \tag{1}$$

Such an equation is said to be exact if

$$\frac{\partial p}{\partial y} = \frac{\partial q}{\partial x}. \tag{2}$$

This statement is equivalent to the requirement that a CONSERVATIVE FIELD exists, so that a scalar potential can be defined. For an exact equation, the solution is

$$\int_{(x_0,y_0)}^{(x,y)} p(x,y)\,dx + q(x,y)\,dy = c, \tag{3}$$

where c is a constant.

A first-order ODE (1) is said to be inexact if

$$\frac{\partial p}{\partial y} \neq \frac{\partial q}{\partial x}. \tag{4}$$

For a nonexact equation, the solution may be obtained by defining an INTEGRATING FACTOR μ of (6) so that the new equation

$$\mu p(x,y)\,dx + \mu q(x,y)\,dy = 0 \tag{5}$$

satisfies

$$\frac{\partial}{\partial y}(\mu p) = \frac{\partial}{\partial x}(\mu q), \tag{6}$$

or, written out explicitly,

$$p\frac{\partial \mu}{\partial y} + \mu\frac{\partial p}{\partial y} = q\frac{\partial \mu}{\partial x} + \mu\frac{\partial p}{\partial x}. \tag{7}$$

This transforms the nonexact equation into an exact one. Solving (7) for μ gives

$$\mu = \frac{q\frac{\partial \mu}{\partial x} - p\frac{\partial \mu}{\partial y}}{\frac{\partial p}{\partial y} - \frac{\partial q}{\partial x}}. \tag{8}$$

Therefore, if a function μ satisfying (8) can be found, then writing

$$P(x,y) = \mu p \tag{9}$$

$$Q(x,y) = \mu q \tag{10}$$

in equation (5) then gives

$$P(x,y)\,dx + Q(x,y)\,dy = 0, \tag{11}$$

which is then an exact ODE. Special cases in which μ can be found include x-dependent, xy-dependent, and y-dependent integrating factors.

Given an inexact first-order ODE, we can also look for an INTEGRATING FACTOR $\mu(x)$ so that

$$\frac{\partial \mu}{\partial y} = 0. \tag{12}$$

For the equation to be exact in μp and μq, the equation for a first-order nonexact ODE

$$p\frac{\partial \mu}{\partial y} + \mu\frac{\partial p}{\partial y} = q\frac{\partial \mu}{\partial x} + \mu\frac{\partial p}{\partial x} \tag{13}$$

becomes

$$\mu\frac{\partial p}{\partial y} = q\frac{\partial \mu}{\partial x} + \mu\frac{\partial p}{\partial x}. \tag{14}$$

Solving for $\partial\mu/\partial x$ gives

$$\frac{\partial \mu}{\partial x} = \mu(x)\frac{\frac{\partial p}{\partial y} - \frac{\partial q}{\partial x}}{q} \equiv f(x,y)\mu(x), \tag{15}$$

which will be integrable if

$$f(x,y) \equiv \frac{\frac{\partial p}{\partial y} - \frac{\partial q}{\partial x}}{q} = f(x), \tag{16}$$

in which case

$$\frac{d\mu}{\mu} = f(x)\,dx, \tag{17}$$

so that the equation is integrable

$$\mu(x) = e^{\int f(x)\,dx}, \tag{18}$$

and the equation

$$[\mu p(x,y)]dx + [\mu q(x,y)]dy = 0 \tag{19}$$

with known $\mu(x)$ is now exact and can be solved as an exact ODE.

Given in an exact first-order ODE, look for an INTEGRATING FACTOR $\mu(x,y) = g(xy)$. Then

$$\frac{\partial \mu}{\partial x} = \frac{\partial g}{\partial x}y \tag{20}$$

$$\frac{\partial \mu}{\partial y} = \frac{\partial g}{\partial y}x. \tag{21}$$

Combining these two,

$$\frac{\partial \mu}{\partial x} = \frac{y}{x}\frac{\partial \mu}{\partial y}. \tag{22}$$

For the equation to be exact in μp and μq, the equation for a first-order nonexact ODE

$$p\frac{\partial \mu}{\partial y} + \mu\frac{\partial p}{\partial y} = q\frac{\partial \mu}{\partial x} + \mu\frac{\partial p}{\partial x} \tag{23}$$

becomes

$$\frac{\partial \mu}{\partial y}\left(p - \frac{y}{x}q\right) = \left(\frac{\partial p}{\partial x} - \frac{\partial p}{\partial y}\right)\mu. \tag{24}$$

Therefore,

$$\frac{1}{x}\frac{\partial \mu}{\partial y} = \frac{\frac{\partial q}{\partial x} - \frac{\partial p}{\partial y}}{xp - yq}\mu. \tag{25}$$

Define a new variable

$$t(x,y) \equiv xy, \tag{26}$$

then $\partial t/\partial y = x$, so

$$\frac{\partial \mu}{\partial t} = \frac{\partial \mu}{\partial y}\frac{\partial y}{\partial t} = \frac{\frac{\partial q}{\partial x} - \frac{\partial p}{\partial y}}{xp - yq}\mu(t) \equiv f(x,y)\mu(t). \quad (27)$$

Now, if

$$f(x,y) \equiv \frac{\frac{\partial q}{\partial x} - \frac{\partial p}{\partial y}}{xp - yq} = f(xy) = f(t), \quad (28)$$

then

$$\frac{\partial \mu}{\partial t} = f(t)\mu(t), \quad (29)$$

so that

$$\mu = e^{\int f(t)\,dt} \quad (30)$$

and the equation

$$[\mu p(x,y)]\,dx + [\mu q(x,y)]\,dy = 0 \quad (31)$$

is now exact and can be solved as an exact ODE.

Given an inexact first-order ODE, assume there exists an integrating factor

$$\mu = f(y), \quad (32)$$

so $\partial\mu/\partial x = 0$. For the equation to be exact in μp and μq, equation (7) becomes

$$\frac{\partial \mu}{\partial y} = \frac{\frac{\partial q}{\partial x} - \frac{\partial p}{\partial y}}{p}\mu = f(x,y)\mu(y). \quad (33)$$

Now, if

$$f(x,y) \equiv \frac{\frac{\partial q}{\partial x} - \frac{\partial p}{\partial y}}{p} = f(y), \quad (34)$$

then

$$\frac{d\mu}{\mu} = f(y)\,dy, \quad (35)$$

so that

$$\mu(y) = e^{\int f(y)\,dy}, \quad (36)$$

and the equation

$$\mu p(x,y)\,dx + \mu q(x,y)\,dy = 0 \quad (37)$$

is now exact and can be solved as an exact ODE.

Given a first-order ODE of the form

$$yf(xy)\,dx + xg(xy)\,dy = 0, \quad (38)$$

define

$$v \equiv xy. \quad (39)$$

Then the solution is

$$\begin{cases} \ln x = \int \frac{g(v)\,dv}{c[g(v)-f(v)]} + c & \text{for } g(v) \neq f(v) \\ xy = c & \text{for } g(v) = f(v). \end{cases} \quad (40)$$

If

$$\frac{dy}{dx} = F(x,y) = G(v), \quad (41)$$

where

$$v \equiv \frac{y}{x}, \quad (42)$$

then letting

$$y \equiv xv \quad (43)$$

gives

$$\frac{dy}{dx} = x\,dv/dx + v \quad (44)$$

$$x\frac{dv}{dx} + v = G(v). \quad (45)$$

This can be integrated by quadratures, so

$$\ln x = \int \frac{dv}{f(v) - v} + c \qquad \text{for } f(v) \neq v \quad (46)$$

$$y = cx \qquad \text{for} f(v) = v. \quad (47)$$

References

Boyce, W. E. and DiPrima, R. C. *Elementary Differential Equations and Boundary Value Problems, 4th ed.* New York: Wiley, 1986.

Ordinary Differential Equation—Second-Order

An ODE

$$y'' + P(x)y' + Q(x)y = 0 \quad (1)$$

has singularities for finite $x = x_0$ under the following conditions: (a) If either $P(x)$ or $Q(x)$ diverges as $x \to x_0$, but $(x - x_0)P(x)$ and $(x - x_0)^2Q(x)$ remain finite as $x \to x_0$, then x_0 is called a regular or nonessential singular point. (b) If $P(x)$ diverges faster than $(x - x_0)^{-1}$ so that $(x - x_0)P(x) \to \infty$ as $x \to x_0$, or $Q(x)$ diverges faster than $(x - x_0)^{-2}$ so that $(x - x_0)^2Q(x) \to \infty$ as $x \to x_0$, then x_0 is called an irregular or essential singularity.

Singularities of equation (1) at infinity are investigated by making the substitution $x \equiv z^{-1}$, so $dx = -z^{-2}\,dz$, giving

$$\frac{dy}{dx} = -z^2\frac{dy}{dz} \quad (2)$$

$$\frac{d^2y}{dx^2} = -z^2\frac{d}{dz}\left(-z^2\frac{dy}{dz}\right) = -z^2\left(-2z\frac{dy}{dz} - z^2\frac{d^2y}{dz^2}\right)$$
$$= 2z^3\frac{dy}{dz} + z^4\frac{d^2y}{dz^2}. \quad (3)$$

Then (1) becomes

$$z^4\frac{d^2y}{dz^2} + [2z^3 - z^2P(z)]\frac{dy}{dz} + Q(z)y = 0. \quad (4)$$

Case (a): If

$$\alpha(z) \equiv \frac{2z - P(z)}{z^2} \tag{5}$$

$$\beta(z) \equiv \frac{Q(z)}{z^4} \tag{6}$$

remain finite at $x = \pm\infty$ ($y = 0$), then the point is ordinary. Case (b): If either $\alpha(z)$ diverges no more rapidly than $1/z$ or $\beta(z)$ diverges no more rapidly than $1/z^2$, then the point is a regular singular point. Case (c): Otherwise, the point is an irregular singular point.

Morse and Feshbach (1953, pp. 667–674) give the canonical forms and solutions for second-order ODEs classified by types of singular points.

For special classes of second-order linear ordinary differential equations, variable COEFFICIENTS can be transformed into constant COEFFICIENTS. Given a second-order linear ODE with variable COEFFICIENTS

$$\frac{d^2y}{dx^2} + p(x)\frac{dy}{dx} + q(x)y = 0. \tag{7}$$

Define a function $z \equiv y(x)$,

$$\frac{dy}{dx} = \frac{dz}{dx}\frac{dy}{dz} \tag{8}$$

$$\frac{d^2y}{dx^2} = \left(\frac{dz}{dx}\right)^2 \frac{d^2y}{dz^2} + \frac{d^2z}{dx^2}\frac{dy}{dz} \tag{9}$$

$$\left(\frac{dz}{dx}\right)^2 \frac{d^2y}{dz^2} + \left[\frac{d^2z}{dx^2} + p(x)\frac{dz}{dx}\right]\frac{dy}{dz} + q(x)y = 0 \tag{10}$$

$$\frac{d^2y}{dz^2} + \left[\frac{\frac{d^2z}{dx^2} + p(x)\frac{dz}{dx}}{\left(\frac{dz}{dx}\right)^2}\right]\frac{dy}{dz} + \left[\frac{q(x)}{\left(\frac{dz}{dx}\right)^2}\right]y$$
$$\equiv \frac{d^2y}{dz^2} + A\frac{dy}{dz} + By = 0. \tag{11}$$

This will have constant COEFFICIENTS if A and B are not functions of x. But we are free to set B to an arbitrary POSITIVE constant for $q(x) \geq 0$ by defining z as

$$z \equiv B^{-1/2}\int [q(x)]^{1/2}\,dx. \tag{12}$$

Then

$$\frac{dz}{dx} = B^{-1/2}[q(x)]^{1/2} \tag{13}$$

$$\frac{d^2z}{dx^2} = \tfrac{1}{2}B^{-1/2}[q(x)]^{-1/2}q'(x), \tag{14}$$

and

$$A = \frac{\tfrac{1}{2}B^{-1/2}[q(x)]^{-1/2}q'(x) + B^{-1/2}p(x)[q(x)]^{1/2}}{B^{-1}q(x)}$$
$$= \frac{q'(x) + 2p(x)q(x)}{2[q(x)]^{3/2}}B^{1/2}. \tag{15}$$

Equation (11) therefore becomes

$$\frac{d^2y}{dz^2} + \frac{q'(x) + 2p(x)q(x)}{2[q(x)]^{3/2}}B^{1/2}\frac{dy}{dz} + By = 0, \tag{16}$$

which has constant COEFFICIENTS provided that

$$A \equiv \frac{q'(x) + 2p(x)q(x)}{2[q(x)]^{3/2}}B^{1/2} = [\text{constant}]. \tag{17}$$

Eliminating constants, this gives

$$A' \equiv \frac{q'(x) + 2p(x)q(x)}{[q(x)]^{3/2}} = [\text{constant}]. \tag{18}$$

So for an ordinary differential equation in which A' is a constant, the solution is given by solving the second-order linear ODE with constant COEFFICIENTS

$$\frac{d^2y}{dz^2} + A\frac{dy}{dz} + By = 0 \tag{19}$$

for z, where z is defined as above.

A linear second-order homogeneous differential equation of the general form

$$y''(x) + P(x)y' + Q(x)y = 0 \tag{20}$$

can be transformed into standard form

$$z''(x) + q(x)z = 0 \tag{21}$$

with the first-order term eliminated using the substitution

$$\ln y \equiv \ln z - \tfrac{1}{2}\int P(x)\,dx. \tag{22}$$

Then

$$\frac{y'}{y} = \frac{z'}{z} - \tfrac{1}{2}P(x) \tag{23}$$

$$\frac{yy'' - y'^2}{y^2} = \frac{zz'' - z'^2}{z^2} - \tfrac{1}{2}P'(x) \tag{24}$$

$$\frac{y''}{y} - \left(\frac{y'}{y}\right)^2 = \frac{z''}{z} - \frac{z'^2}{z} - \frac{z'^2}{z^2} - \tfrac{1}{2}P'(x) \tag{25}$$

$$\frac{y''}{y} = \left[\frac{z'}{z} - \tfrac{1}{2}P(x)\right]^2 + \frac{z''}{z} - \frac{z'^2}{z^2} - \tfrac{1}{2}P'(x)$$
$$= \frac{z'^2}{z^2} - \frac{z'}{z}P(x) + \tfrac{1}{4}P^2(x) + \frac{z''}{z} - \frac{z'^2}{z^2} - \tfrac{1}{2}P'(x), \tag{26}$$

so

$$\frac{y''}{y} + P(x)\frac{y'}{y} + Q(x) = -\frac{z'}{z}P(x)$$
$$+\tfrac{1}{4}P^2(x) + \frac{z''}{z} - \tfrac{1}{2}P'(x) + P(x)\left[\frac{z'}{z} - \tfrac{1}{2}P(x)\right]$$
$$+Q(x) = \frac{z''}{z} - \tfrac{1}{2}P'(x) - \tfrac{1}{4}P^2(x) + Q(x) = 0. \tag{27}$$

Therefore,

$$z'' + [Q(x) - \tfrac{1}{2}P'(x) - \tfrac{1}{4}P^2(x)]z \equiv z''(x) + q(x)z = 0, \quad (28)$$

where

$$q(x) \equiv Q(x) - \tfrac{1}{2}P'(x) - \tfrac{1}{4}P^2(x). \quad (29)$$

If $Q(x) = 0$, then the differential equation becomes

$$y'' + P(x)y' = 0, \quad (30)$$

which can be solved by multiplying by

$$\exp\left[\int^x P(x')\,dx'\right] \quad (31)$$

to obtain

$$0 = \frac{d}{dx}\left\{\exp\left[\int^x P(x')\,dx'\right]\frac{dy}{dx}\right\} \quad (32)$$

$$c_1 = \exp\left[\int^x P(x')\,dx'\right]\frac{dy}{dx} \quad (33)$$

$$y = c_1 \int^x \frac{dx}{\exp\left[\int^x P(x')\,dx'\right]} + c_2. \quad (34)$$

If one solution (y_1) to a second-order ODE is known, the other (y_2) may be found using the REDUCTION OF ORDER method. From the ABEL'S IDENTITY

$$\frac{dW}{W} = -P(x)\,dx, \quad (35)$$

where

$$W \equiv y_1 y_2' - y_1' y_2 \quad (36)$$

$$\int_a^x \frac{dW}{W} = \int_a^x P(x')\,dx' \quad (37)$$

$$\ln\left[\frac{W(x)}{W(a)}\right] = \int_a^x P(x')\,dx' \quad (38)$$

$$W(x) = W(a)\exp\left[-\int_a^x P(x')\,dx'\right]. \quad (39)$$

But

$$W \equiv y_1 y_2' - y_1' y_2 = y_1^2 \frac{d}{dx}\left(\frac{y_2}{y_1}\right). \quad (40)$$

Combining (39) and (40) yields

$$\frac{d}{dx}\left(\frac{y_2}{y1}\right) = W(a)\frac{\exp[-\int_a^x P(x')\,dx']}{y_1^2} \quad (41)$$

$$y_2(x) = y_1(x)W(a)\int_b^x \frac{\exp[-\int_a^{x'} P(x'')\,dx'']}{[y_1(x')]^2}\,dx'. \quad (42)$$

Disregarding $W(a)$, since it is simply a multiplicative constant, and the constants a and b, which will contribute a solution which is not linearly independent of y_1,

$$y_2(x) = y_1(x)\int^x \frac{\exp\left[-\int^{x'} P(x'')\,dx''\right]}{[y_1(x')]^2}\,dx'. \quad (43)$$

If $P(x) = 0$, this simplifies to

$$y_2(x) = y_1(x)\int^x \frac{dx'}{[y_1(x')]^2}. \quad (44)$$

For a nonhomogeneous second-order ODE in which the x term does not appear in the function $f(x, y, y')$,

$$\frac{d^2y}{dx^2} = f(y, y'), \quad (45)$$

let $v \equiv y'$, then

$$\frac{dv}{dx} = f(v, y) = \frac{dv}{dy}\frac{dy}{dx} = v\frac{dv}{dy}. \quad (46)$$

So the first-order ODE

$$v\frac{dv}{dy} = f(y, v), \quad (47)$$

if linear, can be solved for v as a linear first-order ODE. Once the solution is known,

$$\frac{dy}{dx} = v(y) \quad (48)$$

$$\int \frac{dy}{v(y)} = \int dx. \quad (49)$$

On the other hand, if y is missing from $f(x, y, y')$,

$$\frac{d^2y}{dx^2} = f(x, y'), \quad (50)$$

let $v \equiv y'$, then $v' = y''$, and the equation reduces to

$$v' = f(x, v), \quad (51)$$

which, if linear, can be solved for v as a linear first-order ODE. Once the solution is known,

$$y = \int v(x)\,dx. \quad (52)$$

see also ABEL'S IDENTITY, ADJOINT OPERATOR

References

Arfken, G. "A Second Solution." §8.6 in *Mathematical Methods for Physicists, 3rd ed.* Orlando, FL: Academic Press, pp. 467–480, 1985.

Boyce, W. E. and DiPrima, R. C. *Elementary Differential Equations and Boundary Value Problems, 4th ed.* New York: Wiley, 1986.

Morse, P. M. and Feshbach, H. *Methods of Theoretical Physics, Part I.* New York: McGraw-Hill, pp. 667–674, 1953.

Ordinary Differential Equation—System with Constant Coefficients

To solve the system of differential equations

$$\frac{d\mathbf{x}}{dt} = \mathsf{A}\mathbf{x}(t) + \mathbf{p}(t), \tag{1}$$

where A is a MATRIX and $\mathbf{x}$ and $\mathbf{p}$ are VECTORS, first consider the homogeneous case with $\mathbf{p} = \mathbf{0}$. Then the solutions to

$$\frac{d\mathbf{x}}{dt} = \mathsf{A}\mathbf{x}(t) \tag{2}$$

are given by

$$\mathbf{x}(t) = e^{\mathsf{A}t}\mathbf{x}(\mathbf{t}). \tag{3}$$

But, by the MATRIX DECOMPOSITION THEOREM, the MATRIX EXPONENTIAL can be written as

$$e^{\mathsf{A}t} = \mathsf{u}\mathsf{D}\mathsf{u}^{-1}, \tag{4}$$

where the EIGENVECTOR MATRIX is

$$\mathsf{u} = [\,\mathbf{u}_1 \quad \cdots \quad \mathbf{u}_n\,] \tag{5}$$

and the EIGENVALUE MATRIX is

$$\mathsf{D} = \begin{bmatrix} e^{\lambda_1 t} & 0 & \cdots & 0 \\ 0 & e^{\lambda_2 t} & \cdots & 0 \\ \vdots & \vdots & \ddots & 0 \\ 0 & 0 & \cdots & e^{\lambda_n t} \end{bmatrix}. \tag{6}$$

Now consider

$$\begin{aligned} e^{\mathsf{A}t}\mathsf{u} &= \mathsf{u}\mathsf{D}\mathsf{u}^{-1}\mathsf{u} = \mathsf{u}\mathsf{D} \\ &= \begin{bmatrix} u_{11} & u_{21} & \cdots & u_{n1} \\ u_{12} & u_{22} & \cdots & u_{n2} \\ \vdots & \vdots & \ddots & \vdots \\ u_{1n} & u_{2n} & \cdots & u_{nn} \end{bmatrix} \begin{bmatrix} e^{\lambda_1 t} & 0 & \cdots & 0 \\ 0 & e^{\lambda_2 t} & \cdots & 0 \\ \vdots & \vdots & \ddots & 0 \\ 0 & 0 & \cdots & e^{\lambda_n t} \end{bmatrix} \\ &= \begin{bmatrix} u_{11}e^{\lambda_1 t} & \cdots & u_{n1}e^{\lambda_n t} \\ u_{11}e^{\lambda_1 t} & \cdots & u_{n2}e^{\lambda_n t} \\ \vdots & \ddots & \vdots \\ u_{n1}e^{\lambda_1 t} & \cdots & u_{n2}e^{\lambda_n t} \end{bmatrix}. \end{aligned} \tag{7}$$

The individual solutions are then

$$\mathbf{x}_i = (e^{\mathsf{A}t}\mathsf{u}) \cdot \hat{\mathbf{e}}_i = \mathbf{u}_i e^{\lambda_i t}, \tag{8}$$

so the homogeneous solution is

$$\mathbf{x} = \sum_{i=1}^{n} c_i \mathbf{u}_i e^{\lambda_i t}, \tag{9}$$

where the c_is are arbitrary constants.

The general procedure is therefore

1. Find the EIGENVALUES of the MATRIX A (λ_1, ..., λ_n) by solving the CHARACTERISTIC EQUATION.
2. Determine the corresponding EIGENVECTORS $\mathbf{u}_1$, ..., $\mathbf{u}_n$.
3. Compute
$$\mathbf{x}_i \equiv e^{\lambda_i t}\mathbf{u}_i \tag{10}$$
for $i = 1, \ldots, n$. Then the VECTORS $\mathbf{x}_i$ which are REAL are solutions to the homogeneous equation. If A is a 2×2 matrix, the COMPLEX vectors $\mathbf{x}_j$ correspond to REAL solutions to the homogeneous equation given by $\Re(\mathbf{x}_j)$ and $\Im(\mathbf{x}_j)$.
4. If the equation is nonhomogeneous, find the particular solution given by
$$\mathbf{x}^*(t) = \mathsf{X}(t) \int \mathsf{X}^{-1}(t)\mathbf{p}(t)\,dt, \tag{11}$$
where the MATRIX X is defined by
$$\mathsf{X}(t) \equiv [\,\mathbf{x}_1 \quad \cdots \quad \mathbf{x}_n\,]. \tag{12}$$
If the equation is homogeneous so that $\mathbf{p}(t) = \mathbf{0}$, then look for a solution of the form
$$\mathbf{x} = \boldsymbol{\xi} e^{\lambda t}. \tag{13}$$
This leads to an equation
$$(\mathsf{A} - \lambda\mathsf{I})\boldsymbol{\xi} = \mathbf{0}, \tag{14}$$
so ξ is an EIGENVECTOR and λ an EIGENVALUE.
5. The general solution is
$$\mathbf{x}(t) = \mathbf{x}^*(t) + \sum_{i=1}^{n} c_i \mathbf{x}_i. \tag{15}$$

Ordinary Double Point

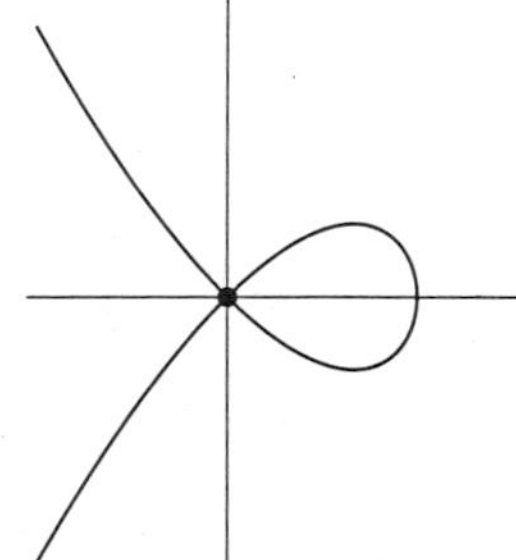

A RATIONAL DOUBLE POINT of CONIC DOUBLE POINT type, known as "A_1." An ordinary DOUBLE POINT is called a NODE. The above plot shows the curve $x^3 - x^2 + y^2 = 0$, which has an ordinary double point at the ORIGIN.

A surface in complex 3-space admits at most finitely many ordinary double points. The maximum possible number of ordinary double points $\mu(d)$ for a surface of degree $d = 1, 2, \ldots$, are 0, 1, 4, 16, 31, 65, $93 \leq \mu(7) \leq 104$, $168 \leq \mu(8) \leq 174$, $216 \leq \mu(8) \leq 246$, $345 \leq \mu(10) \leq 360$, $425 \leq \mu(11) \leq 480$, $576 \leq \mu(12) \leq 645 \ldots$ (Sloane's A046001; Chmutov 1992, Endraß 1995). The fact that $\mu(5) = 31$ was proved by Beauville (1980), and $\mu(6) = 65$ was proved by Jaffe and Ruberman (1994). For $d \geq 3$, the following inequality holds:

$$\mu(d) \leq \tfrac{1}{2}[d(d-1)-3]$$

(Endraß 1995). Examples of ALGEBRAIC SURFACES having the maximum (known) number of ordinary double points are given in the following table.

d	$\mu(d)$	Surface
3	4	Cayley cubic
4	16	Kummer surface
5	31	dervish
6	65	Barth sextic
8	168	Endraß octic
10	345	Barth decic

see also ALGEBRAIC SURFACE, BARTH DECIC, BARTH SEXTIC, CAYLEY CUBIC, CUSP, DERVISH, ENDRASS OCTIC, KUMMER SURFACE, RATIONAL DOUBLE POINT

References

Basset, A. B. "The Maximum Number of Double Points on a Surface." *Nature* **73**, 246, 1906.

Beauville, A. "Sur le nombre maximum de points doubles d'une surface dans $\mathbb{P}^3$ ($\mu(5) = 31$)." *Journées de géométrie algébrique d'Angers (1979).* Sijthoff & Noordhoff, pp. 207–215, 1980.

Chmutov, S. V. "Examples of Projective Surfaces with Many Singularities." *J. Algebraic Geom.* **1**, 191–196, 1992.

Endraß, S. "Surfaces with Many Ordinary Nodes." `http://www.mathematik.uni-mainz.de/AlgebraischeGeometrie/docs/Eflaechen.shtml`.

Endraß, S. "Flächen mit vielen Doppelpunkten." *DMV-Mitteilungen* **4**, 17–20, Apr. 1995.

Endraß, S. *Symmetrische Fläche mit vielen gewöhnlichen Doppelpunkten.* Ph.D. thesis. Erlangen, Germany, 1996.

Fischer, G. (Ed.). *Mathematical Models from the Collections of Universities and Museums.* Braunschweig, Germany: Vieweg, pp. 12–13, 1986.

Jaffe, D. B. and Ruberman, D. "A Sextic Surface Cannot have 66 Nodes." *J. Algebraic Geom.* **6**, 151–168, 1997.

Miyaoka, Y. "The Maximal Number of Quotient Singularities on Surfaces with Given Numerical Invariants." *Math. Ann.* **268**, 159–171, 1984.

Sloane, N. J. A. Sequence A046001 in "An On-Line Version of the Encyclopedia of Integer Sequences."

Togliatti, E. G. "Sulle superficie algebriche col massimo numero di punti doppi." *Rend. Sem. Mat. Torino* **9**, 47–59, 1950.

Varchenko, A. N. "On the Semicontinuity of Spectrum and an Upper Bound for the Number of Singular Points on a Projective Hypersurface." *Dokl. Acad. Nauk SSSR* **270**, 1309–1312, 1983.

Walker, R. J. *Algebraic Curves.* New York: Springer-Verlag, pp. 56–57, 1978.

Ordinary Line

Given an arrangement of $n \geq 3$ points, a LINE containing just two of them is called an ordinary line. Moser (1958) proved that at least $3n/7$ lines must be ordinary (Guy 1989, p. 903).

see also GENERAL POSITION, NEAR-PENCIL, ORDINARY POINT, SPECIAL POINT, SYLVESTER GRAPH

References

Guy, R. K. "Unsolved Problems Come of Age." *Amer. Math. Monthly* **96**, 903–909, 1989.

Ordinary Point

A POINT which lies on at least one ORDINARY LINE.

see also ORDINARY LINE, SPECIAL POINT, SYLVESTER GRAPH

References

Guy, R. K. "Unsolved Problems Come of Age." *Amer. Math. Monthly* **96**, 903–909, 1989.

Ordinate

The y- (vertical) axis of a GRAPH.

see also ABSCISSA, x-AXIS, y-AXIS, z-AXIS

Ore's Conjecture

Define the HARMONIC MEAN of the DIVISORS of n

$$H(n) \equiv \frac{\tau(n)}{\sum_{d|n} \frac{1}{d}},$$

where $\tau(n)$ is the TAU FUNCTION (the number of DIVISORS of n). If n is a PERFECT NUMBER, $H(n)$ is an INTEGER. Ore conjectured that if n is ODD, then $H(n)$ is not an INTEGER. This implies that no ODD PERFECT NUMBERS exist.

see also HARMONIC DIVISOR NUMBER, HARMONIC MEAN, PERFECT NUMBER, TAU FUNCTION

Ore Number

see HARMONIC DIVISOR NUMBER

Ore's Theorem

If a GRAPH G has $n \geq 3$ VERTICES such that every pair of the n VERTICES which are not joined by an EDGE has a sum of VALENCES which is $\geq n$, then G is HAMILTONIAN.

see also HAMILTONIAN GRAPH

Orientable Surface

A REGULAR SURFACE $M \subset \mathbb{R}^n$ is called orientable if each TANGENT SPACE M_p has a COMPLEX STRUCTURE $J_p : M_p \to M_p$ such that $p \to J_p$ is a continuous function.

see also NONORIENTABLE SURFACE, REGULAR SURFACE

References

Gray, A. *Modern Differential Geometry of Curves and Surfaces.* Boca Raton, FL: CRC Press, p. 230, 1993.

Orientation (Plane Curve)

A curve has positive orientation if a region R is on the left when traveling around the outside of R, or on the right when traveling around the inside of R.

Orientation-Preserving

A nonsingular linear MAP $A : \mathbb{R}^n \to \mathbb{R}^n$ is orientation-preserving if $\det(A) > 0$.

see also ORIENTATION-REVERSING, ROTATION

Orientation-Reversing

A nonsingular linear MAP $A : \mathbb{R}^n \to \mathbb{R}^n$ is orientation-reversing if $\det(A) < 0$.

see also ORIENTATION-PRESERVING

Orientation (Vectors)

Let θ be the ANGLE between two VECTORS. If $0 < \theta < \pi$, the VECTORS are positively oriented. If $\pi < \theta < 2\pi$, the vectors are negatively oriented.

Two vectors in the plane

$$\begin{bmatrix} x_1 \\ x_2 \end{bmatrix} \text{ and } \begin{bmatrix} y_1 \\ y_2 \end{bmatrix}$$

are positively oriented IFF the DETERMINANT

$$D \equiv \begin{vmatrix} x_1 & y_1 \\ x_2 & y_2 \end{vmatrix} > 0,$$

and are negatively oriented IFF the DETERMINANT $D < 0$.

Origami

The Japanese art of paper folding to make 3-dimensional objects. CUBE DUPLICATION and TRISECTION of an ANGLE can be solved using origami, although they cannot be solved using the traditional rules for GEOMETRIC CONSTRUCTIONS.

see also FOLDING, GEOMETRIC CONSTRUCTION, STOMACHION, TANGRAM

References

Andersen, E. "Origami on the Web." http://www.netspace.org/users/ema/oriweb.html.

Eppstein, D. "Origami." http://www.ics.uci.edu/~eppstein/junkyard/origami.html.

Geretschläger, R. "Euclidean Constructions and the Geometry of Origami." *Math. Mag.* **68**, 357–371, 1995.

Gurkewitz, R. and Arnstein, B. *3-D Geometric Origami.* New York: Dover, 1996.

Kasahara, K. *Origami Omnibus.* Tokyo: Japan Publications, 1988.

Kasahara, K. and Takahara, T. *Origami for the Connoisseur.* Tokyo: Japan Publications, 1987.

Palacios, V. *Fascinating Origami: 101 Models by Alfredo Cerceda.* New York: Dover, 1997.

Pappas, T. "Mathematics & Paperfolding." *The Joy of Mathematics.* San Carlos, CA: Wide World Publ./Tetra, pp. 48–50, 1989.

Row, T. S. *Geometric Exercises in Paper Folding.* New York: Dover, 1966.

Tomoko, F. *Unit Origami.* Tokyo: Japan Publications, 1990.

Wu, J. "Joseph Wu's Origami Page." http://www.datt.co.jp/Origami.

Origin

The central point ($r = 0$) in POLAR COORDINATES, or the point with all zero coordinates $(0, \ldots, 0)$ in CARTESIAN COORDINATES. In 3-D, the x-AXIS, y-AXIS, and z-AXIS meet at the origin.

see also OCTANT, QUADRANT, x-AXIS, y-AXIS, z-AXIS

Ornstein's Theorem

An important result in ERGODIC THEORY. It states that any two "Bernoulli schemes" with the same MEASURE-THEORETIC ENTROPY are MEASURE-THEORETICALLY ISOMORPHIC.

see also ERGODIC THEORY, ISOMORPHISM, MEASURE THEORY

Orr's Theorem

If

$$(1-z)^{\alpha+\beta+\gamma-1/2}\,{}_2F_1(2\alpha, 2\beta; 2\gamma; z) = \sum a_n z^n, \quad (1)$$

where ${}_2F_1(a, b; c; z)$ is a HYPERGEOMETRIC FUNCTION, then

$${}_2F_1(\alpha, \beta; \gamma; z)\,{}_2F_1(\gamma - \alpha + \tfrac{1}{2}, \gamma - \beta + \tfrac{1}{2}; \gamma + 1; z) = \sum_{(\gamma+\frac{1}{2})_n/(\gamma+1)_n} a_n z^n. \quad (2)$$

Furthermore, if

$$(1-z)^{\alpha+\beta-\gamma-1/2}\,{}_2F_1(2\alpha - 1, 2\beta; 2\gamma - 1; z) = \sum a_n z^n, \quad (3)$$

then

$${}_2F_1(\alpha, \beta; \gamma; z)\Gamma(\gamma - \alpha + \tfrac{1}{2}, \gamma - \beta - \tfrac{1}{2}; \gamma; z) = \sum_{(\gamma-\frac{1}{2})_n/(\gamma)_n} a_n z^n, \quad (4)$$

where $\Gamma(z)$ is the GAMMA FUNCTION.

Orthic Triangle

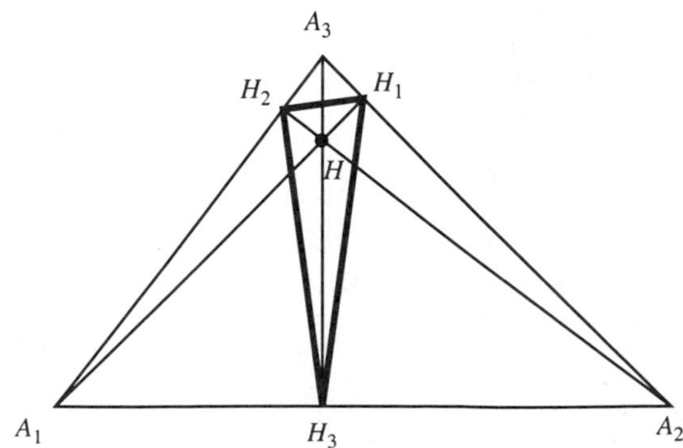

Given a TRIANGLE $\Delta A_1A_2A_3$, the TRIANGLE $\Delta H_1H_2H_3$ with VERTICES at the feet of the ALTITUDES (perpendiculars from a point to the sides) is called the orthic triangle. The three lines A_iH_i are CONCURRENT at the ORTHOCENTER H of $\Delta A_1A_2A_3$.

The centroid of the orthic triangle has TRIANGLE CENTER FUNCTION

$$\alpha = a^2 \cos(B - C)$$

(Casey 1893, Kimberling 1994). The ORTHOCENTER of the orthic triangle has TRIANGLE CENTER FUNCTION

$$\alpha = \cos(2A)\cos(B - C)$$

(Casey 1893, Kimberling 1994). The SYMMEDIAN POINT of the orthic triangle has TRIANGLE CENTER FUNCTION

$$\alpha = \tan A \cos(B - C)$$

(Casey 1893, Kimberling 1994).

see also ALTITUDE, FAGNANO'S PROBLEM, ORTHOCENTER, PEDAL TRIANGLE, SCHWARZ'S TRIANGLE PROBLEM, SYMMEDIAN POINT

References

Casey, J. *A Treatise on the Analytical Geometry of the Point, Line, Circle, and Conic Sections, Containing an Account of Its Most Recent Extensions, with Numerous Examples, 2nd ed., rev. enl.* Dublin: Hodges, Figgis, & Co., p. 9, 1893.

Coxeter, H. S. M. and Greitzer, S. L. *Geometry Revisited.* Washington, DC: Math. Assoc. Amer., pp. 9 and 16–18, 1967.

Kimberling, C. "Central Points and Central Lines in the Plane of a Triangle." *Math. Mag.* **67**, 163–187, 1994.

Orthobicupola

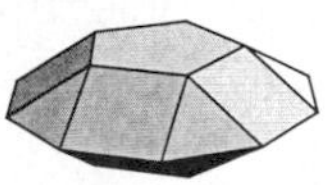
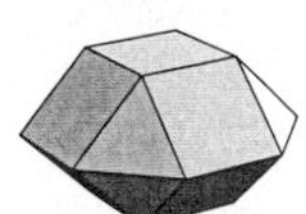
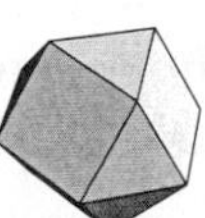

A BICUPOLA in which the bases are in the same orientation.

see also PENTAGONAL ORTHOBICUPOLA, SQUARE ORTHOBICUPOLA, TRIANGULAR ORTHOBICUPOLA

Orthobirotunda

A BIROTUNDA in which the bases are in the same orientation.

Orthocenter

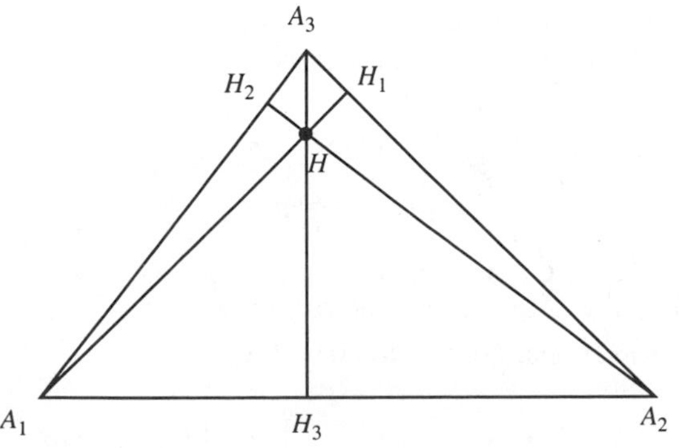

The intersection H of the three ALTITUDES of a TRIANGLE is called the orthocenter. Its TRILINEAR COORDINATES are

$$\cos B \cos C : \cos C \cos A : \cos A \cos B. \quad (1)$$

If the TRIANGLE is not a RIGHT TRIANGLE, then (1) can be divided through by $\cos A \cos B \cos C$ to give

$$\sec A : \sec B : \sec C. \quad (2)$$

If the triangle is ACUTE, the orthocenter is in the interior of the triangle. In a RIGHT TRIANGLE, the orthocenter is the VERTEX of the RIGHT ANGLE.

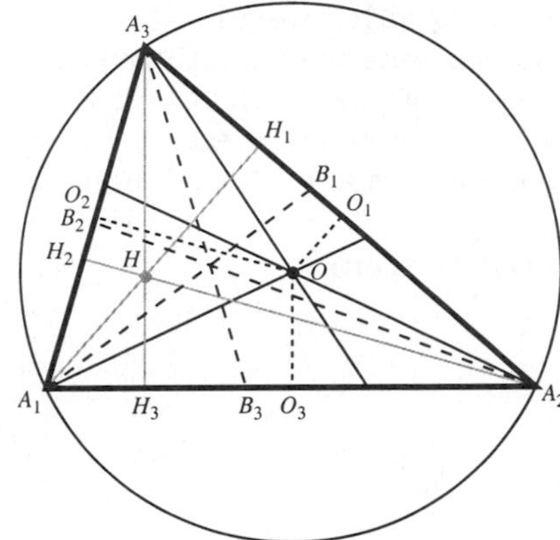

The CIRCUMCENTER O and orthocenter H are ISOGONAL CONJUGATE points. The orthocenter lies on the EULER LINE.

$$a_1{}^2 + a_2{}^2 + a_3{}^2 + \overline{A_1H}^2 + \overline{A_2H}^2 + \overline{A_3H}^2 = 12R^2 \quad (3)$$

$$\overline{A_1H} + \overline{A_2H} + \overline{A_3H} = 2(r + R), \quad (4)$$

$$\overline{A_1H}^2 + \overline{A_2H}^2 + \overline{A_3H}^2 = 4R^2 - 4Rr, \quad (5)$$

where r is the INRADIUS and R is the CIRCUMRADIUS (Johnson 1929, p. 191).

Any HYPERBOLA circumscribed on a TRIANGLE and passing through the orthocenter is RECTANGULAR, and has its center on the NINE-POINT CIRCLE (Falisse 1920, Vandeghen 1965).

see also CENTROID (TRIANGLE), CIRCUMCENTER, EULER LINE, INCENTER, ORTHIC TRIANGLE, ORTHOCENTRIC COORDINATES, ORTHOCENTRIC QUADRILATERAL, ORTHOCENTRIC SYSTEM, POLAR CIRCLE

References

Altshiller-Court, N. *College Geometry: A Second Course in Plane Geometry for Colleges and Normal Schools, 2nd ed.* New York: Barnes and Noble, pp. 165–172, 1952.

Carr, G. S. *Formulas and Theorems in Pure Mathematics, 2nd ed.* New York: Chelsea, p. 622, 1970.

Coxeter, H. S. M. and Greitzer, S. L. *Geometry Revisited.* Washington, DC: Math. Assoc. Amer., pp. 36–40, 1967.

Dixon, R. *Mathographics.* New York: Dover, p. 57, 1991.

Falisse, V. *Cours de géométrie analytique plane.* Brussels, Belgium: Office de Publicité, 1920.

Johnson, R. A. *Modern Geometry: An Elementary Treatise on the Geometry of the Triangle and the Circle.* Boston, MA: Houghton Mifflin, pp. 165–172 and 191, 1929.
Kimberling, C. "Central Points and Central Lines in the Plane of a Triangle." *Math. Mag.* **67**, 163–187, 1994.
Kimberling, C. "Orthocenter." http://www.evansville.edu/~ck6/tcenters/class/orthocn.html.
Vandeghen, A. "Some Remarks on the Isogonal and Cevian Transforms. Alignments of Remarkable Points of a Triangle." *Amer. Math. Monthly* **72**, 1091–1094, 1965.

Orthocentric Coordinates

Coordinates defined by an ORTHOCENTRIC SYSTEM.

see also TRILINEAR COORDINATES

Orthocentric Quadrilateral

If two pairs of opposite sides of a COMPLETE QUADRILATERAL are pairs of PERPENDICULAR lines, the QUADRILATERAL is said to be orthocentric. In such a case, the remaining sides are also PERPENDICULAR.

Orthocentric System

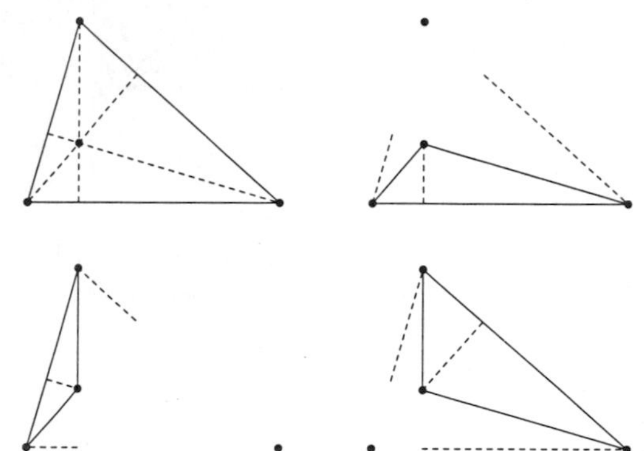

A set of four points, one of which is the ORTHOCENTER of the other three. In an orthocentric system, each point is the ORTHOCENTER of the TRIANGLE of the other three, as illustrated above. The INCENTER and EXCENTERS of a TRIANGLE are an orthocentric system. The centers of the CIRCUMCIRCLES of an orthocentric system form another orthocentric system congruent to the first. The sum of the squares of any nonadjacent pair of connectors of an orthocentric system equals the square of the DIAMETER of the CIRCUMCIRCLE. Orthocentric systems are used to define ORTHOCENTRIC COORDINATES.

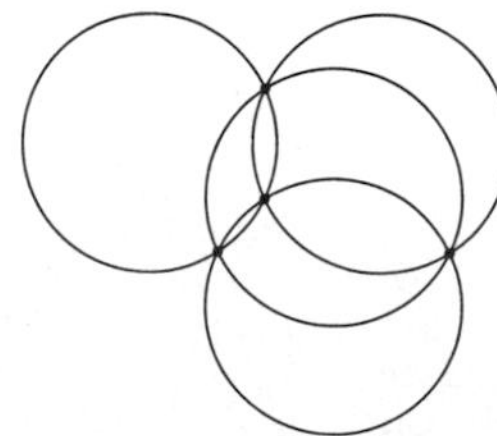

The four CIRCUMCIRCLES of points in an orthocentric system taken three at a time (illustrated above) have equal RADIUS.

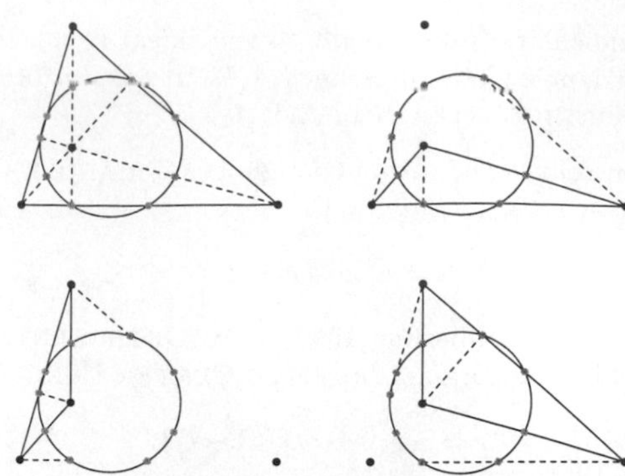

The four triangles of an orthocentric system have a common NINE-POINT CIRCLE, illustrated above.

see also ANGLE BISECTOR, CIRCUMCIRCLE, CYCLIC QUADRANGLE, NINE-POINT CIRCLE, ORTHIC TRIANGLE, ORTHOCENTER, ORTHOCENTRIC SYSTEM, POLAR CIRCLE

References

Altshiller-Court, N. *College Geometry: A Second Course in Plane Geometry for Colleges and Normal Schools, 2nd ed.* New York: Barnes and Noble, pp. 109–114, 1952.
Johnson, R. A. *Modern Geometry: An Elementary Treatise on the Geometry of the Triangle and the Circle.* Boston, MA: Houghton Mifflin, pp. 165–176, 1929.

Orthocupolarotunda

A CUPOLAROTUNDA in which the bases are in the same orientation.

see also GYROCUPOLAROTUNDA, PENTAGONAL ORTHOCUPOLARONTUNDA

Orthodrome

see GREAT CIRCLE

Orthogonal Array

An orthogonal array $OA(k, s)$ is a $k \times s^2$ ARRAY with entries taken from an s-set S having the property that in any two rows, each ordered pair of symbols from S occurs exactly once.

References

Colbourn, C. J. and Dinitz, J. H. (Eds.) *CRC Handbook of Combinatorial Designs.* Boca Raton, FL: CRC Press, p. 111, 1996.

Orthogonal Basis

A BASIS of vectors $\mathbf{x}$ which satisfy

$$x_j x_k = C_{jk}\delta_{jk}$$

$$x^\mu x_\nu = C^\mu_\nu \delta^\mu_\nu,$$

where C_{jk}, C^μ_ν are constants (not necessarily equal to 1) and δ_{jk} is the KRONECKER DELTA.

see also BASIS, ORTHONORMAL BASIS

Orthogonal Circles

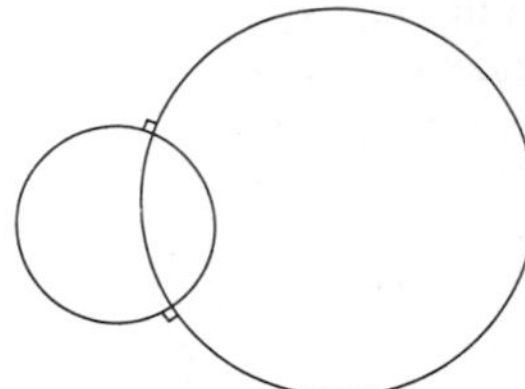

Orthogonal circles are ORTHOGONAL CURVES, i.e., they cut one another at RIGHT ANGLES. Two CIRCLES with equations

$$x^2 + y^2 + 2gx + 2fy + c = 0 \quad (1)$$

$$x^2 + y^2 + 2g'x + 2f'y + c' = 0 \quad (2)$$

are orthogonal if

$$2gg' + 2ff' = c + c'. \quad (3)$$

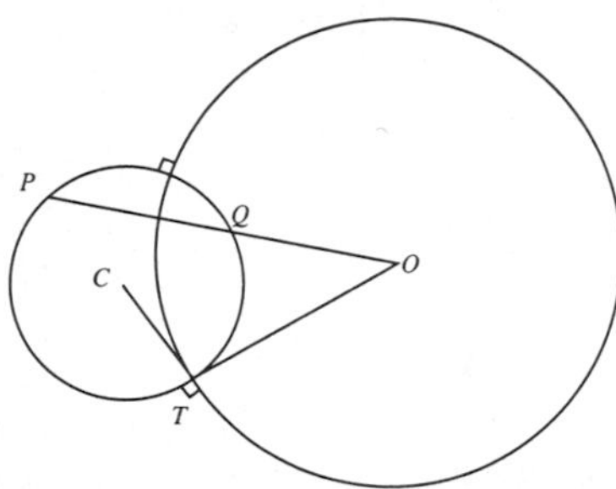

A theorem of Euclid states that, for the orthogonal circles in the above diagram,

$$OP \times OQ = OT^2 \quad (4)$$

(Dixon 1991, p. 65).

References
Dixon, R. *Mathographics.* New York: Dover, pp. 65–66, 1991.
Euclid. *The Thirteen Books of the Elements, 2nd ed. unabridged, Vol. 3: Books X–XIII* New York: Dover, p. 36, 1956.
Pedoe, D. *Circles: A Mathematical View, rev. ed.* Washington, DC: Math. Assoc. Amer., p. xxiv, 1995.

Orthogonal Curves

Two intersecting curves which are PERPENDICULAR at their INTERSECTION are said to be orthogonal.

Orthogonal Functions

Two functions $f(x)$ and $g(x)$ are orthogonal on the interval $a \leq x \leq b$ if

$$\langle f(x)|g(x)\rangle \equiv \int_a^b f(x)g(x)\,dx = 0.$$

see also ORTHOGONAL POLYNOMIALS, ORTHONORMAL FUNCTIONS

Orthogonal Group

see GENERAL ORTHOGONAL GROUP, LIE-TYPE GROUP, ORTHOGONAL ROTATION GROUP, PROJECTIVE GENERAL ORTHOGONAL GROUP, PROJECTIVE SPECIAL ORTHOGONAL GROUP, SPECIAL ORTHOGONAL GROUP

References
Wilson, R. A. "ATLAS of Finite Group Representation." http://for.mat.bham.ac.uk/atlas#orth.

Orthogonal Group Representations

Two representations of a GROUP χ_i and χ_j are said to be orthogonal if

$$\sum_R \chi_i(R)\chi_j(R) = 0$$

for $i \neq j$, where the sum is over all elements R of the representation.

see also GROUP

Orthogonal Lines

Two or more LINES or LINE SEGMENTS which are PERPENDICULAR are said to be orthogonal.

Orthogonal Matrix

Any ROTATION can be given as a composition of rotations about three axes (EULER'S ROTATION THEOREM), and thus can be represented by a 3×3 MATRIX operating on a VECTOR,

$$\begin{bmatrix} x_1' \\ x_2' \\ x_3' \end{bmatrix} = \begin{bmatrix} a_{11} & a_{12} & a_{13} \\ a_{21} & a_{22} & a_{23} \\ a_{31} & a_{32} & a_{33} \end{bmatrix} \begin{bmatrix} x_1 \\ x_2 \\ x_3 \end{bmatrix}. \quad (1)$$

We wish to place conditions on this matrix so that it is consistent with an ORTHOGONAL TRANSFORMATION (basically, a ROTATION or ROTOINVERSION).

In a ROTATION, a VECTOR must keep its original length, so it must be true that

$$x_i'x_i' = x_ix_i \quad (2)$$

for $i = 1, 2, 3$, where EINSTEIN SUMMATION is being used. Therefore, from the transformation equation,

$$(a_{ij}x_j)(a_{ik}x_k) = x_ix_i. \quad (3)$$

This can be rearranged to

$$\begin{aligned} a_{ij}(x_ja_{ik})x_k &= a_{ij}(a_{ik}x_j)x_k \\ &= a_{ij}a_{ik}x_jx_k = x_ix_i. \end{aligned} \quad (4)$$

In order for this to hold, it must be true that

$$a_{ij}a_{ik} = \delta_{jk} \quad (5)$$

for $j, k = 1, 2, 3$, where δ_{ij} is the KRONECKER DELTA. This is known as the ORTHOGONALITY CONDITION, and it guarantees that

$$\mathsf{A}^{-1} = \mathsf{A}^{\mathrm{T}}, \tag{6}$$

and

$$\mathsf{A}^{\mathrm{T}}\mathsf{A} = \mathsf{I}, \tag{7}$$

where A^{T} is the MATRIX TRANSPOSE and I is the IDENTITY MATRIX. Equation (7) is the identity which gives the orthogonal matrix its name. Orthogonal matrices have special properties which allow them to be manipulated and identified with particular ease.

Let A and B be two orthogonal matrices. By the ORTHOGONALITY CONDITION, they satisfy

$$a_{ij}a_{ik} = \delta_{jk}, \tag{8}$$

and

$$b_{ij}b_{ik} = \delta_{jk}, \tag{9}$$

where δ_{ij} is the KRONECKER DELTA. Now

$$\begin{aligned} c_{ij}c_{ik} &= (ab)_{ij}(ab)_{jk} = a_{is}b_{sj}a_{it}b_{tk} = a_{is}a_{it}b_{sj}b_{tk} \\ &= \delta_{st}b_{sj}b_{tk} = b_{tj}b_{tk} = \delta_{jk}, \end{aligned} \tag{10}$$

so the product $\mathsf{C} \equiv \mathsf{AB}$ of two orthogonal matrices is also orthogonal.

The EIGENVALUES of an orthogonal matrix must satisfy one of the following:

1. All EIGENVALUES are 1.
2. One EIGENVALUE is 1 and the other two are -1.
3. One EIGENVALUE is 1 and the other two are COMPLEX CONJUGATES of the form $e^{i\theta}$ and $e^{-i\theta}$.

An orthogonal MATRIX A is classified as proper (corresponding to pure ROTATION) if

$$\det(\mathsf{A}) = 1, \tag{11}$$

where $\det(\mathsf{A})$ is the DETERMINANT of A, or improper (corresponding to inversion with possible rotation; ROTOINVERSION) if

$$\det(\mathsf{A}) = -1. \tag{12}$$

see also EULER'S ROTATION THEOREM, ORTHOGONAL TRANSFORMATION, ORTHOGONALITY CONDITION, ROTATION, ROTATION MATRIX, ROTOINVERSION

References

Arfken, G. "Orthogonal Matrices." *Mathematical Methods for Physicists, 3rd ed.* Orlando, FL: Academic Press, pp. 191–205, 1985.

Goldstein, H. "Orthogonal Transformations." §4–2 in *Classical Mechanics, 2nd ed.* Reading, MA: Addison-Wesley, 132–137, 1980.

Orthogonal Polynomials

Orthogonal polynomials are classes of POLYNOMIALS $\{p_n(x)\}$ over a range $[a, b]$ which obey an ORTHOGONALITY relation

$$\int_a^b w(x)p_m(x)p_n(x)\,dx = \delta_{mn}c_n, \tag{1}$$

where $w(x)$ is a WEIGHTING FUNCTION and δ is the KRONECKER DELTA. If $c_m = 1$, then the POLYNOMIALS are not only orthogonal, but orthonormal.

Orthogonal polynomials have very useful properties in the solution of mathematical and physical problems. Just as FOURIER SERIES provide a convenient method of expanding a periodic function in a series of linearly independent terms, orthogonal polynomials provide a natural way to solve, expand, and interpret solutions to many types of important DIFFERENTIAL EQUATIONS. Orthogonal polynomials are especially easy to generate using GRAM-SCHMIDT ORTHONORMALIZATION. Abramowitz and Stegun (1972, pp. 774–775) give a table of common orthogonal polynomials.

Type	Interval	$w(x)$	c_n
Chebyshev First Kind	$[-1, 1]$	$(1-x^2)^{-1/2}$	$\begin{cases} \frac{1}{2}\pi \\ \pi \end{cases}$ $\begin{cases} \text{for } n = 0 \\ \text{otherwise} \end{cases}$
Chebyshev Second Kind	$[-1, 1]$	$\sqrt{1-x^2}$	$\frac{1}{2}\pi$
Hermite	$(-\infty, \infty)$	e^{-x^2}	$\sqrt{\pi}\,2^n n!$
Jacobi	$(-1, 1)$	$(1-x)^\alpha(1+x)^\beta$	h_n
Laguerre	$[0, \infty)$	e^{-x}	1
Laguerre (Associated)	$[0, \infty)$	$x^k e^{-x}$	$\frac{(n+k)!}{n!}$
Legendre	$[-1, 1]$	1	$\frac{2}{2n+1}$
Ultraspherical	$[-1, 1]$	$(1-x^2)^{\alpha-1/2}$	$\begin{cases} \frac{\pi 2^{1-2\alpha}\Gamma(n+2\alpha)}{n!(n+\alpha)[\Gamma(\alpha)]^2} \\ \frac{2\pi}{n^2} \end{cases}$ $\begin{cases} \text{for } \alpha \neq 0 \\ \text{for } \alpha = 0 \end{cases}$

In the above table, the normalization constant is the value of

$$c_n \equiv \int w(x)[p_n(x)]^2\,dx \tag{2}$$

and

$$h_n \equiv \frac{2^{\alpha+\beta+1}}{2n+\alpha+\beta+1}\,\frac{\Gamma(n+\alpha+1)\Gamma(n+\beta+1)}{n!\Gamma(n+\alpha+\beta+1)}, \tag{3}$$

where $\Gamma(z)$ is a GAMMA FUNCTION.

The ROOTS of orthogonal polynomials possess many rather surprising and useful properties. For instance, let $x_1 < x_2 < \ldots < x_n$ be the ROOTS of the $p_n(x)$ with $x_0 = a$ and $x_{n+1} = b$. Then each interval $[x_\nu, x_{\nu+1}]$ for $\nu = 0, 1, \ldots, n$ contains exactly one ROOT of $p_{n+1}(x)$. Between two ROOTS of $p_n(x)$ there is at least one ROOT of $p_m(x)$ for $m > n$.

Let c be an arbitrary REAL constant, then the POLYNOMIAL

$$p_{n+1}(x) - cp_n(x) \tag{4}$$

has $n+1$ distinct REAL ROOTS. If $c > 0$ ($c < 0$), these ROOTS lie in the interior of $[a,b]$, with the exception of the greatest (least) ROOT which lies in $[a,b]$ only for

$$c \leq \frac{p_{n+1}(b)}{p_n(b)} \qquad \left(c \geq \frac{p_{n+1}(a)}{p_n(a)}\right). \tag{5}$$

The following decomposition into partial fractions holds

$$\frac{p_n(x)}{p_{n+1}(x)} = \sum_{\nu=0}^{n} \frac{l_\nu}{x-\xi}, \tag{6}$$

where $\{\xi_\nu\}$ are the ROOTS of $p_{n+1}(x)$ and

$$\begin{aligned} l_\nu &= \frac{p_n(\xi_\nu)}{p'_{n+1}(\xi_\nu)} \\ &= \frac{p'_{n+1}(\xi_\nu)p_n(\xi_\nu) - p'_n(\xi_\nu)' p_{n+1}(\xi_\nu)}{[p'_{n+1}(\xi_\nu)]^2} > 0. \end{aligned} \tag{7}$$

Another interesting property is obtained by letting $\{p_n(x)\}$ be the orthonormal set of POLYNOMIALS associated with the distribution $d\alpha(x)$ on $[a,b]$. Then the CONVERGENTS R_n/S_n of the CONTINUED FRACTION

$$\frac{1}{A_1x+B_1} - \frac{C_2}{A_2x+B_2} - \frac{C_3}{A_3x+B_3} - \ldots - \frac{C_n}{A_nx+B_n} + \ldots \tag{8}$$

are given by

$$\begin{aligned} R_n &= R_n(x) \\ &= c_0{}^{-3/2}\sqrt{c_0c_2 - c_1{}^2}\int_a^b \frac{p_n(x)-p_n(t)}{x-t}\,d\alpha(t) \quad (9) \\ S_n &= S_n(x) = \sqrt{c_0}\,p_n(x), \quad (10) \end{aligned}$$

where $n = 0, 1, \ldots$ and

$$c_n = \int_a^b x^n\,d\alpha(x). \tag{11}$$

Furthermore, the ROOTS of the orthogonal polynomials $p_n(x)$ associated with the distribution $d\alpha(x)$ on the interval $[a,b]$ are REAL and distinct and are located in the interior of the interval $[a,b]$.

see also CHEBYSHEV POLYNOMIAL OF THE FIRST KIND, CHEBYSHEV POLYNOMIAL OF THE SECOND KIND, GRAM-SCHMIDT ORTHONORMALIZATION, HERMITE POLYNOMIAL, JACOBI POLYNOMIAL, KRAWTCHOUK POLYNOMIAL, LAGUERRE POLYNOMIAL, LEGENDRE POLYNOMIAL, ORTHOGONAL FUNCTIONS, SPHERICAL HARMONIC, ULTRASPHERICAL POLYNOMIAL, ZERNIKE POLYNOMIAL

References

Abramowitz, M. and Stegun, C. A. (Eds.). "Orthogonal Polynomials." Ch. 22 in *Handbook of Mathematical Functions with Formulas, Graphs, and Mathematical Tables, 9th printing.* New York: Dover, pp. 771–802, 1972.

Arfken, G. "Orthogonal Polynomials." *Mathematical Methods for Physicists, 3rd ed.* Orlando, FL: Academic Press, pp. 520–521, 1985.

Iyanaga, S. and Kawada, Y. (Eds.). "Systems of Orthogonal Functions." Appendix A, Table 20 in *Encyclopedic Dictionary of Mathematics.* Cambridge, MA: MIT Press, p. 1477, 1980.

Nikiforov, A. F.; Uvarov, V. B.; and Suslov, S. S. *Classical Orthogonal Polynomials of a Discrete Variable.* New York: Springer-Verlag, 1992.

Sansone, G. *Orthogonal Functions.* New York: Dover, 1991.

Szegő, G. *Orthogonal Polynomials, 4th ed.* Providence, RI: Amer. Math. Soc., pp. 44–47 and 54–55, 1975.

Orthogonal Projection

A PROJECTION of a figure by parallel rays. In such a projection, tangencies are preserved. Parallel lines project to parallel lines. The ratio of lengths of parallel segments is preserved, as is the ratio of areas.

Any TRIANGLE can be positioned such that its shadow under an orthogonal projection is EQUILATERAL. Also, the MEDIANS of a TRIANGLE project to the MEDIANS of the image TRIANGLE. ELLIPSES project to ELLIPSES, and any ELLIPSE can be projected to form a CIRCLE. The center of an ELLIPSE projects to the center of the image ELLIPSE. The CENTROID of a TRIANGLE projects to the CENTROID of its image. Under an ORTHOGONAL TRANSFORMATION, the MIDPOINT ELLIPSE can be transformed into a CIRCLE INSCRIBED in an EQUILATERAL TRIANGLE.

SPHEROIDS project to ELLIPSES (or CIRCLE in the DEGENERATE case).

see also PROJECTION

Orthogonal Rotation Group

Orthogonal rotation groups are LIE GROUPS. The orthogonal rotation group $O_3(n)$ is the set of $n \times n$ REAL ORTHOGONAL MATRICES.

The orthogonal rotation group $O_3^-(n)$ is the set of $n \times n$ REAL ORTHOGONAL MATRICES (having $n(n-1)/2$ independent parameters) with DETERMINANT -1.

The orthogonal rotation group $O_3^+(n)$ is the set of $n \times n$ REAL ORTHOGONAL MATRICES, having $n(n-1)/2$ independent parameters, with DETERMINANT $+1$. $O_3^+(n)$ is

HOMEOMORPHIC with $SU(2)$. Its elements can be written using EULER ANGLES and ROTATION MATRICES as

$$I = \begin{bmatrix} 1 & 0 & 0 \\ 0 & 1 & 0 \\ 0 & 0 & 1 \end{bmatrix} \quad (1)$$

$$R_x(\phi) = \begin{bmatrix} 1 & 0 & 0 \\ 0 & \cos\phi & \sin\phi \\ 0 & -\sin\phi & \cos\phi \end{bmatrix} \quad (2)$$

$$R_y(\theta) = \begin{bmatrix} \cos\theta & 0 & -\sin\theta \\ 0 & 1 & 0 \\ \sin\theta & 0 & \cos\theta \end{bmatrix} \quad (3)$$

$$R_z(\psi) = \begin{bmatrix} \cos\psi & \sin\psi & 0 \\ -\sin\psi & \cos\psi & 0 \\ 0 & 0 & 1 \end{bmatrix}. \quad (4)$$

References

Arfken, G. "Orthogonal Group, O_3^+." *Mathematical Methods for Physicists, 3rd ed.* Orlando, FL: Academic Press, p. 252–253, 1985.

Wilson, R. A. "ATLAS of Finite Group Representation." `http://for.mat.bham.ac.uk/atlas#orth`.

Orthogonal Tensors

Orthogonal CONTRAVARIANT and COVARIANT satisfy

$$g_{ik} g^{ij} = \delta^j_k,$$

where δ^k_j is the KRONECKER DELTA.

see also CONTRAVARIANT TENSOR, COVARIANT TENSOR

Orthogonal Transformation

Any linear transformation

$$\begin{aligned} x_1' &= a_{11}x_1 + a_{12}x_2 + x_{13}x_3 \\ x_2' &= a_{21}x_1 + a_{22}x_2 + a_{23}x_3 \\ x_3' &= a_{31}x_1 + a_{32}x_2 + a_{33}x_3 \end{aligned}$$

satisfying the ORTHOGONALITY CONDITION

$$a_{ij} a_{ik} = \delta_{jk},$$

where EINSTEIN SUMMATION has been used and δ_{ij} is the KRONECKER DELTA, is called an orthogonal transformation.

Orthogonal transformations correspond to rigid ROTATIONS (or ROTOINVERSIONS), and may be represented using ORTHOGONAL MATRICES. If $A : \mathbb{R}^n \to \mathbb{R}^n$ is an orthogonal transformation, then $\det(A) = \pm 1$.

see also AFFINE TRANSFORMATION, ORTHOGONAL MATRIX, ORTHOGONALITY CONDITION, ROTATION, ROTOINVERSION

References

Goldstein, H. "Orthogonal Transformations." §4–2 in *Classical Mechanics, 2nd ed.* Reading, MA: Addison-Wesley, 132–137, 1980.

Gray, A. *Modern Differential Geometry of Curves and Surfaces.* Boca Raton, FL: CRC Press, p. 104, 1993.

Orthogonal Vectors

Two vectors $\mathbf{u}$ and $\mathbf{v}$ whose DOT PRODUCT is $\mathbf{u} \cdot \mathbf{v} = 0$ (i.e., the vectors are PERPENDICULAR) are said to be orthogonal. The definition can be extended to three or more vectors which are mutually PERPENDICULAR.

see also DOT PRODUCT, PERPENDICULAR

Orthogonality Condition

A linear transformation

$$\begin{aligned} x_1' &= a_{11}x_1 + a_{12}x_2 + x_{13}x_3 \\ x_2' &= a_{21}x_1 + a_{22}x_2 + a_{23}x_3 \\ x_3' &= a_{31}x_1 + a_{32}x_2 + a_{33}x_3, \end{aligned}$$

is said to be an ORTHOGONAL TRANSFORMATION if it satisfies the orthogonality condition

$$a_{ij} a_{ik} = \delta_{jk},$$

where EINSTEIN SUMMATION has been used and δ_{ij} is the KRONECKER DELTA.

see also ORTHOGONAL TRANSFORMATION

References

Goldstein, H. "Orthogonal Transformations." §4–2 in *Classical Mechanics, 2nd ed.* Reading, MA: Addison-Wesley, 132–137, 1980.

Orthogonality Theorem

see GROUP ORTHOGONALITY THEOREM

Orthographic Projection

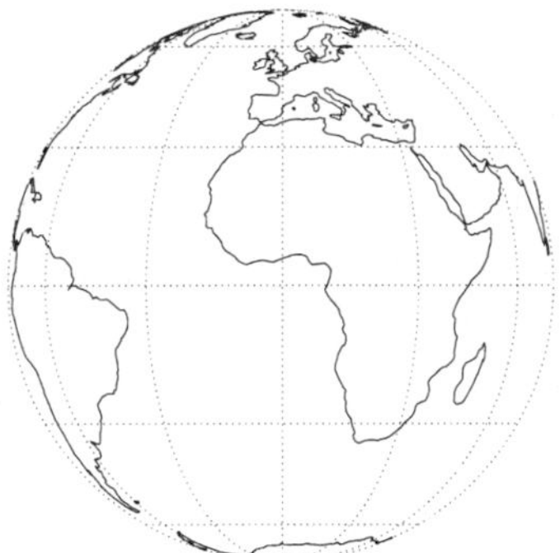

A projection from infinity which preserves neither AREA nor angle.

$$x = \cos\phi \sin(\lambda - \lambda_0) \quad (1)$$

$$y = \cos\phi_1 \sin\phi - \sin\phi_1 \cos\phi \cos(\lambda - \lambda_0). \quad (2)$$

The inverse FORMULAS are

$$\phi = \sin^{-1}\left(\cos c \sin\phi_1 + \frac{y \sin c \cos\phi_1}{\rho}\right) \quad (3)$$

$$\lambda = \lambda_0 + \tan^{-1}\left(\frac{x \sin c}{\rho \cos\phi_1 \cos c - y \sin\phi_1 \sin c}\right), \quad (4)$$

where

$$\rho = \sqrt{x^2 + y^2} \tag{5}$$

$$c = \sin^{-1} \rho. \tag{6}$$

References

Snyder, J. P. *Map Projections—A Working Manual.* U. S. Geological Survey Professional Paper 1395. Washington, DC: U. S. Government Printing Office, pp. 145–153, 1987.

Orthologic

Two TRIANGLES $A_1B_1C_1$ and $A_2B_2C_2$ are orthologic if the perpendiculars from the VERTICES A_1, B_1, C_1 on the sides B_2C_2, A_2C_2, and A_2B_2 pass through one point. This point is known as the orthology center of TRIANGLE 1 with respect to TRIANGLE 2.

Orthonormal Basis

A BASIS of VECTORS **x** which satisfy

$$x_j x_k = \delta_{jk}$$

and

$$x^\mu x_\nu = \delta^\mu_\nu,$$

where δ_{jk} is the KRONECKER DELTA. An orthonormal basis is a normalized ORTHOGONAL BASIS.

see also BASIS, ORTHOGONAL BASIS

Orthonormal Functions

A pair of functions ϕ_i and ϕ_j are orthonormal if they are ORTHOGONAL and each normalized. These two conditions can be succinctly written as

$$\int_a^b \phi_i(x)\phi_j(x)w(x)\,dx = \delta_{ij},$$

where $w(x)$ is a WEIGHTING FUNCTION and δ_{ij} is the KRONECKER DELTA.

see also ORTHOGONAL POLYNOMIALS

Orthonormal Vectors

UNIT VECTORS which are ORTHOGONAL are said to be orthonormal.

see also ORTHOGONAL VECTORS

Orthopole

If perpendiculars are dropped on any line from the vertices of a TRIANGLE, then the perpendiculars to the opposite sides from their FEET are CONCURRENT at a point called the orthopole.

References

Johnson, R. A. *Modern Geometry: An Elementary Treatise on the Geometry of the Triangle and the Circle.* Boston, MA: Houghton Mifflin, p. 247, 1929.

Orthoptic Curve

An ISOPTIC CURVE formed from the locus of TANGENTS meeting at RIGHT ANGLES. The orthoptic of a PARABOLA is its DIRECTRIX. The orthoptic of a central CONIC was investigated by Monge and is a CIRCLE concentric with the CONIC SECTION. The orthoptic of an ASTROID is a CIRCLE.

Curve	Orthoptic
astroid	quadrifolium
cardioid	circle or limaçon
deltoid	circle
logarithmic spiral	equal logarithmic spiral
parabola	directrix

References

Lawrence, J. D. *A Catalog of Special Plane Curves.* New York: Dover, pp. 58 and 207, 1972.

Orthotomic

Given a source S and a curve γ, pick a point on γ and find its tangent T. Then the LOCUS of reflections of S about tangents T is the orthotomic curve (also known as the secondary CAUSTIC). The INVOLUTE of the orthotomic is the CAUSTIC. For a parametric curve $(f(t), g(t))$ with respect to the point (x_0, y_0), the orthotomic is

$$x = x_0 - \frac{2g'[f'(g - y_0) - g'(f - x_0)]}{f'^2 + g'^2}$$

$$y = y_0 + \frac{2f'[f'(g - y_0) - g'(f - x_0)]}{f'^2 + g'^2}.$$

see also CAUSTIC, INVOLUTE

References

Lawrence, J. D. *A Catalog of Special Plane Curves.* New York: Dover, p. 60, 1972.

Orthotope

A PARALLELOTOPE whose edges are all mutually PERPENDICULAR. The orthotope is a generalization of the RECTANGLE and RECTANGULAR PARALLELEPIPED.

see also RECTANGLE, RECTANGULAR PARALLELEPIPED

References

Coxeter, H. S. M. *Regular Polytopes, 3rd ed.* New York: Dover, pp. 122–123, 1973.

Osborne's Rule

The prescription that a TRIGONOMETRY identity can be converted to an analogous identity for HYPERBOLIC FUNCTIONS by expanding, exchanging trigonometric functions with their hyperbolic counterparts, and then flipping the sign of each term involving the product of two HYPERBOLIC SINES. For example, given the identity

$$\cos(x - y) = \cos x \cos y + \sin x \sin y,$$

Osborne's rule gives the corresponding identity

$$\cosh(x - y) = \cosh x \cosh y - \sinh x \sinh y.$$

see also HYPERBOLIC FUNCTIONS, TRIGONOMETRY

Oscillation

The variation of a FUNCTION which exhibits SLOPE changes, also called the SALTUS of a function.

Oscillation Land

see CAROTID-KUNDALINI FUNCTION

Osculating Circle

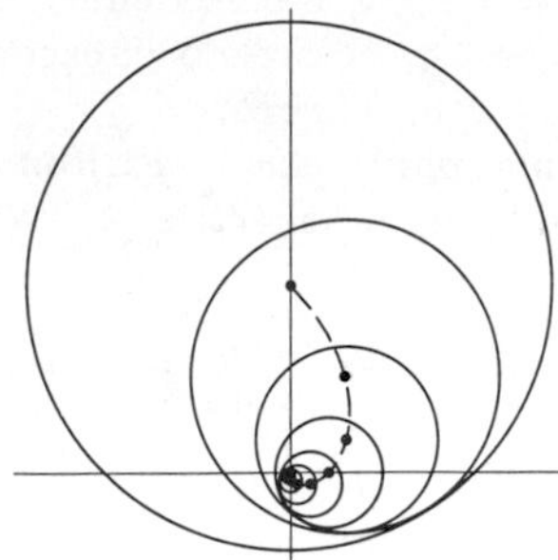

The CIRCLE which shares the same TANGENT as a curve at a given point. The RADIUS OF CURVATURE of the osculating circle is

$$\rho(t) = \frac{1}{|\kappa(t)|}, \tag{1}$$

where κ is the CURVATURE, and the center is

$$x = f - \frac{(f'^2 + g'^2)g'}{f'g'' - f''g'} \tag{2}$$

$$y = g + \frac{(f'^2 + g'^2)g'}{f'g'' - f''g'}, \tag{3}$$

i.e., the centers of the osculating circles to a curve form the EVOLUTE to that curve.

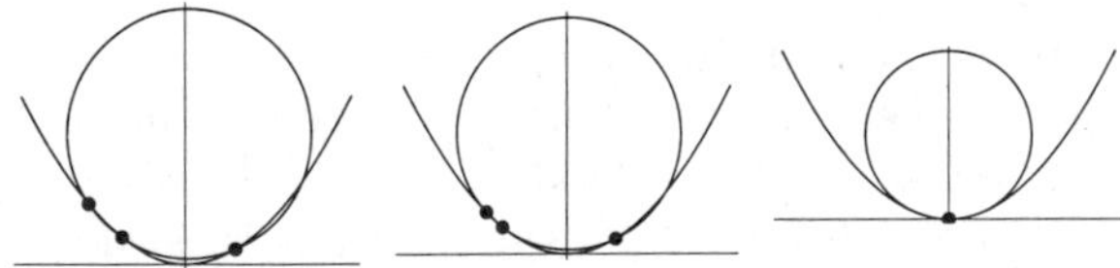

In addition, let $C(t_1, t_2, t_3)$ denote the CIRCLE passing through three points on a curve $(f(t), g(t))$ with $t_1 < t_2 < t_3$. Then the osculating circle C is given by

$$C = \lim_{t_1, t_2, t_3 \to t} C(t_1, t_2, t_3) \tag{4}$$

(Gray 1993).

see also CURVATURE, EVOLUTE, RADIUS OF CURVATURE, TANGENT

References

Gardner, M. "The Game of Life, Parts I–III." Chs. 20–22 in *Wheels, Life, and other Mathematical Amusements.* New York: W. H. Freeman, pp. 221, 237, and 243, 1983.

Gray, A. "Osculating Circles to Plane Curves." §5.6 in *Modern Differential Geometry of Curves and Surfaces.* Boca Raton, FL: CRC Press, pp. 90–95, 1993.

Osculating Curves

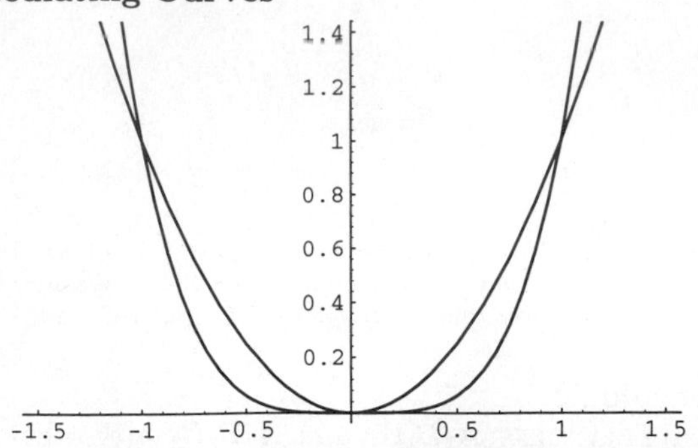

An osculating curve to $f(x)$ at x_0 is tangent at that point and has the same CURVATURE. It therefore satisfies

$$y^{(k)}(x_0) = f^{(k)}(x_0)$$

for $k = 0, 1, 2$. The point of tangency is called a TACNODE. The simplest example of osculating curves are x^2 and x^4, which osculate at the point $x_0 = 0$.

see also TACNODE

Osculating Interpolation

see HERMITE'S INTERPOLATING FUNDAMENTAL POLYNOMIAL

Osculating Plane

The PLANE spanned by the three points $\mathbf{x}(t)$, $\mathbf{x}(t+h_1)$, and $\mathbf{x}(t+h_2)$ on a curve as $h_1, h_2 \to 0$. Let $\mathbf{z}$ be a point on the osculating plane, then

$$[(\mathbf{z} - \mathbf{x}), \mathbf{x}', \mathbf{x}''] = 0,$$

where $[\mathbf{A}, \mathbf{B}, \mathbf{C}]$ denotes the SCALAR TRIPLE PRODUCT. The osculating plane passes through the tangent. The intersection of the osculating plane with the NORMAL PLANE is known as the PRINCIPAL NORMAL VECTOR. The VECTORS $\mathbf{T}$ and $\mathbf{N}$ (TANGENT VECTOR and NORMAL VECTOR) span the osculating plane.

see also NORMAL VECTOR, OSCULATING SPHERE, SCALAR TRIPLE PRODUCT, TANGENT VECTOR

Osculating Sphere

The center of any SPHERE which has a contact of (at least) first-order with a curve C at a point P lies in the normal plane to C at P. The center of any SPHERE which has a contact of (at least) second-order with C at point P, where the CURVATURE $\kappa > 0$, lies on the polar axis of C corresponding to P. All these SPHERES intersect the OSCULATING PLANE of C at P along a circle of curvature at P. The osculating sphere has center

$$\mathbf{a} = \mathbf{x} + \rho\hat{\mathbf{N}} + \frac{\dot{\rho}}{\tau}\hat{\mathbf{B}}$$

where $\hat{\mathbf{N}}$ is the unit NORMAL VECTOR, $\hat{\mathbf{B}}$ is the unit BINORMAL VECTOR, ρ is the RADIUS OF CURVATURE, and τ is the TORSION, and RADIUS

$$R = \sqrt{\rho^2 + \left(\frac{\dot{\rho}}{\tau}\right)^2},$$

and has contact of (at least) third order with C.

see also CURVATURE, OSCULATING PLANE, RADIUS OF CURVATURE, SPHERE, TORSION (DIFFERENTIAL GEOMETRY)

References

Kreyszig, E. *Differential Geometry.* New York: Dover, pp. 54–55, 1991.

Osedelec Theorem

For an n-D MAP, the LYAPUNOV CHARACTERISTIC EXPONENTS are given by

$$\sigma_i = \lim_{N\to\infty} \ln|\lambda_i(N)|$$

for $i = 1, \ldots, n$, where λ_i is the LYAPUNOV CHARACTERISTIC NUMBER.

see also LYAPUNOV CHARACTERISTIC EXPONENT, LYAPUNOV CHARACTERISTIC NUMBER

Ostrowski's Inequality

Let $f(x)$ be a monotonic function integrable on $[a, b]$ and let $f(a), f(b) \leq 0$ and $|f(a)| \geq |f(b)|$, then if g is a REAL function integrable on $[a, b]$,

$$\left|\int_a^b f(x)g(x)\,dx\right| \leq |f(a)| \max_{a\leq\xi\leq b} \left|\int_a^\xi g(x)\,dx\right|.$$

References

Gradshteyn, I. S. and Ryzhik, I. M. *Tables of Integrals, Series, and Products, 5th ed.* San Diego, CA: Academic Press, p. 1100, 1979.

Ostrowski's Theorem

Let $\mathsf{A} = a_{ij}$ be a MATRIX with POSITIVE COEFFICIENTS and λ_0 be the POSITIVE EIGENVALUE in the FROBENIUS THEOREM, then the $n-1$ EIGENVALUES $\lambda_j \neq \lambda_0$ satisfy the INEQUALITY

$$|\lambda_j| \leq \lambda_0 \frac{M^2 - m^2}{M^2 + m^2},$$

where

$$M = \max_{i,j} a_{ij}$$
$$m = \min_{i,j} a_{ij}$$

and $i, j = 1, 2, \ldots, n$.

see also FROBENIUS THEOREM

References

Gradshteyn, I. S. and Ryzhik, I. M. *Tables of Integrals, Series, and Products, 5th ed.* San Diego, CA: Academic Press, p. 1121, 1980.

Otter's Tree Enumeration Constants

see TREE

Outdegree

The number of outward directed EDGES from a given VERTEX in a DIRECTED GRAPH.

see also DIRECTED GRAPH, INDEGREE, LOCAL DEGREE

Outer Automorphism Group

A particular type of AUTOMORPHISM GROUP which exists only for GROUPS. For a GROUP G, the outer automorphism group is the QUOTIENT GROUP $\mathrm{Aut}(G)/\mathrm{Inn}(G)$, which is the AUTOMORPHISM GROUP of G modulo its INNER AUTOMORPHISM GROUP.

see also AUTOMORPHISM GROUP, INNER AUTOMORPHISM GROUP, QUOTIENT GROUP

Outer Product

see DIRECT PRODUCT (TENSOR)

Oval

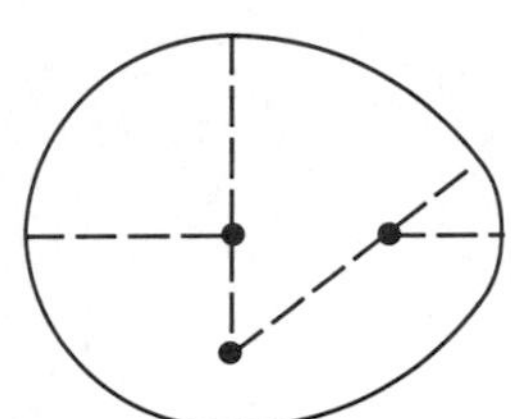

An oval is a curve resembling a squashed CIRCLE but, unlike the ELLIPSE, without a precise mathematical definition. The word oval derived from the Latin word "ovus" for egg. Unlike ellipses, ovals sometimes have only a single axis of reflection symmetry (instead of two).

Ovals can be constructed with a COMPASS by joining together arcs of different radii such that the centers of the arcs lie on a line passing through the join point (Dixon 1991). Albrecht Dürer used this method to design a Roman letter font.

see also CARTESIAN OVALS, CASSINI OVALS, EGG, ELLIPSE, OVOID, SUPERELLIPSE

References

Critchlow, K. *Time Stands Still.* London: Gordon Fraser, 1979.

Cundy, H. and Rollett, A. *Mathematical Models, 3rd ed.* Stradbroke, England: Tarquin Pub., 1989.

Dixon, R. *Mathographics.* New York: Dover, pp. 3–11, 1991.

Dixon, R. "The Drawing Out of an Egg." *New Sci.*, July 29, 1982.

Pedoe, D. *Geometry and the Liberal Arts.* London: Peregrine, 1976.

Oval of Descartes

see CARTESIAN OVALS

Ovals of Cassini

see CASSINI OVALS

Overlapping Resonance Method

see RESONANCE OVERLAP METHOD

Oversampling

A signal sampled at a frequency higher than the NYQUIST FREQUENCY is said to be oversampled β times, where the oversampling ratio is defined as

$$\beta \equiv \frac{\nu_{\text{sampling}}}{\nu_{\text{Nyquist}}}.$$

see also NYQUIST FREQUENCY, NYQUIST SAMPLING

Ovoid

An egg-shaped curve. Lockwood (1967) calls the NEGATIVE PEDAL CURVE of an ELLIPSE with ECCENTRICITY $e \leq 1/2$ an ovoid.

see also OVAL

References

Lockwood, E. H. *A Book of Curves.* Cambridge, England: Cambridge University Press, p. 157, 1967.

P

p-adic Number

A p-adic number is an extension of the FIELD of RATIONAL NUMBERS such that CONGRUENCES MODULO POWERS of a fixed PRIME p are related to proximity in the so called "p-adic metric."

Any NONZERO RATIONAL NUMBER x can be represented by

$$x = \frac{p^a r}{s}, \tag{1}$$

where p is a PRIME NUMBER, r and s are INTEGERS not DIVISIBLE by p, and a is a unique INTEGER. Then define the p-adic absolute value of x by

$$|x|_p = p^{-a}. \tag{2}$$

Also define the p-adic value

$$|0|_p = 0. \tag{3}$$

As an example, consider the FRACTION

$$\tfrac{140}{297} = 2^2 \cdot 3^{-3} \cdot 5 \cdot 7 \cdot 11^{-1}. \tag{4}$$

It has p-adic absolute values given by

$$|\tfrac{140}{297}|_2 = \tfrac{1}{4} \tag{5}$$
$$|\tfrac{140}{297}|_3 = 27 \tag{6}$$
$$|\tfrac{140}{297}|_5 = \tfrac{1}{5} \tag{7}$$
$$|\tfrac{140}{297}|_7 = \tfrac{1}{7} \tag{8}$$
$$|\tfrac{140}{297}|_{11} = 11. \tag{9}$$

The p-adic absolute value satisfies the relations

1. $|x|_p \geq 0$ for all x,
2. $|x|_p = 0$ IFF $x = 0$,
3. $|xy|_p = |x|_p\,|y|_p$ for all x and y,
4. $|x+y|_p \leq |x|_p + |y|_p$ for all x and y (the TRIANGLE INEQUALITY), and
5. $|x+y|_p \leq \max(|x|_p, |y|_p)$ for all x and y (the STRONG TRIANGLE INEQUALITY).

In the above, relation 4 follows trivially from relation 5, but relations 4 and 5 are relevant in the more general VALUATION THEORY.

The p-adics were probably first introduced by Hensel in 1902 in a paper which was concerned with the development of algebraic numbers in POWER SERIES. p-adic numbers were then generalized to VALUATIONS by Kürschák in 1913. In the early 1920s, Hasse formulated the LOCAL-GLOBAL PRINCIPLE (now usually called the HASSE PRINCIPLE), which is one of the chief applications of LOCAL FIELD theory. Skolem's p-adic method, which is used in attacking certain DIOPHANTINE EQUATIONS, is another powerful application of p-adic numbers. Another application is the theorem that the HARMONIC NUMBERS H_n are never INTEGERS (except for H_1). A similar application is the proof of the VON STAUDT-CLAUSEN THEOREM using the p-adic valuation, although the technical details are somewhat difficult. Yet another application is provided by the MAHLER-LECH THEOREM.

Every RATIONAL x has an "essentially" unique p-adic expansion ("essentially" since zero terms can always be added at the beginning)

$$x = \sum_{j=m}^{\infty} a_j p^j, \tag{10}$$

with m an INTEGER, a_j the INTEGERS between 0 and $p-1$ inclusive, and where the sum is convergent with respect to p-adic valuation. If $x \neq 0$ and $a_m \neq 0$, then the expansion is unique. Burger and Struppeck (1996) show that for p a PRIME and n a POSITIVE INTEGER,

$$|n!|_p = p^{-(n-A_p(n))/(p-1)}, \tag{11}$$

where the p-adic expansion of n is

$$n = a_0 + a_1 p + a_2 p^2 + \ldots + a_L P^L, \tag{12}$$

and

$$A_p(n) = a_0 + a_1 + \ldots + a_L. \tag{13}$$

For sufficiently large n,

$$|n!|_p \leq p^{-n/(2p-2)}. \tag{14}$$

The p-adic valuation on $\mathbb{Q}$ gives rise to the p-adic metric

$$d(x,y) = |x-y|_p, \tag{15}$$

which in turn gives rise to the p-adic topology. It can be shown that the rationals, together with the p-adic metric, do not form a COMPLETE METRIC SPACE. The completion of this space can therefore be constructed, and the set of p-adic numbers $\mathbb{Q}_p$ is defined to be this completed space.

Just as the REAL NUMBERS are the completion of the RATIONALS $\mathbb{Q}$ with respect to the usual absolute valuation $|x-y|$, the p-adic numbers are the completion of $\mathbb{Q}$ with respect to the p-adic valuation $|x-y|_p$. The p-adic numbers are useful in solving DIOPHANTINE EQUATIONS. For example, the equation $X^2 = 2$ can easily be shown to have no solutions in the field of 2-adic numbers (we simply take the valuation of both sides). Because the 2-adic numbers contain the rationals as a subset, we can immediately see that the equation has no solutions in the RATIONALS. So we have an immediate proof of the irrationality of $\sqrt{2}$.

This is a common argument that is used in solving these types of equations: in order to show that an equation has no solutions in $\mathbb{Q}$, we show that it has no solutions in a FIELD EXTENSION. For another example, consider $X^2+1=0$. This equation has no solutions in $\mathbb{Q}$ because it has no solutions in the reals $\mathbb{R}$, and $\mathbb{Q}$ is a subset of $\mathbb{R}$.

Now consider the converse. Suppose we have an equation that does have solutions in $\mathbb{R}$ and in all the $\mathbb{Q}_p$. Can we conclude that the equation has a solution in $\mathbb{Q}$? Unfortunately, in general, the answer is no, but there are classes of equations for which the answer is yes. Such equations are said to satisfy the HASSE PRINCIPLE.

see also AX-KOCHEN ISOMORPHISM THEOREM, DIOPHANTINE EQUATION, HARMONIC NUMBER, HASSE PRINCIPLE, LOCAL FIELD, LOCAL-GLOBAL PRINCIPLE, MAHLER-LECH THEOREM, PRODUCT FORMULA, VALUATION, VALUATION THEORY, VON STAUDT-CLAUSEN THEOREM

References

Burger, E. B. and Struppeck, T. "Does $\sum_{n=0}^{\infty}\frac{1}{n!}$ Really Converge? Infinite Series and p-adic Analysis." *Amer. Math. Monthly* **103**, 565–577, 1996.

Cassels, J. W. S. and Scott, J. W. *Local Fields.* Cambridge, England: Cambridge University Press, 1986.

Gouvêa, F. Q. *P-adic Numbers: An Introduction, 2nd ed.* New York: Springer-Verlag, 1997.

Koblitz, N. *P-adic Numbers, P-adic Analysis, and Zeta-Functions, 2nd ed.* New York: Springer-Verlag, 1984.

Mahler, K. *P-adic Numbers and Their Functions, 2nd ed.* Cambridge, England: Cambridge University Press, 1981.

P-Circle

see SPIEKER CIRCLE

p-Element

see SEMISIMPLE

p-Good Path

A LATTICE PATH from one point to another is p-good if it lies completely below the line

$$y=(p-1)x.$$

Hilton and Pederson (1991) show that the number of p-good paths from $(1,\ q-1)$ to $(k,\ n-k)$ under the condition $2\leq k\leq n-p+1\leq p(k-1)$ is

$$\binom{n-q}{k-1}-\sum_{j=1}^{\ell} {}_p d_{qj}\binom{n-pj}{k-j},$$

where $\binom{a}{b}$ is a BINOMIAL COEFFICIENT, and

$$\ell\equiv\left\lfloor\frac{n-k}{p-1}\right\rfloor,$$

where $\lfloor x\rfloor$ is the FLOOR FUNCTION.

see also CATALAN NUMBER, LATTICE PATH, SCHRÖDER NUMBER

References

Hilton, P. and Pederson, J. "Catalan Numbers, Their Generalization, and Their Uses." *Math. Intel.* **13**, 64–75, 1991.

p-Group

A FINITE GROUP of ORDER p^a for p a PRIME is called a p-group. Sylow proved that every GROUP of this form has a POWER-commutator representation on n generators defined by

$$a_i^p=\prod_{k=i+1}^{n} a_k^{\beta(i,k)} \tag{1}$$

for $0\leq\beta(i,k)<p$, $1\leq i\leq n$ and

$$[a_j,a_i]=\prod_{k=j+1}^{n} a_k^{\beta(i,j,k)} \tag{2}$$

for $0\leq\beta(i,j,k)<p$, $1\leq i<j\leq n$. If p is PRIME and $f(p)$ the number of GROUPS of order p^m, then

$$f(p)=p^{Am^2}, \tag{3}$$

where

$$\lim_{m\to\infty} A=\tfrac{2}{27} \tag{4}$$

(Higman 1960a,b).

see also FINITE GROUP

References

Higman, G. "Enumerating p-Groups. I. Inequalities." *Proc. London Math. Soc.* **10**, 24–30, 1960a.

Higman, G. "Enumerating p-Groups. II. Problems Whose Solution is PORC." *Proc. London Math. Soc.* **10**, 566–582, 1960b.

p'-Group

X is a p'-group if p does not divide the ORDER of X.

p-Layer

The p-layer of H, $L_{p'}(H)$ is the unique minimal NORMAL SUBGROUP of H which maps onto $E(H/O_{p'}(H))$.

see also B_p-THEOREM, $L_{p'}$-BALANCE THEOREM, SIGNALIZER FUNCTOR THEOREM

P-Polynomial

see HOMFLY POLYNOMIAL

P-Problem

A problem is assigned to the P (POLYNOMIAL time) class if the number of steps is bounded by a POLYNOMIAL.

see also COMPLEXITY THEORY, NP-COMPLETE PROBLEM, NP-HARD PROBLEM, NP-PROBLEM

References

Borwein, J. M. and Borwein, P. B. *Pi and the AGM: A Study in Analytic Number Theory and Computational Complexity.* New York: Wiley, 1987.

Greenlaw, R.; Hoover, H. J.; and Ruzzo, W. L. *Limits to Parallel Computation: P-Completeness Theory.* Oxford, England: Oxford University Press, 1995.

p-Series

A shorthand name for a POWER SERIES with a NEGATIVE exponent, $\sum_{k=1}^{\infty} k^{-p}$, where $p > 0$.

see also POWER SERIES, RIEMANN ZETA FUNCTION

p-Signature

Diagonalize a form over the rationals to

$$\operatorname{diag}[p^a \cdot A, p^b \cdot B, \ldots],$$

where all the entries are INTEGERS and A, B, ... are RELATIVELY PRIME to p. Then the p-signature of the form (for $p \neq -1, 2$) is

$$p^a + p^b + \ldots + 4k \pmod 8,$$

where k is the number of ANTISQUARES. For $p = -1$, the p-signature is SYLVESTER'S SIGNATURE.

see also SIGNATURE (QUADRATIC FORM)

P-Symbol

A symbol employed in a formal PROPOSITIONAL CALCULUS.

References

Nidditch, P. H. *Propositional Calculus.* New York: Free Press of Glencoe, p. 1, 1962.

P-Value

The PROBABILITY that a variate would assume a value greater than or equal to the observed value strictly by chance: $P(z \geq z_{\text{observed}})$.

see also ALPHA VALUE, SIGNIFICANCE

Paasche's Index

The statistical INDEX

$$P_P \equiv \frac{\sum p_n q_n}{\sum p_0 q_n},$$

where p_n is the price per unit in period n and q_n is the quantity produced in period n.

see also INDEX

References

Kenney, J. F. and Keeping, E. S. *Mathematics of Statistics, Pt. 1, 3rd ed.* Princeton, NJ: Van Nostrand, p. 65, 1962.

Packing

The placement of objects so that they touch in some specified manner, often inside a container with specified properties.

see also BOX-PACKING THEOREM, CIRCLE PACKING, GROEMER PACKING, HYPERSPHERE PACKING, KEPLER PROBLEM, KISSING NUMBER PACKING DENSITY, POLYHEDRON PACKING, SPACE-FILLING POLYHEDRON, SPHERE PACKING

References

Eppstein, D. "Covering and Packing." `http://www.ics.uci.edu/~eppstein/junkyard/cover.html`.

Packing Density

The fraction of a volume filled by a given collection of solids.

see also HYPERSPHERE PACKING, PACKING, SPHERE PACKING

Padé Approximant

Approximants derived by expanding a function as a ratio of two POWER SERIES and determining both the NUMERATOR and DENOMINATOR COEFFICIENTS. Padé approximations are usually superior to TAYLOR EXPANSIONS when functions contain POLES, because the use of RATIONAL FUNCTIONS allows them to be well-represented.

The Padé approximant $R_{L/0}$ corresponds to the MACLAURIN SERIES. When it exists, the $R_{L/M} \equiv [L/M]$ Padé approximant to any POWER SERIES

$$A(x) = \sum_{j=0}^{\infty} a_j x^j \tag{1}$$

is unique. If $A(x)$ is a TRANSCENDENTAL FUNCTION, then the terms are given by the TAYLOR SERIES about x_0

$$a_n = \frac{1}{n!} A^{(n)}(x_0). \tag{2}$$

The COEFFICIENTS are found by setting

$$A(x) - \frac{P_L(x)}{Q_M(x)} = 0 \tag{3}$$

and equating COEFFICIENTS. $Q_M(x)$ can be multiplied by an arbitrary constant which will rescale the other COEFFICIENTS, so an addition constraint can be applied. The conventional normalization is

$$Q_M(0) = 1. \tag{4}$$

Expanding (3) gives

$$P_L(x) = p_0 + p_1 x + \ldots + p_L x^L \tag{5}$$

$$Q_M(x) = 1 + q_1 x + \ldots + q_M x^M. \tag{6}$$

These give the set of equations

$$a_0 = p_0 \tag{7}$$
$$a_1 + a_0 q_1 = p_1 \tag{8}$$
$$a_2 + a_1 q_1 + a_0 q_2 = p_2 \tag{9}$$
$$\vdots$$
$$a_L + a_{L-1} q_1 + \ldots + a_0 q_L = p_L \tag{10}$$
$$a_{L+1} + a_L q_1 + \ldots + a_{L-M+1} q_M = 0 \tag{11}$$
$$\vdots$$
$$q_{L+M} + a_{L+M-1} q_1 + \ldots + a_L q_M = 0, \tag{12}$$

where $a_n = 0$ for $n < 0$ and $q_j = 0$ for $j > M$. Solving these directly gives

$$[L/M] = \frac{\begin{vmatrix} a_{L-m+1} & a_{L-m+2} & \cdots & a_{L+1} \\ \vdots & \vdots & \ddots & \vdots \\ a_L & a_{L+1} & \cdots & a_{L+M} \\ \sum_{j=M}^{L} a_{j-M} x^j & \sum_{j=M-1}^{L} a_{j-M+1} x^j & \cdots & \sum_{j=0}^{L} a_j x^j \end{vmatrix}}{\begin{vmatrix} a_{L-M+1} & a_{L-M+2} & \cdots & a_{L+1} \\ \vdots & \vdots & \ddots & \vdots \\ a_L & a_{L+1} & \cdots & a_{L+M} \\ x^M & x^{M-1} & \cdots & 1 \end{vmatrix}}, \tag{13}$$

where sums are replaced by a zero if the lower index exceeds the upper. Alternate forms are

$$[L/M] = \sum_{j=0}^{L-M} a_j x^j + x^{L-M+1} \mathbf{w}_{L/M}^{\mathrm{T}} \mathsf{W}_{L/M}^{-1} \mathbf{w}_{L/M}$$
$$= \sum_{j=0}^{L+n} a_j x^j + x^{L+n+1} \mathbf{w}_{(L+M)/M}^{\mathrm{T}} \mathsf{W}_{L/M}^{-1} \mathbf{w}_{(L+n)/M}$$

for

$$\mathsf{W}_{L/M} = \begin{bmatrix} a_{L-M+1} - x a_{L-M+2} & \cdots & a_L - x a_{L+1} \\ \vdots & \ddots & \vdots \\ a_L - x a_{L+1} & \cdots & a_{L+M-1} - x a_{L+M} \end{bmatrix} \tag{14}$$

$$\mathbf{w}_{L/M} = \begin{bmatrix} a_{L-M+1} \\ a_{L-M+2} \\ \vdots \\ a_L \end{bmatrix}, \tag{15}$$

and $0 \le n \le M$.

The first few Padé approximants for e^x are

$$\exp_{0/0}(x) = 1$$
$$\exp_{0/1}(x) = \frac{1}{1-x}$$
$$\exp_{0/2}(x) = \frac{2}{2-2x+x^2}$$
$$\exp_{0/3}(x) = \frac{6}{6-6x+3x^2-x^3}$$
$$\exp_{1/0}(x) = 1+x$$
$$\exp_{1/1}(x) = \frac{2+x}{2-x}$$
$$\exp_{1/2}(x) = \frac{6+2x}{6-4x+x^2}$$
$$\exp_{1/3}(x) = \frac{24+6x}{24-18x+6x^2-x^3}$$
$$\exp_{2/0}(x) = \frac{2+2x+x^2}{2}$$
$$\exp_{2/1}(x) = \frac{6+4x+x^2}{6-2x}$$
$$\exp_{2/2}(x) = \frac{12+6x+x^2}{12-6x+x^2}$$
$$\exp_{2/3}(x) = \frac{60+24x+3x^2}{60-36x+9x^2-x^3}$$
$$\exp_{3/0}(x) = \frac{6+6x+3x^2+x^3}{6}$$
$$\exp_{3/1}(x) = \frac{24+18x+16x^2+x^3}{24-6x}$$
$$\exp_{3/2}(x) = \frac{60+36x+9x^2+x^3}{60-24x+3x^2}$$
$$\exp_{3/3}(x) = \frac{120+60x+12x^2+x^3}{120-60x+12x^2-x^3}.$$

Two-term identities include

$$\frac{P_{L+1}(x)}{Q_{M+1}(x)} - \frac{P'_L(x)}{Q'_M(x)} = \frac{C_{(L+1)/(M+1)}{}^2 x^{L+M+1}}{Q_{M+1}(x) Q'_M(x)} \tag{16}$$

$$\frac{P_{L+1}(x)}{Q_M(x)} - \frac{P'_L(x)}{Q'_M(x)} = \frac{C_{(L+1)/M} C_{(L+1)/(M+1)} x^{L+M+1}}{Q_M(x) Q'_M(x)} \tag{17}$$

$$\frac{P_L(x)}{Q_{M+1}(x)} - \frac{P'_L(x)}{Q'_M(x)} = \frac{C_{L/(M+1)} C_{(L+1)/(M+1)} x^{L+M+1}}{Q_M(x) Q'_M(x)} \tag{18}$$

$$\frac{P_L(x)}{Q_{M+1}(x)} - \frac{P'_{L+1}(x)}{Q'_M} = \frac{C_{(L+1)/(M+1)}{}^2 x^{L+M+2}}{Q_{M+1} Q'_M} \tag{19}$$

$$\frac{P_{L+1}}{Q_M(x)} - \frac{P'_{L-1}(x)}{Q'_M(x)} = \frac{C_{L/(M+1)} C_{(L+1)/M} x^{L+M} + C_{L/M} C_{(L+1)/(M+1)} x^{L+M+1}}{Q_M(x) Q'_M(x)} \tag{20}$$

$$\frac{P_L(x)}{Q_{M+1}(x)} - \frac{P'_L(x)}{Q'_{M-1}(x)} = \frac{C_{L/(M+1)}C_{(L+1)/M}x^{L+M} - C_{L/M}C_{(L+1)/(M+1)}x^{L+M+1}}{Q_{M+1}(x)Q'_{M-1}(x)}, \tag{21}$$

where C is the C-DETERMINANT. Three-term identities can be derived using the FROBENIUS TRIANGLE IDENTITIES (Baker 1975, p. 32).

A five-term identity is

$$S_{(L+1)/M}S_{(L-1)/M} - S_{L/(M+1)}S_{L/(M-1)} = S_{L/M}{}^2. \tag{22}$$

Cross ratio identities include

$$\frac{(R_{L/M} - R_{L/(M+1)})(R_{(L+1)/M} - R_{(L+1)/(M+1)})}{(R_{L/M} - R_{(L+1)/M})(R_{L/(M+1)} - R_{(L+1)/(M+1)})} = \frac{C_{L/(M+1)}C_{(L+2)/(M+1)}}{C_{(L+1)/M}C_{(L+1)/(M+2)}} \tag{23}$$

$$\frac{(R_{L/M} - R_{(L+1)/(M+1)})(R_{(L+1)/M} - R_{L/(M+1)})}{(R_{L/M} - R_{L/(M+1)})(R_{(L+1)/M} - R_{(L+1)/(M+1)})} = \frac{C_{(L+1)/(M+1)}{}^2 x}{C_{L/(M+1)}C_{(L+2)/(M+1)}} \tag{24}$$

$$\frac{(R_{L/M} - R_{(L+1)/(M+1)})(R_{(L+1)/M} - R_{L/(M+1)})}{(R_{L/M} - R_{(L+1)/M})(R_{L/(M+1)} - R_{(L+1)/(M+1)})} = \frac{C_{(L+1)/(M+1)}{}^2 x}{C_{(L+1)/M}C_{(L+1)/(M+2)}} \tag{25}$$

$$\frac{(R_{L/M} - R_{(L+1)/(M-1)})(R_{L/(M+1)} - R_{(L+1)/M})}{(R_{L/M} - R_{L/(M+1)})(R_{(L+1)/(M+1)} - R_{(L+1)/M})} = \frac{C_{(L+1)/M}C_{(L+1)/(M+1)}x}{C_{L/(M+1)}C_{(L+2)/M}} \tag{26}$$

$$\frac{(R_{L/M} - R_{(L-1)/(M+1)})(R_{(L+1)/M} - R_{L/(M+1)})}{(R_{L/M} - R_{(L+1)/M})(R_{(L-1)/(M+1)} - R_{L/(M+1)})} = \frac{C_{L/(M+1)}C_{(L+1)/(M+1)}x}{C_{(L+1)/M}C_{L/(M+2)}}. \tag{27}$$

see also C-DETERMINANT, ECONOMIZED RATIONAL APPROXIMATION, FROBENIUS TRIANGLE IDENTITIES

References

Baker, G. A. Jr. "The Theory and Application of The Pade Approximant Method." In *Advances in Theoretical Physics, Vol. 1* (Ed. K. A. Brueckner). New York: Academic Press, pp. 1–58, 1965.

Baker, G. A. Jr. *Essentials of Padé Approximants in Theoretical Physics.* New York: Academic Press, pp. 27–38, 1975.

Baker, G. A. Jr. and Graves-Morris, P. *Padé Approximants.* New York: Cambridge University Press, 1996.

Press, W. H.; Flannery, B. P.; Teukolsky, S. A.; and Vetterling, W. T. "Padé Approximants." §5.12 in *Numerical Recipes in FORTRAN: The Art of Scientific Computing, 2nd ed.* Cambridge, England: Cambridge University Press, pp. 194–197, 1992.

Padé Conjecture

If $P(z)$ is a POWER series which is regular for $|z| \leq 1$ except for m POLES within this CIRCLE and except for $z = +1$, at which points the function is assumed continuous when only points $|z| \leq 1$ are considered, then at least a subsequence of the $[N, N]$ PADÉ APPROXIMANTS are uniformly bounded in the domain formed by removing the interiors of small circles with centers at these POLES and uniformly continuous at $z = +1$ for $|z| \leq 1$.

see also PADÉ APPROXIMANT

References

Baker, G. A. Jr. "The Padé Conjecture and Some Consequences." §II.D in *Advances in Theoretical Physics, Vol. 1* (Ed. K. A. Brueckner). New York: Academic Press, pp. 23–27, 1965.

Padovan Sequence

The INTEGER SEQUENCE defined by the RECURRENCE RELATION

$$P(n) = P(n-2) + P(n-3)$$

with the initial conditions $P(0) = P(1) = P(2) = 1$. The first few terms are 1, 1, 2, 2, 3, 4, 5, 7, 9, 12, ... (Sloane's A000931). The ratio $\lim_{n\to\infty} P(n)/P(n-1)$ is called the PLASTIC CONSTANT.

see also PERRIN SEQUENCE, PLASTIC CONSTANT

References

Sloane, N. J. A. Sequence A000931/M0284 in "An On-Line Version of the Encyclopedia of Integer Sequences."

Stewart, I. "Tales of a Neglected Number." *Sci. Amer.* **274**, 102–103, June 1996.

Painlevé Property

Following the work of Fuchs in classifying first-order ORDINARY DIFFERENTIAL EQUATIONS, Painlevé studied second-order ODEs of the form

$$\frac{d^2y}{dx^2} = F(y', y, x),$$

where F is ANALYTIC in x and rational in y and y'. Painlevé found 50 types whose only movable SINGULARITIES are ordinary POLES. This characteristic is known as the Painlevé property. Six of the transcendents define new transcendents known as PAINLEVÉ TRANSCENDENTS, and the remaining 44 can be integrated in terms of classical transcendents, quadratures, or the PAINLEVÉ TRANSCENDENTS.

see also PAINLEVÉ TRANSCENDENTS

Painlevé Transcendents

$$y'' = 6y^2 + x \tag{1}$$

$$y'' = 2y^3 + xy + \alpha \tag{2}$$

$$y'' = \frac{y'^2}{y} - \frac{1}{xy'} + \alpha y^3 + \frac{\beta}{xy^2} + \frac{\gamma}{x} + \frac{\delta}{y}. \tag{3}$$

Transcendents 4–6 do not have known first integrals, but all transcendents have first integrals for special values of their parameters except (1). Painlevé found the above transcendents (1) to (3), and the rest were investigated by his students. The sixth transcendent was found by Gambier and contains the other five as limiting cases.

see also PAINLEVÉ PROPERTY

Pair

A SET of two numbers or objects linked in some way are said to be a pair. The pair a and b are usually denoted $(a,\ b)$. In certain circumstances, pairs are also called BROTHERS or TWINS.

see also AMICABLE PAIR, AUGMENTED AMICABLE PAIR, BROWN NUMBERS, FRIENDLY PAIR, HEXAD, HOMOGENEOUS NUMBERS, IMPULSE PAIR, IRREGULAR PAIR, LAX PAIR, LONG EXACT SEQUENCE OF A PAIR AXIOM, MONAD, ORDERED PAIR, PERKO PAIR, QUADRUPLET, QUASIAMICABLE PAIR, QUINTUPLET, REDUCED AMICABLE PAIR, SMITH BROTHERS, TRIAD, TRIPLET, TWIN PEAKS, TWIN PRIMES, TWINS, UNITARY AMICABLE PAIR, WILF-ZEILBERGER PAIR

Pair Sum

Given an AMICABLE PAIR (m,n), the quantity

$$\sigma(m)=\sigma(n)=s(m)+s(n)=m+n$$

is called the pair sum, where $\sigma(n)$ is the DIVISOR FUNCTION and $s(n)$ is the RESTRICTED DIVISOR FUNCTION.

see also AMICABLE PAIR

Paired t-Test

Given two paired sets X_i and Y_i of n measured values, the paired t-test determines if they differ from each other in a significant way. Let

$$\hat{X}_i=(X_i-\bar{X}_i)$$
$$\hat{Y}_i=(Y_i-\bar{Y}_i),$$

then define t by

$$t=(\bar{X}-\bar{Y})\sqrt{\frac{n(n-1)}{\sum_{i=1}^n(\hat{X}_i-\hat{Y}_i)^2}}.$$

This statistic has $n-1$ DEGREES OF FREEDOM.

A table of STUDENT'S t-DISTRIBUTION confidence interval can be used to determine the significance level at which two distributions differ.

see also FISHER SIGN TEST, HYPOTHESIS TESTING, STUDENT'S t-DISTRIBUTION, WILCOXON SIGNED RANK TEST

References

Goulden, C. H. *Methods of Statistical Analysis, 2nd ed.* New York: Wiley, pp. 50–55, 1956.

Paley Class

The Paley class of a POSITIVE INTEGER $m\equiv 0 \pmod 4$ is defined as the set of all possible QUADRUPLES (k,e,q,n) where

$$m=2^e(q^n+1),$$

q is an ODD PRIME, and

$$k=\begin{cases}0 & \text{if } q=0\\ 1 & \text{if } q^n-3\equiv 0 \pmod 4\\ 2 & \text{if } q^n-1\equiv 0 \pmod 4\\ \text{undefined} & \text{otherwise.}\end{cases}$$

see also HADAMARD MATRIX, PALEY CONSTRUCTION

Paley Construction

HADAMARD MATRICES H_n can be constructed using GALOIS FIELD GF(p^m) when $p=4l-1$ and m is ODD. Pick a representation r RELATIVELY PRIME to p. Then by coloring white $\lfloor(p-1)/2\rfloor$ (where $\lfloor x\rfloor$ is the FLOOR FUNCTION) distinct equally spaced RESIDUES mod p (r^0, r, r^2, ...; r^0, r^2, r^4, ...; etc.) *in addition to* 0, a HADAMARD MATRIX is obtained if the POWERS of r (mod p) run through $<\lfloor(p-1)/2\rfloor$. For example,

$$n=12=11^1+1=2(5+1)=2^2(2+1)$$

is of this form with $p=11=4\times 3-1$ and $m=1$. Since $m=1$, we are dealing with GF(11), so pick $p=2$ and compute its RESIDUES (mod 11), which are

$$\begin{aligned}p^0&\equiv 1\\ p^1&\equiv 2\\ p^2&\equiv 4\\ p^3&\equiv 8\\ p^4&\equiv 16\equiv 5\\ p^5&\equiv 10\\ p^6&\equiv 20\equiv 9\\ p^7&\equiv 18\equiv 7\\ p^8&\equiv 14\equiv 3\\ p^9&\equiv 6\\ p^{10}&\equiv 12\equiv 1.\end{aligned}$$

Picking the first $\lfloor 11/2\rfloor=5$ RESIDUES and adding 0 gives: 0, 1, 2, 4, 5, 8, which should then be colored in the MATRIX obtained by writing out the RESIDUES increasing to the left and up along the border (0 through $p-1$, followed by ∞), then adding horizontal and vertical coordinates to get the residue to place in each square.

$$\begin{bmatrix} \infty & \infty & \infty & \infty & \infty & \infty & \infty & \infty & \infty & \infty & \infty & \infty \\ 10 & 0 & 1 & 2 & 3 & 4 & 5 & 6 & 7 & 8 & 9 & \infty \\ 9 & 10 & 0 & 1 & 2 & 3 & 4 & 5 & 6 & 7 & 8 & \infty \\ 8 & 9 & 10 & 0 & 1 & 2 & 3 & 4 & 5 & 6 & 7 & \infty \\ 7 & 8 & 9 & 10 & 0 & 1 & 2 & 3 & 4 & 5 & 6 & \infty \\ 6 & 7 & 8 & 9 & 10 & 0 & 1 & 2 & 3 & 4 & 5 & \infty \\ 5 & 6 & 7 & 8 & 9 & 10 & 0 & 1 & 2 & 3 & 4 & \infty \\ 4 & 5 & 6 & 7 & 8 & 9 & 10 & 0 & 1 & 2 & 3 & \infty \\ 3 & 4 & 5 & 6 & 7 & 8 & 9 & 10 & 0 & 1 & 2 & \infty \\ 2 & 3 & 4 & 5 & 6 & 7 & 8 & 9 & 10 & 0 & 1 & \infty \\ 1 & 2 & 3 & 4 & 5 & 6 & 7 & 8 & 9 & 10 & 0 & \infty \\ 0 & 1 & 2 & 3 & 4 & 5 & 6 & 7 & 8 & 9 & 10 & \infty \end{bmatrix}$$

H_{16} can be trivially constructed from $\mathsf{H}_4 \otimes \mathsf{H}_4$. H_{20} cannot be built up from smaller MATRICES, so use $n = 20 = 19 + 1 = 2(3^2 + 1) = 2^2(2^2 + 1)$. Only the first form can be used, with $p = 19 = 4 \times 5 - 1$ and $m = 1$. We therefore use GF(19), and color 9 RESIDUES plus 0 white. H_{24} can be constructed from $\mathsf{H}_2 \otimes \mathsf{H}_{12}$.

Now consider a more complicated case. For $n = 28 = 3^3 + 1 = 2(13 + 1)$, the only form having $p = 4l - 1$ is the first, so use the GF(3^3) field. Take as the modulus the IRREDUCIBLE POLYNOMIAL $x^3 + 2x + 1$, written 1021. A four-digit number can always be written using only three digits, since $1000 - 1021 \equiv 0012$ and $2000 - 2012 \equiv 0021$. Now look at the moduli starting with 10, where each digit is considered separately. Then

$$\begin{array}{lll} x^0 \equiv 1 & x^1 \equiv 10 & x^2 \equiv 100 \\ x^3 \equiv 1000 \equiv 12 & x^4 \equiv 120 & x^5 \equiv 1200 \equiv 212 \\ x^6 \equiv 2120 \equiv 111 & x^7 \equiv 1100 \equiv 122 & x^8 \equiv 1220 \equiv 202 \\ x^9 \equiv 2020 \equiv 11 & x^{10} \equiv 110 & x^{11} \equiv 1100 \equiv 112 \\ x^{12} \equiv 1120 \equiv 102 & x^{13} \equiv 1020 \equiv 2 & x^{14} \equiv 20 \\ x^{15} \equiv 200 & x^{16} \equiv 2000 \equiv 21 & x^{17} \equiv 210 \\ x^{18} \equiv 2100 \equiv 121 & x^{19} \equiv 1210 \equiv 222 & x^{20} \equiv 2220 \equiv 211 \\ x^{21} \equiv 2110 \equiv 101 & x^{22} \equiv 101 \equiv 22 & x^{23} \equiv 220 \\ x^{24} \equiv 2200 \equiv 221 & x^{25} \equiv 2210 \equiv 201 & x^{26} \equiv 2010 \equiv 1 \end{array}$$

Taking the alternate terms gives white squares as 000, 001, 020, 021, 022, 100, 102, 110, 111, 120, 121, 202, 211, and 221.

References

Ball, W. W. R. and Coxeter, H. S. M. *Mathematical Recreations and Essays, 13th ed.* New York: Dover, pp. 107–109 and 274, 1987.

Beth, T.; Jungnickel, D.; and Lenz, H. *Design Theory, 2nd ed. rev.* Cambridge, England: Cambridge University Press, 1998.

Geramita, A. V. *Orthogonal Designs: Quadratic Forms and Hadamard Matrices.* New York: Marcel Dekker, 1979.

Kitis, L. "Paley's Construction of Hadamard Matrices." `http://www.mathsource.com/cgi-bin/MathSource/Applications/Mathematics/0205-760`.

Paley's Theorem

Proved in 1933. If q is an ODD PRIME or $q = 0$ and n is any POSITIVE INTEGER, then there is a HADAMARD MATRIX of order

$$m = 2^e(q^n + 1),$$

where e is any POSITIVE INTEGER such that $m \equiv 0 \pmod 4$. If m is of this form, the matrix can be constructed with a PALEY CONSTRUCTION. If m is divisible by 4 but not of the form (1), the PALEY CLASS is undefined. However, HADAMARD MATRICES have been shown to exist for all $m \equiv 0 \pmod 4$ for $m < 428$.

see also HADAMARD MATRIX, PALEY CLASS, PALEY CONSTRUCTION

Palindrome Number

see PALINDROMIC NUMBER

Palindromic Number

A symmetrical number which is written in some base b as $a_1 a_2 \ldots a_2 a_1$. The first few are 0, 1, 2, 3, 4, 5, 6, 7, 8, 9, 11, 22, 33, 44, 55, 66, 77, 88, 99, 101, 111, 121, ... (Sloane's A002113).

The first few n for which the PRONIC NUMBER P_n is palindromic are 1, 2, 16, 77, 538, 1621, ... (Sloane's A028336), and the first few palindromic numbers which are PRONIC are 2, 6, 272, 6006, 289982, ... (Sloane's A028337). The first few numbers whose squares are palindromic are 1, 2, 3, 11, 22, 26, ... (Sloane's A002778), and the first few palindromic squares are 1, 4, 9, 121, 484, 676, ... (Sloane's A002779).

see also DEMLO NUMBER, PALINDROMIC NUMBER CONJECTURE, REVERSAL

References

de Geest, P. "Palindromic Products of Two Consecutive Integers." `http://www.ping.be/~ping6758/consec.htm`.

de Geest, P. "Palindromic Squares." `http://www.ping.be/~ping6758/square.htm`.

Pappas, T. "Numerical Palindromes." *The Joy of Mathematics.* San Carlos, CA: Wide World Publ./Tetra, p. 146, 1989.

Sloane, N. J. A. Sequences A028336, A028337, A002113/M0484, A0027778/M0807, and A002779/M3371 in "An On-Line Version of the Encyclopedia of Integer Sequences."

Palindromic Number Conjecture

Apply the 196-ALGORITHM, which consists of taking any POSITIVE INTEGER of two digits or more, reversing the digits, and adding to the original number. Now sum the two and repeat the procedure with the sum. Of the first 10,000 numbers, only 251 do not produce a PALINDROMIC NUMBER in ≤ 23 steps (Gardner 1979).

It was therefore conjectured that *all* numbers will eventually yield a PALINDROMIC NUMBER. However, the conjecture has been proven false for bases which are a POWER of 2, and seems to be false for base 10 as well. Among the first 100,000 numbers, 5,996 numbers apparently never generate a PALINDROMIC NUMBER (Gruenberger 1984). The first few are 196, 887, 1675, 7436, 13783, 52514, 94039, 187088, 1067869, 10755470, ... (Sloane's A006960).

It is conjectured, but not proven, that there are an infinite number of palindromic PRIMES. With the exception of 11, palindromic PRIMES must have an ODD number of digits.

see also 196-ALGORITHM

References
Gardner, M. *Mathematical Circus: More Puzzles, Games, Paradoxes and Other Mathematical Entertainments from Scientific American.* New York: Knopf, pp. 242–245, 1979.
Gruenberger, F. "How to Handle Numbers with Thousands of Digits, and Why One Might Want to." *Sci. Amer.* **250**, 19–26, Apr. 1984.
Sloane, N. J. A. Sequence A006960/M5410 in "An On-Line Version of the Encyclopedia of Integer Sequences."

Pancake Cutting

see CIRCLE CUTTING

Pancake Theorem

The 2-D version of the HAM SANDWICH THEOREM.

Pandiagonal Square

see PANMAGIC SQUARE

Pandigital

A decimal INTEGER which contains each of the digits from 0 to 9.

Panmagic Square

8	17	1	15	24
11	25	9	18	2
19	3	12	21	10
22	6	20	4	13
5	14	23	7	16

If all the diagonals (including those obtained by "wrapping around" the edges) of a MAGIC SQUARE, as well as the usual rows, columns, and *main diagonals* sum to the MAGIC CONSTANT, the square is said to be a PANMAGIC SQUARE (also called DIABOLICAL SQUARE, NASIK SQUARE, or PANDIAGONAL SQUARE). No panmagic squares exist of order 3 or any order $4k+2$ for k an INTEGER. The Siamese method for generating MAGIC SQUARES produces panmagic squares for orders $6k \pm 1$ with ordinary vector (2, 1) and break vector (1, −1).

1	15	24	8	17
23	7	16	5	14
20	4	13	22	6
12	21	10	19	3
9	18	2	11	25

The LO SHU is not panmagic, but it is an ASSOCIATIVE MAGIC SQUARE. Order four squares can be panmagic or ASSOCIATIVE, but not both. Order five squares are the smallest which can be both ASSOCIATIVE and panmagic, and 16 distinct ASSOCIATIVE panmagic squares exist, one of which is illustrated above (Gardner 1988).

The number of distinct panmagic squares of order 1, 2, ... are 1, 0, 0, 384, 3600, 0, ... (Sloane's A027567, Hunter and Madachy 1975). Panmagic squares are related to HYPERCUBES.

see also ASSOCIATIVE MAGIC SQUARE, HYPERCUBE, FRANKLIN MAGIC SQUARE, MAGIC SQUARE

References
Gardner, M. *The Second Scientific American Book of Mathematical Puzzles & Diversions: A New Selection.* New York: Simon and Schuster, pp. 135–137, 1961.
Gardner, M. "Magic Squares and Cubes." Ch. 17 in *Time Travel and Other Mathematical Bewilderments.* New York: W. H. Freeman, pp. 213–225, 1988.
Hunter, J. A. H. and Madachy, J. S. "Mystic Arrays." Ch. 3 in *Mathematical Diversions.* New York: Dover, pp. 24–25, 1975.
Kraitchik, M. "Panmagic Squares." §7.9 in *Mathematical Recreations.* New York: W. W. Norton, pp. 174–176, 1942.
Madachy, J. S. *Madachy's Mathematical Recreations.* New York: Dover, p. 87, 1979.
Rosser, J. B. and Walker, R. J. "The Algebraic Theory of Diabolical Squares." *Duke Math. J.* **5**, 705–728, 1939.
Sloane, N. J. A. Sequence A027567 in "An On-Line Version of the Encyclopedia of Integer Sequences."

Pantograph

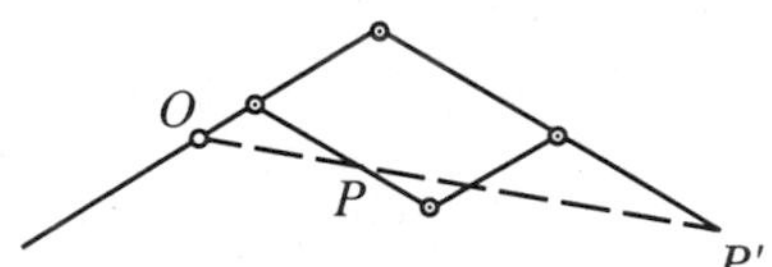

A LINKAGE invented in 1630 by Christoph Scheiner for making a scaled copy of a given figure. The linkage is pivoted at O; hinges are denoted ⊙. By placing a PENCIL at P (or P'), a DILATED image is obtained at P' (or P).

see also LINKAGE

Papal Cross

see also CROSS

Pappus's Centroid Theorem

The SURFACE AREA of a SURFACE OF REVOLUTION is given by

$$S_{\text{solid of rotation}} = [\text{perimenter}] \times [\text{distance traveled by centroid}],$$

and the VOLUME of a SOLID OF REVOLUTION is given by

$$V_{\text{solid of rotation}} = [\text{cross-section area}] \times [\text{distance traveled by centroid}].$$

see also CENTROID (GEOMETRIC), CROSS-SECTION, PERIMETER, SOLID OF REVOLUTION, SURFACE AREA, SURFACE OF REVOLUTION, TOROID, TORUS

References

Beyer, W. H. *CRC Standard Mathematical Tables, 28th ed.* Boca Raton, FL: CRC Press, p. 132, 1987.

Pappus Chain

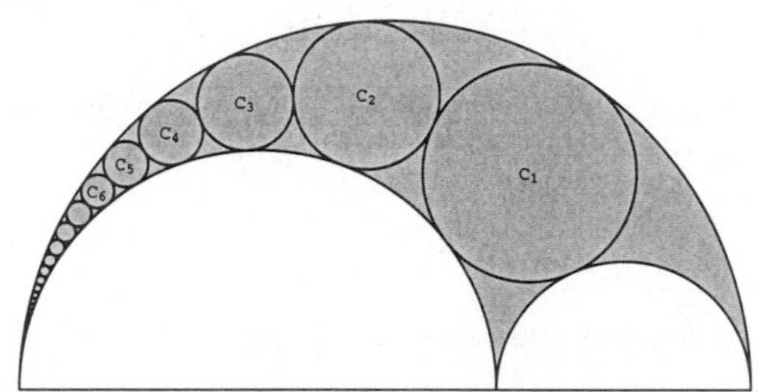

In the ARBELOS, construct a chain of TANGENT CIRCLES starting with the CIRCLE TANGENT to the two small interior semicircles and the large exterior one. Then the distance from the center of the first INSCRIBED CIRCLE to the bottom line is twice the CIRCLE'S RADIUS, from the second CIRCLE is four times the RADIUS, and for the nth CIRCLE is $2n$ times the RADIUS. The centers of the CIRCLES lie on an ELLIPSE, and the DIAMETER of the nth CIRCLE C_n is $(1/n)$th PERPENDICULAR distance to the base of the SEMICIRCLE. This result was known to Pappus, who referred to it as an ancient theorem (Hood 1961, Cadwell 1966, Gardner 1979, Bankoff 1981). The simplest proof is via INVERSIVE GEOMETRY.

If $r \equiv AB/AC$, then the radius of the nth circle in the pappus chain is

$$r_n = \frac{(1-r)r}{2[n^2(1-r)^2+r]}.$$

This equation can be derived by iteratively solving the QUADRATIC FORMULA generated by DESCARTES CIRCLE THEOREM for the radius of the SODDY CIRCLE. This general result simplifies to $r_n = 1/(6+n^2)$ for $r = 2/3$ (Gardner 1979). Further special cases when $AC = 1 + AB$ are considered by Gaba (1940).

If B divides AC in the GOLDEN RATIO ϕ, then the circles in the chain satisfy a number of other special properties (Bankoff 1955).

see also ARBELOS, COXETER'S LOXODROMIC SEQUENCE OF TANGENT CIRCLES, SODDY CIRCLES, STEINER CHAIN

References

Bankoff, L. "The Golden Arbelos." *Scripta Math.* **21**, 70–76, 1955.

Bankoff, L. "Are the Twin Circles of Archimedes Really Twins?" *Math. Mag.* **47**, 214–218, 1974.

Bankoff, L. "How Did Pappus Do It?" In *The Mathematical Gardner* (Ed. D. Klarner). Boston, MA: Prindle, Weber, and Schmidt, pp. 112–118, 1981.

Gaba, M. G. "On a Generalization of the Arbelos." *Amer. Math. Monthly* **47**, 19–24, 1940.

Gardner, M. "Mathematical Games: The Diverse Pleasures of Circles that Are Tangent to One Another." *Sci. Amer.* **240**, 18–28, Jan. 1979.

Hood, R. T. "A Chain of Circles." *Math. Teacher* **54**, 134–137, 1961.

Johnson, R. A. *Advanced Euclidean Geometry: An Elementary Treatise on the Geometry of the Triangle and the Circle.* Boston, MA: Houghton Mifflin, p. 117, 1929.

Pappus-Guldinus Theorem

see PAPPUS'S CENTROID THEOREM

Pappus's Harmonic Theorem

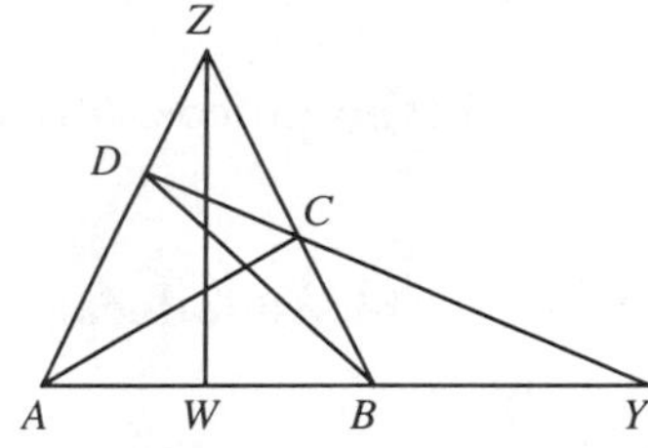

AW, AB, and AY in the above figure are in a HARMONIC RANGE.

see also CEVA'S THEOREM, MENELAUS' THEOREM

References

Coxeter, H. S. M. and Greitzer, S. L. *Geometry Revisited.* Washington, DC: Math. Assoc. Amer., pp. 67–68, 1967.

Pappus's Hexagon Theorem

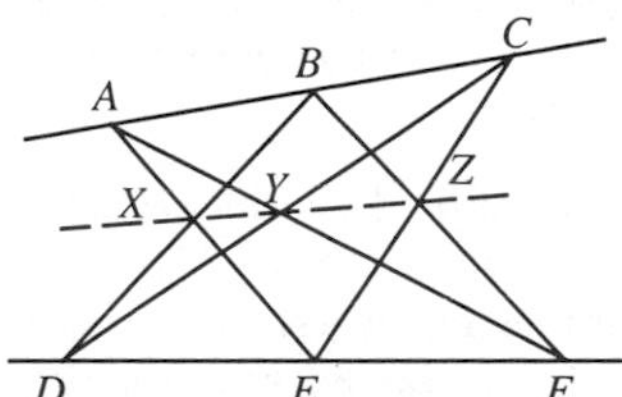

If A, B, and C are three points on one LINE, D, E, and F are three points on another LINE, and AE meets BD at X, AF meets CD at Y, and BF meets CE at Z, then the three points X, Y, and Z are COLLINEAR. Pappus's hexagon theorem is essentially its own dual according to the DUALITY PRINCIPLE of PROJECTIVE GEOMETRY.

see also CAYLEY-BACHARACH THEOREM, DESARGUES' THEOREM, DUALITY PRINCIPLE, PASCAL'S THEOREM, PROJECTIVE GEOMETRY

References

Coxeter, H. S. M. and Greitzer, S. L. *Geometry Revisited.* Washington, DC: Math. Assoc. Amer., pp. 73–74, 1967.
Ogilvy, C. S. *Excursions in Geometry.* New York: Dover, pp. 92–94, 1990.
Pappas, T. "Pappus' Theorem & the Nine Coin Puzzle." *The Joy of Mathematics.* San Carlos, CA: Wide World Publ./ Tetra, p. 163, 1989.

Pappus's Theorem

There are several THEOREMS that generally are known by the generic name "Pappus's Theorem."

see also PAPPUS'S CENTROID THEOREM, PAPPUS CHAIN, PAPPUS'S HARMONIC THEOREM, PAPPUS'S HEXAGON THEOREM

Parabiaugmented Dodecahedron

see JOHNSON SOLID

Parabiaugmented Hexagonal Prism

see JOHNSON SOLID

Parabiaugmented Truncated Dodecahedron

see JOHNSON SOLID

Parabidiminished Rhombicosidodecahedron

see JOHNSON SOLID

Parabigyrate Rhombicosidodecahedron

see JOHNSON SOLID

Parabola

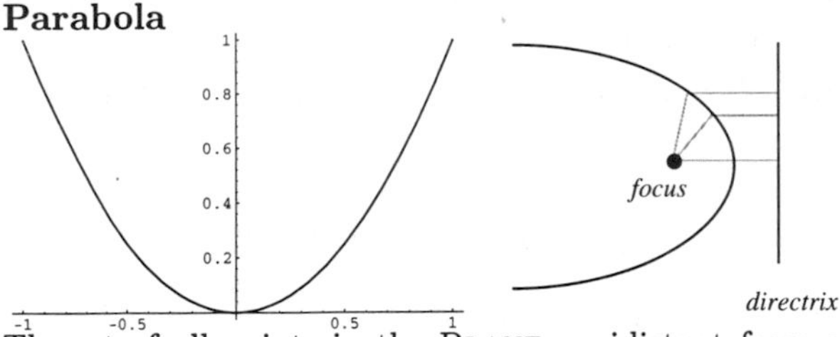

The set of all points in the PLANE equidistant from a given LINE (the DIRECTRIX) and a given point not on the line (the FOCUS).

The parabola was studied by Menaechmus in an attempt to achieve CUBE DUPLICATION. Menaechmus solved the problem by finding the intersection of the two parabolas $x^2 = y$ and $y^2 = 2x$. Euclid wrote about the parabola, and it was given its present name by Apollonius. Pascal considered the parabola as a projection of a CIRCLE, and Galileo showed that projectiles falling under uniform gravity follow parabolic paths. Gregory and Newton considered the CATACAUSTIC properties of a parabola which bring parallel rays of light to a focus (MacTutor Archive).

For a parabola opening to the right, the equation in CARTESIAN COORDINATES is

$$\sqrt{(x-p)^2+y^2} = x+p \tag{1}$$

$$(x-p)^2+y^2 = (x+p)^2 \tag{2}$$

$$x^2-2px+p^2+y^2 = x^2+2px+p^2 \tag{3}$$

$$y^2 = 4px. \tag{4}$$

If the VERTEX is at (x_0, y_0) instead of (0, 0), the equation is

$$(y-y_0)^2 = 4p(x-x_0). \tag{5}$$

If the parabola opens upwards,

$$x^2 = 4py \tag{6}$$

(which is the form shown in the above figure at left). The quantity $4p$ is known as the LATUS RECTUM. In POLAR COORDINATES,

$$r = \frac{2a}{1-\cos\theta}. \tag{7}$$

In PEDAL COORDINATES with the PEDAL POINT at the FOCUS, the equation is

$$p^2 = ar. \tag{8}$$

The parametric equations for the parabola are

$$x = 2at \tag{9}$$

$$y = at^2. \tag{10}$$

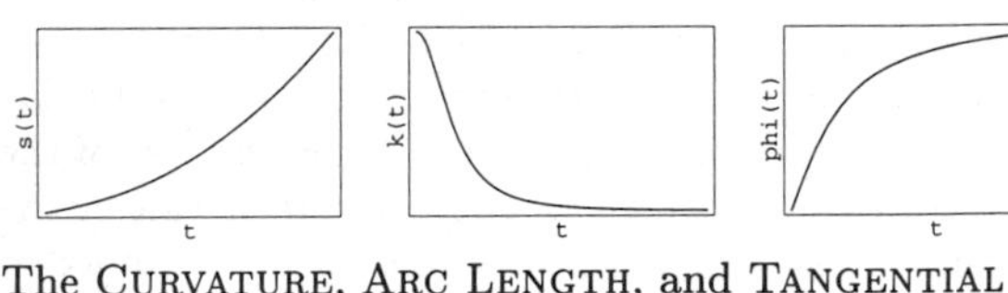

The CURVATURE, ARC LENGTH, and TANGENTIAL ANGLE are

$$\kappa(t) = \frac{1}{2(1+t^2)^{3/2}} \tag{11}$$

$$s(t) = t\sqrt{1+t^2}+\sinh^{-1}t \tag{12}$$

$$\phi(t) = \tan^{-1}t. \tag{13}$$

The TANGENT VECTOR of the parabola is

$$x_T(t) = \frac{1}{\sqrt{1+t^2}} \tag{14}$$

$$y_T(t) = \frac{t}{\sqrt{1+t^2}}. \tag{15}$$

The plots below show the normal and tangent vectors to a parabola.

A system of CURVILINEAR COORDINATES. There are several different conventions for the orientation and designation of these coordinates. Arfken (1970) defines coordinates (ξ, η, z) such that

$$x = \xi\eta \quad (1)$$

$$y = \tfrac{1}{2}(\eta^2 - \xi^2) \quad (2)$$

$$z = z. \quad (3)$$

In this work, following Morse and Feshbach (1953), the coordinates (u, v, z) are used instead. In this convention, the traces of the coordinate surfaces of the xy-PLANE are confocal PARABOLAS with a common axis. The u curves open into the NEGATIVE x-AXIS; the v curves open into the POSITIVE x-AXIS. The u and v curves intersect along the y-AXIS.

$$x = \tfrac{1}{2}(u^2 - v^2) \quad (4)$$

$$y = uv \quad (5)$$

$$z = z, \quad (6)$$

where $u \in [0, \infty)$, $v \in [0, \infty)$, and $z \in (-\infty, \infty)$. The SCALE FACTORS are

$$h_1 = \sqrt{u^2 + v^2} \quad (7)$$

$$h_2 = \sqrt{u^2 + v^2} \quad (8)$$

$$h_3 = 1. \quad (9)$$

LAPLACE'S EQUATION is

$$\nabla^2 f = \frac{1}{u^2 + v^2}\left(\frac{\partial^2 f}{\partial u^2} + \frac{\partial^2 f}{\partial v^2}\right) + \frac{\partial^2 f}{\partial z^2}. \quad (10)$$

The HELMHOLTZ DIFFERENTIAL EQUATION is SEPARABLE in parabolic cylindrical coordinates.

see also CONFOCAL PARABOLOIDAL COORDINATES, HELMHOLTZ DIFFERENTIAL EQUATION—PARABOLIC CYLINDRICAL COORDINATES, PARABOLIC COORDINATES

References

Arfken, G. "Parabolic Cylinder Coordinates (ξ, η, z)." §2.8 in *Mathematical Methods for Physicists, 2nd ed.* Orlando, FL: Academic Press, p. 97, 1970.

Morse, P. M. and Feshbach, H. *Methods of Theoretical Physics, Part I.* New York: McGraw-Hill, p. 658, 1953.

Parabolic Fixed Point

A FIXED POINT of a LINEAR TRANSFORMATION for which the rescaled variables satisfy

$$(\delta - \alpha)^2 + 4\beta\gamma = 0.$$

see also ELLIPTIC FIXED POINT (MAP), HYPERBOLIC FIXED POINT (MAP), LINEAR TRANSFORMATION

Parabolic Geometry

see EUCLIDEAN GEOMETRY

Parabolic Horn Cyclide

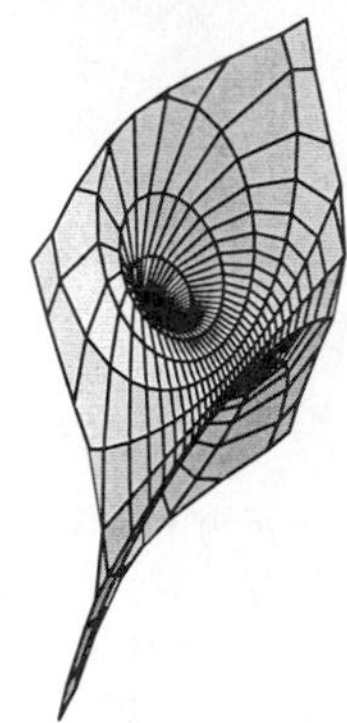

A PARABOLIC CYCLIDE formed by inversion of a HORN TORUS when the inversion sphere is tangent to the TORUS.

see also CYCLIDE, PARABOLIC RING CYCLIDE, PARABOLIC SPINDLE CYCLIDE

Parabolic Partial Differential Equation

A PARTIAL DIFFERENTIAL EQUATION of second-order, i.e., one of the form

$$Au_{xx} + 2Bu_{xy} + Cu_{yy} + Du_x + Eu_y + F = 0, \quad (1)$$

is called parabolic if the MATRIX

$$\mathsf{Z} \equiv \begin{bmatrix} A & B \\ B & C \end{bmatrix} \quad (2)$$

satisfies $\det(\mathsf{Z}) = 0$. The HEAT CONDUCTION EQUATION and other diffusion equations are examples. Initial-boundary conditions are used to give

$$u(x,t) = g(x,t) \quad \text{for } x \in \partial\Omega, t > 0 \quad (3)$$

$$u(x,0) = v(x) \quad \text{for } x \in \Omega, \quad (4)$$

where

$$u_{xx} = f(u_x, u_y, u, x, y) \quad (5)$$

holds in Ω.

see also ELLIPTIC PARTIAL DIFFERENTIAL EQUATION, HYPERBOLIC PARTIAL DIFFERENTIAL EQUATION, PARTIAL DIFFERENTIAL EQUATION

Parabolic Point

A point $\mathbf{p}$ on a REGULAR SURFACE $M \in \mathbb{R}^3$ is said to be parabolic if the GAUSSIAN CURVATURE $K(\mathbf{p}) = 0$ but $S(\mathbf{p}) \neq 0$ (where S is the SHAPE OPERATOR), or equivalently, exactly one of the PRINCIPAL CURVATURES κ_1 and κ_2 is 0.

see also ANTICLASTIC, ELLIPTIC POINT, GAUSSIAN CURVATURE, HYPERBOLIC POINT, PLANAR POINT, SYNCLASTIC

References

Gray, A. *Modern Differential Geometry of Curves and Surfaces.* Boca Raton, FL: CRC Press, p. 280, 1993.

Parabolic Ring Cyclide

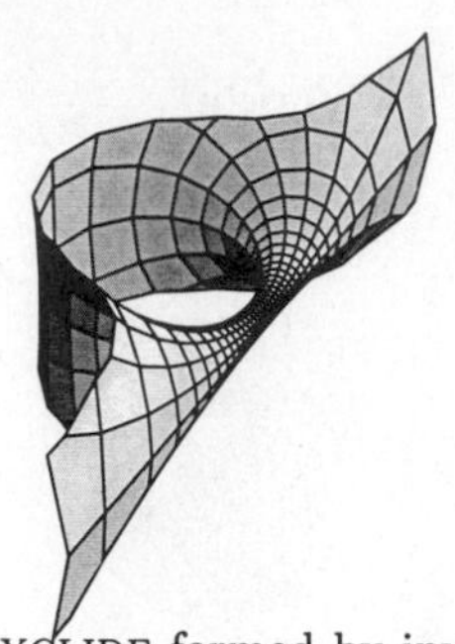

A PARABOLIC CYCLIDE formed by inversion of a RING TORUS when the inversion sphere is tangent to the TORUS.

see also CYCLIDE, PARABOLIC HORN CYCLIDE, PARABOLIC SPINDLE CYCLIDE

Parabolic Rotation

The MAP

$$x' = x + 1 \tag{1}$$
$$y' = 2x + y + 1, \tag{2}$$

which leaves the PARABOLA

$$x'^2 - y' = (x+1)^2 - (2x + y + 1) = x^2 - y \tag{3}$$

invariant.

see also PARABOLA, ROTATION

Parabolic Segment

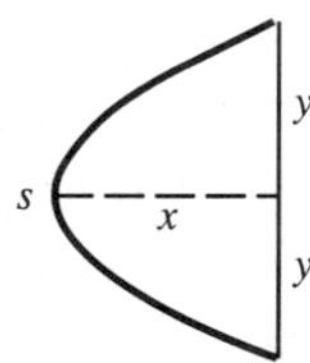

The ARC LENGTH of the parabolic segment shown above is given by

$$s = \sqrt{4x^2 + y^2} + \frac{y^2}{2x} \ln\left(\frac{2x + \sqrt{4x^2 + y^2}}{y}\right). \tag{1}$$

The AREA contained between the curves

$$y = x^2 \tag{2}$$
$$y = ax + b \tag{3}$$

can be found by eliminating y,

$$x^2 - ax - b = 0, \tag{4}$$

so the points of intersection are

$$x_\pm = \tfrac{1}{2}(a \pm \sqrt{a^2 + 4b}). \tag{5}$$

Therefore, for the AREA to be NONNEGATIVE, $a^2 + 4b > 0$, and

$$\begin{aligned} x_\pm &= \tfrac{1}{4}(a^2 \pm 2a\sqrt{a^2 + b^2} + a^2 + 4b) \\ &= \tfrac{1}{4}(2a^2 + 4b \pm 2a\sqrt{a^2 + 4b}) \\ &= \tfrac{1}{2}(a^2 + 2b \pm a\sqrt{a^2 + 4b}), \end{aligned} \tag{6}$$

so the AREA is

$$\begin{aligned} A &= \int_{x_-}^{x_+} [(ax + b) - x^2]\,dx \\ &= \left[\tfrac{1}{2}ax^2 + bx - \tfrac{1}{3}x^3\right]_{(a-\sqrt{a^2+4b})/2}^{(a+\sqrt{a^2+4b})/2}. \end{aligned} \tag{7}$$

Now,

$$\begin{aligned} {x_+}^2 - {x_-}^2 &= \tfrac{1}{4}\left[(a^2 + 2a\sqrt{a^2 + 4b} + a^2 + 4b)\right. \\ &\quad \left. - (a^2 - 2a\sqrt{a^2 + 4b} + a^2 + 4b)\right] \\ &= \tfrac{1}{4}\left[4a\sqrt{a^2 + 4b}\right] = a\sqrt{a^2 + 4b} \end{aligned} \tag{8}$$

$$\begin{aligned} {x_+}^3 - {x_-}^3 &= (x_+ - x_-)({x_+}^2 + x_- x_+ + {x_-}^2) \\ &= \sqrt{a^2 + 4b}\left\{\tfrac{1}{4}(a^2 + 2a\sqrt{a^2 + 4b} + a^2 + 4b)\right. \\ &\quad \left. + \tfrac{1}{4}[a^2 - (a^2 + 4b)] + \tfrac{1}{4}(a^2 - 2a\sqrt{a^2 + 4b} + a^2 + 4b)\right\} \\ &= \tfrac{1}{4}\sqrt{a^2 + 4b}\,(4a^2 + 4b) = \sqrt{a^2 + 4b}\,(a^2 + b). \end{aligned} \tag{9}$$

So

$$\begin{aligned} A &= \tfrac{1}{2}a^2\sqrt{a^2 + 4b} + b\sqrt{a^2 + 4b} = \tfrac{1}{3}(a^2 + b)\sqrt{a^2 + 4b} \\ &= \sqrt{a^2 + 4b}\left[(\tfrac{1}{2} - \tfrac{1}{3})a^2 + b(1 - \tfrac{1}{3})\right] \\ &= (\tfrac{1}{6}a^2 + \tfrac{2}{3}b)\sqrt{a^2 + 4b} \\ &= \tfrac{1}{6}(a^2 + 4b)\sqrt{a^2 + 4b} = \tfrac{1}{6}(a^2 + 4b)^{3/2}. \end{aligned} \tag{10}$$

We now wish to find the maximum AREA of an inscribed TRIANGLE. This TRIANGLE will have two of its VERTICES at the intersections, and AREA

$$A_\Delta = \tfrac{1}{2}(x_- y_* - x_* y_- - x_+ y_* + x_* y_+ + x_+ y_- - x_- y_+). \tag{11}$$

But $y_* = {x_*}^2$, so

$$\begin{aligned} A_\Delta &= \tfrac{1}{2}(x_- {x_*}^2 - x_* y_- - x_+ {x_*}^2 \\ &\quad + x_* y_* + x_+ y_- - x_- y_+) \\ &= \tfrac{1}{2}[-{x_*}^2(x_+ - x_-) + x_*(y_+ - y_-) \\ &\quad + (x_+ y_- - x_- y_+)]. \end{aligned} \tag{12}$$

The maximum AREA will occur when

$$\frac{\partial A_\Delta}{\partial x_*} = \tfrac{1}{2}[-2(x_+ - x_-)x_* + (y_+ - y_-)] = 0. \tag{13}$$

But

$$x_+ - x_- = \sqrt{a^2+4b} \tag{14}$$

$$y_+ - y_- = a\sqrt{a^2+4b}, \tag{15}$$

so

$$x_* = \frac{1}{2}\frac{y_+ - y_-}{x_+ - x_-} = \tfrac{1}{2}a \tag{16}$$

and

$$A_\Delta = \tfrac{1}{2}[-(\tfrac{1}{2}a)^2(x_+ - x_-) + (\tfrac{1}{2}a)(y_+ - y_-) + (x_+y_- - x_-y_+)]. \tag{17}$$

Working on the third term

$$\begin{aligned} x_+y_- &= \tfrac{1}{4}(a+\sqrt{a^2+4b})(a^2+2b-a\sqrt{a^2+4b}) \\ &= \tfrac{1}{4}\Big[a^3+2ab-a^2\sqrt{a^2+4b}+a^2\sqrt{a^2+4b} \\ &\quad +2b\sqrt{a^2+4b}-a(a^2+4b)\Big] \\ &= \tfrac{1}{4}[-2ab+2b\sqrt{a^2+4b}] \end{aligned} \tag{18}$$

$$\begin{aligned} x_-y_+ &= \tfrac{1}{4}(a-\sqrt{a^2+4b})(a^2+2b+a\sqrt{a^2+4b}) \\ &= \tfrac{1}{4}\Big[a^3+2ab+a^2\sqrt{a^2+4b}-a^2\sqrt{a^2+4b} \\ &\quad -2b\sqrt{a^2+4b}-a(a^2+4b)\Big] \\ &= \tfrac{1}{4}[-2ab-2b\sqrt{a^2+4b}], \end{aligned} \tag{19}$$

so

$$x_+y_- - x_-y_+ = \tfrac{1}{4}(4b\sqrt{a^2+4b}) = b\sqrt{a^2+4b} \tag{20}$$

and

$$\begin{aligned} A_\Delta &= \tfrac{1}{2}(-\tfrac{1}{4}a^2\sqrt{a^2+4b}+\tfrac{1}{2}a^2\sqrt{a^2+4b}+b\sqrt{a^2+b^2}) \\ &= \tfrac{1}{2}\sqrt{a^2+4b}\left[(\tfrac{1}{2}-\tfrac{1}{4})a^2+b\right] = \tfrac{1}{2}\sqrt{a^2+4b}\,(\tfrac{1}{4}a^2+b) \\ &= \tfrac{1}{8}\sqrt{a^2+4b}\,(a^2+4b) = \tfrac{1}{8}(a^2+4b)^{3/2}, \end{aligned} \tag{21}$$

which gives the result known to Archimedes in the third century BC that

$$\frac{A}{A_\Delta} = \frac{\frac{1}{6}}{\frac{1}{8}} = \frac{4}{3}. \tag{22}$$

The AREA of the parabolic segment of height h opening upward along the y-AXIS is

$$A = 2\int_0^h \sqrt{y}\,dy = \tfrac{1}{3}h^{3/2}. \tag{23}$$

The weighted mean of y is

$$\langle y\rangle = 2\int_0^h y\sqrt{y}\,dy = 2\int_0^h y^{3/2}\,dy = \tfrac{4}{5}h^{5/2}. \tag{24}$$

The CENTROID is then given by

$$\bar{y} = \frac{\langle y\rangle}{A} = \tfrac{3}{5}h. \tag{25}$$

see also CENTROID (GEOMETRIC), PARABOLA, SEGMENT

References

Beyer, W. H. (Ed.) *CRC Standard Mathematical Tables, 28th ed.* Boca Raton, FL: CRC Press, p. 125, 1987.

Parabolic Spindle Cyclide

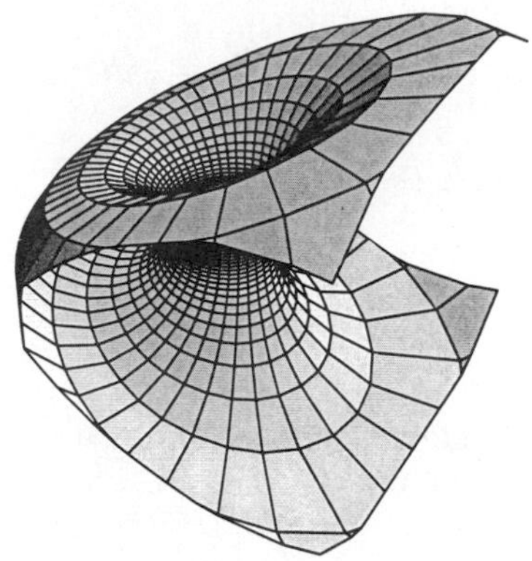

A PARABOLIC CYCLIDE formed by inversion of a SPINDLE TORUS when the inversion sphere is tangent to the TORUS.

see also CYCLIDE, PARABOLIC HORN CYCLIDE, PARABOLIC RING CYCLIDE

Parabolic Spiral

see FERMAT'S SPIRAL

Parabolic Umbilic Catastrophe

A CATASTROPHE which can occur for four control factors and two behavior axes.

Paraboloid

The SURFACE OF REVOLUTION of the PARABOLA. It is a QUADRATIC SURFACE which can be specified by the Cartesian equation

$$z = a(x^2+y^2), \tag{1}$$

or parametrically by

$$x(u,v) = \sqrt{u}\cos v \tag{2}$$

$$y(u,v) = \sqrt{u}\sin v \tag{3}$$

$$z(u,v) = u, \tag{4}$$

where $u \in [0,h]$, $v \in [0,2\pi)$, and h is the height.

The VOLUME of the paraboloid is

$$V = \pi\int_0^h z\,dz = \tfrac{1}{2}\pi h^2. \tag{5}$$

The weighted mean of z over the paraboloid is

$$\langle z \rangle = \pi \int_0^h z^2 \, dz = \tfrac{1}{3}\pi h^3. \qquad (6)$$

The CENTROID is then given by

$$\bar{z} = \frac{\langle z \rangle}{V} = \tfrac{2}{3}h \qquad (7)$$

(Beyer 1987).

see also ELLIPTIC PARABOLOID, HYPERBOLIC PARABOLOID, PARABOLA

References

Beyer, W. H. (Ed.) *CRC Standard Mathematical Tables, 28th ed.* Boca Raton, FL: CRC Press, p. 133, 1987.

Gray, A. "The Paraboloid." §11.5 in *Modern Differential Geometry of Curves and Surfaces.* Boca Raton, FL: CRC Press, pp. 221–222, 1993.

Paraboloid Geodesic

A GEODESIC on a PARABOLOID has differential parameters defined by

$$\begin{aligned} P &\equiv \left(\frac{\partial x}{\partial u}\right)^2 + \left(\frac{\partial y}{\partial u}\right)^2 + \left(\frac{\partial z}{\partial u}\right)^2 \\ &= 1 + \frac{\cos^2 v}{4u} + \frac{\sin^2 v}{4u} = 1 + \frac{1}{4u} \qquad (1) \\ Q &\equiv \frac{\partial^2 x}{\partial u \partial v} + \frac{\partial^2 y}{\partial u \partial v} + \frac{\partial^2 z}{\partial u \partial v} \\ &= 0 + u\cos^2 v + u\sin^2 v = u \qquad (2) \\ R &\equiv 0 - \frac{\sin v}{2\sqrt{u}} + \frac{\cos v}{2\sqrt{u}} = \frac{1}{2\sqrt{u}}(\cos v - \sin v). \qquad (3) \end{aligned}$$

The GEODESIC is then given by solving the EULER-LAGRANGE DIFFERENTIAL EQUATION

$$\frac{\frac{\partial P}{\partial v} + 2v'\frac{\partial Q}{\partial v} + v'^2\frac{\partial R}{\partial v}}{2\sqrt{P + 2Qv' + Rv'^2}} - \frac{d}{du}\left(\frac{Q + Rv'}{\sqrt{P + 2Qv' + Rv'^2}}\right) = 0. \qquad (4)$$

As given by Weinstock (1974), the solution simplifies to

$$\begin{aligned} &u - c^2 \\ &= u(1 + 4c^2)\sin^2\{v - 2c\ln[k(2\sqrt{u - c^2} + \sqrt{4u + 1})]\}. \qquad (5) \end{aligned}$$

see also GEODESIC

References

Weinstock, R. *Calculus of Variations, with Applications to Physics and Engineering.* New York: Dover, p. 45, 1974.

Paraboloidal Coordinates

see CONFOCAL PARABOLOIDAL COORDINATES

Paracompact Space

A paracompact space is a HAUSDORFF SPACE such that every open COVER has a LOCALLY FINITE open REFINEMENT. Paracompactness is a very common property that TOPOLOGICAL SPACES satisfy. Paracompactness is similar to the compactness property, but generalized for slightly "bigger" SPACES. All MANIFOLDS (e.g, second countable and Hausdorff) are paracompact.

see also HAUSDORFF SPACE, LOCALLY FINITE SPACE, MANIFOLD, TOPOLOGICAL SPACE

Paracycle

see ASTROID

Paradox

A statement which appears self-contradictory or contrary to expectations, also known as an ANTINOMY. Bertrand Russell classified known logical paradoxes into seven categories.

Ball and Coxeter (1987) give several examples of geometrical paradoxes.

see also ALIAS' PARADOX, ARISTOTLE'S WHEEL PARADOX, ARROW'S PARADOX, BANACH-TARSKI PARADOX, BARBER PARADOX, BERNOULLI'S PARADOX, BERRY PARADOX, BERTRAND'S PARADOX, CANTOR'S PARADOX, COASTLINE PARADOX, COIN PARADOX, ELEVATOR PARADOX, EPIMENIDES PARADOX, EUBULIDES PARADOX, GRELLING'S PARADOX, HAUSDORFF PARADOX, HEMPEL'S PARADOX, HETEROLOGICAL PARADOX, LEONARDO'S PARADOX, LIAR'S PARADOX, LOGICAL PARADOX, POTATO PARADOX, RICHARD'S PARADOX, RUSSELL'S PARADOX, SAINT PETERSBURG PARADOX, SIEGEL'S PARADOX, SIMPSON'S PARADOX, SKOLEM PARADOX, SMARANDACHE PARADOX, SOCRATES' PARADOX, SORITES PARADOX, THOMSON LAMP PARADOX, UNEXPECTED HANGING PARADOX, ZEEMAN'S PARADOX, ZENO'S PARADOXES

References

Ball, W. W. R. and Coxeter, H. S. M. *Mathematical Recreations and Essays, 13th ed.* New York: Dover, pp. 84–86, 1987.

Bunch, B. *Mathematical Fallacies and Paradoxes.* New York: Dover, 1982.

Carnap, R. *Introduction to Symbolic Logic and Its Applications.* New York: Dover, 1958.

Curry, H. B. *Foundations of Mathematical Logic.* New York: Dover, 1977.

Kasner, E. and Newman, J. R. "Paradox Lost and Paradox Regained." In *Mathematics and the Imagination.* Redmond, WA: Tempus Books, pp. 193–222, 1989.

Northrop, E. P. *Riddles in Mathematics: A Book of Paradoxes.* Princeton, NJ: Van Nostrand, 1944.

O'Beirne, T. H. *Puzzles and Paradoxes.* New York: Oxford University Press, 1965.

Quine, W. V. "Paradox." *Sci. Amer.* **206**, 84–96, Apr. 1962.

Paradromic Rings

Rings produced by cutting a strip that has been given m half twists and been re-attached into n equal strips (Ball and Coxeter 1987, pp. 127–128).

see also MÖBIUS STRIP

References

Ball, W. W. R. and Coxeter, H. S. M. *Mathematical Recreations and Essays, 13th ed.* New York: Dover, pp. 127–128, 1987.

Paragyrate Diminished Rhombicosidodecahedron

see JOHNSON SOLID

Parallel

Two lines in 2-dimensional EUCLIDEAN SPACE are said to be parallel if they do not intersect. In 3-dimensional EUCLIDEAN SPACE, parallel lines not only fail to intersect, but also maintain a constant separation between points closest to each other on the two lines. (Lines in 3-space which are not parallel but do not intersect are called SKEW LINES.)

In a NON-EUCLIDEAN GEOMETRY, the concept of parallelism must be modified from its intuitive meaning. This is accomplished by changing the so-called PARALLEL POSTULATE. While this has counterintuitive results, the geometries so defined are still completely self-consistent.

see also ANTIPARALLEL, HYPERPARALLEL, LINE, NON-EUCLIDEAN GEOMETRY, PARALLEL CURVE, PARALLEL POSTULATE PERPENDICULAR, SKEW LINES

Parallel Axiom

see PARALLEL POSTULATE

Parallel Class

A set of blocks, also called a RESOLUTION CLASS, that partition the set V, where (V, B) is a balanced incomplete BLOCK DESIGN.

see also BLOCK DESIGN, RESOLVABLE

References

Abel, R. J. R. and Furino, S. C. "Resolvable and Near Resolvable Designs." §I.6 in *The CRC Handbook of Combinatorial Designs* (Ed. C. J. Colbourn and J. H. Dinitz). Boca Raton, FL: CRC Press, pp. 87–94, 1996.

Parallel Curve

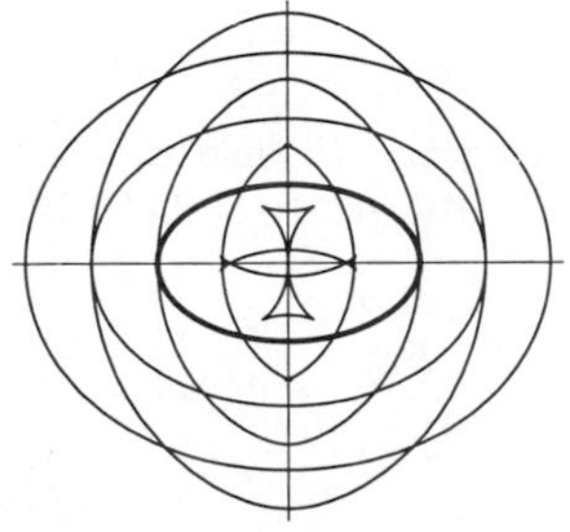

The two branches of the parallel curve a distance k away from a parametrically represented curve $(f(t), g(t))$ are

$$x = f \pm \frac{kg'}{\sqrt{f'^2 + g'^2}}$$

$$y = g \mp \frac{kf'}{\sqrt{f'^2 + g'^2}}.$$

The above figure shows the curves parallel to the ELLIPSE.

References

Gray, A. "Parallel Curves." §5.7 in *Modern Differential Geometry of Curves and Surfaces.* Boca Raton, FL: CRC Press, pp. 95–97, 1993.

Lawrence, J. D. *A Catalog of Special Plane Curves.* New York: Dover, pp. 42–43, 1972.

Lee, X. "Parallel." http://www.best.com/~xah/SpecialPlaneCurves_dir/Parallel_dir/parallel.html.

Yates, R. C. "Parallel Curves." *A Handbook on Curves and Their Properties.* Ann Arbor, MI: J. W. Edwards, pp. 155–159, 1952.

Parallel Postulate

Given any straight line and a point not on it, there "exists one and only one straight line which passes" through that point and never intersects the first line, no matter how far they are extended. This statement is equivalent to the fifth of EUCLID'S POSTULATES, which Euclid himself avoided using until proposition 29 in the *Elements.* For centuries, many mathematicians believed that this statement was not a true postulate, but rather a theorem which could be derived from the first four of EUCLID'S POSTULATES. (That part of geometry which could be derived using only postulates 1–4 came to be known as ABSOLUTE GEOMETRY.)

Over the years, many purported proofs of the parallel postulate were published. However, none were correct, including the 28 "proofs" G. S. Klügel analyzed in his dissertation of 1763 (Hofstadter 1989). In 1823, Janos Bolyai and Lobachevsky independently realized that entirely self-consistent "NON-EUCLIDEAN GEOMETRIES" could be created in which the parallel postulate *did not hold.* (Gauss had also discovered but suppressed the existence of non-Euclidean geometries.)

As stated above, the parallel postulate describes the type of geometry now known as PARABOLIC GEOMETRY. If, however, the phrase "exists one and only one straight line which passes" is replace by "exist no line which passes," or "exist at least two lines which pass," the postulate describes equally valid (though less intuitive) types of geometries known as ELLIPTIC and HYPERBOLIC GEOMETRIES, respectively.

The parallel postulate is equivalent to the EQUIDISTANCE POSTULATE, PLAYFAIR'S AXIOM, PROCLUS' AXIOM, TRIANGLE POSTULATE. There is also a single parallel axiom in HILBERT'S AXIOMS which is equivalent to Euclid's parallel postulate.

see also ABSOLUTE GEOMETRY, EUCLID'S AXIOMS, EUCLIDEAN GEOMETRY, HILBERT'S AXIOMS, NON-EUCLIDEAN GEOMETRY, PLAYFAIR'S AXIOM, TRIANGLE POSTULATE

References

Dixon, R. *Mathographics.* New York: Dover, p. 27, 1991.

Hilbert, D. *The Foundations of Geometry, 2nd ed.* Chicago, IL: Open Court, 1980.

Hofstadter, D. R. *Gödel, Escher, Bach: An Eternal Golden Braid.* New York: Vintage Books, pp. 88–92, 1989.

Iyanaga, S. and Kawada, Y. (Eds.). "Hilbert's System of Axioms." §163B in *Encyclopedic Dictionary of Mathematics.* Cambridge, MA: MIT Press, pp. 544–545, 1980.

Parallel (Surface of Revolution)

A parallel of a SURFACE OF REVOLUTION is the intersection of the surface with a PLANE orthogonal to the axis of revolution.

see also MERIDIAN, SURFACE OF REVOLUTION

References

Gray, A. *Modern Differential Geometry of Curves and Surfaces.* Boca Raton, FL: CRC Press, p. 358, 1993.

Parallelepiped

In 3-D, a parallelepiped is a PRISM whose faces are all PARALLELOGRAMS. The volume of a 3-D parallelepiped is given by the SCALAR TRIPLE PRODUCT

$$\begin{aligned} V_{\text{parallelepiped}} &= |\mathbf{B}\cdot(\mathbf{B}\times\mathbf{C})| \\ &= |\mathbf{C}\cdot(\mathbf{A}\times\mathbf{B})| = |\mathbf{B}\cdot(\mathbf{C}\times\mathbf{A})|. \end{aligned}$$

In n-D, a parallelepiped is the POLYTOPE spanned by n VECTORS $\mathbf{v}_1, \ldots, \mathbf{v}_n$ in a VECTOR SPACE over the reals,

$$\operatorname{span}(\mathbf{v}_1,\ldots,\mathbf{v}_n) = t_1\mathbf{v}_1 + \ldots + t_n\mathbf{v}_n,$$

where $t_i \in [0,1]$ for $i = 1, \ldots, n$. In the usual interpretation, the VECTOR SPACE is taken as EUCLIDEAN SPACE, and the CONTENT of this parallelepiped is given by

$$\operatorname{abs}(\det(\mathbf{v}_1,\ldots,\mathbf{v}_n)),$$

where the sign of the determinant is taken to be the "orientation" of the "oriented volume" of the parallelepiped.

see also PRISMATOID, RECTANGULAR PARALLELEPIPED, ZONOHEDRON

References

Phillips, A. W. and Fisher, I. *Elements of Geometry.* New York: Amer. Book Co., 1896.

Parallelism

see ANGLE OF PARALLELISM

Parallelizable

A sphere $\mathbb{S}^n$ is parallelizable if there exist n cuts containing linearly independent tangent vectors. There exist only three parallelizable spheres: $\mathbb{S}^1$, $\mathbb{S}^3$, and $\mathbb{S}^7$ (Adams 1962, Le Lionnais 1983).

see also SPHERE

References

Adams, J. F. "On the Non-Existence of Elements of Hopf Invariant One." *Bull. Amer. Math. Soc.* **64**, 279–282, 1958.

Adams, J. F. "On the Non-Existence of Elements of Hopf Invariant One." *Ann. Math.* **72**, 20–104, 1960.

Le Lionnais, F. *Les nombres remarquables.* Paris: Hermann, p. 49, 1983.

Parallelogram

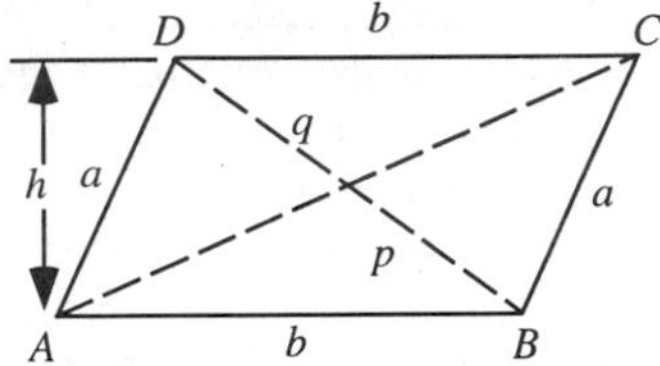

A QUADRILATERAL with opposite sides parallel (and therefore opposite angles equal). A quadrilateral with equal sides is called a RHOMBUS, and a parallelogram whose ANGLES are all RIGHT ANGLES is called a RECTANGLE.

A parallelogram of base b and height h has AREA

$$A = bh = ab\sin A = ab\sin B. \tag{1}$$

The height of a parallelogram is

$$h = a\sin A = a\sin B, \tag{2}$$

and the DIAGONALS are

$$p = \sqrt{a^2+b^2-2ab\cos A} \tag{3}$$

$$q = \sqrt{a^2+b^2-2ab\cos B} \tag{4}$$

$$= \sqrt{a^2+b^2+2ab\cos A} \tag{5}$$

(Beyer 1987).

The AREA of the parallelogram with sides formed by the VECTORS (a,c) and (b,d) is

$$A = \det\left(\begin{bmatrix} a & b \\ c & d \end{bmatrix}\right) = |ad-bc|. \tag{6}$$

Given a parallelogram P with area $A(P)$ and linear transformation T, the AREA of $T(P)$ is

$$A(T(P)) = \begin{vmatrix} a & b \\ c & d \end{vmatrix} A(P). \tag{7}$$

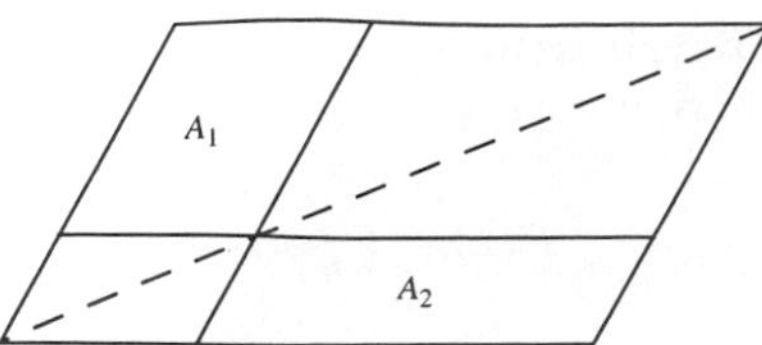

As shown by Euclid, if lines parallel to the sides are drawn through any point on a diagonal of a parallelogram, then the parallelograms not containing segments of that diagonal are equal in AREA (and conversely), so in the above figure, $A_1 = A_2$ (Johnson 1929).

see also DIAMOND, LOZENGE, PARALLELOGRAM ILLUSION, RECTANGLE, RHOMBUS, VARIGNON PARALLELOGRAM, WITTENBAUER'S PARALLELOGRAM

References

Beyer, W. H. (Ed.) *CRC Standard Mathematical Tables, 28th ed.* Boca Raton, FL: CRC Press, p. 123, 1987.

Johnson, R. A. *Modern Geometry: An Elementary Treatise on the Geometry of the Triangle and the Circle.* Boston, MA: Houghton Mifflin, p. 61, 1929.

Parallelogram Illusion

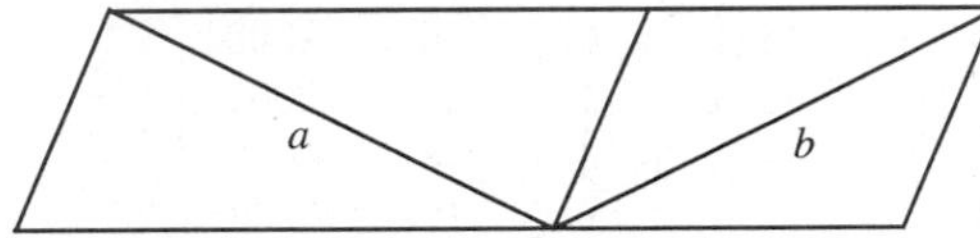

The sides a and b have the same length, appearances to the contrary.

Parallelogram Law

Let $|\cdot|$ denote the NORM of a quantity. Then the quantities x and y satisfy the parallelogram law if

$$\|x+y\|^2 + \|x-y\|^2 = 2\|x\|^2 + 2\|y\|^2.$$

If the NORM is defined as $|f| = \sqrt{\langle f|f\rangle}$ (the so-called L_2-NORM), then the law will always hold.

see also L_2-NORM, NORM

Parallelohedron

A special class of ZONOHEDRON. There are five parallelohedra with an infinity of equal and similarly situated replicas which are SPACE-FILLING POLYHEDRA: the CUBE, ELONGATED DODECAHEDRON, hexagonal PRISM, RHOMBIC DODECAHEDRON, and TRUNCATED OCTAHEDRON.

see also PARALLELOTOPE, SPACE-FILLING POLYHEDRON

References

Coxeter, H. S. M. *Regular Polytopes, 3rd ed.* New York: Dover, p. 29, 1973.

Parallelotope

Move a point Π_0 along a LINE for an initial point to a final point. It traces out a LINE SEGMENT Π_1. When Π_1 is translated from an initial position to a final position, it traces out a PARALLELOGRAM Π_2. When Π_2 is translated, it traces out a PARALLELEPIPED Π_3. The generalization of Π_n to n-D is then called a parallelotope. Π_n has 2^n vertices and

$$N_k = 2^{n-k}\binom{n}{k}$$

Π_ks, where $\binom{n}{k}$ is a BINOMIAL COEFFICIENT and $k = 0$, 1, ..., n (Coxeter 1973). These are also the coefficients of $(2k+1)^n$.

see also HONEYCOMB, HYPERCUBE, ORTHOTOPE, PARALLELOHEDRON

References

Coxeter, H. S. M. *Regular Polytopes, 3rd ed.* New York: Dover, pp. 122–123, 1973.

Klee, V. and Wagon, S. *Old and New Unsolved Problems in Plane Geometry and Number Theory.* Washington, DC: Math. Assoc. Amer., 1991.

Zaks, J. "Neighborly Families of Congruent Convex Polytopes." *Amer. Math. Monthly* **94**, 151–155, 1987.

Paralogic Triangles

At the points where a line cuts the sides of a TRIANGLE $\Delta A_1A_2A_3$, perpendiculars to the sides are drawn, forming a TRIANGLE $\Delta B_1B_2B_3$ similar to the given TRIANGLE. The two triangles are also in perspective. One point of intersection of their CIRCUMCIRCLES is the SIMILITUDE CENTER, and the other is the PERSPECTIVE CENTER. The CIRCUMCIRCLES meet ORTHOGONALLY.

see also CIRCUMCIRCLE, ORTHOGONAL CIRCLES, PERSPECTIVE CENTER, SIMILITUDE CENTER

References

Johnson, R. A. *Modern Geometry: An Elementary Treatise on the Geometry of the Triangle and the Circle.* Boston, MA: Houghton Mifflin, pp. 258–262, 1929.

Parameter

A parameter m used in ELLIPTIC INTEGRALS defined to be $m \equiv k^2$, where k is the MODULUS. An ELLIPTIC INTEGRAL is written $I(\phi|m)$ when the parameter is used. The complementary parameter is defined by

$$m' \equiv 1 - m, \tag{1}$$

where m is the parameter. Let q be the NOME, k the MODULUS, and $m = k^2$ the PARAMETER. Then

$$q(m) = e^{-\pi K'(m)/K(m)} \tag{2}$$

where $K(m)$ is the complete ELLIPTIC INTEGRAL OF THE FIRST KIND. Then the inverse of $q(m)$ is given by

$$m(q) = \frac{{\vartheta_2}^4(q)}{{\vartheta_3}^4(q)}, \tag{3}$$

where ϑ_i is a THETA FUNCTION.

see also AMPLITUDE, CHARACTERISTIC (ELLIPTIC INTEGRAL), ELLIPTIC INTEGRAL, ELLIPTIC INTEGRAL OF THE FIRST KIND, MODULAR ANGLE, MODULUS (ELLIPTIC INTEGRAL), NOME, PARAMETER, THETA FUNCTION

References

Abramowitz, M. and Stegun, C. A. (Eds.). *Handbook of Mathematical Functions with Formulas, Graphs, and Mathematical Tables, 9th printing.* New York: Dover, p. 590, 1972.

Parameter (Quadric)

The number θ in the QUADRIC

$$\frac{x^2}{a^2+\theta}+\frac{y^2}{b^2+\theta}+\frac{z^2}{c^2+\theta}=1$$

is called the parameter.

see also QUADRIC

Parameterization

The specification of a curve, surface, etc., by means of one or more variables which are allowed to take on values in a given specified range.

see also ISOTHERMAL PARAMETERIZATION, REGULAR PARAMETERIZATION, SURFACE PARAMETERIZATION

Parametric Latitude

An AUXILIARY LATITUDE also called the REDUCED LATITUDE and denoted η or θ. It gives the LATITUDE on a SPHERE of RADIUS a for which the parallel has the same radius as the parallel of geodetic latitude ϕ and the ELLIPSOID through a given point. It is given by

$$\eta=\tan^{-1}(\sqrt{1-e^2}\,\tan\phi).$$

In series form,

$$\eta=\phi-e_1\sin(2\phi)+\tfrac{1}{2}{e_1}^2\sin(4\phi)-\tfrac{1}{3}{e_1}^3\sin(6\phi)+\ldots,$$

where

$$e_1\equiv\frac{1-\sqrt{1-e^2}}{1+\sqrt{1-e^2}}.$$

see also AUXILIARY LATITUDE, ELLIPSOID, LATITUDE, SPHERE

References

Adams, O. S. "Latitude Developments Connected with Geodesy and Cartography with Tables, Including a Table for Lambert Equal-Area Meridional Projections." Spec. Pub. No. 67. U. S. Coast and Geodetic Survey, 1921.

Snyder, J. P. *Map Projections—A Working Manual.* U. S. Geological Survey Professional Paper 1395. Washington, DC: U. S. Government Printing Office, p. 18, 1987.

Parametric Test

A STATISTICAL TEST in which assumptions are made about the underlying distribution of observed data.

Pareto Distribution

The distribution

$$P(x)=\left(\frac{x}{b}\right)^{a+2}.$$

References

von Seggern, D. *CRC Standard Curves and Surfaces.* Boca Raton, FL: CRC Press, p. 252, 1993.

Parity

The parity of a number n is the sum of the bits in BINARY representation (mod 2). The parities of the first few integers (starting with 0) are 0, 1, 1, 0, 1, 0, 0, 1, 1, 0, 0, ... (Sloane's A010060) summarized in the following table.

N	Binary	Parity	N	Binary	Parity
1	1	1	11	1011	1
2	10	1	12	1100	0
3	11	0	13	1101	1
4	100	1	14	1110	1
5	101	0	15	1111	0
6	110	0	16	10000	1
7	111	1	17	10001	0
8	1000	1	18	10010	0
9	1001	0	19	10011	1
10	1010	0	20	10100	0

The constant generated by the sequence of parity digits is called the THUE-MORSE CONSTANT.

see also BINARY, THUE-MORSE CONSTANT

References

Sloane, N. J. A. Sequence A010060 in "An On-Line Version of the Encyclopedia of Integer Sequences."

Parity Constant

see THUE-MORSE CONSTANT

Parking Constant

see RÉNYI'S PARKING CONSTANTS

Parodi's Theorem

The EIGENVALUES λ satisfying $P(\lambda)=0$, where $P(\lambda)$ is the CHARACTERISTIC POLYNOMIAL, lie in the unions of the DISKS

$$|z|\leq 1$$

$$|z+b_1|\leq\sum_{j=1}^n|b_j|.$$

References

Gradshteyn, I. S. and Ryzhik, I. M. *Tables of Integrals, Series, and Products, 5th ed.* San Diego, CA: Academic Press, p. 1119, 1979.

Parry Circle

The CIRCLE passing through the ISODYNAMIC POINTS and the CENTROID of a TRIANGLE (Kimberling 1998, pp. 227–228).

see also CENTROID (TRIANGLE), ISODYNAMIC POINTS, PARRY POINT

References

Kimberling, C. "Triangle Centers and Central Triangles." *Congr. Numer.* **129**, 1–295, 1998.

Parry Point

The intersection of the PARRY CIRCLE and the CIRCUMCIRCLE of a TRIANGLE. The TRILINEAR COORDINATES of the Parry point are

$$\frac{a}{2a^2-b^2-c^2}:\frac{b}{2b^2-c^2-a^2}:\frac{c}{2c^2-a^2-b^2}$$

(Kimberling 1998, pp. 227–228).

see also PARRY CIRCLE

References

Kimberling, C. "Parry Point." `http://www.evansville.edu/~ck6/tcenters/recent/parry.html`.

Kimberling, C. "Triangle Centers and Central Triangles." *Congr. Numer.* **129**, 1–295, 1998.

Parseval's Integral

The POISSON INTEGRAL with $n=0$.

$$J_0(z)=\frac{1}{[\Gamma(n+\frac{1}{2})]^2}\int_0^\pi \cos(z\cos\theta)\,d\theta,$$

where $J_0(z)$ is a BESSEL FUNCTION OF THE FIRST KIND and $\Gamma(x)$ is a GAMMA FUNCTION.

Parseval's Relation

Let $F(\nu)$ and $G(\nu)$ be the FOURIER TRANSFORMS of $f(t)$ and $g(t)$, respectively. Then

$$\begin{aligned}&\int_{-\infty}^{\infty} f(t)g^*(t)\,dt\\&=\int_{-\infty}^{\infty}\left[\int_{-\infty}^{\infty} F(\nu)e^{-2\pi i\nu t}\,d\nu\int_{-\infty}^{\infty} G^*(\nu')e^{2\pi i\nu' t}\,d\nu'\right]d\nu'\\&=\int_{-\infty}^{\infty} F(\nu)\int_{-\infty}^{\infty} G^*(\nu')\delta(\nu'-\nu)\,d\nu'\,d\nu\\&=\int_{-\infty}^{\infty} F(\nu)G^*(\nu)\,d\nu.\end{aligned}$$

see also FOURIER TRANSFORM, PARSEVAL'S THEOREM

References

Arfken, G. *Mathematical Methods for Physicists, 3rd ed.* Orlando, FL: Academic Press, p. 425, 1985.

Parseval's Theorem

Let $E(t)$ be a continuous function and $E(t)$ and E_ν be FOURIER TRANSFORM pairs so that

$$E(t)\equiv\int_{-\infty}^{\infty} E_\nu e^{-2\pi i\nu t}\,d\nu \quad (1)$$

$$E^*(t)\equiv\int_{-\infty}^{\infty} E_{\nu'}{}^* e^{2\pi i\nu' t}\,d\nu'. \quad (2)$$

Then

$$\begin{aligned}\int_{-\infty}^{\infty}|E(t)|^2\,dt&=\int_{-\infty}^{\infty} E(t)E^*(t)\,dt\\&=\int_{-\infty}^{\infty}\left[\int_{-\infty}^{\infty} E_\nu e^{-2\pi i\nu t}\,d\nu\int_{-\infty}^{\infty} E_{\nu'}{}^* e^{2\pi i\nu' t}\,d\nu'\right]dt\\&=\int_{-\infty}^{\infty}\int_{-\infty}^{\infty}\int_{-\infty}^{\infty} E_\nu E_{\nu'}{}^* e^{2\pi it(\nu'-\nu)}\,d\nu\,d\nu'\,dt\\&=\int_{-\infty}^{\infty}\int_{-\infty}^{\infty}\int_{-\infty}^{\infty} E_\nu E_{\nu'}{}^* e^{2\pi it(\nu'-\nu)}\,dt\,d\nu\,d\nu'\\&=\int_{-\infty}^{\infty}\int_{-\infty}^{\infty}\delta(\nu'-\nu)E_\nu E_{\nu'}{}^*\,d\nu\,d\nu'\\&=\int_{-\infty}^{\infty} E_\nu E_\nu{}^*\,d\nu=\int_{-\infty}^{\infty}|E_\nu|^2\,d\nu. \quad (3)\end{aligned}$$

where $\delta(x-x_0)$ is the DELTA FUNCTION.

For finite FOURIER TRANSFORM pairs h_k and H_n,

$$\sum_{k=0}^{N-1}|h_k|^2=\frac{1}{N}\sum_{n=0}^{N-1}|H_n|^2. \quad (4)$$

If a function has a FOURIER SERIES given by

$$f(x)=\tfrac{1}{2}a_0+\sum_{n=1}^{\infty}a_n\cos(nx)+\sum_{n=1}^{\infty}b_n\sin(nx), \quad (5)$$

then BESSEL'S INEQUALITY becomes an equality known as Parseval's theorem. From (5),

$$\begin{aligned}[f(x)]^2&=\tfrac{1}{4}{a_0}^2+a_0\sum_{n=1}^{\infty}[a_n\cos(nx)+b_n\sin(nx)]\\&+\sum_{n=1}^{\infty}\sum_{m=1}^{\infty}[a_na_m\cos(nx)\cos(mx)\\&+a_nb_m\cos(nx)\sin(mx)+a_mb_n\sin(nx)\cos(mx)\\&+b_nb_m\sin(nx)\sin(mx)]. \quad (6)\end{aligned}$$

Integrating

$$\int_{-\pi}^{\pi}[f(x)]^2\,dx = \tfrac{1}{4}{a_0}^2\int_{-\pi}^{\pi}dx$$
$$+a_0\int_{-\pi}^{\pi}\sum_{n=1}^{\infty}[a_n\cos(nx)+b_n\sin(nx)]\,dx$$
$$+\int_{-\pi}^{\pi}\sum_{n=1}^{\infty}\sum_{m=1}^{\infty}[a_na_m\cos(nx)\cos(mx)$$
$$+a_nb_m\cos(nx)\sin(mx)+a_mb_n\sin(nx)\cos(mx)$$
$$+b_nb_m\sin(nx)\sin(mx)]\,dx = \tfrac{1}{4}{a_0}^2(2\pi)+0$$
$$+\sum_{n=1}^{\infty}\sum_{m=1}^{\infty}[a_na_m\pi\delta_{nm}+0+0+b_nb_m\pi\delta_{nm}], \tag{7}$$

so

$$\frac{1}{\pi}\int_{-\pi}^{\pi}[f(x)]^2\,dx = \tfrac{1}{2}{a_0}^2+\sum_{n=1}^{\infty}({a_n}^2+{b_n}^2). \tag{8}$$

For a generalized FOURIER SERIES with a COMPLETE BASIS $\{\phi_i\}_{i=1}^{\infty}$, an analogous relationship holds. For a COMPLEX FOURIER SERIES,

$$\frac{1}{2\pi}\int_{-\pi}^{\pi}|f(x)|^2\,dx = \sum_{n=-\infty}^{\infty}|a_n|^2. \tag{9}$$

References

Gradshteyn, I. S. and Ryzhik, I. M. *Tables of Integrals, Series, and Products, 5th ed.* San Diego, CA: Academic Press, p. 1101, 1979.

Part Metric

A METRIC defined by

$$d(z,w) = \sup\left[\left|\frac{\ln u(z)}{u(w)}\right| : u\in H^+\right],$$

where H^+ denotes the POSITIVE HARMONIC FUNCTIONS on a DOMAIN. The part metric is invariant under CONFORMAL MAPS for any DOMAIN.

References

Bear, H. S. "Part Metric and Hyperbolic Metric." *Amer. Math. Monthly* **98**, 109–123, 1991.

Partial Derivative

Partial derivatives are defined as derivatives of a function of multiple variables when all but the variable of interest are held fixed during the differentiation.

$$\frac{\partial f}{\partial x_m} \equiv \lim_{h\to 0}\frac{f(x_1,\ldots,x_m+h,\ldots,x_n)-f(x_1,\ldots,x_m,\ldots,x_n)}{h}. \tag{1}$$

The above partial derivative is sometimes denoted f_{x_m} for brevity. For a "nice" 2-D function $f(x,y)$ (i.e., one for which f, f_x, f_y, f_{xy}, f_{yx} exist and are continuous in a NEIGHBORHOOD (a,b)), then $f_{xy}(a,b)=f_{yx}(a,b)$. Partial derivatives involving more than one variable are called MIXED PARTIAL DERIVATIVES.

For nice functions, mixed partial derivatives must be equal regardless of the order in which the differentiation is performed so, for example,

$$f_{xy} = f_{yx} \tag{2}$$
$$f_{xxy} = f_{xyx} = f_{yxx}. \tag{3}$$

For an EXACT DIFFERENTIAL,

$$df = \left(\frac{\partial f}{\partial x}\right)_y dx + \left(\frac{\partial f}{\partial y}\right)_x dy, \tag{4}$$

so

$$\left(\frac{\partial y}{\partial x}\right)_f = -\frac{\left(\frac{\partial f}{\partial x}\right)_y}{\left(\frac{\partial f}{\partial y}\right)_x}. \tag{5}$$

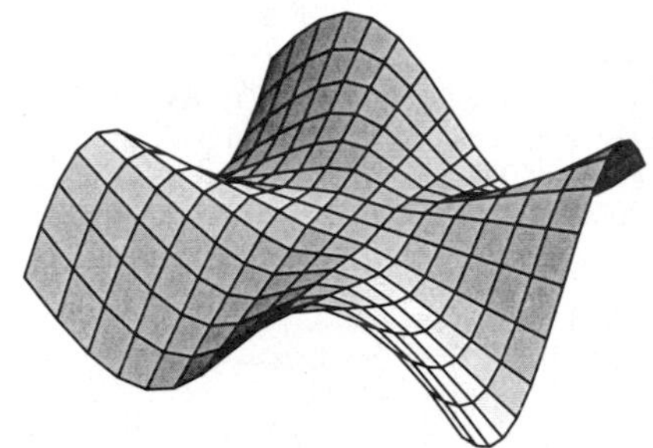

If the continuity requirement for MIXED PARTIALS is dropped, it is possible to construct functions for which MIXED PARTIALS are *not* equal. An example is the function

$$f(x,y) = \begin{cases}\frac{xy(x^2-y^2)}{x^2+y^2} & \text{for } (x,y)=0\\ 0 & \text{for } (x,y)=0,\end{cases} \tag{6}$$

which has $f_{xy}(0,0)=-1$ and $f_{yx}(0,0)=1$ (Wagon 1991). This function is depicted above and by Fischer (1986).

Abramowitz and Stegun (1972) give FINITE DIFFERENCE versions for partial derivatives.

see also ABLOWITZ-RAMANI-SEGUR CONJECTURE, DERIVATIVE, MIXED PARTIAL DERIVATIVE, MONKEY SADDLE

References

Abramowitz, M. and Stegun, C. A. (Eds.). *Handbook of Mathematical Functions with Formulas, Graphs, and Mathematical Tables, 9th printing.* New York: Dover, pp. 883–885, 1972.

Fischer, G. (Ed.). Plate 121 in *Mathematische Modelle/Mathematical Models, Bildband/Photograph Volume.* Braunschweig, Germany: Vieweg, p. 118, 1986.

Thomas, G. B. and Finney, R. L. §16.8 in *Calculus and Analytic Geometry, 9th ed.*0201531747 Reading, MA: Addison-Wesley, 1996.

Wagon, S. *Mathematica in Action.* New York: W. H. Freeman, pp. 83–85, 1991.

Partial Differential Equation

A partial differential equation (PDE) is an equation involving functions and their PARTIAL DERIVATIVES; for example, the WAVE EQUATION

$$\frac{\partial^2\psi}{\partial x^2}+\frac{\partial^2\psi}{\partial y^2}+\frac{\partial^2\psi}{\partial z^2}=\frac{1}{v^2}\frac{\partial^2\psi}{\partial t^2}. \tag{1}$$

In general, partial differential equations are much more difficult to solve analytically than are ORDINARY DIFFERENTIAL EQUATIONS. They may sometimes be solved using a BÄCKLUND TRANSFORMATION, CHARACTERISTIC, GREEN'S FUNCTION, INTEGRAL TRANSFORM, LAX PAIR, SEPARATION OF VARIABLES, or—when all else fails (which it frequently does)—numerical methods.

Fortunately, partial differential equations of second-order are often amenable to analytical solution. Such PDEs are of the form

$$Au_{xx}+2Bu_{xy}+Cu_{yy}+Du_x+Eu_y+F=0. \tag{2}$$

Second-order PDEs are then classified according to the properties of the MATRIX

$$\mathsf{Z}\equiv\begin{bmatrix} A & B\\ B & C\end{bmatrix} \tag{3}$$

as ELLIPTIC, HYPERBOLIC, or PARABOLIC.

If Z is a POSITIVE DEFINITE MATRIX, i.e., $\det(\mathsf{Z})>0$, the PDE is said to be ELLIPTIC. LAPLACE'S EQUATION and POISSON'S EQUATION are examples. Boundary conditions are used to give the constraint $u(x,y)=g(x,y)$ on $\partial\Omega$, where

$$u_{xx}+u_{yy}=f(u_x,u_y,u,x,y) \tag{4}$$

holds in Ω.

If $\det(\mathsf{Z})<0$, the PDE is said to be HYPERBOLIC. The WAVE EQUATION is an example of a hyperbolic partial differential equation. Initial-boundary conditions are used to give

$$u(x,y,t)=g(x,y,t)\quad\text{for } x\in\partial\Omega, t>0 \tag{5}$$

$$u(x,y,0)=v_0(x,y)\quad\text{in }\Omega \tag{6}$$

$$u_t(x,y,0)=v_1(x,y)\quad\text{in }\Omega, \tag{7}$$

where

$$u_{xy}=f(u_x,u_t,x,y) \tag{8}$$

holds in Ω.

If $\det(\mathsf{Z})=0$, the PDE is said to be parabolic. The HEAT CONDUCTION EQUATION equation and other diffusion equations are examples. Initial-boundary conditions are used to give

$$u(x,t)=g(x,t)\quad\text{for } x\in\partial\Omega, t>0 \tag{9}$$

$$u(x,0)=v(x)\quad\text{for } x\in\Omega, \tag{10}$$

where

$$u_{xx}=f(u_x,u_y,u,x,y) \tag{11}$$

holds in Ω.

see also BÄCKLUND TRANSFORMATION, BOUNDARY CONDITIONS, CHARACTERISTIC (PARTIAL DIFFERENTIAL EQUATION), ELLIPTIC PARTIAL DIFFERENTIAL EQUATION, GREEN'S FUNCTION, HYPERBOLIC PARTIAL DIFFERENTIAL EQUATION, INTEGRAL TRANSFORM, JOHNSON'S EQUATION, LAX PAIR, MONGE-AMPÈRE DIFFERENTIAL EQUATION, PARABOLIC PARTIAL DIFFERENTIAL EQUATION, SEPARATION OF VARIABLES

References

Arfken, G. "Partial Differential Equations of Theoretical Physics." §8.1 in *Mathematical Methods for Physicists, 3rd ed.* Orlando, FL: Academic Press, pp. 437–440, 1985.

Bateman, H. *Partial Differential Equations of Mathematical Physics.* New York: Dover, 1944.

Press, W. H.; Flannery, B. P.; Teukolsky, S. A.; and Vetterling, W. T. "Partial Differential Equations." Ch. 19 in *Numerical Recipes in FORTRAN: The Art of Scientific Computing, 2nd ed.* Cambridge, England: Cambridge University Press, pp. 818–880, 1992.

Sobolev, S. L. *Partial Differential Equations of Mathematical Physics.* New York: Dover, 1989.

Sommerfeld, A. *Partial Differential Equations in Physics.* New York: Academic Press, 1964.

Webster, A. G. *Partial Differential Equations of Mathematical Physics, 2nd corr. ed.* New York: Dover, 1955.

Partial Fraction Decomposition

A RATIONAL FUNCTION $P(x)/Q(x)$ can be rewritten using what is known as partial fraction decomposition. This procedure often allows integration to be performed on each term separately by inspection. For each factor of $Q(x)$ the form $(ax+b)^m$, introduce terms

$$\frac{A_1}{ax+b}+\frac{A_2}{(ax+b)^2}+\ldots+\frac{A_m}{(ax+b)^m}. \tag{1}$$

For each factor of the form $(ax^2+bx+c)^m$, introduce terms

$$\frac{A_1x+B_1}{ax^2+bx+c}+\frac{A_2x+B_2}{(ax^2+bx+c)^2}+\ldots+\frac{A_mx+B_m}{(ax^2+bx+c)^m}. \tag{2}$$

Then write

$$\frac{P(x)}{Q(x)}=\frac{A_1}{ax+b}+\ldots+\frac{A_2x+B_2}{ax^2+bx+c}+\ldots \tag{3}$$

and solve for the A_is and B_is.

References

Beyer, W. H. *CRC Standard Mathematical Tables, 28th ed.* Boca Raton, FL: CRC Press, pp. 13–15, 1987.

Partial Latin Square

In a normal $n \times n$ LATIN SQUARE, the entries in each row and column are chosen from a "global" set of n objects. Like a Latin square, a partial Latin square has no two rows or columns which contain the same two symbols. However, in a partial Latin square, each cell is assigned one of its own set of n possible "local" (and distinct) symbols, chosen from an overall set of more than three distinct symbols, and these symbols may vary from location to location. For example, given the possible symbols $\{1, 2, \ldots, 6\}$ which must be arranged as

$$\begin{matrix} \{1,2,3\} & \{1,3,4\} & \{2,5,6\} \\ \{2,3,5\} & \{1,2,3\} & \{4,5,6\} \\ \{4,3,6\} & \{3,5,6\} & \{2,3,5\}, \end{matrix}$$

the 3×3 partial Latin square

$$\begin{matrix} 1 & 3 & 2 \\ 2 & 4 & 5 \\ 6 & 5 & 3 \end{matrix}$$

can be constructed.

see also DINITZ PROBLEM, LATIN SQUARE

References

Cipra, B. "Quite Easily Done." In *What's Happening in the Mathematical Sciences* **2**, pp. 41–46, 1994.

Partial Order

A RELATION "$\leq$" is a partial order on a SET S if it has:

1. Reflexivity: $a \leq a$ for all $a \in S$.
2. Antisymmetry: $a \leq b$ and $b \leq a$ implies $a = b$.
3. Transitivity: $a \leq b$ and $b \leq c$ implies $a \leq c$.

For a partial order, the size of the longest CHAIN (ANTICHAIN) is called the LENGTH (WIDTH). A partially ordered set is also called a POSET.

see also ANTICHAIN, CHAIN, FENCE POSET, IDEAL (PARTIAL ORDER), LENGTH (PARTIAL ORDER), LINEAR EXTENSION, PARTIALLY ORDERED SET, TOTAL ORDER, WIDTH (PARTIAL ORDER)

References

Ruskey, F. "Information on Linear Extension." `http://sue.csc.uvic.ca/~cos/inf/pose/LinearExt.html`.

Partial Quotient

If the SIMPLE CONTINUED FRACTION of a REAL NUMBER x is given by

$$x = a_0 + \cfrac{1}{a_1 + \cfrac{1}{a_2 + \cfrac{1}{a_3 + \ldots}}},$$

then the quantities a_i are called partial quotients.

see also CONTINUED FRACTION, CONVERGENT, SIMPLE CONTINUED FRACTION

Partially Ordered Set

A partially ordered set (or POSET) is a SET taken together with a PARTIAL ORDER on it. Formally, a partially ordered set is defined as an ordered pair $P = (X, \leq)$, where X is called the GROUND SET of P and $\leq$ is the PARTIAL ORDER of P.

see also CIRCLE ORDER, COVER RELATION, DOMINANCE, GROUND SET, HASSE DIAGRAM, INTERVAL ORDER, ISOMORPHIC POSETS, PARTIAL ORDER, POSET DIMENSION, REALIZER, RELATION

References

Dushnik, B. and Miller, E. W. "Partially Ordered Sets." *Amer. J. Math.* **63**, 600–610, 1941.

Fishburn, P. C. *Interval Orders and Interval Sets: A Study of Partially Ordered Sets.* New York: Wiley, 1985.

Trotter, W. T. *Combinatorics and Partially Ordered Sets: Dimension Theory.* Baltimore, MD: Johns Hopkins University Press, 1992.

Particularly Well-Behaved Functions

Functions which have DERIVATIVES of all orders at all points and which, together with their DERIVATIVES, fall off at least as rapidly as $|x|^{-n}$ as $|x| \to \infty$, no matter how large n is.

see also REGULAR SEQUENCE

Partisan Game

A GAME for which each player has a different set of moves in any position. Every position in an IMPARTIAL GAME has a NIM-VALUE.

Partition

A partition is a way of writing an INTEGER n as a sum of POSITIVE INTEGERS without regard to order, possibly subject to one or more additional constraints. Particular types of partition functions include the PARTITION FUNCTION P, giving the number of partitions of a number without regard to order, and PARTITION FUNCTION Q, giving the number of ways of writing the INTEGER n as a sum of POSITIVE INTEGERS without regard to order with the constraint that all INTEGERS in each sum are distinct.

see also AMENABLE NUMBER, DURFEE SQUARE, ELDER'S THEOREM, FERRERS DIAGRAM, GRAPHICAL PARTITION, PARTITION FUNCTION P, Partition Function Q, PERFECT PARTITION, PLANE PARTITION, SET PARTITION, SOLID PARTITION, STANLEY'S THEOREM

References

Andrews, G. E. *The Theory of Partitions.* Cambridge, England: Cambridge University Press, 1998.

Dickson, L. E. "Partitions." Ch. 3 in *History of the Theory of Numbers, Vol. 2: Diophantine Analysis.* New York: Chelsea, pp. 101–164, 1952.

Partition Function P

$P(n)$ gives the number of ways of writing the INTEGER n as a sum of POSITIVE INTEGERS without regard to order. For example, since 4 can be written

$$\begin{aligned} 4 &= 4 \\ &= 3+1 \\ &= 2+2 \\ &= 2+1+1 \\ &= 1+1+1+1, \end{aligned} \tag{1}$$

so $P(4) = 5$. $P(n)$ satisfies

$$P(n) \leq \tfrac{1}{2}[P(n+1) + P(n-1)] \tag{2}$$

(Honsberger 1991). The values of $P(n)$ for $n = 1, 2, \ldots$, are 1, 2, 3, 5, 7, 11, 15, 22, 30, 42, ... (Sloane's A000041). The following table gives the value of $P(n)$ for selected small n.

n	$P(n)$
50	204226
100	190569292
200	3972999029388
300	9253082936723602
400	6727090051741041926
500	2300165032574323995027
600	458004788008144308553622
700	60378285202834474611028659
800	5733052172321422504456911979
900	415873681190459054784114365430
1000	24061467864032622473692149727991

n for which $P(n)$ is PRIME are 2, 3, 4, 5, 6, 13, 36, 77, 132, 157, 168, 186, ... (Sloane's A046063). Numbers which cannot be written as a PRODUCT of $P(n)$ are 13, 17, 19, 23, 26, 29, 31, 34, 37, 38, 39, ... (Sloane's A046064), corresponding to numbers of nonisomorphic ABELIAN GROUPS which are not possible for any group order.

When explicitly listing the partitions of a number n, the simplest form is the so-called *natural representation* which simply gives the sequence of numbers in the representation (e.g., (2, 1, 1) for the number $4 = 2+1+1$). The *multiplicity representation* instead gives the number of times each number occurs together with that number (e.g., (2, 1), (1, 2) for $4 = 2 \cdot 1 + 1 \cdot 2$). The FERRERS DIAGRAM is a pictorial representation of a partition.

Euler invented a GENERATING FUNCTION which gives rise to a POWER SERIES in $P(n)$,

$$P(n) = \sum_{m=1}^{\infty} (-1)^{m+1}[P(n - \tfrac{1}{2}m(3m-1)) + P(n - \tfrac{1}{2}m(3m+1))]. \tag{3}$$

A RECURRENCE RELATION is

$$P(n) = \frac{1}{n} \sum_{m=0}^{n-1} \sigma(n-m)P(m), \tag{4}$$

where $\sigma(n)$ is the DIVISOR FUNCTION (Berndt 1994, p. 108). Euler also showed that, for

$$f(x) \equiv \prod_{m=1}^{\infty} (1 - x^m) = \sum_{n=-\infty}^{\infty} (-1)^n x^{n(3n+1)/2} \tag{5}$$

$$= 1 - x - x^2 + x^5 + x^7 - x^{12} - x^{15} + x^{22} + x^{26} + \ldots, \tag{6}$$

where the exponents are generalized PENTAGONAL NUMBERS 0, 1, 2, 5, 7, 12, 15, 22, 26, 35, ... (Sloane's A001318) and the sign of the kth term (counting 0 as the 0th term) is $(-1)^{\lfloor (k+1)/2 \rfloor}$ (with $\lfloor x \rfloor$ the FLOOR FUNCTION), the partition numbers $P(n)$ are given by the GENERATING FUNCTION

$$\frac{1}{f(x)} = \sum_{n=0}^{\infty} P(n)x^n. \tag{7}$$

MacMahon obtained the beautiful RECURRENCE RELATION

$$\begin{aligned} P(n) - P(n-1) - P(n-2) + P(n-5) + P(n-7) \\ -P(n-12) - P(n-15) + \ldots = 0, \end{aligned} \tag{8}$$

where the sum is over generalized PENTAGONAL NUMBERS $\leq n$ and the sign of the kth term is $(-1)^{\lfloor (k+1)/2 \rfloor}$, as above.

In 1916–1917, Hardy and Ramanujan used the CIRCLE METHOD and elliptic MODULAR FUNCTIONS to obtain the approximate solution

$$P(n) \sim \frac{1}{4n\sqrt{3}} e^{\pi\sqrt{2n/3}}. \tag{9}$$

Rademacher (1937) subsequently obtained an exact series solution which yields the Hardy-Ramanujan FORMULA (9) as the first term:

$$P(n) = \sum_{q=1}^{\infty} L_q(n)\psi_q(n), \tag{10}$$

where

$$K = \pi\sqrt{\frac{2}{3}} \tag{11}$$

$$L_q(n) = \sum_p \omega_{p,q} e^{-2np\pi i/q} \tag{12}$$

$$\omega_{p,q} = e^{\pi i s_{p,q}} \tag{13}$$

$$s_{p,q} = \frac{1}{q} \sum_{\mu=1}^{q-1} \mu \left(\frac{\mu p}{q} - \left\lfloor \frac{\mu p}{q} \right\rfloor - \frac{1}{2} \right) \tag{14}$$

$$\lambda_n = \sqrt{n - \tfrac{1}{24}} \tag{15}$$

$$\psi_q(n) = \frac{\sqrt{q}}{\pi\sqrt{2}} \left\{ \frac{d}{dm} \left[\frac{\sinh\left(\frac{K\lambda_m}{q}\right)}{\lambda_m} \right] \right\}_{m=n}, \tag{16}$$

$\lfloor x \rfloor$ is the FLOOR FUNCTION, and p runs through the INTEGERS less than and RELATIVELY PRIME to q (when $q = 1$, $p = 0$). The remainder after Q terms is

$$R(Q) < CQ^{-1/2} + D\sqrt{\frac{Q}{n}} \sinh\left(\frac{K\sqrt{n}}{Q}\right), \quad (17)$$

where C and D are fixed constants.

With $f(x)$ as defined above, Ramanujan also showed that

$$5\frac{f^5(x^5)}{f^6(x)} = \sum_{m=0}^{\infty} P(5m+4)x^m. \quad (18)$$

Ramanujan also found numerous CONGRUENCES such as

$$P(5m+4) \equiv 0 \pmod 5 \quad (19)$$

$$P(7m+5) \equiv 0 \pmod 7 \quad (20)$$

$$P(11m+6) \equiv 0 \pmod{11}. \quad (21)$$

RAMANUJAN'S IDENTITY gives the first of these.

Let $P_O(n)$ be the number of partitions of n containing ODD numbers only and $P_D(n)$ be the number of partitions of n without duplication, then

$$P_O(n) = P_D(n) = \prod_{k=1,3,\ldots}^{\infty} (1 + x^k + x^{2k} + x^{3k} + \ldots)$$

$$= \prod_{k=1}^{\infty}(1+x^k) = 1+x+x^2+2x^3+2x^4+3x^5+\ldots, \quad (22)$$

as discovered by Euler (Honsberger 1985). The first few values of $P_O = P_D$ are 1, 1, 1, 2, 2, 3, 4, 5, 6, 8, 10, ... (Sloane's A000009).

Let $P_E(n)$ be the number of partitions of EVEN numbers only, and let $P_{EO}(n)$ ($P_{DO}(n)$) be the number of partitions in which the parts are all EVEN (ODD) and all different. The first few values of $P_{DO}(n)$ are 1, 1, 0, 1, 1, 1, 1, 1, 2, 2, 2, 2, 3, 3, 3, 4, ... (Sloane's A000700). Some additional GENERATING FUNCTIONS are given by Honsberger (1985, pp. 241–242)

$$\sum_{n=1}^{\infty} P_{\text{no even part repeated}}(n)x^n = \prod_{k=1} (1 - x^{2k-1})^{-1}(1 + x^{2k}) \quad (23)$$

$$\sum_{n=1}^{\infty} P_{\text{no part occurs more than 3 times}}(n)x^n = \prod_{k=1} (1 + x^k + x^{2k} + x^{3k}) \quad (24)$$

$$\sum_{n=1}^{\infty} P_{\text{no part divisible by 4}}(n)x^n = \prod_{k=1} \frac{1 - x^{4k}}{1 - x^k} \quad (25)$$

$$\sum_{n=1}^{\infty} P_{\text{no part occurs more than } d \text{ times}}(n)x^n = \prod_{k=1}\sum_{i=0}^{d} x^{ik} = \prod_{k=1} \frac{1 - x^{(d+1)k}}{1 - x^k} \quad (26)$$

$$\sum_{n=1}^{\infty} P_{\text{every part occurs 2, 3, or 5 times}}(n)x^n = \prod_{k=1}(1 + x^{2k} + x^{3k} + x^{5k})$$

$$= \prod_{k=1}(1 + x^{2k})(1 + x^{3k}) = \prod_{k=1} \frac{1 - x^{4k}}{1 - x^{2k}} \frac{1 - x^{6k}}{1 - x^{3k}} \quad (27)$$

$$\sum_{n=1}^{\infty} P_{\text{no part occurs exactly once}}(n)x^n = (1 + x^{2k} + x^{3k} + \ldots) = \prod_{k} \frac{1 + x^{6k}}{(1 - x^{2k})(1 - x^{3k})}. \quad (28)$$

Some additional interesting theorems following from these (Honsberger 1985, pp. 64–68 and 143–146) are:

1. The number of partitions of n in which no EVEN part is repeated is the same as the number of partitions of n in which no part occurs more than three times and also the same as the number of partitions in which no part is divisible by four.
2. The number of partitions of n in which no part occurs more often than d times is the same as the number of partitions in which no term is a multiple of $d + 1$.
3. The number of partitions of n in which each part appears either 2, 3, or 5 times is the same as the number of partitions in which each part is CONGRUENT mod 12 to either 2, 3, 6, 9, or 10.
4. The number of partitions of n in which no part appears exactly once is the same as the number of partitions of n in which no part is CONGRUENT to 1 or 5 mod 6.
5. The number of partitions in which the parts are all EVEN and different is equal to the absolute difference of the number of partitions with ODD and EVEN parts.

$P(n,k)$, also written $P_k(n)$, is the number of ways of writing n as a sum of k terms, and can be computed from the RECURRENCE RELATION

$$P(n,k) = P(n-1,k-1) + P(n-k,k) \quad (29)$$

(Ruskey). The number of partitions of n with largest part k is the same as $P(n,k)$.

The function $P(n,k)$ can be given explicitly for the first few values of k,

$$P(n,2) = \left\lfloor \tfrac{1}{2}n \right\rfloor \quad (30)$$

$$P(n,3) = [\tfrac{1}{12}n^2], \quad (31)$$

where $\lfloor x \rfloor$ is the FLOOR FUNCTION and $[x]$ is the NINT function (Honsberger 1985, pp. 40–45).

see also ALCUIN'S SEQUENCE, ELDER'S THEOREM, EULER'S PENTAGONAL NUMBER THEOREM, FERRERS DIAGRAM, PARTITION FUNCTION Q, PENTAGONAL NUMBER, $r_k(n)$, ROGERS-RAMANUJAN IDENTITIES, STANLEY'S THEOREM

References

Abramowitz, M. and Stegun, C. A. (Eds.). "Unrestricted Partitions." §24.2.1 in *Handbook of Mathematical Functions with Formulas, Graphs, and Mathematical Tables, 9th printing.* New York: Dover, p. 825, 1972.

Adler, H. "Partition Identities—From Euler to the Present." *Amer. Math. Monthly* **76**, 733–746, 1969.

Adler, H. "The Use of Generating Functions to Discover and Prove Partition Identities." *Two-Year College Math. J.* **10**, 318–329, 1979.

Andrews, G. *Encyclopedia of Mathematics and Its Applications, Vol. 2: The Theory of Partitions.* Cambridge, England: Cambridge University Press, 1984.

Berndt, B. C. *Ramanujan's Notebooks, Part IV.* New York: Springer-Verlag, 1994.

Conway, J. H. and Guy, R. K. *The Book of Numbers.* New York: Springer-Verlag, pp. 94–96, 1996.

Honsberger, R. *Mathematical Gems III.* Washington, DC: Math. Assoc. Amer., pp. 40–45 and 64–68, 1985.

Honsberger, R. *More Mathematical Morsels.* Washington, DC: Math. Assoc. Amer., pp. 237–239, 1991.

Jackson, D. and Goulden, I. *Combinatorial Enumeration.* New York: Academic Press, 1983.

MacMahon, P. A. *Combinatory Analysis.* New York: Chelsea, 1960.

Rademacher, H. "On the Partition Function $p(n)$." *Proc. London Math. Soc.* **43**, 241–254, 1937.

Ruskey, F. "Information of Numerical Partitions." `http://sue.csc.uvic.ca/~cos/inf/nump/NumPartition.html`.

Sloane, N. J. A. Sequences A000009/M0281, A000041/M0663, and A000700/M0217 in "An On-Line Version of the Encyclopedia of Integer Sequences."

Partition Function Q

$Q(n)$ gives the number of ways of writing the INTEGER n as a sum of POSITIVE INTEGERS without regard to order with the constraint that all INTEGERS in each sum are distinct. The values for $n = 1, 2, \ldots$ are 1, 1, 2, 2, 3, 4, 5, 6, 8, 10, ... (Sloane's A000009). The GENERATING FUNCTION for $Q(n)$ is

$$\prod_{m=1}^{\infty}(1+x^m) = \frac{1}{\prod_{m=0}^{\infty}(1-x^{2m+1})} = 1 + x + x^2 + 2x^3 + 2x^4 + 3x^5 + \ldots.$$

The values of n for which $Q(n)$ is PRIME are 3, 4, 5, 7, 22, 70, 100, 495, 1247, 2072, ... (Sloane's A046065), with no others for $n \leq 15{,}000$.

The number of PARTITIONS of n with $\leq k$ summands is denoted $q(n,k)$ or $q_k(n)$. Therefore, $q_n(n) = P(n)$ and

$$q_k(n) = q_{k-1}(n) + q_k(n-k).$$

see also PARTITION FUNCTION P

References

Abramowitz, M. and Stegun, C. A. (Eds.). "Partitions into Distinct Parts." §24.2.2 in *Handbook of Mathematical Functions with Formulas, Graphs, and Mathematical Tables, 9th printing.* New York: Dover, pp. 823–824, 1972.

Sloane, N. J. A. Sequences A046065 and A000009/M0281 in "An On-Line Version of the Encyclopedia of Integer Sequences."

Party Problem

Also known as the MAXIMUM CLIQUE PROBLEM. Find the minimum number of guests that must be invited so that at least m will know each other or at least n will not know each other. The solutions are known as RAMSEY NUMBERS.

see also CLIQUE, RAMSEY NUMBER

Parzen Apodization Function

An APODIZATION FUNCTION similar to the BARTLETT FUNCTION.

see also APODIZATION FUNCTION, BARTLETT FUNCTION

References

Press, W. H.; Flannery, B. P.; Teukolsky, S. A.; and Vetterling, W. T. *Numerical Recipes in FORTRAN: The Art of Scientific Computing, 2nd ed.* Cambridge, England: Cambridge University Press, p. 547, 1992.

Pascal Distribution

see NEGATIVE BINOMIAL DISTRIBUTION

Pascal's Formula

Each subsequent row of PASCAL'S TRIANGLE is obtained by adding the two entries diagonally above. This follows immediately from the BINOMIAL COEFFICIENT identity

$$\begin{aligned}\binom{n}{r} &\equiv \frac{n!}{(n-r)!r!} = \frac{(n-1)!n}{(n-r)!r!} \\ &= \frac{(n-1)!(n-r)}{(n-r)!r!} + \frac{(n-1)!r}{(n-r)!r!} \\ &= \frac{(n-1)!}{(n-r-1)!r!} + \frac{(n-1)!}{(n-r)!(r-1)!} \\ &= \binom{n-1}{r} + \binom{n-1}{r-1}.\end{aligned}$$

see also BINOMIAL COEFFICIENT, PASCAL'S TRIANGLE

Pascal's Hexagrammum Mysticum

see PASCAL'S THEOREM

Pascal's Limaçon

see LIMAÇON

Pascal Line

The line containing the three points of the intersection of the three pairs of opposite sides of two TRIANGLES.

see also PASCAL'S THEOREM

Pascal's Rule

see PASCAL'S FORMULA

Pascal's Theorem

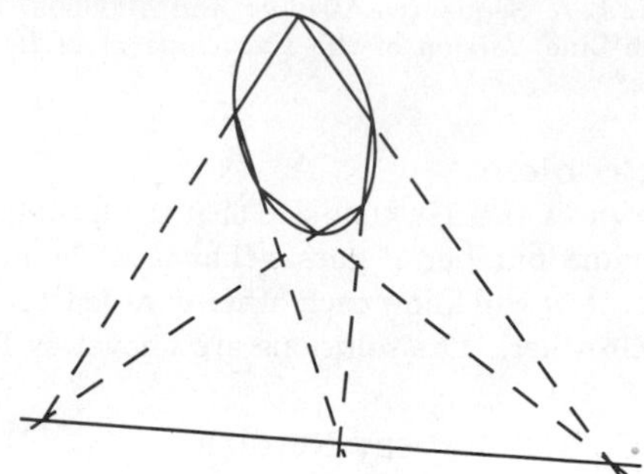

The dual of BRIANCHON'S THEOREM. It states that, given a (not necessarily REGULAR, or even CONVEX) HEXAGON inscribed in a CONIC SECTION, the three pairs of the continuations of opposite sides meet on a straight LINE, called the PASCAL LINE. There are 6! (6! means 6 FACTORIAL, where $6! = 6 \cdot 5 \cdot 4 \cdot 3 \cdot 2 \cdot 1$) possible ways of taking all VERTICES in any order, but among these are six equivalent CYCLIC PERMUTATIONS and two possible orderings, so the total number of different hexagons (not all simple) is

$$\frac{6!}{2 \cdot 6} = \frac{720}{12} = 60.$$

There are therefore a total of 60 PASCAL LINES created by connecting VERTICES in any order. These intersect three by three in 20 STEINER POINTS.

see also BRAIKENRIDGE-MACLAURIN CONSTRUCTION, BRIANCHON'S THEOREM, CAYLEY-BACHARACH THEOREM, CONIC SECTION, DUALITY PRINCIPLE, HEXAGON, PAPPUS'S HEXAGON THEOREM, PASCAL LINE, STEINER POINTS

References

Coxeter, H. S. M. and Greitzer, S. L. *Geometry Revisited.* Washington, DC: Math. Assoc. Amer., pp. 73–76, 1967.

Ogilvy, C. S. *Excursions in Geometry.* New York: Dover, pp. 105–106, 1990.

Pappas, T. "The Mystic Hexagram." *The Joy of Mathematics.* San Carlos, CA: Wide World Publ./Tetra, p. 118, 1989.

Pascal's Triangle

A TRIANGLE of numbers arranged in staggered rows such that

$$a_{nr} \equiv \frac{n!}{r!(n-r)!} \equiv \binom{n}{r}, \qquad (1)$$

where $\binom{n}{r}$ is a BINOMIAL COEFFICIENT. The triangle was studied by B. Pascal, although it had been described centuries earlier by Chinese mathematician Yanghui (about 500 years earlier, in fact) and the Arabian poet-mathematician Omar Khayyám. It is therefore known as the YANGHUI TRIANGLE in China. Starting with $n = 0$, the TRIANGLE is

```
            1
          1   1
        1   2   1
      1   3   3   1
    1   4   6   4   1
  1   5  10  10   5   1
1   6  15  20  15   6   1
```

(Sloane's A007318). PASCAL'S FORMULA shows that each subsequent row is obtained by adding the two entries diagonally above,

$$\binom{n}{r} = \frac{n!}{(n-r)!r!} = \binom{n-1}{r} + \binom{n-1}{r-1}. \qquad (2)$$

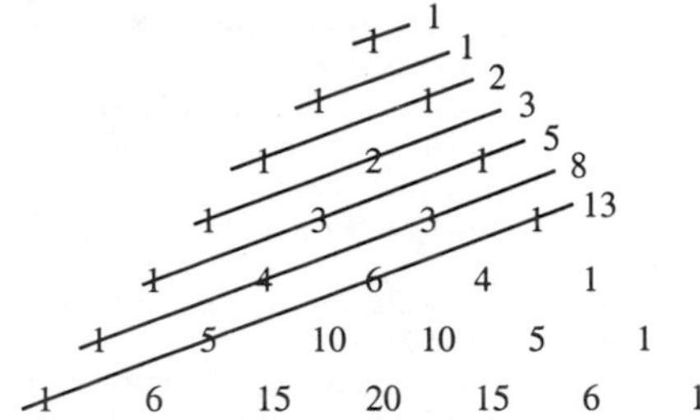

In addition, the "SHALLOW DIAGONALS" of Pascal's triangle sum to FIBONACCI NUMBERS,

$$\sum_{k=1}^{n} \binom{k}{n-k} = \frac{(-1)^n {}_3F_2(1, 2, 1-n; \tfrac{1}{2}(3-n), 2-\tfrac{1}{2}n; -\tfrac{1}{4})}{\pi(2-3n+n^2)} = F_{n+1}, \qquad (3)$$

where ${}_3F_2(a, b, c; d, e; z)$ is a GENERALIZED HYPERGEOMETRIC FUNCTION.

Pascal's triangle contains the FIGURATE NUMBERS along its diagonals. It can be shown that

$$\sum_{i=1}^{n} a_{ij} = \frac{n+1}{j+1} a_{nj} = a_{(n+1),(j+1)} \qquad (4)$$

and

$$\binom{m+1}{1} \sum k^m + \binom{m+1}{2} \sum k^{m-1} + \ldots + \binom{m+1}{m} \sum k = (n+1)[(n+1)^m - 1]. \qquad (5)$$

The "shallow diagonals" sum to the FIBONACCI SEQUENCE, i.e.,

$$\begin{aligned} 1 &= 1 \\ 1 &= 1 \\ 2 &= 1+1 \\ 3 &= 2+1 \\ 5 &= 1+3+1 \\ 8 &= 3+4+1. \end{aligned} \tag{6}$$

In addition,

$$\sum_{j=1}^{i} a_{ij} = 2^i - 1. \tag{7}$$

It is also true that the first number after the 1 in each row divides all other numbers in that row IFF it is a PRIME. If P_n is the number of ODD terms in the first n rows of the Pascal triangle, then

$$0.812\ldots < P_n n^{-\ln 2/\ln 3} < 1 \tag{8}$$

(Harborth 1976, Le Lionnais 1983).

The BINOMIAL COEFFICIENT $\binom{m}{n}$ mod 2 can be computed using the XOR operation n XOR m, making Pascal's triangle mod 2 very easy to construct. Pascal's triangle is unexpectedly connected with the construction of regular POLYGONS and with the SIERPIŃSKI SIEVE.

see also BELL TRIANGLE, BINOMIAL COEFFICIENT, BINOMIAL THEOREM, BRIANCHON'S THEOREM, CATALAN'S TRIANGLE, CLARK'S TRIANGLE, EULER'S TRIANGLE, FIBONACCI NUMBER, FIGURATE NUMBER TRIANGLE, LEIBNIZ HARMONIC TRIANGLE, NUMBER TRIANGLE, PASCAL'S FORMULA, POLYGON, SEIDEL-ENTRINGER-ARNOLD TRIANGLE, SIERPIŃSKI SIEVE, TRINOMIAL TRIANGLE

References

Conway, J. H. and Guy, R. K. "Pascal's Triangle." In *The Book of Numbers*. New York: Springer-Verlag, pp. 68–70, 1996.

Courant, R. and Robbins, H. *What is Mathematics?: An Elementary Approach to Ideas and Methods, 2nd ed.* Oxford, England: Oxford University Press, p. 17, 1996.

Harborth, H. "Number of Odd Binomial Coefficients." *Not. Amer. Math. Soc.* **23**, 4, 1976.

Le Lionnais, F. *Les nombres remarquables.* Paris: Hermann, p. 31, 1983.

Pappas, T. "Pascal's Triangle, the Fibonacci Sequence & Binomial Formula," "Chinese Triangle," and "Probability and Pascal's Triangle." *The Joy of Mathematics.* San Carlos, CA: Wide World Publ./Tetra, pp. 40–41 88, and 184–186, 1989.

Sloane, N. J. A. Sequence A007318/M0082 in "An On-Line Version of the Encyclopedia of Integer Sequences."

Smith, D. E. *A Source Book in Mathematics.* New York: Dover, p. 86, 1984.

Pascal's Wager

"God is or He is not... Let us weigh the gain and the loss in choosing... 'God is.' If you gain, you gain all, if you lose, you lose nothing. Wager, then, unhesitatingly, that He is."

Pasch's Axiom

In the plane, if a line intersects one side of a TRIANGLE and misses the three VERTICES, then it must intersect one of the other two sides. This is a special case of the generalized MENELAUS' THEOREM with $n = 3$.

see also HELLY'S THEOREM, MENELAUS' THEOREM, PASCH'S THEOREM

Pasch's Theorem

A theorem stated in 1882 which cannot be derived from EUCLID'S POSTULATES. Given points a, b, c, and d on a LINE, if it is known that the points are ordered as (a, b, c) and (b, c, d), it is also true that (a, b, d).

see also EUCLID'S POSTULATES, LINE, PASCH'S AXIOM

Pass Equivalent

Two KNOTS are pass equivalent if there exists a sequence of pass moves taking one to the other. Every KNOT is either pass equivalent to the UNKNOT or TREFOIL KNOT. These two knots are not pass equivalent to each other, but the ENANTIOMERS of the TREFOIL KNOT are pass equivalent. A KNOT has ARF INVARIANT 0 if the KNOT is pass equivalent to the UNKNOT and 1 if it is pass equivalent to the TREFOIL KNOT.

see also ARF INVARIANT, KNOT, PASS MOVE, TREFOIL KNOT, UNKNOT

References

Adams, C. C. *The Knot Book: An Elementary Introduction to the Mathematical Theory of Knots.* New York: W. H. Freeman, pp. 223–228, 1994.

Pass Move

A change in a knot projection such that a pair of oppositely oriented strands are passed through another pair of oppositely oriented strands.

see also PASS EQUIVALENT

Patch

A patch (also called a LOCAL SURFACE) is a differentiable mapping $\mathbf{x} : U \to \mathbb{R}^n$, where U is an open subset of $\mathbb{R}^2$. More generally, if A is any SUBSET of $\mathbb{R}^2$, then a map $\mathbf{x} : A \to \mathbb{R}^n$ is a patch provided that $\mathbf{x}$ can be extended to a differentiable map from U into $\mathbb{R}^n$, where U is an open set containing A. Here, $\mathbf{x}(U)$ (or more generally, $\mathbf{x}(A)$) is called the TRACE of $\mathbf{x}$.

see also GAUSS MAP, INJECTIVE PATCH, MONGE PATCH, REGULAR PATCH, TRACE (MAP)

References

Gray, A. "Patches in $\mathbb{R}^3$." §10.2 in *Modern Differential Geometry of Curves and Surfaces.* Boca Raton, FL: CRC Press, pp. 183–184 and 192–193, 1993.

Path

A path γ is a continuous mapping $\gamma : [a,b] \mapsto \mathbb{C}$, where $\gamma(a)$ is the initial point and $\gamma(b)$ is the final point. It is often written parametrically as $\sigma(t)$.

Path Graph

The path P_n is a TREE with two nodes of valency 1, and the other $n-2$ nodes of valency 2. Path graphs P_n are always GRACEFUL for $n > 4$.

see also CHAIN (GRAPH), GRACEFUL GRAPH, HAMILTONIAN PATH, TREE

Path Integral

Let γ be a PATH given parametrically by $\sigma(t)$. Let s denote ARC LENGTH from the initial point. Then

$$\begin{aligned}\int_\gamma f(s)\,ds &= \int_\gamma f(\sigma(t))\,|\sigma'(t)|\,dt \\ &= \int_\gamma f(x(t),y(t),z(t))\,|\sigma'(t)|\,dt.\end{aligned}$$

see also LINE INTEGRAL

References

Press, W. H.; Flannery, B. P.; Teukolsky, S. A.; and Vetterling, W. T. "Evaluation of Functions by Path Integration." §5.14 in *Numerical Recipes in FORTRAN: The Art of Scientific Computing, 2nd ed.* Cambridge, England: Cambridge University Press, pp. 201–204, 1992.

Pathwise-Connected

A TOPOLOGICAL SPACE X is pathwise-connected IFF for every two points $x, y \in X$, there is a CONTINUOUS FUNCTION f from [0,1] to X such that $f(0) = x$ and $f(1) = y$. Roughly speaking, a SPACE X is pathwise-connected if, for every two points in X, there is a path connecting them. For LOCALLY PATHWISE-CONNECTED SPACES (which include most "interesting spaces" such as MANIFOLDS and CW-COMPLEXES), being CONNECTED and being pathwise-connected are equivalent, although there are connected spaces which are not pathwise connected. Pathwise-connected spaces are also called 0-connected.

see also CONNECTED SPACE, CW-COMPLEX, LOCALLY PATHWISE-CONNECTED SPACE, TOPOLOGICAL SPACE

Patriarchal Cross

see GAULLIST CROSS

Pauli Matrices

Matrices which arise in Pauli's treatment of spin in quantum mechanics. They are defined by

$$\sigma_1 = \sigma_x \equiv \mathsf{P}_1 \equiv \begin{bmatrix} 0 & 1 \\ 1 & 0 \end{bmatrix} \tag{1}$$

$$\sigma_2 = \sigma_y \equiv \mathsf{P}_2 \equiv \begin{bmatrix} 0 & i \\ -i & 0 \end{bmatrix} \tag{2}$$

$$\sigma_3 = \sigma_z \equiv \mathsf{P}_3 \equiv \begin{bmatrix} 1 & 0 \\ 0 & -1 \end{bmatrix}. \tag{3}$$

The Pauli matrices plus the 2×2 IDENTITY MATRIX I form a complete set, so any 2×2 matrix A can be expressed as

$$\mathsf{A} = c_0\mathsf{I} + c_1\sigma_1 + c_2\sigma_2 + c_3\sigma_3. \tag{4}$$

The associated matrices

$$\sigma_+ \equiv 2\begin{bmatrix} 0 & 1 \\ 0 & 0 \end{bmatrix} \tag{5}$$

$$\sigma_- \equiv 2\begin{bmatrix} 0 & 0 \\ 1 & 0 \end{bmatrix} \tag{6}$$

$$\sigma^2 \equiv 3\begin{bmatrix} 1 & 0 \\ 0 & 1 \end{bmatrix} \tag{7}$$

can also be defined. The Pauli spin matrices satisfy the identities

$$\sigma_i\sigma_j = \mathsf{I}\delta_{ij} + \epsilon_{ijk}i\sigma_k \tag{8}$$

$$\sigma_i\sigma_j + \sigma_j\sigma_i = 2\sigma_{ij} \tag{9}$$

$$\sigma_x p_x + \sigma_y p_y + \sigma_z p_z = \sqrt{{p_x}^2 + {p_y}^2 + {p_z}^2}. \tag{10}$$

see also DIRAC MATRICES, QUATERNION

References

Arfken, G. *Mathematical Methods for Physicists, 3rd ed.* Orlando, FL: Academic Press, p. 211–212, 1985.

Goldstein, H. "The Cayley-Klein Parameters and Related Quantities." *Classical Mechanics, 2nd ed.* Reading, MA: Addison-Wesley, p. 156, 1980.

Pauli Spin Matrices

see PAULI MATRICES

Payoff Matrix

A $m \times n$ MATRIX which gives the possible outcome of a two-person ZERO-SUM GAME when player A has m possible strategies and player B n strategies. The analysis of the MATRIX in order to determine optimal strategies is the aim of GAME THEORY. The so-called "augmented" payoff matrix is defined as follows:

$$\mathsf{G} = \begin{array}{c} \begin{array}{ccccccccc} P_0 & P_1 & P_2 & \cdots & P_n & P_{n+1} & P_{n+2} & \cdots & P_{n+m} \end{array} \\ \begin{bmatrix} 0 & 1 & 1 & \cdots & 0 & 0 & 0 & \cdots & 0 \\ -1 & a_{11} & a_{12} & \cdots & a_{1n} & 1 & 0 & \cdots & 0 \\ -1 & a_{21} & a_{22} & \cdots & a_{2n} & 0 & 1 & \cdots & 0 \\ \vdots & \vdots & \vdots & \ddots & \vdots & \vdots & \vdots & \ddots & \vdots \\ -1 & a_{m1} & a_{m2} & \cdots & a_{mn} & 0 & 0 & \cdots & 1 \end{bmatrix} \end{array}.$$

see also GAME THEORY, ZERO-SUM GAME

Peacock's Tail

One name for the figure used by Euclid to prove the PYTHAGOREAN THEOREM.

see also BRIDE'S CHAIR, WINDMILL

Peano Arithmetic

The theory of NATURAL NUMBERS defined by the five PEANO'S AXIOMS. Any universal statement which is undecidable in Peano arithmetic is necessarily TRUE. Undecidable statements may be either TRUE or FALSE. Paris and Harrington (1977) gave the first "natural" example of a statement which is true for the integers but unprovable in Peano arithmetic (Spencer 1983).

see also KREISEL CONJECTURE, NATURAL INDEPENDENCE PHENOMENON, NUMBER THEORY, PEANO'S AXIOMS

References

Kirby, L. and Paris, J. "Accessible Independence Results for Peano Arithmetic." *Bull. London Math. Soc.* **14**, 285–293, 1982.

Paris, J. and Harrington, L. "A Mathematical Incompleteness in Peano Arithmetic." In *Handbook of Mathematical Logic* (Ed. J. Barwise). Amsterdam, Netherlands: North-Holland, pp. 1133–1142, 1977.

Spencer, J. "Large Numbers and Unprovable Theorems." *Amer. Math. Monthly* **90**, 669–675, 1983.

Peano's Axioms

1. Zero is a number.
2. If a is a number, the successor of a is a number.
3. ZERO is not the successor of a number.
4. Two numbers of which the successors are equal are themselves equal.
5. (INDUCTION AXIOM.) If a set S of numbers contains ZERO and also the successor of every number in S, then every number is in S.

Peano's axioms are the basis for the version of NUMBER THEORY known as PEANO ARITHMETIC.

see also INDUCTION AXIOM, PEANO ARITHMETIC

Peano Curve

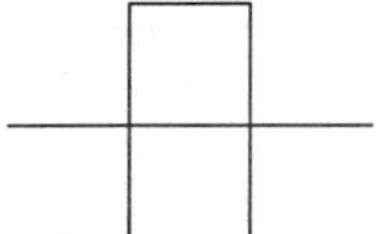

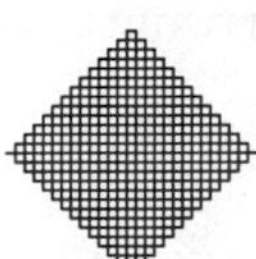

A FRACTAL curve which can be written as a LINDENMAYER SYSTEM.

see also DRAGON CURVE, HILBERT CURVE, LINDENMAYER SYSTEM, SIERPIŃSKI CURVE

References

Dickau, R. M. "Two-Dimensional L-Systems." http://forum.swarthmore.edu/advanced/robertd/lsys2d.html.

Hilbert, D. "Über die stetige Abbildung einer Linie auf ein Flachenstück." *Math. Ann.* **38**, 459–460, 1891.

Peano, G. "Sur une courbe, qui remplit une aire plane." *Math. Ann.* **36**, 157–160, 1890.

Wagon, S. *Mathematica in Action.* New York: W. H. Freeman, p. 207, 1991.

Peano-Gosper Curve

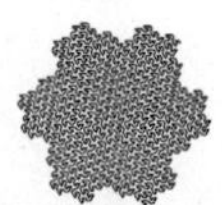

A PLANE-FILLING CURVE originally called a FLOWSNAKE by R. W. Gosper and M. Gardner. Mandelbrot (1977) subsequently coined the name Peano-Gosper curve. The GOSPER ISLAND bounds the space that the Peano-Gosper curve fills.

see also DRAGON CURVE, EXTERIOR SNOWFLAKE, GOSPER ISLAND, HILBERT CURVE, KOCH SNOWFLAKE, PEANO CURVE, SIERPIŃSKI ARROWHEAD CURVE, SIERPIŃSKI CURVE

References

Dickau, R. M. "Two-Dimensional L-Systems." http://forum.swarthmore.edu/advanced/robertd/lsys2d.html.

Mandelbrot, B. B. *Fractals: Form, Chance, & Dimension.* San Francisco, CA: W. H. Freeman, 1977.

Peano Surface

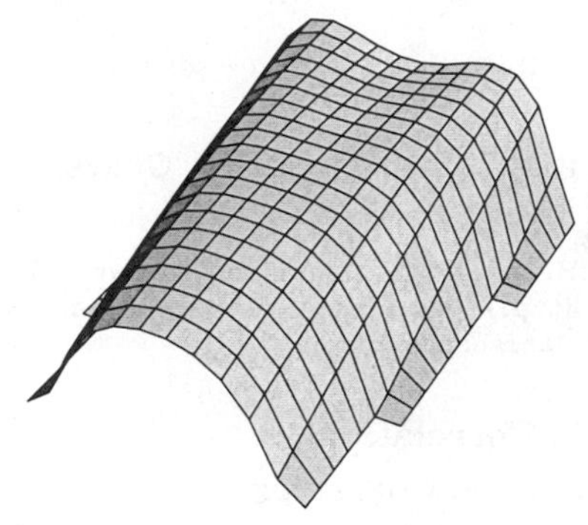

The function

$$f(x,y) = (2x^2 - y)(y - x^2)$$

which does *not* have a LOCAL MAXIMUM at (0, 0), despite criteria commonly touted in the second half of the 1800s which indicated the contrary.

see also LOCAL MAXIMUM

References

Fischer, G. (Ed.). Plate 122 in *Mathematische Modelle/Mathematical Models, Bildband/Photograph Volume.* Braunschweig, Germany: Vieweg, p. 119, 1986.

Leitere, J. "Functions." §7.1.2 in *Mathematical Models from the Collections of Universities and Museums* (Ed. G. Fischer). Braunschweig, Germany: Vieweg, pp. 70–71, 1986.

Pear Curve

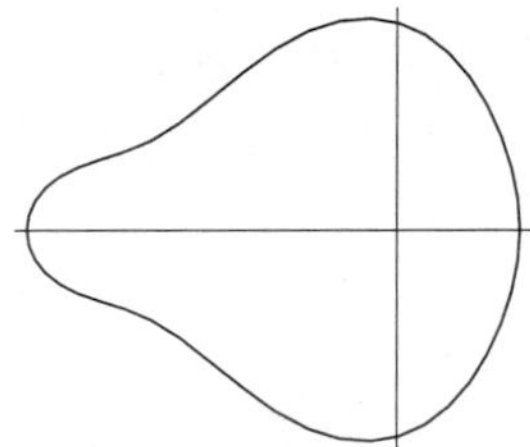

The LEMNISCATE L_3 in the iteration towards the MANDELBROT SET. In CARTESIAN COORDINATES with a constant r, the equation is given by

$$r^2 = (x^2+y^2)(1+2x+5x^2+6x^3+6x^4+4x^5+x^6-3y^2 -2xy^2+8x^2y^2+8x^3y^2+3x^4y^2+2y^4+4xy^4 +3x^2y^4+y^6).$$

see also PEAR-SHAPED CURVE

Pear-Shaped Curve

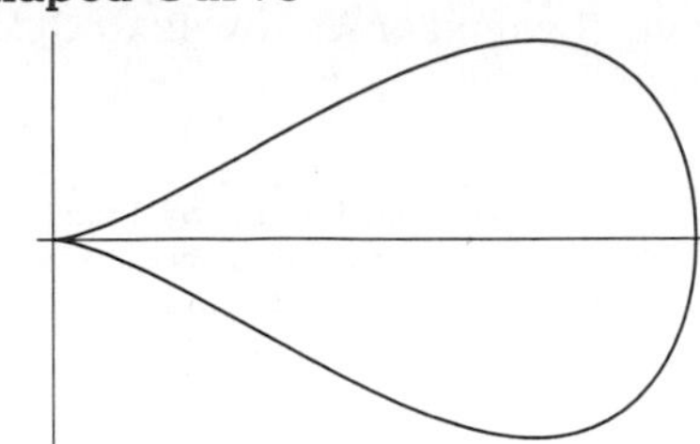

A curve given by the Cartesian equation

$$b^2y^2 = x^3(a-x).$$

see also PEAR CURVE, TEARDROP CURVE

References
MacTutor History of Mathematics Archive. "Pear-Shaped Cubic." http://www-groups.dcs.st-and.ac.uk/~history/Curves/Pearshaped.html.

Pearson's Correlation

see CORRELATION COEFFICIENT

Pearson-Cunningham Function

see CUNNINGHAM FUNCTION

Pearson's Function

$$I\left(\frac{{\chi_s}^2}{\sqrt{2(k-1)}}, \frac{k-3}{2}\right) \equiv \frac{\Gamma\left(\frac{1}{2}{\chi_s}^2, \frac{k-1}{2}\right)}{\Gamma\left(\frac{k-1}{2}\right)},$$

where $\Gamma(x)$ is the GAMMA FUNCTION.

see also CHI-SQUARED TEST, GAMMA FUNCTION

Pearson Kurtosis

Let μ_4 be the fourth MOMENT of a DISTRIBUTION and σ its VARIANCE. Then the Pearson kurtosis is defined by

$$\beta_2 \equiv \frac{\mu_4}{\sigma^4}.$$

see also FISHER KURTOSIS, KURTOSIS

Pearson Mode Skewness

Given a DISTRIBUTION with measured MEAN, MODE, and STANDARD DEVIATION s, the Pearson mode skewness is

$$\frac{\text{mean} - \text{mode}}{s}.$$

see also MEAN, MODE, PEARSON SKEWNESS, PEARSON'S SKEWNESS COEFFICIENTS, SKEWNESS

Pearson Skewness

Let a DISTRIBUTION have third MOMENT μ_3 and STANDARD DEVIATION σ, then the Pearson skewness is defined by

$$\beta_1 = \left(\frac{\mu_3}{\sigma^3}\right)^2.$$

see also FISHER SKEWNESS, PEARSON'S SKEWNESS COEFFICIENTS, SKEWNESS

Pearson's Skewness Coefficients

Given a DISTRIBUTION with measured MEAN, MEDIAN, MODE, and STANDARD DEVIATION s, Pearson's first skewness coefficient is

$$\frac{3[\text{mean}] - [\text{mode}]}{s},$$

and the second coefficient is

$$\frac{3[\text{mean}] - [\text{median}]}{s}.$$

see also FISHER SKEWNESS, PEARSON SKEWNESS, SKEWNESS

Pearson System

Generalizes the differential equation for the GAUSSIAN DISTRIBUTION

$$\frac{dy}{dx} = \frac{y(m-x)}{a} \tag{1}$$

to

$$\frac{dy}{dx} = \frac{y(m-x)}{a+bx+cx^2}. \tag{2}$$

Let c_1, c_2 be the roots of $a+bx+cx^2$. Then the possible types of curves are

0. $b=c=0$, $a>0$. E.g., NORMAL DISTRIBUTION.
I. $b^2/4ac<0$, $c_1 \leq x \leq c_2$. E.g., BETA DISTRIBUTION.
II. $b^2/4ac=0$, $c<0$, $-c_1 \leq x \leq c_1$ where $c_1 \equiv \sqrt{-c/a}$.
III. $b^2/4ac=\infty$, $c=0$, $c_1 \leq x < \infty$ where $c_1 \equiv -a/b$. E.g., GAMMA DISTRIBUTION. This case is intermediate to cases I and VI.
IV. $0<b^2/4ac<1$, $-\infty < x < \infty$.
V. $b^2/4ac=1$, $c_1 \leq x < \infty$ where $c_1 \equiv -b/2a$. Intermediate to cases IV and VI.

VI. $b^2/4ac > 1$, $c_1 \leq x < \infty$ where c_1 is the larger root. E.g., BETA PRIME DISTRIBUTION.

VII. $b^2/4ac = 0$, $c > 0$, $-\infty < x < \infty$. E.g., STUDENT'S t-DISTRIBUTION.

Classes IX–XII are discussed in Pearson (1916). See also Craig (in Kenney and Keeping 1951). If a Pearson curve possesses a MODE, it will be at $x = m$. Let $y(x) = 0$ at c_1 and c_2, where these may be $-\infty$ or ∞. If yx^{r+2} also vanishes at c_1, c_2, then the rth MOMENT and $(r+1)$th MOMENTS exist.

$$\int_{c_1}^{c_2} \frac{dy}{dx}(ax^r+bx^{r+1}+cx^{r+2})\,dx = \int_{c_1}^{c_2} y(mx^r-x^{r+1})\,dx \tag{3}$$

giving

$$[y(ax^r + bx^{r+1} + cx^{r+2})]_{c_1}^{c_2} - \int_{c_1}^{c_2} y[arx^{r-1} + b(r+1)x^r + c(r+2)x^{r+1}]\,dx = \int_{c_1}^{c_2} y(mx^r - x^{r+1})\,dx \tag{4}$$

$$0 - \int_{c_1}^{c_2} y[arx^{r-1} + b(r+1)x^r + c(r+2)x^{r+1}]\,dx = \int_{c_1}^{c_2} y(mx^r - x^{r+1})\,dx \tag{5}$$

also,

$$\nu_r = \int_{c_1}^{c_2} yx^r\,dx, \tag{6}$$

so

$$ar\nu_{r-1} + b(r+1)\nu_r + c(r+2)\nu_{r+1} = -m\nu_r + \nu_{r+1}. \tag{7}$$

For $r = 0$,

$$b + 2c\nu_1 = -m + \nu_1, \tag{8}$$

so

$$\nu_1 = \frac{m+b}{1-2c}. \tag{9}$$

For $r = 1$,

$$a + 2b\nu_1 + 3c\nu_2 = -m\nu_1 + \nu_2, \tag{10}$$

so

$$\nu_2 = \frac{a + (m+2b)\nu_1}{1-3c}. \tag{11}$$

Now let $t \equiv (x - \nu_1)/\sigma$. Then

$$\nu_1 = 0 \tag{12}$$

$$\nu_2 = \mu_2 = 1 \tag{13}$$

$$\alpha_r = \mu_r = \nu_r. \tag{14}$$

Hence $b = -m$, and $a = 1 - c$ so

$$(1-3c)r\alpha_{r-1} - mr\alpha_r + [c(r+2) - 1]\alpha_{r+1} = 0. \tag{15}$$

For $r = 2$,

$$2m + (1-4c)\alpha_3 = 0. \tag{16}$$

For $r = 3$,

$$3(1-3c) - 3m\alpha_3 - (1-5c)\alpha_4 = 0. \tag{17}$$

So the SKEWNESS and KURTOSIS are

$$\gamma_1 = \alpha_3 = \frac{2m}{4c-1} \tag{18}$$

$$\gamma_2 = \alpha_4 - 3 = \frac{6(m^2 - 4c^2 + c)}{(4c-1)(5c-1)}. \tag{19}$$

So the parameters a, b, and c can be written

$$a = 1 - 3c \tag{20}$$

$$b = -m = \frac{\gamma_1}{2(1+2\delta)} \tag{21}$$

$$c = \frac{\delta}{2(1+2\delta)}, \tag{22}$$

where

$$\delta \equiv \frac{2\gamma_2 - 3\gamma_1{}^2}{\gamma_2 + 6}. \tag{23}$$

References

Craig, C. C. "A New Exposition and Chart for the Pearson System of Frequency Curves." *Ann. Math. Stat.* **7**, 16–28, 1936.

Kenney, J. F. and Keeping, E. S. *Mathematics of Statistics, Pt. 2, 2nd ed.* Princeton, NJ: Van Nostrand, p. 107, 1951.

Pearson, K. "Second Supplement to a Memoir on Skew Variation." *Phil. Trans. A* **216**, 429–457, 1916.

Pearson Type III Distribution

A skewed distribution which is similar to the BINOMIAL DISTRIBUTION when $p \neq q$ (Abramowitz and Stegun 1972, p. 930).

$$y = k(t+A)^{A^2-1}e^{-At}, \tag{1}$$

for $t \in [0, \infty)$ where

$$A \equiv 2/\gamma \tag{2}$$

$$K \equiv \frac{A^{A^2}e^{-A^2}}{\Gamma(A^2)}, \tag{3}$$

$\Gamma(z)$ is the GAMMA FUNCTION, and t is a standardized variate. Another form is

$$P(x) = \frac{1}{\beta\Gamma(p)}\left(\frac{x-\alpha}{\beta}\right)^{p-1} \exp\left(\frac{x-\alpha}{\beta}\right). \tag{4}$$

For this distribution, the CHARACTERISTIC FUNCTION is

$$\phi(t) = e^{i\alpha t}(1 - i\beta t)^{-p}, \quad (5)$$

and the MEAN, VARIANCE, SKEWNESS, and KURTOSIS are

$$\mu = \alpha + p\beta \quad (6)$$
$$\sigma^2 = p\beta^2 \quad (7)$$
$$\gamma_1 = \frac{2}{\sqrt{p}} \quad (8)$$
$$\gamma_2 = \frac{6}{p}. \quad (9)$$

References

Abramowitz, M. and Stegun, C. A. (Eds.). *Handbook of Mathematical Functions with Formulas, Graphs, and Mathematical Tables, 9th printing.* New York: Dover, 1972.

Pearls of Sluze

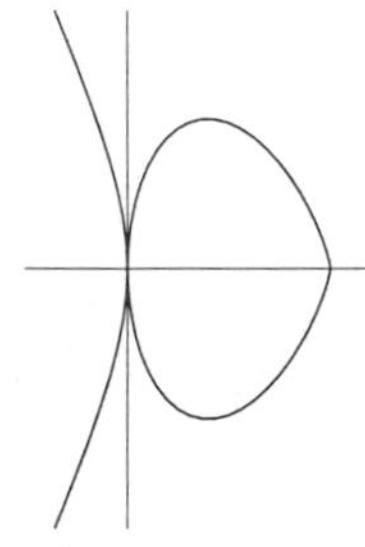

$$y^m = kx^n(a - x)^b.$$

The curves with integral n, p, and m were studied by de Sluze between 1657 and 1698. The name "Pearls of Sluze" was given to these curves by Blaise Pascal (MacTutor Archive).

References

MacTutor History of Mathematics Archive. "Pearls of Sluze." http://www-groups.dcs.st-and.ac.uk/~history /Curves/Pearls.html.

Peaucellier Cell

see PEAUCELLIER INVERSOR

Peaucellier Inversor

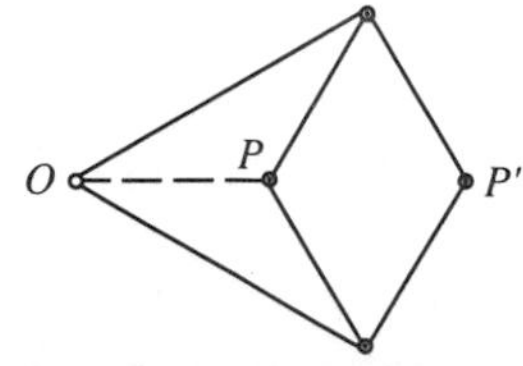

A LINKAGE with six rods which draws the inverse of a given curve. When a pencil is placed at P, the inverse is drawn at P' (or vice versa). If a seventh rod (dashed) is added (with an additional pivot), P is kept on a circle and the locus traced out by P' is a straight line. It therefore converts circular motion to linear motion without sliding, and was discovered in 1864. Another LINKAGE which performs this feat using hinged squares had been published by Sarrus in 1853 but ignored. Coxeter (1969, p. 428) shows that

$$OP \times OP' = OA^2 - PA^2.$$

see also HART'S INVERSOR, LINKAGE

References

Bogomolny, A. "Peaucellier Linkage." http://www.cut-the-knot.com/pythagoras/invert.html.
Courant, R. and Robbins, H. *What is Mathematics?: An Elementary Approach to Ideas and Methods.* Oxford, England: Oxford University Press, p. 156, 1978.
Coxeter, H. S. M. *Introduction to Geometry, 2nd ed.* New York: Wiley, pp. 82–83, 1969.
Ogilvy, C. S. *Excursions in Geometry.* New York: Dover, pp. 46–48, 1990.
Rademacher, H. and Toeplitz, O. *The Enjoyment of Mathematics: Selections from Mathematics for the Amateur.* Princeton, NJ: Princeton University Press, pp. 121–126, 1957.
Smith, D. E. *A Source Book in Mathematics.* New York: Dover, p. 324, 1994.

Peaucellier's Linkage

see PEAUCELLIER INVERSOR

Pedal

The pedal of a curve with respect to a point P is the locus of the foot of the PERPENDICULAR from P to the TANGENT to the curve. When a CLOSED CURVE rolls on a straight line, the AREA between the line and ROULETTE after a complete revolution by any point on the curve is twice the AREA of the pedal (taken with respect to the generating point) of the rolling curve.

Pedal Circle

The pedal CIRCLE of a point P in a TRIANGLE is the CIRCLE through the feet of the perpendiculars from P to the sides of the TRIANGLE (the CIRCUMCIRCLE about the PEDAL TRIANGLE). When P is on a side of the TRIANGLE, the line between the two perpendiculars is called the PEDAL LINE. Given four points, no three of which are COLLINEAR, then the four PEDAL CIRCLES of each point for the TRIANGLE formed by the other three have a common point through which the NINE-POINT CIRCLES of the four TRIANGLES pass. The radius of the pedal circle of a point P is

$$r = \frac{\overline{A_1P} \cdot \overline{A_2P} \cdot \overline{A_3P}}{2(R^2 - \overline{OP}^2)}$$

(Johnson 1929, p. 141).

see also MIQUEL POINT, NINE-POINT CIRCLE, PEDAL TRIANGLE

References

Johnson, R. A. *Modern Geometry: An Elementary Treatise on the Geometry of the Triangle and the Circle.* Boston, MA: Houghton Mifflin, 1929.

Pedal Coordinates

The pedal coordinates of a point P with respect to the curve C and the PEDAL POINT O are the radial distant r from O to P and the PERPENDICULAR distance p from O to the line L tangent to C at P.

References

Lawrence, J. D. *A Catalog of Special Plane Curves.* New York: Dover, pp. 2–3, 1972.

Yates, R. C. "Pedal Equations." *A Handbook on Curves and Their Properties.* Ann Arbor, MI: J. W. Edwards, pp. 166–169, 1952.

Pedal Curve

Given a curve C, the pedal curve of C with respect to a fixed point O (the PEDAL POINT) is the locus of the point P of intersection of the PERPENDICULAR from O to a TANGENT to C. The parametric equations for a curve $(f(t), g(t))$ relative to the PEDAL POINT (x_0, y_0) are

$$x = \frac{x_0 f'^2 + f g'^2 + (y_0 - g) f' g'}{f'^2 + g'^2}$$

$$y = \frac{g f'^2 + y_0 g'^2 + (x_0 - f) f' g'}{f'^2 + g'2^2}.$$

Curve	Pole	Pedal
astroid	center	quadrifolium
cardioid	cusp	Cayley's sextic
central conic	focus	circle
circle	any point	limaçon
circle	on circumference	cardioid
circle involute	center of circle	Archimedean spiral
cissoid of Diocles	focus	cardioid
deltoid	center	trifolium
deltoid	cusp	simple folium
deltoid	on the curve	unsymmetrical double folium
deltoid	vertex	double folium
epicycloid	center	rose
hypocycloid	center	rose
line	any point	point
logarithmic spiral	pole	logarithmic spiral
parabola	focus	line
parabola	foot of directrix	right strophoid
parabola	on directrix	strophoid
parabola	refl. of focus by dir.	Maclaurin trisectrix
parabola	vertex	cissoid of Diocles
sinusoidal spiral	pole	sinusoidal spiral
Tschirnhausen cubic	focus of pedal	parabola

see also NEGATIVE PEDAL CURVE

References

Lawrence, J. D. *A Catalog of Special Plane Curves.* New York: Dover, pp. 46–49 and 204, 1972.

Lee, X. "Pedal." `http://www.best.com/~xah/SpecialPlaneCurves_dir/Pedal_dir/pedal.html`.

Lockwood, E. H. "Pedal Curves." Ch. 18 in *A Book of Curves.* Cambridge, England: Cambridge University Press, pp. 152–155, 1967.

Yates, R. C. "Pedal Curves." *A Handbook on Curves and Their Properties.* Ann Arbor, MI: J. W. Edwards, pp. 160–165, 1952.

Pedal Line

Mark a point P on a side of a TRIANGLE and draw the perpendiculars from the point to the two other sides. The line between the feet of these two perpendiculars is called the pedal line.

see also PEDAL TRIANGLE, SIMSON LINE

Pedal Point

The fixed point with respect to which a PEDAL CURVE is drawn.

Pedal Triangle

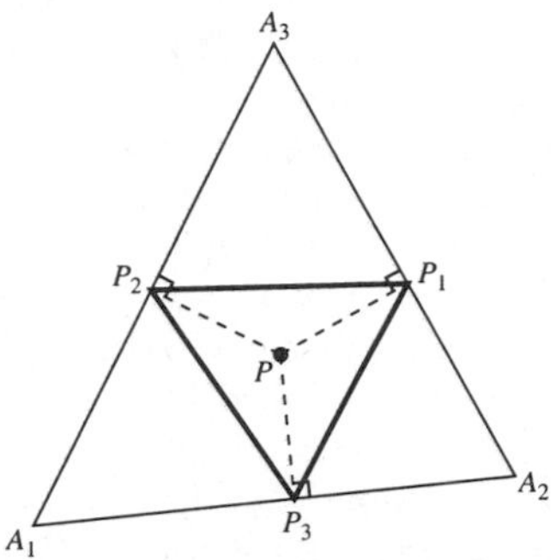

Given a point P, the pedal triangle of P is the TRIANGLE whose VERTICES are the feet of the perpendiculars from P to the side lines. The pedal triangle of a TRIANGLE with TRILINEAR COORDINATES $\alpha : \beta : \gamma$ and angles A, B, and C has VERTICES with TRILINEAR COORDINATES

$$0 : \beta + \alpha \cos C : \gamma + \alpha \cos B \tag{1}$$

$$\alpha + \beta \cos C : 0 : \gamma + \beta \cos A \tag{2}$$

$$\alpha + \gamma \cos B : \beta + \gamma \cos A : 0. \tag{3}$$

The third pedal triangle is similar to the original one. This theorem can be generalized to: the nth pedal n-gon of any n-gon is similar to the original one. It is also true that

$$P_2 P_3 = A_1 P \sin \alpha_1 \tag{4}$$

(Johnson 1929, pp. 135–136). The AREA A of the pedal triangle of a point P is proportional to the POWER of P with respect to the CIRCUMCIRCLE,

$$A = \tfrac{1}{2}(R^2 - \overline{OP}^2) \sin \alpha_1 \sin \alpha_2 \sin \alpha_3 = \frac{R^2 - \overline{OP}^2}{4R^2} \Delta \tag{5}$$

(Johnson 1929, pp. 139–141).

see also ANTIPEDAL TRIANGLE, FAGNANO'S PROBLEM, PEDAL CIRCLE, PEDAL LINE, SCHWARZ'S TRIANGLE PROBLEM

References

Coxeter, H. S. M. and Greitzer, S. L. *Geometry Revisited.* Washington, DC: Math. Assoc. Amer., pp. 22–26, 1967.

Johnson, R. A. *Modern Geometry: An Elementary Treatise on the Geometry of the Triangle and the Circle.* Boston, MA: Houghton Mifflin, 1929.

Peg Knot

see CLOVE HITCH

Peg Solitaire

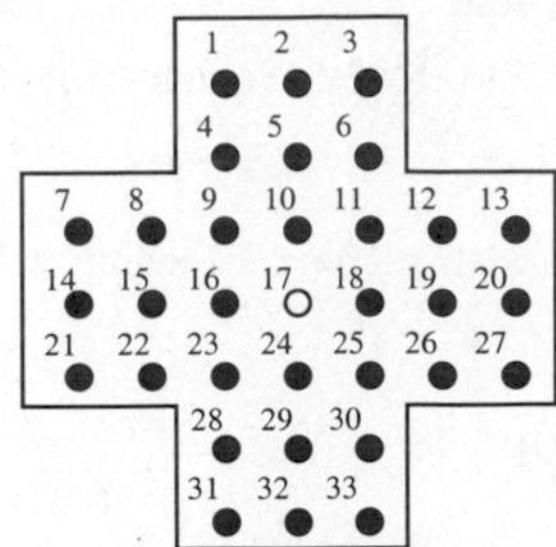

A game played on a cross-shaped board with 33 holes. All holes but the middle one are initially filled with pegs. The goal is to remove all pegs but one by jumping pegs from one side of an occupied peg hole to an empty space, removing the peg which was jumped over. Strategies and symmetries are discussed in Beeler *et al.* (1972, Item 75). A triangular version called HI-Q also exists (Beeler *et al.* 1972, Item 76). Kraitchik (1942) considers a board with one additional hole placed at the vertices of the central right angles.

see also HI-Q

References

Beeler, M.; Gosper, R. W.; and Schroeppel, R. *HAKMEM.* Cambridge, MA: MIT Artificial Intelligence Laboratory, Memo AIM-239, Feb. 1972.

Gardner, M. "Peg Solitaire." Ch. 11 in *The Unexpected Hanging and Other Mathematical Diversions.* New York: Simon and Schuster, pp. 122–135 and 250–251, 1969.

Kraitchik, M. "Peg Solitaire." §12.19 in *Mathematical Recreations.* New York: W. W. Norton, pp. 297–298, 1942.

Peg Top

see PIRIFORM

Peirce's Theorem

The only linear associative algebra in which the coordinates are REAL NUMBERS and products vanish only if one factor is zero are the FIELD of REAL NUMBERS, the FIELD of COMPLEX NUMBERS, and the algebra of QUATERNIONS with REAL COEFFICIENTS.

see also WEIERSTRAß'S THEOREM

Pell Equation

A special case of the quadratic DIOPHANTINE EQUATION having the form

$$x^2 - Dy^2 = 1, \tag{1}$$

where D is a nonsquare NATURAL NUMBER. Dörrie (1965) defines the equation as

$$x^2 - Dy^2 = 4 \tag{2}$$

and calls it the FERMAT DIFFERENCE EQUATION. The general Pell equation was solved by the Indian mathematician Bhaskara.

Pell equations, as well as the analogous equation with a minus sign on the right, can be solved by finding the CONTINUED FRACTION $[a_1, a_2, \ldots]$ for $\sqrt{D}$. (The trivial solution $x = 1$, $y = 0$ is ignored in all subsequent discussion.) Let p_n/q_n denote the nth CONVERGENT $[a_1, a_2, \ldots, a_n]$, then we are looking for a convergent which obeys the identity

$$p_n{}^2 - Dq_n{}^2 = (-1)^n, \tag{3}$$

which turns out to always be possible since the CONTINUED FRACTION of a QUADRATIC SURD always becomes periodic at some term a_{r+1}, where $a_{r+1} = 2a_1$, i.e.,

$$\sqrt{D} = [a_1, \overline{a_2, \ldots, a_r, 2a_1}]. \tag{4}$$

Writing $n = rk$ gives

$$p_{rk}{}^2 - Dq_{rk}{}^2 = (-1)^{rk}, \tag{5}$$

for k anPOSITIVE INTEGER. If r is ODD, solutions to

$$x^2 - Dy^2 = \pm 1 \tag{6}$$

can be obtained if k is chosen to be EVEN or ODD, but if r is EVEN, there are no values of k which can make the exponent ODD.

If r is EVEN, then $(-1)^r$ is POSITIVE and the solution in terms of smallest INTEGERS is $x = p_r$ and $y = q_r$, where p_r/q_r is the rth CONVERGENT. If r is ODD, then $(-1)^r$ is NEGATIVE, but we can take $k = 2$ in this case, to obtain

$$p_{2r}{}^2 - Dq_{2r}{}^2 = 1, \tag{7}$$

so the solution in smallest INTEGERS is $x = p_{2r}$, $y = q_{2r}$. Summarizing,

$$(x, y) = \begin{cases} (p_r, q_r) & \text{for } r \text{ even} \\ (p_{2r}, p_{2r}) & \text{for } r \text{ odd.} \end{cases} \tag{8}$$

Given one solution $(x, y) = (p, q)$ (which can be found as above), a whole family of solutions can be found by taking each side to the nth POWER,

$$x^2 - Dy^2 = (p^2 - Dq^2)^n = 1. \tag{9}$$

Factoring gives

$$(x + \sqrt{D}\,y)(x - \sqrt{D}\,y) = (p + \sqrt{D}\,q)^n (p - \sqrt{D}\,q)^n \tag{10}$$

and

$$x + \sqrt{D}\,y = (p + \sqrt{D}\,q)^n \tag{11}$$

$$x - \sqrt{D}\,y = (p - \sqrt{D}\,q)^n, \tag{12}$$

which gives the family of solutions

$$x = \frac{(p+q\sqrt{D})^n + (p-q\sqrt{D})^n}{2} \tag{13}$$

$$y = \frac{(p+q\sqrt{D})^n - (p-q\sqrt{D})^n}{2\sqrt{D}}. \tag{14}$$

These solutions also hold for

$$x^2 - Dy^2 = -1, \tag{15}$$

except that n can take on only ODD values.

The following table gives the smallest integer solutions (x, y) to the Pell equation with constant $D \leq 102$ (Beiler 1966, p. 254). SQUARE $D = d^2$ are not included, since they would result in an equation of the form

$$x^2 - d^2y^2 = x^2 - (dy)^2 = x^2 - y'^2 = 1, \tag{16}$$

which has no solutions (since the difference of two SQUARES cannot be 1).

D	x	y	D	x	y
2	3	2	54	485	66
3	2	1	55	89	12
5	9	4	56	15	2
6	5	2	57	151	20
7	8	3	58	19603	2574
8	3	1	59	530	69
10	19	6	60	31	4
11	10	3	61	1766319049	226153980
12	7	2	62	63	8
13	649	180	63	8	1
14	15	4	65	129	16
15	4	1	66	65	8
17	33	8	67	48842	5967
18	17	4	68	33	4
19	170	39	69	7775	936
20	9	2	70	251	30
21	55	12	71	3480	413
22	197	42	72	17	2
23	24	5	73	2281249	267000
24	5	1	74	3699	430
26	51	10	75	26	3
27	26	5	76	57799	6630
28	127	24	77	351	40
29	9801	1820	78	53	6
30	11	2	79	80	9
31	1520	273	80	9	1
32	17	3	82	163	18
33	23	4	83	82	9
34	35	6	84	55	6
35	6	1	85	285769	30996
37	73	12	86	10405	1122
38	37	6	87	28	3
39	25	4	88	197	21
40	19	3	89	500001	53000
41	2049	320	90	19	2
42	13	2	91	1574	165
43	3482	531	92	1151	120
44	199	30	93	12151	1260
45	161	24	94	2143295	221064
46	24335	3588	95	39	4
47	48	7	96	49	5
48	7	1	97	62809633	6377352
50	99	14	98	99	10
51	50	7	99	10	1
52	649	90	101	201	20
53	66249	9100	102	101	10

The first few minimal values of x and y for nonsquare D are 3, 2, 9, 5, 8, 3, 19, 10, 7, 649, ... (Sloane's A033313) and 2, 1, 4, 2, 3, 1, 6, 3, 2, 180, ... (Sloane's A033317), respectively. The values of D having $x = 2, 3, \ldots$ are 3, 2, 15, 6, 35, 12, 7, 5, 11, 30, ... (Sloane's A033314) and the values of D having $y = 1, 2, \ldots$ are 3, 2, 7, 5, 23, 10, 47, 17, 79, 26, ... (Sloane's A033318). Values of the incrementally largest minimal x are 3, 9, 19, 649, 9801, 24335, 66249, ... (Sloane's A033315) which occur at $D = 2, 5, 10, 13, 29, 46, 53, 61, 109, 181, \ldots$ (Sloane's A033316). Values of the incrementally largest minimal

y are 2, 4, 6, 180, 1820, 3588, 9100, 226153980, ... (Sloane's A033319), which occur at $D = 2$, 5, 10, 13, 29, 46, 53, 61, ... (Sloane's A033320).

see also DIOPHANTINE EQUATION, DIOPHANTINE EQUATION—QUADRATIC, LAGRANGE NUMBER (DIOPHANTINE EQUATION)

References

Beiler, A. H. "The Pellian." Ch. 22 in *Recreations in the Theory of Numbers: The Queen of Mathematics Entertains.* New York: Dover, pp. 248–268, 1966.

Degan, C. F. *Canon Pellianus.* Copenhagen, Denmark, 1817.

Dörrie, H. *100 Great Problems of Elementary Mathematics: Their History and Solutions.* New York: Dover, 1965.

Lagarias, J. C. "On the Computational Complexity of Determining the Solvability or Unsolvability of the Equation $X^2 - DY^2 = -1$." *Trans. Amer. Math. Soc.* **260**, 485–508, 1980.

Smarandache, F. "Un metodo de resolucion de la ecuacion diofantica." *Gaz. Math.* **1**, 151–157, 1988.

Smarandache, F. " Method to Solve the Diophantine Equation $ax^2 - by^2 + c = 0$." In *Collected Papers, Vol. 1.* Lupton, AZ: Erhus University Press, 1996.

Stillwell, J. C. *Mathematics and Its History.* New York: Springer-Verlag, 1989.

Whitford, E. E. *Pell Equation.* New York: Columbia University Press, 1912.

Pell-Lucas Number

see PELL NUMBER

Pell-Lucas Polynomial

see PELL POLYNOMIAL

Pell Number

The numbers obtained by the U_ns in the LUCAS SEQUENCE with $P = 2$ and $Q = -1$. They and the Pell-Lucas numbers (the V_ns in the LUCAS SEQUENCE) satisfy the recurrence relation

$$P_n = 2P_{n-1} + P_{n-2}. \tag{1}$$

Using P_i to denote a Pell number and Q_i to denote a Pell-Lucas number,

$$P_{m+n} = P_m P_{n+1} + P_{m-1} P_n \tag{2}$$

$$P_{m+n} = 2P_m Q_n - (-1)^n P_{m-n}, \tag{3}$$

$$P_{2^t m} = P_m(2Q_m)(2Q_{2m})(2Q_{4m})\cdots(2Q_{2^{t-1}m}) \tag{4}$$

$$Q_m{}^2 = 2P_m{}^2 + (-1)^m \tag{5}$$

$$Q_{2m} = 2Q_m{}^2 - (-1)^m. \tag{6}$$

The Pell numbers have $P_0 = 0$ and $P_1 = 1$ and are 0, 1, 2, 5, 12, 29, 70, 169, 408, 985, 2378, ... (Sloane's A000129). The Pell-Lucas numbers have $Q_0 = 2$ and $Q_1 = 2$ and are 2, 2, 6, 14, 34, 82, 198, 478, 1154, 2786, 6726, ... (Sloane's A002203).

The only TRIANGULAR Pell number is 1 (McDaniel 1996).

see also BRAHMAGUPTA POLYNOMIAL, PELL POLYNOMIAL

References

McDaniel, W. L. "Triangular Numbers in the Pell Sequence." *Fib. Quart.* **34**, 105–107, 1996.

Sloane, N. J. A. Sequences A000129/M1413 and A002203/M0360 in "An On-Line Version of the Encyclopedia of Integer Sequences."

Pell Polynomial

The Pell polynomials $P(x)$ and Lucas-Pell polynomials $Q(x)$ are generated by a LUCAS POLYNOMIAL SEQUENCE using generator $(2x, 1)$. This gives recursive equations for $P(x)$ from $P_0(x) = P_1(x) = 1$ and

$$P_{n+2}(x) = 2xP_{n+1}(x) + P_n(x). \tag{1}$$

The first few are

$$\begin{aligned} P_1 &= 1 \\ P_2 &= 2x \\ P_3 &= 4x^2 - 1 \\ P_4 &= 8x^3 - 4x \\ P_5 &= 16x^4 - 12x^2 + 1. \end{aligned}$$

The Pell-Lucas numbers are defined recursively by $q_0(x) = 1$, $q_1(x) = x$ and

$$q_{n+2}(x) = 2xq_{n+1}(x) + q_n(x), \tag{2}$$

together with

$$Q_n(x) \equiv 2q_n(x). \tag{3}$$

The first few are

$$\begin{aligned} Q_1 &= 2x \\ Q_2 &= 4x^2 - 2 \\ Q_3 &= 8x^3 - 6x \\ Q_4 &= 16x^4 - 16x^2 + 2 \\ Q_5 &= 32x^5 - 40x^3 + 10x. \end{aligned}$$

see also LUCAS POLYNOMIAL SEQUENCE

References

Horadam, A. F. and Mahon, J. M. "Pell and Pell-Lucas Polynomials." *Fib. Quart.* **23**, 7–20, 1985.

Mahon, J. M. M. A. (Honors) thesis, The University of New England. Armidale, Australia, 1984.

Sloane, N. J. A. Sequence A000129/M1413 in "An On-Line Version of the Encyclopedia of Integer Sequences."

Pell Sequence

see PELL NUMBER

Pencil

The set of all LINES through a point. Woods (1961), however, uses this term as a synonym for RANGE.

see also NEAR-PENCIL, PERSPECTIVITY, RANGE (LINE SEGMENT), SECTION (PENCIL), SHEAF (GEOMETRY)

References

Woods, F. S. *Higher Geometry: An Introduction to Advanced Methods in Analytic Geometry.* New York: Dover, pp. 8 and 11–12, 1961.

Penrose Stairway

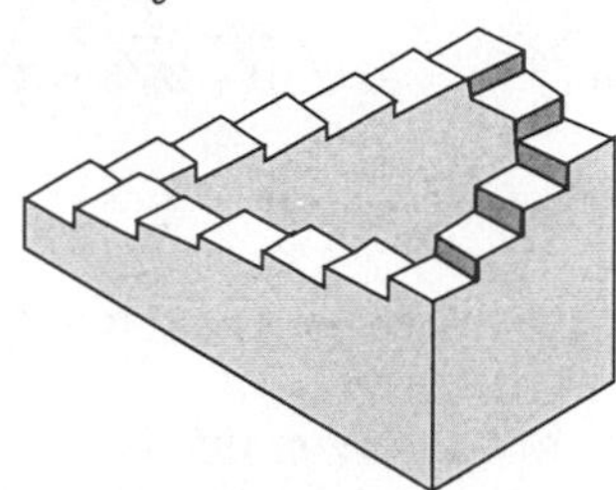

An IMPOSSIBLE FIGURE (also called the SCHROEDER STAIRS) in which a stairway in the shape of a square appears to circulate indefinitely while still possessing normal steps. The Dutch artist M. C. Escher included Penrose stairways in many of his mind-bending illustrations.

see also IMPOSSIBLE FIGURE

References

Hofstadter, D. R. *Gödel, Escher, Bach: An Eternal Golden Braid.* New York: Vintage Books, p. 15, 1989.

Jablan, S. "Impossible Figures." `http://members.tripod.com/~modularity/impos.htm`.

Pappas, T. "Optical Illusions and Computer Graphics." *The Joy of Mathematics.* San Carlos, CA: Wide World Publ./Tetra, p. 5, 1989.

Robinson, J. O. and Wilson, J. A. "The Impossible Colonnade and Other Variations of a Well-Known Figure." *Brit. J. Psych.* **64**, 363–365, 1973.

Penrose Tiles

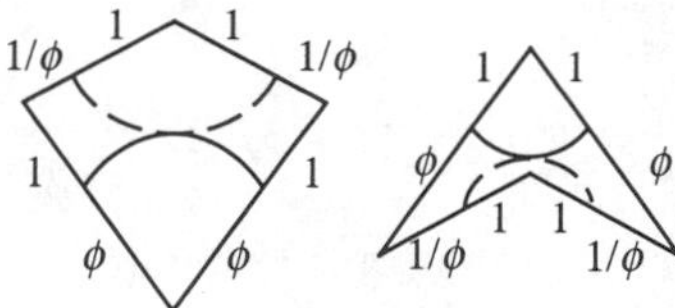

A pair of shapes which tile the plane only aperiodically (when the markings are constrained to match at borders). The two tiles, illustrated above, are called the "KITE" and "DART."

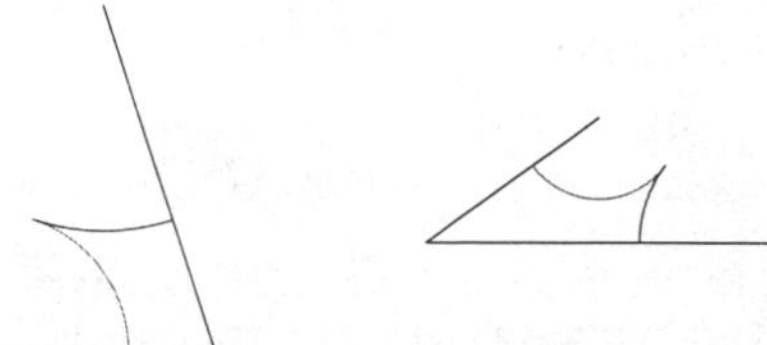

To see how the plane may be tiled aperiodically using the kite and dart, divide the kite into acute and obtuse tiles, shown above. Now define "deflation" and "inflation" operations. The deflation operator takes an acute TRIANGLE to the union of two ACUTE TRIANGLES and one OBTUSE, and the OBTUSE TRIANGLE goes to an ACUTE and an OBTUSE TRIANGLE. These operations are illustrated below.

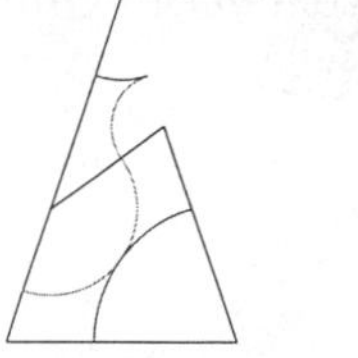

When applied to a collection of tiles, the deflation operator leads to a more refined collection. The operators do not respect tile boundaries, but do respect the half tiles defined above. There are two ways to obtain aperiodic TILINGS with 5-fold symmetry about a single point. They are known an the "star" and "sun" configurations, and are show below.

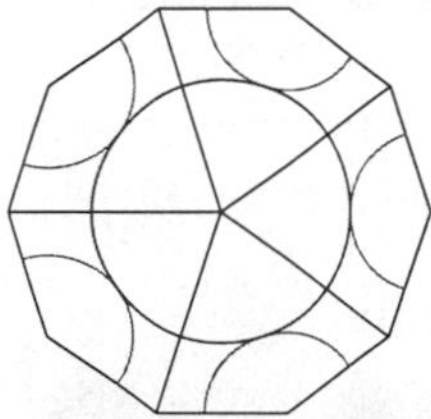

Higher order versions can then be obtained by deflation. For example, the following are third-order deflations:

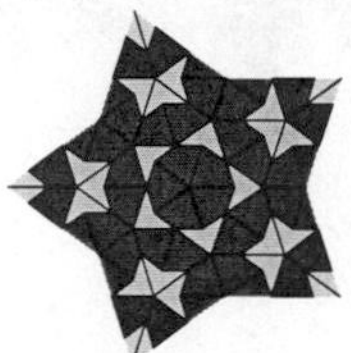

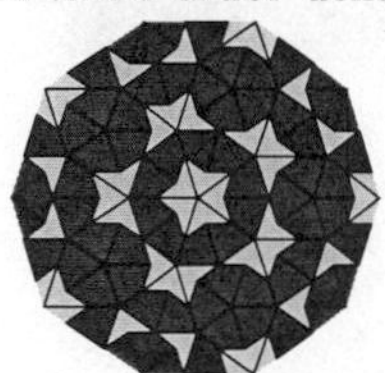

References

Gardner, M. Chs. 1–2 in *Penrose Tiles and Trapdoor Ciphers... and the Return of Dr. Matrix, reissue ed.* New York: W. H. Freeman, pp. 299–300, 1989.

Hurd, L. P. "PenroseTiles." `http://www.mathsource.com/cgi-bin/MathSource/Applications/Graphics/2D/0206-772`.

Peterson, I. *The Mathematical Tourist: Snapshots of Modern Mathematics.* New York: W. H. Freeman, pp. 86–95, 1988.

Wagon, S. "Penrose Tiles." §4.3 in *Mathematica in Action.* New York: W. H. Freeman, pp. 108–117, 1991.

Penrose Triangle

see TRIBAR

Penrose Tribar

see TRIBAR

Pentabolo

A 5-POLYABOLO.

Pentacle

see PENTAGRAM

Pentacontagon

A 50-sided POLYGON.

Pentad

A group of five elements.

see also MONAD, PAIR, QUADRUPLET, QUINTUPLET, TETRAD, TRIAD, TRIPLET, TWINS

Pentadecagon

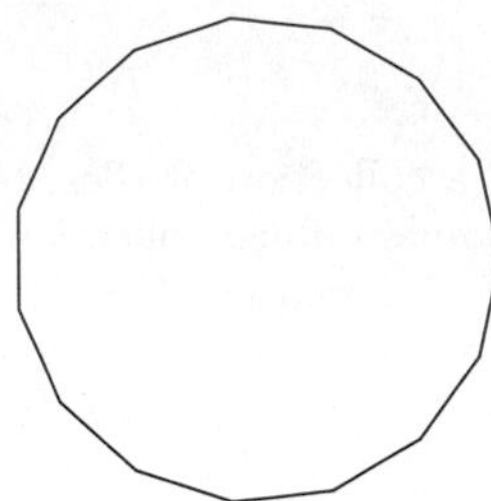

A 15-sided POLYGON, sometimes also called the PENTAKAIDECAGON.

see also POLYGON, REGULAR POLYGON, TRIGONOMETRY VALUES—$\pi/15$

Pentaflake

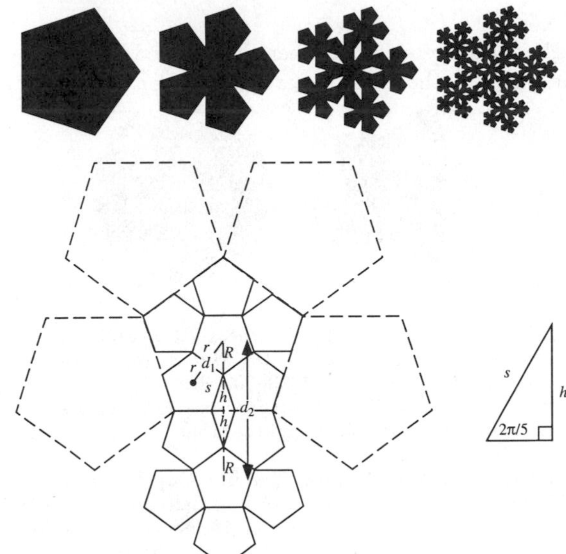

A FRACTAL with 5-fold symmetry. As illustrated above, five PENTAGONS can be arranged around an identical PENTAGON to form the first iteration of the pentaflake. This cluster of six pentagons has the shape of a pentagon with five triangular wedges removed. This construction was first noticed by Albrecht Dürer (Dixon 1991).

For a pentagon of side length 1, the first ring of pentagons has centers at RADIUS

$$d_1 = 2r = \tfrac{1}{2}(1+\sqrt{5})R = \phi R, \tag{1}$$

where ϕ is the GOLDEN RATIO. The INRADIUS r and CIRCUMRADIUS R are related by

$$r = R\cos(\tfrac{1}{5}\pi) = \tfrac{1}{4}(\sqrt{5}+1)R, \tag{2}$$

and these are related to the side length s by

$$s = 2\sqrt{R^2 - r^2} = \tfrac{1}{2}R\sqrt{10-2\sqrt{5}}. \tag{3}$$

The height h is

$$h = s\sin(\tfrac{2}{5}\pi) = \tfrac{1}{4}s\sqrt{10+2\sqrt{5}} = \tfrac{1}{2}\sqrt{5}\,R, \tag{4}$$

giving a RADIUS of the second ring as

$$d_2 = 2(R+h) = (2+\sqrt{5})R = \phi^3 R. \tag{5}$$

Continuing, the nth pentagon ring is located at

$$d_n = \phi^{2n-1}. \tag{6}$$

Now, the length of the side of the first pentagon compound is given by

$$s_2 = 2\sqrt{(2r+R)^2 - (h+R)^2} = R\sqrt{5+2\sqrt{5}}, \tag{7}$$

so the ratio of side lengths of the original pentagon to that of the compound is

$$\frac{s_2}{s} = \frac{R\sqrt{5+2\sqrt{5}}}{\tfrac{1}{2}R\sqrt{10-2\sqrt{5}}} = 1+\phi. \tag{8}$$

We can now calculate the dimension of the pentaflake fractal. Let N_n be the number of black pentagons and L_n the length of side of a pentagon after the n iteration,

$$N_n = 6^n \tag{9}$$

$$L_n = (1+\phi)^{-n}. \tag{10}$$

The CAPACITY DIMENSION is therefore

$$d_{\text{cap}} = -\lim_{n\to\infty}\frac{\ln N_n}{\ln L_n} = \frac{\ln 6}{\ln(1+\phi)} = \frac{\ln 2 + \ln 3}{\ln(1+\phi)} = 1.861715\ldots. \tag{11}$$

see also PENTAGON

References

Dixon, R. *Mathographics.* New York: Dover, pp. 186–188, 1991.

Weisstein, E. W. "Fractals." http://www.astro.virginia.edu/~eww6n/math/notebooks/Fractal.m.

Pentagon

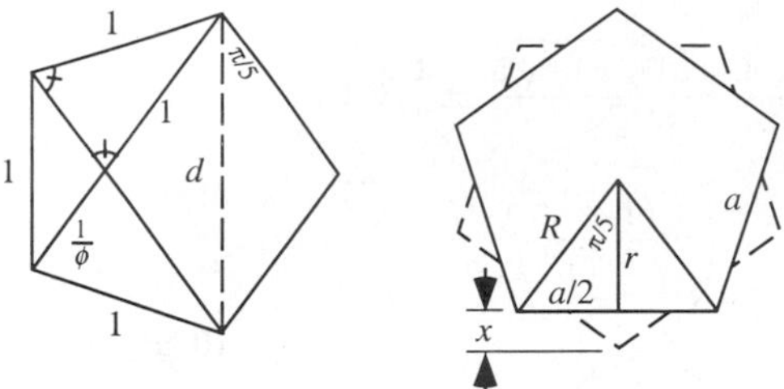

The regular convex 5-gon is called the pentagon. By SIMILAR TRIANGLES in the figure on the left,

$$\frac{d}{1} = \frac{1}{\frac{1}{\phi}} = \phi, \tag{1}$$

where d is the diagonal distance. But the dashed vertical line connecting two nonadjacent VERTICES is the same length as the diagonal one, so

$$\phi = 1 + \frac{1}{\phi} \tag{2}$$

$$\phi^2 - \phi - 1 \tag{3}$$

$$\phi = \frac{1 \pm \sqrt{1+4}}{2} = \frac{1+\sqrt{5}}{2}. \tag{4}$$

This number is the GOLDEN RATIO. The coordinates of the VERTICES relative to the center of the pentagon with unit sides, starting at the right VERTEX and moving clockwise, are $(\cos(\frac{1}{5}n\pi), \sin(\frac{1}{5}n\pi))$ for $n = 0, 1, \ldots, 4$, or

$$(1,0), (c_1, s_1), (c_2, s_2), (c_2, -s_2), (c_1, -s_1), \tag{5}$$

where

$$c_1 = \cos\left(\frac{\pi}{5}\right) = \tfrac{1}{4}(\sqrt{5}+1) \tag{6}$$

$$c_2 = \cos\left(\frac{2\pi}{5}\right) = \tfrac{1}{4}(\sqrt{5}-1) \tag{7}$$

$$s_1 = \sin\left(\frac{\pi}{5}\right) = \tfrac{1}{4}\sqrt{10-2\sqrt{5}} \tag{8}$$

$$s_2 = \sin\left(\frac{2\pi}{5}\right) = \tfrac{1}{4}\sqrt{10+2\sqrt{5}}. \tag{9}$$

For a regular POLYGON, the CIRCUMRADIUS, INRADIUS, SAGITTA, and AREA are given by

$$R_n = \tfrac{1}{2}a\csc\left(\frac{\pi}{n}\right) \tag{10}$$

$$r_n = \tfrac{1}{2}a\cot\left(\frac{\pi}{n}\right) \tag{11}$$

$$x_n = R_n - r_n = \tfrac{1}{2}a\tan\left(\frac{\pi}{2n}\right) \tag{12}$$

$$A_n = \tfrac{1}{4}na^2\cot\left(\frac{\pi}{n}\right). \tag{13}$$

Plugging in $n = 5$ gives

$$R = \tfrac{1}{2}a\csc(\tfrac{1}{5}\pi) = \tfrac{1}{10}a\sqrt{50+10\sqrt{5}} \tag{14}$$

$$r = \tfrac{1}{2}a\cot(\tfrac{1}{5}\pi) = \tfrac{1}{10}a\sqrt{25+10\sqrt{5}} \tag{15}$$

$$x = \tfrac{1}{10}a\sqrt{25-10\sqrt{5}} \tag{16}$$

$$A = \tfrac{5}{4}a^2\sqrt{25+10\sqrt{5}}. \tag{17}$$

Five pentagons can be arranged around an identical pentagon to form the first iteration of the "PENTAFLAKE," which itself has the shape of a pentagon with five triangular wedges removed. For a pentagon of side length 1, the first ring of pentagons has centers at radius ϕ, the second ring at ϕ^3, and the nth at ϕ^{2n-1}.

In proposition IV.11, Euclid showed how to inscribe a regular pentagon in a CIRCLE. Ptolemy also gave a RULER and COMPASS construction for the pentagon in his epoch-making work *The Almagest.* While Ptolemy's construction has a SIMPLICITY of 16, a GEOMETRIC CONSTRUCTION using CARLYLE CIRCLES can be made with GEOMETROGRAPHY symbol $2S_1+S_2+8C_1+0C_2+4C_3$, which has SIMPLICITY 15 (De Temple 1991).

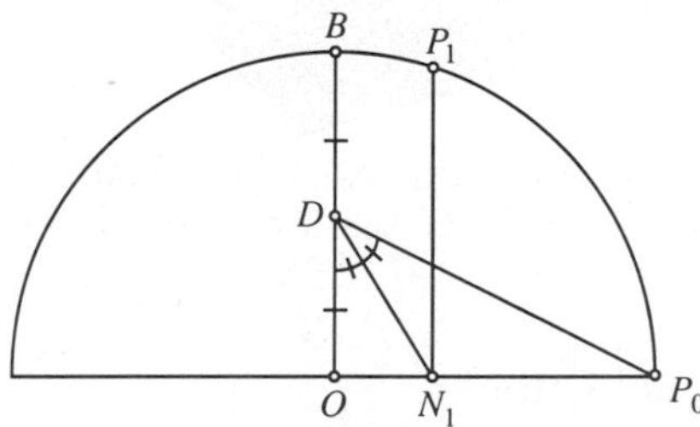

Pentagon

The following elegant construction for the pentagon is due to Richmond (1893). Given a point, a CIRCLE may be constructed of any desired RADIUS, and a DIAMETER drawn through the center. Call the center O, and the right end of the DIAMETER P_0. The DIAMETER PERPENDICULAR to the original DIAMETER may be constructed by finding the PERPENDICULAR BISECTOR. Call the upper endpoint of this PERPENDICULAR DIAMETER B. For the pentagon, find the MIDPOINT of OB and call it D. Draw DP_0, and BISECT $\angle ODP_0$, calling the intersection point with OP_0 N_1. Draw N_1P_1 PARALLEL to OB, and the first two points of the pentagon are P_0 and P_1 (Coxeter 1969).

Madachy (1979) illustrates how to construct a pentagon by folding and knotting a strip of paper.

see also CYCLIC PENTAGON, DECAGON, DISSECTION, FIVE DISKS PROBLEM, HOME PLATE, PENTAFLAKE, PENTAGRAM, POLYGON, TRIGONOMETRY VALUES—$\pi/5$

References

Ball, W. W. R. and Coxeter, H. S. M. *Mathematical Recreations and Essays, 13th ed.* New York: Dover, pp. 95–96, 1987.

Coxeter, H. S. M. *Introduction to Geometry, 2nd ed.* New York: Wiley, pp. 26–28, 1969.

De Temple, D. W. "Carlyle Circles and the Lemoine Simplicity of Polygonal Constructions." *Amer. Math. Monthly* **98**, 97–108, 1991.

Dixon, R. *Mathographics.* New York: Dover, p. 17, 1991.

Dudeney, H. E. *Amusements in Mathematics.* New York: Dover, p. 38, 1970.

Madachy, J. S. *Madachy's Mathematical Recreations.* New York: Dover, p. 59, 1979.

Pappas, T. "The Pentagon, the Pentagram & the Golden Triangle." *The Joy of Mathematics.* San Carlos, CA: Wide World Publ./Tetra, pp. 188–189, 1989.

Richmond, H. W. "A Construction for a Regular Polygon of Seventeen Sides." *Quart. J. Pure Appl. Math.* **26**, 206–207, 1893.

Wantzel, P. L. "Recherches sur les moyens de reconnaître si un Problème de Géométrie peut se résoudre avec la règle et le compas." *J. Math. pures appliq.* **1**, 366–372, 1836.

Pentagonal Antiprism

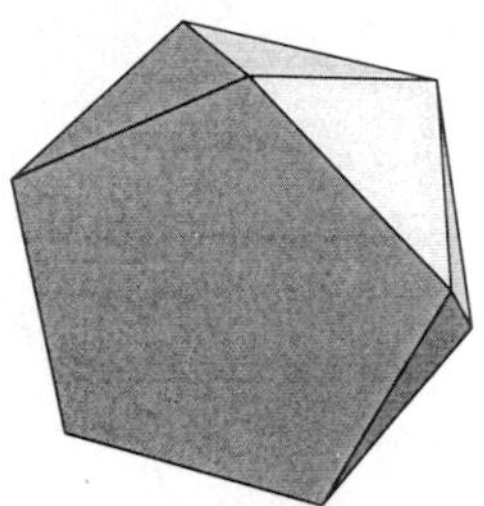

An ANTIPRISM and UNIFORM POLYHEDRON U_{77} whose DUAL POLYHEDRON is the PENTAGONAL DELTAHEDRON.

Pentagonal Cupola

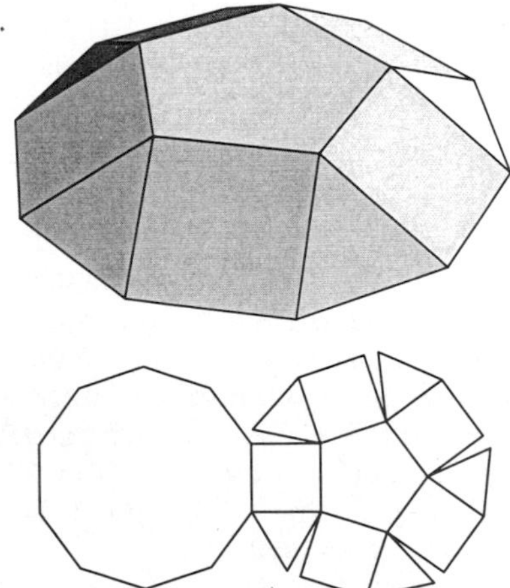

JOHNSON SOLID J_5. The bottom 10 VERTICES are

$$\left(\pm\frac{(1+\sqrt{5})\sqrt{5+\sqrt{5}}}{4\sqrt{2}}, \pm\frac{1}{2}, 0\right),$$
$$\left(\pm\frac{(1+\sqrt{5})\sqrt{5-\sqrt{5}}}{4\sqrt{2}}, \pm\frac{3+\sqrt{5}}{2}, 0\right),$$
$$(0, \pm\tfrac{1}{2}(1+\sqrt{5}), 0)$$

and the top five VERTICES are

$$\left(\frac{\sqrt{5+\sqrt{5}}}{\sqrt{10}}, 0, \frac{\sqrt{5-\sqrt{5}}}{\sqrt{10}}\right),$$
$$\left(\frac{(\sqrt{5}-1)\sqrt{5+\sqrt{5}}}{4\sqrt{10}}, \pm\tfrac{1}{4}(1+\sqrt{5}), \frac{\sqrt{5-\sqrt{5}}}{\sqrt{10}}\right),$$
$$\left(\frac{-(\sqrt{5}+1)\sqrt{5+\sqrt{5}}}{4\sqrt{10}}, \pm\frac{1}{2}, \frac{\sqrt{5-\sqrt{5}}}{\sqrt{10}}\right).$$

Pentagonal Deltahedron

A DELTAHEDRON which is the DUAL POLYHEDRON of the PENTAGONAL ANTIPRISM.

Pentagonal Dipyramid

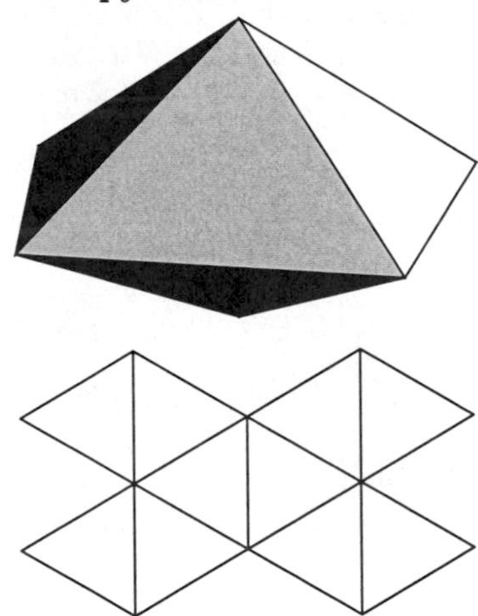

The pentagonal dipyramid is one of the convex DELTAHEDRA, and JOHNSON SOLID J_{13}. It is also the DUAL POLYHEDRON of the PENTAGONAL PRISM. The distance between two adjacent VERTICES on the base of the PENTAGON is

$$\begin{aligned} {d_{12}}^2 &= [1-\cos(\tfrac{2}{5}\pi)]^2 + \sin^2(\tfrac{2}{5}\pi) \\ &= [1-\tfrac{1}{4}(\sqrt{5}-1)]^2 + \left[\frac{(1+\sqrt{5})\sqrt{5-\sqrt{5}}}{4\sqrt{2}}\right]^2 \\ &= \tfrac{1}{2}(5-\sqrt{5}), \end{aligned} \tag{1}$$

and the distance between the apex and one of the base points is

$$d_{1h}^2 = (0-1)^2 + (0-0)^2 + (h-0)^2 = 1 + h^2. \quad (2)$$

But

$$d_{12}^2 = d_{12}^2 \quad (3)$$

$$\tfrac{1}{2}(5-\sqrt{5}) = 1 + h^2 \quad (4)$$

$$h^2 = \tfrac{1}{2}(3-\sqrt{5}), \quad (5)$$

and

$$h = \sqrt{\frac{3-\sqrt{5}}{2}}. \quad (6)$$

This root is of the form $\sqrt{a+b\sqrt{c}}$, so applying SQUARE ROOT simplification gives

$$h = \tfrac{1}{2}(\sqrt{5}-1) \equiv \phi - 1, \quad (7)$$

where ϕ is the GOLDEN MEAN.

see also DELTAHEDRON, DIPYRAMID, GOLDEN MEAN, ICOSAHEDRON, JOHNSON SOLID, TRIANGULAR DIPYRAMID

Pentagonal Gyrobicupola

see JOHNSON SOLID

Pentagonal Gyrocupolarotunda

see JOHNSON SOLID

Pentagonal Hexecontahedron

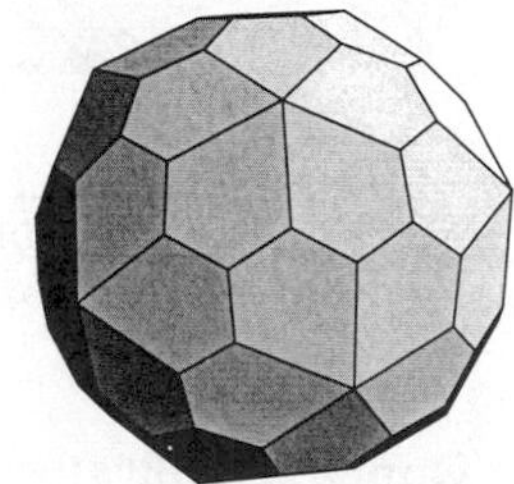

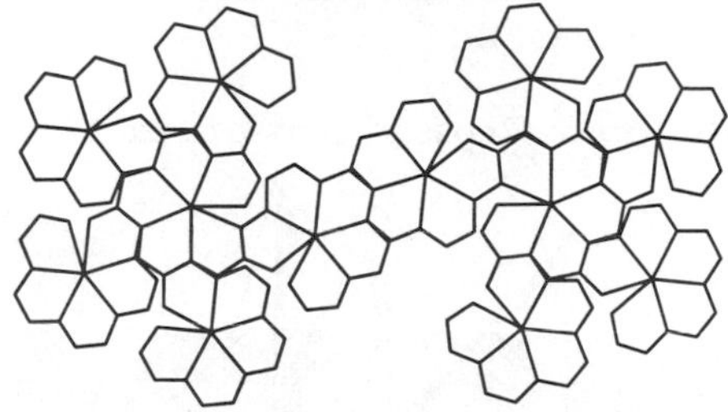

The DUAL POLYHEDRON of the SNUB DODECAHEDRON.

Pentagonal Icositetrahedron

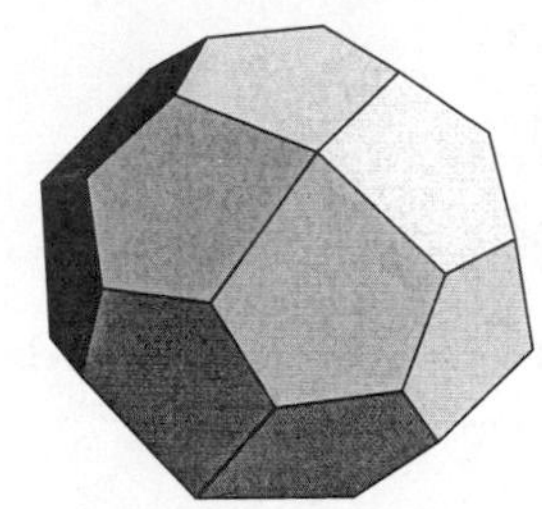

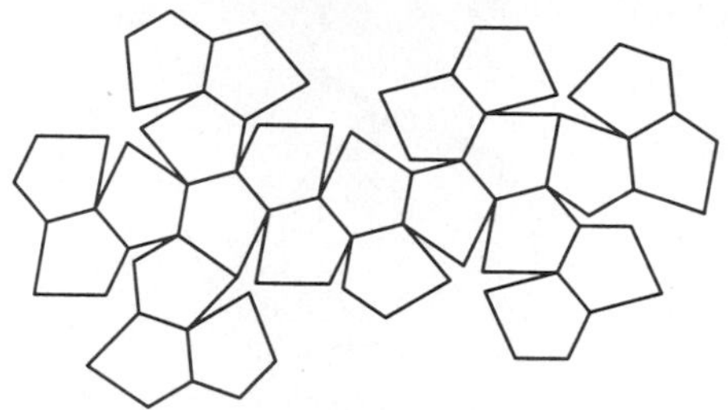

The DUAL POLYHEDRON of the SNUB CUBE.

Pentagonal Number

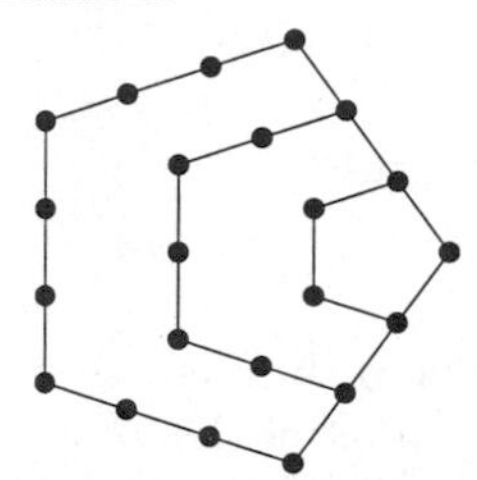

A POLYGONAL NUMBER of the form $n(3n-1)/2$. The first few are 1, 5, 12, 22, 35, 51, 70, ... (Sloane's A000326). The GENERATING FUNCTION for the pentagonal numbers is

$$\frac{x(2x+1)}{(1-x)^3} = x + 5x^2 + 12x^3 + 22x^4 + \ldots.$$

Every pentagonal number is 1/3 of a TRIANGULAR NUMBER.

The so-called generalized pentagonal numbers are given by $n(3n-1)/2$ with $n = 0, \pm1, \pm2, \ldots$, the first few of which are 0, 1, 2, 5, 7, 12, 15, 22, 26, 35, ... (Sloane's A001318).

see also EULER'S PENTAGONAL NUMBER THEOREM, PARTITION FUNCTION P, POLYGONAL NUMBER, TRIANGULAR NUMBER

References

Guy, R. K. "Sums of Squares." §C20 in *Unsolved Problems in Number Theory, 2nd ed.* New York: Springer-Verlag, pp. 136–138, 1994.

Pappas, T. "Triangular, Square & Pentagonal Numbers." *The Joy of Mathematics.* San Carlos, CA: Wide World Publ./Tetra, p. 214, 1989.

Sloane, N. J. A. Sequences A000326/M3818 and A001318/M1336 in "An On-Line Version of the Encyclopedia of Integer Sequences."

Pentagonal Orthobicupola

see JOHNSON SOLID

Pentagonal Orthobirotunda

see JOHNSON SOLID

Pentagonal Orthocupolarontunda

see JOHNSON SOLID

Pentagonal Prism

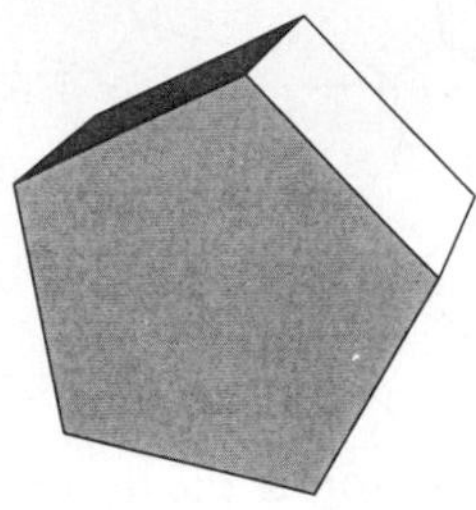

A PRISM and UNIFORM POLYHEDRON U_{76} whose DUAL POLYHEDRON is the PENTAGONAL DIPYRAMID.

see also PENTAGRAMMIC PRISM

Pentagonal Pyramid

see JOHNSON SOLID

Pentagonal Pyramidal Number

A PYRAMIDAL NUMBER of the form $n^2(n+1)/2$. The first few are 1, 6, 18, 40, 75, ... (Sloane's A002411). The GENERATING FUNCTION for the pentagonal pyramidal numbers is

$$\frac{x(2x+1)}{(x-1)^4} = x + 6x^2 + 18x^3 + 40x^4 + \ldots.$$

References

Sloane, N. J. A. Sequence A002411/M4116 in "An On-Line Version of the Encyclopedia of Integer Sequences."

Pentagonal Rotunda

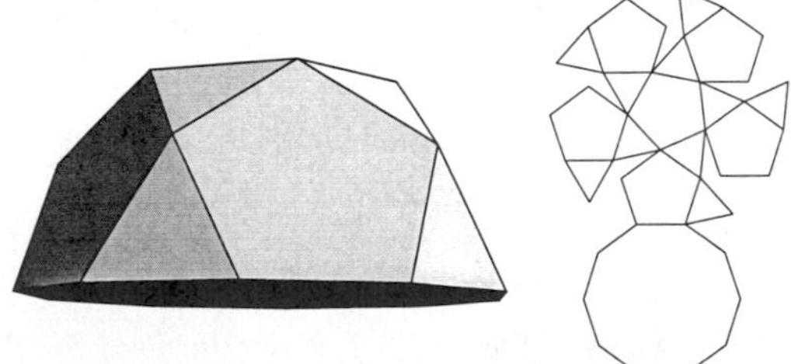

Half of an ICOSIDODECAHEDRON, denoted R_5. It has 10 triangular and five pentagonal faces separating a PENTAGONAL ceiling and a DODECAHEDRAL floor. It is JOHNSON SOLID J_6, and the only true ROTUNDA.

see also ICOSIDODECAHEDRON, JOHNSON SOLID, ROTUNDA

Pentagonal Tiling

see TILING

Pentagram

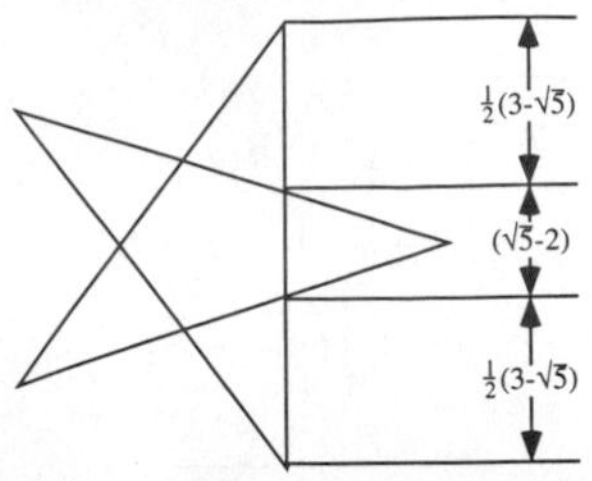

The STAR POLYGON $\{\frac{5}{2}\}$, also called the PENTACLE, PENTALPHA, or PENTANGLE.

see also DISSECTION, HEXAGRAM, HOEHN'S THEOREM, PENTAGON, STAR FIGURE, STAR OF LAKSHMI

References

Ogilvy, C. S. *Excursions in Geometry.* New York: Dover, pp. 122–125, 1990.

Pappas, T. "The Pentagon, the Pentagram & the Golden Triangle." *The Joy of Mathematics.* San Carlos, CA: Wide World Publ./Tetra, pp. 188–189, 1989.

Schwartzman, S. *The Words of Mathematics: An Etymological Dictionary of Mathematical Terms Used in English.* Washington, DC: Math. Assoc. Amer., 1994.

Pentagrammic Antiprism

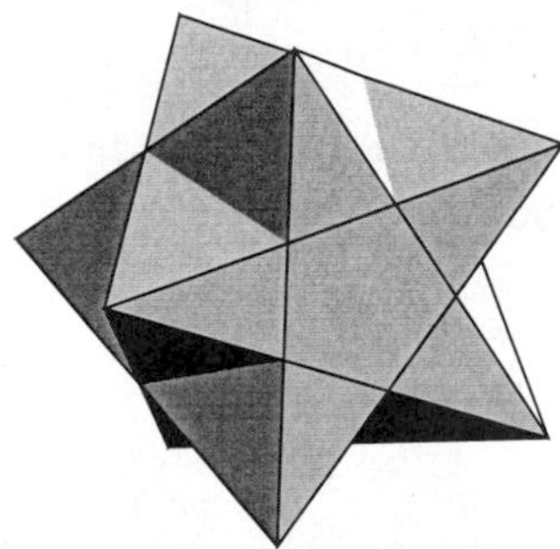

An ANTIPRISM and UNIFORM POLYHEDRON U_{79} whose DUAL POLYHEDRON is the PENTAGRAMMIC DELTAHEDRON.

Pentagrammic Concave Deltahedron

The DUAL POLYHEDRON of the PENTAGRAMMIC CROSSED ANTIPRISM.

Pentagrammic Crossed Antiprism

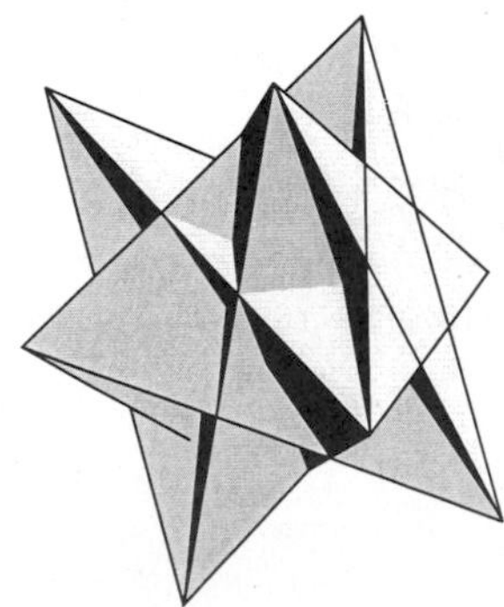

An ANTIPRISM and UNIFORM POLYHEDRON U_{80} whose DUAL POLYHEDRON is the PENTAGRAMMIC CONCAVE DELTAHEDRON.

Pentagrammic Deltahedron

The DUAL POLYHEDRON of the PENTAGRAMMIC ANTIPRISM.

Pentagrammic Dipyramid

The DUAL POLYHEDRON of the PENTAGRAMMIC PRISM.

Pentagrammic Prism

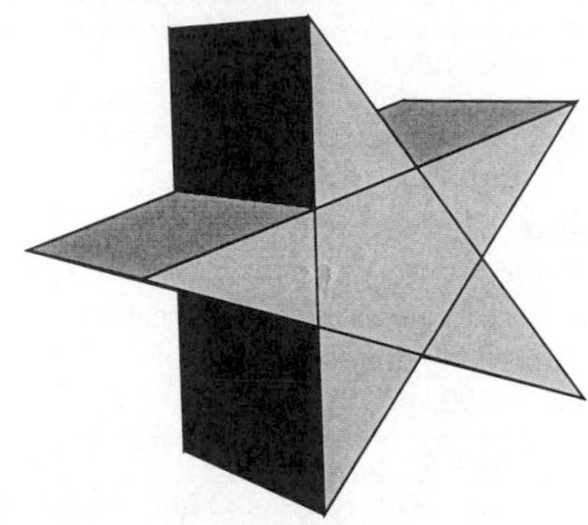

A PRISM and UNIFORM POLYHEDRON U_{78} whose DUAL POLYHEDRON is the PENTAGRAMMIC DIPYRAMID.

see also PENTAGONAL PRISM

Pentakaidecagon

see PENTADECAGON

Pentakis Dodecahedron

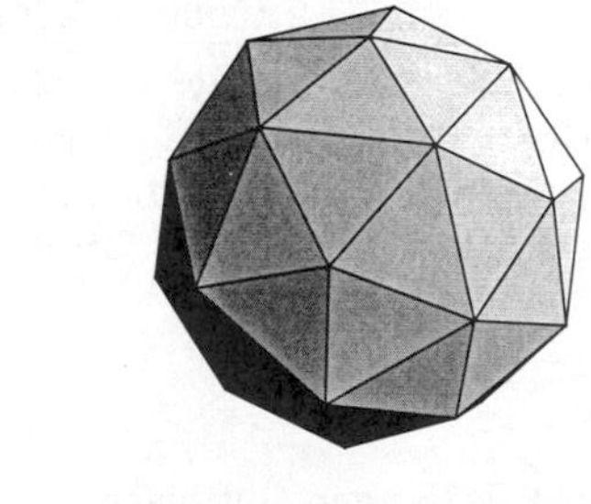

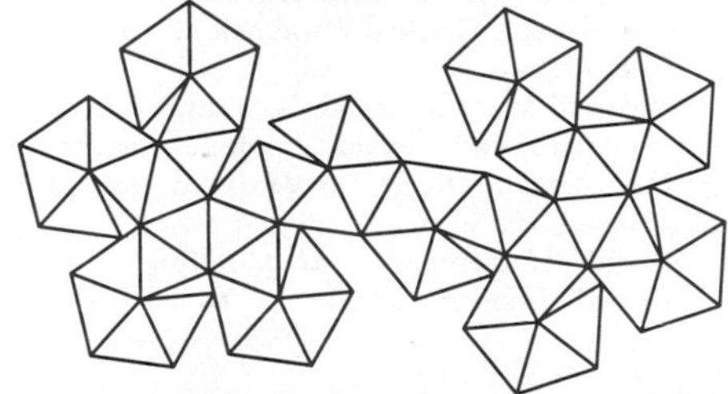

The DUAL POLYHEDRON of the TRUNCATED ICOSAHEDRON.

see also ARCHIMEDEAN SOLID, DUAL POLYHEDRON, TRUNCATED ICOSAHEDRON

Pentalpha

see PENTAGRAM

Pentangle

see PENTAGRAM

Pentatope

The simplest regular figure in 4-D.

Pentatope Number

A FIGURATE NUMBER which is given by

$$Ptop_n = \tfrac{1}{4}\,Te_n(n+3) = \tfrac{1}{24}n(n+1)(n+2)(n+3),$$

where Te_n is the nth TETRAHEDRAL NUMBER. The first few pentatope numbers are 1, 5, 15, 35, 70, 126, ... (Sloane's A000332). The GENERATING FUNCTION for the pentatope numbers is

$$\frac{x}{(1-x)^5} = x + 5x^2 + 15x^3 + 35x^4 + \ldots.$$

see also FIGURATE NUMBER, TETRAHEDRAL NUMBER

References

Conway, J. H. and Guy, R. K. *The Book of Numbers.* New York: Springer-Verlag, pp. 55–57, 1996.

Pentomino

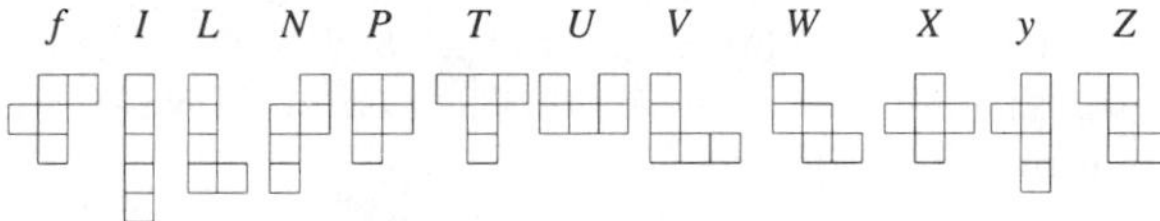

The twelve 5-POLYOMINOES illustrated above and known by the letters of the alphabet they most closely resemble: $f, I, L, N, P, T, U, V, W, X, y, Z$ (Gardner 1960).

References

Ball, W. W. R. and Coxeter, H. S. M. *Mathematical Recreations and Essays, 13th ed.* New York: Dover, pp. 110–111, 1987.

Dudeney, H. E. "The Broken Chessboard." Problem 74 in *The Canterbury Puzzles and Other Curious Problems, 7th ed.* London: Thomas Nelson and Sons, pp. 119–120, 1949.

Gardner, M. "Mathematical Games: More About the Shapes that Can Be Made with Complex Dominoes." *Sci. Amer.* **203**, 186–198, Nov. 1960.

Hunter, J. A. H. and Madachy, J. S. *Mathematical Diversions.* New York: Dover, pp. 80–86, 1975.

Lei, A. "Pentominoes." `http://www.cs.ust.hk/~philipl/omino/pento.html`.

Ruskey, F. "Information on Pentomino Puzzles." `http://sue.csc.uvic.ca/~cos/inf/misc/PentInfo.html`.

Pépin's Test

A test for the PRIMALITY of FERMAT NUMBERS $F_n = 2^{2^n} + 1$, with $n \geq 2$ and $k \geq 2$. Then the two following conditions are equivalent:

1. F_n is PRIME and $k/F_n = -1$.
2. $k^{(F_n - 1)/2} \equiv -1 \pmod{F_n}$.

k is usually taken as 3 as a first test.

see also FERMAT NUMBER, PÉPIN'S THEOREM

References

Ribenboim, P. *The Little Book of Big Primes.* New York: Springer-Verlag, p. 62, 1991.

Shanks, D. *Solved and Unsolved Problems in Number Theory, 4th ed.* New York: Chelsea, pp. 119–120, 1993.

Pépin's Theorem

The FERMAT NUMBER F_n is PRIME IFF

$$3^{2^{2^n-1}} \equiv -1 \pmod{F_n}.$$

see also FERMAT NUMBER, PÉPIN'S TEST, SELFRIDGE-HURWITZ RESIDUE

Percent

The use of percentages is a way of expressing RATIOS in terms of whole numbers. Given a RATIO or FRACTION, it is converted to a percentage by multiplying by 100 and appending a "percentage sign" %. For example, if an investment grows from a number $P = 13.00$ to a number $A = 22.50$, then A is $22.50/13.00 = 1.7308$ times as much as P, or 173.08%, and the investment has grown by 73.08%.

see also PERCENTAGE ERROR, PERMIL

Percentage Error

The percentage error is 100% times the RELATIVE ERROR.

see also ABSOLUTE ERROR, ERROR PROPAGATION, PERCENT, RELATIVE ERROR

References

Abramowitz, M. and Stegun, C. A. (Eds.). *Handbook of Mathematical Functions with Formulas, Graphs, and Mathematical Tables, 9th printing.* New York: Dover, p. 14, 1972.

Percolation Theory

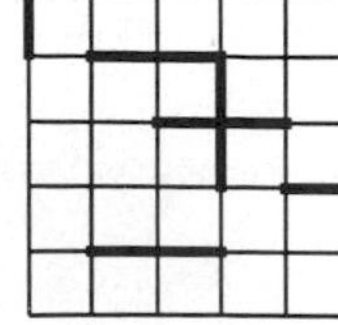

bond percolation

site percolation

Percolation theory deals with fluid flow (or any other similar process) in random media. If the medium is a set of regular LATTICE POINTS, then there are two types of percolation. A SITE PERCOLATION considers the lattice vertices as the relevant entities; a BOND PERCOLATION considers the lattice edges as the relevant entities.

see also BOND PERCOLATION, CAYLEY TREE, CLUSTER, CLUSTER PERIMETER, LATTICE ANIMAL, PERCOLATION THRESHOLD, POLYOMINO, s-CLUSTER, s-RUN, SITE PERCOLATION

References

Deutscher, G.; Zallen, R.; and Adler, J. (Eds.). *Percolation Structures and Processes.* Bristol: Adam Hilger, 1983.

Finch, S. "Favorite Mathematical Constants." http://www.mathsoft.com/asolve/constant/rndprc/rndprc.html.

Grimmett, G. *Percolation.* New York: Springer-Verlag, 1989.

Kesten, H. *Percolation Theory for Mathematicians.* Boston, MA: Birkhäuser, 1982.

Stauffer, D. and Aharony, A. *Introduction to Percolation Theory, 2nd ed.* London: Taylor & Francis, 1992.

Percolation Threshold

The critical fraction of lattice points which must be filled to create a continuous path of nearest neighbors from one side to another. The following table is from Stauffer and Aharony (1992, p. 17).

Lattice	Site	Bond
Cubic (Body-Centered)	0.246	0.1803
Cubic (Face-Centered)	0.198	0.119
Cubic (Simple)	0.3116	0.2488
Diamond	0.43	0.388
Honeycomb	0.6962	0.65271
4-Hypercubic	0.197	0.1601
5-Hypercubic	0.141	0.1182
6-Hypercubic	0.107	0.0942
7-Hypercubic	0.089	0.0787
Square	0.592746	0.50000
Triangular	0.50000	0.34729

The square bond value is 1/2 exactly, as is the triangular site. $p_c = 2\sin(\pi/18)$ for the triangular bond and $p_c = 1 - 2\sin(\pi/18)$ for the honeycomb bond. An exact answer for the square site percolation threshold is not known.

see also PERCOLATION THEORY

References

Essam, J. W.; Gaunt, D. S.; and Guttmann, A. J. "Percolation Theory at the Critical Dimension." *J. Phys. A* **11**, 1983–1990, 1978.

Finch, S. "Favorite Mathematical Constants." http://www.mathsoft.com/asolve/constant/rndprc/rndprc.html.

Kesten, H. *Percolation Theory for Mathematicians.* Boston, MA: Birkhäuser, 1982.

Stauffer, D. and Aharony, A. *Introduction to Percolation Theory, 2nd ed.* London: Taylor & Francis, 1992.

Perfect Box

see EULER BRICK

Perfect Cubic

A perfect cubic POLYNOMIAL can be factored into a linear and a quadratic term,

$$(a^3 - b^3) = (a - b)(a^2 + ab + b^2)$$

$$(a^3 + b^3) = (a + b)(a^2 - ab + b^2).$$

see also CUBIC EQUATION, PERFECT SQUARE, POLYNOMIAL

Perfect Cuboid

see EULER BRICK

Perfect Difference Set

A SET of RESIDUES $\{a_1, a_2, \ldots, a_{k+1}\}$ (mod n) such that every NONZERO RESIDUE can be uniquely expressed in the form $a_i - a_j$. Examples include {1, 2, 4} (mod 7) and {1, 2, 5, 7} (mod 13). A NECESSARY condition for a difference set to exist is that n be of the form $k^2 + k + 1$. A SUFFICIENT condition is that k be a PRIME POWER. Perfect sets can be used in the construction of PERFECT RULERS.

see also PERFECT RULER

References

Guy, R. K. "Modular Difference Sets and Error Correcting Codes." §C10 in *Unsolved Problems in Number Theory, 2nd ed.* New York: Springer-Verlag, pp. 118–121, 1994.

Perfect Digital Invariant

see NARCISSISTIC NUMBER

Perfect Information

A class of GAME in which players move alternately and each player is completely informed of previous moves. FINITE, ZERO-SUM, two-player GAMES with perfect information (including checkers and chess) have a SADDLE POINT, and therefore one or more optimal strategies. However, the optimal strategy may be so difficult to compute as to be effectively impossible to determine (as in the game of CHESS).

see also FINITE GAME, GAME, ZERO-SUM GAME

Perfect Magic Cube

A perfect magic cube is a MAGIC CUBE for which the cross-section diagonals, as well as the space diagonals, sum to the MAGIC CONSTANT.

see also MAGIC CUBE, SEMIPERFECT MAGIC CUBE

References

Gardner, M. "Magic Squares and Cubes." Ch. 17 in *Time Travel and Other Mathematical Bewilderments.* New York: W. H. Freeman, pp. 213–225, 1988.

Perfect Number

Perfect numbers are INTEGERS n such that

$$n = s(n), \tag{1}$$

where $s(n)$ is the RESTRICTED DIVISOR FUNCTION (i.e., the SUM of PROPER DIVISORS of n), or equivalently

$$\sigma(n) = 2n, \tag{2}$$

where $\sigma(n)$ is the DIVISOR FUNCTION (i.e., the SUM of DIVISORS of n including n itself). The first few perfect numbers are 6, 28, 496, 8128, ... (Sloane's A000396). This follows from the fact that

$$\begin{aligned} 6 &= \sum 1, 2, 3 \\ 28 &= \sum 1, 2, 4, 7, 14 \\ 496 &= \sum 1, 2, 4, 8, 16, 31, 62, 124, 248, \end{aligned}$$

etc.

Perfect numbers are intimately connected with a class of numbers known as MERSENNE PRIMES. This can be demonstrated by considering a perfect number P of the form $P = q2^{p-1}$ where q is PRIME. Then

$$\sigma(P) = 2P, \tag{3}$$

and using

$$\sigma(q) = q + 1 \tag{4}$$

for q prime, and

$$\sigma(2^\alpha) = 2^{\alpha+1} - 1 \tag{5}$$

gives

$$\begin{aligned} \sigma(q2^{p-1}) &= \sigma(q)\sigma(2^{p-1}) = (q+1)(2^p - 1) \\ &= 2q2^{p-1} = q2^p \end{aligned} \tag{6}$$

$$q(2^p - 1) + 2^p - 1 = q2^p \tag{7}$$

$$q = 2^p - 1. \tag{8}$$

Therefore, if $M_p \equiv q = 2^p - 1$ is PRIME, then

$$P = \tfrac{1}{2}(M_p + 1)M_p = 2^{p-1}(2^p - 1) \tag{9}$$

is a perfect number, as was stated in Proposition IX.36 of Euclid's *Elements* (Dunham 1990). The first few perfect numbers are summarized in the following table.

#	p	P
1	2	6
2	3	28
3	5	496
4	7	8128
5	13	33550336
6	17	8589869056
7	19	137438691328
8	31	2305843008139952128

All EVEN perfect numbers are of this form, as was proved by Euler in a posthumous paper. The only even perfect number of the form $x^3 + 1$ is 28 (Mąkowski 1962).

It is not known if any ODD perfect numbers exist, although numbers up to 10^{300} have been checked (Brent *et al.* 1991, Guy 1994) without success, improving the result of Tuckerman (1973), who checked odd numbers up to 10^{36}. Euler showed that an ODD perfect number, if it exists, must be of the form

$$m = p^{4a+1}Q^2, \tag{10}$$

where p is an ODD PRIME RELATIVELY PRIME to Q. In 1887, Sylvester conjectured and in 1925, Gradshtein proved that any ODD perfect number must have at least six different prime aliquot factors (or eight if it is not divisible by 3; Ball and Coxeter 1987). Catalan (1888) proved that if an ODD perfect number is not divisible by 3, 5, or 7, it has at least 26 distinct prime aliquot factors. Stuyvaert (1896) proved that an ODD perfect number must be a sum of squares. All EVEN perfect numbers end in 16, 28, 36, 56, or 76 (Lucas 1891) and, with the exception of 6, have DIGITAL ROOT 1.

Every perfect number of the form $2^p(2^{p+1} - 1)$ can be written

$$2^p(2^{p+1} - 1) = \sum_{k=1}^{p/2} (2k-1)^3. \tag{11}$$

All perfect numbers are HEXAGONAL NUMBERS and therefore TRIANGULAR NUMBERS. It therefore follows that perfect numbers are always the sum of consecutive POSITIVE integers starting at 1, for example,

$$6 = \sum_{n=1}^{3} n \tag{12}$$

$$28 = \sum_{n=1}^{7} n \tag{13}$$

$$496 = \sum_{n=1}^{31} n \tag{14}$$

(Singh 1997). All EVEN perfect numbers $P > 6$ are of the form

$$P + 1 + 9T_n, \tag{15}$$

where T_n is a TRIANGULAR NUMBER

$$T_n = \tfrac{1}{2}n(n+1) \tag{16}$$

such that $n = 8j + 2$ (Eaton 1995, 1996). The sum of reciprocals of all the divisors of a perfect number is 2, since

$$\underbrace{n + \ldots + c + b + a}_{n} = 2n \tag{17}$$

$$\frac{n}{a} + \frac{n}{b} + \ldots = 2n \tag{18}$$

$$\frac{1}{a} + \frac{1}{b} + \ldots = 2. \tag{19}$$

If $s(n) > n$, n is said to be an ABUNDANT NUMBER. If $s(n) < n$, n is said to be a DEFICIENT NUMBER. And if $s(n) = kn$ for a POSITIVE INTEGER $k > 1$, n is said to be a MULTIPERFECT NUMBER of order k.

see also ABUNDANT NUMBER, ALIQUOT SEQUENCE, AMICABLE NUMBERS, DEFICIENT NUMBER, DIVISOR FUNCTION, e-PERFECT NUMBER, HARMONIC NUMBER, HYPERPERFECT NUMBER, INFINARY PERFECT NUMBER, MERSENNE NUMBER, MERSENNE PRIME, MULTIPERFECT NUMBER, MULTIPLICATIVE PERFECT NUMBER, PLUPERFECT NUMBER, PSEUDOPERFECT NUMBER, QUASIPERFECT NUMBER, SEMIPERFECT NUMBER, SMITH NUMBER, SOCIABLE NUMBERS, SUBLIME NUMBER, SUPERPERFECT NUMBER, UNITARY PERFECT NUMBER, WEIRD NUMBER

References

Ball, W. W. R. and Coxeter, H. S. M. *Mathematical Recreations and Essays, 13th ed.* New York: Dover, pp. 66–67, 1987.

Brent, R. P.; Cohen, G. L. L.; and te Riele, H. J. J. "Improved Techniques for Lower Bounds for Odd Perfect Numbers." *Math. Comput.* **57**, 857–868, 1991.

Conway, J. H. and Guy, R. K. "Perfect Numbers." In *The Book of Numbers.* New York: Springer-Verlag, pp. 136–137, 1996.

Dickson, L. E. "Notes on the Theory of Numbers." *Amer. Math. Monthly* **18**, 109–111, 1911.

Dickson, L. E. *History of the Theory of Numbers, Vol. 1: Divisibility and Primality.* New York: Chelsea, pp. 3–33, 1952.

Dunham, W. *Journey Through Genius: The Great Theorems of Mathematics.* New York: Wiley, p. 75, 1990.

Eaton, C. F. "Problem 1482." *Math. Mag.* **68**, 307, 1995.

Eaton, C. F. "Perfect Number in Terms of Triangular Numbers." Solution to Problem 1482. *Math. Mag.* **69**, 308–309, 1996.

Gardner, M. "Perfect, Amicable, Sociable." Ch. 12 in *Mathematical Magic Show: More Puzzles, Games, Diversions, Illusions and Other Mathematical Sleight-of-Mind from Scientific American.* New York: Vintage, pp. 160–171, 1978.

Guy, R. K. "Perfect Numbers." §B1 in *Unsolved Problems in Number Theory, 2nd ed.* New York: Springer-Verlag, p. 145, 1994.

Kraitchik, M. "Mersenne Numbers and Perfect Numbers." §3.5 in *Mathematical Recreations.* New York: W. W. Norton, pp. 70–73, 1942.

Madachy, J. S. *Madachy's Mathematical Recreations.* New York: Dover, pp. 145 and 147–151, 1979.

Mąkowski, A. "Remark on Perfect Numbers." *Elemente Math.* **17**, 109, 1962.

Powers, R. E. "The Tenth Perfect Number." *Amer. Math. Monthly* **18**, 195–196, 1911.

Shanks, D. *Solved and Unsolved Problems in Number Theory, 4th ed.* New York: Chelsea, pp. 1–13 and 25–29, 1993.

Singh, S. *Fermat's Enigma: The Epic Quest to Solve the World's Greatest Mathematical Problem.* New York: Walker, pp. 11–13, 1997.

Sloane, N. J. A. Sequence A000396/M4186 in "An On-Line Version of the Encyclopedia of Integer Sequences."

Tuckerman, B. "Odd Perfect Numbers: A Search Procedure, and a New Lower Bound of 10^{36}." *Not. Amer. Math. Soc.* **15**, 226, 1968.

Tuckerman, B. "A Search Procedure and Lower Bound for Odd Perfect Numbers." *Math. Comp.* **27**, 943–949, 1973.
Zachariou, A. and Zachariou, E. "Perfect, Semi-Perfect and Ore Numbers." *Bull. Soc. Math. Grèce (New Ser.)* **13**, 12–22, 1972.

Perfect Partition

A PARTITION of n which can generate any number 1, 2, ..., n.

see also PARTITION

References
Cohen, D. I. A. *Basic Techniques of Combinatorial Theory.* New York: Wiley and Sons, p. 97, 1978.
Honsberger, R. *Mathematical Gems III.* Washington, DC: Math. Assoc. Amer., pp. 140–143, 1985.

Perfect Proportion

Since

$$\frac{2a}{a+b} = \frac{2ab}{(a+b)b}, \tag{1}$$

it follows that

$$\frac{a}{\frac{a+b}{2}} = \frac{\frac{2ab}{a+b}}{b}, \tag{2}$$

so

$$\frac{a}{A} = \frac{H}{b}, \tag{3}$$

where A and H are the ARITHMETIC MEAN and HARMONIC MEAN of a and b. This relationship was purportedly discovered by Pythagoras.

see also ARITHMETIC MEAN, HARMONIC MEAN

Perfect Rectangle

A RECTANGLE which cannot be built up of SQUARES all of different sizes is called an imperfect rectangle. A RECTANGLE which can be built up of SQUARES all of different sizes is called perfect.

order	perfect	imperfect
< 9	0	0
9	2	1
10	6	0
11	22	0
12	67	9
13	213	34
14	744	104
15	2609	282

Perfect Ruler

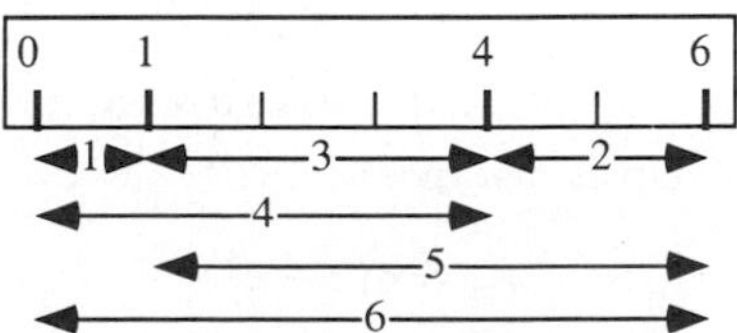

A type of RULER considered by Guy (1994) which has k distinct marks spaced such that the distances between marks can be used to measure all the distances 1, 2, 3, 4, ... up to some maximum distance $n > k$. Such a ruler can be constructed from a PERFECT DIFFERENCE SET by subtracting one from each element. For example, the PERFECT DIFFERENCE SET {1, 2, 5, 7} gives 0, 1, 4, 6, which can be used to measure $1-0=1$, $6-4=2$, $4-1=3$, $4-0=4$, $6-1=5$, $6-0=6$ (so we get 6 distances with only four marks).

see also PERFECT DIFFERENCE SET

References
Guy, R. K. "Modular Difference Sets and Error Correcting Codes." §C10 in *Unsolved Problems in Number Theory, 2nd ed.* New York: Springer-Verlag, pp. 118–121, 1994.

Perfect Set

A SET P is called perfect if $P = P'$, where P' is the DERIVED SET of P.

see also DERIVED SET, SET

Perfect Square

The term perfect square is used to refer to a SQUARE NUMBER, a PERFECT SQUARE DISSECTION, or a factorable quadratic polynomial of the form $a^2 - b^2 = (a-b)(a+b)$.

see also PERFECT SQUARE DISSECTION, QUADRATIC EQUATION, SQUARE NUMBER, SQUAREFREE

Perfect Square Dissection

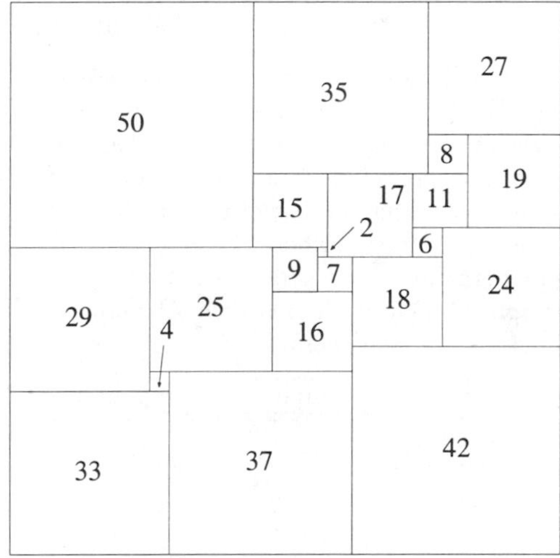

A SQUARE which can be DISSECTED into a number of smaller SQUARES with no two equal is called a PERFECT SQUARE DISSECTION (or a SQUARED SQUARE). Square dissections in which the squares need not be different sizes are called MRS. PERKINS' QUILTS. If no subset of the SQUARES forms a RECTANGLE, then the perfect square is called "simple." Lusin claimed that perfect squares were impossible to construct, but this assertion was proved erroneous when a 55-SQUARE perfect square was published by R. Sprague in 1939 (Wells 1991).

There is a unique simple perfect square of order 21 (the lowest possible order), discovered in 1978 by

A. J. W. Duijvestijn (Bouwkamp and Duijvestijn 1992). It is composed of 21 squares with total side length 112, and is illustrated above. There is a simple notation (sometimes called Bouwkamp code) used to describe perfect squares. In this notation, brackets are used to group adjacent squares with flush tops, and then the groups are sequentially placed in the highest (and left-most) possible slots. For example, the 21-square illustrated above is denoted [50, 35, 27], [8, 19], [15, 17, 11], [6, 24], [29, 25, 9, 2], [7, 18], [16], [42], [4, 37], [33].

The number of simple perfect squares of order n for $n \geq 21$ are 1, 8, 12, 26, 160, 441, ... (Sloane's A006983). Duijvestijn's Table I gives a list of the 441 simple perfect squares of order 26, the smallest with side length 212 and the largest with side length 825. Skinner (1993) gives the smallest possible side length (and smallest order for each) as 110 (22), 112 (21), 120 (24), 139 (22), 140 (23), ... for simple perfect squared squares, and 175 (24), 235 (25), 288 (26), 324 (27), 325 (27), ... for compound perfect squared squares.

There are actually three simple perfect squares having side length 110. They are [60, 50], [23, 27], [24, 22, 14], [7, 16], [8, 6], [12, 15], [13], [2, 28], [26], [4, 21, 3], [18], [17] (order 22; discovered by A. J. W. Duijvestijn); [60, 50], [27, 23], [24, 22, 14], [4, 19], [8, 6], [3, 12, 16], [9], [2, 28], [26], [21], [1, 18], [17] (order 22; discovered by T. H. Willcocks); and [44, 29, 37], [21, 8], [13, 32], [28, 16], [15, 19], [12,4], [3, 1], [2, 14], [5], [10, 41], [38, 7], [31] (order 23; discovered by A. J. W. Duijvestijn).

D. Sleator has developed an efficient ALGORITHM for finding *non*-simple perfect squares using what he calls rectangle and "ell" grow sequences. This algorithm finds a slew of compound perfect squares of orders 24–32. Weisstein gives a partial list of known simple and compound perfect squares (where the number of simple perfect squares is exact for orders less than 27) as well as *Mathematica*® (Wolfram Research, Champaign, IL) algorithms for drawing them.

Order	# Simple	# Compound
21	1	0
22	8	0
23	12	0
24	26	1
25	160	1
26	441	2
27	?	2
28	?	4
29	?	2
30	?	3
31	?	2
32	?	2
38	1	0
69	1	0

see also MRS. PERKINS' QUILT

References

Ball, W. W. R. and Coxeter, H. S. M. *Mathematical Recreations and Essays, 13th ed.* New York: Dover, pp. 115–116, 1987.

Beiler, A. H. *Recreations in the Theory of Numbers: The Queen of Mathematics Entertains.* New York: Dover, pp. 157–161, 1966.

Bouwkamp, C. J. and Duijvestijn, A. J. W. "Catalogue of Simple Perfect Squared Squares of Orders 21 Through 25." Eindhoven Univ. Technology, Dept. Math, Report 92-WSK-03, Nov. 1992.

Brooks, R. L.; Smith, C. A. B.; Stone, A. H.; and Tutte, W. T. "The Dissection of Rectangles into Squares." *Duke Math. J.* **7**, 312–340, 1940.

Duijvestijn, A. J. W. "A Simple Perfect Square of Lowest Order." *J. Combin. Th. Ser. B* **25**, 240–243, 1978.

Duijvestijn, A. J. W. "A Lowest Order Simple Perfect 2 × 1 Squared Rectangle." *J. Combin. Th. Ser. B* **26**, 372–374, 1979.

Duijvestijn, A. J. W. `ftp://ftp.cs.utwente.nl/pub/doc/dvs/TableI`.

Gardner, M. "Squaring the Square." Ch. 17 in *The Second Scientific American Book of Mathematical Puzzles & Diversions: A New Selection.* New York: Simon and Schuster, 1961.

Gardner, M. *Fractal Music, Hypercards, and More: Mathematical Recreations from Scientific American Magazine.* New York: W. H. Freeman, pp. 172–174, 1992.

Kraitchik, M. *Mathematical Recreations.* New York: W. W. Norton, p. 198, 1942.

Madachy, J. S. *Madachy's Mathematical Recreations.* New York: Dover, pp. 15 and 32–33, 1979.

Mauldin, R. D. (Ed.) *The Scottish Book: Math at the Scottish Cafe* Boston, MA: Birkhäuser, 1982.

Moroń, Z. "O rozkładach prostokątów na kwadraty." *Przeglad matematyczno-fizyczny* **3**, 152–153, 1925.

Skinner, J. D. II. *Squares Squares: Who's Who & What's What.* Published by the author, 1993.

Sloane, N. J. A. Sequences A006983/M4482 in "An On-Line Version of the Encyclopedia of Integer Sequences."

Sloane, N. J. A. and Plouffe, S. Extended entry in *The Encyclopedia of Integer Sequences.* San Diego: Academic Press, 1995.

Sprague, R. "Beispiel einer Zerlegung des Quadrats in lauter verschiedene Quadrate." *Math. Z.* **45**, 607–608, 1939.

Weisstein, E. W. "Perfect Squares." `http://www.astro.virginia.edu/~eww6n/math/notebooks/PerfectSquare.m`.

Wells, D. *The Penguin Dictionary of Curious and Interesting Geometry.* London: Penguin, p. 242, 1991.

Periapsis

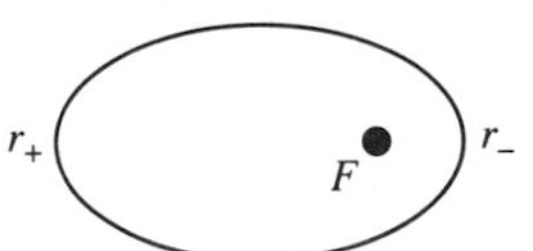

The smallest radial distance of an ELLIPSE as measured from a FOCUS. Taking $v = 0$ in the equation of an ELLIPSE

$$r = \frac{a(1-e^2)}{1+e\cos v}$$

gives the periapsis distance

$$r_- = a(1-e).$$

Periapsis for an orbit around the Earth is called perigee, and periapsis for an orbit around the Sun is called perihelion.

see also APOAPSIS, ECCENTRICITY, ELLIPSE, FOCUS

Perigon

An ANGLE of 2π radians $= 360°$ corresponding to the CENTRAL ANGLE of an entire CIRCLE.

Perimeter

The ARC LENGTH along the boundary of a closed 2-D region. The perimeter of a CIRCLE is called the CIRCUMFERENCE.

see also CIRCUMFERENCE, CLUSTER PERIMETER, SEMIPERIMETER

Perimeter Polynomial

A sum over all CLUSTER PERIMETERS.

Period Doubling

A characteristic of some systems making a transition to CHAOS. Doubling is followed by quadrupling, etc. An example of a map displaying period doubling is the LOGISTIC MAP.

see also CHAOS, LOGISTIC MAP

Period Three Theorem

Li and Yorke (1975) proved that *any* 1-D system which exhibits a regular CYCLE of period three will also display regular CYCLES of every other length as well as completely CHAOTIC CYCLES.

see also CHAOS, CYCLE (MAP)

References

Li, T. Y. and Yorke, J. A. "Period Three Implies Chaos." *Amer. Math. Monthly* **82**, 985–992, 1975.

Periodic Function

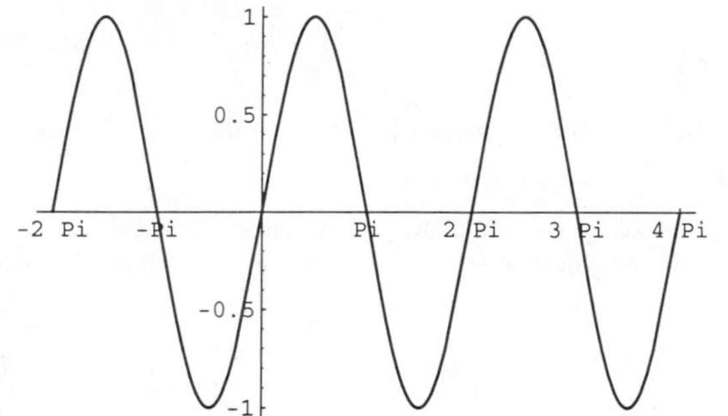

A FUNCTION $f(x)$ is said to be periodic with period p if $f(x) = f(x + np)$ for $n = 1, 2, \ldots$. For example, the SINE function $\sin x$ is periodic with period 2π (as well as with period -2π, 4π, 6π, etc.).

The CONSTANT FUNCTION $f(x) = 0$ is periodic with any period R for all NONZERO REAL NUMBERS R, so there is no concept analogous to the LEAST PERIOD of a PERIODIC POINT for functions.

see also PERIODIC POINT, PERIODIC SEQUENCE

References

Morse, P. M. and Feshbach, H. *Methods of Theoretical Physics, Part I.* New York: McGraw-Hill, pp. 425–427, 1953.
Spanier, J. and Oldham, K. B. "Periodic Functions." Ch. 36 in *An Atlas of Functions.* Washington, DC: Hemisphere, pp. 343–349, 1987.

Periodic Point

A point x_0 is said to be a periodic point of a FUNCTION f of period n if $f^n(x_0) = x_0$, where $f_0(x) = x$ and $f^n(x)$ is defined recursively by $f^n(x) = f(f^{n-1}(x))$.

see also LEAST PERIOD, PERIODIC FUNCTION, PERIODIC SEQUENCE

Periodic Sequence

A SEQUENCE $\{a_i\}$ is said to be periodic with period p with if it satisfies $a_i = a_{i+np}$ for $n = 1, 2, \ldots$. For example, $\{1, 2, 1, 2, 1, 2, 1, 2, 1, 2, 1, 2, 1, 2, \ldots\}$ is a periodic sequence with LEAST PERIOD 2.

see also EVENTUALLY PERIODIC, PERIODIC FUNCTION, PERIODIC POINT

Perkins' Quilt

see MRS. PERKINS' QUILT

Perko Pair

The KNOTS 10_{161} and 10_{162} illustrated above. They are listed as separate knots in the pictorial enumeration of Rolfsen (1976, Appendix C), but were identified as identical by Perko (1974).

References

Perko, K. A. Jr. "On the Classification of Knots." *Proc. Amer. Math. Soc.* **45**, 262–266, 1974.
Rolfsen, D. "Table of Knots and Links." Appendix C in *Knots and Links.* Wilmington, DE: Publish or Perish Press, pp. 280–287, 1976.

Permanence of Algebraic Form

All ELEMENTARY FUNCTIONS can be extended to the COMPLEX PLANE. Such definitions agree with the REAL definitions on the x-AXIS and constitute an ANALYTIC CONTINUATION.

see also ANALYTIC CONTINUATION, ELEMENTARY FUNCTION, PERMANENCE OF MATHEMATICAL RELATIONS PRINCIPLE

References

Arfken, G. *Mathematical Methods for Physicists, 3rd ed.* Orlando, FL: Academic Press, p. 380, 1985.

Permanence of Mathematical Relations Principle

The metric properties discovered for a primitive figure remain applicable, without modifications other than changes of signs, to all correlative figures which can be considered to arise from the first.

This principle was formulated by Poncelet, and amounts to the statement that if an analytic identity in any finite number of variables holds for all real values of the variables, then it also holds by ANALYTIC CONTINUATION for all complex values (Bell 1945). This principle is also called PONCELET'S CONTINUITY PRINCIPLE.

see also ANALYTIC CONTINUATION, CONSERVATION OF NUMBER PRINCIPLE, DUALITY PRINCIPLE, PERMANENCE OF ALGEBRAIC FORM

References

Bell, E. T. *The Development of Mathematics, 2nd ed.* New York: McGraw-Hill, p. 340, 1945.

Permanent

An analog of a DETERMINANT where all the signs in the expansion by MINORS are taken as POSITIVE. The permanent of a MATRIX A is the coefficient of $x_1 \cdots x_n$ in

$$\prod_{i=1}^n (a_{i1}x_1 + a_{i2}x_2 + \ldots + a_{in}x_n)$$

(Vardi 1991). Another equation is the RYSER FORMULA

$$\operatorname{perm}(a_{ij}) = (-1)^n \sum_{s \subseteq \{1,\ldots,n\}} (-1)^{|s|} \prod_{i=1}^n \sum_{j \in s} a_{ij},$$

where the SUM is over all SUBSETS of $\{1, \ldots, n\}$, and $|s$ is the number of elements in s (Vardi 1991).

If M is a UNITARY MATRIX, then

$$|\operatorname{perm}(\mathsf{M})| \leq 1$$

(Minc 1978, p. 25; Vardi 1991).

see also DETERMINANT, FROBENIUS-KÖNIG THEOREM, IMMANANT, RYSER FORMULA, SCHUR MATRIX

References

Borovskikh, Y. V.; Korolyuk, V. S. *Random Permanents.* Philadelphia, PA: Coronet Books, 1994.

Minc, H. *Permanents.* Reading, MA: Addison-Wesley, 1978.

Vardi, I. "Permanents." §6.1 in *Computational Recreations in Mathematica.* Reading, MA: Addison-Wesley, pp. 108 and 110–112, 1991.

Permil

The use of percentages is a way of expressing RATIOS in terms of whole numbers. Given a RATIO or FRACTION, it is converted to a permil-age by multiplying by 1000 and appending a "mil sign" ‰. For example, if an investment grows from a number $P = 13.00$ to a number $A = 22.50$, then A is $22.50/13.00 = 1.7308$ times as much as P, or 1730.8 ‰.

see also PERCENT

Permutation

The rearrangement of elements in a set into a ONE-TO ONE correspondence with itself, also called an ARRANGEMENT or ORDER. The number of ways of obtaining r *ordered* outcomes from a permutation of n elements is

$${}_nP_r \equiv \frac{n!}{(n-r)!} = r!\binom{n}{r}, \tag{1}$$

where $n!$ is n FACTORIAL and $\binom{a}{b}$ is a BINOMIAL COEFFICIENT. The total number of permutations for n elements is given by $n!$.

A representation of a permutation as a product of CYCLES is unique (up to the ordering of the cycles). An example of a cyclic decomposition is $(\{1, 3, 4\}, \{2\})$, corresponding to the permutations $(1 \to 3, 3 \to 4, 4 \to 1)$ and $(2 \to 2)$, which combine to give $\{4, 2, 1, 3\}$.

Any permutation is also a product of TRANSPOSITIONS. Permutations are commonly denoted in LEXICOGRAPHIC or TRANSPOSITION ORDER. There is a correspondence between a PERMUTATION and a pair of YOUNG TABLEAUX known as the SCHENSTED CORRESPONDENCE.

The number of wrong permutations of n objects is $[n!/e]$ where $[x]$ is the NINT function. A permutation of n ordered objects in which no object is in its natural place is called a DERANGEMENT (or sometimes, a COMPLETE PERMUTATION) and the number of such permutations is given by the SUBFACTORIAL $!n$.

Using

$$(x+y)^n = \sum_{r=0}^n \binom{n}{r} x^{n-r} y^r \tag{2}$$

with $x = y = 1$ gives

$$2^n = \sum_{r=0}^n \binom{n}{r}, \tag{3}$$

so the number of ways of choosing 0, 1, ..., or n at a time is 2^n.

The set of all permutations of a set of elements 1, ..., n can be obtained using the following recursive procedure

```
   1 2
    /          (4)
 2 1

   1   2 3
        /
   1 3 2
    /
 3 1   2
 |             (5)
 3 2   1
    \
   2 3 1
        \
   2   1 3
```

Let the set of INTEGERS 1, 2, ..., n be permuted and the resulting sequence be divided into increasing RUNS. As n approaches INFINITY, the average length of the nth RUN is denoted L_n. The first few values are

$$L_1 = e - 1 = 1.7182818\ldots \tag{6}$$
$$L_2 = e^2 - 2e = 1.9524\ldots \tag{7}$$
$$L_3 = e^3 - 3e^2 + \tfrac{3}{2}e = 1.9957\ldots, \tag{8}$$

where e is the base of the NATURAL LOGARITHM (Knuth 1973, Le Lionnais 1983).

see also ALTERNATING PERMUTATION, BINOMIAL COEFFICIENT, CIRCULAR PERMUTATION, COMBINATION, COMPLETE PERMUTATION, DERANGEMENT, DISCORDANT PERMUTATION, EULERIAN NUMBER, LINEAR EXTENSION, PERMUTATION MATRIX, SUBFACTORIAL, TRANSPOSITION

References

Bogomolny, A. "Graphs." http://www.cut-the-knot.com/do_you_know/permutation.html.
Conway, J. H. and Guy, R. K. "Arrangement Numbers." In *The Book of Numbers.* New York: Springer-Verlag, p. 66, 1996.
Dickau, R. M. "Permutation Diagrams." http://forum.swarthmore.edu/advanced/robertd/permutations.html.
Knuth, D. E. *The Art of Computer Programming, Vol. 1: Fundamental Algorithms, 2nd ed.* Reading, MA: Addison-Wesley, 1973.
Kraitchik, M. "The Linear Permutations of n Different Things." §10.1 in *Mathematical Recreations.* New York: W. W. Norton, pp. 239–240, 1942.
Le Lionnais, F. *Les nombres remarquables.* Paris: Hermann, pp. 41–42, 1983.
Ruskey, F. "Information on Permutations." http://sue.csc.uvic.ca/~cos/inf/perm/PermInfo.html.
Sloane, N. J. A. Sequence A000142/M1675 in "An On-Line Version of the Encyclopedia of Integer Sequences."

Permutation Group

A finite GROUP of substitutions of elements for each other. For instance, the order 4 permutation group {4, 2, 1, 3} would rearrange the elements $\{A, B, C, D\}$ in the order $\{D, B, A, C\}$. A SUBSTITUTION GROUP of two elements is called a TRANSPOSITION. Every SUBSTITUTION GROUP with > 2 elements can be written as a product of transpositions. For example,

$$(abc) = (ab)(ac)$$
$$(abcde) = (ab)(ac)(ad)(ae).$$

CONJUGACY CLASSES of elements which are interchanged are called CYCLES (in the above example, the CYCLES are {{1, 3, 4}, {2}}).

see also CAYLEY'S GROUP THEOREM, CYCLE (PERMUTATION), GROUP, SUBSTITUTION GROUP, TRANSPOSITION

Permutation Matrix

A MATRIX p_{ij} obtained by permuting the ith and jth rows of the IDENTITY MATRIX with $i < j$. Every row and column therefore contain precisely a single 1, and every permutation corresponds to a unique permutation matrix. The matrix is nonsingular, so the DETERMINANT is always NONZERO. It satisfies

$$\mathsf{p}_{ij}{}^2 = \mathsf{I},$$

where I is the IDENTITY MATRIX. Applying to another MATRIX, $\mathsf{p}_{ij}\mathsf{A}$ gives A with the ith and jth rows interchanged, and $\mathsf{A}\mathsf{p}_{ij}$ gives A with the ith and jth columns interchanged.

Interpreting the 1s in an $n \times n$ permutation matrix as ROOKS gives an allowable configuration of nonattacking ROOKS on an $n \times n$ CHESSBOARD.

see also ELEMENTARY MATRIX, IDENTITY, PERMUTATION, ROOK NUMBER

Permutation Pseudotensor

see PERMUTATION TENSOR

Permutation Symbol

A three-index object sometimes called the LEVI-CIVITA SYMBOL defined by

$$\epsilon_{ijk} = \begin{cases} 0 & \text{for } i = j, j = k, \text{ or } k = i \\ +1 & \text{for } (i,j,k) \in \{(1,2,3),(2,3,1),(3,1,2)\} \\ -1 & \text{for } (i,j,k) \in \{(1,3,2),(3,2,1),(2,1,3)\}. \end{cases} \tag{1}$$

The permutation symbol satisfies

$$\delta_{ij}\epsilon_{ijk} = 0 \tag{2}$$
$$\epsilon_{ipq}\epsilon_{jpq} = 2\delta_{ij} \tag{3}$$
$$\epsilon_{ijk}\epsilon_{ijk} = 6 \tag{4}$$
$$\epsilon_{ijk}\epsilon_{pqk} = \delta_{ip}\delta_{jq} - \delta_{iq}\delta_{jp}, \tag{5}$$

where δ_{ij} is the KRONECKER DELTA. The symbol can be defined as the SCALAR TRIPLE PRODUCT of unit vectors in a right-handed coordinate system,

$$\epsilon_{ijk} \equiv \hat{\mathbf{x}}_i \cdot (\hat{\mathbf{x}}_j \times \hat{\mathbf{x}}_k). \tag{6}$$

The symbol can also be interpreted as a TENSOR, in which case it is called the PERMUTATION TENSOR.

see also PERMUTATION TENSOR

References

Arfken, G. *Mathematical Methods for Physicists, 3rd ed.* Orlando, FL: Academic Press, pp. 132–133, 1985.

Permutation Tensor

A PSEUDOTENSOR which is ANTISYMMETRIC under the interchange of any two slots. Recalling the definition of the PERMUTATION SYMBOL in terms of a SCALAR TRIPLE PRODUCT of the Cartesian unit vectors,

$$\epsilon_{ijk} \equiv \hat{\mathbf{x}}_i \cdot (\hat{\mathbf{x}}_j \times \hat{\mathbf{x}}_k) = [\hat{\mathbf{x}}_i, \hat{\mathbf{x}}_j, \hat{\mathbf{x}}_k], \tag{1}$$

the pseudotensor is a generalization to an arbitrary BASIS defined by

$$\epsilon_{\alpha\beta\cdots\mu} = \sqrt{|g|}\,[\alpha, \beta, \ldots, \mu] \tag{2}$$

$$\epsilon^{\alpha\beta\cdots\mu} = \frac{[\alpha, \beta, \ldots, \mu]}{\sqrt{|g|}}, \tag{3}$$

where

$$[\alpha, \beta, \ldots, \mu] = \begin{cases} 1 & \text{the arguments are an even permutation} \\ -1 & \text{the arguments are an odd permutation} \\ 0 & \text{two or more arguments are equal,} \end{cases} \tag{4}$$

and $g \equiv \det(g_{\alpha\beta})$, where $g_{\alpha\beta}$ is the METRIC TENSOR. $\epsilon(\mathbf{x}_1, \ldots, \mathbf{x}_n)$ is NONZERO IFF the VECTORS are LINEARLY INDEPENDENT.

see also PERMUTATION SYMBOL, SCALAR TRIPLE PRODUCT

Peron Integral

see DENJOY INTEGRAL

Perpendicular

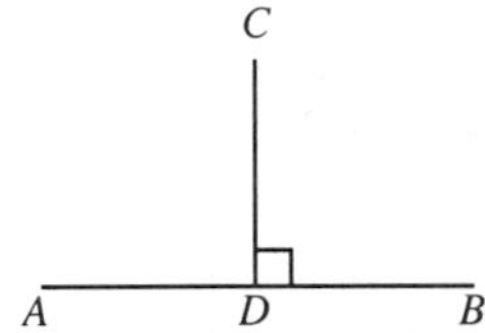

Two lines, vectors, planes, etc., are said to be perpendicular if they meet at a RIGHT ANGLE. In $\mathbb{R}^n$, two VECTORS $\mathbf{A}$ and $\mathbf{B}$ are PERPENDICULAR if their DOT PRODUCT

$$\mathbf{A} \cdot \mathbf{B} = 0.$$

In $\mathbb{R}^2$, a LINE with SLOPE $m_2 = -1/m_1$ is PERPENDICULAR to a LINE with SLOPE m_1. Perpendicular objects are sometimes said to be "orthogonal."

In the above figure, the LINE SEGMENT AB is perpendicular to the LINE SEGMENT CD. This relationship is commonly denoted with a small SQUARE at the vertex where perpendicular objects meet, as shown above.

see also ORTHOGONAL VECTORS, PARALLEL, PERPENDICULAR BISECTOR, PERPENDICULAR FOOT, RIGHT ANGLE

Perpendicular Bisector

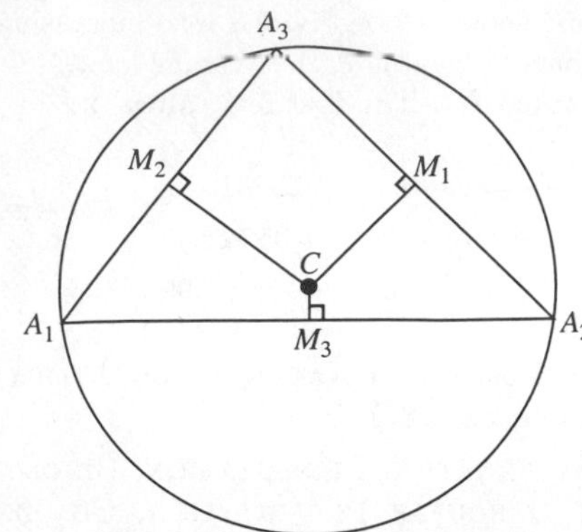

The perpendicular bisectors of a TRIANGLE $\Delta A_1A_2A_3$ are lines passing through the MIDPOINT M_i of each side which are PERPENDICULAR to the given side. A TRIANGLE'S three perpendicular bisectors meet at a point C known as the CIRCUMCENTER (which is also the center of the TRIANGLE'S CIRCUMCIRCLE).

see also CIRCUMCENTER, MIDPOINT, PERPENDICULAR, PERPENDICULAR FOOT

Perpendicular Foot

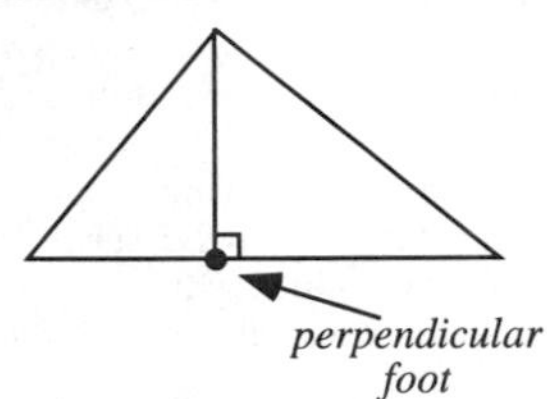

The FOOT of the PERPENDICULAR is the point on the leg opposite a given vertex of a TRIANGLE at which the PERPENDICULAR passing through that vertex intersects the side. The length of the LINE SEGMENT front vertex to perpendicular foot is called the ALTITUDE of the TRIANGLE.

see also ALTITUDE, FOOT, PERPENDICULAR, PERPENDICULAR BISECTOR

Perrin Pseudoprime

If p is PRIME, then $p|P(p)$, where $P(p)$ is a member of the PERRIN SEQUENCE 0, 2, 3, 2, 5, 5, 7, 10, 12, 17, ... (Sloane's A001608). A Perrin pseudoprime is a COMPOSITE NUMBER n such that $n|P(n)$. Several "unrestricted" Perrin pseudoprimes are known, the smallest of which are 271441, 904631, 16532714, 24658561, ... (Sloane's A013998).

Adams and Shanks (1982) discovered the smallest unrestricted Perrin pseudoprime after unsuccessful searches by Perrin (1899), Malo (1900), Escot (1901), and Jarden (1966). (Stewart's 1996 article stating no Perrin pseudoprimes were known was in error.)

Grantham (1996) generalized the definition of Perrin pseudoprime with parameters (r, s) to be an ODD COMPOSITE NUMBER n for which either

1. $(\Delta/n) = 1$ and n has an S-SIGNATURE, or
2. $(\Delta/n) = -1$ and n has a Q-SIGNATURE,

where (a/b) is the JACOBI SYMBOL. All the 55 Perrin pseudoprimes less than 50×10^9 have been computed by Kurtz *et al.* (1986). All have S-SIGNATURE, and form the sequence Sloane calls "restricted" Perrin pseudoprimes: 27664033, 46672291, 102690901, ... (Sloane's A018187).

see also PERRIN SEQUENCE, PSEUDOPRIME

References

Adams, W. W. "Characterizing Pseudoprimes for Third-Order Linear Recurrence Sequences." *Math Comput.* **48**, 1–15, 1987.

Adams, W. and Shanks, D. "Strong Primality Tests that Are Not Sufficient." *Math. Comput.* **39**, 255–300, 1982.

Bach, E. and Shallit, J. *Algorithmic Number Theory, Vol. 1: Efficient Algorithms.* Cambridge, MA: MIT Press, p. 305, 1996.

Escot, E.-B. "Solution to Item 1484." *L'Intermédiare des Math.* **8**, 63–64, 1901.

Grantham, J. "Frobenius Pseudoprimes." `http://www.clark.net/pub/grantham/pseudo/pseudo.ps`

Holzbaur, C. "Perrin Pseudoprimes." `http://ftp.ai.univie.ac.at/perrin.html`.

Jarden, D. *Recurring Sequences.* Jerusalem: Riveon Lematematika, 1966.

Kurtz, G. C.; Shanks, D.; and Williams, H. C. "Fast Primality Tests for Numbers Less than $50 \cdot 10^9$." *Math. Comput.* **46**, 691–701, 1986.

Malo, E. *L'Intermédiare des Math.* **7**, 281 and 312, 1900.

Perrin, R. "Item 1484." *L'Intermédiare des Math.* **6**, 76–77, 1899.

Ribenboim, P. *The New Book of Prime Number Records, 3rd ed.* New York: Springer-Verlag, p. 135, 1996.

Sloane, N. J. A. Sequences A013998, A018187, and A001608/M0429 in "An On-Line Version of the Encyclopedia of Integer Sequences."

Stewart, I. "Tales of a Neglected Number." *Sci. Amer.* **274**, 102–103, June 1996.

Perrin Sequence

The INTEGER SEQUENCE defined by the recurrence

$$P(n) = P(n-2) + P(n-3) \tag{1}$$

with the initial conditions $P(0) = 3$, $P(1) = 0$, $P(2) = 2$. The first few terms are 0, 2, 3, 2, 5, 5, 7, 10, 12, 17, ... (Sloane's A001608). $P(n)$ is the solution of a third-order linear homogeneous DIFFERENCE EQUATION having characteristic equation

$$x^3 - x - 1 = 0, \tag{2}$$

discriminant -23, and ROOTS

$$\alpha \approx 1.324717957 \tag{3}$$

$$\beta \approx -0.6623589786 + 0.5622795121i \tag{4}$$

$$\gamma \approx -0.6623589786 - 0.5622795121i. \tag{5}$$

The solution is then

$$A(n) = \alpha^n + \beta^n + \gamma^n, \tag{6}$$

where

$$A(n) \sim \alpha^n. \tag{7}$$

Perrin (1899) investigated the sequence and noticed that if n is PRIME, then $n|P(n)$. The first statement of this fact is attributed to É. Lucas in 1876 by Stewart (1996). Perrin also searched for but did not find any COMPOSITE NUMBER n in the sequence such that $n|P(n)$. Such numbers are now known as PERRIN PSEUDOPRIMES. Malo (1900), Escot (1901), and Jarden (1966) subsequently investigated the series and also found no PERRIN PSEUDOPRIMES. Adams and Shanks (1982) subsequently found that 271,441 is such a number.

see also PADOVAN SEQUENCE, PERRIN PSEUDOPRIME, SIGNATURE (RECURRENCE RELATION)

References

Adams, W. and Shanks, D. "Strong Primality Tests that Are Not Sufficient." *Math. Comput.* **39**, 255–300, 1982.

Escot, E.-B. "Solution to Item 1484." *L'Intermédiare des Math.* **8**, 63–64, 1901.

Jarden, D. *Recurring Sequences.* Jerusalem: Riveon Lematematika, 1966.

Malo, E. *L'Intermédiare des Math.* **7**, 281 and 312, 1900.

Perrin, R. "Item 1484." *L'Intermédiare des Math.* **6**, 76–77, 1899.

Stewart, I. "Tales of a Neglected Number." *Sci. Amer.* **274**, 102–103, June 1996.

Sloane, N. J. A. Sequence A001608/M0429 in "An On-Line Version of the Encyclopedia of Integer Sequences."

Perron-Frobenius Operator

An OPERATOR which describes the time evolution of densities in PHASE SPACE. The OPERATOR can be defined by

$$\rho_{n+1} = \tilde{L}\rho_n,$$

where ρ_n are the NATURAL DENSITIES after the nth iteration of a map f. This can be explicitly written as

$$\tilde{L}\rho(y) = \sum_{x \in f^{-1}(y)} \frac{\rho(x)}{|f'(x)|}.$$

References

Beck, C. and Schlögl, F. "Transfer Operator Methods." Ch. 17 in *Thermodynamics of Chaotic Systems.* Cambridge, England: Cambridge University Press, pp. 190–203, 1995.

Perron-Frobenius Theorem

If all elements a_{ij} of an IRREDUCIBLE MATRIX A are NONNEGATIVE, then $R = \min M_\lambda$ is an EIGENVALUE of A and all the EIGENVALUES of A lie on the DISK

$$|z| \leq R,$$

where, if $\lambda = (\lambda_1, \lambda_2, \ldots, \lambda_n)$ is a set of NONNEGATIVE numbers (which are not all zero),

$$M_\lambda = \inf\left\{\mu : \mu\lambda_i > \sum_{j=1}^n |a_{ij}|\lambda_j, 1 \leq i \leq n\right\}$$

and $R = \min M_\lambda$. Furthermore, if A has exactly p EIGENVALUES ($p \leq n$) on the CIRCLE $|z| = R$, then the set of all its EIGENVALUES is invariant under rotations by $2\pi/p$ about the ORIGIN.

see also WIELANDT'S THEOREM

References

Gradshteyn, I. S. and Ryzhik, I. M. *Tables of Integrals, Series, and Products, 5th ed.* San Diego, CA: Academic Press, p. 1121, 1979.

Perron's Theorem

If $\boldsymbol{\mu} = (\mu_1, \mu_2, \ldots, \mu_n)$ is an arbitrary set of POSITIVE numbers, then all EIGENVALUES λ of the $n \times n$ MATRIX $\mathsf{A} = a_{ij}$ lie on the DISK $|z| \leq M_\mu$, where

$$M_\mu = \max_{1 \leq i \leq n} \sum_{j=1}^{n} \frac{\mu_j}{\mu_i} |a_{ij}|.$$

References

Gradshteyn, I. S. and Ryzhik, I. M. *Tables of Integrals, Series, and Products, 5th ed.* San Diego, CA: Academic Press, p. 1121, 1979.

Persistence

see ADDITIVE PERSISTENCE, MULTIPLICATIVE PERSISTENCE, PERSISTENT NUMBER, PERSISTENT PROCESS

Persistent Number

An n-persistent number is a POSITIVE INTEGER k which contains the digits 0, 1, ..., 9, and for which $2k$, ..., nk also share this property. No ∞-persistent numbers exist. However, the number $k = 1234567890$ is 2-persistent, since $2k = 2469135780$ but $3k = 3703703670$, and the number $k = 526315789473684210$ is 18-persistent. There exists at least one k-persistent number for each POSITIVE INTEGER k.

see also ADDITIVE PERSISTENCE, MULTIPLICATIVE PERSISTENCE

References

Honsberger, R. *More Mathematical Morsels.* Washington, DC: Math. Assoc. Amer., pp. 15–18, 1991.

Persistent Process

A FRACTAL PROCESS for which $H > 1/2$, so $r > 0$.

see also ANTIPERSISTENT PROCESS, FRACTAL PROCESS

Perspective

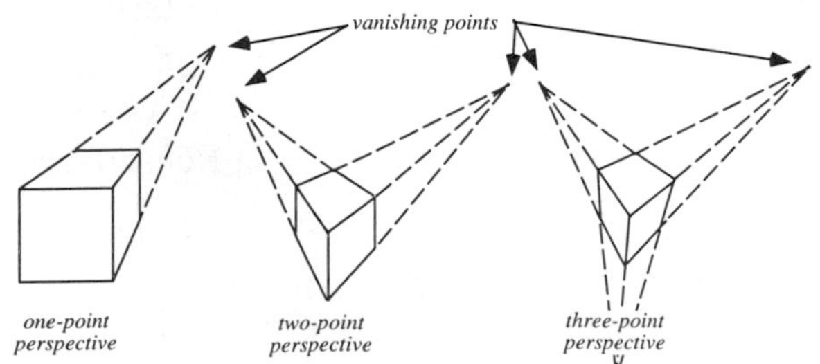

Perspective is the art and mathematics of realistically depicting 3-D objects in a 2-D plane. The study of the projection of objects in a plane is called PROJECTIVE GEOMETRY. The principles of perspective drawing were elucidated by the Florentine architect F. Brunelleschi (1377–1446). These rules are summarized by Dixon (1991):

1. The horizon appears as a line.
2. Straight lines in space appear as straight lines in the image.
3. Sets of PARALLEL lines meet at a VANISHING POINT.
4. Lines PARALLEL to the picture plane appear PARALLEL and therefore have no VANISHING POINT.

There is a graphical method for selecting vanishing points so that a CUBE or box appears to have the correct dimensions (Dixon 1991).

see also LEONARDO'S PARADOX, PERSPECTIVE AXIS, PERSPECTIVE CENTER, PERSPECTIVE COLLINEATION, PERSPECTIVE TRIANGLES, PERSPECTIVITY, PROJECTIVE GEOMETRY, VANISHING POINT, ZEEMAN'S PARADOX

References

de Vries, V. *Perspective.* New York: Dover, 1968.

Dixon, R. "Perspective Drawings." Ch. 3 in *Mathographics.* New York: Dover, pp. 79–88, 1991.

Parramon, J. M. *Perspective—How to Draw.* Barcelona, Spain: Parramon Editions, 1984.

Perspective Axis

The line joining the three collinear points of intersection of the extensions of corresponding sides in PERSPECTIVE TRIANGLES.

see also PERSPECTIVE CENTER, PERSPECTIVE TRIANGLES, SONDAT'S THEOREM

Perspective Center

The point at which the three LINES connecting the VERTICES of PERSPECTIVE TRIANGLES (from a point) CONCUR.

Perspective Collineation

A perspective collineation with center O and axis o is a COLLINEATION which leaves all lines through O and points of o invariant. Every perspective collineation is a PROJECTIVE COLLINEATION.

see also COLLINEATION, ELATION, HOMOLOGY (GEOMETRY), PROJECTIVE COLLINEATION

References

Coxeter, H. S. M. *Introduction to Geometry, 2nd ed.* New York: Wiley, pp. 247–248, 1969.

Perspective Triangles

Two TRIANGLES are perspective from a line if the extensions of their three pairs of corresponding sides meet in COLLINEAR points. The line joining these points is called the PERSPECTIVE AXIS. Two TRIANGLES are perspective from a point if their three pairs of corresponding VERTICES are joined by lines which meet in a point of CONCURRENCE. This point is called the PERSPECTIVE CENTER. DESARGUES' THEOREM guarantees that if two TRIANGLES are perspective from a point, they are perspective from a line.

see also DESARGUES' THEOREM, HOMOTHETIC TRIANGLES, PARALOGIC TRIANGLES, PERSPECTIVE AXIS, PERSPECTIVE CENTER

Perspectivity

A correspondence between two RANGES that are sections of one PENCIL by two distinct lines.

see also PENCIL, PROJECTIVITY, RANGE (LINE SEGMENT)

Pesin Theory

A theory of linear HYPERBOLIC MAPS in which the leading constants do depend on the variable x.

Peter-Weyl Theorem

Establishes completeness for a REPRESENTATION.

References

Knapp, A. W. "Group Representations and Harmonic Analysis, Part II." *Not. Amer. Math. Soc.* **43**, 537–549, 1996.

Peters Projection

A CYLINDRICAL equal-area projection that shifts the standard parallels to 45° or 47°.

see also CYLINDRICAL PROJECTION

References

Dana, P. H. "Map Projections." http://www.utexas.edu/depts/grg/gcraft/notes/mapproj/mapproj.html.

Petersen Graphs

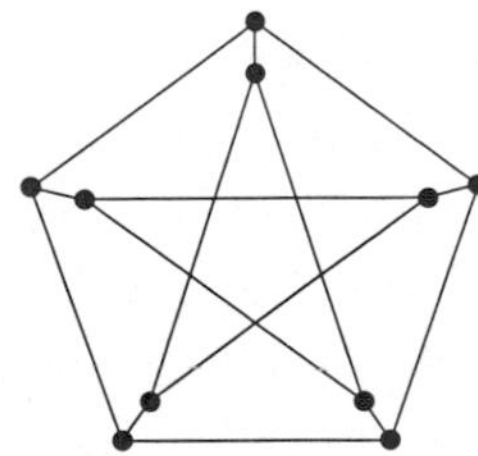

"The" Petersen graph is the GRAPH illustrated above possessing ten VERTICES all of whose nodes have DEGREE 3 (Saaty and Kainen 1986). The Petersen graph is the only smallest-girth graph which has no Tait coloring.

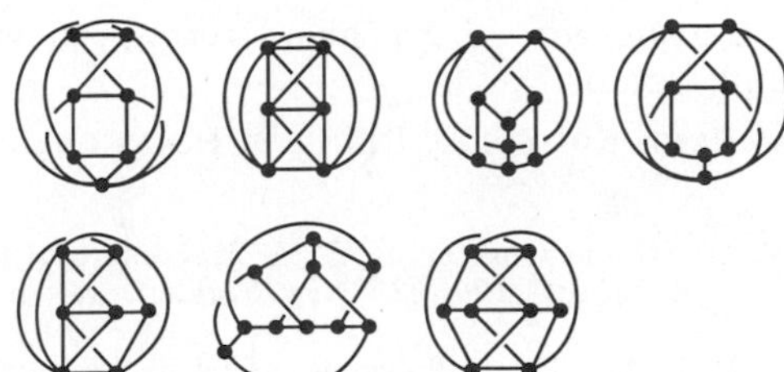

The seven graphs obtainable from the COMPLETE GRAPH K_6 by repeated triangle-Y exchanges are also called Petersen graphs, where the three EDGES forming the TRIANGLE are replaced by three EDGES and a new VERTEX that form a Y, and the reverse operation is also permitted. A GRAPH is intrinsically linked IFF it contains one of the seven Petersen graphs (Robertson *et al.* 1993).

see also HOFFMAN-SINGLETON GRAPH

References

Adams, C. C. *The Knot Book: An Elementary Introduction to the Mathematical Theory of Knots.* New York: W. H. Freeman, pp. 221–222, 1994.

Robertson, N.; Seymour, P. D.; and Thomas, R. "Linkless Embeddings of Graphs in 3-Space." *Bull. Amer. Math. Soc.* **28**, 84–89, 1993.

Saaty, T. L. and Kainen, P. C. *The Four-Color Problem: Assaults and Conquest.* New York: Dover, p. 102, 1986.

Petersen-Shoute Theorem

1. If ΔABC and $\Delta A'B'C'$ are two directly similar triangles, while $\Delta AA'A''$, $\Delta BB'B''$, and $\Delta CC'C''$ are three directly similar triangles, then $\Delta A''B''C''$ is directly similar to ΔABC.
2. When all the points P on AB are related by a SIMILARITY TRANSFORMATION to all the points P' on $A'B'$, the points dividing the segment PP' in a given ratio are distant and collinear, or else they coincide.

References

Coxeter, H. S. M. and Greitzer, S. L. *Geometry Revisited.* Washington, DC: Math. Assoc. Amer., pp. 95–100, 1967.

Petrie Polygon

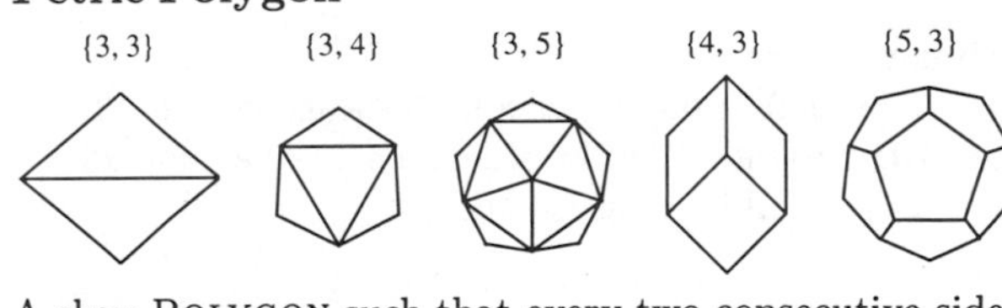

A skew POLYGON such that every two consecutive sides (but no three) belong to a face of a regular POLYHEDRON. Every finite POLYHEDRON can be orthogonally projected onto a plane in such a way that one Petrie polygon becomes a REGULAR POLYGON with the remainder of the projection interior to it. The Petrie polygon of the POLYHEDRON $\{p,q\}$ has h sides, where

$$\cos^2\left(\frac{\pi}{h}\right)=\cos^2\left(\frac{\pi}{p}\right)+\cos^2\left(\frac{\pi}{q}\right).$$

The Petrie polygons shown above correspond to the PLATONIC SOLIDS.

see also PLATONIC SOLID, REGULAR POLYGON

References

Ball, W. W. R. and Coxeter, H. S. M. *Mathematical Recreations and Essays, 13th ed.* New York: Dover, p. 135, 1987.

Coxeter, H. S. M. "Petrie Polygons." §2.6 in *Regular Polytopes, 3rd ed.* New York: Dover, pp.24–25, 1973.

Petrov Notation

A TENSOR notation which considers the RIEMANN TENSOR $R_{\lambda\mu\nu\kappa}$ as a matrix $R_{(\lambda\mu)(\nu\kappa)}$ with indices $\lambda\mu$ and $\nu\kappa$.

References

Weinberg, S. *Gravitation and Cosmology: Principles and Applications of the General Theory of Relativity.* New York: Wiley, p. 142, 1972.

Pfaffian Form

A 1-FORM

$$\omega = \sum_{i=1}^{n} a_i(x)\,dx_i$$

such that

$$\omega = 0.$$

References

Knuth, D. E. "Overlapping Pfaffians." *Electronic J. Combinatorics* **3**, No. 2, R5, 1–13, 1996. `http://www.combinatorics.org/Volume_3/volume3_2.html#R5`.

Phase

The angular position of a quantity. For example, the phase of a function $\cos(\omega t + \phi_0)$ as a function of time is

$$\phi(t) = \omega t + \phi_0.$$

The ARGUMENT of a COMPLEX NUMBER is sometimes also called the phase.

see also ARGUMENT (COMPLEX NUMBER), COMPLEX NUMBER, PHASOR, RETARDANCE

Phase Space

For a function or object with n DEGREES OF FREEDOM, the n-D SPACE which is accessible to the function or object is called its phase space.

see also WORLD LINE

Phase Transition

see RANDOM GRAPH

Phasor

The representation, beloved of engineers and physicists, of a COMPLEX NUMBER in terms of a COMPLEX exponential

$$x + iy = |z|e^{i\phi}, \tag{1}$$

where i (called j by engineers) is the IMAGINARY NUMBER and the MODULUS and ARGUMENT (also called PHASE) are

$$|z| = \sqrt{x^2 + y^2} \tag{2}$$

$$\phi = \tan^{-1}\left(\frac{y}{x}\right). \tag{3}$$

Here, ϕ is the counterclockwise ANGLE from the POSITIVE REAL axis. In the degenerate case when $x = 0$,

$$\phi = \begin{cases} -\frac{1}{2}\pi & \text{if } y < 0 \\ \text{undefined} & \text{if } y = 0 \\ \frac{1}{2}\pi & \text{if } y > 0. \end{cases} \tag{4}$$

It is trivially true that

$$\sum_i \Re[\psi_i] = \Re\left[\sum_i \psi_i\right]. \tag{5}$$

Now consider a SCALAR FUNCTION $\psi \equiv \psi_0 e^{i\phi}$. Then

$$\begin{aligned} I \equiv [\Re(\psi)]^2 &= [\tfrac{1}{2}(\psi + \psi^*)]^2 = \tfrac{1}{4}(\psi + \psi^*)^2 \\ &= \tfrac{1}{4}(\psi^2 + 2\psi\psi^* + \psi^{*2}). \end{aligned} \tag{6}$$

Look at the time averages of each term,

$$\left\langle \psi^2 \right\rangle = \left\langle {\psi_0}^2 e^{2i\phi} \right\rangle = {\psi_0}^2 \left\langle e^{2i\phi} \right\rangle = 0 \tag{7}$$

$$\left\langle \psi\psi^* \right\rangle = \left\langle \psi_0 e^{i\phi} \psi_0 e^{-i\phi} \right\rangle = {\psi_0}^2 = |\psi|^2 \tag{8}$$

$$\left\langle \psi^{*2} \right\rangle = \left\langle {\psi_0}^2 e^{-2i\phi} \right\rangle = {\psi_0}^2 \left\langle e^{-2i\phi} \right\rangle = 0. \tag{9}$$

Therefore,

$$\langle I \rangle = \tfrac{1}{2}|\psi|^2. \tag{10}$$

Consider now two scalar functions

$$\psi_1 \equiv \psi_{1,0} e^{i(kr_1 + \phi_1)} \tag{11}$$

$$\psi_2 \equiv \psi_{2,0} e^{i(kr_2 + \phi_2)}. \tag{12}$$

Then

$$\begin{aligned} I \equiv [\Re(\psi_1) + \Re(\psi_2)]^2 &= \tfrac{1}{4}[(\psi_1 + {\psi_1}^*) + (\psi_2 + {\psi_2}^*)]^2 \\ &= \tfrac{1}{4}[(\psi_1 + {\psi_1}^*)^2 + (\psi_2 + {\psi_2}^*)^2 \\ &\quad + 2(\psi_1\psi_2 + \psi_1{\psi_2}^* + {\psi_1}^*\psi_2 + {\psi_1}^*{\psi_2}^*)] \end{aligned} \tag{13}$$

$$\begin{aligned} \langle I \rangle &= \tfrac{1}{4}[2\psi_1{\psi_1}^* + 2\psi_2{\psi_2}^* + 2\psi_1{\psi_2}^* + 2{\psi_1}^*\psi_2] \\ &= \tfrac{1}{2}[\psi_1({\psi_1}^* + {\psi_2}^*) + \psi_2({\psi_1}^* + {\psi_2}^*)] \\ &= \tfrac{1}{2}(\psi_1 + \psi_2)({\psi_1}^* + {\psi_2}^*) = \tfrac{1}{2}|\psi_1 + \psi_2|^2. \end{aligned} \tag{14}$$

In general,

$$\langle I \rangle = \frac{1}{2}\left|\sum_{i=1}^{n} \psi_i\right|^2. \tag{15}$$

see also AFFIX, ARGUMENT (COMPLEX NUMBER), COMPLEX MULTIPLICATION, COMPLEX NUMBER, MODULUS (COMPLEX NUMBER), PHASE

Phi Curve

An ADJOINT CURVE which bears a special relation to the base curve.

References
Coolidge, J. L. *A Treatise on Algebraic Plane Curves.* New York: Dover, p. 310, 1959.

Phi Number System

For every POSITIVE INTEGER n, there is a corresponding finite sequence of distinct INTEGERS $k_1, \ldots, k_m$ such that

$$n = \phi^{k_1} + \ldots + \phi^{k_m},$$

where ϕ is the GOLDEN MEAN.

References
Bergman, G. "A Number System with an Irrational Base." *Math. Mag.* **31**, 98–110, 1957.
Knuth, D. *The Art of Computer Programming, Vol. 1: Fundamental Algorithms, 2nd ed.* Reading, MA: Addison-Wesley, 1973.
Rousseau, C. "The Phi Number System Revisited." *Math. Mag.* **68**, 283–284, 1995.

Phragmén-Lindelöf Theorem

Let $f(z)$ be an ANALYTIC FUNCTION in an angular domain $W : |\arg z| < \alpha\pi/2$. Suppose there is a constant M such that for each $\epsilon > 0$, each finite boundary point has a NEIGHBORHOOD such that $|f(z)| < M + \epsilon$ on the intersection of D with this NEIGHBORHOOD, and that for some POSITIVE number $\beta > \alpha$ for sufficiently large $|z|$, the INEQUALITY $|f(z)| < \exp(|z|^{1/\beta})$ holds. Then $|f(z)| \leq M$ in D.

References
Iyanaga, S. and Kawada, Y. (Eds.). *Encyclopedic Dictionary of Mathematics.* Cambridge, MA: MIT Press, p. 160, 1980.

Phyllotaxis

The beautiful arrangement of leaves in some plants, called phyllotaxis, obeys a number of subtle mathematical relationships. For instance, the florets in the head of a sunflower form two oppositely directed spirals: 55 of them clockwise and 34 counterclockwise. Surprisingly, these numbers are consecutive FIBONACCI NUMBERS. The ratios of alternate FIBONACCI NUMBERS are given by the convergents to ϕ^{-2}, where ϕ is the GOLDEN RATIO, and are said to measure the fraction of a turn between successive leaves on the stalk of a plant: 1/2 for elm and linden, 1/3 for beech and hazel, 2/5 for oak and apple, 3/8 for poplar and rose, 5/13 for willow and almond, etc. (Coxeter 1969, Ball and Coxeter 1987). A similar phenomenon occurs for DAISIES, pineapples, pinecones, cauliflowers, and so on.

Lilies, irises, and the trillium have three petals; columbines, buttercups, larkspur, and wild rose have five petals; delphiniums, bloodroot, and cosmos have eight petals; corn marigolds have 13 petals; asters have 21 petals; and daisies have 34, 55, or 84 petals—all FIBONACCI NUMBERS.

see also DAISY, FIBONACCI NUMBER, SPIRAL

References
Ball, W. W. R. and Coxeter, H. S. M. *Mathematical Recreations and Essays, 13th ed.* New York: Dover, pp. 56–57, 1987.
Church, A. H. *The Relation of Phyllotaxis to Mechanical Laws.* London: Williams and Norgate, 1904.
Church, A. H. *On the Interpretation of Phenomena of Phyllotaxis.* Riverside, NJ: Hafner, 1968.
Conway, J. H. and Guy, R. K. "Phyllotaxis." In *The Book of Numbers.* New York: Springer-Verlag, pp. 113–125, 1995.
Coxeter, H. S. M. "The Golden Section and Phyllotaxis." Ch. 11 in *Introduction to Geometry, 2nd ed.* New York: Wiley, 1969.
Coxeter, H. S. M. "The Golden Section, Phyllotaxis, and Wythoff's Game." *Scripta Mathematica* **19**, 135–143, 1953.
Dixon, R. *Mathographics.* New York: Dover, 1991.
Douady, S. and Couder, Y. "Phyllotaxis as a Self-Organized Growth Process." In *Growth Patterns in Physical Sciences and Biology* (Ed. Juan M. Garcia-Ruiz *et al.*). Plenum Press, 1993.
Hunter, J. A. H. and Madachy, J. S. *Mathematical Diversions.* New York: Dover, pp. 20–22, 1975.
Jean, R. V. *Phyllotaxis: A Systematic Study in Plant Morphogenesis.* New York: Cambridge University Press, 1994.
Pappas, T. "The Fibonacci Sequence & Nature." *The Joy of Mathematics.* San Carlos, CA: Wide World Publ./Tetra, pp. 222–225, 1989.
Prusinkiewicz, P. and Lindenmayer, A. *The Algorithmic Beauty of Plants.* New York: Springer-Verlag, 1990.
Stewart, I. "Daisy, Daisy, Give Me Your Answer, Do." *Sci. Amer.* **200**, 96–99, Jan. 1995.
Thompson, D. W. *On Growth and Form.* Cambridge, England: Cambridge University Press, 1952.

Pi

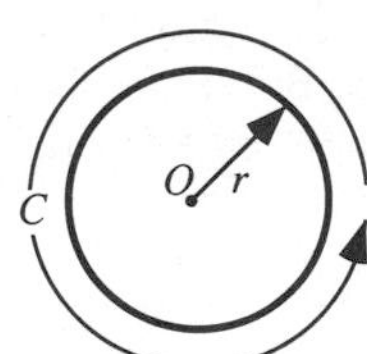

A REAL NUMBER denoted π which is defined as the ratio of a CIRCLE's CIRCUMFERENCE C to its DIAMETER $d = 2r$,

$$\pi \equiv \frac{C}{d} = \frac{C}{2r} \tag{1}$$

It is equal to

$$\pi = 3.141592653589793238462643383279502884197\ldots \tag{2}$$

(Sloane's A000796). π has recently (August 1997) been computed to a world record $51,539,600,000 \approx 3 \cdot 2^{34}$ DECIMAL DIGITS by Y. Kanada. This calculation was done using Borwein's fourth-order convergent algorithm and required 29 hours on a massively parallel 1024-processor Hitachi SR2201 supercomputer. It was checked in 37 hours using the BRENT-SALAMIN FORMULA on the same machine.

The SIMPLE CONTINUED FRACTION for π, which gives the "best" approximation of a given order, is [3, 7, 15,

1, 292, 1, 1, 1, 2, 1, 3, 1, 14, 2, 1, 1, 2, 2, 2, 2, ...] (Sloane's A001203). The very large term 292 means that the CONVERGENT

$$[3,7,15,1]=[3,7,16]=\tfrac{355}{113}=3.14159292\ldots \quad (3)$$

is an extremely good approximation. The first few CONVERGENTS are 22/7, 333/106, 355/113, 103993/33102, 104348/33215, ... (Sloane's A002485 and A002486). The first occurrences of n in the CONTINUED FRACTION are 4, 9, 1, 30, 40, 32, 2, 44, 130, 100, ... (Sloane's A032523).

Gosper has computed 17,001,303 terms of π's CONTINUED FRACTION (Gosper 1977, Ball and Coxeter 1987), although the computer on which the numbers are stored may no longer be functional (Gosper, pers. comm., 1998). According to Gosper, a typical CONTINUED FRACTION term carries only slightly more significance than a decimal DIGIT. The sequence of increasing terms in the CONTINUED FRACTION is 3, 7, 15, 292, 436, 20776, ... (Sloane's A033089), occurring at positions 1, 2, 3, 5, 308, 432, ... (Sloane's A033090). In the first 26,491 terms of the CONTINUED FRACTION (counting 3 as the 0th), the only five-DIGIT terms are 20,776 (the 431st), 19,055 (15,543rd), and 19,308 (23,398th) (Beeler *et al.* 1972, Item 140). The first 6-DIGIT term is 528,210 (the 267,314th), and the first 8-DIGIT term is 12,996,958 (453,294th). The term having the largest known value is the whopping 9-DIGIT 87,878,3625 (the 11,504,931st term).

The SIMPLE CONTINUED FRACTION for π does not show any obvious patterns, but clear patterns *do* emerge in the beautiful non-simple CONTINUED FRACTIONS

$$\frac{4}{\pi}=1+\cfrac{1^2}{2+\cfrac{3^2}{2+\cfrac{5^2}{2+\cfrac{7^2}{2+\ldots}}}} \quad (4)$$

(Brouckner), giving convergents 1, 3/2, 15/13, 105/76, 315/263, ... (Sloane's A025547 and A007509) and

$$\frac{\pi}{2}=1-\cfrac{1}{3-\cfrac{2\cdot 3}{1-\cfrac{1\cdot 2}{3-\cfrac{4\cdot 5}{1-\cfrac{3\cdot 4}{3-\cfrac{6\cdot 7}{1-\cfrac{5\cdot 6}{3-\ldots}}}}}}} \quad (5)$$

(Stern 1833), giving convergents 1, 2/3, 4/3, 16/15, 64/45, 128/105, ... (Sloane's A001901 and A046126).

π crops up in all sorts of unexpected places in mathematics besides CIRCLES and SPHERES. For example, it occurs in the normalization of the GAUSSIAN DISTRIBUTION, in the distribution of PRIMES, in the construction of numbers which are very close to INTEGERS (the RAMANUJAN CONSTANT), and in the probability that a pin dropped on a set of PARALLEL lines intersects a line (BUFFON'S NEEDLE PROBLEM). Pi also appears as the average ratio of the actual length and the direct distance between source and mouth in a meandering river (Støllum 1996, Singh 1997).

A brief history of NOTATION for pi is given by Castellanos (1988). π is sometimes known as LUDOLPH'S CONSTANT after Ludolph van Ceulen (1539–1610), a Dutch π calculator. The symbol π was first used by William Jones in 1706, and subsequently adopted by Euler. In *Measurement of a Circle,* Archimedes (ca. 225 BC) obtained the first rigorous approximation by INSCRIBING and CIRCUMSCRIBING $6\cdot 2^n$-gons on a CIRCLE using the ARCHIMEDES ALGORITHM. Using $n=4$ (a 96-gon), Archimedes obtained

$$3+\tfrac{10}{71}<\pi<3+\tfrac{1}{7} \quad (6)$$

(Shanks 1993, p. 140).

The Bible contains two references (I Kings 7:23 and Chronicles 4:2) which give a value of 3 for π. It should be mentioned, however, that both instances refer to a value obtained from physical measurements and, as such, are probably well within the bounds of experimental uncertainty. I Kings 7:23 states, "Also he made a molten sea of ten Cubits from brim to brim, round in compass, and five cubits in height thereof; and a line thirty cubits did compass it round about." This implies $\pi=C/d=30/10=3$. The Babylonians gave an estimate of π as $3+1/8=3.125$. The Egyptians did better still, obtaining $2^8/3^4=3.1605\ldots$ in the Rhind papyrus, and 22/7 elsewhere. The Chinese geometers, however, did best of all, rigorously deriving π to 6 decimal places.

A method similar to Archimedes' can be used to estimate π by starting with an n-gon and then relating the AREA of subsequent $2n$-gons. Let β be the ANGLE from the center of one of the POLYGON's segments,

$$\beta=\tfrac{1}{4}(n-3)\pi. \quad (7)$$

Then

$$\pi=\frac{\frac{1}{2}n\sin(2\beta)}{\cos\beta\cos\left(\frac{\beta}{2}\right)\cos\left(\frac{\beta}{2^2}\right)\cos\left(\frac{\beta}{2^3}\right)\cdots} \quad (8)$$

(Beckmann 1989, pp. 92–94). Viète (1593) was the first to give an exact expression for π by taking $n=4$ in the above expression, giving

$$\cos\beta=\sin\beta=\frac{1}{\sqrt{2}}=\tfrac{1}{2}\sqrt{2}, \quad (9)$$

which leads to an INFINITE PRODUCT of CONTINUED SQUARE ROOTS,

$$\frac{2}{\pi} = \sqrt{\tfrac{1}{2}}\sqrt{\tfrac{1}{2}+\sqrt{\tfrac{1}{2}}}\sqrt{\tfrac{1}{2}+\sqrt{\tfrac{1}{2}+\sqrt{\tfrac{1}{2}}}}\cdots \tag{10}$$

(Beckmann 1989, p. 95). However, this expression was not rigorously proved to converge until Rudio (1892). Another exact FORMULA is MACHIN'S FORMULA, which is

$$\frac{\pi}{4} = 4\tan^{-1}(\tfrac{1}{5}) - \tan^{-1}(\tfrac{1}{239}). \tag{11}$$

There are three other MACHIN-LIKE FORMULAS, as well as other FORMULAS with more terms. An interesting INFINITE PRODUCT formula due to Euler which relates π and the nth PRIME p_n is

$$\pi = \frac{2}{\prod_{i=n}^{\infty}\left[1+\frac{\sin(\frac{1}{2}\pi p_n)}{p_n}\right]} \tag{12}$$

$$= \frac{2}{\prod_{i=n}^{\infty}\left[1+\frac{(-1)^{(p_n-1)/2}}{p_n}\right]} \tag{13}$$

(Blatner 1997, p. 119), plotted below as a function of the number of terms in the product.

The AREA and CIRCUMFERENCE of the UNIT CIRCLE are given by

$$A = \pi = 4\int_0^1 \sqrt{1-x^2}\,dx \tag{14}$$

$$= \lim_{n\to\infty} \frac{4}{n^2}\sum_{k=0}^{n}\sqrt{n^2-k^2} \tag{15}$$

and

$$C = 2\pi = 4\int_0^1 \frac{dx}{\sqrt{1-x^2}} \tag{16}$$

$$= 4\int_0^1 \sqrt{1+\left(\frac{d}{dx}\sqrt{1-x^2}\right)^2}\,dx. \tag{17}$$

The SURFACE AREA and VOLUME of the unit SPHERE are

$$S = 4\pi \tag{18}$$

$$V = \tfrac{4}{3}\pi. \tag{19}$$

π is known to be IRRATIONAL (Lambert 1761, Legendre 1794) and even TRANSCENDENTAL (Lindemann 1882). Incidentally, Lindemann's proof of the transcendence of π also proved that the GEOMETRIC PROBLEM OF ANTIQUITY known as CIRCLE SQUARING is impossible. A simplified, but still difficult, version of Lindemann's proof is given by Klein (1955).

It is also known that π is not a LIOUVILLE NUMBER (Mahler 1953). In 1974, M. Mignotte showed that

$$\left|\pi - \frac{p}{q}\right| \leq q^{-20} \tag{20}$$

has only a finite number of solutions in INTEGERS (Le Lionnais 1983, p. 50). This result was subsequently improved by Chudnovsky and Chudnovsky (1984) who showed that

$$\left|\pi - \frac{p}{q}\right| > q^{-14.65}, \tag{21}$$

although it is likely that the exponent can be reduced to $2+\epsilon$, where ϵ is an infinitesimally small number (Borwein *et al.* 1989). It is *not* known if π is NORMAL (Wagon 1985), although the first 30 million DIGITS are very UNIFORMLY DISTRIBUTED (Bailey 1988). The following distribution is found for the first n DIGITS of $\pi - 3$. It shows no statistically SIGNIFICANT departure from a UNIFORM DISTRIBUTION (technically, in the CHI-SQUARED TEST, it has a value of ${\chi_s}^2 = 5.60$ for the first 5×10^{10} terms).

digit	1×10^5	1×10^6	6×10^9	5×10^{10}
0	9,999	99,959	599,963,005	5,000,012,647
1	10,137	99,758	600,033,260	4,999,986,263
2	9,908	100,026	599,999,169	5,000,020,237
3	10,025	100,229	600,000,243	4,999,914,405
4	9,971	100,230	599,957,439	5,000,023,598
5	10,026	100,359	600,017,176	4,999,991,499
6	10,029	99,548	600,016,588	4,999,928,368
7	10,025	99,800	600,009,044	5,000,014,860
8	9,978	99,985	599,987,038	5,000,117,637
9	9,902	100,106	600,017,038	4,999,990,486

The digits of $1/\pi$ are also very uniformly distributed (${\chi_s}^2 = 7.04$) shown in the following table.

digit	5×10^{10}
0	4,999,969,955
1	5,000,113,699
2	4,999,987,893
3	5,000,040,906
4	4,999,985,863
5	4,999,977,583
6	4,999,990,916
7	4,999,985,552
8	4,999,881,183
9	5,000,066,450

It is not known if $\pi + e$, π/e, or $\ln \pi$ are IRRATIONAL. However, it is known that they cannot satisfy any POLYNOMIAL equation of degree ≤ 8 with INTEGER COEFFICIENTS of average size 10^9 (Bailey 1988, Borwein *et al.* 1989).

π satisfies the INEQUALITY

$$\left(1 + \frac{1}{\pi}\right)^{\pi+1} \approx 3.14097 < \pi. \tag{22}$$

Beginning with any POSITIVE INTEGER n, round up to the nearest multiple of $n-1$, then up to the nearest multiple of $n-2$, and so on, up to the nearest multiple of 1. Let $f(n)$ denote the result. Then the ratio

$$\lim_{n\to\infty} \frac{n^2}{f(n)} = \pi \tag{23}$$

(Brown). David (1957) credits this result to Jabotinski and Erdős and gives the more precise asymptotic result

$$f(n) = \frac{n^2}{\pi} + \mathcal{O}(n^{4/3}). \tag{24}$$

The first few numbers in the sequence $\{f(n)\}$ are 1, 2, 4, 6, 10, 12, 18, 22, 30, 34, ... (Sloane's A002491).

A particular case of the WALLIS FORMULA gives

$$\frac{\pi}{2} = \prod_{n=1}^{\infty}\left[\frac{(2n)^2}{(2n-1)(2n+1)}\right] = \frac{2\cdot 2}{1\cdot 3}\frac{4\cdot 4}{3\cdot 5}\frac{6\cdot 6}{5\cdot 7}\cdots. \tag{25}$$

This formula can also be written

$$\lim_{n\to\infty}\frac{2^{4n}}{n\binom{2n}{n}^2} = \pi \lim_{n\to\infty}\frac{n[\Gamma(n)]^2}{[\Gamma(\frac{1}{2}+n)]^2} = \pi, \tag{26}$$

where $\binom{n}{k}$ denotes a BINOMIAL COEFFICIENT and $\Gamma(x)$ is the GAMMA FUNCTION (Knopp 1990). Euler obtained

$$\pi = \sqrt{6\left(1 + \frac{1}{2^2} + \frac{1}{3^2} + \frac{1}{4^2} + \ldots\right)}, \tag{27}$$

which follows from the special value of the RIEMANN ZETA FUNCTION $\zeta(2) = \pi^2/6$. Similar FORMULAS follow from $\zeta(2n)$ for all POSITIVE INTEGERS n. Gregory and Leibniz found

$$\frac{\pi}{4} = 1 - \frac{1}{3} + \frac{1}{5} + \ldots, \tag{28}$$

which is sometimes known as GREGORY'S FORMULA. The error after the nth term of this series in GREGORY'S FORMULA is larger than $(2n)^{-1}$ so this sum converges so slowly that 300 terms are not sufficient to calculate π correctly to two decimal places! However, it can be transformed to

$$\pi = \sum_{k=1}^{\infty}\frac{3^k - 1}{4^k}\zeta(k+1), \tag{29}$$

where $\zeta(z)$ is the RIEMANN ZETA FUNCTION (Vardi 1991, pp. 157–158; Flajolet and Vardi 1996), so that the error after k terms is $\approx (3/4)^k$. Newton used

$$\pi = \tfrac{3}{4}\sqrt{3} + 24\int_0^{1/4}\sqrt{x - x^2}\,dx \tag{30}$$

$$= \frac{3\sqrt{3}}{4} + 24\left(\frac{1}{12} - \frac{1}{5\cdot 2^5} - \frac{1}{28\cdot 2^7} - \frac{1}{72\cdot 2^9} - \ldots\right) \tag{31}$$

(Borwein *et al.* 1989). Using Euler's CONVERGENCE IMPROVEMENT transformation gives

$$\frac{\pi}{2} = \frac{1}{2}\sum_{n=0}^{\infty}\frac{(n!)^2 2^{n+1}}{(2n+1)!} = \sum_{n=0}^{\infty}\frac{n!}{(2n+1)!!}$$

$$= 1 + \frac{1}{3} + \frac{1\cdot 2}{3\cdot 5} + \frac{1\cdot 2\cdot 3}{3\cdot 5\cdot 7} + \ldots \tag{32}$$

$$= 1 + \frac{1}{3}\left(1 + \frac{2}{5}\left(1 + \frac{3}{7}\left(1 + \frac{4}{9}(1 + \ldots)\right)\right)\right) \tag{33}$$

(Beeler *et al.* 1972, Item 120). This corresponds to plugging $x = 1/\sqrt{2}$ into the POWER SERIES for the HYPERGEOMETRIC FUNCTION ${}_2F_1(a, b; c; x)$,

$$\frac{\sin^{-1} x}{\sqrt{1-x^2}} = \sum_{i=0}^{\infty}\frac{(2x)^{2i+1}(i!)^2}{2(2i+1)!} = {}_2F_1(1, 1; \tfrac{3}{2}; x^2)x. \tag{34}$$

Despite the convergence improvement, series (33) converges at only one bit/term. At the cost of a SQUARE ROOT, Gosper has noted that $x = 1/2$ gives 2 bits/term,

$$\tfrac{1}{9}\sqrt{3}\pi = \frac{1}{2}\sum_{i=1}^{\infty}\frac{(i!)^2}{(2i+1)!}, \tag{35}$$

and $x = \sin(\pi/10)$ gives almost 3.39 bits/term,

$$\frac{\pi}{5\sqrt{\phi+2}} = \frac{1}{2}\sum_{i=0}^{\infty}\frac{(i!)^2}{\phi^{2i+1}(2i+1)!}, \tag{36}$$

where ϕ is the GOLDEN RATIO. Gosper also obtained

$$\pi = 3 + \frac{1}{60}\left(8 + \frac{2\cdot 3}{7\cdot 8\cdot 3}\left(13 + \frac{3\cdot 5}{10\cdot 11\cdot 3}\right.\right.$$
$$\left.\left.\times\left(18 + \frac{4\cdot 7}{13\cdot 14\cdot 3}(23+\ldots)\right)\right)\right). \quad (37)$$

An infinite sum due to Ramanujan is

$$\frac{1}{\pi} = \sum_{n=0}^{\infty}\binom{2n}{n}^3 \frac{42n+5}{2^{12n+4}} \quad (38)$$

(Borwein *et al.* 1989). Further sums are given in Ramanujan (1913–14),

$$\frac{4}{\pi} = \sum_{n=0}^{\infty} \frac{(-1)^n(1123+21460n)(2n-1)!!(4n-1)!!}{882^{2n+1}32^n(n!)^3} \quad (39)$$

and

$$\frac{1}{\pi} = \sqrt{8}\sum_{n=0}^{\infty}\frac{(1103+26390n)(2n-1)!!(4n-1)!!}{99^{4n+2}32^n(n!)^3}$$
$$= \frac{\sqrt{8}}{9801}\sum_{n=0}^{\infty}\frac{(4n)!(1103+26390n)}{(n!)^4 396^{4n}} \quad (40)$$

(Beeler *et al.* 1972, Item 139; Borwein *et al.* 1989). Equation (40) is derived from a modular identity of order 58, although a first derivation was not presented prior to Borwein and Borwein (1987). The above series both give

$$\pi \approx \frac{2206\sqrt{2}}{9801} = 3.14159273001\ldots \quad (41)$$

as the first approximation and provide, respectively, about 6 and 8 decimal places per term. Such series exist because of the rationality of various modular invariants. The general form of the series is

$$\sum_{n=0}^{\infty}[a(t)+nb(t)]\frac{(6n)!}{(3n)!(n!)^3}\frac{1}{[j(t)]^n} = \frac{\sqrt{-j(t)}}{\pi}, \quad (42)$$

where t is a QUADRATIC FORM DISCRIMINANT, $j(t)$ is the j-FUNCTION,

$$b(t) = \sqrt{t[1728-j(t)]} \quad (43)$$
$$a(t) = \frac{b(t)}{6}\left\{1 - \frac{E_4(t)}{E_6(t)}\left[E_2(t) - \frac{6}{\pi\sqrt{t}}\right]\right\}, \quad (44)$$

and the E_i are RAMANUJAN-EISENSTEIN SERIES. A CLASS NUMBER p field involves pth degree ALGEBRAIC INTEGERS of the constants $A = a(t)$, $B = b(t)$, and $C = c(t)$. The fastest converging series that uses only INTEGER terms corresponds to the largest CLASS NUMBER 1 discriminant of $d = -163$ and was formulated by the Chudnovsky brothers (1987). The 163 appearing here is the same one appearing in the fact that $e^{\pi\sqrt{163}}$ (the RAMANUJAN CONSTANT) is very nearly an INTEGER. The series is given by

$$\frac{1}{\pi} = 12\sum_{n=0}^{\infty}\frac{(-1)^n(6n)!(13591409+545140134n)}{(n!)^3(3n)!(640320^3)^{n+1/2}}$$
$$= \frac{163\cdot 8\cdot 27\cdot 7\cdot 11\cdot 19\cdot 127}{640320^{3/2}}$$
$$\times\sum_{n=0}^{\infty}\left(\frac{13591409}{163\cdot 2\cdot 9\cdot 7\cdot 11\cdot 19\cdot 127}+n\right)$$
$$\times\frac{(6n)!}{(3n)!(n!)^3}\frac{(-1)^n}{640320^{3n}} \quad (45)$$

(Borwein and Borwein 1993). This series gives 14 digits accurately per term. The same equation in another form was given by the Chudnovsky brothers (1987) and is used by *Mathematica*® (Wolfram Research, Champaign, IL) to calculate π (Vardi 1991),

$$\pi = \frac{426880\sqrt{10005}}{A[{}_3F_2(\frac{1}{6},\frac{1}{2},\frac{5}{6};1,1;B) - C\,{}_3F_2(\frac{7}{6},\frac{3}{2},\frac{11}{6};2,2;B)]}, \quad (46)$$

where

$$A \equiv 13591409 \quad (47)$$
$$B \equiv -\frac{1}{151931373056000} \quad (48)$$
$$C \equiv \frac{30285563}{1651969144908540723200}. \quad (49)$$

The best formula for CLASS NUMBER 2 (largest discriminant -427) is

$$\frac{1}{\pi} = 12\sum_{n=0}^{\infty}\frac{(-1)^n(6n)!(A+Bn)}{(n!)^3(3n)!C^{n+1/2}}, \quad (50)$$

where

$$A \equiv 212175710912\sqrt{61} + 1657145277365 \quad (51)$$
$$B \equiv 13773980892672\sqrt{61} + 107578229802750 \quad (52)$$
$$C \equiv [5280(236674 + 30303\sqrt{61})]^3 \quad (53)$$

(Borwein and Borwein 1993). This series adds about 25 digits for each additional term. The fastest converging series for CLASS NUMBER 3 corresponds to $d = -907$ and gives 37–38 digits per term. The fastest converging CLASS NUMBER 4 series corresponds to $d = -1555$ and is

$$\frac{\sqrt{-C^3}}{\pi} = \sum_{n=0}^{\infty}\frac{(6n)!}{(3n)!(n!)^3}\frac{A+nB}{C^{3n}}, \quad (54)$$

where

$$\begin{aligned}A &= 63365028312971999585426220\\&\quad + 28337702140800842046825600\sqrt{5}\\&\quad + 384\sqrt{5}(10891728551171178200467 4\cdots\\&\quad \cdots 3621239520916038565601 7 + 487902908657881022\cdots\\&\quad \cdots 5077338534541688721351255040\sqrt{5}\,)^{1/2}\end{aligned} \tag{55}$$

$$\begin{aligned}B &= 7849910453496627210289749000\\&\quad + 3510586678260932028965606400\sqrt{5}\\&\quad + 2515968\sqrt{3110}(6260208323789001 6\cdots\\&\quad \cdots 3699332265444402088216 1 + 279965027306044429 6\cdots\\&\quad \cdots 5772068907188251902 35\sqrt{5}\,)^{1/2}\end{aligned} \tag{56}$$

$$\begin{aligned}C &= -214772995063512240 - 96049403338648032\sqrt{5}\\&\quad - 1296\sqrt{5}(10985234579463550323713318473\\&\quad + 4912746253692362754607395912\sqrt{5}\,)^{1/2}.\end{aligned} \tag{57}$$

This gives 50 digits per term. Borwein and Borwein (1993) have developed a general ALGORITHM for generating such series for arbitrary CLASS NUMBER. Bellard gives the exotic formula

$$\pi = \frac{1}{740025}\left[\sum_{n=1}^{\infty}\frac{3P(n)}{\binom{7n}{2n}2^{n-1}} - 20379280\right], \tag{58}$$

where

$$P(n) \equiv -885673181n^5 + 3125347237n^4 - 2942969225n^3 + 1031962795n^2 - 196882274n + 10996648. \tag{59}$$

A complete listing of Ramanujan's series for $1/\pi$ found in his second and third notebooks is given by Berndt (1994, pp. 352–354),

$$\frac{4}{\pi} = \sum_{n=0}^{\infty}\frac{(6n+1)(\frac{1}{2})_n{}^3}{4^n(n!)^3} \tag{60}$$

$$\frac{16}{\pi} = \sum_{n=0}^{\infty}\frac{(42n+5)(\frac{1}{2})_n{}^3}{(64)^n(n!)^3} \tag{61}$$

$$\frac{32}{\pi} = \sum_{n=0}^{\infty}\frac{(42\sqrt{5}\,n+5\sqrt{5}+30n-1)(\frac{1}{2})_n{}^3}{(64)^n(n!)^3} \times\left(\frac{\sqrt{5}-1}{2}\right)^{8n} \tag{62}$$

$$\frac{27}{4\pi} = \sum_{n=0}^{\infty}\frac{(15n+2)(\frac{1}{2})_n(\frac{1}{3})_n(\frac{2}{3})_n}{(n!)^3}\left(\frac{2}{27}\right)^n \tag{63}$$

$$\frac{15\sqrt{3}}{2\pi} = \sum_{n=0}^{\infty}\frac{(33n+4)(\frac{1}{2})_n(\frac{1}{3})_n(\frac{2}{3})_n}{(n!)^3}\left(\frac{4}{125}\right)^n \tag{64}$$

$$\frac{5\sqrt{5}}{2\pi\sqrt{3}} = \sum_{n=0}^{\infty}\frac{(11n+1)(\frac{1}{2})_n(\frac{1}{6})_n(\frac{5}{6})_n}{(n!)^3}\left(\frac{4}{125}\right)^n \tag{65}$$

$$\frac{85\sqrt{85}}{18\pi\sqrt{3}} = \sum_{n=0}^{\infty}\frac{(133n+8)(\frac{1}{2})_n(\frac{1}{6})_n(\frac{5}{6})_n}{(n!)^3}\left(\frac{4}{85}\right)^n \tag{66}$$

$$\frac{4}{\pi} = \sum_{n=0}^{\infty}\frac{(-1)^n(20n+3)(\frac{1}{2})_n(\frac{1}{4})_n(\frac{3}{4})_n}{(n!)^3 2^{2n+1}} \tag{67}$$

$$\frac{4}{\pi\sqrt{3}} = \sum_{n=0}^{\infty}\frac{(-1)^n(28n+3)(\frac{1}{2})_n(\frac{1}{4})_n(\frac{3}{4})_n}{(n!)^3 3^n 4^{n+1}} \tag{68}$$

$$\frac{4}{\pi} = \sum_{n=0}^{\infty}\frac{(-1)^n(260n+23)(\frac{1}{2})_n(\frac{1}{4})_n(\frac{3}{4})_n}{(n!)^3(18)^{2n+1}} \tag{69}$$

$$\frac{4}{\pi\sqrt{5}} = \sum_{n=0}^{\infty}\frac{(-1)^n(644n+41)(\frac{1}{2})_n(\frac{1}{4})_n(\frac{3}{4})_n}{(n!)^3 5^n(72)^{2n+1}} \tag{70}$$

$$\frac{4}{\pi} = \sum_{n=0}^{\infty}\frac{(-1)^n(21460n+1123)(\frac{1}{2})_n(\frac{1}{4})_n(\frac{3}{4})_n}{(n!)^3(882)^{2n+1}} \tag{71}$$

$$\frac{2\sqrt{3}}{\pi} = \sum_{n=0}^{\infty}\frac{(8n+1)(\frac{1}{2})_n(\frac{1}{4})_n(\frac{3}{4})_n}{(n!)^3 9^n} \tag{72}$$

$$\frac{1}{2\pi\sqrt{2}} = \sum_{n=0}^{\infty}\frac{(10n+1)(\frac{1}{2})_n(\frac{1}{4})_n(\frac{3}{4})_n}{(n!)^3 9^{2n+1}} \tag{73}$$

$$\frac{1}{3\pi\sqrt{3}} = \sum_{n=0}^{\infty}\frac{(40n+3)(\frac{1}{2})_n(\frac{1}{4})_n(\frac{3}{4})_n}{(n!)^3(49)^{2n+1}} \tag{74}$$

$$\frac{2}{\pi\sqrt{11}} = \sum_{n=0}^{\infty}\frac{(280n+19)(\frac{1}{2})_n(\frac{1}{4})_n(\frac{3}{4})_n}{(n!)^3(99)^{2n+1}} \tag{75}$$

$$\frac{1}{2\pi\sqrt{2}} = \sum_{n=0}^{\infty}\frac{(26390n+1103)(\frac{1}{2})_n(\frac{1}{4})_n(\frac{3}{4})_n}{(n!)^3(99)^{4n+2}}. \tag{76}$$

These equations were first proved by Borwein and Borwein (1987, pp. 177–187). Borwein and Borwein (1987b, 1988, 1993) proved other equations of this type, and Chudnovsky and Chudnovsky (1987) found similar equations for other transcendental constants.

A SPIGOT ALGORITHM for π is given by Rabinowitz and Wagon (1995). Amazingly, a closed form expression giving a *digit extraction* algorithm which produces digits of π (or π^2) in base-16 was recently discovered by Bailey *et al.* (Bailey *et al.* 1995, Adamchik and Wagon 1997),

$$\pi = \sum_{n=0}^{\infty}\left(\frac{4}{8n+1} - \frac{2}{8n+4} - \frac{1}{8n+5} - \frac{1}{8n+6}\right)\left(\frac{1}{16}\right)^n, \tag{77}$$

which can also be written using the shorthand notation

$$\pi = \sum_{i=1}^{\infty}\frac{p_i}{16^{\lfloor i/8\rfloor}i} \qquad \{p_i\} = \{\overline{4,0,0,-2,-1,-1,0,0}\}, \tag{78}$$

where $\{p_i\}$ is given by the periodic sequence obtained by appending copies of $\{4,0,0,-2,-1,-1,0,0\}$ (in other words, $p_i \equiv p_{[(i-1) \pmod 8]+1}$ for $i > 8$) and $\lfloor x \rfloor$ is the FLOOR FUNCTION. This expression was discovered using the PSLQ ALGORITHM and is equivalent to

$$\pi = \int_0^1 \frac{16y-16}{y^4-2y^3+4y-4}\,dy. \tag{79}$$

A similar formula was subsequently discovered by Ferguson, leading to a 2-D lattice of such formulas which can be generated by these two formulas. A related integral is

$$\pi = \tfrac{22}{7} - \int_0^1 \frac{x^4(1-x)^4}{1+x^2}\,dx \tag{80}$$

(Le Lionnais 1983, p. 22). F. Bellard found the more rapidly converging digit-extraction algorithm (in HEXADECIMAL)

$$\pi = \frac{1}{2^6}\sum_{n=0}^{\infty}\frac{(-1)^n}{2^{10n}}\left(-\frac{2^5}{4n+1} - \frac{1}{4n+3} + \frac{2^8}{10n+1} - \frac{2^6}{10n+3} - \frac{2^2}{10n+5} - \frac{2^2}{10n+7} + \frac{1}{10n+9}\right). \tag{81}$$

More amazingly still, S. Plouffe has devised an algorithm to compute the nth DIGIT of π in any base in $\mathcal{O}(n^3(\log n)^3)$ steps.

Another identity is

$$\pi^2 = 36\,\mathrm{Li}_2(\tfrac{1}{2}) - 36\,\mathrm{Li}_2(\tfrac{1}{4}) - 12\,\mathrm{Li}_2(\tfrac{1}{8}) + 6\,\mathrm{Li}_2(\tfrac{1}{64}), \tag{82}$$

where L_n is the POLYLOGARITHM. (82) is equivalent to

$$\frac{\pi^2}{36} = \sum_{i=1}^{\infty}\frac{a_i}{2^i i^2} \qquad \{a_i\} = [\overline{1,-3,-2,-3,1,0}] \tag{83}$$

and

$$\pi^2 = 12L_2(\tfrac{1}{2}) + 6(\ln 2)^2 \tag{84}$$

(Bailey *et al.* 1995). Furthermore

$$\pi^2 = \frac{1}{8}\sum_{k=0}^{\infty}\frac{1}{64^k}\left[\frac{144}{(6k+1)^2} - \frac{216}{(6k+2)^2} - \frac{72}{(6k+3)^2} - \frac{54}{(6k+4)^2} + \frac{9}{(6k+5)^2}\right] \tag{85}$$

and

$$\pi^2 = \sum_{k=0}^{\infty}\frac{1}{16^k}\left[\frac{16}{(8k+1)^2} - \frac{16}{(8k+2)^2} - \frac{8}{(8k+3)^2} - \frac{16}{(8k+4)^2} - \frac{4}{(8k+5)^2} - \frac{4}{(8k+6)^2} + \frac{2}{(8k+7)^2}\right] \tag{86}$$

(Bailey *et al.* 1995, Bailey and Plouffe).

A slew of additional identities due to Ramanujan, Catalan, and Newton are given by Castellanos (1988, pp. 86–88), including several involving sums of FIBONACCI NUMBERS.

Gasper quotes the result

$$\pi = \frac{16}{3}\left[\lim_{x\to\infty} x\,{}_1F_2(\tfrac{1}{2};2,3;-x^2)\right]^{-1}, \tag{87}$$

where ${}_1F_2$ is a GENERALIZED HYPERGEOMETRIC FUNCTION, and transforms it to

$$\pi = \lim_{x\to\infty} 4x\,{}_1F_2(\tfrac{1}{2};\tfrac{3}{2},\tfrac{3}{2};-x^2). \tag{88}$$

Fascinating results due to Gosper include

$$\lim_{n\to\infty}\prod_{i=n}^{2n}\frac{\pi}{2\tan^{-1} i} = 4^{1/\pi} = 1.554682275\ldots \tag{89}$$

and

$$\sum_{n=1}^{\infty}\frac{1}{n^2}\cos\left(\frac{9}{n\pi+\sqrt{n^2\pi^2-9}}\right) = -\frac{\pi^2}{12e^3} = -0.040948222\ldots. \tag{90}$$

Gosper also gives the curious identity

$$\frac{1}{e}\prod_{n=1}^{\infty}\left(\frac{1}{3n}+1\right)^{3n+1/2} = \frac{3\cdot 3^{1/24}\sqrt{(\frac{1}{3})!}}{2^{5/6}\exp\left[\frac{\gamma}{3} - \frac{\pi\sqrt{3}}{18} + \frac{\sqrt{3}\psi_1(\frac{1}{3})}{12\pi} - \frac{2\zeta'(2)}{\pi^2} - 1\right]\pi^{5/6}} = 1.012378552722912\ldots. \tag{91}$$

Another curious fact is the ALMOST INTEGER

$$e^\pi - \pi = 19.999099979\ldots, \tag{92}$$

which can also be written as

$$(\pi+20)^i = -0.9999999992 - 0.0000388927i \approx -1 \tag{93}$$

$$\cos(\ln(\pi+20)) \approx -0.9999999992. \tag{94}$$

Applying COSINE a few more times gives

$$\cos(\pi\cos(\pi\cos(\ln(\pi+20)))) \approx -1 + 3.9321609261\times 10^{-35}. \tag{95}$$

π may also be computed using iterative ALGORITHMS. A quadratically converging ALGORITHM due to Borwein is

$$x_0 = \sqrt{2} \tag{96}$$

$$\pi_0 = 2+\sqrt{2} \tag{97}$$

$$y_1 = 2^{1/4} \tag{98}$$

and

$$x_{n+1} = \frac{1}{2}\left(\sqrt{x_n} + \frac{1}{\sqrt{x_n}}\right) \tag{99}$$

$$y_{n+1} = \frac{y_n\sqrt{x_n} + \frac{1}{\sqrt{x_n}}}{y_n+1} \tag{100}$$

$$\pi_n = \pi_{n-1}\frac{x_n+1}{y_n+1}. \tag{101}$$

π_n decreases monotonically to π with

$$\pi_n - \pi < 10^{-2^{n+1}} \tag{102}$$

for $n \geq 2$. The BRENT-SALAMIN FORMULA is another quadratically converging algorithm which can be used to calculate π. A quadratically convergent algorithm for $\pi/\ln 2$ based on an observation by Salamin is given by defining

$$f(k) = k2^{-k/4}\left[\sum_{n=1}^{\infty} 2^{-k\binom{n}{2}}\right]^2, \tag{103}$$

then writing

$$g_0 \equiv \frac{f(n)}{f(2n)}. \tag{104}$$

Now iterate

$$g_k = \sqrt{\frac{1}{2}\left(g_{k-1} + \frac{1}{g_{k+1}}\right)} \tag{105}$$

to obtain

$$\pi = 2(\ln 2) f(n) \prod_{k=1}^{\infty} g_k. \tag{106}$$

A cubically converging ALGORITHM which converges to the nearest multiple of π to f_0 is the simple iteration

$$f_n = f_{n-1} + \sin(f_{n-1}) \tag{107}$$

(Beeler *et al.* 1972). For example, applying to 23 gives the sequence

$$\{23, 22.1537796, 21.99186453, 21.99114858, \ldots\}, \tag{108}$$

which converges to $7\pi \approx 21.99114858$.

A quartically converging ALGORITHM is obtained by letting

$$y_0 = \sqrt{2} - 1 \tag{109}$$

$$\alpha = 6 - 4\sqrt{2}, \tag{110}$$

then defining

$$y_{n+1} = \frac{1 - (1 - y_n{}^4)^{1/4}}{1 + (1 - y_n{}^4)^{1/4}} \tag{111}$$

$$\alpha_{n+1} = (1 + y_{n+1})^4 \alpha_n - 2^{2n+3} y_{n+1}(1 + y_{n+1} + y_{n+1}{}^2). \tag{112}$$

Then

$$\pi = \lim_{n\to\infty} \frac{1}{\alpha_n} \tag{113}$$

and α_n converges to $1/\pi$ quartically with

$$\alpha_n - \frac{1}{\pi} < 16 \cdot 4^n e^{-2\pi \cdot 4^n} \tag{114}$$

(Borwein and Borwein 1987, Bailey 1988, Borwein *et al.* 1989). This ALGORITHM rests on a MODULAR EQUATION identity of order 4.

A quintically converging ALGORITHM is obtained by letting

$$s_0 = 5(\sqrt{5} - 2) \tag{115}$$

$$\alpha_0 = \tfrac{1}{2}. \tag{116}$$

Then let

$$s_{n+1} = \frac{25}{(z + \frac{x}{z} + 1)^2 s_n}, \tag{117}$$

where

$$x = \frac{5}{s_n} - 1 \tag{118}$$

$$y = (x-1)^2 + 7 \tag{119}$$

$$z = [\tfrac{1}{2}x(y + \sqrt{y^2 - 4x^3})]^{1/5}. \tag{120}$$

Finally, let

$$\alpha_{n+1} = s_n{}^2 \alpha_n - 5^n[\tfrac{1}{2}(s_n{}^2 - 5) + \sqrt{s_n(s_n{}^2 - 2s_n + 5)}], \tag{121}$$

then

$$0 < \alpha_n - \frac{1}{\pi} < 16 \cdot 5^n e^{-\pi 5^n} \tag{122}$$

(Borwein *et al.* 1989). This ALGORITHM rests on a MODULAR EQUATION identity of order 5.

Another ALGORITHM is due to Woon (1995). Define $a(0) \equiv 1$ and

$$a(n) = \sqrt{1 + \left[\sum_{k=0}^{n-1} a(k)\right]^2}. \tag{123}$$

It can be proved by induction that

$$a(n) = \csc\left(\frac{\pi}{2^{n+1}}\right). \tag{124}$$

For $n = 0$, the identity holds. If it holds for $n \leq t$, then

$$a(t+1) = \sqrt{1 + \left[\sum_{k=0}^{t} \csc\left(\frac{\pi}{2^{k+1}}\right)\right]^2}, \tag{125}$$

but

$$\csc\left(\frac{\pi}{2^{k+1}}\right) = \cot\left(\frac{\pi}{2^{k+2}}\right) - \cot\left(\frac{\pi}{2^{k+1}}\right), \tag{126}$$

so

$$\sum_{k=0}^{t} \csc\left(\frac{\pi}{2^{k+1}}\right) = \cot\left(\frac{\pi}{2^{t+2}}\right). \tag{127}$$

Therefore,

$$a(t+1) = \csc\left(\frac{\pi}{2^{t+2}}\right), \tag{128}$$

so the identity holds for $n = t+1$ and, by induction, for all NONNEGATIVE n, and

$$\lim_{n\to\infty} \frac{2^{n+1}}{a(n)} = \lim_{n\to\infty} 2^{n+1} \sin\left(\frac{\pi}{2^{n+1}}\right)$$
$$= \lim_{n\to\infty} 2^{n+1} \frac{\pi}{2^{n+1}} \frac{\sin\left(\frac{\pi}{2^{n+1}}\right)}{\frac{\pi}{2^{n+1}}}$$
$$= \pi \lim_{\theta\to 0} \frac{\sin\theta}{\theta} = \pi. \tag{129}$$

Other iterative ALGORITHMS are the ARCHIMEDES ALGORITHM, which was derived by Pfaff in 1800, and the BRENT-SALAMIN FORMULA. Borwein *et al.* (1989) discuss pth order iterative algorithms.

KOCHANSKY'S APPROXIMATION is the ROOT of

$$9x^4 - 240x^2 + 1492. \tag{130}$$

given by

$$\pi \approx \sqrt{\tfrac{40}{3} - \sqrt{12}} \approx 3.141533. \tag{131}$$

An approximation involving the GOLDEN MEAN is

$$\pi \approx \tfrac{6}{5}\phi^2 = \frac{6}{5}\left(\frac{\sqrt{5}+1}{2}\right)^2 = \tfrac{3}{5}(3+\sqrt{5}) = 3.14164\ldots. \tag{132}$$

Some approximations due to Ramanujan

$$\pi \approx \frac{19\sqrt{7}}{16} \tag{133}$$
$$\approx \tfrac{7}{3}(1+\tfrac{1}{5}\sqrt{3}) \tag{134}$$
$$\approx \tfrac{9}{5} + \sqrt{\tfrac{9}{5}} \tag{135}$$
$$\approx \left(9^2 + \frac{19^2}{22}\right)^{1/4} = \left(102 - \frac{2222}{22^2}\right)^{1/4} \tag{136}$$
$$\approx (97 + \tfrac{1}{2} - \tfrac{1}{11})^{1/4} = (97 + \tfrac{9}{22})^{1/4} \tag{137}$$
$$\approx \frac{63}{25}\left(\frac{17+15\sqrt{5}}{7+15\sqrt{5}}\right) \tag{138}$$
$$\approx \frac{355}{113}\left(1 - \frac{0.0003}{3533}\right) \tag{139}$$
$$\approx \frac{12}{\sqrt{130}} \ln\left[\frac{(3+\sqrt{13})(\sqrt{8}+\sqrt{10})}{2}\right] \tag{140}$$
$$\approx \frac{24}{\sqrt{142}} \ln\left[\frac{\sqrt{10+11\sqrt{2}} + \sqrt{10+7\sqrt{2}}}{2}\right] \tag{141}$$
$$\approx \frac{12}{\sqrt{190}} \ln[(3+\sqrt{10})(\sqrt{8}+\sqrt{10})] \tag{142}$$
$$\approx \frac{12}{\sqrt{310}} \ln\left[\tfrac{1}{4}(3+\sqrt{5})(2+\sqrt{2})\left(5+2\sqrt{10} + \sqrt{61+20\sqrt{10}}\right)\right] \tag{143}$$
$$\approx \frac{4}{\sqrt{522}} \ln\left[\left(\frac{5+\sqrt{29}}{\sqrt{2}}\right)^3 (5\sqrt{29}+11\sqrt{6}) \times \left(\sqrt{\frac{9+3\sqrt{6}}{4}} + \sqrt{\frac{5+3\sqrt{6}}{4}}\right)^6\right], \tag{144}$$

which are accurate to 3, 4, 4, 8, 8, 9, 14, 15, 15, 18, 23, 31 digits, respectively (Ramanujan 1913–1914; Hardy 1952, p. 70; Berndt 1994, pp. 48–49 and 88–89).

Castellanos (1988) gives a slew of curious formulas:

$$\pi \approx (2e^3 + e^8)^{1/7} \tag{145}$$
$$\approx \left(\frac{553}{311+1}\right)^2 \tag{146}$$
$$\approx (\tfrac{3}{14})^4 (\tfrac{193}{5})^2 \tag{147}$$
$$\approx (\tfrac{296}{167})^2 \tag{148}$$
$$\approx \left(\frac{66^3+86^2}{55^3}\right)^2 \tag{149}$$
$$\approx 1.09999901 \cdot 1.19999911 \cdot 1.39999931 \cdot 1.69999961 \tag{150}$$
$$\approx \frac{47^3+20^3}{30^3} - 1 \tag{151}$$
$$\approx 2 + \sqrt{1 + (\tfrac{413}{750})^2} \tag{152}$$
$$\approx (\tfrac{77729}{254})^{1/5} \tag{153}$$
$$\approx \left(31 + \frac{62^2+14}{28^4}\right)^{1/3} \tag{154}$$
$$\approx \frac{1700^3 + 82^3 - 10^3 - 9^3 - 6^3 - 3^3}{69^5} \tag{155}$$
$$\approx \left(95 + \frac{93^4 + 34^4 + 17^4 + 88}{75^4}\right)^{1/4} \tag{156}$$
$$\approx \left(100 - \frac{2125^3 + 214^3 + 30^3 + 37^2}{82^5}\right)^{1/4}, \tag{157}$$

which are accurate to 3, 4, 4, 5, 6, 7, 7, 8, 9, 10, 11, 12, and 13 digits, respectively. An extremely accurate approximation due to Shanks (1982) is

$$\pi \approx \frac{6}{\sqrt{3502}} \ln(2u) + 7.37 \times 10^{-82}, \tag{158}$$

where u is the product of four simple quartic units. A sequence of approximations due to Plouffe includes

$$\pi \approx 43^{7/23} \tag{159}$$
$$\approx \frac{\ln 2198}{\sqrt{6}} \tag{160}$$
$$\approx (\tfrac{13}{4})^{1181/1216} \tag{161}$$
$$\approx \frac{689}{396 \ln\left(\frac{689}{396}\right)} \tag{162}$$
$$\approx (\tfrac{2143}{22})^{1/4} \tag{163}$$
$$\approx \sqrt{\frac{9}{67}} \ln 5280 \tag{164}$$
$$\approx (\tfrac{63023}{30510})^{1/3} + \tfrac{1}{4} + \tfrac{1}{2}(\sqrt{5}+1) \tag{165}$$
$$\approx \tfrac{48}{23} \ln\left(\frac{60318}{13387}\right) \tag{166}$$

$$\approx (228 + \tfrac{16}{1329})^{1/41} + 2 \quad (167)$$

$$\approx \tfrac{125}{124} \ln \left(\frac{28102}{1277} \right) \quad (168)$$

$$\approx \tfrac{276694819753963}{226588}^{1/158} + 2 \quad (169)$$

$$\approx \frac{\ln 262537412640768744}{\sqrt{163}}, \quad (170)$$

which are accurate to 4, 5, 7, 7, 8, 9, 10, 11, 11, 11, 23, and 30 digits, respectively.

Ramanujan (1913–14) and Olds (1963) give geometric constructions for 355/113. Gardner (1966, pp. 92–93) gives a geometric construction for $3 + 16/113 = 3.1415929\ldots$. Dixon (1991) gives constructions for $6/5(1+\phi) = 3.141640\ldots$ and $\sqrt{4 + (3 - \tan(30°))} = 3.141533\ldots$. Constructions for approximations of π are approximations to CIRCLE SQUARING (which is itself impossible).

A short mnemonic for remembering the first eight DECIMAL DIGITS of π is "May I have a large container of coffee?" giving 3.1415926 (Gardner 1959; Gardner 1966, p. 92; Eves 1990, p. 122, Davis 1993, p. 9). A more substantial mnemonic giving 15 digits (3.14159265358979) is "How I want a drink, alcoholic of course, after the heavy lectures involving quantum mechanics," originally due to Sir James Jeans (Gardner 1966, p. 92; Castellanos 1988, p. 152; Eves 1990, p. 122; Davis 1993, p. 9; Blatner 1997, p. 112). A slight extension of this adds the phrase "All of thy geometry, Herr Planck, is fairly hard," giving 24 digits in all (3.14159265358979323846264).

An even more extensive rhyming mnemonic giving 31 digits is "Now I will a rhyme construct, By chosen words the young instruct. Cunningly devised endeavour, Con it and remember ever. Widths in circle here you see, Sketched out in strange obscurity." (Note that the British spelling of "endeavour" is required here.)

The following stanzas are the first part of a poem written by M. Keith based on Edgar Allen Poe's "The Raven." The entire poem gives 740 digits; the fragment below gives only the first 80 (Blatner 1997, p. 113). Words with ten letters represent the digit 0, and those with 11 or more digits are taken to represent two digits.

Poe, E.: Near a Raven.

Midnights so dreary, tired and weary.
Silently pondering volumes extolling all by-now obsolete lore.
During my rather long nap-the weirdest tap!
An ominous vibrating sound disturbing my chamber's antedoor.
'This,' I whispered quietly, 'I ignore.'

Perfectly, the intellect remembers: the ghostly fires, a glittering ember.
Inflamed by lightning's outbursts, windows cast penumbras upon this floor.
Sorrowful, as one mistreated, unhappy thoughts I heeded:
That inimitable lesson in elegance—Lenore—
is delighting, exciting... nevermore.

An extensive collection of π mnemonics in many languages is maintained by A. P. Hatzipolakis. Other mnemonics in various languages are given by Castellanos (1988) and Blatner (1997, pp. 112–118).

In the following, the word "digit" refers to decimal digit after the decimal point. J. H. Conway has shown that there is a sequence of fewer than 40 FRACTIONS F_1, F_2, ... with the property that if you start with 2^n and repeatedly multiply by the first of the F_i that gives an integral answer, then the next POWER of 2 to occur will be the 2^nth decimal digit of π.

The first occurrence of n 0s appear at digits 32, 307, 601, 13390, 17534, ... The sequence 9999998 occurs at decimal 762 (which is sometimes called the FEYNMAN POINT). This is the largest value of any seven digits in the first million decimals. The first time the BEAST NUMBER 666 appears is decimal 2440. The digits 314159 appear at least six times in the first 10 million decimal places of π (Pickover 1995). In the following, "digit" means digit of $\pi - 3$. The sequence 0123456789 occurs beginning at digits 17,387,594,880, 26,852,899,245, 30,243,957,439, 34,549,153,953, 41,952,536,161, and 43,289,964,000. The sequence 9876543210 occurs beginning at digits 21,981,157,633, 29,832,636,867, 39,232,573,648, 42,140,457,481, and 43,065,796,214. The sequence 27182818284 (the digits of e) occur beginning at digit 45,111,908,393. There are also interesting patterns for $1/\pi$. 0123456789 occurs at 6,214,876,462, 9876543210 occurs at 15,603,388,145 and 51,507,034,812, and 999999999999 occurs at 12,479,021,132 of $1/\pi$.

Scanning the decimal expansion of π until all n-digit numbers have occurred, the last 1-, 2-, ... digit numbers appearing are 0, 68, 483, 6716, 33394, 569540, ... (Sloane's A032510). These end at digits 32, 606, 8555, 99849, 1369564, 14118312, ...

see also ALMOST INTEGER, ARCHIMEDES ALGORITHM, BRENT-SALAMIN FORMULA, BUFFON-LAPLACE NEEDLE PROBLEM, BUFFON'S NEEDLE PROBLEM, CIRCLE, DIRICHLET BETA FUNCTION, DIRICHLET ETA FUNCTION, DIRICHLET LAMBDA FUNCTION, e, EULER-MASCHERONI CONSTANT, GAUSSIAN DISTRIBUTION, MACLAURIN SERIES, MACHIN'S FORMULA, MACHIN-LIKE FORMULAS, RELATIVELY PRIME, RIEMANN ZETA FUNCTION, SPHERE, TRIGONOMETRY

References

Adamchik, V. and Wagon, S. "A Simple Formula for π." *Amer. Math. Monthly* **104**, 852–855, 1997.

Adamchik, V. and Wagon, S. "Pi: A 2000-Year Search Changes Direction." `http://www.wolfram.com/~victor/articles/pi/pi.html`.

Almkvist, G. "Many Correct Digits of π, Revisited." *Amer. Math. Monthly* **104**, 351–353, 1997.

Arndt, J. "Cryptic Pi Related Formulas." http://www.jjj.de/hfloat/pise.dvi.

Arndt, J. and Haenel, C. *Pi: Algorithmen, Computer, Arithmetik.* Berlin: Springer-Verlag, 1998.

Assmus, E. F. "Pi." *Amer. Math. Monthly* **92**, 213–214, 1985.

Bailey, D. H. "Numerical Results on the Transcendence of Constants Involving π, e, and Euler's Constant." *Math. Comput.* **50**, 275–281, 1988a.

Bailey, D. H. "The Computation of π to 29,360,00 Decimal Digit using Borwein's' Quartically Convergent Algorithm." *Math. Comput.* **50**, 283–296, 1988b.

Bailey, D.; Borwein, P.; and Plouffe, S. "On the Rapid Computation of Various Polylogarithmic Constants." http://www.cecm.sfu.ca/~pborwein/PAPERS/P123.ps.

Ball, W. W. R. and Coxeter, H. S. M. *Mathematical Recreations and Essays, 13th ed.* New York: Dover, p. 55 and 274, 1987.

Beckmann, P. *A History of Pi, 3rd ed.* New York: Dorset Press, 1989.

Beeler, M.; Gosper, R. W.; and Schroeppel, R. *HAKMEM.* Cambridge, MA: MIT Artificial Intelligence Laboratory, Memo AIM-239, Feb. 1972.

Berggren, L.; Borwein, J.; and Borwein, P. *Pi: A Source Book.* New York: Springer-Verlag, 1997.

Bellard, F. "Fabrice Bellard's Pi Page." http://www-stud.enst.fr/~bellard/pi/.

Berndt, B. C. *Ramanujan's Notebooks, Part IV.* New York: Springer-Verlag, 1994.

Blatner, D. *The Joy of Pi.* New York: Walker, 1997.

Blatner, D. "The Joy of Pi." http://www.joyofpi.com.

Borwein, P. B. "Pi and Other Constants." http://www.cecm.sfu.ca/~pborwein/PISTUFF/Apistuff.html.

Borwein, J. M. "Ramanujan Type Series." http://www.cecm.sfu.ca/organics/papers/borwein/paper/html/local/omlink9/html/node1.html.

Borwein, J. M. and Borwein, P. B. *Pi & the AGM: A Study in Analytic Number Theory and Computational Complexity.* New York: Wiley, 1987a.

Borwein, J. M. and Borwein, P. B. "Ramanujan's Rational and Algebraic Series for $1/\pi$." *Indian J. Math.* **51**, 147–160, 1987b.

Borwein, J. M. and Borwein, P. B. "More Ramanujan-Type Series for $1/\pi$." In *Ramanujan Revisited.* Boston, MA: Academic Press, pp. 359–374, 1988.

Borwein, J. M. and Borwein, P. B. "Class Number Three Ramanujan Type Series for $1/\pi$." *J. Comput. Appl. Math.* **46**, 281–290, 1993.

Borwein, J. M.; Borwein, P. B.; and Bailey, D. H. "Ramanujan, Modular Equations, and Approximations to Pi, or How to Compute One Billion Digits of Pi." *Amer. Math. Monthly* **96**, 201–219, 1989.

Brown, K. S. "Rounding Up to Pi." http://www.seanet.com/~ksbrown/kmath001.htm.

Castellanos, D. "The Ubiquitous Pi. Part I." *Math. Mag.* **61**, 67–98, 1988.

Castellanos, D. "The Ubiquitous Pi. Part II." *Math. Mag.* **61**, 148–163, 1988.

Chan, J. "As Easy as Pi." *Math Horizons,* Winter 1993, pp. 18–19, 1993.

Chudnovsky, D. V. and Chudnovsky, G. V. *Padé and Rational Approximations to Systems of Functions and Their Arithmetic Applications.* Berlin: Springer-Verlag, 1984.

Chudnovsky, D. V. and Chudnovsky, G. V. "Approximations and Complex Multiplication According to Ramanujan." In *Ramanujan Revisited: Proceedings of the Centenary Conference* (Ed. G. E. Andrews, B. C. Berndt, and R. A. Rankin). Boston, MA: Academic Press, pp. 375–472, 1987.

Conway, J. H. and Guy, R. K. "The Number π." In *The Book of Numbers.* New York: Springer-Verlag, pp. 237–239, 1996.

David, Y. "On a Sequence Generated by a Sieving Process." *Riveon Lematematika* **11**, 26–31, 1957.

Davis, D. M. *The Nature and Power of Mathematics.* Princeton, NJ: Princeton University Press, 1993.

Dixon, R. "The Story of Pi (π)." §4.3 in *Mathographics.* New York: Dover, pp. 44–49 and 98–101, 1991.

Dunham, W. "A Gem from Isaac Newton." Ch. 7 in *Journey Through Genius: The Great Theorems of Mathematics.* New York: Wiley, pp. 106–112 and 155–183, 1990.

Eves, H. *An Introduction to the History of Mathematics, 6th ed.* Philadelphia, PA: Saunders, 1990.

Exploratorium. "π Page." http://www.exploratorium.edu/learning_studio/pi.

Finch, S. "Favorite Mathematical Constants." http://www.mathsoft.com/asolve/constant/pi/pi.html.

Flajolet, P. and Vardi, I. "Zeta Function Expansions of Classical Constants." Unpublished manuscript. 1996. http://pauillac.inria.fr/algo/flajolet/Publications/landau.ps.

Gardner, M. "Memorizing Numbers." Ch. 11 in *The Scientific American Book of Mathematical Puzzles and Diversions.* New York: Simon and Schuster, p. 103, 1959.

Gardner, M. "The Transcendental Number Pi." Ch. 8 in *Martin Gardner's New Mathematical Diversions from Scientific American.* New York: Simon and Schuster, 1966.

Gosper, R. W. *Table of Simple Continued Fraction for π and the Derived Decimal Approximation.* Stanford, CA: Artificial Intelligence Laboratory, Stanford University, Oct. 1975. Reviewed in *Math. Comput.* **31**, 1044, 1977.

Hardy, G. H. *A Course of Pure Mathematics, 10th ed.* Cambridge, England: Cambridge University Press, 1952.

Hatzipolakis, A. P. "PiPhilology." http://users.hol.gr/~xpolakis/piphil.html.

Hobsen, E. W. *Squaring the Circle.* New York: Chelsea, 1988.

Johnson-Hill, N. "Extraordinary Pi." http://www.users.globalnet.co.uk/~nickjh/Pi.htm.

Johnson-Hill, N. "The Biggest Selection of Pi Links on the Internet." http://www.users.globalnet.co.uk/~nickjh/pi_links.htm.

Kanada, Y. "New World Record of Pi: 51.5 Billion Decimal Digits." http://www.cecm.sfu.ca/personal/jborwein/Kanada_50b.html.

Klein, F. *Famous Problems.* New York: Chelsea, 1955.

Knopp, K. §32, 136, and 138 in *Theory and Application of Infinite Series.* New York: Dover, p. 238, 1990.

Laczkovich, M. "On Lambert's Proof of the Irrationality of π." *Amer. Math. Monthly* **104**, 439–443, 1997.

Lambert, J. H. "Mémoire sur quelques propriétés remarquables des quantités transcendantes circulaires et logarithmiques." *Mémoires de l'Academie des sciences de Berlin* **17**, 265–322, 1761.

Le Lionnais, F. *Les nombres remarquables.* Paris: Hermann, pp. 22 and 50, 1983.

Lindemann, F. "Über die Zahl π." *Math. Ann.* **20**, 213–225, 1882.

Lopez, A. "Indiana Bill Sets the Value of π to 3." http://daisy.uwaterloo.ca/~alopez-o/math-faq/mathtext/node19.html.

MacTutor Archive. "Pi Through the Ages." http://www-groups.dcs.st-and.ac.uk/~history/HistTopics/Pi_through_the_ages.html.

Mahler, K. "On the Approximation of π." *Nederl. Akad. Wetensch. Proc. Ser. A.* **56**/*Indagationes Math.* **15**, 30–42, 1953.

Ogilvy, C. S. "Pi and Pi-Makers." Ch. 10 in *Excursions in Mathematics.* New York: Dover, pp. 108–120, 1994.

Olds, C. D. *Continued Fractions.* New York: Random House, pp. 59–60, 1963.

Pappas, T. "Probability and π." *The Joy of Mathematics.* San Carlos, CA: Wide World Publ./Tetra, pp. 18–19, 1989.

Peterson, I. *Islands of Truth: A Mathematical Mystery Cruise.* New York: W. H. Freeman, pp. 178–186, 1990.

Pickover, C. A. *Keys to Infinity.* New York: Wiley, p. 62, 1995.

Plouffe, S. "Plouffe's Inverter: Table of Current Records for the Computation of Constants." `http://lacim.uqam.ca/pi/records.html`.

Plouffe, S. "People Who Computed Pi." `http://www.cecm.sfu.ca/projects/ISC/records.html`.

Plouffe, S. "Plouffe's Inverter: A Few Approximations of Pi." `http://www.lacim.uqam.ca/pi/approxpi.html`.

Plouffe, S. "The π Page." `http://www.cecm.sfu.ca/pi/`.

Plouffe, S. "Table of Computation of Pi from 2000 BC to Now." `http://www.cecm.sfu.ca/projects/ISC/Pihistory.html`.

Preston, R. "Mountains of Pi." *New Yorker* **68**, 36–67, Mar. 2, 1992. `http://www.lacim.uqam.ca/plouffe/Chudnovsky.html`.

Project Mathematics! *The Story of Pi.* Videotape (24 minutes). California Institute of Technology. Available from the Math. Assoc. Amer.

Rabinowitz, S. and Wagon, S. "A Spigot Algorithm for the Digits of π." *Amer. Math. Monthly* **102**, 195–203, 1995.

Ramanujan, S. "Modular Equations and Approximations to π." *Quart. J. Pure. Appl. Math.* **45**, 350–372, 1913–1914.

Rudio, F. "Archimedes, Huygens, Lambert, Legendre." In *Vier Abhandlungen über die Kreismessung.* Leipzig, Germany, 1892.

Shanks, D. "Dihedral Quartic Approximations and Series for π." *J. Number. Th.* **14**, 397–423, 1982.

Shanks, D. *Solved and Unsolved Problems in Number Theory, 4th ed.* New York: Chelsea, 1993.

Singh, S. *Fermat's Enigma: The Epic Quest to Solve the World's Greatest Mathematical Problem.* New York: Walker, pp. 17–18, 1997.

Sloane, N. J. A. Sequences A000796/M2218, A001203/M2646, A002486/M4456, and A002491/M1009 in "An On-Line Version of the Encyclopedia of Integer Sequences."

Sloane, N. J. A. Sequences A025547, A001901, A046126, and A007509/M2061 in "An On-Line Version of the Encyclopedia of Integer Sequences."

Sloane, N. J. A. Sequences A032523, A033089, A033090, and A002485/M3097 in "An On-Line Version of the Encyclopedia of Integer Sequences."

Støllum, H.-H. "River Meandering as a Self-Organization Process." *Science* **271**, 1710–1713, 1996.

Vardi, I. *Computational Recreations in Mathematica.* Reading, MA: Addison-Wesley, p. 159, 1991.

Viète, F. *Uriorum de rebus mathematicis responsorum,* liber VIII, 1593.

Wagon, S. "Is π Normal?" *Math. Intel.* **7**, 65–67, 1985.

Whitcomb, C. "Notes on Pi (π)." `http://witcombe.bcpw.sbc.edu/EMPi.html`.

Woon, S. C. "Problem 1441." *Math. Mag.* **68**, 72–73, 1995.

Pi Heptomino

A HEPTOMINO in the shape of the Greek character π.

Piano Mover's Problem

N.B. A detailed on-line essay by S. Finch was the starting point for this entry.

Given an open subset U in n-D space and two compact subsets C_0 and C_1 of U, where C_1 is derived from C_0 by a continuous motion, is it possible to move C_0 to C_1 while remaining entirely inside U?

see also MOVING LADDER CONSTANT, MOVING SOFA CONSTANT

References

Buchberger, B.; Collins, G. E.; and Kutzler, B. "Algebraic Methods in Geometry." *Annual Rev. Comput. Sci.* **3**, 85–119, 1988.

Feinberg, E. B. and Papadimitriou, C. H. "Finding Feasible Points for a Two-point Body." *J. Algorithms* **10**, 109–119, 1989.

Finch, S. "Favorite Mathematical Constants." `http://www.mathsoft.com/asolve/constant/sofa/sofa.html`.

Leven, D. and Sharir, M. "An Efficient and Simple Motion Planning Algorithm for a Ladder Moving in Two-Dimensional Space Amidst Polygonal Barriers." *J. Algorithms* **8**, 192–215, 1987.

Picard's Existence Theorem

If f is a continuous function that satisfies the LIPSCHITZ CONDITION

$$|f(x,t) - f(y,t)| \leq L|x-y|$$

in a surrounding of $(x_0, t_0) \in \Omega \subset \mathbb{R} \times \mathbb{R}^n = \{(x,t) : |x - x_0| < b, |t - t_0| < a\}$, then the differential equation

$$\frac{df}{dx} = f(x,t)$$
$$x(t_0) = x_0$$

has a unique solution $x(t)$ in the interval $|t - t_0| < d$, where $d = \min(a, b/B)$, min denotes the MINIMUM, $B = \sup|f(t,x)|$, and sup denotes the SUPREMUM.

see also ORDINARY DIFFERENTIAL EQUATION

Picard's Little Theorem

Any ENTIRE ANALYTIC FUNCTION whose range omits two points must be a constant.

Picard's Theorem

An ANALYTIC FUNCTION assumes every COMPLEX NUMBER, with possibly one exception, infinitely often in any NEIGHBORHOOD of an ESSENTIAL SINGULARITY.

see also ANALYTIC FUNCTION, ESSENTIAL SINGULARITY, NEIGHBORHOOD

Picard Variety

Let V be a VARIETY, and write $G(V)$ for the set of divisors, $G_l(V)$ for the set of divisors linearly equivalent to 0, and $G_a(V)$ for the group of divisors algebraically equal to 0. Then $G_a(V)/G_l(V)$ is called the Picard variety. The ALBANESE VARIETY is dual to the Picard variety.

see also ALBANESE VARIETY

References

Iyanaga, S. and Kawada, Y. (Eds.). *Encyclopedic Dictionary of Mathematics.* Cambridge, MA: MIT Press, p. 75, 1980.

Pick's Formula

see PICK'S THEOREM

Pick's Theorem

Let A be the AREA of a simply closed POLYGON whose VERTICES are lattice points. Let B denote the number of LATTICE POINTS on the EDGES and I the number of points in the interior of the POLYGON. Then

$$A = I + \tfrac{1}{2}B - 1.$$

The FORMULA has been generalized to 3-D and higher dimensions using EHRHART POLYNOMIALS.

see also BLICHFELDT'S THEOREM, EHRHART POLYNOMIAL, LATTICE POINT, MINKOWSKI CONVEX BODY THEOREM

References

Diaz, R. and Robins, S. "Pick's Formula via the Weierstraß $\wp$-Function." *Amer. Math. Monthly* **102**, 431–437, 1995.

Ewald, G. *Combinatorial Convexity and Algebraic Geometry.* New York: Springer-Verlag, 1996.

Hammer, J. *Unsolved Problems Concerning Lattice Points.* London: Pitman, 1977.

Morelli, R. "Pick's Theorem and the Todd Class of a Toric Variety." *Adv. Math.* **100**, 183–231, 1993.

Pick, G. "Geometrisches zur Zahlentheorie." *Sitzenber. Lotos (Prague)* **19**, 311–319, 1899.

Steinhaus, H. *Mathematical Snapshots, 3rd American ed.* New York: Oxford University Press, pp. 97–98, 1983.

Picone's Theorem

Let $f(x)$ be integrable in $[-1,1]$, let $(1-x^2)f(x)$ be of bounded variation in $[-1,1]$, let M' denote the least upper bound of $|f(x)(1-x^2)|$ in $[-1,1]$, and let V' denote the total variation of $f(x)(1-x^2)$ in $[-1,1]$. Given the function

$$F(x) = F(-1) + \int_1^x f(x)\,dx,$$

then the terms of its LEGENDRE SERIES

$$F(x) \sim \sum_{n=0}^{\infty} a_n P_n(x)$$

$$a_n = \tfrac{1}{2}(2n+1)\int_{-1}^{1} F(x)P_n(x)\,dx,$$

where $P_n(x)$ is a LEGENDRE POLYNOMIAL, satisfy the inequalities

$$|a_n P_n(x)| < \begin{cases} 8\sqrt{\frac{2}{\pi}}\frac{M'+V'}{(1-\delta^2)^{1/4}}n^{-3/2} & \text{for } |x| \leq \delta < 1 \\ 2(M'+V')n^{-1} & \text{for } |x| \leq 1 \end{cases}$$

for $n \geq 1$ (Sansone 1991).

see also JACKSON'S THEOREM, LEGENDRE SERIES

References

Picone, M. *Appunti di Analise Superiore.* Naples, Italy,, p. 260, 1940.

Sansone, G. *Orthogonal Functions, rev. English ed.* New York: Dover, pp. 203–205, 1991.

Pie Cutting

see CIRCLE CUTTING, CYLINDER CUTTING, PANCAKE THEOREM, PIZZA THEOREM

Piecewise Circular Curve

A curve composed exclusively of circular ARCS, e.g., the FLOWER OF LIFE, LENS, REULEAUX TRIANGLE, SEED OF LIFE, and YIN-YANG.

see also ARC, REULEAUX TRIANGLE, YIN-YANG FLOWER OF LIFE, LENS, REULEAUX POLYGON, REULEAUX TRIANGLE, SALINON, SEED OF LIFE, TRIANGLE ARCS, YIN-YANG

References

Banchoff, T. and Giblin, P. "On The Geometry Of Piecewise Circular Curves." *Amer. Math. Monthly* **101**, 403–416, 1994.

Pigeonhole Principle

see DIRICHLET'S BOX PRINCIPLE

Pillai's Conjecture

For every $k > 1$, there exist only finite many pairs of POWERS (p, p') with p and p' PRIME and $k = p' - p$.

References

Ribenboim, P. "Catalan's Conjecture." *Amer. Math. Monthly* **103**, 529–538, 1996.

Pilot Vector

see VECTOR SPHERICAL HARMONIC

Pinch Point

A singular point such that every NEIGHBORHOOD of the point intersects itself. Pinch points are also called Whitney singularities or branch points.

Pinching Theorem

Let $g(x) \leq f(x) \leq h(x)$ for all x in some open interval containing a. If

$$\lim_{\Delta x \to a} g(x) = \lim_{\Delta x \to a} h(x) = L,$$

then $\lim_{\Delta x \to a} f(x) = L$.

Pine Cone Number

see FIBONACCI NUMBER

Piriform

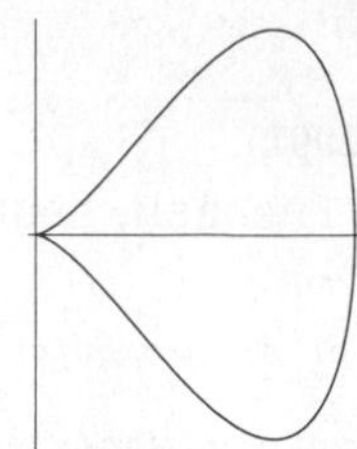

A plane curve also called the PEG TOP and given by the CARTESIAN equation

$$a^4y^2 = b^2x^3(2a-x) \tag{1}$$

and the parametric curves

$$x = a(1+\sin t) \tag{2}$$
$$y = b\cos t(1+\sin t) \tag{3}$$

for $t \in [-\pi/2, \pi/2]$. It was studied by G. de Longchamps in 1886. The generalization to a QUARTIC 3-D surface

$$(x^4 - x^3) + y^2 + z^2 = 0, \tag{4}$$

is shown below (Nordstrand).

see also BUTTERFLY CURVE, DUMBBELL CURVE, EIGHT CURVE, HEART SURFACE, PEAR CURVE

References

Cundy, H. and Rollett, A. *Mathematical Models, 3rd ed.* Stradbroke, England: Tarquin Pub. p. 71, 1989.

Lawrence, J. D. *A Catalog of Special Plane Curves.* New York: Dover, pp. 148–150, 1972.

Nordstrand, T. "Surfaces." `http://www.uib.no/people/nfytn/surfaces.htm`.

Pisot-Vijayaraghavan Constants

Let θ be a number greater than 1, λ a POSITIVE number, and

$$(x) \equiv x - \lfloor x \rfloor \tag{1}$$

denote the fractional part of x. Then for a given λ, the sequence of numbers $(\lambda\theta^n)$ for $n = 1, 2, \ldots$ is uniformly distributed in the interval (0, 1) when θ does not belong to a λ-dependent exceptional set S of MEASURE zero (Koksma 1935). Pisot (1938) and Vijayaraghavan (1941) independently studied the exceptional values of θ, and Salem (1943) proposed calling such values Pisot-Vijayaraghavan numbers.

Pisot (1938) proved that if θ is such that there exists a $\lambda \neq 0$ such that the series $\sum_{n=0}^\infty \sin^2(\pi\lambda\theta)^n$ converges, then θ is an ALGEBRAIC INTEGER whose conjugates all (except for itself) have modulus < 1, and λ is an algebraic INTEGER of the FIELD $K(\theta)$. Vijayaraghavan (1940) proved that the set of Pisot-Vijayaraghavan numbers has infinitely many limit points. Salem (1944) proved that the set of Pisot-Vijayaraghavan constants is closed. The proof of this theorem is based on the LEMMA that for a Pisot-Vijayaraghavan constant θ, there always exists a number λ such that $1 \leq \lambda < \theta$ and the following inequality is satisfied,

$$\sum_{n=0}^\infty \sin^2(\pi\lambda\theta^n) \leq \frac{\pi^2(2\theta+1)^2}{(\theta-1)^2}. \tag{2}$$

The smallest Pisot-Vijayaraghavan constant is given by the POSITIVE ROOT θ_0 of

$$x^3 - x - 1 = 0. \tag{3}$$

This number was identified as the smallest known by Salem (1944), and proved to be the smallest possible by Siegel (1944). Siegel also identified the next smallest Pisot-Vijayaraghavan constant θ_1 as the root of

$$x^4 - x^3 - 1 = 0, \tag{4}$$

showed that θ_1 and θ_2 are isolated in S, and showed that the roots of each POLYNOMIAL

$$x^n(x^2-x-1) + x^2 - 1 \qquad n = 1, 2, 3, \ldots \tag{5}$$

$$x^n - \frac{x^{n+1}-1}{x^2-1} \qquad n = 3, 5, 7, \ldots \tag{6}$$

$$x^n - \frac{x^{n-1}-1}{x-1} \qquad n = 3, 5, 7, \ldots \tag{7}$$

belong to S, where $\theta_0 = \phi$ (the GOLDEN MEAN) is the accumulation point of the set (in fact, the smallest; Le Lionnais 1983, p. 40). Some small Pisot-Vijayaraghavan constants and their POLYNOMIALS are given in the following table. The latter two entries are from Boyd (1977).

k	Number	Order	Polynomial
0	1.3247179572	3	1 0 −1 −1
1	1.3802775691	4	1 −1 0 0 −1
	1.6216584885	16	1 −2 2 −3 2 −2 1 0 0 1 −1 2 −2 2 −2 1 −1
	1.8374664495	20	1 −2 0 1 −1 0 1 −1 0 1 0 −1 0 1 −1 0 1 −1 0 1 −1

All the points in S less than ϕ are known (Dufresnoy and Pisot 1955). Each point of S is a limit point from both sides of the set T of SALEM CONSTANTS (Salem 1945).

see also SALEM CONSTANTS

References

Boyd, D. W. "Small Salem Numbers." *Duke Math. J.* **44**, 315–328, 1977.

Dufresnoy, J. and Pisot, C. "Étude de certaines fonctions méromorphes bornées sur le cercle unité, application à un ensemble fermé d'entiers algébriques." *Ann. Sci. École Norm. Sup.* **72**, 69–92, 1955.

Le Lionnais, F. *Les nombres remarquables.* Paris: Hermann, pp. 38 and 148, 1983.

Koksma, J. F. "Ein mengentheoretischer Satz über die Gleichverteilung modulo Eins." *Comp. Math.* **2**, 250–258, 1935.

Pisot, C. "La répartition modulo 1 et les nombres algébriques." *Annali di Pisa* **7**, 205–248, 1938.

Salem, R. "Sets of Uniqueness and Sets of Multiplicity." *Trans. Amer. Math. Soc.* **54**, 218–228, 1943.

Salem, R. "A Remarkable Class of Algebraic Numbers. Proof of a Conjecture of Vijayaraghavan." *Duke Math. J.* **11**, 103–108, 1944.

Salem, R. "Power Series with Integral Coefficients." *Duke Math. J.* **12**, 153–172, 1945.

Siegel, C. L. "Algebraic Numbers whose Conjugates Lie in the Unit Circle." *Duke Math. J.* **11**, 597–602, 1944.

Vijayaraghavan, T. "On the Fractional Parts of the Powers of a Number, II." *Proc. Cambridge Phil. Soc.* **37**, 349–357, 1941.

Pistol

A 4-POLYHEX.

References

Gardner, M. *Mathematical Magic Show: More Puzzles, Games, Diversions, Illusions and Other Mathematical Sleight-of-Mind from Scientific American.* New York: Vintage, p. 147, 1978.

Pitchfork Bifurcation

Let $f : \mathbb{R} \times \mathbb{R} \to \mathbb{R}$ be a one-parameter family of C^3 map satisfying

$$f(-x,\mu) = -f(x,\mu) \tag{1}$$

$$\left[\frac{\partial f}{\partial x}\right]_{\mu=0,x=0} = 1 \tag{2}$$

$$\left[\frac{\partial f}{\partial x}\right]_{\mu,x} = \left[\frac{\partial f}{\partial x}\right]_{\mu=0,x=\mu} \tag{3}$$

$$\left[\frac{\partial^2 f}{\partial x \partial \mu}\right]_{0,0} > 0 \tag{4}$$

$$\left[\frac{\partial^3 f}{\partial \mu^3}\right]_{\mu=0,x=0} < 0. \tag{5}$$

Then there are intervals having a single stable fixed point and three fixed points (two of which are stable and one of which is unstable). This BIFURCATION is called a pitchfork bifurcation. An example of an equation displaying a pitchfork bifurcation is

$$\dot{x} = \mu x - x^3 \tag{6}$$

(Guckenheimer and Holmes 1997, p. 145).

see also BIFURCATION

References

Guckenheimer, J. and Holmes, P. *Nonlinear Oscillations, Dynamical Systems, and Bifurcations of Vector Fields, 3rd ed.* New York: Springer-Verlag, pp. 145 and 149–150, 1997.

Rasband, S. N. *Chaotic Dynamics of Nonlinear Systems.* New York: Wiley, p. 31, 1990.

Pivot Theorem

If the VERTICES A, B, and C of TRIANGLE ΔABC lie on sides QR, RP, and PQ of the TRIANGLE ΔPQR, then the three CIRCLES CBP, ACQ, and BAR have a common point. In extended form, it is MIQUEL'S THEOREM.

see also MIQUEL'S THEOREM

References

Coxeter, H. S. M. and Greitzer, S. L. *Geometry Revisited.* New York: Random House, pp. 61–62, 1967.

Forder, H. G. *Geometry.* London: Hutchinson, p. 17, 1960.

Pivoting

The element in the diagonal of a matrix by which other elements are divided in an algorithm such as GAUSS-JORDAN ELIMINATION is called the pivot element. Partial pivoting is the interchanging of rows and full pivoting is the interchanging of both rows and columns in order to place a particularly "good" element in the diagonal position prior to a particular operation.

see also GAUSS-JORDAN ELIMINATION

References

Press, W. H.; Flannery, B. P.; Teukolsky, S. A.; and Vetterling, W. T. *Numerical Recipes in FORTRAN: The Art of Scientific Computing, 2nd ed.* Cambridge, England: Cambridge University Press, pp. 29–30, 1992.

Pizza Theorem

If a circular pizza is divided into 8, 12, 16, ... slices by making cuts at equal angles from an arbitrary point, then the sums of the areas of alternate slices are equal.

Place (Digit)

see DIGIT

Place (Field)

A place ν of a number FIELD k is an ISOMORPHISM class of field maps k onto a dense subfield of a nondiscrete locally compact FIELD k_ν.

In the function field case, let F be a function field of algebraic functions of one variable over a FIELD K. Then by a place in F, we mean a subset p of F which is the IDEAL of nonunits of some VALUATION RING O over K.

References
Chevalley, C. *Introduction to the Theory of Algebraic Functions of One Variable.* Providence, RI: Amer. Math. Soc., p. 2, 1951.
Knapp, A. W. "Group Representations and Harmonic Analysis, Part II." *Not. Amer. Math. Soc.* **43**, 537–549, 1996.

Place (Game)

For n players, $n-1$ games are needed to fairly determine first place, and $n-1+\lg(n-1)$ are needed to fairly determine first and second place.

Planar Bubble Problem

see BUBBLE

Planar Distance

For n points in the PLANE, there are at least

$$N_1 = \sqrt{n - \tfrac{3}{4}} - \tfrac{1}{2}$$

different DISTANCES. The minimum DISTANCE can occur only $\leq 3n-6$ times, and the MAXIMUM DISTANCE can occur $\leq n$ times. Furthermore, no DISTANCE can occur as often as

$$N_2 = \tfrac{1}{4}n(1+\sqrt{8n-7}) < \frac{n^{3/2}}{\sqrt{2}} - \frac{n}{4}$$

times. No set of $n > 6$ points in the PLANE can determine only ISOSCELES TRIANGLES.

see also DISTANCE

References
Honsberger, R. "The Set of Distances Determined by n Points in the Plane." Ch. 12 in *Mathematical Gems II.* Washington, DC: Math. Assoc. Amer., pp. 111–135, 1976.

Planar Graph

A GRAPH is planar if it can be drawn in a PLANE without EDGES crossing (i.e., it has CROSSING NUMBER 0). Only planar graphs have DUALS. If G is planar, then G has VERTEX DEGREE ≤ 5. COMPLETE GRAPHS are planar only for $n \leq 4$. The complete BIPARTITE GRAPH $K(3,3)$ in nonplanar. More generally, Kuratowski proved in 1930 that a graph is planar IFF it does not contain within it any graph which can be CONTRACTED to the pentagonal graph $K(5)$ or the hexagonal graph $K(3,3)$. K_5 can be decomposed into a union of two planar graphs, giving it a "DEPTH" of $E(K_5)=2$. Simple CRITERIA for determining the depth of graphs are not known. Beineke and Harary (1964, 1965) have shown that if $n \not\equiv 4 \pmod 6$, then

$$E(K_n) = \left\lfloor \tfrac{1}{6}(n+7) \right\rfloor.$$

The DEPTHS of the graphs K_n for $n = 4$, 10, 22, 28, 34, and 40 are 1, 3, 4, 5, 6, and 7 (Meyer 1970).

see also COMPLETE GRAPH, FABRY IMBEDDING, INTEGRAL DRAWING, PLANAR STRAIGHT LINE GRAPH

References
Beineke, L. W. and Harary, F. "On the Thickness of the Complete Graph." *Bull. Amer. Math. Soc.* **70**, 618–620, 1964.
Beineke, L. W. and Harary, F. "The Thickness of the Complete Graph." *Canad. J. Math.* **17**, 850–859, 1965.
Booth, K. S. and Lueker, G. S. "Testing for the Consecutive Ones Property, Interval Graphs, and Graph Planarity using PQ-Tree Algorithms." *J. Comput. System Sci.* **13**, 335–379, 1976.
Le Lionnais, F. *Les nombres remarquables.* Paris: Hermann, p. 56, 1983.
Meyer, J. "L'épaisseur des graphes completes K_{34} et K_{40}." *J. Comp. Th.* **9**, 1970.

Planar Point

A point $\mathbf{p}$ on a REGULAR SURFACE $M \in \mathbb{R}^3$ is said to be planar if the GAUSSIAN CURVATURE $K(\mathbf{p}) = 0$ and $S(\mathbf{p}) = 0$ (where S is the SHAPE OPERATOR), or equivalently, both of the PRINCIPAL CURVATURES κ_1 and κ_2 are 0.

see also ANTICLASTIC, ELLIPTIC POINT, GAUSSIAN CURVATURE, HYPERBOLIC POINT, PARABOLIC POINT, SYNCLASTIC

References
Gray, A. *Modern Differential Geometry of Curves and Surfaces.* Boca Raton, FL: CRC Press, p. 280, 1993.

Planar Space

Let (ξ_1, ξ_2) be a locally EUCLIDEAN coordinate system. Then

$$ds^2 = d\xi_1{}^2 + d\xi_2{}^2. \tag{1}$$

Now plug in

$$d\xi_1 = \frac{\partial \xi_1}{\partial x_1} dx_1 + \frac{\partial \xi_1}{\partial x_2} dx_2 \tag{2}$$

$$d\xi_2 = \frac{\partial \xi_2}{\partial x_1} dx_1 + \frac{\partial \xi_2}{\partial x_2} dx_2 \tag{3}$$

to obtain

$$\begin{aligned} ds^2 = {} & \left[\left(\frac{\partial \xi_1}{\partial x_1}\right)^2 + \left(\frac{\partial \xi_2}{\partial x_1}\right)^2\right] dx_1{}^2 \\ & + 2\left[\frac{\partial \xi_1}{\partial x_1}\frac{\partial \xi_1}{\partial x_2} + \frac{\partial \xi_2}{\partial x_1}\frac{\partial \xi_2}{\partial x_2}\right] dx_1\, dx_2 \\ & + \left[\left(\frac{\partial \xi_1}{\partial x_2}\right)^2 + \left(\frac{\partial \xi_2}{\partial x_2}\right)^2\right] dx_2{}^2. \end{aligned} \tag{4}$$

Reading off the COEFFICIENTS from

$$ds^2 = g_{11}\, dx_1{}^2 + 2g_{12}\, dx_1\, dx_2 + g_{22}\,(dx_2)^2 \tag{5}$$

gives

$$g_{11} = \left(\frac{\partial \xi_1}{\partial x_1}\right)^2 + \left(\frac{\partial \xi_2}{\partial x_1}\right)^2 \tag{6}$$

$$g_{12} = \frac{\partial \xi_1}{\partial x_1}\frac{\partial \xi_1}{\partial x_2} + \frac{\partial \xi_2}{\partial x_1}\frac{\partial \xi_2}{\partial x_2} \tag{7}$$

$$g_{22} = \left(\frac{\partial \xi_1}{\partial x_2}\right)^2 + \left(\frac{\partial \xi_2}{\partial x_2}\right)^2. \tag{8}$$

Making a change of coordinates $(x_1, x_2) \to (x_1', x_2')$ gives

$$g_{11}' = \left(\frac{\partial \xi_1}{\partial x_1'}\right)^2 + \left(\frac{\partial \xi_2}{\partial x_1'}\right)^2$$
$$= \left(\frac{\partial \xi_1}{\partial x_1}\frac{\partial x_1}{\partial x_1'} + \frac{\partial \xi_1}{\partial x_2}\frac{\partial x_2}{\partial x_1'}\right)^2 + \left(\frac{\partial \xi_2}{\partial x_1}\frac{\partial x_1}{\partial x_1'} + \frac{\partial \xi_2}{\partial x_2}\frac{\partial x_2}{\partial x_1'}\right)^2$$
$$= g_{11}\left(\frac{\partial x_1}{\partial x_1'}\right)^2 + 2g_{12}\frac{\partial x_1}{\partial x_1'}\frac{\partial x_2}{\partial x_1'} + g_{22}\left(\frac{\partial x_2}{\partial x_1'}\right)^2 \quad (9)$$

$$g_{12}' = \frac{\partial \xi_1}{\partial x_1}\frac{\partial x_1}{\partial x_1'}\frac{\partial \xi_1}{\partial x_2}\frac{\partial x_2}{\partial x_2'} + \frac{\partial \xi_2}{\partial x_1}\frac{\partial x_1}{\partial x_1'}\frac{\partial \xi_2}{\partial x_2}\frac{\partial x_2}{\partial x_2'}$$
$$= g_{12}\frac{\partial x_1}{\partial x_1'}\frac{\partial x_2}{\partial x_2'} \quad (10)$$

$$g_{22}' = g_{11}\left(\frac{\partial x_1}{\partial x_2'}\right)^2 + 2g_{12}\frac{\partial x_1}{\partial x_2'}\frac{\partial x_2}{\partial x_2'} + g_{22}\left(\frac{\partial x_2}{\partial x_2'}\right)^2. \quad (11)$$

Planar Straight Line Graph

A PLANAR GRAPH in which only straight line segments are used to connect the VERTICES, where the EDGES may intersect.

see also PLANAR GRAPH

Plancherel's Theorem

$$\int_{-\infty}^{\infty} f(x)g^*(x)\,dx = \int_{-\infty}^{\infty} F(s)G^*(s)\,ds,$$

where $F(s) \equiv \mathcal{F}[f(x)]$ and $\mathcal{F}$ denotes a FOURIER TRANSFORM. If f and g are real

$$\int_{-\infty}^{\infty} f(x)g(-x)\,dx = \int_{-\infty}^{\infty} F(s)G(s)\,ds.$$

see also FOURIER TRANSFORM, PARSEVAL'S THEOREM

Planck's Radiation Function

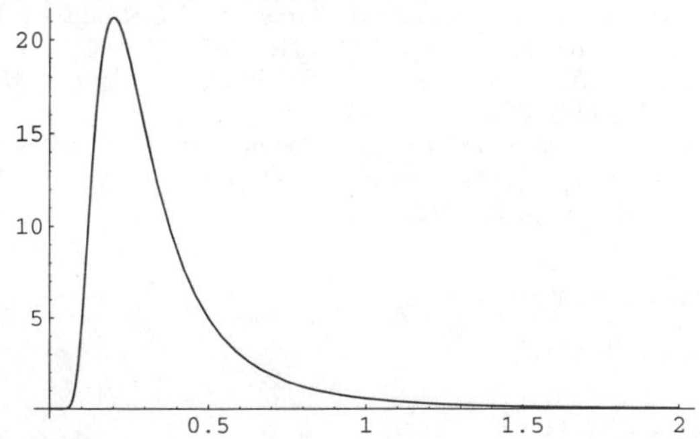

The function

$$f(x) = \frac{1}{x^5(e^{1/x}-1)}.$$

It has a MAXIMUM at $x \approx 0.201405$, where

$$f'(x) = \frac{5x - e^{1/x}(5x-1)}{x^7(e^{1/x}-1)^2} = 0,$$

and inflection points at $x \approx 0.11842$ and $x \approx 0.283757$, where

$$f''(x) = \frac{e^{1/x}(1+e^{1/x}) + 6x(e^{1/x}-1)[e^{1/x}(5x-2)-5x]}{(e^{1/x}-1)^3x^9} = 0.$$

References

Abramowitz, M. and Stegun, C. A. (Eds.). "Planck's Radiation Function." §27.2 in *Handbook of Mathematical Functions with Formulas, Graphs, and Mathematical Tables, 9th printing.* New York: Dover, p. 999, 1972.

Plane

A plane is a 2-D SURFACE spanned by two linearly independent vectors. The generalization of the plane to higher DIMENSIONS is called a HYPERPLANE.

In intercept form, a plane passing through the points $(a, 0, 0)$, $(0, b, 0)$ and $(0, 0, c)$ is given by

$$\frac{x}{a} + \frac{y}{b} + \frac{z}{c} = 1. \quad (1)$$

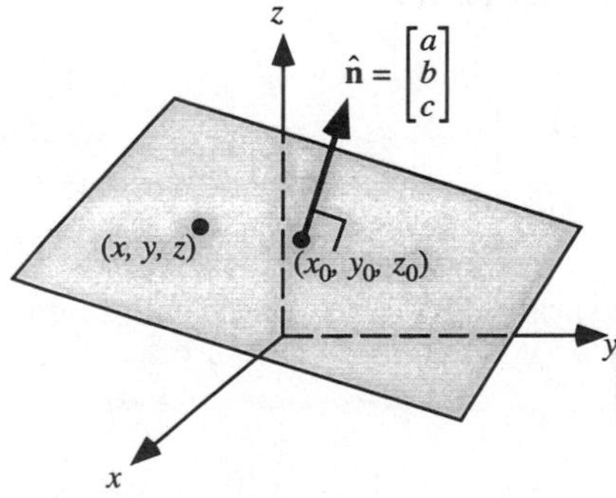

The equation of a plane PERPENDICULAR to the NONZERO VECTOR $\hat{\mathbf{n}} = (a, b, c)$ through the point (x_0, y_0, z_0) is

$$\begin{bmatrix} a \\ b \\ c \end{bmatrix} \cdot \begin{bmatrix} x - x_0 \\ y - y_0 \\ z - z_0 \end{bmatrix} = a(x-x_0) + b(y-y_0) + c(z-z_0) = 0, \quad (2)$$

so

$$ax + by + cz + d = 0, \quad (3)$$

where

$$d \equiv -ax_0 - by_0 - cz_0. \quad (4)$$

A plane specified in this form therefore has x-, y-, and z-intercepts at

$$x = -\frac{d}{a} \quad (5)$$

$$y = -\frac{d}{b} \quad (6)$$

$$z = -\frac{d}{c}, \quad (7)$$

and lies at a DISTANCE

$$h = \frac{|d|}{\sqrt{a^2 + b^2 + c^2}} \quad (8)$$

from the ORIGIN.

The plane through P_1 and parallel to (a_1, b_1, c_1) and (a_2, b_2, c_2) is

$$\begin{vmatrix} x - x_1 & y - y_1 & z - z_1 \\ a_1 & b_1 & c_1 \\ a_2 & b_2 & c_2 \end{vmatrix} = 0. \quad (9)$$

The plane through points P_1 and P_2 parallel to direction (a, b, c) is

$$\begin{vmatrix} x - x_1 & y - y_1 & z - z_1 \\ x_2 - x_1 & y_2 - y_1 & z_2 - z_1 \\ a & b & c \end{vmatrix} = 0. \quad (10)$$

The three-point form is

$$\begin{vmatrix} x & y & z & 1 \\ x_1 & y_1 & z_1 & 1 \\ x_2 & y_2 & z_2 & 1 \\ x_3 & y_3 & z_3 & 1 \end{vmatrix} = \begin{vmatrix} x - x_1 & y - y_1 & z - z_1 \\ x_2 - x_1 & y_2 - y_1 & z_2 - z_1 \\ x_3 - x_1 & y_3 - y_1 & z_3 - z_1 \end{vmatrix} = 0. \quad (11)$$

The DISTANCE from a point (x_1, y_1, z_1) to a plane

$$Ax + By + Cz + D = 0 \quad (12)$$

is

$$d = \frac{Ax_1 + By_1 + Cz_1 + D}{\pm\sqrt{A^2 + B^2 + C^2}}. \quad (13)$$

The DIHEDRAL ANGLE between the planes

$$A_1 x + B_1 y + C_1 z + D_1 = 0 \quad (14)$$

$$A_2 x + B_2 y + C_2 z + D_2 = 0 \quad (15)$$

is

$$\cos\theta = \frac{A_1 A_2 + B_1 B_2 + C_1 C_2}{\sqrt{{A_1}^2 + {B_1}^2 + {C_1}^2}\sqrt{{A_2}^2 + {B_2}^2 + {C_2}^2}}. \quad (16)$$

In order to specify the relative distances of $n > 1$ points in the plane, $1 + 2(n - 2) = 2n - 3$ coordinates are needed, since the first can always be placed at $(0, 0)$ and the second at $(x, 0)$, where it defines the x-AXIS. The remaining $n - 2$ points need two coordinates each. However, the total number of distances is

$${}_nC_2 = \binom{n}{2} = \frac{n!}{2!(n-2)!} = \tfrac{1}{2}n(n-1), \quad (17)$$

where $\binom{n}{k}$ is a BINOMIAL COEFFICIENT, so the distances between points are subject to m relationships, where

$$m \equiv \tfrac{1}{2}n(n-1) - (2n-3) = \tfrac{1}{2}(n-2)(n-3). \quad (18)$$

For $n = 2$ and $n = 3$, there are no relationships. However, for a QUADRILATERAL (with $n = 4$), there is one (Weinberg 1972).

It is impossible to pick random variables which are uniformly distributed in the plane (Eisenberg and Sullivan 1996). In 4-D, it is possible for four planes to intersect in exactly one point. For every set of n points in the plane, there exists a point O in the plane having the property such that *every* straight line through O has at least 1/3 of the points on each side of it (Honsberger 1985).

Every RIGID motion of the plane is one of the following types (Singer 1995):

1. ROTATION about a fixed point P.
2. TRANSLATION in the direction of a line l.
3. REFLECTION across a line l.
4. Glide-reflections along a line l.

Every RIGID motion of the hyperbolic plane is one of the previous types or a

5. Horocycle rotation.

see also ARGAND PLANE, COMPLEX PLANE, DIHEDRAL ANGLE, ELLIPTIC PLANE, FANO PLANE, HYPERPLANE, MOUFANG PLANE, NIRENBERG'S CONJECTURE, NORMAL SECTION, POINT-PLANE DISTANCE, PROJECTIVE PLANE

References

Beyer, W. H. *CRC Standard Mathematical Tables, 28th ed.* Boca Raton, FL: CRC Press, pp. 208–209, 1987.

Eisenberg, B. and Sullivan, R. "Random Triangles *n* Dimensions." *Amer. Math. Monthly* **103**, 308–318, 1996.

Honsberger, R. *Mathematical Gems III.* Washington, DC: Math. Assoc. Amer., pp. 189–191, 1985.

Singer, D. A. "Isometries of the Plane." *Amer. Math. Monthly* **102**, 628–631, 1995.

Weinberg, S. *Gravitation and Cosmology: Principles and Applications of the General Theory of Relativity.* New York: Wiley, p. 7, 1972.

Plane Curve

see CURVE

Plane Cutting

see CIRCLE CUTTING

Plane Division

Consider n intersecting CIRCLES and ELLIPSES. The maximal number of regions in which these divide the PLANE are

$$N_{\text{circle}} = n^2 - n + 2$$
$$N_{\text{ellipse}} = 2n^2 - 2n + 2.$$

see also ARRANGEMENT, CIRCLE, CUTTING, ELLIPSE, SPACE DIVISION

Plane-Filling Curve

see PLANE-FILLING FUNCTION

Plane-Filling Function

A SPACE-FILLING FUNCTION which maps a 1-D INTERVAL into a 2-D area. Plane-filling functions were thought to be impossible until Hilbert discovered the HILBERT CURVE in 1891.

Plane-filling functions are often (imprecisely) defined to be the "limit" of an infinite sequence of specified curves which "fill" the PLANE without "HOLES," hence the more popular term PLANE-FILLING CURVE. The term "plane-filling function" is preferable to "PLANE-FILLING CURVE" because "curve" informally connotes "GRAPH" (i.e., range) of some continuous function, but the GRAPH of a plane-filling function is a solid patch of 2-space with no evidence of the order in which it was traced (and, for a dense set, retraced). Actually, all that is needed to rigorously define a plane-filling function is an arbitrarily refinable correspondence between contiguous subintervals of the domain and contiguous subareas of the range.

True plane-filling functions are not ONE-TO-ONE. In fact, because they map closed intervals onto closed areas, they cannot help but overfill, revisiting at least twice a dense subset of the filled area. Thus, every point in the filled area has *at least* one inverse image.

see also HILBERT CURVE, PEANO CURVE, PEANO-GOSPER CURVE, SIERPIŃSKI CURVE, SPACE-FILLING FUNCTION, SPACE-FILLING POLYHEDRON

References

Bogomolny, A. "Plane Filling Curves." `http://www.cut-the-knot.com/do_you_know/hilbert.html`.

Wagon, S. "A Space-Filling Curve." §6.3 in *Mathematica in Action.* New York: W. H. Freeman, pp. 196–209, 1991.

Plane Geometry

That portion of GEOMETRY dealing with figures in a PLANE, as opposed to SOLID GEOMETRY. Plane geometry deals with the CIRCLE, LINE, POLYGON, etc.

see also CONSTRUCTIBLE POLYGON, GEOMETRIC CONSTRUCTION, GEOMETRY, SOLID GEOMETRY, SPHERICAL GEOMETRY

References

Altshiller-Court, N. *College Geometry: A Second Course in Plane Geometry for Colleges and Normal Schools, 2nd ed., rev. enl.* New York: Barnes and Noble, 1952.

Casey, J. *A Treatise on the Analytical Geometry of the Point, Line, Circle, and Conic Sections, Containing an Account of Its Most Recent Extensions with Numerous Examples, 2nd rev. enl. ed.* Dublin: Hodges, Figgis, & Co., 1893.

Coxeter, H. S. M. and Greitzer, S. L. *Geometry Revisited.* Washington, DC: Math. Assoc. Amer., 1967.

Coxeter, H. S. M. *Introduction to Geometry, 2nd ed.* New York: Wiley, 1969.

Dixon, R. *Mathographics.* New York: Dover, 1991.

Gallatly, W. *The Modern Geometry of the Triangle, 2nd ed.* London: Hodgson, 1913.

Heath, T. L. *The Thirteen Books of the Elements, 2nd ed., Vol. 1: Books I and II.* New York: Dover, 1956.

Heath, T. L. *The Thirteen Books of the Elements, 2nd ed., Vol. 2: Books III–IX.* New York: Dover, 1956.

Heath, T. L. *The Thirteen Books of the Elements, 2nd ed., Vol. 3: Books X–XIII.* New York: Dover, 1956.

Hilbert, D. *The Foundations of Geometry.* Chicago, IL: Open Court, 1980.

Hilbert, D. and Cohn-Vossen, S. *Geometry and the Imagination.* New York: Chelsea, 1952.

Honsberger, R. *Episodes in Nineteenth and Twentieth Century Euclidean Geometry.* Washington, DC: Math. Assoc. Amer., 1995.

Johnson, R. A. *Modern Geometry: An Elementary Treatise on the Geometry of the Triangle and the Circle.* Boston, MA: Houghton Mifflin, 1929.

Kimberling, C. "Triangle Centers and Central Triangles." *Congr. Numer.* **129**, 1–295, 1998.

Klee, V. "Some Unsolved Problems in Plane Geometry." *Math. Mag.* **52**, 131–145, 1979.

Klee, V. and Wagon, S. *Old and New Unsolved Problems in Plane Geometry and Number Theory, rev. ed.* Washington, DC: Math. Assoc. Amer., 1991.

Pedoe, D. *Circles: A Mathematical View, rev. ed.* Washington, DC: Math. Assoc. Amer., 1995.

Plane Partition

A two-dimensional array of INTEGERS nonincreasing both left to right and top to bottom which add up to a given number, i.e., $n_{ij} \geq n_{i(j+1)}$ and $n_{ij} \geq n_{(i+1)j}$. For example, a planar partition of 2 is given by

$$\begin{matrix} 5 & 4 & 2 & 1 & 1 \\ 3 & 2 & & & \\ 2 & 2. & & & \end{matrix}$$

The GENERATING FUNCTION for the number $\mathrm{PL}(n)$ of planar partitions of n is

$$\sum_{n=0}^{\infty} \mathrm{PL}(n) x^n = \frac{1}{\prod_{k=1}^{\infty} (1 - x^k)^k}$$
$$= 1 + x + 3x^2 + 6x^3 + 13x^4 + 24x^5 + \ldots$$

(Sloane's A000219, MacMahon 1912b, Beeler *et al.* 1972, Bender and Knuth 1972). The concept of planar partitions can also be generalized to cubic partitions.

see also PARTITION, SOLID PARTITION

References

Beeler, M.; Gosper, R. W.; and Schroeppel, R. Item 18 in *HAKMEM*. Cambridge, MA: MIT Artificial Intelligence Laboratory, Memo AIM-239, Feb. 1972.

Bender, E. A. and Knuth, D. E. "Enumeration of Plane Partitions." *J. Combin. Theory Ser. A.* **13**, 40–54, 1972.

Knuth, D. E. "A Note on Solid Partitions." *Math. Comput.* **24**, 955–961, 1970.

MacMahon, P. A. "Memoir on the Theory of the Partitions of Numbers. V: Partitions in Two-Dimensional Space." *Phil. Trans. Roy. Soc. London Ser. A* **211**, 75–110, 1912a.

MacMahon, P. A. "Memoir on the Theory of the Partitions of Numbers. VI: Partitions in Two-Dimensional Space, to which is Added an Adumbration of the Theory of Partitions in Three-Dimensional Space." *Phil. Trans. Roy. Soc. London Ser. A* **211**, 345–373, 1912b.

MacMahon, P. A. *Combinatory Analysis, Vol. 2.* New York: Chelsea, 1960.

Sloane, N. J. A. Sequence A000219/M2566 in "An On-Line Version of the Encyclopedia of Integer Sequences."

Plane Symmetry Groups

see WALLPAPER GROUPS

Planted Planar Tree

A planted plane tree (V, E, v, α) is defined as a vertex set V, edges set E, ROOT v, and order relation α on V which satisfies

1. For $x, y \in V$ if $\rho(x) < \rho(y)$, then $x\,\alpha\,y$, where $\rho(x)$ is the length of the path from v to x,
2. If $\{r, s\}, \{x, y\} \in E$, $\rho(r) = \rho(x) = \rho(s)-1 = \rho(y)-1$ and $r\,\alpha\,x$, then $s\,\alpha\,y$

(Klarner 1969, Chorneyko and Mohanty 1975). The CATALAN NUMBERS give the number of planar trivalent planted trees.

see also CATALAN NUMBER, TREE

References

Chorneyko, I. Z. and Mohanty, S. G. "On the Enumeration of Certain Sets of Planted Plane Trees." *J. Combin. Th. Ser. B* **18**, 209–221, 1975.

Harary, F.; Prins, G.; and Tutte, W. T. "The Number of Plane Trees." *Indag. Math.* **26**, 319–327, 1964.

Klarner, D. A. "A Correspondence Between Sets of Trees." *Indag. Math.* **31**, 292–296, 1969.

Plastic Constant

The limiting ratio of the successive terms of the PADOVAN SEQUENCE, $P = 1.32471795\ldots$.

see also PADOVAN SEQUENCE

References

Stewart, I. "Tales of a Neglected Number." *Sci. Amer.* **274**, 102–103, Jun. 1996.

Plat

A BRAID in which strands are intertwined in the center and are free in "handles" on either side of the diagram.

Plateau Curves

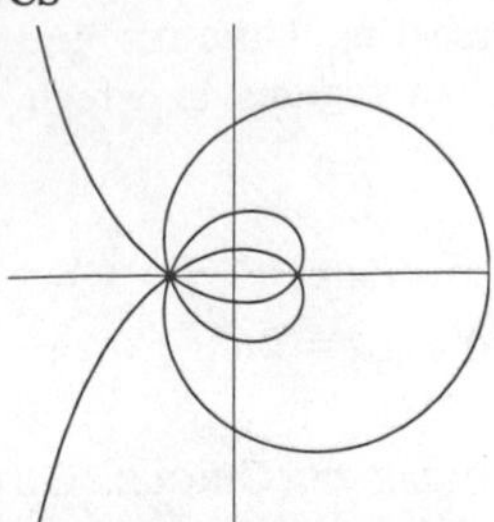

A curve studied by the Belgian physicist and mathematician Joseph Plateau. It has Cartesian equation

$$x = \frac{a\sin[(m+n)t]}{\sin[(m-n)t]}$$

$$y = \frac{2a\sin(mt)\sin(nt)}{\sin[(m-n)t]}.$$

If $m = 2n$, the Plateau curve degenerates to a CIRCLE with center $(1, 0)$ and radius 2.

References

MacTutor History of Mathematics Archive. "Plateau Curves." `http://www-groups.dcs.st-and.ac.uk/~history/Curves/Plateau.html`.

Plateau's Laws

BUBBLES can meet only at ANGLES of 120° (for two BUBBLES) and 109.5° (for three BUBBLES), where the exact value of 109.5° is the TETRAHEDRAL ANGLE. This was proved by Jean Taylor using MEASURE THEORY to study AREA minimization. The DOUBLE BUBBLE is AREA minimizing, but it is not known the triple BUBBLE is also AREA minimizing. It is also unknown if empty chambers trapped inside can minimize AREA for $n \geq 3$ BUBBLES.

see also BUBBLE, CALCULUS OF VARIATIONS, DOUBLE BUBBLE, PLATEAU'S PROBLEM

References

Morgan, F. "Mathematicians, including Undergraduates, Look at Soap Bubbles." *Amer. Math. Monthly* **101**, 343–351, 1994.

Taylor, J. E. "The Structure of Singularities in Soap-Bubble-Like and Soap-Film-Like Minimal Surfaces." *Ann. Math.* **103**, 489–539, 1976.

Plateau's Problem

The problem in CALCULUS OF VARIATIONS to find the MINIMAL SURFACE of a boundary with specified constraints. In general, there may be one, multiple, or no MINIMAL SURFACES spanning a given closed curve in space.

see also CALCULUS OF VARIATIONS, MINIMAL SURFACE

References

Cundy, H. and Rollett, A. *Mathematical Models, 3rd ed.* Stradbroke, England: Tarquin Pub., pp. 48–49, 1989.

Stuwe, M. *Plateau's Problem and the Calculus of Variations.* Princeton, NJ: Princeton University Press, 1989.

Plato's Number

A number appearing in *The Republic* which involves 216 and 12,960,000.

References

Plato. *The Republic.* New York: Oxford University Press, 1994.

Wells, D. G. *The Penguin Dictionary of Curious and Interesting Numbers.* London: Penguin, p. 144, 1986.

Platonic Solid

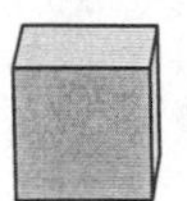 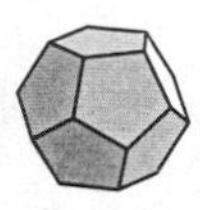 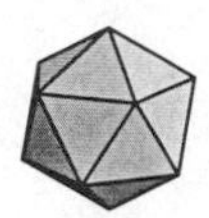 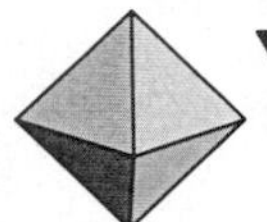 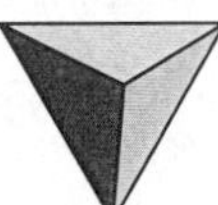

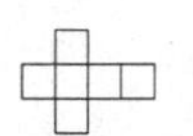 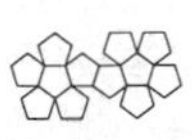 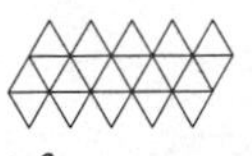 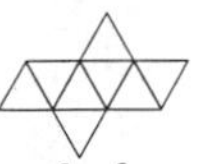 

A solid with equivalent faces composed of congruent regular convex POLYGONS. There are exactly five such solids: the CUBE, DODECAHEDRON, ICOSAHEDRON, OCTAHEDRON, and TETRAHEDRON, as was proved by Euclid in the last proposition of the *Elements*.

The Platonic solids were known to the ancient Greeks, and were described by Plato in his *Timaeus* ca. 350 BC. In this work, Plato equated the TETRAHEDRON with the "element" fire, the CUBE with earth, the ICOSAHEDRON with water, the OCTAHEDRON with air, and the DODECAHEDRON with the stuff of which the constellations and heavens were made (Cromwell 1997).

The Platonic solids are sometimes also known as the REGULAR POLYHEDRA of COSMIC FIGURES (Cromwell 1997), although the former term is sometimes used to refer collectively to both the Platonic solids *and* KEPLER-POINSOT SOLIDS (Coxeter 1973).

If P is a POLYHEDRON with congruent (convex) regular polygonal faces, then Cromwell (1997, pp. 77–78) shows that the following statements are equivalent.

1. The vertices of P all lie on a SPHERE.
2. All the DIHEDRAL ANGLES are equal.
3. All the VERTEX FIGURES are REGULAR POLYGONS.
4. All the SOLID ANGLES are equivalent.
5. All the vertices are surrounded by the same number of FACES.

Let v (sometimes denoted N_0) be the number of VERTICES, e (or N_1) the number of EDGES, and f (or N_2) the number of FACES. The following table gives the SCHLÄFLI SYMBOL, WYTHOFF SYMBOL, and C&R symbol, the number of vertices v, edges e, and faces f, and the POINT GROUPS for the Platonic solids (Wenninger 1989).

Solid	Schläfli	Wyth.	C&R	v	e	f	Grp
cube	$\{4,3\}$	$3 \mid 2\ 4$	4^3	8	12	6	O_h
dodecahedron	$\{5,3\}$	$3 \mid 2\ 5$	5^3	20	30	12	I_h
icosahedron	$\{3,5\}$	$5 \mid 2\ 3$	3^5	12	30	20	I_h
octahedron	$\{3,4\}$	$4 \mid 2\ 3$	3^4	6	12	8	O_h
tetrahedron	$\{3,3\}$	$3 \mid 2\ 3$	3^3	4	6	4	T_d

Let r be the INRADIUS, ρ the MIDRADIUS, and R the CIRCUMRADIUS. The following two tables give the analytic and numerical values of these distances for Platonic solids with unit side length.

Solid	r	ρ	R
cube	$\frac{1}{2}$	$\frac{1}{2}\sqrt{2}$	$\frac{1}{2}\sqrt{3}$
dodecahedron	$\frac{1}{20}\sqrt{250+110\sqrt{5}}$	$\frac{1}{4}(3+\sqrt{5})$	$\frac{1}{4}(\sqrt{15}+\sqrt{3})$
icosahedron	$\frac{1}{12}(3\sqrt{3}+\sqrt{15})$	$\frac{1}{4}(1+\sqrt{5})$	$\frac{1}{4}\sqrt{10+2\sqrt{5}}$
octahedron	$\frac{1}{6}\sqrt{6}$	$\frac{1}{2}$	$\frac{1}{2}\sqrt{2}$
tetrahedron	$\frac{1}{12}\sqrt{6}$	$\frac{1}{4}\sqrt{2}$	$\frac{1}{4}\sqrt{6}$

Solid	r	ρ	R
cube	0.5	0.70711	0.86603
dodecahedron	1.11352	1.30902	1.40126
icosahedron	0.75576	0.80902	0.95106
octahedron	0.40825	0.5	0.70711
tetrahedron	0.20412	0.35355	0.61237

Finally, let A be the AREA of a single FACE, V be the VOLUME of the solid, the EDGES be of unit length on a side, and α be the DIHEDRAL ANGLE. The following table summarizes these quantities for the Platonic solids.

Solid	A	V	α
cube	1	1	$\frac{1}{2}\pi$
dodecahedron	$\frac{1}{4}\sqrt{25+10\sqrt{5}}$	$\frac{1}{4}(15+7\sqrt{5})$	$\cos^{-1}(-\frac{1}{5}\sqrt{5})$
icosahedron	$\frac{1}{4}\sqrt{3}$	$\frac{5}{12}(3+\sqrt{5})$	$\cos^{-1}(-\frac{1}{3}\sqrt{5})$
octahedron	$\frac{1}{4}\sqrt{3}$	$\frac{1}{3}\sqrt{2}$	$\cos^{-1}(-\frac{1}{3})$
tetrahedron	$\frac{1}{4}\sqrt{3}$	$\frac{1}{12}\sqrt{2}$	$\cos^{-1}(\frac{1}{3})$

The number of EDGES meeting at a VERTEX is $2e/v$. The SCHLÄFLI SYMBOL can be used to specify a Platonic solid. For the solid whose faces are p-gons (denoted $\{p\}$), with q touching at each VERTEX, the symbol is $\{p,q\}$. Given p and q, the number of VERTICES, EDGES, and faces are given by

$$N_0 = \frac{4p}{4-(p-2)(q-2)}$$

$$N_1 = \frac{2pq}{4-(p-2)(q-2)}$$

$$N_2 = \frac{4q}{4-(p-2)(q-2)}.$$

MINIMAL SURFACES for Platonic solid frames are illustrated in Isenberg (1992, pp. 82–83).

see also ARCHIMEDEAN SOLID, CATALAN SOLID, JOHNSON SOLID, KEPLER-POINSOT SOLID, QUASIREGULAR POLYHEDRON, UNIFORM POLYHEDRON

References

Artmann, B. "Symmetry Through the Ages: Highlights from the History of Regular Polyhedra." In *In Eves' Circles* (Ed. J. M. Anthony). Washington, DC: Math. Assoc. Amer., pp. 139–148, 1994.

Ball, W. W. R. and Coxeter, H. S. M. "Polyhedra." Ch. 5 in *Mathematical Recreations and Essays, 13th ed.* New York: Dover, pp. 131–136, 1987.

Behnke, H.; Bachman, F.; Fladt, K.; and Kunle, H. (Eds.). *Fundamentals of Mathematics, Vol. 2.* Cambridge, MA: MIT Press, p. 272, 1974.

Beyer, W. H. (Ed.) *CRC Standard Mathematical Tables, 28th ed.* Boca Raton, FL: CRC Press, pp. 128–129, 1987.

Bogomolny, A. "Regular Polyhedra." `http://www.cut-the-knot.com/do_you_know/polyhedra.html`.

Coxeter, H. S. M. *Regular Polytopes, 3rd ed.* New York: Dover, pp. 1–17, 93, and 107–112, 1973.

Critchlow, K. *Order in Space: A Design Source Book.* New York: Viking Press, 1970.

Cromwell, P. R. *Polyhedra.* New York: Cambridge University Press, pp. 51–57, 66–70, and 77–78, 1997.

Dunham, W. *Journey Through Genius: The Great Theorems of Mathematics.* New York: Wiley, pp. 78–81, 1990.

Gardner, M. "The Five Platonic Solids." Ch. 1 in *The Second Scientific American Book of Mathematical Puzzles & Diversions: A New Selection.* New York: Simon and Schuster, pp. 13–23, 1961.

Heath, T. *A History of Greek Mathematics, Vol. 1.* Oxford, England: Oxford University Press, p. 162, 1921.

Isenberg, C. *The Science of Soap Films and Soap Bubbles.* New York: Dover, 1992.

Kepler, J. *Opera Omnia, Vol. 5.* Frankfort, p. 121, 1864.

Ogilvy, C. S. *Excursions in Geometry.* New York: Dover, pp. 129–131, 1990.

Pappas, T. "The Five Platonic Solids." *The Joy of Mathematics.* San Carlos, CA: Wide World Publ./Tetra, pp. 39 and 110–111, 1989.

Rawles, B. A. "Platonic and Archimedean Solids—Faces, Edges, Areas, Vertices, Angles, Volumes, Sphere Ratios." `http://www.intent.com/sg/polyhedra.html`.

Steinhaus, H. "Platonic Solids, Crystals, Bees' Heads, and Soap." Ch. 8 in *Mathematical Snapshots, 3rd American ed.* New York: Oxford University Press, 1960.

Waterhouse, W. "The Discovery of the Regular Solids." *Arch. Hist. Exact Sci.* **9**, 212–221, 1972–1973.

Wenninger, M. J. *Polyhedron Models.* Cambridge, England: Cambridge University Press, 1971.

Platykurtic

A distribution with FISHER KURTOSIS $\gamma_2 < 0$ (and therefore having a flattened shape).

see also FISHER KURTOSIS

Playfair's Axiom

Through any point in space, there is exactly one straight line PARALLEL to a given straight line. This AXIOM is equivalent to the PARALLEL AXIOM.

see also PARALLEL AXIOM

References

Dunham, W. "Hippocrates' Quadrature of the Lune." Ch. 1 in *Journey Through Genius: The Great Theorems of Mathematics.* New York: Wiley, p. 54, 1990.

Plethysm

A group theoretic operation which is useful in the study of complex atomic spectra. A plethysm takes a set of functions of a given symmetry type $\{\mu\}$ and forms from them symmetrized products of a given degree r and other symmetry type $\{\nu\}$. A plethysm

$$\{\mu\}\otimes\{\nu\}=\sum\{\lambda\}$$

satisfies the rules

$$A\otimes(BC)=(A\otimes B)(A\otimes C)=A\otimes BA\otimes C,$$

$$A\otimes(B\pm C)=A\otimes B\pm A\otimes C$$

$$(A\otimes B)\otimes C=A\otimes(B\otimes C)$$

$$(A+B)\otimes\{\lambda\}=\sum\Gamma_{\mu\nu\lambda}(A\otimes\{\mu\})(B\otimes\{\nu\}),$$

where $\Gamma_{\mu\nu\lambda}$ is the coefficient of $\{\lambda\}$ in $\{\mu\}\{\nu\}$,

$$(A-B)\otimes\{\lambda\}=\sum(-1)^r\Gamma_{\mu\nu\lambda}(A\otimes\{\mu\})(B\otimes\{\tilde{\nu}\}),$$

where $\{\tilde{\nu}\}$ is the partition of r conjugate to $\{\nu\}$, and

$$(AB)\otimes\{\lambda\}=\sum g_{\mu\nu\lambda}(A\otimes\{\mu\})(B\otimes\{\nu\}),$$

where $g_{\mu\nu\lambda}$ is the coefficient of $\{\lambda\}$ in the inner product $\{\mu\}\circ\{\nu\}$ (Wybourne 1970).

References

Littlewood, D. E. "Polynomial Concomitants and Invariant Matrices." *J. London Math. Soc.* **11**, 49–55, 1936.

Wybourne, B. G. "The Plethysm of S-Functions" and "Plethysm and Restricted Groups." Chs. 6–7 in *Symmetry Principles and Atomic Spectroscopy.* New York: Wiley, pp. 49–68, 1970.

Plot

see GRAPH (FUNCTION)

Plouffe's Constant

N.B. A detailed on-line essay by S. Finch was the starting point for this entry.

Define the function

$$\rho(x)\equiv\begin{cases}1 & \text{for } x<0\\ 0 & \text{for } x\geq 0.\end{cases}\tag{1}$$

Let

$$a_n=\sin(2^n)=\begin{cases}\sin 1 & \text{for } n=0\\ 2a_0\sqrt{1-{a_0}^2} & \text{for } n=1\\ 2a_{n-1}(1-2{a_{n-2}}^2) & \text{for } n\geq 2,\end{cases}\tag{2}$$

then

$$\sum_{n=0}^{\infty}\frac{\rho(a_n)}{2^{n+1}}=\frac{1}{2\pi}.\tag{3}$$

For

$$b_n = \cos(2^n) = \begin{cases} \cos 1 & \text{for } n = 0 \\ 2b_{n-1}{}^2 - 1 & \text{for } n \geq 1, \end{cases} \tag{4}$$

and

$$\sum_{n=0}^{\infty} \frac{\rho(b_n)}{2^{n+1}} = 0.4756260767\ldots. \tag{5}$$

Letting

$$c_n = \tan(2^n) = \begin{cases} \tan 1 & \text{for } n = 0 \\ \frac{2c_{n-1}}{1-c_{n-1}{}^2} & \text{for } n \geq 1, \end{cases} \tag{6}$$

then

$$\sum_{n=0}^{\infty} \frac{\rho(c_n)}{2^{n+1}} = \frac{1}{\pi}. \tag{7}$$

Plouffe asked if the above processes could be "inverted." He considered

$$\begin{aligned} \alpha_n &= \sin(2^n \sin^{-1} \tfrac{1}{2}) \\ &= \begin{cases} \frac{1}{2} & \text{for } n = 0 \\ \frac{1}{2}\sqrt{3} & \text{for } n = 1 \\ 2\alpha_{n-1}(1 - 2\alpha_{n-2}{}^2) & \text{for } n \geq 2, \end{cases} \end{aligned} \tag{8}$$

giving

$$\sum_{n=0}^{\infty} \frac{\rho(\alpha_n)}{2^{n+1}} = \tfrac{1}{12}, \tag{9}$$

and

$$\beta_n = \cos(2^n \cos^{-1} \tfrac{1}{2}) = \begin{cases} \frac{1}{2} & \text{for } n = 0 \\ 2\beta_{n-1}{}^2 - 1 & \text{for } n \geq 1, \end{cases} \tag{10}$$

giving

$$\sum_{n=0}^{\infty} \frac{\rho(\beta_n)}{2^{n+1}} = \tfrac{1}{2}, \tag{11}$$

and

$$\gamma_n = \tan(2^n \tan^{-1} \tfrac{1}{2}) = \begin{cases} \frac{1}{2} \\ \frac{2\gamma_{n-1}}{1-\gamma_{n-1}{}^2} & \text{for } n \geq 1, \end{cases} \quad \text{for } n = 0 \tag{12}$$

giving

$$\sum_{n=0}^{\infty} \frac{\rho(\alpha_n)}{2^{n+1}} = \frac{1}{\pi} \tan^{-1}(\tfrac{1}{2}). \tag{13}$$

The latter is known as Plouffe's constant (Plouffe 1997). The positions of the 1s in the BINARY expansion of this constant are 3, 6, 8, 9, 10, 13, 21, 23, ... (Sloane's A004715).

Borwein and Girgensohn (1995) extended Plouffe's γ_n to *arbitrary* REAL x, showing that if

$$\xi_n = \tan(2^n \tan^{-1} x) = \begin{cases} x & \text{for } n = 0 \\ \frac{2\xi_{n-1}}{1-\xi_{n-1}{}^2} & \text{for } n \geq 1 \text{ and } |\xi_{n-1}| \neq 1 \\ -\infty & \text{for } n \geq 1 \text{ and } |\xi_{n-1}| = 1, \end{cases} \tag{14}$$

then

$$\sum_{n=0}^{\infty} \frac{\rho(\xi_n)}{2^{n+1}} = \begin{cases} \frac{\tan^{-1} x}{\pi} & \text{for } x \geq 0 \\ 1 + \frac{\tan^{-1} x}{\pi} & \text{for } x < 0. \end{cases} \tag{15}$$

Borwein and Girgensohn (1995) also give much more general recurrences and formulas.

References

Borwein, J. M. and Girgensohn, R. "Addition Theorems and Binary Expansions." *Canad. J. Math.* **47**, 262–273, 1995.

Finch, S. "Favorite Mathematical Constants." `http://www.mathsoft.com/asolve/constant/plff/plff.html`.

Plouffe, S.. "The Computation of Certain Numbers Using a Ruler and Compass." Dec. 12, 1997. `http://www.research.att.com/~njas/sequences/JIS/compass.html`.

Plücker Characteristics

The CLASS m, ORDER n, number of NODES δ, number of CUSPS κ, number of STATIONARY TANGENTS (INFLECTION POINTS) ι, number of BITANGENTS τ, and GENUS p.

see also ALGEBRAIC CURVE, BITANGENT, CUSP, GENUS (SURFACE), INFLECTION POINT, NODE (ALGEBRAIC CURVE), STATIONARY TANGENT

Plücker's Conoid

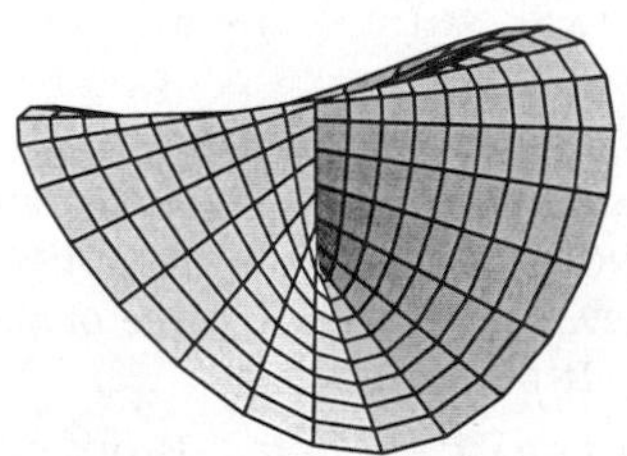

A RULED SURFACE sometimes also called the CYLINDROID. von Seggern (1993) gives the general functional form as

$$ax^2 + by^2 - zx^2 - zy^2 = 0, \tag{1}$$

whereas Fischer (1986) and Gray (1993) give

$$z = \frac{2xy}{(x^2 + y^2)}. \tag{2}$$

A polar parameterization therefore gives

$$x(r, \theta) = r \cos\theta \tag{3}$$

$$y(r, \theta) = r \sin\theta \tag{4}$$

$$z(r, \theta) = 2 \cos\theta \sin\theta. \tag{5}$$

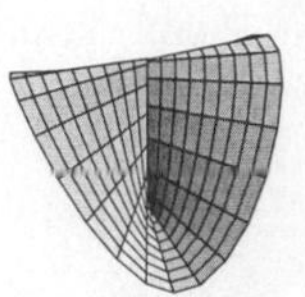
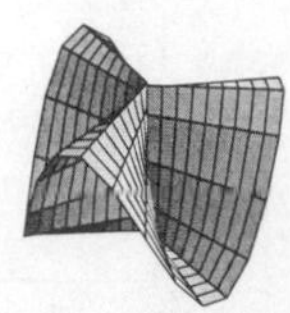
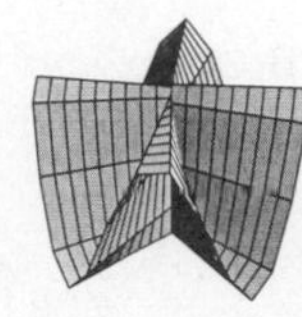

A generalization of Plücker's conoid to n folds is given by

$$x(r,\theta) = r\cos\theta \tag{6}$$
$$y(r,\theta) = r\sin\theta \tag{7}$$
$$z(r,\theta) = \sin(n\theta) \tag{8}$$

(Gray 1993). The cylindroid is the inversion of the CROSS-CAP (Pinkall 1986).

see also CROSS-CAP, RIGHT CONOID, RULED SURFACE

References

Fischer, G. (Ed.). *Mathematical Models from the Collections of Universities and Museums.* Braunschweig, Germany: Vieweg, pp. 4–5, 1986.
Gray, A. *Modern Differential Geometry of Curves and Surfaces.* Boca Raton, FL: CRC Press, pp. 337–339, 1993.
Pinkall, U. *Mathematical Models from the Collections of Universities and Museums* (Ed. G. Fischer). Braunschweig, Germany: Vieweg, p. 64, 1986.
von Seggern, D. *CRC Standard Curves and Surfaces.* Boca Raton, FL: CRC Press, p. 288, 1993.

Plücker's Equations

Relationships between the number of SINGULARITIES of plane algebraic curves. Given a PLANE CURVE,

$$m = n(n-1) - 2\delta - 3\kappa \tag{1}$$
$$n = m(m-1) - 2\tau - 3\iota \tag{2}$$
$$\iota = 3n(n-2) - 6\delta - 8\kappa \tag{3}$$
$$\kappa = 3m(m-2) - 6\tau - 8\iota, \tag{4}$$

where m is the CLASS, n the ORDER, δ the number of NODES, κ the number of CUSPS, ι the number of STATIONARY TANGENTS (INFLECTION POINTS), and τ the number of BITANGENTS. Only three of these equations are LINEARLY INDEPENDENT.

see also ALGEBRAIC CURVE, BIOCHE'S THEOREM, BITANGENT, CUSP, GENUS (SURFACE), INFLECTION POINT, KLEIN'S EQUATION, NODE (ALGEBRAIC CURVE), STATIONARY TANGENT

References

Boyer, C. B. *A History of Mathematics.* New York: Wiley, pp. 581–582, 1968.
Coolidge, J. L. *A Treatise on Algebraic Plane Curves.* New York: Dover, pp. 99–118, 1959.

Plücker Relations

see PLÜCKER'S EQUATIONS

Plumbing

The plumbing of a p-sphere and a q-sphere is defined as the disjoint union of $\mathbb{S}^p \times \mathbb{S}^q$ and $\mathbb{D}^p \times \mathbb{S}^q$ with their common $\mathbb{D}^p \times \mathbb{D}^q$, identified via the identity homeomorphism.

see also HYPERSPHERE

References

Rolfsen, D. *Knots and Links.* Wilmington, DE: Publish or Perish Press, p. 180, 1976.

Pluperfect Number

see MULTIPLY PERFECT NUMBER

Plurisubharmonic Function

An upper semicontinuous function whose restrictions to all COMPLEX lines are subharmonic (where defined). These functions were introduced by P. Lelong and Oka in the early 1940s. Examples of such a function are the logarithms of moduli of holomorphic functions.

References

Range, R. M. and Anderson, R. W. "Hans-Joachim Bremermann, 1926–1996." *Not. Amer. Math. Soc.* **43**, 972–976, 1996.

Plus

The ADDITION of two quantities, i.e., a plus b. The operation is denoted $a+b$, and the symbol $+$ is called the PLUS SIGN. Floating point ADDITION is sometimes denoted $\oplus$.

see also ADDITION, MINUS, PLUS OR MINUS, TIMES

Plus or Minus

The symbol $\pm$ is used to denote a quantity which should be both added and subtracted, as in $a \pm b$. The symbol can be used to denote a range of uncertainty, or to denote a pair of quantities, such as the roots given by the QUADRATIC FORMULA

$$x_\pm = \frac{-b \pm \sqrt{b^2 - 4ac}}{2a}.$$

When order is relevant, the symbol $a \mp b$ is also used, so an expression of the form $x \pm y \mp z$ is interpreted as $x+y-z$ or $x-y+z$. In contrast, the expression $x \pm y \pm z$ is interpreted to mean the set of four quantities $x+y+z$, $x-y+z$, $x+y-z$, and $x-y-z$.

see also MINUS, MINUS SIGN, PLUS, PLUS SIGN, SIGN

Plus Perfect Number

see ARMSTRONG NUMBER

Plus Sign

The symbol "+" which is used to denote a POSITIVE number or to indicate ADDITION.

see also ADDITION, MINUS SIGN, SIGN

Plutarch Numbers

In *Moralia,* the Greek biographer and philosopher Plutarch states "Chrysippus says that the number of compound propositions that can be made from only ten simple propositions exceeds a million. (Hipparchus, to be sure, refuted this by showing that on the affirmative side there are 103,049 compound statements, and on the negative side 310,952.)" These numbers are known as the Plutarch numbers. 103,049 can be interpreted as the number s_{10} of BRACKETINGS on ten letters (Stanley 1997), Habsieger *et al.* 1998). Similarly, Plutarch's second number is given by $(s_{10}+s_{11})/2 = 310,954$ (Habsieger *et al.* 1998).

References

Biermann, K.-R. and Mau, J. "Überprüfung einer frühen Anwendung der Kombinatorik in der Logik." *J. Symbolic Logic* **23**, 129–132, 1958.

Biggs, N. L. "The Roots of Combinatorics." *Historia Mathematica* **6**, 109–136, 1979.

Habsieger, L.; Kazarian, M.; and Lando, S. "On the Second Number of Plutarch." *Amer. Math. Monthly* **105**, 446, 1998.

Heath, T. L. *A History of Greek Mathematics, Vol. 2: From Aristarchus to Diophantus.* New York: Dover, p. 256, 1981.

Kneale, W. and Kneale, M. *The Development of Logic.* Oxford, England: Oxford University Press, p. 162, 1971.

Neugebauer, O. *A History of Ancient Mathematical Astronomy, Vol. 1.* New York: Springer-Verlag, p. 338, 1975.

Plutarch. §VIII.9 in *Moralia, Vol. 9.* Cambridge, MA: Harvard University Press, p. 732, 1961.

Stanley, R. P. *Enumerative Combinatorics, Vol. 1.* Cambridge, England: Cambridge University Press, p. 63, 1996.

Stanley, R. P. "Hipparchus, Plutarch, Schröder, and Hough." *Amer. Math. Monthly* **104**, 344–350, 1997.

Pochhammer Symbol

A.k.a. RISING FACTORIAL. For an INTEGER $n > 0$,

$$(a)_n \equiv \frac{\Gamma(a+k)}{\Gamma(a)} = a(a+1)\cdots(a+n-1), \quad (1)$$

where $\Gamma(z)$ is the GAMMA FUNCTION and

$$(a)_0 \equiv 1. \quad (2)$$

The NOTATION conflicts with both that for q-SERIES and that for GAUSSIAN COEFFICIENTS, so context usually serves to distinguish the three. Additional identities are

$$\frac{d}{da}(a)_n = (a)_n[F(a+n-1) - F(a-1)] \quad (3)$$

$$(a)_{n+k} = (a+n)_k(a)_n, \quad (4)$$

where F is the DIGAMMA FUNCTION. The Pochhammer symbol arises in series expansions of HYPERGEOMETRIC FUNCTIONS and GENERALIZED HYPERGEOMETRIC FUNCTIONS.

see also FACTORIAL, GENERALIZED HYPERGEOMETRIC FUNCTION, HARMONIC LOGARITHM, HYPERGEOMETRIC FUNCTION

References

Abramowitz, M. and Stegun, C. A. (Eds.). *Handbook of Mathematical Functions with Formulas, Graphs, and Mathematical Tables, 9th printing.* New York: Dover, p. 256, 1972.

Spanier, J. and Oldham, K. B. "The Pochhammer Polynomials $(x)_n$." Ch. 18 in *An Atlas of Functions.* Washington, DC: Hemisphere, pp. 149–165, 1987.

Pocklington's Criterion

Let p be an ODD PRIME, k be an INTEGER such that $p \nmid k$ and $1 \leq k \leq 2(p+1)$, and

$$N \equiv 2kp + 1.$$

Then the following are equivalent

1. N is PRIME.
2. $\text{GCD}(a^k + 1, N) = 1$.

This is a modified version of the original theorem due to Lehmer.

References

Pocklington, H. C. "The Determination of the Prime or Composite Nature of Large Numbers by Fermat's Theorem." *Proc. Cambridge Phil. Soc.* **18**, 29–30, 1914/16.

Pocklington-Lehmer Test

see POCKLINGTON'S THEOREM

Pocklington's Theorem

Let $n-1 = FR$ where F is the factored part of a number

$$F = {p_1}^{a_1} \cdots {p_r}^{a_r}, \quad (1)$$

where $(R, F) = 1$, and $R < \sqrt{n}$. If there exists a b_i for $i = 1, \ldots, r$ such that

$${b_i}^{n-1} \equiv 1 \pmod{n} \quad (2)$$

$$\text{GCD}({b_i}^{(n-1)/p_i} - 1, n) = 1, \quad (3)$$

then n is a PRIME.

Poggendorff Illusion

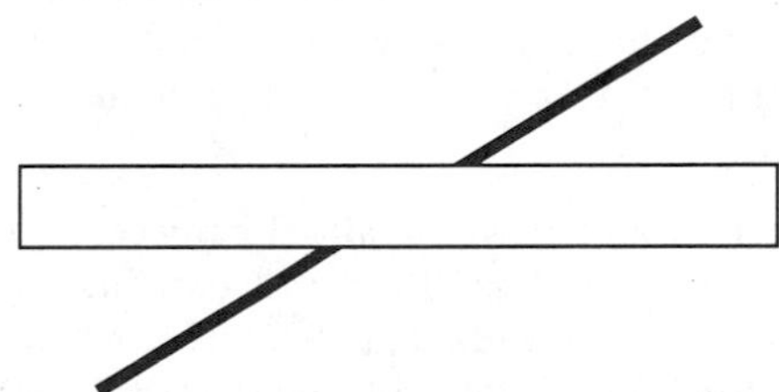

The illusion that the two ends of a straight LINE SEGMENT passing behind an obscuring RECTANGLE are offset when, in fact, they are aligned.

see also ILLUSION, MÜLLER-LYER ILLUSION, PONZO'S ILLUSION, VERTICAL-HORIZONTAL ILLUSION

References

Burmester, E. "Beiträge zu experimentellen Bestimmung geometrisch-optischer Täuschungen." *Z. Psychologie* **12**, 355–394, 1896.

Day, R. H. and Dickenson, R. G. "The Components of the Poggendorff Illusion." *Brit. J. Psychology* **67**, 537–552, 1976.

Fineman, M. "Poggendorff's Illusion." Ch. 19 in *The Nature of Visual Illusion.* New York: Dover, pp. 151–159, 1996.

Pohlke's Theorem

The principal theorem of AXONOMETRY. It states that three segments of arbitrary length $a'x'$, $a'y'$, and $a'z'$ which are drawn in a PLANE from a point a' under arbitrary ANGLES form a parallel projection of three equal segments ax, ay, and az from the ORIGIN of three PERPENDICULAR coordinate axes. However, only one of the segments or one of the ANGLES may vanish.

see also AXONOMETRY

Poincaré-Birkhoff Fixed Point Theorem

For the rational curve of an unperturbed system with ROTATION NUMBER r/s under a map T (for which every point is a FIXED POINT of T^s), only an even number of FIXED POINTS $2ks$ ($k = 1, 2, \ldots$) will remain under perturbation. These FIXED POINTS are alternately stable (ELLIPTIC) and unstable (HYPERBOLIC). Around each elliptic fixed point there is a simultaneous application of the Poincaré-Birkhoff fixed point theorem and the KAM THEOREM, which leads to a self-similar structure on all scales.

The original formulation was: Given a CONFORMAL ONE-TO-ONE transformation from an ANNULUS to itself that advances points on the outer edge positively and on the inner edge negatively, then there are at least two fixed points.

It was conjectured by Poincaré from a consideration of the three-body problem in celestial mechanics and proved by Birkhoff.

Poincaré Conjecture

A SIMPLY CONNECTED 3-MANIFOLD is HOMEOMORPHIC to the 3-SPHERE. The generalized Poincaré conjecture is that a COMPACT n-MANIFOLD is HOMOTOPY equivalent to the n-sphere IFF it is HOMEOMORPHIC to the n-SPHERE. This reduces to the original conjecture for $n = 3$.

The $n = 1$ case of the generalized conjecture is trivial, the $n = 2$ case is classical, $n = 3$ remains open, $n = 4$ was proved by Freedman (1982) (for which he was awarded the 1986 FIELDS MEDAL), $n = 5$ by Zeeman (1961), $n = 6$ by Stallings (1962), and $n \geq 7$ by Smale in 1961 (Smale subsequently extended this proof to include $n \geq 5$.)

see also COMPACT MANIFOLD, HOMEOMORPHIC, HOMOTOPY, MANIFOLD, SIMPLY CONNECTED, SPHERE, THURSTON'S GEOMETRIZATION CONJECTURE

References

Freedman, M. H. "The Topology of Four-Differentiable Manifolds." *J. Diff. Geom.* **17**, 357–453, 1982.

Stallings, J. "The Piecewise-Linear Structure of Euclidean Space." *Proc. Cambridge Philos. Soc.* **58**, 481–488, 1962.

Smale, S. "Generalized Poincaré's Conjecture in Dimensions Greater than Four." *Ann. Math.* **74**, 391–406, 1961.

Zeeman, E. C. "The Generalised Poincaré Conjecture." *Bull. Amer. Math. Soc.* **67**, 270, 1961.

Zeeman, E. C. "The Poincaré Conjecture for $n \geq 5$." In *Topology of 3-Manifolds and Related Topics, Proceedings of the University of Georgia Institute, 1961.* Englewood Cliffs, NJ: Prentice-Hall, pp. 198–204, 1961.

Poincaré Duality

The BETTI NUMBERS of a compact orientable n-MANIFOLD satisfy the relation

$$b_i = b_{n-i}.$$

see also BETTI NUMBER

Poincaré Formula

The POLYHEDRAL FORMULA generalized to a surface of GENUS p.

$$V - E + F = 2 - 2p$$

where V is the number of VERTICES, E is the number of EDGES, F is the number of faces, and

$$\chi \equiv 2 - 2p$$

is called the EULER CHARACTERISTIC.

see also EULER CHARACTERISTIC, GENUS (SURFACE), POLYHEDRAL FORMULA

References

Eppstein, D. "Fourteen Proofs of Euler's Formula: $V - E + F = 2$." http://www.ics.uci.edu/~eppstein/junkyard/euler.

Poincaré-Fuchs-Klein Automorphic Function

$$f(z) = \frac{k}{(cz+d)^r} f\left(\frac{az+b}{cz+d}\right)$$

where $\Im(z) > 0$.

see also AUTOMORPHIC FUNCTION

Poincaré Group

see LORENTZ GROUP

Poincaré's Holomorphic Lemma

Solutions to HOLOMORPHIC differential equations are themselves HOLOMORPHIC FUNCTIONS of time, initial conditions, and parameters.

Poincaré-Hopf Index Theorem

The index of a VECTOR FIELD with finitely many zeros on a compact, oriented MANIFOLD is the same as the EULER CHARACTERISTIC of the MANIFOLD.

see also GAUSS-BONNET FORMULA

Poincaré Hyperbolic Disk

A 2-D space having HYPERBOLIC GEOMETRY defined as the 2-BALL $\{x \in \mathbb{R}^2 : |x| < 1\}$, with HYPERBOLIC METRIC

$$\frac{dx^2 + dy^2}{(1 - r^2)^2}.$$

The Poincaré disk is a model for HYPERBOLIC GEOMETRY, and there is an isomorphism between the Poincaré disk model and the KLEIN-BELTRAMI MODEL.

see also ELLIPTIC PLANE, HYPERBOLIC GEOMETRY, HYPERBOLIC METRIC, KLEIN-BELTRAMI MODEL

Poincaré's Lemma

Let $\wedge$ denote the WEDGE PRODUCT and D the EXTERIOR DERIVATIVE. Then

$$D^2 t = \frac{\partial}{\partial x} \wedge \left(\frac{\partial}{\partial x} \wedge t\right) = \left(\frac{\partial}{\partial x} \wedge \frac{\partial}{\partial x}\right) \wedge t = 0.$$

see also DIFFERENTIAL FORM, EXTERIOR DERIVATIVE, POINCARÉ'S HOLOMORPHIC LEMMA, WEDGE PRODUCT

Poincaré Manifold

A nonsimply connected 3-manifold also called a DODECAHEDRAL SPACE.

References

Rolfsen, D. *Knots and Links.* Wilmington, DE: Publish or Perish Press, pp. 245, 290, and 308, 1976.

Poincaré Metric

The METRIC

$$ds^2 = \frac{dx^2 + dy^2}{(1 - |z|^2)^2}$$

of the POINCARÉ HYPERBOLIC DISK.

see also POINCARÉ HYPERBOLIC DISK

Poincaré Separation Theorem

Let $\{\mathbf{y}^k\}$ be a set of orthonormal vectors with $k = 1, 2, \ldots, K$, such that the INNER PRODUCT $(\mathbf{y}^k, \mathbf{y}^k) = 1$. Then set

$$\mathbf{x} = \sum_{k=1}^{K} u_k \mathbf{y}^k \tag{1}$$

so that for any SQUARE MATRIX A for which the product $\mathsf{A}\mathbf{x}$ is defined, the corresponding QUADRATIC FORM is

$$(\mathbf{x}, \mathsf{A}\mathbf{x}) = \sum_{k,l=1}^{K} u_k u_l (\mathbf{y}^k, \mathsf{A}\mathbf{y}^l). \tag{2}$$

Then if

$$\mathbf{B}_k = (\mathbf{y}^k, \mathsf{A}\mathbf{y}^l) \tag{3}$$

for $k, l = 1, 2, \ldots, K$, it follows that

$$\lambda_i(\mathbf{B}_K) \leq \lambda_1(\mathsf{A}) \tag{4}$$

$$\lambda_{K-j}(\mathbf{B}_K) \geq \lambda_{N-j}(\mathsf{A}) \tag{5}$$

for $i = 1, 2, \ldots, K$ and $j = 0, 1, \ldots, K-1$.

References

Gradshteyn, I. S. and Ryzhik, I. M. *Tables of Integrals, Series, and Products, 5th ed.* San Diego, CA: Academic Press, p. 1120, 1979.

Poinsot Solid

see KEPLER-POINSOT SOLID

Poinsot's Spirals

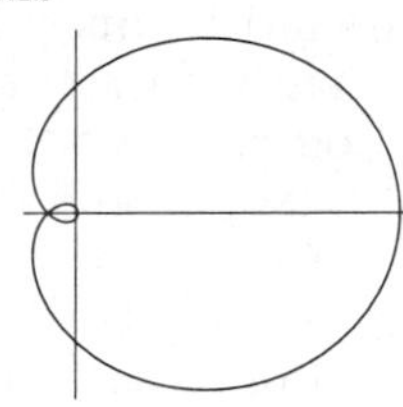

$r \sinh(n\theta) = a.$

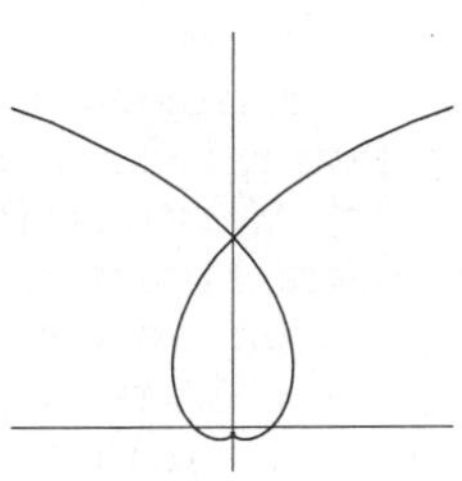

$r \operatorname{csch}(n\theta) = a.$

References

Lawrence, J. D. *A Catalog of Special Plane Curves.* New York: Dover, pp. 192 and 194, 1972.

Point

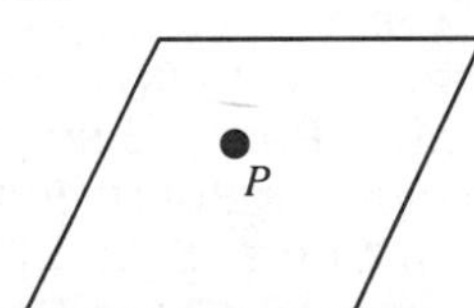

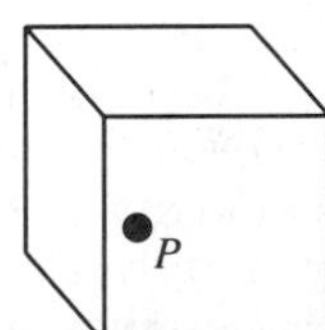

A 0-DIMENSIONAL mathematical object which can be specified in n-D space using n coordinates. Although the notion of a point is intuitively rather clear, the mathematical machinery used to deal with points and point-like objects can be surprisingly slippery. This difficulty was encountered by none other than Euclid himself who, in his *Elements*, gave the vague definition of a point as "that which has no part."

The basic geometric structures of higher DIMENSIONAL geometry—the LINE, PLANE, SPACE, and HYPERSPACE—are all built up of infinite numbers of points arranged in particular ways.

see also ACCUMULATION POINT, ANTIGONAL POINTS, ANTIHOMOLOGOUS POINTS, APOLLONIUS POINT, BOUNDARY POINT, BRANCH POINT, BRIANCHON POINT, BROCARD MIDPOINT, BROCARD POINTS,

Cantor-Dedekind Axiom, Center, Circle Lattice Points, Concur, Concurrent, Congruent Incircles Point, Congruent Isoscelizers Point, Conjugate Points, Critical Point, Crucial Point, Cube Point Picking, Cusp Point, de Longchamps Point, Double Point, Eckardt Point, Elkies Point, Elliptic Fixed Point (Differential Equations), Elliptic Fixed Point (Map), Elliptic Point, Equal Detour Point, Equal Parallelians Point, Equichordal Point, Equilibrium Point, Equiproduct Point, Equireciprocal Point, Evans Point, Exeter Point, Exmedian Point, Fagnano's Point, Far-Out Point, Fejes Tóth's Problem, Fermat Point, Feuerbach Point, Feynman Point, Fixed Point, Fletcher Point, Gergonne Point, Grebe Point, Griffiths Points, Harmonic Conjugate Points, Hermit Point, Hofstadter Point, Homologous Points, Hyperbolic Fixed Point (Differential Equations), Hyperbolic Fixed Point (Map), Hyperbolic Point, Ideal Point, Imaginary Point, Invariant Point, Inverse Points, Isodynamic Points, Isolated Point, Isoperimetric Point, Isotomic Conjugate Point, Lattice Point, Lemoine Point, Limit Point, Malfatti Points, Median Point, Mid-Arc Points, Midpoint, Miquel Point, Nagel Point, Napoleon Points, Nobbs Points, Oldknow Points, Only Critical Point in Town Test, Ordinary Point, Parabolic Point, Parry Point, Pedal Point, Periodic Point, Planar Point, Point at Infinity, Point-Line Distance—2-D, Point-Line Distance—3-D, Point-Quadratic Distance, Point-Plane Distance, Point-Set Topology, Pointwise Dimension, Policeman on Point Duty Curve, Power Point, Radial Point, Radiant Point, Rational Point, Rigby Points, Saddle Point (Game), Saddle Point (Function), Salient Point, Schiffler Point, Self-Homologous Point, Similarity Point, Singular Point (Algebraic Curve), Singular Point (Function), Soddy Points, Special Point, Stationary Point, Steiner Points, Sylvester's Four-Point Problem, Symmedian Point, Symmetric Points, Tarry Point, Torricelli Point, Trisected Perimeter Point, Umbilic Point, Unit Point, Vanishing Point, Visible Point, Weierstraß Point, Wild Point, Yff Points

References

Casey, J. "The Point." Ch. 1 in *A Treatise on the Analytical Geometry of the Point, Line, Circle, and Conic Sections, Containing an Account of Its Most Recent Extensions, with Numerous Examples, 2nd ed., rev. enl.* Dublin: Hodges, Figgis, & Co., pp. 1–29, 1893.

Point Estimator

An Estimator of the actual values of population.

Point Groups

The symmetry groups possible in a crystal lattice without the translation symmetry element. Although an isolated object may have an arbitrary Schönflies Symbol, the requirement that symmetry be present in a lattice requires that only 1, 2, 3, and 6-fold symmetry axes are possible (the Crystallography Restriction), which restricts the number of possible point groups to 32: C_i, C_s, C_1, C_2, C_3, C_4, C_6, C_{2h}, C_{3h}, C_{4h}, C_{6h}, C_{2v}, C_{3v}, C_{4v}, C_{6v}, D_2, D_3, D_4, D_6 (the Dihedral Groups), D_{2h}, D_{3h}, D_{4h}, D_{6h}, D_{2d}, D_{3d}, O, O_h (the Octahedral Group), S_4, S_6, T, T_h, and T_d (the Tetrahedral Group).

see also Crystallography Restriction, Dihedral Group, Group, Group Theory, Hermann-Mauguin Symbol, Lattice Groups, Octahedral Group, Schönflies Symbol, Space Groups, Tetrahedral Group

References

Arfken, G. "Crystallographic Point and Space Groups." *Mathematical Methods for Physicists, 3rd ed.* Orlando, FL: Academic Press, p. 248–249, 1985.

Cotton, F. A. *Chemical Applications of Group Theory, 3rd ed.* New York: Wiley, p. 379, 1990.

Lomont, J. S. "Crystallographic Point Groups." §4.4 in *Applications of Finite Groups.* New York: Dover, pp. 132–146, 1993.

Point at Infinity

P is the point on the line AB such that $\overline{PA}/\overline{PB} = 1$. It can also be thought of as the point of intersection of two Parallel lines.

see also Line at Infinity

References

Behnke, H.; Bachmann, F.; Fladt, K.; and Suss, W. (Eds.). Ch. 7 in *Fundamentals of Mathematics, Vol. 3: Points at Infinity.* Cambridge, MA: MIT Press, 1974.

Point-Line Distance—2-D

Given a line $ax + by + c = 0$ and a point (x_0, y_0), in slope-intercept form, the equation of the line is

$$y = -\frac{a}{b}x - \frac{c}{b}, \tag{1}$$

so the line has Slope $-a/b$. Points on the line have the vector coordinates

$$\begin{bmatrix} x \\ -\frac{a}{b}x - \frac{c}{d} \end{bmatrix} = \begin{bmatrix} 0 \\ -\frac{c}{d} \end{bmatrix} - \frac{1}{b}\begin{bmatrix} -b \\ a \end{bmatrix}. \tag{2}$$

Therefore, the Vector

$$\begin{bmatrix} -b \\ a \end{bmatrix} \tag{3}$$

is Parallel to the line, and the Vector

$$\mathbf{v} = \begin{bmatrix} a \\ b \end{bmatrix} \tag{4}$$

is PERPENDICULAR to it. Now, a VECTOR from the point to the line is given by

$$\mathbf{r} = \begin{bmatrix} x - x_0 \\ y - y_0 \end{bmatrix}. \tag{5}$$

Projecting $\mathbf{r}$ onto $\mathbf{v}$,

$$\begin{aligned} |\text{proj}_{\mathbf{v}}\mathbf{r}| &= \frac{|\mathbf{v}\cdot\mathbf{r}|}{|\mathbf{v}|} = |\hat{\mathbf{v}}\cdot\mathbf{r}| = \frac{|a(x-x_0)+b(y-y_0)|}{\sqrt{a^2+b^2}} \\ &= \frac{|ax+by-ax_0-by_0|}{\sqrt{a^2+b^2}} \\ &= \frac{|ax_0+by_0+c|}{\sqrt{a^2+b^2}}. \end{aligned} \tag{6}$$

If the line is represented by the endpoints of a VECTOR (x_1, y_1) and (x_2, y_2), then the PERPENDICULAR VECTOR is

$$\mathbf{v} = \begin{bmatrix} y_2 - y_1 \\ -(x_2 - x_1) \end{bmatrix} \tag{7}$$

$$\hat{\mathbf{v}} = \frac{1}{s}\begin{bmatrix} y_2 - y_1 \\ -(x_2 - x_1) \end{bmatrix}, \tag{8}$$

where

$$s = |\mathbf{v}| = \sqrt{(x_2-x_1)^2 + (y_2-y_1)^2}, \tag{9}$$

so the distance is

$$d = |\hat{\mathbf{v}}\cdot\mathbf{r}| = \frac{|(y_2-y_1)(x_0-x_1)-(x_2-x_1)(y_0-y_1)|}{s}. \tag{10}$$

The distance from a point (x_1,y_1) to the line $y = a + bx$ can be computed using VECTOR algebra. Let $\mathbf{L}$ be a VECTOR in the same direction as the line

$$\mathbf{L} = \begin{bmatrix} x \\ a+bx \end{bmatrix} - \begin{bmatrix} 0 \\ a \end{bmatrix} = \begin{bmatrix} x \\ bx \end{bmatrix} \tag{11}$$

$$\hat{\mathbf{L}} = \frac{1}{\sqrt{b^2+1}}\begin{bmatrix} 1 \\ b \end{bmatrix}. \tag{12}$$

A given point on the line is

$$\mathbf{x} = \begin{bmatrix} x_1 \\ y_1 \end{bmatrix} - \begin{bmatrix} 0 \\ -a \end{bmatrix} = \begin{bmatrix} x_1 \\ y_1 - a \end{bmatrix}, \tag{13}$$

so the point-line distance is

$$\begin{aligned} \mathbf{r} &= (\mathbf{x}\cdot\hat{\mathbf{L}})\hat{\mathbf{L}} - \mathbf{x} \\ &= \frac{1}{1+b^2}\left(\begin{bmatrix} x_1 \\ y_1 - a \end{bmatrix}\cdot\begin{bmatrix} 1 \\ v \end{bmatrix}\right)\begin{bmatrix} 1 \\ b \end{bmatrix} - \begin{bmatrix} x_1 \\ y_1 - a \end{bmatrix} \\ &= \frac{x_1 + b(y_1-a)}{1+b^2}\begin{bmatrix} 1 \\ b \end{bmatrix} - \begin{bmatrix} x_1 \\ y_1 - a \end{bmatrix} \\ &= \frac{1}{1+b^2}\begin{bmatrix} b(y_1-a) - b^2x_1 \\ bx_1 + b^2y_1 - ab^2 - y_1 + a - b^2y_1 + ab^2 \end{bmatrix} \\ &= \frac{1}{1+b^2}\begin{bmatrix} b[(y_1-a) - bx_1] \\ -[(y_1-a) - bx_1] \end{bmatrix} \\ &= \frac{y_1 - (a+bx_1)}{1+b^2}\begin{bmatrix} b \\ -1 \end{bmatrix}. \end{aligned} \tag{14}$$

Therefore,

$$d = |\mathbf{r}| = \frac{|y_1-(a+bx_1)|}{1+b^2}\sqrt{1+b^2} = \frac{|y_1-(a+bx_1)|}{\sqrt{1+b^2}}. \tag{15}$$

This result can also be obtained much more simply by noting that the PERPENDICULAR distance is just $\cos\theta$ times the vertical distance $|y_1 - (a+bx_1)|$. But the SLOPE b is just $\tan\theta$, so

$$\sin^2\theta + \cos^2\theta = 1 \Rightarrow \tan^2\theta + 1 = \frac{1}{\cos^2\theta}, \tag{16}$$

and

$$\cos\theta = \frac{1}{\sqrt{1+\tan^2\theta}} = \frac{1}{\sqrt{1+b^2}}. \tag{17}$$

The PERPENDICULAR distance is then

$$d = \frac{|y_1-(a+bx_1)|}{\sqrt{1+b^2}}, \tag{18}$$

the same result as before.

see also LINE, POINT, POINT-LINE DISTANCE—3-D

Point-Line Distance—3-D

A line in 3-D is given by the parametric VECTOR

$$\mathbf{v} = \begin{bmatrix} x_0 + at \\ y_0 + bt \\ z_0 + ct \end{bmatrix}. \tag{1}$$

The distance between a point on the line with parameter t and the point (x_1, y_1, z_1) is therefore

$$r^2 = (x_1-x_0-at)^2 + (y_1-y_0-bt)^2 + (z_1-z_0-ct)^2. \tag{2}$$

To minimize the distance, take

$$\begin{aligned} \frac{\partial(r^2)}{\partial t} &= -2a(x_1-x_0-at) - 2b(y_1-y_0-bt) \\ &\quad -2c(z_1-z_0-ct) = 0 \end{aligned} \tag{3}$$

$$a(x_1-x_0)+b(y_1-y_0)+c(z_1-z_0)-t(a^2+b^2+c^2) = 0 \tag{4}$$

$$t = \frac{a(x_1-x_0)+b(y_1-y_0)+c(z_1-z_0)}{a^2+b^2+c^2}, \tag{5}$$

so the minimum distance is found by plugging (5) into (2) and taking the SQUARE ROOT.

see also LINE, POINT, POINT-LINE DISTANCE—2-D

Point Picking

see 18-POINT PROBLEM, BALL TRIANGLE PICKING, CUBE POINT PICKING, CUBE TRIANGLE PICKING, DISCREPANCY THEOREM, ISOSCELES TRIANGLE, OBTUSE TRIANGLE, PLANAR DISTANCE, SYLVESTER'S FOUR-POINT PROBLEM

Point-Plane Distance

Given a PLANE

$$ax + by + cz + d = 0 \tag{1}$$

and a point (x_0, y_0, z_0), the NORMAL to the PLANE is given by

$$\mathbf{v} = \begin{bmatrix} a \\ b \\ c \end{bmatrix}, \tag{2}$$

and a VECTOR from the plane to the point is given by

$$\mathbf{w} = \begin{bmatrix} x - x_0 \\ y - y_0 \\ z - z_0 \end{bmatrix}. \tag{3}$$

Projecting $\mathbf{w}$ onto $\mathbf{v}$,

$$\begin{aligned} |\text{proj}_{\mathbf{v}}\mathbf{w}| &= \frac{|\mathbf{v} \cdot \mathbf{w}|}{|\mathbf{v}|} \\ &= \frac{|a(x-x_0) + b(y-y_0) + c(z-z_0) + d|}{\sqrt{a^2+b^2+c^2}} \\ &= \frac{|ax + by + cz - ax_0 - by_0 - cz_0|}{\sqrt{a^2+b^2+c^2}} \\ &= \frac{|ax_0 + by_0 + cz_0 + d|}{\sqrt{a^2+b^2+c^2}}. \end{aligned} \tag{4}$$

Point-Point Distance—1-D

Given a unit LINE SEGMENT $[0, 1]$, pick two points at random on it. Call the first point x_1 and the second point x_2. Find the distribution of distances d between points. The probability of the points being a (POSITIVE) distance d apart (i.e., without regard to ordering) is given by

$$\begin{aligned} P(d) &= \frac{\int_0^1 \int_0^1 \delta(d - |x_2 - x_1|)\,dx_1\,dx_2}{\int_0^1 \int_0^1 dx_1\,dx_2} \\ &= (1-d)[H(1-d) - H(d-1) + H(d) - H(-d)] \\ &= \begin{cases} 2(1-d) & \text{for } 0 \le d \le 1 \\ 0 & \text{otherwise,} \end{cases} \end{aligned} \tag{1}$$

where δ is the DIRAC DELTA FUNCTION and H is the HEAVISIDE STEP FUNCTION. The MOMENTS are then

$$\begin{aligned} \mu'_m &= \int_0^1 d^m P(d)\,dd = 2\int_0^1 d^m(1-d)\,dd \\ &= 2\left[\frac{d^{m+1}}{m+1} - \frac{d^{m+2}}{m+2}\right]_0^1 \\ &= 2\left(\frac{1}{m+1} - \frac{1}{m+2}\right) = 2\left[\frac{(m+2)-(m+1)}{(m+1)(m+2)}\right] \\ &= \frac{2}{(m+1)(m+2)} \\ &= \begin{cases} \frac{1}{(n+1)(2n+1)} & \text{for } m = 2n \\ \frac{1}{(n+1)(2n+3)} & \text{for } m = 2n+1, \end{cases} \end{aligned} \tag{2}$$

giving MOMENTS about 0

$$\mu'_1 = \tfrac{1}{3} \tag{3}$$

$$\mu'_2 = \tfrac{1}{6} \tag{4}$$

$$\mu'_3 = \tfrac{1}{10} \tag{5}$$

$$\mu'_4 = \tfrac{1}{15}. \tag{6}$$

The MOMENTS can also be computed directly without explicit knowledge of the distribution

$$\begin{aligned} \mu'_1 &= \frac{\int_0^1 \int_0^1 |x_2 - x_1|\,dx_1\,dx_2}{\int_0^1 \int_0^1 dx_1\,dx_2} \\ &= \int_0^1 \int_0^1 |x_2 - x_1|\,dx_1\,dx_2 \\ &= \underset{x_2 - x_1 > 0}{\int_0^1 \int_0^1} (x_2 - x_1)\,dx_1\,dx_2 \\ &\quad + \underset{x_2 - x_1 < 0}{\int_0^1 \int_0^1} (x_1 - x_2)\,dx_1\,dx_2 \\ &= \int_0^1 \int_{x_1}^1 (x_2 - x_1)\,dx_1\,dx_2 \\ &\quad + \int_0^1 \int_0^{x_1} (x_2 - x_1)\,dx_1\,dx_2 \\ &= \int_0^1 \left[\tfrac{1}{2}{x_2}^2 - x_1 x_2\right]_{x_1}^1 dx_1 \\ &\quad + \int_0^1 \left[x_1 x_2 - \tfrac{1}{2}{x_2}^2\right]_0^{x_1} dx_1 \\ &= \int_0^1 \left[(\tfrac{1}{2} - x_1) - (\tfrac{1}{2}{x_1}^2 - {x_1}^2)\right] dx_1 \\ &\quad + \int_0^1 \left[({x_1}^2 - \tfrac{1}{2}{x_1}^2) - (0 - 0)\right] dx_1 \\ &= \int_0^1 (\tfrac{1}{2} - x_1 + {x_1}^2)\,dx_1 = [\tfrac{1}{2}x_1 - \tfrac{1}{2}{x_1}^2 + \tfrac{1}{3}{x_1}^3]_0^1 \\ &= (\tfrac{1}{2} - \tfrac{1}{2} + \tfrac{1}{3}) - (0 - 0 + 0) = \tfrac{1}{3} \end{aligned} \tag{7}$$

$$\begin{aligned} \mu'_2 &= \int_0^1 \int_0^1 (|x_2 - x_1|)^2\,dx_2\,dx_1 \\ &= \int_0^1 \int_0^1 (x_2 - x_1)^2\,dx_1\,dx_2 \\ &= \int_0^1 \int_0^1 ({x_2}^2 - 2x_1x_2 + {x_1}^2)\,dx_1\,dx_2 \\ &= \int_0^1 [\tfrac{1}{3}{x_2}^3 - x_1{x_2}^2 + {x_1}^2 x_2]_0^1\,dx_1 \\ &= \int_0^1 (\tfrac{1}{3} - x_1 + {x_1}^2)\,dx_1 = [\tfrac{1}{3}{x_1}^3 - \tfrac{1}{2}{x_1}^2 + \tfrac{1}{3}x_1]_0^1 \\ &= \tfrac{1}{3} - \tfrac{1}{2} + \tfrac{1}{3} = \tfrac{1}{6}. \end{aligned} \tag{8}$$

The MOMENTS about the MEAN are therefore

$$\mu_2 = \mu_2' - {\mu_1'}^2 = \tfrac{1}{6} - (\tfrac{1}{3})^2 = \tfrac{1}{18} \tag{9}$$

$$\mu_3 = \mu_3' - 3\mu_2'\mu_1' + 2(\mu_1')^3 = \tfrac{1}{135} \tag{10}$$

$$\mu_4 = \mu_4' - 4\mu_3'\mu_1' + 6\mu_2'(\mu_1')^2 - 3(\mu_1')^4 = \tfrac{1}{135}, \tag{11}$$

so the MEAN, VARIANCE, SKEWNESS, and KURTOSIS are

$$\mu = \mu_1' = \tfrac{1}{3} \tag{12}$$

$$\sigma^2 = \mu_2 = \tfrac{1}{18} \tag{13}$$

$$\gamma_1 = \frac{\mu_3}{\sigma^3} = \tfrac{2}{5}\sqrt{2} \tag{14}$$

$$\gamma_2 = \frac{\mu_4}{\sigma^4} - 3 = -\tfrac{3}{5}. \tag{15}$$

The probability distribution of the distance between two points randomly picked on a LINE SEGMENT is germane to the problem of determining the access time of computer hard drives. In fact, the average access time for a hard drive is precisely the time required to seek across 1/3 of the tracks (Benedict 1995).

see also POINT-POINT DISTANCE—2-D, POINT-POINT DISTANCE—3-D, POINT-QUADRATIC DISTANCE, TETRAHEDRON INSCRIBING, TRIANGLE INSCRIBING IN A CIRCLE

References

Arfken, G. *Mathematical Methods for Physicists, 3rd ed.* Orlando, FL: Academic Press, pp. 930–931, 1985.

Benedict, B. *Using Norton Utilities for the Macintosh.* Indianapolis, IN: Que, pp. B-8–B-9, 1995.

Point-Point Distance—2-D

Given two points in the PLANE, find the curve which minimizes the distance between them. The LINE ELEMENT is given by

$$ds = \sqrt{dx^2 + dy^2}, \tag{1}$$

so the ARC LENGTH between the points x_1 and x_2 is

$$L = \int ds = \int_{x_1}^{x_2} \sqrt{1 + y'^2}\, dx, \tag{2}$$

where $y' \equiv dy/dx$ and the quantity we are minimizing is

$$f = \sqrt{1 + y'^2}. \tag{3}$$

Finding the derivatives gives

$$\frac{\partial f}{\partial y} = 0 \tag{4}$$

$$\frac{d}{dx}\frac{\partial f}{\partial y'} = \frac{d}{dx}[(1 + {y'}^2)^{-1/2} y'], \tag{5}$$

so the EULER-LAGRANGE DIFFERENTIAL EQUATION becomes

$$\frac{\partial f}{\partial y} - \frac{d}{dx}\frac{\partial f}{\partial y'} = \frac{d}{dx}\left(\frac{y'}{\sqrt{1 + y'^2}}\right) = 0. \tag{6}$$

Integrating and rearranging,

$$\frac{y'}{\sqrt{1 + y'^2}} = c \tag{7}$$

$$y'^2 = c^2(1 + {y'}^2) \tag{8}$$

$$y'^2(1 - c^2) = c^2 \tag{9}$$

$$y' = \frac{c}{\sqrt{1 - c^2}} \equiv a. \tag{10}$$

The solution is therefore

$$y = ax + b, \tag{11}$$

which is a straight LINE. Now verify that the ARC LENGTH is indeed the straight-line distance between the points. a and b are determined from

$$y_1 = ax_1 + b \tag{12}$$

$$y_2 = ax_2 + b. \tag{13}$$

Writing (12) and (13) as a MATRIX EQUATION gives

$$\begin{bmatrix} y_1 \\ y_2 \end{bmatrix} = \begin{bmatrix} x_1 & 1 \\ x_2 & 1 \end{bmatrix} \begin{bmatrix} a \\ b \end{bmatrix} \tag{14}$$

$$\begin{aligned} \begin{bmatrix} a \\ b \end{bmatrix} &= \begin{bmatrix} x_1 & 1 \\ x_2 & 1 \end{bmatrix}^{-1} \begin{bmatrix} y_1 \\ y_2 \end{bmatrix} \\ &= \frac{1}{x_1 - x_2} \begin{bmatrix} 1 & -1 \\ -x_2 & x_1 \end{bmatrix}^{-1} \begin{bmatrix} y_1 \\ y_2 \end{bmatrix}, \end{aligned} \tag{15}$$

so

$$a = \frac{y_1 - y_2}{x_1 - x_2} = \frac{y_2 - y_1}{x_2 - x_1} \tag{16}$$

$$b = \frac{x_1 y_2 - x_2 y_1}{x_1 - x_2} \tag{17}$$

$$\begin{aligned} L &= \int_{x_1}^{x_2} \sqrt{1 + y'^2}\, dy = (x_2 - x_1)\sqrt{1 + a^2} \\ &= (x_2 - x_1)\sqrt{1 + \left(\frac{y_2 - y_1}{x_2 - x_1}\right)^2} \\ &= \sqrt{(x_2 - x_1)^2 + (y_2 - y_1)^2}, \end{aligned} \tag{18}$$

as expected.

The shortest distance between two points on a SPHERE is the so-called GREAT CIRCLE distance.

see also CALCULUS OF VARIATIONS, GREAT CIRCLE, POINT-POINT DISTANCE—1-D, POINT-POINT DISTANCE—3-D, POINT-QUADRATIC DISTANCE, TETRAHEDRON INSCRIBING, TRIANGLE INSCRIBING IN A CIRCLE

References

Arfken, G. *Mathematical Methods for Physicists, 3rd ed.* Orlando, FL: Academic Press, pp. 930–931, 1985.

Point-Point Distance—3-D

The LINE ELEMENT is

$$ds = \sqrt{dx^2 + dy^2 + dz^2}, \quad (1)$$

so the ARC LENGTH between the points x_1 and x_2 is

$$L = \int ds = \int_{x_1}^{x_2} \sqrt{1 + y'^2 + z'^2}\, dx \quad (2)$$

and the quantity we are minimizing is

$$f = \sqrt{1 + y'^2 + z'^2}. \quad (3)$$

Finding the derivatives gives

$$\frac{\partial f}{\partial y} = 0 \quad (4)$$

$$\frac{\partial f}{\partial z} = 0 \quad (5)$$

and

$$\frac{\partial f}{\partial y'} = \frac{y'}{\sqrt{1 + y'^2 + z'^2}} \quad (6)$$

$$\frac{\partial f}{\partial z'} = \frac{z'}{\sqrt{1 + y'^2 + z'^2}}, \quad (7)$$

so the EULER-LAGRANGE DIFFERENTIAL EQUATIONS become

$$\frac{d}{dx}\left(\frac{y'}{\sqrt{1 + y'^2 + z'^2}}\right) = 0 \quad (8)$$

$$\frac{d}{dx}\left(\frac{z'}{\sqrt{1 + y'^2 + z'^2}}\right) = 0. \quad (9)$$

These give

$$\frac{y'}{\sqrt{1 + y'^2 + z'^2}} = c_1 \quad (10)$$

$$\frac{z'}{\sqrt{1 + y'^2 + z'^2}} = c_2. \quad (11)$$

Taking the ratio,

$$z' = \frac{c_2}{c_1} y' \quad (12)$$

$$\frac{y'}{\sqrt{1 + y'^2 + \left(\frac{c_2}{c_1}\right)^2 y'^2}} = c_1 \quad (13)$$

$$y'^2 = c_1{}^2 \left[1 + y'^2 + \left(\frac{c_2}{c_1}\right)^2 y'^2\right] = c_1{}^2 + y'^2(c_1{}^2 + c_2{}^2), \quad (14)$$

which gives

$$y'^2 = \frac{c_1{}^2}{1 - c_1{}^2 - c_2{}^2} \equiv a_1{}^2 \quad (15)$$

$$z'^2 = \left(\frac{c_2}{c_1}\right)^2 y'^2 = \frac{c_2{}^2}{1 - c_1{}^2 - c_2{}^2} \equiv b_1{}^2. \quad (16)$$

Therefore, $y' = a_1$ and $z' = b_1$, so the solution is

$$\begin{bmatrix} x \\ y \\ z \end{bmatrix} = \begin{bmatrix} x \\ a_1 x + a_0 \\ b_1 x + b_0 \end{bmatrix}, \quad (17)$$

which is the parametric representation of a straight line with parameter $x \in [x_1, x_2]$. Verifying the ARC LENGTH gives

$$L = \sqrt{1 + a_1{}^2 + b_1{}^2}\,(x_2 - x_1) \quad (18)$$

where

$$\begin{bmatrix} y_1 \\ y_2 \end{bmatrix} = \begin{bmatrix} x_1 & 1 \\ x_2 & 1 \end{bmatrix} \begin{bmatrix} a_1 \\ a_0 \end{bmatrix} \quad (19)$$

$$\begin{bmatrix} z_1 \\ z_2 \end{bmatrix} = \begin{bmatrix} x_1 & 1 \\ x_2 & 1 \end{bmatrix} \begin{bmatrix} b_1 \\ b_0 \end{bmatrix}. \quad (20)$$

see also POINT-POINT DISTANCE—1-D, POINT-POINT DISTANCE—2-D, POINT-QUADRATIC DISTANCE

Point Probability

The portion of the probability distribution which has a P-VALUE *equal to* the observed P-VALUE.

see also TAIL PROBABILITY

Point-Quadratic Distance

Find the minimum distance between a point in the plane (x_0, y_0) and a quadratic PLANE CURVE

$$y = a_0 + a_1 x + a_2 x^2. \quad (1)$$

The square of the distance is

$$\begin{aligned} r^2 &= (x - x_0)^2 + (y - y_0)^2 \\ &= (x - x_0)^2 + (a_0 + a_1 x + a_2 x^2 - y_0)^2. \end{aligned} \quad (2)$$

Minimizing the distance squared is the equivalent to minimizing the distance (since r^2 and $|r|$ have minima at the same point), so take

$$\frac{\partial(r^2)}{\partial x} = 2(x - x_0) + 2(a_0 + a_1 x + a_2 x^2 - y_0)(a_1 + 2a_2 x) = 0 \quad (3)$$

$$\begin{aligned} x - x_0 + a_0 a_1 + a_1{}^2 + a_1 a_2 x^2 - a_1 y_0 + 2a_0 a_2 x \\ + 2a_1 a_2 x^2 + 2a_2{}^2 x^3 - 2a_2 y_0 x = 0 \end{aligned} \quad (4)$$

$$\begin{aligned} 2a_2{}^2 x^3 + 3a_1 a_2 x^2 + (a_1{}^2 + 2a_0 a_2 - 2a_2 y_0 + 1)x \\ + (a_0 a_1 - a_1 y_0 - x_0) = 0. \end{aligned} \quad (5)$$

Minimizing the distance therefore requires solution of a CUBIC EQUATION.

see also POINT-POINT DISTANCE—1-D, POINT-POINT DISTANCE—2-D, POINT-POINT DISTANCE—3-D

Point-Set Topology

The low-level language of TOPOLOGY, which is not really considered a separate "branch" of TOPOLOGY. Point-set topology, also called set-theoretic topology or general topology, is the study of the general abstract nature of continuity or "closeness" on SPACES. Basic point-set topological notions are ones like CONTINUITY, DIMENSION, COMPACTNESS, and CONNECTEDNESS. The INTERMEDIATE VALUE THEOREM (which states that if a path in the real line connects two numbers, then it passes over every point between the two) is a basic topological result. Others are that EUCLIDEAN n-space is HOMEOMORPHIC to EUCLIDEAN m-space IFF $m = n$, and that REAL valued functions achieve maxima and minima on COMPACT SETS.

Foundational point-set topological questions are ones like "when can a topology on a space be derived from a metric?" Point-set topology deals with differing notions of continuity and compares them, as well as dealing with their properties. Point-set topology is also the ground-level of inquiry into the geometrical properties of spaces and continuous functions between them, and in that sense, it is the foundation on which the remainder of topology (ALGEBRAIC, DIFFERENTIAL, and LOW-DIMENSIONAL) stands.

see also ALGEBRAIC TOPOLOGY, DIFFERENTIAL TOPOLOGY, LOW-DIMENSIONAL TOPOLOGY, TOPOLOGY

References

Sutherland, W. A. *An Introduction to Metric & Topological Spaces.* New York: Oxford University Press, 1975.

Points Problem

see SHARING PROBLEM

Pointwise Dimension

$$D_P(\mathbf{x}) \equiv \lim_{\epsilon\to 0} \frac{\ln \mu(B_\epsilon(\mathbf{x}))}{\ln \epsilon},$$

where $B_\epsilon(\mathbf{x})$ is an n-D BALL of RADIUS ϵ centered at $\mathbf{x}$ and μ is the PROBABILITY MEASURE.

see also BALL, PROBABILITY MEASURE

References

Nayfeh, A. H. and Balachandran, B. *Applied Nonlinear Dynamics: Analytical, Computational, and Experimental Methods.* New York: Wiley, pp. 541–545, 1995.

Poisson's Bessel Function Formula

For $\Re[\nu] > -1/2$,

$$J_\nu(z) = \left(\frac{z}{2}\right)^\nu \frac{2}{\sqrt{\pi}\,\Gamma(\nu+\frac{1}{2})} \int_0^{\pi/2} \cos(z\cos t)\sin^{2\nu} t\, dt,$$

where $J_\nu(z)$ is a BESSEL FUNCTION OF THE FIRST KIND, and $\Gamma(z)$ is the GAMMA FUNCTION.

References

Iyanaga, S. and Kawada, Y. (Eds.). *Encyclopedic Dictionary of Mathematics.* Cambridge, MA: MIT Press, p. 1472, 1980.

Poisson Bracket

Let F and G be infinitely differentiable functions of x and p. Then the Poisson bracket is defined by

$$(F,G) = \sum_{\nu=1}^{n} \left(\frac{\partial F}{\partial p_\nu}\frac{\partial G}{\partial x_p} - \frac{\partial G}{\partial p_\nu}\frac{\partial F}{\partial x_\nu} \right).$$

If F and G are functions of x and p only, then the LAGRANGE BRACKET $[F,G]$ collapses the Poisson bracket (F,G).

see also LAGRANGE BRACKET, LIE BRACKET

References

Iyanaga, S. and Kawada, Y. (Eds.). *Encyclopedic Dictionary of Mathematics.* Cambridge, MA: MIT Press, p. 1004, 1980.

Poisson-Charlier Function

$$\rho_n(\nu,x) \equiv \frac{(1+\nu-n)_n}{\sqrt{n!x^n}}\, {}_1F_1(-n;1+\nu-n;x),$$

where $(a)_n$ is a POCHHAMMER SYMBOL and ${}_1F_1(a;b;z)$ is a CONFLUENT HYPERGEOMETRIC FUNCTION.

see also POISSON-CHARLIER POLYNOMIAL

Poisson-Charlier Polynomial

Polynomials $p_n(x)$ which belong to the distribution $d\alpha(x)$ where $\alpha(x)$ is a STEP FUNCTION with JUMP

$$j(x) = e^{-a}a^x(x!)^{-1} \tag{1}$$

at $x = 0, 1, \ldots$ for $a > 0$.

$$\begin{aligned} p_n(x) &= a^{n/2}(n!)^{-1/2}\sum_{\nu=0}^{n}(-1)^{n-\nu}\binom{n}{\nu}\nu!a^{-\nu}\binom{x}{\nu} && (2)\\ &= a^{n/2}(n!)^{-1/2}(-1)^n[j(x)]^{-1}\Delta^n j(x-n) && (3)\\ &= a^{-n/2}\sqrt{n!}\,L_n^{x-n}(a), && (4) \end{aligned}$$

where $\binom{n}{k}$ is a BINOMIAL COEFFICIENT, $L_n^k(x)$ is an associated LAGUERRE POLYNOMIAL, and

$$\begin{aligned} \Delta f(x) &= f(x+1) - f(x) && (5)\\ \Delta^n f(x) &= \Delta[\Delta^{n-1} f(x)]\\ &= f(x+n) - \binom{n}{1} f(x+n-1) + \ldots + (-1)^n f(x). && (6) \end{aligned}$$

see also POISSON-CHARLIER FUNCTION

References

Szegő, G. *Orthogonal Polynomials, 4th ed.* Providence, RI: Amer. Math. Soc., pp. 34–35, 1975.

Poisson Distribution

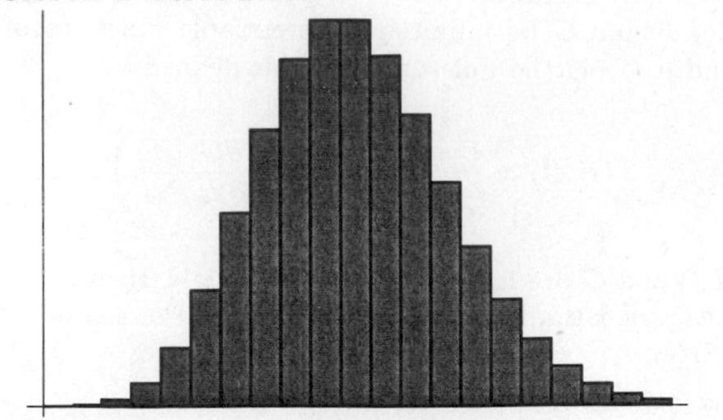

A Poisson distribution is a distribution with the following properties:

1. The number of changes in nonoverlapping intervals are independent for all intervals.
2. The probability of exactly one change in a sufficiently small interval $h \equiv 1/n$ is $P = \nu h \equiv \nu/n$, where ν is the probability of one change and n is the number of TRIALS.
3. The probability of two or more changes in a sufficiently small interval h is essentially 0.

The probability of k changes in a given interval is then given by the limit of the BINOMIAL DISTRIBUTION

$$P(k) = \frac{n!}{k!(n-k)!}\left(\frac{\nu}{n}\right)^k \left(1-\frac{\nu}{n}\right)^{n-k} \quad (1)$$

as the number of trials becomes very large,

$$\lim_{n\to\infty} P(k) =$$

$$\lim_{n\to\infty} \frac{n(n-1)\cdots(n-k-1)}{n^k}\frac{\nu^k}{k!}\left(1-\frac{\nu}{n}\right)^n\left(1-\frac{\nu}{n}\right)^{-k}$$

$$= (1)\left(\frac{\nu^k}{k!}\right)(e^{-\nu})(1) = \frac{\nu^k e^{-\nu}}{k!}. \quad (2)$$

This should be normalized so that the sum of probabilities equals 1. Indeed,

$$\sum_{k=0}^{\infty} P(k) = e^{\nu}\sum_{k=0}^{\infty}\frac{\nu^k}{k!} = e^{\nu}e^{-\nu} = 1, \quad (3)$$

as required. The ratio of probabilities is given by

$$\frac{P(k=i+1)}{P(k=i)} = \frac{\frac{\nu^{i+1}e^{-\nu}}{(i+1)!}}{\frac{i!}{e^{-\nu}\nu^i}} = \frac{\nu}{i+1}. \quad (4)$$

The MOMENT-GENERATING FUNCTION of this distribution is

$$M(t) = \sum_{k=0}^{\infty} e^{tk}\frac{\nu^k e^{-\nu}}{k!} = e^{-\nu}\sum_{k=0}^{\infty}\frac{(\nu e^t)^k}{k!}$$

$$= e^{-\nu}e^{\nu e^t} = e^{\nu(e^t-1)} \quad (5)$$

$$M'(t) = \nu e^t e^{\nu(e^t-1)} \quad (6)$$

$$M''(t) = (\nu e^t)^2 e^{\nu(e^t-1)} + \nu e^t e^{\nu(e^t-1)} \quad (7)$$

$$R(t) \equiv \ln M(t) = \nu(e^t - 1) \quad (8)$$

$$R'(t) = \nu e^t \quad (9)$$

$$R''(t) = \nu e^t, \quad (10)$$

so

$$\mu = R'(0) = \nu \quad (11)$$

$$\sigma^2 = R''(0) = \nu. \quad (12)$$

The MOMENTS about zero can also be computed directly

$$\mu_2' = \nu(1+\nu) \quad (13)$$

$$\mu_3' = \nu(1+3\nu+\nu^2) \quad (14)$$

$$\mu_4' = \nu(1+7\nu+6\nu^2+\nu^3), \quad (15)$$

as can the MOMENTS about the MEAN.

$$\mu_1 = \nu \quad (16)$$

$$\mu_2 = \nu \quad (17)$$

$$\mu_3 = \nu \quad (18)$$

$$\mu_4 = \nu(1+3\nu), \quad (19)$$

so the MEAN, VARIANCE, SKEWNESS, and KURTOSIS are

$$\mu = \nu \quad (20)$$

$$\sigma^2 = \nu \quad (21)$$

$$\gamma_1 \equiv \frac{\mu_3}{\sigma^3} = \frac{\nu}{\nu^{3/2}} = \nu^{-1/2} \quad (22)$$

$$\gamma_2 \equiv \frac{\mu_4}{\sigma^4} - 3 = \frac{\nu(1+3\nu)}{\nu} - 3$$

$$= \frac{\nu+3\nu^2-3\nu^2}{\nu^2} = \nu^{-1}. \quad (23)$$

The CHARACTERISTIC FUNCTION is

$$\phi(t) = e^{m(e^{it}-1)} \quad (24)$$

and the CUMULANT-GENERATING FUNCTION is

$$K(h) = \nu(e^h - 1) = \nu(h + \tfrac{1}{2!}h^2 + \tfrac{1}{3!}h^3 + \ldots), \quad (25)$$

so

$$\kappa_r = \nu. \quad (26)$$

The Poisson distribution can also be expressed in terms of

$$\lambda \equiv \frac{\nu}{x}, \quad (27)$$

the rate of changes, so that

$$P(k) = \frac{(\lambda x)^k e^{-\lambda x}}{k!}. \quad (28)$$

The MOMENT-GENERATING FUNCTION of a Poisson distribution in two variables is given by

$$M(t) = e^{(\nu_1+\nu_2)(e^t-1)}. \quad (29)$$

If the independent variables $x_1, x_2, \ldots, x_N$ have Poisson distributions with parameters $\mu_1, \mu_2, \ldots, \mu_N$, then

$$X = \sum_{j=1}^{N} x_j \tag{30}$$

has a Poisson distribution with parameter

$$\mu = \sum_{j=1}^{N} \mu_j. \tag{31}$$

This can be seen since the CUMULANT-GENERATING FUNCTION is

$$K_j(h) = \mu_j(e^h - 1), \tag{32}$$

$$K \equiv \sum_j K_j(h) = (e^h - 1) \sum_j \mu_j = \mu(e^h - 1). \tag{33}$$

References

Beyer, W. H. *CRC Standard Mathematical Tables, 28th ed.* Boca Raton, FL: CRC Press, p. 532, 1987.

Press, W. H.; Flannery, B. P.; Teukolsky, S. A.; and Vetterling, W. T. "Incomplete Gamma Function, Error Function, Chi-Square Probability Function, Cumulative Poisson Function." §6.2 in *Numerical Recipes in FORTRAN: The Art of Scientific Computing, 2nd ed.* Cambridge, England: Cambridge University Press, pp. 209–214, 1992.

Spiegel, M. R. *Theory and Problems of Probability and Statistics.* New York: McGraw-Hill, p. 111–112, 1992.

Poisson's Equation

A second-order PARTIAL DIFFERENTIAL EQUATION arising in physics:

$$\nabla^2 \psi = -4\pi\rho.$$

If $\rho = 0$, it reduces LAPLACE'S EQUATION. It is also related to the HELMHOLTZ DIFFERENTIAL EQUATION

$$\nabla^2 \psi + k^2 \psi = 0.$$

see also HELMHOLTZ DIFFERENTIAL EQUATION, LAPLACE'S EQUATION

References

Arfken, G. "Gauss's Law, Poisson's Equation." §1.14 in *Mathematical Methods for Physicists, 3rd ed.* Orlando, FL: Academic Press, pp. 74–78, 1985.

Poisson's Harmonic Function Formula

Let $\phi(z)$ be a HARMONIC FUNCTION. Then

$$\phi(z_0) = \frac{1}{2\pi} \int_0^{2\pi} K(r,\theta)\phi(z_0 + re^{i\theta})\, d\theta, \tag{1}$$

where $R = |z_0|$ and $K(r,\theta)$ is the POISSON KERNEL. For a CIRCLE,

$$u(x,y) = \frac{1}{2\pi} \int_0^{2\pi} u(a\cos\phi, a\sin\phi) \frac{a^2 - R^2}{a^2 + R^2 - 2ar\cos(\theta - \phi)}\, d\phi. \tag{2}$$

For a SPHERE,

$$u(x,y,z) = \frac{1}{4\pi a} \iint_S u \frac{a^2 - R^2}{(a^2 + R^2 - 2aR\cos\theta)^{3/2}}\, dS, \tag{3}$$

where

$$\cos\theta \equiv \mathbf{x} \cdot \boldsymbol{\xi}. \tag{4}$$

see also CIRCLE, HARMONIC FUNCTION, POISSON KERNEL, SPHERE

References

Morse, P. M. and Feshbach, H. *Methods of Theoretical Physics, Part I.* New York: McGraw-Hill, pp. 373–374, 1953.

Poisson Integral

A.k.a. BESSEL'S SECOND INTEGRAL.

$$J_n(z) = \frac{(\frac{1}{2})^n}{\Gamma(n + \frac{1}{2})\Gamma(\frac{1}{2})} \int_0^\pi \cos(z\cos\theta)\sin^{2n}\theta\, d\theta,$$

where $J_n(z)$ is a BESSEL FUNCTION OF THE FIRST KIND and $\Gamma(x)$ is a GAMMA FUNCTION. It can be derived from SONINE'S INTEGRAL. With $n = 0$, the integral becomes PARSEVAL'S INTEGRAL.

see also BESSEL FUNCTION OF THE FIRST KIND, PARSEVAL'S INTEGRAL, SONINE'S INTEGRAL

Poisson Integral Representation

$$j_n(z) = \frac{z^n}{2^{n+1} n!} \int_0^\pi \cos(z\cos\theta)\sin^{2n+1}\theta\, d\theta,$$

where $j_n(z)$ is a SPHERICAL BESSEL FUNCTION OF THE FIRST KIND.

Poisson Kernel

In 2-D,

$$\begin{aligned} K(r,\theta) &\equiv \Re\left(\frac{R + re^{i\theta}}{R - re^{i\theta}}\right) \\ &= \Re\left[\frac{(R + re^{i\theta})(R - re^{-i\theta})}{(R - re^{i\theta})(R - re^{-i\theta})}\right] \\ &= \Re\left[\frac{R^2 - rR(e^{i\theta} - e^{-i\theta}) - r^2}{R^2 - rR(e^{i\theta} + e^{-i\theta}) + r^2}\right] \\ &= \Re\left[\frac{R^2 + 2irR\sin\theta - r^2}{R^2 - 2Rr\cos\theta + r^2}\right] \\ &= \frac{R^2 - r^2}{R^2 - 2Rr\cos\theta + r^2}. \end{aligned} \tag{1}$$

In 3-D,

$$u(\mathbf{y}) = \frac{R(R^2 - a^2)}{4\pi} \times \int_0^{2\pi} \int_0^\pi \frac{f(\theta,\phi)\sin\theta\, d\theta\, d\phi}{(R^2 + a^2 - 2aR\cos\gamma)^{3/2}}, \tag{2}$$

where $a = |\mathbf{y}|$ and

$$\cos\gamma = \mathbf{y} \cdot \begin{bmatrix} R\cos\theta\sin\phi \\ R\sin\theta\sin\phi \\ R\cos\phi \end{bmatrix}. \qquad (3)$$

The Poisson kernel for the n-BALL is

$$P(\mathbf{x}, \mathbf{z}) = \frac{1}{2-n}(D_{\mathbf{n}}\mathbf{v})(\mathbf{z}), \qquad (4)$$

where $D_{\mathbf{n}}$ is the outward normal derivative at point $\mathbf{z}$ on a unit n-SPHERE and

$$\mathbf{v}(\mathbf{z}) = |\mathbf{z}-\mathbf{x}|^{2-n} - |\mathbf{x}|^{2-n}\left|\frac{\mathbf{x}}{|\mathbf{x}|^2}\right|^{2-n}. \qquad (5)$$

see also POISSON'S HARMONIC FUNCTION FORMULA

References

Gradshteyn, I. S. and Ryzhik, I. M. *Tables of Integrals, Series, and Products, 5th ed.* San Diego, CA: Academic Press, p. 1090, 1979.

Poisson Manifold

A smooth MANIFOLD with a POISSON BRACKET defined on its FUNCTION SPACE.

Poisson Sum Formula

A special case of the general result

$$\sum_{n=-\infty}^{\infty} f(x+n) = \sum_{k=-\infty}^{\infty} e^{2\pi ikx}\int_{-\infty}^{\infty} f(x_1)e^{-2\pi ikx}\,dx_1 \qquad (1)$$

with $x = 0$, yielding

$$\sum_{n=-\infty}^{\infty} f(n) = \sum_{k=-\infty}^{\infty}\int_{-\infty}^{\infty} f(x_1)e^{-2\pi ikx}\,dx_1. \qquad (2)$$

An alternate form is

$$\frac{1}{2} + \sum_{n=1}^{\infty} e^{-(nx)^2} = \frac{\sqrt{\pi}}{x}\left[\frac{1}{2} + \sum_{n=1}^{\infty} e^{-(n\pi/x)^2}\right]. \qquad (3)$$

Another formula called the Poisson summation formula is

$$\sqrt{\alpha}\,[\tfrac{1}{2}\phi(0) + \phi(\alpha) + \phi(2\alpha) + \ldots] = \sqrt{\beta}[\tfrac{1}{2}\psi(0) + \psi(\beta) + \psi(2\beta) + \ldots], \qquad (4)$$

where

$$\psi(x) = \sqrt{\frac{2}{\pi}}\int_0^{\infty} \psi(t)\cos(xt)\,dt \qquad (5)$$

$$\alpha\beta = 2\pi. \qquad (6)$$

References

Morse, P. M. and Feshbach, H. *Methods of Theoretical Physics, Part I.* New York: McGraw-Hill, pp. 466–467, 1953.

Poisson Trials

A number s TRIALS in which the probability of success p_i varies from trial to trial. Let x be the number of successes, then

$$\operatorname{var}(x) = spq - s{\sigma_p}^2, \qquad (1)$$

where ${\sigma_p}^2$ is the VARIANCE of p_i and $q \equiv (1-p)$. Uspensky has shown that

$$P(s,x) = \beta\frac{m^x e^{-m}}{x!}, \qquad (2)$$

where

$$\beta = [1 - \theta g(x)]e^{h(x)} \qquad (3)$$

$$g(x) = \frac{(s-x)m^3}{3(s-m)^3} + \frac{x^3}{2s(s-x)} \qquad (4)$$

$$h(x) = \frac{mx}{s} - \frac{m^2}{2s^2}(s-x) - \frac{x(x-1)}{2s} = p\left[\frac{x}{2}\left(1+\frac{1}{m}\right) - \frac{(x-m)^2}{2m}\right] \qquad (5)$$

and $\theta \in (0,1)$. The probability that the number of successes is at least x is given by

$$Q_m(x) = \sum_{r=x}^{\infty}\frac{m^r e^{-m}}{r!}. \qquad (6)$$

Uspensky gives the true probability that there are at least x successes in s trials as

$$P_{ms}(x) = Q_m(x) + \Delta, \qquad (7)$$

where

$$|\Delta| < \begin{cases} (e^{\chi}-1)Q_m(x+1) & \text{for } Q_m(x+1) \geq \frac{1}{2} \\ (e^{\chi}-1)[1-Q_m(x+1)] & \text{for } Q_m(x+1) \leq \frac{1}{2} \end{cases} \qquad (8)$$

$$\chi = \frac{m + \frac{1}{4} + \frac{m^3}{s}}{2(s-m)}. \qquad (9)$$

Poke Move

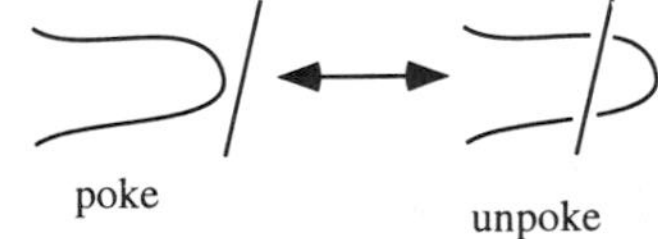

The REIDEMEISTER MOVE of type II.

see also REIDEMEISTER MOVES

Poker

Poker is a CARD game played with a normal deck of 52 CARDS. Sometimes, additional cards called "jokers" are also used. In straight or draw poker, each player is normally dealt a hand of five cards. Depending on the variant, players then discard and redraw CARDS, trying to improve their hands. Bets are placed at each discard step. The number of possible distinct five-card hands is

$$N = \binom{52}{5} = 2{,}598{,}960,$$

where $\binom{n}{k}$ is a BINOMIAL COEFFICIENT.

There are special names for specific types of hands. A royal flush is an ace, king, queen, jack, and 10, all of one suit. A straight flush is five consecutive cards all of the same suit (but not a royal flush), where an ace may count as either high or low. A full house is three-of-a-kind and a pair. A flush is five cards of the same suit (but not a royal flush or straight flush). A straight is five consecutive cards (but not a royal flush or straight flush), where an ace may again count as either high or low.

The probabilities of being dealt five-card poker hands of a given type (before discarding and with no jokers) on the initial deal are given below (Packel 1981). As usual, for a hand with probability P, the ODDS against being dealt it are $(1/r) - 1 : 1$.

Hand	Exact Probability
royal flush	$\frac{4}{N} = \frac{1}{649{,}740}$
straight flush	$\frac{4(10)-4}{N} = \frac{3}{216{,}580}$
four of a kind	$\frac{13(48)}{N} = \frac{1}{4{,}165}$
full house	$\frac{13\binom{4}{3}12\binom{4}{2}}{N} = \frac{6}{4{,}165}$
flush	$\frac{4\binom{13}{5}-36-4}{N} = \frac{1{,}277}{649{,}740}$
straight	$\frac{10(4^5)-36-4}{N} = \frac{5}{1{,}274}$
three of a kind	$\frac{13\binom{4}{3}\frac{(48)(44)}{2!}}{N} = \frac{88}{4{,}165}$
two pair	$\frac{\frac{13\binom{4}{2}12\binom{4}{2}}{2!}44}{N} = \frac{198}{4{,}165}$
one pair	$\frac{13\binom{4}{2}\frac{(48)(44)(40)}{3!}}{N} = \frac{352}{833}$

Hand	Probability	Odds
royal flush	1.54×10^{-6}	649,739.0:1
straight flush	1.39×10^{-5}	72,192.3:1
four of a kind	2.40×10^{-4}	4,164.0:1
full house	1.44×10^{-3}	693.2:1
flush	1.97×10^{-3}	507.8:1
straight	3.92×10^{-3}	253.8:1
three of a kind	0.0211	46.3:1
two pair	0.0475	20.0:1
one pair	0.423	1.366:1

Gadbois (1996) gives probabilities for hands if two jokers are included, and points out that it is *impossible* to rank hands in any single way which is consistent with the relative frequency of the hands.

see also BRIDGE CARD GAME, CARDS

References

Cheung, Y. L. "Why Poker is Played with Five Cards." *Math. Gaz.* **73**, 313–315, 1989.

Conway, J. H. and Guy, R. K. "Choice Numbers with Repetitions." In *The Book of Numbers.* New York: Springer-Verlag, pp. 70–71, 1996.

Gadbois, S. "Poker with Wild Cards—A Paradox?" *Math. Mag.* **69**, 283–285, 1996.

Jacoby, O. *Oswald Jacoby on Poker.* New York: Doubleday, 1981.

Packel, E. W. *The Mathematics of Games and Gambling.* Washington, DC: Math. Assoc. Amer., 1981.

Polar

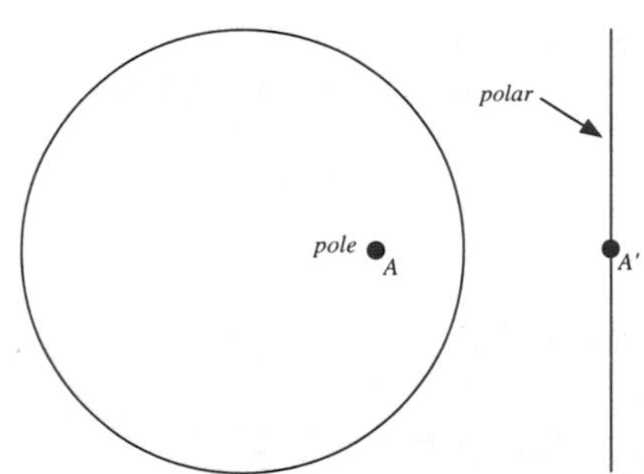

If two points A and A' are INVERSE with respect to a CIRCLE (the INVERSION CIRCLE), then the straight line through A' which is PERPENDICULAR to the line of the points AA' is called the polar of A with respect to the CIRCLE, and A is called the POLE of the polar.

see also APOLLONIUS' PROBLEM, INVERSE POINTS, INVERSION CIRCLE, POLARITY, POLE, TRILINEAR POLAR

References

Dörrie, H. *100 Great Problems of Elementary Mathematics: Their History and Solutions.* New York: Dover, p. 157, 1965.

Johnson, R. A. *Modern Geometry: An Elementary Treatise on the Geometry of the Triangle and the Circle.* Boston, MA: Houghton Mifflin, pp. 100–106, 1929.

Polar Angle

The ANGLE a point makes from the ORIGIN as measured from the x-AXIS.

see also POLAR COORDINATES

Polar Circle

Given a TRIANGLE, the polar circle has center at the ORTHOCENTER H. Call H_i the FEET of the ALTITUDE. Then the RADIUS is

$$r^2 = \overline{HA_1} \cdot \overline{HH_1} = \overline{HA_2} \cdot \overline{HH_2} = \overline{HA_2} \cdot \overline{HH_2} \quad (1)$$

$$= -4R^2 \cos\alpha_1 \cos\alpha_2 \cos\alpha_3 \quad (2)$$

$$= \tfrac{1}{2}({a_1}^2 + {a_2}^2 + {a_3}^2) - 4R^2, \quad (3)$$

where R is the CIRCUMRADIUS, α_i the VERTEX angles, and a_i the corresponding side lengths.

A TRIANGLE is self-conjugate with respect to its polar circle. Also, the RADICAL AXIS of any two polar circles is the ALTITUDE from the third VERTEX. Any two polar circles of an ORTHOCENTRIC SYSTEM are orthogonal. The polar circles of the triangles of a COMPLETE QUADRILATERAL constitute a COAXAL SYSTEM conjugate to that of the circles on the diagonals.

see also COAXAL SYSTEM, ORTHOCENTRIC SYSTEM, POLAR, POLE, RADICAL AXIS

References

Coxeter, H. S. M. and Greitzer, S. L. *Geometry Revisited.* Washington, DC: Math. Assoc. Amer., pp. 136–138, 1967.

Johnson, R. A. *Modern Geometry: An Elementary Treatise on the Geometry of the Triangle and the Circle.* Boston, MA: Houghton Mifflin, pp. 176–181, 1929.

Polar Coordinates

The polar coordinates r and θ are defined by

$$x = r\cos\theta \tag{1}$$
$$y = r\sin\theta. \tag{2}$$

In terms of x and y,

$$r = \sqrt{x^2+y^2} \tag{3}$$
$$\theta = \tan^{-1}\left(\frac{y}{x}\right). \tag{4}$$

The ARC LENGTH of a polar curve given by $r = r(\theta)$ is

$$s = \int_{\theta_1}^{\theta_2}\sqrt{r^2+\left(\frac{dr}{d\theta}\right)^2}\,d\theta. \tag{5}$$

The LINE ELEMENT is given by

$$ds^2 = r^2\,d\theta^2, \tag{6}$$

and the AREA element by

$$dA = r\,dr\,d\theta. \tag{7}$$

The AREA enclosed by a polar curve $r = r(\theta)$ is

$$A = \tfrac{1}{2}\int_{\theta_1}^{\theta_2} r^2\,d\theta. \tag{8}$$

The SLOPE of a polar function $r = r(\theta)$ at the point (r,θ) is given by

$$m = \frac{r+\tan\theta\frac{dr}{d\theta}}{-r\tan\theta+\frac{dr}{d\theta}}. \tag{9}$$

The ANGLE between the tangent and radial line at the point (r,θ) is

$$\psi = \tan^{-1}\left(\frac{r}{\frac{dr}{d\theta}}\right). \tag{10}$$

A polar curve is symmetric about the x-axis if replacing θ by $-\theta$ in its equation produces an equivalent equation, symmetric about the y-axis if replacing θ by $\pi-\theta$ in its equation produces an equivalent equation, and symmetric about the origin if replacing r by $-r$ in its equation produces an equivalent equation.

In Cartesian coordinates, the POSITION VECTOR and its derivatives are

$$\mathbf{r} = \sqrt{x^2+y^2}\,\hat{\mathbf{r}} \tag{11}$$
$$\dot{\mathbf{r}} = \dot{\hat{\mathbf{r}}}\sqrt{x^2+y^2}+\hat{\mathbf{r}}\,(x^2+y^2)^{-1/2}(x\dot{x}+y\dot{y}) \tag{12}$$
$$\hat{\mathbf{r}} = \frac{x\hat{\mathbf{x}}+y\hat{\mathbf{y}}}{\sqrt{x^2+y^2}} \tag{13}$$
$$\begin{aligned}\dot{\hat{\mathbf{r}}} &= \frac{\dot{x}\hat{\mathbf{x}}+\dot{y}\hat{\mathbf{y}}}{\sqrt{x^2+y^2}} \\ &\quad -\tfrac{1}{2}(x^2+y^2)^{-3/2}(2)(x\dot{x}+y\dot{y})(x\hat{\mathbf{x}}+y\hat{\mathbf{y}}) \\ &= \frac{(x\dot{y}-y\dot{x})(x\hat{\mathbf{y}}-y\hat{\mathbf{x}})}{(x^2+y^2)^{3/2}}.\end{aligned} \tag{14}$$

In polar coordinates, the UNIT VECTORS and their derivatives are

$$\mathbf{r} \equiv \begin{bmatrix} r\cos\theta \\ r\sin\theta \end{bmatrix} \tag{15}$$
$$\hat{\mathbf{r}} \equiv \frac{\frac{d\mathbf{r}}{dr}}{\left|\frac{d\mathbf{r}}{dr}\right|} = \begin{bmatrix}\cos\theta \\ \sin\theta\end{bmatrix} \tag{16}$$
$$\hat{\boldsymbol{\theta}} \equiv \frac{\frac{d\boldsymbol{\theta}}{d\theta}}{\left|\frac{d\boldsymbol{\theta}}{d\theta}\right|} = \begin{bmatrix}-\sin\theta \\ \cos\theta\end{bmatrix} \tag{17}$$
$$\dot{\hat{\mathbf{r}}} = \begin{bmatrix}-\sin\theta\dot{\theta} \\ \cos\theta\dot{\theta}\end{bmatrix} = \dot{\theta}\,\hat{\boldsymbol{\theta}} \tag{18}$$
$$\dot{\hat{\boldsymbol{\theta}}} = \begin{bmatrix}-\cos\theta\dot{\theta} \\ -\sin\theta\dot{\theta}\end{bmatrix} = -\dot{\theta}\hat{\mathbf{r}} \tag{19}$$
$$\dot{\mathbf{r}} = \begin{bmatrix}-r\sin\theta\dot{\theta}+\cos\theta\dot{r} \\ r\cos\theta\dot{\theta}+\sin\theta\dot{r}\end{bmatrix} = r\dot{\theta}\,\hat{\boldsymbol{\theta}}+\dot{r}\,\hat{\mathbf{r}} \tag{20}$$

$$\begin{aligned}\ddot{\mathbf{r}} &= \dot{r}\dot{\theta}\,\hat{\boldsymbol{\theta}}+r\ddot{\theta}\hat{\boldsymbol{\theta}}+r\dot{\theta}\dot{\hat{\boldsymbol{\theta}}}+\ddot{r}\hat{\mathbf{r}}+\dot{r}\dot{\hat{\mathbf{r}}} \\ &= \dot{r}\dot{\theta}\hat{\boldsymbol{\theta}}+\dot{r}\ddot{\theta}\hat{\boldsymbol{\theta}}+r\dot{\theta}(-\dot{\theta}\hat{\mathbf{r}})+\ddot{r}\hat{\mathbf{r}}+\dot{r}\dot{\theta}\hat{\boldsymbol{\theta}} \\ &= (\ddot{r}-r\dot{\theta}^2)\hat{\mathbf{r}}+(2\dot{r}\dot{\theta}+r\ddot{\theta})\hat{\boldsymbol{\theta}} \\ &= (\ddot{r}-r\dot{\theta}^2)\hat{\mathbf{r}}+\frac{1}{r}\frac{d}{dt}(r^2\dot{\theta})\hat{\boldsymbol{\theta}}.\end{aligned} \tag{21}$$

see also CARDIOID, CIRCLE, CISSOID, CONCHOID, CURVILINEAR COORDINATES, CYLINDRICAL COORDINATES, EQUIANGULAR SPIRAL, LEMNISCATE, LIMAÇON, ROSE

Polar Line

see POLAR

Polarity

A projective CORRELATION of period two. In a polarity, a is called the POLAR of A, and A the POLE a.

see also CHASLES'S THEOREM, CORRELATION, POLAR, POLE (GEOMETRY)

Pole

A COMPLEX function f has a pole of order m at z_0 if, in the LAURENT SERIES, $a_n = 0$ for $n < -m$ and $a_m \neq 0$. Equivalently, f has a pole of order n at z_0 if n is the smallest POSITIVE INTEGER for which $(z - z_0)^n f(z)$ is differentiable at z_0. If $f(\pm\infty) \neq \pm\infty$, there is no pole at $\pm\infty$. Otherwise, the order of the pole is the greatest POSITIVE COEFFICIENT in the LAURENT SERIES.

This is equivalent to finding the smallest n such that

$$\frac{(z - z_0)^n}{f(z)}$$

is differentiable at 0.

see also LAURENT SERIES, RESIDUE (COMPLEX ANALYSIS)

References

Arfken, G. *Mathematical Methods for Physicists, 3rd ed.* Orlando, FL: Academic Press, pp. 396–397, 1985.

Pole (Geometry)

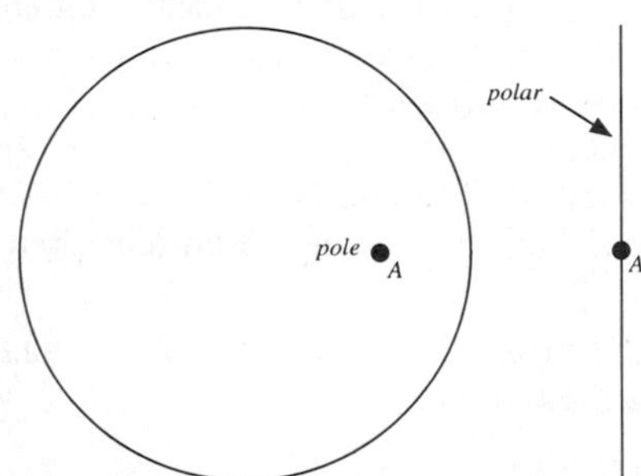

If two points A and A' are INVERSE with respect to a CIRCLE (the INVERSION CIRCLE), then the straight line through A' which is PERPENDICULAR to the line of the points AA' is called the POLAR of the A with respect to the CIRCLE, and A is called the pole of the POLAR.

see also INVERSE POINTS, INVERSION CIRCLE, POLAR, POLARITY, TRILINEAR POLAR

References

Dörrie, H. *100 Great Problems of Elementary Mathematics: Their History and Solutions.* New York: Dover, p. 157, 1965.

Johnson, R. A. *Modern Geometry: An Elementary Treatise on the Geometry of the Triangle and the Circle.* Boston, MA: Houghton Mifflin, pp. 100–106, 1929.

Pole (Origin)

see ORIGIN

Policeman on Point Duty Curve

see CRUCIFORM

Polignac's Conjecture

see DE POLIGNAC'S CONJECTURE

Polish Space

The HOMEOMORPHIC image of a so-called "complete separable" METRIC SPACE. The continuous image of a Polish space is called a SOUSLIN SET.

see also DESCRIPTIVE SET THEORY, STANDARD SPACE

Pollaczek Polynomial

Let $a > |b|$, and write

$$h(\theta) = \frac{a\cos\theta + b}{2\sin\theta}. \tag{1}$$

Then define $P_n(x; a, b)$ by the GENERATING FUNCTION

$$f(x, w) = f(\cos\theta, w) = \sum_{n=0}^{\infty} P_n(x; a, b) w^n$$
$$= (1 - we^{i\theta})^{-1/2+ih(\theta)}(1 - we^{i\theta})^{-1/2-ih(\theta)}. \tag{2}$$

The GENERATING FUNCTION may also be written

$$f(x, w) = (1 - 2xw + w^2)^{-1/2}$$
$$\exp\left[(ax + b)\sum_{m=1}^{\infty} \frac{w^m}{m} U_{m-1}(x)\right], \tag{3}$$

where $U_m(x)$ is a CHEBYSHEV POLYNOMIAL OF THE SECOND KIND. They satisfy the RECURRENCE RELATION

$$nP_n(x; a, b) = [(2n - 1 + 2a)x + 2b]P_{n-1}(x; a, b)$$
$$-(n - 1)P_{n-2}(x; a, b) \tag{4}$$

for $n = 2, 3, \ldots$ with

$$P_0 = 1 \tag{5}$$
$$P_1 = (2a + 1)x + 2b. \tag{6}$$

In terms of the HYPERGEOMETRIC FUNCTION ${}_2F_1(a, b; c; x)$,

$$P_n(\cos\theta; a; b) = e^{in\theta}{}_2F_1(-n, \tfrac{1}{2}+ih(\theta); 1; 1-e^{-2i\theta}). \tag{7}$$

They obey the orthogonality relation

$$\int_{-1}^{1} P_n(x; a, b) P_m(x; a, b) w(x; a, b)\, dx$$
$$= [n + \tfrac{1}{2}(a + 1)]^{-1}\delta_{nm}, \tag{8}$$

where δ_{nm} is the KRONECKER DELTA, for $n, m = 0, 1, \ldots$, with the WEIGHT FUNCTION

$$w(\cos\theta; a, b) = e^{(2\theta-\pi)h(\theta)}\{\cosh[\pi h(\theta)]\}^{-1}. \tag{9}$$

References

Szegő, G. *Orthogonal Polynomials, 4th ed.* Providence, RI: Amer. Math. Soc., pp. 393–400, 1975.

Pollard Monte Carlo Factorization Method

see POLLARD ρ FACTORIZATION METHOD

Pollard $p-1$ Factorization Method

A PRIME FACTORIZATION ALGORITHM which can be implemented in a single-step or double-step form. In the single-step version, PRIMES p are found if $p-1$ is a product of small PRIMES by finding an m such that

$$m \equiv c^q \pmod{n},$$

where $p-1|q$, with q a large number and $(c,n)=1$. Then since $p-1|q$, $m \equiv 1 \pmod{p}$, so $p|m-1$. There is therefore a good chance that $n \nmid m-1$, in which case $\text{GCD}(m-1,n)$ (where GCD is the GREATEST COMMON DIVISOR) will be a nontrivial divisor of n.

In the double-step version, a PRIMES p can be factored if $p-1$ is a product of small PRIMES and a single larger PRIME.

see also PRIME FACTORIZATION ALGORITHMS, WILLIAMS $p+1$ FACTORIZATION METHOD

References

Bressoud, D. M. *Factorization and Prime Testing.* New York: Springer-Verlag, pp. 67–69, 1989.

Pollard, J. M. "Theorems on Factorization and Primality Testing." *Proc. Cambridge Phil. Soc.* **76**, 521–528, 1974.

Pollard ρ Factorization Method

A PRIME FACTORIZATION ALGORITHM also known as POLLARD MONTE CARLO FACTORIZATION METHOD. Let $x_0 = 2$, then compute

$$x_{i+1} = {x_i}^2 - x_i + 1 \pmod{n}.$$

If $\text{GCD}(x_{2i} - x_i, n) > 1$, then n is COMPOSITE and its factors are found. In modified form, it becomes BRENT'S FACTORIZATION METHOD. In practice, almost any unfactorable POLYNOMIAL can be used for the iteration (x^2-2, however, cannot). Under worst conditions, the ALGORITHM can be very slow.

see also BRENT'S FACTORIZATION METHOD, PRIME FACTORIZATION ALGORITHMS

References

Brent, R. P. "Some Integer Factorization Algorithms Using Elliptic Curves." *Austral. Comp. Sci. Comm.* **8**, 149–163, 1986.

Bressoud, D. M. *Factorization and Prime Testing.* New York: Springer-Verlag, pp. 61–67, 1989.

Eldershaw, C. and Brent, R. P. "Factorization of Large Integers on Some Vector and Parallel Computers." ftp://nimbus.anu.edu.au/pub/Brent/156tr.dvi.Z.

Montgomery, P. L. "Speeding the Pollard and Elliptic Curve Methods of Factorization." *Math. Comput.* **48**, 243–264, 1987.

Pollard, J. M. "A Monte Carlo Method for Factorization." *Nordisk Tidskrift for Informationsbehandlung (BIT)* **15**, 331–334, 1975.

Vardi, I. *Computational Recreations in Mathematica.* Reading, MA: Addison-Wesley, pp. 83 and 102–103, 1991.

Poloidal Field

A VECTOR FIELD resembling a magnetic multipole which has a component along the z-AXIS of a SPHERE and continues along lines of LONGITUDE.

see also DIVERGENCELESS FIELD, TOROIDAL FIELD

References

Stacey, F. D. *Physics of the Earth, 2nd ed.* New York: Wiley, p. 239, 1977.

Pólya-Burnside Lemma

see PÓLYA ENUMERATION THEOREM

Pólya Conjecture

Let n be a POSITIVE INTEGER and $r(n)$ the number of (not necessarily distinct) PRIME FACTORS of n (with $r(1)=0$). Let $O(m)$ be the number of POSITIVE INTEGERS $\leq m$ with an ODD number of PRIME factors, and $E(m)$ the number of POSITIVE INTEGERS $\leq m$ with an EVEN number of PRIME factors. Pólya conjectured that

$$L(m) \equiv E(m) - O(m) = \sum_{n=1}^{m} \lambda(n)$$

is ≤ 0, where $\lambda(n)$ is the LIOUVILLE FUNCTION.

The conjecture was made in 1919, and disproven by Haselgrove (1958) using a method due to Ingham (1942). Lehman (1960) found the first explicit counterexample, $L(906{,}180{,}359) = 1$, and the smallest counterexample $m = 906{,}150{,}257$ was found by Tanaka (1980). The first n for which $L(n) = 0$ are $n = 2$, 4, 6, 10, 16, 26, 40, 96, 586, 906150256, ... (Tanaka 1980, Sloane's A028488). It is unknown if $L(x)$ changes sign infinitely often (Tanaka 1980).

see also ANDRICA'S CONJECTURE, LIOUVILLE FUNCTION, PRIME FACTORS

References

Haselgrove, C. B. "A Disproof of a Conjecture of Pólya." *Mathematika* **5**, 141–145, 1958.

Ingham, A. E. "On Two Conjectures in the Theory of Numbers." *Amer. J. Math.* **64**, 313–319, 1942.

Lehman, R. S. "On Liouville's Function." *Math. Comput.* **14**, 311–320, 1960.

Sloane, N. J. A. Sequence A028488 in "An On-Line Version of the Encyclopedia of Integer Sequences."

Tanaka, M. "A Numerical Investigation on Cumulative Sum of the Liouville Function" [sic]. *Tokyo J. Math.* **3**, 187–189, 1980.

Pólya Distribution

see NEGATIVE BINOMIAL DISTRIBUTION

Pólya Enumeration Theorem

A very general theorem which allows the number of discrete combinatorial objects of a given type to be enumerated (counted) as a function of their "order." The most common application is in the counting of the number of GRAPHS of n nodes, TREES and ROOTED TREES with n branches, GROUPS of order n, etc. The theorem is an extension of BURNSIDE'S LEMMA and is sometimes also called the PÓLYA-BURNSIDE LEMMA.

see also BURNSIDE'S LEMMA, GRAPH (GRAPH THEORY), GROUP, ROOTED TREE, TREE

References

Harary, F. "The Number of Linear, Directed, Rooted, and Connected Graphs." *Trans. Amer. Math. Soc.* **78**, 445–463, 1955.

Pólya, G. "Kombinatorische Anzahlbestimmungen für Gruppen, Graphen, und chemische Verbindungen." *Acta Math.* **68**, 145–254, 1937.

Pólya Polynomial

The POLYNOMIAL giving the number of colorings, with m colors, of a structure defined by a PERMUTATION GROUP.

see also PERMUTATION GROUP, PÓLYA ENUMERATION THEOREM

Pólya's Random Walk Constants

N.B. A detailed on-line essay by S. Finch was the starting point for this entry.

Let $p(d)$ be the probability that a RANDOM WALK on a d-D lattice returns to the origin. Pólya (1921) proved that

$$p(1) = p(2) = 1, \tag{1}$$

but

$$p(d) < 1 \tag{2}$$

for $d > 2$. Watson (1939), McCrea and Whipple (1940), Domb (1954), and Glasser and Zucker (1977) showed that

$$p(3) = 1 - \frac{1}{u(3)} = 0.3405373296\ldots, \tag{3}$$

where

$$u(3) = \frac{3}{(2\pi)^3}\int_{-\pi}^{\pi}\int_{-\pi}^{\pi}\int_{-\pi}^{\pi}\frac{dx\,dy\,dz}{3-\cos x-\cos y-\cos z} \tag{4}$$

$$= \frac{12}{\pi^2}(18+12\sqrt{2}-10\sqrt{3}-7\sqrt{6}) \times \{K[(2-\sqrt{3})(\sqrt{3}-\sqrt{2})]\}^2 \tag{5}$$

$$= 3(18+12\sqrt{2}-10\sqrt{3}-7\sqrt{6}) \times \left[1+2\sum_{k=1}^{\infty}\exp(-k^2\pi\sqrt{6})\right]^4 \tag{6}$$

$$= \frac{\sqrt{6}}{32\pi^3}\Gamma(\tfrac{1}{24})\Gamma(\tfrac{5}{24})\Gamma(\tfrac{7}{24})\Gamma(\tfrac{11}{24}) \tag{7}$$

$$= 1.5163860592\ldots, \tag{8}$$

where $K(k)$ is a complete ELLIPTIC INTEGRAL OF THE FIRST KIND and $\Gamma(z)$ is the GAMMA FUNCTION. Closed forms for $d > 3$ are not known, but Montroll (1956) showed that

$$p(d) = 1 - [u(d)]^{-1}, \tag{9}$$

where

$$u(d) = \frac{d}{(2\pi)^d}\underbrace{\int_{-\pi}^{\pi}\int_{-\pi}^{\pi}\cdots\int_{-\pi}^{\pi}}_{d}\left(d-\sum_{k=1}^{d}\cos x_k\right)^{-1} \times dx_1\,dx_2\cdots dx_d$$

$$= \int_0^\infty \left[I_0\left(\frac{t}{d}\right)\right]^d e^{-t}\,dt, \tag{10}$$

and $I_0(z)$ is a MODIFIED BESSEL FUNCTION OF THE FIRST KIND. Numerical values from Montroll (1956) and Flajolet (Finch) are

d	$p(d)$
4	0.20
5	0.136
6	0.105
7	0.0858
8	0.0729

see also RANDOM WALK

References

Finch, S. "Favorite Mathematical Constants." `http://www.mathsoft.com/asolve/constant/polya/polya.html`.

Domb, C. "On Multiple Returns in the Random-Walk Problem." *Proc. Cambridge Philos. Soc.* **50**, 586–591, 1954.

Glasser, M. L. and Zucker, I. J. "Extended Watson Integrals for the Cubic Lattices." *Proc. Nat. Acad. Sci. U.S.A.* **74**, 1800–1801, 1977.

McCrea, W. H. and Whipple, F. J. W. "Random Paths in Two and Three Dimensions." *Proc. Roy. Soc. Edinburgh* **60**, 281–298, 1940.

Montroll, E. W. "Random Walks in Multidimensional Spaces, Especially on Periodic Lattices." *J. SIAM* **4**, 241–260, 1956.

Watson, G. N. "Three Triple Integrals." *Quart. J. Math., Oxford Ser. 2* **10**, 266–276, 1939.

Pólya-Vinogradov Inequality

Let χ be a nonprincipal character (mod q). Then

$$\sum_{n=M+1}^{M+N}\chi(n) \ll \sqrt{q}\ln q,$$

where $\ll$ indicates MUCH LESS than.

References

Davenport, H. "The Pólya-Vinogradov Inequality." Ch. 23 in *Multiplicative Number Theory, 2nd ed.* New York: Springer-Verlag, pp. 135–138, 1980.

Pólya, G. "Über die Verteilung der quadratischen Reste und Nichtreste." *Nachr. Königl. Gesell. Wissensch. Göttingen, Math.-Phys. Klasse,* 21–29, 1918.

Polyabolo

An analog of the POLYOMINO composed of n ISOSCELES RIGHT TRIANGLES joined along edges of the same length. The number of polyaboloes composed of n triangles are 1, 3, 4, 14, 30, 107, 318, 1106, 3671, ... (Sloane's A006074).

see also DIABOLO, HEXABOLO, PENTABOLO, TETRABOLO, TRIABOLO

References

Sloane, N. J. A. Sequence A006074/M2379 in "An On-Line Version of the Encyclopedia of Integer Sequences."

Polyconic Projection

$$x = \cot\phi \sin E \tag{1}$$

$$y = (\phi - \phi_0) + \cot\phi(1 - \cos E), \tag{2}$$

where

$$E = (\lambda - \lambda_0)\sin\phi. \tag{3}$$

The inverse FORMULAS are

$$\lambda = \frac{\sin^{-1}(x\tan\phi)}{\sin\phi} + \lambda_0, \tag{4}$$

and ϕ is determined from

$$\Delta\phi = -\frac{A(\phi\tan\phi + 1) - \phi - \frac{1}{2}(\phi^2 + B)\tan\phi}{\frac{\phi - A}{\tan\phi} - 1}, \tag{5}$$

where $\phi_0 = A$ and

$$A = \phi_0 + y \tag{6}$$

$$B = x^2 + A^2. \tag{7}$$

References

Snyder, J. P. *Map Projections—A Working Manual.* U. S. Geological Survey Professional Paper 1395. Washington, DC: U. S. Government Printing Office, pp. 124–137, 1987.

Polycube

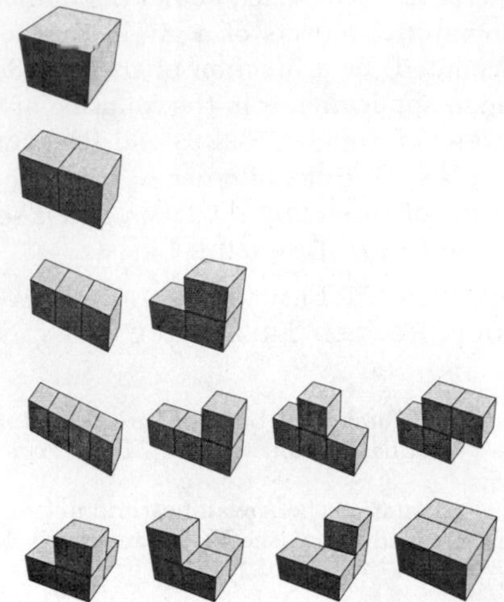

3-D generalization of the POLYOMINOES to n-D. The number of polycubes $N(n)$ composed of n CUBES are 1, 1, 2, 8, 29, 166, 1023, ... (Sloane's A000162, Ball and Coxeter 1987).

see also CONWAY PUZZLE, CUBE DISSECTION, DIABOLICAL CUBE, SLOTHOUBER-GRAATSMA PUZZLE, SOMA CUBE

References

Ball, W. W. R. and Coxeter, H. S. M. *Mathematical Recreations and Essays, 13th ed.* New York: Dover, pp. 112–113, 1987.

Gardner, M. *The Second Scientific American Book of Mathematical Puzzles & Diversions: A New Selection.* New York: Simon and Schuster, pp. 76–77, 1961.

Gardner, M. "Polycubes." Ch. 3 in *Knotted Doughnuts and Other Mathematical Entertainments.* New York: W. H. Freeman, 1986.

Sloane, N. J. A. Sequence A000162/M1845 in "An On-Line Version of the Encyclopedia of Integer Sequences."

Polydisk

Let $\mathbf{c} = (c_1, \ldots, c_n)$ be a point in $\mathbb{C}^n$, then the open polydisk is defined by

$$S = \{z : |z_j - c_j| < |z_j^0 - c_j|\}$$

for $j = 1, \ldots, n$.

see also DISK, OPEN DISK

References

Iyanaga, S. and Kawada, Y. (Eds.). *Encyclopedic Dictionary of Mathematics.* Cambridge, MA: MIT Press, p. 100, 1980.

Polygamma Function

The polygamma function is sometimes denoted $F_m(z)$, and sometimes $\psi_m(z)$. In $F_m(z)$ notation,

$$F_m(z) \equiv \frac{d^{m+1}}{dz^{m+1}} \ln z! \tag{1}$$

$$= (-1)^{m+1} m! \sum_{n=0}^{\infty} \frac{1}{(z+n)^{m+1}} \tag{2}$$

$$= (-1)^{m+1} m! \zeta(m+1, z), \tag{3}$$

where $\zeta(a,z)$ is the HURWITZ ZETA FUNCTION. In the ψ_m NOTATION (the form returned by the `PolyGamma[m,z]` function in *Mathematica*®; Wolfram Research, Champaign, IL),

$$\psi_m(z) = \frac{d^{m+1}}{dz^{m+1}} \ln[\Gamma(z)] = \frac{d^m}{dz^m} \frac{\Gamma'(z)}{\Gamma(z)} = \frac{d^m}{dz^m} \Psi(z), \quad (4)$$

where $\Gamma(z)$ is the GAMMA FUNCTION and $\Psi(z)$ is the DIGAMMA FUNCTION. $\psi_m(z)$ is therefore related to $F_m(z)$ by

$$\psi_m(z) = F_m(z-1). \quad (5)$$

The function $\psi_0(z)$ is equivalent to the DIGAMMA FUNCTION $\Psi(z)$. Note that Morse and Feshbach (1953) adopt a notation no longer in standard use in which Morse and Feshbach's $\psi_m(z)$ is equal to the above $\psi_{m-1}(z)$.

The polygamma function obeys the RECURRENCE RELATION

$$\psi_n(z+1) = \psi_n(z) + (-1)^n n! z^{-n-1}, \quad (6)$$

the reflection FORMULA

$$\psi_n(1-z) + (-1)^{n+1}\psi_n(z) = (-1)^n \pi \frac{d^n}{dz^n} \cot(\pi z), \quad (7)$$

and the multiplication FORMULA,

$$\psi_n(mz) = \delta_{n0} \ln m + \frac{1}{m^{n+1}} \sum_{k=1}^{m-1} \psi_n\left(z + \frac{k}{m}\right), \quad (8)$$

where δ_{mn} is the KRONECKER DELTA.

In general, special values for integral indices are given by

$$\psi_n(1) = (-1)^{n+1} n! \zeta(n+1) \quad (9)$$
$$\psi_n(\tfrac{1}{2}) = (-1)^{n+1} n! (2^{n+1}-1)\zeta(n+1), \quad (10)$$

giving

$$\psi_1(\tfrac{1}{2}) = \tfrac{1}{2}\pi^2 \quad (11)$$
$$\psi_1(1) = \zeta(2) = \tfrac{1}{6}\pi^2 \quad (12)$$
$$\psi_2(1) = -2\zeta(3), \quad (13)$$
$$\psi_3(\tfrac{1}{2}) = \pi^4 \quad (14)$$

and so on.

R. Manzoni has shown that the polygamma function can be expressed in terms of CLAUSEN FUNCTIONS for RATIONAL arguments and integer index. Special cases are given by

$$\psi_1(\tfrac{1}{3}) = \tfrac{2}{3}\pi^2 + \tfrac{3}{2}\sqrt{3}[\mathrm{Cl}_2(\tfrac{2}{3}\pi) - \mathrm{Cl}_2(\tfrac{4}{3}\pi) \quad (15)$$
$$\psi_1(\tfrac{2}{3}) = \tfrac{2}{3}\pi^2 - \tfrac{3}{2}\sqrt{3}[\mathrm{Cl}_2(\tfrac{2}{3}\pi) - \mathrm{Cl}_2(\tfrac{4}{3}\pi) \quad (16)$$
$$\psi_1(\tfrac{1}{4}) = \pi^2 + 4[\mathrm{Cl}_2(\tfrac{1}{2}\pi) - \mathrm{Cl}_2(\tfrac{3}{2}\pi)] \quad (17)$$
$$\psi_1(\tfrac{3}{4}) = \pi^2 - 4[\mathrm{Cl}_2(\tfrac{1}{2}\pi) - \mathrm{Cl}_2(\tfrac{3}{2}\pi)]. \quad (18)$$
$$\psi_2(\tfrac{1}{2}) = -8[\mathrm{Cl}_3(0) - \mathrm{Cl}_3(\pi)]. \quad (19)$$
$$\psi_2(\tfrac{1}{3}) = -\frac{4\pi^3}{3\sqrt{3}} - 18\,\mathrm{Cl}_3(0) + 9[\mathrm{Cl}_3(\tfrac{2}{3}\pi) + \mathrm{Cl}_3(\tfrac{4}{3}\pi)] \quad (20)$$
$$\psi_2(\tfrac{2}{3}) = \frac{4\pi^3}{3\sqrt{3}} - 18\,\mathrm{Cl}_3(0) + 9[\mathrm{Cl}_3(\tfrac{2}{3}\pi) + \mathrm{Cl}_3(\tfrac{4}{3}\pi)] \quad (21)$$
$$\psi_2(\tfrac{1}{4}) = -2\pi^3 - 32[\mathrm{Cl}_3(0) - \mathrm{Cl}_3(\pi)] \quad (22)$$
$$\psi_3(\tfrac{3}{4}) = 2\pi^3 - 32[\mathrm{Cl}_3(0) - \mathrm{Cl}_3(\pi)] \quad (23)$$
$$\psi_3(\tfrac{1}{3}) = \tfrac{8}{3}\pi^4 + 81\sqrt{3}[\mathrm{Cl}_4(\tfrac{2}{3}\pi) - \mathrm{Cl}_4(\tfrac{4}{3}\pi)] \quad (24)$$
$$\psi_3(\tfrac{2}{3}) = \tfrac{8}{3}\pi^4 - 81\sqrt{3}[\mathrm{Cl}_4(\tfrac{2}{3}\pi) - \mathrm{Cl}_4(\tfrac{4}{3}\pi)] \quad (25)$$
$$\psi_3(\tfrac{1}{4}) = 8\pi^4 + 384[\mathrm{Cl}_4(\tfrac{1}{2}\pi) - \mathrm{Cl}_4(\tfrac{3}{2}\pi)] \quad (26)$$
$$\psi_3(\tfrac{3}{4}) = 8\pi^4 - 384[\mathrm{Cl}_4(\tfrac{1}{2}\pi) - \mathrm{Cl}_4(\tfrac{3}{2}\pi)]. \quad (27)$$

see also CLAUSEN FUNCTION, DIGAMMA FUNCTION, GAMMA FUNCTION, STIRLING'S SERIES

References

Abramowitz, M. and Stegun, C. A. (Eds.). "Polygamma Functions." §6.4 in *Handbook of Mathematical Functions with Formulas, Graphs, and Mathematical Tables, 9th printing.* New York: Dover, p. 260, 1972.

Adamchik, V. S. "Polygamma Functions of Negative Order." Submitted to *J. Symb. Comput.* `http://www.wolfram.com/~victor/articles/polyg.html`.

Arfken, G. "Digamma and Polygamma Functions." §10.2 in *Mathematical Methods for Physicists, 3rd ed.* Orlando, FL: Academic Press, pp. 549–555, 1985.

Davis, H. T. *Tables of the Higher Mathematical Functions.* Bloomington, IN: Principia Press, 1933.

Kolbig, V. "The Polygamma Function $\psi_k(x)$ for $x = 1/4$ and $x = 3/4$." *J. Comp. Appl. Math.* **75**, 43–46, 1996.

Morse, P. M. and Feshbach, H. *Methods of Theoretical Physics, Part I.* New York: McGraw-Hill, pp. 422–424, 1953.

Polygenic Function

A function which has infinitely many DERIVATIVES at a point. If a function is not polygenic, it is MONOGENIC.

see also MONOGENIC FUNCTION

References

Newman, J. R. *The World of Mathematics, Vol. 3.* New York: Simon & Schuster, p. 2003, 1956.

Polygon

A closed plane figure with n sides. If all sides and angles are equivalent, the polygon is called regular. Regular polygons can be CONVEX or STAR. The word derives from the Greek *poly* (many) and *gonu* (knee).

The AREA of a polygon with VERTICES (x_1, y_1), ..., (x_n, y_n) is

$$A = \frac{1}{2}\left(\begin{vmatrix} x_1 & y_1 \\ x_2 & y_2 \end{vmatrix} + \begin{vmatrix} x_2 & y_2 \\ x_3 & y_3 \end{vmatrix} + \ldots + \begin{vmatrix} x_n & y_n \\ x_1 & y_1 \end{vmatrix} \right), \quad (1)$$

which can be written

$$A = \tfrac{1}{2}(x_1y_2 + x_2y_1 + \ldots + x_{n-1}y_n + x_ny_1 - y_1x_2 - y_2x_3 - \ldots - y_{n+1}x_n - y_nx_1), \quad (2)$$

where the signs can be found from the following diagram.

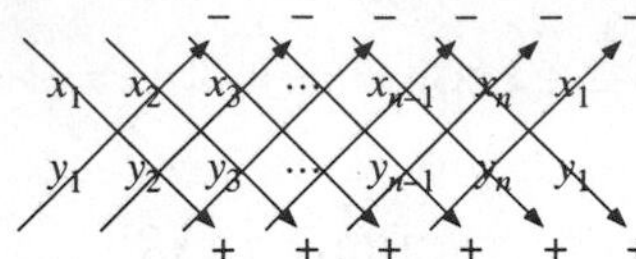

The AREA of a polygon is defined to be POSITIVE if the points are arranged in a counterclockwise order, and NEGATIVE if they are in clockwise order (Beyer 1987).

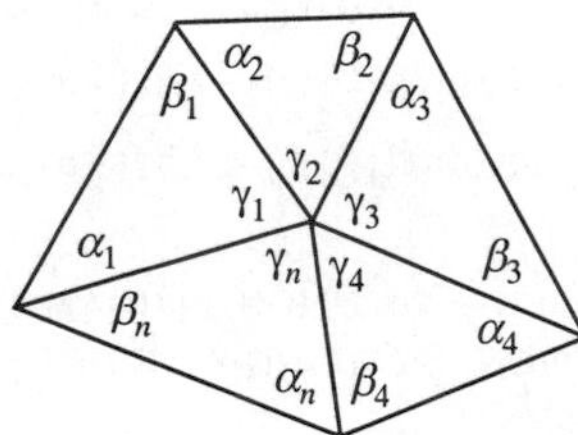

The sum I of internal angles in the above diagram of a dissected PENTAGON is

$$I = \sum_{i=1}^{n}(\alpha_i + \beta_i) = \sum_{i=1}^{n}(\alpha_i + \beta_i + \gamma_i) - \sum_{i=1}^{n}\gamma_i. \quad (3)$$

But

$$\sum_{i=1}^{n}\gamma_i = 360° \quad (4)$$

and the sum of ANGLES of the n TRIANGLES is

$$\sum_{i=1}^{n}(\alpha_i + \beta_i + \gamma_i) = \sum_{i=1}^{n}(180°) = n(180°). \quad (5)$$

Therefore,

$$I = n(180°) - 360° = (n-2)180°. \quad (6)$$

Let n be the number of sides. The *regular* n-gon is then denoted $\{n\}$.

n	$\{n\}$
2	digon
3	equilateral triangle (trigon)
4	square (quadrilateral, tetragon)
5	pentagon
6	hexagon
7	heptagon
8	octagon
9	nonagon (enneagon)
10	decagon
11	undecagon (hendecagon)
12	dodecagon
13	tridecagon (triskaidecagon)
14	tetradecagon (tetrakaidecagon)
15	pentadecagon (pentakaidecagon)
16	hexadecagon (hexakaidecagon)
17	heptadecagon (heptakaidecagon)
18	octadecagon (octakaidecagon)
19	enneadecagon (enneakaidecagon)
20	icosagon
30	triacontagon
40	tetracontagon
50	pentacontagon
60	hexacontagon
70	heptacontagon
80	octacontagon
90	enneacontagon
100	hectogon
10000	myriagon

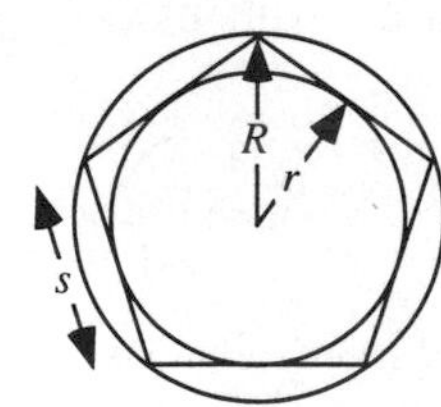

$n = 5$

Let s be the side length, r be the INRADIUS, and R the CIRCUMRADIUS. Then

$$s = 2r\tan\left(\frac{\pi}{n}\right) = 2R\sin\left(\frac{\pi}{n}\right) \quad (7)$$

$$r = \tfrac{1}{2}s\cot\left(\frac{\pi}{n}\right) \quad (8)$$

$$R = \tfrac{1}{2}s\csc\left(\frac{\pi}{n}\right) \quad (9)$$

$$A = \tfrac{1}{4}ns^2\cot\left(\frac{\pi}{n}\right) \quad (10)$$

$$= nr^2\tan\left(\frac{\pi}{n}\right) \quad (11)$$

$$= \tfrac{1}{2}nR^2\sin\left(\frac{2\pi}{n}\right). \quad (12)$$

If the number of sides is doubled, then

$$s_{2n} = \sqrt{2R^2 - R\sqrt{4R^2 - {s_n}^2}} \quad (13)$$

$$A_{2n} = \frac{4rA_n}{2r + \sqrt{4r^2 + {s_n}^2}}. \quad (14)$$

Furthermore, if p_k and P_k are the PERIMETERS of the regular polygons inscribed in and circumscribed around a given CIRCLE and a_k and A_k their areas, then

$$P_{2n} = \frac{2p_n P_n}{p_n + P_n} \tag{15}$$

$$p_{2n} = \sqrt{p_n P_{2n}}, \tag{16}$$

and

$$a_{2n} = \sqrt{a_n A_n} \tag{17}$$

$$A_{2n} = \frac{2a_{2n} A_n}{a_{2n} + A_n} \tag{18}$$

(Beyer 1987, p. 125).

COMPASS and STRAIGHTEDGE constructions dating back to Euclid were capable of inscribing regular polygons of 3, 4, 5, 6, 8, 10, 12, 16, 20, 24, 32, 40, 48, 64, ..., sides. However, this listing is not a complete enumeration of "constructible" polygons. In fact, a regular n-gon is constructible only if $\phi(n)$ is a POWER of 2, where ϕ is the TOTIENT FUNCTION (this is a NECESSARY but not SUFFICIENT condition). More specifically, a regular n-gon ($n \geq 3$) can be constructed by STRAIGHTEDGE and COMPASS (i.e., can have trigonometric functions of its ANGLES expressed in terms of finite SQUARE ROOT extractions) IFF

$$n = 2^k p_1 p_2 \cdots p_s, \tag{19}$$

where k is in INTEGER ≥ 0 and the p_i are distinct FERMAT PRIMES. FERMAT NUMBERS are of the form

$$F_m = 2^{2^m} + 1, \tag{20}$$

where m is an INTEGER ≥ 0. The only known PRIMES of this form are 3, 5, 17, 257, and 65537.

The fact that this condition was SUFFICIENT was first proved by Gauss in 1796 when he was 19 years old, and it relies on the property of IRREDUCIBLE POLYNOMIALS that ROOTS composed of a finite number of SQUARE ROOT extractions exist only if the order of the equation is of the form 2^h. That this condition was also NECESSARY was not explicitly proven by Gauss, and the first proof of this fact is credited to Wantzel (1836).

Constructible values of n for $n < 300$ were given by Gauss (Smith 1994), and the first few are 2, 3, 4, 5, 6, 8, 10, 12, 15, 16, 17, 20, 24, 30, 32, 34, 40, 48, 51, 60, 64, 68, 80, 85, 96, 102, 120, 128, 136, 160, 170, 192, ... (Sloane's A003401). Gardner (1977) and independently Watkins (Conway and Guy 1996) noticed that the number of sides for constructible polygons with an ODD number of sides is given by the first 32 rows of PASCAL'S TRIANGLE (mod 2) interpreted as BINARY numbers, giving 1, 3, 5, 15, 17, 51, 85, 255, ... (Sloane's A004729, Conway and Guy 1996, p. 140).

1	1	1
1 1	1 1	3
1 2 1	1 0 1	5
1 3 3 1	1 1 1 1	15
1 4 6 4 1	1 0 0 0 1	17
1 5 10 10 5 1	1 1 0 0 1 1	51
1 6 15 20 15 6 1	1 0 1 0 1 0 1	85
1 7 21 35 35 21 7 1	1 1 1 1 1 1 1 1	255
1 8 28 56 70 56 28 8 1	1 0 0 0 0 0 0 0 1	257

Although constructions for the regular TRIANGLE, SQUARE, PENTAGON, and their derivatives had been given by Euclid, constructions based on the FERMAT PRIMES $\geq$ 17 were unknown to the ancients. The first explicit construction of a HEPTADECAGON (17-gon) was given by Erchinger in about 1800. Richelot and Schwendenwein found constructions for the 257-GON in 1832, and Hermes spent 10 years on the construction of the 65537-GON at Göttingen around 1900 (Coxeter 1969). Constructions for the EQUILATERAL TRIANGLE and SQUARE are trivial (top figures below). Elegant constructions for the PENTAGON and HEPTADECAGON are due to Richmond (1893) (bottom figures below).

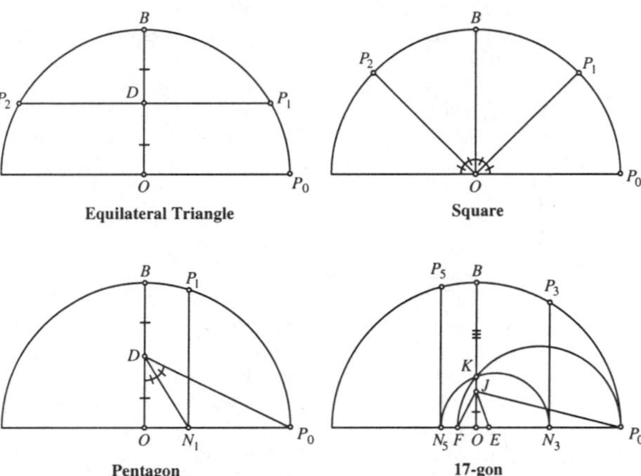

Given a point, a CIRCLE may be constructed of any desired RADIUS, and a DIAMETER drawn through the center. Call the center O, and the right end of the DIAMETER P_0. The DIAMETER PERPENDICULAR to the original DIAMETER may be constructed by finding the PERPENDICULAR BISECTOR. Call the upper endpoint of this PERPENDICULAR DIAMETER B. For the PENTAGON, find the MIDPOINT of OB and call it D. Draw DP_0, and BISECT $\angle ODP_0$, calling the intersection point with OP_0 N_1. Draw N_1P_1 PARALLEL to OB, and the first two points of the PENTAGON are P_0 and P_1. The construction for the HEPTADECAGON is more complicated, but can be accomplished in 17 relatively simple steps. The construction problem has now been automated (Bishop 1978).

see also 257-GON, 65537-GON, ANTHROPOMORPHIC POLYGON, BICENTRIC POLYGON, CARNOT'S POLYGON THEOREM, CHAOS GAME, CONVEX POLYGON, CYCLIC POLYGON, DE MOIVRE NUMBER, DIAGONAL (POLYGON), EQUILATERAL TRIANGLE, EULER'S POLYGON DIVISION PROBLEM, HEPTADECAGON, HEXAGON,

HEXAGRAM, ILLUMINATION PROBLEM, JORDAN POLYGON, LOZENGE, OCTAGON, PARALLELOGRAM, PASCAL'S THEOREM, PENTAGON, PENTAGRAM, PETRIE POLYGON, POLYGON CIRCUMSCRIBING CONSTANT, POLYGON INSCRIBING CONSTANT, POLYGONAL KNOT, POLYGONAL NUMBER, POLYGONAL SPIRAL, POLYGON TRIANGULATION, POLYGRAM, POLYHEDRAL FORMULA, POLYHEDRON, POLYTOPE, QUADRANGLE, QUADRILATERAL, REGULAR POLYGON, REULEAUX POLYGON, RHOMBUS, ROTOR, SIMPLE POLYGON, SIMPLICITY, SQUARE, STAR POLYGON, TRAPEZIUM, TRAPEZOID, TRIANGLE, VISIBILITY, VORONOI POLYGON, WALLACE-BOLYAI-GERWEIN THEOREM

References

Beyer, W. H. *CRC Standard Mathematical Tables, 28th ed.* Boca Raton, FL: CRC Press, pp. 124–125 and 196, 1987.

Bishop, W. "How to Construct a Regular Polygon." *Amer. Math. Monthly* **85**, 186–188, 1978.

Conway, J. H. and Guy, R. K. *The Book of Numbers.* New York: Springer-Verlag, pp. 140 and 197–202, 1996.

Courant, R. and Robbins, H. "Regular Polygons." §3.2 in *What is Mathematics?: An Elementary Approach to Ideas and Methods, 2nd ed.* Oxford, England: Oxford University Press, pp. 122–125, 1996.

Coxeter, H. S.M. *Introduction to Geometry, 2nd ed.* New York: Wiley, 1969.

De Temple, D. W. "Carlyle Circles and the Lemoine Simplicity of Polygonal Constructions." *Amer. Math. Monthly* **98**, 97–108, 1991.

Gardner, M. *Mathematical Carnival: A New Round-Up of Tantalizers and Puzzles from Scientific American.* New York: Vintage Books, p. 207, 1977.

Gauss, C. F. §365 and 366 in *Disquisitiones Arithmeticae.* Leipzig, Germany, 1801. Translated by A. A Clarke. New Haven, CT: Yale University Press, 1965.

The Math Forum. "Naming Polygons and Polyhedra." `http://forum.swarthmore.edu/dr.math/faq/faq.polygon.names.html`.

Rawles, B. *Sacred Geometry Design Sourcebook: Universal Dimensional Patterns.* Nevada City, CA: Elysian Pub., p. 238, 1997.

Richmond, H. W. "A Construction for a Regular Polygon of Seventeen Sides." *Quart. J. Pure Appl. Math.* **26**, 206–207, 1893.

Sloane, N. J. A. Sequences A004729 and A003401/M0505 in "An On-Line Version of the Encyclopedia of Integer Sequences."

Smith, D. E. *A Source Book in Mathematics.* New York: Dover, p. 350, 1994.

Tietze, H. Ch. 9 in *Famous Problems of Mathematics.* New York: Graylock Press, 1965.

Wantzel, P. L. "Recherches sur les moyens de reconnaître si un Problème de Géométrie peut se résoudre avec la règle et le compas." *J. Math. pures appliq.* **1**, 366–372, 1836.

Polygon Circumscribing Constant

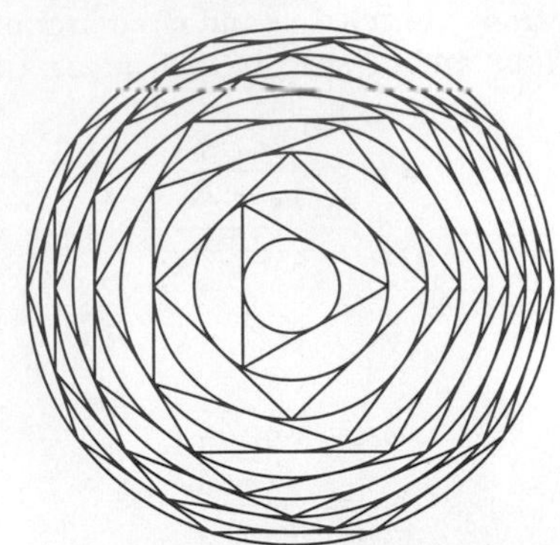

If a TRIANGLE is CIRCUMSCRIBED about a CIRCLE, another CIRCLE around the TRIANGLE, a SQUARE outside the CIRCLE, another CIRCLE outside the SQUARE, and so on. From POLYGONS, the CIRCUMRADIUS and INRADIUS for an n-gon are

$$R = \tfrac{1}{2}s\csc\left(\frac{\pi}{n}\right) \tag{1}$$

$$r = \tfrac{1}{2}s\cot\left(\frac{\pi}{n}\right), \tag{2}$$

where s is the side length. Therefore,

$$\frac{R}{r} = \frac{1}{\cos\left(\frac{\pi}{n}\right)} = \sec\left(\frac{\pi}{n}\right), \tag{3}$$

and an infinitely nested set of circumscribed polygons and circles has

$$K \equiv \frac{r_{\text{final circle}}}{r_{\text{initial circle}}} = \sec\left(\frac{\pi}{3}\right)\sec\left(\frac{\pi}{4}\right)\sec\left(\frac{\pi}{5}\right)\cdots. \tag{4}$$

Kasner and Newman (1989) and Haber (1964) state that $K = 12$, but this is incorrect. Write

$$K = \prod_{n=3}^{\infty} \frac{1}{\cos\left(\frac{\pi}{n}\right)} \tag{5}$$

$$\ln K = -\sum_{n=3}^{\infty} \ln(\cos x). \tag{6}$$

Define

$$y_0(x) \equiv -\ln(\cos x) = \tfrac{1}{2}x^2 + \tfrac{1}{12}x^4 + \tfrac{1}{45}x^6 + \tfrac{17}{2520}x^8 + \ldots. \tag{7}$$

Now define

$$y_1(x) = \tfrac{1}{2}ax^2, \tag{8}$$

with

$$y_1(\tfrac{\pi}{3}) = y_0(\tfrac{\pi}{3}) \tag{9}$$

$$\tfrac{1}{2}a\left(\frac{\pi}{3}\right)^2 = \ln 2, \tag{10}$$

so

$$a = 2\left(\frac{3}{\pi}\right)^2 \ln 2, \tag{11}$$

and

$$y_2(x) = \frac{9 \ln 2}{\pi^2} x^2. \quad (12)$$

But $y_2(x) > y_1(x)$ for $x \in (0, \pi/3)$, so

$$\sum_{n=3}^{\infty} y_2\left(\frac{\pi}{n}\right) > -\sum_{n=3}^{\infty} \ln\left[\cos\left(\frac{\pi}{n}\right)\right] \quad (13)$$

$$\begin{aligned} \ln K &< \sum_{n=3}^{\infty} y_2\left(\frac{\pi}{n}\right) \frac{9\ln 2}{\pi^2} \sum_{n=3}^{\infty} \left(\frac{\pi}{n}\right)^2 = 9 \ln 2 \sum_{n=3}^{\infty} \frac{1}{n^2} \\ &= 9 \ln 2 \left(\sum_{n=1}^{\infty} \frac{1}{n^2} - \sum_{n=1}^{2} \frac{1}{n^2}\right) = 9 \ln 2 [\zeta(2) - \tfrac{5}{4}] \\ &= 9 \ln 2 \left(\frac{\pi^2}{6} - \frac{5}{4}\right) = 2.4637 \end{aligned} \quad (14)$$

$$K < e^{2.4637} = 11.75. \quad (15)$$

If the next term is included,

$$y_2(x) = a(\tfrac{1}{2}x^2 + \tfrac{1}{12}x^4). \quad (16)$$

As before,

$$y_2(\tfrac{\pi}{3}) = y_0(\tfrac{\pi}{3}) \quad (17)$$

$$a = \frac{972 \ln 2}{\pi^2 (54 + \pi^2)}, \quad (18)$$

so

$$y_2(x) = \frac{972 \ln 2}{\pi^2(54 + \pi^2)} (\tfrac{1}{2}x^2 + \tfrac{1}{12}x^4) \quad (19)$$

$$\begin{aligned} \ln K &< \frac{972 \ln 2}{\pi^2(54+\pi^2)} \sum_{n=3}^{\infty} \left[\frac{1}{2}\left(\frac{\pi}{n}\right)^2 + \frac{1}{12}\left(\frac{\pi}{n}\right)^4\right] \\ &= \frac{972 \ln 2}{\pi^2(54+\pi^2)} \left\{\frac{1}{2}\left[\zeta(2) - \frac{5}{4}\right] + \frac{\pi^2}{12}\left[\zeta(4) - 1 - \frac{1}{2^4}\right]\right\} \\ &= \frac{972 \ln 2}{\pi^2(54+\pi^2)} \left[\frac{1}{2}\left(\frac{\pi^2}{6} - \frac{5}{4}\right) + \frac{\pi^2}{12}\left(\frac{\pi^4}{90} - 1 - \frac{1}{2^4}\right)\right] \\ &= \frac{9(8\pi^6 - 45\pi^2 - 5400) \ln 2}{80(\pi^2 + 54)} = 2.255, \end{aligned} \quad (20)$$

and

$$K < e^{2.255} = 9.535. \quad (21)$$

The process can be automated using computer algebra, and the first few bounds are 11.7485, 9.53528, 8.98034, 8.8016, 8.73832, 8.71483, 8.70585, 8.70235, 8.70097, and 8.70042. In order to obtain this accuracy by direct multiplication of the terms, more than 10,000 terms are needed. The limit is

$$K = 8.700036625\ldots. \quad (22)$$

Bouwkamp (1965) produced the following INFINITE PRODUCT formulas

$$K = \frac{2}{\pi} \prod_{m=1}^{\infty} \prod_{n=1}^{\infty} \left[1 - \frac{1}{m^2(n + \frac{1}{2})^2}\right] \quad (23)$$

$$= 6 \exp\left\{\sum_{k=1}^{\infty} \frac{[\lambda(2k) - 1]2^{2k}[\zeta(2k) - 1 - 2^{-2k}]}{k}\right\}, \quad (24)$$

where $\zeta(x)$ is the RIEMANN ZETA FUNCTION and $\lambda(x)$ is the DIRICHLET LAMBDA FUNCTION. Bouwkamp (1965) also produced the formula with accelerated convergence

$$\begin{aligned} K &= \tfrac{1}{12}\sqrt{6}\pi^4(1 - \tfrac{1}{2}\pi^2 + \tfrac{1}{24}\pi^4)(1 - \tfrac{1}{8}\pi^2 + \tfrac{1}{384}\pi^4) \\ &\quad \times \csc\left(\frac{\pi^2}{\sqrt{6+2\sqrt{3}}}\right) \csc\left(\frac{\pi^2}{\sqrt{6-2\sqrt{3}}}\right) B, \end{aligned} \quad (25)$$

where

$$B \equiv \prod_{n=3}^{\infty} \left(1 - \frac{\pi^2}{2n^2} + \frac{\pi^4}{24n^4}\right) \sec\left(\frac{\pi}{n}\right) \quad (26)$$

(cited in Pickover 1995).

see also POLYGON INSCRIBING CONSTANT

References

Bouwkamp, C. "An Infinite Product." *Indag. Math.* **27**, 40–46, 1965.

Finch, S. "Favorite Mathematical Constants." `http://www.mathsoft.com/asolve/constant/infprd/infprd.html`.

Haber, H. "Das Mathematische Kabinett." *Bild der Wissenschaft* **2**, 73, Apr. 1964.

Kasner, E. and Newman, J. R. *Mathematics and the Imagination.* Redmond, WA: Microsoft Press, pp. 311–312, 1989.

Pappas, T. "Infinity & Limits." *The Joy of Mathematics.* San Carlos, CA: Wide World Publ./Tetra, p. 180, 1989.

Pickover, C. A. "Infinitely Exploding Circles." Ch. 18 in *Keys to Infinity.* New York: W. H. Freeman, pp. 147–151, 1995.

Pinkham, R. S. "Mathematics and Modern Technology." *Amer. Math. Monthly* **103**, 539–545, 1996.

Plouffe, S. "Product(cos(Pi/n),n=3..infinity)." `http://lacim.uqam.ca/piDATA/productcos.txt`.

Polygon Construction

see GEOMETRIC CONSTRUCTION, GEOMETROGRAPHY, POLYGON, SIMPLICITY

Polygon Division Problem

see EULER'S POLYGON DIVISION PROBLEM

Polygon Fractal

see CHAOS GAME

Polygon Inscribing Constant

If a TRIANGLE is inscribed in a CIRCLE, another CIRCLE inside the TRIANGLE, a SQUARE inside the CIRCLE, another CIRCLE inside the SQUARE, and so on,

$$K' \equiv \frac{r_{\text{final circle}}}{r_{\text{initial circle}}} = \cos\left(\frac{\pi}{3}\right)\cos\left(\frac{\pi}{4}\right)\cos\left(\frac{\pi}{5}\right)\cdots.$$

Numerically,

$$K' = \frac{1}{K} = \frac{1}{8.7000366252\ldots} = 0.1149420448\ldots,$$

where K is the POLYGON CIRCUMSCRIBING CONSTANT. Kasner and Newman's (1989) assertion that $K = 1/12$ is incorrect.

Let a convex POLYGON be inscribed in a CIRCLE and divided into TRIANGLES from diagonals from one VERTEX. The sum of the RADII of the CIRCLES inscribed in these TRIANGLES is the same independent of the VERTEX chosen (Johnson 1929, p. 193).

see also POLYGON CIRCUMSCRIBING CONSTANT

References

Finch, S. "Favorite Mathematical Constants." `http://www.mathsoft.com/asolve/constant/infprd/infprd.html`.

Johnson, R. A. *Modern Geometry: An Elementary Treatise on the Geometry of the Triangle and the Circle.* Boston, MA: Houghton Mifflin, 1929.

Kasner, E. and Newman, J. R. *Mathematics and the Imagination.* Redmond, WA: Microsoft Press, pp. 311–312, 1989.

Pappas, T. "Infinity & Limits." *The Joy of Mathematics.* San Carlos, CA: Wide World Publ./Tetra, p. 180, 1989.

Plouffe, S. "Product(cos(Pi/n),n=3..infinity)." `http://lacim.uqam.ca/piDATA/productcos.txt`.

Polygon Triangulation

see EULER'S POLYGON DIVISION PROBLEM

Polygonal Knot

A KNOT equivalent to a POLYGON in $\mathbb{R}^3$, also called a TAME KNOT. For a polygonal knot K, there exists a PLANE such that the orthogonal projection π on it satisfies the following conditions:

1. The image $\pi(K)$ has no multiple points other than a FINITE number of double points.
2. The projections of the vertices of K are not double points of $\pi(K)$.

Such a projection $\pi(K)$ is called a regular knot projection.

References

Iyanaga, S. and Kawada, Y. (Eds.). *Encyclopedic Dictionary of Mathematics.* Cambridge, MA: MIT Press, p. 735, 1980.

Polygonal Number

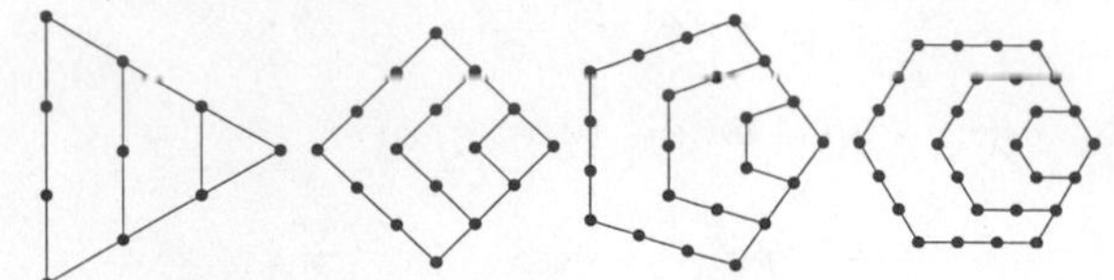

A type of FIGURATE NUMBER which is a generalization of TRIANGULAR, SQUARE, etc., numbers to an arbitrary n-gonal number. The above diagrams graphically illustrate the process by which the polygonal numbers are built up. Starting with the nth TRIANGULAR NUMBER T_n, then

$$n + T_{n-1} = T_n. \tag{1}$$

Now note that

$$n + 2T_{n-1} = n^2 = S_n \tag{2}$$

gives the nth SQUARE NUMBER,

$$n + 3T_{n-1} = \tfrac{1}{2}n(3n-1) = P_n, \tag{3}$$

gives the nth PENTAGONAL NUMBER, and so on. The general polygonal number can be written in the form

$$p_r^n = \tfrac{1}{2}r[(r-1)n - 2(r-2)] = \tfrac{1}{2}r[(n-2)r - (n-4)], \tag{4}$$

where p_r^n is the rth n-gonal number. For example, taking $n = 3$ in (4) gives a TRIANGULAR NUMBER, $n = 4$ gives a SQUARE NUMBER, etc.

Fermat proposed that every number is expressible as *at most k k*-gonal numbers (FERMAT'S POLYGONAL NUMBER THEOREM). Fermat claimed to have a proof of this result, although this proof has never been found. Jacobi, Lagrange (1772), and Euler all proved the square case, and Gauss proved the triangular case in 1796. In 1813, Cauchy proved the proposition in its entirety.

An arbitrary number N can be checked to see if it is a n-gonal number as follows. Note the identity

$$\begin{aligned} 8(n-2)p_n^r + (n-4)^2 &= 4r(n-2)[(r-1)n - 2(r-2)] \\ +(n-4)^2 &= 4r(r-1)n^2 + r[-8(r-1) - 8(r-2)]n \\ &\quad +16r(r-2) + (n^2 - 8n + 16) \\ &= (4r^2 - 4r + 1)n^2 + (-16r^2 + 24r - 8)n \\ &\quad +(16r^2 - 32r + 16) \\ &= (2r-1)^2n^2 - 8(2r^2 - 3r + 1)n + 16(r^2 - 2r + 1) \\ &= (2rn - 4r - n + 4)^2, \end{aligned} \tag{5}$$

so $8(n-2)N + (n-4)^2 = S^2$ must be a PERFECT SQUARE. Therefore, if it is not, the number cannot be n-gonal. If it is a PERFECT SQUARE, then solving

$$S = 2rn - 4r - n + 4 \tag{6}$$

for the rank r gives

$$r = \frac{S+n-4}{2(n-2)}. \quad (7)$$

An n-gonal number is equal to the sum of the $(n-1)$-gonal number of the same RANK and the TRIANGULAR NUMBER of the previous RANK.

see also CENTERED POLYGONAL NUMBER, DECAGONAL NUMBER, FERMAT'S POLYGONAL NUMBER THEOREM, FIGURATE NUMBER, HEPTAGONAL NUMBER, HEXAGONAL NUMBER, NONAGONAL NUMBER, OCTAGONAL NUMBER, PENTAGONAL NUMBER, PYRAMIDAL NUMBER, SQUARE NUMBER, TRIANGULAR NUMBER

References

Beiler, A. H. "Ball Games." Ch. 18 in *Recreations in the Theory of Numbers: The Queen of Mathematics Entertains.* New York: Dover, pp. 184–199, 1966.

Dickson, L. E. *History of the Theory of Numbers, Vol. 1: Divisibility and Primality.* New York: Chelsea, pp. 3–33, 1952.

Guy, K. "Every Number is Expressible as a Sum of How Many Polygonal Numbers?" *Amer. Math. Monthly* **101**, 169–172, 1994.

Pappas, T. "Triangular, Square & Pentagonal Numbers." *The Joy of Mathematics.* San Carlos, CA: Wide World Publ./Tetra, p. 214, 1989.

Sloane, N. J. A. Sequences A000217/M2535 in "An On-Line Version of the Encyclopedia of Integer Sequences."

Sloane, N. J. A. and Plouffe, S. Extended entry in *The Encyclopedia of Integer Sequences.* San Diego: Academic Press, 1995.

Polygonal Spiral

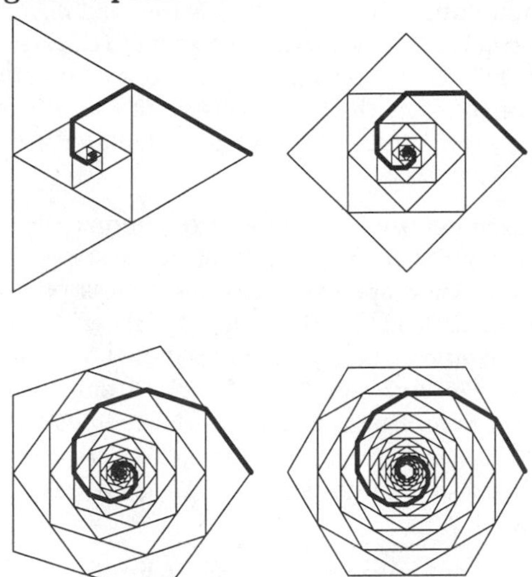

The length of the polygonal spiral is found by noting that the ratio of INRADIUS to CIRCUMRADIUS of a regular POLYGON of n sides is

$$\frac{r}{R} = \frac{\cot\left(\frac{\pi}{n}\right)}{\csc\left(\frac{\pi}{n}\right)} = \cos\left(\frac{\pi}{n}\right). \quad (1)$$

The total length of the spiral for an n-gon with side length s is therefore

$$L = \tfrac{1}{2}s\sum_{k=0}^{\infty}\cos^k\left(\frac{\pi}{n}\right) = \frac{s}{2\left[1-\cos\left(\frac{\pi}{n}\right)\right]}. \quad (2)$$

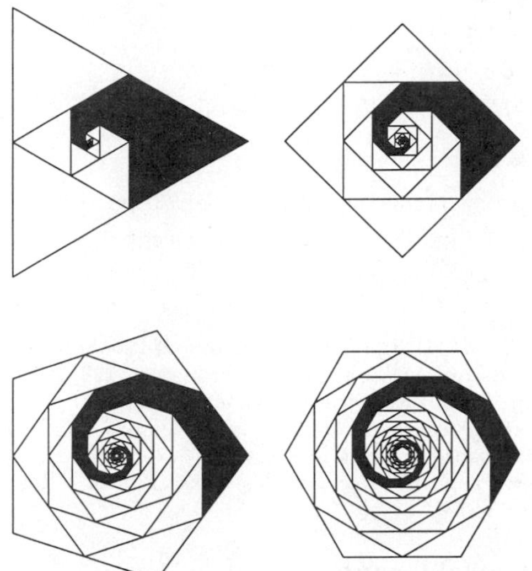

Consider the solid region obtained by filling in subsequent triangles which the spiral encloses. The AREA of this region, illustrated above for n-gons of side length s, is

$$A = \tfrac{1}{4}s^2\cot\left(\frac{\pi}{n}\right). \quad (3)$$

References

Sandefur, J. T. "Using Self-Similarity to Find Length, Area, and Dimension." *Amer. Math. Monthly* **103**, 107–120, 1996.

Polygram

A self-intersecting STAR FIGURE such as the PENTAGRAM or HEXAGRAM.

n	symbol	polygram
5	{5/2}	pentagram
6	{6/2}	hexagram
7	{7/2}	heptagram
8	{8/3}	octagram
	{8/4}	star of Lakshmi
10	{10/3}	decagram

Polyhedral Formula

A formula relating the number of VERTICES, FACES, and EDGES of a POLYHEDRON (or POLYGON). It was discovered independently by Euler and Descartes, so it is also known as the DESCARTES-EULER POLYHEDRAL FORMULA. The polyhedron need not be CONVEX, but the FORMULA does not hold for STELLATED POLYHEDRA.

$$V + F - E = 2, \quad (1)$$

where $V = N_0$ is the number of VERTICES, $E = N_1$ is the number of EDGES, and $F = N_2$ is the number of FACES. For a proof, see Courant and Robbins (1978, pp. 239–240). The FORMULA can be generalized to n-D POLYTOPES.

$$\Pi_1 : N_0 = 2 \tag{2}$$
$$\Pi_2 : N_0 - N_1 = 0 \tag{3}$$
$$\Pi_3 : N_0 - N_1 + N_2 = 2 \tag{4}$$
$$\Pi_4 : N_0 - N_1 + N_2 - N_3 = 0 \tag{5}$$
$$\Pi_n : N_0 - N_1 + N_2 - \ldots + (-1)^{n-1} N_{n-1} = 1 - (-1)^n. \tag{6}$$

For a proof of this, see Coxeter (1973, pp. 166–171).

see also DEHN INVARIANT, DESCARTES TOTAL ANGULAR DEFECT

References
Beyer, W. H. (Ed.) *CRC Standard Mathematical Tables, 28th ed.* Boca Raton, FL: CRC Press, p. 128, 1987.
Courant, R. and Robbins, H. *What is Mathematics?: An Elementary Approach to Ideas and Methods.* Oxford, England: Oxford University Press, 1978.
Coxeter, H. S. M. *Regular Polytopes, 3rd ed.* New York: Dover, 1973.

Polyhedral Graph

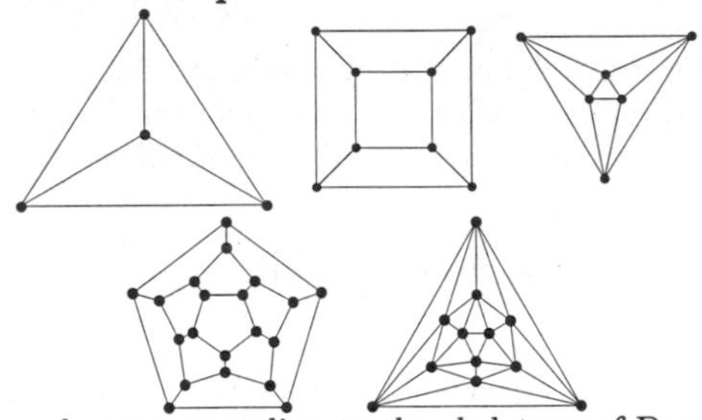

The graphs corresponding to the skeletons of PLATONIC SOLIDS. They are special cases of SCHLEGEL GRAPHS.

see also CUBICAL GRAPH, DODECAHEDRAL GRAPH, ICOSAHEDRAL GRAPH, OCTAHEDRAL GRAPH, SCHLEGEL GRAPH, TETRAHEDRAL GRAPH

Polyhedron

A 3-D solid which consists of a collection of POLYGONS, usually joined at their EDGES. The word derives from the Greek *poly* (many) plus the Indo-European *hedron* (seat). A polyhedron is the 3-D version of the more general POLYTOPE, which can be defined on arbitrary dimensions.

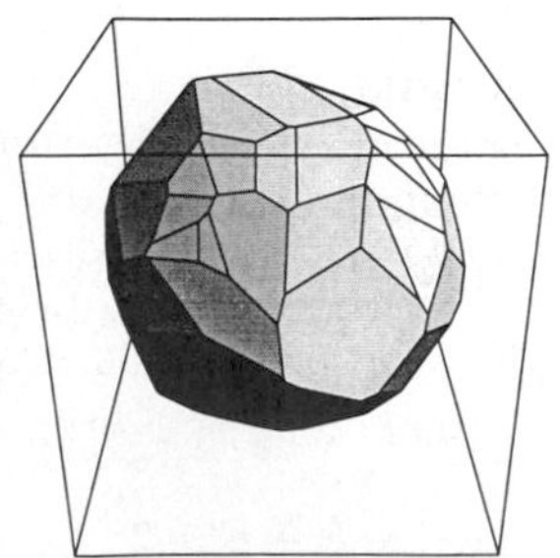

A CONVEX POLYHEDRON can be defined as the set of solutions to a system of linear inequalities

$$\mathbf{mx} \leq \mathbf{b},$$

where $\mathbf{m}$ is a real $s \times 3$ MATRIX and $\mathbf{b}$ is a real s-VECTOR. An example is illustrated above. The more simple DODECAHEDRON is given by a system with $s = 12$. In general, given the MATRICES, the VERTICES (and FACES) can be found using VERTEX ENUMERATION.

A polyhedron is said to be regular if its FACES and VERTEX FIGURES are REGULAR (not necessarily CONVEX) polygons (Coxeter 1973, p. 16). Using this definition, there are a total of nine REGULAR POLYHEDRA, five being the CONVEX PLATONIC SOLIDS and four being the CONCAVE (stellated) KEPLER-POINSOT SOLIDS. However, the term "regular polyhedra" is sometimes also used to refer exclusively to the PLATONIC SOLIDS (Cromwell 1997, p. 53). The DUAL POLYHEDRA of the PLATONIC SOLIDS are not new polyhedra, but are themselves PLATONIC SOLIDS.

A CONVEX polyhedron is called SEMIREGULAR if its FACES have a similar arrangement of nonintersecting regular plane CONVEX polygons of two or more different types about each VERTEX (Holden 1991, p. 41). These solids are more commonly called the ARCHIMEDEAN SOLIDS, and there are 13 of them. The DUAL POLYHEDRA of the ARCHIMEDEAN SOLIDS are 13 new (and beautiful) solids, sometimes called the CATALAN SOLIDS.

A QUASIREGULAR POLYHEDRON is the solid region interior to two DUAL REGULAR POLYHEDRA (Coxeter 1973, pp. 17–20). There are only two CONVEX QUASIREGULAR POLYHEDRA: the CUBOCTAHEDRON and ICOSIDODECAHEDRON. There are also infinite families of PRISMS and ANTIPRISMS.

There exist exactly 92 CONVEX POLYHEDRA with REGULAR POLYGONAL faces (and not necessary equivalent vertices). They are known as the JOHNSON SOLIDS. Polyhedra with identical VERTICES related by a symmetry operation are known as UNIFORM POLYHEDRA. There are 75 such polyhedra in which only two faces may meet at an EDGE, and 76 in which any EVEN number of faces may meet. Of these, 37 were discovered by Badoureau in 1881 and 12 by Coxeter and Miller ca. 1930.

Polyhedra can be superposed on each other (with the sides allowed to pass through each other) to yield additional POLYHEDRON COMPOUNDS. Those made from REGULAR POLYHEDRA have symmetries which are especially aesthetically pleasing. The graphs corresponding to polyhedra skeletons are called SCHLEGEL GRAPHS.

Behnke *et al.* (1974) have determined the symmetry groups of all polyhedra symmetric with respect to their VERTICES.

see also ACOPTIC POLYHEDRON, APEIROGON, ARCHIMEDEAN SOLID, CANONICAL POLYHEDRON, CATALAN SOLID, CUBE, DICE, DIGON, DODECAHEDRON, DUAL POLYHEDRON, ECHIDNAHEDRON, FLEXIBLE POLYHEDRON, HEXAHEDRON, HYPERBOLIC POLYHEDRON, ICOSAHEDRON, ISOHEDRON, JOHNSON SOLID, KEPLER-POINSOT SOLID, NOLID, OCTAHEDRON, PETRIE POLYGON, PLATONIC SOLID, POLYHEDRON COLORING, POLYHEDRON COMPOUND, PRISMATOID, QUADRICORN, QUASIREGULAR POLYHEDRON, RIGIDITY THEOREM, SEMIREGULAR POLYHEDRON, SKELETON, TETRAHEDRON, UNIFORM POLYHEDRON, ZONOHEDRON

References

Ball, W. W. R. and Coxeter, H. S. M. "Polyhedra." Ch. 5 in *Mathematical Recreations and Essays, 13th ed.* New York: Dover, pp. 130–161, 1987.

Behnke, H.; Bachman, F.; Fladt, K.; and Kunle, H. (Eds.). *Fundamentals of Mathematics, Vol. 2.* Cambridge, MA: MIT Press, 1974.

Bulatov, V. "Polyhedra Collection." http://www.physics.orst.edu/~bulatov/polyhedra/.

Coxeter, H. S. M. *Regular Polytopes, 3rd ed.* New York: Dover, 1973.

Critchlow, K. *Order in Space: A Design Source Book.* New York: Viking Press, 1970.

Cromwell, P. R. *Polyhedra.* New York: Cambridge University Press, 1997.

Cundy, H. and Rollett, A. *Mathematical Models, 3rd ed.* Stradbroke, England: Tarquin Pub., 1989.

Davie, T. "Books and Articles about Polyhedra and Polytopes." http://www.dcs.st-andrews.ac.uk/~ad/mathrecs/polyhedra/polyhedrabooks.html.

Davie, T. "The Regular (Platonic) and Semi-Regular (Archimedean) Solids." http://www.dcs.st-andrews.ac.uk/~ad/mathrecs/polyhedra/polyhedratopic.html.

Eppstein, D. "Geometric Models." http://www.ics.uci.edu/~eppstein/junkyard/model.html.

Eppstein, D. "Polyhedra and Polytopes." http://www.ics.uci.edu/~eppstein/junkyard/polytope.html.

Hart, G. W. "Virtual Polyhedra." http://www.li.net/~george/virtual-polyhedra/vp.html.

Hilton, P. and Pedersen, J. *Build Your Own Polyhedra.* Reading, MA: Addison-Wesley, 1994.

Holden, A. *Shapes, Space, and Symmetry.* New York: Dover, 1991.

Lyusternik, L. A. *Convex Figures and Polyhedra.* New York: Dover, 1963.

Malkevitch, J. "Milestones in the History of Polyhedra." In *Shaping Space: A Polyhedral Approach* (Ed. M. Senechal and G. Fleck). Boston, MA: Birkhäuser, pp. 80–92, 1988.

Miyazaki, K. *An Adventure in Multidimensional Space: The Art and Geometry of Polygons, Polyhedra, and Polytopes.* New York: Wiley, 1983.

Paeth, A. W. "Exact Dihedral Metrics for Common Polyhedra." In *Graphic Gems II* (Ed. J. Arvo). New York: Academic Press, 1991.

Pappas, T. "Crystals–Nature's Polyhedra." *The Joy of Mathematics.* San Carlos, CA: Wide World Publ./Tetra, pp. 38–39, 1989.

Pugh, A. *Polyhedra: A Visual Approach.* Berkeley: University of California Press, 1976.

Schaaf, W. L. "Regular Polygons and Polyhedra." Ch. 3, §4 in *A Bibliography of Recreational Mathematics.* Washington, DC: National Council of Teachers of Math., pp. 57–60, 1978.

Virtual Image. "Polytopia I" and "Polytopia II" CD-ROMs. http://ourworld.compuserve.com/homepages/vir_image/html/polytopiai.html and polytopiaii.html.

Polyhedron Coloring

Define a valid "coloring" to occur when no two faces with a common EDGE share the same color. Given two colors, there is a single way to color an OCTAHEDRON. Given three colors, there is one way to color a CUBE and 144 ways to color an ICOSAHEDRON. Given four-colors, there are two distinct ways to color a TETRAHEDRON and 4 ways to color a DODECAHEDRON. Given five colors, there are four ways to color an ICOSAHEDRON.

see also COLORING, POLYHEDRON

References

Ball, W. W. R. and Coxeter, H. S. M. *Mathematical Recreations and Essays, 13th ed.* New York: Dover, 238–242, 1987.

Cundy, H. and Rollett, A. *Mathematical Models, 3rd ed.* Stradbroke, England: Tarquin Pub., pp. 82–83, 1989.

Polyhedron Compound

Solid	Vertices	Symbol
cube-octahedron	both	
dodec.+icos.	both	
two cubes		
three cubes		
four cubes		
five cubes	dodecahedron	$2\{5,3\}[5\{4,3\}]$
five octahedra	icosidodeca.	$[5\{3,4\}]2\{3,5\}$
five tetrahedra	dodecahedron	$\{5,3\}[5\{3,3\}]2\{3,5\}$
two dodecahedra	both	
great dodecahedron-small stellated dodec.		
great icosahedron-great stellated dodec.	both	
stella octangula	cube	$\{4,3\}[2\{3,3\}]\{3,4\}$
ten tetrahedra	dodecahedron	$2\{5,3\}[10\{3,3\}]2\{3,5\}$

The above table gives some common polyhedron compounds. In Coxeter's NOTATION, d distinct VERTICES of $\{m,n\}$ taken c times are denoted

$$c\{m,n\}[d\{p,q\}],$$

or faces of $\{s,t\}$ e times

$$[d\{p,q\}]e\{s,t\},$$

or both

$$c\{m,n\}[d\{p,q\}]e\{s,t\}.$$

The five TETRAHEDRA can be arranged in a laevo or dextro configuration.

see also CUBE-OCTAHEDRON COMPOUND, DODECAHEDRON-ICOSAHEDRON COMPOUND, OCTAHEDRON 5-COMPOUND, STELLA OCTANGULA, TETRAHEDRON 5-COMPOUND

Polyhedron Dissection

A DISSECTION of one or more polyhedra into other shapes.

see also CUBE DISSECTION, DIABOLICAL CUBE, POLYCUBE, SOMA CUBE, WALLACE-BOLYAI-GERWEIN THEOREM

References

Bulatov, V.v "Compounds of Uniform Polyhedra." http://www.physics.orst.edu/~bulatov/polyhedra/uniform_compounds/.

Coffin, S. T. *The Puzzling World of Polyhedral Dissections.* New York: Oxford University Press, 1990.

Polyhedron Dual

see DUAL POLYHEDRON

Polyhedron Hinging

see RIGIDITY THEOREM

Polyhedron Packing

see KELVIN'S CONJECTURE, SPACE-FILLING POLYHEDRON

Polyhex

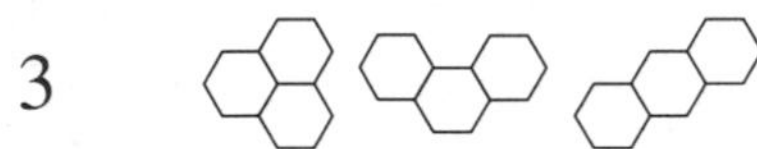

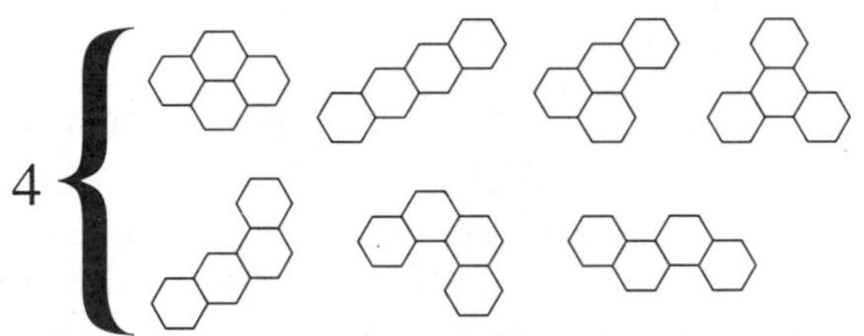

An analog of the POLYOMINOES and POLYIAMONDS in which collections of regular hexagons are arranged with adjacent sides. They are also called HEXES and HEXAS. The number of polyhexes of n hexagons are 1, 1, 2, 7, 22, 82, 333, 1448, 6572, 30490, 143552, 683101, ... (Sloane's A014558). For the 4-hexes (tetrahexes), the possible arrangements are known as the BEE, BAR, PISTOL, PROPELLER, WORM, ARCH, and WAVE.

References

Gardner, M. "Polyhexes and Polyaboloes." Ch. 11 in *Mathematical Magic Show: More Puzzles, Games, Diversions, Illusions and Other Mathematical Sleight-of-Mind from Scientific American.* New York: Vintage, pp. 146–159, 1978.

Gardner, M. "Tiling with Polyominoes, Polyiamonds, and Polyhexes." Ch. 14 in *Time Travel and Other Mathematical Bewilderments.* New York: W. H. Freeman, pp. 175–187, 1988.

Golomb, S. W. *Polyominoes: Puzzles, Patterns, Problems, and Packings, 2nd ed.* Princeton, NJ: Princeton University Press, pp. 92–93, 1994.

Sloane, N. J. A. Sequence A014558 in "An On-Line Version of the Encyclopedia of Integer Sequences."

von Seggern, D. *CRC Standard Curves and Surfaces.* Boca Raton, FL: CRC Press, pp. 342–343, 1993.

Polyiamond

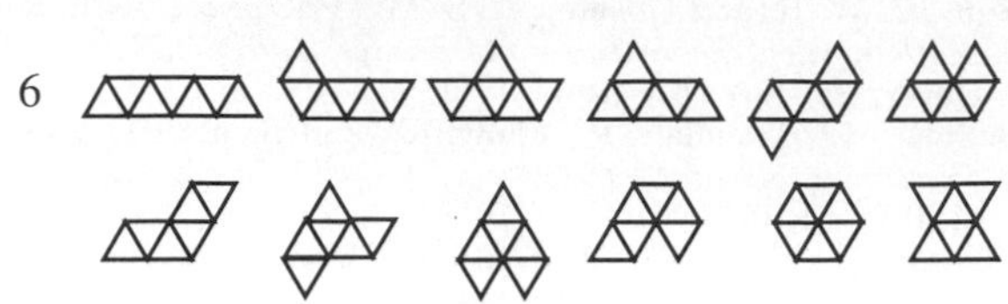

A generalization of the POLYOMINOES using a collection of equal-sized EQUILATERAL TRIANGLES (instead of SQUARES) arranged with coincident sides. Polyiamonds are sometimes simply known as IAMONDS.

The number of two-sided (i.e., can be picked up and flipped, so MIRROR IMAGE pieces are considered identical) polyiamonds made up of n triangles are 1, 1, 1, 3, 4, 12, 24, 66, 160, 448, ... (Sloane's A000577). The number of one-sided polyiamonds composed of n triangles are 1, 1, 1, 4, 6, 19, 43, 121, ... (Sloane's A006534). No HOLES are possible with fewer than seven triangles.

The top row of 6-polyiamonds in the above figure are known as the BAR, CROOK, CROWN, SPHINX, SNAKE, and YACHT. The bottom row of 6-polyiamonds are known as the CHEVRON, SIGNPOST, LOBSTER, HOOK, HEXAGON, and BUTTERFLY.

see also POLYABOLO, POLYHEX, POLYOMINO

References

Beeler, M.; Gosper, R. W.; and Schroeppel, R. *HAKMEM.* Cambridge, MA: MIT Artificial Intelligence Laboratory, Memo AIM-239, Feb. 1972.

Gardner, M. "Mathematical Games." *Sci. Amer.*, Dec. 1964.

Golomb, S. W. *Polyominoes: Puzzles, Patterns, Problems, and Packings, 2nd ed.* Princeton, NJ: Princeton University Press, pp. 90–92, 1994.

Sloane, N. J. A. Sequences A000577/M2374 and A006534/M3287 in "An On-Line Version of the Encyclopedia of Integer Sequences."

von Seggern, D. *CRC Standard Curves and Surfaces.* Boca Raton, FL: CRC Press, pp. 342–343, 1993.

Polyking

see POLYPLET

Polylogarithm

The function

$$\mathrm{Li}_n(z) \equiv \sum_{k=1}^{\infty} \frac{z^k}{k^n}, \quad (1)$$

Also known as JONQUIÈRE'S FUNCTION. (Note that the NOTATION $\mathrm{Li}(z)$ is also used for the LOGARITHMIC INTEGRAL.) The polylogarithm arises in Feynman Diagram integrals, and the special case $n = 2$ is called the DILOGARITHM. The polylogarithm of NEGATIVE INTEGER order arises in sums of the form

$$\sum_{k=1}^{\infty} k^n r^k = \mathrm{Li}_{-n}(r) = \frac{r}{(1-r)^{n+1}} \sum_{i=1}^{n} \left\langle {n \atop i} \right\rangle r^{n-i}, \quad (2)$$

where $\left\langle {n \atop i} \right\rangle$ is an EULERIAN NUMBER.

The polylogarithm satisfies the fundamental identities

$$-\ln(1-2^{-n}) = \mathrm{Li}_1(2^{-n}) \quad (3)$$

$$\mathrm{Li}_s(-1) = -(1-2^{1-s})\zeta(s), \quad (4)$$

where $\zeta(s)$ is the RIEMANN ZETA FUNCTION. The derivative is therefore given by

$$\frac{d}{ds}\mathrm{Li}_s(-1) = -2^{1-s}\zeta(s)\ln 2 - (1-2^{1-s})\zeta'(s), \quad (5)$$

or in the special case $s = 0$, by

$$\left[\frac{d}{ds}\mathrm{Li}_s(-1)\right]_{s=0} = \ln 2 + \zeta'(0) = \ln 2 - \tfrac{1}{2}\ln(2\pi) = \ln\left(\sqrt{\frac{2}{\pi}}\right). \quad (6)$$

This latter fact provides a remarkable proof of the WALLIS FORMULA.

The polylogarithm identities lead to remarkable expressions. Ramanujan gave the polylogarithm identities

$$\mathrm{Li}_2(\tfrac{1}{3}) - \tfrac{1}{6}\mathrm{Li}_2(\tfrac{1}{9}) = \tfrac{1}{18}\pi^2 - \tfrac{1}{6}(\ln 3)^2 \quad (7)$$

$$\mathrm{Li}_2(-\tfrac{1}{2}) + \tfrac{1}{6}\mathrm{Li}_2(\tfrac{1}{9}) = -\tfrac{1}{18}\pi^2 + \ln 2\ln 3 - \tfrac{1}{2}(\ln 2)^2 - \tfrac{1}{3}(\ln 3)^2 \quad (8)$$

$$\mathrm{Li}_2(\tfrac{1}{4}) + \tfrac{1}{3}\mathrm{Li}_2(\tfrac{1}{9}) = \tfrac{1}{18}\pi^2 + 2\ln 2\ln 3 - 2(\ln 2)^2 - \tfrac{2}{3}(\ln 3)^2 \quad (9)$$

$$\mathrm{Li}_2(-\tfrac{1}{3}) - \tfrac{1}{3}\mathrm{Li}_2(\tfrac{1}{9}) = -\tfrac{1}{18}\pi^2 + \tfrac{1}{6}(\ln 3)^2 \quad (10)$$

$$\mathrm{Li}_2(-\tfrac{1}{8}) + \mathrm{Li}_2(\tfrac{1}{9}) = -\tfrac{1}{2}(\ln \tfrac{9}{8})^2 \quad (11)$$

(Berndt 1994), and Bailey *et al.* show that

$$\pi^2 = 36\,\mathrm{Li}_2(\tfrac{1}{2}) - 36\,\mathrm{Li}_2(\tfrac{1}{4}) - 12\,\mathrm{Li}_2(\tfrac{1}{8}) + 6\,\mathrm{Li}_2(\tfrac{1}{64}) \quad (12)$$

$$12\,\mathrm{Li}_2(\tfrac{1}{2}) = \pi^2 - 6(\ln 2)^2 \quad (13)$$

$$\tfrac{35}{2}\zeta(3) - \pi^2\ln 2 = 36\,\mathrm{Li}_3(\tfrac{1}{2}) - 18\,\mathrm{Li}_3(\tfrac{1}{4}) - 4\,\mathrm{Li}_3(\tfrac{1}{8}) + \mathrm{Li}_3(\tfrac{1}{64}) \quad (14)$$

$$2(\ln 2)^3 - 7\zeta(3) = -24\,\mathrm{Li}_3(\tfrac{1}{2}) + 18\,\mathrm{Li}_3(\tfrac{1}{4}) + 4\,\mathrm{Li}_3(\tfrac{1}{8}) - \mathrm{Li}_3(\tfrac{1}{64}) \quad (15)$$

$$10(\ln 2)^3 - 2\pi^2\ln 2 = -48\,\mathrm{Li}_3(\tfrac{1}{2}) + 54\,\mathrm{Li}_3(\tfrac{1}{4}) + 12\,\mathrm{Li}_3(\tfrac{1}{8}) - 3\,\mathrm{Li}_3(\tfrac{1}{64}), \quad (16)$$

and

$$\frac{\mathrm{Li}_m(\frac{1}{64})}{6^{m-1}} - \frac{\mathrm{Li}_m(\frac{1}{8})}{3^{m-1}} - \frac{2\,\mathrm{Li}_m(\frac{1}{4})}{2^{m-1}} + \frac{4\,\mathrm{Li}_m(\frac{1}{2})}{9} - \frac{5(-\ln 2)^m}{9m!} + \frac{\pi^2(-\ln 2)^{m-2}}{54(m-2)!} - \frac{\pi^4(-\ln 2)^{m-4}}{486(m-4)!} - \frac{403\zeta(5)(-\ln 2)^{m-5}}{1296(m-5)!} = 0. \quad (17)$$

No general ALGORITHM is know for the integration of polylogarithms of functions.

see also DILOGARITHM, EULERIAN NUMBER, LEGENDRE'S CHI-FUNCTION, LOGARITHMIC INTEGRAL, NIELSEN-RAMANUJAN CONSTANTS

References

Bailey, D.; Borwein, P.; and Plouffe, S. "On the Rapid Computation of Various Polylogarithmic Constants." `http://www.cecm.sfu.ca/~pborwein/PAPERS/P123.ps`.

Berndt, B. C. *Ramanujan's Notebooks, Part IV.* New York: Springer-Verlag, pp. 323–326, 1994.

Lewin, L. *Polylogarithms and Associated Functions.* New York: North-Holland, 1981.

Lewin, L. *Structural Properties of Polylogarithms.* Providence, RI: Amer. Math. Soc., 1991.

Nielsen, N. *Der Euler'sche Dilogarithms.* Leipzig, Germany: Halle, 1909.

Polymorph

An INTEGER which is expressible in more than one way in the form $x^2 + Dy^2$ or $x^2 - Dy^2$ where x^2 is RELATIVELY PRIME to Dy^2. If the INTEGER is expressible in only one way, it is called a MONOMORPH.

see also ANTIMORPH, IDONEAL NUMBER, MONOMORPH

Polynomial

A POLYNOMIAL is a mathematical expression involving a series of POWERS in one or more variables multiplied by COEFFICIENTS. A POLYNOMIAL in one variable with constant COEFFICIENTS is given by

$$a_n x^n + \ldots + a_2 x^2 + a_1 x + a_0. \quad (1)$$

The highest POWER in a one-variable POLYNOMIAL is called its ORDER. A POLYNOMIAL in two variables with constant COEFFICIENTS is given by

$$a_{nm}x^n y^m + a_{22}x^2y^2 + a_{21}x^2y + a_{12}xy^2 + a_{11}xy + a_{10}x + a_{01}y + a_{00}. \quad (2)$$

Exchanging the COEFFICIENTS of a one-variable POLYNOMIAL end-to-end produces a POLYNOMIAL

$$a_0x^n + a_1x^{n-1} + \ldots + a_{n-1}x + a_n = 0 \quad (3)$$

whose ROOTS are RECIPROCALS $1/x_i$ of the original ROOTS x_i.

The following table gives special names given to polynomials of low orders.

Order	Polynomial Name
1	linear equation
2	quadratic equation
3	cubic equation
4	quartic equation
5	quintic equation
6	sextic equation

Polynomials of fourth degree may be computed using three multiplications and five additions if a few quantities are calculated first (Press *et al.* 1989):

$$a_0 + a_1x + a_2x^2 + a_3x^3 + a_4x^4$$
$$= [(Ax+B)^2 + Ax + C][(Ax+B)^2 + D] + E, \quad (4)$$

where

$$A \equiv (a_4)^{1/4} \quad (5)$$
$$B \equiv \frac{a_3 - A^3}{4A^3} \quad (6)$$
$$D \equiv 3B^2 + 8B^3 + \frac{a_1A - 2a_2B}{A^2} \quad (7)$$
$$C \equiv \frac{a_2}{A^2} - 2B - 6B^2 - D \quad (8)$$
$$E \equiv a_0 - B^4 - B^2(C+D) - CD. \quad (9)$$

Similarly, a POLYNOMIAL of fifth degree may be computed with four multiplications and five additions, and a POLYNOMIAL of sixth degree may be computed with four multiplications and seven additions.

Polynomials of orders 1 to 4 are solvable using only algebraic functions and finite square root extraction. A first-order equation is trivially solvable. A second-order equation is soluble using the QUADRATIC EQUATION. A third-order equation is solvable using the CUBIC EQUATION. A fourth-order equation is solvable using the QUARTIC EQUATION. It was proved by Abel using GROUP THEORY that higher order equations cannot be solved by finite root extraction.

However, the general QUINTIC EQUATION may be given in terms of the THETA FUNCTIONS, or HYPERGEOMETRIC FUNCTIONS in one variable. Hermite and Kronecker proved that higher order POLYNOMIALS are not soluble in the same manner. Klein showed that the work of Hermite was implicit in the GROUP properties of the ICOSAHEDRON. Klein's method of solving the quintic in terms of HYPERGEOMETRIC FUNCTIONS in one variable can be extended to the sextic, but for higher order POLYNOMIALS, either HYPERGEOMETRIC FUNCTIONS in several variables or "Siegel functions" must be used. In the 1880s, Poincaré created functions which give the solution to the nth order POLYNOMIAL equation in finite form. These functions turned out to be "natural" generalizations of the ELLIPTIC FUNCTIONS.

Given an nth degree POLYNOMIAL, the ROOTS can be found by finding the EIGENVALUES of the MATRIX

$$\begin{bmatrix} -a_0/a_n & -a_1/a_n & -a_2/a_n & \cdots & -1 \\ 1 & 0 & 0 & \cdots & 0 \\ 0 & 1 & 0 & \cdots & 0 \\ \vdots & \vdots & 1 & \ddots & 0 \\ 0 & 0 & 0 & \cdots & 0 \end{bmatrix}. \quad (10)$$

This method can be computationally expensive, but is fairly robust at finding close and multiple roots.

Polynomial identities involving sums and differences of like POWERS include

$$x^2 - y^2 = (x-y)(x+y) \quad (11)$$
$$x^3 - y^3 = (x-y)(x^2+xy+y^2) \quad (12)$$
$$x^3 + y^3 = (x+y)(x^2-xy+y^2) \quad (13)$$
$$x^4 - y^4 = (x-y)(x+y)(x^2+y^2) \quad (14)$$
$$x^4 + 4y^4 = (x^2+2xy+2y^2)(x^2-2xy+2y^2) \quad (15)$$
$$x^5 - y^5 = (x-y)(x^4+x^3y+x^2y^2+xy^3+y^4) \quad (16)$$
$$x^5 + y^5 = (x+y)(x^4-x^3y+x^2y^2-xy^3+y^4) \quad (17)$$
$$x^6 - y^6 = (x-y)(x+y)(x^2+xy+y^2)(x^2-xy+y^2) \quad (18)$$
$$x^6 + y^6 = (x^2+y^2)(x^4-x^2y^2+y^4). \quad (19)$$

Further identities include

$$x^4 + x^2y^2 + y^4 = (x^2+xy+y^2)(x^2-xy+y^2) \quad (20)$$
$$({x_1}^2 - D{y_1}^2)({x_2}^2 - D{y_2}^2)$$
$$= (x_1x_2 + Dy_1y_2)^2 - D(x_1y_2 + x_2y_1)^2 \quad (21)$$
$$({x_1}^2 + D{y_1}^2)({x_2}^2 + D{y_2}^2)$$
$$= (x_1x_2 \pm Dy_1y_2)^2 + D(x_1y_2 \mp x_2y_1)^2. \quad (22)$$

The identity

$$(X+Y+Z)^7 - (X^7+Y^7+Z^7) = 7(X+Y)(X+Z)(Y+Z)$$
$$\times[(X^2+Y^2+Z^2+XY+XZ+YZ)^2 + XYZ(X+Y+Z)] \quad (23)$$

was used by Lamé in his proof that FERMAT'S LAST THEOREM was true for $n = 7$.

see also Appell Polynomial, Bernstein Polynomial, Bessel Polynomial, Bezout's Theorem, Binomial, Bombieri Inner Product, Bombieri Norm, Chebyshev Polynomial of the First Kind, Chebyshev Polynomial of the Second Kind, Christoffel-Darboux Formula, Christoffel Number, Complex Number, Cyclotomic Polynomial, Descartes' Sign Rule, Discriminant (Polynomial), Durfee Polynomial, Ehrhart Polynomial, Euler Four-Square Identity, Fibonacci Identity, Fundamental Theorem of Algebra, Fundamental Theorem of Symmetric Functions, Gauss-Jacobi Mechanical Quadrature, Gegenbauer Polynomial, Gram-Schmidt Orthonormalization, Greatest Lower Bound, Hermite Polynomial, Hilbert Polynomial, Irreducible Polynomial, Isobaric Polynomial, Isograph, Jensen Polynomial, Kernel Polynomial, Krawtchouk Polynomial, Laguerre Polynomial, Least Upper Bound, Legendre Polynomial, Liouville Polynomial Identity, Lommel Polynomial, Lukács Theorem, Monomial, Orthogonal Polynomials, Perimeter Polynomial, Poisson-Charlier Polynomial, Pollaczek Polynomial, Polynomial Bar Norm, Quarter Squares Rule, Ramanujan 6-10-8 Identity, Root, Runge-Walsh Theorem, Schläfli Polynomial, Separation Theorem, Stieltjes-Wigert Polynomial, Trinomial, Trinomial Identity, Weierstraß's Polynomial Theorem, Zernike Polynomial

References

Barbeau, E. J. *Polynomials.* New York: Springer-Verlag, 1989.

Bini, D. and Pan, V. Y. *Polynomial and Matrix Computations, Vol. 1: Fundamental Algorithms.* Boston, MA: Birkhäuser, 1994.

Borwein, P. and Erdélyi, T. *Polynomials and Polynomial Inequalities.* New York: Springer-Verlag, 1995.

Cockle, J. "Notes on the Higher Algebra." *Quart. J. Pure Applied Math.* **4**, 49–57, 1861.

Cockle, J. "Notes on the Higher Algebra (Continued)." *Quart. J. Pure Applied Math.* **5**, 1–17, 1862.

Press, W. H.; Flannery, B. P.; Teukolsky, S. A.; and Vetterling, W. T. *Numerical Recipes in C: The Art of Scientific Computing.* Cambridge, England: Cambridge University Press, 1989.

Project Mathematics! *Polynomials.* Videotape (27 minutes). California Institute of Technology. Available from the Math. Assoc. Amer.

Polynomial Bar Norm

For $p = \sum a_j z^j$, define

$$||P||_1 = \int_0^{2\pi} |P(e^{i\theta})| \frac{d\theta}{2\pi} \qquad |P|_1 = \sum_j |a_j|$$

$$||P||_2 = \sqrt{\int_0^{2\pi} |P(e^{\pi\theta})|^2 \frac{d\theta}{2\pi}} \qquad |P|_2 = \sqrt{\sum_j |a_j|^2}$$

$$||P||_\infty = \max_{|z|=1}|P(z)| \qquad |P|_\infty = \max_j |a_j|,$$

where the $||P||_i$ norms are functions on the Unit Circle and the $|P|_i$ norms refer to the Coefficients $a_0, \ldots, a_n$.

see also Bombieri Norm, Norm, Unit Circle

References

Press, W. H.; Flannery, B. P.; Teukolsky, S. A.; and Vetterling, W. T. *Numerical Recipes in FORTRAN: The Art of Scientific Computing, 2nd ed.* Cambridge, England: Cambridge University Press, p. 151, 1989.

Polynomial Bracket Norm

see Bombieri Norm

Polynomial Curve

A curve obtained by fitting Polynomials to each ordinate of an ordered sequence of points. The above plots show Polynomial curves where the order of the fitting Polynomial varies from $p-3$ to $p-1$, where p is the number of points.

Polynomial curves have several undesirable features, including a nonintuitive variation of fitting curve with varying Coefficients, and numerical instability for high orders. Splines such as the Bézier Curve are therefore used more commonly.

see also Bézier Curve, Polynomial, Spline

Polynomial Factor

A Factor of a Polynomial $P(x)$ of degree n is a Polynomial $Q(x)$ of degree less than n which can be multiplied by another Polynomial $R(x)$ of degree less than n to yield $P(x)$, i.e., a Polynomial $Q(x)$ such that

$$P(x) = Q(x)R(x).$$

For example, since

$$x^2 - 1 = (x+1)(x-1),$$

both $x-1$ and $x+1$ are Factors of x^2-1. The Coefficients of factor Polynomials are often required to be Real Numbers or Integers but could, in general, be Complex Numbers.

see also Factor, Factorization, Prime Factorization

Polynomial Norm

see Bombieri Norm, Matrix Norm, Polynomial Bar Norm, Vector Norm

Polynomial Remainder Theorem

If the COEFFICIENTS of the POLYNOMIAL

$$d_n x^n + d_{n-1} x^{n-1} + \ldots + d_0 = 0 \tag{1}$$

are specified to be INTEGERS, then integral ROOTS must have a NUMERATOR which is a factor of d_0 and a DENOMINATOR which is a factor of d_n (with either sign possible). This follows since a POLYNOMIAL of ORDER n with k integral ROOTS can be expressed as

$$(a_1 x + b_1)(a_2 x + b_2) \cdots (a_k x + b_k)(c_{n-k} x^{n-k} + \ldots + c_0) = 0, \tag{2}$$

where the ROOTS are $x_1 = -b_1/a_1$, $x_2 = -b_2/a_2$, ..., and $x_k = -b_k/a_k$. Factoring out the a_is,

$$a_1 a_2 \cdots a_k \left(x - \frac{b_1}{a_1}\right)\left(x - \frac{b_2}{a_2}\right) \cdots \left(x - \frac{b_k}{a_k}\right) \times (c_{n-k} x^{n-k} + \ldots + c_0) = 0. \tag{3}$$

Now, multiplying through,

$$a_1 a_2 \cdots a_k c_{n-k} x^n + \ldots + b_1 b_2 \cdots b_k c_0 = 0, \tag{4}$$

where we have not bothered with the other terms. Since the first and last COEFFICIENTS are d_n and d_0, all the integral roots of (1) are of the form [factors of d_0]/[factors of d_n].

Polynomial Ring

The RING $R[x]$ of POLYNOMIALS in a variable x.

see also POLYNOMIAL, RING

Polynomial Root

If the COEFFICIENTS of the POLYNOMIAL

$$d_n x^n + d_{n-1} x^{n-1} + \ldots + d_0 = 0 \tag{1}$$

are specified to be INTEGERS, then integral roots must have a NUMERATOR which is a factor of d_0 and a DENOMINATOR which is a factor of d_n (with either sign possible). This is known as the POLYNOMIAL REMAINDER THEOREM.

Let the ROOTS of the polynomial

$$P(x) \equiv a_n x^n + a_{n-1} x^{n-1} + \ldots + a_1 x + a_0 \tag{2}$$

be denoted r_1, r_2, ..., r_n. Then NEWTON'S RELATIONS are

$$\sum r_i = -\frac{a_{n-1}}{a_n} \tag{3}$$

$$\sum r_i r_j = \frac{a_{n-2}}{a_n} \tag{4}$$

$$\sum r_1 r_2 \cdots r_k = (-1)^k \frac{a_{n-k}}{a_n}. \tag{5}$$

These can be derived by writing

$$(x-a)(x-b) = 0 \tag{6}$$

$$\left(\frac{x}{a} - 1\right)\left(\frac{x}{b} - 1\right) = 0 \tag{7}$$

$$\frac{1}{ab} x^2 - \left(\frac{1}{a} - \frac{1}{b}\right) x + 1 = 0. \tag{8}$$

Similarly,

$$\left(\frac{x^2}{a^2} - 1\right)\left(\frac{x^2}{b^2} = 1\right) = 0 \tag{9}$$

$$\frac{x^4}{a^2 b^2} - x^2 \left(\frac{1}{a^2} + \frac{1}{b^2}\right) + 1 = 0. \tag{10}$$

Any POLYNOMIAL can be numerically factored, although different ALGORITHMS have different strengths and weaknesses.

If there are no NEGATIVE ROOTS of a POLYNOMIAL (as can be determined by DESCARTES' SIGN RULE), then the GREATEST LOWER BOUND is 0. Otherwise, write out the COEFFICIENTS, let $n = -1$, and compute the next line. Now, if any COEFFICIENTS are 0, set them to minus the sign of the next higher COEFFICIENT, starting with the second highest order COEFFICIENT. If all the signs alternate, n is the greatest lower bound. If not, then subtract 1 from n, and compute another line. For example, consider the POLYNOMIAL

$$y = 2x^4 + 2x^3 - 7x^2 + x - 7. \tag{11}$$

Performing the above ALGORITHM then gives

0	2	2	−7	1	−7
−1	2	0	−7	8	−15
—	2	−1	−7	8	−15
−2	2	−2	−3	7	−21
−3	2	−4	5	−14	35

so the greatest lower bound is −3.

If there are no POSITIVE ROOTS of a POLYNOMIAL (as can be determined by DESCARTES' SIGN RULE), the LEAST UPPER BOUND is 0. Otherwise, write out the COEFFICIENTS of the POLYNOMIALS, including zeros as necessary. Let $n = 1$. On the line below, write the highest order COEFFICIENT. Starting with the second-highest COEFFICIENT, add n times the number just written to the original second COEFFICIENT, and write it below the second COEFFICIENT. Continue through order zero. If all the COEFFICIENTS are NONNEGATIVE, the least upper bound is n. If not, add one to x and repeat the process again. For example, take the POLYNOMIAL

$$y = 2x^4 - x^3 - 7x^2 + x - 7. \tag{12}$$

Performing the above ALGORITHM gives

0	2	−1	−7	1	−7
1	2	1	−6	−5	−12
2	2	3	−1	−1	−9
3	2	5	8	25	68

so the LEAST UPPER BOUND is 3.

see also BAIRSTOW'S METHOD, DESCARTES' SIGN RULE, JENKINS-TRAUB METHOD, LAGUERRE'S METHOD, LEHMER-SCHUR METHOD, MAEHLY'S PROCEDURE, MULLER'S METHOD, ROOT, ZASSENHAUS-BERLEKAMP ALGORITHM

Polynomial Series

see MULTINOMIAL SERIES

Polyomino

A generalization of the DOMINO. An n-omino is defined as a collection of n squares of equal size arranged with coincident sides. FREE polyominoes can be picked up and flipped, so mirror image pieces are considered identical, whereas FIXED polyominoes are distinct if they have different chirality or orientation. FIXED polyominoes are also called LATTICE ANIMALS.

Redelmeier (1981) computed the number of FREE and FIXED polyominoes for $n \leq 24$, and Mertens (1990) gives a simple computer program. The sequence giving the number of FREE polyominoes of each order (Sloane's A000105, Ball and Coxeter 1987) is shown in the second column below, and that for FIXED polyominoes in the third column (Sloane's A014559).

n	Free	Fixed	Pos. Holes
1	1	1	0
2	1	2	0
3	2	6	0
4	5	19	0
5	12	63	0
6	35	216	0
7	108	760	1
8	369	2725	6
9	1285	9910	37
10	4655	39446	384
11	17073	135268	
12	63600	505861	
13	238591	1903890	
14	901971	7204874	
15	3426576	27394666	
16	13079255	104592937	
17	50107909	400795844	
18	192622052	1540820542	
19	742624232	5940738676	
20	2870671950	22964779660	
21	11123060678	88983512783	
22	43191857688	345532572678	
23	168047007728	1344372335524	
24	654999700403	5239988770268	

The best currently known bounds on the number of n-polyominoes are

$$3.72^n < P(n) < 4.65^n$$

(Eden 1961, Klarner 1967, Klarner and Rivest 1973, Ball and Coxeter 1987). For $n = 4$, the quartominoes are called STRAIGHT, L, T, SQUARE, and SKEW. For $n = 5$, the pentominoes are called f, I, L, N, P, T, U, V, W, X, y, and Z (Golomb 1995).

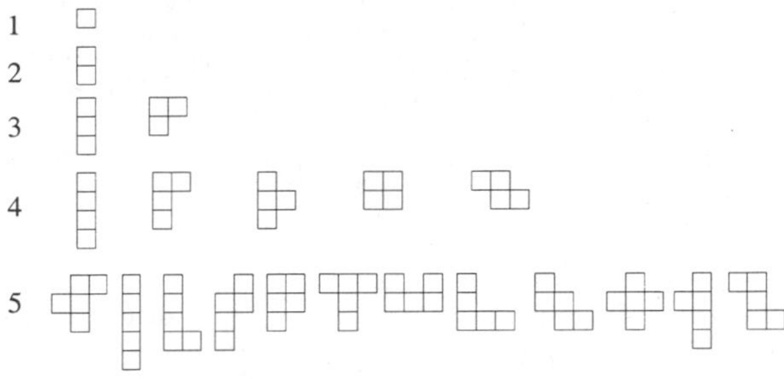

see also DOMINO, HEXOMINO, MONOMINO, PENTOMINO, POLYABOLO, POLYCUBE, POLYHEX, POLYIAMOND, POLYKING, POLYPLET, TETROMINO, TRIOMINO

References

Atkin, A. O. L. and Birch, B. J. (Eds.). *Computers in Number Theory: Proc. Sci. Research Council Atlas Symposium No. 2 Held at Oxford from 18–23 Aug., 1969.* New York: Academic Press, 1971.

Ball, W. W. R. and Coxeter, H. S. M. *Mathematical Recreations and Essays, 13th ed.* New York: Dover, pp. 109–113, 1987.

Beeler, M.; Gosper, R. W.; and Schroeppel, R. Item 77 in *HAKMEM.* Cambridge, MA: MIT Artificial Intelligence Laboratory, Memo AIM-239, pp. 48–50, Feb. 1972.

Eden, M. "A Two-Dimensional Growth Process." *Proc. Fourth Berkeley Symposium Math. Statistics and Probability, Held at the Statistical Laboratory, University of California, June 30–July 30, 1960.* Berkeley, CA: University of California Press, pp. 223–239, 1961.

Finch, S. "Favorite Mathematical Constants." http://www.mathsoft.com/asolve/constant/rndprc/rndprc.html.

Gardner, M. "Polyominoes and Fault-Free Rectangles." Ch. 13 in *Martin Gardner's New Mathematical Diversions from Scientific American.* New York: Simon and Schuster, 1966.

Gardner, M. "Polyominoes and Rectification." Ch. 13 in *Mathematical Magic Show: More Puzzles, Games, Diversions, Illusions and Other Mathematical Sleight-of-Mind from Scientific American.* New York: Vintage, pp. 172–187, 1978.

Golomb, S. W. "Checker Boards and Polyominoes." *Amer. Math. Monthly* **61**, 675–682, 1954.

Golomb, S. W. *Polyominoes: Puzzles, Patterns, Problems, and Packings, rev. enl. 2nd ed.* Princeton, NJ: Princeton University Press, 1995.

Klarner, D. A. "Cell Growth Problems." *Can. J. Math.* **19**, 851–863, 1967.

Klarner, D. A. and Riverst, R. "A Procedure for Improving the Upper Bound for the Number of n-ominoes." *Can. J. Math.* **25**, 585–602, 1973.

Lei, A. "Bigger Polyominoes." http://www.cs.ust.hk/~philipl/omino/bigpolyo.html.

Lei, A. "Polyominoes." http://www.cs.ust.hk/~philipl/omino/omino.html.

Lunnon, W. F. "Counting Polyominoes." In *Computers in Number Theory* (Ed. A. O. L. Atkin and B. J. Brich). London: Academic Press, pp. 347–372, 1971.
Martin, G. *Polyominoes: A Guide to Puzzles and Problems in Tiling.* Washington, DC: Math. Assoc. Amer., 1991.
Mertens, S. "Lattice Animals—A Fast Enumeration Algorithm and New Perimeter Polynomials." *J. Stat. Phys.* **58**, 1095–1108, 1990.
Read, R. C. "Contributions to the Cell Growth Problem." *Canad. J. Math.* **14**, 1–20, 1962.
Redelmeier, D. H. "Counting Polyominoes: Yet Another Attack." *Discrete Math.* **36**, 191–203, 1981.
Ruskey, F. "Information on Polyominoes." `http://sue.csc.uvic.ca/~cos/inf/misc/PolyominoInfo.html`.
Sloane, N. J. A. Sequences A014559 and A000105/M1425 in "An On-Line Version of the Encyclopedia of Integer Sequences."
von Seggern, D. *CRC Standard Curves and Surfaces.* Boca Raton, FL: CRC Press, pp. 342–343, 1993.

Polyomino Tiling

A TILING of the PLANE by specified types of POLYOMINOES. Interestingly, the FIBONACCI NUMBER F_{n+1} gives the number of ways for 2×1 dominoes to cover a $2 \times n$ checkerboard.

see also FIBONACCI NUMBER

References

Gardner, M. "Tiling with Polyominoes, Polyiamonds, and Polyhexes." Ch. 14 in *Time Travel and Other Mathematical Bewilderments.* New York: W. H. Freeman, 1988.

Polyplet

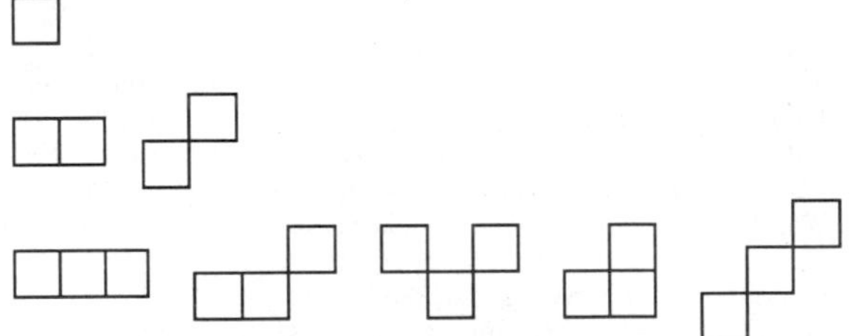

A POLYOMINO-like object made by attaching squares joined either at sides or corners. Because neighboring squares can be in relation to one another as KINGS may move on a CHESSBOARD, polyplets are sometimes also called POLYKINGS. The number of n-polyplets (with holes allowed) are 1, 2, 5, 22, 94, 524, 3031, ... (Sloane's A030222). The number of n-polyplets having bilateral symmetry are 1, 2, 4, 10, 22, 57, 131, ... (Sloane's A030234). The number of n-polyplets not having bilateral symmetry are 0, 0, 1, 12, 72, 467, 2900, ... (Sloane's A030235). The number of fixed n-polyplets are 1, 4, 20, 110, 638, 3832, ... (Sloane's A030232). The number of one-sided n-polyplets are 1, 2, 6, 34, 166, 991, ... (Sloane's A030233).

see also POLYIAMOND, POLYOMINO

References

Sloane, N. J. A. Sequences A030222, A030232, A030233, A030234, and A030235 in "An On-Line Version of the Encyclopedia of Integer Sequences."

Polytope

A convex polytope may be defined as the CONVEX HULL of a finite set of points (which are always bounded), or as the intersection of a finite set of half-spaces. Explicitly, a d-dimensional polytope may be specified as the set of solutions to a system of linear inequalities

$$\mathbf{m}\mathbf{x} \leq \mathbf{b},$$

where $\mathbf{m}$ is a real $s \times d$ MATRIX and $\mathbf{b}$ is a real s-VECTOR. The positions of the vertices given by the above equations may be found using a process called VERTEX ENUMERATION.

A regular polytope is a generalization of the PLATONIC SOLIDS to an arbitrary DIMENSION. The NECESSARY condition for the figure with SCHLÄFLI SYMBOL $\{p, q, r\}$ to be a finite polytope is

$$\cos\left(\frac{\pi}{q}\right) < \sin\left(\frac{\pi}{p}\right) \sin\left(\frac{\pi}{r}\right).$$

SUFFICIENCY can be established by consideration of the six figures satisfying this condition. The table below enumerates the six regular polytopes in 4-D (Coxeter 1969, p. 414).

Name	Schläfli Symbol	N_0	N_1	N_2	N_3
regular simplex	$\{3,3,3\}$	5	10	10	5
hypercube	$\{4,3,3\}$	16	32	24	8
16-cell	$\{3,3,4\}$	8	24	32	16
24-cell	$\{3,4,3\}$	24	96	96	24
120-cell	$\{5,3,3\}$	600	1200	720	120
600-cell	$\{3,3,5\}$	120	720	1200	600

Here, N_0 is the number of VERTICES, N_1 the number of EDGES, N_2 the number of FACES, and N_3 the number of cells. These quantities satisfy the identity

$$N_0 - N_1 + N_2 - N_3 = 0,$$

which is a version of the POLYHEDRAL FORMULA.

For n-D with $n \geq 5$, there are only three regular polytopes, the MEASURE POLYTOPE, CROSS POLYTOPE, and regular SIMPLEX (which are analogs of the CUBE, OCTAHEDRON, and TETRAHEDRON).

see also 16-CELL, 24-CELL, 120-CELL, 600-CELL, CROSS POLYTOPE, EDGE (POLYTOPE), FACE, FACET, HYPERCUBE, INCIDENCE MATRIX, MEASURE POLYTOPE, RIDGE, SIMPLEX, TESSERACT, VERTEX (POLYHEDRON)

References

Coxeter, H. S. M. "Regular and Semi-Regular Polytopes I." *Math. Z.* **46**, 380–407, 1940.
Coxeter, H. S. M. *Introduction to Geometry, 2nd ed.* New York: Wiley, 1969.
Eppstein, D. "Polyhedra and Polytopes." `http://www.ics.uci.edu/~eppstein/junkyard/polytope.html`.

Poncelet's Closure Theorem

If an n-sided PONCELET TRANSVERSE constructed for two given CONIC SECTIONS is closed for one point of origin, it is closed for any position of the point of origin. Specifically, given one ELLIPSE inside another, if there exists one CIRCUMINSCRIBED (simultaneously inscribed in the outer and circumscribed on the inner) n-gon, then any point on the boundary of the outer ELLIPSE is the VERTEX of some CIRCUMINSCRIBED n-gon.

References

Dörrie, H. *100 Great Problems of Elementary Mathematics: Their History and Solutions.* New York: Dover, p. 193, 1965.

Poncelet's Continuity Principle

see PERMANENCE OF MATHEMATICAL RELATIONS PRINCIPLE

Poncelet-Steiner Theorem

All Euclidean GEOMETRIC CONSTRUCTIONS can be carried out with a STRAIGHTEDGE alone if, in addition, one is given the RADIUS of a single CIRCLE and its center. The theorem was suggested by Poncelet in 1822 and proved by Steiner in 1833. A construction using STRAIGHTEDGE alone is called a STEINER CONSTRUCTION.

see also GEOMETRIC CONSTRUCTION, STEINER CONSTRUCTION

References

Dörrie, H. "Steiner's Straight-Edge Problem." §34 in *100 Great Problems of Elementary Mathematics: Their History and Solutions.* New York: Dover, pp. 165–170, 1965.

Steiner, J. *Geometric Constructions with a Ruler, Given a Fixed Circle with Its Center.* New York: Scripta Mathematica, 1950.

Poncelet's Theorem

see PONCELET'S CLOSURE THEOREM

Poncelet Transform

see PONCELET TRANSVERSE

Poncelet Transverse

Let a CIRCLE C_1 lie inside another CIRCLE C_2. From any point on C_2, draw a tangent to C_1 and extend it to C_2. From the point, draw another tangent, etc. For n tangents, the result is called an n-sided PONCELET TRANSFORM.

References

Dörrie, H. *100 Great Problems of Elementary Mathematics: Their History and Solutions.* New York: Dover, p. 192, 1965.

Pong Hau K'i

A Chinese TIC-TAC-TOE-like game.

see also TIC-TAC-TOE

References

Evans, R. "Pong Hau K'i." *Games and Puzzles* **53**, 19, 1976.

Straffin, P. D. Jr. "Position Graphs for Pong Hau K'i and Mu Torere." *Math. Mag.* **68**, 382–386, 1995.

Pons Asinorum

An elementary theorem in geometry whose name means "ass's bridge." The theorem states that the ANGLES at the base of an ISOSCELES TRIANGLE (defined as a TRIANGLE with two legs of equal length) are equal.

see also ISOSCELES TRIANGLE, PYTHAGOREAN THEOREM

References

Dunham, W. *Journey Through Genius: The Great Theorems of Mathematics.* New York: Wiley, p. 38, 1990.

Pontryagin Class

The ith Pontryagin class of a VECTOR BUNDLE is $(-1)^i$ times the ith CHERN CLASS of the complexification of the VECTOR BUNDLE. It is also in the $4i$th cohomology group of the base SPACE involved.

see also CHERN CLASS, STIEFEL-WHITNEY CLASS

Pontryagin Duality

Let G be a locally compact ABELIAN GROUP. Let G^* be the group of all homeomorphisms $G \to R/Z$, in the compact open topology. Then G^* is also a locally compact ABELIAN GROUP, where the asterisk defines a contravariant equivalence of the category of locally compact Abelian groups with itself. The natural mapping $G \to (G^*)^*$, sending g to G, where $G(f) = f(g)$, is an isomorphism and a HOMEOMORPHISM. Under this equivalence, compact groups are sent to discrete groups and vice versa.

see also ABELIAN GROUP, HOMEOMORPHISM

Pontryagin Maximum Principle

A result is CONTROL THEORY. Define

$$H(\psi, x, u) \equiv (\psi, f(x, u)) \equiv \sum_{a=0}^{n} \psi_a f^a(x, u).$$

Then in order for a control $u(t)$ and a trajectory $x(t)$ to be optimal, it is NECESSARY that there exist NONZERO absolutely continuous vector function $\psi(t) = (\psi_0(t), \psi_1(t), \ldots, \psi_n(t))$ corresponding to the functions $u(t)$ and $x(t)$ such that

1. The function $H(\psi(t), x(t), u)$ attains its maximum at the point $u = u(t)$ almost everywhere in the interval $t_0 \leq t \leq t_1$,

$$H(\psi(t), x(t), u(t)) = \max_{u \in U} H(\psi(t), x(t), u).$$

2. At the terminal time t_1, the relations $\psi_0(t_1) \leq 0$ and $H(\psi(t_1), x(t_1), u(t_1)) = 0$ are satisfied.

References

Iyanaga, S. and Kawada, Y. (Eds.). "Pontrjagin's Maximum Principle." §88C in *Encyclopedic Dictionary of Mathematics.* Cambridge, MA: MIT Press, p. 295–296, 1980.

Pontryagin Number

The Pontryagin number is defined in terms of the PONTRYAGIN CLASS of a MANIFOLD as follows. For any collection of PONTRYAGIN CLASSES such that their cup product has the same DIMENSION as the MANIFOLD, this cup product can be evaluated on the MANIFOLD's FUNDAMENTAL CLASS. The resulting number is called the Pontryagin number for that combination of Pontryagin classes. The most important aspect of Pontryagin numbers is that they are COBORDISM invariant. Together, Pontryagin and STIEFEL-WHITNEY NUMBERS determine an oriented manifold's oriented COBORDISM class.

see also CHERN NUMBER, STIEFEL-WHITNEY NUMBER

Ponzo's Illusion

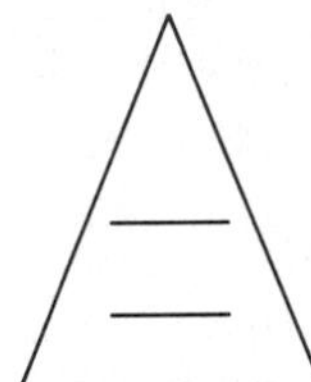

The upper HORIZONTAL line segment in the above figure appears to be longer than the lower line segment despite the fact that both are the same length.

see also ILLUSION, MÜLLER-LYER ILLUSION, POGGENDORFF ILLUSION, VERTICAL-HORIZONTAL ILLUSION

References

Fineman, M. *The Nature of Visual Illusion.* New York: Dover, p. 153, 1996.

Pop

An action which removes a single element from the top of a QUEUE or STACK, turning the LIST $(a_1, a_2, \ldots, a_n)$ into $(a_2, \ldots, a_n)$ and yielding the element a_1.

see also PUSH, STACK

Population Comparison

Let x_1 and x_2 be the number of successes in variates taken from two populations. Define

$$\hat{p}_1 \equiv \frac{x_1}{n_1} \tag{1}$$

$$\hat{p}_2 \equiv \frac{x_2}{n_2}. \tag{2}$$

The ESTIMATOR of the difference is then $\hat{p}_1 - \hat{p}_2$. Doing a z-TRANSFORM,

$$z = \frac{(\hat{p}_1 - \hat{p}_2) - (p_1 - p_2)}{\sigma_{\hat{p}_1 - \hat{p}_2}}, \tag{3}$$

where

$$\sigma_{\hat{p}_1 - \hat{p}_2} \equiv \sqrt{{\sigma_{\hat{p}_1}}^2 - {\sigma_{\hat{p}_2}}^2}. \tag{4}$$

The STANDARD ERROR is

$$\mathrm{SE}_{\hat{p}_1 - \hat{p}_2} = \sqrt{\frac{\hat{p}_1(1-\hat{p}_1)}{n_1} + \frac{\hat{p}_2(1-\hat{p}_2)}{n_2}} \tag{5}$$

$$\mathrm{SE}_{\bar{x}_1 - \bar{x}_2} = \sqrt{\frac{{s_1}^2}{n_1} + \frac{{s_2}^2}{n_2}} \tag{6}$$

$${s_{\text{pool}}}^2 = \frac{(n_1 - 1){s_1}^2 + (n_2 - 1){s_2}^2}{n_1 + n_2 - 2}. \tag{7}$$

see also z-TRANSFORM

Population Growth

The differential equation describing exponential growth is

$$\frac{dN}{dt} = \frac{N}{\tau}. \tag{1}$$

This can be integrated directly

$$\int_{N_0}^{N} \frac{dN}{N} = \int_0^t \frac{dt}{\tau} \tag{2}$$

$$\ln\left(\frac{N}{N_0}\right) = \frac{t}{\tau}. \tag{3}$$

Exponentiating,

$$N(t) = N_0 e^{t/\tau}. \tag{4}$$

Defining $N(t = 1) = N_0 e^{\alpha}$ gives $\tau = 1/\alpha$ in (4), so

$$N(t) = N_0 e^{\alpha t}. \tag{5}$$

The quantity α in this equation is sometimes known as the MALTHUSIAN PARAMETER.

Consider a more complicated growth law

$$\frac{dN}{dt} = \left(\frac{\alpha t - 1}{t}\right) N, \tag{6}$$

where $\alpha > 1$ is a constant. This can also be integrated directly

$$\frac{dN}{N} = \left(\alpha - \frac{1}{t}\right) dt \tag{7}$$

$$\ln N = \alpha t - \ln t + C \tag{8}$$

$$N(t) = \frac{Ce^{\alpha t}}{t}. \tag{9}$$

Note that this expression blows up at $t = 0$. We are given the INITIAL CONDITION that $N(t = 1) = N_0 e^{\alpha}$, so $C = N_0$.

$$N(t) = N_0 \frac{e^{\alpha t}}{t}. \tag{10}$$

The t in the DENOMINATOR of (10) greatly suppresses the growth in the long run compared to the simple growth law.

The LOGISTIC GROWTH CURVE, defined by

$$\frac{dN}{dt} = \frac{r(K-N)}{N} \tag{11}$$

is another growth law which frequently arises in biology. It has a rather complicated solution for $N(t)$.

see also GOMPERTZ CURVE, LIFE EXPECTANCY, LOGISTIC GROWTH CURVE, LOTKA-VOLTERRA EQUATIONS, MAKEHAM CURVE, MALTHUSIAN PARAMETER, SURVIVORSHIP CURVE

Porism

An archaic type of mathematical proposition whose purpose is not entirely known.

see also AXIOM, LEMMA, POSTULATE, PRINCIPLE, STEINER'S PORISM, THEOREM

Porter's Constant

N.B. A detailed on-line essay by S. Finch was the starting point for this entry.

The constant appearing in FORMULAS for the efficiency of the EUCLIDEAN ALGORITHM,

$$\begin{aligned} C &= \frac{6\ln 2}{\pi^2}\left[3\ln 2 + 4\gamma - \frac{24}{\pi^2}\zeta'(2) - 2\right] - \frac{1}{2} \\ &= 1.4670780794\ldots, \end{aligned}$$

where γ is the EULER-MASCHERONI CONSTANT and $\zeta(z)$ is the RIEMANN ZETA FUNCTION.

see also EUCLIDEAN ALGORITHM

References

Finch, S. "Favorite Mathematical Constants." http://www.mathsoft.com/asolve/constant/porter/porter.html.

Porter, J. W. "On a Theorem of Heilbronn." *Mathematika* **22**, 20–28, 1975.

Pósa's Theorem

Let G be a SIMPLE GRAPH with n VERTICES.

1. If, for every k in $1 \leq k < (n-1)/2$, the number of VERTICES of VALENCY not exceeding k is less than k, and
2. If, for n ODD, the number of VERTICES with VALENCY not exceeding $(n-1)/2$ is less than or equal to $(n-1)/2$,

then G contains a HAMILTONIAN CIRCUIT.

see also HAMILTONIAN CIRCUIT

Poset

see PARTIALLY ORDERED SET

Poset Dimension

The DIMENSION of a POSET $P = (X, \leq)$ is the size of the smallest REALIZER of P. Equivalently, it is the smallest INTEGER d such that P is ISOMORPHIC to a DOMINANCE order in $\mathbb{R}^d$.

see also DIMENSION, DOMINANCE, ISOMORPHIC POSETS, REALIZER

References

Dushnik, B. and Miller, E. W. "Partially Ordered Sets." *Amer. J. Math.* **63**, 600–610, 1941.

Trotter, W. T. *Combinatorics and Partially Ordered Sets: Dimension Theory.* Baltimore, MD: Johns Hopkins University Press, 1992.

Position Four-Vector

The CONTRAVARIANT FOUR-VECTOR arising in special and general relativity,

$$x^\mu = \begin{bmatrix} x^0 \\ x^1 \\ x^2 \\ x^3 \end{bmatrix} \equiv \begin{bmatrix} ct \\ x \\ y \\ z \end{bmatrix},$$

where c is the speed of light and t is time. Multiplication of two four-vectors gives the spacetime interval

$$\begin{aligned} I = g_{\mu\nu}x^\mu x^\nu &= (x^0)^2 - (x^1)^2 - (x^2)^2 - (x^3)^2 \\ &= (ct)^2 - (x^1)^2 - (x^2)^2 - (x^3)^2 \end{aligned}$$

see also FOUR-VECTOR, LORENTZ TRANSFORMATION, QUATERNION

Position Vector

see RADIUS VECTOR

Positive

A quantity $x > 0$, which may be written with an explicit PLUS SIGN for emphasis, $+x$.

see also NEGATIVE, NONNEGATIVE, PLUS SIGN, ZERO

Positive Definite Function

A POSITIVE definite FUNCTION f on a GROUP G is a FUNCTION for which the MATRIX $\{f(x_i x_j{}^{-1})\}$ is always POSITIVE SEMIDEFINITE HERMITIAN.

References

Knapp, A. W. "Group Representations and Harmonic Analysis, Part II." *Not. Amer. Math. Soc.* **43**, 537–549, 1996.

Positive Definite Matrix

A MATRIX A is positive definite if

$$(\mathsf{A}\mathbf{v}) \cdot \mathbf{v} > 0 \tag{1}$$

for all VECTORS $\mathbf{v} \neq 0$. All EIGENVALUES of a positive definite matrix are POSITIVE (or, equivalently, the DETERMINANTS associated with *all* upper-left SUBMATRICES are POSITIVE).

The DETERMINANT of a positive definite matrix is POSITIVE, but the converse is not necessarily true (i.e., a matrix with a POSITIVE DETERMINANT is not necessarily positive definite).

A REAL SYMMETRIC MATRIX A is positive definite IFF there exists a REAL nonsingular MATRIX M such that

$$\mathsf{A} = \mathsf{M}\mathsf{M}^{\mathrm{T}}. \tag{2}$$

A 2×2 SYMMETRIC MATRIX

$$\begin{bmatrix} a & b \\ b & c \end{bmatrix} \tag{3}$$

is positive definite if

$$a{v_1}^2 + 2bv_1v_2 + c{v_2}^2 > 0 \tag{4}$$

for all $\mathbf{v} = (v_1, v_2) \neq 0$.

A HERMITIAN MATRIX A is positive definite if

1. $a_{ii} > 0$ for all i,
2. $a_{ii}a_{ij} > |a_{ij}|^2$ for $i \neq j$,
3. The element of largest modulus must lie on the leading diagonal,
4. $|\mathsf{A}| > 0$.

see also DETERMINANT, EIGENVALUE, HERMITIAN MATRIX, MATRIX, POSITIVE SEMIDEFINITE MATRIX

References

Gradshteyn, I. S. and Ryzhik, I. M. *Tables of Integrals, Series, and Products, 5th ed.* San Diego, CA: Academic Press, p. 1106, 1979.

Positive Definite Quadratic Form

A QUADRATIC FORM $Q(\mathbf{x})$ is said to be positive definite if $Q(\mathbf{x}) > 0$ for $\mathbf{x} \neq \mathbf{0}$. A REAL QUADRATIC FORM in n variables is positive definite IFF its canonical form is

$$Q(\mathbf{z}) = {z_1}^2 + {z_2}^2 + \ldots + {z_n}^2. \tag{1}$$

A BINARY QUADRATIC FORM

$$F(x, y) = a_{11}x^2 + 2a_{12}xy + a_{22}y^2 \tag{2}$$

of two REAL variables is positive definite if it is > 0 for any $(x, y) \neq (0, 0)$, therefore if $a_{11} > 0$ and the DISCRIMINANT $a \equiv a_{11}a_{22} - {a_{12}}^2 > 0$. A BINARY QUADRATIC FORM is positive definite if there exist NONZERO x and y such that

$$(ax^2 + 2bxy + cy^2)^2 \leq \tfrac{4}{3}|ac - b^2| \tag{3}$$

(Le Lionnais 1983).

A QUADRATIC FORM $(\mathbf{x}, \mathsf{A}\mathbf{x})$ is positive definite IFF every EIGENVALUE of A is POSITIVE. A QUADRATIC FORM $Q = (\mathbf{x}, \mathsf{A}\mathbf{x})$ with A a HERMITIAN MATRIX is positive definite if all the principal minors in the top-left corner of A are POSITIVE, in other words

$$a_{11} > 0 \tag{4}$$

$$\begin{vmatrix} a_{11} & a_{12} \\ a_{21} & a_{22} \end{vmatrix} > 0 \tag{5}$$

$$\begin{vmatrix} a_{11} & a_{12} & a_{13} \\ a_{21} & a_{22} & a_{23} \\ a_{31} & a_{32} & a_{33} \end{vmatrix} > 0. \tag{6}$$

see also INDEFINITE QUADRATIC FORM, POSITIVE SEMIDEFINITE QUADRATIC FORM

References

Gradshteyn, I. S. and Ryzhik, I. M. *Tables of Integrals, Series, and Products, 5th ed.* San Diego, CA: Academic Press, p. 1106, 1979.

Le Lionnais, F. *Les nombres remarquables.* Paris: Hermann, p. 38, 1983.

Positive Definite Tensor

A TENSOR g whose discriminant satisfies

$$g \equiv g_{11}g_{22} - {g_{12}}^2 > 0.$$

Positive Integer

see $\mathbb{Z}^+$

Positive Semidefinite Matrix

A MATRIX A is positive semidefinite if

$$(\mathsf{A}\mathbf{v}) \cdot \mathbf{v} \geq 0$$

for all $\mathbf{v} \neq 0$.

see also POSITIVE DEFINITE MATRIX

Positive Semidefinite Quadratic Form

A QUADRATIC FORM $Q(\mathbf{x})$ is positive semidefinite if it is never < 0, but is 0 for some $\mathbf{x} \neq \mathbf{0}$. The QUADRATIC FORM, written in the form $(\mathbf{x}, \mathsf{A}\mathbf{x})$, is positive semidefinite IFF every EIGENVALUE of A is NONNEGATIVE.

see also INDEFINITE QUADRATIC FORM, POSITIVE DEFINITE QUADRATIC FORM

References

Gradshteyn, I. S. and Ryzhik, I. M. *Tables of Integrals, Series, and Products, 5th ed.* San Diego, CA: Academic Press, p. 1106, 1979.

Postage Stamp Problem

Consider a SET $A_k = \{a_1, a_2, \ldots, a_k\}$ of INTEGER denomination postage stamps with $1 = a_1 < a_2 < \ldots < a_k$. Suppose they are to be used on an envelope with room for no more than h stamps. The postage stamp problem then consists of determining the smallest INTEGER $N(h, A_k)$ which cannot be represented by a linear combination $\sum_{i=1}^k x_i a_i$ with $x_i \geq 0$ and $\sum_{i=1}^k x_i < h$.

Exact solutions exist for arbitrary A_k for $k = 2$ and 3. The $k = 2$ solution is

$$n(h, A_2) = (h + 3 - a_2)a_2 - 2$$

for $h \geq a_2 - 2$. The general problem consists of finding

$$n(h, k) = \max_{A_k} n(h, A_k).$$

It is known that

$$n(h, 2) = \left\lfloor \tfrac{1}{4}(h^2 + 6h + 1) \right\rfloor,$$

(Stöhr 1955, Guy 1994), where $\lfloor x \rfloor$ is the FLOOR FUNCTION, the first few values of which are 2, 4, 7, 10, 14, 18, 23, 28, 34, 40, ... (Sloane's A014616).

see also HARMONIOUS GRAPH, STAMP FOLDING

References

Guy, R. K. "The Postage Stamp Problem." §C12 in *Unsolved Problems in Number Theory, 2nd ed.* New York: Springer-Verlag, pp. 123–127, 1994.

Sloane, N. J. A. Sequence A014616 in "An On-Line Version of the Encyclopedia of Integer Sequences."

Stöhr, A. "Gelöste und ungelöste Fragen über Basen der natürlichen Zahlenreihe I, II." *J. reine angew. Math.* **194**, 111–140, 1955.

Posterior Distribution

see BAYESIAN ANALYSIS

Postnikov System

An iterated FIBRATION of EILENBERG-MAC LANE SPACES. Every TOPOLOGICAL SPACE has this HOMOTOPY type.

see also EILENBERG-MAC LANE SPACE, FIBRATION, HOMOTOPY

Postulate

A statement, also known as an AXIOM, which is taken to be true without PROOF. Postulates are the basic structure from which LEMMAS and THEOREMS are derived. The whole of EUCLIDEAN GEOMETRY, for example, is based on five postulates known as EUCLID'S POSTULATES.

see also ARCHIMEDES' POSTULATE, AXIOM, BERTRAND'S POSTULATE, CONJECTURE, EQUIDISTANCE POSTULATE, EUCLID'S FIFTH POSTULATE, EUCLID'S POSTULATES, LEMMA, PARALLEL POSTULATE, PORISM, PROOF, THEOREM, TRIANGLE POSTULATE

Potato Paradox

You buy 100 pounds of potatoes and are told that they are 99% water. After leaving them outside, you discover that they are now 98% water. The weight of the dehydrated potatoes is then a surprising 50 pounds!

References

Paulos, J. A. *A Mathematician Reads the Newspaper.* New York: BasicBooks, p. 81, 1995.

Potential Function

The term used in physics and engineering for a HARMONIC FUNCTION. Potential functions are extremely useful, for example, in electromagnetism, where they reduce the study of a 3-component VECTOR FIELD to a 1-component SCALAR FUNCTION.

see also HARMONIC FUNCTION, LAPLACE'S EQUATION, SCALAR POTENTIAL, VECTOR POTENTIAL

Potential Theory

The study of HARMONIC FUNCTIONS (also called POTENTIAL FUNCTIONS).

see also HARMONIC FUNCTION, SCALAR POTENTIAL, VECTOR POTENTIAL

References

Kellogg, O. D. *Foundations of Potential Theory.* New York: Dover, 1953.

MacMillan, W. D. *The Theory of the Potential.* New York: Dover, 1958.

Pothenot Problem

see SNELLIUS-POTHENOT PROBLEM

Poulet Number

A FERMAT PSEUDOPRIME to base 2, denoted psp(2), i.e., a COMPOSITE ODD INTEGER such that

$$2^{n-1} \equiv 1 \pmod{n}.$$

The first few Poulet numbers are 341, 561, 645, 1105, 1387, ... (Sloane's A001567). Pomerance *et al.* (1980) computed all 21,853 Poulet numbers less than 25×10^9.

Pomerance has shown that the number of Poulet numbers less than x for sufficiently large x satisfy

$$\exp[(\ln x)^{5/14}] < P_2(x) < x \exp\left(-\frac{\ln x \ln \ln \ln x}{2 \ln \ln x}\right)$$

(Guy 1994).

A Poulet number all of whose DIVISORS d satisfy $d | 2^d - 2$ is called a SUPER-POULET NUMBER. There are an infinite number of Poulet numbers which are not SUPER-POULET NUMBERS. Shanks (1993) calls *any* integer satisfying $2^{n-1} \equiv 1 \pmod{n}$ (i.e., not limited to ODD composite numbers) a FERMATIAN.

see also FERMAT PSEUDOPRIME, PSEUDOPRIME, SUPER-POULET NUMBER

References

Guy, R. K. *Unsolved Problems in Number Theory, 2nd ed.* New York: Springer-Verlag, pp. 28–29, 1994.

Pomerance, C.; Selfridge, J. L.; and Wagstaff, S. S. Jr. "The Pseudoprimes to $25 \cdot 10^9$." *Math. Comput.* **35**, 1003–1026, 1980. Available electronically from `ftp://sable.ox.ac.uk/pub/math/primes/ps2.Z`.

Shanks, D. *Solved and Unsolved Problems in Number Theory, 4th ed.* New York: Chelsea, pp. 115–117, 1993.

Sloane, N. J. A. Sequence A001567/M5441 in "An On-Line Version of the Encyclopedia of Integer Sequences."

Power

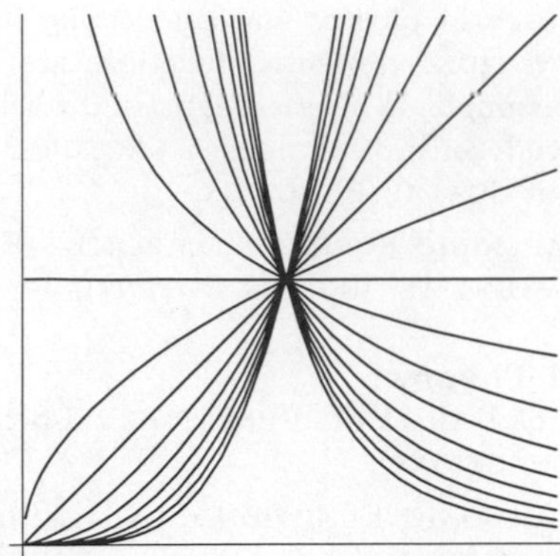

The exponent to which a given quantity is raised is known as its POWER. The expression x^a is therefore known as "x to the ath POWER." The rules for combining quantities containing powers are called the EXPONENT LAWS.

Special names given to various powers are listed in the following table.

Power	Name
1/2	square root
1/3	cube root
2	squared
3	cubed

The SUM of pth POWERS of the first n POSITIVE INTEGERS is given by FAULHABER'S FORMULA,

$$\sum_{k=1}^{n} k^p = \frac{1}{p+1} \sum_{k=1}^{p+1} (-1)^{\delta_{kp}} \binom{p+1}{k} B_{p+1-k} n^k,$$

where δ_{kp} is the KRONECKER DELTA, $\binom{n}{k}$ is a BINOMIAL COEFFICIENT, and B_k is a BERNOULLI NUMBER.

Let s_n be the largest INTEGER that is not the SUM of distinct nth powers of POSITIVE INTEGERS (Guy 1994). The first few values for $n = 2, 3, \ldots$ are 128, 12758, 5134240, 67898771, ... (Sloane's A001661).

CATALAN'S CONJECTURE states that 8 and 9 (2^3 and 3^2) are the only consecutive POWERS (excluding 0 and 1), i.e., the only solution to CATALAN'S DIOPHANTINE PROBLEM. This CONJECTURE has not yet been proved or refuted, although R. Tijdeman has proved that there can be only a finite number of exceptions should the CONJECTURE not hold. It is also known that 8 and 9 are the only consecutive CUBIC and SQUARE NUMBERS (in either order). Hyyrö and Mąkowski proved that there do not exist three consecutive POWERS (Ribenboim 1996).

Very few numbers of the form $n^p \pm 1$ are PRIME (where composite powers $p = kb$ need not be considered, since $n^(kb) \pm 1 = (n^k)^b \pm 1$). The only PRIME NUMBERS of the form $n^p - 1$ for $n \leq 100$ and PRIME $2 \leq p \leq 10$ correspond to $n = 2$, i.e., $2^2 - 1 = 3$, $2^3 - 1 = 7$, $2^5 - 1 = 31$, ... The only PRIME NUMBERS of the form $n^p + 1$ for $n \leq 100$ and PRIME $2 \leq p \leq 10$ correspond to $p = 2$ with $n = 1, 2, 4, 6, 10, 14, 16, 20, 24, 26, \ldots$ (Sloane's A005574).

There are no nontrivial solutions to the equation

$$1^n + 2^n + \ldots + m^n = (m+1)^n$$

for $m \leq 10^{2,000,000}$ (Guy 1994, p. 153).

see also APOCALYPTIC NUMBER, BIQUADRATIC NUMBER, CATALAN'S CONJECTURE, CATALAN'S DIOPHANTINE PROBLEM, CUBE ROOT, CUBED, CUBIC NUMBER, EXPONENT, EXPONENT LAWS, FAULHABER'S FORMULA, FIGURATE NUMBER, MOESSNER'S THEOREM, NARCISSISTIC NUMBER, POWER RULE, SQUARE NUMBER, SQUARE ROOT, SQUARED, SUM, WARING'S PROBLEM

References

Barbeau, E. J. *Power Play: A Country Walk through the Magical World of Numbers.* Washington, DC: Math. Assoc. Amer., 1997.

Beyer, W. H. "Laws of Exponents." *CRC Standard Mathematical Tables, 28th ed.* Boca Raton, FL: CRC Press, p. 158, 1987.

Guy, R. K. "Diophantine Equations." Ch. D in *Unsolved Problems in Number Theory, 2nd ed.* New York: Springer-Verlag, pp. 137, 139–198, and 153–154, 1994.

Ribenboim, P. "Catalan's Conjecture." *Amer. Math. Monthly* **103**, 529–538, 1996.

Sloane, N. J. A. Sequences A001661/M5393 and A005574/M1010 in "An On-Line Version of the Encyclopedia of Integer Sequences."

Spanier, J. and Oldham, K. B. "The Integer Powers $(bx+c)^n$ and x^n" and "The Noninteger Powers x^ν." Ch. 11 and 13 in *An Atlas of Functions.* Washington, DC: Hemisphere, pp. 83–90 and 99–106, 1987.

Power Center

see RADICAL CENTER

Power (Circle)

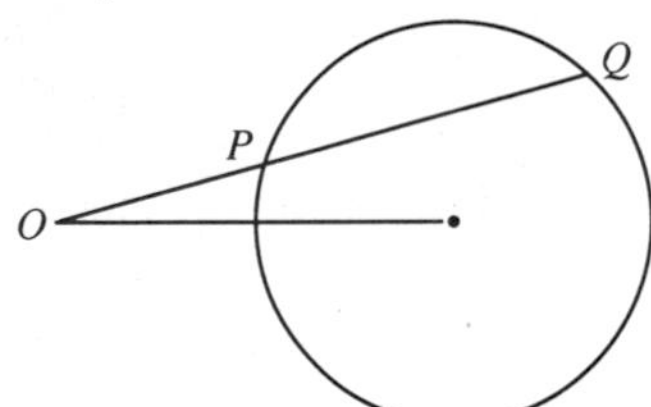

The POWER of the two points P and Q with respect to a CIRCLE is defined by

$$p \equiv OP \times PQ.$$

Let R be the RADIUS of a CIRCLE and d be the distance between a point P and the circle's center. Then the POWER of the point P relative to the circle is

$$p = d^2 - R^2.$$

If P is outside the CIRCLE, its POWER is POSITIVE and equal to the square of the length of the segment from P to the tangent to the CIRCLE through P. If P is inside the CIRCLE, then the POWER is NEGATIVE and equal to the product of the DIAMETERS through P.

The LOCUS of points having POWER k with regard to a fixed CIRCLE of RADIUS r is a CONCENTRIC CIRCLE of RADIUS $\sqrt{r^2+k}$. The CHORDAL THEOREM states that the LOCUS of points having equal POWER with respect to two given nonconcentric CIRCLES is a line called the RADICAL LINE (or CHORDAL; Dörrie 1965).

see also CHORDAL THEOREM, COAXAL CIRCLES, INVERSE POINTS, INVERSION CIRCLE, INVERSION RADIUS, INVERSIVE DISTANCE, RADICAL LINE

References

Coxeter, H. S. M. and Greitzer, S. L. *Geometry Revisited.* Washington, DC: Math. Assoc. Amer., pp. 27–31, 1967.
Dixon, R. *Mathographics.* New York: Dover, p. 68, 1991.
Dörrie, H. *100 Great Problems of Elementary Mathematics: Their History and Solutions.* New York: Dover, p. 153, 1965.
Johnson, R. A. *Modern Geometry: An Elementary Treatise on the Geometry of the Triangle and the Circle.* Boston, MA: Houghton Mifflin, pp. 28–34, 1929.
Pedoe, D. *Circles: A Mathematical View, rev. ed.* Washington, DC: Math. Assoc. Amer., pp. xxii–xxiv, 1995.

Power Curve

The curve with TRILINEAR COORDINATES $a^t : b^t : c^t$ for a given POWER t.

see also POWER

References

Kimberling, C. "Major Centers of Triangles." *Amer. Math. Monthly* **104**, 431–438, 1997.

Power Line

see RADICAL AXIS

Power Point

TRIANGLE centers with TRIANGLE CENTER FUNCTIONS of the form $\alpha = a^n$ are called nth POWER points. The 0th power point is the INCENTER, with TRIANGLE CENTER FUNCTION $\alpha = 1$.

see also INCENTER, TRIANGLE CENTER FUNCTION

References

Groenman, J. T. and Eddy, R. H. "Problem 858 and Solution." *Crux Math.* **10**, 306–307, 1984.
Kimberling, C. "Problem 865." *Crux Math.* **10**, 325–327, 1984.
Kimberling, C. "Central Points and Central Lines in the Plane of a Triangle." *Math. Mag.* **67**, 163–187, 1994.

Power Rule

The DERIVATIVE of the POWER x^n is given by

$$\frac{d}{dx}(x^n) = nx^{n-1}.$$

see also CHAIN RULE, DERIVATIVE, EXPONENT LAWS, PRODUCT RULE

References

Anton, H. *Calculus with Analytic Geometry, 2nd ed.* New York: Wiley, p. 131, 1984.

Power Series

A power series in a variable z is an infinite SUM of the form

$$\sum_n^\infty a_i z^i, \qquad (1)$$

where $n \geq 0$ and a_i are INTEGERS, REAL NUMBERS, COMPLEX NUMBERS, or any other quantities of a given type.

A CONJECTURE of Pólya is that if a FUNCTION has a POWER series with INTEGER COEFFICIENTS and RADIUS OF CONVERGENCE 1, then either the FUNCTION is RATIONAL or the UNIT CIRCLE is a natural boundary.

A generalized POWER sum $a(h)$ for $h = 0, 1, \ldots$ is given by

$$a(h) = \sum_{i=1}^m A_i(h){\alpha_i}^h, \qquad (2)$$

with distinct NONZERO ROOTS α_i, COEFFICIENTS $A_i(h)$ which are POLYNOMIALS of degree $n_i - 1$ for POSITIVE INTEGERS n_i, and $i \in [1, m]$. The generalized POWER sum has order

$$n \equiv \sum_{i=m}^m n_i. \qquad (3)$$

For any power series, one of the following is true:

1. The series converges only for $x = 0$.
2. The series converges absolutely for all x.
3. The series converges absolutely for all x in some finite open interval $(-R, R)$ and diverges if $x < -R$ or $x > R$. At the points $x = R$ and $x = -R$, the series may converge absolutely, converge conditionally, or diverge.

To determine the interval of convergence, apply the RATIO TEST for ABSOLUTE CONVERGENCE and solve for x. A POWER series may be differentiated or integrated within the interval of convergence. Convergent power series may be multiplied and divided (if there is no division by zero).

$$\sum_{k=1}^\infty k^{-p}. \qquad (4)$$

CONVERGES if $p > 1$ and DIVERGES if $0 < p \leq 1$.

see also BINOMIAL SERIES, CONVERGENCE TESTS, LAURENT SERIES, MACLAURIN SERIES, MULTINOMIAL SERIES, p-SERIES, POLYNOMIAL, POWER SET, QUOTIENT-DIFFERENCE ALGORITHM, RECURRENCE SEQUENCE, SERIES, SERIES REVERSION, TAYLOR SERIES

References

Arfken, G. "Power Series." §5.7 in *Mathematical Methods for Physicists, 3rd ed.* Orlando, FL: Academic Press, pp. 313–321, 1985.

Myerson, G. and van der Poorten, A. J. "Some Problems Concerning Recurrence Sequences." *Amer. Math. Monthly* **102**, 698–705, 1995.

Pólya, G. *Mathematics and Plausible Reasoning, Vol. 2: Patterns of Plausible Inference.* Princeton, NJ: Princeton University Press, p. 46, 1954.

Power Set

Given a SET S, the POWER SET of S is the SET of all SUBSETS of S. The order of a POWER set of a SET of order n is 2^n. Power sets are larger than the SETS associated with them.

see also SET, SUBSET

Power Spectrum

For a given signal, the power spectrum gives a plot of the portion of a signal's power (energy per unit time) falling within given frequency bins. The most common way of generating a power spectrum is by using a FOURIER TRANSFORM, but other techniques such as the MAXIMUM ENTROPY METHOD can also be used.

References

Press, W. H.; Flannery, B. P.; Teukolsky, S. A.; and Vetterling, W. T. "Power Spectra Estimation Using the FFT" and "Power Spectrum Estimation by the Maximum Entropy (All Poles) Method." §13.4 and 13.7 in *Numerical Recipes in FORTRAN: The Art of Scientific Computing, 2nd ed.* Cambridge, England: Cambridge University Press, pp. 542–551 and 565–569, 1992.

Power (Statistics)

The probability of getting a positive result for a given test which should produce a positive result.

see also PREDICTIVE VALUE, SENSITIVITY, SPECIFICITY, STATISTICAL TEST

Power Tower

$$a \uparrow\uparrow k \equiv \underbrace{a^{a^{\cdot^{\cdot^{\cdot^{a}}}}}}_{k},$$

where $\uparrow$ is Knuth's (1976) ARROW NOTATION.

$$a \uparrow^k n = a \uparrow^{k-1} [a \uparrow^k (n-1)].$$

The infinite power tower $x \uparrow\uparrow \infty = x^{x^{\cdot^{\cdot^{\cdot}}}}$ converges IFF $e^{-e} \leq x \leq e^{1/e}$ $(0.0659 \leq x \leq 1.4446)$.

see also ACKERMANN FUNCTION, FERMAT NUMBER, MILLS' CONSTANT

References

Knuth, D. E. "Mathematics and Computer Science: Coping with Finiteness. Advances in our Ability to Compute are Bringing us Substantially Closer to Ultimate Limitations." *Science* **194** 1235–1242, 1976.

Vardi, I. *Computational Recreations in Mathematica.* Reading, MA: Addison-Wesley, pp. 11 and 226–229, 1991.

Power (Triangle)

The total POWER of a TRIANGLE is defined by

$$P \equiv \tfrac{1}{2}({a_1}^2 + {a_2}^2 + {a_3}^2), \qquad (1)$$

where a_i are the side lengths, and the "partial power" is defined by

$$p_1 \equiv \tfrac{1}{2}({a_2}^2 + {a_3}^2 - {a_1}^2). \qquad (2)$$

Then

$$P_1 = a_2 a_3 \cos\alpha_1 \qquad (3)$$

$$P = p_1 + p_2 + p_3 \qquad (4)$$

$$P^2 + {p_1}^2 + {p_2}^2 + {p_3}^2 = {a_1}^4 + {a_2}^4 + {a_3}^4 \qquad (5)$$

$$\Delta = \tfrac{1}{2}\sqrt{p_2 p_3 + p_3 p_1 + p_3 p_1} \qquad (6)$$

$$p_1 = \overline{A_1 H_2} \cdot \overline{A_1 A_3} \qquad (7)$$

$$\frac{a_1 p_1}{\cos\alpha_1} = a_1 a_2 a_3 = 4\Delta R \qquad (8)$$

$$p_1 \tan\alpha_1 = p_2 \tan\alpha_2 = p_3 \tan\alpha_3, \qquad (9)$$

where Δ is the AREA of the TRIANGLE and H_i are the FEET of the ALTITUDES. Finally, if a side of the TRIANGLE and the value of any partial power are given, then the LOCUS of the third VERTEX is a CIRCLE or straight line.

see also ALTITUDE, FOOT, TRIANGLE

References

Johnson, R. A. *Modern Geometry: An Elementary Treatise on the Geometry of the Triangle and the Circle.* Boston, MA: Houghton Mifflin, pp. 260–261, 1929.

Powerfree

see BIQUADRATEFREE, CUBEFREE, PRIME NUMBER, SQUAREFREE

Powerful Number

An INTEGER m such that if $p|m$, then $p^2|m$, is called a powerful number. The first few are 1, 4, 8, 9, 16, 25, 27, 32, 36, 49, ... (Sloane's A001694). Powerful numbers are always of the form a^2b^3 for $a, b \geq 1$.

Not every NATURAL NUMBER is the sum of two powerful numbers, but Heath-Brown (1988) has shown that every sufficiently large NATURAL NUMBER is the sum of at most three powerful numbers. There are infinitely many pairs of consecutive powerful numbers, but Erdős has conjectured that there do not exist three consecutive powerful numbers. The CONJECTURE that there are no powerful number triples implies that there are infinitely many Wieferich primes (Granville 1986, Vardi 1991).

A separate usage of the term powerful number is for numbers which are the sums of the positive powers of their digits. The first few are 1, 2, 3, 4, 5, 6, 7, 8, 9, 24, 43, 63, 89, ... (Sloane's A007532).

References
Granville, A. "Powerful Numbers and Fermat's Last Theorem." *C. R. Math. Rep. Acad. Sci. Canada* **8**, 215–218, 1986.
Guy, R. K. "Powerful Numbers." §B16 in *Unsolved Problems in Number Theory, 2nd ed.* New York: Springer-Verlag, pp. 67–73, 1994.
Heath-Brown, D. R. "Ternary Quadratic Forms and Sums of Three Square-Full Numbers." In *Séminaire de Theorie des Nombres, Paris 1986–87* (Ed. C. Goldstein). Boston, MA: Birkhäuser, pp. 137–163, 1988.
Ribenboim, P. "Catalan's Conjecture." *Amer. Math. Monthly* **103**, 529–538, 1996.
Sloane, N. J. A. Sequences A001694/M3325 and A007532/M0487 in "An On-Line Version of the Encyclopedia of Integer Sequences."
Vardi, I. *Computational Recreations in Mathematica.* Reading, MA: Addison-Wesley, pp. 59–62, 1991.

Practical Number

A number n is practical if for all $k \le n$, k is the sum of distinct proper divisors of n. Defined in 1948 by A. K. Srinivasen. All even PERFECT NUMBERS are practical. The number

$$m = 2^{n-1}(2^n - 1)$$

is practical for all $n = 2, 3, \ldots$. The first few practical numbers are 1, 2, 4, 6, 8, 12, 16, 18, 20, 24, 28, 30, 32, 36, 40, 42, 48, 54, 56, ... (Sloane's A005153). G. Melfi has computed twins, triplets, and 5-tuples of practical numbers. The first few 5-tuples are 12, 18, 30, 198, 306, 462, 1482, 2550, 4422, ...

References
Melfi, G. "On Two Conjectures About Practical Numbers." *J. Number Th.* **56**, 205–210, 1996.
Melfi, G. "Practical Numbers." `http://www.dm.unipi.it/gauss-pages/melfi/public_html/pratica.html`.
Sloane, N. J. A. Sequence A005153/M0991 in "An On-Line Version of the Encyclopedia of Integer Sequences."

Pratt Certificate

A primality certificate based on FERMAT'S LITTLE THEOREM CONVERSE. Although the general idea had been well-established for some time, Pratt became the first to prove that the certificate tree was of polynomial size and could also be verified in polynomial time. He was also the first to observe that the tree implies that PRIMES are in the complexity class NP.

To generate a Pratt certificate, assume that n is a POSITIVE INTEGER and $\{p_i\}$ is the set of PRIME FACTORS of $n - 1$. Suppose there exists an INTEGER x (called a "WITNESS") such that $x^{n-1} \equiv 1 \pmod{n}$ but $x^e \not\equiv 1 \pmod{n}$ whenever e is one of $(n-1)/p_i$. Then FERMAT'S LITTLE THEOREM CONVERSE states that n is PRIME (Wagon 1991, pp. 278–279).

By applying FERMAT'S LITTLE THEOREM CONVERSE to n and recursively to each purported factor of $n-1$, a certificate for a given PRIME NUMBER can be generated. Stated another way, the Pratt certificate gives a proof that a number a is a PRIMITIVE ROOT of the multiplicative GROUP (mod p) which, along with the fact that a has order $p-1$, proves that p is a PRIME.

```
7919 —— 7
  2
  37 —— 2
     2
     3 —— 2
        2
  107 —— 2
     2
     53 —— 2
        2
        13 —— 2
           2
           3 —— 2
              2
```

The figure above gives a certificate for the primality of $n = 7919$. The numbers to the right of the dashes are WITNESSES to the numbers to left. The set $\{p_i\}$ for $n - 1 = 7918$ is given by $\{2, 37, 107\}$. Since $7^{7918} \equiv 1 \pmod{7919}$ but $7^{7918/2}$, $7^{7918/37}$, $7^{7918/107} \not\equiv 1 \pmod{7919}$, 7 is a WITNESS for 7919. The PRIME divisors of $7918 = 7919 - 1$ are 2, 37, and 107. 2 is a so-called "self-WITNESS" (i.e., it is recognized as a PRIME without further ado), and the remainder of the witnesses are shown as a nested tree. Together, they certify that 7919 is indeed PRIME. Because it requires the FACTORIZATION of $n - 1$, the METHOD of Pratt certificates is best applied to small numbers (or those numbers n known to have easily factorable $n - 1$).

A Pratt certificate is quicker to generate for small numbers than are other types of primality certificates. The *Mathematica*® (Wolfram Research, Champaign, IL) task `ProvablePrime[n]` therefore generates an ATKIN-GOLDWASSER-KILIAN-MORAIN CERTIFICATE only for numbers above a certain limit (10^{10} by default), and a Pratt certificate for smaller numbers.

see also ATKIN-GOLDWASSER-KILIAN-MORAIN CERTIFICATE, FERMAT'S LITTLE THEOREM CONVERSE, PRIMALITY CERTIFICATE, WITNESS

References
Pratt, V. "Every Prime Has a Succinct Certificate." *SIAM J. Comput.* **4**, 214–220, 1975.
Wagon, S. *Mathematica in Action.* New York: W. H. Freeman, pp. 278–285, 1991.
Wilf, H. §4.10 in *Algorithms and Complexity.* Englewood Cliffs, NJ: Prentice-Hall, 1986.

Pratt-Kasapi Theorem

see HOEHN'S THEOREM

Precedes

The relationship x precedes y is written $x \prec y$. The relation x precedes or is equal to y is written $x \preceq y$.

see also SUCCEEDS

Precession

see CURVE OF CONSTANT PRECESSION

Precisely Unless

If A is true precisely unless B, then B implies not-A and not-B implies A. J. H. Conway has suggested the term "UNLESSS" for this state of affairs, by analogy with IFF.

see also IFF, UNLESS

Predicate

A function whose value is either TRUE or FALSE.

see also AND, FALSE, OR, PREDICATE CALCULUS, TRUE, XOR

Predicate Calculus

The branch of formal LOGIC dealing with representing the logical connections between statements as well as the statements themselves.

see also GÖDEL'S INCOMPLETENESS THEOREM, LOGIC, PREDICATE

Predictability

Predictability at a time τ in the future is defined by

$$\frac{R(x(t), x(t+\tau))}{H(x(t))},$$

and linear predictability by

$$\frac{L(x(t), x(t+\tau))}{H(x(t))},$$

where R and L are the REDUNDANCY and LINEAR REDUNDANCY, and H is the ENTROPY.

Prediction Paradox

see UNEXPECTED HANGING PARADOX

Predictive Value

The POSITIVE predictive value is the probability that a test gives a true result for a true statistic. The negative predictive value is the probability that a test gives a false result for a false statistic.

see also POWER (STATISTICS), SENSITIVITY, SPECIFICITY, STATISTICAL TEST

Predictor-Corrector Methods

A general method of integrating ORDINARY DIFFERENTIAL EQUATIONS. It proceeds by extrapolating a polynomial fit to the derivative from the previous points to the new point (the predictor step), then using this to interpolate the derivative (the corrector step). Press *et al.* (1992) opine that predictor-corrector methods have been largely supplanted by the BULIRSCH-STOER and RUNGE-KUTTA METHODS, but predictor-corrector schemes are still in common use.

see also ADAMS' METHOD, GILL'S METHOD, MILNE'S METHOD, RUNGE-KUTTA METHOD

References

Abramowitz, M. and Stegun, C. A. (Eds.). *Handbook of Mathematical Functions with Formulas, Graphs, and Mathematical Tables, 9th printing.* New York: Dover, pp. 896–897, 1972.

Arfken, G. *Mathematical Methods for Physicists, 3rd ed.* Orlando, FL: Academic Press, pp. 493–494, 1985.

Press, W. H.; Flannery, B. P.; Teukolsky, S. A.; and Vetterling, W. T. "Multistep, Multivalue, and Predictor-Corrector Methods." §16.7 in *Numerical Recipes in FORTRAN: The Art of Scientific Computing, 2nd ed.* Cambridge, England: Cambridge University Press, pp. 740–744, 1992.

Pretzel Curve

see KNOT CURVE

Pretzel Knot

A KNOT obtained from a TANGLE which can be represented by a FINITE sequence of INTEGERS.

see also TANGLE

References

Adams, C. C. *The Knot Book: An Elementary Introduction to the Mathematical Theory of Knots.* New York: W. H. Freeman, p. 48, 1994.

Primality Certificate

A short set of data that proves the primality of a number. A certificate can, in general, be checked much more quickly than the time required to generate the certificate. Varieties of primality certificates include the PRATT CERTIFICATE and ATKIN-GOLDWASSER-KILIAN-MORAIN CERTIFICATE.

see also ATKIN-GOLDWASSER-KILIAN-MORAIN CERTIFICATE, COMPOSITENESS CERTIFICATE, PRATT CERTIFICATE

References

Wagon, S. "Prime Certificates." §8.7 in *Mathematica in Action.* New York: W. H. Freeman, pp. 277–285, 1991.

Primality Test

A test to determine whether or not a given number is PRIME. The RABIN-MILLER STRONG PSEUDOPRIME TEST is a particularly efficient ALGORITHM used by *Mathematica®* version 2.2 (Wolfram Research, Champaign, IL). Like many such algorithms, it is a probabilistic test using PSEUDOPRIMES, and can potentially (although with very small probability) falsely identify a COMPOSITE NUMBER as PRIME (although not vice versa). Unlike PRIME FACTORIZATION, primality testing is believed to be a P-PROBLEM (Wagon 1991). In order to guarantee primality, an almost certainly slower algorithm capable of generating a PRIMALITY CERTIFICATE must be used.

see also ADLEMAN-POMERANCE-RUMELY PRIMALITY TEST, FERMAT'S LITTLE THEOREM CONVERSE, FERMAT'S PRIMALITY TEST, FERMAT'S THEOREM, LUCAS-LEHMER TEST, MILLER'S PRIMALITY TEST, PÉPIN'S TEST, POCKLINGTON'S THEOREM, PROTH'S THEOREM, PSEUDOPRIME, RABIN-MILLER STRONG PSEUDOPRIME TEST, WARD'S PRIMALITY TEST, WILSON'S THEOREM

References

Beauchemin, P.; Brassard, G.; Crépeau, C.; Goutier, C.; and Pomerance, C. "The Generation of Random Numbers that are Probably Prime." *J. Crypt.* **1**, 53–64, 1988.

Brillhart, J.; Lehmer, D. H.; Selfridge, J.; Wagstaff, S. S. Jr.; and Tuckerman, B. *Factorizations of* $b^n \pm 1$, $b = 2, 3, 5, 6, 7, 10, 11, 12$ *Up to High Powers, rev. ed.* Providence, RI: Amer. Math. Soc., pp. lviii–lxv, 1988.

Cohen, H. and Lenstra, A. K. "Primality Testing and Jacobi Sums." *Math. Comput.* **42**, 297–330, 1984.

Knuth, D. E. *The Art of Computer Programming, Vol. 2: Seminumerical Algorithms, 2nd ed.* Reading, MA: Addison-Wesley, 1981.

Riesel, H. *Prime Numbers and Computer Methods for Factorization, 2nd ed.* Boston, MA: Birkhäuser, 1994.

Wagon, S. *Mathematica in Action.* New York: W. H. Freeman, pp. 15–17, 1991.

Primary

Each factor ${p_i}^{\alpha_i}$ in an INTEGER's PRIME DECOMPOSITION is called a primary.

Primary Representation

Let π be a unitary REPRESENTATION of a GROUP G on a separable HILBERT SPACE, and let $R(\pi)$ be the smallest weakly closed algebra of bounded linear operators containing all $\pi(g)$ for $g \in G$. Then π is primary if the center of $R(\pi)$ consists of only scalar operations.

References

Knapp, A. W. "Group Representations and Harmonic Analysis, Part II." *Not. Amer. Math. Soc.* **43**, 537–549, 1996.

Prime

A symbol used to distinguish one quantity x' ("x prime") from another related x. Primes are most commonly used to denote transformed coordinates, conjugate points, and DERIVATIVES.

see also PRIME ALGEBRAIC NUMBER, PRIME NUMBER

Prime Algebraic Number

An irreducible ALGEBRAIC INTEGER which has the property that, if it divides the product of two algebraic INTEGERS, then it DIVIDES at least one of the factors. 1 and -1 are the only INTEGERS which DIVIDE every INTEGER. They are therefore called the PRIME UNITS.

see also ALGEBRAIC INTEGER, PRIME UNIT

Prime Arithmetic Progression

Let the number of PRIMES of the form $mk + n$ less than x be denoted $\pi_{m,n}(x)$. Then

$$\lim_{x\to\infty} \frac{\pi_{a,b}(x)}{\mathrm{Li}(x)} = \frac{1}{\phi(a)},$$

where $\mathrm{Li}(x)$ is the LOGARITHMIC INTEGRAL and $\phi(x)$ is the TOTIENT FUNCTION.

Let P be an increasing arithmetic progression of n PRIMES with minimal difference $d > 0$. If a PRIME $p \leq n$ does not divide d, then the elements of P must assume all residues modulo p, specifically, some element of P must be divisible by p. Whereas P contains only primes, this element must be equal to p.

If $d < n\#$ (where $n\#$ is the PRIMORIAL of n), then some prime $p \leq n$ does not divide d, and that prime p is in P. Thus, in order to determine if P has $d < n\#$, we need only check a finite number of possible P (those with $d < n\#$ and containing prime $p \leq n$) to see if they contain only primes. If not, then $d \geq n\#$. If $d = n\#$, then the elements of P cannot be made to cover all residues of any prime p. The PRIME PATTERNS CONJECTURE then asserts that there are infinitely many arithmetic progressions of primes with difference d.

A computation shows that the smallest possible common difference for a set of n or more PRIMES in arithmetic progression for $n = 1, 2, 3, \ldots$ is 0, 1, 2, 6, 6, 30, 150, 210, 210, 210, 2310, 2310, 30030, 510510, ... (Sloane's A033188, Ribenboim 1989, Dubner and Nelson 1997, Wilson). The values up to $n = 13$ are rigorous, while the remainder are lower bounds which assume the validity of the PRIME PATTERNS CONJECTURE and are simply given by $p_{n-7}\#$, where p_i is the ith PRIME. The smallest first terms of arithmetic progressions of n primes *with minimal differences* are 2, 2, 3, 5, 5, 7, 7, 199, 199, 199, 60858179, 147692845283, 14933623, ... (Sloane's A033189; Wilson).

Smaller first terms are possible for nonminimal n-term progressions. Examples include the 8-term progression $11 + 1210230k$ for $k = 0, 1, \ldots, 7$, the 12-term progression $23143 + 30030k$ for $k = 0, 1, \ldots, 11$ (Golubev 1969, Guy 1994), and the 13-term arithmetic progression $766439 + 510510k$ for $k = 0, 1, \ldots, 12$ (Guy 1994).

The largest known set of primes in ARITHMETIC SEQUENCE is 22,

$$11,410,337,850,553 + 4,609,098,694,200k$$

for $k = 0, 1, \ldots, 21$ (Pritchard *et al.* 1995, UTS School of Mathematical Sciences).

The largest known sequence of *consecutive* PRIMES in ARITHMETIC PROGRESSION (i.e., all the numbers between the first and last term in the progression, except for the members themselves, are composite) is ten, given by

$$\begin{gathered}100{,}996{,}972{,}469{,}714{,}247{,}637{,}786{,}655{,}587{,}969{,}\\840{,}329{,}509{,}324{,}689{,}190{,}041{,}803{,}603{,}417{,}758{,}\\904{,}341{,}703{,}348{,}882{,}159{,}067{,}229{,}719+210k\end{gathered}$$

for $k = 0, 1, \ldots, 9$, discovered by Harvey Dubner, Tony Forbes, Manfred Toplic, *et al.* on March 2, 1998. This beats the record of nine set on January 15, 1998 by the same investigators,

$$\begin{gathered}99{,}679{,}432{,}066{,}701{,}086{,}484{,}490{,}653{,}695{,}853{,}\\561{,}638{,}982{,}364{,}080{,}991{,}618{,}395{,}774{,}048{,}585{,}\\529{,}071{,}475{,}461{,}114{,}799{,}677{,}694{,}651+210k\end{gathered}$$

for $k = 0, 1, \ldots, 8$ (two sequences of nine are now known), the progression of eight consecutive primes given by

$$\begin{gathered}43{,}804{,}034{,}644{,}029{,}893{,}325{,}717{,}710{,}709{,}965{,}\\599{,}930{,}101{,}479{,}007{,}432{,}825{,}862{,}362{,}446{,}333{,}\\961{,}919{,}524{,}977{,}985{,}103{,}251{,}510{,}661+210k\end{gathered}$$

for $k = 0, 1, \ldots, 7$, discovered by Harvey Dubner, Tony Forbes, *et al.* on November 7, 1997 (several are now known), and the progression of seven given by

$$\begin{gathered}1{,}089{,}533{,}431{,}247{,}059{,}310{,}875{,}780{,}378{,}922{,}957{,}732{,}\\908{,}036{,}492{,}993{,}138{,}195{,}385{,}213{,}105{,}561{,}742{,}150{,}\\447{,}308{,}967{,}213{,}141{,}717{,}486{,}151+210k,\end{gathered}$$

for $k = 0, 1, \ldots, 6$, discovered by H. Dubner and H. K. Nelson on Aug. 29, 1995 (Peterson 1995, Dubner and Nelson 1997). The smallest sequence of six consecutive PRIMES in arithmetic progression is

$$121{,}174{,}811+30k$$

for $k = 0, 1, \ldots, 5$ (Lander and Parkin 1967, Dubner and Nelson 1997). According to Dubner *et al.*, a trillion-fold increase in computer speed is needed before the search for a sequence of 11 consecutive primes is practical, so they expect the ten-primes record to stand for a long time to come.

It is conjectured that there are arbitrarily long sequences of PRIMES in ARITHMETIC PROGRESSION (Guy 1994).

see also ARITHMETIC PROGRESSION, CUNNINGHAM CHAIN, DIRICHLET'S THEOREM, LINNIK'S THEOREM, PRIME CONSTELLATION, PRIME-GENERATING POLYNOMIAL, PRIME NUMBER THEOREM, PRIME PATTERNS CONJECTURE, PRIME QUADRUPLET

References

Abel, U. and Siebert, H. "Sequences with Large Numbers of Prime Values." *Amer. Math. Monthly* **100**, 167–169, 1993.

Caldwell, C. K. "Cunningham Chain." `http://www.utm.edu/research/primes/glossary/CunninghamChain.html`.

Courant, R. and Robbins, H. "Primes in Arithmetical Progressions." §1.2b in Supplement to Ch. 1 in *What is Mathematics?: An Elementary Approach to Ideas and Methods, 2nd ed.* Oxford, England: Oxford University Press, pp. 26–27, 1996.

Davenport, H. "Primes in Arithmetic Progression" and "Primes in Arithmetic Progression: The General Modulus." Chs. 1 and 4 in *Multiplicative Number Theory, 2nd ed.* New York: Springer-Verlag, pp. 1–11 and 27–34, 1980.

Dubner, H. and Nelson, H. "Seven Consecutive Primes in Arithmetic Progression." *Math. Comput.* **66**, 1743–1749, 1997.

Forbes, T. "Searching for 9 Consecutive Primes in Arithmetic Progression." `http://www.ltkz.demon.co.uk/ar2/9primes.htm`.

Forman, R. "Sequences with Many Primes." *Amer. Math. Monthly* **99**, 548–557, 1992.

Golubev, V. A. "Faktorisation der Zahlen der Form $x^3 \pm 4x^2 + 3x \pm 1$." *Anz. Österreich. Akad. Wiss. Math.-Naturwiss. Kl.* 184–191, 1969.

Guy, R. K. "Arithmetic Progressions of Primes" and "Consecutive Primes in A.P." §A5 and A6 in *Unsolved Problems in Number Theory, 2nd ed.* New York: Springer-Verlag, pp. 15–17 and 18, 1994.

Lander, L. J. and Parkin, T. R. "Consecutive Primes in Arithmetic Progression." *Math. Comput.* **21**, 489, 1967.

Madachy, J. S. *Madachy's Mathematical Recreations.* New York: Dover, pp. 154–155, 1979.

Nelson, H. L. "There Is a Better Sequence." *J. Recr. Math.* **8**, 39–43, 1975.

Peterson, I. "Progressing to a Set of Consecutive Primes." *Sci. News* **148**, 167, Sep. 9, 1995.

Pritchard, P. A.; Moran, A.; and Thyssen, A. "Twenty-Two Primes in Arithmetic Progression." *Math. Comput.* **64**, 1337–1339, 1995.

Ramaré, O. and Rumely, R. "Primes in Arithmetic Progressions." *Math. Comput.* **65**, 397–425, 1996.

Ribenboim, P. *The Book of Prime Number Records, 2nd ed.* New York: Springer-Verlag, p. 224, 1989.

Shanks, D. "Primes in Some Arithmetic Progressions and a General Divisibility Theorem." §104 in *Solved and Unsolved Problems in Number Theory, 4th ed.* New York: Chelsea, pp. 104–109, 1993.

Sloane, N. J. A. Sequences A033188 and A033189 in "An On-Line Version of the Encyclopedia of Integer Sequences."

Weintraub, S. "Consecutive Primes in Arithmetic Progression." *J. Recr. Math.* **25**, 169–171, 1993.

Zimmerman, P. `http://www.loria.fr/~zimmerma/records/8primes.announce`.

Prime Array

Find the $m \times n$ ARRAY of single digits which contains the maximum possible number of PRIMES, where allowable PRIMES may lie along any horizontal, vertical, or diagonal line. For $m = n = 2$, 11 PRIMES are maximal and are contained in the two distinct arrays

$$A(2,2) = \begin{bmatrix}1 & 3\\4 & 7\end{bmatrix}, \begin{bmatrix}1 & 3\\7 & 9\end{bmatrix},$$

giving the PRIMES (3, 7, 13, 17, 31, 37, 41, 43, 47, 71, 73) and (3, 7, 13, 17, 19, 31, 37, 71, 73, 79, 97), respectively. For the 3×2 array, 18 PRIMES are maximal and are contained in the arrays

$$A(3,2) = \begin{bmatrix} 1 & 1 & 3 \\ 9 & 7 & 4 \end{bmatrix}, \begin{bmatrix} 1 & 7 & 2 \\ 3 & 5 & 9 \end{bmatrix}, \begin{bmatrix} 1 & 7 & 2 \\ 4 & 3 & 9 \end{bmatrix},$$

$$\begin{bmatrix} 1 & 7 & 5 \\ 4 & 3 & 9 \end{bmatrix}, \begin{bmatrix} 1 & 7 & 9 \\ 3 & 2 & 5 \end{bmatrix}, \begin{bmatrix} 1 & 7 & 9 \\ 4 & 3 & 2 \end{bmatrix},$$

$$\begin{bmatrix} 1 & 7 & 9 \\ 4 & 3 & 4 \end{bmatrix}, \begin{bmatrix} 3 & 1 & 6 \\ 4 & 7 & 9 \end{bmatrix}, \begin{bmatrix} 3 & 7 & 6 \\ 4 & 1 & 9 \end{bmatrix}.$$

The best 3×3, 4×4, and 5×5 prime arrays known were found by C. Rivera and J. Ayala in 1998. They are

$$A(3,3) = \begin{bmatrix} 1 & 1 & 3 \\ 7 & 5 & 4 \\ 9 & 3 & 7 \end{bmatrix},$$

which contains 30 PRIMES,

$$A(4,4) = \begin{bmatrix} 1 & 1 & 3 & 9 \\ 6 & 4 & 5 & 1 \\ 7 & 3 & 9 & 7 \\ 3 & 9 & 2 & 9 \end{bmatrix},$$

which contains 63 PRIMES, and

$$A(5,5) = \begin{bmatrix} 1 & 1 & 9 & 3 & 3 \\ 9 & 9 & 5 & 6 & 3 \\ 8 & 9 & 4 & 1 & 7 \\ 3 & 3 & 7 & 3 & 1 \\ 3 & 2 & 9 & 3 & 9 \end{bmatrix},$$

which contains 116 PRIMES. S. C. Root found the a 6×6 array containing 187 primes:

$$A(6,6) = \begin{bmatrix} 3 & 1 & 7 & 3 & 3 & 3 \\ 9 & 9 & 5 & 6 & 3 & 9 \\ 1 & 1 & 8 & 1 & 4 & 2 \\ 1 & 3 & 6 & 3 & 7 & 3 \\ 3 & 4 & 9 & 1 & 9 & 9 \\ 3 & 7 & 9 & 3 & 7 & 9 \end{bmatrix}.$$

In 1998, M. Oswald found five new 6×6 arrays with 187 primes:

$$\begin{bmatrix} 1 & 3 & 9 & 1 & 9 & 9 \\ 3 & 1 & 7 & 2 & 3 & 4 \\ 9 & 9 & 4 & 7 & 9 & 3 \\ 9 & 1 & 5 & 7 & 1 & 3 \\ 9 & 8 & 3 & 6 & 1 & 7 \\ 9 & 1 & 7 & 3 & 3 & 3 \end{bmatrix}, \begin{bmatrix} 1 & 3 & 9 & 1 & 9 & 9 \\ 9 & 1 & 7 & 2 & 3 & 4 \\ 6 & 9 & 4 & 7 & 9 & 3 \\ 7 & 1 & 5 & 7 & 1 & 3 \\ 9 & 8 & 3 & 6 & 1 & 7 \\ 9 & 1 & 7 & 3 & 3 & 3 \end{bmatrix},$$

$$\begin{bmatrix} 3 & 1 & 7 & 3 & 3 & 3 \\ 9 & 9 & 5 & 6 & 3 & 9 \\ 1 & 1 & 8 & 1 & 4 & 2 \\ 1 & 3 & 6 & 3 & 7 & 3 \\ 3 & 4 & 9 & 1 & 9 & 9 \\ 3 & 7 & 9 & 9 & 3 & 9 \end{bmatrix}, \begin{bmatrix} 3 & 1 & 7 & 3 & 3 & 3 \\ 9 & 9 & 5 & 6 & 3 & 9 \\ 1 & 1 & 8 & 1 & 4 & 2 \\ 1 & 3 & 6 & 3 & 7 & 3 \\ 3 & 4 & 9 & 1 & 9 & 9 \\ 9 & 7 & 9 & 3 & 7 & 9 \end{bmatrix},$$

$$\begin{bmatrix} 3 & 1 & 7 & 3 & 3 & 3 \\ 9 & 9 & 5 & 6 & 3 & 9 \\ 1 & 1 & 8 & 1 & 4 & 5 \\ 1 & 3 & 6 & 3 & 7 & 3 \\ 3 & 4 & 9 & 1 & 9 & 9 \\ 9 & 9 & 9 & 2 & 3 & 3 \end{bmatrix}.$$

Rivera and Ayala conjecture that the 30-prime solution for $A(3,3)$ is maximal and unique. The following intervals have been completely searched without finding another 30-prime or better 3×3 array: $[1, 67 \times 10^6]$, $[100 \times 10^6, 133 \times 10^6]$, $[200 \times 10^6, 228 \times 10^6]$, $[300 \times 10^6, 325 \times 10^6]$, and $[400 \times 10^6, 418 \times 10^6]$.

Heuristic arguments by Rivera and Ayala suggest that the maximum possible number of primes in 4×4, 5×5, and 6×6 arrays are 58–63, 112–121, and 205–218, respectively.

see also ARRAY, PRIME ARITHMETIC PROGRESSION, PRIME CONSTELLATION, PRIME STRING

References

Dewdney, A. K. "Computer Recreations: How to Pan for Primes in Numerical Gravel." *Sci. Amer.* **259**, 120–123, July 1988.

Lee, G. "Winners and Losers." *Dragon User.* May 1984.

Lee, G. "Gordon's Paradoxically Perplexing Primesearch Puzzle." `http://www.geocities.com/MotorCity/7983/primesearch.html`.

Rivera, C. "Problems & Puzzles (Puzzles): The Gordon Lee Puzzle." `http://www.sci.net.mx/~crivera/ppp/puzz_001.htm`.

Weisstein, E. W. "Prime Arrays." `http://www.astro.virginia.edu/~eww6n/math/notebooks/PrimeArray.m`.

Prime Circle

A prime circle of order $2m$ is a CIRCULAR PERMUTATION of the numbers from 1 to $2m$ with adjacent PAIRS summing to a PRIME. The number of prime circles for $m = 1, 2, \ldots,$ are 1, 1, 1, 2, 48, ...

References

Filz, A. "Problem 1046." *J. Recr. Math.* **14**, 64, 1982.

Filz, A. "Problem 1046." *J. Recr. Math.* **15**, 71, 1983.

Guy, R. K. *Unsolved Problems in Number Theory, 2nd ed.* New York: Springer-Verlag, pp. 105–106, 1994.

Prime Cluster

see PRIME CONSTELLATION

Prime Constellation

A prime constellation, also called a PRIME k-TUPLE or PRIME k-TUPLET, is a sequence of k consecutive numbers such that the difference between the first and last is, in some sense, the least possible. More precisely, a prime k-tuplet is a sequence of consecutive PRIMES $(p_1, p_2, \ldots, p_k)$ with $p_k - p_1 = s(k)$, where $s(k)$ is the smallest number s for which there exist k integers $b_1 < b_2 < \ldots < b_k$, $b_k - b_1 = s$ and, for every PRIME q, not all the residues modulo q are represented by b_1, b_2, ..., b_k (Forbes). For each k, this definition excludes a finite number of clusters at the beginning of the prime number sequence. For example, (97, 101, 103, 107, 109) satisfies the conditions of the definition of a prime 5-tuplet, but (3, 5, 7, 11, 13) does not because all three residues modulo 3 are represented (Forbes).

A prime double with $s(2) = 2$ is of the form $(p, p+2)$ and is called a pair of TWIN PRIMES. Prime doubles of the form $(p, p+6)$ are called SEXY PRIMES. A prime triplet has $s(3) = 6$. However, the constellation $(p, p+2, p+4)$ cannot exist, since both $p+2$ and $p+4$ cannot be PRIME. However, there are several types of prime triplets which can exist: $(p, p+2, p+6)$, $(p, p+4, p+6)$, $(p, p+6, p+12)$. A PRIME QUADRUPLET is a constellation of four successive PRIMES with minimal distance $s(4) = 8$, and is of the form $(p, p+2, p+6, p+8)$. The sequence $s(n)$ therefore begins 2, 6, 8, and continues 12, 16, 20, 26, 30, ... (Sloane's A008407). Another quadruplet constellation is $(p, p+6, p+12, p+18)$.

The first FIRST HARDY-LITTLEWOOD CONJECTURE states that the number of constellations $\leq x$ are asymptotically given by

$$P_x(p,p+2) \sim 2\prod_{p\geq 3}\frac{p(p-2)}{(p-1)^2}\int_2^x \frac{dx'}{(\ln x')^2}$$

$$= 1.320323632\int_2^x \frac{dx'}{(\ln x')^2} \tag{1}$$

$$P_x(p,p+4) \sim 2\prod_{p\geq 3}\frac{p(p-2)}{(p-1)^2}\int_2^x \frac{dx'}{(\ln x')^2}$$

$$= 1.320323632\int_2^x \frac{dx'}{(\ln x')^2} \tag{2}$$

$$P_x(p,p+6) \sim 4\prod_{p\geq 3}\frac{p(p-2)}{(p-1)^2}\int_2^x \frac{dx'}{(\ln x')^2}$$

$$= 2.640647264\int_2^x \frac{dx'}{(\ln x')^2} \tag{3}$$

$$P_x(p,p+2,p+6) \sim \frac{9}{2}\prod_{p\geq 5}\frac{p^2(p-3)}{(p-1)^3}\int_2^x \frac{dx'}{(\ln x')^3}$$

$$= 2.858248596\int_2^x \frac{dx'}{(\ln x')^3} \tag{4}$$

$$P_x(p,p+4,p+6) \sim \frac{9}{2}\prod_{p\geq 5}\frac{p^2(p-3)}{(p-1)^3}\int_2^x \frac{dx'}{(\ln x')^3}$$

$$= 2.858248596\int_2^x \frac{dx'}{(\ln x')^3} \tag{5}$$

$$P_x(p,p+2,p+6,p+8) \sim \frac{27}{2}\prod_{p\geq 5}\frac{p^3(p-4)}{(p-1)^4}\int_2^x \frac{dx'}{(\ln x')^4}$$

$$= 4.151180864\int_2^x \frac{dx'}{(\ln x')^4} \tag{6}$$

$$P_x(p,p+4,p+6,p+10) \sim 27\prod_{p\geq 5}\frac{p^3(p-4)}{(p-1)^4}\int_2^x \frac{dx'}{(\ln x')^4}$$

$$= 8.302361728\int_2^x \frac{dx'}{(\ln x')^4}. \tag{7}$$

These numbers are sometimes called the HARDY-LITTLEWOOD CONSTANTS. (1) is sometimes called the extended TWIN PRIME CONJECTURE, and

$$C_{p,p+2} = 2\Pi_2, \tag{8}$$

where Π_2 is the TWIN PRIMES CONSTANT. Riesel (1994) remarks that the HARDY-LITTLEWOOD CONSTANTS can be computed to arbitrary accuracy without needing the infinite sequence of primes.

The integrals above have the analytic forms

$$\int_2^x \frac{dx'}{(\ln x')^2} = \mathrm{Li}(x) + \frac{2}{\ln 2} - \frac{n}{\ln n} \tag{9}$$

$$\int_2^x \frac{dx'}{(\ln x')^3} = \tfrac{1}{2}\,\mathrm{Li}(x) - \frac{x(1+\ln x)}{(\ln x)^2} + \frac{1}{\ln 2} + \frac{1}{(\ln 2)^2} \tag{10}$$

$$\int_2^x \frac{dx'}{(\ln x')^4} = \frac{1}{6}\left\{\mathrm{Li}(x) + \frac{2[2+\ln 2+(\ln 2)^2]}{(\ln 2)^3} - \frac{n[2+\ln n+(\ln n)^2]}{(\ln n)^3}\right\}, \tag{11}$$

where $\mathrm{Li}(x)$ is the LOGARITHMIC INTEGRAL.

The following table gives the number of prime constellations $\leq 10^8$, and the second table gives the values predicted by the Hardy-Littlewood formulas.

Count	10^5	10^6	10^7	10^8
$(p,p+2)$	1224	8169	58980	440312
$(p,p+4)$	1216	8144	58622	440258
$(p,p+6)$	2447	16386	117207	879908
$(p,p+2,p+6)$	259	1393	8543	55600
$(p,p+4,p+6)$	248	1444	8677	55556
$(p,p+2,p+6,p+8)$	38	166	899	4768
$(p,p+6,p+12,p+18)$	75	325	1695	9330

Hardy-Littlewood	10^5	10^6	10^7	10^8
$(p,p+2)$	1249	8248	58754	440368
$(p,p+4)$	1249	8248	58754	440368
$(p,p+6)$	2497	16496	117508	880736
$(p,p+2,p+6)$	279	1446	8591	55491
$(p,p+4,p+6)$	279	1446	8591	55491
$(p,p+2,p+6,p+8)$	53	184	863	4735
$(p,p+6,p+12,p+18)$				

Consider prime constellations in which each term is of the form n^2+1. Hardy and Littlewood showed that the number of prime constellations of this form $< x$ is given by

$$P(x) \sim C\sqrt{x}\,(\ln x)^{-1}, \tag{12}$$

where

$$C = \prod_{\substack{p>2\\ p \text{ prime}}}\left[1 - \frac{(-1)^{(p-1)/2}}{p-1}\right] = 1.3727\ldots \tag{13}$$

(Le Lionnais 1983).

Forbes gives a list of the "top ten" prime k-tuples for $2 \leq k \leq 17$. The largest known 14-constellations are (11319107721272355839 + 0, 2, 8, 14, 18, 20, 24, 30, 32, 38, 42, 44, 48, 50), (10756418345074847279 + 0, 2, 8, 14, 18, 20, 24, 30, 32, 38, 42, 44, 48, 50),

(6808488664768715759 + 0, 2, 8, 14, 18, 20, 24, 30, 32, 38, 42, 44, 48, 50), (6120794469172998449 + 0, 2, 8, 14, 18, 20, 24, 30, 32, 38, 42, 44, 48, 50), (5009128141636113611 + 0, 2, 6, 8, 12, 18, 20, 26, 30, 32, 36, 42, 48, 50).

The largest known prime 15-constellations are (84244343639633356306067 + 0, 2, 6, 12, 14, 20, 24, 26, 30, 36, 42, 44, 50, 54, 56), (89852089979514576043337+0, 2, 6, 12, 14, 20, 26, 30, 32, 36, 42, 44, 50, 54, 56), (3594585413466972694697 + 0, 2, 6, 12, 14, 20, 26, 30, 32, 36, 42, 44, 50, 54, 56), (3514383375461541232577+0, 2, 6, 12, 14, 20, 26, 30, 32, 36, 42, 44, 50, 54, 56), (3493864509985912609487 + 0, 2, 6, 12, 14, 20, 24, 26, 30, 36, 42, 44, 50, 54, 56).

The largest known prime 16-constellations are (3259125690557440336637+0, 2, 6, 12, 14, 20, 26, 30, 32, 36, 42, 44, 50, 54, 56, 60), (1522014304823128379267+0, 2, 6, 12, 14, 20, 26, 30, 32, 36, 42, 44, 50, 54, 56, 60), (47710850533373130107 + 0, 2, 6, 12, 14, 20, 26, 30, 32, 36, 42, 44, 50, 54, 56, 60), (13, 17, 19, 23, 29, 31, 37, 41, 43, 47, 53, 59, 61, 67, 71, 73).

The largest known prime 17-constellations are (3259125690557440336631 + 0, 6, 8, 12, 18, 20, 26, 32, 36, 38, 42, 48, 50, 56, 60, 62, 66), (17, 19, 23, 29, 31, 37, 41, 43, 47, 53, 59, 61, 67, 71, 73, 79, 83) (13, 17, 19, 23, 29, 31, 37, 41, 43, 47, 53, 59, 61, 67, 71, 73, 79).

see also COMPOSITE RUNS, PRIME ARITHMETIC PROGRESSION, k-TUPLE CONJECTURE, PRIME k-TUPLES CONJECTURE, PRIME QUADRUPLET, SEXY PRIMES, TWIN PRIMES

References

Forbes, T. "Prime k-tuplets." http://www.ltkz.demon.co.uk/ktuplets.htm.

Guy, R. K. "Patterns of Primes." §A9 in *Unsolved Problems in Number Theory, 2nd ed.* New York: Springer-Verlag, pp. 23–25, 1994.

Le Lionnais, F. *Les nombres remarquables.* Paris: Hermann, p. 38, 1983.

Riesel, H. *Prime Numbers and Computer Methods for Factorization, 2nd ed.* Boston, MA: Birkhäuser, pp. 60–74, 1994.

Sloane, N. J. A. Sequence A008407 in "An On-Line Version of the Encyclopedia of Integer Sequences."

Prime Counting Function

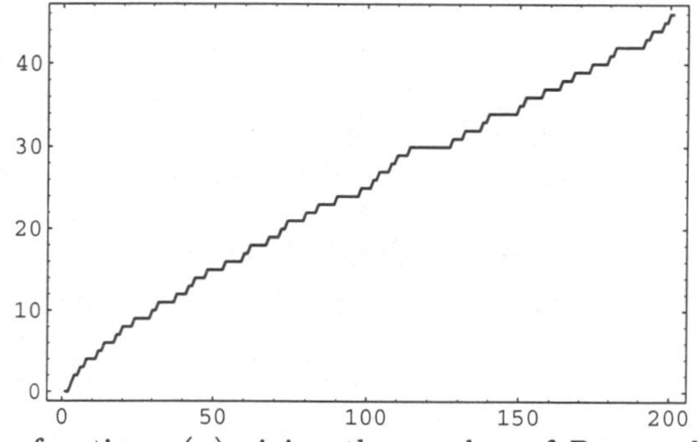

The function $\pi(n)$ giving the number of PRIMES less than n (Shanks 1993, p. 15). The first few values are 0, 1, 2, 2, 3, 3, 4, 4, 4, 4, 5, 5, 6, 6, 6, ... (Sloane's A000720). The following table gives the values of $\pi(n)$ for powers of 10 (Sloane's A006880; Hardy and Wright 1979, p. 4; Shanks 1993, pp. 242–243; Ribenboim 1996, p. 237). Deleglise and Rivat (1996) have computed $\pi(10^{20})$.

$$\begin{aligned}
\pi(10^3) &= 168\\
\pi(10^4) &= 1{,}229\\
\pi(10^5) &= 9{,}592\\
\pi(10^6) &= 78,498\\
\pi(10^7) &= 664,579\\
\pi(10^8) &= 5,761,455\\
\pi(10^9) &= 50,847,534\\
\pi(10^{10}) &= 455,052,511\\
\pi(10^{11}) &= 4,118,054,813\\
\pi(10^{12}) &= 37,607,912,018\\
\pi(10^{13}) &= 346,065,536,839\\
\pi(10^{14}) &= 3,204,941,750,802\\
\pi(10^{15}) &= 29,844,570,422,669\\
\pi(10^{16}) &= 279,238,341,033,925\\
\pi(10^{17}) &= 2,623,557,157,654,233\\
\pi(10^{18}) &= 24,739,954,287,740,860\\
\pi(10^{19}) &= 234,057,667,276,344,607,
\end{aligned}$$

$\pi(10^9)$ is incorrectly given as 50,847,478 in Hardy and Wright (1979). The prime counting function can be expressed by LEGENDRE'S FORMULA, LEHMER'S FORMULA, MAPES' METHOD, or MEISSEL'S FORMULA. A brief history of attempts to calculate $\pi(n)$ is given by Berndt (1994). The following table is taken from Riesel (1994).

Method	Time	Storage
Legendre	$\mathcal{O}(x)$	$\mathcal{O}(x^{1/2})$
Meissel	$\mathcal{O}(x/(\ln x)^3)$	$\mathcal{O}(x^{1/2}/\ln x)$
Lehmer	$\mathcal{O}(x/(\ln x)^4)$	$\mathcal{O}(x^{1/3}/\ln x)$
Mapes'	$\mathcal{O}(x^{0.7})$	$\mathcal{O}(x^{0.7})$
Lagarias-Miller-Odlyzko	$\mathcal{O}(x^{2/3+\epsilon})$	$\mathcal{O}(x^{1/3+\epsilon})$
Lagarias-Odlyzko 1	$\mathcal{O}(x^{3/5+\epsilon})$	$\mathcal{O}(x^{\epsilon})$
Lagarias-Odlyzko 2	$\mathcal{O}(x^{1/2+\epsilon})$	$\mathcal{O}(x^{1/4+\epsilon})$

A modified version of the prime counting function is given by

$$\pi_0(p) \equiv \begin{cases} \pi(p) & \text{for } p \text{ composite} \\ \pi(p) - \frac{1}{2} & \text{for } p \text{ prime} \end{cases}$$

$$\pi_0(p) = \sum_{n=1}^{\infty} \frac{\mu(x) f(x^{1/n})}{n},$$

where $\mu(n)$ is the MÖBIUS FUNCTION and $f(x)$ is the RIEMANN-MANGOLDT FUNCTION.

The notation $\pi_{a,b}$ is also used to denote the number of PRIMES of the form $ak + b$ (Shanks 1993, pp. 21–22).

Groups of EQUINUMEROUS values of $\pi_{a,b}$ include $(\pi_{3,1}, \pi_{3,2})$, $(\pi_{4,1}, \pi_{4,3})$, $(\pi_{5,1}, \pi_{5,2}, \pi_{5,3}, \pi_{5,4})$, $(\pi_{6,1}, \pi_{6,5})$, $(\pi_{7,1}, \pi_{7,2}, \pi_{7,3}, \pi_{7,4}, \pi_{7,5}, \pi_{7,6})$, $(\pi_{8,1}, \pi_{8,3}, \pi_{8,5}, \pi_{8,7})$, $(\pi_{9,1}, \pi_{9,2}, \pi_{9,4}, \pi_{9,5}, \pi_{9,7}, \pi_{9,8})$, and so on. The values of $\pi_{n,k}$ for small n are given in the following table for the first few powers of ten (Shanks 1993).

n	$\pi_{3,1}(n)$	$\pi_{3,2}(n)$	$\pi_{4,1}(n)$	$\pi_{4,3}(n)$
10^1	1	2	1	2
10^2	11	13	11	13
10^3	80	87	80	87
10^4	611	617	609	619
10^5	4784	4807	4783	4808
10^6	39231	39266	39175	39322
10^7	332194	332384	332180	332398

n	$\pi_{5,1}(n)$	$\pi_{5,2}(n)$	$\pi_{5,3}(n)$	$\pi_{5,4}(n)$
10^1	0	2	1	0
10^2	5	7	7	5
10^3	40	47	42	38
10^4	306	309	310	303
10^5	2387	2412	2402	2390
10^6	19617	19622	19665	19593
10^7	166104	166212	166230	166032

n	$\pi_{6,1}(n)$	$\pi_{6,5}(n)$
10^1	1	1
10^2	11	12
10^3	80	86
10^4	611	616
10^5	4784	4806
10^6	39231	39265

n	$\pi_{7,1}$	$\pi_{7,2}$	$\pi_{7,3}$	$\pi_{7,4}$	$\pi_{7,5}$	$\pi_{7,6}$
10^1	0	1	1	0	1	0
10^2	3	4	5	3	5	4
10^3	28	27	30	26	29	27
10^4	203	203	209	202	211	200
10^5	1593	1584	1613	1601	1604	1596
10^6	13063	13065	13105	13069	13105	13090

n	$\pi_{8,1}(n)$	$\pi_{8,3}(n)$	$\pi_{8,5}(n)$	$\pi_{8,7}(n)$
10^1	0	1	1	1
10^2	5	7	6	6
10^3	37	44	43	43
10^4	295	311	314	308
10^5	2384	2409	2399	2399
10^6	19552	19653	19623	19669
10^7	165976	166161	166204	166237

Note that since $\pi_{8,1}(n)$, $\pi_{8,3}(n)$, $\pi_{8,5}(n)$, and $\pi_{8,7}(n)$ are EQUINUMEROUS,

$$\pi_{4,1}(n) = \pi_{8,1}(n) + \pi_{8,5}$$
$$\pi_{4,3}(n) = \pi_{8,3}(n) + \pi_{8,7}$$

are also equinumerous.

Erdős proved that there exist at least one PRIME of the form $4k+1$ and at least one PRIME of the form $4k+3$ between n and $2n$ for all $n > 6$.

The smallest x such that $x \geq n\pi(x)$ for $n = 2, 3, \ldots$ are 2, 27, 96, 330, 1008, ... (Sloane's A038625), and the corresponding $\pi(x)$ are 1, 9, 24, 24, 66, 168, ... (Sloane's A038626). The number of solutions of $x \geq n\pi(x)$ for $n = 2, 3, \ldots$ are 4, 3, 3, 6, 7, 6, ... (Sloane's A038627).

see also BERTELSEN'S NUMBER, EQUINUMEROUS, PRIME ARITHMETIC PROGRESSION, PRIME NUMBER THEOREM, RIEMANN WEIGHTED PRIME-POWER COUNTING FUNCTION

References

Berndt, B. C. *Ramanujan's Notebooks, Part IV.* New York: Springer-Verlag, pp. 134–135, 1994.

Brent, R. P. "Irregularities in the Distribution of Primes and Twin Primes." *Math. Comput.* **29**, 43–56, 1975.

Deleglise, M. and Rivat, J. "Computing $\pi(x)$: The Meissel, Lehmer, Lagarias, Miller, Odlyzko Method." *Math. Comput.* **65**, 235–245, 1996.

Finch, S. "Favorite Mathematical Constants." `http://www.mathsoft.com/asolve/constant/hrdyltl/hrdyltl.html`.

Forbes, T. "Prime k-tuplets." `http://www.ltkz.demon.co.uk/ktuplets.htm`.

Guiasu, S. "Is There Any Regularity in the Distribution of Prime Numbers at the Beginning of the Sequence of Positive Integers?" *Math. Mag.* **68**, 110–121, 1995.

Hardy, G. H. and Wright, E. M. *An Introduction to the Theory of Numbers, 5th ed.* Oxford, England: Clarendon Press, 1979.

Lagarias, J.; Miller, V. S.; and Odlyzko, A. "Computing $\pi(x)$: The Meissel-Lehmer Method." *Math. Comput.* **44**, 537–560, 1985.

Lagarias, J. and Odlyzko, A. "Computing $\pi(x)$: An Analytic Method." *J. Algorithms* **8**, 173–191, 1987.

Mapes, D. C. "Fast Method for Computing the Number of Primes Less than a Given Limit." *Math. Comput.* **17**, 179–185, 1963.

Meissel, E. D. F. "Über die Bestimmung der Primzahlmenge innerhalb gegebener Grenzen." *Math. Ann.* **2**, 636–642, 1870.

Ribenboim, P. *The New Book of Prime Number Records, 3rd ed.* New York: Springer-Verlag, 1996.

Riesel, H. "The Number of Primes Below x." *Prime Numbers and Computer Methods for Factorization, 2nd ed.* Boston, MA: Birkhäuser, pp. 10–12, 1994.

Shanks, D. *Solved and Unsolved Problems in Number Theory, 4th ed.* New York: Chelsea, 1993.

Sloane, N. J. A. Sequences A038625, A038626, A038627, A000720/M2056, and A006880/M3608 in "An On-Line Version of the Encyclopedia of Integer Sequences."

Vardi, I. *Computational Recreations in Mathematica.* Reading, MA: Addison-Wesley, pp. 74–76, 1991.

Wolf, M. "Unexpected Regularities in the Distribution of Prime Numbers." `http://www.ift.uni.wroc.pl/~mwolf`.

Prime Cut

Find two numbers such that $x^2 \equiv y^2 \pmod{n}$. If you know the GREATEST COMMON DIVISOR of n and $x - y$, there exists a high probability of determining a PRIME factor. Taking small numbers x which additionally give small PRIMES $x^2 \equiv p \pmod{n}$ further increases the chances of finding a PRIME factor.

Prime Decomposition

Given an INTEGER n, the prime decomposition is written

$$n = {p_1}^{\alpha_1} {p_2}^{\alpha_2} \cdots {p_n}^{\alpha_n},$$

where p_i are the n PRIME factors, each of order α_i. Each factor ${p_i}^{\alpha_i}$ is called a PRIMARY.

see also PRIMARY, PRIME FACTORIZATION ALGORITHMS, PRIME NUMBER

Prime Difference Function

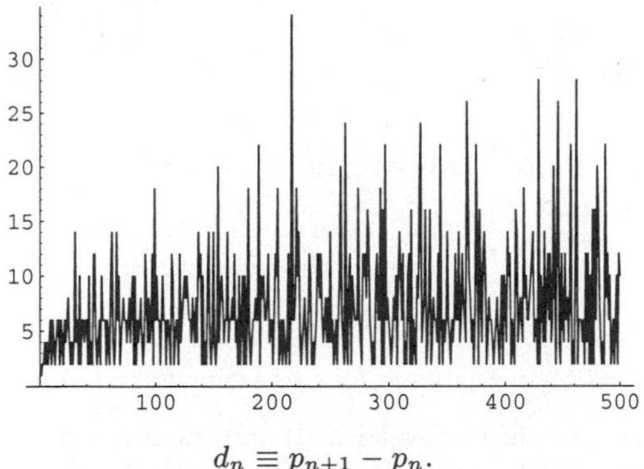

$$d_n \equiv p_{n+1} - p_n.$$

The first few values are 1, 2, 2, 4, 2, 4, 2, 4, 6, 2, 6, 4, 2, 4, 6, 6, ... (Sloane's A001223). Rankin has shown that

$$d_n > \frac{c \ln n \ln \ln n \ln \ln \ln \ln n}{(\ln \ln \ln n)^2}$$

for infinitely many n and for some constant c (Guy 1994).

An integer n is called a JUMPING CHAMPION if n is the most frequently occurring difference between consecutive primes $n \leq N$ for some N (Odlyzko *et al.*).

see also ANDRICA'S CONJECTURE, GOOD PRIME, JUMPING CHAMPION, PÓLYA CONJECTURE, PRIME GAPS, SHANKS' CONJECTURE, TWIN PEAKS

References

Bombieri, E. and Davenport, H. "Small Differences Between Prime Numbers." *Proc. Roy. Soc. A* **293**, 1–18, 1966.

Erdős, P.; and Straus, E. G. "Remarks on the Differences Between Consecutive Primes." *Elem. Math.* **35**, 115–118, 1980.

Guy, R. K. "Gaps between Primes. Twin Primes" and "Increasing and Decreasing Gaps." §A8 and A11 in *Unsolved Problems in Number Theory, 2nd ed.* New York: Springer-Verlag, pp. 19–23 and 26–27, 1994.

Odlyzko, A.; Rubinstein, M.; and Wolf, M. "Jumping Champions." `http://www.research.att.com/~amo/doc/recent.html`.

Riesel, H. "Difference Between Consecutive Primes." *Prime Numbers and Computer Methods for Factorization, 2nd ed.* Boston, MA: Birkhäuser, p. 9, 1994.

Sloane, N. J. A. Sequence A001223/M0296 in "An On-Line Version of the Encyclopedia of Integer Sequences."

Prime Diophantine Equations

$k+2$ is PRIME IFF the 14 DIOPHANTINE EQUATIONS in 26 variables

$$wz + h + j - q = 0 \tag{1}$$
$$(gk + 2g + k + 1)(h + j) + h - z = 0 \tag{2}$$
$$16(k+1)^3(k+2)(n+1)^2 + 1 - f^2 = 0 \tag{3}$$
$$2n + p + q + z - q = 0 \tag{4}$$
$$e^3(e+2)(a+1)^2 + 1 - o^2 = 0 \tag{5}$$
$$(a^2 - 1)y^2 + 1 - x^2 = 0 \tag{6}$$
$$16r^2y^4(a^2 - 1) + 1 - u^2 = 0 \tag{7}$$
$$n + l + v - y = 0 \tag{8}$$
$$(a^2 - 1)l^2 + 1 - m^2 = 0 \tag{9}$$
$$ai + k + 1 - l - i = 0 \tag{10}$$
$$\{[a + u^2(u^2 - a)]^2 - 1\}(n + 4dy)^2 + 1 - (x + cu)^2 = 0 \tag{11}$$
$$p + l(a - n - 1) + b(2an + 2a - n^2 - 2n - 2) - m = 0 \tag{12}$$
$$q + y(a - p - 1) + s(2ap + 2a - p^2 - 2p - 2) - x = 0 \tag{13}$$
$$z + pl(a - p) + t(2ap - p^2 - 1) - pm = 0 \tag{14}$$

have a POSITIVE integral solution.

References

Riesel, H. *Prime Numbers and Computer Methods for Factorization, 2nd ed.* Boston, MA: Birkhäuser, p. 39, 1994.

Prime Factorization

see FACTORIZATION, PRIME DECOMPOSITION, PRIME FACTORIZATION ALGORITHMS, PRIME FACTORS

Prime Factorization Algorithms

Many ALGORITHMS have been devised for determining the PRIME factors of a given number. They vary quite a bit in sophistication and complexity. It is very difficult to build a general-purpose algorithm for this computationally "hard" problem, so any additional information which is known about the number in question or its factors can often be used to save a large amount of time.

The simplest method of finding factors is so-called "DIRECT SEARCH FACTORIZATION" (a.k.a. TRIAL DIVISION). In this method, all possible factors are systematically tested using trial division to see if they actually DIVIDE the given number. It is practical only for very small numbers.

see also BRENT'S FACTORIZATION METHOD, CONTINUED FRACTION FACTORIZATION ALGORITHM, DIRECT SEARCH FACTORIZATION, DIXON'S FACTORIZATION METHOD, ELLIPTIC CURVE FACTORIZATION METHOD, EULER'S FACTORIZATION METHOD, EXCLUDENT FACTORIZATION METHOD, FERMAT'S FACTORIZATION METHOD, LEGENDRE'S FACTORIZATION

METHOD, LENSTRA ELLIPTIC CURVE METHOD, NUMBER FIELD SIEVE FACTORIZATION METHOD, POLLARD $p-1$ FACTORIZATION METHOD, POLLARD ρ FACTORIZATION ALGORITHM, QUADRATIC SIEVE FACTORIZATION METHOD, TRIAL DIVISION, WILLIAMS $p+1$ FACTORIZATION METHOD

References

Bressoud, D. M. *Factorization and Prime Testing.* New York: Springer-Verlag, 1989.

Brillhart, J.; Lehmer, D. H.; Selfridge, J.; Wagstaff, S. S. Jr.; and Tuckerman, B. *Factorizations of $b^n \pm 1$, $b = 2, 3, 5, 6, 7, 10, 11, 12$ Up to High Powers, rev. ed.* Providence, RI: Amer. Math. Soc., liv–lviii, 1988.

Dickson, L. E. "Methods of Factoring." Ch. 14 in *History of the Theory of Numbers, Vol. 1: Divisibility and Primality.* New York: Chelsea, pp. 357–374, 1952.

Herman, P. "The Factoring Page!" http://www.pslc.ucla.edu/~a540pau/factoring.

Lenstra, A. K. and Lenstra, H. W. Jr. "Algorithms in Number Theory." In *Handbook of Theoretical Computer Science, Volume A: Algorithms and Complexity* (Ed. J. van Leeuwen). New York: Elsevier, pp. 673–715, 1990.

Odlyzko, A. M. "The Complexity of Computing Discrete Logarithms and Factoring Integers." §4.5 in *Open Problems in Communication and Computation* (Ed. T. M. Cover and B. Gopinath). New York: Springer-Verlag, pp. 113–116, 1987.

Odlyzko, A. M. "The Future of Integer Factorization." *CryptoBytes: The Technical Newsletter of RSA Laboratories* **1**, No. 2, 5–12, 1995.

Pomerance, C. "A Tale of Two Sieves." *Not. Amer. Math. Soc.* **43**, 1473–1485, 1996.

Riesel, H. *Prime Numbers and Computer Methods for Factorization, 2nd ed.* Boston, MA: Birkhäuser, 1994.

Williams, H. C. and Shallit, J. O. "Factoring Integers Before Computers." In *Mathematics of Computation 1943–1993, Fifty Years of Computational Mathematics* (Ed. W. Gautschi). Providence, RI: Amer. Math. Soc., pp. 481–531, 1994.

Prime Factors

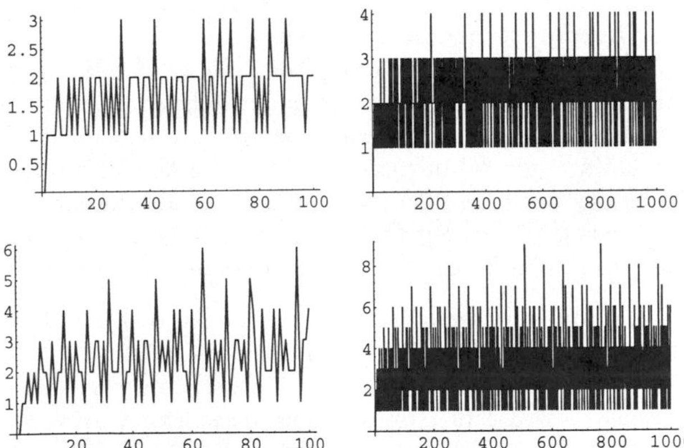

The number of DISTINCT PRIME FACTORS of a number n is denoted $\omega(n)$. The first few values for $n = 1, 2, \ldots$ are 0, 1, 1, 1, 1, 2, 1, 1, 1, 2, 1, 2, 1, 2, 2, 1, 1, 2, 1, 2, ... (Sloane's A001221; top figure). The number of not necessarily distinct prime factors of a number n is denoted $r(n)$. The first few values for $n = 1, 2, \ldots$ are 0, 1, 1, 2, 1, 2, 1, 3, 2, 2, 1, 3, 1, 2, 2, 4, 1, 3, 1, 3, ... (Sloane's A001222; bottom figure).

see also DISTINCT PRIME FACTORS, DIVISOR FUNCTION, GREATEST PRIME FACTOR, LEAST PRIME FACTOR, LIOUVILLE FUNCTION, PÓLYA CONJECTURE, PRIME FACTORIZATION ALGORITHMS

References

Sloane, N. J. A. Sequences A001222/M0094 and A001221/M0056 in "An On-Line Version of the Encyclopedia of Integer Sequences."

Prime Field

A GALOIS FIELD $GF(p)$ where p is PRIME.

Prime Gaps

Letting

$$d_n \equiv p_{n+1} - p_n$$

be the PRIME DIFFERENCE FUNCTION, Rankin has showed that

$$d_n > \frac{c \ln n \ln \ln n \ln \ln \ln \ln n}{(\ln \ln \ln n)^2}$$

for infinitely many n are for some constant c (Guy 1994).

Let $p(d)$ be the smallest PRIME following d or more consecutive COMPOSITE NUMBERS. The largest known is

$$p(804) = 90,874,329,412,297.$$

The largest known prime gap is of length 4247, occurring following $10^{314} - 1929$ (Baugh and O'Hara 1992), although this gap is almost certainly not maximal (i.e., there probably exists a smaller number having a gap of the same length following it).

Let $c(n)$ be the smallest starting INTEGER $c(n)$ for a run of n consecutive COMPOSITE NUMBERS, also called a COMPOSITE RUN. No general method other than exhaustive searching is known for determining the first occurrence for a maximal gap, although arbitrarily large gaps exist (Nicely 1998). Cramér (1937) and Shanks (1964) conjectured that a maximal gap of length n appears at approximately $\exp(\sqrt{n})$. Wolf conjectures that the maximal gap of length n appears approximately at

$$\frac{n}{\pi(n)[2 \ln \pi(n) - \ln n + \ln(2C_2)]},$$

where $\pi(n)$ is the PRIME COUNTING FUNCTION and C_2 is the TWIN PRIMES CONSTANT.

The first few $c(n)$ for $n = 1, 2, \ldots$ are 4, 8, 8, 24, 24, 90, 90, 114, ... (Sloane's A030296). The following table gives the same sequence omitting degenerate runs which are part of a run with greater n, and is a complete list of smallest maximal runs up to 10^{15}. $c(n)$ in this table is given by Sloane's A008950, and n by Sloane's A008996. The ending integers for the run corresponding to $c(n)$ are given by Sloane's A008995. Young and Potler (1989) determined the first occurrences of prime gaps up to 72,635,119,999,997, with all first occurrences found between 1 and 673. Nicely (1998) extended the list of maximal prime gaps to a length of 915, denoting gap lengths by the difference of bounding PRIMES, $c(n) - 1$.

n	$c(n)$	n	$c(n)$
1	4	319	2,300,942,550
3	8	335	3,842,610,774
5	24	353	4,302,407,360
7	90	381	10,726,904,660
13	114	383	20,678,048,298
17	524	393	22,367,084,960
19	888	455	25,056,082,088
21	1,130	463	42,652,618,344
33	1,328	467	127,976,334,672
35	9,552	473	182,226,896,240
43	15,684	485	241,160,024,144
51	19,610	489	297,501,075,800
71	31,398	499	303,371,455,242
85	155,922	513	304,599,508,538
95	360,654	515	416,608,695,822
111	370,262	531	461,690,510,012
113	492,114	533	614,487,453,424
117	1,349,534	539	738,832,927,928
131	1,357,202	581	1,346,294,310,750
147	2,010,734	587	1,408,695,493,610
153	4,652,354	601	1,968,188,556,461
179	17,051,708	651	2,614,941,710,599
209	20,831,324	673	7,177,162,611,713
219	47,326,694	715	13,828,048,559,701
221	122,164,748	765	19,581,334,192,423
233	189,695,660	777	42,842,283,925,352
247	191,912,784	803	90,874,329,411,493
249	387,096,134	805	171,231,342,420,521
281	436,273,010	905	218,209,405,436,543
287	1,294,268,492	915	1,189,459,969,825,483
291	1,453,168,142		

see also JUMPING CHAMPION, PRIME CONSTELLATION, PRIME DIFFERENCE FUNCTION, SHANKS' CONJECTURE

References

Baugh, D. and O'Hara, F. "Large Prime Gaps." *J. Recr. Math.* **24**, 186–187, 1992.

Berndt, B. C. *Ramanujan's Notebooks, Part IV.* New York: Springer-Verlag, pp. 133–134, 1994.

Bombieri, E. and Davenport, H. "Small Differences Between Prime Numbers." *Proc. Roy. Soc. A* **293**, 1–18, 1966.

Brent, R. P. "The First Occurrence of Large Gaps Between Successive Primes." *Math. Comput.* **27**, 959–963, 1973.

Brent, R. P. "The Distribution of Small Gaps Between Successive Primes." *Math. Comput.* **28**, 315–324, 1974.

Brent, R. P. "The First Occurrence of Certain Large Prime Gaps." *Math. Comput.* **35**, 1435–1436, 1980.

Cramér, H. "On the Order of Magnitude of the Difference Between Consecutive Prime Numbers." *Acta Arith.* **2**, 23–46, 1937.

Guy, R. K. "Gaps between Primes. Twin Primes" and "Increasing and Decreasing Gaps." §A8 and A11 in *Unsolved Problems in Number Theory, 2nd ed.* New York: Springer-Verlag, pp. 19–23 and 26–27, 1994.

Lander, L. J. and Parkin, T. R. "On First Appearance of Prime Differences." *Math. Comput.* **21**, 483–488, 1967.

Nicely, T. R. "New Maximal Prime Gaps and First Occurrences." `http://www.lynchburg.edu/public/academic/math/nicely/gaps/gaps.htm`. To Appear in *Math. Comput.*

Shanks, D. "On Maximal Gaps Between Successive Primes." *Math. Comput.* **18**, 646–651, 1964.

Sloane, N. J. A. Sequences A008950, A008995, A008996, and A030296 in "An On-Line Version of the Encyclopedia of Integer Sequences."

Wolf, M. "First Occurrence of a Given Gap Between Consecutive Primes." `http://www.ift.uni.wroc.pl/~mwolf`.

Young, J. and Potler, A. "First Occurrence Prime Gaps." *Math. Comput.* **52**, 221–224, 1989.

Prime-Generating Polynomial

Legendre showed that there is no RATIONAL algebraic function which always gives PRIMES. In 1752, Goldbach showed that no POLYNOMIAL with INTEGER COEFFICIENTS can give a PRIME for all integral values. However, there exists a POLYNOMIAL in 10 variables with INTEGER COEFFICIENTS such that the set of PRIMES equals the set of POSITIVE values of this POLYNOMIAL obtained as the variables run through all NONNEGATIVE INTEGERS, although it is really a set of DIOPHANTINE EQUATIONS in disguise (Ribenboim 1991).

$P(n)$	Range	#	Reference
$36n^2 - 810n + 2753$	[0, 44]	45	Fung and Ruby
$47n^2 - 1701n + 10181$	[0, 42]	43	Fung and Ruby
$n^2 - n + 41$	[0, 39]	40	Euler
$2n^2 + 29$	[0, 28]	29	Legendre
$n^2 + n + 17$	[0, 15]	16	Legendre
$2n^2 + 11$	[0, 10]	11	
$n^3 + n^2 + 17$	[0, 10]	11	

The above table gives some low-order polynomials which generate only PRIMES for the first few NONNEGATIVE values (Mollin and Williams 1990). The best-known of these formulas is that due to Euler (Euler 1772, Ball and Coxeter 1987). Le Lionnais (1983) has christened numbers p such that the Euler-like polynomial

$$n^2 - n + p \tag{1}$$

is PRIME for $p = 0, 1, \ldots, p-2$ as LUCKY NUMBERS OF EULER (where the case $p = 41$ corresponds to Euler's formula). Rabinovitch (1913) showed that for a PRIME $p > 0$, Euler's polynomial represents a PRIME for $n \in [0, p-2]$ (excluding the trivial case $p = 3$) IFF the FIELD $\mathbb{Q}(\sqrt{1-4p})$ has CLASS NUMBER $h = 1$ (Rabinowitz 1913, Le Lionnais 1983, Conway and Guy 1996). As established by Stark (1967), there are only nine numbers $-d$ such that $h(-d) = 1$ (the HEEGNER NUMBERS -2, -3, -7, -11, -19, -43, -67, and -163), and of these, only 7, 11, 19, 43, 67, and 163 are of the required form. Therefore, the only LUCKY NUMBERS OF EULER are 2, 3, 5, 11, 17, and 41 (Le Lionnais 1983, Sloane's A014556), and there does not exist a better prime-generating polynomial of Euler's form.

Euler also considered quadratics of the form

$$2x^2 + p \tag{2}$$

and showed this gives PRIMES for $x \in [0, p-1]$ for PRIME $p > 0$ IFF $\mathbb{Q}(\sqrt{-2p})$ has CLASS NUMBER 2, which permits only $p = 3$, 5, 11, and 29. Baker (1971) and Stark (1971) showed that there are so such FIELDS for $p > 29$. Similar results have been found for POLYNOMIALS of the form

$$px^2 + px + n \tag{3}$$

(Hendy 1974).

see also CLASS NUMBER, HEEGNER NUMBER, LUCKY NUMBER OF EULER, PRIME ARITHMETIC PROGRESSION, PRIME DIOPHANTINE EQUATIONS, SCHINZEL'S HYPOTHESIS

References

Abel, U. and Siebert, H. "Sequences with Large Numbers of Prime Values." *Am. Math. Monthly* **100**, 167–169, 1993.

Baker, A. "Linear Forms in the Logarithms of Algebraic Numbers." *Mathematika* **13**, 204–216, 1966.

Baker, A. "Imaginary Quadratic Fields with Class Number Two." *Ann. Math.* **94**, 139–152, 1971.

Ball, W. W. R. and Coxeter, H. S. M. *Mathematical Recreations and Essays, 13th ed.* New York: Dover, p. 60, 1987.

Boston, N. and Greenwood, M. L. "Quadratics Representing Primes." *Amer. Math. Monthly* **102**, 595–599, 1995.

Conway, J. H. and Guy, R. K. "The Nine Magic Discriminants." In *The Book of Numbers.* New York: Springer-Verlag, pp. 224–226, 1996.

Courant, R. and Robbins, H. *What is Mathematics?: An Elementary Approach to Ideas and Methods, 2nd ed.* Oxford, England: Oxford University Press, p. 26, 1996.

Euler, L. *Nouveaux Mémoires de l'Académie royale des Sciences.* Berlin, p. 36, 1772.

Forman, R. "Sequences with Many Primes." *Amer. Math. Monthly* **99**, 548–557, 1992.

Garrison, B. "Polynomials with Large Numbers of Prime Values." *Amer. Math. Monthly* **97**, 316–317, 1990.

Hendy, M. D. "Prime Quadratics Associated with Complex Quadratic Fields of Class Number 2." *Proc. Amer. Math. Soc.* **43**, 253–260, 1974.

Le Lionnais, F. *Les nombres remarquables.* Paris: Hermann, pp. 88 and 144, 1983.

Mollin, R. A. and Williams, H. C. "Class Number Problems for Real Quadratic Fields." *Number Theory and Cryptology; LMS Lecture Notes Series* **154**, 1990.

Rabinowitz, G. "Eindeutigkeit der Zerlegung in Primzahlfaktoren in quadratischen Zahlkörpern." *Proc. Fifth Internat. Congress Math.* (Cambridge) **1**, 418–421, 1913.

Ribenboim, P. *The Little Book of Big Primes.* New York: Springer-Verlag, 1991.

Sloane, N. J. A. Sequence A014556 in "An On-Line Version of the Encyclopedia of Integer Sequences."

Stark, H. M. "A Complete Determination of the Complex Quadratic Fields of Class Number One." *Michigan Math. J.* **14**, 1–27, 1967.

Stark, H. M. "An Explanation of Some Exotic Continued Fractions Found by Brillhart." In *Computers in Number Theory, Proc. Science Research Council Atlas Symposium No. 2 held at Oxford, from 18–23 August, 1969* (Ed. A. O. L. Atkin and B. J. Birch). London: Academic Press, 1971.

Stark, H. M. "A Transcendence Theorem for Class Number Problems." *Ann. Math.* **94**, 153–173, 1971.

Prime Group

When the ORDER h of a finite GROUP is a PRIME number, there is only one possible GROUP of ORDER h. Furthermore, the GROUP is CYCLIC.

see also p-GROUP

Prime Ideal

An IDEAL I such that if $ab \in I$, then either $a \in I$ or $b \in I$.

see also DEDEKIND RING, IDEAL, KRULL DIMENSION, MAXIMAL IDEAL, STICKELBERGER RELATION, STONE SPACE

Prime Knot

A KNOT other than the UNKNOT which cannot be expressed as a sum of two other KNOTS, neither of which is unknotted. A KNOT which is not prime is called a COMPOSITE KNOT. It is often possible to combine two prime knots to create two different COMPOSITE KNOTS, depending on the orientation of the two.

There is no known FORMULA for giving the number of distinct prime knots as a functions of number of crossings. For the first few n crossings, the numbers of prime knots are 0, 0, 1, 2, 3, 7, 21, 49, 165, 552, 2176, 9988, ... (Sloane's A002863).

see also COMPOSITE KNOT, KNOT

References

Sloane, N. J. A. Sequences A002863/M0851 in "An On-Line Version of the Encyclopedia of Integer Sequences."

Sloane, N. J. A. and Plouffe, S. Extended entry in *The Encyclopedia of Integer Sequences.* San Diego: Academic Press, 1995.

Prime k-Tuple

see PRIME CONSTELLATION

Prime k-Tuples Conjecture

see also k-TUPLE CONJECTURE

Prime k-Tuplet

see PRIME CONSTELLATION

Prime Manifold

An n-MANIFOLD which cannot be "nontrivially" decomposed into other n-MANIFOLDS.

Prime Number

A prime number is a POSITIVE INTEGER p which has no DIVISORS other than 1 and p itself. Although the number 1 used to be considered a prime, it requires special treatment in so many definitions and applications involving primes greater than or equal to 2 that it is usually placed into a class of its own. Since 2 is the only EVEN prime, it is also somewhat special, so the set of all primes excluding 2 is called the "ODD PRIMES." The first few primes are 2, 3, 5, 7, 11, 13, 17, 19, 23, 29,

31, 37, ... (Sloane's A000040, Hardy and Wright 1979, p. 3). POSITIVE INTEGERS other than 1 which are not prime are called COMPOSITE.

The function which gives the number of primes less than a number n is denoted $\pi(n)$ and is called the PRIME COUNTING FUNCTION. The theorem giving an asymptotic form for $\pi(n)$ is called the PRIME NUMBER THEOREM.

Prime numbers can be generated by sieving processes (such as the ERATOSTHENES SIEVE), and LUCKY NUMBERS, which are also generated by sieving, appear to share some interesting asymptotic properties with the primes.

Many PRIME FACTORIZATION ALGORITHMS have been devised for determining the prime factors of a given INTEGER. They vary quite a bit in sophistication and complexity. It is very difficult to build a general-purpose algorithm for this computationally "hard" problem, so any additional information which is known about the number in question or its factors can often be used to save a large amount of time. The simplest method of finding factors is so-called "DIRECT SEARCH FACTORIZATION" (a.k.a. TRIAL DIVISION). In this method, all possible factors are systematically tested using trial division to see if they actually DIVIDE the given number. It is practical only for very small numbers.

Because of their importance in encryption algorithms such as RSA ENCRYPTION, prime numbers can be important commercial commodities. In fact, Roger Schlafly has obtained U.S. Patent 5,373,560 (12/13/94) on the following two primes (expressed in hexadecimal notation):

```
98A3DF52AEAE9799325CB258D767EBD1F4630E9B
9E21732A4AFB1624BA6DF911466AD8DA960586F4
A0D5E3C36AF099660BDDC1577E54A9F402334433
ACB14BCB
```

and

```
93E8965DAFD9DFECFD00B466B68F90EA68AF5DC9
FED915278D1B3A137471E65596C37FED0C7829FF
8F8331F81A2700438ECDCC09447DC397C685F397
294F722BCC484AEDF28BED25AAAB35D35A65DB1F
D62C9D7BA55844FEB1F9401E671340933EE43C54
E4DC459400D7AD61248B83A2624835B31FFF2D95
95A5B90B276E44F9.
```

The FUNDAMENTAL THEOREM OF ARITHMETIC states that any POSITIVE INTEGER can be represented in exactly one way as a PRODUCT of primes. EUCLID'S SECOND THEOREM demonstrated that there are an infinite number of primes. However, it is not known if there are an infinite number of primes of the form x^2+1, whether there are an INFINITE number of TWIN PRIMES, or if a prime can always be found between n^2 and $(n+1)^2$.

Prime numbers satisfy many strange and wonderful properties. For example, there exists a CONSTANT $\theta \approx 1.3064$ known as MILLS' CONSTANT such that

$$\left\lfloor \theta^{3^n} \right\rfloor, \tag{1}$$

where $\lfloor x \rfloor$ is the FLOOR FUNCTION, is prime for all $n \geq 1$. However, it is not known if θ is IRRATIONAL. There also exists a CONSTANT $\omega \approx 1.9287800$ such that

$$\left\lfloor \underbrace{2^{2^{\cdot^{\cdot^{\cdot^{2^\omega}}}}}}_{n} \right\rfloor \tag{2}$$

(Ribenboim 1996, p. 186) is prime for every $n \geq 1$.

Explicit FORMULAS exist for the nth prime both as a function of n and in terms of the primes 2, ..., p_{n-1} (Hardy and Wright 1979, pp. 5–6; Guy 1994, pp. 36–41). Let

$$F(j) = \left\lfloor \cos^2 \left[\pi \frac{(j-1)!+1}{j} \right] \right\rfloor \tag{3}$$

for integral $j > 1$, and define $F(1) = 1$, where $\lfloor x \rfloor$ is again the FLOOR FUNCTION. Then

$$p_n = 1 + \sum_{m=1}^{2^n} \left\lfloor \left\lfloor \frac{n}{\sum_{j=1}^m F(j)} \right\rfloor^{1/n} \right\rfloor \tag{4}$$

$$= 1 + \sum_{m=1}^{2^n} \left\lfloor \left\lfloor \frac{n}{1+\pi(m)} \right\rfloor^{1/n} \right\rfloor, \tag{5}$$

where $\pi(m)$ is the PRIME COUNTING FUNCTION. It is also true that

$$p_{n+1} = 1 + p_n + F(p_n+1) + F(p_n+1)F(p_n+2) + \prod_{j=1}^{p} F(p_n+j) \tag{6}$$

(Ribenboim 1996, pp. 180–182). Note that the number of terms in the summation to obtain the nth prime is 2^n, so these formulas turn out not to be practical in the study of primes. An interesting INFINITE PRODUCT formula due to Euler which relates π and the nth PRIME p_n is

$$\pi = \frac{2}{\prod_{i=n}^{\infty} \left[1 + \frac{\sin(\frac{1}{2}\pi p_n)}{p_n}\right]} \tag{7}$$

$$= \frac{2}{\prod_{i=n}^{\infty} \left[1 + \frac{(-1)^{(p_n-1)/2}}{p_n}\right]} \tag{8}$$

(Blatner 1997). Conway (Guy 1983, Conway and Guy 1996, p. 147) gives an algorithm for generating primes based on 14 fractions, but it is actually just a concealed version of a SIEVE.

Some curious identities satisfied by primes p are

$$\sum_{k=1}^{p-1} \left\lfloor \frac{k^3}{p} \right\rfloor = \frac{(p-2)(p-1)(p+1)}{4} \tag{9}$$

$$\sum_{k=1}^{(p-1)(p-2)} \lfloor (kp)^{1/3} \rfloor = \tfrac{1}{4}(3p-5)(p-2)(p-1) \tag{10}$$

(Doster 1993),

$$\prod_{p \text{ prime}} \frac{p^2+1}{p^2-1} = \frac{5}{2} \tag{11}$$

(Le Lionnais 1983, p. 46),

$$\sum_{k=1}^{\infty} x^k \ln k = \sum_{p \text{ prime}} \sum_{k=1}^{\infty} \frac{x^{p^k}}{1-x^{p^k}}, \tag{12}$$

and

$$\sum_{k=1}^{\infty} (-1)^{k-1} e^{-kx} \ln k$$

$$= -\ln 2 \sum_{k=1}^{\infty} \frac{1}{e^{2^k x}-1} + \sum_{\substack{p \text{ an} \\ \text{odd prime}}} \ln p \sum_{k=1}^{\infty} \frac{1}{e^{p^k x}+1} \tag{13}$$

(Berndt 1994, p. 114).

It has been proven that the set of prime numbers is a DIOPHANTINE SET (Ribenboim 1991, pp. 106–107). Ramanujan also showed that

$$\frac{d\pi(x)}{dx} \sim \frac{1}{x \ln x} \sum_{n=1}^{\infty} \frac{\mu(n)}{n} x^{1/n}, \tag{14}$$

where $\pi(x)$ is the PRIME COUNTING FUNCTION and $\mu(n)$ is the MÖBIUS FUNCTION (Berndt 1994, p. 117). B. M. Bredihin proved that

$$f(x,y) = x^2 + y^2 + 1 \tag{15}$$

takes prime values for infinitely many integral pairs (x, y) (Honsberger 1976, p. 30). In addition, the function

$$f(x,y) = \tfrac{1}{2}(y-1) \left\lfloor |B^2(x,y)-1| - (B^2(x,y)-1) \right\rfloor + 2, \tag{16}$$

where

$$B(x,y) = x(y+1) - (y!+1), \tag{17}$$

$y!$ is the FACTORIAL, and $\lfloor x \rfloor$ is the FLOOR FUNCTION, generates only prime numbers for POSITIVE integral arguments. It not only generates every prime number, but generates ODD primes exactly once each, with all other values being 2 (Honsberger 1976, p. 33). For example,

$$f(1,2) = 3 \tag{18}$$

$$f(5,4) = 5 \tag{19}$$

$$f(103,6) = 7, \tag{20}$$

with no new primes generated for $x, y \leq 1000$.

For n an INTEGER ≥ 2, n is prime IFF

$$\binom{n-1}{k} \equiv (-1)^k \pmod{n} \tag{21}$$

for $k = 0, 1, \ldots, n-1$ (Deutsch 1996).

Cheng (1979) showed that for x sufficiently large, there always exist at least two prime factors between $(x - x^\alpha)$ and x for $\alpha \geq 0.477\ldots$ (Le Lionnais 1983, p. 26). Let $f(n)$ be the number of decompositions of n into two or more consecutive primes. Then

$$\lim_{x \to \infty} \frac{1}{x} \sum_{n=1}^{x} f(n) = \ln 2 \tag{22}$$

(Moser 1963, Le Lionnais 1983, p. 30). Euler showed that the sum of the inverses of primes is infinite

$$\sum_{p \text{ prime}} \frac{1}{p} = \infty \tag{23}$$

(Hardy and Wright 1979, p. 17), although it diverges very slowly. The sum exceeds 1, 2, 3, ... after 3, 59, 361139, ... (Sloane's A046024) primes, and its asymptotic equation is

$$\sum_{\substack{p=2 \\ p \text{ prime}}}^{x} \frac{1}{p} = \ln \ln x + B_1 + o(1), \tag{24}$$

where B_1 is MERTENS CONSTANT (Hardy and Wright 1979, p. 351). Dirichlet showed the even stronger result that

$$\sum_{\substack{\text{prime } p \equiv b \pmod{a} \\ (a,b)=1}} \frac{1}{p} = \infty \tag{25}$$

(Davenport 1980, p. 34).

Despite the fact that $\sum 1/p$ diverges, Brun showed that

$$\sum_{\substack{p \\ p+2 \text{ prime}}} \frac{1}{p} = B < \infty, \tag{26}$$

where B is BRUN'S CONSTANT. The function defined by

$$P(n) \equiv \sum_p \frac{1}{p^n} \tag{27}$$

taken over the primes converges for $n > 1$ and is a generalization of the RIEMANN ZETA FUNCTION known as the PRIME ZETA FUNCTION.

The probability that the largest prime factor of a RANDOM NUMBER x is less than $\sqrt{x}$ is ln 2 (Beeler *et al.* 1972, Item 29). The probability that two INTEGERS picked at random are RELATIVELY PRIME is $[\zeta(2)]^{-1} = 6/\pi^2$, where $\zeta(x)$ is the RIEMANN ZETA FUNCTION (Cesaro and Sylvester 1883). Given three INTEGERS chosen at random, the probability that no common factor will divide them all is

$$[\zeta(3)]^{-1} \approx 1.202^{-1} = 0.832\ldots, \tag{28}$$

where $\zeta(3)$ is APÉRY'S CONSTANT. In general, the probability that n random numbers lack a pth POWER common divisor is $[\zeta(np)]^{-1}$ (Beeler *et al.* 1972, Item 53).

Large primes include the large MERSENNE PRIMES, FERRIER'S PRIME, and $391581(2^{216193}-1)$ (Cipra 1989). The largest known prime as of 1998, is the MERSENNE PRIME $2^{3021377} - 1$.

Primes consisting of consecutive DIGITS (counting 0 as coming after 9) include 2, 3, 5, 7, 23, 67, 89, 4567, 78901, ... (Sloane's A006510).

see also ADLEMAN-POMERANCE-RUMELY PRIMALITY TEST, ALMOST PRIME, ANDRICA'S CONJECTURE, BERTRAND'S POSTULATE, BROCARD'S CONJECTURE, BRUN'S CONSTANT, CARMICHAEL'S CONJECTURE, CARMICHAEL FUNCTION, CARMICHAEL NUMBER, CHEBYSHEV FUNCTION, CHEBYSHEV-SYLVESTER CONSTANT, CHEN'S THEOREM, CHINESE HYPOTHESIS, COMPOSITE NUMBER, COMPOSITE RUNS, COPELAND-ERDŐS CONSTANT, CRAMER CONJECTURE, CUNNINGHAM CHAIN, CYCLOTOMIC POLYNOMIAL, DE POLIGNAC'S CONJECTURE, DIRICHLET'S THEOREM, DIVISOR, ERDŐS-KAC THEOREM, EUCLID'S THEOREMS, FEIT-THOMPSON CONJECTURE, FERMAT NUMBER, FERMAT QUOTIENT, FERRIER'S PRIME, FORTUNATE PRIME, FUNDAMENTAL THEOREM OF ARITHMETIC, GIGANTIC PRIME, GIUGA'S CONJECTURE, GOLDBACH CONJECTURE, GOOD PRIME, GRIMM'S CONJECTURE, HARDY-RAMANUJAN THEOREM, IRREGULAR PRIME, KUMMER'S CONJECTURE, LEHMER'S PROBLEM, LINNIK'S THEOREM, LONG PRIME, MERSENNE NUMBER, MERTENS FUNCTION, MILLER'S PRIMALITY TEST, MIRIMANOFF'S CONGRUENCE, MÖBIUS FUNCTION, PALINDROMIC NUMBER, PÉPIN'S TEST, PILLAI'S CONJECTURE, POULET NUMBER, PRIMARY, PRIME ARRAY, PRIME CIRCLE, PRIME FACTORIZATION ALGORITHMS, PRIME NUMBER OF MEASUREMENT, PRIME NUMBER THEOREM, PRIME POWER SYMBOL, PRIME STRING, PRIME TRIANGLE, PRIME ZETA FUNCTION, PRIMITIVE PRIME FACTOR, PRIMORIAL, PROBABLE PRIME, PSEUDOPRIME, REGULAR PRIME, RIEMANN FUNCTION, ROTKIEWICZ THEOREM, SCHNIRELMANN'S THEOREM, SELFRIDGE'S CONJECTURE, SEMIPRIME, SHAH-WILSON CONSTANT, SIERPIŃSKI'S COMPOSITE NUMBER THEOREM, SIERPIŃSKI'S PRIME SEQUENCE THEOREM, SMOOTH NUMBER, SOLDNER'S CONSTANT, SOPHIE GERMAIN PRIME, TITANIC PRIME, TOTIENT FUNCTION, TOTIENT VALENCE FUNCTION, TWIN PRIMES, TWIN PRIMES CONSTANT, VINOGRADOV'S THEOREM, VON MANGOLDT FUNCTION, WARING'S CONJECTURE, WIEFERICH PRIME, WILSON PRIME, WILSON QUOTIENT, WILSON'S THEOREM, WITNESS, WOLSTENHOLME'S THEOREM, ZSIGMONDY THEOREM

References

Beeler, M.; Gosper, R. W.; and Schroeppel, R. *HAKMEM.* Cambridge, MA: MIT Artificial Intelligence Laboratory, Memo AIM-239, Feb. 1972.

Berndt, B. C. "Ramanujan's Theory of Prime Numbers." Ch. 24 in *Ramanujan's Notebooks, Part IV.* New York: Springer-Verlag, 1994.

Blatner, D. *The Joy of Pi.* New York: Walker, p. 110, 1997.

Caldwell, C. "Largest Primes." `http://www.utm.edu/research/primes/largest.html`.

Cheng, J. R. "On the Distribution of Almost Primes in an Interval II." *Sci. Sinica* **22**, 253–275, 1979.

Cipra, B. A. "Math Team Vaults Over Prime Record." *Science* **245**, 815, 1989.

Conway, J. H. and Guy, R. K. *The Book of Numbers.* New York: Springer-Verlag, p. 130, 1996.

Courant, R. and Robbins, H. "The Prime Numbers." §1 in Supplement to Ch. 1 in *What is Mathematics?: An Elementary Approach to Ideas and Methods, 2nd ed.* Oxford, England: Oxford University Press, pp. 21–31, 1996.

Davenport, H. *Multiplicative Number Theory, 2nd ed.* New York: Springer-Verlag, 1980.

Deutsch, E. "Problem 1494." *Math. Mag.* **69**, 143, 1996.

Dickson, L. E. "Factor Tables, Lists of Primes." Ch. 13 in *History of the Theory of Numbers, Vol. 1: Divisibility and Primality.* New York: Chelsea, pp. 347–356, 1952.

Doster, D. Problem 10346. *Amer. Math. Monthly* **100**, 951, 1993.

Giblin, P. J. *Primes and Programming: Computers and Number Theory.* New York: Cambridge University Press, 1994.

Guy, R. K. "Conway's Prime Producing Machine." *Math. Mag.* **56**, 26–33, 1983.

Guy, R. K. "Prime Numbers," "Formulas for Primes," and "Products Taken Over Primes." Ch. A, §A17, and §B48 in *Unsolved Problems in Number Theory, 2nd ed.* New York: Springer-Verlag, pp. 3–43, 36–41 and 102–103, 1994.

Hardy, G. H. Ch. 2 in *Ramanujan: Twelve Lectures on Subjects Suggested by His Life and Work, 3rd ed.* New York: Chelsea, 1978.

Hardy, G. H. and Wright, E. M. "Prime Numbers" and "The Sequence of Primes." §1.2 and 1.4 in *An Introduction to the Theory of Numbers, 5th ed.* Oxford, England: Clarendon Press, pp. 1–4, 1979.

Honsberger, R. *Mathematical Gems II.* Washington, DC: Math. Assoc. Amer., p. 30, 1976.

Kraitchik, M. "Prime Numbers." §3.9 in *Mathematical Recreations.* New York: W. W. Norton, pp. 78–79, 1942.

Le Lionnais, F. *Les nombres remarquables.* Paris: Hermann, pp. 26, 30, and 46, 1983.

Moser, L. "Notes on Number Theory III. On the Sum of Consecutive Primes." *Can. Math. Bull.* **6**, 159–161, 1963.

Pappas, T. "Prime Numbers." *The Joy of Mathematics.* San Carlos, CA: Wide World Publ./Tetra, pp. 100–101, 1989.
Ribenboim, P. *The Little Book of Big Primes.* New York: Springer-Verlag, 1991.
Ribenboim, P. *The New Book of Prime Number Records.* New York: Springer-Verlag, 1996.
Riesel, H. *Prime Numbers and Computer Methods for Factorization, 2nd ed.* Boston, MA: Birkhäuser, 1994.
Schinzel, A. and Sierpiński, W. "Sur certains hypothèses concernant les nombres premiers." *Acta Arith.* **4**, 185–208, 1958.
Schinzel, A. and Sierpiński, W. Erratum to "Sur certains hypothèses concernant les nombres premiers." *Acta Arith.* **5**, 259, 1959.
Sloane, N. J. A. Sequences A046024, A000040/M0652, and A006510/M0679 in "An On-Line Version of the Encyclopedia of Integer Sequences."
Wagon, S. "Primes Numbers." Ch. 1 in *Mathematica in Action.* New York: W. H. Freeman, pp. 11–37, 1991.
Zaiger, D. "The First 50 Million Prime Numbers." *Math. Intel.* **0**, 221–224, 1977.

Prime Number of Measurement

The set of numbers generated by excluding the SUMS of two or more consecutive earlier members is called the prime numbers of measurement, or sometimes the SEGMENTED NUMBERS. The first few terms are 1, 2, 4, 5, 8, 10, 14, 15, 16, 21, ... (Sloane's A002048). Excluding two *and* three terms gives the sequence 1, 2, 4, 5, 8, 10, 14, 15, 16, 19, 20, 21, ... (Sloane's A005242).

References

Guy, R. K. "MacMahon's Prime Numbers of Measurement." §E30 in *Unsolved Problems in Number Theory, 2nd ed.* New York: Springer-Verlag, pp. 230–231, 1994.
Sloane, N. J. A. Sequence A002048/M0972 in "An On-Line Version of the Encyclopedia of Integer Sequences."0052420971

Prime Number Theorem

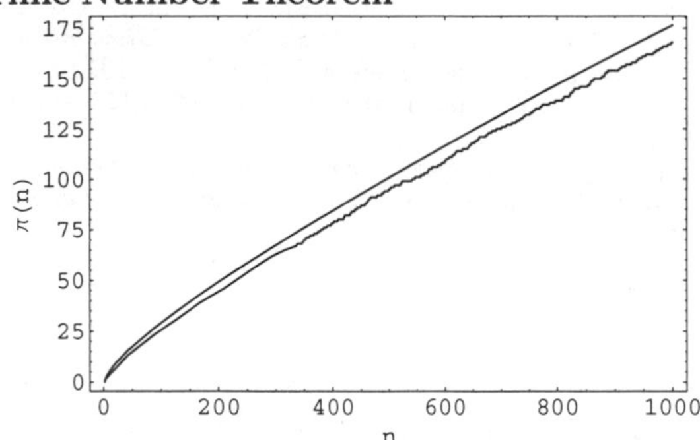

The theorem giving an asymptotic form for the PRIME COUNTING FUNCTION $\pi(n)$ for number of PRIMES less than some INTEGER n. Legendre (1808) suggested that, for large n,

$$\pi(n) \sim \frac{n}{A \ln n + B}, \quad (1)$$

with $A = 1$ and $B = -1.08366$ (where B is sometimes called LEGENDRE'S CONSTANT), a formula which is correct in the leading term only (Wagon 1991, pp. 28–29). In 1791, Gauss became the first to suggest instead

$$\pi(n) \sim \frac{n}{\ln n}. \quad (2)$$

Gauss later refined his estimate to

$$\pi(n) \sim \mathrm{Li}(n), \quad (3)$$

where $\mathrm{Li}(n)$ is the LOGARITHMIC INTEGRAL. This function has $n/\ln n$ as the leading term and has been shown to be a better estimate than $n/\ln n$ alone. The statement (3) is often known as "the" prime number theorem and was proved independently by Hadamard and Vallée Poussin in 1896. A plot of $\pi(n)$ (lower curve) and $\mathrm{Li}(n)$ is shown above for $n \leq 1000$.

For small n, it has been checked and always found that $\pi(n) < \mathrm{Li}(n)$. However, Skewes proved that the first crossing of $\pi(n) - \mathrm{Li}(n) = 0$ occurs before $10^{10^{10^{34}}}$ (the SKEWES NUMBER). The upper bound for the crossing has subsequently been reduced to 10^{371}. Littlewood (1914) proved that the INEQUALITY reverses infinitely often for sufficiently large n (Ball and Coxeter 1987). Lehman (1966) proved that at least 10^{500} reversals occur for numbers with 1166 or 1167 DECIMAL DIGITS.

Chebyshev (Rubinstein and Sarnak 1994) put limits on the RATIO

$$\frac{7}{8} < \frac{\pi(n)}{\frac{n}{\ln n}} < \frac{9}{8}, \quad (4)$$

and showed that if the LIMIT

$$\lim_{n\to\infty} \frac{\pi(n)}{\frac{n}{\ln n}} \quad (5)$$

existed, then it would be 1. This is, in fact, the prime number theorem.

Hadamard and Vallée Poussin proved the prime number theorem by showing that the RIEMANN ZETA FUNCTION $\zeta(z)$ has no zeros of the form $1+it$ (Smith 1994, p. 128). In particular, Vallée Poussin showed that

$$\pi(x) = \mathrm{Li}(x) + \mathcal{O}\left(\frac{x}{\ln x} e^{-a\sqrt{\ln x}}\right) \quad (6)$$

for some constant a. A simplified proof was found by Selberg and Erdős (1949) (Ball and Coxeter 1987, p. 63).

Riemann estimated the PRIME COUNTING FUNCTION with

$$\pi(n) \sim \mathrm{Li}(n) - \tfrac{1}{2}\mathrm{Li}(n^{1/2}), \quad (7)$$

which is a better approximation than $\mathrm{Li}(n)$ for $n < 10^7$. Riemann (1859) also suggested the RIEMANN FUNCTION

$$R(x) = \sum_{n=1}^{\infty} \frac{\mu(n)}{n} \mathrm{Li}(x^{1/n}), \quad (8)$$

where μ is the MÖBIUS FUNCTION (Wagon 1991, p. 29). An even better approximation for small n (by a factor of 10 for $n < 10^9$) is the GRAM SERIES.

The prime number theorem is equivalent to

$$\lim_{x\to\infty} \frac{\psi(x)}{x} = 1, \tag{9}$$

where $\psi(x)$ is the SUMMATORY MANGOLDT FUNCTION.

The RIEMANN HYPOTHESIS is equivalent to the assertion that

$$|\,\mathrm{Li}(x) - \pi(x)| \leq c\sqrt{x}\ln x \tag{10}$$

for some value of c (Ingham 1932, Ball and Coxeter 1987). Some limits obtained without assuming the RIEMANN HYPOTHESIS are

$$\pi(x) = \mathrm{Li}(x) + \mathcal{O}[xe^{-(\ln x)^{1/2}/15}] \tag{11}$$

$$\pi(x) = \mathrm{Li}(x) + \mathcal{O}[xe^{-0.009(\ln x)^{3/5}/(\ln\ln x)^{1/5}}]. \tag{12}$$

Ramanujan showed that for sufficiently large x,

$$\pi^2(x) < \frac{ex}{\ln x}\pi\left(\frac{x}{e}\right). \tag{13}$$

The largest known PRIME for which the inequality fails is 38,358,837,677 (Berndt 1994, pp. 112–113). The related inequality

$$\mathrm{Li}^2(x) < \frac{ex}{\ln x}\,\mathrm{Li}\left(\frac{x}{e}\right) \tag{14}$$

is true for $x \geq 2418$ (Berndt 1994, p. 114).

see also BERTRAND'S POSTULATE, DIRICHLET'S THEOREM, GRAM SERIES, PRIME COUNTING FUNCTION, RIEMANN'S FORMULA, RIEMANN FUNCTION, RIEMANN-MANGOLDT FUNCTION, RIEMANN WEIGHTED PRIME-POWER COUNTING FUNCTION, SKEWES NUMBER

References

Ball, W. W. R. and Coxeter, H. S. M. *Mathematical Recreations and Essays, 13th ed.* New York: Dover, pp. 62–64, 1987.

Berndt, B. C. *Ramanujan's Notebooks, Part IV.* New York: Springer-Verlag, 1994.

Courant, R. and Robbins, H. "The Prime Number Theorem." §1.2c in Supplement to Ch. 1 in *What is Mathematics?: An Elementary Approach to Ideas and Methods, 2nd ed.* Oxford, England: Oxford University Press, pp. 27–30, 1996.

Davenport, H. "Prime Number Theorem." Ch. 18 in *Multiplicative Number Theory, 2nd ed.* New York: Springer-Verlag, pp. 111–114, 1980.

de la Vallée Poussin, C.-J. "Recherches analytiques la théorie des nombres premiers." *Ann. Soc. scient. Bruxelles* **20**, 183–256, 1896.

Hadamard, J. "Sur la distribution des zéros de la fonction $\zeta(s)$ et ses conséquences arithmétiques (')." *Bull. Soc. math. France* **24**, 199–220, 1896.

Hardy, G. H. and Wright, E. M. "Statement of the Prime Number Theorem." §1.8 in *An Introduction to the Theory of Numbers, 5th ed.* Oxford, England: Clarendon Press, pp. 9–10, 1979.

Ingham, A. E. *The Distribution of Prime Numbers.* London: Cambridge University Press, p. 83, 1932.

Legendre, A. M. *Essai sur la Théorie des Nombres.* Paris: Duprat, 1808.

Lehman, R. S. "On the Difference $\pi(x) - \mathrm{li}(x)$." *Acta Arith.* **11**, 397–410, 1966.

Littlewood, J. E. "Sur les distribution des nombres premiers." *C. R. Acad. Sci. Paris* **158**, 1869–1872, 1914.

Nagell, T. "The Prime Number Theorem." Ch. 8 in *Introduction to Number Theory.* New York: Wiley, 1951.

Riemann, G. F. B. "Über die Anzahl der Primzahlen unter einer gegebenen Grösse." *Monatsber. Königl. Preuss. Akad. Wiss. Berlin*, 671, 1859.

Rubinstein, M. and Sarnak, P. "Chebyshev's Bias." *Experimental Math.* **3**, 173–197, 1994.

Selberg, A. and Erdős, P. "An Elementary Proof of the Prime Number Theorem." *Ann. Math.* **50**, 305–313, 1949.

Shanks, D. "The Prime Number Theorem." §1.6 in *Solved and Unsolved Problems in Number Theory, 4th ed.* New York: Chelsea, pp. 15–17, 1993.

Smith, D. E. *A Source Book in Mathematics.* New York: Dover, 1994.

Wagon, S. *Mathematica in Action.* New York: W. H. Freeman, pp. 25–35, 1991.

Prime Pairs

see TWIN PRIMES

Prime Patterns Conjecture

see k-TUPLE CONJECTURE

Prime Polynomial

see PRIME-GENERATING POLYNOMIAL

Prime Power Conjecture

An Abelian planar DIFFERENCE SET of order n exists only for n a PRIME POWER. Gordon (1994) has verified it to be true for $n < 2,000,000$.

see also DIFFERENCE SET

References

Gordon, D. M. "The Prime Power Conjecture is True for $n < 2,000,000$." *Electronic J. Combinatorics* **1**, R6, 1–7, 1994. `http://www.combinatorics.org/Volume_1/volume1.html#R6`.

Prime Power Symbol

The symbol $p^e \| n$ means, for p a PRIME, that $p^e | n$, but $p^{e+1} \nmid n$.

Prime Quadratic Effect

Let $\pi_{m,n}(x)$ denote the number of PRIMES $\leq x$ which are congruent to n modulo m. Then one might expect that

$$\Delta(x) \equiv \pi_{4,3}(x) - \pi_{4,1}(x) \sim \tfrac{1}{2}\pi(x^{1/2}) > 0$$

(Berndt 1994). Although this is true for small numbers, Hardy and Littlewood showed that $\Delta(x)$ changes sign infinitely often. (The first number for which it is false is 26861.) The effect was first noted by Chebyshev in 1853, and is sometimes called the CHEBYSHEV PHENOMENON. It was subsequently studied by Shanks (1959), Hudson (1980), and Bays and Hudson (1977, 1978, 1979). The effect was also noted by Ramanujan, who incorrectly claimed that $\lim_{x\to\infty} \Delta(x) = \infty$ (Berndt 1994).

References

Bays, C. and Hudson, R. H. "The Mean Behavior of Primes in Arithmetic Progressions." *J. Reine Angew. Math.* **296**, 80–99, 1977.

Bays, C. and Hudson, R. H. "On the Fluctuations of Littlewood for Primes of the Form $4n \pm 1$." *Math. Comput.* **32**, 281–286, 1978.

Bays, C. and Hudson, R. H. "Numerical and Graphical Description of All Axis Crossing Regions for the Moduli 4 and 8 which Occur Before 10^{12}." *Internat. J. Math. Math. Sci.* **2**, 111–119, 1979.

Berndt, B. C. *Ramanujan's Notebooks, Part IV.* New York: Springer-Verlag, pp. 135–136, 1994.

Hudson, R. H. "A Common Principle Underlies Riemann's Formula, the Chebyshev Phenomenon, and Other Subtle Effects in Comparative Prime Number Theory. I." *J. Reine Angew. Math.* **313**, 133–150, 1980.

Shanks, D. "Quadratic Residues and the Distribution of Primes." *Math. Comput.* **13**, 272–284, 1959.

Prime Quadruplet

A PRIME CONSTELLATION of four successive PRIMES with minimal distance $(p, p+2, p+6, p+8)$. The quadruplet (2, 3, 5, 7) has smaller minimal distance, but it is an exceptional special case. With the exception of (5, 7, 11, 13), a prime quadruple must be of the form $(30n+11, 30n+13, 30n+17, 30n+19)$. The first few values of n which give prime quadruples are $n = 0$, 3, 6, 27, 49, 62, 69, 108, 115, ... (Sloane's A014561), and the first few values of p are 5 (the exceptional case), 11, 101, 191, 821, 1481, 1871, 2081, 3251, 3461, ... The asymptotic FORMULA for the frequency of prime quadruples is analogous to that for other PRIME CONSTELLATIONS,

$$P_x(p, p+2, p+6, p+8) \sim \frac{27}{2} \prod_{p\geq 5} \frac{p^3(p-4)}{(p-1)^4} \int_2^x \frac{dx}{(\ln x)^4}$$
$$= 4.151180864 \int_2^x \frac{dx}{(\ln x)^4},$$

where $c = 4.15118\ldots$ is the Hardy-Littlewood constant for prime quadruplets. Roonguthai found the large prime quadruplets with

$$p = 10^{99} + 349781731$$
$$p = 10^{199} + 21156403891$$
$$p = 10^{299} + 140159459341$$
$$p = 10^{399} + 34993836001$$
$$p = 10^{499} + 883750143961$$
$$p = 10^{599} + 1394283756151$$
$$p = 10^{699} + 547634621251$$

(Roonguthai).

see also PRIME ARITHMETIC PROGRESSION, PRIME CONSTELLATION, PRIME k-TUPLES CONJECTURE, SEXY PRIMES, TWIN PRIMES

References

Hardy, G. H. and Wright, E. M. *An Introduction to the Theory of Numbers, 5th ed.* New York: Oxford University Press, 1979.

Forbes, T. "Prime k-tuplets." `http://www.ltkz.demon.co.uk/ktuplets.htm`.

Rademacher, H. *Lectures on Elementary Number Theory.* New York: Blaisdell, 1964.

Riesel, H. *Prime Numbers and Computer Methods for Factorization, 2nd ed.* Boston, MA: Birkhäuser, pp. 61–62, 1994.

Roonguthai, W. "Large Prime Quadruplets." `http://www.mathsoft.com/asolve/constant/hrdyltl/roonguth.html`.

Sloane, N. J. A. Sequence A014561 in "An On-Line Version of the Encyclopedia of Integer Sequences."

Prime Representation

Let $a \neq b$, A, and B denote POSITIVE INTEGERS satisfying

$$(a, b) = 1 \qquad (A, B) = 1,$$

(i.e., both pairs are RELATIVELY PRIME), and suppose every PRIME $p \equiv B \pmod{A}$ with $(p, 2ab) = 1$ is expressible if the form $ax^2 - by^2$ for some INTEGERS x and y. Then every PRIME q such that $q \equiv -B \pmod{A}$ and $(q, 2ab) = 1$ is expressible in the form $bX^2 - aY^2$ for some INTEGERS X and Y (Halter-Koch 1993, Williams 1991).

Prime Form	Representation
$4n+1$	x^2+y^2
$8n+1, 8n+3$	x^2+2y^2
$8n\pm 1$	x^2-2y^2
$6n+1$	x^2+3y^2
$12n+1$	x^2-3y^2
$20n+1, 20n+9$	x^2+5y^2
$10n+1, 10n+9$	x^2-5y^2
$14n+1, 14n+9, 14n+25$	x^2+7y^2
$28n+1, 28n+9, 28n+25$	x^2-7y^2
$30n+1, 30n+49$	x^2+15y^2
$60n+1, 60n+49$	x^2-15y^2
$30n-7, 30n+17$	$5x^2+3y^2$
$60n-7, 60n+17$	$5x^2-3y^2$
$24n+1, 24n+7$	x^2+6y^2
$24n+1, 24n+19$	x^2-6y^2
$24n+5, 24n+11$	$2x^2+3y^2$
$24n+5, 24n-1$	$2x^2-3y^2$

References

Berndt, B. C. *Ramanujan's Notebooks, Part IV.* New York: Springer-Verlag, pp. 70–73, 1994.

Halter-Koch, F. "A Theorem of Ramanujan Concerning Binary Quadratic Forms." *J. Number. Theory* **44**, 209–213, 1993.

Williams, K. S. "On an Assertion of Ramanujan Concerning Binary Quadratic Forms." *J. Number Th.* **38**, 118–133, 1991.

Prime Ring

A RING for which the product of any pair of IDEALS is zero only if one of the two IDEALS is zero. All SIMPLE RINGS are prime.

see also IDEAL, RING, SEMIPRIME RING, SIMPLE RING

Prime Sequence

see PRIME ARITHMETIC PROGRESSION, PRIME ARRAY, PRIME-GENERATING POLYNOMIAL, SIERPIŃSKI'S PRIME SEQUENCE THEOREM

Prime Spiral

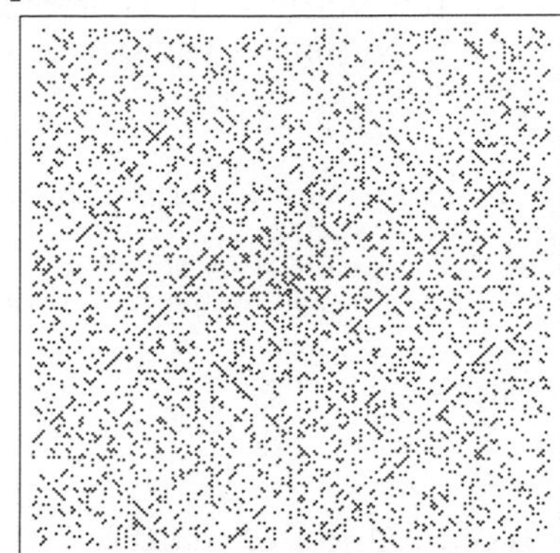

The numbers arranged in a SPIRAL

$$\begin{matrix} 5 & 4 & 3 \\ 6 & 1 & 2 \\ 7 & 8 & 9 \end{matrix}$$

with PRIMES indicated in black, as first drawn by S. Ulam. Unexpected patterns of diagonal lines are apparent in such a plot, as illustrated in the above 199×199 grid.

References

Dewdney, A. K. "Computer Recreations: How to Pan for Primes in Numerical Gravel." *Sci. Amer.* **259**, 120–123, July 1988.

Lane, C. "Prime Spiral." `http://www.best.com/~cdl/PrimeSpiralApplet.html`.

Weisstein, E. W. "Prime Spiral." `http://www.astro.virginia.edu/~eww6n/math/notebooks/PrimeSpiral.m`.

Prime String

Call a number n a prime string from the left if n and all numbers obtained by successively removing the right-most DIGIT are PRIME. There are 83 left prime strings in base 10. The first few are 2, 3, 5, 7, 23, 29, 31, 37, 53, 59, 71, 73, 79, 233, 239, 293, 311, 313, 317, 373, 379, 593, 599, ... (Sloane's A024770), the largest being 73,939,133. Similarly, call a number n a prime string from the right if n and all numbers obtained by successively removing the left-most DIGIT are PRIME. The first few are 2, 3, 5, 7, 13, 17, 23, 37, 43, 47, 53, 67, 73, 83, 97, 103, 107, 113, 137, 167, 173, ... (Sloane's A033664). A large right prime string is 933,739,397.

see also PRIME ARRAY, PRIME NUMBER

References

Beeler, M.; Gosper, R. W.; and Schroeppel, R. *HAKMEM.* Cambridge, MA: MIT Artificial Intelligence Laboratory, Memo AIM-239, Feb. 1972.

Rivera, C. "Problems & Puzzles (Puzzles): Prime Strings." `http://www.sci.net.mx/~crivera/ppp/puzz_002.htm`.

Sloane, N. J. A. Sequence A024770 in "An On-Line Version of the Encyclopedia of Integer Sequences."033664

Prime Sum

Let

$$\Sigma(n) \equiv \sum_{i=1}^{n} p_i$$

be the sum of the first n PRIMES. The first few terms are 2, 5, 10, 17, 28, 41, 58, 77, ... (Sloane's A007504). Bach and Shallit (1996) show that

$$\Sigma(n) \sim \frac{n^2}{2\log n},$$

and provide a general technique for estimating such sums.

see also PRIMORIAL

References

Bach, E. and Shallit, J. §2.7 in *Algorithmic Number Theory, Vol. 1: Efficient Algorithms.* Cambridge, MA: MIT Press, 1996.

Sloane, N. J. A. Sequence A007504/M1370 in "An On-Line Version of the Encyclopedia of Integer Sequences."

Prime Theta Function

The prime theta function is defined as

$$\theta(n) \equiv \sum_{i=1}^{n} \ln p_i,$$

where p_i is the ith PRIME. As shown by Bach and Shallit (1996),

$$\theta(n) \sim n.$$

References

Bach, E. and Shallit, J. *Algorithmic Number Theory, Vol. 1: Efficient Algorithms.* Cambridge, MA: MIT Press, pp. 206 and 233, 1996.

Prime Triangle

$$\begin{matrix} & & & & & * & & & & & \\ & & & & 1 & & 2 & & & & \\ & & & 1 & & 2 & & 3 & & & \\ & & 1 & & 2 & & 3 & & 4 & & \\ & 1 & & 4 & & 3 & & 2 & & 5 & \\ 1 & & 4 & & 3 & & 2 & & 5 & & 6 \end{matrix}$$

This triangle has rows beginning with 1 and ending with n, with the SUM of each two consecutive entries being a PRIME.

see also PASCAL'S TRIANGLE

References

Guy, R. K. *Unsolved Problems in Number Theory, 2nd ed.* New York: Springer-Verlag, p. 106, 1994.

Kenney, M. J. "Student Math Notes." *NCTM News Bulletin.* Nov. 1986.

Prime Unit

1 and -1 are the only INTEGERS which divide every INTEGER. They are therefore called the prime units.

see also INTEGER, PRIME NUMBER, UNIT

Prime Zeta Function

The prime zeta function

$$P(n) \equiv \sum_p \frac{1}{p^n}, \tag{1}$$

where the sum is taken over PRIMES is a generalization of the RIEMANN ZETA FUNCTION

$$\zeta(n) \equiv \sum_{k=1}^{\infty} \frac{1}{k^n}, \tag{2}$$

where the sum is over *all* integers. The prime zeta function can be expressed in terms of the RIEMANN ZETA FUNCTION by

$$\begin{aligned} \ln \zeta(n) &= -\sum_{p\geq 2} \ln(1-p^{-n}) = \sum_{p\geq 2}\sum_{k=1}^{\infty} \frac{p^{-kn}}{k} \\ &= \sum_{k=1}^{\infty} \frac{1}{k} \sum_{p\geq 2} p^{-kn} = \sum_{k=1}^{\infty} \frac{P(kn)}{k}. \end{aligned} \tag{3}$$

Inverting then gives

$$P(n) = \sum_{k=1}^{\infty} \frac{\mu(k)}{k} \ln \zeta(kn), \tag{4}$$

where $\mu(k)$ is the MÖBIUS FUNCTION. The values for the first few integers starting with two are

$$P(2) \approx 0.452247 \tag{5}$$

$$P(3) \approx 0.174763 \tag{6}$$

$$P(4) \approx 0.0769931 \tag{7}$$

$$P(5) \approx 0.035755. \tag{8}$$

see also MÖBIUS FUNCTION, RIEMANN ZETA FUNCTION, ZETA FUNCTION

References

Hardy, G. H. and Weight, E. M. *An Introduction to the Theory of Numbers, 5th ed.* Oxford, England: Oxford University Press, pp. 355–356, 1979.

Primequad

see PRIME QUADRUPLET

Primitive Abundant Number

An ABUNDANT NUMBER for which all PROPER DIVISORS are DEFICIENT is called a primitive abundant number (Guy 1994, p. 46). The first few ODD primitive abundant numbers are 945, 1575, 2205, 3465, ... (Sloane's A006038).

see also ABUNDANT NUMBER, DEFICIENT NUMBER, HIGHLY ABUNDANT NUMBER, SUPERABUNDANT NUMBER, WEIRD NUMBER

References

Guy, R. K. *Unsolved Problems in Number Theory, 2nd ed.* New York: Springer-Verlag, p. 46, 1994.

Sloane, N. J. A. Sequence A006038/M5486 in "An On-Line Version of the Encyclopedia of Integer Sequences."

Primitive Function

see INTEGRAL

Primitive Irreducible Polynomial

An IRREDUCIBLE POLYNOMIAL which generates all elements of an extension field from a base field. For any PRIME or PRIME POWER q and any POSITIVE INTEGER n, there exists a primitive irreducible POLYNOMIAL of degree n over GF(q).

see also GALOIS FIELD, IRREDUCIBLE POLYNOMIAL

Primitive Polynomial Modulo 2

A special type of POLYNOMIAL of which a subclass has COEFFICIENTS of only 0 or 1. Such POLYNOMIALS define a RECURRENCE RELATION which can be used to obtain a new RANDOM bit from the n preceding ones.

Primitive Prime Factor

If $n \geq 1$ is the smallest INTEGER such that $p|a^n - b^n$ (or $a^n + b^n$), then p is a primitive prime factor.

Primitive Pseudoperfect Number

see PRIMITIVE SEMIPERFECT NUMBER

Primitive Recursive Function

For-loops (which have a fixed iteration limit) are a special case of while-loops. A function which can be implemented using only for-loops is called primitive recursive. (In contrast, a COMPUTABLE FUNCTION can be coded using a combination of for- and while-loops, or while-loops only.)

The ACKERMANN FUNCTION is the simplest example of a well-defined TOTAL FUNCTION which is COMPUTABLE but not primitive recursive, providing a counterexample to the belief in the early 1900s that every COMPUTABLE FUNCTION was also primitive recursive (Dötzel 1991).

see also ACKERMANN FUNCTION, COMPUTABLE FUNCTION, TOTAL FUNCTION

References

Dötzel, G. "A Function to End All Functions." *Algorithm: Recreational Programming* **2**, 16–17, 1991.

Primitive Root

A number g is a primitive root of m if

$$g^k \not\equiv 1 \pmod{m} \tag{1}$$

for $1 \leq k < m$ and

$$g^m \equiv 1 \pmod{m}. \tag{2}$$

Only $m = 2$, 4, p^a, and $2p^a$ have primitive roots (where $p > 2$ and a is an INTEGER). For composite m, there may be more than one primitive root (both 3 and 7 are primitive roots mod 10), but for prime p, there is only one primitive root. It is the INTEGER g satisfying $1 \leq g \leq p-1$ such that $g \pmod{p}$ has ORDER $p-1$.

The primitive root of m can also be defined as a cyclic generator of the multiplicative group (mod m) when m is a prime POWER or twice a PRIME POWER. Let p be any ODD PRIME $k \geq 1$, and let

$$s \equiv \sum_{j=1}^{p-1} j^k. \tag{3}$$

Then

$$s = \begin{cases} -1 \pmod{p} & \text{for } p-1 \mid k \\ 0 \pmod{p} & \text{for } p-1 \nmid k. \end{cases} \tag{4}$$

For numbers m with primitive roots, all y satisfying $(p, y) = 1$ are representable as

$$y \equiv g^t \pmod{m}, \tag{5}$$

where $t = 0, 1, \ldots, \phi(m)-1$, t is known as the index, and y is an INTEGER. Kearnes showed that for any POSITIVE INTEGER m, there exist infinitely many PRIMES p such that

$$m < g_p < p - m. \tag{6}$$

Call the least primitive root g_p. Burgess (1962) proved that

$$g_p \leq C p^{1/4+\epsilon} \tag{7}$$

for C and ϵ POSITIVE constants and p sufficiently large.

The table below gives the primitive roots (for prime $m = p$; Sloane's A001918) and least primitive roots (for composite m) for the first few INTEGERS

m	g	m	g	m	g
2	1	53	2	134	7
3	2	54	5	137	3
4	3	58	3	139	2
5	2	59	2	142	7
6	5	61	2	146	5
7	3	62	3	149	2
9	2	67	2	151	6
10	3	71	7	157	5
11	2	73	5	158	3
13	2	74	5	162	5
14	3	79	3	163	2
17	3	81	2	166	5
18	5	82	7	167	5
19	2	83	2	169	2
22	7	86	3	173	2
23	5	89	3	178	3
25	2	94	5	179	2
26	7	97	5	181	2
27	2	98	3	191	19
29	2	101	2	193	5
31	3	103	5	194	5
34	3	106	3	197	2
37	2	107	2	199	3
38	3	109	6	202	3
41	6	113	3	206	5
43	3	118	11	211	2
46	5	121	2	214	5
47	5	122	7	218	11
49	3	125	2	223	3
50	3	127	3	226	3
		131	2	227	2

References

Abramowitz, M. and Stegun, C. A. (Eds.). "Primitive Roots." §24.3.4 in *Handbook of Mathematical Functions with Formulas, Graphs, and Mathematical Tables, 9th printing.* New York: Dover, p. 827, 1972.

Guy, R. K. "Primitive Roots." §F9 in *Unsolved Problems in Number Theory, 2nd ed.* New York: Springer-Verlag, pp. 248–249, 1994.

Sloane, N. J. A. Sequence A001918/M0242 in "An On-Line Version of the Encyclopedia of Integer Sequences."

Primitive Root of Unity

A number r is an nth ROOT OF UNITY if $r^n = 1$ and a primitive nth root of unity if, in addition, n is the smallest INTEGER of $k = 1, \ldots, n$ for which $r^k = 1$.

see also ROOT OF UNITY

Primitive Semiperfect Number

A SEMIPERFECT NUMBER for which none of its PROPER DIVISORS are pseudoperfect (Guy 1994, p. 46). The first few are 6, 20, 28, 88, 104, 272 ... (Sloane's A006036). Primitive pseudoperfect numbers are also called IRREDUCIBLE SEMIPERFECT NUMBERS. There are infinitely many primitive pseudoperfect numbers which are not HARMONIC DIVISOR NUMBERS, and infinitely many ODD primitive semiperfect numbers.

see also HARMONIC DIVISOR NUMBER, SEMIPERFECT NUMBER

References
Guy, R. K. *Unsolved Problems in Number Theory, 2nd ed.* New York: Springer-Verlag, p. 46, 1994.
Sloane, N. J. A. Sequence A006036/M4133 in "An On-Line Version of the Encyclopedia of Integer Sequences."

Primitive Sequence

A SEQUENCE in which no term DIVIDES any other.

References
Guy, R. K. *Unsolved Problems in Number Theory, 2nd ed.* New York: Springer-Verlag, p. 202, 1994.

Primorial

For a PRIME p,

$$\text{primorial}(p_i) = p_i\# \equiv \prod_{j=1}^{i} p_j,$$

where p_i is the ith PRIME. The first few values for $p_i\#$, are 2, 6, 30, 210, 2310, 30030, 510510, ... (Sloane's A002110).

$p\# - 1$ is PRIME for PRIMES $p = 3$, 5, 11, 41, 89, 317, 337, 991, 1873, 2053, 2377, 4093, 4297, ... (Sloane's A014563; Guy 1994), or p_n for $n = 2$, 3, 5, 13, 24, 66, 68, 167, 287, 310, 352, 564, 590, ... $p\# + 1$ is known to be PRIME for the PRIMES $p = 2$, 3, 5, 7, 11, 31, 379, 1019, 1021, 2657, 3229, 4547, 4787, 11549, ... (Sloane's A005234; Guy 1994, Mudge 1997), or p_n for $n = 1$, 2, 3, 4, 5, 11, 75, 171, 172, 384, 457, 616, 643, 1391, ... Both forms have been tested to $p = 25000$ (Caldwell 1995). It is not known if there are an infinite number of PRIMES for which $p\# + 1$ is PRIME or COMPOSITE (Ribenboim 1989).

see also FACTORIAL, FORTUNATE PRIME, PRIME SUM SMARANDACHE NEAR-TO-PRIMORIAL FUNCTION, TWIN PEAKS

References
Borning, A. "Some Results for $k! + 1$ and $2 \cdot 3 \cdot 5 \cdot p + 1$." *Math. Comput.* **26**, 567–570, 1972.
Buhler, J. P.; Crandall, R. E.; and Penk, M. A. "Primes of the form $M! + 1$ and $\cdot 3 \cdot 5 \cdot p + 1$." *Math. Comput.* **38**, 639–643, 1982.
Caldwell, C. "On The Primality of $n! \pm 1$ and $2 \cdot 3 \cdot 5 \cdots p \pm 1$." *Math. Comput.* **64**, 889–890, 1995.
Dubner, H. "Factorial and Primorial Primes." *J. Rec. Math.* **19**, 197–203, 1987.
Dubner, H. "A New Primorial Prime." *J. Rec. Math.* **21**, 276, 1989.
Guy, R. K. *Unsolved Problems in Number Theory, 2nd ed.* New York: Springer-Verlag, pp. 7–8, 1994.
Leyland, P. `ftp://sable.ox.ac.uk/pub/math/factors/primorial-.Z` and `primorial+.Z`.
Mudge, M. "Not Numerology but Numeralogy!" *Personal Computer World,* 279–280, 1997.
Ribenboim, P. *The Book of Prime Number Records, 2nd ed.* New York: Springer-Verlag, p. 4, 1989.
Sloane, N. J. A. Sequences A014563, A002110/M1691, and A005234/M0669 in "An On-Line Version of the Encyclopedia of Integer Sequences."
Temper, M. "On the Primality of $k! + 1$ and $\cdot 3 \cdot 5 \cdots p + 1$." *Math. Comput.* **34**, 303–304, 1980.

Prince Rupert's Cube

The largest CUBE which can be made to pass through a given CUBE. (In other words, the CUBE having a side length equal to the side length of the largest HOLE of a SQUARE CROSS-SECTION which can be cut through a unit CUBE without splitting it into two pieces.) The Prince Rupert's cube has side length $3\sqrt{2}/4 = 1.06065\ldots$, and any CUBE this size or smaller can be made to pass through the original CUBE.

see also CUBE, SQUARE

References
Cundy, H. and Rollett, A. "Prince Rupert's Cubes." §3.15.2 in *Mathematical Models, 3rd ed.* Stradbroke, England: Tarquin Pub., pp. 157–158, 1989.
Schrek, D. J. E. "Prince Rupert's Problem and Its Extension by Pieter Nieuwland." *Scripta Math.* **16**, 73–80 and 261–267, 1950.

Principal

The original amount borrowed or lent on which INTEREST is then paid or given.

see also INTEREST

Principal Curvatures

The MAXIMUM and MINIMUM of the NORMAL CURVATURE κ_1 and κ_2 at a given point on a surface are called the principal curvatures. The principal curvatures measure the MAXIMUM and MINIMUM bending of a REGULAR SURFACE at each point. The GAUSSIAN CURVATURE K and MEAN CURVATURE H are related to κ_1 and κ_2 by

$$K = \kappa_1\kappa_2 \qquad (1)$$

$$H = \tfrac{1}{2}(\kappa_1 + \kappa_2). \qquad (2)$$

This can be written as a QUADRATIC EQUATION

$$\kappa^2 - 2H\kappa + K = 0, \qquad (3)$$

which has solutions

$$\kappa_1 = H + \sqrt{H^2 - K} \qquad (4)$$

$$\kappa_2 = H - \sqrt{H^2 - K}. \qquad (5)$$

see also GAUSSIAN CURVATURE, MEAN CURVATURE, NORMAL CURVATURE, NORMAL SECTION, PRINCIPAL DIRECTION, PRINCIPAL RADIUS OF CURVATURE, RODRIGUES'S CURVATURE FORMULA

References
Geometry Center. "Principal Curvatures." `http://www.geom.umn.edu/zoo/diffgeom/surfspace/concepts/curvatures/prin-curv.html`.
Gray, A. "Normal Curvature." §14.2 in *Modern Differential Geometry of Curves and Surfaces.* Boca Raton, FL: CRC Press, pp. 270–273, 277, and 283, 1993.

Principal Curve

A curve $\boldsymbol{\alpha}$ on a REGULAR SURFACE M is a principal curve IFF the velocity $\boldsymbol{\alpha}'$ always points in a PRINCIPAL DIRECTION, i.e.,

$$S(\boldsymbol{\alpha}') = \kappa_i \boldsymbol{\alpha}',$$

where S is the SHAPE OPERATOR and κ_i is a PRINCIPAL CURVATURE. If a SURFACE OF REVOLUTION generated by a plane curve is a REGULAR SURFACE, then the MERIDIANS and PARALLELS are principal curves.

References

Gray, A. "Principal Curves" and "The Differential Equation for the Principal Curves." §18.1 and 21.1 in *Modern Differential Geometry of Curves and Surfaces.* Boca Raton, FL: CRC Press, pp. 410–413, 1993.

Principal Direction

The directions in which the PRINCIPAL CURVATURES occur.

see also PRINCIPAL DIRECTION

References

Gray, A. *Modern Differential Geometry of Curves and Surfaces.* Boca Raton, FL: CRC Press, p. 270, 1993.

Principal Ideal

An IDEAL I of a RING R is called principal if there is an element a of R such that

$$I = aR = \{ar : r \in R\}.$$

In other words, the IDEAL is generated by the element a. For example, the IDEALS $n\mathbb{Z}$ of the RING of INTEGERS $\mathbb{Z}$ are all principal, and in fact all IDEALS of $\mathbb{Z}$ are principal.

see also IDEAL, RING

Principal Normal Vector

see NORMAL VECTOR

Principal Quintic Form

A general QUINTIC EQUATION

$$a_5x^5 + a_4x^4 + a_3x^3 + a_2x^2 + a_1x + a_0 = 0 \quad (1)$$

can be reduced to one of the form

$$y^5 + b_2y^2 + b_1y + b_0 = 0, \quad (2)$$

called the principal quintic form.

NEWTON'S RELATIONS for the ROOTS y_j in terms of the b_js is a linear system in the b_j, and solving for the b_js expresses them in terms of the POWER sums $s_n(y_j)$. These POWER sums can be expressed in terms of the a_js, so the b_js can be expressed in terms of the a_js. For a quintic to have no quartic or cubic term, the sums of the ROOTS and the sums of the SQUARES of the ROOTS vanish, so

$$s_1(y_j) = 0 \quad (3)$$

$$s_2(y_j) = 0. \quad (4)$$

Assume that the ROOTS y_j of the new quintic are related to the ROOTS x_j of the original quintic by

$$y_j = x_j^2 + \alpha x_j + \beta. \quad (5)$$

Substituting this into (1) then yields two equations for α and β which can be multiplied out, simplified by using NEWTON'S RELATIONS for the POWER sums in the x_j, and finally solved. Therefore, α and β can be expressed using RADICALS in terms of the COEFFICIENTS a_j. Again by substitution into (4), we can calculate $s_3(y_j)$, $s_4(y_j)$ and $s_5(y_j)$ in terms of α and β and the x_j. By the previous solution for α and β and again by using NEWTON'S RELATIONS for the POWER sums in the x_j, we can ultimately express these POWER sums in terms of the a_j.

see also BRING QUINTIC FORM, NEWTON'S RELATIONS, QUINTIC EQUATION

Principal Radius of Curvature

Given a 2-D SURFACE, there are two "principal" RADII OF CURVATURE. The larger is denoted R_1, and the smaller R_2. These are PERPENDICULAR to each other, and both PERPENDICULAR to the tangent PLANE of the surface.

see also GAUSSIAN CURVATURE, MEAN CURVATURE, RADIUS OF CURVATURE

Principal Value

see CAUCHY PRINCIPAL VALUE

Principal Vector

A tangent vector $\mathbf{v_p} = v_1\mathbf{x}_u + v_2\mathbf{x}_v$ is a principal vector IFF

$$\det\begin{bmatrix} {v_2}^2 & -v_1v_2 & {v_1}^2 \\ E & F & G \\ e & f & g \end{bmatrix} = 0,$$

where e, f, and g are coefficients of the first FUNDAMENTAL FORM and E, F, G of the second FUNDAMENTAL FORM.

see also FUNDAMENTAL FORMS, PRINCIPAL CURVE

References

Gray, A. *Modern Differential Geometry of Curves and Surfaces.* Boca Raton, FL: CRC Press, p. 410, 1993.

Principal Vertex

A VERTEX x_i of a SIMPLE POLYGON P is a principal VERTEX if the diagonal $[x_{i-1}, x_{i+1}]$ intersects the boundary of P only at x_{i-1} and x_{i+1}.

see also EAR, MOUTH

References
Meisters, G. H. "Polygons Have Ears." *Amer. Math. Monthly* **82**, 648–751, 1975.
Meisters, G. H. "Principal Vertices, Exposed Points, and Ears." *Amer. Math. Monthly* **87**, 284–285, 1980.
Toussaint, G. "Anthropomorphic Polygons." *Amer. Math. Monthly* **98**, 31–35, 1991.

Principle

A loose term for a true statement which may be a POSTULATE, THEOREM, etc.

see also AREA PRINCIPLE, ARGUMENT PRINCIPLE, AXIOM, CAVALIERI'S PRINCIPLE, CONJECTURE, CONTINUITY PRINCIPLE, COUNTING GENERALIZED PRINCIPLE, DIRICHLET'S BOX PRINCIPLE, DUALITY PRINCIPLE, DUHAMEL'S CONVOLUTION PRINCIPLE, EUCLID'S PRINCIPLE, FUBINI PRINCIPLE, HASSE PRINCIPLE, INCLUSION-EXCLUSION PRINCIPLE, INDIFFERENCE PRINCIPLE, INDUCTION PRINCIPLE, INSUFFICIENT REASON PRINCIPLE, LEMMA, LOCAL-GLOBAL PRINCIPLE, MULTIPLICATION PRINCIPLE, PERMANENCE OF MATHEMATICAL RELATIONS PRINCIPLE, PONCELET'S CONTINUITY PRINCIPLE, PONTRYAGIN MAXIMUM PRINCIPLE, PORISM, POSTULATE, SCHWARZ REFLECTION PRINCIPLE, SUPERPOSITION PRINCIPLE, SYMMETRY PRINCIPLE, THEOREM, THOMSON'S PRINCIPLE, TRIANGLE TRANSFORMATION PRINCIPLE, WELL-ORDERING PRINCIPLE

Pringsheim's Theorem

Let $C^\omega(I)$ be the set of real ANALYTIC FUNCTIONS on I. Then $C^\omega(I)$ is a SUBALGEBRA of $C^\infty(I)$. A NECESSARY and SUFFICIENT condition for a function $f \in C^\infty(I)$ to belong to $C^\omega(I)$ is that

$$|f^{(n)}(x)| \leq k^n n!$$

for $n = 0, 1, \ldots$ for a suitable constant k.

see also ANALYTIC FUNCTION, SUBALGEBRA

References
Iyanaga, S. and Kawada, Y. (Eds.). *Encyclopedic Dictionary of Mathematics.* Cambridge, MA: MIT Press, p. 207, 1980.

Printer's Errors

Typesetting "errors" in which exponents or multiplication signs are omitted but the resulting expression is equivalent to the original one. Examples include

$$2^5 9^2 = 2592$$

$$3^4 425 = 34425$$

$$31^2 325 = 312325$$

$$2^5 \cdot \tfrac{25}{31} = 25\tfrac{25}{31},$$

where a whole number followed by a fraction is interpreted as addition (e.g., $1\frac{1}{2} = 1 + \frac{1}{2} = \frac{3}{2}$).

see also ANOMALOUS CANCELLATION

References
Dudeney, H. E. *Amusements in Mathematics.* New York: Dover, 1970.
Madachy, J. S. *Madachy's Mathematical Recreations.* New York: Dover, pp. 174–175, 1979.

Prior Distribution

see BAYESIAN ANALYSIS

Prism

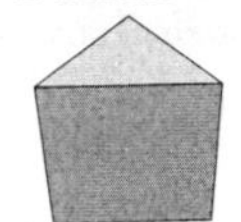

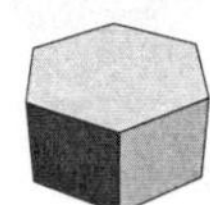
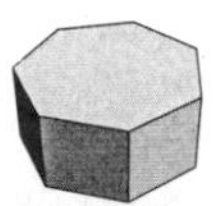
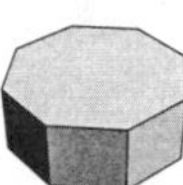
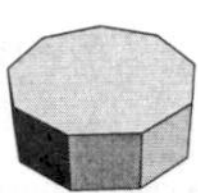
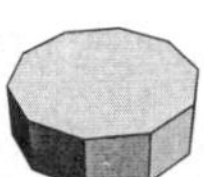

A POLYHEDRON with two congruent POLYGONAL faces and all remaining faces PARALLELOGRAMS. The 3-prism is simply the CUBE. The simple prisms and antiprisms include: decagonal antiprism, decagonal prism, hexagonal antiprism, hexagonal prism, octagonal antiprism, octagonal prism, pentagonal antiprism, pentagonal prism, square antiprism, and triangular prism. The DUAL POLYHEDRON of a simple (Archimedean) prism is a BIPYRAMID.

The triangular prism, square prism (cube), and hexagonal prism are all SPACE-FILLING POLYHEDRA.

see also ANTIPRISM, AUGMENTED HEXAGONAL PRISM, AUGMENTED PENTAGONAL PRISM, AUGMENTED TRIANGULAR PRISM, BIAUGMENTED PENTAGONAL PRISM, BIAUGMENTED TRIANGULAR PRISM, CUBE, METABIAUGMENTED HEXAGONAL PRISM, PARABIAUGMENTED HEXAGONAL PRISM, PRISMATOID, PRISMOID, TRAPEZOHEDRON, TRIAUGMENTED HEXAGONAL PRISM, TRIAUGMENTED TRIANGULAR PRISM

References
Beyer, W. H. (Ed.) *CRC Standard Mathematical Tables, 28th ed.* Boca Raton, FL: CRC Press, p. 127, 1987.
Cromwell, P. R. *Polyhedra.* New York: Cambridge University Press, pp. 85–86, 1997.
Weisstein, E. W. "Prisms and Antiprisms." http://www.astro.virginia.edu/~eww6n/math/notebooks/Prism.m.

Prismatic Ring

A MÖBIUS STRIP with finite width.

see also MÖBIUS STRIP

References
Gardner, M. "Twisted Prismatic Rings." Ch. 5 in *Fractal Music, Hypercards, and More Mathematical Recreations from Scientific American Magazine.* New York: W. H. Freeman, 1992.

Prismatoid

A POLYHEDRON having two POLYGONS in PARALLEL planes as bases and TRIANGULAR or TRAPEZOIDAL lateral faces with one side lying in one base and the opposite VERTEX or side lying in the other base. Examples include the CUBE, PYRAMIDAL FRUSTUM, RECTANGULAR PARALLELEPIPED, PRISM, and PYRAMID. Let A_1 be the AREA of the lower base, A_2 the AREA of the upper base, M the AREA of the midsection, and h the ALTITUDE. Then

$$V = \tfrac{1}{6}h(A_1 + 4M + A_2).$$

see also GENERAL PRISMATOID, PRISMOID

References

Beyer, W. H. *CRC Standard Mathematical Tables, 28th ed.* Boca Raton, FL: CRC Press, pp. 128 and 132, 1987.

Prismoid

A PRISMATOID having planar sides and the same number of vertices in both of its parallel planes. The faces of a prismoid are therefore either TRAPEZOIDS or PARALLELOGRAMS. Ball and Coxeter (1987) use the term to describe an ANTIPRISM.

see also ANTIPRISM, PRISM, PRISMATOID

References

Ball, W. W. R. and Coxeter, H. S. M. *Mathematical Recreations and Essays, 13th ed.* New York: Dover, p. 130, 1987.

Prisoner's Dilemma

A problem in GAME THEORY first discussed by A. Tucker. Suppose each of two prisoners A and B, who are not allowed to communicate with each other, is offered to be set free if he implicates the other. If neither implicates the other, both will receive the usual sentence. However, if the prisoners implicate each other, then both are presumed guilty and granted harsh sentences.

A DILEMMA arises in deciding the best course of action in the absence of knowledge of the other prisoner's decision. Each prisoner's best strategy would appear to be to turn the other in (since if A makes the worst-case assumption that B will turn him in, then B will walk free and A will be stuck in jail if he remains silent). However, if the prisoners turn each other in, they obtain the worst possible outcome for both.

see also DILEMMA, TIT-FOR-TAT

References

Axelrod, R. *The Evolution of Cooperation* New York: Basic-Books, 1985.

Goetz, P. "Phil's Good Enough Complexity Dictionary." http://www.cs.buffalo.edu/~goetz/dict.html.

Prizes

see MATHEMATICS PRIZES

Probability

Probability is the branch of mathematics which studies the possible outcomes of given events together with their relative likelihoods and distributions. In common usage, the word "probability" is used to mean the chance that a particular event (or set of events) will occur expressed on a linear scale from 0 (impossibility) to 1 (certainty), also expressed as a PERCENTAGE between 0 and 100%. The analysis of events governed by probability is called STATISTICS.

There are several competing interpretations of the actual "meaning" of probabilities. Frequentists view probability simply as a measure of the frequency of outcomes (the more conventional interpretation), while BAYESIANS treat probability more subjectively as a statistical procedure which endeavors to estimate parameters of an underlying distribution based on the observed distribution.

A properly normalized function which assigns a probability "density" to each possible outcome within some interval is called a PROBABILITY FUNCTION, and its cumulative value (integral for a continuous distribution or sum for a discrete distribution) is called a DISTRIBUTION FUNCTION.

Probabilities are defined to obey certain assumptions, called the PROBABILITY AXIOMS. Let a SAMPLE SPACE contain the UNION ($\cup$) of all possible events E_i, so

$$S \equiv \left(\bigcup_{i=1}^{N} E_i\right), \tag{1}$$

and let E and F denote subsets of S. Further, let $F' =$ not-F be the complement of F, so that

$$F \cup F' = S. \tag{2}$$

Then the set E can be written as

$$E = E \cap S = E \cap (F \cup F') = (E \cap F) \cup (E \cap F'), \tag{3}$$

where $\cap$ denotes the intersection. Then

$$\begin{aligned} P(E) &= P(E \cap F) + P(E \cap F') - P[(E \cap F) \cap (E \cap F')] \\ &= P(E \cap F) + P(E \cap F') - P[(F \cap F') \cap (E \cap E)] \\ &= P(E \cap F) + P(E \cap F') - P(\varnothing \cap E) \\ &= P(E \cap F) + P(E \cap F') - P(\varnothing) \\ &= P(E \cap F) + P(E \cap F'), \end{aligned} \tag{4}$$

where $\varnothing$ is the EMPTY SET.

Let $P(E|F)$ denote the CONDITIONAL PROBABILITY of E given that F has already occurred, then

$$P(E) = P(E|F)P(F) + P(E|F')P(F') \quad (5)$$
$$= P(E|F)P(F) + P(E|F')[1 - P(F)] \quad (6)$$
$$P(A \cap B) = P(A)P(B|A) \quad (7)$$
$$= P(B)P(A|B) \quad (8)$$
$$P(A' \cap B) = P(A')P(B|A') \quad (9)$$
$$P(E|F) = \frac{P(E \cap F)}{P(F)} \quad (10)$$

A very important result states that

$$P(E \cup F) = P(E) + P(F) - P(E \cap F), \quad (11)$$

which can be generalized to

$$P\left(\bigcup_{i=1}^{n} A_i\right) = \sum_i P(A_i) - \sum_{ij}{}' P(A_i \cup A_j) + \sum_{ijk}{}'' P(A_i \cap A_j \cap A_k) - \ldots + (-1)^{n-1} P\left(\bigcap_{i=1}^{n} A_i\right). \quad (12)$$

see also BAYES' FORMULA, CONDITIONAL PROBABILITY, DISTRIBUTION, DISTRIBUTION FUNCTION, LIKELIHOOD, PROBABILITY AXIOMS, PROBABILITY FUNCTION, PROBABILITY INEQUALITY, STATISTICS

Probability Axioms

Given an event E in a SAMPLE SPACE S which is either finite with N elements or countably infinite with $N = \infty$ elements, then we can write

$$S \equiv \left(\bigcup_{i=1}^{N} E_i\right),$$

and a quantity $P(E_i)$, called the PROBABILITY of event E_i, is defined such that

1. $0 \leq P(E_i) < 1$.
2. $P(S) = 1$.
3. Additivity: $P(E_1 \cup E_2) = P(E_1) + P(E_2)$, where E_1 and E_2 are mutually exclusive.
4. Countable additivity: $P\left(\cup_{i=1}^{n} E_i\right) = \sum_{i=1}^{n} P(E_i)$ for $n = 1, 2, \ldots, N$ where $E_1, E_2, \ldots$ are mutually exclusive (i.e., $E_1 \cap E_2 = \varnothing$).

see also SAMPLE SPACE, UNION

Probability Density Function

see PROBABILITY FUNCTION

Probability Distribution Function

see PROBABILITY FUNCTION

Probability Function

The probability density function $P(x)$ (also called the PROBABILITY DENSITY FUNCTION) of a continuous distribution is defined as the derivative of the (cumulative) DISTRIBUTION FUNCTION $D(x)$,

$$D'(x) = [P(x)]_{-\infty}^{x} = P(x) - P(-\infty) = P(x), \quad (1)$$

so

$$D(x) = P(X \leq x) \equiv \int_{-\infty}^{x} P(y)\,dy. \quad (2)$$

A probability density function satisfies

$$P(x \in B) = \int_B P(x)\,dx \quad (3)$$

and is constrained by the normalization condition,

$$P(-\infty < x < \infty) = \int_{-\infty}^{\infty} P(x)\,dx \equiv 1. \quad (4)$$

Special cases are

$$P(a \leq x \leq b) = \int_a^b P(x)\,dx \quad (5)$$
$$P(a \leq x \leq a + da) = \int_a^{a+da} P(x)\,dx \approx P(a)\,da \quad (6)$$
$$P(x = a) = \int_a^a P(x)\,dx = 0. \quad (7)$$

If $u = u(x, y)$ and $v = v(x, y)$, then

$$P_{u,v}(u, v) = P_{x,y}(x, y) \left|\frac{\partial(x, y)}{\partial(u, v)}\right|. \quad (8)$$

Given the MOMENTS of a distribution (μ, σ, and the GAMMA STATISTICS γ_r), the asymptotic probability function is given by

$$\begin{aligned}P(x) = Z(x) &- [\tfrac{1}{6}\gamma_1 Z^{(3)}(x)] + [\tfrac{1}{24}\gamma_2 Z^{(4)}(x) + \tfrac{1}{72}{\gamma_1}^2 Z^{(6)}(x)] \\ &- [\tfrac{1}{120}\gamma_3 Z^{(5)}(x) + \tfrac{1}{144}\gamma_1\gamma_2 Z^{(7)}(x) + \tfrac{1}{1296}{\gamma_1}^3 Z^{(9)}(x)] \\ &+ [\tfrac{1}{720}\gamma_4 Z^{(6)}(x) + (\tfrac{1}{1152}{\gamma_2}^2 + \tfrac{1}{720}\gamma_1\gamma_3) Z^{(8)}(x) \\ &+ \tfrac{1}{1728}{\gamma_1}^2\gamma_2 Z^{(10)}(x) + \tfrac{1}{31104}{\gamma_1}^4 Z^{(12)}(x)] + \ldots,\end{aligned} \quad (9)$$

where

$$Z(x) = \frac{1}{\sigma\sqrt{2\pi}} e^{-(x-\mu)^2/2\sigma^2} \quad (10)$$

is the NORMAL DISTRIBUTION, and

$$\gamma_r = \frac{\kappa_r}{\sigma^{r+2}} \quad (11)$$

for $r \geq 1$ (with κ_r CUMULANTS and σ the STANDARD DEVIATION; Abramowitz and Stegun 1972, p. 935).

see also CONTINUOUS DISTRIBUTION, CORNISH-FISHER ASYMPTOTIC EXPANSION, DISCRETE DISTRIBUTION, DISTRIBUTION FUNCTION, JOINT DISTRIBUTION FUNCTION

References

Abramowitz, M. and Stegun, C. A. (Eds.). "Probability Functions." Ch. 26 in *Handbook of Mathematical Functions with Formulas, Graphs, and Mathematical Tables, 9th printing.* New York: Dover, pp. 925–964, 1972.

Probability Inequality

If $B \supset A$ (B is a superset of A), then $P(A) \leq P(B)$.

Probability Integral

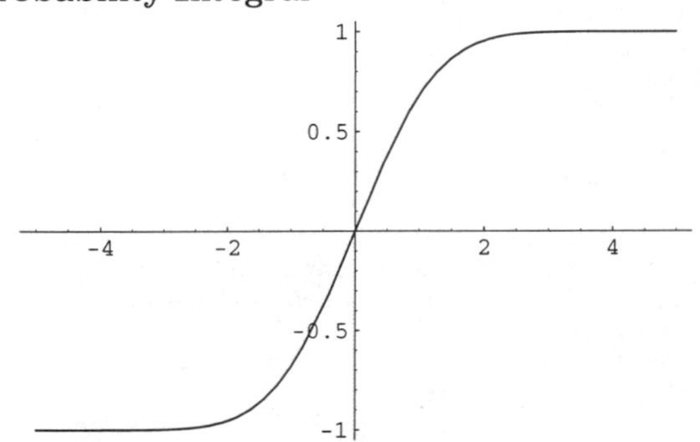

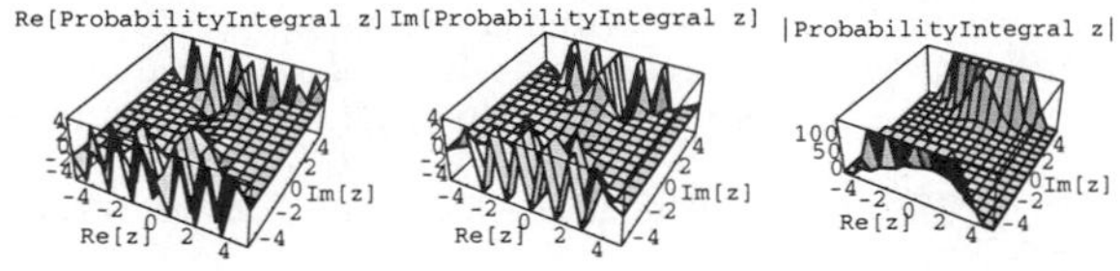

$$\alpha(x) \equiv \frac{1}{\sqrt{2\pi}} \int_{-x}^{x} e^{-t^2/2}\,dt \tag{1}$$

$$= \sqrt{\frac{2}{\pi}} \int_0^x e^{-t^2/2}\,dt \tag{2}$$

$$= 2\Phi(x) \tag{3}$$

$$= \operatorname{erf}\left(\frac{x}{\sqrt{2}}\right), \tag{4}$$

where $\Phi(x)$ is the NORMAL DISTRIBUTION FUNCTION and ERF is the error function.

see also ERF, NORMAL DISTRIBUTION FUNCTION

Probability Measure

Consider a PROBABILITY SPACE $(S, \mathbb{S}, P)$ where $(S, \mathbb{S})$ is a MEASURABLE SPACE and P is a MEASURE on $\mathbb{S}$ with $P(S) = 1$. Then the MEASURE P is said to be a probability measure. Equivalently, P is said to be normalized.

see also MEASURABLE SPACE, MEASURE, PROBABILITY, PROBABILITY SPACE, STATE SPACE

Probability Space

A triple $(S, \mathbb{S}, P)$, where $(S, \mathbb{S})$ is a measurable space and P is a MEASURE on $\mathbb{S}$ with $P(S) = 1$.

see also MEASURABLE SPACE, MEASURE, PROBABILITY, PROBABILITY MEASURE, RANDOM VARIABLE, STATE SPACE

Probable Error

The first QUARTILE of a standard NORMAL DISTRIBUTION occurs when

$$\int_0^t \Phi(z)\,dz = \tfrac{1}{4}.$$

The solution is $t = 0.6745\ldots$. The value of t giving $1/4$ is known as the probable error of a NORMALLY DISTRIBUTED variate. However, the number δ corresponding to the 50% CONFIDENCE INTERVAL,

$$P(\delta) \equiv 1 - 2\int_0^{|\delta|} \phi(t)\,dt = \tfrac{1}{2},$$

is sometimes also called the probable error.

see also SIGNIFICANCE

Probable Prime

A number satisfying FERMAT'S LITTLE THEOREM (or some other primality test) for some nontrivial base. A probable prime which is shown to be COMPOSITE is called a PSEUDOPRIME (otherwise, of course, it is a PRIME).

see also PRIME NUMBER, PSEUDOPRIME

Problem

An exercise whose solution is desired.

see also ALHAZEN'S BILLIARD PROBLEM, ALHAZEN'S PROBLEM, ANDRÉ'S PROBLEM, APOLLONIUS' PROBLEM, APOLLONIUS PURSUIT PROBLEM, ARCHIMEDES' CATTLE PROBLEM, ARCHIMEDES' PROBLEM, BALLOT PROBLEM, BASLER PROBLEM, BERTRAND'S PROBLEM, BILLIARD TABLE PROBLEM, BIRTHDAY PROBLEM, BISHOPS PROBLEM, BOLZA PROBLEM, BOOK STACKING PROBLEM, BOUNDARY VALUE PROBLEM, BOVINUM PROBLEMA, BRACHISTOCHRONE PROBLEM, BRAHMAGUPTA'S PROBLEM, BROCARD'S PROBLEM, BUFFON-LAPLACE NEEDLE PROBLEM, BUFFON'S NEEDLE PROBLEM, BURNSIDE PROBLEM, BUSEMANN-PETTY PROBLEM, CANNONBALL PROBLEM, CASTILLON'S PROBLEM, CATALAN'S DIOPHANTINE PROBLEM, CATALAN'S PROBLEM, CATTLE PROBLEM OF ARCHIMEDES, CAUCHY PROBLEM, CHECKER-JUMPING PROBLEM, CLOSED CURVE PROBLEM, COIN PROBLEM, COLLATZ PROBLEM, CONDOM PROBLEM, CONGRUUM PROBLEM, CONSTANT PROBLEM, COUPON COLLECTOR'S PROBLEM, CROSSED LADDERS PROBLEM, CUBE DOVETAILING PROBLEM, DECISION PROBLEM, DEDEKIND'S PROBLEM, DELIAN PROBLEM, DE

MERE'S PROBLEM, DIAGONALS PROBLEM, DIDO'S PROBLEM, DILEMMA, DINITZ PROBLEM, DIRICHLET DIVISOR PROBLEM, DISK COVERING PROBLEM, EQUICHORDAL PROBLEM, EXTENSION PROBLEM, FAGNANO'S PROBLEM, FEJÉS TÓTH'S PROBLEM, FERMAT'S PROBLEM, FERMAT'S SIGMA PROBLEM, FISHER-BEHRENS PROBLEM, FIVE DISKS PROBLEM, FOUR COINS PROBLEM, FOUR TRAVELERS PROBLEM, FUSS'S PROBLEM, GAUSS'S CIRCLE PROBLEM, GAUSS'S CLASS NUMBER PROBLEM, GLOVE PROBLEM, GUTHRIE'S PROBLEM, HABERDASHER'S PROBLEM, HADWIGER PROBLEM, HALTING PROBLEM, HANSEN'S PROBLEM, HEESCH'S PROBLEM, HEILBRONN TRIANGLE PROBLEM, HILBERT'S PROBLEMS, ILLUMINATION PROBLEM, INDETERMINATE PROBLEMS, INITIAL VALUE PROBLEM, INTERNAL BISECTORS PROBLEM, ISOPERIMETRIC PROBLEM, ISOVOLUME PROBLEM, JEEP PROBLEM, JOSEPHUS PROBLEM, KAKEYA NEEDLE PROBLEM, KAKUTANI'S PROBLEM, KATONA'S PROBLEM, KEPLER PROBLEM, KINGS PROBLEM, KIRKMAN'S SCHOOLGIRL PROBLEM, KISSING CIRCLES PROBLEM, KNAPSACK PROBLEM, KNOT PROBLEM, KÖNIGSBERG BRIDGE PROBLEM, KURATOWSKI'S CLOSURE-COMPONENT PROBLEM, LAM'S PROBLEM, LANGFORD'S PROBLEM, LEBESGUE MEASURABILITY PROBLEM, LEBESGUE MINIMAL PROBLEM, LEHMER'S PROBLEM, LEMOINE'S PROBLEM, LIFTING PROBLEM, LUCAS' MARRIED COUPLES PROBLEM, MALFATTI'S RIGHT TRIANGLE PROBLEM, MALFATTI'S TANGENT TRIANGLE PROBLEM, MARRIED COUPLES PROBLEM, MATCH PROBLEM, MAXIMUM CLIQUE PROBLEM, MÉNAGE PROBLEM, METRIC EQUIVALENCE PROBLEM, MICE PROBLEM, MIKUSIŃSKI'S PROBLEM, MÖBIUS PROBLEM, MONEY-CHANGING PROBLEM, MONKEY AND COCONUT PROBLEM, MONTY HALL PROBLEM, MORTALITY PROBLEM, MOSER'S CIRCLE PROBLEM, NAPOLEON'S PROBLEM, NAVIGATION PROBLEM, NEAREST NEIGHBOR PROBLEM, NP-COMPLETE PROBLEM, NP-PROBLEM, ORCHARD-PLANTING PROBLEM, ORCHARD VISIBILITY PROBLEM, P-PROBLEM, PARTY PROBLEM, PIANO MOVER'S PROBLEM, PLANAR BUBBLE PROBLEM, PLATEAU'S PROBLEM, POINTS PROBLEM, POSTAGE STAMP PROBLEM, POTHENOT PROBLEM, PROUHET'S PROBLEM, QUEENS PROBLEM, RAILROAD TRACK PROBLEM, RIEMANN'S MODULI PROBLEM, SATISFIABILITY PROBLEM, SCHOOLGIRL PROBLEM, SCHUR'S PROBLEM, SCHWARZ'S TRIANGLE PROBLEM, SHARING PROBLEM, SHEPHARD'S PROBLEM, SINCLAIR'S SOAP FILM PROBLEM, SMALL WORLD PROBLEM, SNELLIUS-POTHENOT PROBLEM, STEENROD'S REALIZATION PROBLEM, STEINER'S PROBLEM, STEINER'S SEGMENT PROBLEM, SURVEYING PROBLEMS, SYLVESTER'S FOUR-POINT PROBLEM, SYLVESTER'S LINE PROBLEM, SYLVESTER'S TRIANGLE PROBLEM, SYRACUSE PROBLEM, SYZYGIES PROBLEM, TARRY-ESCOTT PROBLEM, TAUTOCHRONE PROBLEM, THOMSON PROBLEM, THREE JUG PROBLEM, TRAVELING SALESMAN PROBLEM, TRAWLER PROBLEM, ULAM'S PROBLEM, UTILITY PROBLEM, VIBRATION PROBLEM, WALLIS'S PROBLEM, WARING'S PROBLEM

References

Artino, R. A.; Gaglione, A. M.; and Shell, N. *The Contest Problem Book IV: Annual High School Mathematics Examinations 1973–1982.* Washington, DC: Math. Assoc. Amer., 1982.

Alexanderson, G. L.; Klosinski, L.; and Larson, L. *The William Lowell Putnam Mathematical Competition, Problems and Solutions: 1965–1984.* Washington, DC: Math. Assoc. Amer., 1986.

Barbeau, E. J.; Moser, W. O.; and Lamkin, M. S. *Five Hundred Mathematical Challenges.* Washington, DC: Math. Assoc. Amer., 1995.

Brown, K. S. "Most Wanted List of Elementary Unsolved Problems." `http://www.seanet.com/~ksbrown/mwlist.htm`.

Chung, F. and Graham, R. *Erdős on Graphs: His Legacy of Unsolved Problems.* New York: A. K. Peters, 1998.

Cover, T. M. and Gopinath, B. (Eds.). *Open Problems in Communication and Computation.* New York: Springer-Verlag, 1987.

Dörrie, H. *100 Great Problems of Elementary Mathematics: Their History and Solutions.* New York: Dover, 1965.

Dudeney, H. E. *Amusements in Mathematics.* New York: Dover, 1917.

Dudeney, H. E. *The Canterbury Puzzles and Other Curious Problems, 7th ed.* London: Thomas Nelson and Sons, 1949.

Dudeney, H. E. *536 Puzzles & Curious Problems.* New York: Scribner, 1967.

Eppstein, D. "Open Problems." `http://www.ics.uci.edu/~eppstein/junkyard/open.html`.

Erdős, P. "Some Combinatorial Problems in Geometry." In *Geometry and Differential Geometry* (Ed. R. Artzy and I. Vaisman). New York: Springer-Verlag, pp. 46–53, 1980.

Fenchel, W. (Ed.). "Problems." In *Proc. Colloquium on Convexity, 1965.* Københavns Univ. Mat. Inst., pp. 308–325, 1967.

Finch, S. "Unsolved Mathematical Problems." `http://www.mathsoft.com/asolve/`.

Gleason, A. M.; Greenwood, R. E.; and Kelly, L. M. *The William Lowell Putnam Mathematical Competition, Problems and Solutions: 1938–1964.* Washington, DC: Math. Assoc. Amer., 1980.

Graham, L. A. *Ingenious Mathematical Problems and Methods.* New York: Dover, 1959.

Graham, L. A. *The Surprise Attack in Mathematical Problems.* New York: Dover, 1968.

Greitzer, S. L. *International Mathematical Olympiads, 1959–1977.* Providence, RI: Amer. Math. Soc., 1978.

Gruber, P. M. and Schneider, R. "Problems in Geometric Convexity." In *Contributions to Geometry* (Ed. J. Tölke and J. M. Wills.) Boston, MA: Birkhäuser, pp. 255–278, 1979.

Guy, R. K. (Ed.). "Problems." In *The Geometry of Metric and Linear Spaces.* New York: Springer-Verlag, pp. 233–244, 1974.

Halmos, P. R. *Problems for Mathematicians Young and Old.* Washington, DC: Math. Assoc. Amer., 1991.

Honsberger, R. *Mathematical Gems I.* Washington, DC: Math. Assoc. Amer., 1973.

Honsberger, R. *Mathematical Gems II.* Washington, DC: Math. Assoc. Amer., 1976.

Honsberger, R. *Mathematical Morsels.* Washington, DC: Math. Assoc. Amer., 1979.

Honsberger, R. *Mathematical Gems III.* Washington, DC: Math. Assoc. Amer., 1985.

Honsberger, R. *More Mathematical Morsels.* Washington, DC: Math. Assoc. Amer., 1991.

Honsberger, R. *From Erdős to Kiev.* Washington, DC: Math. Assoc. Amer., 1995.
Honsberger, R. *In Pólya's Footsteps: Miscellaneous Problems and Essays.* Washington, DC: Math. Assoc. Amer., 1997.
Honsberger, R. (Ed.). *Mathematical Plums.* Washington, DC: Math. Assoc. Amer., 1979.
Kimberling, C. "Unsolved Problems and Rewards." http://www.evansville.edu/~ck6/integer/unsolved.html.
Klamkin, M. S. *International Mathematical Olympiads, 1978–1985 and Forty Supplementary Problems.* Washington, DC: Math. Assoc. Amer., 1986.
Klamkin, M. S. *U.S.A. Mathematical Olympiads, 1972–1986.* Washington, DC: Math. Assoc. Amer., 1988.
Kordemsky, B. A. *The Moscow Puzzles: 359 Mathematical Recreations.* New York: Dover, 1992.
Kurschak, J. and Hajos, G. *Hungarian Problem Book, Based on the Eőtvős Competitions, Vol. 1: 1894–1905.* New York: Random House, 1963.
Kurschak, J. and Hajos, G. *Hungarian Problem Book, Based on the Eőtvős Competitions, Vol. 2: 1906–1928.* New York: Random House, 1963.
Larson, L. C. *Problem-Solving Through Problems.* New York: Springer-Verlag, 1983.
Mott-Smith, G. *Mathematical Puzzles for Beginners and Enthusiasts.* New York: Dover, 1954.
Ogilvy, C. S. *Tomorrow's Math: Unsolved Problems for the Amateur.* New York: Oxford University Press, 1962.
Ogilvy, C. S. "Some Unsolved Problems of Modern Geometry." Ch. 11 in *Excursions in Geometry.* New York: Dover, pp. 143–153, 1990.
Posamentier, A. S. and Salkind, C. T. *Challenging Problems in Algebra.* New York: Dover, 1997.
Posamentier, A. S. and Salkind, C. T. *Challenging Problems in Geometry.* New York: Dover, 1997.
Rabinowitz, S. (Ed.). *Index to Mathematical Problems 1980–1984.* Westford, MA: MathPro Press, 1992.
Salkind, C. T. *The Contest Problem Book I: Problems from the Annual High School Contests 1950–1960.* New York: Random House, 1961.
Salkind, C. T. *The Contest Problem Book II: Problems from the Annual High School Contests 1961–1965.* Washington, DC: Math. Assoc. Amer., 1966.
Salkind, C. T. and Earl, J. M. *The Contest Problem Book III: Annual High School Contests 1966–1972.* Washington, DC: Math. Assoc. Amer., 1973.
Shanks, D. *Solved and Unsolved Problems in Number Theory, 4th ed.* New York: Chelsea, 1993.
Shkliarskii, D. O.; Chentzov, N. N.; and Yaglom, I. M. *The U.S.S.R. Olympiad Problem Book: Selected Problems and Theorems of Elementary Mathematics.* New York: Dover, 1993.
Sierpiński, W. *A Selection of Problems in the Theory of Numbers.* New York: Pergamon Press, 1964. Sierpiński, W. *Problems in Elementary Number Theory.* New York: Elsevier, 1980.
Smarandache, F. *Only Problems, Not Solutions!, 4th ed.* Phoenix, AZ: Xiquan, 1993.
Steinhaus, H. *One Hundred Problems in Elementary Mathematics.* New York: Dover, 1979.
Tietze, H. *Famous Problems of Mathematics.* New York: Graylock Press, 1965.
Trigg, C. W. *Mathematical Quickies: 270 Stimulating Problems with Solutions.* New York: Dover, 1985.
Ulam, S. M. *A Collection of Mathematical Problems.* New York: Interscience Publishers, 1960.
van Mill, J. and Reed, G. M. (Eds.). *Open Problems in Topology.* New York: Elsevier, 1990.

Procedure

A specific prescription for carrying out a task or solving a problem. Also called an ALGORITHM, METHOD, or TECHNIQUE

see also BISECTION PROCEDURE, MAEHLY'S PROCEDURE

Proclus' Axiom

If a line intersects one of two parallel lines, it must intersect the other also. This AXIOM is equivalent to the PARALLEL AXIOM.

References

Dunham, W. "Hippocrates' Quadrature of the Lune." Ch. 1 in *Journey Through Genius: The Great Theorems of Mathematics.* New York: Wiley, p. 54, 1990.

Procrustian Stretch

see HYPERBOLIC ROTATION

Product

The term "product" refers to the result of one or more MULTIPLICATIONS. For example, the mathematical statement $a \times b = c$ would be read "a TIMES b EQUALS c," where c is the product.

The product symbol is defined by

$$\prod_{i=1}^{n} f_i \equiv f_1 \cdot f_2 \cdots f_n.$$

Useful product identities include

$$\ln\left(\prod_{i=1}^{\infty} f_i\right) = \sum_{i=1}^{\infty} \ln f_i$$

$$\prod_{i=1}^{\infty} f_i = \exp\left(\sum_{i=1}^{\infty} \ln f_i\right).$$

For $0 \leq a_i < 1$, then the products $\prod_{i=1}^{\infty}(1+a_i)$ and $\prod_{i=1}^{\infty}(1-a_i)$ converge and diverge as $\prod_{i=1}^{\infty} a_i$.

see also CROSS PRODUCT, DOT PRODUCT, INNER PRODUCT, MATRIX PRODUCT, MULTIPLICATION, NONASSOCIATIVE PRODUCT, OUTER PRODUCT, SUM, TENSOR PRODUCT, TIMES, VECTOR TRIPLE PRODUCT

References

Guy, R. K. "Products Taken over Primes." §B87 in *Unsolved Problems in Number Theory, 2nd ed.* New York: Springer-Verlag, pp. 102–103, 1994.

Product Formula

Let α be a NONZERO RATIONAL NUMBER $\alpha = \pm p_1{}^{\alpha_1} p_2{}^{\alpha_2} \cdots p_L{}^{\alpha_L}$, where $p_1, \ldots, p_L$ are distinct PRIMES, $\alpha_l \in \mathbb{Z}$ and $\alpha_l \neq 0$. Then

$$|a| \prod_{p \text{ prime}} |\alpha|_p = p_1{}^{\alpha_1} p_2{}^{\alpha_2} \cdots p_L{}^{\alpha_L} \times p_1{}^{-\alpha_1} p_2{}^{-\alpha_2} \cdots p_L{}^{-\alpha_L} = 1.$$

References

Burger, E. B. and Struppeck, T. "Does $\sum_{n=0}^{\infty} \frac{1}{n!}$ Really Converge? Infinite Series and p-adic Analysis." *Amer. Math. Monthly* **103**, 565–577, 1996.

Product-Moment Coefficient of Correlation

see CORRELATION COEFFICIENT

Product Neighborhood

see TUBULAR NEIGHBORHOOD

Product Rule

The DERIVATIVE identity

$$\frac{d}{dx}[f(x)g(x)] = \lim_{h \to 0} \frac{f(x+h)g(x+h) - f(x)g(x)}{h}$$

$$= \lim_{h \to 0} \left[\frac{f(x+h)g(x+h) - f(x+h)g(x)}{h} + \frac{f(x+h)g(x) - f(x)g(x)}{h} \right]$$

$$= \lim_{h \to 0} \left[f(x+h) \frac{g(x+h) - g(x)}{h} + g(x) \frac{f(x+h) - f(x)}{h} \right] = f(x)g'(x) + g(x)f'(x).$$

see also CHAIN RULE, EXPONENT LAWS, QUOTIENT RULE

References

Abramowitz, M. and Stegun, C. A. (Eds.). *Handbook of Mathematical Functions with Formulas, Graphs, and Mathematical Tables, 9th printing.* New York: Dover, p. 11, 1972.

Product Space

A Cartesian product equipped with a "product topology" is called a product space (or product topological space, or direct product).

References

Iyanaga, S. and Kawada, Y. (Eds.). "Product Spaces." §408L *Encyclopedic Dictionary of Mathematics.* Cambridge, MA: MIT Press, pp. 1281–1282, 1980.

Program

A precise sequence of instructions designed to accomplish a given task. The implementation of an ALGORITHM on a computer using a programming language is an example of a program.

see also ALGORITHM

Projection

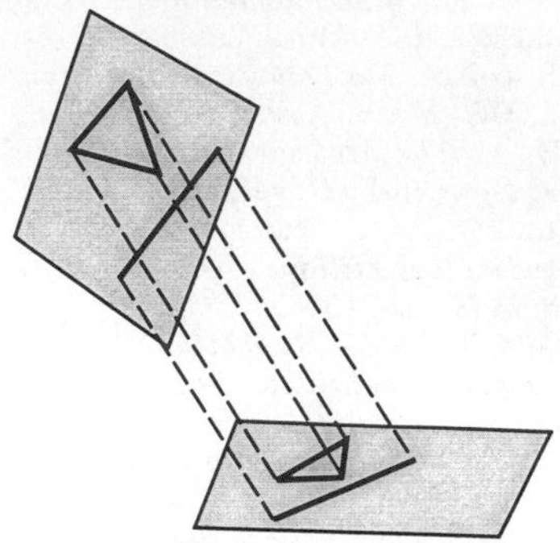

A projection is the transformation of POINTS and LINES in one PLANE onto another PLANE by connecting corresponding points on the two planes with PARALLEL lines. This can be visualized as shining a (point) light source (located at infinity) through a translucent sheet of paper and making an image of whatever is drawn on it on a second sheet of paper. The branch of geometry dealing with the properties and invariants of geometric figures under projection is called PROJECTIVE GEOMETRY.

The projection of a VECTOR $\mathbf{a}$ onto a VECTOR $\mathbf{u}$ is given by

$$\text{proj}_{\mathbf{u}} \mathbf{a} = \frac{\mathbf{a} \cdot \mathbf{u}}{|\mathbf{u}|^2} \mathbf{u},$$

and the length of this projection is

$$|\text{proj}_{\mathbf{u}} \mathbf{a}| = \frac{|\mathbf{a} \cdot \mathbf{u}|}{|\mathbf{u}|}.$$

General projections are considered by Foley and VanDam (1983).

see also MAP PROJECTION, POINT-PLANE DISTANCE, PROJECTIVE GEOMETRY, REFLECTION

References

Casey, J. "Theory of Projections." Ch. 11 in *A Treatise on the Analytical Geometry of the Point, Line, Circle, and Conic Sections, Containing an Account of Its Most Recent Extensions, with Numerous Examples, 2nd ed., rev. enl.* Dublin: Hodges, Figgis, & Co., pp. 349–367, 1893.

Foley, J. D. and VanDam, A. *Fundamentals of Interactive Computer Graphics, 2nd ed.* Reading, MA: Addison-Wesley, 1990.

Projection Operator

$$\tilde{p} \equiv |\phi_i(x)\rangle\langle\phi_i(t)|$$

$$\tilde{p} \sum_j c_j |\phi_j(t)\rangle = c_i |\phi_i(x)\rangle$$

$$\sum_i |\phi_i(x)\rangle\langle\phi_i(x)| = 1.$$

see also BRA, KET

Projective Collineation

A COLLINEATION which transforms every 1-D form projectively. Any COLLINEATION which transforms one range into a projectively related range is a projective collineation. Every PERSPECTIVE COLLINEATION is a projective collineation.

see also COLLINEATION, ELATION, HOMOLOGY (GEOMETRY), PERSPECTIVE COLLINEATION

References

Coxeter, H. S. M. *Introduction to Geometry, 2nd ed.* New York: Wiley, pp. 247–248, 1969.

Projective General Linear Group

The projective general linear group $PGL_n(q)$ is the GROUP obtained from the GENERAL LINEAR GROUP $GL_n(q)$ on factoring the scalar MATRICES contained in that group.

see also GENERAL LINEAR GROUP, PROJECTIVE GENERAL ORTHOGONAL GROUP, PROJECTIVE GENERAL UNITARY GROUP

References

Conway, J. H.; Curtis, R. T.; Norton, S. P.; Parker, R. A.; and Wilson, R. A. "The Groups $GL_n(q)$, $SL_n(q)$, $PGL_n(q)$, and $PSL_n(q) = L_n(q)$." §2.1 in *Atlas of Finite Groups: Maximal Subgroups and Ordinary Characters for Simple Groups.* Oxford, England: Clarendon Press, p. x, 1985.

Projective General Orthogonal Group

The projective general orthogonal group $PGO_n(q)$ is the GROUP obtained from the GENERAL ORTHOGONAL GROUP $GO_n(q)$ on factoring the scalar MATRICES contained in that group.

see also GENERAL ORTHOGONAL GROUP, PROJECTIVE GENERAL LINEAR GROUP, PROJECTIVE GENERAL UNITARY GROUP

References

Conway, J. H.; Curtis, R. T.; Norton, S. P.; Parker, R. A.; and Wilson, R. A. "The Groups $GO_n(q)$, $SO_n(q)$, $PGO_n(q)$, and $PSO_n(q)$, and $O_n(q)$." §2.4 in *Atlas of Finite Groups: Maximal Subgroups and Ordinary Characters for Simple Groups.* Oxford, England: Clarendon Press, pp. xi–xii, 1985.

Projective General Unitary Group

The projective general unitary group $PGU_n(q)$ is the GROUP obtained from the GENERAL UNITARY GROUP $GU_n(q)$ on factoring the scalar MATRICES contained in that group.

see also GENERAL UNITARY GROUP, PROJECTIVE GENERAL LINEAR GROUP, PROJECTIVE GENERAL ORTHOGONAL GROUP, PROJECTIVE GENERAL UNITARY GROUP

References

Conway, J. H.; Curtis, R. T.; Norton, S. P.; Parker, R. A.; and Wilson, R. A. "The Groups $GU_n(q)$, $SU_n(q)$, $PGU_n(q)$, and $PSU_n(q) = U_n(q)$." §2.2 in *Atlas of Finite Groups: Maximal Subgroups and Ordinary Characters for Simple Groups.* Oxford, England: Clarendon Press, p. x, 1985.

Projective Geometry

The branch of geometry dealing with the properties and invariants of geometric figures under PROJECTION. The most amazing result arising in projective geometry is the DUALITY PRINCIPLE, which states that a duality exists between theorems such as PASCAL'S THEOREM and BRIANCHON'S THEOREM which allows one to be instantly transformed into the other. More generally, *all* the propositions in projective geometry occur in dual pairs, which have the property that, starting from either proposition of a pair, the other can be immediately inferred by interchanging the parts played by the words "POINT" and "LINE."

The AXIOMS of projective geometry are:

1. If A and B are distinct points on a PLANE, there is at least one LINE containing both A and B.
2. If A and B are distinct points on a PLANE, there is not more than one LINE containing both A and B.
3. Any two LINES on a PLANE have at least one point of the PLANE in common.
4. There is at least one LINE on a PLANE.
5. Every LINE contains at least three points of the PLANE.
6. All the points of the PLANE do not belong to the same LINE

(Veblin and Young 1910–18, Kasner and Newman 1989).

see also COLLINEATION, DESARGUES' THEOREM, FUNDAMENTAL THEOREM OF PROJECTIVE GEOMETRY, INVOLUTION (LINE), PENCIL, PERSPECTIVITY, PROJECTIVITY, RANGE (LINE SEGMENT), SECTION (PENCIL)

References

Birkhoff, G. and Mac Lane, S. "Projective Geometry." §9.14 in *A Survey of Modern Algebra, 3rd ed.* New York: Macmillan, pp. 275–279, 1965.

Casey, J. "Theory of Projections." Ch. 11 in *A Treatise on the Analytical Geometry of the Point, Line, Circle, and Conic Sections, Containing an Account of Its Most Recent Extensions, with Numerous Examples, 2nd ed., rev. enl.* Dublin: Hodges, Figgis, & Co., pp. 349–367, 1893.

Coxeter, H. S. M. *Projective Geometry, 2nd ed.* New York: Springer-Verlag, 1987.

Kadison, L. and Kromann, M. T. *Projective Geometry and Modern Algebra.* Boston, MA: Birkhäuser, 1996.

Kasner, E. and Newman, J. R. *Mathematics and the Imagination.* Redmond, WA: Microsoft Press, pp. 150–151, 1989.

Ogilvy, C. S. "Projective Geometry." Ch. 7 in *Excursions in Geometry.* New York: Dover, pp. 86–110, 1990.

Pappas, T. "Art & Projective Geometry." *The Joy of Mathematics.* San Carlos, CA: Wide World Publ./Tetra, pp. 66–67, 1989.

Pedoe, D. and Sneddon, I. A. *An Introduction to Projective Geometry.* New York: Pergamon, 1963.
Seidenberg, A. *Lectures in Projective Geometry.* Princeton, NJ: Van Nostrand, 1962.
Struik, D. *Lectures on Projected Geometry.* Reading, MA: Addison-Wesley, 1998.
Veblen, O. and Young, J. W. *Projective Geometry, 2 vols.* Boston, MA: Ginn, 1910–18.
Whitehead, A. N. *The Axioms of Projective Geometry.* New York: Hafner, 1960.

Projective Plane

A projective plane is derived from a usual PLANE by addition of a LINE AT INFINITY. A projective plane of order n is a set of n^2+n+1 POINTS with the properties that:

1. Any two POINTS determine a LINE,
2. Any two LINES determine a POINT,
3. Every POINT has $n+1$ LINES on it, and
4. Every LINE contains $n+1$ POINTS.

(Note that some of these properties are redundant.) A projective plane is therefore a SYMMETRIC $(n^2+n+1, n+1, 1)$ BLOCK DESIGN. An AFFINE PLANE of order n exists IFF a projective plane of order n exists.

A finite projective plane exists when the order n is a POWER of a PRIME, i.e., $n = p^a$ for $a \geq 1$. It is conjectured that these are the *only* possible projective planes, but proving this remains one of the most important unsolved problems in COMBINATORICS. The first few orders which are not of this form are 6, 10, 12, 14, 15, ...

It has been proven analytically that there are no projective planes of order 6. By answering LAM'S PROBLEM in the negative using massive computer calculations on top of some mathematics, it has been proved that there are no finite projective planes of order 10 (Lam 1991). The status of the order 12 projective plane remains open. The remarkable BRUCK-RYSER-CHOWLA THEOREM says that if a projective plane of order n exists, and $n = 1$ or 2 (mod 4), then n is the sum of two SQUARES. This rules out $n = 6$.

The projective plane of order 2, also known as the FANO PLANE, is denoted PG(2, 2). It has INCIDENCE MATRIX

$$\begin{bmatrix} 1 & 1 & 1 & 0 & 0 & 0 & 0 \\ 1 & 0 & 0 & 1 & 1 & 0 & 0 \\ 1 & 0 & 0 & 0 & 0 & 1 & 1 \\ 0 & 1 & 0 & 1 & 0 & 1 & 0 \\ 0 & 1 & 0 & 0 & 1 & 0 & 1 \\ 0 & 0 & 1 & 1 & 0 & 0 & 1 \\ 0 & 0 & 1 & 0 & 1 & 1 & 0 \end{bmatrix}.$$

Every row and column contains 3 1s, and any pair of rows/columns has a single 1 in common.

The projective plane has EULER CHARACTERISTIC 1, and the HEAWOOD CONJECTURE therefore shows that any set of regions on it can be colored using six colors only (Saaty 1986).

see also AFFINE PLANE, BRUCK-RYSER-CHOWLA THEOREM, FANO PLANE, LAM'S PROBLEM, MAP COLORING, MOUFANG PLANE, PROJECTIVE PLANE PK^2, REAL PROJECTIVE PLANE

References

Ball, W. W. R. and Coxeter, H. S. M. *Mathematical Recreations and Essays, 13th ed.* New York: Dover, pp. 281–287, 1987.
Lam, C. W. H. "The Search for a Finite Projective Plane of Order 10." *Amer. Math. Monthly* **98**, 305–318, 1991.
Lindner, C. C. and Rodger, C. A. *Design Theory.* Boca Raton, FL: CRC Press, 1997.
Pinkall, U. "Models of the Real Projective Plane." Ch. 6 in *Mathematical Models from the Collections of Universities and Museums* (Ed. G. Fischer). Braunschweig, Germany: Vieweg, pp. 63–67, 1986.
Saaty, T. L. and Kainen, P. C. *The Four-Color Problem: Assaults and Conquest.* New York: Dover, p. 45, 1986.

Projective Plane PK^2

The 2-D SPACE consisting of the set of TRIPLES

$$\{(a,b,c) : a,b,c \in K, \text{ not all zero}\},$$

where triples which are SCALAR multiples of each other are identified.

Projective Space

A SPACE which is invariant under the GROUP G of all general LINEAR homogeneous transformation in the SPACE concerned, but not under all the transformations of any GROUP containing G as a SUBGROUP.

A projective space is the space of 1-D VECTOR SUBSPACES of a given VECTOR SPACE. For REAL VECTOR SPACES, the NOTATION $\mathbb{RP}^n$ or $\mathbb{P}^n$ denotes the REAL projective space of dimension n (i.e., the SPACE of 1-D VECTOR SUBSPACES of $\mathbb{R}^{n+1}$) and $\mathbb{CP}^n$ denotes the COMPLEX projective space of COMPLEX dimension n (i.e., the space of 1-D COMPLEX VECTOR SUBSPACES of $\mathbb{C}^{n+1}$). $\mathbb{P}^n$ can also be viewed as the set consisting of $\mathbb{R}^n$ together with its POINTS AT INFINITY.

Projective Special Linear Group

The projective special linear group $PSL_n(q)$ is the GROUP obtained from the SPECIAL LINEAR GROUP $SL_n(q)$ on factoring by the SCALAR MATRICES contained in that GROUP. It is SIMPLE for $n \geq 2$ except for

$$PSL_2(2) = S_3$$
$$PSL_2(3) = A_4,$$

and is therefore also denoted $L_n(Q)$.

see also PROJECTIVE SPECIAL ORTHOGONAL GROUP, PROJECTIVE SPECIAL UNITARY GROUP, SPECIAL LINEAR GROUP

References

Conway, J. H.; Curtis, R. T.; Norton, S. P.; Parker, R. A.; and Wilson, R. A. "The Groups $GL_n(q)$, $SL_n(q)$, $PGL_n(q)$, and $PSL_n(q) = L_n(q)$." §2.1 in *Atlas of Finite Groups: Maximal Subgroups and Ordinary Characters for Simple Groups*. Oxford, England: Clarendon Press, p. x, 1985.

Projective Special Orthogonal Group

The projective special orthogonal group $PSO_n(q)$ is the GROUP obtained from the SPECIAL ORTHOGONAL GROUP $SO_n(q)$ on factoring by the SCALAR MATRICES contained in that GROUP. In general, this GROUP is not SIMPLE.

see also PROJECTIVE SPECIAL LINEAR GROUP, PROJECTIVE SPECIAL UNITARY GROUP, SPECIAL ORTHOGONAL GROUP

References

Conway, J. H.; Curtis, R. T.; Norton, S. P.; Parker, R. A.; and Wilson, R. A. "The Groups $GO_n(q)$, $SO_n(q)$, $PGO_n(q)$, and $PSO_n(q)$, and $O_n(q)$." §2.4 in *Atlas of Finite Groups: Maximal Subgroups and Ordinary Characters for Simple Groups*. Oxford, England: Clarendon Press, pp. xi–xii, 1985.

Projective Special Unitary Group

The projective special unitary group $PSU_n(q)$ is the GROUP obtained from the SPECIAL UNITARY GROUP $SU_n(q)$ on factoring by the SCALAR MATRICES contained in that GROUP. $PSU_n(q)$ is SIMPLE except for

$$PSU_2(2) = S_3$$
$$PSU_2(3) = A_4$$
$$PSU_3(2) = 3^2 : Q_8,$$

so it is given the simpler name $U_n(q)$, with $U_2(q) = L_2(q)$.

see also PROJECTIVE SPECIAL LINEAR GROUP, PROJECTIVE SPECIAL ORTHOGONAL GROUP, SPECIAL UNITARY GROUP

References

Conway, J. H.; Curtis, R. T.; Norton, S. P.; Parker, R. A.; and Wilson, R. A. "The Groups $GU_n(q)$, $SU_n(q)$, $PGU_n(q)$, and $PSU_n(q) = U_n(q)$." §2.2 in *Atlas of Finite Groups: Maximal Subgroups and Ordinary Characters for Simple Groups*. Oxford, England: Clarendon Press, p. x, 1985.

Projective Symplectic Group

The projective symplectic group $PSp_n(q)$ is the GROUP obtained from the SYMPLECTIC GROUP $Sp_n(q)$ on factoring by the SCALAR MATRICES contained in that GROUP. $PSp_{2m}(q)$ is SIMPLE except for

$$PSp_2(2) = S_3$$
$$PSp_2(3) = A_4$$
$$PSp_4(2) = S_6,$$

so it is given the simpler name $S_{2m}(q)$, with $S_2(q) = L_2(q)$.

References

Conway, J. H.; Curtis, R. T.; Norton, S. P.; Parker, R. A.; and Wilson, R. A. "The Groups $Sp_n(q)$ and $PSp_n(q) = S_n(q)$." §2.3 in *Atlas of Finite Groups: Maximal Subgroups and Ordinary Characters for Simple Groups*. Oxford, England: Clarendon Press, pp. x–xi, 1985.

Projectivity

The product of any number of PERSPECTIVITIES.

see also INVOLUTION (TRANSFORMATION), PERSPECTIVITY

Prolate Cycloid

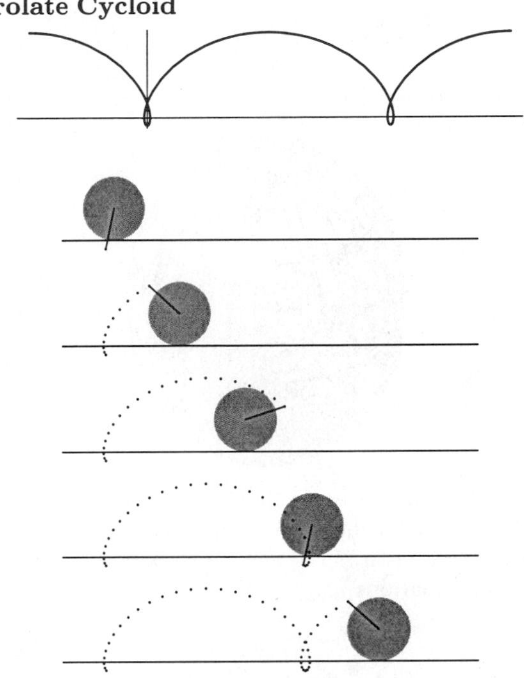

The path traced out by a fixed point at a RADIUS $b > a$, where a is the RADIUS of a rolling CIRCLE, also sometimes called an EXTENDED CYCLOID. The prolate cycloid contains loops, and has parametric equations

$$x = a\phi - b\sin\phi \tag{1}$$
$$y = a - b\cos\phi. \tag{2}$$

The ARC LENGTH from $\phi = 0$ is

$$s = 2(a+b)E(u), \tag{3}$$

where

$$\sin(\tfrac{1}{2}\phi) = \operatorname{sn} u \tag{4}$$
$$k^2 = \frac{4ab}{(a+c)^2}. \tag{5}$$

see also CURTATE CYCLOID, CYCLOID

References

Wagon, S. *Mathematica in Action.* New York: W. H. Freeman, pp. 46–50, 1991.

Prolate Cycloid Evolute

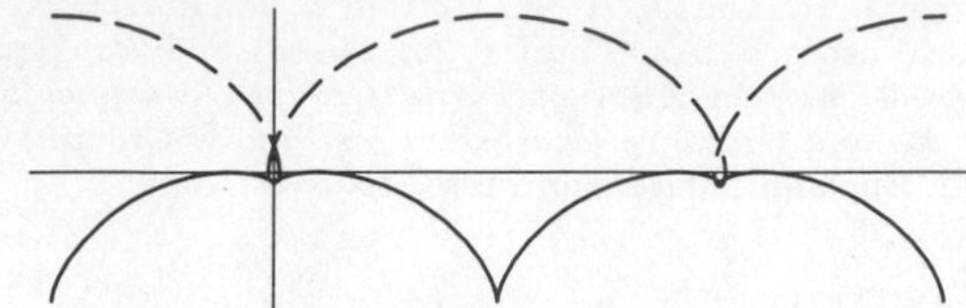

The EVOLUTE of the PROLATE CYCLOID is given by

$$x = \frac{a[-2b\phi + 2a\phi\cos\phi - 2a\sin\phi + b\sin(2\phi)]}{2(a\cos\phi - b)}$$

$$y = \frac{a(a - b\cos\phi)^2}{b(a\cos\phi - b)}.$$

Prolate Spheroid

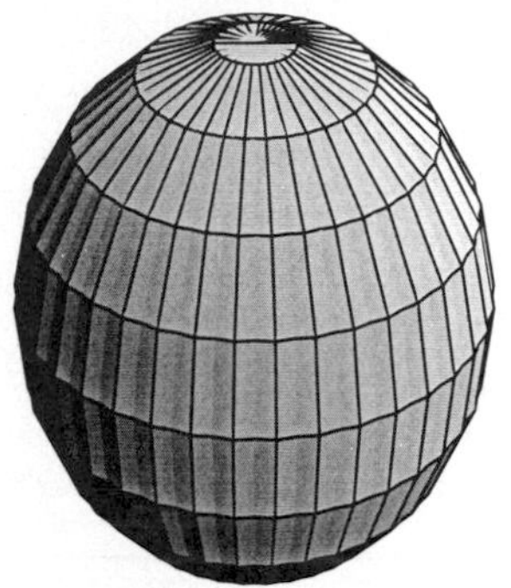

A SPHEROID which is "pointy" instead of "squashed," i.e., one for which the polar radius c is greater than the equatorial radius a, so $c > a$. A prolate spheroid has Cartesian equations

$$\frac{x^2 + y^2}{a^2} + \frac{z^2}{c^2} = 1. \tag{1}$$

The ELLIPTICITY of the prolate spheroid is defined by

$$e \equiv \sqrt{\frac{c^2 - a^2}{c^2}} = \frac{\sqrt{c^2 - a^2}}{c} = \sqrt{1 - \frac{a^2}{c^2}}, \tag{2}$$

so that

$$1 - e^2 = \frac{a^2}{c^2}. \tag{3}$$

Then

$$r = a\left(1 + \frac{e^2}{1 - e^2}\sin^2\delta\right)^{-1/2}. \tag{4}$$

The SURFACE AREA and VOLUME are

$$S = 2\pi a^2 + 2\pi\frac{ac}{e}\sin^{-1} e \tag{5}$$

$$V = \tfrac{4}{3}\pi a^2 c. \tag{6}$$

see also DARWIN-DE SITTER SPHEROID, ELLIPSOID, OBLATE SPHEROID, PROLATE SPHEROIDAL COORDINATES, SPHERE, SPHEROID

References

Beyer, W. H. *CRC Standard Mathematical Tables, 28th ed.* Boca Raton, FL: CRC Press, p. 131, 1987.

Prolate Spheroidal Coordinates

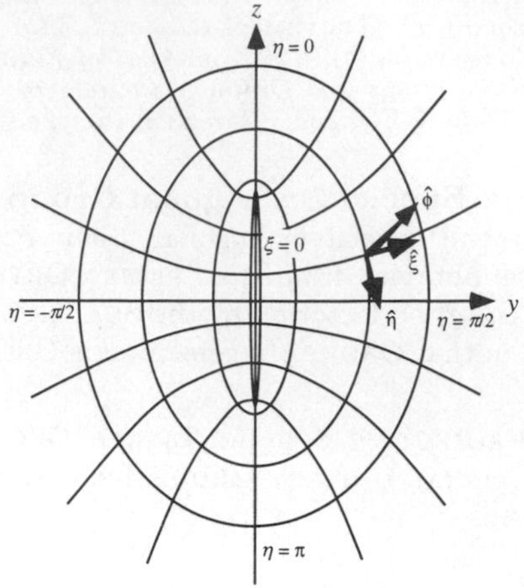

A system of CURVILINEAR COORDINATES in which two sets of coordinate surfaces are obtained by revolving the curves of the ELLIPTIC CYLINDRICAL COORDINATES about the x-AXIS, which is relabeled the z-AXIS. The third set of coordinates consists of planes passing through this axis.

$$x = a\sinh\xi\sin\eta\cos\phi \tag{1}$$

$$y = a\sinh\xi\sin\eta\sin\phi \tag{2}$$

$$z = a\cosh\xi\cos\eta, \tag{3}$$

where $\xi \in [0, \infty)$, $\eta \in [0, \pi)$, and $\phi \in [0, 2\pi)$. Arfken (1970) uses (u, v, φ) instead of (ξ, η, z). The SCALE FACTORS are

$$h_\xi = a\sqrt{\sinh^2\xi + \sin^2\eta} \tag{4}$$

$$h_\eta = a\sqrt{\sinh^2\xi + \sin^2\eta} \tag{5}$$

$$h_\phi = a\sinh\xi\sin\eta. \tag{6}$$

The LAPLACIAN is

$$\nabla^2 f = \frac{1}{a^2(\sinh^2\xi + \sin^2\eta)}\left\{\frac{1}{\sinh\xi}\frac{\partial}{\partial\xi}\left(\sinh\xi\frac{\partial f}{\partial\xi}\right) + \frac{1}{\sin\eta}\frac{\partial}{\partial\eta}\left(\sin\eta\frac{\partial f}{\partial\eta}\right) + \frac{\partial^2 f}{\partial\phi^2}\right\} \tag{7}$$

$$= \frac{1}{a^2(\sin^2\eta + \sinh^2\xi)}\left[(\csc^2\eta + \operatorname{csch}^2\xi)\frac{\partial^2}{\partial\phi^2} + \cot\eta\frac{\partial}{\partial\eta} + \left[\frac{\partial^2}{\partial\eta^2} + \coth\xi\frac{\partial}{\partial\xi} + \frac{\partial^2}{\partial\xi^2}\right]\right]. \tag{8}$$

An alternate form useful for "two-center" problems is defined by

$$\xi_1 = \cosh\xi \tag{9}$$

$$\xi_2 = \cos\eta \tag{10}$$

$$\xi_3 = \phi, \tag{11}$$

where $\xi_1 \in [1, \infty]$, $\xi_2 \in [-1, 1]$, and $\xi_3 \in [0, 2\pi)$ (Abramowitz and Stegun 1972). In these coordinates,

$$z = a\xi_1\xi_2 \tag{12}$$

$$x = a\sqrt{(\xi_1{}^2 - 1)(1 - \xi_2{}^2)} \cos \xi_3 \tag{13}$$

$$y = a\sqrt{(\xi_1{}^2 - 1)(1 - \xi_2{}^2)} \sin \xi_3. \tag{14}$$

In terms of the distances from the two FOCI,

$$\xi_1 = \frac{r_1 + r_2}{2a} \tag{15}$$

$$\xi_2 = \frac{r_1 - r_2}{2a} \tag{16}$$

$$2a = r_{12}. \tag{17}$$

The SCALE FACTORS are

$$h_{\xi_1} = a\sqrt{\frac{\xi_1{}^2 - \xi_2{}^2}{\xi_1{}^2 - 1}} \tag{18}$$

$$h_{\xi_2} = a\sqrt{\frac{\xi_1{}^2 - \xi_2{}^2}{1 - \xi_2{}^2}} \tag{19}$$

$$h_{\xi_3} = a\sqrt{(\xi_1{}^2 - 1)(1 - \xi_2{}^2)}, \tag{20}$$

and the LAPLACIAN is

$$\nabla^2 f = \frac{1}{a^2}\left\{\frac{1}{\xi_1{}^2 - \xi_2{}^2}\frac{\partial}{\partial \xi_1}\left[(\xi_1{}^2 - 1)\frac{\partial f}{\partial \xi_1}\right] + \frac{1}{\xi_1{}^2 - \xi_2{}^2}\frac{\partial}{\partial \xi_2}\left[(1 - \xi_2{}^2)\frac{\partial f}{\partial \xi_2}\right] + \frac{1}{(\xi_1{}^2 - 1)(1 - \xi_2{}^2)}\frac{\partial^2 f}{\partial \xi_3{}^2}\right\}. \tag{21}$$

The HELMHOLTZ DIFFERENTIAL EQUATION is separable in prolate spheroidal coordinates.

see also HELMHOLTZ DIFFERENTIAL EQUATION—PROLATE SPHEROIDAL COORDINATES, LATITUDE, LONGITUDE, OBLATE SPHEROIDAL COORDINATES, SPHERICAL COORDINATES

References

Abramowitz, M. and Stegun, C. A. (Eds.). "Definition of Prolate Spheroidal Coordinates." §21.2 in *Handbook of Mathematical Functions with Formulas, Graphs, and Mathematical Tables, 9th printing.* New York: Dover, p. 752, 1972.

Arfken, G. "Prolate Spheroidal Coordinates (u, v, ϕ)." §2.10 in *Mathematical Methods for Physicists, 2nd ed.* Orlando, FL: Academic Press, pp. 103–107, 1970.

Morse, P. M. and Feshbach, H. *Methods of Theoretical Physics, Part I.* New York: McGraw-Hill, p. 661, 1953.

Prolate Spheroidal Wave Function

The WAVE EQUATION in PROLATE SPHEROIDAL COORDINATES is

$$\nabla^2\Phi + k^2\Phi = \frac{\partial}{\partial \xi_1}\left[(\xi_1{}^2 - 1)\frac{\partial \Phi}{\partial \xi_1}\right] + \frac{\partial}{\partial \xi_2}\left[(1 - \xi_2{}^2)\frac{\partial \Phi}{\partial \xi_2}\right] + \frac{\xi_1{}^2 - \xi_2{}^2}{(\xi_1{}^2 - 1)(1 - x_2{}^2)}\frac{\partial^2 \Phi}{\partial \phi^2} + c^2(\xi_1{}^2 - \xi_2{}^2)\Phi = 0, \tag{1}$$

where

$$c \equiv \tfrac{1}{2}ak. \tag{2}$$

Substitute in a trial solution

$$\Phi = R_{mn}(c, \xi_1)S_{mn}(c, \xi_2)\,{\cos \atop \sin}(m\phi) \tag{3}$$

$$\frac{d}{d\xi_1}\left[(\xi_1{}^2 - 1)\frac{d}{d\xi_1}R_{mn}(c, \xi_1)\right] - \left(\lambda_{mn} - c^2\xi_1{}^2 + \frac{m^2}{\xi_1{}^2 - 1}\right)R_{mn}(c, \xi_1) = 0. \tag{4}$$

The radial differential equation is

$$\frac{d}{d\xi_2}\left[(\xi_2{}^2 - 1)\frac{d}{d\xi_2}S_{mn}(c, \xi_2)\right] - \left(\lambda_{mn} - c^2\xi_2{}^2 + \frac{m^2}{\xi_2{}^2 - 1}\right)R_{mn}(c, \xi_2) = 0, \tag{5}$$

and the angular differential equation is

$$\frac{d}{d\xi_2}\left[(1 - \xi_2{}^2)\frac{d}{d\xi_2}S_{mn}(c, \xi_2)\right] - \left(\lambda_{mn} - c^2\xi_2{}^2 + \frac{m^2}{1 - \xi_2{}^2}\right)R_{mn}(c, \xi_2) = 0. \tag{6}$$

Note that these are identical (except for a sign change). The prolate angular function of the first kind is given by

$$S_{mn}^{(1)} = \begin{cases} \sum_{r=1,3,\ldots}^{\infty} d_r(c)P_{m+r}^m(\eta) & \text{for } n - m \text{ odd} \\ \sum_{r=0,2,\ldots}^{\infty} d_r(c)P_{m+r}^m(\eta) & \text{for } n - m \text{ even,} \end{cases} \tag{7}$$

where $P_m^k(\eta)$ is an associated LEGENDRE POLYNOMIAL. The prolate angular function of the second kind is given by

$$S_{mn}^{(2)} = \begin{cases} \sum\limits_{r=\ldots,-1,1,3,\ldots} d_r(c)Q_{m+r}^m(\eta) & \text{for } n - m \text{ odd} \\ \sum\limits_{r=\ldots,-2,0,2,\ldots} d_r(c)Q_{m+r}^m(\eta) & \text{for } n - m \text{ even,} \end{cases} \tag{8}$$

where $Q_k^m(\eta)$ is an associated LEGENDRE FUNCTION OF THE SECOND KIND and the COEFFICIENTS d_r satisfy the RECURRENCE RELATION

$$\alpha_k d_{k+2} + (\beta_k - \lambda_{mn})d_k + \gamma_k d_{k-2} = 0, \tag{9}$$

with

$$\alpha_k = \frac{(2m + k + 2)(2m + k + 1)c^2}{(2m + 2k + 3)(2m + 2k + 5)} \tag{10}$$

$$\beta_k = (m + k)(m + k + 1) + \frac{2(m + k)(m + k + 1) - 2m^2 - 1}{(2m + 2k - 1)(2m + 2k + 3)}c^2 \tag{11}$$

$$\gamma_k = \frac{k(k - 1)c^2}{(2m + 2k - 3)(2m + 2k - 1)}. \tag{12}$$

Various normalization schemes are used for the ds (Abramowitz and Stegun 1972, p. 758). Meixner and Schäfke (1954) use

$$\int_{-1}^{1} [S_{mn}(c,\eta)]^2 \, d\eta = \frac{2}{2n+1} \frac{(n+m)!}{(n-m)!}. \quad (13)$$

Stratton *et al.* (1956) use

$$\frac{(n+m)!}{(n-m)!} = \begin{cases} \sum_{r=1,3,\ldots}^{\infty} \frac{(r+2m)!}{r!} d_r & \text{for } n-m \text{ odd} \\ \sum_{r=0,2,\ldots}^{\infty} \frac{(r+2m)!}{r!} d_r & \text{for } n-m \text{ even.} \end{cases} \quad (14)$$

Flammer (1957) uses

$$S_{mn}(c,0) = \begin{cases} P_n^{m+1}(0) & \text{for } n-m \text{ odd} \\ P_n^m(0) & \text{for } n-m \text{ even.} \end{cases} \quad (15)$$

see also OBLATE SPHEROIDAL WAVE FUNCTION

References

Abramowitz, M. and Stegun, C. A. (Eds.). "Spheroidal Wave Functions." Ch. 21 in *Handbook of Mathematical Functions with Formulas, Graphs, and Mathematical Tables, 9th printing.* New York: Dover, pp. 751–759, 1972.

Flammer, C. *Spheroidal Wave Functions.* Stanford, CA: Stanford University Press, 1957.

Meixner, J. and Schäfke, F. W. *Mathieusche Funktionen und Sphäroidfunktionen.* Berlin: Springer-Verlag, 1954.

Stratton, J. A.; Morse, P. M.; Chu, L. J.; Little, J. D. C.; and Corbató, F. J. *Spheroidal Wave Functions.* New York: Wiley, 1956.

Pronic Number

A FIGURATE NUMBER of the form $P_n = 2T_n = n(n+1)$, where T_n is the nth TRIANGULAR NUMBER. The first few are 2, 6, 12, 20, 30, 42, 56, 72, 90, 110, ... (Sloane's A002378). The GENERATING FUNCTION of the pronic numbers is

$$\frac{2x}{(1-x)^3} = 2x + 6x^2 + 12x^3 + 20x^4 + \ldots.$$

The first few n for which P_n are PALINDROMIC are 1, 2, 16, 77, 538, 1621, ... (Sloane's A028336), and the first few PALINDROMIC NUMBERS which are pronic are 2, 6, 272, 6006, 289982, ... (Sloane's A028337).

References

De Geest, P. "Palindromic Products of Two Consecutive Integers." `http://www.ping.be/~ping6758/consec.htm`.

Sloane, N. J. A. Sequences A028336, A028337, and A002378/M1581 in "An On-Line Version of the Encyclopedia of Integer Sequences."

Proof

A rigorous mathematical argument which unequivocally demonstrates the truth of a given PROPOSITION. A mathematical statement which has been proven is called a THEOREM.

There is some debate among mathematicians as to just what constitutes a proof. The FOUR-COLOR THEOREM is an example of this debate, since its "proof" relies on an exhaustive computer testing of many individual cases which cannot be verified "by hand." While many mathematicians regard computer-assisted proofs as valid, some purists do not.

see also PARADOX, PROPOSITION, THEOREM

References

Garnier, R. and Taylor, J. *100% Mathematical Proof.* New York: Wiley, 1996.

Solow, D. *How to Read and Do Proofs: An Introduction to Mathematical Thought Process.* New York: Wiley, 1982.

Proofreading Mistakes

If proofreader A finds a mistakes and proofreader B finds b mistakes, c of which were also found by A, how many mistakes were missed by both A and B? Assume there are a total of m mistakes, so proofreader A finds a FRACTION a/m of all mistakes, and also a FRACTION c/b of the mistakes found by B. Assuming these fractions are the same, then solving for m gives

$$m = \frac{ab}{c}.$$

The number of mistakes missed by both is therefore approximately

$$N = m - a - b + c = \frac{(a-c)(b-c)}{c}.$$

References

Pólya, G. "Probabilities in Proofreading." *Amer. Math. Monthly*, **83**, 42, 1976.

Propeller

A 4-POLYHEX.

References

Gardner, M. *Mathematical Magic Show: More Puzzles, Games, Diversions, Illusions and Other Mathematical Sleight-of-Mind from Scientific American.* New York: Vintage, p. 147, 1978.

Proper Cover

see COVER

Proper Divisor

A DIVISOR of a number n excluding n itself.

see also ALIQUANT DIVISOR, ALIQUOT DIVISOR, DIVISOR

Proper Fraction

A FRACTION $p/q < 1$.

see also FRACTION, REDUCED FRACTION

Proper Integral

An INTEGRAL which has neither limit INFINITE and from which the INTEGRAND does not approach INFINITY at any point in the range of integration.

see also IMPROPER INTEGRAL, INTEGRAL

Proper k-Coloring

see k-COLORING

Proper Subset

A SUBSET which is not the entire SET. For example, consider a SET $\{1, 2, 3, 4, 5\}$. Then $\{1, 2, 4\}$ and $\{1\}$ are proper subsets, while $\{1, 2, 6\}$ and $\{1, 2, 3, 4, 5\}$ are not.

see also SET, SUBSET

Proper Superset

A SUPERSET which is not the entire SET.

see also SET, SUPERSET

Proportional

If a is proportional to b, then a/b is a constant. The relationship is written $a \propto b$, which implies

$$a = cb,$$

for some constant c.

Proposition

A statement which is to be proved.

Propositional Calculus

The formal basis of LOGIC dealing with the notion and usage of words such as "NOT," "OR," "AND," and "IMPLIES." Many systems of propositional calculus have been devised which attempt to achieve consistency, completeness, and independence of AXIOMS.

see also LOGIC, P-SYMBOL

References

Cundy, H. and Rollett, A. *Mathematical Models, 3rd ed.* Stradbroke, England: Tarquin Pub., pp. 254–255, 1989.

Nidditch, P. H. *Propositional Calculus.* New York: Free Press of Glencoe, 1962.

Prosthaphaeresis Formulas

TRIGONOMETRY formulas which convert a product of functions into a sum or difference.

Proth's Theorem

For $N = h \cdot 2^n + 1$ with ODD h and $2^n > h$, if there exists an INTEGER a such that

$$a^{(N-1)/2} \equiv -1 \pmod{N},$$

then N is PRIME.

Protractor

A ruled SEMICIRCLE used for measuring and drawing ANGLES.

Prouhet's Problem

A generalization of the TARRY-ESCOTT PROBLEM to three or more sets of INTEGERS.

see also TARRY-ESCOTT PROBLEM

References

Wright, E. M. "Prouhet's 1851 Solution of the Tarry-Escott Problem of 1910." *Amer. Math. Monthly* **102**, 199–210, 1959.

Prüfer Ring

A metric space $\hat{\mathbb{Z}}$ in which the closure of a congruence class $B(j, m)$ is the corresponding congruence class $\{x \in \hat{\mathbb{Z}} | x \equiv j \pmod{m}\}$.

References

Fried, M. D. and Jarden, M. *Field Arithmetic.* New York: Springer-Verlag, pp. 7–11, 1986.

Postnikov, A. G. *Introduction to Analytic Number Theory.* Providence, RI: Amer. Math. Soc., 1988.

Prussian Hat

A device used in the Cornwell smoothness stabilized modification of the CLEAN ALGORITHM.

see also CLEAN ALGORITHM

Pseudoanalytic Function

A pseudoanalytic function is a function defined using generalized CAUCHY-RIEMANN EQUATIONS. Pseudoanalytic functions come as close as possible to having COMPLEX derivatives and are nonsingular "quasiregular" functions.

see also ANALYTIC FUNCTION, SEMIANALYTIC, SUBANALYTIC

Pseudocrosscap

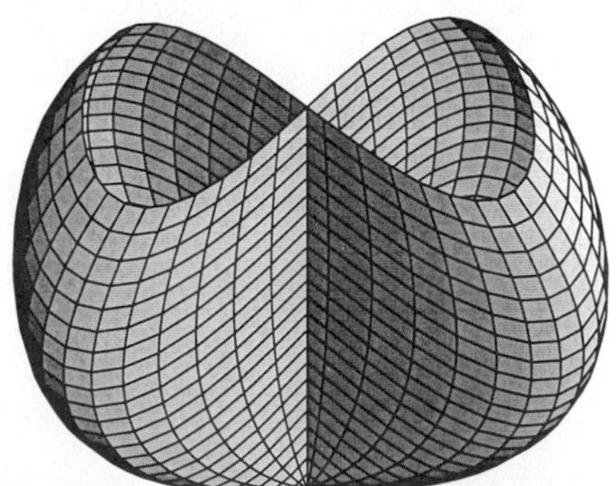

A surface constructed by placing a family of figure-eight curves into $\mathbb{R}^3$ such that the first and last curves reduce to points. The surface has parametric equations

$$x(u,v) = (1-u^2)\sin v$$
$$y(u,v) = (1-u^2)\sin(2v)$$
$$z(u,v) = u.$$

References
Gray, A. *Modern Differential Geometry of Curves and Surfaces.* Boca Raton, FL: CRC Press, pp. 247–248, 1993.

Pseudocylindrical Projection

A projection in which latitude lines are parallel but meridians are curves.

see also CYLINDRICAL PROJECTION, ECKERT IV PROJECTION, ECKERT VI PROJECTION, MOLLWEIDE PROJECTION, ROBINSON PROJECTION, SINUSOIDAL PROJECTION

References
Dana, P. H. "Map Projections." http://www.utexas.edu/depts/grg/gcraft/notes/mapproj/mapproj.html.

Pseudogroup

An algebraic structure whose elements consist of selected HOMEOMORPHISMS between open subsets of a SPACE, with the composition of two transformations defined on the largest possible domain. The "germs" of the elements of a pseudogroup form a GROUPOID (Weinstein 1996).

see also GROUP, GROUPOID, INVERSE SEMIGROUP

References
Weinstein, A. "Groupoids: Unifying Internal and External Symmetry." *Not. Amer. Math. Soc.* **43**, 744–752, 1996.

Pseudolemniscate Case

The case of the WEIERSTRAß ELLIPTIC FUNCTION with invariants $g_2 = -1$ and $g_3 = 0$.

see also EQUIANHARMONIC CASE, LEMNISCATE CASE

References
Abramowitz, M. and Stegun, C. A. (Eds.). "Pseudo-Lemniscate Case ($g_2 = -1$, $g_3 = 0$)." §18.15 in *Handbook of Mathematical Functions with Formulas, Graphs, and Mathematical Tables, 9th printing.* New York: Dover, pp. 662–663, 1972.

Pseudoperfect Number

see SEMIPERFECT NUMBER

Pseudoprime

A pseudoprime is a COMPOSITE number which passes a test or sequence of tests which fail for most COMPOSITE numbers. Unfortunately, some authors drop the "COMPOSITE" requirement, calling any number which passes the specified tests a pseudoprime even if it is PRIME. Pomerance, Selfridge, and Wagstaff (1980) restrict their use of "pseudoprime" to ODD COMPOSITE numbers. "Pseudoprime" used without qualification means FERMAT PSEUDOPRIME.

CARMICHAEL NUMBERS are ODD COMPOSITE numbers which are pseudoprimes to every base; they are sometimes called ABSOLUTE PSEUDOPRIMES. The following table gives the number of FERMAT PSEUDOPRIMES psp, EULER PSEUDOPRIMES epsp, and STRONG PSEUDOPRIMES spsp to the base 2, as well as CARMICHAEL NUMBERS CN which are less the first few powers of 10 (Guy 1994).

	10^3	10^4	10^5	10^6	10^7	10^8	10^9	10^{10}
psp(2)	3	22	78	245	750	2057	5597	14884
epsp(2)	1	12	36	114	375	1071	2939	7706
spsp(2)	0	5	16	46	162	488	1282	3291
CN	1	7	16	43	105	255	646	1547

see also CARMICHAEL NUMBER, ELLIPTIC PSEUDOPRIME, EULER PSEUDOPRIME, EULER-JACOBI PSEUDOPRIME, EXTRA STRONG LUCAS PSEUDOPRIME, FERMAT PSEUDOPRIME, FIBONACCI PSEUDOPRIME, FROBENIUS PSEUDOPRIME, LUCAS PSEUDOPRIME, PERRIN PSEUDOPRIME, PROBABLE PRIME, SOMER-LUCAS PSEUDOPRIME, STRONG ELLIPTIC PSEUDOPRIME, STRONG FROBENIUS PSEUDOPRIME, STRONG LUCAS PSEUDOPRIME, STRONG PSEUDOPRIME

References
Grantham, J. "Frobenius Pseudoprimes." http://www.clark.net/pub/grantham/pseudo/pseudo.ps
Grantham, J. "Pseudoprimes/Probable Primes." http://www.clark.net/pub/grantham/pseudo.
Guy, R. K. "Pseudoprimes. Euler Pseudoprimes. Strong Pseudoprimes." §A12 in *Unsolved Problems in Number Theory, 2nd ed.* New York: Springer-Verlag, pp. 27–30, 1994.
Pomerance, C.; Selfridge, J. L.; and Wagstaff, S. S. "The Pseudoprimes to $25 \cdot 10^9$." *Math. Comput.* **35**, 1003–1026, 1980. Available electronically from ftp://sable.ox.ac.uk/pub/math/primes/ps2.Z.

Pseudorandom Number

A slightly archaic term for a computer-generated RANDOM NUMBER. The prefix pseudo- is used to distinguish this type of number from a "truly" RANDOM NUMBER generated by a random physical process such as radioactive decay.

see also RANDOM NUMBER

References
Luby, M. *Pseudorandomness and Cryptographic Applications.* Princeton, NJ: Princeton University Press, 1996.
Press, W. H.; Flannery, B. P.; Teukolsky, S. A.; and Vetterling, W. T. *Numerical Recipes in FORTRAN: The Art of*

Scientific Computing, 2nd ed. Cambridge, England: Cambridge University Press, p. 266, 1992.

Pseudorhombicuboctahedron

see ELONGATED SQUARE GYROBICUPOLA

Pseudoscalar

A SCALAR which reverses sign under inversion is called a pseudoscalar. The SCALAR TRIPLE PRODUCT

$$\mathbf{A} \cdot (\mathbf{B} \times \mathbf{C})$$

is a pseudoscalar. Given a transformation MATRIX A,

$$S' = \det |\mathsf{A}| S,$$

where det is the DETERMINANT.

see also PSEUDOTENSOR, PSEUDOVECTOR, SCALAR

References
Arfken, G. "Pseudotensors, Dual Tensors." §3.4 in *Mathematical Methods for Physicists, 3rd ed.* Orlando, FL: Academic Press, pp. 128–137, 1985.

Pseudosmarandache Function

The pseudosmarandache function $Z(n)$ is the smallest integer such that

$$\sum_{k=1}^{Z(n)} k = \tfrac{1}{2} Z(n)[Z(n)+1]$$

is divisible by n. The values for $n = 1, 2, \ldots$ are 1, 3, 2, 7, 4, 3, 6, 15, 8, 4, ... (Sloane's A011772).

see also SMARANDACHE FUNCTION

References
Ashbacher, C. "Problem 514." *Pentagon* **57**, 36, 1997.
Kashihara, K. "Comments and Topics on Smarandache Notions and Problems." Vail: Erhus University Press, 1996.
Sloane, N. J. A. Sequence A011772 in "An On-Line Version of the Encyclopedia of Integer Sequences."

Pseudosphere

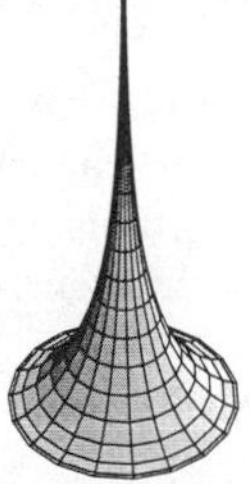

Half the SURFACE OF REVOLUTION generated by a TRACTRIX about its ASYMPTOTE to form a TRACTROID. The Cartesian parametric equations are

$$x = \operatorname{sech} u \cos v \quad (1)$$

$$y = \operatorname{sech} u \sin v \quad (2)$$

$$z = u - \tanh u \quad (3)$$

for $u \geq 0$.

It has constant NEGATIVE CURVATURE, and so is called a pseudosphere by analogy with the SPHERE, which has constant POSITIVE curvature. An equation for the GEODESICS is

$$\cosh^2 u + (v+c)^2 = k^2. \quad (4)$$

see also FUNNEL, GABRIEL'S HORN, TRACTRIX

References
Fischer, G. (Ed.). Plate 82 in *Mathematische Modelle/Mathematical Models, Bildband/Photograph Volume.* Braunschweig, Germany: Vieweg, p. 77, 1986.
Geometry Center. "The Pseudosphere." `http://www.geom.umn.edu/zoo/diffgeom/pseudosphere/`.
Gray, A. *Modern Differential Geometry of Curves and Surfaces.* Boca Raton, FL: CRC Press, pp. 383–384, 1993.

Pseudosquare

Given an ODD PRIME p, a SQUARE NUMBER n satisfies $(n/p) = 0$ or 1 for all $p < n$, where (n/p) is the LEGENDRE SYMBOL. A number $n > 2$ which satisfies this relationship but is not a SQUARE NUMBER is called a pseudosquare. The only pseudoprimes less than 10^8 are 3 and 6.

see also SQUARE NUMBER

Pseudotensor

A TENSOR-like object which reverses sign under inversion. Given a transformation MATRIX A,

$$A_{ij}' = \det |\mathsf{A}| a_{ik} a_{jl} A_{kl},$$

where det is the DETERMINANT. A pseudotensor is sometimes also called a TENSOR DENSITY.

see also PSEUDOSCALAR, PSEUDOVECTOR, SCALAR, TENSOR DENSITY

References
Arfken, G. "Pseudotensors, Dual Tensors." §3.4 in *Mathematical Methods for Physicists, 3rd ed.* Orlando, FL: Academic Press, pp. 128–137, 1985.

Pseudovector

A typical VECTOR is transformed to its NEGATIVE under inversion. A VECTOR which is invariant under inversion is called a pseudovector, also called an AXIAL VECTOR in older literature (Morse and Feshbach 1953). The CROSS PRODUCT

$$\mathbf{A} \times \mathbf{B} \quad (1)$$

is a pseudovector, whereas the VECTOR TRIPLE PRODUCT

$$\mathbf{A} \times (\mathbf{B} \times \mathbf{C}) \quad (2)$$

is a VECTOR.

$$[\text{pseudovector}] \times [\text{pseudovector}] = [\text{pseudovector}] \quad (3)$$

$$[\text{vector}] \times [\text{pseudovector}] = [\text{vector}]. \quad (4)$$

Given a transformation MATRIX A,

$$C_i' = \det|\mathsf{A}|a_{ij}C_j. \quad (5)$$

see also PSEUDOSCALAR, TENSOR, VECTOR

References

Arfken, G. "Pseudotensors, Dual Tensors." §3.4 in *Mathematical Methods for Physicists, 3rd ed.* Orlando, FL: Academic Press, pp. 128–137, 1985.

Morse, P. M. and Feshbach, H. *Methods of Theoretical Physics, Part I.* New York: McGraw-Hill, pp. 46–47, 1953.

Psi Function

$$\Psi(z, s, v) \equiv \sum_{n=0}^{\infty} \frac{z^n}{(v+n)^s}$$

for $|z| < 1$ and $v \neq 0, -1, \ldots$ (Gradshteyn and Ryzhik 1980, pp. 1075–1076).

see also HURWITZ ZETA FUNCTION, RAMANUJAN PSI SUM, THETA FUNCTION

References

Gradshteyn, I. S. and Ryzhik, I. M. *Tables of Integrals, Series, and Products, 5th ed.* San Diego, CA: Academic Press, 1979.

PSLQ Algorithm

An ALGORITHM which finds INTEGER RELATIONS between real numbers $x_1, \ldots, x_n$ such that

$$a_1x_1 + a_2x_2 + \ldots + a_nx_n = 0,$$

with not all $a_i = 0$. This algorithm terminates after a number of iterations bounded by a polynomial in n and uses a numerically stable matrix reduction procedure (Ferguson and Bailey 1992), thus improving upon the FERGUSON-FORCADE ALGORITHM. It is based on a partial sum of squares scheme (like the PSOS ALGORITHM) implemented using LQ decomposition. A much simplified version of the algorithm was developed by Ferguson *et al.* and extended to complex numbers.

see also FERGUSON-FORCADE ALGORITHM, INTEGER RELATION, LLL ALGORITHM, PSOS ALGORITHM

References

Bailey, D. H.; Borwein, J. M.; and Girgensohn, R. "Experimental Evaluation of Euler Sums." *Exper. Math.* **3**, 17–30, 1994.

Bailey, D. and Plouffe, S. "Recognizing Numerical Constants." `http://www.cecm.sfu.ca/organics/papers/bailey`.

Ferguson, H. R. P. and Bailey, D. H. "A Polynomial Time, Numerically Stable Integer Relation Algorithm." RNR Techn. Rept. RNR-91-032, Jul. 14, 1992.

Ferguson, H. R. P.; Bailey, D. H.; and Arno, S. "Analysis of PSLQ, An Integer Relation Finding Algorithm." Unpublished manuscript.

PSOS Algorithm

An INTEGER-RELATION algorithm which is based on a partial sum of squares approach, from which the algorithm takes its name.

see also FERGUSON-FORCADE ALGORITHM, HJLS ALGORITHM, INTEGER RELATION, LLL ALGORITHM, PSLQ ALGORITHM

References

Bailey, D. H. and Ferguson, H. R. P. "Numerical Results on Relations Between Numerical Constants Using a New Algorithm." *Math. Comput.* **53**, 649–656, 1989.

Ptolemy Inequality

For a QUADRILATERAL which is not CYCLIC, PTOLEMY'S THEOREM becomes an INEQUALITY:

$$AB \times CD + BC \times DA > AC \times BD.$$

see also PTOLEMY'S THEOREM, QUADRILATERAL

Ptolemy's Theorem

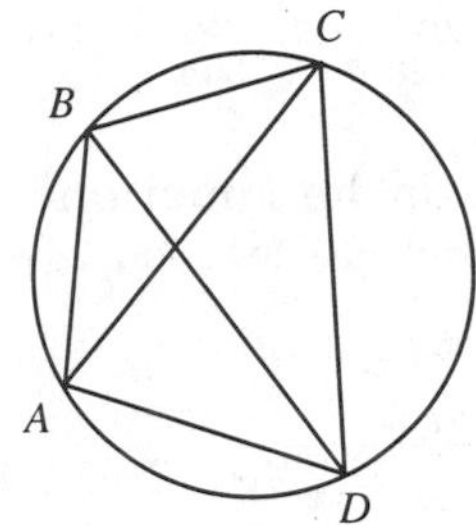

If a QUADRILATERAL is inscribed in a circle (i.e., for a cyclic quadrilateral), the sum of the products of the two pairs of opposite sides equals the product of the diagonals

$$AB \times CD + BC \times DA = AC \times BD.$$

This fact can be used to derive the TRIGONOMETRY addition formulas.

see also FUHRMANN'S THEOREM, PTOLEMY INEQUALITY

References

Coxeter, H. S. M. and Greitzer, S. L. *Geometry Revisited.* Washington, DC: Math. Assoc. Amer., pp. 42–43, 1967.

Public-Key Cryptography

A type of CRYPTOGRAPHY in which the encoding key is revealed without compromising the encoded message. The two best-known methods are the KNAPSACK PROBLEM and RSA ENCRYPTION.

see also KNAPSACK PROBLEM, RSA ENCRYPTION

References

Diffie, W. and Hellman, M. "New Directions in Cryptography." *IEEE Trans. Info. Th.* **22**, 644–654, 1976.

Hellman, M. E. "The Mathematics of Public-Key Cryptography." *Sci. Amer.* **241**, 130–139, Aug. 1979.

Rivest, R.; Shamir, A.; and Adleman, L. "A Method for Obtaining Digital Signatures and Public-Key Cryptosystems." MIT Memo MIT/LCS/TM-82, 1982.

Wagon, S. "Public-Key Encryption." §1.2 in *Mathematica in Action.* New York: W. H. Freeman, pp. 20–22, 1991.

Puiseaux's Theorem

The whole neighborhood of any point y_i of an algebraic PLANE CURVE may be uniformly represented by a certain finite number of convergent developments in POWER SERIES,

$$x_i = \rho_\nu y_i + a_{\nu i1} t_\nu + a_{\nu i2} {t_\nu}^2 + \ldots.$$

References

Coolidge, J. L. *A Treatise on Algebraic Plane Curves.* New York: Dover, p. 207, 1959.

Pullback Map

A pullback is a general CATEGORICAL operation appearing in a number of mathematical contexts, sometimes going under a different name. If $T : V \to W$ is a linear transformation between VECTOR SPACES, then $T^* : W^* \to V^*$ (usually called TRANSPOSE MAP or DUAL MAP because its associated matrix is the MATRIX TRANSPOSE of T) is an example of a pullback map.

In the case of a DIFFEOMORPHISM and DIFFERENTIABLE MANIFOLD, a very explicit definition can be formulated. Given an r-form α on a MANIFOLD M_2, define the r-form $T^*(\alpha)$ on M_1 by its action on an r-tuple of tangent vectors $(X_1, \ldots, X_r)$ as the number $T^*(\alpha)(X_1, \ldots, X_r) = \alpha(T_* X_1, \ldots, T_* X_r)$. This defines a map on r-forms and is the pullback map.

see also CATEGORY

Pulse Function

see RECTANGLE FUNCTION

Purser's Theorem

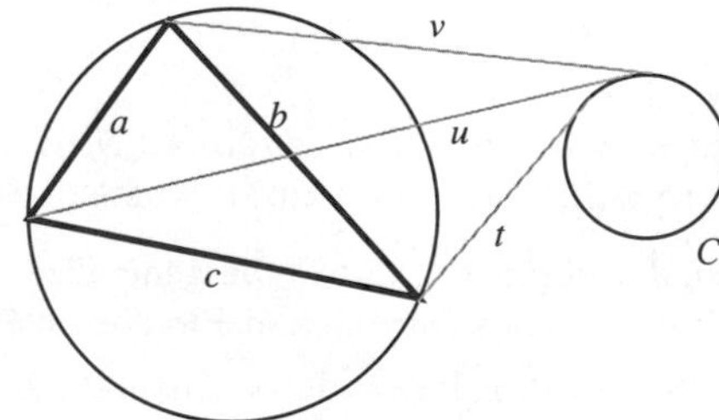

Let t, u, and v be the lengths of the tangents to a CIRCLE C from the vertices of a TRIANGLE with sides of lengths a, b, and c. Then the condition that C is tangent to the CIRCUMCIRCLE of the TRIANGLE is that

$$\pm at \pm bu \pm cv = 0.$$

The theorem was discovered by Casey prior to Purser's independent discovery.

see also CASEY'S THEOREM, CIRCUMCIRCLE

Pursuit Curve

If A moves along a known curve, then P describes a pursuit curve if P is always directed toward A and A and P move with uniform velocities. These were considered in general by the French scientist Pierre Bouguer in 1732. The case restricting A to a straight line was studied by Arthur Bernhart (MacTutor Archive). It has CARTESIAN COORDINATES equation

$$y = cx - \ln x.$$

see also APOLLONIUS PURSUIT PROBLEM, MICE PROBLEM

References

Bernhart, A. "Curves of Pursuit." *Scripta Math.* **20**, 125–141, 1954.

Bernhart, A. "Curves of Pursuit–II." *Scripta Math.* **23**, 49–65, 1957.

Bernhart, A. "Polygons of Pursuit." *Scripta Math.* **24**, 23–50, 1959.

Bernhart, A. "Curves of General Pursuit." *Scripta Math.* **24**, 189–206, 1959.

MacTutor History of Mathematics Archive. "Pursuit Curve." `http://www-groups.dcs.st-and.ac.uk/~history/Curves/Pursuit.html`.

Yates, R. C. "Pursuit Curve." *A Handbook on Curves and Their Properties.* Ann Arbor, MI: J. W. Edwards, pp. 170–171, 1952.

Push

An action which adds a single element to the top of a STACK, turning the STACK (a_1, a_2, ..., a_n) into (a_0, a_1, a_2, ..., a_n).

see also POKE MOVE, POP, STACK

Puzzle

A mathematical PROBLEM, usually not requiring advanced mathematics, to which a solution is desired. Puzzles frequently require the rearrangement of existing pieces (e.g., 15 PUZZLE) or the filling in of blanks (e.g., crossword puzzle).

see also 15 PUZZLE, BAGUENAUDIER, CALIBAN PUZZLE, CONWAY PUZZLE, CRYPTARITHMETIC, DISSECTION PUZZLES, ICOSIAN GAME, PYTHAGOREAN SQUARE PUZZLE, RUBIK'S CUBE, SLOTHOUBER-GRAATSMA PUZZLE, T-PUZZLE

References

Bogomolny, A. "Interactive Mathematics Miscellany and Puzzles." `http://www.cut-the-knot.com/`.

Dudeney, H. E. *Amusements in Mathematics.* New York: Dover, 1917.
Dudeney, H. E. *The Canterbury Puzzles and Other Curious Problems, 7th ed.* London: Thomas Nelson and Sons, 1949.
Dudeney, H. E. *536 Puzzles & Curious Problems.* New York: Scribner, 1967.
Fujii, J. N. *Puzzles and Graphs.* Washington, DC: National Council of Teachers, 1966.

Pyramid

A POLYHEDRON with one face a POLYGON and all the other faces TRIANGLES with a common VERTEX. An n-gonal regular pyramid (denoted Y_n) has EQUILATERAL TRIANGLES, and is possible only for $n = 3, 4, 5$. These correspond to the TETRAHEDRON, SQUARE PYRAMID, and PENTAGONAL PYRAMID, respectively. A pyramid therefore has a single cross-sectional shape in which the length scale of the CROSS-SECTION scales linearly with height. The AREA at a height z is given by

$$A(z) = A_b\left(\frac{z^2}{h^2}\right), \tag{1}$$

where A_b is the base AREA and h is the pyramid height. The VOLUME is therefore given by

$$V = \int_0^h A(z)\,dz = A_b\int_0^h \frac{z^2}{h^2}\,dz = \frac{A_b}{h^2}(\tfrac{1}{3}h^3) = \tfrac{1}{3}A_b h. \tag{2}$$

These results also hold for the CONE, TETRAHEDRON (triangular pyramid), SQUARE PYRAMID, etc.

The CENTROID is the same as for the CONE, given by

$$\bar{z} = \tfrac{1}{4}h. \tag{3}$$

The SURFACE AREA of a pyramid is

$$S = \tfrac{1}{2}ps, \tag{4}$$

where s is the SLANT HEIGHT and p is the base PERIMETER. Joining two PYRAMIDS together at their bases gives a BIPYRAMID, also called a DIPYRAMID.

see also BIPYRAMID, ELONGATED PYRAMID, GYROELONGATED PYRAMID, PENTAGONAL PYRAMID, PYRAMID, PYRAMIDAL FRUSTUM, SQUARE PYRAMID, TETRAHEDRON, TRUNCATED SQUARE PYRAMID

References
Beyer, W. H. (Ed.) *CRC Standard Mathematical Tables, 28th ed.* Boca Raton, FL: CRC Press, p. 128, 1987.
Hart, G. W. "Pyramids, Dipyramids, and Trapezohedra." `http://www.li.net/~george/virtual-polyhedra/pyramids-info.html`.

Pyramidal Frustum

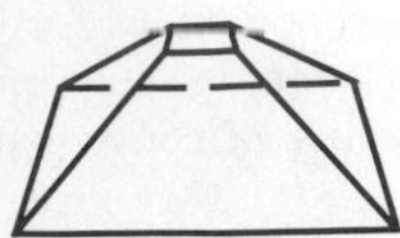

Let s be the slant height, p_i the top and bottom base PERIMETERS, and A_i the top and bottom AREAS. Then the SURFACE AREA and VOLUME of the pyramidal frustum are given by

$$S = \tfrac{1}{2}(p_1 + p_2)s$$
$$V = \tfrac{1}{3}h(A_1 + A_2 + \sqrt{A_1 A_2}\,).$$

see also CONICAL FRUSTUM, FRUSTUM, PYRAMID, SPHERICAL SEGMENT, TRUNCATED SQUARE PYRAMID

References
Beyer, W. H. (Ed.) *CRC Standard Mathematical Tables, 28th ed.* Boca Raton, FL: CRC Press, p. 128, 1987.
Dunham, W. *Journey Through Genius: The Great Theorems of Mathematics.* New York: Wiley, pp. 3–4, 1990.

Pyramidal Number

A FIGURATE NUMBER corresponding to a configuration of points which form a pyramid with r-sided REGULAR POLYGON bases can be thought of as a generalized pyramidal number, and has the form

$$P_n^r = \tfrac{1}{6}(n+1)(2p_n^r + n) = \tfrac{1}{6}n(n+1)[(r-2)n + (5-r)]. \tag{1}$$

The first few cases are therefore

$$P_n^3 = \tfrac{1}{6}n(n+1)(n+2) \tag{2}$$
$$P_n^4 = \tfrac{1}{6}n(n+1)(2n+1) \tag{3}$$
$$P_n^5 = \tfrac{1}{2}n^2(n+1), \tag{4}$$

so $r = 3$ corresponds to a TETRAHEDRAL NUMBER Te_n, and $r = 4$ to a SQUARE PYRAMIDAL NUMBER P_n.

The pyramidal numbers can also be generalized to 4-D and higher dimensions (Sloane and Plouffe 1995).

see also HEPTAGONAL PYRAMIDAL NUMBER, HEXAGONAL PYRAMIDAL NUMBER, PENTAGONAL PYRAMIDAL NUMBER, SQUARE PYRAMIDAL NUMBER, TETRAHEDRAL NUMBER

References
Conway, J. H. and Guy, R. K. "Tetrahedral Numbers" and "Square Pyramidal Numbers" *The Book of Numbers.* New York: Springer-Verlag, pp. 44–49, 1996.
Sloane, N. J. A. and Plouffe, S. "Pyramidal Numbers." Extended entry for sequence M3382 in *The Encyclopedia of Integer Sequences.* San Diego, CA: Academic Press, 1995.

Pyritohedron

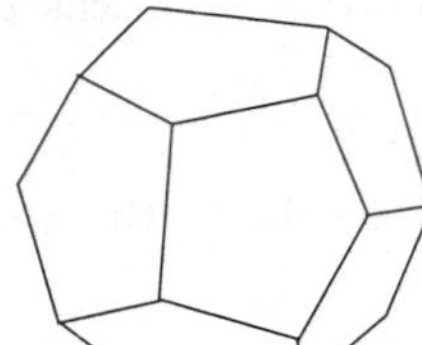

An irregular DODECAHEDRON composed of identical irregular PENTAGONS.

see also DODECAHEDRON, RHOMBIC DODECAHEDRON, TRIGONAL DODECAHEDRON

References

Cotton, F. A. *Chemical Applications of Group Theory, 3rd ed.* New York: Wiley, p. 63, 1990.

Pythagoras's Constant

The number

$$\sqrt{2} = 1.4142135623\ldots,$$

which the Pythagoreans proved to be IRRATIONAL. The Babylonians gave the impressive approximation

$$\sqrt{2} \approx 1 + \frac{24}{60} + \frac{51}{60^2} + \frac{10}{60^3} = 1.41421296296296\ldots$$

(Guy 1990, Conway and Guy 1996, pp. 181–182).

see also IRRATIONAL NUMBER, OCTAGON, PYTHAGORAS'S THEOREM, SQUARE

References

Conway, J. H. and Guy, R. K. *The Book of Numbers.* New York: Springer-Verlag, p. 25 and 181–182, 1996.
Guy, R. K. "Review: The Mathematics of Plato's Academy." *Amer. Math. Monthly* **97**, 440–443, 1990.
Shanks, D. *Solved and Unsolved Problems in Number Theory, 4th ed.* New York: Chelsea, p. 126, 1993.

Pythagoras's Theorem

Proves that the DIAGONAL d of a SQUARE with sides of integral length s cannot be RATIONAL. Assume d/s is rational and equal to p/q where p and q are INTEGERS with no common factors. Then

$$d^2 = s^2 + s^2 = 2s^2,$$

so

$$\left(\frac{d}{s}\right)^2 = \left(\frac{p}{q}\right)^2 = 2,$$

and $p^2 = 2q^2$, so p^2 is even. But if p^2 is EVEN, then p is EVEN. Since p/q is defined to be expressed in lowest terms, q must be ODD; otherwise p and q would have the common factor 2. Since p is EVEN, we can let $p \equiv 2r$, then $4r^2 = 2q^2$. Therefore, $q^2 = 2r^2$, and q^2, so q must be EVEN. But q cannot be both EVEN and ODD, so there are no d and s such that d/s is RATIONAL, and d/s must be IRRATIONAL.

In particular, PYTHAGORAS'S CONSTANT $\sqrt{2}$ is IRRATIONAL. Conway and Guy (1996) give a proof of this fact using paper folding, as well as similar proofs for ϕ (the GOLDEN RATIO) and $\sqrt{3}$ using a PENTAGON and HEXAGON.

see also IRRATIONAL NUMBER, PYTHAGORAS'S CONSTANT, PYTHAGOREAN THEOREM

References

Conway, J. H. and Guy, R. K. *The Book of Numbers.* New York: Springer-Verlag, pp. 183–186, 1996.
Pappas, T. "Irrational Numbers & the Pythagoras Theorem." *The Joy of Mathematics.* San Carlos, CA: Wide World Publ./Tetra, pp. 98–99, 1989.

Pythagoras Tree

A FRACTAL with symmetric

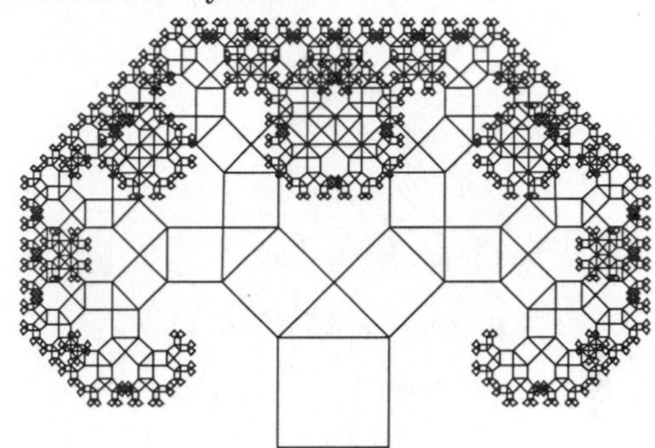

and asymmetric

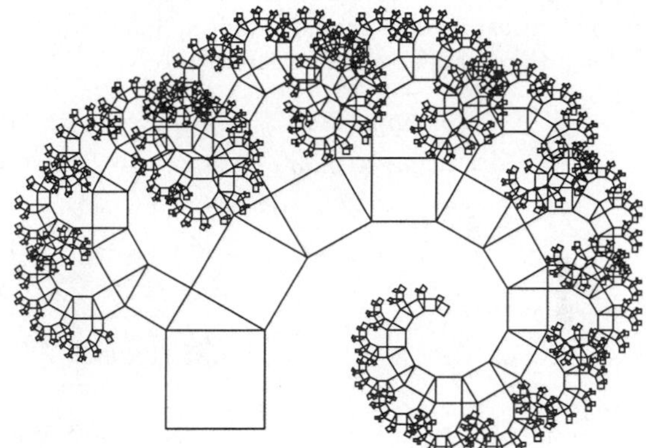

forms.

References

Lauwerier, H. *Fractals: Endlessly Repeated Geometric Figures.* Princeton, NJ: Princeton University Press, pp. 67–77 and 111–113, 1991.
Weisstein, E. W. "Fractals." http://www.astro.virginia.edu/~eww6n/math/notebooks/Fractal.m.

Pythagorean Fraction

Given a PYTHAGOREAN TRIPLE (a, b, c), the fractions a/b and b/a are called Pythagorean fractions. Diophantus showed that the Pythagorean fractions consist precisely of fractions of the form $(p^2 - q^2)/(2pq)$.

References

Conway, J. H. and Guy, R. K. "Pythagorean Fractions." In *The Book of Numbers.* New York: Springer-Verlag, pp. 171–173, 1996.

Pythagorean Quadruple

POSITIVE INTEGERS a, b, c, and d which satisfy

$$a^2 + b^2 + c^2 = d^2. \tag{1}$$

For POSITIVE EVEN a and b, there exist such INTEGERS c and d; for POSITIVE ODD a and b, no such INTEGERS exist (Oliverio 1996). Oliverio (1996) gives the following generalization of this result. Let $S = (a_1, \ldots, a_{n-2})$, where a_i are INTEGERS, and let T be the number of ODD INTEGERS in S. Then IFF $T \not\equiv 2 \pmod 4$, there exist INTEGERS a_{n-1} and a_n such that

$${a_1}^2 + {a_2}^2 + \ldots + {a_{n-1}}^2 = {a_n}^2. \tag{2}$$

A set of Pythagorean quadruples is given by

$$a = 2mp \tag{3}$$
$$b = 2np \tag{4}$$
$$c = p^2 - (m^2 + n^2) \tag{5}$$
$$d = p^2 + (m^2 + n^2), \tag{6}$$

where m, n, and p are INTEGERS,

$$m + n + p \equiv 1 \pmod 2, \tag{7}$$

and

$$(m, n, p) = 1 \tag{8}$$

(Mordell 1969). This does not, however, generate all solutions. For instance, it excludes (36, 8, 3, 37). Another set of solutions can be obtained from

$$a = 2mp + 2nq \tag{9}$$
$$b = 2np - 2mq \tag{10}$$
$$c = p^2 + q^2 - (m^2 + n^2) \tag{11}$$
$$d = p^2 + q^2 + (m^2 + n^2) \tag{12}$$

(Carmichael 1915).

see also EULER BRICK, PYTHAGOREAN TRIPLE

References

Carmichael, R. D. *Diophantine Analysis.* New York: Wiley, 1915.

Mordell, L. J. *Diophantine Equations.* London: Academic Press, 1969.

Oliverio, P. "Self-Generating Pythagorean Quadruples and N-tuples." *Fib. Quart.* **34**, 98–101, 1996.

Pythagorean Square Puzzle

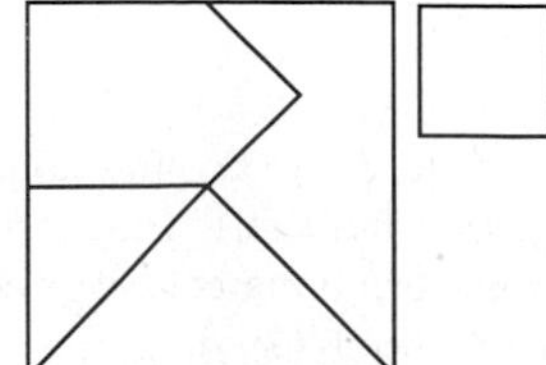

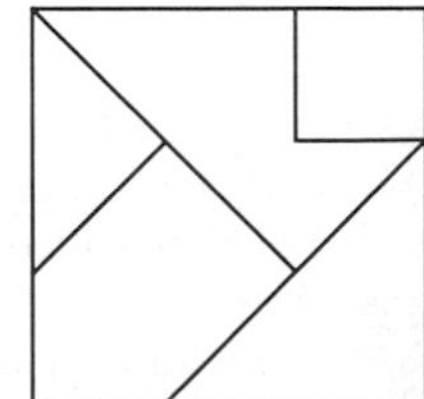

Combine the two above squares on the left into the single large square on the right.

see also DISSECTION, T-PUZZLE

Pythagorean Theorem

For a RIGHT TRIANGLE with legs a and b and HYPOTENUSE c,

$$a^2 + b^2 = c^2. \tag{1}$$

Many different proofs exist for this most fundamental of all geometric theorems.

A clever proof by DISSECTION which reassembles two small squares into one larger one was given by the Arabian mathematician Thabit Ibn Qurra (Ogilvy 1994, Frederickson 1997).

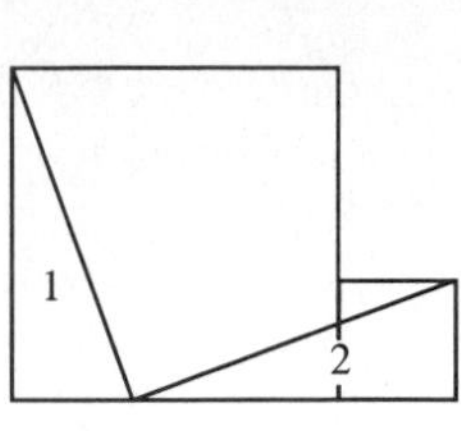

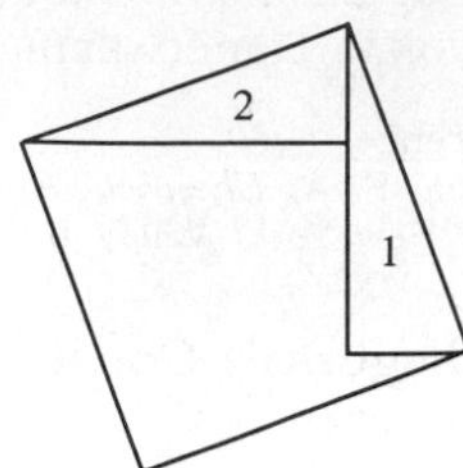

Another proof by DISSECTION is due to Perigal (Pergial 1873, Dudeney 1970, Madachy 1979, Ball and Coxeter 1987).

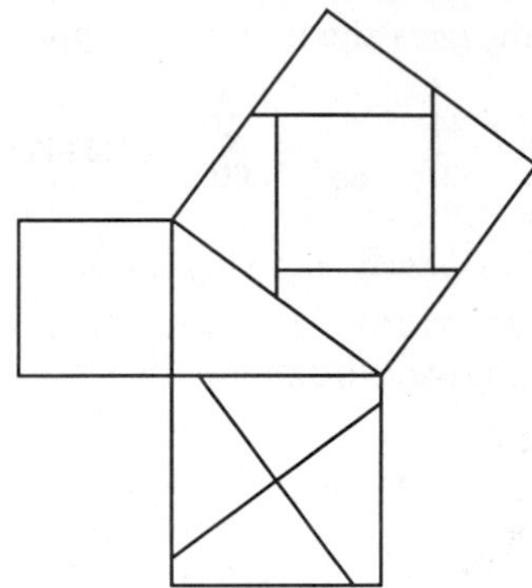

The Indian mathematician Bhaskara constructed a proof using the following figure.

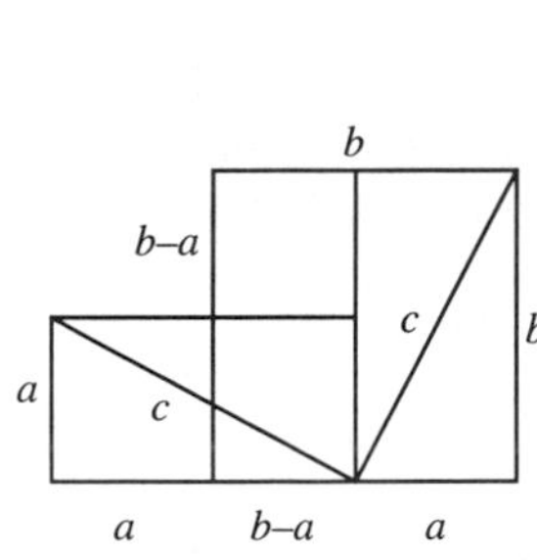

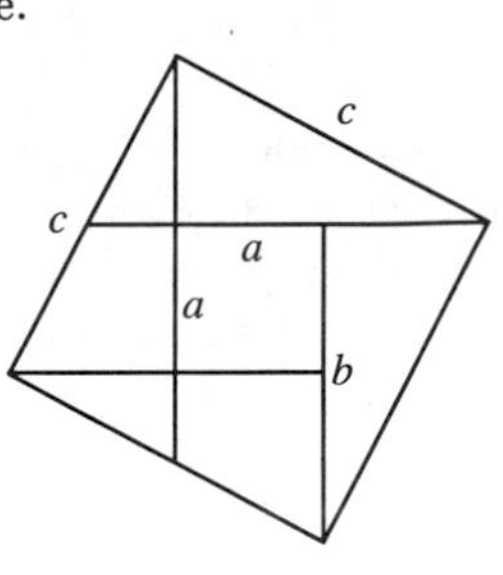

Several similar proofs are shown below.

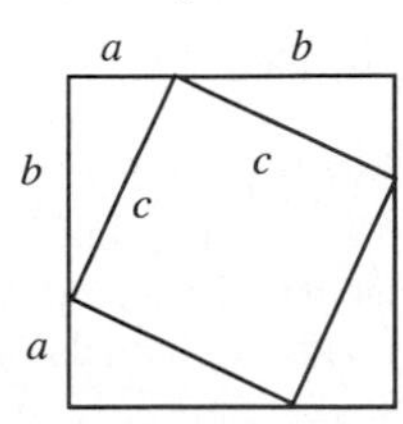

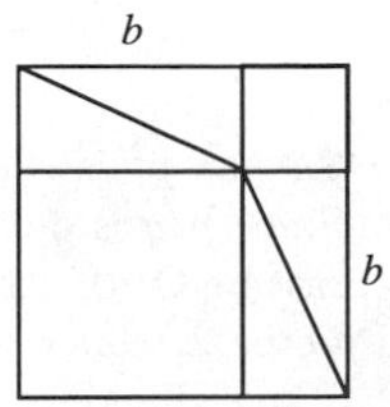

$$c^2 + 4(\tfrac{1}{2}ab) = (a+b)^2 \tag{2}$$

$$c^2 + 2ab = a^2 + 2ab + b^2 \qquad (3)$$

$$c^2 = a^2 + b^2. \qquad (4)$$

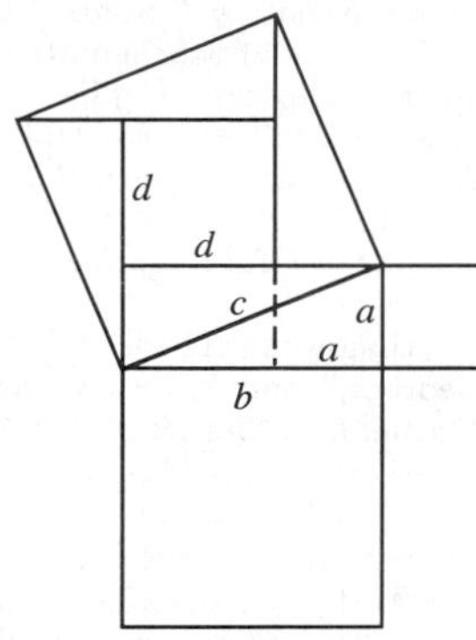

In the above figure, the AREA of the large SQUARE is four times the AREA of one of the TRIANGLES plus the AREA of the interior SQUARE. From the figure, $d = b - a$, so

$$A = 4(\tfrac{1}{2}ab) + d^2 = 2ab + (b-a)^2 = 2ab + b^2 - 2ab + a^2$$
$$= a^2 + b^2 = c^2. \qquad (5)$$

Perhaps the most famous proof of all times is Euclid's geometric proof. Euclid's proof used the figure below, which is sometimes known variously as the BRIDE'S CHAIR, PEACOCK'S TAIL, or WINDMILL.

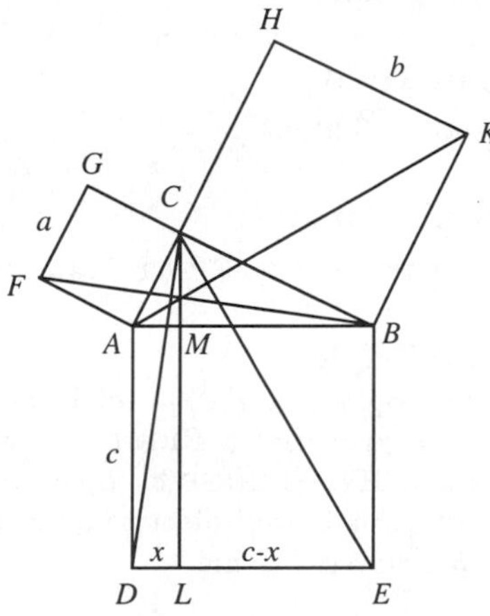

Let ΔABC be a RIGHT TRIANGLE, $\square CAFG$, $\square CBKH$, and $\square ABED$ be squares, and $CL \| BD$. The TRIANGLES ΔFAB and ΔCAD are equivalent except for rotation, so

$$2\Delta FAB = 2\Delta CAD. \qquad (6)$$

Shearing these TRIANGLES gives two more equivalent TRIANGLES

$$2\Delta CAD = \square ADLM. \qquad (7)$$

Therefore,

$$\square ACGF = \square ADLM. \qquad (8)$$

Similarly,

$$\square BC = 2\Delta ABK = 2\Delta BCE = \square BL \qquad (9)$$

so

$$a^2 + b^2 = cx + c(c-x) = c^2. \qquad (10)$$

Heron proved that AK, CL, and BF intersect in a point (Dunham 1990, pp. 48–53).

HERON'S FORMULA for the AREA of the TRIANGLE, contains the Pythagorean theorem implicitly. Using the form

$$K = \tfrac{1}{4}\sqrt{2a^2b^2 + 2a^2c^2 + ab^2c^2 - (a^4 + b^4 + c^4)} \qquad (11)$$

and equating to the AREA

$$K = \tfrac{1}{2}ab \qquad (12)$$

gives

$$\tfrac{1}{4}a^2b^2 = 2a^2b^2 + 2a^2c^2 + ab^2c^2 - (a^4 + b^4 + c^4). \qquad (13)$$

Rearranging and simplifying gives

$$a^2 + b^2 = c^2, \qquad (14)$$

the Pythagorean theorem, where K is the AREA of a TRIANGLE with sides a, b, and c (Dunham 1990, pp. 128–129).

A novel proof using a TRAPEZOID was discovered by James Garfield (1876), later president of the United States, while serving in the House of Representatives (Pappas 1989, pp. 200–201; Bogomolny).

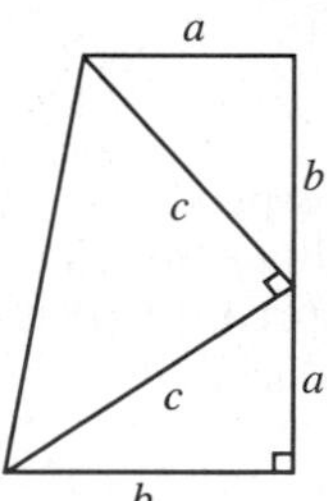

$$A_{\text{trapezoid}} = \tfrac{1}{2}\sum [\text{bases}] \cdot [\text{altitude}]$$
$$= \tfrac{1}{2}(a+b)(a+b)$$
$$= \tfrac{1}{2}ab + \tfrac{1}{2}ab + \tfrac{1}{2}c^2. \qquad (15)$$

Rearranging,

$$\tfrac{1}{2}(a^2 + 2ab + b^2) = ab + \tfrac{1}{2}c^2 \qquad (16)$$

$$a^2 + 2ab + b^2 = 2ab + c^2 \qquad (17)$$

$$a^2 + b^2 = c^2. \qquad (18)$$

An algebraic proof (which would not have been accepted by the Greeks) uses the EULER FORMULA. Let the sides of a TRIANGLE be a, b, and c, and the PERPENDICULAR legs of RIGHT TRIANGLE be aligned along the real and imaginary axes. Then

$$a + bi = ce^{i\theta}. \tag{19}$$

Taking the COMPLEX CONJUGATE gives

$$a - bi = ce^{-i\theta}. \tag{20}$$

Multiplying (19) by (20) gives

$$a^2 + b^2 = c^2. \tag{21}$$

Another algebraic proof proceeds by similarity.

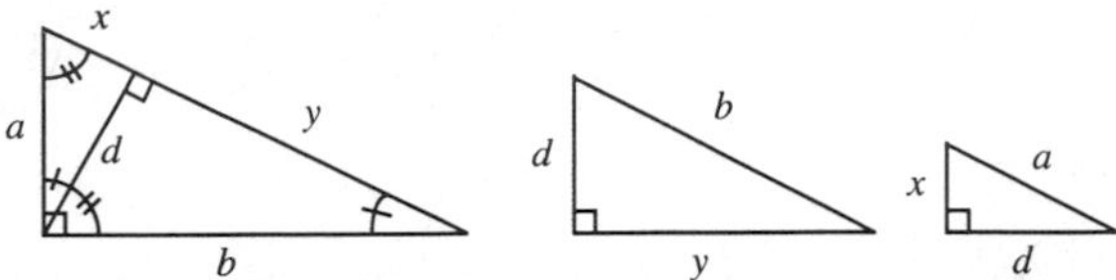

It is a property of RIGHT TRIANGLES, such as the one shown in the above left figure, that the RIGHT TRIANGLE with sides x, a, and d (small triangle in the left figure; reproduced in the right figure) is similar to the RIGHT TRIANGLE with sides d, b, and y (large triangle in the left figure; reproduced in the middle figure), giving

$$\frac{x}{a} = \frac{a}{c} \qquad \frac{y}{b} = \frac{b}{c} \tag{22}$$

$$x = \frac{a^2}{c} \qquad y = \frac{b^2}{c}, \tag{23}$$

so

$$c \equiv x + y = \frac{a^2}{c} + \frac{b^2}{c} = \frac{a^2 + b^2}{c} \tag{24}$$

$$c^2 = a^2 + b^2. \tag{25}$$

Because this proof depends on proportions of potentially IRRATIONAL NUMBERS and cannot be translated directly into a GEOMETRIC CONSTRUCTION, it was not considered valid by Euclid.

see also BRIDE'S CHAIR, COSINES LAW, PEACOCK'S TAIL, PYTHAGORAS'S THEOREM, WINDMILL

References

Ball, W. W. R. and Coxeter, H. S. M. *Mathematical Recreations and Essays, 13th ed.* New York: Dover, pp. 87–88, 1987.

Bogomolny, A. "Pythagorean Theorem." `http://www.cut-the-knot.com/pythagoras/index.html`.

Dixon, R. "The Theorem of Pythagoras." §4.1 in *Mathographics.* New York: Dover, pp. 92–95, 1991.

Dudeney, H. E. *Amusements in Mathematics.* New York: Dover, p. 32, 1958.

Dunham, W. "Euclid's Proof of the Pythagorean Theorem." Ch. 2 in *Journey Through Genius: The Great Theorems of Mathematics.* New York: Wiley, 1990.

Frederickson, G. *Dissections: Plane and Fancy.* New York: Cambridge University Press, pp. 28–29, 1997.

Garfield, J. A. "Pons Asinorum." *New England J. Educ.* **3**, 161, 1876.

Loomis, E. S. *The Pythagorean Proposition: Its Demonstration Analyzed and Classified and Bibliography of Sources for Data of the Four Kinds of "Proofs."* Reston, VA: National Council of Teachers of Mathematics, 1968.

Machover, M. "Euler's Theorem Implies the Pythagorean Proposition." *Amer. Math. Monthly* **103**, 351, 1996.

Madachy, J. S. *Madachy's Mathematical Recreations.* New York: Dover, p. 17, 1979.

Ogilvy, C. S. *Excursions in Mathematics.* New York: Dover, p. 52, 1994.

Pappas, T. "The Pythagorean Theorem," "A Twist to the Pythagorean Theorem," and "The Pythagorean Theorem and President Garfield." *The Joy of Mathematics.* San Carlos, CA: Wide World Publ./Tetra, pp. 4, 30, and 200–201, 1989.

Perigal, H. "On Geometric Dissections and Transformations." *Messenger Math.* **2**, 103–106, 1873.

Project Mathematics! *The Theorem of Pythagoras.* Videotape (22 minutes). California Institute of Technology. Available from the Math. Assoc. Amer.

Shanks, D. *Solved and Unsolved Problems in Number Theory, 4th ed.* New York: Chelsea, pp. 123–127, 1993.

Yancey, B. F. and Calderhead, J. A. "New and Old Proofs of the Pythagorean Theorem." *Amer. Math. Monthly* **3**, 65–67, 110–113, 169–171, and 299–300, 1896.

Yancey, B. F. and Calderhead, J. A. "New and Old Proofs of the Pythagorean Theorem." *Amer. Math. Monthly* **4**, 11–12, 79–81, 168–170, 250–251, and 267–269, 1897.

Yancey, B. F. and Calderhead, J. A. "New and Old Proofs of the Pythagorean Theorem." *Amer. Math. Monthly* **5**, 73–74, 1898.

Yancey, B. F. and Calderhead, J. A. "New and Old Proofs of the Pythagorean Theorem." *Amer. Math. Monthly* **6**, 33–34 and 69–71, 1899.

Pythagorean Triad

see PYTHAGOREAN TRIPLE

Pythagorean Triangle

see PYTHAGOREAN TRIPLE, RIGHT TRIANGLE

Pythagorean Triple

A Pythagorean triple is a TRIPLE of POSITIVE INTEGERS a, b, and c such that a RIGHT TRIANGLE exists with legs a, b and HYPOTENUSE c. By the PYTHAGOREAN THEOREM, this is equivalent to finding POSITIVE INTEGERS a, b, and c satisfying

$$a^2 + b^2 = c^2. \tag{1}$$

The smallest and best-known Pythagorean triple is $(a, b, c) = (3, 4, 5)$.

It is usual to consider only "reduced" (or "primitive") solutions in which a and b are RELATIVELY PRIME, since other solutions can be generated trivially from the primitive ones. For primitive solutions, one of a or b must be EVEN, and the other ODD (Shanks 1993, p. 141), with c always ODD. In addition, in every primitive Pythagorean triple, one side is always DIVISIBLE by 3 and one by 5.

Given a primitive triple (a_0, b_0, c_0), three new primitive triples are obtained from

$$(a_1, b_1, c_1) = (a_0, b_0, c_0)\mathsf{U} \quad (2)$$
$$(a_2, b_2, c_2) = (a_0, b_0, c_0)\mathsf{A} \quad (3)$$
$$(a_3, b_3, c_3) = (a_0, b_0, c_0)\mathsf{D}, \quad (4)$$

where

$$\mathsf{U} \equiv \begin{bmatrix} 1 & 2 & 2 \\ -2 & -1 & -2 \\ 2 & 2 & 3 \end{bmatrix} \quad (5)$$
$$\mathsf{A} \equiv \begin{bmatrix} 1 & 2 & 2 \\ 2 & 1 & 2 \\ 2 & 2 & 3 \end{bmatrix} \quad (6)$$
$$\mathsf{D} \equiv \begin{bmatrix} -1 & -2 & -2 \\ 2 & 1 & 2 \\ 2 & 2 & 3 \end{bmatrix}. \quad (7)$$

Roberts (1977) proves that (a, b, c) is a primitive Pythagorean triple IFF

$$(a, b, c) = (3, 4, 5)\mathsf{M}, \quad (8)$$

where M is a FINITE PRODUCT of the MATRICES U, A, D. It therefore follows that every primitive Pythagorean triple must be a member of the INFINITE array

$$\begin{array}{ccc} & & \begin{array}{l}(7, 24, 25)\\(55, 48, 73)\\(45, 28, 53)\end{array} \\ & (5, 12, 13) & \\ & & \begin{array}{l}(39, 80, 89)\\(119, 120, 169)\\(77, 36, 85)\end{array} \\ (3, 4, 5) & (21, 20, 29) & \\ & & \begin{array}{l}(33, 56, 65)\\(65, 72, 97)\\(35, 12, 37)\end{array} \\ & (15, 8, 17) & \end{array}. \quad (9)$$

For any Pythagorean triple, the PRODUCT of the two nonhypotenuse LEGS (i.e., the two smaller numbers) is always DIVISIBLE by 12, and the PRODUCT of all three sides is DIVISIBLE by 60. It is not known if there are two distinct triples having the same PRODUCT. The existence of two such triples corresponds to a NONZERO solution to the DIOPHANTINE EQUATION

$$xy(x^4 - y^4) = zw(z^4 - w^4) \quad (10)$$

(Guy 1994, p. 188).

Pythagoras and the Babylonians gave a formula for generating (not necessarily primitive) triples:

$$(2m, (m^2 - 1), (m^2 + 1)), \quad (11)$$

and Plato gave

$$(2m^2, (m^2 - 1)^2, (m^2 + 1)^2). \quad (12)$$

A general reduced solution (known to the early Greeks) is

$$(v^2 - u^2, 2uv, u^2 + v^2), \quad (13)$$

where u and v are RELATIVELY PRIME (Shanks 1993, p. 141). Let F_n be a FIBONACCI NUMBER. Then

$$(F_n F_{n+3}, 2F_{n+1}F_{n+2}, {F_{n+1}}^2 + {F_{n+2}}^2) \quad (14)$$

is also a Pythagorean triple.

For a Pythagorean triple (a, b, c),

$$P_3(a) + P_3(b) = P_3(c), \quad (15)$$

where P_3 is the PARTITION FUNCTION P (Garfunkel 1981, Honsberger 1985). Every three-term progression of SQUARES r^2, s^2, t^2 can be associated with a Pythagorean triple (X, Y, Z) by

$$r = X - Y \quad (16)$$
$$s = Z \quad (17)$$
$$t = X + Y \quad (18)$$

(Robertson 1996).

The AREA of a TRIANGLE corresponding to the Pythagorean triple $(u^2 - v^2, 2uv, u^2 + v^2)$ is

$$A = \tfrac{1}{2}(u^2 - v^2)(2uv) = uv(u^2 - v^2). \quad (19)$$

Fermat proved that a number of this form can never be a SQUARE NUMBER.

To find the number $L_p(s)$ of possible *primitive* TRIANGLES which may have a LEG (other than the HYPOTENUSE) of length s, factor s into the form

$$s = p_1^{\alpha_1} \cdots p_n^{\alpha_n}. \quad (20)$$

The number of such TRIANGLES is then

$$L_p(s) = \begin{cases} 0 & \text{for } s \equiv 2 \pmod 4 \\ 2^{n-1} & \text{otherwise,} \end{cases} \quad (21)$$

i.e., 0 for SINGLY EVEN s and 2 to the power one less than the number of distinct prime factors of s otherwise (Beiler 1966, pp. 115–116). The first few numbers for $s = 1, 2, \ldots$, are 0, 0, 1, 1, 1, 0, 1, 1, 1, 0, 1, 2, 1, 0, 2, ... (Sloane's A024361). To find the number of ways $L(s)$ in which a number s can be the LEG (other than the HYPOTENUSE) of a *primitive or nonprimitive* RIGHT TRIANGLE, write the factorization of s as

$$s = 2^{a_0} {p_1}^{\alpha_1} \cdots {p_n}^{\alpha_n}. \quad (22)$$

Then

$$L(s) = \begin{cases} \frac{1}{2}[(2a_1+1)(2a_2+1)\cdots(2a_n+1)-1] \\ \quad \text{for } a_0 = 0 \\ \frac{1}{2}[(2a_0-1)(2a_1+1)(2a_2+1)\cdots(2a_n+1)-1] \\ \quad \text{for } a_0 \geq 2 \end{cases} \tag{23}$$

(Beiler 1966, p. 116). The first few numbers for $s = 1$, 2, ... are 0, 0, 1, 1, 1, 1, 1, 2, 2, 1, 1, 4, 1, ... (Sloane's A046079).

To find the number of ways $H_p(s)$ in which a number s can be the HYPOTENUSE of a *primitive* RIGHT TRIANGLE, write its factorization as

$$s = 2^{a_0}(p_1{}^{a_1}\cdots p_n{}^{a_n})(q_1{}^{b_1}\cdots q_r{}^{b_r}), \tag{24}$$

where the ps are of the form $4x-1$ and the qs are of the form $4x+1$. The number of possible *primitive* RIGHT TRIANGLES is then

$$H_p(s) = \begin{cases} 2^{r-1} & \text{for } n = 0 \text{ and } a_0 = 0 \\ 0 & \text{otherwise,} \end{cases}. \tag{25}$$

The first few PRIMES of the form $4x+1$ are 5, 13, 17, 29, 37, 41, 53, 61, 73, 89, 97, 101, 109, 113, 137, ... (Sloane's A002144), so the smallest side lengths which are the hypotenuses of 1, 2, 4, 8, 16, ... primitive right triangles are 5, 65, 1105, 32045, 1185665, 48612265, ... (Sloane's A006278). The number of possible *primitive or nonprimitive* RIGHT TRIANGLES having s as a HYPOTENUSE is

$$H(s) = \tfrac{1}{2}[(2b_1+1)(2b_2+1)\cdots(2b_r+1)-1] \tag{26}$$

(Beiler 1966, p. 117). The first few numbers for $s = 1$, 2, ... are 0, 0, 0, 0, 1, 0, 0, 0, 0, 1, 0, 0, 1, 0, 1, 0, 1, 0, 0, ... (Sloane's A046080).

Therefore, the total number of ways in which s may be either a LEG or HYPOTENUSE of a RIGHT TRIANGLE is given by

$$T(s) = L(s) + H(s). \tag{27}$$

The values for $s = 1$, 2, ... are 0, 0, 1, 1, 2, 1, 1, 2, 2, 2, 1, 4, 2, 1, 5, 3, ... (Sloane's A046081). The smallest numbers s which may be the sides of T general RIGHT TRIANGLES for $T = 1$, 2, ... are 3, 5, 16, 12, 15, 125, 24, 40, ... (Sloane's A006593; Beiler 1966, p. 114).

There are 50 Pythagorean triples with HYPOTENUSE less than 100, the first few of which, sorted by increasing c, are $(3,4,5)$, $(6,8,10)$, $(5,12,13)$, $(9,12,15)$, $(8,15,17)$, $(12,16,20)$, $(15,20,25)$, $(7,24,25)$, $(10,24,26)$, $(20,21,29)$, $(18,24,30)$, $(16,30,34)$, $(21,28,35)$, ... (Sloane's A046083, A046084, and A046085). Of these, only 16 are primitive triplets with HYPOTENUSE less than 100: $(3,4,5)$, $(5,12,13)$, $(8,15,17)$, $(7,24,25)$, $(20,21,29)$, $(12,35,37)$, $(9,40,41)$, $(28,45,53)$, $(11,60,61)$, $(33,56,65)$, $(16,63,65)$, $(48,55,73)$, $(36,77,85)$, $(13,84,85)$, $(39,80,89)$, and $(65,72,97)$ (Sloane's A046086, A046087, and A046088). Of these 16 primitive triplets, seven are twin triplets (defined as triplets for which two members are consecutive integers). The first few twin triplets, sorted by increasing c, are $(3,4,5)$, $(5,12,13)$, $(7,24,25)$, $(20,21,29)$, $(9,40,41)$, $(11,60,61)$, $(13,84,85)$, $(15,112,113)$, ...

Let the number of triples with HYPOTENUSE less than N be denoted $\Delta(N)$, and the number of twin triplets with HYPOTENUSE less than N be denoted $\Delta_2(N)$. Then, as the following table suggests and Lehmer (1900) proved, the number of primitive solutions with HYPOTENUSE less than N satisfies

$$\lim_{N\to\infty} \frac{\Delta(N)}{N} = \frac{1}{2\pi} = 0.159155\ldots. \tag{28}$$

N	$\Delta(N)$	$\Delta(N)/N$	$\Delta_2(N)$
100	16	0.1600	7
500	80	0.1600	17
1000	158	0.1580	24
2000	319	0.1595	34
3000	477	0.1590	41
4000	639	0.1598	47
5000	792	0.1584	52
10000	1593	0.1593	74

Considering twin triplets in which the LEGS are consecutive, a closed form is available for the rth such pair. Consider the general reduced solution $(u^2-v^2, 2uv, u^2+v^2)$, then the requirement that the LEGS be consecutive integers is

$$u^2 - v^2 = 2uv \pm 1. \tag{29}$$

Rearranging gives

$$(u-v)^2 - 2v^2 = \pm 1. \tag{30}$$

Defining

$$u = x + y \tag{31}$$

$$v = y \tag{32}$$

then gives the PELL EQUATION

$$x^2 - 2y^2 = 1. \tag{33}$$

Solutions to the PELL EQUATION are given by

$$x = \frac{(1+\sqrt{2})^r + (1-\sqrt{2})^r}{2} \tag{34}$$

$$y = \frac{(1+\sqrt{2})^r - (1-\sqrt{2})^r}{2\sqrt{2}}, \tag{35}$$

so the lengths of the legs X_r and Y_r and the HYPOTENUSE Z_r are

$$X_r = u^2 - v^2 = x^2 + 2xy$$
$$= \frac{\sqrt{2}+1)^{2r+1} - (\sqrt{2}-1)^{2r+1}}{4} + \tfrac{1}{2}(-1)^r \quad (36)$$
$$Y_r = 2uv = 2xy + 2y^2$$
$$= \frac{\sqrt{2}+1)^{2r+1} - (\sqrt{2}-1)^{2r+1}}{4} - \tfrac{1}{2}(-1)^r \quad (37)$$
$$Z_r = u^2 + v^2 = x^2 + 2xy + 2y^2$$
$$= \frac{(\sqrt{2}+1)^{2r+1} + (\sqrt{2}-1)^{2r+1}}{2\sqrt{2}}. \quad (38)$$

Denoting the length of the shortest LEG by A_r then gives

$$A_r = \frac{(\sqrt{2}+1)^{2r+1} - (\sqrt{2}-1)^{2r+1}}{4} - \frac{1}{2} \quad (39)$$
$$Z_r = \frac{(\sqrt{2}+1)^{2r+1} + (\sqrt{2}-1)^{2r+1}}{2\sqrt{2}} \quad (40)$$

(Beiler 1966, pp. 124–125 and 256–257), which cannot be solved exactly to give r as a function of Z_r. However, the approximate number of leg-leg twin triplets $\Delta_2^L(N) = r$ less than a given value of $Z_r = N$ can be found by noting that the second term in the DENOMINATOR of Z_r is a small number to the power $1 + 2r$ and can therefore be dropped, leaving

$$N = Z_r > \frac{(\sqrt{2}+1)^{1+2r}}{2\sqrt{2}} \quad (41)$$

$$N > (1+2r)\ln(\sqrt{2}+1) - \ln(2\sqrt{2}). \quad (42)$$

Solving for $r = \Delta_2^L(n)$ gives

$$\Delta_2^L(N) < \frac{\ln N + \ln(2\sqrt{2}) - \ln(\sqrt{2}+1)}{2\ln(\sqrt{2}+1)}$$
$$< \left\lfloor \frac{\ln N}{2\ln(1+\sqrt{2})} \right\rfloor \quad (43)$$
$$\approx 0.567 \ln N. \quad (44)$$

The first few LEG-LEG triplets are (3, 4, 5), (20, 21, 29), (119, 120, 169), (696, 697, 985), ... (Sloane's A046089, A046090, and A046091).

LEG-HYPOTENUSE twin triples $(a, b, c) = (v^2 - u^2, 2uv, u^2 + v^2)$ occur whenever

$$u^2 + v^2 = 2uv + 1 \quad (45)$$

$$(u - v)^2 = 1, \quad (46)$$

that is to say when $v = u + 1$, in which case the HYPOTENUSE exceeds the EVEN LEG by unity and the twin triplet is given by $(1+2u, 2u(1+u), 1+2u(1+u))$. The number of leg-hypotenuse triplets with hypotenuse less than N is therefore given by

$$\Delta_2^L(N) = \left\lfloor \tfrac{1}{2}(\sqrt{2N-1} - 1) \right\rfloor, \quad (47)$$

where $\lfloor x \rfloor$ is the FLOOR FUNCTION. The first few LEG-HYPOTENUSE triples are (3, 4, 5), (5, 12, 13), (7, 24, 25), (9, 40, 41), (11, 60, 61), (13, 84, 85), ... (Sloane's A005408, A046092, and A046093).

The total number of twin triples $\Delta_2(N)$ less than N is therefore approximately given by

$$\Delta_2(N) = \Delta_2^H(N) + \Delta_2^L(N) - 1 \quad (48)$$
$$\approx \left\lfloor \tfrac{1}{2}\sqrt{2N-1} + 0.567 \ln N - 1.5 \right\rfloor, \quad (49)$$

where one has been subtracted to avoid double counting of the leg-leg-hypotenuse double-twin (3,4,5).

There is a general method for obtaining triplets of Pythagorean triangles with equal AREAS. Take the three sets of generators as

$$m_1 = r^2 + rs + s^2 \quad (50)$$
$$n_1 = r^2 - s^2 \quad (51)$$

$$m_2 = r^2 + rs + s^2 \quad (52)$$
$$n_2 = 2rs + s^2 \quad (53)$$

$$m_3 = r^2 + 2rs \quad (54)$$
$$n_3 = r^2 + rs + s^2. \quad (55)$$

Then the RIGHT TRIANGLE generated by each triple $({m_i}^2 - {n_i}^2, 2m_in_i, {m_i}^2 + {n_i}^2)$ has common AREA

$$A = rs(2r+s)(r+2s)(r+s)(r-s)(r^2+rs+s^2) \quad (56)$$

(Beiler 1966, pp. 126–127). The only EXTREMUM of this function occurs at $(r, s) = (0, 0)$. Since $A(r, s) = 0$ for $r = s$, the smallest AREA shared by three nonprimitive RIGHT TRIANGLES is given by $(r, s) = (1, 2)$, which results in an area of 840 and corresponds to the triplets (24, 70, 74), (40, 42, 58), and (15, 112, 113) (Beiler 1966, p. 126). The smallest AREA shared by three *primitive* RIGHT TRIANGLES is 13123110, corresponding to the triples (4485, 5852, 7373), (1380, 19019, 19069), and (3059, 8580, 9109) (Beiler 1966, p. 127).

One can also find quartets of RIGHT TRIANGLES with the same AREA. The QUARTET having smallest known area is (111, 6160, 6161), (231, 2960, 2969), (518, 1320, 1418), (280, 2442, 2458), with AREA 341,880 (Beiler 1966, p. 127). Guy (1994) gives additional information.

It is also possible to find sets of three and four Pythagorean triplets having the same PERIMETER (Beiler 1966, pp. 131–132). Lehmer (1900) showed that the number of primitive triples $N(p)$ with PERIMETER less than p is

$$\lim_{p\to\infty} N(p) = \frac{p\ln 2}{\pi^2} = 0.070230\ldots. \tag{57}$$

In 1643, Fermat challenged Mersenne to find a Pythagorean triplet whose HYPOTENUSE and SUM of the LEGS were SQUARES. Fermat found the smallest such solution:

$$X = 4565486027761 \tag{58}$$
$$Y = 1061652293520 \tag{59}$$
$$Z = 4687298610289, \tag{60}$$

with

$$Z = 2165017^2 \tag{61}$$
$$X + Y = 2372159^2. \tag{62}$$

A related problem is to determine if a specified INTEGER N can be the AREA of a RIGHT TRIANGLE with rational sides. 1, 2, 3, and 4 are not the AREAS of any RATIONAL-sided RIGHT TRIANGLES, but 5 is (3/2, 20/3, 41/6), as is 6 (3, 4, 5). The solution to the problem involves the ELLIPTIC CURVE

$$y^2 = x^3 - N^2 x. \tag{63}$$

A solution (a, b, c) exists if (63) has a RATIONAL solution, in which case

$$x = \tfrac{1}{4}c^2 \tag{64}$$
$$y = \tfrac{1}{8}(a^2 - b^2)c \tag{65}$$

(Koblitz 1993). There is no known general method for determining if there is a solution for arbitrary N, but a technique devised by J. Tunnell in 1983 allows certain values to be ruled out (Cipra 1996).

see also HERONIAN TRIANGLE, PYTHAGOREAN QUADRUPLE, RIGHT TRIANGLE

References

Ball, W. W. R. and Coxeter, H. S. M. *Mathematical Recreations and Essays, 13th ed.* New York: Dover, pp. 57–59, 1987.
Beiler, A H. "The Eternal Triangle." Ch. 14 in *Recreations in the Theory of Numbers: The Queen of Mathematics Entertains.* New York: Dover, 1966.
Cipra, B. "A Proof to Please Pythagoras." *Science* **271**, 1669, 1996.
Courant, R. and Robbins, H. "Pythagorean Numbers and Fermat's Last Theorem." §2.3 in Supplement to Ch. 1 in *What is Mathematics?: An Elementary Approach to Ideas and Methods, 2nd ed.* Oxford, England: Oxford University Press, pp. 40–42, 1996.
Dickson, L. E. "Rational Right Triangles." Ch. 4 in *History of the Theory of Numbers, Vol. 2: Diophantine Analysis.* New York: Chelsea, pp. 165–190, 1952.
Dixon, R. *Mathographics.* New York: Dover, p. 94, 1991.
Garfunkel, J. *Pi Mu Epsilon J.*, p. 31, 1981.
Guy, R. K. "Triangles with Integer Sides, Medians, and Area." §D21 in *Unsolved Problems in Number Theory, 2nd ed.* New York: Springer-Verlag, pp. 188–190, 1994.
Hindin, H. "Stars, Hexes, Triangular Numbers, and Pythagorean Triples." *J. Recr. Math.* **16**, 191–193, 1983–1984.
Honsberger, R. *Mathematical Gems III.* Washington, DC: Math. Assoc. Amer., p. 47, 1985.
Koblitz, N. *Introduction to Elliptic Curves and Modular Forms, 2nd ed.* New York: Springer-Verlag, pp. 1–50, 1993.
Kraitchik, M. *Mathematical Recreations.* New York: W. W. Norton, pp. 95–104, 1942.
Kramer, K. and Tunnell, J. "Elliptic Curves and Local Epsilon Factors." *Comp. Math.* **46**, 307–352, 1982.
Lehmer, D. N. "Asymptotic Evaluation of Certain Totient Sums." *Amer. J. Math.* **22**, 294–335, 1900.
Roberts, J. *Elementary Number Theory: A Problem Oriented Approach.* Cambridge, MA: MIT Press, 1977.
Robertson, J. P. "Magic Squares of Squares." *Math. Mag.* **69**, 289–293, 1996.
Shanks, D. *Solved and Unsolved Problems in Number Theory, 4th ed.* New York: Chelsea, pp. 121 and 141, 1993.
Sloane, N. J. A. Sequences A006278, A046079, A002144/M3823, and A006593/M2499 in "An On-Line Version of the Encyclopedia of Integer Sequences."
Taussky-Todd, O. "The Many Aspects of the Pythagorean Triangles." *Linear Algebra and Appl.* **43**, 285–295, 1982.

Q

$\mathbb{Q}$

The FIELD of RATIONAL NUMBERS.

see also $\mathbb{C}$, $\mathbb{C}^*$, $\mathbb{I}$, $\mathbb{N}$, $\mathbb{R}$, $\mathbb{Z}$

q-Analog

A *q*-analog, also called a *q*-EXTENSION or *q*-GENERALIZATION, is a mathematical expression parameterized by a quantity q which generalizes a known expression and reduces to the known expression in the limit $q \to 1$. There are *q*-analogs of the FACTORIAL, BINOMIAL COEFFICIENT, DERIVATIVE, INTEGRAL, FIBONACCI NUMBERS, and so on. Koornwinder, Suslov, and Bustoz, have even managed some kind of *q*-Fourier analysis.

The *q*-analog of a mathematical object is generally called the "*q*-object", hence *q*-BINOMIAL COEFFICIENT, *q*-FACTORIAL, etc. There are generally several *q*-analogs if there is one, and there is sometimes even a multibasic analog with independent q_1, q_2, ...

see also *d*-ANALOG, *q*-BETA FUNCTION, *q*-BINOMIAL COEFFICIENT, *q*-BINOMIAL THEOREM, *q*-COSINE, *q*-DERIVATIVE, *q*-FACTORIAL, *q*-GAMMA FUNCTION, *q*-SERIES, *q*-SINE, *q*-VANDERMONDE SUM

References

Exton, H. *q-Hypergeometric Functions and Applications.* New York: Halstead Press, 1983.

q-Beta Function

A *q*-ANALOG of the BETA FUNCTION

$$B(a,b) = \int_0^1 t^{a-1}(1-t)^{q-1}\,dt = \frac{\Gamma(a)\Gamma(b)}{\Gamma(a+b)},$$

where $\Gamma(z)$ is a GAMMA FUNCTION, is given by

$$B_q(a,b) \equiv \int_0^1 t^{b-1}(qt;q)_{a-1}\,d(a,t) = \frac{\Gamma_q(b)\Gamma_q(a)}{\Gamma_q(a+b)},$$

where $\Gamma_q(a)$ is a *q*-GAMMA FUNCTION and $(a;q)_n$ is a *q*-SERIES coefficient (Andrews 1986, pp. 11–12).

see also *q*-FACTORIAL, *q*-GAMMA FUNCTION

References

Andrews, G. E. *q-Series: Their Development and Application in Analysis, Number Theory, Combinatorics, Physics, and Computer Algebra.* Providence, RI: Amer. Math. Soc., 1986.

q-Binomial Coefficient

A *q*-ANALOG for the BINOMIAL COEFFICIENT, also called the GAUSSIAN COEFFICIENT. It is given by

$$\binom{n}{m}_q \equiv \frac{(q)_n}{(q)_m(q)_{n-m}} = \prod_{i=0}^{j-1} \frac{1-q^{k-i}}{1-q^{i+1}}, \tag{1}$$

where

$$(q)_k \equiv \prod_{m=1}^{\infty} \frac{1-q^m}{1-q^{k+m}}. \tag{2}$$

For example, the first few *q*-binomial coefficients are

$$\binom{2}{1}_q = \frac{1-q^2}{1-q} = 1+q \tag{3}$$

$$\binom{3}{1}_q = \binom{3}{2}_q = \frac{1-q^3}{1-q} = 1+q+q^2 \tag{4}$$

$$\binom{4}{1}_q = \binom{4}{3}_q = \frac{1-q^4}{1-q} = 1+q+q^2+q^3 \tag{5}$$

$$\binom{4}{2}_q = \frac{(1-q^3)(1-q^4)}{(1-q)(1-q^2)} = (1+q)(1+q+q^2). \tag{6}$$

From the definition, it follows that

$$\binom{n}{1}_q = \binom{n}{n-1}_q = \sum_{i=0}^{n-1} q^n. \tag{7}$$

In the LIMIT $q \to 1$, the *q*-binomial coefficient collapses to the usual BINOMIAL COEFFICIENT.

see also CAUCHY BINOMIAL THEOREM, GAUSSIAN POLYNOMIAL

q-Binomial Theorem

The *q*-ANALOG of the BINOMIAL THEOREM

$$(1-z)^n = 1 - nz + \frac{n(n-1)}{1\cdot 2}z^2 - \frac{n(n-1)(n-2)}{1\cdot 2\cdot 3}z^3 + \ldots$$

is given by

$$\left(1-\frac{z}{q^n}\right)\left(1-\frac{z}{q^{n-1}}\right)\cdots\left(1-\frac{z}{q}\right)$$
$$= 1 - \frac{1-q^n}{1-q}\frac{z}{q^n} + \frac{1-q^n}{1-q}\frac{1-q^{n-1}}{1-q^2}\frac{z^2}{q^{n+(n-1)}} - \ldots \pm \frac{z^n}{q^{n(n+1)/2}}.$$

Written as a *q*-SERIES, the identity becomes

$$\sum_{n=0}^{\infty} \frac{(a;q)_n}{(q;q)_n} z^n = \frac{(az;q)_\infty}{(z;q)_\infty},$$

where

$$(a;q)_n = \prod_{m=0}^{\infty} \frac{(1-aq^m)}{(1-aq^{m+n})}$$

(Heine 1847, p. 303; Andrews 1986). The CAUCHY BINOMIAL THEOREM is a special case of this general theorem.

see also BINOMIAL SERIES, BINOMIAL THEOREM, CAUCHY BINOMIAL THEOREM, HEINE HYPERGEOMETRIC SERIES, RAMANUJAN PSI SUM

References

Andrews, G. E. *q-Series: Their Development and Application in Analysis, Number Theory, Combinatorics, Physics, and Computer Algebra.* Providence, RI: Amer. Math. Soc., p. 10, 1986.

Heine, E. "Untersuchungen über die Reihe $1+\frac{(1-q^\alpha)(1-q^\beta)}{(1-q)(1-q^\gamma)}\cdot x+\frac{(1-q^\alpha)(1-q^{\alpha+1})(1-q^\beta)(1-q^{\beta+1})}{(1-q)(1-q^2)(1-q^\gamma)(1-q^{\gamma+1})}\cdot x^2+\ldots$" *J. reine angew. Math.* **34**, 285–328, 1847.

q-Cosine

The q-ANALOG of the COSINE function, as advocated by R. W. Gosper, is defined by

$$\cos_q(z,q) = \frac{\vartheta_2(z,p)}{\vartheta_2(0,p)},$$

where $\vartheta_2(z,p)$ is a THETA FUNCTION and p is defined via

$$(\ln p)(\ln q) = \pi^2.$$

This is a period 2π, EVEN FUNCTION of unit amplitude with double and triple angle formulas and addition formulas which are analogous to ordinary SINE and COSINE. For example,

$$\cos_q(2z,q) = {\cos_q}^2(z,q^2) - {\sin_q}^2(z,q^2),$$

where $\sin_q(z,a)$ is the q-SINE, and π_q is q-PI. The q-cosine also satisfies

$$\cos_q(\pi a) = \frac{\sum_{n=-\infty}^{\infty}(-1)^n q^{(n+a)^2}}{\sum_{n=-\infty}^{\infty}(-1)^n q^{n^2}}.$$

see also q-FACTORIAL, q-SINE

References

Gosper, R. W. "Experiments and Discoveries in q-Trigonometry." Unpublished manuscript.

q-Derivative

The q-ANALOG of the DERIVATIVE, defined by

$$\left(\frac{d}{dx}\right)_q f(x) = \frac{f(x)-f(qx)}{x-qx}.$$

For example,

$$\left(\frac{d}{dx}\right)_q \sin x = \frac{\sin x - \sin(qx)}{x-qx}$$

$$\left(\frac{d}{dx}\right)_q \ln x = \frac{\ln x - \ln(qx)}{x-qx} = \frac{\ln\left(\frac{1}{q}\right)}{(1-q)x}$$

$$\left(\frac{d}{dx}\right)_q x^2 = \frac{x^2-q^2x^2}{x-qx} = (1+q)x$$

$$\left(\frac{d}{dx}\right)_q x^3 = \frac{x^3-q^3x^3}{x-qx} = (1+q+q^2)x^2.$$

In the LIMIT $q \to 1$, the q-derivative reduces to the usual DERIVATIVE.

see also DERIVATIVE

q-Dimension

$$D_q \equiv \frac{1}{1-q}\lim_{\epsilon\to 0}\frac{\ln I(q,\epsilon)}{\ln\left(\frac{1}{\epsilon}\right)}, \quad (1)$$

where

$$I(q,\epsilon) \equiv \sum_{i=1}^{N} {\mu_i}^q, \quad (2)$$

ϵ is the box size, and μ_i is the NATURAL MEASURE. If $q_1 > q_2$, then

$$D_{q_1} \leq D_{q_2}. \quad (3)$$

The CAPACITY DIMENSION (a.k.a. BOX COUNTING DIMENSION) is given by $q=0$,

$$D_0 = \frac{1}{1-0}\lim_{\epsilon\to 0}\frac{\ln\left(\sum_{i=1}^{N(\epsilon)} 1\right)}{-\ln\epsilon} = -\lim_{\epsilon\to 0}\frac{\ln[N(\epsilon)]}{\ln\epsilon}. \quad (4)$$

If all μ_is are equal, then the CAPACITY DIMENSION is obtained for any q. The INFORMATION DIMENSION is defined by

$$D_1 = \lim_{q\to 1} D_q = \lim_{q\to 1}\frac{\lim_{\epsilon\to 0}\frac{\ln\left[\sum_{i=1}^{N(\epsilon)} {\mu_i}^q\right]}{-\ln\epsilon}}{1-q}$$

$$= \lim_{\epsilon\to 0}\lim_{q\to 1}\frac{\ln\left(\sum_{i=1}^{N(\epsilon)} {\mu_i}^q\right)}{\ln\epsilon(q-1)}. \quad (5)$$

But

$$\lim_{q\to 1}\ln\left(\sum_{i=1}^{N(\epsilon)} {\mu_i}^q\right) = \ln\left(\sum_{i=1}^{N(\epsilon)} \mu_i\right) = \ln 1 = 0, \quad (6)$$

so use L'HOSPITAL'S RULE

$$D_1 = \lim_{\epsilon\to 0}\left(\frac{1}{\ln\epsilon}\lim_{q\to 1}\frac{\sum q{\mu_i}^{q-1}}{\sum {\mu_i}^q}\right). \quad (7)$$

Therefore,

$$D_1 = \lim_{\epsilon\to 0}\frac{\sum_{i=1}^{N(\epsilon)} \mu_i \ln\mu_i}{\ln\epsilon}. \quad (8)$$

D_2 is called the CORRELATION DIMENSION. The q-dimensions satisfy

$$D_{q+1} \leq D_q. \quad (9)$$

see also FRACTAL DIMENSION

Q.E.D.

An abbreviation for the Latin phrase "quod erat demonstrandum" ("that which was to be demonstrated"), a NOTATION which is often placed at the end of a mathematical proof to indicate its completion.

q-Extension

see q-ANALOG

q-Factorial

The q-ANALOG of the FACTORIAL (by analogy with the q-GAMMA FUNCTION). For a an integer, the q-factorial is defined by

$$\operatorname{faq}(a,q) = 1(1+q)(1+q+q^2)\cdots(1+q+\ldots+q^{a-1}).$$

A reflection formula analogous to the GAMMA FUNCTION reflection formula is given by

$$\begin{aligned}\cos_q(\pi a) &= \sin_q[\pi(\tfrac{1}{2}-a)]\\ &= \frac{\pi_q q^{(a-1/2)(a+1/2)}}{\operatorname{faq}(a-\frac{1}{2},q^2)\operatorname{faq}(-(a+\frac{1}{2}),q^2)},\end{aligned}$$

where $\cos_q(z)$ is the q-COSINE, $\sin_q(z)$ is the q-SINE, and π_q is q-PI.

see also q-BETA FUNCTION, q-COSINE, q-GAMMA FUNCTION, q-PI, q-SINE

References

Gosper, R. W. "Experiments and Discoveries in q-Trigonometry." Unpublished manuscript.

Q-Function

Let

$$q = e^{-\pi K'/K} = e^{-i\pi\tau}, \tag{1}$$

then

$$Q_0 \equiv \prod_{n=1}^{\infty}(1-q^{2n}) \tag{2}$$

$$Q_1 \equiv \prod_{n=1}^{\infty}(1+q^{2n}) \tag{3}$$

$$Q_2 \equiv \prod_{n=1}^{\infty}(1+q^{2n-1}) \tag{4}$$

$$Q_3 \equiv \prod_{n=1}^{\infty}(1-q^{2n-1}). \tag{5}$$

The Q-functions are sometimes written using a lowercase q instead of a capital Q. The Q-functions also satisfy the identities

$$Q_0Q_1 = Q_0(q^2) \tag{6}$$

$$Q_0Q_3 = Q_0(q^{1/2}) \tag{7}$$

$$Q_2Q_3 = Q_3(q^2) \tag{8}$$

$$Q_1Q_2 = Q_1(q^{1/2}). \tag{9}$$

see also JACOBI IDENTITIES, q-SERIES

References

Borwein, J. M. and Borwein, P. B. *Pi & the AGM: A Study in Analytic Number Theory and Computational Complexity.* New York: Wiley, pp. 55 and 63–85, 1987.
Tannery, J. and Molk, J. *Elements de la Théorie des Fonctions Elliptiques, 4 vols.* Paris: Gauthier-Villars et fils, 1893–1902.
Whittaker, E. T. and Watson, G. N. *A Course in Modern Analysis, 4th ed.* Cambridge, England: Cambridge University Press, pp. 469–473 and 488–489, 1990.

q-Gamma Function

A q-ANALOG of the GAMMA FUNCTION defined by

$$\Gamma_q(x,q) \equiv \frac{(q;q)_\infty}{(q^x;q)_\infty}(1-q)^{1-x}, \tag{1}$$

where $(x,q)_\infty$ is a q-SERIES. The q-gamma function satisfies

$$\lim_{q\to 1^-}\Gamma_q(x) = \Gamma(x) \tag{2}$$

(Andrews 1986).

A curious identity for the functional equation

$$\begin{aligned}f(a-b)f(a-c)f(a-d)f(a-e) - f(b)f(c)f(d)f(e)\\ = q^b f(a)f(a-b-c)f(a-b-d)f(a-b-e),\end{aligned} \tag{3}$$

where

$$b+c+d+e = 2a \tag{4}$$

is given by

$$f(\alpha) = \begin{cases}\sin(k\alpha) & \text{for } q=1\\ \frac{1}{\Gamma_q(\alpha)\Gamma_q(1-\alpha)} & \text{for } 0<q<1,\end{cases} \tag{5}$$

for any k.

see also q-BETA FUNCTION, q-FACTORIAL

References

Andrews, G. E. "W. Gosper's Proof that $\lim_{q\to 1^-}\Gamma_q(x) = \Gamma(x)$." Appendix A in *q-Series: Their Development and Application in Analysis, Number Theory, Combinatorics, Physics, and Computer Algebra.* Providence, RI: Amer. Math. Soc., p. 11 and 109, 1986.
Wenchang, C. Problem 10226 and Solution. "A q-Trigonometric Identity." *Amer. Math. Monthly* **103**, 175–177, 1996.

q-Generalization

see q-ANALOG

q-Hypergeometric Series

see HEINE HYPERGEOMETRIC SERIES

Q-Matrix

see FIBONACCI Q-MATRIX

Q-Number

see HOFSTADTER'S Q-SEQUENCE

q-Pi

The q-ANALOG of PI π_q can be defined by taking $a = 0$ in the q-FACTORIAL

$$\text{faq}(a,q) = 1(1+q)(1+q+q^2)\cdots(1+q+\ldots+q^{a-1}),$$

giving

$$1 = \sin_q(\tfrac{1}{2}\pi) = \frac{\pi_q}{\text{faq}^2(-\frac{1}{2},q^2)q^{1/4}},$$

where $\sin_q(z)$ is the q-SINE. Gosper has developed an iterative algorithm for computing π based on the algebraic RECURRENCE RELATION

$$\frac{4\pi_{q^4}}{q^4+1} = \frac{(q^2+1)^2{\pi_q}^2}{\pi_{q^2}} - \frac{(q^4+1){\pi_{q^2}}^2}{\pi_{q^4}}.$$

Q-Polynomial

see BLM/HO POLYNOMIAL

q-Product

see Q-FUNCTION

q-Series

A SERIES involving coefficients of the form

$$(a)_n \equiv (a;q)_n \equiv \prod_{k=0}^{\infty} \frac{(1-aq^k)}{(1-aq^{k+n})} \tag{1}$$

$$= \prod_{k=0}^{n-1}(1-aq^k) \tag{2}$$

(Andrews 1986). The symbols

$$[n] \equiv 1+q+q^2+\ldots+q^{n-1} \tag{3}$$

$$[n]! \equiv [n][n-1]\cdots[1] \tag{4}$$

are sometimes also used when discussing q-series.

There are a great many beautiful identities involving q-series, some of which follow directly by taking the q-ANALOG of standard combinatorial identities, e.g., the q-BINOMIAL THEOREM

$$\sum_{n=0}^{\infty} \frac{(a;q)_n z^n}{(q;q)_n} = \frac{(az;q)_\infty}{(z;q)_\infty} \tag{5}$$

($|z| < 1$, $|q| < 1$; Andrews 1986, p. 10) and q-VANDERMONDE SUM

$${}_2\phi_1(a,q^{-n};c;q,q) = \frac{a^n(c/a,q)_n}{(c;q)_n}, \tag{6}$$

where ${}_2\phi_1(a,b;c;q,z)$ is a HEINE HYPERGEOMETRIC SERIES. Other q-series identities, e.g., the JACOBI IDENTITIES, ROGERS-RAMANUJAN IDENTITIES, and HEINE HYPERGEOMETRIC IDENTITY

$${}_2\phi_1(a,b;c;q,z) = \frac{(b;q)_\infty(az;q)_\infty}{(c;q)_\infty(z;q)_\infty}{}_2\phi_1(c/b,a;az;q,b), \tag{7}$$

seem to arise out of the blue.

see also BORWEIN CONJECTURES, FINE'S EQUATION, GAUSSIAN COEFFICIENT, HEINE HYPERGEOMETRIC SERIES, JACKSON'S IDENTITY, JACOBI IDENTITIES, MOCK THETA FUNCTION, q-ANALOG, q-BINOMIAL THEOREM, q-COSINE, q-FACTORIAL, Q-FUNCTION, q-GAMMA FUNCTION, q-SINE, RAMANUJAN PSI SUM, RAMANUJAN THETA FUNCTIONS, ROGERS-RAMANUJAN IDENTITIES

References

Andrews, G. E. *q-Series: Their Development and Application in Analysis, Number Theory, Combinatorics, Physics, and Computer Algebra.* Providence, RI: Amer. Math. Soc., 1986.

Berndt, B. C. "q-Series." Ch. 27 in *Ramanujan's Notebooks, Part IV.* New York: Springer-Verlag, pp. 261–286, 1994.

Gasper, G. and Rahman, M. *Basic Hypergeometric Series.* Cambridge, England: Cambridge University Press, 1990.

Gosper, R. W. "Experiments and Discoveries in q-Trigonometry." Unpublished manuscript.

Q-Signature

see SIGNATURE (RECURRENCE RELATION)

q-Sine

The q-ANALOG of the SINE function, as advocated by R. W. Gosper, is defined by

$$\sin_q(z,q) = \frac{\vartheta_1(z,p)}{\vartheta_1(\frac{1}{2}\pi,p)},$$

where $\vartheta_1(z,p)$ is a THETA FUNCTION and p is defined via

$$(\ln p)(\ln q) = \pi^2.$$

This is a period 2π, ODD FUNCTION of unit amplitude with double and triple angle formulas and addition formulas which are analogous to ordinary SINE and COSINE. For example,

$$\sin_q(2z,q) = (q+1)\frac{\pi_q}{p_{q^2}}\cos_q(z,q^2)\sin_q(z,q^2),$$

where $\cos_q(z,a)$ is the q-COSINE, and π_q is q-PI.

see also q-COSINE, q-FACTORIAL

References

Gosper, R. W. "Experiments and Discoveries in q-Trigonometry." Unpublished manuscript.

q-Vandermonde Sum

$$ {}_2\phi_1(a, q^{-n}; c; q, q) = \frac{a^n (c/a, q)_n}{(c; q)_n}, $$

where ${}_2\phi_1(a, b; c; q, z)$ is a HEINE HYPERGEOMETRIC SERIES.

see also CHU-VANDERMONDE IDENTITY, HEINE HYPERGEOMETRIC SERIES

References

Andrews, G. E. *q-Series: Their Development and Application in Analysis, Number Theory, Combinatorics, Physics, and Computer Algebra.* Providence, RI: Amer. Math. Soc., pp. 15–16, 1986.

QR Decomposition

Given a MATRIX A, its QR-decomposition is of the form

$$ \mathsf{A} = \mathsf{QR}, $$

where R is an upper TRIANGULAR MATRIX and Q is an ORTHOGONAL MATRIX, i.e., one satisfying

$$ \mathsf{Q}^{\mathrm{T}}\mathsf{Q} = \mathsf{I}, $$

where I is the IDENTITY MATRIX. This matrix decomposition can be used to solve linear systems of equations.

see also CHOLESKY DECOMPOSITION, LU DECOMPOSITION, SINGULAR VALUE DECOMPOSITION

References

Householder, A. S. *The Numerical Treatment of a Single Non-Linear Equations.* New York: McGraw-Hill, 1970.

Nash, J. C. *Compact Numerical Methods for Computers: Linear Algebra and Function Minimisation, 2nd ed.* Bristol, England: Adam Hilger, pp. 26–28, 1990.

Press, W. H.; Flannery, B. P.; Teukolsky, S. A.; and Vetterling, W. T. "QR Decomposition." §2.10 in *Numerical Recipes in FORTRAN: The Art of Scientific Computing, 2nd ed.* Cambridge, England: Cambridge University Press, pp. 91–95, 1992.

Stewart, G. W. "A Parallel Implementation of the QR Algorithm." *Parallel Comput.* **5**, 187–196, 1987. `ftp://thales.cs.umd.edu/pub/reports/piqra.ps`.

Quadrable

A plane figure for which QUADRATURE is possible is said to be quadrable.

Quadrangle

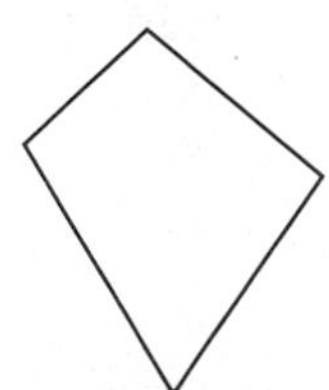

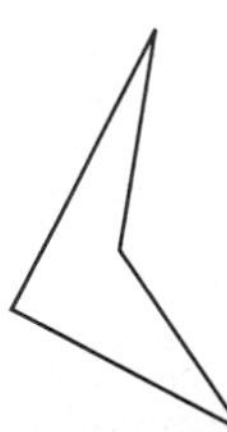

A plane figure consisting of four points, each of which is joined to two other points by a LINE SEGMENT (where the line segments may intersect). A quadrangle may therefore be CONCAVE or CONVEX; if it is CONVEX, it is called a QUADRILATERAL.

see also COMPLETE QUADRANGLE, CYCLIC QUADRANGLE, QUADRILATERAL

Quadrant

$x < 0, y > 0$ Quadrant 2 | Quadrant 1 $x > 0, y > 0$

$x < 0, y < 0$ Quadrant 3 | Quadrant 4 $x > 0, y < 0$

One of the four regions of the PLANE defined by the four possible combinations of SIGNS $(+, +)$, $(+, -)$, $(-, +)$, and $(-, -)$ for (x, y).

see also OCTANT, x-AXIS, y-AXIS

References

Courant, R. and Robbins, H. *What is Mathematics?: An Elementary Approach to Ideas and Methods, 2nd ed.* Oxford, England: Oxford University Press, p. 73, 1996.

Quadratfrei

see SQUAREFREE

Quadratic Congruence

A CONGRUENCE of the form

$$ ax^2 + bx + c \equiv 0 \pmod{m}, $$

where a, b, and c are INTEGERS. A general quadratic congruence can be reduced to the congruence

$$ x^2 \equiv q \pmod{p} $$

and can be solved using EXCLUDENTS, although solution of the general polynomial congruence

$$ a_m x^m + \ldots + a_2 x^2 + a_1 x + a_0 \equiv 0 \pmod{n} $$

is intractable.

see also CONGRUENCE, EXCLUDENT, LINEAR CONGRUENCE

Quadratic Curve

The general 2-variable quadratic equation can be written

$$ ax^2 + 2bxy + cy^2 + 2dx + 2fy + g = 0. \quad (1) $$

Define the following quantities:

$$ \Delta = \begin{vmatrix} a & b & d \\ b & c & f \\ d & f & g \end{vmatrix} \quad (2) $$

$$ J = \begin{vmatrix} a & b \\ b & c \end{vmatrix} \quad (3) $$

$$ I = a + c \quad (4) $$

$$ K = \begin{vmatrix} a & d \\ d & g \end{vmatrix} + \begin{vmatrix} c & f \\ f & g \end{vmatrix}. \quad (5) $$

Then the quadratics are classified into the types summarized in the following table (Beyer 1987). The real (nondegenerate) quadratics (the ELLIPSE, HYPERBOLA, and PARABOLA) correspond to the curves which can be created by the intersection of a PLANE with a (two-NAPPES) CONE, and are therefore known as CONIC SECTIONS.

Curve	Δ	J	Δ/I	K
coincident lines	0	0		0
ellipse (imaginary)	$\neq 0$	>0	>0	
ellipse (real)	$\neq 0$	>0	<0	
hyperbola	$\neq 0$	<0		
intersecting lines (imaginary)	0	>0		
intersecting lines (real)	0	<0		
parabola	$\neq 0$	0		
parallel lines (imaginary)	0	0		>0
parallel lines (real)	0	0		<0

It is always possible to eliminate the xy cross term by a suitable ROTATION of the axes. To see this, consider rotation by an arbitrary angle θ. The ROTATION MATRIX is

$$\begin{bmatrix} x \\ y \end{bmatrix} = \begin{bmatrix} \cos\theta & \sin\theta \\ -\sin\theta & \cos\theta \end{bmatrix} \begin{bmatrix} x' \\ y' \end{bmatrix} = \begin{bmatrix} x'\cos\theta + y'\sin\theta \\ -x'\sin\theta + y'\cos\theta \end{bmatrix}, \tag{6}$$

so

$$x = x'\cos\theta + y'\sin\theta \tag{7}$$

$$y = -x'\sin\theta + y'\cos\theta \tag{8}$$

$$xy = -x'^2\cos\theta\sin\theta + x'y'(\cos^2\theta - \sin^2\theta) + y'^2\cos\theta\sin\theta \tag{9}$$

$$x^2 = x'^2\cos^2\theta + 2x'y'\cos\theta\sin\theta + y'^2\sin^2\theta \tag{10}$$

$$y^2 = -x'^2\sin^2\theta - 2x'y'\sin\theta\cos\theta + y'^2\cos^2\theta. \tag{11}$$

Plugging these into (1) gives

$$\begin{aligned} &a(x'^2\cos^2\theta + 2x'y'\cos\theta + y'^2\sin^2\theta) \\ &+2b(x'\cos\theta + y'\sin\theta)(-x'\sin\theta + y'\cos\theta) \\ &+c(x'^2\sin^2\theta - 2x'y'\cos\theta\sin\theta + y'^2\cos^2\theta) \\ &+2d(x'\cos\theta + y'\sin\theta) \\ &+2f(-x'\sin\theta + y'\cos\theta) + g = 0. \end{aligned} \tag{12}$$

Rewriting,

$$\begin{aligned} &a(x'^2\cos^2\theta + 2x'y'\cos\theta + y'^2\sin^2\theta) \\ &+2b(-x^2\cos^2\theta\sin\theta - xy\sin^2\theta + xy\cos^2\theta + y^2\cos\theta\sin\theta) \\ &+c(x'^2\sin^2\theta - 2x'y'\cos\theta\sin\theta + y'^2\cos^2\theta) \\ &+2d(x'\cos\theta + y'\sin\theta) \\ &+2f(-x'\sin\theta + y'\cos\theta) + g = 0. \end{aligned} \tag{13}$$

Grouping terms,

$$\begin{aligned} &x'^2(a\cos^2\theta + c\sin^2\theta - 2b\cos\theta\sin\theta) \\ &+x'y'[2a\cos\theta\sin\theta - 2c\sin\theta\cos\theta + 2b(\cos^2\theta - \sin^2\theta)] \\ &+y'^2(a\sin^2\theta + c\cos^2\theta + 2b\cos\theta\sin\theta) \\ &+x'(2d\cos\theta - 2f\sin\theta) + y'(-2d\sin\theta + 2f\cos\theta) \\ &+g = 0. \end{aligned} \tag{14}$$

Comparing the COEFFICIENTS with (1) gives an equation of the form

$$a'x'^2 + 2b'x'y' + c'y'^2 + 2d'x' + 2f'y' + g' = 0, \tag{15}$$

where the new COEFFICIENTS are

$$a' = a\cos^2\theta - 2b\cos\theta\sin\theta + c\sin^2\theta \tag{16}$$

$$b' = b(\cos^2\theta - \sin^2\theta) + (a-c)\sin\theta\cos\theta \tag{17}$$

$$c' = a\sin^2\theta + 2b\sin\theta\cos\theta + c\cos^2\theta \tag{18}$$

$$d' = d\cos\theta - f\sin\theta \tag{19}$$

$$f' = -d\sin\theta + f\cos\theta \tag{20}$$

$$g' = g. \tag{21}$$

The cross term $2b'x'y'$ can therefore be made to vanish by setting

$$\begin{aligned} b' &= b(\cos^2\theta - \sin^2\theta) - (c-a)\sin\theta\cos\theta \\ &= b\cos(2\theta) - \tfrac{1}{2}(c-a)\sin(2\theta) = 0. \end{aligned} \tag{22}$$

For b' to be zero, it must be true that

$$\cot(2\theta) = \frac{c-a}{2b} \equiv K. \tag{23}$$

The other components are then given with the aid of the identity

$$\cos[\cot^{-1}(x)] = \frac{x}{\sqrt{1+x^2}} \tag{24}$$

by defining

$$L \equiv \frac{K}{\sqrt{1+K^2}}, \tag{25}$$

so

$$\sin\theta = \sqrt{\frac{1-L}{2}} \tag{26}$$

$$\cos\theta = \sqrt{\frac{1+L}{2}}. \tag{27}$$

Rotating by an angle

$$\theta = \tfrac{1}{2}\cot^{-1}\left(\frac{c-a}{2b}\right) \tag{28}$$

therefore transforms (1) into

$$a'x'^2 + c'y'^2 + 2d'x' + 2f'y' + g' = 0. \tag{29}$$

Completing the Square,

$$a'\left(x'^2+\frac{2d'}{a'}x\right)+c'\left(y'^2+\frac{2f'}{c'}y'\right)+g'=0 \quad (30)$$

$$a'\left(x'+\frac{d'}{a'}\right)^2+c'\left(y'+\frac{f'}{c'}\right)^2=-g'+\frac{d'^2}{a'}+\frac{f'^2}{c'}. \quad (31)$$

Defining $x''\equiv x'+d'/a'$, $y''\equiv y'+f'/c'$, and $g''\equiv -g'+d'^2/a'+f'^2/c'$ gives

$$a'x''^2+c'y''^2=g''. \quad (32)$$

If $g''\neq 0$, then divide both sides by g''. Defining $a''\equiv a'/g''$ and $c''\equiv c'/g''$ then gives

$$a''x''^2+c''y''^2=1. \quad (33)$$

Therefore, in an appropriate coordinate system, the general Conic Section can be written (dropping the primes) as

$$\begin{cases} ax^2+cy^2=1 & a,c,g\neq 0\\ ax^2+cy^2=0 & a,c\neq 0,\ g=0.\end{cases} \quad (34)$$

Consider an equation of the form $ax^2+2bxy+cy^2=1$ where $b\neq 0$. Re-express this using t_1 and t_2 in the form

$$ax^2+2bxy+cy^2=t_1x'^2+t_2y'^2. \quad (35)$$

Therefore, rotate the Coordinate System

$$\begin{bmatrix}x'\\ y'\end{bmatrix}=\begin{bmatrix}\cos\theta & \sin\theta\\ -\sin\theta & \cos\theta\end{bmatrix}\begin{bmatrix}x\\ y\end{bmatrix}, \quad (36)$$

so

$$\begin{aligned}ax^2+2bxy+cy^2&=t_1x'^2+t_2y'^2\\ &=t_1(x^2\cos^2\theta+2xy\cos\theta\sin\theta+y^2\sin^2\theta)\\ &\quad+t_2(x^2\sin^2\theta-2xy\sin\theta\cos\theta+y^2\cos^2\theta)\\ &=x^2(t_1\cos^2\theta+t_2\sin^2\theta)+2xy\cos\theta\sin\theta(t_1-t_2)\\ &\quad+y^2(t_1\sin^2\theta+t_2\cos^2\theta)\end{aligned} \quad (37)$$

and

$$a=t_1\cos^2\theta+t_2\sin^2\theta \quad (38)$$

$$b=(t_1-t_2)\cos\theta\sin\theta=\tfrac{1}{2}(t_1-t_2)\sin(2\theta) \quad (39)$$

$$c=t_1\sin^2\theta+t_2\cos^2\theta. \quad (40)$$

Therefore,

$$\begin{aligned}a+c&=(t_1\cos^2\theta+t_2\sin^2\theta)+(t_1\sin^2\theta+t_2\cos^2\theta)\\ &=t_1+t_2\end{aligned} \quad (41)$$

$$\begin{aligned}a-c&=t_1\cos^2\theta+t_2\sin^2\theta-t_1\sin^2\theta+t_2\cos^2\theta\\ &=(t_1-t_2)(\cos^2\theta-\sin^2\theta)=(t_1-t_2)\cos(2\theta).\end{aligned} \quad (42)$$

From (41) and (42),

$$\frac{a-c}{b}=\frac{(t_1-t_2)\cos(2\theta)}{\frac{1}{2}(t_1-t_2)\sin(2\theta)}=2\cot(2\theta), \quad (43)$$

the same angle as before. But

$$\begin{aligned}\cos(2\theta)&=\cos\left[\cot^{-1}\left(\frac{a-c}{2b}\right)\right]\\ &=\cos\left[\tan^{-1}\left(\frac{2b}{a-c}\right)\right]\\ &=\frac{1}{\sqrt{1+\left(\frac{2b}{a-c}\right)^2}},\end{aligned} \quad (44)$$

so

$$a-c=\frac{t_1-t_2}{\sqrt{1+\left(\frac{2b}{a-c}\right)^2}}. \quad (45)$$

Rewriting and copying (41),

$$\begin{aligned}t_1-t_2&=(a-c)\sqrt{1+\left(\frac{2b}{a-c}\right)^2}\\ &=\sqrt{(a-c)^2+4b^2}\end{aligned} \quad (46)$$

$$t_1+t_2=a+c. \quad (47)$$

Adding (46) and (47) gives

$$t_1=\tfrac{1}{2}[a+c+\sqrt{(a-c)^2+4b^2}\,] \quad (48)$$

$$t_2=a+c-t_1=\tfrac{1}{2}[a+c-\sqrt{(a-c)^2+4b^2}\,]. \quad (49)$$

Note that these Roots can also be found from

$$(t-t_1)(t-t_2)=t^2-t(t_1+t_2)+t_1t_2=0 \quad (50)$$

$$\begin{aligned}&t^2-t(a+c)+\tfrac{1}{4}\{(a+c)^2-[(a-c)^2+4b^2]\}\\ &=t^2-t(a+c)\\ &\quad+\tfrac{1}{4}[a^2+2ac+c^2-a^2+2ac-c^2-4b^2]\\ &=t^2-t(a+c)+(ac-b^2)=(a-t)(c-t)-b^2\\ &=\begin{vmatrix}a-t & b\\ b & c-t\end{vmatrix}=(a-t)(c-t)-b^2=0.\end{aligned} \quad (51)$$

The original problem is therefore equivalent to looking for a solution to

$$\begin{bmatrix}a & b\\ b & c\end{bmatrix}\begin{bmatrix}x\\ y\end{bmatrix}=t\begin{bmatrix}x\\ y\end{bmatrix} \quad (52)$$

$$\begin{bmatrix}ax & bx\\ by & cy\end{bmatrix}\begin{bmatrix}x\\ y\end{bmatrix}=t\begin{bmatrix}x^2\\ y^2\end{bmatrix}, \quad (53)$$

which gives the simultaneous equations

$$\begin{cases}ax^2+bxy=tx^2\\ bxy+cy^2=ty^2.\end{cases} \quad (54)$$

Let $\mathbf{X}$ be any point (x, y) with old coordinates and (x', y') be its new coordinates. Then

$$ax^2 + 2bxy + cy^2 = t_+ x'^2 + t_- y'^2 = 1 \tag{55}$$

and

$$x' = \hat{\mathbf{X}}_+ \cdot \begin{bmatrix} x \\ y \end{bmatrix} \tag{56}$$

$$y' = \hat{\mathbf{X}}_- \cdot \begin{bmatrix} x \\ y \end{bmatrix}. \tag{57}$$

If t_+ and t_- are both > 0, the curve is an ELLIPSE. If t_+ and t_- are both < 0, the curve is empty. If t_+ and t_- have opposite SIGNS, the curve is a HYPERBOLA. If either is 0, the curve is a PARABOLA.

To find the general form of a quadratic curve in POLAR COORDINATES (as given, for example, in Moulton 1970), plug $x = r\cos\theta$ and $y = r\sin\theta$ into (1) to obtain

$$ar^2\cos^2\theta + 2br^2\cos\theta\sin\theta + cr^2\sin^2\theta + 2dr\cos\theta + 2fr\sin\theta + g = 0 \tag{58}$$

$$(a\cos^2\theta + 2b\cos\theta\sin\theta + c\sin^2\theta) + \frac{2}{r}(d\cos\theta + f\sin\theta) + \frac{g}{r^2} = 0. \tag{59}$$

Define $u \equiv 1/r$. For $g \neq 0$,we can divide through by $2g$,

$$\tfrac{1}{2}u^2 + \frac{1}{g}(d\cos\theta + f\sin\theta)u + \frac{1}{2g}(a\cos^2\theta + 2b\cos\theta\sin\theta + c\sin^2\theta) = 0. \tag{60}$$

Applying the QUADRATIC FORMULA gives

$$u = -\frac{d}{g}\cos\theta - \frac{f}{g}\sin\theta \pm \sqrt{R}, \tag{61}$$

where

$$\begin{aligned} R &\equiv \frac{(d\cos\theta + f\sin\theta)^2}{g^2} - 4\left(\frac{1}{2}\right)\left(\frac{1}{2g}\right)(a\cos^2\theta + 2b\cos\theta\sin\theta + c\sin^2\theta) \\ &= \frac{d^2}{g^2}\cos^2\theta + \frac{2df}{g^2}\cos\theta\sin\theta + \frac{f^2}{g^2}\sin^2\theta - \frac{1}{g}(a\cos^2\theta + 2b\cos\theta\sin\theta + c\sin^2\theta). \end{aligned} \tag{62}$$

Using the trigonometric identities

$$\sin^2\theta = 1 - \cos^2\theta \tag{63}$$

$$\sin(2\theta) = 2\sin\theta\cos\theta, \tag{64}$$

it follows that

$$\begin{aligned} R &= \left(\frac{d^2}{g^2} - \frac{a}{g} - \frac{f^2}{g^2} + \frac{c}{g}\right)\cos^2\theta + \left(\frac{df}{g^2} - \frac{b}{g}\right)\sin(2\theta) + \left(\frac{f^2}{g^2} - \frac{c}{g}\right) \\ &= \tfrac{1}{2}[1 + \cos(2\theta)]\frac{d^2 - ag - f^2 + cg}{g^2} + \sin(2\theta)\left(\frac{df - bg}{g^2}\right) + \frac{f^2 - cg}{g^2} \\ &= \frac{d^2 - ag - f^2 + cg}{2g^2}\cos(2\theta) + \frac{df - bg}{g^2}\sin(2\theta) + \frac{d^2 - ag - f^2 + cg + 2f^2 - 2cg}{2g^2}. \end{aligned} \tag{65}$$

Defining

$$A \equiv -\frac{f}{g} \tag{66}$$

$$B \equiv -\frac{d}{g} \tag{67}$$

$$C \equiv \frac{df - bg}{g^2} \tag{68}$$

$$D \equiv \frac{d^2 - f^2 + cg - ag}{2g^2} \tag{69}$$

$$E \equiv \frac{d^2 + f^2 - ag - cg}{2g^2} \tag{70}$$

then gives the equation

$$u \equiv \frac{1}{r} = A\sin\theta + B\cos\theta \pm \sqrt{C\sin(2\theta) + D\cos(2\theta) + E} \tag{71}$$

(Moulton 1970). If $g = 0$, then (59) becomes instead

$$u \equiv \frac{1}{r} = -\frac{a\cos^2\theta + 2b\cos\theta\sin\theta + c\sin^2\theta}{2(d\cos\theta + f\sin\theta)}. \tag{72}$$

Therefore, the general form of a quadratic curve in polar coordinates is given by

$$u = \begin{cases} A\sin\theta + B\cos\theta \pm \sqrt{C\sin(2\theta) + D\cos(2\theta) + E} & \text{for } g \neq 0 \\ -\frac{a\cos^2\theta + 2b\cos\theta\sin\theta + c\sin^2\theta}{2(d\cos\theta + f\sin\theta)} & \text{for } g = 0. \end{cases} \tag{73}$$

see also CONIC SECTION, DISCRIMINANT (QUADRATIC CURVE), ELLIPTIC CURVE

References

Beyer, W. H. *CRC Standard Mathematical Tables, 28th ed.* Boca Raton, FL: CRC Press, pp. 200–201, 1987.

Casey, J. "The General Equation of the Second Degree." Ch. 4 in *A Treatise on the Analytical Geometry of the Point, Line, Circle, and Conic Sections, Containing an Account of Its Most Recent Extensions, with Numerous Examples, 2nd ed., rev. enl.* Dublin: Hodges, Figgis, & Co., pp. 151–172, 1893.

Moulton, F. R. "Law of Force in Binary Stars" and "Geometrical Interpretation of the Second Law." §58 and 59 in *An Introduction to Celestial Mechanics, 2nd rev. ed.* New York: Dover, pp. 86–89, 1970.

Quadratic Effect

see PRIME QUADRATIC EFFECT

Quadratic Equation

A quadratic equation is a second-order POLYNOMIAL

$$ax^2 + bx + c = 0, \tag{1}$$

with $a \neq 0$. The roots x can be found by COMPLETING THE SQUARE:

$$x^2 + \frac{b}{a}x = -\frac{c}{a} \tag{2}$$

$$\left(x + \frac{b}{2a}\right)^2 = -\frac{c}{a} + \frac{b^2}{4a^2} = \frac{b^2 - 4ac}{4a^2} \tag{3}$$

$$x + \frac{b}{2a} = \frac{\pm\sqrt{b^2 - 4ac}}{2a}. \tag{4}$$

Solving for x then gives

$$x = \frac{-b \pm \sqrt{b^2 - 4ac}}{2a}. \tag{5}$$

This is the QUADRATIC FORMULA.

An alternate form is given by dividing (1) through by x^2:

$$a + \frac{b}{x} + \frac{c}{x^2} = 0 \tag{6}$$

$$c\left(\frac{1}{x^2} + \frac{b}{cx}\right) + a = 0 \tag{7}$$

$$c\left(\frac{1}{x} + \frac{b}{2c}\right)^2 = c\left(\frac{b}{2c}\right)^2 - a = \frac{b^2}{4c} - \frac{4ac}{4c} = \frac{b^2 - 4ac}{4c}. \tag{8}$$

Therefore,

$$\frac{1}{x} + \frac{b}{2c} = \pm\frac{\sqrt{b^2 - 4ac}}{2c} \tag{9}$$

$$\frac{1}{x} = \frac{-b \pm \sqrt{b^2 - 4ac}}{2c} \tag{10}$$

$$x = \frac{2c}{-b \pm \sqrt{b^2 - 4ac}}. \tag{11}$$

This form is helpful if $b^2 \gg 4ac$, in which case the usual form of the QUADRATIC FORMULA can give inaccurate numerical results for one of the ROOTS. This can be avoided by defining

$$q \equiv -\tfrac{1}{2}\left[b + \operatorname{sgn}(b)\sqrt{b^2 - 4ac}\right] \tag{12}$$

so that b and the term under the SQUARE ROOT sign always have the same sign. Now, if $b > 0$, then

$$q = -\tfrac{1}{2}(b + \sqrt{b^2 - 4ac}) \tag{13}$$

$$\frac{1}{q} = \frac{-2}{b + \sqrt{b^2 - 4ac}} \frac{b - \sqrt{b^2 - 4ac}}{b - \sqrt{b^2 - 4ac}} = \frac{-2(b - \sqrt{b^2 - 4ac})}{b^2 - (b^2 - 4ac)}$$
$$= \frac{-2(b - \sqrt{b^2 - 4ac})}{4ac} = \frac{-b + \sqrt{b^2 - 4ac}}{2ac}, \tag{14}$$

so

$$x_1 \equiv \frac{q}{a} = \frac{-b - \sqrt{b^2 - 4ac}}{2a} \tag{15}$$

$$x_2 \equiv \frac{c}{q} = \frac{-b + \sqrt{b^2 - 4ac}}{2a}. \tag{16}$$

Similarly, if $b < 0$, then

$$q = -\tfrac{1}{2}(b - \sqrt{b^2 - 4ac}) = \tfrac{1}{2}(-b + \sqrt{b^2 - 4ac}) \tag{17}$$

$$\frac{1}{q} = \frac{2}{-b + \sqrt{b^2 - 4ac}} \frac{b + \sqrt{b^2 - 4ac}}{b + \sqrt{b^2 - 4ac}} = \frac{2(b + \sqrt{b^2 - 4ac})}{-b^2 + (b^2 - 4ac)}$$
$$= \frac{b + \sqrt{b^2 - 4ac}}{-2ac} = \frac{-b - \sqrt{b^2 - 4ac}}{2ac}, \tag{18}$$

so

$$x_1 \equiv \frac{q}{a} = \frac{-b + \sqrt{b^2 - 4ac}}{2a} \tag{19}$$

$$x_2 \equiv \frac{c}{q} = \frac{-b - \sqrt{b^2 - 4ac}}{2a}. \tag{20}$$

Therefore, the ROOTS are always given by $x_1 = q/a$ and $x_2 = c/q$.

see also CARLYLE CIRCLE, CONIC SECTION, CUBIC EQUATION, QUARTIC EQUATION, QUINTIC EQUATION, SEXTIC EQUATION

References

Abramowitz, M. and Stegun, C. A. (Eds.). *Handbook of Mathematical Functions with Formulas, Graphs, and Mathematical Tables, 9th printing.* New York: Dover, p. 17, 1972.

Beyer, W. H. *CRC Standard Mathematical Tables, 28th ed.* Boca Raton, FL: CRC Press, p. 9, 1987.

Courant, R. and Robbins, H. *What is Mathematics?: An Elementary Approach to Ideas and Methods, 2nd ed.* Oxford, England: Oxford University Press, pp. 91–92, 1996.

King, R. B. *Beyond the Quartic Equation.* Boston, MA: Birkhäuser, 1996.

Press, W. H.; Flannery, B. P.; Teukolsky, S. A.; and Vetterling, W. T. "Quadratic and Cubic Equations." §5.6 in *Numerical Recipes in FORTRAN: The Art of Scientific Computing, 2nd ed.* Cambridge, England: Cambridge University Press, pp. 178–180, 1992.

Spanier, J. and Oldham, K. B. "The Quadratic Function $ax^2 + bx + c$ and Its Reciprocal." Ch. 16 in *An Atlas of Functions.* Washington, DC: Hemisphere, pp. 123–131, 1987.

Quadratic Field

An ALGEBRAIC INTEGER of the form $a + b\sqrt{D}$ where D is SQUAREFREE forms a quadratic field and is denoted $\mathbb{Q}(\sqrt{D})$. If $D > 0$, the field is called a REAL QUADRATIC FIELD, and if $D < 0$, it is called an IMAGINARY QUADRATIC FIELD. The integers in $\mathbb{Q}(\sqrt{1})$ are simply called "the" INTEGERS. The integers in $\mathbb{Q}(\sqrt{-1})$ are called GAUSSIAN INTEGERS, and the integers in $\mathbb{Q}(\sqrt{-3})$ are called EISENSTEIN INTEGERS. The ALGEBRAIC INTEGERS in an arbitrary quadratic field do not necessarily have unique factorizations. For example, the fields $\mathbb{Q}(\sqrt{-5})$ and $\mathbb{Q}(\sqrt{6})$ are not uniquely factorable, since

$$21 = 3 \cdot 7 = (1 + 2\sqrt{-5})(1 - 2\sqrt{-5}) \quad (1)$$

$$6 = -\sqrt{6}(\sqrt{-6}) = 2 \cdot 3, \quad (2)$$

although the above factors are all primes within these fields. All other quadratic fields $\mathbb{Q}(\sqrt{D})$ with $|D| \leq 7$ *are* uniquely factorable.

Quadratic fields obey the identities

$$(a + b\sqrt{D}) \pm (c + d\sqrt{D}) = (a \pm c) + (b \pm d)\sqrt{D}, \quad (3)$$

$$(a + b\sqrt{D})(c + d\sqrt{D}) = (ac + bdD) + (ad + bc)\sqrt{D}, \quad (4)$$

and

$$\frac{a + b\sqrt{D}}{c + d\sqrt{D}} = \frac{ac - bdD}{c^2 - d^2D} + \frac{bc - ad}{c^2 - d^2D}\sqrt{D}. \quad (5)$$

The INTEGERS in the real field $\mathbb{Q}(\sqrt{D})$ are of the form $r + s\rho$, where

$$\rho = \begin{cases} \sqrt{D} & \text{for } D \equiv 2 \text{ or } D \equiv 3 \pmod 4 \\ \frac{1}{2}(-1 + \sqrt{D}) & \text{for } D \equiv 1 \pmod 4. \end{cases} \quad (6)$$

There exist 22 quadratic fields in which there is a EUCLIDEAN ALGORITHM (Inkeri 1947).

see also ALGEBRAIC INTEGER, EISENSTEIN INTEGER, GAUSSIAN INTEGER, IMAGINARY QUADRATIC FIELD, INTEGER, NUMBER FIELD, REAL QUADRATIC FIELD

References

Shanks, D. *Solved and Unsolved Problems in Number Theory, 4th ed.* New York: Chelsea, pp. 153–154, 1993.

Quadratic Form

A quadratic form involving n REAL variables $x_1, x_2, \ldots, x_n$ associated with the $n \times n$ MATRIX $\mathsf{A} = a_{ij}$ is given by

$$Q(x_1, x_2, \ldots, x_n) = a_{ij}x_ix_j, \quad (1)$$

where EINSTEIN SUMMATION has been used. Letting $\mathbf{x}$ be a VECTOR made up of $x_1, \ldots, x_n$ and $\mathbf{x}^{\mathrm{T}}$ the TRANSPOSE, then

$$Q(\mathbf{x}) = \mathbf{x}^{\mathrm{T}}\mathsf{A}\mathbf{x}, \quad (2)$$

equivalent to

$$Q(\mathbf{x}) = (\mathbf{x}, \mathsf{A}\mathbf{x}) \quad (3)$$

in INNER PRODUCT notation. A BINARY QUADRATIC FORM has the form

$$Q(x, y) = a_{11}x^2 + 2a_{12}xy + a_{22}y^2. \quad (4)$$

It is always possible to express an arbitrary quadratic form

$$Q(\mathbf{x}) = \alpha_{ij}x_ix_j \quad (5)$$

in the form

$$Q(\mathbf{x}) = (\mathbf{x}, \mathsf{A}\mathbf{x}), \quad (6)$$

where $\mathsf{A} = a_{ii}$ is a SYMMETRIC MATRIX given by

$$a_{ij} = \begin{cases} \alpha_{ii} & i = j \\ \frac{1}{2}(\alpha_{ij} + \alpha_{ji}) & i \neq j. \end{cases} \quad (7)$$

Any REAL quadratic form in n variables may be reduced to the diagonal form

$$Q(\mathbf{x}) = \lambda_1{x_1}^2 + \lambda_2{x_2}^2 + \ldots + \lambda_n{x_n}^2 \quad (8)$$

with $\lambda_1 \geq \lambda_2 \geq \ldots \geq \lambda_n$ by a suitable orthogonal point-transformation. Also, two real quadratic forms are equivalent under the group of linear transformations IFF they have the same RANK and SIGNATURE.

see also DISCONNECTED FORM, INDEFINITE QUADRATIC FORM, INNER PRODUCT, INTEGER-MATRIX FORM, POSITIVE DEFINITE QUADRATIC FORM, POSITIVE SEMIDEFINITE QUADRATIC FORM, RANK (QUADRATIC FORM), SIGNATURE (QUADRATIC FORM), SYLVESTER'S INERTIA LAW

References

Buell, D. A. *Binary Quadratic Forms: Classical Theory and Modern Computations.* New York: Springer-Verlag, 1989.

Conway, J. H. and Fung, F. Y. *The Sensual (Quadratic) Form.* Washington, DC: Math. Assoc. Amer., 1998.

Gradshteyn, I. S. and Ryzhik, I. M. *Tables of Integrals, Series, and Products, 5th ed.* San Diego, CA: Academic Press, pp. 1104–106, 1979.

Lam, T. Y. *The Algebraic Theory of Quadratic Forms.* Reading, MA: W. A. Benjamin, 1973.

Quadratic Formula

The formula giving the ROOTS of a QUADRATIC EQUATION

$$ax^2 + bx + c = 0 \quad (1)$$

as

$$x = \frac{-b \pm \sqrt{b^2 - 4ac}}{2a}. \quad (2)$$

An alternate form is given by

$$x = \frac{2c}{-b \pm \sqrt{b^2 - 4ac}}. \quad (3)$$

see also QUADRATIC EQUATION

Quadratic Integral

To compute integral of the form

$$\int \frac{dx}{a+bx+cx^2}, \tag{1}$$

COMPLETE THE SQUARE in the DENOMINATOR to obtain

$$\int \frac{dx}{a+bx+cx^2} = \frac{1}{c}\int \frac{dx}{\left(x+\frac{b}{2c}\right)^2 + \left(\frac{a}{c}-\frac{b^2}{4c^2}\right)}. \tag{2}$$

Let $u \equiv x + b/2c$. Then define

$$-A^2 \equiv \frac{a}{c} - \frac{b^2}{4c^2} = \frac{1}{4c^2}(4ac-b^2) \equiv \frac{1}{4c^2}q, \tag{3}$$

where

$$q \equiv 4ac - b^2 \tag{4}$$

is the NEGATIVE of the DISCRIMINANT. If $q < 0$, then

$$A = \frac{1}{2c}\sqrt{-q}. \tag{5}$$

Now use PARTIAL FRACTION DECOMPOSITION,

$$\frac{1}{c}\int \frac{du}{(u+A)(u-A)} = \frac{1}{c}\int \left(\frac{A_1}{u+A} + \frac{A_2}{u-A}\right) du \tag{6}$$

$$\begin{aligned}\left(\frac{A_1}{u+A} + \frac{A_2}{u-A}\right) &= \frac{A_1(u-A)+A_2(u+A)}{u^2-A^2}\\ &= \frac{(A_1+A_2)u + A(A_2-A_1)}{u^2-A^2},\end{aligned} \tag{7}$$

so $A_2 + A_1 = 0 \Rightarrow A_2 = -A_1$ and $A(A_2 - A_1) = -2AA_1 = 1 \Rightarrow A_1 = -1/(2A)$. Plugging these in,

$$\begin{aligned}&\frac{1}{c}\int \left(-\frac{1}{2A}\frac{1}{u+A} + \frac{1}{2A}\frac{1}{u-A}\right) du\\ &= \frac{1}{2Ac}\left[-\ln(u+A) + \ln(u-A)\right]\\ &= \frac{1}{2Ac}\ln\left(\frac{u-A}{u+A}\right)\\ &= \frac{1}{2\left(\frac{1}{2c}\right)\sqrt{-q}\,c}\ln\left(\frac{x+\frac{b}{2c}-\frac{1}{2c}\sqrt{-q}}{x+\frac{b}{2c}+\frac{1}{2c}\sqrt{-q}}\right)\\ &= \frac{1}{\sqrt{-q}}\ln\left(\frac{2cx+b-\sqrt{-q}}{2cx+b+\sqrt{-q}}\right)\end{aligned} \tag{8}$$

for $q < 0$. Note that this integral is also tabulated in Gradshteyn and Ryzhik (1979, equation 2.172), where it is given with a sign flipped.

References

Gradshteyn, I. S. and Ryzhik, I. M. *Tables of Integrals, Series, and Products, 5th ed.* San Diego, CA: Academic Press, 1979.

Quadratic Invariant

Given the BINARY QUADRATIC FORM

$$ax^2 + 2bxy + cy^2 \tag{1}$$

with DISCRIMINANT $b^2 - ac$, let

$$x = pX + qY \tag{2}$$

$$y = rX + sY. \tag{3}$$

Then

$$\begin{aligned}a(pX+qY)^2 + 2b(pX+qY)(rX+sY) + c(rX+sY)^2\\ = AX^2 + 2BXY + CY^2,\end{aligned} \tag{4}$$

where

$$A = ap^2 + 2bpr + cr^2 \tag{5}$$

$$B = apq + b(ps+qr) + crs \tag{6}$$

$$C = aq^2 + 2bqs + cs^2, \tag{7}$$

so

$$\begin{aligned}B^2 - AC &= [a^2p^2q^2 + b^2(ps+qr)^2 + c^2r^2s^2\\ &\quad + 2abpq(ps+qr) + 2acpqrs + 2bcrs(ps+qr)]\\ &\quad - (ap^2+2bpr+cr^2)(aq^2+2bqs+cs^2)\\ &= a^2p^2q^2 + b^2p^2s^2 + 2b^2pqrs + b^2q^2r^2 + c^2r^2s^2\\ &\quad + 2abp^2qs + 2abpq^2r + 2acpqrs + 2bcprs^2 + 2bcqr^2s\\ &\quad - a^2p^2q^2 - 2abp^2qs - acp^2s^2 - 2abpq^2r - 4b^2pqrs\\ &\quad - 2bcprs^2 - acq^2r^2 - 2bcqr^2s - c^2r^2s^2\\ &= b^2p^2s^2 - 2b^2pqrs + b^2q^2r^2 + 2acpqrs - acp^2s^2\\ &\quad - acq^2r^2\\ &= p^2s^2(b^2-ac) + q^2r^2(b^2-ac) - 2pqrs(b^2-ac)\\ &= (b^2-ac)(p^2s^2 - 2pqrs + q^2r^2)\\ &= (ps-rq)^2(b^2-ac).\end{aligned} \tag{8}$$

Surprisingly, this is the same discriminant as before, but multiplied by the factor $(ps-rq)^2$. The quantity $ps-rq$ is called the MODULUS.

see also ALGEBRAIC INVARIANT

Quadratic Irrational Number

An IRRATIONAL NUMBER of the form

$$\frac{P \pm \sqrt{D}}{Q},$$

where P and Q are INTEGERS and D is a SQUAREFREE INTEGER. Quadratic irrational numbers are sometimes also called QUADRATIC SURDS. In 1770, Lagrange proved that that any quadratic irrational has a CONTINUED FRACTION which is periodic after some point.

see also CONTINUED FRACTION, QUADRATIC SURD

Quadratic Map

A 1-D MAP often called "the" quadratic map is defined by

$$x_{n+1} = x_n{}^2 + c. \quad (1)$$

This is the real version of the complex map defining the MANDELBROT SET. The quadratic map is called attracting if the JACOBIAN $J < 1$, and repelling if $J > 1$. FIXED POINTS occur when

$$x^{(1)} = [x^{(1)}]^2 + c \quad (2)$$

$$(x^{(1)})^2 - x^{(1)} + c = 0 \quad (3)$$

$$x_\pm^{(1)} = \tfrac{1}{2}(1 \pm \sqrt{1-4c}). \quad (4)$$

Period two FIXED POINTS occur when

$$\begin{aligned} x_{n+2} &= x_{n+1}{}^2 + c = (x_n{}^2 + c)^2 + c \\ &= x_n{}^4 + 2cx_n{}^2 + (c^2 + c) = x_n \end{aligned} \quad (5)$$

$$x^4 + 2x^2 - x + (cx^2 + c) = (x^2 - x + c)(x^2 + x + 1 + c) = 0 \quad (6)$$

$$x_\pm^{(2)} = \tfrac{1}{2}[1 \pm \sqrt{1-4(1+c)}\,] = \tfrac{1}{2}(1 \pm \sqrt{-3-4c}). \quad (7)$$

Period three FIXED POINTS occur when

$$\begin{aligned} x^6 + x^5 + (3c+1)x^4 + (2c+1)x^3 + (c^2+3c+1)x^2 \\ +(c+1)^2 x + (c^3 + 2c^2 + c + 1) = 0. \end{aligned} \quad (8)$$

The most general second-order 2-D MAP with an elliptic fixed point at the origin has the form

$$x' = x\cos\alpha - y\sin\alpha + a_{20}x^2 + a_{11}xy + a_{02}y^2 \quad (9)$$

$$y' = x\sin\alpha + y\cos\alpha + b_{20}x^2 + b_{11}xy + b_{02}y^2. \quad (10)$$

The map must have a DETERMINANT of 1 in order to be AREA preserving, reducing the number of independent parameters from seven to three. The map can then be put in a standard form by scaling and rotating to obtain

$$x' = x\cos\alpha - y\sin\alpha + x^2\sin\alpha \quad (11)$$

$$y' = x\sin\alpha + y\cos\alpha - x^2\cos\alpha. \quad (12)$$

The inverse map is

$$x = x'\cos\alpha + y'\sin\alpha \quad (13)$$

$$y = -x'\sin\alpha + y'\cos\alpha + (x'\cos\alpha + y'\sin\alpha)^2. \quad (14)$$

The FIXED POINTS are given by

$$x_i{}^2\sin\alpha + 2x_i\cos\alpha - x_{i-1} - x_{i+1} = 0 \quad (15)$$

for $i = 0, \ldots, n-1$.

see also BOGDANOV MAP, HÉNON MAP, LOGISTIC MAP, LOZI MAP, MANDELBROT SET

Quadratic Mean

see ROOT-MEAN-SQUARE

Quadratic Reciprocity Law

see QUADRATIC RECIPROCITY THEOREM

Quadratic Reciprocity Relations

$$\left(\frac{-1}{p}\right) = (-1)^{(p-1)/2} \quad (1)$$

$$\left(\frac{2}{p}\right) = (-1)^{(p^2-1)/8} \quad (2)$$

$$\left(\frac{q}{p}\right) = \left(\frac{p}{q}\right)(-1)^{[(p-1)/2][(q-1)/2]}, \quad (3)$$

where $\left(\frac{p}{q}\right)$ is the LEGENDRE SYMBOL.

see also QUADRATIC RECIPROCITY THEOREM

Quadratic Reciprocity Theorem

Also called the AUREUM THEOREMA (GOLDEN THEOREM) by Gauss. If p and q are distinct ODD PRIMES, then the CONGRUENCES

$$x^2 \equiv q \pmod{p}$$
$$x^2 \equiv p \pmod{q}$$

are both solvable or both unsolvable unless both p and q leave the remainder 3 when divided by 4 (in which case one of the CONGRUENCES is solvable and the other is not). Written symbolically,

$$\left(\frac{p}{q}\right)\left(\frac{q}{p}\right) = (-1)^{(p-1)(q-1)/4},$$

where

$$\left(\frac{p}{q}\right) \equiv \begin{cases} 1 & \text{for } x^2 \equiv p \pmod{q} \text{ solvable for } x \\ -1 & \text{for } x^2 \equiv p \pmod{q} \text{ not solvable for } x \end{cases}$$

is known as a LEGENDRE SYMBOL. Legendre was the first to publish a proof, but it was fallacious. Gauss was the first to publish a correct proof. The quadratic reciprocity theorem was Gauss's favorite theorem from NUMBER THEORY, and he devised many proofs of it over his lifetime.

see also JACOBI SYMBOL, KRONECKER SYMBOL, LEGENDRE SYMBOL, QUADRATIC RESIDUE, RECIPROCITY THEOREM

References

Courant, R. and Robbins, H. *What is Mathematics?: An Elementary Approach to Ideas and Methods, 2nd ed.* Oxford, England: Oxford University Press, p. 39, 1996.

Ireland, K. and Rosen, M. "Quadratic Reciprocity." Ch. 5 in *A Classical Introduction to Modern Number Theory, 2nd ed.* New York: Springer-Verlag, pp. 50–65, 1990.

Nagell, T. "Theory of Quadratic Residues." Ch. 4 in *Introduction to Number Theory.* New York: Wiley, 1951.

Riesel, H. "The Law of Quadratic Reciprocity." *Prime Numbers and Computer Methods for Factorization, 2nd ed.* Boston, MA: Birkhäuser, pp. 279–281, 1994.

Shanks, D. *Solved and Unsolved Problems in Number Theory, 4th ed.* New York: Chelsea, pp. 42–49, 1993.

Quadratic Recurrence

N.B. A detailed on-line essay by S. Finch was the starting point for this entry.

A quadratic recurrence is a RECURRENCE RELATION on a SEQUENCE of numbers $\{x_n\}$ expressing x_n as a second degree polynomial in x_k with $k < n$. For example,

$$x_n = x_{n-1}x_{n-2} \tag{1}$$

is a quadratic recurrence. Another simple example is

$$x_n = (x_{n-1})^2 \tag{2}$$

with $x_0 = 2$, which has solution $x_n = 2^{2^n}$. Another example is the number of "strongly" binary trees of height $\leq n$, given by

$$y_n = (y_{n-1})^2 + 1 \tag{3}$$

with $y_0 = 1$. This has solution

$$y_n = \left\lfloor c^{2^n} \right\rfloor, \tag{4}$$

where

$$c = \exp\left[\sum_{j=0}^{\infty} 2^{-j-1} \ln(1 + y_j{}^{-2})\right] = 1.502836801\ldots \tag{5}$$

and $\lfloor x \rfloor$ is the FLOOR FUNCTION (Aho and Sloane 1973). A third example is the closest strict underapproximation of the number 1,

$$S_n = \sum_{i=1}^{n} \frac{1}{z_i}, \tag{6}$$

where $1 < z_1 < \ldots < z_n$ are integers. The solution is given by the recurrence

$$z_n = (z_{n-1})^2 - z_{n-1} + 1, \tag{7}$$

with $z_1 = 2$. This has a closed solution as

$$z_n = \left\lfloor d^{2^n} + \tfrac{1}{2} \right\rfloor \tag{8}$$

where

$$d = \tfrac{1}{2}\sqrt{6}\exp\left\{\sum_{j=1}^{\infty} 2^{-j-1} \ln[1 + (2z_j - 1)^{-2}]\right\} = 1.2640847353\ldots \tag{9}$$

(Aho and Sloane 1973). A final example is the well-known recurrence

$$c_n = (c_{n-1})^2 - \mu \tag{10}$$

with $c_0 = 0$ used to generate the MANDELBROT SET.

see also MANDELBROT SET, RECURRENCE RELATION

References

Aho, A. V. and Sloane, N. J. A. "Some Doubly Exponential Sequences." *Fib. Quart.* **11**, 429–437, 1973.

Finch, S. "Favorite Mathematical Constants." http://www.mathsoft.com/asolve/constant/quad/quad.html.

Quadratic Residue

If there is an INTEGER x such that

$$x^2 \equiv q \pmod{p}, \tag{1}$$

then q is said to be a quadratic residue of x mod p. If not, q is said to be a quadratic nonresidue of x mod p. For example, $4^2 \equiv 6 \pmod{10}$, so 6 is a quadratic residue (mod 10). The entire set of quadratic residues (mod 10) are given by 1, 4, 5, 6, and 9, since

$$1^2 \equiv 1 \pmod{10} \quad 2^2 \equiv 4 \pmod{10} \quad 3^2 \equiv 9 \pmod{10}$$

$$4^2 \equiv 6 \pmod{10} \quad 5^2 \equiv 5 \pmod{10} \quad 6^2 \equiv 6 \pmod{10}$$

$$7^2 \equiv 9 \pmod{10} \quad 8^2 \equiv 4 \pmod{10} \quad 9^2 \equiv 1 \pmod{10}$$

making the numbers 2, 3, 7, and 8 the quadratic nonresidues (mod 10).

A list of quadratic residues for $p \leq 29$ is given below (Sloane's A046071), with those numbers $< p$ not in the list being quadratic nonresidues of p.

p	Quadratic Residues
1	(none)
2	1
3	1
4	1
5	1, 4
6	1, 3, 4
7	1, 2, 4
8	1, 4
9	1, 4, 7
10	1, 4, 5, 6, 9
11	1, 3, 4, 5, 9
12	1, 4, 9
13	1, 3, 4, 9, 10, 12
14	1, 2, 4, 7, 8, 9, 11
15	1, 4, 6, 9, 10
16	1, 4, 9
17	1, 2, 4, 8, 9, 13, 15, 16
18	1, 4, 7, 9, 10, 13, 16
19	1, 4, 5, 6, 7, 9, 11, 16, 17
20	1, 4, 5, 9, 16

The UNITS in the integers (mod n), $\mathbb{Z}_n$, which are SQUARES are the quadratic residues.

Given an ODD PRIME p and an INTEGER a, then the LEGENDRE SYMBOL is given by

$$\left(\frac{a}{p}\right) = \begin{cases} 1 & \text{if } a \text{ is a quadratic residue mod } p \\ -1 & \text{otherwise.} \end{cases} \tag{2}$$

If

$$r^{(p-1)/2} \equiv \pm 1 \pmod{p}, \tag{3}$$

then r is a quadratic residue (+) or nonresidue (−). This can be seen since if r is a quadratic residue of p, then there exists a square x^2 such that $r \equiv x^2 \pmod{p}$, so

$$r^{(p-1)/2} \equiv (x^2)^{(p-1)/2} \equiv x^{p-1} \pmod{p}, \tag{4}$$

and x^{p-1} is congruent to 1 (mod p) by FERMAT'S LITTLE THEOREM. x is given by

$$\begin{cases} q^{k+1} \pmod{p} \\ \quad \text{for } p = 4k+3 \\ q^{k+1} \pmod{p} \\ \quad \text{for } p = 8k+5 \text{ and } q^{2k+1} \equiv 1 \pmod{p} \\ (4q)^{k+1}\left(\frac{p+1}{2}\right) \pmod{p} \\ \quad \text{for } p = 8k+5 \text{ and } q^{2k+1} \equiv -1 \pmod{p}. \end{cases} \tag{5}$$

More generally, let q be a quadratic residue modulo an ODD PRIME p. Choose h such that the LEGENDRE SYMBOL $(h^2 - 4q/p) = -1$. Then defining

$$V_1 = h \tag{6}$$

$$V_2 = h^2 - 2q \tag{7}$$

$$V_i = hV_{i-1} - qV_{i-2} \qquad \text{for } i \geq 3, \tag{8}$$

gives

$$V_{2i} = V_i{}^2 - 2q^i \tag{9}$$

$$V_{2i+1} = V_iV_{i+1} - hn^i, \tag{10}$$

and a solution to the quadratic CONGRUENCE is

$$x = V_{(p+1)/2}\left(\frac{p+1}{2}\right) \pmod{p}. \tag{11}$$

The following table gives the PRIMES which have a given number d as a quadratic residue.

d	Primes
−6	$24k+1, 5, 7, 11$
−5	$20k+1, 3, 7, 9$
−3	$6k+1$
−2	$8k+1, 3$
−1	$4k+1$
2	$8k \pm 1$
3	$12k \pm 1$
5	$10k \pm 1$
6	$24k \pm 1, 5$

Finding the CONTINUED FRACTION of a SQUARE ROOT $\sqrt{D}$ and using the relationship

$$Q_n = \frac{D - P_n{}^2}{Q_{n-1}} \tag{12}$$

for the nth CONVERGENT P_n/Q_n gives

$$P_n{}^2 \equiv -Q_nQ_{n-1} \pmod{D}. \tag{13}$$

Therefore, $-Q_nQ_{n-1}$ is a quadratic residue of D. But since $Q_1 = 1$, $-Q_2$ is a quadratic residue, as must be $-Q_2Q_3$. But since $-Q_2$ is a quadratic residue, so is Q_3, and we see that $(-1)^{n-1}Q_n$ are all quadratic residues of D. This method is not guaranteed to produce all quadratic residues, but can often produce several small ones in the case of large D, enabling D to be factored.

The number of SQUARES $s(n)$ in $\mathbb{Z}_n$ is related to the number $q(n)$ of quadratic residues in $\mathbb{Z}_n$ by

$$q(p^n) = s(p^n) - s(p^{n-2}) \tag{14}$$

for $n \geq 3$ (Stangl 1996). Both q and s are MULTIPLICATIVE FUNCTIONS.

see also EULER'S CRITERION, MULTIPLICATIVE FUNCTION, QUADRATIC RECIPROCITY THEOREM, RIEMANN HYPOTHESIS

References

Burton, D. M. *Elementary Number Theory, 4th ed.* New York: McGraw-Hill, p. 201, 1997.

Courant, R. and Robbins, H. "Quadratic Residues." §2.3 in Supplement to Ch. 1 in *What is Mathematics?: An Elementary Approach to Ideas and Methods, 2nd ed.* Oxford, England: Oxford University Press, pp. 38–40, 1996.

Guy, R. K. "Quadratic Residues. Schur's Conjecture" and "Patterns of Quadratic Residues." §F5 and F6 in *Unsolved Problems in Number Theory, 2nd ed.* New York: Springer-Verlag, pp. 244–248, 1994.

Niven, I. and Zuckerman, H. *An Introduction to the Theory of Numbers, 4th ed.* New York: Wiley, p. 84, 1980.

Rosen, K. H. Ch. 9 in *Elementary Number Theory and Its Applications, 3rd ed.* Reading, MA: Addison-Wesley, 1993.

Shanks, D. *Solved and Unsolved Problems in Number Theory, 4th ed.* New York: Chelsea, pp. 63–66, 1993.

Sloane, N. J. A. Sequence A046071 in "An On-Line Version of the Encyclopedia of Integer Sequences."

Stangl, W. D. "Counting Squares in $\mathbb{Z}_n$." *Math. Mag.* **69**, 285–289, 1996.

Wagon, S. "Quadratic Residues." §9.2 in *Mathematica in Action.* New York: W. H. Freeman, pp. 292–296, 1991.

Quadratic Sieve Factorization Method

A procedure used in conjunction with DIXON'S FACTORIZATION METHOD to factor large numbers. The rs are chosen as

$$\lfloor\sqrt{n}\rfloor + k, \tag{1}$$

where $k = 1, 2, \ldots$ and $\lfloor x \rfloor$ is the FLOOR FUNCTION. We are then looking for factors p such that

$$n \equiv r^2 \pmod{p}, \tag{2}$$

which means that only numbers with LEGENDRE SYMBOL $(n/p) = 1$ (less than $N = \pi(d)$ for trial divisor d) need be considered. The set of PRIMES for which this is true is known as the FACTOR BASE. Next, the CONGRUENCES

$$x^2 \equiv n \pmod{p} \tag{3}$$

must be solved for each p in the FACTOR BASE. Finally, a sieve is applied to find values of $f(r) = r^2 - n$

which can be factored completely using only the FACTOR BASE. GAUSSIAN ELIMINATION is then used as in DIXON'S FACTORIZATION METHOD in order to find a product of the $f(r)$s, yielding a PERFECT SQUARE.

The method requires about $\exp(\sqrt{\log n \log\log n})$ steps, improving on the CONTINUED FRACTION FACTORIZATION ALGORITHM by removing the 2 under the SQUARE ROOT (Pomerance 1996). The use of multiple POLYNOMIALS gives a better chance of factorization, requires a shorter sieve interval, and is well-suited to parallel processing.

see also PRIME FACTORIZATION ALGORITHMS, SMOOTH NUMBER

References

Alford, W. R. and Pomerance, C. "Implementing the Self Initializing Quadratic Sieve on a Distributed Network." In *Number Theoretic and Algebraic Methods in Computer Science, Proc. Internat. Moscow Conf., June–July 1993* (Ed. A. J. van der Poorten, I. Shparlinksi, and H. G. Zimer). World Scientific, pp. 163–174, 1995.

Brent, R. P. "Parallel Algorithms for Integer Factorisation." In *Number Theory and Cryptography* (Ed. J. H. Loxton). New York: Cambridge University Press, 26–37, 1990. `ftp://nimbus.anu.edu.au/pub/Brent/115.dvi.Z`.

Bressoud, D. M. Ch. 8 in *Factorization and Prime Testing.* New York: Springer-Verlag, 1989.

Gerver, J. "Factoring Large Numbers with a Quadratic Sieve." *Math. Comput.* **41**, 287–294, 1983.

Lenstra, A. K. and Manasse, M. S. "Factoring by Electronic Mail." In *Advances in Cryptology—Eurocrypt '89* (Ed. J.-J. Quisquarter and J. Vandewalle). Berlin: Springer-Verlag, pp. 355–371, 1990.

Pomerance, C. "A Tale of Two Sieves." *Not. Amer. Math. Soc.* **43**, 1473–1485, 1996.

Pomerance, C.; Smith, J. W.; and Tuler, R. "A Pipeline Architecture for Factoring Large Integers with the Quadratic Sieve Method." *SIAM J. Comput.* **17**, 387–403, 1988.

Quadratic Surd

see QUADRATIC IRRATIONAL NUMBER

Quadratic Surface

There are 17 standard-form quadratic surfaces. The general quadratic is written

$$ax^2 + by^2 + cz^2 + 2fyz + 2gzx + 2hxy + 2px + 2qy + 2rz + d = 0. \quad (1)$$

Define

$$e = \begin{bmatrix} a & h & g \\ h & b & f \\ g & f & c \end{bmatrix} \quad (2)$$

$$E = \begin{bmatrix} a & h & g & p \\ h & b & f & q \\ g & f & c & r \\ p & q & r & d \end{bmatrix} \quad (3)$$

$$\rho_3 = \text{rank } e \quad (4)$$

$$\rho_4 = \text{rank } E \quad (5)$$

$$\Delta = \det E, \quad (6)$$

and k_1, k_2, as k_3 are the roots of

$$\begin{vmatrix} a-x & h & g \\ h & b-x & f \\ g & f & c-x \end{vmatrix} = 0. \quad (7)$$

Also define

$$k \equiv \begin{cases} 1 & \text{if the signs of nonzero } k\text{s are the same} \\ 0 & \text{otherwise.} \end{cases} \quad (8)$$

Surface	Equation	ρ_3	ρ_4	sgn (Δ)	k
coincident planes	$x^2 = 0$	1	1		
ellipsoid ($\Im$)	$\frac{x^2}{a^2} + \frac{y^2}{b^2} + \frac{z^2}{c^2} = -1$	3	4	+	1
ellipsoid ($\Re$)	$\frac{x^2}{a^2} + \frac{y^2}{b^2} + \frac{z^2}{c^2} = 1$	3	4	−	1
elliptic cone ($\Im$)	$\frac{x^2}{a^2} + \frac{y^2}{b^2} - \frac{z^2}{c^2} = 0$	3	3		1
elliptic cone ($\Re$)	$z^2 = \frac{x^2}{a^2} + \frac{y^2}{b^2}$	3	3		0
elliptic cylinder ($\Im$)	$\frac{x^2}{a^2} + \frac{y^2}{b^2} = -1$	2	3		1
elliptic cylinder ($\Re$)	$\frac{x^2}{a^2} + \frac{y^2}{b^2} = 1$	2	3		1
elliptic paraboloid	$z = \frac{x^2}{a^2} + \frac{y^2}{b^2}$	2	4	−	1
hyperbolic cylinder	$\frac{x^2}{a^2} - \frac{y^2}{b^2} = -1$	2	3		0
hyperbolic paraboloid	$z = \frac{y^2}{b^2} - \frac{x^2}{a^2}$	2	4	+	0
hyperboloid of one sheet	$\frac{x^2}{a^2} + \frac{y^2}{b^2} - \frac{z^2}{c^2} = 1$	3	4	+	0
hyperboloid of two sheets	$\frac{x^2}{a^2} + \frac{y^2}{b^2} - \frac{z^2}{c^2} = -1$	3	4	−	0
intersecting planes ($\Im$)	$\frac{x^2}{a^2} + \frac{y^2}{b^2} = 0$	2	2		1
intersecting planes ($\Re$)	$\frac{x^2}{a^2} - \frac{y^2}{b^2} = 0$	2	2		0
parabolic cylinder	$x^2 + 2rz = 0$	1	3		
parallel planes ($\Im$)	$x^2 = -a^2$	1	2		
parallel planes ($\Re$)	$x^2 = a^2$	1	2		

see also CUBIC SURFACE, ELLIPSOID, ELLIPTIC CONE, ELLIPTIC CYLINDER, ELLIPTIC PARABOLOID, HYPERBOLIC CYLINDER, HYPERBOLIC PARABOLOID, HYPERBOLOID, PLANE, QUARTIC SURFACE, SURFACE

References

Beyer, W. H. *CRC Standard Mathematical Tables, 28th ed.* Boca Raton, FL: CRC Press, pp. 210–211, 1987.

Quadratrix of Hippias

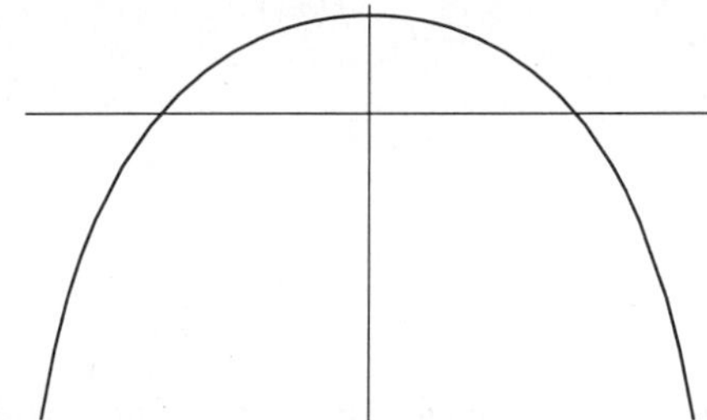

The quadratrix was discovered by Hippias of Elias in 430 BC, and later studied by Dinostratus in 350 BC (MacTutor Archive). It can be used for ANGLE TRISECTION or, more generally, division of an ANGLE into any integral number of equal parts, and CIRCLE SQUARING. In POLAR COORDINATES,

$$\pi\rho = 2r\theta \csc\theta,$$

so

$$r = \frac{\rho\pi \sin\theta}{\theta},$$

which is proportional to the COCHLEOID.

References

Lawrence, J. D. *A Catalog of Special Plane Curves.* New York: Dover, pp. 195 and 198, 1972.

Lee, X. "Quadratrix of Hippias." http://www.best.com/~xah/SpecialPlaneCurves_dir/QuadratrixOfHippias_dir/quadratrixOfHippias.html.

MacTutor History of Mathematics Archive. "Quadratrix of Hippias." http://www-groups.dcs.st-and.ac.uk/~history/Curves/Quadratrix.html.

Quadrature

The word quadrature has (at least) three incompatible meanings. Integration by quadrature either means solving an INTEGRAL analytically (i.e., symbolically in terms of known functions), or solving of an integral numerically (e.g., GAUSSIAN QUADRATURE, QUADRATURE FORMULAS). The word quadrature is also used to mean SQUARING: the construction of a square using only COMPASS and STRAIGHTEDGE which has the same AREA as a given geometric figure. If quadrature is possible for a PLANE figure, it is said to be QUADRABLE.

For a function tabulated at given values x_i (so the ABSCISSAS cannot be chosen at will), write the function ϕ as a sum of ORTHONORMAL FUNCTIONS p_j satisfying

$$\int_a^b p_i(x)p_j(x)W(x)\,dx = \delta_{ij} \tag{1}$$

as

$$\phi(x) = \sum_{j=0}^{\infty} a_j p_j(x), \tag{2}$$

and plug into

$$\int_a^b \phi(x)W(x)\,dx = \int_a^b \sum_{j=1}^{m} \frac{\pi(x)W(x)}{(x-x_j)\pi'(x_j)}\,dx f(x_j)$$
$$\equiv \sum_{j=1}^{m} w_j f(x_j), \tag{3}$$

giving

$$\int_a^b \sum_{j=0}^{\infty} a_j p_j(x)W(x)\,dx = \sum_{i=1}^{n} w_i \left[\sum_{j=0}^{\infty} a_j p_j(x_i)\right]. \tag{4}$$

But we wish this to hold for all degrees of approximation, so

$$a_j \int_a^b p_j(x)W(x)\,dx = a_j \sum_{i=1}^{n} w_i p_j(x_i) \tag{5}$$

$$\int_a^b p_j(x)W(x)\,dx = \sum_{i=1}^{n} w_i p_j(x_i). \tag{6}$$

Setting $i = 0$ in (1) gives

$$\int_a^b p_0(x)p_j(x)W(x)\,dx = \delta_{0j}. \tag{7}$$

The zeroth order orthonormal function can always be taken as $p_0(x) = 1$, so (7) becomes

$$\int_a^b p_j(x)W(x)\,dx = \delta_{0j} \tag{8}$$
$$= \sum_{i=1}^{n} w_i p_j(x_i), \tag{9}$$

where (6) has been used in the last step. We therefore have the MATRIX equation

$$\begin{bmatrix} p_0(x_1) & \cdots & p_0(x_n) \\ p_1(x_1) & \cdots & p_1(x_n) \\ \vdots & \ddots & \vdots \\ p_{n-1}(x_1) & \cdots & p_{n-1}(x_n) \end{bmatrix} \begin{bmatrix} w_1 \\ w_2 \\ \vdots \\ w_n \end{bmatrix} = \begin{bmatrix} 1 \\ 0 \\ \vdots \\ 0 \end{bmatrix}, \tag{10}$$

which can be inverted to solve for the w_is (Press *et al.* 1992).

see also CALCULUS, CHEBYSHEV-GAUSS QUADRATURE, CHEBYSHEV QUADRATURE, DERIVATIVE, FUNDAMENTAL THEOREM OF GAUSSIAN QUADRATURE, GAUSS-JACOBI MECHANICAL QUADRATURE, GAUSSIAN QUADRATURE, HERMITE-GAUSS QUADRATURE, HERMITE QUADRATURE, JACOBI-GAUSS QUADRATURE, JACOBI QUADRATURE, LAGUERRE-GAUSS QUADRATURE, LAGUERRE QUADRATURE, LEGENDRE-GAUSS QUADRATURE, LEGENDRE QUADRATURE, LOBATTO QUADRATURE, MECHANICAL QUADRATURE, MEHLER QUADRATURE, NEWTON-COTES FORMULAS, NUMERICAL INTEGRATION, RADAU QUADRATURE, RECURSIVE MONOTONE STABLE QUADRATURE

References

Abramowitz, M. and Stegun, C. A. (Eds.). "Integration." §25.4 in *Handbook of Mathematical Functions with Formulas, Graphs, and Mathematical Tables, 9th printing.* New York: Dover, pp. 885–897, 1972.

Press, W. H.; Flannery, B. P.; Teukolsky, S. A.; and Vetterling, W. T. *Numerical Recipes in FORTRAN: The Art of Scientific Computing, 2nd ed.* Cambridge, England: Cambridge University Press, pp. 365–366, 1992.

Quadrature Formulas

see NEWTON-COTES FORMULAS

Quadric

An equation of the form

$$\frac{x^2}{a^2+\theta} + \frac{y^2}{b^2+\theta} + \frac{z^2}{c^2+\theta} = 1,$$

where θ is said to be the parameter of the quadric.

Quadricorn

A FLEXIBLE POLYHEDRON due to C. Schwabe (with the appearance of having four horns) which flexes from one totally flat configuration to another, passing through intermediate configurations of positive VOLUME.

see also FLEXIBLE POLYHEDRON

Quadrifolium

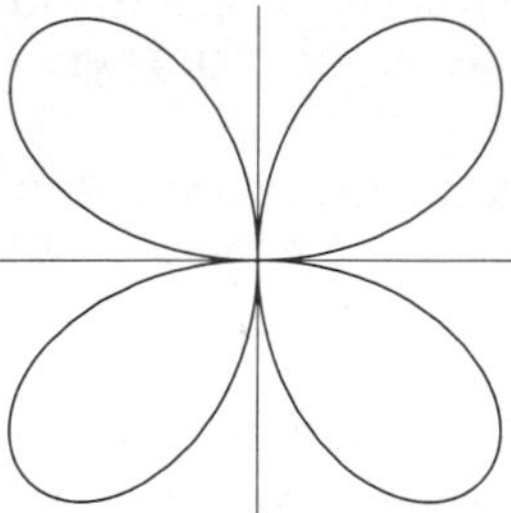

The ROSE with $n = 2$. It has polar equation

$$r = a\sin(2\theta),$$

and Cartesian form

$$(x^2 + y^2)^3 = 4a^2x^2y^2.$$

see also BIFOLIUM, FOLIUM, ROSE, TRIFOLIUM

Quadrilateral

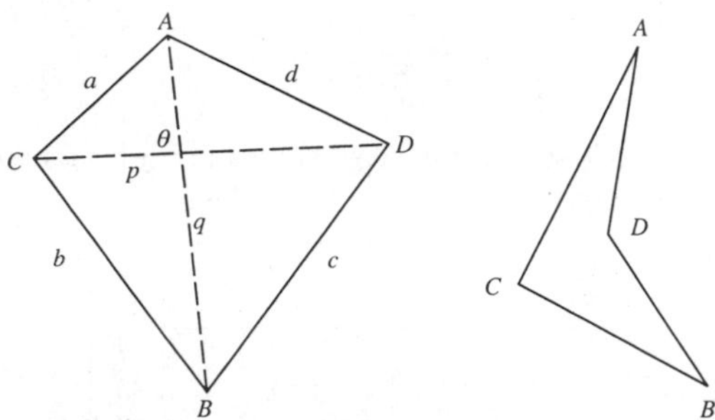

A four-sided POLYGON sometimes (but not very often) also known as a TETRAGON. If not explicitly stated, all four VERTICES are generally taken to lie in a PLANE. If the points do not lie in a PLANE, the quadrilateral is called a SKEW QUADRILATERAL.

For a planar convex quadrilateral (left figure above), let the lengths of the sides be a, b, c, and d, the SEMIPERIMETER s, and the DIAGONALS p and q. The DIAGONALS are PERPENDICULAR IFF $a^2 + c^2 = b^2 + d^2$. An equation for the sum of the squares of side lengths is

$$a^2 + b^2 + c^2 + d^2 = p^2 + q^2 + 4x^2, \tag{1}$$

where x is the length of the line joining the MIDPOINTS of the DIAGONALS. The AREA of a quadrilateral is given by

$$\begin{aligned} K &= \tfrac{1}{2}pq\sin\theta & (2)\\ &= \tfrac{1}{4}(b^2 + d^2 - a^2 - c^2)\tan\theta & (3)\\ &= \tfrac{1}{4}\sqrt{4p^2q^2 - (b^2 + d^2 - a^2 - c^2)^2} & (4)\\ &= \sqrt{(s-a)(s-b)(s-c)(s-d) - abcd\cos^2[\tfrac{1}{2}(A+B)]}, & (5) \end{aligned}$$

where (4) is known as BRETSCHNEIDER'S FORMULA (Beyer 1987).

A special type of quadrilateral is the CYCLIC QUADRILATERAL, for which a CIRCLE can be circumscribed so that it touches each VERTEX. For BICENTRIC quadrilaterals, the CIRCUMCIRCLE and INCIRCLE satisfy

$$2r^2(R^2 - s^2) = (R^2 - s^2)^2 - 4r^2s^2, \tag{6}$$

where R is the CIRCUMRADIUS, r in the INRADIUS, and s is the separation of centers. A quadrilateral with two sides PARALLEL is called a TRAPEZOID.

There is a relationship between the six distances d_{12}, d_{13}, d_{14}, d_{23}, d_{24}, and d_{34} between the four points of a quadrilateral (Weinberg 1972):

$$\begin{aligned} 0 = {} & {d_{12}}^4{d_{34}}^2 + {d_{13}}^4{d_{24}}^2 + {d_{14}}^4{d_{23}}^2 + {d_{23}}^4{d_{14}}^2\\ & + d_{24}^4d_{13}^2 + d_{34}^4d_{12}^2\\ & + d_{12}^2d_{23}^2d_{31}^2 + d_{12}^2d_{24}^2d_{41}^2 + d_{13}^2d_{34}^2d_{41}^2\\ & + d_{23}^2d_{34}^2d_{42}^2 - d_{12}^2d_{23}^2d_{34}^2 - d_{13}^2d_{32}^2d_{24}^2\\ & - d_{12}^2d_{24}^2d_{43}^2 - d_{14}^2d_{42}^2d_{23}^2 - d_{13}^2d_{34}^2d_{42}^2\\ & - d_{14}^2d_{43}^2d_{32}^2 - d_{23}^2d_{31}^2d_{14}^2 - d_{21}^2d_{13}^2d_{34}^2\\ & - d_{24}^2d_{41}^2d_{13}^2 - d_{21}^2d_{14}^2d_{43}^2 - d_{31}^2d_{12}^2d_{24}^2\\ & - d_{32}^2d_{21}^2d_{14}^2. \end{aligned} \tag{7}$$

see also BIMEDIAN, BRAHMAGUPTA'S FORMULA, BRETSCHNEIDER'S FORMULA, COMPLETE QUADRILATERAL, CYCLIC-INSCRIPTABLE QUADRILATERAL, CYCLIC QUADRILATERAL, DIAMOND, EIGHT-POINT CIRCLE THEOREM, EQUILIC QUADRILATERAL, FANO'S AXIOM, LÉON ANNE'S THEOREM, LOZENGE, ORTHOCENTRIC QUADRILATERAL, PARALLELOGRAM, PTOLEMY'S THEOREM, RATIONAL QUADRILATERAL, RHOMBUS, SKEW QUADRILATERAL, TRAPEZOID, VARIGNON'S THEOREM, VON AUBEL'S THEOREM, WITTENBAUER'S PARALLELOGRAM

References

Beyer, W. H. (Ed.) *CRC Standard Mathematical Tables, 28th ed.* Boca Raton, FL: CRC Press, p. 123, 1987.

Routh, E. J. "Moment of Inertia of a Quadrilateral." *Quart. J. Pure Appl. Math.* **11**, 109–110, 1871.

Weinberg, S. *Gravitation and Cosmology: Principles and Applications of the General Theory of Relativity.* New York: Wiley, p. 7, 1972.

Quadrillion

In the American system, 10^{15}.

see also LARGE NUMBER

Quadriplanar Coordinates

The analog of TRILINEAR COORDINATES for TETRAHEDRA.

see also TETRAHEDRON, TRILINEAR COORDINATES

References
Altshiller-Court, N. *Modern Pure Solid Geometry.* New York: Macmillan, 1935.
Mitrinović, D. S.; Pečarić, J. E.; and Volenec, V. Ch. 19 in *Recent Advances in Geometric Inequalities.* Dordrecht, Netherlands: Kluwer, 1989.
Woods, F. S. *Higher Geometry: An Introduction to Advanced Methods in Analytic Geometry.* New York: Dover, pp. 193–196, 1961.

Quadruple

A group of four elements, also called a QUADRUPLET or TETRAD.

see also AMICABLE QUADRUPLE, DIOPHANTINE QUADRUPLE, MONAD, PAIR, PRIME QUADRUPLET, PYTHAGOREAN QUADRUPLE, QUADRUPLET, QUINTUPLET, TETRAD, TRIAD, TRIPLE, TWINS, VECTOR QUADRUPLE PRODUCT

Quadruple Point

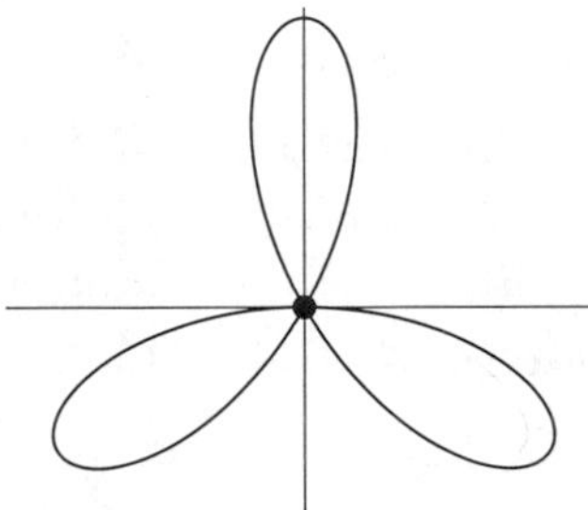

A point where a curve intersects itself along four arcs. The above plot shows the quadruple point at the ORIGIN of the QUADRIFOLIUM $(x^2+y^2)^3 - 4x^2y^2 = 0$.

see also DOUBLE POINT, TRIPLE POINT

References
Walker, R. J. *Algebraic Curves.* New York: Springer-Verlag, pp. 57–58, 1978.

Quadruplet

see QUADRUPLE

Quadtree

A TREE having four branches at each node. Quadtrees are used in the construction of some multidimensional databases (e.g., cartography, computer graphics, and image processing). For a d-D tree, the expected number of comparisons over all pairs of integers for successful and unsuccessful searches are given analytically for $d = 2$ and numerically for $d \geq 3$ by Finch.

References
Finch, S. "Favorite Mathematical Constants." `http://www.mathsoft.com/asolve/constant/qdt/qdt.html`.
Flajolet, P.; Gonnet, G.; Puech, C.; and Robson, J. M. "Analytic Variations on Quadtrees." *Algorithmica* **10**, 473–500, 1993.
Lauwerier, H. *Fractals: Endlessly Repeated Geometric Figures.* Princeton, NJ: Princeton University Press, pp. 11–13, 1991.

Quantic

An m-ary n-ic polynomial (i.e., a HOMOGENEOUS POLYNOMIAL with constant COEFFICIENTS of degree n in m independent variables).

see also ALGEBRAIC INVARIANT, FUNDAMENTAL SYSTEM, p-ADIC NUMBER, SYZYGIES PROBLEM

Quantifier

One of the operations EXISTS $\exists$ or FOR ALL $\forall$.

see also BOUND, EXISTS, FOR ALL, FREE

Quantization Efficiency

Quantization is a nonlinear process which generates additional frequency components (Thompson *et al.* 1986). This means that the signal is no longer band-limited, so the SAMPLING THEOREM no longer holds. If a signal is sampled at the NYQUIST FREQUENCY, information will be lost. Therefore, sampling faster than the NYQUIST FREQUENCY results in detection of more of the signal and a lower signal-to-noise ratio [SNR]. Let β be the OVERSAMPLING ratio and define

$$\eta_Q \equiv \frac{\mathrm{SNR}_{\mathrm{quant}}}{\mathrm{SNR}_{\mathrm{unquant}}}.$$

Then the following table gives values of η_Q for a number of parameters.

Quantization Levels	η_Q $(\beta = 1)$	η_Q $(\beta = 2)$
2	0.64	0.74
3	0.81	0.89
4	0.88	0.94

The Very Large Array of 27 radio telescopes in Socorro, New Mexico uses three-level quantization at $\beta = 1$, so $\eta_Q = 0.81$.

References
Thompson, A. R.; Moran, J. M.; and Swenson, G. W. Jr. Fig. 8.3 in *Interferometry and Synthesis in Radio Astronomy.* New York: Wiley, p. 220, 1986.

Quantum Chaos

The study of the implications of CHAOS for a system in the semiclassical (i.e., between classical and quantum mechanical) regime.

References
Ott, E. "Quantum Chaos." Ch. 10 in *Chaos in Dynamical Systems.* New York: Cambridge University Press, pp. 334–362, 1993.

Quarter

The UNIT FRACTION 1/4, also called one-fourth. It is the value of KOEBE'S CONSTANT.

see also HALF, QUARTILE

Quarter Squares Rule

$$\left(\frac{a+b}{2}\right)^2 - \left(\frac{a-b}{2}\right)^2 = ab.$$

Quartet

A SET of four, also called a TETRAD.

see also HEXAD, MONAD, QUINTET, TETRAD, TRIAD

Quartic Curve

A general plane quartic curve is a curve of the form

$$Ax^4 + By^4 + Cx^3y + Dx^2y^2 + Exy^3 + Fx^3 + Gy^3 + Hx^2y + Ixy^2 + Jx^2 + Ky^2 + Lxy + Mx + Ny + O = 0. \quad (1)$$

The incidence relations of the 28 bitangents of the general quartic curve can be put into a ONE-TO-ONE correspondence with the vertices of a particular POLYTOPE in 7-D space (Coxeter 1928, Du Val 1931). This fact is essentially similar to the discovery by Schoutte (1910) that the 27 SOLOMON'S SEAL LINES on a CUBIC SURFACE can be connected with a POLYTOPE in 6-D space (Du Val 1931). A similar but less complete relation exists between the tritangent planes of the canonical curve of genus 4 and an 8-D POLYTOPE (Du Val 1931).

The maximum number of DOUBLE POINTS for a nondegenerate quartic curve is three.

A quartic curve of the form

$$y^2 = (x-\alpha)(x-\xi)(x-\gamma)(x-\delta) \quad (2)$$

can be written

$$\left(\frac{y}{x-\alpha}\right)^2 = \left(1-\frac{\beta-\alpha}{x-\alpha}\right)\left(1-\frac{\gamma-\alpha}{x-\alpha}\right)\left(1-\frac{\delta-\alpha}{x-\alpha}\right), \quad (3)$$

and so is CUBIC in the coordinates

$$X = \frac{1}{x-\alpha} \quad (4)$$

$$Y = \frac{y}{x-\alpha^2}. \quad (5)$$

This transformation is a BIRATIONAL TRANSFORMATION.

(a)

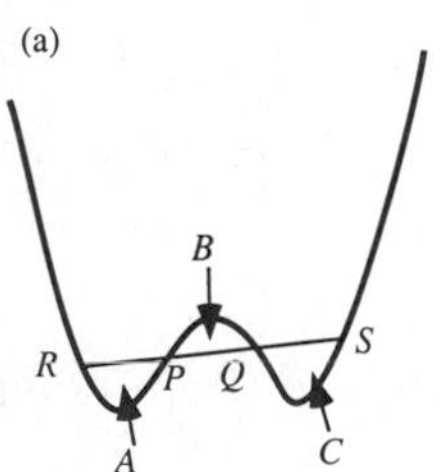

(b)

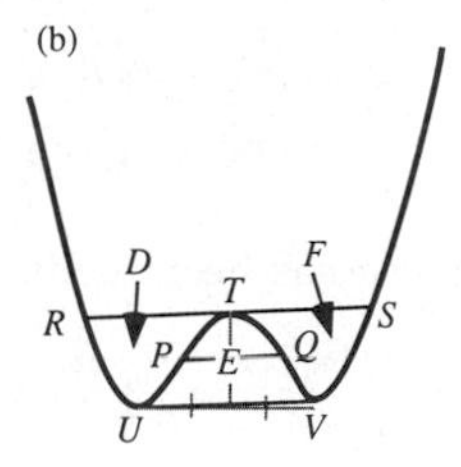

(c)

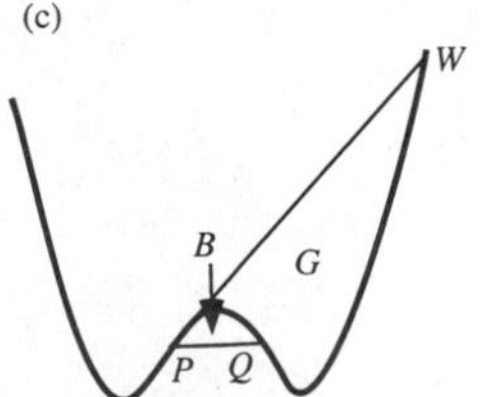

(d)

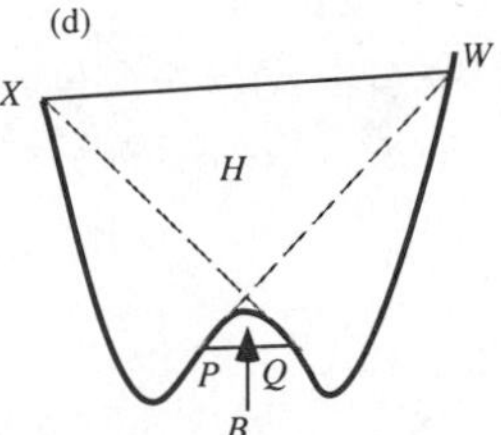

Let P and Q be the INFLECTION POINTS and R and S the intersections of the line PQ with the curve in Figure (a) above. Then

$$A = C \quad (6)$$

$$B = 2A. \quad (7)$$

In Figure (b), let UV be the double tangent, and T the point on the curve whose x coordinate is the average of the x coordinates of U and V. Then $UV||PQ||RS$ and

$$D = F \quad (8)$$

$$E = \sqrt{2}\,D. \quad (9)$$

In Figure (c), the tangent at P intersects the curve at W. Then

$$G = 8B. \quad (10)$$

Finally, in Figure (d), the intersections of the tangents at P and Q are W and X. Then

$$H = 27B \quad (11)$$

(Honsberger 1991).

see also CUBIC SURFACE, PEAR-SHAPED CURVE, SOLOMON'S SEAL LINES

References

Coxeter, H. S. M. "The Pure Archimedean Polytopes in Six and Seven Dimensions." *Proc. Cambridge Phil. Soc.* **24**, 7–9, 1928.

Du Val, P. "On the Directrices of a Set of Points in a Plane." *Proc. London Math. Soc. Ser. 2* **35**, 23–74, 1933.

Honsberger, R. *More Mathematical Morsels.* Washington, DC: Math. Assoc. Amer., pp. 114–118, 1991.

Schoutte, P. H. "On the Relation Between the Vertices of a Definite Sixdimensional Polytope and the Lines of a Cubic Surface." *Proc. Roy. Akad. Acad. Amsterdam* **13**, 375–383, 1910.

Quartic Equation

A general quartic equation (also called a BIQUADRATIC EQUATION) is a fourth-order POLYNOMIAL of the form

$$z^4 + a_3z^3 + a_2z^2 + a_1z + a_0 = 0. \quad (1)$$

The ROOTS of this equation satisfy NEWTON'S RELATIONS:

$$x_1 + x_2 + x_3 + x_4 = -a_3 \quad (2)$$

$$x_1x_2 + x_1x_3 + x_1x_4 + x_2x_3 + x_2x_4 + x_3x_4 = a_2 \quad (3)$$

$$x_1x_2x_3 + x_2x_3x_4 + x_1x_2x_4 + x_1x_3x_4 = -a_1 \quad (4)$$

$$x_1x_2x_3x_4 = a_0, \quad (5)$$

where the denominators on the right side are all $a_4 \equiv 1$.

Ferrari was the first to develop an algebraic technique for solving the general quartic. He applied his technique (which was stolen and published by Cardano) to the equation

$$x^4 + 6x^2 - 60x + 36 = 0 \tag{6}$$

(Smith 1994, p. 207).

The x^3 term can be eliminated from the general quartic (1) by making a substitution of the form

$$z \equiv x - \lambda, \tag{7}$$

so

$$\begin{aligned} x^4 + (a_3 - 4\lambda)x^3 + (a_2 - 3a_3\lambda + 6\lambda^2)x^2 \\ +(a_1 - 2a_2\lambda + 3a_3\lambda^2 - 4\lambda^3)x \\ +(a_0 - a_1\lambda + a_2\lambda^2 - a_3\lambda^3 + \lambda^4). \end{aligned} \tag{8}$$

Letting $\lambda = a_3/4$ so

$$z \equiv x - \tfrac{1}{4}\lambda \tag{9}$$

then gives

$$x^4 + px^2 + qx + r, \tag{10}$$

where

$$p \equiv a_2 - \tfrac{3}{8}a_3{}^2 \tag{11}$$

$$q \equiv a_1 - \tfrac{1}{2}a_2a_3 + \tfrac{1}{8}a_3{}^3 \tag{12}$$

$$r \equiv a_0 - \tfrac{1}{4}a_1a_3 + \tfrac{1}{16}a_2a_3{}^2 - \tfrac{3}{256}a_3{}^4. \tag{13}$$

Adding and subtracting $x^2u + u^2/4$ to (10) gives

$$x^4 + x^2u + \tfrac{1}{4}u^2 - x^2u - \tfrac{1}{4}u^2 + px^2 + qx + r = 0, \tag{14}$$

which can be rewritten

$$(x^2 - \tfrac{1}{2}u)^2 - [(u-p)x^2 - qx + (\tfrac{1}{4}u^2 - r)] = 0 \tag{15}$$

(Birkhoff and Mac Lane 1965). The first term is a perfect square P^2, and the second term is a perfect square Q^2 for those u such that

$$q^2 = 4(u-p)(\tfrac{1}{4}u^2 - r). \tag{16}$$

This is the resolvent CUBIC, and plugging a solution u_1 back in gives

$$P^2 - Q^2 = (P+Q)(P-Q), \tag{17}$$

so (15) becomes

$$(x^2 + \tfrac{1}{2}u_1 + Q)(x^2 + \tfrac{1}{2}u_1 - Q), \tag{18}$$

where

$$Q \equiv Ax - B \tag{19}$$

$$A \equiv \sqrt{u_1 - p} \tag{20}$$

$$B \equiv -\frac{q}{2A}. \tag{21}$$

Let y_1 be a REAL ROOT of the resolvent CUBIC EQUATION

$$y^3 - a_2y^2 + (a_1a_3 - 4a_0)y + (4a_2a_0 - a_1{}^2 - a_3{}^2a_0) = 0. \tag{22}$$

The four ROOTS are then given by the ROOTS of the equation

$$x^2 + \tfrac{1}{2}(a_3 \pm \sqrt{a_3{}^2 - 4a_2 + 4y_1}) + \tfrac{1}{2}(y_1 \mp \sqrt{y_1{}^2 - 4a_0}\,) = 0, \tag{23}$$

which are

$$z_1 = -\tfrac{1}{4}a_3 + \tfrac{1}{2}R + \tfrac{1}{2}D \tag{24}$$

$$z_2 = -\tfrac{1}{4}a_3 + \tfrac{1}{2}R - \tfrac{1}{2}D \tag{25}$$

$$z_3 = -\tfrac{1}{4}a_3 - \tfrac{1}{2}R + \tfrac{1}{2}E \tag{26}$$

$$z_4 = -\tfrac{1}{4}a_3 - \tfrac{1}{2}R - \tfrac{1}{2}E, \tag{27}$$

where

$$R \equiv \sqrt{\tfrac{1}{4}a_3{}^2 - a_2 + y_1} \tag{28}$$

$$D \equiv \begin{cases} \sqrt{\tfrac{3}{4}a_3{}^2 - R^2 - 2a_2 + \tfrac{1}{4}(4a_3a_2 - 8a_1 - a_3{}^3)R^{-1}} & R \neq 0 \\ \sqrt{\tfrac{3}{4}a_3{}^2 - 2a_2 + 2\sqrt{y_1{}^2 - 4a_0}} & R = 0 \end{cases} \tag{29}$$

$$E \equiv \begin{cases} \sqrt{\tfrac{3}{4}a_3{}^2 - R^2 - 2a_2 - \tfrac{1}{4}(4a_3a_2 - 8a_1 - a_3{}^3)R^{-1}} & R \neq 0 \\ \sqrt{\tfrac{3}{4}a_3{}^2 - 2a_2 - 2\sqrt{y_1{}^2 - 4a_0}} & R = 0. \end{cases} \tag{30}$$

Another approach to solving the quartic (10) defines

$$\alpha = (x_1 + x_2)(x_3 + x_4) = -(x_1 + x_2)^2 \tag{31}$$

$$\beta = (x_1 + x_3)(x_2 + x_4) = -(x_1 + x_3)^2 \tag{32}$$

$$\gamma = (x_1 + x_4)(x_2 + x_3) = -(x_2 + x_3)^2, \tag{33}$$

where use has been made of

$$x_1 + x_2 + x_3 + x_4 = 0 \tag{34}$$

(which follows since $a_3 = 0$), and

$$h(x) = (x-\alpha)(x-\beta)(x-\gamma) \tag{35}$$

$$= x^3 - (\alpha+\beta+\gamma)x^2 + (\alpha\beta + \alpha\gamma + \beta\gamma)x - \alpha\beta\gamma. \tag{36}$$

Comparing with

$$\begin{aligned} P(x) &= x^3 + px^2 + qx + r && (37)\\ &= (x - x_1)(x - x_2)(x - x_3)(x - x_4) && (38)\\ &= x^3 + \left(\prod_{i\neq j}^{4} x_i x_j\right) x^2 \\ &\quad + (x_1 + x_2)(x_1 + x_3)(x_2 + x_3)x \\ &\quad - x_1 x_2 x_3 (x_1 + x_2 + x_3), && (39) \end{aligned}$$

gives

$$h(x) = x^3 - 2px^2 = (p^2 - r)z + q^2. \quad (40)$$

Solving this CUBIC EQUATION gives α, β, and γ, which can then be solved for the roots of the quartic x_i (Faucette 1996).

see also CUBIC EQUATION, DISCRIMINANT (POLYNOMIAL), QUINTIC EQUATION

References

Abramowitz, M. and Stegun, C. A. (Eds.). *Handbook of Mathematical Functions with Formulas, Graphs, and Mathematical Tables, 9th printing.* New York: Dover, pp. 17–18, 1972.

Berger, M. §16.4.1–16.4.11.1 in *Geometry I.* New York: Springer-Verlag, 1987.

Beyer, W. H. *CRC Standard Mathematical Tables, 28th ed.* Boca Raton, FL: CRC Press, p. 12, 1987.

Birkhoff, G. and Mac Lane, S. *A Survey of Modern Algebra, 3rd ed.* New York: Macmillan, pp. 107–108, 1965.

Ehrlich, G. §4.16 in *Fundamental Concepts of Abstract Algebra.* Boston, MA: PWS-Kent, 1991.

Faucette, W. M. "A Geometric Interpretation of the Solution of the General Quartic Polynomial." *Amer. Math. Monthly* **103**, 51–57, 1996.

Smith, D. E. *A Source Book in Mathematics.* New York: Dover, 1994.

van der Waerden, B. L. §64 in *Algebra, Vol. 1.* New York: Springer-Verlag, 1993.

Quartic Reciprocity Theorem

Gauss stated the case $n = 4$ using the GAUSSIAN INTEGERS.

see also RECIPROCITY THEOREM

References

Ireland, K. and Rosen, M. "Cubic and Biquadratic Reciprocity." Ch. 9 in *A Classical Introduction to Modern Number Theory, 2nd ed.* New York: Springer-Verlag, pp. 108–137, 1990.

Quartic Surface

An ALGEBRAIC SURFACE of ORDER 4. Unlike CUBIC SURFACES, quartic surfaces have not been fully classified.

see also BOHEMIAN DOME, BURKHARDT QUARTIC, CASSINI SURFACE, CUSHION, CYCLIDE, DESMIC SURFACE, KUMMER SURFACE, MITER SURFACE, PIRIFORM, ROMAN SURFACE, SYMMETROID, TETRAHEDROID, TOOTH SURFACE

References

Fischer, G. (Ed.). *Mathematical Models from the Collections of Universities and Museums.* Braunschweig, Germany: Vieweg, p. 9, 1986.

Fischer, G. (Ed.). Plates 40–41, 45–49, and 52–56 in *Mathematische Modelle/Mathematical Models, Bildband/Photograph Volume.* Braunschweig, Germany: Vieweg, pp. 40–41, 45–49, and 52–56, 1986.

Hunt, B. "Some Quartic Surfaces." Appendix B.5 in *The Geometry of Some Special Arithmetic Quotients.* New York: Springer-Verlag, pp. 310–319, 1996.

Jessop, C. *Quartic Surfaces with Singular Points.* Cambridge, England: Cambridge University Press, 1916.

Quartile

One of the four divisions of observations which have been grouped into four equal-sized sets based on their RANK. The quartile including the top RANKED members is called the first quartile and denoted Q_1. The other quartiles are similarly denoted Q_2, Q_3, and Q_4. For N data points with N of the form $4n+5$ (for $n = 0$, $1, \ldots$), the HINGES are identical to the first and third quartiles.

see also HINGE, INTERQUARTILE RANGE, QUARTILE DEVIATION, QUARTILE VARIATION COEFFICIENT

Quartile Deviation

$$\text{QD} = \tfrac{1}{2}(Q_3 - Q_1),$$

where Q_1 and Q_2 are INTERQUARTILE RANGES.

see also QUARTILE VARIATION COEFFICIENT

Quartile Range

see INTERQUARTILE RANGE

Quartile Skewness Coefficient

see BOWLEY SKEWNESS

Quartile Variation Coefficient

$$V \equiv 100\frac{Q_3 - Q_1}{Q_3 + Q_1},$$

where Q_1 and Q_2 are INTERQUARTILE RANGES.

Quasiamicable Pair

Let $\sigma(m)$ be the DIVISOR FUNCTION of m. Then two numbers m and n are a quasiamicable pair if

$$\sigma(m) = \sigma(n) = m + n + 1.$$

The first few are (48, 75), (140, 195), (1050, 1925), (1575, 1648), ... (Sloane's A005276). Quasiamicable numbers are sometimes called BETROTHED NUMBERS or REDUCED AMICABLE PAIRS.

see also AMICABLE PAIR

References

Beck, W. E. and Najar, R. M. "More Reduced Amicable Pairs." *Fib. Quart.* **15**, 331–332, 1977.

Guy, R. K. "Quasi-Amicable or Betrothed Numbers." §B5 in *Unsolved Problems in Number Theory, 2nd ed.* New York: Springer-Verlag, pp. 59–60, 1994.
Hagis, P. and Lord, G. "Quasi-Amicable Numbers." *Math. Comput.* **31**, 608–611, 1977.
Sloane, N. J. A. Sequence A005276/M5291 in "An On-Line Version of the Encyclopedia of Integer Sequences."

Quasiconformal Map

A generalized CONFORMAL MAP.

see also BELTRAMI DIFFERENTIAL EQUATION

References

Iyanaga, S. and Kawada, Y. (Eds.). "Quasiconformal Mappings." §347 in *Encyclopedic Dictionary of Mathematics.* Cambridge, MA: MIT Press, pp. 1086–1088, 1980.

Quasigroup

A GROUPOID S such that for all $a, b \in S$, there exist unique $x, y \in S$ such that

$$ax = b$$
$$ya = b.$$

No other restrictions are applied; thus a quasigroup need not have an IDENTITY ELEMENT, not be associative, etc. Quasigroups are precisely GROUPOIDS whose multiplication tables are LATIN SQUARES. A quasigroup can be empty.

see also BINARY OPERATOR, GROUPOID, LATIN SQUARE, LOOP (ALGEBRA), MONOID, SEMIGROUP

References

van Lint, J. H. and Wilson, R. M. *A Course in Combinatorics.* New York: Cambridge University Press, 1992.

Quasiperfect Number

A least ABUNDANT NUMBER, i.e., one such that

$$\sigma(n) = 2n + 1.$$

Quasiperfect numbers are therefore the sum of their nontrivial DIVISORS. No quasiperfect numbers are known, although if any exist, they must be greater than 10^{35} and have seven or more DIVISORS. Singh (1997) called quasiperfect numbers SLIGHTLY EXCESSIVE NUMBERS.

see also ABUNDANT NUMBER, ALMOST PERFECT NUMBER, PERFECT NUMBER

References

Guy, R. K. "Almost Perfect, Quasi-Perfect, Pseudoperfect, Harmonic, Weird, Multiperfect and Hyperperfect Numbers." §B2 in *Unsolved Problems in Number Theory, 2nd ed.* New York: Springer-Verlag, pp. 45–53, 1994.
Singh, S. *Fermat's Enigma: The Epic Quest to Solve the World's Greatest Mathematical Problem.* New York: Walker, p. 13, 1997.

Quasiperiodic Function

see WEIERSTRAß SIGMA FUNCTION, WEIERSTRAß ZETA FUNCTION

Quasiperiodic Motion

The type of motion executed by a DYNAMICAL SYSTEM containing two incommensurate frequencies.

Quasirandom Sequence

A sequence of n-tuples that fills n-space more uniformly than uncorrelated random points. Such a sequence is extremely useful in computational problems where numbers are computed on a grid, but it is not known in advance how fine the grid must be to obtain accurate results. Using a quasirandom sequence allows stopping at any point where convergence is observed, whereas the usual approach of halving the interval between subsequent computations requires a huge number of computations between stopping points.

see also PSEUDORANDOM NUMBER, RANDOM NUMBER

References

Press, W. H.; Flannery, B. P.; Teukolsky, S. A.; and Vetterling, W. T. "Quasi- (that is, Sub-) Random Sequences." §7.7 in *Numerical Recipes in FORTRAN: The Art of Scientific Computing, 2nd ed.* Cambridge, England: Cambridge University Press, pp. 299–306, 1992.

Quasiregular Polyhedron

A quasiregular polyhedron is the solid region interior to two DUAL regular polyhedra with SCHLÄFLI SYMBOLS$\{p, q\}$ and $\{q, p\}$. Quasiregular polyhedra are denoted using a SCHLÄFLI SYMBOL of the form $\left\{ {p \atop q} \right\}$, with

$$\left\{ {p \atop q} \right\} = \left\{ {q \atop p} \right\}. \tag{1}$$

Quasiregular polyhedra have two kinds of regular faces with each entirely surrounded by faces of the other kind, equal sides, and equal dihedral angles. They must satisfy the Diophantine inequality

$$\frac{1}{p} + \frac{1}{q} + \frac{1}{r} > 1. \tag{2}$$

But $p, q \geq 3$, so r must be 2. This means that the possible quasiregular polyhedra have symbols $\left\{ {3 \atop 3} \right\}$, $\left\{ {3 \atop 4} \right\}$, and $\left\{ {3 \atop 5} \right\}$. Now

$$\left\{ {3 \atop 3} \right\} = \{3, 4\} \tag{3}$$

is the OCTAHEDRON, which is a regular PLATONIC SOLID and not considered quasiregular. This leaves only two convex quasiregular polyhedra: the CUBOCTAHEDRON $\left\{ {3 \atop 4} \right\}$ and the ICOSIDODECAHEDRON $\left\{ {3 \atop 5} \right\}$.

If nonconvex polyhedra are allowed, then additional quasiregular polyhedra are the GREAT DODECAHEDRON $\{5, \frac{5}{2}\}$ and the GREAT ICOSIDODECAHEDRON $\{3, \frac{5}{2}\}$ (Hart).

For faces to be equatorial $\{h\}$,

$$h = \sqrt{4N_1 + 1} - 1. \tag{4}$$

The EDGES of quasiregular polyhedra form a system of GREAT CIRCLES: the OCTAHEDRON forms three SQUARES, the CUBOCTAHEDRON four HEXAGONS, and the ICOSIDODECAHEDRON six DECAGONS. The VERTEX FIGURES of quasiregular polyhedra are RHOMBUSES (Hart). The EDGES are also all equivalent, a property shared only with the completely regular PLATONIC SOLIDS.

see also CUBOCTAHEDRON, GREAT DODECAHEDRON, GREAT ICOSIDODECAHEDRON, ICOSIDODECAHEDRON, PLATONIC SOLID

References

Coxeter, H. S. M. "Quasi-Regular Polyhedra." §2-3 in *Regular Polytopes, 3rd ed.* New York: Dover, pp. 17–20, 1973.

Hart, G. W. "Quasi-Regular Polyhedra." `http://www.li.net/~george/virtual-polyhedra/quasi-regular-info.html`.

Quasirhombicosidodecahedron

see GREAT RHOMBICOSIDODECAHEDRON (UNIFORM)

Quasirhombicuboctahedron

see GREAT RHOMBICUBOCTAHEDRON (UNIFORM)

Quasisimple Group

A FINITE GROUP L is quasisimple if $L = [L, L]$ and $L/Z(L)$ is a SIMPLE GROUP.

see also COMPONENT, FINITE GROUP, SIMPLE GROUP

Quasithin Theorem

In the classical quasithin case of the QUASI-UNIPOTENT PROBLEM, if G does not have a "strongly embedded" SUBGROUP, then G is a GROUP of LIE-TYPE in characteristic 2 of Lie RANK 2 generated by a pair of parabolic SUBGROUPS P_1 and P_2, or G is one of a short list of exceptions.

see also LIE-TYPE GROUP, QUASI-UNIPOTENT PROBLEM

Quasitruncated Cuboctahedron

see GREAT TRUNCATED CUBOCTAHEDRON

Quasitruncated Dodecadocahedron

see TRUNCATED DODECADODECAHEDRON

Quasitruncated Dodecahedron

see TRUNCATED DODECAHEDRON

Quasitruncated Great Stellated Dodecahedron

see GREAT STELLATED TRUNCATED DODECAHEDRON

Quasitruncated Hexahedron

see STELLATED TRUNCATED HEXAHEDRON

Quasitruncated Small Stellated Dodecahedron

see SMALL STELLATED TRUNCATED DODECAHEDRON

Quasi-Unipotent Group

A GROUP G is quasi-unipotent if every element of G of order p is UNIPOTENT for all PRIMES p such that G has p-RANK ≥ 3.

Quasi-Unipotent Problem

see QUASITHIN THEOREM

Quaternary

The BASE 4 method of counting in which only the DIGITS 0, 1, 2, and 3 are used. These DIGITS have the following multiplication table.

×	0	1	2	3
0	0	0	0	0
1	0	1	2	3
2	0	2	10	12
3	0	3	12	21

see also BASE (NUMBER), BINARY, DECIMAL, HEXADECIMAL, OCTAL, TERNARY

References

Lauwerier, H. *Fractals: Endlessly Repeated Geometric Figures.* Princeton, NJ: Princeton University Press, pp. 9–10, 1991.

Weisstein, E. W. "Bases." `http://www.astro.virginia.edu/~eww6n/math/notebooks/Bases.m`.

Quaternary Tree

see QUADTREE

Quaternion

A member of a *noncommutative* DIVISION ALGEBRA first invented by William Rowan Hamilton. The quaternions are sometimes also known as HYPERCOMPLEX NUMBERS and denoted $\mathbb{H}$. While the quaternions are not commutative, they are associative.

The quaternions can be represented using complex 2×2 MATRICES

$$H = \begin{bmatrix} z & w \\ -w^* & z^* \end{bmatrix} = \begin{bmatrix} a+ib & c+id \\ -c+id & a-ib \end{bmatrix}, \quad (1)$$

where z and w are COMPLEX NUMBERS, a, b, c, and d are REAL, and z^* is the COMPLEX CONJUGATE of z. By analogy with the COMPLEX NUMBERS being representable as a sum of REAL and IMAGINARY PARTS, $a \cdot 1 + bi$, a quaternion can also be written as a linear combination

$$H = a\mathsf{U} + b\mathsf{I} + c\mathsf{J} + d\mathsf{K} \quad (2)$$

of the four matrices

$$\mathsf{U} \equiv \begin{bmatrix} 1 & 0 \\ 0 & 1 \end{bmatrix} \tag{3}$$

$$\mathsf{I} \equiv \begin{bmatrix} i & 0 \\ 0 & -i \end{bmatrix} \tag{4}$$

$$\mathsf{J} \equiv \begin{bmatrix} 0 & 1 \\ -1 & 0 \end{bmatrix} \tag{5}$$

$$\mathsf{K} \equiv \begin{bmatrix} 0 & i \\ i & 0 \end{bmatrix}. \tag{6}$$

(Note that here, U is used to denote the IDENTITY MATRIX, not I.) The matrices are closely related to the PAULI SPIN MATRICES σ_x, σ_y, σ_z, combined with the IDENTITY MATRIX. From the above definitions, it follows that

$$\mathsf{I}^2 = -\mathsf{U} \tag{7}$$

$$\mathsf{J}^2 = -\mathsf{U} \tag{8}$$

$$\mathsf{K}^2 = -\mathsf{U}. \tag{9}$$

Therefore I, J, and K are three essentially different solutions of the matrix equation

$$\mathsf{X}^2 = -\mathsf{U}, \tag{10}$$

which could be considered the square roots of the negative identity matrix.

In $\mathbb{R}^4$, the basis of the quaternions can be given by

$$i \equiv \begin{bmatrix} 0 & 1 & 0 & 0 \\ -1 & 0 & 0 & 0 \\ 0 & 0 & 0 & 1 \\ 0 & 0 & -1 & 0 \end{bmatrix} \tag{11}$$

$$j \equiv \begin{bmatrix} 0 & 0 & 0 & -1 \\ 0 & 0 & -1 & 0 \\ 0 & 1 & 0 & 0 \\ 1 & 0 & 0 & 0 \end{bmatrix} \tag{12}$$

$$k \equiv \begin{bmatrix} 0 & 0 & -1 & 0 \\ 0 & 0 & 0 & 1 \\ 1 & 0 & 0 & 0 \\ 0 & -1 & 0 & 0 \end{bmatrix} \tag{13}$$

$$1 \equiv \begin{bmatrix} 1 & 0 & 0 & 0 \\ 0 & 1 & 0 & 0 \\ 0 & 0 & 1 & 0 \\ 0 & 0 & 0 & 1 \end{bmatrix}. \tag{14}$$

The quaternions satisfy the following identities, sometimes known as HAMILTON'S RULES,

$$i^2 = j^2 = k^2 = -1 \tag{15}$$

$$ij = -ji = k \tag{16}$$

$$jk = -kj = i \tag{17}$$

$$ki = -ik = j. \tag{18}$$

They have the following multiplication table.

	1	i	j	k
1	1	i	j	k
i	i	-1	k	$-j$
j	j	$-k$	-1	i
k	k	j	$-i$	-1

The quaternions ± 1, $\pm i$, $\pm j$, and $\pm k$ form a non-Abelian GROUP of order eight (with multiplication as the group operation) known as Q_8.

The quaternions can be written in the form

$$a = a_1 + a_2 i + a_3 j + a_4 k. \tag{19}$$

The conjugate quaternion is given by

$$a^* = a_1 - a_2 i - a_3 j - a_4 k. \tag{20}$$

The sum of two quaternions is then

$$a+b = (a_1+b_1)+(a_2+b_2)i+(a_3+b_3)j+(a_4+b_4)k, \tag{21}$$

and the product of two quaternions is

$$\begin{aligned} ab = {} & (a_1 b_1 - a_2 b_2 - a_3 b_3 - a_4 b_4) \\ & + (a_1 b_2 + a_2 b_1 + a_3 b_4 - a_4 b_3) i \\ & + (a_1 b_3 - a_2 b_4 + a_3 b_1 + a_4 b_2) j \\ & + (a_1 b_4 + a_2 b_3 - a_3 b_2 + a_4 b_1) k, \end{aligned} \tag{22}$$

so the norm is

$$n(a) = \sqrt{aa^*} = \sqrt{a^* a} = \sqrt{a_1{}^2 + a_2{}^2 + a_3{}^2 + a_4{}^2}. \tag{23}$$

In this notation, the quaternions are closely related to FOUR-VECTORS.

Quaternions can be interpreted as a SCALAR plus a VECTOR by writing

$$a = a_1 + a_2 i + a_3 j + a_4 k = (a_1, \mathbf{a}), \tag{24}$$

where $\mathbf{a} \equiv [a_2\ a_3\ a_4]$. In this notation, quaternion multiplication has the particularly simple form

$$\begin{aligned} q_1 q_2 &= (s_1, \mathbf{v}_1) \cdot (s_2, \mathbf{v}_2) \\ &= (s_1 s_2 - \mathbf{v}_1 \cdot \mathbf{v}_2, s_1 \mathbf{v}_2 + s_2 \mathbf{v}_1 + \mathbf{v}_1 \times \mathbf{v}_2). \end{aligned} \tag{25}$$

Division is uniquely defined (except by zero), so quaternions form a DIVISION ALGEBRA. The inverse of a quaternion is given by

$$a^{-1} = \frac{a^*}{aa^*}, \tag{26}$$

and the norm is multiplicative

$$n(ab) = n(a)n(b). \tag{27}$$

In fact, the product of two quaternion norms immediately gives the EULER FOUR-SQUARE IDENTITY.

A rotation about the UNIT VECTOR $\hat{\mathbf{n}}$ by an angle θ can be computing using the quaternion

$$q = (s, \mathbf{v}) = (\cos(\tfrac{1}{2}\theta), \hat{\mathbf{n}}\sin(\tfrac{1}{2}\theta)) \tag{28}$$

(Arvo 1994, Hearn and Baker 1996). The components of this quaternion are called EULER PARAMETERS. After rotation, a point $p = (0, \mathbf{p})$ is then given by

$$p' = qpq^{-1} = qpq^*, \tag{29}$$

since $n(q) = 1$. A concatenation of two rotations, first q_1 and then q_2, can be computed using the identity

$$q_2(q_1pq_1^*)q_2^* = (q_2q_1)p(q_1^*q_2^*) = (q_2q_1)p(q_2q_1)^* \tag{30}$$

(Goldstein 1980).

see also BIQUATERNION, CAYLEY-KLEIN PARAMETERS, COMPLEX NUMBER, DIVISION ALGEBRA, EULER PARAMETERS, FOUR-VECTOR, OCTONION

References

Altmann, S. L. *Rotations, Quaternions, and Double Groups.* Oxford, England: Clarendon Press, 1986.

Arvo, J. *Graphics Gems 2.* New York: Academic Press, pp. 351–354 and 377–380, 1994.

Baker, A. L. *Quaternions as the Result of Algebraic Operations.* New York: Van Nostrand, 1911.

Beeler, M.; Gosper, R. W.; and Schroeppel, R. *HAKMEM.* Cambridge, MA: MIT Artificial Intelligence Laboratory, Memo AIM-239, Item 107, Feb. 1972.

Conway, J. H. and Guy, R. K. *The Book of Numbers.* New York: Springer-Verlag, pp. 230–234, 1996.

Crowe, M. J. *A History of Vector Analysis: The Evolution of the Idea of a Vectorial System.* New York: Dover, 1994.

Dickson, L. E. *Algebras and Their Arithmetics.* New York: Dover, 1960.

Du Val, P. *Homographies, Quaternions, and Rotations.* Oxford, England: Oxford University Press, 1964.

Ebbinghaus, H. D.; Hirzebruch, F.; Hermes, H.; Prestel, A; Koecher, M.; Mainzer, M.; and Remmert, R. *Numbers.* New York: Springer-Verlag, 1990.

Goldstein, H. *Classical Mechanics, 2nd ed.* Reading, MA: Addison-Wesley, p. 151, 1980.

Hamilton, W. R. *Lectures on Quaternions: Containing a Systematic Statement of a New Mathematical Method.* Dublin: Hodges and Smith, 1853.

Hamilton, W. R. *Elements of Quaternions.* London: Longmans, Green, 1866.

Hamilton, W. R. *The Mathematical Papers of Sir William Rowan Hamilton.* Cambridge, England: Cambridge University Press, 1967.

Hardy, A. S. *Elements of Quaternions.* Boston, MA: Ginn, Heath, & Co., 1881.

Hardy, G. H. and Wright, E. M. *An Introduction to the Theory of Numbers, 5th ed.* Cambridge, England: Clarendon Press, 1965.

Hearn, D. and Baker, M. P. *Computer Graphics: C Version, 2nd ed.* Englewood Cliffs, NJ: Prentice-Hall, pp. 419–420 and 617–618, 1996.

Joly, C. J. *A Manual of Quaternions.* London: Macmillan, 1905.

Julstrom, B. A. "Using Real Quaternions to Represent Rotations in Three Dimensions." *UMAP Modules in Undergraduate Mathematics and Its Applications, Module 652.* Lexington, MA: COMAP, Inc., 1992.

Kelland, P. and Tait, P. G. *Introduction to Quaternions, 3rd ed.* London: Macmillan, 1904.

Nicholson, W. K. *Introduction to Abstract Algebra.* Boston, MA: PWS-Kent, 1993.

Tait, P. G. *An Elementary Treatise on Quaternions, 3rd ed., enl.* Cambridge, England: Cambridge University Press, 1890.

Tait, P. G. "Quaternions." *Encyclopædia Britannica, 9th ed.* ca. 1886. ftp://ftp.netcom.com/pub/hb/hbaker/quaternion/tait/Encyc-Brit.ps.gz.

Quattuordecillion

In the American system, 10^{45}.

see also LARGE NUMBER

Queens Problem

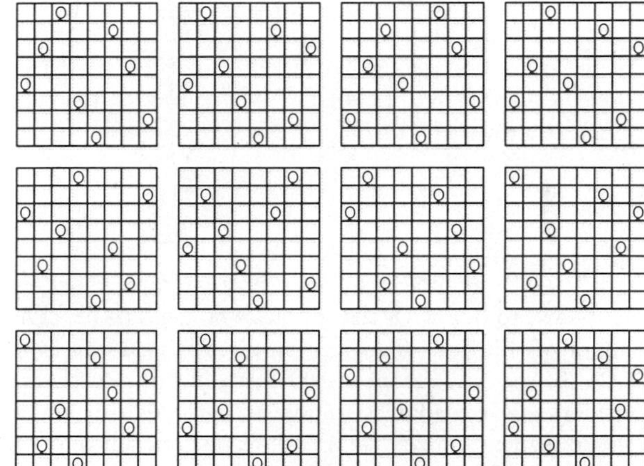

What is the maximum number of queens which can be placed on an $n \times n$ CHESSBOARD such that no two attack one another? The answer is n queens, which gives eight queens for the usual 8×8 board (Madachy 1979). The *number* of different ways the n queens can be placed on an $n \times n$ chessboard so that no two queens may attack each other for the first few n are 1, 0, 0, 2, 10, 4, 40, 92, ... (Sloane's A000170, Madachy 1979). The number of rotationally and reflectively distinct solutions are 1, 0, 0, 1, 2, 1, 6, 12, 46, 92, ... (Sloane's A002562; Dudeney 1970; p. 96). The 12 distinct solutions for $n = 8$ are illustrated above, and the remaining 80 are generated by ROTATION and REFLECTION (Madachy 1979).

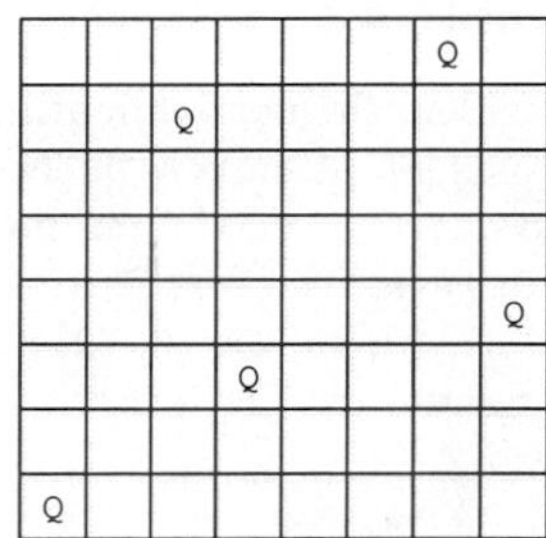

The minimum number of queens needed to occupy or attack all squares of an 8×8 board is 5. Dudeney (1970, pp. 95–96) gave the following results for the number of distinct arrangements $N_p(k,n)$ of k queens attacking or occupying every square of an $n \times n$ board for which every queen is attacked ("protected") by at least one other.

k Queens	$n \times n$	$N_p(k,n)$
2	4	3
3	5	37
3	6	1
4	7	5

Dudeney (1970, pp. 95–96) also gave the following results for the number of distinct arrangements $N_u(k,n)$ of k queens attacking or occupying every square of an $n \times n$ board for which no two queens attack one another (they are "not protected").

k Queens	$n \times n$	$N_u(k,n)$
1	2	1
1	3	1
3	4	2
3	5	2
4	6	17
4	7	1
5	8	91

Vardi (1991) generalizes the problem from a square chessboard to one with the topology of the TORUS. The number of solutions for n queens with n ODD are 1, 0, 10, 28, 0, 88, ... (Sloane's A007705). Vardi (1991) also considers the toroidal "semiqueens" problem, in which a semiqueen can move like a rook or bishop, but only on POSITIVE broken diagonals. The number of solutions to this problem for n queens with n ODD are 1, 3, 15, 133, 2025, 37851, ... (Sloane's A006717), and 0 for EVEN n.

Chow and Velucchi give the solution to the question, "How many different arrangements of k queens are possible on an order n chessboard?" as 1/8th of the COEFFICIENT of $a^k b^{n^2-k}$ in the POLYNOMIAL

$$p(a,b,n) = \begin{cases} (a+b)^{n^2} + 2(a+b)^n (a^2+b^2)^{(n^2-n)/2} \\ \quad +3(a^2+b^2)^{n^2/2} + 2(a^4+b^4)^{n^2/4} & n \text{ even} \\ (a+b)^{n^2} + 2(a+b)(a^4+b^4)^{(n^2-1)/4} \\ \quad +(a+b)(a^2+b^2)^{(n^2-1)/2} \\ \quad +4(a+b)^n(a^2+b^2)^{(n^2-n)/2} & n \text{ odd.} \end{cases}$$

Velucchi also considers the nondominating queens problem, which consists of placing n queens on an order n chessboard to leave a maximum number $U(n)$ of unattacked vacant cells. The first few values are 0, 0, 0, 1, 3, 5, 7, 11, 18, 22, 30, 36, 47, 56, 72, 82, ... (Sloane's A001366). The results can be generalized to k queens on an $n \times n$ board.

see also BISHOPS PROBLEM, CHESS, KINGS PROBLEM, KNIGHTS PROBLEM, KNIGHT'S TOUR, ROOKS PROBLEM

References

Ball, W. W. R. and Coxeter, H. S. M. *Mathematical Recreations and Essays, 13th ed.* New York: Dover, pp. 166–169, 1987.

Campbell, P. J. "Gauss and the 8-Queens Problem: A Study in the Propagation of Historical Error." *Historia Math.* **4**, 397–404, 1977.

Chow, T. and Velucchi, M. "Different Dispositions in the Chessboard." `http://www.cli.di.unipi.it/~velucchi/diff.txt`.

Dudeney, H. E. "The Eight Queens." §300 in *Amusements in Mathematics.* New York: Dover, p. 89, 1970.

Erbas, C. and Tanik, M. M. "Generating Solutions to the N-Queens Problem Using 2-Circulants." *Math. Mag.* **68**, 343–356, 1995.

Erbas, C.; Tanik, M. M.; and Aliyzaicioglu, Z. "Linear Congruence Equations for the Solutions of the N-Queens Problem." *Inform. Proc. Let.* **41**, 301–306, 1992.

Ginsburg, J. "Gauss's Arithmetization of the Problem of n Queens." *Scripta Math.* **5**, 63–66, 1939.

Guy, R. K. "The n Queens Problem." §C18 in *Unsolved Problems in Number Theory, 2nd ed.* New York: Springer-Verlag, pp. 133–135, 1994.

Kraitchik, M. "The Problem of the Queens" and "Domination of the Chessboard." §10.3 and 10.4 in *Mathematical Recreations.* New York: W. W. Norton, pp. 247–256, 1942.

Madachy, J. S. *Madachy's Mathematical Recreations.* New York: Dover, pp. 34–36, 1979.

Pólya, G. "Über die 'doppelt-periodischen' Lösungen des n-Damen-Problems." In *Mathematische Unterhaltungen und Spiele* (Ed. W. Ahrens). 1918.

Riven, I.; Vardi, I.; and Zimmerman, P. "The n-Queens Problem." *Amer. Math. Monthly* **101**, 629–639, 1994.

Riven, I. and Zabih, R. "An Algebraic Approach to Constraint Satisfaction Problems." In *Proc. Eleventh Internat. Joint Conference on Artificial Intelligence, Vol. 1, August 20–25, 1989.* Detroit, MI: IJCAII, pp. 284–289, 1989.

Ruskey, F. "Information on the n Queens Problem." `http://sue.csc.uvic.ca/~cos/inf/misc/Queen.html`.

Sloane, N. J. A. Sequences A001366, A000170/M1958, A006717/M3005, A007705/M4691, and A002562/M0180 in "An On-Line Version of the Encyclopedia of Integer Sequences."Sloane, N. J. A. and Plouffe, S. Extended entry for M0180 in *The Encyclopedia of Integer Sequences.* San Diego: Academic Press, 1995.

Vardi, I. "The n-Queens Problems." Ch. 6 in *Computational Recreations in Mathematica.* Redwood City, CA: Addison-Wesley, pp. 107–125, 1991.

Velucchi, M. "Non-Dominating Queens Problem." `http://www.cli.di.unipi.it/~velucchi/queens.txt`.

Queue

A queue is a special kind of LIST in which elements may only be removed from the bottom by a POP action or added to the top using a PUSH action. Examples of queues include people waiting in line, and submitted jobs waiting to be printed on a printer. The study of queues is called QUEUING THEORY.

see also LIST, QUEUING THEORY, STACK

Queuing Theory

The study of the waiting times, lengths, and other properties of QUEUES.

References

Allen, A. O. *Probability, Statistics, and Queueing Theory with Computer Science Applications.* Orlando, FL: Academic Press, 1978.

Quicksort

The fastest known SORTING ALGORITHM (on average, and for a large number of elements), requiring $\mathcal{O}(n \lg n)$ steps. Quicksort is a recursive algorithm which first partitions an array $\{a_i\}_{i=1}^n$ according to several rules (Sedgewick 1978):

1. Some key ν is in its final position in the array (i.e., if it is the jth smallest, it is in position a_j).
2. All the elements to the left of a_j are less than or equal to a_j. The elements $a_1, a_2, \ldots, a_{j-1}$ are called the "left subfile."
3. All the elements to the right of a_j are greater than or equal to a_j. The elements $a_{j+1}, \ldots, a_n$ are called the "right subfile."

Quicksort was invented by Hoare (1961, 1962), has undergone extensive analysis and scrutiny (Sedgewick 1975, 1977, 1978), and is known to be about twice as fast as the next fastest SORTING algorithm. In the worst case, however, quicksort is a slow n^2 algorithm (and for quicksort, "worst case" corresponds to already sorted).

see also HEAPSORT, SORTING

References

Aho, A. V.; Hopcroft, J. E.; and Ullmann, J. D. *Data Structures and Algorithms.* Reading, MA: Addison-Wesley, pp. 260–270, 1987.

Hoare, C. A. R. "Partition: Algorithm 63," "Quicksort: Algorithm 64," and "Find: Algorithm 65." *Comm. ACM* **4**, 321–322, 1961.

Hoare, C. A. R. "Quicksort." *Computer J.* **5**, 10–15, 1962.

Press, W. H.; Flannery, B. P.; Teukolsky, S. A.; and Vetterling, W. T. "Quicksort." §8.2 in *Numerical Recipes in FORTRAN: The Art of Scientific Computing, 2nd ed.* Cambridge, England: Cambridge University Press, pp. 323–327, 1992.

Sedgewick, R. *Quicksort.* Ph.D. thesis. Stanford Computer Science Report STAN-CS-75-492. Stanford, CA: Stanford University, May 1975.

Sedgewick, R. "The Analysis of Quicksort Programs." *Acta Informatica* **7**, 327–355, 1977.

Sedgewick, R. "Implementing Quicksort Programs." *Comm. ACM* **21**, 847–857, 1978.

Quillen-Lichtenbaum Conjecture

A technical CONJECTURE which connects algebraic k-THEORY to Étale cohomology. The conjecture was made more precise by Dwyer and Friedlander (1982). Thomason (1985) established the first half of this conjecture, but the entire conjecture has not yet been established.

References

Dwyer, W. and Friedlander, E. "Étale K-Theory and Arithmetic." *Bull. Amer. Math. Soc.* **6**, 453–455, 1982.

Thomason, R. W. "Algebraic K-Theory and Étale Cohomology." *Ann. Sci. École Norm. Sup.* **18**, 437–552, 1985.

Weibel, C. A. "The Mathematical Enterprises of Robert Thomason." *Bull. Amer. Math. Soc.* **34**, 1–13, 1996.

Quincunx

The pattern ⁙ of dots on the "5" side of a 6-sided DIE. The word derives from the Latin words for both one and five.

see also DICE

References

Conway, J. H. and Guy, R. K. *The Book of Numbers.* New York: Springer-Verlag, pp. 9 and 22, 1996.

Quindecillion

In the American system, 10^{48}.

see also LARGE NUMBER

Quintet

A SET of five.

see also HEXAD, MONAD, QUARTET, TETRAD, TRIAD

Quintic Equation

A general quintic cannot be solved algebraically in terms of finite additions, multiplications, and root extractions, as rigorously demonstrated by Abel and Galois.

Euler reduced the general quintic to

$$x^5 - 10qx^2 - p = 0. \tag{1}$$

A quintic also can be algebraically reduced to PRINCIPAL QUINTIC FORM

$$x^5 + a_2x^2 + a_1x + a_0 = 0. \tag{2}$$

By solving a quartic, a quintic can be algebraically reduced to the BRING QUINTIC FORM

$$x^5 - x - a = 0, \tag{3}$$

as was first done by Jerrard.

Consider the quintic

$$\prod_{j=0}^{4}[x - (\omega^j u_1 + \omega^{4j} u_2)] = 0, \tag{4}$$

where $\omega = e^{2\pi i/5}$ and u_1 and u_2 are COMPLEX NUMBERS. This is called DE MOIVRE'S QUINTIC. Generalize it to

$$\prod_{j=0}^{4}[x - (\omega^j u_1 + \omega^{2j} u_2 + \omega^{3j} u_3 + \omega^{4j} u_4)] = 0. \tag{5}$$

Expanding,

$$(\omega^j u_1 + \omega^{2j} u_2 + \omega^{3j} u_3 + \omega^{4j} u_4)^5 \\ -5U(\omega^j u_1 + \omega^{2j} u_2 + \omega^{3j} u_3 + \omega^{4j} u_4)^4 \\ -5V(\omega^j u_1 + \omega^{2j} u_2 + \omega^{3j} u_3 + \omega^{4j} u_4)^2 \\ +5W(\omega^j u_1 + \omega^{2j} u_2 + \omega^{3j} u_3 + \omega^{4j} u_4) \\ +[5(X-Y)-Z] = 0, \quad (6)$$

where

$$U = u_1 u_4 + u2u_3 \quad (7)$$
$$V = u_1 {u_2}^2 + u_2 {u_4}^2 + u_3 {u_1}^2 + u_4 {u_3}^2 \quad (8)$$
$$W = {u_1}^2 {u_4}^2 + {u_2}^2 {u_3}^2 - {u_1}^3 u_2 - {u_2}^3 u_4 - {u_3}^3 u_1 - {u_4}^3 u_3 - u_1 u_2 u_3 u_4 \quad (9)$$
$$X = {u_1}^3 u_3 u_4 + {u_2}^3 u_1 u_3 + {u_3}^3 u_2 u_4 + {u_4}^3 u_1 u_2 \quad (10)$$
$$Y = u_1 {u_3}^2 {u_4}^2 + u_2 {u_1}^2 {u_3}^2 + u_3 {u_2}^2 {u_4}^2 + u_4 {u_1}^2 {u_2}^2 \quad (11)$$
$$Z = {u_1}^5 + {u_2}^5 + {u_3}^5 + {u_4}^5. \quad (12)$$

The u_is satisfy

$$u_1 u_4 + u_2 u_3 = 0 \quad (13)$$
$$u_1 {u_2}^2 + u_2 {u_4}^2 + u_3 {u_1}^2 + u_4 {u_3}^2 = 0 \quad (14)$$
$${u_1}^2 {u_4}^2 + {u_2}^2 {u_3}^2 - {u_1}^3 u_2 - {u_2}^3 u_4 - {u_3}^3 u_1 - {u_4}^3 u_3 - u_1 u_2 u_3 u_4 = \tfrac{1}{5} a \quad (15)$$
$$5[({u_1}^3 u_3 u_4 + {u_2}^3 u_1 u_3 + {u_3}^3 u_3 u_4 + {u_4}^3 u_1 u_2) - (u_1 {u_3}^2 {u_4}^2 + u_2 {u_1}^2 {u_3}^2 + u_3 {u_2}^2 {u_4}^2 + u_4 {u_1}^2 {u_2}^2)] - ({u_1}^5 + {u_2}^5 + {u_3}^5 + {u_4}^5) = b. \quad (16)$$

Spearman and Williams (1994) show that an irreducible quintic

$$x^5 + ax + b = 0 \quad (17)$$

with RATIONAL COEFFICIENTS is solvable by radicals IFF there exist rational numbers $\epsilon = \pm 1$, $c \geq 0$, and $e \neq 0$ such that

$$a = \frac{5e^4(3-4\epsilon c)}{c^2+1} \quad (18)$$
$$b = \frac{-4e^5(11\epsilon + 2c)}{c^2+1}. \quad (19)$$

The ROOTS are then

$$x_j = e(\omega^j u_1 + \omega^{2j} u_2 + \omega^{3j} u_3 + \omega^{4j} u_4), \quad (20)$$

where

$$u_1 = \left(\frac{{v_1}^2 v_3}{D^2}\right)^{1/5} \quad (21)$$
$$u_2 = \left(\frac{{v_3}^2 v_4}{D^2}\right)^{1/5} \quad (22)$$
$$u_3 = \left(\frac{{v_2}^2 v_1}{D^2}\right)^{1/5} \quad (23)$$
$$u_4 = \left(\frac{{v_4}^2 v_2}{D^2}\right)^{1/5} \quad (24)$$
$$v_1 = \sqrt{D} + \sqrt{D - \epsilon\sqrt{D}} \quad (25)$$
$$v_2 = -\sqrt{D} - \sqrt{D + \epsilon\sqrt{D}} \quad (26)$$
$$v_3 = -\sqrt{D} + \sqrt{D + \epsilon\sqrt{D}} \quad (27)$$
$$v_4 = \sqrt{D} - \sqrt{D - \epsilon\sqrt{D}} \quad (28)$$
$$D = c^2 + 1. \quad (29)$$

The general quintic can be solved in terms of THETA FUNCTIONS, as was first done by Hermite in 1858. Kronecker subsequently obtained the same solution more simply, and Brioshi also derived the equation. To do so, reduce the general quintic

$$a_5 x^5 + a_4 x^4 + a_3 x^3 + a_2 x^2 + a_1 x + a_0 = 0 \quad (30)$$

into BRING QUINTIC FORM

$$x^5 - x + \rho = 0. \quad (31)$$

Then define

$$k \equiv \tan\left[\tfrac{1}{4}\sin^{-1}\left(\frac{16}{25\sqrt{5}\,\rho^2}\right)\right] \quad (32)$$
$$s \equiv \begin{cases} -\operatorname{sgn}(\Im[\rho]) & \text{for } \Re[\rho] = 0 \\ \operatorname{sgn}(\Re[\rho]) & \text{for } \Re[\rho] \neq 0 \end{cases} \quad (33)$$
$$b = \frac{s(k^2)^{1/8}}{2 \cdot 5^{3/4}\sqrt{k(1-k^2)}} \quad (34)$$
$$q = q(k^2) = e^{i\pi K'(k^2)/K(k^2)}, \quad (35)$$

where k is the MODULUS, $m \equiv k^2$ is the PARAMETER, and q is the NOME. Solving

$$q(m) = e^{i\pi K'(m)/K(m)} \quad (36)$$

for m gives the inverse parameter

$$m(q) = \frac{{\vartheta_2}^4(q)}{{\vartheta_3}^4(q)}. \quad (37)$$

The ROOTS are then given by

$$\begin{aligned} x_1 = {} & (-1)^{3/4} b\{[m(e^{-2\pi i/5} q^{1/5})]^{1/8} \\ & + i[m(e^{2\pi i/5} q^{1/5})]^{1/8}\} \\ & \times\{[m(e^{-4\pi i/5} q^{1/5})]^{1/8} + [m(e^{4\pi i/5} q^{1/5})]^{1/8}\} \\ & \times\{[m(q^{1/5})]^{1/8} + q^{5/8}(q^5)^{-1/8}[m(q^5)]^{1/8}\} \end{aligned} \tag{38}$$

$$\begin{aligned} x_2 = {} & b\{-[m(q^{1/5})]^{1/8} + e^{3\pi i/4}[m(e^{2\pi i/5} q^{1/5})]^{1/8}\} \\ & \times\{e^{-3\pi i/4}[m(e^{-2\pi i/5} q^{1/5})]^{1/8} + i[m(e^{4\pi i/5} q^{1/5})]^{1/8}\} \\ & \times\{i[m(e^{-4\pi i/5} q^{1/5})]^{1/8} + q^{5/8}(q^5)^{-1/8}[m(q^5)]^{1/8}\} \end{aligned} \tag{39}$$

$$\begin{aligned} x_3 = {} & b\{e^{-3\pi i/4}[m(e^{-2\pi i/5} q^{1/5})]^{1/8} \\ & -i[m(e^{-4\pi i/5} q^{1/5})]^{1/8})\}\{-[m(q^{1/5})]^{1/8} \\ & -i[m(e^{4\pi i/5} q^{1/5})]^{1/8}\} \\ & \times\{e^{3\pi i/4}[m(e^{2\pi i/5} q^{1/5})]^{1/8} + q^{5/8}(q^5)^{-1/8}[m(q^5)]^{1/8}\} \end{aligned} \tag{40}$$

$$\begin{aligned} x_4 = {} & b\{[m(q^{1/5})]^{1/8} - i[m(e^{-4\pi i/5} q^{1/5})]^{1/8})\} \\ & \times\{-e^{3\pi i/4}[m(e^{2\pi i/5} q^{1/5})]^{1/8} - i[m(e^{4\pi i/5} q^{1/5})]^{1/8}\} \\ & \times\{e^{-3\pi i/4}[m(e^{-2\pi i/5} q^{1/5})]^{1/8} \\ & + q^{5/8}(q^5)^{-1/8}[m(q^5)]^{1/8}\} \end{aligned} \tag{41}$$

$$\begin{aligned} x_5 = {} & b\{[m(q^{1/5})]^{1/8} - e^{-3\pi i/4}[m(e^{-2\pi i/5} q^{1/5})]^{1/8}\} \\ & \times\{-e^{3\pi i/4}[m(e^{2\pi i/5} q^{1/5})]^{1/8} + i[m(e^{-4\pi i/5} q^{1/5})]^{1/8}\} \\ & \times\{(-i[m(e^{4\pi i/5} q^{1/5})]^{1/8} + q^{5/8}(q^5)^{-1/8}[m(q^5)]^{1/8}\}. \end{aligned} \tag{42}$$

Felix Klein used a TSCHIRNHAUSEN TRANSFORMATION to reduce the general quintic to the form

$$z^5 + 5az^2 + 5bz + c = 0. \tag{43}$$

He then solved the related ICOSAHEDRAL EQUATION

$$\begin{aligned} I(z, 1, Z) = {} & z^5(-1 + 11z^5 + z^{10})^5 \\ & -[1 + z^{30} - 10005(z^{10} + z^{20}) + 522(-z^5 + z^{25})]^2 Z = 0, \end{aligned} \tag{44}$$

where Z is a function of radicals of a, b, and c. The solution of this equation can be given in terms of HYPERGEOMETRIC FUNCTIONS as

$$\frac{Z^{-1/60}{}_2F_1(-\frac{1}{60}, \frac{29}{60}, \frac{4}{5}, 1728Z)}{Z^{11/60}{}_2F_1(\frac{11}{60}, \frac{41}{60}, \frac{6}{5}, 1728Z)}. \tag{45}$$

Another possible approach uses a series expansion, which gives one root (the first one in the list below) of

$$t^5 - t - \rho. \tag{46}$$

All five roots can be derived using differential equations (Cockle 1860, Harley 1862). Let

$$F_1(\rho) = F_2(\rho) \tag{47}$$

$$F_2(\rho) = {}_4F_3(\tfrac{1}{5}, \tfrac{2}{5}, \tfrac{3}{5}, \tfrac{4}{5}; \tfrac{1}{2}, \tfrac{3}{4}, \tfrac{5}{4}; \tfrac{3125}{256}\rho^4) \tag{48}$$

$$F_3(\rho) = {}_4F_3(\tfrac{9}{20}, \tfrac{13}{20}, \tfrac{17}{20}, \tfrac{21}{20}; \tfrac{3}{4}, \tfrac{5}{4}, \tfrac{3}{2}; \tfrac{3125}{256}\rho^4) \tag{49}$$

$$F_4(\rho) = {}_4F_3(\tfrac{7}{10}, \tfrac{9}{10}, \tfrac{11}{10}, \tfrac{13}{10}; \tfrac{5}{4}, \tfrac{3}{2}, \tfrac{7}{4}; \tfrac{3125}{256}\rho^4), \tag{50}$$

then the ROOTS are

$$t_1 = -\rho\, {}_4F_3(\tfrac{1}{5}, \tfrac{2}{5}, \tfrac{3}{5}, \tfrac{4}{5}; \tfrac{1}{2}, \tfrac{3}{4}, \tfrac{5}{4}; \tfrac{3125}{256}\rho^4) \tag{51}$$

$$t_2 = -F_1(\rho) + \tfrac{1}{4}\rho F_2(\rho) + \tfrac{5}{32}\rho^2 F_3(\rho) + \tfrac{5}{32}\rho^3 F_4(\rho) \tag{52}$$

$$t_3 = -F_1(\rho) + \tfrac{1}{4}\rho F_2(\rho) - \tfrac{5}{32}\rho^2 F_3(\rho) + \tfrac{5}{32}\rho^3 F_4(\rho) \tag{53}$$

$$t_4 = -iF_1(\rho) + \tfrac{1}{4}\rho F_2(\rho) - \tfrac{5}{32}i\rho^2 F_3(\rho) - \tfrac{5}{32}\rho^3 F_4(\rho) \tag{54}$$

$$t_5 = iF_1(\rho) + \tfrac{1}{4}\rho F_2(\rho) + \tfrac{5}{32}i\rho^2 F_3(\rho) - \tfrac{5}{32}\rho^3 F_4(\rho). \tag{55}$$

This technique gives closed form solutions in terms of HYPERGEOMETRIC FUNCTIONS in one variable for any POLYNOMIAL equation which can be written in the form

$$x^p + bx^q + c. \tag{56}$$

Cadenhad, Young, and Runge showed in 1885 that all irreducible solvable quintics with COEFFICIENTS of x^4, x^3, and x^2 missing have the following form

$$x^5 + \frac{5\mu^4(4\nu + 3)}{\nu^2 + 1}x + \frac{4\mu^5(2\nu + 1)(4\nu + 3)}{\nu^2 + 1} = 0, \tag{57}$$

where μ and ν are RATIONAL.

see also BRING QUINTIC FORM, BRING-JERRARD QUINTIC FORM, CUBIC EQUATION, DE MOIVRE'S QUINTIC, PRINCIPAL QUINTIC FORM, QUADRATIC EQUATION, QUARTIC EQUATION, SEXTIC EQUATION

References

Birkhoff, G. and Mac Lane, S. *A Survey of Modern Algebra, 3rd ed.* New York: Macmillan, pp. 418–421, 1965.

Chowla, S. "On Quintic Equations Soluble by Radicals." *Math. Student* **13**, 84, 1945.

Cockle, J. "Sketch of a Theory of Transcendental Roots." *Phil. Mag.* **20**, 145–148, 1860.

Cockle, J. " On Transcendental and Algebraic Solution—Supplemental Paper." *Phil. Mag.* **13**, 135–139, 1862.

Davis, H. T. *Introduction to Nonlinear Differential and Integral Equations.* New York: Dover, p. 172, 1960.

Dummit, D. S. "Solving Solvable Quintics." *Math. Comput.* **57**, 387–401, 1991.

Glashan, J. C. "Notes on the Quintic." *Amer. J. Math.* **8**, 178–179, 1885.

Harley, R. "On the Solution of the Transcendental Solution of Algebraic Equations." *Quart. J. Pure Appl. Math.* **5**, 337–361, 1862.

Hermite, C. "Sulla risoluzione delle equazioni del quinto grado." *Annali di math. pura ed appl.* **1**, 256–259, 1858.

King, R. B. *Beyond the Quartic Equation.* Boston, MA: Birkhauser, 1996.
King, R. B. and Cranfield, E. R. "An Algorithm for Calculating the Roots of a General Quintic Equation from Its Coefficients." *J. Math. Phys.* **32**, 823–825, 1991.
Rosen, M. I. "Niels Hendrik Abel and Equations of the Fifth Degree." *Amer. Math. Monthly* **102**, 495–505, 1995.
Shurman, J. *Geometry of the Quintic.* New York: Wiley, 1997.
Spearman, B. K. and Williams, K. S. "Characterization of Solvable Quintics x^5+ax+b." *Amer. Math. Monthly* **101**, 986–992, 1994.
Wolfram Research. "Solving the Quintic." Poster. Champaign, IL: Wolfram Research, 1995. `http://www.wolfram.com/posters/quintic`.
Young, G. P. "Solution of Solvable Irreducible Quintic Equations, Without the Aid of a Resolvent Sextic." *Amer. J. Math.* **7**, 170–177, 1885.

Quintic Surface

A quintic surface is an ALGEBRAIC SURFACE of degree 5. Togliatti (1940, 1949) showed that quintic surfaces having 31 ORDINARY DOUBLE POINTS exist, although he did not explicitly derive equations for such surfaces. Beauville (1978) subsequently proved that 31 double points was the maximum possible, and quintic surfaces having 31 ORDINARY DOUBLE POINTS are therefore sometimes called TOGLIATTI SURFACES. van Straten (1993) subsequently constructed a 3-D family of solutions and in 1994, Barth derived the example known as the DERVISH.

see also ALGEBRAIC SURFACE, DERVISH, KISS SURFACE, ORDINARY DOUBLE POINT

References

Beauville, A. "Surfaces algébriques complexes." *Astérisque* **54**, 1–172, 1978.
Endraß, S. "Togliatti Surfaces." `http://www.mathematik.uni-mainz.de/AlgebraischeGeometrie/docs/Etogliatti.shtml`.
Hunt, B. "Algebraic Surfaces." `http://www.mathematik.uni-kl.de/~wwwagag/Galerie.html`.
Togliatti, E. G. "Una notevole superficie de 5° ordine con soli punti doppi isolati." *Vierteljschr. Naturforsch. Ges. Zürich* **85**, 127–132, 1940.
Togliatti, E. "Sulle superficie monoidi col massimo numero di punti doppi." *Ann. Mat. Pura Appl.* **30**, 201–209, 1949.
van Straten, D. "A Quintic Hypersurface in $\mathbb{P}^4$ with 130 Nodes." *Topology* **32**, 857–864, 1993.

Quintillion

In the American system, 10^{18}.

see also LARGE NUMBER

Quintuple

A group of five elements, also called a QUINTUPLET or PENTAD.

see also MONAD, PAIR, PENTAD, QUADRUPLE, QUADRUPLET, QUINTUPLET, TETRAD, TRIAD, TRIPLET, TWINS

Quintuple Product Identity

A.k.a. the WATSON QUINTUPLE PRODUCT IDENTITY.

$$\prod_{n=1}^{\infty}(1-q^n)(1-zq^n)(1-z^{-1}q^{n-1})(1-z^2q^{2n-1}) \times(1-z^{-2}q^{2n-1}) = \sum_{m=-\infty}^{\infty}(z^{3m}-z^{-3m-1})q^{m(2m+1)/2}. \quad (1)$$

It can also be written

$$\prod_{n=1}^{\infty}(1-q^{2n})(1-q^{2n-1}z)(1-q^{2n-1}z^{-1}) \times(1-q^{4n-3}z^2)(1-q^{4n-4}z^{-2}) = \sum_{n=-\infty}^{\infty} q^{3n^2-2n}[(z^{3n}+z^{-3n})-(z^{3n-2}+z^{-(3n-2)})] \quad (2)$$

or

$$\sum_{k=-\infty}^{\infty}(-1)^k q^{(3k^2-k)/2}x^{3k}(1+zq^k) = \prod_{j=1}^{\infty}(1-q^j)(1+z^{-1}q^j)(1+zq^{j-1}) \times(1+z^{-2}q^{2j-1})(1+z^2q^{2j-1}). \quad (3)$$

Using the NOTATION of the RAMANUJAN THETA FUNCTION (Berndt, p. 83),

$$f(B^3/q,q^5/B^3)-B^2f(q/B^3,B^3q^5) = f(-q^2)\frac{f(-B^2,-q^2/B^2)}{f(Bq,q/B)}. \quad (4)$$

see also JACOBI TRIPLE PRODUCT, RAMANUJAN THETA FUNCTIONS

References

Berndt, B. C. *Ramanujan's Notebooks, Part III.* New York: Springer-Verlag, 1985.
Borwein, J. M. and Borwein, P. B. *Pi & the AGM: A Study in Analytic Number Theory and Computational Complexity.* New York: Wiley, pp. 306–309, 1987.
Gasper, G. and Rahman, M. *Basic Hypergeometric Series.* Cambridge, England: Cambridge University Press, 1990.

Quintuplet

A group of five elements, also called a QUINTUPLE or PENTAD.

see also MONAD, PAIR, PENTAD, QUADRUPLE, QUADRUPLET, QUINTUPLET, TETRAD, TRIAD, TRIPLET, TWINS

Quota Rule

A RECURRENCE RELATION between the function Q arising in QUOTA SYSTEMS,

$$Q(n,r) = Q(n-1,r-1) + Q(n-1,r).$$

References

Young, S. C.; Taylor, A. D.; and Zwicker, W. S. "Counting Quota Systems: A Combinatorial Question from Social Choice Theory." *Math. Mag.* **68**, 331–342, 1995.

Quota System

A generalization of simple majority voting in which a list of quotas $\{q_0, \ldots, q_n\}$ specifies, according to the number of votes, how many votes an alternative needs to win (Taylor 1995). The quota system declares a tie unless for some k, there are exactly k tie votes in the profile and one of the alternatives has at least q_k votes, in which case the alternative is the choice.

Let $Q(n)$ be the number of quota systems for n voters and $Q(n,r)$ the number of quota systems for which $q_0 = r+1$, so

$$Q(n) = \sum_{r=\lfloor n/2 \rfloor}^{n} Q(n,r) = \binom{n+1}{\lfloor \frac{n}{2} \rfloor + 1},$$

where $\lfloor x \rfloor$ is the FLOOR FUNCTION. This produces the sequence of CENTRAL BINOMIAL COEFFICIENTS 1, 2, 3, 6, 10, 20, 35, 70, 126, ... (Sloane's A001405). It may be defined recursively by $Q(0) = 1$ and

$$Q(n+1) = \begin{cases} 2Q(n) & \text{for } n \text{ even} \\ 2Q(n) - C_{(n+1)/2} & \text{for } n \text{ odd,} \end{cases}$$

where C_k is a CATALAN NUMBER (Young *et al.* 1995). The function $Q(n,r)$ satisfies

$$Q(n,r) = \binom{n+1}{r+1} - \binom{n+1}{r+2}$$

for $r > n/2 - 1$ (Young *et al.* 1995). $Q(n,r)$ satisfies the QUOTA RULE.

see also BINOMIAL COEFFICIENT, CENTRAL BINOMIAL COEFFICIENT

References

Sloane, N. J. A. Sequence A001405/M0769 in "An On-Line Version of the Encyclopedia of Integer Sequences."

Taylor, A. *Mathematics and Politics: Strategy, Voting, Power, and Proof.* New York: Springer-Verlag, 1995.

Young, S. C.; Taylor, A. D.; and Zwicker, W. S. "Counting Quota Systems: A Combinatorial Question from Social Choice Theory." *Math. Mag.* **68**, 331–342, 1995.

Quotient

The ratio $q = r/s$ of two quantities r and s, where $s \neq 0$.

see also DIVISION, QUOTIENT GROUP, QUOTIENT RING, QUOTIENT SPACE

Quotient-Difference Algorithm

The ALGORITHM of constructing and interpreting a QUOTIENT-DIFFERENCE TABLE which allows interconversion of CONTINUED FRACTIONS, POWER SERIES, and RATIONAL FUNCTIONS approximations.

see also QUOTIENT-DIFFERENCE TABLE

Quotient-Difference Table

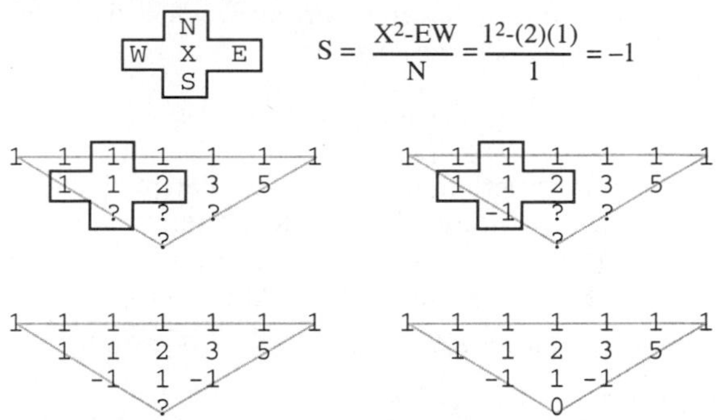

A quotient-difference table is a triangular ARRAY of numbers constructed by drawing a sequence of n numbers in a horizontal row and placing a 1 above each. An additional "1" is then placed at the beginning and end of the row of 1s, and the value of rows underneath the original row is then determined by looking at groups of adjacent numbers

$$\begin{array}{ccc} & N & \\ W & X & E \\ & S & \end{array}$$

and computing

$$S = \frac{X^2 - EW}{N}$$

for the elements falling within a triangle formed by the diagonals extended from the first and last "1," as illustrated above.

0s in quotient-difference tables form square "windows" which are bordered by GEOMETRIC PROGRESSIONS. Quotient-difference tables eventually yield a row of 0s IFF the starting sequence is defined by a linear RECURRENCE RELATION. For example, continuing the above example generated by the FIBONACCI NUMBERS

$$\begin{array}{ccccccccccccc} 1 && 1 && 1 && 1 && 1 && 1 && 1 \\ & 1 && 1 && 2 && 3 && 5 && \\ && && -1 && 1 && -1 && && \\ && && && 0 && && && \end{array}$$

$$\begin{array}{ccccccccccccccc} 1 && 1 && 1 && 1 && 1 && 1 && 1 && 1 \\ & 1 && 1 && 2 && 3 && 5 && 8 & \\ && && -1 && 1 && -1 && 1 && & \\ && && && 0 && 0 && && && \end{array}$$

$$\begin{array}{ccccccccccccccccc} 1 && 1 && 1 && 1 && 1 && 1 && 1 && 1 && 1 \\ & 1 && 1 && 2 && 3 && 5 && 8 && 13 & \\ && && -1 && 1 && -1 && 1 && -1 && && \\ && && && 0 && 0 && 0 && && && \\ && && && && 0 && && && && \end{array}$$

1	1	1	1	1	1	1	1	1	1
	1	1	2	3	5	8	13	21	
		−1	1	−1	1	−1	1		
			0	0	0	0			
				0	0				

and it can be seen that a row of 0s emerges (and furthermore that an attempt to extend the table will result in division by zero). This verifies that the FIBONACCI NUMBERS satisfy a linear recurrence, which is in fact given by the well-known formula

$$F_n = F_{n-1} + F_{n-2}.$$

However, construction of a quotient-difference table for the CATALAN NUMBERS, MOTZKIN NUMBERS, etc., does not lead to a row of zeros, suggesting that these numbers cannot be generated using a linear recurrence.

see also DIFFERENCE TABLE, FINITE DIFFERENCE

References

Conway, J. H. and Guy, R. K. In *The Book of Numbers.* New York: Springer-Verlag, pp. 85–89, 1996.

Quotient Group

The quotient group of G with respect to a SUBGROUP H is denoted G/H and is read "G modulo H." The slash NOTATION conflicts with that for a FIELD EXTENSION, but the meaning can be determined based on context.

see also ABHYANKAR'S CONJECTURE, FIELD EXTENSION, OUTER AUTOMORPHISM GROUP, SUBGROUP

Quotient Ring

The quotient ring of R with respect to a RING modulo some INTEGER n is denoted R/nR and is read "the ring R modulo n." If n is a PRIME p, then $\mathbb{Z}/p\mathbb{Z}$ is the FINITE FIELD $\mathbb{F}_p$. For COMPOSITE

$$n = \prod_{i=1}^{k} p_i$$

with distinct p_i, $\mathbb{Z}/p\mathbb{Z}$ is ISOMORPHIC to the DIRECT SUM

$$\mathbb{Z}/p\mathbb{Z} = \mathbb{F}_{p_1} \otimes \mathbb{F}_{p_2} \otimes \cdots \otimes \mathbb{F}_{p_k}.$$

see also FINITE FIELD, RING

Quotient Rule

The DERIVATIVE rule

$$\frac{d}{dx}\left[\frac{f(x)}{g(x)}\right] = \frac{g(x)f'(x) - f(x)g'(x)}{[g(x)]^2}.$$

see also CHAIN RULE, DERIVATIVE, POWER RULE, PRODUCT RULE

References

Abramowitz, M. and Stegun, C. A. (Eds.). *Handbook of Mathematical Functions with Formulas, Graphs, and Mathematical Tables, 9th printing.* New York: Dover, p. 11, 1972.

Quotient Space

The quotient space $X/\sim$ of a TOPOLOGICAL SPACE X and an EQUIVALENCE RELATION $\sim$ on X is the set of EQUIVALENCE CLASSES of points in X (under the EQUIVALENCE RELATION $\sim$) together with the topology given by a SUBSET U of $X/\sim$. U of $X/\sim$ is open IFF $\cup_{a \in U} a$ is open in X.

This can be stated in terms of MAPS as follows: if $q : X \to X/\sim$ denotes the MAP that sends each point to its EQUIVALENCE CLASS in $X/\sim$, the topology on $X/\sim$ can be specified by prescribing that a subset of $X/\sim$ is open IFF q^{-1}[the set] is open.

In general, quotient spaces are not well behaved, and little is known about them. However, it is known that any compact metrizable space is a quotient of the CANTOR SET, any compact connected n-dimensional MANIFOLD for $n > 0$ is a quotient of any other, and a function out of a quotient space $f : X/\sim \to Y$ is continuous IFF the function $f \circ q : X \to Y$ is continuous.

Let $\mathbb{D}^n$ be the closed n-D DISK and $\mathbb{S}^{n-1}$ its boundary, the $(n-1)$-D sphere. Then $\mathbb{D}^n/\mathbb{S}^{n-1}$ (which is homeomorphic to $\mathbb{S}^n$), provides an example of a quotient space. Here, $\mathbb{D}^n/\mathbb{S}^{n-1}$ is interpreted as the space obtained when the boundary of the n-DISK is collapsed to a point, and is formally the "quotient space by the equivalence relation generated by the relations that all points in $\mathbb{S}^{n-1}$ are equivalent."

see also EQUIVALENCE RELATION, TOPOLOGICAL SPACE

References

Munkres, J. R. *Topology: A First Course.* Englewood Cliffs, NJ: Prentice-Hall, 1975.

R

$\mathbb{R}$

The FIELD of REAL NUMBERS.

see also $\mathbb{C}$, $\mathbb{C}^*$, $\mathbb{I}$, $\mathbb{N}$, $\mathbb{Q}$, $\mathbb{R}^-$, $\mathbb{R}^+$, $\mathbb{Z}$

$\mathbb{R}^-$

The REAL NEGATIVE numbers.

see also $\mathbb{R}$, $\mathbb{R}^+$

$\mathbb{R}^+$

The REAL POSITIVE numbers.

see also $\mathbb{R}$, $\mathbb{R}^-$

$r_k(n)$

The number of representations of n by k squares is denoted $r_k(n)$. The *Mathematica*® (Wolfram Research, Champaign, IL) function `NumberTheory‘NumberTheoryFunctions‘SumOfSquaresR[k,n]` gives $r_k(n)$.

$r_2(n)$ is often simply written $r(n)$. Jacobi solved the problem for $k = 2$, 4, 6, and 8. The first cases $k = 2$, 4, and 6 were found by equating COEFFICIENTS of the THETA FUNCTION $\vartheta_3(x)$, $\vartheta_3{}^2(x)$, and $\vartheta_3{}^4(x)$. The solutions for $k = 10$ and 12 were found by Liouville and Eisenstein, and Glaisher (1907) gives a table of $r_k(n)$ for $k = 2s = 18$. $r_3(n)$ was found as a finite sum involving quadratic reciprocity symbols by Dirichlet. $r_5(n)$ and $r_7(n)$ were found by Eisenstein, Smith, and Minkowski.

$r(n) = r_2(n)$ is 0 whenever n has a PRIME divisor of the form $4k+3$ to an ODD POWER; it doubles upon reaching a new PRIME of the form $4k+1$. It is given explicitly by

$$r(n) = 4 \sum_{d=1,3,\ldots|n} (-1)^{(d-1)/2} = 4[d_1(n) - d_3(n)], \quad (1)$$

where $d_k(n)$ is the number of DIVISORS of n of the form $4m+k$. The first few values are 4, 4, 0, 4, 8, 0, 0, 4, 4, 8, 0, 0, 8, 0, 0, 4, 8, 4, 0, 8, 0, 0, 0, 0, 12, 8, 0, 0, ... (Sloane's A004018). The first few values of the summatory function

$$R(n) \equiv \sum_{k=1}^{n} r(n), \quad (2)$$

where are 0, 4, 8, 8, 12, 20, 20, 20, 24, 28, 36, ... (Sloane's A014198). Shanks (1993) defines instead $R'(n) = 1 + R(n)$, with $R'(0) = 1$. A LAMBERT SERIES for $r(n)$ is

$$\sum_{n=1}^{\infty} \frac{4(-1)^{n+1}x^n}{1-x^n} = \sum_{n=1}^{\infty} r(n)x^n \quad (3)$$

(Hardy and Wright 1979).

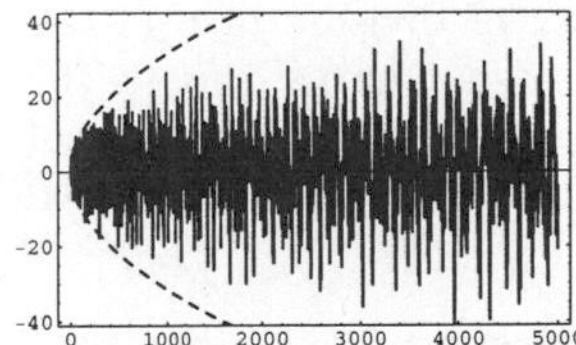

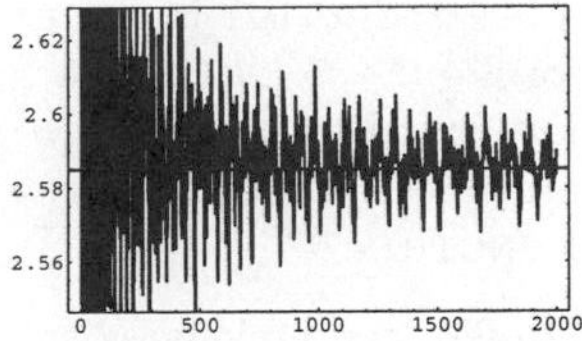

Asymptotic results include

$$\sum_{k=1}^{n} r_2(k) = \pi n + \mathcal{O}(\sqrt{n}) \quad (4)$$

$$\sum_{k=1}^{n} \frac{r_2(k)}{k} = K + \pi \ln n + \mathcal{O}(n^{-1/2}), \quad (5)$$

where K is a constant known as the SIERPIŃSKI CONSTANT. The left plot above

$$\left[\sum_{k=1}^{n} r_2(k)\right] - \pi n, \quad (6)$$

with $\pm\sqrt{n}$ illustrated by the dashed curve, and the right plot shows

$$\left[\sum_{k=1}^{n} \frac{r_2(k)}{k}\right] - \pi \ln n, \quad (7)$$

with the value of K indicated as the solid horizontal line.

The number of solutions of

$$x^2 + y^2 + z^2 = n \quad (8)$$

for a given n without restriction on the signs or relative sizes of x, y, and z is given by $r_3(n)$. If $n > 4$ is SQUAREFREE, then Gauss proved that

$$r_3(n) = \begin{cases} 24h(-n) & \text{for } n \equiv 3 \pmod 8 \\ 12h(-4n) & \text{for } n \equiv 1,2,5,6 \pmod 8 \\ 0 & \text{for } n \equiv 7 \pmod 8 \end{cases} \quad (9)$$

(Arno 1992), where $h(x)$ is the CLASS NUMBER of x.

Additional higher-order identities are given by

$$r_4(n) = 8 \sum_{d|n} d = 8\sigma(n) \quad (10)$$

$$= 24 \sum_{d=1,3,\ldots|n} d = 24\sigma_0(n) \quad (11)$$

$$r_{10}(n) = \tfrac{4}{5}[E_4(n) + 16E_4'(n) + 8\chi_4(n)] \quad (12)$$

$$r_{24}(n) = \rho_{24}(n) + \tfrac{128}{691}[(-1)^{n-1}259\tau(n) - 512\tau(\tfrac{1}{2}n)], \quad (13)$$

where

$$E_4(n) = \sum_{d=1,3,\ldots|n} (-1)^{(d-1)/2} d^4 \quad (14)$$

$$E_4'(n) = \sum_{d'=1,3,\ldots|n} (-1)^{(d'-1)/2} d^4 \quad (15)$$

$$\chi_4(n) = \tfrac{1}{4} \sum_{a^2+b^2=n} (a+bi)^4, \quad (16)$$

$d' \equiv n/d$, $d_k(n)$ is the number of divisors of n of the form $4m+k$, $\rho_{24}(n)$ is a SINGULAR SERIES, $\sigma(n)$ is the DIVISOR FUNCTION, $\sigma_0(n)$ is the DIVISOR FUNCTION of order 0 (i.e., the number of DIVISORS), and τ is the TAU FUNCTION.

Similar expressions exist for larger EVEN k, but they quickly become extremely complicated and can be written simply only in terms of expansions of modular functions.

see also CLASS NUMBER, LANDAU-RAMANUJAN CONSTANT, PRIME FACTORS, SIERPIŃSKI CONSTANT, TAU FUNCTION

References

Arno, S. "The Imaginary Quadratic Fields of Class Number 4." *Acta Arith.* **60**, 321–334, 1992.

Boulyguine. *Comptes Rendus Paris* **161**, 28–30, 1915.

New York: Chelsea, p. 317, 1952.

Glaisher, J. W. L. "On the Numbers of a Representation of a Number as a Sum of $2r$ Squares, where $2r$ Does Not Exceed 18." *Proc. London Math. Soc.* **5**, 479–490, 1907.

Grosswald, E. *Representations of Integers as Sums of Squares.* New York: Springer-Verlag, 1985.

Hardy, G. H. "The Representation of Numbers as Sums of Squares." Ch. 9 in *Ramanujan: Twelve Lectures on Subjects Suggested by His Life and Work, 3rd ed.* New York: Chelsea, 1959.

Hardy, G. H. and Wright, E. M. "The Function $r(n)$," "Proof of the Formula for $r(n)$," "The Generating Function of $r(n)$," and "The Order of $r(n)$." §16.9, 16.10, 17.9 , and 18.7 in *An Introduction to the Theory of Numbers, 5th ed.* Oxford, England: Clarendon Press, pp. 241–243, 256–258, and 270–271, 1979.

Shanks, D. *Solved and Unsolved Problems in Number Theory, 4th ed.* New York: Chelsea, pp. 162–153, 1993.

Sloane, N. J. A. Sequence A004018/M3218 in "An On-Line Version of the Encyclopedia of Integer Sequences."

R-Estimate

A ROBUST ESTIMATION based on RANK tests. Examples include the statistic of the KOLMOGOROV-SMIRNOV TEST, SPEARMAN RANK CORRELATION, and WILCOXON SIGNED RANK TEST.

see also L-ESTIMATE, M-ESTIMATE, ROBUST ESTIMATION

References

Press, W. H.; Flannery, B. P.; Teukolsky, S. A.; and Vetterling, W. T. "Robust Estimation." §15.7 in *Numerical Recipes in FORTRAN: The Art of Scientific Computing, 2nd ed.* Cambridge, England: Cambridge University Press, pp. 694–700, 1992.

Raabe's Test

Given a SERIES of POSITIVE terms u_i and a SEQUENCE of POSITIVE constants $\{a_i\}$, use KUMMER'S TEST

$$\rho' \equiv \lim_{n\to\infty}\left(a_n \frac{u_n}{u_{n+1}} - a_{n+1}\right).$$

with $a_n = n$, giving

$$\begin{aligned}\rho' &\equiv \lim_{n\to\infty}\left(n\frac{u_n}{u_{n+1}} - (n+1)\right)\\ &= \lim_{n\to\infty}\left[n\left(\frac{u_n}{u_{n+1}} - 1\right) - 1\right].\end{aligned}$$

Defining

$$\rho \equiv \rho' + 1 = \lim_{n\to\infty}\left[n\left(\frac{u_n}{u_{n+1}} - 1\right)\right],$$

then gives Raabe's test:

1. If $\rho > 1$, the SERIES CONVERGES.
2. If $\rho < 1$, the SERIES DIVERGES.
3. If $\rho = 1$, the SERIES may CONVERGE or DIVERGE.

see also CONVERGENT SERIES, CONVERGENCE TESTS, DIVERGENT SERIES, KUMMER'S TEST

References

Arfken, G. *Mathematical Methods for Physicists, 3rd ed.* Orlando, FL: Academic Press, pp. 286–287, 1985.

Bromwich, T. J. I'a and MacRobert, T. M. *An Introduction to the Theory of Infinite Series, 3rd ed.* New York: Chelsea, p. 39, 1991.

Rabbit Constant

The limiting RABBIT SEQUENCE written as a BINARY FRACTION $0.1011010110110\ldots_2$ (Sloane's A005614), where b_2 denotes a BINARY number (a number in base-2). The DECIMAL value is

$$R = 0.709803442861291314\,6\ldots$$

(Sloane's A014565).

Amazingly, the rabbit constant is also given by the CONTINUED FRACTION $[0, 2^{F_0}, 2^{F_1}, 2^{F_2}, 2^{F_3}, \ldots]$, where F_n are FIBONACCI NUMBERS with F_0 taken as 0 (Gardner 1989, Schroeder 1991). Another amazing connection was discovered by S. Plouffe. Define the BEATTY SEQUENCE $\{a_i\}$ by

$$a_i \equiv \lfloor i\phi \rfloor,$$

where $\lfloor x \rfloor$ is the FLOOR FUNCTION and ϕ is the GOLDEN RATIO. The first few terms are 1, 3, 4, 6, 8, 9, 11, ... (Sloane's A000201). Then

$$R = \sum_{i=1}^{\infty} 2^{-a_i}.$$

see also RABBIT SEQUENCE, THUE CONSTANT, THUE-MORSE CONSTANT

References

Finch, S. "Favorite Mathematical Constants." `http://www.mathsoft.com/asolve/constant/cntfrc/cntfrc.html`.

Gardner, M. *Penrose Tiles and Trapdoor Ciphers... and the Return of Dr. Matrix, reissue ed.* New York: W. H. Freeman, pp. 21–22, 1989.
Plouffe, S. "The Rabbit Constant to 330 Digits." `http://lacim.uqam.ca/piDATA/rabbit.txt`.
Schroeder, M. *Fractals, Chaos, Power Laws: Minutes from an Infinite Paradise.* New York: W. H. Freeman, p. 55, 1991.
Sloane, N. J. A. Sequences A005614, A014565, and A000201/M2322 in "An On-Line Version of the Encyclopedia of Integer Sequences."

Rabbit-Duck Illusion

A perception ILLUSION in which the brain switches between seeing a rabbit and a duck.

see also YOUNG GIRL-OLD WOMAN ILLUSION

Rabbit Sequence

A SEQUENCE which arises in the hypothetical reproduction of a population of rabbits. Let the SUBSTITUTION MAP $0 \to 1$ correspond to young rabbits growing old, and $1 \to 10$ correspond to old rabbits producing young rabbits. Starting with 0 and iterating using STRING REWRITING gives the terms 1, 10, 101, 10110, 10110101, 1011010110110, ... The limiting sequence written as a BINARY FRACTION $0.1011010110110\ldots_2$ (Sloane's A005614), where b_2 denotes a BINARY number (a number in base-2) is called the RABBIT CONSTANT.

see also RABBIT CONSTANT, THUE-MORSE SEQUENCE

References

Schroeder, M. *Fractals, Chaos, Power Laws: Minutes from an Infinite Paradise.* New York: W. H. Freeman, p. 55, 1991.
Sloane, N. J. A. Sequence A005614 in "An On-Line Version of the Encyclopedia of Integer Sequences."

Rabdology

see NAPIER'S BONES

Rabin-Miller Strong Pseudoprime Test

A PRIMALITY TEST which provides an efficient probabilistic ALGORITHM for determining if a given number is PRIME. It is based on the properties of STRONG PSEUDOPRIMES. Given an ODD INTEGER n, let $n = 2^r s + 1$ with s ODD. Then choose a random integer a with $1 \leq a \leq n-1$. If $a^s \equiv 1 \pmod{n}$ or $a^{2^j s} \equiv -1 \pmod{n}$ for some $0 \leq j \leq r-1$, then n passes the test. A PRIME will pass the test for all a.

The test is very fast and requires no more than $(1 + o(1)) \lg n$ multiplications (mod n), where LG is the LOGARITHM base 2. Unfortunately, a number which passes the test is not necessarily PRIME. Monier (1980) and Rabin (1980) have shown that a COMPOSITE NUMBER passes the test for at most 1/4 of the possible bases a.

The Rabin-Miller test (combined with a LUCAS PSEUDOPRIME test) is the PRIMALITY TEST used by *Mathematica*® versions 2.2 and later (Wolfram Research, Champaign, IL). As of 1991, the combined test had been proven correct for all $n < 2.5 \times 10^{10}$, but not beyond. The test potentially could therefore incorrectly identify a large COMPOSITE NUMBER as PRIME (but not vice versa). STRONG PSEUDOPRIME tests have been subsequently proved valid for every number up to 3.4×10^{14}.

see also LUCAS-LEHMER TEST, MILLER'S PRIMALITY TEST, PSEUDOPRIME, STRONG PSEUDOPRIME

References

Arnault, F. "Rabin-Miller Primality Test: Composite Numbers Which Pass It." *Math. Comput.* **64**, 355–361, 1995.
Miller, G. "Riemann's Hypothesis and Tests for Primality." *J. Comp. Syst. Sci.* **13**, 300–317, 1976.
Monier, L. "Evaluation and Comparison of Two Efficient Probabilistic Primality Testing Algorithms." *Theor. Comput. Sci.* **12**, 97–108, 1980.
Rabin, M. O. "Probabilistic Algorithm for Testing Primality." *J. Number Th.* **12**, 128–138, 1980.
Wagon, S. *Mathematica in Action.* New York: W. H. Freeman, pp. 15–17, 1991.

Rabinovich-Fabrikant Equation

The 3-D MAP

$$\begin{aligned} \dot{x} &= y(z - 1 + x^2) + \gamma x \\ \dot{y} &= x(3z + 1 - x^2) + \gamma y \\ \dot{z} &= -2z(\alpha + xy) \end{aligned}$$

(Rabinovich and Fabrikant 1979). The parameters are most commonly taken as $\gamma = 0.87$ and $\alpha = 1.1$. It has a CORRELATION EXPONENT of 2.19 ± 0.01.

References

Grassberger, P. and Procaccia, I. "Measuring the Strangeness of Strange Attractors." *Physica D* **9**, 189–208, 1983.
Rabinovich, M. I. and Fabrikant, A. L. *Sov. Phys. JETP* **50**, 311-317, 1979.

Racah V-Coefficient

The Racah V-COEFFICIENTS are written

$$V(j_1 j_2 j; m_1 m_2 m) \tag{1}$$

and are sometimes expressed using the related CLEBSCH-GORDON COEFFICIENTS

$$C^j_{m_1 m_2} = (j_1 j_2 m_1 m_2 | j_1 j_2 j m), \tag{2}$$

or WIGNER 3j-SYMBOLS. Connections among the three are

$$(j_1 j_2 m_1 m_2 | j_1 j_2 m) = (-1)^{-j_1+j_2-m}\sqrt{2j+1}\begin{pmatrix} j_1 & j_2 & j \\ m_1 & m_2 & -m \end{pmatrix} \quad (3)$$

$$(j_1 j_2 m_1 m_2 | j_1 j_2 j m) = (-1)^{j+m}\sqrt{2j+1}\,V(j_1 j_2 j; m_1 m_2 - m) \quad (4)$$

$$V(j_1 j_2 j; m_1 m_2 m) = (-1)^{-j_1+j_2+j}\begin{pmatrix} j_1 & j_2 & j_1 \\ m_2 & m_1 & m_2 \end{pmatrix}. \quad (5)$$

see also CLEBSCH-GORDON COEFFICIENT, RACAH W-COEFFICIENT, WIGNER 3j-SYMBOL, WIGNER 6j-SYMBOL, WIGNER 9j-SYMBOL

References

Sobel'man, I. I. "Angular Momenta." Ch. 4 in *Atomic Spectra and Radiative Transitions, 2nd ed.* Berlin: Springer-Verlag, 1992.

Racah W-Coefficient

Related to the CLEBSCH-GORDON COEFFICIENTS by

$$(J_1 J_2[J']J_3 | J_1, J_2 J_3[J'']) = \sqrt{(2J'+1)(2J''+1)}\,W(J_1 J_2 J J_3; J' J'')$$

and

$$(J_1 J_2[J']J_3 | J_1 J_3[J'']J_2) = \sqrt{(2J'+1)(2J''+1)}\,W(J_1' J_3 J_2 J''; J J_1).$$

see also CLEBSCH-GORDON COEFFICIENT, RACAH V-COEFFICIENT, WIGNER 3j-SYMBOL, WIGNER 6j-SYMBOL, WIGNER 9j-SYMBOL

References

Messiah, A. "Racah Coefficients and '6j' Symbols." Appendix C.II in *Quantum Mechanics, Vol. 2.* Amsterdam, Netherlands: North-Holland, pp. 1061–1066, 1962.

Sobel'man, I. I. "Angular Momenta." Ch. 4 in *Atomic Spectra and Radiative Transitions, 2nd ed.* Berlin: Springer-Verlag, 1992.

Radau Quadrature

A GAUSSIAN QUADRATURE-like formula for numerical estimation of integrals. It requires $m+1$ points and fits all POLYNOMIALS to degree $2m$, so it effectively fits exactly all POLYNOMIALS of degree $2m-1$. It uses a WEIGHTING FUNCTION $W(x) = 1$ in which the endpoint -1 in the interval $[-1, 1]$ is included in a total of n ABSCISSAS, giving $r = n-1$ free abscissas. The general formula is

$$\int_{-1}^{1} f(x)\,dx = w_1 f(-1) + \sum_{i=2}^{n} w_i f(x_i). \quad (1)$$

The free abscissas x_i for $i = 2, \ldots, n$ are the roots of the POLYNOMIAL

$$\frac{P_{n-1}(x) + P_n(x)}{1+x}, \quad (2)$$

where $P(x)$ is a LEGENDRE POLYNOMIAL. The weights of the free abscissas are

$$w_i = \frac{1-x_i}{n^2[P_{n-1}(x_i)]^2} = \frac{1}{(1-x_i)[P'_{n-1}(x_i)]^2}, \quad (3)$$

and of the endpoint

$$w_1 = \frac{2}{n^2}. \quad (4)$$

The error term is given by

$$E = \frac{2^{2n-1}n[(n-1)!]^4}{[(2n-1)!]^3} f^{(2n-1)}(\xi), \quad (5)$$

for $\xi \in (-1, 1)$.

n	x_i	w_i
2	−1	0.5
	0.333333	1.5
3	−1	0.222222
	−0.289898	1.02497
	0.689898	0.752806
4	−1	0.125
	−0.575319	0.657689
	0.181066	0.776387
	0.822824	0.440924
5	−1	0.08
	−0.72048	0.446208
	−0.167181	0.623653
	0.446314	0.562712
	0.885792	0.287427

The ABSCISSAS and weights can be computed analytically for small n.

n	x_i	w_i
2	-1	$\frac{1}{2}$
	$\frac{1}{3}$	$\frac{3}{2}$
3	-1	$\frac{2}{9}$
	$\frac{1}{5}(1-\sqrt{6})$	$\frac{1}{18}(16+\sqrt{6})$
	$\frac{1}{5}(1+\sqrt{6})$	$\frac{1}{18}(16-\sqrt{6})$

see also CHEBYSHEV QUADRATURE, LOBATTO QUADRATURE

References

Abramowitz, M. and Stegun, C. A. (Eds.). *Handbook of Mathematical Functions with Formulas, Graphs, and Mathematical Tables, 9th printing.* New York: Dover, p. 888, 1972.

Chandrasekhar, S. *Radiative Transfer.* New York: Dover, p. 61, 1960.

Hildebrand, F. B. *Introduction to Numerical Analysis.* New York: McGraw-Hill, pp. 338–343, 1956.

Rademacher Function

see SQUARE WAVE

Radial Curve

Let C be a curve and let O be a fixed point. Let P be on C and let Q be the CURVATURE CENTER at P. Let P_1 be the point with P_1O a line segment PARALLEL and of equal length to PQ. Then the curve traced by P_1 is the radial curve of C. It was studied by Robert Tucker in 1864. The parametric equations of a curve (f, g) with RADIAL POINT (x_0, y_0) are

$$x = x_0 - \frac{g'(f'^2 + g'^2)}{f'g'' - f''g'}$$
$$y = y_0 + \frac{f'(f'^2 + g'^2)}{f'g'' - f''g'}.$$

Curve	Radial Curve
astroid	quadrifolium
catenary	kampyle of Eudoxus
cycloid	circle
deltoid	trifolium
logarithmic spiral	logarithmic spiral
tractrix	kappa curve

References

Lawrence, J. D. *A Catalog of Special Plane Curves.* New York: Dover, pp. 40 and 202, 1972.

Yates, R. C. "Radial Curves." *A Handbook on Curves and Their Properties.* Ann Arbor, MI: J. W. Edwards, pp. 172–174, 1952.

Radial Point

The point with respect to which a RADIAL CURVE is computed.

see also RADIANT POINT

Radian

A unit of angular measure in which the ANGLE of an entire CIRCLE is 2π radians. There are therefore 360° per 2π radians, equal to $180/\pi$ or $57.29577951°$/radian. A RIGHT ANGLE is $\pi/2$ radians.

see also ANGLE, ARC MINUTE, ARC SECOND, DEGREE, GRADIAN, STERADIAN

Radiant Point

The point of illumination for a CAUSTIC.

see also CAUSTIC, RADIAL POINT

Radical

The symbol $\sqrt[n]{x}$ used to indicate a root is called a radical. The expression $\sqrt[n]{x}$ is therefore read "x radical n," or "the nth ROOT of x." $n = 2$ is written $\sqrt{x}$ and is called the SQUARE ROOT of x. $n = 3$ corresponds to the CUBE ROOT. The quantity under the root is called the RADICAND.

Some interesting radical identities are due to Ramanujan, and include the equivalent forms

$$(2^{1/3} + 1)(2^{1/3} - 1)^{1/3} = 3^{1/3}$$

and

$$(2^{1/3} - 1)^{1/3} = (\tfrac{1}{9})^{1/3} - (\tfrac{2}{9})^{1/3} + (\tfrac{4}{9})^{1/3}.$$

Another such identity is

$$(5^{1/3} - 4^{1/3})^{1/2} = \tfrac{1}{3}(2^{1/3} + 20^{1/3} - 25^{1/3}).$$

see also CUBE ROOT, NESTED RADICAL, POWER, RADICAL INTEGER, RADICAND, ROOT (RADICAL), SQUARE ROOT, VINCULUM

Radical Axis

see RADICAL LINE

Radical Center

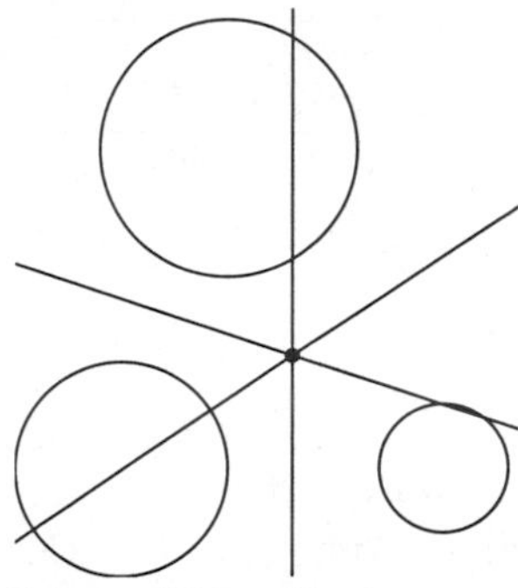

The RADICAL LINES of three CIRCLES are CONCURRENT in a point known as the radical center (also called the POWER CENTER). This theorem was originally demonstrated by Monge (Dörrie 1965, p. 153).

see also APOLLONIUS' PROBLEM, CONCURRENT, MONGE'S PROBLEM, RADICAL LINE

References

Dörrie, H. *100 Great Problems of Elementary Mathematics: Their History and Solutions.* New York: Dover, 1965.

Johnson, R. A. *Modern Geometry: An Elementary Treatise on the Geometry of the Triangle and the Circle.* Boston, MA: Houghton Mifflin, p. 32, 1929.

Radical Integer

A radical integer is a number obtained by closing the INTEGERS under ADDITION, DIVISION, MULTIPLICATION, SUBTRACTION, and ROOT extraction. An example of such a number is $\sqrt[3]{7} + \sqrt{-2} - \sqrt{3 + \sqrt[4]{1} + \sqrt{2}}$. The radical integers are a subring of the ALGEBRAIC INTEGERS. If there are ALGEBRAIC INTEGERS which are not radical integers, they must at least be cubic.

see also ALGEBRAIC INTEGER, ALGEBRAIC NUMBER, EUCLIDEAN NUMBER

Radical Line

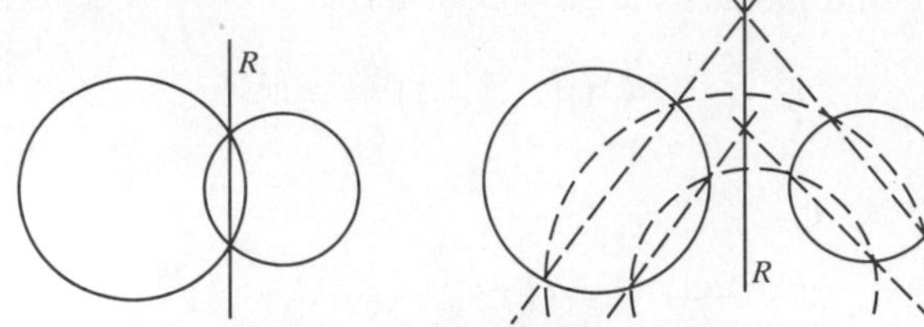

The LOCUS of points of equal POWER with respect to two nonconcentric CIRCLES which is PERPENDICULAR to the line of centers (the CHORDAL THEOREM; Dörrie 1965). Let the circles have RADII r_1 and r_2 and their centers be separated by a distance d. If the CIRCLES intersect in two points, then the radical line is the line passing through the points of intersection. If not, then draw any two CIRCLES which cut each original CIRCLE twice. Draw lines through each pair of points of intersection of each CIRCLE. The line connecting their two points of intersection is then the radical line.

The radical line is located at distances

$$d_1 = \frac{d^2 + {r_1}^2 - {r_2}^2}{2d} \tag{1}$$

$$d_2 = -\frac{d^2 + {r_2}^2 - {r_1}^2}{2d} \tag{2}$$

along the line of centers from C_1 and C_2, respectively, where

$$d \equiv d_1 - d_2. \tag{3}$$

The radical line of any two POLAR CIRCLES is the ALTITUDE from the third vertex.

see also CHORDAL THEOREM, COAXAL CIRCLES, INVERSE POINTS, INVERSION, POWER (CIRCLE), RADICAL CENTER

References

Coxeter, H. S. M. and Greitzer, S. L. *Geometry Revisited.* Washington, DC: Math. Assoc. Amer., pp. 31–34, 1967.

Dixon, R. *Mathographics.* New York: Dover, p. 68, 1991.

Dörrie, H. *100 Great Problems of Elementary Mathematics: Their History and Solutions.* New York: Dover, p. 153, 1965.

Johnson, R. A. *Modern Geometry: An Elementary Treatise on the Geometry of the Triangle and the Circle.* Boston, MA: Houghton Mifflin, pp. 28–34 and 176–177, 1929.

Radicand

The quantity under a RADICAL sign.

see also RADICAL, VINCULUM

Radius

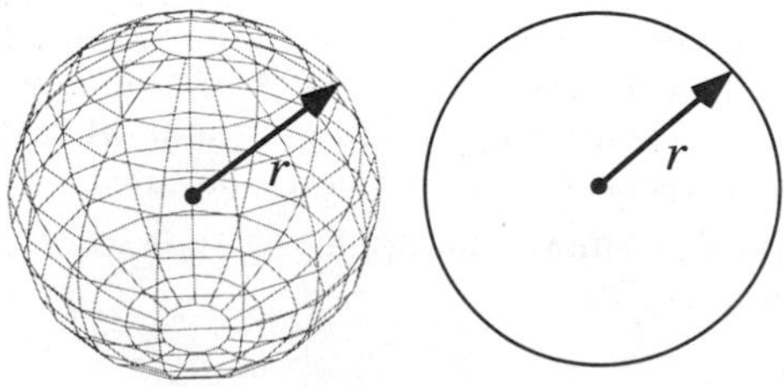

The distance from the center of a CIRCLE to its PERIMETER, or from the center of a SPHERE to its surface. The radius is equal to half the DIAMETER.

see also BERTRAND'S PROBLEM, CIRCLE, CIRCUMFERENCE, DIAMETER, EXTENT, INVERSION RADIUS, KINNEY'S SET, PI, RADIUS OF CONVERGENCE, RADIUS OF CURVATURE, RADIUS (GRAPH), RADIUS OF GYRATION, RADIUS OF TORSION, RADIUS VECTOR, SPHERE

Radius of Convergence

The RADIUS (or 1-D distance in the 1-D case) over which series expansion CONVERGES.

Radius of Curvature

The radius of curvature is given by

$$R \equiv \frac{1}{\kappa}, \tag{1}$$

where κ is the CURVATURE. At a given point on a curve, R is the radius of the OSCULATING CIRCLE. The symbol ρ is sometimes used instead of R to denote the radius of curvature.

Let x and y be given parametrically by

$$x = x(t) \tag{2}$$

$$y = y(t), \tag{3}$$

then

$$R = \frac{(x'^2 + y'^2)^{3/2}}{x'y'' - y'x''}, \tag{4}$$

where $x' = dx/dt$ and $y' = dy/dt$. Similarly, if the curve is written in the form $y = f(x)$, then the radius of curvature is given by

$$R = \frac{\left[1 + \left(\frac{dy}{dx}\right)^2\right]^{3/2}}{\frac{d^2y}{dx^2}}. \tag{5}$$

see also BEND (CURVATURE), CURVATURE, OSCULATING CIRCLE, TORSION (DIFFERENTIAL GEOMETRY)

References

Kreyszig, E. *Differential Geometry.* New York: Dover, p. 34, 1991.

Radius (Graph)

The minimum ECCENTRICITY of any VERTEX of a GRAPH.

Radius of Gyration

A function quantifying the spatial extent of the structure of a curve. It is defined by

$$R_g = \frac{\sqrt{\int_0^\infty r^2 p(r)\,dr}}{2\int_0^\infty p(r)\,dr},$$

where $p(r)$ is the LENGTH DISTRIBUTION FUNCTION. Small compact patterns have small R_g.

References

Pickover, C. A. *Keys to Infinity.* New York: W. H. Freeman, pp. 204–206, 1995.

Radius of Torsion

$$\sigma \equiv \frac{1}{\tau},$$

where τ is the TORSION. The symbol ϕ is also sometimes used instead of σ.

see also TORSION (DIFFERENTIAL GEOMETRY)

References
Kreyszig, E. *Differential Geometry.* New York: Dover, p. 39, 1991.

Radius Vector

The VECTOR $\mathbf{r}$ from the ORIGIN to the current position. It is also called the POSITION VECTOR. The derivative of $\mathbf{r}$ satisfies

$$\mathbf{r}\cdot\frac{d\mathbf{r}}{dt} = \frac{1}{3}\frac{d}{dt}(\mathbf{r}\cdot\mathbf{r}) = \frac{1}{2}\frac{d}{dt}(r^2) = r\frac{dr}{dt} = rv,$$

where v is the magnitude of the VELOCITY (i.e., the SPEED).

Radix

The BASE of a number system, i.e., 2 for BINARY, 8 for OCTAL, 10 for DECIMAL, and 16 for HEXADECIMAL. The radix is sometimes called the BASE or SCALE.

see also BASE (NUMBER)

Rado's Sigma Function

see BUSY BEAVER

Radon-Nikodym Theorem

A THEOREM which gives NECESSARY and SUFFICIENT conditions for a countably additive function of sets can be expressed as an integral over the set.

References
Doob, J. L. "The Development of Rigor in Mathematical Probability (1900–1950)." *Amer. Math. Monthly* **103**, 586–595, 1996.

Radon Transform

An INTEGRAL TRANSFORM whose inverse is used to reconstruct images from medical CT scans. A technique for using Radon transforms to reconstruct a map of a planet's polar regions using a spacecraft in a polar orbit has also been devised (Roulston and Muhleman 1997).

The Radon transform can be defined by

$$R(p,\tau)[f(x,y)] = \int_{-\infty}^{\infty} f(x,\tau+px)\,dx$$

$$= \int_{-\infty}^{\infty}\int_{-\infty}^{\infty} f(x,y)\delta[y-(\tau+px)]\,dy\,dx \equiv U(p,\tau), \quad (1)$$

where p is the SLOPE of a line and τ is its intercept. The inverse Radon transform is

$$f(x,y) = \frac{1}{2\pi}\int_{-\infty}^{\infty} \frac{d}{dy} H[U(p,y-px)]\,dp, \quad (2)$$

where H is a HILBERT TRANSFORM. The transform can also be defined by

$$R'(r,\alpha)[f(x,y)]$$

$$= \int_{-\infty}^{\infty}\int_{-\infty}^{\infty} f(x,y)\delta(r-x\cos\alpha-y\sin\alpha)\,dx\,dy, \quad (3)$$

where r is the PERPENDICULAR distance from a line to the origin and α is the ANGLE formed by the distance VECTOR.

Using the identity

$$\mathcal{F}[R[f(\omega,\alpha)]] = \mathcal{F}^2[f(u,v)], \quad (4)$$

where $\mathcal{F}$ is the FOURIER TRANSFORM, gives the inversion formula

$$f(x,y) =$$

$$c\int_0^{\pi}\int_{-\infty}^{\infty} \mathcal{F}[R[f(\omega,\alpha)]]|\omega|e^{i\omega(x\cos\alpha+y\sin\alpha)}\,d\omega\,d\alpha. \quad (5)$$

The FOURIER TRANSFORM can be eliminated by writing

$$f(x,y) = \int_0^{\pi}\int_{-\infty}^{\infty} R[f(r,\alpha)]W(r,\alpha,x,y)\,dr\,d\alpha, \quad (6)$$

where W is a WEIGHTING FUNCTION such as

$$W(r,\alpha,x,y) = h(x\cos\alpha+y\sin\alpha-r) = \mathcal{F}^{-1}[|\omega|]. \quad (7)$$

Nievergelt (1986) uses the inverse formula

$$f(x,y) = \frac{1}{\pi}\lim_{c\to 0}$$

$$\int_0^{\pi}\int_{-\infty}^{\infty} R[f(r+x\cos\alpha+y\sin\alpha,\alpha)]G_c(r)\,dr\,d\alpha, \quad (8)$$

where

$$G_c(r) = \begin{cases} \frac{1}{\pi c^2} & \text{for } |r|\le c \\ \frac{1}{\pi c^2}\left(1-\frac{1}{\sqrt{1-c^2/r^2}}\right) & \text{for } |r|>c. \end{cases} \quad (9)$$

LUDWIG'S INVERSION FORMULA expresses a function in terms of its Radon transform. $R'(r,\alpha)$ and $R(p,\tau)$ are related by

$$p = \cot\alpha \qquad \tau = r\csc\alpha \quad (10)$$

$$r = \frac{\tau}{1+p^2} \qquad \alpha = \cot^{-1}p. \quad (11)$$

The Radon transform satisfies superposition

$$R(p,\tau)[f_1(x,y)+f_2(x,y)] = U_1(p,\tau)+U_2(p,\tau), \quad (12)$$

linearity

$$R(p,\tau)[af(x,y)] = aU(p,\tau), \tag{13}$$

scaling

$$R(p,\tau)\left[f\left(\frac{x}{a},\frac{y}{b}\right)\right] = |a|U\left(p\frac{a}{b},\frac{\tau}{b}\right), \tag{14}$$

ROTATION, with R_ϕ ROTATION by ANGLE ϕ

$$R(p,\tau)[R_\phi f(x,y)] = \frac{1}{|\cos\phi + p\sin\phi|} U\left(\frac{p-\tan\phi}{1+p\tan\phi}, \frac{\tau}{\cos\phi + p\sin\phi}\right), \tag{15}$$

and skewing

$$R(p,\tau)[f(ax+by,cx+dy)] = \frac{1}{|a+bp|}U\left[\frac{c+dp}{a+bp}, \tau\frac{d-b(c+bd)}{a+bp}\right] \tag{16}$$

(Durrani and Bisset 1984).

The line integral along p,τ is

$$I = \sqrt{1+p^2}\,U(p,\tau). \tag{17}$$

The analog of the 1-D CONVOLUTION THEOREM is

$$R(p,\tau)[f(x,y)*g(y)] = U(p,\tau)*g(\tau), \tag{18}$$

the analog of PLANCHEREL'S THEOREM is

$$\int_{-\infty}^{\infty} U(p,\tau)\,d\tau = \int_{-\infty}^{\infty}\int_{-\infty}^{\infty} f(x,y)\,dx\,dy, \tag{19}$$

and the analog of PARSEVAL'S THEOREM is

$$\int_{-\infty}^{\infty} R(p,\tau)[f(x,y)]^2\,d\tau = \int_{-\infty}^{\infty}\int_{-\infty}^{\infty} f^2(x,y)\,dx\,dy. \tag{20}$$

If f is a continuous function on $\mathbb{C}$, integrable with respect to a plane LEBESGUE MEASURE, and

$$\int_l f\,ds = 0 \tag{21}$$

for every (doubly) infinite line l where s is the length measure, then f must be identically zero. However, if the global integrability condition is removed, this result fails (Zalcman 1982, Goldstein 1993).

see also TOMOGRAPHY

References

Anger, B. and Portenier, C. *Radon Integrals.* Boston, MA: Birkhäuser, 1992.

Armitage, D. H. and Goldstein, M. "Nonuniqueness for the Radon Transform." *Proc. Amer. Math. Soc.* **117**, 175–178, 1993.

Deans, S. R. *The Radon Transform and Some of Its Applications.* New York: Wiley, 1983.

Durrani, T. S. and Bisset, D. "The Radon Transform and its Properties." *Geophys.* **49**, 1180–1187, 1984.

Esser, P. D. (Ed.). *Emission Computed Tomography: Current Trends.* New York: Society of Nuclear Medicine, 1983.

Gindikin, S. (Ed.). *Applied Problems of Radon Transform.* Providence, RI: Amer. Math. Soc., 1994.

Gradshteyn, I. S. and Ryzhik, I. M. *Tables of Integrals, Series, and Products, 5th ed.* San Diego, CA: Academic Press, 1979.

Helgason, S. *The Radon Transform.* Boston, MA: Birkhäuser, 1980.

Kunyansky, L. A. "Generalized and Attenuated Radon Transforms: Restorative Approach to the Numerical Inversion." *Inverse Problems* **8**, 809–819, 1992.

Nievergelt, Y. "Elementary Inversion of Radon's Transform." *SIAM Rev.* **28**, 79–84, 1986.

Rann, A. G. and Katsevich, A. I. *The Radon Transform and Local Tomography.* Boca Raton, FL: CRC Press, 1996.

Robinson, E. A. "Spectral Approach to Geophysical Inversion Problems by Lorentz, Fourier, and Radon Transforms." *Proc. Inst. Electr. Electron. Eng.* **70**, 1039–1053, 1982.

Roulston, M. S. and Muhleman, D. O. "Synthesizing Radar Maps of Polar Regions with a Doppler-Only Method." *Appl. Opt.* **36**, 3912–3919, 1997.

Shepp, L. A. and Kruskal, J. B. "Computerized Tomography: The New Medical X-Ray Technology." *Amer. Math. Monthly* **85**, 420–439, 1978.

Strichartz, R. S. "Radon Inversion—Variation on a Theme." *Amer. Math. Monthly* **89**, 377–384 and 420–423, 1982.

Zalcman, L. "Uniqueness and Nonuniqueness for the Radon Transform." *Bull. London Math. Soc.* **14**, 241–245, 1982.

Radon Transform—Cylinder

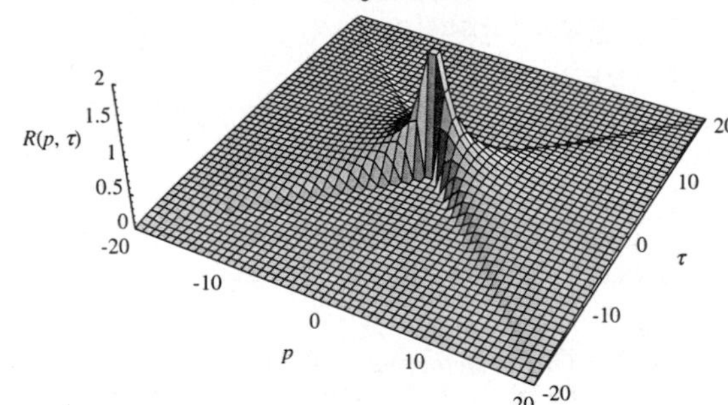

Let the 2-D cylinder function be defined by

$$f(x,y) \equiv \begin{cases} 1 & \text{for } r < R \\ 0 & \text{for } r > R. \end{cases} \tag{1}$$

Then the Radon transform is given by

$$R(p,\tau) = \int_{-\infty}^{\infty}\int_{-\infty}^{\infty} f(x,y)\delta[y-(\tau+px)]\,dy\,dx, \tag{2}$$

where

$$\delta(x) = \frac{1}{2\pi}\int_{-\infty}^{\infty} e^{-ikx} \tag{3}$$

is the DELTA FUNCTION.

$$\begin{aligned} R(p,\tau) &= \frac{1}{2\pi}\int_0^{2\pi}\int_0^R\int_{-\infty}^{\infty} e^{-ik(r\sin\theta - pr\cos\theta)} r\,dr\,d\theta\,dk \\ &= \frac{1}{2\pi}\int_{-\infty}^{\infty} e^{ik\tau}\int_0^{2\pi}\int_0^R e^{-ikr(\sin\theta - p\cos\theta)} r\,dr\,d\theta\,dk. \end{aligned} \tag{4}$$

Now write

$$\sin\theta - p\cos\theta = \sqrt{1+p^2}\,\cos(\theta+\phi) \equiv \sqrt{1+p^2}\,\cos\theta', \tag{5}$$

with ϕ a phase shift. Then

$$\begin{aligned} R(p,\tau) &= \\ \frac{1}{2\pi}\int_{-\infty}^{\infty} e^{ik\tau} \int_0^R &\left(\int_0^{2\pi} e^{-ik\sqrt{1+p^2}\,r\cos\theta'}\,d\theta'\right) r\,dr\,dk \\ &= \frac{1}{2\pi}\int_{-\infty}^{\infty} e^{ik\tau}\int_0^R 2\pi J_0(k\sqrt{1+p^2}\,r) r\,dr\,dk \\ &= \int_{-\infty}^{\infty} e^{ik\tau}\int_0^R J_0(k\sqrt{1+p^2}\,r) r\,dr\,dk. \end{aligned} \tag{6}$$

Then use

$$\int_0^z t^{n+1} J_n(t)\,dt = z^{n+1} J_{n+1}(z), \tag{7}$$

which, with $n = 0$, becomes

$$\int_0^z tJ_0(t)\,dt = zJ_1(z). \tag{8}$$

Define

$$t \equiv k\sqrt{1+p^2}\,r \tag{9}$$

$$dt = k\sqrt{1+p^2}\,dr \tag{10}$$

$$r\,dr = \frac{t\,dt}{k^2(1+p^2)}, \tag{11}$$

so the inner integral is

$$\begin{aligned} \int_0^{R\sqrt{1+p^2}} & J_0(t)\frac{t\,dt}{k^2(1+p^2)} \\ &= \frac{1}{k^2(1+p^2)} kR\sqrt{1+p^2} J_1(kR\sqrt{1+p^2}\,) \\ &= \frac{J_1(kR\sqrt{1+p^2}\,)}{k\sqrt{1+p^2}} R, \end{aligned} \tag{12}$$

and the Radon transform becomes

$$\begin{aligned} R(p,\tau) &= \frac{R}{\sqrt{1+p^2}}\int_{-\infty}^{\infty} \frac{e^{ik\tau} J_1(kR\sqrt{1+p^2}\,)}{k}\,dk \\ &= \frac{2R}{\sqrt{1+p^2}}\int_0^{\infty} \frac{\cos(k\tau) J_1(kR\sqrt{1+p^2}\,)}{k}\,dk \\ &= \begin{cases} \frac{2}{1+p^2}\sqrt{R^2(1+p^2)-\tau^2} & \text{for } \tau^2 < R^2(1+p^2) \\ 0 & \text{for } \tau^2 \geq R^2(1+p^2). \end{cases} \end{aligned} \tag{13}$$

Converting to R' using $p = \cot\alpha$,

$$\begin{aligned} R'(r,\alpha) &= \frac{2}{\sqrt{1+\cot^2\alpha}}\sqrt{(1+\cot^2\alpha)R^2 - r^2\csc^2\alpha} \\ &= \frac{2}{\csc\alpha}\sqrt{\csc^2\alpha R^2 - r^2\csc^2\alpha} \\ &= 2\sqrt{R^2-r^2}, \end{aligned} \tag{14}$$

which could have been derived more simply by

$$R'(r,\alpha) = \int_{-\sqrt{R^2-r^2}}^{\sqrt{R^2-r^2}} dy. \tag{15}$$

Radon Transform—Delta Function

For a Delta Function at (x_0, y_0),

$$\begin{aligned} R(p,\tau) &= \int_{-\infty}^{\infty}\int_{-\infty}^{\infty} \delta(x-x_0)\delta(y-y_0)\delta[y-(\tau+px)]\,dy\,dx \\ &= \frac{1}{2\pi}\int_{-\infty}^{\infty}\int_{-\infty}^{\infty}\int_{-\infty}^{\infty} e^{-ik[y-(\tau+px)]}\delta(x-x_0)\delta(y-y_0) \\ &\qquad \times dk\,dy\,dx \\ &= \frac{1}{2\pi}\int_{-\infty}^{\infty} e^{ik\tau}\left[\int_{-\infty}^{\infty} e^{-iky}\delta(y-y_0)\,dy \right. \\ &\qquad \left. \times \int_{-\infty}^{\infty} e^{ikpx}\delta(x-x_0)\,dx\right] dk \\ &= \frac{1}{2\pi}\int_{-\infty}^{\infty} e^{ik\tau}e^{-iky_0}e^{ikpx_0}\,dk \\ &= \frac{1}{2\pi}\int_{-\infty}^{\infty} e^{ik(\tau+px_0-y_0)}\,dk = \delta(\tau+px_0-y_0). \end{aligned}$$

Radon Transform—Gaussian

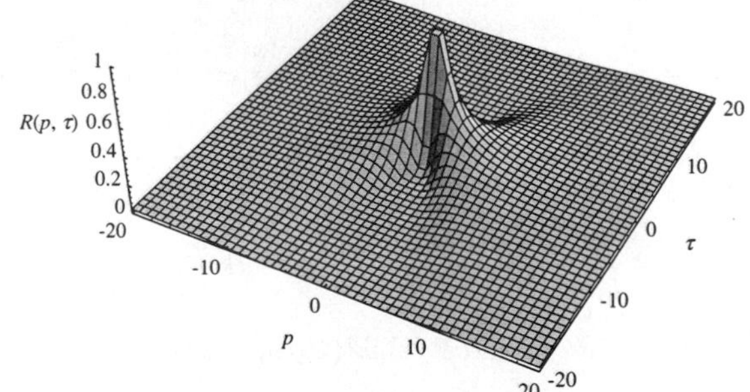

$$\begin{aligned} R(p,\tau) &= \int_{-\infty}^{\infty}\int_{-\infty}^{\infty}\left[\frac{1}{\sigma\sqrt{2\pi}}e^{-(x^2+y^2)/2\sigma^2}\right] \\ &\qquad \times\delta[y-(\tau+px)]\,dy\,dx \\ &= \frac{1}{\sigma\sqrt{2\pi}}\int_{-\infty}^{\infty} e^{-[x^2+(\tau+px)^2]/2\sigma^2}\,dx \\ &= \frac{1}{\sqrt{1+p^2}}e^{-t^2/[2(1+p^2)\sigma^2]}. \end{aligned}$$

Radon Transform—Square

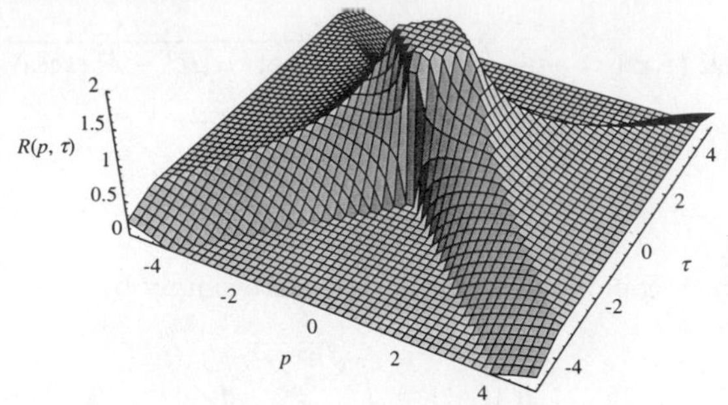

$$R(p,\tau)=\int_{-\infty}^{\infty}\int_{-\infty}^{\infty}f(x,y)\delta[y-(\tau+px)]\,dy\,dx, \quad (1)$$

where

$$f(x,y)\equiv\begin{cases}1 & \text{for } x,y\in[-a,a]\\ 0 & \text{otherwise}\end{cases} \quad (2)$$

and

$$\delta(x)=\frac{1}{2\pi}\int_{-\infty}^{\infty}e^{-ikx} \quad (3)$$

is the DELTA FUNCTION.

$$\begin{aligned}R(p,\tau)&=\frac{1}{2\pi}\int_{-a}^{a}\int_{-a}^{a}\int_{-\infty}^{\infty}e^{-ik[y-(\tau+px)]}\,dk\,dy\,dx\\&=\frac{1}{2\pi}\int_{-\infty}^{\infty}e^{ik\tau}\left[\int_{-a}^{a}e^{-ky}\,dy\int_{-a}^{a}e^{ikpx}\,dx\right]dk\\&=\frac{1}{2\pi}e^{ik\tau}\frac{1}{-ik}[e^{-iky}]_{-a}^{a}\frac{1}{ikp}[e^{ikpx}]_{-a}^{a}\,dk\\&=\frac{1}{2\pi}\int_{-\infty}^{\infty}e^{ik\tau}\frac{1}{k^2p}[-2i\sin(ka)][2i\sin(kpa)]\,dk\\&=\frac{2}{\pi p}\int_{-\infty}^{\infty}\frac{\sin(ka)\sin(kpa)e^{ik\tau}}{k^2}\,dk\\&=\frac{4}{\pi p}\int_{0}^{\infty}\frac{\sin(ka)\sin(kpa)\cos(k\tau)}{k^2}\,dk\\&=\frac{2}{\pi p}\int_{0}^{\infty}\frac{\sin[k(\tau+a)]-\sin[k(\tau-a)]}{k^2}\sin(kpa)\,dk\\&=\frac{2}{\pi p}\left\{\int_{0}^{\infty}\frac{\sin[k(\tau+a)]\sin(kpa)}{k^2}\,dk\right.\\&\quad\left.-\int_{0}^{\infty}\frac{\sin[k(\tau-a)]\sin(kpa)}{k^2}\,dk\right\}. \quad (4)\end{aligned}$$

From Gradshteyn and Ryzhik (1979, equation 3.741.3),

$$\int_{0}^{\infty}\frac{\sin(ax)\sin(bx)}{x^2}\,dx=\tfrac{1}{2}\pi\operatorname{sgn}(ab)\min(|a|,|b|), \quad (5)$$

so

$$\begin{aligned}R(p,\tau)=\frac{1}{p}\{&\operatorname{sgn}[(\tau+a)pa]\min(|\tau+a|,|pa|)\\&-\operatorname{sgn}[(\tau-a)pa]\min(|\tau-a|,|pa|)\}. \quad (6)\end{aligned}$$

References

Gradshteyn, I. S. and Ryzhik, I. M. *Tables of Integrals, Series, and Products, 5th ed.* San Diego, CA: Academic Press, 1979.

Railroad Track Problem

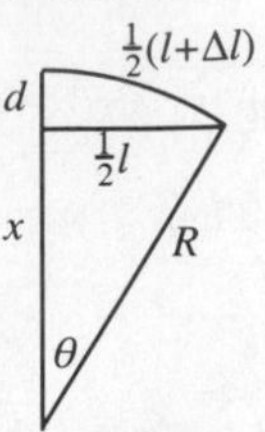

Given a straight segment of track of length l, add a small segment Δl so that the track bows into a circular ARC. Find the maximum displacement d of the bowed track. The PYTHAGOREAN THEOREM gives

$$R^2=x^2+(\tfrac{1}{2}l)^2, \quad (1)$$

But R is simply $x+d$, so

$$R^2=(x+d)^2=x^2+2xd+d^2. \quad (2)$$

Solving (1) and (2) for x gives

$$x=\frac{\frac{1}{4}l^2-d^2}{2d}. \quad (3)$$

Expressing the length of the ARC in terms of the central angle,

$$\begin{aligned}\tfrac{1}{2}(l+\Delta l)&=\theta(d+x)=\theta\left(d+\frac{\frac{1}{4}l^2-d^2}{2d}\right)\\&=\theta\left(\frac{2d^2+\frac{1}{4}l^2-d^2}{2d}\right)=\theta\left(\frac{d^2+\frac{1}{4}l^2}{2d}\right). \quad (4)\end{aligned}$$

But θ is given by

$$\tan\theta=\frac{\frac{1}{2}l}{x}=\frac{\frac{1}{2}l(2d)}{\frac{1}{4}l^2-d^2}=\frac{dl}{\frac{1}{4}l^2-d^2}, \quad (5)$$

so plugging θ in gives

$$\tfrac{1}{2}(l+\Delta l)=\left(\frac{d^2+\frac{1}{4}l^2}{2d}\right)\tan^{-1}\left(\frac{dl}{\frac{1}{4}l^2-d^2}\right) \quad (6)$$

$$d(l+\Delta l)=(d^2+\tfrac{1}{4}l^2)\tan^{-1}\left(\frac{dl}{\frac{1}{4}l^2-d^2}\right). \quad (7)$$

For $l\gg d$,

$$\frac{dl}{\frac{1}{4}l^2\left(1-\frac{d^2}{4l^2}\right)}=\frac{4d}{l}\left(1-\frac{4d^2}{l^2}\right)^{-1}\approx\frac{4d}{l}\left(1+\frac{4d}{l^2}\right). \quad (8)$$

Therefore,

$$\begin{aligned}&d(l+\Delta l)\\&\approx(d^2+\tfrac{1}{4}l^2)\left\{\frac{4d}{l}\left(1+\frac{4d^2}{l^2}\right)-\frac{1}{3}\left[\frac{4d}{l}\left(1+\frac{4d^2}{l^2}\right)\right]^3\right\}\\&\approx(d^2+\tfrac{1}{4}l^2)\left[\frac{4d}{l}+\frac{16d^3}{l^3}-\frac{1}{3}\left(\frac{4d}{l}\right)^3\left(1+3\frac{4d^2}{l^2}\right)\right]. \quad (9)\end{aligned}$$

Keeping only terms to order $(d/l)^3$,

$$dl + \Delta l \approx \frac{4d^3}{l} + dl + \frac{4d^3}{l} - \frac{16}{3}\frac{d^3}{l} \tag{10}$$

$$\Delta l \approx \left(8 - \tfrac{16}{3}\right)\frac{d^3}{l} = \frac{24-16}{3}\frac{d^3}{l} = \frac{8}{3}\frac{d^3}{l}, \tag{11}$$

so

$$d^2 = \tfrac{3}{8} l \Delta l \tag{12}$$

and

$$d \approx \tfrac{1}{2}\sqrt{\tfrac{3}{2} l \Delta l} = \tfrac{1}{4}\sqrt{6 l \Delta l}. \tag{13}$$

If we take $l = 1$ mile $= 5280$ feet and $\Delta l = 1$ foot, then $d \approx 44.450$ feet.

Ramanujan 6-10-8 Identity

Let $ad = bc$, then

$$\begin{aligned} &64[(a+b+c)^6 + (b+c+d)^6 - (c+d+a)^6 \\ &\qquad -(d+a+b)^6 + (a-d)^6 - (b-c)^6] \\ &\times[(a+b+c)^{10} + (b+c+d)^{10} - (c+d+a)^{10} \\ &\qquad -(d+a+b)^{10} + (a-d)^{10} - (b-c)^{10}] \\ &= 45[(a+b+c)^8 + (b+c+d)^8 - (c+d+a)^8 \\ &\qquad -(d+a+b)^8 + (a-d)^8 - (b-c)^8]^2. \end{aligned} \tag{1}$$

This can also be expressed by defining

$$F_{2m}(a,b,c,d) = (a+b+c)^{2m} + (b+c+d)^{2m} - (c+d+a)^{2m} - (d+a+b)^{2m} + (a-d)^{2m} - (b-c)^{2m} \tag{2}$$

$$f_{2m}(x,y) = (1+x+y)^{2m} + (x+y+xy)^{2m} - (y+xy+1)^{2m} - (xy+1+x)^{2m} + (1-xy)^{2m} - (x-y)^{2m}. \tag{3}$$

Then

$$F_{2m}(a,b,c,d) = a^{2m} f_{2m}(x,y), \tag{4}$$

and identity (1) can then be written

$$64 f_6(x,y) f_{10}(x,y) = 45 {f_8}^2(x,y). \tag{5}$$

Incidentally,

$$f_2(x,y) = 0 \tag{6}$$

$$f_4(x,y) = 0. \tag{7}$$

References

Berndt, B. C. *Ramanujan's Notebooks, Part IV.* New York: Springer-Verlag, pp. 3 and 102–106, 1994.

Berndt, B. C. and Bhargava, S. "A Remarkable Identity Found in Ramanujan's Third Notebook." *Glasgow Math. J.* **34**, 341–345, 1992.

Berndt, B. C. and Bhargava, S. "Ramanujan—For Lowbrows." *Amer. Math. Monthly* **100**, 644–656, 1993.

Bhargava, S. "On a Family of Ramanujan's Formulas for Sums of Fourth Powers." *Ganita* **43**, 63–67, 1992.

Hirschhorn, M. D. "Two or Three Identities of Ramanujan." *Amer. Math. Monthly* **105**, 52–55, 1998.

Nanjundiah, T. S. "A Note on an Identity of Ramanujan." *Amer. Math. Monthly* **100**, 485–487, 1993.

Ramanujan, S. *Notebooks.* New York: Springer-Verlag, pp. 385–386, 1987.

Ramanujan Constant

The IRRATIONAL constant

$$R \equiv e^{\pi\sqrt{163}} = 262537412640768743.99999999999925\ldots$$

which is very close to an INTEGER. Numbers such as the Ramanujan constant can be found using the theory of MODULAR FUNCTIONS. A few rather spectacular examples are given by Ramanujan (1913–14), including the one above, and can be generated using some amazing properties of the j-FUNCTION.

M. Gardner (Apr. 1975) played an April Fool's joke on the readers of *Scientific American* by claiming that this number was exactly an INTEGER. He admitted the hoax a few months later (Gardner, July 1975).

see also ALMOST INTEGER, CLASS NUMBER, j-FUNCTION

References

Ball, W. W. R. and Coxeter, H. S. M. *Mathematical Recreations and Essays, 13th ed.* New York: Dover, p. 387, 1987.

Castellanos, D. "The Ubiquitous Pi. Part I." *Math. Mag.* **61**, 67–98, 1988.

Gardner, M. "Mathematical Games: Six Sensational Discoveries that Somehow or Another have Escaped Public Attention." *Sci. Amer.* **232**, 127–131, Apr. 1975.

Gardner, M. "Mathematical Games: On Tessellating the Plane with Convex Polygons." *Sci. Amer.* **232**, 112–117, Jul. 1975.

Good, I. J. "What is the Most Amazing Approximate Integer in the Universe?" *Pi Mu Epsilon J.* **5**, 314–315, 1972.

Plouffe, S. "$e^{\pi\sqrt{163}}$, the Ramanujan Number." `http://lacim.uqam.ca/piDATA/ramanujan.txt`.

Ramanujan, S. "Modular Equations and Approximations to π." *Quart. J. Pure Appl. Math.* **45**, 350–372, 1913–1914.

Wolfram, S. *The Mathematica Book, 3rd ed.* New York: Cambridge University Press, p. 52, 1996.

Ramanujan Continued Fraction

Let $f(a,b)$ be a RAMANUJAN THETA FUNCTION. Then

$$\frac{f(-q,-q^4)}{f(-q^2,-q^3)} = \frac{1}{1+}\,\frac{q}{1+}\,\frac{q^2}{1+}\,\frac{q^3}{1+\ldots},$$

where the quantity on the right is a CONTINUED FRACTION.

see also RAMANUJAN THETA FUNCTIONS

Ramanujan Cos/Cosh Identity

$$\left[1 + 2\sum_{n=1}^\infty \frac{\cos(n\theta)}{\cosh(n\pi)}\right]^{-2} + \left[1 + 2\sum_{n=1}^\infty \frac{\cosh(n\theta)}{\cosh(n\pi)}\right]^{-2} = \frac{2\Gamma^4(\frac{3}{4})}{\pi},$$

where $\Gamma(z)$ is the GAMMA FUNCTION.

Ramanujan-Eisenstein Series

Let t be a discriminant,

$$q \equiv -e^{-\pi\sqrt{t}}, \quad (1)$$

then

$$E_2(q) \equiv L(q) \equiv 1 - 24\sum_{k=1}^{\infty}\frac{(2k+1)q^{2k+1}}{1-q^{2k+1}} = \left(\frac{2K}{\pi}\right)^2 (1-2k^2) \quad (2)$$

$$E_4(q) \equiv M(q) \equiv 1 + 240\sum_{k=1}^{\infty}\frac{k^3 q^{2k}}{1-q^{2k}} = \left(\frac{2K}{\pi}\right)^4 (1-k^2k'^2) \quad (3)$$

$$E_6(q) \equiv N(q) \equiv 1 - 504\sum_{k=1}^{\infty}\frac{k^5 q^{2k}}{1-q^{2k}} = \left(\frac{2K}{\pi}\right)^6 (1-2k^2)(1+\tfrac{1}{2}k^2k'^2). \quad (4)$$

see also KLEIN'S ABSOLUTE INVARIANT, PI

References

Borwein, J. M. and Borwein, P. B. "Class Number Three Ramanujan Type Series for $1/\pi$." *J. Comput. Appl. Math.* **46**, 281–290, 1993.

Ramanujan, S. "Modular Equations and Approximations to π." *Quart. J. Pure Appl. Math.* **45**, 350–372, 1913–1914.

Ramanujan Function

$$\phi(a,n) \equiv 1 + 2\sum_{k=1}^{n}\frac{1}{(ak)^3 - ak}$$

$$\phi(a) \equiv \lim_{n\to\infty}\phi(a,n) = 1 + 2\sum_{k=1}^{\infty}\frac{1}{(ak)^3 - ak}.$$

The values of $\phi(n)$ for $n = 2, 3, \ldots$ are

$$\phi(2) = 2\ln 2$$
$$\phi(3) = \ln 3$$
$$\phi(4) = \tfrac{3}{2}\ln 2$$
$$\phi(6) = \tfrac{1}{2}\ln 3 + \tfrac{1}{3}\ln 4.$$

Ramanujan g- and G-Functions

Following Ramanujan (1913–14), write

$$\prod_{k=1,3,5,\ldots}^{\infty}(1+e^{-k\pi\sqrt{n}}) = 2^{1/4}e^{-\pi\sqrt{n}/24}G_n \quad (1)$$

$$\prod_{k=1,3,5,\ldots}^{\infty}(1-e^{-k\pi\sqrt{n}}) = 2^{1/4}e^{-\pi\sqrt{n}/24}g_n. \quad (2)$$

These satisfy the equalities

$$g_{4n} = 2^{1/4}g_nG_n \quad (3)$$
$$G_n = G_{1/n} \quad (4)$$
$$g_n{}^{-1} = g_{4/n} \quad (5)$$
$$\tfrac{1}{4} = (g_nG_n)^8(G_n{}^8 - g_n{}^8). \quad (6)$$

G_n and g_n can be derived using the theory of MODULAR FUNCTIONS and can always be expressed as roots of algebraic equations when n is RATIONAL. For simplicity, Ramanujan tabulated g_n for n EVEN and G_n for n ODD. However, (6) allows G_n and g_n to be solved for in terms of g_n and G_n, giving

$$g_n = \tfrac{1}{2}\left(G_n{}^8 + \sqrt{G_n{}^{16} - G_n{}^{-8}}\right)^{1/8} \quad (7)$$

$$G_n = \tfrac{1}{2}\left(g_n{}^8 + \sqrt{g_n{}^{16} + Gg_n{}^{-8}}\right)^{1/8}. \quad (8)$$

Using (3) and the above two equations allows g_{4n} to be computed in terms of g_n or G_n

$$g_{4n} = \begin{cases} 2^{1/8}g_n\left(g_n{}^8 + \sqrt{g_n{}^{16} + g_n{}^{-8}}\right)^{1/8} & \text{for } n \text{ even} \\ 2^{1/8}G_n\left(G_n{}^8 + \sqrt{G_n{}^{16} - G_n{}^{-8}}\right)^{1/8} & \text{for } n \text{ odd.} \end{cases} \quad (9)$$

In terms of the PARAMETER k and complementary PARAMETER k',

$$G_n = (2k_nk'_n)^{-1/12} \quad (10)$$

$$g_n = \left(\frac{k'^2_n}{2k}\right)^{1/12}. \quad (11)$$

Here,

$$k_n = \lambda^*(n) \quad (12)$$

is the ELLIPTIC LAMBDA FUNCTION, which gives the value of k for which

$$\frac{K'(k)}{K(k)} = \sqrt{n}. \quad (13)$$

Solving for $\lambda^*(n)$ gives

$$\lambda^*(n) = \tfrac{1}{2}[\sqrt{1+G_n{}^{-12}} - \sqrt{1-G_n{}^{-12}}] \quad (14)$$

$$\lambda^*(n) = g_n{}^6[\sqrt{g_n{}^{12} + g_n{}^{-12}} - g_n{}^6]. \quad (15)$$

Analytic values for small values of n can be found in Ramanujan (1913–1914) and Borwein and Borwein (1987), and have been compiled in Weisstein (1996). Ramanujan (1913–1914) contains a typographical error labeling G_{465} as G_{265}.

see also G-FUNCTION

References

Borwein, J. M. and Borwein, P. B. *Pi & the AGM: A Study in Analytic Number Theory and Computational Complexity.* New York: Wiley, pp. 139 and 298, 1987.

Ramanujan, S. "Modular Equations and Approximations to π." *Quart. J. Pure. Appl. Math.* **45**, 350–372, 1913–1914.

Weisstein, E. W. "Elliptic Singular Values." `http://www.astro.virginia.edu/~eww6n/math/notebooks/EllipticSingular`.

Ramanujan's Hypergeometric Identity

$$1-\left(\frac{1}{2}\right)^3+\left(\frac{1\cdot 3}{2\cdot 4}\right)^3+\ldots = {}_3F_2\left(\begin{matrix}\frac{1}{2},\frac{1}{2},\frac{1}{2}\\ 1,1\end{matrix};-1\right)$$
$$= \left[{}_2F_1\left(\begin{matrix}\frac{1}{4},\frac{1}{4}\\ 1\end{matrix};-1\right)\right]^2 = \frac{\Gamma^2(\frac{9}{8})}{\Gamma^2(\frac{5}{4})\Gamma^2(\frac{7}{8})},$$

where ${}_2F_1(a,b;c;x)$ is a HYPERGEOMETRIC FUNCTION, ${}_3F_2(a,b,c;d;e;x)$ is a GENERALIZED HYPERGEOMETRIC FUNCTION, and $\Gamma(z)$ is a GAMMA FUNCTION.

References

Hardy, G. H. *Ramanujan: Twelve Lectures on Subjects Suggested by His Life and Work, 3rd ed.* New York: Chelsea, p. 106, 1959.

Ramanujan's Hypothesis

see TAU CONJECTURE

Ramanujan's Identity

$$5\frac{\phi^5(x^5)}{\phi^6(x)} = \sum_{m=0}^{\infty} P(5m+4)x^m,$$

where

$$\phi(x) = \prod_{m=1}^{\infty}(1-x^m)$$

and $P(n)$ is the PARTITION FUNCTION P.

see also RAMANUJAN'S SUM IDENTITY

Ramanujan's Integral

$$\int_{-\infty}^{\infty} \frac{J_{\mu+\xi}(x)}{x^{\mu+\xi}}\frac{J_{\nu-\xi}(y)}{y^{\nu-\xi}}e^{it\xi}\,d\xi$$
$$= \left[\frac{2\cos\left(\frac{1}{2}t\right)}{x^2e^{-it/2}+y^2e^{it/2}}\right]^{(\mu+\nu)/2}$$
$$\times J_{\mu+\nu}\left[\sqrt{2\cos\left(\tfrac{1}{2}t\right)(x^2e^{-it/2}+y^2e^{it/2})}\right]e^{it(\nu-\mu)/2},$$

where $J_n(z)$ is a BESSEL FUNCTION OF THE FIRST KIND.

References

Watson, G. N. *A Treatise on the Theory of Bessel Functions, 2nd ed.* Cambridge, England: Cambridge University Press, 1966.

Ramanujan's Interpolation Formula

$$\int_0^{\infty} x^{s-1}\sum_{k=0}^{\infty}(-1)^k x^k \phi(k)\,dx = \frac{\pi\phi(-s)}{\sin(s\pi)} \qquad (1)$$
$$\int_0^{\infty} x^{s-1}\sum_{k=0}^{\infty}(-1)^k \frac{x^k}{k!}\lambda(k)\,dx = \Gamma(s)\lambda(-s), \qquad (2)$$

where $\lambda(z)$ is the DIRICHLET LAMBDA FUNCTION and $\Gamma(z)$ is the GAMMA FUNCTION. Equation (2) is obtained from (1) by defining

$$\phi(u) = \frac{\lambda(u)}{\Gamma(1+u)}. \qquad (3)$$

These formulas give valid results only for certain classes of functions.

References

Hardy, G. H. *Ramanujan: Twelve Lectures on Subjects Suggested by His Life and Work, 3rd ed.* New York: Chelsea, pp. 15 and 186–195, 1959.

Ramanujan's Master Theorem

Suppose that in some NEIGHBORHOOD of $x=0$,

$$F(x) = \sum_{k=0}^{\infty}\frac{\phi(k)(-x)^k}{k!}.$$

Then

$$\int_0^{\infty} x^{n-1}F(x)\,dx = \Gamma(n)\phi(-n).$$

References

Berndt, B. C. *Ramanujan's Notebooks: Part I.* New York: Springer-Verlag, p. 298, 1985.

Ramanujan-Petersson Conjecture

A CONJECTURE for the EIGENVALUES of modular forms under HECKE OPERATORS.

Ramanujan Psi Sum

A sum which includes both the JACOBI TRIPLE PRODUCT and the q-BINOMIAL THEOREM as special cases. Ramanujan's sum is

$$\sum_{n=-\infty}^{\infty}\frac{(a)_n}{(b)_n}x^n = \frac{(ax)_\infty(q/ax)_\infty(q)_\infty(b/a)_\infty}{(x)_\infty(b/ax)_\infty(b)_\infty(q/a)_\infty},$$

where the NOTATION $(q)_k$ denotes q-SERIES. For $b=q$, this becomes the q-BINOMIAL THEOREM.

see also JACOBI TRIPLE PRODUCT, q-BINOMIAL THEOREM, q-SERIES

Ramanujan's Square Equation

It has been proved that the only solutions to the DIOPHANTINE EQUATION

$$2^n - 7 = x^2$$

are $n = 3$, 4, 5, 7, and 15 (Beeler *et al.* 1972, Item 31).

References

Beeler, M.; Gosper, R. W.; and Schroeppel, R. *HAKMEM.* Cambridge, MA: MIT Artificial Intelligence Laboratory, Memo AIM-239, Feb. 1972.

Ramanujan's Sum

The sum

$$c_q(m) = \sum_{h^*(q)} e^{2\pi i h m/q}, \quad (1)$$

where h runs through the residues RELATIVELY PRIME to q, which is important in the representation of numbers by the sums of squares. If $(q, q') = 1$ (i.e., q and q' are RELATIVELY PRIME), then

$$c_{qq'}(m) = c_q(m) c_{q'}(m). \quad (2)$$

For argument 1,

$$c_b(1) = \mu(b), \quad (3)$$

where μ is the MÖBIUS FUNCTION, and for general m,

$$c_b(m) = \mu\left(\frac{b}{(b,m)}\right) \frac{\phi(b)}{\phi\left(\frac{b}{(b,m)}\right)}. \quad (4)$$

see also MÖBIUS FUNCTION, WEYL'S CRITERION

References

Vardi, I. *Computational Recreations in Mathematica.* Redwood City, CA: Addison-Wesley, p. 254, 1991.

Ramanujan's Sum Identity

If

$$\frac{1 + 53x + 9x^2}{1 - 82x - 82x^2 + x^3} = \sum_{n=1}^{\infty} a_n x^n \quad (1)$$

$$\frac{2 - 26x - 12x^2}{1 - 82x - 82x^2 + x^3} = \sum_{n=0}^{\infty} b_n x^n \quad (2)$$

$$\frac{2 + 8x - 10x^2}{1 - 82x - 82x^2 + x^3} = \sum_{n=0}^{\infty} c_n x^n, \quad (3)$$

then

$$a_n{}^3 + b_n{}^3 = c_n{}^3 + (-1)^n. \quad (4)$$

Hirschhorn (1995) showed that

$$a_n = \tfrac{1}{85}[(64 + 8\sqrt{85})\alpha^n + (64 - 8\sqrt{85})\beta^n - 43(-1)^n] \quad (5)$$

$$b_n = \tfrac{1}{85}[(77 + 7\sqrt{85})\alpha^n + (77 - 7\sqrt{85})\beta^n + 16(-1)^n] \quad (6)$$

$$c_n = \tfrac{1}{85}[(93 + 9\sqrt{85})\alpha^n + (93 - 9\sqrt{85})\beta^n - 16(-1)^n], \quad (7)$$

where

$$\alpha = \tfrac{1}{2}(83 + 9\sqrt{85}) \quad (8)$$

$$\beta = \tfrac{1}{2}(83 - 9\sqrt{85}). \quad (9)$$

Hirschhorn (1996) showed that checking the first seven cases $n = 0$ to 6 is sufficient to prove the result.

References

Hirschhorn, M. D. "An Amazing Identity of Ramanujan." *Math. Mag.* **68**, 199–201, 1995.

Hirschhorn, M. D. "A Proof in the Spirit of Zeilberger of an Amazing Identity of Ramanujan." *Math. Mag.* **69**, 267–269, 1996.

Ramanujan's Tau-Dirichlet Series

see TAU-DIRICHLET SERIES

Ramanujan's Tau Function

see TAU FUNCTION

Ramanujan Theta Functions

Ramanujan's one-variable theta function is defined by

$$\varphi(x) \equiv \sum_{m=-\infty}^{\infty} x^{m^2}. \quad (1)$$

It is equal to the function in the JACOBI TRIPLE PRODUCT with $z = 1$,

$$\varphi(x) = G(1) = \prod_{n=1}^{\infty} (1 + x^{2n-1})^2 (1 - x^{2n})$$

$$= \sum_{m=-\infty}^{\infty} x^{m^2} = 1 + 2\sum_{m=0}^{\infty} x^{m^2}. \quad (2)$$

Special values include

$$\varphi(e^{-\pi\sqrt{2}}) = \frac{\Gamma(\frac{9}{8})}{\Gamma(\frac{5}{4})} \sqrt{\frac{\Gamma(\frac{1}{4})}{2^{1/4}\pi}} \quad (3)$$

$$\varphi(e^{-\pi}) = \frac{\pi^{1/4}}{\Gamma(\frac{3}{4})} \quad (4)$$

$$\varphi(e^{-2\pi}) = \frac{\sqrt{2 + \sqrt{2}}}{2} \frac{\pi}{\Gamma(\frac{3}{4})}. \quad (5)$$

Ramanujan's two-variable theta function is defined by

$$f(a, b) \equiv \sum_{n=-\infty}^{\infty} a^{n(n+1)/2} b^{n(n-1)/2} \quad (6)$$

for $|ab| < 1$. It satisfies

$$f(-1, a) = 0 \quad (7)$$

$$f(a, b) = f(b, a) = (-a; ab)_\infty (-b; ab)_\infty (ab; ab)_\infty \quad (8)$$

$$f(-q) \equiv f(-q, -q^2) = \sum_{k=0}^{\infty} (-1)^k q^{k(2k-1)/2} + \sum_{k=1}^{\infty} (-1)^k q^{k(2k+1)/2} = (q;q)_\infty, \quad (9)$$

where $(q)_\infty$ are q-SERIES.

see also JACOBI TRIPLE PRODUCT, SCHRÖTER'S FORMULA, q-SERIES

Ramp Function

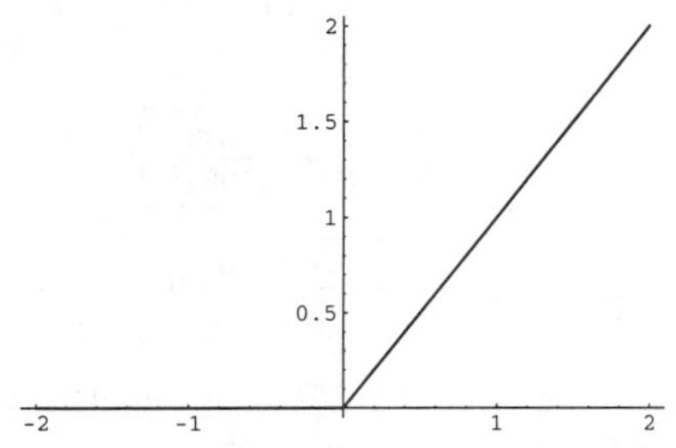

$$R(x) \equiv xH(x) \quad (1)$$
$$= \int_{-\infty}^{x} H(x')\,dx' \quad (2)$$
$$= \int_{-\infty}^{\infty} H(x')H(x-x')\,dx' \quad (3)$$
$$= H(x) * H(x), \quad (4)$$

where $H(x)$ is the HEAVISIDE STEP FUNCTION and $*$ is the CONVOLUTION. The DERIVATIVE is

$$R'(x) = -H(x). \quad (5)$$

The FOURIER TRANSFORM of the ramp function is given by

$$\mathcal{F}[R(x)] = \int_{-\infty}^{\infty} e^{-2\pi ikx} R(x)\,dx = \pi i\delta'(2\pi k) - \frac{1}{4\pi^2 k^2}, \quad (6)$$

where $\delta(x)$ is the DELTA FUNCTION and $\delta'(x)$ its DERIVATIVE.

see also FOURIER TRANSFORM—RAMP FUNCTION, HEAVISIDE STEP FUNCTION, RECTANGLE FUNCTION, SGN, SQUARE WAVE

Ramphoid Cusp

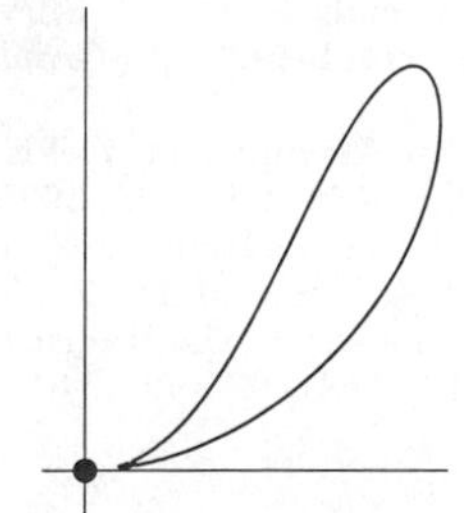

A type of CUSP as illustrated above for the curve $x^4 + x^2y^2 - 2x^2y - xy^2 + y^2 = 0$.

see also CUSP

References

Walker, R. J. *Algebraic Curves.* New York: Springer-Verlag, pp. 57–58, 1978.

Ramsey Number

The Ramsey number $R(m,n)$ gives the solution to the PARTY PROBLEM, which asks the minimum number of guests $R(m,n)$ that must be invited so that at least m will know each other (i.e., there exists a CLIQUE of order m) or at least n will not know each other (i.e., there exists an independent set of order n). By symmetry, it is true that

$$R(m,n) = R(n,m). \quad (1)$$

It also must be true that

$$R(2,m) = m. \quad (2)$$

A generalized Ramsey number is written

$$R(m_1, \ldots, m_k; n) \quad (3)$$

and is the smallest INTEGER R such that, no matter how each n-element SUBSET of an r-element SET are colored with k colors, there exists an i such that there is a SUBSET of size m_i, all of whose n-element SUBSETS are color i. The usual Ramsey numbers are then equivalent to $R(m,n) = R(m,n;2)$.

Bounds are given by

$$R(k,l) \leq \begin{cases} R(k-1,l) + R(k,l-1) - 1 & \text{for } R(k-1,l) \text{ and } R(k,l-1) \text{ even} \\ R(k-1,l) + R(k,l-1) & \text{otherwise} \end{cases} \quad (4)$$

and

$$R(k,k) \leq 4R(k-2,k) + 2 \quad (5)$$

(Chung and Grinstead 1983). Erdős proved that for diagonal Ramsey numbers $R(k,k)$,

$$\frac{k2^{k/2}}{e\sqrt{2}} < R(k,k). \quad (6)$$

This result was subsequently improved by a factor of 2 by Spencer (1975). $R(3,k)$ was known since 1980 to be bounded from above by $c_2k^2/\ln k$, and Griggs (1983) showed that $c_2 = 5/12$ was an acceptable limit. J.-H. Kim (Cipra 1995) subsequently bounded $R(3,k)$ by a similar expression from below, so

$$c_1 \frac{k^2}{\ln k} \leq R(3,k) \leq c_2 \frac{k^2}{\ln k}. \quad (7)$$

Burr (1983) gives Ramsey numbers for all 113 graphs with no more than 6 EDGES and no isolated points.

A summary of known results up to 1983 for $R(m,n)$ is given in Chung and Grinstead (1983). Radziszowski maintains an up-to-date list of the best current bounds, reproduced in part in the following table for $R(m,n;2)$.

m	n	$R(m,n)$
3	3	6
3	4	9
3	5	14
3	6	18
3	7	23
3	8	28
3	9	36
3	10	[40, 43]
3	11	[46, 51]
3	12	[52, 60]
3	13	[60, 69]
3	14	[66, 78]
3	15	[73, 89]
3	16	[79, ∞]
3	17	[92, ∞]
3	18	[98, ∞]
3	19	[106, ∞]
3	20	[109, ∞]
3	21	[122, ∞]
3	22	[125, ∞]
3	23	[136, ∞]

m	n	$R(m,n)$
4	4	18
4	5	25
4	6	[35, 41]
4	7	[49, 62]
4	8	[55, 85]
4	9	[69, 116]
4	10	[80, 151]
4	11	[93, 191]
4	12	[98, 238]
4	13	[112, 291]
4	14	[119, 349]
4	15	[128, 417]

m	n	$R(m,n)$
5	5	[43, 49]
5	6	[58, 87]
5	7	[80, 143]
5	8	[95, 216]
5	9	[116, 371]
5	10	[1, 445]

m	n	$R(m,n)$
6	6	[102, 165]
6	7	[109, 300]
6	8	[122, 497]
6	9	[153, 784]
6	10	[167, 1180]

m	n	$R(m,n)$
7	7	[205, 545]
7	8	[1, 1035]
7	9	[1, 1724]
7	10	[1, 2842]

m	n	$R(m,n)$
8	8	[282, 1874]
8	9	[1, 3597]
8	10	[1, 6116]

m	n	$R(m,n)$
9	9	[565, 6680]
9	10	[1, 12795]

m	n	$R(m,n)$
10	10	[798, 23981]

m	n	$R(m,n)$
11	11	[522, ∞]

Known values for generalized Ramsey numbers are given in the following table.

$R(\ldots;2)$	Bounds
$R(3,3,3;2)$	17
$R(3,3,4;2)$	[30, 32]
$R(3,3,5;2)$	[45, 59]
$R(3,4,4;2)$	[55, 81]
$R(3,4,5;2)$	≥ 80
$R(4,4,4;2)$	[128, 242]
$R(3,3,3,3;2)$	[51, 64]
$R(3,3,3,4;2)$	[87, 159]
$R(3,3,3,3,3;2)$	[162, 317]
$R(3,3,3,3,3,3;2)$	[1, 500]

$R(\ldots;3)$	Bounds
$R(4,4;3)$	[14, 15]

see also CLIQUE, COMPLETE GRAPH, EXTREMAL GRAPH, IRREDUNDANT RAMSEY NUMBER, SCHUR NUMBER

References

Burr, S. A. "Generalized Ramsey Theory for Graphs—A Survey." In *Graphs and Combinatorics* (Ed. R. A. Bari and F. Harary). New York: Springer-Verlag, pp. 52–75, 1964.

Burr, S. A. "Diagonal Ramsey Numbers for Small Graphs." *J. Graph Th.* **7**, 57–69, 1983.

Chartrand, G. "The Problem of the Eccentric Hosts: An Introduction to Ramsey Numbers." §5.1 in *Introductory Graph Theory.* New York: Dover, pp. 108–115, 1985.

Chung, F. R. K. "On the Ramsey Numbers $N(3,3,\ldots,3;2)$." *Discrete Math.* **5**, 317–321, 1973.

Chung, F. and Grinstead, C. G. "A Survey of Bounds for Classical Ramsey Numbers." *J. Graph. Th.* **7**, 25–37, 1983.

Cipra, B. "A Visit to Asymptopia Yields Insights into Set Structures." *Science* **267**, 964–965, 1995.

Exoo, G. "On Two Classical Ramsey Numbers of the Form $R(3,n)$." *SIAM J. Discrete Math.* **2**, 488–490, 1989.

Exoo, G. "Announcement: On the Ramsey Numbers $R(4,6)$, $R(5,6)$ and $R(3,12)$." *Ars Combin.* **35**, 85, 1993.

Exoo, G. "Some New Ramsey Colorings." *Electronic J. Combinatorics* **5**, No. 1, R29, 1–5, 1998. http://www.combinatorics.org/Volume_5/v5i1toc.html.
Folkmann, J. "Notes on the Ramsey Number $N(3,3,3,3)$." *J. Combinat. Theory. Ser. A* **16**, 371–379, 1974.
Gardner, M. "Mathematical Games: In Which Joining Sets of Points by Lines Leads into Diverse (and Diverting) Paths." *Sci. Amer.* **237**, 18–28, 1977.
Gardner, M. *Penrose Tiles and Trapdoor Ciphers... and the Return of Dr. Matrix, reissue ed.* New York: W. H. Freeman, pp. 240–241, 1989.
Giraud, G. "Une minoration du nombre de quadrangles unicolores et son application a la majoration des nombres de Ramsey binaires bicolors." *C. R. Acad. Sci. Paris A* **276**, 1173–1175, 1973.
Graham, R. L.; Rothschild, B. L.; and Spencer, J. H. *Ramsey Theory, 2nd ed.* New York: Wiley, 1990.
Graver, J. E. and Yackel, J. "Some Graph Theoretic Results Associated with Ramsey's Theorem." *J. Combin. Th.* **4**, 125–175, 1968.
Greenwood, R. E. and Gleason, A. M. "Combinatorial Relations and Chromatic Graphs." *Canad. J. Math.* **7**, 1–7, 1955.
Griggs, J. R. "An Upper Bound on the Ramsey Numbers $R(3,k)$." *J. Comb. Th. A* **35**, 145–153, 1983.
Grinstead, C. M. and Roberts, S. M. "On the Ramsey Numbers $R(3,8)$ and $R(3,9)$." *J. Combinat. Th. Ser. B* **33**, 27–51, 1982.
Guldan, F. and Tomasta, P. "New Lower Bounds of Some Diagonal Ramsey Numbers." *J. Graph. Th.* **7**, 149–151, 1983.
Hanson, D. "Sum-Free Sets and Ramsey Numbers." *Discrete Math.* **14**, 57–61, 1976.
Harary, F. "Recent Results on Generalized Ramsey Theory for Graphs." *Graph Theory and Applications* (Ed. Y. Alai, D. R. Lick, and A. T. White). New York: Springer-Verlag, pp. 125–138, 1972.
Hill, R. and Irving, R. W. "On Group Partitions Associated with Lower Bounds for Symmetric Ramsey Numbers." *European J. Combin.* **3**, 35–50, 1982.
Kalbfleisch, J. G. *Chromatic Graphs and Ramsey's Theorem.* Ph.D. thesis, University of Waterloo, January 1966.
McKay, B. D. and Min, Z. K. "The Value of the Ramsey Number $R(3,8)$." *J. Graph Th.* **16**, 99–105, 1992.
McKay, B. D. and Radziszowski, S. P. "$R(4,5)=25$." *J. Graph. Th* **19**, 309–322, 1995.
Piwakowski, K. "Applying Tabu Search to Determine New Ramsey Numbers." *Electronic J. Combinatorics* **3**, R6, 1–4, 1996. http://www.combinatorics.org/Volume_3/volume3.html#R6.
Radziszowski, S. P. "Small Ramsey Numbers." *Electronic J. Combin.* **1**, DS1 1–29, Rev. Mar. 25, 1996. http://ejc.math.gatech.edu:8080/Journal/Surveys/ds1.ps.
Radziszowski, S. and Kreher, D. L. "Upper Bounds for Some Ramsey Numbers $R(3,k)$." *J. Combinat. Math. Combin. Comput.* **4**, 207–212, 1988.
Spencer, J. H. "Ramsey's Theorem—A New Lower Bound." *J. Combinat. Theory Ser. A* **18**, 108–115, 1975.
Wang, Q. and Wang, G. "New Lower Bounds for the Ramsey Numbers $R(3,q)$." *Beijing Daxue Xuebao* **25**, 117–121, 1989.
Whitehead, E. G. "The Ramsey Number $N(3,3,3,3;2)$." *Discrete Math.* **4**, 389–396, 1973.

Ramsey's Theorem

A generalization of DILWORTH'S LEMMA. For each $m,n \in \mathbb{N}$ with $m,n \geq 2$, there exists a least INTEGER $R(m,n)$ (the RAMSEY NUMBER) such that no matter how the COMPLETE GRAPH $K_{R(m,n)}$ is two-colored, it will contain a green SUBGRAPH K_m or a red subgroup K_n. Furthermore,

$$R(m,n) \leq R(m-1,n) + R(m,n-1)$$

if $m,n \geq 3$. The theorem can be equivalently stated that, for all $\in \mathbb{N}$, there exists an $n \in \mathbb{N}$ such that any complete DIGRAPH on n VERTICES contains a complete transitive SUBGRAPH of m VERTICES. Ramsey's theorem is a generalization of the PIGEONHOLE PRINCIPLE since

$$R(\underbrace{2,2,\ldots,2}_{t}) = t+1.$$

see also DILWORTH'S LEMMA, NATURAL INDEPENDENCE PHENOMENON, PIGEONHOLE PRINCIPLE, RAMSEY NUMBER

References

Graham, R. L.; Rothschild, B. L.; and Spencer, J. H. *Ramsey Theory, 2nd ed.* New York: Wiley, 1990.
Spencer, J. "Large Numbers and Unprovable Theorems." *Amer. Math. Monthly* **90**, 669–675, 1983.

Randelbrot Set

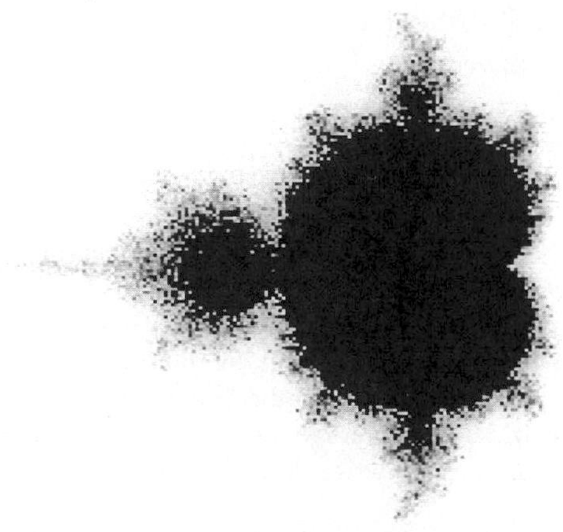

The FRACTAL-like figure obtained by performing the same iteration as for the MANDELBROT SET, but adding a random component R,

$$z_{n+1} = {z_n}^2 + c + R.$$

In the above plot, $R \equiv R_x + iR_y$, where $R_x, R_y \in [-0.05, 0.05]$.

see also MANDELBROT SET

References

Dickau, R. M. "Randelbrot Set." http://forum.swarthmore.edu/advanced/robertd/randelbrot.html.

Random Distribution

A DISTRIBUTION in which the variates occur with PROBABILITIES asymptotically matching their "true" underlying DISTRIBUTION is said to be random.

see also DISTRIBUTION, RANDOM NUMBER

Random Dot Stereogram

see STEREOGRAM

Random Graph

A random graph is a GRAPH in which properties such as the number of NODES, EDGES, and connections between them are determined in some random way. Erdős and Rényi showed that for many monotone-increasing properties of random graphs, graphs of a size slightly less than a certain threshold are very unlikely to have the property, whereas graphs with a few more EDGES are almost certain to have it. This is known as a PHASE TRANSITION.

see also GRAPH (GRAPH THEORY), GRAPH THEORY

References

Bollobás, B. *Random Graphs.* London: Academic Press, 1985.

Steele, J. M. "Gibbs' Measures on Combinatorial Objects and the Central Limit Theorem for an Exponential Family of Random Trees." *Prob. Eng. Inform. Sci.* **1**, 47–59, 1987.

Random Matrix

A random matrix is a MATRIX of given type and size whose entries consist of random numbers from some specified distribution.

see also MATRIX

Random Number

Computer-generated random numbers are sometimes called PSEUDORANDOM NUMBERS, while the term "random" is reserved for the output of unpredictable physical processes. It is impossible to produce an arbitrarily long string of random digits and prove it is random. Strangely, it is very difficult for humans to produce a string of random digits, and computer programs can be written which, on average, actually predict some of the digits humans will write down based on previous ones.

The LINEAR CONGRUENCE METHOD is one algorithm for generating PSEUDORANDOM NUMBERS. The initial number used as the starting point in a random number generating algorithm is known as the SEED. The goodness of random numbers generated by a given ALGORITHM can be analyzed by examining its NOISE SPHERE.

see also BAYS' SHUFFLE, CLIFF RANDOM NUMBER GENERATOR, QUASIRANDOM SEQUENCE, SCHRAGE'S ALGORITHM, STOCHASTIC

References

Bassein, S. "A Sampler of Randomness." *Amer. Math. Monthly* **103**, 483–490, 1996.

Bratley, P.; Fox, B. L.; and Schrage, E. L. *A Guide to Simulation, 2nd ed.* New York: Springer-Verlag, 1996.

Dahlquist, G. and Bjorck, A. Ch. 11 in *Numerical Methods.* Englewood Cliffs, NJ: Prentice-Hall, 1974.

Deak, I. *Random Number Generators and Simulation.* New York: State Mutual Book & Periodical Service, 1990.

Forsythe, G. E.; Malcolm, M. A.; and Moler, C. B. Ch. 10 in *Computer Methods for Mathematical Computations.* Englewood Cliffs, NJ: Prentice-Hall, 1977.

Gardner, M. "Random Numbers." Ch. 13 in *Mathematical Carnival: A New Round-Up of Tantalizers and Puzzles from Scientific American.* New York: Vintage, 1977.

James, F. "A Review of Pseudorandom Number Generators." *Computer Physics Comm.* **60**, 329–344, 1990.

Kac, M. "What is Random?" *Amer. Sci.* **71**, 405–406, 1983.

Kenney, J. F. and Keeping, E. S. *Mathematics of Statistics, Pt. 1, 3rd ed.* Princeton, NJ: Van Nostrand, pp. 200–201 and 205–207, 1962.

Kenney, J. F. and Keeping, E. S. *Mathematics of Statistics, Pt. 2, 2nd ed.* Princeton, NJ: Van Nostrand, pp. 151–154, 1951.

Knuth, D. E. Ch. 3 in *The Art of Computer Programming, Vol. 2: Seminumerical Algorithms, 2nd ed.* Reading, MA: Addison-Wesley, 1981.

Marsaglia, G. "A Current View of Random Number Generators." In *Computer Science and Statistics: Proceedings of the Symposium on the Interface, 16th, Atlanta, Georgia, March 1984* (Ed. L. Billard). New York: Elsevier, 1985.

Park, S. and Miller, K. "Random Number Generators: Good Ones are Hard to Find." *Comm. ACM* **31**, 1192–1201, 1988.

Peterson, I. *The Jungles of Randomness: A Mathematical Safari.* New York: Wiley, 1997.

Pickover, C. A. "Computers, Randomness, Mind, and Infinity." Ch. 31 in *Keys to Infinity.* New York: W. H. Freeman, pp. 233–247, 1995.

Press, W. H.; Flannery, B. P.; Teukolsky, S. A.; and Vetterling, W. T. "Random Numbers." Ch. 7 in *Numerical Recipes in FORTRAN: The Art of Scientific Computing, 2nd ed.* Cambridge, England: Cambridge University Press, pp. 266–306, 1992.

Schrage, L. "A More Portable Fortran Random Number Generator." *ACM Trans. Math. Software* **5**, 132–138, 1979.

Schroeder, M. "Random Number Generators." In *Number Theory in Science and Communication, with Applications in Cryptography, Physics, Digital Information, Computing and Self-Similarity, 3rd ed.* New York: Springer-Verlag, pp. 289–295, 1990.

Random Percolation

see PERCOLATION THEORY

Random Polynomial

A POLYNOMIAL having random COEFFICIENTS.

see also KAC FORMULA

Random Variable

A random variable is a measurable function from a PROBABILITY SPACE $(S, \mathbb{S}, P)$ into a MEASURABLE SPACE $(S', \mathbb{S}')$ known as the STATE SPACE.

see also PROBABILITY SPACE, RANDOM DISTRIBUTION, RANDOM NUMBER, STATE SPACE, VARIATE

References

Gikhman, I. I. and Skorokhod, A. V. *Introduction to the Theory of Random Processes.* New York: Dover, 1997.

Random Walk

A random process consisting of a sequence of discrete steps of fixed length. The random thermal perturbations in a liquid are responsible for a random walk phenomenon known as Brownian motion, and the collisions of molecules in a gas are a random walk responsible for diffusion. Random walks have interesting mathematical properties that vary greatly depending on the dimension in which the walk occurs and whether it is confined to a lattice.

see also RANDOM WALK—1-D, RANDOM WALK—2-D, RANDOM WALK—3-D, SELF-AVOIDING WALK

References

Barber, M. N. and Ninham, B. W. *Random and Restricted Walks: Theory and Applications.* New York: Gordon and Breach, 1970.

Chandrasekhar, S. In *Selected Papers on Noise and Stochastic Processes* (Ed. N. Wax). New York: Dover, 1954.

Doyle, P. G. and Snell, J. L. *Random Walks and Electric Networks.* Washington, DC: Math. Assoc. Amer, 1984.

Dykin, E. B. and Uspenskii, V. A. *Random Walks.* New York: Heath, 1963.

Feller, W. *An Introduction to Probability Theory and Its Applications, Vol. 1, 3rd ed.* New York: Wiley, 1968.

Gardner, M. "Random Walks." Ch. 6–7 in *Mathematical Circus: More Puzzles, Games, Paradoxes, and Other Mathematical Entertainments.* Washington, DC: Math. Assoc. Amer., 1992.

Hughes, B. D. *Random Walks and Random Environments, Vol. 1: Random Walks.* New York: Oxford University Press, 1995.

Hughes, B. D. *Random Walks and Random Environments, Vol. 2: Random Environments.* New York: Oxford University Press, 1996.

Lawler, G. F. *Intersections of Random Walks.* Boston, MA: Birkhäuser, 1996.

Spitzer, F. *Principles of Random Walk, 2nd ed.* New York: Springer-Verlag, 1976.

Random Walk—1-D

Let N steps of equal length be taken along a LINE. Let p be the probability of taking a step to the right, q the probability of taking a step to the left, n_1 the number of steps taken to the right, and n_2 the number of steps taken to the left. The quantities p, q, n_1, n_2, and N are related by

$$p+q=1 \tag{1}$$

and

$$n_1+n_2=N. \tag{2}$$

Now examine the probability of taking exactly n_1 steps out of N to the right. There are $\binom{N}{n_1}=\binom{n_1+n_2}{n_1}$ ways of taking n_1 steps to the right and n_2 to the left, where $\binom{n}{m}$ is a BINOMIAL COEFFICIENT. The probability of taking a particular ordered sequence of n_1 and n_2 steps is $p^{n_1}q^{n_2}$. Therefore,

$$P(n_1)=\frac{(n_1+n_2)!}{n_1!n_2!}p^{n_1}q^{n_2}=\frac{N!}{n_1!(N-n_1)!}p^{n_1}q^{N-n_1}, \tag{3}$$

where $n!$ is a FACTORIAL. This is a BINOMIAL DISTRIBUTION and satisfies

$$\sum_{n_1=0}^{N} P(n_1)=(p+q)^N=1^N=1. \tag{4}$$

The MEAN number of steps n_1 to the right is then

$$\langle n_1\rangle \equiv \sum_{n_1=0}^{N} n_1P(n_1)=\sum_{n_1=0}^{N}\frac{N!}{n_1!(N-n_1)!}p^{n_1}q^{N-n_1}n_1, \tag{5}$$

but

$$n_1p^{n_1}=p\frac{\partial}{\partial p}p^{n_1}, \tag{6}$$

so

$$\begin{aligned}\langle n_1\rangle &= \sum_{n_1=0}^{N}\frac{N!}{n_1!(N-n_1)!}\left(p\frac{\partial}{\partial p}p^{n_1}\right)q^{N-n_1}\\ &= p\frac{\partial}{\partial p}\sum_{n_1=0}^{N}\frac{N!}{n_1!(N-n_1)!}p^{n_1}q^{N-n_1}\\ &= p\frac{\partial}{\partial p}(p+q)^N=pN(p+q)^{N-1}=pN.\end{aligned} \tag{7}$$

From the BINOMIAL THEOREM,

$$\langle n_2\rangle = N-\langle n_1\rangle = N(1-p)=qN. \tag{8}$$

The VARIANCE is given by

$$\sigma_{n_1}{}^2=\left\langle {n_1}^2\right\rangle-\langle n_1\rangle^2. \tag{9}$$

But

$$\left\langle {n_1}^2\right\rangle=\sum_{n_1=0}^{N}\frac{N!}{n_1!(N-n_1)!}p^{n_1}q^{N-n_1}{n_1}^2, \tag{10}$$

so

$$\begin{aligned}{n_1}^2p^{n_1} &= n_1\left(p\frac{\partial}{\partial p}\right)p^{n_1}=\left(p\frac{\partial}{\partial p}\right)^2p^{n_1}\\ &= \sum_{n_1=0}^{N}\frac{N!}{n_1!(N-n_1)!}\left(p\frac{\partial}{\partial p}\right)^2p^{n_1}q^{N-n_1}\\ &= \left(p\frac{\partial}{\partial p}\right)^2\sum_{n_1=0}^{N}\frac{N!}{n_1!(N-n_1)!}p^{n_1}q^{N-n_1}\\ &= \left(p\frac{\partial}{\partial p}\right)^2(p+q)^N=\frac{\partial}{\partial p}[pN(p+q)N-1]\\ &= p[N(p+q)^{N-1}+pN(N-1)(p+q)^{N-2}]\\ &= p[N+pN(N-1)]\\ &= pN[1+pN-p]=(Np)^2+Npq\\ &= \langle n_1\rangle^2+Npq.\end{aligned} \tag{11}$$

Therefore,

$$\sigma_{n_1}{}^2=\left\langle {n_1}^2\right\rangle-\langle n_1\rangle^2=Npq, \tag{12}$$

and the ROOT-MEAN-SQUARE deviation is

$$\sigma_{n_1}=\sqrt{Npq}. \tag{13}$$

For a large number of total steps N, the BINOMIAL DISTRIBUTION characterizing the distribution approaches a GAUSSIAN DISTRIBUTION.

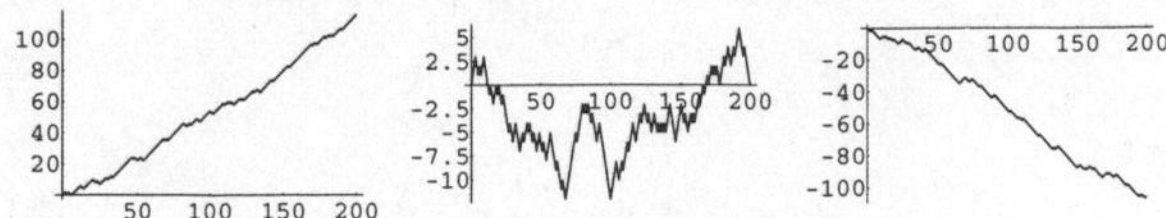

Consider now the distribution of the distances d_N traveled after a given number of steps,

$$d_N \equiv n_1 - n_2 = 2n_1 - N, \tag{14}$$

as opposed to the *number* of steps in a given direction. The above plots show $d_N(p)$ for $N = 200$ and three values $p = 0.1$, $p = 0.5$, and $p = 0.9$, respectively. Clearly, weighting the steps toward one direction or the other influences the overall trend, but there is still a great deal of random scatter, as emphasized by the plot below, which shows three random walks all with $p = 0.5$.

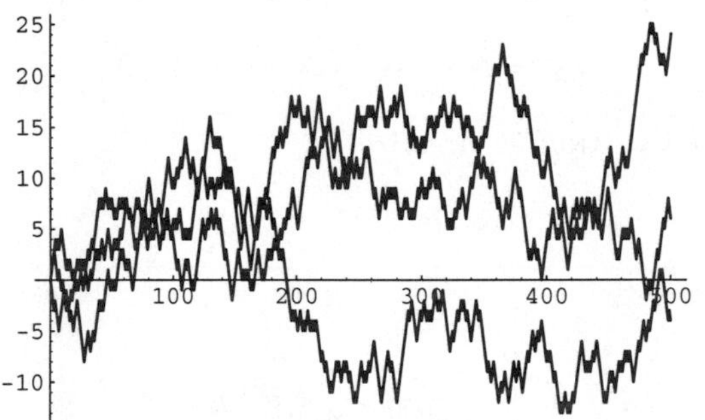

Surprisingly, the most probable number of sign changes in a walk is 0, followed by 1, then 2, etc.

For a random walk with $p = 1/2$, the probability $P_N(d)$ of traveling a given distance d after N steps is given in the following table.

steps	−5	−4	−3	−2	−1	0	1	2	3	4	5
0						1					
1					$\frac{1}{2}$	0	$\frac{1}{2}$				
2				$\frac{1}{4}$	0	$\frac{2}{4}$	0	$\frac{1}{4}$			
3			$\frac{1}{8}$	0	$\frac{3}{8}$	0	$\frac{3}{8}$	0	$\frac{1}{8}$		
4		$\frac{1}{16}$	0	$\frac{4}{16}$	0	$\frac{6}{16}$	0	$\frac{4}{16}$	0	$\frac{1}{16}$	
5	$\frac{1}{32}$	0	$\frac{5}{32}$	0	$\frac{10}{32}$	0	$\frac{10}{32}$	0	$\frac{5}{32}$	0	$\frac{1}{32}$

In this table, subsequent rows are found by adding HALF of each cell in a given row to each of the two cells diagonally below it. In fact, it is simply PASCAL'S TRIANGLE padded with intervening zeros and with each row multiplied by an additional factor of 1/2. The COEFFICIENTS in this triangle are given by

$$P_N(d) = \frac{1}{2^N}\binom{N}{\frac{d+N}{2}}. \tag{15}$$

The expectation value of the distance after N steps is therefore

$$\begin{aligned}\langle d_N\rangle &= \sum_{d=-N,-(N-2),\ldots}^{N} |d|\, P_N(d)\\ &= \frac{1}{2^N}\sum_{d=-N,-(N-2),\ldots}^{N} \frac{|d|\, N!}{\left(\frac{N+d}{2}\right)!\left(\frac{N-d}{2}\right)!}. \end{aligned}\tag{16}$$

This sum can be done symbolically by separately considering the cases N EVEN and N ODD. First, consider EVEN N so that $N \equiv 2J$. Then

$$\begin{aligned}\langle d_{2J}\rangle &= \frac{N!}{2^N}\left[\sum_{\substack{d=-2J,\\ -2(J-1),\ldots}}^{-2} \frac{|d|}{\left(\frac{2J+d}{2}\right)!\left(\frac{2J-d}{2}\right)!}\right.\\ &\quad \left. + \sum_{d=0} \frac{|d|}{\left(\frac{2J+d}{2}\right)!\left(\frac{2J-d}{2}\right)!} + \sum_{d=2,4,\ldots}^{2J} \frac{|d|}{\left(\frac{2J+d}{2}\right)!\left(\frac{2J-d}{2}\right)!}\right]\\ &= \frac{N!}{2^N}\left[\sum_{d=-J,-(J-1),\ldots}^{-1} \frac{|2d|}{\left(\frac{2J+2d}{2}\right)!\left(\frac{2J-2d}{2}\right)!}\right.\\ &\quad \left. + \sum_{d=1,2,\ldots}^{J} \frac{|2d|}{\left(\frac{2J+2d}{2}\right)!\left(\frac{2J-2d}{2}\right)!}\right]\\ &= \frac{N!}{2^N}\left[2\sum_{d=1}^{J}\frac{2d}{(J+d)!(J-d)!}\right]\\ &= \frac{N!}{2^{N-2}}\sum_{d=1}^{J}\frac{d}{(J+d)!(J-d)!}. \end{aligned}\tag{17}$$

But this sum can be evaluated analytically as

$$\sum_{d=1}^{J}\frac{d}{(J+d)!(J-d)!} = \frac{J}{2\Gamma^2(1+J)}, \tag{18}$$

which, when combined with $N = 2J$ and plugged back in, gives

$$\langle d_{2J}\rangle = \frac{\Gamma(2J+1)J}{2^{2J-1}\Gamma^2(1+J)} = \frac{\Gamma(2J)}{2^{2J-2}\Gamma^2(J)}. \tag{19}$$

But the LEGENDRE DUPLICATION FORMULA gives

$$\Gamma(2J) = \frac{2^{2J-1/2}\Gamma(J)\Gamma(J+\frac{1}{2})}{\sqrt{2\pi}}, \tag{20}$$

so

$$\langle d_{2J}\rangle = \frac{\frac{1}{\sqrt{2\pi}}2^{2J-1/2}\Gamma(J)\Gamma(J+\frac{1}{2})}{2^{2J-2}\Gamma^2(J)} = \frac{2}{\sqrt{\pi}}\frac{\Gamma(J+\frac{1}{2})}{\Gamma(J)}. \tag{21}$$

Now consider N ODD, so $N \equiv 2J-1$. Then

$$\begin{aligned}\langle d_{2J-1}\rangle &= \frac{N!}{2^N}\left[\sum_{\substack{d=-(2J-1),\\-(2J+1),\ldots}}^{-1} \frac{|d|}{\left(\frac{2J-1+d}{2}\right)!\left(\frac{2J-1-d}{2}\right)!}\right.\\
&\quad\left.+\sum_{d=1,3,\ldots}^{2J-1} \frac{|d|}{\left(\frac{2J-1+d}{2}\right)!\left(\frac{2J-1-d}{2}\right)!}\right]\\
&= \frac{N!}{2^{N-1}}\left[\sum_{d=1,3,\ldots}^{2J-1} \frac{d}{\left(\frac{2J-1+d}{2}\right)!\left(\frac{2J-1-d}{2}\right)!}\right]\\
&= \frac{N!}{2^{N-1}}\left[\sum_{d=2,4,\ldots}^{2J} \frac{d-1}{\left(\frac{2J-2+d}{2}\right)!\left(\frac{2J-d}{2}\right)!}\right]\\
&= \frac{\Gamma(2J)}{2^{2J-2}}\left[\sum_{d=1}^{J} \frac{2d-1}{(J+d-1)!(J-d)!}\right]\\
&= \Gamma(2J)\left[\frac{1+J-{}_2F_1(1,-J;J;1)}{2^{2J-2}\Gamma(J)\Gamma(1+J)}+\frac{1}{\Gamma(2J)}\right]\\
&= \frac{\frac{2^{2J-1/2}}{\sqrt{2\pi}}\Gamma(J)\Gamma(J+1/2)}{2^{2J-2}\Gamma^2(J)J}[1+J-{}_2F_1(1,-J;J;-1)]+1\\
&= \frac{2}{\sqrt{\pi}}\frac{\Gamma(J+\frac{1}{2})}{J\Gamma(J)}[1+J-{}_2F_1(1,-J;J;-1)]+1. \qquad (22)\end{aligned}$$

But the HYPERGEOMETRIC FUNCTION ${}_2F_1$ has the special value

$${}_2F_1(1,-J;J;-1) = \frac{\sqrt{\pi}}{2}\frac{J\Gamma(J)}{\Gamma(J+\frac{1}{2})}+1, \qquad (23)$$

so

$$\langle d_{2J-1}\rangle = \frac{2}{\sqrt{\pi}}\frac{\Gamma(J+\frac{1}{2})}{\Gamma(J)}. \qquad (24)$$

Summarizing the EVEN and ODD solutions,

$$\langle d_N\rangle = \frac{2}{\sqrt{\pi}}\frac{\Gamma(J+\frac{1}{2})}{\Gamma(J)}, \qquad (25)$$

where

$$\begin{cases} J = \frac{1}{2}N & \text{for } N \text{ even}\\ J = \frac{1}{2}(N+1) & \text{for } N \text{ odd.}\end{cases} \qquad (26)$$

Written explicitly in terms of N,

$$\langle d_N\rangle = \begin{cases} \frac{2}{\sqrt{\pi}}\frac{\Gamma(\frac{1}{2}N+\frac{1}{2})}{\Gamma(\frac{1}{2}N)} & \text{for } N \text{ even}\\ \frac{2}{\sqrt{\pi}}\frac{\Gamma(\frac{1}{2}N+1)}{\Gamma(\frac{1}{2}N+\frac{1}{2})} & \text{for } N \text{ odd.}\end{cases} \qquad (27)$$

The first few values of $\langle d_N\rangle$ are then

$$\begin{aligned}\langle d_0\rangle &= 0\\ \langle d_1\rangle = \langle d_2\rangle &= 1\\ \langle d_3\rangle = \langle d_4\rangle &= \tfrac{3}{2}\\ \langle d_5\rangle = \langle d_6\rangle &= \tfrac{15}{8}\\ \langle d_7\rangle = \langle d_8\rangle &= \tfrac{35}{16}\\ \langle d_9\rangle = \langle d_{10}\rangle &= \tfrac{315}{128}\\ \langle d_{11}\rangle = \langle d_{12}\rangle &= \tfrac{693}{256}\\ \langle d_{13}\rangle = \langle d_{14}\rangle &= \tfrac{3003}{1024}.\end{aligned}$$

Now, examine the asymptotic behavior of $\langle d_N\rangle$. The asymptotic expansion of the GAMMA FUNCTION ratio is

$$\frac{\Gamma(J+\frac{1}{2})}{\Gamma(J)} = \sqrt{J}\left(1-\frac{1}{8J}+\frac{1}{128J^2}+\ldots\right) \qquad (28)$$

(Graham *et al.* 1994), so plugging in the expression for $\langle d_N\rangle$ gives the asymptotic series

$$\langle d_N\rangle = \sqrt{\frac{2N}{\pi}}\left(1\mp\frac{1}{4N}+\frac{1}{32N^2}\pm\frac{5}{128N^3}-\frac{21}{2048N^4}\mp\ldots\right), \qquad (29)$$

where the top signs are taken for N EVEN and the bottom signs for N ODD. Therefore, for large N,

$$\langle d_N\rangle \sim \sqrt{\frac{2N}{\pi}}, \qquad (30)$$

which is also shown in Mosteller *et al.* (1961, p. 14).

see also BINOMIAL DISTRIBUTION, CATALAN NUMBER, p-GOOD PATH, PÓLYA'S RANDOM WALK CONSTANTS, RANDOM WALK—2-D, RANDOM WALK—3-D, SELF-AVOIDING WALK

References

Chandrasekhar, S. "Stochastic Problems in Physics and Astronomy." *Rev. Modern Phys.* **15**, 1–89, 1943. Reprinted in *Noise and Stochastic Processes* (Ed. N. Wax). New York: Dover, pp. 3–91, 1954.

Feller, W. Ch. 3 in *An Introduction to Probability Theory and Its Applications, Vol. 1, 3rd ed., rev. printing.* New York: Wiley, 1968.

Gardner, M. Chs. 6–7 in *Mathematical Carnival: A New Round-Up of Tantalizers and Puzzles from Scientific American.* New York: Vintage Books, 1977.

Graham, R. L.; Knuth, D. E.; and Patashnik, O. Answer to problem 9.60 in *Concrete Mathematics: A Foundation for Computer Science, 2nd ed.* Reading, MA: Addison-Wesley, 1994.

Hersh, R. and Griego, R. J. "Brownian Motion and Potential Theory." *Sci. Amer.* **220**, 67–74, 1969.

Kac, M. "Random Walk and the Theory of Brownian Motion." *Amer. Math. Monthly* **54**, 369–391, 1947. Reprinted in *Noise and Stochastic Processes* (Ed. N. Wax). New York: Dover, pp. 295–317, 1954.

Mosteller, F.; Rourke, R. E. K.; and Thomas, G. B. *Probability and Statistics.* Reading, MA: Addison-Wesley, 1961.

Random Walk—2-D

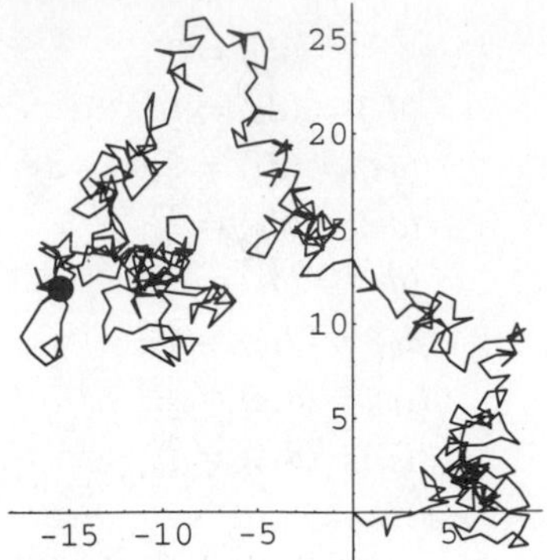

In a PLANE, consider a sum of N 2-D VECTORS with random orientations. Use PHASOR notation, and let the phase of each VECTOR be RANDOM. Assume N unit steps are taken in an arbitrary direction (i.e., with the angle θ uniformly distributed in $[0, 2\pi)$ and *not* on a LATTICE), as illustrated above. The position z in the COMPLEX PLANE after N steps is then given by

$$z = \sum_{j=1}^{N} e^{i\theta_j}, \tag{1}$$

which has ABSOLUTE SQUARE

$$|z|^2 = \sum_{j=1}^{N} e^{i\theta_j} \sum_{k=1}^{N} e^{-i\theta_k} = \sum_{j=1}^{N}\sum_{k=1}^{N} e^{i(\theta_j-\theta_k)}$$

$$= N + \sum_{\substack{j,k=1\\k\neq j}}^{N} e^{i(\theta_j-\theta_k)}. \tag{2}$$

Therefore,

$$\langle |z|^2 \rangle = N + \left\langle \sum_{\substack{j,k=1\\k\neq j}}^{N} e^{i(\theta_j-\theta_k)} \right\rangle. \tag{3}$$

Each step is likely to be in any direction, so both θ_j and θ_k are RANDOM VARIABLES with identical MEANS of zero, and their difference is also a random variable. Averaging over this distribution, which has equally likely POSITIVE and NEGATIVE values yields an expectation value of 0, so

$$\langle |z|^2 \rangle = N. \tag{4}$$

The root-mean-square distance after N unit steps is therefore

$$|z|_{\mathrm{rms}} = \sqrt{N}, \tag{5}$$

so with a step size of l, this becomes

$$d_{\mathrm{rms}} = l\sqrt{N}. \tag{6}$$

In order to travel a distance d

$$N \approx \left(\frac{d}{l}\right)^2 \tag{7}$$

steps are therefore required.

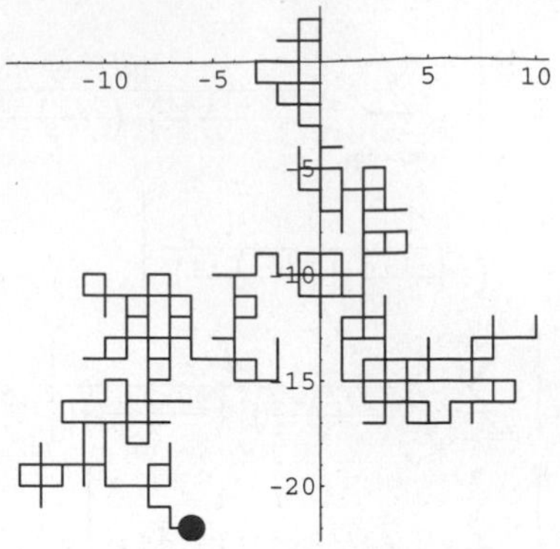

Amazingly, it has been proven that on a 2-D LATTICE, a random walk has unity probability of reaching any point (including the starting point) as the number of steps approaches INFINITY.

see also PÓLYA'S RANDOM WALK CONSTANTS, RANDOM WALK—1-D, RANDOM WALK—3-D

Random Walk—3-D

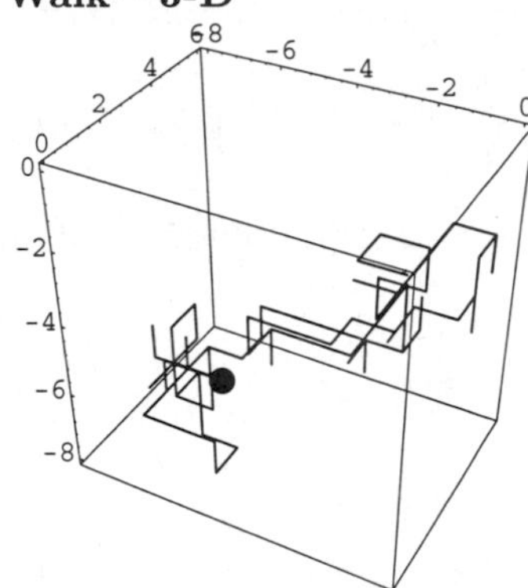

On a 3-D LATTICE, a random walk has *less than* unity probability of reaching any point (including the starting point) as the number of steps approaches infinity. The probability of reaching the starting point again is 0.3405373296.... This is one of PÓLYA'S RANDOM WALK CONSTANTS.

see also PÓLYA'S RANDOM WALK CONSTANTS, RANDOM WALK—1-D, RANDOM WALK—2-D

Range (Image)

If T is MAP over a DOMAIN D, then the range of T is defined as

$$\mathrm{Range}(T) = T(D) = \{T(\mathbf{X}) : \mathbf{X} \in D\}.$$

The range $T(D)$ is also called the IMAGE of D under T.

see also DOMAIN, MAP

Range (Line Segment)

The set of all points on a LINE SEGMENT, also called a PENCIL.

see also PERSPECTIVITY, SECTION (PENCIL)

References

Woods, F. S. *Higher Geometry: An Introduction to Advanced Methods in Analytic Geometry.* New York: Dover, p. 8, 1961.

Range (Statistics)

$$R \equiv \max(x_i) - \min(x_i). \quad (1)$$

For small samples, the range is a good estimator of the population STANDARD DEVIATION (Kenney and Keeping 1962, pp. 213–214). For a continuous UNIFORM DISTRIBUTION

$$P(x) = \begin{cases} \frac{1}{C} & \text{for } 0 < x < C \\ 0 & \text{for } |x| < C, \end{cases} \quad (2)$$

the distribution of the range is given by

$$D(R) = N\left(\frac{R}{C}\right)^{N-1} - (N-1)\left(\frac{R}{C}\right)^{N}. \quad (3)$$

Given two samples with sizes m and n and ranges R_1 and R_2, let $u \equiv R_1/R_2$. Then

$$D(u) = \begin{cases} \frac{m(m-1)n(n-1)}{(m+n)(m+n-1)(m+n-2)} \\ \times[(m+n)u^{m-2} - (m+n-2)u^{m-1}] \\ \qquad \text{for } 0 \leq u \leq 1 \\ \frac{m(m-1)n(n-1)}{(m+n)(m+n-1)(m+n-2)} \\ \times[(m+n)u^{-n} - (m+n-2)u^{-n-1}] \\ \qquad \text{for } 1 \leq u < \infty. \end{cases} \quad (4)$$

The MEAN is

$$\mu_u = \frac{(m-1)n}{(m+1)(n-2)}, \quad (5)$$

and the MODE is

$$\hat{u} = \begin{cases} \frac{(m-2)(m+n)}{(m-1)(m+n-2)} & \text{for } m-n \leq 2 \\ \frac{(n+1)(m+n-2)}{n(m+n)} & \text{for } m-n \geq 2. \end{cases} \quad (6)$$

References

Kenney, J. F. and Keeping, E. S. *Mathematics of Statistics, Pt. 1, 3rd ed.* Princeton, NJ: Van Nostrand, pp. 213–214, 1962.

Rank

In a total generality, the "rank" of a mathematical object is defined whenever that object is FREE. In general, the rank of a FREE object is the CARDINALITY of the FREE generating SUBSET G. The word "rank" also refers to several unrelated concepts in mathematics involving groups, quadratic forms, sequences, statistics, and tensors.

see also RANK (GROUP), RANK (QUADRATIC FORM), RANK (SEQUENCE), RANK (STATISTICS), RANK (TENSOR)

Rank (Group)

For an arbitrary finitely generated ABELIAN GROUP G, the rank of G is defined to be the rank of the FREE generating SUBSET G modulo its TORSION SUBGROUP. For a finitely generated GROUP, the rank is defined to be the rank of its "Abelianization."

see also ABELIAN GROUP, BETTI NUMBER, BURNSIDE PROBLEM, QUASITHIN THEOREM, QUASI-UNIPOTENT GROUP, TORSION (GROUP THEORY)

Rank (Quadratic Form)

For a QUADRATIC FORM Q in the canonical form

$$Q = {y_1}^2 + {y_2}^2 + \ldots + {y_p}^2 - {y_{p+1}}^2 - {y_{p+2}}^2 - \ldots - {y_r}^2,$$

the rank is the total number r of square terms (both POSITIVE and NEGATIVE).

see also SIGNATURE (QUADRATIC FORM)

References

Gradshteyn, I. S. and Ryzhik, I. M. *Tables of Integrals, Series, and Products, 5th ed.* San Diego, CA: Academic Press, p. 1105, 1979.

Rank (Sequence)

The position of a RATIONAL NUMBER in the SEQUENCE 1, $\frac{1}{2}$, 2, $\frac{1}{3}$, 3, $\frac{1}{4}$, $\frac{2}{3}$, $\frac{3}{2}$, 4, $\frac{1}{5}$, ..., ordered in terms of increasing NUMERATOR+DENOMINATOR.

see also ENCODING, FAREY SERIES

Rank (Statistics)

The ORDINAL NUMBER of a value in a list arranged in a specified order (usually decreasing).

see also SPEARMAN RANK CORRELATION, WILCOXON RANK SUM TEST, WILCOXON SIGNED RANK TEST, ZIPF'S LAW

Rank (Tensor)

The total number of CONTRAVARIANT and COVARIANT indices of a TENSOR. The rank of a TENSOR is independent of the number of DIMENSIONS of the SPACE.

Rank	Object
0	scalar
1	vector
≥ 2	tensor

see also CONTRAVARIANT TENSOR, COVARIANT TENSOR, SCALAR, TENSOR, VECTOR

Ranunculoid

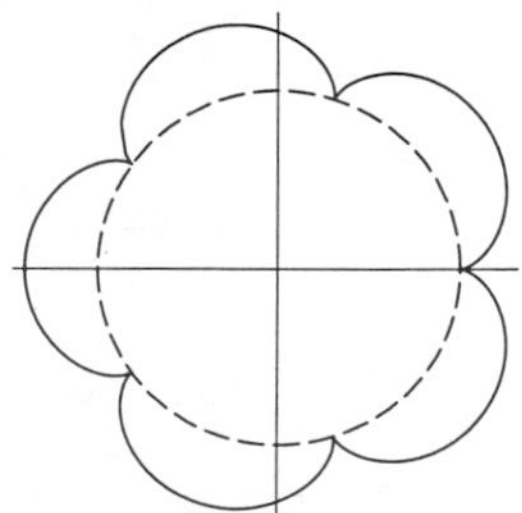

An EPICYCLOID with $n = 5$ cusps, named after the buttercup genus *Ranunculus* (Madachy 1979).

see also EPICYCLOID.

References

Madachy, J. S. *Madachy's Mathematical Recreations.* New York: Dover, p. 223, 1979.

Pickover, C. A. *Keys to Infinity.* New York: Wiley, pp. 79–80, 1995.

RAT-Free Set

A RAT-free set is a set of points, no three of which determine a RIGHT TRIANGLE. Let $f(n)$ be the smallest RAT-free subset guaranteed to be contained in a planar set of n points, then the function $f(n)$ is bounded by

$$\sqrt{n} \leq f(n) \leq 2\sqrt{n}.$$

References

Abbott, H. L. "On a Conjecture of Erdős and Silverman in Combinatorial Geometry." *J. Combin. Th. A* **29**, 380–381, 1980.

Chan, W. K. "On the Largest RAT-FREE Subset of a Finite Set of Points." *Pi Mu Epsilon*, Spring 1987.

Honsberger, R. *More Mathematical Morsels.* Washington, DC: Math. Assoc. Amer., pp. 250–251, 1991.

Seidenberg, A. "A Simple Proof of a Theorem of Erdős and Szekeres." *J. London Math. Soc.* **34**, 352, 1959.

Ratio

The ratio of two numbers r and s is written r/s, where r is the NUMERATOR and s is the DENOMINATOR. The ratio of r to s is equivalent to the QUOTIENT r/s. Betting ODDS written as $r : s$ correspond to $s/(r+s)$. A number which can be expressed as a ratio of INTEGERS is called a RATIONAL NUMBER.

see also DENOMINATOR, FRACTION, NUMERATOR, ODDS, QUOTIENT, RATIONAL NUMBER

Ratio Distribution

Given two distributions Y and X with joint probability density function $f(x,y)$, let $U = Y/X$ be the ratio distribution. Then the distribution function of u is

$$\begin{aligned} D(u) &= P(U \leq u) \\ &= P(Y \leq uX | X > 0) + P(Y \geq uX | X < 0) \\ &= \int_0^\infty \int_0^{ux} f(x,y)\,dy\,dx + \int_{-\infty}^0 \int_{ux}^0 f(x,y)\,dy\,dx. \end{aligned} \tag{1}$$

The probability function is then

$$\begin{aligned} P(u) = D'(u) &= \int_0^\infty x f(x,ux)\,dx - \int_{-\infty}^0 x f(x,ux)\,dx \\ &= \int_{-\infty}^\infty |x| f(x,ux)\,dx. \end{aligned} \tag{2}$$

For variates with a standard NORMAL DISTRIBUTION, the ratio distribution is a CAUCHY DISTRIBUTION. For a UNIFORM DISTRIBUTION

$$f(x,y) = \begin{cases} 1 & \text{for } x, y \in [0,1] \\ 0 & \text{otherwise,} \end{cases} \tag{3}$$

$$P(u) = \begin{cases} 0 & u < 0 \\ \int_0^1 x\,dx = [\tfrac{1}{2}x^2] = \tfrac{1}{2} & \text{for } 0 \leq u \leq 1 \\ \int_0^{1/u} x\,dx = [\tfrac{1}{2}x^2]_0^{1/u} = \tfrac{1}{2u^2} & \text{for } u > 1. \end{cases} \tag{4}$$

see also CAUCHY DISTRIBUTION

Ratio Test

Let u_k be a SERIES with POSITIVE terms and suppose

$$\rho \equiv \lim_{k\to\infty} \frac{u_{k+1}}{u_k}.$$

Then

1. If $\rho < 1$, the SERIES CONVERGES.
2. If $\rho > 1$ or $\rho = \infty$, the SERIES DIVERGES.
3. If $\rho = 1$, the SERIES may CONVERGE or DIVERGE.

The test is also called the CAUCHY RATIO TEST or D'ALEMBERT RATIO TEST.

see also CONVERGENCE TESTS

References

Arfken, G. *Mathematical Methods for Physicists, 3rd ed.* Orlando, FL: Academic Press, pp. 282–283, 1985.

Bromwich, T. J. I'a. and MacRobert, T. M. *An Introduction to the Theory of Infinite Series, 3rd ed.* New York: Chelsea, p. 28, 1991.

Rational Approximation

If r is any number and n is any INTEGER, then there is a RATIONAL NUMBER m/n for which

$$0 \leq r - \frac{m}{n} < \frac{1}{n}. \tag{1}$$

If r is IRRATIONAL and k is any WHOLE NUMBER, there is a FRACTION m/n with $n \leq k$ and for which

$$0 \leq r - \frac{m}{n} < \frac{1}{nk}. \tag{2}$$

Furthermore, there are an infinite number of FRACTIONS m/n for which

$$0 \leq r - \frac{m}{n} < \frac{1}{n^2}. \tag{3}$$

Hurwitz has shown that for an IRRATIONAL NUMBER ζ

$$\left|\zeta - \frac{h}{k}\right| < \frac{1}{ck^2}, \tag{4}$$

there are infinitely RATIONAL NUMBERS h/k if $0 < c \leq \sqrt{5}$, but if $c > \sqrt{5}$, there are some ζ for which this approximation holds for only finitely many h/k.

Rational Canonical Form

There is an invertible matrix Q such that

$$\mathsf{Q}^{-1}\mathsf{T}\mathsf{Q} = \text{diag}[L(\psi_1), L(\psi_2), \ldots, L(\psi_s)],$$

where $L(f)$ is the companion MATRIX for any MONIC POLYNOMIAL

$$f(\lambda) = f_0 + f_1\lambda + \ldots + f_n\lambda^n$$

with $f_n = 1$. The POLYNOMIALS ψ_i are called the "invariant factors" of T, and satisfy $\psi_{i+1}|\psi_i$ for $i = s - 1$, ..., 1 (Hartwig 1996).

References

Gantmacher, F. R. *The Theory of Matrices, Vol. 1.* New York: Chelsea, 1960.

Hartwig, R. E. "Roth's Removal Rule and the Rational Canonical Form." *Amer. Math. Monthly* **103**, 332–335, 1996.

Herstein, I. N. *Topics in Algebra, 2nd ed.* New York: Springer-Verlag, p. 162, 1975.

Hoffman, K. and Kunze, K. *Linear Algebra, 3rd ed.* Englewood Cliffs, NJ: Prentice-Hall, 1996.

Lancaster, P. and Tismenetsky, M. *The Theory of Matrices, 2nd ed.* New York: Academic Press, 1985.

Turnbull, H. W. and Aitken, A. C. *An Introduction to the Theory of Canonical Matrices, 2nd impression.* New York: Blackie and Sons, 1945.

Rational Cuboid

see EULER BRICK

Rational Distances

It is possible to find six points in the PLANE, no three on a LINE and no four on a CIRCLE (i.e., none of which are COLLINEAR or CONCYCLIC), such that all the mutual distances are RATIONAL. An example is illustrated by Guy (1994, p. 185).

It is not known if a TRIANGLE with INTEGER sides, MEDIANS, and AREA exists (although there are incorrect PROOFS of the impossibility in the literature). However, R. L. Rathbun, A. Kemnitz, and R. H. Buchholz have showed that there are infinitely many triangles with RATIONAL sides (HERONIAN TRIANGLES) with *two* RATIONAL MEDIANS (Guy 1994, p. 188).

see also COLLINEAR, CONCYCLIC, CYCLIC QUADRILATERAL, EQUILATERAL TRIANGLE, EULER BRICK, HERONIAN TRIANGLE, RATIONAL QUADRILATERAL, RATIONAL TRIANGLE, SQUARE, TRIANGLE

References

Guy, R. K. "Six General Points at Rational Distances" and "Triangles with Integer Sides, Medians, and Area." §D20 and D21 in *Unsolved Problems in Number Theory, 2nd ed.* New York: Springer-Verlag, pp. 185–190 and 188–190, 1994.

Rational Domain

see FIELD

Rational Double Point

There are nine possible types of ISOLATED SINGULARITIES on a CUBIC SURFACE, eight of them rational double points. Each type of ISOLATED SINGULARITY has an associated normal form and COXETER-DYNKIN DIAGRAM (A_1, A_2, A_3, A_4, A_5, D_4, D_5, E_6 and $\tilde{E}_6$).

The eight types of rational double points (the $\tilde{E}_6$ type being the one excluded) can occur in only 20 combinations on a CUBIC SURFACE (of which Fischer 1986 gives 19): A_1, $2A_1$, $3A_1$, $4A_1$, A_2, (A_2, A_1), $2A_2$, $(2A_2, A_1)$, $3A_2$, A_3, (A_3, A_1), $(A_3, 2A_1)$, A_4, (A_4, A_1), A_5, (A_5, A_1), D_4, D_5, and E_6 (Looijenga 1978, Bruce and Wall 1979, Fischer 1986).

In particular, on a CUBIC SURFACE, precisely those configurations of rational double points occur for which the disjoint union of the COXETER-DYNKIN DIAGRAM is a SUBGRAPH of the COXETER-DYNKIN DIAGRAM $\tilde{E}_6$. Also, a surface specializes to a more complicated one precisely when its graph is contained in the graph of the other one (Fischer 1986).

see also COXETER-DYNKIN DIAGRAM, CUBIC SURFACE, ISOLATED SINGULARITY

References

Bruce, J. and Wall, C. T. C. "On the Classification of Cubic Surfaces." *J. London Math. Soc.* **19**, 245–256, 1979.

Fischer, G. (Ed.). *Mathematical Models from the Collections of Universities and Museums.* Braunschweig, Germany: Vieweg, p. 13, 1986.

Fischer, G. (Ed.). Plates 14–31 in *Mathematische Modelle/Mathematical Models, Bildband/Photograph Volume.* Braunschweig, Germany: Vieweg, pp. 17–31, 1986.

Looijenga, E. "On the Semi-Universal Deformation of a Simple Elliptic Hypersurface Singularity. Part II: The Discriminant." *Topology* **17**, 23–40, 1978.

Rodenberg, C. "Modelle von Flächen dritter Ordnung." In *Mathematische Abhandlungen aus dem Verlage Mathematischer Modelle von Martin Schilling.* Halle a. S., 1904.

Rational Function

A quotient of two polynomials $P(z)$ and $Q(z)$,

$$R(z) \equiv \frac{P(z)}{Q(z)},$$

is called a rational function. More generally, if P and Q are POLYNOMIALS in multiple variables, their quotient is a rational function.

see also ABEL'S CURVE THEOREM, CLOSED FORM, FUNDAMENTAL THEOREM OF SYMMETRIC FUNCTIONS, QUOTIENT-DIFFERENCE ALGORITHM, RATIONAL INTEGER, RATIONAL NUMBER, RIEMANN CURVE THEOREM

Rational Integer

A synonym for INTEGER. The word "rational" is sometimes used for emphasis to distinguish it from other types of "integers" such as CYCLOTOMIC INTEGERS, EISENSTEIN INTEGERS, and GAUSSIAN INTEGERS.

see also CYCLOTOMIC INTEGER, EISENSTEIN INTEGER, GAUSSIAN INTEGER, INTEGER, RATIONAL NUMBER

References
Hardy, G. H. and Wright, E. M. *An Introduction to the Theory of Numbers, 5th ed.* Oxford, England: Clarendon Press, p. 1, 1979.

Rational Number

A number that can be expressed as a FRACTION p/q where p and q are INTEGERS, is called a rational number with NUMERATOR p and DENOMINATOR q. Numbers which are not rational are called IRRATIONAL NUMBERS. Any rational number is trivially also an ALGEBRAIC NUMBER.

For a, b, and c any different rational numbers, then

$$\frac{1}{(a-b)^2}+\frac{1}{(b-c)^2}+\frac{1}{(c-a)^2}$$

is the SQUARE of a rational number (Honsberger 1991). The probability that a random rational number has an EVEN DENOMINATOR is 1/3 (Beeler *et al.* 1972, Item 54).

see also ALGEBRAIC INTEGER, ALGEBRAIC NUMBER, ANOMALOUS CANCELLATION, DENOMINATOR, DIRICHLET FUNCTION, FRACTION, INTEGER, IRRATIONAL NUMBER, NUMERATOR, QUOTIENT, TRANSCENDENTAL NUMBER

References
Beeler, M.; Gosper, R. W.; and Schroeppel, R. *HAKMEM.* Cambridge, MA: MIT Artificial Intelligence Laboratory, Memo AIM-239, Feb. 1972.
Courant, R. and Robbins, H. "The Rational Numbers." §2.1 in *What is Mathematics?: An Elementary Approach to Ideas and Methods, 2nd ed.* Oxford, England: Oxford University Press, pp. 52–58,, 1996.
Honsberger, R. *More Mathematical Morsels.* Washington, DC: Math. Assoc. Amer., pp. 52–53, 1991.

Rational Point

A K-rational point is a point (X, Y) on an ALGEBRAIC CURVE, where X and Y are in a FIELD K.

The rational point may also be a POINT AT INFINITY. For example, take the ELLIPTIC CURVE

$$Y^2 = X^3 + X + 42$$

and homogenize it by introducing a third variable Z so that each term has degree 3 as follows:

$$ZY^2 = X^3 + XZ^2 + 42Z^3.$$

Now, find the points at infinity by setting $Z = 0$, obtaining

$$0 = X^3.$$

Solving gives $X = 0$, Y equal to any value, and (by definition) $Z = 0$. Despite freedom in the choice of Y, there is only a single POINT AT INFINITY because the two triples (X_1, Y_1, Z_1), (X_2, Y_2, Z_2) are considered to be equivalent (or identified) only if one is a scalar multiple of the other. Here, (0, 0, 0) is not considered to be a valid point. The triples $(a, b, 1)$ correspond to the ordinary points (a, b), and the triples $(a, b, 0)$ correspond to the POINTS AT INFINITY, usually called the LINE AT INFINITY.

The rational points on ELLIPTIC CURVES over the GALOIS FIELD GF(q) are 5, 7, 9, 10, 13, 14, 16, ... (Sloane's A005523).

see also ELLIPTIC CURVE, LINE AT INFINITY, POINT AT INFINITY

References
Sloane, N. J. A. Sequence A005523/M3757 in "An On-Line Version of the Encyclopedia of Integer Sequences."

Rational Quadrilateral

A rational quadrilateral is a QUADRILATERAL for which the sides, DIAGONALS, and AREA are RATIONAL. The simplest case has sides $a = 52$, $b = 25$, $c = 39$, and $d = 60$ and DIAGONALS of length $p = 63$ and $q = 56$.

see also AREA, DIAGONAL (POLYGON), RATIONAL QUADRILATERAL

Rational Triangle

A rational triangle is a TRIANGLE all of whose sides are RATIONAL NUMBERS and all of whose ANGLES are RATIONAL numbers of DEGREES. The only such triangle is the EQUILATERAL TRIANGLE (Conway and Guy 1996).

see also EQUILATERAL TRIANGLE, FERMAT'S RIGHT TRIANGLE THEOREM, RIGHT TRIANGLE

References
Conway, J. H. and Guy, R. K. "The Only Rational Triangle." In *The Book of Numbers.* New York: Springer-Verlag, pp. 201 and 228–239, 1996.

RATS Sequence

A sequence produced by the instructions "reverse, add, then sort the digits," where zeros are suppressed. For example, after 668 we get

$$668 + 866 = 1534,$$

so the next term is 1345. Applied to 1, the sequence gives 1, 2, 4, 8, 16, 77, 145, 668, 1345, 6677, 13444, 55778, ... (Sloane's A004000)

see also 196-ALGORITHM, KAPREKAR ROUTINE, REVERSAL, SORT-THEN-ADD SEQUENCE

References
Sloane, N. J. A. Sequence A004000/M1137 in "An On-Line Version of the Encyclopedia of Integer Sequences."

Ray

A VECTOR $\overrightarrow{AB}$ from a point A to a point B. In GEOMETRY, a ray is usually taken as a half-infinite LINE with one of the two points A and B taken to be at INFINITY.

see also LINE, VECTOR

Rayleigh Distribution

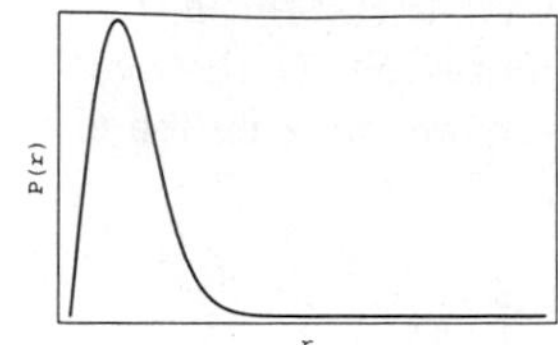

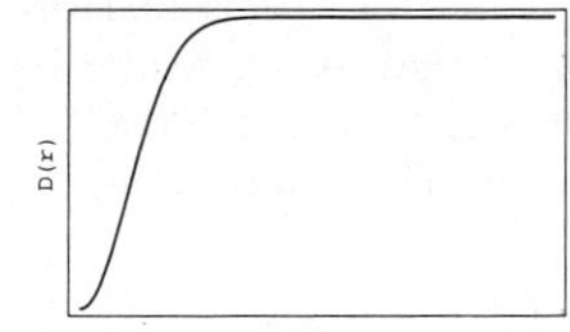

The distribution with PROBABILITY FUNCTION

$$P(r) = \frac{re^{-r^2/2s^2}}{s^2} \tag{1}$$

for $r \in [0, \infty)$. The MOMENTS about 0 are given by

$$\begin{aligned} \mu'_m &\equiv \int_0^\infty r^m P(r)\,dr = s^{-2} \int_0^\infty r^{m+1} e^{-r^2/2s^2}\,dr \\ &= s^{-2} I_{m+1}\left(\frac{1}{2s^2}\right), \end{aligned} \tag{2}$$

where $I(x)$ is a GAUSSIAN INTEGRAL. The first few of these are

$$I_1(a^{-1}) = \tfrac{1}{2}a \tag{3}$$
$$I_2(a^{-1}) = \tfrac{1}{4}a\sqrt{a\pi} \tag{4}$$
$$I_3(a^{-1}) = \tfrac{1}{2}a^2 \tag{5}$$
$$I_4(a^{-1}) = \tfrac{3}{8}a^2\sqrt{a\pi} \tag{6}$$
$$I_5(a^{-1}) = a^3, \tag{7}$$

so

$$\mu'_0 = s^{-2}\tfrac{1}{2}(2s^2) = 1 \tag{8}$$
$$\mu'_1 = s^{-2}\tfrac{1}{4}(2s^2)\sqrt{2s^2\pi} = \tfrac{1}{2}s\sqrt{2\pi} = s\sqrt{\frac{\pi}{2}} \tag{9}$$
$$\mu'_2 = s^{-2}\tfrac{1}{2}(2s^2)^2 = 2s^2 \tag{10}$$
$$\mu'_3 = s^{-2}\tfrac{3}{8}(2s^2)^2\sqrt{2s^2\pi} = \tfrac{3}{2}s^3\sqrt{2\pi} = 3s^3\sqrt{\frac{\pi}{2}} \tag{11}$$
$$\mu'_4 = s^{-2}(2s^2)^3 = 8s^4. \tag{12}$$

The MOMENTS about the MEAN are

$$\mu_2 = \mu'_2 - (\mu'_1)^2 = \frac{4-\pi}{2}s^2 \tag{13}$$
$$\mu_3 = \mu'_3 - 3\mu'_2\mu'_1 + 2(\mu'_1)^3 = \sqrt{\frac{\pi}{2}}(\pi-3)s^3 \tag{14}$$
$$\begin{aligned} \mu_4 &= \mu'_4 - 4\mu'_3\mu'_1 + 6\mu'_2(\mu'_1)^2 - 3(\mu - 1')^4 \\ &= \frac{32-3\pi^2}{4}s^4, \end{aligned} \tag{15}$$

so the MEAN, VARIANCE, SKEWNESS, and KURTOSIS are

$$\mu = \mu'_1 = s\sqrt{\frac{\pi}{2}} \tag{16}$$
$$\sigma^2 = \mu_2 = \frac{4-\pi}{2}s^2 \tag{17}$$
$$\gamma_1 = \frac{\mu_3}{\sigma^3} = \frac{2(\pi-3)\sqrt{\pi}}{(4-\pi)^{3/2}} \tag{18}$$
$$\gamma_2 = \frac{\mu_4}{\sigma^4} - 3 = \frac{2(-3\pi^2+12\pi-8)}{(\pi-4)^2}. \tag{19}$$

Rayleigh Differential Equation

$$y'' - \mu(1 - \tfrac{1}{3}y'^2)y' + y = 0,$$

where $\mu > 0$. Differentiating and setting $y = y'$ gives the VAN DER POL EQUATION.

see also VAN DER POL EQUATION

Rayleigh's Formulas

The formulas

$$j_n(z) = z^n\left(-\frac{1}{z}\frac{d}{dz}\right)^n \frac{\sin z}{z}$$
$$y_n(z) = -z^n\left(-\frac{1}{z}\frac{d}{dz}\right)^n \frac{\cos z}{z}$$

for $n = 0, 1, 2, \ldots$, where $j_n(z)$ is a SPHERICAL BESSEL FUNCTION OF THE FIRST KIND and $y_n(z)$ is a SPHERICAL BESSEL FUNCTION OF THE SECOND KIND.

References

Abramowitz, M. and Stegun, C. A. (Eds.). *Handbook of Mathematical Functions with Formulas, Graphs, and Mathematical Tables, 9th printing.* New York: Dover, p. 439, 1972.

Rayleigh-Ritz Variational Technique

A technique for computing EIGENFUNCTIONS and EIGENVALUES. It proceeds by requiring

$$J = \int_a^b [p(x){y_x}^2 - q(x)y^2]\,dx \tag{1}$$

to have a STATIONARY VALUE subject to the normalization condition

$$\int_a^b y^2 w(x)\,dx = 1 \tag{2}$$

and the boundary conditions

$$py_x y|_a^b = 0. \tag{3}$$

This leads to the STURM-LIOUVILLE EQUATION

$$\frac{d}{dx}\left(p\frac{dy}{dx}\right) + qy + \lambda wy = 0, \tag{4}$$

which gives the stationary values of

$$F[y(x)] = \frac{\int_a^b (p{y_x}^2 - qy^2)\,dx}{\int_a^b y^2 w\,dx} \tag{5}$$

as

$$F[y_n(x)] = \lambda_n, \tag{6}$$

where λ_n are the EIGENVALUES corresponding to the EIGENFUNCTION y_n.

References

Arfken, G. "Rayleigh-Ritz Variational Technique." §17.8 in *Mathematical Methods for Physicists, 3rd ed.* Orlando, FL: Academic Press, pp. 957–961, 1985.

Rayleigh's Theorem

see PARSEVAL'S THEOREM

Re-Entrant Circuit

A CYCLE in a GRAPH which terminates at the starting point.

see also CYCLE (GRAPH), EULERIAN CIRCUIT, HAMILTONIAN CYCLE

Real Analysis

That portion of mathematics dealing with functions of real variables. While this includes some portions of TOPOLOGY, it is most commonly used to distinguish that portion of CALCULUS dealing with real as opposed to COMPLEX NUMBERS.

Real Axis

see REAL LINE

Real Function

A FUNCTION whose RANGE is in the REAL NUMBERS is said to be a real function.

see also COMPLEX FUNCTION, SCALAR FUNCTION, VECTOR FUNCTION

Real Line

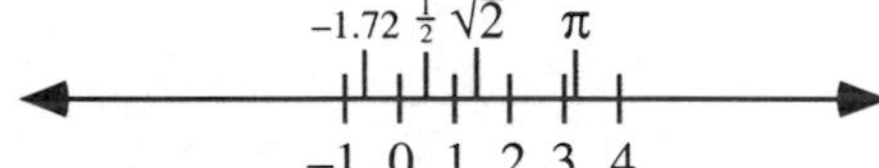

A LINE with a fixed scale so that every REAL NUMBER corresponds to a unique POINT on the LINE. The generalization of the real line to 2-D is called the COMPLEX PLANE.

see also ABSCISSA, COMPLEX PLANE

References

Courant, R. and Robbins, H. *What is Mathematics?: An Elementary Approach to Ideas and Methods, 2nd ed.* Oxford, England: Oxford University Press, p. 57, 1996.

Real Matrix

A MATRIX whose elements consist entirely of REAL NUMBERS.

Real Number

The set of all RATIONAL and IRRATIONAL numbers is called the real numbers, or simply the "reals," and denoted $\mathbb{R}$. The set of real numbers is also called the CONTINUUM, denoted C.

The real numbers can be extended with the addition of the IMAGINARY NUMBER i, equal to $\sqrt{-1}$. Numbers of the form $x + iy$, where x and y are both real, are then called COMPLEX NUMBERS. Another extension which includes both the real numbers and the infinite ORDINAL NUMBERS of Georg Cantor is the SURREAL NUMBERS.

Pick two real numbers x and y at random in $(0,1)$ with a UNIFORM DISTRIBUTION. What is the PROBABILITY $P_{\rm even}$ that $[x/y]$, where $[r]$ denotes NINT, the nearest INTEGER to r, is EVEN? The answer may be found as follows (Putnam Exam).

$$P\left(a<\frac{x}{y}<b\right)=\begin{cases}P(ay<x<by)\\ P\left(\frac{x}{b}<y<\frac{x}{a}\right)\end{cases}$$
$$=\begin{cases}\int_0^1\int_{ay}^{by}dx\,dy=\frac{1}{2}(b-a) & \text{for } 0\leq a<b<1\\ \int_0^1\int_{x/b}^{x/a}dy\,dx=\frac{1}{2a}-\frac{1}{2b} & \text{for } 1<a<b\end{cases}\tag{1}$$

$$\begin{aligned}P_{\rm even}&=P\left(0<\frac{x}{y}<\tfrac{1}{2}\right)+\sum_{n=1}^{\infty}P\left(2n-\tfrac{1}{2}<\frac{x}{y}<2n+\tfrac{1}{2}\right)\\&=\tfrac{1}{2}(\tfrac{1}{2}-0)+\sum_{n=1}^{\infty}\left[\frac{1}{2(2n-\frac{1}{2})}-\frac{1}{2(2n+\frac{1}{2})}\right]\\&=\frac{1}{4}+\sum_{n=1}^{\infty}\left(\frac{1}{4n-1}+\frac{1}{4n-1}\right)\\&=\tfrac{1}{4}+(\tfrac{1}{3}-\tfrac{1}{5}+\tfrac{1}{7}-\tfrac{1}{9}+\ldots)=\tfrac{1}{4}+(1-\tan^{-1}1)\\&=\frac{5}{4}-\frac{\pi}{4}=\tfrac{1}{4}(5-\pi)\approx 46.460\%.\end{aligned}\tag{2}$$

Plouffe's "Inverse Symbolic Calculator" includes a huge database of 54 million real numbers which are algebraically related to fundamental mathematical constants and functions.

see also COMPLEX NUMBER, CONTINUUM, i, IMAGINARY NUMBER, INTEGER RELATION, RATIONAL NUMBER, REAL PART, SURREAL NUMBER

References

Plouffe, S. "Inverse Symbolic Calculator." http://www.cecm.sfu.ca/projects/ISC/.
Plouffe, S. "Plouffe's Inverter." http://www.lacim.uqam.ca/pi/.
Putnam Exam. Problem B-3 in the 54th Putnam Exam.

Real Part

The real part $\Re$ of a COMPLEX NUMBER $z = x + iy$ is the REAL NUMBER *not* multiplying i, so $\Re[x + iy] = x$. In terms of z itself,

$$\Re[z]=\tfrac{1}{2}(z+z^*),$$

where z^* is the COMPLEX CONJUGATE of z.

see also ABSOLUTE SQUARE, COMPLEX CONJUGATE, IMAGINARY PART

References

Abramowitz, M. and Stegun, C. A. (Eds.). *Handbook of Mathematical Functions with Formulas, Graphs, and Mathematical Tables, 9th printing.* New York: Dover, p. 16, 1972.

Real Polynomial

A POLYNOMIAL having only REAL NUMBERS as COEFFICIENTS.

see also POLYNOMIAL

Real Projective Plane

The closed topological MANIFOLD, denoted $\mathbb{R}P^2$, which is obtained by projecting the points of a plane E from a fixed point P (not on the plane), with the addition of the LINE AT INFINITY, is called the real projective plane. There is then a one-to-one correspondence between points in E and lines through P. Since each line through P intersects the sphere $\mathbb{S}^2$ centered at P and tangent to E in two ANTIPODAL POINTS, $\mathbb{R}P^2$ can be described as a QUOTIENT SPACE of $\mathbb{S}^2$ by identifying any two such points. The real projective plane is a NONORIENTABLE SURFACE.

The BOY SURFACE, CROSS-CAP, and ROMAN SURFACE are all homeomorphic to the real projective plane and, because $\mathbb{R}P^2$ is nonorientable, these surfaces contain self-intersections (Kuiper 1961, Pinkall 1986).

see also BOY SURFACE, CROSS-CAP, NONORIENTABLE SURFACE, PROJECTIVE PLANE, ROMAN SURFACE

References

Geometry Center. "The Projective Plane." `http://www.geom.umn.edu/zoo/toptype/pplane/`.

Gray, A. "Realizations of the Real Projective Plane." §12.5 in *Modern Differential Geometry of Curves and Surfaces.* Boca Raton, FL: CRC Press, pp. 241–245, 1993.

Klein, F. §1.2 in *Vorlesungen über nicht-euklidische Geometrie.* Berlin, 1928.

Kuiper, N. H. "Convex Immersion of Closed Surfaces in E^3." *Comment. Math. Helv.* **35**, 85–92, 1961.

Pinkall, U. *Mathematical Models from the Collections of Universities and Museums* (Ed. G. Fischer). Braunschweig, Germany: Vieweg, pp. 64–65, 1986.

Real Quadratic Field

A QUADRATIC FIELD $\mathbb{Q}(\sqrt{D})$ with $D > 0$.

see also QUADRATIC FIELD

Realizer

A SET of R of LINEAR EXTENSIONS of a POSET $P = (X, \leq)$ is a realizer of P (and is said to realize P) provided that for all $x, y \in X$, $x \leq y$ IFF x is below y in every member of R.

see also DOMINANCE, LINEAR EXTENSION, PARTIALLY ORDERED SET, POSET DIMENSION

Rearrangement Theorem

Each row and each column in the GROUP multiplication table lists each of the GROUP elements once and only once. From this, it follows that no two elements may be in the identical location in two rows or two columns. Thus, each row and each column is a rearranged list of the GROUP elements. Stated otherwise, given a GROUP of n distinct elements $(I, a, b, c, \ldots, n)$, the set of products $(aI, a^2, ab, ac, \ldots, an)$ reproduces the n original distinct elements in a new order.

see also GROUP

Reciprocal

The reciprocal of a REAL or COMPLEX NUMBER z is its MULTIPLICATIVE INVERSE $1/z$. The reciprocal of a COMPLEX NUMBER $z = x + iy$ is given by

$$\frac{1}{x+iy} = \frac{x-iy}{x^2+y^2} = \frac{x}{x^2+y^2} - \frac{y}{x^2+y^2}i.$$

Reciprocal Difference

The reciprocal differences are closely related to the DIVIDED DIFFERENCE. The first few are explicitly given by

$$\rho(x_0, x_1) = \frac{x_0 - x_1}{f_0 - f_1} \tag{1}$$

$$\rho_2(x_0, x_1, x_2) = \frac{x_0 - x_2}{\rho(x_0, x_1) - \rho(x_1, x_2)} + f_1 \tag{2}$$

$$\rho_3(x_0, x_1, x_2, x_3) = \frac{x_0 - x_3}{\rho_2(x_0, x_1, x_2) - \rho_2(x_1, x_2, x_3)} + \rho(x_1, x_2) \tag{3}$$

$$\rho_n(x_0, x_1, \ldots, x_n) = \frac{x_0 - x_n}{\rho_{n-1}(x_0, \ldots, x_{n-1}) - \rho_{n-1}(x_1, \ldots, x_n)} + \rho_{n-2}(x_1, \ldots, x_{n-1}). \tag{4}$$

see also BACKWARD DIFFERENCE, CENTRAL DIFFERENCE, DIVIDED DIFFERENCE, FINITE DIFFERENCE, FORWARD DIFFERENCE

References

Abramowitz, M. and Stegun, C. A. (Eds.). *Handbook of Mathematical Functions with Formulas, Graphs, and Mathematical Tables, 9th printing.* New York: Dover, p. 878, 1972.

Beyer, W. H. (Ed.) *CRC Standard Mathematical Tables, 28th ed.* Boca Raton, FL: CRC Press, p. 443, 1987.

Reciprocal Polyhedron

see DUAL POLYHEDRON

Reciprocating Sphere

see MIDSPHERE

Reciprocation

An incidence-preserving transformation in which points and lines are transformed into their poles and polars. A PROJECTIVE GEOMETRY-like DUALITY PRINCIPLE holds for reciprocation.

References

Coxeter, H. S. M. and Greitzer, S. L. "Reciprocation." §6.1 in *Geometry Revisited.* Washington, DC: Math. Assoc. Amer., pp. 132–136, 1967.

Reciprocity Theorem

If there exists a RATIONAL INTEGER x such that, when n, p, and q are POSITIVE INTEGERS,

$$x^n \equiv q \pmod{p},$$

then q is the n-adic reside of p, i.e., q is an n-adic residue of p IFF $x^n \equiv q \pmod{p}$ is solvable for x.

The first case to be considered was $n = 2$ (the QUADRATIC RECIPROCITY THEOREM), of which Gauss gave the first correct proof. Gauss also solved the case $n = 3$ (CUBIC RECIPROCITY THEOREM) using INTEGERS of the form $a + b\rho$, when ρ is a root if $x^2 + x + 1 = 0$ and a, b are rational INTEGERS. Gauss stated the case $n = 4$ (QUARTIC RECIPROCITY THEOREM) using the GAUSSIAN INTEGERS.

Proof of n-adic reciprocity for PRIME n was given by Eisenstein in 1844–50 and by Kummer in 1850–61. In the 1920s, Artin formulated ARTIN'S RECIPROCITY THEOREM, a general reciprocity law for all orders.

see also ARTIN RECIPROCITY, CUBIC RECIPROCITY THEOREM, LANGLANDS RECIPROCITY, QUADRATIC RECIPROCITY THEOREM, QUARTIC RECIPROCITY THEOREM, ROOK RECIPROCITY THEOREM

Rectangle

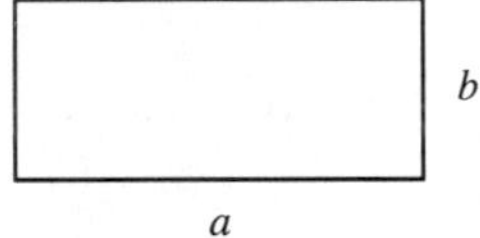

A closed planar QUADRILATERAL with opposite sides of equal lengths a and b, and with four RIGHT ANGLES. The AREA of the rectangle is

$$A = ab,$$

and its DIAGONALS are of length

$$p, q = \sqrt{a^2 + b^2}.$$

A SQUARE is a degenerate rectangle with $a = b$.

see also GOLDEN RECTANGLE, PERFECT RECTANGLE, SQUARE

References

Beyer, W. H. (Ed.) *CRC Standard Mathematical Tables, 28th ed.* Boca Raton, FL: CRC Press, p. 122, 1987.

Eppstein, D. "Rectilinear Geometry." http://www.ics.uci.edu/~eppstein/junkyard/rect.html.

Rectangle Function

The rectangle function $\Pi(x)$ is a function which is 0 outside the interval $[-1, 1]$ and unity inside it. It is also called the GATE FUNCTION, PULSE FUNCTION, or WINDOW FUNCTION, and is defined by

$$\Pi(x) \equiv \begin{cases} 0 & \text{for } |x| > \frac{1}{2} \\ \frac{1}{2} & \text{for } |x| = \frac{1}{2} \\ 1 & \text{for } |x| < \frac{1}{2}. \end{cases} \qquad (1)$$

The function $f(x) = h\Pi((x - c)/b)$ has height h, center c, and full-width b. Identities satisfied by the rectangle function include

$$\Pi(x) = H(x + \tfrac{1}{2}) - H(x - \tfrac{1}{2}) \qquad (2)$$
$$= H(\tfrac{1}{2} + x) + H(\tfrac{1}{2} - x) - 1 \qquad (3)$$
$$= H(\tfrac{1}{4} - x^2) \qquad (4)$$
$$= \tfrac{1}{2}[\operatorname{sgn}(x + \tfrac{1}{2}) - \operatorname{sgn}(x - \tfrac{1}{2})], \qquad (5)$$

where $H(x)$ is the HEAVISIDE STEP FUNCTION. The FOURIER TRANSFORM of the rectangle function is given by

$$\mathcal{F}[\Pi(x)] = \int_{-\infty}^{\infty} e^{-2\pi ikx} \Pi(x)\, dx = \operatorname{sinc}(\pi k), \qquad (6)$$

where $\operatorname{sinc}(x)$ is the SINC FUNCTION.

see also FOURIER TRANSFORM—RECTANGLE FUNCTION, HEAVISIDE STEP FUNCTION, RAMP FUNCTION

Rectangle Squaring

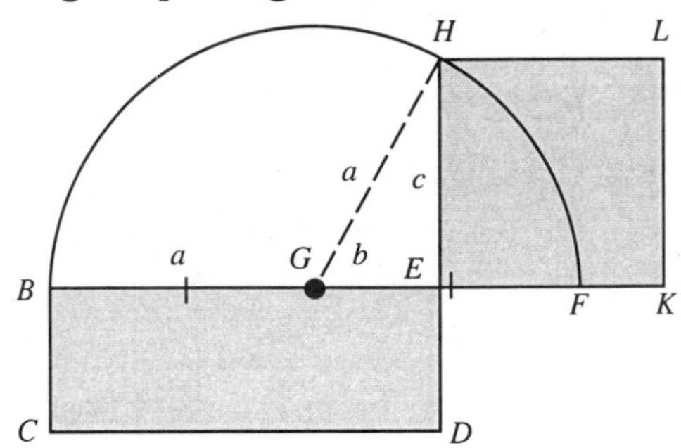

Given a RECTANGLE $\square BCDE$, draw $EF = DE$ on an extension of BE. Bisect BF and call the MIDPOINT G. Now draw a SEMICIRCLE centered at G, and construct the extension of ED which passes through the SEMICIRCLE at H. Then $\square EKLH$ has the same AREA as $\square BCDE$. This can be shown as follows:

$$A(\square BCDE) = BE \cdot ED = BE \cdot EF$$
$$= (a + b)(a - b) = a^2 - b^2 = c^2.$$

References

Dunham, W. "Hippocrates' Quadrature of the Lune." Ch. 1 in *Journey Through Genius: The Great Theorems of Mathematics.* New York: Wiley, pp. 13–14, 1990.

Rectangular Coordinates

see CARTESIAN COORDINATES

Rectangular Distribution

see UNIFORM DISTRIBUTION

Rectangular Hyperbola

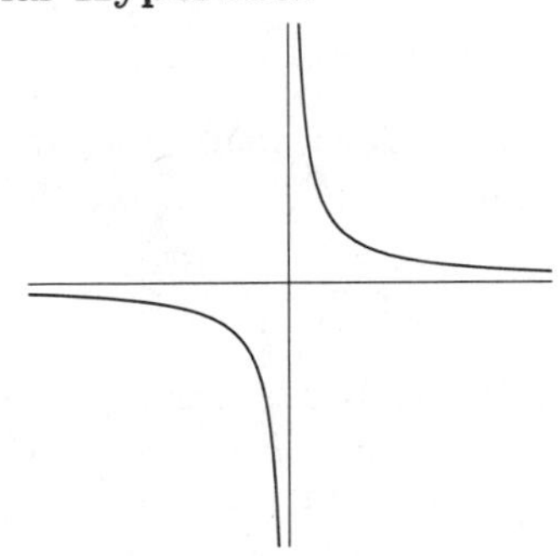

A RIGHT HYPERBOLA of the special form

$$xy = ab,$$

so that the ASYMPTOTES are the lines $x = 0$ and $y = 0$. The rectangular hyperbola is sometimes also called an EQUILATERAL HYPERBOLA.

see also HYPERBOLA, RIGHT HYPERBOLA

References

Courant, R. and Robbins, H. *What is Mathematics?: An Elementary Approach to Ideas and Methods, 2nd ed.* Oxford, England: Oxford University Press, pp. 76–77, 1996.

Rectangular Parallelepiped

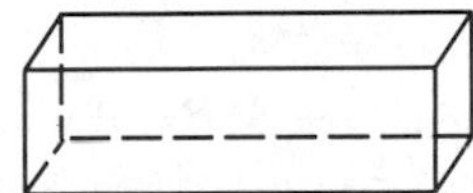

A closed box composed of 3 pairs of rectangular faces placed opposite each other and joined at RIGHT ANGLES to each other. This PARALLELEPIPED therefore corresponds to a rectangular "box." If the lengths of the sides are denoted a, b, and c, then the VOLUME is

$$V = abc, \tag{1}$$

the total SURFACE AREA is

$$A = 2(ab + bc + ca), \tag{2}$$

and the length of the "space" DIAGONAL is

$$d_{abc} = \sqrt{a^2 + b^2 + c^2}. \tag{3}$$

If $a = b = c$, then the rectangular parallelepiped is a CUBE.

see also CUBE, EULER BRICK, PARALLELEPIPED

References

Beyer, W. H. (Ed.) *CRC Standard Mathematical Tables, 28th ed.* Boca Raton, FL: CRC Press, p. 127, 1987.

Rectangular Projection

see EQUIRECTANGULAR PROJECTION

Rectifiable Current

The space of currents arising from rectifiable sets by integrating a differential form is called the space of 2-D rectifiable currents. For C a closed bounded rectifiable curve of a number of components in $\mathbb{R}^3$, C bounds a rectifiable current of least AREA. The theory of rectifiable currents generalizes to m-D surfaces in $\mathbb{R}^n$.

see also INTEGRAL CURRENT, REGULARITY THEOREM

References

Morgan, F. "What is a Surface?" *Amer. Math. Monthly* **103**, 369–376, 1996.

Rectifiable Set

The rectifiable sets include the image of any LIPSCHITZ FUNCTION f from planar domains into $\mathbb{R}^3$. The full set is obtained by allowing arbitrary measurable subsets of countable unions of such images of Lipschitz functions as long as the total AREA remains finite. Rectifiable sets have an "approximate" tangent plane at almost every point.

References

Morgan, F. "What is a Surface?" *Amer. Math. Monthly* **103**, 369–376, 1996.

Rectification

Rectification is the determination of the length of a curve.

see also QUADRABLE, SQUARING

Rectifying Latitude

An AUXILIARY LATITUDE which gives a sphere having correct distances along the meridians. It is denoted μ (or ω) and is given by

$$\mu = \frac{\pi M}{2M_p}. \tag{1}$$

M_p is evaluated for M at the north pole ($\phi = 90°$), and M is given by

$$\begin{aligned} M &= a(1-e^2)\int_0^\phi \frac{d\phi}{(1-e^2\sin^2\phi)^{3/2}} \\ &= a\left[\int_0^\phi \sqrt{1-e^2\sin^2\phi}\,d\phi - \frac{e^2\sin\phi\cos\phi}{\sqrt{1-e^2\sin^2\phi}}\right]. \end{aligned} \tag{2}$$

A series for M is

$$\begin{aligned} M = a[&(1 - \tfrac{1}{4}e^2 - \tfrac{3}{64}e^4 - \tfrac{5}{256}e^6 - \ldots)\phi \\ &- (\tfrac{3}{8}e^2 + \tfrac{3}{32}e^4 + \tfrac{45}{1024}e^6 + \ldots)\sin(2\phi) \\ &+ (\tfrac{15}{256}e^4 + \tfrac{45}{1024}e^6 + \ldots)\sin(4\phi) \\ &- (\tfrac{35}{3072}e^6 + \ldots)\sin(6\phi) + \ldots], \end{aligned} \tag{3}$$

and a series for μ is

$$\mu = \phi - (\tfrac{3}{2}e_1 - \tfrac{9}{16}{e_1}^3 + \ldots)\sin(2\phi) + (\tfrac{15}{16}{e_1}^2 - \tfrac{15}{32}{e_1}^4 + \ldots)\sin(4\phi) - (\tfrac{35}{48}{e_1}^3 - \ldots)\sin(6\phi) + (\tfrac{315}{512}{e_1}^4 - \ldots)\sin(8\phi) + \ldots, \quad (4)$$

where

$$e_1 \equiv \frac{1-\sqrt{1-e^2}}{1+\sqrt{1-e^2}}. \quad (5)$$

The inverse formula is

$$\phi = \mu + (\tfrac{3}{2}e_1 - \tfrac{27}{32}{e_1}^3 + \ldots)\sin(2\mu) + (\tfrac{21}{16}{e_1}^2 - \tfrac{55}{32}{e_1}^4 + \ldots)\sin(4\mu) + (\tfrac{151}{96}{e_1}^3 - \ldots)\sin(6\mu) + (\tfrac{1097}{512}{e_1}^4 - \ldots)\sin(8\mu) + \ldots. \quad (6)$$

see also LATITUDE

References

Adams, O. S. "Latitude Developments Connected with Geodesy and Cartography with Tables, Including a Table for Lambert Equal-Area Meridional Projections." Spec. Pub. No. 67. U. S. Coast and Geodetic Survey, pp. 125–128, 1921.

Snyder, J. P. *Map Projections—A Working Manual.* U. S. Geological Survey Professional Paper 1395. Washington, DC: U. S. Government Printing Office, pp. 16–17, 1987.

Rectifying Plane

The PLANE spanned by the TANGENT VECTOR **T** and BINORMAL VECTOR **B** .

see also BINORMAL VECTOR, TANGENT VECTOR

Recurrence Relation

A mathematical relationship expressing f_n as some combination of f_i with $i < n$. The solutions to linear recurrence can be computed straightforwardly, but QUADRATIC RECURRENCES are not so well understood. The sequence generated by a recurrence relation is called a RECURRENCE SEQUENCE. Perhaps the most famous example of a recurrence relation is the one defining the FIBONACCI NUMBERS,

$$F_n = F_{n-2} + F_{n-1}$$

for $n \geq 3$ and with $F_1 = F_2 = 1$.

see also ARGUMENT ADDITION RELATION, ARGUMENT MULTIPLICATION RELATION, CLENSHAW RECURRENCE FORMULA, QUADRATIC RECURRENCE, RECURRENCE SEQUENCE, REFLECTION RELATION, TRANSLATION RELATION

References

Press, W. H.; Flannery, B. P.; Teukolsky, S. A.; and Vetterling, W. T. "Recurrence Relations and Clenshaw's Recurrence Formula." §5.5 in *Numerical Recipes in FORTRAN: The Art of Scientific Computing, 2nd ed.* Cambridge, England: Cambridge University Press, pp. 172–178, 1992.

Recurrence Sequence

A sequence of numbers generated by a RECURRENCE RELATION is called a recurrence sequence. Perhaps the most famous recurrence sequence is the FIBONACCI NUMBERS.

If a sequence $\{x_n\}$ with $x_1 = x_2 = 1$ is described by a two-term linear recurrence relation of the form

$$x_n = Ax_{n-1} + Bx_{n-2} \quad (1)$$

for $n \geq 3$ and A and B constants, then the closed form for x_n is given by

$$x_n = \frac{\alpha^n - \beta^n}{\alpha - \beta} \quad (2)$$

where α and β are the ROOTS of the QUADRATIC EQUATION

$$x^2 - Ax - B = 0, \quad (3)$$

$$\alpha = \tfrac{1}{2}(A + \sqrt{A^2+4B}) \quad (4)$$

$$\beta = \tfrac{1}{2}(A - \sqrt{A^2+4B}). \quad (5)$$

The general second-order linear recurrence

$$x_n = Ax_{n-1} + Bx_{n-2} \quad (6)$$

for constants A and B with arbitrary x_1 and x_2 has terms

$$\begin{aligned} x_1 &= x_1 \\ x_2 &= x_2 \\ x_3 &= Bx_1 + Ax_2 \\ x_4 &= Bx_2 + ABx_1 + A^2x_2 \\ x_5 &= B^2x_1 + 2ABx_2 + A^2Bx_1 + A^3x_2 \\ x_6 &= B^2x_2 + 2AB^2x_1 + 3A^2Bx_2 + A^3Bx_1 + A^4x_2. \end{aligned}$$

Dropping x_1, x_2, and A, this can be written

$$\begin{array}{lllll} 1 \\ 1 \\ B & 1 \\ B & B & 1 \\ B^2 & 2B & B & 1 \\ B^2 & 2B^2 & 3B & B & 1, \end{array}$$

which is simply PASCAL'S TRIANGLE on its side. An arbitrary term can therefore be written as

$$x_n = \sum_{k=0}^{n-2} \binom{\lfloor \frac{1}{2}(n+k-2) \rfloor}{k} A^k B^{\lfloor (n-k-1)/2 \rfloor} \times {x_1}^{[n+k \pmod 2]} {x_2}^{[n+k+1 \pmod 2]}. \quad (7)$$

$$= -(Ax_1 - x_2)\sum_{k=0}^{n-2} A^{2k-n+2}B^{-k+n-2}\binom{k}{n-k-2} + x_1 \sum_{k=0}^{n-1} A^{2k-n+1}B^{-k+n-1}\binom{k}{n-k-1}. \quad (8)$$

The general linear third-order recurrence

$$x_n = Ax_{n-1} + Bx_{n-2} + Cx_{n-3} \quad (9)$$

has solution

$$x_n = x_1\left(\frac{\alpha^{-n}}{A+2\alpha B+3\alpha^2C}+\frac{\beta^{-n}}{A+2\beta B+3\beta^2C}+\frac{\gamma^{-n}}{A+2\gamma B+3\gamma^2C}\right)$$
$$-(Ax_1-x_2)\left(\frac{\alpha^{1-n}}{A+2\alpha B+3\alpha^2C}+\frac{\beta^{1-n}}{A+2\beta B+3\beta^2B}+\frac{\gamma^{1-n}}{A+2\gamma C+3\gamma^2C}\right)$$
$$-(Bx_1+Ax_2-x_3)\left(\frac{\alpha^{2-n}}{A+2\alpha B+3\alpha^2C}+\frac{\beta^{2-n}}{A+2\beta B+3\beta^2C}+\frac{\gamma^{2-n}}{A+2\gamma B+3\gamma^2C}\right), \quad (10)$$

where α, β, and γ are the roots of the polynomial

$$Cx^3 + Bx^2 + Ax = 1. \quad (11)$$

A QUOTIENT-DIFFERENCE TABLE eventually yields a line of 0s IFF the starting sequence is defined by a linear recurrence relation.

A linear second-order recurrence

$$f_{n+1} = xf_n + yf_{n-1} \quad (12)$$

can be solved rapidly using a "rate doubling,"

$$f_{n+2} = (x^2+2y)f_n - y^2f_{n-2}, \quad (13)$$

"rate tripling"

$$f_{n+3} = (x^3+3xy)f_n + y^3f_{n-3}, \quad (14)$$

or in general, "rate k-tupling" formula

$$f_{n+k} = p_kf_n + q_kf_{n-k}, \quad (15)$$

where

$$p_0 = 2 \quad (16)$$
$$p_1 = x \quad (17)$$
$$p_k = 2(-y)^{k/2}T_k(x/(2i\sqrt{y})) \quad (18)$$
$$p_{k+1} = xp_k + yp_{k-1} \quad (19)$$

(here, $T_k(x)$ is a CHEBYSHEV POLYNOMIAL OF THE FIRST KIND) and

$$q_0 = -1 \quad (20)$$
$$q_1 = y \quad (21)$$
$$q_k = -(-y)^k \quad (22)$$
$$q_{k+1} = -yq_k \quad (23)$$

(Beeler *et al.* 1972, Item 14).

Let

$$s(X) = \prod_{i=1}^{m}(1-\alpha_iX)^{n_i} = 1 - s_1X - \ldots - s_nX^n, \quad (24)$$

where the generalized POWER sum $a(h)$ for $h = 0, 1, \ldots$ is given by

$$a(h) = \sum_{i=1}^{m} A_i(h){\alpha_i}^h, \quad (25)$$

with distinct NONZERO roots α_i, COEFFICIENTS $A_i(h)$ which are POLYNOMIALS of degree $n_i - 1$ for POSITIVE INTEGERS n_i, and $i \in [1,m]$. Then the sequence $\{a_h\}$ with $a_h = a(h)$ satisfies the RECURRENCE RELATION

$$a_{h+n} = s_ia_{h+n-1} + \ldots + s_na_h \quad (26)$$

(Meyerson and van der Poorten 1995).

The terms in a general recurrence sequence belong to a finitely generated RING over the INTEGERS, so it is impossible for every RATIONAL NUMBER to occur in any finitely generated recurrence sequence. If a recurrence sequence vanishes infinitely often, then it vanishes on an arithmetic progression with a common difference 1 that depends only on the roots. The number of values that a recurrence sequence can take on infinitely often is bounded by some INTEGER l that depends only on the roots. There is no recurrence sequence in which each INTEGER occurs infinitely often, or in which every GAUSSIAN INTEGER occurs (Myerson and van der Poorten 1995).

Let $\mu(n)$ be a bound so that a nondegenerate INTEGER recurrence sequence of order n takes the value zero at least $\mu(n)$ times. Then $\mu(2) = 1$, $\mu(3) = 6$, and $\mu(4) \geq 9$ (Myerson and van der Poorten 1995). The maximal case for $\mu(3)$ is

$$a_{n+3} = 2a_{n+2} - 4a_{n+1} + 4a_n \quad (27)$$

with

$$a_0 = a_1 = 0 \quad (28)$$
$$a_2 = 1. \quad (29)$$

The zeros are

$$a_0 = a_1 = a_4 = a_6 = a_{13} = a_{52} = 0 \quad (30)$$

(Beukers 1991).

see also BINET FORMS, BINET'S FORMULA, FAST FIBONACCI TRANSFORM, FIBONACCI SEQUENCE, LUCAS SEQUENCE, QUOTIENT-DIFFERENCE TABLE, SKOLEM-MAHLER-LERCH THEOREM

References

Beeler, M.; Gosper, R. W.; and Schroeppel, R. *HAKMEM.* Cambridge, MA: MIT Artificial Intelligence Laboratory, Memo AIM-239, Feb. 1972.

Beukers, F. "The Zero-Multiplicity of Ternary Recurrences." *Composito Math.* **77**, 165–177, 1991.

Myerson, G. and van der Poorten, A. J. "Some Problems Concerning Recurrence Sequences." *Amer. Math. Monthly* **10?**, 698–705, 1995.

Recurring Digital Invariant

To define a recurring digital invariant of order k, compute the sum of the kth powers of the digits of a number n. If this number n' is equal to the original number n, then $n = n'$ is called a k-NARCISSISTIC NUMBER. If not, compute the sums of the kth powers of the digits of n', and so on. If this process eventually leads back to the original number n, the *smallest number* in the sequence $\{n, n', n'', \ldots\}$ is said to be a k-recurring digital invariant. For example,

$$55 : 5^3 + 5^3 = 250$$
$$250 : 2^3 + 5^3 + 0^3 = 133$$
$$133 : 1^3 + 3^3 + 3^3 = 55,$$

so 55 is an order 3 recurring digital invariant. The following table gives recurring digital invariants of orders 2 to 10 (Madachy 1979).

Order	RDIs	Cycle Lengths
2	4	8
3	55, 136, 160, 919	3, 2, 3, 2
4	1138, 2178	7, 2
5	244, 8294, 8299, 9044, 9045, 10933,24584, 58618, 89883	28, 10, 6, 10, 22, 4, 12, 2, 2
6	17148, 63804, 93531, 239459, 282595	30, 2, 4, 10, 3
7	80441, 86874, 253074, 376762, 922428, 982108, five more	92, 56, 27, 30, 14, 21
8	6822, 7973187, 8616804	
9	322219, 2274831, 20700388, eleven more	
10	20818070, five more	

see also 196-ALGORITHM, ADDITIVE PERSISTENCE, DIGITAL ROOT, DIGITADITION, HAPPY NUMBER, KAPREKAR NUMBER, NARCISSISTIC NUMBER, VAMPIRE NUMBER

References

Madachy, J. S. *Madachy's Mathematical Recreations.* New York: Dover, pp. 163–165, 1979.

Recursion

A recursive process is one in which objects are defined in terms of other objects of the same type. Using some sort of RECURRENCE RELATION, the entire class of objects can then be built up from a few initial values and a small number of rules. The FIBONACCI NUMBERS are most commonly defined recursively. Care, however, must be taken to avoid SELF-RECURSION, in which an object is defined in terms of itself, leading to an infinite nesting.

see also ACKERMANN FUNCTION, PRIMITIVE RECURSIVE FUNCTION, RECURRENCE RELATION, RECURRENCE SEQUENCE, RICHARDSON'S THEOREM, SELF-RECURSION, SELF-SIMILARITY, TAK FUNCTION

References

Buck, R. C. "Mathematical Induction and Recursive Definitions." *Amer. Math. Monthly* **70**, 128–135, 1963.

Knuth, D. E. "Textbook Examples of Recursion." In *Artificial Intelligence and Mathematical Theory of Computation, Papers in Honor of John McCarthy* (Ed. V. Lifschitz). Boston, MA: Academic Press, pp. 207–229, 1991.

Péter, R. *Rekursive Funktionen.* Budapest: Akad. Kiado, 1951.

Recursive Function

A recursive function is a function generated by (1) ADDITION, (2) MULTIPLICATION, (3) selection of an element from a list, and (4) determination of the truth or falsity of the INEQUALITY $a < b$ according to the technical rules:

1. If F and the sequence of functions $G_1, \ldots, G_n$ are recursive, then so is $F(G_1, \ldots, G_n)$.
2. If F is a recursive function such that there is an x for each a with $H(a,x) = 0$, then the smallest x can be obtained recursively.

A TURING MACHINE is capable of computing recursive functions.

see also TURING MACHINE

References

Kleene, S. C. *Introduction to Metamathematics.* Princeton, NJ: Van Nostrand, 1952.

Recursive Monotone Stable Quadrature

A QUADRATURE (NUMERICAL INTEGRATION) algorithm which has a number of desirable properties.

References

Favati, P.; Lotti, G.; and Romani, F. "Interpolary Integration Formulas for Optimal Composition." *ACM Trans. Math. Software* **17**, 207–217, 1991.

Favati, P.; Lotti, G.; and Romani, F. "Algorithm 691: Improving QUADPACK Automatic Integration Routines." *ACM Trans. Math. Software* **17**, 218–232, 1991.

Red-Black Tree

An extended BINARY TREE satisfying the following conditions:

1. Every node has two CHILDREN, each colored either red or black.
2. Every LEAF node is colored black.
3. Every red node has both of its CHILDREN colored black.
4. Every path from the ROOT to a LEAF contains the same number (the "black-height") of black nodes.

Let n be the number of internal nodes of a red-black tree. Then the number of red-black trees for $n = 1$, 2, ... is 2, 2, 3, 8, 14, 20, 35, 64, 122, ... (Sloane's A001131). The number of trees with black roots and red roots are given by Sloane's A001137 and Sloane's A001138, respectively.

Let T_h be the GENERATING FUNCTION for the number of red-black trees of black-height h indexed by the number of LEAVES. Then

$$T_{h+1}(x) = [T_h(x)]^2 + [T_h(x)]^4, \tag{1}$$

where $T_1(x) = x+x^2$. If $T(x)$ is the GENERATING FUNCTION for the number of red-black trees, then

$$T(x) = x + x^2 + T(x^2(1+x)^2) \quad (2)$$

(Ruskey). Let $rb(n)$ be the number of red-black trees with n LEAVES, $r(n)$ the number of red-rooted trees, and $b(n)$ the number of black-rooted trees. All three of the quantities satisfy the RECURRENCE RELATION

$$R(n) = \sum_{n/4\leq n\leq n/2} \binom{2m}{n-2m} R(m), \quad (3)$$

where $\binom{n}{k}$ is a BINOMIAL COEFFICIENT, $\mathrm{rb}(1) = 1$, $\mathrm{rb}(2) = 2$ for $R(n) = \mathrm{rb}(n)$, $r(1) = r(3) = 0$, $r(2) = 1$ for $R(n) = r(n)$, and $b(1) = 1$ for $R(n) = b(n)$ (Ruskey).

References

Beyer, R. "Symmetric Binary *B*-Trees: Data Structures and Maintenance Algorithms." *Acta Informat.* **1**, 290–306, 1972.

Rivest, R. L.; Leiserson, C. E.; and Cormen, R. H. *Introduction to Algorithms.* New York: McGraw-Hill, 1990.

Ruskey, F. "Information on Red-Black Trees." `http://sue.csc.uvic.ca/~cos/inf/tree/RedBlackTree.html`.

Sloane, N. J. A. Sequences A001131, A001137, and A001138 in "An On-Line Version of the Encyclopedia of Integer Sequences."

Red Net

The coloring red of two COMPLETE SUBGRAPHS of $n/2$ points (for EVEN n) in order to generate a BLUE-EMPTY GRAPH.

see also BLUE-EMPTY GRAPH, COMPLETE GRAPH

Reduced Amicable Pair

see QUASIAMICABLE PAIR

Reduced Fraction

A FRACTION a/b written in lowest terms, i.e., by dividing NUMERATOR and DENOMINATOR through by their GREATEST COMMON DIVISOR (a, b). For example, $2/3$ is the reduced fraction of $8/12$.

see also FRACTION, PROPER FRACTION

Reduced Latitude

see PARAMETRIC LATITUDE

Reducible Crossing

A crossing in a LINK projection which can be removed by rotating part of the LINK, also called REMOVABLE CROSSING.

see also ALTERNATING KNOT

Reducible Representation

see IRREDUCIBLE REPRESENTATION

Reducible Matrix

A SQUARE $n\times n$ matrix $\mathsf{A} = a_{ij}$ is called reducible if the indices 1, 2, ..., n can be divided into two disjoint nonempty sets $i_1, i_2, \ldots, i_\mu$ and $j_1, j_2, \ldots, j_\nu$ (with $\mu+\nu = n$) such that

$$a_{i_\alpha j_\beta} = 0$$

for $\alpha = 1, 2, \ldots, \mu$ and $\beta = 1, 2, \ldots, \nu$. A SQUARE MATRIX which is not reducible is said to be IRREDUCIBLE.

see also SQUARE MATRIX

References

Gradshteyn, I. S. and Ryzhik, I. M. *Tables of Integrals, Series, and Products, 5th ed.* San Diego, CA: Academic Press, p. 1103, 1979.

Reduction of Order

see ORDINARY DIFFERENTIAL EQUATION—SECOND-ORDER

Reduction Theorem

If a fixed point is added to each group of a special complete series, then the resulting series is complete.

References

Coolidge, J. L. *A Treatise on Algebraic Plane Curves.* New York: Dover, p. 253, 1959.

Redundancy

$$R(X_1,\ldots X_n) \equiv \sum_{i=1}^n H(X_i) - H(X_1,\ldots,X_n),$$

where $H(x_i)$ is the ENTROPY and $H(X_1,\ldots,X_n)$ is the joint ENTROPY. Linear redundancy is defined as

$$L(X_1,\ldots,X_n) \equiv -\tfrac{1}{2}\sum_{i=1}^n \ln\sigma_i,$$

where σ_i are EIGENVALUES of the correlation matrix.

see also PREDICTABILITY

References

Fraser, A. M. "Reconstructing Attractors from Scalar Time Series: A Comparison of Singular System and Redundancy Criteria." *Phys. D* **34**, 391–404, 1989.

Paluš, M. "Identifying and Quantifying Chaos by Using Information-Theoretic Functionals." In *Time Series Prediction: Forecasting the Future and Understanding the Past* (Ed. A. S. Weigend and N. A. Gerschenfeld). Proc. NATO Advanced Research Workshop on Comparative Time Series Analysis held in Sante Fe, NM, May 14–17, 1992. Reading, MA: Addison-Wesley, pp. 387–413, 1994.

Reeb Foliation

The Reeb foliation of the HYPERSPHERE $\mathbb{S}^3$ is a FOLIATION constructed as the UNION of two solid TORI with common boundary.

see also FOLIATION

References

Rolfsen, D. *Knots and Links.* Wilmington, DE: Publish or Perish Press, pp. 287–288, 1976.

Reef Knot

see SQUARE KNOT

Refinement

A refinement X of a COVER Y is a COVER such that every element $x \in X$ is a SUBSET of an element $y \in Y$.

see also COVER

Reflection

The operation of exchanging all points of a mathematical object with their MIRROR IMAGES (i.e., reflections in a mirror). Objects which do not change HANDEDNESS under reflection are said to be AMPHICHIRAL; those that do are said to be CHIRAL.

If the PLANE of reflection is taken as the yz-PLANE, the reflection in 2- or 3-D SPACE consists of making the transformation $x \to -x$ for each point. Consider an arbitrary point $\mathbf{x}_0$ and a PLANE specified by the equation

$$ax + by + xz + d = 0. \tag{1}$$

This PLANE has NORMAL VECTOR

$$\mathbf{n} = \begin{bmatrix} a \\ b \\ c \end{bmatrix}, \tag{2}$$

and the POINT-PLANE DISTANCE is

$$D = \frac{|ax_0 + by_0 + cz_0 + d|}{\sqrt{a^2 + b^2 + c^2}}. \tag{3}$$

The position of the point reflected in the given plane is therefore given by

$$\begin{aligned} \mathbf{x}_0' &= \mathbf{x}_0 - 2D\hat{\mathbf{n}} \\ &= \begin{bmatrix} x_0 \\ y_0 \\ z_0 \end{bmatrix} - 2|ax_0 + by_0 + cz_0 + d| \begin{bmatrix} a \\ b \\ c \end{bmatrix}. \end{aligned} \tag{4}$$

see also AMPHICHIRAL, CHIRAL, DILATION, ENANTIOMER, EXPANSION, GLIDE, HANDEDNESS, IMPROPER ROTATION, INVERSION OPERATION, MIRROR IMAGE, PROJECTION, REFLECTION PROPERTY, REFLECTION RELATION, REFLEXIBLE, ROTATION, ROTOINVERSION, TRANSLATION

Reflection Property

In the plane, the reflection property can be stated as three theorems (Ogilvy 1990, pp. 73–77):

1. The LOCUS of the center of a variable CIRCLE, tangent to a fixed CIRCLE and passing through a fixed point inside that CIRCLE, is an ELLIPSE.
2. If a variable CIRCLE is tangent to a fixed CIRCLE and also passes through a fixed point outside the CIRCLE, then the LOCUS of its moving center is a HYPERBOLA.
3. If a variable CIRCLE is tangent to a fixed straight line and also passes through a fixed point not on the line, then the LOCUS of its moving center is a PARABOLA.

Let $\alpha : I \to \mathbb{R}^2$ be a smooth regular parameterized curve in $\mathbb{R}^2$ defined on an OPEN INTERVAL I, and let F_1 and F_2 be points in $\mathbb{P}^2\backslash\alpha(I)$, where $\mathbb{P}^n$ is an n-D PROJECTIVE SPACE. Then α has a reflection property with FOCI F_1 and F_2 if, for each point $P \in \alpha(I)$,

1. Any vector normal to the curve α at P lies in the SPAN of the vectors $\overrightarrow{F_1P}$ and $\overrightarrow{F_2P}$.
2. The line normal to α at P bisects one of the pairs of opposite ANGLES formed by the intersection of the lines joining F_1 and F_2 to P.

A smooth connected plane curve has a reflection property IFF it is part of an ELLIPSE, HYPERBOLA, PARABOLA, CIRCLE, or straight LINE.

Foci	Sign	Both foci finite	One focus finite	Both foci ∞
distinct	+	confocal ellipses	confocal parabolas	$\parallel$ lines
distinct	−	confocal hyperbola and $\perp$ bisector of interfoci line segment	confocal parabolas	$\parallel$ lines
equal		concentric circles		$\parallel$ lines

Let $S \in \mathbb{R}^3$ be a smooth connected surface, and let F_1 and F_2 be points in $\mathbb{P}^3\backslash S$, where $\mathbb{P}^n$ is an n-D PROJECTIVE SPACE. Then S has a reflection property with FOCI F_1 and F_2 if, for each point $P \in S$,

1. Any vector normal to S at P lies in the SPAN of the vectors $\overrightarrow{F_1P}$ and $\overrightarrow{F_2P}$.
2. The line normal to S at P bisects one of the pairs of opposite angles formed by the intersection of the lines joining F_1 and F_2 to P.

A smooth connected surface has a reflection property IFF it is part of an ELLIPSOID of revolution, a HYPERBOLOID of revolution, a PARABOLOID of revolution, a SPHERE, or PLANE.

Foci	Sign	Both foci finite	One focus finite	Both foci ∞
distinct	+	confocal ellipsoids	confocal paraboloids	$\parallel$ planes
distinct	−	confocal hyperboloids and plane $\perp$ bisector of interfoci line segment	confocal paraboloids	$\parallel$ planes
equal		concentric spheres		$\parallel$ planes

see also BILLIARDS

References

Drucker, D. "Euclidean Hypersurfaces with Reflective Properties." *Geometrica Dedicata* **33**, 325–329, 1990.

Drucker, D. "Reflective Euclidean Hypersurfaces." *Geometrica Dedicata* **39**, 361–362, 1991.

Drucker, D. "Reflection Properties of Curves and Surfaces." *Math. Mag.* **65**, 147–157, 1992.
Drucker, D. and Locke, P. "A Natural Classification of Curves and Surfaces with Reflection Properties." *Math. Mag.* **69**, 249–256, 1996.
Ogilvy, C. S. *Excursions in Geometry.* New York: Dover, pp. 73–77, 1990.
Wegner, B. "Comment on 'Euclidean Hypersurfaces with Reflective Properties'." *Geometrica Dedicata* **39**, 357–359, 1991.

Reflection Relation

A mathematical relationship relating $f(-x)$ to $f(x)$.

see also ARGUMENT ADDITION RELATION, ARGUMENT MULTIPLICATION RELATION, RECURRENCE RELATION, TRANSLATION RELATION

Reflexible

An object is reflexible if it is superposable with its image in a plane mirror. Also called AMPHICHIRAL.

see also AMPHICHIRAL, CHIRAL, ENANTIOMER, HANDEDNESS, MIRROR IMAGE, REFLECTION

References

Ball, W. W. R. and Coxeter, H. S. M. "Polyhedra." Ch. 5 in *Mathematical Recreations and Essays, 13th ed.* New York: Dover, p. 130, 1987.

Reflexible Map

An AUTOMORPHISM which interchanges the two vertices of a regular map at each edge without interchanging the vertices.

see also EDMONDS' MAP

Reflexive Closure

The reflexive closure of a binary RELATION R on a SET X is the minimal REFLEXIVE RELATION R' on X that contains R. Thus $aR'a$ for every element a of X and $aR'b$ for distinct elements a and b, provided that aRb.

see also REFLEXIVE REDUCTION, REFLEXIVE RELATION, RELATION, TRANSITIVE CLOSURE

Reflexive Graph

see DIRECTED GRAPH

Reflexive Reduction

The reflexive reduction of a binary RELATION R on a SET X is the minimum relation R' on X with the same REFLEXIVE CLOSURE as R. Thus $aR'b$ for any elements a and b of X, provided that a and b are distinct and aRb.

see also REFLEXIVE CLOSURE, RELATION, TRANSITIVE REDUCTION

Reflexive Relation

A RELATION R on a SET S is reflexive provided that xRx for every x in S.

see also RELATION

Reflexivity

A REFLEXIVE RELATION.

Region

An open connected set is called a region (sometimes also called a DOMAIN).

Regression

A method for fitting a curve (not necessarily a straight line) through a set of points using some goodness-of-fit criterion. The most common type of regression is LINEAR REGRESSION.

see also LEAST SQUARES FITTING, LINEAR REGRESSION, MULTIPLE REGRESSION, NONLINEAR LEAST SQUARES FITTING, REGRESSION COEFFICIENT

References

Kleinbaum, D. G. and Kupper, L. L. *Applied Regression Analysis and Other Multivariable Methods.* North Scituate, MA: Duxbury Press, 1978.

Regression Coefficient

The slope b of a line obtained using linear LEAST SQUARES FITTING is called the regression coefficient.

see also CORRELATION COEFFICIENT, LEAST SQUARES FITTING

References

Kenney, J. F. and Keeping, E. S. *Mathematics of Statistics, Pt. 2, 2nd ed.* Princeton, NJ: Van Nostrand, p. 254, 1951.

Regula Falsi

see FALSE POSITION METHOD

Regular Function

see HOLOMORPHIC FUNCTION

Regular Graph

A GRAPH is said to be regular of degree r if all LOCAL DEGREES are the same number r. Then

$$E = \tfrac{1}{2}nr,$$

where E is the number of EDGES. The connected 3-regular graphs have been determined by G. Brinkman up to 24 VERTICES.

see also COMPLETE GRAPH, COMPLETELY REGULAR GRAPH, LOCAL DEGREE, SUPERREGULAR GRAPH

References

Chartrand, G. *Introductory Graph Theory.* New York: Dover, p. 29, 1985.

Regular Isotopy

The equivalence of MANIFOLDS under continuous deformation within the embedding space. KNOTS of opposite CHIRALITY have AMBIENT ISOTOPY, but not regular isotopy.

see also AMBIENT ISOTOPY

Regular Isotopy Invariant

see BRACKET POLYNOMIAL

Regular Local Ring

A regular local ring is a LOCAL RING R with MAXIMAL IDEAL m so that m can be generated with exactly d elements where d is the KRULL DIMENSION of the RING R. Equivalently, R is regular if the VECTOR SPACE m/m^2 has dimension d.

see also KRULL DIMENSION, LOCAL RING, REGULAR RING, RING

References

Eisenbud, D. *Commutative Algebra with a View Toward Algebraic Geometry.* New York: Springer-Verlag, p. 242, 1995.

Regular Number

A number which has a finite DECIMAL expansion. A number which is not regular is said to be nonregular.

see also DECIMAL EXPANSION, REPEATING DECIMAL

Regular Parameterization

A parameterization of a SURFACE $\mathbf{x}(u,v)$ in u and v is regular if the TANGENT VECTORS

$$\frac{\partial \mathbf{x}}{\partial u} \quad \text{and} \quad \frac{\partial \mathbf{x}}{\partial v}$$

are always LINEARLY INDEPENDENT.

Regular Patch

A regular patch is a PATCH $\mathbf{x} : U \to \mathbb{R}^n$ for which the JACOBIAN $J(\mathbf{x})(u,v)$ has rank 2 for all $(u,v) \in U$. A PATCH is said to be regular at a point $(u_0, v_0) \in U$ providing that its JACOBIAN has rank 2 at (u_0, v_0). For example, the points at $\phi = \pm\pi/2$ in the standard parameterization of the SPHERE $(\cos\theta\sin\phi, \sin\theta\sin\phi, \cos\phi)$ are not regular.

An example of a PATCH which is regular but not INJECTIVE is the CYLINDER defined parametrically by $(\cos u, \sin u, v)$ with $u \in (-\infty, \infty)$ and $v \in (-2, 2)$. However, if $\mathbf{x} : U \to \mathbb{R}^n$ is an injective regular patch, then $\mathbf{x}$ maps U diffeomorphically onto $\mathbf{x}(U)$.

see also INJECTIVE PATCH, PATCH, REGULAR SURFACE

References

Gray, A. *Modern Differential Geometry of Curves and Surfaces.* Boca Raton, FL: CRC Press, p. 187, 1993.

Regular Point

see ORDINARY POINT

Regular Polygon

An n-sided POLYGON in which the sides are all the same length and are symmetrically placed about a common center. The sum of PERPENDICULARS from any point to the sides of a regular polygon of n sides is n times the APOTHEM. Only certain regular polygons are "CONSTRUCTIBLE" with RULER and STRAIGHTEDGE.

n	Regular Polygon
3	equilateral triangle
4	square
5	pentagon
6	hexagon
7	heptagon
8	octagon
9	nonagon
10	decagon
12	dodecagon
15	pentadecagon
16	hexadecagon
17	heptadecagon
18	octadecagon
20	icosagon
30	triacontagon

see also CONSTRUCTIBLE POLYGON, GEOMETROGRAPHY, HEPTADECAGON, POLYGON

References

Bishop, W. "How to Construct a Regular Polygon." *Amer. Math. Monthly* **85**, 186–188, 1978.

Regular Polyhedron

A polyhedron is said to be regular if its FACES and VERTEX FIGURES are REGULAR (not necessarily CONVEX) polygons (Coxeter 1973, p. 16). Using this definition, there are a total of nine regular polyhedra, five being the CONVEX PLATONIC SOLIDS and four being the CONCAVE (stellated) KEPLER-POINSOT SOLIDS. However, the term "regular polyhedra" is sometimes used to refer exclusively to the CONVEX PLATONIC SOLIDS.

It can be proven that only nine regular solids (in the Coxeter sense) exist by noting that a possible regular polyhedron must satisfy

$$\cos^2\left(\frac{\pi}{p}\right) + \cos^2\left(\frac{\pi}{q}\right) + \cos^2\left(\frac{\pi}{r}\right) = 1.$$

Gordon showed that the only solutions to

$$1 + \cos\phi_1 + \cos\phi_2 + \cos\phi_3 = 0$$

of the form $\phi_i = \pi m_i/n_i$ are the permutations of $(\frac{2}{3}\pi, \frac{2}{3}\pi, \frac{1}{3}\pi)$ and $(\frac{2}{3}\pi, \frac{2}{5}\pi, \frac{4}{5}\pi)$. This gives three permutations of (3, 3, 4) and six of (3, 5, $\frac{5}{3}$) as possible solutions to the first equation. Plugging back in gives the SCHLÄFLI SYMBOLS of possible regular polyhedra as $\{3,3\}$, $\{3,4\}$, $\{4,3\}$, $\{3,5\}$, $\{5,3\}$, $\{3,\frac{5}{2}\}$, $\{\frac{5}{2},3\}$, $\{5,\frac{5}{2}\}$, and $\{\frac{5}{2},5\}$ (Coxeter 1973, pp. 107–109). The first five of these are the PLATONIC SOLIDS and the remaining four the KEPLER-POINSOT SOLIDS.

Every regular polyhedron has $e+1$ axes of symmetry, where e is the number of EDGES, and $3h/2$ PLANES of symmetry, where h is the number of sides of the corresponding PETRIE POLYGON.

see also CONVEX POLYHEDRON, KEPLER-POINSOT SOLID, PETRIE POLYGON, PLATONIC SOLID, POLYHEDRON, POLYHEDRON COMPOUND, SPONGE, VERTEX FIGURE

References
Coxeter, H. S. M. "Regular and Semi-Regular Polytopes I." *Math. Z.* **46**, 380–407, 1940.
Coxeter, H. S. M. *Regular Polytopes, 3rd ed.* New York: Dover, pp. 1–17, 93, and 107–112, 1973.
Cromwell, P. R. *Polyhedra.* New York: Cambridge University Press, pp. 85–86, 1997.

Regular Prime

A PRIME which does not DIVIDE the CLASS NUMBER $h(p)$ of the CYCLOTOMIC FIELD obtained by adjoining a PRIMITIVE pTH ROOT of unity to the rational FIELD. A PRIME p is regular IFF p does not divide the NUMERATORS of the BERNOULLI NUMBERS B_{10}, B_{12}, ..., B_{2p-2}. A PRIME which is not regular is said to be an IRREGULAR PRIME.

In 1915, Jensen proved that there are infinitely many IRREGULAR PRIMES. It has not yet been proven that there are an INFINITE number of regular primes (Guy 1994, p. 145). Of the 283,145 PRIMES $< 4\times 10^6$, 171,548 (or 60.59%) are regular (the conjectured FRACTION is $e^{-1/2} \approx 60.65\%$). The first few are 3, 5, 7, 11, 13, 17, 19, 23, 29, 31, 41, 43, 47, ... (Sloane's A007703).

see also BERNOULLI NUMBER, FERMAT'S THEOREM, IRREGULAR PRIME

References
Buhler, J.; Crandall, R. Ernvall, R.; and Metsankyla, T. "Irregular Primes and Cyclotomic Invariants to Four Million." *Math. Comput.* **61**, 151–153, 1993.
Guy, R. K. *Unsolved Problems in Number Theory, 2nd ed.* New York: Springer-Verlag, p. 145, 1994.
Ribenboim, P. "Regular Primes." §5.1 in *The New Book of Prime Number Records.* New York: Springer-Verlag, pp. 323–329, 1996.
Shanks, D. *Solved and Unsolved Problems in Number Theory, 4th ed.* New York: Chelsea, p. 153, 1993.
Sloane, N. J. A. Sequence A007703/M2411 in "An On-Line Version of the Encyclopedia of Integer Sequences."

Regular Ring

In the sense of von Neumann, a regular ring is a RING R such that for all $a \in R$, there exists a $b \in R$ satisfying $a = aba$.

see also REGULAR LOCAL RING, RING

References
Jacobson, N. *Basic Algebra II, 2nd ed.* New York: W. H. Freeman, p. 196, 1989.

Regular Sequence

Let there be two PARTICULARLY WELL-BEHAVED FUNCTIONS $F(x)$ and $p_\tau(x)$. If the limit

$$\lim_{\tau\to 0}\int_{-\infty}^{\infty} p_\tau(x)F(x)\,dx$$

exists, then $p_\tau(x)$ is a regular sequence of PARTICULARLY WELL-BEHAVED FUNCTIONS.

Regular Singular Point

Consider a second-order ORDINARY DIFFERENTIAL EQUATION

$$y'' + P(x)y' + Q(x)y = 0.$$

If $P(x)$ and $Q(x)$ remain FINITE at $x = x_0$, then x_0 is called an ORDINARY POINT. If either $P(x)$ or $Q(x)$ diverges as $x \to x_0$, then x_0 is called a singular point. If either $P(x)$ or $Q(x)$ diverges as $x \to x_0$ but $(x-x_0)P(x)$ and $(x-x_0)^2Q(x)$ remain FINITE as $x \to x_0$, then $x = x_0$ is called a regular singular point (or NONESSENTIAL SINGULARITY).

see also IRREGULAR SINGULARITY, SINGULAR POINT (DIFFERENTIAL EQUATION)

References
Arfken, G. "Singular Points." §8.4 in *Mathematical Methods for Physicists, 3rd ed.* Orlando, FL: Academic Press, pp. 451–453 and 461–463, 1985.

Regular Singularity

see REGULAR SINGULAR POINT

Regular Surface

A SUBSET $M \subset \mathbb{R}^n$ is called a regular surface if for each point $p \in M$, there exists a NEIGHBORHOOD V of p in $\mathbb{R}^n$ and a MAP $x : U \to \mathbb{R}^n$ of a OPEN SET $U \subset \mathbb{R}^2$ onto $V \cap M$ such that

1. x is differentiable,
2. $x : U \to V \cap M$ is a HOMEOMORPHISM,
3. Each map $x : U \to M$ is a REGULAR PATCH.

Any open subset of a regular surface is also a regular surface.

see also REGULAR PATCH

References
Gray, A. "The Definition of a Regular Surface in $\mathbb{R}^n$." §10.4 in *Modern Differential Geometry of Curves and Surfaces.* Boca Raton, FL: CRC Press, pp. 195–200, 1993.

Regular Triangle Center

A TRIANGLE CENTER is regular IFF there is a TRIANGLE CENTER FUNCTION which is a POLYNOMIAL in Δ, a, b, and c (where Δ is the AREA of the TRIANGLE) such that the TRILINEAR COORDINATES of the center are

$$f(a,b,c) : f(b,c,a) : f(c,a,b).$$

The ISOGONAL CONJUGATE of a regular center is a regular center. Furthermore, given two regular centers, any two of their HARMONIC CONJUGATE POINTS are also regular centers.

see also ISOGONAL CONJUGATE, TRIANGLE CENTER, TRIANGLE CENTER FUNCTION

Regularity Theorem

An AREA-minimizing surface (RECTIFIABLE CURRENT) bounded by a smooth curve in $\mathbb{R}^3$ is a smooth submanifold with boundary.

see also MINIMAL SURFACE, RECTIFIABLE CURRENT

References

Morgan, F. "What is a Surface?" *Amer. Math. Monthly* **103**, 369–376, 1996.

Regularized Beta Function

The regularized beta function is defined by

$$I(z;a,b) = \frac{B(z;a,b)}{B(a,b)},$$

where $B(z;a,b)$ is the incomplete BETA FUNCTION and $B(a,b)$ is the complete BETA FUNCTION.

see also BETA FUNCTION, REGULARIZED GAMMA FUNCTION

Regularized Gamma Function

The regularized gamma functions are defined by

$$P(a,z) = 1 - Q(a,z) \equiv \frac{\gamma(a,z)}{\Gamma(a)}$$

and

$$Q(a,z) = 1 - P(a,z) \equiv \frac{\Gamma(a,z)}{\Gamma(a)},$$

where $\gamma(a,z)$ and $\Gamma(a,z)$ are incomplete GAMMA FUNCTIONS and $\Gamma(a)$ is a complete GAMMA FUNCTION. Their derivatives are

$$\frac{d}{dz}P(a,z) = e^{-z}z^{a-1}$$
$$\frac{d}{dz}Q(a,z) = -e^{-z}z^{a-1}.$$

see also GAMMA FUNCTION, REGULARIZED BETA FUNCTION

References

Press, W. H.; Flannery, B. P.; Teukolsky, S. A.; and Vetterling, W. T. *Numerical Recipes in FORTRAN: The Art of Scientific Computing, 2nd ed.* Cambridge, England: Cambridge University Press, pp. 160–161, 1992.

Regulus

The locus of lines meeting three given SKEW LINES. ("Regulus" is also the name of the brightest star in the constellation Leo.)

Reidemeister Moves

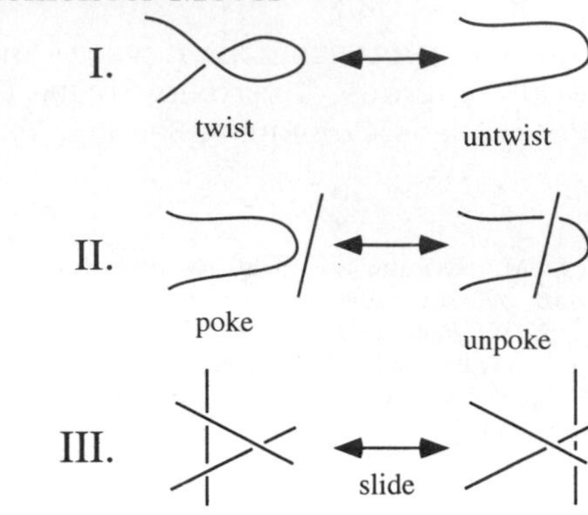

In the 1930s, Reidemeister first rigorously proved that KNOTS exist which are distinct from the UNKNOT. He did this by showing that all KNOT deformations can be reduced to a sequence of three types of "moves," called the (I) TWIST MOVE, (II) POKE MOVE, and (III) SLIDE MOVE.

REIDEMEISTER'S THEOREM guarantees that moves I, II, and III correspond to AMBIENT ISOTOPY (moves II and III alone correspond to REGULAR ISOTOPY). He then defined the concept of COLORABILITY, which is invariant under Reidemeister moves.

see also AMBIENT ISOTOPY, COLORABLE, MARKOV MOVES, REGULAR ISOTOPY, UNKNOT

Reidemeister's Theorem

Two LINKS can be continuously deformed into each other IFF any diagram of one can be transformed into a diagram of the other by a sequence of REIDEMEISTER MOVES.

see also REIDEMEISTER MOVES

Reinhardt Domain

A Reinhardt domain with center **c** is a DOMAIN D in C^n such that whenever D contains z_0, the DOMAIN D also contains the closed POLYDISK.

References

Iyanaga, S. and Kawada, Y. (Eds.). *Encyclopedic Dictionary of Mathematics.* Cambridge, MA: MIT Press, p. 101, 1980.

Relation

A relation is any SUBSET of a CARTESIAN PRODUCT. For instance, a SUBSET of $A \times B$, called a (binary) "relation from A to B," is a collection of ORDERED PAIRS (a,b) with first components from A and second components from B, and, in particular, a SUBSET of $A \times A$ is called a "relation on A." For a binary relation R, one often writes aRb to mean that (a,b) is in R.

see also ADJACENCY RELATION, ANTISYMMETRIC RELATION, ARGUMENT ADDITION RELATION, ARGUMENT MULTIPLICATION RELATION, COVER RELATION, EQUIVALENCE RELATION, IRREFLEXIVE, PARTIAL ORDER, RECURRENCE RELATION, REFLECTION RELATION, REFLEXIVE RELATION, SYMMETRIC RELATION, TRANSITIVE, TRANSLATION RELATION

Relative Error

Let the true value of a quantity be x and the measured or inferred value x_0. Then the relative error is defined by

$$\delta x = \frac{\Delta x}{x} = \frac{x_0 - x}{x} = \frac{x_0}{x} - 1,$$

where Δx is the ABSOLUTE ERROR. The relative error of the QUOTIENT or PRODUCT of a number of quantities is less than or equal to the SUM of their relative errors. The PERCENTAGE ERROR is 100% times the relative error.

see also ABSOLUTE ERROR, ERROR PROPAGATION, PERCENTAGE ERROR

References

Abramowitz, M. and Stegun, C. A. (Eds.). *Handbook of Mathematical Functions with Formulas, Graphs, and Mathematical Tables, 9th printing.* New York: Dover, p. 14, 1972.

Relative Extremum

A RELATIVE MAXIMUM or RELATIVE MINIMUM, also called a LOCAL EXTREMUM.

see also EXTREMUM, GLOBAL EXTREMUM, RELATIVE MAXIMUM, RELATIVE MINIMUM

Relative Maximum

A MAXIMUM within some NEIGHBORHOOD which need not be a GLOBAL MAXIMUM.

see also GLOBAL MAXIMUM, MAXIMUM, RELATIVE MINIMUM

Relative Minimum

A MINIMUM within some NEIGHBORHOOD which need not be a GLOBAL MINIMUM.

see also GLOBAL MINIMUM, MINIMUM, RELATIVE MAXIMUM

Relatively Prime

Two integers are relatively prime if they share no common factors (divisors) except 1. Using the notation (m,n) to denote the GREATEST COMMON DIVISOR, two integers m and n are relatively prime if $(m,n) = 1$. Relatively prime integers are sometimes also called STRANGERS or COPRIME and are denoted $m \perp n$.

The probability that two INTEGERS picked at random are relatively prime is $[\zeta(2)]^{-1} = 6/\pi^2$, where $\zeta(z)$ is the RIEMANN ZETA FUNCTION. This result is related to the fact that the GREATEST COMMON DIVISOR of m and n, $(m,n) = k$, can be interpreted as the number of LATTICE POINTS in the PLANE which lie on the straight LINE connecting the VECTORS $(0,0)$ and (m,n) (excluding (m,n) itself). In fact $6/\pi^2$ the fractional number of LATTICE POINTS VISIBLE from the ORIGIN (Castellanos 1988, pp. 155–156).

Given three INTEGERS chosen at random, the probability that no common factor will divide them all is

$$[\zeta(3)]^{-1} \approx 1.202^{-1} = 0.832\ldots,$$

where $\zeta(3)$ is APÉRY'S CONSTANT. This generalizes to k random integers (Schoenfeld 1976).

see also DIVISOR, GREATEST COMMON DIVISOR, VISIBILITY

References

Castellanos, D. "The Ubiquitous Pi." *Math. Mag.* **61**, 67–98, 1988.

Guy, R. K. *Unsolved Problems in Number Theory, 2nd ed.* New York: Springer-Verlag, pp. 3–4, 1994.

Schoenfeld, L. "Sharper Bounds for the Chebyshev Functions $\theta(x)$ and $\psi(x)$, II." *Math. Comput.* **30**, 337–360, 1976.

Relaxation Methods

Methods of solving an ORDINARY DIFFERENTIAL EQUATION by replacing it with a FINITE DIFFERENCE equation on a regular grid spanning the domain of interest. The finite difference equations are then solved using an n-D NEWTON'S METHOD or other similar algorithm.

References

Press, W. H.; Flannery, B. P.; Teukolsky, S. A.; and Vetterling, W. T. "Richardson Extrapolation and the Bulirsch-Stoer Method." §17.3 in *Numerical Recipes in FORTRAN: The Art of Scientific Computing, 2nd ed.* Cambridge, England: Cambridge University Press, pp. 753–763, 1992.

Remainder

In general, a remainder is a quantity "left over" after performing a particular algorithm. The term is most commonly used to refer to the number left over when two integers are divided by each other in INTEGER DIVISION. For example, $55\backslash 7 = 7$, with a remainder of 6. Of course in real division, there is no such thing as a remainder since, for example, $55/7 = 7 + 6/7$.

The term remainder is also sometimes to the RESIDUE of a CONGRUENCE.

see also DIVISION, INTEGER DIVISION, RESIDUE (CONGRUENCE)

Remainder Theorem

see POLYNOMIAL REMAINDER THEOREM

Rembs' Surfaces

A special class of ENNEPER'S SURFACES which can be given parametrically by

$$x = a(U \cos u - U' \sin u) \quad (1)$$
$$y = -a(U \sin u + U' \cos u) \quad (2)$$
$$z = v - aV', \quad (3)$$

where

$$U \equiv \frac{\cosh(u\sqrt{C})}{\sqrt{C}} \quad (4)$$
$$V \equiv \frac{\cos(v\sqrt{C+1})}{\sqrt{C+1}} \quad (5)$$
$$a \equiv \frac{2V}{(C+1)(U^2 - V^2)}. \quad (6)$$

The value of v is restricted to

$$|v| \leq v_0 \equiv \frac{\pi}{2\sqrt{C+1}} \quad (7)$$

(Reckziegel 1986), and the values $v = \pm v_0$ correspond to the ends of the cleft in the surface.

see also ENNEPER'S SURFACES, KUEN SURFACE, SIEVERT'S SURFACE

References

Fischer, G. (Ed.). Plate 88 in *Mathematische Modelle/Mathematical Models, Bildband/Photograph Volume.* Braunschweig, Germany: Vieweg, p. 84, 1986.

Reckziegel, H. "Sievert's Surface." §3.4.4.3 in *Mathematical Models from the Collections of Universities and Museums* (Ed. G. Fischer). Braunschweig, Germany: Vieweg, pp. 39–40, 1986.

Rembs, E. "Enneper'sche Flächen konstanter positiver Krümmung und Hazzidakissche Transformationen." *Jahrber. DMV* **39**, 278–283, 1930.

Removable Crossing

see REDUCIBLE CROSSING

Removable Singularity

A SINGULAR POINT z_0 of a FUNCTION $f(z)$ for which it is possible to assign a COMPLEX NUMBER in such a way that $f(z)$ becomes ANALYTIC. A more precise way of defining a removable singularity is as a SINGULARITY z_0 of a function $f(z)$ about which the function $f(z)$ is bounded. For example, the point $x_0 = 0$ is a removable singularity in the SINC FUNCTION $\operatorname{sinc} x = \sin x / x$, since this function satisfies $\operatorname{sinc} 0 = 1$.

Rencontres Number

see DERANGEMENT, SUBFACTORIAL

Rendezvous Values

see MAGIC GEOMETRIC CONSTANTS

Rényi's Parking Constants

N.B. A detailed on-line essay by S. Finch was the starting point for this entry.

Given the CLOSED INTERVAL $[0, x]$ with $x > 1$, let 1-D "cars" of unit length be parked randomly on the interval. The MEAN number $M(x)$ of cars which can fit (without overlapping!) satisfies

$$M(x) = \begin{cases} 0 & \text{for } 0 \leq x < 1 \\ 1 + \frac{2}{x-1}\int_0^{x-1} M(y)\, dy & \text{for } x \geq 1. \end{cases} \quad (1)$$

The mean density of the cars for large x is

$$m \equiv \lim_{x\to\infty} \frac{M(x)}{x} = \int_0^\infty \exp\left(-2\int_0^x \frac{1-e^{-y}}{y}\, dy\right) dx$$
$$= 0.7475979203\ldots. \quad (2)$$

Furthermore,

$$M(x) = mx + m - 1 + \mathcal{O}(x^{-n}) \quad (3)$$

for all n (Rényi 1958), which was strengthened by Dvoretzky and Robbins (1964) to

$$M(x) = mx + m - 1 + \mathcal{O}\left[\left(\frac{2e}{x}\right)^{x-3/2}\right]. \quad (4)$$

Dvoretzky and Robbins (1964) also proved that

$$\inf_{x\leq t\leq x+1} \frac{M(t)+1}{t+1} \leq m \leq \sup_{x\leq t\leq x+1} \frac{M(t)+1}{t+1}. \quad (5)$$

Let $V(x)$ be the variance of the number of cars, then Dvoretzky and Robbins (1964) and Mannion (1964) showed that

$$v \equiv \lim_{x\to\infty} \frac{V(x)}{x}$$
$$= 2\int_0^\infty \left\{x\int_0^1 e^{-xy} R_2(y)\, dy + x^2\left[\int_0^\infty e^{-xy} R_1(y)\, dy\right]^2\right\}$$
$$\times \exp\left(-2\int_0^x \frac{1-e^{-y}}{y}\, dy\right) dx = 0.038156\ldots, \quad (6)$$

where

$$R_1(x) = M(x) - mx - m + 1 \quad (7)$$

$$R_2(x) = \begin{cases} (1-m-mx)^2 & \text{for } 0 \leq x \leq 1 \\ 4(1-m)^2 & \text{for } x = 1 \\ \frac{2}{x-1}\left[\int_0^{x-1} R_2(y)\, dy\right. & \text{for } x > 1 \\ \quad \left. + \int_0^{x-1} R_1(y) R_1(x-y-1)\, dy\right], & \end{cases}$$
$$(8)$$

and the numerical value is due to Blaisdell and Solomon (1970). Dvoretzky and Robbins (1964) also proved that

$$\inf_{x\leq t\leq x+1} \frac{V(t)}{t+1} \leq v \leq \sup_{x\leq t\leq x+1} \frac{V(t)}{t+1}, \tag{9}$$

and that

$$V(x) = vx + v + \mathcal{O}\left[\left(\frac{4e}{x}\right)^{x-4}\right]. \tag{10}$$

Palasti (1960) conjectured that in 2-D,

$$\lim_{x,y\to\infty} \frac{M(x,y)}{xy} = m^2, \tag{11}$$

but this has not yet been proven or disproven (Finch).

References

Blaisdell, B. E. and Solomon, H. "On Random Sequential Packing in the Plane and a Conjecture of Palasti." *J. Appl. Prob.* **7**, 667–698, 1970.

Dvoretzky, A. and Robbins, H. "On the Parking Problem." *Publ. Math. Inst. Hung. Acad. Sci.* **9**, 209–224, 1964.

Finch, S. "Favorite Mathematical Constants." `http://www.mathsoft.com/asolve/constant/renyi/renyi.html`.

Mannion, D. "Random Space-Filling in One Dimension." *Publ. Math. Inst. Hung. Acad. Sci.* **9**, 143–154, 1964.

Palasti, I. "On Some Random Space Filling Problems." *Publ. Math. Inst. Hung. Acad. Sci.* **5**, 353–359, 1960.

Rényi, A. "On a One-Dimensional Problem Concerning Random Space-Filling." *Publ. Math. Inst. Hung. Acad. Sci.* **3**, 109–127, 1958.

Solomon, H. and Weiner, H. J. "A Review of the Packing Problem." *Comm. Statist. Th. Meth.* **15**, 2571–2607, 1986.

Rep-Tile

A POLYGON which can be divided into smaller copies of itself.

see also DISSECTION

References

Gardner, M. Ch. 19 in *The Unexpected Hanging and Other Mathematical Diversions.* Chicago, IL: Chicago University Press, 1991.

Repartition

see ADÉLE

Repdigit

A number composed of a single digit is called a repdigit. If the digits are all 1s, the repdigit is called a REPUNIT. The BEAST NUMBER 666 is a repdigit.

see also KEITH NUMBER, REPUNIT

Repeating Decimal

A number whose decimal representation eventually becomes periodic (i.e., the same sequence of digits repeats indefinitely) is called a repeating decimal. Numbers such as 0.5 can be regarded as repeating decimals since $0.5 = 0.5000\ldots = 0.4999\ldots$. All RATIONAL NUMBERS have repeating decimals, e.g., $1/11 = 0.\overline{09}$. However, TRANSCENDENTAL NUMBERS, such as $\pi = 3.141592\ldots$ do not.

see also CYCLIC NUMBER, DECIMAL EXPANSION, FULL REPTEND PRIME, IRRATIONAL NUMBER, MIDY'S THEOREM, RATIONAL NUMBER, REGULAR NUMBER

References

Ball, W. W. R. and Coxeter, H. S. M. *Mathematical Recreations and Essays, 13th ed.* New York: Dover, pp. 53–54, 1987.

Courant, R. and Robbins, H. "Rational Numbers and Periodic Decimals." §2.2.4 in *What is Mathematics?: An Elementary Approach to Ideas and Methods, 2nd ed.* Oxford, England: Oxford University Press, pp. 66–68, 1996.

Repfigit Number

see KEITH NUMBER

Replicate

One out of a set of identical observations in a given experiment under identical conditions.

Reptend Prime

see FULL REPTEND PRIME

Representation

The representation of a GROUP G on a COMPLEX VECTOR SPACE V is a group action of G on V by linear transformations. Two finite dimensional representations π on V and π' on V' are equivalent if there is an invertible linear map $E : V \mapsto V'$ such that $\pi'(g)E = E\pi(g)$ for all $g \in G$. π is said to be irreducible if it has no proper NONZERO invariant SUBSPACES.

see also CHARACTER (MULTIPLICATIVE), PETER-WEYL THEOREM, PRIMARY REPRESENTATION, SCHUR'S LEMMA

References

Knapp, A. W. "Group Representations and Harmonic Analysis, Part II." *Not. Amer. Math. Soc.* **43**, 537–549, 1996.

Repunit

A (generalized) repunit to the base b is a number of the form

$$M_n^b = \frac{b^n - 1}{b - 1}.$$

The term "repunit" was coined by Beiler (1966), who also gave the first tabulation of known factors. Repunits $M_n = M_n^2 = 2^n - 1$ with $b = 2$ are called MERSENNE

NUMBERS. If $b = 10$, the number is called a repunit (since the digits are all 1s). A number of the form

$$R_n = \frac{10^n - 1}{10 - 1} = R_n = \frac{10^n - 1}{9}$$

is therefore a (decimal) repunit of order n.

b	Sloane	b-Repunits
2	000225	1, 3, 7, 15, 31, 63, 127, ...
3	003462	1, 4, 13, 40, 121, 364, ...
4	002450	1, 5, 21, 85, 341, 1365, ...
5	003463	1, 6, 31, 156, 781, 3906, ...
6	003464	1, 7, 43, 259, 1555, 9331, ...
7	023000	1, 8, 57, 400, 2801, 19608, ...
8	023001	1, 9, 73, 585, 4681, 37449, ...
9	002452	1, 10, 91, 820, 7381, 66430, ...
10	002275	1, 11, 111, 1111, 11111, ...
11	016123	1, 12, 133, 1464, 16105, 177156, ...
12	016125	1, 13, 157, 1885, 22621, 271453, ...

Williams and Seah (1979) factored generalized repunits for $3 \leq b \leq 12$ and $2 \leq n \leq 1000$. A (base-10) repunit can be PRIME only if n is PRIME, since otherwise $10^{ab} - 1$ is a BINOMIAL NUMBER which can be factored algebraically. In fact, if $n = 2a$ is EVEN, then $10^{2a} - 1 = (10^a - 1)(10^a + 1)$. The only base-10 repunit PRIMES R_n for $n \leq 16,500$ are $n = 2$, 19, 23, 317, and 1031 (Sloane's A004023; Madachy 1979, Williams and Dubner 1986, Ball and Coxeter 1987). The number of factors for the base-10 repunits for $n = 1$, 2, ... are 1, 1, 2, 2, 2, 5, 2, 4, 4, 4, 2, 7, 3, ... (Sloane's A046053).

b	Sloane	n of Prime b-Repunits
2	000043	2, 3, 5, 7, 13, 17, 19, 31, 61, 89, 107, ...
3	028491	3, 7, 13, 71, 103, 541, 1091, 1367, ...
5	004061	3, 7, 11, 13, 47, 127, 149, 181, 619, ...
6	004062	2, 3, 7, 29, 71, 127, 271, 509, 1049, ...
7	004063	5, 13, 131, 149, 1699, ...
10	004023	2, 19, 23, 317, 1031, ...
11	005808	17, 19, 73, 139, 907, 1907, 2029, 4801, ...
12	004064	2, 3, 5, 19, 97, 109, 317, 353, 701, ...

A table of the factors not obtainable algebraically for generalized repunits (a continuously updated version of Brillhart *et al.* 1988) is maintained online. These tables include factors for $10^n - 1$ (with $n \leq 209$ odd) and $10^n + 1$ (for $n \leq 210$ EVEN and ODD) in the files `ftp://sable.ox.ac.uk/pub/math/cunningham/10-` and `ftp://sable.ox.ac.uk/pub/math/cunningham/10+`. After algebraically factoring R_n, these are sufficient for complete factorizations. Yates (1982) published all the repunit factors for $n \leq 1000$, a portion of which are reproduced in the *Mathematica*® notebook by Weisstein.

A SMITH NUMBER can be constructed from every factored repunit.

see also CUNNINGHAM NUMBER, FERMAT NUMBER, MERSENNE NUMBER, REPDIGIT, SMITH NUMBER

References

Ball, W. W. R. and Coxeter, H. S. M. *Mathematical Recreations and Essays, 13th ed.* New York: Dover, p. 66, 1987.

Beiler, A. H. "11111...111." Ch. 11 in *Recreations in the Theory of Numbers: The Queen of Mathematics Entertains.* New York: Dover, 1966.

Brillhart, J.; Lehmer, D. H.; Selfridge, J.; Wagstaff, S. S. Jr.; and Tuckerman, B. *Factorizations of $b^n \pm 1$, $b = 2, 3, 5, 6, 7, 10, 11, 12$ Up to High Powers, rev. ed.* Providence, RI: Amer. Math. Soc., 1988. Updates are available electronically from `ftp://sable.ox.ac.uk/pub/math/cunningham`.

Dubner, H. "Generalized Repunit Primes." *Math. Comput.* **61**, 927–930, 1993.

Guy, R. K. "Mersenne Primes. Repunits. Fermat Numbers. Primes of Shape $k \cdot 2^n + 2$." §A3 in *Unsolved Problems in Number Theory, 2nd ed.* New York: Springer-Verlag, pp. 8–13, 1994.

Madachy, J. S. *Madachy's Mathematical Recreations.* New York: Dover, pp. 152–153, 1979.

Ribenboim, P. "Repunits and Similar Numbers." §5.5 in *The New Book of Prime Number Records.* New York: Springer-Verlag, pp. 350–354, 1996.

Snyder, W. M. "Factoring Repunits." *Am. Math. Monthly* **89**, 462–466, 1982.

Sorli, R. "Factorization Tables." `http://www.maths.uts.edu.au/staff/ron/fact/fact.html`.

Weisstein, E. W. "Repunits." `http://www.astro.virginia.edu/~eww6n/math/notebooks/Repunit.m`.

Williams, H. C. and Dubner, H. "The Primality of R1031." *Math. Comput.* **47**, 703–711, 1986.

Williams, H. C. and Seah, E. "Some Primes of the Form $(a^n - 1)/(a - 1)$. *Math. Comput.* **33**, 1337–1342, 1979.

Yates, S. "Prime Divisors of Repunits." *J. Recr. Math.* **8**, 33–38, 1975.

Yates, S. "The Mystique of Repunits." *Math. Mag.* **51**, 22–28, 1978.

Yates, S. *Repunits and Reptends.* Delray Beach, FL: S. Yates, 1982.

Residual

The residual is the sum of deviations from a best-fit curve of arbitrary form.

$$R \equiv \sum [y_i - f(x_i, a_1, \ldots, a_n)]^2.$$

The residual should not be confused with the CORRELATION COEFFICIENT.

Residual vs. Predictor Plot

A plot of y_i vs. $e_i \equiv \hat{y}_i - y_i$. Random scatter indicates the model is probably good. A pattern indicates a problem with the model. If the spread in e_i increases as y_i increases, the errors are called HETEROSCEDASTIC.

Residue Class

The residue classes of a function $f(x)$ mod n are all possible values of the RESIDUE $f(x)$ (mod n). For example, the residue classes of x^2 (mod 6) are $\{0, 1, 3, 4\}$, since

$$\begin{aligned} 0^2 &\equiv 0 \pmod 6 \\ 1^2 &\equiv 1 \pmod 6 \\ 3^2 &\equiv 3 \pmod 6 \\ 4^2 &\equiv 4 \pmod 6 \end{aligned}$$

are all the possible residues. A COMPLETE RESIDUE SYSTEM is a set of integers containing one element from each class, so in this case, $\{0, 1, 9, 4\}$ would be a COMPLETE RESIDUE SYSTEM.

The $\phi(m)$ residue classes prime to m form a GROUP under the binary multiplication operation (mod m), where $\phi(m)$ is the TOTIENT FUNCTION (Shanks 1993) and the GROUP is classed a MODULO MULTIPLICATION GROUP.

see also COMPLETE RESIDUE SYSTEM, CONGRUENCE, CUBIC NUMBER, QUADRATIC RECIPROCITY THEOREM, QUADRATIC RESIDUE, RESIDUE (CONGRUENCE), SQUARE NUMBER

References

Shanks, D. *Solved and Unsolved Problems in Number Theory, 4th ed.* New York: Chelsea, p. 56 and 59–63, 1993.

Residue (Complex Analysis)

The constant a_{-1} in the LAURENT SERIES

$$f(z) = \sum_{n=-\infty}^{\infty} a_n (z - z_0)^n$$

of $f(z)$ is called the residue of $f(z)$. The residue is a very important property of a complex function and appears in the amazing RESIDUE THEOREM of CONTOUR INTEGRATION.

see also CONTOUR INTEGRATION, LAURENT SERIES, RESIDUE THEOREM

References

Arfken, G. "Calculus of Residues." §7.2 in *Mathematical Methods for Physicists, 3rd ed.* Orlando, FL: Academic Press, pp. 400–421, 1985.

Residue (Congruence)

The number b in the CONGRUENCE $a \equiv b \pmod m$ is called the residue of a (mod m). The residue of large numbers can be computed quickly using CONGRUENCES. For example, to find 37^{13} (mod 17), note that

$$\begin{aligned} 37 &\equiv 3 \\ 37^2 &\equiv 3^2 \equiv 9 \equiv -8 \\ 37^4 &\equiv 81 \equiv -4 \\ 37^8 &\equiv 16 \equiv -1, \end{aligned}$$

so

$$37^{13} \equiv 37^{1+4+8} \equiv 3(-4)(-1) \equiv 12 \pmod{17}.$$

see also COMMON RESIDUE, CONGRUENCE, MINIMAL RESIDUE

References

Shanks, D. *Solved and Unsolved Problems in Number Theory, 4th ed.* New York: Chelsea, pp. 55–56, 1993.

Residue Index

$p-1$ divided by the HAUPT-EXPONENT of a base b mod p for a given PRIME p.

see also HAUPT-EXPONENT

Residue Theorem (Complex Analysis)

Given a complex function $f(z)$, consider the LAURENT SERIES

$$f(z) = \sum_{n=-\infty}^{\infty} a_n (z - z_0)^n. \tag{1}$$

Integrate term by term using a closed contour γ encircling z_0,

$$\begin{aligned} \int_\gamma f(z)\,dz &= \sum_{n=-\infty}^{\infty} a_n \int_\gamma (z - z_0)^n\,dz \\ &= \sum_{n=-\infty}^{-2} a_n \int_\gamma (z - z_0)^n\,dz \\ &\quad + a_{-1} \int_\gamma \frac{dz}{z - z_0} + \sum_{n=0}^{\infty} a_n \int_\gamma (z - z_0)^n\,dz. \end{aligned} \tag{2}$$

The CAUCHY INTEGRAL THEOREM requires that the first and last terms vanish, so we have

$$\int_\gamma f(z)\,dz = a_{-1} \int_\gamma \frac{dz}{z - z_0}. \tag{3}$$

But we can evaluate this function (which has a POLE at z_0) using the CAUCHY INTEGRAL FORMULA,

$$f(z_0) = \frac{1}{2\pi i} \int_\gamma \frac{f(z)\,dz}{z - z_0}. \tag{4}$$

This equation must also hold for the constant function $f(z) = 1$, in which case it is also true that $f(z_0) = 1$, so

$$1 = \frac{1}{2\pi i} \int_\gamma \frac{dz}{z - z_0}, \tag{5}$$

$$\int_\gamma \frac{dz}{z - z_0} = 2\pi i, \tag{6}$$

and (3) becomes

$$\int_\gamma f(z)\,dz = 2\pi i a_{-1}. \quad (7)$$

The quantity a_{-1} is known as the RESIDUE of $f(z)$ at z_0. Generalizing to a curve passing through multiple poles, (7) becomes

$$\int_\gamma f(z)\,dz = 2\pi i \sum_{i=1}^{\text{poles in } \gamma} n(\gamma, z_0^{(i)}) a_{-1}^{(i)}, \quad (8)$$

where n is the WINDING NUMBER and the (i) superscript denotes the quantity corresponding to POLE i.

If the path does not completely encircle the RESIDUE, take the CAUCHY PRINCIPAL VALUE to obtain

$$\int f(z)\,dz = (\theta_2 - \theta_1) i a_{-1}. \quad (9)$$

If f has only ISOLATED SINGULARITIES, then

$$\sum_{z_0^{(i)} \in \mathbb{C}^*} a_{-1}{}^{(i)} = 0. \quad (10)$$

The residues may be found without explicitly expanding into a LAURENT SERIES as follows.

$$f(z) = \sum_{n=-\infty}^{\infty} a_n (z - z_0)^n. \quad (11)$$

If $f(z)$ has a POLE of order m at z_0, then $a_n = 0$ for $n < -m$ and $a_{-m} \neq 0$. Therefore,

$$f(z) = \sum_{n=-m}^{\infty} a_n (z - z_0)^n = \sum_{n=0}^{\infty} a_{-m+n} (z - z_0)^{-m+n} \quad (12)$$

$$(z - z_0)^m f(z) = \sum_{n=0}^{\infty} a_{-m+n} (z - z_0)^n \quad (13)$$

$$\frac{d}{dz}[(z - z_0)^m f(z)] = \sum_{n=0}^{\infty} n a_{-m+n} (z - z_0)^{n-1}$$

$$= \sum_{n=1}^{\infty} n a_{-m+n} (z - z_0)^{n-1}$$

$$= \sum_{n=0}^{\infty} (n+1) a_{-m+n+1} (z - z_0)^n \quad (14)$$

$$\frac{d^2}{dz^2}[(z - z_0)^m f(z)] = \sum_{n=0}^{\infty} n(n+1) a_{-m+n+1} (z - z_0)^{n-1}$$

$$= \sum_{n=1}^{\infty} n(n+1) a_{-m+n+1} (z - z_0)^{n-1}$$

$$= \sum_{n=0}^{\infty} (n+1)(n+2) a_{-m+n+2} (z - z_0)^n. \quad (15)$$

Iterating,

$$\frac{d^{m-1}}{dz^{m-1}}[(z - z_0)^m f(z)]$$

$$= \sum_{n=0}^{\infty} (n+1)(n+2)(n+m-1) a_{n-1} (z - z_0)^n$$

$$= (m-1)! a_{-1}$$

$$+ \sum_{n=1}^{\infty} (n+1)(n+2)(n+m-1) a_{n-1} (z - z_0)^{n-1}. \quad (16)$$

So

$$\lim_{z \to z_0} \frac{d^{m-1}}{dz^{m-1}}[(z - z_0)^m f(z)]$$

$$= \lim_{z \to z_0} (m-1)! a_{-1} + 0 = (m-1)! a_{-1}, \quad (17)$$

and the RESIDUE is

$$a_{-1} = \frac{1}{(m-1)!} \frac{d^{m-1}}{dz^{m-1}}[(z - z_0)^m f(z)]_{z=z_0}. \quad (18)$$

This amazing theorem says that the value of a CONTOUR INTEGRAL in the COMPLEX PLANE depends *only* on the properties of a few special points *inside* the contour.

see also CAUCHY INTEGRAL FORMULA, CAUCHY INTEGRAL THEOREM, CONTOUR INTEGRAL, LAURENT SERIES, POLE, RESIDUE (COMPLEX ANALYSIS)

Residue Theorem (Group)

If two groups are residual to a third, every group residual to one is residual to the other. The Gambier extension of this theorem states that if two groups are pseudoresidual to a third, then every group pseudoresidual to the first with an excess greater than or equal to the excess of the first minus the excess of the second is pseudoresidual to the second, with an excess ≥ 0.

References

Coolidge, J. L. *A Treatise on Algebraic Plane Curves.* New York: Dover, pp. 30–31, 1959.

Resolution

Resolution is a widely used word with many different meanings. It can refer to resolution of equations, resolution of singularities (in ALGEBRAIC GEOMETRY), resolution of modules or more sophisticated structures, etc. In a BLOCK DESIGN, a PARTITION R of a BIBD's set of blocks B into PARALLEL CLASSES, each of which in turn partitions the set V, is called a resolution (Abel and Furino 1996).

A resolution of the MODULE M over the RING R is a complex of R-modules C_i and morphisms d_i and a MORPHISM ϵ such that

$$\cdots \to C_i \to^{d_i} C_{i-1} \to \cdots \to C_0 \to^{\epsilon} M \to 0$$

satisfying the following conditions:

1. The composition of any two consecutive morphisms is the zero map,
2. For all i, $(\ker d_i)/(\operatorname{im} d_{i+1}) = 0$,
3. $C_0/(\ker \epsilon) \simeq M$,

where ker is the kernel and im is the image. Here, the quotient

$$\frac{(\ker d_i)}{(\operatorname{im} d_{i+1})}$$

is the ith HOMOLOGY GROUP.

If all modules C_i are projective (free), then the resolution is called projective (free). There is a similar concept for resolutions "to the right" of M, which are called injective resolutions.

see also HOMOLOGY GROUP, MODULE, MORPHISM, RING

References

Abel, R. J. R. and Furino, S. C. "Resolvable and Near Resolvable Designs." §I.6 in *The CRC Handbook of Combinatorial Designs* (Ed. C. J. Colbourn and J. H. Dinitz). Boca Raton, FL: CRC Press, p. 4 and 87–94, 1996.

Jacobson, N. *Basic Algebra II, 2nd ed.* New York: W. H. Freeman, p. 339, 1989.

Resolution Class

see PARALLEL CLASS

Resolution Modulus

The least POSITIVE INTEGER m^* with the property that $\chi(y) = 1$ whenever $y \equiv 1 \pmod{m^*}$ and $(y, m) = 1$.

Resolvable

A balanced incomplete BLOCK DESIGN (B, V) is called resolvable if there exists a PARTITION R of its set of blocks B into PARALLEL CLASSES, each of which in turn partitions the set V. The partition R is called a RESOLUTION.

see also BLOCK DESIGN, PARALLEL CLASS

References

Abel, R. J. R. and Furino, S. C. "Resolvable and Near Resolvable Designs." §I.6 in *The CRC Handbook of Combinatorial Designs* (Ed. C. J. Colbourn and J. H. Dinitz). Boca Raton, FL: CRC Press, p. 4 and 87–94, 1996.

Resolving Tree

A tree of LINKS obtained by repeatedly choosing a crossing, applying the SKEIN RELATIONSHIP to obtain two simpler LINKS, and repeating the process. The DEPTH of a resolving tree is the number of levels of links, not including the top. The DEPTH of the LINK is the minimal depth for any resolving tree of that LINK.

Resonance Overlap

Isolated resonances in a DYNAMICAL SYSTEM can cause considerable distortion of preserved TORI in their NEIGHBORHOOD, but they do not introduce any CHAOS into a system. However, when two or more resonances are simultaneously present, they will render a system nonintegrable. Furthermore, if they are sufficiently "close" to each other, they will result in the appearance of widespread (large-scale) CHAOS.

To investigate this problem, Walker and Ford (1969) took the integrable Hamiltonian

$$H_0(I_1, I_2) = I_1 + I_2 - I_1^2 - 3I_1I_2 + {I_2}^2$$

and investigated the effect of adding a 2:2 resonance and a 3:2 resonance

$$H(\mathbf{I}, \theta) = H_0(\mathbf{I}) + \alpha I_1 I_2 \cos(2\theta_1 - 2\theta_2) + \beta {I_1}^{3/2} I_2 \cos(2\theta_1 - 3\theta_2).$$

At low energies, the resonant zones are well-separated. As the energy increases, the zones overlap and a "macroscopic zone of instability" appears. When the overlap starts, many higher-order resonances are also involved so fairly large areas of PHASE SPACE have their TORI destroyed and the ensuing CHAOS is "widespread" since trajectories are now free to wander between regions that previously were separated by nonresonant TORI.

Walker and Ford (1969) were able to numerically predict the energy at which the overlap of the resonances first occurred. They plotted the θ_2-axis intercepts of the inner 2:2 and the outer 2:3 separatrices as a function of total energy. The energy at which they crossed was found to be identical to that at which 2:2 and 2:3 resonance zones began to overlap.

see also CHAOS, RESONANCE OVERLAP METHOD

References

Walker, G. H. and Ford, J. "Amplitude Instability and Ergodic Behavior for Conservative Nonlinear Oscillator Systems." *Phys. Rev.* **188**, 416–432, 1969.

Resonance Overlap Method

A method for predicting the onset of widespread CHAOS.

see also GREENE'S METHOD

References

Chirikov, B. V. "A Universal Instability of Many-Dimensional Oscillator Systems." *Phys. Rep.* **52**, 264–379, 1979.

Tabor, M. *Chaos and Integrability in Nonlinear Dynamics: An Introduction.* New York: Wiley, pp. 154–163, 1989.

Restricted Divisor Function

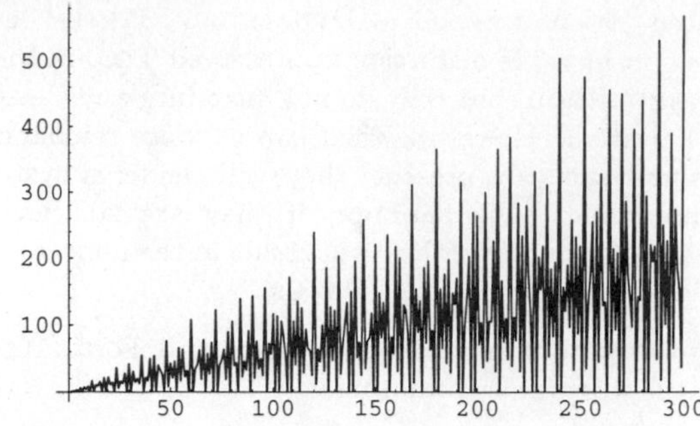

The sum of the ALIQUOT DIVISORS of n, given by

$$s(n) \equiv \sigma(n) - n,$$

where $\sigma(n)$ is the DIVISOR FUNCTION. The first few values are 0, 1, 1, 3, 1, 6, 1, 7, 4, 8, 1, 16, ... (Sloane's A001065).

see also DIVISOR FUNCTION

References

Sloane, N. J. A. Sequence A001065/M2226 in "An On-Line Version of the Encyclopedia of Integer Sequences."

Restricted Growth Function

see RESTRICTED GROWTH STRING

Restricted Growth String

For a SET PARTITION of n elements, the n-character string $a_1a_2\ldots a_n$ in which each character gives the BLOCK ($\mathbf{B}_0$, $\mathbf{B}_1$, ...) into which the corresponding element belongs is called the restricted growth string (or sometimes the RESTRICTED GROWTH FUNCTION). For example, for the SET PARTITION $\{\{1\},\{2\},\{3,4\}\}$, the restricted growth string would be 0122. If the BLOCKS are "sorted" so that $a_1 = 0$, then the restricted growth string satisfies the INEQUALITY

$$a_{i+1} \leq 1 + \max\{a_1, a_2, \ldots, a_i\}$$

for $i = 1, 2, \ldots, n-1$.

References

Ruskey, F. "Info About Set Partitions." `http://sue.csc.uvic.ca/~cos/inf/setp/SetPartitions.html`.

Resultant

Given a POLYNOMIAL $p(x)$ of degree n with roots α_i, $i = 1, \ldots, n$ and a POLYNOMIAL $q(x)$ of degree m with roots β_j, $j = 1, \ldots, m$, the resultant is defined by

$$R(p,q) = \prod_{i=1}^{n}\prod_{j=1}^{m}(\beta_j - \alpha_i).$$

There exists an ALGORITHM similar to the EUCLIDEAN ALGORITHM for computing resultants (Pohst and Zassenhaus 1989). The resultant is the DETERMINANT of the corresponding SYLVESTER MATRIX. Given p and q, then

$$h(x) = R(q(t), p(x-t))$$

is a POLYNOMIAL of degree mn, having as its roots all sums of the form $\alpha_i + \beta_j$.

see also DISCRIMINANT (POLYNOMIAL), SYLVESTER MATRIX

References

Pohst, M. and Zassenhaus, H. *Algorithmic Algebraic Number Theory.* Cambridge, England: Cambridge University Press, 1989.

Wagon, S. *Mathematica in Action.* New York: W. H. Freeman, p. 348, 1991.

Retardance

A shift in PHASE.

see also PHASE

Reuleaux Polygon

A curvilinear polygon built up of circular ARCS. The Reuleaux polygon is a generalization of the REULEAUX TRIANGLE.

see also CURVE OF CONSTANT WIDTH, REULEAUX TRIANGLE

References

Wagon, S. *Mathematica in Action.* New York: W. H. Freeman, pp. 52–54, 1991.

Reuleaux Triangle

A CURVE OF CONSTANT WIDTH constructed by drawing arcs from each VERTEX of an EQUILATERAL TRIANGLE between the other two VERTICES. It is the basis for the Harry Watt square drill bit. It has the smallest AREA for a given width of any CURVE OF CONSTANT WIDTH.

The AREA of each meniscus-shaped portion is

$$A = \tfrac{1}{6}\pi r^2 - \tfrac{1}{2}r\left(\frac{\sqrt{3}}{2}r\right) = \left(\frac{\pi}{6} - \frac{\sqrt{3}}{4}\right)r^2, \quad (1)$$

where we have subtracted the AREA of the wedge from that of the EQUILATERAL TRIANGLE. The total AREA is then

$$A = 3\left(\frac{\pi}{6} - \frac{\sqrt{3}}{4}\right)r^2 + \frac{\sqrt{3}}{4}r^2 = \frac{\pi - \sqrt{3}}{2}r^2. \quad (2)$$

When rotated in a square, the fractional AREA covered is

$$A_{\text{covered}} = 2\sqrt{3} + \tfrac{1}{6}\pi = 0.9877700392\ldots. \quad (3)$$

The center does *not* stay fixed as the TRIANGLE is rotated, but moves along a curve composed of four arcs of an ELLIPSE (Wagon 1991).

see also CURVE OF CONSTANT WIDTH, FLOWER OF LIFE, PIECEWISE CIRCULAR CURVE, REULEAUX POLYGON

References

Bogomolny, A. "Shapes of Constant Width." http://www.cut-the-knot.com/do_you_know/cwidth.html.

Eppstein, D. "Reuleaux Triangles." http://www.ics.uci.edu/~eppstein/junkyard/reuleaux.html.

Reuleaux, F. *The Kinematics of Machinery.* New York: Dover, 1963.

Wagon, S. *Mathematica in Action.* New York: W. H. Freeman, pp. 52–54 and 381–383, 1991.

Yaglom, I. M. and Boltyansky, B. G. *Convex Shapes.* Moscow: Nauka, 1951.

Reversal

The reversal of a decimal number $abc\cdots$ is $\cdots cba$. Ball and Coxeter (1987) consider numbers whose reversals are integral multiples of themselves. PALINDROMIC NUMBER and numbers ending with a ZERO are trivial examples. The first few nontrivial examples are 8712, 9801, 87912, 98901, 879912, 989901, 8799912, 9899901, 87128712, 87999912, 98019801, 98999901, ... (Sloane's A031877). The pattern continues for large numbers, with numbers of the form $87\underbrace{9\cdots9}12$ equal to 4 times their reversals and numbers of the form $98\underbrace{9\cdots9}01$ equal to 9 times their reversals. In addition, runs of numbers of either of these forms can be concatenated to yield numbers of the form $87\underbrace{9\cdots9}12\cdots87\underbrace{9\cdots9}12$, equal to 4 times their reversals, and $98\underbrace{9\cdots9}01\cdots98\underbrace{9\cdots9}01$, equal to 9 times their reversals.

The product of a 2-digit number and its reversal is never a SQUARE NUMBER except when the digits are the same (Ogilvy 1988). Numbers whose product is the reversal of the products of their reversals include

$$312 \times 221 = 68952$$

$$213 \times 122 = 25986$$

(Ball and Coxeter 1987, p. 14).

see also RATS SEQUENCE

References

Ball, W. W. R. and Coxeter, H. S. M. *Mathematical Recreations and Essays, 13th ed.* New York: Dover, pp. 14–15, 1987.

Ogilvy, C. S. and Anderson, J. T. *Excursions in Number Theory.* New York: Dover, pp. 88–89, 1988.

Sloane, N. J. A. Sequence A031877 in "An On-Line Version of the Encyclopedia of Integer Sequences."

Reverse Greedy Algorithm

An algorithm for computing a UNIT FRACTION.

see also GREEDY ALGORITHM, UNIT FRACTION

Reversion of Series

see SERIES REVERSION

Reverse-Then-Add Sequence

An integer sequence produced by the 196-ALGORITHM.

see also 196-ALGORITHM, SORT-THEN-ADD SEQUENCE

Reznik's Identity

For P and Q POLYNOMIALS in n variables,

$$|P\cdot Q|_2{}^2 = \sum_{i_1,\ldots,i_n\geq 0} \frac{|P^{i_1,\ldots,i_n)}(D_1,\ldots,D_n)Q(x_1,\ldots,x_n)|_2{}^2}{i_1!\cdots i_n!},$$

where $D_i \equiv \partial/\partial x_i$, $|X|_2$ is the BOMBIERI NORM, and

$$P^{(i_1,\ldots,i_n)} = D_1^{i_1}\cdots D_n^{i_n}P.$$

BOMBIERI'S INEQUALITY follows from this identity.

see also BEAUZAMY AND DÉGOT'S IDENTITY

Rhodonea

see ROSE

Rhomb

see RHOMBUS

Rhombic Dodecahedral Number

A FIGURATE NUMBER which is constructed as a centered CUBE with a SQUARE PYRAMID appended to each face,

$$RhoDod_n = CCub_n + 6P_{n-1} = (2n-1)(2n^2-2n+1),$$

where $CCub_n$ is a CENTERED CUBE NUMBER and P_n is a PYRAMIDAL NUMBER. The first few are 1, 15, 65, 175, 369, 671, ... (Sloane's A005917). The GENERATING FUNCTION of the rhombic dodecahedral numbers is

$$\frac{x(1+11x+11x^2+x^3)}{(x-1)^4} = x+15x^2+65x^3+175x^4+\ldots.$$

see also OCTAHEDRAL NUMBER

References

Conway, J. H. and Guy, R. K. *The Book of Numbers.* New York: Springer-Verlag, pp. 53–54, 1996.

Sloane, N. J. A. Sequence A005917/M4968 in "An On-Line Version of the Encyclopedia of Integer Sequences."

Rhombic Dodecahedron

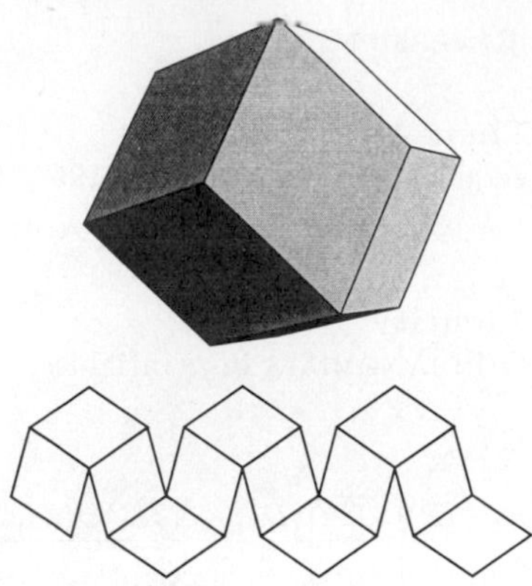

The DUAL POLYHEDRON of the CUBOCTAHEDRON, also sometimes called the RHOMBOIDAL DODECAHEDRON (Cotton 1990). Its 14 vertices are joined by 12 RHOMBUSES, and one possible way to construct it is known as the BAUSPIEL. The rhombic dodecahedron is a ZONOHEDRON and a SPACE-FILLING POLYHEDRON. The vertices are given by $(\pm 1, \pm 1, \pm 1)$, $(\pm 2, 0, 0)$, $(0, \pm 2, 0)$, $(0, 0, \pm 2)$.

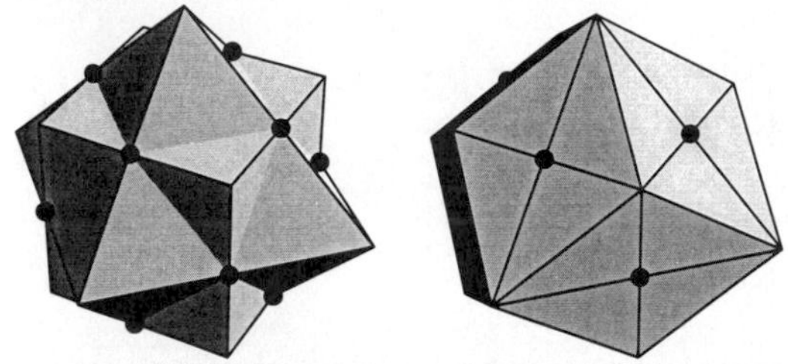

The edges of the CUBE-OCTAHEDRON COMPOUND intersecting in the points plotted above are the diagonals of RHOMBUSES, and the 12 RHOMBUSES form a rhombic dodecahedron (Ball and Coxeter 1987).

see also BAUSPIEL, CUBE-OCTAHEDRON COMPOUND, DODECAHEDRON, PYRITOHEDRON, RHOMBIC TRIACONTAHEDRON, RHOMBUS, TRIGONAL DODECAHEDRON, ZONOHEDRON

References

Ball, W. W. R. and Coxeter, H. S. M. *Mathematical Recreations and Essays, 13th ed.* New York: Dover, p. 137, 1987.

Cotton, F. A. *Chemical Applications of Group Theory, 3rd ed.* New York: Wiley, p. 62, 1990.

Rhombic Icosahedron

A ZONOHEDRON which can be derived from the TRIACONTAHEDRON by removing any one of the zones and bringing together the two pieces into which the remainder of the surface is thereby divided.

References

Ball, W. W. R. and Coxeter, H. S. M. *Mathematical Recreations and Essays, 13th ed.* New York: Dover, p. 143, 1987.

Bilinski, S. "Über die Rhombenisoeder." *Glasnik Mat.-Fiz. Astron. Društro Mat. Fiz. Hrvatske Ser. II* **15**, 251–263, 1960.

Rhombic Polyhedron

A POLYHEDRON with extra square faces, given by the SCHLÄFLI SYMBOL $r\left\{ {p \atop q} \right\}$.

see also RHOMBIC DODECAHEDRON, RHOMBIC ICOSAHEDRON, RHOMBIC TRIACONTAHEDRON, SNUB POLYHEDRON, TRUNCATED POLYHEDRON

Rhombic Triacontahedron

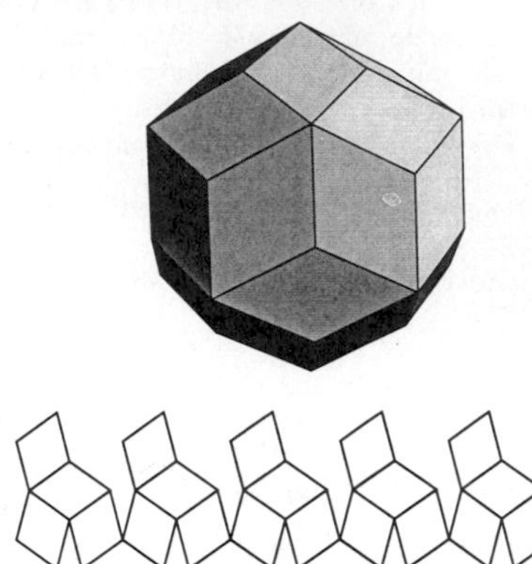

A ZONOHEDRON which is the DUAL POLYHEDRON of the ICOSIDODECAHEDRON. It is composed of 30 RHOMBUSES joined at 32 vertices. Ede (1958) enumerates 13 basic series of stellations of the rhombic triacontahedron, the total number of which is extremely large. Messer (1995) describes 226 stellations. The intersecting edges of the DODECAHEDRON-ICOSAHEDRON COMPOUND form the diagonals of 30 RHOMBUSES which comprise the TRIACONTAHEDRON. The CUBE 5-COMPOUND has the 30 facial planes of the rhombic triacontahedron (Ball and Coxeter 1987).

see also CUBE 5-COMPOUND, DODECAHEDRON-ICOSAHEDRON COMPOUND, ICOSIDODECAHEDRON, RHOMBIC DODECAHEDRON, RHOMBUS, ZONOHEDRON

References

Ball, W. W. R. and Coxeter, H. S. M. *Mathematical Recreations and Essays, 13th ed.* New York: Dover, p. 137, 1987.

Bulatov, V.v "Stellations of Rhombic Triacontahedron." `http://www.physics.orst.edu/~bulatov/polyhedra/rtc/`.

Cundy, H. and Rollett, A. *Mathematical Models, 3rd ed.* Stradbroke, England: Tarquin Pub., p. 127, 1989.

Ede, J. D. "Rhombic Triacontahedra." *Math. Gazette* **42**, 98–100, 1958.

Messer, P. W. "Les étoilements du rhombitricontaèdre et plus." *Structural Topology* No. 21, 25–46, 1995.

Rhombicosacron

The DUAL POLYHEDRON of the RHOMBICOSAHEDRON.

Rhombicosahedron

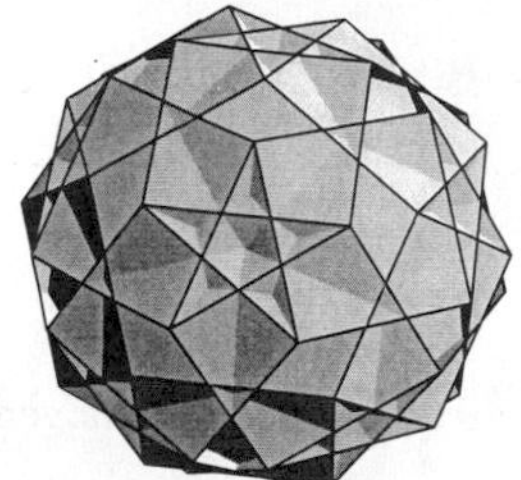

The UNIFORM POLYHEDRON U_{56} whose DUAL POLYHEDRON is the RHOMBICOSACRON. It has WYTHOFF SYMBOL $2\,3\,\frac{\frac{5}{4}}{\frac{5}{2}}$. Its faces are $20\{6\} + 30\{4\}$. The CIRCUMRADIUS for unit edge length is

$$R = \tfrac{1}{2}\sqrt{7}.$$

References

Wenninger, M. J. *Polyhedron Models.* Cambridge, England: Cambridge University Press, pp. 149–150, 1971.

Rhombicosidodecahedron

see BIGYRATE DIMINISHED RHOMBICOSIDODECAHEDRON, DIMINISHED RHOMBICOSIDODECAHEDRON, GREAT RHOMBICOSIDODECAHEDRON (ARCHIMEDEAN), GREAT RHOMBICOSIDODECAHEDRON (UNIFORM), GYRATE BIDIMINISHED RHOMBICOSIDODECAHEDRON, GYRATE RHOMBICOSIDODECAHEDRON, METABIDIMINISHED RHOMBICOSIDODECAHEDRON, METABIGYRATE RHOMBICOSIDODECAHEDRON, METAGYRATE DIMINISHED RHOMBICOSIDODECAHEDRON, PARABIDIMINISHED RHOMBICOSIDODECAHEDRON, PARABIGYRATE RHOMBICOSIDODECAHEDRON, PARAGYRATE DIMINISHED RHOMBICOSIDODECAHEDRON, SMALL RHOMBICOSIDODECAHEDRON, TRIDIMINISHED RHOMBICOSIDODECAHEDRON, TRIGYRATE RHOMBICOSIDODECAHEDRON

Rhombicuboctahedron

see GREAT RHOMBICUBOCTAHEDRON (ARCHIMEDEAN), GREAT RHOMBICUBOCTAHEDRON (UNIFORM), SMALL RHOMBICUBOCTAHEDRON

Rhombidodecadodecahedron

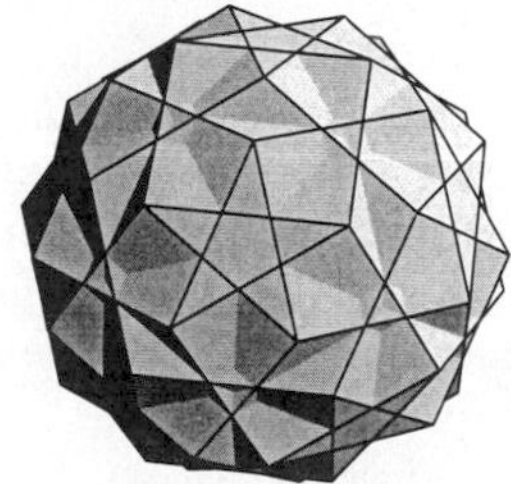

The UNIFORM POLYHEDRON U_{38} whose DUAL POLYHEDRON is the MEDIAL DELTOIDAL HEXECONTAHEDRON. It has SCHLÄFLI SYMBOL $r\left\{\begin{smallmatrix}\frac{5}{2}\\ 5\end{smallmatrix}\right\}$ and WYTHOFF SYMBOL $\frac{5}{2}\,5\,|\,2$. Its faces are $12\{\frac{5}{2}\} + 30\{4\} + 12\{5\}$. The CIRCUMRADIUS for unit edge length is

$$R = \tfrac{1}{2}\sqrt{7}.$$

References

Wenninger, M. J. *Polyhedron Models.* Cambridge, England: Cambridge University Press, pp. 116–117, 1989.

Rhombihexacron

see GREAT RHOMBIHEXACRON, SMALL RHOMBIHEXACRON

Rhombihexahedron

see GREAT RHOMBIHEXAHEDRON, SMALL RHOMBIHEXAHEDRON

Rhombitruncated Cuboctahedron

see GREAT RHOMBICUBOCTAHEDRON (ARCHIMEDEAN)

Rhombitruncated Icosidodecahedron

see GREAT RHOMBICOSIDODECAHEDRON (ARCHIMEDEAN)

Rhombohedron

A PARALLELEPIPED bounded by six congruent RHOMBS.

see also PARALLELEPIPED, RHOMB

References

Ball, W. W. R. and Coxeter, H. S. M. *Mathematical Recreations and Essays, 13th ed.* New York: Dover, pp. 142 and 161, 1987.

Rhomboid

A PARALLELOGRAM in which angles are oblique and adjacent sides are of unequal length.

see also DIAMOND, LOZENGE, PARALLELOGRAM, QUADRILATERAL, RHOMBUS, SKEW QUADRILATERAL, TRAPEZIUM, TRAPEZOID

Rhomboidal Dodecahedron

see RHOMBIC DODECAHEDRON

Rhombus

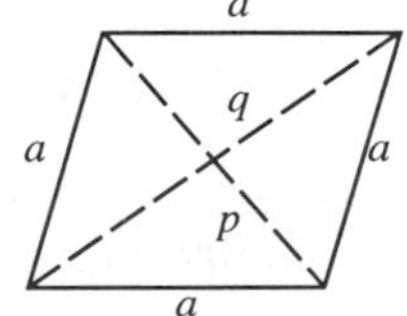

A QUADRILATERAL with both pairs of opposite sides PARALLEL and all sides the same length, i.e., an equilateral PARALLELOGRAM. The word RHOMB is sometimes used instead of rhombus. The DIAGONALS p and q of a rhombus satisfy

$$p^2 + q^2 = 4a^2,$$

and the AREA is

$$A = \tfrac{1}{2}pq.$$

A rhombus whose ACUTE ANGLES are 45° is called a LOZENGE.

see also DIAMOND, LOZENGE, PARALLELOGRAM, QUADRILATERAL, RHOMBIC DODECAHEDRON, RHOMBIC ICOSAHEDRON, RHOMBIC TRIACONTAHEDRON, RHOMBOID, SKEW QUADRILATERAL, TRAPEZIUM, TRAPEZOID

References

Beyer, W. H. (Ed.) *CRC Standard Mathematical Tables, 28th ed.* Boca Raton, FL: CRC Press, p. 123, 1987.

Rhumb Line

see LOXODROME

Ribbon Knot

If the KNOT K is the boundary $K = f(\mathbb{S}^1)$ of a singular disk $f : \mathbb{D} \to \mathbb{S}^3$ which has the property that each self-intersecting component is an arc $A \subset f(\mathbb{D}^2)$ for which $f^{-1}(A)$ consists of two arcs in $\mathbb{D}^2$, one of which is interior, then K is said to be a ribbon knot. Every ribbon knot is a SLICE KNOT, and it is conjectured that every SLICE KNOT is a ribbon knot.

see also SLICE KNOT

References

Rolfsen, D. *Knots and Links.* Wilmington, DE: Publish or Perish Press, p. 225, 1976.

Ribet's Theorem

If the TANIYAMA-SHIMURA CONJECTURE holds for all semistable ELLIPTIC CURVES, then FERMAT'S LAST THEOREM is true. Before its proof by Ribet in 1986, the theorem had been called the epsilon conjecture. It had its roots in a surprising result of G. Frey.

see also ELLIPTIC CURVE, FERMAT'S LAST THEOREM, MODULAR FORM, MODULAR FUNCTION, TANIYAMA-SHIMURA CONJECTURE

Riccati-Bessel Functions

$$S_n(z) \equiv zj_n(z) = \sqrt{\frac{\pi z}{2}} J_{n+1/2}(z)$$

$$C_n(z) \equiv -zn_n(z) = -\sqrt{\frac{\pi z}{2}} N_{n+1/2}(z),$$

where $j_n(z)$ and $n_n(z)$ are SPHERICAL BESSEL FUNCTIONS OF THE FIRST and SECOND KIND.

References

Abramowitz, M. and Stegun, C. A. (Eds.). "Riccati-Bessel Functions." §10.3 in *Handbook of Mathematical Functions with Formulas, Graphs, and Mathematical Tables, 9th printing.* New York: Dover, p. 445, 1972.

Riccati Differential Equation

$$y' = P(z) + Q(z)y + R(z)y^2, \tag{1}$$

where $y' \equiv dy/dz$. The transformation

$$w \equiv -\frac{y'}{yR(z)} \tag{2}$$

leads to the second-order linear homogeneous equation

$$R(z)y'' - [R'(z) + Q(z)R(z)]y' + [R(z)]^2 P(z)y = 0. \tag{3}$$

Another equation sometimes called the Riccati differential equation is

$$z^2 w'' + [z^2 - n(n+1)]w = 0, \tag{4}$$

which has solutions

$$w = Azj_n(z) + Bzy_n(z). \tag{5}$$

Yet another form of "the" Riccati differential equation is

$$\frac{dy}{dz} = az^n + by^2, \tag{6}$$

which is solvable by algebraic, exponential, and logarithmic functions only when $n = -4m/(2m \pm 1)$, for $m = 0$, 1, 2, ...

References

Abramowitz, M. and Stegun, C. A. (Eds.). "Riccati-Bessel Functions." §10.3 in *Handbook of Mathematical Functions with Formulas, Graphs, and Mathematical Tables, 9th printing.* New York: Dover, p. 445, 1972.

Glaisher, J. W. L. "On Riccati's Equation." *Quart. J. Pure Appl. Math.* **11**, 267–273, 1871.

Ricci Curvature

The mathematical object which controls the growth rate of the volume of metric balls in a MANIFOLD.

see also BISHOP'S INEQUALITY, MILNOR'S THEOREM

Ricci Tensor

$$R_{\mu\kappa} \equiv R^{\lambda}{}_{\mu\lambda\kappa},$$

where $R^{\lambda}{}_{\mu\lambda\kappa}$ is the RIEMANN TENSOR.

see also CURVATURE SCALAR, RIEMANN TENSOR

Rice Distribution

$$P(Z) = \frac{Z}{\sigma^2} \exp\left(-\frac{Z^2 + |V|^2}{2\sigma^2}\right) I_0\left(\frac{Z|V|}{\sigma^2}\right),$$

where $I_0(z)$ is a MODIFIED BESSEL FUNCTION OF THE FIRST KIND and $Z > 0$. For a derivation, see Papoulis (1962). For $|V| = 0$, this reduces to the RAYLEIGH DISTRIBUTION.

see also RAYLEIGH DISTRIBUTION

References

Papoulis, A. *The Fourier Integral and Its Applications.* New York: McGraw-Hill, 1962.

Richard's Paradox

It is possible to describe a set of POSITIVE INTEGERS that cannot be listed in a book containing a set of counting numbers on each consecutively numbered page.

Richardson Extrapolation

The consideration of the result of a numerical calculation as a function of an adjustable parameter (usually the step size). The function can then be fitted and evaluated at $h = 0$ to yield very accurate results. Press *et al.* (1992) describe this process as turning lead into gold. Richardson extrapolation is one of the key ideas used in the popular and robust BULIRSCH-STOER ALGORITHM of solving ORDINARY DIFFERENTIAL EQUATIONS.

see also BULIRSCH-STOER ALGORITHM

References

Acton, F. S. *Numerical Methods That Work, 2nd printing.* Washington, DC: Math. Assoc. Amer., p. 106, 1990.

Press, W. H.; Flannery, B. P.; Teukolsky, S. A.; and Vetterling, W. T. "Richardson Extrapolation and the Bulirsch-Stoer Method." §16.4 in *Numerical Recipes in FORTRAN: The Art of Scientific Computing, 2nd ed.* Cambridge, England: Cambridge University Press, pp. 718–725, 1992.

Richardson's Theorem

Let R be the class of expressions generated by

1. The RATIONAL NUMBERS and the two REAL NUMBERS π and $\ln 2$,
2. The variable x,
3. The operations of ADDITION, MULTIPLICATION, and composition, and
4. The SINE, EXPONENTIAL, and ABSOLUTE VALUE functions.

Then if $E \in R$, the predicate "$E = 0$" is recursively UNDECIDABLE.

see also RECURSION, UNDECIDABLE

References

Caviness, B. F. "On Canonical Forms and Simplification." *J. Assoc. Comp. Mach.* **17**, 385–396, 1970.

Petkovšek, M.; Wilf, H. S.; and Zeilberger, D. *A=B.* Wellesley, MA: A. K. Peters, 1996.

Richardson, D. "Some Unsolvable Problems Involving Elementary Functions of a Real Variable." *J. Symbolic Logic* **33**, 514–520, 1968.

Ridders' Method

A variation of the FALSE POSITION METHOD for finding ROOTS which fits the function in question with an exponential.

see also FALSE POSITION METHOD

References

Ostrowski, A. M. Ch. 12 in *Solutions of Equations and Systems of Equations, 2nd ed.* New York: Academic Press, 1966.

Press, W. H.; Flannery, B. P.; Teukolsky, S. A.; and Vetterling, W. T. "Secant Method, False Position Method, and Ridders' Method." §9.2 in *Numerical Recipes in FORTRAN: The Art of Scientific Computing, 2nd ed.* Cambridge, England: Cambridge University Press, pp. 347–352, 1992.

Ralston, A. and Rabinowitz, P. §8.3 in *A First Course in Numerical Analysis, 2nd ed.* New York: McGraw-Hill, 1978.

Ridders, C. F. J. "A New Algorithm for Computing a Single Root of a Real Continuous Function." *IEEE Trans. Circuits Systems* **26**, 979–980, 1979.

Ridge

An $(n-2)$-D FACE of an n-D POLYTOPE.

see also POLYTOPE

Riemann-Christoffel Tensor

see RIEMANN TENSOR

Riemann Curve Theorem

If two algebraic plane curves with only ordinary singular points and CUSPS are related such that the coordinates of a point on either are RATIONAL FUNCTIONS of a corresponding point on the other, then the curves have the same GENUS (CURVE). This can be stated equivalently as the GENUS of a curve is unaltered by a BIRATIONAL TRANSFORMATION.

References

Coolidge, J. L. *A Treatise on Algebraic Plane Curves.* New York: Dover, p. 120, 1959.

Riemann Differential Equation

see RIEMANN P-DIFFERENTIAL EQUATION

Riemann's Formula

$$J(x) = \mathrm{Li}(x) - \sum \mathrm{Li}(x^\rho) + \ln 2 \int_x^\infty \frac{dt}{t(t^2-1)\ln t},$$

where $\mathrm{Li}(x)$ is the LOGARITHMIC INTEGRAL, the sum is taken over all nontrivial zeros ρ (i.e., those other than $-2, -4, \ldots$) of the RIEMANN ZETA FUNCTION $\zeta(s)$, and $J(x)$ is RIEMANN WEIGHTED PRIME-POWER COUNTING FUNCTION.

see also LOGARITHMIC INTEGRAL, PRIME NUMBER THEOREM, RIEMANN WEIGHTED PRIME-POWER COUNTING FUNCTION, RIEMANN ZETA FUNCTION

Riemann Function

The function obtained by approximating the RIEMANN WEIGHTED PRIME-POWER COUNTING FUNCTION J_2 in

$$\pi(x) = \sum_{n=1}^{\infty} \frac{\mu(n)}{n} J_2(x^{1/n}) \tag{1}$$

by the LOGARITHMIC INTEGRAL $\mathrm{Li}(x)$. This gives

$$R(n) \equiv 1 + \sum_{k=1}^{\infty} \frac{1}{k\zeta(k+1)} \frac{(\ln n)^k}{k!} \quad (2)$$

$$= \sum_{m=1}^{\infty} \frac{\mu(m)}{m} \mathrm{Li}(n^{1/m}), \quad (3)$$

where $\zeta(z)$ is the RIEMANN ZETA FUNCTION, $\mu(n)$ is the MÖBIUS FUNCTION, and $\mathrm{Li}(x)$ is the LOGARITHMIC INTEGRAL. Then

$$\pi(x) = R(x) - \sum_{\rho} R(x^{\rho}), \quad (4)$$

where π is the PRIME COUNTING FUNCTION. Ramanujan independently derived the formula for $R(n)$, but nonrigorously (Berndt 1994, p. 123).

see also MANGOLDT FUNCTION, PRIME NUMBER THEOREM, RIEMANN-MANGOLDT FUNCTION, RIEMANN ZETA FUNCTION

References
Berndt, B. C. *Ramanujan's Notebooks, Part IV.* New York: Springer-Verlag, 1994.
Conway, J. H. and Guy, R. K. *The Book of Numbers.* New York: Springer-Verlag, pp. 144–145, 1996.
Riesel, H. and Göhl, G. "Some Calculations Related to Riemann's Prime Number Formula." *Math. Comput.* **24**, 969–983, 1970.
Wagon, S. *Mathematica in Action.* New York: W. H. Freeman, pp. 28–29 and 362–372, 1991.

Riemann Hypothesis

First published in Riemann (1859), the Riemann hypothesis states that the nontrivial ROOTS of the RIEMANN ZETA FUNCTION

$$\zeta(s) \equiv \sum_{n=1}^{\infty} \frac{1}{n^s}, \quad (1)$$

where $s \in \mathbb{C}$ (the COMPLEX NUMBERS), all lie on the "CRITICAL LINE" $\Re[s] = 1/2$, where $\Re[z]$ denotes the REAL PART of z. The Riemann hypothesis is also known as ARTIN'S CONJECTURE.

In 1914, Hardy proved that an INFINITE number of values for s can be found for which $\zeta(s) = 0$ and $\Re[s] = 1/2$. However, it is not known if *all* nontrivial roots s satisfy $\Re[s] = 1/2$, so the conjecture remains open. André Weil proved the Riemann hypothesis to be true for field functions (Weil 1948, Eichler 1966, Ball and Coxeter 1987). In 1974, Levin showed that at least 1/3 of the ROOTS must lie on the CRITICAL LINE (Le Lionnais 1983), a result which has since been sharpened to 40% (Vardi 1991, p. 142). It is known that the zeros are symmetrical placed about the line $\Im[s] = 0$.

The Riemann hypothesis is equivalent to $\Lambda \leq 0$, where Λ is the DE BRUIJN-NEWMAN CONSTANT (Csordas *et al.* 1994). It is also equivalent to the assertion that for some constant c,

$$|\mathrm{Li}(x) - \pi(x)| \leq c\sqrt{x} \ln x, \quad (2)$$

where $\mathrm{Li}(x)$ is the LOGARITHMIC INTEGRAL and π is the PRIME COUNTING FUNCTION (Wagon 1991).

The hypothesis was computationally tested and found to be true for the first 2×10^8 zeros by Brent *et al.* (1979), a limit subsequently extended to the first $1.5 \times 10^9 + 1$ zeros by Brent *et al.* (1979). Brent's calculation covered zeros $\sigma + it$ in the region $0 < t < 81,702,130.19$.

There is also a finite analog of the Riemann hypothesis concerning the location of zeros for function fields defined by equations such as

$$ay^l + bz^m + c = 0. \quad (3)$$

This hypothesis, developed by Weil, is analogous to the usual Riemann hypothesis. The number of solutions for the particular cases $(l, m) = (2, 2)$, (3,3), (4,4), and (2,4) were known to Gauss.

see also CRITICAL LINE, EXTENDED RIEMANN HYPOTHESIS, GRONWALL'S THEOREM, MERTENS CONJECTURE, MILLS' CONSTANT, RIEMANN ZETA FUNCTION

References
Ball, W. W. R. and Coxeter, H. S. M. *Mathematical Recreations and Essays, 13th ed.* New York: Dover, p. 75, 1987.
Brent, R. P.; Vandelune, J.; te Riele, H. J. J.; and Winter, D. T. "On the Zeros of the Riemann Zeta Function in the Critical Strip. I." *Math. Comput.* **33**, 1361–1372, 1979.
Brent, R. P.; Vandelune, J.; te Riele, H. J. J.; and Winter, D. T. "On the Zeros of the Riemann Zeta Function in the Critical Strip. II." *Math. Comput.* **39**, 681–688, 1982. Abstract available at `ftp://nimbus.anu.edu.au/pub/Brent/rpb070a.dvi.Z`.
Csordas, G.; Smith, W.; and Varga, R. S. "Lehmer Pairs of Zeros, the de Bruijn-Newman Constant and the Riemann Hypothesis." *Constr. Approx.* **10**, 107–129, 1994.
Eichler, M. *Introduction to the Theory of Algebraic Numbers and Functions.* New York: Academic Press, 1966.
Le Lionnais, F. *Les nombres remarquables.* Paris: Hermann, p. 25, 1983.
Odlyzko, A. "The 10^{20}th Zero of the Riemann Zeta Function and 70 Million of Its Neighbors."
Riemann, B. "Über die Anzahl der Primzahlen unter einer gegebenen Grösse," *Mon. Not. Berlin Akad.*, pp. 671–680, Nov. 1859.
Sloane, N. J. A. Sequence A002410/M4924 in "An On-Line Version of the Encyclopedia of Integer Sequences."
Vandelune, J. and te Riele, H. J. J. "On The Zeros of the Riemann Zeta-Function in the Critical Strip. III." *Math. Comput.* **41**, 759–767, 1983.
Vandelune, J.; te Riele, H. J. J.; and Winter, D. T. "On the Zeros of the Riemann Zeta Function in the Critical Strip. IV." *Math. Comput.* **46**, 667–681, 1986.
Wagon, S. *Mathematica in Action.* New York: W. H. Freeman, p. 33, 1991.
Weil, A. *Sur les courbes algébriques et les variét'es qui s'en déduisent.* Paris, 1948.

Riemann Integral

The Riemann integral is the INTEGRAL normally encountered in CALCULUS texts and used by physicists and engineers. Other types of integrals exist (e.g., the LEBESGUE INTEGRAL), but are unlikely to be encountered outside the confines of advanced mathematics texts.

The Riemann integral is based on the JORDAN MEASURE, and defined by taking a limit of a RIEMANN SUM,

$$\int_b^a f(x)\,dx \equiv \lim_{\max \Delta x_k \to 0} \sum_{k=1}^n f(x_k^*)\Delta x_k \qquad (1)$$

$$\iint f(x,y)\,dA \equiv \lim_{\max \Delta A_k \to 0} \sum_{k=1}^n f(x_k^*, y_k^*)\Delta A_k \qquad (2)$$

$$\iiint f(x,y,z)\,dV \equiv \lim_{\max \Delta V_k \to 0} \sum_{k=1}^n f(x_k^*, y_k^*, z_k^*)\Delta V_k, \qquad (3)$$

where $a \leq x \leq b$ and x_k^*, y_k^*, and z_k^* are arbitrary points in the intervals Δx_k, Δy_k, and Δz_k, respectively. The value $\max \Delta x_k$ is called the MESH SIZE of a partition of the interval $[a,b]$ into subintervals Δx_k.

As an example of the application of the Riemann integral definition, find the AREA under the curve $y = x^r$ from 0 to a. Divide (a,b) into n segments, so $\Delta x_k = \frac{b-a}{n} \equiv h$, then

$$f(x_1) = f(0) = 0 \qquad (4)$$

$$f(x_2) = f(\Delta x_k) = h^r \qquad (5)$$

$$f(x_3) = f(2\Delta x_k) = (2h)^r. \qquad (6)$$

By induction

$$f(x_k) = f([k-1]\Delta x_k) = [(k-1)h]^r = h^r(k-1)^r, \qquad (7)$$

so

$$f(x_k)\Delta x_k = h^{r+1}(k-1)^r \qquad (8)$$

$$\sum_{k=1}^n f(x_k)\Delta x_k = h^{r+1}\sum_{k=1}^n (k-1)^r. \qquad (9)$$

For example, take $r = 2$.

$$\sum_{k=1}^n f(x_k)\Delta x_k = h^3 \sum_{k=1}^n (k-1)^2$$

$$= h^3\left(\sum_{k=1}^n k^2 - 2\sum_{k=1}^n k + \sum_{k=1}^n 1\right)$$

$$= h^3\left[\frac{n(n+1)(2n+1)}{6} - 2\frac{n(n+1)}{2} + n\right], \qquad (10)$$

so

$$I \equiv \lim_{n\to\infty} \sum_{k=1}^n f(x_k{}^*)\Delta x_k = \lim_{n\to\infty} \sum_{k=1}^n f(x_k)\Delta x_k$$

$$= \lim_{n\to\infty} h^3\left[\frac{n(n+1)(2n+1)}{6} - 2\frac{n(n+1)}{2} + n\right]$$

$$= a^3 \lim_{n\to\infty}\left[\frac{n(n+1)(2n+1)}{6n^3} - \frac{n(n+1)}{n^3} + \frac{n}{n^3}\right]$$

$$= \tfrac{1}{3}a^3. \qquad (11)$$

see also INTEGRAL, RIEMANN SUM

References

Kestelman, H. "Riemann Integration." Ch. 2 in *Modern Theories of Integration, 2nd rev. ed.* New York: Dover, pp. 33–66, 1960.

Riemann's Integral Theorem

Associated with an irreducible curve of GENUS (CURVE) p, there are p LINEARLY INDEPENDENT integrals of the first sort. The ROOTS of the integrands are groups of the canonical series, and every such group will give rise to exactly one integral of the first sort.

References

Coolidge, J. L. *A Treatise on Algebraic Plane Curves.* New York: Dover, p. 274, 1959.

Riemann-Lebesgue Lemma

Sometimes also called MERCER'S THEOREM.

$$\lim_{n\to\infty} \int_a^b K(\lambda, z) C \sin(nz)\,dz = 0$$

for arbitrarily large C and "nice" $K(\lambda, z)$. Gradshteyn and Ryzhik (1979) state the lemma as follows. If $f(x)$ is integrable on $[\pi, \pi]$, then

$$\lim_{t\to\infty} \int_{-\pi}^{\pi} f(x)\sin(tx)\,dx \to 0$$

and

$$\lim_{t\to\infty} \int_{-\pi}^{\pi} f(x)\cos(tx)\,dx \to 0.$$

References

Gradshteyn, I. S. and Ryzhik, I. M. *Tables of Integrals, Series, and Products, 5th ed.* San Diego, CA: Academic Press, p. 1101, 1979.

Riemann-Mangoldt Function

$$f(x) = \sum_{n>1}^{\infty} \frac{\pi_0(x^{1/n})}{n}$$

$$= \mathrm{Li}(x) - \sum_{\substack{\text{nontrivial } \rho \\ \zeta(\rho)=0}} \mathrm{ei}(\rho \ln x) - \ln 2$$

$$+ \int_x^\infty \frac{dt}{t(t^2-1)\ln t}, \qquad (1)$$

where $\zeta(z)$ is the RIEMANN ZETA FUNCTION, $\mathrm{Li}(x)$ is the LOGARITHMIC INTEGRAL and $\mathrm{ei}(x)$ is the EXPONENTIAL INTEGRAL. The MANGOLDT FUNCTION is given by

$$\Lambda(n) = \begin{cases} \ln p & \text{if } n = p^m \text{ for } (m \geq 1) \text{ and } p \text{ prime} \\ 0 & \text{otherwise} \end{cases} \tag{2}$$

$$-\frac{\zeta'(x)}{\zeta(s)} = \sum_{n=1}^{\infty} \frac{\Lambda(n)}{n^s} \tag{3}$$

for $\Re[s] > 1$.

$$J(x) = \sum_{n \leq x} \frac{\Lambda(n)}{\ln n}. \tag{4}$$

The SUMMATORY Riemann-Mangoldt function is defined by

$$\psi(x) = \sum_{n \leq x} \Lambda(n) = \theta(x) + \theta(x^{1/2}) + \ldots. \tag{5}$$

see also PRIME NUMBER THEOREM, RIEMANN FUNCTION

References

Wagon, S. *Mathematica in Action.* New York: W. H. Freeman, pp. 364–365, 1991.

Riemann Mapping Theorem

Let z_0 be a point in a simply connected region $R \neq \mathbb{C}$. Then there is a unique ANALYTIC FUNCTION $w = f(z)$ mapping R one-to-one onto the DISK $|w| < 1$ such that $f(z_0) = 0$ and $f'(z_0) = 0$. The COROLLARY guarantees that any two simply connected regions except $\mathbb{R}^2$ can be mapped CONFORMALLY onto each other.

Riemann's Moduli Problem

Find an ANALYTIC parameterization of the compact RIEMANN SURFACES in a fixed HOMOMORPHISM class. The AHLFORS-BERS THEOREM proved that RIEMANN'S MODULI SPACE gives the solution.

see also AHLFORS-BERS THEOREM, RIEMANN'S MODULI SPACE

Riemann's Moduli Space

Riemann's moduli space R_p is the space of ANALYTIC EQUIVALENCE CLASSES of RIEMANN SURFACES of fixed GENUS p.

see also AHLFORS-BERS THEOREM, RIEMANN'S MODULI PROBLEM, RIEMANN SURFACE

Riemann P-Differential Equation

$$\frac{d^2u}{dz^2} + \left[\frac{1-\alpha-\alpha'}{z-a} + \frac{1-\beta-\beta'}{z-b} + \frac{1-\gamma-\gamma'}{z-c}\right]\frac{du}{dz} + \left[\frac{\alpha\alpha'(a-b)(a-c)}{z-a} + \frac{\beta\beta'(b-c)(b-a)}{z-b} + \frac{\gamma\gamma'(c-a)(c-b)}{z-c}\right]\frac{u}{(z-a)(z-b)(z-c)} = 0,$$

where

$$\alpha + \alpha' + \beta + \beta' + \gamma + \gamma' = 1.$$

Solutions are RIEMANN P-SERIES (Abramowitz and Stegun 1972, pp. 564–565).

References

Abramowitz, M. and Stegun, C. A. (Eds.). "Riemann's Differential Equation." §15.6 in *Handbook of Mathematical Functions with Formulas, Graphs, and Mathematical Tables, 9th printing.* New York: Dover, pp. 564–565, 1972.

Morse, P. M. and Feshbach, H. *Methods of Theoretical Physics, Part I.* New York: McGraw-Hill, pp. 541–543, 1953.

Riemann P-Series

The solutions to the RIEMANN P-DIFFERENTIAL EQUATION

$$z = P\left\{\begin{matrix} a & b & c \\ \alpha & \beta & \gamma \ ; \ z \\ \alpha' & \beta' & \gamma' \end{matrix}\right\}.$$

Solutions are given in terms of the HYPERGEOMETRIC FUNCTION by

$$u_1 = \left(\frac{z-a}{z-b}\right)^{\alpha} \left(\frac{z-c}{z-b}\right)^{\gamma} \times {}_2F_1(\alpha+\beta+\gamma, \alpha+\beta'+\gamma; 1+\alpha-\alpha'; \lambda)$$

$$u_2 = \left(\frac{z-a}{z-b}\right)^{\alpha'} \left(\frac{z-c}{z-b}\right)^{\gamma} \times {}_2F_1(\alpha'+\beta+\gamma, \alpha'+\beta'+\gamma; 1+\alpha'-\alpha; \lambda)$$

$$u_3 = \left(\frac{z-a}{z-b}\right)^{\alpha} \left(\frac{z-c}{z-b}\right)^{\gamma'} \times {}_2F_1(\alpha+\beta+\gamma', \alpha+\beta'+\gamma'; 1+\alpha-\alpha'; \lambda)$$

$$u_4 = \left(\frac{z-a}{z-b}\right)^{\alpha'} \left(\frac{z-c}{z-b}\right)^{\gamma'} \times {}_2F_1(\alpha'+\beta+\gamma', \alpha'+\beta'+\gamma'; 1+\alpha'-\alpha; \lambda),$$

where

$$\lambda \equiv \frac{(z-a)(c-b)}{(z-b)(c-a)}.$$

References

Abramowitz, M. and Stegun, C. A. (Eds.). "Riemann's Differential Equation." §15.6 in *Handbook of Mathematical Functions with Formulas, Graphs, and Mathematical Tables, 9th printing.* New York: Dover, pp. 564–565, 1972.

Morse, P. M. and Feshbach, H. *Methods of Theoretical Physics, Part I.* New York: McGraw-Hill, pp. 541–543, 1953.

Whittaker, E. T. and Watson, G. N. *A Course in Modern Analysis, 4th ed.* Cambridge, England: Cambridge University Press, pp. 283–284, 1990.

Riemann-Roch Theorem

The dimension of a complete series is equal to the sum of the order and index of specialization of any group, less the GENUS of the base curve

$$r = N + i + p.$$

References

Coolidge, J. L. *A Treatise on Algebraic Plane Curves.* New York: Dover, p. 261, 1959.

Riemann Series Theorem

By a suitable rearrangement of terms, a conditionally convergent SERIES may be made to converge to any desired value, or to DIVERGE.

References

Bromwich, T. J. I'a. and MacRobert, T. M. *An Introduction to the Theory of Infinite Series, 3rd ed.* New York: Chelsea, p. 74, 1991.

Riemann-Siegel Functions

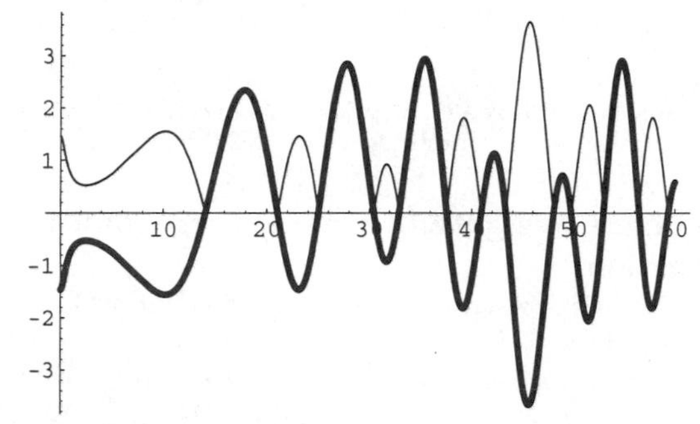

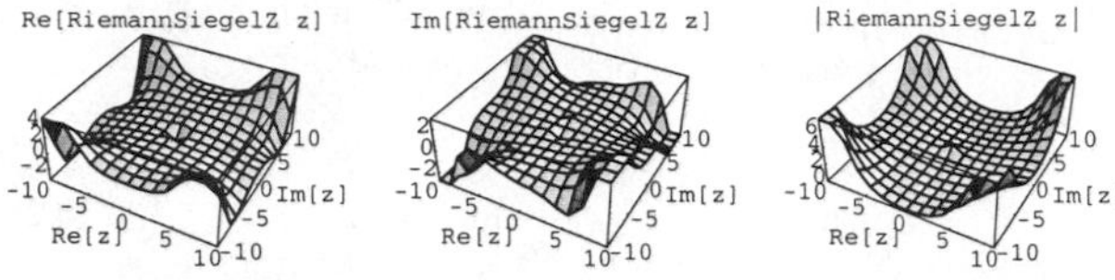

For a REAL POSITIVE t, the Riemann-Siegel Z function is defined by

$$Z(t) \equiv e^{i\vartheta(t)}\zeta(\tfrac{1}{2}+it).$$

The top plot superposes $Z(t)$ (thick line) on $|\zeta(\frac{1}{2}+it)|$, where $\zeta(z)$ is the RIEMANN ZETA FUNCTION.

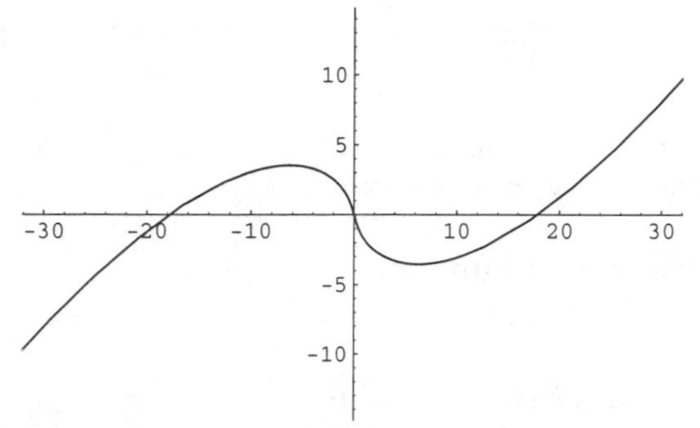

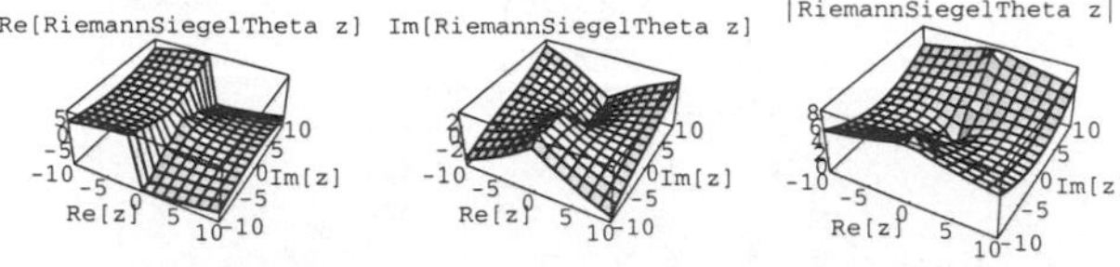

The Riemann-Siegel theta function appearing above is defined by

$$\begin{aligned}\vartheta &\equiv \Im[\ln\Gamma(\tfrac{1}{4}+\tfrac{1}{2}it)-\tfrac{1}{2}t\ln\pi]\\ &= \arg[\Gamma(\tfrac{1}{4}+\tfrac{1}{2}it)]-\tfrac{1}{2}t\ln\pi.\end{aligned}$$

These functions are implemented in *Mathematica*® (Wolfram Research, Champaign, IL) as `RiemannSiegelZ[z]` and `RiemannSiegelTheta[z]`, illustrated above.

see also RIEMANN ZETA FUNCTION

References

Vardi, I. *Computational Recreations in Mathematica.* Reading, MA: Addison-Wesley, p. 143, 1991.

Riemann Space

see METRIC SPACE

Riemann Sphere

A 1-D COMPLEX MANIFOLD $\mathbb{C}^*$, which is the one-point compactification of the COMPLEX numbers $\mathbb{C}\cup\{\infty\}$, together with two charts. For all points in the COMPLEX PLANE, the chart is the IDENTITY MAP from the SPHERE (with infinity removed) to the COMPLEX PLANE. For the point at infinity, the chart neighborhood is the sphere (with the ORIGIN removed), and the chart is given by sending infinity to 0 and all other points z to $1/z$.

Riemann-Stieltjes Integral

see STIELTJES INTEGRAL

Riemann Sum

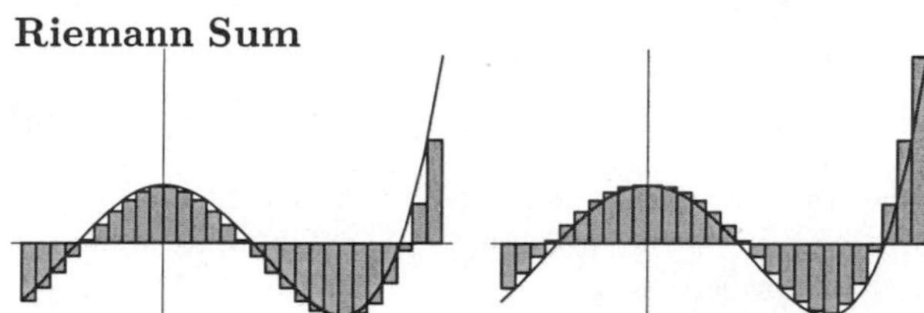

Let a CLOSED INTERVAL $[a,b]$ be partitioned by points $a<x_1<x_2<\ldots<x_{n-1}<b$, the lengths of the resulting intervals between the points are denoted Δx_1, Δx_2, ..., Δx_n. Then the quantity

$$\sum_{k=1}^{n} f(x_k^*)\Delta x_k$$

is called a Riemann sum for a given function $f(x)$ and partition. The value $\max\Delta x_k$ is called the MESH SIZE of the partition. If the LIMIT $\max\Delta x_k\to 0$ exists, this limit is known as the Riemann INTEGRAL of $f(x)$ over the interval $[a,b]$. The shaded areas in the above plots show the LOWER and UPPER SUMS for a constant MESH SIZE.

see also LOWER SUM, RIEMANN INTEGRAL, UPPER SUM

Riemann Surface

The Riemann surface S of the ALGEBRAIC FUNCTION FIELD K is the set of nontrivial discrete valuations on K. Here, the set S corresponds to the IDEALS of the RING A of INTEGERS of K over $\mathbb{C}(z)$. (A consists of the elements of K that are ROOTS of MONIC POLYNOMIALS over $\mathbb{C}[z]$.)

see also ALGEBRAIC FUNCTION FIELD, IDEAL, RING

References

Fischer, G. (Ed.). Plates 123–126 in *Mathematische Modelle/Mathematical Models, Bildband/Photograph Volume*. Braunschweig, Germany: Vieweg, pp. 120–123, 1986.

Riemann Tensor

A TENSOR sometimes known as the RIEMANN-CHRISTOFFEL TENSOR. Let

$$\tilde{D}_s \equiv \frac{\partial}{\partial x^s} - \sum_l \left\{ \begin{matrix} s & u \\ & l \end{matrix} \right\}, \tag{1}$$

where the quantity inside the $\left\{ \begin{matrix} s & u \\ & l \end{matrix} \right\}$ is a CHRISTOFFEL SYMBOL OF THE SECOND KIND. Then

$$R_{pqrs} \equiv \tilde{D}_q \left\{ \begin{matrix} p & r \\ & s \end{matrix} \right\} - \tilde{D}_r \left\{ \begin{matrix} r & q \\ & s \end{matrix} \right\}. \tag{2}$$

Broken down into its simplest decomposition in N-D,

$$R_{\lambda\mu\nu\kappa} = \frac{1}{N-2}(g_{\lambda\nu}R_{\mu\kappa} - g_{\lambda\kappa}R_{\mu\nu} - g_{\mu\nu}R_{\lambda\kappa} + g_{\mu\kappa}R_{\lambda\nu}) - \frac{R}{(N-1)(N-2)}(g_{\lambda\nu}g_{\mu\kappa} - g_{\lambda\kappa}g_{\mu\nu}) + C_{\lambda\mu\nu\kappa}. \tag{3}$$

Here, $R_{\mu\nu}$ is the RICCI TENSOR, R is the CURVATURE SCALAR, and $C_{\lambda\mu\nu\kappa}$ is the WEYL TENSOR. In terms of the JACOBI TENSOR $J^\mu{}_{\nu\alpha\beta}$,

$$R^\mu{}_{\alpha\nu\beta} = \tfrac{2}{3}(J^\mu_{\nu\alpha\beta} - J^\mu_{\beta\alpha\nu}). \tag{4}$$

The Riemann tensor is the only tensor that can be constructed from the METRIC TENSOR and its first and second derivatives,

$$R^\alpha{}_{\beta\gamma\delta} = \Gamma^\alpha_{\beta\delta,\gamma} - \Gamma^\alpha_{\beta\gamma,\delta} + \Gamma^\alpha_{\mu\gamma}\Gamma^\mu_{\beta\delta} - \Gamma^\alpha_{\mu\delta} - \Gamma^\alpha_{\beta\mu}c_{\gamma\delta}{}^\mu, \tag{5}$$

where Γ are CONNECTION COEFFICIENTS and c are COMMUTATION COEFFICIENTS. The number of independent coordinates in n-D is

$$C_n \equiv \tfrac{1}{12}n^2(n^2-1), \tag{6}$$

and the number of SCALARS which can be constructed from $R_{\lambda\mu\nu\kappa}$ and $g_{\mu\nu}$ is

$$S_n \equiv \begin{cases} 1 & \text{for } n=2 \\ \frac{1}{12}n(n-1)(n-2)(n+3) & \text{for } n=1, n>2. \end{cases} \tag{7}$$

In 1-D, $R_{1111} = 0$.

n	C_n	S_n
1	0	0
2	1	1
3	6	3
4	20	14

see also BIANCHI IDENTITIES, CHRISTOFFEL SYMBOL OF THE SECOND KIND, COMMUTATION COEFFICIENT, CONNECTION COEFFICIENT, CURVATURE SCALAR, GAUSSIAN CURVATURE, JACOBI TENSOR, PETROV NOTATION, RICCI TENSOR, WEYL TENSOR

Riemann Theta Function

Let the IMAGINARY PART of a $g \times g$ MATRIX F be POSITIVE DEFINITE, and $\mathbf{m} = (m_1, \ldots, m_g)$ be a row VECTOR with coefficients in $\mathbb{Z}$. Then the Riemann theta function is defined by

$$\vartheta(u) = \sum_{\mathbf{m}} \exp[2\pi i(\mathbf{m}^{\mathrm{T}}u + \tfrac{1}{2}\mathsf{F}^{\mathrm{T}}\mathbf{m})].$$

see also RAMANUJAN THETA FUNCTIONS, THETA FUNCTION

References

Iyanaga, S. and Kawada, Y. (Eds.). *Encyclopedic Dictionary of Mathematics*. Cambridge, MA: MIT Press, p. 9, 1980.

Riemann Weighted Prime-Power Counting Function

The Riemann weighted prime-power counting function is defined by

$$J_2(x) \equiv \begin{cases} \pi(x) + \frac{1}{2}\pi(x^{1/2}) + \frac{1}{3}\pi(x^{1/3}) + \ldots - \frac{1}{2m} & \text{for } p^m \text{ with } p \text{ a prime} \\ \pi(x) + \frac{1}{2}\pi(x^{1/2}) + \frac{1}{3}\pi(x^{1/3}) + \ldots & \text{otherwise} \end{cases} \tag{1}$$

$$= \lim_{t\to\infty} \frac{1}{2\pi i}\int_{2-iT}^{2+iT} \frac{x^s}{s} \ln\zeta(s)\,ds. \tag{2}$$

The PRIME COUNTING FUNCTION is given in terms of $J_2(x)$ by

$$\pi(x) = \sum_{n=1}^\infty \frac{\mu(n)}{n} J_2(x^{1/n}). \tag{3}$$

The function also satisfies the identity

$$\frac{\ln\zeta(s)}{s} = \int_1^\infty J_2(x)x^{-s-1}\,dx. \tag{4}$$

see also MANGOLDT FUNCTION, PRIME COUNTING FUNCTION, RIEMANN'S FORMULA

Riemann Xi Function

see XI FUNCTION

Riemann Zeta Function

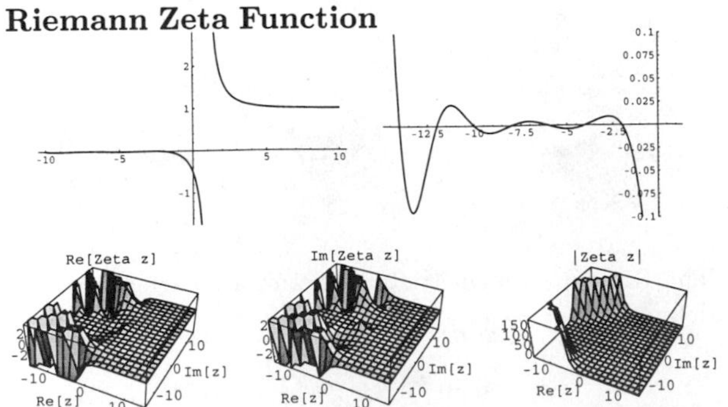

The Riemann zeta function can be defined by the integral

$$\zeta(x) \equiv \frac{1}{\Gamma(x)} \int_0^\infty \frac{u^{x-1}}{e^u - 1}\, du, \tag{1}$$

where $x > 1$. If x is an INTEGER n, then

$$\frac{u^{n-1}}{e^u - 1} = \frac{e^{-u} u^{n-1}}{1 - e^{-u}} = e^{-u} u^{n-1} \sum_{k=1}^\infty e^{-ku} u^{n-1}, \tag{2}$$

so

$$\int_0^\infty \frac{u^{n-1}}{e^u - 1}\, du = \sum_{k=1}^\infty \int_0^\infty e^{-ku} u^{n-1}\, du. \tag{3}$$

Let $y \equiv ku$, then $dy = k\, du$ and

$$\begin{aligned} \zeta(n) &= \frac{1}{\Gamma(n)} \sum_{k=1}^\infty \int_0^\infty e^{-ku} u^{n-1}\, du \\ &= \frac{1}{\Gamma(n)} \sum_{k=1}^\infty \int_0^\infty e^{-y} \left(\frac{y}{k}\right)^{n-1} \frac{dy}{k} \\ &= \frac{1}{\Gamma(n)} \sum_{k=1}^\infty \frac{1}{k^n} \int_0^\infty e^{-y} y^{n-1}\, dy, \end{aligned} \tag{4}$$

where $\Gamma(n)$ is the GAMMA FUNCTION. Integrating the final expression in (4) gives $\Gamma(n)$, which cancels the factor $1/\Gamma(n)$ and gives the most common form of the Riemann zeta function,

$$\zeta(n) = \sum_{k=1}^\infty \frac{1}{k^n}. \tag{5}$$

At $n = 1$, the zeta function reduces to the HARMONIC SERIES (which diverges), and therefore has a singularity. In the COMPLEX PLANE, trivial zeros occur at -2, -4, -6, ..., and nontrivial zeros at

$$s \equiv \sigma + it \tag{6}$$

for $0 \leq \sigma \leq 1$. The figures below show the structure of $\zeta(z)$ by plotting $|\zeta(z)|$ and $1/|\zeta(z)|$.

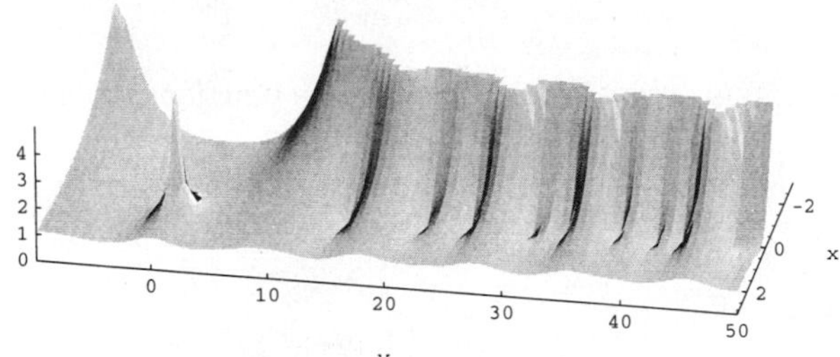

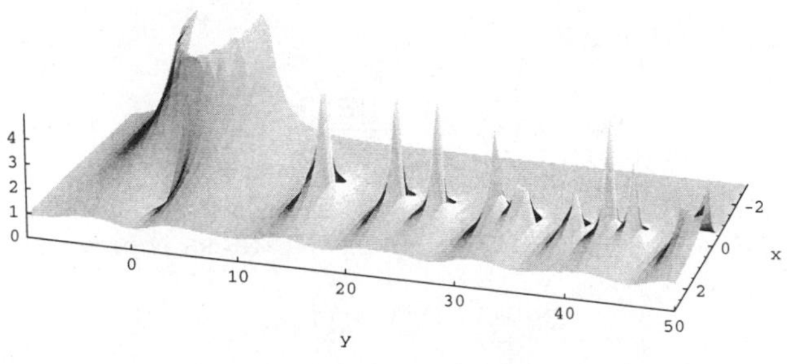

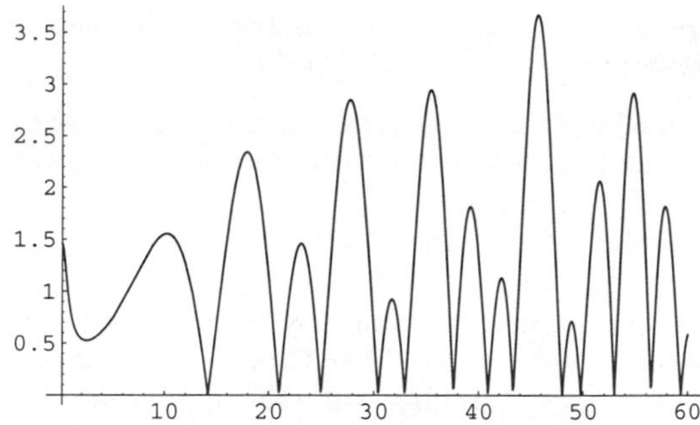

The RIEMANN HYPOTHESIS asserts that the nontrivial ROOTS of $\zeta(s)$ all have REAL PART $\sigma = \Re[s] = 1/2$, a line called the "CRITICAL STRIP." This is known to be true for the first 1.5×10^{12} roots (Brent *et al.* 1979). The above plot shows $|\zeta(1/2+it)|$ for t between 0 and 60. As can be seen, the first few nontrivial zeros occur at $t =$ 14.134725, 21.022040, 25.010858, 30.424876, 32.935062, 37.586178, ... (Wagon 1991, pp. 361–362 and 367–368).

The Riemann zeta function can also be defined in terms of MULTIPLE INTEGRALS by

$$\zeta(n) = \underbrace{\int_0^1 \cdots \int_0^1}_{n} \frac{\prod_{i=1}^n dx_i}{1 - \prod_{i=1}^n x_i}. \tag{7}$$

The Riemann zeta function can be split up into

$$\zeta(\tfrac{1}{2} + it) = z(t) e^{-i\vartheta(t)}, \tag{8}$$

where $z(t)$ and $\vartheta(t)$ are the RIEMANN-SIEGEL FUNCTIONS. An additional identity is

$$\lim_{s \to 1} \zeta(s) - \frac{1}{s-1} = \gamma, \tag{9}$$

where γ is the EULER-MASCHERONI CONSTANT.

The Riemann zeta function is related to the DIRICHLET LAMBDA FUNCTION $\lambda(\nu)$ and DIRICHLET ETA FUNCTION $\eta(\nu)$ by

$$\frac{\zeta(\nu)}{2^\nu} = \frac{\lambda(\nu)}{2^\nu - 1} = \frac{\eta(\nu)}{2^\nu - 2} \tag{10}$$

and

$$\zeta(\nu) + \eta(\nu) = 2\lambda(\nu) \tag{11}$$

(Spanier and Oldham 1987). It is related to the LIOUVILLE FUNCTION $\lambda(n)$ by

$$\frac{\zeta(2s)}{\zeta(s)} = \sum_{n=1}^\infty \frac{\lambda(n)}{n^s} \tag{12}$$

(Lehman 1960, Hardy and Wright 1979). Furthermore,

$$\frac{\zeta^2(s)}{\zeta(2s)} = \sum_{n=1}^\infty \frac{2^{\omega(n)}}{n^s}, \tag{13}$$

where $\omega(n) = \sigma_0(n)$ is the number of different prime factors of n (Hardy and Wright 1979).

A generalized Riemann zeta function $\zeta(s,a)$ known as the HURWITZ ZETA FUNCTION can also be defined such that

$$\zeta(s) \equiv \zeta(s,0). \tag{14}$$

The Riemann zeta function may be computed analytically for EVEN n using either CONTOUR INTEGRATION or PARSEVAL'S THEOREM with the appropriate FOURIER SERIES. An interesting formula involving the product of PRIMES was first discovered by Euler in 1737,

$$\zeta(x)(1-2^{-x}) = \left(1+\frac{1}{2^x}+\frac{1}{3^x}+\ldots\right)\left(1-\frac{1}{2^x}\right)$$
$$= \left(1+\frac{1}{2^x}+\frac{1}{3^x}+\ldots\right)-\left(\frac{1}{2^x}+\frac{1}{4^x}+\frac{1}{6^x}+\ldots\right) \tag{15}$$

$$\zeta(x)(1-2^{-x})(1-3^{-x})$$
$$= \left(1+\frac{1}{3^x}+\frac{1}{5^x}+\frac{1}{7^x}+\ldots\right)-\left(\frac{1}{3^x}+\frac{1}{9^x}+\frac{1}{15^x}+\ldots\right) \tag{16}$$

$$\zeta(x)(1-2^{-x})(1-3^{-x})\cdots(1-p^{-x})\cdots$$
$$= \zeta(x)\prod_{n=2}^{\infty}(1-p^{-x}) = 1. \tag{17}$$

Here, each subsequent multiplication by the next PRIME p leaves only terms which are POWERS of p^{-x}. Therefore,

$$\zeta(x) = \left[\prod_{p=2}^{\infty}(1-p^{-x})\right]^{-1}, \tag{18}$$

where p runs over all PRIMES. Euler's product formula can also be written

$$\zeta(s) = (1-2^{-s})^{-1}\prod_{\substack{q\equiv 1\\ (\text{mod } 4)}}(1-q^{-s})^{-1}\prod_{\substack{r\equiv 3\\ (\text{mod } 4)}}(1-r^{-s})^{-1}. \tag{19}$$

A few sum identities involving $\zeta(n)$ are

$$\sum_{n=2}^{\infty}[\zeta(n)-1] = 1 \tag{20}$$

$$\sum_{n=2}^{\infty}(-1)^n[\zeta(n)-1] = \tfrac{1}{2}. \tag{21}$$

The Riemann zeta function is related to the GAMMA FUNCTION $\Gamma(z)$ by

$$\Gamma\left(\frac{s}{2}\right)\pi^{-z/2}\zeta(s) = \Gamma\left(\frac{1-s}{2}\right)\pi^{-(1-s)/2}\zeta(1-s). \tag{22}$$

$\zeta(n)$ was proved to be transcendental for all even n by Euler. Apéry (1979) proved $\zeta(3)$ to be IRRATIONAL with the aid of the k^{-3} sum formula below. As a result, $\zeta(3)$ is sometimes called APÉRY'S CONSTANT.

$$\zeta(2) = 3\sum_{k=1}^{\infty}\frac{1}{k^2\binom{2k}{k}} \tag{23}$$

$$\zeta(3) = \frac{5}{2}\sum_{k=1}^{\infty}\frac{(-1)^{k-1}}{k^3\binom{2k}{k}} \tag{24}$$

$$\zeta(4) = \frac{36}{17}\sum_{k=1}^{\infty}\frac{1}{k^4\binom{2k}{k}} \tag{25}$$

(Guy 1994, p. 257). A relation of the form

$$\zeta(5) = Z_5\sum_{k=1}^{\infty}\frac{(-1)^{k-1}}{k^5\binom{2k}{k}} \tag{26}$$

has been searched for with Z_5 a RATIONAL or ALGEBRAIC NUMBER, but if Z_5 is a ROOT of a POLYNOMIAL of degree 25 or less, then the Euclidean norm of the coefficients must be larger than 2×10^{37} (Bailey, Bailey and Plouffe). Therefore, no such sums are known for $\zeta(n)$ are known for $n \geq 5$.

The zeta function is defined for $\Re[s] > 1$, but can be analytically continued to $\Re[s] > 0$ as follows

$$\sum_{n=1}^{\infty}(-1)^n n^{-s} + \sum_{n=1}^{\infty}n^{-s} = 2\sum_{n=2,4,\ldots}^{\infty}n^{-s}$$
$$= 2\sum_{k=1}^{\infty}(2k)^{-s} = 2^{1-s}\sum_{k=1}^{\infty}k^{-s} \tag{27}$$

$$\sum_{n=1}^{\infty}(-1)^n n^{-s} + \zeta(s) = 2^{1-s}\zeta(s) \tag{28}$$

$$\zeta(s) = \frac{1}{1-2^{1-s}}\sum_{n=1}^{\infty}(-1)^n n^{-s}. \tag{29}$$

The DERIVATIVE of the Riemann zeta function is defined by

$$\zeta'(s) = -s\sum_{k=1}^{\infty}k^{-s}\ln k = -\sum_{k=2}^{\infty}\frac{\ln k}{k^s}. \tag{30}$$

As $s \to 0$,

$$\zeta'(0) = -\tfrac{1}{2}\ln(2\pi). \tag{31}$$

For EVEN $n \equiv 2k$,

$$\zeta(n) = \frac{2^{n-1}|B_n|\pi^n}{n!}, \tag{32}$$

where B_n is a BERNOULLI NUMBER. Another intimate connection with the BERNOULLI NUMBERS is provided by

$$B_n = (-1)^{n+1}n\zeta(1-n). \tag{33}$$

No analytic form for $\zeta(n)$ is known for ODD $n \equiv 2k+1$, but $\zeta(2k+1)$ can be expressed as the sum limit

$$\zeta(2k+1) = (\tfrac{1}{2}\pi)^{2k+1} \lim_{t\to\infty} \frac{1}{t^{2k+1}} \sum_{i=1}^{\infty} \left[\cot\left(\frac{i}{2t+1}\right)\right]^{2k+1} \quad (34)$$

(Stark 1974). The values for the first few integral arguments are

$$\begin{aligned} \zeta(0) &\equiv -\tfrac{1}{2} \\ \zeta(1) &= \infty \\ \zeta(2) &= \frac{\pi^2}{6} \\ \zeta(3) &= 1.2020569032\ldots \\ \zeta(4) &= \frac{\pi^4}{90} \\ \zeta(5) &= 1.0369277551\ldots \\ \zeta(6) &= \frac{\pi^6}{945} \\ \zeta(7) &= 1.0083492774\ldots \\ \zeta(8) &= \frac{\pi^8}{9450} \\ \zeta(9) &= 1.0020083928\ldots \\ \zeta(10) &= \frac{\pi^{10}}{93{,}555}. \end{aligned}$$

Euler gave $\zeta(2)$ to $\zeta(26)$ for EVEN n, and Stieltjes (1993) determined the values of $\zeta(2), \ldots, \zeta(70)$ to 30 digits of accuracy in 1887. The denominators of $\zeta(2n)$ for $n = 1, 2, \ldots$ are 6, 90, 945, 9450, 93555, 638512875, ... (Sloane's A002432).

Using the LLL ALGORITHM, Plouffe (inspired by Zucker 1979, Zucker 1984, and Berndt 1988) has found some beautiful infinite sums for $\zeta(n)$ with ODD n. Let

$$S_\pm(n) \equiv \sum_{k=1}^{\infty} \frac{1}{k^n(e^{2\pi k} \pm 1)}, \quad (35)$$

then

$$\zeta(3) = \tfrac{7}{180}\pi^3 - 2S_-(3) \quad (36)$$

$$\zeta(5) = \tfrac{1}{294}\pi^5 - \tfrac{72}{35}S_-(5) - \tfrac{2}{35}S_+(5) \quad (37)$$

$$\zeta(7) = \tfrac{19}{56700}\pi^7 - 2S_-(7) \quad (38)$$

$$\zeta(9) = \tfrac{125}{3704778}\pi^9 - \tfrac{992}{495}S_-(9) - \tfrac{2}{495}S_+(9) \quad (39)$$

$$\zeta(11) = \tfrac{1453}{425675250}\pi^{11} - 2S_-(11) \quad (40)$$

$$\zeta(13) = \tfrac{89}{257432175}\pi^{13} - \tfrac{16512}{8255}S_-(13) - \tfrac{2}{8255}S_+(13) \quad (41)$$

$$\zeta(15) = \tfrac{13687}{390769879500}\pi^{15} - 2S_-(15) \quad (42)$$

$$\zeta(17) = \tfrac{397549}{112024529867250}\pi^{17} - \tfrac{261632}{130815}S_-(17) - \tfrac{2}{130815}S_+(17) \quad (43)$$

$$\zeta(19) = \tfrac{7708537}{21438612514068750}\pi^{19} - 2S_-(19) \quad (44)$$

$$\zeta(21) = \tfrac{68529640373}{1881063815762259253125}\pi^{21} - \tfrac{4196352}{2098175}S_-(21) - \tfrac{2}{2098175}S_+(21) \quad (45)$$

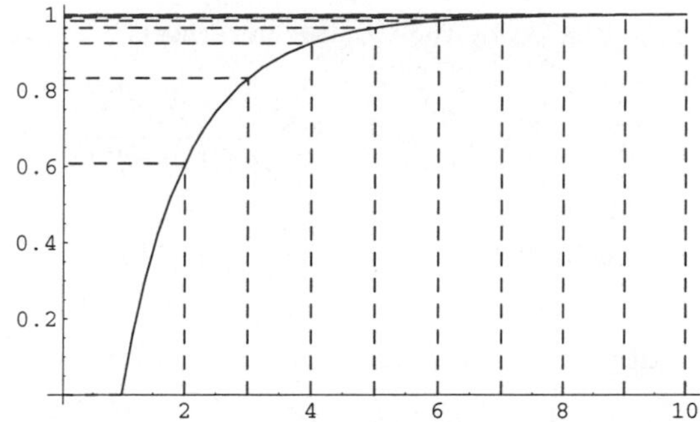

The inverse of the RIEMANN ZETA FUNCTION $1/\zeta(p)$ is the asymptotic density of pth-powerfree numbers (i.e., SQUAREFREE numbers, CUBEFREE numbers, etc.). The following table gives the number $Q_p(n)$ of pth-powerfree numbers $\leq n$ for several values of n.

p	$1/\zeta(p)$	10	100	10^3	10^4	10^5	10^6
2	0.607927	7	61	608	6083	60794	607926
3	0.831907	9	85	833	8319	83190	831910
4	0.923938	10	93	925	9240	92395	923939
5	0.964387	10	97	965	9645	96440	964388
6	0.982953	10	99	984	9831	98297	982954

The value for $\zeta(2)$ can be found using a number of different techniques (Apostol 1983, Choe 1987, Giesy 1972, Holme 1970, Kimble 1987, Knopp and Schur 1918, Kortram 1996, Matsuoka 1961, Papadimitriou 1973, Simmons 1992, Stark 1969, Stark 1970, Yaglom and Yaglom 1987). The problem of finding this value analytically is sometimes known as the BASLER PROBLEM (Castellanos 1988). Yaglom and Yaglom (1987), Holme (1970), and Papadimitrou (1973) all derive the result from DE MOIVRE'S IDENTITY or related identities.

Consider the FOURIER SERIES of $f(x) = x^{2n}$

$$f(x) = \tfrac{1}{2}a_0 + \sum_{m=1}^{\infty} a_m \cos(mx) + \sum_{m=1}^{\infty} b_m \sin(mx), \quad (46)$$

which has coefficients given by

$$\begin{aligned} a_0 &= \frac{1}{\pi}\int_{-\pi}^{\pi} f(x)\,dx = \frac{2}{\pi}\int_0^{\pi} x^{2n}\,dx \\ &= \frac{2}{\pi}\left[\frac{x^{2n+1}}{2n+1}\right]_0^{\pi} = \frac{2\pi^{2n}}{2n+1} \end{aligned} \quad (47)$$

$$\begin{aligned} a_m &= \frac{1}{\pi}\int_{\pi}^{\pi} x^{2n}\cos(mx)\,dx \\ &= \frac{2}{\pi}\int_0^{\pi} x^{2n}\cos(mx)\,dx \end{aligned} \quad (48)$$

$$b_m = \frac{1}{\pi}\int_{-\pi}^{\pi} x^{2n}\sin(mx)\,dx = 0, \quad (49)$$

where the latter is true since the integrand is ODD. Therefore, the FOURIER SERIES is given explicitly by

$$x^{2n} = \frac{\pi^{2n}}{2n+1} + \sum_{m=1}^{\infty} a_m \cos(mx). \quad (50)$$

Now, a_m is given by the COSINE INTEGRAL

$$a_m = \frac{2}{\pi}(-1)^{n+1}(2n)! \left[\sin(mx) \sum_{k=0}^{n} \frac{(-1)^k}{(2k)!m^{2n-2k+1}} x^{2k} \right.$$

$$\left. + \cos(mx) \sum_{k=1}^{n} \frac{(-1)^{k+1}}{(2k-3)!m^{2n-2k+2}} x^{2k-1} \right]_0^\pi . \quad (51)$$

But $\cos(m\pi) = (-1)^m$, and $\sin(m\pi) = \sin 0 = 0$, so

$$a_m = \frac{2}{\pi}(-1)^{n+1}(2n)!(-1)^m \sum_{k=1}^{n} \frac{(-1)^{k+1}}{(2k-3)!m^{2n-2k+2}} \pi^{2k-1}$$

$$= (-1)^{m+n} 2(2n)! \sum_{k=1}^{n} \frac{(-1)^k}{(2k-3)!m^{2n-2k+2}} \pi^{2k-2}. \quad (52)$$

Now, if $n = 1$,

$$a_m = (-1)^{m+1} 2(2!) \sum_{k=1}^{1} \frac{(-1)^k}{(2k-3)!m^{4-2k}} \pi^{2k-2}$$

$$= 4(-1)^{m+1} \frac{(-1)}{(-1)!m^2} \pi^0 = \frac{4(-1)^m}{m^2}, \quad (53)$$

so the FOURIER SERIES is

$$x^2 = \frac{\pi^2}{3} + 4 \sum_{m=1}^{\infty} \frac{(-1)^m \cos(mx)}{m^2}. \quad (54)$$

Letting $x \equiv \pi$ gives $\cos(m\pi) = (-1)^m$, so

$$\pi^2 = \frac{\pi^2}{3} + 4 \sum_{m=1}^{\infty} \frac{1}{m^2}, \quad (55)$$

and we have

$$\zeta(2) = \sum_{m=1}^{\infty} \frac{1}{m^2} = \frac{\pi^2}{6}. \quad (56)$$

Higher values of n can be obtained by finding a_m and proceeding as above.

The value $\zeta(2)$ can also be found simply using the ROOT LINEAR COEFFICIENT THEOREM. Consider the equation $\sin z = 0$ and expand sin in a MACLAURIN SERIES

$$\sin z = z - \frac{z^3}{3!} + \frac{z^5}{5!} + \ldots = 0 \quad (57)$$

$$0 = 1 - \frac{z^2}{3!} + \frac{z^4}{5!} + \ldots = 1 - \frac{w}{3!} + \frac{w^2}{5!} + \ldots, \quad (58)$$

where $w \equiv z^2$. But the zeros of $\sin(z)$ occur at π, 2π, 3π, $\ldots$, so the zeros of $\sin w = \sin\sqrt{z}$ occur at π^2, $(2\pi)^2$, $\ldots$. Therefore, the sum of the roots equals the COEFFICIENT of the leading term

$$\frac{1}{\pi^2} + \frac{1}{(2\pi)^2} + \frac{1}{(3\pi^2)} + \ldots = \frac{1}{3!} = \frac{1}{6}, \quad (59)$$

which can be rearranged to yield

$$\zeta(2) = \frac{\pi^2}{6}. \quad (60)$$

Yet another derivation (Simmons 1992) evaluates the integral using the integral

$$I = \int_0^1 \int_0^1 \frac{dx\,dy}{1-xy} = \int_0^1 \int_0^1 (1 + xy + x^2y^2 + \ldots)\,dx\,dy$$

$$= \int_0^1 [(x + \tfrac{1}{2}x^2y + \tfrac{1}{3}x^3y^2 + \ldots)]_0^1\,dy$$

$$= \int_0^1 (1 + \tfrac{1}{2}y + \tfrac{1}{3}y^2 + \ldots)\,dy$$

$$= \left[y + \frac{y^2}{2^2} + \frac{y^3}{3^2} + \ldots \right]_0^1 = 1 + \frac{1}{2^2} + \frac{1}{3^2} + \ldots. \quad (61)$$

To evaluate the integral, rotate the coordinate system by $\pi/4$ so

$$x = u\cos\theta - v\sin\theta = \tfrac{1}{2}\sqrt{2}\,(u - v) \quad (62)$$

$$y = u\sin\theta + v\cos\theta = \tfrac{1}{2}\sqrt{2}\,(u + v) \quad (63)$$

and

$$xy = \tfrac{1}{2}(u^2 - v^2) \quad (64)$$

$$1 - xy = \tfrac{1}{2}(2 - u^2 + v^2). \quad (65)$$

Then

$$I = 4 \int_0^{\sqrt{2}/2} \int_0^u \frac{du\,dv}{2 - u^2 + v^2}$$

$$+ 4 \int_{\sqrt{2}/2}^{\sqrt{2}} \int_0^{\sqrt{2}-u} \frac{du\,dv}{2 - u^2 + v^2} \equiv I_1 + I_2. \quad (66)$$

Now compute the integrals I_1 and I_2.

$$I_1 = 4 \int_0^{\sqrt{2}/2} \left[\int_0^u \frac{dv}{2 - u^2 + v^2} \right] du$$

$$= 4 \int_0^{\sqrt{2}\,2} \left[\frac{1}{\sqrt{2-u^2}} \tan^{-1}\left(\frac{v}{\sqrt{2-u^2}} \right) \right]_0^u du$$

$$= 4 \int_0^{\sqrt{2}/2} \frac{1}{\sqrt{2-u^2}} \tan^{-1}\left(\frac{u}{\sqrt{2-u^2}} \right) du. \quad (67)$$

Make the substitution

$$u = \sqrt{2}\sin\theta \quad (68)$$

$$\sqrt{2-u^2} = \sqrt{2}\cos\theta \quad (69)$$

$$du = \sqrt{2}\cos\theta\,d\theta, \quad (70)$$

so

$$\tan^{-1}\left(\frac{u}{\sqrt{2-u^2}}\right) = \tan^{-1}\left(\frac{\sqrt{2}\sin\theta}{\sqrt{2}\cos\theta}\right) = \theta \quad (71)$$

and

$$I_1 = 4\int_0^{\pi/6} \frac{1}{\sqrt{2}\cos\theta}\theta\sqrt{2}\cos\theta\,d\theta = 2[\theta^2]_0^{\pi/6} = \frac{\pi^2}{18}. \quad (72)$$

I_2 can also be computed analytically,

$$\begin{aligned} I_2 &= 4\int_{\sqrt{2}/2}^{\sqrt{2}} \left[\int_0^{\sqrt{2}-u} \frac{dv}{2-u^2+v^2}\right] du \\ &= 4\int_{\sqrt{2}/2}^{\sqrt{2}} \left[\frac{1}{\sqrt{2-u^2}}\tan^{-1}\left(\frac{v}{\sqrt{2-u^2}}\right)\right]_0^{\sqrt{2}-u} du \\ &= 4\int_{\sqrt{2}/2}^{\sqrt{2}} \frac{1}{\sqrt{2-u^2}}\tan^{-1}\left(\frac{\sqrt{2}-u}{\sqrt{2-u^2}}\right) du. \quad (73) \end{aligned}$$

But

$$\begin{aligned} &\tan^{-1}\left(\frac{\sqrt{2}-u}{\sqrt{2-u^2}}\right) = \tan^{-1}\left(\frac{\sqrt{2}-\sqrt{2}\sin\theta}{\sqrt{2}\cos\theta}\right) \\ &= \tan\left(\frac{1-\sin\theta}{\cos\theta}\right) = \tan^{-1}\left(\frac{\cos\theta}{1+\sin\theta}\right) \\ &= \tan^{-1}\left[\frac{\sin(\frac{1}{2}\pi-\theta)}{1+\cos(\frac{1}{2}\pi-\theta)}\right] \\ &= \tan^{-1}\left\{\frac{2\sin[\frac{1}{2}(\frac{1}{2}\pi-\theta)]\cos[\frac{1}{2}(\frac{1}{2}\pi-\theta)]}{2\cos^2[\frac{1}{2}(\frac{1}{2}\pi-\theta)]}\right\} \\ &= \tfrac{1}{2}(\tfrac{1}{2}\pi-\theta), \quad (74) \end{aligned}$$

so

$$\begin{aligned} I_2 &= 4\int_{\pi/6}^{\pi/2} \frac{1}{\sqrt{2}\cos\theta}(\tfrac{1}{4}\pi-\tfrac{1}{2}\theta)\sqrt{2}\cos\theta\,d\theta \\ &= 4\left[\tfrac{1}{4}\pi\theta-\tfrac{1}{4}\theta^2\right]_{\pi/6}^{\pi/2} \\ &= 4\left[\left(\frac{\pi^2}{8}-\frac{\pi^2}{16}\right)-\left(\frac{\pi^2}{24}-\frac{\pi^2}{144}\right)\right] = \frac{\pi^2}{9}. \quad (75) \end{aligned}$$

Combining I_1 and I_2 gives

$$\zeta(2) = I_1 + I_2 = \frac{\pi^2}{18} + \frac{\pi^2}{9} = \frac{\pi^2}{6}. \quad (76)$$

see also ABEL'S FUNCTIONAL EQUATION, DEBYE FUNCTIONS, DIRICHLET BETA FUNCTION, DIRICHLET ETA FUNCTION, DIRICHLET LAMBDA FUNCTION, HARMONIC SERIES, HURWITZ ZETA FUNCTION, KHINTCHINE'S CONSTANT, LEHMER'S PHENOMENON, PSI FUNCTION, RIEMANN HYPOTHESIS, RIEMANN *P*-SERIES, RIEMANN-SIEGEL FUNCTIONS, STIELTJES CONSTANTS, XI FUNCTION

References

Abramowitz, M. and Stegun, C. A. (Eds.). "Riemann Zeta Function and Other Sums of Reciprocal Powers." §23.2 in *Handbook of Mathematical Functions with Formulas, Graphs, and Mathematical Tables, 9th printing.* New York: Dover, pp. 807–808, 1972.

Apéry, R. "Irrationalité de $\zeta(2)$ et $\zeta(3)$." *Astérisque* **61**, 11–13, 1979.

Apostol, T. M. "A Proof that Euler Missed: Evaluating $\zeta(2)$ the Easy Way." *Math. Intel.* **5**, 59–60, 1983.

Arfken, G. *Mathematical Methods for Physicists, 3rd ed.* Orlando, FL: Academic Press, pp. 332–335, 1985.

Ayoub, R. "Euler and the Zeta Function." *Amer. Math. Monthly* **71**, 1067–1086, 1974.

Bailey, D. H. "Multiprecision Translation and Execution of Fortran Programs." *ACM Trans. Math. Software.* To appear.

Bailey, D. and Plouffe, S. "Recognizing Numerical Constants." `http://www.cecm.sfu.ca/organics/papers/bailey`.

Berndt, B. C. Ch. 14 in *Ramanujan's Notebooks, Part II.* New York: Springer-Verlag, 1988.

Borwein, D. and Borwein, J. "On an Intriguing Integral and Some Series Related to $\zeta(4)$." *Proc. Amer. Math. Soc.* **123**, 1191–1198, 1995.

Brent, R. P.; van der Lune, J.; te Riele, H. J. J.; and Winter, D. T. "On the Zeros of the Riemann Zeta Function in the Critical Strip 1." *Math. Comput.* **33**, 1361–1372, 1979.

Castellanos, D. "The Ubiquitous Pi. Part I." *Math. Mag.* **61**, 67–98, 1988.

Choe, B. R. "An Elementary Proof of $\sum_{n=1}^{\infty}\frac{1}{n^2} = \frac{\pi^2}{6}$." *Amer. Math. Monthly* **94**, 662–663, 1987.

Davenport, H. *Multiplicative Number Theory, 2nd ed.* New York: Springer-Verlag, 1980.

Edwards, H. M. *Riemann's Zeta Function.* New York: Academic Press, 1974.

Farmer, D. W. "Counting Distinct Zeros of the Riemann Zeta-Function." *Electronic J. Combinatorics* **2**, R1, 1–5, 1995. `http://www.combinatorics.org/Volume_2/volume2.html#R1`.

Giesy, D. P. "Still Another Proof that $\sum 1/k^2 = \pi^2/6$." *Math. Mag.* **45**, 148–149, 1972.

Guy, R. K. "Series Associated with the ζ-Function." §F17 in *Unsolved Problems in Number Theory, 2nd ed.* New York: Springer-Verlag, pp. 257–258, 1994.

Hardy, G. H. and Wright, E. M. *An Introduction to the Theory of Numbers, 5th ed.* Oxford, England: Clarendon Press, p. 255, 1979.

Holme, F. "Ein enkel beregning av $\sum_{k=1}^{\infty}\frac{1}{k^2}$." *Nordisk Mat. Tidskr.* **18**, 91–92 and 120, 1970.

Ivic, A. A. *The Riemann Zeta-Function.* New York: Wiley, 1985.

Ivic, A. A. *Lectures on Mean Values of the Riemann Zeta Function.* Berlin: Springer-Verlag, 1991.

Karatsuba, A. A. and Voronin, S. M. *The Riemann Zeta-Function.* Hawthorne, NY: De Gruyter, 1992.

Katayama, K. "On Ramanujan's Formula for Values of Riemann Zeta-Function at Positive Odd Integers." *Acta Math.* **22**, 149–155, 1973.

Kimble, G. "Euler's Other Proof." *Math. Mag.* **60**, 282, 1987.

Knopp, K. and Schur, I. "Über die Herleitug der Gleichung $\sum_{n=1}^{\infty}\frac{1}{n^2} = \frac{\pi^2}{6}$." *Archiv der Mathematik u. Physik* **27**, 174–176, 1918.

Kortram, R. A. "Simple Proofs for $\sum_{k=1}^{\infty}\frac{1}{k^2} = \frac{\pi^2}{6}$ and $\sin x = x\prod_{k=1}^{\infty}\left(1-\frac{x^2}{k^2\pi^2}\right)$." *Math. Mag.* **69**, 122–125, 1996.

Le Lionnais, F. *Les nombres remarquables.* Paris: Hermann, p. 35, 1983.

Lehman, R. S. "On Liouville's Function." *Math. Comput.* **14**, 311–320, 1960.
Matsuoka, Y. "An Elementary Proof of the Formula $\sum_{k=1}^{\infty} \frac{1}{k^2} = \frac{\pi^2}{6}$." *Amer. Math. Monthly* **68**, 486–487, 1961.
Papadimitriou, I. "A Simple Proof of the Formula $\sum_{k=1}^{\infty} \frac{1}{k^2} = \frac{\pi^2}{6}$." *Amer. Math. Monthly* **80**, 424–425, 1973.
Patterson, S. J. *An Introduction to the Theory of the Riemann Zeta-Function.* New York: Cambridge University Press, 1988.
Plouffe, S. "Identities Inspired from Ramanujan Notebooks." `http://www.lacim.uqam.ca/plouffe/identities.html`.
Simmons, G. F. "Euler's Formula $\sum_1^{\infty} 1/n^2 = \pi^2/6$ by Double Integration." Ch. B. 24 in *Calculus Gems: Brief Lives and Memorable Mathematics.* New York: McGraw-Hill, 1992.
Sloane, N. J. A. Sequence A002432/M4283 in "An On-Line Version of the Encyclopedia of Integer Sequences."
Spanier, J. and Oldham, K. B. "The Zeta Numbers and Related Functions." Ch. 3 in *An Atlas of Functions.* Washington, DC: Hemisphere, pp. 25–33, 1987.
Stark, E. L. "Another Proof of the Formula $\sum_{k=1}^{\infty} \frac{1}{k^2} = \frac{\pi^2}{6}$." *Amer. Math. Monthly* **76**, 552–553, 1969.
Stark, E. L. "$1 - \frac{1}{4} + \frac{1}{9} - \frac{1}{16} + \ldots = \frac{\pi^2}{12}$." *Praxis Math.* **12**, 1–3, 1970.
Stark, E. L. "The Series $\sum_{k=1}^{\infty} k^{-s}$ $s = 2, 3, 4, \ldots$, Once More." *Math. Mag.* **47**, 197–202, 1974.
Stieltjes, T. J. *Oeuvres Complètes, Vol. 2* (Ed. G. van Dijk.) New York: Springer-Verlag, p. 100, 1993.
Titchmarsh, E. C. *The Zeta-Function of Riemann.* New York: Stechert-Hafner Service Agency, 1964.
Titchmarsh, E. C. and Heath-Brown, D. R. *The Theory of the Riemann Zeta-Function, 2nd ed.* Oxford, England: Oxford University Press, 1986.
Vardi, I. "The Riemann Zeta Function." Ch. 8 in *Computational Recreations in Mathematica.* Reading, MA: Addison-Wesley, pp. 141–174, 1991.
Wagon, S. "The Evidence: Where Are the Zeros of Zeta of s?" *Math. Intel.* **8**, 57–62, 1986.
Wagon, S. "The Riemann Zeta Function." §10.6 in *Mathematica in Action.* New York: W. H. Freeman, pp. 353–362, 1991.
Yaglom, A. M. and Yaglom, I. M. Problem 145 in *Challenging Mathematical Problems with Elementary Solutions, Vol. 2.* New York: Dover, 1987.
Zucker, I. J. "The Summation of Series of Hyperbolic Functions." *SIAM J. Math. Anal.* **10**, 192–206, 1979.
Zucker, I. J. "Some Infinite Series of Exponential and Hyperbolic Functions." *SIAM J. Math. Anal.* **15**, 406–413, 1984.

Riemannian Geometry

The study of MANIFOLDS having a complete RIEMANNIAN METRIC. Riemannian geometry is a general space based on the LINE ELEMENT

$$ds = F(x^1, \ldots, x^n; dx^1, \ldots, dx^n),$$

with $F(x, y) > 0$ for $y \neq 0$ a function on the TANGENT BUNDLE TM. In addition, F is homogeneous of degree 1 in y and of the form

$$F^2 = g_{ij}(x)\, dx^i\, dx^j$$

(Chern 1996). If this restriction is dropped, the resulting geometry is called FINSLER GEOMETRY.

References

Besson, G.; Lohkamp, J.; Pansu, P.; and Petersen, P. *Riemannian Geometry.* Providence, RI: Amer. Math. Soc., 1996.
Buser, P. *Geometry and Spectra of Compact Riemann Surfaces.* Boston, MA: Birkhäuser, 1992.
Chavel, I. *Eigenvalues in Riemannian Geometry.* New York: Academic Press, 1984.
Chavel, I. *Riemannian Geometry: A Modern Introduction.* New York: Cambridge University Press, 1994.
Chern, S.-S. "Finsler Geometry is Just Riemannian Geometry without the Quadratic Restriction." *Not. Amer. Math. Soc.* **43**, 959–963, 1996.
do Carmo, M. P. *Riemannian Geometry.* Boston, MA: Birkhäuser, 1992.

Riemannian Geometry (Non-Euclidean)

see ELLIPTIC GEOMETRY

Riemannian Manifold

A MANIFOLD possessing a METRIC TENSOR. For a complete Riemannian manifold, the METRIC $d(x, y)$ is defined as the length of the shortest curve (GEODESIC) between x and y.

see also BISHOP'S INEQUALITY, CHEEGER'S FINITENESS THEOREM

Riemannian Metric

Suppose for every point x in a COMPACT MANIFOLD M, an INNER PRODUCT $\langle \cdot, \cdot \rangle_x$ is defined on a TANGENT SPACE $T_x M$ of M at x. Then the collection of all these INNER PRODUCTS is called the Riemannian metric. In 1870, Christoffel and Lipschitz showed how to decide when two Riemannian metrics differ by only a coordinate transformation.

see also COMPACT MANIFOLD, LINE ELEMENT, METRIC TENSOR

Riesel Number

There exist infinitely many ODD INTEGERS k such that $k \cdot 2^n - 1$ is COMPOSITE for every $n \geq 1$. Numbers k with this property are called RIESEL NUMBERS, and analogous numbers with the minus sign replaced by a plus are called SIERPIŃSKI NUMBERS OF THE SECOND KIND. The smallest known Riesel number is $k = 509{,}203$, but there remain 963 smaller candidates (the smallest of which is 659) which generate only composite numbers for all n which have been checked (Ribenboim 1996, p. 358).

Let $a(k)$ be smallest n for which $(2k-1) \cdot 2^n - 1$ is PRIME, then the first few values are 2, 0, 2, 1, 1, 2, 3, 1, 2, 1, 1, 4, 3, 1, 4, 1, 2, 2, 1, 3, 2, 7, ... (Sloane's A046069), and second smallest n are 3, 1, 4, 5, 3, 26, 7, 2, 4, 3, 2, 6, 9, 2, 16, 5, 3, 6, 2553, ... (Sloane's A046070).

see also CUNNINGHAM NUMBER, MERSENNE NUMBER, SIERPIŃSKI'S COMPOSITE NUMBER THEOREM, SIERPIŃSKI NUMBER OF THE SECOND KIND

References

Ribenboim, P. *The New Book of Prime Number Records.* New York: Springer-Verlag, p. 357, 1996.

Riesel, H. "Några stora primtal." *Elementa* **39**, 258–260, 1956.
Sloane, N. J. A. Sequence A046068 in "An On-Line Version of the Encyclopedia of Integer Sequences."

Riesz-Fischer Theorem

A function is L_2- (square-) integrable IFF its FOURIER SERIES is L_2-convergent. The application of this theorem requires use of the LEBESGUE INTEGRAL.

see also LEBESGUE INTEGRAL

Riesz Representation Theorem

Let f be a bounded linear FUNCTIONAL on a HILBERT SPACE H. Then there exists exactly one $x_0 \in H$ such that $f(x) = \langle x, x_0 \rangle$ for all $x \in H$. Also, $\|f\| = \|x_0\|$.

see also FUNCTIONAL, HILBERT SPACE

References
Debnath, L. and Mikusiński, P. *Introduction to Hilbert Spaces with Applications.* San Diego, CA: Academic Press, 1990.

Riesz's Theorem

Every continuous linear functional $U[f]$ for $f \in C[a, b]$ can be expressed as a STIELTJES INTEGRAL

$$U[f] = \int_a^b f(x)\,dw(x),$$

where $w(x)$ is determined by U and is of bounded variation on $[a, b]$.

see also STIELTJES INTEGRAL

References
Kestelman, H. "Riesz's Theorem." §11.5 in *Modern Theories of Integration, 2nd rev. ed.* New York: Dover, pp. 265–269, 1960.

Riffle Shuffle

A SHUFFLE, also called a FARO SHUFFLE, in which a deck of $2n$ cards is divided into two HALVES which are then alternatively interleaved from the left and right hands (an "in-shuffle") or from the right and left hands (an "out-shuffle"). Using an "in-shuffle," a deck originally arranged as 1 2 3 4 5 6 7 8 would become 5 1 6 2 7 3 8 4. Using an "out-shuffle," the deck order would become 1 5 2 6 3 7 4 8. Riffle shuffles are used in card tricks (Marlo 1958ab, Adler 1973), and also in the theory of parallel processing (Stone 1971, Chen *et al.* 1981).

In general, card k moves to the position originally occupied by the $2k$th card (mod $2n + 1$). Therefore, in-shuffling $2n$ cards $2n$ times (where $2n + 1$ is PRIME) results in the original card order. Similarly, out-shuffling $2n$ cards $2n - 2$ times (where $2n - 1$ is PRIME) results in the original order (Diaconis *et al.* 1983, Conway and Guy 1996). Amazingly, this means that an ordinary deck of 52 cards is returned to its original order after 8 out-shuffles.

Morris (1994) further discusses aspects of the perfect riffle shuffle (in which the deck is cut exactly in half and cards are perfectly interlaced). Ramnath and Scully (1996) give an algorithm for the shortest sequence of in- and out-shuffles to move a card from arbitrary position i to position j. This algorithm works for any deck with an EVEN number of cards and is $\mathcal{O}(\log n)$.

see also CARDS, SHUFFLE

References
Adler, I. "Make Up Your Own Card Tricks." *J. Recr. Math.* **6**, 87–91, 1973.
Ball, W. W. R. and Coxeter, H. S. M. *Mathematical Recreations and Essays, 13th ed.* New York: Dover, pp. 323–325, 1987.
Chen, P. Y.; Lawrie, D. H.; Yew, P.-C.; and Padua, D. A. "Interconnection Networks Using Shuffles." *Computer* **33**, 55–64, Dec. 1981.
Conway, J. H. and Guy, R. K. "Fractions Cycle into Decimals." In *The Book of Numbers.* New York: Springer-Verlag, pp. 163–165, 1996.
Diaconis, P.; Graham, R. L.; and Kantor, W. M. "The Mathematics of Perfect Shuffles." *Adv. Appl. Math.* **4**, 175–196, 1983.
Gardner, M. *Mathematical Carnival: A New Round-Up of Tantalizers and Puzzles from Scientific American.* Washington, DC: Math. Assoc. Amer., 1989.
Herstein, I. N. and Kaplansky, I. *Matters Mathematical.* New York: Harper & Row, 1974.
Mann, B. "How Many Times Should You Shuffle a Deck of Cards." *UMAP J.* **15**, 303–332, 1994.
Marlo, E. *Faro Notes.* Chicago, IL: Ireland Magic Co., 1958a.
Marlo, E. *Faro Shuffle.* Chicago, IL: Ireland Magic Co., 1958b.
Medvedoff, S. and Morrison, K. "Groups of Perfect Shuffles." *Math. Mag.* **60**, 3–14, 1987.
Morris, S. B. and Hartwig, R. E. "The Generalized Faro Shuffle." *Discrete Math.* **15**, 333–346, 1976.
Peterson, I. *Islands of Truth: A Mathematical Mystery Cruise.* New York: W. H. Freeman, pp. 240–244, 1990.
Ramnath, S. and Scully, D. "Moving Card i to Position j with Perfect Shuffles." *Math. Mag.* **69**, 361–365, 1996.
Stone, H. S. "Parallel Processing with the Perfect Shuffle." *IEEE Trans. Comput.* **2**, 153–161, 1971.

Rigby Points

The PERSPECTIVE CENTERS of the TANGENTIAL and CONTACT TRIANGLES of the inner and outer SODDY POINTS. The Rigby points are given by

$$Ri = I + \tfrac{4}{3}\,Ge$$

$$Ri' = I - \tfrac{4}{3}\,Ge,$$

where I is the INCENTER and Ge is the GERGONNE POINT.

see also CONTACT TRIANGLE, GERGONNE POINT, GRIFFITHS POINTS, INCENTER, OLDKNOW POINTS, SODDY POINTS, TANGENTIAL TRIANGLE

References
Oldknow, A. "The Euler-Gergonne-Soddy Triangle of a Triangle." *Amer. Math. Monthly* **103**, 319–329, 1996.

Right Angle

An ANGLE equal to half the ANGLE from one end of a line segment to the other. A right angle is $\pi/2$ radians or $90°$. A TRIANGLE containing a right angle is called a RIGHT TRIANGLE. However, a TRIANGLE cannot contain more than one right angle, since the sum of the two right angles plus the third angle would exceed the $180°$ total possessed by a TRIANGLE.

see also ACUTE ANGLE, OBLIQUE ANGLE, OBTUSE ANGLE, RIGHT TRIANGLE, SEMICIRCLE, STRAIGHT ANGLE, THALES' THEOREM

Right Conoid

A RULED SURFACE is called a right conoid if it can be generated by moving a straight LINE intersecting a fixed straight LINE such that the LINES are always PERPENDICULAR (Kreyszig 1991, p. 87). Taking the PERPENDICULAR plane as the xy-plane and the line to be the x-AXIS gives the parametric equations

$$x(u,v) = v\cos\vartheta(u)$$
$$y(u,v) = v\sin\vartheta(u)$$
$$z(u,v) = h(u)$$

(Gray 1993). Taking $h(u) = 2u$ and $\vartheta(u) = u$ gives the HELICOID.

see also HELICOID, PLÜCKER'S CONOID, WALLIS'S CONICAL EDGE

References

Dixon, R. *Mathographics.* New York: Dover, p. 20, 1991.

Gray, A. *Modern Differential Geometry of Curves and Surfaces.* Boca Raton, FL: CRC Press, pp. 351–352, 1993.

Kreyszig, E. *Differential Geometry.* New York: Dover, 1991.

Right Hyperbola

A HYPERBOLA for which the ASYMPTOTES are PERPENDICULAR. This occurs when the SEMIMAJOR and SEMIMINOR AXES are equal. Taking $a = b$ in the equation of a HYPERBOLA with SEMIMAJOR AXIS parallel to the x-AXIS and SEMIMINOR AXIS parallel to the y-AXIS (i.e., vertical DIRECTRIX),

$$\frac{(x-x_0)^2}{a^2} - \frac{(y-y_0)^2}{b^2} = 1$$

therefore gives

$$(x-x_0)^2 - (y-y_0)^2 = a^2.$$

A special type of right hyperbola is the so-called RECTANGULAR HYPERBOLA, which has equation $xy = ab$.

see also HYPERBOLA, RECTANGULAR HYPERBOLA

Right Line

see LINE

Right Strophoid

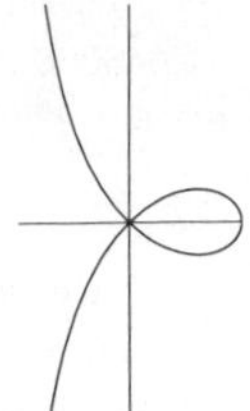

The STROPHOID of a line L with pole O not on L and fixed point O' being the point where the PERPENDICULAR from O to L cuts L is called a right strophoid. It is therefore a general STROPHOID with $a = \pi/2$.

The right strophoid is given by the Cartesian equation

$$y^2 = \frac{c-x}{c+x}x^2, \tag{1}$$

or the polar equation

$$r = c\cos(2\theta)\sec\theta. \tag{2}$$

The parametric form of the strophoid is

$$x(t) = \frac{1-t^2}{t^2+1} \tag{3}$$

$$y(t) = \frac{t(t^2-1)}{t^2+1}. \tag{4}$$

The right strophoid has CURVATURE

$$\kappa(t) = -\frac{4(1+3t^2)}{(1+6t^2+t^4)^{3/2}} \tag{5}$$

and TANGENTIAL ANGLE

$$\phi(t) = -2\tan^{-1}t - \tan^{-1}\left(\frac{2t}{1+t^2}\right). \tag{6}$$

The right strophoid first appears in work by Isaac Barrow in 1670, although Torricelli describes the curve in his letters around 1645 and Roberval found it as the LOCUS of the focus of the conic obtained when the plane cutting the CONE rotates about the tangent at its vertex (MacTutor Archive). The AREA of the loop is

$$A_{\text{loop}} = \tfrac{1}{2}c^2(4-\pi) \tag{7}$$

(MacTutor Archive).

Let C be the CIRCLE with center at the point where the right strophoid crosses the x-axis and radius the distance of that point from the origin. Then the right strophoid is invariant under inversion in the CIRCLE C and is therefore an ANALLAGMATIC CURVE.

see also STROPHOID, TRISECTRIX

References

Gray, A. *Modern Differential Geometry of Curves and Surfaces.* Boca Raton, FL: CRC Press, p. 71, 1993.

Lawrence, J. D. *A Catalog of Special Plane Curves.* New York: Dover, pp. 100–104, 1972.

Lockwood, E. H. "The Right Strophoid." Ch. 10 in *A Book of Curves.* Cambridge, England: Cambridge University Press, pp. 90–97, 1967.

MacTutor History of Mathematics Archive. "Right Strophoid." `http://www-groups.dcs.st-and.ac.uk/~history/Curves/Right.html`.

Right Strophoid Inverse Curve

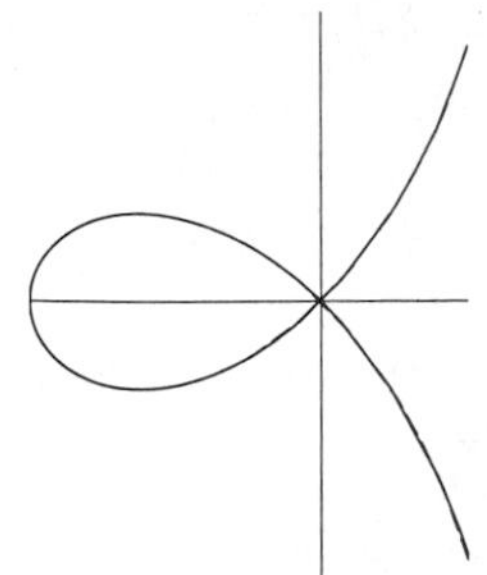

The INVERSE CURVE of a right strophoid is the same strophoid.

Right Triangle

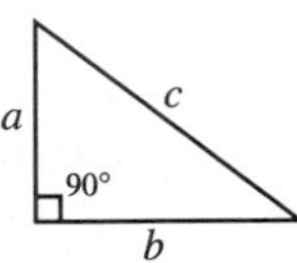

A TRIANGLE with an ANGLE of 90° ($\pi/2$ radians). The sides a, b, and c of such a TRIANGLE satisfy the PYTHAGOREAN THEOREM. The largest side is conventionally denoted c and is called the HYPOTENUSE.

For any three similar shapes on the sides of a right triangle,

$$A_1 + A_2 = A_3, \tag{1}$$

which is equivalent to the PYTHAGOREAN THEOREM. For a right triangle with sides a, b, and HYPOTENUSE c, let r be the INRADIUS. Then

$$\tfrac{1}{2}ab = \tfrac{1}{2}ra + \tfrac{1}{2}rb + \tfrac{1}{2}rc = \tfrac{1}{2}r(a+b+c). \tag{2}$$

Solving for r gives

$$r = \frac{ab}{a+b+c}. \tag{3}$$

But any PYTHAGOREAN TRIPLE can be written

$$a = m^2 - n^2 \tag{4}$$

$$b = 2mn \tag{5}$$

$$c = m^2 + n^2, \tag{6}$$

so (5) becomes

$$r = \frac{(m^2 - n^2)2mn}{m^2 - n^2 + 2mn + m^2 + n^2} = n(m-n), \tag{7}$$

which is an INTEGER when m and n are integers.

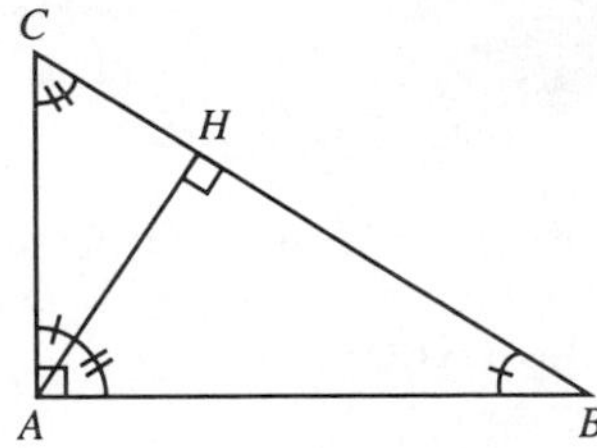

Given a right triangle ΔABC, draw the ALTITUDE AH from the RIGHT ANGLE A. Then the triangles ΔAHC and ΔBHA are similar.

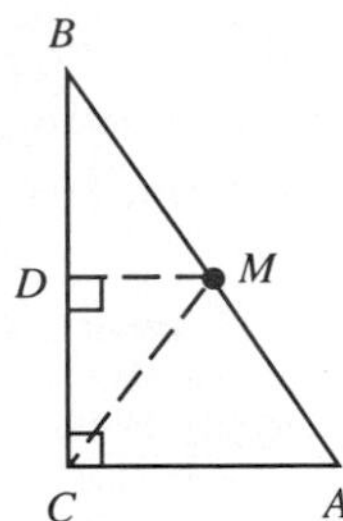

In a right triangle, the MIDPOINT of the HYPOTENUSE is equidistant from the three VERTICES (Dunham 1990). This can be proved as follows. Given ΔABC, let M be the MIDPOINT of AB (so that $AM = BM$). Draw $DM \| CA$, then since ΔBDM is similar to ΔBCA, it follows that $BD = DC$. Since both ΔBDM and ΔCDM are right triangles and the corresponding legs are equal, the HYPOTENUSES are also equal, so we have $AM = BM = CM$ and the theorem is proved.

see also ACUTE TRIANGLE, ARCHIMEDES' MIDPOINT THEOREM, BROCARD MIDPOINT, CIRCLE-POINT MIDPOINT THEOREM, FERMAT'S RIGHT TRIANGLE THEOREM, ISOSCELES TRIANGLE, MALFATTI'S RIGHT TRIANGLE PROBLEM, OBTUSE TRIANGLE, PYTHAGOREAN TRIPLE, QUADRILATERAL, RAT-FREE SET, TRIANGLE

References

Beyer, W. H. (Ed.) *CRC Standard Mathematical Tables, 28th ed.* Boca Raton, FL: CRC Press, p. 121, 1987.

Dunham, W. *Journey Through Genius: The Great Theorems of Mathematics.* New York: Wiley, pp. 120–121, 1990.

Rigid

A FRAMEWORK is rigid IFF continuous motion of the points of the configuration maintaining the bar constraints comes from a family of motions of all EUCLIDEAN SPACE which are distance-preserving. A GRAPH G is (generically) d-rigid if, for almost all (i.e., an open dense set of) CONFIGURATIONS of p, the FRAMEWORK $G(p)$ is rigid in $\mathbb{R}^d$.

One of the first results in rigidity theory was the RIGIDITY THEOREM by Cauchy in 1813. Although rigidity problems were of immense interest to engineers, intensive mathematical study of these types of problems has occurred only relatively recently (Connelly 1993, Graver *et al.* 1993).

see also BAR (EDGE), FLEXIBLE POLYHEDRON, FRAMEWORK, LAMAN'S THEOREM, LIEBMANN'S THEOREM, RIGIDITY THEOREM

References

Connelly, R. "Rigidity." Ch. 1.7 in *Handbook of Convex Geometry, Vol. A* (Ed. P. M. Gruber and J. M. Wills). Amsterdam, Netherlands: North-Holland, pp. 223–271, 1993.

Crapo, H. and Whiteley, W. "Statics of Frameworks and Motions of Panel Structures, A Projective Geometry Introduction." *Structural Topology* **6**, 43–82, 1982.

Graver, J.; Servatius, B.; and Servatius, H. *Combinatorial Rigidity.* Providence, RI: Amer. Math. Soc., 1993.

Rigid Motion

A transformation consisting of ROTATIONS and TRANSLATIONS which leaves a given arrangement unchanged.

see also EUCLIDEAN MOTION, PLANE, ROTATION

References

Courant, R. and Robbins, H. *What is Mathematics?: An Elementary Approach to Ideas and Methods, 2nd ed.* Oxford, England: Oxford University Press, p. 141, 1996.

Rigidity Theorem

If the faces of a *convex* POLYHEDRON were made of metal plates and the EDGES were replaced by hinges, the POLYHEDRON would be RIGID. The theorem was stated by Cauchy (1813), although a mistake in this paper went unnoticed for more than 50 years. An example of a *concave* "FLEXIBLE POLYHEDRON" (with 18 triangular faces) for which this is not true was given by Connelly (1978), and a FLEXIBLE POLYHEDRON with only 14 triangular faces was subsequently found by Steffen (Mackenzie 1998).

see also FLEXIBLE POLYHEDRON, RIGID

References

Cauchy, A. L. "Sur les polygons et le polyhéders." *XVIe Cahier* **IX**, 87–89, 1813.
Connelly, R. "A Flexible Sphere." *Math. Intel.* **1**, 130–131, 1978.
Graver, J.; Servatius, B.; and Servatius, H. *Combinatorial Rigidity.* Providence, RI: Amer. Math. Soc., 1993.
Mackenzie, D. "Polyhedra Can Bend But Not Breathe." *Science* **279**, 1637, 1998.

Ring

A ring is a set together with two BINARY OPERATORS $S(+,*)$ satisfying the following conditions:

1. Additive associativity: For all $a, b, c \in S$, $(a+b)+c = a+(b+c)$,
2. Additive commutativity: For all $a, b \in S$, $a+b = b+a$,
3. Additive identity: There exists an element $0 \in S$ such that for all $a \in S$, $0+a = a+0 = a$,
4. Additive inverse: For every $a \in S$ there exists $-ainS$ such that $a+(-a) = (-a)+a = 0$,
5. Multiplicative associativity: For all $a, b, c \in S$, $(a*b)*c = a*(b*c)$,
6. Left and right distributivity: For all $a, b, c \in S$, $a*(b+c) = (a*b)+(a*c)$ and $(b+c)*a = (b*a)+(c*a)$.

A ring is therefore an ABELIAN GROUP under addition and a SEMIGROUP under multiplication. A ring must contain at least one element, but need not contain a multiplicative identity or be commutative. The number of finite rings of n elements for $n = 1, 2, \ldots$, are 1, 2, 2, 11, 2, 4, 2, 52, 11, 4, 2, 22, 2, 4, 4, ... (Sloane's A027623 and A037234; Fletcher 1980). In general, the number of rings of order p^3 for p an ODD PRIME is $3p+50$ and 52 for $p=2$ (Ballieu 1947, Gilmer and Mott 1973).

A ring with a multiplicative identity is sometimes called a UNIT RING. Fraenkel (1914) gave the first abstract definition of the ring, although this work did not have much impact.

A ring that is COMMUTATIVE under multiplication, has a unit element, and has no divisors of zero is called an INTEGRAL DOMAIN. A ring which is also a COMMUTATIVE multiplication group is called a FIELD. The simplest rings are the INTEGERS $\mathbb{Z}$, POLYNOMIALS $\mathbb{R}[x]$ and $\mathbb{R}[x,y]$ in one and two variables, and SQUARE $n\times n$ REAL MATRICES.

Rings which have been investigated and found to be of interest are usually named after one or more of their investigators. This practice unfortunately leads to names which give very little insight into the relevant properties of the associated rings.

see also ABELIAN GROUP, ARTINIAN RING, CHOW RING, DEDEKIND RING, DIVISION ALGEBRA, FIELD, GORENSTEIN RING, GROUP, GROUP RING, IDEAL, INTEGRAL DOMAIN, MODULE, NILPOTENT ELEMENT, NOETHERIAN RING, NUMBER FIELD, PRIME RING, PRÜFER RING, QUOTIENT RING, REGULAR RING, RINGOID, SEMIPRIME RING, SEMIRING, SEMISIMPLE RING, SIMPLE RING, UNIT RING, ZERO DIVISOR

References

Ballieu, R. "Anneaux finis; systèmes hypercomplexes de rang trois sur un corps commutatif." *Ann. Soc. Sci. Bruxelles. Sér. I* **61**, 222–227, 1947.
Fletcher, C. R. "Rings of Small Order." *Math. Gaz.* **64**, 9–22, 1980.
Fraenkel, A. "Über die Teiler der Null und die Zerlegung von Ringen." *J. Reine Angew. Math.* **145**, 139–176, 1914.
Gilmer, R. and Mott, J. "Associative Rings of Order p^3." *Proc. Japan Acad.* **49**, 795–799, 1973.
Kleiner, I. "The Genesis of the Abstract Ring Concept." *Amer. Math. Monthly* **103**, 417–424, 1996.
Sloane, N. J. A. Sequences A027623 and A037234 in "An On-Line Version of the Encyclopedia of Integer Sequences."
van der Waerden, B. L. *A History of Algebra.* New York: Springer-Verlag, 1985.

Ring Cyclide

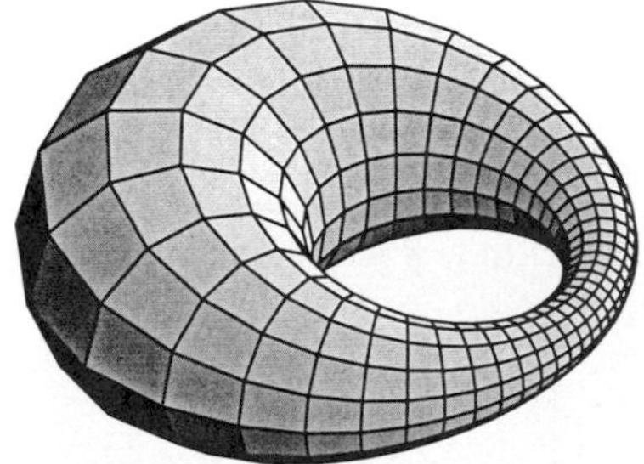

The inversion of a RING TORUS. If the inversion center lies on the torus, then the ring cyclide degenerates to a PARABOLIC RING CYCLIDE.

see also CYCLIDE, PARABOLIC CYCLIDE, RING CYCLIDE, RING TORUS, SPINDLE CYCLIDE, TORUS

Ring Function

see TOROIDAL FUNCTION

Ring Torus

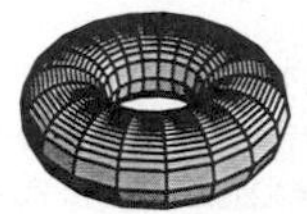
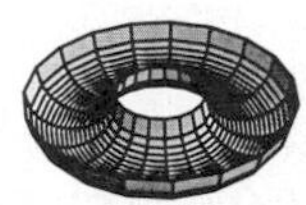
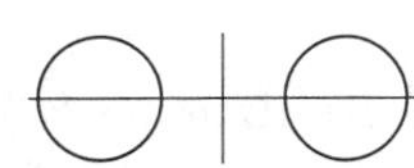

One of the three STANDARD TORI given by the parametric equations

$$x = (c + a\cos v)\cos u$$
$$y = (c + a\cos v)\sin u$$
$$z = a\sin v$$

with $c > a$. This is the TORUS which is generally meant when the term "torus" is used without qualification. The inversion of a ring torus is a RING CYCLIDE if the INVERSION CENTER does not lie on the torus and a PARABOLIC RING CYCLIDE if it does. The above left figure shows a ring torus, the middle a cutaway, and the right figure shows a CROSS-SECTION of the ring torus through the xz-plane.

see also CYCLIDE, HORN TORUS, PARABOLIC RING CYCLIDE, RING CYCLIDE, SPINDLE TORUS, STANDARD TORI, TORUS

References

Gray, A. "Tori." §11.4 in *Modern Differential Geometry of Curves and Surfaces.* Boca Raton, FL: CRC Press, pp. 218–220, 1993.

Pinkall, U. "Cyclides of Dupin." §3.3 in *Mathematical Models from the Collections of Universities and Museums* (Ed. G. Fischer). Braunschweig, Germany: Vieweg, pp. 28–30, 1986.

Ringoid

A ringoid R is a set $(R, +, \times)$ with two binary operators, conventionally denoted addition $(+)$ and multiplication $(\times)$, where $\times$ distributes over $+$ left and right:

$$a(b+c) = ab + acand(b+c)a = ba + ca.$$

A ringoid can be empty.

see also BINARY OPERATOR, RING, SEMIRING

References

Rosenfeld, A. *An Introduction to Algebraic Structures.* New York: Holden-Day, 1968.

Risch Algorithm

An ALGORITHM for indefinite integration.

see also INDEFINITE INTEGRAL

Rising Factorial

see POCHHAMMER SYMBOL

Rivest-Shamir-Adleman Number

see RSA NUMBER

RMS

see ROOT-MEAN-SQUARE

Robbin Constant

$$R = \tfrac{4}{105} + \tfrac{17}{105}\sqrt{2} - \tfrac{2}{35}\sqrt{3} + \tfrac{1}{5}\ln(1+\sqrt{2}) + \tfrac{2}{5}\ln(2+\sqrt{3}) - \tfrac{1}{15}\pi = 0.661707182\ldots.$$

see also TRANSFINITE DIAMETER

References

Plouffe, S. "The Robbin Constant." http://lacim.uqam.ca/piDATA/robbin.txt.

Robbin's Inequality

If the fourth MOMENT $\mu_4 \neq 0$, then

$$P(|\bar{x} - \mu_4| \geq \lambda) \leq \frac{\mu_4 + 3(N-1)\sigma^4}{N^3\lambda^4},$$

where σ^2 is the VARIANCE.

Robbins Algebra

Building on work of Huntington (1933), Robbins conjectured that the equations for a Robbins algebra, commutivity, associativity, and the ROBBINS EQUATION

$$n(n(x+y) + n(x + n(y))) = x,$$

imply those for a BOOLEAN ALGEBRA. The conjecture was finally proven using a computer (McCune 1997).

References

Huntington, E. V. "New Sets of Independent Postulates for the Algebra of Logic, with Special Reference to Whitehead and Russell's *Principia Mathematica.*" *Trans. Amer. Math. Soc.* **35**, 274–304, 1933.

Huntington, E. V. "Boolean Algebra. A Correction." *Trans. Amer. Math. Soc.* **35**, 557–558, 1933.

McCune, W. "Solution of the Robbins Problem." *J. Automat. Reason.* **19**, 263–276, 1997.

McCune, W. "Robbins Algebras are Boolean." http://www.mcs.anl.gov/~mccune/papers/robbins/.

Nelson, E. "Automated Reasoning." http://www.math.princeton.edu/~nelson/ar.html.

Robbins Equation

$$n(n(x+y) + n(x + n(y))) = x.$$

see also ROBBINS ALGEBRA

Robertson Condition

For the HELMHOLTZ DIFFERENTIAL EQUATION to be SEPARABLE in a coordinate system, the SCALE FACTORS h_i in the LAPLACIAN

$$\nabla^2 = \sum_{i=1}^{3} \frac{1}{h_1h_2h_3} \frac{\partial}{\partial u_i} \left(\frac{h_1h_2h_3}{{h_i}^2} \frac{\partial}{\partial u_i} \right) \quad (1)$$

and the functions $f_i(u_i)$ and Φ_{ij} defined by

$$\frac{1}{f_n} \frac{\partial}{\partial u_n} \left(f_n \frac{\partial X_n}{\partial u_n} \right) + ({k_1}^2\Phi_{n1} + {k_2}^2\Phi_{n2} + {k_3}^2\Phi_{n3})X_n = 0 \quad (2)$$

must be of the form of a STÄCKEL DETERMINANT

$$S = |\Phi_{mn}| = \begin{vmatrix} \Phi_{11} & \Phi_{12} & \Phi_{13} \\ \Phi_{21} & \Phi_{22} & \Phi_{23} \\ \Phi_{31} & \Phi_{32} & \Phi_{33} \end{vmatrix} = \frac{h_1h_2h_3}{f_1(u_1)f_2(u_2)f_3(u_3)}. \quad (3)$$

see also HELMHOLTZ DIFFERENTIAL EQUATION, LAPLACE'S EQUATION, SEPARATION OF VARIABLES, STÄCKEL DETERMINANT

References

Morse, P. M. and Feshbach, H. *Methods of Theoretical Physics, Part 1.* New York: McGraw-Hill, p. 510, 1953.

Robertson Conjecture

A conjecture due to M. S. Robertson (1936) which treats a UNIVALENT POWER SERIES containing only ODD powers within the UNIT DISK. This conjecture IMPLIES the BIEBERBACH CONJECTURE and follows in turn from the MILIN CONJECTURE. de Branges' proof of the BIEBERBACH CONJECTURE proceeded by proving the MILIN CONJECTURE, thus establishing the Robertson conjecture and hence implying the truth of the BIEBERBACH CONJECTURE.

see also BIEBERBACH CONJECTURE, MILIN CONJECTURE

References

Stewart, I. *From Here to Infinity: A Guide to Today's Mathematics.* Oxford, England: Oxford University Press, p. 165, 1996.

Robertson-Seymour Theorem

A generalization of the KURATOWSKI REDUCTION THEOREM by Robertson and Seymour, which states that the collection of finite graphs is well-quasi-ordered by minor embeddability, from which it follows that Kuratowski's "forbidden minor" embedding obstruction generalizes to higher genus surfaces.

Formally, for a fixed INTEGER $g \geq 0$, there is a finite list of graphs $L(g)$ with the property that a graph C embeds on a surface of genus g IFF it does not contain, as a minor, any of the graphs on the list L.

References

Fellows, M. R. "The Robertson-Seymour Theorems: A Survey of Applications." *Comtemp. Math.* **89**, 1–18, 1987.

Robin Boundary Conditions

PARTIAL DIFFERENTIAL EQUATION BOUNDARY CONDITIONS which, for an elliptic partial differential equation in a region Ω, specify that the sum of αu and the normal derivative of $u = f$ at all points of the boundary of Ω, α and f being prescribed.

Robin's Constant

see TRANSFINITE DIAMETER

Robinson Projection

A PSEUDOCYLINDRICAL MAP PROJECTION which distorts shape, AREA, scale, and distance to create attraction average projection properties.

References

Dana, P. H. "Map Projections." `http://www.utexas.edu/depts/grg/gcraft/notes/mapproj/mapproj.html`.

Robust Estimation

An estimation technique which is insensitive to small departures from the idealized assumptions which have been used to optimize the algorithm. Classes of such techniques include M-ESTIMATES (which follow from maximum likelihood considerations), L-ESTIMATES (which are linear combinations of ORDER STATISTICS), and R-ESTIMATES (based on RANK tests).

see also L-ESTIMATE, M-ESTIMATE, R-ESTIMATE

References

Press, W. H.; Flannery, B. P.; Teukolsky, S. A.; and Vetterling, W. T. "Robust Estimation." §15.7 in *Numerical Recipes in FORTRAN: The Art of Scientific Computing, 2nd ed.* Cambridge, England: Cambridge University Press, pp. 694–700, 1992.

Rodrigues's Curvature Formula

$$d\hat{\mathbf{N}} + \kappa_i \, d\mathbf{r} = \mathbf{0},$$

where $\hat{\mathbf{N}}$ is the unit NORMAL VECTOR and κ_i is one of the two PRINCIPAL CURVATURES.

see also NORMAL VECTOR, PRINCIPAL CURVATURES

Rodrigues Formula

An operator definition of a function. A Rodrigues formula may be converted into a SCHLÄFLI INTEGRAL.

see also SCHLÄFLI INTEGRAL

Rogers-Ramanujan Continued Fraction

see RAMANUJAN CONTINUED FRACTION

Rogers-Ramanujan Identities

For $|q| < 1$ and using the NOTATION of the RAMANUJAN THETA FUNCTION, the Rogers-Ramanujan identities are

$$\frac{f(-q^5)}{f(-q,-q^4)} = \sum_{k=0}^{\infty} \frac{q^{k^2}}{(q)_k} \tag{1}$$

$$\frac{f(-q^5)}{f(-q^2,-q^3)} = \sum_{k=0}^{\infty} \frac{q^{k(k+1)}}{(q)_k}, \tag{2}$$

where $(q)_k$ are q-SERIES. Written out explicitly (Hardy 1959, p. 13),

$$1+\frac{q}{1-q}+\frac{q^4}{(1-q)(1-q^2)}+\frac{q^9}{(1-q)(1-q^2)(1-q^3)}+\ldots = \frac{1}{(1-q)(1-q^6)\cdots(1-q^4)(1-q^9)\cdots} \tag{3}$$

$$1+\frac{q^2}{1-q}+\frac{q^6}{(1-q)(1-q^2)}+\frac{q^{12}}{(1-q)(1-q^2)(1-q^3)}+\ldots = \frac{1}{(1-q^2)(1-q^7)\cdots(1-q^3)(1-q^8)\cdots}. \tag{4}$$

The identities can also be written succinctly as

$$1+\sum_{k=1}^{\infty} \frac{q^{k^2+ak}}{(1-q)(1-q^2)\cdots(1-q^k)} = \prod_{j=0}^{\infty} \frac{1}{(1-q^{5j+a+1})(1-q^{5j-a+4})}, \tag{5}$$

where $a = 0, 1$.

Other forms of the Rogers-Ramanujan identities include

$$\sum_k \frac{q^{k^2}}{(q;q)_k(q;q)_{n-k}} = \sum_k \frac{(-1)^k q^{(5k^2-k)/2}}{(q;q)_{n-k}(q;q)_{n+k}} \tag{6}$$

and

$$\sum_k \frac{2q^{k^2}}{(q;q)_k(q;q)_{n-k}} = \sum_k \frac{(-1)^k(1+q^k)q^{(5k^2-k)/2}}{(q;q)_{n-k}(q;q)_{n+k}} \tag{7}$$

(Petkovšek *et al.* 1996).

see also ANDREWS-SCHUR IDENTITY

References

Andrews, G. E. *The Theory of Partitions.* Cambridge, England: Cambridge University Press, 1985.

Andrews, G. E. *q-Series: Their Development and Application in Analysis, Number Theory, Combinatorics, Physics, and Computer Algebra.* Providence, RI: Amer. Math. Soc., pp. 17–20, 1986.

Andrews, G. E. and Baxter, R. J. "A Motivated Proof of the Rogers-Ramanujan Identities." *Amer. Math. Monthly* **96**, 401–409, 1989.

Bressoud, D. M. *Analytic and Combinatorial Generalizations of the Rogers-Ramanujan Identities.* Providence, RI: Amer. Math. Soc., 1980.

Hardy, G. H. *Ramanujan: Twelve Lectures on Subjects Suggested by His Life and Work, 3rd ed.* New York: Chelsea, p. 13, 1959.

Paule, P "Short and Easy Computer Proofs of the Rogers-Ramanujan Identities and of Identities of Similar Type." *Electronic J. Combinatorics* **1**, R10, 1–9, 1994. `http://www.combinatorics.org/Volume_1/volume1.html#R10`.

Petkovšek, M.; Wilf, H. S.; and Zeilberger, D. *A=B.* Wellesley, MA: A. K. Peters, p. 117, 1996.

Robinson, R. M. "Comment to: 'A Motivated Proof of the Rogers-Ramanujan Identities.'" *Amer. Math. Monthly* **97**, 214–215, 1990.

Rogers, L. J. "Second Memoir on the Expansion of Certain Infinite Products." *Proc. London Math. Soc.* **25**, 318–343, 1894.

Sloane, N. J. A. Sequence A006141/M0260 in "An On-Line Version of the Encyclopedia of Integer Sequences."

Rolle's Theorem

Let f be differentiable on (a, b) and continuous on $[a, b]$. If $f(a) = f(b) = 0$, then there is at least one point $c \in (a, b)$ where $f'(c) = 0$.

see also FIXED POINT THEOREM, MEAN-VALUE THEOREM

Roman Coefficient

A generalization of the BINOMIAL COEFFICIENT whose NOTATION was suggested by Knuth,

$$\left\lfloor \begin{matrix} n \\ k \end{matrix} \right\rceil = \frac{\lfloor n\rceil!}{\lfloor k\rceil!\,\lfloor n-k\rceil!}. \tag{1}$$

The above expression is read "Roman n choose k." Whenever the BINOMIAL COEFFICIENT is defined (i.e., $n \geq k \geq 0$ or $k \geq 0 > n$), the Roman coefficient agrees with it. However, the Roman coefficients are defined for values for which the BINOMIAL COEFFICIENTS are not, e.g.,

$$\left\lfloor \begin{matrix} n \\ -1 \end{matrix} \right\rceil = \frac{1}{\lfloor n+1\rceil} \tag{2}$$

$$\left\lfloor \begin{matrix} 0 \\ k \end{matrix} \right\rceil = \frac{(-1)^{k+(k>0)}}{\lfloor k\rceil}, \tag{3}$$

where

$$n<0 \equiv \begin{cases} 1 & \text{for } n<0 \\ 0 & \text{for } n \geq 0. \end{cases} \tag{4}$$

The Roman coefficients also satisfy properties like those of the BINOMIAL COEFFICIENT,

$$\left\lfloor \begin{matrix} n \\ k \end{matrix} \right\rceil = \left\lfloor \begin{matrix} n \\ n-k \end{matrix} \right\rceil \tag{5}$$

$$\left\lfloor \begin{matrix} n \\ k \end{matrix} \right\rceil \left\lfloor \begin{matrix} k \\ r \end{matrix} \right\rceil = \left\lfloor \begin{matrix} n \\ r \end{matrix} \right\rceil \left\lfloor \begin{matrix} n-r \\ k-r \end{matrix} \right\rceil, \tag{6}$$

an analog of PASCAL'S FORMULA

$$\left\lfloor \begin{matrix} n \\ k \end{matrix} \right\rceil = \left\lfloor \begin{matrix} n-1 \\ k \end{matrix} \right\rceil + \left\lfloor \begin{matrix} n-1 \\ k-1 \end{matrix} \right\rceil, \tag{7}$$

and a curious rotation/reflection law due to Knuth

$$(-1)^{k+(k>0)} \left\lfloor \begin{matrix} -n \\ k-1 \end{matrix} \right\rceil = (-1)^{n+(n>0)} \left\lfloor \begin{matrix} -k \\ n-1 \end{matrix} \right\rceil \quad (8)$$

(Roman 1992).

see also BINOMIAL COEFFICIENT, ROMAN FACTORIAL

References

Roman, S. "The Logarithmic Binomial Formula." *Amer. Math. Monthly* **99**, 641–648, 1992.

Roman Factorial

$$\lfloor n \rceil ! \equiv \begin{cases} n! & \text{for } n \geq 0 \\ \frac{(-1)^{-n-1}}{(-n-1)!} & \text{for } n < 0. \end{cases} \quad (1)$$

The Roman factorial arises in the definition of the HARMONIC LOGARITHM and ROMAN COEFFICIENT. It obeys the identities

$$\lfloor n \rceil ! = \lfloor n \rceil \lfloor n-1 \rceil ! \quad (2)$$

$$\frac{\lfloor n \rceil !}{\lfloor n-k \rceil !} = \lfloor n \rceil \lfloor n-1 \rceil \cdots \lfloor n-k+1 \rceil \quad (3)$$

$$\lfloor n \rceil ! \lfloor -n-1 \rceil ! = (-1)^{n+(n<0)}, \quad (4)$$

where

$$\lfloor n \rceil \equiv \begin{cases} n & \text{for } n \neq 0 \\ 1 & \text{for } n = 0 \end{cases} \quad (5)$$

and

$$n < 0 \equiv \begin{cases} 1 & \text{for } n < 0 \\ 0 & \text{for } n \geq 0. \end{cases} \quad (6)$$

see also HARMONIC LOGARITHM, HARMONIC NUMBER, ROMAN COEFFICIENT

References

Loeb, D. and Rota, G.-C. "Formal Power Series of Logarithmic Type." *Advances Math.* **75**, 1–118, 1989.
Roman, S. "The Logarithmic Binomial Formula." *Amer. Math. Monthly* **99**, 641–648, 1992.

Roman Numeral

A system of numerical notations used by the Romans. It is an additive (and subtractive) system in which letters are used to denote certain "base" numbers, and arbitrary numbers are then denoted using combinations of symbols.

Character	Numerical Value
I	1
V	5
X	10
L	50
C	100
D	500
M	1000

For example, the number 1732 would be denoted MDCCXXXII. One additional rule states that, instead of using four symbols to represent a 4, 40, 9, 90, etc., such numbers are instead denoted by preceding the symbol for 5, 50, 10, 100, etc., with a symbol indicating *subtraction*. For example, 4 is denoted IV, 9 as IX, 40 as XL, etc. However, this rule is generally *not* followed on the faces of clocks, where IIII is usually encountered instead of IV.

Roman numerals are encountered in the release year for movies and occasionally on the numerals on the faces of watches and clocks, but in few other modern instances. They do have the advantage that ADDITION can be done "symbolically" (and without worrying about the "place" of a given DIGIT) by simply combining all the symbols together, grouping, writing groups of 5 Is as V, groups of 2 Vs as X, etc.

Roman Surface

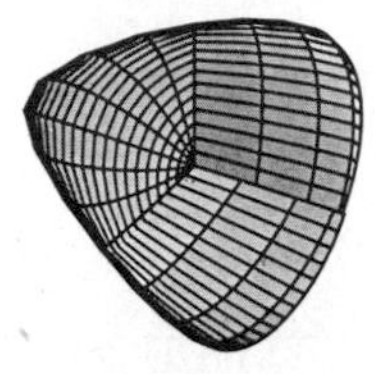
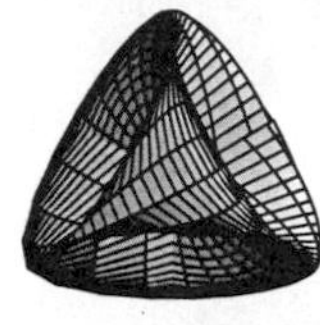

A QUARTIC NONORIENTABLE SURFACE, also known as the STEINER SURFACE. The Roman surface is one of the three possible surfaces obtained by sewing a MÖBIUS STRIP to the edge of a DISK. The other two are the BOY SURFACE and CROSS-CAP, all of which are homeomorphic to the REAL PROJECTIVE PLANE (Pinkall 1986).

The center point of the Roman surface is an ordinary TRIPLE POINT with $(\pm 1, 0, 0) = (0, \pm 1, 0) = (0, 0, \pm 1)$, and the six endpoints of the three lines of self-intersection are singular PINCH POINTS, also known as WHITNEY SINGULARITIES. The Roman surface is essentially six CROSS-CAPS stuck together and contains a double INFINITY of CONICS.

The Roman surface can given by the equation

$$(x^2+y^2+z^2-k^2)^2 = [(z-k)^2-2x^2][(z+k)^2-2y^2]. \quad (1)$$

Solving for z gives the pair of equations

$$z = \frac{k(y^2-x^2) \pm (x^2-y^2)\sqrt{k^2-x^2-y^2}}{2(x^2+y^2)}. \quad (2)$$

If the surface is rotated by 45° about the z-AXIS via the ROTATION MATRIX

$$\mathsf{R}_z(45°) = \frac{1}{\sqrt{2}} \begin{bmatrix} 1 & 1 & 0 \\ -1 & 1 & 0 \\ 0 & 0 & 1 \end{bmatrix} \quad (3)$$

to give

$$\begin{bmatrix} x' \\ y' \\ z' \end{bmatrix} = \mathsf{R}_z(45°) \begin{bmatrix} x \\ y \\ z \end{bmatrix}, \quad (4)$$

then the simple equation

$$x^2y^2 + x^2z^2 + y^2z^2 + 2kxyz = 0 \tag{5}$$

results. The Roman surface can also be generated using the general method for NONORIENTABLE SURFACES using the polynomial function

$$\mathbf{f}(x, y, z) = (xy, yz, zx) \tag{6}$$

(Pinkall 1986). Setting

$$x = \cos u \sin v \tag{7}$$
$$y = \sin u \sin v \tag{8}$$
$$z = \cos v \tag{9}$$

in the former gives

$$x(u, v) = \tfrac{1}{2}\sin(2u)\sin^2 v \tag{10}$$
$$y(u, v) = \tfrac{1}{2}\sin u \cos(2v) \tag{11}$$
$$z(u, v) = \tfrac{1}{2}\cos u \sin(2v) \tag{12}$$

for $u \in [0, 2\pi)$ and $v \in [-\pi/2, \pi/2]$. Flipping $\sin v$ and $\cos v$ and multiplying by 2 gives the form shown by Wang.

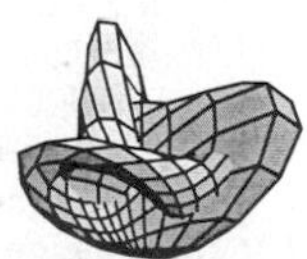

A HOMOTOPY (smooth deformation) between the Roman surface and BOY SURFACE is given by the equations

$$x(u, v) = \frac{\sqrt{2}\cos(2u)\cos^2 v + \cos u \sin(2v)}{2 - \alpha\sqrt{2}\sin(3u)\sin(2v)} \tag{13}$$

$$y(u, v) = \frac{\sqrt{2}\sin(2u)\cos^2 v - \sin u \sin(2v)}{2 - \alpha\sqrt{2}\sin(3u)\sin(2v)} \tag{14}$$

$$z(u, v) = \frac{3\cos^2 v}{2 - \alpha\sqrt{2}\sin(3u)\sin(2v)} \tag{15}$$

for $u \in [-\pi/2, \pi/2]$ and $v \in [0, \pi]$ as α varies from 0 to 1. $\alpha = 0$ corresponds to the Roman surface and $\alpha = 1$ to the BOY SURFACE (Wang).

see also BOY SURFACE, CROSS-CAP, HEPTAHEDRON, MÖBIUS STRIP, NONORIENTABLE SURFACE, QUARTIC SURFACE, STEINER SURFACE

References

Fischer, G. (Ed.). *Mathematical Models from the Collections of Universities and Museums.* Braunschweig, Germany: Vieweg, p. 19, 1986.

Fischer, G. (Ed.). Plates 42–44 and 108–114 in *Mathematische Modelle/Mathematical Models, Bildband/Photograph Volume.* Braunschweig, Germany: Vieweg, pp. 42–44 and 108–109, 1986.

Geometry Center. "The Roman Surface." `http://www.geom.umn.edu/zoo/toptype/pplane/roman/`.

Gray, A. *Modern Differential Geometry of Curves and Surfaces.* Boca Raton, FL: CRC Press, pp. 242–243, 1993.

Nordstrand, T. "Steiner's Roman Surface." `http://www.uib.no/people/nfytn/steintxt.htm`.

Pinkall, U. *Mathematical Models from the Collections of Universities and Museums* (Ed. G. Fischer). Braunschweig, Germany: Vieweg, p. 64, 1986.

Wang, P. "Renderings." `http://www.ugcs.caltech.edu/~peterw/portfolio/renderings/`.

Roman Symbol

$$\lfloor n \rceil \equiv \begin{cases} n & \text{for } n \neq 0 \\ 1 & \text{for } n = 0. \end{cases}$$

see also ROMAN FACTORIAL, HARMONIC LOGARITHM

References

Roman, S. "The Logarithmic Binomial Formula." *Amer. Math. Monthly* **99**, 641–648, 1992.

Romberg Integration

A powerful NUMERICAL INTEGRATION technique which uses k refinements of the extended TRAPEZOIDAL RULE to remove error terms less than order $\mathcal{O}(N^{-2k})$. The routine advocated by Press *et al.* (1992) makes use of NEVILLE'S ALGORITHM.

References

Acton, F. S. *Numerical Methods That Work, 2nd printing.* Washington, DC: Math. Assoc. Amer., pp. 106–107, 1990.

Dahlquist, G. and Bjorck, A. §7.4.1–7.4.2 in *Numerical Methods.* Englewood Cliffs, NJ: Prentice-Hall, 1974.

Press, W. H.; Flannery, B. P.; Teukolsky, S. A.; and Vetterling, W. T. "Romberg Integration." §4.3 in *Numerical Recipes in FORTRAN: The Art of Scientific Computing, 2nd ed.* Cambridge, England: Cambridge University Press, pp. 134–135, 1992.

Ralston, A. and Rabinowitz, P. §4.10 in *A First Course in Numerical Analysis, 2nd ed.* New York: McGraw-Hill, 1978.

Stoer, J.; and Burlisch, R. §3.4–3.5 in *Introduction to Numerical Analysis.* New York: Springer-Verlag, 1980.

Rook Number

The rook numbers r_n^B of an $n \times n$ BOARD B are the number of subsets of size n such that no two elements have the same first or second coordinate. In other word, it is the number of ways of placing n rooks on B such that none attack each other. The rook numbers of a board determine the rook numbers of the complementary board $\overline{B}$, defined to be $\mathbf{d} \times \mathbf{d} \backslash B$. This is known as the ROOK RECIPROCITY THEOREM. The first few rook numbers are 1, 2, 7, 23, 115, 694, 5282, 46066, ... (Sloane's A000903). For an $n \times n$ board, each $n \times n$

PERMUTATION MATRIX corresponds to an allowed configuration of rooks.

see also ROOK RECIPROCITY THEOREM

References

Sloane, N. J. A. Sequence A000903/M1761 in "An On-Line Version of the Encyclopedia of Integer Sequences."

Rook Reciprocity Theorem

$$\sum_{k=0}^{d} r_k^B (d-k)! x^k = \sum_{k=0}^{d} (-1)^k r_k^{\bar{B}} (d-k)! x^k (x+1)^{d-k}.$$

References

Chow, T. Y. "The Path-Cycle Symmetric Function of a Digraph." *Adv. Math.* **118**, 71–98, 1996.

Chow, T. "A Short Proof of the Rook Reciprocity Theorem." *Electronic J. Combinatorics* **3**, R10, 1–2, 1996. `http://www.combinatorics.org/Volume_3/volume3.html#R10`.

Goldman, J. R.; Joichi, J. T.; and White, D. E. "Rook Theory I. Rook Equivalence of Ferrers Boards." *Proc. Amer. Math. Soc.* **52**, 485–492, 1975.

Riordan, J. *An Introduction to Combinatorial Analysis.* New York: Wiley, 1958.

Rooks Problem

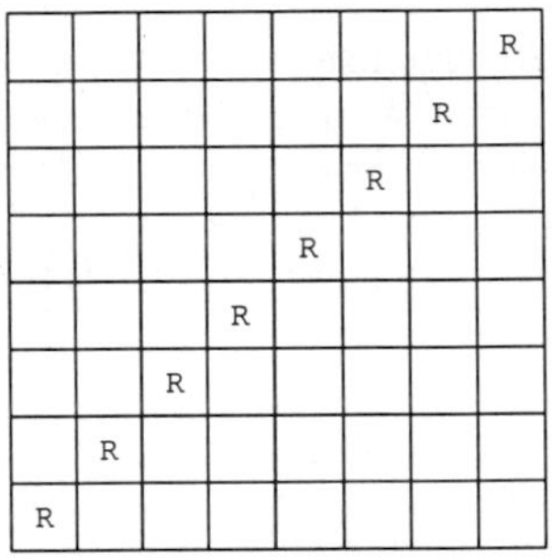

							R
						R	
					R		
				R			
			R				
		R					
	R						
R							

The rook is a CHESS piece which may move any number of spaces either horizontally or vertically per move. The maximum number of nonattacking rooks which may be placed on an $n \times n$ CHESSBOARD is n. This arrangement is achieved by placing the rooks along the diagonal (Madachy 1979). The total number of ways of placing n nonattacking rooks on an $n \times n$ board is $n!$ (Madachy 1979, p. 47). The number of rotationally and reflectively inequivalent ways of placing n nonattacking rooks on an $n \times n$ board are 1, 2, 7, 23, 115, 694, ... (Sloane's A000903; Dudeney 1970, p. 96; Madachy 1979, pp. 46–54).

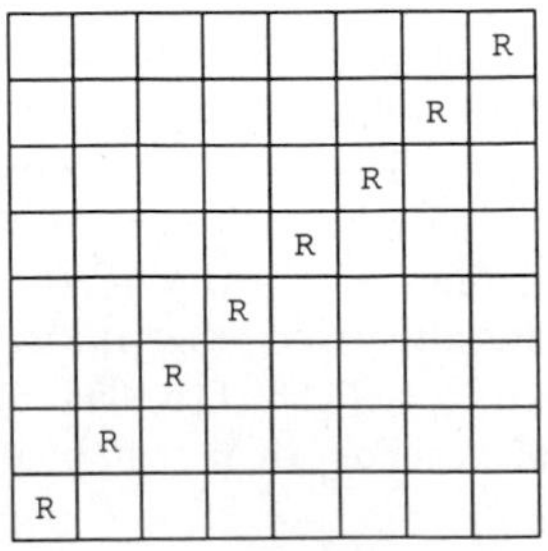

							R
						R	
					R		
				R			
			R				
		R					
	R						
R							

The minimum number of rooks needed to occupy or attack all spaced on an 8×8 CHESSBOARD is 8, illustrated above (Madachy 1979).

Consider an $n \times n$ chessboard with the restriction that, for every subset of $\{1, \ldots, n\}$, a rook may not be put in column $s + j \pmod{n}$ when on row j, where the rows are numbered 0, 1, ..., $n - 1$. Vardi (1991) denotes the number of rook solutions so restricted as $\mathrm{rook}(s, n)$. $\mathrm{rook}(\{1\}, n)$ is simply the number of DERANGEMENTS on n symbols, known as a SUBFACTORIAL. The first few values are 1, 2, 9, 44, 265, 1854, ... (Sloane's A000166). $\mathrm{rook}(\{1, 2\}, n)$ is a solution to the MARRIED COUPLES PROBLEM, sometimes known as MÉNAGE NUMBERS. The first few MÉNAGE NUMBERS are -1, 1, 0, 2, 13, 80, 579, ... (Sloane's A000179).

Although simple formulas are not known for general $\{1, \ldots, p\}$, RECURRENCE RELATIONS can be used to compute $\mathrm{rook}(\{1, \ldots, p\}, n)$ in polynomial time for $p = 3$, ..., 6 (Metropolis *et al.* 1969, Minc 1978, Vardi 1991).

see also CHESS, MÉNAGE NUMBER, ROOK NUMBER, ROOK RECIPROCITY THEOREM

References

Dudeney, H. E. "The Eight Rooks." §295 in *Amusements in Mathematics.* New York: Dover, p. 88, 1970.

Kraitchik, M. "The Problem of the Rooks" and "Domination of the Chessboard." §10.2 and 10.4 in *Mathematical Recreations.* New York: W. W. Norton, pp. 240–247 and 255–256, 1942.

Madachy, J. S. *Madachy's Mathematical Recreations.* New York: Dover, pp. 36–37, 1979.

Metropolis, M.; Stein, M. L.; and Stein, P. R. "Permanents of Cyclic (0, 1) Matrices." *J. Combin. Th.* **7**, 291–321, 1969.

Minc, H. §3.1 in *Permanents.* Reading, MA: Addison-Wesley, 1978.

Riordan, J. Chs. 7–8 in *An Introduction to Combinatorial Analysis.* Princeton, NJ: Princeton University Press, 1978.

Sloane, N. J. A. Sequences A000903/M1761, A000166/M1937, and A000179/M2062 in "An On-Line Version of the Encyclopedia of Integer Sequences." `http://www.research.att.com/~njas/sequences/eisonline.html`.

Sloane, N. J. A. and Plouffe, S. Extended entry for M2062 in *The Encyclopedia of Integer Sequences.* San Diego: Academic Press, 1995.

Vardi, I. *Computational Recreations in Mathematica.* Reading, MA: Addison-Wesley, pp. 123–124, 1991.

Room Square

A Room square (named after T. G. Room) of order n (for n ODD) is an arrangement in an $n \times n$ SQUARE MATRIX of $n + 1$ objects such that each cell is either empty or holds exactly two different objects. Furthermore, each object appears once in each row and column and each unordered pair occupies exactly one cell. The Room square of order 2 is shown below.

1, 2

The Room square of order 8 is

1, 8			5, 7		3, 4	2, 6
3, 7	2, 8			6, 1		4, 5
5, 6	4, 1	3, 8			7, 2	
	6, 7	5, 2	4, 8			1, 3
2, 4		7, 1	6, 3	5 ,8		
	3, 5		1, 2	7, 4	6, 8	
		4, 6		2, 3	1, 5	7, 8

References

Dinitz, J. H. and Stinson, D. R. In *Contemporary Design Theory: A Collection of Surveys* (Ed. J. H. Dinitz and D. R. Stinson). New York: Wiley, 1992.

Gardner, M. *Time Travel and Other Mathematical Bewilderments.* New York: W. H. Freeman, pp. 146–147 and 151–152, 1988.

Mullin, R. C. and Nemeth, E. "On Furnishing Room Squares." *J. Combin. Th.* **7**, 266–272, 1969.

Mullin, R. D. and Wallis, W. D. "The Existence of Room Squares." *Aequationes Math.* **13**, 1–7, 1975.

O'Shaughnessy, C. D. "On Room Squares of Order $6m+2$." *J. Combin. Th.* **13**, 306–314, 1972.

Room, T. G. "A New Type of Magic Square" (Note 2569). *Math. Gaz.* **39**, 307, 1955.

Wallis, W. D. "Solution of the Room Square Existence Problem." *J. Combin. Th.* **17**, 379–383, 1974.

Root

The roots of an equation

$$f(x) = 0 \tag{1}$$

are the values of x for which the equation is satisfied. The FUNDAMENTAL THEOREM OF ALGEBRA states that every POLYNOMIAL equation of degree n has exactly n roots, where some roots may have a multiplicity greater than 1 (in which case they are said to be degenerate).

To find the nth roots of a COMPLEX NUMBER, solve the equation $z^n = w$. Then

$$z^n = |z|^n[\cos(n\theta) + i\sin(n\theta)] = |w|\,(\cos\phi + i\sin\phi), \tag{2}$$

so

$$|z| = |w|^{1/n} \tag{3}$$

and

$$\arg(z) = \frac{\phi}{n}. \tag{4}$$

Rolle proved that any number has n nth roots (Boyer 1968, p. 476). Householder (1970) gives an algorithm for constructing root-finding algorithms with an arbitrary order of convergence. Special root-finding techniques can often be applied when the function in question is a POLYNOMIAL.

see also BAILEY'S METHOD, BISECTION PROCEDURE, BRENT'S METHOD, CROUT'S METHOD, DESCARTES' SIGN RULE, FALSE POSITION METHOD, FUNDAMENTAL THEOREM OF SYMMETRIC FUNCTIONS, GRAEFFE'S METHOD, HALLEY'S IRRATIONAL FORMULA, HALLEY'S METHOD, HALLEY'S RATIONAL FORMULA, HORNER'S METHOD, HOUSEHOLDER'S METHOD, HUTTON'S METHOD, ISOGRAPH, JENKINS-TRAUB METHOD, LAGUERRE'S METHOD, LAMBERT'S METHOD, LEHMER-SCHUR METHOD, LIN'S METHOD, MAEHLY'S PROCEDURE, MULLER'S METHOD, NEWTON'S METHOD, RIDDERS' METHOD, ROOT DRAGGING THEOREM, SCHRÖDER'S METHOD, POLYNOMIAL, SECANT METHOD, STURM FUNCTION, STURM THEOREM, TANGENT HYPERBOLAS METHOD, WEIERSTRAß APPROXIMATION THEOREM

References

Arfken, G. "Appendix 1: Real Zeros of a Function." *Mathematical Methods for Physicists, 3rd ed.* Orlando, FL: Academic Press, pp. 963–967, 1985.

Boyer, C. B. *A History of Mathematics.* New York: Wiley, 1968.

Householder, A. S. *The Numerical Treatment of a Single Nonlinear Equation.* New York: McGraw-Hill, 1970.

Press, W. H.; Flannery, B. P.; Teukolsky, S. A.; and Vetterling, W. T. "Roots of Polynomials." §9.5 in *Numerical Recipes in FORTRAN: The Art of Scientific Computing, 2nd ed.* Cambridge, England: Cambridge University Press, pp. 362–372, 1992.

Root Dragging Theorem

If any of the ROOTS of a POLYNOMIAL are increased, then *all* of the critical points increase.

References

Anderson, B. "Polynomial Root Dragging." *Amer. Math. Monthly* **100**, 864–866, 1993.

Root Linear Coefficient Theorem

The sum of the reciprocals of ROOTS of an equation equals the NEGATIVE COEFFICIENT of the linear term in the MACLAURIN SERIES.

see also NEWTON'S RELATIONS

Root-Mean-Square

The root-mean-square (RMS) of a variate x, sometimes called the QUADRATIC MEAN, is the SQUARE ROOT of the mean squared value of x:

$$R(x) \equiv \sqrt{\langle x^2 \rangle} \tag{1}$$

$$= \begin{cases} \sqrt{\dfrac{\sum_{i=1}^n x_i^2}{n}} & \text{for a discrete distribution} \\ \sqrt{\dfrac{\int P(x)x^2\,dx}{\int P(x)\,dx}} & \text{for a continuous distribution.} \end{cases} \tag{2}$$

Hoehn and Niven (1985) show that

$$R(a_1+c, a_2+c, \ldots, a_n+c) < c + R(a_1, a_2, \ldots, a_n)$$

for any POSITIVE constant c.

Physical scientists often use the term root-mean-square as a synonym for STANDARD DEVIATION when they refer to the SQUARE ROOT of the mean squared deviation of a signal from a given baseline or fit.

see also ARITHMETIC-GEOMETRIC MEAN, ARITHMETIC-HARMONIC MEAN, GENERALIZED MEAN, GEOMETRIC MEAN, HARMONIC MEAN, HARMONIC-GEOMETRIC MEAN, MEAN, MEDIAN (STATISTICS), STANDARD DEVIATION, VARIANCE

References

Hoehn, L. and Niven, I. "Averages on the Move." *Math. Mag.* **58**, 151–156, 1985.

Root (Radical)

The nth root (or "RADICAL") of a quantity z is a value r such that $z = r^n$, and therefore is the INVERSE FUNCTION to the taking of a POWER. The nth root is denoted $r = \sqrt[n]{z}$ or, using POWER notation, $r = z^{1/n}$. The special case of the SQUARE ROOT is denoted $\sqrt{z}$. The quantities for which a general FUNCTION equals 0 are also called ROOTS, or sometimes ZEROS.

see also CUBE ROOT, ROOT, SQUARE ROOT, VINCULUM

Root Test

Let u_k be a SERIES with POSITIVE terms, and let

$$\rho \equiv \lim_{k\to\infty} {u_k}^{1/k}.$$

1. If $\rho < 1$, the SERIES CONVERGES.
2. If $\rho > 1$ or $\rho = \infty$, the SERIES DIVERGES.
3. If $\rho = 1$, the SERIES may CONVERGE or DIVERGE.

This test is also called the CAUCHY ROOT TEST.

see also CONVERGENCE TESTS

References

Arfken, G. *Mathematical Methods for Physicists, 3rd ed.* Orlando, FL: Academic Press, pp. 281–282, 1985.

Bromwich, T. J. I'a and MacRobert, T. M. *An Introduction to the Theory of Infinite Series, 3rd ed.* New York: Chelsea, pp. 31–39, 1991.

Root (Tree)

A special node which is designated to turn a TREE into a ROOTED TREE. The root is sometimes also called "EVE," and each of the nodes which is one EDGE further away from a given EDGE is called a CHILD. Nodes connected to the same node are then called SIBLINGS.

see also CHILD, ROOTED TREE, SIBLING, TREE

Root of Unity

The nth ROOTS of UNITY are ROOTS $\zeta_k = e^{2\pi ik/p}$ of the CYCLOTOMIC EQUATION

$$x^p = 1,$$

which are known as the DE MOIVRE NUMBERS.

see also CYCLOTOMIC EQUATION, DE MOIVRE'S IDENTITY, DE MOIVRE NUMBER, UNITY

References

Courant, R. and Robbins, H. "De Moivre's Formula and the Roots of Unity." §5.3 in *What is Mathematics?: An Elementary Approach to Ideas and Methods, 2nd ed.* Oxford, England: Oxford University Press, pp. 98–100, 1996.

Rooted Tree

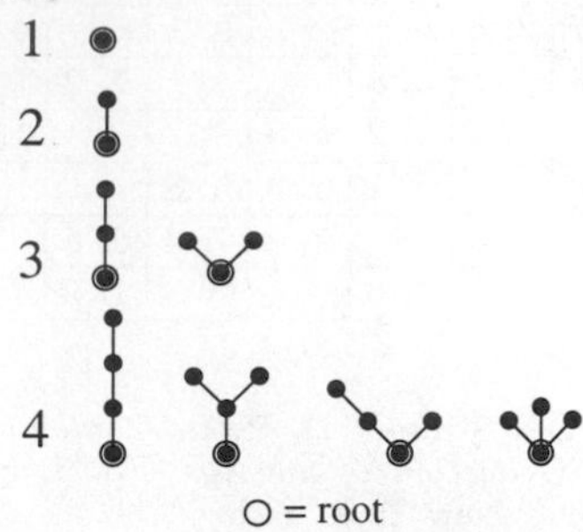

A TREE with a special node called the "ROOT" or "EVE." Denote the number of rooted trees with n nodes by T_n, then the GENERATING FUNCTION is

$$T(x) \equiv \sum_{n=0}^{\infty} T_n x^n = x + x^2 + 2x^3 + 4x^4 + 9x^5 + 20x^6 + 48x^7 + 115x^8 + 286x^9 + 719x^{10} + \ldots \quad (1)$$

(Sloane's A000081). This POWER SERIES satisfies

$$T(x) = x \exp\left[\sum_{r=1}^{\infty} \frac{1}{r} T(x^r)\right] \quad (2)$$

$$t(x) = T(x) - \tfrac{1}{2}[T^2(x) - T(x^2)], \quad (3)$$

where $t(x)$ is the GENERATING FUNCTION for unrooted TREES. A GENERATING FUNCTION for T_n can be written using a product involving *the sequence itself* as

$$x \prod_{n=1}^{\infty} \frac{1}{(1-x^n)^{T_n}} = \sum_{n=1}^{\infty} T_n x^n. \quad (4)$$

The number of rooted trees can also be calculated from the RECURRENCE RELATION

$$T_{i+1} = \frac{1}{i} \sum_{j=1}^{i} \left(\sum_{d|j} dT_d\right) T_{i-j+1}, \quad (5)$$

with $T_0 = 0$ and $T_1 = 1$, where the second sum is over all d which DIVIDE j (Finch).

see also ORDERED TREE, RED-BLACK TREE, WEAKLY BINARY TREE

References

Finch, S. "Favorite Mathematical Constants." `http://www.mathsoft.com/asolve/constant/otter/otter.html`.

Ruskey, F. "Information on Rooted Trees." `http://sue.csc.uvic.ca/~cos/inf/tree/RootedTree.html`.

Sloane, N. J. A. Sequence A000081/M1180 in "An On-Line Version of the Encyclopedia of Integer Sequences."

Rosatti's Theorem

There is a one-to-one correspondence between the sets of equivalent correspondences (not of value 0) on an irreducible curve of GENUS (CURVE) p, and the rational COLLINEATIONS of a projective space of $2p-1$ dimensions which leave invariant a space of $p-1$ dimensions. The number of linearly independent correspondences will be that of linearly independent COLLINEATIONS.

References
Coolidge, J. L. *A Treatise on Algebraic Plane Curves.* New York: Dover, p. 339, 1959.

Rose

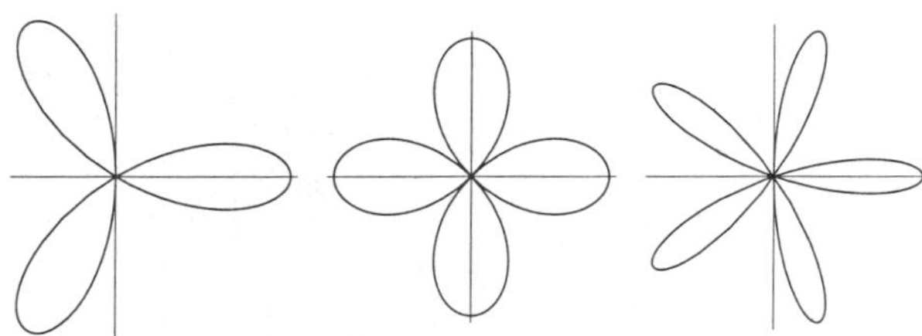

A curve which has the shape of a petalled flower. This curve was named RHODONEA by the Italian mathematician Guido Grandi between 1723 and 1728 because it resembles a rose (MacTutor Archive). The polar equation of the rose is

$$r = a\sin(n\theta),$$

or

$$r = a\cos(n\theta).$$

If n is ODD, the rose is n-petalled. If n is EVEN, the rose is $2n$-petalled. If n is IRRATIONAL, then there are an infinite number of petals.

The QUADRIFOLIUM is the rose with $n = 2$. The rose is the RADIAL CURVE of the EPICYCLOID.

see also DAISY, MAURER ROSE, STARR ROSE

References
Lawrence, J. D. *A Catalog of Special Plane Curves.* New York: Dover, pp. 175–177, 1972.
Lee, X. "Rose." http://www.best.com/~xah/SpecialPlaneCurves_dir/Rose_dir/rose.html.
MacTutor History of Mathematics Archive. "Rhodonea Curves." http://www-groups.dcs.st-and.ac.uk/~history/Curves/Rhodonea.html.
Wagon, S. "Roses." §4.1 in *Mathematica in Action.* New York: W. H. Freeman, pp. 96–102, 1991.

Rosenbrock Methods

A generalization of the RUNGE-KUTTA METHOD for solution of ORDINARY DIFFERENTIAL EQUATIONS, also called KAPS-RENTROP METHODS.

see also RUNGE-KUTTA METHOD

References
Press, W. H.; Flannery, B. P.; Teukolsky, S. A.; and Vetterling, W. T. *Numerical Recipes in FORTRAN: The Art of Scientific Computing, 2nd ed.* Cambridge, England: Cambridge University Press, pp. 730–735, 1992.

Rössler Model

The nonlinear 3-D MAP

$$\dot{X} = -(Y+Z)$$
$$\dot{Y} = X + 0.2Y$$
$$\dot{Z} = 0.2 + XZ - cZ.$$

see also LORENZ SYSTEM

References
Dickau, R. M. "Rössler Attractor." http://www.prairienet.org/~pops/rossler.html.
Peitgen, H.-O.; Jürgens, H.; and Saupe, D. §12.3 in *Chaos and Fractals: New Frontiers of Science.* New York: Springer-Verlag, pp. 686–696, 1992.

Rotation

The turning of an object or coordinate system by an ANGLE about a fixed point. A rotation is an ORIENTATION-PRESERVING ORTHOGONAL TRANSFORMATION. EULER'S ROTATION THEOREM states that an arbitrary rotation can be parameterized using three parameters. These parameters are commonly taken as the EULER ANGLES. Rotations can be implemented using ROTATION MATRICES.

The rotation SYMMETRY OPERATION for rotation by $360°/n$ is denoted "n." For periodic arrangements of points ("crystals"), the CRYSTALLOGRAPHY RESTRICTION gives the only allowable rotations as 1, 2, 3, 4, and 6.

see also DILATION, EUCLIDEAN GROUP, EULER'S ROTATION THEOREM, EXPANSION, IMPROPER ROTATION, INFINITESIMAL ROTATION, INVERSION OPERATION, MIRROR PLANE, ORIENTATION-PRESERVING, ORTHOGONAL TRANSFORMATION, REFLECTION, ROTATION FORMULA, ROTATION GROUP, ROTATION MATRIX, ROTATION OPERATOR, ROTOINVERSION, SHIFT, TRANSLATION

References
Beyer, W. H. (Ed.) *CRC Standard Mathematical Tables, 28th ed.* Boca Raton, FL: CRC Press, p. 211, 1987.
Yates, R. C. "Instantaneous Center of Rotation and the Construction of Some Tangents." *A Handbook on Curves and Their Properties.* Ann Arbor, MI: J. W. Edwards, pp. 119–122, 1952.

Rotation Formula

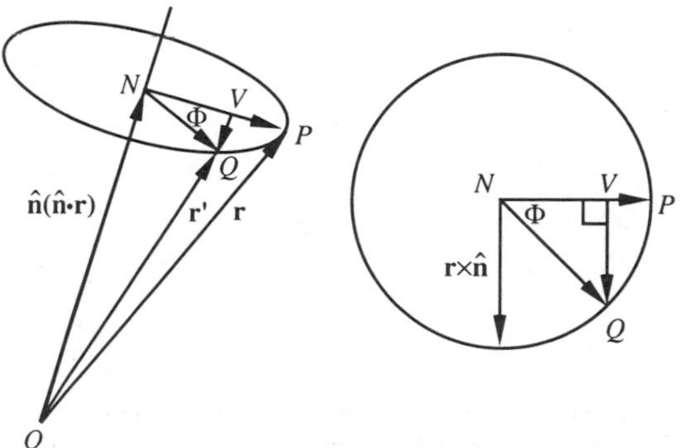

A formula which relates the VECTOR $\mathbf{r}'$ to the ANGLE Φ in the above figure (Goldstein 1980). Referring to the figure,

$$\begin{aligned}\mathbf{r}' &= \overrightarrow{ON} + \overrightarrow{NV} + \overrightarrow{VQ}\\ &= \hat{\mathbf{n}}(\hat{\mathbf{n}}\cdot\mathbf{r}) + [\mathbf{r} - \hat{\mathbf{n}}(\hat{\mathbf{n}}\cdot\mathbf{r})]\cos\Phi + (\mathbf{r}\times\hat{\mathbf{n}})\sin\Phi\\ &= \mathbf{r}\cos\Phi + \hat{\mathbf{n}}(\hat{\mathbf{n}}\cdot\mathbf{r})(1-\cos\Phi) + (\mathbf{r}\times\hat{\mathbf{n}})\sin\Phi.\end{aligned}$$

The ANGLE Φ and unit normal $\hat{\mathbf{n}}$ may also be expressed as EULER ANGLES. In terms of the EULER PARAMETERS,

$$\mathbf{r}' = \mathbf{r}({e_0}^2 - {e_1}^2 - {e_2}^2 - {e_3}^2) + 2\mathbf{e}(\mathbf{e}\cdot\mathbf{r}) + 2(\mathbf{r}\times\hat{\mathbf{n}})\sin\Phi.$$

see also EULER ANGLES, EULER PARAMETERS

References

Goldstein, H. *Classical Mechanics, 2nd ed.* Reading, MA: Addison-Wesley, 1980.

Rotation Group

There are three representations of the rotation groups, corresponding to EXPANSION/DILATION, ROTATION, and SHEAR.

Rotation Matrix

When discussing a ROTATION, there are two possible conventions: rotation of the *axes* and rotation of the *object* relative to fixed axes.

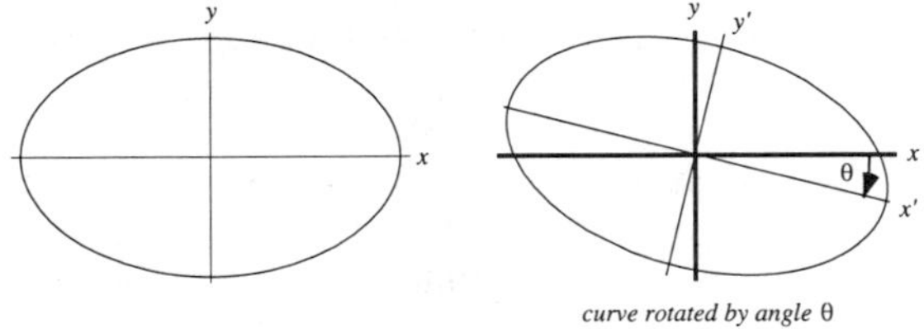

curve rotated by angle θ

In $\mathbb{R}^2$, let a curve be rotated by a clockwise ANGLE θ, so that the original axes of the curve are $\hat{\mathbf{x}}$ and $\hat{\mathbf{y}}$, and the new axes of the curve are $\hat{\mathbf{x}}'$ and $\hat{\mathbf{y}}'$. The MATRIX transforming the original curve to the rotated curve, referred to the original $\hat{\mathbf{x}}$ and $\hat{\mathbf{y}}$ axes, is

$$\mathsf{R}_\theta = \begin{bmatrix}\cos\theta & \sin\theta\\ -\sin\theta & \cos\theta\end{bmatrix}, \quad (1)$$

i.e.,

$$\mathbf{x} = \mathsf{R}_\theta\mathbf{x}'. \quad (2)$$

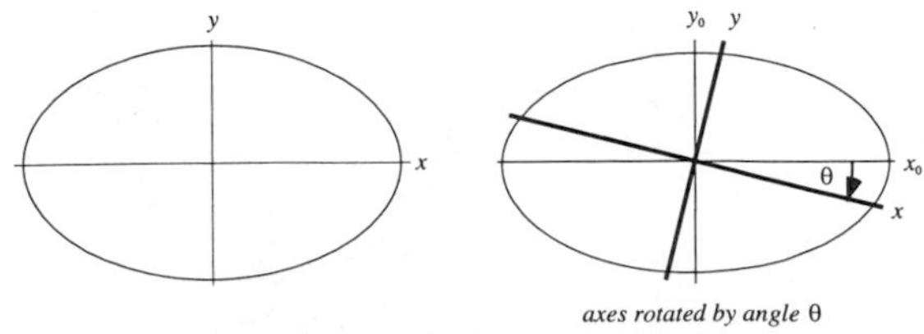

axes rotated by angle θ

On the other hand, let the *axes* with respect to which a curve is measured be rotated by a clockwise ANGLE θ, so that the original axes are $\hat{\mathbf{x}}_0$ and $\hat{\mathbf{y}}_0$, and the new axes are $\hat{\mathbf{x}}$ and $\hat{\mathbf{y}}$. Then the MATRIX transforming the coordinates of the curve with respect to $\hat{\mathbf{x}}$ and $\hat{\mathbf{y}}$ is given by the MATRIX TRANSPOSE of the above matrix:

$$\mathsf{R}'_\theta = \begin{bmatrix}\cos\theta & -\sin\theta\\ \sin\theta & \cos\theta\end{bmatrix}, \quad (3)$$

i.e.,

$$\mathbf{x} = \mathsf{R}'_\theta\mathbf{x}_0. \quad (4)$$

In $\mathbb{R}^3$, rotations of the x-, y-, and z-axes give the matrices

$$\mathsf{R}_x(\alpha) = \begin{bmatrix}1 & 0 & 0\\ 0 & \cos\alpha & \sin\alpha\\ 0 & -\sin\alpha & \cos\alpha\end{bmatrix} \quad (5)$$

$$\mathsf{R}_y(\beta) = \begin{bmatrix}\cos\beta & 0 & -\sin\beta\\ 0 & 1 & 0\\ \sin\beta & 0 & \cos\beta\end{bmatrix} \quad (6)$$

$$\mathsf{R}_z(\gamma) = \begin{bmatrix}\cos\gamma & \sin\gamma & 0\\ -\sin\gamma & \cos\gamma & 0\\ 0 & 0 & 1\end{bmatrix}. \quad (7)$$

see also EULER ANGLES, EULER'S ROTATION THEOREM, ROTATION

Rotation Number

The period for a QUASIPERIODIC trajectory to pass through the same point in a SURFACE OF SECTION. If the rotation number is IRRATIONAL, the trajectory will densely fill out a curve in the SURFACE OF SECTION. If the rotation number is RATIONAL, it is called the WINDING NUMBER, and only a finite number of points in the SURFACE OF SECTION will be visited by the trajectory.

see also QUASIPERIODIC FUNCTION, SURFACE OF SECTION, WINDING NUMBER (MAP)

Rotation Operator

The rotation operator can be derived from examining an INFINITESIMAL ROTATION

$$\left(\frac{d}{dt}\right)_{\text{space}} = \left(\frac{d}{dt}\right)_{\text{body}} + \boldsymbol{\omega}\times,$$

where d/dt is the time derivative, $\boldsymbol{\omega}$ is the ANGULAR VELOCITY, and $\times$ is the CROSS PRODUCT operator.

see also ACCELERATION, ANGULAR ACCELERATION, INFINITESIMAL ROTATION

Roth's Removal Rule

If the matrices A, X, B, and C satisfy

$$\mathsf{AX} - \mathsf{XB} = \mathsf{C},$$

then

$$\begin{bmatrix}\mathsf{I} & \mathsf{X}\\ 0 & \mathsf{I}\end{bmatrix}\begin{bmatrix}\mathsf{A} & \mathsf{C}\\ 0 & \mathsf{B}\end{bmatrix}\begin{bmatrix}\mathsf{I} & -\mathsf{X}\\ 0 & \mathsf{I}\end{bmatrix} = \begin{bmatrix}\mathsf{A} & 0\\ 0 & \mathsf{B}\end{bmatrix},$$

where I is the IDENTITY MATRIX.

References
Roth, W. E. "The Equations $AX - YB = C$ and $AX - XB = C$ in Matrices." *Proc. Amer. Math. Soc.* **3**, 392–396, 1952.
Turnbull, H. W. and Aitken, A. C. *An Introduction to the Theory of Canonical Matrices.* New York: Dover, p. 422, 1961.

Roth's Theorem

For ALGEBRAIC α

$$\left|\alpha - \frac{p}{q}\right| < \frac{1}{q^{2+\epsilon}},$$

with $\epsilon > 0$, has finitely many solutions. Klaus Roth received a FIELDS MEDAL for this result.

see also HURWITZ EQUATION, HURWITZ'S IRRATIONAL NUMBER THEOREM, LAGRANGE NUMBER (RATIONAL APPROXIMATION), LIOUVILLE'S RATIONAL APPROXIMATION THEOREM, LIOUVILLE-ROTH CONSTANT, MARKOV NUMBER, SEGRE'S THEOREM, THUE-SIEGEL-ROTH THEOREM

References
Davenport, H. and Roth, K. F. "Rational Approximations to Algebraic Numbers." *Mathematika* **2**, 160–167, 1955.
Roth, K. F. "Rational Approximations to Algebraic Numbers." *Mathematika* **2**, 1–20, 1955.
Roth, K. F. "Corrigendum to 'Rational Approximations to Algebraic Numbers'." *Mathematika* **2**, 168, 1955.

Rotkiewicz Theorem

If $n > 19$, there exists a base-2 PSEUDOPRIME between n and n^2. The theorem was proved in 1965.

see also PSEUDOPRIME

References
Rotkiewicz, A. "Les intervalles contenants les nombres pseudoprimiers." *Rend. Circ. Mat. Palermo Ser. 2* **14**, 278–280, 1965.
Rotkiewicz, A. "Sur les nombres de Mersenne dépourvus de diviseurs carrés er sur les nombres naturels n, tel que $n^2 - 2^n - 2$." *Mat. Vesnik* **2 (17)**, 78–80, 1965.
Rotkiewicz, A. "Sur les nombres pseudoprimiers carrés." *Elem. Math.* **20**, 39–40, 1965.

Rotoinversion

see IMPROPER ROTATION

Rotor

A convex figure that can be rotated inside a POLYGON (or POLYHEDRON) while always touching every side (or face). The least AREA rotor in a SQUARE is the REULEAUX TRIANGLE. The least AREA rotor in an EQUILATERAL TRIANGLE is a LENS with two 60° ARCS of CIRCLES and RADIUS equal to the TRIANGLE ALTITUDE.

There exist nonspherical rotors for the TETRAHEDRON, OCTAHEDRON, and CUBE, but not for the DODECAHEDRON and ICOSAHEDRON.

see also LENS, REULEAUX TRIANGLE

References
Gardner, M. *The Unexpected Hanging and Other Mathematical Diversions.* Chicago, IL: Chicago University Press, p. 219, 1991.

Rotunda

A class of solids whose only true member is the PENTAGONAL ROTUNDA.

see also ELONGATED ROTUNDA, GYROELONGATED ROTUNDA, PENTAGONAL ROTUNDA, TRIANGULAR HEBESPHENOROTUNDA

References
Johnson, N. W. "Convex Polyhedra with Regular Faces." *Canad. J. Math.* **18**, 169–200, 1966.

Rouché's Theorem

Given two functions f and g ANALYTIC in A with γ a simple loop HOMOTOPIC to a point in A, if $|g(z)| < |f(z)|$ for all z on γ, then f and $f + g$ have the same number of ROOTS inside γ.

References
Szegő, G. *Orthogonal Polynomials, 4th ed.* Providence, RI: Amer. Math. Soc., p. 22, 1975.

Roulette

The curve traced by a fixed point on a closed convex curve as that curve rolls without slipping along a second curve. The roulettes described by the FOCI of CONICS when rolled upon a line are sections of MINIMAL SURFACES (i.e., they yield MINIMAL SURFACES when revolved about the line) known as UNDULOIDS.

Curve 1	Curve 2	Pole	Roulette
circle	exterior circle	on c.	epicycloid
circle	interior circle	on c.	hypocycloid
circle	line	on c.	cycloid
circle	same circle	any point	rose
circle involute	line	center	parabola
cycloid	line	center	ellipse
ellipse	line	focus	elliptic catenary
hyperbola	line	focus	hyperbolic catenary
hyperbolic spiral	line	origin	tractrix
line	any curve	on line	involute of curve
logarithmic spiral	line	any point	line
parabola	equal parabola	vertex	cissoid of Diocles
parabola	line	focus	catenary

see also GLISSETTE, UNDULOID

References
Besant, W. H. *Notes on Roulettes and Glissettes, 2nd enl. ed.* Cambridge, England: Deighton, Bell & Co., 1890.

Cundy, H. and Rollett, A. "Roulettes and Involutes." §2.6 in *Mathematical Models, 3rd ed.* Stradbroke, England: Tarquin Pub., pp. 46–55, 1989.
Lawrence, J. D. *A Catalog of Special Plane Curves.* New York: Dover, pp. 56–58 and 206, 1972.
Lockwood, E. H. "Roulettes." Ch. 17 in *A Book of Curves.* Cambridge, England: Cambridge University Press, pp. 138–151, 1967.
Yates, R. C. "Roulettes." *A Handbook on Curves and Their Properties.* Ann Arbor, MI: J. W. Edwards, pp. 175–185, 1952.
Zwillinger, D. (Ed.). "Roulettes (Spirograph Curves)." §8.2 in *CRC Standard Mathematical Tables and Formulae, 3rd ed.* Boca Raton, FL: CRC Press, 1996. http://www.geom.umn.edu/docs/reference/CRC-formulas/node34.html.

Round

see NINT

Rounding

The process of approximating a quantity, usually done for convenience or, in the case of numerical computations, of necessity. If rounding is performed on each of a series of numbers in a long computation, round-off errors can become important, especially if division by a small number ever occurs.

see also SHADOWING THEOREM

References

Wilkinson, J. H. *Rounding Errors in Algebraic Processes.* New York: Dover, 1994.

Routh-Hurwitz Theorem

Consider the CHARACTERISTIC EQUATION

$$|\lambda\mathsf{I} - \mathsf{A}| = \lambda^n + b_1\lambda^{n-1} + \ldots + b_{n-1}\lambda + b_n = 0$$

determining the n EIGENVALUES λ of a REAL $n \times n$ MATRIX A, where I is the IDENTITY MATRIX. Then the EIGENVALUES λ all have NEGATIVE REAL PARTS if

$$\Delta_1 > 0, \Delta_2 > 0, \ldots, \Delta_n > 0,$$

where

$$\Delta_k = \begin{vmatrix} b_1 & 1 & 0 & 0 & 0 & 0 & \cdots & 0 \\ b_3 & b_2 & b_1 & 1 & 0 & 0 & \cdots & 0 \\ b_5 & b_4 & b_3 & b_2 & b_1 & 0 & \cdots & 0 \\ \vdots & \vdots & \vdots & \vdots & \vdots & \vdots & \ddots & \vdots \\ b_{2k-1} & b_{2k-2} & b_{2k-3} & b_{2k-4} & b_{2k-5} & b_{2k-6} & \cdots & b_k \end{vmatrix}.$$

References

Gradshteyn, I. S. and Ryzhik, I. M. *Tables of Integrals, Series, and Products, 5th ed.* San Diego, CA: Academic Press, p. 1119, 1979.

Routh's Theorem

If the sides of a TRIANGLE are divided in the ratios $\lambda : 1$, $\mu : 1$, and $\nu : 1$, the CEVIANS form a central TRIANGLE whose AREA is

$$A = \frac{(\lambda\mu\nu - 1)^2}{(\lambda\mu + \lambda + 1)(\mu\nu + \mu + 1)(\nu\lambda + \nu + 1)}\Delta, \quad (1)$$

where Δ is the AREA of the original TRIANGLE. For $\lambda = \mu = \nu \equiv n$,

$$A = \frac{(n-1)^2}{n^2 + n + 1}\Delta. \quad (2)$$

For $n = 2, 3, 4, 5$, the areas are $\frac{1}{7}$, $\frac{3}{7}$, and $\frac{16}{31}$. The AREA of the TRIANGLE formed by connecting the division points on each side is

$$A' = \frac{\lambda\mu\nu}{(\lambda+1)(\mu+1)(\nu+1)}\Delta. \quad (3)$$

Routh's theorem gives CEVA'S THEOREM and MENELAUS' THEOREM as special cases.

see also CEVA'S THEOREM, CEVIAN, MENELAUS' THEOREM

References

Coxeter, H. S. M. *Introduction to Geometry, 2nd ed.* New York: Wiley, pp. 211–212, 1969.

RSA Encryption

A PUBLIC-KEY CRYPTOGRAPHY ALGORITHM which uses PRIME FACTORIZATION as the TRAPDOOR FUNCTION. Define

$$n \equiv pq \quad (1)$$

for p and q PRIMES. Also define a private key d and a public key e such that

$$de \equiv 1 \pmod{\phi(n)} \quad (2)$$

$$(e, \phi(n)) = 1, \quad (3)$$

where $\phi(n)$ is the TOTIENT FUNCTION.

Let the message be converted to a number M. The sender then makes n and e public and sends

$$E = M^e \pmod{n}. \quad (4)$$

To decode, the receiver (who knows d) computes

$$E^d \equiv (M^e)^d \equiv M^{ed} \equiv M^{N\phi(n)+1} \equiv M \pmod{n}, \quad (5)$$

since N is an INTEGER. In order to crack the code, d must be found. But this requires factorization of n since

$$\phi(n) = (p-1)(q-1). \quad (6)$$

Both p and q should be picked so that $p \pm 1$ and $q \pm 1$ are divisible by large PRIMES, since otherwise the POLLARD $p-1$ FACTORIZATION METHOD or WILLIAMS $p+1$ FACTORIZATION METHOD potentially factor n easily. It is also desirable to have $\phi(\phi(pq))$ large and divisible by large PRIMES.

It is possible to break the cryptosystem by repeated encryption if a unit of $\mathbb{Z}/\phi(n)\mathbb{Z}$ has small ORDER (Simmons and Norris 1977, Meijer 1996), where $\mathbb{Z}/s\mathbb{Z}$ is the RING of INTEGERS between 0 and $s-1$ under addition and multiplication (mod s). Meijer (1996) shows that "almost" every encryption exponent e is safe from breaking using repeated encryption for factors of the form

$$p = 2p_1 + 1 \tag{7}$$

$$q = 2q_1 + 1, \tag{8}$$

where

$$p_1 = 2p_2 + 1 \tag{9}$$

$$q_1 = 2q_2 + 1, \tag{10}$$

and p, p_1, p_2, q, q_1, and q_2 are all PRIMES. In this case,

$$\phi(n) = 4p_1q_1 \tag{11}$$

$$\phi(\phi(n)) = 8p_2q_2. \tag{12}$$

Meijer (1996) also suggests that p_2 and q_2 should be of order 10^{75}.

Using the RSA system, the identity of the sender can be identified as genuine without revealing his private code.

see also PUBLIC-KEY CRYPTOGRAPHY

References

Honsberger, R. *Mathematical Gems III.* Washington, DC: Math. Assoc. Amer., pp. 166–173, 1985.

Meijer, A. R. "Groups, Factoring, and Cryptography." *Math. Mag.* **69**, 103–109, 1996.

Rivest, R. L. "Remarks on a Proposed Cryptanalytic Attack on the MIT Public-Key Cryptosystem." *Cryptologia* **2**, 62–65, 1978.

Rivest, R.; Shamir, A.; and Adleman, L. "A Method for Obtaining Digital Signatures and Public Key Cryptosystems." *Comm. ACM* **21**, 120–126, 1978.

RSA Data Security.® A Security Dynamics Company. `http://www.rsa.com`.

Simmons, G. J. and Norris, M. J. "Preliminary Comments on the MIT Public-Key Cryptosystem." *Cryptologia* **1**, 406–414, 1977.

RSA Number

Numbers contained in the "factoring challenge" of RSA Data Security, Inc. An additional number which is not part of the actual challenge is the RSA-129 number. The RSA numbers which have been factored are RSA-100, RSA-110, RSA-120, RSA-129, and RSA-130 (Cowie *et al.* 1996).

RSA-129 is a 129-digit number used to encrypt one of the first public-key messages. This message was published by R. Rivest, A. Shamir, and L. Adleman (Gardner 1977), along with the number and a $100 reward for its decryption. Despite belief that the message encoded by RSA-129 "would take millions of years of break," RSA-129 was factored in 1994 using a distributed computation which harnessed networked computers spread around the globe performing a multiple polynomial QUADRATIC SIEVE FACTORIZATION METHOD. The effort was coordinated by P. Leylad, D. Atkins, and M. Graff. They received 112,011 full factorizations, 1,431,337 single partial factorizations, and 8,881,138 double partial factorizations out of a factor base of 524,339 PRIMES. The final MATRIX obtained was $188,346 \times 188,346$ square.

The text of the message was "The magic words are squeamish ossifrage" (an ossifrage is a rare, predatory vulture found in the mountains of Europe), and the FACTORIZATION (into a 64-DIGIT number and a 65-DIGIT number) is

$$\begin{aligned}
&114381625757888867669235779976146612010218296\cdots\\
&\cdots 7212423625625618429357069352457338978305971\cdots\\
&\cdots 235639587050589890751475992900268795435 41\\
&= 3490529510847650949147849619903898133417764\cdots\\
&\cdots 638493387843990820577 \cdot 3276913299326\cdots\\
&\cdots 670954996198819083446141317764296799 2\cdots\\
&\cdots 9425397982885 33
\end{aligned}$$

(Leutwyler 1994, Cipra 1995).

References

Cipra, B. "The Secret Life of Large Numbers." *What's Happening in the Mathematical Sciences, 1995–1996, Vol. 3.* Providence, RI: Amer. Math. Soc., pp. 90–99, 1996.

Cowie, J.; Dodson, B.; Elkenbracht-Huizing, R. M.; Lenstra, A. K.; Montgomery, P. L.; Zayer, J. A. "World Wide Number Field Sieve Factoring Record: On to 512 Bits." In *Advances in Cryptology—ASIACRYPT '96 (Kyongju)* (Ed. K. Kim and T. Matsumoto.) New York: Springer-Verlag, pp. 382–394, 1996.

Gardner, M. "Mathematical Games: A New Kind of Cipher that Would Take Millions of Years to Break." *Sci. Amer.* **237**, 120–124, Aug. 1977.

Klee, V. and Wagon, S. *Old and New Unsolved Problems in Plane Geometry and Number Theory, rev. ed.* Washington, DC: Math. Assoc. Amer., p. 223, 1991.

Leutwyler, K. "Superhack: Forty Quadrillion Years Early, a 129-Digit Code is Broken." *Sci. Amer.* **271**, 17–20, 1994.

Leyland, P. `ftp://sable.ox.ac.uk/pub/math/rsa129`.

RSA Data Security.® A Security Dynamics Company. `http://www.rsa.com`.

Taubes, G. "Small Army of Code-breakers Conquers a 129-Digit Giant." *Science* **264**, 776–777, 1994.

Weisstein, E. W. "RSA Numbers." `http://www.astro.virginia.edu/~eww6n/math/notebooks/RSANumbers.m`.

Rubber-Sheet Geometry

see ALGEBRAIC TOPOLOGY

Rubik's Clock

A puzzle consisting of 18 small clocks. There are 12^{18} possible configurations, although not all are realizable.

see also RUBIK'S CUBE

References

Dénes, J. and Mullen, G. L. "Rubik's Clock and Its Solution." *Math. Mag.* **68**, 378–381, 1995.

Zeilberger, D. "Doron Zeilberger's Maple Packages and Programs: RubikClock." http://www.math.temple.edu/~zeilberg/programs.html.

Rubik's Cube

A $3\times3\times3$ CUBE in which the 26 subcubes on the outside are internally hinged in such a way that rotation (by a quarter turn in either direction or a half turn) is possible in any plane of cubes. Each of the six sides is painted a distinct color, and the goal of the puzzle is to return the cube to a state in which each side has a single color after it has been randomized by repeated rotations. The PUZZLE was invented in the 1970s by the Hungarian Erno Rubik and sold millions of copies worldwide over the next decade.

The number of possible positions of Rubik's cube is

$$\frac{8!12!3^8 2^{12}}{2\cdot3\cdot2} = 43{,}252{,}003{,}274{,}489{,}856{,}000$$

(Turner and Gold 1985). Hoey showed using the PÓLYA-BURNSIDE LEMMA that there are 901,083,404,981,813,-616 positions up to conjugacy by whole-cube symmetries.

Algorithms exist for solving a cube from an arbitrary initial position, but they are not necessarily optimal (i.e., requiring a minimum number of turns). The maximum number of turns required for an arbitrary starting position is still not known, although it is bounded from above. Michael Reid (1995) produced the best proven bound of 29 turns (or 42 "quarter-turns"). The proof involves large tables of "subroutines" generated by computer.

However, Dik Winter has produced a program based on work by Kociemba which has solved each of millions of cubes in at most 21 turns. Recently, Richard Korf (1997) has produced a different algorithm which is practical for cubes up to 18 moves away from solved. Out of 10 randomly generated cubes, one was solved in 16 moves, three required 17 moves, and six required 18 moves.

see also RUBIK'S CLOCK

References

Hofstadter, D. R. "Metamagical Themas: The Magic Cube's Cubies are Twiddled by Cubists and Solved by Cubemeisters." *Sci. Amer.* **244**, 20–39, Mar. 1981.

Larson, M. E. "Rubik's Revenge: The Group Theoretical Solution." *Amer. Math. Monthly* **92**, 381–390, 1985.

Miller, D. L. W. "Solving Rubik's Cube Using the 'Bestfast' Search Algorithm and 'Profile' Tables." http://www.sunyit.edu/~millerd1/RUBIK.HTM.

Schubart, M. "Rubik's Cube Resource List." http://www.best.com/~schubart/rc/resources.html.

Singmaster, D. *Notes on Rubik's 'Magic Cube.'* Hillside, NJ: Enslow Pub., 1981.

Taylor, D. *Mastering Rubik's Cube.* New York: Holt, Rinehart, and Winston, 1981.

Taylor, D. and Rylands, L. *Cube Games: 92 Puzzles & Solutions* New York: Holt, Rinehart, and Winston, 1981.

Turner, E. C. and Gold, K. F. "Rubik's Groups." *Amer. Math. Monthly* **92**, 617–629, 1985.

Rudin-Shapiro Sequence

The sequence of numbers given by

$$a_n = (-1)^{\sum_{i=1}^{k-1} \epsilon_i\epsilon_{i+1}}, \tag{1}$$

where n is written in binary

$$n = \epsilon_1\epsilon_2\ldots\epsilon_k. \tag{2}$$

It is therefore the parity of the number of pairs of consecutive 1s in the BINARY expansion of n. The SUMMATORY sequence is

$$s_n \equiv \sum_{j=0}^{n} a_j, \tag{3}$$

which gives

$$s_n = \begin{cases} 2^{k/2}+1 & \text{if } k \text{ is even} \\ 2^{(k-1)/2}+1 & \text{if } k \text{ is odd} \end{cases} \tag{4}$$

(Blecksmith and Laud 1995).

References

Blecksmith, R. and Laud, P. W. "Some Exact Number Theory Computations via Probability Mechanisms." *Amer. Math. Monthly* **102**, 893–903, 1995.

Brillhart, J.; Erdős, P.; and Morton, P. "On the Sums of the Rudin-Shapiro Coefficients II." *Pac. J. Math.* **107**, 39–69, 1983.

Brillhart, J. and Morton, P. "Über Summen von Rudin-Shapiroschen Koeffizienten." *Ill. J. Math.* **22**, 126–148, 1978.

France, M. M. and van der Poorten, A. J. "Arithmetic and Analytic Properties of Paper Folding Sequences." *Bull. Austral. Math. Soc.* **24**, 123–131, 1981.

Rudvalis Group

The SPORADIC GROUP *Ru*.

see also SPORADIC GROUP

References

Wilson, R. A. "ATLAS of Finite Group Representation." http://for.mat.bham.ac.uk/atlas/Ru.html.

Rule

A usually simple ALGORITHM or IDENTITY. The term is frequently applied to specific orders of NEWTON-COTES FORMULAS.

see also ALGORITHM, BAC-CAB RULE, BODE'S RULE, CHAIN RULE, CRAMER'S RULE, DESCARTES' SIGN RULE, DURAND'S RULE, ESTIMATOR, EULER'S RULE, EULER'S TOTIENT RULE, GOLDEN RULE, HARDY'S RULE, HORNER'S RULE, IDENTITY, L'HOSPITAL'S RULE, LEIBNIZ INTEGRAL RULE, METHOD, OSBORNE'S RULE, PASCAL'S RULE, POWER RULE, PRODUCT RULE, QUARTER SQUARES RULE, QUOTA RULE, QUOTIENT RULE, ROTH'S REMOVAL RULE, RULE OF 72, SIMPSON'S RULE, SLIDE RULE, SUM RULE, TRAPEZOIDAL RULE, WEDDLE'S RULE, ZEUTHEN'S RULE

Rule of 72

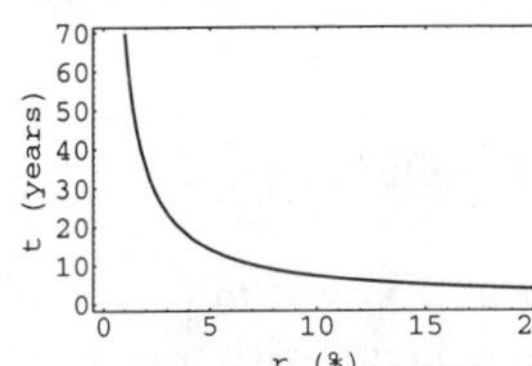

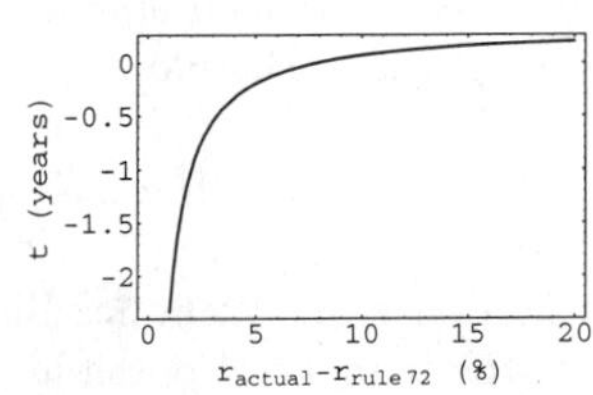

The time required for a given PRINCIPAL to double (assuming $n = 1$ CONVERSION PERIOD) for COMPOUND INTEREST is given by solving

$$2P = P(1+r)^t, \tag{1}$$

or

$$t = \frac{\ln 2}{\ln(1+r)}, \tag{2}$$

where LN is the NATURAL LOGARITHM. This function can be approximated by the so-called "rule of 72":

$$t \approx \frac{0.72}{r}. \tag{3}$$

The above plots show the actual doubling time t (left plot) and difference between actual and time calculated using the rule of 72 (right plot) as a function of the interest rate r.

see also COMPOUND INTEREST, INTEREST

References

Avanzini, J. F. *Rapid Debt-Reduction Strategies.* Fort Worth, TX: HIS Pub., 1990.

Ruled Surface

A SURFACE which can be swept out by a moving LINE in space and therefore has a parameterization of the form

$$\mathbf{x}(u,v) = \mathbf{b}(u) + v\boldsymbol{\delta}(u), \tag{1}$$

where $\mathbf{b}$ is called the DIRECTRIX (also called the BASE CURVE) and $\boldsymbol{\delta}$ is the DIRECTOR CURVE. The straight lines themselves are called RULINGS. The rulings of a ruled surface are ASYMPTOTIC CURVES. Furthermore, the GAUSSIAN CURVATURE on a ruled REGULAR SURFACE is everywhere NONPOSITIVE.

Examples of ruled surfaces include the elliptic HYPERBOLOID of one sheet (a doubly ruled surface)

$$\begin{bmatrix} a(cosu \mp v\sin u) \\ b(\sin u \pm \cos u) \\ \pm cv \end{bmatrix} = \begin{bmatrix} a\cos u \\ b\sin u \\ 0 \end{bmatrix} \pm v \begin{bmatrix} -a\cos u \\ b\sin u \\ c \end{bmatrix}, \tag{2}$$

the HYPERBOLIC PARABOLOID (a doubly ruled surface)

$$\begin{bmatrix} a(u+v) \\ \pm bv \\ u^2 + 2uv \end{bmatrix} = \begin{bmatrix} au \\ 0 \\ u^2 \end{bmatrix} + v \begin{bmatrix} a \\ \pm b \\ 2u \end{bmatrix}, \tag{3}$$

PLÜCKER'S CONOID

$$\begin{bmatrix} r\cos\theta \\ r\sin\theta \\ 2\cos\theta\sin\theta \end{bmatrix} = \begin{bmatrix} 0 \\ 0 \\ 2\cos\theta\sin\theta \end{bmatrix} + r \begin{bmatrix} \cos\theta \\ \sin\theta \\ 0 \end{bmatrix}, \tag{4}$$

and the MÖBIUS STRIP

$$a \begin{bmatrix} \cos u + v\cos(\tfrac{1}{2}u)\cos u \\ \sin u + v\cos(\tfrac{1}{2}u)\sin u \\ v\sin(\tfrac{1}{2}u) \end{bmatrix} = a \begin{bmatrix} \cos u \\ \sin u \\ 0 \end{bmatrix} + av \begin{bmatrix} \cos(\tfrac{1}{2}u)\cos u \\ \cos(\tfrac{1}{2}u)\sin u \\ \sin(\tfrac{1}{2}u) \end{bmatrix} \tag{5}$$

(Gray 1993).

The only ruled MINIMAL SURFACES are the PLANE and HELICOID (Catalan 1842, do Carmo 1986).

see also ASYMPTOTIC CURVE, CAYLEY'S RULED SURFACE, DEVELOPABLE SURFACE, DIRECTOR CURVE, DIRECTRIX (RULED SURFACE), GENERALIZED CONE, GENERALIZED CYLINDER, HELICOID, NONCYLINDRICAL RULED SURFACE, PLANE, RIGHT CONOID, RULING

References

Catalan E. "Sur les surfaces régléess dont l'aire est un minimum." *J. Math. Pure. Appl.* **7**, 203–211, 1842.

do Carmo, M. P. "The Helicoid." §3.5B in *Mathematical Models from the Collections of Universities and Museums* (Ed. G. Fischer). Braunschweig, Germany: Vieweg, pp. 44–45, 1986.

Fischer, G. (Ed.). Plates 32–33 in *Mathematische Modelle/Mathematical Models, Bildband/Photograph Volume.* Braunschweig, Germany: Vieweg, pp. 32–33, 1986.

Gray, A. "Ruled Surfaces." Ch. 17 in *Modern Differential Geometry of Curves and Surfaces.* Boca Raton, FL: CRC Press, pp. 333–355, 1993.

Ruler

A STRAIGHTEDGE with markings to indicate distances. Although GEOMETRIC CONSTRUCTIONS are sometimes said to be performed with a ruler and COMPASS, the term STRAIGHTEDGE is preferable to ruler since markings are not allowed by the classical Greek rules.

see also COASTLINE PARADOX, COMPASS, GEOMETRIC CONSTRUCTION, GEOMETROGRAPHY, GOLOMB RULER, PERFECT RULER, SIMPLICITY, SLIDE RULE, STRAIGHTEDGE

Ruler Function

The exponent of the largest POWER of 2 which DIVIDES a given number k. The values of the ruler function are 1, 2, 1, 3, 1, 2, 1, 4, 1, 2, ... (Sloane's A001511).

References

Guy, R. K. "Cycles and Sequences Containing All Permutations as Subsequences." §E22 in *Unsolved Problems in Number Theory, 2nd ed.* New York: Springer-Verlag, p. 224, 1994.

Sloane, N. J. A. Sequence A001511/M0127 in "An On-Line Version of the Encyclopedia of Integer Sequences."

Ruling

One of the straight lines sweeping out a RULED SURFACE. The rulings on a ruled surface are ASYMPTOTIC CURVES.

see also ASYMPTOTIC CURVE, DIRECTOR CURVE, DIRECTRIX (RULED SURFACE), RULED SURFACE

Run

A run is a sequence of more than one consecutive identical outcomes, also known as a CLUMP. Given n BERNOULLI TRIALS (say, in the form of COIN TOSSINGS), the probability $P_t(n)$ of a run of t consecutive heads or tails is given by the RECURRENCE RELATION

$$P_t(n) = P_t(n-1) + 2^{-t}[1 - P_t(n-t)], \qquad (1)$$

where $P_t(n) = 0$ for $n < t$ and $P_t(t) = 2^{1-t}$ (Bloom 1996).

Let $C_t(m,k)$ denote the number of sequences of m indistinguishable objects of type A and k indistinguishable objects of type B in which *no* t-run occurs. The probability that a t-run *does* occur is then given by

$$P_t(m,k) = 1 - \frac{C_t(m,k)}{\binom{m+k}{k}}, \qquad (2)$$

where $\binom{a}{b}$ is a BINOMIAL COEFFICIENT. Bloom (1996) gives the following recurrence sequence for $C_t(m,k)$,

$$C_t(m,k) = \sum_{i=0}^{t-1} C_t(m-1, k-i) - \sum_{i=1}^{t-1} C_t(m-t, k-i) + e_t(m,k), \qquad (3)$$

where

$$e_t(m,k) \equiv \begin{cases} 1 & \text{if } m = 0 \text{ and } 0 \le k < t \\ -1 & \text{if } m = t \text{ and } 0 \le k < t \\ 0 & \text{otherwise.} \end{cases} \qquad (4)$$

Another recurrence which has only a fixed number of terms is given by

$$C_t(m,k) = C_t(m-1,k) + C_t(m,k-1) - C_t(m-t,k-1) - C_t(m-1,k-t) + C_t(m-t,k-t) + e_t^*(m,k), \qquad (5)$$

where

$$e_t^*(m,k) \equiv \begin{cases} 1 & \text{if } (m,k) = (0,0) \text{ or } (t,t) \\ -1 & \text{if } (m,k) = (0,t) \text{ or } (t,0) \\ 0 & \text{otherwise} \end{cases} \qquad (6)$$

(Goulden and Jackson 1983, Bloom 1996). These formulas disprove the assertion of Gardner (1982) that "there will almost always be a clump of six or seven CARDS of the same color" in a normal deck of cards by giving $P_6(26,26) = 0.46424$.

Given n BERNOULLI TRIALS with a probability of success (heads) p, the expected number of tails is $n(1-p)$, so the expected number of tail runs ≥ 1 is $\approx n(1-p)p$. Continuing,

$$N_R = n(1-p)p^R \qquad (7)$$

is the expected number of runs $\ge R$. The longest expected run is therefore given by

$$R = \log_{1/p}[n(1-p)] \qquad (8)$$

(Gordon *et al.* 1986, Schilling 1990). Given m 0s and n 1s, the number of possible arrangements with u runs is

$$f_u = \begin{cases} 2\binom{m-1}{k-1}\binom{n-1}{k-1} & u \equiv 2k \\ \binom{m-1}{k-1}\binom{n-1}{k-2} + \binom{m-1}{k-2}\binom{n-1}{k-1} & u \equiv 2k+1 \end{cases} \qquad (9)$$

for k an INTEGER, where $\binom{n}{k}$ is a BINOMIAL COEFFICIENT. Then

$$P(u \le u') = \sum_{u=2}^{u'} \frac{f_u}{\binom{m+n}{m}}. \qquad (10)$$

Bloom (1996) gives the expected number of noncontiguous t-runs in a sequence of m 0s and n 1s as

$$E(n,m,t) = \frac{(m+1)(n)_t + (n+1)(m)_t}{(m+n)_t}, \qquad (11)$$

where $(a)_n$ is the POCHHAMMER SYMBOL. For $m > 10$, u has an approximately NORMAL DISTRIBUTION with MEAN and VARIANCE

$$\mu_u = 1 + \frac{2mn}{m+n} \qquad (12)$$

$$\sigma_u{}^2 = \frac{2mn(2mn - m - n)}{(m+n)^2(m+n-1)}. \qquad (13)$$

see also COIN TOSSING, EULERIAN NUMBER, PERMUTATION, s-RUN

References

Bloom, D. M. "Probabilities of Clumps in a Binary Sequence (and How to Evaluate Them Without Knowing a Lot)." *Math. Mag.* **69**, 366–372, 1996.

Gardner, M. *Aha! Gotcha: Paradoxes to Puzzle and Delight.* New York: W. H. Freeman, p. 124, 1982.

Godbole, A. P. "On Hypergeometric and Related Distributions of Order *k*." *Commun. Stat.: Th. and Meth.* **19**, 1291–1301, 1990.
Godbole, A. P. and Papastavnidis, G. (Eds.). *Runs and Patterns in Probability: Selected Papers.* New York: Kluwer, 1994.
Gordon, L.; Schilling, M. F.; and Waterman, M. S. "An Extreme Value Theory for Long Head Runs." *Prob. Th. and Related Fields* **72**, 279–287, 1986.
Goulden, I. P. and Jackson, D. M. *Combinatorial Enumeration.* New York: Wiley, 1983.
Mood, A. M. "The Distribution Theory of Runs." *Ann. Math. Statistics* **11**, 367–392, 1940.
Philippou, A. N. and Makri, F. S. "Successes, Runs, and Longest Runs." *Stat. Prob. Let.* **4**, 211–215, 1986.
Schilling, M. F. "The Longest Run of Heads." *Coll. Math. J.* **21**, 196–207, 1990.
Schuster, E. F. In *Runs and Patterns in Probability: Selected Papers* (Ed. A. P. Godbole and S. Papastavridis). Boston, MA: Kluwer, pp. 91–111, 1994.

Runge-Kutta Method

A method of integrating ORDINARY DIFFERENTIAL EQUATIONS by using a trial step at the midpoint of an interval to cancel out lower-order error terms. The second-order formula is

$$\begin{aligned} k_1 &= hf(x_n, y_n) \\ k_2 &= hf(x_n + \tfrac{1}{2}h, y_n + \tfrac{1}{2}k_1) \\ y_{n+1} &= y_n + k_2 + \mathcal{O}(h^3), \end{aligned}$$

and the fourth-order formula is

$$\begin{aligned} k_1 &= hf(x_n, y_n) \\ k_2 &= hf(x_n + \tfrac{1}{2}h, y_n + \tfrac{1}{2}k_1) \\ k_3 &= hf(x_n + \tfrac{1}{2}h, y_n + \tfrac{1}{2}k_2) \\ k_4 &= hf(x_n + h, y_n + k_3) \\ y_{n+1} &= y_n + \tfrac{1}{6}k_1 + \tfrac{1}{3}k_2 + \tfrac{1}{3}k_3 + \tfrac{1}{6}k_4 + \mathcal{O}(h^5). \end{aligned}$$

(Press *et al.* 1992). This method is reasonably simple and robust and is a good general candidate for numerical solution of differential equations when combined with an intelligent adaptive step-size routine.

see also ADAMS' METHOD, GILL'S METHOD, MILNE'S METHOD, ORDINARY DIFFERENTIAL EQUATION, ROSENBROCK METHODS

References

Abramowitz, M. and Stegun, C. A. (Eds.). *Handbook of Mathematical Functions with Formulas, Graphs, and Mathematical Tables, 9th printing.* New York: Dover, pp. 896–897, 1972.
Arfken, G. *Mathematical Methods for Physicists, 3rd ed.* Orlando, FL: Academic Press, pp. 492–493, 1985.
Cartwright, J. H. E. and Piro, O. "The Dynamics of Runge-Kutta Methods." *Int. J. Bifurcations Chaos* **2**, 427–449, 1992. `http://formentor.uib.es/~julyan/TeX/rkpaper/root/root.html`.
Lambert, J. D. and Lambert, D. Ch. 5 in *Numerical Methods for Ordinary Differential Systems: The Initial Value Problem.* New York: Wiley, 1991.
Press, W. H.; Flannery, B. P.; Teukolsky, S. A.; and Vetterling, W. T. "Runge-Kutta Method" and "Adaptive Step Size Control for Runge-Kutta." §16.1 and 16.2 in *Numerical Recipes in FORTRAN: The Art of Scientific Computing, 2nd ed.* Cambridge, England: Cambridge University Press, pp. 704–716, 1992.

Runge-Walsh Theorem

Let $f(x)$ be an ANALYTIC FUNCTION which is REGULAR in the interior of a JORDAN CURVE C and continuous in the closed DOMAIN bounded by C. Then $f(x)$ can be approximated with an arbitrary accuracy by POLYNOMIALS.

see also ANALYTIC FUNCTION, JORDAN CURVE

References

Szegő, G. *Orthogonal Polynomials, 4th ed.* Providence, RI: Amer. Math. Soc., p. 7, 1975.

Running Knot

A KNOT which tightens around an object when strained but slackens when the strain is removed. Running knots are sometimes also known as slip knots or nooses.

References

Owen, P. *Knots.* Philadelphia, PA: Courage, p. 60, 1993.

Russell's Antinomy

Let R be the set of all sets which are not members of themselves. Then R is neither a member of itself nor not a member of itself. Symbolically, let $R = \{x : x \notin x\}$. Then $R \in R$ IFF $R \notin R$.

Bertrand Russell discovered this PARADOX and sent it in a letter to G. Frege just as Frege was completing *Grundlagen der Arithmetik.* This invalidated much of the rigor of the work, and Frege was forced to add a note at the end stating, "A scientist can hardly meet with anything more undesirable than to have the foundation give way just as the work is finished. I was put in this position by a letter from Mr. Bertrand Russell when the work was nearly through the press."

see also GRELLING'S PARADOX

References

Courant, R. and Robbins, H. "The Paradoxes of the Infinite." §2.4.5 in *What is Mathematics?: An Elementary Approach to Ideas and Methods, 2nd ed.* Oxford, England: Oxford University Press, p. 78, 1996.
Frege, G. *Foundations of Arithmetic.* Evanston, IL: Northwestern University Press, 1968.
Hofstadter, D. R. *Gödel, Escher, Bach: An Eternal Golden Braid.* New York: Vintage Books, pp. 20–21, 1989.

Russell's Paradox

see RUSSELL'S ANTINOMY

Russian Multiplication

Also called ETHIOPIAN MULTIPLICATION. To multiply two numbers a and b, write $a_0 \equiv a$ and $b_0 \equiv b$ in two columns. Under a_0, write $\lfloor a_0/2 \rfloor$, where $\lfloor x \rfloor$ is the FLOOR FUNCTION, and under b_0, write $2b_0$. Continue until $a_i = 1$. Then cross out any entries in the b column which are opposite an EVEN NUMBER in the a column and add the b column. The result is the desired product. For example, for $a = 27, b = 35$

$$\begin{array}{rr} 27 & 35 \\ 13 & 70 \\ 6 & \not{140} \\ 3 & 280 \\ 1 & \underline{560} \\ & 945 \end{array}$$

Russian Roulette

Russian roulette is a GAME of chance in which one or more of the six chambers of a gun are filled with bullets, the magazine is rotated at random, and the gun is shot. The shooter bets on whether the chamber which rotates into place will be loaded. If it is, he loses not only his bet but his life.

A modified version is considered by Blom *et al.* (1996) and Blom (1989). In this variant, the revolver is loaded with a single bullet, and two duelists alternately spin the chamber and fire at themselves until one is killed. The probability that the first duelist is killed is then 6/11.

References

Blom, G. *Probabilities and Statistics: Theory and Applications.* New York: Springer-Verlag, p. 32, 1989.

Blom, G.; Englund, J,.-E.; and Sandell, D. "General Russian Roulette." *Math. Mag.* **69**, 293–297, 1996.

Ryser Formula

A formula for the PERMANENT of a MATRIX

$$\operatorname{perm}(a_{ij}) = (-1)^n \sum_{s \subseteq \{1,\ldots,n\}} (-1)^{|s|} \prod_{i=1}^{n} \sum_{j \in s} a_{ij},$$

where the SUM is over all SUBSETS of $\{1, \ldots, n\}$, and $|s|$ is the number of elements in s. The formula can be optimized by picking the SUBSETS so that only a single element is changed at a time (which is precisely a GRAY CODE), reducing the number of additions from n^2 to n.

It turns out that the number of disks moved after the kth step in the TOWERS OF HANOI is the same as the element which needs to be added or deleted in the kth ADDEND of the RYSER FORMULA (Gardner 1988, Vardi 1991, p. 111)

see also DETERMINANT, GRAY CODE, PERMANENT, TOWERS OF HANOI

References

Gardner, M. "The Icosian Game and the Tower of Hanoi." Ch. 6 in *The Scientific American Book of Mathematical Puzzles & Diversions.* New York: Simon and Schuster, 1959.

Knuth, D. E. *The Art of Computer Programming, Vol. 2: Seminumerical Algorithms, 2nd ed.* Reading, MA: Addison-Wesley, p. 497, 1981.

Nijenhuis, A. and Wilf, H. Chs. 7–8 in *Combinatorial Algorithms.* New York: Academic Press, 1975.

Vardi, I. *Computational Recreations in Mathematica.* Reading, MA: Addison-Wesley, p. 111, 1991.

S

s-Additive Sequence

A generalization of an ULAM SEQUENCE in which each term is the SUM of two earlier terms in exactly s ways. (s,t)-additive sequences are a further generalization in which each term has exactly s representations as the SUM of t distinct earlier numbers. It is conjectured that 0-additive sequences ultimately have periodic differences of consecutive terms (Guy 1994, p. 233).

see also GREEDY ALGORITHM, STÖHR SEQUENCE, ULAM SEQUENCE

References

Finch, S. R. "Conjectures about s-Additive Sequences." *Fib. Quart.* **29**, 209–214, 1991.
Finch, S. R. "Are 0-Additive Sequences Always Regular?" *Amer. Math. Monthly* **99**, 671–673, 1992.
Finch, S. R. "On the Regularity of Certain 1-Additive Sequences." *J. Combin. Th. Ser. A.* **60**, 123–130, 1992.
Finch, S. R. "Patterns in 1-Additive Sequences." *Experiment. Math.* **1**, 57–63, 1992.
Guy, R. K. *Unsolved Problems in Number Theory, 2nd ed.* New York: Springer-Verlag, pp. 110 and 233, 1994.
Ulam, S. M. *Problems in Modern Mathematics.* New York: Interscience, p. ix, 1964.

s-Cluster

N.B. A detailed on-line essay by S. Finch was the starting point for this entry.

Let an $n \times n$ MATRIX have entries which are either 1 (with probability p) or 0 (with probability $q = 1 - p$). An s-cluster is an isolated group of s adjacent (i.e., horizontally or vertically connected) 1s. Let C_n be the total number of "SITE" clusters. Then the value

$$K_S(p) = \lim_{n\to\infty} \frac{\langle C_n\rangle}{n^2}, \tag{1}$$

called the MEAN CLUSTER COUNT PER SITE or MEAN CLUSTER DENSITY, exists. Numerically, it is found that $K_S(1/2) \approx 0.065770\ldots$ (Ziff *et al.* 1997).

Considering instead "BOND" clusters (where numbers are assigned to the edges of a grid) and letting C_n be the total number of bond clusters, then

$$K_B(p) \equiv \lim_{n\to\infty} \frac{\langle C_n\rangle}{n^2} \tag{2}$$

exists. The analytic value is known for $p = 1/2$,

$$K_B(\tfrac{1}{2}) = \tfrac{3}{2}\sqrt{3} - \tfrac{41}{16} \tag{3}$$

(Ziff *et al.* 1997).

see also BOND PERCOLATION, PERCOLATION THEORY, s-RUN, SITE PERCOLATION

References

Finch, S. "Favorite Mathematical Constants." `http://www.mathsoft.com/asolve/constant/rndprc/rndprc.html`.
Temperley, H. N. V. and Lieb, E. H. "Relations Between the 'Percolation' and 'Colouring' Problem and Other Graph-Theoretical Problems Associated with Regular Planar Lattices; Some Exact Results for the 'Percolation' Problem." *Proc. Roy. Soc. London A* **322**, 251–280, 1971.
Ziff, R.; Finch, S.; and Adamchik, V. "Universality of Finite-Sized Corrections to the Number of Percolation Clusters." *Phys. Rev. Let.* To appear, 1998.

s-Run

N.B. A detailed on-line essay by S. Finch was the starting point for this entry.

Let $\mathbf{v}$ be a n-VECTOR whose entries are each 1 (with probability p) or 0 (with probability $q = 1 - p$). An s-run is an isolated group of s consecutive 1s. Ignoring the boundaries, the total number of runs R_n satisfies

$$K_n = \frac{\langle R_n\rangle}{n} = (1-p)^2 \sum_{s=1}^{n} p^s = p(1-p)(1-p^n),$$

so

$$K(p) \equiv \lim_{n\to\infty} K_n = p(1-p),$$

which is called the MEAN RUN COUNT PER SITE or MEAN RUN DENSITY in PERCOLATION THEORY.

see also PERCOLATION THEORY, s-CLUSTER

References

Finch, S. "Favorite Mathematical Constants." `http://www.mathsoft.com/asolve/constant/rndprc/rndprc.html`.

S-Signature

see SIGNATURE (RECURRENCE RELATION)

Saalschützian

For a GENERALIZED HYPERGEOMETRIC FUNCTION

$${}_{p+1}F_p\left[\begin{matrix}\alpha_1, \alpha_2, \ldots, \alpha_{p+1} \\ \beta_1, \beta_2, \ldots, \beta_p\end{matrix}; z\right],$$

the Saalschützian S is defined if

$$\sum \beta = \sum \alpha + 1.$$

see also GENERALIZED HYPERGEOMETRIC FUNCTION

Saalschütz's Theorem

$${}_3F_2\left[\begin{matrix}-x, -y, -z \\ n+1, -x-y-z\end{matrix}\right] = \frac{\Gamma(n+1)\Gamma(x+y+n+1)}{\Gamma(x+n+1)\Gamma(y+n+1)} \times \frac{\Gamma(y+z+n+1)\Gamma(z+x+n+1)}{\Gamma(z+n+1)\Gamma(x+y+z+n+1)},$$

where ${}_3F_2(a,b,c;d,e;z)$ is a GENERALIZED HYPERGEOMETRIC FUNCTION and $\Gamma(z)$ is the GAMMA FUNCTION.

It can be derived from the DOUGALL-RAMANUJAN IDENTITY and written in the symmetric form

$$ {}_3F_2(a,b,c;d,e;1) = \frac{(d-a)_{|c|}(d-b)_{|c|}}{d_{|c|}(d-a-b)_{|c|}} $$

for $d+e = a+b+c+1$ with c a negative integer and $(a)_n$ the POCHHAMMER SYMBOL (Petkovšek *et al.* 1996).

see also DOUGALL-RAMANUJAN IDENTITY, GENERALIZED HYPERGEOMETRIC FUNCTION

References

Petkovšek, M.; Wilf, H. S.; and Zeilberger, D. *A=B.* Wellesley, MA: A. K. Peters, pp. 43 and 126, 1996.

Saddle

A SURFACE possessing a SADDLE POINT.

see also HYPERBOLIC PARABOLOID, MONKEY SADDLE, SADDLE POINT (FUNCTION)

Saddle-Node Bifurcation

see FOLD BIFURCATION

Saddle Point (Fixed Point)

see HYPERBOLIC FIXED POINT (DIFFERENTIAL EQUATIONS), HYPERBOLIC FIXED POINT (MAP)

Saddle Point (Function)

A POINT of a FUNCTION or SURFACE which is a STATIONARY POINT but not an EXTREMUM. An example of a 1-D FUNCTION with a saddle point is $f(x) = x^3$, which has

$$ \begin{aligned} f'(x) &= 3x^2 \\ f''(x) &= 6x \\ f'''(x) &= 6. \end{aligned} $$

This function has a saddle point at $x_0 = 0$ by the EXTREMUM TEST since $f''(x_0) = 0$ and $f'''(x_0) = 6 \neq 0$. An example of a SURFACE with a saddle point is the MONKEY SADDLE.

Saddle Point (Game)

For a general two-player ZERO-SUM GAME,

$$ \min_{i\leq m}\min_{j\leq n} a_{ij} \leq \min_{j\leq n}\max_{i\leq m} a_{ij}. $$

If the two are equal, then write

$$ \min_{i\leq m}\min_{j\leq n} a_{ij} = \min_{j\leq n}\max_{i\leq m} a_{ij} \equiv v, $$

where v is called the VALUE of the GAME. In this case, there exist optimal strategies for the first and second players.

A NECESSARY and SUFFICIENT condition for a saddle point to exist is the presence of a PAYOFF MATRIX element which is both a minimum of its row and a maximum of its column. A GAME may have more than one saddle point, but all must have the same VALUE.

see also GAME, PAYOFF MATRIX, VALUE

References

Dresher, M. "Saddle Points." §1.5 in *The Mathematics of Games of Strategy: Theory and Applications.* New York: Dover, pp. 12–14, 1981.

Llewellyn, D. C.; Tovey, C.; and Trick, M. "Finding Saddlepoints of Two-Person, Zero Sum Games." *Amer. Math. Monthly* **95**, 912–918, 1988.

Safarevich Conjecture

see SHAFAREVICH CONJECTURE

Safe

A position in a GAME is safe if the person who plays next will lose.

see also GAME, UNSAFE

Sagitta

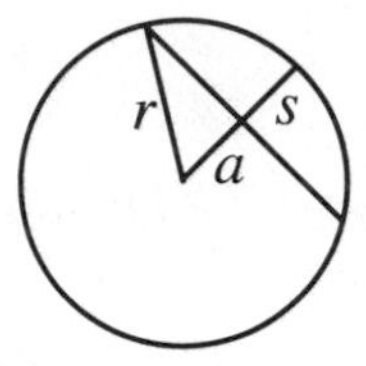

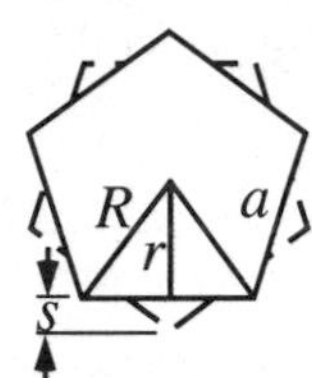

The PERPENDICULAR distance s from an ARC's MIDPOINT to the CHORD across it, equal to the RADIUS r minus the APOTHEM a,

$$ s = r - a. \tag{1} $$

For a regular POLYGON of side length a,

$$ \begin{aligned} s \equiv R - r &= \tfrac{1}{2}a\left[\csc\left(\frac{\pi}{n}\right) - \cot\left(\frac{\pi}{n}\right)\right] \\ &= \tfrac{1}{2}a\tan\left(\frac{\pi}{2n}\right) && (2) \\ &= r\tan\left(\frac{\pi}{n}\right)\tan\left(\frac{\pi}{2n}\right) && (3) \\ &= 2R\sin^2\left(\frac{\pi}{2n}\right), && (4) \end{aligned} $$

where R is the CIRCUMRADIUS, r the INRADIUS, a is the side length, and n is the number of sides.

see also APOTHEM, CHORD, SECTOR, SEGMENT

Saint Andrew's Cross

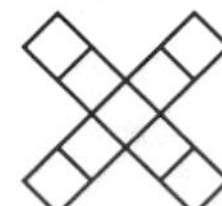

A GREEK CROSS rotated by 45°, also called the crux decussata. The MULTIPLICATION SIGN × is based on Saint Andrew's cross (Bergamini 1969).

see also CROSS, GREEK CROSS, MULTIPLICATION SIGN

References

Bergamini, D. *Mathematics.* New York: Time-Life Books, p. 11, 1969.

Saint Anthony's Cross

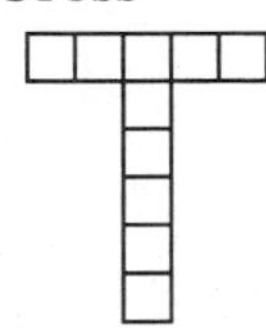

A CROSS also called the tau cross or crux commissa.

see also CROSS

Saint Petersburg Paradox

Consider a game in which a player bets on whether a given TOSS of a COIN will turn up heads or tails. If he bets \$1 that heads will turn up on the first throw, \$2 that heads will turn up on the second throw (if it did not turn up on the first), \$4 that heads will turn up on the third throw, etc., his expected *payoff* is

$$\tfrac{1}{2}(1)+\tfrac{1}{4}(2)+\tfrac{1}{8}(4)+\ldots=\tfrac{1}{2}+\tfrac{1}{2}+\tfrac{1}{2}+\ldots=\infty.$$

Apparently, the first player can be in the hole by any amount of money and still come out ahead in the end. This PARADOX was first proposed by Daniel Bernoulli.

The paradox arises as a result of muddling the distinction between the amount of the final payoff and the net amount won in the game. It is misleading to consider the payoff without taking into account the amount lost on previous bets, as can be shown as follows. At the time the player first wins (say, on the nth toss), he will have lost

$$\sum_{k=1}^{n-1} 2^{k-1} = 2^{n-1}-1$$

dollars. In this toss, however, he wins 2^{n-1} dollars. This means that the net gain for the player is a whopping \$1, no matter how many tosses it takes to finally win. As expected, the large payoff after a long run of tails is exactly balanced by the large amount that the player has to invest.

In fact, by noting that the probability of winning on the nth toss is $1/2^n$, it can be seen that the probability distribution for the number of tosses needed to win is simply a GEOMETRIC DISTRIBUTION with $p=1/2$.

see also COIN TOSSING, GAMBLER'S RUIN, GEOMETRIC DISTRIBUTION, MARTINGALE

References

Ball, W. W. R. and Coxeter, H. S. M. *Mathematical Recreations and Essays, 13th ed.* New York: Dover, pp. 44–45, 1987.

Gardner, M. *The Scientific American Book of Mathematical Puzzles & Diversions.* New York: Simon and Schuster, pp. 51–52, 1959.

Kamke, E. *Einführung in die Wahrscheinlichkeitstheorie.* Leipzig, Germany, pp. 82–89, 1932.

Keynes, J. M. K. "The Application of Probability to Conduct." In *The World of Mathematics, Vol. 2* (Ed. K. Newman). Redmond, WA: Microsoft Press, 1988.

Kraitchik, M. "The Saint Petersburg Paradox." §6.18 in *Mathematical Recreations.* New York: W. W. Norton, pp. 138–139, 1942.

Todhunter, I. §391 in *History of the Mathematical Theory of Probability.* New York: Chelsea, p. 221, 1949.

Sal

see WALSH FUNCTION

Salamin Formula

see BRENT-SALAMIN FORMULA

Salem Constants

Each point of the PISOT-VIJAYARAGHAVAN CONSTANTS S is a LIMIT POINT from both sides of a set T known as the Salem constants (Salem 1945). The Salem constants are algebraic INTEGERS >1 in which one or more of the conjugates is on the UNIT CIRCLE with the others inside (Le Lionnais 1983, p. 150). The smallest known Salem number was found by Lehmer (1933) as the largest REAL ROOT of

$$x^{10}+x^9-x^7-x^6-x^5-x^4-x^3+x+1=0,$$

which is

$$\sigma_1 = 1.176280818\ldots$$

(Le Lionnais 1983, p. 35). Boyd (1977) found the following table of small Salem numbers, and suggested that σ_1, σ_2, σ_3, and σ_4 are the smallest Salem numbers. The NOTATION 1 1 0 −1 −1 −1 is short for 1 1 0 −1 −1 −1 −1 −1 0 1 1, the coefficients of the above polynomial.

k	σ_k	°	polynomial
1	1.1762808183	10	1 1 0 −1 −1 −1
2	1.1883681475	18	1 −1 1 −1 0 0 −1 1 −1 1
3	1.2000265240	14	1 0 0 −1 −1 0 0 1
4	1.2026167437	14	1 0 −1 0 0 0 0 −1
5	1.2163916611	10	1 0 0 0 −1 −1
6	1.2197208590	18	1 −1 0 0 0 0 0 0 −1 1
7	1.2303914344	10	1 0 0 −1 0 −1
8	1.2326135486	20	1 −1 0 0 0 −1 1 0 0 −1 1
9	1.2356645804	22	1 0 −1 −1 0 0 0 1 1 0 −1 −1
10	1.2363179318	16	1 −1 0 0 0 0 0 0 −1
11	1.2375048212	26	1 0 −1 0 0 −1 0 0 −1 0 1 0 0 1
12	1.2407264237	12	1 −1 1 −1 0 0 −1
13	1.2527759374	18	1 0 0 0 0 0 −1 −1 −1 −1
14	1.2533306502	20	1 0 −1 0 0 −1 0 0 0 0 0
15	1.2550935168	14	1 0 −1 −1 0 1 0 −1
16	1.2562211544	18	1 −1 0 0 −1 1 0 0 0 −1
17	1.2601035404	24	1 −1 0 0 −1 1 0 −1 1 −1 0 1 −1
18	1.2602842369	22	1 −1 0 −1 1 0 0 0 −1 1 −1 1
19	1.2612309611	10	1 0 −1 0 0 −1
20	1.2630381399	26	1 −1 0 0 0 0 −1 0 0 0 0 0 0 1
21	1.2672964425	14	1 −1 0 0 0 0 −1 1
22	1.2806381563	8	1 0 0 −1 −1
23	1.2816913715	26	1 0 0 0 0 0 −1 −1 −1 −1 −1 −1 −1 −1
24	1.2824955606	20	1 −2 2 −2 2 −2 1 0 −1 1 −1
25	1.2846165509	18	1 0 0 0 −1 0 −1 −1 0 −1
26	1.2847468215	26	1 −2 1 1 −2 1 0 0 −1 1 0 −1 1 −1
27	1.2850993637	30	1 0 0 0 0 −1 −1 −1 −1 −1 −1 0 0 0 0 1
28	1.2851215202	30	1 −2 2 −2 1 0 −1 2 −2 1 0 −1 1 −1 1 −1
29	1.2851856708	30	1 −1 0 0 0 0 0 0 −1 0 0 0 −1 0 0 −1
30	1.2851967268	26	1 0 −1 −1 0 0 0 1 0 −1 −1 0 1 1
31	1.2851991792	44	1 −1 0 0 0 0 0 −1 0 0 0 −1 0 0 0 0 0 0 0 1 0 0 1
32	1.2852354362	30	1 0 −1 0 0 −1 −1 0 0 0 1 0 0 1 0 −1
33	1.2854090648	34	1 −1 0 0 −1 1 −1 0 1 −1 1 0 −1 1 −1 0 1 −1
34	1.2863959668	18	1 −2 2 −2 2 −2 2 −3 3 −3
35	1.2867301820	26	1 −1 0 0 −1 1 −1 0 1 −1 1 0 −1 1
36	1.2917414257	24	1 −1 0 0 0 0 −1 0 0 0 0 0 0
37	1.2920391602	20	1 0 −1 0 0 −1 0 0 −1 0 1
38	1.2934859531	10	1 0 −1 −1 0 1
39	1.2956753719	18	1 −1 0 0 −1 1 −1 0 1 −1

see also PISOT-VIJAYARAGHAVAN CONSTANTS

References

Boyd, D. W. "Small Salem Numbers." *Duke Math. J.* **44**, 315–328, 1977.

Le Lionnais, F. *Les nombres remarquables.* Paris: Hermann, 1983.

Lehmer, D. H. "Factorization of Certain Cyclotomic Functions." *Ann. Math., Ser. 2* **34**, 461–479, 1933.

Salem, R. "Power Series with Integral Coefficients." *Duke Math. J.* **12**, 153–172, 1945.

Stewart, C. L. "Algebraic Integers whose Conjugates Lie Near the Unit Circle." *Bull. Soc. Math. France* **106**, 169–176, 1978.

Salesman Problem

see TRAVELING SALESMAN PROBLEM

Salient Point

A point at which two noncrossing branches of a curve meet with different tangents.

see also CUSP

Salinon

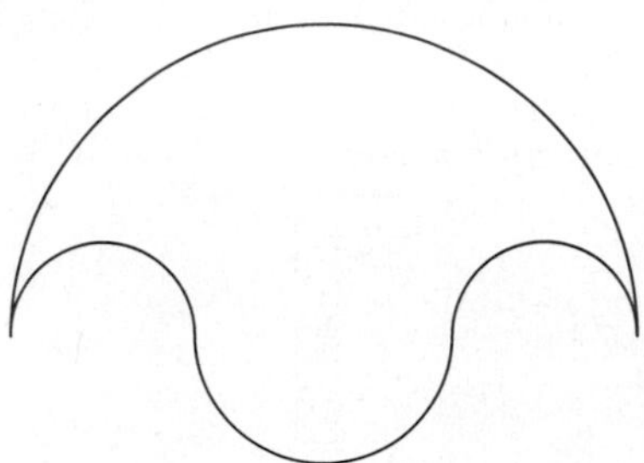

The above figure formed from four connected SEMICIRCLES. The word salinon is Greek for salt cellar, which the figure resembles.

see also ARBELOS, PIECEWISE CIRCULAR CURVE, SEMICIRCLE

Salmon's Theorem

Given a track bounded by two confocal ELLIPSES, if a ball is rolled so that its trajectory is tangent to the inner ELLIPSE, the ball's trajectory will be tangent to the inner ELLIPSE following all subsequent caroms as well.

References

Salmon, G. *A Treatise on Conic Sections.* New York: Chelsea, p. 182, 1954.

Saltus

The word saltus has two different meanings: either a jump or an oscillation of a function.

Sample Proportion

Let there be x successes out of n BERNOULLI TRIALS. The sample proportion is the fraction of samples which were successes, so

$$\hat{p} = \frac{x}{n}. \tag{1}$$

For large n, $\hat{p}$ has an approximately NORMAL DISTRIBUTION. Let RE be the RELATIVE ERROR and SE the STANDARD ERROR, then

$$\langle p \rangle = p \tag{2}$$

$$\mathrm{SE}(\hat{p}) \equiv \sigma(\hat{p}) = \sqrt{\frac{p(1-p)}{n}} \tag{3}$$

$$\mathrm{RE}(\hat{p}) = \sqrt{\frac{2\hat{p}(1-\hat{p})}{n}}\,\mathrm{erf}^{-1}(\mathrm{CI}), \tag{4}$$

where CI is the CONFIDENCE INTERVAL and $\mathrm{erf}\,x$ is the ERF function. The number of tries needed to determine p with RELATIVE ERROR RE and CONFIDENCE INTERVAL CI is

$$n = \frac{2[\mathrm{erf}^{-1}(\mathrm{CI})]^2 \hat{p}(1-\hat{p})}{(\mathrm{RE})^2}. \tag{5}$$

Sample Space

Informally, the sample space for a given set of events is the set of all possible values the events may assume. Formally, the set of possible events for a given variate forms a SIGMA ALGEBRA, and sample space is defined as the largest set in the SIGMA ALGEBRA.

see also PROBABILITY SPACE, RANDOM VARIABLE, SIGMA ALGEBRA, STATE SPACE

Sample Variance

To estimate the population VARIANCE from a sample of N elements with a priori *unknown* MEAN (i.e., the MEAN is estimated from the sample itself), we need an unbiased ESTIMATOR for σ. This is the k-STATISTIC k_2, where

$$k_2 = \frac{N}{N-1} m_2 \tag{1}$$

and $m_2 \equiv s^2$ is the sample variance

$$s^2 \equiv \frac{1}{N} \sum_{i=1}^{N} (x_i - \bar{x})^2. \tag{2}$$

Note that some authors prefer the definition

$$s'^2 \equiv \frac{1}{N-1} \sum_{i=1}^{N} (x_i - \bar{x})^2, \tag{3}$$

since this makes the sample variance an UNBIASED ESTIMATOR for the population variance.

see also k-STATISTIC, VARIANCE

Sampling

For infinite precision sampling of a band-limited signal at the NYQUIST FREQUENCY, the signal-to-noise ratio after N_q samples is

$$\mathrm{SNR} = \frac{\langle r_\infty \rangle}{\sigma_\infty} = \frac{\rho\sigma^2}{\sigma^2 {N_q}^{-1/2} \sqrt{1+\rho^2}} = \frac{\rho}{\sqrt{1+\rho^2}} \sqrt{N_q}, \tag{1}$$

where ρ is the normalized cross-correlation COEFFICIENT

$$\rho \equiv \frac{\langle x(t) \rangle \langle y(t) \rangle}{\sqrt{\langle x^2(t) \rangle \langle y^2(t) \rangle}}. \tag{2}$$

For $\rho \ll 1$,

$$\mathrm{SNR} \approx \rho \sqrt{N_q}. \tag{3}$$

The identical result is obtained for oversampling. For undersampling, the SNR decreases (Thompson *et al.* 1986).

see also NYQUIST SAMPLING, OVERSAMPLING, QUANTIZATION EFFICIENCY, SAMPLING FUNCTION, SHANNON SAMPLING THEOREM, SINC FUNCTION

References

Feuer, A. *Sampling in Digital Signal Processing and Control.* Boston, MA: Birkhäuser, 1996.

Thompson, A. R.; Moran, J. M.; and Swenson, G. W. Jr. *Interferometry and Synthesis in Radio Astronomy.* New York: Wiley, pp. 214–216, 1986.

Sampling Function

The 1-D sampling function is given by

$$S(x) = \sum_{n=-\infty}^{\infty} \delta(x - n\Delta x),$$

where δ is the DIRAC DELTA FUNCTION. The 2-D version is

$$S(u,v) = \sum \delta(u - u_n, v - v_n),$$

which can be weighted to

$$S(u,v) = \sum R_n T_n D_n \delta(u - u_n, v - v_n),$$

where R_n is a reliability weight, D_n is a density weight (WEIGHTING FUNCTION), and T_n is a taper.

see also SHAH FUNCTION, SINC FUNCTION

Sampling Theorem

In order for a band-limited (i.e., one with a zero POWER SPECTRUM for frequencies $f > B$) baseband ($f > 0$) signal to be reconstructed fully, it must be sampled at a rate $f \geq 2B$. A signal sampled at $f = 2B$ is said to be NYQUIST SAMPLED, and $f = 2B$ is called the NYQUIST FREQUENCY. No information is lost if a signal is sampled at the NYQUIST FREQUENCY, and no additional information is gained by sampling faster than this rate.

see also ALIASING, NYQUIST FREQUENCY, NYQUIST SAMPLING, OVERSAMPLING

San Marco Fractal

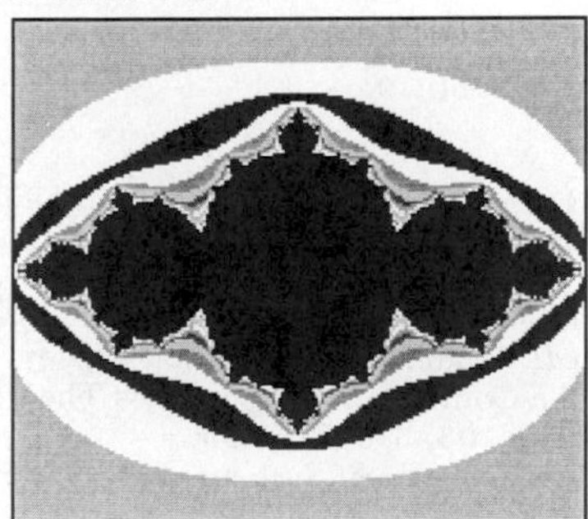

The FRACTAL $J(-3/4, 0)$, where J is the JULIA SET. It slightly resembles the MANDELBROT SET.

see also DOUADY'S RABBIT FRACTAL, JULIA SET, MANDELBROT SET

References

Wagon, S. *Mathematica in Action.* New York: W. H. Freeman, p. 173, 1991.

Sandwich Theorem

see HAM SANDWICH THEOREM, SQUEEZING THEOREM

Sard's Theorem

The set of "critical values" of a MAP $u : \mathbb{R}^n \to \mathbb{R}^n$ of CLASS C^1 has LEBESGUE MEASURE 0 in $\mathbb{R}^n$.

see also CLASS (MAP), LEBESGUE MEASURE

References

Iyanaga, S. and Kawada, Y. (Eds.). *Encyclopedic Dictionary of Mathematics*. Cambridge, MA: MIT Press, p. 682, 1980.

Šarkovskii's Theorem

Order the NATURAL NUMBERS as follows:

$$\begin{aligned}3 \prec 5 \prec 7 \prec 9 \prec 11 \prec 13 \prec 15 \prec \ldots \prec 2\cdot 3 \prec 2\cdot 5 \prec 2\cdot 7 \\ \prec 2\cdot 9 \prec \ldots \prec 2\cdot 2\cdot 3 \prec 2\cdot 2\cdot 5 \prec 2\cdot 2\cdot 7 \\ \prec 2\cdot 2\cdot 9 \prec \ldots \prec 2\cdot 2\cdot 2\cdot 3 \prec \ldots \\ \prec 2^5 \prec 2^4 \prec 2^3 \prec 2^2 \prec 2 \prec 1.\end{aligned}$$

Now let F be a CONTINUOUS FUNCTION from the REALS to the REALS and suppose $p \prec q$ in the above ordering. Then if F has a point of LEAST PERIOD p, then F also has a point of LEAST PERIOD q.

A special case of this general result, also known as Šarkovskii's theorem, states that if a CONTINUOUS REAL function has a PERIODIC POINT with period 3, then there is a PERIODIC POINT of period n for every INTEGER n.

A converse to Šarkovskii's theorem says that if $p \prec q$ in the above ordering, then we can find a CONTINUOUS FUNCTION which has a point of LEAST PERIOD q, but does not have any points of LEAST PERIOD p (Elaydi 1996). For example, there is a CONTINUOUS FUNCTION with no points of LEAST PERIOD 3 but having points of all other LEAST PERIODS.

see also LEAST PERIOD

References

Conway, J. H. and Guy, R. K. "Periodic Points." In *The Book of Numbers*. New York: Springer-Verlag, pp. 207–208, 1996.

Devaney, R. L. *An Introduction to Chaotic Dynamical Systems, 2nd ed.* Reading, MA: Addison-Wesley, 1989.

Elaydi, S. "On a Converse of Sharkovsky's Theorem." *Amer. Math. Monthly* **103**, 386–392, 1996.

Ott, E. *Chaos in Dynamical Systems*. New York: Cambridge University Press, p. 49, 1993.

Sharkovsky, A. N. "Co-Existence of Cycles of a Continuous Mapping of a Line onto Itself." *Ukranian Math. Z.* **16**, 61–71, 1964.

Stefan, P. "A Theorem of Sharkovsky on the Existence of Periodic Orbits of Continuous Endomorphisms of the Real Line." *Comm. Math. Phys.* **54**, 237–248, 1977.

Sárközy's Theorem

A partial solution to the ERDŐS SQUAREFREE CONJECTURE which states that the BINOMIAL COEFFICIENT $\binom{2n}{n}$ is never SQUAREFREE for all sufficiently large $n \geq n_0$. Sárközy (1985) showed that if $s(n)$ is the square part of the BINOMIAL COEFFICIENT $\binom{2n}{n}$, then

$$\ln s(n) \sim (\sqrt{2} - 2)\zeta(\tfrac{1}{2})\sqrt{n},$$

where $\zeta(z)$ is the RIEMANN ZETA FUNCTION. An upper bound on n_0 of $2^{8,000}$ has been obtained.

see also BINOMIAL COEFFICIENT, ERDŐS SQUAREFREE CONJECTURE

References

Erdős, P. and Graham, R. L. *Old and New Problems and Results in Combinatorial Number Theory*. Geneva, Switzerland: L'Enseignement Mathématique Université de Genève, Vol. 28, 1980.

Sander, J. W. "A Story of Binomial Coefficients and Primes." *Amer. Math. Monthly* **102**, 802–807, 1995.

Sárközy, A. "On the Divisors of Binomial Coefficients, I." *J. Number Th.* **20**, 70–80, 1985.

Vardi, I. "Applications to Binomial Coefficients." *Computational Recreations in Mathematica*. Reading, MA: Addison-Wesley, pp. 25–28, 1991.

Sarrus Linkage

A LINKAGE which converts circular to linear motion using a hinged square.

see also HART'S INVERSOR, LINKAGE, PEAUCELLIER INVERSOR

Sarrus Number

see POULET NUMBER

SAS Theorem

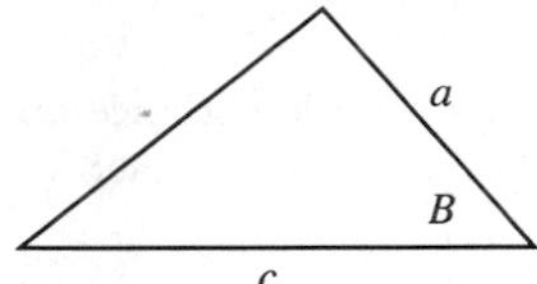

Specifying two sides and the ANGLE between them uniquely determines a TRIANGLE. Let b be the base length and h be the height. Then the AREA is

$$K = \tfrac{1}{2}ch = \tfrac{1}{2}ac\sin B. \tag{1}$$

The length of the third side is given by the LAW OF COSINES,

$$b^2 = a^2 + c^2 - 2ac\cos B,$$

so

$$b = \sqrt{a^2 + c^2 - 2ac\cos B}. \tag{2}$$

Using the LAW OF SINES

$$\frac{a}{\sin A} = \frac{b}{\sin B} = \frac{c}{\sin C} \tag{3}$$

then gives the two other ANGLES as

$$A = \sin^{-1}\left(\frac{a\sin B}{\sqrt{a^2 + c^2 - 2ac\cos B}}\right) \tag{4}$$

$$C = \sin^{-1}\left(\frac{c\sin B}{\sqrt{a^2 + c^2 - 2ac\cos B}}.\right) \tag{5}$$

see also AAA THEOREM, AAS THEOREM, ASA THEOREM, ASS THEOREM, SSS THEOREM, TRIANGLE

Satellite Knot

Let K_1 be a knot inside a TORUS. Now knot the TORUS in the shape of a second knot (called the COMPANION KNOT) K_2. Then the new knot resulting from K_1 is called the satellite knot K_3. COMPOSITE KNOTS are special cases of satellite knots. The only KNOTS which are not HYPERBOLIC KNOTS are TORUS KNOTS and satellite knots (including COMPOSITE KNOTS). No satellite knot is an ALMOST ALTERNATING KNOT.

see also ALMOST ALTERNATING KNOT, COMPANION KNOT, COMPOSITE KNOT, HYPERBOLIC KNOT, TORUS KNOT

References

Adams, C. C. *The Knot Book: An Elementary Introduction to the Mathematical Theory of Knots.* New York: W. H. Freeman, pp. 115–118, 1994.

Satisfiability Problem

Deciding whether a given Boolean formula in conjunctive normal form has an assignment that makes the formula "true." In 1971, Cook showed that the problem is NP-COMPLETE.

see also BOOLEAN ALGEBRA

References

Cook, S. A. and Mitchell, D. G. "Finding Hard Instances of the Satisfiability Problem: A Survey." In *Satisfiability problem: theory and applications (Piscataway, NJ, 1996). Theoret. Comput. Sci., Vol. 35.* Providence, RI: Amer. Math. Soc., pp. 1–17, 1997.

Sausage Conjecture

In n-D for $n \geq 5$ the arrangement of HYPERSPHERES whose CONVEX HULL has minimal CONTENT is always a "sausage" (a set of HYPERSPHERES arranged with centers along a line), independent of the number of n-spheres. The CONJECTURE was proposed by Fejes Tóth, and solved for dimensions ≥ 42 by Betke *et al.* (1994) and Betke and Henk (1998).

see also CONTENT, CONVEX HULL, HYPERSPHERE, HYPERSPHERE PACKING, SPHERE PACKING

References

Betke, U.; Henk, M.; and Wills, J. M. "Finite and Infinite Packings." *J. Reine Angew. Math.* **453**, 165–191, 1994.

Betke, U. and Henk, M. "Finite Packings of Spheres." *Discrete Comput. Geom.* **19**, 197–227, 1998.

Croft, H. T.; Falconer, K. J.; and Guy, R. K. Problem D9 in *Unsolved Problems in Geometry.* New York: Springer-Verlag, 1991.

Fejes Tóth, L. "Research Problems." *Periodica Methematica Hungarica* **6**, 197–199, 1975.

Savitzky-Golay Filter

A low-pass filter which is useful for smoothing data.

see also FILTER

References

Press, W. H.; Flannery, B. P.; Teukolsky, S. A.; and Vetterling, W. T. *Numerical Recipes in FORTRAN: The Art of Scientific Computing, 2nd ed.* Cambridge, England: Cambridge University Press, pp. 183 and 644–645, 1992.

Savoy Knot

see FIGURE-OF-EIGHT KNOT

Scalar

A one-component quantity which is invariant under ROTATIONS of the coordinate system.

see also PSEUDOSCALAR, SCALAR FIELD, SCALAR FUNCTION, SCALAR POTENTIAL, SCALAR TRIPLE PRODUCT, TENSOR, VECTOR

Scalar Curvature

see CURVATURE SCALAR

Scalar Field

A MAP $f : \mathbb{R}^n \mapsto \mathbb{R}$ which assigns each $\mathbf{x}$ a SCALAR FUNCTION $f(\mathbf{x})$.

see also VECTOR FIELD

References

Morse, P. M. and Feshbach, H. "Scalar Fields." §1.1 in *Methods of Theoretical Physics, Part I.* New York: McGraw-Hill, pp. 4–8, 1953.

Scalar Function

A function $f(x_1, \ldots, x_n)$ of one or more variables whose RANGE is one-dimensional, as compared to a VECTOR FUNCTION, whose RANGE is three-dimensional (or, in general, n-dimensional).

see also COMPLEX FUNCTION, REAL FUNCTION, VECTOR FUNCTION

Scalar Potential

A conservative VECTOR FIELD (for which the CURL $\nabla \times \mathbf{F} = \mathbf{0}$) may be assigned a scalar potential

$$\begin{aligned}\phi(x,y,z) - \phi(0,0,0) &\equiv -\int_C \mathbf{F} \cdot d\mathbf{s} \\ &= -\int_{(0,0,0)}^{(x,0,0)} F_1(t,0,0)\,dt + \int_{(x,0,0)}^{(x,y,0)} F_2(x,t,0)\,dt \\ &\quad + \int_{(x,y,0)}^{x,y,z} F_3(x,y,t)\,dt,\end{aligned}$$

where $\int_C \mathbf{F} \cdot d\mathbf{s}$ is a LINE INTEGRAL.

see also POTENTIAL FUNCTION, VECTOR POTENTIAL

Scalar Triple Product

The VECTOR product

$$\begin{aligned}[\mathbf{A}, \mathbf{B}, \mathbf{C}] &\equiv \mathbf{A} \cdot (\mathbf{B} \times \mathbf{C}) = \mathbf{B} \cdot (\mathbf{C} \times \mathbf{A}) \\ &= \mathbf{C} \cdot (\mathbf{A} \times \mathbf{B}) = \begin{vmatrix} A_1 & A_2 & A_3 \\ B_1 & B_2 & B_3 \\ C_1 & C_2 & C_3 \end{vmatrix},\end{aligned}$$

which yields a SCALAR (actually, a PSEUDOSCALAR).

The VOLUME of a PARALLELEPIPED whose sides are given by the vectors **A**, **B**, and **C** is

$$V_{\text{parallelepiped}} = |\mathbf{A} \cdot (\mathbf{B} \times \mathbf{C})|.$$

see also CROSS PRODUCT, DOT PRODUCT, PARALLELEPIPED, VECTOR TRIPLE PRODUCT

References

Arfken, G. "Triple Scalar Product, Triple Vector Product." §1.5 in *Mathematical Methods for Physicists, 3rd ed.* Orlando, FL: Academic Press, pp. 26–33, 1985.

Scale

see BASE (NUMBER)

Scale Factor

For a diagonal METRIC TENSOR $g_{ij} = g_{ii}\delta_{ij}$, where δ_{ij} is the KRONECKER DELTA, the scale factor is defined by

$$h_i \equiv \sqrt{g_{ii}}. \tag{1}$$

The LINE ELEMENT (first FUNDAMENTAL FORM) is then given by

$$\begin{aligned} ds^2 &= g_{11}\,dx_{11}{}^2 + g_{22}\,dx_{22}{}^2 + g_{33}\,dx_{33}{}^2 && (2)\\ &= h_1{}^2\,dx_{11}{}^2 + h_2{}^2\,dx_{22}{}^2 + h_3{}^2\,dx_{33}{}^2. && (3) \end{aligned}$$

The scale factor appears in vector derivatives of coordinates in CURVILINEAR COORDINATES.

see also CURVILINEAR COORDINATES, FUNDAMENTAL FORMS, LINE ELEMENT

Scalene Triangle

A TRIANGLE with three unequal sides.

see also ACUTE TRIANGLE, EQUILATERAL TRIANGLE, ISOSCELES TRIANGLE, OBTUSE TRIANGLE, TRIANGLE

Scaling

Increasing a plane figure's linear dimensions by a scale factor s increases the PERIMETER $p' \to sp$ and the AREA $A' \to s^2 A$.

see also DILATION, EXPANSION, FRACTAL, SELF-SIMILARITY

Scattering Operator

An OPERATOR relating the past asymptotic state of a DYNAMICAL SYSTEM governed by the Schrödinger equation

$$i\frac{d}{dt}\psi(t) = H\psi(t)$$

to its future asymptotic state.

see also WAVE OPERATOR

Scattering Theory

The mathematical study of the SCATTERING OPERATOR and Schrödinger equation.

see also SCATTERING OPERATOR

References

Yafaev, D. R. *Mathematical Scattering Theory: General Theory.* Providence, RI: Amer. Math. Soc., 1996.

Schaar's Identity

A generalization of the GAUSSIAN SUM. For p and q of opposite PARITY (i.e., one is EVEN and the other is ODD), Schaar's identity states

$$\frac{1}{\sqrt{q}}\sum_{r=0}^{q-1} e^{-\pi i r^2 p/q} = \frac{e^{-\pi i/4}}{\sqrt{p}}\sum_{r=0}^{p-1} e^{\pi i r^2 q/p}.$$

see also GAUSSIAN SUM

References

Evans, R. and Berndt, B. "The Determination of Gauss Sums." *Bull. Amer. Math. Soc.* **5**, 107–129, 1981.

Schanuel's Conjecture

Let $\lambda_1, \ldots, \lambda_n \in \mathbb{C}$ be linearly independent over the RATIONALS $\mathbb{Q}$, then

$$\mathbb{Q}(\lambda_1, \ldots, \lambda_n, e^{\lambda_1}, \ldots, e^{\lambda_n})$$

has TRANSCENDENCE degree at least n over $\mathbb{Q}$. Schanuel's conjecture is a generalization of the LINDEMANN-WEIERSTRAß THEOREM. If the conjecture is true, then it follows that e and π are algebraically independent. Mcintyre (1991) proved that the truth of Schanuel's conjecture also guarantees that there are no unexpected exponential-algebraic relations on the INTEGERS $\mathbb{Z}$ (Marker 1996).

see also CONSTANT PROBLEM

References

Macintyre, A. "Schanuel's Conjecture and Free Exponential Rings." *Ann. Pure Appl. Logic* **51**, 241–246, 1991.

Marker, D. "Model Theory and Exponentiation." *Not. Amer. Math. Soc.* **43**, 753–759, 1996.

Schauder Fixed Point Theorem

Let A be a closed convex subset of a BANACH SPACE and assume there exists a continuous MAP T sending A to a countably compact subset $T(A)$ of A. Then T has fixed points.

References

Iyanaga, S. and Kawada, Y. (Eds.). *Encyclopedic Dictionary of Mathematics.* Cambridge, MA: MIT Press, p. 543, 1980.

Schauder, J. "Der Fixpunktsatz in Funktionalräumen." *Studia Math.* **2**, 171–180, 1930.

Zeidler, E. *Applied Functional Analysis: Applications to Mathematical Physics.* New York: Springer-Verlag, 1995.

Scheme

A local-ringed SPACE which is locally isomorphic to an AFFINE SCHEME.

see also AFFINE SCHEME

References

Iyanaga, S. and Kawada, Y. (Eds.). "Schemes." §18E in *Encyclopedic Dictionary of Mathematics.* Cambridge, MA: MIT Press, p. 69, 1980.

Schensted Correspondence

A correspondence between a PERMUTATION and a pair of YOUNG TABLEAUX.

see also PERMUTATION, YOUNG TABLEAU

References

Knuth, D. E. *The Art of Computer Programming, Vol. 3: Sorting and Searching, 2nd ed.* Reading, MA: Addison-Wesley, 1973.

Stanton, D. W. and White, D. E. §3.6 in *Constructive Combinatorics.* New York: Springer-Verlag, pp. 85–87, 1986.

Scherk's Minimal Surfaces

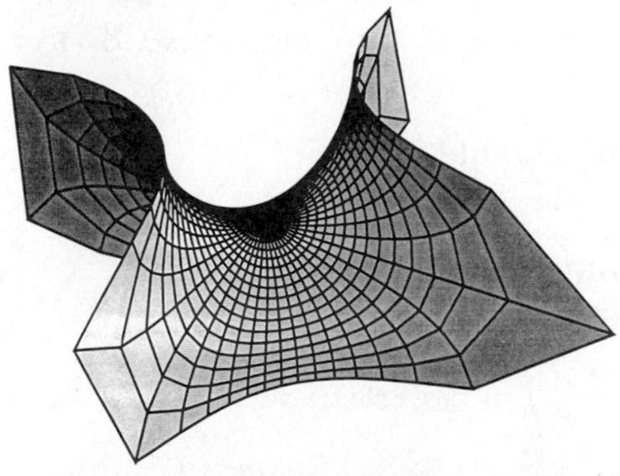

A class of MINIMAL SURFACES discovered by Scherk (1834) which were the first new surfaces discovered since Meusnier in 1776. Scherk's first surface is doubly periodic. Scherk's second surface, illustrated above, can be written parametrically as

$$x = 2\Re[\ln(1+re^{i\theta}) - \ln(1-re^{i\theta})]$$
$$y = \Re[4i\tan^{-1}(re^{i\theta})]$$
$$z = \Re\left\{2i(-\ln[1-r^2e^{2i\theta}] + \ln[1+r^2e^{2i\theta}])\right\}$$

for $\theta \in [0, 2\pi)$, and $r \in (0, 1)$. Scherk's first surface has been observed to form in layers of block copolymers (Peterson 1988).

von Seggern (1993) calls

$$z = c\ln\left[\frac{\cos(2\pi y)}{\cos(2\pi x)}\right]$$

"Scherk's surface." Beautiful images of wood sculptures of Scherk surfaces are illustrated by Séquin.

References

Dickson, S. "Minimal Surfaces." *Mathematica J.* **1**, 38–40, 1990.

do Carmo, M. P. *Mathematical Models from the Collections of Universities and Museums* (Ed. G. Fischer). Braunschweig, Germany: Vieweg, p. 41, 1986.

Meusnier, J. B. "Mémoire sur la courbure des surfaces." *Mém. des savans étrangers* **10** (lu 1776), 477–510, 1785.

Peterson, I. "Geometry for Segregating Polymers." *Sci. News*, 151, Sep. 3, 1988.

Scherk, H. F. "Bemerkung über der kleinste Fläche innerhalb gegebener Grenzen." *J. Reine. angew. Math.* **13**, 185–208, 1834.

Thomas, E. L.; Anderson, D. M.; Henkee, C. S.; and Hoffman, D. "Periodic Area-Minimizing Surfaces in Block Copolymers." *Nature* **334**, 598–601, 1988.

von Seggern, D. *CRC Standard Curves and Surfaces.* Boca Raton, FL: CRC Press, p. 304, 1993.

Wolfram Research "Mathematica Version 2.0 Graphics Gallery." `http://www.mathsource.com/cgi-bin/MathSource/Applications/Graphics/3D/0207-155`.

Schiffler Point

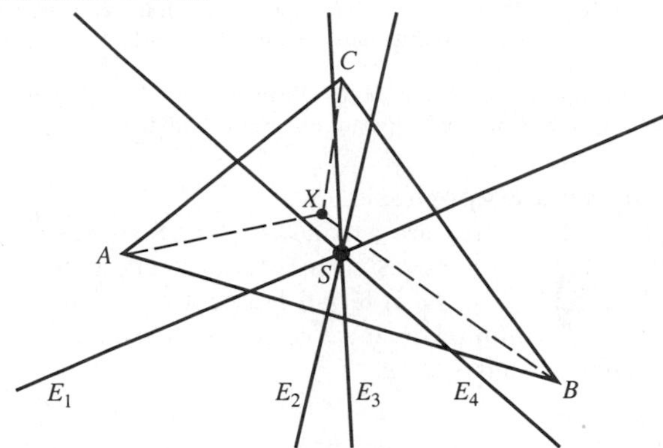

The CONCURRENCE S of the EULER LINES E_n of the TRIANGLES ΔXBC, ΔXCA, ΔXAB, and ΔABC where X is the INCENTER. The TRIANGLE CENTER FUNCTION is

$$\alpha = \frac{1}{\cos B + \cos C} = \frac{b+c-a}{b+c}.$$

References

Kimberling, C. "Central Points and Central Lines in the Plane of a Triangle." *Math. Mag.* **67**, 163–187, 1994.

Kimberling, C. "Schiffler Point." `http://www.evansville.edu/~ck6/tcenters/recent/schiff.html`.

Schiffler, K.; Veldkamp, G. R.; and van der Spek, W. A. "Problem 1018 and Solution." *Crux Math.* **12**, 176–179, 1986.

Schinzel Circle

A CIRCLE having a given number of LATTICE POINTS on its CIRCUMFERENCE. The Schinzel circle halving n lattice points is given by the equation

$$\begin{cases} (x-\frac{1}{2})^2 + y^2 = \frac{1}{4}5^{k-1} & \text{for } n = 2k \text{ even} \\ (x-\frac{1}{3})^2 + y^2 = \frac{1}{9}5^{2k} & \text{for } n = 2k+1 \text{ odd.} \end{cases}$$

Note that these solutions do not necessarily have the smallest possible RADIUS. For example, while the Schinzel circle centered at (1/3, 0) and with radius 625/3 has nine lattice points on its circumference, so does the circle centered at (1/3, 0) with radius 65/3.

see also CIRCLE, CIRCLE LATTICE POINTS, KULIKOWSKI'S THEOREM, LATTICE POINT, SCHINZEL'S THEOREM, SPHERE

References

Honsberger, R. "Circles, Squares, and Lattice Points." Ch. 11 in *Mathematical Gems I.* Washington, DC: Math. Assoc. Amer., pp. 117–127, 1973.

Kulikowski, T. "Sur l'existence d'une sphère passant par un nombre donné aux coordonnées entières." *L'Enseignement Math. Ser. 2* **5**, 89–90, 1959.

Schinzel, A. "Sur l'existence d'un cercle passant par un nombre donné de points aux coordonnées entières." *L'Enseignement Math. Ser. 2* **4**, 71–72, 1958.

Sierpiński, W. "Sur quelques problèmes concernant les points aux coordonnées entières." *L'Enseignement Math. Ser. 2* **4**, 25–31, 1958.

Sierpiński, W. "Sur un problème de H. Steinhaus concernant les ensembles de points sur le plan." *Fund. Math.* **46**, 191–194, 1959.

Sierpiński, W. *A Selection of Problems in the Theory of Numbers.* New York: Pergamon Press, 1964.

Schinzel's Hypothesis

If $f_1(x), \ldots, f_s(x)$ are irreducible POLYNOMIALS with INTEGER COEFFICIENTS such that no INTEGER $n > 1$ divides $f_1(x), \ldots, f_s(x)$ for all INTEGERS x, then there should exist infinitely many x such that $f_1(x), \ldots, f_s(x)$ are simultaneous PRIME.

References

Schinzel, A. and Sierpiński, W. "Sur certaines hypothéses concernant les nombres premiers. Remarque." *Acta Arithm.* **4**, 185–208, 1958.

Schinzel's Theorem

For every POSITIVE INTEGER n, there exists a CIRCLE in the plane having exactly n LATTICE POINTS on its CIRCUMFERENCE. The theorem is based on the number $r(n)$ of integral solutions (x, y) to the equation

$$x^2 + y^2 = n, \tag{1}$$

given by

$$r(n) = 4(d_1 - d_3), \tag{2}$$

where d_1 is the number of divisors of n of the form $4k+1$ and d_3 is the number of divisors of the form $4k+3$. It explicitly identifies such circles (the SCHINZEL CIRCLES) as

$$\begin{cases} (x - \frac{1}{2})^2 + y^2 = \frac{1}{4}5^{k-1} & \text{for } n = 2k \\ (x - \frac{1}{3})^2 + y^2 = \frac{1}{9}5^{2k} & \text{for } n = 2k+1. \end{cases} \tag{3}$$

Note, however, that these solutions do not necessarily have the smallest possible radius.

see also BROWKIN'S THEOREM, KULIKOWSKI'S THEOREM, SCHINZEL CIRCLE

References

Honsberger, R. "Circles, Squares, and Lattice Points." Ch. 11 in *Mathematical Gems I.* Washington, DC: Math. Assoc. Amer., pp. 117–127, 1973.

Kulikowski, T. "Sur l'existence d'une sphère passant par un nombre donné aux coordonnées entières." *L'Enseignement Math. Ser. 2* **5**, 89–90, 1959.

Schinzel, A. "Sur l'existence d'un cercle passant par un nombre donné de points aux coordonnées entières." *L'Enseignement Math. Ser. 2* **4**, 71–72, 1958.

Sierpiński, W. "Sur quelques problèmes concernant les points aux coordonnées entières." *L'Enseignement Math. Ser. 2* **4**, 25–31, 1958.

Sierpiński, W. "Sur un problème de H. Steinhaus concernant les ensembles de points sur le plan." *Fund. Math.* **46**, 191–194, 1959.

Sierpiński, W. *A Selection of Problems in the Theory of Numbers.* New York: Pergamon Press, 1964.

Schisma

The musical interval by which eight fifths and a major third exceed five octaves,

$$\frac{(\frac{3}{2})^8(\frac{5}{4})}{2^5} = \frac{3^8 \cdot 5}{2^{15}} = \frac{32805}{32768} = 1.00112915\ldots.$$

see also COMMA OF DIDYMUS, COMMA OF PYTHAGORAS, DIESIS

Schläfli Double Six

see DOUBLE SIXES

Schläfli's Formula

For $\Re[z] > 0$,

$$J_\nu(z) = \frac{1}{\pi}\int_0^{\pi/2} \cos(z \sin t - \nu t)\,dt - \frac{\sin(\nu\pi)}{\pi}\int_0^\infty e^{-z\sinh t} e^{-\nu t}\,dt,$$

where $J_\nu(z)$ is a BESSEL FUNCTION OF THE FIRST KIND.

References

Iyanaga, S. and Kawada, Y. (Eds.). *Encyclopedic Dictionary of Mathematics.* Cambridge, MA: MIT Press, p. 1472, 1980.

Schläfli Function

The function giving the VOLUME of the spherical quadrectangular TETRAHEDRON:

$$V = \frac{\pi^2}{8} f\left(\frac{\pi}{p}, \frac{\pi}{q}, \frac{\pi}{r}\right),$$

where

$$\frac{\pi^2}{2} f\left(\frac{\pi}{2} - x, y, \frac{\pi}{2} - z\right) = \sum_{m=1}^\infty \left(\frac{D - \sin x \sin z}{D + \sin x \sin z}\right)^m \times \frac{\cos(2mx) - \cos(2my) + \cos(2mz) - 1}{m^2} - x^2 - y^2 - z^2,$$

and

$$D \equiv \sqrt{\cos^2 x \cos^2 z - \cos^2 y}.$$

see also TETRAHEDRON

Schläfli Integral

A definition of a function using a CONTOUR INTEGRAL. Schläfli integrals may be converted into RODRIGUES FORMULAS.

see also RODRIGUES FORMULA

Schläfli's Modular Form

The MODULAR EQUATION of degree 5 can be written

$$\left(\frac{u}{v}\right)^3 + \left(\frac{v}{u}\right)^3 = 2\left(u^2v^2 - \frac{1}{u^2v^2}\right).$$

see also MODULAR EQUATION

Schläfli Polynomial

A polynomial given in terms of the NEUMANN POLYNOMIALS $O_n(x)$ by

$$S_n(x) = \frac{2xO_n(x) - 2\cos^2(\tfrac{1}{2}n\pi)}{n}.$$

see also NEUMANN POLYNOMIAL

References

Iyanaga, S. and Kawada, Y. (Eds.). *Encyclopedic Dictionary of Mathematics.* Cambridge, MA: MIT Press, p. 1477, 1980.

von Seggern, D. *CRC Standard Curves and Surfaces.* Boca Raton, FL: CRC Press, p. 196, 1993.

Schläfli Symbol

The symbol $\{p,q\}$ is used to denote a TESSELLATION of regular p-gons, with q of them surrounding each VERTEX. The Schläfli symbol can be used to describe PLATONIC SOLIDS, and a generalized version describes QUASIREGULAR POLYHEDRA and ARCHIMEDEAN SOLIDS.

see also ARCHIMEDEAN SOLID, PLATONIC SOLID, QUASIREGULAR POLYHEDRON, TESSELLATION

Schlegel Graph

A GRAPH corresponding to POLYHEDRA skeletons. The POLYHEDRAL GRAPHS are special cases.

References

Gardner, M. *Wheels, Life, and Other Mathematical Amusements.* New York: W. H. Freeman, p. 158, 1983.

Schlömilch's Function

$$S(\nu,z) \equiv \int_0^\infty (1+t)^{-\nu}e^{-zt}\,dt = z^{\nu-1}e^z\int_z^\infty u^{-\nu}e^{-u}\,du$$
$$= z^{\nu/2-1}e^{z/2}W_{-\nu/2,(1-\nu)/2}(z),$$

where $W_{k,m}(z)$ is the WHITTAKER FUNCTION.

Schlömilch's Series

A FOURIER SERIES-like expansion of a twice continuously differentiable function

$$f(x) = \tfrac{1}{2}a_0 + \sum_{n=1}^\infty a_nJ_0(nx)$$

for $0 < x < \pi$, where $J_0(x)$ is a zeroth order BESSEL FUNCTION OF THE FIRST KIND and

$$a_0 \equiv 2f(0) + \frac{2}{\pi}\int_0^\pi du\int_0^{\pi/2} f'(u\sin\phi)\,d\phi$$
$$a_n \equiv \frac{2}{\pi}\int_0^\pi du\int_0^{\pi/2} uf'(u\sin\phi)\cos(n\pi)\,d\phi.$$

A special case gives the amazing identity

$$1 = J_0(z) + 2\sum_{n=1}^\infty J_{2n}(z) = [J_0(z)]^2 + 2\sum_{n=1}^\infty [J_n(z)]^2.$$

see also BESSEL FUNCTION OF THE FIRST KIND, BESSEL FUNCTION FOURIER EXPANSION, FOURIER SERIES

References

Iyanaga, S. and Kawada, Y. (Eds.). *Encyclopedic Dictionary of Mathematics.* Cambridge, MA: MIT Press, p. 1473, 1980.

Schmitt-Conway Biprism

A CONVEX POLYHEDRON which is SPACE-FILLING, but only aperiodically, was found by Conway in 1993.

see also CONVEX POLYHEDRON, SPACE-FILLING POLYHEDRON

Schnirelmann Constant

The constant s_0 in SCHNIRELMANN'S THEOREM.

see also SCHNIRELMANN'S THEOREM

Schnirelmann Density

The Schnirelmann density of a sequence of natural numbers is the greatest lower bound of the fractions $A(n)/n$ where $A(n)$ is the number of terms in the sequence $\leq n$.

References

Khinchin, A. Y. "The Landau-Schnirelmann Hypothesis and Mann's Theorem." Ch. 2 in *Three Pearls of Number Theory.* New York: Dover, pp. 18–36, 1998.

Schnirelmann's Theorem

There exists a POSITIVE INTEGER s such that every sufficiently large INTEGER is the sum of at most s PRIMES. It follows that there exists a POSITIVE INTEGER $s_0 \geq s$ such that every INTEGER > 1 is a sum of at most s_0 PRIMES, where s_0 is the SCHNIRELMANN CONSTANT. The best current estimate is $s_0 = 19$.

see also PRIME NUMBER, SCHNIRELMANN DENSITY, WARING'S PROBLEM

References

Khinchin, A. Y. "The Landau-Schnirelmann Hypothesis and Mann's Theorem." Ch. 2 in *Three Pearls of Number Theory.* New York: Dover, pp. 18–36, 1998.

Schoenemann's Theorem

If the integral COEFFICIENTS $C_0, C_1, \ldots, C_{N-1}$ of the POLYNOMIAL

$$f(x) = C_0 + C_1 x + C_2 x^2 + \ldots + C_{N-1} x^{N-1} + x^N$$

are divisible by a PRIME NUMBER p, while the free term C_0 is not divisible by p^2, then $f(x)$ is irreducible in the natural rationality domain.

see also ABEL'S IRREDUCIBILITY THEOREM, ABEL'S LEMMA, GAUSS'S POLYNOMIAL THEOREM, KRONECKER'S POLYNOMIAL THEOREM

References

Dörrie, H. *100 Great Problems of Elementary Mathematics: Their History and Solutions.* New York: Dover, p. 118, 1965.

Scholz Conjecture

Let the minimal length of an ADDITION CHAIN for a number n be denoted $l(n)$. Then the Scholz conjecture states that

$$l(2^n - 1) \leq n - 1 + l(n).$$

The conjecture has been proven for a variety of special cases but not in general.

see also ADDITION CHAIN

References

Guy, R. K. *Unsolved Problems in Number Theory, 2nd ed.* New York: Springer-Verlag, p. 111, 1994.

Schönflies Symbol

One of the set of symbols C_i, C_s, C_1, C_2, C_3, C_4, C_5, C_6, C_7, C_8, C_{2h}, C_{3h}, C_{4h}, C_{5h}, C_{6h}, C_{2v}, C_{3v}, C_{4v}, C_{5v}, C_{6v}, $C_{\infty v}$, D_2, D_3, D_4, D_5, D_6, D_{2h}, D_{3h}, D_{4h}, D_{5h}, D_{6h}, D_{8h}, $D_{\infty h}$, D_{2d}, D_{3d}, D_{4d}, D_{5d}, D_{6d}, I, I_h, O, O_h, S_4, S_6, S_8, T, T_d, and T_h used to identify crystallographic symmetry GROUPS.

Cotton (1990), gives a table showing the translations between Schönflies symbols and HERMANN-MAUGUIN SYMBOLS. Some of the Schönflies symbols denote different sets of symmetry operations but correspond to the same abstract GROUP and so have the same CHARACTER TABLE.

see also CHARACTER TABLE, HERMANN-MAUGUIN SYMBOL, POINT GROUPS, SPACE GROUPS, SYMMETRY OPERATION

References

Cotton, F. A. *Chemical Applications of Group Theory, 3rd ed.* New York: Wiley, p. 379, 1990.

Schönflies Theorem

If J is a simple closed curve in $\mathbb{R}^2$, the closure of one of the components of $\mathbb{R}^2 - J$ is HOMEOMORPHIC with the unit 2-BALL. This theorem may be proved using the RIEMANN MAPPING THEOREM, but the easiest proof is via MORSE THEORY.

The generalization to n-D is called MAZUR'S THEOREM. It follows from the Schönflies theorem that any two KNOTS of $\mathbb{S}^1$ in $\mathbb{S}^2$ or $\mathbb{R}^2$ are equivalent.

see also JORDAN CURVE THEOREM, MAZUR'S THEOREM, RIEMANN MAPPING THEOREM

References

Rolfsen, D. *Knots and Links.* Wilmington, DE: Publish or Perish Press, p. 9, 1976.

Thomassen, C. "The Jordan-Schönflies Theorem and the Classification of Surfaces." *Amer. Math. Monthly* **99**, 116–130, 1992.

Schoolgirl Problem

see KIRKMAN'S SCHOOLGIRL PROBLEM

Schoute Coaxal System

The CIRCUMCIRCLE, BROCARD CIRCLE, LEMOINE LINE, and ISODYNAMIC POINTS belong to a COAXAL SYSTEM orthogonal to the the APOLLONIUS CIRCLES, called the Schoute coaxal system. In general, there are 12 points whose PEDAL TRIANGLES with regard to a given TRIANGLE have a given form. They lie six by six on two CIRCLES of the Schoute coaxal system.

References

Johnson, R. A. *Modern Geometry: An Elementary Treatise on the Geometry of the Triangle and the Circle.* Boston, MA: Houghton Mifflin, pp. 297–299, 1929.

Schoute's Theorem

In any TRIANGLE, the LOCUS of a point whose PEDAL TRIANGLE has a constant BROCARD ANGLE and is described in a given direction is a CIRCLE of the SCHOUTE COAXAL SYSTEM.

References

Johnson, R. A. *Modern Geometry: An Elementary Treatise on the Geometry of the Triangle and the Circle.* Boston, MA: Houghton Mifflin, pp. 297–299, 1929.

Schoute, P. H. *Proc. Amsterdam Acad.,* 39–62, 1887–1888.

Schrage's Algorithm

An algorithm for multiplying two 32-bit integers modulo a 32-bit constant without using any intermediates larger than 32 bits. It is also useful in certain types of RANDOM NUMBER generators.

References

Bratley, P.; Fox, B. L.; and Schrage, E. L. *A Guide to Simulation, 2nd ed.* New York: Springer-Verlag, 1996.

Press, W. H.; Flannery, B. P.; Teukolsky, S. A.; and Vetterling, W. T. "Random Numbers." Ch. 7 in *Numerical Recipes in FORTRAN: The Art of Scientific Computing, 2nd ed.* Cambridge, England: Cambridge University Press, p. 269, 1992.

Schrage, L. "A More Portable Fortran Random Number Generator." *ACM Trans. Math. Software* **5**, 132–138, 1979.

Schröder-Bernstein Theorem

The Schröder-Bernstein theorem for numbers states that if

$$n \le m \le n,$$

then $m = n$. For SETS, the theorem states that if there are INJECTIONS of the SET A into the SET B and of B into A, then there is a BIJECTIVE correspondence between A and B (i.e., they are EQUIPOLLENT).

see also BIJECTION, EQUIPOLLENT, INJECTION

Schröder's Equation

$$f(\lambda z) = R(z),$$

where $R(z) = \lambda x + a_2 x^2 + \ldots$, $\lambda \equiv R'(0)$, $|\lambda| = 1$, and $\lambda^n \ne 1$ for all $n \in \mathbb{N}$.

Schröder's Method

Two families of equations used to find roots of nonlinear functions of a single variable. The "B" family is more robust and can be used in the neighborhood of degenerate multiple roots while still providing a guaranteed convergence rate. Almost all other root-finding methods can be considered as special cases of Schröder's method. Householder humorously claimed that papers on root-finding could be evaluated quickly by looking for a citation of Schröder's paper; if the reference were missing, the paper probably consisted of a rediscovery of a result due to Schröder (Stewart 1993).

One version of the "A" method is obtained by applying NEWTON'S METHOD to f/f',

$$x_{n+1} = x_n - \frac{f(x_n)f'(x_n)}{[f'(x_n)]^2 - f(x_n)f''(x_n)}$$

(Scavo and Thoo 1995).

see also NEWTON'S METHOD

References

Householder, A. S. *The Numerical Treatment of a Single Nonlinear Equation.* New York: McGraw-Hill, 1970.

Scavo, T. R. and Thoo, J. B. "On the Geometry of Halley's Method." *Amer. Math. Monthly* **102**, 417–426, 1995.

Schröder, E. "Über unendlich viele Algorithmen zur Auflösung der Gleichungen." *Math. Ann.* **2**, 317–365, 1870.

Stewart, G. W. "On Infinitely Many Algorithms for Solving Equations." English translation of Schröder's original paper. College Park, MD: University of Maryland, Institute for Advanced Computer Studies, Department of Computer Science, 1993. `ftp://thales.cs.umd.edu/pub/reports/imase.ps`.

Schröder Number

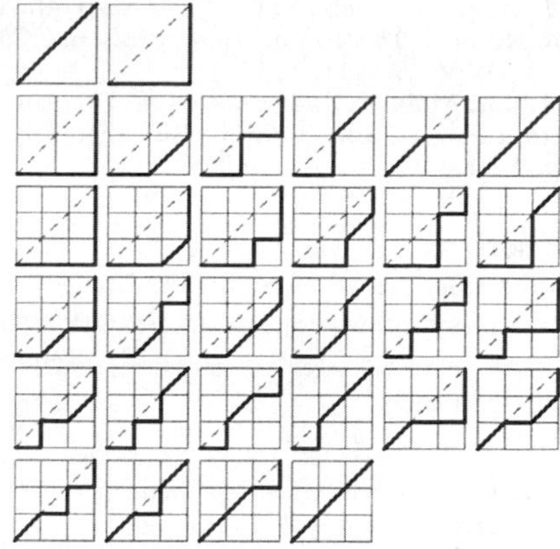

The Schröder number S_n is the number of LATTICE PATHS in the Cartesian plane that start at (0, 0), end at (n, n), contain no points above the line $y = x$, and are composed only of steps (0, 1), (1, 0), and (1, 1), i.e., $\rightarrow$, $\uparrow$, and $\nearrow$. The diagrams illustrating the paths generating S_1, S_2, and S_3 are illustrated above. The numbers S_n are given by the RECURRENCE RELATION

$$S_n = S_{n-1} + \sum_{k=0}^{n-1} S_k S_{n-1-k},$$

where $S_0 = 1$, and the first few are 2, 6, 22, 90, ... (Sloane's A006318). The Schröder Numbers bear the same relation to the DELANNOY NUMBERS as the CATALAN NUMBERS do to the BINOMIAL COEFFICIENTS.

see also BINOMIAL COEFFICIENT, CATALAN NUMBER, DELANNOY NUMBER, LATTICE PATH, MOTZKIN NUMBER, p-GOOD PATH

References

Sloane, N. J. A. Sequence A006318/M1659 in "An On-Line Version of the Encyclopedia of Integer Sequences."

Schroeder Stairs

see PENROSE STAIRWAY

Schröter's Formula

Let a general THETA FUNCTION be defined as

$$T(x, q) \equiv \sum_{n=-\infty}^{\infty} x^n q^{n^2},$$

then

$$T(x, q^a)T(x, q^b) = \sum_{k=0}^{a+b-1} y^k q^{bk^2} T(xyq^{2bk}, q^{a+b}) T(y^q x^{-b} q^{2abk}, q^{ab(1+b)}).$$

see also BLECKSMITH-BRILLHART-GERST THEOREM, JACOBI TRIPLE PRODUCT, RAMANUJAN THETA FUNCTIONS

References

Borwein, J. M. and Borwein, P. B. *Pi & the AGM: A Study in Analytic Number Theory and Computational Complexity.* New York: Wiley, p. 111, 1987.

Tannery, J. and Molk, J. *Elements de la Théorie des Fonctions Elliptiques, 4 vols.* Paris: Gauthier-Villars et fils, 1893–1902.

Schur Algebra

An Auslander algebra which connects the representation theories of the symmetric group of PERMUTATIONS and the GENERAL LINEAR GROUP $GL(n,\mathbb{C})$. Schur algebras are "quasihereditary."

References

Martin, S. *Schur Algebras and Representation Theory.* New York: Cambridge University Press, 1993.

Schur Functor

A FUNCTOR which defines an equivalence of module CATEGORIES.

References

Martin, S. *Schur Algebras and Representation Theory.* New York: Cambridge University Press, 1993.

Schur's Inequalities

Let $\mathsf{A} = a_{ij}$ be an $n \times n$ MATRIX with COMPLEX (or REAL) entries and EIGENVALUES λ_1, λ_2, ..., λ_n, then

$$\sum_{i=1}^n |\lambda_i|^2 \leq \sum_{i,j=1}^n |a_{ij}|^2$$

$$\sum_{i=1}^n |\Re[\lambda_i]|^2 \leq \sum_{i,j=1}^n \left|\frac{a_{ij}+a_{ji}^*}{2}\right|^2$$

$$\sum_{i=1}^n |\Im[\lambda_i]|^2 \leq \sum_{i,j=1}^n \left|\frac{a_{ij}-a_{ji}^*}{2}\right|^2.$$

References

Gradshteyn, I. S. and Ryzhik, I. M. *Tables of Integrals, Series, and Products, 5th ed.* San Diego, CA: Academic Press, p. 1120, 1979.

Schur's Lemma

For each $k \in \mathbb{N}$ there exists a largest INTEGER $s(k)$ (known as the SCHUR NUMBER) such that no matter how the set of INTEGERS less than $\lfloor n!e \rfloor$ (where $\lfloor x \rfloor$ is the FLOOR FUNCTION) is partitioned into k classes, one class must contain INTEGERS x, y, z such that $x+y=z$, where x and y are not necessarily distinct. The upper bound has since been slightly improved to $\lfloor n!(e-1/24) \rfloor$.

see also COMBINATORICS, SCHUR NUMBER, SCHUR'S THEOREM

References

Guy, R. K. "Schur's Problem. Partitioning Integers into Sum-Free Classes" and "The Modular Version of Schur's Problem." §E11 and E12 in *Unsolved Problems in Number Theory, 2nd ed.* New York: Springer-Verlag, pp. 209–212, 1994.

Schur Matrix

The $p \times p$ SQUARE MATRIX formed by setting $s_{ij} = \xi^{ij}$, where ξ is an pth ROOT OF UNITY. The Schur matrix has a particularly simple DETERMINANT given by

$$\det \mathsf{S} = \epsilon_p p^{p/2},$$

where p is an ODD PRIME and

$$\epsilon_p \equiv \begin{cases} 1 & \text{if } p \equiv 1 \pmod 4 \\ i & \text{if } p \equiv 3 \pmod 4. \end{cases}$$

This determinant has been used to prove the QUADRATIC RECIPROCITY LAW (Landau 1958, Vardi 1991). The ABSOLUTE VALUES of the PERMANENTS of the Schur matrix of order $2p+1$ are given by 1, 3, 5, 105, 81, 6765, ... (Sloane's A003112, Vardi 1991).

Denote the Schur matrix S_p with the first row and first row column omitted by S'_p. Then

$$\operatorname{perm} \mathsf{S}_p = p \operatorname{perm} \mathsf{S}'_p,$$

where perm denoted the PERMANENT (Vardi 1991).

References

Graham, R. L. and Lehmer, D. H. "On the Permanent of Schur's Matrix." *J. Austral. Math. Soc.* **21**, 487–497, 1976.

Landau, E. *Elementary Number Theory.* New York: Chelsea, 1958.

Sloane, N. J. A. Sequence A003112/M2509 in "An On-Line Version of the Encyclopedia of Integer Sequences."

Vardi, I. *Computational Recreations in Mathematica.* Reading, MA: Addison-Wesley, pp. 119–122 and 124, 1991.

Schur Multiplier

A property of FINITE SIMPLE GROUPS which is known for all such GROUPS.

see also FINITE GROUP, SIMPLE GROUP

Schur Number

The Schur numbers are the numbers in the partitioning of a set which are guaranteed to exist by SCHUR'S LEMMA. Schur numbers satisfy the inequality

$$s(k) \geq c(315)^{k/5}$$

for $k > 5$ and some constant c. SCHUR'S THEOREM also shows that

$$s(n) \leq R(n),$$

where $R(n)$ is a RAMSEY NUMBER. The first few Schur numbers are 1, 4, 13, 44, ($\geq$ 157), ... (Sloane's A045652).

see also RAMSEY NUMBER, RAMSEY'S THEOREM, SCHUR'S LEMMA, SCHUR'S THEOREM

References

Frederickson, H. "Schur Numbers and the Ramsey Numbers $N(3,3,\ldots,3;2)$." *J. Combin. Theory Ser. A* **27**, 376–377, 1979.

Guy, R. K. "Schur's Problem. Partitioning Integers into Sum-Free Classes" and "The Modular Version of Schur's Problem." §E11 and E12 in *Unsolved Problems in Number Theory, 2nd ed.* New York: Springer-Verlag, pp. 209–212, 1994.

Sloane, N. J. A. Sequence A045652 in "An On-Line Version of the Encyclopedia of Integer Sequences."

Schur's Problem

see SCHUR'S LEMMA

Schur's Representation Lemma

If π on V and π' on V' are irreducible representations and $E : V \mapsto V'$ is a linear map such that $\pi'(g)E = E\pi(g)$ for all $g \in$ and group G, then $E = 0$ or E is invertible. Furthermore, if $V = V'$, then E is a SCALAR.

References

Knapp, A. W. "Group Representations and Harmonic Analysis, Part II." *Not. Amer. Math. Soc.* **43**, 537–549, 1996.

Schur's Theorem

As shown by Schur in 1916, the SCHUR NUMBER $s(n)$ satisfies

$$s(n) \leq R(n)$$

for $n = 1, 2, \ldots$, where $R(n)$ is a RAMSEY NUMBER.

see also RAMSEY NUMBER, SCHUR'S LEMMA, SCHUR NUMBER

Schwarz's Inequality

$$|\langle\psi_1|\psi_2\rangle|^2 \leq \langle\psi_1|\psi_1\rangle\langle\psi_2|\psi_2\rangle. \quad (1)$$

Written out explicitly

$$\left[\int_a^b \psi_1(x)\psi_2(x)\,dx\right]^2 \leq \int_a^b [\psi_1(x)]^2\,dx \int_a^b [\psi_2(x)]^2\,dx, \quad (2)$$

with equality IFF $g(x) = \alpha f(x)$ with α a constant. To derive, let $\psi(x)$ be a COMPLEX function and λ a COMPLEX constant such that $\psi(x) \equiv f(x) + \lambda g(x)$ for some f and g. Then

$$\int \psi^*\psi\,dx = \int f^* f\,dx + \lambda \int f^* g\,dx + \lambda^* \int g^* f\,dx + \lambda\lambda^* \int g^* g\,dx \geq 0, \quad (3)$$

with equality when $\psi(x) = 0$. Now, note that λ and λ^* are LINEARLY INDEPENDENT (they are ORTHOGONAL), so differentiate with respect to one of them (say λ^*) and set to zero to minimize $\int \psi^*\psi\,dx$.

$$\frac{\partial}{\partial} \int \psi^*\psi\,dx = \int g^* f\,dx + \lambda \int g^* g\,dx = 0 \quad (4)$$

$$\lambda = -\frac{\int g^* f\,dx}{\int g^* g\,dx}, \quad (5)$$

which means that

$$\lambda^* = -\frac{\int f^* g\,dx}{\int g^* g\,dx}. \quad (6)$$

Plugging back in,

$$\int \psi^*\psi\,dx = \int f^* f\,dx - \frac{\int g^* f\,dx}{\int g^* g\,dx}\int f^* g\,dx - \frac{\int f^* g\,dx}{\int g^* g\,dx}\int g^* f\,dx + \frac{\int g^* f\,dx \int f^* g\,dx}{(\int g^* g\,dx)^2}\int g^* g\,dx \geq 0. \quad (7)$$

Multiplying through by $\int g^* g\,dx$ gives

$$\int f^* f\,dx \int g^* g\,dx - \int g^* f\,dx \int f^* g\,dx - \int f^* g\,dx \int g^* f\,dx + \int g^* f\,dx \int f^* g\,dx \geq 0 \quad (8)$$

$$\int g^* f\,dx \int f^* g\,dx \leq \int f^* f\,dx \int g^* g\,dx \quad (9)$$

$$\left|\int g^* f\,dx\right| = \left|\int f^* g\,dx\right| \leq \int f^* f\,dx \int g^* g\,dx \quad (10)$$

or

$$|\langle f|g\rangle|^2 \leq \langle f|f\rangle\langle g|g\rangle. \quad (11)$$

BESSEL'S INEQUALITY can be derived from this.

References

Abramowitz, M. and Stegun, C. A. (Eds.). *Handbook of Mathematical Functions with Formulas, Graphs, and Mathematical Tables, 9th printing.* New York: Dover, p. 11, 1972.

Arfken, G. *Mathematical Methods for Physicists, 3rd ed.* Orlando, FL: Academic Press, pp. 527–529, 1985.

Schwarz-Pick Lemma

If f is an analytic map of the DISK $\mathbb{D}$ into $\mathbb{D}$ and f preserves the hyperbolic distance between any two points, then f is a disk map and preserves all distance.

References

Busemann, H. *The Geometry of Geodesics.* New York: Academic Press, p. 41, 1955.

Schwarz Reflection Principle

Let

$$g(z) \equiv \sum_{n=0}^{\infty}(z-z_0)^n \frac{f^{(n)}(z_0)}{n!}, \quad (1)$$

then

$$g^*(z) = \left[\sum_{n=0}^{\infty}(z-z_0)^n \frac{f^{(n)}(z_0)}{n!}\right]^* = \sum_{n=0}^{\infty}(z^*-z_0{}^*)n\frac{f^{(n)}(z_0{}^*)}{n!}. \quad (2)$$

If z_0 is pure real, then $z_0 = z_0{}^*$, so

$$g^*(z) = \sum_{n=0}^{\infty}(z^*-z_0)^n \frac{f^{(n)}(z_0)}{n!} = g(z^*). \quad (3)$$

Therefore, if a function $f(z)$ is ANALYTIC over some region including the REAL LINE and $f(z)$ is REAL when z is real, then $f^*(z) = f(z^*)$.

Schwarz Triangle

The Schwarz triangles are SPHERICAL TRIANGLES which, by repeated reflection in their indices, lead to a set of congruent SPHERICAL TRIANGLES covering the SPHERE a finite number of times.

Schwarz triangles are specified by triples of numbers (p,q,r). There are four "families" of Schwarz triangles, and the largest triangles from each of these families are

$$(2\ 2\ n'), (\tfrac{3}{2}\ \tfrac{3}{2}\ \tfrac{3}{2}), (\tfrac{3}{2}\ \tfrac{4}{3}\ \tfrac{4}{3}), (\tfrac{5}{4}\ \tfrac{5}{4}\ \tfrac{5}{4}).$$

The others can be derived from

$$(p\ q\ r) = (p\ x\ r_1) + (x\ q\ r_2),$$

where

$$\frac{1}{r_1} + \frac{1}{r_2} = \frac{1}{r}$$

and

$$\cos\left(\frac{\pi}{x}\right) = -\cos\left(\frac{\pi}{x'}\right) = \frac{\cos\left(\frac{\pi}{q}\right)\sin\left(\frac{\pi}{r_1}\right) - \cos\left(\frac{\pi}{p}\right)\sin\left(\frac{\pi}{r_2}\right)}{\sin\left(\frac{\pi}{r}\right)}.$$

see also COLUNAR TRIANGLE, SPHERICAL TRIANGLE

References

Coxeter, H. S. M. *Regular Polytopes, 3rd ed.* New York: Dover, pp. 112–113 and 296, 1973.

Schwarz, H. A. "Zur Theorie der hypergeometrischen Reihe." *J. reine angew. Math.* **75**, 292–335, 1873.

Schwarz's Triangle Problem

see FAGNANO'S PROBLEM

Schwarzian Derivative

The Schwarzian derivative is defined by

$$D_{\text{Schwarzian}} \equiv \frac{f'''(x)}{f'(x)} - \frac{3}{2}\left[\frac{f''(x)}{f'(x)}\right]^2.$$

The FEIGENBAUM CONSTANT is universal for 1-D MAPS if its Schwarzian derivative is NEGATIVE in the bounded interval (Tabor 1989, p. 220).

see also FEIGENBAUM CONSTANT

References

Tabor, M. *Chaos and Integrability in Nonlinear Dynamics: An Introduction.* New York: Wiley, 1989.

Schwenk's Formula

Let $R+B$ be the number of MONOCHROMATIC FORCED TRIANGLES (where R and B are the number of red and blue TRIANGLES) in an EXTREMAL GRAPH. Then

$$R+B = \binom{n}{3} - \left\lfloor \tfrac{1}{2}n \left\lfloor \tfrac{1}{4}(n-1)^2 \right\rfloor \right\rfloor,$$

where $\binom{n}{k}$ is a BINOMIAL COEFFICIENT and $\lfloor x \rfloor$ is the FLOOR FUNCTION (Schwenk 1972).

see also EXTREMAL GRAPH, MONOCHROMATIC FORCED TRIANGLE

References

Schwenk, A. J. "Acquaintance Party Problem." *Amer. Math. Monthly* **79**, 1113–1117, 1972.

Scientific Notation

Scientific notation is the expression of a number n in the form $a \times 10^p$, where

$$p \equiv \lfloor \log_{10} |n| \rfloor$$

is the FLOOR of the base-10 LOGARITHM of n (the "order of magnitude"), and

$$a \equiv \frac{n}{10^p}$$

is a REAL NUMBER satisfying $1 \leq |a| < 10$. For example, in scientific notation, the number $n = 101,325$ has order of magnitude

$$p = \lfloor \log_{10} 101,325 \rfloor = \lfloor 5.00572 \rfloor = 5,$$

so n would be written 1.01325×10^5. The special case of 0 does not have a unique representation in scientific notation, i.e., $0 = 0 \times 10^0 = 0 \times 10^1 = \ldots$.

see also CHARACTERISTIC (REAL NUMBER), FIGURES, MANTISSA, SIGNIFICANT FIGURES

Score Sequence

The score sequence of a TOURNAMENT is a monotonic nondecreasing sequence of the OUTDEGREES of the VERTICES. The score sequences for $n = 1, 2, \ldots$ are 1, 1, 2,4, 9, 22, 59, 167, ... (Sloane's A000571).

see also TOURNAMENT

References

Ruskey, F. "Information on Score Sequences." `http://sue.csc.uvic.ca/~cos/inf/nump/ScoreSequence.html`.

Ruskey, F.; Cohen, R.; Eades, P.; and Scott, A. "Alley CATs in Search of Good Homes." *Congres. Numer.* **102**, 97–110, 1994.

Sloane, N. J. A. Sequence A000571/M1189 in "An On-Line Version of the Encyclopedia of Integer Sequences."

Screw

A TRANSLATION along a straight line L and a ROTATION about L such that the angle of ROTATION is proportional to the TRANSLATION at each instant. Also known as a TWIST.

see also DINI'S SURFACE, HELICOID, ROTATION, SCREW THEOREM, SEASHELL, TRANSLATION

Screw Theorem

Any motion of a rigid body in space at every instant is a SCREW motion. This theorem was proved by Mozzi and Cauchy.

see also SCREW

Scruple

An archaic UNIT FRACTION variously defined as 1/200 (of an hour), 1/10 or 1/12 (of an inch), 1/12 (of a celestial body's angular diameter), or 1/60 (of an hour or DEGREE).

see also CALCUS, UNCIA

Sea Horse Valley

A portion of the MANDELBROT SET centered around $-1.25 + 0.047i$ with width approximately $0.009 + 0.005i$.

see also MANDELBROT SET

Searching

Searching refers to locating a given element or an element satisfying certain conditions from some (usually ordered or partially ordered) table, list, TREE, etc.

see also SORTING, TABU SEARCH, TREE SEARCHING

References

Knuth, D. E. *The Art of Computer Programming, 2nd ed, Vol. 3: Sorting and Searching.* Reading, MA: Addison-Wesley, 1973.

Press, W. H.; Flannery, B. P.; Teukolsky, S. A.; and Vetterling, W. T. "How to Search an Ordered Table." §3.4 in *Numerical Recipes in FORTRAN: The Art of Scientific Computing, 2nd ed.* Cambridge, England: Cambridge University Press, pp. 110–113, 1992.

Search Tree

see TREE SEARCHING

Seashell

see CONICAL SPIRAL

Secant

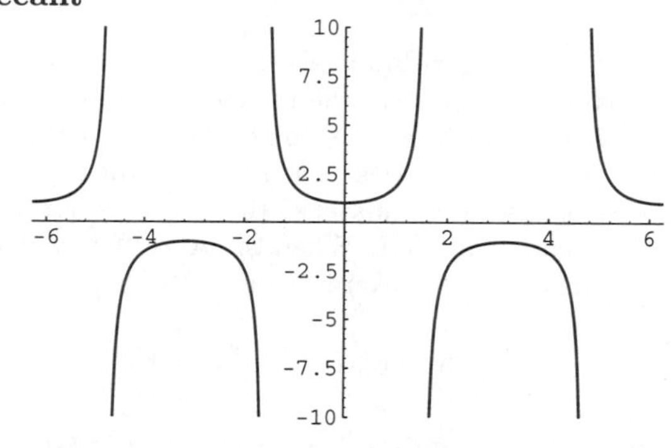

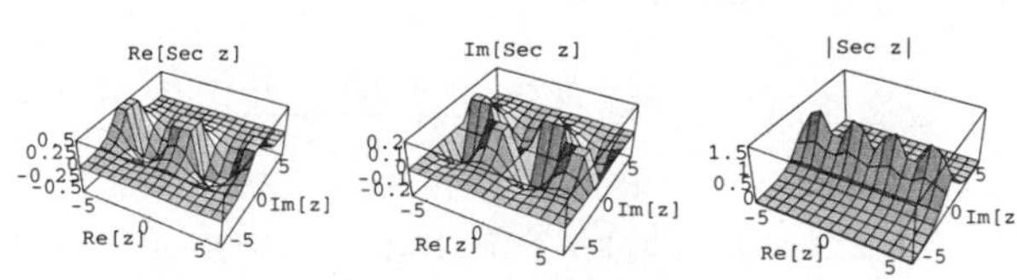

The function defined by $\sec x \equiv 1/\cos x$, where $\cos x$ is the COSINE. The MACLAURIN SERIES of the secant is

$$\begin{aligned}\sec x &= \frac{(-1)^n E_{2n}}{(2n)!} x^{2n} \\ &= 1 + \tfrac{1}{2}x^2 + \tfrac{5}{24}x^4 + \tfrac{61}{720}x^6 + \tfrac{277}{8064}x^8 + \ldots,\end{aligned}$$

where E_{2n} is an EULER NUMBER.

see also ALTERNATING PERMUTATION, COSECANT, COSINE, EULER NUMBER, EXSECANT, INVERSE SECANT

References

Abramowitz, M. and Stegun, C. A. (Eds.). "Circular Functions." §4.3 in *Handbook of Mathematical Functions with Formulas, Graphs, and Mathematical Tables, 9th printing.* New York: Dover, pp. 71–79, 1972.

Spanier, J. and Oldham, K. B. "The Secant sec(x) and Cosecant csc(x) Functions." Ch. 33 in *An Atlas of Functions.* Washington, DC: Hemisphere, pp. 311–318, 1987.

Secant Line

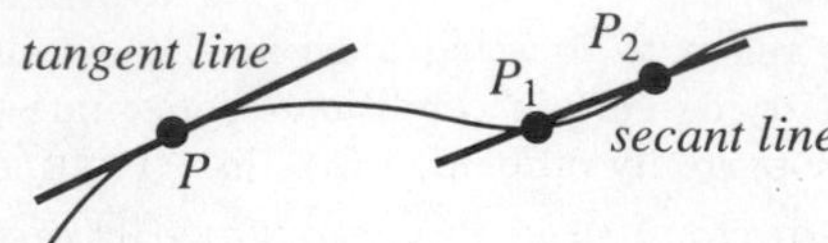

A line joining two points of a curve. In abstract mathematics, the points which a secant line connects can be either REAL or COMPLEX CONJUGATE IMAGINARY.

see also BITANGENT, TANGENT LINE, TRANSVERSAL LINE

Secant Method

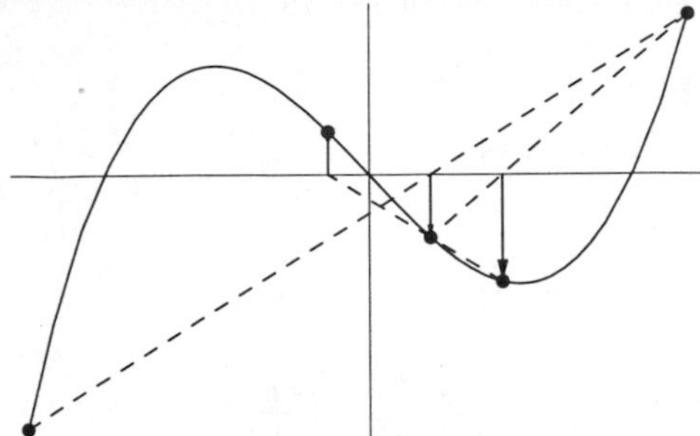

A ROOT-finding algorithm which assumes a function to be approximately linear in the region of interest. Each improvement is taken as the point where the approximating line crosses the axis. The secant method retains only the most recent estimate, so the root does not necessarily remain bracketed. When the ALGORITHM does converge, its order of convergence is

$$\lim_{k\to\infty} |\epsilon_{k+1}| \approx C|\epsilon|^{\phi}, \tag{1}$$

where C is a constant and ϕ is the GOLDEN MEAN.

$$f'(x_{n-1}) \approx \frac{f(x_{n-1}) - f(x_{n-2})}{x_{n-1} - x_{n-2}} \tag{2}$$

$$f(x_n) \approx f(x_{n-1}) + f'(x_n)(x_n - x_{n-1}) = 0 \tag{3}$$

$$f(x_{n-1}) + \frac{f(x_{n-1}) - f(x_{n-2})}{x_{n-1} - x_{n-2}}(x_n - x_{n-1}) = 0, \tag{4}$$

so

$$x_n = x_{n-1} - \frac{f(x_{n-1})(x_{n-1} - x_{n-2})}{f(x_{n-1}) - f(x_{n-2})}. \tag{5}$$

see also FALSE POSITION METHOD

References

Press, W. H.; Flannery, B. P.; Teukolsky, S. A.; and Vetterling, W. T. "Secant Method, False Position Method, and Ridders' Method." §9.2 in *Numerical Recipes in FORTRAN: The Art of Scientific Computing, 2nd ed.* Cambridge, England: Cambridge University Press, pp. 347–352, 1992.

Secant Number

A number, more commonly called an EULER NUMBER, giving the number of ODD ALTERNATING PERMUTATIONS. The term ZAG NUMBER is sometimes also used.

see also ALTERNATING PERMUTATION, EULER NUMBER, EULER ZIGZAG NUMBER, TANGENT NUMBER

Sech

see HYPERBOLIC SECANT

Second

see ARC SECOND

Second Curvature

see TORSION (DIFFERENTIAL GEOMETRY)

Second Derivative Test

Suppose $f(x)$ is a FUNCTION of x which is twice DIFFERENTIABLE at a STATIONARY POINT x_0.

1. If $f''(x_0) > 0$, then f has a RELATIVE MINIMUM at x_0.
2. If $f''(x_0) < 0$, then f has a RELATIVE MAXIMUM at x_0.

The EXTREMUM TEST gives slightly more general conditions under which functions with $f''(x_0) = 0$.

If $f(x,y)$ is a 2-D FUNCTION which has a RELATIVE EXTREMUM at a point (x_0, y_0) and has CONTINUOUS PARTIAL DERIVATIVES at this point, then $f_x(x_0, y_0) = 0$ and $f_y(x_0, y_0) = 0$. The second PARTIAL DERIVATIVES test classifies the point as a MAXIMUM or MINIMUM. Define the DISCRIMINANT as

$$D \equiv f_{xx}f_{yy} - f_{xy}f_{yx} = f_{xx}f_{yy} - {f_{xy}}^2.$$

1. If $D > 0$, $f_{xx}(x_0, y_0) > 0$ and $f_{xx}(x_0, y_0) + f_{yy}(x_0, y_0) > 0$, the point is a RELATIVE MINIMUM.
2. If $D > 0$, $f_{xx}(x_0, y_0) < 0$, and $f_{xx}(x_0, y_0) + f_{yy}(x_0, y_0) < 0$, the point is a RELATIVE MAXIMUM.
3. If $D < 0$, the point is a SADDLE POINT.
4. If $D = 0$, higher order tests must be used.

see also DISCRIMINANT (SECOND DERIVATIVE TEST), EXTREMUM, EXTREMUM TEST, FIRST DERIVATIVE TEST, GLOBAL MAXIMUM, GLOBAL MINIMUM, HESSIAN DETERMINANT, MAXIMUM, MINIMUM, RELATIVE MAXIMUM, RELATIVE MINIMUM, SADDLE POINT (FUNCTION)

References

Abramowitz, M. and Stegun, C. A. (Eds.). *Handbook of Mathematical Functions with Formulas, Graphs, and Mathematical Tables, 9th printing.* New York: Dover, p. 14, 1972.

Second Fundamental Tensor

see WEINGARTEN MAP

Section (Graph)

A section of a GRAPH is obtained by finding its intersection with a PLANE.

Section (Pencil)

The lines of a PENCIL joining the points of a RANGE to another POINT.

see also PENCIL, RANGE (LINE SEGMENT)

Section (Tangent Bundle)

A VECTOR FIELD is a section of its TANGENT BUNDLE, meaning that to every point x in a MANIFOLD M, a VECTOR $X(x) \in T_x M$ is associated, where T_x is the TANGENT SPACE.

see also TANGENT BUNDLE, TANGENT SPACE

Sectional Curvature

The mathematical object κ which controls the rate of geodesic deviation.

see also BISHOP'S INEQUALITY, CHEEGER'S FINITENESS THEOREM, GEODESIC

Sector

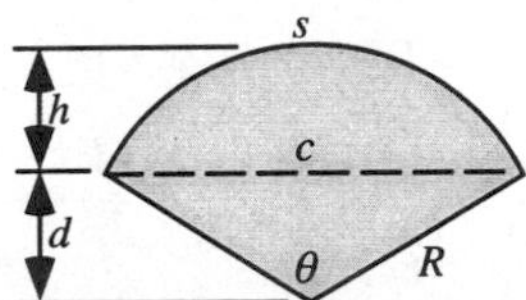

A WEDGE obtained by taking a portion of a CIRCLE with CENTRAL ANGLE $\theta < \pi$ radians (180°), illustrated above as the shaded region. A sector of π radians would be a SEMICIRCLE. Let R be the radius of the CIRCLE, c the CHORD length, s the ARC LENGTH, h the height of the arced portion, and d the height of the triangular portion. Then

$$R = h + d \tag{1}$$

$$s = R\theta \tag{2}$$

$$d = R\cos(\tfrac{1}{2}\theta) \tag{3}$$

$$= \tfrac{1}{2}c\cot(\tfrac{1}{2}\theta) \tag{4}$$

$$= \tfrac{1}{2}\sqrt{4R^2 - c^2} \tag{5}$$

$$c = 2R\sin(\tfrac{1}{2}\theta) \tag{6}$$

$$= 2d\tan(\tfrac{1}{2}\theta) \tag{7}$$

$$= 2\sqrt{R^2 - d^2} \tag{8}$$

$$= 2\sqrt{h(2R - h)}. \tag{9}$$

The ANGLE θ obeys the relationships

$$\theta = \frac{s}{R} = 2\cos^{-1}\left(\frac{d}{R}\right) = 2\tan^{-1}\left(\frac{c}{2d}\right) = 2\sin^{-1}\left(\frac{c}{2R}\right). \tag{10}$$

The AREA of the sector is

$$A = \tfrac{1}{2}Rs = \tfrac{1}{2}R^2\theta \tag{11}$$

(Beyer 1987).

see also CIRCLE-CIRCLE INTERSECTION, LENS, OBTUSE TRIANGLE, SEGMENT

References

Beyer, W. H. (Ed.) *CRC Standard Mathematical Tables, 28th ed.* Boca Raton, FL: CRC Press, p. 125, 1987.

Sectorial Harmonic

A SPHERICAL HARMONIC of the form

$$\sin(m\theta)P_m^m(\cos\phi)$$

or

$$\cos(m\theta)P_m^m(\cos\phi).$$

see also SPHERICAL HARMONIC

Secular Equation

see CHARACTERISTIC EQUATION

Seed

The initial number used as the starting point in a RANDOM NUMBER generating ALGORITHM.

Seed of Life

One of the beautiful arrangements of CIRCLES found at the Temple of Osiris at Abydos, Egypt (Rawles 1997). The CIRCLES are placed with 6-fold symmetry, forming a mesmerizing pattern of CIRCLES and LENSES.

see also CIRCLE, FIVE DISKS PROBLEM, FLOWER OF LIFE, VENN DIAGRAM

References

Rawles, B. *Sacred Geometry Design Sourcebook: Universal Dimensional Patterns.* Nevada City, CA: Elysian Pub., p. 15, 1997.

Weisstein, E. W. "Flower of Life." `http://www.astro.virginia.edu/~eww6n/math/notebooks/FlowerOfLife.m.`

Seek Time

see POINT-POINT DISTANCE—1-D

Segment

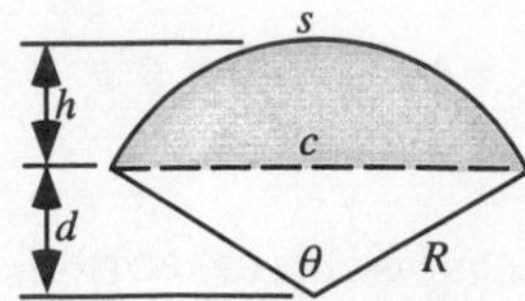

A portion of a CIRCLE whose upper boundary is a circular ARC and whose lower boundary is a CHORD making a CENTRAL ANGLE $\theta < \pi$ radians (180°), illustrated above as the shaded region. Let R be the radius of the CIRCLE, c the CHORD length, s the ARC LENGTH, h the height of the arced portion, and d the height of the triangular portion. Then

$$R = h + d \tag{1}$$
$$s = R\theta \tag{2}$$
$$d = R\cos(\tfrac{1}{2}\theta) \tag{3}$$
$$= \tfrac{1}{2}c\cot(\tfrac{1}{2}\theta) \tag{4}$$
$$= \tfrac{1}{2}\sqrt{4R^2 - c^2} \tag{5}$$
$$c = 2R\sin(\tfrac{1}{2}\theta) \tag{6}$$
$$= 2d\tan(\tfrac{1}{2}\theta) \tag{7}$$
$$= 2\sqrt{R^2 - d^2} \tag{8}$$
$$= 2\sqrt{h(2R - h)}. \tag{9}$$

The ANGLE θ obeys the relationships

$$\theta = \frac{s}{R} = 2\cos^{-1}\left(\frac{d}{R}\right) = 2\tan^{-1}\left(\frac{c}{2d}\right) = 2\sin^{-1}\left(\frac{c}{2R}\right). \tag{10}$$

The AREA of the segment is then

$$A = A_{\text{sector}} - A_{\text{isosceles triangle}} \tag{11}$$
$$= \tfrac{1}{2}R^2(\theta - \sin\theta) \tag{12}$$
$$= \tfrac{1}{2}(Rs - cd) \tag{13}$$
$$= R^2\cos^{-1}\left(\frac{d}{R}\right) - d\sqrt{R^2 - d^2} \tag{14}$$
$$= R^2\cos^{-1}\left(\frac{R-h}{R}\right) - (R-h)\sqrt{2Rh - h^2}, \tag{15}$$

where the formula for the ISOSCELES TRIANGLE in terms of the VERTEX angle has been used (Beyer 1987).

see also CHORD, CIRCLE-CIRCLE INTERSECTION, CYLINDRICAL SEGMENT, LENS, PARABOLIC SEGMENT, SAGITTA, SECTOR, SPHERICAL SEGMENT

References

Beyer, W. H. (Ed.) *CRC Standard Mathematical Tables, 28th ed.* Boca Raton, FL: CRC Press, p. 125, 1987.

Segmented Number

see PRIME NUMBER OF MEASUREMENT

Segner's Recurrence Formula

The recurrence FORMULA

$$E_n = E_2E_{n-1} + E_3E_{n-2} + \ldots + E_{n-1}E_2$$

which gives the solution to EULER'S POLYGON DIVISION PROBLEM.

see also CATALAN NUMBER, EULER'S POLYGON DIVISION PROBLEM

Segre's Theorem

For any REAL NUMBER $r \geq 0$, an IRRATIONAL number α can be approximated by infinitely many RATIONAL fractions p/q in such a way that

$$-\frac{1}{\sqrt{1+4r}\,q^2} < \frac{p}{q} - \alpha < \frac{r}{\sqrt{1+4r}\,q^2}.$$

If $r = 1$, this becomes HURWITZ'S IRRATIONAL NUMBER THEOREM.

see also HURWITZ'S IRRATIONAL NUMBER THEOREM

Seiberg-Witten Equations

$$D_A\psi = 0$$
$$F_A^+ = -\tau(\psi, \psi),$$

where τ is the sesquilinear map $\tau : W^+ \times W^+ \to \Lambda^+ \otimes \mathbb{C}$.

see also WITTEN'S EQUATIONS

References

Donaldson, S. K. "The Seiberg-Witten Equations and 4-Manifold Topology." *Bull. Amer. Math. Soc.* **33**, 45–70, 1996.

Morgan, J. W. *The Seiberg-Witten Equations and Applications to the Topology of Smooth Four-Manifolds.* Princeton, NJ: Princeton University Press, 1996.

Seiberg-Witten Invariants

see WITTEN'S EQUATIONS

Seidel-Entringer-Arnold Triangle

The NUMBER TRIANGLE consisting of the ENTRINGER NUMBERS $E_{n,k}$ arranged in "ox-plowing" order,

$$\begin{array}{c} E_{00} \\ E_{10} \to E_{11} \\ E_{22} \leftarrow E_{21} \leftarrow E_{20} \\ E_{30} \to E_{31} \to E_{32} \to E_{33} \\ E_{44} \leftarrow E_{43} \leftarrow E_{42} \leftarrow E_{41} \leftarrow E_{40} \end{array}$$

giving

$$\begin{array}{c} 1 \\ 0 \to 1 \\ 1 \leftarrow 1 \leftarrow 0 \\ 0 \to 1 \to 2 \to 2 \\ 5 \leftarrow 5 \leftarrow 4 \leftarrow 2 \leftarrow 0 \end{array}$$

see also BELL NUMBER, BOUSTROPHEDON TRANSFORM, CLARK'S TRIANGLE, ENTRINGER NUMBER, EULER'S TRIANGLE, LEIBNIZ HARMONIC TRIANGLE, NUMBER TRIANGLE, PASCAL'S TRIANGLE

References

Arnold, V. I. "Bernoulli-Euler Updown Numbers Associated with Function Singularities, Their Combinatorics, and Arithmetics." *Duke Math. J.* **63**, 537–555, 1991.

Arnold, V. I. "Snake Calculus and Combinatorics of Bernoulli, Euler, and Springer Numbers for Coxeter Groups." *Russian Math. Surveys* **47**, 3–45, 1992.

Conway, J. H. and Guy, R. K. In *The Book of Numbers.* New York: Springer-Verlag, 1996.

Dumont, D. "Further Triangles of Seidel-Arnold Type and Continued Fractions Related to Euler and Springer Numbers." *Adv. Appl. Math.* **16**, 275–296, 1995.

Entringer, R. C. "A Combinatorial Interpretation of the Euler and Bernoulli Numbers." *Nieuw. Arch. Wisk.* **14**, 241–246, 1966.

Millar, J.; Sloane, N. J. A.; and Young, N. E. "A New Operation on Sequences: The Boustrophedon Transform." *J. Combin. Th. Ser. A* **76**, 44–54, 1996.

Seidel, I. "Über eine einfache Entstehungsweise der Bernoullischen Zahlen und einiger verwandten Reihen." *Sitzungsber. Münch. Akad.* **4**, 157–187, 1877.

Seifert Circle

Eliminate each knot crossing by connecting each of the strands coming into the crossing to the adjacent strand leaving the crossing. The resulting strands no longer cross but form instead a set of nonintersecting CIRCLES called Seifert circles.

References

Adams, C. C. *The Knot Book: An Elementary Introduction to the Mathematical Theory of Knots.* New York: W. H. Freeman, p. 96, 1994.

Seifert Conjecture

Every smooth NONZERO VECTOR FIELD on the 3-SPHERE has at least one closed orbit. The conjecture was proposed in 1950, proved true for Hopf fibrations, but proved false in general by Kuperberg (1994).

References

Kuperberg, G. "A Volume-Preserving Counterexample to the Seifert Conjecture." *Comment. Math. Helv.* **71**, 70–97, 1996.

Kuperberg, G. and Kuperberg, K. "Generalized counterexamples to the Seifert Conjecture." *Ann. Math.* **143**, 547–576, 1996.

Kuperberg, G. and Kuperberg, K. "Generalized Counterexamples to the Seifert Conjecture." *Ann. Math.* **144**, 239–268, 1996.

Kuperberg, K. "A Smooth Counterexample to the Seifert Conjecture." *Ann. Math.* **140**, 723–732, 1994.

Seifert Form

For K a given KNOT in $\mathbb{S}^3$, choose a SEIFERT SURFACE M^2 in $\mathbb{S}^3$ for K and a bicollar $\hat{M} \times [-1,1]$ in $\mathbb{S}^3 - K$. If $x \in H_1(\hat{M})$ is represented by a 1-cycle in $\hat{M}$, let x^+ denote the homology cycle carried by $x \times 1$ in the bicollar. Similarly, let x^- denote $x \times -1$. The function $f : H_1(\hat{M}) \times H_1(\hat{M}) \to Z$ defined by

$$f(x,y) = \operatorname{lk}(x, y^+),$$

where lk denotes the LINKING NUMBER, is called a Seifert form for K.

see also SEIFERT MATRIX

References

Rolfsen, D. *Knots and Links.* Wilmington, DE: Publish or Perish Press, pp. 200–201, 1976.

Seifert Matrix

Given a SEIFERT FORM $f(x,y)$, choose a basis e_1, ..., e_{2g} for $H_1(\hat{M})$ as a Z-module so every element is uniquely expressible as

$$n_1 e_1 + \ldots + n_{2g} e_{2g}$$

with n_i integer, define the Seifert matrix V as the $2g \times 2g$ integral MATRIX with entries

$$v_{ij} = \operatorname{lk}(e_i, e_j^+).$$

The right-hand TREFOIL KNOT has Seifert matrix

$$V = \begin{bmatrix} -1 & 1 \\ 0 & -1 \end{bmatrix}.$$

A Seifert matrix is not a knot invariant, but it can be used to distinguish between different SEIFERT SURFACES for a given knot.

see also ALEXANDER MATRIX

References

Rolfsen, D. *Knots and Links.* Wilmington, DE: Publish or Perish Press, pp. 200–203, 1976.

Seifert's Spherical Spiral

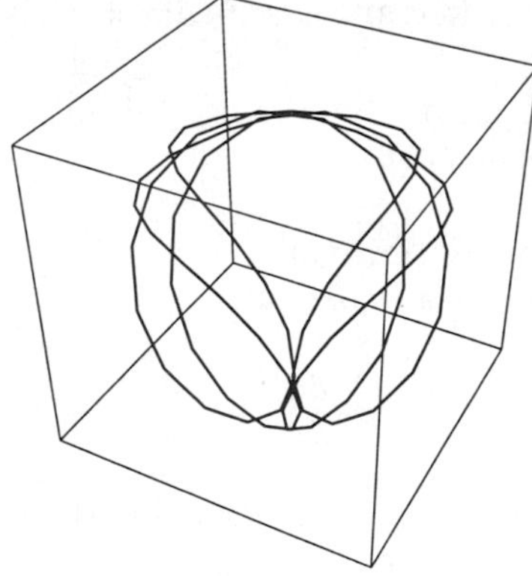

Is given by the CYLINDRICAL COORDINATES parametric equation

$$\begin{aligned} r &= \operatorname{sn}(s) \\ \theta &= ks \\ z &= \operatorname{cn}(s), \end{aligned}$$

where k is a POSITIVE constant and $\operatorname{sn}(s)$ and $\operatorname{cn}(s)$ are JACOBI ELLIPTIC FUNCTIONS (Whittaker and Watson 1990, pp. 527–528).

References
Bowman, F. *Introduction to Elliptic Functions, with Applications.* New York: Dover, p. 34, 1961.
Whittaker, E. T. and Watson, G. N. *A Course in Modern Analysis, 4th ed.* Cambridge, England: Cambridge University Press, 1990.

Seifert Surface

An orientable surface with one boundary component such that the boundary component of the surface is a given KNOT K. In 1934, Seifert proved that such a surface can be constructed for any KNOT. The process of generating this surface is known as Seifert's algorithm. Applying Seifert's algorithm to an alternating projection of an alternating knot yields a Seifert surface of minimal GENUS.

There are KNOTS for which the minimal genus Seifert surface cannot be obtained by applying Seifert's algorithm to any projection of that KNOT, as proved by Morton in 1986 (Adams 1994, p. 105).

see also GENUS (KNOT), SEIFERT MATRIX

References
Adams, C. C. *The Knot Book: An Elementary Introduction to the Mathematical Theory of Knots.* New York: W. H. Freeman, pp. 95–106, 1994.
Seifert, H. "Über das Geschlecht von Knotten." *Math. Ann.* **110**, 571–592, 1934.

Self-Adjoint Matrix

A MATRIX A for which

$$\mathsf{A}^\dagger \equiv (\mathsf{A}^{\mathrm{T}})^* = \mathsf{A},$$

where the ADJOINT OPERATOR is denoted $\mathsf{A}^\dagger$, A^{T} is the MATRIX TRANSPOSE, and $*$ is the COMPLEX CONJUGATE. If a MATRIX is self-adjoint, it is said to be HERMITIAN.

see also ADJOINT OPERATOR, HERMITIAN MATRIX, MATRIX TRANSPOSE

Self-Adjoint Operator

Given a differential equation

$$\tilde{\mathcal{L}}u(x) \equiv p_0 \frac{du^2}{dx^2} + p_1 \frac{du}{dx} + p_2 u, \tag{1}$$

where $p_i \equiv p_i(x)$ and $u \equiv u(x)$, the ADJOINT OPERATOR $\tilde{\mathcal{L}}^\dagger$ is defined by

$$\begin{aligned}\tilde{\mathcal{L}}^\dagger u &\equiv \frac{d}{dx^2}(p_0 u) - \frac{d}{dx}(p_1 u) + p_2 u \qquad (2)\\ &= p_0 \frac{d^2 u}{dx^2} + (2p_0{}' - p_1)\frac{du}{dx} + (p_0{}'' - p_1{}' + p_2)u. \qquad (3)\end{aligned}$$

In order for the operator to be self-adjoint, i.e.,

$$\tilde{\mathcal{L}} = \tilde{\mathcal{L}}^\dagger, \tag{4}$$

the second terms in (1) and (3) must be equal, so

$$p_0{}'(x) = p_1(x). \tag{5}$$

This also guarantees that the third terms are equal, since

$$p_0{}'(x) = p_1(x) \Rightarrow p_0{}''(x) = p_1{}'(x), \tag{6}$$

so (3) becomes

$$\begin{aligned}\tilde{\mathcal{L}}u = \tilde{\mathcal{L}}^\dagger u &= p_0 \frac{d^2}{dx^2} + p_0{}' \frac{du}{dx} + p_2 u \qquad (7)\\ &= \frac{d}{dx}\left(p_0 \frac{du}{dx}\right) + p_2 u = 0. \qquad (8)\end{aligned}$$

The LEGENDRE DIFFERENTIAL EQUATION and the equation of SIMPLE HARMONIC MOTION are self-adjoint, but the LAGUERRE DIFFERENTIAL EQUATION and HERMITE DIFFERENTIAL EQUATION are not.

A nonself-adjoint second-order linear differential operator can always be transformed into a self-adjoint one using STURM-LIOUVILLE THEORY. In the special case $p_2(x) = 0$, (8) gives

$$\frac{d}{dx}\left[p_0(x)\frac{du}{dx}\right] = 0 \tag{9}$$

$$p_0(x)\frac{du}{dx} = C \tag{10}$$

$$du = C\frac{dx}{p_0(x)} \tag{11}$$

$$u = C\int \frac{dx}{p_0(x)}, \tag{12}$$

where C is a constant of integration.

A self-adjoint operator which satisfies the BOUNDARY CONDITIONS

$$v^* pU'|_{x=a} = v^* pU'|_{x=b} \tag{13}$$

is automatically a HERMITIAN OPERATOR.

see also ADJOINT OPERATOR, HERMITIAN OPERATOR, STURM-LIOUVILLE THEORY

References
Arfken, G. "Self-Adjoint Differential Equations." §9.1 in *Mathematical Methods for Physicists, 3rd ed.* Orlando, FL: Academic Press, pp. 497–509, 1985.

Self-Avoiding Walk

N.B. A detailed on-line essay by S. Finch was the starting point for this entry.

Let the number of RANDOM WALKS on a d-D lattice starting at the ORIGIN which never land on the same lattice point twice in n steps be denoted $c(n)$. The first few values are

$$c_d(0) = 1 \quad (1)$$

$$c_d(1) = 2d \quad (2)$$

$$c_d(2) = 2d(2d-1). \quad (3)$$

The connective constant

$$\mu_d \equiv \lim_{n\to\infty} [c_d(n)]^{1/n} \quad (4)$$

is known to exist and be FINITE. The best ranges for these constants are

$$\mu_2 \in [2.62002, 2.6939] \quad (5)$$

$$\mu_3 \in [4.572140, 4.7476] \quad (6)$$

$$\mu_4 \in [6.742945, 6.8179] \quad (7)$$

$$\mu_5 \in [8.828529, 8.8602] \quad (8)$$

$$\mu_6 \in [10.874038, 10.8886] \quad (9)$$

(Finch).

For the triangular lattice in the plane, $\mu < 4.278$ (Alm 1993), and for the hexagonal planar lattice, it is conjectured that

$$\mu = \sqrt{2+\sqrt{2}} \quad (10)$$

(Madras and Slade 1993).

The following limits are also believed to exist and to be FINITE:

$$\begin{cases} \lim_{n\to\infty} \frac{c(n)}{\mu^n n^{\gamma-1}} & \text{for } d \neq 4 \\ \lim_{n\to\infty} \frac{c(n)}{\mu^n n^{\gamma-1}(\ln n)^{1/4}} & \text{for } d = 4, \end{cases} \quad (11)$$

where the critical exponent $\gamma = 1$ for $d > 4$ (Madras and Slade 1993) and it has been conjectured that

$$\gamma = \begin{cases} \frac{43}{32} & \text{for } d = 2 \\ 1.162\ldots & \text{for } d = 3 \\ 1 & \text{for } d = 4. \end{cases} \quad (12)$$

Define the mean square displacement over all n-step self-avoiding walks ω as

$$s(n) \equiv \langle |\omega(n)|^2 \rangle = \frac{1}{c(n)} \sum_\omega |\omega(n)|^2. \quad (13)$$

The following limits are believed to exist and be FINITE:

$$\begin{cases} \lim_{n\to\infty} \frac{s(n)}{n^{2\nu}} & \text{for } d \neq 4 \\ \lim_{n\to\infty} \frac{s(n)}{n^{2\nu}(\ln n)^{1/4}} & \text{for } d = 4, \end{cases} \quad (14)$$

where the critical exponent $\nu = 1/2$ for $d > 4$ (Madras and Slade 1993), and it has been conjectured that

$$\nu = \begin{cases} \frac{3}{4} & \text{for } d = 2 \\ 0.59\ldots & \text{for } d = 3 \\ \frac{1}{2} & \text{for } d = 4. \end{cases} \quad (15)$$

see also RANDOM WALK

References

Alm, S. E. "Upper Bounds for the Connective Constant of Self-Avoiding Walks." *Combin. Prob. Comput.* **2**, 115–136, 1993.

Finch, S. "Favorite Mathematical Constants." `http://www.mathsoft.com/asolve/constant/cnntv/cnntv.html`.

Madras, N. and Slade, G. *The Self-Avoiding Walk.* Boston, MA: Birkhäuser, 1993.

Self-Conjugate Subgroup

see INVARIANT SUBGROUP

Self-Descriptive Number

A 10-DIGIT number satisfying the following property. Number the DIGITS 0 to 9, and let DIGIT n be the number of ns in the number. There is exactly one such number: 6210001000.

References

Pickover, C. A. "Chaos in Ontario." Ch. 28 in *Keys to Infinity.* New York: W. H. Freeman, pp. 217–219, 1995.

Self-Homologous Point

see SIMILITUDE CENTER

Self Number

A number (usually base 10 unless specified otherwise) which has no GENERATOR. Such numbers were originally called COLUMBIAN NUMBERS (S. 1974). There are infinitely many such numbers, since an infinite sequence of self numbers can be generated from the RECURRENCE RELATION

$$C_k = 8 \cdot 10^{k-1} + C_{k-1} + 8, \quad (1)$$

for $k = 2, 3, \ldots$, where $C_1 = 9$. The first few self numbers are 1, 3, 5, 7, 9, 20, 31, 42, 53, 64, 75, 86, 97, ... (Sloane's A003052).

An infinite number of 2-self numbers (i.e., base-2 self numbers) can be generated by the sequence

$$C_k = 2^j + C_{k-1} + 1 \quad (2)$$

for $k = 1, 2, \ldots$, where $C_1 = 1$ and j is the number of digits in C_{k-1}. An infinite number of n-self numbers can be generated from the sequence

$$C_k = (n-2)n^{k-1} + C_{k-1} + (n-2) \quad (3)$$

for $k = 2, 3, \ldots$, and

$$C_1 = \begin{cases} n-1 & \text{for } n \text{ even} \\ n-2 & \text{for } n \text{ odd.} \end{cases} \tag{4}$$

Joshi (1973) proved that if k is ODD, then m is a k-self number IFF m is ODD. Patel (1991) proved that $2k$, $4k+2$, and k^2+2k+1 are k-self numbers in every EVEN base $k \geq 4$.

see also DIGITADITION

References

Cai, T. "On k-Self Numbers and Universal Generated Numbers." *Fib. Quart.* **34**, 144–146, 1996.

Gardner, M. *Time Travel and Other Mathematical Bewilderments.* New York: W. H. Freeman, pp. 115–117, 122, 1988.

Joshi, V. S. Ph.D. dissertation. Gujarat University, Ahmadabad, 1973.

Kaprekar, D. R. *The Mathematics of New Self-Numbers.* Devaiali, pp. 19–20, 1963.

Patel, R. B. "Some Tests for k-Self Numbers." *Math. Student* **56**, 206–210, 1991.

S., B. R. Solution to Problem E 2048. *Amer. Math. Monthly* **81**, 407, 1974.

Sloane, N. J. A. Sequence A003052/M2404 in "An On-Line Version of the Encyclopedia of Integer Sequences."

Self-Reciprocating Property

Let h be the number of sides of certain skew POLYGONS (Coxeter 1973, p. 15). Then

$$h = \frac{2(p+q+2)}{10-p-q}.$$

References

Coxeter, H. S. M. *Regular Polytopes, 3rd ed.* New York: Dover, 1973.

Self-Recursion

Self-recursion is a RECURSION which is defined in terms of itself, resulting in an ill-defined infinite regress.

see SELF-RECURSION

Self-Similarity

An object is said to be self-similar if it looks "roughly" the same on any scale. FRACTALS are a particularly interesting class of self-similar objects.

see also FRACTAL

References

Hutchinson, J. "Fractals and Self-Similarity." *Indiana Univ. J. Math.* **30**, 713–747, 1981.

Self-Transversality Theorem

Let j, r, and s be distinct INTEGERS (mod n), and let W_i be the point of intersection of the side or diagonal V_iV_{i+j} of the n-gon $P = [V_1, \ldots, V_n]$ with the transversal $V_{i+r}V_{i+s}$. Then a NECESSARY and SUFFICIENT condition for

$$\prod_{i=1}^{n} \left[\frac{V_iW_i}{W_iV_{i+j}}\right] = (-1)^n,$$

where $AB||CD$ and

$$\left[\frac{AB}{CD}\right],$$

is the ratio of the lengths $[A, B]$ and $[C, D]$ with a plus or minus sign depending on whether these segments have the same or opposite direction, is that

1. $n = 2m$ is EVEN with $j \equiv m \pmod n$ and $s \equiv r + m \pmod n$,
2. n is arbitrary and either $s \equiv 2r$ and $j \equiv 3r$, or
3. $r \equiv 2s \pmod n$ and $j \equiv 3s \pmod n$.

References

Grünbaum, B. and Shepard, G. C. "Ceva, Menelaus, and the Area Principle." *Math. Mag.* **68**, 254–268, 1995.

Selfridge's Conjecture

There exist infinitely many $n > 0$ with ${p_n}^2 > p_{n-i}p_{n+i}$ for all $i < n$. Also, there exist infinitely many $n > 0$ such that $2p_n < p_{n-i} + p_{n-i}$ for all $i < n$.

Selfridge-Hurwitz Residue

Let the RESIDUE from PÉPIN'S THEOREM be

$$R_n \equiv 3^{(F_n-1)/2} \pmod{F_n},$$

where F_n is a FERMAT NUMBER. Selfridge and Hurwitz use

$$R_n (\text{mod } 2^{35} - 1, 2^{36}, 2^{36} - 1).$$

A nonvanishing $R_n \pmod{2^{36}}$ indicates that F_n is COMPOSITE for $n > 5$.

see also FERMAT NUMBER, PÉPIN'S THEOREM

References

Crandall, R.; Doenias, J.; Norrie, C.; and Young, J. "The Twenty-Second Fermat Number is Composite." *Math. Comput.* **64**, 863–868, 1995.

Selmer Group

A GROUP which is related to the TANIYAMA-SHIMURA CONJECTURE.

see also TANIYAMA-SHIMURA CONJECTURE

Semi-Integral

An INTEGRAL of order 1/2. The semi-integral of the CONSTANT FUNCTION $f(x) = c$ is

$$\frac{d^{-1/2}c}{dx^{-1/2}} = 2c\sqrt{\frac{x}{\pi}}.$$

see also SEMIDERIVATIVE

References

Spanier, J. and Oldham, K. B. *An Atlas of Functions.* Washington, DC: Hemisphere, pp. 8 and 14, 1987.

Semialgebraic Number

A subset of $\mathbb{R}^n$ which is a finite Boolean combination of sets of the form $\{\bar{x} = (x_1, \ldots, x_m) : f(\bar{x}) > 0\}$ and $\{\bar{x} : g(\bar{x}) = 0\}$, where $f, g \in \mathbb{R}[X_1, \ldots, X_n]$.

References

Bierstone, E. and Milman, P. "Semialgebraic and Subanalytic Sets." *IHES Pub. Math.* **67**, 5–42, 1988.
Marker, D. "Model Theory and Exponentiation." *Not. Amer. Math. Soc.* **43**, 753–759, 1996.

Semianalytic

$X \subseteq \mathbb{R}^n$ is semianalytic if, for all $x \in \mathbb{R}^n$, there is an open neighborhood U of x such that $X \cap U$ is a finite Boolean combination of sets $\{\bar{x} \in U : f(\bar{x}) = 0\}$ and $\{\bar{x} \in U : g(\bar{x}) > 0\}$, where $f, g : U \to \mathbb{R}$ are ANALYTIC.

see also ANALYTIC FUNCTION, PSEUDOANALYTIC FUNCTION, SUBANALYTIC

References

Marker, D. "Model Theory and Exponentiation." *Not. Amer. Math. Soc.* **43**, 753–759, 1996.

Semicircle

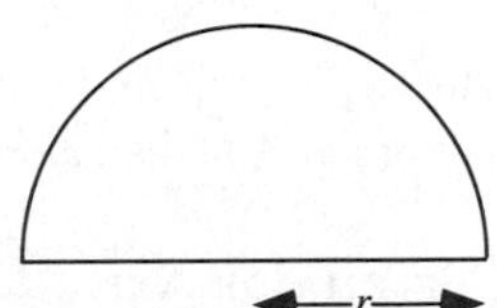

Half a CIRCLE. The PERIMETER of the semicircle of RADIUS r is

$$L = 2r + \pi r = r(2 + \pi), \tag{1}$$

and the AREA is

$$A = 2\int_0^r \sqrt{r^2 - y^2}\,dy = \tfrac{1}{2}\pi r^2. \tag{2}$$

The weighted mean of y is

$$\langle y \rangle = 2\int_0^r y\sqrt{r^2 - y^2}\,dy = \tfrac{2}{3}r^3. \tag{3}$$

The CENTROID is then given by

$$\bar{y} = \frac{\langle y \rangle}{A} = \frac{4r}{3\pi}. \tag{4}$$

The semicircle is the CROSS-SECTION of a HEMISPHERE for any PLANE through the z-AXIS.

see also ARBELOS, ARC, CIRCLE, DISK, HEMISPHERE, LENS, RIGHT ANGLE, SALINON, THALES' THEOREM, YIN-YANG

Semicolon Derivative

see COVARIANT DERIVATIVE

Semiconvergent Series

see ASYMPTOTIC SERIES

Semicubical Parabola

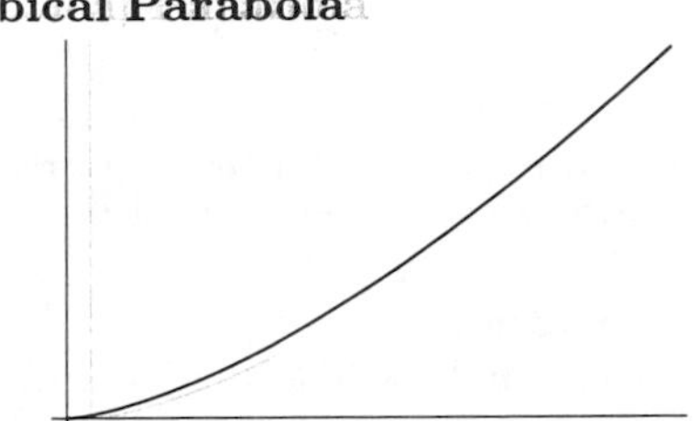

A PARABOLA-like curve with Cartesian equation

$$y = ax^{3/2}, \tag{1}$$

parametric equations

$$x = t^2 \tag{2}$$

$$y = at^3, \tag{3}$$

and POLAR COORDINATES,

$$r = \frac{\tan^2\theta \sec\theta}{a}. \tag{4}$$

The semicubical parabola is the curve along which a particle descending under gravity describes equal vertical spacings within equal times, making it an ISOCHRONOUS CURVE. The problem of finding the curve having this property was posed by Leibniz in 1687 and solved by Huygens (MacTutor Archive).

The ARC LENGTH, CURVATURE, and TANGENTIAL ANGLE are

$$s(t) = \tfrac{1}{27}(4 + 9t^2)^{3/2} - \tfrac{8}{27} \tag{5}$$

$$\kappa(t) = \frac{6}{t(4 + 9t^2)^{3/2}} \tag{6}$$

$$\phi(t) = \tan^{-1}(\tfrac{3}{2}t). \tag{7}$$

see also NEILE'S PARABOLA, PARABOLA INVOLUTE

References

Gray, A. "The Semicubical Parabola." §1.7 in *Modern Differential Geometry of Curves and Surfaces.* Boca Raton, FL: CRC Press, pp. 15–16, 1993.
Lawrence, J. D. *A Catalog of Special Plane Curves.* New York: Dover, pp. 85–87, 1972.
Lee, X. "Semicubic Parabola." http://www.best.com/~xah/Special Plane Curves _ dir / Semicubic Parabola _ dir / semicubicParabola.html.
MacTutor History of Mathematics Archive. "Neile's Parabola." http://www-groups.dcs.st-and.ac.uk/~history/Curves/Neiles.html.
Yates, R. C. "Semi-Cubic Parabola." *A Handbook on Curves and Their Properties.* Ann Arbor, MI: J. W. Edwards, pp. 186–187, 1952.

Semiderivative

A DERIVATIVE of order 1/2. The semiderivative of the CONSTANT FUNCTION $f(x) = c$ is

$$\frac{d^{1/2}c}{dx^{1/2}} = \frac{c}{\sqrt{\pi x}}.$$

see also DERIVATIVE, SEMI-INTEGRAL

References

Spanier, J. and Oldham, K. B. *An Atlas of Functions.* Washington, DC: Hemisphere, pp. 8 and 14, 1987.

Semidirect Product

The "split" extension G of GROUPS N and F which contains a SUBGROUP $\bar{F}$ isomorphic to F with $G = \bar{F}\bar{N}$ and $\bar{F} \cap \bar{N} = \{e\}$.

References

Iyanaga, S. and Kawada, Y. (Eds.). *Encyclopedic Dictionary of Mathematics.* Cambridge, MA: MIT Press, p. 613, 1980.

Semiflow

An ACTION with $G = \mathbb{R}^+$.

see also FLOW

Semigroup

A mathematical object defined for a set and a BINARY OPERATOR in which the multiplication operation is ASSOCIATIVE. No other restrictions are placed on a semigroup; thus a semigroup need not have an IDENTITY ELEMENT and its elements need not have inverses within the semigroup. A semigroup is an ASSOCIATIVE GROUPOID.

A semigroup can be empty. The total number of semigroups of order n are 1, 4, 18, 126, 1160, 15973, 836021, ... (Sloane's A001423). The number of semigroups of order n with one IDEMPOTENT are 1, 2, 5, 19, 132, 3107, 623615, ... (Sloane's A002786), and with two IDEMPOTENTS are 2, 7, 37, 216, 1780, 32652, ... (Sloane's A002787). The number $a(n)$ of semigroups having n IDEMPOTENTS are 1, 2, 6, 26, 135, 875, ... (Sloane's A002788).

see also ASSOCIATIVE, BINARY OPERATOR, FREE SEMIGROUP, GROUPOID, INVERSE SEMIGROUP, MONOID, QUASIGROUP

References

Clifford, A. H. and Preston, G. B. *The Algebraic Theory of Semigroups.* Providence, RI: Amer. Math. Soc., 1961.

Sloane, N. J. A. Sequences A001423/M3550, A002786/M1522, A002787/M1802, and A002788/M1679 in "An On-Line Version of the Encyclopedia of Integer Sequences."

Semilatus Rectum

Given an ELLIPSE, the semilatus rectum is defined as the distance L measured from a FOCUS such that

$$\frac{1}{L} \equiv \frac{1}{2}\left(\frac{1}{r_+} + \frac{1}{r_-}\right), \tag{1}$$

where $r_+ = a(1+e)$ and $r_- = a(1-e)$ are the APOAPSIS and PERIAPSIS, and e is the ELLIPSE's ECCENTRICITY. Plugging in for r_+ and r_- then gives

$$\frac{1}{L} = \frac{1}{2a}\left(\frac{1}{1-e} + \frac{1}{1+e}\right) = \frac{1}{2a}\frac{(1+e)+(1-e)}{1-e^2} = \frac{1}{a}\frac{1}{1-e^2}, \tag{2}$$

so

$$L = a(1-e^2). \tag{3}$$

see also ECCENTRICITY, ELLIPSE, FOCUS, LATUS RECTUM, SEMIMAJOR AXIS, SEMIMINOR AXIS

Semimagic Square

A square that fails to be a MAGIC SQUARE only because one or both of the main diagonal sums do not equal the MAGIC CONSTANT is called a SEMIMAGIC SQUARE.

see also MAGIC SQUARE

Semimajor Axis

HALF the distance across an ELLIPSE along its long principal axis.

see also ELLIPSE, SEMIMINOR AXIS

Semiminor Axis

Half the distance across an ELLIPSE along its short principal axis.

see also ELLIPSE, SEMIMAJOR AXIS

Semiperfect Magic Cube

A semiperfect magic cube, also called an ANDREWS CUBE, is a MAGIC CUBE for which the cross-section diagonals do not sum to the MAGIC CONSTANT.

see also MAGIC CUBE, PERFECT MAGIC CUBE

References

Gardner, M. "Magic Squares and Cubes." Ch. 17 in *Time Travel and Other Mathematical Bewilderments.* New York: W. H. Freeman, pp. 213–225, 1988.

Semiperfect Number

A number such as $20 = 1 + 4 + 5 + 10$ which is the SUM of some (or all) its PROPER DIVISORS. A semiperfect number which is the SUM of *all* its PROPER DIVISORS is called a PERFECT NUMBER. The first few semiperfect numbers are 6, 12, 18, 20, 24, 28, 30, 36, 40, ... (Sloane's A005835). Every multiple of a semiperfect number is semiperfect, as are all numbers $2^m p$ for $m \geq 1$ and p a PRIME between 2^m and 2^{m+1} (Guy 1994, p. 47).

A semiperfect number cannot be DEFICIENT. Rare ABUNDANT NUMBERS which are not semiperfect are called WEIRD NUMBERS. Semiperfect numbers are sometimes also called PSEUDOPERFECT NUMBERS.

see also ABUNDANT NUMBER, DEFICIENT NUMBER, PERFECT NUMBER, PRIMITIVE SEMIPERFECT NUMBER, WEIRD NUMBER

References

Guy, R. K. "Almost Perfect, Quasi-Perfect, Pseudoperfect, Harmonic, Weird, Multiperfect and Hyperperfect Numbers." §B2 in *Unsolved Problems in Number Theory, 2nd ed.* New York: Springer-Verlag, pp. 45–53, 1994.

Sloane, N. J. A. Sequence A005835/M4094 in "An On-Line Version of the Encyclopedia of Integer Sequences."

Zachariou, A. and Zachariou, E. "Perfect, Semi-Perfect and Ore Numbers." *Bull. Soc. Math. Gréce (New Ser.)* **13**, 12–22, 1972.

Semiperimeter

The semiperimeter on a figure is defined as

$$s \equiv \tfrac{1}{2}p, \tag{1}$$

where p is the PERIMETER. The semiperimeter of POLYGONS appears in unexpected ways in the computation of their AREAS. The most notable cases are in the ALTITUDE, EXRADIUS, and INRADIUS of a TRIANGLE, the SODDY CIRCLES, HERON'S FORMULA for the AREA of a TRIANGLE in terms of the legs a, b, and c

$$A_\Delta = \sqrt{s(s-a)(s-b)(s-c)}, \tag{2}$$

and BRAHMAGUPTA'S FORMULA for the AREA of a QUADRILATERAL

$$A_{\text{quadrilateral}} = \sqrt{(s-a)(s-b)(s-c)(s-d) - abcd\cos^2\left(\frac{A+B}{2}\right)}. \tag{3}$$

The semiperimeter also appears in the beautiful L'HUILIER'S THEOREM about SPHERICAL TRIANGLES.

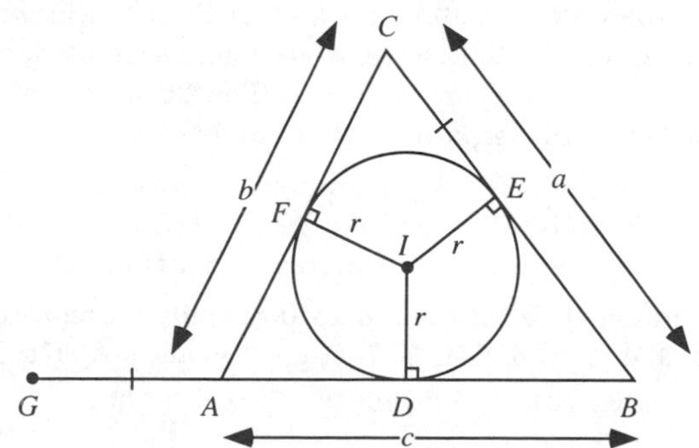

For a TRIANGLE, the following identities hold,

$$s - a = \tfrac{1}{2}(-a+b+c) \tag{4}$$
$$s - b = \tfrac{1}{2}(a+b-c) \tag{5}$$
$$s - c = \tfrac{1}{2}(a+b-c). \tag{6}$$

Now consider the above figure. Let I be the INCENTER of the TRIANGLE ΔABC, with D, E, and F the tangent points of the INCIRCLE. Extend the line BA with $GA = CE$. Note that the pairs of triangles (ADI, AFI), (BDI, BEI), (CFI, CEI) are congruent. Then

$$\begin{aligned} BG &= BD + AD + AG = BD + AD + CE \\ &= \tfrac{1}{2}(2BD + 2AD + 2CE) \\ &= \tfrac{1}{2}[(BD+BE)+(AD+AF)+(CE+CF)] \\ &= \tfrac{1}{2}[(BD+AD)+(BE+CE)+(AF+CF)] \\ &= \tfrac{1}{2}(AB+BC+AC) = \tfrac{1}{2}(a+b+c) = s. \end{aligned} \tag{7}$$

Furthermore,

$$\begin{aligned} s-a &= BG - BC \\ &= (BD+AD+AG)-(BE+CE) \\ &= (BD+AD+CE)-(BD+CE) = AC \quad (8) \\ s-b &= BG - AC \\ &= (BD+AD+AG)-(AF+CF) \\ &= (BD+AD+CE)-(AD+CE) = BD \quad (9) \\ s-c &= BG - AB = AG \quad (10) \end{aligned}$$

(Dunham 1990). These equations are some of the building blocks of Heron's derivation of HERON'S FORMULA.

References

Dunham, W. "Heron's Formula for Triangular Area." Ch. 5 in *Journey Through Genius: The Great Theorems of Mathematics.* New York: Wiley, pp. 113–132, 1990.

Semiprime

A COMPOSITE number which is the PRODUCT of two PRIMES (possibly equal). They correspond to the 2-ALMOST PRIMES. The first few are 4, 6, 9, 10, 14, 15, 21, 22, ... (Sloane's A001358).

see also ALMOST PRIME, CHEN'S THEOREM, COMPOSITE NUMBER, PRIME NUMBER

References

Sloane, N. J. A. Sequence A001358/M3274 in "An On-Line Version of the Encyclopedia of Integer Sequences."

Semiprime Ring

Given an IDEAL A, a semiprime ring is one for which $A^n = 0$ IMPLIES $A = 0$ for any POSITIVE n. Every PRIME RING is semiprime.

see also PRIME RING

Semiregular Polyhedron

A POLYHEDRON or plane TESSELLATION is called semiregular if its faces are all REGULAR POLYGONS and its corners are alike (Walsh 1972; Coxeter 1973, pp. 4 and 58; Holden 1991, p. 41). The usual name for a semiregular polyhedron is an ARCHIMEDEAN SOLIDS, of which there are exactly 13.

see also ARCHIMEDEAN SOLID, POLYHEDRON, TESSELLATION

References
Coxeter, H. S. M. "Regular and Semi-Regular Polytopes I." *Math. Z.* **46**, 380–407, 1940.
Coxeter, H. S. M. *Regular Polytopes, 3rd ed.* New York: Dover, 1973.
Holden, A. *Shapes, Space, and Symmetry.* New York: Dover, 1991.
Walsh, T. R. S. "Characterizing the Vertex Neighbourhoods of Semi-Regular Polyhedra." *Geometriae Dedicata* **1**, 117–123, 1972.

Semiring

A semiring is a set together with two BINARY OPERATORS $S(+,*)$ satisfying the following conditions:

1. Additive associativity: For all $a, b, c \in S$, $(a+b)+c = a+(b+c)$,
2. Additive commutativity: For all $a, b \in S$, $a+b = b+a$,
3. Multiplicative associativity: For all $a, b, c \in S$, $(a*b)*c = a*(b*c)$,
4. Left and right distributivity: For all $a, b, c \in S$, $a*(b+c) = (a*b)+(a*c)$ and $(b+c)*a = (b*a)+(c*a)$.

Thus a semiring is therefore a commutative SEMIGROUP under addition and a SEMIGROUP under multiplication. A semiring can be empty.

see also BINARY OPERATOR, RING, RINGOID, SEMIGROUP

References
Rosenfeld, A. *An Introduction to Algebraic Structures.* New York: Holden-Day, 1968.

Semisecant

see TRANSVERSAL LINE

Semisimple

A p-ELEMENT x of a GROUP G is semisimple if $E(C_G(x)) \neq 1$, where $E(H)$ is the commuting product of all components of H and $C_G(x)$ is the CENTRALIZER of G.

see also CENTRALIZER, p-ELEMENT

Semisimple Algebra

An ALGEBRA with no nontrivial nilpotent IDEALS. In the 1890s, Cartan, Frobenius, and Molien independently proved that any finite-dimensional semisimple algebra over the REAL or COMPLEX numbers is a finite and unique DIRECT SUM of SIMPLE ALGEBRAS. This result was then extended to algebras over arbitrary fields by Wedderburn in 1907 (Kleiner 1996).

see also IDEAL, NILPOTENT ELEMENT, SIMPLE ALGEBRA

References
Kleiner, I. "The Genesis of the Abstract Ring Concept." *Amer. Math. Monthly* **103**, 417–424, 1996.

Semisimple Lie Group

A LIE GROUP which has a simply connected covering group HOMEOMORPHIC to $\mathbb{R}^n$. The prototype is any connected closed subgroup of upper TRIANGULAR COMPLEX MATRICES. The HEISENBERG GROUP is such a group.

see also HEISENBERG GROUP, LIE GROUP

References
Knapp, A. W. "Group Representations and Harmonic Analysis, Part II." *Not. Amer. Math. Soc.* **43**, 537–549, 1996.

Semisimple Ring

A SEMIPRIME RING which is also an ARTINIAN RING.

see also ARTINIAN RING

Semistable

When a PRIME l divides the DISCRIMINANT of a ELLIPTIC CURVE E, two or all three roots of E become congruent mod l. An ELLIPTIC CURVE is semistable if, for all such PRIMES l, only two roots become CONGRUENT mod l (with more complicated definitions for $p = 2$ or 3).

see also DISCRIMINANT (ELLIPTIC CURVE), ELLIPTIC CURVE

Sensitivity

The probability that a STATISTICAL TEST will be positive for a true statistic.

see also SPECIFICITY, STATISTICAL TEST, TYPE I ERROR, TYPE II ERROR

Sentence

A LOGIC FORMULA with no FREE variables.

Separating Edge

An EDGE of a GRAPH is separating if a path from a point A to a point B must pass over it. Separating EDGES can therefore be viewed as either bridges or dead ends.

see also EDGE (GRAPH)

Separating Family

A SEPARATING FAMILY is a SET of SUBSETS in which each pair of adjacent elements are found separated, each in one of two disjoint subsets. The 26 letters of the alphabet can be separated by a family of 9,

$$\begin{matrix} (abcdefghi) & (jklmnopqr) & (stuvwxyz) \\ (abcjklstu) & (defmnovwx) & (ghipqryz) \\ (adgjmpsvy) & (behknqtwz) & (cfilorux) \end{matrix}.$$

The minimal size of the separating family for an n-set is 0, 2, 3, 4, 5, 5, 6, 6, 6, 7, 7, 7, ... (Sloane's A007600).

see also KATONA'S PROBLEM

References
Honsberger, R. "Cai Mao-Cheng's Solution to Katona's Problem on Families of Separating Subsets." Ch. 18 in *Mathematical Gems III.* Washington, DC: Math. Assoc. Amer., pp. 224–239, 1985.
Sloane, N. J. A. Sequence A007600/M0456 in "An On-Line Version of the Encyclopedia of Integer Sequences."

Separation

Two distinct point pairs AC and BD separate each other if A, B, C, and D lie on a CIRCLE (or line) in such order that either of the arcs (or the line segment AC) contains one but not both of B and D. In addition, the point pairs separate each other if every CIRCLE through A and C intersects (or coincides with) every CIRCLE through B and D. If the point pairs separate each other, then the symbol $AC//BD$ is used.

Separation of Variables

A method of solving partial differential equations in a function Φ and variables x, y, ... by making a substitution of the form

$$\Phi(x,y,\ldots) \equiv X(x)Y(y)\cdots,$$

breaking the resulting equation into a set of independent ordinary differential equations, solving these for $X(x)$, $Y(y)$, ..., and then plugging them back into the original equation.

This technique works because if the product of functions of independent variables is a constant, each function must separately be a constant. Success requires choice of an appropriate coordinate system and may not be attainable at all depending on the equation. Separation of variables was first used by L'Hospital in 1750. It is especially useful in solving equations arising in mathematical physics, such as LAPLACE'S EQUATION, the HELMHOLTZ DIFFERENTIAL EQUATION, and the Schrödinger equation.

see also HELMHOLTZ DIFFERENTIAL EQUATION, LAPLACE'S EQUATION

References

Arfken, G. "Separation of Variables" and "Separation of Variables—Ordinary Differential Equations." §2.6 and §8.3 in *Mathematical Methods for Physicists, 3rd ed.* Orlando, FL: Academic Press, pp. 111–117 and 448–451, 1985.

Morse, P. M. and Feshbach, H. "Separable Coordinates" and "Table of Separable Coordinates in Three Dimensions." §5.1 in *Methods of Theoretical Physics, Part I.* New York: McGraw-Hill, pp. 464–523 and 655–666, 1953.

Separation Theorem

There exist numbers $y_1 < y_2 < \ldots < x_{n-1}$, $a < y_{n-1}$, $y_{n-1} < b$, such that

$$\lambda_\nu = \alpha(y_\nu) - \alpha(y_{\nu-1}),$$

where $\nu = 1, 2, \ldots, n$, $y_0 = a$ and $y_n = b$. Furthermore, the zeros $x_1, \ldots, x_n$, arranged in increasing order, alternate with the numbers $y_1, \ldots y_{n-1}$, so

$$x_\nu < y_\nu < x_{\nu+1}.$$

More precisely,

$$\alpha(x_\nu + \epsilon) - \alpha(a) < \alpha(y_\nu) - \alpha(a)$$
$$= \lambda_1 + \ldots + \lambda_\nu < \alpha(x_{\nu+1} - \epsilon) - \alpha(a)$$

for $\nu = 1, \ldots, n-1$.

see also POINCARÉ SEPARATION THEOREM, STURMIAN SEPARATION THEOREM

References

Szegő, G. *Orthogonal Polynomials, 4th ed.* Providence, RI: Amer. Math. Soc., p. 50, 1975.

Separatrix

A phase curve (invariant MANIFOLD) which meets a HYPERBOLIC FIXED POINT (intersection of a stable and an unstable invariant MANIFOLD). A separatrix marks a boundary between phase curves with different properties. For example, the separatrix in the equation of motion for the pendulum occurs at the angular momentum where oscillation gives way to rotation.

Septendecillion

In the American system, 10^{54}.

see also LARGE NUMBER

Septillion

In the American system, 10^{24}.

see also LARGE NUMBER

Sequence

A sequence is an ordered set of mathematical objects which is denoted using braces. For example, the symbol $\{2n\}_{n=1}^\infty$ denotes the infinite sequence of EVEN NUMBERS $\{2, 4, \ldots, 2n, \ldots\}$.

see also 196-ALGORITHM, A-SEQUENCE, ALCUIN'S SEQUENCE, $B2$-SEQUENCE, BEATTY SEQUENCE, CARMICHAEL SEQUENCE, CAUCHY SEQUENCE, CONVERGENT SEQUENCE, DEGREE SEQUENCE, DENSITY (SEQUENCE), FRACTAL SEQUENCE, GIUGA SEQUENCE, INFINITIVE SEQUENCE, INTEGER SEQUENCE, ITERATION SEQUENCE, LIST, NONAVERAGING SEQUENCE, PRIMITIVE SEQUENCE, REVERSE-THEN-ADD SEQUENCE, SCORE SEQUENCE, SERIES, SIGNATURE SEQUENCE, SORT-THEN-ADD SEQUENCE, ULAM SEQUENCE

Sequency

The sequency k of a WALSH FUNCTION is defined as half the number of zero crossings in the time base.

see also WALSH FUNCTION

Sequency Function

see WALSH FUNCTION

Sequential Graph

A CONNECTED GRAPH having e EDGES is said to be sequential if it is possible to label the nodes i with distinct INTEGERS f_i in$\{0, 1, 2, \ldots, e-1\}$ such that when EDGE ij is labeled $f_i + f_j$, the set of EDGE labels is a block of e consecutive integers (Grace 1983, Gallian 1990). No HARMONIOUS GRAPH is known which cannot also be labeled sequentially.

see also CONNECTED GRAPH, HARMONIOUS GRAPH

References
Gallian, J. A. "Open Problems in Grid Labeling." *Amer. Math. Monthly* **97**, 133–135, 1990.
Grace, T. "On Sequential Labelings of Graphs." *J. Graph Th.* **7**, 195–201, 1983.

Series

A series is a sum of terms specified by some rule. If each term increases by a constant amount, it is said to be an ARITHMETIC SERIES. If each term equals the previous multiplied by a constant, it is said to be a GEOMETRIC SERIES. A series usually has an INFINITE number of terms, but the phrase INFINITE SERIES is sometimes used for emphasis or clarity.

If the sum of partial sequences comprising the first few terms of the series does not converge to a LIMIT (e.g., it oscillates or approaches $\pm\infty$), it is said to diverge. An example of a convergent series is the GEOMETRIC SERIES

$$\sum_{n=0}^{\infty}(\tfrac{1}{2})^n = 2,$$

and an example of a divergent series is the HARMONIC SERIES

$$\sum_{n=1}^{\infty}\frac{1}{n} = \infty.$$

A number of methods known as CONVERGENCE TESTS can be used to determine whether a given series converges. Although terms of a series can have either sign, convergence properties can often be computed in the "worst case" of all terms being POSITIVE, and then applied to the particular series at hand. A series of terms u_n is said to be ABSOLUTELY CONVERGENT if the series formed by taking the absolute values of the u_n,

$$\sum_n |u_n|,$$

converges.

An especially strong type of convergence is called UNIFORM CONVERGENCE, and series which are uniformly convergent have particularly "nice" properties. For example, the sum of a UNIFORMLY CONVERGENT series of continuous functions is continuous. A CONVERGENT SERIES can be DIFFERENTIATED term by term, provided that the functions of the series have continuous derivatives and that the series of DERIVATIVES is UNIFORMLY CONVERGENT. Finally, a UNIFORMLY CONVERGENT series of continuous functions can be INTEGRATED term by term.

For a table listing the COEFFICIENTS for various series operations, see Abramowitz and Stegun (1972, p. 15).

While it can be difficult to calculate analytical expressions for arbitrary convergent infinite series, many algorithms can handle a variety of common series types. The program *Mathematica*® (Wolfram Research, Champaign, IL) implements many of these algorithms. General techniques also exist for computing the numerical values to any but the most pathological series (Braden 1992).

see also ALTERNATING SERIES, ARITHMETIC SERIES, ARTISTIC SERIES, ASYMPTOTIC SERIES, BIAS (SERIES), CONVERGENCE IMPROVEMENT, CONVERGENCE TESTS, EULER-MACLAURIN INTEGRATION FORMULAS, GEOMETRIC SERIES, HARMONIC SERIES, INFINITE SERIES, MELODIC SERIES, q-SERIES, RIEMANN SERIES THEOREM, SEQUENCE, SERIES EXPANSION, SERIES REVERSION

References
Abramowitz, M. and Stegun, C. A. (Eds.). "Infinite Series." §3.6 in *Handbook of Mathematical Functions with Formulas, Graphs, and Mathematical Tables, 9th printing.* New York: Dover, p. 14, 1972.
Arfken, G. "Infinite Series." Ch. 5 in *Mathematical Methods for Physicists, 3rd ed.* Orlando, FL: Academic Press, pp. 277–351, 1985.
Boas, R. P. Jr. "Partial Sums of Infinite Series, and How They Grow." *Amer. Math. Monthly* **84**, 237–258, 1977.
Boas, R. P. Jr. "Estimating Remainders." *Math. Mag.* **51**, 83–89, 1978.
Borwein, J. M. and Borwein, P. B. "Strange Series and High Precision Fraud." *Amer. Math. Monthly* **99**, 622–640, 1992.
Braden, B. "Calculating Sums of Infinite Series." *Amer. Math. Monthly* **99**, 649–655, 1992.
Bromwich, T. J. I'a. and MacRobert, T. M. *An Introduction to the Theory of Infinite Series, 3rd ed.* New York: Chelsea, 1991.
Hansen, E. R. *A Table of Series and Products.* Englewood Cliffs, NJ: Prentice-Hall, 1975.
Hardy, G. H. *Divergent Series.* Oxford, England: Clarendon Press, 1949.
Jolley, L. B. W. *Summation of Series, 2nd rev. ed.* New York: Dover, 1961.
Knopp, K. *Theory and Application of Infinite Series.* New York: Dover, 1990.
Mangulis, V. *Handbook of Series for Scientists and Engineers.* New York: Academic Press, 1965.
Press, W. H.; Flannery, B. P.; Teukolsky, S. A.; and Vetterling, W. T. "Series and Their Convergence." §5.1 in *Numerical Recipes in FORTRAN: The Art of Scientific Computing, 2nd ed.* Cambridge, England: Cambridge University Press, pp. 159–163, 1992.
Rainville, E. D. *Infinite Series.* New York: Macmillan, 1967.

Series Expansion

see LAURENT SERIES, MACLAURIN SERIES, POWER SERIES, SERIES REVERSION, TAYLOR SERIES

Series Inversion

see SERIES REVERSION

Series Multisection

If

$$f(x) = f_0 + f_1x + f_2x^2 + \ldots + f_nx^n + \ldots,$$

then

$$S(n,j) = f_jx^j + f_{j+n}x^{j+n} + f_{j+2n}x^{j+2n} + \ldots$$

is given by

$$S(n,j) = \frac{1}{n}\sum_{t=0}^{n-1} w^{-jt} f(w^t x),$$

where $w = e^{2\pi i/n}$.

see also SERIES REVERSION

References

Honsberger, R. *Mathematical Gems III.* Washington, DC: Math. Assoc. Amer., pp. 210–214, 1985.

Series Reversion

Series reversion is the computation of the COEFFICIENTS of the inverse function given those of the forward function. For a function expressed in a series as

$$y = a_1x + a_2x^2 + a_3x^3 + \ldots, \tag{1}$$

the series expansion of the inverse series is given by

$$x = A_1y + A_2y^2 + A_3y^3 + \ldots. \tag{2}$$

By plugging (2) into (1), the following equation is obtained

$$\begin{aligned} y = a_1A_1y &+ (a_2{A_1}^2 + a_1A_2)y^2 \\ &+(a_3{A_1}^3 + 2a_2A_1A_2 + a_1A_3)y^3 \\ &+(3a_3{A_1}^2A_2 + a_2{A_2}^2 + a_2A_1A_3) + \ldots. \end{aligned} \tag{3}$$

Equating COEFFICIENTS then gives

$$A_1 = {a_1}^{-1} \tag{4}$$

$$A_2 = -\frac{a_2}{a_1}{A_1}^2 = -{a_1}^{-3}a_2 \tag{5}$$

$$A_3 = {a_1}^{-5}(2{a_2}^2 - a_1a_3) \tag{6}$$

$$A_4 = {a_1}^{-7}(5a_1a_2a_3 - {a_1}^2a_4 - 5{a_2}^3) \tag{7}$$

$$\begin{aligned} A_5 = {a_1}^{-9}&(6{a_1}^2a_2a_4 + 3{a_1}^2a_2a_3 + 14{a_2}^4 - {a_1}^3a_5 \\ &- 21a_1{a_2}^2a_3) \end{aligned} \tag{8}$$

$$\begin{aligned} A_6 = {a_1}^{-11}&(7{a_1}^3a_2a_5 + 7{a_1}^3a_3a_4 + 84a_1{a_2}^3a_3 \\ &- {a_1}^4a_6 - 28{a_1}^2a_2{a_3}^2 - 42{a_2}^5 - 28{a_1}^2{a_2}^2a_4) \end{aligned} \tag{9}$$

$$\begin{aligned} A_7 = {a_1}^{-13}&(8{a_1}^4a_2a_6 + 8{a_1}^4a_3a_5 + 4{a_1}^4{a_4}^2 \\ &+ 120{a_1}^2{a_2}^3a_4 + 180{a_1}^2{a_2}^2{a_3}^2 + 132{a_2}^6 \\ &- {a_1}^5a_7 - 36{a_1}^3{a_2}^2a_5 - 72{a_1}^3a_2a_3a_4 \\ &- 12{a_1}^3{a_3}^3 - 330a_1{a_2}^4a_3) \end{aligned} \tag{10}$$

(Dwight 1961, Abramowitz and Stegun 1972, p. 16). A derivation of the explicit formula for the nth term is given by Morse and Feshbach (1953),

$$\begin{aligned} A_n = \frac{1}{n{a_1}^n} &\sum_{s,t,u,\ldots} (-1)^{s+t+u+\ldots} \\ &\times \frac{n(n+1)\cdots(n-1+s+t+u+\ldots)}{s!t!u!\cdots}\left(\frac{a_2}{a_1}\right)^s\left(\frac{a_3}{a_1}\right)^t\cdots, \end{aligned} \tag{11}$$

where

$$s + 2t + 3u + \ldots = n - 1. \tag{12}$$

References

Abramowitz, M. and Stegun, C. A. (Eds.). *Handbook of Mathematical Functions with Formulas, Graphs, and Mathematical Tables, 9th printing.* New York: Dover, 1972.

Arfken, G. *Mathematical Methods for Physicists, 3rd ed.* Orlando, FL: Academic Press, pp. 316–317, 1985.

Beyer, W. H. *CRC Standard Mathematical Tables, 28th ed.* Boca Raton, FL: CRC Press, p. 297, 1987.

Dwight, H. B. *Table of Integrals and Other Mathematical Data, 4th ed.* New York: Macmillan, 1961.

Morse, P. M. and Feshbach, H. *Methods of Theoretical Physics, Part I.* New York: McGraw-Hill, pp. 411–413, 1953.

Serpentine Curve

A curve named and studied by Newton in 1701 and contained in his classification of CUBIC CURVES. It had been studied earlier by L'Hospital and Huygens in 1692 (MacTutor Archive).

The curve is given by the CARTESIAN equation

$$y(x) = \frac{abx}{x^2 + a^2} \tag{1}$$

and parametric equations

$$x(t) = a\cot t \tag{2}$$

$$y(t) = b\sin t\cos t. \tag{3}$$

The curve has a MAXIMUM at $x = a$ and a MINIMUM at $x = -a$, where

$$y'(x) = \frac{ab(a-x)(a+x)}{(a^2+x^2)^2} = 0, \tag{4}$$

and inflection points at $x = \pm\sqrt{3}\,a$, where

$$y''(x) = \frac{2abx(x^2 - 3a^2)}{(x^2+a^2)^3} = 0. \tag{5}$$

The CURVATURE is given by

$$\kappa(x) = \frac{2abx(x^2-3a^2)}{(x^2+a^2)^3\left[1 + \frac{(a^3b - abx^2)^2}{(x^2+a^2)^4}\right]^{3/2}} \tag{6}$$

$$\kappa(t) = -\frac{4\sqrt{2}\,ab[2\cos(2t) - 1]\cot t\csc^2 t}{\{b^2[1+\cos(4t)] + 2a^2\csc^4 t\}^{3/2}}. \tag{7}$$

References

Lawrence, J. D. *A Catalog of Special Plane Curves.* New York: Dover, pp. 111–112, 1972.

MacTutor History of Mathematics Archive. "Serpentine." `http://www-groups.dcs.st-and.ac.uk/~history/Curves/Serpentine.html`.

Serret-Frenet Formulas

see FRENET FORMULAS

Set

A set is a FINITE or INFINITE collection of objects. Older words for set include AGGREGATE and CLASS. Russell also uses the term MANIFOLD to refer to a set. The study of sets and their properties is the object of SET THEORY. Symbols used to operate on sets include $\wedge$ (which denotes the EMPTY SET $\varnothing$), $\vee =$ (which denotes the POWER SET of a set), $\cap$ (which means "and" or INTERSECTION), and $\cup$ (which means "or" or UNION).

The NOTATION A^B, where A and B are arbitrary sets, is used to denote the set of MAPS from B to A. For example, an element of $X^{\mathbb{N}}$ would be a MAP from the NATURAL NUMBERS $\mathbb{N}$ to the set X. Call such a function f, then $f(1)$, $f(2)$, etc., are elements of X, so call them x_1, x_2, etc. This now looks like a SEQUENCE of elements of X, so sequences are really just functions from $\mathbb{N}$ to X. This NOTATION is standard in mathematics and is frequently used in symbolic dynamics to denote sequence spaces.

Let E, F, and G be sets. Then operation on these sets using the $\cap$ and $\cup$ operators is COMMUTATIVE

$$E \cap F = F \cap E \tag{1}$$

$$E \cup F = F \cup E, \tag{2}$$

ASSOCIATIVE

$$(E \cap F) \cap G = E \cap (F \cap G) \tag{3}$$

$$A \cap \left(\bigcup_{i=1}^{n} B_i \right) = \bigcup_{i=1}^{n} (A \cap B_i) \tag{4}$$

$$(E \cup F) \cup G = E \cup (F \cup G), \tag{5}$$

and DISTRIBUTIVE

$$(E \cap F) \cup G = (E \cup G) \cap (F \cup G) \tag{6}$$

$$(E \cup F) \cap G = (E \cap G) \cup (F \cap G). \tag{7}$$

The proofs follow trivially using VENN DIAGRAMS.

$$P\left(\bigcup_{i=1}^{n} A_i \right) = \sum_{i=1}^{n} P(A_i). \tag{8}$$

The table below gives symbols for some common sets in mathematics.

Symbol	Set
$\mathbb{B}^n$	n-ball
$\mathbb{C}$	complex numbers
C^n, $C^{(n)}$	n-differentiable functions
$\mathbb{D}^n$	n-disk
$\mathbb{H}$	quaternions
$\mathbb{I}$	integers
$\mathbb{N}$	natural numbers
$\mathbb{Q}$	rational numbers
$\mathbb{R}^n$	real numbers in n-D
$\mathbb{S}^n$	n-sphere
$\mathbb{Z}$	integers
$\mathbb{Z}_n$	integers (mod n)
$\mathbb{Z}^-$	negative integers
$\mathbb{Z}^+$	positive integers
$\mathbb{Z}^*$	nonnegative integers

see also AGGREGATE, ANALYTIC SET, BOREL SET, $\mathbb{C}$, CLASS (SET), COANALYTIC SET, DEFINABLE SET, DERIVED SET, DOUBLE-FREE SET, EXTENSION, GROUND SET, $\mathbb{I}$, INTENSION, INTERSECTION, KINNEY'S SET, MANIFOLD, $\mathbb{N}$, PERFECT SET, POSET, $\mathbb{Q}$, $\mathbb{R}$, SET DIFFERENCE, SET THEORY, TRIPLE-FREE SET, UNION, VENN DIAGRAM, WELL-ORDERED SET, $\mathbb{Z}$, $\mathbb{Z}^-$, $\mathbb{Z}^+$

References

Courant, R. and Robbins, H. "The Algebra of Sets." Supplement to Ch. 2 in *What is Mathematics?: An Elementary Approach to Ideas and Methods, 2nd ed.* Oxford, England: Oxford University Press, pp. 108–116, 1996.

Set Difference

The set difference $A\backslash B$ is defined by

$$A\backslash B = \{x : x \in A \text{ and } x \notin B\}.$$

The same symbol is also used for QUOTIENT GROUPS.

Set Partition

A set partition of a SET S is a collection of disjoint SUBSETS $\mathbf{B}_0$, $\mathbf{B}_1$, ... of S whose UNION is S, where each $\mathbf{B}_i$ is called a BLOCK. The number of partitions of the SET $\{k\}_{k=1}^{n}$ is called a BELL NUMBER.

see also BELL NUMBER, BLOCK, RESTRICTED GROWTH STRING, STIRLING NUMBER OF THE SECOND KIND

References

Ruskey, F. "Info About Set Partitions." `http://sue.csc.uvic.ca/~cos/inf/setp/SetPartitions.html`.

Set Theory

The mathematical theory of SETS. Set theory is closely associated with the branch of mathematics known as LOGIC.

There are a number of different versions of set theory, each with its own rules and AXIOMS. In order of increasing CONSISTENCY STRENGTH, several versions of set theory include PEANO ARITHMETIC (ordinary ALGEBRA), second-order arithmetic (ANALYSIS), ZERMELO-FRAENKEL SET THEORY, Mahlo, weakly compact, hyper-Mahlo, ineffable, measurable, Ramsey, supercompact, huge, and n-huge set theory.

Given a set of REAL NUMBERS, there are 14 versions of set theory which can be obtained using only closure and complement (Beeler *et al.* 1972, Item 105).

see also AXIOMATIC SET THEORY, CONSISTENCY STRENGTH, CONTINUUM HYPOTHESIS, DESCRIPTIVE SET THEORY, IMPREDICATIVE, NAIVE SET THEORY, PEANO ARITHMETIC, SET, ZERMELO-FRAENKEL SET THEORY

References
Beeler, M.; Gosper, R. W.; and Schroeppel, R. *HAKMEM.* Cambridge, MA: MIT Artificial Intelligence Laboratory, Memo AIM-239, pp. 36–44, Feb. 1972.
Brown, K. S. "Set Theory and Foundations." `http://www.seanet.com/~ksbrown/ifoundat.htm`.
Courant, R. and Robbins, H. "The Algebra of Sets." Supplement to Ch. 2 in *What is Mathematics?: An Elementary Approach to Ideas and Methods, 2nd ed.* Oxford, England: Oxford University Press, pp. 108–116, 1996.
Devlin, K. *The Joy of Sets: Fundamentals of Contemporary Set Theory, 2nd ed.* New York: Springer-Verlag, 1993.
Halmos, P. R. *Naive Set Theory.* New York: Springer-Verlag, 1974.
MacTutor History of Mathematics Archive. "The Beginnings of Set Theory." `http://www-groups.dcs.st-and.ac.uk/~history/HistTopics/Beginnings_of_set_theory.html`.
Stewart, I. *The Problems of Mathematics, 2nd ed.* Oxford: Oxford University Press, p. 96, 1987.

Sexagesimal

The base-60 notational system for representing REAL NUMBERS. A base-60 number system was used by the Babylonians and is preserved in the modern measurement of time (hours, minutes, and seconds) and ANGLES (DEGREES, ARC MINUTES, and ARC SECONDS).

see also BASE (NUMBER), BINARY, DECIMAL, HEXADECIMAL, OCTAL, QUATERNARY, SCRUPLE, TERNARY, VIGESIMAL

References
Bergamini, D. *Mathematics.* New York: Time-Life Books, pp. 16–17, 1969.
Weisstein, E. W. "Bases." `http://www.astro.virginia.edu/~eww6n/math/notebooks/Bases.m`.

Sexdecillion

In the American system, 10^{51}.

see also LARGE NUMBER

Sextic Equation

The general sextic polynomial equation

$$x^6 + a_5x^5 + a_4x^4 + a_3x^3 + a_2x^2 + a_1x + a_0 = 0$$

can be solved in terms of HYPERGEOMETRIC FUNCTIONS in one variable using Klein's approach to solving the QUINTIC EQUATION.

see also CUBIC EQUATION, QUADRATIC EQUATION, QUARTIC EQUATION, QUINTIC EQUATION

References
Coble, A. B. "The Reduction of the Sextic Equation to the Valentiner Form—Problem." *Math. Ann.* **70**, 337–350, 1911a.
Coble, A. B. "An Application of Moore's Cross-ratio Group to the Solution of the Sextic Equation." *Trans. Amer. Math. Soc.* **12**, 311–325, 1911b.
Cole, F. N. "A Contribution to the Theory of the General Equation of the Sixth Degree." *Amer. J. Math.* **8**, 265–286, 1886.

Sextic Surface

An ALGEBRAIC SURFACE which can be represented implicitly by a polynomial of degree six in x, y, and z. Examples are the BARTH SEXTIC and BOY SURFACE.

see also ALGEBRAIC SURFACE, BARTH SEXTIC, BOY SURFACE, CUBIC SURFACE, DECIC SURFACE, QUADRATIC SURFACE, QUARTIC SURFACE

References
Catanese, F. and Ceresa, G. "Constructing Sextic Surfaces with a Given Number of Nodes." *J. Pure Appl. Algebra* **23**, 1–12, 1982.
Hunt, B. "Algebraic Surfaces." `http://www.mathematik.uni-kl.de/~wwwagag/Galerie.html`.

Sextillion

In the American system, 10^{21}.

see also LARGE NUMBER

Sexy Primes

Since a PRIME NUMBER cannot be divisible by 2 or 3, it must be true that, for a PRIME p, $p \equiv 6 \pmod{1,5}$. This motivates the definition of sexy primes as a pair of primes (p, q) such that $p - q = 6$ ("sexy" since "sex" is the Latin word for "six."). The first few sexy prime pairs are (5, 11), (7, 13), (11, 17), (13, 19), (17, 23), (23, 29), (31, 37), (37, 43), (41, 47), (47, 53), ... (Sloane's A023201 and A046117).

Sexy constellations also exist. The first few sexy triplets (i.e., numbers such that each of $(p, p + 6, p + 12)$ is PRIME but $p + 18$ is *not* PRIME) are (7, 13, 19), (17, 23, 29), (31, 37, 43), (47, 53, 59), ... (Sloane's A046118, A046119, and A046120). The first few sexy quadruplets are (11, 17, 23, 29), (41, 47, 53, 59), (61, 67, 73, 79), (251, 257, 263, 269), ... (Sloane's A046121, A046122, A046123, A046124). Sexy quadruplets can only begin with a PRIME ending in a "1." There is only a single sexy quintuplet, (5, 11, 17, 23, 29), since every fifth number of the form $6n \pm 1$ is divisible by 5, and therefore cannot be PRIME.

see also PRIME CONSTELLATION, PRIME QUADRUPLET, TWIN PRIMES

References
Sloane, N. J. A. Sequences A023201, A046117, A046118, A046119, A046120, A046121, A046122, A046123, and A046124 in "An On-Line Version of the Encyclopedia of Integer Sequences."
Trotter, T. "Sexy Primes." `http://www.geocities.com/CapeCanaveral/Launchpad/8202/sexyprim.html`.

Seydewitz's Theorem

If a TRIANGLE is inscribed in a CONIC SECTION, any line conjugate to one side meets the other two sides in conjugate points.

see also CONIC SECTION, TRIANGLE

Sgn

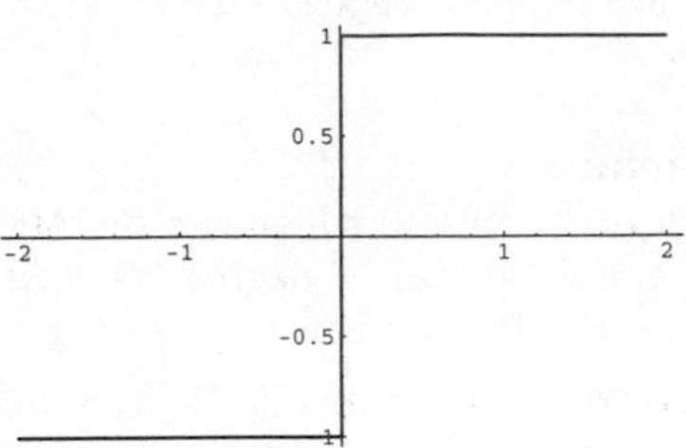

Also called SIGNUM. It can be defined as

$$\operatorname{sgn} \equiv \begin{cases} -1 & x < 0 \\ 0 & x = 0 \\ 1 & x > 0 \end{cases} \quad (1)$$

or

$$\operatorname{sgn}(x) = 2H(x) - 1, \quad (2)$$

where $H(x)$ is the HEAVISIDE STEP FUNCTION. For $x \neq 0$, this can be written

$$\operatorname{sgn}(x) \equiv \frac{x}{|x|} \text{ for } x \neq 0. \quad (3)$$

see also HEAVISIDE STEP FUNCTION, RAMP FUNCTION

Shadow

The SURFACE corresponding to the region of obscuration when a solid is illuminated from a point light source (located at the RADIANT POINT). A DISK is the SHADOW of a SPHERE on a PLANE perpendicular to the SPHERE-RADIANT POINT line. If the PLANE is tilted, the shadow can be the interior of an ELLIPSE or a PARABOLA.

see also PROJECTIVE GEOMETRY

Shadowing Theorem

Although a numerically computed CHAOTIC trajectory diverges exponentially from the true trajectory with the same initial coordinates, there exists an errorless trajectory with a slightly different initial condition that stays near ("shadows") the numerically computed one. Therefore, the FRACTAL structure of chaotic trajectories seen in computer maps is real.

References

Ott, E. *Chaos in Dynamical Systems.* New York: Cambridge University Press, pp. 18–19, 1993.

Shafarevich Conjecture

A conjecture which implies the MORDELL CONJECTURE, as proved in 1968 by A. N. Parshin.

see also MORDELL CONJECTURE

References

Stewart, I. *The Problems of Mathematics, 2nd ed.* Oxford, England: Oxford University Press, p. 45, 1987.

Shah Function

$$\text{Ш}(x) \equiv \sum_{n=-\infty}^{\infty} \delta(x-n) \quad (1)$$

where $\delta(x)$ is the DELTA FUNCTION, so $\text{Ш}(x) = 0$ for $x \notin \mathbb{Z}$ (i.e., x not an INTEGER). The shah function obeys the identities

$$\text{Ш}(ax) = \frac{1}{a} \sum_{n=-\infty}^{\infty} \delta\left(x - \frac{n}{a}\right) \quad (2)$$

$$\text{Ш}(-x) = \text{Ш}(x) \quad (3)$$

$$\text{Ш}(x+n) = \text{Ш}(x), \quad (4)$$

for $2n \in \mathbb{Z}$ (i.e., n a half-integer).

It is normalized so that

$$\int_{n-1/2}^{n+1/2} \text{Ш}(x)\, dx = 1. \quad (5)$$

The "sampling property" is

$$\text{Ш}(x) f(x) = \sum_{n=-\infty}^{\infty} f(n)\delta(x-n) \quad (6)$$

and the "replicating property" is

$$\text{Ш}(x) * f(x) = \sum_{n=-\infty}^{\infty} f(x-n), \quad (7)$$

where $*$ denotes CONVOLUTION.

see also CONVOLUTION, DELTA FUNCTION, IMPULSE PAIR

Shah-Wilson Constant

see TWIN PRIMES CONSTANT

Shallit Constant

Define $f(x_1, x_2, \ldots, x_n)$ with x_i POSITIVE as

$$f(x_1, x_2, \ldots, x_n) \equiv \sum_{i=1}^{n} x_i + \sum_{1 \leq i \leq k \leq n} \prod_{j=i}^{k} \frac{1}{x_j}.$$

Then

$$\min f = 3n - C + o(1)$$

as n increases, where the Shallit constant is

$$C = 1.369451403937\ldots$$

(Shallit 1995). In their solution, Grosjean and De Meyer (quoted in Shallit 1995) reduced the complexity of the problem.

References

MacLeod, A. `http://www.mathsoft.com/asolve/constant/shapiro/macleod.html`.

Shallit, J. Solution by C. C. Grosjean and H. E. De Meyer. "A Minimization Problem." Problem 94-15 in *SIAM Review* **37**, 451–458, 1995.

Shallow Diagonal

see PASCAL'S TRIANGLE

Shanks' Algorithm

An ALGORITHM which finds the least NONNEGATIVE value of $\sqrt{a} \pmod{p}$ for given a and PRIME p.

Shanks' Conjecture

Let $p(g)$ be the first PRIME which follows a PRIME GAP of g between consecutive PRIMES. Shanks' conjecture holds that

$$\ln[p(g)] \sim \sqrt{g}.$$

see also PRIME DIFFERENCE FUNCTION, PRIME GAPS

References

Guy, R. K. *Unsolved Problems in Number Theory, 2nd ed.* New York: Springer-Verlag, p. 21, 1994.

Rivera, C. "Problems & Puzzles (Conjectures): Shanks' Conjecture." http://www.sci.net.mx/~crivera/ppp/conj_009.htm.

Shanks, D. "On Maximal Gaps Between Successive Primes." *Math. Comput.* **18**, 646–651, 1964.

Shannon Entropy

see ENTROPY

Shannon Sampling Theorem

see SAMPLING THEOREM

Shape Operator

The negative derivative

$$S(\mathbf{v}) = -D_{\mathbf{v}}\mathbf{N} \tag{1}$$

of the unit normal $\mathbf{N}$ vector field of a SURFACE is called the shape operator (or WEINGARTEN MAP or SECOND FUNDAMENTAL TENSOR). The shape operator S is an EXTRINSIC CURVATURE, and the GAUSSIAN CURVATURE is given by the DETERMINANT of S. If $\mathbf{x} : U \to \mathbb{R}^3$ is a REGULAR PATCH, then

$$S(\mathbf{x}_u) = -\mathbf{N}_u \tag{2}$$

$$S(\mathbf{x}_v) = -\mathbf{N}_v. \tag{3}$$

At each point $\mathbf{p}$ on a REGULAR SURFACE $M \subset \mathbb{R}^3$, the shape operator is a linear map

$$S : M_{\mathbf{p}} \to M_{\mathbf{p}}. \tag{4}$$

The shape operator for a surface is given by the WEINGARTEN EQUATIONS.

see also CURVATURE, FUNDAMENTAL FORMS, WEINGARTEN EQUATIONS

References

Gray, A. "The Shape Operator," "Calculation of the Shape Operator," and "The Eigenvalues of the Shape Operator." §14.1, 14.3, and 14.4 in *Modern Differential Geometry of Curves and Surfaces.* Boca Raton, FL: CRC Press, pp. 268–269, 274–279, 1993.

Reckziegel, H. In *Mathematical Models from the Collections of Universities and Museums* (Ed. G. Fischer). Braunschweig, Germany: Vieweg, p. 30, 1986.

Shapiro's Cyclic Sum Constant

N.B. A detailed on-line essay by S. Finch was the starting point for this entry.

Consider the sum

$$f_n(x_1, x_2, \ldots, x_n) = \frac{x_1}{x_2 + x_3} + \frac{x_2}{x_3 + x_4} + \ldots + \frac{x_{n-1}}{x_n + x_1} + \frac{x_n}{x_1 + x_2}, \tag{1}$$

where the x_js are NONNEGATIVE and the DENOMINATORS are POSITIVE. Shapiro (1954) asked if

$$f_n(x_1, x_2, \ldots, x_n) \geq \tfrac{1}{2}n \tag{2}$$

for all n. It turns out (Mitrinovic *et al.* 1993) that this INEQUALITY is true for all EVEN $n \leq 12$ and ODD $n \leq 23$. Ranikin (1958) proved that for

$$f(n) = \inf_{x \geq 0} f_n(x_1, x_2, \ldots, x_n), \tag{3}$$

$$\lambda = \lim_{n\to\infty} \frac{f(n)}{n} = \inf_{n \geq 1} \frac{f(n)}{n} < \tfrac{1}{2} - 7 \times 10^{-8}. \tag{4}$$

λ can be computed by letting $\phi(x)$ be the CONVEX HULL of the functions

$$y_1 = e^{-x} \tag{5}$$

$$y_2 = \frac{2}{e^x + e^{x/2}}. \tag{6}$$

Then

$$\lambda = \tfrac{1}{2}\phi(0) = 0.4945668\ldots \tag{7}$$

(Drinfeljd 1971).

A modified sum was considered by Elbert (1973):

$$g_n(x_1, x_1, \ldots, x_n) = \frac{x_1 + x_3}{x_1 + x_2} + \frac{x_2 + x_4}{x_2 + x_3} + \ldots + \frac{x_{x-1} + x_1}{x_{n-1} + x_n} + \frac{x_n + x_2}{x_n + x_1}. \tag{8}$$

Consider

$$\mu = \lim_{n\to\infty} \frac{g(n)}{n}, \tag{9}$$

where

$$g(n) = \inf_{x \geq 0} g_n(x_1, x_2, \ldots, x_n), \tag{10}$$

and let $\psi(x)$ be the CONVEX HULL of

$$y_1 = \tfrac{1}{2}(1 + e^x) \tag{11}$$

$$y_2 = \frac{1 + e^x}{1 + e^{x/2}}. \tag{12}$$

Then

$$\mu = \psi(0) = 0.978012\ldots. \tag{13}$$

see also CONVEX HULL

References

Drinfeljd, V. G. "A Cyclic Inequality." *Math. Notes. Acad. Sci. USSR* **9**, 68–71, 1971.

Elbert, A. "On a Cyclic Inequality." *Period. Math. Hungar.* **4**, 163–168, 1973.

Finch, S. "Favorite Mathematical Constants." http://www.mathsoft.com/asolve/constant/shapiro/shapiro.html.

Mitrinovic, D. S.; Pecaric, J. E.; and Fink, A. M. *Classical and New Inequalities in Analysis.* New York: Kluwer, 1993.

Sharing Problem

A problem also known as the POINTS PROBLEM or UNFINISHED GAME. Consider a tournament involving k players playing the same game repetitively. Each game has a single winner, and denote the number of games won by player i at some juncture w_i. The games are independent, and the probability of the ith player winning a game is p_i. The tournament is specified to continue until one player has won n games. If the tournament is discontinued before any player has won n games so that $w_i < n$ for $i = 1, \ldots, k$, how should the prize money be shared in order to distribute it proportionally to the players' chances of winning?

For player i, call the number of games left to win $r_i \equiv n - w_i > 0$ the "quota." For two players, let $p \equiv p_1$ and $q \equiv p_2 = 1 - p$ be the probabilities of winning a single game, and $a \equiv r_1 = n - w_1$ and $b \equiv r_2 = n - w_2$ be the number of games needed for each player to win the tournament. Then the stakes should be divided in the ratio $m : n$, where

$$m = p^a \left[1 + \frac{a}{1}q + \frac{a(a+1)}{2!}q^2 + \ldots + \frac{a(a+1)\cdots(a+b-2)}{(b-1)!}q^{b-1} \right] \quad (1)$$

$$n = q^b \left[1 + \frac{b}{1}p + \frac{b(b+1)}{2!}p^2 + \ldots + \frac{b(b+1)\cdots(b+a-2)}{(a-1)!}p^{a-1} \right] \quad (2)$$

(Kraitchik 1942).

If i players have equal probability of winning ("cell probability"), then the chance of player i winning for quotas $r_1, \ldots, r_k$ is

$$W_i = D_1^{k-1}(r_1, \ldots, r_{i-1}, r_{i+1}, \ldots, r_k; r_i), \quad (3)$$

where D is the DIRICHLET INTEGRAL of type 2D. Similarly, the chance of player i losing is

$$L_i = C_1^{k-1}(r_1, \ldots, r_{i-1}, r_{i+1}, \ldots, r_k; r_i), \quad (4)$$

where C is the DIRICHLET INTEGRAL of type 2C. If the cell quotas are not equal, the general Dirichlet integral $D_{\mathbf{a}}$ must be used, where

$$a_i = \frac{p_i}{1 - \sum_{i=1}^{k-1} p_i}. \quad (5)$$

If $r_i = r$ and $a_i = 1$, then W_i and L_i reduce to $1/k$ as they must. Let $P(r_1, \ldots, r_k)$ be the joint probability that the players would be RANKED in the order of the r_is in the argument list if the contest were completed. For $k = 3$,

$$P(r_1, r_2, r_3) = CD_1^{(1,1)}(r_1, r_2, r_3). \quad (6)$$

For $k = 4$ with quota vector $\mathbf{r} = (r_1, r_2, r_3, r_4)$ and $\Delta = p_2 + p_3 + p_4$,

$$P(\mathbf{r}) = \sum_{i=0}^{r_3-1} \sum_{j=0}^{r_4-1} \binom{r_2 - 1 + i + j}{r_2 - 1, i, j} \left(\frac{p_2}{\Delta}\right)^{r_2} \left(\frac{p_3}{\Delta}\right)^i \left(\frac{p_4}{\Delta}\right)^j \times C_{p_1/\Delta}^{(1)}(r_1, r_2 + i + j) D_{p_4/p_3}^{(1)}(r_4 - j, r_3 - i). \quad (7)$$

An expression for $k = 5$ is given by Sobel and Frankowski (1994, p. 838).

see also DIRICHLET INTEGRALS

References

Kraitchik, M. "The Unfinished Game." §6.1 in *Mathematical Recreations.* New York: W. W. Norton, pp. 117–118, 1942.

Sobel, M. and Frankowski, K. "The 500th Anniversary of the Sharing Problem (The Oldest Problem in the Theory of Probability)." *Amer. Math. Monthly* **101**, 833–847, 1994.

Sharkovsky's Theorem

see ŠARKOVSKII'S THEOREM

Sharpe's Differential Equation

A generalization of the BESSEL DIFFERENTIAL EQUATION for functions of order 0, given by

$$zy'' + y' + (z + A)y = 0.$$

Solutions are

$$y = e^{\pm iz}{}_1F_1\left(\tfrac{1}{2} \mp \tfrac{1}{2}iA; 1; \mp 2iz\right),$$

where ${}_1F_1(a; b; x)$ is a CONFLUENT HYPERGEOMETRIC FUNCTION.

see also BESSEL DIFFERENTIAL EQUATION, CONFLUENT HYPERGEOMETRIC FUNCTION

Sharpe Ratio

A risk-adjusted financial measure developed by Nobel Laureate William Sharpe. It uses a fund's standard deviation and excess return to determine the reward per unit of risk. The higher a fund's Sharpe ratio, the better the fund's "risk-adjusted" performance.

see also ALPHA, BETA

Sheaf (Geometry)

The set of all PLANES through a LINE.

see also LINE, PENCIL, PLANE

References

Woods, F. S. *Higher Geometry: An Introduction to Advanced Methods in Analytic Geometry.* New York: Dover, p. 12, 1961.

Sheaf (Topology)

A topological GADGET related to families of ABELIAN GROUPS and MAPS.

References

Iyanaga, S. and Kawada, Y. (Eds.). "Sheaves." §377 in *Encyclopedic Dictionary of Mathematics.* Cambridge, MA: MIT Press, p. 1171–1174, 1980.

Shear

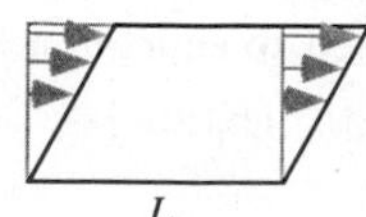

A transformation in which all points along a given LINE L remain fixed while other points are shifted parallel to L by a distance proportional to their PERPENDICULAR distance from L. Shearing a plane figure does not change its AREA. The shear can also be generalized to 3-D, in which PLANES are translated instead of lines.

Shear Matrix

The shear matrix e_{ij}^s is obtained from the IDENTITY MATRIX by inserting s at (i,j), e.g.,

$$\mathsf{e}_{12}^s = \begin{bmatrix} 1 & s & 0 \\ 0 & 1 & 0 \\ 0 & 0 & 1 \end{bmatrix}.$$

see also ELEMENTARY MATRIX

Shephard's Problem

Measurements of a centered convex body in Euclidean n-space (for $n \geq 3$) show that its brightness function (the volume of each projection) is smaller than that of another such body. Is it true that its VOLUME is also smaller? C. M. Petty and R. Schneider showed in 1967 that the answer is yes if the body with the larger brightness function is a projection body, but no in general for every n.

References

Gardner, R. J. "Geometric Tomography." *Not. Amer. Math. Soc.* **42**, 422–429, 1995.

Sheppard's Correction

A correction which must be applied to the MOMENTS computed from NORMALLY DISTRIBUTED data which have been binned. The corrected versions of the second, third, and fourth moments are

$$\mu_2 = {\mu_2}^{(0)} - \tfrac{1}{12}c^2 \qquad (1)$$

$$\mu_3 = {\mu_3}^{(0)} \qquad (2)$$

$$\mu_4 = {\mu_4}^{(0)} - \tfrac{1}{2}{\mu_2}^{(0)} + \tfrac{7}{240}c^2, \qquad (3)$$

where c is the CLASS INTERVAL. If κ_r' is the rth CUMULANT of an ungrouped distribution and κ_r the rth CUMULANT of the grouped distribution with CLASS INTERVAL c, the corrected cumulants (under rather restrictive conditions) are

$$\kappa_r' = \begin{cases} \kappa_r & \text{for } r \text{ odd} \\ \kappa_r - \frac{B_r}{r}c^r & \text{for } r \text{ even,} \end{cases} \qquad (4)$$

where B_r is the rth BERNOULLI NUMBER, giving

$$\kappa_1' = \kappa_1 \qquad (5)$$

$$\kappa_2' = \kappa_2 - \tfrac{1}{12}c^2 \qquad (6)$$

$$\kappa_3' = \kappa_3 \qquad (7)$$

$$\kappa_4' = \kappa_4 + \tfrac{1}{120}c^4 \qquad (8)$$

$$\kappa_5' = \kappa_5 \qquad (9)$$

$$\kappa_6' = \kappa_6 - \tfrac{1}{252}c^6. \qquad (10)$$

For a proof, see Kendall *et al.* (1987).

References

Kendall, M. G.; Stuart, A.; and Ord, J. K. *Kendall's Advanced Theory of Statistics, Vol. 1: Distribution Theory, 6th ed.* New York: Oxford University Press, 1987.

Kenney, J. F. and Keeping, E. S. "Sheppard's Correction." §4.12 in *Mathematics of Statistics, Pt. 2, 2nd ed.* Princeton, NJ: Van Nostrand, pp. 80–82, 1951.

Sherman-Morrison Formula

A formula which allows the new MATRIX to be computed for a small change to a MATRIX A. If the change can be written in the form

$$\mathbf{u} \otimes \mathbf{v}$$

for two vectors $\mathbf{u}$ and $\mathbf{v}$, then the Sherman-Morrison formula is

$$(\mathsf{A} + \mathbf{u} \otimes \mathbf{v})^{-1} = \mathsf{A}^{-1} - \frac{(\mathsf{A}^{-1}\mathbf{u}) \otimes (\mathbf{v} \cdot \mathsf{A}^{-1})}{1 + \lambda},$$

where

$$\lambda \equiv \mathbf{v} \cdot \mathsf{A}^{-1}\mathbf{u}.$$

see also WOODBURY FORMULA

References

Press, W. H.; Flannery, B. P.; Teukolsky, S. A.; and Vetterling, W. T. "Sherman-Morrison Formula." In *Numerical Recipes in FORTRAN: The Art of Scientific Computing, 2nd ed.* Cambridge, England: Cambridge University Press, pp. 65–67, 1992.

Shi

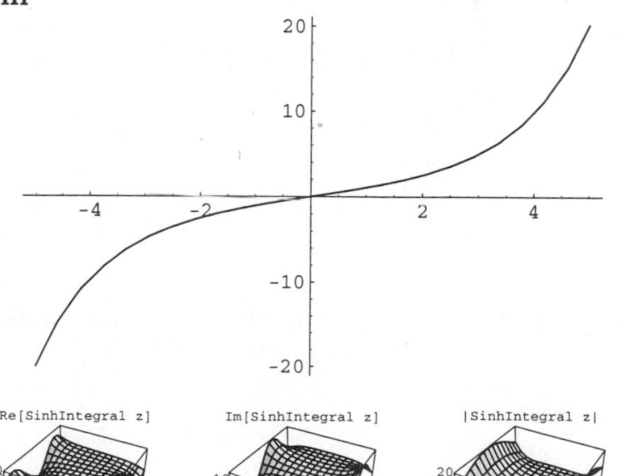

$$\mathrm{Shi}(z) = \int_0^z \frac{\sinh t}{t}\,dt.$$

The function is given by the *Mathematica*® (Wolfram Research, Champaign, IL) command `SinhIntegral[z]`.

see also CHI, COSINE INTEGRAL, SINE INTEGRAL

References
Abramowitz, M. and Stegun, C. A. (Eds.). "Sine and Cosine Integrals." §5.2 in *Handbook of Mathematical Functions with Formulas, Graphs, and Mathematical Tables, 9th printing.* New York: Dover, pp. 231–233, 1972.

Shift

A TRANSLATION without ROTATION or distortion.

see also DILATION, EXPANSION, ROTATION, TRANSLATION, TWIRL

Shift Property

see DELTA FUNCTION

Shimura-Taniyama Conjecture

see TANIYAMA-SHIMURA CONJECTURE

Shimura-Taniyama-Weil Conjecture

see TANIYAMA-SHIMURA CONJECTURE

Shoe Surface

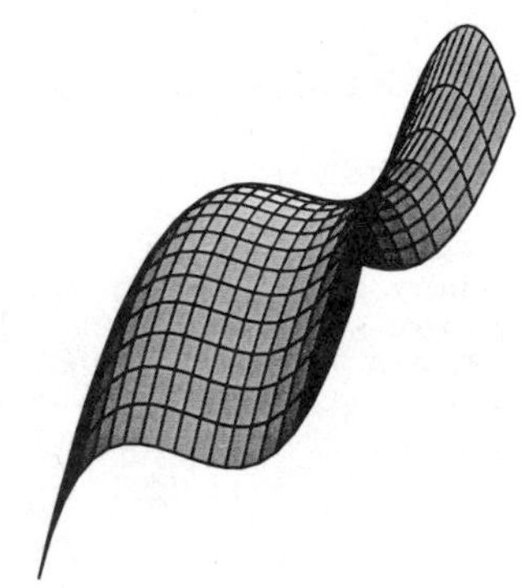

A surface given by the parametric equations

$$x(u,v) = u$$
$$y(u,v) = v$$
$$z(u,v) = \tfrac{1}{3}u^3 - \tfrac{1}{2}v^2.$$

References
Gray, A. *Modern Differential Geometry of Curves and Surfaces.* Boca Raton, FL: CRC Press, p. 634, 1993.

Shoemaker's Knife

see ARBELOS

Shortening

A KNOT used to shorten a long rope.

see also BEND (KNOT)

References
Owen, P. *Knots.* Philadelphia, PA: Courage, p. 65, 1993.

Shuffle

The randomization of a deck of CARDS by repeated interleaving. More generally, a shuffle is a rearrangement of the elements in an ordered list. Shuffling by exactly interleaving two halves of a deck is called a RIFFLE SHUFFLE. Normal shuffling leaves gaps of different lengths between the two layers of cards and so randomizes the order of the cards.

A deck of 52 CARDS must be shuffled seven times for it to be randomized (Aldous and Diaconis 1986, Bayer and Diaconis 1992). This is intermediate between too few shuffles and the decreasing effectiveness of many shuffles. One of Bayer and Diaconis's randomness CRITERIA, however, gives $3\lg k/2$ shuffles for a k-card deck, yielding 11–12 shuffles for 52 CARDS. Keller (1995) shows that roughly $\ln k$ shuffles are needed just to randomize the bottom card.

see also BAYS' SHUFFLE, CARDS, FARO SHUFFLE, MONGE'S SHUFFLE, RIFFLE SHUFFLE

References
Aldous, D. and Diaconis, P. "Shuffling Cards and Stopping Times." *Amer. Math. Monthly* **93**, 333–348, 1986.
Bayer, D. and Diaconis, P. "Trailing the Dovetail Shuffle to Its Lair." *Ann. Appl. Probability* **2**, 294–313, 1992.
Keller, J. B. "How Many Shuffles to Mix a Deck?" *SIAM Review* **37**, 88–89, 1995.
Morris, S. B. "Practitioner's Commentary: Card Shuffling." *UMAP J.* **15**, 333–338, 1994.
Rosenthal, J. W. "Card Shuffling." *Math. Mag.* **54**, 64–67, 1981.

Siamese Dodecahedron

see SNUB DISPHENOID

Siamese Method

A method for constructing MAGIC SQUARES of ODD order, also called DE LA LOUBERE'S METHOD.

see also MAGIC SQUARE

Sibling

Two nodes connected to the same node in a ROOTED TREE are called siblings.

see also CHILD, ROOTED TREE

Sicherman Dice

2
4 2 1 3
3

6
8 4 1 5
3

A pair of DICE which have the same ODDS for throwing every number as a normal pair of 6-sided DICE. They are the only such alternate arrangement.

see also DICE, EFRON'S DICE

Sici Spiral

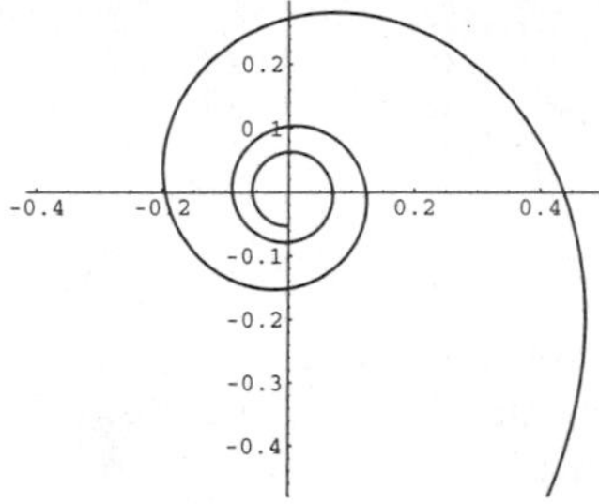

The spiral

$$x = c\,\mathrm{ci}\,t$$
$$y = c(\mathrm{si}\,t - \tfrac{1}{2}\pi),$$

where $\mathrm{ci}(t)$ and $\mathrm{si}(t)$ are the COSINE INTEGRAL and SINE INTEGRAL and c is a constant.

see also COSINE INTEGRAL, SINE INTEGRAL, SPIRAL

References

von Seggern, D. *CRC Standard Curves and Surfaces.* Boca Raton, FL: CRC Press, pp. 204 and 270, 1993.

Side

The edge of a POLYGON and face of a POLYHEDRON are sometimes called sides.

Sidon Sequence

see $B2$-SEQUENCE

Siegel Disk Fractal

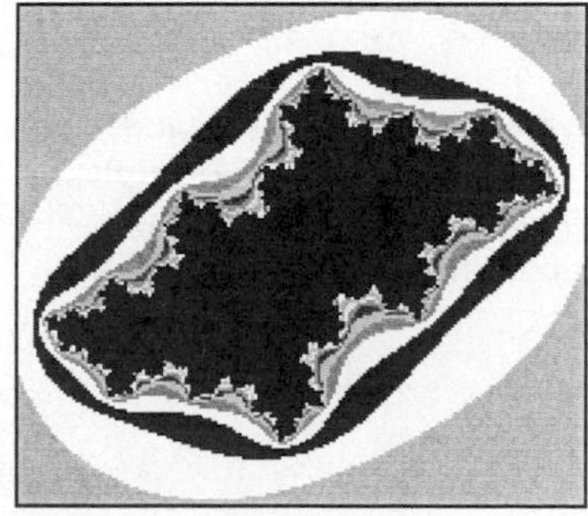

A JULIA SET with $c = -0.390541 - 0.586788i$. The FRACTAL somewhat resembles the better known MANDELBROT SET.

see also DOUADY'S RABBIT FRACTAL, JULIA SET, MANDELBROT SET, SAN MARCO FRACTAL

References

Wagon, S. *Mathematica in Action.* New York: W. H. Freeman, p. 176, 1991.

Siegel Modular Function

A Γ_n-invariant meromorphic function on the space of all $n \times n$ complex symmetric matrices with POSITIVE IMAGINARY PART. In 1984, H. Umemura expressed the ROOTS of an arbitrary POLYNOMIAL in terms of elliptic Siegel functions.

References

Iyanaga, S. and Kawada, Y. (Eds.). "Siegel Modular Functions." §34F in *Encyclopedic Dictionary of Mathematics.* Cambridge, MA: MIT Press, pp. 131–132, 1980.

Siegel's Paradox

If a fixed FRACTION x of a given amount of money P is lost, and then the same FRACTION x of the remaining amount is gained, the result is less than the original and equal to the final amount if a FRACTION x is first gained, then lost. This can easily be seen from the fact that

$$[P(1-x)](1+x) = P(1-x^2) < P$$
$$[P(1+x)](1-x) = P(1-x^2) < P.$$

Siegel's Theorem

An ELLIPTIC CURVE can have only a finite number of points with INTEGER coordinates.

see also ELLIPTIC CURVE

References

Davenport, H. "Siegel's Theorem." Ch. 21 in *Multiplicative Number Theory, 2nd ed.* New York: Springer-Verlag, pp. 126–125, 1980.

Sierpiński Arrowhead Curve

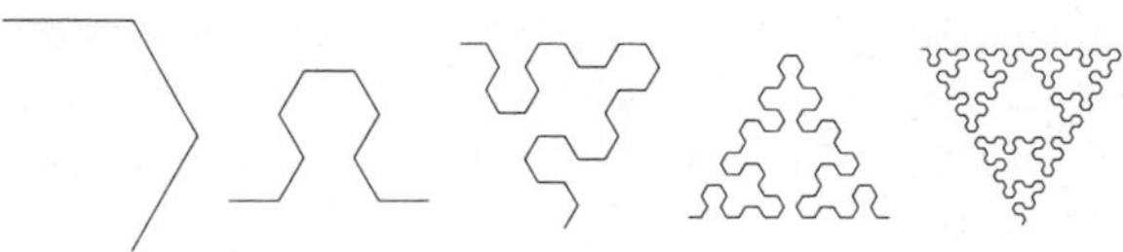

A FRACTAL which can be written as a LINDENMAYER SYSTEM with initial string `"YF"`, STRING REWRITING rules `"X" -> "YF+XF+Y"`, `"Y" -> "XF-YF-X"`, and angle 60°.

see also DRAGON CURVE, HILBERT CURVE, KOCH SNOWFLAKE, LINDENMAYER SYSTEM, PEANO CURVE, PEANO-GOSPER CURVE, SIERPIŃSKI CURVE, SIERPIŃSKI SIEVE

References

Dickau, R. M. "Two-Dimensional L-Systems." `http://forum.swarthmore.edu/advanced/robertd/lsys2d.html`.

Sierpiński Carpet

A FRACTAL which is constructed analogously to the SIERPIŃSKI SIEVE, but using squares instead of triangles. Let N_n be the number of black boxes, L_n the length of a side of a white box, and A_n the fractional AREA of black boxes after the nth iteration. Then

$$N_n = 8^n \tag{1}$$

$$L_n = (\tfrac{1}{3})^n = 3^{-n} \tag{2}$$

$$A_n = {L_n}^2 N_n = (\tfrac{8}{9})^n. \tag{3}$$

The CAPACITY DIMENSION is therefore

$$d_{\text{cap}} = -\lim_{n\to\infty} \frac{\ln N_n}{\ln L_n} = -\lim_{n\to\infty} \frac{\ln(8^n)}{\ln(3^{-n})} = \frac{\ln 8}{\ln 3}$$

$$= \frac{3\ln 2}{\ln 3} = 1.892789261\ldots. \tag{4}$$

see also MENGER SPONGE, SIERPIŃSKI SIEVE

References

Dickau, R. M. "The Sierpinski Carpet." `http://forum.swarthmore.edu/advanced/robertd/carpet.html`.

Peitgen, H.-O.; Jürgens, H.; and Saupe, D. *Chaos and Fractals: New Frontiers of Science.* New York: Springer-Verlag, pp. 112–121, 1992.

Weisstein, E. W. "Fractals." `http://www.astro.virginia.edu/~eww6n/math/notebooks/Fractal.m`.

Sierpiński's Composite Number Theorem

There exist infinitely many ODD INTEGERS k such that $k \cdot 2^n + 1$ is COMPOSITE for every $n \geq 1$. Numbers k with this property are called SIERPIŃSKI NUMBERS OF THE SECOND KIND, and analogous numbers with the plus sign replaced by a minus are called RIESEL NUMBERS. It is conjectured that the smallest SIERPIŃSKI NUMBER OF THE SECOND KIND is $k = 78{,}557$ and the smallest RIESEL NUMBER is $k = 509{,}203$.

see also CUNNINGHAM NUMBER, SIERPIŃSKI NUMBER OF THE SECOND KIND

References

Buell, D. A. and Young, J. "Some Large Primes and the Sierpiński Problem." SRC Tech. Rep. 88004, Supercomputing Research Center, Lanham, MD, 1988.

Jaeschke, G. "On the Smallest k such that $k \cdot 2^N + 1$ are Composite." *Math. Comput.* **40**, 381–384, 1983.

Jaeschke, G. Corrigendum to "On the Smallest k such that $k \cdot 2^N + 1$ are Composite." *Math. Comput.* **45**, 637, 1985.

Keller, W. "Factors of Fermat Numbers and Large Primes of the Form $k \cdot 2^n + 1$." *Math. Comput.* **41**, 661–673, 1983.

Keller, W. "Factors of Fermat Numbers and Large Primes of the Form $k \cdot 2^n + 1$, II." In prep.

Ribenboim, P. *The New Book of Prime Number Records.* New York: Springer-Verlag, pp. 357–359, 1996.

Riesel, H. "Några stora primtal." *Elementa* **39**, 258–260, 1956.

Sierpiński, W. "Sur un problème concernant les nombres $k \cdot 2^n + 1$." *Elem. d. Math.* **15**, 73–74, 1960.

see also COMPOSITE NUMBER, SIERPIŃSKI NUMBERS OF THE SECOND KIND, SIERPIŃSKI'S PRIME SEQUENCE THEOREM

Sierpiński Constant

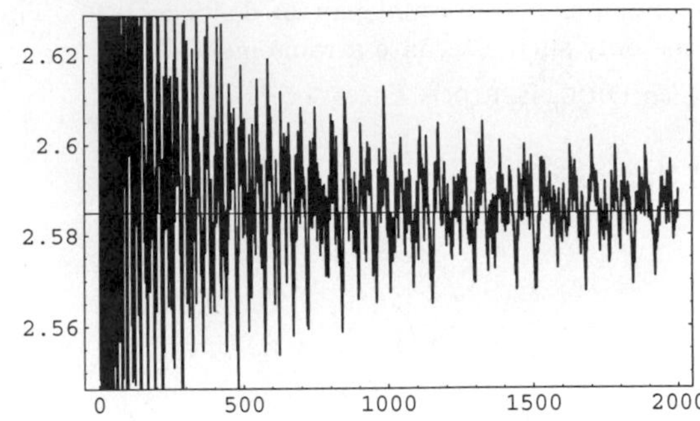

Let $r_k(n)$ denote the number of representations of n by k squares, then the SUMMATORY FUNCTION of $r_2(k)/k$ has the ASYMPTOTIC expansion

$$\sum_{k=1}^{n} \frac{r_2(k)}{k} = K + \pi \ln n + \mathcal{O}(n^{-1/2}),$$

where $K = 2.5849817596$ is the Sierpiński constant. The above plot shows

$$\left[\sum_{k=1}^{n} \frac{r_2(k)}{k}\right] - \pi \ln n,$$

with the value of K indicated as the solid horizontal line.

see also $r_k(n)$

References

Sierpiński, W. *Oeuvres Choiseies, Tome 1.* Editions Scientifiques de Pologne, 1974.

Sierpiński Curve

There are several FRACTAL curves associated with Sierpiński. The above curve is one example, and the SIERPIŃSKI ARROWHEAD CURVE is another. The limit of the curve illustrated above has AREA

$$A = \tfrac{5}{12}.$$

The AREA for a related curve illustrated by Cundy and Rollett (1989) is

$$A = \tfrac{1}{3}(7 - 4\sqrt{2}).$$

see also EXTERIOR SNOWFLAKE, GOSPER ISLAND, HILBERT CURVE, KOCH ANTISNOWFLAKE, KOCH SNOWFLAKE, PEANO CURVE, PEANO-GOSPER CURVE, SIERPIŃSKI ARROWHEAD CURVE

References
Cundy, H. and Rollett, A. *Mathematical Models, 3rd ed.* Stradbroke, England: Tarquin Pub., pp. 67–68, 1989.
Dickau, R. M. "Two-Dimensional L-Systems." `http://forum.swarthmore.edu/advanced/robertd/lsys2d.html`.
Gardner, M. *Penrose Tiles and Trapdoor Ciphers... and the Return of Dr. Matrix, reissue ed.* New York: W. H. Freeman, p. 34, 1989.
Wagon, S. *Mathematica in Action.* New York: W. H. Freeman, p. 207, 1991.

Sierpiński Gasket

see SIERPIŃSKI SIEVE

Sierpiński-Menger Sponge

see MENGER SPONGE

Sierpiński Number of the First Kind

Numbers of the form $S_n \equiv n^n + 1$. The first few are 2, 5, 28, 257, 3126, 46657, 823544, 16777217, ... (Sloane's A014566). Sierpiński proved that if S_n is PRIME with $n \geq 2$, then $S_n = F_{m+2^m}$, where F_m is a FERMAT NUMBER with $m \geq 0$. The first few such numbers are $F_1 = 5$, $F_3 = 257$, F_6, F_{11}, F_{20}, and F_{37}. Of these, 5 and 257 are PRIME, and the first unknown case is $F_{37} > 10^{3\times 10^{10}}$.

see also CULLEN NUMBER, CUNNINGHAM NUMBER, FERMAT NUMBER, WOODALL NUMBER

References
Madachy, J. S. *Madachy's Mathematical Recreations.* New York: Dover, p. 155, 1979.
Ribenboim, P. *The Book of Prime Number Records, 2nd ed.* New York: Springer-Verlag, p. 74, 1989.
Sloane, N. J. A. Sequence A014566 in "An On-Line Version of the Encyclopedia of Integer Sequences."

Sierpiński Number of the Second Kind

A number k satisfying SIERPIŃSKI'S COMPOSITE NUMBER THEOREM, i.e., such that $k \cdot 2^n + 1$ is COMPOSITE for every $n \geq 1$. The smallest known is $k = 78,557$, but there remain 35 smaller candidates (the smallest of which is 4847) which are known to generate only composite numbers for $n \leq 18,000$ or more (Ribenboim 1996, p. 358).

Let $a(k)$ be smallest n for which $(2k-1)\cdot 2^n + 1$ is PRIME, then the first few values are 0, 1, 1, 2, 1, 1, 2, 1, 3, 6, 1, 1, 2, 2, 1, 8, 1, 1, 2, 1, 1, 2, 2, 583, ... (Sloane's A046067). The second smallest n are given by 1, 2, 3, 4, 2, 3, 8, 2, 15, 10, 4, 9, 4, 4, 3, 60, 6, 3, 4, 2, 11, 6, 9, 1483, ... (Sloane's A046068). Quite large n can be required to obtain the first prime even for small k. For example, the smallest prime of the form $383 \cdot 2^n + 1$ is $383 \cdot 2^{6393} + 1$. There are an infinite number of Sierpiński numbers which are PRIME.

The smallest odd k such that $k + 2^n$ is COMPOSITE for all $n < k$ are 773, 2131, 2491, 4471, 5101, ...

see also MERSENNE NUMBER, RIESEL NUMBER, SIERPIŃSKI'S COMPOSITE NUMBER THEOREM

References
Buell, D. A. and Young, J. "Some Large Primes and the Sierpiński Problem." SRC Tech. Rep. 88004, Supercomputing Research Center, Lanham, MD, 1988.
Jaeschke, G. "On the Smallest k such that $k \cdot 2^N + 1$ are Composite." *Math. Comput.* **40**, 381–384, 1983.
Jaeschke, G. Corrigendum to "On the Smallest k such that $k \cdot 2^N + 1$ are Composite." *Math. Comput.* **45**, 637, 1985.
Keller, W. "Factors of Fermat Numbers and Large Primes of the Form $k \cdot 2^n + 1$." *Math. Comput.* **41**, 661–673, 1983.
Keller, W. "Factors of Fermat Numbers and Large Primes of the Form $k \cdot 2^n + 1$, II." In prep.
Ribenboim, P. *The New Book of Prime Number Records.* New York: Springer-Verlag, pp. 357–359, 1996.
Sierpiński, W. "Sur un problème concernant les nombres $k \cdot 2^n + 1$." *Elem. d. Math.* **15**, 73–74, 1960.
Sloane, N. J. A. Sequence A046067 in "An On-Line Version of the Encyclopedia of Integer Sequences."046068

Sierpiński's Prime Sequence Theorem

For any M, there exists a t' such that the sequence

$$n^2 + t',$$

where $n = 1, 2, \ldots$ contains at least M PRIMES.

see also DIRICHLET'S THEOREM, FERMAT $4n+1$ THEOREM, SIERPIŃSKI'S COMPOSITE NUMBER THEOREM

References
Abel, U. and Siebert, H. "Sequences with Large Numbers of Prime Values." *Amer. Math. Monthly* **100**, 167–169, 1993.
Ageev, A. A. "Sierpinski's Theorem is Deducible from Euler and Dirichlet." *Amer. Math. Monthly* **101**, 659–660, 1994.
Forman, R. "Sequences with Many Primes." *Amer. Math. Monthly* **99**, 548–557, 1992.
Garrison, B. "Polynomials with Large Numbers of Prime Values." *Amer. Math. Monthly* **97**, 316–317, 1990.
Sierpiński, W. "Les binômes $x^2 + n$ et les nombres premiers." *Bull. Soc. Roy. Sci. Liege* **33**, 259–260, 1964.

Sierpiński Sieve

A FRACTAL described by Sierpiński in 1915. It is also called the SIERPIŃSKI GASKET or SIERPIŃSKI TRIANGLE. The curve can be written as a LINDENMAYER SYSTEM with initial string `"FXF--FF--FF"`, STRING REWRITING rules `"F" -> "FF"`, `"X"->"--FXF++FXF++FXF--"`, and angle 60°.

Let N_n be the number of black triangles after iteration n, L_n the length of a side of a triangle, and A_n the fractional AREA which is black after the nth iteration. Then

$$N_n = 3^n \tag{1}$$

$$L_n = (\tfrac{1}{2})^n = 2^{-n} \tag{2}$$

$$A_n = {L_n}^2 N_n = (\tfrac{3}{4})^n. \tag{3}$$

The CAPACITY DIMENSION is therefore

$$d_{\text{cap}} = -\lim_{n\to\infty}\frac{\ln N_n}{\ln L_n} = -\lim_{n\to\infty}\frac{\ln(3^n)}{\ln(2^{-n})} = \frac{\ln 3}{\ln 2}$$
$$= 1.584962501\ldots. \qquad (4)$$

In PASCAL'S TRIANGLE, coloring all ODD numbers black and EVEN numbers white produces a Sierpiński sieve.

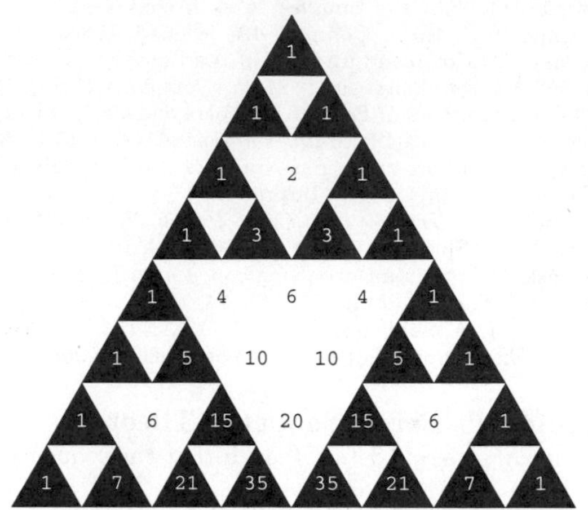

see also LINDENMAYER SYSTEM, SIERPIŃSKI ARROWHEAD CURVE, SIERPIŃSKI CARPET, TETRIX

References

Crownover, R. M. *Introduction to Fractals and Chaos.* Sudbury, MA: Jones & Bartlett, 1995.

Dickau, R. M. "Two-Dimensional L-Systems." `http://forum.swarthmore.edu/advanced/robertd/lsys2d.html`.

Dickau, R. M. "Typeset Fractals." *Mathematica J.* **7**, 15, 1997.

Dickau, R. "Sierpinski-Menger Sponge Code and Graphic." `http://www.mathsource.com/cgi-bin/MathSource/Applications/Graphics/0206-110`.

Lauwerier, H. *Fractals: Endlessly Repeated Geometric Figures.* Princeton, NJ: Princeton University Press, pp. 13–14, 1991.

Peitgen, H.-O.; Jürgens, H.; and Saupe, D. *Chaos and Fractals: New Frontiers of Science.* New York: Springer-Verlag, pp. 78–88, 1992.

Peitgen, H.-O. and Saupe, D. (Eds.). *The Science of Fractal Images.* New York: Springer-Verlag, p. 282, 1988.

Wagon, S. *Mathematica in Action.* New York: W. H. Freeman, pp. 108 and 151–153, 1991.

Wang, P. "Renderings." `http://www.ugcs.caltech.edu/~peterw/portfolio/renderings/`.

Weisstein, E. W. "Fractals." `http://www.astro.virginia.edu/~eww6n/math/notebooks/Fractal.m`.

Sierpiński Sponge

see TETRIX

Sierpiński Tetrahedron

see TETRIX

Sierpiński's Theorem

see SIERPIŃSKI'S COMPOSITE NUMBER THEOREM, SIERPIŃSKI'S PRIME SEQUENCE THEOREM

Sierpiński Triangle

see SIERPIŃSKI SIEVE

Sieve

A process of successively crossing out members of a list according to a set of rules such that only some remain. The best known sieve is the ERATOSTHENES SIEVE for generating PRIME NUMBERS. In fact, numbers generated by sieves seem to share a surprisingly large number of properties with the PRIME NUMBERS.

see also HAPPY NUMBER, NUMBER FIELD SIEVE FACTORIZATION METHOD, PRIME NUMBER, QUADRATIC SIEVE FACTORIZATION METHOD, SIERPIŃSKI SIEVE, SIEVE OF ERATOSTHENES, WALLIS SIEVE

References

Halberstam, H. and Richert, H.-E. *Sieve Methods.* New York: Academic Press, 1974.

Pomerance, C. "A Tale of Two Sieves." *Not. Amer. Math. Soc.* **43**, 1473–1485, 1996.

Sieve of Eratosthenes

1	2	3	~~4~~ 2	5	~~6~~ 2	7	~~8~~ 2	9	~~10~~ 2
11	~~12~~ 2	13	~~14~~ 2	15	~~16~~ 2	17	~~18~~ 2	19	~~20~~ 2
21	~~22~~ 2	23	~~24~~ 2	25	~~26~~ 2	27	~~28~~ 2	29	~~30~~ 2
31	~~32~~ 2	33	~~34~~ 2	35	~~36~~ 2	37	~~38~~ 2	39	~~40~~ 2
41	~~42~~ 2	43	~~44~~ 2	45	~~46~~ 2	47	~~48~~ 2	49	~~50~~ 2

1	2	3	~~4~~ 2	5	~~6~~ 23	7	~~8~~ 2	~~9~~ 3	~~10~~ 2
11	~~12~~ 23	13	~~14~~ 2	~~15~~ 3	~~16~~ 2	17	~~18~~ 23	19	~~20~~ 2
~~21~~ 3	~~22~~ 2	23	~~24~~ 23	25	~~26~~ 2	~~27~~ 3	~~28~~ 2	29	~~30~~ 23
31	~~32~~ 2	~~33~~ 3	~~34~~ 2	35	~~36~~ 23	37	~~38~~ 2	~~39~~ 3	~~40~~ 2
41	~~42~~ 23	43	~~44~~ 2	~~45~~ 3	~~46~~ 2	47	~~48~~ 23	49	~~50~~ 2

1	2	3	~~4~~ 2	5	~~6~~ 23	7	~~8~~ 2	~~9~~ 3	~~10~~ 25
11	~~12~~ 23	13	~~14~~ 2	~~15~~ 35	~~16~~ 2	17	~~18~~ 23	19	~~20~~ 25
~~21~~ 3	~~22~~ 2	23	~~24~~ 23	~~25~~ 5	~~26~~ 2	~~27~~ 3	~~28~~ 2	29	~~30~~ 235
31	~~32~~ 2	~~33~~ 3	~~34~~ 2	~~35~~ 5	~~36~~ 23	37	~~38~~ 2	~~39~~ 3	~~40~~ 25
41	~~42~~ 23	43	~~44~~ 2	~~45~~ 35	~~46~~ 2	47	~~48~~ 23	49	~~50~~ 25

1	2	3	~~4~~ 2	5	~~6~~ 23	7	~~8~~ 2	~~9~~ 3	~~10~~ 25
11	~~12~~ 23	13	~~14~~ 27	~~15~~ 35	~~16~~ 2	17	~~18~~ 23	19	~~20~~ 25
~~21~~ 37	~~22~~ 2	23	~~24~~ 23	~~25~~ 5	~~26~~ 2	~~27~~ 3	~~28~~ 27	29	~~30~~ 235
31	~~32~~ 2	~~33~~ 3	~~34~~ 2	~~35~~ 57	~~36~~ 23	37	~~38~~ 2	~~39~~ 3	~~40~~ 25
41	~~42~~ 237	43	~~44~~ 2	~~45~~ 35	~~46~~ 2	47	~~48~~ 23	~~49~~ 7	~~50~~ 25

An ALGORITHM for making tables of PRIMES. Sequentially write down the INTEGERS from 2 to the highest number n you wish to include in the table. Cross out all numbers > 2 which are divisible by 2 (every second number). Find the smallest remaining number > 2. It is 3. So cross out all numbers > 3 which are divisible by 3 (every third number). Find the smallest remaining number > 3. It is 5. So cross out all numbers > 5 which are divisible by 5 (every fifth number).

Continue until you have crossed out all numbers divisible by $\lfloor\sqrt{n}\rfloor$, where $\lfloor x\rfloor$ is the FLOOR FUNCTION. The numbers remaining are PRIME. This procedure is illustrated in the above diagram which sieves up to 50, and therefore crosses out PRIMES up to $\lfloor\sqrt{50}\rfloor = 7$. If the procedure is then continued up to n, then the number of cross-outs gives the number of distinct PRIME factors of each number.

References

Conway, J. H. and Guy, R. K. *The Book of Numbers.* New York: Springer-Verlag, pp. 127–130, 1996.

Pappas, T. *The Joy of Mathematics.* San Carlos, CA: Wide World Publ./Tetra, pp. 100–101, 1989.

Ribenboim, P. *The New Book of Prime Number Records.* New York: Springer-Verlag, pp. 20–21, 1996.

Sievert Integral

The integral

$$\int_0^{\theta} e^{-x \sec\phi}\, d\phi.$$

References

Abramowitz, M. and Stegun, C. A. (Eds.). "Sievert Integral." §27.4 in *Handbook of Mathematical Functions with Formulas, Graphs, and Mathematical Tables, 9th printing.* New York: Dover, pp. 1000–1001, 1972.

Sievert's Surface

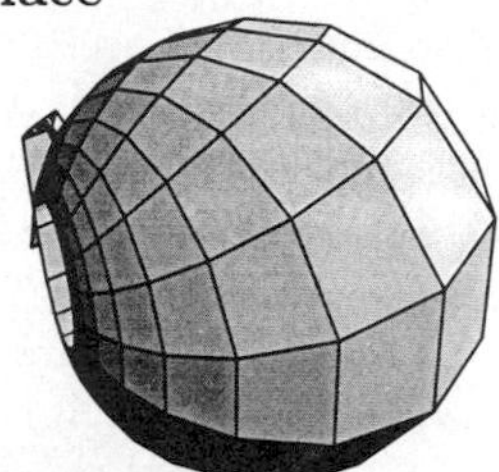

A special case of ENNEPER'S SURFACES which can be given parametrically by

$$x = r\cos\phi \tag{1}$$

$$y = r\sin\phi \tag{2}$$

$$z = \frac{\ln[\tan(\frac{1}{2}v)] + a(C+1)\cos v}{\sqrt{C}}, \tag{3}$$

where

$$\phi \equiv -\frac{u}{\sqrt{C+1}} + \tan^{-1}(\tan u\sqrt{C+1}) \tag{4}$$

$$a \equiv \frac{2}{C+1-C\sin^2 v\cos^2 u} \tag{5}$$

$$r \equiv \frac{a\sqrt{(C+1)(1+C\sin^2 u)}\,\sin v}{\sqrt{C}}, \tag{6}$$

with $|u| < \pi/2$ and $0 < v < \pi$ (Reckziegel 1986).

see also ENNEPER'S SURFACES, KUEN SURFACE, REMBS' SURFACES

References

Fischer, G. (Ed.). Plate 87 in *Mathematische Modelle/Mathematical Models, Bildband/Photograph Volume.* Braunschweig, Germany: Vieweg, p. 83, 1986.

Reckziegel, H. "Sievert's Surface." §3.4.4.3 in *Mathematical Models from the Collections of Universities and Museums* (Ed. G. Fischer). Braunschweig, Germany: Vieweg, pp. 38–39, 1986.

Sievert, H. *Über die Zentralflächen der Enneperschen Flačhen konstanten Krümmungsmaßes.* Dissertation, Tübingen, 1886.

Sifting Property

The property

$$\int f(\mathbf{y})\delta(\mathbf{x}-\mathbf{y})\, d\mathbf{y} = f(\mathbf{x})$$

obeyed by the DELTA FUNCTION $\delta(\mathbf{x})$.

see also DELTA FUNCTION

Sigma Algebra

Let X be a SET. Then a σ-algebra F is a nonempty collection of SUBSETS of X such that the following hold:

1. The EMPTY SET is in F.
2. If A is in F, then so is the complement of A.
3. If A_n is a SEQUENCE of elements of F, then the UNION of the A_ns is in F.

If S is any collection of subsets of X, then we can always find a σ-algebra containing S, namely the POWER SET of X. By taking the INTERSECTION of all σ-algebras containing S, we obtain the smallest such σ-algebra. We call the smallest σ-algebra containing S the σ-algebra generated by S.

see also BOREL SIGMA ALGEBRA, BOREL SPACE, MEASURABLE SET, MEASURABLE SPACE, MEASURE ALGEBRA, STANDARD SPACE

Sigma Function

see DIVISOR FUNCTION

Sigmoid Curve

see SIGMOID FUNCTION

Sigmoid Function

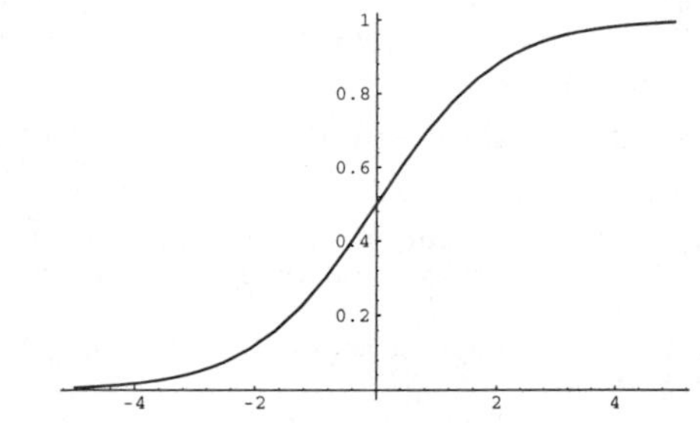

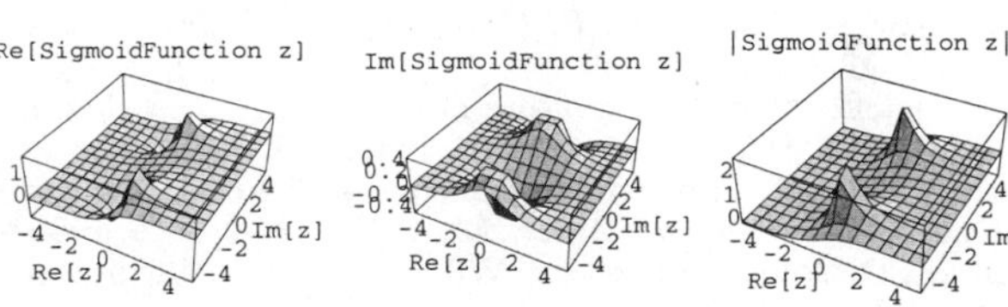

The function

$$y = \frac{1}{1+e^{-x}}$$

which is the solution to the ORDINARY DIFFERENTIAL EQUATION

$$\frac{dy}{dx} = y(1-y).$$

It has an inflection point at $x = 0$, where

$$y''(x) = -\frac{e^x(e^x-1)}{(e^x+1)^3} = 0.$$

see also EXPONENTIAL FUNCTION, EXPONENTIAL RAMP

References

von Seggern, D. *CRC Standard Curves and Surfaces.* Boca Raton, FL: CRC Press, p. 124, 1993.

Sign

The sign of a number, also called SGN, is -1 for a NEGATIVE number (i.e., one with a MINUS SIGN "$-$"), 0 for the number ZERO, or $+1$ for a POSITIVE number (i.e., one with a PLUS SIGN "+").

see also ABSOLUTE VALUE, MINUS SIGN, NEGATIVE, PLUS SIGN, POSITIVE, SGN, ZERO

Signalizer Functor Theorem

$$\Theta(G;A) = \langle \theta(a) : a \in A - 1 \rangle$$

is an A-invariant solvable p'-subgroup of G.

Signature (Knot)

The signature $s(K)$ of a KNOT K can be defined using the SKEIN RELATIONSHIP

$$s(\text{unknot}) = 0$$

$$s(K_+) - s(K_-) \in \{0, 2\},$$

and

$$4|s(K) \leftrightarrow \nabla(K)(2i) > 0,$$

where $\nabla(K)$ is the ALEXANDER-CONWAY POLYNOMIAL and $\nabla(K)(2i)$ is an ODD NUMBER.

Many UNKNOTTING NUMBERS can be determined using a knot's signature.

see also SKEIN RELATIONSHIP, UNKNOTTING NUMBER

References

Gordon, C. McA.; Litherland, R. A.; and Murasugi, K. "Signatures of Covering Links." *Canad. J. Math.* **33**, 381–394, 1981.

Murasugi, K. "On the Signature of Links." *Topology* **9**, 283–298, 1970.

Murasugi, K. "Signatures and Alexander Polynomials of Two-Bridge Knots." *C. R. Math. Rep. Acad. Sci. Canada* **5**, 133–136, 1983.

Murasugi, K. "On the Signature of a Graph." *C. R. Math. Rep. Acad. Sci. Canada* **10**, 107–111, 1988.

Murasugi, K. "On Invariants of Graphs with Applications to Knot Theory." *Trans. Amer. Math. Soc.* **314**, 1–49, 1989.

Rolfsen, D. *Knots and Links.* Wilmington, DE: Publish or Perish Press, 1976.

Stoimenow, A. "Signatures." http://www.informatik.hu-berlin.de/~stoimeno/ptab/sig10.html.

Signature (Quadratic Form)

The signature of the QUADRATIC FORM

$$Q = {y_1}^2 + {y_2}^2 + \ldots + {y_p}^2 - {y_{p+1}}^2 - {y_{p+2}}^2 - \ldots - {y_r}^2$$

is the number s of POSITIVE squared terms in the reduced form. (The signature is sometimes defined as $2s - r$.)

see also p-SIGNATURE, RANK (QUADRATIC FORM), SYLVESTER'S INERTIA LAW, SYLVESTER'S SIGNATURE

References

Gradshteyn, I. S. and Ryzhik, I. M. *Tables of Integrals, Series, and Products, 5th ed.* San Diego, CA: Academic Press, p. 1105, 1979.

Signature (Recurrence Relation)

Let a sequence be defined by

$$\begin{aligned} A_{-1} &= s \\ A_0 &= 3 \\ A_1 &= r \\ A_n &= rA_{n-1} - sA_{n-2} + A_{n-3}. \end{aligned}$$

Also define the associated POLYNOMIAL

$$f(x) = x^3 - rx^2 + sx + 1,$$

and let Δ be its discriminant. The PERRIN SEQUENCE is a special case corresponding to $A_n(0,-1)$. The signature mod m of an INTEGER n with respect to the sequence $A_k(r,s)$ is then defined as the 6-tuple (A_{-n-1}, A_{-n}, A_{-n+1}, A_{n-1}, A_n, A_{n+1}) (mod m).

1. An INTEGER n has an S-signature if its signature (mod n) is (A_{-2}, A_{-1}, A_0, A_1, A_2).
2. An INTEGER n has a Q-signature if its signature (mod n) is CONGRUENT to (A, s, B, B, r, C) where, for some INTEGER a with $f(a) \equiv 0 \pmod{n}$, $A \equiv a^{-2} + 2a$, $B \equiv -ra^2 + (r^2 - s)a$, and $C \equiv a^2 + 2a^{-1}$.
3. An INTEGER n has an I-signature if its signature (mod n) is CONGRUENT to (r, s, D', D, r, s), where $D' + D \equiv rs - 3$ and $(D' - D)^2 \equiv \Delta$.

see also PERRIN PSEUDOPRIME

References

Adams, W. and Shanks, D. "Strong Primality Tests that Are Not Sufficient." *Math. Comput.* **39**, 255–300, 1982.

Grantham, J. "Frobenius Pseudoprimes." http://www.clark.net/pub/grantham/pseudo/pseudo.ps

Signature Sequence

Let θ be an IRRATIONAL NUMBER, define $S(\theta) = \{c + d\theta : c, d \in \mathbb{N}\}$, and let $c_n(\theta) + d_n\theta(\theta)$ be the sequence obtained by arranging the elements of $S(\theta)$ in increasing order. A sequence x is said to be a signature sequence if there EXISTS a POSITIVE IRRATIONAL NUMBER θ such that $x = \{c_n(\theta)\}$, and x is called the signature of θ.

The signature of an IRRATIONAL NUMBER is a FRACTAL SEQUENCE. Also, if x is a signature sequence, then the LOWER-TRIMMED SUBSEQUENCE is $V(x) = x$.

References

Kimberling, C. "Fractal Sequences and Interspersions." *Ars Combin.* **45**, 157–168, 1997.

Signed Deviation

The signed deviation is defined by

$$\Delta u_i \equiv (u_i - \bar{u}),$$

so the average deviation is

$$\overline{\Delta u} = \overline{u_i - \bar{u}} = \overline{u_i} - \bar{u} = 0.$$

see also ABSOLUTE DEVIATION, DEVIATION, DISPERSION (STATISTICS), MEAN DEVIATION, QUARTILE DEVIATION, STANDARD DEVIATION, VARIANCE

Significance

Let $\delta \equiv z \leq z_{\text{observed}}$. A value $0 \leq \alpha \leq 1$ such that $P(\delta) \leq \alpha$ is considered "significant" (i.e., is not simply due to chance) is known as an ALPHA VALUE. The PROBABILITY that a variate would assume a value greater than or equal to the observed value strictly by chance, $P(\delta)$, is known as a P-VALUE.

Depending on the type of data and conventional practices of a given field of study, a variety of different alpha values may be used. One commonly used terminology takes $P(\delta) \geq 5\%$ as "not significant," $1\% < P(\delta) < 5\%$, as "significant" (sometimes denoted *), and $P(\delta) < 1\%$ as "highly significant" (sometimes denoted **). Some authors use the term "almost significant" to refer to $5\% < P(\delta) < 10\%$, although this practice is not recommended.

see also ALPHA VALUE, CONFIDENCE INTERVAL, P-VALUE, PROBABLE ERROR, SIGNIFICANCE TEST, STATISTICAL TEST

Significance Test

A test for determining the probability that a given result could not have occurred by chance (its SIGNIFICANCE).

see also SIGNIFICANCE, STATISTICAL TEST

References

Beyer, W. H. *CRC Standard Mathematical Tables, 28th ed.* Boca Raton, FL: CRC Press, pp. 491–492, 1987.

Significant Digits

When a number is expressed in SCIENTIFIC NOTATION, the number of significant figures is the number of DIGITS needed to express the number to within the uncertainty of measurement. For example, if a quantity had been measured to be 1.234 ± 0.002, four figures would be significant. No more figures should be given than are allowed by the uncertainty. For example, a quantity written as 1.234 ± 0.1 is incorrect; it should really be written as 1.2 ± 0.1.

The number of significant figures of a MULTIPLICATION or DIVISION of two or more quantities is equal to the smallest number of significant figures for the quantities involved. For ADDITION or MULTIPLICATION, the number of significant figures is determined with the smallest significant figure of all the quantities involved. For example, the sum $10.234 + 5.2 + 100.3234$ is 115.7574, but should be written 115.8 (with rounding), since the quantity 5.2 is significant only to ± 0.1.

see also NINT, ROUND, TRUNCATE

Significant Figures

see SIGNIFICANT DIGITS

Signpost

A 6-POLYIAMOND.

References

Golomb, S. W. *Polyominoes: Puzzles, Patterns, Problems, and Packings, 2nd ed.* Princeton, NJ: Princeton University Press, p. 92, 1994.

Signum

see SGN

Silver Constant

The REAL ROOT of the equation

$$x^3 - 5x^2 + 6x - 1 = 0,$$

which is 3.2469.... It is the seventh BERAHA CONSTANT.

see also BERAHA CONSTANTS

References

Le Lionnais, F. *Les nombres remarquables.* Paris: Hermann, pp. 51 and 143, 1983.

Silver Mean

see SILVER RATIO

Silver Ratio

The quantity defined by the CONTINUED FRACTION

$$\delta_S \equiv [2, 2, 2, \ldots] = 2 + \cfrac{1}{2 + \cfrac{1}{2 + \cfrac{1}{2 + \cdots}}}.$$

It follows that

$$(\delta_S - 1)^2 = 2,$$

so

$$\delta_S = \sqrt{2} + 1 = 2.41421\ldots.$$

see also GOLDEN RATIO, GOLDEN RATIO CONJUGATE

Silverman Constant

$$\sum_{n=1}^{\infty} \frac{1}{\phi(n)\sigma(n)} = \prod_{p \text{ prime}} \left(1 + \sum_{k=1}^{\infty} \frac{1}{p^{2k} - p^{k-1}}\right) = 1.786576459\ldots,$$

where $\phi(n)$ is the TOTIENT FUNCTION and $\sigma(n)$ is the DIVISOR FUNCTION.

References

Finch, S. "Favorite Mathematical Constants." `http://www.mathsoft.com/asolve/constant/totient/totient.html`.

Zimmerman, P. `http://www.mathsoft.com/asolve/constant/totient/zimmermn.html`.

Silverman's Sequence

Let $f(1) = 1$, and let $f(n)$ be the number of occurrences of n in a nondecreasing sequence of INTEGERS. Then the first few values of $f(n)$ are 1, 2, 2, 3, 3, 4, 4, 4, 5, 5, 5, ... (Sloane's A001462). The asymptotic value of the nth term is $\phi^{2-\phi}n^{\phi-1}$, where ϕ is the GOLDEN RATIO.

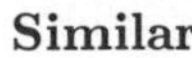

Guy, R. K. "Silverman's Sequences." §E25 in *Unsolved Problems in Number Theory, 2nd ed.* New York: Springer-Verlag, pp. 225–226, 1994.
Sloane, N. J. A. Sequence A001462/M0257 in "An On-Line Version of the Encyclopedia of Integer Sequences."

Similar

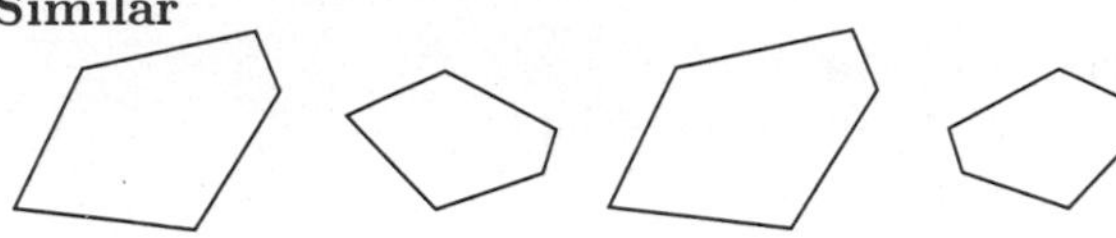

directly similar *inversely similar*

Two figures are said to be similar when all corresponding ANGLES are equal. Two figures are DIRECTLY SIMILAR when all corresponding ANGLES are equal and described in the same rotational sense. This relationship is written $A \sim B$. (The symbol $\sim$ is also used to mean "is the same order of magnitude as" and "is ASYMPTOTIC to.") Two figures are INVERSELY SIMILAR when all corresponding ANGLES are equal and described in the opposite rotational sense.

see also DIRECTLY SIMILAR, INVERSELY SIMILAR, SIMILARITY TRANSFORMATION

References

Project Mathematics! *Similarity.* Videotape (27 minutes). California Institute of Technology. Available from the Math. Assoc. Amer.

Similarity Axis

see D'ALEMBERT'S THEOREM

Similarity Dimension

To multiply the size of a d-D object by a factor a, $c \equiv a^d$ copies are required, and the quantity

$$d = \frac{\ln c}{\ln a}$$

is called the similarity dimension.

Similarity Point

External (or positive) and internal (or negative) similarity points of two CIRCLES with centers C and C' and RADII r and r' are the points E and I on the lines CC' such that

$$\frac{CE}{C'E} = \frac{r}{r'},$$

or

$$\frac{CI}{C'I} = -\frac{r}{r'}.$$

Similarity Transformation

An ANGLE-preserving transformation. A similarity transformation has a transformation MATRIX A' of the form

$$\mathsf{A}' \equiv \mathsf{BAB}^{-1}.$$

If A is an ANTISYMMETRIC MATRIX ($a_{ij} = -a_{ji}$) and B is an ORTHOGONAL MATRIX, then

$$(bab^{-1})_{ij} = b_{ik}a_{kl}b_{lj}^{-1} = -b_{ik}a_{lk}b_{lj}^{-1} = -b^\dagger{}_{ki}a_{lk}(b^\dagger)^{-1}{}_{jl}$$
$$= -b^{-1}{}_{ki}a_{ki}b_{jl} = b_{jl}a_{lk}b_{ki}^{-1} = -(bab^{-1})_{ji}.$$

Similarity transformations and the concept of SELF-SIMILARITY are important foundations of FRACTALS and ITERATED FUNCTION SYSTEMS.

see also CONFORMAL TRANSFORMATION

References

Lauwerier, H. *Fractals: Endlessly Repeated Geometric Figures.* Princeton, NJ: Princeton University Press, pp. 83–103, 1991.

Similitude Center

Also called a SELF-HOMOLOGOUS POINT. If two SIMILAR figures lie in the plane but do not have parallel sides (they are not HOMOTHETIC), there exists a center of similitude which occupies the same homologous position with respect to the two figures. The LOCUS of similitude centers of two nonconcentric circles is another circle having the line joining the two homothetic centers as its DIAMETER.

There are a number of interesting theorems regarding three CIRCLES (Johnson 1929, pp. 151–152).

1. The external similitude centers of three circles are COLLINEAR.
2. Any two internal similitude centers are COLLINEAR with the third external one.
3. If the center of each circle is connected with the internal similitude center of the other three [sic], the connectors are CONCURRENT.
4. If one center is connected with the internal similitude center of the other two, the others with the corresponding external centers, the connectors are CONCURRENT.

References

Johnson, R. A. *Modern Geometry: An Elementary Treatise on the Geometry of the Triangle and the Circle.* Boston, MA: Houghton Mifflin, pp. 19–27 and 151–153, 1929.

Similitude Ratio

Two figures are HOMOTHETIC if they are related by a DILATION (a dilation is also known as a HOMOTHECY). This means that the connectors of corresponding points are CONCURRENT at a point which divides each connector in the same ratio k, known as the similitude ratio.

see also CONCURRENT, DILATION, HOMOTHECY, HOMOTHETIC

Simple Algebra

An ALGEBRA with no nontrivial IDEALS.

see also ALGEBRA, IDEAL, SEMISIMPLE ALGEBRA

Simple Continued Fraction

A CONTINUED FRACTION

$$\sigma = b_0 + \cfrac{a_1}{b_1 + \cfrac{a_2}{b_2 + \cfrac{a_3}{b_3 + \ldots}}} \tag{1}$$

in which the b_is are all unity, leaving a continued fraction of the form

$$\sigma = a_0 + \cfrac{1}{a_1 + \cfrac{1}{a_2 + \cfrac{1}{a_3 + \ldots}}}. \tag{2}$$

A simple continued fraction can be written in a compact abbreviated NOTATION as

$$\sigma = [a_0, a_1, a_2, a_3, \ldots]. \tag{3}$$

Bach and Shallit (1996) show how to compute the JACOBI SYMBOL in terms of the simple continued fraction of a RATIONAL NUMBER a/b.

see also CONTINUED FRACTION

References

Bach, E. and Shallit, J. *Algorithmic Number Theory, Vol. 1: Efficient Algorithms.* Cambridge, MA: MIT Press, pp. 343–344, 1996.

Simple Curve

A curve is simple closed if it does not cross itself.

see also JORDAN CURVE

Simple Graph

A GRAPH for which at most one EDGE connects any two nodes.

see also ADJACENCY MATRIX, EDGE (GRAPH)

Simple Group

A simple group is a GROUP whose NORMAL SUBGROUPS (INVARIANT SUBGROUPS) are ORDER one or the whole of the original GROUP. Simple groups include ALTERNATING GROUPS, CYCLIC GROUPS, LIE-TYPE GROUPS (five varieties), and SPORADIC GROUPS (26 varieties, including the MONSTER GROUP). The CLASSIFICATION THEOREM of finite simple groups states that such groups can be classified completely into the three types:

1. CYCLIC GROUPS of PRIME ORDER,
2. ALTERNATING GROUPS of degree at least five
3. LIE-TYPE CHEVALLEY GROUPS,
4. LIE-TYPE (TWISTED CHEVALLEY GROUPS or the TITS GROUP), and
5. SPORADIC GROUPS.

BURNSIDE'S CONJECTURE states that every non-ABELIAN SIMPLE GROUP has EVEN ORDER.

see also ALTERNATING GROUP, BURNSIDE'S CONJECTURE, CHEVALLEY GROUPS, CLASSIFICATION THEOREM, CYCLIC GROUP, FEIT-THOMPSON THEOREM, FINITE GROUP, GROUP, INVARIANT SUBGROUP, LIE-TYPE GROUP, MONSTER GROUP, SCHUR MULTIPLIER, SPORADIC GROUP, TITS GROUP, TWISTED CHEVALLEY GROUPS

Simple Harmonic Motion

Simple harmonic motion refers to the periodic sinusoidal oscillation of an object or quantity. Simple harmonic motion is executed by any quantity obeying the DIFFERENTIAL EQUATION

$$\ddot{x} + {\omega_0}^2 x = 0, \tag{1}$$

where $\ddot{x}$ denotes the second DERIVATIVE of x with respect to t, and ω_0 is the angular frequency of oscillation. This ORDINARY DIFFERENTIAL EQUATION has an irregular SINGULARITY at ∞. The general solution is

$$x = A\sin(\omega_0 t) + B\cos(\omega_0 t) \tag{2}$$

$$= C\cos(\omega_0 t + \phi), \tag{3}$$

where the two constants A and B (or C and ϕ) are determined from the initial conditions.

Many physical systems undergoing small displacements, including any objects obeying Hooke's law, exhibit simple harmonic motion. This equation arises, for example, in the analysis of the flow of current in an electronic CL circuit (which contains a capacitor and an inductor). If a damping force such as Friction is present, an additional term $\beta\dot{x}$ must be added to the DIFFERENTIAL EQUATION and motion dies out over time.

Adding a damping force proportional to $\dot{x}$, the first derivative of x with respect to time, the equation of motion for *damped* simple harmonic motion is

$$\ddot{x} + \beta\dot{x} + {\omega_0}^2 x = 0, \tag{4}$$

where β is the damping constant. This equation arises, for example, in the analysis of the flow of current in an electronic CLR circuit, (which contains a capacitor, an inductor, and a resistor). This ORDINARY DIFFERENTIAL EQUATION can be solved by looking for trial solutions of the form $x = e^{rt}$. Plugging this into (4) gives

$$(r^2 + \beta r + {\omega_0}^2)e^{rt} = 0 \tag{5}$$

$$r^2 + \beta r + {\omega_0}^2 = 0. \tag{6}$$

This is a QUADRATIC EQUATION with solutions

$$r = \tfrac{1}{2}(-\beta \pm \sqrt{\beta^2 - 4\omega_0{}^2}\,). \tag{7}$$

There are therefore three solution regimes depending on the SIGN of the quantity inside the SQUARE ROOT,

$$\alpha \equiv \beta^2 - 4\omega_0{}^2. \tag{8}$$

The three regimes are

1. $\alpha > 0$ is POSITIVE: overdamped,
2. $\alpha = 0$ is ZERO: critically damped,
3. $\alpha < 0$ is NEGATIVE: underdamped.

If a periodic (sinusoidal) forcing term is added at angular frequency ω, the same three solution regimes are again obtained. Surprisingly, the resulting motion is still periodic (after an initial transient response, corresponding to the solution to the unforced case, has died out), but it has an amplitude different from the forcing amplitude.

The "particular" solution $x_p(t)$ to the forced second-order nonhomogeneous ORDINARY DIFFERENTIAL EQUATION

$$\ddot{x} + p(t)\dot{x} + q(t)x = A\cos(\omega t) \tag{9}$$

due to forcing is given by the equation

$$x_p(t) = -x_1(t)\int \frac{x_2(t)g(t)}{W(t)}\,dt + x_2(t)\int \frac{x_1(t)g(t)}{W(t)}\,dt, \tag{10}$$

where x_1 and x_2 are the homogeneous solutions to the unforced equation

$$\ddot{x} + p(t)\dot{x} + q(t)x = 0 \tag{11}$$

and $W(t)$ is the WRONSKIAN of these two functions. Once the sinusoidal case of forcing is solved, it can be generalized to *any* periodic function by expressing the periodic function in a FOURIER SERIES.

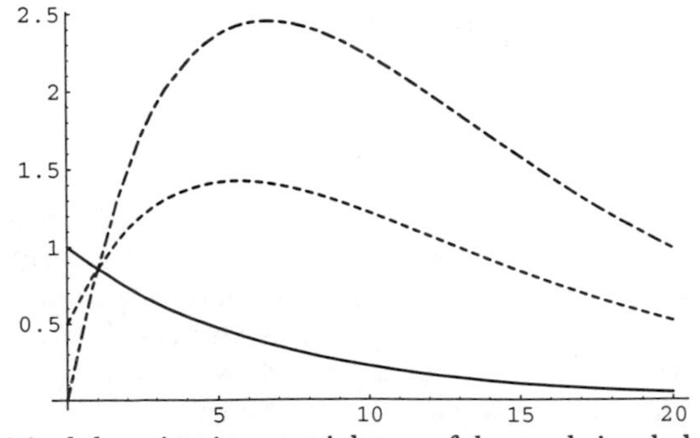

Critical damping is a special case of damped simple harmonic motion in which

$$\alpha \equiv \beta^2 - 4\omega_0{}^2 = 0, \tag{12}$$

so

$$\beta = 2\omega_0. \tag{13}$$

The above plot shows an underdamped simple harmonic oscillator with $\omega = 0.3$, $\beta = 0.15$. The solid curve is for $(A, B) = (1, 0)$, the dot-dashed for $(0, 1)$, and the dotted for $(1/2, 1/2)$. In this case, $\alpha = 0$ so the solutions of the form $x = e^{rt}$ satisfy

$$r_\pm = \tfrac{1}{2}(-\beta) = -\tfrac{1}{2}\beta = -\omega_0. \tag{14}$$

One of the solutions is therefore

$$x_1 = e^{-\omega_0 t}. \tag{15}$$

In order to find the other linearly independent solution, we can make use of the identity

$$x_2(t) = x_1(t)\int \frac{e^{-\int p(t)\,dt}}{[x_1(t)]^2}\,dt. \tag{16}$$

Since we have $p(t) = 2\omega_0$, $e^{-\int p(t)\,dt}$ simplifies to $e^{-2\omega_0 t}$. Equation (16) therefore becomes

$$x_2(t) = e^{-\omega_0 t}\int \frac{e^{-2\omega_0 t}}{[e^{-\omega_0 t}]^2}\,dt = e^{-\omega_0 t}\int dt = te^{-\omega_0 t}. \tag{17}$$

The general solution is therefore

$$x = (A + Bt)e^{-\omega_0 t}. \tag{18}$$

In terms of the constants A and B, the initial values are

$$x(0) = A \tag{19}$$
$$x'(0) = B - A\omega, \tag{20}$$

so

$$A = x(0) \tag{21}$$
$$B = x'(0) + \omega_0 x(0). \tag{22}$$

For sinusoidally forced simple harmonic motion with critical damping, the equation of motion is

$$\ddot{x} + 2\omega_0\dot{x} + \omega_0{}^2 x = A\cos(\omega t), \tag{23}$$

and the WRONSKIAN is

$$\begin{aligned} W(t) &\equiv x_1\dot{x}_2 - \dot{x}_1 x_2 \\ &= e^{-\omega_0 t}(e^{-\omega_0 t} - \omega_0 te^{-\omega_0 t}) + \omega_0 e^{-\omega_0 t}te^{-\omega_0 t} \\ &= e^{-2\omega_0 t}(1 - \omega_0 t + \omega_0 t) = e^{-2\omega_0 t}. \end{aligned} \tag{24}$$

Plugging this into the equation for the particular solution gives

$$\begin{aligned} x_p(t) &= -e^{-\omega_0 t} \int \frac{te^{-\omega_0 t} A\cos(\omega t)}{e^{-2\omega_0 t}}\, dt \\ &\quad + te^{-\omega_0 t} \int \frac{e^{-\omega_0 t} A\cos(\omega t)}{e^{-2\omega_0 t}}\, dt \\ &= Ae^{-\omega_0 t}\left[-\int te^{\omega_0 t}\cos(\omega t)\, dt + t\int e^{\omega_0 t}\cos(\omega t)\, dt\right] \\ &= Ae^{-\omega_0 t}\left\{-\frac{e^{\omega_0 t}}{(\omega^2+\omega_0{}^2)^2}[(\omega^2 + t\omega^2\omega_0 - \omega_0{}^2 + t\omega_0{}^3) \right. \\ &\quad \times \cos(\omega t) + \omega(t\omega^2 - 2\omega_0 + t\omega_0{}^2)\sin(\omega t)] \\ &\quad \left. + t\frac{e^{\omega_0 t}}{\omega^2+\omega_0{}^2}[\omega_0\cos(\omega t) + \omega\sin(\omega t)\right\} \\ &= \frac{A}{(\omega^2+\omega_0{}^2)^2}[(\omega_0{}^2 - \omega^2)\cos(\omega t) + 2\omega\omega_0\sin(\omega t)]. \end{aligned} \tag{25}$$

In order to put this in the desired form, note that we want to equate

$$\begin{aligned} C\cos\theta + S\sin\theta &= Q\cos(\theta+\delta) \\ &= Q(\cos\theta\cos\delta - \sin\theta\sin\delta). \end{aligned} \tag{26}$$

This means

$$C \equiv Q\cos\delta = \omega_0{}^2 - \omega^2 \tag{27}$$
$$S \equiv -Q\sin\delta = 2\omega\omega_0, \tag{28}$$

so

$$Q = \sqrt{C^2+S^2} \tag{29}$$
$$\delta = \tan^{-1}\left(-\frac{S}{C}\right). \tag{30}$$

Plugging in,

$$\begin{aligned} Q &= \sqrt{\omega_0{}^4 - 2\omega_0{}^2\omega^2 + \omega^4 + 4\omega_0{}^2\omega^2} \\ &= \sqrt{\omega_0{}^4 + 2\omega_0{}^2\omega^2 + \omega^4} = \omega_0{}^2 + \omega^2. \end{aligned} \tag{31}$$
$$\delta = \tan^{-1}\left(\frac{2\omega\omega_0}{\omega^2 - \omega_0{}^2}\right). \tag{32}$$

The solution in the requested form is therefore

$$\begin{aligned} x_p &= \frac{A}{(\omega^2+\omega_0{}^2)^2}(\omega_0{}^2 + \omega^2)\cos(\omega t + \delta) \\ &= \frac{A}{\omega^2+\omega_0{}^2}\cos(\omega t+\delta), \end{aligned} \tag{33}$$

where δ is defined by (32).

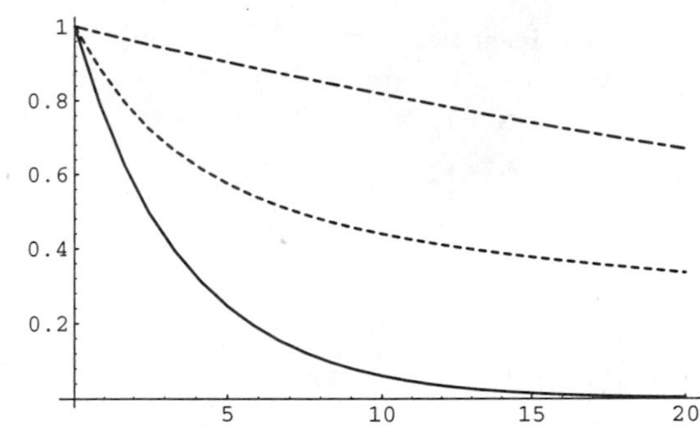

Overdamped simple harmonic motion occurs when

$$\beta^2 - 4\omega_0{}^2 > 0, \tag{34}$$

so

$$\alpha \equiv \beta^2 - 4\omega_0{}^2 > 0. \tag{35}$$

The above plot shows an overdamped simple harmonic oscillator with $\omega = 0.3$, $\beta = 0.075$. The solid curve is for $(A, B) = (1, 0)$, the dot-dashed for (0, 1), and the dotted for (1/2, 1/2). The solutions are

$$x_1 = e^{r_- t} \tag{36}$$
$$x_2 = e^{r_+ t}, \tag{37}$$

where

$$r_\pm \equiv \tfrac{1}{2}(-\beta \pm \sqrt{\beta^2 - 4\omega_0{}^2}\,). \tag{38}$$

The general solution is therefore

$$x = Ae^{r_- t} + Be^{r_+ t}, \tag{39}$$

where A and B are constants. The initial values are

$$x(0) = A + B \tag{40}$$
$$x'(0) = Ar_- + Br_+, \tag{41}$$

so

$$A = x(0) + \frac{r_+ x(0) - x'(0)}{r_- - r_+} \tag{42}$$
$$B = -\frac{r_+ x(0) - x'(0)}{r_- - r_+}. \tag{43}$$

For a cosinusoidally forced overdamped oscillator with forcing function $g(t) = C\cos(\omega t)$, the particular solutions are

$$y_1(t) = e^{r_1 t} \tag{44}$$
$$y_2(t) = e^{r_2 t}, \tag{45}$$

where

$$r_1 \equiv \tfrac{1}{2}(-\beta + \sqrt{\beta^2 - 4\omega_0{}^2}\,) \tag{46}$$
$$r_2 \equiv \tfrac{1}{2}(-\beta - \sqrt{\beta^2 - 4\omega_0{}^2}\,). \tag{47}$$

These give the identities

$$r_1 + r_2 = -\beta \tag{48}$$
$$r_1 - r_2 = \sqrt{\beta^2 - 4{\omega_0}^2} \tag{49}$$

and

$$\begin{aligned} {\omega_0}^2 &= \tfrac{1}{4}[\beta - (r_1 - r_2)^2] = \tfrac{1}{4}[(r_1 + r_2)^2 - (r_1 - r_2)^2] \\ &= \tfrac{1}{4}[2r_1r_2 + 2r_1r_2] = r_1r_2. \end{aligned} \tag{50}$$

The WRONSKIAN is

$$\begin{aligned} W(t) &\equiv y_1y_2' - y_1'y_2 = e^{r_1t}r_2e^{r_2t} - r_1e^{r_1t}e^{r_2t} \\ &= (r_2 - r_1)e^{(r_1+r_2)t}. \end{aligned} \tag{51}$$

The particular solution is

$$y_p = -y_1v_1 + y_2v_2, \tag{52}$$

where

$$v_1 \equiv \int \frac{y_2 g(t)}{W(t)} = \frac{C}{r_2 - r_1}\frac{\omega\sin(\omega t) - r_2\cos(\omega t)}{e^{r_2t}({r_2}^2 + \omega^2)} \tag{53}$$

$$v_2 \equiv \int \frac{y_2 g(t)}{W(t)} = \frac{C}{r_2 - r_1}\frac{\omega\sin(\omega t) - r_1\cos(\omega t)}{e^{r_1t}({r_2}^2 + \omega^2)}. \tag{54}$$

Therefore,

$$\begin{aligned} y_p &= C\frac{\cos(\omega t)(r_1r_2 - \omega^2) - \sin(\omega t)\omega(r_1 + r_2)}{({r_1}^2 + \omega^2)({r_2}^2 + \omega^2)} \\ &= C\frac{({\omega_0}^2 - \omega^2)\cos(\omega t) + \beta\omega\sin(\omega t)}{\omega^2\beta^2 + (\omega^2 - {\omega_0}^2)} \\ &= \frac{C}{\omega^2\beta^2 + (\omega^2 - {\omega_0}^2)^2}\sqrt{(\omega^2 - {\omega_0}^2)^2 + \beta^2\omega^2} \\ &\quad \times \cos(\omega t + \delta) \\ &= \frac{C}{\sqrt{\beta^2\omega^2 + (\omega^2 - {\omega_0}^2)^2}}\cos(\omega t + \delta), \end{aligned} \tag{55}$$

where

$$\delta = \tan^{-1}\left(\frac{\beta\omega}{\omega^2 - {\omega_0}^2}\right). \tag{56}$$

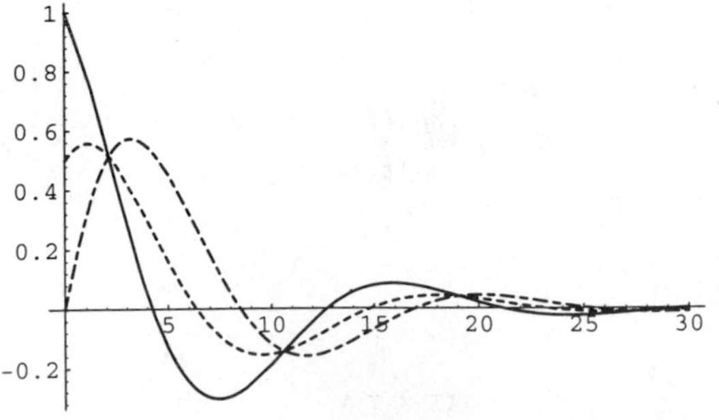

Underdamped simple harmonic motion occurs when

$$\beta^2 - 4{\omega_0}^2 < 0, \tag{57}$$

so

$$\alpha \equiv \beta^2 - 4{\omega_0}^2 < 0. \tag{58}$$

The above plot shows an underdamped simple harmonic oscillator with $\omega = 0.3$, $\beta = 0.4$. The solid curve is for $(A, B) = (1, 0)$, the dot-dashed for $(0, 1)$, and the dotted for $(1/2, 1/2)$. Define

$$\gamma \equiv \sqrt{-\alpha} = \tfrac{1}{2}\sqrt{4{\omega_0}^2 - \beta^2}, \tag{59}$$

then solutions satisfy

$$r_\pm = -\tfrac{1}{2}\beta \pm i\gamma, \tag{60}$$

where

$$r_\pm \equiv \tfrac{1}{2}(-\beta \pm \sqrt{\beta^2 - 4{\omega_0}^2}\,), \tag{61}$$

and are of the form

$$x = e^{-(\beta/2 \pm i\gamma)t}. \tag{62}$$

Using the EULER FORMULA

$$e^{ix} = \cos x + i\sin x, \tag{63}$$

this can be rewritten

$$x = e^{-(\beta/2)t}\left[\cos(\gamma t) \pm i\sin(\gamma t)\right]. \tag{64}$$

We are interested in the *real* solutions. Since we are dealing here with a *linear homogeneous* ODE, linear sums of LINEARLY INDEPENDENT solutions are also solutions. Since we have a sum of such solutions in (64), it follows that the IMAGINARY and REAL PARTS separately satisfy the ODE and are therefore the solutions we seek. The constant in front of the sine term is arbitrary, so we can identify the solutions as

$$x_1 = e^{-(\beta/2)t}\cos(\gamma t) \tag{65}$$
$$x_2 = e^{-(\beta/2)t}\sin(\gamma t), \tag{66}$$

so the general solution is

$$x = e^{-(\beta/2)t}[A\cos(\gamma t) + B\sin(\gamma t)]. \tag{67}$$

The initial values are

$$x(0) = A \tag{68}$$
$$x'(0) = -\tfrac{1}{2}\beta A + B, \gamma \tag{69}$$

so A and B can be expressed in terms of the initial conditions by

$$A = x(0) \tag{70}$$
$$B = \frac{\beta x(0)}{2\gamma} + \frac{x'(0)}{\gamma}. \tag{71}$$

For a cosinusoidally forced underdamped oscillator with forcing function $g(t) = C\cos(\omega t)$, use

$$\gamma \equiv \tfrac{1}{2}\sqrt{4\omega_0{}^2 - \beta^2} \tag{72}$$
$$\alpha \equiv \tfrac{1}{2}\beta \tag{73}$$

to obtain

$$4\omega_0{}^2 - \beta^2 = 4\gamma^2 \tag{74}$$
$$\omega_0{}^2 = \gamma^2 + \tfrac{1}{4}\beta^2 = \gamma^2 + \alpha^2 \tag{75}$$
$$\beta = 2\alpha. \tag{76}$$

The particular solutions are

$$y_1(t) = e^{-\alpha t}\cos(\gamma t) \tag{77}$$
$$y_2(t) = e^{-\alpha t}\sin(\gamma t). \tag{78}$$

The WRONSKIAN is

$$\begin{aligned} W(t) &\equiv y_1y_2' - y_1'y_2 \\ &= e^{-\alpha t}\cos(\gamma t)[-\alpha e^{-\alpha t}\sin(\gamma t) + e^{-\alpha t}\gamma\cos(\gamma t)] \\ &\quad - e^{-\alpha t}\sin(\gamma t)[-\alpha e^{-\alpha t}\cos(\gamma t) - e^{-\alpha t}\gamma\sin(\gamma t)] \\ &= e^{-2\alpha t}\{\alpha[-\sin(\gamma t)\cos(\gamma t) + \sin(\gamma t)\cos(\gamma t)] \\ &\quad + \gamma[\cos^2(\gamma t) + \sin^2(\gamma t)]\} \\ &= \gamma e^{-2\alpha t}. \end{aligned} \tag{79}$$

The particular solution is given by

$$y_p = -y_1v_1 + y_2v_2, \tag{80}$$

where

$$v_1 \equiv \int \frac{y_2 g(t)}{W(t)} = \frac{C}{\gamma}\int e^{\alpha t}\cos(\gamma t)\cos(\omega t)\,dt \tag{81}$$
$$v_2 \equiv \int \frac{y_2 g(t)}{W(t)} = \frac{C}{\gamma}\int e^{\alpha t}\cos(\gamma t)\cos(\omega t)\,dt. \tag{82}$$

Using computer algebra to perform the algebra, the particular solution is

$$\begin{aligned} y_p(t) &= C\frac{(\alpha^2+\gamma^2-\omega^2)\cos(\omega t) + 2\alpha\omega\sin(\omega t)}{[\alpha^2 + (\gamma-\omega)^2][\alpha^2 + (\gamma+\omega)^2]} \\ &= C\frac{(\omega_0{}^2 - \omega^2)\cos(\omega t) + \beta\omega\sin(\omega t)}{(\alpha^2+\gamma^2+\omega^2)^2 - 4\gamma^2\omega^2} \\ &= C\frac{(\omega_0{}^2 - \omega^2)\cos(\omega t) + \beta\omega\sin(\omega t)}{(\omega_0{}^2+\omega^2)^2 - 4\frac{1}{4}(4\omega_0{}^2 - \beta^2)\omega^2} \\ &= C\frac{(\omega_0{}^2 - \omega^2)\cos(\omega t) + \beta\omega\sin(\omega t)}{(\omega_0{}^2-\omega^2)^2 - \omega^2(4\omega_0{}^2 - \beta^2)} \\ &= \frac{C\sqrt{(\omega_0{}^2-\omega^2)^2 + \beta^2\omega^2}}{(\omega_0{}^2-\omega^2)^2 - \omega^2(4\omega_0{}^2 - \beta^2)}\cos(\omega t + \delta) \\ &= C\frac{\sqrt{(\omega_0{}^2-\omega^2)^2 + \beta^2\omega^2}}{(\omega_0{}^2-\omega^2)^2 - \omega^2(4\omega_0{}^2 - \beta^2)}\cos(\omega t + \delta), \end{aligned} \tag{83}$$

where

$$\delta = \tan^{-1}\left(\frac{\beta\omega}{\omega^2 - \omega_0{}^2}\right). \tag{84}$$

If the forcing function is sinusoidal instead of cosinusoidal, then

$$\delta' = \delta - \tfrac{1}{2}\pi = \tan^{-1}x - \tfrac{1}{2}\pi = \tan^{-1}\left(-\frac{1}{x}\right), \tag{85}$$

so

$$\delta' = \tan^{-1}\left(\frac{\omega_0{}^2 - \omega^2}{\beta\omega}\right). \tag{86}$$

Simple Harmonic Motion Quadratic Perturbation

Given a simple harmonic oscillator with a quadratic perturbation ϵx^2,

$$\ddot{x} + \omega_0{}^2 x - \alpha\epsilon x^2 = 0, \tag{1}$$

find the first-order solution using a perturbation method. Write

$$x \equiv x_0 + \epsilon x_1 + \ldots, \tag{2}$$

so

$$\ddot{x} = \ddot{x}_0 + \epsilon\ddot{x}_1 + \ldots. \tag{3}$$

Plugging (2) and (3) back into (1) gives

$$(\ddot{x}_0 + \epsilon\ddot{x}_1) + (\omega_0{}^2x_0 + \omega_0{}^2\epsilon x_1) - \alpha\epsilon(x_0 + 2x_0x_1\epsilon + \ldots) = 0. \tag{4}$$

Keeping only terms of order ϵ and lower and grouping, we obtain

$$(\ddot{x}_0 + \omega_0{}^2x_0) + (\ddot{x}_1 + \omega_0{}^2x_1 - \alpha x_0{}^2)\epsilon = 0. \tag{5}$$

Since this equation must hold for all POWERS of ϵ, we can separate it into the two differential equations

$$\ddot{x}_0 + \omega_0{}^2x_0 = 0 \tag{6}$$

$$\ddot{x}_1 + \omega_0{}^2x_1 = \alpha x_0{}^2. \tag{7}$$

The solution to (6) is just

$$x_0 = A\cos(\omega_0 t + \phi). \tag{8}$$

Setting our clock so that $\phi = 0$ gives

$$x_0 = A\cos(\omega_0 t). \tag{9}$$

Plugging this into (7) then gives

$$\ddot{x}_1 + \omega_0{}^2x_1 = \alpha A^2\cos^2(\omega_0 t). \tag{10}$$

The two homogeneous solutions to (10) are

$$x_1 = \cos(\omega_0 t) \tag{11}$$
$$x_2 = \sin(\omega_0 t). \tag{12}$$

The particular solution to (10) is therefore given by

$$x_p(t) = -x_1(t)\int \frac{x_2(t)g(t)}{W(t)}\,dt + x_2(t)\int \frac{x_1(t)g(t)}{W(t)}\,dt, \tag{13}$$

where

$$g(t) = \alpha A^2 \cos^2(\omega_0 t), \tag{14}$$

and the WRONSKIAN is

$$\begin{aligned} W &\equiv x_1\dot{x}_2 - \dot{x}_1 x_2 \\ &= \cos(\omega_0 t)\omega_0\cos(\omega_0 t) - [-\omega_0\sin(\omega_0 t)]\sin(\omega_0 t) \\ &= \omega_0. \end{aligned} \tag{15}$$

Plugging everything into (13),

$$\begin{aligned} x_p &= \alpha A^2\left[-\cos(\omega_0 t)\int \frac{\sin(\omega_0 t)\cos^2(\omega_0 t)}{\omega_0}\,dt \right. \\ &\quad \left. + \sin(\omega_0 t)\int \frac{\cos^3(\omega_0 t)}{\omega_0}\,dt\right] \\ &= \frac{\alpha A^2}{\omega_0}\left\{\sin(\omega_0 t)\int [1-\sin^2(\omega_0 t)]\cos(\omega_0 t)\,dt \right. \\ &\quad \left. - \cos(\omega_0 t)\int \sin(\omega_0 t)\cos^2(\omega_0 t)\,dt\right\}. \end{aligned} \tag{16}$$

Now let

$$u \equiv \sin(\omega_0 t) \tag{17}$$

$$du = \omega_0\cos(\omega_0 t)\,dt \tag{18}$$

$$v \equiv \cos(\omega_0 t) \tag{19}$$

$$dv = -\omega_0\sin(\omega_0 t)\,dt. \tag{20}$$

Then

$$\begin{aligned} x_p &= \frac{\alpha A^2}{{\omega_0}^2}\left[\sin(\omega_0 t)\int (1-u^2)\,du + \cos(\omega_0 t)\int v^2\,dv\right] \\ &= \frac{\alpha A^2}{{\omega_0}^2}\left[\sin(\omega_0 t)(1-\tfrac{1}{3}u^3) + \cos(\omega_0 t)\tfrac{1}{3}v^3\right] \\ &= \frac{\alpha A^2}{{\omega_0}^2}\{\sin(\omega_0 t)[1-\tfrac{1}{3}\sin^3(\omega_0 t)] \\ &\quad + \tfrac{1}{3}\cos(\omega_0 t)\cos^3(\omega_0 t)\} \\ &= \frac{\alpha A^2}{{\omega_0}^2}\left\{\tfrac{1}{3}[\cos^4(\omega_0 t) - \sin^4(\omega_0 t)] + \sin^2(\omega_0 t)\right\} \\ &= \frac{\alpha A^2}{{\omega_0}^2}\left\{\tfrac{1}{3}[\cos^2(\omega_0 t) - \sin^2(\omega_0 t)] + \sin^2(\omega_0 t)\right\} \\ &= \frac{\alpha A^2}{{\omega_0}^2}\tfrac{1}{3}[\cos^2(\omega_0 t) + 2\sin^2(\omega_0 t)] \\ &= \frac{\alpha A^2}{3{\omega_0}^2}[2-\cos^2(\omega_0 t)] = \frac{\alpha A^2}{3{\omega_0}^2}\{2-\tfrac{1}{2}[1+\cos(2\omega_0 t)]\} \\ &= \frac{\alpha A^2}{6{\omega_0}^2}[3-\cos(2\omega_0 t)]. \end{aligned} \tag{21}$$

Plugging $x_0(t)$ and (21) into (2), we obtain the solution

$$x(t) = A\cos(\omega_0 t) - \frac{\alpha A^2}{6{\omega_0}^2}\epsilon[\cos(2\omega_0 t) - 3]. \tag{22}$$

Simple Harmonic Oscillator

see SIMPLE HARMONIC MOTION

Simple Interest

INTEREST which is paid only on the PRINCIPAL and not on the additional amount generated by previous INTEREST payments. A formula for computing simple interest is

$$a(t) = a(0)(1+rt),$$

where $a(t)$ is the sum of PRINCIPAL and INTEREST at time t for a constant interest rate r.

see also COMPOUND INTEREST, INTEREST

References

Kellison, S. G. *Theory of Interest, 2nd ed.* Burr Ridge, IL: Richard D. Irwin, 1991.

Simple Polygon

A POLYGON P is said to be simple (or JORDAN) if the only points of the plane belonging to two EDGES of P are the VERTICES of P. Such a polygon has a well-defined interior and exterior.

see also POLYGON, REGULAR POLYGON, TWO-EARS THEOREM

References

Toussaint, G. "Anthropomorphic Polygons." *Amer. Math. Monthly* **122**, 31–35, 1991.

Simple Ring

A NONZERO RING S whose only (two-sided) IDEALS are S itself and zero. Every commutative simple ring is a FIELD. Every simple ring is a PRIME RING.

see also FIELD, IDEAL, PRIME RING, RING

Simplex

The generalization of a tetrahedral region of space to n-D. The boundary of a k-simplex has $k+1$ 0-faces (VERTICES), $k(k+1)/2$ 1-faces (EDGES), and $\binom{k+1}{i+1}$ i-faces, where $\binom{n}{k}$ is a BINOMIAL COEFFICIENT.

The simplex in 4-D is a regular TETRAHEDRON $ABCD$ in which a point E along the fourth dimension through the center of $ABCD$ is chosen so that $EA = EB = EC = ED = AB$. The 4-D simplex has SCHLÄFLI SYMBOL $\{3,3,3\}$.

n	Simplex
0	point
1	line segment
2	equilateral triangular plane region
3	tetrahedral region
4	4-simplex

The only irreducible spherical simplexes generated by reflection are A_n ($n \geq 1$), B_n ($n \geq 4$), C_n ($n \geq 2$), D_2^p ($p \geq 5$), E_6, E_7, E_8, F_4, G_3, and G_4. The only irreducible Euclidean simplexes generated by reflection are W_2, P_m ($m \geq 3$), Q_m ($m \geq 5$), R_m ($m \geq 3$), S_m ($m \geq 4$), V_3, T_7, T_8, T_9, and U_5.

The regular simplex in n-D with $n \geq 5$ is denoted α_n and has SCHLÄFLI SYMBOL $\{\underbrace{3, \ldots, 3}_{3^{n-1}}\}$.

see also COMPLEX, CROSS POLYTOPE, EQUILATERAL TRIANGLE, LINE SEGMENT, MEASURE POLYTOPE, NERVE, POINT, SIMPLEX METHOD, TETRAHEDRON

References

Eppstein, D. "Triangles and Simplices." http://www.ics.uci.edu/~eppstein/junkyard/triangulation.html.

Simplex Method

A method for solving problems in LINEAR PROGRAMMING. This method, invented by G. B. Dantzig in 1947, runs along EDGES of the visualization SOLID to find the best answer. In 1970, Klee and Minty constructed examples in which the simplex method required an exponential number of steps, but such cases seem never to be encountered in practical applications.

A much more efficient (POLYNOMIAL-time) ALGORITHM was found in 1984 by N. Karmarkar. This method goes through the middle of the SOLID and then transforms and warps. It offers many advantages over the simplex method (Nemirovsky and Yudin 1994).

see also LINEAR PROGRAMMING

References

Nemirovsky, A. and Yudin, N. *Interior-Point Polynomial Methods in Convex Programming.* Philadelphia, PA: SIAM, 1994.

Press, W. H.; Flannery, B. P.; Teukolsky, S. A.; and Vetterling, W. T. "Downhill Simplex Method in Multidimensions" and "Linear Programming and the Simplex Method." §10.4 and 10.8 in *Numerical Recipes in FORTRAN: The Art of Scientific Computing, 2nd ed.* Cambridge, England: Cambridge University Press, pp. 402–406 and 423–436, 1992.

Tokhomirov, V. M. "The Evolution of Methods of Convex Optimization." *Amer. Math. Monthly* **103**, 65–71, 1996.

Simplicial Complex

A simplicial complex is a SPACE with a TRIANGULATION. Objects in the space made up of only the simplices in the triangulation of the space are called simplicial subcomplexes. When only simplicial complexes and subcomplexes are considered, defining HOMOLOGY is particularly easy (and, in fact, combinatorial because of its finite/counting nature). This kind of homology is called SIMPLICIAL HOMOLOGY.

see also HOMOLOGY (TOPOLOGY), NERVE, SIMPLICIAL HOMOLOGY, SPACE, TRIANGULATION

Simplicial Homology

The type of HOMOLOGY which results when the spaces being studied are restricted to SIMPLICIAL COMPLEXES and subcomplexes.

see also SIMPLICIAL COMPLEX

Simplicity

The number of operations needed to effect a GEOMETRIC CONSTRUCTION as determined in GEOMETROGRAPHY. If the number of operations of the five GEOMETROGRAPHIC types are denoted m_1, m_2, n_1, n_2, and n_3, respectively, then the simplicity is $m_1+m_2+n_1+n_2+n_3$ and the symbol $m_1S_1 + m_2S_2 + n_1C_1 + n_2C_2 + n_3C_3$. It is apparently an unsolved problem to determine if a given GEOMETRIC CONSTRUCTION is of smallest possible simplicity.

see also GEOMETRIC CONSTRUCTION, GEOMETROGRAPHY

References

De Temple, D. W. "Carlyle Circles and the Lemoine Simplicity of Polygonal Constructions." *Amer. Math. Monthly* **98**, 97–108, 1991.

Eves, H. *An Introduction to the History of Mathematics, 6th ed.* New York: Holt, Rinehart, and Winston, 1976.

Simply Connected

A CONNECTED DOMAIN is said to be simply connected (also called 1-connected) if any simple closed curve can be shrunk to a point continuously in the set. If the domain is CONNECTED but not simply, it is said to be MULTIPLY CONNECTED.

A SPACE S is simply connected if it is 0-connected and if every MAP from the 1-SPHERE to S extends continuously to a MAP from the 2-DISK. In other words, every loop in the SPACE is contractible.

see also CONNECTED SPACE, MULTIPLY CONNECTED

Simpson's Paradox

It is not necessarily true that averaging the averages of different populations gives the average of the combined population.

References

Paulos, J. A. *A Mathematician Reads the Newspaper.* New York: BasicBooks, p. 135, 1995.

Simpson's Rule

Let $h \equiv (b-a)/n$, and assume a function $f(x)$ is defined at points $f(a+kh) = y_k$ for $k = 0, \ldots, n$. Then

$$\int_a^b f(x)\,dx = \tfrac{1}{3}h(y_1 + 4y_2 + 2y_3 + 4y_4 + \ldots + 2y_{n-2} + 4y_{n-1} + y_n) - R_n,$$

where the remainder is

$$R_n = \tfrac{1}{90}(b-a)^4 f^{(4)}(x^*)$$

for some $x^* \in [a, b]$.

see also BODE'S RULE, NEWTON-COTES FORMULAS, SIMPSON'S 3/8 RULE, TRAPEZOIDAL RULE

References

Abramowitz, M. and Stegun, C. A. (Eds.). *Handbook of Mathematical Functions with Formulas, Graphs, and Mathematical Tables, 9th printing.* New York: Dover, p. 886, 1972.

Simpson's 3/8 Rule

$$\int_{x_1}^{x_4} f(x)\,dx = \tfrac{3}{8}h(f_1 + 3f_2 + 3f_3 + f_4) - \tfrac{3}{80}h^5 f^{(4)}(\xi).$$

see also BODE'S RULE, NEWTON-COTES FORMULAS, SIMPSON'S RULE

References

Abramowitz, M. and Stegun, C. A. (Eds.). *Handbook of Mathematical Functions with Formulas, Graphs, and Mathematical Tables, 9th printing.* New York: Dover, p. 886, 1972.

Simson Line

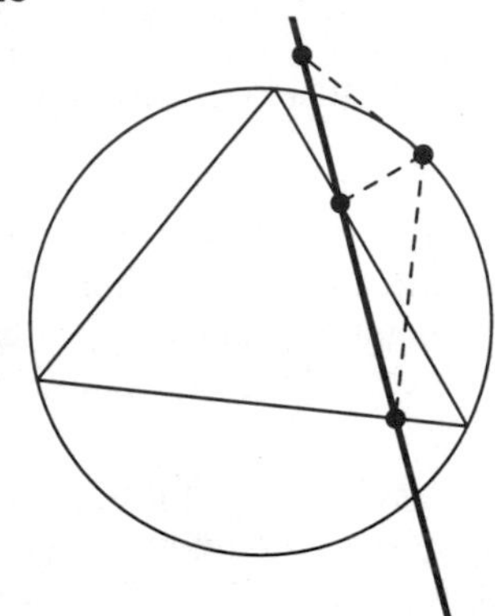

The Simson line is the LINE containing the feet of the perpendiculars from a point on the CIRCUMCIRCLE of a TRIANGLE to the sides (or their extensions) of the TRIANGLE. The Simson line is sometimes known as the WALLACE-SIMSON LINE, since it does not appear in any work of Simson (Johnson 1929, p. 137).

The ANGLE between the Simson lines of two points P and P' is half the ANGLE of the arc PP'. The Simson line of any VERTEX is the ALTITUDE through that VERTEX. The Simson line of a point opposite a VERTEX is the corresponding side. If $T_1T_2T_3$ is the Simson line of a point T of the CIRCUMCIRCLE, then the triangles TT_1T_2 and TA_2A_1 are directly similar.

see also CIRCUMCIRCLE

References

Coxeter, H. S. M. and Greitzer, S. L. *Geometry Revisited.* Washington, DC: Math. Assoc. Amer., pp. 40–41 and 43–45, 1967.

Johnson, R. A. *Modern Geometry: An Elementary Treatise on the Geometry of the Triangle and the Circle.* Boston, MA: Houghton Mifflin, pp. 137–139, 1929.

Sinc Function

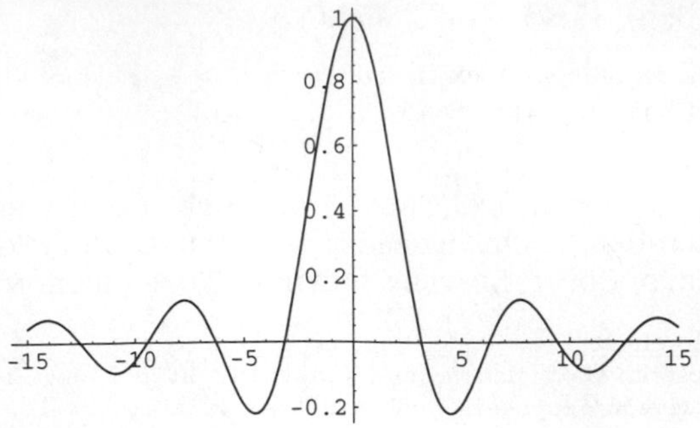

A function also called the SAMPLING FUNCTION and defined by

$$\operatorname{sinc}(x) \equiv \begin{cases} 1 & \text{for } x = 0 \\ \frac{\sin x}{x} & \text{otherwise,} \end{cases} \tag{1}$$

where $\sin x$ is the SINE function. Let $\Pi(x)$ be the RECTANGLE FUNCTION, then the FOURIER TRANSFORM of $\Pi(x)$ is the sinc function

$$\mathcal{F}[\Pi(x)] = \operatorname{sinc}(\pi k). \tag{2}$$

The sinc function therefore frequently arises in physical applications such as Fourier transform spectroscopy as the so-called INSTRUMENT FUNCTION, which gives the instrumental response to a DELTA FUNCTION input. Removing the instrument functions from the final spectrum requires use of some sort of DECONVOLUTION algorithm.

The sinc function can be written as a complex INTEGRAL by noting that

$$\begin{aligned} \operatorname{sinc}(nx) &\equiv \frac{\sin(nx)}{nx} = \frac{1}{nx}\frac{e^{inx} - e^{-inx}}{2i} \\ &= \frac{1}{2inx}[e^{itx}]_{-n}^{n} = \frac{1}{2n}\int_{-n}^{n} e^{ixt}\,dt. \end{aligned} \tag{3}$$

The sinc function can also be written as the INFINITE PRODUCT

$$\frac{\sin x}{x} = \prod_{k=1}^{\infty} \cos\left(\frac{x}{2^k}\right). \tag{4}$$

Definite integrals involving the sinc function include

$$\int_0^\infty \operatorname{sinc}(x)\,dx = \tfrac{1}{2}\pi \tag{5}$$

$$\int_0^\infty \operatorname{sinc}^2(x)\,dx = \tfrac{1}{2}\pi \tag{6}$$

$$\int_0^\infty \operatorname{sinc}^3(x)\,dx = \tfrac{3}{8}\pi \tag{7}$$

$$\int_0^\infty \operatorname{sinc}^4(x)\,dx = \tfrac{1}{3}\pi \tag{8}$$

$$\int_0^\infty \operatorname{sinc}^5(x)\,dx = \tfrac{115}{384}\pi. \tag{9}$$

These are all special cases of the amazing general result

$$\int_0^\infty \frac{\sin^a x}{x^b}\,dx = \frac{\pi^{1-c}(-1)^{\lfloor (a-b)/2\rfloor}}{2^{a-c}(b-1)!} \times \sum_{k=0}^{\lfloor a/2\rfloor - c} (-1)^k \binom{a}{k}(a-2k)^{b-1}[\ln(a-2k)]^c, \quad (10)$$

where a and b are POSITIVE integers such that $a \geq b > c$, $c \equiv a - b \pmod 2$, $\lfloor x\rfloor$ is the FLOOR FUNCTION, and 0^0 is taken to be equal to 1 (Kogan). This spectacular formula simplifies in the special case when n is a POSITIVE EVEN integer to

$$\int_0^\infty \frac{\sin^{2n} x}{x^{2n}}\,dx = \frac{\pi}{2(2n-1)!}\left\langle {2n-1 \atop n-1} \right\rangle, \quad (11)$$

where $\left\langle {n \atop k} \right\rangle$ is an EULERIAN NUMBER (Kogan). The solution of the integral can also be written in terms of the RECURRENCE RELATION for the coefficients

$$c(a,b) = \begin{cases} \frac{\pi}{2^{a+1-b}}\binom{a-1}{\frac{1}{2}(a-1)} & \text{for } b = 1 \text{ or } b = 2 \\ \frac{a}{(b-1)(b-2)}[(a-1)c(a-2,b-2) \\ \quad -a \cdot c(a,b-2)] & \text{otherwise} \end{cases} \quad (12)$$

(Zimmerman).

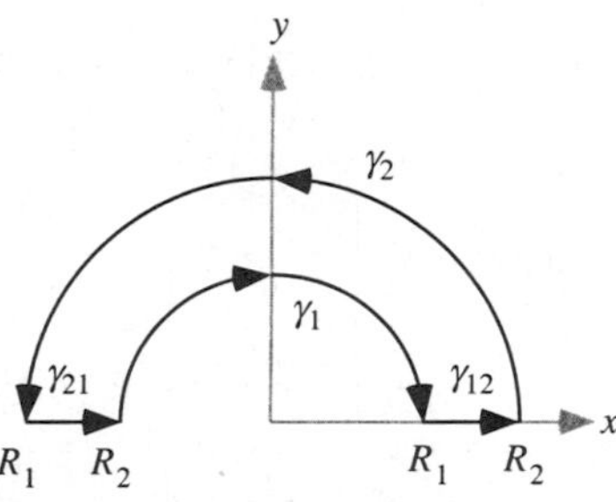

The half-infinite integral of $\operatorname{sinc}(x)$ can be derived using CONTOUR INTEGRATION. In the above figure, consider the path $\gamma \equiv \gamma_1 + \gamma_{12} + \gamma_2 + \gamma_{21}$. Now write $z = Re^{i\theta}$. On an arc, $dz = iRe^{i\theta}\,d\theta$ and on the x-AXIS, $dz = e^{i\theta}\,dR$. Write

$$\int_{-\infty}^\infty \frac{\sin x}{x}\,dx = \Im \int_\gamma \frac{e^{iz}}{z}\,dx, \quad (13)$$

where $\Im$ denotes the IMAGINARY POINT. Now define

$$\begin{aligned} I &\equiv \int_\gamma \frac{e^{iz}}{z}\,dz \\ &= \lim_{R_1\to 0}\int_\pi^0 \frac{\exp(iR_1e^{i\theta})}{R_1e^{i\theta}}\, i\theta R_1 e^{i\theta}\,d\theta \\ &\quad + \lim_{R_1\to 0}\lim_{R_2\to\infty}\int_{R_1}^{R_2}\frac{e^{iR}}{R}\,dR \\ &\quad + \lim_{R_2\to\infty}\int_0^\pi \frac{\exp(iz)}{z}\,dx + \lim_{R_1\to 0}\int_{R_2}^{R_1}\frac{e^{-iR}}{-R}\,(-dR), \end{aligned} \quad (14)$$

where the second and fourth terms use the identities $e^{i0} = 1$ and $e^{i\pi} = -1$. Simplifying,

$$\begin{aligned} I &= \lim_{R_1\to 0}\int_\pi^0 \exp(iR_1e^{i\theta})i\theta\,d\theta + \int_{0+}^\infty \frac{e^{iR}}{R}\,dR \\ &\quad + \lim_{R_2\to\infty}\int_0^\pi \frac{\exp(iz)}{z}\,dz + \int_\infty^{0+}\frac{e^{-iR}}{-R}\,(-dR) \\ &= -\int_0^\pi i\theta\,d\theta + \int_{0+}^\infty \frac{e^{iR}}{R}\,dR + 0 + \int_{-\infty}^{0-}\frac{e^{iR}}{R}\,dR, \end{aligned} \quad (15)$$

where the third term vanishes by JORDAN'S LEMMA. Performing the integration of the first term and combining the others yield

$$I = -i\pi + \int_{-\infty}^\infty \frac{e^{iz}}{z}\,dz = 0. \quad (16)$$

Rearranging gives

$$\int_{-\infty}^\infty \frac{e^{iz}}{z}\,dz = i\pi, \quad (17)$$

so

$$\int_{-\infty}^\infty \frac{\sin z}{z}\,dz = \pi. \quad (18)$$

The same result is arrived at using the method of RESIDUES by noting

$$\begin{aligned} I &= 0 + \tfrac{1}{2}2\pi i \operatorname{Res}[f(z)]_{z=0} \\ &= i\pi\left[(z-0)\frac{e^{iz}}{z}\right]_{z=0} = i\pi[e^{iz}]_{z=0} \\ &= i\pi, \end{aligned} \quad (19)$$

so

$$\Im(I) = \pi. \quad (20)$$

Since the integrand is symmetric, we therefore have

$$\int_0^\infty \frac{\sin x}{x}\,dx = \tfrac{1}{2}\pi, \quad (21)$$

giving the SINE INTEGRAL evaluated at 0 as

$$\operatorname{si}(0) = -\int_0^\infty \frac{\sin x}{x}\,dx = -\tfrac{1}{2}\pi. \quad (22)$$

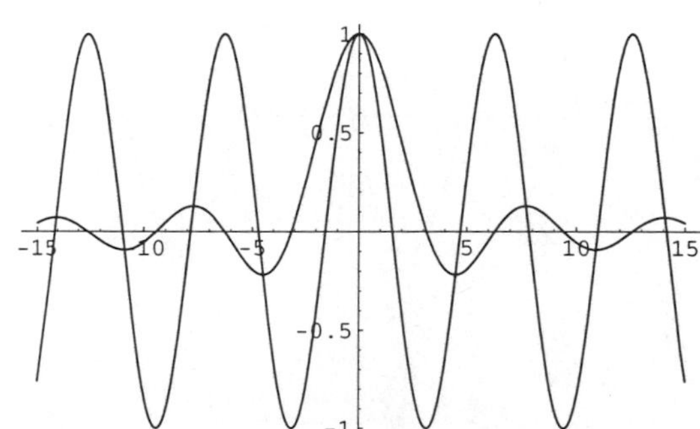

An interesting property of $\operatorname{sinc}(x)$ is that the set of LOCAL EXTREMA of $\operatorname{sinc}(x)$ corresponds to its intersections with the COSINE function $\cos(x)$, as illustrated above.

see also FOURIER TRANSFORM, FOURIER TRANSFORM—RECTANGLE FUNCTION, INSTRUMENT FUNCTION, JINC FUNCTION, SINE, SINE INTEGRAL

References

Kogan, S. "A Note on Definite Integrals Involving Trigonometric Functions." http://www.mathsoft.com/asolve/constant/pi/sin/sin.html.

Morrison, K. E. "Cosine Products, Fourier Transforms, and Random Sums." *Amer. Math. Monthly* **102**, 716–724, 1995.

Sinclair's Soap Film Problem

Find the shape of a soap film (i.e., MINIMAL SURFACE) which will fill two inverted conical FUNNELS facing each other is known as Sinclair's soap film problem (Bliss 1925, p. 121). The soap film will assume the shape of a CATENOID.

see also CATENOID, FUNNEL, MINIMAL SURFACE

References

Bliss, G. A. *Calculus of Variations.* Chicago, IL: Open Court, pp. 121–122, 1925.

Isenberg, C. *The Science of Soap Films and Soap Bubbles.* New York: Dover, p. 81, 1992.

Sinclair, M. E. "On the Minimum Surface of Revolution in the Case of One Variable End Point." *Ann. Math.* **8**, 177–188, 1907.

Sine

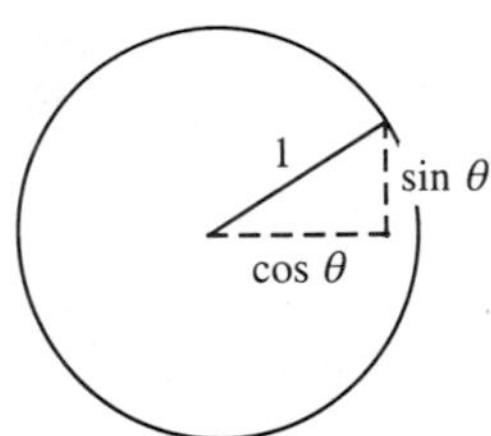

Let θ be an ANGLE measured counterclockwise from the x-AXIS along the arc of the UNIT CIRCLE. Then $\sin\theta$ is the vertical coordinate of the arc endpoint. As a result of this definition, the sine function is periodic with period 2π. By the PYTHAGOREAN THEOREM, $\sin\theta$ also obeys the identity

$$\sin^2\theta + \cos^2\theta = 1. \tag{1}$$

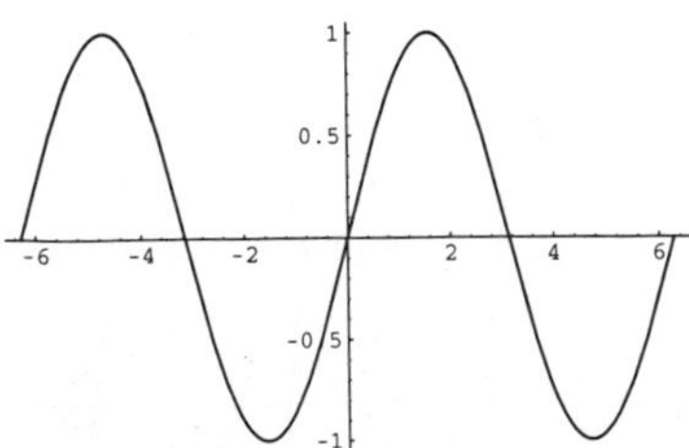

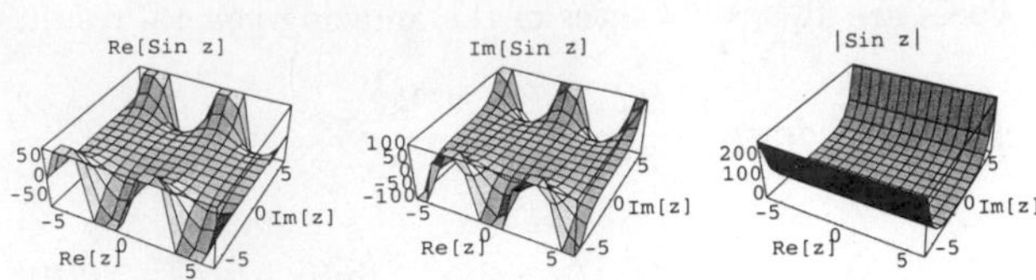

The sine function can be defined algebraically by the infinite sum

$$\sin x = \sum_{n=1}^{\infty} \frac{(-1)^{n-1}}{(2n-1)!} x^{2n-1} \tag{2}$$

and INFINITE PRODUCT

$$\sin x = x \prod_{n=1}^{\infty} \left(1 - \frac{x^2}{n^2\pi^2}\right). \tag{3}$$

It is also given by the IMAGINARY PART of the complex exponential

$$\sin x = \Im[e^{ix}]. \tag{4}$$

The multiplicative inverse of the sine function is the COSECANT, defined as

$$\csc x \equiv \frac{1}{\sin x}. \tag{5}$$

Using the results from the EXPONENTIAL SUM FORMULAS

$$\begin{aligned}\sum_{n=0}^{\infty} \sin(nx) &= \Im\left[\sum_{n=0}^{\infty} e^{inx}\right] \\ &= \Im\left[\frac{\sin(\frac{1}{2}Nx)}{\sin(\frac{1}{2}x)} e^{i(N-1)x/2}\right] \\ &= \frac{\sin(\frac{1}{2}Nx)}{\sin(\frac{1}{2}x)} \sin[\tfrac{1}{2}x(N-1)].\end{aligned} \tag{6}$$

Similarly,

$$\begin{aligned}\sum_{n=0}^{\infty} p^n \sin(nx) &= \Im\left[\sum_{n=0}^{\infty} p^n e^{inx}\right] \\ &= \Im\left[\frac{1-pe^{-ix}}{1-2p\cos x + p^2}\right] = \frac{p\sin x}{1-2p\cos x+p^2}.\end{aligned} \tag{7}$$

Other identities include

$$\sin(n\theta) = 2\cos\theta\sin[(n-1)\theta] - \sin[(n-2)\theta] \tag{8}$$

$$\begin{aligned}\sin(nx) = \binom{n}{1}\cos^{n-1}x\sin x - \binom{n}{3}\cos^{n-3}x\sin^3 x \\ + \binom{n}{5}\cos^{n-5}x\sin^5 x - \ldots,\end{aligned} \tag{9}$$

where $\binom{n}{k}$ is a BINOMIAL COEFFICIENT.

Cvijović and Klinowski (1995) show that the sum

$$S_\nu(\alpha) = \sum_{k=0}^{\infty} \frac{\sin(2k+1)\alpha}{(2k+1)^\nu} \tag{10}$$

has closed form for $\nu = 2n+1$,

$$S_{2n+1}(\alpha) = \frac{(-1)^n}{4(2n)!}\pi^{2n+1} E_{2n}\left(\frac{\alpha}{\pi}\right), \tag{11}$$

where $E_n(x)$ is an EULER POLYNOMIAL.

A CONTINUED FRACTION representation of $\sin x$ is

$$\sin x = \cfrac{x}{1 + \cfrac{x^2}{(2\cdot 3 - x^2) + \cfrac{2\cdot 3x^2}{(4\cdot 5 - x^2) + \cfrac{4\cdot 5x^2}{(6\cdot 7 - x^2) + \ldots}}}}. \tag{12}$$

The value of $\sin(2\pi/n)$ is IRRATIONAL for all n except 4 and 12, for which $\sin(\pi/2) = 1$ and $\sin(\pi/6) = 1/2$.

The FOURIER TRANSFORM of $\sin(2\pi k_0 x)$ is given by

$$\begin{aligned} \mathcal{F}[\sin(2\pi k_0 x)] &= \int_{-\infty}^{\infty} e^{-2\pi i k_0 x} \sin(2\pi k_0 x)\,dx \\ &= \tfrac{1}{2} i[\delta(k+k_0) - \delta(k-k_0)]. \end{aligned} \tag{13}$$

Definite integrals involving $\sin x$ include

$$\int_0^\infty \sin(x^2)\,dx = \tfrac{1}{4}\sqrt{2\pi} \tag{14}$$

$$\int_0^\infty \sin(x^3)\,dx = \tfrac{1}{6}\Gamma(\tfrac{1}{3}) \tag{15}$$

$$\int_0^\infty \sin(x^4)\,dx = -\cos(\tfrac{5}{8}\pi)\Gamma(\tfrac{5}{4}) \tag{16}$$

$$\int_0^\infty \sin(x^5)\,dx = \tfrac{1}{4}(\sqrt{5}-1)\Gamma(\tfrac{6}{5}), \tag{17}$$

where $\Gamma(x)$ is the GAMMA FUNCTION.

see also ANDREW'S SINE, COSECANT, COSINE, FOURIER TRANSFORM—SINE, HYPERBOLIC SINE, SINC FUNCTION, TANGENT, TRIGONOMETRY

References

Abramowitz, M. and Stegun, C. A. (Eds.). "Circular Functions." §4.3 in *Handbook of Mathematical Functions with Formulas, Graphs, and Mathematical Tables, 9th printing.* New York: Dover, pp. 71–79, 1972.

Cvijović, D. and Klinowski, J. "Closed-Form Summation of Some Trigonometric Series." *Math. Comput.* **64**, 205–210, 1995.

Hansen, E. R. *A Table of Series and Products.* Englewood Cliffs, NJ: Prentice-Hall, 1975.

Project Mathematics! *Sines and Cosines, Parts I–III.* Videotapes (28, 30, and 30 minutes). California Institute of Technology. Available from the Math. Assoc. Amer.

Spanier, J. and Oldham, K. B. "The Sine sin(x) and Cosine cos(x) Functions." Ch. 32 in *An Atlas of Functions.* Washington, DC: Hemisphere, pp. 295–310, 1987.

Sine-Gordon Equation

A PARTIAL DIFFERENTIAL EQUATION which appears in differential geometry and relativistic field theory. Its name is a pun on its similar form to the KLEIN-GORDON EQUATION. The sine-Gordon equation is

$$v_{tt} - v_{xx} + \sin v = 0, \tag{1}$$

where v_{tt} and v_{xx} are PARTIAL DERIVATIVES. The equation can be transformed by defining

$$\xi \equiv \tfrac{1}{2}(x-t) \tag{2}$$

$$\eta \equiv \tfrac{1}{2}(x+t), \tag{3}$$

giving

$$v_{\xi\eta} = \sin v. \tag{4}$$

Traveling wave analysis gives

$$z - z_0 = \sqrt{c^2-1}\int \frac{df}{\sqrt{2[d - 2\sin^2(\frac{1}{2}f)]}}. \tag{5}$$

For $d = 0$,

$$z - z_0 = \pm\sqrt{1-c^2}\,\ln[\pm\tan(\tfrac{1}{4}f)] \tag{6}$$

$$f(z) = \pm 4\tan^{-1}[e^{\pm(z-z_0)/(1-c^2)^{1/2}}]. \tag{7}$$

Letting $z \equiv \xi\eta$ then gives

$$zf'' + f' = \sin f. \tag{8}$$

Letting $g \equiv e^{if}$ gives

$$g'' - \frac{g'^2}{f} + \frac{2g' - g^2 + 1}{2z} = 0, \tag{9}$$

which is the third PAINLEVÉ TRANSCENDENT. Look for a solution of the form

$$v(x,t) = 4\tan^{-1}\left[\frac{\phi(x)}{\psi(t)}\right]. \tag{10}$$

Taking the partial derivatives gives

$$\phi_{xx} = -k^2\phi^4 + m^2\phi^2 + n^2 \tag{11}$$

$$\psi_{tt} = k^2\psi^4 + (m^2-1)\psi^2 - n^2, \tag{12}$$

which can be solved in terms of ELLIPTIC FUNCTIONS. A single SOLITON solution exists with $k = n = 0$, $m > 1$:

$$v = 4\tan^{-1}\left[\exp\left(\frac{\pm x - \beta t}{\sqrt{1-\beta^2}}\right)\right], \tag{13}$$

where

$$\beta \equiv \frac{\sqrt{m^2-1}}{m}. \tag{14}$$

A two-SOLITON solution exists with $k = 0$, $m > 1$:

$$v = 4\tan^{-1}\left[\frac{\sinh(\beta mx)}{\beta\cosh(\beta mt)}\right]. \quad (15)$$

A SOLITON-antisoliton solution exists with $k \neq 0$, $n = 0$, $m^2 > 1$:

$$v = -4\tan^{-1}\left[\frac{\sinh(\beta mx)}{\beta\cosh(mt)}\right]. \quad (16)$$

A "breather" solution is

$$v = -4\tan^{-1}\left[\frac{m}{\sqrt{1-m^2}}\frac{\sin(\sqrt{1-m^2}t)}{\cosh(mx)}\right]. \quad (17)$$

References

Infeld, E. and Rowlands, G. *Nonlinear Waves, Solitons, and Chaos.* Cambridge, England: Cambridge University Press, pp. 199–200, 1990.

Sine Integral

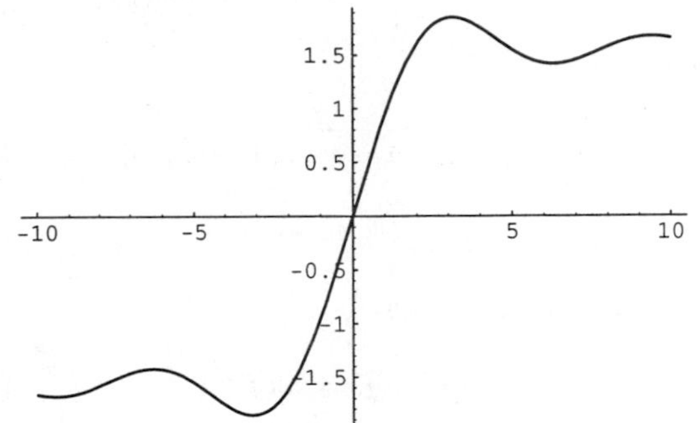

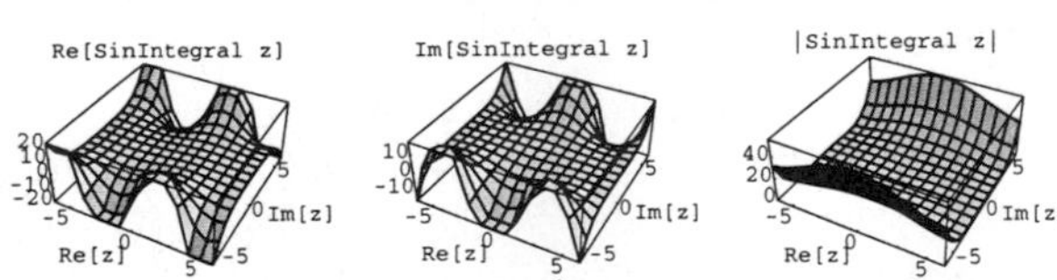

There are two types of "sine integrals" commonly defined,

$$\mathrm{Si}(x) \equiv \int_0^z \frac{\sin t}{t}\,dt \quad (1)$$

and

$$\mathrm{si}(x) \equiv -\int_x^\infty \frac{\sin t}{t}\,dt \quad (2)$$

$$= \frac{1}{2i}[\mathrm{ei}(ix) - \mathrm{ei}(-ix)]$$

$$= \frac{1}{2i}[\mathrm{e}_1(ix) - \mathrm{e}_1(-ix)] \quad (3)$$

$$= \mathrm{Si}(z) - \tfrac{1}{2}\pi, \quad (4)$$

where $\mathrm{ei}(x)$ is the EXPONENTIAL INTEGRAL and

$$\mathrm{e}_1(x) \equiv -\mathrm{ei}(-x). \quad (5)$$

$\mathrm{Si}(x)$ is the function returned by the *Mathematica*® (Wolfram Research, Champaign, IL) command `SinIntegral[x]` and displayed above. The half-infinite integral of the SINC FUNCTION is given by

$$\mathrm{si}(0) = -\int_0^\infty \frac{\sin x}{x}\,dx = -\tfrac{1}{2}\pi. \quad (6)$$

To compute the integral of a sine function times a power

$$I \equiv \int x^{2n}\sin(mx)\,dx, \quad (7)$$

use INTEGRATION BY PARTS. Let

$$u = x^{2n} \qquad dv = \sin(mx)\,dx \quad (8)$$

$$du = 2nx^{2n-1}\,dx \qquad v = -\frac{1}{m}\cos(mx), \quad (9)$$

so

$$I = -\frac{1}{m}x^{2n}\cos(mx) + \frac{2n}{m}\int x^{2n-1}\cos(mx)\,dx. \quad (10)$$

Using INTEGRATION BY PARTS again,

$$u = x^{2n-1} \qquad dv = \cos(mx)\,dx \quad (11)$$

$$du = (2n-1)x^{2n-2}\,dx \qquad v = \frac{1}{m}\sin(mx) \quad (12)$$

$$\begin{aligned}\int x^{2n}\sin(mx)\,dx &= -\frac{1}{m}x^{2n}\cos(mx) \\ &+ \frac{2n}{m}\left[\frac{1}{m}x^{2n-1}\cos(mx) - \frac{2n-1}{m}\int x^{2n-2}\sin(mx)\,dx\right] \\ &= -\frac{1}{m}x^{2n}\sin(mx) + \frac{2n}{m^2}x^{2n-1}\sin(mx) \\ &- \frac{(2n)(2n-1)}{m^2}\int x^{2n-2}\sin(mx)\,dx \\ &= -\frac{1}{m}x^{2n}\cos(mx) + \frac{2n}{m^2}x^{2n-1}\sin(mx) + \ldots \\ &+ \frac{(2n)!}{m^{2n}}\int x^0\sin(mx)\,dx \\ &= -\frac{1}{m}x^{2n}\cos(mx) + \frac{2n}{m^2}x^{2n-1}\sin(mx) + \ldots \\ &- \frac{(2n)!}{m^{2n+1}}\cos(mx) \\ &= \cos(mx)\sum_{k=0}^{n}(-1)^{k+1}\frac{(2n)!}{(2n-2k)!m^{2k+1}}x^{2n-2k} \\ &+ \sin(mx)\sum_{k=1}^{n}(-1)^{k+1}\frac{(2n)!}{(2k-2n-1)!m^{2k}}x^{2n-2k+1}.\end{aligned} \quad (13)$$

Letting $k' \equiv n-k$, so

$$\int x^{2n}\sin(mx)\,dx$$
$$= \cos(mx)\sum_{k=0}^{n}(-1)^{n-k+1}\frac{(2n)!}{(2k)!m^{2n-2k+1}}x^{2k}$$
$$+\sin(mx)\sum_{k=0}^{n-1}(-1)^{n-k+1}\frac{(2n)!}{(2k-1)!m^{2n-2k}}x^{2k+1}$$
$$=(-1)^{n+1}(2n)!\left[\cos(mx)\sum_{k=0}^{n}\frac{(-1)^k}{(2k)!m^{2n-2k+1}}x^{2k}\right.$$
$$\left.+\sin(mx)\sum_{k=1}^{n}\frac{(-1)^{k+1}}{(2k-3)!m^{2n-2k+2}}x^{2k-1}\right]. \tag{14}$$

General integrals of the form

$$I(k,l)=\int_0^\infty \frac{\sin^k x}{x^l}\,dx \tag{15}$$

are related to the SINC FUNCTION and can be computed analytically.

see also CHI, COSINE INTEGRAL, EXPONENTIAL INTEGRAL, NIELSEN'S SPIRAL, SHI, SICI SPIRAL, SINC FUNCTION

References

Abramowitz, M. and Stegun, C. A. (Eds.). "Sine and Cosine Integrals." §5.2 in *Handbook of Mathematical Functions with Formulas, Graphs, and Mathematical Tables, 9th printing.* New York: Dover, pp. 231–233, 1972.

Arfken, G. *Mathematical Methods for Physicists, 3rd ed.* Orlando, FL: Academic Press, pp. 342–343, 1985.

Press, W. H.; Flannery, B. P.; Teukolsky, S. A.; and Vetterling, W. T. "Fresnel Integrals, Cosine and Sine Integrals." §6.79 in *Numerical Recipes in FORTRAN: The Art of Scientific Computing, 2nd ed.* Cambridge, England: Cambridge University Press, pp. 248–252, 1992.

Spanier, J. and Oldham, K. B. "The Cosine and Sine Integrals." Ch. 38 in *An Atlas of Functions.* Washington, DC: Hemisphere, pp. 361–372, 1987.

Sine-Tangent Theorem

If

$$\frac{\sin\alpha}{\sin\beta}=\frac{m}{n},$$

then

$$\frac{\tan[\frac{1}{2}(\alpha-\beta)]}{\tan[\frac{1}{2}(\alpha+\beta)]}=\frac{m-n}{m+n}.$$

Sines Law

see LAW OF SINES

Singly Even Number

An EVEN NUMBER of the form $4n+2$ (i.e., an INTEGER which is DIVISIBLE by 2 but not by 4). The first few for $n = 0, 1, 2, \ldots$ are 2, 6, 10, 14, 18, ... (Sloane's A016825)

see also DOUBLY EVEN NUMBER, EVEN NUMBER, ODD NUMBER

References

Conway, J. H. and Guy, R. K. *The Book of Numbers.* New York: Springer-Verlag, p. 30, 1996.

Sloane, N. J. A. Sequence A016825 in "An On-Line Version of the Encyclopedia of Integer Sequences."

Singular Homology

The general type of HOMOLOGY which is what mathematicians generally mean when they say "homology." Singular homology is a more general version than Poincaré's original SIMPLICIAL HOMOLOGY.

see also HOMOLOGY (TOPOLOGY), SIMPLICIAL HOMOLOGY

Singular Point (Algebraic Curve)

A singular point of an ALGEBRAIC CURVE is a point where the curve has "nasty" behavior such as a CUSP or a point of self-intersection (when the underlying field K is taken as the REALS). More formally, a point (a,b) on a curve $f(x,y)=0$ is singular if the x and y PARTIAL DERIVATIVES of f are both zero at the point (a,b). (If the field K is not the REALS or COMPLEX NUMBERS, then the PARTIAL DERIVATIVE is computed formally using the usual rules of CALCULUS.)

Consider the following two examples. For the curve

$$x^3-y^2=0,$$

the CUSP at (0, 0) is a singular point. For the curve

$$x^2+y^2=-1,$$

$(0,i)$ is a nonsingular point and this curve is nonsingular.

see also ALGEBRAIC CURVE, CUSP

Singular Point (Differential Equation)

Consider a second-order ORDINARY DIFFERENTIAL EQUATION

$$y''+P(x)y'+Q(x)y=0.$$

If $P(x)$ and $Q(x)$ remain FINITE at $x=x_0$, then x_0 is called an ORDINARY POINT. If either $P(x)$ or $Q(x)$ diverges as $x\to x_0$, then x_0 is called a singular point. Singular points are further classified as follows:

1. If either $P(x)$ or $Q(x)$ diverges as $x\to x_0$ but $(x-x_0)P(x)$ and $(x-x_0)^2Q(x)$ remain FINITE as $x\to x_0$, then $x=x_0$ is called a REGULAR SINGULAR POINT (or NONESSENTIAL SINGULARITY).

2. If $P(x)$ diverges more quickly than $1/(x-x_0)$, so $(x-x_0)P(x)$ approaches INFINITY as $x \to x_0$, or $Q(x)$ diverges more quickly than $1/(x-x_0)^2 Q$ so that $(x-x_0)^2 Q(x)$ goes to INFINITY as $x \to x_0$, then x_0 is called an IRREGULAR SINGULARITY (or ESSENTIAL SINGULARITY).

see also IRREGULAR SINGULARITY, REGULAR SINGULAR POINT, SINGULARITY

References

Arfken, G. "Singular Points." §8.4 in *Mathematical Methods for Physicists, 3rd ed.* Orlando, FL: Academic Press, pp. 451–454, 1985.

Singular Point (Function)

Singular points (also simply called "singularities") are points z_0 in the DOMAIN of a FUNCTION f where f fails to be ANALYTIC. ISOLATED SINGULARITIES may be classified as ESSENTIAL SINGULARITIES, POLES, or REMOVABLE SINGULARITIES.

ESSENTIAL SINGULARITIES are POLES of INFINITE order.

A POLE of order n is a singularity z_0 of $f(z)$ for which the function $(z-z_0)^n f(z)$ is nonsingular and for which $(z-z_0)^k f(z)$ is singular for $k = 0, 1, \ldots, n-1$.

REMOVABLE SINGULARITIES are singularities for which it is possible to assign a COMPLEX NUMBER in such a way that $f(z)$ becomes ANALYTIC. For example, the function $f(z) = z^2/z$ has a REMOVABLE SINGULARITY at 0, since $f(z) = z$ everywhere but 0, and $f(z)$ can be set equal to 0 at $z = 0$. REMOVABLE SINGULARITIES are not POLES.

The function $f(z) = \csc(1/z)$ has POLES at $z = 1/(2\pi n)$, and a nonisolated singularity at 0.

see also ESSENTIAL SINGULARITY, IRREGULAR SINGULARITY, ORDINARY POINT, POLE, REGULAR SINGULAR POINT, REMOVABLE SINGULARITY, SINGULAR POINT (DIFFERENTIAL EQUATION)

References

Arfken, G. "Singularities." §7.1 in *Mathematical Methods for Physicists, 3rd ed.* Orlando, FL: Academic Press, pp. 396–400, 1985.

Singular Series

$$\rho_{2s}(n) = \frac{\pi^s}{\Gamma(s)} n^{s-1} \sum_{p,q} \left(\frac{S_{p,q}}{q}\right)^{2s} e^{2np\pi i/q},$$

where $S_{p,q}$ is a GAUSSIAN SUM, and $\Gamma(s)$ is the GAMMA FUNCTION.

Singular System

A system is singular if the CONDITION NUMBER is INFINITE and ILL-CONDITIONED if it is too large.

see also CONDITION NUMBER, ILL-CONDITIONED

Singular Value

A MODULUS k_r such that

$$\frac{K'(k_r)}{K(k_r)} = \sqrt{r},$$

where $K(k)$ is a complete ELLIPTIC INTEGRAL OF THE FIRST KIND, and $K'(k_r) \equiv K(\sqrt{1-k_r{}^2})$. The ELLIPTIC LAMBDA FUNCTION $\lambda^*(r)$ gives the value of k_r.

Abel (quoted in Whittaker and Watson 1990, p. 525) proved that if r is an INTEGER, or more generally whenever

$$\frac{K'(k)}{K(k)} = \frac{a+b\sqrt{n}}{c+d\sqrt{n}},$$

where a, b, c, d, and n are INTEGERS, then the MODULUS k is the ROOT of an algebraic equation with INTEGER COEFFICIENTS.

see also ELLIPTIC INTEGRAL SINGULAR VALUE, ELLIPTIC INTEGRAL OF THE FIRST KIND, ELLIPTIC LAMBDA FUNCTION, MODULUS (ELLIPTIC INTEGRAL)

References

Whittaker, E. T. and Watson, G. N. *A Course in Modern Analysis, 4th ed.* Cambridge, England: Cambridge University Press, pp. 524–528, 1990.

Singular Value Decomposition

An expansion of a REAL $M \times N$ MATRIX by ORTHOGONAL OUTER PRODUCTS according to

$$A = \sum_{k=1}^{K} s_k \mathbf{u}_k \mathbf{v}_k^{\mathrm{T}}, \tag{1}$$

where $s_1 \geq s_2 \geq \ldots \geq 0$,

$$K \equiv \min\{M, N\} \tag{2}$$

and

$$\mathbf{u}_k^{\mathrm{T}} \mathbf{u}_{k'} = \mathbf{v}_l^{\mathrm{T}} \mathbf{v}_{k'} = \delta_{kk'}. \tag{3}$$

Here δ_{ij} is the KRONECKER DELTA and A^{T} is the MATRIX TRANSPOSE.

see also CHOLESKY DECOMPOSITION, LU DECOMPOSITION, QR DECOMPOSITION

References

Nash, J. C. "The Singular-Value Decomposition and Its Use to Solve Least-Squares Problems." Ch. 3 in *Compact Numerical Methods for Computers: Linear Algebra and Function Minimisation, 2nd ed.* Bristol, England: Adam Hilger, pp. 30–48, 1990.

Press, W. H.; Flannery, B. P.; Teukolsky, S. A.; and Vetterling, W. T. "Singular Value Decomposition." §2.6 in *Numerical Recipes in FORTRAN: The Art of Scientific Computing, 2nd ed.* Cambridge, England: Cambridge University Press, pp. 51–63, 1992.

Singularity
In general, a point at which an equation, surface, etc., blows up or becomes DEGENERATE.

see also ESSENTIAL SINGULARITY, ISOLATED SINGULARITY, SINGULAR POINT (ALGEBRAIC CURVE), SINGULAR POINT (DIFFERENTIAL EQUATION), SINGULAR POINT (FUNCTION), WHITNEY SINGULARITY

Sinh
see HYPERBOLIC SINE

Sink (Directed Graph)
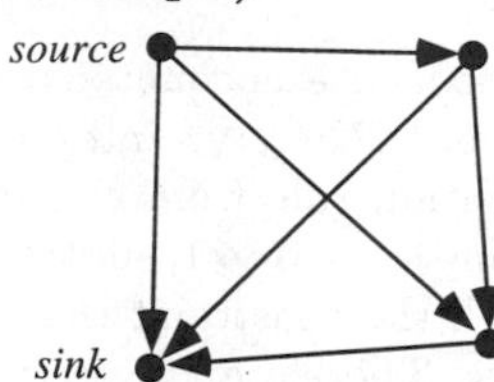

A vertex of a DIRECTED GRAPH with no exiting edges, also called a TERMINAL.

see also DIRECTED GRAPH, NETWORK, SOURCE

Sink (Map)
A stable fixed point of a MAP which, in a dissipative DYNAMICAL SYSTEM, is an ATTRACTOR.

see also ATTRACTOR, DYNAMICAL SYSTEM

Sinusoidal Projection
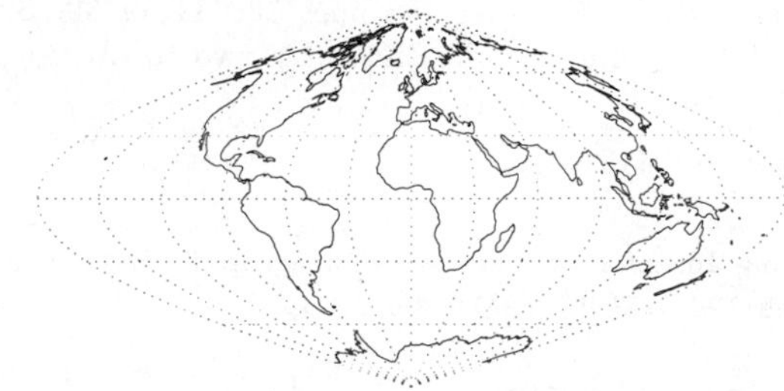

An equal AREA MAP PROJECTION.

$$x = (\lambda - \lambda_0)\cos\phi \quad (1)$$
$$y = \phi. \quad (2)$$

The inverse FORMULAS are

$$\phi = y \quad (3)$$
$$\lambda = \lambda_0 + \frac{x}{\cos\phi}. \quad (4)$$

References
Snyder, J. P. *Map Projections—A Working Manual.* U. S. Geological Survey Professional Paper 1395. Washington, DC: U. S. Government Printing Office, pp. 243–248, 1987.

Sinusoidal Spiral
A curve of the form

$$r^n = a^n \cos(n\theta)$$

with n RATIONAL, which is not a true SPIRAL. Sinusoidal spirals were first studied by Maclaurin. Special cases are given in the following table.

n	Curve
-2	hyperbola
-1	line
$-\frac{1}{2}$	parabola
$-\frac{1}{3}$	Tschirnhausen cubic
0	logarithmic spiral
$\frac{1}{3}$	Cayley sextic
$\frac{1}{2}$	cardioid
1	circle
2	Bernoulli lemniscate

References
Lawrence, J. D. *A Catalog of Special Plane Curves.* New York: Dover, p. 184, 1972.
Lee, X. "Sinusoid." http://www.best.com/~xah/SpecialPlaneCurves_dir/Sinusoid_dir/sinusoid.html.
Lockwood, E. H. *A Book of Curves.* Cambridge, England: Cambridge University Press, p. 175, 1967.
MacTutor History of Mathematics Archive. "Sinusoidal Spirals." http://www-groups.dcs.st-and.ac.uk/~history/Curves/Sinusoidal.html.

Sinusoidal Spiral Inverse Curve
The INVERSE CURVE of a SINUSOIDAL SPIRAL

$$r = a^{(1/n)}[\cos(nt)]^{1/n}$$

with INVERSION CENTER at the origin and inversion radius k is another SINUSOIDAL SPIRAL

$$r = ka^{(1/n)}[\cos(nt)]^{1/n}.$$

Sinusoidal Spiral Pedal Curve
The PEDAL CURVE of a SINUSOIDAL SPIRAL

$$r = a^{(1/n)}[\cos(nt)]^{1/n}$$

with PEDAL POINT at the center is another SINUSOIDAL SPIRAL

$$x = \cos^{1+1/n}(nt)\cos[(n+1)t]$$
$$y = \cos^{1+1/n}(nt)\sin[(n+1)t].$$

Sister Celine's Method

A method for finding RECURRENCE RELATIONS for hypergeometric polynomials directly from the series expansions of the polynomials. The method is effective and easily implemented, but usually slower than ZEILBERGER'S ALGORITHM. Given a sum $f(n) = \sum_k F(n,k)$, the method operates by finding a recurrence of the form

$$\sum_{i=0}^{I}\sum_{j=0}^{J} a_{ij}(n)F(n-j,k-i) = 0$$

by proceeding as follows (Petkovšek *et al.* 1996, p. 59):

1. Fix trial values of I and J.
2. Assume a recurrence formula of the above form where $a_{ij}(n)$ are to be solved for.
3. Divide each term of the assumed recurrence by $F(n,k)$ and reduce every ratio $F(n-j,k-i)/F(n,k)$ by simplifying the ratios of its constituent factorials so that only RATIONAL FUNCTIONS in n and k remain.
4. Put the resulting expression over a common DENOMINATOR, then collect the numerator as a POLYNOMIAL in k.
5. Solve the system of linear equations that results after setting the coefficients of each power of k in the NUMERATOR to 0 for the unknown coefficients a_{ij}.
6. If no solution results, start again with larger I or J.

Under suitable hypotheses, a "fundamental theorem" (Verbaten 1974, Wilf and Zeilberger 1992, Petkovšek *et al.* 1996) guarantees that this algorithm always succeeds for large enough I and J (which can be estimated in advance). The theorem also generalizes to multivariate sums and to q- and multi-q-sums (Wilf and Zeilberger 1992, Petkovšek *et al.* 1996).

see also GENERALIZED HYPERGEOMETRIC FUNCTION, GOSPER'S ALGORITHM, HYPERGEOMETRIC IDENTITY, HYPERGEOMETRIC SERIES, ZEILBERGER'S ALGORITHM

References
Fasenmyer, Sister M. C. *Some Generalized Hypergeometric Polynomials.* Ph.D. thesis. University of Michigan, Nov. 1945.
Fasenmyer, Sister M. C. "Some Generalized Hypergeometric Polynomials." *Bull. Amer. Math. Soc.* **53**, 806–812, 1947.
Fasenmyer, Sister M. C. "A Note on Pure Recurrence Relations." *Amer. Math. Monthly* **56**, 14–17, 1949.
Petkovšek, M.; Wilf, H. S.; and Zeilberger, D. "Sister Celine's Method." Ch. 4 in *A=B.* Wellesley, MA: A. K. Peters, pp. 55–72, 1996.
Rainville, E. D. Chs. 14 and 18 in *Special Functions.* New York: Chelsea, 1971.
Verbaten, P. "The Automatic Construction of Pure Recurrence Relations." *Proc. EUROSAM '74, ACM-SIGSAM Bull.* **8**, 96–98, 1974.
Wilf, H. S. and Zeilberger, D. "An Algorithmic Proof Theory for Hypergeometric (Ordinary and "q") Multisum/Integral Identities." *Invent. Math.* **108**, 575–633, 1992.

Site Percolation

site percolation *bond percolation*

A PERCOLATION which considers the lattice vertices as the relevant entities (left figure).

see also BOND PERCOLATION, PERCOLATION THEORY

Siteswap

A siteswap is a sequence encountered in JUGGLING in which each term is a POSITIVE integer, encoded in BINARY. The transition rule from one term to the next consists of changing some 0 to 1, subtracting 1, and then dividing by 2, with the constraint that the DIVISION by two must be exact. Therefore, if a term is EVEN, the bit to be changed must be the units bit. In siteswaps, the number of 1-bits is a constant.

Each transition is characterized by the bit position of the toggled bit (denoted here by the numeral on top of the arrow). For example,

$$111 \xrightarrow{5} 10011 \xrightarrow{2} 1011 \xrightarrow{5} 10101 \xrightarrow{1} 1011 \xrightarrow{2} 111$$
$$\xrightarrow{6} 100011 \xrightarrow{3} 10101 \xrightarrow{3} 1110 \xrightarrow{0} 111 \xrightarrow{4} 1011 \ldots$$

The second term is given from the first as follows: 000111 with bit 5 flipped becomes 100111, or 39. Subtract 1 to obtain 38 and divide by two to obtain 19, which is 10011.

see also JUGGLING

References
Juggling Information Service. "Siteswaps." http://www.juggling.org/help/siteswap.

Six-Color Theorem

To color any map on the SPHERE or the PLANE requires at most six-colors. This number can be easily be reduced to five, and the FOUR-COLOR THEOREM demonstrates that the NECESSARY number is, in fact, four.

see also FOUR-COLOR THEOREM, HEAWOOD CONJECTURE, MAP COLORING

References
Franklin, P. "A Six Colour Problem." *J. Math. Phys.* **13**, 363–369, 1934.
Hoffman, I. and Soifer, A. "Another Six-Coloring of the Plane." *Disc. Math.* **150**, 427–429, 1996.
Saaty, T. L. and Kainen, P. C. *The Four-Color Problem: Assaults and Conquest.* New York: Dover, 1986.

Skein Relationship

A relationship between KNOT POLYNOMIALS for links in different orientations (denoted below as L_+, L_0, and L_-). J. H. Conway was the first to realize that the ALEXANDER POLYNOMIAL could be defined by a relationship of this type.

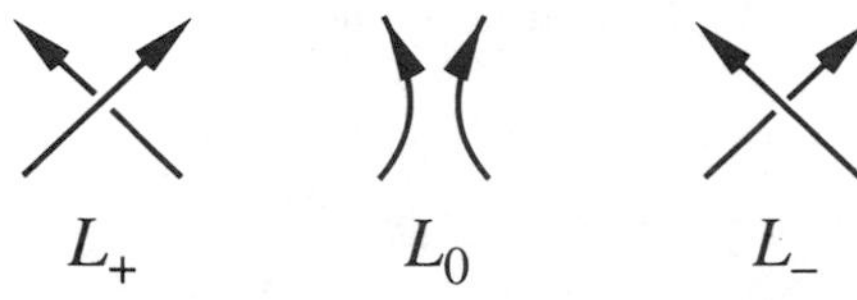

see also ALEXANDER POLYNOMIAL, HOMFLY POLYNOMIAL, SIGNATURE (KNOT)

Skeleton

The GRAPH obtained by collapsing a POLYHEDRON into the PLANE. The number of topologically distinct skeletons $N(n)$ with n VERTICES is given in the following table.

n	$N(n)$
4	1
5	2
6	7

References

Gardner, M. *Martin Gardner's New Mathematical Diversions from Scientific American.* New York: Simon and Schuster, p. 233, 1966.

Skeleton Division

A LONG DIVISION in which most or all of the digits are replaced by a symbol (usually asterisks) to form a CRYPTARITHM.

see also CRYPTARITHM

Skew Conic

Also known as a GAUCHE CONIC, SPACE CONIC, TWISTED CONIC, or CUBICAL CONIC SECTION. A third-order SPACE CURVE having up to three points in common with a plane and having three points in common with the plane at infinity. A skew cubic is determined by six points, with no four of them COPLANAR. A line is met by up to four tangents to a skew cubic.

A line joining two points of a skew cubic (REAL or conjugate imaginary) is called a SECANT of the curve, and a line having one point in common with the curve is called a SEMISECANT or TRANSVERSAL. Depending on the nature of the roots, the skew conic is classified as follows:

1. The three ROOTS are REAL and distinct (CUBICAL HYPERBOLA).
2. One root is REAL and the other two are COMPLEX CONJUGATES (CUBICAL ELLIPSE).
3. Two of the ROOTS coincide (CUBICAL PARABOLIC HYPERBOLA).
4. All three ROOTS coincide (CUBICAL PARABOLA).

see also CONIC SECTION, CUBICAL ELLIPSE, CUBICAL HYPERBOLA, CUBICAL PARABOLA, CUBICAL PARABOLIC HYPERBOLA

Skew Field

A FIELD in which the commutativity of multiplication is not required, more commonly called a DIVISION ALGEBRA.

see also DIVISION ALGEBRA, FIELD

Skew Lines

Two or more LINES which have no intersections but are not PARALLEL, also called AGONIC LINES. Since two LINES in the PLANE must intersect or be PARALLEL, skew lines can exist only in three or more DIMENSIONS.

see also GALLUCCI'S THEOREM, REGULUS

Skew Polyomino

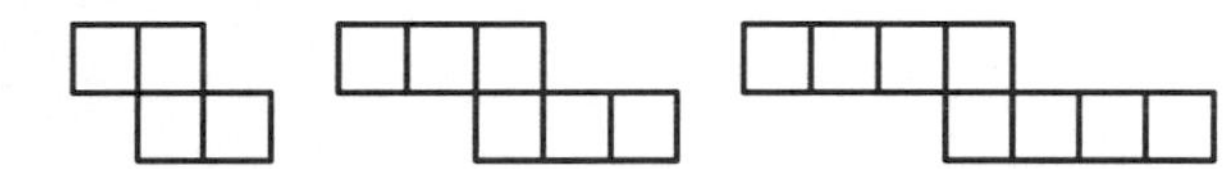

see also L-POLYOMINO, SQUARE POLYOMINO, STRAIGHT POLYOMINO, T-POLYOMINO

Skew Quadrilateral

A four-sided QUADRILATERAL not contained in a plane. The problem of finding the minimum bounding surface of a skew quadrilateral was solved by Schwarz (1890) in terms of ABELIAN INTEGRALS and has the shape of a SADDLE. It is given by solving

$$(1+{f_y}^2)f_{xx} - 2f_xf_yf_{xy} + (1+{f_x}^2)f_{yy} = 0.$$

see also QUADRILATERAL

References

Isenberg, C. *The Science of Soap Films and Soap Bubbles.* New York: Dover, p. 81, 1992.

Forsyth, A. R. *Calculus of Variations.* New York: Dover, p. 503, 1960.

Schwarz, H. A. *Gesammelte Mathematische Abhandlungen, 2nd ed.* New York: Chelsea.

Skew Symmetric Matrix

A MATRIX A where

$$\mathsf{A}^{\mathrm{T}} = -\mathsf{A},$$

with A^{T} denoting the MATRIX TRANSPOSE.

see also MATRIX TRANSPOSE, SYMMETRIC MATRIX

Skewes Number

The Skewes number (or first Skewes number) is the number Sk_1 above which $\pi(n) < \mathrm{Li}(n)$ must fail (assuming that the RIEMANN HYPOTHESIS is true), where $\pi(n)$ is the PRIME COUNTING FUNCTION and $\mathrm{Li}(n)$ is the LOGARITHMIC INTEGRAL.

$$\mathrm{Sk}_1 = e^{e^{e^{79}}} \approx 10^{10^{10^{34}}}.$$

The Skewes number has since been reduced to $e^{e^{27/4}} \approx 8.185 \times 10^{370}$ by te Riele (1987), although Conway and Guy (1996) claim that the best current limit is 10^{1167}. In 1914, Littlewood proved that the inequality must, in fact, fail infinitely often.

The second Skewes number Sk_2 is the number above which $\pi(n) < \mathrm{Li}(n)$ must fail (assuming that the RIEMANN HYPOTHESIS is false). It is much larger than the Skewes number Sk_1,

$$\mathrm{Sk}_2 = 10^{10^{10^{10^3}}}.$$

see also GRAHAM'S NUMBER, RIEMANN HYPOTHESIS

References

Asimov, I. "Skewered!" *Of Matters Great and Small.* New York: Ace Books, 1976. Originally published in *Magazine of Fantasy and Science Fiction,* Nov. 1974.

Ball, W. W. R. and Coxeter, H. S. M. *Mathematical Recreations and Essays, 13th ed.* New York: Dover, p. 63, 1987.

Boas, R. P. "The Skewes Number." In *Mathematical Plums* (Ed. R. Honsberger). Washington, DC: Math. Assoc. Amer., 1979.

Conway, J. H. and Guy, R. K. *The Book of Numbers.* New York: Springer-Verlag, p. 61, 1996.

Lehman, R. S. "On the Difference $\pi(x) - \mathrm{li}(x)$." *Acta Arith.* **11**, 397–410, 1966.

te Riele, H. J. J. "On the Sign of the Difference $\pi(x) - \mathrm{li}(x)$." *Math. Comput.* **48**, 323–328, 1987.

Wagon, S. *Mathematica in Action.* New York: W. H. Freeman, p. 30, 1991.

Skewness

The degree of asymmetry of a distribution. If the distribution has a longer tail less than the maximum, the function has NEGATIVE skewness. Otherwise, it has POSITIVE skewness. Several types of skewness are defined. The FISHER SKEWNESS is defined by

$$\gamma_1 = \frac{\mu_3}{{\mu_2}^{3/2}} = \frac{\mu_3}{\sigma^3}, \tag{1}$$

where μ_3 is the third MOMENT, and ${\mu_2}^{1/2} \equiv \sigma$ is the STANDARD DEVIATION. The PEARSON SKEWNESS is defined by

$$\beta_1 = \left(\frac{\mu_3}{\sigma^3}\right)^2 = {\gamma_1}^2. \tag{2}$$

The MOMENTAL SKEWNESS is defined by

$$\alpha^{(m)} \equiv \tfrac{1}{2}\gamma_1. \tag{3}$$

The PEARSON MODE SKEWNESS is defined by

$$\frac{[\text{mean}] - [\text{mode}]}{\sigma}. \tag{4}$$

PEARSON'S SKEWNESS COEFFICIENTS are defined by

$$\frac{3[\text{mean}] - [\text{mode}]}{s} \tag{5}$$

and

$$\frac{3[\text{mean}] - [\text{median}]}{s}. \tag{6}$$

The BOWLEY SKEWNESS (also known as QUARTILE SKEWNESS COEFFICIENT) is defined by

$$\frac{(Q_3 - Q_2) - (Q_2 - Q_1)}{Q_3 - Q_1} = \frac{Q_1 - 2Q_2 + Q_3}{Q_3 - Q_1}, \tag{7}$$

where the Qs denote the INTERQUARTILE RANGES. The MOMENTAL SKEWNESS is

$$\alpha^{(m)} \equiv \tfrac{1}{2}\gamma = \frac{\mu_3}{2\sigma^3}. \tag{8}$$

An ESTIMATOR for the FISHER SKEWNESS γ_1 is

$$g_1 = \frac{k_3}{{k_2}^{3/2}}, \tag{9}$$

where the ks are k-STATISTICS. The STANDARD DEVIATION of g_1 is

$${\sigma_{g_1}}^2 \approx \frac{6}{N}. \tag{10}$$

see also BOWLEY SKEWNESS, FISHER SKEWNESS, GAMMA STATISTIC, KURTOSIS, MEAN, MOMENTAL SKEWNESS, PEARSON SKEWNESS, STANDARD DEVIATION

References

Abramowitz, M. and Stegun, C. A. (Eds.). *Handbook of Mathematical Functions with Formulas, Graphs, and Mathematical Tables, 9th printing.* New York: Dover, p. 928, 1972.

Press, W. H.; Flannery, B. P.; Teukolsky, S. A.; and Vetterling, W. T. "Moments of a Distribution: Mean, Variance, Skewness, and So Forth." §14.1 in *Numerical Recipes in FORTRAN: The Art of Scientific Computing, 2nd ed.* Cambridge, England: Cambridge University Press, pp. 604–609, 1992.

Sklar's Theorem

Let H be a 2-D distribution function with marginal distribution functions F and G. Then there exists a COPULA C such that

$$H(x, y) = C(F(x), G(y)).$$

Conversely, for any univariate distribution functions F and G and any COPULA C, the function H is a two-dimensional distribution function with marginals F and G. Furthermore, if F and G are continuous, then C is unique.

Skolem-Mahler-Lerch Theorem

If $\{a_0, a_1, \ldots\}$ is a RECURRENCE SEQUENCE, then the set of all k such that $a_k = 0$ is the union of a finite (possibly EMPTY) set and a finite number (possibly zero) of full arithmetical progressions, where a full arithmetic progression is a set of the form $\{r, r+d, r+2d, \ldots\}$ with $r \in [0, d)$.

References

Myerson, G. and van der Poorten, A. J. "Some Problems Concerning Recurrence Sequences." *Amer. Math. Monthly* **102**, 698–705, 1995.

Skolem Paradox

Even though ARITHMETIC is uncountable, it possesses a countable "model."

Skolem Sequence

A Skolem sequence of order n is a sequence $S = \{s_1, s_2, \ldots, s_{2n}\}$ of $2n$ integers such that

1. For every $k \in \{1, 2, \ldots, n\}$, there exist exactly two elements $s_i, s_j \in S$ such that $s_i = s_j = k$, and
2. If $s_i = s_j = k$ with $i < j$, then $j - i = k$.

References

Colbourn, C. J. and Dinitz, J. H. (Eds.) "Skolem Sequences." Ch. 43 in *CRC Handbook of Combinatorial Designs.* Boca Raton, FL: CRC Press, pp. 457–461, 1996.

Slant Height

The height of an object (such as a CONE) measured along a side from the edge of the base to the apex.

Slice Knot

A KNOT K in $\mathbb{S}^3 = \partial\mathbb{D}^4$ is a slice knot if it bounds a DISK Δ^2 in $\mathbb{D}^4$ which has a TUBULAR NEIGHBORHOOD $\Delta^2 \times \mathbb{D}^2$ whose intersection with $\mathbb{S}^3$ is a TUBULAR NEIGHBORHOOD $K \times \mathbb{D}^2$ for K.

Every RIBBON KNOT is a slice knot, and it is conjectured that every slice knot is a RIBBON KNOT.

see also RIBBON KNOT, TUBULAR NEIGHBORHOOD

References

Rolfsen, D. *Knots and Links.* Wilmington, DE: Publish or Perish Press, p. 218, 1976.

Slide Move

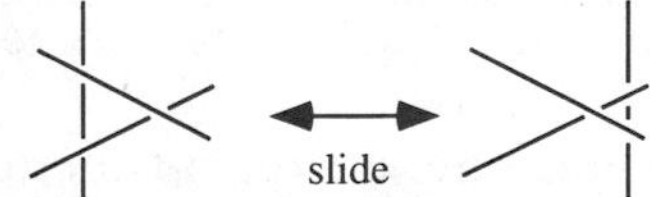

The REIDEMEISTER MOVE of type III.

see also REIDEMEISTER MOVES

Slide Rule

A mechanical device consisting of a sliding portion and a fixed case, each marked with logarithmic axes. By lining up the ticks, it is possible to do MULTIPLICATION by taking advantage of the additive property of LOGARITHMS. More complicated slide rules also allow the extraction of roots and computation of trigonometric functions. The development of the desk calculator (and subsequently pocket calculator) rendered slide rules largely obsolete beginning in the 1960s.

see also ABACUS, RULER, STRAIGHTEDGE

References

Electronic Teaching Laboratories. *Simplify Math: Learn to Use the Slide Rule.* New Augusta, IN: Editors and Engineers, 1966.

Saffold, R. *The Slide Rule.* Garden City, NY: Doubleday, 1962.

Slightly Defective Number

see ALMOST PERFECT NUMBER

Slightly Excessive Number

see QUASIPERFECT NUMBER

Slip Knot

see RUNNING KNOT

Slope

A quantity which gives the inclination of a curve or line with respect to another curve or line. For a LINE in the PLANE making an ANGLE θ with the x-AXIS, the SLOPE m is a constant given by

$$m \equiv \frac{\Delta y}{\Delta x} = \tan\theta,$$

where Δx and Δy are changes in the two coordinates over some distance. It is meaningless to talk about the slope in 3-D unless the slope *with respect to what* is specified.

Slothouber-Graatsma Puzzle

Assemble six $1 \times 2 \times 2$ blocks and three $1 \times 1 \times 1$ blocks into a $3 \times 3 \times 3$ CUBE.

see also BOX-PACKING THEOREM, CONWAY PUZZLE, CUBE DISSECTION, DE BRUIJN'S THEOREM, KLARNER'S THEOREM, POLYCUBE

References

Honsberger, R. *Mathematical Gems II.* Washington, DC: Math. Assoc. Amer., pp. 75–77, 1976.

Slutzky-Yule Effect

A MOVING AVERAGE may generate an irregular oscillation even if none exists in the original data.

see also MOVING AVERAGE

Sluze Pearls

see PEARLS OF SLUZE

Smale-Hirsch Theorem

The SPACE of IMMERSIONS of a MANIFOLD in another MANIFOLD is HOMOTOPICALLY equivalent to the space of bundle injections from the TANGENT SPACE of the first to the TANGENT BUNDLE of the second.

see also HOMOTOPY, IMMERSION, MANIFOLD, TANGENT BUNDLE, TANGENT SPACE

Smale Horseshoe Map

The basic topological operations for constructing an ATTRACTOR consist of stretching (which gives sensitivity to initial conditions) and folding (which gives the attraction). Since trajectories in PHASE SPACE cannot cross, the repeated stretching and folding operations result in an object of great topological complexity.

The Smale horseshoe map consists of a sequence of operations on the unit square. First, stretch by a factor of 2 in the x direction, then compress by $2a$ in the y direction. Then, fold the rectangle and fit it back into the square. Repeating this generates the horseshoe attractor. If one looks at a cross-section of the final structure, it is seen to correspond to a CANTOR SET.

see also ATTRACTOR, CANTOR SET

References
Gleick, J. *Chaos: Making a New Science.* New York: Penguin, pp. 50–51, 1988.
Rasband, S. N. *Chaotic Dynamics of Nonlinear Systems.* New York: Wiley, p. 77, 1990.
Tabor, M. *Chaos and Integrability in Nonlinear Dynamics: An Introduction.* New York: Wiley, 1989.

Small Cubicuboctahedron

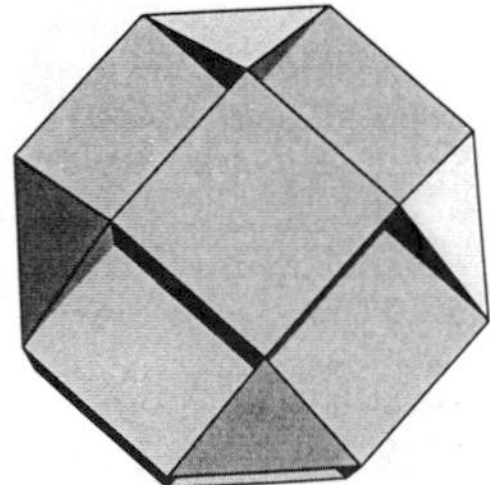

UNIFORM POLYHEDRON U_{13} whose DUAL POLYHEDRON is the SMALL HEXACRONIC ICOSITETRAHEDRON. It has WYTHOFF SYMBOL $\frac{3}{2}\,4\,|\,4$. Its faces are $8\{3\}+6\{4\}+6\{8\}$. The CIRCUMRADIUS for the solid with unit edge length is

$$R=\tfrac{1}{2}\sqrt{5+2\sqrt{2}}\,.$$

FACETED versions include the GREAT RHOMBICUBOCTAHEDRON (UNIFORM) and SMALL RHOMBIHEXAHEDRON.

References
Wenninger, M. J. *Polyhedron Models.* Cambridge, England: Cambridge University Press, pp. 104–105, 1971.

Small Ditrigonal Dodecacronic Hexecontahedron

The DUAL POLYHEDRON of the SMALL DITRIGONAL DODECICOSIDODECAHEDRON.

Small Ditrigonal Dodecicosidodecahedron

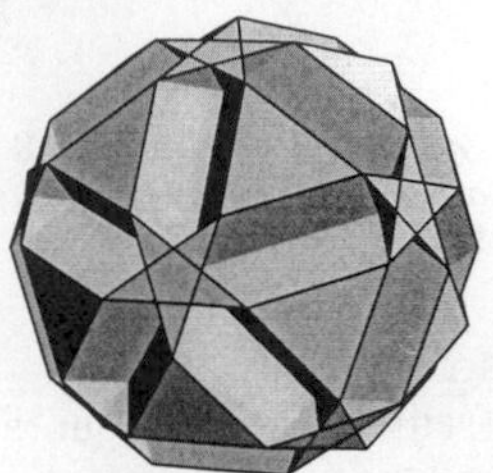

The UNIFORM POLYHEDRON U_{43} whose DUAL POLYHEDRON is the SMALL DITRIGONAL DODECACRONIC HEXECONTAHEDRON. It has WYTHOFF SYMBOL $3\,\frac{5}{3}\,|\,5$. Its faces are $20\{3\}+12\{\frac{5}{2}\}+12\{10\}$. Its CIRCUMRADIUS with $a=1$ is

$$R=\tfrac{1}{4}\sqrt{34+6\sqrt{5}}\,.$$

References
Wenninger, M. J. *Polyhedron Models.* Cambridge, England: Cambridge University Press, pp. 126–127, 1971.

Small Ditrigonal Icosidodecahedron

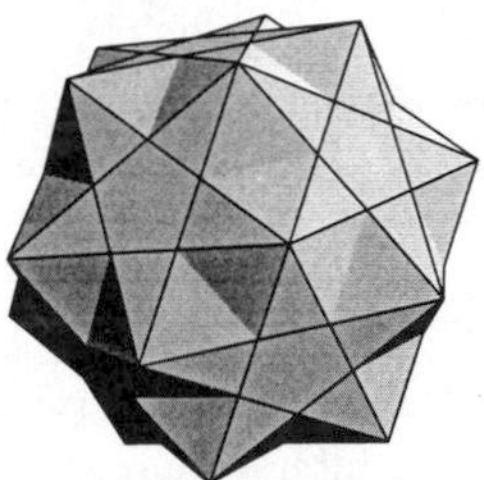

The UNIFORM POLYHEDRON U_{30} whose DUAL POLYHEDRON is the SMALL TRIAMBIC ICOSAHEDRON. It has WYTHOFF SYMBOL $3\,|\,3\,\frac{5}{2}$. Its faces are $20\{3\}+12\{\frac{5}{2}\}$. A FACETED version is the DITRIGONAL DODECADODECAHEDRON. Its CIRCUMRADIUS with $a=1$ is

$$R=\tfrac{1}{2}\sqrt{3}\,.$$

References
Wenninger, M. J. *Polyhedron Models.* Cambridge, England: Cambridge University Press, pp. 106–107, 1971.

Small Dodecacronic Hexecontahedron

The DUAL POLYHEDRON of the SMALL DODECICOSIDODECAHEDRON.

Small Dodecahemicosacron

The DUAL POLYHEDRON of the SMALL DODECAHEMICOSAHEDRON.

Small Dodecahemicosahedron

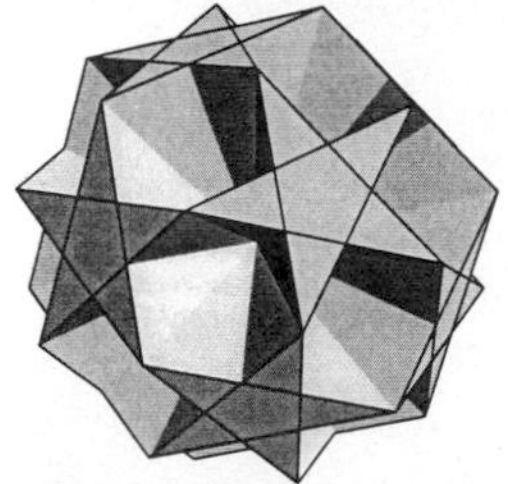

The UNIFORM POLYHEDRON U_{62} whose DUAL POLYHEDRON is the SMALL DODECAHEMICOSACRON. It has WYTHOFF SYMBOL $\frac{5}{3}\,\frac{5}{2}\,|\,3$. Its faces are $10\{6\}+12\{\frac{5}{2}\}$. It is a FACETED version of the ICOSIDODECAHEDRON. Its CIRCUMRADIUS with unit edge length is

$$R=1.$$

References

Wenninger, M. J. *Polyhedron Models.* Cambridge, England: Cambridge University Press, p. 155, 1971.

Small Dodecahemidodecacron

The DUAL POLYHEDRON of the SMALL DODECAHEMIDODECAHEDRON.

Small Dodecahemidodecahedron

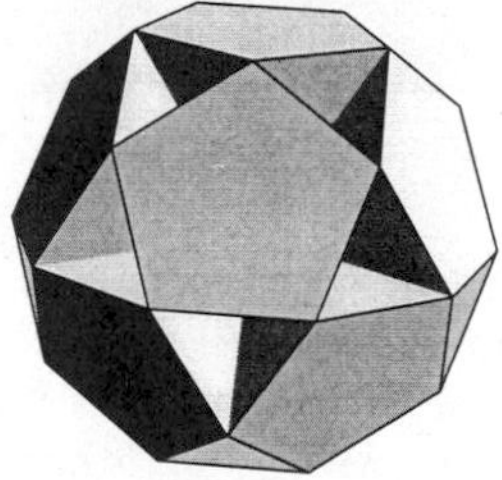

The UNIFORM POLYHEDRON U_{51} whose DUAL POLYHEDRON is the SMALL DODECAHEMIDODECACRON. It has WYTHOFF SYMBOL $2\,5\,\frac{3}{2}\,\frac{5}{2}$. Its faces are $30\{4\}+12\{10\}$. Its CIRCUMRADIUS with $a=1$ is

$$R=\tfrac{1}{2}\sqrt{11+4\sqrt{5}}.$$

References

Wenninger, M. J. *Polyhedron Models.* Cambridge, England: Cambridge University Press, pp. 113–114, 1971.

Small Dodecicosacron

The DUAL POLYHEDRON of the SMALL DODECICOSAHEDRON.

Small Dodecicosahedron

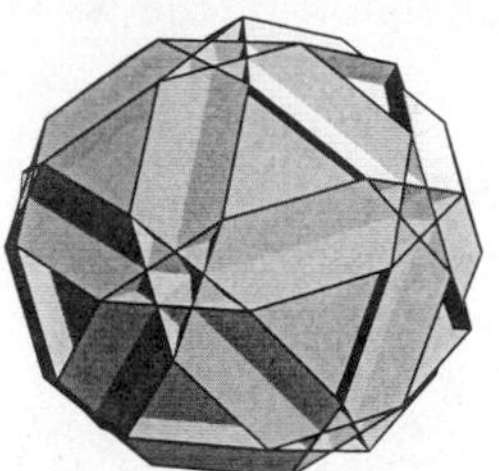

The UNIFORM POLYHEDRON U_{50} whose DUAL POLYHEDRON is the SMALL DODECICOSACRON. It has WYTHOFF SYMBOL $3\,5\,\frac{3}{2}\,\frac{5}{4}\,|$. Its faces are $20\{6\}+12\{10\}$. Its CIRCUMRADIUS with $a=1$ is

$$R=\tfrac{1}{4}\sqrt{34+6\sqrt{5}}.$$

References

Wenninger, M. J. *Polyhedron Models.* Cambridge, England: Cambridge University Press, pp. 141–142, 1971.

Small Dodecicosidodecahedron

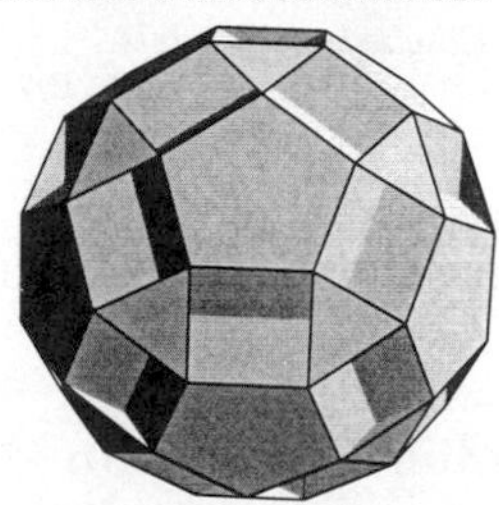

The UNIFORM POLYHEDRON U_{33} whose DUAL POLYHEDRON is the SMALL DODECACRONIC HEXECONTAHEDRON. It has WYTHOFF SYMBOL $\frac{3}{2}\,5\,|\,5$. Its faces are $20\{3\}+12\{5\}+12\{10\}$. It is a FACETED version of the SMALL RHOMBICOSIDODECAHEDRON. Its CIRCUMRADIUS with $a=1$ is

$$R=\tfrac{1}{2}\sqrt{11+4\sqrt{5}}.$$

References

Wenninger, M. J. *Polyhedron Models.* Cambridge, England: Cambridge University Press, pp. 110–111, 1971.

Small Hexacronic Icositetrahedron

The DUAL POLYHEDRON of the SMALL CUBICUBOCTAHEDRON.

Small Hexagonal Hexecontahedron

The DUAL POLYHEDRON of the SMALL SNUB ICOSICOSIDODECAHEDRON.

Small Hexagrammic Hexecontahedron

The DUAL POLYHEDRON of the SMALL RETROSNUB ICOSICOSIDODECAHEDRON.

Small Icosacronic Hexecontahedron

The DUAL POLYHEDRON of the SMALL ICOSICOSIDODECAHEDRON.

Small Icosicosidodecahedron

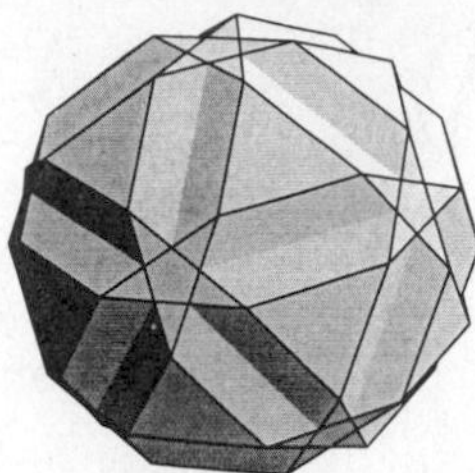

The UNIFORM POLYHEDRON U_{31} whose DUAL POLYHEDRON is the SMALL ICOSACRONIC HEXECONTAHEDRON. It has WYTHOFF SYMBOL $\frac{5}{4}\,5\,|\,5$. Its faces are $12\{5\}+6\{10\}$. Its CIRCUMRADIUS with $a=1$ is

$$R=\phi=\tfrac{1}{2}(1+\sqrt{5}\,),$$

where ϕ is the GOLDEN RATIO.

References
Wenninger, M. J. *Polyhedron Models.* Cambridge, England: Cambridge University Press, p. 143, 1971.

Small Icosihemidodecacron

The DUAL POLYHEDRON of the SMALL ICOSIHEMIDODECAHEDRON.

Small Icosihemidodecahedron

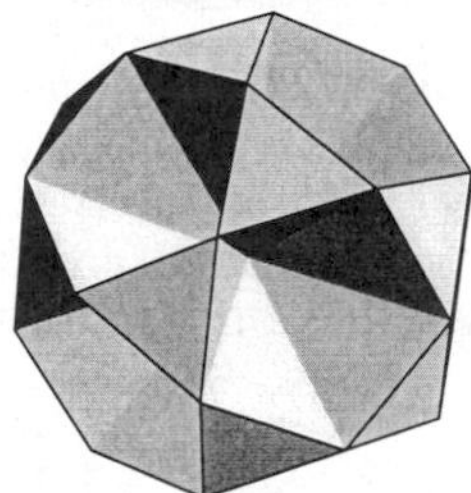

The UNIFORM POLYHEDRON U_{49} whose DUAL POLYHEDRON is the SMALL ICOSIHEMIDODECACRON. It has WYTHOFF SYMBOL $\frac{3}{2}\,3\,|\,5$. Its faces are $20\{3\}+6\{10\}$. It is a FACETED version of the ICOSIDODECAHEDRON. Its CIRCUMRADIUS with $a=1$ is

$$R=\phi=\tfrac{1}{2}(1+\sqrt{5}\,).$$

References
Wenninger, M. J. *Polyhedron Models.* Cambridge, England: Cambridge University Press, p. 140, 1971.

Small Inverted Retrosnub Icosicosidodecahedron

see SMALL RETROSNUB ICOSICOSIDODECAHEDRON

Small Multiple Method

An algorithm for computing a UNIT FRACTION.

Small Number

Guy's "STRONG LAW OF SMALL NUMBERS" states that there aren't enough small numbers to meet the many demands made of them. Guy (1988) also gives several interesting and misleading facts about small numbers:

1. 10% of the first 100 numbers are SQUARE NUMBERS.
2. A QUARTER of the numbers < 100 are PRIMES.
3. All numbers less than 10, except for 6, are PRIME POWERS.
4. Half the numbers less than 10 are FIBONACCI NUMBERS.

see also LARGE NUMBER, STRONG LAW OF SMALL NUMBERS

References
Guy, R. K. "The Strong Law of Small Numbers." *Amer. Math. Monthly* **95**, 697–712, 1988.

Small Retrosnub Icosicosidodecahedron

The UNIFORM POLYHEDRON U_{72} also called the SMALL INVERTED RETROSNUB ICOSICOSIDODECAHEDRON whose DUAL POLYHEDRON is the SMALL HEXAGRAMMIC HEXECONTAHEDRON. It has WYTHOFF SYMBOL $|\,\frac{3}{2}\,\frac{3}{2}\,\frac{5}{2}$. Its faces are $100\{3\}+12\{\frac{5}{2}\}$. It has CIRCUMRADIUS with $a=1$

$$\begin{aligned}R&=\tfrac{1}{4}\sqrt{13+3\sqrt{5}-\sqrt{102+46\sqrt{5}}}\\&\approx 0.580694800133921.\end{aligned}$$

References
Wenninger, M. J. *Polyhedron Models.* Cambridge, England: Cambridge University Press, pp. 194–199, 1971.

Small Rhombicosidodecahedron

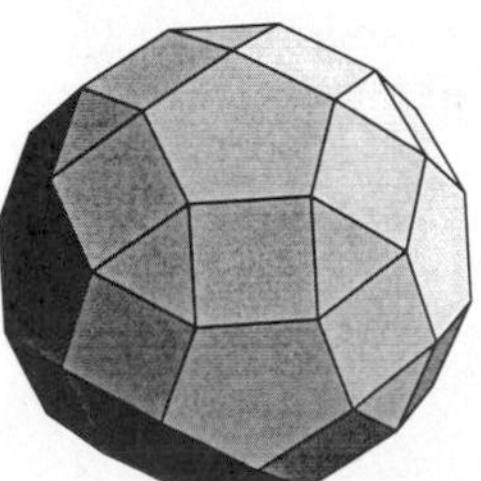

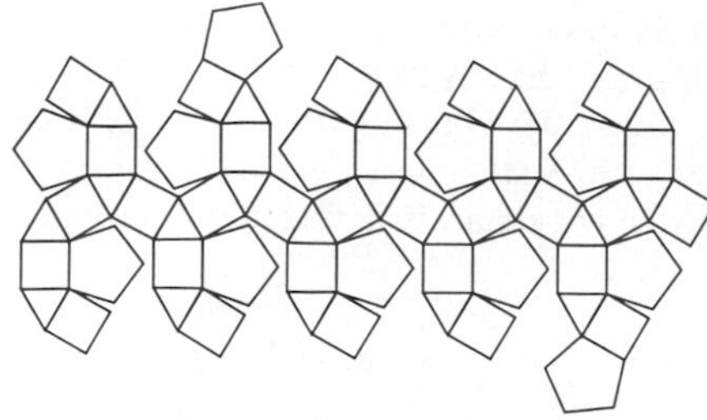

An ARCHIMEDEAN SOLID whose DUAL POLYHEDRON is the DELTOIDAL HEXECONTAHEDRON. It has SCHLÄFLI SYMBOL $r\left\{\begin{smallmatrix}3\\5\end{smallmatrix}\right\}$. It is also UNIFORM POLYHEDRON U_{27} with WYTHOFF SYMBOL $3\,5\,|\,2$. Its faces are $20\{3\} + 30\{4\} + 12\{5\}$. The SMALL DODECICOSIDODECAHEDRON and SMALL RHOMBIDODECAHEDRON are FACETED versions. The INRADIUS, MIDRADIUS, and CIRCUMRADIUS for $a = 1$ are

$$r = \tfrac{1}{41}(15 + 2\sqrt{5})\sqrt{11 + 4\sqrt{5}} = 2.12099\ldots$$

$$\rho = \tfrac{1}{2}\sqrt{10 + 4\sqrt{5}} = 2.17625\ldots$$

$$R = \tfrac{1}{2}\sqrt{11 + 4\sqrt{5}} = 2.23295\ldots.$$

see also GREAT RHOMBICOSIDODECAHEDRON (ARCHIMEDEAN), GREAT RHOMBICOSIDODECAHEDRON (UNIFORM)

Small Rhombicuboctahedron

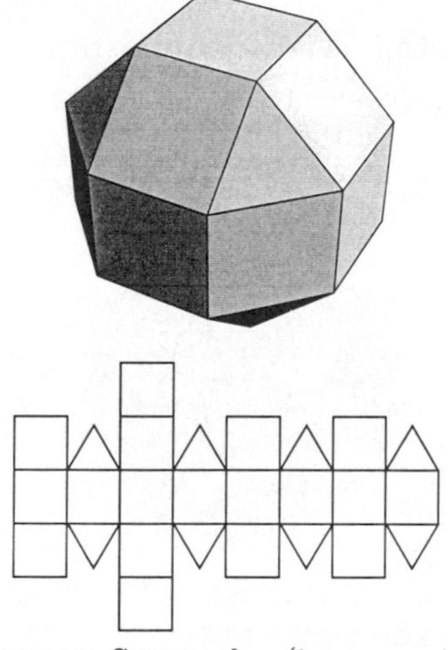

An ARCHIMEDEAN SOLID also (inappropriately) called the TRUNCATED ICOSIDODECAHEDRON. This name is inappropriate since truncation would yield rectangular instead of square faces. Its DUAL POLYHEDRON is the DELTOIDAL ICOSITETRAHEDRON, also called the TRAPEZOIDAL ICOSITETRAHEDRON. It has SCHLÄFLI SYMBOL $r\left\{\begin{smallmatrix}3\\4\end{smallmatrix}\right\}$. It is also UNIFORM POLYHEDRON U_{10} and has WYTHOFF SYMBOL $3\,4\,|\,2$. Its INRADIUS, MIDRADIUS, and CIRCUMRADIUS for $a = 1$ are

$$r = \tfrac{1}{17}(6 + \sqrt{2})\sqrt{5 + 2\sqrt{2}} = 1.22026\ldots$$

$$\rho = \tfrac{1}{2}\sqrt{4 + 2\sqrt{2}} = 1.30656\ldots$$

$$R = \tfrac{1}{2}\sqrt{5 + 2\sqrt{2}} = 1.39897\ldots.$$

A version in which the top and bottom halves are rotated with respect to each other is known as the ELONGATED SQUARE GYROBICUPOLA.

see also ELONGATED SQUARE GYROBICUPOLA, GREAT RHOMBICUBOCTAHEDRON (ARCHIMEDEAN), GREAT RHOMBICUBOCTAHEDRON (UNIFORM)

References

Ball, W. W. R. and Coxeter, H. S. M. *Mathematical Recreations and Essays, 13th ed.* New York: Dover, pp. 137–138, 1987.

Small Rhombidodecacron

The DUAL POLYHEDRON of the SMALL RHOMBIDODECAHEDRON.

Small Rhombidodecahedron

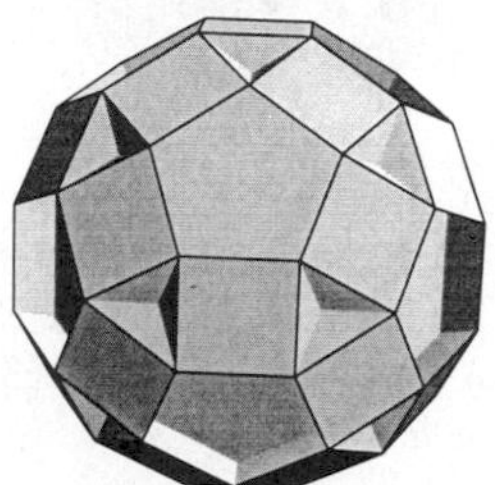

The UNIFORM POLYHEDRON U_{39} whose DUAL POLYHEDRON is the SMALL RHOMBIDODECACRON. It has WYTHOFF SYMBOL $2\,5\,\frac{\frac{3}{2}}{\frac{5}{2}}$. Its faces are $30\{4\} + 12\{10\}$. It is a FACETED version of the SMALL RHOMBICOSIDODECAHEDRON. Its CIRCUMRADIUS with $a = 1$ is

$$R = \tfrac{1}{2}\sqrt{11 + 4\sqrt{5}}.$$

References

Wenninger, M. J. *Polyhedron Models.* Cambridge, England: Cambridge University Press, pp. 113–114, 1971.

Small Rhombihexacron

The DUAL POLYHEDRON of the SMALL RHOMBIHEXAHEDRON.

Small Rhombihexahedron

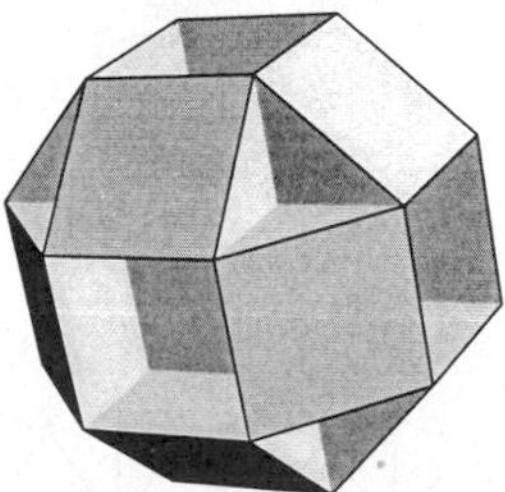

The UNIFORM POLYHEDRON U_{18} whose DUAL POLYHEDRON is the SMALL RHOMBIHEXACRON. It has WYTHOFF SYMBOL $2\,4\,\frac{\frac{3}{2}}{\frac{4}{2}}\,\Big|$. Its faces are $12\{4\} + 6\{8\}$. It is a FACETED version of the SMALL RHOMBICUBOCTAHEDRON. Its CIRCUMRADIUS with $a = 1$ is

$$R = \tfrac{1}{2}\sqrt{5 + 2\sqrt{2}}.$$

References
Wenninger, M. J. *Polyhedron Models.* Cambridge, England: Cambridge University Press, p. 134, 1971.

Small Snub Icosicosidodecahedron

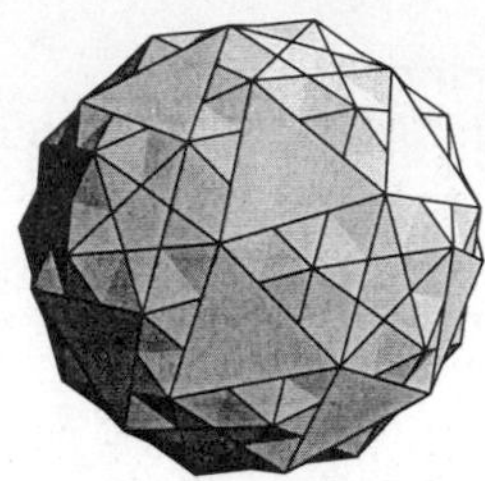

The UNIFORM POLYHEDRON U_{32} whose DUAL POLYHEDRON is the SMALL HEXAGONAL HEXECONTAHEDRON. It has WYTHOFF SYMBOL $|\,3\,3\,\frac{5}{2}$ (Har'El 1993 gives the symbol as $|\,\frac{5}{2}\,3\,3$.) Its faces are $100\{3\} + 12\{\frac{5}{2}\}$. Its CIRCUMRADIUS for $a = 1$ is

$$R = \tfrac{1}{4}\sqrt{13 + 3\sqrt{5} + \sqrt{102 + 46\sqrt{5}}}$$
$$= 1.4581903307387\ldots.$$

References
Har'El, Z. "Uniform Solution for Uniform Polyhedra." *Geometriae Dedicata* **47**, 57–110, 1993.
Wenninger, M. J. *Polyhedron Models.* Cambridge, England: Cambridge University Press, pp. 172–173, 1971.

Small Stellapentakis Dodecahedron

The DUAL POLYHEDRON of the TRUNCATED GREAT DODECAHEDRON.

Small Stellated Dodecahedron

One of the KEPLER-POINSOT SOLIDS whose DUAL POLYHEDRON is the GREAT DODECAHEDRON. Its SCHLÄFLI SYMBOL is $\{\frac{5}{2}, 5\}$. It is also UNIFORM POLYHEDRON U_{34} and has WYTHOFF SYMBOL $5\,|\,2\,\frac{5}{2}$. It was originally called the URCHIN by Kepler. It is composed of 12 PENTAGRAMMIC faces. Its faces are $12\{\frac{5}{2}\}$. The easiest way to construct it is to build twelve pentagonal PYRAMIDS

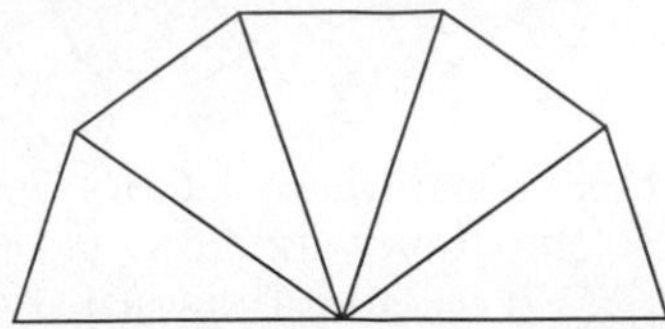

and attach them to the faces of a DODECAHEDRON. The CIRCUMRADIUS of the small stellated dodecahedron with $a = 1$ is

$$R = \tfrac{1}{2}5^{1/4}\phi^{-1/2} = \tfrac{1}{4}5^{1/4}\sqrt{2(\sqrt{5} - 1)}.$$

see also GREAT DODECAHEDRON, GREAT ICOSAHEDRON, GREAT STELLATED DODECAHEDRON, KEPLER-POINSOT SOLID

References
Fischer, G. (Ed.). Plate 103 in *Mathematische Modelle/Mathematical Models, Bildband/Photograph Volume.* Braunschweig, Germany: Vieweg, p. 102, 1986.
Rawles, B. *Sacred Geometry Design Sourcebook: Universal Dimensional Patterns.* Nevada City, CA: Elysian Pub., p. 219, 1997.

Small Stellated Triacontahedron

see MEDIAL RHOMBIC TRIACONTAHEDRON

Small Stellated Truncated Dodecahedron

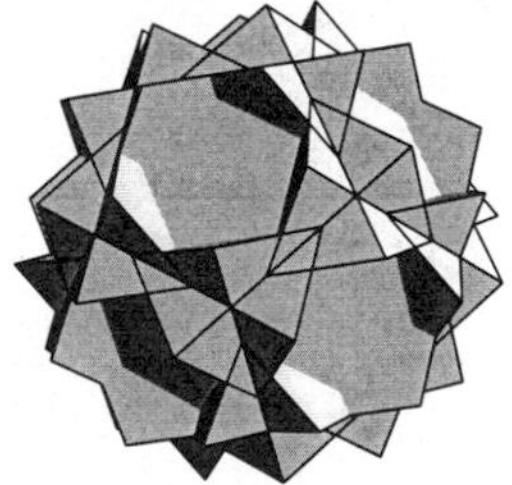

The UNIFORM POLYHEDRON U_{58} also called the QUASITRUNCATED SMALL STELLATED DODECAHEDRON whose DUAL POLYHEDRON is the GREAT PENTAKIS DODECAHEDRON. It has SCHLÄFLI SYMBOL $t'\{\frac{5}{2}, 5\}$ and WYTHOFF SYMBOL $2\,5\,|\,\frac{5}{3}$. Its faces are $12\{5\} + 12\{\frac{10}{3}\}$. Its CIRCUMRADIUS with $a = 1$ is

$$R = \tfrac{1}{4}\sqrt{34 - 10\sqrt{5}}.$$

References
Wenninger, M. J. *Polyhedron Models.* Cambridge, England: Cambridge University Press, p. 151, 1971.

Small Triakis Octahedron

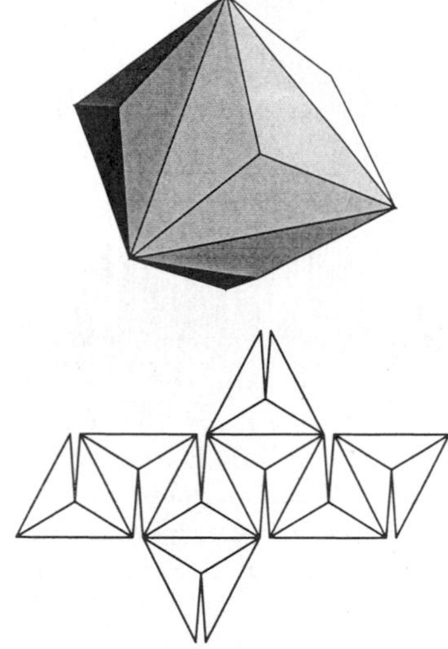

The Dual Polyhedron of the Truncated Cube.

see also Great Triakis Octahedron

Small Triambic Icosahedron

The Dual Polyhedron of the Small Ditrigonal Icosidodecahedron.

Small World Problem

The small world problem asks for the probability that two people picked at random have at least one acquaintance in common.

see also Birthday Problem

Smarandache Ceil Function

A Smarandache-like function which is defined where $S_k(n)$ is defined as the smallest integer for which $n|S_k(n)^k$. The Smarandache $S_k(n)$ function can therefore be obtained by replacing any factors which are kth powers in n by their k roots. The functions $S_k(n)$ for $k = 2, 3, \ldots, 6$ for values such that $S_k(n) \neq n$ are tabulated by Begay (1997).

$S_1(n) = n$, so the first few values of $S_1(n)$ are 1, 2, 3, 4, 5, 6, 7, 8, 9, 10, 11, 12, 13, 14, 15, 16, 17, 18, 19, 20, ... (Sloane's A000027). The first few values of $S_2(n)$ are 1, 2, 3, 2, 5, 6, 7, 4, 3, 10, 11, 6, 13, 14, 15, 4, 17, 6, 19, 10, ... (Sloane's A019554) The first few values of $S_3(n)$ are 1, 2, 3, 2, 5, 6, 7, 2, 3, 10, 11, 6, 13, 14, 15, 4, 17, 6, 19, 10, ... (Sloane's A019555) The first few values of $S_4(n)$ are 1, 2, 3, 2, 5, 6, 7, 2, 3, 10, 11, 6, 13, 14, 15, 2, 17, 6, 19, 10, ... (Sloane's A007947).

see also Pseudosmarandache Function, Smarandache Function, Smarandache-Kurepa Function, Smarandache Near-to-Primorial Function, Smarandache Sequences, Smarandache-Wagstaff Function, Smarandache Function

References

Begay, A. "Smarandache Ceil Functions." *Bull. Pure Appl. Sci.* **16E**, 227–229, 1997.

"Functions in Number Theory." `http://www.gallup.unm.edu/~smarandache/FUNCT1.TXT`.

Sloane, N. J. A. Sequences A007947, A019554, A019555, and A0472/M000027 in "An On-Line Version of the Encyclopedia of Integer Sequences."

Smarandache, F. *Collected Papers, Vol. 2.* Kishinev, Moldova: Kishinev University Press, 1997.

Smarandache, F. *Only Problems, Not Solutions!, 4th ed.* Phoenix, AZ: Xiquan, 1993.

Smarandache Constants

The first Smarandache constant is defined as

$$S_1 \equiv \sum_{n=2}^{\infty} \frac{1}{[S(n)]!} > 1.093111,$$

where $S(n)$ is the Smarandache Function. Cojocaru and Cojocaru (1996a) prove that S_1 exists and is bounded by $0.717 < S_1 < 1.253$. The lower limit given above is obtained by taking 40,000 terms of the sum.

Cojocaru and Cojocaru (1996b) prove that the second Smarandache constant

$$S_2 \equiv \sum_{n=2}^{\infty} \frac{S(n)}{n!} \approx 1.71400629359162$$

is an Irrational Number.

Cojocaru and Cojocaru (1996c) prove that the series

$$S_3 \equiv \sum_{n=2}^{\infty} \frac{1}{\prod_{i=2}^{n} S(i)} \approx 0.719960700043708$$

converges to a number $0.71 < S_3 < 1.01$, and that

$$S_4(a) \equiv \sum_{n=2}^{\infty} \frac{n^a}{\prod_{i=2}^{n} S(i)}$$

converges for a fixed Real Number $a \geq 1$. The values for small a are

$$S_4(1) \approx 1.72875760530223$$
$$S_4(2) \approx 4.50251200619297$$
$$S_4(3) \approx 13.0111441949445$$
$$S_4(4) \approx 42.4818449849626$$
$$S_4(5) \approx 158.105463729329.$$

Sandor (1997) shows that the series

$$S_5 \equiv \sum_{n=1}^{\infty} \frac{(-1)^{n-1} S(n)}{n!}$$

converges to an IRRATIONAL. Burton (1995) and Dumitrescu and Seleacu (1996) show that the series

$$S_6 \equiv \sum_{n=2}^{\infty} \frac{S(n)}{(n+1)!}$$

converges. Dumitrescu and Seleacu (1996) show that the series

$$S_7 \equiv \sum_{n=r}^{\infty} \frac{S(n)}{(n+r)!}$$

and

$$S_8 \equiv \sum_{n=r}^{\infty} \frac{S(n)}{(n-r)!}$$

converge for r a natural number (which must be nonzero in the latter case). Dumitrescu and Seleacu (1996) show that

$$S_9 \equiv \sum_{n=2}^{\infty} \frac{1}{\sum_{i=2}^{n} \frac{S(i)}{i!}}$$

converges. Burton (1995) and Dumitrescu and Seleacu (1996) show that the series

$$S_{10} \sum_{n=2}^{\infty} \frac{1}{[S(n)]^{\alpha} \sqrt{S(n)!}}$$

and

$$S_{11} \sum_{n=2}^{\infty} \frac{1}{[S(n)]^{\alpha} \sqrt{[S(n)+1]!}}$$

converge for $\alpha > 1$.

see also SMARANDACHE FUNCTION

References

Burton, E. "On Some Series Involving the Smarandache Function." *Smarandache Notions J.* **6**, 13–15, 1995.

Burton, E. "On Some Convergent Series." *Smarandache Notions J.* **7**, 7–9, 1996.

Cojocaru, I. and Cojocaru, S. "The First Constant of Smarandache." *Smarandache Notions J.* **7**, 116–118, 1996a.

Cojocaru, I. and Cojocaru, S. "The Second Constant of Smarandache." *Smarandache Notions J.* **7**, 119–120, 1996b.

Cojocaru, I. and Cojocaru, S. "The Third and Fourth Constants of Smarandache." *Smarandache Notions J.* **7**, 121–126, 1996c.

"Constants Involving the Smarandache Function." `http://www.gallup.unm.edu/~smarandache/CONSTANT.TXT`.

Dumitrescu, C. and Seleacu, V. "Numerical Series Involving the Function S." *The Smarandache Function.* Vail: Erhus University Press, pp. 48–61, 1996.

Ibstedt, H. *Surfing on the Ocean of Numbers—A Few Smarandache Notions and Similar Topics.* Lupton, AZ: Erhus University Press, pp. 27–30, 1997.

Sandor, J. 'On The Irrationality Of Certain Alternative Smarandache Series." *Smarandache Notions J.* **8**, 143–144, 1997.

Smarandache, F. *Collected Papers, Vol. 1.* Bucharest, Romania: Tempus, 1996.

Smarandache, F. *Collected Papers, Vol. 2.* Kishinev, Moldova: Kishinev University Press, 1997.

Smarandache Function

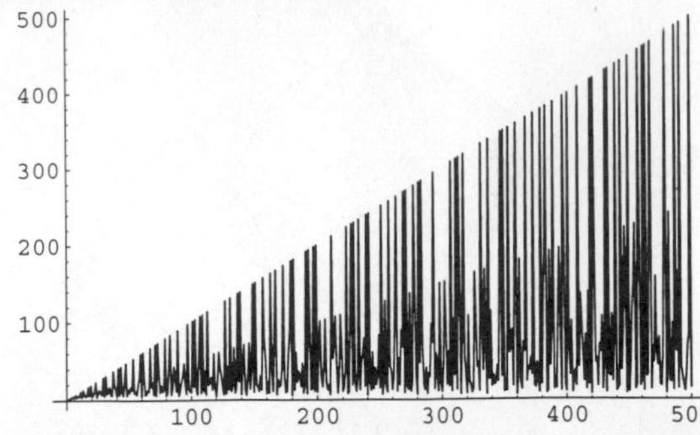

The smallest value $S(n)$ for a given n for which $n|S(n)!$ (n divides $S(n)$ FACTORIAL). For example, the number 8 does not divide 1!, 2!, 3!, but does divide $4! = 4\cdot3\cdot2\cdot1 = 8\cdot3$, so $S(8) = 4$. For a PRIME p, $S(p) = p$, and for an EVEN PERFECT NUMBER r, $S(r)$ is PRIME (Ashbacher 1997).

The Smarandache numbers for $n = 1, 2, \ldots$ are 1, 2, 3, 4, 5, 3, 7, 4, 6, 5, 11, ... (Sloane's A002034). Letting $a(n)$ denote the smallest value of n for which $S(n) = 1, 2, \ldots$, then $a(n)$ is given by 1, 2, 3, 4, 5, 9, 7, 32, 27, 25, 11, 243, ... (Sloane's A046021). Some values of $S(n)$ first occur only for very large n, for example, $S(59,049) = 24$, $S(177,147) = 27$, $S(134,217,728) = 30$, $S(43,046,721) = 36$, and $S(9,765,625) = 45$. D. Wilson points out that if we let

$$I(n,p) = \frac{n - \Sigma(n,p)}{p-1},$$

be the power of the PRIME p in $n!$, where $\Sigma(n,p)$ is the sum of the base-p digits of n, then it follows that

$$a(n) = \min p^{I(n-1,p)+1},$$

where the minimum is taken over the PRIMES p dividing n. This minimum appears to always be achieved when p is the GREATEST PRIME FACTOR of n.

The incrementally largest values of $S(n)$ are 1, 2, 3, 4, 5, 7, 11, 13, 17, 19, 23, 29, ... (Sloane's A046022), which occur for $n = 1, 2, 3, 4, 5, 7, 11, 13, 17, 19, 23, 29, \ldots$, i.e., the values where $S(n) = n$.

Tutescu (1996) conjectures that the DIOPHANTINE EQUATION $S(n) = S(n+1)$ has no solution.

see also FACTORIAL, GREATEST PRIME FACTOR, PSEUDOSMARANDACHE FUNCTION, SMARANDACHE CEIL FUNCTION, SMARANDACHE CONSTANTS, SMARANDACHE-KUREPA FUNCTION, SMARANDACHE NEAR-TO-PRIMORIAL FUNCTION, SMARANDACHE-WAGSTAFF FUNCTION

References

Ashbacher, C. *An Introduction to the Smarandache Function.* Cedar Rapids, IA: Decisionmark, 1995.

Ashbacher, C. "Problem 4616." *School Sci. Math.* **97**, 221, 1997.

Begay, A. "Smarandache Ceil Functions." *Bulletin Pure Appl. Sci. India* **16E**, 227–229, 1997.
Dumitrescu, C. and Seleacu, V. *The Smarandache Function.* Vail, AZ: Erhus University Press, 1996.
"Functions in Number Theory." `http://www.gallup.unm.edu/~smarandache/FUNCT1.TXT`.
Ibstedt, H. *Surfing on the Ocean of Numbers—A Few Smarandache Notions and Similar Topics.* Lupton, AZ: Erhus University Press, pp. 27–30, 1997.
Sandor, J. "On Certain Inequalities Involving the Smarandache Function." *Abstracts of Papers Presented to the Amer. Math. Soc.* **17**, 583, 1996.
Sloane, N. J. A. Sequences A046021, A046022, A046023, and A002034/M0453 in "An On-Line Version of the Encyclopedia of Integer Sequences."
Smarandache, F. *Collected Papers, Vol. 1.* Bucharest, Romania: Tempus, 1996.
Smarandache, F. *Collected Papers, Vol. 2.* Kishinev, Moldova: Kishinev University Press, 1997.
Tutescu, L. "On a Conjecture Concerning the Smarandache Function." *Abstracts of Papers Presented to the Amer. Math. Soc.* **17**, 583, 1996.

Smarandache-Kurepa Function

Given the sum-of-factorials function

$$\Sigma(n) = \sum_{k=1}^{n} k!,$$

SK(p) is the smallest integer for p PRIME such that $1 + \Sigma[\mathrm{SK}(p-1)]$ is divisible by p. The first few known values of SK(p) are 2, 4, 6, 6, 5, 7, 7, 12, 22, 16, 55, 54, 42, 24, ... for $p = 2$, 5, 7, 11, 17, 19, 23, 31, 37, 41, 61, 71, 73, 89, ... The values for $p = 3$, 13, 29, 43, 47, 53, 67, 79, 83, ..., if they are finite, must be very large (e.g., $\mathrm{SK}(3) > 100{,}000$).

see also PSEUDOSMARANDACHE FUNCTION, SMARANDACHE CEIL FUNCTION, SMARANDACHE FUNCTION, SMARANDACHE-WAGSTAFF FUNCTION, SMARANDACHE FUNCTION

References

Ashbacher, C. "Some Properties of the Smarandache-Kurepa and Smarandache-Wagstaff Functions." *Math. Informatics Quart.* **7**, 114–116, 1997.
Mudge, M. "Introducing the Smarandache-Kurepa and Smarandache-Wagstaff Functions." *Smarandache Notions J.* **7**, 52–53, 1996.
Mudge, M. "Introducing the Smarandache-Kurepa and Smarandache-Wagstaff Functions." *Abstracts of Papers Presented to the Amer. Math. Soc.* **17**, 583, 1996.

Smarandache Near-to-Primorial Function

$SNTP(n)$ is the smallest PRIME such that $p\# - 1$, $p\#$, or $p\# + 1$ is divisible by n, where $p\#$ is the PRIMORIAL of p. Ashbacher (1996) shows that $SNTP(n)$ only exists

1. If there are no square or higher powers in the factorization of n, or
2. If there exists a PRIME $q < p$ such that $n|(q\# \pm 1)$, where p is the smallest power contained in the factorization of n.

Therefore, $SNTP(n)$ does not exist for the SQUAREFUL numbers $n = 4$, 8, 9, 12, 16, 18, 20, 24, 25, 27, 28, ... (Sloane's A002997) The first few values of $SNTP(n)$, where defined, are 2, 2, 2, 3, 3, 3, 5, 7, ... (Sloane's A046026).

see also PRIMORIAL, SMARANDACHE FUNCTION

References

Ashbacher, C. "A Note on the Smarandache Near-To-Primordial Function." *Smarandache Notions J.* **7**, 46–49, 1996.
Mudge, M. R. "The Smarandache Near-To-Primorial Function." *Abstracts of Papers Presented to the Amer. Math. Soc.* **17**, 585, 1996.
Sloane, N. J. A. Sequence A002997 in "An On-Line Version of the Encyclopedia of Integer Sequences."

Smarandache Paradox

Let A be some attribute (e.g., possible, present, perfect, etc.). If all is A, then the non-A must also be A. For example, "All is possible, the impossible too," and "Nothing is perfect, not even the perfect."

References

Le, C. T. "The Smarandache Class of Paradoxes." *Bull. Transylvania Univ. Brasov* **36**, 7–8, 1994.
Le, C. T. "The Smarandache Class of Paradoxes." *Bull. Pure Appl. Sci.* **14E**, 109–110, 1995.
Le, C. T. "The Smarandache Class of Paradoxes." *J. Indian Acad. Math.* **18**, 53–55, 1996.
Mitroiescu, I. *The Smarandache Class of Paradoxes.* Glendale, AZ: Erhus University Press, 1994.
Mitroiescu, I. "The Smarandache's Class of Paradoxes Applied in Computer Science." *Abstracts of Papers Presented to the Amer. Math. Soc.* **16**, 651, 1995.

Smarandache Sequences

Smarandache sequences are any of a number of simply generated INTEGER SEQUENCES resembling those considered in published works by Smarandache such as the CONSECUTIVE NUMBER SEQUENCES and EUCLID NUMBERS (Iacobescu 1997). Other Smarandache-type sequences are given below.

1. The concatenation of n copies of the INTEGER n: 1, 22, 333, 4444, 55555, ... (Sloane's A000461; Marimutha 1997),
2. The concatenation of the first n FIBONACCI NUMBERS: 1, 11, 112, 1123, 11235, ... (Sloane's A019523; Marimutha 1997),
3. The smallest number that is the sum of squares of *two* distinct earlier terms: 1, 2, 5, 26, 29, 677, ... (Sloane's A008318, Bencze 1997),
4. The smallest number that is the sum of squares of any number of distinct earlier terms: 1, 1, 2, 4, 5, 6, 16, 17, ... (Sloane's A008319, Bencze 1997),
5. The smallest number that is *not* the sum of squares of *two* distinct earlier terms: 1, 2, 3, 4, 6, 7, 8, 9, 11, ... (Sloane's A008320, Bencze 1997),
6. The smallest number that is *not* the sum of squares of any number of distinct earlier terms: 1, 2, 3, 6, 7, 8, 11, ... (Sloane's A008321, Bencze 1997),

7. The smallest number that is a sum of cubes of *two* distinct earlier terms: 1, 2, 9, 730, 737, ... (Sloane's A008322, Bencze 1997),
8. The smallest number that is a sum of cubes of any number of distinct earlier terms: 1, 1, 2, 8, 9, 512, 513, 514, ... (Sloane's A008323, Bencze 1997),
9. The smallest number that is *not* a sum of cubes of *two* of distinct earlier terms: 1, 2, 3, 4, 5, 6, 7, 8, 10, ... (Sloane's A008380, Bencze 1997),
10. The smallest number that is *not* a sum of cubes of any number of distinct earlier terms: 1, 2, 3, 4, 5, 6, 7, 10, 11, ... (Sloane's A008381, Bencze 1997),
11. The number of PARTITIONS of a number $n = 1$, 2, ... into SQUARE NUMBERS: 1, 1, 1, 1, 2, 2, 2, 2, 3, 4, 4, 4, 5, 6, 6, 6, 8, 9, 10, 10, 12, 13, ... (Sloane's A001156, Iacobescu 1997),
12. The number of PARTITIONS of a number $n = 1$, 2, ... into CUBIC NUMBERS: 1, 1, 1, 1, 1, 1, 1, 1, 2, 2, 2, 2, 2, 2, 2, 2, 3, 3, 3, 3, 3, 3, 3, ... (Sloane's A003108, Iacobescu 1997),
13. Two copies of the first n POSITIVE integers: 11, 1212, 123123, 12341234, ... (Sloane's A019524, Iacobescu 1997),
14. Numbers written in base of triangular numbers: 1, 2, 10, 11, 12, 100, 101, 102, 110, 1000, 1001, 1002, ... (Sloane's A000462, Iacobescu 1997),
15. Numbers written in base of double factorial numbers: 1, 10, 100, 101, 110, 200, 201, 1000, 1001, 1010, ... (Sloane's A019513, Iacobescu 1997),
16. Sequences starting with terms $\{a_1, a_2\}$ which contain no three-term arithmetic progressions starting with $\{1, 2\}$: 1, 2, 4, 5, 10, 11, 13, 14, 28, ... (Sloane's A033155, Iacobescu 1997, Mudge 1997, Weisstein),
17. Numbers of the form $(n!)^2 + 1$: 2, 5, 37, 577, 14401, 518401, 25401601, 1625702401, 131681894401, ... (Sloane's A020549, Iacobescu 1997),
18. Numbers of the form $(n!)^3 + 1$: 2, 9, 217, 13825, 1728001, 373248001, 128024064001, ... (Sloane's A019514, Iacobescu 1997),
19. Numbers of the form $1 + 1!2!3!\cdots n!$: 2, 3, 13, 289, 34561, 24883201, 125411328001, 5056584744960001, ... (Sloane's A019515, Iacobescu 1997),
20. Sequences starting with terms $\{a_1, a_2\}$ which contain no three-term geometric progressions starting with $\{1, 2\}$: 1, 2, 3, 5, 6, 7, 8, 10, 11, 13, 14, 15, 16, ... (Sloane's A000452, Iacobescu 1997),
21. Numbers repeating the digit 1 p_n times, where p_n is the nth prime: 11, 111, 11111, 1111111, ... (Sloane's A031974, Iacobescu 1997). These are a subset of the REPUNITS,
22. Integers with all 2s, 3s, 5s, and 7s (prime digits) removed: 1, 4, 6, 8, 9, 10, 11, 1, 1, 14, 1, 16, 1, 18, 19, 0, ... (Sloane's A019516, Iacobescu 1997),
23. Integers with all 0s, 1s, 4s, and 9s (square digits) removed: 2, 3, 5, 6, 7, 8, 2, 3, 5, 6, 7, 8, 2, 2, 22, 23, ... (Sloane's A031976, Iacobescu 1997).
24. (Smarandache-Fibonacci triples) Integers n such that $S(n) = S(n-1) + S(n-2)$, where $S(k)$ is the SMARANDACHE FUNCTION: 3, 11, 121, 4902, 26245, ... (Sloane's A015047; Aschbacher and Mudge 1995; Ibstedt 1997, pp. 19–23; Begay 1997). The largest known is 19,448,047,080,036,
25. (Smarandache-Radu triplets) Integers n such that there are no primes between the smaller and larger of $S(n)$ and $S(n+1)$: 224, 2057, 265225, ... (Sloane's A015048; Radu 1994/1995, Begay 1997, Ibstedt 1997). The largest known is 270,329,975,921, 205,253,634,707,051,822,848,570,391,313,
26. (Smarandache crescendo sequence): Integers obtained by concatenating strings of the first $n+1$ integers for $n = 0, 1, 2, \ldots$: 1, 1, 2, 1, 2, 3, 1, 2, 3, 4, ... (Sloane's A002260; Brown 1997, Brown and Castillo 1997). The nth term is given by $n - m(m+1)/2 + 1$, where $m = \lfloor(\sqrt{8n+1} - 1)/2\rfloor$, with $\lfloor x \rfloor$ the FLOOR FUNCTION (Hamel 1997),
27. (Smarandache descrescendo sequence): Integers obtained by concatenating strings of the first n integers for $n = \ldots, 2, 1$: 1, 2, 1, 3, 2, 1, 4, 3, 2, 1, ... (Sloane's A004736; Smarandache 1997, Brown 1997),
28. (Smarandache crescendo pyramidal sequence): Integers obtained by concatenating strings of rising and falling integers: 1, 1, 2, 1, 1, 2, 3, 2, 1, 1, 2, 3, 4, 3, 2, 1, ... (Sloane's A004737; Brown 1997, Brown and Castillo 1997, Smarandache 1997),
29. (Smarandache descrescendo pyramidal sequence): Integers obtained by concatenating strings of falling and rising integers: 1, 2, 1, 2, 3, 2, 1, 2, 3, 4, 3, 2, 1, 2, 3, 4, ... (Brown 1997),
30. (Smarandache crescendo symmetric sequence): 1, 1, 1, 2, 2, 1, 1, 2, 3, 3, 2, 1, ... (Sloane's A004739, Brown 1997, Smarandache 1997),
31. (Smarandache descrescendo symmetric sequence): 1, 1, 2, 1, 1, 2, 3, 2, 1, 1, 2, 3, ... (Sloane's A004740; Brown 1997, Smarandache 1997),
32. (Smarandache permutation sequence): Numbers obtained by concatenating sequences of increasing length of increasing ODD NUMBERS and decreasing EVEN NUMBERS: 1, 2, 1, 3, 4, 2, 1, 3, 5, 6, 4, 2, ... (Sloane's A004741; Brown 1997, Brown and Castillo 1997),
33. (Smarandache pierced chain sequence): Numbers of the form $c(n) = 101\underbrace{0101}_{n}$ for $n = 0, 1, \ldots$: 101, 1010101, 10101010101, ... (Sloane's A031982; Ashbacher 1997). In addition, $c(n)/101$ contains no PRIMES (Ashbacher 1997),

34. (Smarandache symmetric sequence): 1, 11, 121, 1221, 12321, 123321, ... (Sloane's A007907; Smarandache 1993, Dumitrescu and Seleacu 1994, sequence 3; Mudge 1995),
35. (Smarandache square-digital sequence): square numbers all of whose digits are also squares: 1, 4, 9, 49, 100, 144, ... (Sloane's A019544; Mudge 1997),
36. (Square-digits): numbers composed of digits which are squares: 1, 4, 9, 10, 14, 19, 40, 41, 44, 49, ... (Sloane's A066030),
37. (Smarandache square-digital sequence): square-digit numbers which are themselves squares: 1, 4, 9, 49, 100, 144, ... (Sloane's A019544; Mudge 1997),
38. (Cube-digits): numbers composed of digits which are cubes: 1, 4, 10, 11, 14, 40, 41, 44, 100, 101, ... (Sloane's A046031),
39. (Smarandache cube-digital sequence): cube-digit numbers which are themselves cubes: 1, 8, 1000, 8000, 1000000, ... (Sloane's A019545; Mudge 1997),
40. (Prime-digits): numbers composed of digits which are primes: 2, 3, 5, 7, 22, 23, 25, 27, 32, 33, 35, ... (Sloane's A046034),
41. (Smarandache prime-digital sequence): prime-digit numbers which are themselves prime: 2, 3, 5, 7, 23, 37, 53, ... (Smith 1996, Mudge 1997).

see also ADDITION CHAIN, CONSECUTIVE NUMBER SEQUENCES, CUBIC NUMBER, EUCLID NUMBER, EVEN NUMBER, FIBONACCI NUMBER, INTEGER SEQUENCE, ODD NUMBER, PARTITION, SMARANDACHE FUNCTION, SQUARE NUMBER

References

Aschbacher, C. *Collection of Problems On Smarandache Notions.* Vail, AZ: Erhus University Press, 1996.
Aschbacher, C. and Mudge, M. *Personal Computer World.* pp. 302, Oct. 1995.
Begay, A. "Smarandache Ceil Functions." *Bull. Pure Appl. Sci.* **16E**, 227–229, 1997.
Bencze, M. "Smarandache Recurrence Type Sequences." *Bull. Pure Appl. Sci.* **16E**, 231–236, 1997.
Bencze, M. and Tutescu, L. (Eds.). *Some Notions and Questions in Number Theory, Vol. 2.* `http://www.gallup.unm.edu/~smarandache/SNAQINT2.TXT`.
Brown, J. "Crescendo & Descrescendo." In *Richard Henry Wilde: An Anthology in Memoriam (1789–1847)* (Ed. M. Myers). Bristol, IN: Bristol Banner Books, p. 19, 1997.
Brown, J. and Castillo, J. "Problem 4619." *School Sci. Math.* **97**, 221–222, 1997.
Dumitrescu, C. and Seleacu, V. (Ed.). *Some Notions and Questions in Number Theory, 4th ed.* Glendale, AZ: Erhus University Press, 1994. `http://www.gallup.unm.edu/~smarandache/SNAQINT.TXT`.
Dumitrescu, C. and Seleacu, V. (Ed.). *Proceedings of the First International Conference on Smarandache Type Notions in Number Theory.* Lupton, AZ: American Research Press, 1997.
Hamel, E. Solution to Problem 4619. *School Sci. Math.* **97**, 221–222, 1997.
Iacobescu, F. "Smarandache Partition Type and Other Sequences." *Bull. Pure Appl. Sci.* **16E**, 237–240, 1997.
Ibstedt, H. *Surfing on the Ocean of Numbers—A Few Smarandache Notions and Similar Topics.* Lupton, AZ: Erhus University Press, 1997.
Kashihara, K. *Comments and Topics on Smarandache Notions and Problems.*ail, AZ: Erhus University Press, 1996.
Mudge, M. "Top of the Class." *Personal Computer World,* 674–675, June 1995.
Mudge, M. "Not Numerology but Numeralogy!" *Personal Computer World,* 279–280, 1997.
Programs and the Abstracts of the First International Conference on Smarandache Notions in Number Theory. Craiova, Romania, Aug. 21–23, 1997.
Radu, I. M. *Mathematical Spectrum* **27**, 43, 1994/1995.
Sloane, N. J. A. Sequences A001156/M0221 and A003108/M0209 in "An On-Line Version of the Encyclopedia of Integer Sequences."
Smarandache, F. "Properties of the Numbers." Tempe, AZ: Arizona State University Special Collection, 1975.
Smarandache, F. *Only Problems, Not Solutions!, 4th ed.* Phoenix, AZ: Xiquan, 1993.
Smarandache, F. *Collected Papers, Vol. 2.* Kishinev, Moldova: Kishinev University Press, 1997.
Smith, S. "A Set of Conjectures on Smarandache Sequences." *Bull. Pure Appl. Sci.* **15E**, 101–107, 1996.

Smarandache-Wagstaff Function

Given the sum-of-FACTORIALS function

$$\Sigma(n) = \sum_{k=1}^{n} k!,$$

$SW(p)$ is the smallest integer for p PRIME such that $\Sigma[SW(p)]$ is divisible by p. The first few known values are 2, 4, 5, 12, 19, 24, 32, 19, 20, 20, 20, 7, 57, 6, ... for $p = 3, 11, 17, 23, 29, 37, 41, 43, 53, 67, 73, 79, 97, \ldots$. The values for 5, 7, 13, 31, ..., if they are finite, must be very large.

see also FACTORIAL, SMARANDACHE FUNCTION

References

Ashbacher, C. "Some Properties of the Smarandache-Kurepa and Smarandache-Wagstaff Functions." *Math. Informatics Quart.* **7**, 114–116, 1997.
"Functions in Number Theory." `http://www.gallup.unm.edu/~smarandache/FUNCT1.TXT`.
Mudge, M. "Introducing the Smarandache-Kurepa and Smarandache-Wagstaff Functions." *Smarandache Notions J.* **7**, 52–53, 1996.
Mudge, M. "Introducing the Smarandache-Kurepa and Smarandache-Wagstaff Functions." *Abstracts of Papers Presented to the Amer. Math. Soc.* **17**, 583, 1996.

Smith Brothers

Consecutive SMITH NUMBERS. The first two brothers are (728, 729) and (2964, 2965).

see also SMITH NUMBER

Smith Conjecture

The set of fixed points which do not move as a knot is transformed into itself is not a KNOT. The conjecture was proved in 1978.

References

Rolfsen, D. *Knots and Links.* Wilmington, DE: Publish or Perish Press, pp. 350–351, 1976.

Smith's Markov Process Theorem

Consider

$$P_2(y_1, t|y_3, t_3) = \int P_2(y_1, t_1|y_2, t_1) P_3(y_1, t_1; y_2, t_2|y_3, t_3)\, dy_2. \quad (1)$$

If the probability distribution is governed by a MARKOV PROCESS, then

$$P_3(y_1, t_1; y_2, t_2|y_3, t_3) = P_2(y_2, t_2|y_3, t_3) = P_2(y_2|y_3, t_3 - t_2). \quad (2)$$

Assuming no time dependence, so $t_1 \equiv 0$,

$$P_2(y_1|y_3, t_3) = \int P_2(y_1|y_2, t_2) P_2(y_2|y_3, t_3 - t_2)\, dy_2. \quad (3)$$

see also MARKOV PROCESS

Smith's Network Theorem

In a NETWORK with three EDGES at each VERTEX, the number of HAMILTONIAN CIRCUITS through a specified EDGE is 0 or EVEN.

see also EDGE (GRAPH), HAMILTONIAN CIRCUIT, NETWORK

Smith Normal Form

A form for INTEGER matrices.

Smith Number

A COMPOSITE NUMBER the SUM of whose DIGITS is the sum of the DIGITS of its PRIME factors (excluding 1). (The PRIMES are excluded since they trivially satisfy this condition). One example of a Smith number is the BEAST NUMBER

$$666 = 2 \cdot 3 \cdot 3 \cdot 37,$$

since

$$6 + 6 + 6 = 2 + 3 + 3 + (3 + 7) = 18.$$

Another Smith number is

$$4937775 = 3 \cdot 5 \cdot 5 \cdot 65837,$$

since

$$4+9+3+7+7+7+5 = 3+5+5+(6+5+8+3+7) = 42.$$

The first few Smith numbers are 4, 22, 27, 58, 85, 94, 121, 166, 202, 265, 274, 319, 346, ... (Sloane's A006753). There are 360 Smith numbers less than 10^4 and 29,928 $\leq 10^6$. McDaniel (1987a) showed that an infinite number exist.

A generalized k-Smith number can also be defined as a number m satisfying $S_p(m) = kS(m)$, where S_p is the sum of prime factors and S is the sum of digits. There are 47 1-Smith numbers, 21 2-Smith numbers, three 3-Smith numbers, and one 7-Smith, 9-Smith, and 14-Smith number < 1000.

A Smith number can be constructed from every factored REPUNIT R_n. The largest known Smith number is

$$9 \times R_{1031}(10^{4594} + 3 \times 10^{2297} + 1)^{1476} \times 10^{3913210}.$$

see also MONICA SET, PERFECT NUMBER, REPUNIT, SMITH BROTHERS, SUZANNE SET

References

Gardner, M. *Penrose Tiles and Trapdoor Ciphers... and the Return of Dr. Matrix, reissue ed.* New York: W. H. Freeman, pp. 99–300, 1989.

Guy, R. K. "Smith Numbers." §B49 in *Unsolved Problems in Number Theory, 2nd ed.* New York: Springer-Verlag, pp. 103–104, 1994.

McDaniel, W. L. "The Existence of Infinitely Many k-Smith Numbers." *Fib. Quart.*, **25**, 76–80, 1987a.

McDaniel, W. L. "Powerful K-Smith Numbers." *Fib. Quart.* **25**, 225-228, 1987b.

Oltikar, S. and Weiland, K. "Construction of Smith Numbers." *Math. Mag.* **56**, 36–37, 1983.

Sloane, N. J. A. Sequence A006753/M3582 in "An On-Line Version of the Encyclopedia of Integer Sequences."

Wilansky, A. "Smith Numbers." *Two-Year College Math. J.* **13**, 21, 1982.

Yates, S. "Special Sets of Smith Numbers." *Math. Mag.* **59**, 293–296, 1986.

Yates, S. "Smith Numbers Congruent to 4 (mod 9)." *J. Recr. Math.* **19**, 139–141, 1987.

Smooth Manifold

Another word for a C^∞ (infinitely differentiable) MANIFOLD. A smooth manifold is a TOPOLOGICAL MANIFOLD together with its "functional structure" (Bredon 1995) and so differs from a TOPOLOGICAL MANIFOLD because the notion of differentiability exists on it. Every smooth manifold is a TOPOLOGICAL MANIFOLD, but not necessarily vice versa. (The first nonsmooth TOPOLOGICAL MANIFOLD occurs in 4-D.) In 1959, Milnor showed that a 7-D HYPERSPHERE can be made into a smooth manifold in 28 ways.

see also DIFFERENTIABLE MANIFOLD, HYPERSPHERE, MANIFOLD, TOPOLOGICAL MANIFOLD

References

Bredon, G. E. *Topology & Geometry.* New York: Springer-Verlag, p. 69, 1995.

Smooth Number

An INTEGER is k-smooth if it has no PRIME FACTORS $> k$. The probability that a random POSITIVE INTEGER $\leq n$ is k-smooth is $\psi(n,k)/n$, where $\psi(n,k)$ is the number of k-smooth numbers $\leq n$. This fact is important in application of Kraitchik's extension of FERMAT'S FACTORIZATION METHOD because it is related to the number of random numbers which must be examined to find a suitable subset whose product is a square.

Since about $\pi(k)$ k-smooth numbers must be found (where $\pi(k)$ is the PRIME COUNTING FUNCTION), the number of random numbers which must be examined is about $\pi(k)n/\psi(n,k)$. But because it takes about $\pi(k)$ steps to determine if a number is k-smooth using TRIAL DIVISION, the expected number of steps needed to find a subset of numbers whose product is a square is $\sim [\pi(k)]^2 n/\psi(n,k)$ (Pomerance 1996). Canfield *et al.* (1983) showed that this function is minimized when

$$k \sim \exp(\tfrac{1}{2}\sqrt{\ln n \ln \ln n})$$

and that the minimum value is about

$$\exp(2\sqrt{\ln n \ln \ln n}).$$

In the CONTINUED FRACTION FACTORIZATION ALGORITHM, n can be taken as $2\sqrt{n}$, but in FERMAT'S FACTORIZATION METHOD, it is $n^{1/2+\epsilon}$. k is an estimate for the largest PRIME in the FACTOR BASE (Pomerance 1996).

References
Canfield, E. R.; Erdős, P.; and Pomerance, C. "On a Problem of Oppenheim Concerning 'Factorisation Numerorum.'" *J. Number Th.* **17**, 1–28, 1983.
Pomerance, C. "On the Role of Smooth Numbers in Number Theoretic Algorithms." In *Proc. Internat. Congr. Math., Zürich, Switzerland, 1994, Vol. 1* (Ed. S. D. Chatterji). Basel: Birkhäuser, pp. 411–422, 1995.
Pomerance, C. "A Tale of Two Sieves." *Not. Amer. Math. Soc.* **43**, 1473–1485, 1996.

Smooth Surface

A surface PARAMETERIZED in variables u and v is called smooth if the TANGENT VECTORS in the u and v directions satisfy

$$\mathbf{T}_u \times \mathbf{T}_v \neq \mathbf{0},$$

where $\mathbf{A} \times \mathbf{B}$ is a CROSS PRODUCT.

Snake

A simple circuit in the d-HYPERCUBE which has no chords (i.e., for which all snake edges are edges of the HYPERCUBE). Klee (1970) asked for the maximum length $s(d)$ of a d-snake. Klee (1970) gave the bounds

$$\frac{7}{4(d-1)} \leq \frac{s(d)}{2^d}\frac{1}{2} - \frac{1-12\cdots 2^{-d}}{7d(d-1)^2+2} \tag{1}$$

for $d \geq 6$ (Danzer and Klee 1967, Douglas 1969), as well as numerous references. Abbott and Katchalski (1988) show

$$s(d) \geq 77 \cdot 2^{d-8}, \tag{2}$$

and Snevily (1994) showed that

$$s(d) \leq 2^{n-1}\left(1 - \frac{1}{20n-41}\right) \tag{3}$$

for $n \leq 12$, and conjectured

$$s(d) \leq 3 \cdot 2^{n-3} + 2 \tag{4}$$

for $n \leq 5$. The first few values for $s(d)$ for $d = 1, 2, \ldots$, are 2, 4, 6, 8, 14, 26, ... (Sloane's A000937).

see also HYPERCUBE

References
Abbott, H. L. and Katchalski, M. "On the Snake in the Box Problem." *J. Combin. Th. Ser. B* **44**, 12–24, 1988.
Danzer, L. and Klee, V. "Length of Snakes in Boxes." *J. Combin. Th.* **2**, 258–265, 1967.
Douglas, R. J. "Some Results on the Maximum Length of Circuits of Spread k in the d-Cube." *J. Combin. Th.* **6**, 323–339, 1969.
Evdokimov, A. A. "Maximal Length of a Chain in a Unit n-Dimensional Cube." *Mat. Zametki* **6**, 309–319, 1969.
Guy, R. K. "Unsolved Problems Come of Age." *Amer. Math. Monthly* **96**, 903–909, 1989.
Kautz, W. H. "Unit-Distance Error-Checking Codes." *IRE Trans. Elect. Comput.* **7**, 177–180, 1958.
Klee, V. "What is the Maximum Length of a d-Dimensional Snake?" *Amer. Math. Monthly* **77**, 63–65, 1970.
Sloane, N. J. A. Sequence A000937/M0995 in "An On-Line Version of the Encyclopedia of Integer Sequences."
Snevily, H. S. "The Snake-in-the-Box Problem: A New Upper Bound." *Disc. Math.* **133**, 307–314, 1994.

Snake Eyes

A roll of two 1s (the lowest roll possible) on a pair of six-sided DICE. The probability of rolling snake eyes is 1/36, or 2.777...%.

see also BOXCARS

Snake Oil Method

The expansion of the two sides of a sum equality in terms of POLYNOMIALS in x^m and y^k, followed by closed form summation in terms of x and y. For an example of the technique, see Bloom (1995).

References
Bloom, D. M. "A Semi-Unfriendly Identity." Problem 10206. Solution by R. J. Chapman. *Amer. Math. Monthly* **102**, 657–658, 1995.
Wilf, H. S. *Generatingfunctionology, 2nd ed.* New York: Academic Press, 1993.

Snake Polyiamond

A 6-POLYIAMOND.

References
Golomb, S. W. *Polyominoes: Puzzles, Patterns, Problems, and Packings, 2nd ed.* Princeton, NJ: Princeton University Press, p. 92, 1994.

Snedecor's F-Distribution

If a random variable X has a CHI-SQUARED DISTRIBUTION with m degrees of freedom ($\chi_m{}^2$) and a random variable Y has a CHI-SQUARED DISTRIBUTION with n degrees of freedom ($\chi_n{}^2$), and X and Y are independent, then

$$F \equiv \frac{X/m}{Y/n} \tag{1}$$

is distributed as Snedecor's F-distribution with m and n degrees of freedom

$$f(F(m,n)) = \frac{\Gamma\left(\frac{m+n}{2}\right)\left(\frac{m}{n}\right)^{m/2} F^{(m-2)/2}}{\Gamma\left(\frac{m}{2}\right)\Gamma\left(\frac{n}{2}\right)\left(1+\frac{m}{n}F\right)^{(m+n)/2}} \tag{2}$$

for $0 < F < \infty$. The MOMENTS about 0 are

$$\mu_1' = \frac{n}{n-2} \tag{3}$$

$$\mu_2' = \frac{n^2(m+2)}{m(n-2)(n-4)} \tag{4}$$

$$\mu_3' = \frac{n^3(m+2)(m+4)}{m^2(n-2)(n-4)(n-6)} \tag{5}$$

$$\mu_4' = \frac{n^4(m+2)(m+4)(m+6)}{m^3(n-2)(n-4)(n-6)(n-8)}, \tag{6}$$

so the MOMENTS about the MEAN are given by

$$\mu_2 = \frac{2n^2(m+n-2)}{m(n-2)^2(n-4)} \tag{7}$$

$$\mu_3 = \frac{8n^3(m+n-2)(2m+n-2)}{m^2(n-2)^3(n-4)(n-6)} \tag{8}$$

$$\mu_4 = \frac{12n^4(m+n-2)}{m^3(n-2)^4(n-4)(n-6)(n-8)} g(m,n), \tag{9}$$

where

$$g(m,n) = mn^2 + 4n^2 + m^2n + 8mn - 16n + 10m^2 - 20m + 16, \tag{10}$$

and the MEAN, VARIANCE, SKEWNESS, and KURTOSIS are

$$\mu = \mu_1' = \frac{n}{n-2} \tag{11}$$

$$\sigma^2 = \frac{2n^2(m+n-2)}{m(n-2)^2(n-4)} \tag{12}$$

$$\gamma_1 = \frac{\mu_3}{\sigma^3} = 2\sqrt{\frac{2(n-4)}{m(m+n-2)}}\,\frac{2m+n-2}{n-6} \tag{13}$$

$$\gamma_2 = \frac{\mu_4}{\sigma^4} - 3 = \frac{12h(m,n)}{m(m+n-2)(n-6)(n-8)}, \tag{14}$$

where

$$h(m,n) \equiv n^3 + 5mn^2 - 8n^2 + 5m^2n - 32mn + 20n - 22m^2 + 44m - 16. \tag{15}$$

Letting

$$w \equiv \frac{\frac{mF}{n}}{1+\frac{mF}{n}} \tag{16}$$

gives a BETA DISTRIBUTION.

see also BETA DISTRIBUTION, CHI-SQUARED DISTRIBUTION, STUDENT'S t-DISTRIBUTION

References

Beyer, W. H. *CRC Standard Mathematical Tables, 28th ed.* Boca Raton, FL: CRC Press, p. 536, 1987.

Snellius-Pothenot Problem

A SURVEYING PROBLEM which asks: Determine the position of an unknown accessible point P by its bearings from three inaccessible known points A, B, and C.

see also SURVEYING PROBLEMS

References

Dörrie, H. "Annex to a Survey." §40 in *100 Great Problems of Elementary Mathematics: Their History and Solutions.* New York: Dover, pp. 193–197, 1965.

Snowflake

see EXTERIOR SNOWFLAKE, KOCH ANTISNOWFLAKE, KOCH SNOWFLAKE, PENTAFLAKE

Snub Cube

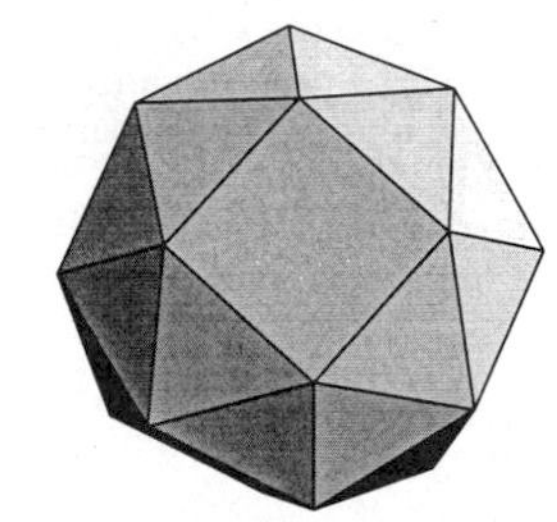

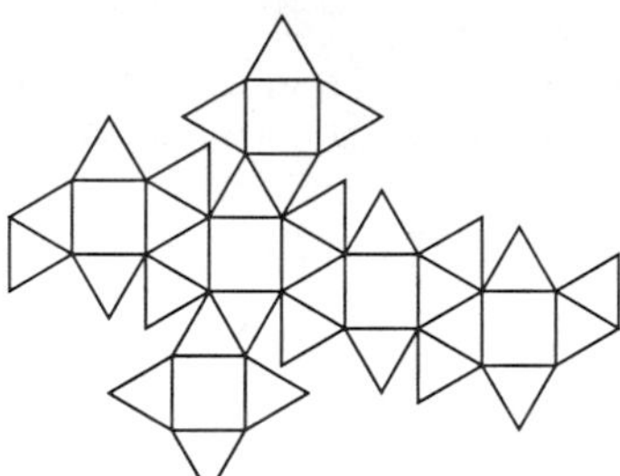

An ARCHIMEDEAN SOLID also called the SNUB CUBOCTAHEDRON whose VERTICES are the 24 points on the surface of a SPHERE for which the smallest distance between any two is as great as possible. It has two ENANTIOMERS, and its DUAL POLYHEDRON is the PENTAGONAL ICOSITETRAHEDRON. It has SCHLÄFLI SYMBOL $s\left\{ {3 \atop 4} \right\}$. It is also UNIFORM POLYHEDRON U_{12} and has WYTHOFF SYMBOL $|\,2\,3\,4$. Its faces are $32\{3\} + 6\{4\}$.

The INRADIUS, MIDRADIUS, and CIRCUMRADIUS for $a = 1$ are

$$r = 1.157661791\ldots$$
$$\rho = 1.247223168\ldots$$
$$R = \tfrac{1}{2}\sqrt{\frac{x^2 - 8x + 4}{x^2 - 5x + 4}} = 1.3437133737446\ldots,$$

where

$$x \equiv (19 + 3\sqrt{33}\,)^{1/3},$$

and the exact expressions for r and ρ can be computed using

$$r = \frac{R^2 - \frac{1}{4}a^2}{R}$$
$$\rho = \sqrt{R^2 - \tfrac{1}{4}a^2}.$$

see also SNUB DODECAHEDRON

References

Ball, W. W. R. and Coxeter, H. S. M. *Mathematical Recreations and Essays, 13th ed.* New York: Dover, p. 139, 1987.

Coxeter, H. S. M.; Longuet-Higgins, M. S.; and Miller, J. C. P. "Uniform Polyhedra." *Phil. Trans. Roy. Soc. London Ser. A* **246**, 401–450, 1954.

Snub Cuboctahedron

see SNUB CUBE

Snub Disphenoid

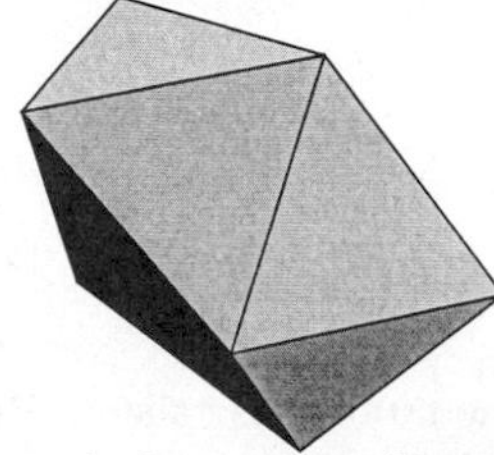

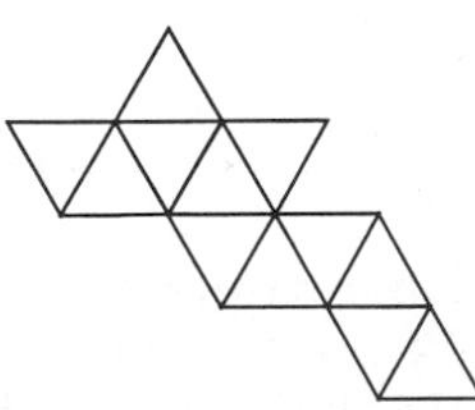

One of the convex DELTAHEDRA also known as the SIAMESE DODECAHEDRON. It is JOHNSON SOLID J_{84}.

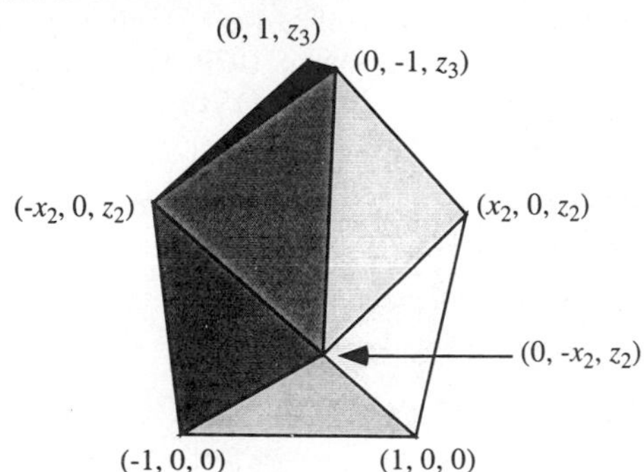

The coordinates of the VERTICES may be found by solving the set of four equations

$$1 + {x_2}^2 + {z_1}^2 = 4$$
$$(x_2 - 1)^2 + (z_3 - z_1)^2 = 4$$
$${x_2}^2 + (z_3 - z_2)^2 = 4$$
$${x_2}^2 + {x_2}^2 + (z_2 - z_1)^2 = 4$$

for the four unknowns x_2, z_1, z_2, and z_3. Numerically,

$$x_2 = 1.28917$$
$$z_1 = 1.15674$$
$$z_2 = 1.97898$$
$$z_3 = 3.13572.$$

The analytic solution requires solving the CUBIC EQUATION and gives

$$x_2 = 1 - 7\cdot 2^{-2/3}(1 - i\sqrt{3})\alpha^{-1} - \tfrac{1}{6}\cdot 2^{-1/3}(1 + i\sqrt{3})\alpha$$
$$z_1 = \tfrac{1}{3}\cdot 2^{-1/2}[-48 + 6\beta(1 + i\sqrt{3}) + \beta^2(1 - i\sqrt{3}) + 147\beta\gamma(\sqrt{3} - i) + 42\beta^2\gamma(\sqrt{3} + i)]^{1/2},$$

where

$$\alpha \equiv (12i\sqrt{237} - 54)^{1/3}$$
$$\beta \equiv 3^{1/3}(2i\sqrt{237} - 9)^{1/3}$$
$$\gamma \equiv (9i + 2\sqrt{237}\,)^{-1}.$$

Snub Dodecadodecahedron

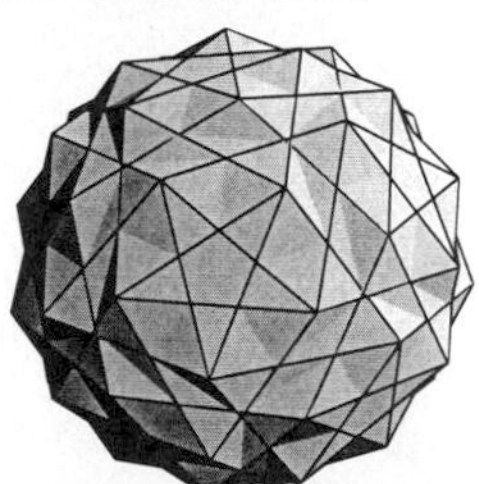

The UNIFORM POLYHEDRON U_{40} whose DUAL POLYHEDRON is the MEDIAL PENTAGONAL HEXECONTAHEDRON. It has WYTHOFF SYMBOL $|\,2\,\frac{5}{2}\,5$. Its faces are $12\{\frac{5}{2}\} + 60\{3\} + 12\{5\}$. It has CIRCUMRADIUS for $a = 1$ of

$$R = 1.27443994.$$

see also SNUB CUBE

References

Ball, W. W. R. and Coxeter, H. S. M. *Mathematical Recreations and Essays, 13th ed.* New York: Dover, p. 139, 1987.

Coxeter, H. S. M.; Longuet-Higgins, M. S.; and Miller, J. C. P. "Uniform Polyhedra." *Phil. Trans. Roy. Soc. London Ser. A* **246**, 401–450, 1954.

Wenninger, M. J. *Polyhedron Models.* Cambridge, England: Cambridge University Press, pp. 174–176, 1971.

Snub Dodecahedron

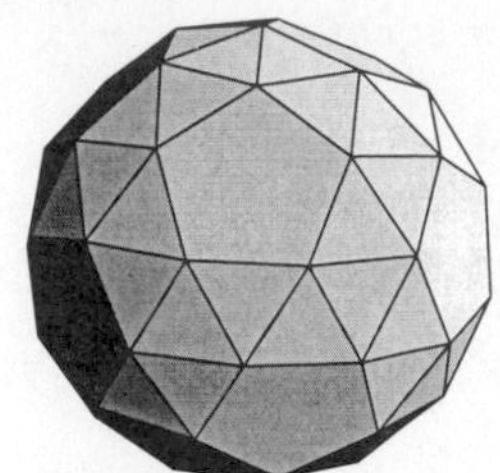

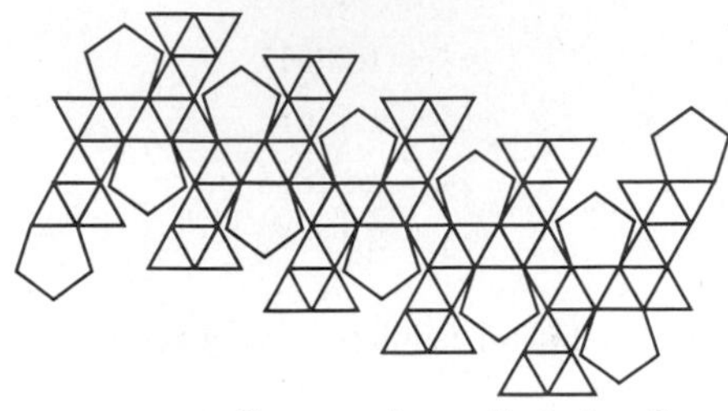

An ARCHIMEDEAN SOLID, also called the SNUB ICOSIDODECAHEDRON, whose DUAL POLYHEDRON is the PENTAGONAL HEXECONTAHEDRON. It has SCHLÄFLI SYMBOL $s\left\{ {3 \atop 5} \right\}$. It is also UNIFORM POLYHEDRON U_{29} and has WYTHOFF SYMBOL $|\,2\,3\,5$. Its faces are $80\{3\}+12\{5\}$. For $a=1$, it has INRADIUS, MIDRADIUS, and CIRCUMRADIUS

$$\begin{aligned} r &= 2.039873155\ldots \\ \rho &= 2.097053835\ldots \\ R &= \frac{1}{2}\sqrt{\frac{8\cdot 2^{2/3}-16x+2^{1/3}x^2}{8\cdot 2^{2/3}-10x+2^{1/3}x^2}} \\ &= 2.15583737511564\ldots, \end{aligned}$$

where

$$x \equiv \left(49+27\sqrt{5}+3\sqrt{6}\sqrt{93+49\sqrt{5}}\right)^{1/3},$$

and the exact expressions for r and ρ can be computed using

$$\begin{aligned} r &= \frac{R^2-\frac{1}{4}a^2}{R} \\ \rho &= \sqrt{R^2-\tfrac{1}{4}a^2}. \end{aligned}$$

References

Coxeter, H. S. M.; Longuet-Higgins, M. S.; and Miller, J. C. P. "Uniform Polyhedra." *Phil. Trans. Roy. Soc. London Ser. A* **246**, 401–450, 1954.

Snub Icosidodecadodecahedron

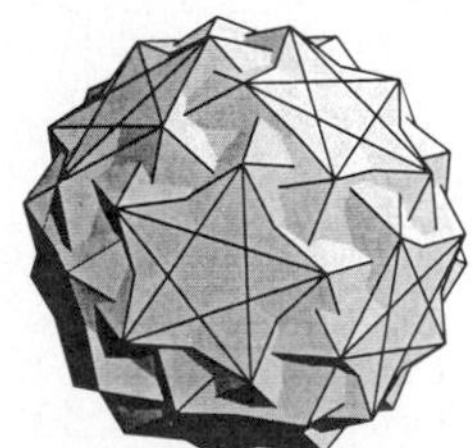

The UNIFORM POLYHEDRON U_{46} whose DUAL POLYHEDRON is the MEDIAL HEXAGONAL HEXECONTAHEDRON. It has WYTHOFF SYMBOL $|\,3\,\frac{5}{3}\,5$. Its faces are $12\{\frac{4}{2}\}+80\{3\}+12\{5\}$. It has CIRCUMRADIUS for $a=1$ of

$$\begin{aligned} R &= \frac{1}{2}\sqrt{\frac{2^{4/3}-14x+2^{2/3}x^2}{2^{4/3}-8x+2^{2/3}x^2}} \\ &= 1.12689791279994\ldots, \end{aligned}$$

where

$$x = (25+3\sqrt{69})^{1/3}.$$

References

Wenninger, M. J. *Polyhedron Models.* Cambridge, England: Cambridge University Press, pp. 177–178, 1971.

Snub Icosidodecahedron

see SNUB DODECAHEDRON

Snub Polyhedron

A polyhedron with extra triangular faces, given by the SCHLÄFLI SYMBOL $s\left\{ {p \atop q} \right\}$.

see also RHOMBIC POLYHEDRON, TRUNCATED POLYHEDRON

Snub Square Antiprism

see JOHNSON SOLID

Soap Bubble

see BUBBLE

Soccer Ball

see TRUNCATED ICOSAHEDRON

Sociable Numbers

Numbers which result in a periodic ALIQUOT SEQUENCE. If the period is 1, the number is called a PERFECT NUMBER. If the period is 2, the two numbers are called an AMICABLE PAIR. If the period is $t \geq 3$, the number is called sociable of order t. Only two sociable numbers were known prior to 1970, the sets of orders 5 and 28 discovered by Poulet (1918). In 1970, Cohen discovered nine groups of order 4.

The table below summarizes the number of sociable cycles known as given in the compilation by Moews (1995).

order	known
3	0
4	38
5	1
6	2
8	2
9	1
28	1

see also ALIQUOT SEQUENCE, PERFECT NUMBER, UNITARY SOCIABLE NUMBERS

References

Beeler, M.; Gosper, R. W.; and Schroeppel, R. *HAKMEM.* Cambridge, MA: MIT Artificial Intelligence Laboratory, Memo AIM-239, Item 61, Feb. 1972.

Borho, W. "Über die Fixpunkte der k-fach iterierten Teilerersummenfunktion." *Mitt. Math. Gesellsch. Hamburg* **9**, 34–48, 1969.

Cohen, H. "On Amicable and Sociable Numbers." *Math. Comput.* **24**, 423–429, 1970.

Devitt, J. S.; Guy, R. K.; and Selfridge, J. L. *Third Report on Aliquot Sequences,* Congr. Numer. XVIII, Proc. 6th Manitoba Conf. Numerical Math, pp. 177–204, 1976.

Flammenkamp, A. "New Sociable Numbers." *Math. Comput.* **56**, 871–873, 1991.

Gardner, M. "Perfect, Amicable, Sociable." Ch. 12 in *Mathematical Magic Show: More Puzzles, Games, Diversions, Illusions and Other Mathematical Sleight-of-Mind from Scientific American.* New York: Vintage, pp. 160–171, 1978.

Guy, R. K. "Aliquot Cycles or Sociable Numbers." §B7 in *Unsolved Problems in Number Theory, 2nd ed.* New York: Springer-Verlag, pp. 62–63, 1994.

Madachy, J. S. *Madachy's Mathematical Recreations.* New York: Dover, pp. 145–146, 1979.

Moews, D. and Moews, P. C. "A Search for Aliquot Cycles Below 10^{10}." *Math. Comput.* **57**, 849–855, 1991.

Moews, D. and Moews, P. C. "A Search for Aliquot Cycles and Amicable Pairs." *Math. Comput.* **61**, 935–938, 1993.

Moews, D. "A List of Aliquot Cycles of Length Greater than 2." Rev. Dec. 18, 1995. `http://xraysgi.ims.uconn.edu:8080/sociable.txt`.

Poulet, P. Question 4865. *L'interméd. des Math.* **25**, 100–101, 1918.

te Riele, H. J. J. "Perfect Numbers and Aliquot Sequences." In *Computational Methods in Number Theory, Part I.* (Eds. H. W. Lenstra Jr. and R. Tijdeman). Amsterdam, Netherlands: Mathematisch Centrum, pp. 141–157, 1982.

Weisstein, E. W. "Sociable and Amicable Numbers." `http://www.astro.virginia.edu/~eww6n/math/notebooks/Sociable.m`.

Social Choice Theory

The theory of analyzing a decision between a collection of alternatives made by a collection of n voters with separate opinions. Any choice for the entire group should reflect the desires of the individual voters to the extent possible.

Fair choice procedures usually satisfy ANONYMITY (invariance under permutation of voters), DUALITY (each alternative receives equal weight for a single vote), and MONOTONICITY (a change favorable for X does not hurt X). Simple majority vote is anonymous, dual, and monotone. MAY'S THEOREM states a stronger result.

see also ANONYMOUS, DUAL VOTING, MAY'S THEOREM, MONOTONIC VOTING, VOTING

References

Taylor, A. *Mathematics and Politics: Strategy, Voting, Power, and Proof.* New York: Springer-Verlag, 1995.

Young, S. C.; Taylor, A. D.; and Zwicker, W. S. "Counting Quota Systems: A Combinatorial Question from Social Choice Theory." *Math. Mag.* **68**, 331–342, 1995.

Socrates' Paradox

Socrates is reported to have stated: "One thing I know is that I know nothing."

see also LIAR'S PARADOX

References

Pickover, C. A. *Keys to Infinity.* New York: W. H. Freeman, p. 134, 1995.

Soddy Circles

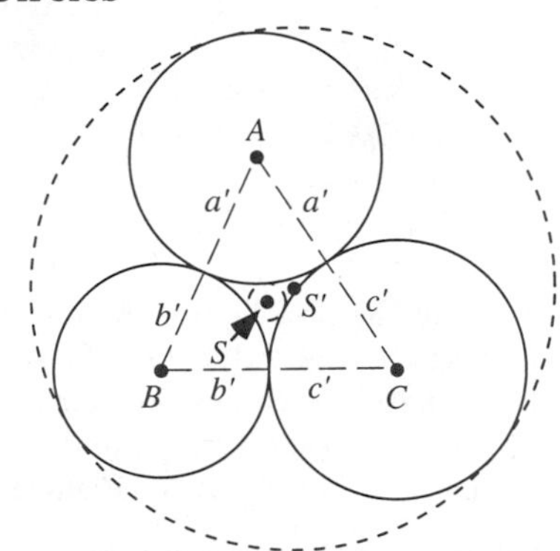

Given three distinct points A, B, and C, let three CIRCLES be drawn, one centered about each point and each one tangent to the other two. Call the RADII r_i ($r_3 = a'$, $r_1 = b'$, $r_2 = c'$). Then the CIRCLES satisfy

$$a' + b' = c \tag{1}$$

$$a' + c' = b \tag{2}$$

$$b' + c' = a, \tag{3}$$

as shown in the diagram below.

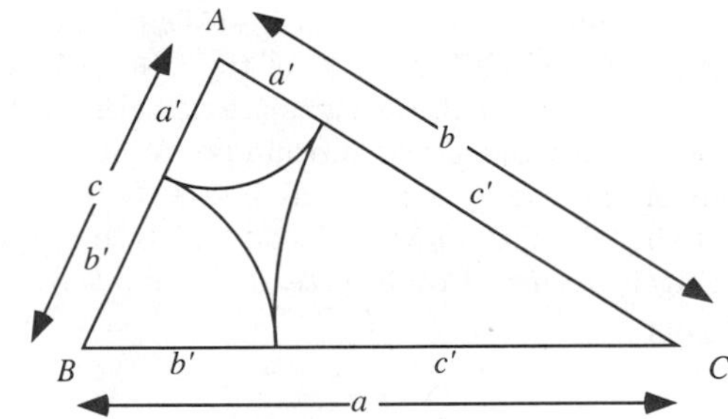

Solving for the RADII then gives

$$a' = \tfrac{1}{2}(b + c - a) \tag{4}$$

$$b' = \tfrac{1}{2}(a + c - b) \tag{5}$$

$$c' = \tfrac{1}{2}(a + b - c). \tag{6}$$

The above TRIANGLE has sides a, b, and c, and SEMIPERIMETER

$$s \equiv \tfrac{1}{2}(a + b + c). \tag{7}$$

Plugging in,

$$2s = (a' + b') + (a' + c') + (b' + c') = 2(a' + b' + c'), \tag{8}$$

giving

$$a' + b' + c' = s. \tag{9}$$

In addition,

$$a = b' + c' = a' + b' + c' - a' = s - a'. \tag{10}$$

Switching a and a' to opposite sides of the equation and noting that the above argument applies equally well to b' and c' then gives

$$a' = s - a \tag{11}$$
$$b' = s - b \tag{12}$$
$$c' = s - c. \tag{13}$$

As can be seen from the first figure, there exist exactly two nonintersecting CIRCLES which are TANGENT to all three CIRCLES. These are called the inner and outer Soddy circles (S and S', respectively), and their centers are called the inner and outer SODDY POINTS.

The inner Soddy circle is the solution to the FOUR COINS PROBLEM. The center S of the inner Soddy circle is the EQUAL DETOUR POINT, and the center of the outer Soddy circle S' is the ISOPERIMETRIC POINT (Kimberling 1994).

Frederick Soddy (1936) gave the FORMULA for finding the RADII of the Soddy circles (r_4) given the RADII r_i ($i = 1, 2, 3$) of the other three. The relationship is

$$2({\epsilon_1}^2 + {\epsilon_2}^2 + {\epsilon_3}^2 + {\epsilon_4}^2) = (\epsilon_1 + \epsilon_2 + \epsilon_3 + \epsilon_4)^2, \tag{14}$$

where $\epsilon_i \equiv \pm\kappa_i = \pm 1/r_i$ are the so-called BENDS, defined as the signed CURVATURES of the CIRCLES. If the contacts are all external, the signs are all taken as POSITIVE, whereas if one circle surrounds the other three, the sign of this circle is taken as NEGATIVE (Coxeter 1969). Using the QUADRATIC FORMULA to solve for ϵ_4, expressing in terms of radii instead of curvatures, and simplifying gives

$$r_4^\pm = \frac{r_1 r_2 r_3}{r_2 r_3 + r_1(r_2 + r_3) \pm 2\sqrt{r_1 r_2 r_3 (r_1 + r_2 + r_3)}}. \tag{15}$$

Here, the NEGATIVE solution corresponds to the outer Soddy circle and the POSITIVE one to the inner Soddy circle.

This FORMULA is called the DESCARTES CIRCLE THEOREM since it was known to Descartes. However, Soddy also extended it to SPHERES. Gosper has further extended the result to $n + 2$ mutually tangent n-D HYPERSPHERES, whose CURVATURES satisfy

$$\left(\sum_{i=0}^{n+1} \kappa_i\right)^2 - n \sum_{i=0}^{n+1} {\kappa_i}^2 = 0. \tag{16}$$

Solving for κ_{n+1} gives

$$\kappa_{n+1} = \frac{\sqrt{n}\sqrt{\left(\sum_{i=0}^n \kappa_i\right)^2 - (n-1)\sum_{i=0}^n {\kappa_i}^2} + \sum_{i=0}^n \kappa_i}{n-1}. \tag{17}$$

For (at least) $n = 2$ and 3, the RADICAL equals

$$f(n) V \kappa_0 \kappa_1 \cdots \kappa_n, \tag{18}$$

where V is the CONTENT of the SIMPLEX whose vertices are the centers of the $n+1$ independent HYPERSPHERES. The RADICAND can also become NEGATIVE, yielding an IMAGINARY κ_{n+1}. For $n = 3$, this corresponds to a sphere touching three large bowling balls and a small BB, all mutually tangent, which is an impossibility.

Bellew has derived a generalization applicable to a CIRCLE surrounded by n CIRCLES which are, in turn, circumscribed by another CIRCLE. The relationship is

$$[n(c_n - 1)^2 + 1] \sum_{i=1}^n {\kappa_i}^2 + n(3n{c_n}^2 - 2n - 6){c_n}^2 (c_n - 1)^2 =$$
$$\frac{1}{[n(c_n - 1) + 1]^2} \times \{n(c_n - 1)^2 + 1] \sum_{i=1}^n \kappa_i$$
$$+ n c_n (c_n - 1)(n{c_n}^2 + (3 - n)c_n - 4])\}, \tag{19}$$

where

$$c_n \equiv \csc\left(\frac{\pi}{n}\right). \tag{20}$$

For $n = 3$, this simplifies to the Soddy formula.

see also APOLLONIUS CIRCLES, APOLLONIUS' PROBLEM, ARBELOS, BEND (CURVATURE), CIRCUMCIRCLE, DESCARTES CIRCLE THEOREM, FOUR COINS PROBLEM, HART'S THEOREM, PAPPUS CHAIN, SPHERE PACKING, STEINER CHAIN

References

Coxeter, H. S. M. *Introduction to Geometry, 2nd ed.* New York: Wiley, pp. 13–14, 1969.

Elkies, N. D. and Fukuta, J. "Problem E3236 and Solution." *Amer. Math. Monthly* **97**, 529–531, 1990.

Kimberling, C. "Central Points and Central Lines in the Plane of a Triangle." *Math. Mag.* **67**, p. 181, 1994.

"The Kiss Precise." *Nature* **139**, 62, 1937.

Soddy, F. "The Kiss Precise." *Nature* **137**, 1021, 1936.

Vandeghen, A. "Soddy's Circles and the De Longchamps Point of a Triangle." *Amer. Math. Monthly* **71**, 176–179, 1964.

Soddy's Hexlet

see HEXLET

Soddy Line

A LINE on which the INCENTER I, GERGONNE POINT Ge, and inner and outer SODDY POINTS S and S' lie (the latter two of which are the EQUAL DETOUR POINT and the ISOPERIMETRIC POINT). The Soddy line can be given parametrically by

$$I + \lambda Ge,$$

where λ is a parameter. It is also given by

$$\sum (f - e)\alpha = 0,$$

where cyclic permutations of d, e, and f are taken and the sum is over TRILINEAR COORDINATES α, β, and γ.

λ	Center
-4	outer Griffiths point Gr'
-2	outer Oldknow point Ol'
$-\frac{4}{3}$	outer Rigby point Ri'
-1	outer Soddy center S'
0	incenter I
1	inner Soddy center S
$\frac{4}{3}$	inner Rigby point Ri
2	inner Oldknow point Ol
4	inner Griffiths point Gr
∞	Gergonne point

S', I, S, and Ge are COLLINEAR and form a HARMONIC RANGE (Vandeghen 1964, Oldknow 1996). There are a total of 22 HARMONIC RANGES for sets of four points out of these 10 (Oldknow 1996).

The Soddy line intersects the EULER LINE in the DE LONGCHAMPS POINT, and the GERGONNE LINE in the FLETCHER POINT.

see also DE LONGCHAMPS POINT, EULER LINE, FLETCHER POINT, GERGONNE POINT, GRIFFITHS POINTS, HARMONIC RANGE, INCENTER, OLDKNOW POINTS, RIGBY POINTS, SODDY POINTS

References

Oldknow, A. "The Euler-Gergonne-Soddy Triangle of a Triangle." *Amer. Math. Monthly* **103**, 319–329, 1996.

Vandeghen, A. "Soddy's Circles and the De Longchamps Point of a Triangle." *Amer. Math. Monthly* **71**, 176–179, 1964.

Soddy Points

Given three mutually tangent CIRCLES, there exist exactly two nonintersecting CIRCLES TANGENT to all three CIRCLES. These are called the inner and outer SODDY CIRCLES, and their centers are called the inner and outer Soddy points. The outer Soddy circle is the solution to the FOUR COINS PROBLEM. The center S of the inner Soddy circle is the EQUAL DETOUR POINT, and the center of the outer Soddy circle S' is the ISOPERIMETRIC POINT (Kimberling 1994).

see also EQUAL DETOUR POINT, ISOPERIMETRIC POINT, SODDY CIRCLES

References

Kimberling, C. "Central Points and Central Lines in the Plane of a Triangle." *Math. Mag.* **67**, p. 181, 1994.

Sofa Constant

see MOVING SOFA CONSTANT

Sol Geometry

The GEOMETRY of the LIE GROUP R SEMIDIRECT PRODUCT with R^2, where R acts on R^2 by $(t, (x, y)) \to (e^t x, e^{-t} y)$.

see also THURSTON'S GEOMETRIZATION CONJECTURE

Soldner's Constant

Consider the following formulation of the PRIME NUMBER THEOREM,

$$\pi(x) = \sum \frac{\mu(m)}{m} \int_c^x \frac{dt}{\ln t},$$

where $\mu(m)$ is the MÖBIUS FUNCTION and c (sometimes also denoted μ) is Soldner's constant. Ramanujan found $c = 1.45136380\ldots$ (Hardy 1969, Le Lionnais 1983, Berndt 1994). Soldner (cited in Nielsen 1965) derived the correct value of c as $1.4513692346\ldots$, where c is the root of

$$L(x) = \lim_{\epsilon \to 0} \int_0^{1-\epsilon} \frac{dt}{\ln t} + \int_{1+\epsilon}^{\infty} \frac{dt}{\ln t}$$

(Le Lionnais 1983).

References

Berndt, B. C. *Ramanujan's Notebooks, Part IV.* New York: Springer-Verlag, pp. 123–124, 1994.

Hardy, G. H. *Ramanujan: Twelve Lectures on Subjects Suggested by His Life and Work, 3rd ed.* New York: Chelsea, p. 45, 1959.

Le Lionnais, F. *Les nombres remarquables.* Paris: Hermann, p. 39, 1983.

Nielsen, N. *Théorie des Integrallogarithms.* New York: Chelsea, p. 88, 1965.

Solenoidal Field

A solenoidal VECTOR FIELD satisfies

$$\nabla \cdot \mathbf{B} = 0 \tag{1}$$

for every VECTOR $\mathbf{B}$, where $\nabla \cdot \mathbf{B}$ is the DIVERGENCE. If this condition is satisfied, there exists a vector $\mathbf{A}$, known as the VECTOR POTENTIAL, such that

$$\mathbf{B} \equiv \nabla \times \mathbf{A}, \tag{2}$$

where $\nabla \times \mathbf{A}$ is the CURL. This follows from the vector identity

$$\nabla \cdot \mathbf{B} = \nabla \cdot (\nabla \times \mathbf{A}) = 0. \tag{3}$$

If $\mathbf{A}$ is an IRROTATIONAL FIELD, then

$$\mathbf{A} \times \mathbf{r} \tag{4}$$

is solenoidal. If $\mathbf{u}$ and $\mathbf{v}$ are irrotational, then

$$\mathbf{u} \times \mathbf{v} \tag{5}$$

is solenoidal. The quantity

$$(\nabla u) \times (\nabla v), \tag{6}$$

where ∇u is the GRADIENT, is always solenoidal. For a function ϕ satisfying LAPLACE'S EQUATION

$$\nabla^2 \phi = 0, \tag{7}$$

it follows that $\nabla \phi$ is solenoidal (and also IRROTATIONAL).

see also BELTRAMI FIELD, CURL, DIVERGENCE, DIVERGENCELESS FIELD, GRADIENT, IRROTATIONAL FIELD, LAPLACE'S EQUATION, VECTOR FIELD

References
Gradshteyn, I. S. and Ryzhik, I. M. *Tables of Integrals, Series, and Products, 5th ed.* San Diego, CA: Academic Press, pp. 1084, 1980.

Solid

A closed 3-D figure (which may, according to some terminology conventions, be self-intersecting). Among the simplest solids are the SPHERE, CUBE, CONE, CYLINDER, and more generally, the POLYHEDRA.

see also APPLE, ARCHIMEDEAN SOLID, CATALAN SOLID, CONE, CORK PLUG, CUBE, CUBOCTAHEDRON, CYLINDER, CYLINDRICAL HOOF, CYLINDRICAL WEDGE, DODECAHEDRON, GEODESIC DOME, GREAT DODECAHEDRON, GREAT ICOSAHEDRON, GREAT RHOMBICOSIDODECAHEDRON (ARCHIMEDEAN), GREAT RHOMBICUBOCTAHEDRON (ARCHIMEDEAN), GREAT STELLATED DODECAHEDRON, ICOSAHEDRON, ICOSIDODECAHEDRON, JOHNSON SOLID, KEPLER-POINSOT SOLID, LEMON, MÖBIUS STRIP, OCTAHEDRON, PLATONIC SOLID, POLYHEDRON, PSEUDOSPHERE, RHOMBICOSIDODECAHEDRON, RHOMBICUBOCTAHEDRON, SMALL STELLATED DODECAHEDRON, SNUB CUBE, SNUB DODECAHEDRON, SOLID OF REVOLUTION, SPHERE, STEINMETZ SOLID, STELLA OCTANGULA, SURFACE, TETRAHEDRON, TORUS, TRUNCATED CUBE, TRUNCATED DODECAHEDRON, TRUNCATED ICOSAHEDRON, TRUNCATED OCTAHEDRON, TRUNCATED TETRAHEDRON, UNIFORM POLYHEDRON, WULFF SHAPE

Solid Angle

Defined as the SURFACE AREA Ω of a UNIT SPHERE which is subtended by a given object S. Writing the SPHERICAL COORDINATES as ϕ for the COLATITUDE (angle from the pole) and θ for the LONGITUDE (azimuth),

$$\Omega \equiv A_{\text{projected}} = \iint_S \sin\phi \, d\theta \, d\phi.$$

Solid angle is measured in STERADIANS, and the solid angle corresponding to all of space being subtended is 4π STERADIANS.

see also SPHERE, STERADIAN

Solid Geometry

That portion of GEOMETRY dealing with SOLIDS, as opposed to PLANE GEOMETRY. Solid geometry is concerned with POLYHEDRA, SPHERES, 3-D SOLIDS, lines in 3-space, PLANES, and so on.

see also GEOMETRY, PLANE GEOMETRY, SPHERICAL GEOMETRY

References
Altshiller-Court, N. *Modern Pure Solid Geometry.* New York: Chelsea, 1979.
Bell, R. J. T. *An Elementary Treatise on Coordinate Geometry of Three Dimensions.* London: Macmillan, 1926.
Cohn, P. M. *Solid Geometry.* New York: Routledge, 1968.
Frost, P. *Solid Geometry, 3rd ed.* London: Macmillan, 1886.
Lines, L. *Solid Geometry.* New York: Dover, 1965.
Salmon, G. *Treatise on the Analytic Geometry of Three Dimensions, 6th ed.* London: Longmans Green, 1914.
Shute, W. G.; Shirk, W. W.; and Porter, G. F. *Solid Geometry.* New York: American Book Co., 1960.
Wentworth, G. A. and Smith, D. E. *Solid Geometry.* Boston, MA: Ginn and Company, 1913.

Solid Partition

Solid partitions are generalizations of PLANE PARTITIONS. MacMohan (1960) conjectured the GENERATING FUNCTION for the number of solid partitions was

$$f(z) = \frac{1}{(1-z)(1-z^2)^3(1-z^3)^6(1-z^4)^{10}\cdots},$$

but this was subsequently shown to disagree at $n = 6$ (Atkin *et al.* 1967). Knuth (1970) extended the tabulation of values, but was unable to find a correct generating function. The first few values are 1, 4, 10, 26, 59, 140, ... (Sloane's A000293).

References
Atkin, A. O. L.; Bratley, P.; MacDonald, I. G.; and McKay, J. K. S. "Some Computations for m-Dimensional Partitions." *Proc. Cambridge Philos. Soc.* **63**, 1097–1100, 1967.
Knuth, D. E. "A Note on Solid Partitions." *Math. Comput.* **24**, 955–961, 1970.
MacMahon, P. A. "Memoir on the Theory of the Partitions of Numbers. VI: Partitions in Two-Dimensional Space, to which is Added an Adumbration of the Theory of Partitions in Three-Dimensional Space." *Phil. Trans. Roy. Soc. London Ser. A* **211**, 345–373, 1912b.
MacMahon, P. A. *Combinatory Analysis, Vol. 2.* New York: Chelsea, pp. 75–176, 1960.
Sloane, N. J. A. Sequence A3392/M000293 in "An On-Line Version of the Encyclopedia of Integer Sequences."

Solid of Revolution

To find the VOLUME of a solid of rotation by adding up a sequence of thin cylindrical shells, consider a region bounded above by $y = f(x)$, below by $y = g(x)$, on the left by the LINE $x = a$, and on the right by the LINE $x = b$. When the region is rotated about the y-AXIS, the resulting VOLUME is given by

$$V = 2\pi \int_b^a x[f(x) - g(x)]\, dx.$$

To find the volume of a solid of rotation by adding up a sequence of thin flat disks, consider a region bounded above by $y = f(x)$, below by $y = g(x)$, on the left by the LINE $x = a$, and on the right by the LINE $x = b$. When the region is rotated about the x-AXIS, the resulting VOLUME is

$$V = \pi \int_b^a \{[f(x)]^2 - [g(x)]^2\}\, dx.$$

see also SURFACE OF REVOLUTION, VOLUME

Solidus

The diagonal slash "/" used to denote DIVISION for in-line equations such as a/b, $1/(x-1)^2$, etc. The solidus is also called a DIAGONAL.

see also DIVISION, OBELUS

Solitary Number

A number which does not have any FRIENDS. Solitary numbers include all PRIMES and POWERS of PRIMES. More generally, numbers for which $(n, \sigma(n)) = 1$ are solitary, where (a, b) is the GREATEST COMMON DIVISOR of a and b and $\sigma(n)$ is the DIVISOR FUNCTION. The first few solitary numbers are 1, 2, 3, 4, 5, 7, 8, 9, 11, 13, 16, 17, 19, 21, ... (Sloane's A014567).

see also FRIEND

References

Anderson, C. W. and Hickerson, D. Problem 6020. "Friendly Integers." *Amer. Math. Monthly* **84**, 65–66, 1977.

Sloane, N. J. A. Sequence A014567 in "An On-Line Version of the Encyclopedia of Integer Sequences."

Soliton

A stable isolated (i.e., solitary) traveling wave solution to a set of equations.

see also LAX PAIR, SINE-GORDON EQUATION

References

Bullough, R. K. and Caudrey, P. J. (Eds.). *Solitons.* Berlin: Springer-Verlag, 1980.

Dodd, R. K. *Solitons and Nonlinear Equations.* London: Academic Press, 1984.

Drazin, P. G. and Johnson, R. S. *Solitons: An Introduction.* Cambridge, England: Cambridge University Press, 1988.

Filippov, A. *The Versatile Solitons.* Boston, MA: Birkhäuser, 1996.

Gu, C. H. *Soliton Theory and Its Applications.* New York: Springer-Verlag, 1995.

Infeld, E. and Rowlands, G. *Nonlinear Waves, Solitons, and Chaos.* Cambridge, England: Cambridge University Press, 1990.

Lamb, G. L. Jr. *Elements of Soliton Theory.* New York: Wiley, 1980.

Makhankov, V. G.; Fedyann, V. K.; and Pashaev, O. K. (Eds.). *Solitons and Applications.* Singapore: World Scientific, 1990.

Newell, A. C. *Solitons in Mathematics and Physics.* Philadelphia, PA: SIAM, 1985.

Olver, P. J. and Sattinger, D. H. (Eds.). *Solitons in Physics, Mathematics, and Nonlinear Optics.* New York: Springer-Verlag, 1990.

Remoissent, M. *Waves Called Solitons, 2nd ed.* New York: Springer-Verlag, 1996.

Solomon's Seal Knot

The (5,2) TORUS KNOT 05_{001} with BRAID WORD $\sigma_1{}^5$.

Solomon's Seal Lines

The 27 REAL or IMAGINARY straight LINES which lie on the general CUBIC SURFACE and the 45 triple tangent PLANES to the surface. All are related to the 28 BITANGENTS of the general QUARTIC CURVE.

Schoutte (1910) showed that the 27 lines can be put into a ONE-TO-ONE correspondence with the vertices of a particular POLYTOPE in 6-D space in such a manner that all incidence relations between the lines are mirrored in the connectivity of the POLYTOPE and conversely (Du Val 1931). A similar correspondence can be made between the 28 bitangents and a 7-D POLYTOPE (Coxeter 1928) and between the tritangent planes of the canonical curve of genus four and an 8-D POLYTOPE (Du Val 1933).

see also BRIANCHON'S THEOREM, CUBIC SURFACE, DOUBLE SIXES, PASCAL'S THEOREM, QUARTIC SURFACE, STEINER SET

References

Bell, E. T. *The Development of Mathematics, 2nd ed.* New York: McGraw-Hill, pp. 322–325, 1945.

Coxeter, H. S. M. "The Pure Archimedean Polytopes in Six and Seven Dimensions." *Proc. Cambridge Phil. Soc.* **24**, 7–9, 1928.

Du Val, P. "On the Directrices of a Set of Points in a Plane." *Proc. London Math. Soc. Ser. 2* **35**, 23–74, 1933.

Schoutte, P. H. "On the Relation Between the Vertices of a Definite Sixdimensional Polytope and the Lines of a Cubic Surface." *Proc. Roy. Akad. Acad. Amsterdam* **13**, 375–383, 1910.

Solomon's Seal Polygon

see HEXAGRAM

Solvable Congruence

A CONGRUENCE that has a solution.

Solvable Group

A solvable group is a group whose composition indices are all PRIME NUMBERS. Equivalently, a solvable is a GROUP having a "normal series" such that each "normal factor" is ABELIAN. The term solvable derives from this type of group's relationship to GALOIS'S THEOREM, namely that the SYMMETRIC GROUP S_n is insoluble for $n \geq 5$ while it is solvable for $n = 1$, 2, 3, and 4. As a result, the POLYNOMIAL equations of degree ≥ 5 are not solvable using finite additions, multiplications, divisions, and root extractions.

Every FINITE GROUP of order < 60, every ABELIAN GROUP, and every SUBGROUP of a solvable group is solvable.

see also ABELIAN GROUP, COMPOSITION SERIES, GALOIS'S THEOREM, SYMMETRIC GROUP

References

Lomont, J. S. *Applications of Finite Groups.* New York: Dover, p. 26, 1993.

Solvable Lie Group

The connected closed SUBGROUPS (up to an ISOMORPHISM) of COMPLEX MATRICES that are closed under conjugate transpose and have a discrete finite center. Examples include SPECIAL LINEAR GROUPS, SYMPLECTIC GROUPS, and certain isometry groups of QUADRATIC FORMS.

see also LIE GROUP

References

Knapp, A. W. "Group Representations and Harmonic Analysis, Part II." *Not. Amer. Math. Soc.* **43**, 537–549, 1996.

Soma Cube

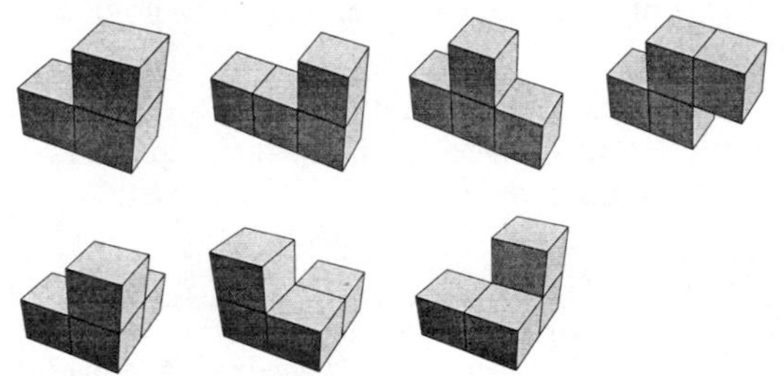

A solid DISSECTION puzzle invented by Piet Hein during a lecture on Quantum Mechanics by Werner Heisenberg. There are seven soma pieces composed of all the *irregular* face-joined cubes (POLYCUBES) with ≤ 4 cubes. The object is to assemble the pieces into a CUBE. There are 240 essentially distinct ways of doing so (Beeler *et al.* 1972, Berlekamp *et al.* 1982), as first enumerated one rainy afternoon in 1961 by J. H. Conway and Mike Guy.

A commercial version of the cube colors the pieces black, green, orange, white, red, and blue. When the 48 symmetries of the cube, three ways of assembling the black piece, and 2^5 ways of assembling the green, orange, white, red, and blue pieces are counted, the total number of solutions rises to 1,105,920.

see also CUBE DISSECTION, POLYCUBE

References

Albers, D. J. and Alexanderson, G. L. (Eds.). *Mathematical People: Profiles and Interviews.* Boston, MA: Birkhäuser, p. 43, 1985.

Ball, W. W. R. and Coxeter, H. S. M. *Mathematical Recreations and Essays, 13th ed.* New York: Dover, pp. 112–113, 1987.

Beeler, M.; Gosper, R. W.; and Schroeppel, R. *HAKMEM.* Cambridge, MA: MIT Artificial Intelligence Laboratory, Memo AIM-239, Item 112, Feb. 1972.

Berlekamp, E. R.; Conway, J. H.; and Guy, R. K. Ch. 24 in *Winning Ways, For Your Mathematical Plays, Vol. 2: Games in Particular.* London: Academic Press, 1982.

Cundy, H. and Rollett, A. *Mathematical Models, 3rd ed.* Stradbroke, England: Tarquin Pub., pp. 203–205, 1989.

Gardner, M. Ch. 6 in *The Second Scientific American Book of Mathematical Puzzles & Diversions: A New Selection.* New York: Simon and Schuster, pp. 65–77, 1961.

Steinhaus, H. *Mathematical Snapshots, 3rd American ed.* New York: Oxford University Press, pp. 168–169, 1983.

Somer-Lucas Pseudoprime

An ODD COMPOSITE NUMBER N is called a Somer-Lucas d-pseudoprime (with $d \geq 1$) if there EXISTS a nondegenerate LUCAS SEQUENCE $U(P,Q)$ with $U_0 = 0$, $U_1 = 1$, $D = P^2 - 4Q$, such that $(N,D) = 1$ and the rank appearance of N in the sequence $U(P,Q)$ is $(1/a)(N - (D/N))$, where (D/N) denotes the JACOBI SYMBOL.

see also LUCAS SEQUENCE, PSEUDOPRIME

References

Ribenboim, P. "Somer-Lucas Pseudoprimes." §2.X.D in *The New Book of Prime Number Records, 3rd ed.* New York: Springer-Verlag, pp. 131–132, 1996.

Sommerfeld's Formula

There are (at least) two equations known as Sommerfeld's formula. The first is

$$J_\nu(z) = \frac{1}{2\pi}\int_{-\eta+i\infty}^{2\pi-\eta+i\infty} e^{iz\cos t}e^{i\nu(t-\pi/2)}\,dt,$$

where $J_\nu(z)$ is a BESSEL FUNCTION OF THE FIRST KIND. The second states that under appropriate restrictions,

$$\int_0^\infty J_0(\tau r)e^{-|x|\sqrt{\tau^2-k^2}}\frac{\tau\,d\tau}{\sqrt{\tau^2-k^2}} = \frac{e^{ik\sqrt{r^2+k^2}}}{\sqrt{r^2+x^2}}.$$

see also WEYRICH'S FORMULA

References

Iyanaga, S. and Kawada, Y. (Eds.). *Encyclopedic Dictionary of Mathematics.* Cambridge, MA: MIT Press, pp. 1472 and 1474, 1980.

Somos Sequence

The Somos sequences are a set of related symmetrical RECURRENCE RELATIONS which, surprisingly, always give integers. The Somos sequence of order k is defined by

$$a_n = \frac{\sum_{j=1}^{\lfloor k/2\rfloor} a_{n-j}a_{n-(k-j)}}{a_{n-k}},$$

where $\lfloor x \rfloor$ is the FLOOR FUNCTION and $a_j = 1$ for $j = 0$, ..., $k-1$. The 2- and 3-Somos sequences consist entirely of 1s. The k-Somos sequences for $k = 4$, 5, 6, and 7 are

$$a_n = \frac{a_{n-1}a_{n-3} + {a_{n-2}}^2}{a_{n-4}}$$
$$a_n = \frac{a_{n-1}a_{n-4} + a_{n-2}a_{n-3}}{a_{n-5}}$$
$$a_n = \frac{1}{a_{n-6}}[a_{n-1}a_{n-5} + a_{n-2}a_{n-4} + {a_{n-3}}^2]$$
$$a_n = \frac{1}{a_{n-7}}[a_{n-1}a_{n-6} + a_{n-2}a_{n-5} + a_{n-3}a_{n-4}],$$

giving 1, 1, 1, 2, 3, 7, 23, 59, 314, 1529, ... (Sloane's A006720), 1, 1, 1, 1, 2, 3, 5, 11, 37, 83, 274, 1217, ... (Sloane's A006721), 1, 1, 1, 1, 1, 3, 5, 9, 23, 75, 421, 1103, ... (Sloane's A006722), 1, 1, 1, 1, 1, 1, 3, 5, 9, 17, 41, 137, 769, ... (Sloane's A006723). Gale (1991) gives simple proofs of the integer-only property of the 4-Somos and 5-Somos sequences. Hickerson proved 6-Somos generates only integers using computer algebra, and empirical evidence suggests 7-Somos is also integer-only.

However, the k-Somos sequences for $k \geq 8$ do not give integers. The values of n for which a_n first becomes nonintegral for the k-Somos sequence for $k = 8$, 9, ... are 17, 19, 20, 22, 24, 27, 28, 30, 33, 34, 36, 39, 41, 42, 44, 46, 48, 51, 52, 55, 56, 58, 60, ... (Sloane's A030127).

see also GÖBEL'S SEQUENCE, HERONIAN TRIANGLE

References

Buchholz, R. H. and Rathbun, R. L. "An Infinite Set of Heron Triangles with Two Rational Medians." *Amer. Math. Monthly* **104**, 107–115, 1997.

Gale, D. "Mathematical Entertainments: The Strange and Surprising Saga of the Somos Sequences." *Math. Intel.* **13**, 40–42, 1991.

Sloane, N. J. A. Sequences A030127, A006720/M0857, A006721/M0735, A006722/M2457, and A006723/M2456 in "An On-Line Version of the Encyclopedia of Integer Sequences."

Sondat's Theorem

The PERSPECTIVE AXIS bisects the line joining the two ORTHOCENTERS.

see also ORTHOCENTER, PERSPECTIVE AXIS

References

Johnson, R. A. *Modern Geometry: An Elementary Treatise on the Geometry of the Triangle and the Circle.* Boston, MA: Houghton Mifflin, p. 259, 1929.

Sonine's Integral

$$J_m(x) = \frac{2x^{m-n}}{2^{m-n}\Gamma(m-n)} \int_0^1 J_n(xt)t^{n+1}(1-t^2)^{m-n-1}\,dt,$$

where $J_m(x)$ is a BESSEL FUNCTION OF THE FIRST KIND and $\Gamma(x)$ is the GAMMA FUNCTION.

see also HANKEL'S INTEGRAL, POISSON INTEGRAL

Sonine Polynomial

A polynomial which differs from the associated LAGUERRE POLYNOMIAL by only a normalization constant,

$$\begin{aligned} S_r^s(x) &= \frac{1}{s!}e^x x^{-r}\frac{d^s}{dx^s}(e^{-x}x^{r+s}) = \frac{(-1)^r}{(r+s)!}L_{r+s}^r(x) \\ &= \frac{x^s}{s!(r+s)!0!} - \frac{x^{s-1}}{(s-1)!(r+s-1)!1!} \\ &\quad + \frac{x^{r-2}}{(r-2)!(r+s-2)!2!} - \ldots \\ &= \frac{1}{s!(r+s)!}x^{-(r+1)/2}e^{x/2}W_{s+r/2+1/2,r/2}(x), \end{aligned}$$

where $W_{k,m}(z)$ is a WHITTAKER FUNCTION.

see also LAGUERRE POLYNOMIAL, WHITTAKER FUNCTION

Sonine-Schafheitlin Formula

$$\begin{aligned} &\int_0^\infty J_\mu(at)J_\nu(bt)t^{-\lambda}\,dt \\ &= \frac{a^\mu\Gamma[(\mu+\nu-\lambda+1)/2]}{2^\lambda b^{\mu-\lambda+1}\Gamma[(-\mu+\nu+\lambda+1)/2]\Gamma(\mu+1)} \\ &\times {}_2F_1((\mu+\nu-\lambda+1)/2, (\mu-\nu-\lambda+1)/2; \mu+1; a^2/b^2), \end{aligned}$$

where $\Re[\mu+\nu-\lambda+1] > 0$, $\Re[\lambda] > -1$, $0 < a < b$, $J_\nu(x)$ is a BESSEL FUNCTION OF THE FIRST KIND, $\Gamma(x)$ is the GAMMA FUNCTION, and ${}_2F_1(a,b;c;x)$ is a HYPERGEOMETRIC FUNCTION.

References

Iyanaga, S. and Kawada, Y. (Eds.). *Encyclopedic Dictionary of Mathematics.* Cambridge, MA: MIT Press, p. 1474, 1980.

Sophie Germain Prime

A PRIME p is said to be a Sophie Germain prime if both p and $2p+1$ are PRIME. The first few Sophie Germain primes are 2, 3, 5, 11, 23, 29, 41, 53, 83, 89, 113, 131, ... (Sloane's A005384).

Around 1825, Sophie Germain proved that the first case of FERMAT'S LAST THEOREM is true for such primes, i.e., if p is a Sophie Germain prime, there do not exist INTEGERS x, y, and z different from 0 and not multiples of p such that

$$x^p + y^p = z^p.$$

Sophie Germain primes p of the form $p = k \cdot 2^n - 1$ (which makes $2p+1$ a PRIME) are COMPOSITE MERSENNE NUMBERS. Since the largest known COMPOSITE MERSENNE NUMBER is M_p with $p = 39051 \times 2^{6001} - 1$, p is the largest known Sophie Germain prime.

see also CUNNINGHAM CHAIN, FERMAT'S LAST THEOREM, MERSENNE NUMBER, TWIN PRIMES

References
Dubner, H. "Large Sophie Germain Primes." *Math. Comput.* **65**, 393–396, 1996.
Ribenboim, P. "Sophie Germane Primes." §5.2 in *The New Book of Prime Number Records.* New York: Springer-Verlag, pp. 329–332, 1996.
Shanks, D. *Solved and Unsolved Problems in Number Theory, 4th ed.* New York: Chelsea, pp. 154–157, 1993.
Sloane, N. J. A. Sequence A005384 in "An On-Line Version of the Encyclopedia of Integer Sequences."

Sorites Paradox

Sorites paradoxes are a class of paradoxical arguments also known as little-by-little arguments. The name "sorites" derives from the Greek word *soros*, meaning "pile" or "heap". Sorites paradoxes are exemplified by the problem that a single grain of wheat does not comprise a heap, nor do two grains of wheat, three grains of wheat, etc. However, at some point, the collection of grains becomes large enough to be called a heap, but there is apparently no definite point where this occurs.

see also UNEXPECTED HANGING PARADOX

Sort-Then-Add Sequence

A sequence produced by sorting the digits of a number and adding them to the previous number. The algorithm terminates when a sorted number is obtained. For $n = 1, 2, \ldots$, the algorithm terminates on 1, 2, 3, 4, 5, 6, 7, 8, 9, 11, 11, 12, 13, 14, 15, 16, 17, 18, 19, 22, 33, ... (Sloane's A033862). The first few numbers not known to terminate are 316, 452, 697, 1376, 2743, 5090, ... (Sloane's A033861). The least numbers of sort-then-add persistence $n = 1, 2, \ldots$, are 1, 10, 65, 64, 175, 98, 240, 325, 302, 387, 198, 180, 550, ... (Sloane's A033863).

see also 196-ALGORITHM, RATS SEQUENCE

References
Sloane, N. J. A. Sequences A033861, A033862, and A033863 in "An On-Line Version of the Encyclopedia of Integer Sequences."

Sorting

Sorting is the rearrangement of numbers (or other orderable objects) in a list into their correct lexographic order. Alphabetization is therefore a form of sorting. Because of the extreme importance of sorting in almost all database applications, a great deal of effort has been expended in the creation and analysis of efficient sorting algorithms.

see also HEAPSORT, ORDERING, QUICKSORT

References
Knuth, D. E. *The Art of Computer Programming, Vol. 3: Sorting and Searching, 2nd ed.* Reading, MA: Addison-Wesley, 1973.
Press, W. H.; Flannery, B. P.; Teukolsky, S. A.; and Vetterling, W. T. "Sorting." Ch. 8 in *Numerical Recipes in FORTRAN: The Art of Scientific Computing, 2nd ed.* Cambridge, England: Cambridge University Press, pp. 320–339, 1992.

Source

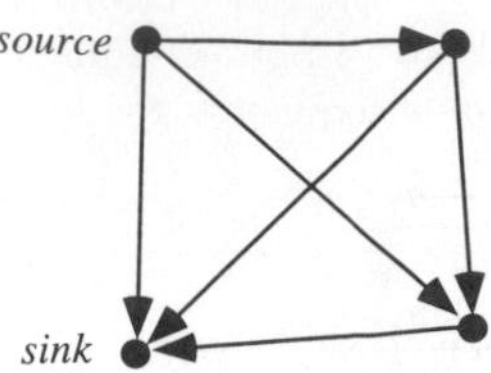

A vertex of a DIRECTED GRAPH with no entering edges.

see also DIRECTED GRAPH, NETWORK, SINK (DIRECTED GRAPH)

Sous-Double

A MULTIPERFECT NUMBER P_3. Six sous-doubles are known, and these are believed to comprise all sous-doubles.

see also MULTIPERFECT NUMBER, SOUS-TRIPLE

Souslin's Hypothesis

Every dense linear order complete set without endpoints having at most ω disjoint intervals is order isomorphic to the CONTINUUM of REAL NUMBERS, where ω is the set of NATURAL NUMBERS.

References
Iyanaga, S. and Kawada, Y. (Eds.). "Souslin's Hypothesis." §35E.4 in in *Encyclopedic Dictionary of Mathematics.* Cambridge, MA: MIT Press, p. 137, 1980.

Souslin Set

The continuous image of a POLISH SPACE, also called an ANALYTIC SET.

see also ANALYTIC SET, POLISH SPACE

Sous-Triple

A MULTIPERFECT NUMBER P_4. 36 sous-triples are known, and these are believed to comprise all sous-triples.

see also MULTIPERFECT NUMBER, SOUS-DOUBLE

Space

The concept of a space is an extremely general and important mathematical construct. Members of the space obey certain addition properties. Spaces which have been investigated and found to be of interest are usually named after one or more of their investigators. This practice unfortunately leads to names which give very little insight into the relevant properties of a given space.

One of the most general type of mathematical spaces is the TOPOLOGICAL SPACE.

see also AFFINE SPACE, BAIRE SPACE, BANACH SPACE, BASE SPACE, BERGMAN SPACE, BESOV SPACE, BOREL SPACE, CALABI-YAU SPACE, CELLULAR SPACE, CHU SPACE, DODECAHEDRAL SPACE, DRINFELD'S SYMMETRIC SPACE, EILENBERG-MAC LANE SPACE, EUCLIDEAN SPACE, FIBER SPACE, FINSLER SPACE,

FIRST-COUNTABLE SPACE, FRÉCHET SPACE, FUNCTION SPACE, G-SPACE, GREEN SPACE, HAUSDORFF SPACE, HEISENBERG SPACE, HILBERT SPACE, HYPERBOLIC SPACE, INNER PRODUCT SPACE, L_2-SPACE, LENS SPACE, LINE SPACE, LINEAR SPACE, LIOUVILLE SPACE, LOCALLY CONVEX SPACE, LOCALLY FINITE SPACE, LOOP SPACE, MAPPING SPACE, MEASURE SPACE, METRIC SPACE, MINKOWSKI SPACE, MÜNTZ SPACE, NON-EUCLIDEAN GEOMETRY, NORMED SPACE, PARACOMPACT SPACE, PLANAR SPACE, POLISH SPACE, PROBABILITY SPACE, PROJECTIVE SPACE, QUOTIENT SPACE, RIEMANN'S MODULI SPACE, RIEMANN SPACE, SAMPLE SPACE, STANDARD SPACE, STATE SPACE, STONE SPACE, TEICHMÜLLER SPACE, TENSOR SPACE, TOPOLOGICAL SPACE, TOPOLOGICAL VECTOR SPACE, TOTAL SPACE, VECTOR SPACE

Space of Closed Paths

see LOOP SPACE

Space Conic

see SKEW CONIC

Space Curve

A curve which may pass through any region of 3-D space, as contrasted to a PLANE CURVE which must lie in a single PLANE. Von Staudt (1847) classified space curves geometrically by considering the curve

$$\phi : I \to \mathbb{R}^3 \tag{1}$$

at $t_0 = 0$ and assuming that the parametric functions $\phi_i(t)$ for $i = 1, 2, 3$ are given by POWER SERIES which converge for small t. If the curve is contained in no PLANE for small t, then a coordinate transformation puts the parametric equations in the normal form

$$\phi_1(t) = t^{1+k_1} + \ldots \tag{2}$$

$$\phi_2(t) = t^{2+k_1+k_2} + \ldots \tag{3}$$

$$\phi_3(t) = t^{3+k_1+k_2+k_3} + \ldots \tag{4}$$

for integers $k_1, k_2, k_3 \geq 0$, called the local numerical invariants.

see also CURVE, CYCLIDE, FUNDAMENTAL THEOREM OF SPACE CURVES, HELIX, PLANE CURVE, SEIFERT'S SPHERICAL SPIRAL, SKEW CONIC, SPACE-FILLING FUNCTION, SPHERICAL SPIRAL, SURFACE, VIVIANI'S CURVE

References

do Carmo, M.; Fischer, G.; Pinkall, U.; and Reckziegel, H. "Singularities of Space Curves." §3.1 in *Mathematical Models from the Collections of Universities and Museums* (Ed. G. Fischer). Braunschweig, Germany: Vieweg, pp. 24–25, 1986.

Fine, H. B. "On the Singularities of Curves of Double Curvature." *Amer. J. Math.* **8**, 156–177, 1886.

Fischer, G. (Ed.). Plates 57–64 in *Mathematische Modelle/Mathematical Models, Bildband/Photograph Volume.* Braunschweig, Germany: Vieweg, pp. 58–59, 1986.

Gray, A. "Curves in $\mathbb{R}^n$" and "Curves in Space." §1.2 and Ch. 7 in *Modern Differential Geometry of Curves and Surfaces.* Boca Raton, FL: CRC Press, pp. 4–6 and 123–151, 1993.

Griffiths, P. and Harris, J. *Principles of Algebraic Geometry.* New York: Wiley, 1978.

Saurel, P. "On the Singularities of Tortuous Curves." *Ann. Math.* **7**, 3–9, 1905.

Staudt, C. von. *Geometrie der Lage.* Nürnberg, Germany, 1847.

Wiener, C. "Die Abhängigkeit der Rückkehrelemente der Projektion einer unebenen Curve von deren der Curve selbst." *Z. Math. & Phys.* **25**, 95–97, 1880.

Space Diagonal

The LINE SEGMENT connecting opposite VERTICES (i.e., two VERTICES which do not share a common face) in a PARALLELEPIPED or other similar solid.

see also DIAGONAL (POLYGON), DIAGONAL (POLYHEDRON), EULER BRICK

Space Distance

The maximum distance in 3-D can occur no more than $2n-2$ times. Also, there exists a fixed number c such that no distance determined by a set of n points in 3-D space occurs more than $cn^{5/3}$ times. The maximum distance can occur no more than $\lfloor \frac{1}{4}n^2 \rfloor$ times in 4-D, where $\lfloor x \rfloor$ is the FLOOR FUNCTION.

References

Honsberger, R. *Mathematical Gems II.* Washington, DC: Math. Assoc. Amer., pp. 122–123, 1976.

Space Division

The number of regions into which space can be divided by n SPHERES is

$$N = \tfrac{1}{3}n(n^2 - 3n + 8),$$

giving 2, 4, 8, 16, 30, 52, 84, ... (Sloane's A046127).

see also PLANE DIVISION

References

Sloane, N. J. A. Sequence A046127 in "An On-Line Version of the Encyclopedia of Integer Sequences."

Space-Filling Curve

see SPACE-FILLING FUNCTION

Space-Filling Function

A "CURVE" (i.e., a continuous map of a 1-D INTERVAL) into a 2-D area (a PLANE-FILLING FUNCTION) or a 3-D volume.

see also HILBERT CURVE, PEANO CURVE, PEANO-GOSPER CURVE, PLANE-FILLING CURVE, SIERPIŃSKI CURVE, SPACE-FILLING POLYHEDRON

References

Pappas, T. "Paradoxical Curve–Space-Filling Curve." *The Joy of Mathematics.* San Carlos, CA: Wide World Publ./ Tetra, p. 208, 1989.

Platzman, L. K. and Bartholdi, J. J. "Spacefilling Curves and the Planar Travelling Salesman Problem." *J. Assoc. Comput. Mach.* **46**, 719–737, 1989.

Wagon, S. "A Spacefilling Curve." §6.3 in *Mathematica in Action.* New York: W. H. Freeman, pp. 196–209, 1991.

Space-Filling Polyhedron

A space-filling polyhedron is a POLYHEDRON which can be used to generate a TESSELLATION of space. There exists one 16-sided space-filling POLYHEDRON, but it is unknown if this is the unique 16-sided space-filler. The CUBE, RHOMBIC DODECAHEDRON, and TRUNCATED OCTAHEDRON are space-fillers, as are the ELONGATED DODECAHEDRON and hexagonal PRISM. These five solids are all "primary" PARALLELOHEDRA (Coxeter 1973).

P. Schmitt discovered a nonconvex aperiodic polyhedral space-filler around 1990, and a convex POLYHEDRON known as the SCHMITT-CONWAY BIPRISM which fills space only aperiodically was found by J. H. Conway in 1993 (Eppstein).

see also CUBE, ELONGATED DODECAHEDRON, KELLER'S CONJECTURE, PARALLELOHEDRON, PRISM, RHOMBIC DODECAHEDRON, SCHMITT-CONWAY BIPRISM, TESSELLATION, TILING, TRUNCATED OCTAHEDRON

References

Coxeter, H. S. M. *Regular Polytopes, 3rd ed.* New York: Dover, pp. 29–30, 1973.

Critchlow, K. *Order in Space: A Design Source Book.* New York: Viking Press, 1970.

Devlin, K. J. "An Aperiodic Convex Space-filler is Discovered." *Focus: The Newsletter of the Math. Assoc. Amer.* **13**, 1, Dec. 1993.

Eppstein, D. "Re: Aperiodic Space-Filling Tile?." `http://www.ics.uci.edu/~eppstein/junkyard/biprism.html`.

Holden, A. *Shapes, Space, and Symmetry.* New York: Dover, pp. 154–163, 1991.

Thompson, D'A. W. *On Growth and Form, 2nd ed., compl. rev. ed.* New York: Cambridge University Press, 1992.

Tutton, A. E. H. *Crystallography and Practical Crystal Measurement, 2nd ed.* London: Lubrecht & Cramer, pp. 567 and 723, 1964.

Space Groups

The space groups in 2-D are called WALLPAPER GROUPS. In 3-D, the space groups are the symmetry GROUPS possible in a crystal lattice with the translation symmetry element. There are 230 space groups in $\mathbb{R}^3$, although 11 are MIRROR IMAGES of each other. They are listed by HERMANN-MAUGUIN SYMBOL in Cotton (1990).

see also HERMANN-MAUGUIN SYMBOL, LATTICE GROUPS, POINT GROUPS, WALLPAPER GROUPS

References

Arfken, G. "Crystallographic Point and Space Groups." *Mathematical Methods for Physicists, 3rd ed.* Orlando, FL: Academic Press, p. 248–249, 1985.

Buerger, M. J. *Elementary Crystallography.* New York: Wiley, 1956.

Cotton, F. A. *Chemical Applications of Group Theory, 3rd ed.* New York: Wiley, pp. 250–251, 1990.

Span (Geometry)

The largest possible distance between two points for a finite set of points.

see also JUNG'S THEOREM

Span (Link)

The span of an unoriented LINK diagram (also called the SPREAD) is the difference between the highest and lowest degrees of its BRACKET POLYNOMIAL. The span is a topological invariant of a knot. If a KNOT K has a reduced alternating projection of n crossings, then the span of K is $4n$.

see also LINK

Span (Polynomial)

The difference between the highest and lowest degrees of a POLYNOMIAL.

Span (Set)

For a SET S, the span is defined by $\max S - \min S$, where max is the MAXIMUM and min is the MINIMUM.

References

Guy, R. K. *Unsolved Problems in Number Theory, 2nd ed.* New York: Springer-Verlag, p. 207, 1994.

Span (Vectors)

The span of SUBSPACE generated by VECTORS $\mathbf{v}_1$ and $\mathbf{v}_2 \in \mathbb{V}$ is

$$\text{Span}(\mathbf{v}_1, \mathbf{v}_2) \equiv \{r\mathbf{v}_1 + s\mathbf{v}_2 : r, s \in \mathbb{R}\}.$$

Sparse Matrix

A MATRIX which has only a small number of NONZERO elements.

References

Press, W. H.; Flannery, B. P.; Teukolsky, S. A.; and Vetterling, W. T. "Sparse Linear Systems." §2.7 in *Numerical Recipes in FORTRAN: The Art of Scientific Computing, 2nd ed.* Cambridge, England: Cambridge University Press, pp. 63–82, 1992.

Spearman Rank Correlation

The Spearman rank correlation is defined by

$$r' \equiv \frac{\sum xy}{\sqrt{\sum x^2 \sum y^2}} = 1 - 6\sum \frac{d^2}{N(N^2-1)}. \quad (1)$$

The VARIANCE, KURTOSIS, and higher order MOMENTS are

$$\sigma^2 = \frac{1}{N-1} \quad (2)$$

$$\gamma_2 = -\frac{114}{25N} - \frac{6}{5N^2} - \ldots \quad (3)$$

$$\gamma_3 = \gamma_5 = \ldots = 0. \quad (4)$$

Student was the first to obtain the VARIANCE. The Spearman rank correlation is an R-ESTIMATE.

References

Press, W. H.; Flannery, B. P.; Teukolsky, S. A.; and Vetterling, W. T. *Numerical Recipes in FORTRAN: The Art of Scientific Computing, 2nd ed.* Cambridge, England: Cambridge University Press, pp. 634–637, 1992.

Special Curve

see PLANE CURVE, SPACE CURVE

Special Function

see FUNCTION

Special Linear Group

The special linear group $SL_n(q)$ is the MATRIX GROUP corresponding to the set of $n \times n$ COMPLEX MATRICES having DETERMINANT +1. It is a SUBGROUP of the GENERAL LINEAR GROUP $GL_n(q)$ and is also a LIE GROUP.

see also GENERAL LINEAR GROUP, SPECIAL ORTHOGONAL GROUP, SPECIAL UNITARY GROUP

References

Conway, J. H.; Curtis, R. T.; Norton, S. P.; Parker, R. A.; and Wilson, R. A. "The Groups $GL_n(q)$, $SL_n(q)$, $PGL_n(q)$, and $PSL_n(q) = L_n(q)$." §2.1 in *Atlas of Finite Groups: Maximal Subgroups and Ordinary Characters for Simple Groups.* Oxford, England: Clarendon Press, p. x, 1985.

Special Matrix

A matrix whose entries satisfy

$$a_{ij} = \begin{cases} 0 & \text{if } j > i+1 \\ -1 & \text{if } j = i+1 \\ 0 \text{ or } 1 & \text{if } j \leq i. \end{cases}$$

There are 2^{n-1} special MINIMAL MATRICES of size $n \times n$.

References

Knuth, D. E. "Problem 10470." *Amer. Math. Monthly* **102**, 655, 1995.

Special Orthogonal Group

The special orthogonal group $SO_n(q)$ is the SUBGROUP of the elements of GENERAL ORTHOGONAL GROUP $GO_n(q)$ with DETERMINANT 1.

see also GENERAL ORTHOGONAL GROUP, SPECIAL LINEAR GROUP, SPECIAL UNITARY GROUP

References

Conway, J. H.; Curtis, R. T.; Norton, S. P.; Parker, R. A.; and Wilson, R. A. "The Groups $GO_n(q)$, $SO_n(q)$, $PGO_n(q)$, and $PSO_n(q)$, and $O_n(q)$." §2.4 in *Atlas of Finite Groups: Maximal Subgroups and Ordinary Characters for Simple Groups.* Oxford, England: Clarendon Press, pp. xi–xii, 1985.

Special Point

A POINT which does not lie on at least one ORDINARY LINE.

see also ORDINARY POINT

References

Guy, R. K. "Unsolved Problems Come of Age." *Amer. Math. Monthly* **96**, 903–909, 1989.

Special Series Theorem

If the difference between the order and the dimension of a series is less than the GENUS (CURVE), then the series is special.

References

Coolidge, J. L. *A Treatise on Algebraic Plane Curves.* New York: Dover, p. 253, 1959.

Special Unitary Group

The special unitary group $SU_n(q)$ is the set of $n \times n$ UNITARY MATRICES with DETERMINANT +1 (having $n^2 - 1$ independent parameters). $SU(2)$ is HOMEOMORPHIC with the ORTHOGONAL GROUP $O_3^+(2)$. It is also called the UNITARY UNIMODULAR GROUP and is a LIE GROUP. The special unitary group can be represented by the MATRIX

$$U(a,b) = \begin{bmatrix} a & b \\ -b^* & a^* \end{bmatrix}, \quad (1)$$

where $a^*a + b^*b = 1$ and a, b are the CAYLEY-KLEIN PARAMETERS. The special unitary group may also be represented by the MATRIX

$$U(\xi,\eta,\zeta) = \begin{bmatrix} e^{i\xi}\cos\eta & e^{i\zeta}\sin\eta \\ -e^{-i\zeta}\sin\eta & e^{-i\xi}\cos\eta \end{bmatrix}, \quad (2)$$

or the matrices

$$U_x(\tfrac{1}{2}\phi) = \begin{bmatrix} \cos(\frac{1}{2}\phi) & i\sin(\frac{1}{2}\phi) \\ i\sin(\frac{1}{2}\phi) & \cos(\frac{1}{2}\phi) \end{bmatrix} \quad (3)$$

$$U_y(\tfrac{1}{2}\beta) = \begin{bmatrix} \cos(\frac{1}{2}\beta) & \sin(\frac{1}{2}\beta) \\ -\sin(\frac{1}{2}\beta) & \cos(\frac{1}{2}\beta) \end{bmatrix} \quad (4)$$

$$U_z(\xi) = \begin{bmatrix} e^{i\xi} & 0 \\ 0 & e^{-i\xi} \end{bmatrix}. \quad (5)$$

The order $2j+1$ representation is

$$U_{p,q}{}^{(j)}(\alpha,\beta,\gamma) = \sum_m \frac{(-1)^{m-q-p}\sqrt{(j+p)!(j-p)!(j+q)!(j-q)!}}{(j-p-m)!(j+q-m)!(m+p-q)!m!} \times e^{iq\alpha}\cos^{2j+q-p-2m}(\tfrac{1}{2}\beta)\sin^{p+2m-q}(\tfrac{1}{2}\beta)e^{ip\gamma}. \quad (6)$$

The summation is terminated by putting $1/(-N)! = 0$. The CHARACTER is given by

$$\chi^{(j)}(\alpha) = \begin{cases} 1+2\cos\alpha+\ldots+2\cos(j\alpha) \\ 2[\cos(\tfrac{1}{2}\alpha)+\cos(\tfrac{3}{2}\alpha)+\ldots+\cos(j\alpha)] \end{cases} = \begin{cases} \frac{\sin[(j+\frac{1}{2})\alpha]}{\sin(\frac{1}{2}\alpha)} & \text{for } j=0,1,2,\ldots \\ \frac{\sin[(j+\frac{1}{2})\alpha]}{\sin(\frac{1}{2}\alpha)} & \text{for } j=\frac{1}{2},\frac{3}{2},\ldots. \end{cases} \quad (7)$$

see also ORTHOGONAL GROUP, SPECIAL LINEAR GROUP, SPECIAL ORTHOGONAL GROUP

References

Arfken, G. "Special Unitary Group, $SU(2)$ and $SU(2)$–O_3^+ Homomorphism." *Mathematical Methods for Physicists, 3rd ed.* Orlando, FL: Academic Press, pp. 253–259, 1985.

Conway, J. H.; Curtis, R. T.; Norton, S. P.; Parker, R. A.; and Wilson, R. A. "The Groups $GU_n(q)$, $SU_n(q)$, $PGU_n(q)$, and $PSU_n(q) = U_n(q)$." §2.2 in *Atlas of Finite Groups: Maximal Subgroups and Ordinary Characters for Simple Groups.* Oxford, England: Clarendon Press, p. x, 1985.

Species

A species of structures is a rule F which

1. Produces, for each finite set U, a finite set $F[U]$,
2. Produces, for each bijection $\sigma : U \to V$, a function

$$F[\sigma] : F[U] \to F[V].$$

The functions $F[\sigma]$ should further satisfy the following functorial properties:

1. For all bijections $\sigma : U \to V$ and $\tau : V \to W$,

$$F[\tau\circ\sigma] = F[\tau]\circ F[\sigma],$$

2. For the IDENTITY MAP $\mathrm{Id}_U : U \to U$,

$$F[\underset{U}{\mathrm{Id}}] = \underset{F[U]}{\mathrm{Id}}.$$

An element $\sigma \in F[U]$ is called an F-structure on U (or a structure of species F on U). The function $F[\sigma]$ is called the transport of F-structures along σ.

References

Bergeron, F.; Labelle, G.; and Leroux, P. *Combinatorial Species and Tree-Like Structures.* Cambridge, England: Cambridge University Press, p. 5, 1998.

Specificity

The probability that a STATISTICAL TEST will be negative for a negative statistic.

see also SENSITIVITY, STATISTICAL TEST, TYPE I ERROR, TYPE II ERROR

Spectral Norm

The NATURAL NORM induced by the L_2-NORM. Let $\mathsf{A}^\dagger$ be the ADJOINT of the SQUARE MATRIX A, so that $\mathsf{A}^\dagger = a_{ji}^*$, then

$$||\mathsf{A}||_2 = (\text{maximum eigenvalue of } \mathsf{A}^\dagger\mathsf{A})^{1/2} = \max_{||x||_2\neq 0} \frac{||\mathsf{A}\mathbf{x}||_2}{||\mathbf{x}||_2}.$$

see also L_2-NORM, MATRIX NORM

References

Gradshteyn, I. S. and Ryzhik, I. M. *Tables of Integrals, Series, and Products, 5th ed.* San Diego, CA: Academic Press, pp. 1115, 1979.

Strang, G. §6.2 and 7.2 in *Linear Algebra and Its Applications, 4th ed.* New York: Academic Press, 1980.

Spectral Power Density

$$P_y(\nu) \equiv \lim_{T\to\infty} \frac{2}{T}\left|\int_{-T/2}^{T/2} [y(t)-\bar{y}]e^{-2\pi i\nu t}\,dt\right|^2,$$

so

$$\int_0^\infty P_y(\nu)\,d\nu = \lim_{T\to\infty}\frac{1}{T}\int_{-T/2}^{T/2}[y(t)-\bar{y}]^2\,dt = \langle (y-\bar{y})^2\rangle = \sigma_y{}^2.$$

see also POWER SPECTRUM

Spectral Radius

Let A be an $n\times n$ MATRIX with COMPLEX or REAL elements with EIGENVALUES $\lambda_1, \ldots, \lambda_n$. Then the spectral radius $\rho(\mathsf{A})$ of A is

$$\rho(\mathsf{A}) = \max_{1\leq i\leq n} |\lambda_i|.$$

References

Gradshteyn, I. S. and Ryzhik, I. M. *Tables of Integrals, Series, and Products, 5th ed.* San Diego, CA: Academic Press, pp. 1115–1116, 1979.

Spectral Rigidity

The mean square deviation of the best local fit straight line to a staircase cumulative spectral density over a normalized energy scale.

References

Ott, E. *Chaos in Dynamical Systems.* New York: Cambridge University Press, p. 341, 1993.

Spectral Theorem

Let H be a HILBERT SPACE, $B(H)$ the set of BOUNDED linear operators from H to itself, and $\sigma(T)$ the SPECTRUM of T. Then if $T \in B(H)$ and T is normal, there exists a unique resolution of the identity E on the Borel subsets of $\sigma(T)$ which satisfies

$$T = \operatorname*{int}_{\sigma(T)} \lambda \, dE(\lambda).$$

Furthermore, every projection $E(\omega)$ COMMUTES with every $S \in B(H)$ which COMMUTES with T.

References

Rudin, W. Theorem 12.23 in *Functional Analysis, 2nd ed.* New York: McGraw-Hill, 1991.

Spectrum (Operator)

Let T be an OPERATOR on a HILBERT SPACE. The spectrum $\sigma(T)$ of T is the set of λ such that $(T - \lambda I)$ is not invertible on all of the HILBERT SPACE, where the λs are COMPLEX NUMBERS and I is the IDENTITY OPERATOR. The definition can also be stated in terms of the resolvent of an operator

$$\rho(T) = \{\lambda : (T - \lambda I) \text{ is invertible}\},$$

and then the spectrum is defined to be the complement of $\rho(T)$ in the COMPLEX PLANE. The reason for doing this is that it is easy to demonstrate that $\rho(T)$ is an OPEN SET, which shows that the spectrum is closed.

see also HILBERT SPACE

Spectrum Sequence

A spectrum sequence is a SEQUENCE formed by successive multiples of a REAL NUMBER a rounded down to the nearest INTEGER $s_n = \lfloor na \rfloor$. If a is IRRATIONAL, the spectrum is called a BEATTY SEQUENCE.

see also BEATTY SEQUENCE, LAGRANGE SPECTRUM, MARKOV SPECTRUM

Speed

The SCALAR $|\mathbf{v}|$ equal to the magnitude of the VELOCITY $\mathbf{v}$.

see also ANGULAR VELOCITY, VELOCITY

Spence's Function

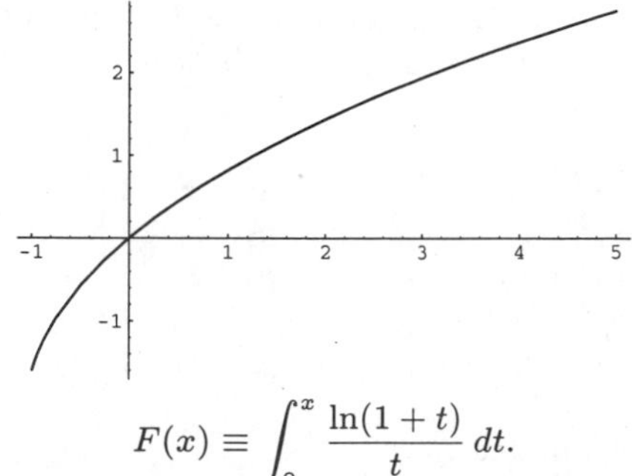

$$F(x) \equiv \int_0^x \frac{\ln(1+t)}{t} \, dt.$$

see also SPENCE'S INTEGRAL

References

Berestetskii, V. B.; Lifschitz, E. M.; and Ditaevskii, L. P. *Quantum Electrodynamics, 2nd ed.* Oxford, England: Pergamon Press, p. 596, 1982.

Spence's Integral

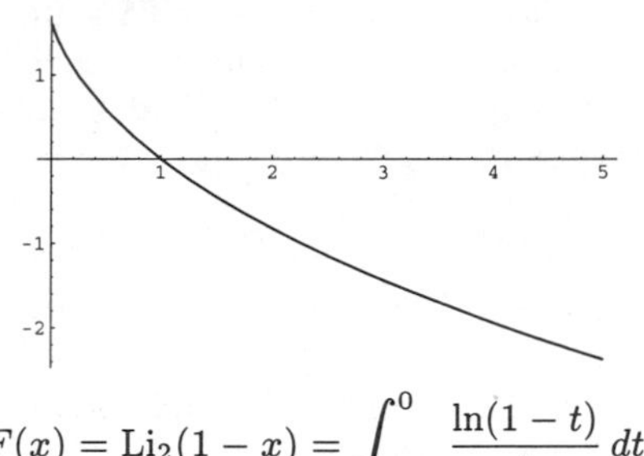

$$F(x) = \mathrm{Li}_2(1-x) = \int_{1-x}^0 \frac{\ln(1-t)}{t} \, dt,$$

where $\mathrm{Li}_2(x)$ is the DILOGARITHM.

see also SPENCE'S FUNCTION

Spencer's 15-Point Moving Average

A MOVING AVERAGE using 15 points having weights -3, -6, -5, 3, 21, 46, 67, 74, 67, 46, 21, 3, -5, -6, and -3. It is sometimes used by actuaries.

see also MOVING AVERAGE

References

Kenney, J. F. and Keeping, E. S. *Mathematics of Statistics, Pt. 1, 3rd ed.* Princeton, NJ: Van Nostrand, p. 223, 1962.

Sperner's Theorem

The MAXIMUM CARDINALITY of a collection of SUBSETS of a t-element SET T, none of which contains another, is the BINOMIAL COEFFICIENT $\binom{t}{\lfloor t/2 \rfloor}$, where $\lfloor x \rfloor$ is the FLOOR FUNCTION.

see also CARDINALITY

Sphenocorona

see JOHNSON SOLID

Sphenoid

see DISPHENOID

Sphenomegacorona

see JOHNSON SOLID

Sphere

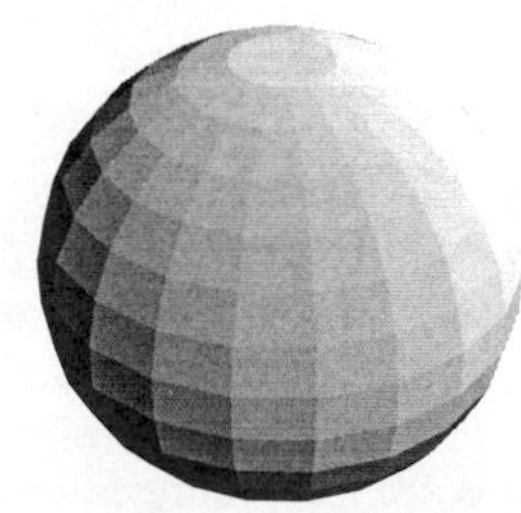

A sphere is defined as the set of all points in $\mathbb{R}^3$ which are a distance r (the "RADIUS") from a given point (the "CENTER"). Twice the RADIUS is called the DIAMETER, and pairs of points on opposite sides of a DIAMETER are called ANTIPODES. The term "sphere" technically refers to the outer surface of a "BUBBLE," which is denoted $\mathbb{S}^2$. However, in common usage, the word *sphere* is also used to mean the UNION of a sphere and its INTERIOR (a "solid sphere"), where the INTERIOR is called a BALL. The SURFACE AREA of the sphere and VOLUME of the BALL of RADIUS r are given by

$$S = 4\pi r^2 \tag{1}$$
$$V = \tfrac{4}{3}\pi r^3 \tag{2}$$

(Beyer 1987, p. 130). In *On the Sphere and Cylinder* (ca. 225 BC), Archimedes became the first to derive these equations (although he expressed π in terms of the sphere's circular cross-section). The fact that

$$\frac{V_{\text{sphere}}}{V_{\text{circumscribed cylinder}} - V_{\text{sphere}}} = \frac{\frac{4}{3}}{2 - \frac{4}{3}} = \frac{\frac{4}{3}}{\frac{2}{3}} = 2 \tag{3}$$

was also known to Archimedes.

Any cross-section through a sphere is a CIRCLE (or, in the degenerate case where the slicing PLANE is tangent to the sphere, a point). The size of the CIRCLE is maximized when the PLANE defining the cross-section passes through a DIAMETER.

The equation of a sphere of RADIUS r is given in CARTESIAN COORDINATES by

$$x^2 + y^2 + z^2 = r^2, \tag{4}$$

which is a special case of the ELLIPSOID

$$\frac{x^2}{a^2} + \frac{y^2}{b^2} + \frac{z^2}{c^2} = 1 \tag{5}$$

and SPHEROID

$$\frac{x^2 + y^2}{a^2} + \frac{z^2}{c^2} = 1. \tag{6}$$

A sphere may also be specified in SPHERICAL COORDINATES by

$$x = \rho \cos\theta \sin\phi \tag{7}$$
$$y = \rho \sin\theta \sin\phi \tag{8}$$
$$z = \rho \cos\phi, \tag{9}$$

where θ is an azimuthal coordinate running from 0 to 2π (LONGITUDE), ϕ is a polar coordinate running from 0 to π (COLATITUDE), and ρ is the RADIUS. Note that there are several other notations sometimes used in which the symbols for θ and ϕ are interchanged or where r is used instead of ρ. If ρ is allowed to run from 0 to a given RADIUS r, then a solid BALL is obtained. Converting to "standard" parametric variables $a = \rho$, $u = \theta$, and $v = \phi$ gives the first FUNDAMENTAL FORMS

$$E = a \sin^2 v \tag{10}$$
$$F = 0 \tag{11}$$
$$G = a, \tag{12}$$

second FUNDAMENTAL FORMS

$$e = a^2 \sin^2 v \tag{13}$$
$$f = 0 \tag{14}$$
$$g = a^2, \tag{15}$$

AREA ELEMENT

$$dA = a \sin v, \tag{16}$$

GAUSSIAN CURVATURE

$$K = \frac{1}{a^2}, \tag{17}$$

and MEAN CURVATURE

$$H = \frac{1}{a}. \tag{18}$$

A sphere may also be represented parametrically by letting $u \equiv r\cos\phi$, so

$$x = \sqrt{r^2 - u^2}\cos\theta \tag{19}$$
$$y = \sqrt{r^2 - u^2}\sin\theta \tag{20}$$
$$z = u, \tag{21}$$

where θ runs from 0 to 2π and u runs from $-r$ to r.

Given two points on a sphere, the shortest path on the surface of the sphere which connects them (the SPHERE GEODESIC) is an ARC of a CIRCLE known as a GREAT CIRCLE. The equation of the sphere with points (x_1, y_1, z_1) and (x_2, y_2, z_2) lying on a DIAMETER is given by

$$(x - x_1)(x - x_2) + (y - y_1)(y - y_2) + (z - z_1)(z - z_2) = 0. \tag{22}$$

Four points are sufficient to uniquely define a sphere. Given the points (x_i, y_i, z_i) with $i = 1$, 2, 3, and 4, the sphere containing them is given by the beautiful DETERMINANT equation

$$\begin{vmatrix} x^2 + y^2 + z^2 & x & y & z & 1 \\ {x_1}^2 + {y_1}^2 + {z_1}^2 & x_1 & y_1 & z_1 & 1 \\ {x_2}^2 + {y_2}^2 + {z_2}^2 & x_2 & y_2 & z_2 & 1 \\ {x_3}^2 + {y_3}^2 + {z_3}^2 & x_3 & y_3 & z_3 & 1 \\ {x_4}^2 + {y_4}^2 + {z_4}^2 & x_4 & y_4 & z_4 & 1 \end{vmatrix} = 0 \tag{23}$$

(Beyer 1987, p. 210).

The generalization of a sphere in n dimensions is called a HYPERSPHERE. An n-D HYPERSPHERE can be specified by the equation

$$x_1{}^2 + x_2{}^2 + \ldots + x_n{}^2 = r^2. \quad (24)$$

The distribution of ANGLES for random rotation of a sphere is

$$P(\theta) = \frac{2}{\pi}\sin^2(\tfrac{1}{2}\theta), \quad (25)$$

giving a MEAN of $\pi/2 + 2/\pi$.

To pick a random point on the surface of a sphere, let u and v be random variates on $[0, 1]$. Then

$$\theta = 2\pi u \quad (26)$$

$$\phi = \cos^{-1}(2v - 1). \quad (27)$$

This works since the SOLID ANGLE is

$$d\Omega = \sin\phi\, d\theta\, d\phi = d\theta\, d(\cos\phi). \quad (28)$$

Another easy way to pick a random point on a SPHERE is to generate three gaussian random variables x, y, and z. Then the distribution of the vectors

$$\frac{1}{\sqrt{x^2 + y^2 + z^2}}\begin{bmatrix} x \\ y \\ z \end{bmatrix} \quad (29)$$

is uniform over the surface $\mathbb{S}^2$. Another method is to pick z from a UNIFORM DISTRIBUTION over $[-1, 1]$ and θ from a UNIFORM DISTRIBUTION over $[0, 2\pi)$. Then the points

$$\begin{bmatrix} \sqrt{1-z^2}\cos\theta \\ \sqrt{1-z^2}\sin\theta \\ z \end{bmatrix} \quad (30)$$

are uniformly distributed over $\mathbb{S}^2$.

Pick four points on a sphere. What is the probability that the TETRAHEDRON having these points as VERTICES contains the CENTER of the sphere? In the 1-D case, the probability that a second point is on the opposite side of 1/2 is 1/2. In the 2-D case, pick two points. In order for the third to form a TRIANGLE containing the CENTER, it must lie in the quadrant bisected by a LINE SEGMENT passing through the center of the CIRCLE and the bisector of the two points. This happens for one QUADRANT, so the probability is 1/4. Similarly, for a sphere the probability is one OCTANT, or 1/8.

Pick two points at random on a unit sphere. The first one can be assigned the coordinate (0, 0, 1) without loss of generality. The second point can be given the coordinates $(\sin\phi, 0\cos\phi)$ with $\theta \equiv 0$ since all points with the same ϕ are rotationally identical. The distance between the two points is then

$$r = \sqrt{\sin^2\phi + (1-\cos\phi)^2} = \sqrt{2 - \cos\phi} = 2\sin(\tfrac{1}{2}\phi). \quad (31)$$

Because the surface AREA element is

$$d\Omega = \sin\phi\, d\theta\, d\phi, \quad (32)$$

the probability that two points are a distance r apart is

$$\begin{aligned} P_\phi(r) &= \frac{\int_0^\pi \delta(\phi - r)\sin\phi\, d\phi}{\int_0^\pi \sin\phi\, d\phi} \\ &= \tfrac{1}{2}\int_0^\pi \delta[r - 2\sin(\tfrac{1}{2}\phi)]\sin\phi\, d\phi. \quad (33) \end{aligned}$$

The DELTA FUNCTION contributes when

$$\tfrac{1}{2}r = \sin(\tfrac{1}{2}\phi) \quad (34)$$

$$\phi = 2\sin^{-1}(\tfrac{1}{2}r), \quad (35)$$

so

$$\begin{aligned} P_\phi(r) &= \tfrac{1}{2}\sin[2\sin^{-1}(\tfrac{1}{2}r)] = \sin[\sin^{-1}(\tfrac{1}{2}r)]\cos[\sin^{-1}(\tfrac{1}{2}r)] \\ &= \tfrac{1}{2}r\sqrt{1 - (\tfrac{1}{2}r)^2} = \tfrac{1}{4}r\sqrt{4 - r^2}. \quad (36) \end{aligned}$$

However, we need

$$P_r(r)\, dr = P_\phi(r)\frac{d\phi}{dr}\, dr, \quad (37)$$

and

$$\begin{aligned} \tfrac{1}{2}dr &= \tfrac{1}{2}\cos(\tfrac{1}{2}\phi)\, d\phi = \tfrac{1}{2}\sqrt{1 - \sin^2(\tfrac{1}{2}\phi)}\, d\phi \\ &= \tfrac{1}{2}\sqrt{1 - (\tfrac{1}{2}r)^2}\, d\phi = \tfrac{1}{4}\sqrt{4 - r^2}\, d\phi \quad (38) \end{aligned}$$

so

$$\frac{d\phi}{dr} = \frac{2}{\sqrt{4 - r^2}}, \quad (39)$$

and

$$P_r(r) = \tfrac{1}{4}r\sqrt{4 - r^2}\frac{2}{\sqrt{4 - r^2}} = \tfrac{1}{2}r \quad (40)$$

for $r \in [0, 2]$. Somewhat surprisingly, the largest distances are the most common, contrary to most people's intuition. A plot of 15 random lines is shown below.

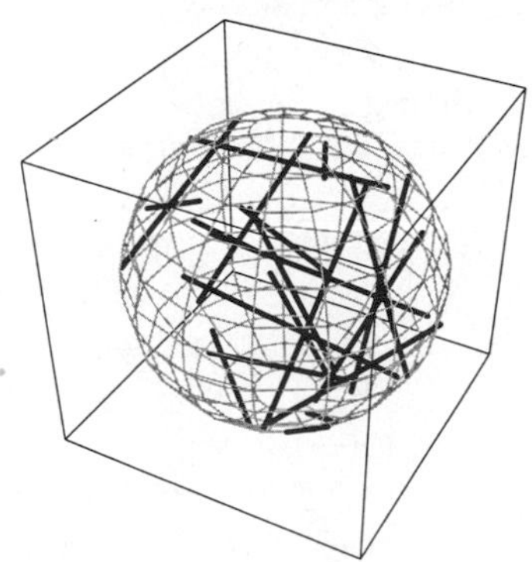

The MOMENTS about zero are

$$\mu'_n = \langle r^n \rangle = \int_0^2 r^n\,dr = \frac{2^{n+1}}{2+n}, \tag{41}$$

giving the first few as

$$\mu'_1 = \tfrac{4}{3} \tag{42}$$
$$\mu'_2 = 2 \tag{43}$$
$$\mu'_3 = \tfrac{16}{5} \tag{44}$$
$$\mu'_4 = \tfrac{16}{3}. \tag{45}$$

Moments about the MEAN are

$$\mu = \tfrac{4}{3} \tag{46}$$
$$\mu_2 = \sigma^2 = \tfrac{2}{9} \tag{47}$$
$$\mu_3 = -\tfrac{8}{135} \tag{48}$$
$$\mu_4 = \tfrac{16}{135}, \tag{49}$$

so the SKEWNESS and KURTOSIS are

$$\gamma_1 = \tfrac{4}{5}\sqrt{2} \tag{50}$$
$$\gamma_2 = -\tfrac{5}{3}. \tag{51}$$

see also BALL, BING'S THEOREM, BUBBLE, CIRCLE, DANDELIN SPHERES, DIAMETER, ELLIPSOID, EXOTIC SPHERE, FEJES TÓTH'S PROBLEM, HYPERSPHERE, LIEBMANN'S THEOREM, LIOUVILLE'S SPHERE-PRESERVING THEOREM, MIKUSIŃSKI'S PROBLEM, NOISE SPHERE, OBLATE SPHEROID, OSCULATING SPHERE, PARALLELIZABLE, PROLATE SPHEROID, RADIUS, SPACE DIVISION, SPHERE PACKING, TENNIS BALL THEOREM

References

Beyer, W. H. (Ed.) *CRC Standard Mathematical Tables, 28th ed.* Boca Raton, FL: CRC Press, 1987.

Eppstein, D. "Circles and Spheres." http://www.ics.uci.edu/~eppstein/junkyard/sphere.html.

Geometry Center. "The Sphere." http://www.geom.umn.edu/zoo/toptype/sphere/.

Sphere-Cylinder Intersection

see CYLINDER-SPHERE INTERSECTION

Sphere Embedding

A 4-sphere has POSITIVE CURVATURE, with

$$R^2 = x^2 + y^2 + z^2 + w^2 \tag{1}$$

$$2x\frac{dx}{dw} + 2y\frac{dy}{dw} + 2z\frac{dz}{dw} + 2w = 0. \tag{2}$$

Since

$$\mathbf{r} \equiv x\hat{\mathbf{x}} + y\hat{\mathbf{y}} + z\hat{\mathbf{z}}, \tag{3}$$

$$dw = -\frac{x\,dx + y\,dy + z\,dz}{w} = -\frac{\mathbf{r}\cdot d\mathbf{r}}{\sqrt{R^2 - r^2}}. \tag{4}$$

To stay on the surface of the sphere,

$$\begin{aligned} ds^2 &= dx^2 + dy^2 + dz^2 + dw^2 \\ &= dx^2 + dy^2 + dz^2 + \frac{r^2\,dr^2}{R^2 - r^2} \\ &= dr^2 + r^2\,d\Omega^2 + \frac{dr^2}{\frac{R^2}{r^2} - 1} \\ &= dr^2\left(1 + \frac{1}{\frac{R^2}{r^2} - 1}\right) + r^2\,d\Omega^2 \\ &= dr^2\left(\frac{\frac{R^2}{r^2}}{\frac{R^2}{r^2} - 1}\right) + r^2\,d\Omega^2 \\ &= \frac{dr^2}{1 - \frac{r^2}{R^2}} + r^2\,d\Omega^2. \end{aligned} \tag{5}$$

With the addition of the so-called expansion parameter, this is the Robertson-Walker line element.

Sphere Eversion

Smale (1958) proved that it is mathematically possible to turn a SPHERE inside-out without introducing a sharp crease at any point. This means there is a regular homotopy from the standard embedding of the 2-SPHERE in EUCLIDEAN 3-space to the mirror-reflection embedding such that at every stage in the homotopy, the sphere is being IMMERSED in EUCLIDEAN SPACE. This result is so counterintuitive and the proof so technical that the result remained controversial for a number of years.

In 1961, Arnold Shapiro devised an explicit eversion but did not publicize it. Phillips (1966) heard of the result and, in trying to reproduce it, actually devised an independent method of his own. Yet another eversion was devised by Morin, which became the basis for the movie by Max (1977). Morin's eversion also produced explicit algebraic equations describing the process. The original method of Shapiro was subsequently published by Francis and Morin (1979).

see also EVERSION, SPHERE

References

Francis, G. K. Ch. 6 in *A Topological Picturebook.* New York: Springer-Verlag, 1987.

Francis, G. K. and Morin, B. "Arnold Shapiro's Eversion of the Sphere." *Math. Intell.* **2**, 200–203, 1979.

Levy, S. *Making Waves: A Guide to the Ideas Behind Outside In.* Wellesley, MA: A. K. Peters, 1995.

Levy, S. "A Brief History of Sphere Eversions." http://www.geom.umn.edu/docs/outreach/oi/history.html.

Levy, S.; Maxwell, D.; and Munzner, T. *Outside-In.* 22 minute videotape. http://www.geom.umn.edu/docs/outreach/oi/.

Max, N. "Turning a Sphere Inside Out." Videotape. Chicago, IL: International Film Bureau, 1977.

Peterson, I. *Islands of Truth: A Mathematical Mystery Cruise.* New York: W. H. Freeman, pp. 240–244, 1990.

Petersen, I. "Forging Links Between Mathematics and Art." *Science News* **141**, 404–405, June 20, 1992.

Phillips, A. "Turning a Surface Inside Out." *Sci. Amer.* **214**, 112–120, Jan. 1966.

Smale, S. "A Classification of Immersions of the Two-Sphere." *Trans. Amer. Math. Soc.* **90**, 281–290, 1958.

Sphere Geodesic

see GREAT CIRCLE

Sphere Packing

Let η denote the PACKING DENSITY, which is the fraction of a VOLUME filled by identical packed SPHERES. In 2-D (CIRCLE PACKING), there are two periodic packings for identical CIRCLES: square lattice and hexagonal lattice. Fejes Tóth (1940) proved that the hexagonal lattice is indeed the densest of *all* possible plane packings (Conway and Sloane 1993, pp. 8–9).

In 3-D, there are three periodic packings for identical spheres: cubic lattice, face-centered cubic lattice, and hexagonal lattice. It was hypothesized by Kepler in 1611 that close packing (cubic or hexagonal) is the densest possible (has the greatest η), and this assertion is known as the KEPLER CONJECTURE. The problem of finding the densest packing of spheres (not necessarily periodic) is therefore known as the KEPLER PROBLEM. The KEPLER CONJECTURE is intuitively obvious, but the proof still remains elusive. However, Gauss (1831) did prove that the face-centered cubic is the densest *lattice* packing in 3-D (Conway and Sloane 1993, p. 9). This result has since been extended to HYPERSPHERE PACKING.

In 3-D, face-centered cubic close packing and hexagonal close packing (which is distinct from hexagonal lattice), both give

$$\eta = \frac{\pi}{3\sqrt{2}} \approx 74.048\%. \tag{1}$$

For packings in 3-D, C. A. Rogers (1958) showed that

$$\eta < \sqrt{18}\,(\cos^{-1}\tfrac{1}{3} - \tfrac{1}{3}\pi) \approx 77.96355700\% \tag{2}$$

(Le Lionnais 1983). This was subsequently improved to 77.844% (Lindsey 1986), then 77.836% (Muder 1988). However, Rogers (1958) remarks that "many mathematicians believe, and all physicists know" that the actual answer is 74.05% (Conway and Sloane 1993, p. 3).

"Random" close packing in 3-D gives only $\eta \approx 64\%$ (Jaeger and Nagel 1992).

The PACKING DENSITIES for several packing types are summarized in the following table.

Packing	η (exact)	η (approx.)
square lattice (2-D)	$\frac{\pi}{4}$	0.7854
hexagonal lattice (2-D)	$\frac{\pi}{2\sqrt{3}}$	0.9069
cubic lattice	$\frac{\pi}{6}$	0.5236
hexagonal lattice	$\frac{\pi}{3\sqrt{3}}$	0.6046
face-centered cubic lattice	$\frac{\pi}{3\sqrt{2}}$	0.7405
random	—	0.6400

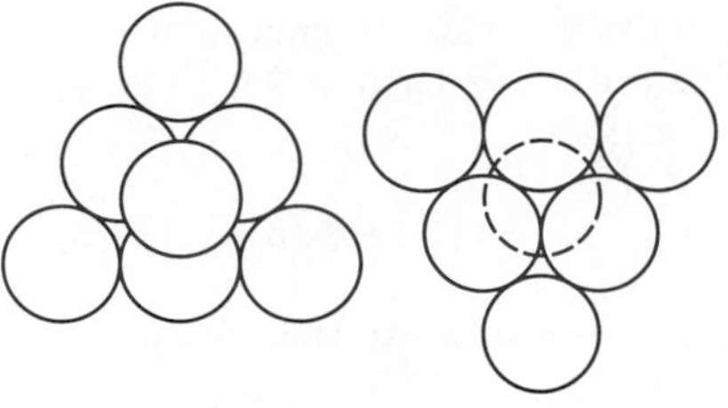

For cubic close packing, pack six SPHERES together in the shape of an EQUILATERAL TRIANGLE and place another SPHERE on top to create a TRIANGULAR PYRAMID. Now create another such grouping of seven SPHERES and place the two PYRAMIDS together facing in opposite directions. A CUBE emerges. Consider a face of the CUBE, illustrated below.

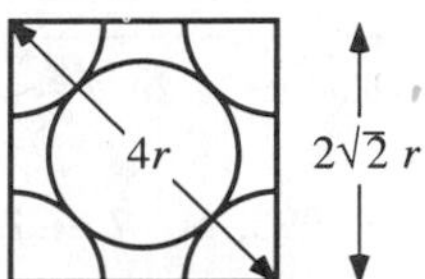

The "unit cell" cube contains eight 1/8-spheres (one at each VERTEX) and six HEMISPHERES. The total VOLUME of SPHERES in the unit cell is

$$\begin{aligned} V_{\text{spheres in unit cell}} &= (8\cdot\tfrac{1}{8} + 6\cdot\tfrac{1}{2})\frac{4\pi}{3}r^3 \\ &= 4\cdot\frac{4\pi}{3}r^3 = \tfrac{16}{3}\pi r^3. \end{aligned} \tag{3}$$

The diagonal of the face is $4r$, so each side is $2\sqrt{2}\,r$. The VOLUME of the unit cell is therefore

$$V_{\text{unit cell}} = (2\sqrt{2}\,r)^3 = 16\sqrt{2}\,r^3. \tag{4}$$

The PACKING DENSITY is therefore

$$\eta_{\text{CCP}} = \frac{\frac{16}{3}\pi r^2}{16\sqrt{2}r^3} = \frac{\pi}{3\sqrt{2}} \tag{5}$$

(Conway and Sloane 1993, p. 2).

Hexagonal close packing must give the same values, since sliding one sheet of SPHERES cannot affect the volume they occupy. To verify this, construct a 3-D diagram containing a hexagonal unit cell with three layers. Both the top and the bottom contain six 1/6-SPHERES and one HEMISPHERE. The total number of spheres in these two rows is therefore

$$2(6\tfrac{1}{6} + 1\tfrac{1}{2}) = 3. \tag{6}$$

The VOLUME of SPHERES in the middle row cannot be simply computed using geometry. However, symmetry requires that the piece of the SPHERE which is cut off is exactly balanced by an extra piece on the other side. There are therefore three SPHERES in the middle layer, for a total of six, and a total VOLUME

$$V_{\text{spheres in unit cell}} = 6\cdot\frac{4\pi}{3}r^3(3+3) = 8\pi r^3. \tag{7}$$

The base of the HEXAGON is made up of 6 EQUILATERAL TRIANGLES with side lengths $2r$. The unit cell base AREA is therefore

$$A_{\text{unit cell}} = 6[\tfrac{1}{2}(2r)(\sqrt{3}\,r)] = 6\sqrt{3}\,r^2. \tag{8}$$

The height is the same as that of two TETRAHEDRA length $2r$ on a side, so

$$h_{\text{unit cell}} = 2\left(2r\sqrt{\frac{2}{3}}\right), \tag{9}$$

giving

$$\eta_{\text{HCP}} = \frac{8\pi r^3}{(6\sqrt{3}\,r^2)\left(4r\sqrt{\frac{2}{3}}\right)} = \frac{\pi}{3\sqrt{2}} \tag{10}$$

(Conway and Sloane 1993, pp. 7 and 9).

If we had actually wanted to compute the VOLUME of SPHERE inside and outside the HEXAGONAL PRISM, we could use the SPHERICAL CAP equation to obtain

$$V_\subset = \tfrac{1}{3}\pi h^2(3r-h) = \frac{1}{3}\pi r^3\frac{1}{3}\left(3-\frac{1}{\sqrt{3}}\right)$$
$$= \frac{1}{9}\pi r^3\left(3-\frac{\sqrt{3}}{3}\right) = \tfrac{1}{27}\pi r^3(9-\sqrt{3}) \tag{11}$$
$$V_\supset = \pi r^3\left[\tfrac{4}{3}-\tfrac{1}{27}(9-\sqrt{3})\right] = \tfrac{1}{27}\pi r^3(36-9+\sqrt{3})$$
$$= \tfrac{1}{27}\pi r^3(27+\sqrt{3}). \tag{12}$$

The rigid packing with *lowest* density known has $\eta \approx 0.0555$ (Gardner 1966). To be RIGID, each SPHERE must touch at least four others, and the four contact points cannot be in a single HEMISPHERE or all on one equator.

If spheres packed in a cubic lattice, face-centered cubic lattice, and hexagonal lattice are allowed to expand, they form cubes, hexagonal prisms, and rhombic dodecahedra. Compressing a random packing gives polyhedra with an average of 13.3 faces (Coxeter 1958, 1961).

For sphere packing inside a CUBE, see Goldberg (1971) and Schaer (1966).

see also CANNONBALL PROBLEM, CIRCLE PACKING, DODECAHEDRAL CONJECTURE, HEMISPHERE, HERMITE CONSTANTS, HYPERSPHERE, HYPERSPHERE PACKING, KEPLER CONJECTURE, KEPLER PROBLEM, KISSING NUMBER, LOCAL DENSITY, LOCAL DENSITY CONJECTURE, SPHERE

References

Conway, J. H. and Sloane, N. J. A. *Sphere Packings, Lattices, and Groups, 2nd ed.* New York: Springer-Verlag, 1993.

Coxeter, H. S. M. "Close-Packing and so Forth." *Illinois J. Math.* **2**, 746–758, 1958.

Coxeter, H. S. M. "Close Packing of Equal Spheres." Section 22.4 in *Introduction to Geometry, 2nd ed.* New York: Wiley, pp. 405–411, 1961.

Coxeter, H. S. M. "The Problem of Packing a Number of Equal Nonoverlapping Circles on a Sphere." *Trans. New York Acad. Sci.* **24**, 320–331, 1962.

Critchlow, K. *Order in Space: A Design Source Book.* New York: Viking Press, 1970.

Cundy, H. and Rollett, A. *Mathematical Models, 3rd ed.* Stradbroke, England: Tarquin Pub., pp. 195–197, 1989.

Eppstein, D. "Covering and Packing." `http://www.ics.uci.edu/~eppstein/junkyard/cover.html`.

Fejes Tóth, G. "Über einen geometrischen Satz." *Math. Z.* **46**, 78–83, 1940.

Fejes Tóth, G. *Lagerungen in der Ebene, auf der Kugel und in Raum, 2nd ed.* Berlin: Springer-Verlag, 1972.

Gardner, M. "Packing Spheres." Ch. 7 in *Martin Gardner's New Mathematical Diversions from Scientific American.* New York: Simon and Schuster, 1966.

Gauss, C. F. "Besprechung des Buchs von L. A. Seeber: Intersuchungen über die Eigenschaften der positiven ternären quadratischen Formen usw." *Göttingsche Gelehrte Anzeigen (1831, July 9)* **2**, 188–196, 1876.

Goldberg, M. "On the Densest Packing of Equal Spheres in a Cube." *Math. Mag.* **44**, 199–208, 1971.

Hales, T. C. "The Sphere Packing Problem." *J. Comput. Appl. Math* **44**, 41–76, 1992.

Jaeger, H. M. and Nagel, S. R. "Physics of Granular States." *Science* **255**, 1524, 1992.

Le Lionnais, F. *Les nombres remarquables.* Paris: Hermann, p. 31, 1983.

Lindsey, J. H. II. "Sphere Packing in $\mathbb{R}^3$." *Math.* **33**, 137–147, 1986.

Muder, D. J. "Putting the the Best Face of a Voronoi Polyhedron." *Proc. London Math. Soc.* **56**, 329–348, 1988.

Rogers, C. A. "The Packing of Equal Spheres." *Proc. London Math. Soc.* **8**, 609–620, 1958.

Rogers, C. A. *Packing and Covering.* Cambridge, England: Cambridge University Press, 1964.

Schaer, J. "On the Densest Packing of Spheres in a Cube." *Can. Math. Bul.* **9**, 265–270, 1966.

Sloane, N. J. A. "The Packing of Spheres." *Sci. Amer.* **250**, 116–125, 1984.

Stewart, I. *The Problems of Mathematics, 2nd ed.* Oxford, England: Oxford University Press, pp. 69–82, 1987.

Thompson, T. M. *From Error-Correcting Codes Through Sphere Packings to Simple Groups.* Washington, DC: Math. Assoc. Amer., 1984.

Sphere Point Picking

see FEJES TÓTH'S PROBLEM

Sphere-Sphere Intersection

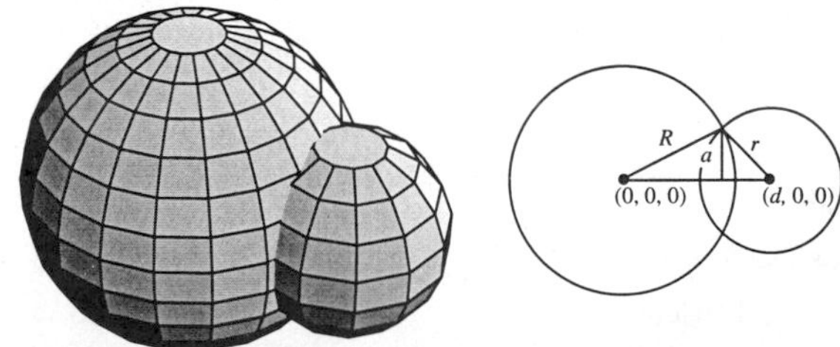

Let two spheres of RADII R and r be located along the x-AXIS centered at $(0,0,0)$ and $(d,0,0)$, respectively. Not surprisingly, the analysis is very similar to the case of the CIRCLE-CIRCLE INTERSECTION. The equations of the two SPHERES are

$$x^2 + y^2 + z^2 = R^2 \quad (1)$$
$$(x-d)^2 + y^2 + z^2 = r^2. \quad (2)$$

Combining (1) and (2) gives

$$(x-d)^2 + (R^2 - x^2) = r^2. \quad (3)$$

Multiplying through and rearranging give

$$x^2 - 2dx + d^2 - x^2 = r^2 - R^2. \quad (4)$$

Solving for x gives

$$x = \frac{d^2 - r^2 + R^2}{2d}. \quad (5)$$

The intersection of the SPHERES is therefore a curve lying in a PLANE parallel to the yz-plane at a single x-coordinate. Plugging this back into (1) gives

$$\begin{aligned} y^2 + z^2 = R^2 - x^2 &= R^2 - \left(\frac{d^2 - r^2 + R^2}{2d}\right)^2 \\ &= \frac{4d^2R^2 - (d^2 - r^2 + R^2)^2}{4d^2}, \end{aligned} \quad (6)$$

which is a CIRCLE with RADIUS

$$\begin{aligned} a &= \frac{1}{2d}\sqrt{4d^2R^2 - (d^2 - r^2 + R^2)^2} \\ &= \frac{1}{2d}[(-d + r - R)(-d - r + R) \\ &\quad \times [(-d + r + R)(d + r + R)]^{1/2}. \end{aligned} \quad (7)$$

The VOLUME of the 3-D LENS common to the two spheres can be found by adding the two SPHERICAL CAPS. The distances from the SPHERES' centers to the bases of the caps are

$$d_1 = x \quad (8)$$
$$d_2 = d - x, \quad (9)$$

so the heights of the caps are

$$h_1 = R - d_1 = \frac{(r - R + d)(r + R - d)}{2d} \quad (10)$$
$$h_2 = r - d_2 = \frac{(R - r + d)(R + r - d)}{2d}. \quad (11)$$

The VOLUME of a SPHERICAL CAP of height h' for a SPHERE of RADIUS R' is

$$V(R', h') = \tfrac{1}{3}\pi h'^2(3R' - h'). \quad (12)$$

Letting $R_1 = R$ and $R_2 = r$ and summing the two caps gives

$$\begin{aligned} V &= V(R_1, h_1) + V(R_2, h_2) \\ &= \frac{\pi(R + r - d)^2(d^2 + 2dr - 3r^2 + 2dR + 6rR - 3R^2)}{12d}. \end{aligned} \quad (13)$$

This expression gives $V = 0$ for $d = r + R$ as it must. In the special case $r = R$, the VOLUME simplifies to

$$V = \tfrac{1}{12}\pi(4R + d)(2R - d)^2. \quad (14)$$

see also APPLE, CIRCLE-CIRCLE INTERSECTION, DOUBLE BUBBLE, LENS, SPHERE

Sphere with Tunnel

Find the tunnel between two points A and B on a gravitating SPHERE which gives the shortest transit time under the force of gravity. Assume the SPHERE to be nonrotating, of RADIUS a, and with uniform density ρ. Then the standard form EULER-LAGRANGE DIFFERENTIAL EQUATION in polar coordinates is

$$r_{\phi\phi}(r^3 - ra^2) + r_\phi{}^2(2a^2 - r^2) + a^2r^2 = 0, \quad (1)$$

along with the boundary conditions $r(\phi = 0) = r_0$, $r_\phi(\phi = 0) = 0$, $r(\phi = \phi_A) = a$, and $r(\phi = \phi_B) = a$. Integrating once gives

$$r_\phi{}^2 = \frac{a^2r^2}{r_0{}^2}\frac{r^2 - r_0{}^2}{a^2 - r^2}. \quad (2)$$

But this is the equation of a HYPOCYCLOID generated by a CIRCLE of RADIUS $\frac{1}{2}(a - r_0)$ rolling inside the CIRCLE of RADIUS a, so the tunnel is shaped like an arc of a HYPOCYCLOID. The transit time from point A to point B is

$$T = \pi\sqrt{\frac{a^2 - r_0{}^2}{ag}}, \quad (3)$$

where

$$g = \frac{GM}{a^2} = \tfrac{4}{3}\pi\rho Ga \quad (4)$$

is the surface gravity with G the universal gravitational constant.

Spherical Bessel Differential Equation

Take the HELMHOLTZ DIFFERENTIAL EQUATION

$$\nabla^2 F + k^2 F = 0 \quad (1)$$

in SPHERICAL COORDINATES. This is just LAPLACE'S EQUATION in SPHERICAL COORDINATES with an additional term,

$$\begin{aligned} &\frac{d^2R}{dr^2}\Phi\Theta + \frac{2}{r}\frac{dR}{dr} + \frac{1}{r^2\sin^2\phi}\frac{d^2\Theta}{d\theta^2}\Phi R \\ &\quad + \frac{\cos\phi}{r^2\sin\phi}\frac{d\Phi}{d\phi}\Theta R + \frac{1}{r^2}\frac{d^2\Phi}{d\phi^2} + k^2R\Phi\Theta = 0. \end{aligned} \quad (2)$$

Multiply through by $r^2/R\Phi\Theta$,

$$\frac{r^2}{R}\frac{d^2R}{dr^2}+\frac{2r}{R}\frac{dR}{dr}+k^2r^2+\frac{1}{\Theta\sin^2\phi}\frac{d^2\Theta}{d\theta^2}+\frac{\cos\phi}{\Phi\sin\phi}\frac{d\Phi}{d\phi}+\frac{1}{\Phi}\frac{d^2\Phi}{d\phi^2}=0. \quad (3)$$

This equation is separable in R. Call the separation constant $n(n+1)$,

$$\frac{r^2}{R}\frac{d^2R}{dr^2}+\frac{2r}{R}\frac{dR}{dr}+k^2r^2=n(n+1). \quad (4)$$

Now multiply through by R,

$$r^2\frac{d^2R}{dr^2}+2r\frac{dR}{dr}+[k^2r^2-n(n+1)]R=0. \quad (5)$$

This is the SPHERICAL BESSEL DIFFERENTIAL EQUATION. It can be transformed by letting $x\equiv kr$, then

$$r\frac{dR(r)}{dr}=kr\frac{dR(r)}{k\,dr}=kr\frac{dR(r)}{d(kr)}=x\frac{dR(r)}{dx}. \quad (6)$$

Similarly,

$$r^2\frac{d^2R(r)}{dr^2}=x^2\frac{d^2R(r)}{dx^2}, \quad (7)$$

so the equation becomes

$$x^2\frac{d^2R}{dx^2}+2x\frac{dR}{dx}+[x^2-n(n+1)]R=0. \quad (8)$$

Now look for a solution of the form $R(r)=Z(x)x^{-1/2}$, denoting a derivative with respect to x by a prime,

$$R'=Z'x^{-1/2}-\tfrac{1}{2}Zx^{-3/2} \quad (9)$$

$$\begin{aligned}R''&=Z''x^{-1/2}-\tfrac{1}{2}Z'x^{-3/2}-\tfrac{1}{2}Z'x^{-3/2}\\&\quad-\tfrac{1}{2}(-\tfrac{3}{2})Zx^{-5/2}\\&=Z''x^{-1/2}-Z'x^{-3/2}+\tfrac{3}{4}Zx^{-5/2},\end{aligned} \quad (10)$$

so

$$\begin{aligned}&x^2(Z''x^{-1/2}-Z'x^{-3/2}+\tfrac{3}{4}Zx^{-5/2})\\&+2x(Z'x^{-1/2}-\tfrac{1}{2}Zx^{-3/2})+[x^2-n(n+1)]Zx^{-1/2}=0\end{aligned} \quad (11)$$

$$x^2(Z''-Z'x^{-1}+\tfrac{3}{4}Zx^{-2})+2x(Z'-\tfrac{1}{2}Zx^{-1})+[x^2-n(n+1)]Z=0 \quad (12)$$

$$x^2Z''+(-x+2x)Z'+[\tfrac{3}{4}-1+x^2-n(n+1)]Z=0 \quad (13)$$

$$\begin{aligned}&x^2Z''+xZ'+[x^2-(n^2+n+\tfrac{1}{4})]Z=0\\&x^2Z''+xZ'+[x^2-(n+\tfrac{1}{2})^2]Z=0.\end{aligned} \quad (14)$$

But the solutions to this equation are BESSEL FUNCTIONS of half integral order, so the normalized solutions to the original equation are

$$R(r)\equiv A\frac{J_{n+1/2}(kr)}{\sqrt{kr}}+B\frac{Y_{n+1/2}(kr)}{\sqrt{kr}} \quad (15)$$

which are known as SPHERICAL BESSEL FUNCTIONS. The two types of solutions are denoted $j_n(x)$ (SPHERICAL BESSEL FUNCTION OF THE FIRST KIND) or $n_n(x)$ (SPHERICAL BESSEL FUNCTION OF THE SECOND KIND), and the general solution is written

$$R(r)=A'j_n(kr)+B'n_n(kr), \quad (16)$$

where

$$j_n(z)\equiv\sqrt{\frac{\pi}{2}}\frac{J_{n+1/2}(z)}{\sqrt{z}} \quad (17)$$

$$n_n(z)\equiv\sqrt{\frac{\pi}{2}}\frac{Y_{n+1/2}(z)}{\sqrt{z}}. \quad (18)$$

see also SPHERICAL BESSEL FUNCTION, SPHERICAL BESSEL FUNCTION OF THE FIRST KIND, SPHERICAL BESSEL FUNCTION OF THE SECOND KIND

References

Abramowitz, M. and Stegun, C. A. (Eds.). *Handbook of Mathematical Functions with Formulas, Graphs, and Mathematical Tables, 9th printing.* New York: Dover, p. 437, 1972.

Spherical Bessel Function

A solution to the SPHERICAL BESSEL DIFFERENTIAL EQUATION. The two types of solutions are denoted $j_n(x)$ (SPHERICAL BESSEL FUNCTION OF THE FIRST KIND) or $n_n(x)$ (SPHERICAL BESSEL FUNCTION OF THE SECOND KIND).

see also SPHERICAL BESSEL FUNCTION OF THE FIRST KIND, SPHERICAL BESSEL FUNCTION OF THE SECOND KIND

References

Abramowitz, M. and Stegun, C. A. (Eds.). "Spherical Bessel Functions." §10.1 in *Handbook of Mathematical Functions with Formulas, Graphs, and Mathematical Tables, 9th printing.* New York: Dover, pp. 437–442, 1972.

Arfken, G. "Spherical Bessel Functions." §11.7 in *Mathematical Methods for Physicists, 3rd ed.* Orlando, FL: Academic Press, pp. 622–636, 1985.

Press, W. H.; Flannery, B. P.; Teukolsky, S. A.; and Vetterling, W. T. "Bessel Functions of Fractional Order, Airy Functions, Spherical Bessel Functions." §6.7 in *Numerical Recipes in FORTRAN: The Art of Scientific Computing, 2nd ed.* Cambridge, England: Cambridge University Press, pp. 234–245, 1992.

Spherical Bessel Function of the First Kind

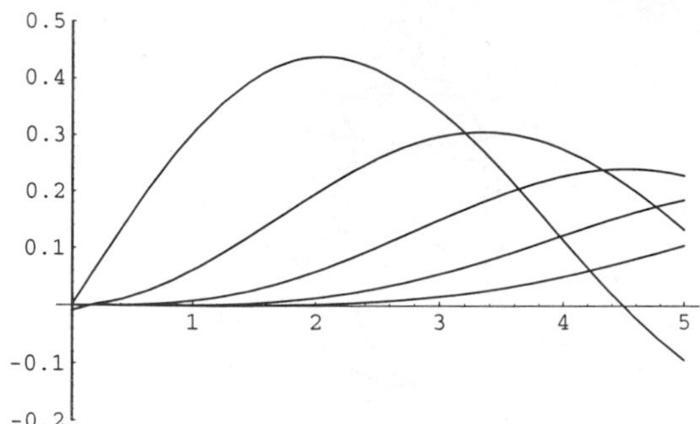

$$\begin{aligned} j_n(x) &\equiv \sqrt{\frac{\pi}{2x}} J_{n+1/2}(x) \\ &= 2^n x^n \sum_{s=0}^{\infty} \frac{(-1)^s (s-n)!}{s!(2s+2n+1)!} x^{2s} \\ &= \frac{x^n}{(2n+1)!!} \left[1 - \frac{\frac{1}{2}x^2}{1!(2n+3)} \right. \\ &\quad \left. + \frac{\left(\frac{1}{2}x^2\right)^2}{2!(2n+3)(2n+5)} + \ldots \right] \\ &= (-1)^n x^n \left(\frac{d}{x\,dx} \right)^n \frac{\sin x}{x}. \end{aligned}$$

The first few functions are

$$\begin{aligned} j_0(x) &= \frac{\sin x}{x} \\ j_1(x) &= \frac{\sin x}{x^2} - \frac{\cos x}{x} \\ j_2(x) &= \left(\frac{3}{x^3} - \frac{1}{x} \right) \sin x - \frac{3}{x^2} \cos x. \end{aligned}$$

see also POISSON INTEGRAL REPRESENTATION, RAYLEIGH'S FORMULAS

References

Abramowitz, M. and Stegun, C. A. (Eds.). "Spherical Bessel Functions." §10.1 in *Handbook of Mathematical Functions with Formulas, Graphs, and Mathematical Tables, 9th printing.* New York: Dover, pp. 437–442, 1972.

Arfken, G. "Spherical Bessel Functions." §11.7 in *Mathematical Methods for Physicists, 3rd ed.* Orlando, FL: Academic Press, pp. 622–636, 1985.

Spherical Bessel Function of the Second Kind

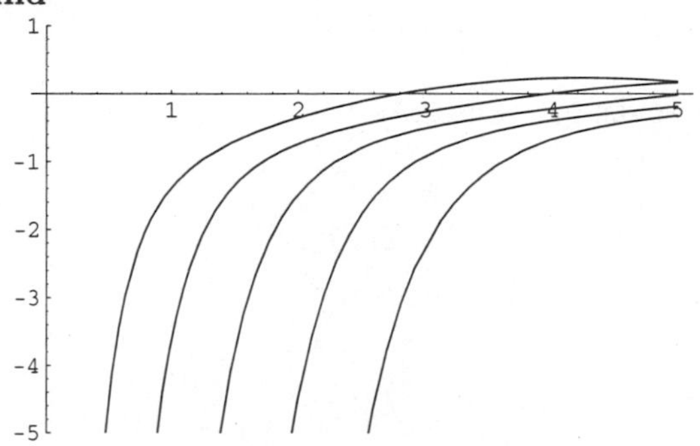

$$\begin{aligned} n_n(x) &\equiv \sqrt{\frac{\pi}{2x}} Y_{n+1/2}(x) \\ &= \frac{(-1)^{n+1}}{2^n x^{n+1}} \sum_{n=0}^{\infty} \frac{(-1)^s (s-n)!}{s!(2s-2n)!} x^{2s} \\ &= -\frac{(2n-1)!!}{x^{n+1}} \left[1 - \frac{\frac{1}{2}x^2}{1!(1-2n)} \right. \\ &\quad \left. + \frac{\left(\frac{1}{2}x^2\right)^2}{2!(1-2n)(3-2n)} + \ldots \right] \\ &= (-1)^{n+1} \sqrt{\frac{\pi}{2x}} J_{-n-1/2}(x). \end{aligned}$$

The first few functions are

$$\begin{aligned} n_0(x) &= -\frac{\cos x}{x} \\ n_1(x) &= -\frac{\cos x}{x^2} - \frac{\sin x}{x} \\ n_2(x) &= -\left(\frac{3}{x^3} - \frac{1}{x} \right) \cos x - \frac{3}{x^2} \sin x. \end{aligned}$$

see also RAYLEIGH'S FORMULAS

References

Abramowitz, M. and Stegun, C. A. (Eds.). "Spherical Bessel Functions." §10.1 in *Handbook of Mathematical Functions with Formulas, Graphs, and Mathematical Tables, 9th printing.* New York: Dover, pp. 437–442, 1972.

Arfken, G. "Spherical Bessel Functions." §11.7 in *Mathematical Methods for Physicists, 3rd ed.* Orlando, FL: Academic Press, pp. 622–636, 1985.

Spherical Bessel Function of the Third Kind

see SPHERICAL HANKEL FUNCTION OF THE FIRST KIND, SPHERICAL HANKEL FUNCTION OF THE SECOND KIND

Spherical Cap

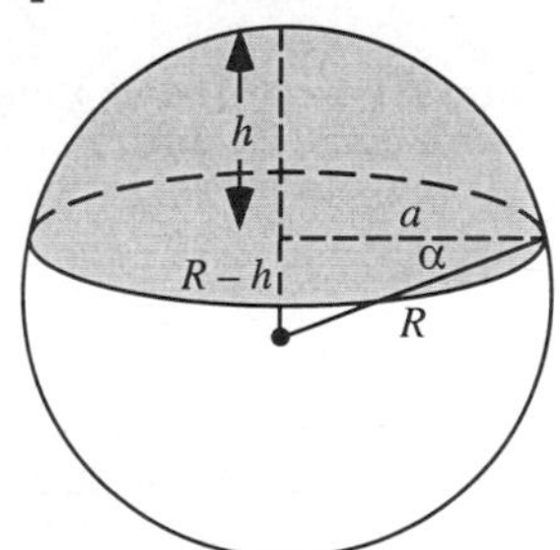

A spherical cap is the region of a SPHERE which lies above (or below) a given PLANE. If the PLANE passes through the CENTER of the SPHERE, the cap is a HEMISPHERE. Let the SPHERE have RADIUS R, then the VOLUME of a spherical cap of height h and base RADIUS a is

given by the equation of a SPHERICAL SEGMENT (which is a spherical cut by a second PLANE)

$$V_{\text{spherical segment}} = \tfrac{1}{6}\pi h(3a^2 + 3b^2 + h^2) \quad (1)$$

with $b = 0$, giving

$$V_{\text{cap}} = \tfrac{1}{6}\pi h(3a^2 + h^2). \quad (2)$$

Using the PYTHAGOREAN THEOREM gives

$$(R-h)^2 + a^2 = R^2, \quad (3)$$

which can be solved for a^2 as

$$a^2 = 2Rh - h^2, \quad (4)$$

and plugging this in gives the equivalent formula

$$V_{\text{cap}} = \tfrac{1}{3}\pi h^2(3R - h). \quad (5)$$

In terms of the so-called CONTACT ANGLE (the angle between the normal to the sphere at the bottom of the cap and the base plane)

$$R - h = R\sin\theta \quad (6)$$

$$\alpha \equiv \sin^{-1}\left(\frac{R-h}{R}\right), \quad (7)$$

so

$$V_{\text{cap}} = \tfrac{1}{3}\pi R^3(2 - 3\sin\alpha + \sin^3\alpha). \quad (8)$$

Consider a cylindrical box enclosing the cap so that the top of the box is tangent to the top of the SPHERE. Then the enclosing box has VOLUME

$$\begin{aligned} V_{\text{box}} &= \pi a^2 h = \pi(R\cos\alpha)[R(1-\sin\alpha)] \\ &= \pi R^3(1 - \sin\alpha - \sin^2\alpha + \sin^3\alpha), \end{aligned} \quad (9)$$

so the hollow volume between the cap and box is given by

$$V_{\text{box}} - V_{\text{cap}} = \tfrac{1}{3}\pi R^3(1 - 3\sin^2\alpha + 2\sin^3\alpha). \quad (10)$$

If a second PLANE cuts the cap, the resulting SPHERICAL FRUSTUM is called a SPHERICAL SEGMENT. The SURFACE AREA of the spherical cap is given by the same equation as for a general ZONE:

$$S_{\text{cap}} = 2\pi Rh. \quad (11)$$

see also CONTACT ANGLE, DOME, FRUSTUM, HEMISPHERE, SOLID OF REVOLUTION, SPHERE, SPHERICAL SEGMENT, TORISPHERICAL DOME, ZONE

Spherical Coordinates

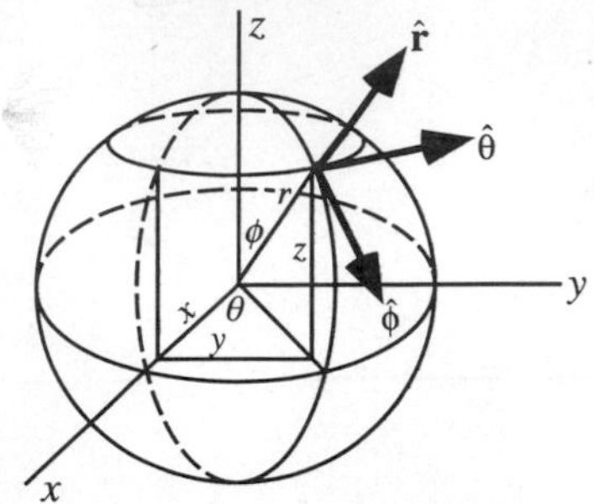

A system of CURVILINEAR COORDINATES which is natural for describing positions on a SPHERE or SPHEROID. Define θ to be the azimuthal ANGLE in the xy-PLANE from the x-AXIS with $0 \leq \theta < 2\pi$ (denoted λ when referred to as the LONGITUDE), ϕ to be the polar ANGLE from the z-AXIS with $0 \leq \phi \leq \pi$ (COLATITUDE, equal to $\phi = 90° - \delta$ where δ is the LATITUDE), and r to be distance (RADIUS) from a point to the ORIGIN.

Unfortunately, the convention in which the symbols θ and ϕ are reversed is frequently used, especially in physics, leading to unnecessary confusion. The symbol ρ is sometimes also used in place of r. Arfken (1985) uses (r, ϕ, θ), whereas Beyer (1987) uses (ρ, θ, ϕ). Be very careful when consulting the literature.

In this work, the symbols for the azimuthal, polar, and radial coordinates are taken as θ, ϕ, and r, respectively. Note that this definition provides a logical extension of the usual POLAR COORDINATES notation, with θ remaining the ANGLE in the xy-PLANE and ϕ becoming the ANGLE out of the PLANE.

$$r = \sqrt{x^2 + y^2 + z^2} \quad (1)$$

$$\theta = \tan^{-1}\left(\frac{y}{x}\right) \quad (2)$$

$$\phi = \sin^{-1}\left(\frac{\sqrt{x^2+y^2}}{r}\right) = \cos^{-1}\left(\frac{z}{r}\right), \quad (3)$$

where $r \in [0, \infty)$, $\theta \in [0, 2\pi)$, and $\phi \in [0, \pi]$. In terms of CARTESIAN COORDINATES,

$$x = r\cos\theta\sin\phi \quad (4)$$

$$y = r\sin\theta\sin\phi \quad (5)$$

$$z = r\cos\phi. \quad (6)$$

The SCALE FACTORS are

$$h_r = 1 \quad (7)$$

$$h_\theta = r\sin\phi \quad (8)$$

$$h_\phi = r, \quad (9)$$

so the METRIC COEFFICIENTS are

$$g_{rr} = 1 \tag{10}$$

$$g_{\theta\theta} = r^2 \sin^2\phi \tag{11}$$

$$g_{\phi\phi} = r^2. \tag{12}$$

The LINE ELEMENT is

$$d\mathbf{s} = dr\,\hat{\mathbf{r}} + r\,d\phi\,\hat{\boldsymbol{\phi}} + r\sin\phi\,d\theta\,\hat{\boldsymbol{\theta}}, \tag{13}$$

the AREA element

$$d\mathbf{a} = r^2 \sin\phi\,d\theta\,d\phi\,\hat{\mathbf{r}}, \tag{14}$$

and the VOLUME ELEMENT

$$dV = r^2 \sin\phi\,d\theta\,d\phi\,dr. \tag{15}$$

The JACOBIAN is

$$\left|\frac{\partial(x,y,z)}{\partial(r,\theta,\phi)}\right| = r^2|\sin\phi|. \tag{16}$$

The POSITION VECTOR is

$$\mathbf{r} \equiv \begin{bmatrix} r\cos\theta\sin\phi \\ r\sin\theta\sin\phi \\ r\cos\phi \end{bmatrix}, \tag{17}$$

so the UNIT VECTORS are

$$\hat{\mathbf{r}} \equiv \frac{\frac{d\mathbf{r}}{dr}}{\left|\frac{d\mathbf{r}}{dr}\right|} = \begin{bmatrix} \cos\theta\sin\phi \\ \sin\theta\sin\phi \\ \cos\phi \end{bmatrix} \tag{18}$$

$$\hat{\boldsymbol{\theta}} \equiv \frac{\frac{d\mathbf{r}}{d\theta}}{\left|\frac{d\mathbf{r}}{d\theta}\right|} = \begin{bmatrix} -\sin\theta \\ \cos\theta \\ 0 \end{bmatrix} \tag{19}$$

$$\hat{\boldsymbol{\phi}} \equiv \frac{\frac{d\mathbf{r}}{d\phi}}{\left|\frac{d\mathbf{r}}{d\phi}\right|} = \begin{bmatrix} \cos\theta\cos\phi \\ \sin\theta\cos\phi \\ -\sin\phi \end{bmatrix}. \tag{20}$$

Derivatives of the UNIT VECTORS are

$$\frac{\partial\hat{\mathbf{r}}}{\partial r} = \mathbf{0} \tag{21}$$

$$\frac{\partial\hat{\boldsymbol{\theta}}}{\partial r} = \mathbf{0} \tag{22}$$

$$\frac{\partial\hat{\boldsymbol{\phi}}}{\partial r} = \mathbf{0} \tag{23}$$

$$\frac{\partial\hat{\mathbf{r}}}{\partial\theta} = \begin{bmatrix} -\sin\theta\sin\phi \\ \cos\theta\sin\phi \\ 0 \end{bmatrix} = \sin\phi\,\hat{\boldsymbol{\theta}} \tag{24}$$

$$\frac{\partial\hat{\boldsymbol{\theta}}}{\partial\theta} = \begin{bmatrix} -\cos\theta \\ -\sin\theta \\ 0 \end{bmatrix} = -\cos\phi\,\hat{\boldsymbol{\phi}} - \sin\phi\,\hat{\mathbf{r}} \tag{25}$$

$$\frac{\partial\hat{\boldsymbol{\phi}}}{\partial\theta} = \begin{bmatrix} -\sin\theta\cos\phi \\ \cos\theta\cos\phi \\ 0 \end{bmatrix} = \cos\phi\,\hat{\boldsymbol{\theta}} \tag{26}$$

$$\frac{\partial\hat{\mathbf{r}}}{\partial\phi} = \begin{bmatrix} \cos\theta \\ \sin\theta\cos\phi \\ -\sin\phi \end{bmatrix} = \hat{\boldsymbol{\phi}} \tag{27}$$

$$\frac{\partial\hat{\boldsymbol{\theta}}}{\partial\phi} = \mathbf{0} \tag{28}$$

$$\frac{\partial\hat{\boldsymbol{\phi}}}{\partial\phi} = \begin{bmatrix} -\cos\theta\sin\phi \\ -\sin\theta\sin\phi \\ -\cos\phi \end{bmatrix} = -\hat{\mathbf{r}}. \tag{29}$$

The GRADIENT is

$$\nabla = \hat{\mathbf{r}}\frac{\partial}{\partial r} + \frac{1}{r}\hat{\boldsymbol{\phi}}\frac{\partial}{\partial\phi} + \frac{1}{r\sin\phi}\hat{\boldsymbol{\theta}}\frac{\partial}{\partial\theta}, \tag{30}$$

so

$$\nabla_r\hat{\mathbf{r}} = \mathbf{0} \tag{31}$$

$$\nabla_r\hat{\boldsymbol{\theta}} = \mathbf{0} \tag{32}$$

$$\nabla_r\hat{\boldsymbol{\phi}} = \hat{\mathbf{0}} \tag{33}$$

$$\nabla_\theta\hat{\mathbf{r}} = \frac{\sin\phi\,\hat{\boldsymbol{\theta}}}{r\sin\phi} = \frac{1}{r}\hat{\boldsymbol{\theta}} \tag{34}$$

$$\nabla_\theta\hat{\boldsymbol{\theta}} = -\frac{\cos\phi\,\hat{\boldsymbol{\phi}} + \sin\phi\,\hat{\mathbf{r}}}{r\sin\phi} = -\frac{\cot\phi}{r}\hat{\boldsymbol{\phi}} - \frac{1}{r}\hat{\mathbf{r}} \tag{35}$$

$$\nabla_\theta\hat{\boldsymbol{\phi}} = \frac{\cos\phi\,\hat{\boldsymbol{\phi}}}{r\sin\phi} = \frac{1}{r}\cot\phi\,\hat{\boldsymbol{\theta}}. \tag{36}$$

Now, since the CONNECTION COEFFICIENTS are given by $\Gamma^i_{jk} = \hat{\mathbf{x}}_i \cdot (\nabla_k \hat{\mathbf{x}}_j)$,

$$\Gamma^\theta = \begin{bmatrix} 0 & \frac{1}{r} & 0 \\ 0 & 0 & 0 \\ 0 & \frac{\cot\phi}{r} & 0 \end{bmatrix} \tag{37}$$

$$\Gamma^\phi = \begin{bmatrix} 0 & 0 & \frac{1}{r} \\ 0 & -\frac{\cot\phi}{r} & 0 \\ 0 & 0 & 0 \end{bmatrix} \tag{38}$$

$$\Gamma^r = \begin{bmatrix} 0 & 0 & 0 \\ 0 & -\frac{1}{r} & 0 \\ 0 & 0 & -\frac{1}{r} \end{bmatrix}. \tag{39}$$

The DIVERGENCE is

$$\begin{aligned} \nabla\cdot\mathbf{F} &= A^k{}_{,k} + \Gamma^k_{jk}A^j \\ &= [A^r{}_{,r} + (\Gamma^r_{rr}A^r + \Gamma^r_{\theta r}A^\theta + \Gamma^r_{\phi r}A^\phi] \\ &\quad + [A^\theta{}_{,\theta} + (\Gamma^\theta_{r\theta}A^r + \Gamma^\theta_{\theta\theta}A^\theta + \Gamma^\theta_{\phi\theta}A^\phi)] \\ &\quad + [A^\phi{}_{,\phi} + (\Gamma^\phi_{r\phi}A^r + \Gamma^\phi_{\theta\phi}A^\theta + \Gamma^\phi_{\phi\phi}A^\phi)] \\ &= \frac{1}{g_r}\frac{\partial A^r}{\partial r} + \frac{1}{g_\theta}\frac{\partial A^\theta}{\partial\theta} + \frac{1}{g_\phi}\frac{\partial A^\phi}{\partial\phi} + (0+0+0) \\ &\quad + \left(\frac{1}{r}A^r + 0 + \frac{\cot\phi}{r}A^\phi\right) + \left(\frac{1}{r}A^r + 0 + 0\right) \\ &= \frac{\partial}{\partial r}A^r + \frac{2}{r}A^r + \frac{1}{r\sin\phi}\frac{\partial}{\partial\theta}A^\theta + \frac{1}{r}\frac{\partial}{\partial\phi}A^\phi + \frac{\cot\phi}{r}A^\phi, \end{aligned} \tag{40}$$

or, in VECTOR notation,

$$\begin{aligned} \nabla\cdot\mathbf{F} &= \left(\frac{2}{r} + \frac{\partial}{\partial r}\right)F_r + \left(\frac{1}{r}\frac{\partial}{\partial\phi} + \frac{\cot\phi}{r}\right)F_\phi + \frac{1}{\sin\phi}\frac{\partial F_\theta}{\partial\theta} \\ &= \frac{1}{r^2}\frac{\partial}{\partial r}(r^2F_r) + \frac{1}{r\sin\phi}\frac{\partial}{\partial\phi}(\sin\phi F_\phi) + \frac{1}{r\sin\phi}\frac{\partial F_\theta}{\partial\theta}. \end{aligned} \tag{41}$$

The COVARIANT DERIVATIVES are given by

$$A_{j;k} = \frac{1}{g_{kk}}\frac{\partial A_j}{\partial x_k} - \Gamma^i_{jk}A_i, \tag{42}$$

Spherical Coordinates

so

$$A_{r;r} = \frac{\partial A_r}{\partial r} - \Gamma^i_{rr} A_i = \frac{\partial A_r}{\partial r} \tag{43}$$

$$A_{r;\theta} = \frac{1}{r\sin\phi}\frac{\partial A_r}{\partial\theta} - \Gamma^i_{r\theta} = \frac{1}{r\sin\phi}\frac{\partial A_r}{\partial\theta} - \Gamma_{r\theta}A_\theta$$
$$= \frac{1}{r\sin\phi}\frac{\partial A_r}{\partial\phi} - \frac{A_\theta}{r} \tag{44}$$

$$A_{r;\phi} = \frac{1}{r}\frac{\partial A_r}{\partial\phi} - \Gamma^i_{r\phi}A_i = \frac{1}{r}\frac{\partial A_r}{\partial\phi} - \Gamma^\phi_{r\phi}A_\phi$$
$$= \frac{1}{r}\left(\frac{\partial A_r}{\partial\phi} - A_\phi\right) \tag{45}$$

$$A_{\theta;r} = \frac{\partial A_\theta}{\partial r} - \Gamma^i_{\theta r}A_i = \frac{\partial A_\theta}{\partial r} \tag{46}$$

$$A_{\theta;\theta} = \frac{1}{r\sin\phi}\frac{\partial A_\phi}{\partial\theta} - \Gamma^i_{\theta\theta}A_i$$
$$= \frac{1}{r\sin\phi}\partial A_\theta \partial\theta - \Gamma^\phi_{\theta\theta}A_\phi - \Gamma^r_{\theta\theta}A_r$$
$$= \frac{1}{r\sin\phi}\frac{\partial A_\theta}{\partial\theta} + \frac{\cot\phi}{r}A_\phi + \frac{A_r}{r} \tag{47}$$

$$A_{\theta;\phi} = \frac{1}{r}\frac{\partial A_\theta}{\partial r} - \Gamma^i_{\phi r}A_i\frac{\partial A_\theta}{\partial\phi} \tag{48}$$

$$A_{\phi;r} = \frac{\partial A_\phi}{\partial r} - \Gamma^i_{\phi r}A_i = \frac{\partial A_\phi}{r} \tag{49}$$

$$A_{\phi;\theta} = \frac{1}{r\sin\phi}\frac{\partial A_\phi}{\partial\theta} - \Gamma^i_{\phi\theta}A_i = \frac{1}{r\sin\phi}\frac{\partial A_\phi}{\partial\theta} - \Gamma^\theta_{\phi\theta}$$
$$= \frac{1}{r\sin\phi}\frac{\partial A_\phi}{\partial\theta} - \frac{\cot\phi}{r}A_\theta \tag{50}$$

$$A_{\phi;\phi} = \frac{1}{r}\frac{\partial A_\phi}{\partial\phi} - \Gamma^i_{\phi\phi}A_i = \frac{1}{r}\frac{\partial A_\phi}{\partial\phi} - \Gamma^r_{\phi\phi}A_r$$
$$= \frac{1}{r}\frac{\partial A_\phi}{\partial\phi} + \frac{A_r}{r}. \tag{51}$$

The COMMUTATION COEFFICIENTS are given by

$$c^\mu_{\alpha\beta}\vec{e}_\mu = [\vec{e}_\alpha, \vec{e}_\beta] = \nabla_\alpha\vec{e}_\beta - \nabla_\beta\vec{e}_\alpha \tag{52}$$

$$[\hat{\mathbf{r}}, \hat{\mathbf{r}}] = [\hat{\boldsymbol{\theta}}, \hat{\boldsymbol{\theta}}] = [\hat{\boldsymbol{\phi}}, \hat{\boldsymbol{\phi}}] = \mathbf{0}, \tag{53}$$

so $c^\alpha_{rr} = c^\alpha_{\theta\theta} = c^\alpha_{\phi\phi} = 0$, where $\alpha = r, \theta, \phi$.

$$[\hat{\mathbf{r}}, \hat{\boldsymbol{\theta}}] = -[\hat{\boldsymbol{\theta}}, \hat{\mathbf{r}}] = \nabla_r\hat{\boldsymbol{\theta}} - \nabla_\theta\hat{\mathbf{r}} = \mathbf{0} - \frac{1}{r}\hat{\boldsymbol{\theta}} = -\frac{1}{r}\hat{\boldsymbol{\theta}}, \tag{54}$$

so $c^\theta_{r\theta} = -c^\theta_{\theta r} = -\frac{1}{r}$, $c^r_{r\theta} = c^\phi_{r\theta} = 0$.

$$[\hat{\mathbf{r}}, \hat{\boldsymbol{\phi}}] = -[\hat{\boldsymbol{\phi}}, \hat{\mathbf{r}}] = \mathbf{0} - \frac{1}{r}\hat{\boldsymbol{\phi}} = -\frac{1}{r}\hat{\boldsymbol{\phi}}, \tag{55}$$

so $c^\phi_{r\phi} = -c^\phi_{\phi r} = \frac{1}{r}$.

$$[\hat{\boldsymbol{\theta}}, \hat{\boldsymbol{\phi}}] = -[\hat{\boldsymbol{\phi}}, \hat{\boldsymbol{\theta}}] = \frac{1}{r}\cot\phi\hat{\boldsymbol{\theta}} - \mathbf{0} = \frac{1}{r}\cot\phi\hat{\boldsymbol{\theta}}, \tag{56}$$

so

$$c^\theta_{\theta\phi} = -c^\theta_{\phi\theta} = \frac{1}{r}\cot\phi. \tag{57}$$

Summarizing,

$$c^r = \begin{bmatrix} 0 & 0 & 0 \\ 0 & 0 & 0 \\ 0 & 0 & 0 \end{bmatrix} \tag{58}$$

$$c^\theta = \begin{bmatrix} 0 & -\frac{1}{r} & 0 \\ \frac{1}{r} & 0 & \frac{1}{r}\cot\phi \\ 0 & -\frac{1}{r}\cot\phi & 0 \end{bmatrix} \tag{59}$$

$$c^\phi = \begin{bmatrix} 0 & 0 & -\frac{1}{r} \\ 0 & 0 & 0 \\ \frac{1}{r} & 0 & 0 \end{bmatrix}. \tag{60}$$

Time derivatives of the POSITION VECTOR are

$$\dot{\mathbf{r}} = \begin{bmatrix} \cos\theta\sin\phi\,\dot{r} - r\sin\theta\sin\phi\,\dot{\theta} + r\cos\theta\cos\phi\,\dot{\phi} \\ \sin\theta\sin\phi\,\dot{r} + r\cos\theta\sin\phi\,\dot{\theta} + r\sin\theta\cos\phi\,\dot{\phi} \\ \cos\phi\,\dot{r} - r\sin\phi\,\dot{\phi} \end{bmatrix}$$
$$= \begin{bmatrix} \cos\theta\sin\phi \\ \sin\theta\sin\phi \\ \cos\phi \end{bmatrix}\dot{r} + r\sin\phi\begin{bmatrix} -\sin\theta \\ \cos\theta \\ 0 \end{bmatrix}\dot{\theta}$$
$$+ r\begin{bmatrix} \cos\theta\cos\phi \\ \sin\theta\cos\phi \\ -\sin\phi \end{bmatrix}\dot{\phi}$$
$$= \dot{r}\,\hat{\mathbf{r}} + r\sin\phi\,\dot{\theta}\,\hat{\boldsymbol{\theta}} + r\,\dot{\phi}\,\hat{\boldsymbol{\phi}}. \tag{61}$$

The SPEED is therefore given by

$$v \equiv |\dot{\mathbf{r}}| = \sqrt{\dot{r}^2 + r^2\sin^2\phi\dot{\theta}^2 + r^2\dot{\phi}^2}. \tag{62}$$

The ACCELERATION is

$$\begin{aligned}\ddot{x} &= (-\sin\theta\sin\phi\dot{\theta}\dot{r} + \cos\theta\cos\phi\dot{r}\dot{\phi} + \cos\theta\sin\phi\ddot{r}) \\ &\quad - (\sin\theta\sin\phi\dot{r}\dot{\theta} + r\cos\theta\sin\phi\dot{\theta}^2 + r\sin\theta\cos\phi\dot{\theta}\dot{\phi} \\ &\quad + r\sin\theta\sin\phi\ddot{\theta}) + (\cos\theta\cos\phi\dot{r}\dot{\phi} - r\sin\theta\cos\dot{\theta}\dot{\phi} \\ &\quad - r\cos\theta\sin\phi\dot{\phi}^2 + r\cos\theta\cos\phi\ddot{\phi}) \\ &= -2\sin\theta\sin\phi\dot{\theta}\dot{r} + 2\cos\theta\cos\phi\dot{r}\dot{\phi} - 2r\sin\theta\cos\phi\dot{\theta}\dot{\phi} \\ &\quad + \cos\theta\sin\phi\ddot{r} - r\sin\theta\sin\phi\ddot{\theta} + r\cos\theta\cos\phi\ddot{\phi} \\ &\quad - r\cos\theta\sin\phi(\dot{\theta}^2 + \dot{\phi}^2)\end{aligned} \tag{63}$$

$$\begin{aligned}\ddot{y} &= (\sin\theta\sin\phi\ddot{r} + r\cos\theta\sin\phi\dot{\theta} + r\cos\phi\sin\theta\dot{\phi}) \\ &\quad + (\cos\theta\sin\phi\dot{r}\dot{\theta} - r\sin\theta\sin\phi\dot{\theta}^2 + r\cos\theta\cos\phi\dot{\theta}\dot{\phi} \\ &\quad + r\cos\theta\sin\phi\ddot{\theta}) + (\sin\theta\cos\phi\dot{r}\dot{\phi} + r\cos\theta\cos\phi\dot{\theta}\dot{\phi} \\ &\quad - r\sin\theta\sin\phi\dot{\phi}^2 + r\sin\theta\cos\phi\ddot{\phi}) \\ &= 2\cos\theta\sin\phi\dot{\theta}\dot{r} + 2\sin\theta\cos\phi\dot{r}\dot{\phi} + 2r\cos\theta\cos\phi\dot{\theta}\dot{\phi} \\ &\quad + \sin\theta\sin\phi\ddot{r} + r\cos\theta\sin\phi\ddot{\theta} + r\sin\theta\cos\phi\ddot{\phi} \\ &\quad - r\sin\theta\sin\phi(\dot{\theta}^2 + \dot{\phi}^2)\end{aligned} \tag{64}$$

$$\begin{aligned}\ddot{z} &= (\cos\phi\ddot{r} - \sin\phi\dot{r}\dot{\phi}) - (\dot{r}\sin\phi\dot{\phi} + r\cos\phi\dot{\phi}^2 + r\sin\phi\ddot{\phi}) \\ &= -r\cos\phi\dot{\phi}^2 + \cos\phi\ddot{r} - 2\sin\phi\dot{\phi}\dot{r} - r\sin\phi\ddot{\phi}.\end{aligned} \tag{65}$$

Plugging these in gives

$$\ddot{\mathbf{r}} = (\ddot{r} - r\dot{\phi}^2)\begin{bmatrix}\cos\theta\sin\phi\\ \sin\theta\sin\phi\\ \cos\phi\end{bmatrix} + (2r\cos\phi\dot{\theta}\dot{\phi} + r\sin\phi\ddot{\theta})\begin{bmatrix}-\sin\theta\\ \cos\theta\\ 0\end{bmatrix} + (2\dot{r}\dot{\phi} + r\ddot{\phi})\begin{bmatrix}\cos\theta\cos\phi\\ \sin\theta\cos\phi\\ -\sin\phi\end{bmatrix} - r\sin\phi\dot{\theta}^2\begin{bmatrix}\cos\theta\\ \sin\theta\\ 0\end{bmatrix}, \tag{66}$$

but

$$\sin\phi\hat{\mathbf{r}} + \cos\phi\hat{\boldsymbol{\phi}} = \begin{bmatrix}\cos\theta\sin^2\phi + \cos\theta\cos^2\phi\\ \sin\theta\sin^2\phi + \sin\theta\cos^2\phi\\ 0\end{bmatrix} = \begin{bmatrix}\cos\theta\\ \sin\theta\\ 0\end{bmatrix}, \tag{67}$$

so

$$\begin{aligned}\ddot{\mathbf{r}} &= (\ddot{r} - r\dot{\phi}^2)\hat{\mathbf{r}} + (2r\cos\phi\dot{\theta}\dot{\phi} + 2\sin\phi\dot{\theta}\dot{r} + r\sin\phi\ddot{\theta})\hat{\boldsymbol{\theta}}\\ &\quad + (2\dot{r}\dot{\phi} + r\ddot{\phi})\hat{\boldsymbol{\phi}} - r\sin\phi\dot{\theta}^2(\sin\phi\hat{\mathbf{r}} + \cos\phi\hat{\boldsymbol{\phi}})\\ &= (\ddot{r} - r\dot{\phi}^2 - r\sin^2\phi\dot{\theta}^2)\hat{\mathbf{r}}\\ &\quad + (2\sin\phi\dot{\theta}\dot{r} + 2r\cos\phi\dot{\theta}\dot{\phi} + r\sin\phi\ddot{\theta})\hat{\boldsymbol{\theta}}\\ &\quad + (2\dot{r}\dot{\phi} + r\ddot{\phi} - r\sin\phi\cos\phi\dot{\theta}^2)\hat{\boldsymbol{\phi}}.\end{aligned} \tag{68}$$

Time DERIVATIVES of the UNIT VECTORS are

$$\dot{\hat{\mathbf{r}}} = \begin{bmatrix}-\sin\theta\sin\phi\,\dot{\theta} + \cos\theta\cos\phi\,\dot{\phi}\\ \cos\theta\sin\phi\,\dot{\theta} + \sin\theta\cos\phi\,\dot{\phi}\\ -\sin\phi\,\dot{\phi}\end{bmatrix} = \sin\phi\,\dot{\theta}\hat{\boldsymbol{\theta}} + \dot{\phi}\hat{\boldsymbol{\phi}} \tag{69}$$

$$\dot{\hat{\boldsymbol{\theta}}} = \begin{bmatrix}-\cos\theta\dot{\theta}\\ -\sin\theta\dot{\theta}\\ 0\end{bmatrix} = -\dot{\theta}\begin{bmatrix}\cos\theta\\ \sin\theta\\ 0\end{bmatrix} = -\dot{\theta}(\sin\phi\hat{\mathbf{r}} + \cos\phi\hat{\boldsymbol{\phi}}) \tag{70}$$

$$\dot{\hat{\boldsymbol{\phi}}} = \begin{bmatrix}-\sin\theta\cos\phi\,\dot{\theta} - \cos\theta\sin\phi\,\dot{\phi}\\ \cos\theta\cos\phi\,\dot{\theta} - \sin\theta\sin\phi\,\dot{\phi}\\ -\cos\phi\,\dot{\phi}\end{bmatrix} = -\dot{\phi}\hat{\mathbf{r}} + \cos\phi\dot{\theta}\hat{\boldsymbol{\theta}}. \tag{71}$$

The CURL is

$$\nabla\times\mathbf{F} = \frac{1}{r\sin\phi}\left[\frac{\partial}{\partial\phi}(\sin\phi F_\theta) - \frac{\partial F_\phi}{\partial\theta}\right]\hat{\mathbf{r}} + \frac{1}{r}\left[\frac{1}{\sin\phi}\frac{\partial F_r}{\partial\theta} - \frac{\partial}{\partial r}(rF_\theta)\right]\hat{\boldsymbol{\phi}} + \frac{1}{r}\left[\frac{\partial}{\partial r}(rF_\phi) - \frac{\partial F_r}{\partial\phi}\right]\hat{\boldsymbol{\theta}}. \tag{72}$$

The LAPLACIAN is

$$\begin{aligned}\nabla^2 &\equiv \frac{1}{r^2}\frac{\partial}{\partial r}\left(r^2\frac{\partial}{\partial r}\right) + \frac{1}{r^2\sin^2\phi}\frac{\partial^2}{\partial\theta^2} + \frac{1}{r^2\sin\phi}\frac{\partial}{\partial\phi}\left(\sin\phi\frac{\partial}{\partial\phi}\right)\\ &= \frac{1}{r^2}\left(r^2\frac{\partial^2}{\partial r^2} + 2r\frac{\partial}{\partial r}\right) + \frac{1}{r^2\sin^2\phi}\frac{\partial^2}{\partial\theta^2} + \frac{1}{r^2\sin\phi}\left(\cos\phi\frac{\partial}{\partial\phi} + \sin\phi\frac{\partial^2}{\partial\phi^2}\right)\\ &= \frac{\partial^2}{\partial r^2} + \frac{2}{r}\frac{\partial}{\partial r} + \frac{1}{r^2\sin^2\phi}\frac{\partial^2}{\partial\theta^2} + \frac{\cos\phi}{r^2\sin\phi}\frac{\partial}{\partial\phi} + \frac{1}{r^2}\frac{\partial^2}{\partial\phi^2}.\end{aligned} \tag{73}$$

The vector LAPLACIAN is

$$\nabla^2\mathbf{v} = \begin{bmatrix}\frac{1}{r}\frac{\partial^2(rv_r)}{\partial r^2} + \frac{1}{r^2}\frac{\partial^2 v_r}{\partial\theta^2} + \frac{1}{r^2\sin^2\theta}\frac{\partial^2 v_r}{\partial\phi^2} + \frac{\cot\theta}{r^2}\frac{\partial v_r}{\theta} - \frac{2}{r^2}\frac{\partial v_\theta}{\partial\theta} - \frac{2}{r^2\sin\theta}\frac{\partial v_\phi}{\partial\phi} - \frac{2v_r}{r^2} - \frac{2\cot\theta}{r^2}v_\theta\\ \frac{1}{r}\frac{\partial^2(rv_\theta)}{\partial r^2} + \frac{1}{r^2}\frac{\partial^2 v_\theta}{\partial\theta^2} + \frac{1}{r^2\sin^2\theta}\frac{\partial^2 v_\theta}{\partial\phi^2} + \frac{\cot\theta}{r^2}\frac{\partial v_\theta}{\partial\theta} - \frac{2}{r^2}\frac{2\cot\theta}{r^2\sin\theta}\frac{\partial v_\phi}{\partial\phi} + \frac{2}{r^2}\frac{\partial v_r}{\partial\theta} - \frac{v_\theta}{r^2\sin^2\theta}\\ \frac{1}{r}\frac{\partial^2(rv_\phi)}{\partial r^2} + \frac{1}{r^2}\frac{\partial^2 v_\phi a}{\partial\theta^2} + \frac{1}{r^2\sin^2\theta}\frac{\partial^2 v_\phi}{\partial\phi^2} + \frac{\cot\theta}{r^2}\frac{\partial v_\phi}{\partial\theta} + \frac{2}{r^2}\frac{\partial v_r}{\partial\phi} + \frac{2\cot\theta}{r^2\sin\theta}\frac{\partial v_\theta}{\partial\phi} - \frac{v_\phi}{r^2\sin^2\theta}\end{bmatrix}. \tag{74}$$

To express PARTIAL DERIVATIVES with respect to Cartesian axes in terms of PARTIAL DERIVATIVES of the spherical coordinates,

$$\begin{bmatrix}x\\ y\\ z\end{bmatrix} = \begin{bmatrix}r\cos\theta\sin\phi\\ r\sin\theta\sin\phi\\ r\cos\phi\end{bmatrix} \tag{75}$$

$$\begin{bmatrix}dx\\ dy\\ dz\end{bmatrix} = \begin{bmatrix}\cos\theta\sin\phi\,dr - r\sin\theta\sin\phi\,d\theta + r\cos\theta\cos\phi\,d\phi\\ \sin\theta\sin\phi\,dr + r\sin\phi\cos\theta\,d\theta + r\sin\theta\cos\phi\,d\phi\\ \cos\phi\,dr - r\sin\phi\,d\phi\end{bmatrix} = \begin{bmatrix}\cos\theta\sin\phi & -r\sin\theta\sin\phi & r\cos\theta\cos\phi\\ \sin\theta\sin\phi & r\sin\phi\cos\theta & r\sin\theta\cos\phi\\ \cos\phi & 0 & -r\sin\phi\end{bmatrix}\begin{bmatrix}dr\\ d\theta\\ d\phi\end{bmatrix}. \tag{76}$$

Upon inversion, the result is

$$\begin{bmatrix}dr\\ d\theta\\ d\phi\end{bmatrix} = \begin{bmatrix}\cos\theta\sin\phi & \sin\theta\sin\phi & \cos\phi\\ -\frac{\sin\theta}{r\sin\phi} & \frac{\cos\theta}{r\sin\phi} & 0\\ \frac{\cos\theta\cos\phi}{r} & \frac{\sin\theta\cos\phi}{r} & -\frac{\sin\phi}{r}\end{bmatrix}\begin{bmatrix}dx\\ dy\\ dz\end{bmatrix}. \tag{77}$$

The Cartesian PARTIAL DERIVATIVES in spherical coordinates are therefore

$$\frac{\partial}{\partial x} = \frac{\partial r}{\partial x}\frac{\partial}{\partial r} + \frac{\partial \theta}{\partial x}\frac{\partial}{\partial \theta} + \frac{\partial \phi}{\partial x}\frac{\partial}{\partial \phi}$$
$$= \cos\theta\sin\phi\frac{\partial}{\partial r} - \frac{\sin\theta}{r\sin\phi}\frac{\partial}{\partial\theta} + \frac{\cos\theta\cos\phi}{r}\frac{\partial}{\partial\phi} \tag{78}$$

$$\frac{\partial}{\partial y} = \frac{\partial r}{\partial y}\frac{\partial}{\partial r} + \frac{\partial \theta}{\partial y}\frac{\partial}{\partial \theta} + \frac{\partial \phi}{\partial y}\frac{\partial}{\partial \phi}$$
$$= \sin\theta\sin\phi\frac{\partial}{\partial r} + \frac{\cos\theta}{r\sin\phi}\frac{\partial}{\partial\theta} + \frac{\sin\theta\cos\phi}{r}\frac{\partial}{\partial\phi} \tag{79}$$

$$\frac{\partial}{\partial z} = \frac{\partial r}{\partial z}\frac{\partial}{\partial r} + \frac{\partial \theta}{\partial z}\frac{\partial}{\partial \theta} + \frac{\partial \phi}{\partial z}\frac{\partial}{\partial \phi}$$
$$= \cos\phi\frac{\partial}{\partial r} - \frac{\sin\phi}{r}\frac{\partial}{\partial\phi} \tag{80}$$

(Gasiorowicz 1974, pp. 167–168).

The HELMHOLTZ DIFFERENTIAL EQUATION is separable in spherical coordinates.

see also COLATITUDE, GREAT CIRCLE, HELMHOLTZ DIFFERENTIAL EQUATION—SPHERICAL COORDINATES, LATITUDE, LONGITUDE, OBLATE SPHEROIDAL COORDINATES, PROLATE SPHEROIDAL COORDINATES

References

Arfken, G. "Spherical Polar Coordinates." §2.5 in *Mathematical Methods for Physicists, 3rd ed.* Orlando, FL: Academic Press, pp. 102–111, 1985.

Beyer, W. H. *CRC Standard Mathematical Tables, 28th ed.* Boca Raton, FL: CRC Press, p. 212, 1987.

Gasiorowicz, S. *Quantum Physics.* New York: Wiley, 1974.

Morse, P. M. and Feshbach, H. *Methods of Theoretical Physics, Part I.* New York: McGraw-Hill, p. 658, 1953.

Spherical Design

X is a spherical t-design in E IFF it is possible to exactly determine the average value on E of any POLYNOMIAL f of degree at most t by sampling f at the points of X. In other words,

$$\frac{1}{\text{volume } E}\int_E f(\xi)\,d\xi = \frac{1}{|X|}\sum_{x\in X} f(x).$$

References

Colbourn, C. J. and Dinitz, J. H. (Eds.) "Spherical t-Designs." Ch. 44 in *CRC Handbook of Combinatorial Designs.* Boca Raton, FL: CRC Press, pp. 462–466, 1996.

Spherical Excess

The difference between the sum of the angles of a SPHERICAL TRIANGLE and 180°.

see also ANGULAR DEFECT, DESCARTES TOTAL ANGULAR DEFECT, GIRARD'S SPHERICAL EXCESS FORMULA, L'HUILIER'S THEOREM, SPHERICAL TRIANGLE

Spherical Frustum

see SPHERICAL SEGMENT

Spherical Geometry

The study of figures on the surface of a SPHERE (such as the SPHERICAL TRIANGLE and SPHERICAL POLYGON), as opposed to the type of geometry studied in PLANE GEOMETRY or SOLID GEOMETRY.

see also PLANE GEOMETRY, SOLID GEOMETRY, SPHERICAL TRIGONOMETRY, THURSTON'S GEOMETRIZATION CONJECTURE

Spherical Hankel Function of the First Kind

$$h_n^{(1)}(x) \equiv \sqrt{\frac{\pi}{2x}}H_{n+1/2}^{(1)}(x) = j_n(x) + in_n(x),$$

where $H^{(1)}(x)$ is the HANKEL FUNCTION OF THE FIRST KIND and $j_n(x)$ and $n_n(x)$ are the SPHERICAL BESSEL FUNCTIONS OF THE FIRST and SECOND KINDS. Explicitly, the first few are

$$h_0^{(1)}(x) = \frac{1}{x}(\sin x - i\cos x) = -\frac{i}{x}e^{ix}$$
$$h_1^{(1)}(x) = e^{ix}\left(-\frac{1}{x} - \frac{i}{x^2}\right)$$
$$h_2^{(1)}(x) = e^{ix}\left(\frac{i}{x} - \frac{3}{x^2} - \frac{3i}{x^3}\right).$$

References

Abramowitz, M. and Stegun, C. A. (Eds.). "Spherical Bessel Functions." §10.1 in *Handbook of Mathematical Functions with Formulas, Graphs, and Mathematical Tables, 9th printing.* New York: Dover, pp. 437–442, 1972.

Spherical Hankel Function of the Second Kind

$$h_n^{(2)}(x) \equiv \sqrt{\frac{\pi}{2x}}H_{n+1/2}^{(2)}(x) = j_n(x) - in_n(x),$$

where $H^{(2)}(x)$ is the HANKEL FUNCTION OF THE SECOND KIND and $j_n(x)$ and $n_n(x)$ are the SPHERICAL BESSEL FUNCTIONS OF THE FIRST and SECOND KINDS. Explicitly, the first is

$$h_0^{(2)}(x) = \frac{1}{x}(\sin x + i\cos x) = \frac{i}{x}e^{-ix}.$$

References

Abramowitz, M. and Stegun, C. A. (Eds.). "Spherical Bessel Functions." §10.1 in *Handbook of Mathematical Functions with Formulas, Graphs, and Mathematical Tables, 9th printing.* New York: Dover, pp. 437–442, 1972.

Spherical Harmonic

The spherical harmonics $Y_l^m(\theta,\phi)$ are the angular portion of the solution to LAPLACE'S EQUATION in SPHERICAL COORDINATES where azimuthal symmetry is not present. Some care must be taken in identifying the notational convention being used. In the below equations, θ is taken as the azimuthal (longitudinal) coordinate, and ϕ as the polar (latitudinal) coordinate (opposite the notation of Arfken 1985).

$$Y_l^m(\theta,\phi) \equiv \sqrt{\frac{2l+1}{4\pi}\frac{(l-m)!}{(l+m)!}} P_l^m(\cos\phi)e^{im\theta}, \tag{1}$$

where $m = -l, -1+1, \ldots, 0, \ldots, l$ and the normalization is chosen such that

$$\int_0^{2\pi}\int_0^{\pi} Y_l^m Y_{l'}^{m'*} \sin\phi\,d\phi\,d\theta = \int_0^{2\pi}\int_{-1}^{1} Y_l^m Y_{l'}^{m'*}\,d(\cos\phi)\,d\theta = \delta_{mm'}\delta_{ll'}, \tag{2}$$

where δ_{mn} is the KRONECKER DELTA. Sometimes, the CONDON-SHORTLEY PHASE $(-1)^m$ is prepended to the definition of the spherical harmonics.

Integrals of the spherical harmonics are given by

$$\int Y_{l_1}^{m_1} Y_{l_2}^{m_2} Y_{l_3}^{m_3}\,d\Omega = \sqrt{\frac{(2l_1+1)(2l_2+1)(2l_3+1)}{4\pi}} \times \begin{pmatrix} l_1 & l_2 & l_3 \\ 0 & 0 & 0 \end{pmatrix}\begin{pmatrix} l_1 & l_2 & l_3 \\ m_1 & m_2 & m_3 \end{pmatrix}, \tag{3}$$

where $\begin{pmatrix} l_1 & l_2 & l_3 \\ m_1 & m_2 & m_3 \end{pmatrix}$ is a WIGNER $3j$-SYMBOL (which is related to the CLEBSCH-GORDON COEFFICIENTS). The spherical harmonics obey

$$Y_l^{-l} = \frac{1}{2^l l!}\sqrt{\frac{(2l+1)!}{4\pi}} \sin^l\phi e^{-il\theta} \tag{4}$$

$$Y_l^0 = \sqrt{\frac{2l+1}{4\pi}} P_l(\cos\phi) \tag{5}$$

$$Y_l^{-m} = (-1)^m Y_l^{m*}, \tag{6}$$

where $P_l(x)$ is a LEGENDRE POLYNOMIAL.

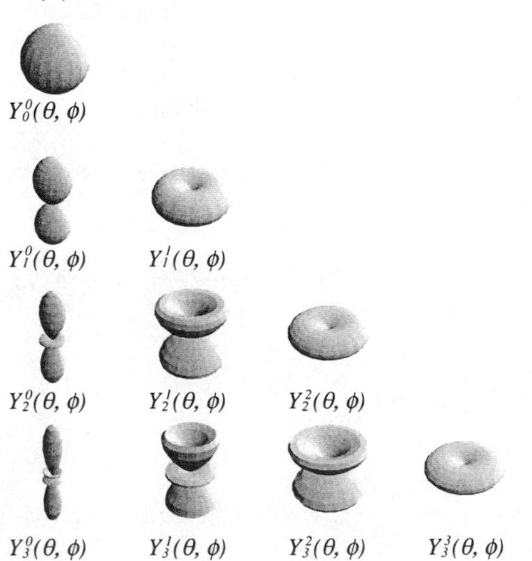

The above illustrations show $|Y_l^m(\theta,\phi)|$ (top) and $\Re[Y_l^m(\theta,\phi)]$ and $\Im[Y_l^m(\theta,\phi)]$ (bottom). The first few spherical harmonics are

$$Y_0^0 = \frac{1}{2}\frac{1}{\sqrt{\pi}}$$

$$Y_1^{-1} = \frac{1}{2}\sqrt{\frac{3}{2\pi}} \sin\phi\,e^{-i\theta}$$

$$Y_1^0 = \frac{1}{2}\sqrt{\frac{3}{\pi}} \cos\phi$$

$$Y_1^1 = -\frac{1}{2}\sqrt{\frac{3}{2\pi}} \sin\phi\,e^{i\theta}$$

$$Y_2^{-2} = \frac{1}{4}\sqrt{\frac{15}{2\pi}} \sin^2\phi\,e^{-2i\theta}$$

$$Y_2^{-1} = \frac{1}{2}\sqrt{\frac{15}{2\pi}} \sin\phi\cos\phi\,e^{-i\theta}$$

$$Y_2^0 = \frac{1}{4}\sqrt{\frac{5}{\pi}} (3\cos^2\phi - 1)$$

$$Y_2^1 = -\frac{1}{2}\sqrt{\frac{15}{2\pi}} \sin\phi\cos\phi\,e^{i\theta}$$

$$Y_2^2 = \frac{1}{4}\sqrt{\frac{15}{2\pi}} \sin^2\phi\,e^{2i\theta}$$

$$Y_3^{-3} = \frac{1}{8}\sqrt{\frac{35}{\pi}} \sin^3\phi\,e^{-3i\theta}$$

$$Y_3^{-2} = \frac{1}{4}\sqrt{\frac{105}{2\pi}} \sin^2\phi\cos\phi\,e^{-2i\theta}$$

$$Y_3^{-1} = \frac{1}{8}\sqrt{\frac{21}{\pi}} \sin\phi(5\cos^2\phi - 1)e^{-i\theta}$$

$$Y_3^0 = \frac{1}{4}\sqrt{\frac{7}{\pi}} (5\cos^3\phi - 3\cos\phi)$$

$$Y_3^1 = -\frac{1}{8}\sqrt{\frac{21}{\pi}} \sin\phi(5\cos^2\phi - 1)e^{i\theta}$$

$$Y_3^2 = \frac{1}{4}\sqrt{\frac{105}{2\pi}} \sin^2\phi\cos\phi\,e^{2i\theta}$$

$$Y_3^3 = -\frac{1}{8}\sqrt{\frac{35}{\pi}} \sin^3\phi\,e^{3i\theta}.$$

Written in terms of CARTESIAN COORDINATES,

$$e^{i\theta} = \frac{x+iy}{\sqrt{x^2+y^2}} \tag{7}$$

$$\phi = \sin^{-1}\left(\sqrt{\frac{x^2+y^2}{x^2+y^2+z^2}}\right) \tag{8}$$

$$= \cos^{-1}\left(\frac{z}{\sqrt{x^2+y^2+z^2}}\right), \tag{9}$$

so

$$Y_0^0 = \frac{1}{2}\frac{1}{\sqrt{\pi}} \tag{10}$$

$$Y_1^0 = \frac{1}{2}\sqrt{\frac{3}{\pi}}\frac{z}{\sqrt{x^2+y^2+z^2}} \tag{11}$$

$$Y_1^1 = -\frac{1}{2}\sqrt{\frac{3}{2\pi}}\frac{x+iy}{\sqrt{x^2+y^2+z^2}} \tag{12}$$

$$Y_2^0 = \frac{1}{4}\sqrt{\frac{5}{\pi}}\left(\frac{3z^2}{x^2+y^2+z^2}-1\right) \tag{13}$$

$$Y_2^1 = -\frac{1}{2}\sqrt{\frac{15}{2\pi}}\frac{z(x+iy)}{x^2+y^2+z^2} \tag{14}$$

$$Y_2^2 = \frac{1}{4}\sqrt{\frac{15}{2\pi}}\frac{(x+iy)^2}{x^2+y^2+z^2}. \tag{15}$$

These can be separated into their REAL and IMAGINARY PARTS

$$Y_l^{ms}(\theta,\phi) \equiv P_l^m(\cos\phi)\sin(m\theta) \tag{16}$$

$$Y_l^{mc}(\theta,\phi) \equiv P_l^m(\cos\phi)\cos(m\theta). \tag{17}$$

The ZONAL HARMONICS are defined to be those of the form

$$P_n^m(\cos\theta). \tag{18}$$

The TESSERAL HARMONICS are those of the form

$$\sin(m\phi)P_n^m(\cos\theta) \tag{19}$$

$$\cos(m\phi)P_n^m(\cos\theta) \tag{20}$$

for $n \neq m$. The SECTORIAL HARMONICS are of the form

$$\sin(m\phi)P_m^m(\cos\theta) \tag{21}$$

$$\cos(m\phi)P_m^m(\cos\theta). \tag{22}$$

The spherical harmonics form a COMPLETE ORTHONORMAL BASIS, so an arbitrary REAL function $f(\theta,\phi)$ can be expanded in terms of COMPLEX spherical harmonics

$$f(\theta,\phi) \equiv \sum_{l=0}^{\infty}\sum_{m=-1}^{l} A_l^m Y_l^m(\theta,\phi), \tag{23}$$

or REAL spherical harmonics

$$f(\theta,\phi) \equiv \sum_{l=0}^{\infty}\sum_{m=0}^{l}[C_l^m Y_l^{mc}(\theta,\phi)\sin(m\theta) + S_l^m Y_l^{ms}(\theta,\phi)]. \tag{24}$$

see also CORRELATION COEFFICIENT, SPHERICAL HARMONIC ADDITION THEOREM, SPHERICAL HARMONIC CLOSURE RELATIONS, SPHERICAL VECTOR HARMONIC

References

Arfken, G. "Spherical Harmonics." §12.6 in *Mathematical Methods for Physicists, 3rd ed.* Orlando, FL: Academic Press, pp. 680–685, 1985.

Ferrers, N. M. *An Elementary Treatise on Spherical Harmonics and Subjects Connected with Them.* London: Macmillan, 1877.

Groemer, H. *Geometric Applications of Fourier Series and Spherical Harmonics.* New York: Cambridge University Press, 1996.

Hobson, E. W. *The Theory of Spherical and Ellipsoidal Harmonics.* New York: Chelsea, 1955.

MacRobert, T. M. and Sneddon, I. N. *Spherical Harmonics: An Elementary Treatise on Harmonic Functions, with Applications, 3rd ed. rev.* Oxford, England: Pergamon Press, 1967.

Press, W. H.; Flannery, B. P.; Teukolsky, S. A.; and Vetterling, W. T. "Spherical Harmonics." §6.8 in *Numerical Recipes in FORTRAN: The Art of Scientific Computing, 2nd ed.* Cambridge, England: Cambridge University Press, pp. 246–248, 1992.

Sansone, G. "Harmonic Polynomials and Spherical Harmonics," "Integral Properties of Spherical Harmonics and the Addition Theorem for Legendre Polynomials," and "Completeness of Spherical Harmonics with Respect to Square Integrable Functions." §3.18–3.20 in *Orthogonal Functions, rev. English ed.* New York: Dover, pp. 253–272, 1991.

Sternberg, W. and Smith, T. L. *The Theory of Potential and Spherical Harmonics, 2nd ed.* Toronto: University of Toronto Press, 1946.

Spherical Harmonic Addition Theorem

A FORMULA also known as the LEGENDRE ADDITION THEOREM which is derived by finding GREEN'S FUNCTIONS for the SPHERICAL HARMONIC expansion and equating them to the generating function for LEGENDRE POLYNOMIALS. When γ is defined by

$$\cos\gamma \equiv \cos\theta_1\cos\theta_2 + \sin\theta_1\sin\theta_2\cos\phi_1 - \phi_2,$$

$$P_n(\cos\gamma) = \frac{4\pi}{2n+1}\sum_{m=-n}^{n}(-1)^m Y_m^n(\theta_1,\phi_1)Y_{-m}^n(\theta_2,\phi_2)$$

$$= \frac{4\pi}{2n+1}\sum_{m=-n}^{n} Y_m^n(\theta_1,\phi_1)Y_m^{n*}(\theta_2,\phi_2)$$

$$= P_n(\cos\theta_1)P_n(\cos\theta_2)$$

$$+2\sum_{m=-n}^{n}\frac{(n-m)!}{(n+m)!}P_m^n(\cos\theta_1)P_m^n(\cos\theta_2)\cos[m(\phi_1-\phi_2)].$$

References
Arfken, G. "The Addition Theorem for Spherical Harmonics." §12.8 in *Mathematical Methods for Physicists, 3rd ed.* Orlando, FL: Academic Press, pp. 693–695, 1985.

Spherical Harmonic Closure Relations

The sum of the absolute squares of the SPHERICAL HARMONICS $Y_l^m(\theta, \phi)$ over all values of m is

$$\sum_{m=-l}^{l} |Y_l^m(\theta,\phi)|^2 = \frac{2l+1}{4\pi}.$$

The double sum over m and l is given by

$$\sum_{l=0}^{\infty}\sum_{m=-l}^{l} Y_l^m(\theta_1,\phi_1)Y_l^{m*}(\theta_2,\phi_2) = \frac{1}{\sin\theta_1}\delta(\theta_1-\theta_2)\delta(\phi_1-\phi_2) = \delta(\cos\theta_1-\cos\theta_2)\delta(\phi_1-\phi_2),$$

where $\delta(x)$ is the DELTA FUNCTION.

Spherical Harmonic Tensor

A tensor defined in terms of the TENSORS which satisfy the DOUBLE CONTRACTION RELATION.

see also DOUBLE CONTRACTION RELATION, SPHERICAL HARMONIC

Spherical Helix

The TANGENT INDICATRIX of a CURVE OF CONSTANT PRECESSION is a spherical helix. The equation of a spherical helix on a SPHERE with RADIUS r making an ANGLE θ with the z-axis is

$$x(\psi) = \tfrac{1}{2}r(1+\cos\theta)\cos\psi - \tfrac{1}{2}r(1-\cos\theta)\cos\left(\frac{1+\cos\theta}{1-\cos\theta}\psi\right) \tag{1}$$

$$y(\psi) = \tfrac{1}{2}r(1+\cos\theta)\sin\psi - \tfrac{1}{2}r(1-\cos\theta)\sin\left(\frac{1+\cos\theta}{1-\cos\theta}\psi\right) \tag{2}$$

$$z(\psi) = r\sin\theta\cos\left(\frac{\cos\theta}{1-\cos\theta}\psi\right). \tag{3}$$

The projection on the xy-plane is an EPICYCLOID with RADII

$$a = r\cos\theta \tag{4}$$

$$b = r\sin^2(\tfrac{1}{2}\theta). \tag{5}$$

see also HELIX, LOXODROME, SPHERICAL SPIRAL

References
Scofield, P. D. "Curves of Constant Precession." *Amer. Math. Monthly* **102**, 531–537, 1995.

Spherical Point System

How can n points be distributed on a SPHERE such that they maximize the minimum distance between any pair of points? This is FEJES TÓTH'S PROBLEM.

see also FEJES TÓTH'S PROBLEM

Spherical Polygon

A closed geometric figure on the surface of a SPHERE which is formed by the ARCS of GREAT CIRCLES. The spherical polygon is a generalization of the SPHERICAL TRIANGLE. If θ is the sum of the RADIAN ANGLES of a spherical polygon on a SPHERE of RADIUS r, then the AREA is

$$S = [\theta - (n-2)\pi]r^2.$$

see also GREAT CIRCLE, SPHERICAL TRIANGLE

References
Beyer, W. H. *CRC Standard Mathematical Tables, 28th ed.* Boca Raton, FL: CRC Press, p. 131, 1987.

Spherical Ring

A SPHERE with a CYLINDRICAL HOLE cut so that the centers of the CYLINDER and SPHERE coincide, also called a NAPKIN RING.

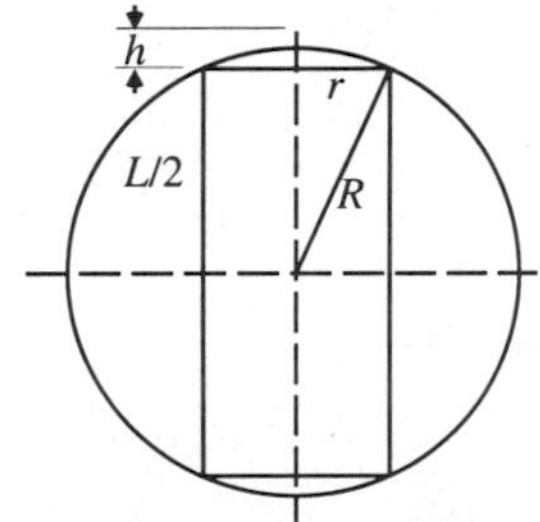

The volume of the entire CYLINDER is

$$V_{\rm cyl} = \pi LR^2, \tag{1}$$

and the VOLUME of the upper segment is

$$V_{\rm seg} = \tfrac{1}{6}\pi h(3R^2+h^2), \tag{2}$$

where

$$R = \sqrt{r^2 - \tfrac{1}{4}L^2} \tag{3}$$

$$h = r - \tfrac{1}{2}L, \tag{4}$$

so the VOLUME removed upon drilling of a CYLINDRICAL hole is

$$\begin{aligned} V_{\rm rem} &= V_{\rm cyl} + 2V_{\rm seg} = \pi[LR^2 + \tfrac{1}{3}h(3R^2+h^2)] \\ &= \pi(LR^2 + hR^2 + \tfrac{1}{3}h^3) \\ &= \pi[L(r^2-\tfrac{1}{4}L^2) + (r-\tfrac{1}{2}L)(r^2-\tfrac{1}{4}L^2) \\ &\quad + \tfrac{1}{3}(r-\tfrac{1}{2}L)^3] \\ &= \pi[Lr^2 - \tfrac{1}{4}L^3 + (r^3 - \tfrac{1}{2}r^2L - \tfrac{1}{4}RL^2 + \tfrac{1}{8}L^3) \\ &\quad + \tfrac{1}{3}(r^3 - \tfrac{3}{2}r^2L + \tfrac{3}{4}rL^2 - \tfrac{1}{8}L^3)] \\ &= \pi[\tfrac{4}{3}r^3 + (1-\tfrac{1}{2}-\tfrac{1}{2})r^2L + (-\tfrac{1}{4}+\tfrac{1}{4})RL^2 \\ &\quad + L^3(-\tfrac{1}{4}+\tfrac{1}{8}-\tfrac{1}{24})] \\ &= \tfrac{4}{3}\pi r^3 - \tfrac{1}{6}\pi L^3 = \tfrac{1}{6}\pi(8r^3 - L^3), \end{aligned} \tag{5}$$

so

$$V_{\text{left}} = V_{\text{sphere}} - V_{\text{rem}} = \tfrac{4}{3}\pi r^3 - (\tfrac{4}{3}\pi r^3 - \tfrac{1}{6}\pi L^3) = \tfrac{1}{6}\pi L^3. \tag{6}$$

Spherical Sector

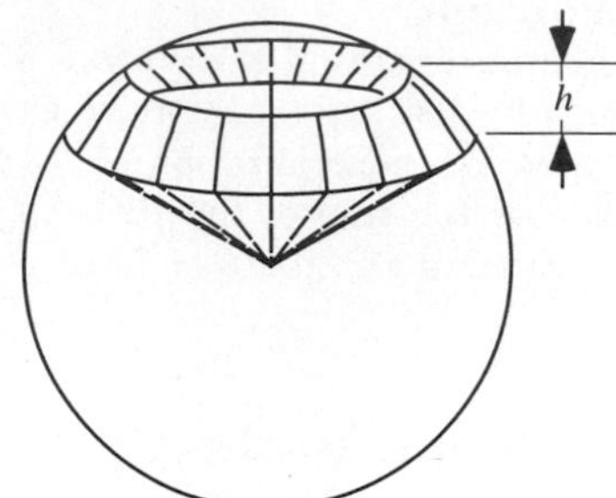

The VOLUME of a spherical sector, depicted above, is given by

$$V = \tfrac{2}{3}\pi R^2 h,$$

where h is the vertical height of the upper and lower curves.

see also CYLINDRICAL SEGMENT, SPHERE, SPHERICAL CAP, SPHERICAL SEGMENT, ZONE

References
Beyer, W. H. *CRC Standard Mathematical Tables, 28th ed.* Boca Raton, FL: CRC Press, p. 131, 1987.

Spherical Segment

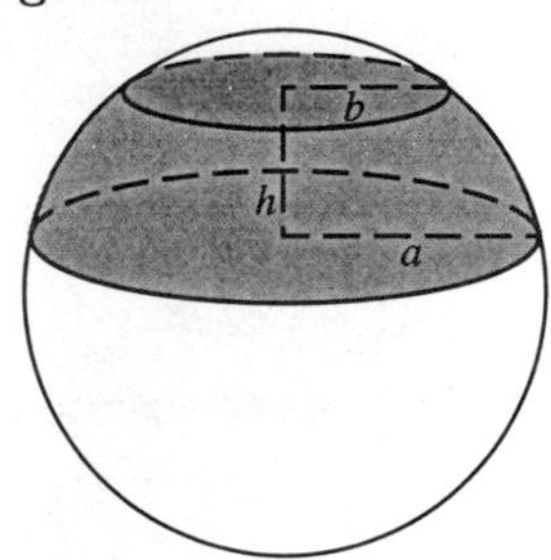

A spherical segment is the solid defined by cutting a SPHERE with a pair of PARALLEL PLANES. It can be thought of as a SPHERICAL CAP with the top truncated, and so it corresponds to a SPHERICAL FRUSTUM. The surface of the spherical segment (excluding the bases) is called a ZONE.

Call the RADIUS of the SPHERE R and the height of the segment (the distance from the plane to the top of SPHERE) h. Let the RADII of the lower and upper bases be denoted a and b, respectively. Call the distance from the center to the start of the segment d, and the height from the bottom to the top of the segment h. Call the RADIUS parallel to the segment r, and the height above the center y. Then $r^2 = R^2 - y^2$,

$$\begin{aligned} V &= \int_d^{d+h} \pi r^2\,dy = \pi \int_d^{d+h} (R^2 - y^2)\,dy \\ &= \pi \left[R^2 y - \tfrac{1}{3}y^3\right]_d^{d+h} = \pi\{R^2 h - \tfrac{1}{3}[(d+h)^3 - d^3]\} \\ &= \pi[R^2 h - \tfrac{1}{3}(d^3 + 3d^2 h + 3h^2 d + h^3 - d^3)] \\ &= \pi(R^2 h - d^2 h - h^2 d - \tfrac{1}{3}h^3) \\ &= \pi h(R^2 - d^2 - hd - \tfrac{1}{3}h^2). \end{aligned} \tag{1}$$

Using

$$a^2 = R^2 - d^2 \tag{2}$$

$$b^2 = R^2 - (d+h)^2 = R^2 - d^2 - 2dh - h^2, \tag{3}$$

gives

$$a^2 + b^2 = 2R^2 - 2d^2 - 2dh - h^2 \tag{4}$$

$$R^2 - d^2 - dh = \tfrac{1}{2}(a^2 + b^2 + h^2), \tag{5}$$

so

$$\begin{aligned} V &= \pi h[\tfrac{1}{2}(a^2 + b^2 + h^2) - \tfrac{1}{3}h^2] = \pi h(\tfrac{1}{2}a^2 + \tfrac{1}{2}b^2 + \tfrac{1}{6}h^2) \\ &= \tfrac{1}{6}\pi h(3a^2 + 3b^2 + h^2). \end{aligned} \tag{6}$$

The surface area of the ZONE (which excludes the top and bottom bases) is given by

$$S = 2\pi R h. \tag{7}$$

see also ARCHIMEDES' PROBLEM, FRUSTUM, HEMISPHERE, SPHERE, SPHERICAL CAP, SPHERICAL SECTOR, SURFACE OF REVOLUTION, ZONE

References
Beyer, W. H. *CRC Standard Mathematical Tables, 28th ed.* Boca Raton, FL: CRC Press, p. 130, 1987.

Spherical Shell

A generalization of an ANNULUS to 3-D. A spherical shell is the intersection of two concentric BALLS of differing RADII.

see also ANNULUS, BALL, CHORD, SPHERE, SPHERICAL HELIX

Spherical Spiral

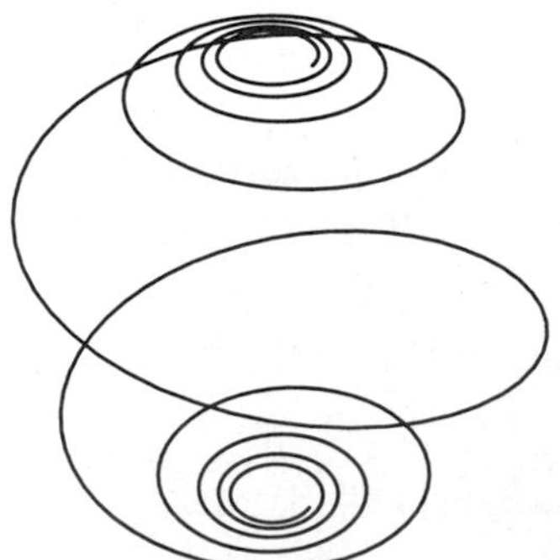

The path taken by a ship which travels from the south pole to the north pole of a SPHERE while keeping a fixed (but not RIGHT) ANGLE with respect to the meridians. The curve has an infinite number of loops since the separation of consecutive revolutions gets smaller and smaller near the poles. It is given by the parametric equations

$$x = \cos t \cos c$$
$$y = \sin t \cos c$$
$$z = -\sin c,$$

where

$$c \equiv \tan^{-1}(at)$$

and a is a constant.

see also MERCATOR PROJECTION, SEIFERT'S SPHERICAL SPIRAL

References

Gray, A. *Modern Differential Geometry of Curves and Surfaces.* Boca Raton, FL: CRC Press, p. 162, 1993.

Lauwerier, H. "Spherical Spiral." In *Fractals: Endlessly Repeated Geometric Figures.* Princeton, NJ: Princeton University Press, pp. 64–66, 1991.

Spherical Symmetry

Let **A** and **B** be constant VECTORS. Define

$$Q \equiv 3(\mathbf{A}\cdot\hat{\mathbf{r}})(\mathbf{B}\cdot\hat{\mathbf{r}}) - \mathbf{A}\cdot\mathbf{B}.$$

Then the average of Q over a spherically symmetric surface or volume is

$$\langle Q\rangle = \left\langle 3\cos^2\theta - 1\right\rangle(\mathbf{A}\cdot\mathbf{B}) = 0,$$

since $\left\langle 3\cos^2\theta - 1\right\rangle = 0$ over the sphere.

Spherical Tessellation

see TRIANGULAR SYMMETRY GROUP

Spherical Triangle

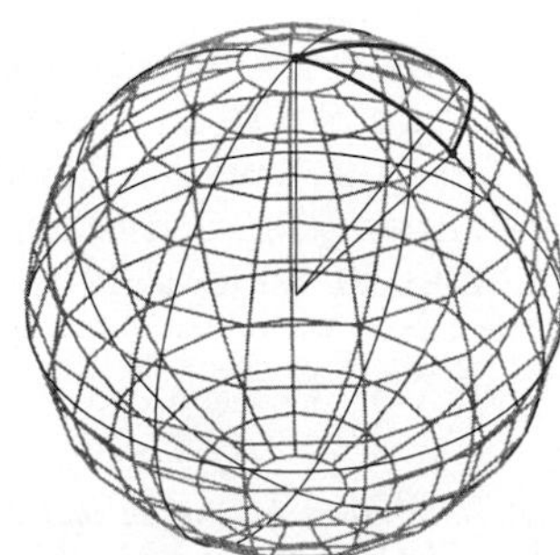

A spherical triangle is a figure formed on the surface of a sphere by three great circular arcs intersecting pairwise in three vertices. The spherical triangle is the spherical analog of the planar TRIANGLE. Let a spherical triangle have ANGLES α, β, and γ and RADIUS r. Then the AREA of the spherical triangle is

$$K = r^2[(\alpha+\beta+\gamma)-\pi].$$

The sum of the angles of a spherical triangle is between 180° and 540°. The amount by which it exceeds 180° is called the SPHERICAL EXCESS and is denoted E or Δ.

The study of angles and distances of figures on a sphere is known as SPHERICAL TRIGONOMETRY.

see also COLUNAR TRIANGLE, GIRARD'S SPHERICAL EXCESS FORMULA, L'HUILIER'S THEOREM, SPHERICAL POLYGON, SPHERICAL TRIGONOMETRY

References

Abramowitz, M. and Stegun, C. A. (Eds.). *Handbook of Mathematical Functions with Formulas, Graphs, and Mathematical Tables, 9th printing.* New York: Dover, p. 79, 1972.

Beyer, W. H. *CRC Standard Mathematical Tables, 28th ed.* Boca Raton, FL: CRC Press, pp. 131 and 147–150, 1987.

Spherical Trigonometry

Define a SPHERICAL TRIANGLE on the surface of a unit SPHERE, centered at a point O, with vertices A, B, and C. Define ANGLES $a \equiv \angle BOC$, $b \equiv \angle COA$, and $c \equiv \angle AOB$. Let the ANGLE between PLANES AOB and AOC be α, the ANGLE between PLANES BOC and AOB be β, and the ANGLE between PLANES BOC and AOC be γ. Define the VECTORS

$$\mathbf{a} \equiv \overrightarrow{OA} \quad (1)$$
$$\mathbf{b} \equiv \overrightarrow{OB} \quad (2)$$
$$\mathbf{c} \equiv \overrightarrow{OC}. \quad (3)$$

Then

$$(\hat{\mathbf{a}}\times\hat{\mathbf{b}})\cdot(\hat{\mathbf{a}}\times\hat{\mathbf{c}}) = (|\hat{\mathbf{a}}|\,|\hat{\mathbf{b}}|\sin c)(|\hat{\mathbf{a}}|\,|\hat{\mathbf{c}}|\sin b)\cos\alpha$$
$$= \sin b\sin c\cos\alpha. \quad (4)$$

Equivalently,

$$(\hat{\mathbf{a}}\times\hat{\mathbf{b}})\cdot(\hat{\mathbf{a}}\times\hat{\mathbf{c}}) = \hat{\mathbf{a}}\cdot[\hat{\mathbf{b}}\times(\hat{\mathbf{a}}\times\hat{\mathbf{c}})]$$
$$= \hat{\mathbf{a}}\cdot[\hat{\mathbf{a}}(\hat{\mathbf{b}}\cdot\hat{\mathbf{c}}) - \hat{\mathbf{c}}(\hat{\mathbf{a}}\cdot\hat{\mathbf{b}})]$$
$$= (\hat{\mathbf{b}}\cdot\hat{\mathbf{c}}) - (\hat{\mathbf{a}}\cdot\hat{\mathbf{c}})(\hat{\mathbf{a}}\cdot\hat{\mathbf{b}})$$
$$= \cos a - \cos c\cos b. \quad (5)$$

Since these two expressions are equal, we obtain the identity

$$\cos a = \cos b\cos c + \sin b\sin c\cos\alpha \quad (6)$$

The identity

$$\sin\alpha = \frac{|(\hat{\mathbf{a}}\times\hat{\mathbf{b}})\times(\hat{\mathbf{a}}\times\hat{\mathbf{c}})|}{|\hat{\mathbf{a}}\times\hat{\mathbf{b}}||\hat{\mathbf{a}}\times\hat{\mathbf{c}}|} = -\frac{|\hat{\mathbf{a}}[\hat{\mathbf{b}},\hat{\mathbf{a}},\hat{\mathbf{c}}] + \hat{\mathbf{b}}[\hat{\mathbf{a}},\hat{\mathbf{a}},\hat{\mathbf{c}}]|}{\sin b\sin c}$$
$$= \frac{[\hat{\mathbf{a}},\hat{\mathbf{b}},\hat{\mathbf{c}}]}{\sin b\sin c}, \quad (7)$$

where $[\mathbf{a}, \mathbf{b}, \mathbf{c}]$ is the SCALAR TRIPLE PRODUCT, gives a spherical analog of the LAW OF SINES,

$$\frac{\sin\alpha}{\sin a} = \frac{\sin\beta}{\sin b} = \frac{\sin\gamma}{\sin c} = \frac{6\,\mathrm{Vol}(OABC)}{\sin a \sin b \sin c}, \tag{8}$$

where $\mathrm{Vol}(OABC)$ is the VOLUME of the TETRAHEDRON. From (7) and (8), it follows that

$$\sin a \cos\beta = \cos b \sin c - \sin b \cos c \cos\alpha \tag{9}$$

$$\cos a \cos\gamma = \sin a \cot b - \sin\gamma \cot\beta. \tag{10}$$

These are the fundamental equalities of spherical trigonometry.

There are also spherical analogs of the LAW OF COSINES for the sides of a spherical triangle,

$$\cos a = \cos b \cos c + \sin b \sin c \cos A \tag{11}$$

$$\cos b = \cos c \cos a + \sin c \sin a \cos B \tag{12}$$

$$\cos c = \cos a \cos b + \sin a \sin b \cos C, \tag{13}$$

and the angles of a spherical triangle,

$$\cos A = -\cos B \cos C + \sin B \sin C \cos a \tag{14}$$

$$\cos B = -\cos C \cos A + \sin C \sin A \cos b \tag{15}$$

$$\cos C = -\cos A \cos B + \sin A \sin B \cos c \tag{16}$$

(Beyer 1987), as well as the LAW OF TANGENTS

$$\frac{\tan[\frac{1}{2}(a-b)]}{\tan[\frac{1}{2}(a+b)]} = \frac{\tan[\frac{1}{2}(A-B)]}{\tan[\frac{1}{2}(A+B)]}. \tag{17}$$

Let

$$s \equiv \tfrac{1}{2}(a+b+c) \tag{18}$$

$$S \equiv \tfrac{1}{2}(A+B+C), \tag{19}$$

then the half-angle formulas are

$$\tan(\tfrac{1}{2}A) = \frac{k}{\sin(s-a)} \tag{20}$$

$$\tan(\tfrac{1}{2}B) = \frac{k}{\sin(s-b)} \tag{21}$$

$$\tan(\tfrac{1}{2}C) = \frac{k}{\sin(s-c)}, \tag{22}$$

where

$$k^2 = \frac{\sin(s-a)\sin(s-b)\sin(s-c)}{\sin s} = \tan^2 r, \tag{23}$$

and the half-side formulas are

$$\tan(\tfrac{1}{2}a) = K\cos(S-A) \tag{24}$$

$$\tan(\tfrac{1}{2}b) = K\cos(S-B) \tag{25}$$

$$\tan(\tfrac{1}{2}c) = K\cos(S-C), \tag{26}$$

where

$$K^2 = -\frac{\cos S}{(\cos(S-A)\cos(S-B)\cos(S-C)} = \tan^2 R, \tag{27}$$

where R is the RADIUS of the SPHERE on which the spherical triangle lies.

Additional formulas include the HAVERSINE formulas

$$\operatorname{hav} a = \operatorname{hav}(b-c) + \sin b \sin c \sin(s-c) \tag{28}$$

$$\operatorname{hav} A = \frac{\sin(s-b)\sin(s-c)}{\sin b \sin c} \tag{29}$$

$$= \frac{\operatorname{hav} a - \operatorname{hav}(b-c)}{\sin b \sin c} \tag{30}$$

$$= \operatorname{hav}[\pi - (B+C)] + \sin B \sin C \operatorname{hav} a, \tag{31}$$

GAUSS'S FORMULAS

$$\frac{\sin[\frac{1}{2}(a-b)]}{\sin(\frac{1}{2}c)} = \frac{\sin[\frac{1}{2}(A-B)]}{\cos(\frac{1}{2}C)} \tag{32}$$

$$\frac{\sin[\frac{1}{2}(a+b)]}{\sin(\frac{1}{2}c)} = \frac{\cos[\frac{1}{2}(A-B)]}{\sin(\frac{1}{2}C)} \tag{33}$$

$$\frac{\cos[\frac{1}{2}(a-b)]}{\cos(\frac{1}{2}c)} = \frac{\sin[\frac{1}{2}(A+B)]}{\cos(\frac{1}{2}C)} \tag{34}$$

$$\frac{\cos[\frac{1}{2}(a+b)]}{\cos(\frac{1}{2}c)} = \frac{\cos[\frac{1}{2}(A+B)]}{\sin(\frac{1}{2}C)}, \tag{35}$$

and NAPIER'S ANALOGIES

$$\frac{\sin[\frac{1}{2}(A-B)]}{\sin[\frac{1}{2}(A+B)]} = \frac{\tan[\frac{1}{2}(a-b)]}{\tan(\frac{1}{2}c)} \tag{36}$$

$$\frac{\cos[\frac{1}{2}(A-B)]}{\cos[\frac{1}{2}(A+B)]} = \frac{\tan[\frac{1}{2}(a+b)]}{\tan(\frac{1}{2}c)} \tag{37}$$

$$\frac{\sin[\frac{1}{2}(a-b)]}{\sin[\frac{1}{2}(a+b)]} = \frac{\tan[\frac{1}{2}(A-B)]}{\cot(\frac{1}{2}C)} \tag{38}$$

$$\frac{\cos[\frac{1}{2}(a-b)]}{\cos[\frac{1}{2}(a+b)]} = \frac{\tan[\frac{1}{2}(A+B)]}{\cot(\frac{1}{2}C)} \tag{39}$$

(Beyer 1987).

see also ANGULAR DEFECT, DESCARTES TOTAL ANGULAR DEFECT, GAUSS'S FORMULAS, GIRARD'S SPHERICAL EXCESS FORMULA, LAW OF COSINES, LAW OF SINES, LAW OF TANGENTS, L'HUILIER'S THEOREM, NAPIER'S ANALOGIES, SPHERICAL EXCESS, SPHERICAL GEOMETRY, SPHERICAL POLYGON, SPHERICAL TRIANGLE

References

Beyer, W. H. *CRC Standard Mathematical Tables, 28th ed.* Boca Raton, FL: CRC Press, pp. 131 and 147–150, 1987.

Danby, J. M. *Fundamentals of Celestial Mechanics, 2nd ed., rev. ed.* Richmond, VA: Willmann-Bell, 1988.

Smart, W. M. *Text-Book on Spherical Astronomy, 6th ed.* Cambridge, England: Cambridge University Press, 1960.

Spherical Vector Harmonic

see VECTOR SPHERICAL HARMONIC

Spheroid

A spheroid is an ELLIPSOID

$$\frac{r^2\cos^2\theta\sin^2\phi}{a^2}+\frac{r^2\sin^2\theta\sin^2\phi}{b^2}+\frac{r^2\cos^2\phi}{c^2}=1 \quad (1)$$

with two SEMIMAJOR AXES equal. Orient the ELLIPSE so that the a and b axes are equal, then

$$\frac{r^2\cos^2\theta\sin^2\phi}{a^2}+\frac{r^2\sin^2\theta\sin^2\phi}{a^2}+\frac{r^2\cos^2\phi}{c^2}=1 \quad (2)$$

$$\frac{r^2\sin^2\phi}{a^2}+\frac{r^2\cos^2\phi}{c^2}=1, \quad (3)$$

where a is the equatorial RADIUS and c is the polar RADIUS. Here ϕ is the colatitude, so take $\delta\equiv\pi/2-\phi$ to express in terms of latitude.

$$\frac{r^2\cos^2\delta}{a^2}+\frac{r^2\sin^2\delta}{c^2}=1. \quad (4)$$

Rewriting $\cos^2\delta=1-\sin^2\delta$ gives

$$\frac{r^2}{a^2}+r^2\sin^2\delta\left(\frac{1}{c^2}-\frac{1}{a^2}\right)=1 \quad (5)$$

$$r^2\left(1+a^2\sin^2\delta\frac{a^2-c^2}{c^2a^2}\right)=r^2\left(1+\sin^2\delta\frac{a^2-c^2}{c^2}\right)=a^2, \quad (6)$$

so

$$r=a\left(1+\sin^2\delta\frac{a^2-c^2}{c^2}\right)^{-1/2}. \quad (7)$$

If $a>c$, the spheroid is OBLATE. If $a<c$, the spheroid is PROLATE. If $a=c$, the spheroid degenerates to a SPHERE.

see also DARWIN-DE SITTER SPHEROID, ELLIPSOID, OBLATE SPHEROID, PROLATE SPHEROID

Spheroidal Harmonic

A spheroidal harmonic is a special case of the ELLIPSOIDAL HARMONIC which satisfies the differential equation

$$\frac{d}{dx}\left[(1-x^2)\frac{dS}{dx}\right]+\left(\lambda-c^2x^2-\frac{m^2}{1-x^2}\right)S=0$$

on the interval $-1\leq x\leq 1$.

see also ELLIPSOIDAL HARMONIC

References

Press, W. H.; Flannery, B. P.; Teukolsky, S. A.; and Vetterling, W. T. "A Worked Example: Spheroidal Harmonics." §17.4 in *Numerical Recipes in FORTRAN: The Art of Scientific Computing, 2nd ed.* Cambridge, England: Cambridge University Press, pp. 764–773, 1992.

Spheroidal Wavefunction

Whittaker and Watson (1990, p. 403) define the internal and external spheroidal wavefunctions as

$$S_{mn}^{(1)}=2\pi\frac{(n-m)!}{(n+m)!}P_n^m(ir)P_n^m(\cos\theta)\begin{matrix}\cos\\ \sin\end{matrix}(m\phi)$$

$$S_{mn}^{(2)}=2\pi\frac{(n-m)!}{(n+m)!}Q_n^m(ir)Q_n^m(\cos\theta)\begin{matrix}\cos\\ \sin\end{matrix}(m\phi).$$

see also ELLIPSOIDAL HARMONIC, OBLATE SPHEROIDAL WAVE FUNCTION, PROLATE SPHEROIDAL WAVE FUNCTION, SPHERICAL HARMONIC

References

Abramowitz, M. and Stegun, C. A. (Eds.). "Spheroidal Wave Functions." Ch. 21 in *Handbook of Mathematical Functions with Formulas, Graphs, and Mathematical Tables, 9th printing.* New York: Dover, pp. 751–759, 1972.

Morse, P. M. and Feshbach, H. *Methods of Theoretical Physics, Part I.* New York: McGraw-Hill, pp. 642–644, 1953.

Whittaker, E. T. and Watson, G. N. *A Course in Modern Analysis, 4th ed.* Cambridge, England: Cambridge University Press, 1990.

Sphinx

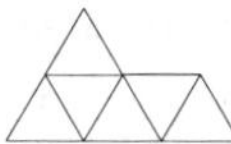

A 6-POLYIAMOND named for its resemblance to the Great Sphinx of Egypt.

References

Golomb, S. W. *Polyominoes: Puzzles, Patterns, Problems, and Packings, 2nd ed.* Princeton, NJ: Princeton University Press, p. 92, 1994.

Spider and Fly Problem

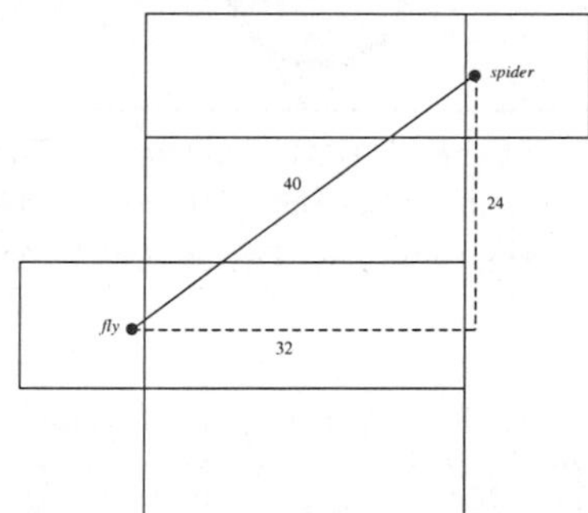

In a rectangular room (a CUBOID) with dimensions $30'\times 12'\times 12'$, a spider is located in the middle of one $12'\times 12'$ wall one foot away from the ceiling. A fly is in the middle of the opposite wall one foot away from the floor. If the fly remains stationary, what is the shortest distance the spider must crawl to capture the fly? The answer, $40'$, can be obtained by "flattening" the walls as illustrated above.

References

Pappas, T. "The Spider & the Fly Problem." *The Joy of Mathematics.* San Carlos, CA: Wide World Publ./Tetra, pp. 218 and 233, 1989.

Spider Lines

see EPITROCHOID

Spiegeldrieck

see FUHRMANN TRIANGLE

Spieker Center

The center of the SPIEKER CIRCLE. It is the CENTROID of the PERIMETER of the original TRIANGLE. The third BROCARD POINT is COLLINEAR with the Spieker center and the ISOTOMIC CONJUGATE POINT of its INCENTER.

see also BROCARD POINTS, CENTROID (TRIANGLE), INCENTER, ISOTOMIC CONJUGATE POINT, PERIMETER, SPIEKER CIRCLE, TAYLOR CENTER

References

Casey, J. *A Treatise on the Analytical Geometry of the Point, Line, Circle, and Conic Sections, Containing an Account of Its Most Recent Extensions, with Numerous Examples, 2nd ed., rev. enl.* Dublin: Hodges, Figgis, & Co., p. 81, 1893.

Johnson, R. A. *Modern Geometry: An Elementary Treatise on the Geometry of the Triangle and the Circle.* Boston, MA: Houghton Mifflin, pp. 226–229 and 249, 1929.

Kimberling, C. "Central Points and Central Lines in the Plane of a Triangle." *Math. Mag.* **67**, 163–187, 1994.

Spieker Circle

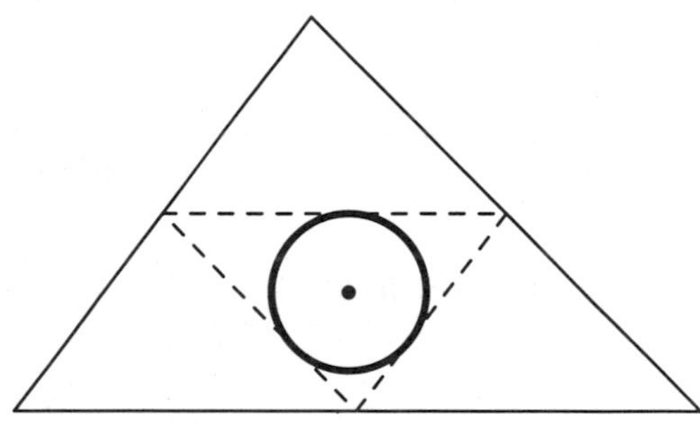

The INCIRCLE of the MEDIAL TRIANGLE. The center of the Spieker circle is called the SPIEKER CENTER.

see also INCIRCLE, MEDIAL TRIANGLE, SPIEKER CENTER

References

Johnson, R. A. *Modern Geometry: An Elementary Treatise on the Geometry of the Triangle and the Circle.* Boston, MA: Houghton Mifflin, pp. 226–228, 1929.

Spigot Algorithm

An ALGORITHM which generates digits of a quantity one at a time without using or requiring previously computed digits. Amazingly, spigot ALGORITHMS are known for both PI and e.

Spijker's Lemma

The image on the RIEMANN SPHERE of any CIRCLE under a COMPLEX rational mapping with NUMERATOR and DENOMINATOR having degrees no more than n has length no longer than $2n\pi$.

References

Edelman, A. and Kostlan, E. "How Many Zeros of a Random Polynomial are Real?" *Bull. Amer. Math. Soc.* **32**, 1–37, 1995.

Spindle Cyclide

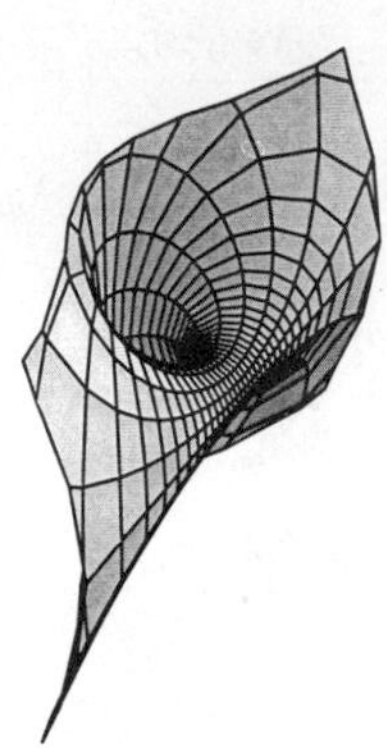

The inversion of a SPINDLE TORUS. If the inversion center lies on the torus, then the spindle cyclide degenerates to a PARABOLIC SPINDLE CYCLIDE.

see also CYCLIDE, HORN CYCLIDE, PARABOLIC CYCLIDE, RING CYCLIDE, SPINDLE TORUS, TORUS

Spindle Torus

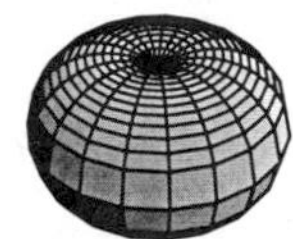  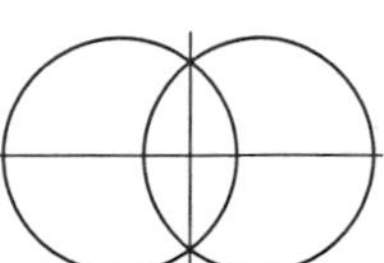

One of the three STANDARD TORI given by the parametric equations

$$x = (c + a\cos v)\cos u$$
$$y = (c + a\cos v)\sin u$$
$$z = a\sin v$$

with $c < a$. The exterior surface is called an APPLE and the interior surface a LEMON. The above left figure shows a spindle torus, the middle a cutaway, and the right figure shows a cross-section of the spindle torus through the xz-plane.

see also APPLE, CYCLIDE, HORN TORUS, LEMON, PARABOLIC SPINDLE CYCLIDE, RING TORUS, SPINDLE CYCLIDE, STANDARD TORI, TORUS

References

Gray, A. "Tori." §11.4 in *Modern Differential Geometry of Curves and Surfaces.* Boca Raton, FL: CRC Press, pp. 218–220, 1993.

Pinkall, U. "Cyclides of Dupin." §3.3 in *Mathematical Models from the Collections of Universities and Museums* (Ed. G. Fischer). Braunschweig, Germany: Vieweg, pp. 28–30, 1986.

Spinode

see also ACNODE, CRUNODE, CUSP, TACNODE

Spinor

A two-component COMPLEX column VECTOR. Spinors are used in physics to represent particles with half-integral spin (i.e., Fermions).

References

Lounesto, P. "Counterexamples to Theorems Published and Proved in Recent Literature on Clifford Algebras, Spinors, Spin Groups, and the Exterior Algebra." http://www.hit.fi/~lounesto/counterexamples.htm.

Morse, P. M. and Feshbach, H. "The Lorentz Transformation, Four-Vectors, Spinors." §1.7 in *Methods of Theoretical Physics, Part I.* New York: McGraw-Hill, pp. 93–107, 1953.

Spira Mirabilis

see LOGARITHMIC SPIRAL

Spiral

In general, a spiral is a curve with $\tau(s)/\kappa(s)$ equal to a constant for all s, where τ is the TORSION and κ is the CURVATURE.

see also ARCHIMEDES' SPIRAL, CIRCLE INVOLUTE, CONICAL SPIRAL, CORNU SPIRAL, COTES' SPIRAL, DAISY, EPISPIRAL, FERMAT'S SPIRAL, HYPERBOLIC SPIRAL, LOGARITHMIC SPIRAL, MICE PROBLEM, NIELSEN'S SPIRAL, PHYLLOTAXIS, POINSOT'S SPIRALS, POLYGONAL SPIRAL, SPHERICAL SPIRAL

References

Eppstein, D. "Spirals." http://www.ics.uci.edu/~eppstein/junkyard/spiral.html.

Lauwerier, H. *Fractals: Endlessly Repeated Geometric Figures.* Princeton, NJ: Princeton University Press, pp. 54–66, 1991.

Lockwood, E. H. "Spirals." Ch. 22 in *A Book of Curves.* Cambridge, England: Cambridge University Press, pp. 172–175, 1967.

Yates, R. C. "Spirals." *A Handbook on Curves and Their Properties.* Ann Arbor, MI: J. W. Edwards, pp. 206–216, 1952.

Spiral Point

A FIXED POINT for which the EIGENVALUES are COMPLEX CONJUGATES.

see also STABLE SPIRAL POINT, UNSTABLE SPIRAL POINT

References

Tabor, M. "Classification of Fixed Points." §1.4.b in *Chaos and Integrability in Nonlinear Dynamics: An Introduction.* New York: Wiley, pp. 22–25, 1989.

Spiric Section

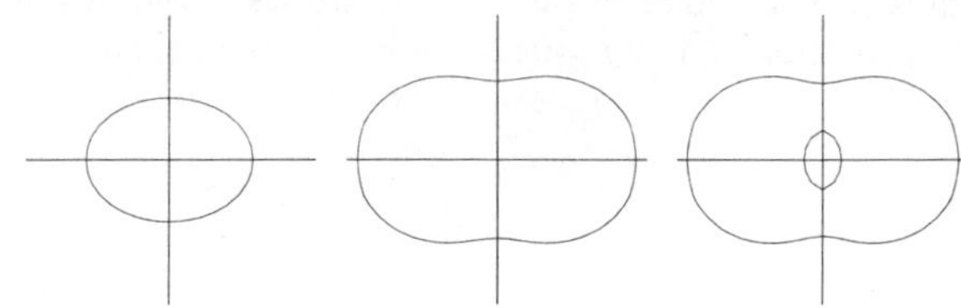

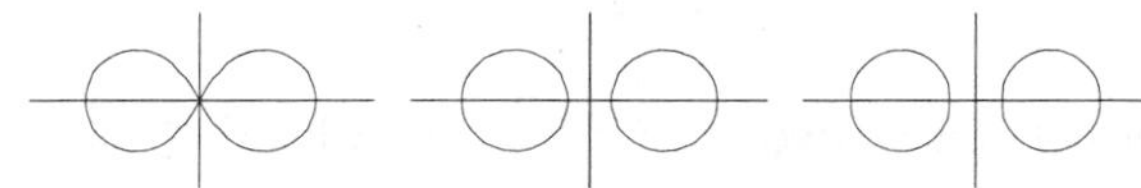

A curve with Cartesian equation

$$(r^2 - a^2 + c^2 + x^2 + y^2) = 4r^2(x^2 + c^2).$$

Around 150 BC, Menaechmus constructed CONIC SECTIONS by cutting a CONE by a PLANE. Two hundred years later, the Greek mathematician Perseus investigated the curves obtained by cutting a TORUS by a PLANE which is PARALLEL to the line through the center of the HOLE of the TORUS (MacTutor).

In the FORMULA of the curve given above, the TORUS is formed from a CIRCLE of RADIUS a whose center is rotated along a CIRCLE of RADIUS r. The value of c gives the distance of the cutting PLANE from the center of the TORUS.

When $c = 0$, the curve consists of two CIRCLES of RADIUS a whose centers are at $(r, 0)$ and $(-r, 0)$. If $c = r + a$, the curve consists of one point (the origin), while if $c > r + a$, no point lies on the curve. The above curves have $(a, b, r) = (3, 4, 2)$, (3, 1, 2) (3, 0.8, 2), (3, 1, 4), (3, 1, 4.5), and (3, 0, 4.5).

References

MacTutor History of Mathematics Archive. "Spiric Sections." http://www-groups.dcs.st-and.ac.uk/~history/Curves/Spiric.html.

Spirograph

A HYPOTROCHOID generated by a fixed point on a CIRCLE rolling inside a fixed CIRCLE. It has parametric equations,

$$x = (R + r)\cos\theta - (r + \rho)\cos\left(\frac{R + r}{r}\theta\right) \quad (1)$$

$$y = (R + r)\sin\theta - (r + \rho)\sin\left(\frac{R + r}{r}\theta\right), \quad (2)$$

where R is the radius of the fixed circle, r is the radius of the rotating circle, and ρ is the offset of the edge of the rotating circle. The figure closes only if R, r, and ρ are RATIONAL. The equations can also be written

$$x = x_0[m\cos t + a\cos(nt)] - y_0[m\sin t - a\sin(nt)] \quad (3)$$

$$y = y_0[m\cos t + a\cos(nt)] + x_0[m\sin t - a\sin(nt)], \quad (4)$$

where the outer wheel has radius 1, the inner wheel a radius p/q, the pen is placed a units from the center, the beginning is at θ radians above the x-axis, and

$$m \equiv \frac{q-p}{q} \quad (5)$$

$$n \equiv \frac{q-p}{p} \quad (6)$$

$$x_0 \equiv \cos\theta \quad (7)$$

$$y_0 \equiv \sin\theta. \quad (8)$$

The following curves are for $a = i/10$, with $i = 1, 2, \ldots, 10$, and $\theta = 0$.

$(p,q) = (1,3)$

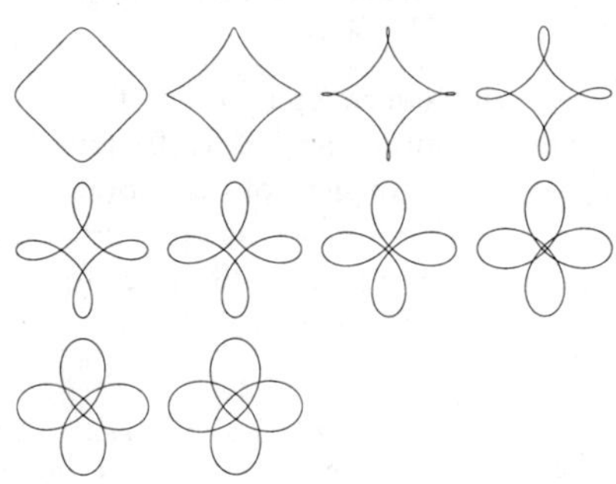

$(p,q) = (1,4)$

$(p,q) = (1,5)$

$(p,q) = (2,5)$

$(p,q) = (2,7)$

$(p,q) = (3,7)$

Additional attractive designs such as the following can also be made by superposing individual spirographs.

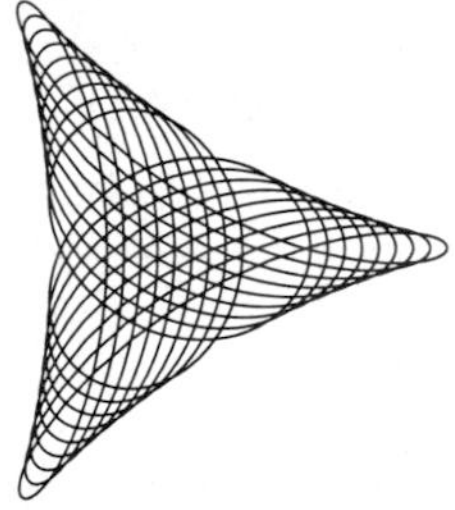

see also EPITROCHOID, HYPOTROCHOID, MAURER ROSE, SPIROLATERAL

Spirolateral

A figure formed by taking a series of steps of length 1, 2, ..., n, with an angle θ turn after each step. The symbol for a spirolateral is ${}^{a_1,\ldots,a_k}n_\theta$, where the a_is indicate that turns are in the $-\theta$ direction for these steps.

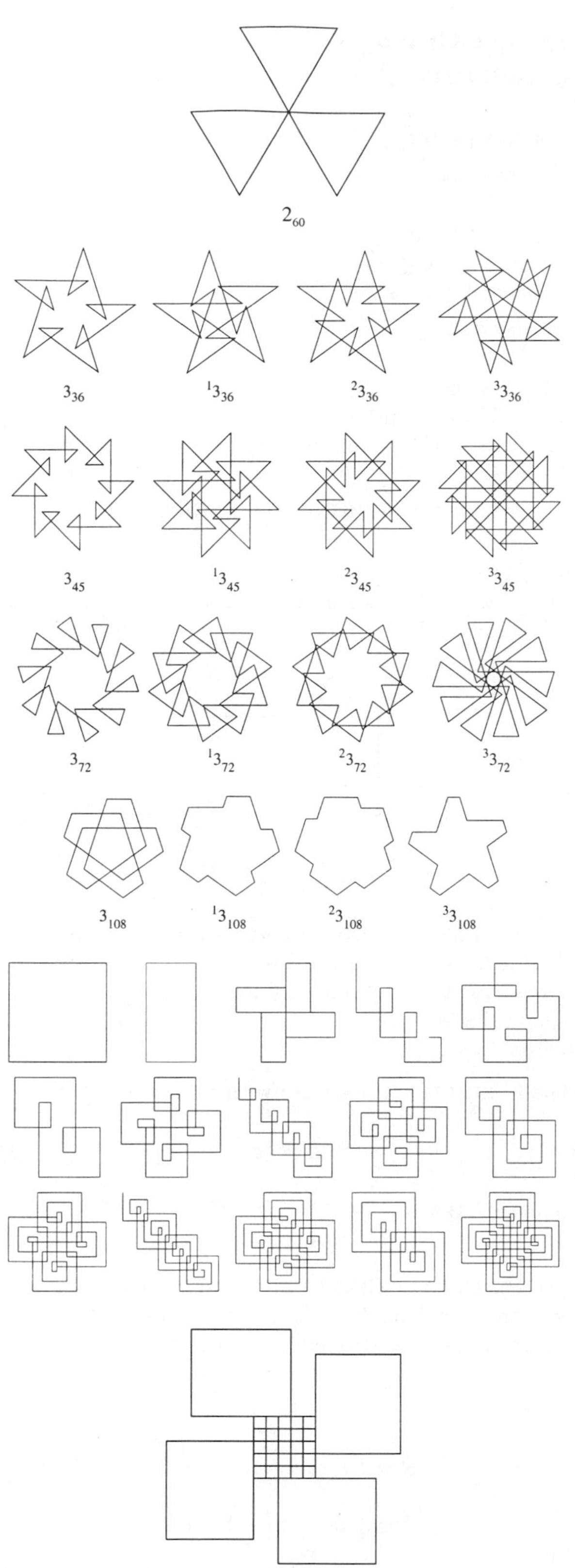

see also MAURER ROSE, SPIROGRAPH

References

Gardner, M. "Worm Paths." Ch. 17 in *Knotted Doughnuts and Other Mathematical Entertainments.* New York: W. H. Freeman, 1986.

Odds, F. C. "Spirolaterals." *Math. Teacher* **66**, 121–124, 1973.

Spline

An interpolating POLYNOMIAL which uses information from neighboring points to obtain a degree of global smoothness.

see also B-SPLINE, BÉZIER SPLINE, CUBIC SPLINE, NURBS CURVE

References

Bartels, R. H.; Beatty, J. C.; and Barsky, B. A. *An Introduction to Splines for Use in Computer Graphics and Geometric Modelling.* San Francisco, CA: Morgan Kaufmann, 1987.

de Boor, C. *A Practical Guide to Splines.* New York: Springer-Verlag, 1978.

Press, W. H.; Flannery, B. P.; Teukolsky, S. A.; and Vetterling, W. T. "Interpolation and Extrapolation." Ch. 3 in *Numerical Recipes in FORTRAN: The Art of Scientific Computing, 2nd ed.* Cambridge, England: Cambridge University Press, pp. 99–122, 1992.

Späth, H. *One Dimensional Spline Interpolation Algorithms.* Wellesley, MA: A. K. Peters, 1995.

Splitting

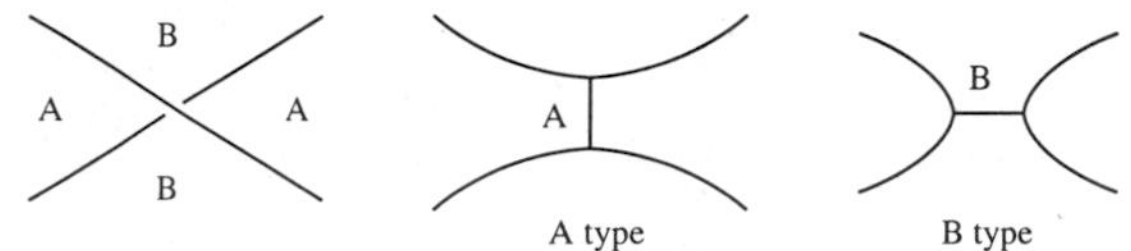

Splitting Algorithm

A method for computing a UNIT FRACTION. This method always terminates (Beeckmans 1993).

References

Beeckmans, L. "The Splitting Algorithm for Egyptian Fractions." *J. Number Th.* **43**, 173–185, 1993.

Sponge

A sponge is a solid which can be parameterized by INTEGERS p, q, and n which satisfy the equation

$$2\sin\left(\frac{\pi}{p}\right)\sin\left(\frac{\pi}{q}\right)=\cos\left(\frac{\pi}{k}\right).$$

The possible sponges are $\{p,q|k\}=\{6,6|3\}$, $\{6,4|4\}$, $\{4,6|4\}$, $\{3,6|6\}$, and $\{4,4|\infty\}$ (Ball and Coxeter 1987).

see also HONEYCOMB, MENGER SPONGE, SIERPIŃSKI SPONGE, TETRIX

References

Ball, W. W. R. and Coxeter, H. S. M. *Mathematical Recreations and Essays, 13th ed.* New York: Dover, p. 152, 1987.

Cromwell, P. R. *Polyhedra.* New York: Cambridge University Press, p. 79, 1997.

Sporadic Group

One of the 26 finite SIMPLE GROUPS. The most complicated is the MONSTER GROUP. A summary, as given by Conway *et al.* (1985), is given below.

Sym	Name	Order	M	A
M_{11}	Mathieu	$2^4 \cdot 3^2 \cdot 5 \cdot 11$	1	1
M_{12}	Mathieu	$2^6 \cdot 3^3 \cdot 5 \cdot 11$	2	2
M_{22}	Mathieu	$2^7 \cdot 3^2 \cdot 5 \cdot 7 \cdot 11$	12	2
M_{23}	Mathieu	$2^7 \cdot 3^2 \cdot 5 \cdot 7 \cdot 11 \cdot 23$	1	1
M_{24}	Mathieu	$2^{10} \cdot 3^3 \cdot 5 \cdot 7 \cdot 11 \cdot 23$	1	1
$J_2 = HJ$	Janko	$2^7 \cdot 3^3 \cdot 5^2 \cdot 7$	2	2
Suz	Suzuki	$2^{13} \cdot 3^7 \cdot 5^2 \cdot 7 \cdot 11 \cdot 13$	6	2
HS	Higman-Sims	$2^9 \cdot 3^2 \cdot 5^3 \cdot 7 \cdot 11$	2	2
McL	McLaughlin	$2^7 \cdot 3^6 \cdot 5^3 \cdot 7 \cdot 11$	3	2
Co_3	Conway	$2^{10} \cdot 3^7 \cdot 5^3 \cdot 7 \cdot 11 \cdot 23$	1	1
Co_2	Conway	$2^{18} \cdot 3^6 \cdot 5^3 \cdot 7 \cdot 11 \cdot 23$	1	1
Co_1	Conway	$2^{21} \cdot 3^9 \cdot 5^4 \cdot 7^2 \cdot 11 \cdot 13 \cdot 23$	2	1
He	Held	$2^{10} \cdot 3^3 \cdot 5^2 \cdot 7^3 \cdot 17$	1	2
Fi_{22}	Fischer	$2^{17} \cdot 3^9 \cdot 5^2 \cdot 7 \cdot 11 \cdot 13$	6	2
Fi_{23}	Fischer	$2^{18} \cdot 3^{13} \cdot 5^2 \cdot 7 \cdot 11 \cdot 13 \cdot 17 \cdot 23$	1	1
Fi'_{24}	Fischer	$2^{21} \cdot 3^{16} \cdot 5^2 \cdot 7^3 \cdot 11 \cdot 13 \cdot 17$ $\cdot 23 \cdot 29$	3	2
HN	Harada-Norton	$2^{14} \cdot 3^6 \cdot 5^6 \cdot 7 \cdot 11 \cdot 19$	1	2
Th	Thompson	$2^{15} \cdot 3^{10} \cdot 5^3 \cdot 7^2 \cdot 13 \cdot 19 \cdot 31$	1	1
B	Baby Monster	$2^{41} \cdot 3^{13} \cdot 5^6 \cdot 7^2 \cdot 11 \cdot 13 \cdot 17 \cdot 19$ $\cdot 23 \cdot 31 \cdot 47$	2	1
M	Monster	$2^{46} \cdot 3^{20} \cdot 5^9 \cdot 7^6 \cdot 11^2 \cdot 13^3 \cdot 17 \cdot 19$ $\cdot 23 \cdot \cdot 29 \cdot 31 \cdot 41 \cdot 47 \cdot 59 \cdot 71$	1	1
J_1	Janko	$2^3 \cdot 3 \cdot 5 \cdot 7 \cdot 11 \cdot 19$	1	1
$O'N$	O'Nan	$2^9 \cdot 3^4 \cdot 7^3 \cdot 5 \cdot 11 \cdot 19 \cdot 31$	3	2
J_3	Janko	$2^7 \cdot 3^5 \cdot 5 \cdot 17 \cdot 19$	3	2
Ly	Lyons	$2^8 \cdot 3^7 \cdot 5^6 \cdot 7 \cdot 11 \cdot 31 \cdot 37 \cdot 67$	1	1
Ru	Rudvalis	$2^{14} \cdot 3^3 \cdot 5^3 \cdot 7 \cdot 13 \cdot 29$	2	1
J_4	Janko	$2^{21} \cdot 3^3 \cdot 5 \cdot 7 \cdot 11^3 \cdot 23 \cdot 29 \cdot 31$ $\cdot 37 \cdot 43$	1	1

see also BABY MONSTER GROUP, CONWAY GROUPS, FISCHER GROUPS, HARADA-NORTON GROUP, HELD GROUP, HIGMAN-SIMS GROUP, JANKO GROUPS, LYONS GROUP, MATHIEU GROUPS, MCLAUGHLIN GROUP, MONSTER GROUP, O'NAN GROUP, RUDVALIS GROUP, SUZUKI GROUP, THOMPSON GROUP

References

Aschbacher, M. *Sporadic Groups.* New York: Cambridge University Press, 1994.

Conway, J. H.; Curtis, R. T.; Norton, S. P.; Parker, R. A.; and Wilson, R. A. *Atlas of Finite Groups: Maximal Subgroups and Ordinary Characters for Simple Groups.* Oxford, England: Clarendon Press, p. viii, 1985.

Math. Intell. Cover of volume **2**, 1980.

Wilson, R. A. "ATLAS of Finite Group Representation." http://for.mat.bham.ac.uk/atlas#spo.

Sports

see also BASEBALL, BOWLING, CHECKERS, CHESS, GO

Sprague-Grundy Function

see NIM-VALUE

Sprague-Grundy Number

see NIM-VALUE

Sprague-Grundy Value

see NIM-VALUE

Spread (Link)

see SPAN (LINK)

Spread (Tree)

A TREE having an infinite number of branches and whose nodes are sequences generated by a set of rules.

see also FAN

Spun Knot

A 3-D KNOT spun about a plane in 4-D. Unlike SUSPENDED KNOTS, spun knots are smoothly embedded at the poles.

see also SUSPENDED KNOT, TWIST-SPUN KNOT

Squarable

An object which can be constructed by SQUARING is called squarable.

Square

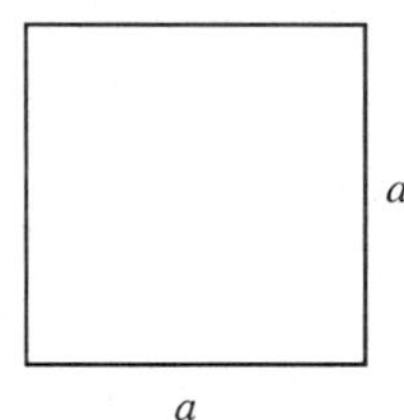

The term square is sometimes used to mean SQUARE NUMBER. When used in reference to a geometric figure, however, it means a convex QUADRILATERAL with four equal sides at RIGHT ANGLES to each other, illustrated above.

The PERIMETER of a square with side length a is

$$L = 4a \tag{1}$$

and the AREA is

$$A = a^2. \tag{2}$$

The INRADIUS r, CIRCUMRADIUS R, and AREA A can be computed directly from the formulas for a general regular POLYGON with side length a and $n = 4$ sides,

$$r = \tfrac{1}{2} a \cot\left(\frac{\pi}{4}\right) = \tfrac{1}{2} a \tag{3}$$

$$R = \tfrac{1}{2} a \csc\left(\frac{\pi}{4}\right) = \tfrac{1}{2}\sqrt{2}\, a \tag{4}$$

$$A = \tfrac{1}{4} n a^2 \cot\left(\frac{\pi}{4}\right) = a^2. \tag{5}$$

The length of the DIAGONAL of the UNIT SQUARE is $\sqrt{2}$, sometimes known as PYTHAGORAS'S CONSTANT.

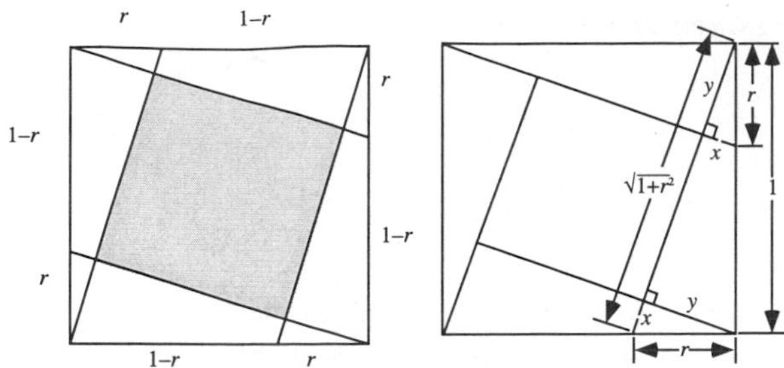

The AREA of a square inscribed inside a UNIT SQUARE as shown in the above diagram can be found as follows. Label x and y as shown, then

$$x^2 + y^2 = r^2 \tag{6}$$

$$(\sqrt{1+r^2} - x)^2 + y^2 = 1. \tag{7}$$

Plugging (6) into (7) gives

$$(\sqrt{1+r^2} - x)^2 + (r^2 - x^2) = 1. \tag{8}$$

Expanding

$$x^2 - 2x\sqrt{1+r^2} + 1 + r^2 + r^2 - x^2 = 1 \tag{9}$$

and solving for x gives

$$x = \frac{r^2}{\sqrt{1+r^2}}. \tag{10}$$

Plugging in for y yields

$$y = \sqrt{r^2 - x^2} = \frac{r}{\sqrt{1+r^2}}. \tag{11}$$

The area of the shaded square is then

$$A = (\sqrt{1+r^2} - x - y)^2 = \frac{(1-r)^2}{1+r^2} \tag{12}$$

(Detemple and Harold 1996).

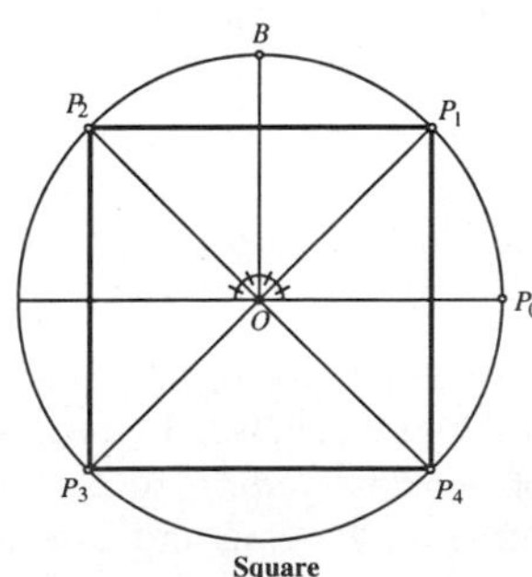

Square

The STRAIGHTEDGE and COMPASS construction of the square is simple. Draw the line OP_0 and construct a circle having OP_0 as a radius. Then construct the perpendicular OB through O. Bisect P_0OB and $P_0'OB$ to locate P_1 and P_2, where P_0' is opposite P_0. Similarly, construct P_3 and P_4 on the other SEMICIRCLE. Connecting $P_1P_2P_3P_4$ then gives a square.

As shown by Schnirelmann, a square can be INSCRIBED in any closed convex planar curve (Steinhaus 1983). A square can also be CIRCUMSCRIBED about any closed curve (Steinhaus 1983).

An infinity of points in the interior of a square are known whose distances from three of the corners of a square are RATIONAL NUMBERS. Calling the distances a, b, and c where s is the side length of the square, these solutions satisfy

$$(s^2 + b^2 - a^2)^2 + (s^2 + b^2 - c^2)^2 = (2bs)^2 \tag{13}$$

(Guy 1994). In this problem, one of a, b, c, and s is DIVISIBLE by 3, one by 4, and one by 5. It is not known if there are points having distances from *all four* corners RATIONAL, but such a solution requires the additional condition

$$a^2 + c^2 = b^2 + d^2. \tag{14}$$

In this problem, s is DIVISIBLE by 4 and a, b, c, and d are ODD. If s is not DIVISIBLE by 3 (5), then two of a, b, c, and d are DIVISIBLE by 3 (5) (Guy 1994).

see also BROWKIN'S THEOREM, DISSECTION, DOUGLAS-NEUMANN THEOREM, FINSLER-HADWIGER THEOREM, LOZENGE, PERFECT SQUARE DISSECTION, PYTHAGORAS'S CONSTANT, PYTHAGOREAN SQUARE PUZZLE, RECTANGLE, SQUARE CUTTING, SQUARE NUMBER, SQUARE PACKING, SQUARE QUADRANTS, UNIT SQUARE, VON AUBEL'S THEOREM

References

Detemple, D. and Harold, S. "A Round-Up of Square Problems." *Math. Mag.* **69**, 15–27, 1996.

Dixon, R. *Mathographics.* New York: Dover, p. 16, 1991.

Eppstein, D. "Rectilinear Geometry." `http://www.ics.uci.edu/~eppstein/junkyard/rect.html`.

Guy, R. K. "Rational Distances from the Corners of a Square." §D19 in *Unsolved Problems in Number Theory, 2nd ed.* New York: Springer-Verlag, pp. 181–185, 1994.

Steinhaus, H. *Mathematical Snapshots, 3rd American ed.* New York: Oxford University Press, p. 104, 1983.

Square Bracket Polynomial

A POLYNOMIAL which is not necessarily an invariant of a LINK. It is related to the DICHROIC POLYNOMIAL. It is defined by the SKEIN RELATIONSHIP

$$B_{L_+} = q^{-1/2} v B_{L_0} + B_{L_\infty}, \tag{1}$$

and satisfies

$$B_{\text{unknot}} = q^{1/2} \tag{2}$$

and

$$B_{L \cup \text{unknot}} = q^{1/2} B_L. \tag{3}$$

References

Adams, C. C. *The Knot Book: An Elementary Introduction to the Mathematical Theory of Knots.* New York: W. H. Freeman, pp. 235–241, 1994.

Square Cupola

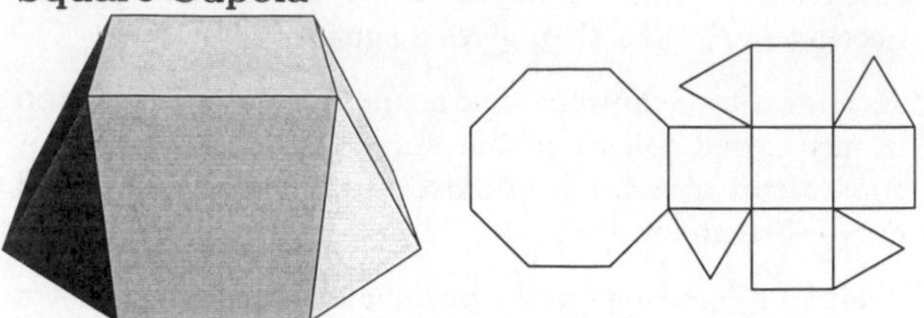

JOHNSON SOLID J_4. The bottom eight VERTICES are

$$(\pm\tfrac{1}{2}(1+\sqrt{2}), \pm\tfrac{1}{2}, 0), (\pm\tfrac{1}{2}, \pm\tfrac{1}{2}(1+\sqrt{2}), 0),$$

and the top four VERTICES are

$$\left(\pm\frac{1}{\sqrt{2}}, 0, \frac{1}{\sqrt{2}}\right), \left(0, \pm\frac{1}{\sqrt{2}}, \frac{1}{\sqrt{2}}\right).$$

Square Curve

see SIERPIŃSKI CURVE

Square Cutting

The average number of regions into which N lines divide a SQUARE is

$$\tfrac{1}{16}N(N-1)\pi + N + 1$$

(Santaló 1976).

see also CIRCLE CUTTING

References

Finch, S. "Favorite Mathematical Constants." http://www.mathsoft.com/asolve/constant/geom/geom.html.
Santaló, L. A. *Integral Geometry and Geometric Probability.* Reading, MA: Addison-Wesley, 1976.

Square-Free

see SQUAREFREE

Square Gyrobicupola

see JOHNSON SOLID

Square Integrable

A function $f(x)$ is said to be square integrable if

$$\int_{-\infty}^{\infty} |f(x)|^2 \, dx$$

is finite.

see also INTEGRABLE, L_2-NORM, TITCHMARSH THEOREM

References

Sansone, G. "Square Integrable Functions." §1.1 in *Orthogonal Functions, rev. English ed.* New York: Dover, pp. 1–2, 1991.

Square Knot

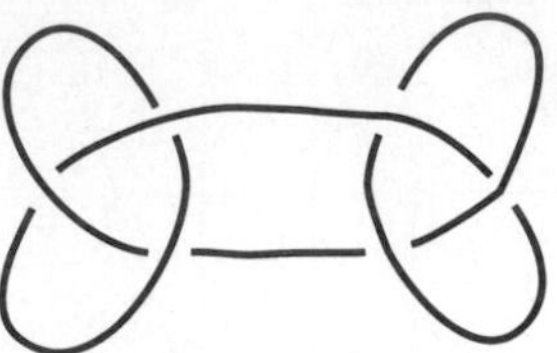

A composite KNOT of six crossings consisting of a KNOT SUM of a TREFOIL KNOT and its MIRROR IMAGE. The GRANNY KNOT has the same ALEXANDER POLYNOMIAL $(x^2-x+1)^2$ as the square knot. The square knot is also called the REEF KNOT.

see also GRANNY KNOT, MIRROR IMAGE, TREFOIL KNOT

References

Owen, P. *Knots.* Philadelphia, PA: Courage, p. 50, 1993.

Square Matrix

A MATRIX for which horizontal and vertical dimensions are the same (i.e., an $n \times n$ MATRIX).

see also MATRIX

Square Number

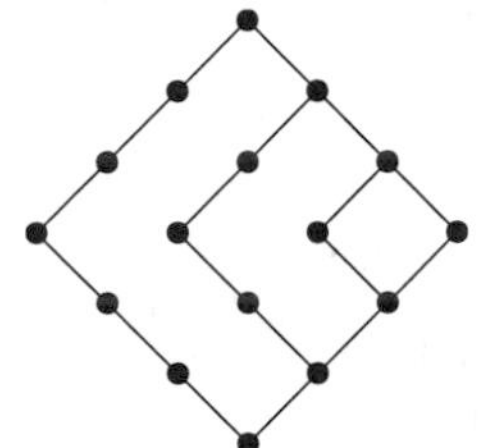

A FIGURATE NUMBER of the form $m = n^2$, where n is an INTEGER. A square number is also called a PERFECT SQUARE. The first few square numbers are 1, 4, 9, 25, 36, 49, ... (Sloane's A000290). The GENERATING FUNCTION giving the square numbers is

$$\frac{x(x+1)}{(1-x)^3} = x + 4x^2 + 9x^3 + 16x^4 + \ldots. \tag{1}$$

The kth nonsquare number a_k is given by

$$a_n = n + \left\lfloor \tfrac{1}{2} + \sqrt{n} \right\rfloor, \tag{2}$$

where $\lfloor x \rfloor$ is the FLOOR FUNCTION, and the first few are 2, 3, 5, 6, 7, 8, 10, 11, ... (Sloane's A000037).

The only numbers which are simultaneously square and PYRAMIDAL (the CANNONBALL PROBLEM) are $P_1 = 1$ and $P_{24} = 4900$, corresponding to $S_1 = 1$ and $S_{70} = 4900$ (Dickson 1952, p. 25; Ball and Coxeter 1987, p. 59; Ogilvy 1988), as conjectured by Lucas (1875, 1876) and proved by Watson (1918). The CANNONBALL PROBLEM is equivalent to solving the DIOPHANTINE EQUATION

$$y^2 = \tfrac{1}{6}x(x+1)(2x+1) \tag{3}$$

(Guy 1994, p. 147).

The only numbers which are square and TETRAHEDRAL are $Te_1 = 1$, $Te_2 = 4$, and $Te_{48} = 19600$ (giving $S_1 = 1$, $S_2 = 4$, and $S_{140} = 19600$), as proved by Meyl (1878; cited in Dickson 1952, p. 25; Guy 1994, p. 147). In general, proving that only certain numbers are simultaneously figurate in two different ways is far from elementary.

To find the possible last digits for a square number, write $n = 10a+b$ for the number written in decimal NOTATION as ab_{10} (a, $b = 0, 1, \ldots, 9$). Then

$$n^2 = 100a^2 + 20ab + b^2, \tag{4}$$

so the last digit of n^2 is the same as the last digit of b^2. The following table gives the last digit of b^2 for $b = 0$, $1, \ldots, 9$. As can be seen, the last digit can be only 0, 1, 4, 5, 6, or 9.

0	1	2	3	4	5	6	7	8	9
0	1	4	9	_6	_5	_6	_9	_4	_1

We can similarly examine the allowable last two digits by writing abc_{10} as

$$n = 100a + 10b + c, \tag{5}$$

so

$$\begin{aligned} n^2 &= (100a + 10b + c)^2 \\ &= 10^4a^2 + 2(1000ab + 100ac + 10bc) + 100b^2 + c^2 \\ &= (10^4a^2 + 2000ab + 100ac + 100b^2) + 20bc + c^2, \end{aligned} \tag{6}$$

so the last two digits are given by $20bc+c^2 = c(20b+c)$. But since the last digit must be 0, 1, 4, 5, 6, or 9, the following table exhausts all possible last two digits.

c	*b*									
	0	1	2	3	4	5	6	7	8	9
1	01	21	41	61	81	_01	_21	_41	_61	_81
4	16	96	_76	_56	_36	_16	_96	_76	_56	_36
5	25	_25	_25	_25	_25	_25	_25	_25	_25	_25
6	36	_56	_76	_96	_16	_36	_56	_76	_96	_16
9	81	_61	_41	_21	_01	_81	_61	_41	_21	_01

The only possibilities are 00, 01, 04, 09, 16, 21, 24, 25, 29, 36, 41, 44, 49, 56, 61, 64, 69, 76, 81, 84, 89, and 96, which can be summarized succinctly as 00, $e1$, $e4$, 25, $o6$, and $e9$, where e stands for an EVEN NUMBER and o for an ODD NUMBER. Additionally, unless the sum of the digits of a number is 1, 4, 7, or 9, it cannot be a square number.

The following table gives the possible residues mod n for square numbers for $n = 1$ to 20. The quantity $s(n)$ gives the number of distinct residues for a given n.

n	$s(n)$	$x^2 \pmod n$
2	2	0, 1
3	2	0, 1
4	2	0, 1
5	3	0, 1, 4
6	4	0, 1, 3, 4
7	4	0, 1, 2, 4
8	3	0, 1, 4
9	4	0, 1, 4, 7
10	6	0, 1, 4, 5, 6, 9
11	6	0, 1, 3, 4, 5, 9
12	4	0, 1, 4, 9
13	7	0, 1, 3, 4, 9, 10, 12
14	8	0, 1, 2, 4, 7, 8, 9, 11
15	6	0, 1, 4, 6, 9, 10
16	4	0, 1, 4, 9
17	9	0, 1, 2, 4, 8, 9, 13, 15, 16
18	8	0, 1, 4, 7, 9, 10, 13, 16
19	10	0, 1, 4, 5, 6, 7, 9, 11, 16, 17
20	6	0, 1, 4, 5, 9, 16

In general, the ODD squares are congruent to 1 (mod 8) (Conway and Guy 1996). Stangl (1996) gives an explicit formula by which the number of squares $s(n)$ in $\mathbb{Z}_n$ (i.e., mod n) can be calculated. Let p be an ODD PRIME. Then $s(n)$ is the MULTIPLICATIVE FUNCTION given by

$$s(2) = 2 \tag{7}$$

$$s(p) = \tfrac{1}{2}(p+1) \qquad (p \neq 2) \tag{8}$$

$$s(p^2) = \tfrac{1}{2}(p^2 - p + 2) \qquad (p \neq 2) \tag{9}$$

$$s(2^n) = \begin{cases} \frac{1}{3}(2^{n-1}+4) & \text{for } n \text{ even} \\ \frac{1}{3}(2^{n-1}+5) & \text{for } n \text{ odd} \end{cases} \tag{10}$$

$$s(p^n) = \begin{cases} \frac{p^{n+1}+p+2}{2(p+1)} & \text{for } n \geq 3 \text{ even} \\ \frac{p^{n+1}+2p+1}{2(p+1)} & \text{for } n \geq 3 \text{ odd.} \end{cases} \tag{11}$$

$s(n)$ is related to the number $q(n)$ of QUADRATIC RESIDUES in $\mathbb{Z}_n$ by

$$q(p^n) = s(p^n) - s(p^{n-2}) \tag{12}$$

for $n \geq 3$ (Stangl 1996).

For a perfect square n, $(n/p) = 0$ or 1 for all ODD PRIMES $p < n$ where (n/p) is the LEGENDRE SYMBOL. A number n which is not a perfect square but which satisfies this relationship is called a PSEUDOSQUARE.

The minimum number of squares needed to represent the numbers 1, 2, 3, ... are 1, 2, 3, 1, 2, 3, 4, 2, 1, 2, ... (Sloane's A002828), and the number of distinct ways to represent the numbers 1, 2, 3, ... in terms of squares are 1, 1, 1, 2, 2, 2, 2, 3, 4, 4, ... (Sloane's A001156). A brute-force algorithm for enumerating the square permutations of n is repeated application of the GREEDY ALGORITHM. However, this approach rapidly becomes impractical since the number of representations grows extremely rapidly with n, as shown in the following table.

n	Square Partitions
10	4
50	104
100	1116
150	6521
200	27482

Every POSITIVE integer is expressible as a SUM of (at most) $g(2) = 4$ square numbers (WARING'S PROBLEM). (Actually, the basis set is {0, 1, 4, 9, 16, 25, 36, 64, 81, 100, ...}, so 49 need never be used.) Furthermore, an infinite number of n require four squares to represent them, so the related quantity $G(2)$ (the least INTEGER n such that every POSITIVE INTEGER beyond a certain point requires $G(2)$ squares) is given by $G(2) = 4$.

Numbers expressible as the sum of two squares are those whose PRIME FACTORS are of the form $4k - 1$ taken to an EVEN POWER. Numbers expressible as the sum of three squares are those not of the form $4^k(8l + 7)$ for $k, l \geq 0$. The following table gives the first few numbers which require N = 1, 2, 3, and 4 squares to represent them as a sum.

N	Sloane	Numbers
1	000290	1, 4, 9, 16, 25, 36, 49, 64, 81, ...
2	000415	2, 5, 8, 10, 13, 17, 18, 20, 26, 29, ...
3	000419	3, 6, 11, 12, 14, 19, 21, 22, 24, 27, ...
4	004215	7, 15, 23, 28, 31, 39, 47, 55, 60, 63, ...

The FERMAT $4n + 1$ THEOREM guarantees that every PRIME of the form $4n+1$ is a sum of two SQUARE NUMBERS in only one way.

There are only 31 numbers which cannot be expressed as the sum of *distinct* squares: 2, 3, 6, 7, 8, 11, 12, 15, 18, 19, 22, 23, 24, 27, 28, 31, 32, 33, 43, 44, 47, 48, 60, 67, 72, 76, 92, 96, 108, 112, 128 (Sloane's A001422; Guy 1994). All numbers > 188 can be expressed as the sum of at most five distinct squares, and only

$$124 = 1 + 4 + 9 + 25 + 36 + 49 \tag{13}$$

and

$$188 = 1 + 4 + 9 + 25 + 49 + 100 \tag{14}$$

require six distinct squares (Bohman *et al.* 1979; Guy 1994, p. 136). In fact, 188 can also be represented using seven distinct squares:

$$188 = 1 + 4 + 9 + 25 + 36 + 49 + 64. \tag{15}$$

The following table gives the numbers which can be represented in W different ways as a sum of S squares. For example,

$$50 = 1^2 + 7^2 = 5^2 + 5^2$$

can be represented in two ways ($W = 2$) by two squares ($S = 2$).

S	W	Sloane	Numbers
1	1	000290	1, 4, 9, 16, 25, 36, 49, 64, 81, 100, ...
2	1	025284	2, 5, 8, 10, 13, 17, 18, 20, 25, 26, 29, ...
2	2	025285	50, 65, 85, 125, 130, 145, 170, 185, ...
3	1	025321	3, 6, 9, 11, 12, 14, 17, 18, 19, 21, 22, ...
3	2	025322	27, 33, 38, 41, 51, 57, 59, 62, 69, 74, ...
3	3	025323	54, 66, 81, 86, 89, 99, 101, 110, 114, ...
3	4	025324	129, 134, 146, 153, 161, 171, 189, ...
4	1	025357	4, 7, 10, 12, 13, 15, 16, 18, 19, 20, ...
4	2	025358	31, 34, 36, 37, 39, 43, 45, 47, 49, ...
4	3	025359	28, 42, 55, 60, 66, 67, 73, 75, 78, ...
4	4	025360	52, 58, 63, 70, 76, 84, 87, 91, 93, ...

The number of INTEGERS $< x$ which are squares or sums of two squares is

$$N(x) \sim kx(\ln x)^{-1/2}, \tag{16}$$

where

$$k = \sqrt{\frac{1}{2} \prod_{\substack{r=4n+3 \\ r \text{ prime}}} (1 - r^{-2})^{-1}} \tag{17}$$

(Landau 1908; Le Lionnais 1983, p. 31). The product of four distinct NONZERO INTEGERS in ARITHMETIC PROGRESSION is square only for (−3, −1, 1, 3), giving $(-3)(-1)(1)(3) = 9$ (Le Lionnais 1983, p. 53). It is possible to have three squares in ARITHMETIC PROGRESSION, but not four (Dickson 1952, pp. 435–440). If these numbers are r^2, s^2, and t^2, there are POSITIVE INTEGERS p and q such that

$$r = |p^2 - 2pq - q^2| \tag{18}$$

$$s = p^2 + q^2 \tag{19}$$

$$t = p^2 + 2pq - q^2, \tag{20}$$

where $(p, q) = 1$ and one of r, s, or t is EVEN (Dickson 1952, pp. 437–438). Every three-term progression of squares can be associated with a PYTHAGOREAN TRIPLE (X, Y, Z) by

$$X = \tfrac{1}{2}(r + t) \tag{21}$$

$$Y = \tfrac{1}{2}(t - r) \tag{22}$$

$$Z = s \tag{23}$$

(Robertson 1996).

CATALAN'S CONJECTURE states that 8 and 9 (2^3 and 3^2) are the only consecutive POWERS (excluding 0 and 1), i.e., the only solution to CATALAN'S DIOPHANTINE PROBLEM. This CONJECTURE has not yet been proved or refuted, although R. Tijdeman has proved that there can be only a finite number of exceptions should the CONJECTURE not hold. It is also known that 8 and 9 are the only consecutive CUBIC and square numbers (in either order).

A square number can be the concatenation of two squares, as in the case $16 = 4^2$ and $9 = 3^2$ giving $169 = 13^2$.

It is conjectured that, other than 10^{2n}, 4×10^{2n} and 9×10^{2n}, there are only a FINITE number of squares n^2 having exactly two distinct NONZERO DIGITS (Guy 1994, p. 262). The first few such n are 4, 5, 6, 7, 8, 9, 11, 12, 15, 21, ... (Sloane's A016070), corresponding to n^2 of 16, 25, 36, 49, 64, 81, 121, ... (Sloane's A016069).

The following table gives the first few numbers which, when squared, give numbers composed of only certain digits. The only known square number composed only of the digits 7, 8, and 9 is 9. Vardi (1991) considers numbers composed only of the square digits: 1, 4, and 9.

Digits	Sloane	n, n^2
1, 2, 3	030175	1, 11, 111, 36361, 363639, ...
	030174	1, 121, 12321, 1322122321, ...
1, 4, 6	027677	1, 2, 4, 8, 12, 31, 38, 108, ...
	027676	1, 4, 16, 64, 144, 441, 1444, ...
1, 4, 9	027675	1, 2, 3, 7, 12, 21, 38, 107, ...
	006716	1, 4, 9, 49, 144, 441, 1444, 11449, ...
2, 4, 8	027679	2, 22, 168, 478, 2878, 210912978, ...
	027678	4, 484, 28224, 228484, 8282884, ...
4, 5, 6	030177	2, 8, 216, 238, 258, 738, 6742, ...
	030176	4, 64, 46656, 56644, 66564, ...

BROWN NUMBERS are pairs (m,n) of INTEGERS satisfying the condition of BROCARD'S PROBLEM, i.e., such that

$$n! + 1 = m^2, \tag{24}$$

where $n!$ is a FACTORIAL. Only three such numbers are known: (5,4), (11,5), (71,7). Erdős conjectured that these are the only three such pairs.

Either $5x^2 + 4 = y^2$ or $5x^2 - 4 = y^2$ has a solution in POSITIVE INTEGERS IFF, for some n, $(x, y) = (F_n, L_n)$, where F_n is a FIBONACCI NUMBER and L_n is a LUCAS NUMBER (Honsberger 1985, pp. 114–118).

The smallest and largest square numbers containing the digits 1 to 9 are

$$11,826^2 = 139,854,276, \tag{25}$$

$$30,384^2 = 923,187,456. \tag{26}$$

The smallest and largest square numbers containing the digits 0 to 9 are

$$32,043^2 = 1,026,753,849, \tag{27}$$

$$99,066^2 = 9,814,072,356 \tag{28}$$

(Madachy 1979, p. 159). The smallest and largest square numbers containing the digits 1 to 9 twice each are

$$335,180,136^2 = 112,345,723,568,978,496 \tag{29}$$

$$999,390,432^2 = 998,781,235,573,146,624, \tag{30}$$

and the smallest and largest containing 1 to 9 three times are

$$10,546,200,195,312^2 = 111,222,338,559,598,866,946,777,344 \tag{31}$$

$$31,621,017,808,182^2 = 999,888,767,225,363,175,346,145,124 \tag{32}$$

(Madachy 1979, p. 159).

Madachy (1979, p. 165) also considers number which are equal to the sum of the squares of their two "halves" such as

$$1233 = 12^2 + 33^2 \tag{33}$$

$$8833 = 88^2 + 33^2 \tag{34}$$

$$10100 = 10^2 + 100^2 \tag{35}$$

$$5882353 = 588^2 + 2353^2, \tag{36}$$

in addition to a number of others.

see also ANTISQUARE NUMBER, BIQUADRATIC NUMBER, BROCARD'S PROBLEM, BROWN NUMBERS, CANNONBALL PROBLEM, CATALAN'S CONJECTURE, CENTERED SQUARE NUMBER, CLARK'S TRIANGLE, CUBIC NUMBER, DIOPHANTINE EQUATION, FERMAT $4n + 1$ THEOREM, GREEDY ALGORITHM, GROSS, LAGRANGE'S FOUR-SQUARE THEOREM, LANDAU-RAMANUJAN CONSTANT, PSEUDOSQUARE, PYRAMIDAL NUMBER, $r_k(n)$, SQUAREFREE, SQUARE TRIANGULAR NUMBER, WARING'S PROBLEM

References

Ball, W. W. R. and Coxeter, H. S. M. *Mathematical Recreations and Essays, 13th ed.* New York: Dover, p. 59, 1987.

Bohman, J.; Fröberg, C.-E.; and Riesel, H. "Partitions in Squares." *BIT* **19**, 297–301, 1979.

Conway, J. H. and Guy, R. K. *The Book of Numbers.* New York: Springer-Verlag, pp. 30–32, 1996.

Dickson, L. E. *History of the Theory of Numbers, Vol. 2: Diophantine Analysis.* New York: Chelsea, 1952.

Grosswald, E. *Representations of Integers as Sums of Squares.* New York: Springer-Verlag, 1985.

Guy, R. K. "Sums of Squares" and "Squares with Just Two Different Decimal Digits." §C20 and F24 in *Unsolved Problems in Number Theory, 2nd ed.* New York: Springer-Verlag, pp. 136–138 and 262, 1994.

Honsberger, R. "A Second Look at the Fibonacci and Lucas Numbers." Ch. 8 in *Mathematical Gems III.* Washington, DC: Math. Assoc. Amer., 1985.

Le Lionnais, F. *Les nombres remarquables.* Paris: Hermann, 1983.

Lucas, É. Question 1180. *Nouv. Ann. Math. Ser. 2* **14**, 336, 1875.

Lucas, É. Solution de Question 1180. *Nouv. Ann. Math. Ser. 2* **15**, 429–432, 1876.

Madachy, J. S. *Madachy's Mathematical Recreations.* New York: Dover, pp. 159 and 165, 1979.

Meyl, A.-J.-J. Solution de Question 1194. *Nouv. Ann. Math.* **17**, 464–467, 1878.

Ogilvy, C. S. and Anderson, J. T. *Excursions in Number Theory.* New York: Dover, pp. 77 and 152, 1988.

Pappas, T. "Triangular, Square & Pentagonal Numbers." *The Joy of Mathematics.* San Carlos, CA: Wide World Publ./Tetra, p. 214, 1989.
Pietenpol, J. L. "Square Triangular Numbers." *Amer. Math. Monthly* **69**, 168–169, 1962.
Robertson, J. P. "Magic Squares of Squares." *Math. Mag.* **69**, 289-293, 1996.
Stangl, W. D. "Counting Squares in $\mathbb{Z}_n$." *Math. Mag.* **69**, 285–289, 1996.
Taussky-Todd, O. "Sums of Squares." *Amer. Math. Monthly* **77**, 805–830, 1970.
Vardi, I. *Computational Recreations in Mathematica.* Reading, MA: Addison-Wesley, pp. 20 and 234–237, 1991.
Watson, G. N. "The Problem of the Square Pyramid." *Messenger. Math.* **48**, 1–22, 1918.

Square Orthobicupola

see JOHNSON SOLID

Square Packing

Find the minimum size SQUARE capable of bounding n equal SQUARES arranged in any configuration. The only packings which have been proven optimal are 2, 3, 5, and SQUARE NUMBERS (4, 9, ...). If $n = a^2 - a$ for some a, it is CONJECTURED that the size of the minimum bounding square is a for small n. The smallest n for which the CONJECTURE is known to be violated is 1560. The size is known to scale as k^b, where

$$\tfrac{1}{2}(3-\sqrt{3}) < b < \tfrac{1}{2}.$$

n	Exact	Decimal
1	1	1
2	2	2
3	2	2
4	2	2
5	$2+\frac{1}{2}\sqrt{2}$	2.707...
6	3	3
7	3	3
8	3	3
9	3	3
10	$3+\frac{1}{2}\sqrt{2}$	3.707...
11		3.877...
12	4	4
13	4	4
14	4	4
15	4	4
16	4	4
17	$4+\frac{1}{2}\sqrt{2}$	4.707...
18	$2(7+\sqrt{7})$	4.822...
19	$3+\frac{4}{3}\sqrt{2}$	4.885...
20	5	5
21	5	5
22	5	5
23	5	5
24	5	5
25	5	5
26		5.650...

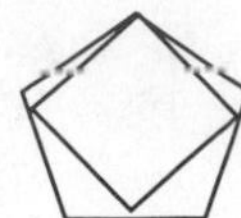

The best packing of a SQUARE inside a PENTAGON, illustrated above, is 1.0673....

References

Erdős, P. and Graham, R. L. "On Packing Squares with Equal Squares." *J. Combin. Th. Ser. A* **19**, 119–123, 1975.
Friedman, E. "Packing Unit Squares in Squares." *Elec. J. Combin.* DS7, 1–24, Mar. 5, 1998. http://www.combinatorics.org/Surveys/.
Gardner, M. "Packing Squares." Ch. 20 in *Fractal Music, Hypercards, and More Mathematical Recreations from Scientific American Magazine.* New York: W. H. Freeman, 1992.
Göbel, F. "Geometrical Packing and Covering Problems." In *Packing and Covering in Combinatorics* (Ed. A. Schrijver). Amsterdam: Tweede Boerhaavestraat, 1979.
Roth, L. F. and Vaughan, R. C. "Inefficiency in Packing Squares with Unit Squares." *J. Combin. Th. Ser. A* **24**, 170–186, 1978.

Square Polyomino

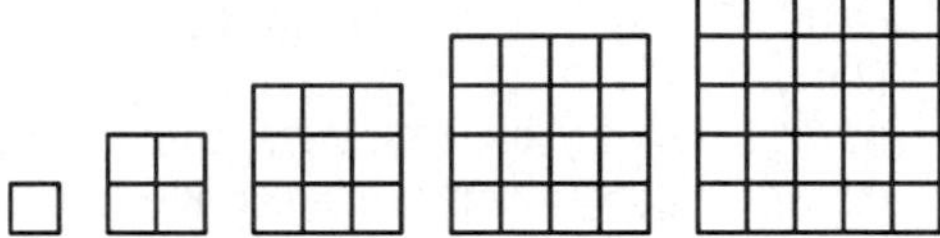

see also L-POLYOMINO, SKEW POLYOMINO, STRAIGHT POLYOMINO, T-POLYOMINO

Square Pyramid

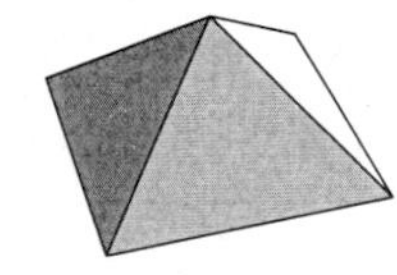
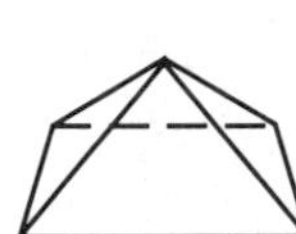
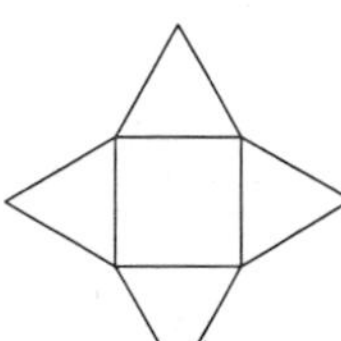

A square pyramid is a PYRAMID with a SQUARE base. If the top of the pyramid is cut off by a PLANE, a square PYRAMIDAL FRUSTUM is obtained. If the four TRIANGLES of the square pyramid are EQUILATERAL, the square pyramid is the "regular" POLYHEDRON known as JOHNSON SOLID J_1 and, for side length a, has height

$$h = \tfrac{1}{2}\sqrt{2}\,a. \quad (1)$$

Using the equation for a general PYRAMID, the VOLUME of the "regular" is therefore

$$V = \tfrac{1}{3}hA_b = \tfrac{1}{6}\sqrt{2}\,a^3. \quad (2)$$

If the apex of the pyramid does not lie atop the center of the base, then the SLANT HEIGHT is given by

$$s = \sqrt{h^2 + \tfrac{1}{2}a^2}, \quad (3)$$

where h is the height and a is the length of a side of the base.

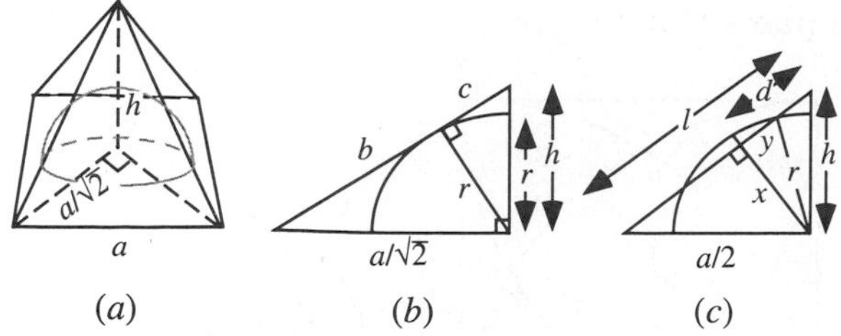

(a) (b) (c)

Consider a HEMISPHERE placed on the base of a square pyramid (having side lengths a and height h). Further, let the hemisphere be tangent to the four apex edges. Then what is the volume of the HEMISPHERE which is interior the pyramid (Cipra 1993)?

From Fig. (a), the CIRCUMRADIUS of the base is $a/\sqrt{2}$. Now find h in terms of r and a. Fig. (b) shows a CROSS-SECTION cut by the plane through the pyramid's apex, one of the base's vertices, and the base center. This figure gives

$$b = \sqrt{\tfrac{1}{2}a^2 - r^2} \tag{4}$$

$$c = \sqrt{h^2 - r^2}, \tag{5}$$

so the SLANT HEIGHT is

$$s = \sqrt{h^2 + \tfrac{1}{2}a^2} = b + c = \sqrt{\tfrac{1}{2}a^2 - r^2} + \sqrt{h^2 - r^2}. \tag{6}$$

Solving for h gives

$$h = \frac{ra}{\sqrt{a^2 - 2r^2}}. \tag{7}$$

We know, however, that the HEMISPHERE must be tangent to the sides, so $r = a/2$, and

$$h = \frac{\tfrac{1}{2}a}{\sqrt{a^2 - \tfrac{1}{2}a^2}}a = \frac{\tfrac{1}{2}}{\sqrt{\tfrac{1}{2}}}a = \tfrac{1}{2}\sqrt{2}\,a. \tag{8}$$

Fig. (c) shows a CROSS-SECTION through the center, apex, and midpoints of opposite sides. The PYTHAGOREAN THEOREM once again gives

$$l = \sqrt{\tfrac{1}{4}a^2 + h^2} = \sqrt{\tfrac{1}{4}a^2 + \tfrac{1}{2}a^2} = \tfrac{1}{2}\sqrt{3}\,a. \tag{9}$$

We now need to find x and y.

$$\sqrt{\tfrac{1}{4}a^2 - x^2} + d = l. \tag{10}$$

But we know l and h, and d is given by

$$d = \sqrt{h^2 - x^2}, \tag{11}$$

so

$$\sqrt{\tfrac{1}{4}a^2 - x^2} + \sqrt{\tfrac{1}{2}a^2 - x^2} = \tfrac{1}{2}\sqrt{3}\,a. \tag{12}$$

Solving gives

$$x = \tfrac{1}{6}\sqrt{6}\,a, \tag{13}$$

so

$$y = \sqrt{r^2 - x^2} = \sqrt{\tfrac{1}{4} - \tfrac{1}{6}}\,a = \sqrt{\frac{3-2}{12}}\,a = \frac{a}{2\sqrt{3}}. \tag{14}$$

We can now find the AREA of the SPHERICAL CAP as

$$V_{\text{cap}} = \tfrac{1}{6}\pi H(3A^2 + H^2), \tag{15}$$

where

$$A \equiv y = \frac{a}{2\sqrt{3}} \tag{16}$$

$$H \equiv r - x = \tfrac{1}{2}a - \frac{a}{\sqrt{6}} = a\left(\frac{1}{2} - \frac{1}{\sqrt{6}}\right), \tag{17}$$

so

$$\begin{aligned} V_{\text{cap}} &= \tfrac{1}{6}\pi a^3 \left[3\left(\frac{1}{12}\right) + \left(\frac{1}{2} - \frac{1}{\sqrt{6}}\right)^2\right]\left(\frac{1}{2} - \frac{1}{\sqrt{6}}\right) \\ &= \tfrac{1}{6}\pi a^3 \left[\frac{1}{4} + \left(\frac{1}{4} + \frac{1}{6} - \frac{1}{\sqrt{6}}\right)\right]\left(\frac{1}{2} - \frac{1}{\sqrt{6}}\right) \\ &= \tfrac{1}{6}\pi a^3 \left(\frac{2}{3} - \frac{1}{\sqrt{6}}\right)\left(\frac{1}{2} - \frac{1}{\sqrt{6}}\right) \\ &= \tfrac{1}{6}\pi a^3 \left(\frac{1}{3} - \frac{1}{2\sqrt{6}} - \frac{2}{3\sqrt{6}} + \frac{1}{6}\right) \\ &= \tfrac{1}{6}\pi a^3 \left(\frac{1}{2} - \frac{7}{6\sqrt{6}}\right). \end{aligned} \tag{18}$$

Therefore, the volume within the pyramid is

$$\begin{aligned} V_{\text{inside}} &= \tfrac{2}{3}\pi r^3 - 4V_{\text{cap}} = \tfrac{2}{3}\pi\tfrac{1}{8}a^3 - \tfrac{2}{3}\pi a^3\left(\frac{1}{2} - \frac{7}{6\sqrt{6}}\right) \\ &= \tfrac{2}{3}\pi a^3\left(\frac{1}{8} - \frac{1}{2} + \frac{7}{6\sqrt{6}}\right) = \tfrac{2}{3}\pi a^3\left(\frac{7}{6\sqrt{6}} - \frac{3}{8}\right) \\ &= \pi a^3\left(\frac{7}{9\sqrt{6}} - \frac{1}{4}\right). \end{aligned} \tag{19}$$

This problem appeared in the Japanese scholastic aptitude test (Cipra 1993).

see also SQUARE PYRAMIDAL NUMBER

References

Cipra, B. "An Awesome Look at Japan Math SAT." *Science* **259**, 22, 1993.

Square Pyramidal Number

A FIGURATE NUMBER of the form

$$P_n = \frac{1}{6}n(n+1)(2n+1), \tag{1}$$

corresponding to a configuration of points which form a SQUARE PYRAMID, is called a square pyramidal number (or sometimes, simply a PYRAMIDAL NUMBER). The first few are 1, 5, 14, 30, 55, 91, 140, 204, ... (Sloane's A000330). They are sums of consecutive pairs of TETRAHEDRAL NUMBERS and satisfy

$$P_n = \frac{1}{3}(2n+1)T_n, \tag{2}$$

where T_n is the nth TRIANGULAR NUMBER.

The only numbers which are simultaneously SQUARE and pyramidal (the CANNONBALL PROBLEM) are $P_1 = 1$ and $P_{24} = 4900$, corresponding to $S_1 = 1$ and $S_{70} = 4900$ (Dickson 1952, p. 25; Ball and Coxeter 1987, p. 59; Ogilvy 1988), as conjectured by Lucas (1875, 1876) and proved by Watson (1918). The proof is far from elementary, and is equivalent to solving the DIOPHANTINE EQUATION

$$y^2 = \frac{1}{6}x(x+1)(2x+1) \tag{3}$$

(Guy 1994, p. 147). However, an elementary proof has also been given by a number of authors.

Numbers which are simultaneously TRIANGULAR and square pyramidal satisfy the DIOPHANTINE EQUATION

$$3(2y+1)^2 = 8x^3 + 12x^2 + 4x + 3. \tag{4}$$

The only solutions are $x = -1$, 0, 1, 5, 6, and 85 (Guy 1994, p. 147). Beukers (1988) has studied the problem of finding numbers which are simultaneously TETRAHEDRAL and square pyramidal via INTEGER points on an ELLIPTIC CURVE. He finds that the only solution is the trivial $Te_1 = P_1 = 1$.

see also TETRAHEDRAL NUMBER

References

Ball, W. W. R. and Coxeter, H. S. M. *Mathematical Recreations and Essays, 13th ed.* New York: Dover, p. 59, 1987.

Beukers, F. "On Oranges and Integral Points on Certain Plane Cubic Curves." *Nieuw Arch. Wisk.* **6**, 203–210, 1988.

Conway, J. H. and Guy, R. K. *The Book of Numbers.* New York: Springer-Verlag, pp. 47–50, 1996.

Dickson, L. E. *History of the Theory of Numbers, Vol. 2: Diophantine Analysis.* New York: Chelsea, 1952.

Guy, R. K. "Figurate Numbers." §D3 in *Unsolved Problems in Number Theory, 2nd ed.* New York: Springer-Verlag, pp. 147–150, 1994.

Lucas, É. Question 1180. *Nouvelles Ann. Math. Ser. 2* **14**, 336, 1875.

Lucas, É. Solution de Question 1180. *Nouvelles Ann. Math. Ser. 2* **15**, 429–432, 1876.

Ogilvy, C. S. and Anderson, J. T. *Excursions in Number Theory.* New York: Dover, pp. 77 and 152, 1988.

Sloane, N. J. A. Sequence A000330/M3844 in "An On-Line Version of the Encyclopedia of Integer Sequences."

Watson, G. N. "The Problem of the Square Pyramid." *Messenger. Math.* **48**, 1–22, 1918.

Square Quadrants

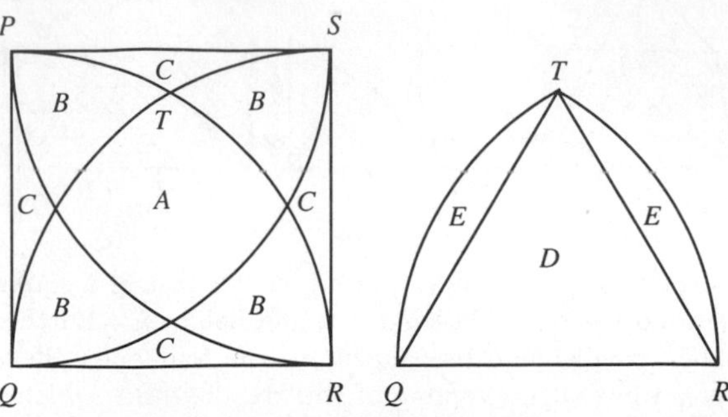

The areas of the regions illustrated above can be found from the equations

$$A + 4B + 4C = 1 \tag{1}$$

$$A + 3B + 2C = \frac{1}{4}\pi. \tag{2}$$

Since we want to solve for three variables, we need a third equation. This can be taken as

$$A + 2B + C = 2E + D, \tag{3}$$

where

$$D = \frac{1}{4}\sqrt{3} \tag{4}$$

$$D + E = \frac{1}{6}\pi, \tag{5}$$

leading to

$$A+2B+C = D+2E = 2(D+E) - D = \frac{1}{3}\pi - \frac{1}{4}\sqrt{3}. \tag{6}$$

Combining the equations (1), (2), and (6) gives the matrix equation

$$\begin{bmatrix} 1 & 4 & 4 \\ 1 & 3 & 2 \\ 1 & 2 & 1 \end{bmatrix} \begin{bmatrix} A \\ B \\ C \end{bmatrix} = \begin{bmatrix} 1 \\ \frac{1}{4}\pi \\ \frac{1}{3}\pi - \frac{1}{4}\sqrt{3} \end{bmatrix}, \tag{7}$$

which can be inverted to yield

$$A = 1 - \sqrt{3} - \frac{1}{3}\pi \tag{8}$$

$$B = -1 + \frac{1}{2}\sqrt{3} + \frac{1}{12}\pi \tag{9}$$

$$C = 1 - \frac{1}{4}\sqrt{3} + \frac{1}{6}\pi. \tag{10}$$

References

Honsberger, R. *More Mathematical Morsels.* Washington, DC: Math. Assoc. Amer., pp. 67–69, 1991.

Square Root

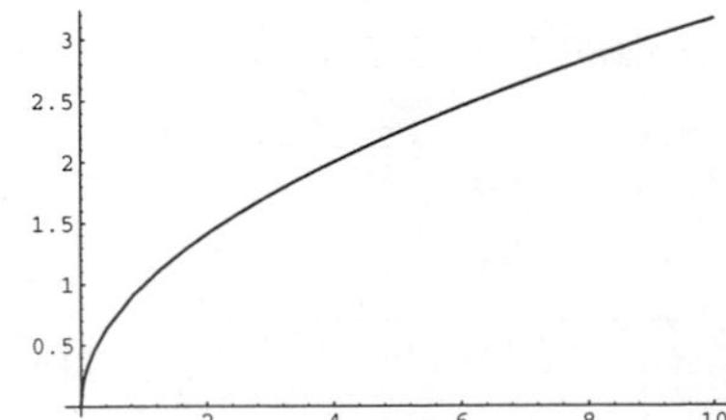

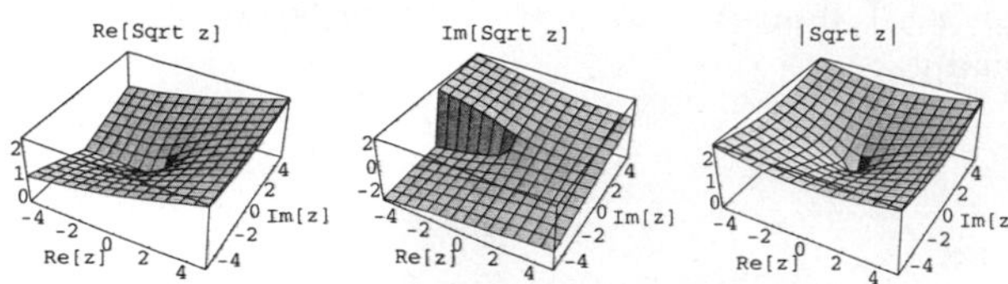

A square root of x is a number r such that $r^2 = x$. This is written $r = x^{1/2}$ (x to the 1/2 POWER) or $r = \sqrt{x}$. The square root function $f(x) = \sqrt{x}$ is the INVERSE FUNCTION of $f(x) = x^2$. Square roots are also called RADICALS or SURDS. A general COMPLEX NUMBER z has *two* square roots. For example, for the real POSITIVE number $x = 9$, the two square roots are $\sqrt{9} = \pm 3$, since $3^2 = (-3)^2 = 9$. Similarly, for the real NEGATIVE number $x = -9$, the two square roots are $\sqrt{-9} = \pm 3i$, where i is the IMAGINARY NUMBER defined by $i^2 = -1$. In common usage, unless otherwise specified, "the" square root is generally taken to mean the POSITIVE square root.

The square root of 2 is the IRRATIONAL NUMBER $\sqrt{2} \approx 1.41421356$ (Sloane's A002193), which has the simple periodic CONTINUED FRACTION 1, 2, 2, 2, 2, 2, ... The square root of 3 is the IRRATIONAL NUMBER $\sqrt{3} \approx 1.73205081$ (Sloane's A002194), which has the simple periodic CONTINUED FRACTION 1, 1, 2, 1, 2, 1, 2, ... In general, the CONTINUED FRACTIONS of the square roots of all POSITIVE integers are periodic.

The square roots of a COMPLEX NUMBER are given by

$$\sqrt{x+iy} = \pm\sqrt{x^2+y^2}\left\{\cos\left[\frac{1}{2}\tan^{-1}\left(\frac{y}{x}\right)\right] + i\sin\left[\frac{1}{2}\tan^{-1}\left(\frac{y}{x}\right)\right]\right\}. \quad (1)$$

As can be seen in the above figure, the IMAGINARY PART of the complex square root function has a BRANCH CUT along the NEGATIVE real axis.

A NESTED RADICAL of the form $\sqrt{a \pm b\sqrt{c}}$ can sometimes be simplified into a simple square root by equating

$$\sqrt{a \pm b\sqrt{c}} = \sqrt{d} \pm \sqrt{e}. \quad (2)$$

Squaring gives

$$a \pm b\sqrt{c} = d + e \pm 2\sqrt{de}, \quad (3)$$

so

$$a = d + e \quad (4)$$

$$b^2 c = 4de. \quad (5)$$

Solving for d and e gives

$$d, e = \frac{a \pm \sqrt{a^2 - b^2 c}}{2}. \quad (6)$$

A sequence of approximations a/b to $\sqrt{n}$ can be derived by factoring

$$a^2 - nb^2 = \pm 1 \quad (7)$$

(where -1 is possible only if -1 is a QUADRATIC RESIDUE of n). Then

$$(a + b\sqrt{n})(a - b\sqrt{n}) = \pm 1 \quad (8)$$

$$(a + b\sqrt{n})^k (a - b\sqrt{n})^k = (\pm 1)^k = \pm 1, \quad (9)$$

and

$$(1 + \sqrt{n})^1 = 1 + \sqrt{n} \quad (10)$$

$$(1 + \sqrt{n})^2 = (1 + n) + 2\sqrt{n} \quad (11)$$

$$(1 + \sqrt{n})(a + b\sqrt{n}) = (a + bn) + \sqrt{n}(a + b). \quad (12)$$

Therefore, a and b are given by the RECURRENCE RELATIONS

$$a_i = a_{i-1} + b_{i-1} n \quad (13)$$

$$b_i = a_{i-1} + b_{i-1} \quad (14)$$

with $a_1 = b_1 = 1$. The error obtained using this method is

$$\left|\frac{a}{b} - \sqrt{n}\right| = \frac{1}{b(a + b\sqrt{n})} < \frac{1}{2b^2}. \quad (15)$$

The first few approximants to $\sqrt{n}$ are therefore given by

$$1, \tfrac{1}{2}(1+n), \frac{1+3n}{3+n}, \frac{1+6n+n^2}{4(n+1)}, \frac{1+10n+5n^2}{5+10n+n^2}, \ldots. \quad (16)$$

This ALGORITHM is sometimes known as the BHASKARA-BROUCKNER ALGORITHM. For the case $n = 2$, this gives the convergents to $\sqrt{2}$ as 1, 3/2, 7/5, 17/12, 41/29, 99/70, ...

Another general technique for deriving this sequence, known as NEWTON'S ITERATION, is obtained by letting $x = \sqrt{n}$. Then $x = n/x$, so the SEQUENCE

$$x_k = \frac{1}{2}\left(x_{k-1} + \frac{n}{x_{k-1}}\right) \quad (17)$$

converges quadratically to the root. The first few approximants to $\sqrt{n}$ are therefore given by

$$1, \tfrac{1}{2}(1+n), \frac{1+6n+n^2}{4(n+1)}, \frac{1+26n+70n^2+28n^3+n^4}{8(1+n)(1+6n+n^2)}, \ldots. \quad (18)$$

For $\sqrt{2}$, this gives the convergents 1, 3/2, 17/12, 577/408, 665857/470832, ...

see also CONTINUED SQUARE ROOT, CUBE ROOT, NESTED RADICAL, NEWTON'S ITERATION, QUADRATIC SURD, ROOT OF UNITY, SQUARE NUMBER, SQUARE TRIANGULAR NUMBER, SURD

References

Sloane, N. J. A. Sequences A002193/M3195 and A002194/M4326 in "An On-Line Version of the Encyclopedia of Integer Sequences."

Spanier, J. and Oldham, K. B. "The Square-Root Function $\sqrt{bx+c}$ and Its Reciprocal," "The $b\sqrt{a^2-x^2}$ Function and Its Reciprocal," and "The $b\sqrt{x^2+a}$ Function." Chs. 12, 14, and 15 in *An Atlas of Functions*. Washington, DC: Hemisphere, pp. 91–99, 107–115, and 115–122, 1987.

Williams, H. C. "A Numerical Investigation into the Length of the Period of the Continued Fraction Expansion of $\sqrt{D}$." *Math. Comp.* **36**, 593–601, 1981.

Square Root Inequality

$$2\sqrt{n+1}-2\sqrt{n}<\frac{1}{\sqrt{n}}<2\sqrt{n}-2\sqrt{n-1}.$$

Square Root Method

The square root method is an algorithm which solves the MATRIX EQUATION

$$\mathsf{A}\mathbf{u}=\mathbf{g} \tag{1}$$

for $\mathbf{u}$, with A a $p\times p$ SYMMETRIC MATRIX and $\mathbf{g}$ a given VECTOR. Convert A to a TRIANGULAR MATRIX such that

$$\mathsf{T}^{\mathrm{T}}\mathsf{T}=\mathsf{A}, \tag{2}$$

where T^{T} is the MATRIX TRANSPOSE. Then

$$\mathsf{T}^{\mathrm{T}}\mathbf{k}=\mathbf{g} \tag{3}$$

$$\mathsf{T}\mathbf{u}=\mathbf{k}, \tag{4}$$

so

$$\mathsf{T}=\begin{bmatrix} s_{11} & s_{12} & \cdots & \cdots \\ 0 & s_{22} & \cdots & \cdots \\ \vdots & \vdots & \ddots & \vdots \\ 0 & 0 & \cdots & s_{pp}\end{bmatrix}, \tag{5}$$

giving the equations

$$\begin{aligned} {s_{11}}^2 &= a_{11} \\ s_{11}s_{12} &= a_{12} \\ {s_{12}}^2+{s_{22}}^2 &= a_{22} \\ {s_{1j}}^2+{s_{2j}}^2+\ldots+{s_{jj}}^2 &= a_{jj} \\ s_{1j}+s_{2j}s_{2k}+\ldots+s_{jj}s_{jk} &= a_{jk}. \end{aligned} \tag{6}$$

These give

$$\begin{aligned} s_{11} &= \sqrt{a_{11}} \\ s_{12} &= \frac{a_{12}}{s_{11}} \\ s_{22} &= \sqrt{a_{22}-{s_{12}}^2} \\ s_{jj} &= \sqrt{a_{jj}-{s_{ij}}^2-{s_{2j}}^2-\ldots-{s_{j-1,j}}^2} \\ s_{jk} &= \frac{a_{jk}-s_{1j}s_{1k}-s_{2j}s_{2k}-\ldots-s_{j-1,j}s_{j-1,k}}{s_{jj}}, \end{aligned} \tag{7}$$

giving T from A. Now solve for $\mathbf{k}$ in terms of the s_{ij}s and $\mathbf{g}$,

$$\begin{aligned} s_{11}k_1 &= g_1 \\ s_{12}k_1+s_{22}k_2 &= g_2 \\ s_{1j}k_1+s_{2j}k_2+\ldots+s_{jj}k_j &= g_j, \end{aligned} \tag{8}$$

which gives

$$\begin{aligned} k_1 &= \frac{g_1}{s_{11}} \\ k_2 &= \frac{g_2-s_{12}k_1}{s_{22}} \\ k_j &= \frac{g_j-s_{1j}k_1-s_{2j}k_2-\ldots-s_{j-1,j}k_{j-1}}{s_{jj}}. \end{aligned} \tag{9}$$

Finally, find $\mathbf{u}$ from the s_{ij}s and $\mathbf{k}$,

$$\begin{aligned} s_{11}u_1+s_{12}u_2\ldots+s_{1p}u_p &= k_1 \\ s_{22}u_2+\ldots+s_{2p}u_p &= k_2 \\ s_{pp}u_p &= k_p, \end{aligned} \tag{10}$$

giving the desired solution,

$$\begin{aligned} u_p &= \frac{k_p}{s_{pp}} \\ u_{p-1} &= \frac{k_{p-1}-s_{p-1,p}u_p}{s_{p-1,p-1}} \\ u_j &= \frac{k_j-s_{j,j+1}u_{j+1}-s_{j,j+2}u_{j+2}-\ldots-s_{jp}u_p}{s_{jj}}. \end{aligned} \tag{11}$$

see also LU DECOMPOSITION

References

Kenney, J. F. and Keeping, E. S. *Mathematics of Statistics, Pt. 2, 2nd ed.* Princeton, NJ: Van Nostrand, pp. 298–300, 1951.

Square Triangular Number

A number which is simultaneously SQUARE and TRIANGULAR. The first few are 1, 36, 1225, 41616, 1413721, 48024900, ... (Sloane's A001110), corresponding to $T_1=S_1$, $T_8=S_6$, $T_{49}=S_{35}$, $T_{288}=S_{204}$, $T_{1681}=S_{1189}$, ... (Pietenpol 1962), but there are an infinite number, as first shown by Euler in 1730 (Dickson 1952).

The general FORMULA for a square triangular number ST_n is b^2c^2, where b/c is the nth convergent to the CONTINUED FRACTION of $\sqrt{2}$ (Ball and Coxeter 1987, p. 59; Conway and Guy 1996). The first few are

$$\frac{1}{1},\frac{3}{2},\frac{7}{5},\frac{17}{12},\frac{41}{29},\frac{99}{70},\frac{239}{169},\ldots. \tag{1}$$

The NUMERATORS and DENOMINATORS give solutions to the PELL EQUATION

$$x^2-2y^2=\pm1, \tag{2}$$

but can also be obtained by doubling the previous FRACTION and adding to the FRACTION before that. The connection with the PELL EQUATION can be seen by letting N denote the Nth TRIANGULAR NUMBER and M the Mth SQUARE NUMBER, then

$$\tfrac{1}{2}N(N+1) = M^2. \tag{3}$$

Defining

$$x \equiv 2N+1 \tag{4}$$
$$y \equiv 2M \tag{5}$$

then gives the equation

$$x^2 - 2y^2 = 1 \tag{6}$$

(Conway and Guy 1996). Numbers which are simultaneously TRIANGULAR and SQUARE PYRAMIDAL also satisfy the DIOPHANTINE EQUATION

$$3(2y+1)^2 = 8x^3 + 12x^2 + 4x + 3. \tag{7}$$

The only solutions are $x = -1$, 0, 1, 5, 6, and 85 (Guy 1994, p. 147).

A general FORMULA for square triangular numbers is

$$ST_n = \left[\frac{(1+\sqrt{2})^{2n} - (1-\sqrt{2})^{2n}}{4\sqrt{2}}\right]^2 \tag{8}$$
$$= \tfrac{1}{32}[(17+12\sqrt{2})^n + (17-12\sqrt{2})^n - 2]. \tag{9}$$

The square triangular numbers also satisfy the RECURRENCE RELATION

$$ST_n = 34ST_{n-1} - ST_{n-2} + 2 \tag{10}$$
$$u_{n+2} = 6u_{n+1} - u_n, \tag{11}$$

with $u_0 = 0$, $u_1 = 1$, where $ST_n \equiv u_n{}^2$. A curious product formula for ST_n is given by

$$ST_n = 2^{2n-5}\prod_{k=1}^{2n}\left[3 + \cos\left(\frac{k\pi}{n}\right)\right]. \tag{12}$$

An amazing GENERATING FUNCTION is

$$f(x) = \frac{1+x}{(1-x)(1-34x+x^2)} = 1 + 36x + 1225x^2 + \ldots \tag{13}$$

(Sloane and Plouffe 1995).

see also SQUARE NUMBER, SQUARE ROOT, TRIANGULAR NUMBER

References

Allen, B. M. "Squares as Triangular Numbers." *Scripta Math.* **20**, 213–214, 1954.

Ball, W. W. R. and Coxeter, H. S. M. *Mathematical Recreations and Essays, 13th ed.* New York: Dover, 1987.

Conway, J. H. and Guy, R. K. *The Book of Numbers.* New York: Springer-Verlag, pp. 203–205, 1996.

Dickson, L. E. *A History of the Theory of Numbers, Vol. 2: Diophantine Analysis.* New York: Chelsea, pp. 10, 16, and 27, 1952.

Guy, R. K. "Sums of Squares" and "Figurate Numbers." §C20 and §D3 in *Unsolved Problems in Number Theory, 2nd ed.* New York: Springer-Verlag, pp. 136–138 and 147–150, 1994.

Khatri, M. N. "Triangular Numbers Which are Also Squares." *Math. Student* **27**, 55–56, 1959.

Pietenpol, J. L. "Square Triangular Numbers." Problem E 1473. *Amer. Math. Monthly* **69**, 168–169, 1962.

Sierpiński, W. *Teoria Liczb, 3rd ed.* Warsaw, Poland: Monografie Matematyczne t. 19, p. 517, 1950.

Sierpiński, W. "Sur les nombres triangulaires carrés." *Pub. Faculté d'Électrotechnique l'Université Belgrade*, No. 65, 1–4, 1961.

Sierpiński, W. "Sur les nombres triangulaires carrés." *Bull. Soc. Royale Sciences Liège*, 30 ann., 189–194, 1961.

Sloane, N. J. A. Sequence A001110/M5259 in "An On-Line Version of the Encyclopedia of Integer Sequences."

Walker, G. W. "Triangular Squares." Problem E 954. *Amer. Math. Monthly* **58**, 568, 1951.

Square Wave

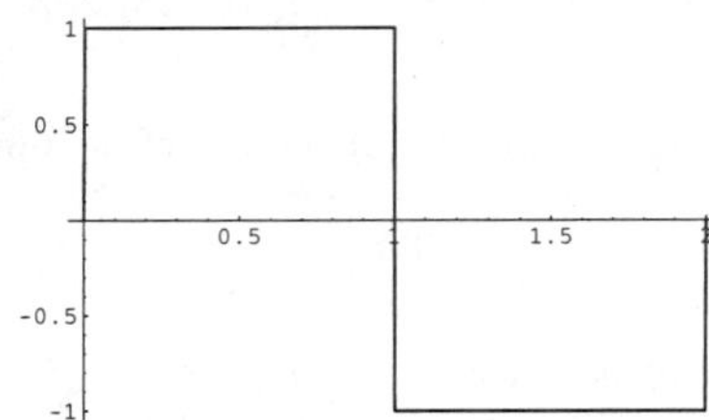

The square wave is a periodic waveform consisting of instantaneous transitions between two levels which can be denoted ± 1. The square wave is sometimes also called the RADEMACHER FUNCTION. Let the square wave have period $2L$. The square wave function is ODD, so the FOURIER SERIES has $a_0 = a_n = 0$ and

$$b_n = \frac{2}{L}\int_0^L \sin\left(\frac{n\pi x}{L}\right)dx$$
$$= \frac{4}{n\pi}\sin^2(\tfrac{1}{2}n\pi) = \frac{4}{n\pi}\begin{cases}0 & n \text{ even}\\ 1 & n \text{ odd.}\end{cases}$$

The FOURIER SERIES for the square wave is therefore

$$f(x) = \frac{4}{\pi}\sum_{n=1,3,5,\ldots}^{\infty}\frac{1}{n}\sin\left(\frac{n\pi x}{L}\right).$$

see also HADAMARD MATRIX, WALSH FUNCTION

References

Thompson, A. R.; Moran, J. M.; and Swenson, G. W. Jr. *Interferometry and Synthesis in Radio Astronomy.* New York: Wiley, p. 203, 1986.

Squared

A number to the POWER 2 is said to be squared, so that x^2 is called "x squared."

see also CUBED, SQUARE ROOT

Squared Square

see PERFECT SQUARE DISSECTION

Squarefree

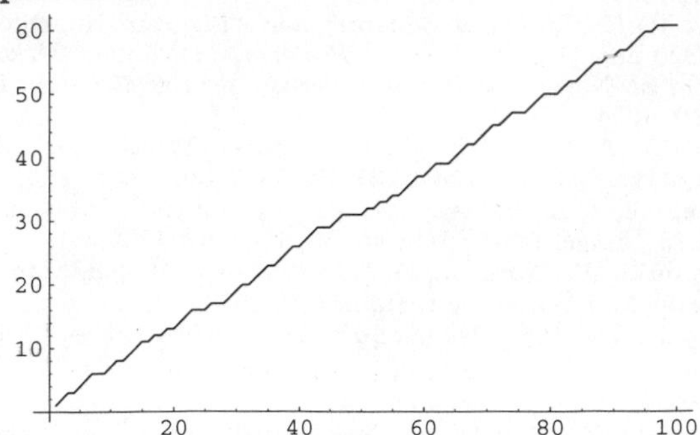

A number is said to be squarefree (or sometimes QUADRATFREI; Shanks 1993) if its PRIME decomposition contains no repeated factors. All PRIMES are therefore trivially squarefree. The squarefree numbers are 1, 2, 3, 5, 6, 7, 10, 11, 13, 14, 15, ... (Sloane's A005117). The SQUAREFUL numbers (i.e., those that contain at least one square) are 4, 8, 9, 12, 16, 18, 20, 24, 25, ... (Sloane's A013929).

The asymptotic number $Q(n)$ of squarefree numbers $\leq n$ is given by

$$Q(n) = \frac{6n}{\pi^2} + \mathcal{O}(\sqrt{n}) \qquad (1)$$

(Hardy and Wright 1979, pp. 269–270). $Q(n)$ for $n = 10, 100, 1000, \ldots$ are 7, 61, 608, 6083, 60794, 607926, ..., while the asymptotic density is $1/\zeta(2) = 6/\pi^2 \approx 0.607927$, where $\zeta(n)$ is the RIEMANN ZETA FUNCTION.

The MÖBIUS FUNCTION is given by

$$\mu(n) \equiv \begin{cases} 0 & \text{if } n \text{ has one or more repeated prime factors} \\ 1 & \text{if } n = 1 \\ (-1)^k & \text{if } n \text{ is product of } k \text{ distinct primes,} \end{cases} \qquad (2)$$

so $\mu(n) \neq 0$ indicates that n is squarefree. The asymptotic formula for $Q(x)$ is equivalent to the formula

$$\sum_{n=1}^{x} |\mu(n)| = \frac{6x}{\pi^2} + \mathcal{O}(\sqrt{x}) \qquad (3)$$

(Hardy and Wright 1979, p. 270)

There is no known polynomial-time algorithm for recognizing squarefree INTEGERS or for computing the squarefree part of an INTEGER. In fact, this problem may be no easier than the general problem of integer factorization (obviously, if an integer n can be factored completely, n is squarefree IFF it contains no duplicated factors). This problem is an important unsolved problem in NUMBER THEORY because computing the RING of integers of an algebraic number field is reducible to computing the squarefree part of an INTEGER (Lenstra 1992, Pohst and Zassenhaus 1997). The *Mathematica*® (Wolfram Research, Champaign, IL) function `NumberTheory'NumberTheoryFunctions' SquareFreeQ[n]` determines whether a number is squarefree.

The largest known SQUAREFUL FIBONACCI NUMBER is F_{336}, and no SQUAREFUL FIBONACCI NUMBERS F_p are known with p PRIME. All numbers less than 2.5×10^{15} in SYLVESTER'S SEQUENCE are squarefree, and no SQUAREFUL numbers in this sequence are known (Vardi 1991). Every CARMICHAEL NUMBER is squarefree. The BINOMIAL COEFFICIENTS $\binom{2n-1}{n}$ are squarefree only for $n = 2, 3, 4, 6, 9, 10, 12, 36, \ldots$, with no others less than $n = 1500$. The CENTRAL BINOMIAL COEFFICIENTS are SQUAREFREE only for $n = 1, 2, 3, 4, 5, 7, 8, 11, 17, 19, 23, 71, \ldots$ (Sloane's A046098), with no others less than 1500.

see also BINOMIAL COEFFICIENT, BIQUADRATEFREE, COMPOSITE NUMBER, CUBEFREE, ERDŐS SQUAREFREE CONJECTURE, FIBONACCI NUMBER, KORSELT'S CRITERION, MÖBIUS FUNCTION, PRIME NUMBER, RIEMANN ZETA FUNCTION, SÁRKÖZY'S THEOREM, SQUARE NUMBER, SQUAREFUL, SYLVESTER'S SEQUENCE

References

Bellman, R. and Shapiro, H. N. "The Distribution of Squarefree Integers in Small Intervals." *Duke Math. J.* **21**, 629–637, 1954.

Hardy, G. H. and Wright, E. M. "The Number of Squarefree Numbers." §18.6 in *An Introduction to the Theory of Numbers, 5th ed.* Oxford, England: Clarendon Press, pp. 269–270, 1979.

Lenstra, H. W. Jr. "Algorithms in Algebraic Number Theory." *Bull. Amer. Math. Soc.* **26**, 211–244, 1992.

Pohst, M. and Zassenhaus, H. *Algorithmic Algebraic Number Theory.* Cambridge, England: Cambridge University Press, p. 429, 1997.

Shanks, D. *Solved and Unsolved Problems in Number Theory, 4th ed.* New York: Chelsea, p. 114, 1993.

Sloane, N. J. A. Sequences A013929 and A005117/M0617 in "An On-Line Version of the Encyclopedia of Integer Sequences."

Vardi, I. "Are All Euclid Numbers Squarefree?" §5.1 in *Computational Recreations in Mathematica.* Reading, MA: Addison-Wesley, pp. 7–8, 82–85, and 223–224, 1991.

Squareful

A number is squareful, also called NONSQUAREFREE, if it contains at least one SQUARE in its prime factorization. Such a number is also called SQUAREFUL. The first few are 4, 8, 9, 12, 16, 18, 20, 24, 25, ... (Sloane's A013929). The greatest multiple prime factors for the squareful integers are 2, 2, 3, 2, 2, 3, 2, 2, 5, 3, 2, 2, 3, ... (Sloane's A046028). The least multiple prime factors for squareful integers are 2, 2, 3, 2, 2, 3, 2, 2, 5, 3, 2, 2, 2, ... (Sloane's A046027).

see also GREATEST PRIME FACTOR, LEAST PRIME FACTOR, SMARANDACHE NEAR-TO-PRIMORIAL FUNCTION, SQUAREFREE

References
Sloane, N. J. A. Sequences A013929, A046027, and A046028 in "An On-Line Version of the Encyclopedia of Integer Sequences."

Squaring

Squaring is the GEOMETRIC CONSTRUCTION, using only COMPASS and STRAIGHTEDGE, of a SQUARE which has the same area as a given geometric figure. Squaring is also called QUADRATURE. An object which can be constructed by squaring is called SQUARABLE.

see also CIRCLE SQUARING, COMPASS, CONSTRUCTIBLE NUMBER, GEOMETRIC CONSTRUCTION, RECTANGLE SQUARING, STRAIGHTEDGE, TRIANGLE SQUARING

Squeezing Theorem

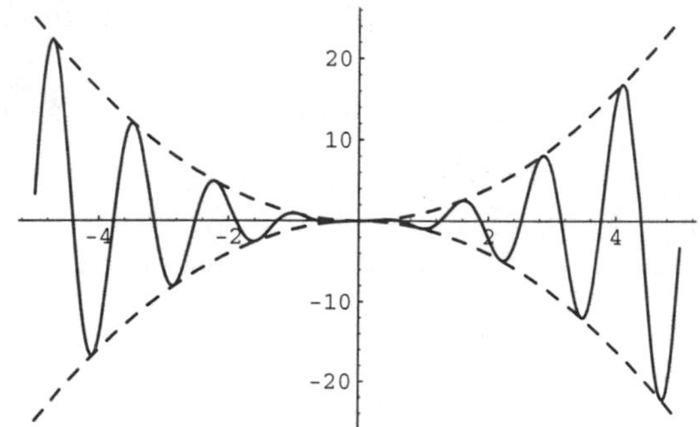

Let there be two functions $f_-(x)$ and $f_+(x)$ such that $f(x)$ is "squeezed" between the two,

$$f_-(x) \leq f(x) \leq f_+(x).$$

If

$$r = \lim_{x\to a} f_-(x) = \lim_{x\to a} f_+(x),$$

then $\lim_{x\to a} f(x) = r$. In the above diagram the functions $f_-(x) = -x^2$ and $f_+(x) = x^2$ "squeeze" $x^2 \sin(cx)$ at 0, so $\lim_{x\to 0} x^2 \sin(cx) = 0$. The squeezing theorem is also called the SANDWICH THEOREM.

SSS Theorem

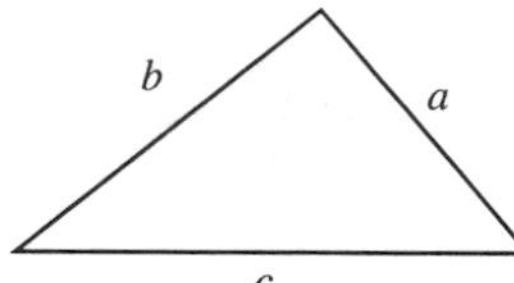

Specifying three sides uniquely determines a TRIANGLE whose AREA is given by HERON'S FORMULA,

$$A = \sqrt{s(s-a)(s-b)(s-c)}, \tag{1}$$

where

$$s \equiv \tfrac{1}{2}(a+b+c) \tag{2}$$

is the SEMIPERIMETER of the TRIANGLE. Let R be the CIRCUMRADIUS, then

$$A = \frac{abc}{4R}. \tag{3}$$

Using the LAW OF COSINES

$$a^2 = b^2 + c^2 - 2bc\cos A \tag{4}$$
$$b^2 = a^2 + c^2 - 2ac\cos B \tag{5}$$
$$c^2 = a^2 + b^2 - 2ab\cos C \tag{6}$$

gives the three ANGLES as

$$A = \cos^{-1}\left(\frac{b^2+c^2-a^2}{2bc}\right) \tag{7}$$
$$B = \cos^{-1}\left(\frac{b^2+c^2-b^2}{2ac}\right) \tag{8}$$
$$C = \cos^{-1}\left(\frac{a^2+b^2-c^2}{2ab}\right). \tag{9}$$

see also AAA THEOREM, AAS THEOREM, ASA THEOREM, ASS THEOREM, HERON'S FORMULA, SAS THEOREM, SEMIPERIMETER, TRIANGLE

Stability

The robustness of a given outcome to small changes in initial conditions or small random fluctuations. CHAOS is an example of a process which is not stable.

see also STABILITY MATRIX

Stability Matrix

Given a system of two ordinary differential equations

$$\dot{x} = f(x,y) \tag{1}$$
$$\dot{y} = g(x,y), \tag{2}$$

let x_0 and y_0 denote FIXED POINTS with $\dot{x} = \dot{y} = 0$, so

$$f(x_0, y_0) = 0 \tag{3}$$
$$g(x_0, y_0) = 0. \tag{4}$$

Then expand about (x_0, y_0) so

$$\delta\dot{x} = f_x(x_0,y_0)\delta x + f_y(x_0,y_0)\delta y + f_{xy}(x_0,y_0)\delta x\delta y + \ldots \tag{5}$$
$$\delta\dot{y} = g_x(x_0,y_0)\delta x + g_y(x_0,y_0)\delta y + g_{xy}(x_0,y_0)\delta x\delta y + \ldots. \tag{6}$$

To first-order, this gives

$$\frac{d}{dt}\begin{bmatrix}\delta x\\ \delta y\end{bmatrix} = \begin{bmatrix} f_x(x_0,y_0) & f_y(x_0,y_0)\\ g_x(x_0,y_0) & g_y(x_0,y_0)\end{bmatrix}\begin{bmatrix}\delta x\\ \delta y\end{bmatrix}, \tag{7}$$

where the 2×2 MATRIX, or its generalization to higher dimension, is called the stability matrix. Analysis of the EIGENVALUES (and EIGENVECTORS) of the stability matrix characterizes the type of FIXED POINT.

see also ELLIPTIC FIXED POINT (DIFFERENTIAL EQUATIONS), FIXED POINT, HYPERBOLIC FIXED POINT

(DIFFERENTIAL EQUATIONS), LINEAR STABILITY, STABLE IMPROPER NODE, STABLE NODE, STABLE SPIRAL POINT, STABLE STAR, UNSTABLE IMPROPER NODE, UNSTABLE NODE, UNSTABLE SPIRAL POINT, UNSTABLE STAR

References
Tabor, M. "Linear Stability Analysis." §1.4 in *Chaos and Integrability in Nonlinear Dynamics: An Introduction.* New York: Wiley, pp. 20–31, 1989.

Stabilization

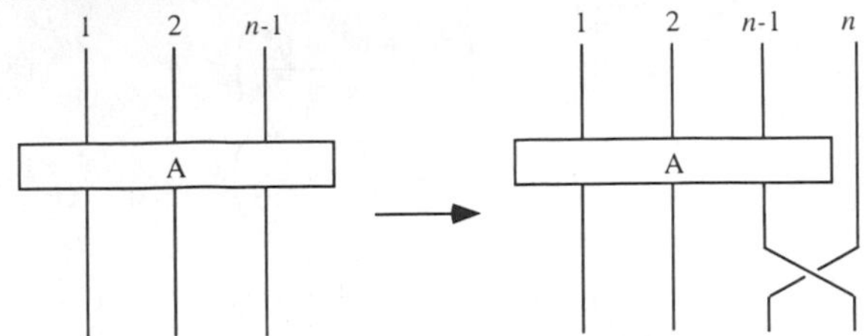

A type II MARKOV MOVE.

see also MARKOV MOVES

Stable Equivalence

Two VECTOR BUNDLES are stably equivalent IFF ISOMORPHIC VECTOR BUNDLES are obtained upon WHITNEY SUMMING each VECTOR BUNDLE with a trivial VECTOR BUNDLE.

see also VECTOR BUNDLE, WHITNEY SUM

Stable Improper Node

A FIXED POINT for which the STABILITY MATRIX has equal NEGATIVE EIGENVALUES.

see also ELLIPTIC FIXED POINT (DIFFERENTIAL EQUATIONS), FIXED POINT, HYPERBOLIC FIXED POINT (DIFFERENTIAL EQUATIONS), STABLE NODE, STABLE SPIRAL POINT, UNSTABLE IMPROPER NODE, UNSTABLE NODE, UNSTABLE SPIRAL POINT, UNSTABLE STAR

References
Tabor, M. "Classification of Fixed Points." §1.4.b in *Chaos and Integrability in Nonlinear Dynamics: An Introduction.* New York: Wiley, pp. 22–25, 1989.

Stable Node

A FIXED POINT for which the STABILITY MATRIX has both EIGENVALUES NEGATIVE, so $\lambda_1 < \lambda_2 < 0$.

see also ELLIPTIC FIXED POINT (DIFFERENTIAL EQUATIONS), FIXED POINT, HYPERBOLIC FIXED POINT (DIFFERENTIAL EQUATIONS), STABLE IMPROPER NODE, STABLE SPIRAL POINT, STABLE STAR, UNSTABLE IMPROPER NODE, UNSTABLE NODE, UNSTABLE SPIRAL POINT, UNSTABLE STAR

References
Tabor, M. "Classification of Fixed Points." §1.4.b in *Chaos and Integrability in Nonlinear Dynamics: An Introduction.* New York: Wiley, pp. 22–25, 1989.

Stable Spiral Point

A FIXED POINT for which the STABILITY MATRIX has EIGENVALUES of the form $\lambda_\pm = -\alpha \pm i\beta$ (with $\alpha, \beta > 0$).

see also ELLIPTIC FIXED POINT (DIFFERENTIAL EQUATIONS), FIXED POINT, HYPERBOLIC FIXED POINT (DIFFERENTIAL EQUATIONS), STABLE IMPROPER NODE, STABLE NODE, STABLE STAR, UNSTABLE IMPROPER NODE, UNSTABLE NODE, UNSTABLE SPIRAL POINT, UNSTABLE STAR

References
Tabor, M. "Classification of Fixed Points." §1.4.b in *Chaos and Integrability in Nonlinear Dynamics: An Introduction.* New York: Wiley, pp. 22–25, 1989.

Stable Star

A FIXED POINT for which the STABILITY MATRIX has one zero EIGENVECTOR with NEGATIVE EIGENVALUE $\lambda < 0$.

see also ELLIPTIC FIXED POINT (DIFFERENTIAL EQUATIONS), FIXED POINT, HYPERBOLIC FIXED POINT (DIFFERENTIAL EQUATIONS), STABLE IMPROPER NODE, STABLE NODE, STABLE SPIRAL POINT, UNSTABLE IMPROPER NODE, UNSTABLE NODE, UNSTABLE SPIRAL POINT, UNSTABLE STAR

References
Tabor, M. "Classification of Fixed Points." §1.4.b in *Chaos and Integrability in Nonlinear Dynamics: An Introduction.* New York: Wiley, pp. 22–25, 1989.

Stable Type

A POLYNOMIAL equation whose ROOTS all have NEGATIVE REAL PARTS. For a REAL QUADRATIC EQUATION

$$z^2 + Bz + C = 0,$$

the stability conditions are $B, C > 0$. For a REAL CUBIC EQUATION

$$z^3 + Az^2 + Bz + C = 0,$$

the stability conditions are $A, B, C > 0$ and $AB > C$.

References
Birkhoff, G. and Mac Lane, S. *A Survey of Modern Algebra, 3rd ed.* New York: Macmillan, pp. 108–109, 1965.

Stack

A DATA STRUCTURE which is a special kind of LIST in which elements may be added to or removed from the top only. These actions are called a PUSH or a POP, respectively. Actions may be taken by popping one or more values, operating on them, and then pushing the result back onto the stack.

Stacks are used as the basis for computer languages such as FORTH, PostScript® (Adobe Systems), and the RPN language used in Hewlett-Packard® programmable calculators.

see also LIST, POP, PUSH, QUEUE

Stäckel Determinant

A DETERMINANT used to determine in which coordinate systems the HELMHOLTZ DIFFERENTIAL EQUATION is separable (Morse and Feshbach 1953). A determinant

$$S = |\Phi_{mn}| = \begin{vmatrix} \Phi_{11} & \Phi_{12} & \Phi_{13} \\ \Phi_{21} & \Phi_{22} & \Phi_{23} \\ \Phi_{31} & \Phi_{32} & \Phi_{33} \end{vmatrix} \tag{1}$$

in which Φ_{ni} are functions of u_i alone is called a Stäckel determinant. A coordinate system is separable if it obeys the ROBERTSON CONDITION, namely that the SCALE FACTORS h_i in the LAPLACIAN

$$\nabla^2 = \sum_{i=1}^{3} \frac{1}{h_1 h_2 h_3} \frac{\partial}{\partial u_i} \left(\frac{h_1 h_2 h_3}{{h_i}^2} \frac{\partial}{\partial u_i} \right) \tag{2}$$

can be rewritten in terms of functions $f_i(u_i)$ defined by

$$\begin{aligned} \frac{1}{h_1 h_2 h_3} \frac{\partial}{\partial u_i} \left(\frac{h_1 h_2 h_3}{{h_i}^2} \frac{\partial}{\partial u_i} \right) &= \frac{g(u_{i+1}, u_{i+2})}{h_1 h_2 h_3} \frac{\partial}{\partial u_i} \left[f_i(u_i) \frac{\partial}{\partial u_i} \right] \\ &= \frac{1}{{h_i}^2 f_i} \frac{\partial}{\partial u_i} \left(f_i \frac{\partial}{\partial u_i} \right) \end{aligned} \tag{3}$$

such that S can be written

$$S = \frac{h_1 h_2 h_3}{f_1(u_1) f_2(u_2) f_3(u_3)}. \tag{4}$$

When this is true, the separated equations are of the form

$$\frac{1}{f_n} \frac{\partial}{\partial u_n} \left(f_n \frac{\partial X_n}{\partial u_n} \right) + ({k_1}^2 \Phi_{n1} + {k_2}^2 \Phi_{n2} + {k_3}^2 \Phi_{n3}) X_n = 0 \tag{5}$$

The Φ_{ij}s obey the minor equations

$$M_1 = \Phi_{22}\Phi_{33} - \Phi_{23}\Phi_{32} = \frac{S}{h_1^2} \tag{6}$$

$$M_2 = \Phi_{13}\Phi_{31} - \Phi_{12}\Phi_{33} = \frac{S}{h_2^2} \tag{7}$$

$$M_3 = \Phi_{12}\Phi_{23} - \Phi_{13}\Phi_{22} = \frac{S}{h_3^2}, \tag{8}$$

which are equivalent to

$$M_1\Phi_{11} + M_2\Phi_{21} + M_3\Phi_{31} = S \tag{9}$$

$$M_1\Phi_{12} + M_2\Phi_{22} + M_3\Phi_{32} = 0 \tag{10}$$

$$M_1\Phi_{13} + M_2\Phi_{23} + M_3\Phi_{33} = 0. \tag{11}$$

This gives a total of four equations in nine unknowns. Morse and Feshbach (1953, pp. 655–666) give not only the Stäckel determinants for common coordinate systems, but also the elements of the determinant (although it is not clear how these are derived).

see also HELMHOLTZ DIFFERENTIAL EQUATION, LAPLACE'S EQUATION, POISSON'S EQUATION, ROBERTSON CONDITION, SEPARATION OF VARIABLES

References

Morse, P. M. and Feshbach, H. "Tables of Separable Coordinates in Three Dimensions." *Methods of Theoretical Physics, Part I.* New York: McGraw-Hill, pp. 509–511 and 655–666, 1953.

Stamp Folding

The number of ways of folding a strip of stamps has several possible variants. Considering only positions of the hinges for unlabeled stamps without regard to orientation of the stamps, the number of foldings is denoted $U(n)$. If the stamps are labelled and orientation is taken into account, the number of foldings is denoted $N(n)$. Finally, the number of symmetric foldings is denoted $S(n)$. The following table summarizes these values for the first n.

n	$S(n)$	$U(n)$	$N(n)$
1	1	1	1
2	1	1	1
3	2	2	6
4	4	5	16
5	6	14	50
6	8	39	144
7	18	120	462
8	20	358	1392
9	56	1176	4536
10		3572	

see also MAP FOLDING

References

Gardner, M. "The Combinatorics of Paper-Folding." In *Wheels, Life, and Other Mathematical Amusements.* New York: W. H. Freeman, pp. 60–73, 1983.

Ruskey, F. "Information of Stamp Folding." `http://sue.csc.uvic.ca/~cos/inf/perm/StampFolding.html`.

Sloane, N. J. A. *A Handbook of Integer Sequences.* Boston, MA: Academic Press, p. 22, 1973.

Standard Deviation

The standard deviation is defined as the SQUARE ROOT of the VARIANCE,

$$\sigma = \sqrt{\langle x^2 \rangle - \langle x \rangle^2} = \sqrt{\mu_2' - \mu^2}, \tag{1}$$

where $\mu = \langle x \rangle$ is the MEAN and $\mu_2' = \left\langle x^2 \right\rangle$ is the second MOMENT about 0. The variance σ^2 is equal to the second MOMENT about the MEAN,

$$\sigma^2 = \mu_2. \tag{2}$$

The square root of the SAMPLE VARIANCE is the "sample" standard deviation,

$$s_N = \sqrt{\frac{1}{N} \sum_{i=1}^{N} (x_i - \bar{x})^2}. \tag{3}$$

It is a BIASED ESTIMATOR of the population standard deviation. As unbiased ESTIMATOR is given by

$$s_{N-1} = \sqrt{\frac{1}{N-1}\sum_{i=1}^{N}(x_i - \bar{x})^2}. \quad (4)$$

Physical scientists often use the term ROOT-MEAN-SQUARE as a synonym for standard deviation when they refer to the SQUARE ROOT of the mean squared deviation of a signal from a given baseline or fit.

see also MEAN, MOMENT, ROOT-MEAN-SQUARE, SAMPLE VARIANCE, STANDARD ERROR, VARIANCE

Standard Error

The square root of the ESTIMATED VARIANCE of a quantity. The standard error is also sometimes used to mean

$$\operatorname{var}(\bar{x}) = \sum_{i=1}^{n}\left(\frac{1}{n}\right)^2 {\sigma_i}^2 = \sum_{i=1}^{n}\left(\frac{1}{n}\right)^2 \sigma^2 = \frac{\sigma^2}{n}.$$

see also STANDARD DEVIATION

Standard Map

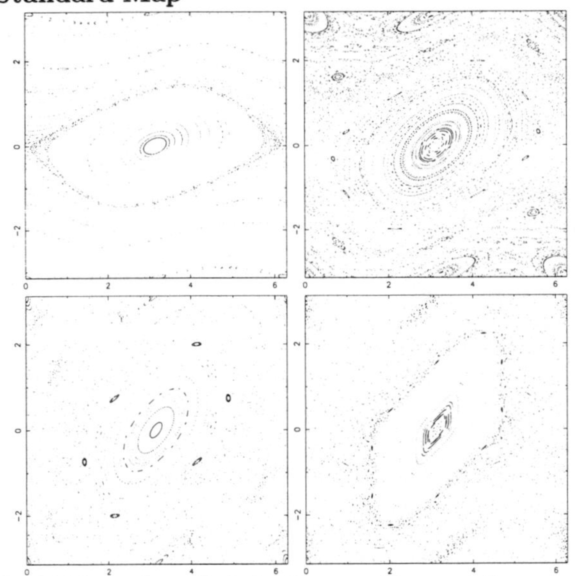

A 2-D MAP, also called the TAYLOR-GREENE-CHIRIKOV MAP in some of the older literature.

$$I_{n+1} = I_n + K\sin\theta_n \quad (1)$$
$$\theta_{n+1} = \theta_n + I_{n+1} = I_n + \theta_n + K\sin\theta_n, \quad (2)$$

where I and θ are computed mod 2π and K is a POSITIVE constant. An analytic estimate of the width of the CHAOTIC zone (Chirikov 1979) finds

$$\delta I = Be^{-AK^{-1/2}}. \quad (3)$$

Numerical experiments give $A \approx 5.26$ and $B \approx 240$. The value of K at which global CHAOS occurs has been bounded by various authors. GREENE'S METHOD is the most accurate method so far devised.

Author	Bound	Fraction	Decimal
Hermann	>	$\frac{1}{34}$	0.029411764
Italians	>	–	0.65
Greene	≈	–	0.971635406
MacKay and Pearson	<	$\frac{63}{64}$	0.984375000
Mather	<	$\frac{4}{3}$	1.333333333

FIXED POINTS are found by requiring that

$$I_{n+1} = I_n \quad (4)$$
$$\theta_{n+1} = \theta_n. \quad (5)$$

The first gives $K\sin\theta_n = 0$, so $\sin\theta_n = 0$ and

$$\theta_n = 0, \pi. \quad (6)$$

The second requirement gives

$$I_n + K\sin\theta_n = I_n = 0. \quad (7)$$

The FIXED POINTS are therefore $(I,\theta) = (0,0)$ and $(0,\pi)$. In order to perform a LINEAR STABILITY analysis, take differentials of the variables

$$dI_{n+1} = dI_n + K\cos\theta_n\, d\theta_n \quad (8)$$
$$d\theta_{n+1} = dI_n + (1 + K\cos\theta_n)\, d\theta_n. \quad (9)$$

In MATRIX form,

$$\begin{bmatrix}\delta I_{n+1}\\ \delta\theta_{n+1}\end{bmatrix} = \begin{bmatrix}1 & K\cos\theta_n\\ 1 & 1+K\cos\theta_n\end{bmatrix}\begin{bmatrix}\delta I_n\\ \delta\theta_n\end{bmatrix}. \quad (10)$$

The EIGENVALUES are found by solving the CHARACTERISTIC EQUATION

$$\begin{vmatrix}1-\lambda & K\cos\theta_n\\ 1 & 1+K\cos\theta_n-\lambda\end{vmatrix} = 0, \quad (11)$$

so

$$\lambda^2 - \lambda(K\cos\theta_n + 2) + 1 = 0 \quad (12)$$

$$\lambda_\pm = \tfrac{1}{2}[K\cos\theta_n + 2 \pm \sqrt{(K\cos\theta_n + 2)^2 - 4}]. \quad (13)$$

For the FIXED POINT $(0,\pi)$,

$$\lambda_\pm^{(0,\pi)} = \tfrac{1}{2}[2 - K \pm \sqrt{(2-K)^2 - 4}]$$
$$= \tfrac{1}{2}(2 - K \pm \sqrt{K^2 - 4K}). \quad (14)$$

The FIXED POINT will be stable if $|\Re(\lambda^{(0,\pi)})| < 2$. Here, that means

$$\tfrac{1}{2}|2-K| < 1 \quad (15)$$
$$|2-K| < 2 \quad (16)$$
$$-2 < 2-K < 2 \quad (17)$$
$$-4 < -K < 0 \quad (18)$$

so $K \in [0,4)$. For the FIXED POINT (0, 0), the EIGENVALUES are

$$\lambda_{\pm}^{(0,0)} = \tfrac{1}{2}[2 + K \pm \sqrt{(K+2)^2 - 4}]$$
$$= \tfrac{1}{2}(2 + K \pm \sqrt{K^2 + 4K}). \quad (19)$$

If the map is unstable for the larger EIGENVALUE, it is unstable. Therefore, examine $\lambda_{+}^{(0,0)}$. We have

$$\frac{1}{2}\left|2 + K + \sqrt{K^2 + 4K}\right| < 1, \quad (20)$$

so

$$-2 < 2 + K + \sqrt{K^2 + 4K} < 2 \quad (21)$$
$$-4 - K < \sqrt{K^2 + 4K} < -K. \quad (22)$$

But $K > 0$, so the second part of the inequality cannot be true. Therefore, the map is unstable at the FIXED POINT (0, 0).

References

Chirikov, B. V. "A Universal Instability of Many-Dimensional Oscillator Systems." *Phys. Rep.* **52**, 264–379, 1979.

Standard Normal Distribution

A NORMAL DISTRIBUTION with zero MEAN ($\mu = 0$) and unity STANDARD DEVIATION ($\sigma^2 = 1$).

see also NORMAL DISTRIBUTION

Standard Space

A SPACE which is ISOMORPHIC to a BOREL SUBSET B of a POLISH SPACE equipped with its SIGMA ALGEBRA of BOREL SETS.

see also BOREL SET, POLISH SPACE, SIGMA ALGEBRA

Standard Tori

full view *cutaway* *cross-section*

ring torus

horn torus

spindle torus

One of the three classes of TORI illustrated above and given by the parametric equations

$$x = (c + a\cos v)\cos u \quad (1)$$
$$y = (c + a\cos v)\sin u \quad (2)$$
$$z = a\sin v. \quad (3)$$

The three different classes of standard tori arise from the three possible relative sizes of a and c. $c > a$ corresponds to the RING TORUS shown above, $c = a$ corresponds to a HORN TORUS which touches itself at the point (0, 0, 0), and $c < a$ corresponds to a self-intersecting SPINDLE TORUS (Pinkall 1986). If no specification is made, "torus" is taken to mean RING TORUS.

The standard tori and their inversions are CYCLIDES.

see also APPLE, CYCLIDE, HORN TORUS, LEMON, RING TORUS, SPINDLE TORUS, TORUS

References

Pinkall, U. "Cyclides of Dupin." §3.3 in *Mathematical Models from the Collections of Universities and Museums* (Ed. G. Fischer). Braunschweig, Germany: Vieweg, pp. 28–30, 1986.

Standardized Moment

Defined for samples x_i, $i = 1, \ldots, N$ by

$$\alpha_r \equiv \frac{1}{N}\sum_{i=1}^{N} {z_i}^r = \frac{\mu_r}{\sigma^r}, \quad (1)$$

where

$$z_i \equiv \frac{x_i - \bar{x}}{s_x}. \quad (2)$$

The first few are

$$\alpha_1 = 0 \quad (3)$$
$$\alpha_2 = 1 \quad (4)$$
$$\alpha_3 = \frac{\mu_3}{s^3} \quad (5)$$
$$\alpha_4 = \frac{\mu_4}{s^4}. \quad (6)$$

see also KURTOSIS, MOMENT, SKEWNESS

Standardized Score

see z-SCORE

Stanley's Theorem

The total number of 1s that occur among all unordered PARTITIONS of a POSITIVE INTEGER is equal to the sum of the numbers of distinct parts of (i.e., numbers in) those PARTITIONS.

see also ELDER'S THEOREM, PARTITION

References

Honsberger, R. *Mathematical Gems III.* Washington, DC: Math. Assoc. Amer, pp. 6–8, 1985.

Star

In formal geometry, a star is a set of $2n$ VECTORS $\pm\mathbf{a}_1$, ..., $\pm\mathbf{a}_n$ which form a fixed center in EUCLIDEAN 3-SPACE. In common usage, a star is a STAR POLYGON (i.e., regular convex polygon) such as the PENTAGRAM or HEXAGRAM

see also CROSS, EUTACTIC STAR, STAR OF GOLIATH, STAR POLYGON

Star of David

see HEXAGRAM

Star Figure

A STAR POLYGON-like figure $\left\{\frac{p}{q}\right\}$ for which p and q are not RELATIVELY PRIME.

see also STAR POLYGON

Star (Fixed Point)

A FIXED POINT which has one zero EIGENVECTOR.

see STABLE STAR, UNSTABLE STAR

Star Fractal

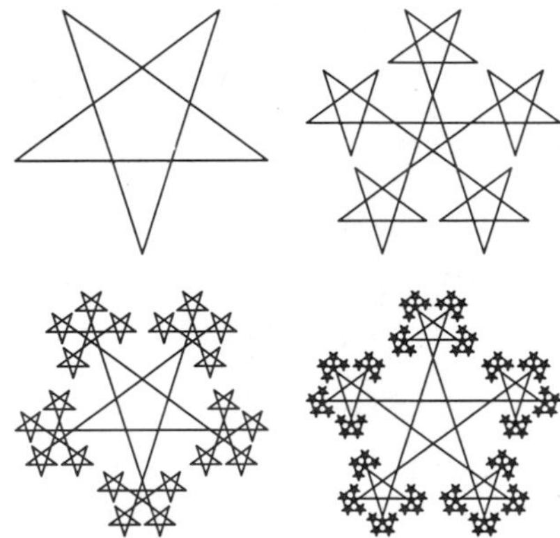

A FRACTAL composed of repeated copies of a PENTAGRAM or other polygon.

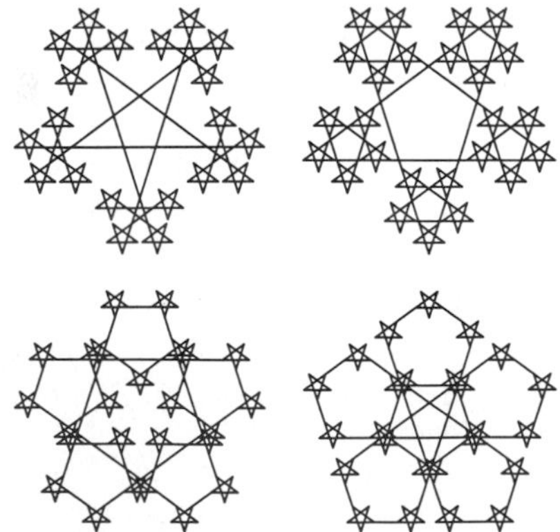

The above figure shows a generalization to different offsets from the center.

References

Lauwerier, H. *Fractals: Endlessly Repeated Geometric Figures.* Princeton, NJ: Princeton University Press, pp. 72–77, 1991.

Weisstein, E. W. "Fractals." `http://www.astro.virginia.edu/~eww6n/math/notebooks/Fractal.m`.

Star of Goliath

see NONAGRAM

Star Graph

The k-star graph is a TREE on $k+1$ nodes with one node having valency k and the others having valency 1. Star graphs S_n are always GRACEFUL.

Star of Lakshmi

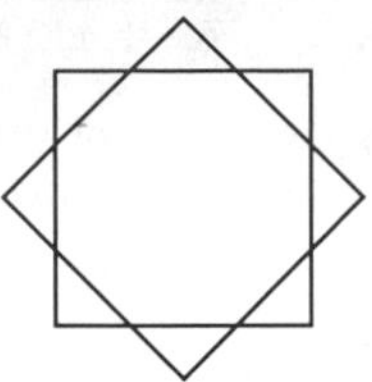

The STAR FIGURE {8/2}, which is used by Hindus to symbolize *Ashtalakshmi*, the eight forms of wealth. This symbol appears prominently in the Lugash national museum portrayed in the fictional film *Return of the Pink Panther.*

see also DISSECTION, HEXAGRAM, PENTAGRAM, STAR FIGURE, STAR POLYGON

References

Savio, D. Y. and Suryanaroyan, E. R. "Chebyshev Polynomials and Regular Polygons." *Amer. Math. Monthly* **100**, 657–661, 1993.

Star Number

The number of cells in a generalized Chinese checkers board (or "centered" HEXAGRAM).

$$S_n = 6n(n+1)+1 = S_{n-1}+12(n-1). \quad (1)$$

The first few are 1, 13, 37, 73, 121, ... (Sloane's A003154). Every star number has DIGITAL ROOT 1 or 4, and the final digits must be one of: 01, 21, 41, 61, 81, 13, 33, 53, 73, 93, or 37.

The first TRIANGULAR star numbers are 1, 253, 49141, 9533161, ... (Sloane's A006060), and can be computed using

$$\begin{aligned} TS_n &= \frac{3[(7+4\sqrt{3})^{2n-1}+(7-4\sqrt{3})^{2n-1}]-10}{32} \\ &= 194\,TS_{n-1}+60-TS_{n-2}. \end{aligned} \quad (2)$$

The first few SQUARE star numbers are 1, 121, 11881, 1164241, 114083761, ... (Sloane's A006061). SQUARE star numbers are obtained by solving the DIOPHANTINE EQUATION

$$2x^2+1=3y^2 \quad (3)$$

and can be computed using

$$SS_n = \frac{[(5+2\sqrt{6})^n(\sqrt{6}-2)-(5-2\sqrt{6})^n(\sqrt{6}+2)]^2}{4}. \quad (4)$$

see also HEX NUMBER, SQUARE NUMBER, TRIANGULAR NUMBER

References

Gardner, M. "Hexes and Stars." Ch. 2 in *Time Travel and Other Mathematical Bewilderments.* New York: W. H. Freeman, 1988.

Hindin, H. "Stars, Hexes, Triangular Numbers, and Pythagorean Triples." *J. Recr. Math.* **16**, 191–193, 1983–1984.

Sloane, N. J. A. Sequences A003154/M4893, A006060/M5425, and A006061/M5385 in "An On-Line Version of the Encyclopedia of Integer Sequences."

Star Polygon

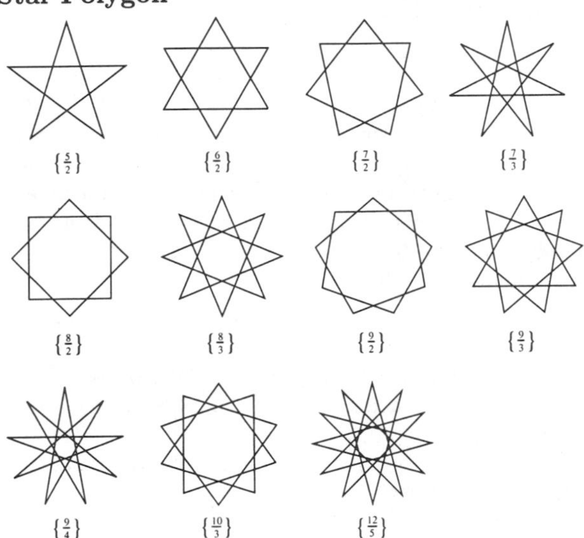

A star polygon $\{p/q\}$, with p, q POSITIVE INTEGERS, is a figure formed by connecting with straight lines every qth point out of p regularly spaced points lying on a CIRCUMFERENCE. The number q is called the DENSITY of the star polygon. Without loss of generality, take $q < p/2$.

The usual definition (Coxeter 1969) requires p and q to be RELATIVELY PRIME. However, the star polygon can also be generalized to the STAR FIGURE (or "improper" star polygon) when p and q share a common divisor (Savio and Suryanaroyan 1993). For such a figure, if all points are not connected after the first pass, i.e., if $(p,q) \neq 1$, then start with the first unconnected point and repeat the procedure. Repeat until all points are connected. For $(p,q) \neq 1$, the $\{p/q\}$ symbol can be factored as

$$\left\{\frac{p}{q}\right\} = n\left\{\frac{p}{q'}\right\}, \quad (1)$$

where

$$p' \equiv \frac{p}{n} \quad (2)$$

$$q' \equiv \frac{q}{n}, \quad (3)$$

to give $n\ \{p'/q'\}$ figures, each rotated by $2\pi/p$ radians, or $360°/p$.

If $q = 1$, a REGULAR POLYGON $\{p\}$ is obtained. Special cases of $\{p/q\}$ include $\{5/2\}$ (the PENTAGRAM), $\{6/2\}$ (the HEXAGRAM, or STAR OF DAVID), $\{8/2\}$ (the STAR OF LAKSHMI), $\{8/3\}$ (the OCTAGRAM), $\{10/3\}$ (the DECAGRAM), and $\{12/5\}$ (the DODECAGRAM).

The star polygons were first systematically studied by Thomas Bradwardine.

see also DECAGRAM, HEXAGRAM, NONAGRAM, OCTAGRAM, PENTAGRAM, REGULAR POLYGON, STAR OF LAKSHMI, STELLATED POLYHEDRON

References

Coxeter, H. S. M. "Star Polygons." §2.8 in *Introduction to Geometry, 2nd ed.* New York: Wiley, pp. 36–38, 1969.

Frederickson, G. "Stardom." Ch. 16 in *Dissections: Plane and Fancy.* New York: Cambridge University Press, pp. 172–186, 1997.

Savio, D. Y. and Suryanaroyan, E. R. "Chebyshev Polynomials and Regular Polygons." *Amer. Math. Monthly* **100**, 657–661, 1993.

Star Polyhedron

see KEPLER-POINSOT SOLID

Starr Rose

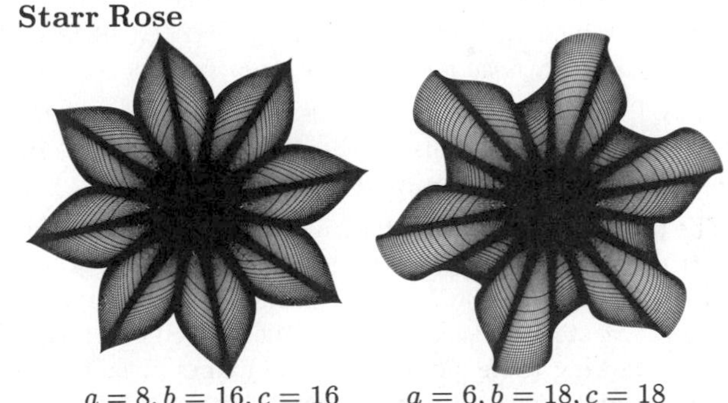

$a = 8, b = 16, c = 16$ $\quad$ $a = 6, b = 18, c = 18$

see also MAURER ROSE

References

Wagon, S. "Variations of Circular Motion." §4.5 in *Mathematica in Action.* New York: W. H. Freeman, pp. 137–140, 1991.

State Space

The measurable space $(S', \mathbb{S}')$ into which a RANDOM VARIABLE from a PROBABILITY SPACE is a measurable function.

see also PROBABILITY SPACE, RANDOM VARIABLE

Stationary Point

stationary point $\quad$ *minimum* $\quad$ *maximum*

A point x_0 at which the DERIVATIVE of a FUNCTION $f(x)$ vanishes,

$$f'(x_0) = 0.$$

A stationary point may be a MINIMUM, MAXIMUM, or INFLECTION POINT.

see also CRITICAL POINT, DERIVATIVE, EXTREMUM, FIRST DERIVATIVE TEST, INFLECTION POINT, MAXIMUM, MINIMUM, SECOND DERIVATIVE TEST

Stationary Tangent

see INFLECTION POINT

Stationary Value

The value at a STATIONARY POINT.

Statistic

A function of one or more random variables.

see also ANDERSON-DARLING STATISTIC, KUIPER STATISTIC, VARIATE

Statistical Test

A test used to determine the statistical SIGNIFICANCE of an observation. Two main types of error can occur:

1. A TYPE I ERROR occurs when a false negative result is obtained in terms of the NULL HYPOTHESIS by obtaining a *false positive measurement.*
2. A TYPE II ERROR occurs when a false positive result is obtained in terms of the NULL HYPOTHESIS by obtaining a *false negative measurement.*

The probability that a statistical test will be positive for a true statistic is sometimes called the test's SENSITIVITY, and the probability that a test will be negative for a negative statistic is sometimes called the SPECIFICITY. The following table summarizes the names given to the various combinations of the actual state of affairs and observed test results.

result	name
true positive result	sensitivity
false negative result	1 − sensitivity
true negative result	specificity
false positive result	1 − specificity

Multiple-comparison corrections to statistical tests are used when several statistical tests are being performed simultaneously. For example, let's suppose you were measuring leg length in eight different lizard species and wanted to see whether the MEANS of any pair were different. Now, there are $8!/2!6! = 28$ pairwise comparisons possible, so even if all of the *population* means are equal, it's quite likely that at least one pair of sample means would differ significantly at the 5% level. An ALPHA VALUE of 0.05 is therefore appropriate for each individual comparison, but not for the set of *all* comparisons.

In order to avoid a lot of spurious positives, the ALPHA VALUE therefore needs to be lowered to account for the number of comparisons being performed. This is a correction for multiple comparisons. There are *many* different ways to do this. The simplest, and the most conservative, is the BONFERRONI CORRECTION. In practice, more people are more willing to accept false positives (false rejection of NULL HYPOTHESIS) than false negatives (false acceptance of NULL HYPOTHESIS), so less conservative comparisons are usually used.

see also ANOVA, BONFERRONI CORRECTION, CHI-SQUARED TEST, FISHER'S EXACT TEST, FISHER SIGN TEST, KOLMOGOROV-SMIRNOV TEST, LIKELIHOOD RATIO, LOG LIKELIHOOD PROCEDURE, NEGATIVE LIKELIHOOD RATIO, PAIRED t-TEST, PARAMETRIC TEST, PREDICTIVE VALUE, SENSITIVITY, SIGNIFICANCE TEST, SPECIFICITY, TYPE I ERROR, TYPE II ERROR, WILCOXON RANK SUM TEST, WILCOXON SIGNED RANK TEST

Statistics

The mathematical study of the LIKELIHOOD and PROBABILITY of events occurring based on known information and inferred by taking a limited number of samples. Statistics plays an extremely important role in many aspects of economics and science, allowing educated guesses to be made with a minimum of expensive or difficult-to-obtain data.

see also BOX-AND-WHISKER PLOT, BUFFON-LAPLACE NEEDLE PROBLEM, BUFFON'S NEEDLE PROBLEM, CHERNOFF FACE, COIN FLIPPING, DE MERE'S PROBLEM, DICE, DISTRIBUTION, GAMBLER'S RUIN, INDEX, LIKELIHOOD, MOVING AVERAGE, P-VALUE, POPULATION COMPARISON, POWER (STATISTICS), PROBABILITY, RESIDUAL VS. PREDICTOR PLOT, RUN, SHARING PROBLEM, STATISTICAL TEST, TAIL PROBABILITY

References

Brown, K. S. "Probability." `http://www.seanet.com/~ksbrown/iprobabi.htm`.

Babu, G. and Feigelson, E. *Astrostatistics.* New York: Chapman & Hall, 1996.

Dixon, W. J. and Massey, F. J. *Introduction to Statistical Analysis, 4th ed.* New York: McGraw-Hill, 1983.

Doob, J. L. *Stochastic Processes.* New York: Wiley, 1953.

Feller, W. *An Introduction to Probability Theory and Its Applications, Vol. 1, 3rd ed.* New York: Wiley, 1968.

Feller, W. *An Introduction to Probability Theory and Its Applications, Vol. 2, 2nd ed.* New York: Wiley, 1968.

Fisher, N. I.; Lewis, T.; and Embleton, B. J. J. *Statistical Analysis of Spherical Data.* Cambridge, England: Cambridge University Press, 1987.

Fisher, R. A. and Prance, G. T. *The Design of Experiments, 9th ed. rev.* New York: Hafner, 1974.

Fisher, R. A. *Statistical Methods for Research Workers, 14th ed., rev. and enl.* Darien, CO: Hafner, 1970.

Goldberg, S. *Probability: An Introduction.* New York: Dover, 1986.

Gonick, L. and Smith, W. *The Cartoon Guide to Statistics.* New York: Harper Perennial, 1993.

Goulden, C. H. *Methods of Statistical Analysis, 2nd ed.* New York: Wiley, 1956.

Hoel, P. G.; Port, S. C.; and Stone, C. J. *Introduction to Statistical Theory.* New York: Houghton Mifflin, 1971.

Hogg, R. V. and Tanis, E. A. *Probability and Statistical Inference, 3rd ed.* New York: Macmillan, 1988.
Keeping, E. S. *Introduction to Statistical Inference.* New York: Dover, 1995.
Kenney, J. F. and Keeping, E. S. *Mathematics of Statistics, Pt. 1, 3rd ed.* Princeton, NJ: Van Nostrand, 1962.
Kenney, J. F. and Keeping, E. S. *Mathematics of Statistics, Pt. 2, 2nd ed.* Princeton, NJ: Van Nostrand, 1951.
Kendall, M. G.; Stuart, A.; and Ord, J. K. *Kendall's Advanced Theory of Statistics, Vol. 1: Distribution Theory, 6th ed.*0340614307 New York: Oxford University Press, 1987.
Kendall, M. G.; Stuart, A.; and Ord, J. K. *Kendall's Advanced Theory of Statistics, Vol. 2A: 5th ed.* New York: Oxford University Press, 1987.
Kendall, M. G.; Stuart, A.; and Ord, J. K. *Kendall's Advanced Theory of Statistics, Vol. 2B: Bayesian Inference.* New York: Oxford University Press, 1987.
Keynes, J. M. *A Treatise on Probability.* London: Macmillan, 1921.
Mises, R. von *Mathematical Theory of Probability and Statistics.* New York: Academic Press, 1964.
Mises, R. von *Probability, Statistics, and Truth, 2nd rev. English ed.* New York: Dover, 1981.
Mood, A. M. *Introduction to the Theory of Statistics.* New York: McGraw-Hill, 1950.
Mosteller, F. *Fifty Challenging Problems in Probability with Solutions.* New York: Dover, 1987.
Mosteller, F.; Rourke, R. E. K.; and Thomas, G. B. *Probability: A First Course, 2nd ed.* Reading, MA: Addison-Wesley, 1970.
Neyman, J. *First Course in Probability and Statistics.* New York: Holt, 1950.
Ostle, B. *Statistics in Research: Basic Concepts and Techniques for Research Workers, 4th ed.* Ames, IA: Iowa State University Press, 1988.
Papoulis, A. *Probability, Random Variables, and Stochastic Processes, 2nd ed.* New York: McGraw-Hill, 1984.
Press, W. H.; Flannery, B. P.; Teukolsky, S. A.; and Vetterling, W. T. "Statistical Description of Data." Ch. 14 in *Numerical Recipes in FORTRAN: The Art of Scientific Computing, 2nd ed.* Cambridge, England: Cambridge University Press, pp. 603–649, 1992.
Pugh, E. M. and Winslow, G. H. *The Analysis of Physical Measurements.* Reading, MA: Addison-Wesley, 1966.
Rényi, A. *Foundations of Probability.* San Francisco, CA: Holden-Day, 1970.
Robbins, H. and van Ryzin, J. *Introduction to Statistics.* Chicago, IL: Science Research Associates, 1975.
Ross, S. M. *A First Course in Probability.* New York: Macmillan, 1976.
Ross, S. M. *Introduction to Probability and Statistics for Engineers and Scientists.* New York: Wiley, 1987.
Ross, S. M. *Applied Probability Models with Optimization Applications.* New York: Dover, 1992.
Ross, S. M. *Introduction to Probability Models, 5th ed.* New York: Academic Press, 1993.
Snedecor, G. W. *Statistical Methods Applied to Experiments in Agriculture and Biology, 5th ed.* Ames, IA: State College Press, 1956.
Tippett, L. H. C. *The Methods of Statistics: An Introduction Mainly for Experimentalists, 3rd rev. ed.* London: Williams and Norgate, 1941.
Todhunter, I. *A History of the Mathematical Theory of Probability from the Time of Pascal to that of Laplace.* New York: Chelsea, 1949.
Tukey, J. W. *Explanatory Data Analysis.* Reading, MA: Addison-Wesley, 1977.
Uspensky, J. V. *Introduction to Mathematical Probability.* New York: McGraw-Hill, 1937.
Weaver, W. *Lady Luck: The Theory of Probability.* New York: Dover, 1963.
Whittaker, E. T. and Robinson, G. *The Calculus of Observations: A Treatise on Numerical Mathematics, 4th ed.* New York: Dover, 1967.
Young, H. D. *Statistical Treatment of Experimental Data.* New York: McGraw-Hill, 1962.
Yule, G. U. and Kendall, M. G. *An Introduction to the Theory of Statistics, 14th ed., rev. and enl.* New York: Hafner, 1950.

Staudt-Clausen Theorem

see VON STAUDT-CLAUSEN THEOREM

Steenrod Algebra

The Steenrod algebra has to do with the COHOMOLOGY operations in singular COHOMOLOGY with INTEGER mod 2 COEFFICIENTS. For every $n \in \mathbb{Z}$ and $i \in \{0, 1, 2, 3, \ldots\}$ there are natural transformations of FUNCTORS

$$Sq^i : H^n(\bullet; \mathbb{Z}_2) \to H^{n+i}(\bullet; \mathbb{Z}_2)$$

satisfying:

1. $Sq^i = 0$ for $i > n$.
2. $Sq^n(x) = x \smile x$ for all $x \in H^n(X, A; \mathbb{Z}_2)$ and all pairs (X, A).
3. $Sq^0 = id_{H^n(\bullet; \mathbb{Z}_2)}$.
4. The Sq^i maps commute with the coboundary maps in the long exact sequence of a pair. In other words,
$$Sq^i : H^*(\bullet; \mathbb{Z}_2) \to H^{*+i}(\bullet; \mathbb{Z}_2)$$
is a degree i transformation of cohomology theories.
5. (CARTAN RELATION)
$$Sq^i(x \smile y) = \Sigma_{j+k=i} Sq^j(x) \smile Sq^k(y).$$
6. (ADEM RELATIONS) For $i < 2j$,
$$Sq^i \circ Sq^j(x) = \Sigma_{k=0}^{\lfloor i \rfloor} \binom{j-k-1}{i-2k} Sq^{i+j-k} \circ Sq^k(x).$$
7. $Sq^i \circ \Sigma = \Sigma \circ Sq^i$ where Σ is the cohomology suspension isomorphism.

The existence of these cohomology operations endows the cohomology ring with the structure of a MODULE over the Steenrod algebra $\mathcal{A}$, defined to be $T(F_{\mathbb{Z}_2}\{Sq^i : i \in \{0, 1, 2, 3, \ldots\}\})/R$, where $F_{\mathbb{Z}_2}(\bullet)$ is the free module functor that takes any set and sends it to the free $\mathbb{Z}_2$ module over that set. We think of $F_{\mathbb{Z}_2}\{Sq^i : i \in \{0, 1, 2, \ldots\}\}$ as being a graded $\mathbb{Z}_2$ module, where the i-th gradation is given by $\mathbb{Z}_2 \cdot Sq^i$. This makes the tensor algebra $T(F_{\mathbb{Z}_2}\{Sq^i : i \in \{0, 1, 2, 3, \ldots\}\})$ into a GRADED ALGEBRA over $\mathbb{Z}_2$. R is the IDEAL generated by the elements $Sq^i Sq^j + \Sigma_{k=0}^{\lfloor i \rfloor} \binom{j-k-1}{i-2k} Sq^{i+j-k} Sq^k$ and

$1 + Sq^0$ for $0 < i < 2j$. This makes $\mathcal{A}$ into a graded $\mathbb{Z}_2$ algebra.

By the definition of the Steenrod algebra, for any SPACE (X, A), $H^*(X, A; \mathbb{Z}_2)$ is a MODULE over the Steenrod algebra $\mathcal{A}$, with multiplication induced by $Sq^i \cdot x \equiv Sq^i(x)$. With the above definitions, cohomology with COEFFICIENTS in the RING $\mathbb{Z}_2$, $H^*(\bullet; \mathbb{Z}_2)$ is a FUNCTOR from the category of pairs of TOPOLOGICAL SPACES to graded modules over $\mathcal{A}$.

see also ADEM RELATIONS, CARTAN RELATION, COHOMOLOGY, GRADED ALGEBRA, IDEAL, MODULE, TOPOLOGICAL SPACE

Steenrod-Eilenberg Axioms

see EILENBERG-STEENROD AXIOMS

Steenrod's Realization Problem

When can homology classes be realized as the image of fundamental classes of MANIFOLDS? The answer is known, and singular BORDISM GROUPS provide insight into this problem.

see also BORDISM GROUP, MANIFOLD

Steepest Descent Method

An ALGORITHM for calculating the GRADIENT $\nabla f(\mathbf{P})$ of a function at an n-D point $\mathbf{P}$. The steepest descent method starts at a point $\mathbf{P}_0$ and, as many times as needed, moves from $\mathbf{P}_i$ to $\mathbf{P}_{i+1}$ by minimizing along the line extending from $\mathbf{P}_i$ in the direction of $-\nabla f(\mathbf{P}_i)$, the local downhill gradient. This method has the severe drawback of requiring a great many iterations for functions which have long, narrow valley structures. In such cases, a CONJUGATE GRADIENT METHOD is preferable.

see also CONJUGATE GRADIENT METHOD, GRADIENT

References

Arfken, G. "The Method of Steepest Descents." §7.4 in *Mathematical Methods for Physicists, 3rd ed.* Orlando, FL: Academic Press, pp. 428–436, 1985.

Menzel, D. (Ed.). *Fundamental Formulas of Physics, Vol. 2, 2nd ed.* New York: Dover, p. 80, 1960.

Morse, P. M. and Feshbach, H. "Asymptotic Series; Method of Steepest Descent." §4.6 in *Methods of Theoretical Physics, Part I.* New York: McGraw-Hill, pp. 434–443, 1953.

Press, W. H.; Flannery, B. P.; Teukolsky, S. A.; and Vetterling, W. T. *Numerical Recipes in FORTRAN: The Art of Scientific Computing, 2nd ed.* Cambridge, England: Cambridge University Press, p. 414, 1992.

Steffenson's Formula

$$f_p = f_0 + \tfrac{1}{2}p(p+1)\delta_{1/2} - \tfrac{1}{2}(p-1)p\delta_{-1/2} + (S_3 + S_4)\delta_{1/2}^3 + (S_3 - S_4)\delta_{-1/2}^3 + \ldots, \quad (1)$$

for $p \in [-\frac{1}{2}, \frac{1}{2}]$, where δ is the CENTRAL DIFFERENCE and

$$S_{2n+1} = \frac{1}{2}\binom{p+n}{2n+1} \quad (2)$$

$$S_{2n+2} = \frac{p}{2n+2}\binom{p+n}{2n+1} \quad (3)$$

$$S_{2n+1} - S_{2n+2} = \binom{p+n+1}{2n+2} \quad (4)$$

$$S_{2n+1} - S_{2n+2} = -\binom{p+n}{2n+2}, \quad (5)$$

where $\binom{n}{k}$ is a BINOMIAL COEFFICIENT.

see also CENTRAL DIFFERENCE, STIRLING'S FINITE DIFFERENCE FORMULA

References

Beyer, W. H. *CRC Standard Mathematical Tables, 28th ed.* Boca Raton, FL: CRC Press, p. 433, 1987.

Steffensen's Inequality

Let $f(x)$ be a NONNEGATIVE and monotonic decreasing function in $[a, b]$ and $g(x)$ satisfy such that $0 \leq g(x) \leq 1$ in $[a, b]$, then

$$\int_{b-k}^{b} f(x)\,dx \leq \int_a^b f(x)g(x)\,dx \leq \int_a^{a+k} f(x)\,dx,$$

where

$$k = \int_a^b g(x)\,dx.$$

References

Gradshteyn, I. S. and Ryzhik, I. M. *Tables of Integrals, Series, and Products, 5th ed.* San Diego, CA: Academic Press, p. 1099, 1979.

Steinbach Screw

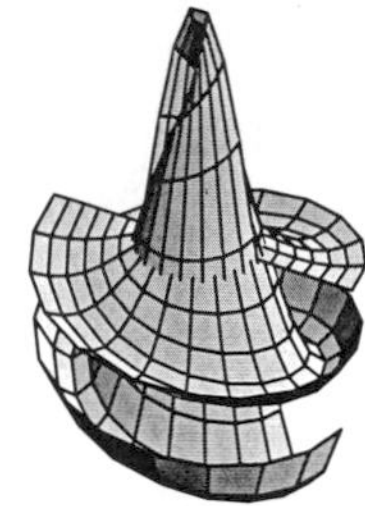

A SURFACE generated by the parametric equations

$$x(u, v) = u \cos v$$

$$y(u, v) = u \sin v$$

$$z(u, v) = v \cos u.$$

The above image uses $u \in [-4, 4]$ and $v \in [0, 6.25]$.

References

Naylor, B. "Steinbach Screw 1." `http://www.garlic.com/~bnaylor/rtstein1.html`.

Pickover, C. A. *Mazes for the Mind: Computers and the Unexpected.* New York: St. Martin's Press, 1992.

Wang, P. "Renderings." `http://www.ugcs.caltech.edu/~peterw/portfolio/renderings/`.

Steiner Chain

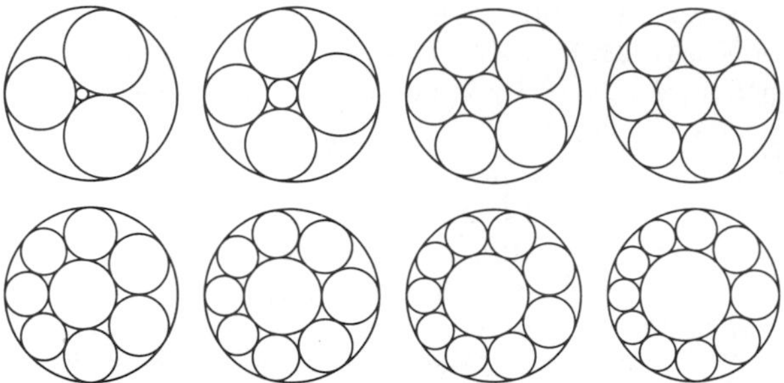

Given two nonconcentric CIRCLES with one interior to the other, if small TANGENT CIRCLES can be inscribed around the region between the two CIRCLES such that the final CIRCLE is TANGENT to the first, the CIRCLES form a Steiner chain.

The simplest way to construct a Steiner chain is to perform an INVERSION on a symmetrical arrangement on n circles packed between a central circle of radius b and an outer concentric circle of radius a. In this arrangement,

$$\sin\left(\frac{\pi}{n}\right) = \frac{a-b}{a+b}, \tag{1}$$

so the ratio of the radii for the small and large circles is

$$\frac{b}{a} = \frac{1-\sin\left(\frac{\pi}{n}\right)}{1+\sin\left(\frac{\pi}{n}\right)}. \tag{2}$$

To transform the symmetrical arrangement into a Steiner chain, find an INVERSION CENTER which transforms two centers initially offset by a fixed distance c to the same point. This can be done by equating

$$\frac{k^2 x}{x^2 - a^2} = \frac{k^2(x-c)}{(x-c)^2 - b^2}, \tag{3}$$

giving the offset of the inversion center from the large circle's center as

$$x = \frac{a^2 - b^2 + c^2 \pm \sqrt{(a^2-b^2+c^2)^2 - 4a^2c^2}}{2c}. \tag{4}$$

Plugging in a fixed value of a fixes b, which therefore determines x for a given c. Equivalently, a Steiner chain results whenever the INVERSIVE DISTANCE between the two original circles is given by

$$\delta = 2\ln\left[\sec\left(\frac{\pi}{n}\right) + \tan\left(\frac{\pi}{n}\right)\right] \tag{5}$$

$$= 2\ln\left[\tan\left(\frac{\pi}{4} + \frac{\pi}{2n}\right)\right] \tag{6}$$

(Coxeter and Greitzer 1967). The centers of the circles in a Steiner chain lie on an ELLIPSE (Ogilvy 1990, p. 57).

STEINER'S PORISM states that if a Steiner chain is formed from one starting circle, then a Steiner chain is also formed from any other starting circle.

see also ARBELOS, COXETER'S LOXODROMIC SEQUENCE OF TANGENT CIRCLES, HEXLET, PAPPUS CHAIN, STEINER'S PORISM

References

Coxeter, H. S. M. "Interlocking Rings of Spheres." *Scripta Math.* **18**, 113–121, 1952.

Coxeter, H. S. M. *Introduction to Geometry, 2nd ed.* New York: Wiley, p. 87, 1969.

Coxeter, H. S. M. and Greitzer, S. L. *Geometry Revisited.* Washington, DC: Math. Assoc. Amer., pp. 124–126, 1967.

Forder, H. G. *Geometry, 2nd ed.* London: Hutchinson's University Library, p. 23, 1960.

Gardner, M. "Mathematical Games: The Diverse Pleasures of Circles that Are Tangent to One Another." *Sci. Amer.* **240**, 18–28, Jan. 1979.

Johnson, R. A. *Modern Geometry: An Elementary Treatise on the Geometry of the Triangle and the Circle.* Boston, MA: Houghton Mifflin, pp. 113–115, 1929.

Ogilvy, C. S. *Excursions in Geometry.* New York: Dover, pp. 51–54, 1990.

Weisstein, E. W. "Plane Geometry." `http://www.astro.virginia.edu/~eww6n/math/notebooks/PlaneGeometry.m`.

Steiner Construction

A construction done using only a STRAIGHTEDGE. The PONCELET-STEINER THEOREM proves that all constructions possible using a COMPASS and STRAIGHTEDGE are possible using a STRAIGHTEDGE alone, as long as a fixed CIRCLE and its center, two intersecting CIRCLES without their centers, or three nonintersecting CIRCLES are drawn beforehand.

see also GEOMETRIC CONSTRUCTION, MASCHERONI CONSTRUCTION, PONCELET-STEINER THEOREM, STRAIGHTEDGE

References

Dörrie, H. "Steiner's Straight-Edge Problem." §34 in *100 Great Problems of Elementary Mathematics: Their History and Solutions.* New York: Dover, pp. 165–170, 1965.

Steiner, J. *Geometric Constructions with a Ruler, Given a Fixed Circle with Its Center.* Translated from the first German ed. (1833). New York: Scripta Mathematica, 1950.

Steiner's Ellipse

Let $\alpha' : \beta' : \gamma'$ be the ISOTOMIC CONJUGATE POINT of a point with TRILINEAR COORDINATES $\alpha : \beta : \gamma$. The isotomic conjugate of the LINE AT INFINITY having trilinear equation

$$a\alpha + b\beta + c\gamma = 0$$

is

$$\frac{\beta'\gamma'}{a} + \frac{\gamma'\alpha'}{b} + \frac{\alpha'\beta'}{c} = 0,$$

known as Steiner's ellipse (Vandeghen 1965).

see also ISOTOMIC CONJUGATE POINT, LINE AT INFINITY

References

Vandeghen, A. "Some Remarks on the Isogonal and Cevian Transforms. Alignments of Remarkable Points of a Triangle." *Amer. Math. Monthly* **72**, 1091–1094, 1965.

Steiner's Hypocycloid

see DELTOID

Steiner-Lehmus Theorem

Any TRIANGLE that has two equal ANGLE BISECTORS (each measured from a VERTEX to the opposite sides) is an ISOSCELES TRIANGLE. This theorem is also called the INTERNAL BISECTORS PROBLEM and LEHMUS' THEOREM.

see also ISOSCELES TRIANGLE

References

Altshiller-Court, N. *College Geometry: A Second Course in Plane Geometry for Colleges and Normal Schools, 2nd ed., rev. enl.* New York: Barnes and Noble, p. 72, 1952.

Coxeter, H. S. M. *Introduction to Geometry, 2nd ed.* New York: Wiley, p. 9, 1969.

Coxeter, H. S. M. and Greitzer, S. L. *Geometry Revisited.* Washington, DC: Math. Assoc. Amer., pp. 14–16, 1967.

Gardner, M. *Martin Gardner's New Mathematical Diversions from Scientific American.* New York: Simon and Schuster, pp. 198–199 and 206–207, 1966.

Henderson, A. "The Lehmus-Steiner-Terquem Problem in Global Survey." *Scripta Math.* **21**, 223–232 and 309-312, 1955.

Hunter, J. A. H. and Madachy, J. S. *Mathematical Diversions.* New York: Dover, pp. 72–73, 1975.

Steiner Points

There are two different types of points known as Steiner points.

The point of CONCURRENCE of the three lines drawn through the VERTICES of a TRIANGLE PARALLEL to the corresponding sides of the first BROCARD TRIANGLE. It lies on the CIRCUMCIRCLE opposite the TARRY POINT and has TRIANGLE CENTER FUNCTION

$$\alpha = bc(a^2 - b^2)(a^2 - c^2).$$

The BRIANCHON POINT for KIEPERT'S PARABOLA is the Steiner point. The LEMOINE POINT K is the Steiner point of the first BROCARD TRIANGLE. The SIMSON LINE of the Steiner point is PARALLEL to the line OK, when O is the CIRCUMCENTER and K is the LEMOINE POINT.

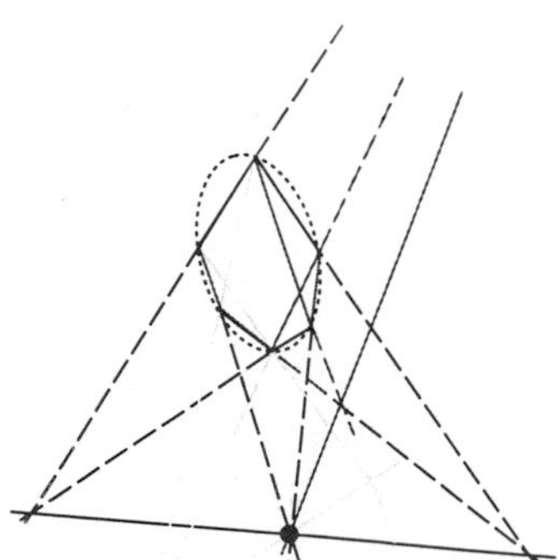

If triplets of opposites sides on a CONIC SECTION in PASCAL'S THEOREM are extended for all permutations of VERTICES, 60 PASCAL LINES are produced. The 20 points of their 3 by 3 intersections are called Steiner points.

see also BRIANCHON POINT, BROCARD TRIANGLES, CIRCUMCIRCLE, CONIC SECTION, KIEPERT'S PARABOLA, LEMOINE POINT, PASCAL LINE, PASCAL'S THEOREM, STEINER SET, STEINER TRIPLE SYSTEM, TARRY POINT

References

Casey, J. *A Treatise on the Analytical Geometry of the Point, Line, Circle, and Conic Sections, Containing an Account of Its Most Recent Extensions, with Numerous Examples, 2nd ed., rev. enl.* Dublin: Hodges, Figgis, & Co., p. 66, 1893.

Gallatly, W. *The Modern Geometry of the Triangle, 2nd ed.* London: Hodgson, p. 102, 1913.

Johnson, R. A. *Modern Geometry: An Elementary Treatise on the Geometry of the Triangle and the Circle.* Boston, MA: Houghton Mifflin, pp. 281–282, 1929.

Kimberling, C. "Central Points and Central Lines in the Plane of a Triangle." *Math. Mag.* **67**, 163–187, 1994.

Steiner's Porism

If a STEINER CHAIN is formed from one starting circle, then a STEINER CHAIN is formed from any other starting circle. In other words, given two nonconcentric CIRCLES, draw CIRCLES successively touching them and each other. If the last touches the first, this will also happen for any position of the first CIRCLE.

see also HEXLET, STEINER CHAIN

References

Coxeter, H. S. M. "Interlocking Rings of Spheres." *Scripta Math.* **18**, 113–121, 1952.

Coxeter, H. S. M. *Introduction to Geometry, 2nd ed.* New York: Wiley, p. 87, 1969.

Coxeter, H. S. M. and Greitzer, S. L. *Geometry Revisited.* Washington, DC: Math. Assoc. Amer., pp. 124–126, 1967.

Forder, H. G. *Geometry, 2nd ed.* London: Hutchinson's University Library, p. 23, 1960.

Gardner, M. "Mathematical Games: The Diverse Pleasures of Circles that Are Tangent to One Another." *Sci. Amer.* **240**, 18–28, Jan. 1979.

Johnson, R. A. *Modern Geometry: An Elementary Treatise on the Geometry of the Triangle and the Circle.* Boston, MA: Houghton Mifflin, pp. 113–115, 1929.

Ogilvy, C. S. *Excursions in Geometry.* New York: Dover, pp. 53–54, 1990.

Steiner's Problem

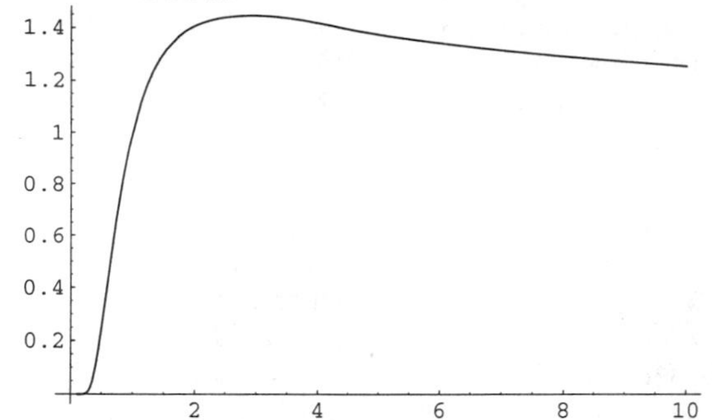

For what value of x is $f(x) = x^{1/x}$ a MAXIMUM? The maximum occurs at $x = e$, where

$$f'(x) = x^{-2+1/x}(1 - \ln x) = 0,$$

which gives a maximum of

$$e^{1/e} = 1.444667861\ldots.$$

The function has an inflection point at $x = 0.581933\ldots$, where

$$f''(x) = x^{-4+1/x}[1 - 3x + (\ln x)(2x - 2 + \ln x)] = 0.$$

see also FERMAT'S PROBLEM

Steiner Quadruple System

A Steiner quadruple system is a STEINER SYSTEM $S(t = 3, k = 4, v)$, where S is a v-set and B is a collection of k-sets of S such that every t-subset of S is contained in exactly one member of B. Barrau (1908) established the uniqueness of $S(3,4,8)$,

1	2	4	8	3	5	6	7
2	3	5	8	1	4	6	7
3	4	6	8	1	2	5	7
4	5	7	8	1	2	3	6
1	5	6	8	2	3	4	7
2	6	7	8	1	3	4	5
1	3	7	8	2	4	5	6

and $S(3,4,10)$

1	2	4	5	1	2	3	7	1	3	5	8
2	3	5	6	2	3	4	8	2	4	6	9
3	4	6	7	3	4	5	9	3	5	7	0
4	5	7	8	4	5	6	0	1	4	6	8
5	6	8	9	1	5	6	7	2	5	7	9
6	7	9	0	2	6	7	8	3	6	8	0
1	7	8	0	3	7	8	9	1	4	7	9
1	2	8	9	4	8	9	0	2	5	8	0
2	3	9	0	1	5	9	0	1	3	6	9
1	3	4	0	1	2	6	0	2	4	7	0

Fitting (1915) subsequently constructed the cyclic systems $S(3,4,26)$ and $S(3,4,34)$, and Bays and de Weck (1935) showed the existence of at least one $S(3,4,14)$. Hanani (1960) proved that a NECESSARY and SUFFICIENT condition for the existence of an $S(3,4,v)$ is that $v \equiv 2$ or 4 (mod 6).

The number of nonisomorphic steiner quadruple systems of orders 8, 10, 14, and 16 are 1, 1, 4 (Mendelsohn and Hung 1972), and at least 31,021 (Lindner and Rosa 1976).

see also STEINER SYSTEM, STEINER TRIPLE SYSTEM

References

Barrau, J. A. "On the Combinatory Problem of Steiner." *K. Akad. Wet. Amsterdam Proc. Sect. Sci.* **11**, 352–360, 1908.

Bays, S. and de Weck, E. "Sur les systémes de quadruples." *Comment. Math. Helv.* **7**, 222–241, 1935.

Fitting, F. "Zyklische Lösungen des Steiner'schen Problems." *Nieuw. Arch. Wisk.* **11**, 140–148, 1915.

Hanani, M. "On Quadruple Systems." *Canad. J. Math.* **12**, 145–157, 1960.

Lindner, C. L. and Rosa, A. "There are at Least 31,021 Nonisomorphic Steiner Quadruple Systems of Order 16." *Utilitas Math.* **10**, 61–64, 1976.

Lindner, C. L. and Rosa, A. "Steiner Quadruple Systems—A Survey." *Disc. Math.* **22**, 147–181, 1978.

Mendelsohn, N. S. and Hung, S. H. Y. "On the Steiner Systems $S(3,4,14)$ and $S(4,5,15)$." *Utilitas Math.* **1**, 5–95, 1972.

Steiner's Segment Problem

Given n points, find the line segments with the shortest possible total length which connect the points. The segments need not necessarily be straight from one point to another.

For three points, if all ANGLES are less than 120°, then the line segments are those connecting the three points to a central point P which makes the ANGLES $\langle A\rangle PB$, $\langle B\rangle PC$, and $\langle C\rangle PA$ all 120°. If one ANGLE is greater that 120°, then P coincides with the offending ANGLE.

For four points, P is the intersection of the two diagonals, but the required minimum segments are not necessarily these diagonals.

A modified version of the problem is, given two points, to find the segments with the shortest total length connecting the points such that each branch point may be connected to only three segments. There is no general solution to this version of the problem.

Steiner Set

Three sets of three LINES such that each line is incident with two from both other sets.

see also SOLOMON'S SEAL LINES, STEINER POINTS, STEINER TRIPLE SYSTEM

Steiner Surface

A projection of the VERONESE SURFACE into 3-D (which must contain singularities) is called a Steiner surface. A classification of Steiner surfaces allowing complex parameters and projective transformations was accomplished in the 19th century. The surfaces obtained by restricting to real parameters and transformations were classified into 10 types by Coffman *et al.* (1996). Examples of Steiner surfaces include the ROMAN SURFACE (Coffman type 1) and CROSS-CAP (type 3).

The Steiner surface of type 2 is given by the implicit equation

$$x^2y^2 - x^2z^2 + y^2z^2 - xyz = 0,$$

and can be transformed into the ROMAN SURFACE or CROSS-CAP by a complex projective change of coordinates (but not by a real transformation). It has two pinch points and three double lines and, unlike the ROMAN SURFACE or CROSS-CAP, is not compact in any affine neighborhood.

The Steiner surface of type 4 has the implicit equation

$$y^2 - 2xy^2 - xz^2 + x^2y^2 + x^2z^2 - z^4 = 0,$$

and two of the three double lines of surface 2 coincide along a line where the two noncompact "components" are tangent.

see also CROSS-CAP, ROMAN SURFACE, VERONESE VARIETY

References

Coffman, A. "Steiner Surfaces." http://www.ipfw.edu/math/Coffman/steinersurface.html.

Coffman, A.; Schwartz, A.; and Stanton, C. "The Algebra and Geometry of Steiner and Other Quadratically Parametrizable Surfaces." *Computer Aided Geom. Design* **13**, 257–286, 1996.

Nordstrand, T. "Steiner Relative." http://www.uib.no/people/nfytn/stmtxt.htm.

Nordstrand, T. "Steiner Relative [2]." http://www.uib.no/people/nfytn/stm2txt.htm.

Steiner System

A Steiner system is a set X of v points, and a collection of subsets of X of size k (called blocks), such that any t points of X are in exactly one of the blocks. The special case $t = 2$ and $k = 3$ corresponds to a so-called STEINER TRIPLE SYSTEM. For a PROJECTIVE PLANE, $v = n^2 + n + 1$, $k = n + 1$, $t = 2$, and the blocks are simply lines.

see also STEINER QUADRUPLE SYSTEM, STEINER TRIPLE SYSTEM.

References

Colbourn, C. J. and Dinitz, J. H. (Eds.) *CRC Handbook of Combinatorial Designs.* Boca Raton, FL: CRC Press, 1996.

Woolhouse, W. S. B. "Prize Question 1733." *Lady's and Gentleman's Diary.* 1844.

Steiner's Theorem

Let LINES x and y join a variable point on a CONIC SECTION to two fixed points on the same CONIC SECTION. Then x and y are PROJECTIVELY related.

see also CONIC SECTION, PROJECTION

Steiner Triple System

Let X be a set of $v \geq 3$ elements together with a set B of 3-subset (triples) of X such that every 2-SUBSET of X occurs in exactly one triple of B. Then B is called a Steiner triple system and is a special case of a STEINER SYSTEM with $t = 2$ and $k = 3$. A Steiner triple system $S(v) = S(v, k = 3, \lambda = 1)$ of order v exists IFF $v \equiv 1, 3 \pmod 6$ (Kirkman 1847). In addition, if Steiner triple systems S_1 and S_2 of orders v_1 and v_2 exist, then so does a Steiner triple system S of order v_1v_2 (Ryser 1963, p. 101).

Examples of Steiner triple systems $S(v)$ of small orders v are

$$\begin{aligned} S_3 &= \{\{1,2,3\}\} \\ S_7 &= \{\{1,2,4\},\{2,3,5\},\{3,4,6\},\{4,5,7\}, \\ &\quad \{5,6,1\},\{6,7,2\},\{7,1,3\}\} \\ S_9 &= \{\{1,2,3\},\{4,5,6\},\{7,8,9\},\{1,4,7\}, \\ &\quad \{2,5,8\},\{3,6,9\},\{1,5,9\},\{2,6,7\}\}. \end{aligned}$$

The number of nonisomorphic Steiner triple systems $S(v)$ of orders v = 7, 9, 13, 15, 19, ... (i.e., $6k + 1, 3$) are 1, 1, 20, 80, $> 1.1 \times 10^9$, ... (Colbourn and Dinitz 1996, pp. 14–15; Sloane's A030129). $S(7)$ is the same as the finite PROJECTIVE PLANE of order 2. $S(9)$ is a finite AFFINE PLANE which can be constructed from the array

$$\begin{matrix} a & b & c \\ d & e & f \\ g & h & i \end{matrix}.$$

One of the two $S(13)$s is a finite HYPERBOLIC PLANE. The 80 Steiner triple systems $S(15)$ have been studied by Tonchev and Weishaar (1997). There are more than 1.1×10^9 Steiner triple systems of order 19 (Stinson and Ferch 1985; Colbourn and Dinitz 1996, p. 15).

see also HADAMARD MATRIX, KIRKMAN TRIPLE SYSTEM, STEINER QUADRUPLE SYSTEM, STEINER SYSTEM

References

Colbourn, C. J. and Dinitz, J. H. (Eds.) "Steiner Triple Systems." §4.5 in *CRC Handbook of Combinatorial Designs.* Boca Raton, FL: CRC Press, pp.14–15 and 70, 1996.

Kirkman, T. P. "On a Problem in Combinatorics." *Cambridge Dublin Math. J.* **2**, 191–204, 1847.

Lindner, C. C. and Rodger, C. A. *Design Theory.* Boca Raton, FL: CRC Press, 1997.

Ryser, H. J. *Combinatorial Mathematics.* Buffalo, NY: Math. Assoc. Amer., pp. 99–102, 1963.

Sloane, N. J. A. Sequence A030129 in "An On-Line Version of the Encyclopedia of Integer Sequences."

Stinson, D. R. and Ferch, H. "2000000 Steiner Triple Systems of Order 19." *Math. Comput.* **44**, 533–535, 1985.

Tonchev, V. D. and Weishaar, R. S. "Steiner Triple Systems of Order 15 and Their Codes." *J. Stat. Plan. Inference* **58**, 207–216, 1997.

Steinerian Curve

The LOCUS of points whose first POLARS with regard to the curves of a linear net have a common point. It is also the LOCUS of points of CONCURRENCE of line POLARS of points of the JACOBIAN CURVE. It passes through all points common to all curves of the system and is of order $3(n-1)^2$.

see also CAYLEYIAN CURVE, JACOBIAN CURVE

References

Coolidge, J. L. *A Treatise on Algebraic Plane Curves.* New York: Dover, p. 150, 1959.

Steinhaus-Moser Notation

A NOTATION for LARGE NUMBERS defined by Steinhaus (1983, pp. 28–29). In this notation, $\triangle n$ denotes n^n, $\boxed{n}$ denotes "n in n TRIANGLES," and $\textcircled{n}$ denotes "n in n SQUARES." A modified version due to Moser eliminates the circle notation, continuing instead with POLYGONS of ever increasing size, so n in a PENTAGON is n with n SQUARES around it, etc.

see also CIRCLE NOTATION, LARGE NUMBER, MEGA, MOSER

References

Steinhaus, H. *Mathematical Snapshots, 3rd American ed.* New York: Oxford University Press, 1983.

Steinitz's Theorem

A GRAPH G is the edge graph of a POLYHEDRON IFF G is a SIMPLE, PLANAR GRAPH which is 3-connected.

see also PLANAR GRAPH, SIMPLE GRAPH

Steinmetz Solid

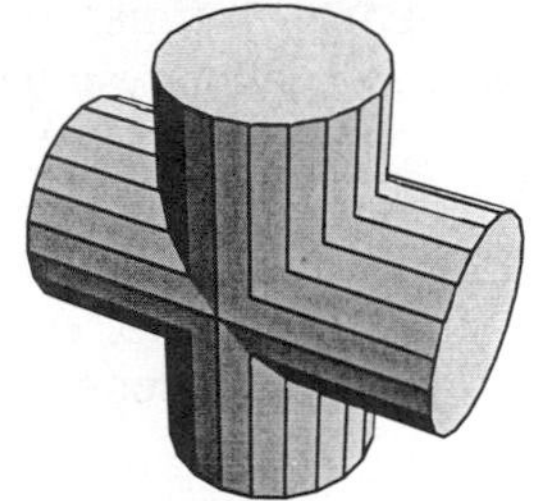

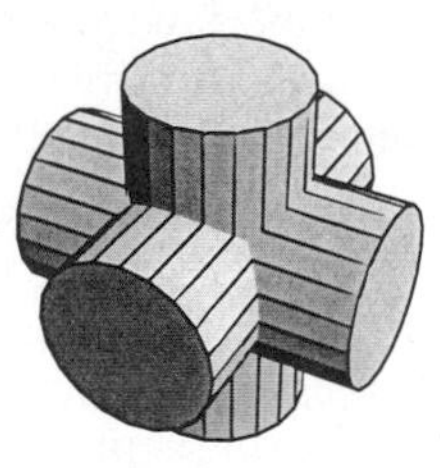

The solid common to two (or three) right circular CYLINDERS of equal RADII intersecting at RIGHT ANGLES is called the Steinmetz solid. (Two CYLINDERS intersecting at RIGHT ANGLES are sometimes called a BICYLINDER, and three intersecting CYLINDERS a TRICYLINDER.)

The VOLUME common to two intersecting right CYLINDERS of RADIUS r is

$$V_2(r,r) = \tfrac{16}{3}r^3. \quad (1)$$

If the two right CYLINDERS are of *different* RADII a and b with $a > b$, then the VOLUME common to them is

$$V_2(a,b) = \tfrac{8}{3}a[(a^2+b^2)E(k) - (a^2-b^2)K(k)], \quad (2)$$

where $K(k)$ is the complete ELLIPTIC INTEGRAL OF THE FIRST KIND, $E(k)$ is the complete ELLIPTIC INTEGRAL OF THE SECOND KIND, and $k \equiv b/a$ is the MODULUS.

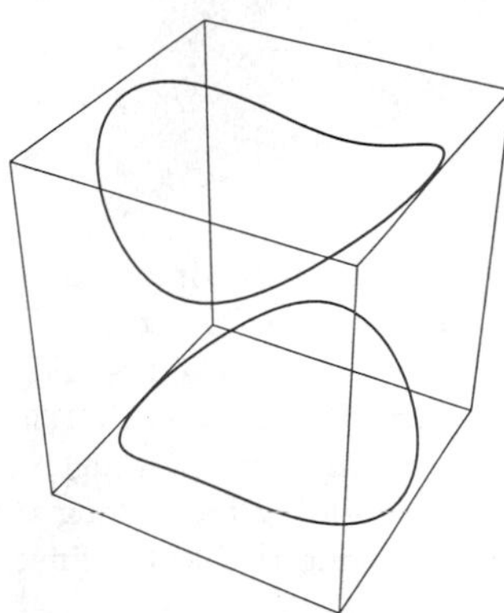

The curves of intersection of two cylinders of RADII a and b, shown above, are given by the parametric equations

$$x(t) = a\cos t \quad (3)$$

$$y(t) = a\sin t \quad (4)$$

$$z(t) = \pm\sqrt{b^2 - a^2\sin^2 t} \quad (5)$$

(Gray 1993).

The VOLUME common to two ELLIPTIC CYLINDERS

$$\frac{x^2}{a^2} + \frac{z^2}{c^2} = 1 \qquad \frac{y^2}{b^2} + \frac{z^2}{c'^2} = 1 \quad (6)$$

with $c < c'$ is

$$V_2(a,c;b,c') = \frac{8ab}{3c}[(c'^2+c^2)E(k) - (c'^2-c^2)K(k)], \quad (7)$$

where $k = c/c'$ (Bowman 1961, p. 34).

For three CYLINDERS of RADII r intersecting at RIGHT ANGLES, the VOLUME of intersection is

$$V_3(r,r,r) = 8(2-\sqrt{2})r^3. \quad (8)$$

see also BICYLINDER, CYLINDER

References

Bowman, F. *Introduction to Elliptic Functions, with Applications.* New York: Dover, 1961.

Gardner, M. *The Unexpected Hanging and Other Mathematical Diversions.* Chicago, IL: Chicago University Press, pp. 183–185, 1991.

Gray, A. *Modern Differential Geometry of Curves and Surfaces.* Boca Raton, FL: CRC Press, pp. 149–150, 1993.

Wells, D. G. #555 in *The Penguin Book of Curious and Interesting Puzzles.* London: Penguin Books, 1992.

Stella Octangula

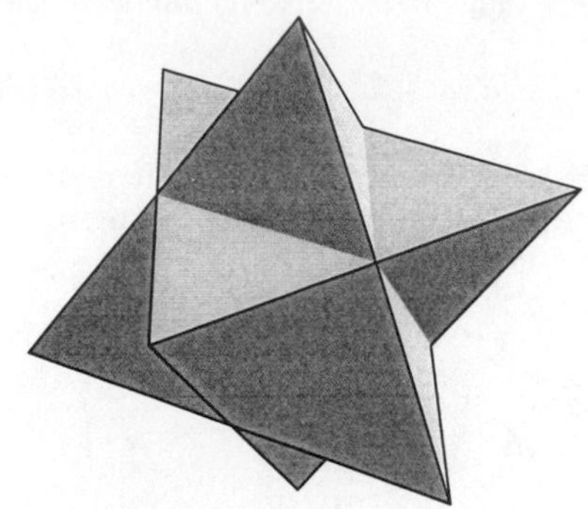

A POLYHEDRON COMPOUND composed of a TETRAHEDRON and its RECIPROCAL (a second TETRAHEDRON rotated 180° with respect to the first). The stella octangula is also called a STELLATED TETRAHEDRON. It can be constructed using the following NET by cutting along the solid lines, folding back along the plain lines, and folding forward along the dotted lines.

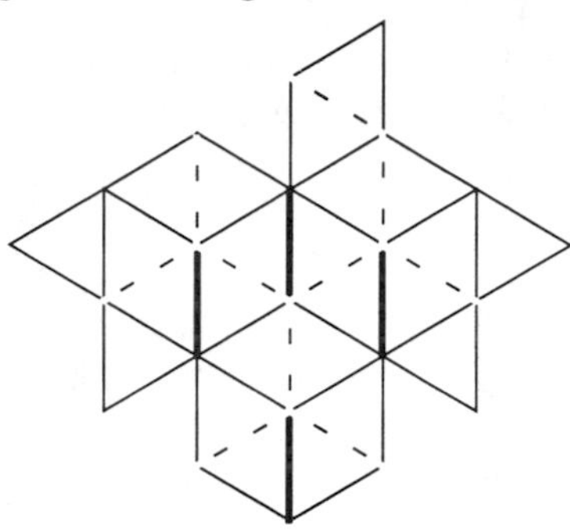

Another construction builds a single TETRAHEDRON, then attaches four tetrahedral caps, one to each face.

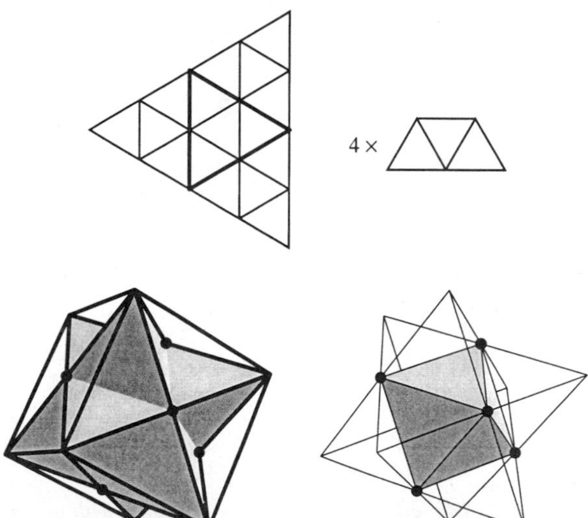

The edges of the two tetrahedra form the 12 DIAGONALS of a CUBE. The solid common to both tetrahedra is an OCTAHEDRON (Ball and Coxeter 1987).

see also CUBE, OCTAHEDRON, POLYHEDRON COMPOUND, TETRAHEDRON

References

Ball, W. W. R. and Coxeter, H. S. M. *Mathematical Recreations and Essays, 13th ed.* New York: Dover, pp. 135–137, 1987.

Coxeter, H. S. M. *Introduction to Geometry, 2nd ed.* New York: Wiley, p. 158, 1969.

Cundy, H. and Rollett, A. *Mathematical Models, 3rd ed.* Stradbroke, England: Tarquin Pub., p. 129, 1989.

Stella Octangula Number

A FIGURATE NUMBER of the form,

$$StOct_n = O_n + 8T_{n-1} = n(2n^2 - 1).$$

The first few are 1, 14, 51, 124, 245, ... (Sloane's A007588). The GENERATING FUNCTION for the stella octangula numbers is

$$\frac{x(x^2 + 10x + 1)}{(x-1)^4} = x + 14x^2 + 51x^3 + 124x^4 + \ldots.$$

References

Conway, J. H. and Guy, R. K. *The Book of Numbers.* New York: Springer-Verlag, p. 51, 1996.

Sloane, N. J. A. Sequence A007588/M4932 in "An On-Line Version of the Encyclopedia of Integer Sequences."

Stellated Polyhedron

A convex regular POLYHEDRON. Stellated polyhedra include the KEPLER-POINSOT SOLIDS, which consist of three DODECAHEDRON STELLATIONS and one of the ICOSAHEDRON STELLATIONS. Coxeter (1982) shows that 59 ICOSAHEDRON STELLATIONS exist. The CUBE and the TETRAHEDRON cannot be stellated. The OCTAHEDRON has only one stellation, the STELLA OCTANGULA which is a compound of two TETRAHEDRA.

There are therefore a total of $3 + 1 + (59 - 1) + 1 = 63$ stellated POLYHEDRA, although some are COMPOUND POLYHEDRA and therefore not UNIFORM POLYHEDRA. The set of all possible EDGES of the stellations can be obtained by finding all intersections on the facial planes.

see also ARCHIMEDEAN SOLID STELLATION, DODECAHEDRON STELLATIONS, ICOSAHEDRON STELLATIONS, KEPLER-POINSOT SOLID, POLYHEDRON, STELLA OCTANGULA, STELLATED TRUNCATED HEXAHEDRON, STELLATION, UNIFORM POLYHEDRON

References

Coxeter, H. S. M. *The Fifty-Nine Icosahedra.* New York: Springer-Verlag, 1982.

Cundy, H. and Rollett, A. *Mathematical Models, 3rd ed.* Stradbroke, England: Tarquin Publications, 1989.

Wenninger, M. J. *Polyhedron Models.* Cambridge, England: University Press, 1974.

Stellated Tetrahedron

see STELLA OCTANGULA

Stellated Truncated Hexahedron

The UNIFORM POLYHEDRON U_{19}, also called the QUASITRUNCATED HEXAHEDRON, whose DUAL POLYHEDRON is the GREAT TRIAKIS OCTAHEDRON. It has SCHLÄFLI SYMBOL $t'\{4,3\}$ and WYTHOFF SYMBOL $2\,3\,|\,\frac{4}{3}$. Its faces are $8\{3\}+6\{\frac{8}{3}\}$. For $a=1$, its CIRCUMRADIUS is

$$R=\tfrac{1}{2}\sqrt{7-4\sqrt{2}}\,.$$

References

Wenninger, M. J. *Polyhedron Models.* Cambridge, England: Cambridge University Press, p. 144, 1989.

Stellation

The process of constructing POLYHEDRA by extending the facial PLANES past the EDGES of a given POLYHEDRON.

see also ARCHIMEDEAN SOLID STELLATION, DODECAHEDRON STELLATIONS, FACETING, ICOSAHEDRON STELLATIONS, KEPLER-POINSOT SOLID, POLYHEDRON, STELLA OCTANGULA, STELLATED POLYHEDRON, STELLATED TRUNCATED HEXAHEDRON, STELLATION TRUNCATION, UNIFORM POLYHEDRON

References

Fleurent, G. M. "Symmetry and Polyhedral Stellation Ia and Ib. Symmetry 2: Unifying Human Understanding, Part 1." *Comput. Math. Appl.* **17**, 167–193, 1989.

Messer, P. W. "Les étoilements du rhombitricontaèdre et plus." *Structural Topology* **21**, 25–46, 1995.

Messer, P. W. and Wenninger, M. J. "Symmetry and Polyhedral Stellation. II. Symmetry 2: Unifying Human Understanding, Part 1." *Comput. Math. Appl.* **17**, 195–201, 1989.

Stem-and-Leaf Diagram

The "stem" is a column of the data with the last digit removed. The final digits of each column are placed next to each other in a row next to the appropriate column. Then each row is sorted in numerical order. This diagram was invented by John Tukey.

References

Tukey, J. W. *Explanatory Data Analysis.* Reading, MA: Addison-Wesley, pp. 7–16, 1977.

Step

1.5 times the H-SPREAD.

see also FENCE, H-SPREAD

References

Tukey, J. W. *Explanatory Data Analysis.* Reading, MA: Addison-Wesley, p. 44, 1977.

Step Function

A function on the REALS $\mathbb{R}$ is a step function if it can be written as a finite linear combination of semi-open intervals $[a,b)\subseteq\mathbb{R}$. Therefore, a step function f can be written as

$$f(x)=\alpha_1 f_1(x)+\cdots+\alpha_n f_n(x),$$

where $\alpha_i\in\mathbb{R}$, $f_i(x)=1$ if $x\in[a_i,b_i)$ and 0 otherwise, for $i=1,\ldots,n$.

see also HEAVISIDE STEP FUNCTION

Step Polynomial

see HERMITE'S INTERPOLATING FUNDAMENTAL POLYNOMIAL

Steradian

The unit of SOLID ANGLE. The SOLID ANGLE corresponding to all of space being subtended is 4π steradian.

see also RADIAN, SOLID ANGLE

Stereogram

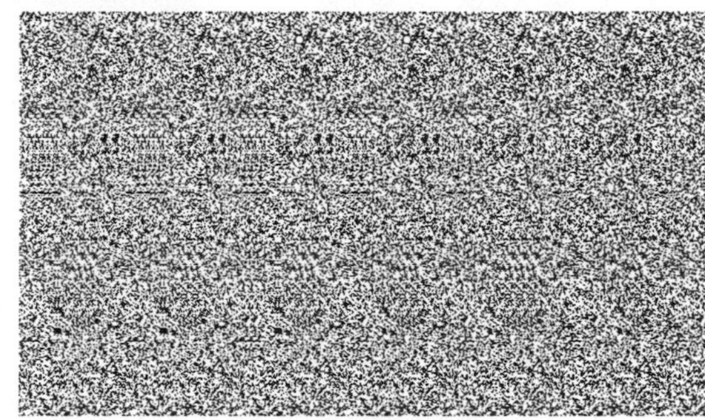

A plane image or pair of 2-D images which, when appropriately viewed using both eyes, produces an image which appears to be three-dimensional. By taking a pair of photographs from slightly different angles and then allowing one eye to view each image, a stereogram is not difficult to produce.

Amazingly, it turns out that the 3-D effect can be produced by both eyes looking at a *single* image by defocusing the eyes at a certain distance. Such stereograms are called "random-dot stereograms."

References

Bar-Natan, D. "Random-Dot Stereograms." *Math. J.* **1**, 69–71, 1991.

Fineman, M. *The Nature of Visual Illusion.* New York: Dover, pp. 89–93, 1996.

Julesz, B. *Foundations of Cyclopean Perception.* Chicago, IL: University of Chicago Press, 1971.

Julesz, B. "Stereoscopic Vision." *Vision Res.* **26**, 1601–1611, 1986.

Terrell, M. S. and Terrell, R. E. "Behind the Scenes of a Random Dot Stereogram." *Amer. Math. Monthly* **101**, 715–724, 1994.

Tyler, C. "Sensory Processing of Binocular Disparity." In *Vergence Eye Movements: Basic and Clinical Aspects.* Boston, MA: Butterworth, pp. 199–295, 1983.

Stereographic Projection

A MAP PROJECTION in which GREAT CIRCLES are CIRCLES and LOXODROMES are LOGARITHMIC SPIRALS.

$$x = k\cos\phi\sin(\lambda - \lambda_0) \quad (1)$$
$$y = k[\cos\phi_1\sin\phi - \sin\phi_1\cos\phi\cos(\lambda - \lambda_0)], \quad (2)$$

where

$$k = \frac{2}{1 + \sin\phi_1\sin\phi + \cos\phi_1\cos\phi\cos(\lambda - \lambda_0)}. \quad (3)$$

The inverse FORMULAS are given by

$$\phi = \sin^{-1}\left(\cos c\sin\phi_1 + \frac{y\sin c\cos\phi_1}{\rho}\right) \quad (4)$$
$$\lambda = \lambda_0 + \tan^{-1}\left(\frac{x\sin c}{\rho\cos\phi_1\cos c - y\sin\phi_1\sin c}\right), \quad (5)$$

where

$$\rho = \sqrt{x^2 + y^2} \quad (6)$$
$$c = 2\tan^{-1}(\tfrac{1}{2}\rho). \quad (7)$$

see also GALL'S STEREOGRAPHIC PROJECTION

References

Coxeter, H. S. M. and Greitzer, S. L. *Geometry Revisited.* Washington, DC: Math. Assoc. Amer., pp. 150–153, 1967.

Snyder, J. P. *Map Projections—A Working Manual.* U. S. Geological Survey Professional Paper 1395. Washington, DC: U. S. Government Printing Office, pp. 154–163, 1987.

Stereology

The exploration of 3-D space from 2-D sections of PROJECTIONS of solid bodies.

see also AXONOMETRY, CORK PLUG, CROSS-SECTION, PROJECTION, TRIP-LET

Stern-Brocot Tree

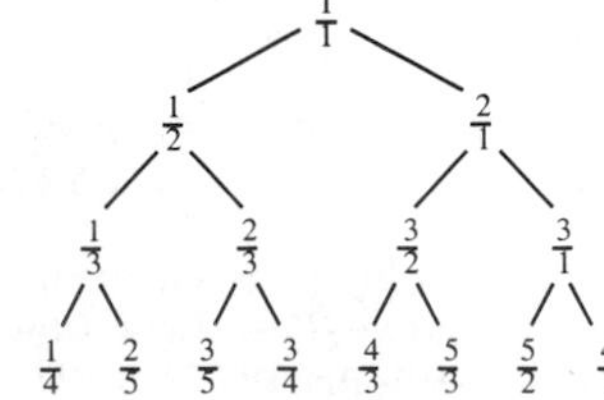

A special type of BINARY TREE obtained by starting with the fractions $\frac{0}{1}$ and $\frac{1}{0}$ and iteratively inserting $(m+m')/(n+n')$ between each two adjacent fractions m/n and m'/n'. The result can be arranged in tree form as illustrated above. The FAREY SEQUENCE F_n defines a subtree of the Stern-Brocot tree obtained by pruning off unwanted branches (Vardi 1991, Graham *et al.* 1994).

see also BINARY TREE, FAREY SEQUENCE, FORD CIRCLE

References

Brocot, A. "Calcul des rouages par approximation, nouvelle méthode." *Revue Chonométrique* **6**, 186–194, 1860.

Graham, R. L.; Knuth, D. E.; and Patashnik, O. *Concrete Mathematics: A Foundation for Computer Science, 2nd ed.* Reading, MA: Addison-Wesley, pp. 116–117, 1994.

Stern, M. A. "Über eine zahlentheoretische Funktion." *J. reine angew. Math.* **55**, 193–220, 1858.

Vardi, I. *Computational Recreations in Mathematica.* Redwood City, CA: Addison-Wesley, p. 253, 1991.

Stevedore's Knot

The 6-crossing KNOT 06_{001} having CONWAY-ALEXANDER POLYNOMIAL

$$\Delta(t) = 2t^2 - 5t + 2.$$

References

Rolfsen, D. *Knots and Links.* Wilmington, DE: Publish or Perish Press, pp. 225, 1976.

Stewart's Theorem

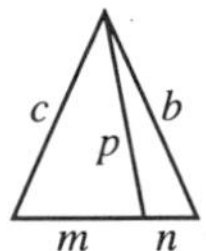

$$a(p^2 + mn) = b^2m + c^2n,$$

where

$$a \equiv m + n.$$

References

Altshiller-Court, N. "Stewart's Theorem." §6B in *College Geometry: A Second Course in Plane Geometry for Colleges and Normal Schools, 2nd ed., rev. enl.* New York: Barnes and Noble, pp. 152–153, 1952.

Coxeter, H. S. M. and Greitzer, S. L. *Geometry Revisited.* Washington, DC: Math. Assoc. Amer., p. 6, 1967.

Stick Number

Let the stick number $s(K)$ of a KNOT K be the least number of straight sticks needed to make a KNOT K. The smallest stick number of any KNOT is $s(T) = 6$, where T is the TREFOIL KNOT. If J and K are KNOTS, then

$$s(J + K) \le s(J) + s(K) + 1.$$

For a nontrivial KNOT K, let $c(K)$ be the CROSSING NUMBER (i.e., the least number of crossings in any projection of K). Then

$$\tfrac{1}{2}[5+\sqrt{25+8(c(K)-2)}] \leq s(K) \leq 2c(K).$$

The following table gives the stick number for some common knots.

Knot	s
trefoil knot	6
Whitehead link	8

see also CROSSING NUMBER (LINK), TRIANGLE COUNTING

References

Adams, C. C. *The Knot Book: An Elementary Introduction to the Mathematical Theory of Knots.* New York: W. H. Freeman, pp. 27–30, 1994.

Stickelberger Relation

Let P be a PRIME IDEAL in D_m not containing m. Then

$$(\Phi(P)) = P^{\sum t\sigma_t^{-1}},$$

where the sum is over all $1 \leq t < m$ which are RELATIVELY PRIME to m. Here D_m is the RING of integers in $\mathbb{Q}(\zeta_m)$, $\Phi(P) = g(P)^m$, and other quantities are defined by Ireland and Rosen (1990).

see also PRIME IDEAL

References

Ireland, K. and Rosen, M. "The Stickelberger Relation and the Eisenstein Reciprocity Law." Ch. 14 in *A Classical Introduction to Modern Number Theory, 2nd ed.* New York: Springer-Verlag, pp. 203–227, 1990.

Stiefel Manifold

The Stiefel manifold of ORTHONORMAL k-frames in $\mathbb{R}^n$ is the collection of vectors $(v_1, \ldots, v_k)$ where v_i is in $\mathbb{R}^n$ for all i, and the k-tuple $(v_1, \ldots, v_k)$ is ORTHONORMAL. This is a submanifold of $\mathbb{R}^{nk}$, having DIMENSION $nk - (k+1)k/2$.

Sometimes the "orthonormal" condition is dropped in favor of the mildly weaker condition that the k-tuple $(v_1, \ldots, v_k)$ is linearly independent. Usually, this does not affect the applications since Stiefel manifolds are usually considered only during HOMOTOPY THEORETIC considerations. With respect to HOMOTOPY THEORY, the two definitions are more or less equivalent since GRAM-SCHMIDT ORTHONORMALIZATION gives rise to a smooth deformation retraction of the second type of Stiefel manifold onto the first.

see also GRASSMANN MANIFOLD

Stiefel-Whitney Class

The ith Stiefel-Whitney class of a REAL VECTOR BUNDLE (or TANGENT BUNDLE or a REAL MANIFOLD) is in the ith cohomology group of the base SPACE involved. It is an OBSTRUCTION to the existence of $(n-i+1)$ REAL linearly independent VECTOR FIELDS on that VECTOR BUNDLE, where n is the dimension of the FIBER. Here, OBSTRUCTION means that the ith Stiefel-Whitney class being NONZERO implies that there do *not* exist $(n-i+1)$ everywhere linearly dependent VECTOR FIELDS (although the Stiefel-Whitney classes are not always the OBSTRUCTION).

In particular, the nth Stiefel-Whitney class is the obstruction to the existence of an everywhere NONZERO VECTOR FIELD, and the first Stiefel-Whitney class of a MANIFOLD is the obstruction to orientability.

see also CHERN CLASS, OBSTRUCTION, PONTRYAGIN CLASS, STIEFEL-WHITNEY NUMBER

Stiefel-Whitney Number

The Stiefel-Whitney number is defined in terms of the STIEFEL-WHITNEY CLASS of a MANIFOLD as follows. For any collection of STIEFEL-WHITNEY CLASSES such that their cup product has the same DIMENSION as the MANIFOLD, this cup product can be evaluated on the MANIFOLD'S FUNDAMENTAL CLASS. The resulting number is called the PONTRYAGIN NUMBER for that combination of Pontryagin classes.

The most important aspect of Stiefel-Whitney numbers is that they are COBORDISM invariant. Together, PONTRYAGIN and Stiefel-Whitney numbers determine an oriented MANIFOLD'S COBORDISM class.

see also CHERN NUMBER, PONTRYAGIN NUMBER, STIEFEL-WHITNEY CLASS

Stieltjes Constants

N.B. A detailed on-line essay by S. Finch was the starting point for this entry.

Expanding the RIEMANN ZETA FUNCTION about $z=1$ gives

$$\zeta(z) = \frac{1}{z-1} + \sum_{n=0}^{\infty} \frac{(-1)^n}{n!}\gamma_n(z-1)^n, \tag{1}$$

where

$$\gamma_n \equiv \lim_{m\to\infty}\left[\sum_{k=1}^{m} \frac{(\ln k)^n}{k} - \frac{(\ln m)^{n+1}}{n+1}\right]. \tag{2}$$

An alternative definition is given by

$$\gamma_n' \equiv \frac{(-1)^n}{n!}\gamma_n. \tag{3}$$

The case $n = 0$ gives the EULER-MASCHERONI CONSTANT γ. The first few numerical values are given in the following table.

n	γ_n
0	0.5772156649
1	−0.07281584548
2	−0.009690363192
3	0.002053834420
4	0.002325370065
5	0.0007933238173

Briggs (1955–1956) proved that there infinitely many γ_n of each SIGN. Berndt (1972) gave upper bounds of

$$|\gamma_n| < \begin{cases} \frac{4(n-1)!}{\pi^n} & \text{for } n \text{ even} \\ \frac{2(n-1)!}{\pi^n} & \text{for } n \text{ odd.} \end{cases} \tag{4}$$

Vacca (1910) proves that the EULER-MASCHERONI CONSTANT may be expressed as

$$\gamma = \sum_{k=1}^{\infty} \frac{(-1)^k}{k} \lfloor \lg k \rfloor, \tag{5}$$

where $\lfloor x \rfloor$ is the FLOOR FUNCTION. Hardy (1912) gave the FORMULA

$$\frac{2\gamma_1}{\ln 2} = \sum_{k=1}^{\infty} \frac{(-1)^k}{k} [2 \lg k - \lfloor \lg(2k) \rfloor] \lfloor \lg k \rfloor. \tag{6}$$

Kluyver (1927) gave similar series for γ_n with $n > 1$.

A set of constants related to γ_n is

$$\delta_n \equiv \lim_{m\to\infty} \left[\sum_{k=1}^{m} (\ln k)^n - \int_1^m (\ln x)^n \, dx - \tfrac{1}{2}(\ln m)^n \right] \tag{7}$$

(Sitaramachandrarao 1986, Lehmer 1988).

References

Berndt, B. C. "On the Hurwitz Zeta-Function." *Rocky Mountain J. Math.* **2**, 151–157, 1972.

Bohman, J. and Fröberg, C.-E. "The Stieltjes Function—Definitions and Properties." *Math. Comput.* **51**, 281–289, 1988.

Briggs, W. E. "Some Constants Associated with the Riemann Zeta-Function." *Mich. Math. J.* **3**, 117–121, 1955–1956.

Finch, S. "Favorite Mathematical Constants." `http://www.mathsoft.com/asolve/constant/stltjs/stltjs.html`.

Hardy, G. H. "Note on Dr. Vacca's Series for γ." *Quart. J. Pure Appl. Math.* **43**, 215-216, 1912.

Kluyver, J. C. "On Certain Series of Mr. Hardy." *Quart. J. Pure Appl. Math.* **50**, 185–192, 1927.

Knopfmacher, J. "Generalised Euler Constants." *Proc. Edinburgh Math. Soc.* **21**, 25–32, 1978.

Lehmer, D. H. "The Sum of Like Powers of the Zeros of the Riemann Zeta Function." *Math. Comput.* **50**, 265–273, 1988.

Liang, J. J. Y. and Todd, J. "The Stieltjes Constants." *J. Res. Nat. Bur. Standards—Math. Sci.* **76B**, 161–178, 1972.

Sitaramachandrarao, R. "Maclaurin Coefficients of the Riemann Zeta Function." *Abstracts Amer. Math. Soc.* **7**, 280, 1986.

Vacca, G. "A New Series for the Eulerian Constant." *Quart. J. Pure Appl. Math.* **41**, 363–368, 1910.

Stieltjes Integral

The Stieltjes integral is a generalization of the RIEMANN INTEGRAL. Let $f(x)$ and $\alpha(x)$ be real-values bounded functions defined on a CLOSED INTERVAL $[a,b]$. Take a partition of the INTERVAL

$$a = x_0 < x_1 < x_2, \ldots < x_{n-1} < x_n = b, \tag{1}$$

and consider the Riemann sum

$$\sum_{i=0}^{n-1} f(\xi_i)[\alpha(x_{i+1}) - \alpha(x_i)] \tag{2}$$

with $\xi_i \in [x_i, x_{i+1}]$. If the sum tends to a fixed number I as $\max(x_{i+1} - x_i) \to 0$, then I is called the Stieltjes integral, or sometimes the RIEMANN-STIELTJES INTEGRAL. The Stieltjes integral of P with respect to F is denoted

$$\int P(x)\, dF(x), \tag{3}$$

where

$$\int P(x)\, dF(x) = \begin{cases} \int f(x)\, dx & \text{for } x \text{ continuous} \\ \sum_x f(x) & \text{for } x \text{ discrete.} \end{cases} \tag{4}$$

If P and F have a common point of discontinuity, then the integral does not exist. However, if the Stieltjes integral exists and F has a derivative F', then

$$\int P(x)\, dF(x) = \int P(x) F'(x)\, dx. \tag{5}$$

For enumeration of many of the integral's properties, see Dresher (1981, p. 105).

see also RIEMANN INTEGRAL

References

Dresher, M. *The Mathematics of Games of Strategy: Theory and Applications.* New York: Dover, 1981.

Hardy, G. H.; Littlewood, J. E.; and Pólya, G. *Inequalities, 2nd ed.* Cambridge, England: Cambridge University Press, pp. 152–155, 1988.

Kestelman, H. "Riemann-Stieltjes Integration." Ch. 11 in *Modern Theories of Integration, 2nd rev. ed.* New York: Dover, pp. 247–269, 1960.

Stieltjes' Theorem

The $m+1$ ELLIPSOIDAL HARMONICS when κ_1, κ_2, and κ_3 are given can be arranged in such a way that the rth function has $r-1$ zeros between $-a^2$ and $-b^2$ and the remaining $m+r-1$ zeros between $-b^2$ and $-c^2$ (Whittaker and Watson 1990).

see also ELLIPSOIDAL HARMONIC

References

Whittaker, E. T. and Watson, G. N. *A Course in Modern Analysis, 4th ed.* Cambridge, England: Cambridge University Press, pp. 560–562, 1990.

Stieltjes-Wigert Polynomial

Orthogonal POLYNOMIALS associated with WEIGHTING FUNCTION

$$w(x) = \pi^{-1/2} k \exp(-k^2 \ln^2 x) = \pi^{-1/2} k x^{-k^2 \ln x} \quad (1)$$

for $x \in (0, \infty)$ and $k > 0$. Using

$$\begin{bmatrix} n \\ \nu \end{bmatrix} = \frac{(1-q^n)(1-q^{n-1})\cdots(1-q^{n-\nu+1})}{(1-q)(1-q^2)\cdots(1-q^\nu)} \quad (2)$$

where $0 < \nu < n$,

$$\begin{bmatrix} n \\ 0 \end{bmatrix} = \begin{bmatrix} n \\ n \end{bmatrix} = 1, \quad (3)$$

and

$$q = \exp[-(2k^2)^{-1}]. \quad (4)$$

Then

$$p_n(x) = (-1)^n q^{n/2+1/4}[(1-q)(1-q^2)\cdots(1-q^n)]^{-1/2} \times \sum_{\nu=0}^{n} \begin{bmatrix} n \\ \nu \end{bmatrix} q^{\nu^2}(-q^{1/2}x)^\nu \quad (5)$$

for $n > 0$ and

$$p_0(x) = q^{1/4}. \quad (6)$$

References

Szegő, G. *Orthogonal Polynomials, 4th ed.* Providence, RI: Amer. Math. Soc., p. 33, 1975.

Stirling's Approximation

Stirling's approximation gives an approximate value for the FACTORIAL function $n!$ or the GAMMA FUNCTION $\Gamma(n)$ for $n \gg 1$. The approximation can most simply be derived for n an INTEGER by approximating the sum over the terms of the FACTORIAL with an INTEGRAL, so that

$$\ln n! = \ln 1 + \ln 2 + \ldots + \ln n = \sum_{k=1}^{n} \ln k \approx \int_1^n \ln x \, dx$$
$$= [x \ln x - x]_1^n = n \ln n - n + 1 \approx n \ln n - n. \quad (1)$$

The equation can also be derived using the integral definition of the FACTORIAL,

$$n! = \int_0^\infty e^{-x} x^n \, dx. \quad (2)$$

Note that the derivative of the LOGARITHM of the integrand can be written

$$\frac{d}{dx} \ln(e^{-x} x^n) = \frac{d}{dx}(n \ln x - x) = \frac{n}{x} - 1. \quad (3)$$

The integrand is sharply peaked with the contribution important only near $x = n$. Therefore, let $x \equiv n + \xi$ where $\xi \ll n$, and write

$$\ln(x^n e^{-x}) = n \ln x - x = n \ln(n+\xi) - (n+\xi). \quad (4)$$

Now,

$$\ln(n+\xi) = \ln\left[n\left(1+\frac{\xi}{n}\right)\right] = \ln n + \ln\left(1+\frac{\xi}{n}\right)$$
$$= \ln n + \frac{\xi}{n} - \frac{1}{2}\frac{\xi^2}{n^2} + \ldots, \quad (5)$$

so

$$\ln(x^n e^{-x}) = n \ln(n+\xi) - (n+\xi)$$
$$= n \ln n + \xi - \frac{1}{2}\frac{\xi^2}{n} - n - \xi + \ldots$$
$$= n \ln n - n - \frac{\xi^2}{2n} + \ldots. \quad (6)$$

Taking the EXPONENTIAL of each side then gives

$$x^n e^{-x} \approx e^{n \ln n} e^{-n} e^{-\xi^2/2n} = n^n e^{-n} e^{-\xi^2/2n}. \quad (7)$$

Plugging into the integral expression for $n!$ then gives

$$n! \approx \int_{-n}^{\infty} n^n e^{-n} e^{-\xi^2/2n} \, d\xi \approx n^n e^{-n} \int_{-\infty}^{\infty} e^{-\xi^2/2n} \, d\xi$$
$$= n^n. \quad (8)$$

Evaluating the integral gives

$$n! \approx n^n e^{-n} \sqrt{2\pi n}, \quad (9)$$
$$\approx \sqrt{2\pi}\, n^{n+1/2} e^{-n}. \quad (10)$$

Taking the LOGARITHM of both sides then gives

$$\ln n! \approx n \ln n - n + \tfrac{1}{2}\ln(2\pi n) = (n+\tfrac{1}{2})\ln n - n + \tfrac{1}{2}\ln(2\pi). \quad (11)$$

This is STIRLING'S SERIES with only the first term retained and, for large n, it reduces to Stirling's approximation

$$\ln n! \approx n \ln n - n. \quad (12)$$

Gosper notes that a better approximation to $n!$ (i.e., one which approximates the terms in STIRLING'S SERIES instead of truncating them) is given by

$$n! \approx \sqrt{(2n+\tfrac{1}{3})\pi}\, n^n e^{-n}. \quad (13)$$

This also gives a much closer approximation to the FACTORIAL of 0, $0! = 1$, yielding $\sqrt{\pi/3} \approx 1.02333$ instead of 0 obtained with the conventional Stirling approximation.

see also STIRLING'S SERIES

Stirling Cycle Number

see STIRLING NUMBER OF THE FIRST KIND

Stirling's Finite Difference Formula

$$f_p = f_0 + \tfrac{1}{2}p(\delta_{1/2} + \delta_{-1/2}) + \tfrac{1}{2}p^2\delta_0^2 + S_3(\delta_{1/2}^2 + \delta_{-1/2}^2) + S_4\delta_0^4 + \ldots,$$

for $p \in [-1/2, 1/2]$, where δ is the CENTRAL DIFFERENCE and

$$S_{2n+1} = \frac{1}{2}\binom{p+n}{2n+1}$$

$$S_{2n+2} = \frac{p}{2n+2}\binom{p+n}{2n+1},$$

with $\binom{n}{k}$ a BINOMIAL COEFFICIENT.

see also CENTRAL DIFFERENCE, STEFFENSON'S FORMULA

References

Beyer, W. H. *CRC Standard Mathematical Tables, 28th ed.* Boca Raton, FL: CRC Press, p. 433, 1987.

Stirling's Formula

see STIRLING'S SERIES

Stirling Number of the First Kind

The definition of the (signed) Stirling number of the first kind is a number $S_n^{(m)}$ such that the number of permutations of n elements which contain exactly m CYCLES is

$$(-1)^{n-m}S_n^{(m)}. \tag{1}$$

This means that $S_n^{(m)} = 0$ for $m > n$ and $S_n^{(n)} = 1$. The GENERATING FUNCTION is

$$x(x-1)\cdots(x-n+1) = \sum_{m=0}^{n} S_n^{(m)}x^m. \tag{2}$$

This is the Stirling number of the first kind returned by the *Mathematica*® (Wolfram Research, Champaign, IL) command `StirlingS1[n,m]`. The triangle of signed Stirling numbers of the first kind is

```
        1
      -1  1
    2  -3  1
  -6  11  6  1
24 -50 35 -10 1
```

(Sloane's A008275).

The NONNEGATIVE version simply gives the number of PERMUTATIONS of n objects having m CYCLES (with cycles in opposite directions counted as distinct) and is obtained by taking the ABSOLUTE VALUE of the signed version. The nonnegative Stirling number of the first kind is denoted $S_1(n,m) = |S_n^{(m)}|$ or $\begin{bmatrix} n \\ m \end{bmatrix}$. Diagrams illustrating $S_1(5,1) = 24$, $S_1(5,3) = 35$, $S_1(5,4) = 10$, and $S_1(5,5) = 1$ (Dickau) are shown below.

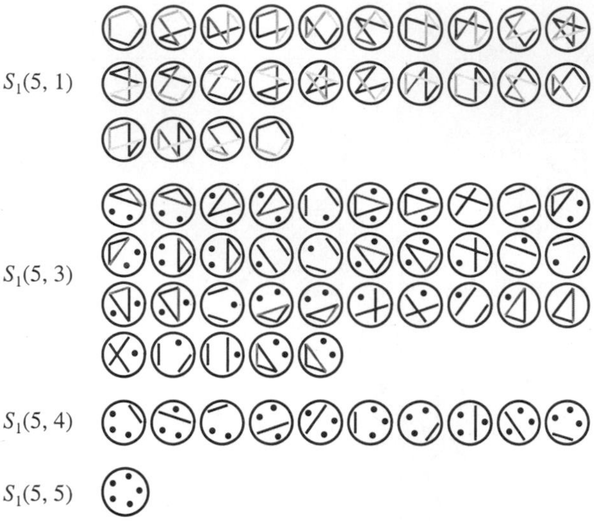

$S_1(5,1)$

$S_1(5,3)$

$S_1(5,4)$

$S_1(5,5)$

The nonnegative Stirling numbers of the first kind satisfy the curious identity

$$\sum_{n=1}^{\infty}\left[\sum_{k=0}^{n-2}\frac{(e^x - x - 1)^{k+1}S_1(n, n-k)}{(k+1)!}\right]e^{-xn} = \ln(x+1) \tag{3}$$

(Gosper) and have the GENERATING FUNCTION

$$(1+x)(1+2x)\cdots(1+nx) = \sum_{k=1}^{n} S_1(n,m)x^k \tag{4}$$

and satisfy

$$_1(n+1,k) = nS_1(n,k) + S_1(n,k-1). \tag{5}$$

The Stirling numbers can be generalized to nonintegral arguments (a sort of "Stirling polynomial") using the identity

$$\frac{\Gamma(j+h)}{j^h\Gamma(j)} = \sum_{k=0}^{\infty}\frac{S_1(h,h-k)}{j^k} = 1 + \frac{(h-1)h}{2j} + \frac{(h-2)(3h-1)(h-1)h}{24j^2} + \frac{(h-3)(h-2)(h-1)^2h^2}{48j^3} + \ldots, \tag{6}$$

which is a generalization of an ASYMPTOTIC SERIES for a ratio of GAMMA FUNCTIONS $\Gamma(j+1/2)/\Gamma(j)$ (Gosper).

see also CYCLE (PERMUTATION), HARMONIC NUMBER, PERMUTATION, STIRLING NUMBER OF THE SECOND KIND

References

Abramowitz, M. and Stegun, C. A. (Eds.). "Stirling Numbers of the First Kind." §24.1.3 in *Handbook of Mathematical Functions with Formulas, Graphs, and Mathematical Tables, 9th printing.* New York: Dover, p. 824, 1972.

Adamchik, V. "On Stirling Numbers and Euler Sums." *J. Comput. Appl. Math.* **79**, 119–130, 1997. `http://www.wolfram.com/~victor/articles/stirling.html`.

Conway, J. H. and Guy, R. K. In *The Book of Numbers.* New York: Springer-Verlag, pp. 91–92, 1996.

Dickau, R. M. "Stirling Numbers of the First Kind." `http://forum.swarthmore.edu/advanced/robertd/stirling1.html`.

Knuth, D. E. "Two Notes on Notation." *Amer. Math. Monthly* **99**, 403–422, 1992.

Sloane, N. J. A. Sequence A008275 in "An On-Line Version of the Encyclopedia of Integer Sequences."

Stirling Number of the Second Kind

The number of ways of partitioning a set of n elements into m nonempty SETS (i.e., m BLOCKS), also called a STIRLING SET NUMBER. For example, the SET $\{1,2,3\}$ can be partitioned into three SUBSETS in one way: $\{\{1\},\{2\},\{3\}\}$; into two SUBSETS in three ways: $\{\{1,2\},\{3\}\}$, $\{\{1,3\},\{2\}\}$, and $\{\{1\},\{2,3\}\}$; and into one SUBSET in one way: $\{\{1,2,3\}\}$.

The Stirling numbers of the second kind are denoted $s_n^{(m)}$, $S_2(n,m)$, $s(n,m)$, or $\left\{ n \atop m \right\}$, so the Stirling numbers of the second kind for three elements are

$$s(3,1) = 1 \quad (1)$$
$$s(3,2) = 3 \quad (2)$$
$$s(3,3) = 1. \quad (3)$$

Since a set of n elements can only be partitioned in a single way into 1 or n SUBSETS,

$$s(n,1) = s(n,n) = 1. \quad (4)$$

The triangle of Stirling numbers of the second kind is

$$\begin{array}{c} 1 \\ 1 \quad 1 \\ 1 \quad 3 \quad 1 \\ 1 \quad 7 \quad 6 \quad 1 \\ 1 \quad 15 \quad 25 \quad 10 \quad 1 \\ 1 \quad 31 \quad 90 \quad 65 \quad 15 \quad 1 \end{array}$$

(Sloane's A008277).

The Stirling numbers of the second kind can be computed from the sum

$$s(n,k) = \frac{1}{k!}\sum_{i=0}^{k-1}(-1)^i \binom{k}{i}(k-i)^n, \quad (5)$$

with $\binom{n}{k}$ a BINOMIAL COEFFICIENT, or the GENERATING FUNCTIONS

$$x^n = \sum_{m=0}^{n} s(n,m)x(x-1)\cdots(x-m+1), \quad (6)$$

$$\sum_{n\geq k} s(n,k)\frac{x^n}{n!} = \frac{1}{k!}(e^x-1)^k, \quad (7)$$

and

$$\frac{1}{(1-x)(1-2x)\cdots(1-kx)} = \sum_{n=1}^{k} s(n,k)x^n. \quad (8)$$

The following diagrams (Dickau) illustrate the definition of the Stirling numbers of the second kind $s(n,m)$ for $n=3$ and 4.

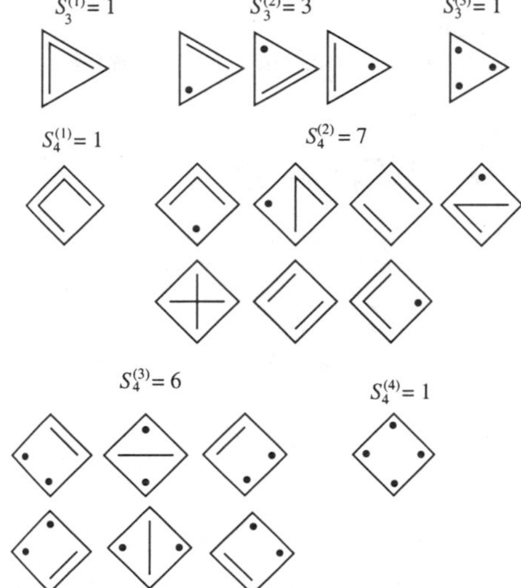

Stirling numbers of the second kind obey the RECURRENCE RELATIONS

$$s(n,k) = s(n-1,k-1) + ks(n-1,k). \quad (9)$$

An identity involving Stirling numbers of the second kind is

$$f(m,n) \equiv \sum_{k=1}^{\infty} k^n \left(\frac{m}{m+1}\right)^l = (m+1)\sum_{k=1}^{m} k!s(n,k)m^k. \quad (10)$$

It turns out that $f(1,n)$ can have only 0, 2, or 6 as a last DIGIT (Riskin 1995).

see also BELL NUMBER, COMBINATION LOCK, LENGYEL'S CONSTANT, MINIMAL COVER, STIRLING NUMBER OF THE FIRST KIND

References

Abramowitz, M. and Stegun, C. A. (Eds.). "Stirling Numbers of the Second Kind." §24.1.4 in *Handbook of Mathematical Functions with Formulas, Graphs, and Mathematical Tables, 9th printing.* New York: Dover, pp. 824–825, 1972.

Comtet, L. *Advanced Combinatorics.* Boston, MA: Reidel, 1974.
Conway, J. H. and Guy, R. K. In *The Book of Numbers.* New York: Springer-Verlag, pp. 91–92, 1996.
Dickau, R. M. "Stirling Numbers of the Second Kind." http://forum.swarthmore.edu/advanced/robertd/stirling2.html.
Graham, R. L.; Knuth, D. E.; and Patashnik, O. *Concrete Mathematics: A Foundation for Computer Science, 2nd ed.* Reading, MA: Addison-Wesley, 1994.
Knuth, D. E. "Two Notes on Notation." *Amer. Math. Monthly* **99**, 403–422, 1992.
Riordan, J. *An Introduction to Combinatorial Analysis.* New York: Wiley, 1958.
Riordan, J. *Combinatorial Identities.* New York: Wiley, 1968.
Riskin, A. "Problem 10231." *Amer. Math. Monthly* **102**, 175–176, 1995.
Sloane, N. J. A. Sequence A008277 in "An On-Line Version of the Encyclopedia of Integer Sequences."
Stanley, R. P. *Enumerative Combinatorics, Vol. 1.* Cambridge, England: Cambridge University Press, 1997.

Stirling's Series

The ASYMPTOTIC SERIES for the GAMMA FUNCTION is given by

$$\Gamma(z) = e^{-z} z^{z-1/2} \sqrt{2\pi} \left(1 + \frac{1}{12z} + \frac{1}{288z^2} - \frac{139}{51840z^3} - \frac{571}{2488320z^4} + \ldots\right) \quad (1)$$

(Sloane's A001163 and A001164). The series for $z!$ is obtained by adding an additional factor of z,

$$z! = \Gamma(z+1) = e^{-z} z^{z+1/2} \sqrt{2\pi} \left(1 + \frac{1}{12z} + \frac{1}{288z^2} - \frac{139}{51840z^3} - \frac{571}{2488320z^4} + \ldots\right). \quad (2)$$

The expansion of $\ln\Gamma(z)$ is what is usually called Stirling's series. It is given by the simple analytic expression

$$\ln\Gamma(z) = \sum_{n=1}^{\infty} \frac{B_{2n}}{2n(2n-1)z^{2n-1}} \quad (3)$$

$$= \tfrac{1}{2}\ln(2\pi) + (z+\tfrac{1}{2})\ln z - z + \frac{1}{12z} - \frac{1}{360z^3} + \frac{1}{1260z^5} - \ldots, \quad (4)$$

where B_n is a BERNOULLI NUMBER.

see also BERNOULLI NUMBER, K-FUNCTION, STIRLING'S APPROXIMATION

References

Abramowitz, M. and Stegun, C. A. (Eds.). *Handbook of Mathematical Functions with Formulas, Graphs, and Mathematical Tables, 9th printing.* New York: Dover, p. 257, 1972.
Arfken, G. "Stirling's Series." §10.3 in *Mathematical Methods for Physicists, 3rd ed.* Orlando, FL: Academic Press, pp. 555–559, 1985.
Conway, J. H. and Guy, R. K. "Stirling's Formula." In *The Book of Numbers.* New York: Springer-Verlag, pp. 260–261, 1996.
Morse, P. M. and Feshbach, H. *Methods of Theoretical Physics, Part I.* New York: McGraw-Hill, p. 443, 1953.
Sloane, N. J. A. Sequences A001163/M5400 and A001164/M4878 in "An On-Line Version of the Encyclopedia of Integer Sequences."

Stirling Set Number

see STIRLING NUMBER OF THE SECOND KIND

Stirrup Curve

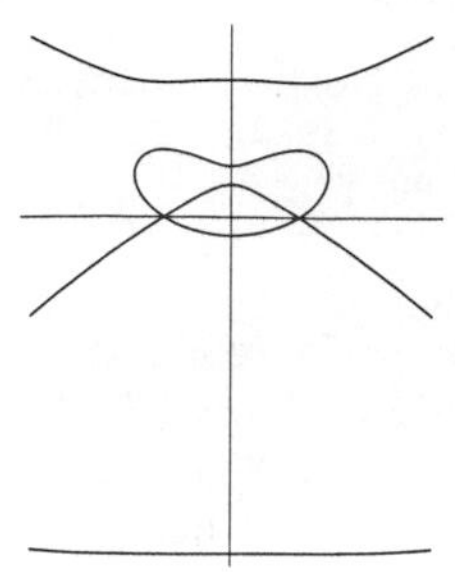

A plane curve given by the equation

$$(x^2-1)^2 = y^2(y-1)(y-2)(y+5).$$

References

Cundy, H. and Rollett, A. *Mathematical Models, 3rd ed.* Stradbroke, England: Tarquin Pub., p. 72, 1989.

Stochastic

see RANDOM VARIABLE

Stochastic Calculus of Variations

see MALLIAVIN CALCULUS

Stochastic Group

The GROUP of all nonsingular $n \times n$ STOCHASTIC MATRICES over a FIELD F. It is denoted $S(n, F)$. If p is PRIME and F is the GALOIS FIELD of ORDER $q = p^m$, $S(n, q)$ is written instead of $S(n, F)$. Particular examples include

$$S(2,2) = \mathbb{Z}_2$$
$$S(2,3) = S_3$$
$$S(2,4) = A_4$$
$$S(3,2) = S_4$$
$$S(2,5) = \mathbb{Z}_4 \times_\theta \mathbb{Z}_5,$$

where $\mathbb{Z}_2$ is an ABELIAN GROUP, S_n are SYMMETRIC GROUPS on n elements, and $\times_\theta$ denotes the semidirect product with $\theta : \mathbb{Z}_4 \to \mathrm{Aut}(\mathbb{Z}_5)$ (Poole 1995).

see also STOCHASTIC MATRIX

References

Poole, D. G. "The Stochastic Group." *Amer. Math. Monthly* **102**, 798–801, 1995.

Stochastic Matrix

A Stochastic matrix is the transition matrix for a finite MARKOV CHAIN, also called a MARKOV MATRIX. Elements of the matrix must be REAL NUMBERS in the CLOSED INTERVAL [0, 1].

A completely independent type of stochastic matrix is defined as a SQUARE MATRIX with entries in a FIELD F such that the sum of elements in each column equals 1. There are two nonsingular 2×2 STOCHASTIC MATRICES over $\mathbb{Z}_2$ (i.e., the integers mod 2),

$$\begin{bmatrix} 1 & 0 \\ 0 & 1 \end{bmatrix} \quad \text{and} \quad \begin{bmatrix} 0 & 1 \\ 1 & 0 \end{bmatrix}.$$

There are six nonsingular stochastic 3×3 MATRICES over $\mathbb{Z}_3$,

$$\begin{bmatrix} 1 & 0 \\ 0 & 1 \end{bmatrix}, \begin{bmatrix} 0 & 1 \\ 1 & 0 \end{bmatrix}, \begin{bmatrix} 2 & 1 \\ 2 & 0 \end{bmatrix}, \begin{bmatrix} 2 & 0 \\ 2 & 1 \end{bmatrix}, \begin{bmatrix} 0 & 2 \\ 1 & 2 \end{bmatrix}, \begin{bmatrix} 1 & 2 \\ 0 & 2 \end{bmatrix}.$$

In fact, the set S of all nonsingular stochastic $n \times n$ matrices over a FIELD F forms a GROUP under MATRIX MULTIPLICATION. This GROUP is called the STOCHASTIC GROUP.

see also MARKOV CHAIN, STOCHASTIC GROUP

References

Poole, D. G. "The Stochastic Group." *Amer. Math. Monthly* **102**, 798–801, 1995.

Stochastic Process

A stochastic process is a family of RANDOM VARIABLES $\{x(t,\bullet), t \in \mathcal{J}\}$ from some PROBABILITY SPACE $(S, \mathbb{S}, P)$ into a STATE SPACE $(S', \mathbb{S}')$. Here, $\mathcal{J}$ is the INDEX SET of the process.

see also INDEX SET, PROBABILITY SPACE, RANDOM VARIABLE, STATE SPACE

References

Doob, J. L. "The Development of Rigor in Mathematical Probability (1900–1950)." *Amer. Math. Monthly* **103**, 586–595, 1996.

Stochastic Resonance

A stochastic resonance is a phenomenon in which a nonlinear system is subjected to a periodic modulated signal so weak as to be normally undetectable, but it becomes detectable due to resonance between the weak deterministic signal and stochastic NOISE. The earliest definition of stochastic resonance was the maximum of the output signal strength as a function of NOISE (Bulsara and Gammaitoni 1996).

see also KRAMERS RATE, NOISE

References

Benzi, R.; Sutera, A.; and Vulpiani, A. "The Mechanism of Stochastic Resonance." *J. Phys. A* **14**, L453–L457, 1981.

Bulsara, A. R. and Gammaitoni, L. "Tuning in to Noise." *Phys. Today* **49**, 39–45, March 1996.

Stöhr Sequence

Let $a_1 = 1$ and define a_{n+1} to be the least INTEGER greater than a_n for $n \geq k$ which cannot be written as the SUM of at most h addends among the terms a_1, a_2, ..., a_n.

see also GREEDY ALGORITHM, s-ADDITIVE SEQUENCE, ULAM SEQUENCE

References

Guy, R. K. *Unsolved Problems in Number Theory, 2nd ed.* New York: Springer-Verlag, p. 233, 1994.

Stokes Phenomenon

The asymptotic expansion of the AIRY FUNCTION $\mathrm{Ai}(z)$ (and other similar functions) has a different form in different sectors of the COMPLEX PLANE.

see also AIRY FUNCTIONS

References

Morse, P. M. and Feshbach, H. *Methods of Theoretical Physics, Part I.* New York: McGraw-Hill, pp. 609–611, 1953.

Stokes' Theorem

For w a DIFFERENTIAL $(n-1)$-FORM with compact support on an oriented n-dimensional MANIFOLD M,

$$\int_M dw = \int_{\partial M} w, \tag{1}$$

where dw is the EXTERIOR DERIVATIVE of the differential form w. This connects to the "standard" GRADIENT, CURL, and DIVERGENCE THEOREMS by the following relations. If f is a function on $\mathbb{R}^3$,

$$\mathrm{grad}(f) = c^{-1}\, df, \tag{2}$$

where $c : \mathbb{R}^3 \to \mathbb{R}^{3*}$ (the dual space) is the duality isomorphism between a VECTOR SPACE and its dual, given by the Euclidean INNER PRODUCT on $\mathbb{R}^3$. If f is a VECTOR FIELD on a $\mathbb{R}^3$,

$$\mathrm{div}(f) = {}^*d^*c(f), \tag{3}$$

where $*$ is the HODGE STAR operator. If f is a VECTOR FIELD on $\mathbb{R}^3$,

$$\mathrm{curl}(f) = c^{-1*}dc(f). \tag{4}$$

With these three identities in mind, the above Stokes' theorem in the three instances is transformed into the GRADIENT, CURL, and DIVERGENCE THEOREMS respectively as follows. If f is a function on $\mathbb{R}^3$ and γ is a curve in $\mathbb{R}^3$, then

$$\int_\gamma \mathrm{grad}(f) \cdot d\mathbf{l} = \int_\gamma df = f(\gamma(1)) - f(\gamma(0)), \tag{5}$$

which is the GRADIENT THEOREM. If $f : \mathbb{R}^3 \to \mathbb{R}^3$ is a VECTOR FIELD and M an embedded compact 3-manifold with boundary in $\mathbb{R}^3$, then

$$\int_{\partial M} f \cdot dA = \int_{\partial M} {}^*cf = \int_M d*cf = \int_M \operatorname{div}(f)\, dV, \quad (6)$$

which is the DIVERGENCE THEOREM. If f is a VECTOR FIELD and M is an oriented, embedded, compact 2-MANIFOLD with boundary in $\mathbb{R}^3$, then

$$\int_{\partial M} f\, dl = \int_{\partial M} cf = \int_M dc(f) = \int_M \operatorname{curl}(f) \cdot dA, \quad (7)$$

which is the CURL THEOREM.

Physicists generally refer to the CURL THEOREM

$$\int_S (\nabla \times \mathbf{F}) \cdot d\mathbf{a} = \int_{\partial S} \mathbf{F} \cdot d\mathbf{s} \quad (8)$$

as Stokes' theorem.

see also CURL THEOREM, DIVERGENCE THEOREM, GRADIENT THEOREM

Stolarsky Array

A INTERSPERSION array given by

1	2	3	5	8	13	21	34	55	⋯
4	6	10	16	26	42	68	110	178	⋯
7	11	18	29	47	76	123	199	322	⋯
9	15	24	39	63	102	165	267	432	⋯
12	19	31	50	81	131	212	343	555	⋯
14	23	37	60	97	157	254	411	665	⋯
17	28	45	73	118	191	309	500	809	⋯
20	32	52	84	136	220	356	576	932	⋯
22	36	58	94	152	246	398	644	1042	⋯
⋮	⋮	⋮	⋮	⋮	⋮	⋮	⋮	⋮	⋱

the first row of which is the FIBONACCI NUMBERS.

see also INTERSPERSION, WYTHOFF ARRAY

References

Kimberling, C. "Interspersions and Dispersions." *Proc. Amer. Math. Soc.* **117**, 313–321, 1993.

Morrison, D. R. "A Stolarsky Array and Wythoff Pairs." In *A Collection of Manuscripts Related to the Fibonacci Sequence.* Santa Clara, CA: Fibonacci Assoc., pp. 134–136, 1980.

Stolarsky-Harborth Constant

N.B. A detailed on-line essay by S. Finch was the starting point for this entry.

Let $b(k)$ be the number of 1s in the BINARY expression of k. Then the number of ODD BINOMIAL COEFFICIENTS $\binom{k}{j}$ where $0 \le j \le k$ is $2^{b(k)}$ (Glaisher 1899, Fine 1947). The number of ODD elements in the first n rows of PASCAL'S TRIANGLE is

$$f(n) = \sum_{k=0}^{n-1} 2^{b(k)}. \quad (1)$$

This function is well approximated by n^θ, where

$$\theta \equiv \frac{\ln 3}{\ln 2} = 1.58496\ldots. \quad (2)$$

Stolarsky and Harborth showed that

$$0.812556 \le \liminf_{n\to\infty} \frac{f(n)}{n^\theta} < 0.812557 < \limsup_{n\to\infty} \frac{f(n)}{n^\theta} = 1. \quad (3)$$

The value

$$SH = \liminf_{n\to\infty} \frac{f(n)}{n^\theta} \quad (4)$$

is called the Stolarsky-Harborth constant.

References

Finch, S. "Favorite Mathematical Constants." `http://www.mathsoft.com/asolve/constant/stlrsky/stlrsky.html`.

Fine, N. J. "Binomial Coefficients Modulo a Prime." *Amer. Math. Monthly* **54**, 589–592, 1947.

Wolfram, S. "Geometry of Binomial Coefficients." *Amer. Math. Monthly* **91**, 566–571, 1984.

Stolarsky's Inequality

If $0 \le g(x) \le 1$ and g is nonincreasing on the INTERVAL [0,1], then for all possible values of a and b,

$$\int_0^1 g(x^{1/(a+b)})\, dx \ge \int_0^1 g(x^{1/a})\, dx \int_0^1 g(x^{1/b})\, dx.$$

Stomachion

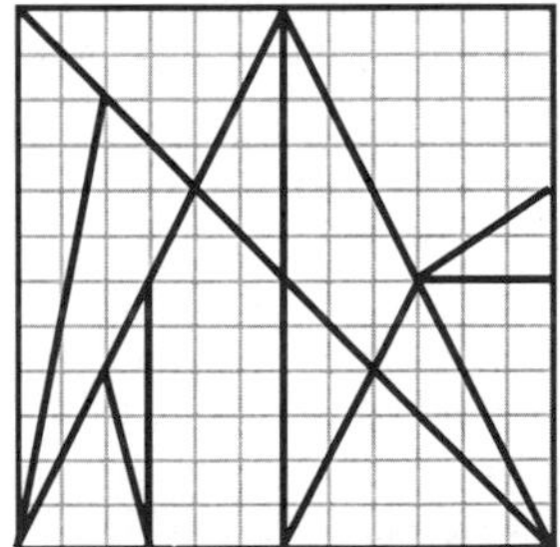

A DISSECTION game similar to TANGRAMS described in fragmentary manuscripts attributed to Archimedes and was referred to as the LOCULUS OF ARCHIMEDES (Archimedes' box) in Latin texts. The word Stomachion has as its root the Greek word for stomach. The game consisted of 14 flat pieces of various shapes arranged in the shape of a square. Like TANGRAMS, the object is to rearrange the pieces to form interesting shapes.

see also DISSECTION, TANGRAM

References

Rorres, C. "Stomachion Introduction." http://www.mcs.drexel.edu/~crorres/Archimedes/Stomachion/intro.html.

Rorres, C. "Stomachion Construction." http://www.mcs.drexel.edu/~crorres/Archimedes/Stomachion/construction.html.

Stone Space

Let $P(L)$ be the set of all PRIME IDEALS of L, and define $r(a) = \{P|a \notin P\}$. Then the Stone space of L is the TOPOLOGICAL SPACE defined on $P(L)$ by postulating that the sets of the form $r(a)$ are a subbase for the open sets.

see also PRIME IDEAL, TOPOLOGICAL SPACE

References

Grätzer, G. *Lattice Theory: First Concepts and Distributive Lattices.* San Francisco, CA: W. H. Freeman, p. 119, 1971.

Stone-von Neumann Theorem

A theorem which specifies the structure of the generic unitary representation of the Weyl relations and thus establishes the equivalence of Heisenberg's matrix mechanics and Schrödinger's wave mechanics formulations of quantum mechanics.

References

Neumann, J. von. "Die Eindeutigkeit der Schrödingerschen Operationen." *Math. Ann.* **104**, 570–578, 1931.

Stopper Knot

A KNOT used to prevent the end of a string from slipping through a hole.

References

Owen, P. *Knots.* Philadelphia, PA: Courage, p. 11, 1993.

Størmer Number

A Størmer number is a POSITIVE INTEGER n for which the largest PRIME factor p of n^2+1 is at least $2n$. Every GREGORY NUMBER t_x can be expressed uniquely as a sum of t_ns where the ns are Størmer numbers. Conway and Guy (1996) give a table of Størmer numbers reproduced below (Sloane's A005529). In a paper on INVERSE TANGENT relations, Todd (1949) gives a similar compilation.

n	p	n	p	n	p	n	p	n	p
1	2	10	101	19	181	26	617	35	613
2	5	11	61	20	401	27	73	36	1297
4	17	12	29	22	97	28	157	37	137
5	13	14	197	23	53	29	421	39	761
6	37	15	113	24	577	33	109	40	1601
9	41	16	257	25	313	34	89	42	353

see also GREGORY NUMBER, INVERSE TANGENT

References

Conway, J. H. and Guy, R. K. "Størmer's Numbers." *The Book of Numbers.* New York: Springer-Verlag, pp. 245–248, 1996.

Sloane, N. J. A. Sequence A005529/M1505 in "An On-Line Version of the Encyclopedia of Integer Sequences."

Todd, J. "A Problem on Arc Tangent Relations." *Amer. Math. Monthly* **56**, 517–528, 1949.

Straight Angle

An ANGLE of $180° = \pi$ RADIANS.

see also DIGON, RIGHT ANGLE

Straight Line

see LINE

Straight Polyomino

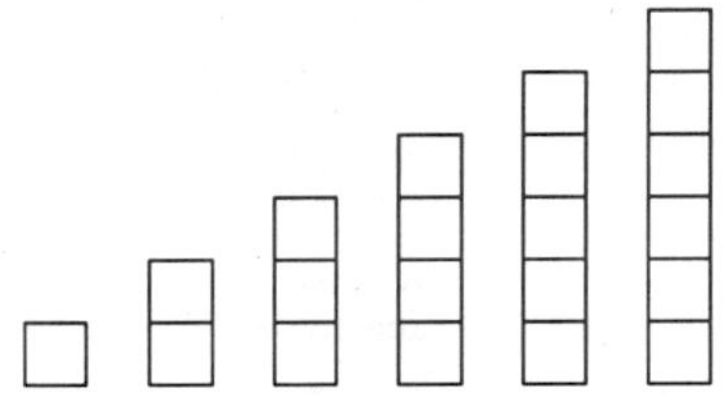

The straight polyomino of order n is the n-POLYOMINO in which all squares are placed along a line.

see also L-POLYOMINO, SKEW POLYOMINO, SQUARE POLYOMINO, T-POLYOMINO

Straightedge

An idealized mathematical object having a rigorously straight edge which can be used to draw a LINE SEGMENT. Although GEOMETRIC CONSTRUCTIONS are sometimes said to be performed with a RULER and COMPASS, the term straightedge is preferable to RULER since markings on the straightedge (usually assumed to be present on a RULER) are not allowed by the classical Greek rules.

see also COMPASS, GEOMETRIC CONSTRUCTION, GEOMETROGRAPHY, MASCHERONI CONSTANT, POLYGON, PONCELET-STEINER THEOREM, RULER, SIMPLICITY, STEINER CONSTRUCTION

Strange Attractor

An attracting set that has zero MEASURE in the embedding PHASE SPACE and has FRACTAL dimension. Trajectories within a strange attractor appear to skip around randomly.

see also CORRELATION EXPONENT, FRACTAL

References

Benmizrachi, A.; Procaccia, I.; and Grassberger, P. "Characterization of Experimental (Noisy) Strange Attractors." *Phys. Rev. A* **29**, 975–977, 1984.

Grassberger, P. "On the Hausdorff Dimension of Fractal Attractors." *J. Stat. Phys.* **26**, 173–179, 1981.

Grassberger, P. and Procaccia, I. "Measuring the Strangeness of Strange Attractors." *Physica D* **9**, 189–208, 1983a.

Grassberger, P. and Procaccia, I. "Characterization of Strange Attractors." *Phys. Rev. Let.* **50**, 346–349, 1983b.

Lauwerier, H. *Fractals: Endlessly Repeated Geometric Figures.* Princeton, NJ: Princeton University Press, pp. 137–138, 1991.
Sprott, J. C. *Strange Attractors: Creating Patterns in Chaos.* New York: Henry Holt, 1993.

Strange Loop

A phenomenon in which, whenever movement is made upwards or downwards through the levels of some heirarchial system, the system unexpectedly arrives back where it started. Hofstadter (1987) uses the strange loop as a paradigm in which to interpret paradoxes in logic (such as GRELLING'S PARADOX and RUSSELL'S PARADOX) and calls a system in which a strange loop appears a TANGLED HIERARCHY.

see also GRELLING'S PARADOX, RUSSELL'S PARADOX, TANGLED HIERARCHY

References

Hofstadter, D. R. *Gödel, Escher, Bach: An Eternal Golden Braid.* New York: Vintage Books, p. 10, 1989.

Strangers

Two numbers which are RELATIVELY PRIME.

References

Le Lionnais, F. *Les nombres remarquables.* Paris: Hermann, p. 145, 1983.

Strassen Formulas

The usual number of scalar operations (i.e., the total number of additions and multiplications) required to perform $n \times n$ MATRIX MULTIPLICATION is

$$M(n) = 2n^3 - n^2 \tag{1}$$

(i.e., n^3 multiplications and $n^3 - n^2$ additions). However, Strassen (1969) discovered how to multiply two MATRICES in

$$S(n) = 7 \cdot 7^{\lg n} - 6 \cdot 4^{\lg n} \tag{2}$$

scalar operations, where lg is the LOGARITHM to base 2, which is less than $M(n)$ for $n > 654$. For n a power of two ($n = 2^k$), the two parts of (2) can be written

$$7 \cdot 7^{\lg n} = 7 \cdot 7^{\lg 2^k} = 7 \cdot 7^k = 7 \cdot 2^{k \lg 7} = 7(2^k)^{\lg 7} = 7n^{\lg 7} \tag{3}$$

$$6 \cdot 4^{\lg n} = 6 \cdot 4^{\lg 2^k} = 6 \cdot 4^{k \lg 2} = 6 \cdot 4^k = 6(2^k)^2 = 6n^2, \tag{4}$$

so (2) becomes

$$S(2^k) = 7n^{\lg 7} - 6n^2. \tag{5}$$

Two 2×2 matrices can therefore be multiplied

$$\mathsf{C} = \mathsf{AB} \tag{6}$$

$$\begin{bmatrix} c_{11} & c_{12} \\ c_{21} & c_{22} \end{bmatrix} = \begin{bmatrix} a_{11} & a_{12} \\ a_{21} & a_{22} \end{bmatrix} \begin{bmatrix} b_{11} & b_{12} \\ b_{21} & b_{22} \end{bmatrix} \tag{7}$$

with only

$$S(2) = 7 \cdot 2^{\lg 7} - 6 \cdot 2^2 = 49 - 24 = 25 \tag{8}$$

scalar operations (as it turns out, seven of them are multiplications and 18 are additions). Define the seven products (involving a total of 10 additions) as

$$Q_1 \equiv (a_{11} + a_{22})(b_{11} + b_{22}) \tag{9}$$
$$Q_2 \equiv (a_{21} + a_{22})b_{11} \tag{10}$$
$$Q_3 \equiv a_{11}(b_{12} - b_{22}) \tag{11}$$
$$Q_4 \equiv a_{22}(-b_{11} + b_{21}) \tag{12}$$
$$Q_5 \equiv (a_{11} + a_{12})b_{22} \tag{13}$$
$$Q_6 \equiv (-a_{11} + a_{12})(b_{11} + b_{12}) \tag{14}$$
$$Q_7 \equiv (a_{12} - a_{22})(b_{21} + b_{22}). \tag{15}$$

Then the matrix product is given using the remaining eight additions as

$$c_{11} = Q_1 + Q_4 - Q_5 + Q_7 \tag{16}$$
$$c_{21} = Q_2 + Q_4 \tag{17}$$
$$c_{12} = Q_3 + Q_5 \tag{18}$$
$$c_{22} = Q_1 + Q_3 - Q_2 + Q_6 \tag{19}$$

(Strassen 1969, Press *et al.* 1989).

Matrix inversion of a 2×2 matrix A to yield $\mathsf{C} = \mathsf{A}^{-1}$ can also be done in fewer operations than expected using the formulas

$$R_1 \equiv {a_{11}}^{-1} \tag{20}$$
$$R_2 \equiv a_{21} R_1 \tag{21}$$
$$R_3 \equiv R_1 a_{12} \tag{22}$$
$$R_4 \equiv a_{21} R_3 \tag{23}$$
$$R_5 \equiv R_4 - a_{22} \tag{24}$$
$$R_6 \equiv {R_5}^{-1} \tag{25}$$
$$c_{12} = R_3 R_6 \tag{26}$$
$$c_{21} = R_6 R_2 \tag{27}$$
$$R_7 = R_3 c_{21} \tag{28}$$
$$c_{11} = R_1 - R_7 \tag{29}$$
$$c_{22} = -R_6 \tag{30}$$

(Strassen 1969, Press *et al.* 1989). The leading exponent for Strassen's algorithm for a POWER of 2 is $\lg 7 \approx 2.808$. The best leading exponent currently known is 2.376 (Coppersmith and Winograd 1990). It has been shown that the exponent must be at least 2.

see also COMPLEX MULTIPLICATION, KARATSUBA MULTIPLICATION

References

Coppersmith, D. and Winograd, S. "Matrix Multiplication via Arithmetic Programming." *J. Symb. Comput.* **9**, 251–280, 1990.

Pan, V. *How to Multiply Matrices Faster.* New York: Springer-Verlag, 1982.

Press, W. H.; Flannery, B. P.; Teukolsky, S. A.; and Vetterling, W. T. "Is Matrix Inversion an N^3 Process?" §2.11 in *Numerical Recipes in FORTRAN: The Art of Scientific Computing, 2nd ed.* Cambridge, England: Cambridge University Press, pp. 95–98, 1989.

Strassen, V. "Gaussian Elimination is Not Optimal." *Numerische Mathematik* **13**, 354–356, 1969.

Strassman's Theorem

Let $(K, |\cdot|)$ be a complete non-ARCHIMEDEAN VALUATED FIELD, with VALUATION RING R, and let $f(X)$ be a POWER series with COEFFICIENTS in R. Suppose at least one of the COEFFICIENTS is NONZERO (so that f is not identically zero) and the sequence of COEFFICIENTS converges to 0 with respect to $|\cdot|$. Then $f(X)$ has only finitely many zeros in R.

see also ARCHIMEDEAN VALUATION, MAHLER-LECH THEOREM, VALUATION, VALUATION RING

Strassnitzky's Formula

The MACHIN-LIKE FORMULA

$$\tfrac{1}{4}\pi = \cot^{-1} 2 + \cot^{-1} 5 + \cot^{-1} 8.$$

see also MACHIN'S FORMULA, MACHIN-LIKE FORMULAS

Strategy

A set of moves which a player plans to follow while playing a GAME.

see also GAME, MIXED STRATEGY

Stratified Manifold

A set that is a smooth embedded 2-D MANIFOLD except for a subset that consists of smooth embedded curves, except for a set of ISOLATED POINTS.

References

Morgan, F. "What is a Surface?" *Amer. Math. Monthly* **103**, 369–376, 1996.

Strehl Identity

The sum identity

$$\sum_{j=0}^{\infty} \binom{n}{j}^3 = \sum_{k=0}^{\infty} \binom{n}{k}^2 \binom{2(n-k)}{n},$$

where $\binom{n}{k}$ is a BINOMIAL COEFFICIENT.

see also BINOMIAL COEFFICIENT

Striction Curve

A NONCYLINDRICAL RULED SURFACE always has a parameterization of the form

$$\mathbf{x}(u,v) = \boldsymbol{\sigma}(u) + v\boldsymbol{\delta}(u), \tag{1}$$

where $|\boldsymbol{\delta}| = 1$, $\boldsymbol{\sigma}' \cdot \boldsymbol{\delta}' = 0$, and $\boldsymbol{\sigma}$ is called the striction curve of $\mathbf{x}$. Furthermore, the striction curve does not depend on the choice of the base curve. The striction and DIRECTOR CURVES of the HELICOID

$$\mathbf{x}(u,v) = \begin{bmatrix} 0 \\ 0 \\ bu \end{bmatrix} + av \begin{bmatrix} \cos u \\ \sin u \\ 0 \end{bmatrix} \tag{2}$$

are

$$\boldsymbol{\sigma}(u) = \begin{bmatrix} 0 \\ 0 \\ bu \end{bmatrix} \tag{3}$$

$$\boldsymbol{\delta}(u) = \begin{bmatrix} a\cos u \\ a\sin u \\ 0 \end{bmatrix}. \tag{4}$$

For the HYPERBOLIC PARABOLOID

$$\mathbf{x}(u,v) = \begin{bmatrix} u \\ 0 \\ 0 \end{bmatrix} + v \begin{bmatrix} 0 \\ 1 \\ u \end{bmatrix}, \tag{5}$$

the striction and DIRECTOR CURVES are

$$\boldsymbol{\sigma}(u) = \begin{bmatrix} u \\ 0 \\ 0 \end{bmatrix} \tag{6}$$

$$\boldsymbol{\delta}(u) = \begin{bmatrix} 0 \\ 1 \\ u \end{bmatrix}. \tag{7}$$

see also DIRECTOR CURVE, DISTRIBUTION PARAMETER, NONCYLINDRICAL RULED SURFACE, RULED SURFACE,

References

Gray, A. "Noncylindrical Ruled Surfaces" and "Examples of Striction Curves of Noncylindrical Ruled Surfaces." §17.3 and 17.4 in *Modern Differential Geometry of Curves and Surfaces.* Boca Raton, FL: CRC Press, pp. 345–350, 1993.

String Rewriting

A SUBSTITUTION MAP in which rules are used to operate on a string consisting of letters of a certain alphabet. String rewriting is a particularly useful technique for generating successive iterations of certain types of FRACTALS, such as the BOX FRACTAL, CANTOR DUST, CANTOR SQUARE FRACTAL, and SIERPIŃSKI CARPET.

see also RABBIT SEQUENCE, SUBSTITUTION MAP

References

Peitgen, H.-O. and Saupe, D. (Eds.). "String Rewriting Systems." §C.1 in *The Science of Fractal Images.* New York: Springer-Verlag, pp. 273–275, 1988.

Wagon, S. "Recursion via String Rewriting." §6.2 in *Mathematica in Action.* New York: W. H. Freeman, pp. 190–196, 1991.

Strip

see CRITICAL STRIP, MÖBIUS STRIP

Strong Convergence

Strong convergence is the type of convergence usually associated with convergence of a SEQUENCE. More formally, a SEQUENCE $\{x_n\}$ of VECTORS in an INNER PRODUCT SPACE E is called convergent to a VECTOR x in E if

$$\|x_n - x\| \to 0 \quad \text{as } n \to \infty.$$

see also CONVERGENT SEQUENCE, INNER PRODUCT SPACE, WEAK CONVERGENCE

Strong Elliptic Pseudoprime

Let n be an ELLIPTIC PSEUDOPRIME associated with (E, P), and let $n+1 = 2^s k$ with k ODD and $s \geq 0$. Then n is a strong elliptic pseudoprime when either $kP \equiv 0 \pmod{n}$ or $2^r kP \equiv 0 \pmod{n}$ for some r with $1 \leq r < s$.

see also ELLIPTIC PSEUDOPRIME

References

Ribenboim, P. *The New Book of Prime Number Records, 3rd ed.* New York: Springer-Verlag, pp. 132–134, 1996.

Strong Frobenius Pseudoprime

A PSEUDOPRIME which obeys an additional restriction beyond that required for a FROBENIUS PSEUDOPRIME. A number n with $(n, 2a) = 1$ is a strong Frobenius pseudoprime with respect to $x - a$ IFF n is a STRONG PSEUDOPRIME with respect to $f(x)$. Every strong Frobenius pseudoprime with respect to $x - a$ is an EULER PSEUDOPRIME to the base a.

Every strong Frobenius pseudoprime with respect to $f(x) = x^2 - bx - c$ such that $((b^2 + 4c)/n) = -1$ is a STRONG LUCAS PSEUDOPRIME with parameters (b, c). Every strong Frobenius pseudoprime n with respect to $x^2 - bx + 1$ is an EXTRA STRONG LUCAS PSEUDOPRIME to the base b.

see also FROBENIUS PSEUDOPRIME

References

Grantham, J. "Frobenius Pseudoprimes." 1996. `http://www.clark.net/pub/grantham/pseudo/pseudo.ps`

Strong Law of Large Numbers

For a set of random variates x_i from a distribution having unit MEAN,

$$P\left(\lim_{n\to\infty} \frac{x_1 + \ldots + x_n}{n}\right) = P\left(\lim_{n\to\infty} \langle x \rangle\right) = 1.$$

This result is due to Kolmogorov.

see also LAW OF TRULY LARGE NUMBERS, STRONG LAW OF SMALL NUMBERS, WEAK LAW OF LARGE NUMBERS

Strong Law of Small Numbers

There aren't enough small numbers to meet the many demands made of them.

References

Gardner, M. "Patterns in Primes are a Clue to the Strong Law of Small Numbers." *Sci. Amer.* **243**, 18–28, Dec. 1980.

Guy, R. K. "The Strong Law of Small Numbers." *Amer. Math. Monthly* **95**, 697–712, 1988.

Strong Lucas Pseudoprime

Let $U(P, Q)$ and $V(P, Q)$ be LUCAS SEQUENCES generated by P and Q, and define

$$D \equiv P^2 - 4Q.$$

Let n be an ODD COMPOSITE NUMBER with $(n, D) = 1$, and $n - (D/n) = 2^s d$ with d ODD and $s \geq 0$, where (a/b) is the LEGENDRE SYMBOL. If

$$U_d \equiv 0 \pmod{n}$$

or

$$V_{2^r d} \equiv 0 \pmod{n}$$

for some r with $0 \leq r < s$, then n is called a strong Lucas pseudoprime with parameters (P, Q).

A strong Lucas pseudoprime is a LUCAS PSEUDOPRIME to the same base. Arnault (1997) showed that any COMPOSITE NUMBER n is a strong Lucas pseudoprime for at most 4/15 of possible bases (unless n is the PRODUCT of TWIN PRIMES having certain properties).

see also EXTRA STRONG LUCAS PSEUDOPRIME, LUCAS PSEUDOPRIME

References

Arnault, F. "The Rabin-Monier Theorem for Lucas Pseudoprimes." *Math. Comput.* **66**, 869–881, 1997.

Ribenboim, P. "Euler-Lucas Pseudoprimes (elpsp(P, Q)) and Strong Lucas Pseudoprimes (slpsp(P, Q))." §2.X.C in *The New Book of Prime Number Records, 3rd ed.* New York: Springer-Verlag, pp. 130–131, 1996.

Strong Pseudoprime

A strong pseudoprime to a base a is an ODD COMPOSITE NUMBER n with $n - 1 = d \cdot 2^s$ (for d ODD) for which either

$$a^d \equiv 1 \pmod{n} \tag{1}$$

or

$$a^{d\cdot 2^r} \equiv -1 \pmod{n} \tag{2}$$

for some $r \in [0, s)$.

The definition is motivated by the fact that a FERMAT PSEUDOPRIME n to the base b satisfies

$$b^{n-1} - 1 \equiv 0 \pmod{n}. \tag{3}$$

But since n is ODD, it can be written $n = 2m + 1$, and

$$b^{2m} - 1 = (b^m - 1)(b^m + 1) \equiv 0 \pmod{n}. \tag{4}$$

If n is PRIME, it must DIVIDE at least one of the FACTORS, but can't DIVIDE both because it would then DIVIDE their difference

$$(b^m + 1) - (b^m - 1) = 2. \quad (5)$$

Therefore,

$$b^m \equiv \pm 1 \pmod{n}, \quad (6)$$

so write $n = 2^a t + 1$ to obtain

$$b^{n-1} - 1 = (b^t - 1)(b^t + 1)(b^{2t} + 1) \cdots (b^{2^{a-1}t} + 1). \quad (7)$$

If n DIVIDES exactly one of these FACTORS but is COMPOSITE, it is a strong pseudoprime. A COMPOSITE number is a strong pseudoprime to at most 1/4 of all bases less than itself (Monier 1980, Rabin 1980). The strong pseudoprimes provide the basis for MILLER'S PRIMALITY TEST and RABIN-MILLER STRONG PSEUDOPRIME TEST.

A strong pseudoprime to the base a is also an EULER PSEUDOPRIME to the base a (Pomerance *et al.* 1980). The strong pseudoprimes include some EULER PSEUDOPRIMES, FERMAT PSEUDOPRIMES, and CARMICHAEL NUMBERS.

There are 4842 strong psp(2) less than 2.5×10^{10}, where a psp(2) is also known as a POULET NUMBER. The strong k-pseudoprime test for $k = 2$, 3, 5 correctly identifies all PRIMES below 2.5×10^{10} with only 13 exceptions, and if 7 is added, then the only exception less than 2.5×10^{10} is 315031751. Jaeschke (1993) showed that there are only 101 strong pseudoprimes for the bases 2, 3, and 5 less than 10^{12}, nine if 7 is added, and none if 11 is added. Also, the bases 2, 13, 23, and 1662803 have no exceptions up to 10^{12}.

If n is COMPOSITE, then there is a base for which n is not a strong pseudoprime. There are therefore no "strong CARMICHAEL NUMBERS." Let ψ_k denote the smallest strong pseudoprime to all of the first k PRIMES taken as bases (i.e, the smallest ODD NUMBER for which the RABIN-MILLER STRONG PSEUDOPRIME TEST on bases less than or equal to k fails). Jaeschke (1993) computed ψ_k from $k = 5$ to 8 and gave upper bounds for $k = 9$ to 11.

$$\begin{aligned}
\psi_1 &= 2047 \\
\psi_2 &= 1373653 \\
\psi_3 &= 25326001 \\
\psi_4 &= 3215031751 \\
\psi_5 &= 2152302898747 \\
\psi_6 &= 3474749660383 \\
\psi_7 &= 34155071728321 \\
\psi_8 &= 34155071728321 \\
\psi_9 &\leq 41234316135705689041 \\
\psi_{10} &\leq 1553360566073143205541002401 \\
\psi_{11} &\leq 56897193526942024370326972321
\end{aligned}$$

(Sloane's A014233). A seven-step test utilizing these results (Riesel 1994) allows all numbers less than 3.4×10^{14} to be tested.

Pomerance *et al.* (1980) have proposed a test based on a combination of STRONG PSEUDOPRIMES and LUCAS PSEUDOPRIMES. They offer a $620 reward for discovery of a COMPOSITE NUMBER which passes their test (Guy 1994, p. 28).

see also CARMICHAEL NUMBER, MILLER'S PRIMALITY TEST, POULET NUMBER, RABIN-MILLER STRONG PSEUDOPRIME TEST, ROTKIEWICZ THEOREM, STRONG ELLIPTIC PSEUDOPRIME, STRONG LUCAS PSEUDOPRIME

References

Baillie, R. and Wagstaff, S. "Lucas Pseudoprimes." *Math. Comput.* **35**, 1391–1417, 1980.

Guy, R. K. "Pseudoprimes. Euler Pseudoprimes. Strong Pseudoprimes." §A12 in *Unsolved Problems in Number Theory, 2nd ed.* New York: Springer-Verlag, pp. 27–30, 1994.

Jaeschke, G. "On Strong Pseudoprimes to Several Bases." *Math. Comput.* **61**, 915–926, 1993.

Monier, L. "Evaluation and Comparison of Two Efficient Probabilistic Primality Testing Algorithms." *Theor. Comput. Sci.* **12**, 97–108, 1980.

Pomerance, C.; Selfridge, J. L.; and Wagstaff, S. S. Jr. "The Pseudoprimes to $25 \cdot 10^9$." *Math. Comput.* **35**, 1003–1026, 1980. Available electronically from ftp://sable.ox.ac.uk/pub/math/primes/ps2.Z.

Rabin, M. O. "Probabilistic Algorithm for Testing Primality." *J. Number Th.* **12**, 128–138, 1980.

Riesel, H. *Prime Numbers and Computer Methods for Factorization, 2nd ed.* Basel: Birkhäuser, p. 92, 1994.

Sloane, N. J. A. Sequence A014233 in "An On-Line Version of the Encyclopedia of Integer Sequences."

Strong Pseudoprime Test

see RABIN-MILLER STRONG PSEUDOPRIME TEST

Strong Subadditivity Inequality

$$\phi(A) + \phi(B) - \phi(A \cup B) \geq \phi(A \cap B).$$

References

Doob, J. L. "The Development of Rigor in Mathematical Probability (1900–1950)." *Amer. Math. Monthly* **103**, 586–595, 1996.

Strong Triangle Inequality

$$|x + y|_p \leq \max(|x|_p, |y|_p)$$

for all x and y.

see also p-ADIC NUMBER, TRIANGLE INEQUALITY

Strongly Connected Component

A maximal subgraph of a DIRECTED GRAPH such that for every pair of vertices u, v in the SUBGRAPH, there is a directed path from u to v and a directed path from v to u.

see also BI-CONNECTED COMPONENT

Strongly Embedded Theorem

The strongly embedded theorem identifies all SIMPLE GROUPS with a strongly 2-embedded SUBGROUP. In particular, it asserts that no SIMPLE GROUP has a strongly 2-embedded 2'-local SUBGROUP.

see also SIMPLE GROUP, SUBGROUP

Strongly Independent

An infinite sequence $\{a_i\}$ of POSITIVE INTEGERS is called strongly independent if any relation $\sum \epsilon_i a_i$, with $\epsilon_i = 0, \pm 1$, or ± 2 and $\epsilon_i = 0$ except finitely often, IMPLIES $\epsilon_i = 0$ for all i.

see also WEAKLY INDEPENDENT

References

Guy, R. K. *Unsolved Problems in Number Theory, 2nd ed.* New York: Springer-Verlag, p. 136, 1994.

Strongly Triple-Free Set

see TRIPLE-FREE SET

Strophoid

Let C be a curve, let O be a fixed point (the POLE), and let O' be a second fixed point. Let P and P' be points on a line through O meeting C at Q such that $P'Q = QP = QO'$. The LOCUS of P and P' is called the strophoid of C with respect to the POLE O and fixed point O'. Let C be represented parametrically by $(f(t), g(t))$, and let $O = (x_0, y_0)$ and $O' = (x_1, y_1)$. Then the equation of the strophoid is

$$x = f \pm \sqrt{\frac{(x_1 - f)^2 + (y_1 - g)^2}{1 + m^2}} \qquad (1)$$

$$y = g \pm \sqrt{\frac{(x_1 - f)^2 + (y_1 - g)^2}{1 + m^2}}, \qquad (2)$$

where

$$m \equiv \frac{g - y_0}{f - x_0}. \qquad (3)$$

The name strophoid means "belt with a twist," and was proposed by Montucci in 1846 (MacTutor Archive). The polar form for a general strophoid is

$$r = \frac{b \sin(a - 2\theta)}{\sin(a - \theta)}. \qquad (4)$$

If $a = \pi/2$, the curve is a RIGHT STROPHOID. The following table gives the strophoids of some common curves.

Curve	Pole	Fixed Point	Strophoid
line	not on line	on line	oblique strophoid
line	not on line	foot of ⊥ origin to line	right strophoid
circle	center	on circumf.	Freeth's nephroid

see also RIGHT STROPHOID

References

Lawrence, J. D. *A Catalog of Special Plane Curves.* New York: Dover, pp. 51–53 and 205, 1972.
Lockwood, E. H. "Strophoids." Ch. 16 in *A Book of Curves.* Cambridge, England: Cambridge University Press, pp. 134–137, 1967.
MacTutor History of Mathematics Archive. "Right." http://www-groups.dcs.st-and.ac.uk/~history/Curves/Right.html.
Yates, R. C. "Strophoid." *A Handbook on Curves and Their Properties.* Ann Arbor, MI: J. W. Edwards, pp. 217–220, 1952.

Structurally Stable

A MAP $\phi : M \to M$ where M is a MANIFOLD is C^r structurally stable if any C^r perturbation is TOPOLOGICALLY CONJUGATE to ϕ. Here, C^r perturbation means a FUNCTION ψ such that ψ is close to ϕ and the first r derivatives of ψ are close to those of ϕ.

see also TOPOLOGICALLY CONJUGATE

Structure

see LATTICE

Structure Constant

The structure constant is defined as $i\epsilon_{ijk}$, where ϵ_{ijk} is the PERMUTATION SYMBOL. The structure constant forms the starting point for the development of LIE ALGEBRA.

see also LIE ALGEBRA, PERMUTATION SYMBOL

Structure Factor

The structure factor S_Γ of a discrete set Γ is the FOURIER TRANSFORM of δ-scatterers of equal strengths on all points of Γ,

$$S_\Gamma(k) = \int \sum_{x \in \Gamma} \delta(x' - x) e^{-2\pi i k x'}\, dx' = \sum_{x \in \Gamma} e^{-2\pi i k x}.$$

References

Baake, M.; Grimm, U.; and Warrington, D. H. "Some Remarks on the Visible Points of a Lattice." *J. Phys. A: Math. General* **27**, 2669–2674, 1994.

Struve Differential Equation

The ordinary differential equation

$$z^2 y'' + z y' + (z^2 - \nu^2) y = \frac{4(\frac{1}{2} z)^{\nu+1}}{\sqrt{\pi}\, \Gamma(\nu + \frac{1}{2})},$$

where $\Gamma(z)$ is the GAMMA FUNCTION. The solution is

$$y = a J_\nu(z) + b Y_\nu(z) + \mathcal{H}_\nu(z),$$

where $J_\nu(z)$ and $Y_\nu(z)$ are BESSEL FUNCTIONS OF THE FIRST and SECOND KINDS, and $\mathcal{H}_\nu(z)$ is a STRUVE FUNCTION (Abramowitz and Stegun 1972).

see also BESSEL FUNCTION OF THE FIRST KIND, BESSEL FUNCTION OF THE SECOND KIND, STRUVE FUNCTION

References

Abramowitz, M. and Stegun, C. A. (Eds.). *Handbook of Mathematical Functions with Formulas, Graphs, and Mathematical Tables, 9th printing.* New York: Dover, pp. 496, 1972.

Struve Function

Abramowitz and Stegun (1972, pp. 496–499) define the Struve function as

$$\mathcal{H}_\nu(z) = (\tfrac{1}{2}z)^{\nu+1} \sum_{k=0}^{\infty} \frac{(-1)^k (\frac{1}{2}z)^{2k}}{\Gamma(k+\frac{3}{2})\Gamma(k+\nu+\frac{3}{2})}, \qquad (1)$$

where $\Gamma(z)$ is the GAMMA FUNCTION. Watson (1966, p. 338) defines the Struve function as

$$\mathcal{H}_\nu(z) \equiv \frac{2(\frac{1}{2}z)^\nu}{\Gamma(\nu+\frac{1}{2})\Gamma(\frac{1}{2})} \int_0^1 (1-t^2)^{\nu-1/2} \sin(zt)\, dt. \qquad (2)$$

The series expansion is

$$\mathcal{H}_\nu(z) = \sum_{m=0}^{\infty} (-1)^m \frac{(\frac{1}{2}z)^{2m+\nu+1}}{\Gamma(m+\frac{3}{2})\Gamma(\nu+m+\frac{3}{2})}. \qquad (3)$$

For half integral orders,

$$\mathcal{H}_{n+1/2}(z) = Y_{n+1/2}(z) + \frac{1}{\pi} \sum_{m=0}^{n} \frac{\Gamma(m+\frac{1}{2})(\frac{1}{2}z)^{-2m+n-1/2}}{\Gamma(n+1-m)} \qquad (4)$$

$$\mathcal{H}_{-(n+1/2)}(z) = (-1)^n J_{n+1/2}(z). \qquad (5)$$

The Struve function and its derivatives satisfy

$$\mathcal{H}_{\nu-1}(z) - \mathcal{H}_{\nu+1}(z) = 2\mathcal{H}'_\nu(z) - \frac{(\frac{1}{2}z)^\nu}{\sqrt{\pi}\,\Gamma(\nu+\frac{3}{2})}. \qquad (6)$$

see also ANGER FUNCTION, BESSEL FUNCTION, MODIFIED STRUVE FUNCTION, WEBER FUNCTIONS

References

Abramowitz, M. and Stegun, C. A. (Eds.). "Struve Function $\mathbf{H}_\nu(x)$." §12.1 in *Handbook of Mathematical Functions with Formulas, Graphs, and Mathematical Tables, 9th printing.* New York: Dover, pp. 496–498, 1972.

Spanier, J. and Oldham, K. B. "The Struve Function." Ch. 57 in *An Atlas of Functions.* Washington, DC: Hemisphere, pp. 563–571, 1987.

Watson, G. N. *A Treatise on the Theory of Bessel Functions, 2nd ed.* Cambridge, England: Cambridge University Press, 1966.

Student's t-Distribution

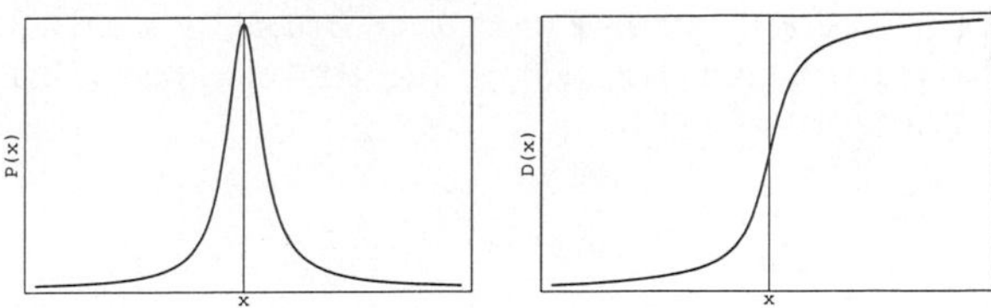

A DISTRIBUTION published by William Gosset in 1908. His employer, Guinness Breweries, required him to publish under a pseudonym, so he chose "Student." Given n independent measurements x_i, let

$$t \equiv \frac{\bar{x}-\mu}{s/\sqrt{n}}, \qquad (1)$$

where μ is the population MEAN, $\bar{x}$ is the sample MEAN, and s is the ESTIMATOR for population STANDARD DEVIATION (i.e., the SAMPLE VARIANCE) defined by

$$s^2 \equiv \frac{1}{N-1} \sum_{i=1}^{n} (x_i - \bar{x})^2. \qquad (2)$$

Student's t-distribution is defined as the distribution of the random variable t which is (very loosely) the "best" that we can do not knowing σ. If $\sigma = s$, $t = z$ and the distribution becomes the NORMAL DISTRIBUTION. As N increases, Student's t-distribution approaches the NORMAL DISTRIBUTION.

Student's t-distribution is arrived at by transforming to STUDENT'S z-DISTRIBUTION with

$$z \equiv \frac{\bar{x}-\mu}{s}. \qquad (3)$$

Then define

$$t \equiv z\sqrt{n-1}. \qquad (4)$$

The resulting probability and cumulative distribution functions are

$$f_r(t) = \frac{\Gamma\left(\frac{r+1}{2}\right)}{\sqrt{r\pi}\,\Gamma\left(\frac{r}{2}\right)\left(1+\frac{t^2}{r}\right)^{(r+1)/2}} = \frac{\left(\frac{r}{r+t^2}\right)^{(1+r)/2}}{\sqrt{r}\,B(\frac{1}{2}r,\frac{1}{2})} \qquad (5)$$

$$F_r(t) = \int_{-\infty}^{t} \frac{\Gamma\left(\frac{r+1}{2}\right)}{\sqrt{r\pi}\,\Gamma\left(\frac{r}{2}\right)\left(1+\frac{t^2}{r}\right)^{(r+1)/2}}\,dt = \frac{1}{\sqrt{n}B\left(\frac{r}{2},\frac{1}{2}\right)\left(1+\frac{t^2}{r}\right)^{(r+1)/2}} = \frac{1}{2}+\frac{1}{2}\left[I(1;\tfrac{1}{2}n,\tfrac{1}{2}) - I\left(\frac{r}{r+t^2},\tfrac{1}{2}n,\tfrac{1}{2}\right)\right], \qquad (6)$$

where

$$r \equiv n-1 \qquad (7)$$

is the number of DEGREES OF FREEDOM, $-\infty < t < \infty$, $\Gamma(z)$ is the GAMMA FUNCTION, $B(a,b)$ is the BETA FUNCTION, and $I(z;a,b)$ is the REGULARIZED BETA FUNCTION defined by

$$I(z;a,b) = \frac{B(z;a,b)}{B(a,b)}. \quad (8)$$

The MEAN, VARIANCE, SKEWNESS, and KURTOSIS of Student's t-distribution are

$$\mu = 0 \quad (9)$$

$$\sigma^2 = \frac{r}{r-2} \quad (10)$$

$$\gamma_1 = 0 \quad (11)$$

$$\gamma_2 = \frac{6}{r-4}. \quad (12)$$

Beyer (1987, p. 514) gives 60%, 70%, 90%, 95%, 97.5%, 99%, 99.5%, and 99.95% confidence intervals, and Goulden (1956) gives 50%, 90%, 95%, 98%, 99%, and 99.9% confidence intervals. A partial table is given below for small r and several common confidence intervals.

r	80%	90%	95%	99%
1	3.08	6.31	12.71	63.66
2	1.89	2.92	4.30	9.92
3	1.64	2.35	3.18	5.84
4	1.53	2.13	2.78	4.60
5	1.48	2.01	2.57	4.03
10	1.37	1.81	2.23	4.14
30	1.31	1.70	2.04	2.75
100	1.29	1.66	1.98	2.63
∞	1.28	1.65	1.96	2.58

The so-called $A(t|n)$ distribution is useful for testing if two observed distributions have the same MEAN. $A(t|n)$ gives the probability that the difference in two observed MEANS for a certain statistic t with n DEGREES OF FREEDOM would be smaller than the observed value purely by chance:

$$A(t|n) = \frac{1}{\sqrt{n}\,B(\tfrac{1}{2},\tfrac{1}{2}n)} \int_{-t}^{t} \left(1+\frac{x^2}{n}\right)^{-(1+n)/2} dx. \quad (13)$$

Let X be a NORMALLY DISTRIBUTED random variable with MEAN 0 and VARIANCE σ^2, let Y^2/σ^2 have a CHI-SQUARED DISTRIBUTION with n DEGREES OF FREEDOM, and let X and Y be independent. Then

$$t \equiv \frac{X\sqrt{n}}{Y} \quad (14)$$

is distributed as Student's t with n DEGREES OF FREEDOM.

The noncentral Student's t-distribution is given by

$$P(x) = \frac{n^{n/2} n!}{2^n e^{\lambda^2/2}\Gamma(\tfrac{1}{2}n)} \times \left\{ \frac{\sqrt{2}\,\lambda x(n+x^2)^{-(1+n/2)}\,{}_1F_1\left(1+\tfrac{1}{2}n;\tfrac{3}{2};\tfrac{\lambda^2x^2}{2(n+x^2)}\right)}{\Gamma[\tfrac{1}{2}(1+n)]} + \frac{e^{(\lambda^2x^2)/[2(n+x^2)]}\sqrt{\pi}(n+x^2)^{-(n+1)/2} L_{n/2}^{-1/2}\left(-\tfrac{\lambda^2x^2}{2(n+x^2)}\right)}{\Gamma[\tfrac{1}{2}(1+n)]} \right\}, \quad (15)$$

where $\Gamma(z)$ is the GAMMA FUNCTION, ${}_1F_1(a;b;z)$ is a CONFLUENT HYPERGEOMETRIC FUNCTION, and $L_n^m(x)$ is an associated LAGUERRE POLYNOMIAL.

see also PAIRED t-TEST, STUDENT'S z-DISTRIBUTION

References

Abramowitz, M. and Stegun, C. A. (Eds.). *Handbook of Mathematical Functions with Formulas, Graphs, and Mathematical Tables, 9th printing.* New York: Dover, pp. 948–949, 1972.

Beyer, W. H. *CRC Standard Mathematical Tables, 28th ed.* Boca Raton, FL: CRC Press, p. 536, 1987.

Fisher, R. A. "Applications of 'Student's' Distribution." *Metron* **5**, 3–17, 1925.

Fisher, R. A. "Expansion of 'Student's' Integral in Powers of $n-1$." *Metron* **5**, 22–32, 1925.

Fisher, R. A. *Statistical Methods for Research Workers, 10th ed.* Edinburgh: Oliver and Boyd, 1948.

Goulden, C. H. Table A–3 in *Methods of Statistical Analysis, 2nd ed.* New York: Wiley, p. 443, 1956.

Press, W. H.; Flannery, B. P.; Teukolsky, S. A.; and Vetterling, W. T. "Incomplete Beta Function, Student's Distribution, F-Distribution, Cumulative Binomial Distribution." §6.2 in *Numerical Recipes in FORTRAN: The Art of Scientific Computing, 2nd ed.* Cambridge, England: Cambridge University Press, pp. 219–223, 1992.

Spiegel, M. R. *Theory and Problems of Probability and Statistics.* New York: McGraw-Hill, pp. 116–117, 1992.

Student. "The Probable Error of a Mean." *Biometrika* **6**, 1–25, 1908.

Student's z-Distribution

The probability density function and cumulative distribution functions for Student's z-distribution are given by

$$f(z) = \frac{\Gamma\left(\frac{n}{2}\right)}{\sqrt{\pi}\,\Gamma\left(\frac{n-1}{2}\right)}(1+z^2)^{-n/2} \quad (1)$$

$$D(z) = \frac{-z^{1-n}\Gamma(\tfrac{1}{2}n)\,{}_2F_1(\tfrac{1}{2}(n-1),\tfrac{1}{2}n;\tfrac{1}{2}(n+1);-z^{-2})}{2\sqrt{\pi}\,\Gamma[\tfrac{1}{2}(n+1)]}. \quad (2)$$

The MEAN is 0, so the MOMENTS are

$$\mu_1 = 0 \quad (3)$$

$$\mu_2 = \frac{1}{n-3} \quad (4)$$

$$\mu_3 = 0 \quad (5)$$

$$\mu_4 = \frac{3}{(n-3)(n-5)}. \quad (6)$$

The MEAN, VARIANCE, SKEWNESS, and KURTOSIS are

$$\mu = 0 \tag{7}$$
$$\sigma^2 = \frac{1}{n-3} \tag{8}$$
$$\gamma_1 = 0 \tag{9}$$
$$\gamma_2 = \frac{6}{n-5}. \tag{10}$$

Letting

$$z \equiv \frac{(\bar{x}-\mu)}{s}, \tag{11}$$

where x is the sample MEAN and μ is the population MEAN gives STUDENT'S t-DISTRIBUTION.

see also STUDENT'S t-DISTRIBUTION

Study's Theorem

Given three curves ϕ_1, ϕ_2, ϕ_3 with the common group of ordinary points G (which may be empty), let their remaining groups of intersections g_{23}, g_{31}, and g_{12} also be ordinary points. If ϕ_1' is any other curve through g_{23}, then there exist two other curves ϕ_2', ϕ_3' such that the three combined curves $\phi_i\phi_i'$ are of the same order and LINEARLY DEPENDENT, each curve ϕ_k' contains the corresponding group g_{ij}, and every intersection of ϕ_i or ϕ_i' with ϕ_j or ϕ_j' lies on ϕ_k or ϕ_k'.

References

Coolidge, J. L. *A Treatise on Algebraic Plane Curves.* New York: Dover, p. 34, 1959.

Sturm Chain

The series of STURM FUNCTIONS arising in application of the STURM THEOREM.

see also STURM FUNCTION, STURM THEOREM

Sturm Function

Given a function $f(x) \equiv f_0(x)$, write $f_1 \equiv f'(x)$ and define the Sturm functions by

$$f_n(x) = -\left\{f_{n-2}(x) - f_{n-1}(x)\left[\frac{f_{n-2}(x)}{f_{n-1}(x)}\right]\right\}, \tag{1}$$

where $[P(x)/Q(x)]$ is a polynomial quotient. Then construct the following chain of Sturm functions,

$$\begin{aligned} f_0 &= q_0 f_1 - f_2 \\ f_1 &= q_1 f_2 - f_3 \\ f_2 &= q_2 f_3 - f_4 \\ &\vdots \\ f_{s-2} &= q_{s-2} f_{s-1} - f_s, \end{aligned} \tag{2}$$

known as a STURM CHAIN. The chain is terminated when a constant $-f_s(x)$ is obtained.

Sturm functions provide a convenient way for finding the number of real roots of an algebraic equation with real coefficients over a given interval. Specifically, the difference in the number of sign changes between the Sturm functions evaluated at two points $x = a$ and $x = b$ gives the number of real roots in the interval (a, b). This powerful result is known as the STURM THEOREM.

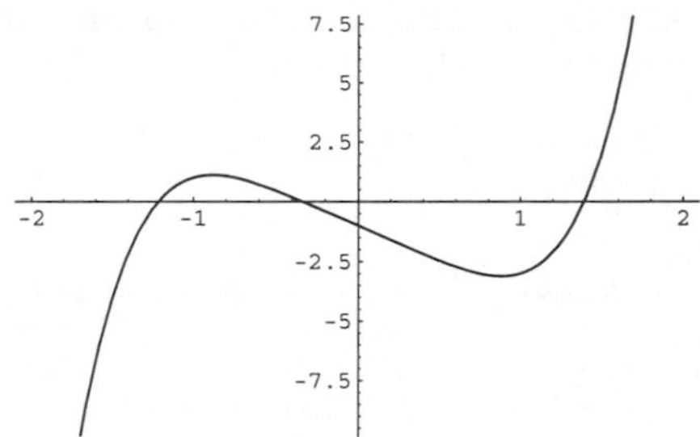

As a specific application of Sturm functions toward finding POLYNOMIAL ROOTS, consider the function $f_0(x) = x^5 - 3x - 1$, plotted above, which has roots -1.21465, -0.334734, $0.0802951 \pm 1.32836i$, and 1.38879 (three of which are real). The DERIVATIVE is given by $f'(x) = 5x^4 - 3$, and the STURM CHAIN is then given by

$$f_0 = x^5 - 3x - 1 \tag{3}$$
$$f_1 = 5x^4 - 3 \tag{4}$$
$$f_2 = \tfrac{1}{5}(12x + 5) \tag{5}$$
$$f_3 = \tfrac{59083}{20736}. \tag{6}$$

The following table shows the signs of f_i and the number of sign changes Δ obtained for points separated by $\Delta x = 2$.

x	f_0	f_1	f_2	f_3	Δ
-2	-1	1	-1	1	3
0	-1	-1	1	1	1
2	1	1	1	1	0

This shows that $3 - 1 = 2$ real roots lie in $(-2, 0)$, and $1 - 0 = 1$ real root lies in $(0, 2)$. Reducing the spacing to $\Delta x = 0.5$ gives the following table.

x	f_0	f_1	f_2	f_3	Δ
-2.0	-1	1	-1	1	3
-1.5	-1	1	-1	1	3
-1.0	1	1	-1	1	2
-0.5	1	-1	-1	1	2
0.0	-1	-1	1	1	1
0.5	-1	-1	1	1	1
1.0	-1	1	1	1	1
1.5	1	1	1	1	0
2.0	1	1	1	1	0

This table isolates the three real roots and shows that they lie in the intervals $(-1.5, -1.0)$, $(-0.5, 0.0)$, and $(1.0, 1.5)$. If desired, the intervals in which the roots fall could be further reduced.

The Sturm functions satisfy the following conditions:

1. Two neighboring functions do not vanish simultaneously at any point in the interval.
2. At a null point of a Sturm function, its two neighboring functions are of different signs.
3. Within a sufficiently small AREA surrounding a zero point of $f_0(x)$, $f_1(x)$ is everywhere greater than zero or everywhere smaller than zero.

see also DESCARTES' SIGN RULE, STURM CHAIN, STURM THEOREM

References

Acton, F. S. *Numerical Methods That Work, 2nd printing.* Washington, DC: Math. Assoc. Amer., p. 334, 1990.
Dörrie, H. "Sturm's Problem of the Number of Roots." §24 in *100 Great Problems of Elementary Mathematics: Their History and Solutions.* New York: Dover, pp. 112–116, 1965.
Press, W. H.; Flannery, B. P.; Teukolsky, S. A.; and Vetterling, W. T. *Numerical Recipes in FORTRAN: The Art of Scientific Computing, 2nd ed.* Cambridge, England: Cambridge University Press, p. 469, 1992.
Rusin, D. "Known Math." http://www.math.niu.edu./~rusin/known-math/polynomials/sturm.
Sturm, C. "Mémoire sur la résolution des équations numériques." *Bull. des sciences de Férussac* **11**, 1929.

Sturm-Liouville Equation

A second-order ORDINARY DIFFERENTIAL EQUATION

$$\frac{d}{dx}\left[p(x)\frac{dy}{dx}\right]+[\lambda w(x)-q(x)]y=0,$$

where λ is a constant and $w(x)$ is a known function called either the density or WEIGHTING FUNCTION. The solutions (with appropriate boundary conditions) of λ are called EIGENVALUES and the corresponding $u_\lambda(x)$ EIGENFUNCTIONS. The solutions of this equation satisfy important mathematical properties under appropriate boundary conditions (Arfken 1985).

see also ADJOINT OPERATOR, SELF-ADJOINT OPERATOR

References

Arfken, G. "Sturm-Liouville Theory—Orthogonal Functions." Ch. 9 in *Mathematical Methods for Physicists, 3rd ed.* Orlando, FL: Academic Press, pp. 497–538, 1985.

Sturm-Liouville Theory

see STURM-LIOUVILLE EQUATION

Sturm Theorem

The number of REAL ROOTS of an algebraic equation with REAL COEFFICIENTS whose REAL ROOTS are simple over an interval, the endpoints of which are not ROOTS, is equal to the difference between the number of sign changes of the STURM CHAINS formed for the interval ends.

see also STURM CHAIN, STURM FUNCTION

References

Dörrie, H. "Sturm's Problem of the Number of Roots." §24 in *100 Great Problems of Elementary Mathematics: Their History and Solutions.* New York: Dover, pp. 112–116, 1965.
Rusin, D. "Known Math." http://www.math.niu.edu./~rusin/known-math/polynomials/sturm.

Sturmian Separation Theorem

Let $\mathsf{A}_r = a_{ij}$ be a SEQUENCE of N SYMMETRIC MATRICES of increasing order with $i, j = 1, 2, \ldots, r$ and $r = 1, 2, \ldots, N$. Let $\lambda_k(\mathsf{A}_r)$ be the kth EIGENVALUE of A_r for $k = 1, 2, \ldots, r$, where the ordering is given by

$$\lambda_1(\mathsf{A}_r) \geq \lambda_2(\mathsf{A}_r) \geq \ldots \geq \lambda_r(\mathsf{A}_r).$$

Then it follows that

$$\lambda_{k+1}(\mathsf{A}_{i+1}) \leq \lambda_k(\mathsf{A}_i) \leq \lambda_k(\mathsf{A}_{i+1}).$$

References

Gradshteyn, I. S. and Ryzhik, I. M. *Tables of Integrals, Series, and Products, 5th ed.* San Diego, CA: Academic Press, p. 1121, 1979.

Sturmian Sequence

If a SEQUENCE has the property that the BLOCK GROWTH function $B(n) = n + 1$ for all n, then it is said to have minimal block growth, and the sequence is called a Sturmian sequence. An example of this is the sequence arising from the SUBSTITUTION MAP

$$0 \to 01$$
$$1 \to 0,$$

yielding $0 \to 01 \to 010 \to 01001 \to 01001010 \to \ldots$, which gives us the Sturmian sequence 01001010....

STURM FUNCTIONS are sometimes also said to form a Sturmian sequence.

see also STURM FUNCTION, STURM THEOREM

Subalgebra

An ALGEBRA S' which is part of a large ALGEBRA S and shares its properties.

see also ALGEBRA

Subanalytic

$X \subseteq \mathbb{R}^n$ is subanalytic if, for all $x \in \mathbb{R}^n$, there is an open U and $Y \subset \mathbb{R}^{n+m}$ a bounded SEMIANALYTIC set such that $X \cap U$ is the projection of Y into U.

see also SEMIANALYTIC

References

Bierstone, E. and Milman, P. "Semialgebraic and Subanalytic Sets." *IHES Pub. Math.* **67**, 5–42, 1988.
Marker, D. "Model Theory and Exponentiation." *Not. Amer. Math. Soc.* **43**, 753–759, 1996.

Subfactorial

The number of PERMUTATIONS of n objects in which no object appear in its natural place (i.e., so-called "DERANGEMENTS").

$$!n \equiv n! \sum_{k=0}^{n} \frac{(-1)^k}{k!} \tag{1}$$

or

$$!n \equiv \left[\frac{n!}{e}\right], \tag{2}$$

where $k!$ is the usual FACTORIAL and $[x]$ is the NINT function. The first few values are $!1 = 0$, $!2 = 1$, $!3 = 2$, $!4 = 9$, $!5 = 44$, $!6 = 265$, $!7 = 1854$, $!8 = 14833$, ... (Sloane's A000166). For example, the only DERANGEMENTS of $\{1,2,3\}$ are $\{2,3,1\}$ and $\{3,1,2\}$, so $!3 = 2$. Similarly, the DERANGEMENTS of $\{1,2,3,4\}$ are $\{2,1,4,3\}$, $\{2,3,4,1\}$, $\{2,4,1,3\}$, $\{3,1,4,2\}$, $\{3,4,1,2\}$, $\{3,4,2,1\}$, $\{4,1,2,3\}$, $\{4,3,1,2\}$, and $\{4,3,2,1\}$, so $!4 = 9$.

The subfactorials are also called the RENCONTRES NUMBERS and satisfy the RECURRENCE RELATIONS

$$!n = n{\cdot}!(n-1) + (-1)^n \tag{3}$$

$$!(n+1) = n[!n + !(n-1)]. \tag{4}$$

The subfactorial can be considered a special case of a restricted ROOKS PROBLEM.

The only number equal to the sum of subfactorials of its digits is

$$148,349 = !1 + !4 + !8 + !3 + !4 + !9 \tag{5}$$

(Madachy 1979).

see also DERANGEMENT, FACTORIAL, MARRIED COUPLES PROBLEM, ROOKS PROBLEM, SUPERFACTORIAL

References

Dörrie, H. §6 in *100 Great Problems of Elementary Mathematics: Their History and Solutions.* New York: Dover, pp. 19–21, 1965.

Madachy, J. S. *Madachy's Mathematical Recreations.* New York: Dover, p. 167, 1979.

Sloane, N. J. A. Sequences A000166/M1937 in "An On-Line Version of the Encyclopedia of Integer Sequences."

Sloane, N. J. A. and Plouffe, S. Extended entry in *The Encyclopedia of Integer Sequences.* San Diego: Academic Press, 1995.

Stanley, R. P. *Enumerative Combinatorics, Vol. 1.* Cambridge, England: Cambridge University Press, p. 67, 1997.

Subfield

If a subset S of the elements of a FIELD F satisfies the FIELD AXIOMS with the same operations of F, then S is called a subfield of F. Let F be a FINITE FIELD of order p^n, then there exists a subfield of ORDER p^m for PRIME p IFF m DIVIDES n.

see also FIELD, SUBMANIFOLD, SUBSPACE

Subgraph

A GRAPH G' whose VERTICES and EDGES form subsets of the VERTICES and EDGES of a given GRAPH G. If G' is a subgraph of G, then G is said to be a SUPERGRAPH of G'.

see also GRAPH (GRAPH THEORY), SUPERGRAPH

Subgroup

A subset of GROUP elements which satisfies the four GROUP requirements. The ORDER of any subgroup of a GROUP ORDER h must be a DIVISOR of h.

see also CARTAN SUBGROUP, COMPOSITION SERIES, FITTING SUBGROUP, GROUP

Sublime Number

Let $\tau(n)$ and $\sigma(n)$ denote the number and sum of the divisors of n, respectively (i.e., the zeroth- and first-order DIVISOR FUNCTIONS). A number N is called sublime if $\tau(N)$ and $\sigma(N)$ are both PERFECT NUMBERS. The only two known sublime numbers are 12 and

$$6086555670238378989670371734243169 6\cdots$$
$$\cdots 22657830773351885970528324860512791691264.$$

It is not known if any ODD sublime number exists.

see also DIVISOR FUNCTION, PERFECT NUMBER

Submanifold

A C^∞ (infinitely differentiable) MANIFOLD is said to be a submanifold of a C^∞ MANIFOLD M' if M is a SUBSET of M' and the IDENTITY MAP of M into M' is an embedding.

see also MANIFOLD, SUBFIELD, SUBSPACE

Submatrix

An $p \times q$ submatrix of an $m \times n$ MATRIX (with $p \leq m$, $n \leq q$) is a $p \times q$ MATRIX formed by taking a block of the entries of this size from the original matrix.

see also MATRIX

Subnormal

L is a subnormal SUBGROUP of H if there is a a "normal series" (in the sense of Jordan-Holder) from L to H.

Subordinate Norm

see NATURAL NORM

Subscript

A quantity displayed below the normal line of text (and generally in a smaller point size), as the "i" in a_i, is called a subscript. Subscripts are commonly used to indicate indices (a_{ij} is the entry in the ith row and jth column of a MATRIX A), partial differentiation (y_x is an abbreviation for $\partial y/\partial x$), and a host of other operations and notations in mathematics.

see also SUPERSCRIPT

Subsequence

A subsequence of a SEQUENCE $S = \{x_i\}_{i=1}^n$ is a derived sequence $\{y_i\}_{i=1}^N = \{x_{i+j}\}$ for some $j \geq 0$ and $N \leq n - j$. More generally, the word subsequence is sometimes used to mean a sequence derived from a sequence S by discarding some of its terms.

see also LOWER-TRIMMED SUBSEQUENCE, UPPER-TRIMMED SUBSEQUENCE

Subset

A portion of a SET. B is a subset of A (written $B \subseteq A$) IFF every member of B is a member of A. If B is a PROPER SUBSET of A (i.e., a subset other than the set itself), this is written $B \subset A$.

A SET of n elements has 2^n subsets (including the set itself and the EMPTY SET). For sets of $n = 1, 2, \ldots$ elements, the numbers of subsets are therefore 2, 4, 8, 16, 32, 64, ... (Sloane's A000079). For example, the set $\{1\}$ has the two subsets $\varnothing$ and $\{1\}$. Similarly, the set $\{1,2\}$ has subsets $\varnothing$ (the EMPTY SET, $\{1\}$, $\{2\}$, and $\{1,2\}$.

see also EMPTY SET, IMPLIES, k-SUBSET, PROPER SUBSET, SUPERSET, VENN DIAGRAM

References

Courant, R. and Robbins, H. *What is Mathematics?: An Elementary Approach to Ideas and Methods, 2nd ed.* Oxford, England: Oxford University Press, p. 109, 1996.

Ruskey, F. "Information of Subsets of a Set." http://sue.csc.uvic.ca/~cos/inf/comb/SubsetInfo.html.

Sloane, N. J. A. Sequence A000079/M1129 in "An On-Line Version of the Encyclopedia of Integer Sequences."

Subspace

Let $\mathbb{V}$ be a REAL VECTOR SPACE (e.g., the real continuous functions $C(I)$ on a CLOSED INTERVAL I, 2-D EUCLIDEAN SPACE $\mathbb{R}^2$, the twice differentiable real functions $C^{(2)}(I)$ on I, etc.). Then $\mathbb{W}$ is a real SUBSPACE of $\mathbb{V}$ if $\mathbb{W}$ is a SUBSET of $\mathbb{V}$ and, for every $\mathbf{w}_1, \mathbf{w}_2 \in \mathbb{W}$ and $t \in \mathbb{R}$ (the REALS), $\mathbf{w}_1 + \mathbf{w}_2 \in \mathbb{W}$ and $t\mathbf{w}_1 \in \mathbb{W}$. Let (H) be a homogeneous system of linear equations in $x_1, \ldots, x_n$. Then the SUBSET S of $\mathbb{R}^n$ which consists of all solutions of the system (H) is a subspace of $\mathbb{R}^n$.

More generally, let F_q be a FIELD with $q = p^\alpha$, where p is PRIME, and let $F_{q,n}$ denote the n-D VECTOR SPACE over F_q. The number of k-D linear subspaces of $F_{q,n}$ is

$$N(F_{q,n}) = \binom{n}{k}_q,$$

where this is the q-BINOMIAL COEFFICIENT (Aigner 1979, Exton 1983). The asymptotic limit is

$$N(F_{q,n}) = \begin{cases} c_e q^{n^2/4}[1+o(1)] & \text{for } n \text{ even} \\ c_o q^{n^2/4}[1+o(1)] & \text{for } n \text{ odd,} \end{cases}$$

where

$$c_e = \frac{\sum_{k=-\infty}^{\infty} q^{-k^2}}{\prod_{j=1}^{\infty}(1-q^{-j})}$$

$$c_o = \frac{\sum_{k=-\infty}^{\infty} q^{-(k+1/2)^2}}{\prod_{j=1}^{\infty}(1-q^{-j})}$$

(Finch). The case $q = 2$ gives the q-ANALOG of the WALLIS FORMULA.

see also q-BINOMIAL COEFFICIENT, SUBFIELD, SUBMANIFOLD

References

Aigner, M. *Combinatorial Theory.* New York: Springer-Verlag, 1979.

Exton, H. *q-Hypergeometric Functions and Applications.* New York: Halstead Press, 1983.

Finch, S. "Favorite Mathematical Constants." http://www.mathsoft.com/asolve/constant/dig/dig.html.

Substitution Group

see PERMUTATION GROUP

Substitution Map

A MAP which uses a set of rules to transform elements of a sequence into a new sequence using a set of rules which "translate" from the original sequence to its transformation. For example, the substitution map $\{1 \to 0, 0 \to 11\}$ would take 10 to 011.

see also GOLDEN RATIO, MORSE-THUE SEQUENCE, STRING REWRITING, THUE CONSTANT

Subtend

Given a geometric object O in the PLANE and a point P, let A be the ANGLE from one edge of O to the other with VERTEX at P. Then O is said to subtend an ANGLE A from P.

see also ANGLE, VERTEX ANGLE

Subtraction

Subtraction is the operation of taking the DIFFERENCE $x - y$ of two numbers x and y. Here, the symbol between the x and y is called the MINUS SIGN and $x - y$ is read "x MINUS y."

see also ADDITION, DIVISION, MINUS, MINUS SIGN, MULTIPLICATION

Succeeds

The relationship x succeeds (or FOLLOWS) y is written $x \succ y$. The relation x succeeds or is equal to y is written $x \succeq y$.

see also PRECEDES

Successes

see DIFFERENCE OF SUCCESSES

Sufficient

A CONDITION which, if true, guarantees that a result is also true. (However, the result may also be true if the CONDITION is not met.) If a CONDITION is both NECESSARY and SUFFICIENT, then the result is said to be true IFF ("if and only if") the CONDITION holds.

For example, the condition that a decimal number n end in the DIGIT 2 is a sufficient but not NECESSARY condition that n be EVEN.

see also IFF, IMPLIES, NECESSARY

Suitable Number

see IDONEAL NUMBER

Sum

A sum is the result of an ADDITION. For example, adding 1, 2, 3, and 4 gives the sum 10, written

$$1+2+3+4=10. \qquad (1)$$

The numbers being summed are called ADDENDS, or sometimes SUMMANDS. The summation operation can also be indicated using a capital sigma with upper and lower limits written above and below, and the index indicated below. For example, the above sum could be written

$$\sum_{k=1}^{4} k = 10. \qquad (2)$$

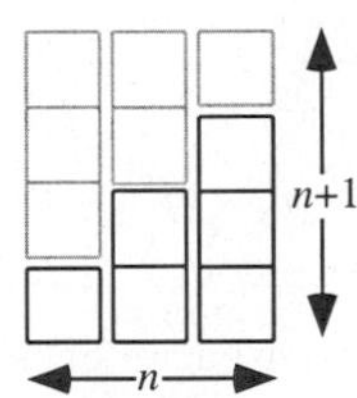

A simple graphical proof of the sum $\sum_{k=1}^{n} k = n(n+1)/2$ can also be given. Construct a sequence of stacks of boxes, each 1 unit across and k units high, where $k = 1$, 2, ..., n. Now add a rotated copy on top, as in the above figure. Note that the resulting figure has WIDTH n and HEIGHT $n+1$, and so has AREA $n(n+1)$. The desired sum is half this, so the AREA of the boxes in the sum is $n(n+1)/2$. Since the boxes are of unit width, this is also the value of the sum.

The sum can also be computed using the first EULER-MACLAURIN INTEGRATION FORMULA

$$\sum_{k=1}^{n} f(k) = \int_1^n f(x)\,dx + \tfrac{1}{2}f(1) + \tfrac{1}{2}f(n) + \tfrac{1}{2!}B_2[f'(n) - f'(1)] + \ldots \qquad (3)$$

with $f(k) = k$. Then

$$\sum_{k=1}^{n} k = \int_1^n x\,dx + \tfrac{1}{2}\cdot 1 + \tfrac{1}{2}\cdot n + \tfrac{1}{6}(1-1) + \ldots = \tfrac{1}{2}(n^2-1) - \tfrac{1}{2} + h + \tfrac{1}{2}n = \tfrac{1}{2}n(n+1). \qquad (4)$$

The general finite sum of integral POWERS can be given by the expression

$$\sum_{k=1}^{n} k^p = \frac{(B+n+1)^{[p+1]} - B^{[p+1]}}{p+1}, \qquad (5)$$

where the NOTATION $B^{[k]}$ means the quantity in question is raised to the appropriate POWER k and all terms of the form B^m are replaced with the corresponding BERNOULLI NUMBERS B_m. It is also true that the COEFFICIENTS of the terms in such an expansion sum to 1, as stated by Bernoulli without proof (Boyer 1943).

An analytic solution for a sum of POWERS of integers is

$$\sum_{k=1}^{n} k^p = \zeta(-p) - \zeta(-p, 1+n), \qquad (6)$$

where $\zeta(z)$ is the RIEMANN ZETA FUNCTION and $\zeta(z;a)$ is the HURWITZ ZETA FUNCTION. For the special case of p a POSITIVE integer, FAULHABER'S FORMULA gives the SUM explicitly as

$$\sum_{k=1}^{n} k^p = \frac{1}{p+1}\sum_{k=1}^{p+1}(-1)^{\delta_{kp}}\binom{p+1}{k}B_{p+1-k}n^k, \qquad (7)$$

where δ_{kp} is the KRONECKER DELTA, $\binom{n}{k}$ is a BINOMIAL COEFFICIENT, and B_k is a BERNOULLI NUMBER. Written explicitly in terms of a sum of POWERS,

$$\sum_{k=1}^{n} k^p = \frac{B_k p!}{k!(p-k+1)!}n^{p-k+1}. \qquad (8)$$

Computing the sums for $p = 1, \ldots, 10$ gives

$$\sum_{k=1}^{n} k = \tfrac{1}{2}(n^2+n) \qquad (9)$$

$$\sum_{k=1}^{n} k^2 = \tfrac{1}{6}(2n^3+3n^2+n) \qquad (10)$$

$$\sum_{k=1}^{n} k^3 = \tfrac{1}{4}(n^4+2n^3+n^2) \qquad (11)$$

$$\sum_{k=1}^{n} k^4 = \tfrac{1}{30}(6n^5+15n^4+10n^3-n) \qquad (12)$$

$$\sum_{k=1}^{n} k^5 = \tfrac{1}{12}(2n^6+6n^5+5n^4-n^2) \qquad (13)$$

$$\sum_{k=1}^{n} k^6 = \tfrac{1}{42}(6n^7 + 21n^6 + 21n^5 - 7n^3 + n) \quad (14)$$

$$\sum_{k=1}^{n} k^7 = \tfrac{1}{24}(3n^8 + 12n^7 + 14n^6 - 7n^4 + 2n^2) \quad (15)$$

$$\sum_{k=1}^{n} k^8 = \tfrac{1}{90}(10n^9 + 45n^8 + 60n^7 - 42n^5 + 20n^3 - 3n) \quad (16)$$

$$\sum_{k=1}^{n} k^9 = \tfrac{1}{20}(2n^{10} + 10n^9 + 15n^8 - 14n^6 + 10n^4 - 3n^2) \quad (17)$$

$$\sum_{k=1}^{n} k^{10} = \tfrac{1}{66}(6n^{11} + 33n^{10} + 55n^9 - 66n^7 + 66n^5 - 33n^3 + 5n). \quad (18)$$

Factoring the above equations results in

$$\sum_{k=1}^{n} k = \tfrac{1}{2}n(n+1) \quad (19)$$

$$\sum_{k=1}^{n} k^2 = \tfrac{1}{6}n(n+1)(2n+1) \quad (20)$$

$$\sum_{k=1}^{n} k^3 = \tfrac{1}{4}n^2(n+1)^2 \quad (21)$$

$$\sum_{k=1}^{n} k^4 = \tfrac{1}{30}n(n+1)(2n+1)(3n^2+3n-1) \quad (22)$$

$$\sum_{k=1}^{n} k^5 = \tfrac{1}{12}n^2(n+1)^2(2n^2+2n-1) \quad (23)$$

$$\sum_{k=1}^{n} k^6 = \tfrac{1}{42}n(n+1)(2n+1)(3n^4+6n^3-3n+1) \quad (24)$$

$$\sum_{k=1}^{n} k^7 = \tfrac{1}{24}n^2(n+1)^2(3n^4+6n^3-n^2-4n+2) \quad (25)$$

$$\sum_{k=1}^{n} k^8 = \tfrac{1}{90}n(n+1)(2n+1)(5n^6+15n^5+5n^4 - 15n^3 - n^2 + 9n - 3) \quad (26)$$

$$\sum_{k=1}^{n} k^9 = \tfrac{1}{20}n^2(n+1)^2(n^2+n-1) \times (2n^4+4n^3-n^2-3n+3) \quad (27)$$

$$\sum_{k=1}^{n} k^{10} = \tfrac{1}{66}n(n+1)(2n+1)(n^2+n-1) \times (3n^6+9n^5+2n^4-11n^3+3n^2+10n-5). \quad (28)$$

From the above, note the interesting identity

$$\sum_{k=1}^{n} k^3 = \left(\sum_{k=1}^{n} k\right)^2. \quad (29)$$

Sums of the following type can also be done analytically.

$$\left(\sum_{k=0}^{\infty} x^k\right)^2 = \sum_{n=0}^{\infty}\left(\sum_{k=0}^{n} 1\right)x^n = \sum_{n=0}^{\infty}(n+1)x^n \quad (30)$$

$$\left(\sum_{k=0}^{\infty} x^k\right)^3 = \sum_{n=0}^{\infty}\left(\sum_{k=0}^{n} k\right)x^n = \frac{1}{2}\sum_{n=0}^{\infty}(n+1)(n+2)x^n \quad (31)$$

$$\begin{aligned}\left(\sum_{k=0}^{\infty} x^k\right)^4 &= \sum_{n=0}^{\infty}\left[\sum_{k=0}^{n}\tfrac{1}{2}(k+1)(k+2)\right]x^n \\ &= \frac{1}{2}\sum_{n=0}^{\infty}\left(\sum_{k=0}^{n} k^2+3k+2\right)x^n \\ &= \frac{1}{2}\sum_{n=0}^{\infty}[\tfrac{1}{6}n(n+1)(2n+1) + 3\tfrac{1}{2}n(n+1) + 2(n+1)]x^n \\ &= \frac{1}{12}\sum_{n=0}^{\infty}(n+1)[n(2n+1)+9n+12]x^n \\ &= \frac{1}{12}\sum_{n=0}^{\infty}(n+1)(2n^2+10n+12)x^n \\ &= \frac{1}{6}\sum_{n=0}^{\infty}(n+1)(n+2)(n+3)x^n. \end{aligned} \quad (32)$$

By INDUCTION, the sum for an arbitrary POWER p is

$$\left(\sum_{k=0}^{\infty} x^k\right)^p = \frac{1}{(p-1)!}\sum_{n=0}^{\infty}\frac{(n+p-1)!}{n!}x^n. \quad (33)$$

Other analytic sums include

$$\left(\sum_{k=0}^{n} x^k\right)^2 = \frac{1}{(p-1)!}\sum_{k=0}^{2n}\frac{(n-|n-k|+p-1)!}{(n-|n-k|)!}x^k \quad (34)$$

$$\left(\sum_{n=0}^{\infty} a_n x^n\right)^2 = \sum_{n=0}^{\infty} {a_n}^2 x^{2n} + 2\sum_{\substack{n=1\\ i+j=n\\ i<j}}^{\infty} a_i a_j x^n. \quad (35)$$

$$\begin{aligned}\sum xy &= x_1y_1 + x_1y_2 + \ldots + x_2y_1 + x_2y_2 + \ldots \\ &= (x_1+x_2+\ldots)y_1 + (x_1+x_2+\ldots)y_2 \\ &= \left(\sum x\right)(y_1+y_2+\ldots) = \sum x \sum y, \end{aligned} \quad (36)$$

so

$$\sum_{i=1}^{m}\sum_{j=1}^{n} x_i y_j = \left(\sum_{i=1}^{m} x_i\right)\left(\sum_{j=1}^{n} y_j\right). \quad (37)$$

$$\sum_{j=0}^{n} jx^j = \frac{nx^{n+2}-(n+1)x^{n+1}+x}{(x-1)^2} \quad (38)$$

$$\sum_{j=1}^{n} \frac{{x_j}^r}{\prod_{\substack{k=1\\k\neq j}}^{n}(x_j-x_k)} = \begin{cases} 0 & \text{for } 0 \leq r < n-1 \\ 1 & \text{for } r = n-1 \\ \sum_{j=1}^{n} x_j & \text{for } r = n \end{cases} \quad (39)$$

$$\sum_{k=1}^{n} \frac{\prod_{\substack{r=1\\r\neq k}}^{n}(x+k-r)}{\prod_{\substack{r=1\\r\neq k}}^{n}(k-r)} = 1 \quad (40)$$

$$(n+1)\sum_{m=1}^{n} m^k = \sum_{m=1}^{n}\left[m^{k+1} + \sum_{p=1}^{n}\left(\sum_{m=1}^{p} m^k\right)\right]. \quad (41)$$

To minimize the sum of a set of squares of numbers $\{x_i\}$ about a given number x_0

$$S \equiv \sum_i (x_i - x_0)^2 = \sum_i {x_i}^2 - 2x_0 \sum x_i + N{x_0}^2, \quad (42)$$

take the DERIVATIVE.

$$\frac{d}{dx_0} S = -2\sum_i x_i + 2Nx_0 = 0. \quad (43)$$

Solving for x_0 gives

$$x_0 \equiv \bar{x} = \frac{1}{N}\sum_i x_i, \quad (44)$$

so S is maximized when x_0 is set to the MEAN.

see also ARITHMETIC SERIES, BERNOULLI NUMBER, CLARK'S TRIANGLE, CONVERGENCE IMPROVEMENT, DEDEKIND SUM, DOUBLE SUM, EULER SUM, FACTORIAL SUM, FAULHABER'S FORMULA, GABRIEL'S STAIRCASE, GAUSSIAN SUM, GEOMETRIC SERIES, GOSPER'S METHOD, HURWITZ ZETA FUNCTION, INFINITE PRODUCT, KLOOSTERMAN'S SUM, LEGENDRE SUM, LERCH TRANSCENDENT, PASCAL'S TRIANGLE, PRODUCT, RAMANUJAN'S SUM, RIEMANN ZETA FUNCTION, WHITNEY SUM

References

Boyer, C. B. "Pascal's Formula for the Sums of the Powers of the Integers." *Scripta Math.* **9**, 237–244, 1943.

Courant, R. and Robbins, H. "The Sum of the First n Squares." §1.4 in *What is Mathematics?: An Elementary Approach to Ideas and Methods, 2nd ed.* Oxford, England: Oxford University Press, pp. 14–15, 1996.

Petkovšek, M.; Wilf, H. S.; and Zeilberger, D. *A=B*. Wellesley, MA: A. K. Peters, 1996.

Sum-Product Number

A sum-product number is a number n such that the sum of n's digits times the product of n's digit is n itself, for example

$$135 = (1+3+5)(1 \cdot 3 \cdot 5).$$

The only sum-product numbers less than 10^7 are 1, 135, and 144.

see also AMENABLE NUMBER

Sum Rule

$$\frac{d}{dx}[f(x)+g(x)] = f'(x)+g'(x),$$

where d/dx denotes a derivative and $f'(x)$ and $g'(x)$ are the derivatives of f and g, respectively.

see also DERIVATIVE

Summand

see ADDEND

Summatory Function

For an discrete function $f(n)$, the summatory function is defined by

$$F(n) \equiv \sum_{k \in D}^{n} f(k),$$

where D is the DOMAIN of the function.

see also DIVISOR FUNCTION, MANGOLDT FUNCTION, MERTENS FUNCTION, RUDIN-SHAPIRO SEQUENCE, TAU FUNCTION, TOTIENT FUNCTION

Sup

see SUPREMUM, SUPREMUM LIMIT

Super-3 Number

An INTEGER n such that $3n^3$ contains three consecutive 3s in its DECIMAL representation. The first few super-3 numbers are 261, 462, 471, 481, 558, 753, 1036, ... (Sloane's A014569). A. Anderson has conjectured that all numbers ending in 471, 4710, or 47100 are super-3 (Pickover 1995).

For a digit d, super-3 numbers can be generalized to super-d numbers n such that dn^d contains d ds in its DECIMAL representation. The following table gives the first few super-d numbers for small d.

d	Sloane	Super-d numbers
2	032743	19, 31, 69, 81, 105, 106, 107, 119, ...
3	014569	261, 462, 471, 481, 558, 753, 1036, ...
4	032744	1168, 4972, 7423, 7752, 8431, 10267, ...
5	032745	4602, 5517, 7539, 12955, 14555, 20137, ...
6	032746	27257, 272570, 302693, 323576, ...
7	032747	140997, 490996, 1184321, 1259609, ...
8	032748	185423, 641519, 1551728, 1854230, ...
9	032749	17546133, 32613656, 93568867, ...

References
Pickover, C. A. *Keys to Infinity.* New York: Wiley, p. 7, 1995.
Sloane, N. J. A. Sequence A014569 in "An On-Line Version of the Encyclopedia of Integer Sequences."

Super Catalan Number

While the CATALAN NUMBERS are the number of p-GOOD PATHS from (n, n) to (0,0) which do not cross the diagonal line, the super Catalan numbers count the number of LATTICE PATHS with diagonal steps from (n, n) to (0,0) which do not touch the diagonal line $x = y$.

The super Catalan numbers are given by the RECURRENCE RELATION

$$S(n) = \frac{3(2n-3)S(n-1) - (n-3)S(n-2)}{n}$$

(Comtet 1974), with $S(1) = S(2) = 1$. (Note that the expression in Vardi (1991, p. 198) contains *two* errors.) A closed form expression in terms of LEGENDRE POLYNOMIALS $P_n(x)$ is

$$S(n) = \frac{3P_{n-1}(3) - P_{n-2}(3)}{4n}$$

(Vardi 1991, p. 199). The first few super Catalan numbers are 1, 1, 3, 11, 45, 197, ... (Sloane's A001003).

see also CATALAN NUMBER

References
Comtet, L. *Advanced Combinatorics.* Dordrecht, Netherlands: Reidel, p. 56, 1974.
Graham, R. L.; Knuth, D. E.; and Patashnik, O. Exercise 7.50 in *Concrete Mathematics: A Foundation for Computer Science, 2nd ed.* Reading, MA: Addison-Wesley, 1994.
Motzkin, T. "Relations Between Hypersurface Cross Ratios and a Combinatorial Formula for Partitions of a Polygon for Permanent Preponderance and for Non-Associative Products." *Bull. Amer. Math. Soc.* **54**, 352–360, 1948.
Schröder, E. "Vier combinatorische Probleme." *Z. Math. Phys.* **15**, 361–376, 1870.
Sloane, N. J. A. Sequence A001003/M2898 in "An On-Line Version of the Encyclopedia of Integer Sequences."
Vardi, I. *Computational Recreations in Mathematica.* Reading, MA: Addison-Wesley, pp. 198–199, 1991.

Super-Poulet Number

A POULET NUMBER whose DIVISORS d all satisfy $d|2^d - 2$.

see also POULET NUMBER

Superabundant Number

see HIGHLY COMPOSITE NUMBER

Superegg

A superegg is a solid described by the equation

$$\left|\sqrt{\frac{x^2+y^2}{a^2}}\right|^n + \left|\frac{z}{b}\right|^n = 1.$$

Supereggs will balance on either end for any a, b, and n.

see also EGG, SUPERELLIPSE

References
Gardner, M. "Pier Hein's Superellipse." Ch. 18 in *Mathematical Carnival: A New Round-Up of Tantalizers and Puzzles from Scientific American.* New York: Vintage, 1977.

Superellipse

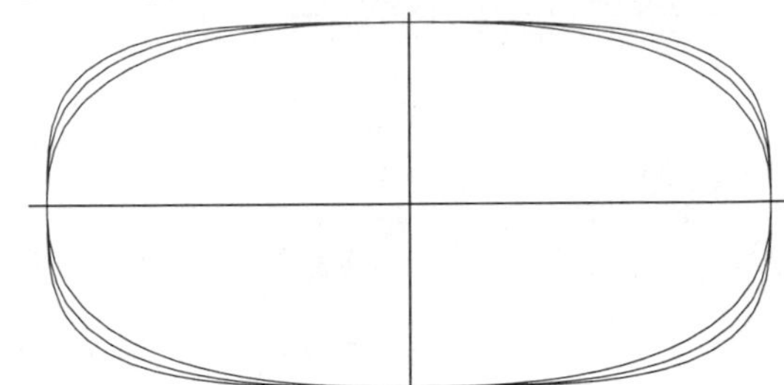

A curve of the form

$$\left|\frac{x}{a}\right|^r + \left|\frac{y}{b}\right|^r = 1.$$

where $r > 2$. "The" superellipse is sometimes taken as the curve of the above form with $r = 5/2$. Superellipses with $a = b$ are also known as LAMÉ CURVES. The above curves are for $a = 1$, $b = 2$, and $r = 2.5$, 3.0, and 3.5.

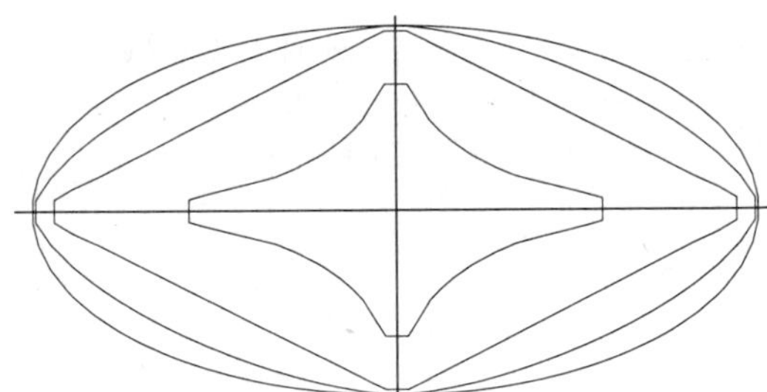

A degenerate superellipse is a superellipse with $r \leq 2$. The above curves are for $a = 1$, $b = 2$, and $r = 0.5$, 1.0, 1.5, and 2.0.

see also ELLIPSE, LAMÉ CURVE, SUPEREGG

References
Gardner, M. "Piet Hein's Superellipse." Ch. 18 in *Mathematical Carnival: A New Round-Up of Tantalizers and Puzzles from Scientific American.* New York: Vintage, 1977.

Superfactorial
The superfactorial of n is defined by Pickover (1995) as

$$n\$ \equiv \underbrace{n!^{n!^{\cdot^{\cdot^{\cdot^{n!}}}}}}_{n!}.$$

The first two values are 1 and 4, but subsequently grow so rapidly that 3$ already has a huge number of digits.

Sloane and Plouffe (1995) define the superfactorial by

$$n\$ \equiv \prod_{i=1}^{n} i!,$$

which is equivalent to the integral values of the G-FUNCTION. The first few values are 1, 1, 2, 12, 288, 34560, ... (Sloane's A000178).

see also FACTORIAL, G-FUNCTION, LARGE NUMBER, SUBFACTORIAL

References
Pickover, C. A. *Keys to Infinity.* New York: Wiley, p. 102, 1995.
Sloane, N. J. A. Sequence A000178/M2049 in "An On-Line Version of the Encyclopedia of Integer Sequences."

Supergraph
If G' is a SUBGRAPH of G, then G is said to be a supergraph of G'.

see also GRAPH (GRAPH THEORY), SUBGRAPH

Supernormal
Trials for which the LEXIS RATIO

$$L \equiv \frac{\sigma}{\sigma_B},$$

satisfies $L > 1$, where σ is the VARIANCE in a set of s LEXIS TRIALS and σ_B is the VARIANCE assuming BERNOULLI TRIALS.

see also BERNOULLI TRIAL, LEXIS TRIALS, SUBNORMAL

Superperfect Number
A number n such that

$$\sigma^2(n) = \sigma(\sigma(n)) = 2n,$$

where $\sigma(n)$ is the DIVISOR FUNCTION. EVEN superperfect numbers are just 2^{p-1}, where $M_p = 2^p - 1$ is a MERSENNE PRIME. If any ODD superperfect numbers exist, they are SQUARE NUMBERS and either n or $\sigma(n)$ is DIVISIBLE by at least three distinct PRIMES.

More generally, an m-superperfect number is a number for which $\sigma^m(n) = 2n$. For $m \geq 3$, there are no EVEN m-superperfect numbers.

see also MERSENNE NUMBER

References
Guy, R. K. "Superperfect Numbers." §B9 in *Unsolved Problems in Number Theory, 2nd ed.* New York: Springer-Verlag, pp. 65–66, 1994.
Kanold, H.-J. "Über 'Super Perfect Numbers.'" *Elem. Math.* **24**, 61–62, 1969.
Lord, G. "Even Perfect and Superperfect Numbers." *Elem. Math.* **30**, 87–88, 1975.
Suryanarayana, D. "Super Perfect Numbers." *Elem. Math.* **20**, 16–17, 1969.
Suryanarayana, D. "There is No Odd Super Perfect Number of the Form $p^{2\alpha}$." *Elem. Math.* **24**, 148–150, 1973.

Superposition Principle
For a linear homogeneous ORDINARY DIFFERENTIAL EQUATION, if $y_1(x)$ and $y_2(x)$ are solutions, then so is $y_1(x) + y_2(x)$.

Superregular Graph
For a VERTEX x of a GRAPH, let Γ_x and Δ_x denote the SUBGRAPHS of $\Gamma - x$ induced by the VERTICES adjacent to and nonadjacent to x, respectively. The empty graph is defined to be superregular, and Γ is said to be superregular if Γ is a REGULAR GRAPH and both Γ_x and Δ_x are superregular for all x.

The superregular graphs are precisely C_5, mK_n ($m, n \geq 1$), G_n ($n \geq 1$), and the complements of these graphs, where C_n is a CYCLIC GRAPH, K_n is a COMPLETE GRAPH and mK_n is m disjoint copies of K_n, and G_n is the Cartesian product of K_n with itself (the graph whose VERTEX set consists of n^2 VERTICES arranged in an $n \times n$ square with two VERTICES adjacent IFF they are in the same row or column).

see also COMPLETE GRAPH, CYCLIC GRAPH, REGULAR GRAPH

References
Vince, A. "The Superregular Graph." Problem 6617. *Amer. Math. Monthly* **103**, 600–603, 1996.
West, D. B. "The Superregular Graphs." *J. Graph Th.* **23**, 289–295, 1996.

Superscript
A quantity displayed above the normal line of text (and generally in a smaller point size), as the "i" in x^i, is called a superscript. Superscripts are commonly used to indicate raising to a POWER (x^3 means $x \cdot x \cdot x$ or x CUBED), multiple differentiation ($f^{(3)}(x)$ is an abbreviation for $f'''(x) = d^3f/dx^3$), and a host of other operations and notations in mathematics.

see also SUBSCRIPT

Superset
A SET containing all elements of a smaller SET. If B is a SUBSET of A, then A is a superset of B, written $A \supseteq B$. If A is a PROPER SUPERSET of B, this is written $A \supset B$.

see also PROPER SUBSET, PROPER SUPERSET, SUBSET

Supplementary Angle

Two ANGLES α and $\pi - \alpha$ which together form a STRAIGHT ANGLE are said to be supplementary.

see also ANGLE, COMPLEMENTARY ANGLE, DIGON, STRAIGHT ANGLE

Support

The CLOSURE of the SET of arguments of a FUNCTION f for which f is not zero.

see also CLOSURE

Support Function

Let M be an oriented REGULAR SURFACE in $\mathbb{R}^3$ with normal $\mathbf{N}$. Then the support function of M is the function $h : M \to \mathbb{R}$ defined by

$$h(\mathbf{p}) = \mathbf{p} \cdot \mathbf{N}(\mathbf{p}).$$

References

Gray, A. *Modern Differential Geometry of Curves and Surfaces.* Boca Raton, FL: CRC Press, p. 293, 1993.

Supremum

The supremum of a set is the least upper bound of the set. It is denoted

$$\sup_S .$$

On the REAL LINE, the supremum of a set is the same as the supremum of its CLOSURE.

see also INFIMUM, SUPREMUM LIMIT

Supremum Limit

The limit supremum is used for sequences and nets (as opposed to sets) and is denoted

$$\limsup_S .$$

see also SUPREMUM

Surd

An archaic term for a SQUARE ROOT.

see also QUADRATIC SURD, SQUARE ROOT

Surface

The word "surface" is an important term in mathematics and is used in many ways. The most common and straightforward use of the word is to denote a 2-D SUBMANIFOLD of 3-D EUCLIDEAN SPACE. Surfaces can range from the very complicated (e.g., FRACTALS such as the MANDELBROT SET) to the very simple (such as the PLANE). More generally, the word "surface" can be used to denote an $(n-1)$-D SUBMANIFOLD of an n-D MANIFOLD, or in general, any co-dimension 1 subobject in an object (like a BANACH SPACE or an infinite-dimensional MANIFOLD).

Even simple surfaces can display surprisingly counterintuitive properties. For example, the SURFACE OF REVOLUTION of $y = 1/x$ around the x-AXIS for $x \geq 1$ (called GABRIEL'S HORN) has FINITE VOLUME but INFINITE SURFACE AREA.

see also ALGEBRAIC SURFACE, BARTH DECIC, BARTH SEXTIC, BERNSTEIN MINIMAL SURFACE THEOREM, BOHEMIAN DOME, BOY SURFACE, CATALAN'S SURFACE, CAYLEY'S RULED SURFACE, CHAIR, CLEBSCH DIAGONAL CUBIC, COMPACT SURFACE, CONE, CONICAL WEDGE, CONOCUNEUS OF WALLIS, CORK PLUG, CORKSCREW SURFACE, CORNUCOPIA, COSTA MINIMAL SURFACE, CROSS-CAP, CROSSED TROUGH, CUBIC SURFACE, CYCLIDE, CYLINDER, CYLINDROID, DARWIN-DE SITTER SPHEROID, DECIC SURFACE, DEL PEZZO SURFACE, DERVISH, DESMIC SURFACE, DEVELOPABLE SURFACE, DINI'S SURFACE, EIGHT SURFACE, ELLIPSOID, ELLIPTIC CONE, ELLIPTIC CYLINDER, ELLIPTIC HELICOID, ELLIPTIC HYPERBOLOID, ELLIPTIC PARABOLOID, ELLIPTIC TORUS, ENNEPER'S SURFACES, ENRIQUES SURFACES, ETRUSCAN VENUS SURFACE, FLAT SURFACE, FRESNEL'S ELASTICITY SURFACE, GABRIEL'S HORN, HANDKERCHIEF SURFACE, HELICOID, HENNEBERG'S MINIMAL SURFACE, HOFFMAN'S MINIMAL SURFACE, HORN CYCLIDE, HORN TORUS, HUNT'S SURFACE, HYPERBOLIC CYLINDER, HYPERBOLIC PARABOLOID, HYPERBOLOID, IDA SURFACE, IMMERSED MINIMAL SURFACE, KISS SURFACE, KLEIN BOTTLE, KUEN SURFACE, KUMMER SURFACE, LICHTENFELS SURFACE, MAEDER'S OWL MINIMAL SURFACE, MANIFOLD, MENN'S SURFACE, MINIMAL SURFACE, MITER SURFACE, MÖBIUS STRIP, MONGE'S FORM, MONKEY SADDLE, NONORIENTABLE SURFACE, NORDSTRAND'S WEIRD SURFACE, NURBS SURFACE, OBLATE SPHEROID, OCTIC SURFACE, ORIENTABLE SURFACE, PARABOLIC CYLINDER, PARABOLIC HORN CYCLIDE, PARABOLIC RING CYCLIDE, PARABOLIC SPINDLE CYCLIDE, PARABOLOID, PEANO SURFACE, PIRIFORM, PLANE, PLÜCKER'S CONOID, POLYHEDRON, PRISM, PRISMATOID, PROLATE SPHEROID, PSEUDOCROSSCAP, QUADRATIC SURFACE, QUARTIC SURFACE, QUINTIC SURFACE, REGULAR SURFACE, REMBS' SURFACES, RIEMANN SURFACE, RING CYCLIDE, RING TORUS, ROMAN SURFACE, RULED SURFACE, SCHERK'S MINIMAL SURFACES, SEIFERT SURFACE, SEXTIC SURFACE, SHOE SURFACE, SIEVERT'S SURFACE, SMOOTH SURFACE, SOLID, SPHERE, SPHEROID, SPINDLE CYCLIDE, SPINDLE TORUS, STEINBACH SCREW, STEINER SURFACE, SWALLOWTAIL CATASTROPHE, SYMMETROID, TANGLECUBE, TETRAHEDRAL SURFACE, TOGLIATTI SURFACE, TOOTH SURFACE, TRINOID, UNDULOID, VERONESE SURFACE, VERONESE VARIETY, WALLIS'S CONICAL EDGE, WAVE SURFACE, WEDGE, WHITNEY UMBRELLA

References

Endraß, S. "Home Page of S. Endraß." `http://www.mathematik.uni-mainz.de/~endrass/`.

Fischer, G. (Ed.). *Mathematical Models from the Collections of Universities and Museums.* Braunschweig, Germany: Vieweg, 1986.
Francis, G. K. *A Topological Picturebook.* New York: Springer-Verlag, 1987.
Geometry Center. "The Topological Zoo." http://www.geom.umn.edu/zoo/.
Gray, A. *Modern Differential Geometry of Curves and Surfaces.* Boca Raton, FL: CRC Press, 1993.
Hunt, B. "Algebraic Surfaces." http://www.mathematik.uni-kl.de/~wwwagag/Galerie.html.
Morgan, F. "What is a Surface?" *Amer. Math. Monthly* **103**, 369–376, 1996.
Nordstrand, T. "Gallery." http://www.uib.no/people/nfytn/mathgal.htm.
Nordstrand, T. "Surfaces." http://www.uib.no/people/nfytn/surfaces.htm.
von Seggern, D. *CRC Standard Curves and Surfaces.* Boca Raton, FL: CRC Press, 1993.
Wagon, S. "Surfaces." Ch. 3 in *Mathematica in Action.* New York: W. H. Freeman, pp. 67–91, 1991.
Yamaguchi, F. *Curves and Surfaces in Computer Aided Geometric Design.* New York: Springer-Verlag, 1988.

Surface Area

Surface area is the AREA of a given surface. Roughly speaking, it is the "amount" of a surface, and has units of distance squares. It is commonly denoted S for a surface in 3-D, or A for a region of the plane (in which case it is simply called "the" AREA).

If the surface is PARAMETERIZED using u and v, then

$$S = \int_S |\mathbf{T}_u \times \mathbf{T}_v|\,du\,dv, \tag{1}$$

where $\mathbf{T}_u$ and $\hat{\mathbf{T}}_v$ are tangent vectors and $\mathbf{a}\times\mathbf{b}$ is the CROSS PRODUCT.

The surface area given by rotating the curve $y = f(x)$ from $x = a$ to $x = b$ about the x-axis is

$$S = \int_b^a 2\pi f(x)\sqrt{1+[f'(x)]^2}\,dx. \tag{2}$$

If $z = f(x,y)$ is defined over a region R, then

$$S = \iint_R \sqrt{\left(\frac{\partial z}{\partial x}\right)^2 + \left(\frac{\partial z}{\partial y}\right)^2 + 1}\,dA, \tag{3}$$

where the integral is taken over the entire surface.

The following tables gives surface areas for some common SURFACES. In the first table, S denotes the lateral surface, and in the second, T denotes the total surface. In both tables, r denotes the RADIUS, h the height, p the base PERIMETER, and s the SLANT HEIGHT (Beyer 1987).

Surface	S
cone	$\pi r\sqrt{r^2+h^2}$
conical frustum	$\pi(R_1+R_2)\sqrt{(R_1-R_2)^2+h^2}$
cube	$6a^2$
cylinder	$2\pi rh$
lune	$2r^2\theta$
oblate spheroid	$2\pi a^2 + \frac{\pi b^2}{e}\ln\left(\frac{1+e}{1-e}\right)$
prolate spheroid	$2\pi b^2 + \frac{2\pi ab}{e}\sin^{-1} e$
pyramid	$\frac{1}{2}ps$
pyramidal frustum	$\frac{1}{2}ps$
sphere	$4\pi r^2$
torus	$4\pi^2 Rr$
zone	$2\pi rh$

Surface	T
cone	$\pi r(r+\sqrt{r^2+h^2})$
conical frustum	$\pi[{R_1}^2+{R_2}^2$ $+(R_1+R_2)\sqrt{(R_1-R_2)^2+h^2}\,]$
cylinder	$2\pi r(r+h)$

Even simple surfaces can display surprisingly counterintuitive properties. For instance, the surface of revolution of $y = 1/x$ around the x-AXIS for $x \geq 1$ is called GABRIEL'S HORN, and has FINITE VOLUME but INFINITE surface AREA.

see also AREA, SURFACE INTEGRAL, SURFACE OF REVOLUTION, VOLUME

References
Beyer, W. H. *CRC Standard Mathematical Tables, 28th ed.* Boca Raton, FL: CRC Press, pp. 127–132, 1987.

Surface Integral

For a SCALAR FUNCTION f over a surface parameterized by u and v, the surface integral is given by

$$\Phi = \int_S f\,da = \int_S f(u,v)\,|\mathbf{T}_u \times \mathbf{T}_v|\,du\,dv, \tag{1}$$

where $\mathbf{T}_u$ and $\hat{\mathbf{T}}_v$ are tangent vectors and $\mathbf{a}\times\mathbf{b}$ is the CROSS PRODUCT.

For a VECTOR FUNCTION over a surface, the surface integral is given by

$$\Phi = \int_S \mathbf{F}\cdot d\mathbf{a} = \int_S (\mathbf{F}\cdot\hat{\mathbf{n}})\,da \tag{2}$$

$$= \int_S f_x\,dy\,dz + f_y\,dz\,dx + f_z\,dx\,dy, \tag{3}$$

where $\mathbf{a}\cdot\mathbf{b}$ is a DOT PRODUCT and $\hat{\mathbf{n}}$ is a unit NORMAL VECTOR. If $z = f(x,y)$, then $d\mathbf{a}$ is given explicitly by

$$d\mathbf{a} = \pm\left(-\frac{\partial z}{\partial x}\hat{\mathbf{x}} - \frac{\partial z}{\partial y}\hat{\mathbf{y}} + \hat{\mathbf{z}}\right)dx\,dy. \tag{4}$$

If the surface is SURFACE PARAMETERIZED using u and v, then

$$\Phi = \int_S \mathbf{F}\cdot(\mathbf{T}_u \times \mathbf{T}_v)\,du\,dv. \tag{5}$$

see also SURFACE PARAMETERIZATION

Surface Parameterization

A surface in 3-SPACE can be parameterized by two variables (or coordinates) u and v such that

$$x = x(u,v) \tag{1}$$
$$y = y(u,v) \tag{2}$$
$$z = z(u,v). \tag{3}$$

If a surface is parameterized as above, then the tangent VECTORS

$$\mathbf{T}_u = \frac{\partial x}{\partial u}\hat{\mathbf{x}} + \frac{\partial y}{\partial u}\hat{\mathbf{y}} + \frac{\partial z}{\partial u}\hat{\mathbf{z}} \tag{4}$$

$$\mathbf{T}_v = \frac{\partial x}{\partial v}\hat{\mathbf{x}} + \frac{\partial y}{\partial v}\hat{\mathbf{y}} + \frac{\partial z}{\partial v}\hat{\mathbf{z}} \tag{5}$$

are useful in computing the SURFACE AREA and SURFACE INTEGRAL.

see also SMOOTH SURFACE, SURFACE AREA, SURFACE INTEGRAL

Surface of Revolution

A surface of revolution is a SURFACE generated by rotating a 2-D CURVE about an axis. The resulting surface therefore always has azimuthal symmetry. Examples of surfaces of revolution include the APPLE, CONE (excluding the base), CONICAL FRUSTUM (excluding the ends), CYLINDER (excluding the ends), DARWIN-DE SITTER SPHEROID, GABRIEL'S HORN, HYPERBOLOID, LEMON, OBLATE SPHEROID, PARABOLOID, PROLATE SPHEROID, PSEUDOSPHERE, SPHERE, SPHEROID, and TORUS (and its generalization, the TOROID).

The standard parameterization of a surface of revolution is given by

$$x(u,v) = \phi(v)\cos u \tag{1}$$
$$y(u,v) = \phi(v)\sin u \tag{2}$$
$$z(u,v) = \psi(v). \tag{3}$$

For a curve so parameterized, the first FUNDAMENTAL FORM has

$$E = \psi^2 \tag{4}$$
$$F = 0 \tag{5}$$
$$G = \phi'^2 + \psi'^2. \tag{6}$$

Wherever ϕ and $\phi'^2 + \psi'^2$ are nonzero, then the surface is regular and the second FUNDAMENTAL FORM has

$$e = -\frac{|\phi|\psi'}{\sqrt{\phi'^2+\psi'^2}} \tag{7}$$
$$f = 0 \tag{8}$$
$$g = \frac{\operatorname{sgn}(\phi)(\phi''\psi' - \phi'\psi'')}{\sqrt{\phi'^2+\psi'^2}}. \tag{9}$$

Furthermore, the unit NORMAL VECTOR is

$$\hat{\mathbf{N}}(u,v) = \frac{\operatorname{sgn}(\phi)}{\sqrt{\phi'^2+\psi'^2}}\begin{bmatrix}\phi'\cos u\\ \psi'\sin u\\ \phi'\end{bmatrix}, \tag{10}$$

and the PRINCIPAL CURVATURES are

$$\kappa_1 = \frac{g}{G} = \frac{\operatorname{sgn}(\phi)(\phi''\psi' - \phi'\psi'')}{(\phi'^2+\psi'^2)^{3/2}} \tag{11}$$

$$\kappa_2 = \frac{e}{E} = -\frac{\psi'}{|\phi|\sqrt{\phi'^2+\psi'^2}}. \tag{12}$$

The GAUSSIAN and MEAN CURVATURES are

$$K = \frac{-\psi'^2\phi'' + \phi'\psi'\psi''}{\phi(\phi'^2+\psi'^2)^2} \tag{13}$$

$$H = \frac{\phi(\phi''\psi' - \phi'\psi'') - \psi'(\phi'^2+\psi'^2)}{2|\phi|(\phi'^2+\psi'^2)^{3/2}} \tag{14}$$

(Gray 1993).

PAPPUS'S CENTROID THEOREM gives the VOLUME of a solid of rotation as the cross-sectional AREA times the distance traveled by the centroid as it is rotated.

CALCULUS OF VARIATIONS can be used to find the curve from a point (x_1, y_1) to a point (x_2, y_2) which, when revolved around the x-AXIS, yields a surface of smallest SURFACE AREA A (i.e., the MINIMAL SURFACE). This is equivalent to finding the MINIMAL SURFACE passing through two circular wire frames. The AREA element is

$$dA = 2\pi y\,ds = 2\pi y\sqrt{1+y'^2}\,dx, \tag{15}$$

so the SURFACE AREA is

$$A = 2\pi \int y\sqrt{1+y'^2}\,dx, \tag{16}$$

and the quantity we are minimizing is

$$f = y\sqrt{1+y'^2}. \tag{17}$$

This equation has $f_x = 0$, so we can use the BELTRAMI IDENTITY

$$f - y_x\frac{\partial f}{\partial y_x} = a \tag{18}$$

to obtain

$$y\sqrt{1+y'^2} - y'\frac{yy'}{\sqrt{1+y'^2}} = a \tag{19}$$

$$y(1+y'^2) - yy'^2 = a\sqrt{1+y'^2} \tag{20}$$

$$y = a\sqrt{1+y'^2} \tag{21}$$

$$\frac{y}{\sqrt{1+y'^2}} = a \tag{22}$$

$$\frac{y^2}{a} - 1 = y'^2 \tag{23}$$

$$\frac{dx}{dy} = \frac{1}{y'} = \frac{a}{\sqrt{y^2 - a^2}} \tag{24}$$

$$x = a \int \frac{dy}{\sqrt{y^2 - a^2}} = a\cosh^{-1}\left(\frac{y}{a}\right) + b \tag{25}$$

$$y = a\cosh\left(\frac{x-b}{a}\right), \tag{26}$$

which is called a CATENARY, and the surface generated by rotating it is called a CATENOID. The two constants a and b are determined from the two implicit equations

$$y_1 = a\cosh\left(\frac{x_1 - b}{a}\right) \tag{27}$$

$$y_2 = a\cosh\left(\frac{x_2 - b}{a}\right), \tag{28}$$

which cannot be solved analytically.

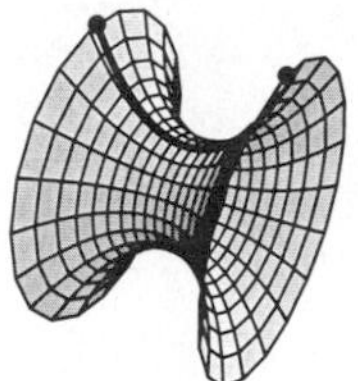

The general case is somewhat more complicated than this solution suggests. To see this, consider the MINIMAL SURFACE between two rings of equal RADIUS y_0. Without loss of generality, take the origin at the midpoint of the two rings. Then the two endpoints are located at $(-x_0, y_0)$ and (x_0, y_0), and

$$y_0 = a\cosh\left(\frac{-x_0 - b}{a}\right) = a\cosh\left(\frac{x_0 - b}{a}\right). \tag{29}$$

But $\cosh(-x) = \cosh(x)$, so

$$\cosh\left(\frac{-x_0 - b}{a}\right) = \cosh\left(\frac{-x_0 + b}{a}\right). \tag{30}$$

Inverting each side

$$-x_0 - b = -x_0 + b, \tag{31}$$

so $b = 0$ (as it must by symmetry, since we have chosen the origin between the two rings), and the equation of the MINIMAL SURFACE reduces to

$$y = a\cosh\left(\frac{x}{a}\right). \tag{32}$$

At the endpoints

$$y_0 = a\cosh\left(\frac{x_0}{a}\right), \tag{33}$$

but for certain values of x_0 and y_0, this equation has no solutions. The physical interpretation of this fact is that the surface breaks and forms circular disks in each ring to minimize AREA. CALCULUS OF VARIATIONS cannot be used to find such discontinuous solutions (known in this case as GOLDSCHMIDT SOLUTIONS). The minimal surfaces for several choices of endpoints are shown above. The first two cases are CATENOIDS, while the third case is a GOLDSCHMIDT SOLUTION.

To find the maximum value of x_0/y_0 at which CATENARY solutions can be obtained, let $p \equiv 1/a$. Then (31) gives

$$y_0 p = \cosh(px_0). \tag{34}$$

Now, denote the maximum value of x_0 as x_0^*. Then it will be true that $dx_0/dp = 0$. Take d/dp of (34),

$$y_0 = \sinh(px_0)\left(x_0 + p\frac{dx_0}{dp}\right). \tag{35}$$

Now set $dx_0/dp = 0$

$$y_0 = x_0 \sinh(px_0^*). \tag{36}$$

From (34),

$$py_0{}^* = \cosh(px_0{}^*). \tag{37}$$

Take (37) ÷ (36),

$$px_0^* = \coth(px_0^*). \tag{38}$$

Defining $u \equiv px_0{}^*$,

$$u = \coth u. \tag{39}$$

This has solution $u = 1.1996789403\ldots$. From (36), $y_0 p = \cosh u$. Divide this by (39) to obtain $y_0/x_0 = \sinh u$, so the maximum possible value of x_0/y_0 is

$$\frac{x_0}{y_0} = \operatorname{csch} u = 0.6627434193\ldots. \tag{40}$$

Therefore, only Goldschmidt ring solutions exist for $x_0/y_0 > 0.6627\ldots$.

The SURFACE AREA of the minimal CATENOID surface is given by

$$A = 2(2\pi)\int_0^{x_0} y\sqrt{1 + y'^2}\,dx, \tag{41}$$

but since

$$y = \sqrt{1 + y'^2}\,a \tag{42}$$

$$y = a\cosh\left(\frac{x}{a}\right), \tag{43}$$

$$A = \frac{4\pi}{a}\int_0^{x_0} y^2\,dx = 4\pi a\int_0^{x_0}\cosh^2\left(\frac{x}{a}\right)dx$$
$$= 4\pi a\int_0^{x_0}\tfrac{1}{2}\left[\cosh\left(\frac{2x}{a}\right)+1\right]dx$$
$$= 2\pi a\left[\int_0^{x_0}\cosh\left(\frac{2x}{a}\right)dx + \int_0^{x_0}dx\right]$$
$$= 2\pi a\left[\frac{a}{2}\sinh\left(\frac{2x}{a}\right)+x\right]_0^{x_0}$$
$$= \pi a^2\left[\sinh\left(\frac{2x}{a}\right)+\frac{2x}{a}\right]_0^{x_0}$$
$$= \pi a^2\left[\sinh\left(\frac{2x_0}{a}\right)+\frac{2x_0}{a}\right]. \quad (44)$$

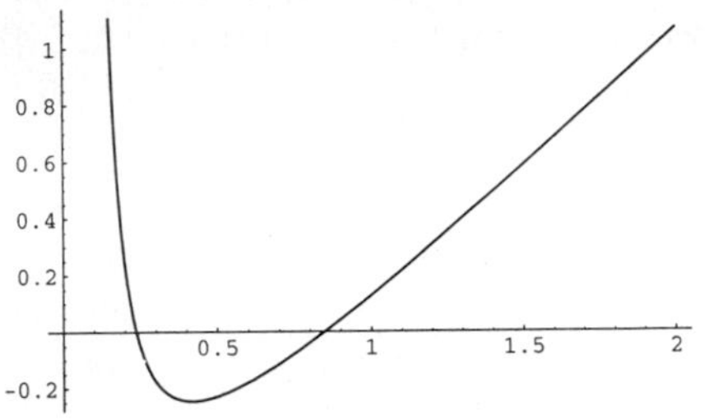

Some caution is needed in solving (33) for a. If we take $x_0 = 1/2$ and $y_0 = 1$ then (33) becomes

$$1 = a\cosh\left(\frac{1}{2a}\right), \quad (45)$$

which has *two* solutions: $a_1 = 0.2350\ldots$ ("deep"), and $a_2 = 0.8483\ldots$ ("flat"). However, upon plugging these into (44) with $x_0 = 1/2$, we find $A_1 = 6.8456\ldots$ and $A_2 = 5.9917\ldots$. So A_1 is *not,* in fact, a local minimum, and A_2 is the only *true* minimal solution.

The SURFACE AREA of the CATENOID solution equals that of the GOLDSCHMIDT SOLUTION when (44) equals the AREA of two disks,

$$\pi a^2\left[\sinh\left(\frac{2x_0}{a}\right)+\frac{2x_0}{a}\right] = 2\pi {y_0}^2 \quad (46)$$

$$a^2\left[2\sinh\left(\frac{x_0}{a}\right)\cosh\left(\frac{x_0}{a}\right)+\frac{2x_0}{a}\right]-2{y_0}^2 = 0 \quad (47)$$

$$a^2\left[\cosh\left(\frac{x_0}{a}\right)\sqrt{\cosh^2\left(\frac{x_0}{a}\right)-1}+\frac{x_0}{a}\right]-{y_0}^2 = 0. \quad (48)$$

Plugging in

$$\frac{y_0}{a} = \cosh\left(\frac{x_0}{a}\right), \quad (49)$$

$$\frac{y_0}{a}\sqrt{\left(\frac{y_0}{a}\right)^2-1}+\cosh^{-1}\left(\frac{y_0}{a}\right)-\left(\frac{y_0}{a}\right)^2 = 0. \quad (50)$$

Defining

$$u \equiv \frac{y_0}{a} \quad (51)$$

gives

$$u\sqrt{u^2-1}+\cosh^{-1}u-u^2 = 0. \quad (52)$$

This has a solution $u = 1.2113614259$. The value of x_0/y_0 for which

$$A_{\text{catenary}} = A_{2\text{ disks}} \quad (53)$$

is therefore

$$\frac{x_0}{y_0} = \frac{\frac{x_0}{a}}{\frac{y_0}{a}} = \frac{\cosh^{-1}\left(\frac{y_0}{a}\right)}{\frac{y_0}{a}} = \frac{\cosh^{-1}u}{u} = 0.5276973967. \quad (54)$$

For $x_0/y_0 \in (0.52770, 0.6627)$, the CATENARY solution has larger AREA than the two disks, so it exists only as a RELATIVE MINIMUM.

There also exist solutions with a disk (of radius r) between the rings supported by two CATENOIDS of revolution. The AREA is larger than that for a simple CATENOID, but it is a RELATIVE MINIMUM. The equation of the POSITIVE half of this curve is

$$y = c_1\cosh\left(\frac{x}{c_1}+c_3\right). \quad (55)$$

At $(0, r)$,

$$r = c_1\cosh(c_3). \quad (56)$$

At (x_0, y_0),

$$y_0 = c_1\cosh\left(\frac{x_0}{c_1}+c_3\right). \quad (57)$$

The AREA of the two CATENOIDS is

$$A_{\text{catenoids}} = 2(2\pi)\int_0^{x_0} y\sqrt{1+y'^2}\,dx = \frac{4\pi}{c_1}\int_0^{x_0} y^2\,dx$$
$$= 4\pi c_1\int_0^{x_0}\cosh^2\left(\frac{x}{c_1}+c_3\right)dx. \quad (58)$$

Now let $u \equiv x/c_1 + c_3$, so $du = dx/c_1$

$$A = 4\pi {c_1}^2\int_{c_3}^{x_0/x_1+c_3}\cosh^2 u\,du$$
$$= 4\pi {c_1}^2\tfrac{1}{2}\int_{c_3}^{x_0/x_1+c_3}[\cosh(2u)+1]\,du$$
$$= 2\pi {c_1}^2\left[\tfrac{1}{2}\sinh(2u)+u\right]_{c_3}^{x_0/x_1+c_3}$$
$$= 2\pi {c_1}^2\left\{\tfrac{1}{2}\sinh\left[2\left(\frac{x_0}{c_1}+c_3\right)\right]-\tfrac{1}{2}\sinh(2c_3)+\frac{x_0}{c_1}\right\}$$
$$= \pi {c_1}^2\left\{\sinh\left[2\left(\frac{x_0}{c_1}+c_3\right)\right]-\sinh(2c_3)+\frac{2x_0}{c_1}\right\}. \quad (59)$$

The AREA of the central DISK is

$$A_{\text{disk}} = \pi r^2 = \pi {c_1}^2\cosh^2 c_3, \quad (60)$$

so the total AREA is

$$A = \pi {c_1}^2 \left\{ \sinh\left[2\left(\frac{x_0}{c_1}+c_3\right)\right] + \left[\cosh^2 c_3 - \sinh(2c_3)\right] + \frac{2x_0}{c_1} \right\}. \tag{61}$$

By PLATEAU'S LAWS, the CATENOIDS meet at an ANGLE of 120°, so

$$\tan 30^\circ = \left[\frac{dy}{dx}\right]_{x=0} = \left[\sinh\left(\frac{x}{c_1}+c_3\right)\right]_{x=0} = \sinh c_3 = \frac{1}{\sqrt{3}} \tag{62}$$

and

$$c_3 = \sinh^{-1}\left(\frac{1}{\sqrt{3}}\right). \tag{63}$$

This means that

$$\begin{aligned} &\cosh^2 c_3 - \sinh(2c_3) \\ &= [1+\sinh^2 c_3] - 2\sinh c_3 \sqrt{1+\sinh^2 c_3} \\ &= (1+\tfrac{1}{3}) - 2\left(\frac{1}{\sqrt{3}}\right)\sqrt{1+\tfrac{1}{3}} \\ &= \frac{4}{3} - \frac{2}{\sqrt{3}}\frac{2}{\sqrt{3}} = 0, \end{aligned} \tag{64}$$

so

$$A = \pi {c_1}^2 \left\{ \sinh\left[2\left(\frac{x_0}{c_1}+c_3\right)\right] + \frac{2x_0}{c_1} \right\}. \tag{65}$$

Now examine x_0/y_0,

$$\frac{x_0}{y_0} = \frac{\frac{x_0}{c_1}}{\frac{y_0}{c_1}} = \frac{\frac{x_0}{c_1}}{\cosh\left(\frac{x_0}{c_1}+c_3\right)} = u \operatorname{sech}(u+c_3), \tag{66}$$

where $u \equiv x_0/c_1$. Finding the maximum ratio of x_0/y_0 gives

$$\frac{d}{du}\left(\frac{x_0}{y_0}\right) = \operatorname{sech}(u+c_3) - u\tanh(u+c_3)\operatorname{sech}(u+c_3) = 0 \tag{67}$$

$$u\tanh(u+c_3) = 1, \tag{68}$$

with $c_3 = \sinh^{-1}(1/\sqrt{3})$ as given above. The solution is $u = 1.0799632187$, so the maximum value of x_0/y_0 for two CATENOIDS with a central disk is $y_0 = 0.4078241702$.

If we are interested instead in finding the curve from a point (x_1, y_1) to a point (x_2, y_2) which, when revolved around the y-AXIS (as opposed to the x-AXIS), yields a surface of smallest SURFACE AREA A, we proceed as above. Note that the solution is physically equivalent to that for rotation about the x-AXIS, but takes on a different mathematical form. The AREA element is

$$dA = 2\pi x\, ds = 2\pi x\sqrt{1+y'^2}\, dx \tag{69}$$

$$A = 2\pi \int x\sqrt{1+y'^2}\, dx, \tag{70}$$

and the quantity we are minimizing is

$$f = x\sqrt{1+y'^2}. \tag{71}$$

Taking the derivatives gives

$$\frac{\partial f}{\partial y} = 0 \tag{72}$$

$$\frac{d}{dx}\frac{\partial f}{\partial y'} = \frac{d}{dx}\left(\frac{xy'}{\sqrt{1+y'^2}}\right), \tag{73}$$

so the EULER-LAGRANGE DIFFERENTIAL EQUATION becomes

$$\frac{\partial f}{\partial y} - \frac{d}{dx}\frac{\partial f}{\partial y'} = \frac{d}{dx}\left(\frac{xy'}{\sqrt{1+y'^2}}\right) = 0. \tag{74}$$

$$\frac{xy'}{\sqrt{1+y'^2}} = a \tag{75}$$

$$x^2 y'^2 = a^2(1+y'^2) \tag{76}$$

$$y'^2(x^2-a^2) = a^2 \tag{77}$$

$$\frac{dy}{dx} = \frac{a}{\sqrt{x^2-a^2}} \tag{78}$$

$$y = a\int \frac{dx}{\sqrt{x^2-a^2}} + b = a\cosh^{-1}\left(\frac{x}{a}\right) + b. \tag{79}$$

Solving for x then gives

$$x = a\cosh\left(\frac{y-b}{a}\right), \tag{80}$$

which is the equation for a CATENARY. The SURFACE AREA of the CATENOID product by rotation is

$$\begin{aligned} A &= 2\pi\int x\sqrt{1+y'^2}\, dx = 2\pi\int x\sqrt{1+\frac{a^2}{x^2-a^2}}\, dx \\ &= 2\pi\int \frac{x}{\sqrt{x^2-a^2}}\sqrt{(x^2-a^2)+a^2}\, dx \\ &= 2\pi\int \frac{x^2\, dx}{\sqrt{x^2-a^2}} \\ &= \left[\frac{x}{2}\sqrt{x^2-a^2} + \frac{a^2}{2}\ln\left(x+\sqrt{x^2-a^2}\right)\right]_{x_1}^{x_2} \\ &= \frac{1}{2}\left[x_2\sqrt{{x_2}^2-a^2} - x_1\sqrt{{x_1}^2-a^2} + a^2\ln\left(\frac{x_2+\sqrt{{x_2}^2-a^2}}{x_1+\sqrt{{x_1}^2-a^2}}\right)\right]. \end{aligned} \tag{81}$$

Isenberg (1992, p. 80) discusses finding the MINIMAL SURFACE passing through two rings with axes offset from each other.

see also APPLE, CATENOID, CONE CONICAL FRUSTUM, CYLINDER, DARWIN-DE SITTER SPHEROID, EIGHT SURFACE, GABRIEL'S HORN, HYPERBOLOID, LEMON, MERIDIAN, OBLATE SPHEROID, PAPPUS'S CENTROID THEOREM, PARABOLOID, PARALLEL (SURFACE OF REVOLUTION), PROLATE SPHEROID, PSEUDOSPHERE, SINCLAIR'S SOAP FILM PROBLEM, SOLID OF REVOLUTION, SPHERE, SPHEROID, TOROID, TORUS

References

Arfken, G. *Mathematical Methods for Physicists, 3rd ed.* Orlando, FL: Academic Press, pp. 931–937, 1985.

Goldstein, H. *Classical Mechanics, 2nd ed.* Reading, MA: Addison-Wesley, p. 42, 1980.

Gray, A. "Surfaces of Revolution." Ch. 18 in *Modern Differential Geometry of Curves and Surfaces.* Boca Raton, FL: CRC Press, pp. 357–375, 1993.

Isenberg, C. *The Science of Soap Films and Soap Bubbles.* New York: Dover, pp. 79–80 and Appendix III, 1992.

Surface of Section

A surface (or "space") of section is a way of presenting a trajectory in n-D PHASE SPACE in an $(n-1)$-D SPACE. By picking one phase element constant and plotting the values of the other elements each time the selected element has the desired value, an intersection surface is obtained. If the equations of motion can be formulated as a MAP in which an explicit FORMULA gives the values of the other elements at successive passages through the selected element value, the time required to compute the surface of section is greatly reduced.

see also PHASE SPACE

Surgery

In the process of attaching a k-HANDLE to a MANIFOLD M, the BOUNDARY of M is modified by a process called $(k-1)$-surgery. Surgery consists of the removal of a TUBULAR NEIGHBORHOOD of a $(k-1)$-SPHERE $\mathbb{S}^{k-1}$ from the BOUNDARIES of M and the $\dim(M)-1$ standard SPHERE, and the gluing together of these two scarred-up objects along their common BOUNDARIES.

see also BOUNDARY, DEHN SURGERY, HANDLE, MANIFOLD, SPHERE, TUBULAR NEIGHBORHOOD

Surjection

An ONTO (SURJECTIVE) MAP.

see also BIJECTION, INJECTION, ONTO

Surjective

see ONTO

Surprise Examination Paradox

see UNEXPECTED HANGING PARADOX

Surreal Number

The most natural collection of numbers which includes both the REAL NUMBERS and the infinite ORDINAL NUMBERS of Georg Cantor. They were invented by John H. Conway in 1969. Every REAL NUMBER is surrounded by surreals, which are closer to it than any REAL NUMBER. Knuth (1974) describes the surreal numbers in a work of fiction.

The surreal numbers are written using the NOTATION $\{a|b\}$, where $\{|\}=0$, $\{0|\}=1$ is the simplest number greater than 0, $\{1|\}=2$ is the simplest number greater than 1, etc. Similarly, $\{|0\}=-1$ is the simplest number less than 1, etc. However, 2 can also be represented by $\{1|3\}$, $\{3/2|4\}$, $\{1|\omega\}$, etc.

see also OMNIFIC INTEGER, ORDINAL NUMBER, REAL NUMBER

References

Berlekamp, E. R.; Conway, J. H.; and Guy, R. K. *Winning Ways, For Your Mathematical Plays, Vol. 1: Games in General.* London: Academic Press, 1982.

Conway, J. H. *On Numbers and Games.* New York: Academic Press, 1976.

Conway, J. H. and Guy, R. K. *The Book of Numbers.* New York: Springer-Verlag, pp. 283–284, 1996.

Conway, J. H. and Jackson, A. "Budding Mathematician Wins Westinghouse Competition." *Not. Amer. Math. Soc.* **43**, 776–779, 1996.

Gonshor, H. *An Introduction to Surreal Numbers.* Cambridge: Cambridge University Press, 1986.

Knuth, D. *Surreal Numbers: How Two Ex-Students Turned on to Pure Mathematics and Found Total Happiness.* Reading, MA: Addison-Wesley, 1974. http://www-cs-faculty.stanford.edu/~knuth/sn.html.

Surrogate

Surrogate data are artificially generated data which mimic statistical properties of real data. Isospectral surrogates have identical POWER SPECTRA as real data but with randomized phases. Scrambled surrogates have the same probability distribution as real data, but with white noise POWER SPECTRA.

see also POWER SPECTRUM

Surveying Problems

see HANSEN'S PROBLEM, SNELLIUS-POTHENOT PROBLEM

Survivorship Curve

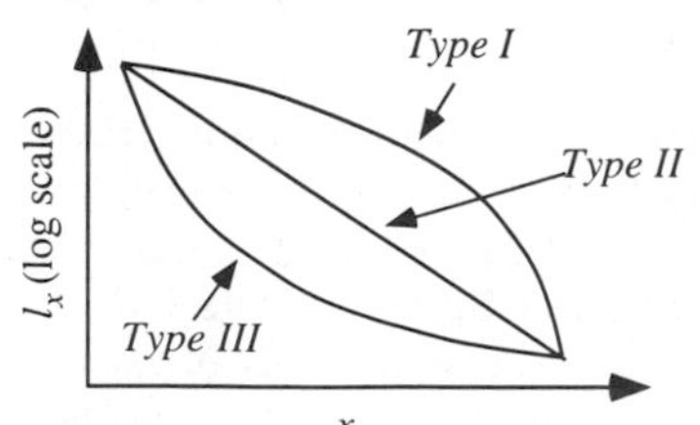

Plotting l_x from a LIFE EXPECTANCY table on a logarithmic scale versus x gives a curve known as a survivorship curve. There are three general classes of survivorship curves, illustrated above.

1. Type I curves are typical of populations in which most mortality occurs among the elderly (e.g., humans in developed countries).
2. Type II curves occur when mortality is not dependent on age (e.g., many species of large birds and fish). For an infinite type II population, $e_0 = e_1 = \ldots$, but this cannot hold for a finite population.
3. Type III curves occur when juvenile mortality is extremely high (e.g., plant and animal species producing many offspring of which few survive). In type III populations, it is often true that $e_{i+1} > e_i$ for small i. In other words, life expectancy increases for individuals who survive their risky juvenile period.

see also LIFE EXPECTANCY

Suslin's Theorem

A SET in a POLISH SPACE is a BOREL SET IFF it is both ANALYTIC and COANALYTIC. For subsets of w, a set is δ_1^1 IFF it is "hyperarithmetic."

see also ANALYTIC SET, BOREL SET, COANALYTIC SET, POLISH SPACE

Suspended Knot

An ordinary KNOT in 3-D suspended in 4-D to create a knotted 2-sphere. Suspended knots are not smooth at the poles.

see also SPUN KNOT, TWIST-SPUN KNOT

Suspension

The JOIN of a TOPOLOGICAL SPACE X and a pair of points S^0, $\Sigma(X) = X * S^0$.

see also JOIN (SPACES), TOPOLOGICAL SPACE

References
Rolfsen, D. *Knots and Links.* Wilmington, DE: Publish or Perish Press, p. 6, 1976.

Suzanne Set

The nth Suzanne set S_n is defined as the set of COMPOSITE NUMBERS x for which $n|S(x)$ and $n|S_p(x)$, where

$$x = a_0 + a_1(10^1) + \ldots + a_d(10^d) = p_1 p_2 \cdots p_n,$$

and

$$S(x) = \sum_{j=0}^{d} a_j$$
$$S_p(x) = \sum_{i=1}^{m} S(p_i).$$

Every Suzanne set has an infinite number of elements. The Suzanne set S_n is a superset of the MONICA SET M_n.

see also MONICA SET

References
Smith, M. "Cousins of Smith Numbers: Monica and Suzanne Sets." *Fib. Quart.* **34**, 102–104, 1996.

Suzuki Group

The SPORADIC GROUP *Suz.*

References
Wilson, R. A. "ATLAS of Finite Group Representation." `http://for.mat.bham.ac.uk/atlas/Suz.html`.

Swallowtail Catastrophe

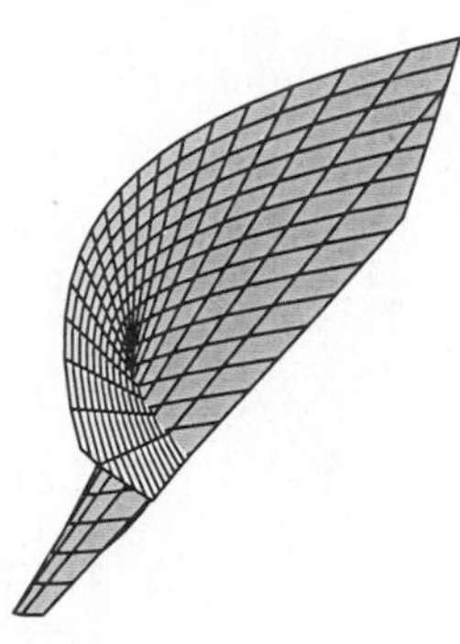

A CATASTROPHE which can occur for three control factors and one behavior axis. The equations

$$x = uv^2 + 3v^4$$
$$y = -2uv - 4v^3$$
$$z = u$$

display such a catastrophe (von Seggern 1993, Nordstrand). The above surface uses $u \in [-2, 2]$ and $v \in [-0.8, 0.8]$.

References
Nordstrand, T. "Swallowtail." `http://www.uib.no/people/nfytn/stltxt.htm`.
von Seggern, D. *CRC Standard Curves and Surfaces.* Boca Raton, FL: CRC Press, p. 94, 1993.

Swastika

An irregular ICOSAGON, also called the gammadion or fylfot, which symbolized good luck in ancient Arabic and Indian cultures. In more recent times, it was adopted as the symbol of the Nazi Party in Hitler's Germany and has thence come to symbolize anti-Semitism.

see also CROSS, DISSECTION

Swastika Curve

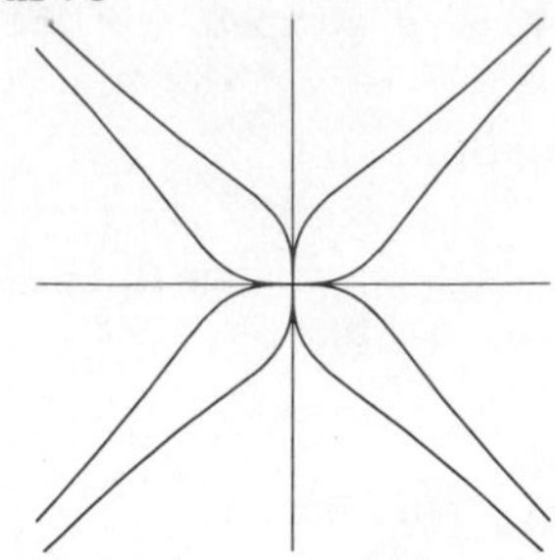

The plane curve with Cartesian equation

$$y^4 - x^4 = xy$$

and polar equation

$$r^2 = \frac{\sin\theta\cos\theta}{\sin^4\theta - \cos^4\theta}.$$

References

Cundy, H. and Rollett, A. *Mathematical Models, 3rd ed.* Stradbroke, England: Tarquin Pub., p. 71, 1989.

Sweep Signal

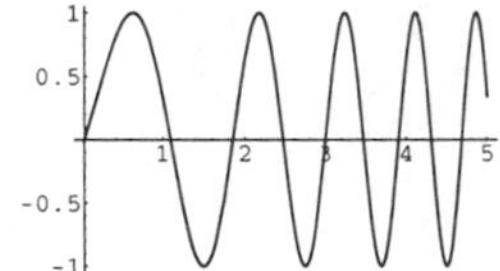

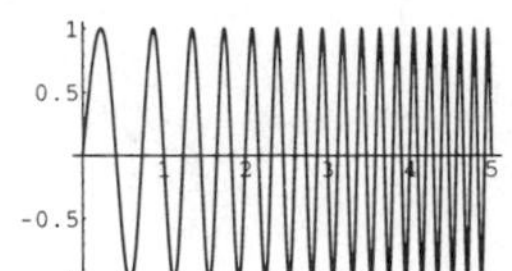

The general function

$$y(a,b,c,d) = c\sin\left\{\frac{\pi}{b-a}\left[\left((b-a)\frac{x}{d}+a\right)^2 - a^2\right]\right\}.$$

References

von Seggern, D. *CRC Standard Curves and Surfaces.* Boca Raton, FL: CRC Press, p. 160, 1993.

Swinnerton-Dyer Conjecture

In the early 1960s, B. Birch and H. P. F. Swinnerton-Dyer conjectured that if a given ELLIPTIC CURVE has an infinite number of solutions, then the associated L-function has value 0 at a certain fixed point. In 1976, Coates and Wiles showed that elliptic curves with COMPLEX multiplication having an infinite number of solutions have L-functions which are zero at the relevant fixed point (COATES-WILES THEOREM), but they were unable to prove the converse. V. Kolyvagin extended this result to modular curves.

see also COATES-WILES THEOREM, ELLIPTIC CURVE

References

Cipra, B. "Fermat Prover Points to Next Challenges." *Science* **271**, 1668–1669, 1996.

Ireland, K. and Rosen, M. "New Results on the Birch–Swinnerton-Dyer Conjecture." §20.5 in *A Classical Introduction to Modern Number Theory, 2nd ed.* New York: Springer-Verlag, pp. 353–357, 1990.

Mazur, B. and Stevens, G. (Eds.). *p-Adic Monodromy and the Birch and Swinnerton-Dyer Conjecture.* Providence, RI: Amer. Math. Soc., 1994.

Swinnerton-Dyer Polynomial

The minimal POLYNOMIAL $S_n(x)$ whose ROOTS are sums and differences of the SQUARE ROOTS of the first n PRIMES,

$$S_n(x) = \prod(x \pm \sqrt{2} \pm \sqrt{3} \pm \sqrt{5} \pm \ldots \pm \sqrt{p_n}).$$

References

Vardi, I. *Computational Recreations in Mathematica.* Redwood City, CA: Addison-Wesley, pp. 11 and 225–226, 1991.

Swirl

A swirl is a generic word to describe a function having arcs which double back swirl around each other. The plots above correspond to the function

$$f(r,\theta) = \sin(6\cos r - n\theta)$$

for $n = 0, 1, \ldots, 5$.

see also DAISY, WHIRL

Sylow p-Subgroup

If p^k is the highest POWER of a PRIME p dividing the ORDER of a finite GROUP G, then a SUBGROUP of G of ORDER p^k is called a Sylow p-subgroup of G.

see also ABHYANKAR'S CONJECTURE, SUBGROUP, SYLOW THEOREMS

Sylow Theorems

Let p be a PRIME NUMBER, G a GROUP, and $|G|$ the order of G.

1. If p divides $|G|$, then G has a SYLOW p-SUBGROUP.
2. In a FINITE GROUP, all the SYLOW p-SUBGROUPS are isomorphic for some fixed p.
3. The number of SYLOW p-SUBGROUPS for a fixed p is CONGRUENT to 1 (mod p).

Sylvester Cyclotomic Number

Given a LUCAS SEQUENCE with parameters P and Q, discriminant $D \neq 0$, and roots α and β, the Sylvester cyclotomic numbers are

$$Q_n = \prod_r (\alpha - \zeta^r \beta),$$

where

$$\zeta \equiv \cos\left(\frac{2\pi}{n}\right) + i \sin\left(\frac{2\pi}{n}\right)$$

is a PRIMITIVE ROOT OF UNITY and the product is over all exponents r RELATIVELY PRIME to n such that $r \in [1, n)$.

see also LUCAS SEQUENCE

References

Ribenboim, P. *The Book of Prime Number Records, 2nd ed.* New York: Springer-Verlag, p. 69, 1989.

Sylvester's Determinant Identity

$$|A|\,|A_{rs,pq}| = |A_{r,p}|\,|A_{s,q}| - |A_{r,q}|\,|A_{s,p}|,$$

where $A_{u,w}$ is the submatrix of A formed by the intersection of the subset w of columns and u of rows.

Sylvester's Four-Point Problem

Let $q(R)$ be the probability that four points chosen at random in a region R have a CONVEX HULL which is a QUADRILATERAL. For an open, convex subset of the PLANE of finite AREA,

$$0.667 \approx \tfrac{2}{3} \leq q(R) \leq 1 - \frac{35}{12\pi^2} \approx 0.704.$$

References

Schneinerman, E. and Wilf, H. S. "The Rectilinear Crossing Number of a Complete Graph and Sylvester's 'Four Point' Problem of Geometric Probability." *Amer. Math. Monthly* **101**, 939–943, 1994.

Sylvester Graph

The Sylvester graph of a configuration is the set of ORDINARY POINTS and ORDINARY LINES.

see also ORDINARY LINE, ORDINARY POINT

References

Guy, R. K. "Monthly Unsolved Problems, 1969–1987." *Amer. Math. Monthly* **94**, 961–970, 1987.

Guy, R. K. "Unsolved Problems Come of Age." *Amer. Math. Monthly* **96**, 903–909, 1989.

Sylvester's Inertia Law

The numbers of EIGENVALUES that are POSITIVE, NEGATIVE, or 0 do not change under a congruence transformation. Gradshteyn and Ryzhik (1979) state it as follows: when a QUADRATIC FORM Q in n variables is reduced by a nonsingular linear transformation to the form

$$Q = {y_1}^2 + {y_2}^2 + \ldots + {y_p}^2 - {p_{p+1}}^2 - {y_{p2}}^2 - \ldots - {y_r}^2,$$

the number p of POSITIVE SQUARES appearing in the reduction is an invariant of the QUADRATIC FORM Q and does not depend on the method of reduction.

see also EIGENVALUE, QUADRATIC FORM

References

Gradshteyn, I. S. and Ryzhik, I. M. *Tables of Integrals, Series, and Products, 5th ed.* San Diego, CA: Academic Press, p. 1105, 1979.

Sylvester's Line Problem

It is not possible to arrange a finite number of points so that a LINE through every two of them passes through a third, unless they are all on a single LINE.

see also COLLINEAR, SYLVESTER'S FOUR-POINT PROBLEM

Sylvester Matrix

For POLYNOMIALS of degree m and n, the Sylvester matrix is an $(m+n) \times (m+n)$ matrix whose DETERMINANT is the RESULTANT of the two POLYNOMIALS.

see also RESULTANT

Sylvester's Sequence

The sequence defined by $e_0 = 2$ and the RECURRENCE RELATION

$$e_n = 1 + \prod_{i=0}^{n-1} e_i = {e_{n-1}}^2 - e_{n-1} + 1. \tag{1}$$

This sequence arises in Euclid's proof that there are an INFINITE number of PRIMES. The proof proceeds by constructing a sequence of PRIMES using the RECURRENCE RELATION

$$e_{n+1} = e_0 e_1 \cdots e_n + 1 \tag{2}$$

(Vardi 1991). Amazingly, there is a constant

$$E \approx 1.264084735306 \tag{3}$$

such that

$$e_n = \left\lfloor E^{2^{n+1}} + \tfrac{1}{2} \right\rfloor \tag{4}$$

(Vardi 1991, Graham *et al.* 1994). The first few numbers in Sylvester's sequence are 2, 3, 7, 43, 1807, 3263443, 10650056950807, ... (Sloane's A000058). The e_n satisfy

$$\sum_{n=0}^{\infty} \frac{1}{e_n} = 1. \quad (5)$$

In addition, if $0 < x < 1$ is an IRRATIONAL NUMBER, then the nth term of an infinite sum of unit fractions used to represent x as computed using the GREEDY ALGORITHM must be smaller than $1/e_n$.

The n of the first few PRIME e_n are 0, 1, 2, 3, 5, ... Vardi (1991) gives a lists of factors less than 5×10^7 of e_n for $n \leq 200$ and shows that e_n is COMPOSITE for $6 \leq n \leq 17$. Furthermore, all numbers less than 2.5×10^{15} in Sylvester's sequence are SQUAREFREE, and no SQUAREFUL numbers in this sequence are known (Vardi 1991).

see also EUCLID'S THEOREMS, GREEDY ALGORITHM, SQUAREFREE, SQUAREFUL

References

Graham, R. L.; Knuth, D. E.; and Patashnik, O. Research problem 4.65 in *Concrete Mathematics: A Foundation for Computer Science, 2nd ed.* Reading, MA: Addison-Wesley, 1994.

Sloane, N. J. A. Sequence A000058/M0865 in "An On-Line Version of the Encyclopedia of Integer Sequences."

Vardi, I. "Are All Euclid Numbers Squarefree?" and "PowerMod to the Rescue." §5.1 and 5.2 in *Computational Recreations in Mathematica.* Reading, MA: Addison-Wesley, pp. 82–89, 1991.

Sylvester's Signature

Diagonalize a form over the RATIONALS to

$$\mathrm{diag}[p^a \cdot A, p^b \cdot B, \ldots],$$

where all the entries are INTEGERS and A, B, ... are RELATIVELY PRIME to p. Then Sylvester's signature is the sum of the -1-parts of the entries.

see also p-SIGNATURE

Sylvester's Triangle Problem

The resultant of the vectors represented by the three RADII from the center of a TRIANGLE'S CIRCUMCIRCLE to its VERTICES is the segment extending from the CIRCUMCENTER to the ORTHOCENTER.

see also CIRCUMCENTER, CIRCUMCIRCLE, ORTHOCENTER, TRIANGLE

References

Dörrie, H. *100 Great Problems of Elementary Mathematics: Their History and Solutions.* New York: Dover, p. 142, 1965.

Symbolic Logic

The study of the meaning and relationships of statements used to represent precise mathematical ideas. Symbolic logic is also called FORMAL LOGIC.

see also FORMAL LOGIC, LOGIC, METAMATHEMATICS

References

Carnap, R. *Introduction to Symbolic Logic and Its Applications.* New York: Dover, 1958.

Symmedian Line

The lines ISOGONAL to the MEDIANS of a TRIANGLE are called the triangle's symmedian lines. The symmedian lines are concurrent in a point called the LEMOINE POINT.

see also ISOGONAL CONJUGATE, LEMOINE POINT, MEDIAN (TRIANGLE)

References

Johnson, R. A. *Modern Geometry: An Elementary Treatise on the Geometry of the Triangle and the Circle.* Boston, MA: Houghton Mifflin, pp. 213–218, 1929.

Symmedian Point

see LEMOINE POINT

Symmetric

A quantity which remains unchanged in SIGN when indices are reversed. For example, $A_{ij} \equiv a_i + a_j$ is symmetric since $A_{ij} = A_{ji}$.

see also ANTISYMMETRIC

Symmetric Block Design

A symmetric design is a BLOCK DESIGN (v, k, λ, r, b) with the same number of blocks as points, so $b = v$ (or, equivalently, $r = k$). An example of a symmetric block design is a PROJECTIVE PLANE.

see also BLOCK DESIGN, PROJECTIVE PLANE

References

Dinitz, J. H. and Stinson, D. R. "A Brief Introduction to Design Theory." Ch. 1 in *Contemporary Design Theory: A Collection of Surveys* (Ed. J. H. Dinitz and D. R. Stinson). New York: Wiley, pp. 1–12, 1992.

Symmetric Design

see SYMMETRIC BLOCK DESIGN

Symmetric Function

A symmetric function on n variables x_1, ..., x_n is a function that is unchanged by any PERMUTATION of its variables. In most contexts, the term "symmetric function" refers to a polynomial on n variables with this feature (more properly called a "symmetric polynomial"). Another type of symmetric functions is symmetric rational functions, which are the RATIONAL FUNCTIONS that are unchanged by PERMUTATION of variables.

The symmetric polynomials (respectively, symmetric rational functions) can be expressed as polynomials (respectively, rational functions) in the ELEMENTARY SYMMETRIC FUNCTIONS. This is called the FUNDAMENTAL THEOREM OF SYMMETRIC FUNCTIONS.

A function $f(x)$ is sometimes said to be symmetric about the y-AXIS if $f(-x) = f(x)$. Examples of such functions include $|x|$ (the ABSOLUTE VALUE) and x^2 (the PARABOLA).

see also ELEMENTARY SYMMETRIC FUNCTION, FUNDAMENTAL THEOREM OF SYMMETRIC FUNCTIONS, RATIONAL FUNCTION

References
Macdonald, I. G. *Symmetric Functions and Hall Polynomials, 2nd ed.* Oxford, England: Oxford University Press, 1995.
Macdonald, I. G. *Symmetric Funtions and Orthogonal Polynomials.* Providence, RI: Amer. Math. Soc., 1997.
Petkovšek, M.; Wilf, H. S.; and Zeilberger, D. "Symmetric Function Identities." §1.7 in *A=B.* Wellesley, MA: A. K. Peters, pp. 12–13, 1996.

Symmetric Group

The symmetric group S_n of DEGREE n is the GROUP of all PERMUTATIONS on n symbols. S_n is therefore of ORDER $n!$ and contains as SUBGROUPS every GROUP of ORDER n. The number of CONJUGACY CLASSES of S_n is given by the PARTITION FUNCTION P.

NETTO'S CONJECTURE states that the probability that two elements P_1 and P_2 of a symmetric group generate the entire group tends to 3/4 as $n \to \infty$. This was proven by Dixon in 1967.

see also ALTERNATING GROUP, CONJUGACY CLASS, FINITE GROUP, NETTO'S CONJECTURE, PARTITION FUNCTION P, SIMPLE GROUP

References
Lomont, J. S. "Symmetric Groups." Ch. 7 in *Applications of Finite Groups.* New York: Dover, pp. 258–273, 1987.
Wilson, R. A. "ATLAS of Finite Group Representation." `http://for.mat.bham.ac.uk/atlas#alt`.

Symmetric Matrix

A symmetric matrix is a SQUARE MATRIX which satisfies $\mathsf{A}^{\mathrm{T}} = \mathsf{A}$ where A^{T} denotes the TRANSPOSE, so $a_{ij} = a_{ji}$. This also implies

$$\mathsf{A}^{-1}\mathsf{A}^{\mathrm{T}} = \mathsf{I}, \tag{1}$$

where I is the IDENTITY MATRIX. Written explicitly,

$$\begin{bmatrix} a_{11} & a_{12} & \cdots & a_{1n} \\ a_{21} & a_{22} & \cdots & a_{2n} \\ \vdots & \vdots & \ddots & \vdots \\ a_{n1} & a_{n2} & \cdots & a_{nn} \end{bmatrix}. \tag{2}$$

The symmetric part of any MATRIX may be obtained from

$$\mathsf{A}_s = \tfrac{1}{2}(\mathsf{A} + \mathsf{A}^{\mathrm{T}}). \tag{3}$$

A MATRIX A is symmetric if it can be expressed in the form

$$\mathsf{A} = \mathsf{QDQ}^{\mathrm{T}}, \tag{4}$$

where Q is an ORTHOGONAL MATRIX and D is a DIAGONAL MATRIX. This is equivalent to the MATRIX equation

$$\mathsf{AQ} = \mathsf{QD}, \tag{5}$$

which is equivalent to

$$\mathsf{A}\mathbf{Q}_n = \lambda_n \mathbf{Q}_n \tag{6}$$

for all n, where $\lambda_n = D_{nn}$. Therefore, the diagonal elements of D are the EIGENVALUES of A, and the columns of Q are the corresponding EIGENVECTORS.

see also ANTISYMMETRIC MATRIX, SKEW SYMMETRIC MATRIX

References
Nash, J. C. "Real Symmetric Matrices." Ch. 10 in *Compact Numerical Methods for Computers: Linear Algebra and Function Minimisation, 2nd ed.* Bristol, England: Adam Hilger, pp. 119–134, 1990.

Symmetric Points

Two points z and $z^S \in \mathbb{C}^*$ are symmetric with respect to a CIRCLE or straight LINE L if all CIRCLES and straight LINES passing through z and z^S are orthogonal to L. MÖBIUS TRANSFORMATIONS preserve symmetry. Let a straight line be given by a point z_0 and a unit VECTOR $e^{i\theta}$, then

$$z^S = e^{2i\theta}(z - z_0)^* + z_0.$$

Let a CIRCLE be given by center z_0 and RADIUS r, then

$$z^S = z_0 + \frac{r^2}{(z - z_0)^*}.$$

see also MÖBIUS TRANSFORMATION

Symmetric Relation

A RELATION R on a SET S is symmetric provided that for every x and y in S we have xRy IFF yRx.

see also RELATION

Symmetric Tensor

A second-RANK symmetric TENSOR is defined as a TENSOR A for which

$$A^{mn} = A^{nm}. \tag{1}$$

Any TENSOR can be written as a sum of symmetric and ANTISYMMETRIC parts

$$\begin{aligned} A^{mn} &= \tfrac{1}{2}(A^{mn} + A^{nm}) + \tfrac{1}{2}(A^{mn} - A^{nm}) \\ &= \tfrac{1}{2}({B_S}^{mn} + {B_A}^{mn}). \end{aligned} \tag{2}$$

The symmetric part of a TENSOR is denoted by parentheses as follows:

$$T_{(a,b)} \equiv \tfrac{1}{2}(T_{ab} + T_{ba}) \tag{3}$$

$$T_{(a_1,a_2,\ldots,a_n)} \equiv \frac{1}{n!} \sum_{\text{permutations}} T_{a_1 a_2 \cdots a_n}. \tag{4}$$

The product of a symmetric and an ANTISYMMETRIC TENSOR is 0. This can be seen as follows. Let $a^{\alpha\beta}$ be ANTISYMMETRIC, so

$$a^{11} = a^{22} = 0 \tag{5}$$

$$a^{21} = -a^{12}. \tag{6}$$

Let $b_{\alpha\beta}$ be symmetric, so

$$b_{12} = b_{21}. \tag{7}$$

Then

$$\begin{aligned} a^{\alpha\beta} b_{\alpha\beta} &= a^{11} b_{11} + a^{12} b_{12} + a^{21} b_{21} + a^{22} b_{22} \\ &= 0 + a^{12} b_{12} - a^{12} b_{12} + 0 = 0. \end{aligned} \tag{8}$$

A symmetric second-RANK TENSOR A_{mn} has SCALAR invariants

$$s_1 = A_{11} + A_{22} + A_{22} \tag{9}$$

$$\begin{aligned} s_2 = A_{22}A_{33} + A_{33}A_{11} + A_{11}A_{22} - {A_{23}}^2 \\ - {A_{31}}^2 - {A_{12}}^2. \end{aligned} \tag{10}$$

Symmetroid

A QUARTIC SURFACE which is the locus of zeros of the DETERMINANT of a SYMMETRIC 4×4 matrix of linear forms. A general symmetroid has 10 ORDINARY DOUBLE POINTS (Jessop 1916, Hunt 1996).

References

Hunt, B. "Algebraic Surfaces." http://www.mathematik.uni-kl.de/~wwwagag/Galerie.html.

Hunt, B. "Symmetroids and Weddle Surfaces." §B.5.3 in *The Geometry of Some Special Arithmetic Quotients.* New York: Springer-Verlag, pp. 315–319, 1996.

Jessop, C. *Quartic Surfaces with Singular Points.* Cambridge, England: Cambridge University Press, p. 166, 1916.

Symmetry

An intrinsic property of a mathematical object which causes it to remain invariant under certain classes of transformations (such as ROTATION, REFLECTION, INVERSION, or more abstract operations). The mathematical study of symmetry is systematized and formalized in the extremely powerful and beautiful AREA of mathematics called GROUP THEORY.

Symmetry can be present in the form of coefficients of equations as well as in the physical arrangement of objects. By classifying the symmetry of polynomial equations using the machinery of GROUP THEORY, for example, it is possible to prove the unsolvability of the general QUINTIC EQUATION.

In physics, an extremely powerful theorem of Noether states that each symmetry of a system leads to a physically conserved quantity. Symmetry under TRANSLATION corresponds to momentum conservation, symmetry under ROTATION to angular momentum conservation, symmetry in time to energy conservation, etc.

see also GROUP THEORY

References

Eppstein, D. "Symmetry and Group Theory." http://www.ics.uci.edu/~eppstein/junkyard/sym.html.

Farmer, D. *Groups and Symmetry.* Providence, RI: Amer. Math. Soc., 1995.

Pappas, T. "Art & Dynamic Symmetry." *The Joy of Mathematics.* San Carlos, CA: Wide World Publ./Tetra, pp. 154–155, 1989.

Rosen, J. *Symmetry in Science: An Introduction to the General Theory.* New York: Springer-Verlag, 1995.

Schattschneider, D. *Visions of Symmetry: Notebooks, Periodic Drawings, and Related Work of M. C. Escher.* New York: W. H. Freeman, 1990.

Stewart, I. and Golubitsky, M. *Fearful Symmetry.* New York: Viking Penguin, 1993.

Symmetry Group

see GROUP

Symmetry Operation

Symmetry operations include the IMPROPER ROTATION, INVERSION OPERATION, MIRROR PLANE, and ROTATION. Together, these operations create 32 crystal classes corresponding to the 32 POINT GROUPS.

The INVERSION OPERATION takes

$$(x, y, z) \to (-x, -y, -z)$$

and is denoted i. When used in conjunction with a ROTATION, it becomes an IMPROPER ROTATION. An IMPROPER ROTATION by $360°/n$ is denoted $\bar{n}$ (or S_n). For periodic crystals, the CRYSTALLOGRAPHY RESTRICTION allows only the IMPROPER ROTATIONS $\bar{1}$, $\bar{2}$, $\bar{3}$, $\bar{4}$, and $\bar{6}$.

The MIRROR PLANE symmetry operation takes

$$(x, y, z) \to (x, y, -z), (x, -y, z) \to (x, -y, z),$$

etc., which is equivalent to $\bar{2}$. Invariance under reflection can be denoted $n\sigma_v$ or $n\sigma_h$. The ROTATION symmetry operation for $360°/n$ is denoted n (or C_n). For periodic crystals, CRYSTALLOGRAPHY RESTRICTION allows only 1, 2, 3, 4, and 6.

Symmetry operations can be indicated with symbols such as C_n, S_n, E, i, $n\sigma_v$, and $n\sigma_h$.

1. C_n indicates ROTATION about an n-fold symmetry axis.
2. S_n indicates IMPROPER ROTATION about an n-fold symmetry axis.
3. E (or I) indicates invariance under TRANSLATION.
4. i indicates a center of symmetry under INVERSION.

5. $n\sigma_v$ indicates invariance under n vertical REFLECTIONS.
6. $n\sigma_h$ indicates invariance under n horizontal REFLECTIONS.

see also CRYSTALLOGRAPHY RESTRICTION, IMPROPER ROTATION, INVERSION OPERATION, MIRROR PLANE, POINT GROUPS, ROTATION, SYMMETRY

Symmetry Principle

SYMMETRIC POINTS are preserved under a MÖBIUS TRANSFORMATION.

see also MÖBIUS TRANSFORMATION, SYMMETRIC POINTS

Symplectic Diffeomorphism

A MAP $T : (M_1, \omega_1) \to (M_2, \omega_2)$ between the SYMPLECTIC MANIFOLDS (M_1, ω_1) and (M_2, ω_2) which is a DIFFEOMORPHISM and $T^*(\omega_2) = \omega_1$ (where T^* is the PULLBACK MAP induced by T, i.e., the derivative of the DIFFEOMORPHISM T acting on tangent vectors). A symplectic diffeomorphism is also known as a SYMPLECTOMORPHISM or CANONICAL TRANSFORMATION.

see also DIFFEOMORPHISM, PULLBACK MAP, SYMPLECTIC MANIFOLD

References

Guillemin, V. and Sternberg, S. *Symplectic Techniques in Physics.* New York: Cambridge University Press, p. 34, 1984.

Symplectic Form

A symplectic form on a SMOOTH MANIFOLD M is a smooth closed 2-FORM ω on M which is nondegenerate such that at every point m, the alternating bilinear form ω_m on the TANGENT SPACE $T_m M$ is nondegenerate.

A symplectic form on a VECTOR SPACE V over F_q is a function $f(x, y)$ (defined for all $x, y \in V$ and taking values in F_q) which satisfies

$$f(\lambda_1 x_1 + \lambda_2 x_2, y) = \lambda_1 f(x_1, y) + \lambda_2 f(x_2, y),$$

$$f(y, x) = -f(x, y),$$

and

$$f(x, x) = 0.$$

Symplectic forms can exist on M (or V) only if M (or V) is EVEN-dimensional.

Symplectic Group

The symplectic group $Sp_n(q)$ for n EVEN is the GROUP of elements of the GENERAL LINEAR GROUP GL_n that preserve a given nonsingular SYMPLECTIC FORM. Any such MATRIX has DETERMINANT 1.

see also GENERAL LINEAR GROUP, LIE-TYPE GROUP, PROJECTIVE SYMPLECTIC GROUP, SYMPLECTIC FORM

References

Conway, J. H.; Curtis, R. T.; Norton, S. P.; Parker, R. A.; and Wilson, R. A. "The Groups $Sp_n(q)$ and $PSp_n(q) = S_n(q)$." §2.3 in *Atlas of Finite Groups: Maximal Subgroups and Ordinary Characters for Simple Groups.* Oxford, England: Clarendon Press, pp. x–xi, 1985.

Wilson, R. A. "ATLAS of Finite Group Representation." `http://for.mat.bham.ac.uk/atlas#symp`.

Symplectic Manifold

A pair (M, ω), where M is a MANIFOLD and ω is a SYMPLECTIC FORM on M. The PHASE SPACE $\mathbb{R}^{2n} = \mathbb{R}^n \times \mathbb{R}^n$ is a symplectic manifold. Near every point on a symplectic manifold, it is possible to find a set of local "Darboux coordinates" in which the SYMPLECTIC FORM has the simple form

$$\omega = \sum_k dq_k \wedge dp_k$$

(Sjamaar 1996), where $dq_k \wedge dp_k$ is a WEDGE PRODUCT.

see also MANIFOLD, SYMPLECTIC DIFFEOMORPHISM, SYMPLECTIC FORM

References

Sjamaar, R. "Symplectic Reduction and Riemann-Roch Formulas for Multiplicities." *Bull. Amer. Math. Soc.* **33**, 327–338, 1996.

Symplectic Map

A MAP which preserves the sum of AREAS projected onto the set of (p_i, q_i) planes. It is the generalization of an AREA-PRESERVING MAP.

see also AREA-PRESERVING MAP, LIOUVILLE'S PHASE SPACE THEOREM

Symplectomorphism

see SYMPLECTIC DIFFEOMORPHISM

Synclastic

A surface on which the GAUSSIAN CURVATURE K is everywhere POSITIVE. When K is everywhere NEGATIVE, a surface is called ANTICLASTIC. A point at which the GAUSSIAN CURVATURE is POSITIVE is called an ELLIPTIC POINT.

see also ANTICLASTIC, ELLIPTIC POINT, GAUSSIAN QUADRATURE, HYPERBOLIC POINT, PARABOLIC POINT, PLANAR POINT

Synergetics

Synergetics deals with systems composed of many subsystems which may each be of a very different nature. In particular, synergetics treats systems in which cooperation among subsystems creates organized structure on macroscopic scales (Haken 1993). Examples of problems treated by synergetics include BIFURCATIONS, phase transitions in physics, convective instabilities, coherent oscillations in lasers, nonlinear oscillations in electrical circuits, population dynamics, etc.

see also BIFURCATION, CHAOS, DYNAMICAL SYSTEM

References

Haken, H. *Synergetics, an Introduction: Nonequilibrium Phase Transitions and Self-Organization in Physics, Chemistry, and Biology, 3rd rev. enl. ed.* New York: Springer-Verlag, 1983.

Haken, H. *Advanced Synergetics: Instability Hierarchies of Self-Organizing Systems and Devices.* New York: Springer-Verlag, 1993.

Mikhailov, A. S. *Foundations of Synergetics: Distributed Active Systems, 2nd ed.* New York: Springer-Verlag, 1994.

Mikhailov, A. S. and Loskutov, A. Y. *Foundations of Synergetics II: Complex Patterns, 2nd ed., enl. rev.* New York: Springer-Verlag, 1996.

Synthesized Beam

see DIRTY BEAM

Syntonic Comma

see COMMA OF DIDYMUS

Syracuse Algorithm

see COLLATZ PROBLEM

Syracuse Problem

see COLLATZ PROBLEM

System of Differential Equations

see ORDINARY DIFFERENTIAL EQUATION

System of Equations

Let a linear system of equations be denoted

$$\mathsf{A}\mathbf{X} = \mathbf{Y}, \tag{1}$$

where A is a MATRIX and $\mathbf{X}$ and $\mathbf{Y}$ are VECTORS. As shown by CRAMER'S RULE, there is a unique solution if A has a MATRIX INVERSE A^{-1}. In this case,

$$\mathbf{X} = \mathsf{A}^{-1}\mathbf{Y}. \tag{2}$$

If $\mathbf{Y} = \mathbf{0}$, then the solution is $\mathbf{X} = \mathbf{0}$. If A has no MATRIX INVERSE, then the solution SUBSPACE is either a LINE or the EMPTY SET. If two equations are multiples of each other, solutions are of the form

$$\mathbf{X} = \mathbf{A} + t\mathbf{B}, \tag{3}$$

for t a REAL NUMBER.

see also CRAMER'S RULE, MATRIX INVERSE

Syzygies Problem

The problem of finding all independent irreducible algebraic relations among any finite set of QUANTICS.

see also QUANTIC

Syzygy

A technical mathematical object defined in terms of a POLYNOMIAL RING of n variables over a FIELD k.

see also FUNDAMENTAL SYSTEM, HILBERT BASIS THEOREM, SYZYGIES PROBLEM

References

Hilbert, D. "Über die Theorie der algebraischen Formen." *Math. Ann.* **36**, 473–534, 1890.

Iyanaga, S. and Kawada, Y. (Eds.). "Syzygy Theory." §364F in *Encyclopedic Dictionary of Mathematics.* Cambridge, MA: MIT Press, p. 1140, 1980.

Szilassi Polyhedron

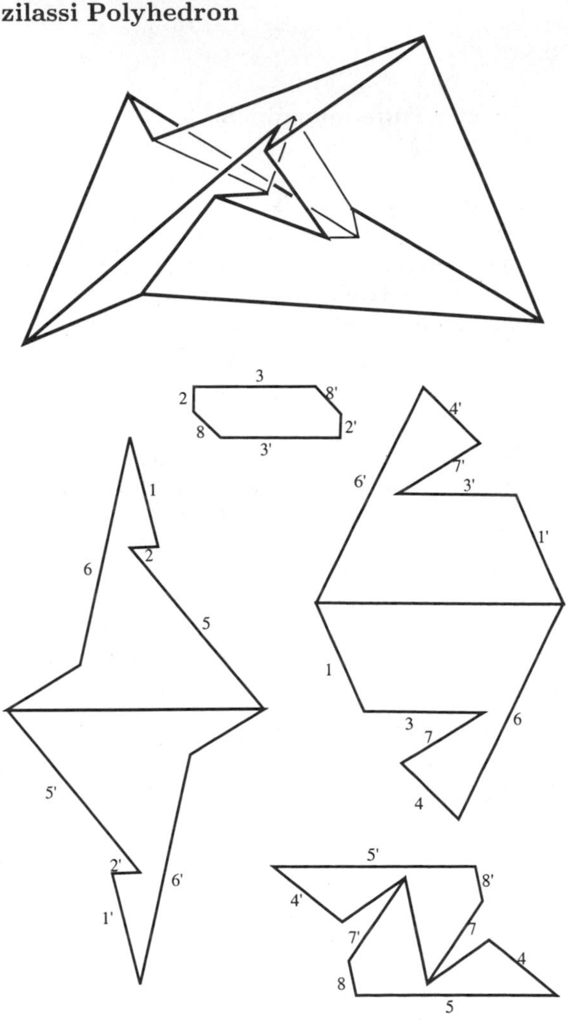

A POLYHEDRON which is topologically equivalent to a TORUS and for which every pair of faces has an EDGE in common. This polyhedron was discovered by L. Szilassi in 1977. Its SKELETON is equivalent to the seven-color torus map illustrated below.

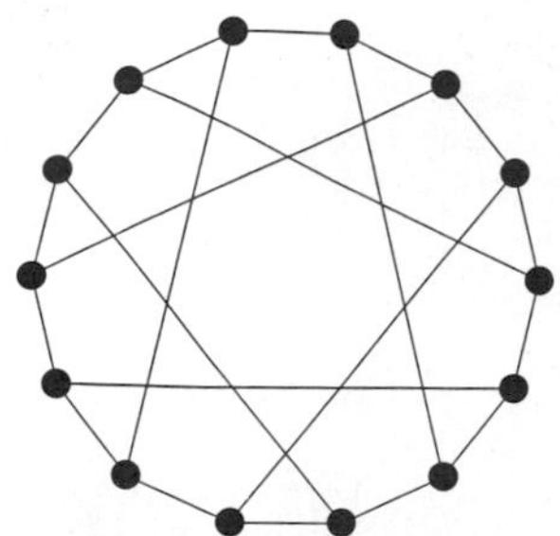

The Szilassi polyhedron has 14 VERTICES, seven faces, and 21 EDGES, and is the DUAL POLYHEDRON of the CSÁSZÁR POLYHEDRON.

see also CSÁSZÁR POLYHEDRON, TOROIDAL POLYHEDRON

References

Eppstein, D. "Polyhedra and Polytopes." http://www.ics.uci.edu/~eppstein/junkyard/polytope.html.

Gardner, M. *Fractal Music, Hypercards, and More Mathematical Recreations from Scientific American Magazine.* New York: W. H. Freeman, pp. 118–120, 1992.

Hart, G. "Toroidal Polyhedra." http://www.li.net/~george/virtual-polyhedra/toroidal.html.

Szpiro's Conjecture

A conjecture which relates the minimal DISCRIMINANT of an ELLIPTIC CURVE to the CONDUCTOR. If true, it would imply FERMAT'S LAST THEOREM for sufficiently large exponents.

see also CONDUCTOR, DISCRIMINANT (ELLIPTIC CURVE), ELLIPTIC CURVE

References

Cox, D. A. "Introduction to Fermat's Last Theorem." *Amer. Math. Monthly* **101**, 3–14, 1994.

T

t-Distribution

see STUDENT'S *t*-DISTRIBUTION

T-Polyomino

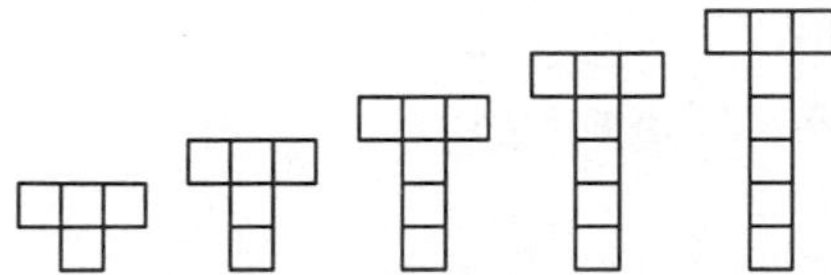

The order n T-polyomino consists of a vertical line of $n-3$ squares capped by a horizontal line of three squares centered on the line.

see also L-POLYOMINO, SKEW POLYOMINO, SQUARE POLYOMINO, STRAIGHT POLYOMINO

T-Puzzle

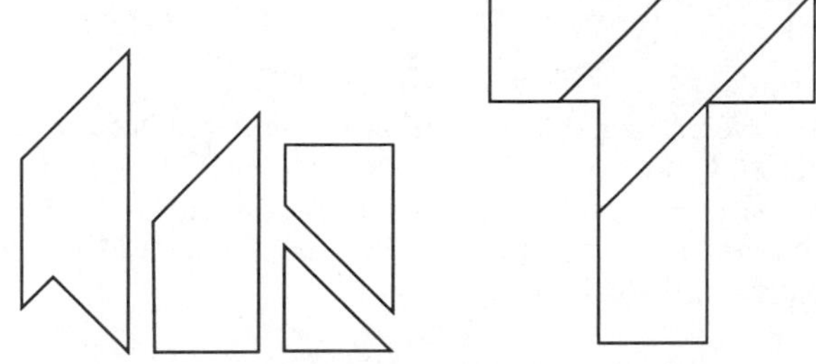

The DISSECTION of the four pieces shown at left into the capital letter "T" shown at right.

see also DISSECTION

References

Pappas, T. "The T Problem." *The Joy of Mathematics.* San Carlos, CA: Wide World Publ./Tetra, pp. 35 and 230, 1989.

T2-Separation Axiom

Finite SUBSETS are CLOSED.

see also CLOSURE

Tableau

see YOUNG TABLEAU

Tabu Search

A heuristic procedure which has proven efficient at solving COMBINATORIAL optimization problems.

References

Glover, F.; Taillard, E.; and De Werra, D. "A User's Guide to Tabu Search." *Ann. Oper. Res.* **41**, 3–28, 1993.

Piwakowski, K. "Applying Tabu Search to Determine New Ramsey Numbers." *Electronic J. Combinatorics* **3**, R6, 1–4, 1996. `http://www.combinatorics.org/Volume_3/volume3.html#R6`.

Tacnode

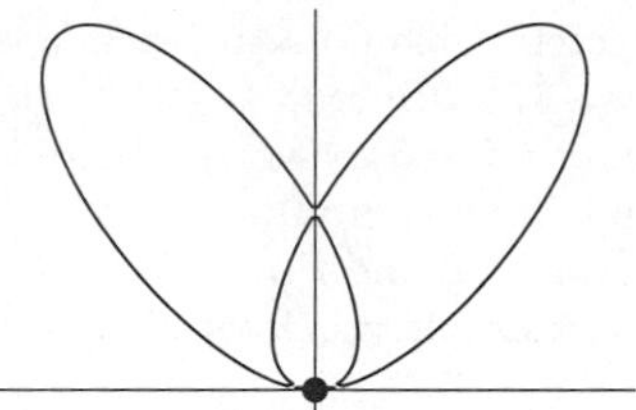

A DOUBLE POINT at which two OSCULATING CURVES are tangent. The above plot shows the tacnode of the curve $2x^4 - 3x^2y + y^2 - 2y^3 + y^4 = 0$. The LINKS CURVE also has a tacnode at the origin.

see also ACNODE, CRUNODE, DOUBLE POINT SPINODE

References

Walker, R. J. *Algebraic Curves.* New York: Springer-Verlag, pp. 57–58, 1978.

Tacpoint

A tangent point of two similar curves.

Tactix

see NIM

Tail Probability

Define T as the set of all points t with probabilities $P(x)$ such that $a > t \Rightarrow P(a \leq x \leq a + da) < P_0$ or $a < t \Rightarrow P(a \leq x \leq a + da < P_0$, where P_0 is a POINT PROBABILITY (often, the likelihood of an observed event). Then the associated tail probability is given by $\int_T P(x)\,dx$.

see also P-VALUE, POINT PROBABILITY

Tait Coloring

A 3-coloring of GRAPH EDGES so that no two EDGES of the same color meet at a VERTEX (Ball and Coxeter 1987, pp. 265–266).

see also EDGE (GRAPH), TAIT CYCLE, VERTEX (GRAPH)

References

Ball, W. W. R. and Coxeter, H. S. M. *Mathematical Recreations and Essays, 13th ed.* New York: Dover, 1987.

Tait Cycle

A set of circuits going along the EDGES of a GRAPH, each with an EVEN number of EDGES, such that just one of the circuits passes through each VERTEX (Ball and Coxeter 1987, pp. 265–266).

see also EDGE (GRAPH), EULERIAN CYCLE, HAMILTONIAN CYCLE, TAIT COLORING, VERTEX (GRAPH)

References

Ball, W. W. R. and Coxeter, H. S. M. *Mathematical Recreations and Essays, 13th ed.* New York: Dover, 1987.

Tait Flyping Conjecture

see FLYPING CONJECTURE

Tait's Hamiltonian Graph Conjecture

Every 3-connected cubic GRAPH (each VERTEX has VALENCY 3) has a HAMILTONIAN CIRCUIT. Proposed by Tait in 1880 and refuted by W. T. Tutte in 1946 with a counterexample, TUTTE'S GRAPH. If it had been true, it would have implied the FOUR-COLOR THEOREM. A simpler counterexample was later given by Kozyrev and Grinberg.

see also HAMILTONIAN CIRCUIT, TUTTE'S GRAPH, VERTEX (GRAPH)

References
Honsberger, R. *Mathematical Gems I.* Washington, DC: Math. Assoc. Amer., pp. 82–89, 1973.

Tait's Knot Conjectures

P. G. Tait undertook a study of KNOTS in response to Kelvin's conjecture that the atoms were composed of knotted vortex tubes of ether (Thomson 1869). He categorized KNOTS in terms of the number of crossings in a plane projection. He also made some conjectures which remained unproven until the discovery of JONES POLYNOMIALS.

Tait's FLYPING CONJECTURE states that the number of crossings is the same for any diagram of an ALTERNATING KNOT. This was proved true in 1986.

see also ALTERNATING KNOT, FLYPING CONJECTURE, JONES POLYNOMIAL, KNOT

References
Tait, P. G. "On Knots I, II, III." *Scientific Papers, Vol. 1.* London: Cambridge University Press, pp. 273–347, 1900.
Thomson, W. H. "On Vortex Motion." *Trans. Roy. Soc. Edinburgh* **25**, 217–260, 1869.

TAK Function

A RECURSIVE FUNCTION devised by I. Takeuchi. For INTEGERS x, y, and z, and a function h, it is

$$\mathrm{TAK}_h(x,y,z) = \begin{cases} h(x,y,z) & \text{for } x \leq y \\ h(h(x-1,y,z),h(y-1,z,x), & \text{for } x > y. \\ \quad h(z-1,x,y)) & \end{cases}$$

The number of function calls $F_0(a,b)$ required to compute $\mathrm{TAK}_0(a,b,0)$ for $a > b > 0$ is

$$F_0(a,b) = 4\sum_{k=0}^{b} \frac{a-b}{a+b-2k}\binom{a+b-2k}{b-k} - 3$$
$$= 1 + 4\sum_{k=0}^{b-1} \frac{a-b}{a+b-2k}\binom{a+b-2k}{b-k}$$

(Vardi 1991).

The TAK function is also connected with the BALLOT PROBLEM (Vardi 1991).

see also ACKERMANN FUNCTION, BALLOT PROBLEM

References
Gabriel, R. P. *Performance and Implementation of Lisp Systems.* Cambridge, MA: MIT Press, 1985.
Knuth, D. E. *Textbook Examples of Recursion.* Preprint 1990.
Vardi, I. "The Running Time of TAK." Ch. 9 in *Computational Recreations in Mathematica.* Redwood City, CA: Addison-Wesley, pp. 179–199, 1991.

Takagi Fractal Curve

see BLANCMANGE FUNCTION

Take-Away Game

see NIM-HEAP

Takeuchi Function

see TAK FUNCTION

Talbot's Curve

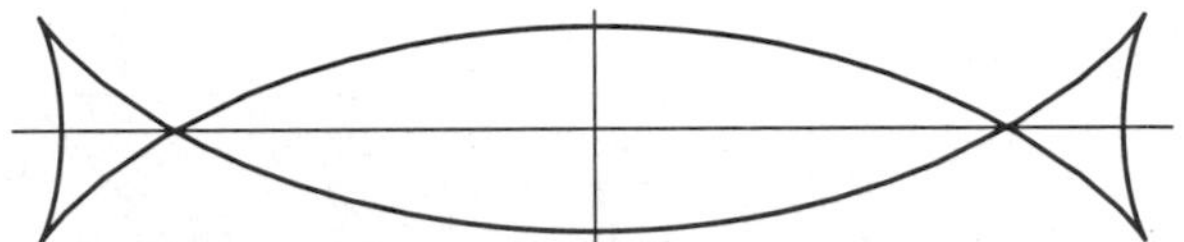

A curve investigated by Talbot which is the NEGATIVE PEDAL CURVE of an ELLIPSE with respect to its center. It has four CUSPS and two NODES, provided the ECCENTRICITY of the ELLIPSE is greater than $1/\sqrt{2}$. Its CARTESIAN EQUATION is

$$x = \frac{(a^2 + f^2 \sin^2 t)\cos t}{a}$$
$$y = \frac{(a^2 - 2f^2 + f^2 \sin^2 t)\sin t}{b},$$

where f is a constant.

References
Lockwood, E. H. *A Book of Curves.* Cambridge, England: Cambridge University Press, p. 157, 1967.
MacTutor History of Mathematics Archive. "Talbot's Curve." `http://www-groups.dcs.st-and.ac.uk/~history/Curves/Talbots.html`.

Talisman Hexagon

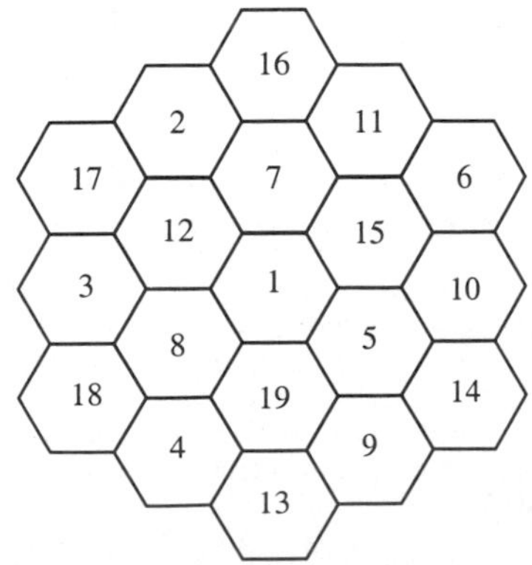

An (n,k)-talisman hexagon is an arrangement of nested hexagons containing the integers 1, 2, ..., $H_n = 3n(n-$

1) + 1, where H_n is the nth HEX NUMBER, such that the difference between all adjacent hexagons is at least as large as k. The hexagon illustrated above is a (3, 5)-talisman hexagon.

see also HEX NUMBER, MAGIC SQUARE, TALISMAN SQUARE

References

Madachy, J. S. *Madachy's Mathematical Recreations.* New York: Dover, pp. 111–112, 1979.

Talisman Square

1	5	3	7
9	11	13	15
2	6	4	8
10	12	14	16

5	15	9	12
10	1	6	3
13	16	11	14
2	8	4	7

15	1	12	4	9
20	7	22	18	24
16	2	13	5	10
21	8	23	19	25
17	3	14	6	11

28	10	31	13	34	16
19	1	22	4	25	7
29	11	32	14	35	17
20	2	23	5	26	8
30	12	33	15	36	18
21	3	24	6	27	9

An $n \times n$ ARRAY of the integers from 1 to n^2 such that the difference between any one integer and its neighbor (horizontally, vertically, or diagonally, without wrapping around) is greater than or equal to some value k is called a (n, k)-talisman square. The above illustrations show (4, 2)-, (4, 3)-, (5, 4)-, and (6, 8)-talisman squares.

see also ANTIMAGIC SQUARE, HETEROSQUARE, MAGIC SQUARE, TALISMAN HEXAGON

References

Madachy, J. S. *Madachy's Mathematical Recreations.* New York: Dover, pp. 110–113, 1979.

Weisstein, E. W. "Magic Squares." http://www.astro.virginia.edu/~eww6n/math/notebooks/MagicSquares.m.

Tame Algebra

Let A denote an $\mathbb{R}$-algebra, so that A is a VECTOR SPACE over R and

$$A \times A \to A$$

$$(x, y) \mapsto x \cdot y,$$

where $x \cdot y$ is vector multiplication which is assumed to be BILINEAR. Now define

$$Z \equiv \{x \in a : x \cdot y = 0 \text{ for some nonzero } y \in A\},$$

where $0 \in Z$. A is said to be tame if Z is a finite union of SUBSPACES of A. A 2-D 0-ASSOCIATIVE algebra is tame, but a 4-D 4-ASSOCIATIVE algebra and a 3-D 1-ASSOCIATIVE algebra need not be tame. It is conjectured that a 3-D 2-ASSOCIATIVE algebra is tame, and proven that a 3-D 3-ASSOCIATIVE algebra is tame if it possesses a multiplicative IDENTITY ELEMENT.

References

Finch, S. "Zero Structures in Real Algebras." http://www.mathsoft.com/asolve/zerodiv/zerodiv.html.

Tame Knot

A KNOT equivalent to a POLYGONAL KNOT. Knots which are not tame are called WILD KNOTS.

References

Rolfsen, D. *Knots and Links.* Wilmington, DE: Publish or Perish Press, p. 49, 1976.

Tangency Theorem

The external (internal) SIMILARITY POINT of two fixed CIRCLES is the point at which all the CIRCLES homogeneously (nonhomogeneously) tangent to the fixed CIRCLES have the same POWER and at which all the tangency secants intersect.

References

Dörrie, H. *100 Great Problems of Elementary Mathematics: Their History and Solutions.* New York: Dover, p. 157, 1965.

Tangent

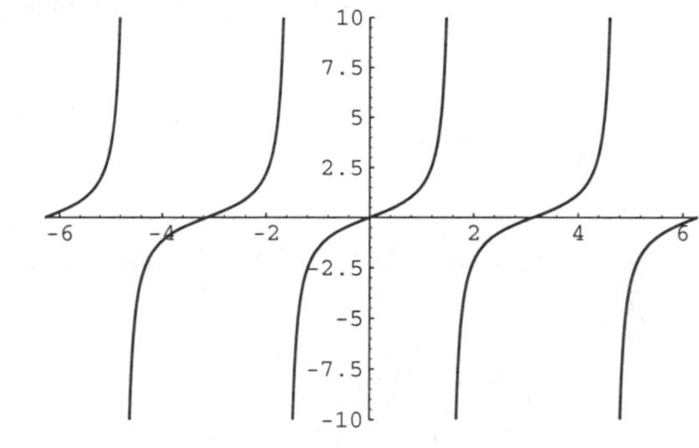

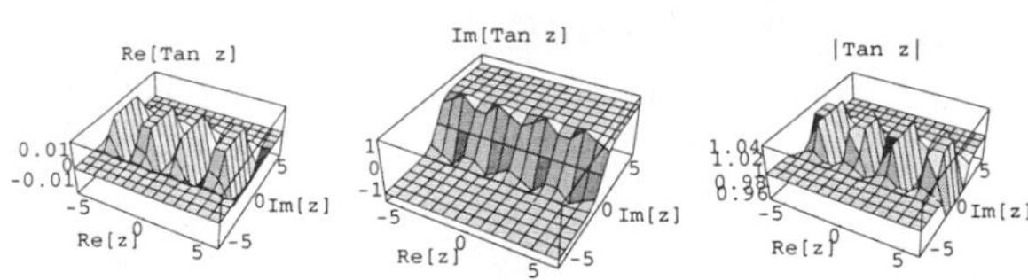

The tangent function is defined by

$$\tan\theta \equiv \frac{\sin\theta}{\cos\theta}, \tag{1}$$

where $\sin x$ is the SINE function and $\cos x$ is the COSINE function. The word "tangent," however, also has an important related meaning as a LINE or PLANE which touches a given curve or solid at a single point. These geometrical objects are then called a TANGENT LINE or TANGENT PLANE, respectively.

The MACLAURIN SERIES for the tangent function is

$$\tan x = \sum \frac{(-1)^{n-1}2^{2n}(2^{2n}-1)B_{2n}}{(2n)!}x^{2n-1}+\ldots$$
$$= x+\tfrac{1}{3}x^3+\tfrac{2}{15}x^5+\tfrac{17}{315}x^7+\tfrac{62}{2835}x^9+\ldots, \quad (2)$$

where B_n is a BERNOULLI NUMBER.

$\tan x$ is IRRATIONAL for any RATIONAL $x \neq 0$, which can be proved by writing $\tan x$ as a CONTINUED FRACTION

$$\tan x = \cfrac{x}{1-\cfrac{x^2}{3-\cfrac{x^2}{5-\cfrac{x^2}{7-\ldots}}}}. \quad (3)$$

Lambert derived another CONTINUED FRACTION expression for the tangent,

$$\tan x = \cfrac{1}{\frac{1}{x}-\cfrac{1}{\frac{3}{x}-\cfrac{1}{\frac{5}{x}-\cfrac{1}{\frac{7}{x}-\ldots}}}}. \quad (4)$$

An interesting identity involving the PRODUCT of tangents is

$$\prod_{k=1}^{\lfloor(n-1)/2\rfloor}\tan\left(\frac{k\pi}{n}\right)=\begin{cases}\sqrt{n} & \text{for } n \text{ odd}\\ 1 & \text{for } n \text{ even,}\end{cases} \quad (5)$$

where $\lfloor x \rfloor$ is the FLOOR FUNCTION. Another tangent identity is

$$\tan(n\tan^{-1}x)=\frac{1}{i}\frac{(1+ix)^n-(1-ix)^n}{(1+ix)^n+(1-ix)^m} \quad (6)$$

(Beeler *et al.* 1972, Item 16).

see also ALTERNATING PERMUTATION, COSINE, COTANGENT, INVERSE TANGENT, MORRIE'S LAW, SINE, TANGENT LINE, TANGENT PLANE

References

Abramowitz, M. and Stegun, C. A. (Eds.). "Circular Functions." §4.3 in *Handbook of Mathematical Functions with Formulas, Graphs, and Mathematical Tables, 9th printing.* New York: Dover, pp. 71–79, 1972.

Beeler, M.; Gosper, R. W.; and Schroeppel, R. *HAKMEM.* Cambridge, MA: MIT Artificial Intelligence Laboratory, Memo AIM-239, Feb. 1972.

Spanier, J. and Oldham, K. B. "The Tangent tan(x) and Cotangent cot(x) Functions." Ch. 34 in *An Atlas of Functions.* Washington, DC: Hemisphere, pp. 319–330, 1987.

Tangent Bifurcation

see FOLD BIFURCATION

Tangent Bundle

The tangent bundle TM of a SMOOTH MANIFOLD M is the SPACE of TANGENT VECTORS to points in the manifold, i.e., it is the set (x, v) where $x \in M$ and v is tangent to $x \in M$. For example, the tangent bundle to the CIRCLE is the CYLINDER.

see also COTANGENT BUNDLE, TANGENT VECTOR

Tangent Developable

A RULED SURFACE M is a tangent developable of a curve $\mathbf{y}$ if M can be parameterized by $\mathbf{x}(u,v)=\mathbf{y}(u)+v\mathbf{y}'(u)$. A tangent developable is a FLAT SURFACE.

see also BINORMAL DEVELOPABLE, NORMAL DEVELOPABLE

References

Gray, A. *Modern Differential Geometry of Curves and Surfaces.* Boca Raton, FL: CRC Press, pp. 341–343, 1993.

Tangent Hyperbolas Method

see HALLEY'S METHOD

Tangent Indicatrix

Let the SPEED $\boldsymbol{\sigma}$ of a closed curve on the unit sphere S^2 never vanish. Then the tangent indicatrix

$$\boldsymbol{\tau} \equiv \frac{\dot{\boldsymbol{\sigma}}}{|\dot{\boldsymbol{\sigma}}|}$$

is another closed curve on S^2. It is sometimes called the TANTRIX. If $\boldsymbol{\sigma}$ IMMERSES in S^2, then so will $\boldsymbol{\tau}$.

References

Solomon, B. "Tantrices of Spherical Curves." *Amer. Math. Monthly* **103**, 30–39, 1996.

Tangent Line

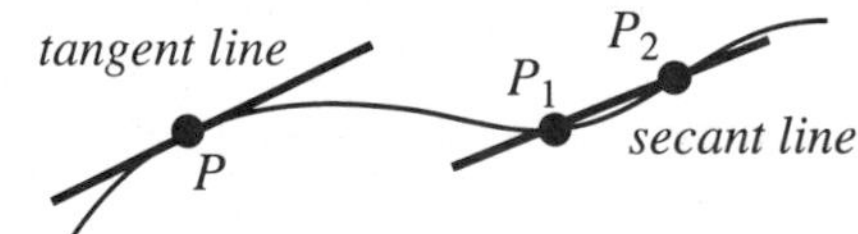

A tangent line is a LINE which meets a given curve at a single POINT.

see also CIRCLE TANGENTS, SECANT LINE, TANGENT, TANGENT PLANE, TANGENT SPACE, TANGENT VECTOR

References

Yates, R. C. "Instantaneous Center of Rotation and the Construction of Some Tangents." *A Handbook on Curves and Their Properties.* Ann Arbor, MI: J. W. Edwards, pp. 119–122, 1952.

Tangent Map

If $f : M \to N$, then the tangent map Tf associated to f is a VECTOR BUNDLE HOMEOMORPHISM $Tf : TM \to TN$ (i.e., a MAP between the TANGENT BUNDLES of M and N respectively). The tangent map corresponds to DIFFERENTIATION by the formula

$$Tf(v) = (f \circ \phi)'(0), \tag{1}$$

where $\phi'(0) = v$ (i.e., ϕ is a curve passing through the base point to v in TM at time 0 with velocity v). In this case, if $f : M \to N$ and $g : N \to O$, then the CHAIN RULE is expressed as

$$T(f \circ g) = Tf \circ Tg. \tag{2}$$

In other words, with this way of formalizing differentiation, the CHAIN RULE can be remembered by saying that "the process of taking the tangent map of a map is functorial." To a topologist, the form

$$(f \circ g)'(a) = f'(g(a)) \circ g'(a), \tag{3}$$

for all a, is more intuitive than the usual form of the CHAIN RULE.

see also DIFFEOMORPHISM

References

Gray, A. "Tangent Maps." §9.3 in *Modern Differential Geometry of Curves and Surfaces.* Boca Raton, FL: CRC Press, pp. 168–171, 1993.

Tangent Number

A number also called a ZAG NUMBER giving the number of EVEN ALTERNATING PERMUTATIONS. The first few are 1, 2, 16, 272, 7936, ... (Sloane's A000182).

see also ALTERNATING PERMUTATION, EULER ZIGZAG NUMBER, SECANT NUMBER

References

Knuth, D. E. and Buckholtz, T. J. "Computation of Tangent, Euler, and Bernoulli Numbers." *Math. Comput.* **21**, 663–688, 1967.

Sloane, N. J. A. Sequence A000182/M2096 in "An On-Line Version of the Encyclopedia of Integer Sequences."

Tangent Plane

A tangent plane is a PLANE which meets a given SURFACE at a single POINT. Let (x_0, y_0) be any point of a surface function $z = f(x, y)$. The surface has a nonvertical tangent plane at (x_0, y_0) with equation

$$z = f(x_0, y_0) + f_x(x_0, y_0)(x - x_0) + f_y(x_0, y_0)(y - y_0).$$

see also NORMAL VECTOR, TANGENT, TANGENT LINE, TANGENT SPACE, TANGENT VECTOR

Tangent Space

Let x be a point in an n-dimensional COMPACT MANIFOLD M, and attach at x a copy of $\mathbb{R}^n$ tangential to M. The resulting structure is called the TANGENT SPACE of M at x and is denoted $T_x M$. If γ is a smooth curve passing through x, then the derivative of γ at x is a VECTOR in $T_x M$.

see also TANGENT, TANGENT BUNDLE, TANGENT PLANE, TANGENT VECTOR

Tangent Vector

For a curve with POSITION VECTOR $\mathbf{r}(t)$, the unit tangent vector $\hat{\mathbf{T}}(t)$ is defined by

$$\hat{\mathbf{T}}(t) \equiv \frac{\mathbf{r}'(t)}{|\mathbf{r}'(t)|} = \frac{\frac{d\mathbf{r}}{dt}}{\left|\frac{d\mathbf{r}}{dt}\right|} \tag{1}$$

$$= \frac{\frac{d\mathbf{r}}{dt}}{\frac{ds}{dt}} \tag{2}$$

$$= \frac{d\mathbf{r}}{ds}, \tag{3}$$

where t is a parameterization variable and s is the ARC LENGTH. For a function given parametrically by $(f(t), g(t))$, the tangent vector relative to the point $(f(t), g(t))$ is therefore given by

$$x(t) = \frac{f'}{\sqrt{f'^2 + g'^2}} \tag{4}$$

$$y(t) = \frac{g'}{\sqrt{f'^2 + g'^2}}. \tag{5}$$

To actually place the vector tangent to the curve, it must be displaced by $(f(t), g(t))$. It is also true that

$$\frac{d\hat{\mathbf{T}}}{ds} = \kappa \hat{\mathbf{N}} \tag{6}$$

$$\frac{d\hat{\mathbf{T}}}{dt} = \kappa \frac{ds}{dt} \hat{\mathbf{N}} \tag{7}$$

$$[\dot{\mathbf{T}}, \ddot{\mathbf{T}}, \dddot{\mathbf{T}}] = \kappa^5 \frac{d}{ds}\left(\frac{\tau}{\kappa}\right), \tag{8}$$

where $\mathbf{N}$ is the NORMAL VECTOR, κ is the CURVATURE, and τ is the TORSION.

see also CURVATURE, NORMAL VECTOR, TANGENT, TANGENT BUNDLE, TANGENT PLANE, TANGENT SPACE, TORSION (DIFFERENTIAL GEOMETRY)

References

Gray, A. "Tangent and Normal Lines to Plane Curves." §5.5 in *Modern Differential Geometry of Curves and Surfaces.* Boca Raton, FL: CRC Press, pp. 85–90, 1993.

Tangential Angle

For a PLANE CURVE, the tangential angle ϕ is defined by

$$\rho\, d\phi = ds, \tag{1}$$

where s is the ARC LENGTH and ρ is the RADIUS OF CURVATURE. The tangential angle is therefore given by

$$\phi = \int_0^t s'(t)\kappa(t)\, dt, \tag{2}$$

where $\kappa(t)$ is the CURVATURE. For a plane curve $\mathbf{r}(t)$, the tangential angle $\phi(t)$ can also be defined by

$$\frac{\mathbf{r}'(t)}{|\mathbf{r}'(t)|} = \begin{bmatrix} \cos[\phi(t)] \\ \sin[\phi(t)] \end{bmatrix}. \tag{3}$$

Gray (1993) calls ϕ the TURNING ANGLE instead of the tangential angle.

see also ARC LENGTH, CURVATURE, PLANE CURVE, RADIUS OF CURVATURE, TORSION (DIFFERENTIAL GEOMETRY)

References

Gray, A. "The Turning Angle." §1.6 in *Modern Differential Geometry of Curves and Surfaces.* Boca Raton, FL: CRC Press, pp. 13–14, 1993.

Tangential Triangle

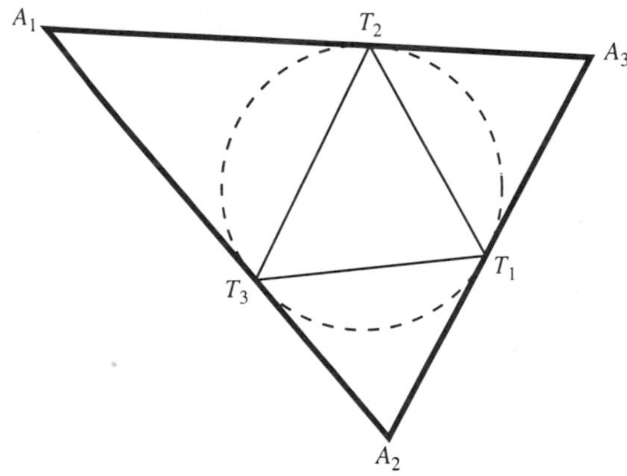

The TRIANGLE $\Delta T_1T_2T_3$ formed by the lines tangent to the CIRCUMCIRCLE of a given TRIANGLE $\Delta A_1A_2A_3$ at its VERTICES. It is the PEDAL TRIANGLE of $\Delta A_1A_2A_3$ with the CIRCUMCENTER as the PEDAL POINT. The TRILINEAR COORDINATES of the VERTICES of the tangential triangle are

$$\begin{aligned} A' &= -a : b : c \\ B' &= a : -b : c \\ C' &= a : b : -c. \end{aligned}$$

The CONTACT TRIANGLE and tangential triangle are perspective from the GERGONNE POINT.

see also CIRCUMCIRCLE, CONTACT TRIANGLE, GERGONNE POINT, PEDAL TRIANGLE, PERSPECTIVE

Tangential Triangle Circumcenter

A POINT with TRIANGLE CENTER FUNCTION

$$\alpha = a[b^2\cos(2B) + c^2\cos(2C) - a^2\cos(2A)].$$

It lies on the EULER LINE.

References

Kimberling, C. "Central Points and Central Lines in the Plane of a Triangle." *Math. Mag.* **67**, 163–187, 1994.

Tangents Law

see LAW OF TANGENTS

Tangle

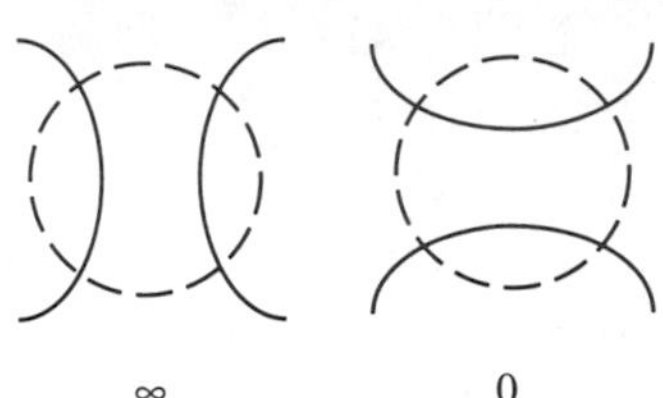

A region in a KNOT or LINK projection plane surrounded by a CIRCLE such that the KNOT or LINK crosses the circle exactly four times. Two tangles are equivalent if a sequence of REIDEMEISTER MOVES can be used to transform one into the other while keeping the four string endpoints fixed and not allowing strings to pass outside the CIRCLE.

The simplest tangles are the ∞-tangle and 0-tangle, shown above. A tangle with n left-handed twists is called an n-tangle, and one with n right-handed twists is called a $-n$-tangle. By placing tangles side by side, more complicated tangles can be built up such as (−2, 3, 2), etc. The link created by connecting the ends of the tangles is now described by the sequence of tangle symbols, known as CONWAY'S KNOT NOTATION. If tangles are multiplied by 0 and then added, the resulting tangle symbols are separated by commas. Additional symbols which are used are the period, colon, and asterisk.

Amazingly enough, two tangles described in this NOTATION are equivalent IFF the CONTINUED FRACTIONS of the form

$$2 + \cfrac{1}{3 + \cfrac{1}{-2}}$$

are equal (Burde and Zieschang 1985)! An ALGEBRAIC TANGLE is any tangle obtained by ADDITIONS and MULTIPLICATIONS of rational tangles (Adams 1994). Not all tangles are ALGEBRAIC.

see also ALGEBRAIC LINK, FLYPE, PRETZEL KNOT

References

Adams, C. C. *The Knot Book: An Elementary Introduction to the Mathematical Theory of Knots.* New York: W. H. Freeman pp. 41–51, 1994.

Burde, G. and Zieschang, H. *Knots.* Berlin: de Gruyter, 1985.

Tanglecube

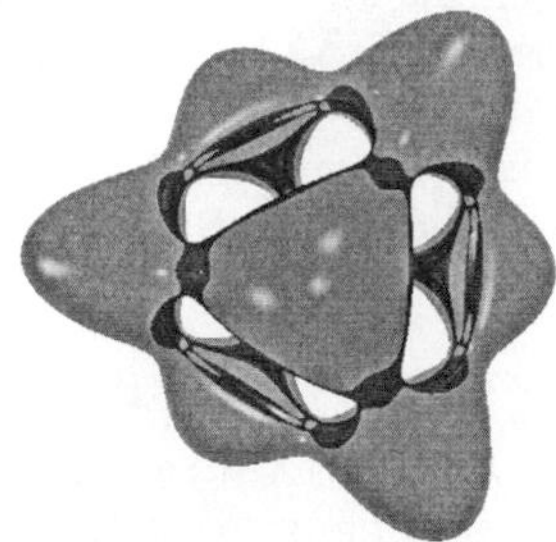

A QUARTIC SURFACE given by the implicit equation

$$x^4 - 5x^2 + y^4 - 5y^2 + z^4 - 5z^2 + 11.8 = 0.$$

References

Banchoff, T. "The Best Homework Ever?" http://www.brown.edu/Administration/Brown_Alumni_Monthly/12-96/features/homework.html.

Nordstrand, T. "Tangle." http://www.uib.no/people/nfytn/tangltxt.htm.

Tangled Hierarchy

A system in which a STRANGE LOOP appears.

see also STRANGE LOOP

References

Hofstadter, D. R. *Gödel, Escher, Bach: An Eternal Golden Braid.* New York: Vintage Books, p. 10, 1989.

Tangram

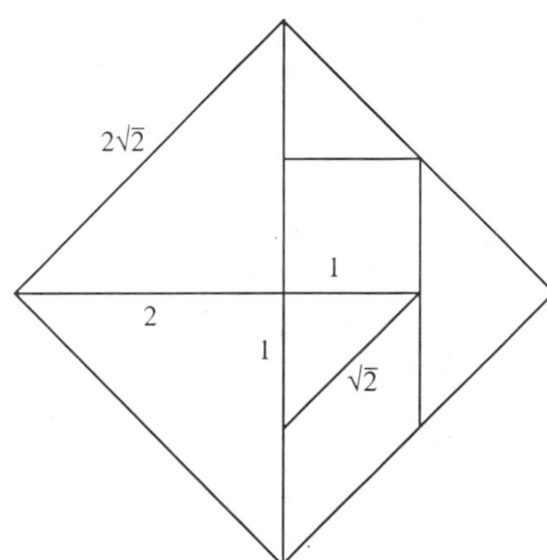

A combination of the above plane polygonal pieces such that the EDGES are coincident. There are 13 convex tangrams (where a "convex tangram" is a set of tangram pieces arranged into a CONVEX POLYGON).

see also ORIGAMI, STOMACHION

References

Cundy, H. and Rollett, A. *Mathematical Models, 3rd ed.* Stradbroke, England: Tarquin Pub., pp. 19–20, 1989.

Gardner, M. "Tangrams, Parts 1 and 2." Ch. 3-4 in *Time Travel and Other Mathematical Bewilderments.* New York: W. H. Freeman, 1988.

Johnston, S. *Fun with Tangrams Kit: 120 Puzzles with Two Complete Sets of Tangram Pieces.* New York: Dover, 1977.

Pappas, T. "Tangram Puzzle." *The Joy of Mathematics.* San Carlos, CA: Wide World Publ./Tetra, p. 212, 1989.

Tanh

see HYPERBOLIC TANGENT

Taniyama Conjecture

see TANIYAMA-SHIMURA CONJECTURE

Taniyama-Shimura Conjecture

A conjecture which arose from several problems proposed by Taniyama in an international mathematics symposium in 1955. Let E be an ELLIPTIC CURVE whose equation has INTEGER COEFFICIENTS, let N be the CONDUCTOR of E and, for each n, let a_n be the number appearing in the L-function of E. Then there exists a MODULAR FORM of weight two and level N which is an eigenform under the HECKE OPERATORS and has a FOURIER SERIES $\sum a_n q^n$.

The conjecture says, in effect, that every rational ELLIPTIC CURVE is a MODULAR FORM in disguise. Stated formally, the conjecture suggests that, for every ELLIPTIC CURVE $y^2 = Ax^3 + Bx^2 + Cx + D$ over the RATIONALS, there exist nonconstant MODULAR FUNCTIONS $f(z)$ and $g(z)$ of the same level N such that

$$[f(z)]^2 = A[g(z)]^2 + Cg(z) + D.$$

Equivalently, for every ELLIPTIC CURVE, there is a MODULAR FORM with the same DIRICHLET L-SERIES.

In 1985, starting with a fictitious solution to FERMAT'S LAST THEOREM, G. Frey showed that he could create an unusual ELLIPTIC CURVE which appeared not to be modular. If the curve were not modular, then this would show that if FERMAT'S LAST THEOREM were false, then the Taniyama-Shimura conjecture would also be false. Furthermore, if the Taniyama-Shimura conjecture were true, then so would be FERMAT'S LAST THEOREM!

However, Frey did not actually prove whether his curve was modular. The conjecture that Frey's curve *was* modular came to be called the "epsilon conjecture," and was quickly proved by Ribet (RIBET'S THEOREM) in 1986, establishing a very close link between two mathematical structures (the Taniyama-Shimura conjecture and FERMAT'S LAST THEOREM) which appeared previously to be completely unrelated.

As of the early 1990s, most mathematicians believed that the Taniyama-Shimura conjecture was not accessible to proof. However, A. Wiles was not one of these. He attempted to establish the correspondence between the set of ELLIPTIC CURVES and the set of modular elliptic curves by showing that the number of each was the same. Wiles accomplished this by "counting" Galois representations and comparing them with the number of modular forms. In 1993, after a monumental seven-year effort, Wiles (almost) proved the Taniyama-Shimura conjecture for special classes of curves called SEMISTABLE ELLIPTIC CURVES.

Wiles had tried to use horizontal Iwasawa theory to create a so-called CLASS NUMBER formula, but was initially unsuccessful and therefore used instead an extension of a result of Flach based on ideas from Kolyvagin. However, there was a problem with this extension which was discovered during review of Wiles' manuscript in September 1993. Former student Richard Taylor came to Princeton in early 1994 to help Wiles patch up this error. After additional effort, Wiles discovered the reason that the Flach/Kolyvagin approach was failing, and also discovered that it was precisely what had prevented Iwasawa theory from working.

With this additional insight, he was able to successfully complete the erroneous portion of the proof using Iwasawa theory, proving the SEMISTABLE case of the Taniyama-Shimura conjecture (Taylor and Wiles 1995, Wiles 1995) and, at the same time, establishing FERMAT'S LAST THEOREM as a true theorem.

see also ELLIPTIC CURVE, FERMAT'S LAST THEOREM, MODULAR FORM, MODULAR FUNCTION, RIBET'S THEOREM

References

Lang, S. "Some History of the Shimura-Taniyama Conjecture." *Not. Amer. Math. Soc.* **42**, 1301–1307, 1995.

Taylor, R. and Wiles, A. "Ring-Theoretic Properties of Certain Hecke Algebras." *Ann. Math.* **141**, 553–572, 1995.

Wiles, A. "Modular Elliptic-Curves and Fermat's Last Theorem." *Ann. Math.* **141**, 443–551, 1995.

Tank

see CYLINDRICAL SEGMENT

Tantrix

see TANGENT INDICATRIX

Tapering Function

see APODIZATION FUNCTION

Tarry-Escott Problem

For each POSITIVE INTEGER l, there exists a POSITIVE INTEGER n and a PARTITION of $\{1, \ldots, n\}$ as a disjoint union of two sets A and B, such that for $1 \leq i \leq l$,

$$\sum_{a \in A} a^i = \sum_{b \in B} b^i.$$

The results extended to three or more sets of INTEGERS are called PROUHET'S PROBLEM.

see also PROUHET'S PROBLEM

References

Dickson, L. E. *History of the Theory of Numbers, Vol. 2: Diophantine Analysis.* New York: Chelsea, pp. 709–710, 1971.

Hahn, L. "The Tarry-Escott Problem." Problem 10284. *Amer. Math. Monthly* **102**, 843–844, 1995.

Tarry Point

The point at which the lines through the VERTICES of a TRIANGLE PERPENDICULAR to the corresponding sides of the first BROCARD TRIANGLE, are CONCURRENT. The Tarry point lies on the CIRCUMCIRCLE opposite the STEINER POINT. It has TRIANGLE CENTER FUNCTION

$$\alpha = \frac{bc}{b^4 + c^4 - a^2b^2 - a^2c^2} = \sec(A + \omega),$$

where ω is the BROCARD ANGLE. The SIMSON LINE of the Tarry point is PERPENDICULAR to the line OK, when O is the CIRCUMCENTER and K is the LEMOINE POINT.

see also BROCARD ANGLE, BROCARD TRIANGLES, CIRCUMCIRCLE, LEMOINE POINT, SIMSON LINE, STEINER POINTS

References

Gallatly, W. *The Modern Geometry of the Triangle, 2nd ed.* London: Hodgson, p. 102, 1913.

Johnson, R. A. *Modern Geometry: An Elementary Treatise on the Geometry of the Triangle and the Circle.* Boston, MA: Houghton Mifflin, pp. 281–282, 1929.

Kimberling, C. "Central Points and Central Lines in the Plane of a Triangle." *Math. Mag.* **67**, 163–187, 1994.

Tarski's Theorem

Tarski's theorem states that the first-order theory of the FIELD of REAL NUMBERS is DECIDABLE. However, the best-known ALGORITHM for eliminating QUANTIFIERS is doubly exponential in the number of QUANTIFIER blocks (Heintz *et al.* 1989).

References

Heintz, J.; Roy, R.-F.; and Solerno, P. "Complexité du principe de Tarski-Seidenberg." *C. R. Acad. Sci. Paris Sér. I Math.* **309**, 825–830, 1989.

Marker, D. "Model Theory and Exponentiation." *Not. Amer. Math. Soc.* **43**, 753–759, 1996.

Tarski, A. "Sur les ensembles définissables de nombres réels." *Fund. Math.* **17**, 210–239, 1931.

Tarski, A. "A Decision Method for Elementary Algebra and Geometry." RAND Corp. monograph, 1948.

Tau Conjecture

Also known as RAMANUJAN'S HYPOTHESIS. Ramanujan proposed that

$$\tau(n) \sim \mathcal{O}(n^{11/2+\epsilon}),$$

where $\tau(n)$ is the TAU FUNCTION, defined by

$$\sum_{n=1}^{\infty} \tau(n)x^n = x(1 - 3x + 5x^3 - 7x^6 + \ldots)^8.$$

This was proven by Deligne (1974), who was subsequently awarded the FIELDS MEDAL for his proof.

see also TAU FUNCTION

References

Deligne, P. "La conjecture de Weil. I." *Inst. Hautes Études Sci. Publ. Math.* **43**, 273–307, 1974.

Deligne, P. "La conjecture de Weil. II." *Inst. Hautes Études Sci. Publ. Math.* **52**, 137–252, 1980.
Hardy, G. H. *Ramanujan: Twelve Lectures on Subjects Suggested by His Life and Work, 3rd ed.* New York: Chelsea, p. 169, 1959.

Tau-Dirichlet Series

$$\tau_{DS}(s) \equiv \sum_{n=1}^{\infty} \frac{\tau(n)}{n^s},$$

where $\tau(n)$ is the TAU FUNCTION. Ramanujan conjectured that all nontrivial zeros of $f(z)$ lie on the line $\Re[s] = 6$, where

$$f(s) \equiv \sum_{n=1}^{\infty} \tau(n) n^{-z}$$

and $\tau(n)$ is the TAU FUNCTION.

see also TAU FUNCTION

References

Spira, R. "Calculation of the Ramanujan Tau-Dirichlet Series." *Math. Comput.* **27**, 379–385, 1973.
Yoshida, H. "On Calculations of Zeros of L-Functions Related with Ramanujan's Discriminant Function on the Critical Line." *J. Ramanujan Math. Soc.* **3**, 87–95, 1988.

Tau Function

A function $\tau(n)$ related to the DIVISOR FUNCTION $\sigma_k(n)$, also sometimes called RAMANUJAN'S TAU FUNCTION. It is given by the GENERATING FUNCTION

$$\sum_{n=1}^{\infty} \tau(n) x^n = \prod_{n=1}^{\infty} (1 - x^n)^{24}, \tag{1}$$

and the first few values are 1, −24, 252, −1472, 4380, ... (Sloane's A000594). $\tau(n)$ is also given by

$$g(-x) = \sum_{n=1}^{\infty} (-1)^n \tau(n) x^n \tag{2}$$

$$g(x^2) = \sum_{n=1}^{\infty} \tau(\tfrac{1}{2}n) x^n \tag{3}$$

$$\sum_{n=1}^{\infty} \tau(n) x^n = x(1 - 3x + 5x^3 - 7x^6 + \ldots)^8. \tag{4}$$

In ORE'S CONJECTURE, the tau function appears as the number of DIVISORS of n. Ramanujan conjectured and Mordell proved that if (n, n'), then

$$\tau(nn') = \tau(n)\tau(n'). \tag{5}$$

Ramanujan conjectured and Watson proved that $\tau(n)$ is divisible by 691 for almost all n. If

$$\tau(p) \equiv 0 \pmod{p}, \tag{6}$$

then

$$\tau(pn) \equiv 0 \pmod{p}. \tag{7}$$

Values of p for which the first equation holds are $p = 2$, 3, 5, 7, 23.

Ramanujan also studied

$$f(x) \equiv \sum_{n=1}^{\infty} \tau(n) n^{-s}, \tag{8}$$

which has properties analogous to the RIEMANN ZETA FUNCTION. It satisfies

$$\frac{f(s)\Gamma(s)}{(2\pi)^s} = \frac{f(12-s)}{(2\pi)^{12-s}}, \tag{9}$$

and Ramanujan's TAU-DIRICHLET SERIES conjecture alleges that all nontrivial zeros of $f(s)$ lie on the line $\Re[s] = 6$. f can be split up into

$$f(6 + it) = z(t) e^{-i\theta(t)}, \tag{10}$$

where

$$z(t) = \Gamma(6 + it) f(6 + it)(2\pi)^{-it} \times \sqrt{\frac{\sinh(\pi t)}{\pi t(1+t^2)(4+t^2)(9+t^2)(16+t^2)(25+t^2)}} \tag{11}$$

$$\theta(t) = -\tfrac{1}{2} i \ln\left[\frac{\Gamma(6+it)}{\Gamma(6-it)}\right] - t\ln(2\pi). \tag{12}$$

The SUMMATORY tau function is given by

$$T(n) = \sum_{n \le x}{}' \tau(n). \tag{13}$$

Here, the prime indicates that when x is an INTEGER, the last term $\tau(x)$ should be replaced by $\frac{1}{2}\tau(x)$.

Ramanujan's tau theta function $Z(t)$ is a REAL function for REAL t and is analogous to the RIEMANN-SIEGEL FUNCTION Z. The number of zeros in the critical strip from $t = 0$ to T is given by

$$N(t) = \frac{\Theta(T) + \Im\{\ln[\tau_{DS}(6 + iT)]\}}{\pi}, \tag{14}$$

where Θ is the RIEMANN THETA FUNCTION and τ_{DS} is the TAU-DIRICHLET SERIES, defined by

$$\tau_{DS}(s) \equiv \sum_{n=1}^{\infty} \frac{\tau(n)}{n^s}. \tag{15}$$

Ramanujan conjectured that the nontrivial zeros of the function are all real.

Ramanujan's τ_z function is defined by

$$\tau_z(t) = \frac{\Gamma(6+it)(2\pi)^{-it}}{\tau_{DS}(6+it)\sqrt{\frac{\sinh(\pi t)}{\pi t \prod_{k=1}^{5} k^2+t^2}}}, \tag{16}$$

where $\tau_{DS}(z)$ is the TAU-DIRICHLET SERIES.

see also ORE'S CONJECTURE, TAU CONJECTURE, TAU-DIRICHLET SERIES

References

Hardy, G. H. "Ramanujan's Function $\tau(n)$." Ch. 10 in *Ramanujan: Twelve Lectures on Subjects Suggested by His Life and Work, 3rd ed.* New York: Chelsea, 1959.

Sloane, N. J. A. Sequence A000594/M5153 in "An On-Line Version of the Encyclopedia of Integer Sequences."

Tauberian Theorem

A Tauberian theorem is a theorem which deduces the convergence of an INFINITE SERIES on the basis of the properties of the function it defines and any kind of auxiliary HYPOTHESIS which prevents the general term of the series from converging to zero too slowly.

see also HARDY-LITTLEWOOD TAUBERIAN THEOREM

Tautochrone Problem

Find the curve down which a bead placed anywhere will fall to the bottom in the same amount of time. The solution is a CYCLOID, a fact first discovered and published by Huygens in *Horologium oscillatorium* (1673). Huygens also constructed the first pendulum clock with a device to ensure that the pendulum was isochronous by forcing the pendulum to swing in an arc of a CYCLOID.

The parametric equations of the CYCLOID are

$$x = a(\theta - \sin\theta) \tag{1}$$

$$y = a(1 - \cos\theta). \tag{2}$$

To see that the CYCLOID satisfies the tautochrone property, consider the derivatives

$$x' = a(1 - \cos\theta) \tag{3}$$

$$y' = a\sin\theta, \tag{4}$$

and

$$\begin{aligned} x'^2 + y'^2 &= a^2[(1 - 2\cos\theta + \cos^2\theta) + \sin^2\theta] \\ &= 2a^2(1 - \cos\theta). \end{aligned} \tag{5}$$

Now

$$\tfrac{1}{2}mv^2 = mgy \tag{6}$$

$$v = \frac{ds}{dt} = \sqrt{2gy} \tag{7}$$

$$\begin{aligned} dt &= \frac{ds}{\sqrt{2gy}} = \frac{\sqrt{dx^2+dy^2}}{\sqrt{2gy}} \\ &= \frac{a\sqrt{2(1-\cos\theta)}\,d\theta}{\sqrt{2ga(1-\cos\theta)}} = \sqrt{\frac{a}{g}}\,d\theta, \end{aligned} \tag{8}$$

so the time required to travel from the top of the CYCLOID to the bottom is

$$T = \int_0^\pi dt = \sqrt{\frac{a}{g}}\,\pi. \tag{9}$$

However, from an intermediate point θ_0,

$$v = \frac{ds}{dt} = \sqrt{2g(y - y_0)}, \tag{10}$$

so

$$\begin{aligned} T &= \int_{\theta_0}^{\pi} \frac{\sqrt{2a^2(1-\cos\theta)}}{2ag(\cos\theta_0 - \cos\theta)}\,d\theta \\ &= \sqrt{\frac{a}{g}} \int_{\theta_0}^{\pi} \sqrt{\frac{1-\cos\theta}{\cos\theta_0 - \cos\theta}}\,d\theta \\ &= \sqrt{\frac{a}{g}} \int_{\theta_0}^{\pi} \frac{\sin(\frac{1}{2}\theta)\,d\theta}{\sqrt{\cos^2(\frac{1}{2}\theta_0) - \cos^2(\frac{1}{2}\theta)}}. \end{aligned} \tag{11}$$

Now let

$$u = \frac{\cos(\frac{1}{2}\theta)}{\cos(\frac{1}{2}\theta_0)} \tag{12}$$

$$du = -\frac{\sin(\frac{1}{2}\theta)d\theta}{2\cos(\theta_0)}, \tag{13}$$

so

$$T = -2\sqrt{\frac{a}{g}} \int_1^0 \frac{du}{\sqrt{1-u^2}} = 2\sqrt{\frac{a}{g}}\,[\sin^{-1} u]_0^1 = \pi\sqrt{\frac{a}{g}}, \tag{14}$$

and the amount of time is the same from any point!

see also BRACHISTOCHRONE PROBLEM, CYCLOID

References

Muterspaugh, J.; Driver, T.; and Dick, J. E. "The Cycloid and Tautochronism." http://ezinfo.ucs.indiana.edu/~jedick/project/intro.html.

Muterspaugh, J.; Driver, T.; and Dick, J. E. "P221 Tautochrone Problem." http://ezinfo.ucs.indiana.edu/~jedick/project/project.html.

Wagon, S. *Mathematica in Action.* New York: W. H. Freeman, pp. 54–60 and 384–385, 1991.

Tautology

A logical statement in which the conclusion is equivalent to the premise. If p is a tautology, it is written $\models p$.

Taxicab Number

The nth taxicab number $\mathrm{Ta}(n)$ is the smallest number representable in n ways as a sum of POSITIVE CUBES. The numbers derive their name from the HARDY-RAMANUJAN NUMBER

$$\begin{aligned}\mathrm{Ta}(2) &= 1729\\ &= 1^3 + 12^3\\ &= 9^3 + 10^3, \qquad (1)\end{aligned}$$

which is associated with the following story told about Ramanujan by G. H. Hardy. "Once, in the taxi from London, Hardy noticed its number, 1729. He must have thought about it a little because he entered the room where Ramanujan lay in bed and, with scarcely a hello, blurted out his disappointment with it. It was, he declared, 'rather a dull number,' adding that he hoped that wasn't a bad omen. 'No, Hardy,' said Ramanujan, 'it is a very interesting number. It is the smallest number expressible as the sum of two [POSITIVE] cubes in two different ways'" (Hofstadter 1989, Kanigel 1991, Snow 1993).

However, this property was also known as early as 1657 by F. de Bessy (Berndt and Bhargava 1993, Guy 1994). Leech (1957) found

$$\begin{aligned}\mathrm{Ta}(3) &= 87539319\\ &= 167^3 + 436^3\\ &= 228^3 + 423^3\\ &= 255^3 + 414^3. \qquad (2)\end{aligned}$$

Rosenstiel *et al.* (1991) recently found

$$\begin{aligned}\mathrm{Ta}(4) &= 6963472309248\\ &= 2421^3 + 19083^3\\ &= 5436^3 + 18948^3\\ &= 10200^3 + 18072^3\\ &= 13322^3 + 16630^3. \qquad (3)\end{aligned}$$

D. Wilson found

$$\begin{aligned}\mathrm{Ta}(5) &= 48988659276962496\\ &= 38787^3 + 365757^3\\ &= 107839^3 + 362753^3\\ &= 205292^3 + 342952^3\\ &= 221424^3 + 336588^3\\ &= 231518^3 + 331954^3. \qquad (4)\end{aligned}$$

The first few taxicab numbers are therefore 2, 1729, 87539319, 6963472309248, ... (Sloane's A011541).

Hardy and Wright (Theorem 412, 1979) show that the number of such sums can be made arbitrarily large but, updating Guy (1994) with Wilson's result, the least example is not known for six or more equal sums.

Sloane defines a slightly different type of taxicab numbers, namely numbers which are sums of two cubes in two or more ways, the first few of which are 1729, 4104, 13832, 20683, 32832, 39312, 40033, 46683, 64232, ... (Sloane's A001235).

see also DIOPHANTINE EQUATION—CUBIC, HARDY-RAMANUJAN NUMBER

References

Berndt, B. C. and Bhargava, S. "Ramanujan—For Lowbrows." *Am. Math. Monthly* **100**, 645–656, 1993.

Guy, R. K. "Sums of Like Powers. Euler's Conjecture." §D1 in *Unsolved Problems in Number Theory, 2nd ed.* New York: Springer-Verlag, pp. 139–144, 1994.

Hardy, G. H. *Ramanujan: Twelve Lectures on Subjects Suggested by His Life and Work, 3rd ed.* New York: Chelsea, p. 68, 1959.

Hardy, G. H. and Wright, E. M. *An Introduction to the Theory of Numbers, 5th ed.* Oxford, England: Clarendon Press, 1979.

Hofstadter, D. R. *Gödel, Escher, Bach: An Eternal Golden Braid.* New York: Vintage Books, p. 564, 1989.

Kanigel, R. *The Man Who Knew Infinity: A Life of the Genius Ramanujan.* New York: Washington Square Press, p. 312, 1991.

Leech, J. "Some Solutions of Diophantine Equations." *Proc. Cambridge Phil. Soc.* **53**, 778–780, 1957.

Plouffe, S. "Taxicab Numbers." `http://www.lacim.uqam.ca/pi/problem.html`.

Rosenstiel, E.; Dardis, J. A.; and Rosenstiel, C. R. "The Four Least Solutions in Distinct Positive Integers of the Diophantine Equation $s = x^3 + y^3 = z^3 + w^3 = u^3 + v^3 = m^3 + n^3$." *Bull. Inst. Math. Appl.* **27**, 155–157, 1991.

Silverman, J. H. "Taxicabs and Sums of Two Cubes." *Amer. Math. Monthly* **100**, 331–340, 1993.

Sloane, N. J. A. Sequences A001235 and A011541 in "An On-Line Version of the Encyclopedia of Integer Sequences."

Snow, C. P. Foreword to *A Mathematician's Apology, reprinted with a foreword by C. P. Snow* (by G. H. Hardy). New York: Cambridge University Press, p. 37, 1993.

Wooley, T. D. "Sums of Two Cubes." *Internat. Math. Res. Not.*, 181–184, 1995.

Taylor Center

The center of the TAYLOR CIRCLE, which is the SPIEKER CENTER of $\Delta H_1H_2H_3$, where H_i are the ALTITUDES.

References

Johnson, R. A. *Modern Geometry: An Elementary Treatise on the Geometry of the Triangle and the Circle.* Boston, MA: Houghton Mifflin, p. 277, 1929.

Taylor Circle

From the feet of each ALTITUDE of a TRIANGLE, draw lines PERPENDICULAR to the adjacent sides. Then the feet of these perpendiculars lie on a CIRCLE called the TAYLOR CIRCLE.

see also TUCKER CIRCLES

References

Johnson, R. A. *Modern Geometry: An Elementary Treatise on the Geometry of the Triangle and the Circle.* Boston, MA: Houghton Mifflin, p. 277, 1929.

Taylor's Condition

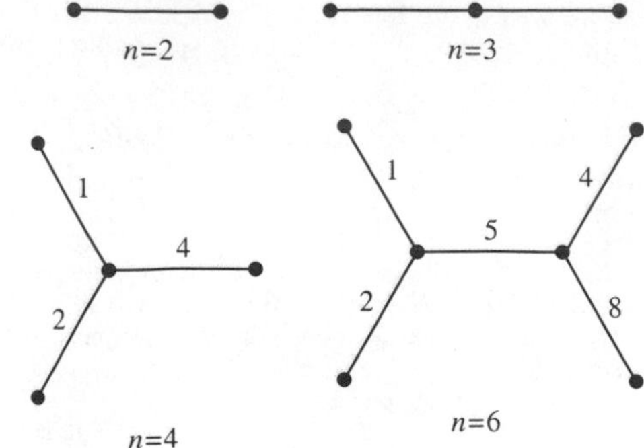

For a given POSITIVE INTEGER n, does there exist a WEIGHTED TREE with n VERTICES whose paths have weights 1, 2, ..., $\binom{n}{2}$, where $\binom{n}{2}$ is a BINOMIAL COEFFICIENT? Taylor showed that no such TREE can exist unless it is a PERFECT SQUARE or a PERFECT SQUARE plus 2. No such TREES are known except $n = 2, 3, 4$, and 6.

see also GOLOMB RULER, PERFECT DIFFERENCE SET

References

Honsberger, R. *Mathematical Gems III.* Washington, DC: Math. Assoc. Amer., pp. 56–60, 1985.

Leech, J. "Another Tree Labeling Problem." *Amer. Math. Monthly* **82**, 923–925, 1975.

Taylor, H. "Odd Path Sums in an Edge-Labeled Tree." *Math. Mag.* **50**, 258–259, 1977.

Taylor Expansion

see TAYLOR SERIES

Taylor-Greene-Chirikov Map

see STANDARD MAP

Taylor Polynomial

see TAYLOR SERIES

Taylor Series

A Taylor series is a series expansion of a FUNCTION about a point. A 1-D Taylor series is an expansion of a SCALAR FUNCTION $f(x)$ about a point $x = a$. If $a = 0$, the expansion is known as a MACLAURIN SERIES.

$$\int_a^x f^{(n)}(x)\,dx = [f^{(n-1)}(x)]_a^x = f^{(n-1)}(x) - f^{(n-1)}(a) \tag{1}$$

$$\int_a^x \left[\int_a^x f^{(n)}(x)\,dx\right] dx = \int_a^x [f^{(n-1)}(x) - f^{(n-1)}(a)]\,dx$$
$$= f^{(n-2)}(x) - f^{(n-2)}(a) - (x-a)f^{(n-1)}(a). \tag{2}$$

Continuing,

$$\iiint_a^x f^{(n)}(x)\,(dx)^3 = f^{(n-3)}(a) - (x-a)f^{(n-2)}(a)$$
$$-\tfrac{1}{2!}(x-a)^2 f^{(n-1)}(a) \tag{3}$$

$$\underbrace{\int \cdots \int_a^x}_{n} f^{(n)}(x)\,(dx)^n = f(x) - f(a) - (x-a)f'(a)$$
$$-\tfrac{1}{2!}(x-a)^2 f''(a) - \ldots - \tfrac{1}{(n-1)!}(x-a)^{n-1} f^{(n-1)}(a). \tag{4}$$

Therefore, we obtain the 1-D Taylor series

$$f(x) = f(a) + (x-a)f'(a) + \tfrac{1}{2!}(x-a)^2 f''(a) + \ldots$$
$$+ \tfrac{1}{(n-1)!}(x-a)^{n-1} f^{(n-1)}(a) + R_n, \tag{5}$$

where R_n is a remainder term defined by

$$R_n = \underbrace{\int \cdots \int_a^x}_{n} f^{(n)}(x)\,(dx)^n. \tag{6}$$

Using the MEAN-VALUE THEOREM for a function g, it must be true that

$$\int_a^x g(x)\,dx = (x-a)g(x^*) \tag{7}$$

for some $x^* \in [a, x]$. Therefore, integrating n times gives the result

$$R_n = \frac{(x-a)^n}{n!} f^{(n)}(x). \tag{8}$$

The maximum error is then the maximum value of (8) for all possible $x^* \in [a, x]$.

An alternative form of the 1-D Taylor series may be obtained by letting

$$x - a \equiv \Delta x \tag{9}$$

so that

$$x = a + \Delta x \equiv x_0 + \Delta x. \tag{10}$$

Substitute this result into (5) to give

$$f(x_0 + \Delta x) = f(x_0) + \Delta x f'(x_0) + \tfrac{1}{2!}(\Delta x)^2 f''(x_0) + \ldots. \tag{11}$$

A Taylor series of a FUNCTION in two variables $f(x, y)$ is given by

$$f(x+\Delta x, y+\Delta y) = f(x,y) + [f_x(x,y)\Delta x + f_y(x,y)\Delta y]$$
$$+ \tfrac{1}{2!}[(\Delta x)^2 f_{xx}(x,y) + 2\Delta x \Delta y f_{xy}(x,y) + (\Delta y)^2 f_{yy}(x,y)]$$
$$+ \tfrac{1}{3!}[(\Delta x)^3 f_{xxx}(x,y) + 3(\Delta x)^2 \Delta y f_{xxy}(x,y)$$
$$+ 3\Delta x(\Delta y)^2 f_{xyy}(x,y) + (\Delta y)^3 f_{yyy}(x,y)] + \ldots. \tag{12}$$

This can be further generalized for a FUNCTION in n variables,

$$f(x_1, \ldots, x_n)$$
$$= \sum_{j=0}^{\infty} \left\{ \frac{1}{j!} \left[\sum_{k=1}^{n} (x'_k - a_k) \frac{\partial}{\partial x'_k} \right]^j f(x'_1, \ldots, x'_n) \right\}_{x'_1 = a_1, \ldots, x'_n = a_n}. \tag{13}$$

Rewriting,

$$f(x_1+a_1,\ldots,x_n+a_n) = \sum_{j=0}^{\infty}\left\{\frac{1}{j!}\left(\sum_{k=1}^{n} a_k\frac{\partial}{\partial x'_k}\right)^j f(x'_1,\ldots,x'_n)\right\}_{x'_1=x_1,\ldots,x'_n=x_n}. \tag{14}$$

Taking $n=2$ in (13) gives

$$\begin{aligned} f(x_1,x_2) &= \sum_{j=0}^{\infty}\left\{\frac{1}{j!}\left[(x'_1-a_1)\frac{\partial}{\partial x'_1} + (x'_2-a_2)\frac{\partial}{\partial x'_2}\right]^j f(x'_1,x'_2)\right\}_{x'_1=x_1,x'_2=x_2} \\ &= f(a_1,a_2) + \left[(x_1-a_1)\frac{\partial f}{\partial x_1} + (x_2-a_2)\frac{\partial f}{\partial x_2}\right] \\ &\quad + \frac{1}{2!}\left[(x_1-a_1)^2\frac{\partial^2 f}{\partial {x_1}^2} + 2(x_1-a_1)(x_2-a_2)\frac{\partial^2 f}{\partial x_1 \partial x_2} + (x_2-a_2)^2\frac{\partial^2 f}{\partial {x_2}^2}\right] + \ldots. \end{aligned} \tag{15}$$

Taking $n=3$ in (14) gives

$$f(x_1+a_1,x_2+a_2,x_3+a_3) = \sum_{j=0}^{\infty}\left\{\frac{1}{j!}\left(a_1\frac{\partial}{\partial x'_1} + a_2\frac{\partial}{\partial x'_2} + a_3\frac{\partial}{\partial x'_3}\right)^j \times f(x'_1,x'_2,x'_3)\right\}_{x'_1=x_1,x'_2=x_2,x'_3=x_3}, \tag{16}$$

or, in VECTOR form

$$f(\mathbf{r}+\mathbf{a}) = \sum_{j=0}^{\infty}\left[\frac{1}{j!}(\mathbf{a}\cdot\nabla_{\mathbf{r}'})^j f(\mathbf{r}')\right]_{\mathbf{r}'=\mathbf{r}}. \tag{17}$$

The zeroth- and first-order terms are

$$f(\mathbf{r}) \tag{18}$$

and

$$(\mathbf{a}\cdot\nabla_{\mathbf{r}'})f(\mathbf{r}')|_{\mathbf{r}'=\mathbf{r}}, \tag{19}$$

respectively. The second-order term is

$$\begin{aligned} \tfrac{1}{2}(\mathbf{a}\cdot\nabla_{\mathbf{r}'})(\mathbf{a}\cdot\nabla_{\mathbf{r}'})f(\mathbf{r}')|_{\mathbf{r}'=\mathbf{r}} &= \tfrac{1}{2}\mathbf{a}\cdot\nabla_{\mathbf{r}'}[\mathbf{a}\cdot(\nabla f(\mathbf{r}'))]_{\mathbf{r}'=\mathbf{r}} \\ &= \tfrac{1}{2}\mathbf{a}\cdot[\mathbf{a}\cdot\nabla_{\mathbf{r}'}(\nabla_{\mathbf{r}'}f(\mathbf{r}'))]|_{\mathbf{r}'=\mathbf{r}}, \end{aligned} \tag{20}$$

so the first few terms of the expansion are

$$f(\mathbf{r}+\mathbf{a}) = f(\mathbf{r}) + (\mathbf{a}\cdot\nabla_{\mathbf{r}'})f(\mathbf{r}')|_{\mathbf{r}'=\mathbf{r}} + \tfrac{1}{2}\mathbf{a}\cdot[\mathbf{a}\cdot\nabla_{\mathbf{r}'}(\nabla_{\mathbf{r}'}f(\mathbf{r}'))]|_{\mathbf{r}'=\mathbf{r}}. \tag{21}$$

Taylor series can also be defined for functions of a COMPLEX variable. By the CAUCHY INTEGRAL FORMULA,

$$\begin{aligned} f(z) &= \frac{1}{2\pi i}\int_C \frac{f(z')\,dz}{z'-z} = \frac{1}{2\pi i}\int_C \frac{f(z')\,dz'}{(z'-z_0)-(z-z_0)} \\ &= \frac{1}{2\pi i}\int_C \frac{f(z')\,dz'}{(z'-z_0)\left(1-\frac{z-z_0}{z'-z_0}\right)}. \end{aligned} \tag{22}$$

In the interior of C,

$$\frac{|z-z_0|}{|z'-z_0|} < 1 \tag{23}$$

so, using

$$\frac{1}{1-t} = \sum_{n=0}^{\infty} t^n, \tag{24}$$

it follows that

$$\begin{aligned} f(z) &= \frac{1}{2\pi i}\int_C \sum_{n=0}^{\infty} \frac{(z-z_0)^n f(z')\,dz'}{(z'-z_0)^{n+1}} \\ &= \frac{1}{2\pi i}\sum_{n=0}^{\infty}(z-z_0)^n \int_C \frac{f(z')\,dz}{(z'-z_0)^{n+1}}. \end{aligned} \tag{25}$$

Using the the CAUCHY INTEGRAL FORMULA for derivatives,

$$f(z) = \sum_{n=0}^{\infty}(z-z_0)^n \frac{f^{(n)}(z_0)}{n!}. \tag{26}$$

see also CAUCHY REMAINDER FORM, LAGRANGE EXPANSION, LAURENT SERIES, LEGENDRE SERIES, MACLAURIN SERIES, NEWTON'S FORWARD DIFFERENCE FORMULA

References

Abramowitz, M. and Stegun, C. A. (Eds.). *Handbook of Mathematical Functions with Formulas, Graphs, and Mathematical Tables, 9th printing.* New York: Dover, p. 880, 1972.

Arfken, G. "Taylor's Expansion." §5.6 in *Mathematical Methods for Physicists, 3rd ed.* Orlando, FL: Academic Press, pp. 303–313, 1985.

Morse, P. M. and Feshbach, H. "Derivatives of Analytic Functions, Taylor and Laurent Series." §4.3 in *Methods of Theoretical Physics, Part I.* New York: McGraw-Hill, pp. 374–398, 1953.

Tchebycheff

see CHEBYSHEV APPROXIMATION FORMULA, CHEBYSHEV CONSTANTS, CHEBYSHEV DEVIATION, CHEBYSHEV DIFFERENTIAL EQUATION, CHEBYSHEV FUNCTION, CHEBYSHEV-GAUSS QUADRATURE, CHEBYSHEV INEQUALITY, CHEBYSHEV INEQUALITY, CHEBYSHEV INTEGRAL, CHEBYSHEV PHENOMENON, CHEBYSHEV POLYNOMIAL OF THE FIRST KIND, CHEBYSHEV POLYNOMIAL OF THE SECOND KIND, CHEBYSHEV QUADRATURE, CHEBYSHEV-RADAU QUADRATURE, CHEBYSHEV-SYLVESTER CONSTANT

Teardrop Curve

A plane curve given by the parametric equations

$$x = \cos t$$
$$y = \sin t \sin^m(\tfrac{1}{2}t).$$

see also PEAR-SHAPED CURVE

References

von Seggern, D. *CRC Standard Curves and Surfaces.* Boca Raton, FL: CRC Press, p. 174, 1993.

Technique

A specific method of performing an operation. The terms ALGORITHM, METHOD, and PROCEDURE are also used interchangeably.

see also ALGORITHM, METHOD, PROCEDURE

Teichmüller Space

TEICHMÜLLER'S THEOREM asserts the EXISTENCE and UNIQUENESS of the extremal quasiconformal map between two compact RIEMANN SURFACES of the same GENUS modulo an EQUIVALENCE RELATION. The equivalence classes form the Teichmüller space T_p of compact RIEMANN SURFACES of GENUS p.

see also RIEMANN'S MODULI PROBLEM

Teichmüller's Theorem

Asserts the EXISTENCE and UNIQUENESS of the extremal quasiconformal map between two compact RIEMANN SURFACES of the same GENUS modulo an EQUIVALENCE RELATION.

see also TEICHMÜLLER SPACE

Telescoping Sum

A sum in which subsequent terms cancel each other, leaving only initial and final terms. For example,

$$\begin{aligned} S &= \sum_{i=1}^{n-1}\left(\frac{1}{a_i}-\frac{1}{a_{i+1}}\right) \\ &= \left(\frac{1}{a_1}-\frac{1}{a_2}\right)+\left(\frac{1}{a_2}-\frac{1}{a_3}\right)+\ldots \\ &\quad +\left(\frac{1}{a_{n-2}}-\frac{1}{a_{n-1}}\right)+\left(\frac{1}{a_{n-1}}-\frac{1}{a_n}\right) \\ &= \frac{1}{a_1}-\frac{1}{a_n} \end{aligned}$$

is a telescoping sum.

see also ZEILBERGER'S ALGORITHM

Temperature

The "temperature" of a curve Γ is defined as

$$T \equiv \frac{1}{\ln\left(\frac{2l}{2l-h}\right)},$$

where l is the length of Γ and h is the length of the PERIMETER of the CONVEX HULL. The temperature of a curve is 0 only if the curve is a straight line, and increases as the curve becomes more "wiggly."

see also CURLICUE FRACTAL

References

Pickover, C. A. *Keys to Infinity.* New York: W. H. Freeman, pp. 164–165, 1995.

Templar Magic Square

S	A	T	O	R
A	R	E	P	O
T	E	N	E	T
O	P	E	R	A
R	O	T	A	S

A MAGIC SQUARE-type arrangement of the words in the Latin sentence "Sator Arepo tenet opera rotas" ("the farmer Arepo keeps the world rolling"). This square has been found in excavations of ancient Pompeii.

see also MAGIC SQUARE

References

Bouisson, S. M. *La Magie: Ses Grands Rites, Son Histoire.* Paris, pp. 147–148, 1958.

Grosser, F. "Ein neuer Vorschlag zur Deutung der Sator-Formel." *Archiv. f. Relig.* **29**, 165–169, 1926.

Heietala, H. "The Templar Magic Square." `http://www.trantex.fi/staff/heikkih/knights/pubsator.htm`.

Hocke, G. R. *Manierismus in der Literatur: Sprach-Alchimie und esoterische Kombinationskunst.* Hamburg, Germany: Rowohlt, p. 24, 1967.

Tennis Ball Theorem

A closed simple smooth spherical curve dividing the SPHERE into two parts of equal areas has at least four inflection points.

see also BALL, BASEBALL COVER

References

Arnold, V. I. *Topological Invariants of Plane Curves and Caustics.* Providence, RI: Amer. Math. Soc., 1994.

Martinez-Maure, Y. "A Note on the Tennis Ball Theorem." *Amer. Math. Monthly* **103**, 338–340, 1996.

Tensor

An nth-RANK tensor of order m is a mathematical object in m-dimensional space which has n indices and m^n components and obeys certain transformation rules. Each index of a tensor ranges over the number of dimensions of SPACE. If the components of any tensor of any RANK vanish in one particular coordinate system, they vanish in all coordinate systems.

Zeroth-RANK tensors are called SCALARS, and first-RANK tensors are called VECTORS. In tensor notation, a vector $\mathbf{v}$ would be written v_i, where $i = 1, \ldots, m$. Tensor notation can provide a very concise way of writing vector and more general identities. For example, in tensor notation, the DOT PRODUCT $\mathbf{u} \cdot \mathbf{v}$ is simply written

$$\mathbf{u} \cdot \mathbf{v} = u_i v_i, \tag{1}$$

where repeated indices are summed over (EINSTEIN SUMMATION) so that $u_i v_i$ stands for $u_1 v_1 + \ldots + u_m v_m$. Similarly, the CROSS PRODUCT can be concisely written as

$$\mathbf{u} \times \mathbf{v} = \epsilon_{ijk} u^j v^k, \tag{2}$$

where ϵ_{ijk} is the LEVI-CIVITA TENSOR.

Second-RANK tensors resemble square MATRICES. CONTRAVARIANT second-RANK tensors are objects which transform as

$$A'^{ij} = \frac{\partial x'_i}{\partial x_k} \frac{\partial x'_j}{\partial x_l} A^{kl}. \tag{3}$$

COVARIANT second-RANK tensors are objects which transform as

$$C'_{ij} = \frac{\partial x_k}{\partial x'_i} \frac{\partial x_l}{\partial x'_i} C_{kl}. \tag{4}$$

MIXED second-RANK tensors are objects which transform as

$$B'^{j}_{i} = \frac{\partial x'_i}{\partial x_k} \frac{\partial x_l}{\partial x'_j} B^k_l. \tag{5}$$

If two tensors A and B have the same RANK and the same COVARIANT and CONTRAVARIANT indices, then

$$A^{ij} + B^{ij} = C^{ij} \tag{6}$$

$$A_{ij} + B_{ij} = C_{ij} \tag{7}$$

$$A^i_j + B^i_j = C^i_j. \tag{8}$$

A transformation of the variables of a tensor changes the tensor into another whose components are linear HOMOGENEOUS FUNCTIONS of the components of the original tensor.

see also ANTISYMMETRIC TENSOR, CURL, DIVERGENCE, GRADIENT, IRREDUCIBLE TENSOR, ISOTROPIC TENSOR, JACOBI TENSOR, RICCI TENSOR, RIEMANN TENSOR, SCALAR, SYMMETRIC TENSOR, TORSION TENSOR, VECTOR, WEYL TENSOR

References

Abraham, R.; Marsden, J. E.; and Ratiu, T. S. *Manifolds, Tensor Analysis, and Applications.* New York: Springer-Verlag, 1991.

Akivis, M. A. and Goldberg, V. V. *An Introduction to Linear Algebra and Tensors.* New York: Dover, 1972.

Arfken, G. "Tensor Analysis." Ch. 3 in *Mathematical Methods for Physicists, 3rd ed.* Orlando, FL: Academic Press, pp. 118–167, 1985.

Aris, R. *Vectors, Tensors, and the Basic Equations of Fluid Mechanics.* New York: Dover, 1989.

Bishop, R. and Goldberg, S. *Tensor Analysis on Manifolds.* New York: Dover, 1980.

Jeffreys, H. *Cartesian Tensors.* Cambridge, England: Cambridge University Press, 1931.

Joshi, A. W. *Matrices and Tensors in Physics, 3rd ed.* New York: Wiley, 1995.

Lass, H. *Vector and Tensor Analysis.* New York: McGraw-Hill, 1950.

Lawden, D. F. *An Introduction to Tensor Calculus, Relativity, and Cosmology, 3rd ed.* Chichester, England: Wiley, 1982.

McConnell, A. J. *Applications of Tensor Analysis.* New York: Dover, 1947.

Morse, P. M. and Feshbach, H. "Vector and Tensor Formalism." §1.5 in *Methods of Theoretical Physics, Part I.* New York: McGraw-Hill, pp. 44–54, 1953.

Simmonds, J. G. *A Brief on Tensor Analysis, 2nd ed.* New York: Springer-Verlag, 1994.

Sokolnikoff, I. S. *Tensor Analysis—Theory and Applications, 2nd ed.* New York: Wiley, 1964.

Synge, J. L. and Schild, A. *Tensor Calculus.* New York: Dover, 1978.

Wrede, R. C. *Introduction to Vector and Tensor Analysis.* New York: Wiley, 1963.

Tensor Calculus

The set of rules for manipulating and calculating with TENSORS.

Tensor Density

A quantity which transforms like a TENSOR except for a scalar factor of a JACOBIAN.

Tensor Dual

see DUAL TENSOR

Tensor Product

see DIRECT PRODUCT (TENSOR)

Tensor Space

Let E be a linear space over a FIELD K. Then the DIRECT PRODUCT $\bigotimes_{\lambda=1}^{k} E$ is called a tensor space of degree k.

References

Yokonuma, T. *Tensor Spaces and Exterior Algebra.* Providence, RI: Amer. Math. Soc., 1992.

Tensor Spherical Harmonic

see DOUBLE CONTRACTION RELATION

Tensor Transpose

see TRANSPOSE

Tent Map

A piecewise linear, 1-D MAP on the interval $[0,1]$ exhibiting CHAOTIC dynamics and given by

$$x_{n+1} = \mu(1 - 2|x_n - \tfrac{1}{2}|).$$

The case $\mu = 1$ is equivalent to the LOGISTIC EQUATION WITH $r = 4$, so the NATURAL INVARIANT in this case is

$$\rho(x) = \frac{1}{\pi\sqrt{x(1-x)}}.$$

see also $2x$ MOD 1 MAP, LOGISTIC EQUATION, LOGISTIC EQUATION WITH $r = 4$

Terminal

see SINK (DIRECTED GRAPH)

Ternary

The BASE 3 method of counting in which only the digits 0, 1, and 2 are used. These digits have the following multiplication table.

×	0	1	2
0	0	0	0
1	0	1	2
2	0	2	11

Erdős and Graham (1980) conjectured that no POWER of 2, 2^n, is a SUM of distinct powers of 3 for $n > 8$. This is equivalent to the requirement that the ternary expansion of 2^n always contains a 2. This has been verified by Vardi (1991) up to $n = 2 \cdot 3^{20}$. N. J. A. Sloane has conjectured that any POWER of 2 has a 0 in its ternary expansion (Vardi 1991, p. 28).

see also BASE (NUMBER), BINARY, DECIMAL, HEXADECIMAL, OCTAL, QUATERNARY

References

Erdős, P. and Graham, R. L. *Old and New Problems and Results in Combinatorial Number Theory.* Geneva, Switzerland: L'Enseignement Mathématique Université de Genève, Vol. 28, 1980.

Lauwerier, H. *Fractals: Endlessly Repeated Geometric Figures.* Princeton, NJ: Princeton University Press, pp. 10–11, 1991.

Vardi, I. "The Digits of 2^n in Base Three." *Computational Recreations in Mathematica.* Reading, MA: Addison-Wesley, pp. 20–25, 1991.

Weisstein, E. W. "Bases." http://www.astro.virginia.edu/~eww6n/math/notebooks/Bases.m.

Tessellation

A regular TILING of POLYGONS (in 2-D), POLYHEDRA (3-D), or POLYTOPES (n-D) is called a tessellation. Tessellations can be specified using a SCHLÄFLI SYMBOL.

Consider a 2-D tessellation with q regular p-gons at each VERTEX. In the PLANE,

$$\left(1 - \frac{2}{p}\right)\pi = \frac{2\pi}{q} \tag{1}$$

$$\frac{1}{p} + \frac{1}{q} = \frac{1}{2}, \tag{2}$$

so

$$(p-2)(q-2) = 4 \tag{3}$$

(Ball and Coxeter 1987), and the only factorizations are

$$4 = 4 \cdot 1 = (6-2)(3-2) \Rightarrow \{6,3\} \tag{4}$$

$$= 2 \cdot 2 = (4-2)(4-2) \Rightarrow \{4,4\} \tag{5}$$

$$= 1 \cdot 4 = (3-2)(6-2) \Rightarrow \{3,6\}. \tag{6}$$

Therefore, there are only three regular tessellations (composed of the HEXAGON, SQUARE, and TRIANGLE), illustrated as follows.

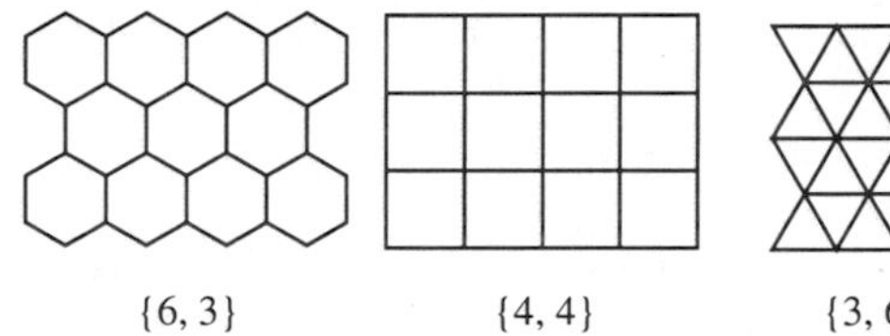

{6, 3} {4, 4} {3, 6}

There do not exist any regular STAR POLYGON tessellations in the PLANE. Regular tessellations of the SPHERE by SPHERICAL TRIANGLES are called TRIANGULAR SYMMETRY GROUPS.

Regular tilings of the plane by *two or more* convex regular POLYGONS such that the same POLYGONS in the same order surround each VERTEX are called semiregular tilings. In the plane, there are eight such tessellations, illustrated below.

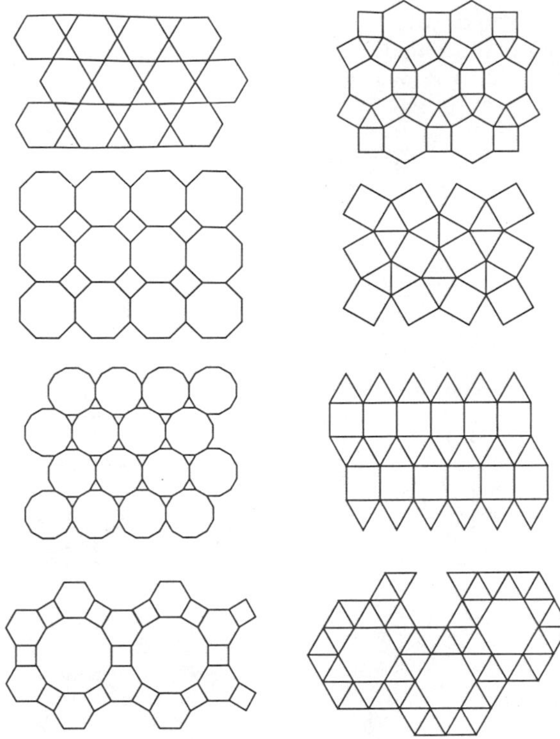

In 3-D, a POLYHEDRON which is capable of tessellating space is called a SPACE-FILLING POLYHEDRON. Examples include the CUBE, RHOMBIC DODECAHEDRON, and TRUNCATED OCTAHEDRON. There is also a 16-sided space-filler and a convex POLYHEDRON known as the SCHMITT-CONWAY BIPRISM which fills space only aperiodically.

A tessellation of n-D polytopes is called a HONEYCOMB.

see also ARCHIMEDEAN SOLID, CELL, HONEYCOMB, SCHLÄFLI SYMBOL, SEMIREGULAR POLYHEDRON, SPACE-FILLING POLYHEDRON, TILING, TRIANGULAR SYMMETRY GROUP

References

Ball, W. W. R. and Coxeter, H. S. M. *Mathematical Recreations and Essays, 13th ed.* New York: Dover, pp. 105–107, 1987.
Cundy, H. and Rollett, A. *Mathematical Models, 3rd ed.* Stradbroke, England: Tarquin Pub., pp. 60–63, 1989.
Gardner, M. *Martin Gardner's New Mathematical Diversions from Scientific American.* New York: Simon and Schuster, pp. 201–203, 1966.
Gardner, M. "Tilings with Convex Polygons." Ch. 13 in *Time Travel and Other Mathematical Bewilderments.* New York: W. H. Freeman, pp. 162–176, 1988.
Kraitchik, M. "Mosaics." §8.2 in *Mathematical Recreations.* New York: W. W. Norton, pp. 199–207, 1942.
Lines, L. *Solid Geometry.* New York: Dover, pp. 199 and 204–207 1965.
Pappas, T. "Tessellations." *The Joy of Mathematics.* San Carlos, CA: Wide World Publ./Tetra, pp. 120–122, 1989.
Peterson, I. *The Mathematical Tourist: Snapshots of Modern Mathematics.* New York: W. H. Freeman, p. 75, 1988.
Rawles, B. *Sacred Geometry Design Sourcebook: Universal Dimensional Patterns.* Nevada City, CA: Elysian Pub., 1997.
Walsh, T. R. S. "Characterizing the Vertex Neighbourhoods of Semi-Regular Polyhedra." *Geometriae Dedicata* **1**, 117–123, 1972.

Tesseract

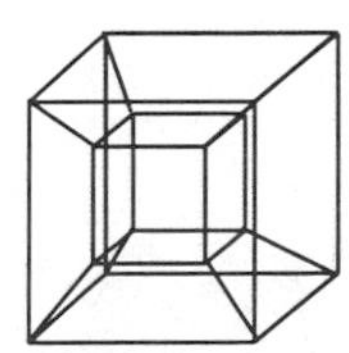

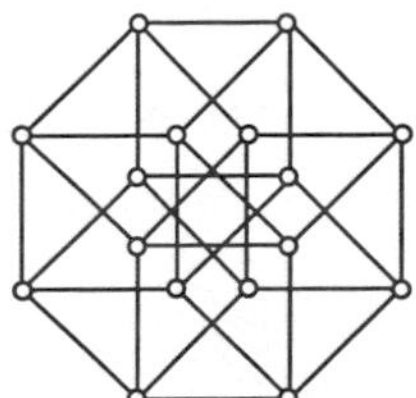

The HYPERCUBE in $\mathbb{R}^4$ is called a tesseract. It has the SCHLÄFLI SYMBOL $\{4,3,3\}$, and VERTICES $(\pm1,\pm1,\pm1,\pm1)$. The above figures show two visualizations of the TESSERACT. The figure on the left is a projection of the TESSERACT in 3-space (Gardner 1977), and the figure on the right is the GRAPH of the TESSERACT symmetrically projected into the PLANE (Coxeter 1973). A TESSERACT has 16 VERTICES, 32 EDGES, 4 SQUARES, and 8 CUBES.

see also HYPERCUBE, POLYTOPE

References

Coxeter, H. S. M. *Regular Polytopes, 3rd ed.* New York: Dover, p. 123, 1973.
Gardner, M. "Hypercubes." Ch. 4 in *Mathematical Carnival: A New Round-Up of Tantalizers and Puzzles from Scientific American.* New York: Vintage Books, 1977.
Geometry Center. "The Tesseract (or Hypercube)." `http://www.geom.umn.edu/docs/outreach/4-cube/`.

Tesseral Harmonic

A SPHERICAL HARMONIC which is expressible as products of factors linear in x^2, y^2, and z^2 multiplied by one of 1, x, y, z, yz, zx, xy, and xyz.

see also ZONAL HARMONIC

Tethered Bull Problem

Let a bull be tethered to a silo whose horizontal CROSS-SECTION is a CIRCLE of RADIUS R by a leash of length L. Then the AREA which the bull can graze if $L \leq R\pi$ is

$$A = \frac{\pi L^2}{2} + \frac{L^3}{3R}.$$

References

Hoffman, M. E. "The Bull and the Silo: An Application of Curvature." *Amer. Math. Monthly* **105**, 55–58, 1998.

Tetrabolo

A 4-POLYABOLO.

Tetrachoric Function

The function defined by

$$T_n \equiv \frac{(-1)^{n-1}}{\sqrt{n!}} Z^{(n-1)}(x),$$

where

$$Z(x) = \frac{1}{\sqrt{2\pi}} e^{-x^2/2}.$$

see also NORMAL DISTRIBUTION

References

Kenney, J. F. and Keeping, E. S. "Tetrachoric Correlation." §8.5 in *Mathematics of Statistics, Pt. 2, 2nd ed.* Princeton, NJ: Van Nostrand, pp. 205–207, 1951.

Tetracontagon

A 40-sided POLYGON.

Tetracuspid

see HYPOCYCLOID—4-CUSPED

Tetrad

A SET of four, also called a QUARTET.

see also HEXAD, MONAD, PAIR, QUARTET, QUINTET, TRIAD, TRIPLE, TWINS

Tetradecagon

A 14-sided POLYGON, sometimes called a TETRAKAIDECAGON.

Tetradecahedron

A 14-sided POLYHEDRON, sometimes called a TETRAKAIDECAHEDRON.

see also CUBOCTAHEDRON, TRUNCATED OCTAHEDRON

References

Ghyka, M. *The Geometry of Art and Life.* New York: Dover, p. 54, 1977.

Tetradic

Tetradics transform DYADICS in much the same way that DYADICS transform VECTORS. They are represented using Hebrew characters and have 81 components (Morse and Feshbach 1953, pp. 72–73). The use of tetradics is archaic, since TENSORS perform the same function but are notationally simpler.

References

Morse, P. M. and Feshbach, H. *Methods of Theoretical Physics, Vol. 1.* New York: McGraw-Hill, 1953.

Tetradyakis Hexahedron

The DUAL POLYHEDRON of the CUBITRUNCATED CUBOCTAHEDRON.

Tetraflexagon

A FLEXAGON made with SQUARE faces. Gardner (1961) shows how to construct a tri-tetraflexagon,

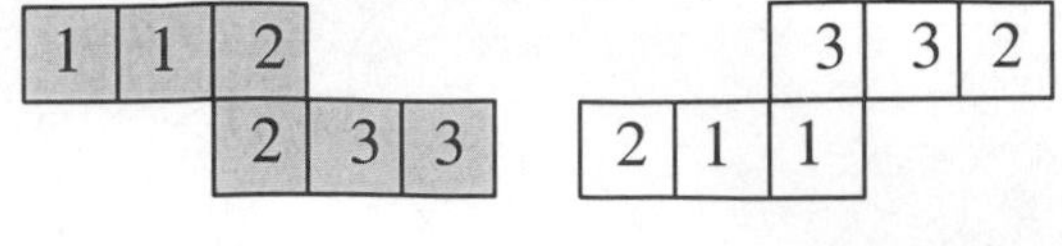

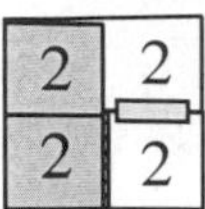

tetra-tetraflexagon,

1	1	2	3
3	2	1	1
1	1	2	3

4	4	3	3
2	3	4	4
4	4	3	2

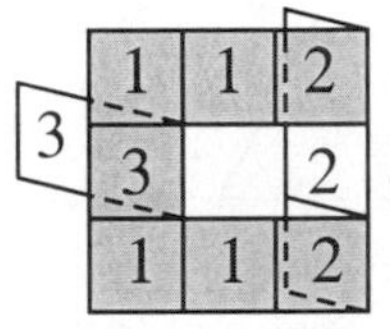

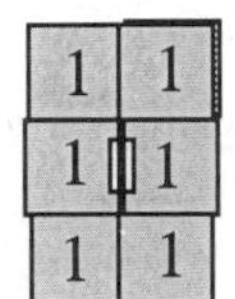

and hexa-tetraflexagon.

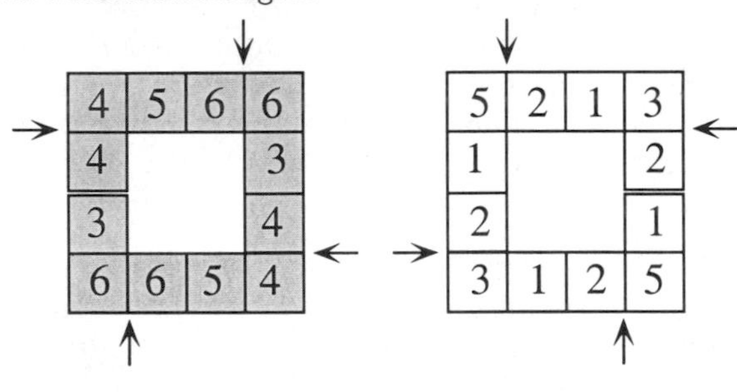

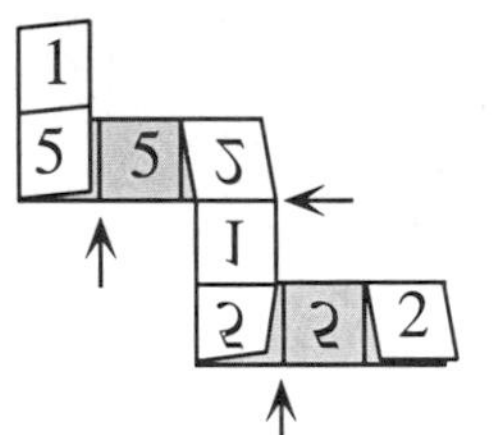

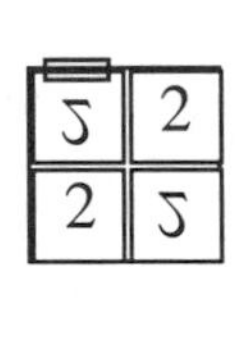

see also FLEXAGON, FLEXATUBE, HEXAFLEXAGON

References

Chapman, P. B. "Square Flexagons." *Math. Gaz.* **45**, 192–194, 1961.
Cundy, H. and Rollett, A. *Mathematical Models, 3rd ed.* Stradbroke, England: Tarquin Pub., p. 207, 1989.
Gardner, M. Ch. 1 in *The Scientific American Book of Mathematical Puzzles & Diversions.* New York: Simon and Schuster, 1959.
Gardner, M. Ch. 2 in *The Second Scientific American Book of Mathematical Puzzles & Diversions: A New Selection.* New York: Simon and Schuster, 1961.
Pappas, T. "Making a Tri-Tetra Flexagon." *The Joy of Mathematics.* San Carlos, CA: Wide World Publ./Tetra, p. 107, 1989.

Tetragon

see QUADRILATERAL

Tetrahedral Coordinates

Coordinates useful for plotting projective 3-D curves of the form $f(x_0, x_1, x_2, x_3) = 0$ which are defined by

$$x_0 = 1 - z - \sqrt{2}\,x$$
$$x_1 = 1 - z + \sqrt{2}\,x$$
$$x_2 = 1 + z + \sqrt{2}\,y$$
$$x_3 = 1 + z - \sqrt{2}\,y.$$

see also CAYLEY CUBIC, KUMMER SURFACE

Tetrahedral Graph

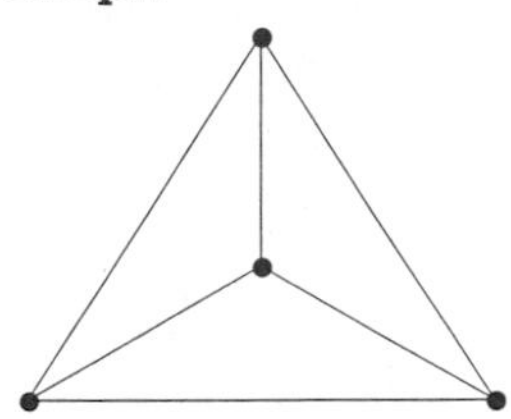

A POLYHEDRAL GRAPH which is also the COMPLETE GRAPH K_4.

see also CUBICAL GRAPH, DODECAHEDRAL GRAPH, ICOSAHEDRAL GRAPH, OCTAHEDRAL GRAPH, TETRAHEDRON

Tetrahedral Group

The POINT GROUP of symmetries of the TETRAHEDRON, denoted T_d. The tetrahedral group has symmetry operations E, $8C_3$, $3C_2$, $6S_4$, and $6\sigma_d$ (Cotton 1990).

see also ICOSAHEDRAL GROUP, OCTAHEDRAL GROUP, POINT GROUPS, TETRAHEDRON

References

Cotton, F. A. *Chemical Applications of Group Theory, 3rd ed.* New York: Wiley, p. 47, 1990.

Lomont, J. S. "Icosahedral Group." §3.10.C in *Applications of Finite Groups.* New York: Dover, p. 81, 1987.

Tetrahedral Number

A FIGURATE NUMBER Te_n of the form

$$Te_n = \sum_{i=1}^{n} T_n = \tfrac{1}{6}n(n+1)(n+2) = \binom{n+2}{3}, \quad (1)$$

where T_n is the nth TRIANGULAR NUMBER and $\binom{n}{m}$ is a BINOMIAL COEFFICIENT. These numbers correspond to placing discrete points in the configuration of a TETRAHEDRON (triangular base pyramid). Tetrahedral numbers are PYRAMIDAL NUMBERS with $r = 3$, and are the sum of consecutive TRIANGULAR NUMBERS. The first few are 1, 4, 10, 20, 35, 56, 84, 120, ... (Sloane's A000292). The GENERATING FUNCTION of the tetrahedral numbers is

$$\frac{x}{(x-1)^4} = x + 4x^2 + 10x^3 + 20x^4 + \ldots. \quad (2)$$

Tetrahedral numbers are EVEN, except for every fourth tetrahedral number, which is ODD (Conway and Guy 1996).

The only numbers which are simultaneously SQUARE and TETRAHEDRAL are $Te_1 = 1$, $Te_2 = 4$, and $Te_{48} = 19600$ (giving $S_1 = 1$, $S_2 = 4$, and $S_{140} = 19600$), as proved by Meyl (1878; cited in Dickson 1952, p. 25). Numbers which are simultaneously TRIANGULAR and tetrahedral satisfy the BINOMIAL COEFFICIENT equation

$$\binom{n}{2} = \binom{m}{3}, \quad (3)$$

the only solutions of which are $(m, n) = (10, 16)$, (22, 56), and (36, 120) (Guy 1994, p. 147). Beukers (1988) has studied the problem of finding numbers which are simultaneously tetrahedral and PYRAMIDAL via INTEGER points on an ELLIPTIC CURVE, and finds that the only solution is the trivial $Te_1 = P_1 = 1$.

see also PYRAMIDAL NUMBER, TRUNCATED TETRAHEDRAL NUMBER

References

Ball, W. W. R. and Coxeter, H. S. M. *Mathematical Recreations and Essays, 13th ed.* New York: Dover, p. 59, 1987.

Beukers, F. "On Oranges and Integral Points on Certain Plane Cubic Curves." *Nieuw Arch. Wisk.* **6**, 203–210, 1988.

Conway, J. H. and Guy, R. K. *The Book of Numbers.* New York: Springer-Verlag, pp. 44–46, 1996.

Dickson, L. E. *History of the Theory of Numbers, Vol. 2: Diophantine Analysis.* New York: Chelsea, 1952.

Guy, R. K. "Figurate Numbers." §D3 in *Unsolved Problems in Number Theory, 2nd ed.* New York: Springer-Verlag, pp. 147–150, 1994.

Meyl, A.-J.-J. "Solution de Question 1194." *Nouv. Ann. Math.* **17**, 464–467, 1878.

Sloane, N. J. A. Sequence A000292/M3382 in "An On-Line Version of the Encyclopedia of Integer Sequences."

Tetrahedral Surface

A SURFACE given by the parametric equations

$$x = A(u-a)^m(v-a)^n$$
$$y = B(u-b)^m(v-b)^n$$
$$z = C(u-c)^m(v-c)^n.$$

References

Eisenhart, L. P. *A Treatise on the Differential Geometry of Curves and Surfaces.* New York: Dover, p. 267, 1960.

Tetrahedroid

A special case of a quartic KUMMER SURFACE.

References

Fischer, G. (Ed.). *Mathematical Models from the Collections of Universities and Museums.* Braunschweig, Germany: Vieweg, pp. 17–19, 1986.

Guy, R. K. *Unsolved Problems in Number Theory, 2nd ed.* New York: Springer-Verlag, p. 183, 1994.

Tetrahedron

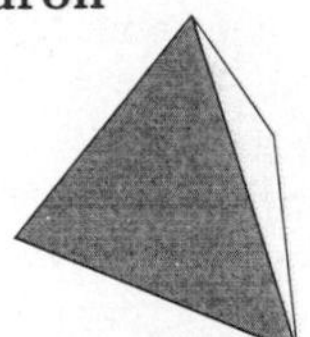
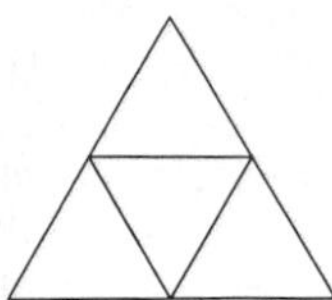

The regular tetrahedron, often simply called "the" tetrahedron, is the PLATONIC SOLID P_1 with four VERTICES, six EDGES, and four equivalent EQUILATERAL TRIANGULAR faces ($4\{3\}$). It is also UNIFORM POLYHEDRON U_1. It is described by the SCHLÄFLI SYMBOL $\{3,3\}$ and the WYTHOFF SYMBOL is $3\,|\,2\,3$. It is the prototype of the TETRAHEDRAL GROUP T_d,

The tetrahedron is its own DUAL POLYHEDRON. It is the only simple POLYHEDRON with no DIAGONALS, and cannot be STELLATED. The VERTICES of a tetrahedron are given by $(0,0,\sqrt{3})$, $(0,\frac{2}{3}\sqrt{6},-\frac{1}{3}\sqrt{3})$, $(-\sqrt{2},-\frac{1}{3}\sqrt{6},-\frac{1}{3}\sqrt{3})$, and $(\sqrt{2},-\frac{1}{3}\sqrt{6},-\frac{1}{3}\sqrt{3})$, or by (0, 0, 0), (0, 1, 1), (1, 0, 1), (1, 1, 0). In the latter case, the face planes are

$$x+y+z=2 \tag{1}$$

$$x-y-z=0 \tag{2}$$

$$-x+y-z=0 \tag{3}$$

$$x+y-z=0. \tag{4}$$

Let a tetrahedron be length a on a side. The VERTICES are located at $(x,\ 0,\ 0)$, $(-d,\ \pm a/2,\ 0)$, and $(0,\ 0,\ h)$. From the figure,

Perspective View

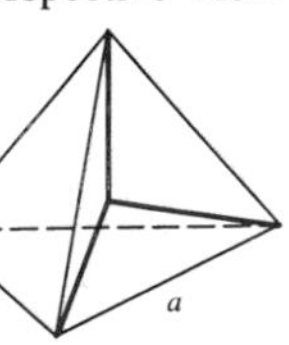

Bottom View

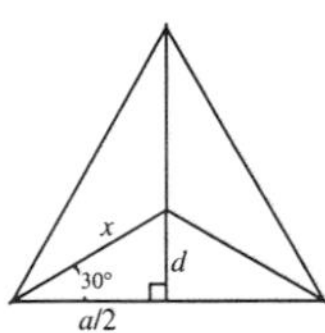

Side View

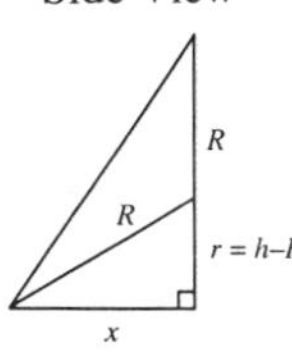

$$x=\frac{\frac{a}{2}}{\cos(\frac{\pi}{6})}=\frac{1}{2}\frac{2}{\sqrt{3}}a=\tfrac{1}{3}\sqrt{3}\,a. \tag{5}$$

d is then

$$d=\sqrt{x^2-(\tfrac{1}{2}a)^2}=a\sqrt{\tfrac{1}{3}-\tfrac{1}{4}}=a\sqrt{\frac{4-3}{12}}=\frac{a}{\sqrt{12}}$$
$$=\tfrac{1}{6}\sqrt{3}\,a. \tag{6}$$

This gives the AREA of the base as

$$A=\tfrac{1}{2}a(R+x)=\tfrac{1}{2}a\left(\frac{\sqrt{3}}{6}a+\frac{1}{\sqrt{3}}a\right)$$
$$=\tfrac{1}{2}a^2\left(\frac{\sqrt{3}}{6}+\frac{2\sqrt{3}}{6}\right)$$
$$=\tfrac{1}{2}a^2\frac{3\sqrt{3}}{6}=\tfrac{1}{4}\sqrt{3}\,a^2. \tag{7}$$

The height is

$$h=\sqrt{a^2-x^2}=a\sqrt{1-\tfrac{1}{3}}=\tfrac{1}{3}\sqrt{6}\,a. \tag{8}$$

The CIRCUMRADIUS R is found from

$$x^2+(h-R)^2=R^2 \tag{9}$$

$$x^2+h^2-2hR+R^2=R^2. \tag{10}$$

Solving gives

$$R=\frac{x^2+h^2}{2h}=\frac{\frac{1}{3}+\frac{2}{3}}{2\sqrt{\frac{2}{3}}}=\frac{1}{2}\sqrt{\frac{3}{2}}=\tfrac{1}{4}\sqrt{6}\,a\approx 0.61237a. \tag{11}$$

The INRADIUS r is

$$r\equiv h-R=\sqrt{\frac{2}{3}}\,a-\frac{\sqrt{3}}{6}\,a=\tfrac{1}{12}\sqrt{6}\,a\approx 0.20412a, \tag{12}$$

which is also

$$r=\tfrac{1}{4}h=\tfrac{1}{3}R. \tag{13}$$

The MIDRADIUS is

$$\rho=\sqrt{r^2+d^2}=a\sqrt{\tfrac{6}{144}+\tfrac{3}{36}}=\sqrt{\tfrac{1}{8}}\,a=\tfrac{1}{4}\sqrt{2}\,a$$
$$\approx 0.35355a. \tag{14}$$

Plugging in for the VERTICES gives

$$(a\sqrt{3},0,0),(-\tfrac{1}{6}\sqrt{3}\,a,\pm\tfrac{1}{2}a,0),\ \text{and}\ (0,0,\tfrac{1}{3}\sqrt{6}\,a). \tag{15}$$

Since a tetrahedron is a PYRAMID with a triangular base, $V=\frac{1}{3}A_bh$, and

$$V=\tfrac{1}{3}\left(\tfrac{1}{4}\sqrt{3}\,a^2\right)\left(\sqrt{\frac{2}{3}}\,a\right)=\tfrac{1}{12}\sqrt{2}\,a^3. \tag{16}$$

The DIHEDRAL ANGLE is

$$\theta=\tan^{-1}(2\sqrt{2})=2\sin^{-1}(\tfrac{1}{3}\sqrt{6})=\cos^{-1}(\tfrac{1}{3}). \tag{17}$$

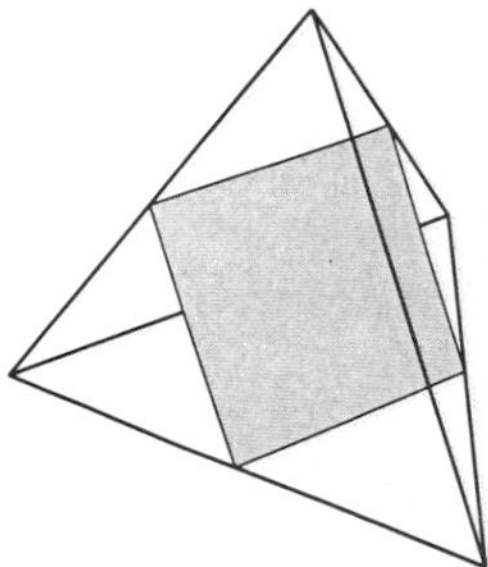

By slicing a tetrahedron as shown above, a SQUARE can be obtained. This cut divides the tetrahedron into two congruent solids rotated by 90°.

Now consider a general (not necessarily regular) tetrahedron, defined as a convex POLYHEDRON consisting of four (not necessarily identical) TRIANGULAR faces. Let the tetrahedron be specified by its VERTICES at (x_i, y_i) where $i = 1, \ldots, 4$. Then the VOLUME is given by

$$V = \frac{1}{3!} \begin{vmatrix} x_1 & y_1 & z_1 & 1 \\ x_2 & y_2 & z_2 & 1 \\ x_3 & y_3 & z_3 & 1 \\ x_4 & y_4 & z_4 & 1 \end{vmatrix}. \tag{18}$$

Specifying the tetrahedron by the three EDGE vectors **a**, **b**, and **c** from a given VERTEX, the VOLUME is

$$V = \tfrac{1}{3!} |\mathbf{a} \cdot (\mathbf{b} \times \mathbf{c})|. \tag{19}$$

If the faces are congruent and the sides have lengths a, b, and c, then

$$V = \sqrt{\frac{(a^2+b^2-c^2)(a^2+c^2-b^2)(b^2+c^2-a^2)}{72}} \tag{20}$$

(Klee and Wagon 1991, p. 205). Let a, b, c, and d be the areas of the four faces, and define

$$B \equiv \angle cd \tag{21}$$

$$C \equiv \angle bd \tag{22}$$

$$D \equiv \angle bc, \tag{23}$$

where $\angle jk$ means here the ANGLE between the PLANES formed by the FACES j and k, with VERTEX along their intersecting EDGE. Then

$$a^2 = b^2 + c^2 + d^2 - 2cd\cos B - 2bd\cos C - 2bc\cos D. \tag{24}$$

The analog of GAUSS'S CIRCLE PROBLEM can be asked for tetrahedra: how many LATTICE POINTS lie within a tetrahedron centered at the ORIGIN with a given INRADIUS (Lehmer 1940, Granville 1991, Xu and Yau 1992, Guy 1994).

see also AUGMENTED TRUNCATED TETRAHEDRON, BANG'S THEOREM, EHRHART POLYNOMIAL, HERONIAN TETRAHEDRON, HILBERT'S 3RD PROBLEM, ISOSCELES TETRAHEDRON, SIERPIŃSKI TETRAHEDRON, STELLA OCTANGULA, TETRAHEDRON 5-COMPOUND, TETRAHEDRON 10-COMPOUND, TRUNCATED TETRAHEDRON

References

Davie, T. "The Tetrahedron." http://www.dcs.st-and.ac.uk/~ad/mathrecs/polyhedra/tetrahedron.html.

Granville, A. "The Lattice Points of an n-Dimensional Tetrahedron." *Aequationes Math.* **41**, 234–241, 1991.

Guy, R. K. "Gauß's Lattice Point Problem." §F1 in *Unsolved Problems in Number Theory, 2nd ed.* New York: Springer-Verlag, pp. 240–241, 1994.

Klee, V. and Wagon, S. *Old and New Unsolved Problems in Plane Geometry and Number Theory, rev. ed.* Washington, DC: Math. Assoc. Amer., 1991.

Lehmer, D. H. "The Lattice Points of an n-Dimensional Tetrahedron." *Duke Math. J.* **7**, 341–353, 1940.

Xu, Y. and Yau, S. "A Sharp Estimate of the Number of Integral Points in a Tetrahedron." *J. reine angew. Math.* **423**, 199–219, 1992.

Tetrahedron 5-Compound

A POLYHEDRON COMPOUND composed of 5 TETRAHEDRA. Two tetrahedron 5-compounds of opposite CHIRALITY combine to make a TETRAHEDRON 10-COMPOUND. The following diagram shows pieces which can be assembled to form a tetrahedron 5-compound (Cundy and Rollett 1989).

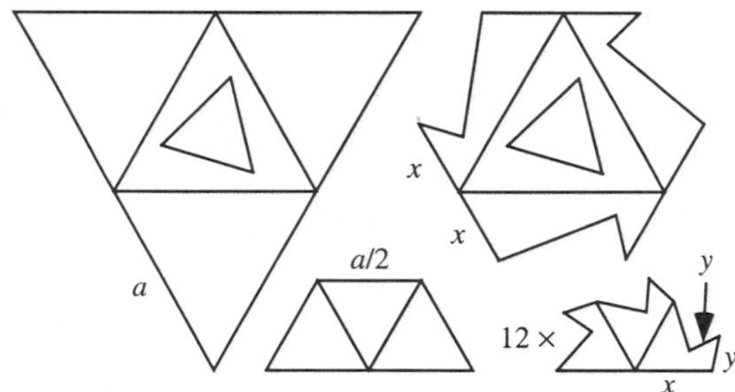

see also POLYHEDRON COMPOUND, TETRAHEDRON 10-COMPOUND

References

Ball, W. W. R. and Coxeter, H. S. M. *Mathematical Recreations and Essays, 13th ed.* New York: Dover, p. 135, 1987.

Cundy, H. and Rollett, A. *Mathematical Models, 3rd ed.* Stradbroke, England: Tarquin Pub., pp. 139–141, 1989.

Wang, P. "Renderings." http://www.ugcs.caltech.edu/~peterw/portfolio/renderings/.

Wenninger, M. J. *Polyhedron Models.* New York: Cambridge University Press, p. 44, 1989.

Tetrahedron 10-Compound

Two TETRAHEDRON 5-COMPOUNDS of opposite CHIRALITY combined.

see also POLYHEDRON COMPOUND, TETRAHEDRON 5-COMPOUND

References

Ball, W. W. R. and Coxeter, H. S. M. *Mathematical Recreations and Essays, 13th ed.* New York: Dover, p. 135, 1987.

Cundy, H. and Rollett, A. *Mathematical Models, 3rd ed.* Stradbroke, England: Tarquin Pub., pp. 141–142, 1989.

Wenninger, M. J. *Polyhedron Models.* New York: Cambridge University Press, p. 45, 1989.

Tetrahedron Inscribing

Pick four points at random on the surface of a unit SPHERE. Find the distribution of possible volumes of (nonregular) TETRAHEDRA. Without loss of generality, the first point can be chosen as (1, 0, 0). Designate the other points **a**, **b**, and **c**. Then the distances from the first VERTEX are

$$\mathbf{a} = \begin{bmatrix} \cos\theta_1 - 1 \\ \sin\theta_1 \\ 0 \end{bmatrix} \tag{1}$$

$$\mathbf{b} = \begin{bmatrix} \cos\theta_2 \sin\phi_2 - 1 \\ \sin\theta_2 \sin\phi_2 \\ \cos\phi_2 \end{bmatrix} \tag{2}$$

$$\mathbf{c} = \begin{bmatrix} \cos\theta_3 \sin\phi_3 - 1 \\ \sin\theta_3 \sin\phi_3 \\ \cos\phi_3 \end{bmatrix}. \tag{3}$$

The average volume is then

$$\bar{V} = \frac{1}{C}\int_0^{2\pi}\int_0^{2\pi}\int_0^{2\pi}\int_{-\pi/2}^{\pi/2}\int_{-\pi/2}^{\pi/2} \tfrac{1}{3!}|\mathbf{a}\cdot(\mathbf{b}\times\mathbf{c})|\, d\phi_3\, d\phi_2\, d\theta_3\, d\theta_2\, d\theta_1, \tag{4}$$

where

$$C = \int_0^{2\pi}\int_0^{2\pi}\int_0^{2\pi}\int_{-\pi/2}^{\pi/2} d\phi_3\, d\phi_2\, d\theta_3\, d\theta_2\, d\theta_1 = 8\pi^5 \tag{5}$$

and

$$\begin{aligned}\mathbf{a}\cdot(\mathbf{b}\times\mathbf{c}) = &-\cos\phi_2\sin\theta_1 + \cos\phi_3\sin\theta_1 \\ &-\cos\phi_3\cos\theta_2\sin\phi_2\sin\theta_1 + \cos\phi_2\cos\theta_3\sin\phi_3\sin\theta_1 \\ &-\cos\phi_3\sin\phi_2\sin\theta_2 + \cos\phi_3\cos\theta_1\sin\phi_2\sin\theta_2 \\ &+\cos\phi_2\sin\phi_3\sin\theta_3 - \cos\phi_2\cos\theta_1\sin\phi_3\sin\theta_3.\end{aligned} \tag{6}$$

The integrals are difficult to compute analytically, but 10^7 computer TRIALS give

$$\langle V\rangle \approx 0.1080 \tag{7}$$

$$\langle V^2\rangle \approx 0.02128 \tag{8}$$

$$\sigma_V{}^2 = \langle V^2\rangle - \langle V\rangle^2 \approx 0.009937. \tag{9}$$

see also POINT-POINT DISTANCE—1-D, TRIANGLE INSCRIBING IN A CIRCLE, TRIANGLE INSCRIBING IN AN ELLIPSE

References

Buchta, C. "A Note on the Volume of a Random Polytope in a Tetrahedron." *Ill. J. Math.* **30**, 653–659, 1986.

Tetrahemihexacron

The DUAL POLYHEDRON of the TETRAHEMIHEXAHEDRON.

Tetrahemihexahedron

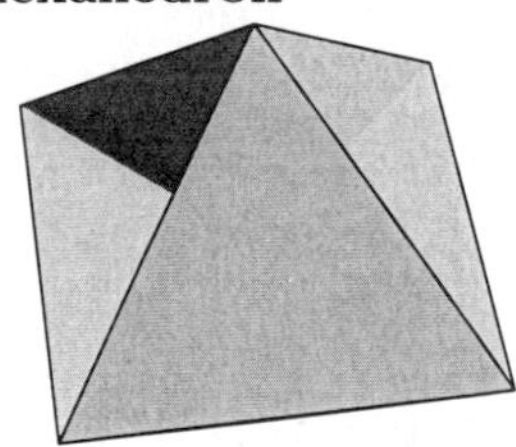

The UNIFORM POLYHEDRON U_4 whose DUAL POLYHEDRON is the TETRAHEMIHEXACRON. It has SCHLÄFLI SYMBOL $r'\{{}^3_3\}$ and WYTHOFF SYMBOL $\frac{3}{2}\,3\,|\,2$. Its faces are $4\{3\}+3\{4\}$. It is a faceted form of the OCTAHEDRON. Its CIRCUMRADIUS is

$$R = \tfrac{1}{2}\sqrt{2}.$$

References

Wenninger, M. J. *Polyhedron Models.* Cambridge, England: Cambridge University Press, pp. 101–102, 1971.

Tetrakaidecagon

see TETRADECAGON

Tetrakaidecahedron

see TETRADECAHEDRON

Tetrakis Hexahedron

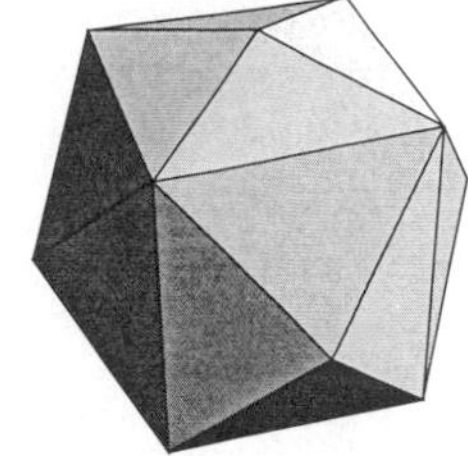
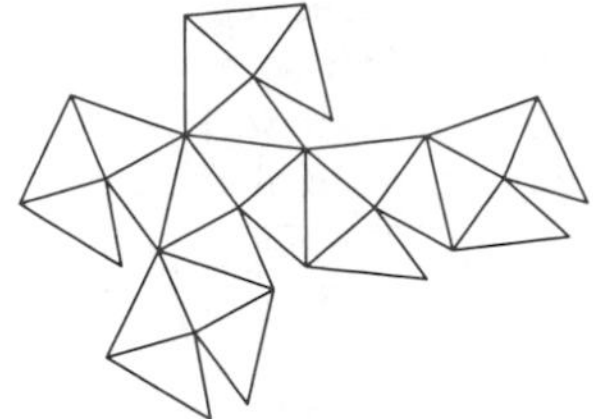

The DUAL POLYHEDRON of the TRUNCATED OCTAHEDRON.

Tetranacci Number

The tetranacci numbers are a generalization of the FIBONACCI NUMBERS defined by $T_0 = 0$, $T_1 = 1$, $T_2 = 1$, $T_3 = 2$, and the RECURRENCE RELATION

$$T_n = T_{n-1} + T_{n-2} + T_{n-3} + T_{n-4}$$

for $n \geq 4$. They represent the $n = 4$ case of the FIBONACCI n-STEP NUMBERS. The first few terms are 1, 1, 2, 4, 8, 15, 29, 56, 108, 208, ... (Sloane's A000078). The ratio of adjacent terms tends to 1.92756, which is the REAL ROOT of $x^5 - 2x^4 + 1 = 0$.

see also FIBONACCI n-STEP NUMBER, FIBONACCI NUMBER, TRIBONACCI NUMBER

References

Sloane, N. J. A. Sequence A000078/M1108 in "An On-Line Version of the Encyclopedia of Integer Sequences."

Tetrix

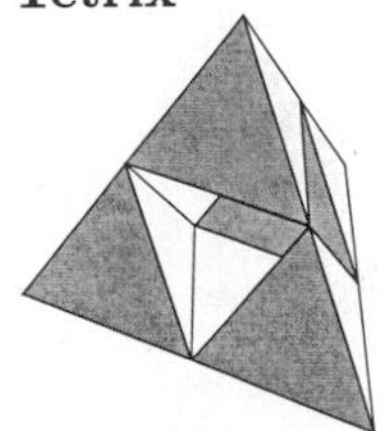
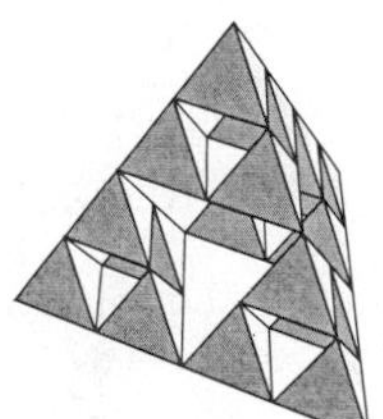
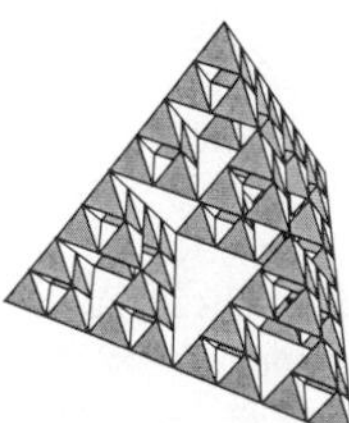

The 3-D analog of the SIERPIŃSKI SIEVE illustrated above, also called the SIERPIŃSKI SPONGE or SIERPIŃSKI TETRAHEDRON. Let N_n be the number of tetrahedra, L_n the length of a side, and A_n the fractional VOLUME of tetrahedra after the nth iteration. Then

$$N_n = 4^n \tag{1}$$

$$L_n = (\tfrac{1}{2})^n = 2^{-n} \tag{2}$$

$$A_n = {L_n}^3 N_n = (\tfrac{1}{2})^n. \tag{3}$$

The CAPACITY DIMENSION is therefore

$$d_{\text{cap}} = -\lim_{n\to\infty} \frac{\ln N_n}{\ln L_n} = -\lim_{n\to\infty} \frac{\ln(4^n)}{\ln(2^{-n})} = \frac{\ln 4}{\ln 2} = \frac{2\ln 2}{\ln 2} = 2, \tag{4}$$

so the tetrix has an INTEGRAL CAPACITY DIMENSION (albeit one less than the DIMENSION of the 3-D TETRAHEDRA from which it is built), despite the fact that it is a FRACTAL.

The following illustration demonstrates how this counterintuitive fact can be true by showing three stages of the rotation of a tetrix, viewed along one of its edges. In the last frame, the tetrix "looks" like the 2-D PLANE.

see also MENGER SPONGE, SIERPIŃSKI SIEVE

References

Dickau, R. M. "Sierpinski Tetrahedron." `http://forum.swarthmore.edu/advanced/robertd/tetrahedron.html`.

Eppstein, D. "Sierpinski Tetrahedra and Other Fractal Sponges." `http://www.ics.uci.edu/~eppstein/junkyard/sierpinski.html`.

Tetromino

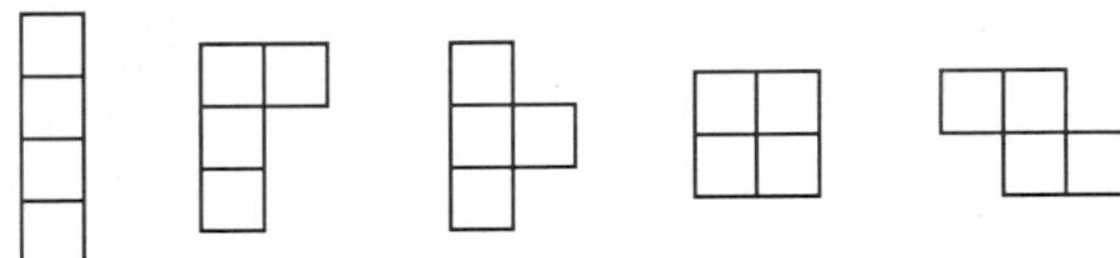

The five 4-POLYOMINOES, known as STRAIGHT, L-, T-, SQUARE, and SKEW.

References

Gardner, M. "Polyominoes." Ch. 13 in *The Scientific American Book of Mathematical Puzzles & Diversions.* New York: Simon and Schuster, pp. 124–140, 1959.

Hunter, J. A. H. and Madachy, J. S. *Mathematical Diversions.* New York: Dover, pp. 80–81, 1975.

Thales' Theorem

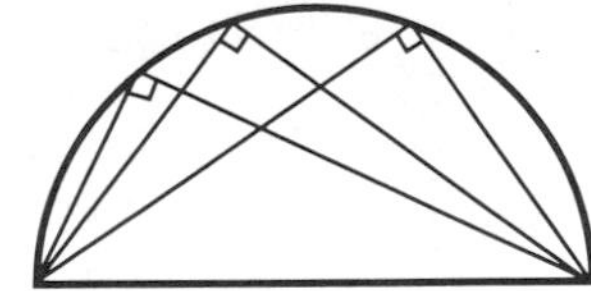

An ANGLE inscribed in a SEMICIRCLE is a RIGHT ANGLE.

see also RIGHT ANGLE, SEMICIRCLE

Theorem

A statement which can be demonstrated to be true by accepted mathematical operations and arguments. In general, a theorem is an embodiment of some general principle that makes it part of a larger theory.

According to the Nobel Prize-winning physicist Richard Feynman (1985), any theorem, no matter how difficult to prove in the first place, is viewed as "TRIVIAL" by mathematicians once it has been proven. Therefore, there are exactly two types of mathematical objects: TRIVIAL ones, and those which have not yet been proven.

see also AXIOM, AXIOMATIC SYSTEM, COROLLARY, DEEP THEOREM, PORISM, LEMMA, POSTULATE, PRINCIPLE, PROPOSITION

References

Feynman, R. P. and Leighton, R. *Surely You're Joking, Mr. Feynman!* New York: Bantam Books, 1985.

Theorema Egregium

see GAUSS'S THEOREMA EGREGIUM

Theta Function

The theta functions are the elliptic analogs of the EXPONENTIAL FUNCTION, and may be used to express the JACOBI ELLIPTIC FUNCTIONS. Let t be a constant COMPLEX NUMBER with $\Im[t] > 0$. Define the NOME

$$q \equiv e^{i\pi t} = e^{\pi K'(k)/K(k)}, \tag{1}$$

where

$$t \equiv -i\frac{K'(k)}{K(k)}, \tag{2}$$

and $K(k)$ is a complete ELLIPTIC INTEGRAL OF THE FIRST KIND, k is the MODULUS, and k' is the complementary MODULUS. Then the theta functions are, in the NOTATION of Whittaker and Watson,

$$\vartheta_1(z,q) \equiv 2\sum_{n=0}^{\infty}(-1)^n q^{(n+1/2)^2}\sin[(2n+1)z]$$
$$= zq^{1/4}\sum_{n=0}^{\infty}(-1)^n q^{n(n+1)}\sin[(2n+1)z] \tag{3}$$

$$\vartheta_2(z,q) \equiv 2\sum_{n=0}^{\infty} q^{(n+1/2)^2}\cos[(2n+1)z]$$
$$= 2q^{1/4}\sum_{n=0}^{\infty} q^{n(n+1)}\cos[(2n+1)z] \tag{4}$$

$$\vartheta_3(z,q) \equiv 1+2\sum_{n=1}^{\infty} q^{n^2}\cos(2nz) \tag{5}$$

$$\vartheta_4(z,q) \equiv \sum_{n=-\infty}^{\infty}(-1)^n q^{n^2} e^{2niz}$$
$$= 1+2\sum_{n=1}^{\infty}(-1)^n q^{n^2}\cos(2nz). \tag{6}$$

Written in terms of t,

$$\vartheta_2(t,q) = \sum_{n=-\infty}^{\infty} q^{(n+1/2)^2} e^{\Im[t]} \tag{7}$$

$$\vartheta_3(t,q) = \sum_{n=-\infty}^{\infty} q^{n^2} e^{\Im[t]}. \tag{8}$$

These functions are sometimes denoted Θ_i or θ_i, and a number of indexing conventions have been used. For a summary of these notations, see Whittaker and Watson (1990). The theta functions are quasidoubly periodic, as illustrated in the following table.

ϑ_i	$\vartheta_i(z+\pi)/\vartheta_i(z)$	$\vartheta_i(z+t\pi)/\vartheta_i(z)$
ϑ_1	-1	$-N$
ϑ_2	-1	N
ϑ_3	1	N
ϑ_4	1	$-N$

Here,

$$N \equiv q^{-1}e^{-2iz}. \tag{9}$$

The quasiperiodicity can be established as follows for the specific case of ϑ_4,

$$\vartheta_4(z+\pi,q) = \sum_{n=-\infty}^{\infty}(-1)^n q^{n^2} e^{2niz} e^{2ni\pi}$$
$$= \sum_{n=-\infty}^{\infty}(-1)^n q^{n^2} e^{2niz} = \vartheta_4(z,q) \tag{10}$$

$$\vartheta_4(z+\pi t,q) = \sum_{n=-\infty}^{\infty}(-1)^n q^{n^2} e^{2ni\pi t} e^{2niz}$$
$$= \sum_{n=-\infty}^{\infty}(-1)^n q^{n^2} q^{2n} e^{2niz}$$
$$= -q^{-1}e^{-2iz}\sum_{n=-\infty}^{\infty}(-1)^{n+1} q^{(n+1)^2} q^{2(n+1)iz}$$
$$= -q^{-1}e^{-2iz}\sum_{n=-\infty}^{\infty}(-1)^n q^{n^2} q^{2niz}$$
$$= -q^{-1}e^{-2iz}\vartheta_4(z,q). \tag{11}$$

The theta functions can be written in terms of each other:

$$\vartheta_1(z,q) = -ie^{iz+\pi it/4}\vartheta_4(z+\tfrac{1}{4}\pi t,q) \tag{12}$$
$$\vartheta_2(z,q) = \vartheta_1(z+\tfrac{1}{2}\pi,q) \tag{13}$$
$$\vartheta_3(z,q) = \vartheta_4(z+\tfrac{1}{2}\pi,q). \tag{14}$$

Any theta function of given arguments can be expressed in terms of any other two theta functions with the same arguments.

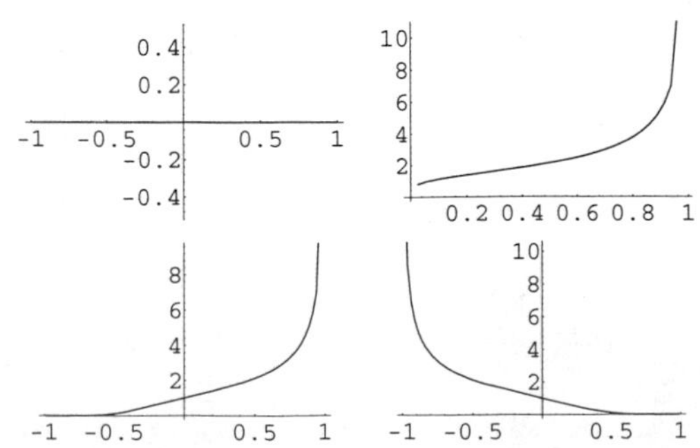

Define

$$\vartheta_i \equiv \vartheta_i(z=0), \tag{15}$$

which are plotted above. Then we have the identities

$$\vartheta_1{}^2(z)\vartheta_4{}^2 = \vartheta_3{}^2(z)\vartheta_2{}^2 - \vartheta_2{}^2(z)\vartheta_3{}^2 \tag{16}$$

$$\vartheta_2{}^2(z)\vartheta_4{}^2 = \vartheta_4{}^2(z)\vartheta_2{}^2 - \vartheta_1{}^2(z)\vartheta_3{}^2 \tag{17}$$

$$\vartheta_3{}^2(z)\vartheta_4{}^2 = \vartheta_4{}^2(z)\vartheta_3{}^2 - \vartheta_1{}^2(z)\vartheta_2{}^2 \tag{18}$$

$$\vartheta_4{}^2(z)\vartheta_4{}^2 = \vartheta_3{}^2(z)\vartheta_3{}^2 - \vartheta_2{}^2(z)\vartheta_2{}^2. \tag{19}$$

Taking $z=0$ in the last gives the special case

$$\vartheta_4{}^4 = \vartheta_3{}^4 - \vartheta_2{}^4. \tag{20}$$

In addition,

$$\vartheta_3(x) = \sum_{n=-\infty}^{\infty} x^{n^2} = 1 + 2x + 2x^4 + 2x^9 + \ldots \tag{21}$$

$$\vartheta_3{}^2(x) = 1 + 4\left(\frac{x}{1-x} - \frac{x^3}{1-x^3} + \frac{x^5}{1-x^5} - \frac{x^7}{1-x^7} + \ldots\right) \tag{22}$$

$$\vartheta_3{}^4(x) = 1 + 8\left(\frac{x}{1-x} + \frac{2x^2}{1+x^2} + \frac{3x^3}{1-x^3} + \frac{4x^4}{1+x^4} + \ldots\right). \tag{23}$$

The theta functions obey addition rules such as

$$\vartheta_3(z+y)\vartheta_3(z-y)\vartheta_3{}^2 = \vartheta_3{}^2(y)\vartheta_3{}^2(z) + \vartheta_1{}^2(y)\vartheta_1{}^2(z). \tag{24}$$

Letting $y = z$ gives a duplication FORMULA

$$\vartheta_3(2z)\vartheta_3{}^3 = \vartheta_3{}^4(z) + \vartheta_1{}^4(z). \tag{25}$$

For more addition FORMULAS, see Whittaker and Watson (1990, pp. 487–488). Ratios of theta function derivatives to the functions themselves have the simple forms

$$\frac{\vartheta_1'(z)}{\vartheta_1(z)} = \cot z + 4\sum_{n=1}^{\infty} \frac{q^{2n}}{1-q^{2n}}\sin(2nz) \tag{26}$$

$$\frac{\vartheta_2'(z)}{\vartheta_2(z)} = -\tan z + 4\sum_{n=1}^{\infty}(-1)^n \frac{q^{2n}}{1-q^{2n}}\sin(2nz) \tag{27}$$

$$\frac{\vartheta_3'(z)}{\vartheta_3(z)} = 4\sum_{n=1}^{\infty}(-1)^n \frac{q^n}{1-q^{2n}}\sin(2nz) \tag{28}$$

$$\frac{\vartheta_4'(z)}{\vartheta_4(z)} = \sum_{n=1}^{\infty} \frac{q^{2n-1}\sin(2z)}{1-2q^{2n-1}\cos(2z)+q^{4n-2}} = \sum_{n=1}^{\infty}\frac{4q^n\sin(2nz)}{1-q^{2n}}. \tag{29}$$

The theta functions can be expressed as products instead of sums by

$$\vartheta_1(z) = 2Gq^{1/4}\sin z \prod_{n=1}^{\infty}[1-2q^{2n}\cos(2z)+q^{4n}] \tag{30}$$

$$\vartheta_2(z) = 2Gq^{1/4}\cos z \prod_{n=1}^{\infty}[1+2q^{2n}\cos(2z)+q^{4n}] \tag{31}$$

$$\vartheta_3(z) = G\prod_{n=1}^{\infty}[1+2q^{2n-1}\cos(2z)+q^{4n-2}] \tag{32}$$

$$\vartheta_4(z) = G\prod_{n=1}^{\infty}[1-2q^{2n-1}\cos(2z)+q^{4n-2}], \tag{33}$$

where

$$G \equiv \prod_{n=1}^{\infty}(1-q^{2n}) \tag{34}$$

(Whittaker and Watson 1990, pp. 469–470).

The theta functions satisfy the PARTIAL DIFFERENTIAL EQUATION

$$\tfrac{1}{4}\pi i\frac{\partial^2 y}{\partial z^2} + \frac{\partial y}{\partial t} = 0, \tag{35}$$

where $y \equiv \vartheta_j(z|t)$. Ratios of the theta functions with ϑ_4 in the DENOMINATOR also satisfy differential equations

$$\frac{d}{dz}\left[\frac{\vartheta_1(z)}{\vartheta_4(z)}\right] = \vartheta_4{}^2\frac{\vartheta_2(z)\vartheta_3(z)}{\vartheta_4{}^2(z)} \tag{36}$$

$$\frac{d}{dz}\left[\frac{\vartheta_2(z)}{\vartheta_4(z)}\right] = -\vartheta_3{}^2\frac{\vartheta_1(z)\vartheta_3(z)}{\vartheta_4{}^2(z)} \tag{37}$$

$$\frac{d}{dz}\left[\frac{\vartheta_3(z)}{\vartheta_4(z)}\right] = -\vartheta_2{}^2\frac{\vartheta_1(z)\vartheta_2(z)}{\vartheta_4{}^2(z)}. \tag{38}$$

Some additional remarkable identities are

$$\vartheta_1' = \vartheta_2\vartheta_3\vartheta_4 \tag{39}$$

$$\vartheta_3(z,t) = -(it)^{1/2}e^{z^2/\pi it}\vartheta_3\left(\frac{2}{t}, -\frac{1}{t}\right), \tag{40}$$

which were discovered by Poisson in 1827 and are equivalent to

$$\sum_{n=-\infty}^{\infty} e^{-t(x+n)^2} = \sqrt{\frac{\pi}{t}}\sum_{k=-\infty}^{\infty} 2^{2\pi ikx-(\pi^2k^2/t)}. \tag{41}$$

Another amazing identity is

$$\begin{aligned}2\vartheta_1[\tfrac{1}{2}(-b+c+d+e)]\vartheta_2[\tfrac{1}{2}(b-c+d+e)]\vartheta_3[\tfrac{1}{2}(b+c-d+e)]\\ \times\vartheta_4[\tfrac{1}{2}(b+c+d-e)] = \vartheta_3(b)\vartheta_4(c)\vartheta_1(d)\vartheta_2(e)\\ +\vartheta_2(b)\vartheta_1(c)\vartheta_4(d)\vartheta_3(e) - \vartheta_1(b)\vartheta_2(c)\vartheta_3(d)\vartheta_4(e)\\ +\vartheta_4(b)\vartheta_3(c)\vartheta_2(d)\vartheta_1(e)\end{aligned} \tag{42}$$

(Whittaker and Watson 1990, p. 469).

The complete ELLIPTIC INTEGRALS OF THE FIRST and SECOND KINDS can be expressed using theta functions. Let

$$\xi \equiv \frac{\vartheta_1(z)}{\vartheta_4(z)}, \tag{43}$$

and plug into (36)

$$\left(\frac{d\xi}{dz}\right)^2 = ({\vartheta_2}^2 - \xi^2{\vartheta_3}^2)({\vartheta_3}^2 - \xi^2{\vartheta_2}^2). \tag{44}$$

Now write

$$\xi\frac{\vartheta_3}{\vartheta_2} \equiv y \tag{45}$$

and

$$z{\vartheta_3}^2 \equiv u. \tag{46}$$

Then

$$\left(\frac{dy}{du}\right)^2 = (1-y^2)(1-k^2y^2), \tag{47}$$

where the MODULUS is defined by

$$k = k(q) = \frac{{\vartheta_2}^2(q)}{{\vartheta_3}^2(q)}. \tag{48}$$

Define also the complementary MODULUS

$$k' = k'(q) = \frac{{\vartheta_4}^2(-q)}{{\vartheta_3}^2(q)}. \tag{49}$$

Now, since

$${\vartheta_2}^4 + {\vartheta_4}^4 = {\vartheta_3}^4, \tag{50}$$

we have shown

$$k^2 + k'^2 = 1. \tag{51}$$

The solution to the equation is

$$y = \frac{\vartheta_3}{\vartheta_2}\frac{\vartheta_1(u{\vartheta_3}^{-2}|t)}{\vartheta_4(u{\vartheta_3}^{-2}|t)} \equiv \operatorname{sn}(u,k), \tag{52}$$

which is a JACOBI ELLIPTIC FUNCTION with periods

$$4K(k) = 2\pi{\vartheta_3}^2(q) \tag{53}$$

and

$$2iK'(k) = \pi t{\vartheta_3}^2(q). \tag{54}$$

Here, K is the complete ELLIPTIC INTEGRAL OF THE FIRST KIND,

$$K(k) = \tfrac{1}{2}\pi{\vartheta_3}^2(q). \tag{55}$$

see also BLECKSMITH-BRILLHART-GERST THEOREM, ELLIPTIC FUNCTION, ETA FUNCTION, EULER'S PENTAGONAL NUMBER THEOREM, JACOBI ELLIPTIC FUNCTIONS, JACOBI TRIPLE PRODUCT, LANDEN'S FORMULA, MOCK THETA FUNCTION, MODULAR EQUATION, MODULAR TRANSFORMATION, MORDELL INTEGRAL, NEVILLE THETA FUNCTION, NOME, POINCARÉ-FUCHS-KLEIN AUTOMORPHIC FUNCTION, PRIME THETA FUNCTION, QUINTUPLE PRODUCT IDENTITY, RAMANUJAN THETA FUNCTIONS, SCHRÖTER'S FORMULA, WEBER FUNCTIONS

References

Abramowitz, M. and Stegun, C. A. (Eds.). *Handbook of Mathematical Functions with Formulas, Graphs, and Mathematical Tables, 9th printing.* New York: Dover, p. 577, 1972.

Bellman, R. E. *A Brief Introduction to Theta Functions.* New York: Holt, Rinehart and Winston, 1961.

Berndt, B. C. "Theta-Functions and Modular Equations." Ch. 25 in *Ramanujan's Notebooks, Part IV.* New York: Springer-Verlag, pp. 138–244, 1994.

Morse, P. M. and Feshbach, H. *Methods of Theoretical Physics, Part I.* New York: McGraw-Hill, pp. 430–432, 1953.

Whittaker, E. T. and Watson, G. N. *A Course in Modern Analysis, 4th ed.* Cambridge, England: Cambridge University Press, 1990.

Theta Operator

In the NOTATION of Watson (1966),

$$\vartheta \equiv z\frac{d}{dz}.$$

References

Watson, G. N. *A Treatise on the Theory of Bessel Functions, 2nd ed.* Cambridge, England: Cambridge University Press, 1966.

Theta Subgroup

see LAMBDA GROUP

Thiele's Interpolation Formula

Let ρ be a RECIPROCAL DIFFERENCE. Then Thiele's interpolation formula is the CONTINUED FRACTION

$$f(x) = f(x_1) + \frac{x-x_1}{\rho(x_1,x_2)+}\,\frac{x-x_2}{\rho_2(x_1,x_2,x_3)-f(x_1)+}\,\frac{x-x_3}{\rho_3(x_1,x_2,x_3,x_4)-\rho(x_1,x_2)+\ldots}.$$

References

Abramowitz, M. and Stegun, C. A. (Eds.). *Handbook of Mathematical Functions with Formulas, Graphs, and Mathematical Tables, 9th printing.* New York: Dover, p. 881, 1972.

Milne-Thomson, L. M. *The Calculus of Finite Differences.* London: Macmillan, 1951.

Thiessen Polytope

see VORONOI POLYGON

Third Curvature

Also known as the TOTAL CURVATURE. The linear element of the INDICATRIX

$$ds_P = \sqrt{{ds_T}^2 + {ds_B}^2}.$$

see also LANCRET EQUATION

Thirteenth

see FRIDAY THE THIRTEENTH

Thom's Eggs

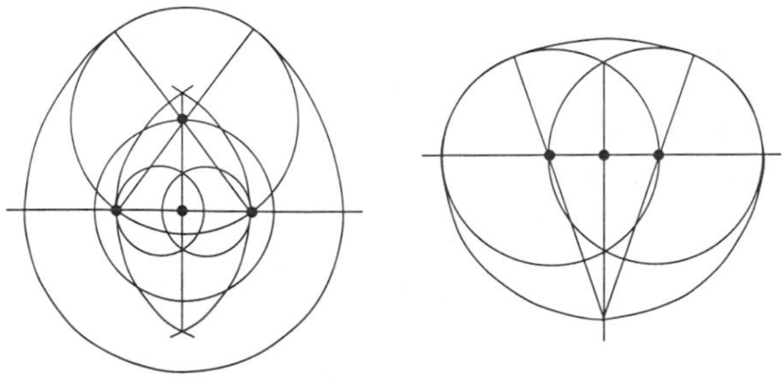

EGG-shaped curves constructed using multiple CIRCLES which Thom (1967) used to model Megalithic stone rings in Britain.

see also EGG, OVAL

References
Dixon, R. *Mathographics.* New York: Dover, p. 6, 1991.
Thom, A. "Mathematical Background." Ch. 4 in *Megalithic Sites in Britain.* Oxford, England: Oxford University Press, pp. 27–33, 1967.

Thomae's Theorem

$$\frac{\Gamma(x+y+s+1)}{\Gamma(x+s+1)\Gamma(y+s+1)}\,{}_3F_2\left(\begin{matrix}-a,-b,x+y+s+1\\ x+s+1,y+s+1\end{matrix};1\right)$$
$$=\frac{\Gamma(a+b+s+1)}{\Gamma(a+s+1)\Gamma(b+s+1)}\,{}_3F_2\left(\begin{matrix}-x,-y,a+b+s+1\\ a+s+1,b+s+1\end{matrix};1\right),$$

where $\Gamma(z)$ is the GAMMA FUNCTION and the function ${}_3F_2(a,b,c;d,e;z)$ is a GENERALIZED HYPERGEOMETRIC FUNCTION.

see also GENERALIZED HYPERGEOMETRIC FUNCTION

References
Hardy, G. H. *Ramanujan: Twelve Lectures on Subjects Suggested by His Life and Work, 3rd ed.* New York: Chelsea, pp. 104–105, 1959.

Thomassen Graph

The GRAPH illustrated above.

see also THOMSEN GRAPH

Thompson's Functions

see BEI, BER, KELVIN FUNCTIONS

Thompson Group

The SPORADIC GROUP *Th.*

References
Wilson, R. A. "ATLAS of Finite Group Representation." `http://for.mat.bham.ac.uk/atlas/Th.html`.

Thomsen's Figure

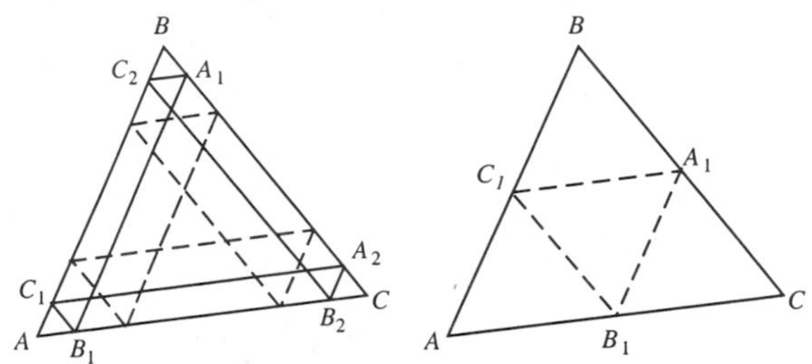

Take any TRIANGLE with VERTICES A, B, and C. Pick a point A_1 on the side opposite A, and draw a line PARALLEL to AB. Upon reaching the side AC at B_1, draw the line PARALLEL to BC. Continue (left figure). Then $A_3 = A_1$ for any TRIANGLE. If A_1 is the MIDPOINT of BC, then $A_2 = A_1$ (right figure).

see also MIDPOINT, TRIANGLE

References
Madachy, J. S. *Madachy's Mathematical Recreations.* New York: Dover, pp. 234, 1979.

Thomsen Graph

The COMPLETE BIPARTITE GRAPH $K_{3,3}$, which is equivalent to the UTILITY GRAPH. It has a CROSSING NUMBER 1.

see also COMPLETE BIPARTITE GRAPH, CROSSING NUMBER (GRAPH), THOMASSEN GRAPH, UTILITY GRAPH

Thomson Lamp Paradox

A lamp is turned on for 1/2 minute, off for 1/4 minute, on for 1/8 minute, etc. At the end of one minute, the lamp switch will have been moved $\aleph_0$ times, where $\aleph_0$ is ALEPH-0. Will the lamp be on or off? This PARADOX is actually nonsensical, since it is equivalent to asking if the "last" INTEGER is EVEN or ODD.

References
Pickover, C. A. *Keys to Infinity.* New York: Wiley, pp. 19–23, 1995.

Thomson's Principle

see DIRICHLET'S PRINCIPLE

Thomson Problem

Determine the stable equilibrium positions of N classical electrons constrained to move on the surface of a SPHERE and repelling each other by an inverse square law. Exact solutions for $N = 2$ to 8 are known, but $N = 9$ and 11 are still unknown.

In reality, Earnshaw's theorem guarantees that no system of discrete electric charges can be held in stable equilibrium under the influence of their electrical interaction alone (Aspden 1987).

see also FEJES TÓTH'S PROBLEM

References

Altschuler, E. L.; Williams, T. J.; Ratner, E. R.; Dowla, F.; and Wooten, F. "Method of Constrained Global Optimization." *Phys. Rev. Let.* **72**, 2671–2674, 1994.

Altschuler, E. L.; Williams, T. J.; Ratner, E. R.; Dowla, F.; and Wooten, F. "Method of Constrained Global Optimization—Reply." *Phys. Rev. Let.* **74**, 1483, 1995.

Ashby, N. and Brittin, W. E. "Thomson's Problem." *Amer. J. Phys.* **54**, 776–777, 1986.

Aspden, H. "Earnshaw's Theorem." *Amer. J. Phys.* **55**, 199–200, 1987.

Berezin, A. A. "Spontaneous Symmetry Breaking in Classical Systems." *Amer. J. Phys.* **53**, 1037, 1985.

Calkin, M. G.; Kiang, D.; and Tindall, D. A. "Minimum Energy Configurations." *Nature* **319**, 454, 1986.

Erber, T. and Hockney, G. M. "Comment on 'Method of Constrained Global Optimization.'" *Phys. Rev. Let.* **74**, 1482–1483, 1995.

Marx, E. "Five Charges on a Sphere." *J. Franklin Inst.* **290**, 71–74, Jul. 1970.

Melnyk, T. W.; Knop, O.; and Smith, W. R. "Extremal Arrangements of Points and Unit Charges on a Sphere: Equilibrium Configurations Revisited." *Canad. J. Chem.* **55**, 1745–1761, 1977.

Whyte, L. L. "Unique Arrangement of Points on a Sphere." *Amer. Math. Monthly* **59**, 606–611, 1952.

Thousand

$1{,}000 = 10^3$. The word "thousand" appears in common expressions in a number of languages, for example, "a thousand pardons" in English and "tusen takk" ("a thousand thanks") in Norwegian.

see also HUNDRED, LARGE NUMBER, MILLION

Three

see 3

Three-Colorable

see COLORABLE

Three-In-A-Row

see TIC-TAC-TOE

Three Jug Problem

Given three jugs with x pints in the first, y in the second, and z in the third, obtain a desired amount in one of the vessels by completely filling up and/or emptying vessels into others. This problem can be solved with the aid of TRILINEAR COORDINATES.

References

Coxeter, H. S. M. and Greitzer, S. L. *Geometry Revisited.* Washington, DC: Math. Assoc. Amer., pp. 89–93, 1967.

Three-Valued Logic

A logical structure which does not assume the EXCLUDED MIDDLE LAW. Three possible truth values are possible: true, false, or undecided. There are 3072 such logics.

see also EXCLUDED MIDDLE LAW, FUZZY LOGIC, LOGIC

Threefoil Knot

see TREFOIL KNOT

Thue Constant

The base-2 TRANSCENDENTAL NUMBER

$$0.110110111110110111111\ldots_2,$$

where the nth bit is 1 if n is not divisible by 3 and is the complement of the $(n/3)$th bit if n is divisible by 3. It is also given by the SUBSTITUTION MAP

$$0 \to 111$$
$$1 \to 110.$$

In decimal, the Thue constant equals 0.8590997969....

see also RABBIT CONSTANT, THUE-MORSE CONSTANT

References

Thue-Morse Constant

The constant also called the PARITY CONSTANT and defined by

$$P \equiv \tfrac{1}{2}\sum_{n=0}^{\infty} P(n)2^{-n} = 0.4124540336401075977\ldots \quad (1)$$

(Sloane's A014571), where $P(n)$ is the PARITY of n. Dekking (1977) proved that the Thue-Morse constant is TRANSCENDENTAL, and Allouche and Shallit give a complete proof correcting a minor error of Dekking.

The Thue-Morse constant can be written in base 2 by stages by taking the previous iteration a_n, taking the complement $\overline{a_n}$, and appending, producing

$$\begin{aligned} a_0 &= 0.0_2 \\ a_1 &= 0.01_2 \\ a_2 &= 0.0110_2 \\ a_3 &= 0.01101001_2 \\ a_4 &= 0.0110100110010110_2. \end{aligned} \quad (2)$$

This can be written symbolically as

$$a_{n+1} = a_n + \overline{a_n} \cdot 2^{-2^n} \quad (3)$$

with $a_0 = 0$. Here, the complement is the number $\overline{a_n}$ such that $a_n + \overline{a_n} = 0.11\ldots_2$, which can be found from

$$a_n + \overline{a_n} = \sum_{k=1}^{2^n} (\tfrac{1}{2})^k = \frac{1-(\frac{1}{2})^{2^n}}{1-\frac{1}{2}} - 1 = 1 - 2^{-2^n}. \quad (4)$$

Therefore,

$$\overline{a_n} = 1 - a_n - 2^{-2^n}, \tag{5}$$

and

$$a_{n+1} = a_n + (1 - 2^{-2^n} - a_n)2^{-2^n}. \tag{6}$$

The regular CONTINUED FRACTION for the Thue-Morse constant is [0 2 2 2 1 4 3 5 2 1 4 2 1 5 44 1 4 1 2 4 1 1 1 5 14 1 50 15 5 1 1 1 4 2 1 4 1 43 1 4 1 2 1 3 16 1 2 1 2 1 50 1 2 424 1 2 5 2 1 1 1 5 5 2 22 5 1 1 1 1274 3 5 2 1 1 1 4 1 1 15 154 7 2 1 2 2 1 2 1 1 50 1 4 1 2 867374 1 1 1 5 5 1 1 6 1 2 7 2 1650 23 3 1 1 1 2 5 3 84 1 1 1 1284 ...] (Sloane's A014572), and seems to continue with sporadic large terms in suspicious-looking patterns. A nonregular CONTINUED FRACTION is

$$P = \cfrac{1}{3 - \cfrac{1}{2 - \cfrac{1}{4 - \cfrac{3}{16 - \cfrac{15}{256 - \cfrac{255}{65536 - \ldots}}}}}}. \tag{7}$$

A related infinite product is

$$4P = 2 - \frac{1 \cdot 3 \cdot 15 \cdot 255 \cdot 65535 \cdots}{2 \cdot 4 \cdot 16 \cdot 256 \cdot 65536 \cdots}. \tag{8}$$

The SEQUENCE a_∞ = 0110100110010110100101100... (Sloane's A010060) is known as the THUE-MORSE SEQUENCE.

see also RABBIT CONSTANT, THUE CONSTANT

References

Allouche, J. P.; Arnold, A.; Berstel, J.; Brlek, S.; Jockusch, W.; Plouffe, S.; and Sagan, B. "A Relative of the Thue-Morse Sequence." *Discr. Math.* **139**, 455–461, 1995.

Allouche, J. P. and Shallit, J. In preparation.

Beeler, M.; Gosper, R. W.; and Schroeppel, R. *HAKMEM.* Cambridge, MA: MIT Artificial Intelligence Laboratory, Memo AIM-239, Item 122, Feb. 1972.

Dekking, F. M. "Transcendence du nombre de Thue-Morse." *Comptes Rendus de l'Academie des Sciences de Paris* **285**, 157–160, 1977.

Finch, S. "Favorite Mathematical Constants." `http://www.mathsoft.com/asolve/constant/cntfrc/cntfrc.html`.

Sloane, N. J. A. Sequences A010060, A014571, and A014572 in "An On-Line Version of the Encyclopedia of Integer Sequences."

Thue-Morse Sequence

The INTEGER SEQUENCE (also called the MORSE-THUE SEQUENCE)

$$0110100110010110100101100110 1001\ldots \tag{1}$$

(Sloane's A010060) which arises in the THUE-MORSE CONSTANT. It can be generated from the SUBSTITUTION MAP

$$0 \to 01 \tag{2}$$

$$1 \to 10 \tag{3}$$

starting with 0 as follows:

$$0 \to 01 \to 0110 \to 01101001 \to \ldots. \tag{4}$$

Writing the sequence as a POWER SERIES over the GALOIS FIELD GF(2),

$$F(x) = 0 + 1x + 1x^2 + 0x^3 + 1x^4 + \ldots, \tag{5}$$

then F satisfies the quadratic equation

$$(1 + x)F^2 + F = \frac{x}{1 + x^2} \pmod 2. \tag{6}$$

This equation has two solutions, F and F', where F' is the complement of F, i.e.,

$$F + F' = 1 + x + x^2 + x^3 + \ldots = \frac{1}{1 + x}, \tag{7}$$

which is consistent with the formula for the sum of the roots of a quadratic. The equality (6) can be demonstrated as follows. Let $(abcdef\ldots)$ be a shorthand for the POWER series

$$a + bx + cx^2 + dx^3 + \ldots, \tag{8}$$

so $F(x)$ is (0110100110010110...). To get F^2, simply use the rule for squaring POWER SERIES over GF(2)

$$(A + B)^2 = A^2 + B^2 \pmod 2, \tag{9}$$

which extends to the simple rule for squaring a POWER SERIES

$$(a_0 + a_1 x + a_2 x^2 + \ldots)^2 = a_0 + a_1 x^2 + a_2 x^4 + \ldots \pmod 2, \tag{10}$$

i.e., space the series out by a factor of 2, (0 1 1 0 1 0 0 1 ...), and insert zeros in the ODD places to get

$$F^2 = (0010100010000010\ldots). \tag{11}$$

Then multiply by x (which just adds a zero at the front) to get

$$xF^2 = (00010100010000010\ldots). \tag{12}$$

Adding to F^2 gives

$$(1 + x)F^2 = (0011110011000011\ldots). \tag{13}$$

This is the first term of the quadratic equation, which is the Thue-Morse sequence with each term doubled up. The next term is F, so we have

$$(1 + x)F^2 = (0011110011000011\ldots) \tag{14}$$

$$F = (0110100110010110\ldots). \tag{15}$$

The sum is the above two sequences XORed together (there are no CARRIES because we're working over GF(2)), giving

$$(1+x)F^2 + F = (0101010101010101\ldots). \quad (16)$$

We therefore have

$$(1+x)F^2 + F = \frac{x}{1+x^2}$$
$$= x + x^3 + x^5 + x^7 + x^9 + x^{11} + \ldots \pmod 2. \quad (17)$$

The Thue-Morse sequence is an example of a cube-free sequence on two symbols (Morse and Hedlund 1944), i.e., it contains no substrings of the form WWW, where W is any WORD. For example, it does not contain the WORDS 000, 010101 or 010010010. In fact, the following stronger statement is true: the Thue-Morse sequence does not contain any substrings of the form WWa, where a is the first symbol of W. We can obtain a SQUAREFREE sequence on three symbols by doing the following: take the Thue-Morse sequence 0110100110010110... and look at the sequence of WORDS of length 2 that appear: 01 11 10 01 10 00 01 11 10 .... Replace 01 by 0, 10 by 1, 00 by 2 and 11 by 2 to get the following: 021012021.... Then this SEQUENCE is SQUAREFREE (Morse and Hedlund 1944).

The Thue-Morse sequence has important connections with the GRAY CODE. Kindermann generates fractal music using the SELF-SIMILARITY of the Thue-Morse sequence.

see also GRAY CODE, PARITY CONSTANT, RABBIT SEQUENCE, THUE SEQUENCE

References
Kindermann, L. "MusiNum—The Music in the Numbers." http://www.forwiss.uni-erlangen.de/~kinderma/musinum/.
Morse, M. and Hedlund, G. A. "Unending Chess, Symbolic Dynamics, and a Problem in Semigroups." *Duke Math. J.* **11**, 1–7, 1944.
Schroeder, M. R. *Fractals, Chaos, and Power Laws: Minutes from an Infinite Paradise.* New York: W. H. Freeman, 1991.
Sloane, N. J. A. Sequence A010060 in "An On-Line Version of the Encyclopedia of Integer Sequences."

Thue Sequence

The SEQUENCE of BINARY DIGITS of the THUE CONSTANT, $0.110110111110110111110110110\ldots_2$ (Sloane's A014578).

see also RABBIT CONSTANT, THUE CONSTANT

References
Guy, R. K. "Thue Sequences." §E21 in *Unsolved Problems in Number Theory, 2nd ed.* New York: Springer-Verlag, pp. 223–224, 1994.
Sloane, N. J. A. Sequence A014578 in "An On-Line Version of the Encyclopedia of Integer Sequences."

Thue-Siegel-Roth Theorem

If α is a TRANSCENDENTAL NUMBER, it can be approximated by infinitely many RATIONAL NUMBERS m/n to within n^{-r}, where r is any POSITIVE number.

see also LIOUVILLE'S RATIONAL APPROXIMATION THEOREM, LIOUVILLE-ROTH CONSTANT, ROTH'S THEOREM

Thue-Siegel-Schneider-Roth Theorem

see THUE-SIEGEL-ROTH THEOREM

Thue's Theorem

If $n > 1$, $(a, n) = 1$ (i.e., a and n are RELATIVELY PRIME), and m is the least integer $> \sqrt{n}$, then there exist an x and y such that

$$ay \equiv \pm x \pmod n$$

where $0 < x < m$ and $0 < y < m$.

References
Shanks, D. *Solved and Unsolved Problems in Number Theory, 4th ed.* New York: Chelsea, p. 161, 1993.

Thurston's Geometrization Conjecture

Thurston's conjecture has to do with geometric structures on 3-D MANIFOLDS. Before stating Thurston's conjecture, some background information is useful. 3-dimensional MANIFOLDS possess what is known as a standard 2-level DECOMPOSITION. First, there is the CONNECTED SUM DECOMPOSITION, which says that every COMPACT 3-MANIFOLD is the CONNECTED SUM of a unique collection of PRIME 3-MANIFOLDS.

The second DECOMPOSITION is the JACO-SHALEN-JOHANNSON TORUS DECOMPOSITION, which states that irreducible orientable COMPACT 3-MANIFOLDS have a canonical (up to ISOTOPY) minimal collection of disjointly EMBEDDED incompressible TORI such that each component of the 3-MANIFOLD removed by the TORI is either "atoroidal" or "Seifert-fibered."

Thurston's conjecture is that, after you split a 3-MANIFOLD into its CONNECTED SUM and then JACO-SHALEN-JOHANNSON TORUS DECOMPOSITION, the remaining components each admit exactly one of the following geometries:

1. EUCLIDEAN GEOMETRY,
2. HYPERBOLIC GEOMETRY,
3. SPHERICAL GEOMETRY,
4. the GEOMETRY of $\mathbb{S}^2 \times \mathbb{R}$,
5. the GEOMETRY of $\mathbb{H}^2 \times \mathbb{R}$,
6. the GEOMETRY of SL_2R,
7. NIL GEOMETRY, or
8. SOL GEOMETRY.

Here, $\mathbb{S}^2$ is the 2-SPHERE and $\mathbb{H}^2$ is the HYPERBOLIC PLANE. If Thurston's conjecture is true, the truth of the POINCARÉ CONJECTURE immediately follows.

see also CONNECTED SUM DECOMPOSITION, EUCLIDEAN GEOMETRY, HYPERBOLIC GEOMETRY, JACO-SHALEN-JOHANNSON TORUS DECOMPOSITION, NIL GEOMETRY, POINCARÉ CONJECTURE, SOL GEOMETRY, SPHERICAL GEOMETRY

Thwaites Conjecture

see COLLATZ PROBLEM

Tic-Tac-Toe

The usual game of tic-tac-toe (also called TICKTACKTOE) is 3-in-a-row on a 3×3 board. However, a generalized n-IN-A-ROW on an $n \times m$ board can also be considered. For $n = 1$ and 2 the first player can always win. If the board is at least 3×4, the first player can win for $n = 3$.

However, for TIC-TAC-TOE which uses a 3×3 board, a draw can always be obtained. If the board is at least 4×30, the first player can win for $n = 4$. For $n = 5$, a draw can always be obtained on a 5×5 board, but the first player can win if the board is at least 15×15. The cases $n = 6$ and 7 have not yet been fully analyzed for an $n \times n$ board, although draws can always be forced for $n = 8$ and 9. On an $\infty \times \infty$ board, the first player can win for $n = 1$, 2, 3, and 4, but a tie can always be forced for $n \geq 8$. For $3 \times 3 \times 3$ and $4 \times 4 \times 4$, the first player can always win (Gardner 1979).

see also PONG HAU K'I

References

Ball, W. W. R. and Coxeter, H. S. M. *Mathematical Recreations and Essays, 13th ed.* New York: Dover, pp. 103–104, 1987.

de Fouquières, B. Ch. 18 in *Les Jeux des Anciens, 2nd ed.*. Paris, 1873.

Gardner, M. "Mathematical Games: The Diverse Pleasures of Circles that Are Tangent to One Another." *Sci. Amer.* **240**, 18–28, Jan. 1979a.

Gardner, M. "Ticktacktoe Games." Ch. 9 in *Wheels, Life, and Other Mathematical Amusements.* New York: W. H. Freeman, 1983.

Stewart, I. "A Shepherd Takes A Sheep Shot." *Sci. Amer.* **269**, 154–156, 1993.

Ticktacktoe

see TIC-TAC-TOE

Tight Closure

The application of characteristic p methods in COMMUTATIVE ALGEBRA, which is a synthesis of some areas of COMMUTATIVE ALGEBRA and ALGEBRAIC GEOMETRY.

see also ALGEBRAIC GEOMETRY, COMMUTATIVE ALGEBRA

References

Bruns, W. "Tight Closure." *Bull. Amer. Math. Soc.* **33**, 447–457, 1996.

Huneke, C. "An Algebraist Commuting in Berkeley." *Math. Intell.* **11**, 40–52, 1989.

Tightly Embedded

Q is said to be tightly embedded if $|Q \cap Q^g|$ is ODD for all $g \in G - N_G(Q)$, where $N_G(Q)$ is the NORMALIZER of Q in G.

Tiling

A plane-filling arrangement of plane figures or its generalization to higher dimensions. Formally, a tiling is a collection of disjoint open sets, the closures of which cover the plane. Given a single tile, the so-called first CORONA is the set of all tiles that have a common boundary point with the tile (including the original tile itself).

WANG'S CONJECTURE (1961) stated that if a set of tiles tiled the plane, then they could always be arranged to do so periodically. A periodic tiling of the PLANE by POLYGONS or SPACE by POLYHEDRA is called a TESSELLATION. The conjecture was refuted in 1966 when R. Berger showed that an aperiodic set of 20,426 tiles exists. By 1971, R. Robinson had reduced the number to six and, in 1974, R. Penrose discovered an aperiodic set (when color-matching rules are included) of two tiles: the so-called PENROSE TILES. (Penrose also sued the Kimberly Clark Corporation over their quilted toilet paper, which allegedly resembles a Penrose aperiodic tiling; Mirsky 1997.)

It is not known if there is a single aperiodic tile.

n-gon	tilings
3	any
4	any
5	14
6	3

The number of tilings possible for convex irregular POLYGONS are given in the above table. Any TRIANGLE or convex QUADRILATERAL tiles the plane. There are at least 14 classes of convex PENTAGONAL tilings. There are at least three aperiodic tilings of HEXAGONS, given by the following types:

$$\begin{array}{cc} A+B+C=360^\circ & a=d \\ A+B+D=360^\circ & a=d, c=e \\ A=C=E & a=b, c=d, e=f \end{array} \quad (1)$$

(Gardner 1988). Note that the periodic hexagonal TESSELLATION is a degenerate case of all three tilings with

$$A=B=C=D=E=F \qquad a=b=c=d=e=f. \quad (2)$$

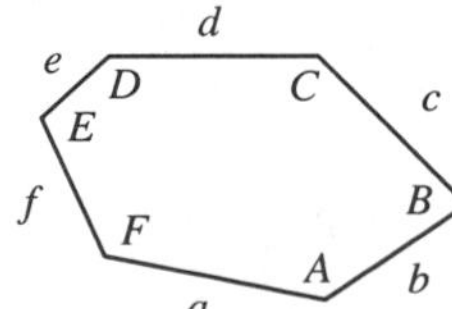

There are no tilings for convex n-gons for $n \geq 7$.

see also ANISOHEDRAL TILING, CORONA (TILING), GOSPER ISLAND, HEESCH'S PROBLEM, ISOHEDRAL TILING, KOCH SNOWFLAKE, MONOHEDRAL TILING, PENROSE TILES, POLYOMINO TILING, SPACE-FILLING POLYHEDRON, TILING THEOREM, TRIANGLE TILING

References

Eppstein, D. "Tiling." http://www.ics.uci.edu/~eppstein/junkyard/tiling.html.

Gardner, M. "Tilings with Convex Polygons." Ch. 13 in *Time Travel and Other Mathematical Bewilderments.* New York: W. H. Freeman, pp. 162–176, 1988.

Gardner, M. Chs. 1–2 in *Penrose Tiles to Trapdoor Ciphers... and the Return of Dr. Matrix, reissue ed.* Washington, DC: Math. Assoc. Amer.

Grünbaum, B. and Shepard, G. C. "Some Problems on Plane Tilings." In *The Mathematical Gardner* (Ed. D. Klarner). Boston, MA: Prindle, Weber, and Schmidt, pp. 167–196, 1981.

Grünbaum, B. and Sheppard, G. C. *Tilings and Patterns.* New York: W. H. Freeman, 1986.

Lee, X. "Visual Symmetry." http://www.best.com/~xah/MathGraphicsGallery_dir/Tiling_dir/tiling.html.

Mirsky, S. "The Emperor's New Toilet Paper." *Sci. Amer.* **277**, 24, July 1997.

Pappas, T. "Mathematics & Moslem Art." *The Joy of Mathematics.* San Carlos, CA: Wide World Publ./Tetra, p. 178, 1989.

Peterson, I. *The Mathematical Tourist: Snapshots of Modern Mathematics.* New York: W. H. Freeman, pp. 82–85, 1988.

Rawles, B. *Sacred Geometry Design Sourcebook: Universal Dimensional Patterns.* Nevada City, CA: Elysian Pub., 1997. http://www.oro.net/~elysian/bruce_rawles_books.html.

Schattschneider, D. "In Praise of Amateurs." In *The Mathematical Gardner* (Ed. D. Klarner). Boston, MA: Prindle, Weber, and Schmidt, pp. 140–166, 1981.

Seyd, J. A. and Salman, A. S. *Symmetries of Islamic Geometrical Patterns.* River Edge, NJ: World Scientific, 1995.

Stein, S. and Szabó, S. *Algebra and Tiling.* Washington, DC: Math. Assoc. Amer., 1994.

Tiling Theorem

Due to Lebesgue and Brouwer. If an n-D figure is covered in any way by sufficiently small subregions, then there will exist points which belong to at least $n+1$ of these subareas. Moreover, it is always possible to find a covering by arbitrarily small regions for which no point will belong to more than $n+1$ regions.

see also TESSELLATION, TILING

Times

The operation of MULTIPLICATION, i.e., a times b. Various notations are $a \times b$, $a \cdot b$, ab, and $(a)(b)$. The "multiplication sign" $\times$ is based on SAINT ANDREW'S CROSS (Bergamini 1969). Floating point MULTIPLICATION is sometimes denoted $\otimes$.

see also CROSS PRODUCT, DOT PRODUCT, MINUS, MULTIPLICATION, PLUS, PRODUCT

References

Bergamini, D. *Mathematics.* New York: Time-Life Books, p. 11, 1969.

Tit-for-Tat

A strategy for the iterated PRISONER'S DILEMMA in which a prisoner cooperates on the first move, and thereafter copies the previous move of the other prisoner. Any better strategy has more complicated rules.

see also PRISONER'S DILEMMA

References

Goetz, P. "Phil's Good Enough Complexity Dictionary." http://www.cs.buffalo.edu/~goetz/dict.html.

Titanic Prime

A PRIME with ≥ 1000 DIGITS. As of 1990, there were more than 1400 known (Ribenboim 1990). The table below gives the number of known titanic primes as a function of year end.

Year	Titanic Primes
1992	2254
1993	9166
1994	9779
1995	12391

References

Caldwell, C. "The Ten Largest Known Primes." http://www.utm.edu/research/primes/largest.html#largest.

Morain, F. "Elliptic Curves, Primality Proving and Some Titanic Primes." *Astérique* **198–200**, 245–251, 1992.

Ribenboim, P. *The Little Book of Big Primes.* Berlin: Springer-Verlag, p. 97, 1990.

Yates, S. "Titanic Primes." *J. Recr. Math.* **16**, 250–262, 1983-84.

Yates, S. "Sinkers of the Titanics." *J. Recr. Math.* **17**, 268–274, 1984–85.

Titchmarsh Theorem

If $f(\omega)$ is SQUARE INTEGRABLE over the REAL ω axis, then any one of the following implies the other two:

1. The FOURIER TRANSFORM of $f(\omega)$ is 0 for $t < 0$.
2. Replacing ω by z, the function $f(z)$ is analytic in the COMPLEX PLANE z for $y > 0$ and approaches $f(x)$ almost everywhere as $y \to 0$. Furthermore, $\int_{-\infty}^{\infty} |f(x+iy)|^2\,dx < k$ for some number k and $y > 0$ (i.e., the integral is bounded).
3. The REAL and IMAGINARY PARTS of $f(z)$ are HILBERT TRANSFORMS of each other.

Tits Group

A finite SIMPLE GROUP which is a SUBGROUP of the TWISTED CHEVALLEY GROUP ${}^2F_4(2)$.

Toeplitz Matrix

Given $2N-1$ numbers r_k where $k = -N+1, \ldots, -1, 0, 1, \ldots, N-1$, a MATRIX of the form

$$\begin{bmatrix} r_0 & r_{-1} & r_{-2} & \cdots & r_{-n+1} \\ r_1 & r_0 & r_{-1} & \cdots & r_{-n+2} \\ \vdots & \vdots & \vdots & \ddots & \vdots \\ r_{n-1} & r_{n-2} & r_{n-3} & \cdots & r_0 \end{bmatrix}$$

is called a Toeplitz matrix. MATRIX equations of the form

$$\sum_{j=1}^{N} r_{i-j}x_j = y_i$$

can be solved with $\mathcal{O}(N^2)$ operations.

see also VANDERMONDE MATRIX

References

Press, W. H.; Flannery, B. P.; Teukolsky, S. A.; and Vetterling, W. T. "Vandermonde Matrices and Toeplitz Matrices." §2.8 in *Numerical Recipes in FORTRAN: The Art of Scientific Computing, 2nd ed.* Cambridge, England: Cambridge University Press, pp. 82–89, 1992.

Togliatti Surface

Togliatti (1940, 1949) showed that QUINTIC SURFACES having 31 ORDINARY DOUBLE POINTS exist, although he did not explicitly derive equations for such surfaces. Beauville (1978) subsequently proved that 31 double points are the maximum possible, and quintic surfaces having 31 ORDINARY DOUBLE POINTS are therefore sometimes called Togliatti surfaces. van Straten (1993) subsequently constructed a 3-D family of solutions and in 1994, Barth derived the example known as the DERVISH.

see also DERVISH, ORDINARY DOUBLE POINT, QUINTIC SURFACE

References

Beauville, A. "Surfaces algébriques complexes." *Astérisque* **54**, 1–172, 1978.

Endraß, S. "Togliatti Surfaces." `http://www.mathematik.uni-mainz.de/Algebraische Geometrie/docs/Etogliatti.shtml`.

Hunt, B. "Algebraic Surfaces." `http://www.mathematik.uni-kl.de/~wwwagag/Galerie.html`.

Togliatti, E. G. "Una notevole superficie de 5° ordine con soli punti doppi isolati." *Vierteljschr. Naturforsch. Ges. Zürich* **85**, 127–132, 1940.

Togliatti, E. "Sulle superficie monoidi col massimo numero di punti doppi." *Ann. Mat. Pura Appl.* **30**, 201–209, 1949.

van Straten, D. "A Quintic Hypersurface in $\mathbb{P}^4$ with 130 Nodes." *Topology* **32**, 857–864, 1993.

Tomography

Tomography is the study of the reconstruction of 2- and 3-dimensional objects from 1-dimensional slices. The RADON TRANSFORM is an important tool in tomography.

Rather surprisingly, there exist certain sets of four directions in Euclidean n-space such that X-rays of a convex body in these directions distinguish it from all other convex bodies.

see also ALEKSANDROV'S UNIQUENESS THEOREM, BRUNN-MINKOWSKI INEQUALITY, BUSEMANN-PETTY PROBLEM, DVORETZKY'S THEOREM, RADON TRANSFORM, STEREOLOGY

References

Gardner, R. J. "Geometric Tomography." *Not. Amer. Math. Soc.* **42**, 422–429, 1995.

Gardner, R. J. *Geometric Tomography.* New York: Cambridge University Press, 1995.

Tooth Surface

The QUARTIC SURFACE given by the equation

$$x^4 + y^4 + z^4 - (x^2 + y^2 + z^2) = 0.$$

References

Nordstrand, T. "Surfaces." `http://www.uib.no/people/nfytn/surfaces.htm`.

Topological Basis

A topological basis is a SUBSET B of a SET T in which all other OPEN SETS can be written as UNIONS or finite INTERSECTIONS of B. For the REAL NUMBERS, the SET of all OPEN INTERVALS is a basis.

Topological Completion

The topological completion C of a FIELD F with respect to the ABSOLUTE VALUE $|\cdot|$ is the smallest FIELD containing F for which all CAUCHY SEQUENCES or rationals converge.

References

Burger, E. B. and Struppeck, T. "Does $\sum_{n=0}^{\infty} \frac{1}{n!}$ Really Converge? Infinite Series and p-adic Analysis." *Amer. Math. Monthly* **103**, 565–577, 1996.

Topologically Conjugate

Two MAPS $\phi, \psi : M \to M$ are said to be topologically conjugate if there EXISTS a HOMEOMORPHISM $h : M \to M$ such that $\phi \circ h = h \circ \psi$, i.e., h maps ψ-orbits onto ϕ-orbits. Two maps which are topologically conjugate cannot be distinguished topologically.

see also ANOSOV DIFFEOMORPHISM, STRUCTURALLY STABLE

Topological Dimension

see LEBESGUE COVERING DIMENSION

Topological Entropy

The topological entropy of a MAP M is defined as

$$h_T(M) = \sup_{\{W_i\}} h(M, \{W_i\}),$$

where $\{W_i\}$ is a partition of a bounded region W containing a probability measure which is invariant under M, and sup is the SUPREMUM.

References

Ott, E. *Chaos in Dynamical Systems.* New York: Cambridge University Press, pp. 143–144, 1993.

Topological Groupoid

A topological groupoid over B is a GROUPOID G such that B and G are TOPOLOGICAL SPACES and α, β, and multiplication are continuous maps. Here, α and β are maps from G onto $\mathbb{R}^2$ with $\alpha : (x, \gamma, y) \mapsto x$ and $\beta : (x, \gamma, y) \mapsto y$.

References

Weinstein, A. "Groupoids: Unifying Internal and External Symmetry." *Not. Amer. Math. Soc.* **43**, 744–752, 1996.

Topological Manifold

A TOPOLOGICAL SPACE M satisfying some separability (i.e., it is a HAUSDORFF SPACE) and countability (i.e., it is a PARACOMPACT SPACE) conditions such that every point $p \in M$ has a NEIGHBORHOOD homeomorphic to an OPEN SET in $\mathbb{R}^n$ for some $n \geq 0$. Every SMOOTH MANIFOLD is a topological manifold, but not necessarily vice versa. The first nonsmooth topological manifold occurs in 4-D.

Nonparacompact manifolds are of little use in mathematics, but non-Hausdorff manifolds do occasionally arise in research (Hawking and Ellis 1975). For manifolds, Hausdorff and second countable are equivalent to Hausdorff and paracompact, and both are equivalent to the manifold being embeddable in some large-dimensional Euclidean space.

see also HAUSDORFF SPACE, MANIFOLD, PARACOMPACT SPACE, SMOOTH MANIFOLD, TOPOLOGICAL SPACE

References

Hawking, S. W. and Ellis, G. F. R. *The Large Scale Structure of Space-Time.* New York: Cambridge University Press, 1975.

Topological Space

A SET X for which a TOPOLOGY T has been specified is called a topological space (Munkres 1975, p. 76).

see also KURATOWSKI'S CLOSURE-COMPONENT PROBLEM, OPEN SET, TOPOLOGICAL VECTOR SPACE

References

Berge, C. *Topological Spaces Including a Treatment of Multi-Valued Functions, Vector Spaces and Convexity.* New York: Dover, 1997.

Munkres, J. R. *Topology: A First Course.* Englewood Cliffs, NJ: Prentice-Hall, 1975.

Topological Vector Space

A TOPOLOGICAL SPACE such that the two algebraic operations of VECTOR SPACE are continuous in the topology.

References

Köthe, G. *Topological Vector Spaces.* New York: Springer-Verlag, 1979.

Topologically Transitive

A FUNCTION f is topologically transitive if, given any two intervals U and V, there is some POSITIVE INTEGER k such that $f^k(U) \cap V = \varnothing$. Vaguely, this means that neighborhoods of points eventually get flung out to "big" sets so that they don't necessarily stick together in one localized clump.

see also CHAOS

Topology

Topology is the mathematical study of properties of objects which are preserved through deformations, twistings, and stretchings. (Tearing, however, is not allowed.) A CIRCLE is topologically equivalent to an ELLIPSE (into which it can be deformed by stretching) and a SPHERE is equivalent to an ELLIPSOID. Continuing along these lines, the SPACE of all positions of the minute hand on a clock is topologically equivalent to a CIRCLE (where SPACE of all positions means "the collection of all positions"). Similarly, the SPACE of all positions of the minute and hour hands is equivalent to a TORUS. The SPACE of all positions of the hour, minute and second hands form a 4-D object that cannot be visualized quite as simply as the former objects since it cannot be placed in our 3-D world, although it can be visualized by other means.

There is more to topology, though. Topology began with the study of curves, surfaces, and other objects in the plane and 3-space. One of the central ideas in topology is that spatial objects like CIRCLES and SPHERES can be treated as objects in their own right, and knowledge of objects is independent of how they are "represented" or "embedded" in space. For example, the statement "if you remove a point from a CIRCLE, you get a line segment" applies just as well to the CIRCLE as to an ELLIPSE, and even to tangled or knotted CIRCLES, since the statement involves only topological properties.

Topology has to do with the study of spatial objects such as curves, surfaces, the space we call our universe, the space-time of general relativity, fractals, knots, manifolds (objects with some of the same basic spatial properties as our universe), phase spaces that are encountered in physics (such as the space of hand-positions of a clock), symmetry groups like the collection of ways of rotating a top, etc.

The "objects" of topology are often formally defined as TOPOLOGICAL SPACES. If two objects have the same topological properties, they are said to be HOMEOMORPHIC (although, strictly speaking, properties that are not destroyed by stretching and distorting an object are really properties preserved by ISOTOPY, not HOMEOMORPHISM; ISOTOPY has to do with distorting embedded objects, while HOMEOMORPHISM is intrinsic).

Topology is divided into ALGEBRAIC TOPOLOGY (also called COMBINATORIAL TOPOLOGY), DIFFERENTIAL TOPOLOGY, and LOW-DIMENSIONAL TOPOLOGY.

There is also a formal definition for a topology defined in terms of set operations. A SET X along with a collection T of SUBSETS of it is said to be a topology if the SUBSETS in T obey the following properties:

1. The (trivial) subsets X and the EMPTY SET $\varnothing$ are in T.
2. Whenever sets A and B are in T, then so is $A \cap B$.
3. Whenever two or more sets are in T, then so is their UNION

(Bishop and Goldberg 1980).

A SET X for which a topology T has been specified is called a TOPOLOGICAL SPACE (Munkres 1975, p. 76). For example, the SET $X = \{0, 1, 2, 3\}$ together with the SUBSETS $T = \{0\}, \{1, 2, 3\}, \varnothing, \{0, 1, 2, 3\}\}$ comprises a topology, and X is a TOPOLOGICAL SPACE.

Topologies can be built up from TOPOLOGICAL BASES. For the REAL NUMBERS, the topology is the UNION of OPEN INTERVALS.

see also ALGEBRAIC TOPOLOGY, DIFFERENTIAL TOPOLOGY, GENUS, KLEIN BOTTLE, KURATOWSKI REDUCTION THEOREM, LEFSHETZ TRACE FORMULA, LOW-DIMENSIONAL TOPOLOGY, POINT-SET TOPOLOGY, ZARISKI TOPOLOGY

References

Adamson, I. *A General Topology Workbook.* Boston, MA: Birkhäuser, 1996.

Armstrong, M. A. *Basic Topology, rev.* New York: Springer-Verlag, 1997.

Barr, S. *Experiments in Topology.* New York: Dover, 1964.

Berge, C. *Topological Spaces Including a Treatment of Multi-Valued Functions, Vector Spaces and Convexity.* New York: Dover, 1997.

Bishop, R. and Goldberg, S. *Tensor Analysis on Manifolds.* New York: Dover, 1980.

Blackett, D. W. *Elementary Topology: A Combinatorial and Algebraic Approach.* New York: Academic Press, 1967.

Bloch, E. *A First Course in Geometric Topology and Differential Geometry.* Boston, MA: Birkhäuser, 1996.

Chinn, W. G. and Steenrod, N. E. *First Concepts of Topology: The Geometry of Mappings of Segments, Curves, Circles, and Disks.* Washington, DC: Math. Assoc. Amer., 1966.

Eppstein, D. "Geometric Topology." `http://www.ics.uci.edu/~eppstein/junkyard/topo.html`.

Francis, G. K. *A Topological Picturebook.* New York: Springer-Verlag, 1987.

Gemignani, M. C. *Elementary Topology.* New York: Dover, 1990.

Greever, J. *Theory and Examples of Point-Set Topology.* Belmont, CA: Brooks/Cole, 1967.

Hirsch, M. W. *Differential Topology.* New York: Springer-Verlag, 1988.

Hocking, J. G. and Young, G. S. *Topology.* New York: Dover, 1988.

Kahn, D. W. *Topology: An Introduction to the Point-Set and Algebraic Areas.* New York: Dover, 1995.

Kelley, J. L. *General Topology.* New York: Springer-Verlag, 1975.

Kinsey, L. C. *Topology of Surfaces.* New York: Springer-Verlag, 1993.

Lipschutz, S. *Theory and Problems of General Topology.* New York: Schaum, 1965.

Mendelson, B. *Introduction to Topology.* New York: Dover, 1990.

Munkres, J. R. *Elementary Differential Topology.* Princeton, NJ: Princeton University Press, 1963.

Munkres, J. R. *Topology: A First Course.* Englewood Cliffs, NJ: Prentice-Hall, 1975.

Praslov, V. V. and Sossinsky, A. B. *Knots, Links, Braids and 3-Manifolds: An Introduction to the New Invariants in Low-Dimensional Topology.* Providence, RI: Amer. Math. Soc., 1996.

Shakhmatv, D. and Watson, S. "Topology Atlas." `http://www.unipissing.ca/topology/`.

Steen, L. A. and Seebach, J. A. Jr. *Counterexamples in Topology.* New York: Dover, 1996.

Thurston, W. P. *Three-Dimensional Geometry and Topology, Vol. 1.* Princeton, NJ: Princeton University Press, 1997.

van Mill, J. and Reed, G. M. (Eds.). *Open Problems in Topology.* New York: Elsevier, 1990.

Veblen, O. *Analysis Situs, 2nd ed.* New York: Amer. Math. Soc., 1946.

Topos

A CATEGORY modeled after the properties of the CATEGORY of sets.

see also CATEGORY, LOGOS

References

Freyd, P. J. and Scedrov, A. *Categories, Allegories.* Amsterdam, Netherlands: North-Holland, 1990.

McLarty, C. *Elementary Categories, Elementary Toposes.* New York: Oxford University Press, 1992.

Toric Variety

Let $m_1, m_2, \ldots, m_n$ be distinct primitive elements of a 2-D LATTICE M such that $\det(m_i, m_{i+1}) > 0$ for $i = 1, \ldots, n$. Each collection $\Gamma = \{m_1, m_2, \ldots, m_n\}$ then forms a set of rays of a unique complete fan in M, and therefore determines a 2-D toric variety X_Γ.

References

Danilov, V. I. "The Geometry of Toric Varieties." *Russ. Math. Surv.* **33**, 97–154, 1978.

Fulton, W. *Introduction to Toric Varieties.* Princeton, NJ: Princeton University Press, 1993.

Morelli, R. "Pick's Theorem and the Todd Class of a Toric Variety." *Adv. Math.* **100**, 183–231, 1993.

Oda, T. *Convex Bodies and Algebraic Geometry.* New York: Springer-Verlag, 1987.

Pommersheim, J. E. "Toric Varieties, Lattice Points, and Dedekind Sums." *Math. Ann.* **295**, 1–24, 1993.

Torispherical Dome

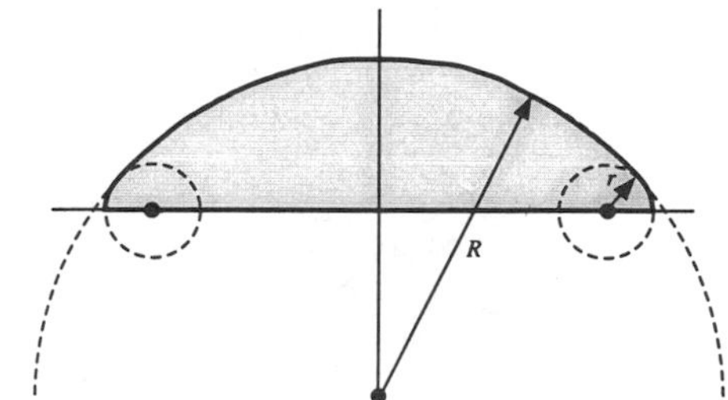

A torispherical dome is the surface obtained from the intersection of a SPHERICAL CAP with a tangent TORUS, as illustrated above. The radius of the sphere R is called the "crown radius," and the radius of the torus is called the "knuckle radius." Torispherical domes are used to construct pressure vessels.

see also DOME, SPHERICAL CAP

Torn Square Fractal

see CESÀRO FRACTAL

Toroid

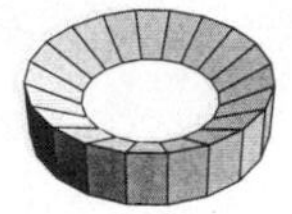
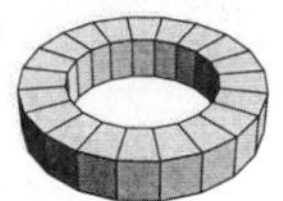
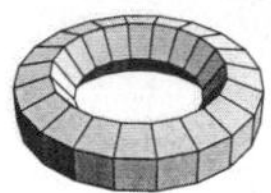

A SURFACE OF REVOLUTION obtained by rotating a closed PLANE CURVE about an axis parallel to the plane which does not intersect the curve. The simplest toroid is the TORUS.

see also PAPPUS'S CENTROID THEOREM, SURFACE OF REVOLUTION, TORUS

Toroidal Coordinates

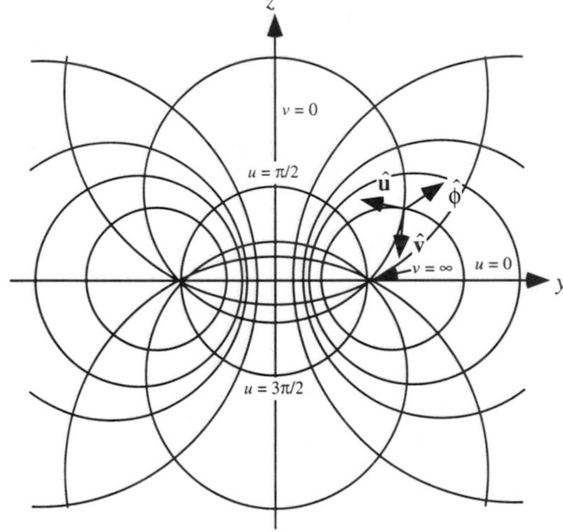

A system of CURVILINEAR COORDINATES for which several different notations are commonly used. In this work (u, v, ϕ) is used, whereas Arfken (1970) uses (ξ, η, φ). The toroidal coordinates are defined by

$$x = \frac{a \sinh v \cos\phi}{\cosh v - \cos u} \tag{1}$$

$$y = \frac{a \sinh v \sin\phi}{\cosh v - \cos u} \tag{2}$$

$$z = \frac{a \sin u}{\cosh v - \cos u}, \tag{3}$$

where $\sinh z$ is the HYPERBOLIC SINE and $\cosh z$ is the HYPERBOLIC COSINE. The SCALE FACTORS are

$$h_u = \frac{a}{\cosh v - \cos u} \tag{4}$$

$$h_v = \frac{a}{\cosh v - \cos u} \tag{5}$$

$$h_\phi = \frac{a \sinh v}{\cosh v - \cos u}. \tag{6}$$

The LAPLACIAN is

$$\nabla^2 f = \frac{(\cosh v - \cos u)^3}{a^2}\frac{\partial}{\partial u}\left(\frac{1}{\cosh v - \cos u}\frac{\partial f}{\partial u}\right) + \frac{(\cosh v - \cos u)^3}{a^2 \sinh v}\frac{\partial}{\partial v}\left(\frac{\sinh v}{\cosh v - \cos u}\frac{\partial f}{\partial v}\right) + \frac{(\cosh v - \cos u)^2}{a^2 \sinh v}\frac{\partial^2 f}{\partial \phi^2} \tag{7}$$

$$= \left(\frac{-3\cos\coth^2 v + \cosh v \coth^2 v}{\cosh v - \cos u} + \frac{+3\cos^2 u \coth v \operatorname{csch} v - \cos^3 u \operatorname{csch}^2 v}{\cosh v - \cos u}\right)\frac{\partial^2}{\partial\phi^2} + (\cos u - \cosh v)\sin u \frac{\partial}{\partial u} + (\cosh v - \cos u)^2 \frac{\partial^2}{\partial u^2} + (\cosh v - \cos u)(\cosh v \coth v - \sinh v - \cos u \coth v)\frac{\partial}{\partial v} + (\cosh^2 v - \cos u)^2 \frac{\partial^2}{\partial v^2}. \tag{8}$$

The HELMHOLTZ DIFFERENTIAL EQUATION is not separable in toroidal coordinates, but LAPLACE'S EQUATION is.

see also BISPHERICAL COORDINATES, LAPLACE'S EQUATION—TOROIDAL COORDINATES

References

Arfken, G. "Toroidal Coordinates (ξ, η, ϕ)." §2.13 in *Mathematical Methods for Physicists, 2nd ed.* Orlando, FL: Academic Press, pp. 112–115, 1970.

Morse, P. M. and Feshbach, H. *Methods of Theoretical Physics, Part I.* New York: McGraw-Hill, p. 666, 1953.

Toroidal Field

A VECTOR FIELD resembling a TORUS which is purely circular about the z-AXIS of a SPHERE (i.e., follows lines of LATITUDE). A toroidal field takes the form

$$\mathbf{T} = \begin{bmatrix} 0 \\ \frac{1}{\sin\theta}\frac{\partial T}{\partial\phi} \\ -\frac{\partial T}{\partial\theta} \end{bmatrix}.$$

see also DIVERGENCELESS FIELD, POLOIDAL FIELD

References

Stacey, F. D. *Physics of the Earth, 2nd ed.* New York: Wiley, p. 239, 1977.

Toroidal Function

A class of functions also called RING FUNCTIONS which appear in systems having toroidal symmetry. Toroidal functions can be expressed in terms of the LEGENDRE FUNCTIONS and SECOND KINDS (Abramowitz and Stegun 1972, p. 336):

$$P_{\nu-1/2}^{\mu}(\cosh\eta) = [\Gamma(1-\mu)]^{-1}2^{2\mu}(1-e^{-2\eta})^{-\mu}e^{-(\nu+1/2)\eta}$$
$$\times {}_2F_1(\tfrac{1}{2}-\mu, \tfrac{1}{2}+\nu-\mu; 1-2\mu; 1-e^{-2\eta})$$
$$P_{n-1/2}^{m}(\cosh\eta) = \frac{\Gamma(n+m+\frac{1}{2})(\sinh\eta)^m}{\Gamma(n-m+\frac{1}{2})2^m\sqrt{\pi}\,\Gamma(m+\frac{1}{2})}$$
$$\times \int_0^\pi \frac{sin^{2m}\phi\,d\phi}{(\cosh\eta+\cos\phi\sinh\eta)^{n+m+1/2}}$$
$$Q_{\nu-1/2}^{\mu}(\cosh\eta) = [\Gamma(1+\nu)]^{-1}\sqrt{\pi}\,e^{i\mu\pi}\Gamma(\tfrac{1}{2}+\nu+\mu)$$
$$\times (1-e^{-2\eta})^{\mu}e^{-(\nu+1/2)\eta}{}_2F_1(\tfrac{1}{2}-\mu, \tfrac{1}{2}+\nu+\mu; 1+\mu; 1-e^{-2\eta})$$
$$Q_{n-1/2}^{m}(\cosh\eta) = \frac{(-1)^m\Gamma(n+\frac{1}{2})}{\Gamma(n-m+\frac{1}{2})}$$
$$\times \int_0^\infty \frac{\cosh(mt)\,dt}{(\cosh\eta+\cosh t\sinh\eta)^{n+1/2}}$$

for $n > m$. Byerly (1959) identifies

$$\frac{1}{i^{n/2}}P_m^n(\coth x) = \operatorname{csch}^n x\frac{d^n P_m(\coth x)}{d(\coth x)^n}$$

as a TOROIDAL HARMONIC.

see also CONICAL FUNCTION

References

Abramowitz, M. and Stegun, C. A. (Eds.). "Toroidal Functions (or Ring Functions)." §8.11 in *Handbook of Mathematical Functions with Formulas, Graphs, and Mathematical Tables, 9th printing.* New York: Dover, p. 336, 1972.

Byerly, W. E. *An Elementary Treatise on Fourier's Series, and Spherical, Cylindrical, and Ellipsoidal Harmonics, with Applications to Problems in Mathematical Physics.* New York: Dover, p. 266, 1959.

Iyanaga, S. and Kawada, Y. (Eds.). *Encyclopedic Dictionary of Mathematics.* Cambridge, MA: MIT Press, p. 1468, 1980.

Toroidal Harmonic

see TOROIDAL FUNCTION

Toroidal Polyhedron

A toroidal polyhedron is a POLYHEDRON with GENUS $g \geq 1$ (i.e., having one or more HOLES). Examples of toroidal polyhedra include the CSÁSZÁR POLYHEDRON and SZILASSI POLYHEDRON, both of which have GENUS 1 (i.e., the TOPOLOGY of a TORUS).

The only known TOROIDAL POLYHEDRON with no DIAGONALS is the CSÁSZÁR POLYHEDRON. If another exists, it must have 12 or more VERTICES and GENUS $g \geq 6$. The smallest known single-hole toroidal POLYHEDRON made up of only EQUILATERAL TRIANGLES is composed of 48 of them.

see also CSÁSZÁR POLYHEDRON, SZILASSI POLYHEDRON

References

Gardner, M. *Time Travel and Other Mathematical Bewilderments.* New York: W. H. Freeman, p. 141, 1988.

Hart, G. "Toroidal Polyhedra." `http://www.li.net/~george/virtual-polyhedra/toroidal.html`.

Stewart, B. M. *Adventures Among the Toroids, 2nd rev. ed.* Okemos, MI: B. M. Stewart, 1984.

Toronto Function

$$T(m,n,r) \equiv \frac{\Gamma(\frac{1}{2}m+\frac{1}{2})}{n!r^{-2n+m-1}}\,{}_1F_1(\tfrac{1}{2}; i+n; r^2),$$

where ${}_1F_1(a;b;z)$ is a CONFLUENT HYPERGEOMETRIC FUNCTION and $\Gamma(z)$ is the GAMMA FUNCTION (Abramowitz and Stegun 1972).

References

Abramowitz, M. and Stegun, C. A. (Eds.). *Handbook of Mathematical Functions with Formulas, Graphs, and Mathematical Tables, 9th printing.* New York: Dover, p. 509, 1972.

Torricelli Point

see FERMAT POINT

Torsion (Differential Geometry)

The rate of change of the OSCULATING PLANE of a SPACE CURVE. The torsion τ is POSITIVE for a right-handed curve, and NEGATIVE for a left-handed curve. A curve with CURVATURE $\kappa \neq 0$ is planar IFF $\tau = 0$.

The torsion can be defined by

$$\tau \equiv -\mathbf{N}\cdot\mathbf{B}',$$

where $\mathbf{N}$ is the unit NORMAL VECTOR and $\mathbf{B}$ is the unit BINORMAL VECTOR. Written explicitly in terms of a parameterized VECTOR FUNCTION $\mathbf{x}$,

$$\tau = \frac{|\dot{\mathbf{x}}\,\ddot{\mathbf{x}}\,\dddot{\mathbf{x}}|}{\ddot{\mathbf{x}}\cdot\ddot{\mathbf{x}}} = \rho^2|\dot{\mathbf{x}}\,\ddot{\mathbf{x}}\,\dddot{\mathbf{x}}|,$$

where $|\mathbf{a}\,\mathbf{b}\,\mathbf{c}|$ denotes a SCALAR TRIPLE PRODUCT and ρ is the RADIUS OF CURVATURE. The quantity $1/\tau$ is called the RADIUS OF TORSION and is denoted σ or ϕ.

see also CURVATURE, RADIUS OF CURVATURE, RADIUS OF TORSION

References

Gray, A. "Drawing Space Curves with Assigned Curvature." §7.8 in *Modern Differential Geometry of Curves and Surfaces.* Boca Raton, FL: CRC Press, pp. 145–147, 1993.

Kreyszig, E. "Torsion." §14 in *Differential Geometry.* New York: Dover, pp. 37–40, 1991.

Torsion (Group Theory)

If G is a GROUP, then the torsion elements $\mathrm{Tor}(G)$ of G (also called the torsion of G) are defined to be the set of elements g in G such that $g^n = e$ for some NATURAL NUMBER n, where e is the IDENTITY ELEMENT of the GROUP G.

In the case that G is ABELIAN, $\mathrm{Tor}(G)$ is a SUBGROUP and is called the torsion subgroup of G. If $\mathrm{Tor}(G)$ consists only of the IDENTITY ELEMENT, the GROUP G is called torsion-free.

see also ABELIAN GROUP, GROUP, IDENTITY ELEMENT

Torsion Number

One of a set of numbers defined in terms of an invariant generated by the finite cyclic covering spaces of a KNOT complement. The torsion numbers for KNOTS up to 9 crossings were cataloged by Reidemeister (1948).

References

Reidemeister, K. *Knotentheorie.* New York: Chelsea, 1948.

Rolfsen, D. "Torsion Numbers." §6A in *Knots and Links.* Wilmington, DE: Publish or Perish Press, pp. 145–146, 1976.

Torsion Tensor

The TENSOR defined by

$$T^l{}_{jk} \equiv -(\Gamma^l{}_{jk} - \Gamma^l{}_{kj}),$$

where $\Gamma^l{}_{jk}$ are CONNECTION COEFFICIENTS.

see also CONNECTION COEFFICIENT

Torus

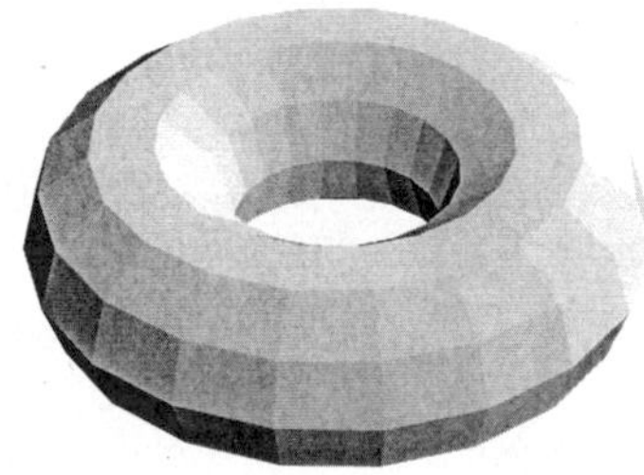

A torus is a surface having GENUS 1, and therefore possessing a single "HOLE." The usual torus in 3-D space is shaped like a donut, but the concept of the torus is extremely useful in higher dimensional space as well. One of the more common uses of n-D tori is in DYNAMICAL SYSTEMS. A fundamental result states that the PHASE SPACE trajectories of a HAMILTONIAN SYSTEM with n DEGREES OF FREEDOM and possessing n INTEGRALS OF MOTION lie on an n-D MANIFOLD which is topologically equivalent to an n-torus (Tabor 1989).

The usual 3-D "ring" torus is known in older literature as an "ANCHOR RING." Let the radius from the center of the hole to the center of the torus tube be c, and the radius of the tube be a. Then the equation in CARTESIAN COORDINATES is

$$\left(c - \sqrt{x^2 + y^2}\,\right)^2 + z^2 = a^2. \tag{1}$$

The parametric equations of a torus are

$$x = (c + a\cos v)\cos u \tag{2}$$

$$y = (c + a\cos v)\sin u \tag{3}$$

$$z = a\sin v \tag{4}$$

for $u, v \in [0, 2\pi)$. Three types of torus, known as the STANDARD TORI, are possible, depending on the relative sizes of a and c. $c > a$ corresponds to the RING TORUS (shown above), $c = a$ corresponds to a HORN TORUS which is tangent to itself at the point (0, 0, 0), and $c < a$ corresponds to a self-intersecting SPINDLE TORUS (Pinkall 1986).

If no specification is made, "torus" is taken to mean RING TORUS. The three STANDARD TORI are illustrated below, where the first image shows the full torus, the second a cut-away of the bottom half, and the third a CROSS-SECTION of a plane passing through the z-AXIS.

	full view	*cutaway*	*cross-section*
ring torus			
horn torus			
spindle torus			

The STANDARD TORI and their inversions are CYCLIDES. If the coefficient of $\sin v$ in the formula for z is changed to $b \neq a$, an ELLIPTIC TORUS results.

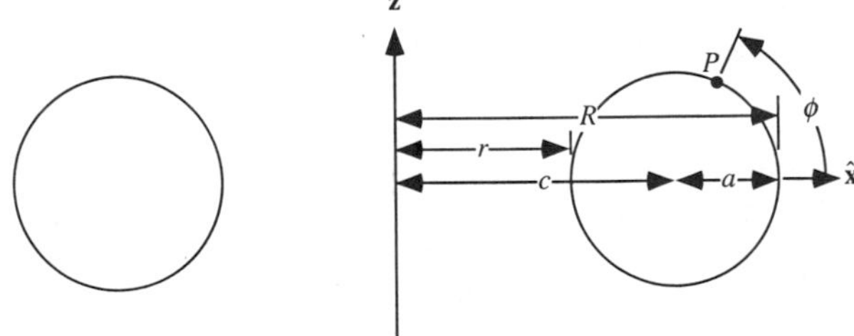

To compute the metric properties of the ring torus, define the inner and outer radii by

$$r \equiv c - a \tag{5}$$

$$R \equiv c + a. \tag{6}$$

Solving for a and c gives

$$a = \tfrac{1}{2}(R - r) \quad (7)$$
$$c = \tfrac{1}{2}(R + r). \quad (8)$$

Then the SURFACE AREA of this torus is

$$S = (2\pi a)(2\pi c) = 4\pi^2 ac \quad (9)$$
$$= \pi^2(R + r)(R - r), \quad (10)$$

and the VOLUME can be computed from PAPPUS'S CENTROID THEOREM

$$V = (\pi a^2)^2 \pi c = 2\pi^2 a^2 c \quad (11)$$
$$= \tfrac{1}{4}\pi^2(R + r)(R - r)^2. \quad (12)$$

The coefficients of the first and second FUNDAMENTAL FORMS of the torus are given by

$$e = -(c + a\cos v)\cos v \quad (13)$$
$$f = 0 \quad (14)$$
$$g = -a \quad (15)$$
$$E = (c + a\cos v)^2 \quad (16)$$
$$F = 0 \quad (17)$$
$$G = a^2, \quad (18)$$

giving RIEMANNIAN METRIC

$$ds^2 = (c + a\cos v)^2\, du^2 + a^2\, dv^2, \quad (19)$$

AREA ELEMENT

$$dA = a(c + a\cos v)\, du \wedge dv \quad (20)$$

(where $du \wedge dv$ is a WEDGE PRODUCT), and GAUSSIAN and MEAN CURVATURES as

$$K = \frac{\cos v}{a(c + a\cos v)} \quad (21)$$
$$H = -\frac{c + 2a\cos v}{2a(c + a\cos v)} \quad (22)$$

(Gray 1993, pp. 289–291).

A torus with a HOLE in *its surface* can be turned inside out to yield an identical torus. A torus can be knotted externally or internally, but not both. These two cases are AMBIENT ISOTOPIES, but not REGULAR ISOTOPIES. There are therefore three possible ways of embedding a torus with zero or one KNOT.

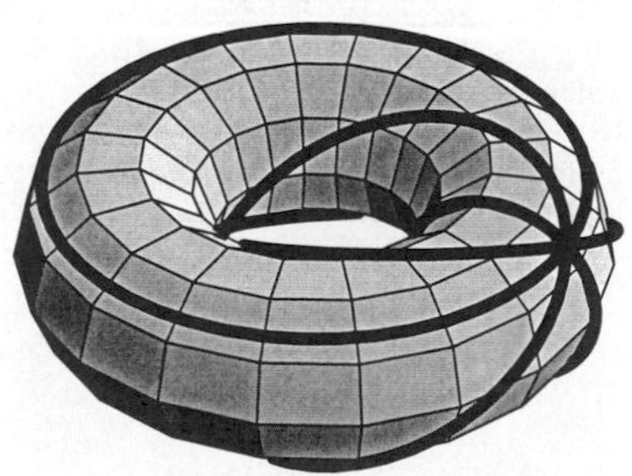

An arbitrary point P on a torus (not lying in the xy-plane) can have four CIRCLES drawn through it. The first circle is in the plane of the torus and the second is PERPENDICULAR to it. The third and fourth CIRCLES are called VILLARCEAU CIRCLES (Villarceau 1848, Schmidt 1950, Coxeter 1969, Melnick 1983).

To see that two additional CIRCLES exist, consider a coordinate system with origin at the center of torus, with $\hat{\mathbf{z}}$ pointing up. Specify the position of P by its ANGLE ϕ measured around the tube of the torus. Define $\phi = 0$ for the circle of points farthest away from the center of the torus (i.e., the points with $x^2 + y^2 = R^2$), and draw the x-AXIS as the intersection of a plane through the z-axis and passing through P with the xy-plane. Rotate about the y-AXIS by an ANGLE θ, where

$$\theta = \sin^{-1}\left(\frac{a}{c}\right). \quad (23)$$

In terms of the old coordinates, the new coordinates are

$$x = x_1 \cos\theta - z_1 \sin\theta \quad (24)$$
$$z = x_1 \sin\theta + z_1 \cos\theta. \quad (25)$$

So in (x_1, y_1, z_1) coordinates, equation (1) of the torus becomes

$$[\sqrt{(x_1\cos\theta - z_1\sin\theta)^2 + {y_1}^2} - c]^2 + (x_1\sin\theta + z_1\cos\theta)^2 = a^2. \quad (26)$$

Squaring both sides gives

$$(x_1\cos\theta - z_1\sin\theta)^2 + {y_1}^2 + c^2 - 2c\sqrt{(x_1\cos\theta - z_1\sin\theta)^2 + {y_1}^2} + (x_1\sin\theta + z_1\cos\theta)^2 = a^2. \quad (27)$$

But

$$(x_1\cos\theta - z_1\sin\theta)^2 + (x_1\sin\theta + z_1\cos\theta)^2 = {x_1}^2 + {z_1}^2, \quad (28)$$

so

$${x_1}^2 + {y_1}^2 + {z_1}^2 + c^2 - 2c\sqrt{(x_1\cos\theta - z_1\sin\theta)^2 + {y_1}^2} = a^2. \quad (29)$$

In the $z_1 = 0$ plane, plugging in (23) and factoring gives

$$[{x_1}^2 + (y_1 - a)^2 - c^2][{x_1}^2 + (y_1 + a) - c^2] = 0. \quad (30)$$

This gives the CIRCLES

$${x_1}^2 + (y_1 - a)^2 = c^2 \quad (31)$$

and

$${x_1}^2 + (y_1 + a)^2 = c^2 \quad (32)$$

in the z_1 plane. Written in MATRIX form with parameter $t \in [0, 2\pi)$, these are

$$C_1 = \begin{bmatrix} c\cos t \\ c\sin t + a \\ 0 \end{bmatrix} \quad (33)$$

$$C_2 = \begin{bmatrix} c\cos t \\ c\sin t - a \\ 0 \end{bmatrix}. \quad (34)$$

In the original (x, y, z) coordinates,

$$\begin{aligned} C_1 &= \begin{bmatrix} \cos\theta & 0 & -\sin\theta \\ 0 & 1 & 0 \\ -\sin\theta & 0 & \cos\theta \end{bmatrix} \begin{bmatrix} c\cos t \\ c\sin t + a \\ 0 \end{bmatrix} \\ &= \begin{bmatrix} c\cos\theta\cos t \\ c\sin t + a \\ -c\sin\theta\cos t \end{bmatrix} \end{aligned} \quad (35)$$

$$\begin{aligned} C_2 &= \begin{bmatrix} \cos\theta & 0 & \sin\theta \\ 0 & 1 & 0 \\ -\sin\theta & 0 & \cos\theta \end{bmatrix} \begin{bmatrix} c\cos t \\ c\sin t - a \\ 0 \end{bmatrix} \\ &= \begin{bmatrix} c\cos\theta\cos t \\ c\sin t - a \\ -c\sin\theta\cos t \end{bmatrix}. \end{aligned} \quad (36)$$

The point P must satisfy

$$z = a\sin\phi = c\sin\theta\cos t, \quad (37)$$

so

$$\cos t = \frac{a\sin\phi}{c\sin\theta}. \quad (38)$$

Plugging this in for x_1 and y_1 gives the ANGLE ψ by which the CIRCLE must be rotated about the z-AXIS in order to make it pass through P,

$$\psi = \tan^{-1}\left(\frac{y}{x}\right) = \frac{c\sin t + a}{c\cos\theta\cos t} = \frac{c\sqrt{1-\cos^2 t} + a}{c\cos\theta\cos t}. \quad (39)$$

The four CIRCLES passing through P are therefore

$$C_1 = \begin{bmatrix} \cos\psi & \sin\psi & 0 \\ -\sin\psi & \cos\psi & 0 \\ 0 & 0 & 1 \end{bmatrix} \begin{bmatrix} c\cos\theta\cos t \\ c\sin t + a \\ -c\sin\theta\cos t \end{bmatrix} \quad (40)$$

$$C_2 = \begin{bmatrix} \cos\psi & \sin\psi & 0 \\ -\sin\psi & \cos\psi & 0 \\ 0 & 0 & 1 \end{bmatrix} \begin{bmatrix} c\cos\theta\cos t \\ c\sin t - a \\ -c\sin\theta\cos t \end{bmatrix} \quad (41)$$

$$C_3 = \begin{bmatrix} (c + a\cos\phi)\cos t \\ (c + a\cos\phi)\sin t \\ a\sin\phi \end{bmatrix} \quad (42)$$

$$C_4 = \begin{bmatrix} c + a\cos t \\ 0 \\ a\sin t \end{bmatrix}. \quad (43)$$

see also APPLE, CYCLIDE, ELLIPTIC TORUS, GENUS (SURFACE), HORN TORUS, KLEIN QUARTIC, LEMON, RING TORUS, SPINDLE TORUS, SPIRIC SECTION, STANDARD TORI, TOROID, TORUS COLORING, TORUS CUTTING

References

Beyer, W. H. *CRC Standard Mathematical Tables, 28th ed.* Boca Raton, FL: CRC Press, pp. 131–132, 1987.

Coxeter, H. S. M. *Introduction to Geometry, 2nd ed.* New York: Wiley, pp. 132–133, 1969.

Geometry Center. "The Torus." `http://www.geom.umn.edu/zoo/toptype/torus/`.

Gray, A. "Tori." §11.4 in *Modern Differential Geometry of Curves and Surfaces.* Boca Raton, FL: CRC Press, pp. 218–220 and 289–290, 1993.

Melzak, Z. A. *Invitation to Geometry.* New York: Wiley, pp. 63–72, 1983.

Pinkall, U. "Cyclides of Dupin." §3.3 in *Mathematical Models from the Collections of Universities and Museums* (Ed. G. Fischer). Braunschweig, Germany: Vieweg, pp. 28–30, 1986.

Schmidt, H. *Die Inversion und ihre Anwendungen.* Munich: Oldenbourg, p. 82, 1950.

Tabor, M. *Chaos and Integrability in Nonlinear Dynamics: An Introduction.* New York: Wiley, pp. 71–74, 1989.

Villarceau, M. "Théorème sur le tore." *Nouv. Ann. Math.* **7**, 345–347, 1848.

Torus Coloring

The number of colors SUFFICIENT for MAP COLORING on a surface of GENUS g is given by the HEAWOOD CONJECTURE,

$$\chi(g) = \left\lfloor \tfrac{1}{2}(7 + \sqrt{48g+1}) \right\rfloor,$$

where $\lfloor x \rfloor$ is the FLOOR FUNCTION. The fact that $\chi(g)$ (which is called the CHROMATIC NUMBER) is also NECESSARY was proved by Ringel and Youngs (1968) with two exceptions: the SPHERE (which requires the same number of colors as the PLANE) and the KLEIN BOTTLE. A g-holed TORUS therefore requires $\chi(g)$ colors. For $g = 0, 1, \ldots$, the first few values of $\chi(g)$ are 4, 7, 8, 9, 10, 11, 12, 12, 13, 13, 14, 15, 15, 16, ... (Sloane's A000934).

see also CHROMATIC NUMBER, FOUR-COLOR THEOREM, HEAWOOD CONJECTURE, KLEIN BOTTLE, MAP COLORING

References

Gardner, M. "Mathematical Games: The Celebrated Four-Color Map Problem of Topology." *Sci. Amer.* **203**, 218–222, Sep. 1960.

Ringel, G. *Map Color Theorem.* New York: Springer-Verlag, 1974.

Ringel, G. and Youngs, J. W. T. "Solution of the Heawood Map-Coloring Problem." *Proc. Nat. Acad. Sci. USA* **60**, 438–445, 1968.

Sloane, N. J. A. Sequence A000934/M3292 in "An On-Line Version of the Encyclopedia of Integer Sequences."

Wagon, S. "Map Coloring on a Torus." §7.5 in *Mathematica in Action.* New York: W. H. Freeman, pp. 232–237, 1991.

Torus Cutting

With n cuts of a TORUS of GENUS 1, the maximum number of pieces which can be obtained is

$$N(n) = \tfrac{1}{6}(n^3 + 3n^2 + 8n).$$

The first few terms are 2, 6, 13, 24, 40, 62, 91, 128, 174, 230, ... (Sloane's A003600).

see also CAKE CUTTING, CIRCLE CUTTING, CYLINDER CUTTING, PANCAKE CUTTING, PLANE CUTTING, PIE CUTTING, SQUARE CUTTING

References

Gardner, M. *Mathematical Magic Show: More Puzzles, Games, Diversions, Illusions and Other Mathematical Sleight-of-Mind from Scientific American.* New York: Vintage, pp. 149–150, 1978.

Sloane, N. J. A. Sequence A003600/M1594 in "An On-Line Version of the Encyclopedia of Integer Sequences."

Torus Knot

A (p, q)-torus KNOT is obtained by looping a string through the HOLE of a TORUS p times with q revolutions before joining its ends, where p and q are RELATIVELY PRIME. A (p, q)-torus knot is equivalent to a (q, p)-torus knot. The CROSSING NUMBER of a (p, q)-torus knot is

$$c = \min\{p(q-1), q(p-1)\} \tag{1}$$

(Murasugi 1991). The UNKNOTTING NUMBER of a (p, q)-torus knot is

$$u = \tfrac{1}{2}(p-1)(q-1) \tag{2}$$

(Adams 1991).

Torus knots with fewer than 11 crossings are the TREFOIL KNOT 03_{001} (3, 2), SOLOMON'S SEAL KNOT 05_{001} (5, 2), 07_{001} (7, 2), 08_{019} (4, 3), 09_{001} (9, 2), and 10_{124} (5, 3) (Adams *et al.* 1991). The only KNOTS which are not HYPERBOLIC KNOTS are torus knots and SATELLITE KNOTS (including COMPOSITE KNOTS). The $(2, q)$, $(3, 4)$, and $(3, 5)$-torus knots are ALMOST ALTERNATING KNOTS.

The JONES POLYNOMIAL of an (m, n)-TORUS KNOT is

$$\frac{t^{(m-1)(n-1)/2}(1 - t^{m+1} - t^{n+1} + t^{m+n}}{1-t^2}. \tag{3}$$

The BRACKET POLYNOMIAL for the torus knot $K_n = (2, n)$ is given by the RECURRENCE RELATION

$$\langle K_n \rangle = A\langle K_{n-1}\rangle + (-1)^{n-1}A^{-3n+2}, \tag{4}$$

where

$$\langle K_1 \rangle = -A^3. \tag{5}$$

see also ALMOST ALTERNATING KNOT, HYPERBOLIC KNOT, KNOT, SATELLITE KNOT, SOLOMON'S SEAL KNOT, TREFOIL KNOT

References

Adams, C.; Hildebrand, M.; and Weeks, J. "Hyperbolic Invariants of Knots and Links." *Trans. Amer. Math. Soc.* **326**, 1–56, 1991.

Gray, A. "Torus Knots." §8.2 in *Modern Differential Geometry of Curves and Surfaces.* Boca Raton, FL: CRC Press, pp. 155–161, 1993.

Murasugi, K. "On the Braid Index of Alternating Links." *Trans. Amer. Math. Soc.* **326**, 237–260, 1991.

Total Angular Defect

see DESCARTES TOTAL ANGULAR DEFECT

Total Curvature

The total curvature of a curve is the quantity $\sqrt{\tau^2 + \kappa^2}$, where τ is the TORSION and κ is the CURVATURE. The total curvature is also called the THIRD CURVATURE.

see also CURVATURE, TORSION (DIFFERENTIAL GEOMETRY)

Total Differential

see EXACT DIFFERENTIAL

Total Function

A FUNCTION defined for all possible input values.

Total Intersection Theorem

If one part of the total intersection group of a curve of order n with a curve of order $n_1 + n_2$ constitutes the total intersection with a curve of order n_1, then the other part will constitute the total intersection with a curve of order n_2.

References

Coolidge, J. L. *A Treatise on Algebraic Plane Curves.* New York: Dover, p. 32, 1959.

Total Order

A total order satisfies the conditions for a PARTIAL ORDER plus the comparability condition. A RELATION $\leq$ is a partial order on a SET S if

1. Reflexivity: $a \leq a$ for all $a \in S$
2. Antisymmetry: $a \leq b$ and $b \leq a$ implies $a = b$
3. Transitivity: $a \leq b$ and $b \leq c$ implies $a \leq c$,

and is a total order if, in addition,

4. Comparability: For any $a, b \in S$, either $a \leq b$ or $b \leq a$.

see also PARTIAL ORDER, RELATION

Total Space

The SPACE E of a FIBER BUNDLE given by the MAP $f : E \to B$, where B is the BASE SPACE of the FIBER BUNDLE.

see also BASE SPACE, FIBER BUNDLE, SPACE

Totative

A POSITIVE INTEGER less than or equal to a number n which is also RELATIVELY PRIME to n, where 1 is counted as being RELATIVELY PRIME to all numbers. The number of totatives of n is the value of the TOTIENT FUNCTION $\phi(n)$.

see also RELATIVELY PRIME, TOTIENT FUNCTION

Totient Function

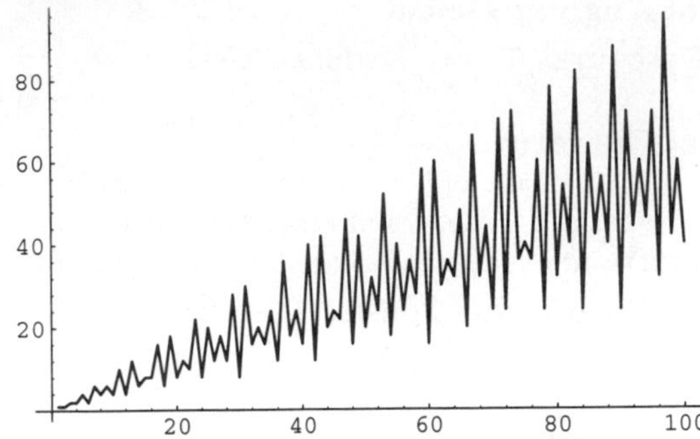

The totient function $\phi(n)$, also called Euler's totient function, is defined as the number of POSITIVE INTEGERS $\leq n$ which are RELATIVELY PRIME to (i.e., do not contain any factor in common with) n, where 1 is counted as being RELATIVELY PRIME to all numbers. Since a number less than or equal to and RELATIVELY PRIME to a given number is called a TOTATIVE, the totient function $\phi(n)$ can be simply defined as the number of TOTATIVES of n. For example, there are eight TOTATIVES of 24 (1, 5, 7, 11, 13, 17, 19, and 23), so $\phi(24) = 8$.

By convention, $\phi(0) = 1$. The first few values of $\phi(n)$ for $n = 1, 2, \ldots$ are 1, 1, 2, 2, 4, 2, 6, 4, 6, 4, 10, ... (Sloane's A000010). $\phi(n)$ is plotted above for small n.

For a PRIME p,

$$\phi(p) = p - 1, \tag{1}$$

since all numbers less than p are RELATIVELY PRIME to p. If $m = p^\alpha$ is a POWER of a PRIME, then the numbers which have a common factor with m are the multiples of p: $p, 2p, \ldots, (p^{\alpha-1})p$. There are $p^{\alpha-1}$ of these multiples, so the number of factors RELATIVELY PRIME to p^α is

$$\phi(p^\alpha) = p^\alpha - p^{\alpha-1} = p^{\alpha-1}(p-1) = p^\alpha\left(1 - \frac{1}{p}\right). \tag{2}$$

Now take a general m divisible by p. Let $\phi_p(m)$ be the number of POSITIVE INTEGERS $\leq m$ not DIVISIBLE by p. As before, $p, 2p, \ldots, (m/p)p$ have common factors, so

$$\phi_p(m) = m - \frac{m}{p} = m\left(1 - \frac{1}{p}\right). \tag{3}$$

Now let q be some other PRIME dividing m. The INTEGERS divisible by q are $q, 2q, \ldots, (m/q)q$. But these duplicate $pq, 2pq, \ldots, (m/pq)pq$. So the number of terms which must be subtracted from ϕ_p to obtain ϕ_{pq} is

$$\Delta\phi_q(m) = \frac{m}{q} - \frac{m}{pq} = \frac{m}{q}\left(1 - \frac{1}{p}\right), \tag{4}$$

and

$$\begin{aligned}\phi_{pq}(m) &\equiv \phi_q(m) - \Delta\phi_q(m) \\ &= m\left(1 - \frac{1}{p}\right) - \frac{m}{q}\left(1 - \frac{1}{p}\right) \\ &= m\left(1 - \frac{1}{p}\right)\left(1 - \frac{1}{q}\right).\end{aligned} \tag{5}$$

By induction, the general case is then

$$\phi(n) = n\left(1 - \frac{1}{p_1}\right)\left(1 - \frac{1}{p_2}\right)\cdots\left(1 - \frac{1}{p_r}\right). \tag{6}$$

An interesting identity relates $\phi(n^2)$ to $\phi(n)$,

$$\phi(n^2) = n\phi(n). \tag{7}$$

Another identity relates the DIVISORS d of n to n via

$$\sum_d \phi(d) = n. \tag{8}$$

The DIVISOR FUNCTION satisfies the CONGRUENCE

$$n\sigma(n) \equiv 2 \pmod{\phi(n)} \tag{9}$$

for all PRIMES and no COMPOSITE with the exceptions of 4, 6, and 22 (Subbarao 1974), where $\sigma(n)$ is the DIVISOR FUNCTION. No COMPOSITE solution is currently known to

$$n - 1 \equiv 0 \pmod{\phi(n)} \tag{10}$$

(Honsberger 1976, p. 35).

Walfisz (1963), building on the work of others, showed that

$$\sum_{n=1}^N \phi(n) = \frac{3N^2}{\pi^2} + \mathcal{O}[N(\ln N)^{2/3}(\ln\ln N)^{4/3}], \tag{11}$$

and Landau (1900, quoted in Dickson 1952) showed that

$$\sum_{n=1}^N \frac{1}{\phi(n)} = A\ln N + B + \mathcal{O}\left(\frac{\ln N}{N}\right), \tag{12}$$

where

$$\begin{aligned}A &= \sum_{k=1}^\infty \frac{[\mu(k)]^2}{k\phi(k)} = \frac{\zeta(2)\zeta(3)}{\zeta(6)} = \frac{315}{2\pi^4}\zeta(3) \\ &= 1.9435964368\ldots\end{aligned} \tag{13}$$

$$\begin{aligned}B &= \gamma\frac{315}{2\pi^4}\zeta(3) - \sum_{k=1}^\infty \frac{[\mu(k)]^2 \ln k}{k\phi(k)} \\ &= -0.0595536246\ldots,\end{aligned} \tag{14}$$

$\mu(k)$ is the MÖBIUS FUNCTION, $\zeta(z)$ is the RIEMANN ZETA FUNCTION, and γ is the EULER-MASCHERONI CONSTANT (Dickson). A can also be written

$$A = \prod_{k=1}^{\infty} \frac{1 - {p_k}^6}{(1 - {p_k}^{-2})(1 - {p_k}^{-3})} = \prod_{k=1}^{\infty} \left[1 + \frac{1}{p_k(p_k - 1)}\right]. \quad (15)$$

Note that this constant is similar to ARTIN'S CONSTANT.

If the GOLDBACH CONJECTURE is true, then for every number m, there are PRIMES p and q such that

$$\phi(p) + \phi(q) = 2m \quad (16)$$

(Guy 1994, p. 105).

Curious equalities of consecutive values include

$$\phi(5186) = \phi(5187) = \phi(5188) = 2^5 3^4 \quad (17)$$

$$\phi(25930) = \phi(25935) = \phi(25940) = \phi(25942) = 2^7 3^4 \quad (18)$$

$$\phi(404471) = \phi(404473) = \phi(404477) = 2^8 3^2 5^2 7 \quad (19)$$

(Guy 1994, p. 91).

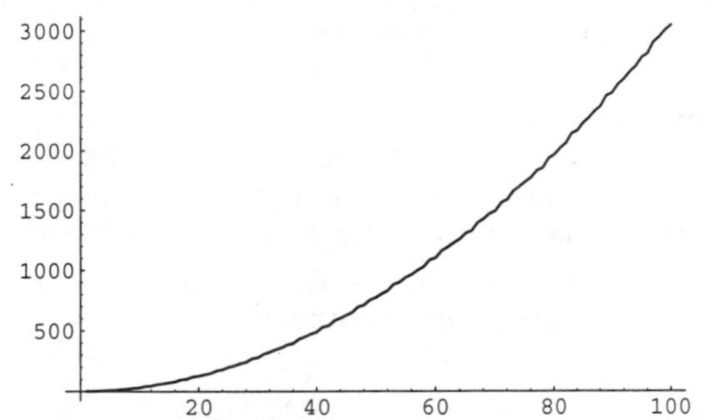

The SUMMATORY totient function, plotted above, is defined by

$$\Phi(n) \equiv \sum_{k=1}^{n} \phi(k) \quad (20)$$

and has the asymptotic series

$$\Phi(x) \sim \frac{1}{2\zeta(2)} x^2 + (x \ln x) \quad (21)$$

$$\sim \frac{3}{\pi^2} x^2 + \mathcal{O}(x \ln x), \quad (22)$$

where $\zeta(z)$ is the RIEMANN ZETA FUNCTION (Perrot 1881). The first values of $\Phi(n)$ are 1, 2, 4, 6, 10, 12, 18, 22, 28, ... (Sloane's A002088).

see also DEDEKIND FUNCTION, EULER'S TOTIENT RULE, FERMAT'S LITTLE THEOREM, LEHMER'S PROBLEM, LEUDESDORF THEOREM, NONCOTOTIENT, NONTOTIENT, SILVERMAN CONSTANT, TOTATIVE, TOTIENT VALENCE FUNCTION

References

Abramowitz, M. and Stegun, C. A. (Eds.). "The Euler Totient Function." §24.3.2 in *Handbook of Mathematical Functions with Formulas, Graphs, and Mathematical Tables, 9th printing.* New York: Dover, p. 826, 1972.

Beiler, A. H. Ch. 12 in *Recreations in the Theory of Numbers: The Queen of Mathematics Entertains.* New York: Dover, 1966.

Conway, J. H. and Guy, R. K. "Euler's Totient Numbers." *The Book of Numbers.* New York: Springer-Verlag, pp. 154–156, 1996.

Courant, R. and Robbins, H. "Euler's φ Function. Fermat's Theorem Again." §2.4.3 in Supplement to Ch. 1 in *What is Mathematics?: An Elementary Approach to Ideas and Methods, 2nd ed.* Oxford, England: Oxford University Press, pp. 48–49, 1996.

DeKoninck, J.-M. and Ivic, A. *Topics in Arithmetical Functions: Asymptotic Formulae for Sums of Reciprocals of Arithmetical Functions and Related Fields.* Amsterdam, Netherlands: North-Holland, 1980.

Dickson, L. E. *History of the Theory of Numbers, Vol. 1: Divisibility and Primality.* New York: Chelsea, pp. 113–158, 1952.

Finch, S. "Favorite Mathematical Constants." `http://www.mathsoft.com/asolve/constant/totient/totient.html.`

Guy, R. K. "Euler's Totient Function," "Does $\phi(n)$ Properly Divide $n-1$," "Solutions of $\phi(m) = \sigma(n)$," "Carmichael's Conjecture," "Gaps Between Totatives," "Iterations of ϕ and σ," "Behavior of $\phi(\sigma(n))$ and $\sigma(\phi(n))$." §B36–B42 in *Unsolved Problems in Number Theory, 2nd ed.* New York: Springer-Verlag, pp. 90-99, 1994.

Halberstam, H. and Richert, H.-E. *Sieve Methods.* New York: Academic Press, 1974.

Honsberger, R. *Mathematical Gems II.* Washington, DC: Math. Assoc. Amer., p. 35, 1976.

Perrot, J. 1811. Quoted in Dickson, L. E. *History of the Theory of Numbers, Vol. 1: Divisibility and Primality.* New York: Chelsea, p. 126, 1952.

Shanks, D. "Euler's ϕ Function." §2.27 in *Solved and Unsolved Problems in Number Theory, 4th ed.* New York: Chelsea, pp. 68–71, 1993.

Sloane, N. J. A. Sequences A000010/M0299 and A002088/M1008 in "An On-Line Version of the Encyclopedia of Integer Sequences."

Subbarao, M. V. "On Two Congruences for Primality." *Pacific J. Math.* **52**, 261–268, 1974.

Totient Function Constants

see SILVERMAN CONSTANT, TOTIENT FUNCTION

Totient Valence Function

$N_\phi(m)$ is the number of INTEGERS n for which $\phi(n) = m$, also called the MULTIPLICITY of m (Guy 1994). The table below lists values for $\phi(N) \leq 50$.

$\phi(N)$	m	N
1	2	1, 2
2	3	3, 4, 6
4	4	5, 8, 10, 12
6	4	7, 9, 14, 18
8	5	15, 16, 20, 24, 30
10	2	11, 22
12	6	13, 21, 26, 28, 36, 42
16	6	17, 32, 34, 40, 48, 60
18	4	19, 27, 38, 54
20	5	25, 33, 44, 50, 66
22	2	23, 46
24	10	35, 39, 45, 52, 56, 70, 72, 78, 84, 90
28	2	29, 58
30	2	31, 62
32	7	51, 64, 68, 80, 96, 102, 120
36	8	37, 57, 63, 74, 76, 108, 114, 126
40	9	41, 55, 75, 82, 88, 100, 110, 132, 150
42	4	43, 49, 86, 98
44	3	69, 92, 138
46	2	47, 94
48	11	65, 104, 105, 112, 130, 140, 144, 156, 168, 180, 210

A table listing the first value of $\phi(N)$ with multiplicities up to 100 follows (Sloane's A014573).

M	ϕ	M	ϕ	M	ϕ	M	ϕ
0	3	26	2560	51	4992	76	21840
2	1	27	384	52	17640	77	9072
3	2	28	288	53	2016	78	38640
4	4	29	1320	54	1152	79	9360
5	8	30	3696	55	6000	80	81216
6	12	31	240	56	12288	81	4032
7	32	32	768	57	4752	82	5280
8	36	33	9000	58	2688	83	4800
9	40	34	432	59	3024	84	4608
10	24	35	7128	60	13680	85	16896
11	48	36	4200	61	9984	86	3456
12	160	37	480	62	1728	87	3840
13	396	38	576	63	1920	88	10800
14	2268	39	1296	64	2400	89	9504
15	704	40	1200	65	7560	90	18000
16	312	41	15936	66	2304	91	23520
17	72	42	3312	67	22848	92	39936
18	336	43	3072	68	8400	93	5040
19	216	44	3240	69	29160	94	26208
20	936	45	864	70	5376	95	27360
21	144	46	3120	71	3360	96	6480
22	624	47	7344	72	1440	97	9216
23	1056	48	3888	73	13248	98	2880
24	1760	49	720	74	11040	99	26496
25	360	50	1680	75	27720	100	34272

It is thought that $N_\phi(m) \geq 2$ (i.e., the totient valence function never takes on the value 1), but this has not been proven. This assertion is called CARMICHAEL'S TOTIENT FUNCTION CONJECTURE and is equivalent to the statement that for all n, there exists $m \neq n$ such that $\phi(n) = \phi(m)$ (Ribenboim 1996, pp. 39–40). Any counterexample must have more than 10,000,000 DIGITS (Schlafly and Wagon 1994, Conway and Guy 1996).

see also CARMICHAEL'S TOTIENT FUNCTION CONJECTURE, TOTIENT FUNCTION

References

Conway, J. H. and Guy, R. K. *The Book of Numbers.* New York: Springer-Verlag, p. 155, 1996.

Guy, R. K. *Unsolved Problems in Number Theory, 2nd ed.* New York: Springer-Verlag, p. 94, 1994.

Ribenboim, P. *The New Book of Prime Number Records.* New York: Springer-Verlag, 1996.

Schlafly, A. and Wagon, S. "Carmichael's Conjecture on the Euler Function is Valid Below $10^{10,000,000}$." *Math. Comput.* **63**, 415–419, 1994.

Sloane, N. J. A. Sequence A014573 in "An On-Line Version of the Encyclopedia of Integer Sequences."

Touchard's Congruence

$$B_{p+k} \equiv B_k + B_{k+1} \pmod{p},$$

when p is PRIME and B_n is a BELL NUMBER.

see also BELL NUMBER

Tour

A sequence of moves on a chessboard by a CHESS piece in which each square of a CHESSBOARD is visited exactly once.

see also CHESS, KNIGHT'S TOUR, MAGIC TOUR, TRAVELING SALESMAN CONSTANTS

Tournament

A COMPLETE DIRECTED GRAPH. A so-called SCORE SEQUENCE can be associated with every tournament. Every tournament contains a HAMILTONIAN PATH.

see also COMPLETE GRAPH, DIRECTED GRAPH, HAMILTONIAN PATH, SCORE SEQUENCE

References

Chartrand, G. "Tournaments." §27.2 in *Introductory Graph Theory.* New York: Dover, pp. 155–161, 1985.

Moon, J. W. *Topics on Tournaments.* New York: Holt, Rinehart, and Winston, 1968.

Ruskey, F. "Information on Score Sequences." `http://sue.csc.uvic.ca/~cos/inf/nump/ScoreSequence.html`.

Tournament Matrix

A matrix for a round-robin tournament involving n players competing in $n(n-1)/2$ matches (no ties allowed) having entries

$$a_{ij} = \begin{cases} 1 & \text{if player } i \text{ defeats player } j \\ -1 & \text{if player } i \text{ loses to player } j \\ 0 & \text{if } i = j. \end{cases}$$

The MATRIX satisfies

$$\mathsf{A} + \mathsf{A}^{\mathrm{T}} + \mathsf{I} = \mathsf{J},$$

where I is the IDENTITY MATRIX, J is an $n \times n$ MATRIX of all 1s, and A^{T} is the MATRIX TRANSPOSE of A.

The tournament matrix for n players has zero DETERMINANT IFF n is ODD (McCarthy and Benjamin 1996). The dimension of the NULLSPACE of an n-player tournament matrix is

$$\dim[\text{nullspace}] = \begin{cases} 0 & \text{for } n \text{ even} \\ 1 & \text{for } n \text{ odd} \end{cases}$$

(McCarthy 1996).

References
McCarthy, C. A. and Benjamin, A. T. "Determinants of the Tournaments." *Math. Mag.* **69**, 133–135, 1996.
Michael, T. S. "The Ranks of Tournament Matrices." *Amer. Math. Monthly* **102**, 637–639, 1995.

Tower of Power

see POWER TOWER

Towers of Hanoi

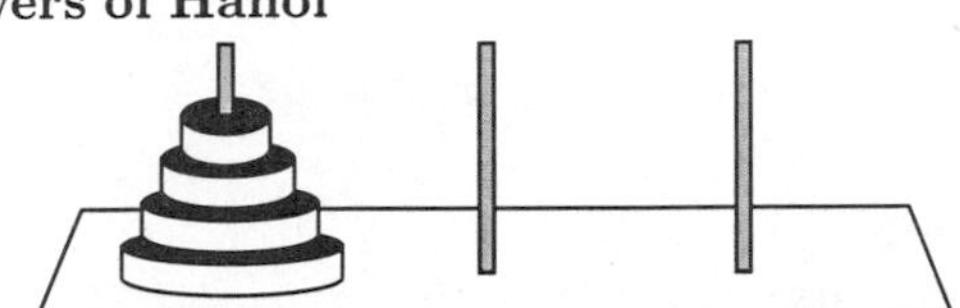

A PUZZLE invented by E. Lucas in 1883. Given a stack of n disks arranged from largest on the bottom to smallest on top placed on a rod, together with two empty rods, the towers of Hanoi puzzle asks for the minimum number of moves required to reverse the order of the stack (where moves are allowed only if they place smaller disks on top of larger disks). The problem is ISOMORPHIC to finding a HAMILTONIAN PATH on an n-HYPERCUBE.

For n disks, the number of moves h_n required is given by the RECURRENCE RELATION

$$h_n = 2h_{n-1} + 1.$$

Solving gives

$$h_n = 2^n - 1.$$

The number of disks moved after the kth step is the same as the element which needs to be added or deleted in the kth ADDEND of the RYSER FORMULA (Gardner 1988, Vardi 1991).

A HANOI GRAPH can be constructed whose VERTICES correspond to legal configurations of n towers of Hanoi, where the VERTICES are adjacent if the corresponding configurations can be obtained by a legal move. It can be solved using a binary GRAY CODE.

Poole (1994) gives *Mathematica*® (Wolfram Research, Champaign, IL) routines for solving an arbitrary disk configuration in the fewest possible moves. The proof of minimality is achieved using the LUCAS CORRESPONDENCE which relates PASCAL'S TRIANGLE to the HANOI GRAPH. ALGORITHMS are known for transferring disks for four pegs, but none has been proved minimal. For additional references, see Poole (1994).

see also GRAY CODE, RYSER FORMULA

References
Bogomolny, A. "Towers of Hanoi." http://www.cut-the-knot.com/recurrence/hanoi.html.
Chartrand, G. "The Tower of Hanoi Puzzle." §6.3 in *Introductory Graph Theory.* New York: Dover, pp. 135–139, 1985.
Dubrovsky, V. "Nesting Puzzles, Part I: Moving Oriental Towers." *Quantum* **6**, 53–57 (Jan.) and 49–51 (Feb.), 1996.
Gardner, M. "The Icosian Game and the Tower of Hanoi." Ch. 6 in *The Scientific American Book of Mathematical Puzzles & Diversions.* New York: Simon and Schuster, 1959.
Kasner, E. and Newman, J. R. *Mathematics and the Imagination.* Redmond, WA: Tempus Books, pp. 169–171, 1989.
Kolar, M. "Towers of Hanoi." http://www.pangea.ca/kolar/javascript/Hanoi/Hanoi.html.
Poole, D. G. "The Towers and Triangles of Professor Claus (or, Pascal Knows Hanoi)." *Math. Mag.* **67**, 323–344, 1994.
Poole, D. G. "Towers of Hanoi." http://www.astro.virginia.edu/~eww6n/math/notebooks/Hanoi.m.
Ruskey, F. "Towers of Hanoi." http://sue.csc.uvic.ca/~cos/inf/comb/SubsetInfo.html#Hanoi.
Schoutte, P. H. "De Ringen van Brahma." *Eigen Haard* **22**, 274–276, 1884.
Kraitchik, M. "The Tower of Hanoi." §3.12.4 in *Mathematical Recreations.* New York: W. W. Norton, pp. 91–93, 1942.
Vardi, I. *Computational Recreations in Mathematica.* Reading, MA: Addison-Wesley, pp. 111–112, 1991.

Trace (Complex)

The image of the path γ in $\mathbb{C}$ under the FUNCTION f is called the trace. This term is unrelated to that applied to MATRICES and TENSORS.

Trace (Group)

see CHARACTER (GROUP)

Trace (Map)

Let a PATCH be given by the map $\mathbf{x} : U \to \mathbb{R}^n$, where U is an open subset of $\mathbb{R}^2$, or more generally by $\mathbf{x} : A \to \mathbb{R}^n$, where A is any SUBSET of $\mathbb{R}^2$. Then $\mathbf{x}(U)$ (or more generally, $\mathbf{x}(A)$) is called the trace of $\mathbf{x}$.

see also PATCH

References
Gray, A. *Modern Differential Geometry of Curves and Surfaces.* Boca Raton, FL: CRC Press, pp. 183–184, 1993.

Trace (Matrix)

The trace of an $n \times n$ SQUARE MATRIX A is defined by

$$\text{Tr}(\mathsf{A}) \equiv a_{ii}, \tag{1}$$

where EINSTEIN SUMMATION is used (i.e., the a_{ii} is summed over $i = 1, \ldots, n$). For SQUARE MATRICES A and B, it is true that

$$\text{Tr}(\mathsf{A}) = \text{Tr}(\mathsf{A}^{\text{T}}) \tag{2}$$

$$\text{Tr}(\mathsf{A} + \mathsf{B}) = \text{Tr}(\mathsf{A}) + \text{Tr}(\mathsf{B}) \tag{3}$$

$$\text{Tr}(\alpha\mathsf{A}) = \alpha\,\text{Tr}(\mathsf{A}) \tag{4}$$

(Lange 1987, p. 40). The trace is invariant under a SIMILARITY TRANSFORMATION

$$A' \equiv BAB^{-1} \tag{5}$$

(Lange 1987, p. 64). Since

$$(bab^{-1})_{ij} = b_{il}a_{lk}b_{kj}^{-1}, \tag{6}$$

$$\begin{aligned} \mathrm{Tr}(BAB^{-1}) &= b_{il}a_{lk}b^{-1}{}_{ki} \\ &= (b^{-1}b)_{kl}a_{lk} = \delta_{kl}a_{lk} \\ &= a_{kk} = \mathrm{Tr}(A), \end{aligned} \tag{7}$$

where δ_{ij} is the KRONECKER DELTA.

The trace of a product of square matrices is independent of the order of the multiplication since

$$\begin{aligned} \mathrm{Tr}(AB) &= (ab)_{ii} = a_{ij}b_{ji} = b_{ji}a_{ij} \\ &= (ba)_{jj} = \mathrm{Tr}(BA). \end{aligned} \tag{8}$$

Therefore, the trace of the COMMUTATOR of A and B is given by

$$\mathrm{Tr}([A, B]) \equiv \mathrm{Tr}(AB) - \mathrm{Tr}(BA) = 0. \tag{9}$$

The product of a SYMMETRIC and an ANTISYMMETRIC MATRIX has zero trace,

$$\mathrm{Tr}(A_S B_A) = 0. \tag{10}$$

The value of the trace can be found using the fact that the matrix can always be transformed to a coordinate system where the z-AXIS lies along the axis of rotation. In the new coordinate system, the MATRIX is

$$A' = \begin{bmatrix} \cos\phi & \sin\phi & 0 \\ -\sin\phi & \cos\phi & 0 \\ 0 & 0 & 1 \end{bmatrix}, \tag{11}$$

so the trace is

$$\mathrm{Tr}(A') = \mathrm{Tr}(A) \equiv a_{ii} = 1 + 2\cos\phi. \tag{12}$$

References

Lang, S. *Linear Algebra, 3rd ed.* New York: Springer-Verlag, pp. 40 and 64, 1987.

Trace (Tensor)

The trace of a second-RANK TENSOR T is a SCALAR given by the CONTRACTED mixed TENSOR equal to T_i^i. The trace satisfies

$$\mathrm{Tr}\left[M^{-1}(x)\frac{\partial}{\partial x^\lambda}M(x)\right] = \frac{\partial}{\partial x^\lambda}\ln[\det(x)],$$

and

$$\begin{aligned} \delta\ln[\det M] &= \ln[\det(M+\delta M)] - \ln(\det M) \\ &= \ln\left[\frac{\det(M+\delta M)}{\det M}\right] \\ &= \ln[\det M^{-1}(M+\delta M)] \\ &= \ln[\det(1+M^{-1}\delta M)] \\ &\approx \ln[1+\mathrm{Tr}(M^{-1}\delta M)] \\ &\approx \mathrm{Tr}(M^{-1}\delta M). \end{aligned}$$

see also CONTRACTION (TENSOR)

Tractory

see TRACTRIX

Tractrix

The tractrix is the CATENARY INVOLUTE described by a point initially on the vertex. It has a constant NEGATIVE CURVATURE and is sometimes called the TRACTORY or EQUITANGENTIAL CURVE. The tractrix was first studied by Huygens in 1692, who gave it the name "tractrix." Later, Leibniz, Johann Bernoulli, and others studied the curve.

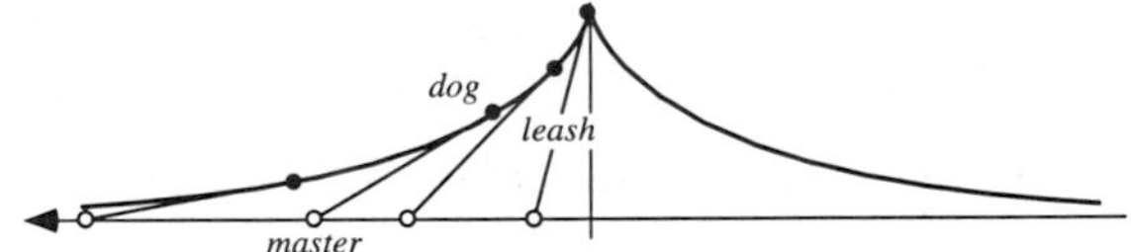

The tractrix arises from the following problem posed to Leibniz: What is the path of an object starting off with a vertical offset when it is dragged along by a string of constant length being pulled along a straight horizontal line? By associating the object with a dog, the string with a leash, and the pull along a horizontal line with the dog's master, the curve has the descriptive name HUNDKURVE (hound curve) in German. Leibniz found the curve using the fact that the axis is an asymptote to the tractrix (MacTutor Archive).

In CARTESIAN COORDINATES the tractrix has equation

$$x = a\,\mathrm{sech}^{-1}\left(\frac{y}{a}\right) - \sqrt{a^2 - y^2}. \tag{1}$$

One parametric form is

$$x(t) = a(t - \tanh t) \tag{2}$$

$$y(t) = a \operatorname{sech} t. \tag{3}$$

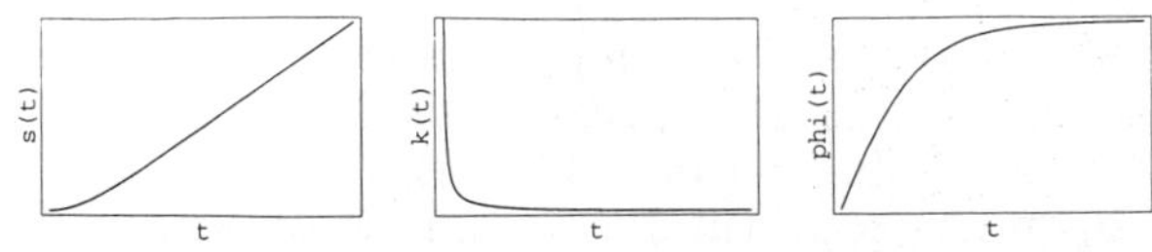

The ARC LENGTH, CURVATURE, and TANGENTIAL ANGLE are

$$s(t) = \ln(\cosh t) \tag{4}$$

$$\kappa(t) = \operatorname{csch} t \tag{5}$$

$$\phi(t) = 2\tan^{-1}[\tanh(\tfrac{1}{2}t)]. \tag{6}$$

A second parametric form in terms of the ANGLE ϕ of the straight line tangent to the tractrix is

$$x = a\{\ln[\tan(\tfrac{1}{2}\phi)] + \cos\phi\} \tag{7}$$

$$y = a \sin\phi \tag{8}$$

(Gray 1993). This parameterization has CURVATURE

$$\kappa(\phi) = |\tan\phi|. \tag{9}$$

A parameterization which traverses the tractrix with constant speed a is given by

$$x(t) = \begin{cases} ae^{-v/a} & \text{for } v \in [0,\infty) \\ ae^{v/a} & \text{for } v \in (-\infty, 0] \end{cases} \tag{10}$$

$$y(t) = \begin{cases} a[\tanh^{-1}(\sqrt{1-e^{-2v/a}}) - \sqrt{1-e^{-2v/a}}] \\ \qquad \text{for } v \in [0,\infty) \\ a[-\tanh^{-1}(\sqrt{1-e^{2v/a}}) + \sqrt{1-e^{2v/a}}] \\ \qquad \text{for } v \in (-\infty, 0]. \end{cases} \tag{11}$$

When a tractrix is rotated around its asymptote, a PSEUDOSPHERE results. This is a surface of constant NEGATIVE CURVATURE. For a tractrix, the length of a TANGENT from its point of contact to an asymptote is constant. The AREA between the tractrix and its asymptote is finite.

see also CURVATURE, DINI'S SURFACE, MICE PROBLEM, PSEUDOSPHERE, PURSUIT CURVE, TRACTROID

References

Geometry Center. "The Tractrix." `http://www.geom.umn.edu/zoo/diffgeom/pseudosphere/tractrix.html`.

Gray, A. "The Tractrix" and "The Evolute of a Tractrix is a Catenary." §3.5 and 5.3 in *Modern Differential Geometry of Curves and Surfaces.* Boca Raton, FL: CRC Press, pp. 46–50 and 80–81, 1993.

Lawrence, J. D. *A Catalog of Special Plane Curves.* New York: Dover, pp. 199–200, 1972.

Lee, X. "Tractrix." `http://www.best.com/~xah/SpecialPlaneCurves_dir/Tractrix_dir/tractrix.html`.

Lockwood, E. H. "The Tractrix and Catenary." Ch. 13 in *A Book of Curves.* Cambridge, England: Cambridge University Press, pp. 118–124, 1967.

MacTutor History of Mathematics Archive. "Tractrix." `http://www-groups.dcs.st-and.ac.uk/~history/Curves/Tractrix.html`.

Yates, R. C. "Tractrix." *A Handbook on Curves and Their Properties.* Ann Arbor, MI: J. W. Edwards, pp. 221–224, 1952.

Tractrix Evolute

The EVOLUTE of the TRACTRIX is the CATENARY.

Tractrix Radial Curve

The RADIAL CURVE of the TRACTRIX is the KAPPA CURVE.

Tractroid

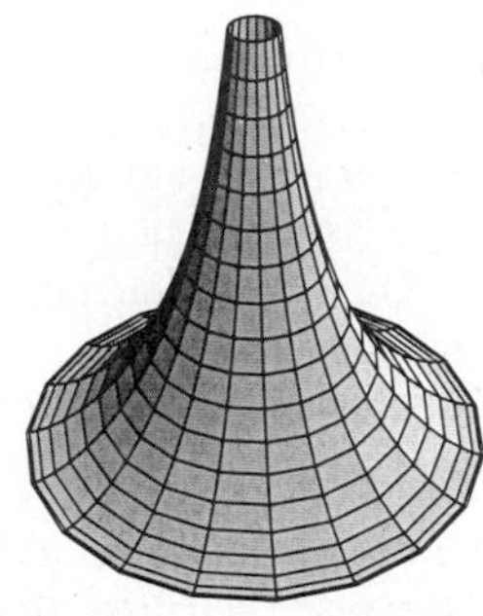

The SURFACE OF REVOLUTION produced by revolving the TRACTRIX

$$x = \operatorname{sech} u \tag{1}$$

$$z = u - \tanh u \tag{2}$$

about the z-AXIS is a tractroid given by

$$x = \operatorname{sech} u \cos v \tag{3}$$

$$y = \operatorname{sech} u \sin v \tag{4}$$

$$z = u - \tanh u. \tag{5}$$

see also PSEUDOSPHERE, SURFACE OF REVOLUTION, TRACTRIX

Transcendental Curve

A curve which intersects some straight line in an infinity of points (but for which not every point lies on this curve).

References

Borwein, J. M.; Borwein, P. B.; and Bailey, D. H. "Ramanujan, Modular Equations, and Approximations to Pi or How to Compute One Billion Digits of Pi." *Amer. Math. Monthly* **96**, 201–219, 1989.

Transcendental Equation

An equation or formula involving TRANSCENDENTAL FUNCTIONS.

Transcendental Function

A function which "transcends," i.e., cannot be expressed in terms of, the usual ELEMENTARY FUNCTIONS. Define

$$l_1(z) \equiv l(z) \equiv \ln(z)$$
$$e_1(z) \equiv e(z) \equiv e^z$$
$$\varsigma_1 f(z) \equiv \varsigma f(z) \equiv \int f(z)\,dz,$$

and let $l_2 \equiv l(l(z))$, etc. These are called the "elementary" transcendental functions (Watson 1966, p. 111).

see also ALGEBRAIC FUNCTION, ELEMENTARY FUNCTION

References

Watson, G. N. *A Treatise on the Theory of Bessel Functions, 2nd ed.* Cambridge, England: Cambridge University Press, 1966.

Transcendental Number

A number which is not the ROOT of *any* POLYNOMIAL equation with INTEGER COEFFICIENTS, meaning that it not an ALGEBRAIC NUMBER of any degree, is said to be transcendental. This definition guarantees that every transcendental number must also be IRRATIONAL, since a RATIONAL NUMBER is, by definition, an ALGEBRAIC NUMBER of degree one.

Transcendental numbers are important in the history of mathematics because their investigation provided the first proof that CIRCLE SQUARING, one of the GEOMETRIC PROBLEMS OF ANTIQUITY which had baffled mathematicians for more than 2000 years was, in fact, insoluble. Specifically, in order for a number to be produced by a GEOMETRIC CONSTRUCTION using the ancient Greek rules, it must be either RATIONAL or a very special kind of ALGEBRAIC NUMBER known as a EUCLIDEAN NUMBER. Because the number π is transcendental, the construction cannot be done according to the Greek rules.

Georg Cantor was the first to prove the EXISTENCE of transcendental numbers. Liouville subsequently showed how to construct special cases (such as LIOUVILLE'S CONSTANT) using LIOUVILLE'S RATIONAL APPROXIMATION THEOREM. In particular, he showed that any number which has a rapidly converging sequence of rational approximations must be transcendental. For many years, it was only known how to determine if special classes of numbers were transcendental. The determination of the status of more general numbers was considered an important enough unsolved problem that it was one of HILBERT'S PROBLEMS.

Great progress was subsequently made by GELFOND'S THEOREM, which gives a general rule for determining if special cases of numbers of the form α^β are transcendental. Baker produced a further revolution by proving the transcendence of sums of numbers of the form $\alpha \ln \beta$ for ALGEBRAIC NUMBERS α and β.

The number e was proven to be transcendental by Hermite in 1873, and PI (π) by Lindemann in 1882. e^π is transcendental by GELFOND'S THEOREM since

$$(-1)^{-i} = (e^{i\pi})^{-i} = e^\pi.$$

The GELFOND-SCHNEIDER CONSTANT $2^{\sqrt{2}}$ is also transcendental. Other known transcendentals are $\sin 1$ where $\sin x$ is the SINE function, $J_0(1)$ where $J_0(x)$ is a BESSEL FUNCTION OF THE FIRST KIND (Hardy and Wright 1985), $\ln 2$, $\ln 3/\ln 2$, the first zero $x_0 = 2.4048255\ldots$ of the BESSEL FUNCTION $J_0(x_0)$ (Le Lionnais 1983, p. 46), $\pi + \ln 2 + \sqrt{2}\ln 3$ (Borwein *et al.* 1989), the THUE-MORSE CONSTANT $P = 0.4124540336\ldots$ (Dekking 1977, Allouche and Shallit), the CHAMPERNOWNE CONSTANT $0.1234567891011\ldots$, the THUE CONSTANT

$$0.110110111110110111110110110\ldots,$$

$\Gamma(\frac{1}{3})$ (Le Lionnais 1983, p. 46), $\Gamma(\frac{1}{4})\pi^{-1/4}$ (Davis 1959), and $\Gamma(\frac{1}{4})$ (Chudnovsky, Waldschmidt), where $\Gamma(x)$ is the GAMMA FUNCTION. At least one of πe and $\pi + e$ (and probably both) are transcendental, but transcendence has not been proven for either number on its own.

It is not known if e^e, π^π, π^e, γ (the EULER-MASCHERONI CONSTANT), $I_0(2)$, or $I_1(2)$ (where $I_n(x)$ is a MODIFIED BESSEL FUNCTION OF THE FIRST KIND) are transcendental.

The "degree" of transcendence of a number can be characterized by a so-called LIOUVILLE-ROTH CONSTANT. There are still many fundamental and outstanding problems in transcendental number theory, including the CONSTANT PROBLEM and SCHANUEL'S CONJECTURE.

see also ALGEBRAIC NUMBER, CONSTANT PROBLEM, GELFOND'S THEOREM, IRRATIONAL NUMBER, LINDEMANN-WEIERSTRAß THEOREM, LIOUVILLE-ROTH CONSTANT, ROTH'S THEOREM, SCHANUEL'S CONJECTURE, THUE-SIEGEL-ROTH THEOREM

References

Allouche, J. P. and Shallit, J. In preparation.

Baker, A. "Approximations to the Logarithm of Certain Rational Numbers." *Acta Arith.* **10**, 315–323, 1964.

Baker, A. "Linear Forms in the Logarithms of Algebraic Numbers I." *Mathematika* **13**, 204–216, 1966.

Baker, A. "Linear Forms in the Logarithms of Algebraic Numbers II." *Mathematika* **14**, 102–107, 1966.

Baker, A. "Linear Forms in the Logarithms of Algebraic Numbers III." *Mathematika* **14**, 220–228, 1966.

Baker, A. "Linear Forms in the Logarithms of Algebraic Numbers IV." *Mathematika* **15**, 204–216, 1966.

Borwein, J. M.; Borwein, P. B.; and Bailey, D. H. "Ramanujan, Modular Equations, and Approximations to Pi or How to Compute One Billion Digits of Pi." *Amer. Math. Monthly* **96**, 201–219, 1989.

Chudnovsky, G. V. *Contributions to the Theory of Transcendental Numbers.* Providence, RI: Amer. Math. Soc., 1984.

Courant, R. and Robbins, H. "Algebraic and Transcendental Numbers." §2.6 in *What is Mathematics?: An Elementary*

Approach to Ideas and Methods, 2nd ed. Oxford, England: Oxford University Press, pp. 103–107, 1996.
Davis, P. J. "Leonhard Euler's Integral: A Historical Profile of the Gamma Function." *Amer. Math. Monthly* **66**, 849–869, 1959.
Dekking, F. M. "Transcendence du nombre de Thue-Morse." *Comptes Rendus de l'Academie des Sciences de Paris* **285**, 157–160, 1977.
Gray, R. "Georg Cantor and Transcendental Numbers." *Amer. Math. Monthly* **101**, 819–832, 1994.
Hardy, G. H. and Wright, E. M. *An Introduction to the Theory of Numbers, 5th ed.* Oxford, England: Oxford University Press, 1985.
Le Lionnais, F. *Les nombres remarquables.* Paris: Hermann, p. 46, 1983.
Siegel, C. L. *Transcendental Numbers.* New York: Chelsea, 1965.

Transcritical Bifurcation

Let $f : \mathbb{R} \times \mathbb{R} \to \mathbb{R}$ be a one-parameter family of C^2 maps satisfying

$$f(0,\mu) = 0 \tag{1}$$
$$\left[\frac{\partial f}{\partial x}\right]_{\mu=0,x=0} = 1 \tag{2}$$
$$\left[\frac{\partial f}{\partial x}\right]_{\mu,x} = \left[\frac{\partial f}{\partial x}\right]_{\mu=0,x=\mu} \tag{3}$$
$$\left[\frac{\partial^2 f}{\partial x \partial \mu}\right]_{0,0} > 0 \tag{4}$$
$$\left[\frac{\partial^2 f}{\partial \mu^2}\right]_{\mu=0,x=0} > 0. \tag{5}$$

Then there are two branches, one stable and one unstable. This BIFURCATION is called a transcritical bifurcation. An example of an equation displaying a transcritical bifurcation is

$$\dot{x} = \mu x - x^2. \tag{6}$$

(Guckenheimer and Holmes 1997, p. 145).

see also BIFURCATION

References
Guckenheimer, J. and Holmes, P. *Nonlinear Oscillations, Dynamical Systems, and Bifurcations of Vector Fields, 3rd ed.* New York: Springer-Verlag, pp. 145 and 149–150, 1997.
Rasband, S. N. *Chaotic Dynamics of Nonlinear Systems.* New York: Wiley, pp. 27–28, 1990.

Transfer Function

The engineering terminology for one use of FOURIER TRANSFORMS. By breaking up a wave pulse into its frequency spectrum

$$f_\nu = F(\nu)e^{2\pi i\nu t}, \tag{1}$$

the entire signal can be written as a sum of contributions from each frequency,

$$f(t) = \int_{-\infty}^{\infty} f_\nu \, d\nu = \int_{-\infty}^{\infty} F(\nu)e^{2\pi i\nu t} \, d\nu. \tag{2}$$

If the signal is modified in some way, it will become

$$g_\nu(t) = \phi(\nu)f_\nu(t) = \phi(\nu)F(\nu)e^{2\pi i\nu t} \tag{3}$$
$$g(t) = \int_{-\infty}^{\infty} g_\nu(t)\, dt = \int_{-\infty}^{\infty} \phi(\nu)F(\nu)e^{2\pi i\nu t} \, d\nu, \tag{4}$$

where $\phi(\nu)$ is known as the "transfer function." FOURIER TRANSFORMING ϕ and F,

$$\phi(\nu) = \int_{-\infty}^{\infty} \Phi(t)e^{-2\pi i\nu t}\, dt \tag{5}$$

$$F(\nu) = \int_{-\infty}^{\infty} f(t)e^{-2\pi i\nu t}\, dt. \tag{6}$$

From the CONVOLUTION THEOREM,

$$g(t) = f(t) * \Phi(t) = \int_{-\infty}^{\infty} f(t)\Phi(t-\tau)\, d\tau. \tag{7}$$

see also CONVOLUTION THEOREM, FOURIER TRANSFORM

Transfinite Diameter

Let

$$\phi(z) = cz + c_0 + c_1 z^{-1} + c_2 z^{-2} + \ldots$$

be an ANALYTIC FUNCTION, REGULAR and UNIVALENT for $|z| > 1$, which maps $|z| > 1$ CONFORMALLY onto the region T preserving the POINT AT INFINITY and its direction. Then the function $\phi(z)$ is uniquely determined and c is called the transfinite diameter, sometimes also known as ROBIN'S CONSTANT or the CAPACITY of $\phi(z)$.

see also ANALYTIC FUNCTION, REGULAR FUNCTION, UNIVALENT FUNCTION

Transfinite Number

One of Cantor's ORDINAL NUMBERS ω, $\omega+1$, $\omega+2$, ..., $\omega+\omega$, $\omega+\omega+1$, ... which is "larger" than any WHOLE NUMBER.

see also $\aleph_0$, $\aleph_1$, CARDINAL NUMBER, CONTINUUM, ORDINAL NUMBER, WHOLE NUMBER

References
Pappas, T. "Transfinite Numbers." *The Joy of Mathematics.* San Carlos, CA: Wide World Publ./Tetra, pp. 156–158, 1989.

Transform

A shortened term for INTEGRAL TRANSFORM.

Geometrically, if S and T are two transformations, then the SIMILARITY TRANSFORMATION TST^{-1} is sometimes called the transform (Woods 1961).

see also ABEL TRANSFORM, BOUSTROPHEDON TRANSFORM, DISCRETE FOURIER TRANSFORM, FAST FOURIER TRANSFORM, FOURIER TRANSFORM, FRACTIONAL FOURIER TRANSFORM, HANKEL TRANSFORM, HARTLEY TRANSFORM, HILBERT TRANSFORM, LAPLACE-STIELTJES TRANSFORM, LAPLACE TRANSFORM, MELLIN TRANSFORM, NUMBER THEORETIC TRANSFORM, PONCELET TRANSFORM, RADON TRANSFORM, WAVELET TRANSFORM, z-TRANSFORM, Z-TRANSFORM

References

Woods, F. S. *Higher Geometry: An Introduction to Advanced Methods in Analytic Geometry.* New York: Dover, p. 5, 1961.

Transformation

see FUNCTION, MAP

Transitive

A RELATION R on a SET S is transitive provided that for all x, y and z in S such that xRy and yRz, we also have xRz.

see also ASSOCIATIVE, COMMUTATIVE, RELATION

Transitive Closure

The transitive closure of a binary RELATION R on a SET X is the minimal TRANSITIVE relation R' on X that contains R. Thus $aR'b$ for any elements a and b of X, provided either that aRb or that there exists some element c of X such that aRc and cRb.

see also REFLEXIVE CLOSURE, TRANSITIVE REDUCTION

Transitive Reduction

The transitive reduction of a binary RELATION R on a SET X is the minimum relation R' on X with the same TRANSITIVE CLOSURE as R. Thus $aR'b$ for any elements a and b of X, provided that aRb and there exists no element c of X such that aRc and cRb.

see also REFLEXIVE REDUCTION, TRANSITIVE CLOSURE

Transitivity Class

Let $S(T)$ be the group of symmetries which map a MONOHEDRAL TILING T onto itself. The TRANSITIVITY CLASS of a given tile T is then the collection of all tiles to which T can be mapped by one of the symmetries of $S(T)$.

see also MONOHEDRAL TILING

References

Berglund, J. "Is There a k-Anisohedral Tile for $k \geq 5$?" *Amer. Math. Monthly* **100**, 585–588, 1993.

Translation

A transformation consisting of a constant offset with no ROTATION or distortion. In n-D EUCLIDEAN SPACE, a translation may be specified simply as a VECTOR giving the offset in each of the n coordinates.

see also AFFINE GROUP, DILATION, EUCLIDEAN GROUP, EXPANSION, GLIDE, IMPROPER ROTATION, INVERSION OPERATION, MIRROR IMAGE, REFLECTION, ROTATION

References

Beyer, W. H. (Ed.) *CRC Standard Mathematical Tables, 28th ed.* Boca Raton, FL: CRC Press, p. 211, 1987.

Translation Relation

A mathematical relationship transforming a function $f(x)$ to the form $f(x+a)$.

see also ARGUMENT ADDITION RELATION, ARGUMENT MULTIPLICATION RELATION, RECURRENCE RELATION, REFLECTION RELATION

Transpose

The object obtained by replacing all elements a_{ij} with a_{ji}. For a second-RANK TENSOR a_{ij}, the tensor transpose is simply a_{ji}. The matrix transpose, written A^{T}, is the MATRIX obtained by exchanging A's rows and columns, and satisfies the identity

$$(\mathsf{A}^{\mathrm{T}})^{-1} = (\mathsf{A}^{-1})^{\mathrm{T}}.$$

The product of two transposes satisfies

$$(\mathsf{B}^{\mathrm{T}}\mathsf{A}^{\mathrm{T}})_{ij} = (b^{\mathrm{T}})_{ik}(a^{\mathrm{T}})_{kj} = b_{ki}a_{jk} = a_{jk}b_{ki} = (\mathsf{AB})_{ji} = (\mathsf{AB})^{\mathrm{T}}_{ij}.$$

Therefore,

$$(\mathsf{AB})^{\mathrm{T}} = \mathsf{B}^{\mathrm{T}}\mathsf{A}^{\mathrm{T}}.$$

Transpose Map

see PULLBACK MAP

Transposition

An exchange of two elements of a SET with all others staying the same. A transposition is therefore a PERMUTATION of two elements. For example, the swapping of 2 and 5 to take the list 123456 to 153426 is a transposition.

see also PERMUTATION, TRANSPOSITION ORDER

Transposition Group

A PERMUTATION GROUP in which the PERMUTATIONS are limited to TRANSPOSITIONS.

see also PERMUTATION GROUP

Transposition Order

An ordering of PERMUTATIONS in which each two adjacent permutations differ by the TRANSPOSITION of two elements. For the permutations of $\{1, 2, 3\}$ there are two listings which are in transposition order. One is 123, 132, 312, 321, 231, 213, and the other is 123, 321, 312, 213, 231, 132.

see also LEXICOGRAPHIC ORDER, PERMUTATION

References

Ruskey, F. "Information on Combinations of a Set." http://sue.csc.uvic.ca/~cos/inf/comb/CombinationsInfo.html.

Transversal Array

A set of n cells in an $n \times n$ SQUARE such that no two come from the same row and no two come from the same column. The number of transversals of an $n \times n$ SQUARE is $n!$ (n FACTORIAL).

Transversal Design

A transversal design $\mathrm{TD}_\lambda(k, n)$ of order n, block size k, and index λ is a triple (V, G, B) such that

1. V is a set of kn elements,
2. G is a partition of V into k classes, each of size n (the "groups"),
3. B is a collection of k-subsets of V (the "blocks"), and
4. Every unordered pair of elements from V is contained in either exactly one group or in exactly λ blocks, but not both.

References

Colbourn, C. J. and Dinitz, J. H. (Eds.) *CRC Handbook of Combinatorial Designs.* Boca Raton, FL: CRC Press, p. 112, 1996.

Transversal Line

A transversal line is a LINE which intersects each of a given set of other lines. It is also called a SEMISECANT.

see also LINE

Transylvania Lottery

A lottery in which three numbers are picked at random from the INTEGERS 1–14.

see also FANO PLANE

Trapdoor Function

An easily computed function whose inverse is extremely difficult to compute. An example is the multiplication of two large PRIMES. Finding and verifying two large PRIMES is easy, as is their multiplication. But factorization of the resultant product is very difficult.

see also RSA ENCRYPTION

References

Gardner, M. Chs. 13–14 in *Penrose Tiles and Trapdoor Ciphers... and the Return of Dr. Matrix, reissue ed.* New York: W. H. Freeman, pp. 299–300, 1989.

Trapezium

There are two common definitions of the trapezium. The American definition is a QUADRILATERAL with no PARALLEL sides. The British definition for a trapezium is a QUADRILATERAL *with* two sides PARALLEL. Such a trapezium is equivalent to a TRAPEZOID and therefore has AREA

$$A = \tfrac{1}{2}(a + b)h.$$

see also DIAMOND, LOZENGE, PARALLELOGRAM, QUADRILATERAL, RHOMBOID, RHOMBUS, SKEW QUADRILATERAL, TRAPEZOID

Trapezohedron

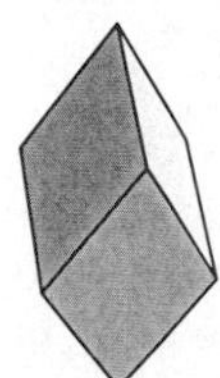
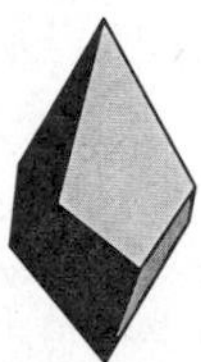
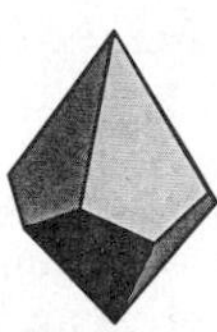
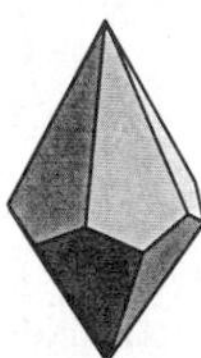

The trapezohedra are the DUAL POLYHEDRA of the Archimedean ANTIPRISMS. However, their faces are not TRAPEZOIDS.

see also ANTIPRISM, DIPYRAMID, HEXAGONAL SCALENOHEDRON, PRISM, TRAPEZOID

References

Cundy, H. and Rollett, A. *Mathematical Models, 3rd ed.* Stradbroke, England: Tarquin Pub., p. 117, 1989.

Trapezoid

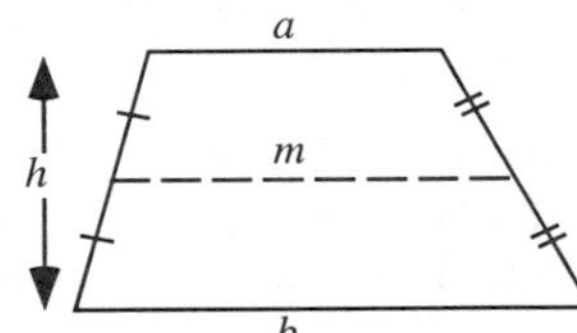

A QUADRILATERAL with two sides PARALLEL. The trapezoid depicted above satisfies

$$m = \tfrac{1}{2}(a + b)$$

and has AREA

$$A = \tfrac{1}{2}(a + b)h = mh.$$

The trapezoid is equivalent to the British definition of TRAPEZIUM.

see also PYRAMIDAL FRUSTUM, TRAPEZIUM

References

Beyer, W. H. (Ed.) *CRC Standard Mathematical Tables, 28th ed.* Boca Raton, FL: CRC Press, p. 123, 1987.

Trapezoidal Hexecontahedron

see DELTOIDAL HEXECONTAHEDRON

Trapezoidal Icositetrahedron

see Deltoidal Icositetrahedron

Trapezoidal Rule

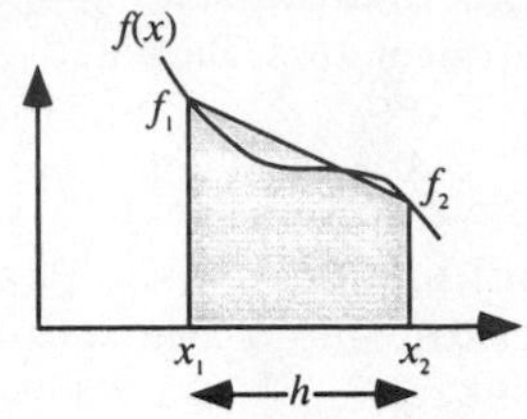

The 2-point Newton-Cotes Formula

$$\int_{x_1}^{x_2} f(x)\,dx = \tfrac{1}{2}h(f_1+f_2) - \tfrac{1}{2}h^3 f''(\xi),$$

where $f_i \equiv f(x_i)$, h is the separation between the points, and ξ is a point satisfying $x_1 \leq \xi \leq x_2$. Picking ξ to maximize $f''(\xi)$ gives an upper bound for the error in the trapezoidal approximation to the Integral.

see also Bode's Rule, Hardy's Rule, Newton-Cotes Formulas, Simpson's 3/8 Rule, Simpson's Rule, Weddle's Rule

References

Abramowitz, M. and Stegun, C. A. (Eds.). *Handbook of Mathematical Functions with Formulas, Graphs, and Mathematical Tables, 9th printing.* New York: Dover, p. 885, 1972.

Traveling Salesman Constants

N.B. A detailed on-line essay by S. Finch was the starting point for this entry.

Let $L(n,d)$ be the smallest Tour length for n points in a d-D Hypercube. Then there exists a smallest constant $\alpha(d)$ such that for all optimal Tours in the Hypercube,

$$\limsup_{n\to\infty} \frac{L(n,d)}{n^{(d-1)/d}\sqrt{d}} \leq \alpha(d), \tag{1}$$

and a constant $\beta(d)$ such that for *almost* all optimal tours in the Hypercube,

$$\lim_{n\to\infty} \frac{L(n,d)}{n^{(d-1)/d}\sqrt{d}} = \beta(d). \tag{2}$$

These constants satisfy the inequalities

$$\begin{aligned}0.44194 < \gamma_2 = \tfrac{5}{16}\sqrt{2} \leq \beta(2) \\ \leq \delta < 0.6508 < 0.75983 < 3^{-1/4} \leq \alpha(2) \\ \leq \phi < 0.98398 \end{aligned}\tag{3}$$

$$\begin{aligned}0.37313 < \gamma_3 \leq \beta(3) \leq 12^{1/6}6^{-1/2} < 0.61772 < 0.64805 \\ < 2^{1/6}3^{-1/2} \leq \alpha(3) \leq 0.90422\end{aligned}\tag{4}$$

$$\begin{aligned}0.34207 < \gamma_4 \leq \beta(4) \leq 12^{1/8}6^{-1/2} < 0.55696 \\ < 0.59460 < 2^{-3/4} \leq \alpha(4) \leq 0.8364\end{aligned}\tag{5}$$

(Fejes Tóth 1940, Verblunsky 1951, Few 1955, Beardwood *et al.* 1959), where

$$\gamma_d \equiv \frac{\Gamma\left(3+\frac{1}{d}\right)[\Gamma(\frac{1}{2}d+1)]^{1/d}}{2\sqrt{\pi}(d^{1/2}+d^{-1/2})}, \tag{6}$$

$\Gamma(z)$ is the Gamma Function, δ is an expression involving Struve Functions and Neumann Functions,

$$\phi \equiv \frac{280(3-\sqrt{3})}{840-280\sqrt{3}+4\sqrt{5}-\sqrt{10}} \tag{7}$$

(Karloff 1989), and

$$\psi \equiv \tfrac{1}{2}3^{-2/3}(4+\ln 3)^{2/3} \tag{8}$$

(Goddyn 1990). In the Limit $d\to\infty$,

$$\begin{aligned}0.24197 < \lim_{d\to\infty}\gamma_d = \frac{1}{\sqrt{2\pi e}} \leq \liminf_{d\to\infty}\beta(d) \\ \leq \limsup_{d\to\infty}\beta(d) \leq \lim_{d\to\infty} 12^{1/(2d)}6^{-1/2} \\ = \frac{1}{\sqrt{6}} < 0.40825\end{aligned}\tag{9}$$

and

$$\begin{aligned}0.24197 < \frac{1}{\sqrt{2\pi e}} \leq \lim_{d\to\infty}\alpha(d) \\ \leq \frac{2(3-\sqrt{3})\theta}{\sqrt{2\pi e}} < 0.4052,\end{aligned}\tag{10}$$

where

$$\tfrac{1}{2} \leq \theta = \lim_{d\to\infty}[\theta(d)]^{1/d} \leq 0.6602, \tag{11}$$

and $\theta(d)$ is the best Sphere Packing density in d-D space (Goddyn 1990, Moran 1984, Kabatyanskii and Levenshtein 1978). Steele and Snyder (1989) proved that the limit $\alpha(d)$ exists.

Now consider the constant

$$\kappa \equiv \lim_{n\to\infty}\frac{L(n,2)}{\sqrt{n}} = \beta(2)\sqrt{2}, \tag{12}$$

so

$$\tfrac{5}{8} = \gamma_2\sqrt{2} \leq \kappa \leq \delta\sqrt{2} < 0.9204. \tag{13}$$

The best current estimate is $\kappa \approx 0.7124$.

A certain self-avoiding Space-Filling Curve is an optimal Tour through a set of n points, where n can be arbitrarily large. It has length

$$\lambda \equiv \lim_{m\to\infty}\frac{L_m}{\sqrt{n_m}} = \frac{4(1+2\sqrt{2})\sqrt{51}}{153} = 0.7147827\ldots, \tag{14}$$

where L_m is the length of the curve at the mth iteration and n_m is the point-set size (Moscato and Norman).

References
Beardwood, J.; Halton, J. H.; and Hammersley, J. M. "The Shortest Path Through Many Points." *Proc. Cambridge Phil. Soc.* **55**, 299–327, 1959.
Chartrand, G. "The Salesman's Problem: An Introduction to Hamiltonian Graphs." §3.2 in *Introductory Graph Theory.* New York: Dover, pp. 67–76, 1985.
Fejes Tóth, L. "Über einen geometrischen Satz." *Math. Zeit.* **46**, 83–85, 1940.
Few, L. "The Shortest Path and the Shortest Road Through n Points." *Mathematika* **2**, 141–144, 1955.
Finch, S. "Favorite Mathematical Constants." http://www.mathsoft.com/asolve/constant/sales/sales.html.
Flood, M. "The Travelling Salesman Problem." *Operations Res.* **4**, 61–75, 1956.
Goddyn, L. A. "Quantizers and the Worst Case Euclidean Traveling Salesman Problem." *J. Combin. Th. Ser. B* **50**, 65–81, 1990.
Kabatyanskii, G. A. and Levenshtein, V. I. "Bounds for Packing on a Sphere and in Space." *Problems Inform. Transm.* **14**, 1–17, 1978.
Karloff, H. J. "How Long Can a Euclidean Traveling Salesman Tour Be?" *SIAM J. Disc. Math.* **2**, 91–99, 1989.
Moran, S. "On the Length of Optimal TSP Circuits in Sets of Bounded Diameter." *J. Combin. Th. Ser. B* **37**, 113–141, 1984.
Moscato, P. "Fractal Instances of the Traveling Salesman Constant." http://www.ing.unlp.edu.ar/cetad/mos/FRACTAL_TSP_home.html
Steele, J. M. and Snyder, T. L. "Worst-Case Growth Rates of Some Classical Problems of Combinatorial Optimization." *SIAM J. Comput.* **18**, 278–287, 1989.
Verblunsky, S. "On the Shortest Path Through a Number of Points." *Proc. Amer. Math. Soc.* **2**, 904–913, 1951.

Traveling Salesman Problem

A problem in GRAPH THEORY requiring the most efficient (i.e., least total distance) TOUR (i.e., closed path) a salesman can take through each of n cities. No general method of solution is known, and the problem is NP-HARD.

see also TRAVELING SALESMAN CONSTANTS

References
Platzman, L. K. and Bartholdi, J. J. "Spacefilling Curves and the Planar Travelling Salesman Problem." *J. Assoc. Comput. Mach.* **46**, 719–737, 1989.

Trawler Problem

A fast boat is overtaking a slower one when fog suddenly sets in. At this point, the boat being pursued changes course, but not speed. How should the pursuing vessel proceed in order to be sure of catching the other boat?

The amazing answer is that the pursuing boat should continue to the point where the slow boat would be if it had set its course directly for the pursuing boat when the fog set in. If the boat is not there, it should proceed in a SPIRAL whose origin is the point where the slow boat was when the fog set in. The SPIRAL can be constructed in such a way that the two boats will intersect before a complete turn is made.

References
Ogilvy, C. S. *Excursions in Mathematics.* New York: Dover, pp. 84 and 148, 1994.

Trebly Magic Square

see TRIMAGIC SQUARE

Tredecillion

In the American system, 10^{42}.

see also LARGE NUMBER

Tree

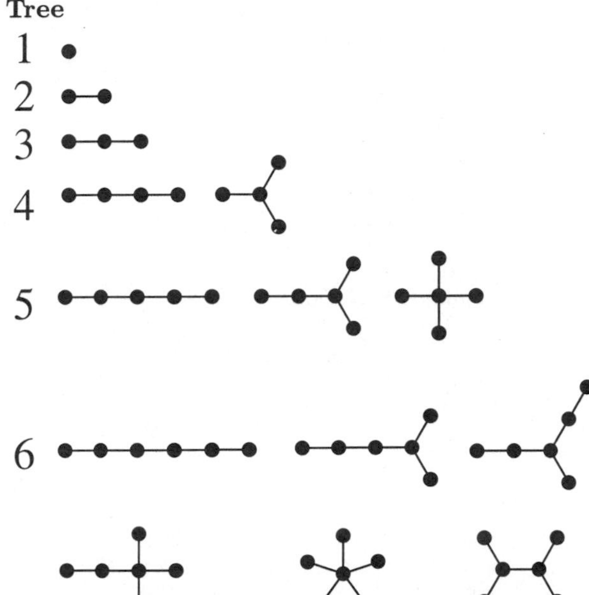

A tree is a mathematical structure which can be viewed as either a GRAPH or as a DATA STRUCTURE. The two views are equivalent, since a tree DATA STRUCTURE contains not only a set of elements, but also connections between elements, giving a tree graph.

A tree graph is a set of straight line segments connected at their ends containing no closed loops (cycles). A tree with n nodes has $n-1$ EDGES. The points of connection are known as FORKS and the segments as BRANCHES. Final segments and the nodes at their ends are called LEAVES. A tree with two BRANCHES at each FORK and with one or two LEAVES at the end of each branch is called a BINARY TREE.

When a special node is designated to turn a tree into a ROOTED TREE, it is called the ROOT (or sometimes "EVE.") In such a tree, each of the nodes which is one EDGE further away from a given EDGE is called a CHILD, and nodes connected to the same node are then called SIBLINGS.

Note that two BRANCHES placed end-to-end are equivalent to a single BRANCH which means, for example, that there is only *one* tree of order 3. The number $t(n)$ of nonisomorphic trees of order $n = 1, 2, \ldots$ (where trees of orders 1, 2, ..., 6 are illustrated above), are 1, 1, 1, 2, 3, 6, 11, 23, 47, 106, 235, ... (Sloane's A000055).

Otter showed that

$$\lim_{n\to\infty} \frac{t(n)n^{5/2}}{\alpha^n} = \beta, \qquad (1)$$

(Otter 1948, Harary and Palmer 1973, Knuth 1969), where the constants α and β are sometimes called OTTER'S TREE ENUMERATION CONSTANTS. Write the GENERATING FUNCTION for ROOTED TREES as

$$f(z) = \sum_{i=0}^{\infty} f_i z^i, \qquad (2)$$

where the COEFFICIENTS are

$$f_{i+1} = \frac{1}{i} \sum_{j=1}^{i} \left(\sum_{d|j} d f_d \right) f_{i-j+1}, \qquad (3)$$

with $f_0 = 0$ and $f_1 = 1$. Then

$$\alpha = 2.955765\ldots \qquad (4)$$

is the unique POSITIVE ROOT of

$$f\left(\frac{1}{x}\right) = 1, \qquad (5)$$

and

$$\beta = \frac{1}{\sqrt{2\pi}} \left[1 + \sum_{k=2}^{\infty} f'\left(\frac{1}{\alpha_k}\right) \frac{1}{\alpha_k} \right]^{3/2} = 0.5349485\ldots. \qquad (6)$$

see also B-TREE, BINARY TREE, CATERPILLAR GRAPH, CAYLEY TREE, CHILD, DIJKSTRA TREE, EVE, FOREST, KRUSKAL'S ALGORITHM, KRUSKAL'S TREE THEOREM, LEAF (TREE), ORCHARD-PLANTING PROBLEM, ORDERED TREE, PATH GRAPH, PLANTED PLANAR TREE, PÓLYA ENUMERATION THEOREM, QUADTREE, RED-BLACK TREE, ROOT (TREE), ROOTED TREE, SIBLING, STAR GRAPH, STERN-BROCOT TREE, WEAKLY BINARY TREE, WEIGHTED TREE

References

Finch, S. "Favorite Mathematical Constants." http://www.mathsoft.com/asolve/constant/otter/otter.html.

Chauvin, B.; Cohen, S.; and Rouault, A. (Eds.). *Trees: Workshop in Versailles, June 14–16, 1995.* Basel, Switzerland: Birkhäuser, 1996.

Gardner, M. "Trees." Ch. 17 in *Mathematical Magic Show: More Puzzles, Games, Diversions, Illusions and Other Mathematical Sleight-of-Mind from Scientific American.* New York: Vintage, pp. 240–250, 1978.

Harary, F. *Graph Theory.* Reading, MA: Addison-Wesley, 1994.

Harary, F. and Manvel, B. "Trees." *Scripta Math.* **28**, 327–333, 1970.

Harary, F. and Palmer, E. M. *Graphical Enumeration.* New York: Academic Press, 1973.

Knuth, D. E. *The Art of Computer Programming, Vol. 1: Fundamental Algorithms, 2nd ed.* Reading, MA: Addison-Wesley, 1973.

Otter, R. "The Number of Trees." *Ann. Math.* **49**, 583–599, 1948.

Sloane, N. J. A. Sequences A000055/M0791 in "An On-Line Version of the Encyclopedia of Integer Sequences."

Sloane, N. J. A. and Plouffe, S. Extended entry in *The Encyclopedia of Integer Sequences.* San Diego: Academic Press, 1995.

Tree-Planting Problem

see ORCHARD-PLANTING PROBLEM

Tree Searching

N.B. A detailed on-line essay by S. Finch was the starting point for this entry.

In database structures, two quantities are generally of interest: the average number of comparisons required to

1. Find an existing random record, and
2. Insert a new random record into a data structure.

Some constants which arise in the theory of digital tree searching are

$$\alpha \equiv \sum_{k=1}^{\infty} \frac{1}{2^k - 1} = 1.6066951524\ldots \qquad (1)$$

$$\beta \equiv \sum_{n=1}^{\infty} \frac{1}{(2^n - 1)^2} = 1.1373387363\ldots. \qquad (2)$$

Erdős (1948) proved that α is IRRATIONAL. The expected number of comparisons for a successful search is

$$E = \frac{\ln n}{\ln 2} + \frac{\gamma - 1}{\ln 2} - \alpha + \tfrac{3}{2} + \delta(n) + \mathcal{O}(n^{-1/2}) \qquad (3)$$

$$\sim \lg n - 0.716644\ldots + \delta(n), \qquad (4)$$

and for an unsuccessful search is

$$E = \frac{\ln n}{\ln 2} + \frac{\gamma}{\ln 2} - \alpha + \tfrac{1}{2} + \delta(n) + \mathcal{O}(n^{-1/2}) \qquad (5)$$

$$\sim \lg n - 0.273948\ldots + \delta(n). \qquad (6)$$

Here $\delta(n)$, $\epsilon(s)$, and $\rho(n)$ are small-amplitude periodic functions, and LG is the base 2 LOGARITHM. The VARIANCE for searching is

$$V \sim \frac{1}{12} + \frac{\pi^2 + 6}{6(\ln 2)^2} - \alpha - \beta + \epsilon(s) \sim 2.844383\ldots + \epsilon(s) \qquad (7)$$

and for inserting is

$$V \sim \frac{1}{12} + \frac{\pi^2}{6(\ln 2)^2} - \alpha - \beta + \epsilon(s) \sim 0.763014\ldots + \epsilon(s). \qquad (8)$$

The expected number of pairs of twin vacancies in a digital search tree is

$$\langle A_n\rangle = \left[\theta + 1 - \frac{1}{Q}\left(\frac{1}{\ln 2} + \alpha^2 - \alpha\right) + \rho(n)\right] n + \mathcal{O}(\sqrt{n}), \tag{9}$$

where

$$Q \equiv \prod_{k=1}^{\infty}\left(1 - \frac{1}{2^k}\right) = 0.2887880950\ldots \tag{10}$$

$$= \frac{1}{3} - \frac{1}{3\cdot 7} + \frac{1}{3\cdot 5\cdot 15} - \frac{1}{3\cdot 5\cdot 15\cdot 21} + \ldots \tag{11}$$

$$= \exp\left[-\sum_{n=1}^{\infty}\frac{1}{n(2^n-1)}\right] \tag{12}$$

$$= \sqrt{\frac{2\pi}{\ln 2}}\exp\left(\frac{\ln 2}{24} - \frac{\pi^2}{6\ln 2}\right) \times \prod_{n=1}^{\infty}\left[1 - \exp\left(-\frac{4\pi^2 n}{\ln 2}\right)\right] \tag{13}$$

and

$$\theta = \sum_{k=1}^{\infty}\frac{k2^{k+1}}{1\cdot 3\cdot 7\cdot 16\cdots(2^k-1)}\sum_{j=1}^{k}\frac{1}{2^j-1} = 7.7431319855\ldots. \tag{14}$$

(Flajolet and Sedgewick 1986). The linear COEFFICIENT of $\langle A_n\rangle$ fluctuates around

$$c = \theta + 1 - \frac{1}{Q}\left(\frac{1}{\ln 2} + \alpha^2 - \alpha\right) = 0.3720486812\ldots, \tag{15}$$

which can also be written

$$c = \frac{1}{\ln 2}\int_0^{\infty}\frac{x}{1+x} \times \frac{dx}{(1+x)(1+\frac{1}{2}x)(1+\frac{1}{4}x)(1+\frac{1}{8}x)\cdots}. \tag{16}$$

(Flajolet and Richmond 1992).

References

Finch, S. "Favorite Mathematical Constants." `http://www.mathsoft.com/asolve/constant/bin/bin.html`.

Finch, S. "Favorite Mathematical Constants." `http://www.mathsoft.com/asolve/constant/dig/dig.html`.

Finch, S. "Favorite Mathematical Constants." `http://www.mathsoft.com/asolve/constant/qdt/qdt.html`.

Flajolet, P. and Richmond, B. "Generalized Digital Trees and their Difference-Differential Equations." *Random Structures and Algorithms* **3**, 305–320, 1992.

Flajolet, P. and Sedgewick, R. "Digital Search Trees Revisited." *SIAM Review* **15**, 748–767, 1986.

Knuth, D. E. *The Art of Computer Programming, Vol. 3: Sorting and Searching, 2nd ed.* Reading, MA: Addison-Wesley, pp. 21, 134, 156, 493–499, and 580, 1973.

Trefoil Curve

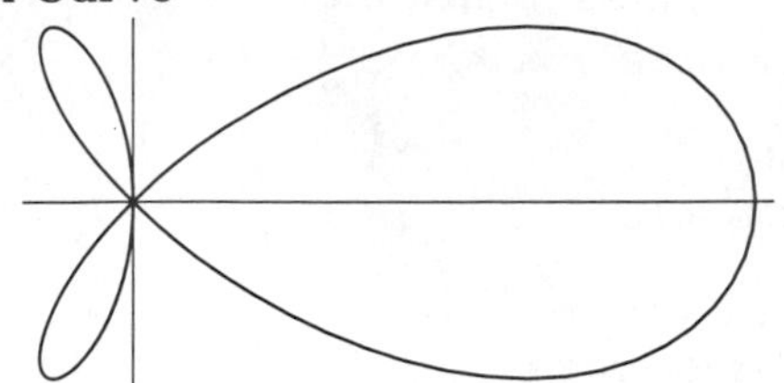

The plane curve given by the equation

$$x^4 + x^2y^2 + y^4 = x(x^2 - y^2).$$

Trefoil Knot

The knot 03_{001}, also called the THREEFOIL KNOT, which is the unique PRIME KNOT of three crossings. It has BRAID WORD $\sigma_1{}^3$. The trefoil and its MIRROR IMAGE are not equivalent. The trefoil has ALEXANDER POLYNOMIAL $-x^2 + x - 1$ and is a (3, 2)-TORUS KNOT. The BRACKET POLYNOMIAL can be computed as follows.

$$\begin{aligned}\langle L\rangle &= A^3d^{2-1} + A^2Bd^{1-1} + A^2Bd^{1-1} + AB^2d^{2-1} \\ &\quad + A^2Bd^{1-1} + AB^2d^{2-1} + AB^2d^{2-1} + B^3d^{3-1} \\ &= A^3d^1 + 3A^2Bd^0 + 3AB^2d^1 + B^3d^2.\end{aligned}$$

Plugging in

$$B = A^{-1}$$
$$d = -A^2 - A^{-2}$$

gives

$$\langle L\rangle = A^{-7} - A^{-3} - A^5.$$

The normalized one-variable KAUFFMAN POLYNOMIAL X is then given by

$$\begin{aligned}X_L &= (-A^3)^{-w(L)}\langle L\rangle = (-A^3)^{-3}(A^{-7} - A^{-3} - A^5) \\ &= A^{-4} + A^{-12} - A^{-16},\end{aligned}$$

where the WRITHE $w(L) = 3$. The JONES POLYNOMIAL is therefore

$$V(t) = L(A = t^{-1/4}) = t + t^3 - t^4 = t(1 + t^2 - t^3).$$

Since $V(t^{-1}) \neq V(t)$, we have shown that the mirror images are not equivalent.

References

Claremont High School. "Trefoil_Knot Movie." Binary encoded QuickTime movie. `ftp://chs.cusd.claremont.edu/pub/knot/trefoil.cptbin`.

Crandall, R. E. *Mathematica for the Sciences.* Redwood City, CA: Addison-Wesley, 1993.

Kauffman, L. H. *Knots and Physics.* Singapore: World Scientific, pp. 29–35, 1991.

Nordstrand, T. "Threefoil Knot." `http://www.uib.no/people/nfytn/tknottxt.htm`.

Pappas, T. "The Trefoil Knot." *The Joy of Mathematics.* San Carlos, CA: Wide World Publ./Tetra, p. 96, 1989.

Trench Diggers' Constant

see BEAM DETECTOR

Triabolo

A 3-POLYABOLO.

Triacontagon

A 30-sided POLYGON.

Triacontahedron

A 30-sided POLYHEDRON such as the RHOMBIC TRIACONTAHEDRON.

Triad

A SET with three elements.

see also HEXAD, MONAD, QUARTET, QUINTET, TETRAD

Triakis Icosahedron

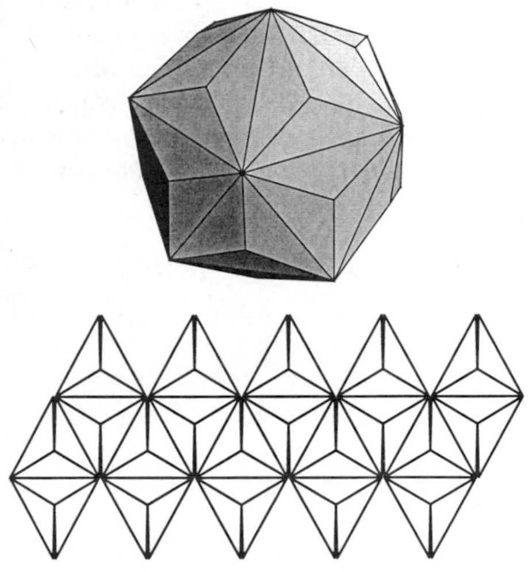

The DUAL POLYHEDRON of the TRUNCATED DODECAHEDRON ARCHIMEDEAN SOLID. The triakis icosahedron is also ICOSAHEDRON STELLATION #2.

References

Wenninger, M. J. *Polyhedron Models.* New York: Cambridge University Press, p. 46, 1989.

Triakis Octahedron

see GREAT TRIAKIS OCTAHEDRON, SMALL TRIAKIS OCTAHEDRON

Triakis Tetrahedron

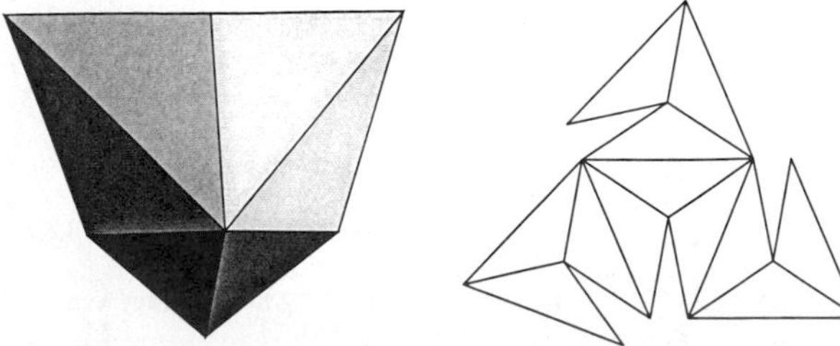

The DUAL POLYHEDRON of the TRUNCATED TETRAHEDRON ARCHIMEDEAN SOLID.

Trial

In statistics, a trial is a single measurable random event, such as the flipping of a COIN, the generation of a RANDOM NUMBER, the dropping of a ball down the apex of a triangular lattice and having it fall into a single bin at the bottom, etc.

see also BERNOULLI TRIAL, LEXIS TRIALS, POISSON TRIALS

Trial Division

A brute-force method of finding a DIVISOR of an INTEGER n by simply plugging in one or a set of INTEGERS and seeing if they DIVIDE n. Repeated application of trial division to obtain the complete PRIME FACTORIZATION of a number is called DIRECT SEARCH FACTORIZATION. An individual integer being tested is called a TRIAL DIVISOR.

see also DIRECT SEARCH FACTORIZATION, DIVISION, PRIME FACTORIZATION

Trial Divisor

An INTEGER n which is tested to see if it divides a given number.

see also TRIAL DIVISION

Triamond

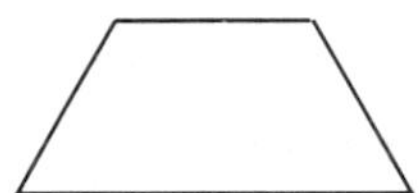

The unique 3-POLYIAMOND, illustrated above.

see also POLYIAMOND, TRAPEZOID

Triangle

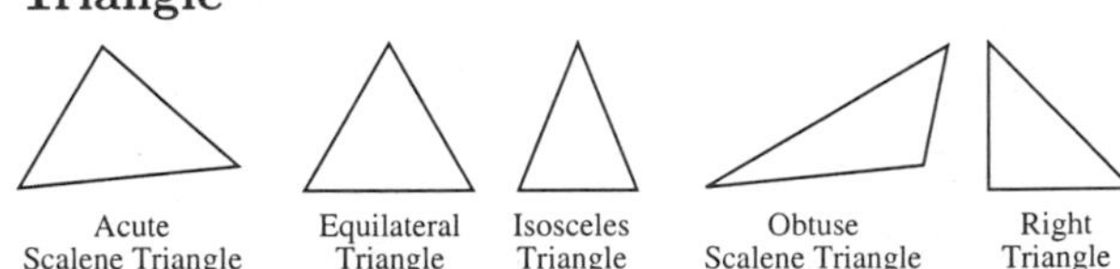

A triangle is a 3-sided POLYGON sometimes (but not very commonly) called the TRIGON. All triangles are convex. An ACUTE TRIANGLE is a triangle whose three angles are all ACUTE. A triangle with all sides equal is called EQUILATERAL. A triangle with two sides equal is called ISOSCELES. A triangle having an OBTUSE ANGLE is called an OBTUSE TRIANGLE. A triangle with a RIGHT ANGLE is called RIGHT. A triangle with all sides a different length is called SCALENE.

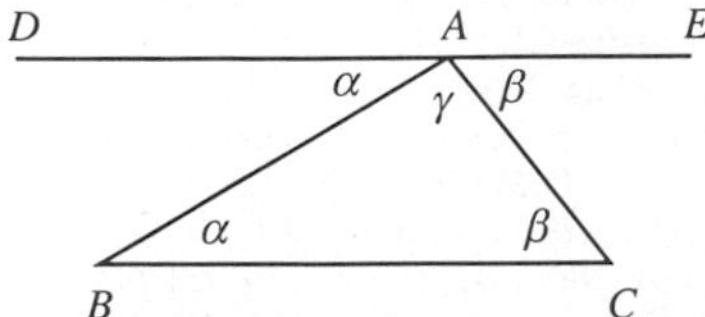

The sum of ANGLES in a triangle is 180°. This can be established as follows. Let $DAE||BC$ (DAE be PARALLEL to BC) in the above diagram, then the angles α and β

satisfy $\alpha = \angle DAB = \angle ABC$ and $\beta = \angle EAC = \angle BCE$, as indicated. Adding γ, it follows that

$$\alpha + \beta + \gamma = 180°, \tag{1}$$

since the sum of angles for the line segment must equal two RIGHT ANGLES. Therefore, the sum of angles in the triangle is also 180°.

Let S stand for a triangle side and A for an angle, and let a set of Ss and As be concatenated such that adjacent letters correspond to adjacent sides and angles in a triangle. Triangles are uniquely determined by specifying three sides (SSS THEOREM), two angles and a side (AAS THEOREM), or two sides with an adjacent angle (SAS THEOREM). In each of these cases, the unknown three quantities (there are three sides and three angles total) can be uniquely determined. Other combinations of sides and angles do not uniquely determine a triangle: three angles specify a triangle only modulo a scale size (AAA THEOREM), and one angle and two sides not containing it may specify one, two, or no triangles (ASS THEOREM).

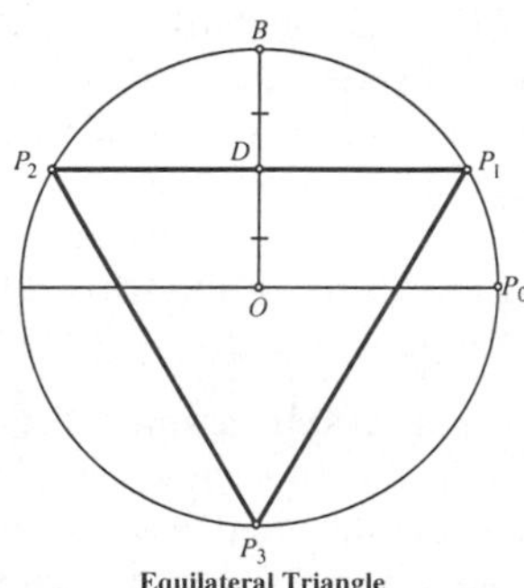

Equilateral Triangle

The RULER and COMPASS construction of the triangle can be accomplished as follows. In the above figure, take OP_0 as a RADIUS and draw $OB \perp OP_0$. Then bisect OB and construct $P_2P_1 || OP_0$. Extending BO to locate P_3 then gives the EQUILATERAL TRIANGLE $\Delta P_1P_2P_3$.

In Proposition IV.4 of the *Elements*, Euclid showed how to inscribe a CIRCLE (the INCIRCLE) in a given triangle by locating the CENTER as the point of intersection of ANGLE BISECTORS. In Proposition IV.5, he showed how to circumscribe a CIRCLE (the CIRCUMCIRCLE) about a given triangle by locating the CENTER as the point of intersection of the perpendicular bisectors.

If the coordinates of the triangle VERTICES are given by (x_i, y_i) where $i = 1, 2, 3$, then the AREA Δ is given by the DETERMINANT

$$\Delta = \frac{1}{2!}\begin{vmatrix} x_1 & y_1 & 1 \\ x_2 & y_2 & 1 \\ x_3 & y_3 & 1 \end{vmatrix}. \tag{2}$$

If the coordinates of the triangle VERTICES are given in 3-D by (x_i, y_i, z_i) where $i = 1, 2, 3$, then

$$\Delta = \frac{1}{2}\sqrt{\begin{vmatrix} y_1 & z_1 & 1 \\ y_2 & z_2 & 1 \\ y_3 & z_3 & 1 \end{vmatrix}^2 + \begin{vmatrix} z_1 & x_1 & 1 \\ z_2 & x_2 & 1 \\ z_3 & x_3 & 1 \end{vmatrix}^2 + \begin{vmatrix} x_1 & y_1 & 1 \\ x_2 & y_2 & 1 \\ x_3 & y_3 & 1 \end{vmatrix}^2}. \tag{3}$$

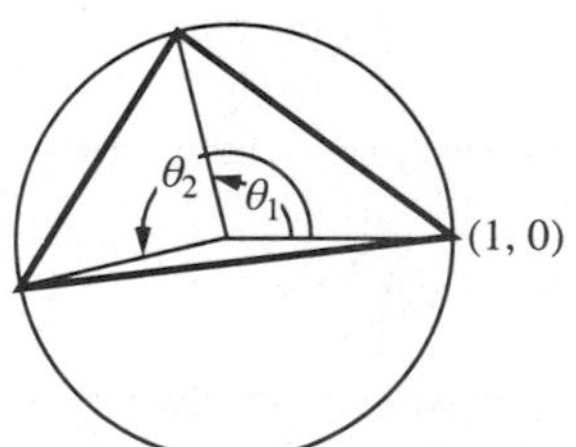

In the above figure, let the CIRCUMCIRCLE passing through a triangle's VERTICES have RADIUS r, and denote the CENTRAL ANGLES from the first point to the second θ_1, and to the third point by θ_2. Then the AREA of the triangle is given by

$$\Delta = 2r^2 \left|\sin(\tfrac{1}{2}\theta_1)\sin(\tfrac{1}{2}\theta_2)\sin[\tfrac{1}{2}(\theta_1 - \theta_2)]\right|. \tag{4}$$

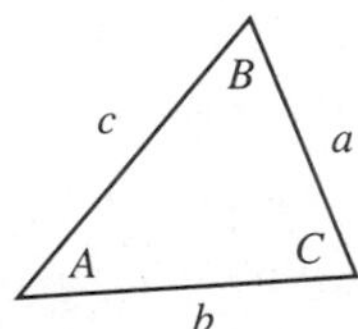

If a triangle has sides a, b, c, call the angles opposite these sides A, B, and C, respectively. Also define the SEMIPERIMETER s as HALF the PERIMETER:

$$s \equiv \tfrac{1}{2}p = \tfrac{1}{2}(a + b + c). \tag{5}$$

The AREA of a triangle is then given by HERON'S FORMULA

$$\Delta = \sqrt{s(s-a)(s-b)(s-c)}, \tag{6}$$

as well by the FORMULAS

$$\Delta = \tfrac{1}{4}\sqrt{(a+b+c)(b+c-a)(c+a-b)(a+b-c)} \tag{7}$$

$$= \tfrac{1}{4}\sqrt{2(a^2b^2 + a^2c^2 + b^2c^2) - (a^4 + b^4 + c^4)} \tag{8}$$

$$= \tfrac{1}{4}\sqrt{[(a+b)^2 - c^2][c^2 - (a-b)^2]} \tag{9}$$

$$= \tfrac{1}{4}\sqrt{p(p-2a)(p-2b)(p-2c)}, \tag{10}$$

$$= 2R^2 \sin A \sin B \sin C \tag{11}$$

$$= \frac{abc}{4R} = rs \tag{12}$$

$$= \tfrac{1}{2}ah_a \tag{13}$$

$$= \tfrac{1}{2}bc\sin A. \tag{14}$$

In the above formulas, h_i is the ALTITUDE on side i, R is the CIRCUMRADIUS, and r is the INRADIUS (Johnson 1929, p. 11). Expressing the side lengths a, b, and c in terms of the radii a', b', and c' of the mutually tangent circles centered on the TRIANGLE vertices (which define the SODDY CIRCLES),

$$a = b' + c' \tag{15}$$

$$b = a' + c' \tag{16}$$

$$c = a' + b', \tag{17}$$

gives the particularly pretty form

$$\Delta = \sqrt{a'b'c'(a' + b' + c')}. \tag{18}$$

For additional FORMULAS, see Beyer (1987) and Baker (1884), who gives 110 FORMULAS for the AREA of a triangle.

The ANGLES of a triangle satisfy

$$\cot A = \frac{b^2 + c^2 - a^2}{4\Delta} \tag{19}$$

where Δ is the AREA (Johnson 1929, p. 11, with missing squared symbol added). This gives the pretty identity

$$\cot A + \cot B + \cot C = \frac{a^2 + b^2 + c^2}{4\Delta}. \tag{20}$$

Let a triangle have ANGLES A, B, and C. Then

$$\sin A \sin B \sin C \leq kABC, \tag{21}$$

where

$$k = \left(\frac{3\sqrt{3}}{2\pi}\right)^3 \tag{22}$$

(Abi-Khuzam 1974, Le Lionnais 1983). This can be used to prove that

$$8\omega^3 < ABC, \tag{23}$$

where ω is the BROCARD ANGLE.

TRIGONOMETRIC FUNCTIONS of half angles can be expressed in terms of the triangle sides:

$$\cos(\tfrac{1}{2}A) = \sqrt{\frac{s(s-a)}{bc}} \tag{24}$$

$$\sin(\tfrac{1}{2}A) = \sqrt{\frac{(s-b)(s-c)}{bc}} \tag{25}$$

$$\tan(\tfrac{1}{2}A) = \sqrt{\frac{(s-b)(s-c)}{s(s-a)}}, \tag{26}$$

where s is the SEMIPERIMETER.

The number of different triangles which have INTEGRAL sides and PERIMETER n is

$$\begin{aligned} T(n) &= P_3(n) - \sum_{1 \leq j \leq \lfloor n/2 \rfloor} P_2(j) \\ &= \left[\frac{n^2}{12}\right] - \left\lfloor \frac{n}{4} \right\rfloor \left\lfloor \frac{n+2}{4} \right\rfloor \\ &= \begin{cases} \left[\frac{n^2}{48}\right] & \text{for } n \text{ even} \\ \left[\frac{(n+3)^2}{48}\right] & \text{for } n \text{ odd,} \end{cases} \end{aligned} \tag{27}$$

where P_2 and P_3 are PARTITION FUNCTIONS P, $[x]$ is the NINT function, and $\lfloor x \rfloor$ is the FLOOR FUNCTION (Jordan *et al.* 1979, Andrews 1979, Honsberger 1985). The values of $T(n)$ for $n = 1, 2, \ldots$ are 0, 0, 1, 0, 1, 1, 2, 1, 3, 2, 4, 3, 5, 4, 7, 5, 8, 7, 10, 8, 12, 10, 14, 12, 16, ... (Sloane's A005044), which is also ALCUIN'S SEQUENCE padded with two initial 0s. $T(n)$ also satisfies

$$T(2n) = T(2n - 3) = P_3(n). \tag{28}$$

It is not known if a triangle with INTEGER sides, MEDIANS, and AREA exists (although there are incorrect PROOFS of the impossibility in the literature). However, R. L. Rathbun, A. Kemnitz, and R. H. Buchholz have shown that there are infinitely many triangles with RATIONAL sides (HERONIAN TRIANGLES) with *two* RATIONAL MEDIANS (Guy 1994).

In the following paragraph, assume the specified sides and angles are adjacent to each other. Specifying three ANGLES does not uniquely define a triangle, but any two triangles with the same ANGLES are similar (the AAA THEOREM). Specifying two ANGLES A and B and a side a uniquely determines a triangle with AREA

$$\Delta = \frac{a^2 \sin B \sin C}{2 \sin A} = \frac{a^2 \sin B \sin(\pi - A - B)}{2 \sin A} \tag{29}$$

(the AAS THEOREM). Specifying an ANGLE A, a side c, and an ANGLE B uniquely specifies a triangle with AREA

$$\Delta = \frac{c^2}{2(\cot A + \cot B)} \tag{30}$$

(the ASA THEOREM). Given a triangle with two sides, a the smaller and c the larger, and one known ANGLE A, ACUTE and opposite a, if $\sin A < a/c$, there are two possible triangles. If $\sin A = a/c$, there is one possible triangle. If $\sin A > a/c$, there are no possible triangles. This is the ASS THEOREM. Let a be the base length and h be the height. Then

$$\Delta = \tfrac{1}{2}ah = \tfrac{1}{2}ac \sin B \tag{31}$$

(the SAS THEOREM). Finally, if all three sides are specified, a unique triangle is determined with AREA given by HERON'S FORMULA or by

$$\Delta = \frac{abc}{4R}, \tag{32}$$

where R is the CIRCUMRADIUS. This is the SSS THEOREM.

There are four CIRCLES which are tangent to the sides of a triangle, one internal and the rest external. Their centers are the points of intersection of the ANGLE BISECTORS of the triangle.

Any triangle can be positioned such that its shadow under an orthogonal projection is EQUILATERAL.

see also AAA THEOREM, AAS THEOREM, ACUTE TRIANGLE, ALCUIN'S SEQUENCE, ALTITUDE, ANGLE BISECTOR, ANTICEVIAN TRIANGLE, ANTICOMPLEMENTARY TRIANGLE, ANTIPEDAL TRIANGLE, ASS THEOREM, BELL TRIANGLE, BRIANCHON POINT, BROCARD ANGLE, BROCARD CIRCLE, BROCARD MIDPOINT, BROCARD POINTS, BUTTERFLY THEOREM, CENTROID (TRIANGLE), CEVA'S THEOREM, CEVIAN, CEVIAN TRIANGLE, CHASLES'S THEOREM, CIRCUMCENTER, CIRCUMCIRCLE, CIRCUMRADIUS, CONTACT TRIANGLE, CROSSED LADDERS PROBLEM, CRUCIAL POINT, D-TRIANGLE, DE LONGCHAMPS POINT, DESARGUES' THEOREM, DISSECTION, ELKIES POINT, EQUAL DETOUR POINT, EQUILATERAL TRIANGLE, EULER LINE, EULER'S TRIANGLE, EULER TRIANGLE FORMULA, EXCENTER, EXCENTRAL TRIANGLE, EXCIRCLE, EXETER POINT, EXMEDIAN, EXMEDIAN POINT, EXRADIUS, EXTERIOR ANGLE THEOREM, FAGNANO'S PROBLEM, FAR-OUT POINT, FERMAT POINT, FERMAT'S PROBLEM, FEUERBACH POINT, FEUERBACH'S THEOREM, FUHRMANN TRIANGLE, GERGONNE POINT, GREBE POINT, GRIFFITHS POINTS, GRIFFITHS' THEOREM, HARMONIC CONJUGATE POINTS, HEILBRONN TRIANGLE PROBLEM, HERON'S FORMULA, HERONIAN TRIANGLE, HOFSTADTER TRIANGLE, HOMOTHETIC TRIANGLES, INCENTER, INCIRCLE, INRADIUS, ISODYNAMIC POINTS, ISOGONAL CONJUGATE, ISOGONIC CENTERS, ISOPERIMETRIC POINT, ISOSCELES TRIANGLE, KABON TRIANGLES, KANIZSA TRIANGLE, KIEPERT'S HYPERBOLA, KIEPERT'S PARABOLA, LAW OF COSINES, LAW OF SINES, LAW OF TANGENTS, LEIBNIZ HARMONIC TRIANGLE, LEMOINE CIRCLE, LEMOINE POINT, LINE AT INFINITY, MALFATTI POINTS, MEDIAL TRIANGLE, MEDIAN (TRIANGLE), MEDIAN TRIANGLE, MENELAUS' THEOREM, MID-ARC POINTS, MITTENPUNKT, MOLLWEIDE'S FORMULAS, MORLEY CENTERS, MORLEY'S THEOREM, NAGEL POINT, NAPOLEON'S THEOREM, NAPOLEON TRIANGLES, NEWTON'S FORMULAS, NINE-POINT CIRCLE, NUMBER TRIANGLE, OBTUSE TRIANGLE, ORTHIC TRIANGLE, ORTHOCENTER, ORTHOLOGIC, PARALOGIC TRIANGLES, PASCAL'S TRIANGLE, PASCH'S AXIOM, PEDAL TRIANGLE, PERPENDICULAR BISECTOR, PERSPECTIVE TRIANGLES, PETERSEN-SHOUTE THEOREM, PIVOT THEOREM, POWER POINT, POWER (TRIANGLE), PRIME TRIANGLE, PURSER'S THEOREM, QUADRILATERAL, RATIONAL TRIANGLE, ROUTH'S THEOREM, SAS THEOREM, SCALENE TRIANGLE, SCHIFFLER POINT, SCHWARZ TRIANGLE, SCHWARZ'S TRIANGLE PROBLEM, SEIDEL-ENTRINGER-ARNOLD TRIANGLE, SEYDEWITZ'S THEOREM, SIMSON LINE, SPIEKER CENTER, SSS THEOREM, STEINER-LEHMUS THEOREM, STEINER POINTS, STEWART'S THEOREM, SYMMEDIAN POINT, TANGENTIAL TRIANGLE, TANGENTIAL TRIANGLE CIRCUMCENTER, TARRY POINT, THOMSEN'S FIGURE, TORRICELLI POINT, TRIANGLE TILING, TRIANGLE TRANSFORMATION PRINCIPLE, YFF POINTS, YFF TRIANGLES

References

Abi-Khuzam, F. "Proof of Yff's Conjecture on the Brocard Angle of a Triangle." *Elem. Math.* **29**, 141–142, 1974.

Andrews, G. "A Note on Partitions and Triangles with Integer Sides." *Amer. Math. Monthly* **86**, 477, 1979.

Baker, M. "A Collection of Formulæ for the Area of a Plane Triangle." *Ann. Math.* **1**, 134–138, 1884.

Berkhan, G. and Meyer, W. F. "Neuere Dreiecksgeometrie." In *Encyklopaedie der Mathematischen Wissenschaften, Vol. 3AB 10* (Ed. F. Klein). Leipzig: Teubner, pp. 1173–1276, 1914.

Beyer, W. H. (Ed.) *CRC Standard Mathematical Tables, 28th ed.* Boca Raton, FL: CRC Press, pp. 123–124, 1987.

Coxeter, H. S. M. *Introduction to Geometry, 2nd ed.* New York: Wiley, 1969.

Davis, P. "The Rise, Fall, and Possible Transfiguration of Triangle Geometry: A Mini-History." *Amer. Math. Monthly* **102**, 204–214, 1995.

Eppstein, D. "Triangles and Simplices." `http://www.ics.uci.edu/~eppstein/junkyard/triangulation.html`.

Feuerbach, K. W. *Eigenschaften einiger merkwürdingen Punkte des geradlinigen Dreiecks, und mehrerer durch die bestimmten Linien und Figuren.* Nürnberg, Germany, 1822.

Guy, R. K. "Triangles with Integer Sides, Medians, and Area." §D21 in *Unsolved Problems in Number Theory, 2nd ed.* New York: Springer-Verlag, pp. 188–190, 1994.

Honsberger, R. *Mathematical Gems III.* Washington, DC: Math. Assoc. Amer., pp. 39–47, 1985.

Johnson, R. A. *Modern Geometry: An Elementary Treatise on the Geometry of the Triangle and the Circle.* Boston, MA: Houghton Mifflin, 1929.

Jordan, J. H.; Walch, R.; and Wisner, R. J. "Triangles with Integer Sides." *Amer. Math. Monthly* **86**, 686–689, 1979.

Kimberling, C. "Central Points and Central Lines in the Plane of a Triangle." *Math. Mag.* **67**, 163–187, 1994.

Kimberling, C. "Triangle Centers and Central Triangles." *Congr. Numer.* **129**, 1–295, 1998.

Le Lionnais, F. *Les nombres remarquables.* Paris: Hermann, p. 28, 1983.

Schroeder. *Das Dreieck und seine Beruhungskreise.*

Sloane, N. J. A. Sequence A005044/M0146 in "An On-Line Version of the Encyclopedia of Integer Sequences."

Vandeghen, A. "Some Remarks on the Isogonal and Cevian Transforms. Alignments of Remarkable Points of a Triangle." *Amer. Math. Monthly* **72**, 1091–1094, 1965.

Weisstein, E. W. "Plane Geometry." `http://www.astro.virginia.edu/~eww6n/math/notebooks/PlaneGeometry.m`.

Triangle Arcs

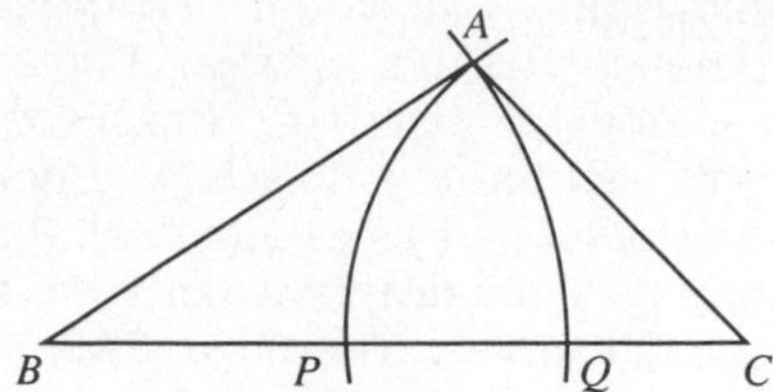

In the above figure, the curves are arcs of a CIRCLE and

$$a = BC \quad (1)$$
$$b = CA = CP \quad (2)$$
$$c = BA = BQ. \quad (3)$$

Then

$$PQ^2 = 2BP \cdot QC. \quad (4)$$

The figure also yields the algebraic identity

$$(b + c - \sqrt{b^2 + c^2})^2 = 2(\sqrt{b^2 + c^2} - b)(\sqrt{b^2 + c^2} - c). \quad (5)$$

see also ARC

References
Berndt, B. C. *Ramanujan's Notebooks, Part IV.* New York: Springer-Verlag, pp. 8–9, 1994.
Dharmarajan, T. and Srinivasan, P. K. *An Introduction to Creativity of Ramanujan, Part III.* Madras: Assoc. Math. Teachers, pp. 11–13, 1987.

Triangle Center

A triangle center is a point whose TRILINEAR COORDINATES are defined in terms of the side lengths and angles of a TRIANGLE. The function giving the coordinates $\alpha : \beta : \gamma$ is called the TRIANGLE CENTER FUNCTION. The four ancient centers are the CENTROID, INCENTER, CIRCUMCENTER, and ORTHOCENTER. For a listing of these and other triangle centers, see Kimberling (1994).

A triangle center is said to be REGULAR IFF there is a TRIANGLE CENTER FUNCTION which is a POLYNOMIAL in Δ, a, b, and c (where Δ is the AREA of the TRIANGLE) such that the TRILINEAR COORDINATES of the center are

$$f(a, b, c) : f(b, c, a) : f(c, a, b).$$

A triangle center is said to be a MAJOR TRIANGLE CENTER if the TRIANGLE CENTER FUNCTION α is a function of ANGLE A alone, and therefore β and γ of B and C alone, respectively.

see also MAJOR TRIANGLE CENTER, REGULAR TRIANGLE CENTER, TRIANGLE, TRIANGLE CENTER FUNCTION, TRILINEAR COORDINATES, TRILINEAR POLAR

References
Davis, P. J. "The Rise, Fall, and Possible Transfiguration of Triangle Geometry: A Mini-History." *Amer. Math. Monthly* **102**, 204–214, 1995.
Dixon, R. "The Eight Centres of a Triangle." §1.5 in *Mathographics.* New York: Dover, pp. 55–61, 1991.
Gale, D. "From Euclid to Descartes to Mathematica to Oblivion?" *Math. Intell.* **14**, 68–69, 1992.
Kimberling, C. "Central Points and Central Lines in the Plane of a Triangle." *Math. Mag.* **67**, 163–167, 1994.
Kimberling, C. "Triangle Centers and Central Triangles." *Congr. Numer.* **129**, 1–295, 1998.

Triangle Center Function

A HOMOGENEOUS FUNCTION $f(a, b, c)$, i.e., a function f such that

$$f(ta, tb, tc) = t^n f(a, b, c),$$

which gives the TRILINEAR COORDINATES of a TRIANGLE CENTER as

$$\alpha : \beta : \gamma = f(a, b, c) : f(b, c, a) : f(c, a, b).$$

The variables may correspond to angles (A, B, C) or side lengths (a, b, c), since these can be interconverted using the LAW OF COSINES.

see also MAJOR TRIANGLE CENTER, REGULAR TRIANGLE CENTER, TRIANGLE CENTER, TRILINEAR COORDINATES

References
Kimberling, C. "Triangle Centers as Functions." *Rocky Mtn. J. Math.* **23**, 1269–1286, 1993.
Kimberling, C. "Triangle Centers." http://www.evansville.edu/~ck6/tcenters/.
Kimberling, C. "Triangle Centers and Central Triangles." *Congr. Numer.* **129**, 1–295, 1998.

Triangle Coefficient

A function of three variables written $\Delta(abc) \equiv \Delta(a, b, c)$ and defined by

$$\Delta(abc) \equiv \sqrt{\frac{(a+b-c)!(a-b+c)!(-a+b+c)!}{(a+b+c+1)!}}.$$

References
Shore, B. W. and Menzel, D. H. *Principles of Atomic Spectra.* New York: Wiley, p. 273, 1968.

Triangle Condition

The condition that j takes on the values

$$j = j_1 + j_2, j_1 + j_2 - 1, \ldots, |j_1 - j_2|,$$

denoted $\Delta(j_1 j_2 j)$.

References
Sobelman, I. I. *Atomic Spectra and Radiative Transitions, 2nd ed.* Berlin: Springer-Verlag, p. 60, 1992.

Triangle Counting

Given rods of length 1, 2, ..., n, how many distinct triangles $T(n)$ can be made? Lengths for which

$$l_i = l_j + l_k$$

obviously do not give triangles, but all other combinations of three rods do. The answer is

$$T(n) = \begin{cases} \frac{1}{24}n(n-2)(2n-5) & \text{for } n \text{ even} \\ \frac{1}{24}(n-1)(n-3)(2n-1) & \text{for } n \text{ odd.} \end{cases}$$

The values for $n = 1, 2, \ldots$ are 0, 0, 0, 1, 3, 7, 13, 22, 34, 50, ... (Sloane's A002623). Somewhat surprisingly, this sequence is also given by the GENERATING FUNCTION

$$f(x) = \frac{x^4}{(1-x)^3(1-x^2)} = x^4 + 3x^5 + 7x^6 + 13x^7 + \ldots.$$

References

Honsberger, R. *More Mathematical Morsels.* Washington, DC: Math. Assoc. Amer., pp. 278–282, 1991.

Sloane, N. J. A. Sequence A002623/M2640 in "An On-Line Version of the Encyclopedia of Integer Sequences."

Triangle of Figurate Numbers

see FIGURATE NUMBER TRIANGLE

Triangle Function

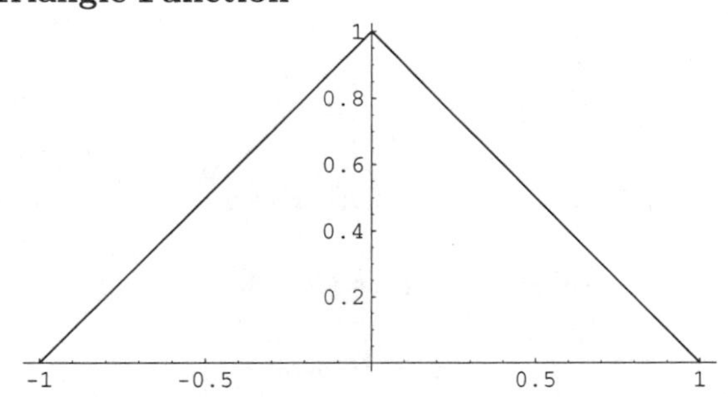

$$\Lambda(x) \equiv \begin{cases} 0 & |x| > 1 \\ 1 - |x| & |x| < 1 \end{cases} \quad (1)$$

$$= \Pi(x) * \Pi(x) \quad (2)$$

$$= \Pi(x) * H(x + \tfrac{1}{2}) - \Pi(x) * H(x - \tfrac{1}{2}), \quad (3)$$

where Π is the RECTANGLE FUNCTION and H is the HEAVISIDE STEP FUNCTION. An obvious generalization used as an APODIZATION FUNCTION goes by the name of the BARTLETT FUNCTION.

There is also a three-argument function known as the triangle function:

$$\lambda(x, y, z) \equiv x^2 + y^2 + z^2 - 2xy - 2xz - 2yz. \quad (4)$$

It follows that

$$\lambda(a^2, b^2, c^2) = (a+b+c)(a+b-c)(a-b+c)(a-b-c). \quad (5)$$

see also ABSOLUTE VALUE, BARTLETT FUNCTION, HEAVISIDE STEP FUNCTION, RAMP FUNCTION, SGN, TRIANGLE COEFFICIENT

Triangle Inequality

Let $\mathbf{x}$ and $\mathbf{y}$ be vectors

$$|\mathbf{x}| - |\mathbf{y}| \leq |\mathbf{x} + \mathbf{y}| \leq |\mathbf{x}| + |\mathbf{y}|. \quad (1)$$

Equivalently, for COMPLEX NUMBERS z_1 and z_2,

$$|z_1| - |z_2| \leq |z_1 + z_2| \leq |z_1| + |z_2|. \quad (2)$$

A generalization is

$$\left| \sum_{k=1}^{n} a_k \right| \leq \sum_{k=1}^{n} |a_k|. \quad (3)$$

see also p-ADIC NUMBER, STRONG TRIANGLE INEQUALITY

References

Abramowitz, M. and Stegun, C. A. (Eds.). *Handbook of Mathematical Functions with Formulas, Graphs, and Mathematical Tables, 9th printing.* New York: Dover, p. 11, 1972.

Triangle Inscribing in a Circle

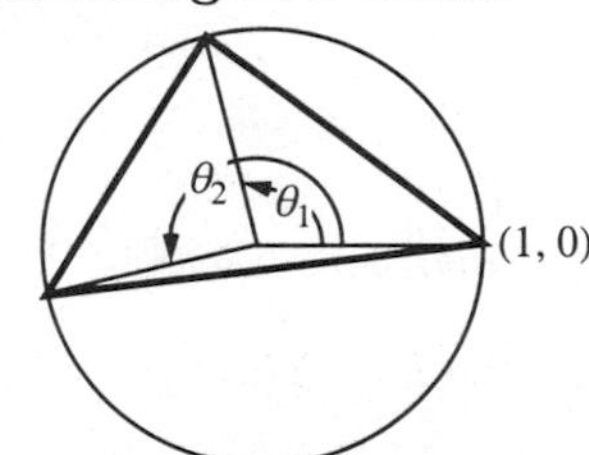

Select three points at random on a unit CIRCLE. Find the distribution of possible areas. The first point can be assigned coordinates (1, 0) without loss of generality. Call the central angles from the first point to the second and third θ_1 and θ_2. The range of θ_1 can be restricted to $[0, \pi]$ because of symmetry, but θ_2 can range from $[0, 2\pi)$. Then

$$A(\theta_1, \theta_2) = 2\left|\sin(\tfrac{1}{2}\theta_1)\sin(\tfrac{1}{2}\theta_2)\sin[\tfrac{1}{2}(\theta_1 - \theta_2)]\right|, \quad (1)$$

so

$$\bar{A} = \frac{\int_0^\pi \int_0^{2\pi} A(\theta_1, \theta_2)\, d\theta_2\, d\theta_1}{C}, \quad (2)$$

where

$$C \equiv \int_0^\pi \int_0^{2\pi} d\theta_2\, d\theta_1 = 2\pi^2. \quad (3)$$

Therefore,

$$\bar{A} = \frac{2}{2\pi^2}\int_0^\pi \int_0^{2\pi} \left|\sin(\tfrac{1}{2}\theta_1)\sin(\tfrac{1}{2}\theta_2)\sin[\tfrac{1}{2}(\theta_1-\theta_2)]\right| d\theta_2\, d\theta_1$$
$$= \frac{1}{\pi^2}\int_0^\pi \sin(\tfrac{1}{2}\theta_1)\left[\int_0^{2\pi} \sin(\tfrac{1}{2}\theta_2)\left|\sin[\tfrac{1}{2}(\theta_2-\theta_1)]\right| d\theta_2\right] d\theta_1$$
$$= \frac{1}{\pi^2}\int_0^\pi \int_{0\ \theta_2-\theta_1>0}^{2\pi} \sin(\tfrac{1}{2}\theta_1)\sin(\tfrac{1}{2}\theta_2)\sin[\tfrac{1}{2}(\theta_1-\theta_2)]\, d\theta_2\, d\theta_1$$
$$+ \frac{1}{\pi^2}\int_0^\pi \int_{0\ \theta_2-\theta_1<0}^{2\pi} \sin(\tfrac{1}{2}\theta_1)\sin(\tfrac{1}{2}\theta_2)\sin[\tfrac{1}{2}(\theta_1-\theta_2)]\, d\theta_2\, d\theta_1$$
$$= \frac{1}{\pi^2}\int_0^\pi \sin(\tfrac{1}{2}\theta_1)\left[\int_{\theta_1}^{2\pi} \sin(\tfrac{1}{2}\theta_2)\sin[\tfrac{1}{2}(\theta_2-\theta_1)]\, d\theta_2\right] d\theta_1$$
$$+ \frac{1}{\pi^2}\int_0^\pi \sin(\tfrac{1}{2}\theta_1)\left[\int_0^{\theta_1} \sin(\tfrac{1}{2}\theta_2)\sin[\tfrac{1}{2}(\theta_2-\theta_1)]\, d\theta_2\right] d\theta_1. \quad (4)$$

But

$$\int (\tfrac{1}{2}\theta_2)\sin[\tfrac{1}{2}(\theta_2-\theta_1)]\, d\theta_2$$
$$= \int \sin(\tfrac{1}{2}\theta_2)\left[\sin(\tfrac{1}{2}\theta_2)\cos(\tfrac{1}{2}\theta_2) - \sin(\tfrac{1}{2}\theta_1)\cos(\tfrac{1}{2}\theta_2)\right] d\theta_2$$
$$= \cos(\tfrac{1}{2}\theta_1)\int \sin^2(\tfrac{1}{2}\theta_2)\, d\theta_2 - \sin(\tfrac{1}{2}\theta_1)\int \sin(\tfrac{1}{2}\theta_1)\cos(\tfrac{1}{2}\theta_2)\, d\theta_2$$
$$= \tfrac{1}{2}\cos(\tfrac{1}{2}\theta_1)\int (1-\cos\theta_2)\, d\theta_2 - \tfrac{1}{2}\sin(\tfrac{1}{2}\theta_2)\int \sin\theta_2\, d\theta_2$$
$$= \tfrac{1}{2}\cos(\tfrac{1}{2}\theta_1)(\theta_2 - \sin\theta_2) + \tfrac{1}{2}\sin(\tfrac{1}{2}\theta_1)\cos(\theta_2). \quad (5)$$

Write (4) as

$$\bar{A} = \frac{1}{\pi^2}\left[\int_0^\pi \sin(\tfrac{1}{2}\theta_1) I_1\, d\theta_1 + \int_0^\pi \sin(\tfrac{1}{2}\theta_1) I_2\, d\theta_1\right], \quad (6)$$

then

$$I_1 \equiv \int_{\theta_1}^{2\pi} \sin(\tfrac{1}{2}\theta_2)\sin[\tfrac{1}{2}(\theta_2-\theta_1)]\, d\theta_2, \quad (7)$$

and

$$I_2 \equiv \int_0^{\theta_1} \sin(\tfrac{1}{2}\theta_2)\sin[\tfrac{1}{2}(\theta_1-\theta_2)]\, d\theta_2. \quad (8)$$

From (6),

$$I_1 = \tfrac{1}{2}\cos(\tfrac{1}{2}\theta_2)[\theta_2 - \sin\theta_2]_{\theta_1}^{2\pi} + \tfrac{1}{2}\sin(\tfrac{1}{2}\theta_1)[\cos\theta_2]_{\theta_1}^{2\pi}$$
$$= \tfrac{1}{2}\cos(\tfrac{1}{2}\theta_1)(2\pi - \theta_1 + \sin\theta_1)$$
$$+ \tfrac{1}{2}\sin(\tfrac{1}{2}\theta_1)(1-\cos\theta_1)$$
$$= \pi\cos(\tfrac{1}{2}\theta_1) - \tfrac{1}{2}\theta_1\cos(\tfrac{1}{2}\theta_1) + \tfrac{1}{2}[\cos(\tfrac{1}{2}\theta_1)\sin\theta_1$$
$$- \cos\theta_1\sin(\tfrac{1}{2}\theta_1)] + \tfrac{1}{2}\sin(\tfrac{1}{2}\theta_1)$$
$$= \pi\cos(\tfrac{1}{2}\theta_1) - \tfrac{1}{2}\theta_1\cos(\tfrac{1}{2}\theta_1) + \tfrac{1}{2} + \tfrac{1}{2}\sin(\theta_1 - \tfrac{1}{2}\theta_1)$$
$$+ \tfrac{1}{2}\sin(\tfrac{1}{2}\theta_1)$$
$$= \pi\cos(\tfrac{1}{2}\theta_1) - \tfrac{1}{2}\theta_1\cos(\tfrac{1}{2}\theta_1) + \sin(\tfrac{1}{2}\theta_1), \quad (9)$$

so

$$\int_0^\pi I_1 \sin(\tfrac{1}{2}\theta_1)\, d\theta_1 = \tfrac{5}{4}\pi. \quad (10)$$

Also,

$$I_2 = \tfrac{1}{2}\cos(\tfrac{1}{2}\theta_1)[\sin\theta_2 - \theta_2]_0^{\theta_1} - \tfrac{1}{2}\sin(\tfrac{1}{2}\theta_1)[\cos\theta_2]_0^{\theta_1}$$
$$= \tfrac{1}{2}\cos(\tfrac{1}{2}\theta_2)(\sin\theta_1 - \theta_1) - \tfrac{1}{2}\sin(\tfrac{1}{2}\theta_1)(\cos\theta_1 - 1)$$
$$= -\tfrac{1}{2}\theta_1\cos(\tfrac{1}{2}\theta_1)$$
$$+ \tfrac{1}{2}[\sin\theta_1\cos(\tfrac{1}{2}\theta_1) - \cos\theta_1\sin(\tfrac{1}{2}\theta_2)]$$
$$+ \tfrac{1}{2}\sin(\tfrac{1}{2}\theta_1)$$
$$= -\tfrac{1}{2}\theta_1\cos(\tfrac{1}{2}\theta_1) + \sin(\tfrac{1}{2}\theta_1), \quad (11)$$

so

$$\int_0^\pi I_2 \sin(\tfrac{1}{2}\theta_1)\, d\theta_1 = \tfrac{1}{4}\pi. \quad (12)$$

Combining (10) and (12) gives

$$\bar{A} = \frac{1}{\pi^2}\left(\frac{5\pi}{4} + \frac{\pi}{4}\right) = \frac{3}{2\pi} \approx 0.4775. \quad (13)$$

The VARIANCE is

$$\sigma_A{}^2 = \frac{1}{2\pi^2}\int_0^\pi \int_0^{2\pi} [A(\theta_1,\theta_2) - \tfrac{3}{2\pi}]^2\, d\theta_2\, d\theta_1$$
$$= \frac{1}{2\pi^2}\int_0^\pi \int_0^{2\pi} \left[2\left|\sin(\tfrac{1}{2}\theta_1)\sin(\tfrac{1}{2}\theta_2)\sin[\tfrac{1}{2}(\theta_1-\theta_2)]\right|\right.$$
$$\left. - \frac{3}{2\pi}\right]^2 d\theta_2\, d\theta_1$$
$$= \frac{1}{2\pi^2}\int_0^\pi \int_0^{2\pi} \left\{4\sin^2(\tfrac{1}{2}\theta_1)\sin^2(\tfrac{1}{2}\theta_2)\sin^2[\tfrac{1}{2}(\theta_2-\theta_1)]\right.$$
$$\left. - \frac{6}{\pi}\left|\sin(\tfrac{1}{2}\theta_1)\sin(\tfrac{1}{2}\theta_2)\sin[\tfrac{1}{2}(\theta_1-\theta_2)]\right| + \frac{9}{4\pi^2}\right\} d\theta_2\, d\theta_1$$
$$= \frac{1}{2\pi^2}\left[\int_0^\pi \pi(2+\theta_1)\sin^2(\tfrac{1}{2}\theta_1)\, d\theta_1\right.$$
$$\left. - \frac{6}{\pi}\left(\frac{5\pi}{4} + \frac{\pi}{4}\right) + \frac{9}{4\pi^2}(2\pi^2)\right]$$
$$= \frac{1}{2\pi^2}\left(\frac{3\pi^2}{4} - 9 + \frac{9}{2}\right) = \frac{1}{2\pi^2}\left(\frac{3\pi^2}{4} - \frac{9}{2}\right)$$
$$= \frac{3(\pi^2-6)}{8\pi^2} \approx 0.1470. \quad (14)$$

see also POINT-POINT DISTANCE—1-D, TETRAHEDRON INSCRIBING

Triangle Inscribing in an Ellipse

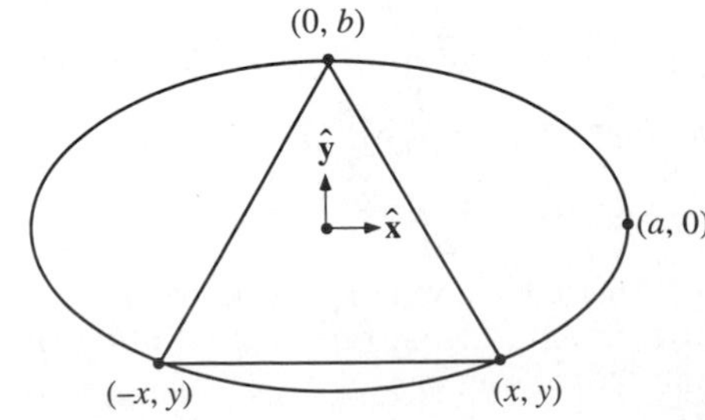

To inscribe an EQUILATERAL TRIANGLE in an ELLIPSE, place the top VERTEX at $(0, b)$, then solve to find the (x, y) coordinate of the other two VERTICES.

$$\sqrt{x^2 + (b-y)^2} = 2x \tag{1}$$

$$x^2 + (b-y)^2 = 4x^2 \tag{2}$$

$$3x^2 = (b-y)^2. \tag{3}$$

Now plugging in the equation of the ELLIPSE

$$\frac{x^2}{a^2} + \frac{y^2}{b^2} = 1, \tag{4}$$

gives

$$3a^2\left(1 - \frac{y^2}{b^2}\right) = b^2 - 2by + y^2 \tag{5}$$

$$y^2\left(1 + 3\frac{a^2}{b^2}\right) - 2by + (b^2 - 3a^2) = 0 \tag{6}$$

$$\begin{aligned} y &= \frac{2b - \sqrt{4b^2 - 4(b^2 - 3a^2)\left(1 + 3\frac{a^2}{b^2}\right)}}{2\left(1 + 3\frac{a^2}{b^2}\right)} \\ &= \frac{1 - \sqrt{1 - \left(1 - 3\frac{a^2}{b^2}\right)\left(1 + 3\frac{a^2}{b^2}\right)}}{1 + 3\frac{a^2}{b^2}} b, \end{aligned} \tag{7}$$

and

$$x = \pm a\sqrt{1 - \frac{y^2}{b^2}}. \tag{8}$$

Triangle Postulate

The sum of the ANGLES of a TRIANGLE is two RIGHT ANGLES. This POSTULATE is equivalent to the PARALLEL AXIOM.

References

Dunham, W. "Hippocrates' Quadrature of the Lune." Ch. 1 in *Journey Through Genius: The Great Theorems of Mathematics.* New York: Wiley, p. 54, 1990.

Triangle Squaring

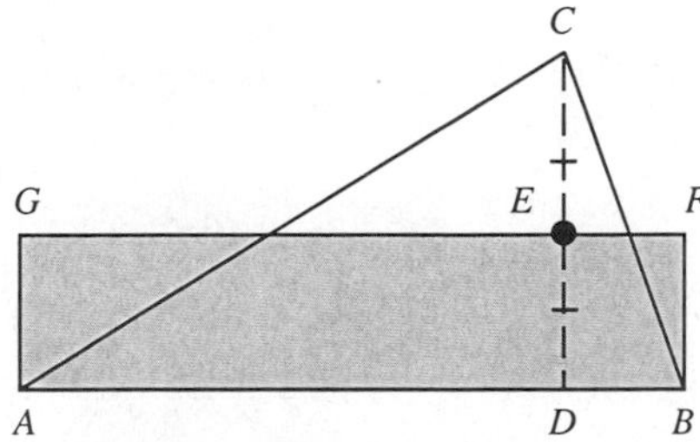

Let CD be the ALTITUDE of a TRIANGLE ΔABC and let E be its MIDPOINT. Then

$$\text{area}(\Delta ABC) = \tfrac{1}{2} AB \cdot CD = AB \cdot DE,$$

and $\square ABFG$ can be squared by RECTANGLE SQUARING. The general POLYGON can be treated by drawing diagonals, squaring the constituent triangles, and then combining the squares together using the PYTHAGOREAN THEOREM.

see also PYTHAGOREAN THEOREM, RECTANGLE SQUARING

References

Dunham, W. "Hippocrates' Quadrature of the Lune." Ch. 1 in *Journey Through Genius: The Great Theorems of Mathematics.* New York: Wiley, pp. 14–15, 1990.

Triangle Tiling

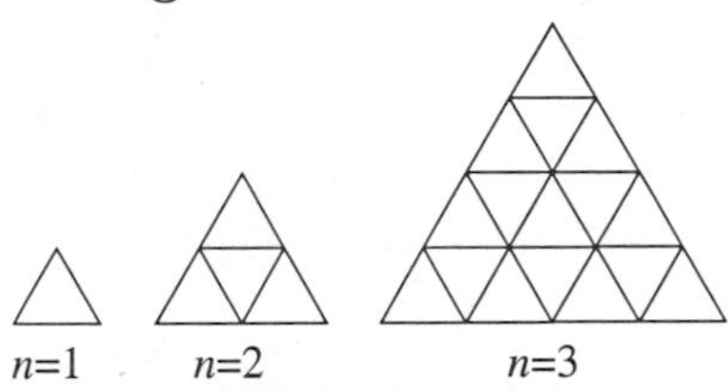

The total number of triangle (including inverted ones) in the above figures is given by

$$N(n) = \begin{cases} \frac{1}{8}n(n+2)(2n+1) & \text{for } n \text{ even} \\ \frac{1}{8}[n(n+2)(2n+1) - 1] & \text{for } n \text{ odd.} \end{cases}$$

The first few values are 1, 5, 13, 27, 48, 78, 118, 170, 235, 315, 411, 525, 658, 812, 988, 1188, 1413, 1665, ... (Sloane's A002717).

References

Conway, J. H. and Guy, R. K. "How Many Triangles." In *The Book of Numbers.* New York: Springer-Verlag, pp. 83–84, 1996.

Sloane, N. J. A. Sequence A002717/M3827 in "An On-Line Version of the Encyclopedia of Integer Sequences."

Triangle Transformation Principle

The triangle transformation principle gives rules for transforming equations involving an INCIRCLE to equations about EXCIRCLES.

see also EXCIRCLE, INCIRCLE

References

Johnson, R. A. *Modern Geometry: An Elementary Treatise on the Geometry of the Triangle and the Circle.* Boston, MA: Houghton Mifflin, pp. 191–192, 1929.

Triangular Cupola

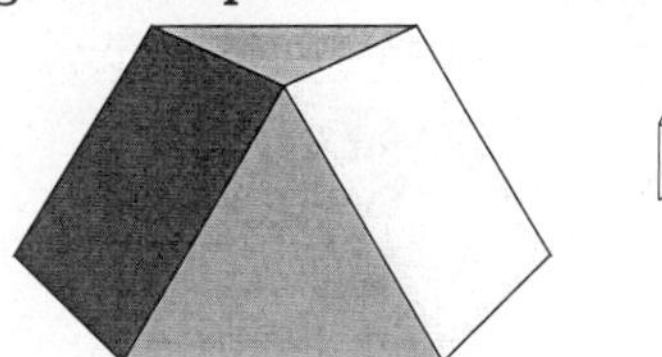

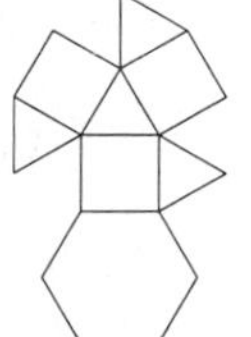

JOHNSON SOLID J_3. The bottom six VERTICES are

$$(\pm\tfrac{1}{2}\sqrt{3}, \pm\tfrac{1}{2}, 0), (0, \pm 1, 0),$$

and the top three VERTICES are

$$\left(\frac{1}{\sqrt{3}}, 0, \sqrt{\frac{2}{3}}\right), -\left(\frac{1}{2\sqrt{3}}, \pm\frac{1}{2}, \sqrt{\frac{2}{3}}\right).$$

see also JOHNSON SOLID

Triangular Dipyramid

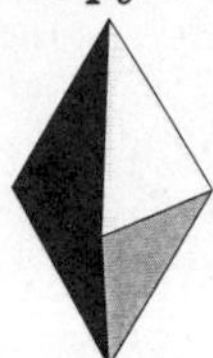
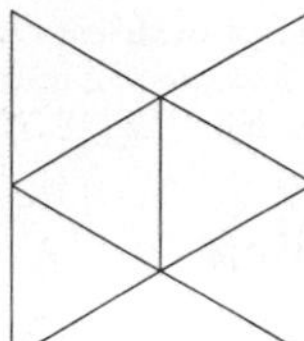

The triangular (or TRIGONAL) dipyramid is one of the convex DELTAHEDRA, and JOHNSON SOLID J_{12}.

see also DELTAHEDRON, DIPYRAMID, JOHNSON SOLID, PENTAGONAL DIPYRAMID

Triangular Graph

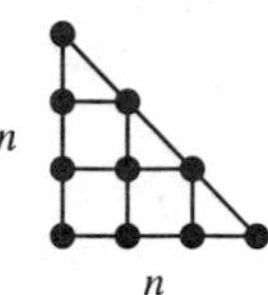

The triangular graph with n nodes on a side is denoted $T(n)$. Tutte (1970) showed that the CHROMATIC POLYNOMIALS of planar triangular graphs possess a ROOT close to $\phi^2 = 2.618033\ldots$, where ϕ is the GOLDEN MEAN. More precisely, if n is the number of VERTICES of G, then

$$P_G(\phi^2) \leq \phi^{5-n}$$

(Le Lionnais 1983, p. 46). Every planar triangular graph possesses a VERTEX of degree 3, 4, or 5 (Le Lionnais 1983, pp. 49 and 53).

see also LATTICE GRAPH

References

Le Lionnais, F. *Les nombres remarquables.* Paris: Hermann, 1983.

Tutte, W. T. "On Chromatic Polynomials and the Golden Ratio." *J. Combin. Theory* **9**, 289–296, 1970.

Triangular Hebesphenorotunda

see JOHNSON SOLID

Triangular Matrix

An upper triangular MATRIX U is defined by

$$U_{ij} = \begin{cases} a_{ij} & \text{for } i \leq j \\ 0 & \text{for } i > j. \end{cases} \quad (1)$$

Written explicitly,

$$\mathsf{U} = \begin{bmatrix} a_{11} & a_{12} & \cdots & a_{1n} \\ 0 & a_{22} & \cdots & a_{2n} \\ \vdots & \vdots & \ddots & \vdots \\ 0 & 0 & \cdots & a_{nn} \end{bmatrix}. \quad (2)$$

A lower triangular MATRIX L is defined by

$$L_{ij} = \begin{cases} a_{ij} & \text{for } i \geq j \\ 0 & \text{for } i < j. \end{cases} \quad (3)$$

Written explicitly,

$$\mathsf{L} = \begin{bmatrix} a_{11} & 0 & \cdots & 0 \\ a_{21} & a_{22} & \cdots & 0 \\ \vdots & \vdots & \ddots & 0 \\ a_{n1} & a_{n2} & \cdots & a_{nn} \end{bmatrix}. \quad (4)$$

see also HESSENBERG MATRIX, HILBERT MATRIX, MATRIX, VANDERMONDE MATRIX

Triangular Number

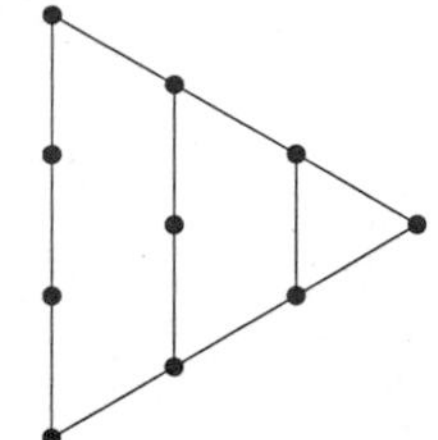

A FIGURATE NUMBER of the form $T_n \equiv n(n+1)/2$ obtained by building up regular triangles out of dots. The first few triangle numbers are 1, 3, 6, 10, 15, 21, ... (Sloane's A000217). $T_4 = 10$ gives the number and arrangement of BOWLING pins, while $T_5 = 15$ gives the number and arrangement of balls in BILLIARDS. Triangular numbers satisfy the RECURRENCE RELATION

$$T_{n+1}{}^2 - T_n{}^2 = (n+1)^3, \quad (1)$$

as well as

$$3T_n + T_{n-1} = T_{2n} \quad (2)$$

$$3T_n + T_{n+1} = T_{2n+1} \quad (3)$$

$$1 + 3 + 5 + \ldots + (2n-1) = T_n + T_{n-1} \quad (4)$$

and

$$(2n+1)^2 = 8T + 1 = T_{n-1} + 6T_n + T_{n+1} \quad (5)$$

(Conway and Guy 1996). They have the simple GENERATING FUNCTION

$$f(x) = \frac{x}{(1-x)^3} = x + 3x^2 + 6x^3 + 10x^4 + 15x^5 + \ldots. \quad (6)$$

Every triangular number is also a HEXAGONAL NUMBER, since

$$\tfrac{1}{2}r(r+1) = \begin{cases} \left(\frac{r+1}{2}\right)\left[2\left(\frac{r+1}{2}\right) - 1\right] & \text{for } r \text{ odd} \\ \left(-\frac{r}{2}\right)\left[2\left(-\frac{r}{2}\right) - 1\right] & \text{for } r \text{ even.} \end{cases} \quad (7)$$

Also, every PENTAGONAL NUMBER is 1/3 of a triangular number. The sum of consecutive triangular numbers is a SQUARE NUMBER, since

$$\begin{aligned} T_r + T_{r-1} &= \tfrac{1}{2}r(r+1) + \tfrac{1}{2}(r-1)r \\ &= \tfrac{1}{2}r[(r+1)+(r-1)] = r^2. \end{aligned} \quad (8)$$

Interesting identities involving triangular numbers and SQUARE NUMBERS are

$$\sum_{k=1}^{2n-1} (-1)^{k+1} T_k = n^2 \quad (9)$$

$$T_n{}^2 = \sum_{k=1}^{n} k^3 = \tfrac{1}{4}n^2(n+1)^2 \quad (10)$$

$$\sum_{k=1,3,\ldots,q} k^3 = T_n \quad (11)$$

for q ODD and

$$n = \tfrac{1}{2}(q^2 + 2q - 1). \quad (12)$$

All EVEN PERFECT NUMBERS are triangular T_p with PRIME p. Furthermore, every EVEN PERFECT NUMBER $P > 6$ is of the form

$$P = 1 + 9T_n = T_{3n+1}, \quad (13)$$

where T_n is a triangular number with $n = 8j+2$ (Eaton 1995, 1996). Therefore, the nested expression

$$9(9\cdots(9(9(9(9T_n+1)+1)+1)+1)\ldots+1)+1 \quad (14)$$

generates triangular numbers for any T_n. An INTEGER k is a triangular number IFF $8k+1$ is a SQUARE NUMBER > 1.

The numbers 1, 36, 1225, 41616, 1413721, 48024900, ... (Sloane's A001110) are SQUARE TRIANGULAR NUMBERS, i.e., numbers which are simultaneously triangular and SQUARE (Pietenpol 1962). Numbers which are simultaneously triangular and TETRAHEDRAL satisfy the BINOMIAL COEFFICIENT equation

$$\binom{n}{2} = \binom{m}{3}, \quad (15)$$

the only solutions of which are $(m,n) = (10,16)$, (22, 56), and (36, 120) (Guy 1994, p. 147).

The smallest of two INTEGERS for which $n^3 - 13$ is four times a triangular number is 5 (Cesaro 1886; Le Lionnais 1983, p. 56). The only FIBONACCI NUMBERS which are triangular are 1, 3, 21, and 55 (Ming 1989), and the only PELL NUMBER which is triangular is 1 (McDaniel 1996). The BEAST NUMBER 666 is triangular, since

$$T_{6\cdot 6} = T_{36} = 666. \quad (16)$$

In fact, it is the largest REPDIGIT triangular number (Bellew and Weger 1975–76).

FERMAT'S POLYGONAL NUMBER THEOREM states that every POSITIVE INTEGER is a sum of *most* three TRIANGULAR NUMBERS, four SQUARE NUMBERS, five PENTAGONAL NUMBERS, and n n-POLYGONAL NUMBERS. Gauss proved the triangular case, and noted the event in his diary on July 10, 1796, with the notation

$$**\,E\Upsilon RHKA \qquad num = \Delta + \Delta + \Delta. \quad (17)$$

This case is equivalent to the statement that every number of the form $8m+3$ is a sum of three ODD SQUARES (Duke 1997). Dirichlet derived the number of ways in which an INTEGER m can be expressed as the sum of three triangular numbers (Duke 1997). The result is particularly simple for a PRIME of the form $8m+3$, in which case it is the number of squares mod $8m+3$ minus the number of nonsquares mod $8m+3$ in the INTERVAL $4m+1$ (Deligne 1973).

The only triangular numbers which are the PRODUCT of three consecutive INTEGERS are 6, 120, 210, 990, 185136, 258474216 (Guy 1994, p. 148).

see also FIGURATE NUMBER, PRONIC NUMBER, SQUARE TRIANGULAR NUMBER

References

Ball, W. W. R. and Coxeter, H. S. M. *Mathematical Recreations and Essays, 13th ed.* New York: Dover, p. 59, 1987.

Bellew, D. W. and Weger, R. C. "Repdigit Triangular Numbers." *J. Recr. Math.* **8**, 96–97, 1975–76.

Conway, J. H. and Guy, R. K. *The Book of Numbers.* New York: Springer-Verlag, pp. 33–38, 1996.

Deligne, P. "La Conjecture de Weil." *Inst. Hautes Études Sci. Pub. Math.* **43**, 273–308, 1973.

Dudeney, H. E. *Amusements in Mathematics.* New York: Dover, pp. 67 and 167, 1970.

Duke, W. "Some Old Problems and New Results about Quadratic Forms." *Not. Amer. Math. Soc.* **44**, 190–196, 1997.

Eaton, C. F. "Problem 1482." *Math. Mag.* **68**, 307, 1995.

Eaton, C. F. "Perfect Number in Terms of Triangular Numbers." Solution to Problem 1482. *Math. Mag.* **69**, 308–309, 1996.

Guy, R. K. "Sums of Squares" and "Figurate Numbers." §C20 and §D3 in *Unsolved Problems in Number Theory, 2nd ed.* New York: Springer-Verlag, pp. 136–138 and 147–150, 1994.

Hindin, H. "Stars, Hexes, Triangular Numbers and Pythagorean Triples." *J. Recr. Math.* **16**, 191–193, 1983–1984.

Le Lionnais, F. *Les nombres remarquables.* Paris: Hermann, p. 56, 1983.

McDaniel, W. L. "Triangular Numbers in the Pell Sequence." *Fib. Quart.* **34**, 105–107, 1996.

Ming, L. "On Triangular Fibonacci Numbers." *Fib. Quart.* **27**, 98–108, 1989.

Pappas, T. "Triangular, Square & Pentagonal Numbers." *The Joy of Mathematics.* San Carlos, CA: Wide World Publ./Tetra, p. 214, 1989.

Pietenpol, J. L "Square Triangular Numbers." *Amer. Math. Monthly* **109**, 168–169, 1962.
Satyanarayana, U. V. "On the Representation of Numbers as the Sum of Triangular Numbers." *Math. Gaz.* **45**, 40–43, 1961.
Sloane, N. J. A. Sequences A000217/M2535 and A001110/M5259 in "An On-Line Version of the Encyclopedia of Integer Sequences."

Triangular Orthobicupola

see JOHNSON SOLID

Triangular Pyramid

see TETRAHEDRON

Triangular Square Number

see SQUARE TRIANGULAR NUMBER

Triangular Symmetry Group

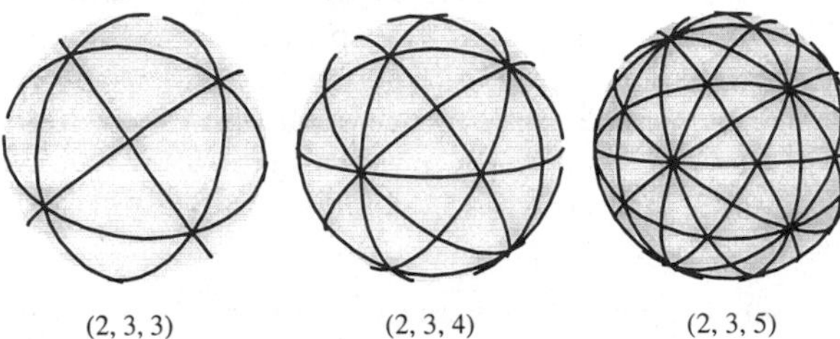

(2, 3, 3) (2, 3, 4) (2, 3, 5)

Given a TRIANGLE with angles $(\pi/p,\ \pi/q,\ \pi/r)$, the resulting symmetry GROUP is called a (p,q,r) triangle group (also known as a SPHERICAL TESSELLATION). In 3-D, such GROUPS must satisfy

$$\frac{1}{p}+\frac{1}{q}+\frac{1}{r}>1,$$

and so the only solutions are $(2,2,n)$, $(2,3,3)$, $(2,3,4)$, and $(2,3,5)$ (Ball and Coxeter 1987). The group $(2,3,6)$ gives rise to the semiregular planar TESSELLATIONS of types 1, 2, 5, and 7. The group $(2,3,7)$ gives hyperbolic tessellations.

see also GEODESIC DOME

References

Ball, W. W. R. and Coxeter, H. S. M. *Mathematical Recreations and Essays, 13th ed.* New York: Dover, pp. 155–161, 1987.
Coxeter, H. S. M. "The Partition of a Sphere According to the Icosahedral Group." *Scripta Math* **4**, 156–157, 1936.
Coxeter, H. S. M. *Regular Polytopes, 3rd ed.* New York: Dover, 1973.
Kraitchik, M. "A Mosaic on the Sphere." §7.3 in *Mathematical Recreations.* New York: W. W. Norton, pp. 208–209, 1942.

Triangulation

Triangulation is the division of a surface into a set of TRIANGLES, usually with the restriction that each TRIANGLE side is entirely shared by two adjacent TRIANGLES. It was proved in 1930 that every surface has a triangulation, but it might require an infinite number of TRIANGLES. A surface with a finite number of triangles in its triangulation is called COMPACT. B. Chazelle showed that an arbitrary SIMPLE POLYGON can be triangulated in linear time.

see also COMPACT SURFACE, DELAUNAY TRIANGULATION, JAPANESE TRIANGULATION THEOREM, SIMPLE POLYGON

Triaugmented Dodecahedron

see JOHNSON SOLID

Triaugmented Hexagonal Prism

see JOHNSON SOLID

Triaugmented Triangular Prism

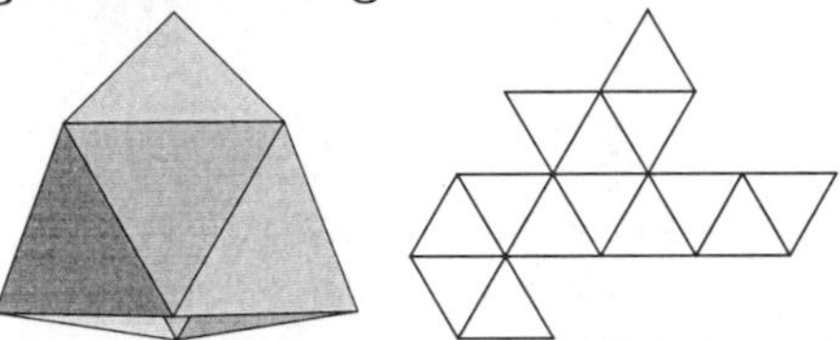

One of the convex DELTAHEDRA and JOHNSON SOLID J_{51}. The VERTICES are $(\pm1,\pm1,0)$, $(0,0,\sqrt{2})$, $(0,\pm1,-\sqrt{3})$, $(\pm(1+\sqrt{6})/2,0,-(\sqrt{2}+\sqrt{3})/2)$, where the x and z coordinates of the last are found by solving

$$x^2+1^2+(z+\sqrt{3})^2=2^2$$
$$(x-1)^2+1^2+z^2=2^2.$$

see also DELTAHEDRON, JOHNSON SOLID

Triaugmented Truncated Dodecahedron

see JOHNSON SOLID

Triaxial Ellipsoid

see ELLIPSOID

Tribar

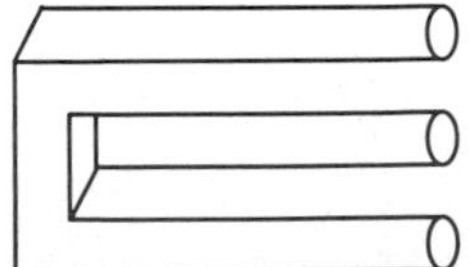

An IMPOSSIBLE FIGURE published by R. Penrose (1958). It also exists as a TRIBOX.

References

Draper, S. W. "The Penrose Triangle and a Family of Related Figures." *Perception* **7**, 283–296, 1978.
Fineman, M. *The Nature of Visual Illusion.* New York: Dover, p. 119, 1996.

Jablan, S. "Set of Modular Elements 'Space Tiles'." http://members.tripod.com/~modularity/space.htm.
Pappas, T. "The Impossible Tribar." *The Joy of Mathematics.* San Carlos, CA: Wide World Publ./Tetra, p. 13, 1989.
Penrose, R. "Impossible Objects: A Special Type of Visual Illusion." *Brit. J. Psychology* **49**, 31–33, 1958.

Tribox

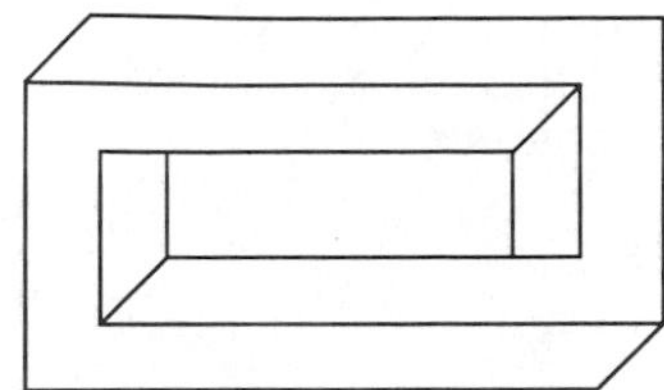

An IMPOSSIBLE FIGURE.

see also IMPOSSIBLE FIGURE, TRIBAR

References

Jablan, S. "Are Impossible Figures Possible?" http://members.tripod.com/~modularity/kulpa.htm.

Tribonacci Number

The tribonacci numbers are a generalization of the FIBONACCI NUMBERS defined by $T_1 = 1$, $T_2 = 1$, $T_3 = 2$, and the RECURRENCE RELATION

$$T_n = T_{n-1} + T_{n-2} + T_{n-3} \tag{1}$$

for $n \geq 4$. The represent the $n = 3$ case of the FIBONACCI n-STEP NUMBERS. The first few terms are 1, 1, 2, 4, 7, 13, 24, 44, 81, 149, ... (Sloane's A000073). The ratio of adjacent terms tends to 1.83929, which is the REAL ROOT of $x^4 - 2x^3 + 1 = 0$. The Tribonacci numbers can also be computed using the GENERATING FUNCTION

$$\frac{1}{1-z-z^2-z^3} = 1 + z + 2z^2 + 4z^3 + 7z^4 + 13z^5 + 24z^6 + 44z^7 + 81z^8 + 149z^9 + \ldots. \tag{2}$$

An explicit FORMULA for T_n is also given by

$$\left[3\frac{\{\frac{1}{3}(19+3\sqrt{33})^{1/3}+\frac{1}{3}(19-3\sqrt{33})^{1/3}+\frac{1}{3}\}^n(586+102\sqrt{33})^{1/3}}{(586+102\sqrt{33})^{2/3}+4-2(586+102\sqrt{33})^{1/3}}\right], \tag{3}$$

where $[x]$ denotes the NINT function (Plouffe). The first part of a NUMERATOR is related to the REAL root of $x^3 - x^2 - x - 1$, but determination of the DENOMINATOR requires an application of the LLL ALGORITHM. The numbers increase asymptotically to

$$T_n \sim c^n, \tag{4}$$

where

$$c = \left(\tfrac{19}{27} + \tfrac{1}{9}\sqrt{33}\right)^{1/3} + \tfrac{4}{9}\left(\tfrac{19}{27} + \tfrac{1}{9}\sqrt{33}\right)^{-1/3} + \tfrac{1}{3} = 1.83928675521\ldots \tag{5}$$

(Plouffe).

see also FIBONACCI n-STEP NUMBER, FIBONACCI NUMBER, TETRANACCI NUMBER

References

Plouffe, S. "Tribonacci Constant." http://lacim.uqam.ca/piDATA/tribo.txt.
Sloane, N. J. A. Sequence A000073/M1074 in "An On-Line Version of the Encyclopedia of Integer Sequences."

Trichotomy Law

Every REAL NUMBER is NEGATIVE, 0, or POSITIVE.

Tricolorable

A projection of a LINK is tricolorable if each of the strands in the projection can be colored in one of three different colors such that, at each crossing, all three colors come together or only one does and at least two different colors are used. The TREFOIL KNOT and trivial 2-link are tricolorable, but the UNKNOT, WHITEHEAD LINK, and FIGURE-OF-EIGHT KNOT are not.

If the projection of a knot is tricolorable, then REIDEMEISTER MOVES on the knot preserve tricolorability, so either every projection of a knot is tricolorable or none is.

Tricomi Function

see CONFLUENT HYPERGEOMETRIC FUNCTION OF THE SECOND KIND, GORDON FUNCTION

Tricuspoid

see DELTOID

Tricylinder

see STEINMETZ SOLID

Tridecagon

A 13-sided POLYGON, sometimes also called the TRISKAIDECAGON.

Trident

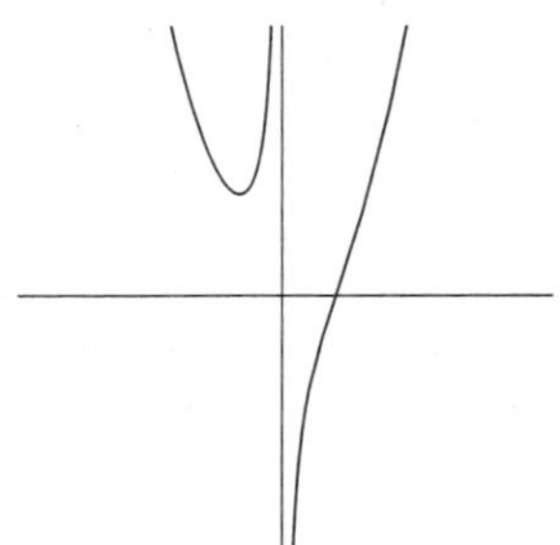

The plane curve given by the equation

$$xy = x^3 - a^3.$$

see also TRIDENT OF DESCARTES, TRIDENT OF NEWTON

Trident of Descartes

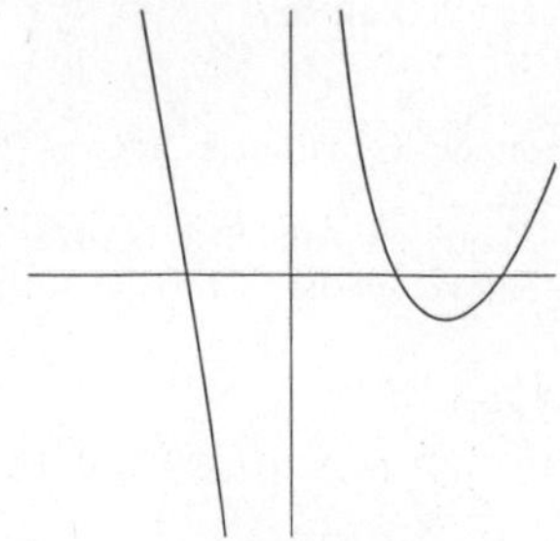

The plane curve given by the equation

$$(a+x)(a-x)(2a-x) = x^3 - 2ax^2 - a^2x + 2a^3 = axy$$

$$y = \frac{(a+x)(a-x)(2a-x)}{ax}$$

The above plot has $a = 2$.

Trident of Newton

The CUBIC CURVE defined by

$$ax^3 + bx^2 + cx + d = xy$$

with $a \neq 0$. The curve cuts the axis in either one or three points. It was the 66th curve in Newton's classification of CUBICS. Newton stated that the curve has four infinite legs and that the y-axis is an ASYMPTOTE to two tending toward contrary parts.

References

Lawrence, J. D. *A Catalog of Special Plane Curves.* New York: Dover, pp. 109–110, 1972.

MacTutor History of Mathematics Archive. "Trident of Newton." `http://www-groups.dcs.st-and.ac.uk/~history/Curves/Trident.html`.

Tridiagonal Matrix

A MATRIX with NONZERO elements only on the diagonal and slots horizontally or vertically adjacent the diagonal. A general 4×4 tridiagonal MATRIX has the form

$$\begin{bmatrix} a_{11} & a_{12} & 0 & 0 \\ a_{21} & a_{22} & a_{23} & 0 \\ 0 & a_{32} & a_{33} & a_{34} \\ 0 & 0 & a_{43} & a_{44} \end{bmatrix}.$$

Inversion of such a matrix requires only n (as opposed to n^3) arithmetic operations (Acton 1990).

see also DIAGONAL MATRIX, JACOBI ALGORITHM

References

Acton, F. S. *Numerical Methods That Work, 2nd printing.* Washington, DC: Math. Assoc. Amer., p. 103, 1990.

Press, W. H.; Flannery, B. P.; Teukolsky, S. A.; and Vetterling, W. T. "Tridiagonal and Band Diagonal Systems of Equations." §2.4 in *Numerical Recipes in FORTRAN: The Art of Scientific Computing, 2nd ed.* Cambridge, England: Cambridge University Press, pp. 42–47, 1992.

Tridiminished Icosahedron

see JOHNSON SOLID

Tridiminished Rhombicosidodecahedron

see JOHNSON SOLID

Tridyakis Icosahedron

The DUAL POLYHEDRON of the ICOSITRUNCATED DODECADODECAHEDRON.

Trifolium

Lawrence (1972) defines a trifolium as a FOLIUM with $b \in (0, 4a)$. However, the term "the" trifolium is sometimes applied to the FOLIUM with $b = a$, which is then the 3-petalled ROSE with Cartesian equation

$$(x^2 + y^2)[y^2 + x(x+a)] = 4axy^2$$

and polar equation

$$r = a\cos\theta(4\sin^2\theta - 1) = -a\cos(3\theta).$$

The trifolium with $b = a$ is the RADIAL CURVE of the DELTOID.

see also BIFOLIUM, FOLIUM, QUADRIFOLIUM

References

Lawrence, J. D. *A Catalog of Special Plane Curves.* New York: Dover, pp. 152–153, 1972.

MacTutor History of Mathematics Archive. "Trifolium." `http://www-groups.dcs.st-and.ac.uk/~history/Curves/Trifolium.html`.

Trigon

see TRIANGLE

Trigonal Dipyramid

see TRIANGULAR DIPYRAMID

Trigonal Dodecahedron

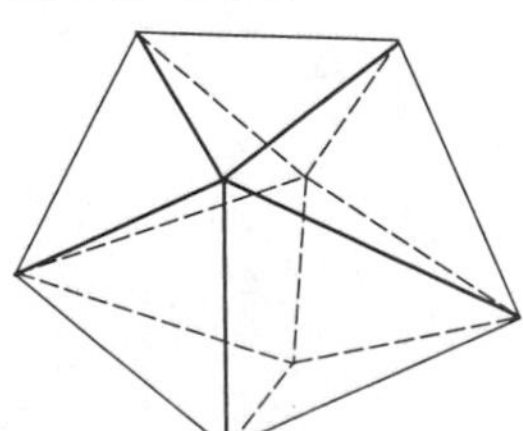

An irregular DODECAHEDRON.

see also DODECAHEDRON, PYRITOHEDRON, RHOMBIC DODECAHEDRON

References
Cotton, F. A. *Chemical Applications of Group Theory, 3rd ed.* New York: Wiley, p. 62, 1990.

Trigonometric Functions

see TRIGONOMETRY

Trigonometric Series

$$\begin{aligned} &A\sin(2\phi)+B\sin(4\phi)+C\sin(6\phi)+D\sin(8\phi)\\ &=\sin(2\phi)(A'+\cos(2\phi)(B'+\cos(2\phi)(C'+D'\cos(2\phi)))), \end{aligned}$$

where

$$\begin{aligned} A' &\equiv A-C\\ B' &\equiv 2B-4D\\ C' &\equiv 4C\\ D' &\equiv 8D. \end{aligned}$$

$$\begin{aligned} &A\sin\phi+B\sin(3\phi)+C\sin(5\phi)+D\sin(7\phi)\\ &\quad=\sin\phi(A'+\sin^2\phi(B'+\sin^2\phi(C'+D'\sin^2\phi))), \end{aligned}$$

where

$$\begin{aligned} A' &\equiv A+3B+5C+7D\\ B' &\equiv -4B-20C-56D\\ C' &\equiv 16C+112D\\ D' &\equiv -64D. \end{aligned}$$

$$\begin{aligned} &A+B\cos(2\phi)+C\cos(4\phi)+D\cos(6\phi)+E\cos(8\phi)\\ &\quad=A'+\cos(2\phi)(B'+\cos(2\phi)(C'+\cos(2\phi)\\ &\qquad\times(D'+E'\cos(2\phi)))), \end{aligned}$$

where

$$\begin{aligned} A' &\equiv A-C+E\\ B' &\equiv B-3D\\ C' &\equiv 2C-8E\\ D' &\equiv 4D\\ E' &\equiv 8E. \end{aligned}$$

References
Snyder, J. P. *Map Projections—A Working Manual.* U. S. Geological Survey Professional Paper 1395. Washington, DC: U. S. Government Printing Office, p. 19, 1987.

Trigonometric Substitution

INTEGRALS of the form

$$\int f(\cos\theta,\sin\theta)\,d\theta \tag{1}$$

can be solved by making the substitution $z=e^{i\theta}$ so that $dz=ie^{i\theta}\,d\theta$ and expressing

$$\cos\theta=\frac{e^{i\theta}+e^{-i\theta}}{2}=\frac{z+z^{-1}}{2} \tag{2}$$

$$\sin\theta=\frac{e^{i\theta}-e^{-i\theta}}{2i}=\frac{z-z^{-1}}{2i}. \tag{3}$$

The integral can then be solved by CONTOUR INTEGRATION.

Alternatively, making the substitution $t\equiv\tan(\theta/2)$ transforms (1) into

$$\int f\left(\frac{2t}{1+t^2},\frac{1-t^2}{1+t^2}\right)\frac{2\,dt}{1+t^2}. \tag{4}$$

The following table gives trigonometric substitutions which can be used to transform integrals involving square roots.

Form	Substitution
$\sqrt{a^2-x^2}$	$x=a\sin\theta$
$\sqrt{a^2+x^2}$	$x=a\tan\theta$
$\sqrt{x^2-a^2}$	$x=a\sec\theta$

see also HYPERBOLIC SUBSTITUTION

Trigonometry

The study of ANGLES and of the angular relationships of planar and 3-D figures is known as trigonometry. The trigonometric functions (also called the CIRCULAR FUNCTIONS) comprising trigonometry are the COSECANT $\csc x$, COSINE $\cos x$, COTANGENT $\cot x$, SECANT $\sec x$, SINE $\sin x$, and TANGENT $\tan x$. The inverses of these functions are denoted $\csc^{-1}x$, $\cos^{-1}x$, $\cot^{-1}x$, $\sec^{-1}x$, $\sin^{-1}x$, and $\tan^{-1}x$. Note that the f^{-1} NOTATION here means INVERSE FUNCTION, *not* f to the -1 POWER.

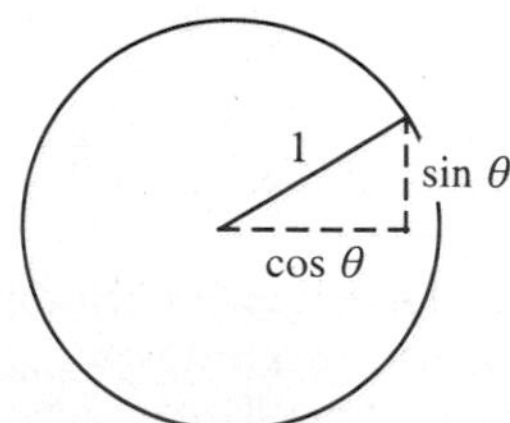

The trigonometric functions are most simply defined using the UNIT CIRCLE. Let θ be an ANGLE measured counterclockwise from the x-AXIS along an ARC of the CIRCLE. Then $\cos\theta$ is the horizontal coordinate of the ARC endpoint, and $\sin\theta$ is the vertical component. The RATIO $\sin\theta/\cos\theta$ is defined as $\tan\theta$. As a result of this definition, the trigonometric functions are periodic with period 2π, so

$$\text{func}(2\pi n + \theta) = \text{func}(\theta), \quad (1)$$

where n is an INTEGER and func is a trigonometric function.

From the PYTHAGOREAN THEOREM,

$$\sin^2\theta + \cos^2\theta = 1. \quad (2)$$

Therefore, it is also true that

$$\tan^2\theta + 1 = \sec^2\theta \quad (3)$$

$$1 + \cot^2\theta = \csc^2\theta. \quad (4)$$

The trigonometric functions can be defined algebraically in terms of COMPLEX EXPONENTIALS (i.e., using the EULER FORMULA) as

$$\sin z \equiv \frac{e^{iz} - e^{-iz}}{2i} \quad (5)$$

$$\csc z \equiv \frac{1}{\sin z} = \frac{2i}{e^{iz} - e^{-iz}} \quad (6)$$

$$\cos z \equiv \frac{e^{iz} + e^{-iz}}{2} \quad (7)$$

$$\sec z \equiv \frac{1}{\cos z} = \frac{2}{e^{iz} + e^{-iz}} \quad (8)$$

$$\tan z \equiv \frac{\sin z}{\cos z} = \frac{e^{iz} - e^{-iz}}{i(e^{iz} + e^{-iz})} \quad (9)$$

$$\cot z \equiv \frac{1}{\tan z} = \frac{i(e^{iz} + e^{-iz})}{e^{iz} - e^{-iz}} = \frac{i(1 + e^{-2iz})}{1 - e^{-2iz}}. \quad (10)$$

OSBORNE'S RULE gives a prescription for converting trigonometric identities to analogous identities for HYPERBOLIC FUNCTIONS.

The ANGLES $n\pi/m$ (with m, n integers) for which the trigonometric function may be expressed in terms of finite root extraction of *real numbers* are limited to values of m which are precisely those which produce constructible POLYGONS. Gauss showed these to be of the form

$$m = 2^k p_1 p_2 \cdots p_s, \quad (11)$$

where k is an INTEGER ≥ 0 and the p_i are distinct FERMAT PRIMES. The first few values are m = 1, 2, 3, 4, 5, 6, 8, 10, 12, 15, 16, 17, 20, ... (Sloane's A003401). Although formulas for trigonometric functions may be found analytically for other m as well, the expressions involve ROOTS of COMPLEX NUMBERS obtained by solving a CUBIC, QUARTIC, or higher order equation. The cases $m = 7$ and $m = 9$ involve the CUBIC EQUATION and QUARTIC EQUATION, respectively. A partial table of the analytic values of SINE, COSINE, and TANGENT for arguments π/m is given below. Derivations of these formulas appear in the following entries.

°	rad	$\sin x$	$\cos x$	$\tan x$
0.0	0	0	1	0
15.0	$\frac{1}{12}\pi$	$\frac{1}{4}(\sqrt{6}-\sqrt{2})$	$\frac{1}{4}(\sqrt{6}+\sqrt{2})$	$2-\sqrt{3}$
18.0	$\frac{1}{10}\pi$	$\frac{1}{4}(\sqrt{5}-1)$	$\frac{1}{4}\sqrt{10+2\sqrt{5}}$	$\frac{1}{5}\sqrt{25-10\sqrt{5}}$
22.5	$\frac{1}{8}\pi$	$\frac{1}{2}\sqrt{2-\sqrt{2}}$	$\frac{1}{2}\sqrt{2+\sqrt{2}}$	$\sqrt{2}-1$
30.0	$\frac{1}{6}\pi$	$\frac{1}{2}$	$\frac{1}{2}\sqrt{3}$	$\frac{1}{3}\sqrt{3}$
36.0	$\frac{1}{5}\pi$	$\frac{1}{4}\sqrt{10-2\sqrt{5}}$	$\frac{1}{4}(1+\sqrt{5})$	$\sqrt{5-2\sqrt{5}}$
45.0	$\frac{1}{4}\pi$	$\frac{1}{2}\sqrt{2}$	$\frac{1}{2}\sqrt{2}$	1
60.0	$\frac{1}{3}\pi$	$\frac{1}{2}\sqrt{3}$	$\frac{1}{2}$	$\sqrt{3}$
90.0	$\frac{1}{2}\pi$	1	0	∞
180.0	π	0	-1	0

The INVERSE TRIGONOMETRIC FUNCTIONS are generally defined on the following domains.

Function	Domain
$\sin^{-1} x$	$-\frac{1}{2}\pi \leq y \leq \frac{1}{2}\pi$
$\cos^{-1} x$	$0 \leq y \leq \pi$
$\tan^{-1} x$	$-\frac{1}{2}\pi < y < \frac{1}{2}\pi$
$\csc^{-1} x$	$0 \leq y \leq \frac{1}{2}\pi$ or $\pi \leq y \leq \frac{3\pi}{2}$
$\sec^{-1} x$	$0 \leq y \leq \pi$
$\cot^{-1} x$	$0 \leq y \leq \frac{1}{2}\pi$ or $-\pi \leq y \leq -\frac{1}{2}\pi$

Inverse-forward identities are

$$\tan^{-1}(\cot x) = \tfrac{1}{2}\pi - x \quad (12)$$

$$\sin^{-1}(\cos x) = \tfrac{1}{2}\pi - x \quad (13)$$

$$\sec^{-1}(\csc x) = \tfrac{1}{2}\pi - x, \quad (14)$$

and forward-inverse identities are

$$\cos(\sin^{-1} x) = \sqrt{1-x^2} \quad (15)$$

$$\cos(\tan^{-1} x) = \frac{1}{\sqrt{1+x^2}} \quad (16)$$

$$\sin(\cos^{-1} x) = \sqrt{1-x^2}$$

$$\sin(\tan^{-1} x) = \frac{x}{\sqrt{1+x^2}} \quad (17)$$

$$\tan(\cos^{-1} x) = \frac{\sqrt{1-x^2}}{x} \quad (18)$$

$$\tan(\sin^{-1} x) = \frac{x}{\sqrt{1-x^2}}. \quad (19)$$

Inverse sum identities include

$$\sin^{-1} x + \cos^{-1} x = \tfrac{1}{2}\pi \quad (20)$$

$$\tan^{-1} x + \cot^{-1} x = \tfrac{1}{2}\pi \quad (21)$$

$$\sec^{-1} x + \csc^{-1} x = \tfrac{1}{2}\pi, \quad (22)$$

where (20) follows from

$$x = \sin(\sin^{-1} x) = \cos(\tfrac{1}{2}\pi - \sin^{-1} x). \quad (23)$$

Complex inverse identities in terms of LOGARITHMS include

$$\sin^{-1}(z) = -i\ln(iz \pm \sqrt{1-z^2}) \quad (24)$$

$$\cos^{-1}(z) = -i\ln(z \pm i\sqrt{1-z^2}) \quad (25)$$

$$\tan^{-1}(z) = -i\ln\left(\frac{1+iz}{\sqrt{1+z^2}}\right) \quad (26)$$

$$= \tfrac{1}{2}i\ln\left(\frac{1-iz}{1+iz}\right). \quad (27)$$

For IMAGINARY arguments,

$$\sin(iz) = i\sinh z \quad (28)$$

$$\cos(iz) = \cosh z. \quad (29)$$

For COMPLEX arguments,

$$\sin(x+iy) = \sin x\cosh y + i\cos x\sinh y \quad (30)$$

$$\cos(x+iy) = \cos x\cosh y - i\sin x\sinh y. \quad (31)$$

For the ABSOLUTE SQUARE of COMPLEX arguments $z = x+iy$,

$$|\sin(x+iy)|^2 = \sin^2 x + \sinh^2 y \quad (32)$$

$$|\cos(x+iy)|^2 = \cos^2 x + \sinh^2 y. \quad (33)$$

The MODULUS also satisfies the curious identity

$$|\sin(x+iy)| = |\sin x + \sin(iy)|. \quad (34)$$

The only functions satisfying identities of this form,

$$|f(x+iy)| = |f(x)+f(iy)| \quad (35)$$

are $f(z) = Az$, $f(z) = A\sin(bz)$, and $f(z) = A\sinh(bz)$ (Robinson 1957).

Trigonometric product formulas can be derived using the following figure (Kung 1996).

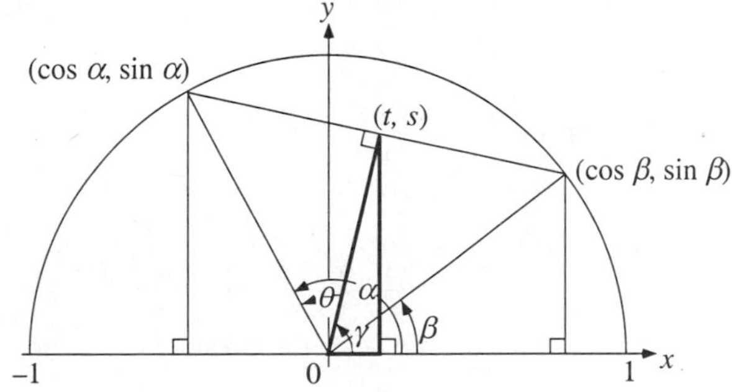

In the figure,

$$\theta = \tfrac{1}{2}(\alpha-\beta) \quad (36)$$

$$\gamma = \tfrac{1}{2}(\alpha+\beta), \quad (37)$$

so

$$s = \tfrac{1}{2}(\sin\alpha+\sin\beta) = \cos[\tfrac{1}{2}(\alpha-\beta)]\sin[\tfrac{1}{2}(\alpha+\beta)] \quad (38)$$

$$t = \tfrac{1}{2}(\cos\alpha+\cos\beta) = \cos[\tfrac{1}{2}(\alpha-\beta)]\cos[\tfrac{1}{2}(\alpha+\beta)]. \quad (39)$$

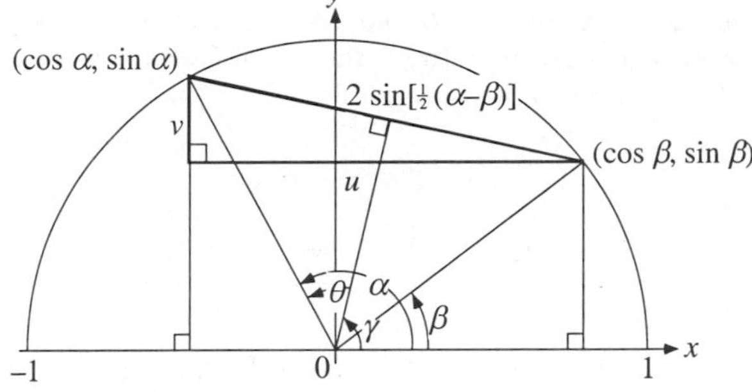

With θ and γ as previously defined, the above figure (Kung 1996) gives

$$u = \cos\beta - \cos\alpha = 2\sin[\tfrac{1}{2}(\alpha-\beta)]\sin[\tfrac{1}{2}(\alpha+\beta)] \quad (40)$$

$$v = \sin\alpha - \sin\beta = 2\sin[\tfrac{1}{2}(\alpha-\beta)]\cos[\tfrac{1}{2}(\alpha+\beta)]. \quad (41)$$

Angle addition FORMULAS express trigonometric functions of sums of angles $\alpha \pm \beta$ in terms of functions of α and β. They can be simply derived used COMPLEX exponentials and the EULER FORMULA,

$$\begin{aligned}\sin(\alpha+\beta) &= \frac{e^{i(\alpha+\beta)} - e^{-i(\alpha+\beta)}}{2i} = \frac{e^{i\alpha}e^{i\beta} - e^{-i\alpha}e^{-i\beta}}{2i} \\ &= \frac{(\cos\alpha + i\sin\alpha)(\cos\beta + i\sin\beta)}{2i} \\ &\quad - \frac{(\cos\alpha - i\sin\alpha)(\cos\beta - i\sin\beta)}{2i} \\ &= \frac{\cos\alpha\cos\beta + i\sin\beta\cos\alpha + i\sin\alpha\cos\beta - \sin\alpha\sin\beta}{2i} \\ &\quad + \frac{-\cos\alpha\cos\beta + i\cos\alpha\sin\beta + i\sin\alpha\cos\beta + \sin\alpha\sin\beta}{2i} \\ &= \sin\alpha\cos\beta + \sin\beta\cos\alpha \end{aligned} \quad (42)$$

$$\begin{aligned}\cos(\alpha+\beta) &= \frac{e^{i(\alpha+\beta)} + e^{-i(\alpha+\beta)}}{2} = \frac{e^{i\alpha}e^{i\beta} + e^{-i\alpha}e^{-i\beta}}{2} \\ &= \frac{(\cos\alpha + i\sin\alpha)(\cos\beta + i\sin\beta)}{2} \\ &\quad + \frac{(\cos\alpha - i\sin\alpha)(\cos\beta - i\sin\beta)}{2} \\ &= \frac{\cos\alpha\cos\beta + i\cos\alpha\sin\beta + i\sin\alpha\cos\beta - \sin\alpha\sin\beta}{2} \\ &\quad + \frac{\cos\alpha\cos\beta - i\cos\alpha\sin\beta - i\sin\alpha\cos\beta - \sin\alpha\sin\beta}{2} \\ &= \cos\alpha\cos\beta - \sin\alpha\sin\beta. \end{aligned} \quad (43)$$

Taking the ratio gives the tangent angle addition FORMULA

$$\tan(\alpha+\beta) \equiv \frac{\sin(\alpha+\beta)}{\cos(\alpha+\beta)} = \frac{\sin\alpha\cos\beta+\sin\beta\cos\alpha}{\cos\alpha\cos\beta-\sin\alpha\sin\beta}$$

$$= \frac{\frac{\sin\alpha}{\cos\alpha}+\frac{\sin\beta}{\cos\beta}}{1-\frac{\sin\alpha\sin\beta}{\cos\alpha\cos\alpha\beta}} = \frac{\tan\alpha+\tan\beta}{1-\tan\alpha\tan\beta}. \quad (44)$$

The angle addition FORMULAS can also be derived purely algebraically without the use of COMPLEX NUMBERS. Consider the following figure.

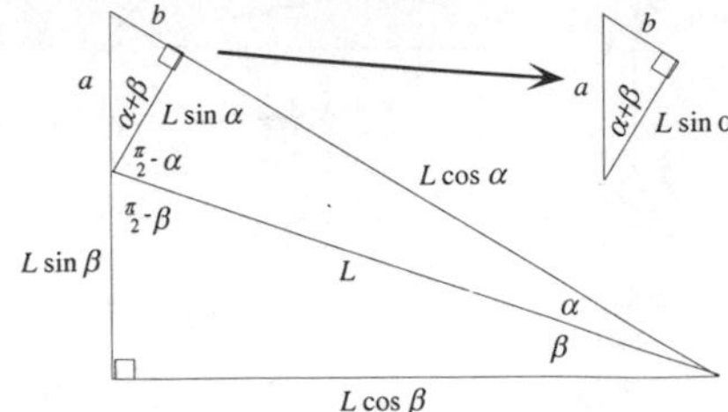

From the large RIGHT TRIANGLE,

$$\sin(\alpha+\beta) = \frac{L\sin\beta+a}{L\cos\alpha+b} \quad (45)$$

$$\cos(\alpha+\beta) = \frac{L\cos\beta}{L\cos\alpha+b}. \quad (46)$$

But, from the small triangle (inset at upper right),

$$a = \frac{L\sin\alpha}{\cos(\alpha+\beta)} \quad (47)$$

$$b = L\sin\alpha\tan(\alpha+\beta). \quad (48)$$

Plugging a and b from (47) and (48) into (45) and (46) gives

$$\sin(\alpha+\beta) = \frac{L\sin\beta+\frac{L\sin\alpha}{\cos(\alpha+\beta)}}{L\cos\alpha+\frac{L\sin\alpha\sin(\alpha+\beta)}{\cos(\alpha+\beta)}}$$

$$= \frac{\sin\beta\cos(\alpha+\beta)+\sin\alpha}{\cos\alpha\cos(\alpha+\beta)+\sin\alpha\sin(\alpha+\beta)}, \quad (49)$$

and

$$\cos(\alpha+\beta) = \frac{L\cos\beta}{L\cos\alpha+\frac{L\sin\alpha\sin(\alpha+\beta)}{\cos(\alpha+\beta)}}$$

$$= \frac{\cos\beta}{\cos\alpha+\frac{\sin\alpha\sin(\alpha+\beta)}{\cos(\alpha+\beta)}}. \quad (50)$$

Now solve (50) for $\cos(\alpha+\beta)$,

$$\cos(\alpha+\beta)\cos\alpha+\sin\alpha\sin(\alpha+\beta) = \cos\beta \quad (51)$$

to obtain

$$\cos(\alpha+\beta) = \frac{\cos\beta-\sin\alpha\sin(\alpha+\beta)}{\cos\alpha}. \quad (52)$$

Plugging (52) into (49) gives

$$\sin(\alpha+\beta) = \frac{\sin\beta\left[\frac{\cos\beta-\sin\alpha\sin(\alpha+\beta)}{\cos\alpha}\right]+\sin\alpha}{\cos\alpha\left[\frac{\cos\beta-\sin\alpha\sin(\alpha+\beta)}{\cos\alpha}\right]+\sin\alpha\sin(\alpha+\beta)}$$

$$= \frac{\sin\beta\cos\beta-\sin\alpha\sin\beta\sin(\alpha+\beta)+\sin\alpha\cos\alpha}{\cos\alpha\cos\beta-\sin\alpha\cos\alpha\sin(\alpha+\beta)+\sin\alpha\cos\alpha\sin(\alpha+\beta)}$$

$$= \frac{\sin\beta\cos\beta-\sin\alpha\sin\beta\sin(\alpha+\beta)+\sin\alpha\cos\alpha}{\cos\alpha\cos\beta}$$

$$= \frac{\sin\alpha\cos\alpha+\sin\beta\cos\beta}{\cos\alpha\cos\beta} - \frac{\sin\alpha\sin\beta}{\cos\alpha\cos\beta}\sin(\alpha+\beta), \quad (53)$$

so

$$\sin(\alpha+\beta)\left(1+\frac{\sin\alpha\sin\beta}{\cos\alpha\cos\beta}\right) = \frac{\sin\alpha\cos\alpha+\sin\beta\cos\beta}{\cos\alpha\cos\beta} \quad (54)$$

$$\sin(\alpha+\beta)(\cos\alpha\cos\beta+\sin\alpha\sin\beta)$$
$$= \sin\alpha\cos\alpha+\sin\beta\cos\beta, \quad (55)$$

and

$$\sin(\alpha+\beta) = \frac{\sin\alpha\cos\alpha+\sin\beta\cos\beta}{\sin\alpha\sin\beta+\cos\alpha\cos\beta}$$

$$= \frac{\sin\alpha\cos\alpha+\sin\beta\cos\beta}{\sin\alpha\sin\beta+\cos\alpha\cos\beta}\frac{\sin\alpha\cos\beta+\sin\beta\cos\alpha}{\sin\alpha\cos\beta+\sin\beta\cos\alpha}. \quad (56)$$

Multiplying out the DENOMINATOR gives

$$(\cos\alpha\cos\beta+\sin\alpha\sin\beta)(\sin\alpha\cos\beta+\sin\beta\cos\alpha)$$
$$= \sin\alpha\cos\alpha\cos^2\beta+\cos^2\alpha\sin\beta\cos\beta$$
$$+\sin^2\alpha\sin\beta\cos\beta+\sin\alpha\cos\alpha\sin^2\beta$$
$$= \sin\alpha\cos\alpha+\sin\beta\cos\beta, \quad (57)$$

so

$$\sin(\alpha+\beta) = \sin\alpha\cos\beta+\sin\beta\cos\alpha. \quad (58)$$

Multiplying out (50),

$$\cos(\alpha+\beta)\cos\alpha+\sin\alpha\sin(\alpha+\beta) = \cos\beta \quad (59)$$

$$\cos(\alpha+\beta) = \frac{\cos\beta-\sin\alpha\sin(\alpha+\beta)}{\cos\alpha}$$

$$= \frac{\cos\beta-\sin\alpha(\sin\alpha\cos\beta+\sin\beta\cos\alpha)}{\cos\alpha}$$

$$= \frac{\cos\beta(1-\sin^2\alpha)+\sin\alpha\cos\alpha\sin\beta}{\cos\alpha}$$

$$= \frac{\cos^2\alpha\cos\beta+\sin\alpha\cos\alpha\sin\beta}{\cos\alpha}$$

$$= \cos\alpha\cos\beta+\sin\alpha\sin\beta. \quad (60)$$

Summarizing,

$$\sin(\alpha+\beta) = \sin\alpha\cos\beta + \sin\beta\cos\alpha \quad (61)$$
$$\sin(\alpha-\beta) = \sin\alpha\cos\beta - \sin\beta\cos\alpha \quad (62)$$
$$\cos(\alpha+\beta) = \cos\alpha\cos\beta - \sin\alpha\sin\beta \quad (63)$$
$$\cos(\alpha-\beta) = \cos\alpha\cos\beta + \sin\alpha\sin\beta \quad (64)$$
$$\tan(\alpha+\beta) = \frac{\tan\alpha+\tan\beta}{1-\tan\alpha\tan\beta} \quad (65)$$
$$\tan(\alpha-\beta) = \frac{\tan\alpha-\tan\beta}{1+\tan\alpha\tan\beta}. \quad (66)$$

The sine and cosine angle addition identities can be summarized by the MATRIX EQUATION

$$\begin{bmatrix} \cos x & \sin x \\ -\sin x & \cos x \end{bmatrix} \begin{bmatrix} \cos y & \sin y \\ -\sin y & \cos y \end{bmatrix} = \begin{bmatrix} \cos(x+y) & \sin(x+y) \\ -\sin(x+y) & \cos(x+y) \end{bmatrix}. \quad (67)$$

The double angle formulas are

$$\sin(2\alpha) = 2\sin\alpha\cos\alpha \quad (68)$$
$$\cos(2\alpha) = \cos^2\alpha - \sin^2\alpha \quad (69)$$
$$= 2\cos^2\alpha - 1 \quad (70)$$
$$= 1 - 2\sin^2\alpha \quad (71)$$
$$\tan(2\alpha) = \frac{2\tan\alpha}{1-\tan^2\alpha}. \quad (72)$$

General multiple angle formulas are

$$\sin(n\alpha) = 2\sin[(n-1)\alpha]\cos\alpha - \sin[(n-2)\alpha] \quad (73)$$
$$\sin(nx) = n\cos^{n-1}x\sin x - \frac{n(n-1)(n-2)}{1\cdot 2\cdot 3}\cos^{n-3}x\sin^3 x + \ldots \quad (74)$$
$$\cos(n\alpha) = 2\cos[(n-1)\alpha]\cos\alpha - \cos[(n-2)\alpha] \quad (75)$$
$$\cos(nx) = \cos^n x - \frac{n(n-1)}{1\cdot 2}\cos^{n-2}x\sin^2 x + \frac{n(n-1)(n-2)(n-3)}{1\cdot 2\cdot 3\cdot 4}\cos^{n-4}x\sin^4 x - \ldots \quad (76)$$
$$\tan(n\alpha) = \frac{\tan[(n-1)\alpha]+\tan\alpha}{1-\tan[(n-1)\alpha]\tan\alpha}. \quad (77)$$

Therefore, any trigonometric function of a sum can be broken up into a sum of trigonometric functions with $\sin\alpha\cos\alpha$ cross terms. Particular cases for multiple angle formulas up to $n = 4$ are given below.

$$\sin(3\alpha) = 3\sin\alpha - 4\sin^3\alpha \quad (78)$$
$$\cos(3\alpha) = 4\cos^3\alpha - 3\cos\alpha \quad (79)$$
$$\tan(3\alpha) = \frac{3\tan\alpha - \tan^3\alpha}{1-3\tan^2\alpha} \quad (80)$$
$$\sin(4\alpha) = 4\sin\alpha\cos\alpha - 8\sin^3\alpha\cos\alpha \quad (81)$$
$$\cos(4\alpha) = 8\cos^4\alpha - 8\cos^2\alpha + 1 \quad (82)$$
$$\tan(4\alpha) = \frac{4\tan\alpha - 4\tan^3\alpha}{1-6\tan^2\alpha+\tan^4\alpha}. \quad (83)$$

Beyer (1987, p. 139) gives formulas up to $n = 6$.

Sum identities include

$$\frac{\tan(\alpha-\beta)}{\tan(\alpha+\beta)} = \frac{\sin(\alpha-\beta)\cos(\alpha+\beta)}{\cos(\alpha-\beta)\sin(\alpha+\beta)} = \frac{(\sin\alpha\cos\beta-\sin\beta\cos\alpha)(\cos\alpha\cos\beta-\sin\alpha\sin\beta)}{(\cos\alpha\cos\beta+\sin\alpha\sin\beta)(\sin\alpha\cos\beta+\sin\beta\cos\alpha)} = \frac{\sin\alpha\cos\alpha-\sin\beta\cos\beta}{\sin\alpha\cos\alpha+\sin\beta\cos\beta}. \quad (84)$$

Infinite sum identities include

$$\sum_{k=1,\,3,\,5,\ldots}^{\infty} \frac{e^{-kx}\sin(ky)}{k} = \frac{1}{2}\tan^{-1}\left(\frac{\sin y}{\sinh x}\right). \quad (85)$$

Trigonometric half-angle formulas include

$$\sin\left(\frac{\alpha}{2}\right) = \sqrt{\frac{1-\cos\alpha}{2}} \quad (86)$$
$$\cos\left(\frac{\alpha}{2}\right) = \sqrt{\frac{1+\cos\alpha}{2}} \quad (87)$$
$$\tan\left(\frac{\alpha}{2}\right) = \frac{\sin\alpha}{1+\cos\alpha} \quad (88)$$
$$= \frac{1-\cos\alpha}{\sin\alpha} \quad (89)$$
$$= \frac{1\pm\sqrt{1+\tan^2\alpha}}{\tan\alpha} \quad (90)$$
$$= \frac{\tan\alpha\sin\alpha}{\tan\alpha+\sin\alpha}. \quad (91)$$

The PROSTHAPHAERESIS FORMULAS are

$$\sin\alpha+\sin\beta = 2\sin[\tfrac{1}{2}(\alpha+\beta)]\cos[\tfrac{1}{2}(\alpha-\beta)] \quad (92)$$
$$\sin\alpha-\sin\beta = 2\cos[\tfrac{1}{2}(\alpha+\beta)]\sin[\tfrac{1}{2}(\alpha-\beta)] \quad (93)$$
$$\sin\alpha+\cos\beta = 2\cos[\tfrac{1}{2}(\alpha+\beta)]\cos[\tfrac{1}{2}(\alpha-\beta)] \quad (94)$$
$$\cos\alpha-\cos\beta = -2\sin[\tfrac{1}{2}(\alpha+\beta)]\sin[\tfrac{1}{2}(\alpha-\beta)]. \quad (95)$$

Related formulas are

$$\sin\alpha\cos\beta = \tfrac{1}{2}[\sin(\alpha-\beta)+\sin(\alpha+\beta)] \quad (96)$$
$$\cos\alpha\cos\beta = \tfrac{1}{2}[\cos(\alpha-\beta)+\cos(\alpha+\beta)] \quad (97)$$
$$\cos\alpha\sin\beta = \tfrac{1}{2}[\sin(\alpha+\beta)-\sin(\alpha-\beta)] \quad (98)$$
$$\sin\alpha\sin\beta = \tfrac{1}{2}[\cos(\alpha-\beta)-\cos(\alpha+\beta)]. \quad (99)$$

Multiplying both sides by 2 gives the equations sometimes known as the WERNER FORMULAS.

Trigonometric product/sum formulas are

$$\sin(\alpha+\beta)\sin(\alpha-\beta) = \sin^2\alpha - \sin^2\beta = \cos^2\beta - \cos^2\alpha \quad (100)$$

$$\cos(\alpha+\beta)\cos(\alpha-\beta) = \cos^2\alpha - \sin^2\beta = \cos^2\beta - \sin^2\alpha. \tag{101}$$

Power formulas include

$$\sin^2 x = \tfrac{1}{2}[1-\cos(2x)] \tag{102}$$
$$\sin^3 x = \tfrac{1}{4}[3\sin x - \sin(3x)] \tag{103}$$
$$\sin^4 x = \tfrac{1}{8}[3 - 4\cos(2x) + \cos(4x)] \tag{104}$$

and

$$\cos^2 x = \tfrac{1}{2}[1+\cos(2x)] \tag{105}$$
$$\cos^3 x = \tfrac{1}{4}[3\cos x + \cos(3x)] \tag{106}$$
$$\cos^4 x = \tfrac{1}{8}[3 + 4\cos(2x) + \cos(4x)] \tag{107}$$

(Beyer 1987, p. 140). Formulas of these types can also be given analytically as

$$\sin^{2n} x = \frac{1}{2^{2n}}\binom{2n}{n} + \frac{(-1)^n}{2^{2n-1}}\sum_{k=0}^{n-1}(-1)^k\binom{2n}{k}\cos[2(n-k)x] \tag{108}$$

$$\sin^{2n+1} = \frac{(-1)^n}{4^n}\sum_{k=0}^{n}(-1)^k\binom{2n+1}{k}\sin[(2n+1-2k)x] \tag{109}$$

$$\cos^{2n} x = \frac{(-1)^n}{2^{2n}}\binom{2n}{n} + \frac{1}{2^{2n-1}}\sum_{k=0}^{n-1}\binom{2n}{k}\cos[2(n-k)x] \tag{110}$$

$$\cos^{2n+1} x = \frac{1}{4^n}\sum_{k=0}^{n}\binom{2n+1}{k}\cos[(2n+1-2k)x] \tag{111}$$

(Kogan), where $\binom{n}{m}$ is a BINOMIAL COEFFICIENT.

see also COSECANT, COSINE, COTANGENT, EUCLIDEAN NUMBER, INVERSE COSECANT, INVERSE COSINE, INVERSE COTANGENT, INVERSE SECANT, INVERSE SINE, INVERSE TANGENT, INVERSE TRIGONOMETRIC FUNCTIONS, OSBORNE'S RULE, POLYGON, SECANT, SINE, TANGENT, TRIGONOMETRY VALUES: π, $\pi/2$, $\pi/3$, $\pi/4$, $\pi/5$, $\pi/6$, $\pi/7$, $\pi/8$, $\pi/9$, $\pi/10$, $\pi/11$, $\pi/12$, $\pi/15$, $\pi/16$, $\pi/17$, $\pi/18$, $\pi/20$, 0, WERNER FORMULAS

References

Abramowitz, M. and Stegun, C. A. (Eds.). "Circular Functions." §4.3 in *Handbook of Mathematical Functions with Formulas, Graphs, and Mathematical Tables, 9th printing.* New York: Dover, pp. 71–79, 1972.

Bahm, L. B. *The New Trigonometry on Your Own.* Patterson, NJ: Littlefield, Adams & Co., 1964.

Beyer, W. H. "Trigonometry." *CRC Standard Mathematical Tables, 28th ed.* Boca Raton, FL: CRC Press, pp. 134–152, 1987.

Dixon, R. "The Story of Sine and Cosine." § 4.4 in *Mathographics.* New York: Dover, pp. 102–106, 1991.

Hobson, E. W. *A Treatise on Plane Trigonometry.* London: Cambridge University Press, 1925.

Kells, L. M.; Kern, W. F.; and Bland, J. R. *Plane and Spherical Trigonometry.* New York: McGraw-Hill, 1940.

Kogan, S. "A Note on Definite Integrals Involving Trigonometric Functions." `http://www.mathsoft.com/asolve/constant/pi/sin/sin.html`.

Kung, S. H. "Proof Without Words: The Difference-Product Identities" and "Proof Without Words: The Sum-Product Identities." *Math. Mag.* **69**, 269, 1996.

Maor, E. *Trigonometric Delights.* Princeton, NJ: Princeton University Press, 1998.

Morrill, W. K. *Plane Trigonometry, rev. ed.* Dubuque, IA: Wm. C. Brown, 1964.

Robinson, R. M. "A Curious Mathematical Identity." *Amer. Math. Monthly* **64**, 83–85, 1957.

Sloane, N. J. A. Sequence A003401/M0505 in "An On-Line Version of the Encyclopedia of Integer Sequences."

Thompson, J. E. *Trigonometry for the Practical Man.* Princeton, NJ: Van Nostrand.

Weisstein, E. W. "Exact Values of Trigonometric Functions." `http://www.astro.virginia.edu/~eww6n/math/notebooks/TrigExact.m`.

Yates, R. C. "Trigonometric Functions." *A Handbook on Curves and Their Properties.* Ann Arbor, MI: J. W. Edwards, pp. 225–232, 1952.

Zill, D. G. and Dewar, J. M. *Trigonometry.* New York: McGraw-Hill 1990.

Trigonometry Values—π

By the definition of the trigonometric functions,

$$\sin\pi = 0 \tag{1}$$
$$\cos\pi = -1 \tag{2}$$
$$\tan\pi = 0 \tag{3}$$
$$\csc\pi = \infty \tag{4}$$
$$\sec\pi = -1 \tag{5}$$
$$\cot\pi = \infty. \tag{6}$$

Trigonometry Values—$\pi/2$

By the definition of the trigonometric functions,

$$\sin\left(\frac{\pi}{2}\right) = 1 \tag{1}$$
$$\cos\left(\frac{\pi}{2}\right) = 0 \tag{2}$$
$$\tan\left(\frac{\pi}{2}\right) = \infty \tag{3}$$
$$\csc\left(\frac{\pi}{2}\right) = 1 \tag{4}$$
$$\sec\left(\frac{\pi}{2}\right) = \infty \tag{5}$$
$$\cot\left(\frac{\pi}{2}\right) = 0. \tag{6}$$

see also DIGON

Trigonometry Values—$\pi/3$

From TRIGONOMETRY VALUES: $\pi/6$

$$\sin\left(\frac{\pi}{6}\right) = \tfrac{1}{2} \tag{1}$$

$$\cos\left(\frac{\pi}{6}\right) = \tfrac{1}{2}\sqrt{3} \tag{2}$$

together with the trigonometric identity

$$\sin(2\alpha) = 2\sin\alpha\cos\alpha, \tag{3}$$

the identity

$$\sin\left(\frac{\pi}{3}\right) = 2\sin\left(\frac{\pi}{6}\right)\cos\left(\frac{\pi}{6}\right) = 2(\tfrac{1}{2})(\tfrac{1}{2}\sqrt{3}) = \tfrac{1}{2}\sqrt{3} \tag{4}$$

is obtained. Using the identity

$$\cos(2\alpha) = 1 - 2\sin^2\alpha, \tag{5}$$

then gives

$$\cos\left(\frac{\pi}{3}\right) = 1 - 2\sin^2\left(\frac{\pi}{6}\right) = 1 - 2(\tfrac{1}{2})^2 = \tfrac{1}{2}. \tag{6}$$

Summarizing,

$$\sin\left(\frac{\pi}{3}\right) = \tfrac{1}{2}\sqrt{3} \tag{7}$$

$$\cos\left(\frac{\pi}{3}\right) = \tfrac{1}{2} \tag{8}$$

$$\tan\left(\frac{\pi}{3}\right) = \sqrt{3}. \tag{9}$$

see also EQUILATERAL TRIANGLE

Trigonometry Values—$\pi/4$

For a RIGHT ISOSCELES TRIANGLE, symmetry requires that the angle at each VERTEX be given by

$$\tfrac{1}{2}\pi + 2\alpha = \pi, \tag{1}$$

so $\alpha = \pi/4$. The sides are equal, so

$$\sin^2\alpha + \cos^2\alpha = 2\sin^2\alpha = 1. \tag{2}$$

Solving,

$$\sin\left(\frac{\pi}{4}\right) = \tfrac{1}{2}\sqrt{2} \tag{3}$$

$$\cos\left(\frac{\pi}{4}\right) = \tfrac{1}{2}\sqrt{2} \tag{4}$$

$$\tan\left(\frac{\pi}{4}\right) = 1. \tag{5}$$

see also SQUARE

Trigonometry Values—$\pi/5$

Use the identity

$$\sin(5\alpha) = 5\sin\alpha - 20\sin^3\alpha + 16\sin^5\alpha. \tag{1}$$

Now, let $\alpha \equiv \pi/5$ and $x \equiv \sin\alpha$. Then

$$\sin\pi = 0 = 5x - 20x^3 + 16x^5 \tag{2}$$

$$16x^4 - 20x^2 + 5 = 0. \tag{3}$$

Solving the QUADRATIC EQUATION for x^2 gives

$$\sin^2\left(\frac{\pi}{5}\right) = x^2 = \frac{20 \pm \sqrt{(-20)^2 - 4\cdot 16\cdot 5}}{2\cdot 16} = \frac{20 \pm \sqrt{80}}{32} = \tfrac{1}{8}(5 \pm \sqrt{5}). \tag{4}$$

Now, $\sin(\pi/5)$ must be less than

$$\sin\left(\frac{\pi}{4}\right) = \tfrac{1}{2}\sqrt{2}, \tag{5}$$

so taking the MINUS SIGN and simplifying gives

$$\sin\left(\frac{\pi}{5}\right) = \sqrt{\frac{5-\sqrt{5}}{8}} = \tfrac{1}{4}\sqrt{10 - 2\sqrt{5}}. \tag{6}$$

$\cos(\pi/5)$ can be computed from

$$\cos\left(\frac{\pi}{5}\right) = \sqrt{1 - \sin^2\left(\frac{\pi}{5}\right)} = \tfrac{1}{4}(1 + \sqrt{5}). \tag{7}$$

Summarizing,

$$\sin\left(\frac{\pi}{5}\right) = \tfrac{1}{4}\sqrt{10 - 2\sqrt{5}} \tag{8}$$

$$\sin\left(\frac{2\pi}{5}\right) = \tfrac{1}{4}\sqrt{10 + 2\sqrt{5}} \tag{9}$$

$$\sin\left(\frac{3\pi}{5}\right) = \tfrac{1}{4}\sqrt{10 + 2\sqrt{5}} \tag{10}$$

$$\sin\left(\frac{4\pi}{5}\right) = \tfrac{1}{4}\sqrt{10 - 2\sqrt{5}} \tag{11}$$

$$\cos\left(\frac{\pi}{5}\right) = \tfrac{1}{4}(1 + \sqrt{5}) \tag{12}$$

$$\cos\left(\frac{2\pi}{5}\right) = \tfrac{1}{4}(-1 + \sqrt{5}) \tag{13}$$

$$\cos\left(\frac{3\pi}{5}\right) = \tfrac{1}{4}(1 - \sqrt{5}) \tag{14}$$

$$\cos\left(\frac{4\pi}{5}\right) = -\tfrac{1}{4}(1 + \sqrt{5}) \tag{15}$$

$$\tan\left(\frac{\pi}{5}\right) = \sqrt{5 - 2\sqrt{5}} \tag{16}$$

$$\tan\left(\frac{2\pi}{5}\right) = \sqrt{5 + 2\sqrt{5}} \tag{17}$$

$$\tan\left(\frac{3\pi}{5}\right) = -\sqrt{5 + 2\sqrt{5}} \tag{18}$$

$$\tan\left(\frac{4\pi}{5}\right) = -\sqrt{5 - 2\sqrt{5}}. \tag{19}$$

see also DODECAHEDRON, ICOSAHEDRON, PENTAGON, PENTAGRAM

Trigonometry Values—$\pi/6$

Given a RIGHT TRIANGLE with angles defined to be α and 2α, it must be true that

$$\alpha + 2\alpha + \tfrac{1}{2}\pi = \pi, \tag{1}$$

so $\alpha = \pi/6$. Define the hypotenuse to have length 1 and the side opposite α to have length x, then the side opposite 2α has length $\sqrt{1-x^2}$. This gives $\sin\alpha \equiv x$ and

$$\sin(2\alpha) = \sqrt{1-x^2}. \tag{2}$$

But

$$\sin(2\alpha) = 2\sin\alpha\cos\alpha = 2x\sqrt{1-x^2}, \tag{3}$$

so we have

$$\sqrt{1-x^2} = 2x\sqrt{1-x^2}. \tag{4}$$

This gives $2x = 1$, or

$$\sin\left(\frac{\pi}{6}\right) = \tfrac{1}{2}. \tag{5}$$

$\cos(\pi/6)$ is then computed from

$$\cos\left(\frac{\pi}{6}\right) = \sqrt{1-\sin^2\left(\frac{\pi}{6}\right)} = \sqrt{1-(\tfrac{1}{2})^2} = \tfrac{1}{2}\sqrt{3}. \tag{6}$$

Summarizing,

$$\sin\left(\frac{\pi}{6}\right) = \tfrac{1}{2} \tag{7}$$

$$\cos\left(\frac{\pi}{6}\right) = \tfrac{1}{2}\sqrt{3} \tag{8}$$

$$\tan\left(\frac{\pi}{6}\right) = \tfrac{1}{3}\sqrt{3}. \tag{9}$$

see also HEXAGON, HEXAGRAM

Trigonometry Values—$\pi/7$

Trigonometric functions of $n\pi/7$ for n an integer cannot be expressed in terms of sums, products, and finite root extractions on *real* rational numbers because 7 is not a FERMAT PRIME. This also means that the HEPTAGON is not a CONSTRUCTIBLE POLYGON.

However, exact expressions involving roots of *complex* numbers can still be derived using the trigonometric identity

$$\sin(n\alpha) = 2\sin[(n-1)\alpha]\cos\alpha - \sin[(n-2)\alpha]. \tag{1}$$

The case $n = 7$ gives

$$\begin{aligned}\sin(7\alpha) &= 2\sin(6\alpha)\cos\alpha - \sin(5\alpha)\\ &= 2(32\cos^5\alpha\sin\alpha - 32\cos^3\alpha\sin\alpha + 6\cos\alpha\sin\alpha)\cos\alpha\\ &\quad -(5\sin\alpha - 20\sin^3\alpha + 16\sin^5\alpha)\\ &= 64\cos^6\alpha\sin\alpha - 64\cos^4\alpha\sin\alpha + 12\cos^2\alpha\sin\alpha\\ &\quad -5\sin\alpha + 20(1-\cos^2\alpha)\sin\alpha\\ &\quad -16(1-2\cos^2\alpha+\cos^4\alpha)\sin\alpha\\ &= \sin\alpha(64\cos^6\alpha - 80\cos^4\alpha + 24\cos^2\alpha - 1).\end{aligned} \tag{2}$$

Rewrite this using the identity $\cos^2\alpha = 1 - \sin^2\alpha$,

$$\begin{aligned}\sin\left(\frac{\pi}{7}\right) &= \sin\alpha(7 - 56\sin^2\alpha + 112\sin^4\alpha - 64\sin^6\alpha)\\ &= -64\sin\alpha(\sin^6\alpha - \tfrac{112}{64}\sin^4\alpha + \tfrac{56}{64}\sin^2\alpha - \tfrac{7}{64}).\end{aligned} \tag{3}$$

Now, let $\alpha \equiv \pi/7$ and $x \equiv \sin^2\alpha$, then

$$\sin(\pi) = 0 = x^3 - \tfrac{7}{4}x^2 + \tfrac{7}{8}x - \tfrac{7}{64}, \tag{4}$$

which is a CUBIC EQUATION in x. The ROOTS are numerically found to be $x \approx 0.188255$, $0.611260\ldots$, $0.950484\ldots$. But $\sin\alpha = \sqrt{x}$, so these ROOTS correspond to $\sin\alpha \approx 0.4338$, $\sin(2\alpha) \approx 0.7817$, $\sin(3\alpha) \approx 0.9749$. By NEWTON'S RELATION

$$\prod_i r_i = -a_0, \tag{5}$$

we have

$$x_1x_2x_3 = \tfrac{7}{64}, \tag{6}$$

or

$$\sin\left(\frac{\pi}{7}\right)\sin\left(\frac{2\pi}{7}\right)\sin\left(\frac{3\pi}{7}\right) = \sqrt{\frac{7}{64}} = \tfrac{1}{8}\sqrt{7}. \tag{7}$$

Similarly,

$$\cos\left(\frac{\pi}{7}\right)\cos\left(\frac{2\pi}{7}\right)\cos\left(\frac{3\pi}{7}\right) = \frac{1}{8}. \tag{8}$$

The constants of the CUBIC EQUATION are given by

$$Q \equiv \tfrac{1}{9}(3a_1 - a_2{}^2) = \tfrac{1}{9}[3\cdot\tfrac{7}{8} - (-\tfrac{7}{4})^2] = -\tfrac{7}{144} \tag{9}$$

$$\begin{aligned}R &\equiv \tfrac{1}{54}(9a_2a_1 - 2a_2^3 - 27a_0)\\ &= \tfrac{1}{54}[9(-\tfrac{7}{4})(\tfrac{1}{7}8) - 2(-\tfrac{7}{4})^3 - 27(-\tfrac{7}{64})]\\ &= -\tfrac{7}{3456}.\end{aligned} \tag{10}$$

The DISCRIMINANT is then

$$\begin{aligned}D \equiv Q^3 + R^2 &= -\tfrac{343}{2{,}985{,}984} + \tfrac{49}{11{,}943{,}936}\\ &= -\tfrac{49}{442{,}368} < 0,\end{aligned} \tag{11}$$

so there are three distinct REAL ROOTS. Finding the first one,

$$x = \sqrt[3]{R+\sqrt{D}} + \sqrt[3]{R-\sqrt{D}} - \tfrac{1}{3}a_2. \tag{12}$$

Writing

$$\sqrt{D} = 3^{-3/2}\tfrac{7}{128}i, \tag{13}$$

plugging in from above, and anticipating that the solution we have picked corresponds to $\sin(3\pi/7)$,

$$\begin{aligned}\sin\left(\frac{3\pi}{7}\right) &= \sqrt{x} = \\ &\sqrt{\sqrt[3]{-\frac{7}{3456}+3^{-3/2}\frac{7}{128}i}+\sqrt[3]{-\frac{7}{3456}-3^{-3/2}\frac{7}{128}i}-\frac{1}{3}\left(-\frac{7}{4}\right)} \\ &= \sqrt{\sqrt[3]{-\frac{7}{3456}+3^{-3/2}\frac{7}{128}i}+\sqrt[3]{-\frac{7}{3456}-3^{-3/2}\frac{7}{128}i}+\frac{7}{12}} \\ &= \sqrt{\sqrt[3]{\frac{7}{3456}(-1+3^{3/2}i)}-\sqrt[3]{\frac{7}{3456}(1+3^{3/2}i)}+\frac{7}{12}} \\ &= \sqrt{\frac{1}{12}\left[\sqrt[3]{\frac{7}{2}(-1+3^{3/2}i)}-\sqrt[3]{\frac{7}{2}(1+3^{3/2}i)}+7\right]}. \quad (14)\end{aligned}$$

see also HEPTAGON

Trigonometry Values—$\pi/8$

$$\begin{aligned}\sin\left(\frac{\pi}{8}\right) &= \sin\left(\frac{1}{2}\cdot\frac{\pi}{4}\right) = \sqrt{\frac{1}{2}\left(1-\cos\frac{\pi}{4}\right)} \\ &= \sqrt{\tfrac{1}{2}(1-\tfrac{1}{2}\sqrt{2})} = \tfrac{1}{2}\sqrt{2-\sqrt{2}}. \quad (1)\end{aligned}$$

Now, checking to see if the SQUARE ROOT can be simplified gives

$$a^2 - b^2c = 2^2 - 1^2\cdot 2 = 4-2 = 2, \quad (2)$$

which is not a PERFECT SQUARE, so the above expression cannot be simplified. Similarly,

$$\begin{aligned}\cos\left(\frac{\pi}{8}\right) &= \cos\left(\frac{1}{2}\frac{\pi}{4}\right) = \sqrt{\frac{1}{2}\left(1+\cos\frac{\pi}{4}\right)} \\ &= \sqrt{\frac{1}{2}\left(1+\frac{\sqrt{2}}{2}\right)} = \tfrac{1}{2}\sqrt{2+\sqrt{2}} \quad (3)\end{aligned}$$

$$\begin{aligned}\tan\left(\frac{\pi}{8}\right) &= \sqrt{\frac{2-\sqrt{2}}{2+\sqrt{2}}} = \sqrt{\frac{(2-\sqrt{2})^2}{4-2}} = \sqrt{\frac{4+2-4\sqrt{2}}{2}} \\ &= \sqrt{\frac{6-4\sqrt{2}}{2}} = \sqrt{3-2\sqrt{2}}. \quad (4)\end{aligned}$$

But

$$a^2 - b^2c = 3^2 - 2^2 2 = 9-8 = 1 \quad (5)$$

is a PERFECT SQUARE, so we can find

$$d = \tfrac{1}{2}(3\pm 1) = 1, 2.$$

Rewrite the above as

$$\tan\left(\frac{\pi}{8}\right) = \sqrt{2}-1 \quad (6)$$

$$\cot\left(\frac{\pi}{8}\right) = \frac{1}{\sqrt{2}-1} = \frac{\sqrt{2}+1}{2-1} = \sqrt{2}+1. \quad (7)$$

Summarizing,

$$\sin\left(\frac{\pi}{8}\right) = \tfrac{1}{2}\sqrt{2-\sqrt{2}} \quad (8)$$

$$\sin\left(\frac{3\pi}{8}\right) = \tfrac{1}{2}\sqrt{2+\sqrt{2}} \quad (9)$$

$$\cos\left(\frac{\pi}{8}\right) = \tfrac{1}{2}\sqrt{2+\sqrt{2}} \quad (10)$$

$$\cos\left(\frac{3\pi}{8}\right) = \tfrac{1}{2}\sqrt{2-\sqrt{2}} \quad (11)$$

$$\tan\left(\frac{\pi}{8}\right) = \sqrt{2}-1 \quad (12)$$

$$\tan\left(\frac{3\pi}{8}\right) = \sqrt{2}+1. \quad (13)$$

see also OCTAGON

Trigonometry Values—$\pi/9$

Trigonometric functions of $n\pi/9$ radians for n an integer not divisible by 3 (e.g., 40° and 80°) cannot be expressed in terms of sums, products, and finite root extractions on *real* rational numbers because 9 is not a product of distinct FERMAT PRIMES. This also means that the NONAGON is not a CONSTRUCTIBLE POLYGON.

However, exact expressions involving roots of *complex* numbers can still be derived using the trigonometric identity

$$\sin(3\alpha) = 3\sin\alpha - 4\sin^3\alpha. \quad (1)$$

Let $\alpha \equiv \pi/9$ and $x \equiv \sin\alpha$. Then the above identity gives the CUBIC EQUATION

$$4x^3 - 3x + \tfrac{1}{2}\sqrt{3} = 0 \quad (2)$$

$$x^3 - \tfrac{3}{4}x = -\tfrac{1}{8}\sqrt{3}. \quad (3)$$

This cubic is of the form

$$x^3 + px = q, \quad (4)$$

where

$$p = -\tfrac{3}{4} \quad (5)$$

$$q = -\tfrac{1}{8}\sqrt{3}. \quad (6)$$

The DISCRIMINANT is then

$$\begin{aligned}D &\equiv \left(\frac{p}{3}\right)^3 + \left(\frac{q}{2}\right)^2 \\ &= \left(-\frac{1}{4}\right)^3 + \left(\frac{\sqrt{3}}{16}\right)^2 = -\frac{1}{16\cdot 4} + \frac{3}{16\cdot 16} = \frac{-4+3}{256} \\ &= -\frac{1}{256} < 0. \quad (7)\end{aligned}$$

There are therefore three REAL distinct roots, which are approximately -0.9848, 0.3240, and 0.6428. We want the one in the first QUADRANT, which is 0.3240.

$$\begin{aligned}\sin\left(\frac{\pi}{9}\right) &= \sqrt[3]{-\frac{\sqrt{3}}{16}+\sqrt{-\frac{1}{256}}}+\sqrt[3]{-\frac{\sqrt{3}}{16}-\sqrt{-\frac{1}{256}}}\\ &= \sqrt[3]{-\frac{\sqrt{3}}{16}+\frac{1}{16}i}-\sqrt[3]{\frac{\sqrt{3}}{16}+\frac{1}{16}i}\\ &= 2^{-4/3}(\sqrt[3]{i-\sqrt{3}}-\sqrt[3]{i+\sqrt{3}})\\ &\approx 0.3240\ldots. \end{aligned} \tag{8}$$

Similarly,

$$\begin{aligned}\cos\left(\frac{\pi}{9}\right) &= 2^{-4/3}(\sqrt[3]{1+i\sqrt{3}}+\sqrt[3]{1-i\sqrt{3}})\\ &\approx 0.7660\ldots. \end{aligned} \tag{9}$$

Because of the NEWTON'S RELATIONS, we have the identities

$$\sin\left(\frac{\pi}{9}\right)\sin\left(\frac{2\pi}{9}\right)\sin\left(\frac{4\pi}{9}\right)=\tfrac{1}{8} \tag{10}$$

$$\cos\left(\frac{\pi}{9}\right)\cos\left(\frac{2\pi}{9}\right)\cos\left(\frac{4\pi}{9}\right)=\tfrac{1}{8}\sqrt{3} \tag{11}$$

$$\tan\left(\frac{\pi}{9}\right)\tan\left(\frac{2\pi}{9}\right)\tan\left(\frac{4\pi}{9}\right)=\sqrt{3}. \tag{12}$$

see also NONAGON, STAR OF GOLIATH

Trigonometry Values—$\pi/10$

$$\begin{aligned}\sin\left(\frac{\pi}{10}\right) &= \sin\left(\frac{1}{2}\cdot\frac{\pi}{5}\right)=\sqrt{\tfrac{1}{2}\left[1-\cos\left(\frac{\pi}{5}\right)\right]}\\ &= \sqrt{\tfrac{1}{2}[1-\tfrac{1}{4}(1+\sqrt{5})]}=\tfrac{1}{4}(\sqrt{5}-1). \end{aligned} \tag{1}$$

So we have

$$\begin{aligned}\cos\left(\frac{\pi}{10}\right) &= \cos\left(\frac{1}{2}\cdot\frac{\pi}{5}\right)=\sqrt{\frac{1}{2}\left[1+\cos\left(\frac{\pi}{5}\right)\right]}\\ &= \sqrt{\tfrac{1}{2}[1+\tfrac{1}{4}(1+\sqrt{5})]}\\ &= \tfrac{1}{4}\sqrt{10+2\sqrt{5}}, \end{aligned} \tag{2}$$

and

$$\tan\left(\frac{\pi}{10}\right)=\sqrt{\frac{3-\sqrt{5}}{5+\sqrt{5}}}=\tfrac{1}{5}\sqrt{25-10\sqrt{5}}. \tag{3}$$

Summarizing,

$$\sin\left(\frac{\pi}{10}\right)=\tfrac{1}{4}(\sqrt{5}-1) \tag{4}$$

$$\cos\left(\frac{\pi}{10}\right)=\tfrac{1}{4}\sqrt{10+2\sqrt{5}} \tag{5}$$

$$\tan\left(\frac{\pi}{10}\right)=\tfrac{1}{5}\sqrt{25-10\sqrt{5}} \tag{6}$$

$$\sin\left(\frac{3\pi}{10}\right)=\tfrac{1}{4}(1+\sqrt{5}) \tag{7}$$

$$\cos\left(\frac{3\pi}{10}\right)=\tfrac{1}{4}(10-2\sqrt{5}) \tag{8}$$

$$\tan\left(\frac{3\pi}{10}\right)=\tfrac{1}{5}\sqrt{25+10\sqrt{5}}. \tag{9}$$

An interesting near-identity is given by

$$\frac{1}{4}\left[\cos(\tfrac{1}{10})+\cosh(\tfrac{1}{10})+2\cos(\tfrac{1}{20}\sqrt{2})\cosh(\tfrac{1}{20}\sqrt{2})\right]\approx 1. \tag{10}$$

In fact, the left-hand side is approximately equal to $1+2.480\times 10^{-13}$.

see also DECAGON, DECAGRAM

Trigonometry Values—$\pi/11$

Trigonometric functions of $n\pi/11$ for n an integer cannot be expressed in terms of sums, products, and finite root extractions on *real* rational numbers because 11 is not a FERMAT PRIME. This also means that the UNDECAGON is not a CONSTRUCTIBLE POLYGON.

However, exact expressions involving roots of *complex* numbers can still be derived using the trigonometric identity

$$\begin{aligned}\sin(11\alpha) &= \sin(12\alpha-\alpha)\cos\alpha-\cos(12\alpha)\sin\alpha\\ &= 2\sin(6\alpha)\cos(6\alpha)\cos\alpha-[1-2\sin^2(6\alpha)]\sin\alpha. \end{aligned} \tag{1}$$

Using the identities from Beyer (1987, p. 139),

$$\sin(6\alpha)=\sin\alpha\cos\alpha[32\cos^4\alpha-32\cos^2\alpha+6] \tag{2}$$

$$\cos(6\alpha)=32\cos^6\alpha-48\cos^4\alpha+18\cos^2\alpha-1 \tag{3}$$

gives

$$\begin{aligned}\sin(11\alpha) &= 2\cos^2\alpha\sin\alpha(32\cos^4\alpha-32\cos^2\alpha+6)\\ &\quad\times(32\cos^6\alpha-48\cos^4\alpha+18\cos^2\alpha-1)\\ &\quad-\sin\alpha[1-2\sin^2\alpha\cos^2\alpha(32\cos^4\alpha-32\cos^2\alpha+6)^2]\\ &= \sin\alpha(11-220\sin^2\alpha+1232\sin^4\alpha\alpha\\ &\quad-2816\sin^6\alpha+2816\sin^8-1024\sin^{10}\alpha). \end{aligned} \tag{4}$$

Now, let $\alpha\equiv\pi/11$ and $x\equiv\sin^2\alpha$, then

$$\begin{aligned}\sin\pi &= 0\\ &= 11-220x+1232x^2-2816x^3+2816x^4-1024x^5. \end{aligned} \tag{5}$$

This equation is an irreducible QUINTIC EQUATION, so an analytic solution involving FINITE ROOT extractions does not exist. The numerical ROOTS are $x = 0.07937$, 0.29229, 0.57115, 0.82743, 0.97974. So $\sin\alpha = 0.2817$, $\sin(2\alpha) = 0.5406$, $\sin(3\alpha) = 0.7557$, $\sin(4\alpha) = 0.9096$, $\sin(5\alpha) = 0.9898$. From one of NEWTON'S IDENTITIES,

$$\sin\left(\frac{\pi}{11}\right)\sin\left(\frac{2\pi}{11}\right)\sin\left(\frac{3\pi}{11}\right)\sin\left(\frac{4\pi}{11}\right)\sin\left(\frac{5\pi}{11}\right) = \sqrt{\frac{11}{1024}} = \frac{\sqrt{11}}{32} \quad (6)$$

$$\cos\left(\frac{\pi}{11}\right)\cos\left(\frac{2\pi}{11}\right)\cos\left(\frac{3\pi}{11}\right)\cos\left(\frac{4\pi}{11}\right)\cos\left(\frac{5\pi}{11}\right) = \frac{1}{32} \quad (7)$$

$$\tan\left(\frac{\pi}{11}\right)\tan\left(\frac{2\pi}{11}\right)\tan\left(\frac{3\pi}{11}\right)\tan\left(\frac{4\pi}{11}\right)\tan\left(\frac{5\pi}{11}\right) = \sqrt{11}. \quad (8)$$

The trigonometric functions of $\pi/11$ also obey the identity

$$\tan\left(\frac{3\pi}{11}\right) + 4\sin\left(\frac{2\pi}{11}\right) = \sqrt{11}. \quad (9)$$

see also UNDECAGON

References

Beyer, W. H. "Trigonometry." *CRC Standard Mathematical Tables, 28th ed.* Boca Raton, FL: CRC Press, 1987.

Trigonometry Values—$\pi/12$

$$\begin{aligned}\sin\left(\frac{\pi}{12}\right) &= \sin\left(\frac{\pi}{3} - \frac{\pi}{4}\right)\\ &= -\sin\left(\frac{\pi}{4}\right)\cos\left(\frac{\pi}{3}\right) + \sin\left(\frac{\pi}{3}\right)\cos\left(\frac{\pi}{4}\right)\\ &= -\tfrac{1}{2}\sqrt{2}(\tfrac{1}{2}) + \tfrac{1}{2}\sqrt{3}(\tfrac{1}{2}\sqrt{2})\\ &= \tfrac{1}{4}(\sqrt{6} - \sqrt{2}). \quad (1)\end{aligned}$$

Similarly,

$$\begin{aligned}\cos\left(\frac{\pi}{12}\right) &= \cos\left(\frac{\pi}{3} - \frac{\pi}{4}\right)\\ &= \cos\left(\frac{\pi}{4}\right)\cos\left(\frac{\pi}{3}\right) - \sin\left(\frac{\pi}{3}\right)\sin\left(\frac{\pi}{4}\right)\\ &= \tfrac{1}{2}(\tfrac{1}{2}\sqrt{2}) + \tfrac{1}{2}\sqrt{3}(-\tfrac{1}{2}\sqrt{2})\\ &= \tfrac{1}{4}(\sqrt{6} + \sqrt{2}). \quad (2)\end{aligned}$$

Summarizing,

$$\sin\left(\frac{\pi}{12}\right) = \tfrac{1}{4}(\sqrt{6} - \sqrt{2}) \approx 0.25881 \quad (3)$$

$$\cos\left(\frac{\pi}{12}\right) = \tfrac{1}{4}(\sqrt{6} + \sqrt{2}) \approx 0.96592 \quad (4)$$

$$\tan\left(\frac{\pi}{12}\right) = 2 - \sqrt{3} \approx 0.26794 \quad (5)$$

$$\csc\left(\frac{\pi}{12}\right) = \sqrt{6} + \sqrt{2} \approx 3.86370 \quad (6)$$

$$\sec\left(\frac{\pi}{12}\right) = \sqrt{6} - \sqrt{2} \approx 1.03527 \quad (7)$$

$$\cot\left(\frac{\pi}{12}\right) = 2 + \sqrt{3} \approx 3.73205. \quad (8)$$

Trigonometry Values—$\pi/15$

$$\begin{aligned}\sin\left(\frac{\pi}{15}\right) &= \sin\left(\frac{\pi}{6} - \frac{\pi}{10}\right)\\ &= \sin\left(\frac{\pi}{6}\right)\cos\left(\frac{\pi}{10}\right) - \sin\left(\frac{\pi}{10}\right)\cos\left(\frac{\pi}{6}\right)\\ &= \frac{1}{2}\sqrt{\frac{1}{8}\left(5 + \sqrt{5}\right)} - \frac{\sqrt{3}}{2}\frac{1}{4}(\sqrt{5} - 1)\\ &= \tfrac{1}{16}(2\sqrt{3} - 2\sqrt{15} + \sqrt{40 + 8\sqrt{5}}) \quad (1)\end{aligned}$$

and

$$\begin{aligned}\cos\left(\frac{\pi}{15}\right) &= \cos\left(\frac{\pi}{6} - \frac{\pi}{10}\right)\\ &= \cos\left(\frac{\pi}{6}\right)\cos\left(\frac{\pi}{10}\right) + \sin\left(\frac{\pi}{6}\right)\sin\left(\frac{\pi}{10}\right)\\ &= \frac{\sqrt{3}}{2}\sqrt{\frac{1}{8}\left(5 + \sqrt{5}\right)} + \frac{1}{2}\frac{1}{4}(\sqrt{5} - 1)\\ &= \tfrac{1}{8}(\sqrt{30 + 6\sqrt{5}} + \sqrt{5} - 1). \quad (2)\end{aligned}$$

Summarizing,

$$\sin\left(\frac{\pi}{15}\right) = \tfrac{1}{16}(2\sqrt{3} - 2\sqrt{15} + \sqrt{40 + 8\sqrt{5}}) \approx 0.20791 \quad (3)$$

$$\sin\left(\frac{2\pi}{15}\right) = \tfrac{1}{8}(\sqrt{3} + \sqrt{15} - \sqrt{10 - 2\sqrt{5}}) \approx 0.40673 \quad (4)$$

$$\cos\left(\frac{\pi}{15}\right) = \tfrac{1}{8}(\sqrt{30 + 6\sqrt{5}} + \sqrt{5} - 1) \approx 0.97814 \quad (5)$$

$$\cos\left(\frac{2\pi}{15}\right) = \tfrac{1}{8}(\sqrt{30 - 6\sqrt{5}} + 1) \approx 0.91354 \quad (6)$$

$$\tan\left(\frac{\pi}{15}\right) = \tfrac{1}{2}(3\sqrt{3} - \sqrt{15} - \sqrt{50 - 22\sqrt{5}}) \approx 0.21255. \quad (7)$$

Trigonometry Values—$\pi/16$

$$\begin{aligned}\sin\left(\frac{\pi}{16}\right) &= \sin\left(\frac{1}{2}\cdot\frac{\pi}{8}\right)\\ &= \sqrt{\tfrac{1}{2}\left(1 - \cos\frac{\pi}{8}\right)} = \sqrt{\tfrac{1}{2}\left(1 - \tfrac{1}{2}\sqrt{2 + \sqrt{2}}\right)}\\ &= \sqrt{\tfrac{1}{2} - \tfrac{1}{4}\sqrt{2 + \sqrt{2}}} = \tfrac{1}{2}\sqrt{2 - \sqrt{2 + \sqrt{2}}} \quad (1)\end{aligned}$$

$$\begin{aligned}\cos\left(\frac{\pi}{16}\right) &= \cos\left(\frac{1}{2}\cdot\frac{\pi}{8}\right)\\ &= \sqrt{\frac{1}{2}\left(1 + \cos\frac{\pi}{8}\right)} = \sqrt{\tfrac{1}{2}\left(1 + \tfrac{1}{2}\sqrt{2 + \sqrt{2}}\right)}\\ &= \sqrt{\tfrac{1}{2} + \tfrac{1}{4}\sqrt{2 + \sqrt{2}}} = \tfrac{1}{2}\sqrt{2 + \sqrt{2 + \sqrt{2}}} \quad (2)\end{aligned}$$

$$\begin{aligned}\tan\left(\frac{\pi}{16}\right) &= \sqrt{\frac{2 - \sqrt{2 + \sqrt{2}}}{2 + \sqrt{2 + \sqrt{2}}}}\\ &= \sqrt{4 + 2\sqrt{2}} - \sqrt{2} - 1. \quad (3)\end{aligned}$$

Summarizing,

$$\sin\left(\frac{\pi}{16}\right) = \tfrac{1}{2}\sqrt{2-\sqrt{2+\sqrt{2}}} \approx 0.19509 \quad (4)$$

$$\sin\left(\frac{3\pi}{16}\right) = \tfrac{1}{2}\sqrt{2-\sqrt{2-\sqrt{2}}} \approx 0.55557 \quad (5)$$

$$\cos\left(\frac{\pi}{16}\right) = \tfrac{1}{2}\sqrt{2+\sqrt{2+\sqrt{2}}} \approx 0.98079 \quad (6)$$

$$\cos\left(\frac{3\pi}{16}\right) = \tfrac{1}{2}\sqrt{2+\sqrt{2-\sqrt{2}}} \approx 0.83147 \quad (7)$$

$$\tan\left(\frac{\pi}{16}\right) = \sqrt{4+2\sqrt{2}} - \sqrt{2} - 1 \approx 0.19891. \quad (8)$$

Trigonometry Values—$\pi/17$

Rather surprisingly, trigonometric functions of $n\pi/17$ for n an integer can be expressed in terms of sums, products, and finite root extractions because 17 is a FERMAT PRIME. This makes the HEPTADECAGON a CONSTRUCTIBLE, as first proved by Gauss. Although Gauss did not actually explicitly provide a construction, he did derive the trigonometric formulas below using a series of intermediate variables from which the final expressions were then built up.

Let

$$\epsilon \equiv \sqrt{17+\sqrt{17}}$$
$$\epsilon^* \equiv \sqrt{17-\sqrt{17}}$$
$$\alpha \equiv \sqrt{\sqrt{34}+6\sqrt{17}+(\sqrt{34}-\sqrt{2})\epsilon^* - 8\sqrt{2}\,\epsilon}$$
$$\beta \equiv 2\sqrt{17+3\sqrt{17}-2\sqrt{2}\,\epsilon-\sqrt{2}\,\epsilon^*}\,.$$

Then

$$\sin\left(\frac{\pi}{17}\right) = \tfrac{1}{8}[34-2\sqrt{17}-2\sqrt{2}\,\epsilon^* - 2\sqrt{68+12\sqrt{17}+2\sqrt{2}(\sqrt{17}-1)\epsilon^*-16\sqrt{2}\,\epsilon}\,]^{1/2} \approx 0.18375$$

$$\cos\left(\frac{\pi}{17}\right) = \tfrac{1}{8}[30+2\sqrt{17}+2\sqrt{2}\,\epsilon^* + 2\sqrt{68+12\sqrt{17}+2\sqrt{2}(\sqrt{17}-1)\epsilon^*-16\sqrt{2}\,\epsilon}\,]^{1/2} \approx 0.98297$$

$$\sin\left(\frac{2\pi}{17}\right) = \tfrac{1}{16}[136-8\sqrt{17}+4\sqrt{2}(1-\sqrt{17})\epsilon^*+16\sqrt{2}\,\epsilon + 2(\sqrt{2}-\sqrt{34}-2\epsilon^*)\sqrt{34+6\sqrt{17}+(\sqrt{34}-\sqrt{2})\epsilon^*-8\sqrt{2}\epsilon}\,]^{1/2} \approx 0.36124$$

$$\cos\left(\frac{2\pi}{17}\right) = \tfrac{1}{16}[-1+\sqrt{17}+\sqrt{2}\epsilon^* + \sqrt{68+12\sqrt{17}-2\sqrt{2}(1-\sqrt{17})\epsilon^*-16\sqrt{2}\,\epsilon}\,] \approx 0.0.93247$$

$$\sin\left(\frac{4\pi}{17}\right) = \tfrac{1}{128}(-\sqrt{2}+\sqrt{34}+2\epsilon^*+2\alpha) \times\sqrt{68-4\sqrt{17}-2(\sqrt{34}-\sqrt{2})\epsilon^*+8\sqrt{2}\,\epsilon+\alpha(\sqrt{2}-\sqrt{34}-2\epsilon^*)} \approx 0.0.67370$$

$$\sin\left(\frac{8\pi}{17}\right) = \tfrac{1}{16}[136-8\sqrt{17}+8\sqrt{2}\epsilon-2(\sqrt{34}-3\sqrt{2})\epsilon^* - 2\beta(1-\sqrt{17}-\sqrt{2}\epsilon^*)\,]^{1/2} \approx 0.99573$$

$$\cos\left(\frac{8\pi}{17}\right) = \tfrac{1}{16}(-1+\sqrt{17}+\sqrt{2}\,\epsilon^* - 2\sqrt{17+3\sqrt{17}-\sqrt{2}\,\epsilon^*-2\sqrt{2}\,\epsilon}\,). \approx 0.09227$$

There are some interesting analytic formulas involving the trigonometric functions of $n\pi/17$. Define

$$P(x) \equiv (x-1)(x-2)(x^2+1)$$
$$g_1(x) \equiv \frac{2+\sqrt{P(x)}}{1-x}$$
$$g_4(x) \equiv \frac{2-\sqrt{P(x)}}{1-x}$$
$$f_i(x) \equiv \tfrac{1}{4}[g_i(x)-1]$$
$$a \equiv \tfrac{1}{4}\tan^{-1}4,$$

where $i = 1$ or 4. Then

$$f_1(\tan a) = \cos\left(\frac{2\pi}{17}\right)$$
$$f_4(\tan a) = \cos\left(\frac{8\pi}{17}\right).$$

see also CONSTRUCTIBLE POLYGON, FERMAT PRIME, HEPTADECAGON

References

Casey, J. *Plane Trigonometry.* Dublin: Hodges, Figgis, & Co., p. 220, 1888.

Conway, J. H. and Guy, R. K. *The Book of Numbers.* New York: Springer-Verlag, pp. 192–194 and 229–230, 1996.

Dörrie, H. "The Regular Heptadecagon." §37 in *100 Great Problems of Elementary Mathematics: Their History and Solutions.* New York: Dover, pp. 177–184, 1965.

Ore, Ø. *Number Theory and Its History.* New York: Dover, 1988.

Smith, D. E. *A Source Book in Mathematics.* New York: Dover, p. 348, 1994.

Trigonometry Values—$\pi/18$

The exact values of $\cos(\pi/18)$ and $\sin(\pi/18)$ are given by infinite NESTED RADICALS.

$$\sin\left(\frac{\pi}{18}\right) = \tfrac{1}{2}\sqrt{2-\sqrt{2+\sqrt{2+\sqrt{2-\ldots}}}} \approx 0.17365,$$

where the sequence of signs $+$, $+$, $-$ repeats with period 3, and

$$\cos\left(\frac{\pi}{18}\right) = \tfrac{1}{6}\sqrt{3}\left(\sqrt{8-\sqrt{8-\sqrt{8+\sqrt{8-\ldots}}}}+1\right) \approx 0.98481,$$

where the sequence of signs $-$, $-$, $+$ repeats with period 3.

Trigonometry Values—$\pi/20$

$$\sin\left(\frac{\pi}{20}\right) = \sin\left(\frac{1}{2}\frac{\pi}{10}\right) = \sqrt{\frac{1}{2}\left(1-\cos\frac{\pi}{10}\right)}$$
$$= \tfrac{1}{4}\sqrt{8-2\sqrt{10+2\sqrt{5}}}$$
$$\approx 0.15643\ldots \quad (1)$$
$$\cos\left(\frac{\pi}{20}\right) = \cos\left(\frac{1}{2}\frac{\pi}{10}\right) = \sqrt{\frac{1}{2}\left(1+\cos\frac{\pi}{10}\right)}$$
$$= \tfrac{1}{4}\sqrt{8+2\sqrt{10+2\sqrt{5}}}$$
$$\approx 0.98768\ldots \quad (2)$$
$$\tan\left(\frac{\pi}{20}\right) = 1+\sqrt{5}-\sqrt{5+2\sqrt{5}}$$
$$\approx 0.15838. \quad (3)$$

An interesting near-identity is given by

$$\frac{1}{4}\left[\cos(\tfrac{1}{10}) + \cosh(\tfrac{1}{10}) + 2\cos(\tfrac{1}{20}\sqrt{2})\cosh(\tfrac{1}{20}\sqrt{2})\right] \approx 1. \quad (4)$$

In fact, the left-hand side is approximately equal to $1 + 2.480 \times 10^{-13}$.

Trigonometry Values—0

By the definition of the trigonometric functions,

$$\sin 0 = 0$$
$$\cos 0 = 1$$
$$\tan 0 = 0$$
$$\csc 0 = \infty$$
$$\sec 0 = 1$$
$$\cot 0 = \infty.$$

Trigyrate Rhombicosidodecahedron

see JOHNSON SOLID

Trihedron

The TRIPLE of unit ORTHOGONAL VECTORS **T**, **N**, and **B** (TANGENT VECTOR, NORMAL VECTOR, and BINORMAL VECTOR).

see also BINORMAL VECTOR, NORMAL VECTOR, TANGENT VECTOR

Trilinear Coordinates

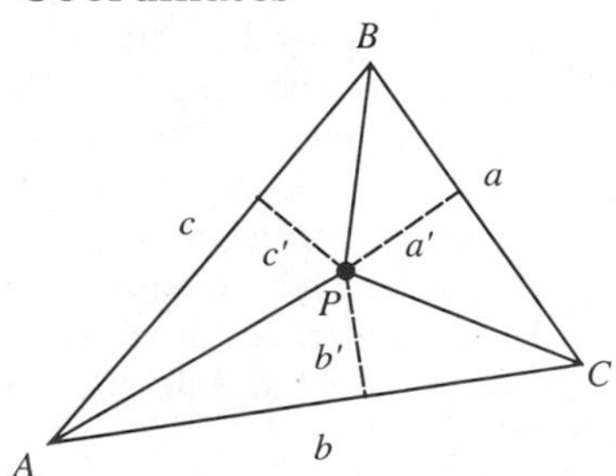

Given a TRIANGLE ΔABC, the trilinear coordinates of a point P with respect to ΔABC are an ordered TRIPLE of numbers, each of which is PROPORTIONAL to the directed distance from P to one of the side lines. Trilinear coordinates are denoted $\alpha:\beta:\gamma$ or (α,β,γ) and also are known as BARYCENTRIC COORDINATES, HOMOGENEOUS COORDINATES, or "trilinears."

In trilinear coordinates, the three VERTICES A, B, and C are given by $1:0:0$, $0:1:0$, and $0:0:1$. Let the point P in the above diagram have trilinear coordinates $\alpha:\beta:\gamma$ and lie at distances a', b', and c' from the sides BC, AC, and AB, respectively. Then the distances $a'=k\alpha$, $b'=k\beta$, and $c'=k\gamma$ can be found by writing Δ_a for the AREA of ΔBPC, and similarly for Δ_b and Δ_c. We then have

$$\Delta = \Delta_a+\Delta_b+\Delta_c = \tfrac{1}{2}aa' + \tfrac{1}{2}bb' + \tfrac{1}{2}cc'$$
$$= \tfrac{1}{2}(ak\alpha + bk\beta + ck\gamma) = \tfrac{1}{2}k(a\alpha+b\beta+c\gamma). \quad (1)$$

so

$$k \equiv \frac{2\Delta}{a\alpha+b\beta+c\gamma}, \quad (2)$$

where Δ is the AREA of ΔABC and a, b, and c are the lengths of its sides. When the values of the coordinates are taken as the actual lengths (i.e., the trilinears are chosen so that $k=1$), the coordinates are known as EXACT TRILINEAR COORDINATES.

Trilinear coordinates are unchanged when each is multiplied by any constant μ, so

$$t_1 : t_2 : t_3 = \mu t_1 : \mu t_2 : \mu t_3. \quad (3)$$

When normalized so that

$$t_1+t_2+t_3 = 1, \quad (4)$$

trilinear coordinates are called AREAL COORDINATES. The trilinear coordinates of the line

$$ux+vy+wz=0 \quad (5)$$

are

$$u:v:w = ad_A : bd_B : cd_C, \quad (6)$$

where d_i is the POINT-LINE DISTANCE from VERTEX A to the LINE.

Trilinear coordinates for some common POINTS are summarized in the following table, where A, B, and C are the angles at the corresponding vertices and a, b, and c are the opposite side lengths.

Point	Triangle Center Function
centroid M	$\csc A$, $1/a$
circumcenter O	$\cos A$
de Longchamps point	$\cos A - \cos B \cos C$
equal detour point	$\sec(\frac{1}{2}A)\cos(\frac{1}{2}B)\cos(\frac{1}{2}C)+1$
Feuerbach point F	$1-\cos(B-C)$
incenter I	1
isoperimetric point	$\sec(\frac{1}{2}A)\cos(\frac{1}{2}B)\cos(\frac{1}{2}C)-1$
Lemoine point	a
nine-point center N	$\cos(B-C)$
orthocenter H	$\cos B \cos C$
vertex A	$1:0:0$
vertex B	$0:1:0$
vertex C	$0:0:1$

To convert trilinear coordinates to a vector position for a given triangle specified by the x- and y-coordinates of its axes, pick two UNIT VECTORS along the sides. For instance, pick

$$\hat{\mathbf{a}} = \begin{bmatrix} a_1 \\ a_2 \end{bmatrix} \quad (7)$$

$$\hat{\mathbf{c}} = \begin{bmatrix} c_1 \\ c_2 \end{bmatrix}, \quad (8)$$

where these are the UNIT VECTORS BC and AB. Assume the TRIANGLE has been labeled such that $A = \mathbf{x}_1$ is the lower rightmost VERTEX and $C = \mathbf{x}_2$. Then the VECTORS obtained by traveling l_a and l_c along the sides and then inward PERPENDICULAR to them must meet

$$\begin{bmatrix} x_1 \\ y_1 \end{bmatrix} + l_c \begin{bmatrix} c_1 \\ c_2 \end{bmatrix} - k\gamma \begin{bmatrix} c_2 \\ -c_1 \end{bmatrix} = \begin{bmatrix} x_2 \\ y_2 \end{bmatrix} + l_a \begin{bmatrix} a_1 \\ a_2 \end{bmatrix} - k\alpha \begin{bmatrix} a_2 \\ -a_1 \end{bmatrix}. \quad (9)$$

Solving the two equations

$$x_1 + l_c c_1 - k\gamma c_2 = x_2 l_a a_1 - k\alpha a_2 \quad (10)$$

$$y_1 + l_c c_2 + k\gamma c_1 = y_2 l_a a_2 + k\alpha a_1, \quad (11)$$

gives

$$l_c = \frac{k\alpha(a_1c_1+a_2c_2) - \gamma k({c_1}^2+{c_2}^2) + c_2(x_1-x_2) + c_1(y_2-y_1)}{a_1c_2 - a_2c_1} \quad (12)$$

$$l_a = \frac{k\alpha({a_1}^2+{a_2}^2) - \gamma k(a_1c_1+a_2c_2) + a_2(x_1-x_2) + a_1(y_2-y_1)}{a_1c_2 - a_2c_1}. \quad (13)$$

But $\hat{\mathbf{a}}$ and $\hat{\mathbf{c}}$ are UNIT VECTORS, so

$$l_c = \frac{k\alpha(a_1c_1+a_2c_2) - \gamma k + c_2(x_1-x_2) + c_1(y_2-y_1)}{a_1c_2 - a_2c_1} \quad (14)$$

$$l_a = \frac{k\alpha - \gamma k(a_1c_1+a_2c_2) + a_2(x_1-x_2) + a_1(y_2-y_1)}{a_1c_2 - a_2c_1}. \quad (15)$$

And the VECTOR coordinates of the point $\alpha:\beta:\gamma$ are then

$$\mathbf{x} = \mathbf{x}_1 + l_c \begin{bmatrix} c_1 \\ c_2 \end{bmatrix} - k\gamma \begin{bmatrix} c_2 \\ -c_1 \end{bmatrix}. \quad (16)$$

see also AREAL COORDINATES, EXACT TRILINEAR COORDINATES, ORTHOCENTRIC COORDINATES, POWER CURVE, QUADRIPLANAR COORDINATES, TRIANGLE, TRILINEAR POLAR

References

Boyer, C. B. *History of Analytic Geometry.* New York: Yeshiva University, 1956.

Casey, J. "The General Equation—Trilinear Co-Ordinates." Ch. 10 in *A Treatise on the Analytical Geometry of the Point, Line, Circle, and Conic Sections, Containing an Account of Its Most Recent Extensions, with Numerous Examples, 2nd ed., rev. enl.* Dublin: Hodges, Figgis, & Co., pp. 333–348, 1893.

Coolidge, J. L. *A Treatise on Algebraic Plane Curves.* New York: Dover, pp. 67–71, 1959.

Coxeter, H. S. M. *Introduction to Geometry, 2nd ed.* New York: Wiley, 1969.

Coxeter, H. S. M. "Some Applications of Trilinear Coordinates." *Linear Algebra Appl.* **226–228**, 375–388, 1995.

Kimberling, C. "Triangle Centers and Central Triangles." *Congr. Numer.* **129**, 1–295, 1998.

Trilinear Line

A LINE is given in TRILINEAR COORDINATES by

$$l\alpha + m\beta + n\gamma = 0.$$

see also LINE, TRILINEAR COORDINATES

Trilinear Polar

Given a TRIANGLE CENTER $X = l:m:n$, the line

$$l\alpha + m\beta + n\gamma = 0$$

is called the trilinear polar of X^{-1} and is denoted L.

see also CHASLES'S POLARS THEOREM

Trillion

The word trillion denotes different numbers in American and British usage. In the American system, one trillion equals 10^{12}. In the British, French, and German systems, one trillion equals 10^{18}.

see also BILLION, LARGE NUMBER, MILLION

Trimagic Square

If replacing each number by its square or cube in a MAGIC SQUARE produces another MAGIC SQUARE, the square is said to be a trimagic square. Trimagic squares of order 32, 64, 81, and 128 are known. Tarry gave a method for constructing a trimagic square of order 128, Cazalas a method for trimagic squares of orders 64 and 81, and R. V. Heath a method for constructing an order 64 trimagic square which is different from Cazalas's (Kraitchik 1942).

Trimagic squares are also called TREBLY MAGIC SQUARES, and are 3-MULTIMAGIC SQUARES.

see also BIMAGIC SQUARE, MAGIC SQUARE, MULTIMAGIC SQUARE

References

Ball, W. W. R. and Coxeter, H. S. M. *Mathematical Recreations and Essays, 13th ed.* New York: Dover, pp. 212–213, 1987.

Kraitchik, M. "Multimagic Squares." §7.10 in *Mathematical Recreations.* New York: W. W. Norton, pp. 176–178, 1942.

Trimean

The trimean is defined to be

$$TM \equiv \tfrac{1}{4}(H_1 + 2M + H_2),$$

where H_i are the HINGES and M is the MEDIAN. Press *et al.* (1992) call this TUKEY'S TRIMEAN. It is an L-ESTIMATE.

see also HINGE, L-ESTIMATE, MEAN, MEDIAN (STATISTICS)

References

Press, W. H.; Flannery, B. P.; Teukolsky, S. A.; and Vetterling, W. T. *Numerical Recipes in FORTRAN: The Art of Scientific Computing, 2nd ed.* Cambridge, England: Cambridge University Press, p. 694, 1992.

Tukey, J. W. *Explanatory Data Analysis.* Reading, MA: Addison-Wesley, pp. 46–47, 1977.

Trimorphic Number

A number n such that the last digits of n^3 are the same as n. 49 is trimorphic since $49^3 = 117649$ (Wells 1986, p. 124). The first few are 1, 4, 5, 6, 9, 24, 25, 49, 51, 75, 76, 99, 125, 249, 251, 375, 376, 499, ...

see also AUTOMORPHIC NUMBER, NARCISSISTIC NUMBER, SUPER-3 NUMBER

References

Wells, D. *The Penguin Dictionary of Curious and Interesting Numbers.* Middlesex, England: Penguin Books, 1986.

Trinoid

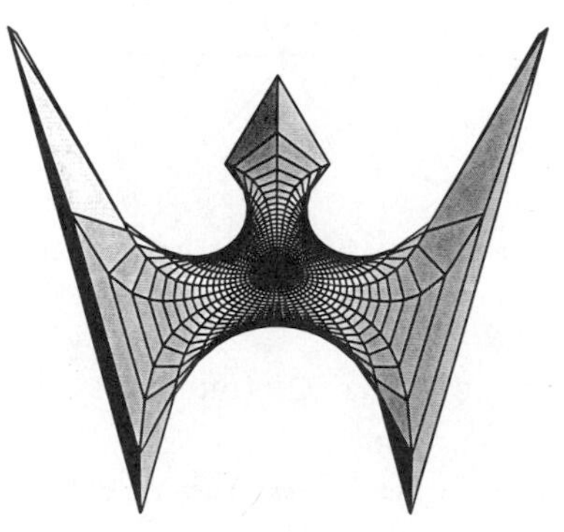

A MINIMAL SURFACE discovered by L. P. M. Jorge and W. Meeks III in 1983 with ENNEPER-WEIERSTRAß PARAMETERIZATION

$$f = \frac{1}{(\zeta^3 - 1)^2} \tag{1}$$

$$g = \zeta^2 \tag{2}$$

(Dickson 1990). Explicitly, it is given by

$$x = \Re\left[\frac{re^{i\theta}}{3(1 + re^{i\theta} + r^2e^{2i\theta})} - \frac{4\ln(re^{i\theta} - 1)}{9} + \frac{2\ln(1 + re^{i\theta} + r^2e^{2i\theta})}{9}\right] \tag{3}$$

$$y = -\tfrac{1}{9}\Im\left[\frac{-3re^{i\theta}(1 + re^{i\theta})}{r^3e^{3i\theta} - 1} + \frac{4\sqrt{3}\,(r^3e^{3i\theta} - 1)\tan^{-1}\left(\frac{1+2re^{i\theta}}{\sqrt{3}}\right)}{r^3e^{3i\theta} - 1}\right] \tag{4}$$

$$z = \Re\left[-\frac{2}{3} - \frac{2}{3(r^3e^{3i\theta} - 1)}\right], \tag{5}$$

for $\theta \in [0, 2\pi)$ and $r \in [0, 4]$.

see also MINIMAL SURFACE

References

Dickson, S. "Minimal Surfaces." *Mathematica J.* **1**, 38–40, 1990.

Wolfram Research "Mathematica Version 2.0 Graphics Gallery." `http://www.mathsource.com/cgi-bin/MathSource/Applications/Graphics/3D/0207-155`.

Trinomial

A POLYNOMIAL with three terms.

see also BINOMIAL, MONOMIAL, POLYNOMIAL

Trinomial Identity

$$(x^2 + axy + by^2)(t^2 + atu + bu^2) = r^2 + ars + bs^2, \tag{1}$$

where

$$r = xt - byu \tag{2}$$

$$s = yt + xu + ayu. \tag{3}$$

Trinomial Triangle

The NUMBER TRIANGLE obtained by starting with a row containing a single "1" and the next row containing three 1s and then letting subsequent row elements be computed by summing the elements above to the left, directly above, and above to the right:

$$\begin{array}{ccccccccc} &&&&1&&&& \\ &&&1&1&1&&& \\ &&1&2&3&2&1&& \\ &1&3&6&7&6&3&1& \\ 1&4&10&16&19&16&10&4&1 \end{array}$$

(Sloane's A027907). The nth row can also be obtained by expanding $(1+x+x^2)^n$ and taking coefficients:

$$(1+x+x^2)^0 = 1$$
$$(1+x+x^2)^1 = 1+x+x^2$$
$$(1+x+x^2)^2 = 1+2x+3x^2+2x^3+x^4$$

and so on.

see also PASCAL'S TRIANGLE

References

Sloane, N. J. A. Sequence A027907 in "An On-Line Version of the Encyclopedia of Integer Sequences."

Triomino

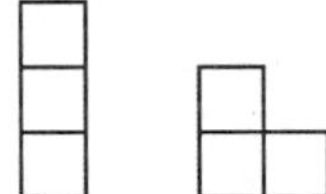

The two 3-POLYOMINOES are called triominoes, and are also known as the TROMINOES. The left triomino above is "STRAIGHT," while the right triomino is called "right" or L-.

see also L-POLYOMINO, POLYOMINO, STRAIGHT POLYOMINO

References

Gardner, M. "Polyominoes." Ch. 13 in *The Scientific American Book of Mathematical Puzzles & Diversions.* New York: Simon and Schuster, pp. 124–140, 1959.

Hunter, J. A. H. and Madachy, J. S. *Mathematical Diversions.* New York: Dover, pp. 80–81, 1975.

Lei, A. "Tromino." `http://www.cs.ust.hk/~philipl/omino/tromino.html`

Trip-Let

A 3-dimensional solid which is shaped in such a way that its projections along three mutually perpendicular axes are three different letters of the alphabet. Hofstadter (1989) has constructed such a solid for the letters G, E, and B.

see also CORK PLUG

References

Hofstadter, D. R. *Gödel, Escher, Bach: An Eternal Golden Braid.* New York: Vintage Books, cover and pp. xiv, 1, and 273, 1989.

Triple

A group of three elements, also called a TRIAD.

see also AMICABLE TRIPLE, MONAD, PAIR, PYTHAGOREAN TRIPLE, QUADRUPLET, QUINTUPLET, TETRAD, TRIAD, TWINS

Triple-Free Set

A SET of POSITIVE integers is called weakly triple-free if, for any integer x, the SET $\{x, 2x, 3x\} \not\subset S$. It is called strongly triple-free if $x \in S$ IMPLIES $2x \notin S$ and $3x \notin S$. Define

$$p(n) = \max\{|S| : S \subset \{1, 2, \ldots, n\} \text{ is weakly triple-free}\}$$
$$q(n) = \max\{|S| : S \subset \{1, 2, \ldots, n\} \text{ is strongly triple-free}\},$$

where $|S|$ denotes the CARDINALITY of S, then

$$\lim_{n\to\infty} \frac{p(n)}{n} \geq \tfrac{4}{5}$$

and

$$\lim_{n\to\infty} \frac{q(n)}{n} = 0.6134752692\ldots$$

(Finch).

see also DOUBLE-FREE SET

References

Finch, S. "Favorite Mathematical Constants." `http://www.mathsoft.com/asolve/constant/triple/triple.html`.

Triple Jacobi Product

see JACOBI TRIPLE PRODUCT

Triple Point

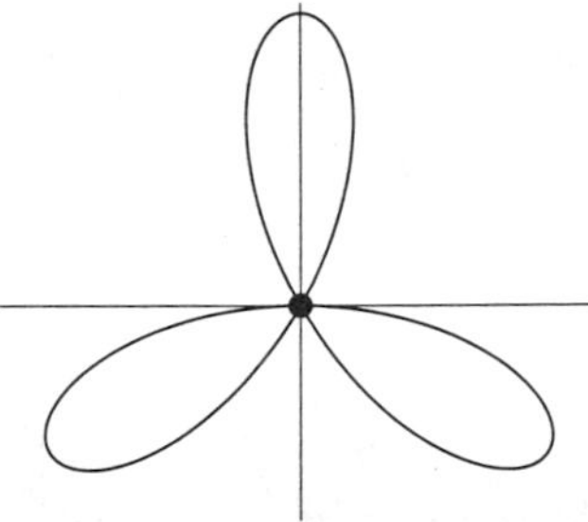

A point where a curve intersects itself along three arcs. The above plot shows the triple point at the ORIGIN of the TRIFOLIUM $(x^2+y^2)^2 + 3x^2y - y^3 = 0$.

see also DOUBLE POINT, QUADRUPLE POINT

References

Walker, R. J. *Algebraic Curves.* New York: Springer-Verlag, pp. 57–58, 1978.

Triple Scalar Product

see SCALAR TRIPLE PRODUCT

Triple Vector Product

see VECTOR TRIPLE PRODUCT

Triplet

see TRIPLE

Triplicate-Ratio Circle

see LEMOINE CIRCLE

Trisected Perimeter Point

A triangle center which has a TRIANGLE CENTER FUNCTION

$$\alpha = bc(v - c + a)(v - a + b),$$

where v is the unique REAL ROOT of

$$2x^3 - 3(a+b+c)x^2 + (a^2+b^2+c^2+8bc+8ca+8ab)x - (b^2c+c^2a+a^2b+5bc^2+5ca^2+5ab^2+9abc) = 0.$$

References

Kimberling, C. "Central Points and Central Lines in the Plane of a Triangle." *Math. Mag.* **67**, 163–187, 1994.

Trisection

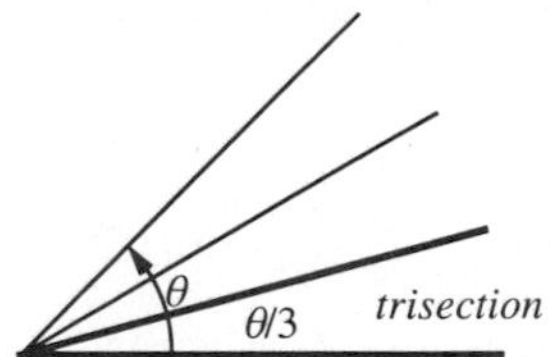

Angle trisection is the division of an *arbitrary* ANGLE into three equal ANGLES. It was one of the three GEOMETRIC PROBLEMS OF ANTIQUITY for which solutions using only COMPASS and STRAIGHTEDGE were sought. The problem was algebraically proved impossible by Wantzel (1836).

Although trisection is not possible for a general ANGLE using a Greek construction, there are some specific angles, such as $\pi/2$ and π radians (90° and 180°, respectively), which *can* be trisected. Furthermore, some ANGLES are geometrically trisectable, but cannot be constructed in the first place, such as $3\pi/7$ (Honsberger 1991). In addition, trisection of an arbitrary angle *can* be accomplished using a *marked* RULER (a NEUSIS CONSTRUCTION) as illustrated below (Courant and Robbins 1996).

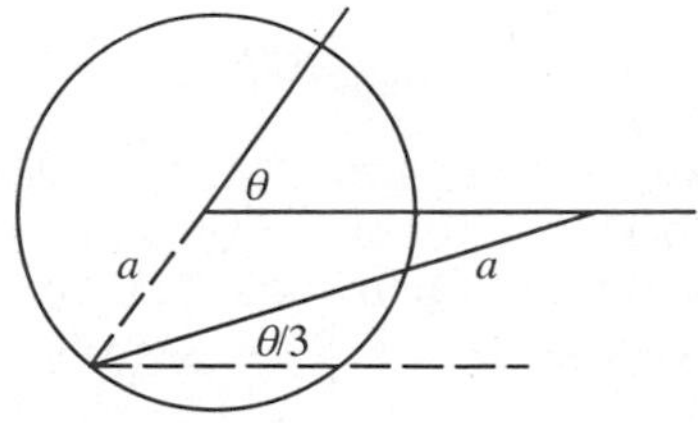

An ANGLE can also be divided into three (or any WHOLE NUMBER) of equal parts using the QUADRATRIX OF HIPPIAS or TRISECTRIX.

see also ANGLE BISECTOR, MACLAURIN TRISECTRIX, QUADRATRIX OF HIPPIAS, TRISECTRIX

References

Bogomolny, A. "Angle Trisection." `http://www.cut-the-knot.com/pythagoras/archi.html`.

Conway, J. H. and Guy, R. K. *The Book of Numbers.* New York: Springer-Verlag, pp. 190–191, 1996.

Courant, R. and Robbins, H. "Trisecting the Angle." §3.3.3 in *What is Mathematics?: An Elementary Approach to Ideas and Methods, 2nd ed.* Oxford, England: Oxford University Press, pp. 137–138, 1996.

Coxeter, H. S.M. "Angle Trisection." §2.2 in *Introduction to Geometry, 2nd ed.* New York: Wiley, p. 28, 1969.

Dixon, R. *Mathographics.* New York: Dover, pp. 50–51, 1991.

Dörrie, H. "Trisection of an Angle." §36 in *100 Great Problems of Elementary Mathematics: Their History and Solutions.* New York: Dover, pp. 172–177, 1965.

Dudley, U. *The Trisectors.* Washington, DC: Math. Assoc. Amer., 1994.

Geometry Center. "Angle Trisection." `http://www.geom.umn.edu:80/docs/forum/angtri/`.

Honsberger, R. *More Mathematical Morsels.* Washington, DC: Math. Assoc. Amer., pp. 25–26, 1991.

Ogilvy, C. S. "Angle Trisection." *Excursions in Geometry.* New York: Dover, pp. 135–141, 1990.

Wantzel, P. L. "Recherches sur les moyens de reconnaître si un Problème de Géométrie peut se résoudre avec la règle et le compas." *J. Math. pures appliq.* **1**, 366–372, 1836.

Trisectrix

see CATALAN'S TRISECTRIX, MACLAURIN TRISECTRIX

Trisectrix of Catalan

see CATALAN'S TRISECTRIX

Trisectrix of Maclaurin

see MACLAURIN TRISECTRIX

Triskaidecagon

see TRIDECAGON

Triskaidekaphobia

The number 13 is traditionally associated with bad luck. This superstition leads some people to fear or avoid anything involving this number, a condition known as triskaidekaphobia. Triskaidekaphobia leads to interesting practices such as the numbering of floors as 1, 2, ..., 11, 12, 14, 15, ..., *omitting the number 13,* in many high-rise hotels.

see also 13, BAKER'S DOZEN, FRIDAY THE THIRTEENTH, TRISKAIDEKAPHOBIA

Tritangent

The tritangent of a CUBIC SURFACE is a PLANE which intersects the surface in three mutually intersecting lines. Each intersection of two lines is then a tangent point of the surface.

see also CUBIC SURFACE

References

Hunt, B. "Algebraic Surfaces." `http://www.mathematik.uni-kl.de/~wwwagag/Galerie.html`.

Tritangent Triangle

see EXCENTRAL TRIANGLE

Trivial

According to the Nobel Prize-winning physicist Richard Feynman (1985), mathematicians designate any THEOREM as "trivial" once a proof has been obtained—no matter how difficult the theorem was to prove in the first place. There are therefore exactly two types of true mathematical propositions: trivial ones, and those which have not yet been proven.

see also PROOF, THEOREM

References

Feynman, R. P. and Leighton, R. *Surely You're Joking, Mr. Feynman!* New York: Bantam Books, 1985.

Trivialization

In the definition of a FIBER BUNDLE $f : E \to B$, the homeomorphisms $g_U : f^{-1}(U) \to U \times F$ that commute with projection are called local trivializations for the FIBER BUNDLE f.

see also FIBER BUNDLE

Trochoid

The curve described by a point at a distance b from the center of a rolling CIRCLE of RADIUS a.

$$x = a\phi - b\sin\phi$$
$$y = a - b\cos\phi.$$

If $b < a$, the curve is a CURTATE CYCLOID. If $b = a$, the curve is a CYCLOID. If $b > a$, the curve is a PROLATE CYCLOID.

see also CURTATE CYCLOID, CYCLOID, PROLATE CYCLOID

References

Lee, X. "Trochoid." `http://www.best.com/~xah/SpecialPlaneCurves_dir/Trochoid_dir/trochoid.html`.

Wagon, S. *Mathematica in Action.* New York: W. H. Freeman, pp. 46–50, 1991.

Yates, R. C. "Trochoids." *A Handbook on Curves and Their Properties.* Ann Arbor, MI: J. W. Edwards, pp. 233–236, 1952.

Tromino

see TRIOMINO

True

A statement which is rigorously known to be correct. A statement which is not true is called FALSE, although certain statements can be proved to be rigorously UNDECIDABLE within the confines of a given set of assumptions and definitions. Regular two-valued LOGIC allows statements to be only true or FALSE, but FUZZY LOGIC treats "truth" as a continuum which can have any value between 0 and 1.

see also ALETHIC, FALSE, FUZZY LOGIC, LOGIC, TRUTH TABLE, UNDECIDABLE

Truncate

To truncate a REAL NUMBER is to remove its nonintegral part. Truncation of a number x therefore corresponds to taking the FLOOR FUNCTION $\lfloor x \rfloor$.

see also CEILING FUNCTION, FLOOR FUNCTION, NINT, ROUND

Truncated Cone

see CONICAL FRUSTUM

Truncated Cube

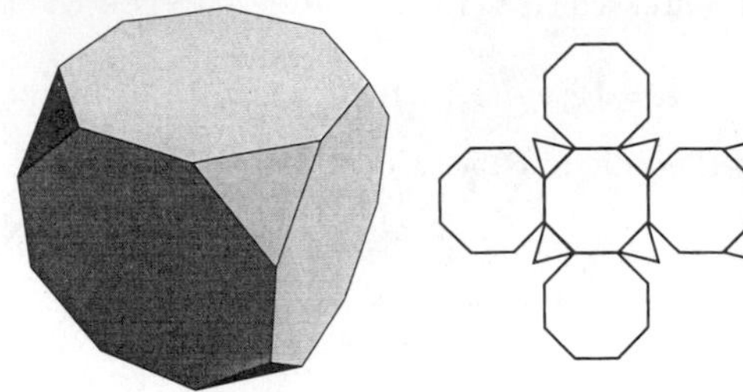

An ARCHIMEDEAN SOLID whose DUAL POLYHEDRON is the TRIAKIS OCTAHEDRON. It has SCHLÄFLI SYMBOL t{4,3}. It is also UNIFORM POLYHEDRON U_9 and has WYTHOFF SYMBOL $2\,3\,|\,4$. Its faces are $8\{3\}+6\{8\}$. The INRADIUS, MIDRADIUS, and CIRCUMRADIUS for $a = 1$ are

$$r = \tfrac{1}{17}(5+2\sqrt{2})\sqrt{7+4\sqrt{2}} \approx 1.63828$$
$$\rho = \tfrac{1}{2}(2+\sqrt{2}) \approx 1.70711$$
$$R = \tfrac{1}{2}\sqrt{7+4\sqrt{2}} \approx 1.77882.$$

References

Ball, W. W. R. and Coxeter, H. S. M. *Mathematical Recreations and Essays, 13th ed.* New York: Dover, p. 138, 1987.

Truncated Cuboctahedron

see GREAT RHOMBICUBOCTAHEDRON (ARCHIMEDEAN)

Truncated Dodecadodecahedron

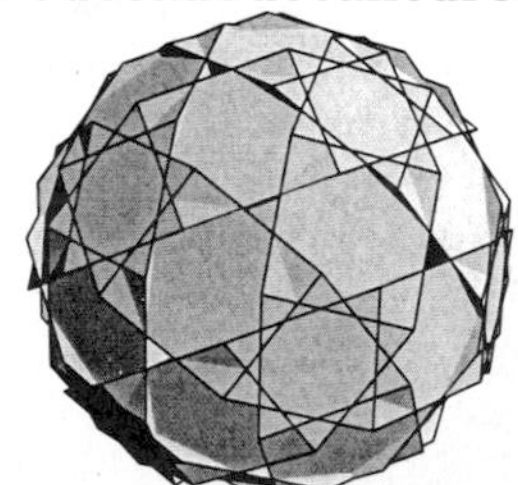

The UNIFORM POLYHEDRON U_{59}, also called the QUASITRUNCATED DODECAHEDRON, whose DUAL POLYHEDRON is the MEDIAL DISDYAKIS TRIACONTAHEDRON. It has SCHLÄFLI SYMBOL $t'\left\{\begin{matrix}\frac{5}{2}\\ 5\end{matrix}\right\}$ and WYTHOFF SYMBOL $2\,\frac{5}{3}\,|\,5$. Its faces are $12\{10\}+30\{4\}+12\{\frac{10}{3}\}$. Its CIRCUMRADIUS for $a = 1$ is

$$R = \tfrac{1}{2}\sqrt{11}.$$

References
Wenninger, M. J. *Polyhedron Models.* Cambridge, England: Cambridge University Press, pp. 152–153, 1989.

Truncated Dodecahedron

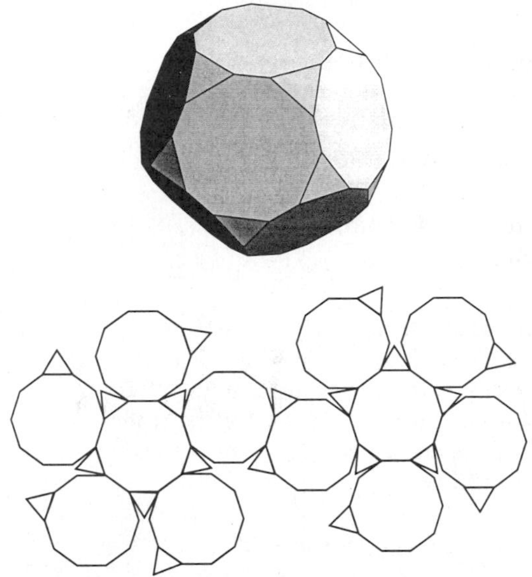

An ARCHIMEDEAN SOLID whose DUAL POLYHEDRON is the TRIAKIS ICOSAHEDRON. It has SCHLÄFLI SYMBOL t{5,3}. It is also UNIFORM POLYHEDRON U_{26} and has WYTHOFF SYMBOL $2\,3\,|\,5$. Its faces are $20\{3\}+12\{10\}$. The INRADIUS, MIDRADIUS, and CIRCUMRADIUS for $a=1$ are

$$r = \tfrac{5}{488}(17\sqrt{2}+3\sqrt{10})\sqrt{37+15\sqrt{5}} \approx 2.88526$$
$$\rho = \tfrac{1}{4}(5+3\sqrt{5}) \approx 2.92705$$
$$R = \tfrac{1}{4}\sqrt{74+30\sqrt{5}} \approx 2.96945.$$

Truncated Great Dodecahedron

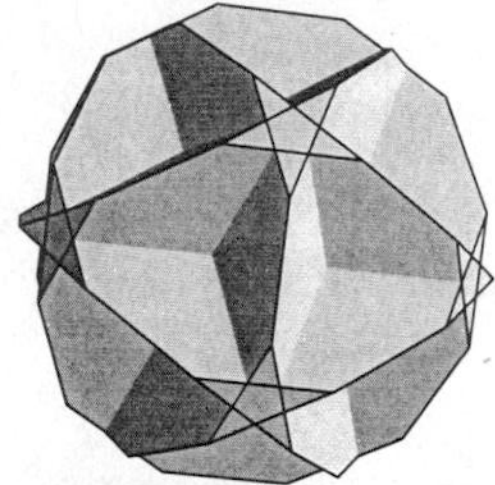

The UNIFORM POLYHEDRON U_{37} whose DUAL POLYHEDRON is the SMALL STELLAPENTAKIS DODECAHEDRON. It has SCHLÄFLI SYMBOL $t\{5,\frac{5}{2}\}$. It has WYTHOFF SYMBOL $2\,\frac{5}{2}\,5$. Its faces are $12\{\frac{5}{2}\}+12\{10\}$. Its CIRCUMRADIUS for $a=1$ is

$$R = \tfrac{1}{4}\sqrt{34+10\sqrt{5}}.$$

see also GREAT ICOSAHEDRON

References
Wenninger, M. J. *Polyhedron Models.* Cambridge, England: Cambridge University Press, p. 115, 1971.

Truncated Great Icosahedron

see GREAT TRUNCATED ICOSAHEDRON

Truncated Hexahedron

see TRUNCATED CUBE

Truncated Icosahedron

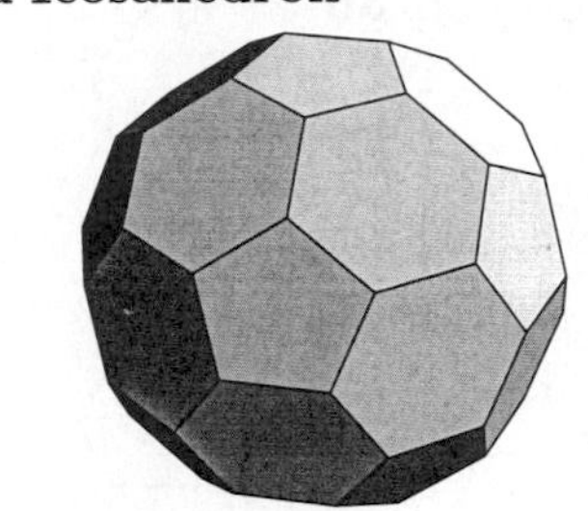

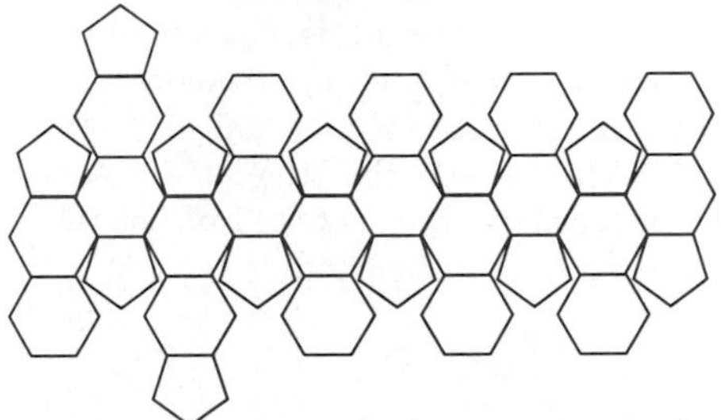

An ARCHIMEDEAN SOLID used in the construction of SOCCER BALLS. Its DUAL POLYHEDRON is the PENTAKIS DODECAHEDRON. It has SCHLÄFLI SYMBOL t{3,5}. It is also UNIFORM POLYHEDRON U_{25} and has WYTHOFF SYMBOL $2\,5\,|\,3$. Its faces are $20\{6\}+12\{5\}$. The INRADIUS, MIDRADIUS, and CIRCUMRADIUS for $a=1$ are

$$r = \tfrac{9}{872}(21+\sqrt{5})\sqrt{58+18\sqrt{5}} \approx 2.37713$$
$$\rho = \tfrac{3}{4}(1+\sqrt{5}) \approx 2.42705$$
$$R = \tfrac{1}{4}\sqrt{58+18\sqrt{5}} \approx 2.47802.$$

Truncated Icosidodecahedron

see GREAT RHOMBICOSIDODECAHEDRON (ARCHIMEDEAN)

Truncated Octahedral Number

A FIGURATE NUMBER which is constructed as an OCTAHEDRAL NUMBER with a SQUARE PYRAMID removed from each of the six VERTICES,

$$TO_n = O_{3n-2} - 6P_{n-1} = \tfrac{1}{3}(3n-2)[2(3n-2)^2+1],$$

where O_n is an OCTAHEDRAL NUMBER and P_n is a PYRAMIDAL NUMBER. The first few are 1, 38, 201, 586, ... (Sloane's A005910). The GENERATING FUNCTION for the truncated octahedral numbers is

$$\frac{x(6x^3+55x^2+34x+1)}{(x-1)^4} = x+38x^2+201x^3+\ldots.$$

see also OCTAHEDRAL NUMBER

References

Conway, J. H. and Guy, R. K. *The Book of Numbers.* New York: Springer-Verlag, p. 52, 1996.

Sloane, N. J. A. Sequence A005910/M5266 in "An On-Line Version of the Encyclopedia of Integer Sequences."

Truncated Octahedron

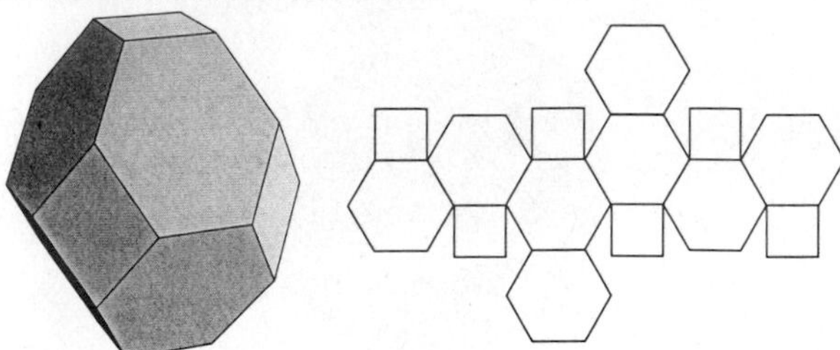

An ARCHIMEDEAN SOLID, also known as the MECON, whose DUAL POLYHEDRON is the TETRAKIS HEXAHEDRON. It is also UNIFORM POLYHEDRON U_8 and has SCHLÄFLI SYMBOL t$\{3,4\}$ and WYTHOFF SYMBOL $2\,4\,|\,3$. The faces of the truncated octahedron are $8\{6\}+6\{4\}$. The truncated octahedron has the O_h OCTAHEDRAL GROUP of symmetries.

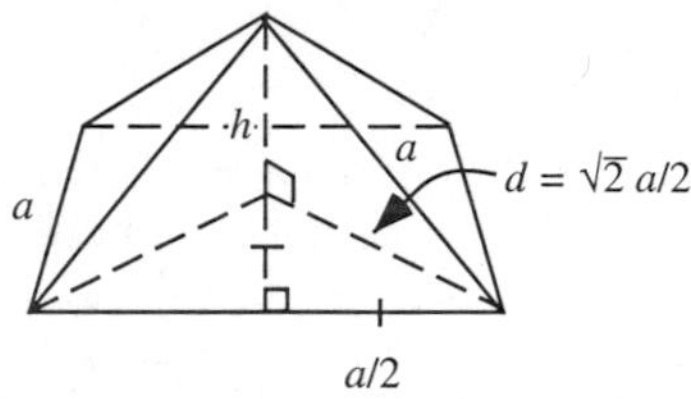

The solid can be formed from an OCTAHEDRON via TRUNCATION by removing six SQUARE PYRAMIDS, each with edge slant height $a = s/3$ and height h, where s is the side length of the original OCTAHEDRON. From the above diagram, the height and base area of the SQUARE PYRAMID are

$$h = \sqrt{a^2 - d^2} = \tfrac{1}{2}\sqrt{2}\,a \qquad (1)$$

$$A_b = a^2. \qquad (2)$$

The VOLUME of the truncated octahedron is then given by the VOLUME of the OCTAHEDRON

$$V_{\rm octahedron} = \tfrac{1}{3}\sqrt{2}\,s^3 = 9\sqrt{2}\,a^3 \qquad (3)$$

minus six times the volume of the SQUARE PYRAMID,

$$V = V_{\rm octahedron} - 6(\tfrac{1}{3}A_b h) = (9\sqrt{2} - \sqrt{2})a^3 = 8\sqrt{2}\,a^3. \qquad (4)$$

The truncated octahedron is a SPACE-FILLING POLYHEDRON. The INRADIUS, MIDRADIUS, and CIRCUMRADIUS for $a = 1$ are

$$r = \tfrac{9}{20}\sqrt{10} \approx 1.42302 \qquad (5)$$

$$\rho = \tfrac{3}{2} = 1.5 \qquad (6)$$

$$R = \tfrac{1}{2}\sqrt{10} \approx 1.58114. \qquad (7)$$

see also OCTAHEDRON, SQUARE PYRAMID, TRUNCATION

References

Coxeter, H. S. M. *Regular Polytopes, 3rd ed.* New York: Dover, pp. 29–30 and 257, 1973.

Truncated Polyhedron

A polyhedron with truncated faces, given by the SCHLÄFLI SYMBOL t$\left\{ {p \atop q} \right\}$.

see also RHOMBIC POLYHEDRON, SNUB POLYHEDRON

Truncated Pyramid

see PYRAMIDAL FRUSTUM

Truncated Square Pyramid

The truncated square pyramid is a special case of a PYRAMIDAL FRUSTUM for a SQUARE PYRAMID. Let the base and top side lengths of the truncated pyramid be a and b, and let the height be h. Then the VOLUME of the solid is

$$V = \tfrac{1}{3}(a^2 + ab + b^2)h.$$

This FORMULA was known to the Egyptians ca. 1850 BC. The Egyptians cannot have proved it without calculus, however, since Dehn showed in 1900 that no proof of this equation exists which does not rely on the concept of continuity (and therefore some form of INTEGRATION).

see also FRUSTUM, PYRAMID, PYRAMIDAL FRUSTUM, SQUARE PYRAMID

Truncated Tetrahedral Number

A FIGURATE NUMBER constructed by taking the $(3n-2)$th TETRAHEDRAL NUMBER and removing the $(n-1)$th TETRAHEDRAL NUMBER from each of the four corners,

$$\mathrm{Ttet}_n \equiv \mathrm{Te}_{3n-3} - 4\mathrm{Te}_{n-1} = \tfrac{1}{6}n(23n^2 - 27n + 10).$$

The first few are 1, 16, 68, 180, 375, ... (Sloane's A005906). The GENERATING FUNCTION for the truncated tetrahedral numbers is

$$\frac{x(10x^2 + 12x + 1)}{(x-1)^4} = x + 16x^2 + 89x^3 + 180x^4 + \ldots.$$

References

Conway, J. H. and Guy, R. K. *The Book of Numbers.* New York: Springer-Verlag, pp. 46–47, 1996.

Sloane, N. J. A. Sequence A005906/M5002 in "An On-Line Version of the Encyclopedia of Integer Sequences."

Truncated Tetrahedron

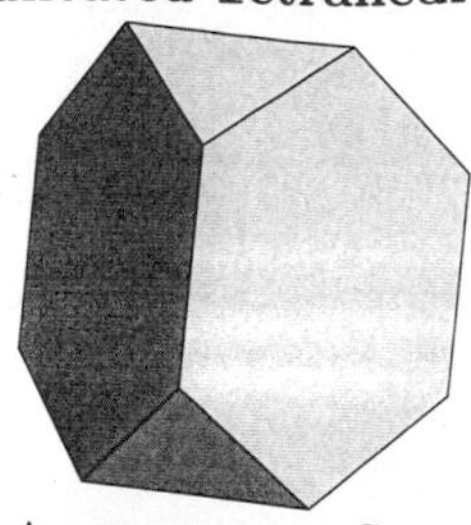

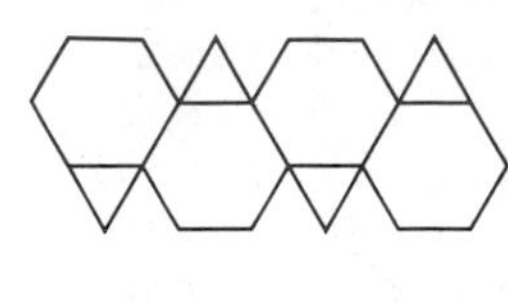

An ARCHIMEDEAN SOLID whose dual is the TRIAKIS TETRAHEDRON. It has SCHLÄFLI SYMBOL t{3,3}. It is also UNIFORM POLYHEDRON U_2 and has WYTHOFF SYMBOL $2\,3\,|\,3$. Its faces are $4\{3\}+4\{6\}$. The INRADIUS, MIDRADIUS, and CIRCUMRADIUS for a truncated tetrahedron with $a=1$ are

$$r = \tfrac{9}{44}\sqrt{22} \approx 0.95940$$
$$\rho = \tfrac{3}{4}\sqrt{2} \approx 1.06066$$
$$R = \tfrac{1}{4}\sqrt{22} \approx 1.17260.$$

Truncation

The removal of portions of SOLIDS falling outside a set of symmetrically placed planes. The five PLATONIC SOLIDS belong to one of the following three truncation series (which, in the first two cases, carry the solid to its DUAL POLYHEDRON).

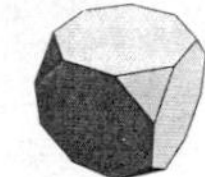
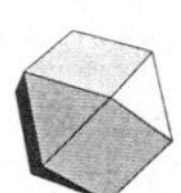
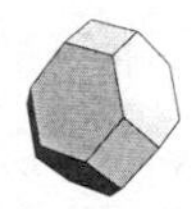
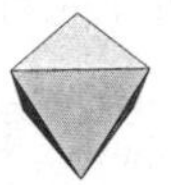

Cube, Truncated Cube, Cuboctahedron, Truncated Octahedron, Octahedron

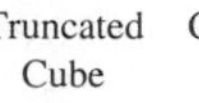
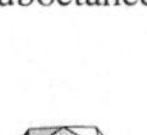
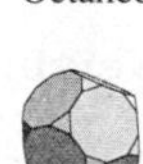
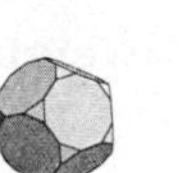

Icosahedron, Truncated Icosahedron, Icosidodecahedron, Truncated Dodecahedron, Dodecahedron

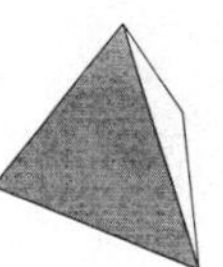
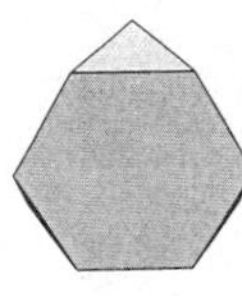
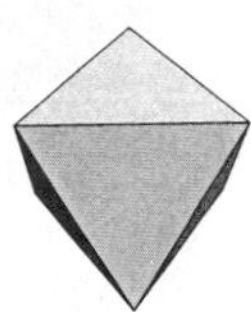

Tetrahedron, Truncated Tetrahedron, Octahedron

see also STELLATION, TRUNCATED CUBE, TRUNCATED DODECAHEDRON, TRUNCATED ICOSAHEDRON, TRUNCATED OCTAHEDRON, TRUNCATED TETRAHEDRON, VERTEX FIGURE

Truth Table

A truth table is a 2-D array with $n+1$ columns. The first n columns correspond to the possible values of n inputs, and the last column to the operation being performed. The rows list all possible combinations of inputs together with the corresponding outputs. For example, the following truth table shows the result of the binary AND operator acting on two inputs A and B, each of which may be true or false.

A	B	$A \wedge B$
F	F	F
F	T	F
T	F	F
T	T	T

see also AND, MULTIPLICATION TABLE, OR, XOR

Tschebyshev

An alternative spelling of the name "Chebyshev."

see also CHEBYSHEV APPROXIMATION FORMULA, CHEBYSHEV CONSTANTS, CHEBYSHEV DEVIATION, CHEBYSHEV DIFFERENTIAL EQUATION, CHEBYSHEV FUNCTION, CHEBYSHEV-GAUSS QUADRATURE, CHEBYSHEV INEQUALITY, CHEBYSHEV INTEGRAL, CHEBYSHEV PHENOMENON, CHEBYSHEV POLYNOMIAL OF THE FIRST KIND, CHEBYSHEV POLYNOMIAL OF THE SECOND KIND, CHEBYSHEV QUADRATURE, CHEBYSHEV-RADAU QUADRATURE, CHEBYSHEV-SYLVESTER CONSTANT

Tschirnhausen Cubic

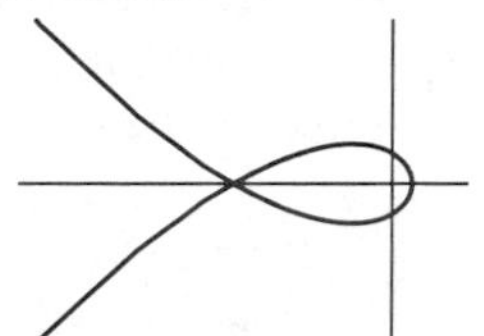

The Tschirnhausen cubic is a plane curve given by

$$a = r\cos^3\left(\tfrac{1}{3}\theta\right),$$

and is also known as CATALAN'S TRISECTRIX and L'HOSPITAL'S CUBIC. The name Tschirnhaus's cubic is given in R. C. Archibald's 1900 paper attempting to classify curves (MacTutor Archive). Tschirnhaus's cubic is the NEGATIVE PEDAL CURVE of a PARABOLA with respect to the FOCUS.

References

Lawrence, J. D. *A Catalog of Special Plane Curves.* New York: Dover, pp. 87–90, 1972.

MacTutor History of Mathematics Archive. "Tschirnhaus's Cubic." `http://www-groups.dcs.st-and.ac.uk/~history/Curves/Tschirnhaus.html`.

Tschirnhausen Cubic Caustic

The CAUSTIC of the TSCHIRNHAUSEN CUBIC taking the RADIANT POINT as the pole is NEILE'S PARABOLA.

Tschirnhausen Cubic Pedal Curve

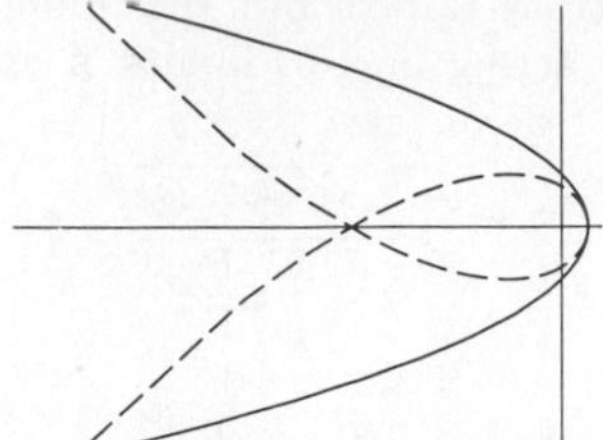

The PEDAL CURVE to the TSCHIRNHAUSEN CUBIC for PEDAL POINT at the origin is the PARABOLA

$$x = 1 - t^2$$
$$y = 2t.$$

see also PARABOLA, PEDAL CURVE, PEDAL POINT, TSCHIRNHAUSEN CUBIC

Tschirnhausen Transformation

A transformation of a POLYNOMIAL equation $f(x) = 0$ which is of the form $y = g(x)/h(x)$ where g and h are POLYNOMIALS and $h(x)$ does not vanish at a root of $f(x) = 0$. The CUBIC EQUATION is a special case of such a transformation. Tschirnhaus (1683) showed that a POLYNOMIAL of degree $n > 2$ can be reduced to a form in which the x^{n-1} and x^{n-2} terms have 0 COEFFICIENTS. In 1786, E. S. Bring showed that a general QUINTIC EQUATION can be reduced to the form

$$x^5 + px + q = 0.$$

In 1834, G. B. Jerrard showed that a Tschirnhaus transformation can be used to eliminate the x^{n-1}, x^{n-2}, *and* x^{n-3} terms for a general POLYNOMIAL equation of degree $n > 3$.

see also BRING QUINTIC FORM, CUBIC EQUATION

References

Boyer, C. B. *A History of Mathematics.* New York: Wiley, pp. 472–473, 1968.

Tschirnhaus. *Acta Eruditorum.* 1683.

Tubular Neighborhood

The tubular embedding of a SUBMANIFOLD $M^m \subset N^n$ of another MANIFOLD N^n is an EMBEDDING $t : M \times \mathbb{B}^{n-m} \to N$ such that $t(x,0) = x$ whenever $x \in M$, where $\mathbb{B}^{n-m}$ is the unit BALL in $\mathbb{R}^{n-m}$ centered at 0. The tubular neighborhood is also called the PRODUCT NEIGHBORHOOD.

see also BALL, EMBEDDING, PRODUCT NEIGHBORHOOD

References

Rolfsen, D. *Knots and Links.* Wilmington, DE: Publish or Perish Press, pp. 34–35, 1976.

Tucker Circles

Let three equal lines P_1Q_1, P_2Q_2, and P_3Q_3 be drawn ANTIPARALLEL to the sides of a triangle so that two (say P_2Q_2 and P_3Q_3) are on the same side of the third line as $A_2P_2Q_3A_3$. Then $P_2Q_3P_3Q_2$ is an isosceles TRAPEZOID, i.e., P_3Q_2, P_1Q_3, and P_2Q_1 are parallel to the respective sides. The MIDPOINTS C_1, C_2, and C_3 of the antiparallels are on the respective symmedians and divide them proportionally.

If T divides KO in the same ratio, TC_1, TC_2, TC_3 are parallel to the radii OA_1, OA_2, and OA_3 and equal. Since the antiparallels are perpendicular to the symmedians, they are equal chords of a circle with center T which passes through the six given points. This circle is called the Tucker circle.

If

$$c \equiv \frac{\overline{KC_1}}{\overline{KA_1}} = \frac{\overline{KC_2}}{\overline{KA_2}} = \frac{\overline{KC_3}}{\overline{KA_3}} = \frac{\overline{KT}}{\overline{KO}},$$

then the radius of the Tucker circle is

$$R\sqrt{c^2 + (1-c)^2 \tan\omega},$$

where ω is the BROCARD ANGLE.

The COSINE CIRCLE, LEMOINE CIRCLE, and TAYLOR CIRCLE are Tucker circles.

see also ANTIPARALLEL, BROCARD ANGLE, COSINE CIRCLE, LEMOINE CIRCLE, TAYLOR CIRCLE

References

Johnson, R. A. *Modern Geometry: An Elementary Treatise on the Geometry of the Triangle and the Circle.* Boston, MA: Houghton Mifflin, pp. 271–277 and 300–301, 1929.

Tukey's Biweight

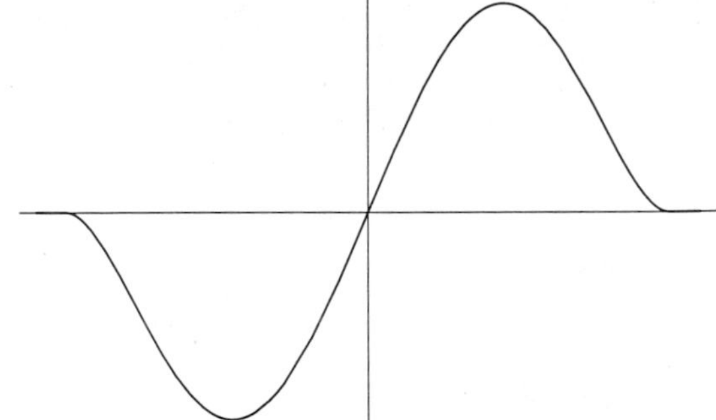

The function

$$\psi(z) = \begin{cases} z\left(1 - \frac{z^2}{c^2}\right)^2 & \text{for } |z| < c \\ 0 & \text{for } |z| > c \end{cases}$$

sometimes used in ROBUST ESTIMATION. It has a minimum at $z = -c/\sqrt{3}$ and a maximum at $z = c/\sqrt{3}$, where

$$\psi'(z) = 1 - \frac{3x^2}{c^2} = 0,$$

and an inflection point at $z = 0$, where

$$\psi''(z) = -\frac{6z}{c^2} = 0.$$

References
Press, W. H.; Flannery, B. P.; Teukolsky, S. A.; and Vetterling, W. T. *Numerical Recipes in FORTRAN: The Art of Scientific Computing, 2nd ed.* Cambridge, England: Cambridge University Press, p. 697, 1992.

Tukey's Trimean

see TRIMEAN

Tunnel Number

Let a KNOT K be n-EMBEDDABLE. Then its tunnel number is a KNOT invariant which is related to n.

see also EMBEDDABLE KNOT

References
Adams, C. C. *The Knot Book: An Elementary Introduction to the Mathematical Theory of Knots.* New York: W. H. Freeman, p. 114, 1994.

Turán Graph

The (n, k)-Turán graph is the EXTREMAL GRAPH on n VERTICES which contains no k-CLIQUE. In other words, the Turán graph has the maximum possible number of EDGES of any n-vertex graph not containing a COMPLETE GRAPH K_k. TURÁN'S THEOREM gives the maximum number of edges $t(n, k)$ for the (n, k)-Turán graph. For $k = 3$,

$$t(n,3) = \tfrac{1}{4}n^4,$$

so the Turán graph is given by the COMPLETE BIPARTITE GRAPHS

$$\begin{cases} K_{n/2,n/2} & n \text{ even} \\ K_{(n-1)/2,(n+1)/2} & n \text{ odd.} \end{cases}$$

see also CLIQUE, COMPLETE BIPARTITE GRAPH, TURÁN'S THEOREM

References
Aigner, M. "Turán's Graph Theorem." *Amer. Math. Monthly* **102**, 808–816, 1995.

Turán's Inequalities

For a set of POSITIVE γ_k, $k = 0, 1, 2\ldots$, Turán's inequalities are given by

$$\gamma_k{}^2 - \gamma_{k-1}\gamma_{k+1} \geq 0$$

for $k = 1, 2, \ldots$.

see also JENSEN POLYNOMIAL

References
Csordas, G.; Varga, R. S.; and Vincze, I. "Jensen Polynomials with Applications to the Riemann ζ-Function." *J. Math. Anal. Appl.* **153**, 112–135, 1990.
Szegő, G. *Orthogonal Polynomials, 4th ed.* Providence, RI: Amer. Math. Soc., p. 388, 1975.

Turán's Theorem

Let $G(V, E)$ be a GRAPH with VERTICES V and EDGES E on n VERTICES without a k-CLIQUE. Then

$$t(n,k) \leq \frac{(k-2)n^2}{2(k-1)},$$

where $t(n,k) = |E|$ is the EDGE NUMBER. More precisely, the K-GRAPH $K_{n_1,\ldots,n_{k-1}}$ with $|n_i - n_j| \leq 1$ for $i \neq j$ is the unique GRAPH without a k-CLIQUE with the maximal number of EDGES $t(n, k)$.

see also CLIQUE, K-GRAPH, TURÁN GRAPH

References
Aigner, M. "Turán's Graph Theorem." *Amer. Math. Monthly* **102**, 808–816, 1995.

Turbine

A VECTOR FIELD on a CIRCLE in which the directions of the VECTORS are all at the same ANGLE to the CIRCLE.

see also CIRCLE, VECTOR FIELD

Turing Machine

A theoretical computing machine which consists of an infinitely long magnetic tape on which instructions can be written and erased, a single-bit register of memory, and a processor capable of carrying out the following instructions: move the tape right, move the tape left, change the state of the register based on its current value and a value on the tape, and write or erase a value on the tape. The machine keeps processing instructions until it reaches a particular state, causing it to halt. Determining whether a Turing machine will halt for a given input and set of rules is called the HALTING PROBLEM.

see also BUSY BEAVER, CELLULAR AUTOMATON, CHAITIN'S OMEGA, CHURCH-TURING THESIS, COMPUTABLE NUMBER, HALTING PROBLEM, UNIVERSAL TURING MACHINE

References
Penrose, R. "Algorithms and Turning Machines." Ch. 2 in *The Emperor's New Mind: Concerning Computers, Minds, and the Laws of Physics.* Oxford, England: Oxford University Press, pp. 30–73, 1989.
Turing, A. M. "On Computable Numbers, with an Application to the Entscheidungsproblem." *Proc. London Math. Soc. Ser. 2* **42**, 230–265, 1937.
Turing, A. M. "Correction to: On Computable Numbers, with an Application to the Entscheidungsproblem." *Proc. London Math. Soc. Ser. 2* **43**, 544–546, 1938.

Turning Angle

see TANGENTIAL ANGLE

Tutte's Graph

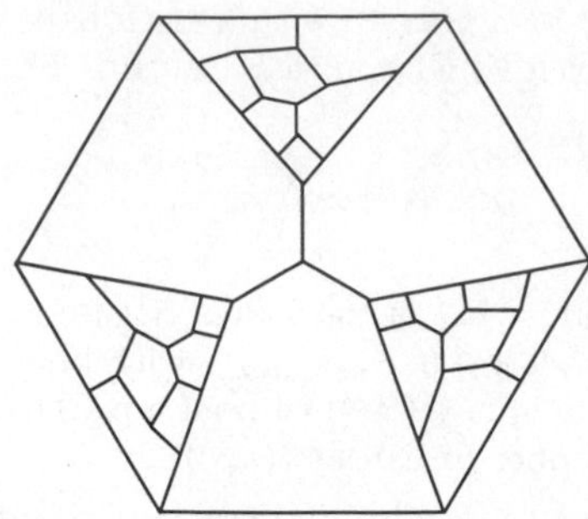

A counterexample to TAIT'S HAMILTONIAN GRAPH CONJECTURE given by Tutte (1946). A simpler counterexample was later given by Kozyrev and Grinberg.

see also HAMILTONIAN CIRCUIT, TAIT'S HAMILTONIAN GRAPH CONJECTURE

References

Honsberger, R. *Mathematical Gems I.* Washington, DC: Math. Assoc. Amer., pp. 82–89, 1973.

Saaty, T. L. and Kainen, P. C. *The Four-Color Problem: Assaults and Conquest.* New York: Dover, p. 112, 1986.

Tutte, W. T. "On Hamiltonian Circuits." *J. London Math. Soc.* **21**, 98–101, 1946.

Tutte Polynomial

Let G be a GRAPH, and let $\mathrm{ea}(T)$ denote the cardinality of the set of externally active edges of a spanning tree T of G and $\mathrm{ia}(T)$ denote the cardinality of the set of internally active edges of T. Then

$$t_G(x,y) = \sum_{T\subseteq G} x^{\mathrm{ia}(T)} y^{\mathrm{ea}(T)}.$$

References

Gessel, I. M. and Sagan, B. E. "The Tutte Polynomial of a Graph, Depth-First Search, and Simplicial Complex Partitions." *Electronic J. Combinatorics* **3**, No. 2, R9, 1–36, 1996. `http://www.combinatorics.org/Volume_3/volume3_2.html#R9`.

Tutte, W. T. "A Contribution to the Theory of Chromatic Polynomials." *Canad. J. Math.* **6**, 80–91, 1953.

Tutte's Theorem

Let G be a GRAPH and S a SUBGRAPH of G. Let the number of ODD components in $G - S$ be denoted S', and $|S|$ the number of VERTICES of S. The condition $|S| \geq S'$ for every SUBSET of VERTICES is NECESSARY and SUFFICIENT for G to have a 1-FACTOR.

see also FACTOR (GRAPH)

References

Honsberger, R. "Lovász' Proof of a Theorem of Tutte." Ch. 14 in *Mathematical Gems II.* Washington, DC: Math. Assoc. Amer., pp. 147–157, 1976.

Tutte, W. T. "The Factorization of Linear Graphs." *J. London Math. Soc.* **22**, 107–111, 1947.

Twin Peaks

For an INTEGER $n \geq 2$, let $\mathrm{lpf}(x)$ denote the LEAST PRIME FACTOR of n. A PAIR of INTEGERS (x, y) is called a twin peak if

1. $x < y$,
2. $\mathrm{lpf}(x) = \mathrm{lpf}(y)$,
3. For all z, $x < z < y$ IMPLIES $\mathrm{lpf}(z) < \mathrm{lpf}(x)$.

A broken-line graph of the least prime factor function resembles a jagged terrain of mountains. In terms of this terrain, a twin peak consists of two mountains of equal height with no mountain of equal or greater height between them. Denote the height of twin peak (x, y) by $p = \mathrm{lpf}(x) = \mathrm{lpf}(y)$. By definition of the LEAST PRIME FACTOR function, p must be PRIME.

Call the distance between two twin peaks (x, y)

$$s \equiv y - x.$$

Then s must be an EVEN multiple of p; that is, $s = kp$ where k is EVEN. A twin peak with $s = kp$ is called a kp-twin peak. Thus we can speak of $2p$-twin peaks, $4p$-twin peaks, etc. A kp-twin peak is fully specified by k, p, and x, from which we can easily compute $y \equiv x + kp$.

The set of kp-twin peaks is periodic with period $q = p\#$, where $p\#$ is the PRIMORIAL of p. That is, if (x, y) is a kp-twin peak, then so is $(x + q, y + q)$. A fundamental kp-twin peak is a twin peak having x in the fundamental period $[0, q)$. The set of fundamental kp-twin peaks is symmetric with respect to the fundamental period; that is, if (x, y) is a twin peak on $[0, q)$, then so is $(q-y, q-x)$.

The question of the EXISTENCE of twin peaks was first raised by David Wilson in the math-fun mailing list on Feb. 10, 1997. Wilson already had privately showed the EXISTENCE of twin peaks of height $p \leq 13$ to be unlikely, but was unable to rule them out altogether. Later that same day, John H. Conway, Johan de Jong, Derek Smith, and Manjul Bhargava collaborated to discover the first twin peak. Two hours at the blackboard revealed that $p = 113$ admits the $2p$-twin peak

$$x = 126972592296404970720882679404584182254788131,$$

which settled the EXISTENCE question. Immediately thereafter, Fred Helenius found the smaller $2p$-twin peak with $p = 89$ and

$$x = 9503844926749390990454854843625839.$$

The effort now shifted to finding the least PRIME p admitting a $2p$-twin peak. On Feb. 12, 1997, Fred Helenius found $p = 71$, which admits 240 fundamental $2p$-twin peaks, the least being

$$x = 7310131732015251470110369.$$

Helenius's results were confirmed by Dan Hoey, who also computed the least $2p$-twin peak $L(2p)$ and number of fundamental $2p$-twin peaks $N(2p)$ for $p = 73$, 79, and 83. His results are summarized in the following table.

p	$L(2p)$	$N(2p)$
71	7310131732015251470110369	240
73	2061519317176132799110061	40296
79	3756800873017263196139951	164440
83	6316254452384500173544921	6625240

The $2p$-twin peak of height $p = 73$ is the smallest known twin peak. Wilson found the smallest known $4p$-twin peak with $p = 1327$, as well as another very large $4p$-twin peak with $p = 3203$. Richard Schroeppel noted that the latter twin peak is at the high end of its fundamental period and that its reflection within the fundamental period $[0, p\#)$ is smaller.

Many open questions remain concerning twin peaks, e.g.,

1. What is the smallest twin peak (smallest n)?
2. What is the least PRIME p admitting a $4p$-twin peak?
3. Do $6p$-twin peaks exist?
4. Is there, as Conway has argued, an upper bound on the span of twin peaks?
5. Let $p < q < r$ be PRIME. If p and r each admit kp-twin peaks, does q then necessarily admit a kp-twin peak?

see also ANDRICA'S CONJECTURE, DIVISOR FUNCTION, LEAST COMMON MULTIPLE, LEAST PRIME FACTOR

Twin Prime Conjecture

Adding a correction proportional to $1/\ln p$ to a computation of BRUN'S CONSTANT ending with $\ldots + 1/p + 1/(p+2)$ will give an estimate with error less than $c(\sqrt{p}\ln p)^{-1}$. An extended form of the conjecture states that

$$P_x(p, p+2) \sim 2\Pi_2 \int_2^x \frac{dx}{(\ln x)^2},$$

where Π_2 is the TWIN PRIMES CONSTANT. The twin prime conjecture is a special case of the more general PRIME PATTERNS CONJECTURE corresponding to the set $S = \{0, 2\}$.

see also BRUN'S CONSTANT, PRIME ARITHMETIC PROGRESSION, PRIME CONSTELLATION, PRIME PATTERNS CONJECTURE, TWIN PRIMES

Twin Primes

Twin primes are PRIMES (p, q) such that $p - q = 2$. The first few twin primes are $n \pm 1$ for $n = 4$, 6, 12, 18, 30, 42, 60, 72, 102, 108, 138, 150, 180, 192, 198, 228, 240, 270, 282, ... (Sloane's A014574). Explicitly, these are (3, 5), (5, 7), (11, 13), (17, 19), (29, 31), (41, 43), ... (Sloane's A001359 and A006512).

Let $\pi_2(n)$ be the number of twin primes p and $p+2$ such that $p \leq n$. It is not known if there are an infinite number of such PRIMES (Shanks 1993), but all twin primes except (3, 5) are of the form $6n \pm 1$. J. R. Chen has shown there exists an INFINITE number of PRIMES p such that $p + 2$ has at most two factors (Le Lionnais 1983, p. 49). Bruns proved that there exists a computable INTEGER x_0 such that if $x \geq x_0$, then

$$\pi_2(x) < \frac{100x}{(\ln x)^2} \tag{1}$$

(Ribenboim 1989, p. 201). It has been shown that

$$\pi_2(x) \leq c \prod_{p>2} \left[1 - \frac{1}{(p-1)^2}\right] \frac{x}{(\ln x)^2} \left[1 + \mathcal{O}\left(\frac{\ln \ln x}{\ln x}\right)\right], \tag{2}$$

where c has been reduced to $68/9 \approx 7.5556$ (Fouvry and Iwaniec 1983), $128/17 \approx 7.5294$ (Fouvry 1984), 7 (Bombieri *et al.* 1986), 6.9075 (Fouvry and Grupp 1986), and 6.8354 (Wu 1990). The bound on c is further reduced to 6.8324107886 in a forthcoming thesis by Haugland (1998). This calculation involved evaluation of 7-fold integrals and fitting of three different parameters. Hardy and Littlewood conjectured that $c = 2$ (Ribenboim 1989, p. 202).

Define

$$E \equiv \liminf_{n\to\infty} \frac{p_{n+1} - p_n}{\ln p_n}. \tag{3}$$

If there are an infinite number of twin primes, then $E = 0$. The best upper limit to date is $E \leq \frac{1}{4} + \pi/16 = 0.44634\ldots$ (Huxley 1973, 1977). The best previous values were $15/16$ (Ricci), $(2+\sqrt{3})/8 = 0.46650\ldots$ (Bombieri and Davenport 1966), and $(2\sqrt{2} - 1)/4 = 0.45706\ldots$ (Pil'Tai 1972), as quoted in Le Lionnais (1983, p. 26).

Some large twin primes are $10,006,428 \pm 1$, $1,706,595 \times 2^{11235} \pm 1$, and $571,305 \times 2^{7701} \pm 1$. An up-to-date table of known twin primes with 2000 or more digits follows. An extensive list is maintained by `Caldwell`.

$(p, p+1)$	dig.	Reference
$260,497,545 \times 2^{6625} \pm 1$	2003	Atkin & Rickert 1984
$43,690,485,351,513 \times 10^{1995} \pm 1$	2009	Dubner, Atkin 1985
$2,846!!!! \pm 1$	2151	Dubner 1992
$10,757,0463 \times 10^{2250} \pm 1$	2259	Dubner, Atkin 1985
$663,777 \times 2^{7650} \pm 1$	2309	Brown *et al.* 1989
$75,188,117,004 \times 10^{2298} \pm 1$	2309	Dubner 1989
$571305 \times 2^{7701} \pm 1$	2324	Brown *et al.* 1989
$1,171,452,282 \times 10^{2490} \pm 1$	2500	Dubner 1991
$459 \cdot 2^{8529} \pm 1$	2571	Dubner 1993
$1,706,595 \cdot 2^{11235} \pm 1$	3389	Noll *et al.* 1989
$4,655,478,828 \cdot 10^{3429} \pm 1$	3439	Dubner 1993
$1,692,923,232 \cdot 10^{4020} \pm 1$	4030	Dubner 1993
$6,797,727 \cdot 2^{15328} \pm 1$	4622	Forbes 1995
$697,053,813\dot{2}^{16352} \pm 1$	4932	Indlekofer & Ja'rai 1994
$570,918,348 \cdot 10^{5120} \pm 1$	5129	Dubner 1995
$242,206,083 \cdot 2^{38880} \pm 1$	11713	Indlekofer & Ja'rai 1995

The last of these is the largest known twin prime pair. In 1995, Nicely discovered a flaw in the Intel® Pentium™ microprocessor by computing the reciprocals of 824,633,702,441 and 824,633,702,443, which should have been accurate to 19 decimal places but were incorrect from the tenth decimal place on (Cipra 1995, 1996; Nicely 1996).

If $n \geq 2$, the INTEGERS n and $n+2$ form a pair of twin primes IFF

$$4[(n-1)!+1]+n \equiv 0 \pmod{n(n+2)}. \tag{4}$$

$n = pp'$ where (p, p') is a pair of twin primes IFF

$$\phi(n)\sigma(n) = (n-3)(n+1) \tag{5}$$

(Ribenboim 1989). The values of $\pi_2(n)$ were found by Brent (1976) up to $n = 10^{11}$. T. Nicely calculated them up to 10^{14} in his calculation of BRUN'S CONSTANT. The following table gives the number less than increasing powers of 10 (Sloane's A007508).

n	$\pi_2(n)$
10^3	35
10^4	205
10^5	1224
10^6	8,169
10^7	58,980
10^8	440,312
10^9	3,424,506
10^{10}	27,412,679
10^{11}	224,376,048
10^{12}	1,870,585,220
10^{13}	15,834,664,872
10^{14}	135,780,321,665

see also BRUN'S CONSTANT, DE POLIGNAC'S CONJECTURE PRIME CONSTELLATION, SEXY PRIMES, TWIN PRIME CONJECTURE, TWIN PRIMES CONSTANT

References

Bombieri, E. and Davenport, H. "Small Differences Between Prime Numbers." *Proc. Roy. Soc. Ser. A* **293**, 1–8, 1966.
Bombieri, E.; Friedlander, J. B.; and Iwaniec, H. "Primes in Arithmetic Progression to Large Moduli." *Acta Math.* **156**, 203–251, 1986.
Bradley, C. J. "The Location of Twin Primes." *Math. Gaz.* **67**, 292–294, 1983.
Brent, R. P. "Irregularities in the Distribution of Primes and Twin Primes." *Math. Comput.* **29**, 43–56, 1975.
Brent, R. P. "UMT 4." *Math. Comput.* **29**, 221, 1975.
Brent, R. P. "Tables Concerning Irregularities in the Distribution of Primes and Twin Primes to 10^{11}." *Math. Comput.* **30**, 379, 1976.
Caldwell, C. `http://www.utm.edu/cgi-bin/caldwell/primes.cgi/twin`.
Cipra, B. "How Number Theory Got the Best of the Pentium Chip." *Science* **267**, 175, 1995.
Cipra, B. "Divide and Conquer." *What's Happening in the Mathematical Sciences, 1995–1996, Vol. 3.* Providence, RI: Amer. Math. Soc., pp. 38–47, 1996.
Fouvry, É. "Autour du theoreme de Bombieri-Vinogradov." *Acta. Math.* **152**, 219–244, 1984.
Fouvry, É. and Grupp, F. "On the Switching Principle in Sieve Theory." *J. Reine Angew. Math.* **370**, 101–126, 1986.
Fouvey, É. and Iwaniec, H. "Primes in Arithmetic Progression." *Acta Arith.* **42**, 197–218, 1983.
Guy, R. K. "Gaps between Primes. Twin Primes." §A8 in *Unsolved Problems in Number Theory, 2nd ed.* New York: Springer-Verlag, pp. 19–23, 1994.
Haugland, J. K. *Topics in Analytic Number Theory.* Ph.D. thesis. Oxford, England: Oxford University, Oct. 1998.
Huxley, M. N. "Small Differences between Consecutive Primes." *Mathematica* **20**, 229–232, 1973.
Huxley, M. N. "Small Differences between Consecutive Primes. II." *Mathematica* **24**, 142–152, 1977.
Le Lionnais, F. *Les nombres remarquables.* Paris: Hermann, 1983.
Nicely, T. R. "The Pentium Bug.' `http://www.lynchburg.edu/public/academic/math/nicely/pentbug/pentbug.htm`.
Nicely, T. "Enumeration to 10^{14} of the Twin Primes and Brun's Constant." *Virginia J. Sci.* **46**, 195–204, 1996. `http://www.lynchburg.edu/public/academic/math/nicely/twins/twins.htm`.
Parady, B. K.; Smith, J. F.; and Zarantonello, S. E. "Largest Known Twin Primes." *Math. Comput.* **55**, 381–382, 1990.
Ribenboim, P. *The Book of Prime Number Records, 2nd ed.* New York: Springer-Verlag, pp. 199–204, 1989.
Shanks, D. *Solved and Unsolved Problems in Number Theory, 4th ed.* New York: Chelsea, p. 30, 1993.
Sloane, N. J. A. Sequences A014574, A001359/M2476, A006512/M3763, and A007508/M1855 in "An On-Line Version of the Encyclopedia of Integer Sequences."
Weintraub, S. "A Prime Gap of 864." *J. Recr. Math.* **25**, 42–43, 1993.
Wu, J. "Sur la suite des nombres premiers jumeaux." *Acta. Arith.* **55**, 365–394, 1990.

Twin Primes Constant

The twin primes constant Π_2 is defined by

$$\Pi_2 \equiv \prod_{\substack{p>2 \\ p \text{ prime}}} \left[1 - \frac{1}{(p-1)^2}\right] \tag{1}$$

$$\begin{aligned} \ln(\tfrac{1}{2}\Pi_2) &= \sum_{\substack{p\geq 3 \\ p \text{ prime}}} \ln\left[\frac{p(p-2)}{(p-1)^2}\right] \\ &= \sum_{\substack{p\geq 3 \\ p \text{ prime}}} \left[\ln\left(1-\frac{2}{p}\right) - 2\ln\left(1-\frac{1}{p}\right)\right] \\ &= -\sum_{j=2}^{\infty} \frac{2^j-2}{j} \sum_{\substack{p\geq 3 \\ p \text{ prime}}} p^{-j}, \end{aligned} \tag{2}$$

where the ps in sums and products are taken over PRIMES only. Flajolet and Vardi (1996) give series with accelerated convergence

$$\Pi_2 = \prod_{n=2}^{\infty} [\zeta(n)(1-2^{-n})]^{-I_n} \tag{3}$$

$$= \tfrac{3}{4}\tfrac{15}{16}\tfrac{35}{36} \prod_{n=2}^{\infty} [\zeta(n)(1-2^{-n})(1-3^{-n})(1-5^{-n}) \times(1-7^{-n})]^{-I_n}, \tag{4}$$

with

$$I_n \equiv \frac{1}{n} \sum_{d|n} \mu(d) 2^{n/d}, \qquad (5)$$

where $\mu(x)$ is the MÖBIUS FUNCTION. (4) has convergence like $\sim (11/2)^{-n}$.

The most accurately known value of Π_2 is

$$\Pi_2 = 0.6601618158\ldots. \qquad (6)$$

Le Lionnais (1983, p. 30) calls C_2 the SHAH-WILSON CONSTANT, and $2C_2$ the twin prime constant (Le Lionnais 1983, p. 37).

see also BRUN'S CONSTANT, GOLDBACH CONJECTURE, MERTENS CONSTANT

References

Finch, S. "Favorite Mathematical Constants." http://www.mathsoft.com/asolve/constant/hrdyltl/hrdyltl.html.

Flajolet, P. and Vardi, I. "Zeta Function Expansions of Classical Constants." Unpublished manuscript. 1996. http://pauillac.inria.fr/algo/flajolet/Publications/landau.ps.

Le Lionnais, F. *Les nombres remarquables.* Paris: Hermann, 1983.

Ribenboim, P. *The Book of Prime Number Records, 2nd ed.* New York: Springer-Verlag, p. 202, 1989.

Ribenboim, P. *The Little Book of Big Primes.* New York: Springer-Verlag, p. 147, 1991.

Riesel, H. *Prime Numbers and Computer Methods for Factorization, 2nd ed.* Boston, MA: Birkhäuser, pp. 61–66, 1994.

Shanks, D. *Solved and Unsolved Problems in Number Theory, 4th ed.* New York: Chelsea, p. 30, 1993.

Wrench, J. W. "Evaluation of Artin's Constant and the Twin Prime Constant." *Math. Comput.* **15**, 396–398, 1961.

Twins

see BROTHERS, PAIR

Twirl

A ROTATION combined with an EXPANSION or DILATION.

see also SCREW, SHIFT

Twist

The twist of a ribbon measures how much it twists around its axis and is defined as the integral of the incremental twist around the ribbon. Letting Lk be the linking number of the two components of a ribbon, Tw be the twist, and Wr be the WRITHE, then

$$\mathrm{Lk}(R) = \mathrm{Tw}(R) + \mathrm{Wr}(R)$$

(Adams 1994, p. 187).

see also SCREW, WRITHE

References

Adams, C. C. *The Knot Book: An Elementary Introduction to the Mathematical Theory of Knots.* New York: W. H. Freeman, 1994.

Twist Map

A class of AREA-PRESERVING MAPS of the form

$$\theta_{i+1} = \theta_i + 2\pi\alpha(r_i)$$
$$r_{i+1} = r_i,$$

which maps CIRCLES into CIRCLES but with a twist resulting from the $\alpha = \alpha(r_i)$ term.

Twist Move

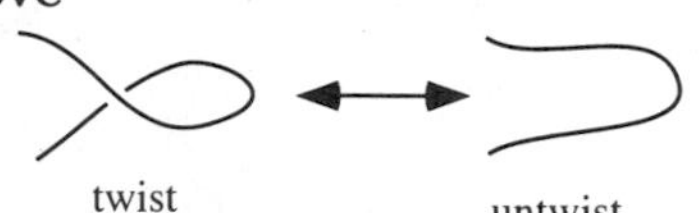

The REIDEMEISTER MOVE of type II.

see also REIDEMEISTER MOVES

Twist Number

see WRITHE

Twist-Spun Knot

A generalization of SPUN KNOTS due to Zeeman. This method produces 4-D KNOT types that cannot be produced by ordinary spinning.

see also SPUN KNOT

Twisted Chevalley Groups

FINITE SIMPLE GROUPS of LIE-TYPE of ORDERS 14, 52, 78, 133, and 248. They are denoted ${}^3D_4(q)$, $E_6(q)$, $E_7(q)$, $E_8(q)$, $F_4(q)$, ${}^2F_4(2^n)'$, $G_2(q)$,${}^2G_2(3^n)$, ${}^2B(2^n)$.

see also CHEVALLEY GROUPS, FINITE GROUP, SIMPLE GROUP, TITS GROUP

References

Wilson, R. A. "ATLAS of Finite Group Representation." http://for.mat.bham.ac.uk/atlas#twi.

Twisted Conic

see SKEW CONIC

Twisted Sphere

see CORKSCREW SURFACE

Two

see 2

Two-Ears Theorem

Except for TRIANGLES, every SIMPLE POLYGON has at least two nonoverlapping EARS.

see also EAR, ONE-MOUTH THEOREM, PRINCIPAL VERTEX

References

Meisters, G. H. "Principal Vertices, Exposed Points, and Ears." *Amer. Math. Monthly* **87**, 284–285, 1980.

Toussaint, G. "Anthropomorphic Polygons." *Amer. Math. Monthly* **122**, 31–35, 1991.

Two-Point Distance

see POINT-POINT DISTANCE—1-D, POINT-POINT DISTANCE—2-D, POINT-POINT DISTANCE—3-D

Two Triangle Theorem

see DESARGUES' THEOREM

Tychonof Compactness Theorem

The topological product of any number of COMPACT SPACES is COMPACT.

Type

Whitehead and Russell (1927) devised a hierarchy of "types" in order to eliminate self-referential statements from *Principia Mathematica*, which purported to derive all of mathematics from logic. A set of the lowest type contained only objects (not sets), a set of the next higher type could contain only objects or sets of the lower type, and so on. Unfortunately, GÖDEL'S INCOMPLETENESS THEOREM showed that both *Principia Mathematica* and all consistent formal systems must be incomplete.

see also GÖDEL'S INCOMPLETENESS THEOREM

References

Hofstadter, D. R. *Gödel, Escher, Bach: An Eternal Golden Braid.* New York: Vintage Books, pp. 21–22, 1989.

Whitehead, A. N. and Russell, B. *Principia Mathematica.* New York: Cambridge University Press, 1927.

Type I Error

An error in a STATISTICAL TEST which occurs when a true hypothesis is rejected (a false negative in terms of the NULL HYPOTHESIS).

see also NULL HYPOTHESIS, SENSITIVITY, SPECIFICITY, STATISTICAL TEST, TYPE II ERROR

Type II Error

An error in a STATISTICAL TEST which occurs when a false hypothesis is accepted (a false positive in terms of the NULL HYPOTHESIS).

see also NULL HYPOTHESIS, SENSITIVITY, SPECIFICITY, STATISTICAL TEST, TYPE I ERROR

U

U-Number

see ULAM SEQUENCE

Ulam Map

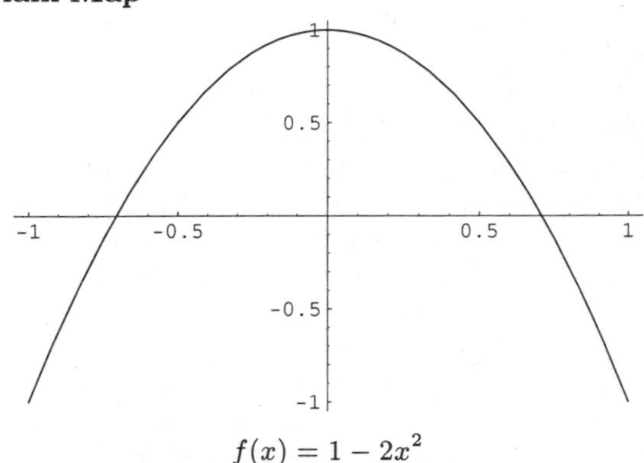

$$f(x) = 1 - 2x^2$$

for $x \in [-1, 1]$. Fixed points occur at $x = -1$, $1/2$, and order 2 fixed points at $x = (1 \pm \sqrt{5})/4$. The INVARIANT DENSITY of the map is

$$\rho(y) = \frac{1}{\pi\sqrt{1-y^2}}.$$

References

Beck, C. and Schlögl, F. *Thermodynamics of Chaotic Systems: An Introduction.* Cambridge, England: Cambridge University Press, p. 194, 1995.

Ulam Number

see ULAM SEQUENCE

Ulam's Problem

see COLLATZ PROBLEM

Ulam Sequence

The Ulam sequence $\{a_i\} = (u, v)$ is defined by $a_1 = u$, $a_2 = v$, with the general term a_n for $n > 2$ given by the least INTEGER expressible uniquely as the SUM of two distinct earlier terms. The numbers so produced are sometimes called U-NUMBERS or ULAM NUMBERS.

The first few numbers in the (1, 2) Ulam sequence are 1, 2, 3, 4, 6, 8, 11, 13, 16, ... (Sloane's A002858). Here, the first term after the initial 1, 2 is obviously 3 since $3 = 1 + 2$. The next term is $4 = 1 + 3$. (We don't have to worry about $4 = 2 + 2$ since it is a sum of a single term instead of *unique* terms.) 5 is not a member of the sequence since it is representable in *two* ways, $5 = 1 + 4 = 2 + 3$, but $6 = 2 + 4$ is a member.

Proceeding in the manner, we can generate Ulam sequences for any (u, v), examples of which are given below.

$$\begin{aligned}
(1,2) &= \{1,2,3,4,6,8,11,13,16,18,\ldots\}\\
(1,3) &= \{1,3,4,5,6,8,10,12,17,21,\ldots\}\\
(1,4) &= \{1,4,5,6,7,8,10,16,18,19,\ldots\}\\
(1,5) &= \{1,5,6,7,8,9,10,12,20,22,\ldots\}\\
(2,3) &= \{2,3,5,7,8,9,13,14,18,19,\ldots\}\\
(2,4) &= \{2,4,6,8,12,16,22,26,32,36,\ldots\}\\
(2,5) &= \{2,5,7,9,11,12,13,15,19,23,\ldots\}.
\end{aligned}$$

Schmerl and Spiegel (1994) proved that Ulam sequences $(2, v)$ for ODD $v \geq 5$ have exactly two EVEN terms. Ulam sequences with only finitely many EVEN terms eventually must have periodic successive differences (Finch 1991, 1992abc). Cassaigne and Finch (1995) proved that the Ulam sequences $(4, v)$ for $5 \leq v \equiv 1 \pmod 4$ have exactly three EVEN terms.

The Ulam sequence can be generalized by the s-ADDITIVE SEQUENCE.

see also GREEDY ALGORITHM, s-ADDITIVE SEQUENCE, STÖHR SEQUENCE

References

Cassaigne, J. and Finch, S. "A Class of 1-Additive Sequences and Quadratic Recurrences." *Exper. Math* **4**, 49–60, 1995.

Finch, S. "Conjectures About 1-Additive Sequences." *Fib. Quart.* **29**, 209–214, 1991.

Finch, S. "Are 0-Additive Sequences Always Regular?" *Amer. Math. Monthly* **99**, 671–673, 1992a.

Finch, S. "On the Regularity of Certain 1-Additive Sequences." *J. Combin. Th. Ser. A* **60**, 123–130, 1992b.

Finch, S. "Patterns in 1-Additive Sequences." *Exper. Math.* **1**, 57–63, 1992c.

Finch, S. "Ulam s-Additive Sequences." `http://www.mathsoft.com/asolve/sadd/sadd.html`.

Guy, R. K. "A Quarter Century of *Monthly* Unsolved Problems, 1969–1993." *Amer. Math. Monthly* **100**, 945–949, 1993.

Guy, R. K. "Ulam Numbers." §C4 in *Unsolved Problems in Number Theory, 2nd ed.* New York: Springer-Verlag, pp. 109–110, 1994.

Guy, R. K. and Nowakowski, R. J. "*Monthly* Unsolved Problems, 1969–1995." *Amer. Math. Monthly* **102**, 921–926, 1995.

Recaman, B. "Questions on a Sequence of Ulam." *Amer. Math. Monthly* **80**, 919–920, 1973.

Schmerl, J. and Spiegel, E. "The Regularity of Some 1-Additive Sequences." *J. Combin. Theory Ser. A* **66**, 172–175, 1994.

Sloane, N. J. A. Sequence A002858/M0557 in "An On-Line Version of the Encyclopedia of Integer Sequences."

Ultrametric

An ultrametric is a METRIC which satisfies the following strengthened version of the TRIANGLE INEQUALITY,

$$d(x, z) \leq \max(d(x, y), d(y, z))$$

for all x, y, z. At least two of $d(x,y)$, $d(y,z)$, and $d(x,z)$ are the same.

Let X be a SET, and let $X^{\mathbb{N}}$ (where $\mathbb{N}$ is the SET of NATURAL NUMBERS) denote the collection of sequences of elements of X (i.e., all the possible sequences x_1, x_2, x_3, ...). For sequences $a = (a_1, a_2, \ldots)$, $b = (b_1, b_2, \ldots)$, let n be the number of initial places where the sequences agree, i.e., $a_1 = b_1$, $a_2 = b_2$, ..., $a_n = b_n$, but $a_{n+1} \neq b_{n+1}$. Take $n = 0$ if $a_1 \neq b_1$. Then defining $d(a,b) = 2^{-n}$ gives an ultrametric.

The p-ADIC NUMBER metric is another example of an ultrametric.

see also METRIC, p-ADIC NUMBER

Ultraradical

A symbol which can be used to express solutions not obtainable by finite ROOT extraction. The solution to the irreducible QUINTIC EQUATION

$$x^5 + x = a$$

is written $\sqrt[\int]{a}$.

see also RADICAL

Ultraspherical Differential Equation

$$(1-x^2)y'' - (2\alpha+1)xy' + n(n+2\alpha)y = 0. \quad (1)$$

Alternate forms are

$$(1-x^2)Y'' + (2\lambda-3)xY' + (n+1)(n+2\lambda-1)Y = 0, \quad (2)$$

where

$$Y = (1-x^2)^{\lambda-1/2} P_n^{(\lambda)}(x), \quad (3)$$

$$\frac{d^2u}{dx^2} + \left[\frac{(n+\lambda)^2}{1-x^2} + \frac{\frac{1}{2}+\lambda-\lambda^2+\frac{1}{4}x^2}{(1-x^2)^2}\right]u = 0, \quad (4)$$

where

$$u = (1-x^2)^{\lambda/2+1/4} P_n^{(\lambda)}(x), \quad (5)$$

and

$$\frac{d^2u}{d\theta^2} + \left[(n+\lambda)^2 + \frac{\lambda(1-\lambda)}{\sin^2\theta}\right]u = 0, \quad (6)$$

where

$$u = \sin^\lambda\theta\, P_n^{(\lambda)}(\cos\theta). \quad (7)$$

The solutions are the ULTRASPHERICAL FUNCTIONS $P_n^{(\lambda)}(x)$. For integral n with $\alpha < 1/2$, the function converges to the ULTRASPHERICAL POLYNOMIALS $C_n^{(\alpha)}(x)$.

References

Morse, P. M. and Feshbach, H. *Methods of Theoretical Physics, Part I.* New York: McGraw-Hill, pp. 547–549, 1953.

Ultraspherical Function

A function defined by a POWER SERIES whose coefficients satisfy the RECURRENCE RELATION

$$a_{j+2} = a_j \frac{(k+j)(k+j+2\alpha) - n(n+2\alpha)}{(k+j+1)(k+j+2)}.$$

For $x \neq -1$, the function converges for $\alpha < 1/2$ and diverges for $\alpha > 1/2$.

Ultraspherical Polynomial

The ultraspherical polynomials are solutions $P_n^{(\lambda)}(x)$ to the ULTRASPHERICAL DIFFERENTIAL EQUATION for INTEGER n and $\alpha < 1/2$. They are generalizations of LEGENDRE POLYNOMIALS to $(n+2)$-D space and are proportional to (or, depending on the normalization, equal to) the GEGENBAUER POLYNOMIALS $C_n^{(\lambda)}(x)$, denoted in *Mathematica*® (Wolfram Research, Champaign, IL) `GegenbauerC[n,lambda,x]`. The ultraspherical polynomials are also JACOBI POLYNOMIALS with $\alpha = \beta$. They are given by the GENERATING FUNCTION

$$\frac{1}{(1-2xt+t^2)^\lambda} = \sum_{n=0}^\infty P_n^{(\lambda)}(x)t^n, \quad (1)$$

and can be given explicitly by

$$P_n^{(\lambda)}(x) = \frac{\Gamma(\lambda+\frac{1}{2})}{\Gamma(2\lambda)} \frac{\Gamma(n+2\lambda)}{\Gamma(n+\lambda+\frac{1}{2})} P_n^{(\lambda-1/2,\lambda-1/2)}(x), \quad (2)$$

where $P_n^{(\lambda-1/2,\lambda-1/2)}$ is a JACOBI POLYNOMIAL (Szegő 1975, p. 80). The first few ultraspherical polynomials are

$$P_0^{(\lambda)}(x) = 1 \quad (3)$$

$$P_1^{(\lambda)}(x) = 2\lambda x \quad (4)$$

$$P_2^{(\lambda)}(x) = -\lambda + 2\lambda(1+\lambda)x^2 \quad (5)$$

$$P_3^{(\lambda)}(x) = -2\lambda(1+\lambda)x + \tfrac{4}{3}\lambda(1+\lambda)(2+\lambda)x^3. \quad (6)$$

In terms of the HYPERGEOMETRIC FUNCTIONS,

$$P_n^{(\lambda)}(x) = \binom{n+2\lambda-1}{n} \times {}_2F_1(-n, n+2\lambda; \lambda+\tfrac{1}{2}; \tfrac{1}{2}(1-x)) \quad (7)$$

$$= 2^n \binom{n+\lambda-1}{n}(x-1)^n \times {}_2F_1\left(-n, -n-\lambda+\tfrac{1}{2}; -2n-2\lambda+1; \frac{2}{1-x}\right) \quad (8)$$

$$= \binom{n+2\lambda+1}{n}\left(\frac{x+1}{2}\right)^n \times {}_2F_1\left(-n, -n-\lambda+\tfrac{1}{2}; \lambda+\tfrac{1}{2}; \frac{x-1}{x+1}\right). \quad (9)$$

They are normalized by

$$\int_{-1}^{1}(1-x^2)^{\lambda-1/2}[P_n^{(\lambda)}]^2\,dx = 2^{1-2\lambda}\pi\frac{\Gamma(n+2\lambda)}{(n+\lambda)\Gamma^2(\lambda)\Gamma(n+1)}. \quad (10)$$

Derivative identities include

$$\frac{d}{dx}P_n^{(\lambda)}(x) = 2\lambda P_{n-1}^{(\lambda+1)}(x) \quad (11)$$

$$(1-x^2)\frac{d}{dx}[P_n^{(\lambda)}] = [2(n+\lambda)]^{-1}[(n+2\lambda-1) \times(n+2\lambda)P_{n-1}^{(\lambda)}(x) - n(n+1)P_{n+1}^{(\lambda)}(x)] \quad (12)$$

$$= -nxP_n^{(\lambda)}(x) + (n+2\lambda-1)P_{n-1}^{(\lambda)}(x) \quad (13)$$

$$= (n+2\lambda)xP_n^{(\lambda)}(x) - (n+1)P_{n+1}^{(\lambda)}(x) \quad (14)$$

$$nP_n^{(\lambda)}(x) = x\frac{d}{dx}[P_n^{(\lambda)}(x)] - \frac{d}{dx}[P_{n-1}^{(\lambda)}(x)] \quad (15)$$

$$(n+2\lambda)P_n^{(\lambda)}(x) = \frac{d}{dx}[P_{n+1}^{(\lambda)}(x)] - x\frac{d}{dx}[P_n^{(\lambda)}(x)] \quad (16)$$

$$\frac{d}{dx}[P_{n+1}^{(\lambda)}(x) - P_{n-1}^{(\lambda)}(x)] = 2(n+\lambda)P_n^{(\lambda)}P_n^{(\lambda)}(x) \quad (17)$$

$$= 2\lambda[P_n^{(\lambda+1)}(x) - P_{n-2}^{(\lambda+1)}(x)] \quad (18)$$

(Szegő 1975, pp. 80–83).

A RECURRENCE RELATION is

$$nP_n^{(\lambda)}(x) = 2(n+\lambda-1)xP_{n-1}^{(\lambda)}(x) - (n+2\lambda-2)P_{n-2}^{(\lambda)}(x) \quad (19)$$

for $n = 2, 3, \ldots$.

Special double-ν FORMULAS also exist

$$P_{2\nu}^{(\lambda)}(x) = \binom{2\nu+2\lambda-1}{2\nu}{}_2F_1(-\nu,\nu+\lambda;\lambda+\tfrac{1}{2};1-x^2) \quad (20)$$

$$= (-1)^\nu\binom{\nu+\lambda-1}{\nu}{}_2F_1(-\nu,\nu+\lambda;\tfrac{1}{2};x^2) \quad (21)$$

$$P_{2\nu+1}^{(\lambda)}(x) = \binom{2\nu+2\lambda}{2\nu+1}x\,{}_2F_1(-\nu,\nu+\lambda+1;\lambda+\tfrac{1}{2};1-x^2) \quad (22)$$

$$= (-1)^\nu 2\lambda\binom{\nu+\lambda}{\nu}x\,{}_2F_1(-\nu,\nu+\lambda+1;\tfrac{3}{2};x^2). \quad (23)$$

Special values are given in the following table.

λ	Special Polynomial
$\frac{1}{2}$	Legendre
1	Chebyshev polynomial of the second kind

Koschmieder (1920) gives representations in terms of ELLIPTIC FUNCTIONS for $\alpha = -3/4$ and $\alpha = -2/3$.

see also BIRTHDAY PROBLEM, CHEBYSHEV POLYNOMIAL OF THE SECOND KIND, ELLIPTIC FUNCTION, HYPERGEOMETRIC FUNCTION, JACOBI POLYNOMIAL

References

Abramowitz, M. and Stegun, C. A. (Eds.). "Orthogonal Polynomials." Ch. 22 in *Handbook of Mathematical Functions with Formulas, Graphs, and Mathematical Tables, 9th printing.* New York: Dover, pp. 771–802, 1972.

Arfken, G. *Mathematical Methods for Physicists, 3rd ed.* Orlando, FL: Academic Press, p. 643, 1985.

Iyanaga, S. and Kawada, Y. (Eds.). "Gegenbauer Polynomials (Gegenbauer Functions)." Appendix A, Table 20.I in *Encyclopedic Dictionary of Mathematics.* Cambridge, MA: MIT Press, pp. 1477–1478, 1980.

Koschmieder, L. "Über besondere Jacobische Polynome." *Math. Zeitschrift* **8**, 123–137, 1920.

Morse, P. M. and Feshbach, H. *Methods of Theoretical Physics, Part I.* New York: McGraw-Hill, pp. 547–549 and 600–604, 1953.

Szegő, G. *Orthogonal Polynomials, 4th ed.* Providence, RI: Amer. Math. Soc., 1975.

Umbilic Point

A point on a surface at which the CURVATURE is the same in any direction.

Umbral Calculus

The study of certain properties of FINITE DIFFERENCES. The term was coined by Sylvester from the word "umbra" (meaning "shadow" in Latin), and reflects the fact that for many types of identities involving sequences of polynomials with POWERS a^n, "shadow" identities are obtained when the polynomials are changed to discrete values and the exponent in a^n is changed to the POCHHAMMER SYMBOL $(a)_n \equiv a(a-1)\cdots(a-n+1)$.

For example, NEWTON'S FORWARD DIFFERENCE FORMULA written in the form

$$f(x+a) = \sum_{n=0}^{\infty}\frac{(a)_n\Delta^n f(x)}{n!} \quad (1)$$

with $f(x+a) \equiv f_{x+a}$ looks suspiciously like a finite analog of the TAYLOR SERIES expansion

$$f(x+a) = \sum_{n=0}^{\infty}\frac{a^n\tilde{D}^n f(x)}{n!}, \quad (2)$$

where $\tilde{D}$ is the DIFFERENTIAL OPERATOR. Similarly, the CHU-VANDERMONDE IDENTITY

$$(x+a)_n = \sum_{k=0}^{\infty}\binom{n}{k}(a)_k(x)_{n-k} \quad (3)$$

with $\binom{n}{k}$ a BINOMIAL COEFFICIENT, looks suspiciously like an analog of the BINOMIAL THEOREM

$$(x+a)^n = \sum_{k=0}^{\infty}\binom{n}{k}a^k x^{n-k} \quad (4)$$

(Di Bucchianico and Loeb).

see also BINOMIAL THEOREM, CHU-VANDERMONDE IDENTITY, FINITE DIFFERENCE

References

Roman, S. and Rota, G.-C. "The Umbral Calculus." *Adv. Math.* **27**, 95–188, 1978.

Roman, S. *The Umbral Calculus.* New York: Academic Press, 1984.

Umbrella

see WHITNEY UMBRELLA

Unambiguous

see WELL-DEFINED

Unbiased

A quantity which does not exhibit BIAS. An ESTIMATOR $\hat{\theta}$ is an UNBIASED ESTIMATOR of θ if

$$\langle\hat{\theta}\rangle = \theta.$$

see also BIAS (ESTIMATOR), ESTIMATOR

Uncia

$$1 \text{ uncia} \equiv \tfrac{1}{12}.$$

The word *uncia* was Latin for a unit equal to 1/12 of another unit called the *as*. The words "inch" (1/12 of a foot) and "ounce" (originally 1/12 of a pound and still 1/12 of a "Troy pound," now used primarily to weigh precious metals) are derived from the word *uncia*.

see also CALCUS, HALF, QUARTER, SCRUPLE, UNIT FRACTION

References

Conway, J. H. and Guy, R. K. *The Book of Numbers.* New York: Springer-Verlag, p. 4, 1996.

Uncorrelated

Variables x_i and x_j are said to be uncorrelated if their COVARIANCE is zero:

$$\operatorname{cov}(x_i, x_j) = 0.$$

INDEPENDENT STATISTICS are always uncorrelated, but the converse is not necessarily true.

see also COVARIANCE, INDEPENDENT STATISTICS

Uncountable Set

see UNCOUNTABLY INFINITE SET

Uncountably Infinite Set

An INFINITE SET which is not a COUNTABLY INFINITE SET.

see also ALEPH-0, ALEPH-1, COUNTABLE SET, COUNTABLY INFINITE SET, FINITE, INFINITE

Undecagon

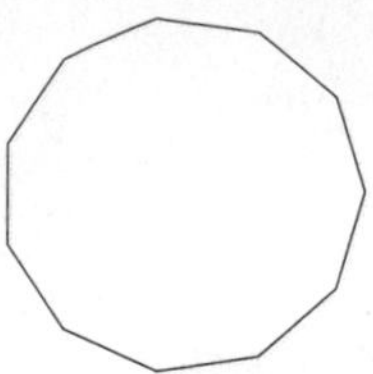

The unconstructible 11-sided POLYGON with SCHLÄFLI SYMBOL {11}.

see also DECAGON, DODECAGON, TRIGONOMETRY VALUES—$\pi/11$

Undecidable

Not DECIDABLE as a result of being neither formally provable nor unprovable.

see also GÖDEL'S INCOMPLETENESS THEOREM, RICHARDSON'S THEOREM

Undecillion

In the American system, 10^{36}.

see also LARGE NUMBER

Undetermined Coefficients Method

Given a nonhomogeneous ORDINARY DIFFERENTIAL EQUATION, select a differential operator which will annihilate the right side, and apply it to both sides. Find the solution to the homogeneous equation, plug it into the left side of the original equation, and solve for constants by setting it equal to the right side. The solution is then obtained by plugging the determined constants into the homogeneous equation.

see also ORDINARY DIFFERENTIAL EQUATION

Undulating Number

A number of the form $aba\cdots$, $abab\cdots$, etc. The first few nontrivial undulants (with the stipulation that $a \neq b$) are 101, 121, 131, 141, 151, 161, 171, 181, 191, 202, 212, ... (Sloane's A046075). Including the trivial 1- and 2-digit undulants and dropping the requirement that $a \neq b$ gives Sloane's A033619.

The first few undulating SQUARES are 121, 484, 676, 69696, ... (Sloane's A016073), with no larger such numbers of fewer than a million digits (Pickover 1995). Several tricks can be used to speed the search for square undulating numbers, especially by examining the possible patterns of ending digits. For example, the only possible sets of four trailing digits for undulating SQUARES are 0404, 1616, 2121, 2929, 3636, 6161, 6464, 6969, 8484, and 9696.

The only undulating POWER $n^p = aba\cdots$ for $3 \leq p \leq 31$ and up to 100 digits is $7^3 = 343$ (Pickover 1995). A large undulating prime is given by $7+720(100^{49}-1)/99$ (Pickover 1995).

A binary undulant is a POWER of 2 whose base-10 representation contains one or both of the sequences $010\cdots$ and $101\cdots$. The first few are 2^n for $n = 103$, 107, 138, 159, 179, 187, 192, 199, 205, ... (Sloane's A046076). The smallest n for which an undulating sequence of *exactly* d-digit occurs for $d = 3, 4, \ldots$ are $n = 103$, 138, 875, 949, 6617, 1802, 14545, ... (Sloane's A046077). An undulating binary sequence of length 10 occurs for $n = 1,748,219$ (Pickover 1995).

References

Pickover, C. A. "Is There a Double Smoothly Undulating Integer?" In *Computers, Pattern, Chaos and Beauty.* New York: St. Martin's Press, 1990.

Pickover, C. A. "The Undulation of the Monks." Ch. 20 in *Keys to Infinity.* New York: W. H. Freeman, pp. 159–161 1995.

Sloane, N. J. A. Sequences A016073, A033619, A046075, A046076, and A046077 in "An On-Line Version of the Encyclopedia of Integer Sequences."

Unduloid

A SURFACE OF REVOLUTION with constant NONZERO MEAN CURVATURE also called an ONDULOID. It is a ROULETTE obtained from the path described by the FOCI of a CONIC SECTION when rolled on a LINE. This curve then generates an unduloid when revolved about the LINE. These curves are special cases of the shapes assumed by soap film spanning the gap between prescribed boundaries. The unduloid of a PARABOLA gives a CATENOID.

see also CALCULUS OF VARIATIONS, CATENOID, ROULETTE

References

Cundy, H. and Rollett, A. *Mathematical Models, 3rd ed.* Stradbroke, England: Tarquin Pub., p. 48, 1989.

Delaunay, C. "Sur la surface de révolution dont la courbure moyenne est constante." *J. math. pures appl.* **6**, 309–320, 1841.

do Carmo, M. P. "The Onduloid." §3.5G in *Mathematical Models from the Collections of Universities and Museums* (Ed. G. Fischer). Braunschweig, Germany: Vieweg, pp. 47–48, 1986.

Fischer, G. (Ed.). Plate 97 in *Mathematische Modelle/Mathematical Models, Bildband/Photograph Volume.* Braunschweig, Germany: Vieweg, p. 93, 1986.

Thompson, D'A. W. *On Growth and Form, 2nd ed., compl. rev. ed.* New York: Cambridge University Press, 1992.

Yates, R. C. *A Handbook on Curves and Their Properties.* Ann Arbor, MI: J. W. Edwards, p. 184, 1952.

Unexpected Hanging Paradox

A PARADOX also known as the SURPRISE EXAMINATION PARADOX or PREDICTION PARADOX.

A prisoner is told that he will be hanged on some day between Monday and Friday, but that he will not know on which day the hanging will occur before it happens. He cannot be hanged on Friday, because if he were still alive on Thursday, he would know that the hanging will occur on Friday, but he has been told he will not know the day of his hanging in advance. He cannot be hanged Thursday for the same reason, and the same argument shows that he cannot be hanged on any other day. Nevertheless, the executioner unexpectedly arrives on some day other than Friday, surprising the prisoner.

This PARADOX is similar to that in Robert Louis Stevenson's "The Imp in the Bottle," in which you are offered the opportunity to buy, for whatever price you wish, a bottle containing a genie who will fulfill your every desire. The only catch is that the bottle must thereafter be resold for a price smaller than what you paid for it, or you will be condemned to live out the rest of your days in excrutiating torment. Obviously, no one would buy the bottle for 1¢ since he would have to give the bottle away, but no one would accept the bottle knowing he would be unable to get rid of it. Similarly, no one would buy it for 2¢, and so on. However, for some reasonably large amount, it will always be possible to find a next buyer, so the bottle will be bought (Paulos 1995).

see also SORITES PARADOX

References

Chow, T. Y. "The Surprise Examination or Unexpected Hanging Paradox." *Amer. Math. Monthly* **105**, 41–51, 1998.

Clark, D. "How Expected is the Unexpected Hanging?" *Math. Mag.* **67**, 55–58, 1994.

Gardner, M. "The Paradox of the Unexpected Hanging." Ch. 1 in *The Unexpected Hanging and Other Mathematical Diversions.* Chicago, IL: Chicago University Press, 1991.

Margalit, A. and Bar-Hillel, M. "Expecting the Unexpected." *Philosophia* **13**, 263–288, 1983.

Pappas, T. "The Paradox of the Unexpected Exam." *The Joy of Mathematics.* San Carlos, CA: Wide World Publ./ Tetra, p. 147, 1989.

Paulos, J. A. *A Mathematician Reads the Newspaper.* New York: BasicBooks, p. 97, 1995.

Quine, W. V. O. "On a So-Called Paradox." *Mind* **62**, 65–67, 1953.

Unfinished Game

see SHARING PROBLEM

Unhappy Number

A number which is not HAPPY is said to be unhappy.

see also HAPPY NUMBER

Unicursal Circuit

A CIRCUIT in which an entire GRAPH is traversed in one route. An example of a curve which can be traced unicursally is the MOHAMMED SIGN.

Uniform Apodization Function

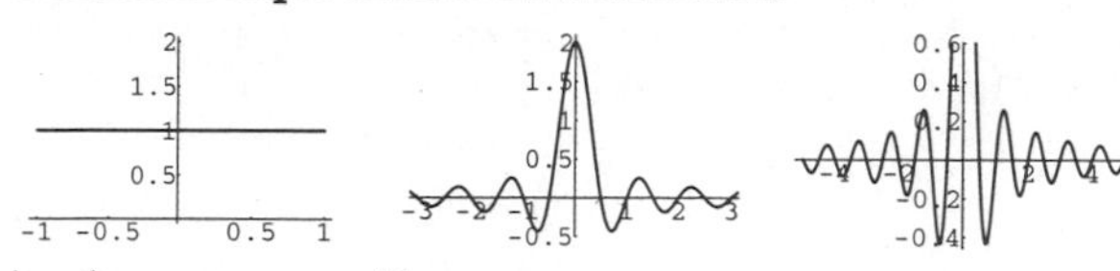

An APODIZATION FUNCTION

$$f(x) = 1, \tag{1}$$

having INSTRUMENT FUNCTION

$$I(x) = \int_{-a}^{a} e^{-2\pi ikx}\,dx = -\frac{1}{2\pi ik}(e^{-2\pi ika} - e^{2\pi ikx})$$
$$= \frac{\sin(2\pi ka)}{\pi k} = 2a\,\mathrm{sinc}(2\pi ka). \quad (2)$$

The peak (in units of a) is 2. The extrema are given by letting $\beta \equiv 2\pi ka$ and solving

$$\frac{d}{d\beta}(\beta \sin\beta) = \frac{\sin\beta - \beta\cos\beta}{\beta^2} = 0 \quad (3)$$

$$\sin\beta - \beta\cos\beta = 0 \quad (4)$$

$$\tan\beta = \beta. \quad (5)$$

Solving this numerically gives $\beta_0 = 0$, $\beta_1 = 4.49341$, $\beta_2 = 7.72525$, ... for the first few solutions. The second of these is the peak POSITIVE sidelobe, and the third is the peak NEGATIVE sidelobe. As a fraction of the peak, they are 0.128375 and -0.217234. The FULL WIDTH AT HALF MAXIMUM is found by setting $I(x) = 1$

$$\mathrm{sinc}(x) = \tfrac{1}{2}, \quad (6)$$

and solving for $x_{1/2}$, yielding

$$x_{1/2} = 2\pi k_{1/2} a = 1.89549. \quad (7)$$

Therefore, with $L \equiv 2a$,

$$\mathrm{FWHM} = 2k_{1/2} = \frac{0.603353}{a} = \frac{1.20671}{L}. \quad (8)$$

see also APODIZATION FUNCTION

Uniform Boundedness Principle

If a "pointwise-bounded" family of continuous linear OPERATORS from a BANACH SPACE to a NORMED SPACE is "uniformly bounded." Symbolically, if $\sup ||T_i(x)||$ is FINITE for each x in the unit BALL, then $\sup ||T_i||$ is FINITE. The theorem is also called the BANACH-STEINHAUS THEOREM.

References

Zeidler, E. *Applied Functional Analysis: Applications to Mathematical Physics.* New York: Springer-Verlag, 1995.

Uniform Convergence

A SERIES $\sum_{n=1}^{\infty} u_n(x)$ is uniformly convergent to $S(x)$ for a set E of values of x if, for each $\epsilon > 0$, an INTEGER N can be found such that

$$|S_n(x) - S(x)| < \epsilon \quad (1)$$

for $n \geq N$ and all $x \in E$. To test for uniform convergence, use ABEL'S UNIFORM CONVERGENCE TEST or the WEIERSTRAß M-TEST. If individual terms $u_n(x)$ of a uniformly converging series are continuous, then

1. The series sum

$$f(x) = \sum_{n=1}^{\infty} u_n(x) \quad (2)$$

is continuous,

2. The series may be integrated term by term

$$\int_a^b f(x)\,dx = \sum_{n=1}^{\infty} \int_a^b u_n(x)\,dx, \quad (3)$$

and

3. The series may be differentiated term by term

$$\frac{d}{dx} f(x) = \sum_{n=1}^{\infty} n \frac{d}{dx} u_n(x). \quad (4)$$

see also ABEL'S THEOREM, ABEL'S UNIFORM CONVERGENCE TEST, WEIERSTRAß M-TEST

References

Arfken, G. *Mathematical Methods for Physicists, 3rd ed.* Orlando, FL: Academic Press, pp. 299–301, 1985.

Uniform Distribution

A distribution which has constant probability is called a uniform distribution, sometimes also called a RECTANGULAR DISTRIBUTION. The probability density function and cumulative distribution function for a *continuous* uniform distribution are

$$P(x) = \begin{cases} \frac{1}{b-a} & \text{for } a < x < b \\ 0 & \text{for } x < a,\ x > b \end{cases} \quad (1)$$

$$D(x) = \begin{cases} 0 & \text{for } x < a \\ \frac{x-a}{b-a} & \text{for } a \leq x < b \\ 1 & \text{for } x \geq b. \end{cases} \quad (2)$$

With $a = 0$ and $b = 1$, these can be written

$$P(x) = \tfrac{1}{2}\,\mathrm{sgn}(x) - \mathrm{sgn}(x-1) \quad (3)$$
$$D(x) = \tfrac{1}{2}[1 - (1-x)^2\,\mathrm{sgn}(1-x) + x\,\mathrm{sgn}(x)]. \quad (4)$$

The CHARACTERISTIC FUNCTION is

$$\phi(t) = \frac{2}{ht}\sin(\tfrac{1}{2}ht)e^{imt}, \quad (5)$$

where

$$a = m - \tfrac{1}{2}h \quad (6)$$
$$b = m + \tfrac{1}{2}h. \quad (7)$$

The MOMENT-GENERATING FUNCTION is

$$M(t) = \langle e^{xt} \rangle = \int_a^b \frac{e^{xt}}{b-a}\,dx = \left[\frac{e^{xt}}{t(b-a)}\right]_a^b, \quad (8)$$

so

$$M(t) = \begin{cases} \frac{e^{tb}-e^{ta}}{t(b-a)} & \text{for } t \neq 0 \\ 0 & \text{for } t = 0, \end{cases} \quad (9)$$

and

$$M'(t) = \frac{1}{b-a}\left[\frac{1}{t}(be^{bt} - ae^{at}) - \frac{1}{t^2}(e^{bt} - e^{at})\right]$$
$$= \frac{e^{bt}(bt-1) - e^{at}(at-1)}{(b-a)t^2}. \quad (10)$$

The function is not differentiable at zero, so the MOMENTS cannot be found using the standard technique. They can, however, be found by direct integration. The MOMENTS about 0 are

$$\mu_1' = \tfrac{1}{2}(a+b) \quad (11)$$
$$\mu_2' = \tfrac{1}{3}(a^2+ab+b^2) \quad (12)$$
$$\mu_3' = \tfrac{1}{4}(a+b)(a^2+b^2) \quad (13)$$
$$\mu_4' = \tfrac{1}{5}(a^4+a^3b+a^2b^2+ab^3+b^4). \quad (14)$$

The MOMENTS about the MEAN are

$$\mu_1 = 0 \quad (15)$$
$$\mu_2 = \tfrac{1}{12}(b-a)^2 \quad (16)$$
$$\mu_3 = 0 \quad (17)$$
$$\mu_4 = \tfrac{1}{80}(b-a)^4, \quad (18)$$

so the MEAN, VARIANCE, SKEWNESS, and KURTOSIS are

$$\mu = \tfrac{1}{2}(a+b) \quad (19)$$
$$\sigma^2 = \mu_2 = \tfrac{1}{12}(b-a)^2 \quad (20)$$
$$\gamma_1 = \frac{\mu_3}{\sigma^{3/2}} = 0 \quad (21)$$
$$\gamma_2 = -\tfrac{6}{5}. \quad (22)$$

The probability distribution function and cumulative distributions function for a *discrete* uniform distribution are

$$P(n) = \frac{1}{N} \quad (23)$$
$$D(n) = \frac{n}{N} \quad (24)$$

for $n = 1, \ldots, N$. The MOMENT-GENERATING FUNCTION is

$$M(t) = \langle e^{nt} \rangle = \sum_{n=1}^{N} \frac{1}{N} e^{nt} = \frac{1}{N}\frac{e^t - e^{t(N+1)}}{1-e^t}$$
$$= \frac{e^t(1-e^{Nt})}{N(1-e^t)}. \quad (25)$$

The MOMENTS about 0 are

$$\mu_m' = \frac{1}{N}\sum_{n=1}^{N} n^m, \quad (26)$$

so

$$\mu_1' = \tfrac{1}{2}(N+1) \quad (27)$$
$$\mu_2' = \tfrac{1}{6}(N+1)(2N+1) \quad (28)$$
$$\mu_3' = \tfrac{1}{4}N(N+1)^2 \quad (29)$$
$$\mu_4' = \tfrac{1}{30}(N+1)(2N+1)(3N^2+3N-1), \quad (30)$$

and the MOMENTS about the MEAN are

$$\mu_2 = \tfrac{1}{12}(N-1)(N+1) \quad (31)$$
$$\mu_3 = 0 \quad (32)$$
$$\mu_4 = \tfrac{1}{240}(N-1)(N+1)(3N^2-7). \quad (33)$$

The MEAN, VARIANCE, SKEWNESS, and KURTOSIS are

$$\mu = \tfrac{1}{2}(N+1) \quad (34)$$
$$\sigma^2 = \mu_2 = \tfrac{1}{12}(N-1)(N+1) \quad (35)$$
$$\gamma_1 = \frac{\mu_3}{\sigma^{3/2}} = 0 \quad (36)$$
$$\gamma_2 = \frac{6(N^2+1)}{5(N-1)(N+1)}. \quad (37)$$

References

Beyer, W. H. *CRC Standard Mathematical Tables, 28th ed.* Boca Raton, FL: CRC Press, pp. 531 and 533, 1987.

Uniform Polyhedron

The uniform polyhedra are POLYHEDRA with identical VERTICES. Coxeter *et al.* (1954) conjectured that there are 75 such polyhedra in which only two faces are allowed to meet at an EDGE, and this was subsequently proven. (However, when any EVEN number of faces may meet, there are 76 polyhedra.) If the five pentagonal PRISMS are included, the number rises to 80.

The VERTICES of a uniform polyhedron all lie on a SPHERE whose center is their CENTROID. The VERTICES joined to another VERTEX lie on a CIRCLE.

Source code and binary programs for generating and viewing the uniform polyhedra are also available at `http://www.math.technion.ac.il/~rl/kaleido/`. The following depictions of the polyhedra were produced by R. Maeder's `UniformPolyhedra.m` package for *Mathematica*® (Wolfram Research, Champaign, IL). Due to a limitation in *Mathematica*'s renderer, uniform polyhedra 69, 72, 74, and 75 cannot be displayed using this package.

n	Name/Dual
1	tetrahedron
	tetrahedron
2	truncated tetrahedron
	triakis tetrahedron
3	octahemioctahedron
	octahemioctacron
4	tetrahemihexahedron
	tetrahemihexacron
5	octahedron
	cube
6	cube
	octahedron
7	cuboctahedron
	rhombic dodecahedron
8	truncated octahedron
	tetrakis hexahedron
9	truncated cube
	triakis octahedron
10	small rhombicuboctahedron
	deltoidal icositetrahedron
11	truncated cuboctahedron
	disdyakis dodecahedron
12	snub cube
	pentagonal icositetrahedron
13	small cubicuboctahedron
	small hexacronic icositetrahedron
14	great cubicuboctahedron
	great hexacronic icositetrahedron
15	cubohemioctahedron
	hexahemioctahedron
16	cubitruncated cuboctahedron
	tetradyakis hexahedron
17	great rhombicuboctahedron
	great deltoidal icositetrahedron
18	small rhombihexahedron
	small rhombihexacron
19	stellated truncated hexahedron
	great triakis octahedron
20	great truncated cuboctahedron
	great disdyakis dodecahedron
21	great rhombihexahedron
	great rhombihexacron
22	icosahedron
	dodecahedron
23	dodecahedron
	icosahedron
24	icosidodecahedron
	rhombic triacontahedron
25	truncated icosahedron
	pentakis dodecahedron
26	truncated dodecahedron
	triakis icosahedron
27	small rhombicosidodecahedron
	deltoidal hexecontahedron
28	truncated icosidodecahedron
	disdyakis triacontahedron
29	snub dodecahedron
	pentagonal hexecontahedron
30	small ditrigonal icosidodecahedron
	small triambic icosahedron
31	small icosicosidodecahedron
	small icosacronic hexecontahedron
32	small snub icosicosidodecahedron
	small hexagonal hexecontahedron
33	small dodecicosidodecahedron
	small dodecacronic hexecontahedron
34	small stellated dodecahedron
	great dodecahedron
35	great dodecahedron
	small stellated dodecahedron
36	dodecadodecahedron
	medial rhombic triacontahedron
37	truncated great dodecahedron
	small stellapentakis dodecahedron
38	rhombidodecadodecahedron
	medial deltoidal hexecontahedron
39	small rhombidodecahedron
	small rhombidodecacron
40	snub dodecadodecahedron
	medial pentagonal hexecontahedron
41	ditrigonal dodecadodecahedron
	medial triambic icosahedron
42	great ditrigonal dodecicosidodecahedron
	great ditrigonal dodecacronic hexecontahedron
43	small ditrigonal dodecicosidodecahedron
	small ditrigonal dodecacronic hexecontahedron
44	icosidodecadodecahedron
	medial icosacronic hexecontahedron
45	icositruncated dodecadodecahedron
	tridyakis icosahedron
46	snub icosidodecadodecahedron
	medial hexagonal hexecontahedron
47	great ditrigonal icosidodecahedron
	great triambic icosahedron
48	great icosicosidodecahedron
	great icosacronic hexecontahedron
49	small icosihemidodecahedron
	small icosihemidodecacron
50	small dodecicosahedron
	small dodecicosacron

n	Name/Dual
51	small dodecahemidodecahedron
	small dodecahemidodecacron
52	great stellated dodecahedron
	great icosahedron
53	great icosahedron
	great stellated dodecahedron
54	great icosidodecahedron
	great rhombic triacontahedron
55	great truncated icosahedron
	great stellapentakis dodecahedron
56	rhombicosahedron
	rhombicosacron
57	great snub icosidodecahedron
	great pentagonal hexecontahedron
58	small stellated truncated dodecahedron
	great pentakis dodecahedron
59	truncated dodecadodecahedron
	medial disdyakis triacontahedron
60	inverted snub dodecadodecahedron
	medial inverted pentagonal hexecontahedron
61	great dodecicosidodecahedron
	great dodecacronic hexecontahedron
62	small dodecahemicosahedron
	small dodecahemicosacron
63	great dodecicosahedron
	great dodecicosacron
64	great snub dodecicosidodecahedron
	great hexagonal hexecontahedron
65	great dodecahemicosahedron
	great dodecahemicosacron
66	great stellated truncated dodecahedron
	great triakis icosahedron
67	great rhombicosidodecahedron
	great deltoidal hexecontahedron
68	great truncated icosidodecahedron
	great disdyakis triacontahedron
69	great inverted snub icosidodecahedron
	great inverted pentagonal hexecontahedron
70	great dodecahemidodecahedron
	great dodecahemidodecacron
71	great icosihemidodecahedron
	great icosihemidodecacron
72	small retrosnub icosicosidodecahedron
	small hexagrammic hexecontahedron
73	great rhombidodecahedron
	great rhombidodecacron
74	great retrosnub icosidodecahedron
	great pentagrammic hexecontahedron
75	great dirhombicosidodecahedron
	great dirhombicosidodecacron

n	Name/Dual
76	pentagonal prism
	pentagonal dipyramid
77	pentagonal antiprism
	pentagonal deltahedron
78	pentagrammic prism
	pentagrammic dipyramid
79	pentagrammic antiprism
	pentagrammic deltahedron
80	pentagrammic crossed antiprism
	pentagrammic concave deltahedron

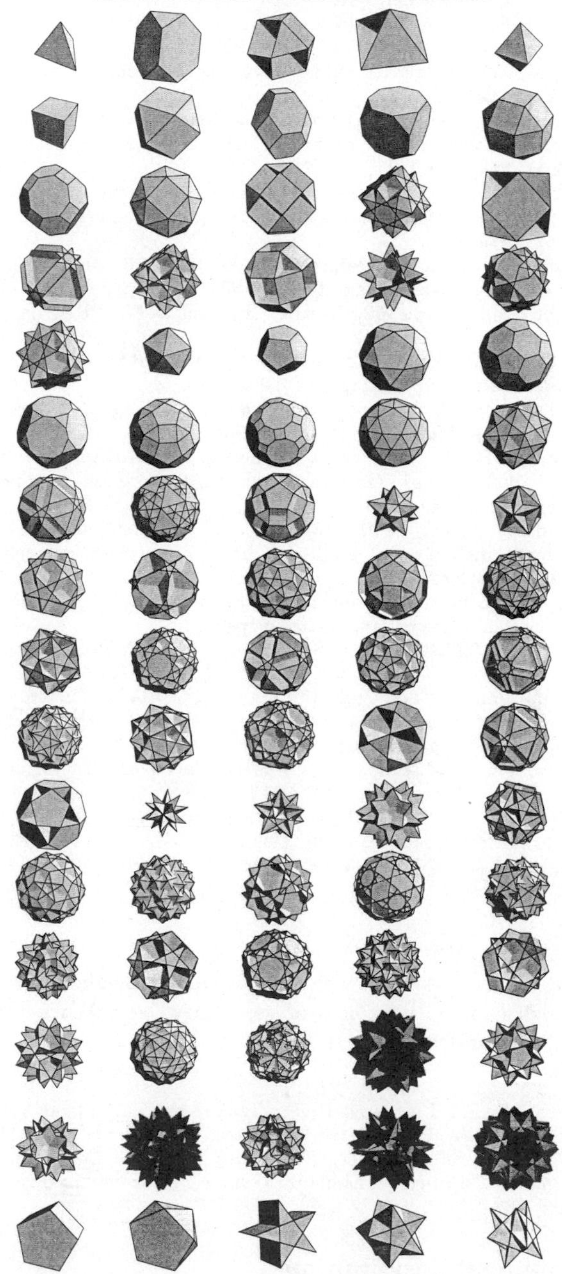

see also ARCHIMEDEAN SOLID, AUGMENTED POLYHEDRON, JOHNSON SOLID, KEPLER-POINSOT SOLID, PLATONIC SOLID, POLYHEDRON, VERTEX FIGURE, WYTHOFF SYMBOL

References

Ball, W. W. R. and Coxeter, H. S. M. *Mathematical Recreations and Essays, 13th ed.* New York: Dover, p. 136, 1987.

Bulatov, V.v "Compounds of Uniform Polyhedra." `http://www.physics.orst.edu/~bulatov/polyhedra/uniform_compounds/`.

Bulatov, V. "Dual Uniform Polyhedra." `http://www.physics.orst.edu/~bulatov/polyhedra/dual/`.

Bulatov, V. "Uniform Polyhedra." `http://www.physics.orst.edu/~bulatov/polyhedra/uniform/`.

Coxeter, H. S. M.; Longuet-Higgins, M. S.; and Miller, J. C. P. "Uniform Polyhedra." *Phil. Trans. Roy. Soc. London Ser. A* **246**, 401–450, 1954.

Har'El, Z. "Uniform Solution for Uniform Polyhedra." *Geometriae Dedicata* **47**, 57–110, 1993.

Har'El, Z. "Kaleido." `http://www.math.technion.ac.il/~rl/kaleido/`.

Har'El, Z. "Eighty Dual Polyhedra Generated by Kaleido." `http://www.math.technion.ac.il/~rl/kaleido/dual.html`.

Har'El, Z. "Eighty Uniform Polyhedra Generated by Kaleido." `http://www.math.technion.ac.il/~rl/kaleido/poly.html`.

Hume, A. "Exact Descriptions of Regular and Semi-Regular Polyhedra and Their Duals." Computing Science Tech. Rept. No. 130. Murray Hill, NJ: AT&T Bell Lab., 1986.

Hume, A. Information files on polyhedra. `http://netlib.bell-labs.com/netlib/polyhedra/`.

Johnson, N. W. "Convex Polyhedra with Regular Faces." *Canad. J. Math.* **18**, 169–200, 1966.

Maeder, R. E. "Uniform Polyhedra." *Mathematica J.* **3**, 1993. `ftp://ftp.inf.ethz.ch/doc/papers/ti/scs/unipoly.ps.gz`.

Maeder, R. E. `Polyhedra.m` and `PolyhedraExamples` *Mathematica®* notebooks. `http://www.inf.ethz.ch/department/TI/rm/programs.html`.

Maeder, R. E. "The Uniform Polyhedra." `http://www.inf.ethz.ch/department/TI/rm/unipoly/`.

Skilling, J. "The Complete Set of Uniform Polyhedron." *Phil. Trans. Roy. Soc. London, Ser. A* **278**, 111–136, 1975.

Virtual Image. "The Uniform Polyhedra CD-ROM." `http://ourworld.compuserve.com/homepages/vir_image/html/uniformpolyhedra.html`.

Wenninger, M. J. *Polyhedron Models.* New York: Cambridge University Press, pp. 1–10 and 98, 1989.

Zalgaller, V. *Convex Polyhedra with Regular Faces.* New York: Consultants Bureau, 1969.

Ziegler, G. M. *Lectures on Polytopes.* Berlin: Springer-Verlag, 1995.

Uniform Variate

A RANDOM NUMBER which lies within a specified range (which can, without loss of generality, be taken as [0, 1]), with a UNIFORM DISTRIBUTION.

References

Press, W. H.; Flannery, B. P.; Teukolsky, S. A.; and Vetterling, W. T. "Uniform Deviates." §7.1 in *Numerical Recipes in FORTRAN: The Art of Scientific Computing, 2nd ed.* Cambridge, England: Cambridge University Press, pp. 267–277, 1992.

Unimodal Distribution

A DISTRIBUTION such as the GAUSSIAN DISTRIBUTION which has a single "peak."

see also BIMODAL DISTRIBUTION

Unimodal Sequence

A finite SEQUENCE which first increases and then decreases. A SEQUENCE $\{s_1, s_2, \ldots, s_n\}$ is unimodal if there exists a t such that

$$s_1 \leq s_2 \leq \ldots \leq s_t$$

and

$$s_t \geq s_{t+1} \geq \ldots \geq s_n.$$

Unimodular Group

A group whose left HAAR MEASURE equals its right HAAR MEASURE.

see also HAAR MEASURE

References

Knapp, A. W. "Group Representations and Harmonic Analysis, Part II." *Not. Amer. Math. Soc.* **43**, 537–549, 1996.

Unimodular Matrix

A MATRIX A with INTEGER elements and DETERMINANT $\det(\mathsf{A}) = \pm 1$, also called a UNIT MATRIX.

The inverse of a unimodular matrix is another unimodular matrix. A POSITIVE unimodular matrix has $\det(\mathsf{A}) = +1$. The nth POWER of a POSITIVE UNIMODULAR MATRIX

$$\mathsf{M} \equiv \begin{bmatrix} m_{11} & m_{12} \\ m_{21} & m_{22} \end{bmatrix} \tag{1}$$

is

$$\mathsf{M}^n = \begin{bmatrix} m_{11}U_{n-1}(a) - U_{n-2}(a) & m_{12}U_{n-1}(a) \\ m_{21}U_{n-1}(a) & m_{22}U_{n-1}(a) - U_{n-2}(a) \end{bmatrix}, \tag{2}$$

where

$$a \equiv \tfrac{1}{2}(m_{11} + m_{22}) \tag{3}$$

and the U_n are CHEBYSHEV POLYNOMIALS OF THE SECOND KIND,

$$U_m(x) = \frac{\sin[(m+1)\cos^{-1} x]}{\sqrt{1-x^2}}. \tag{4}$$

see also CHEBYSHEV POLYNOMIAL OF THE SECOND KIND

References

Born, M. and Wolf, E. *Principles of Optics: Electromagnetic Theory of Propagation, Interference, and Diffraction of Light, 6th ed.* New York: Pergamon Press, p. 67, 1980.

Goldstein, H. *Classical Mechanics, 2nd ed.* Reading, MA: Addison-Wesley, p. 149, 1980.

Unimodular Transformation

A transformation $\mathbf{x}' = \mathsf{A}\mathbf{x}$ is unimodular if the DETERMINANT of the MATRIX A satisfies

$$\det(\mathsf{A}) = \pm 1.$$

A NECESSARY and SUFFICIENT condition that a linear transformation transform a lattice to itself is that the transformation be unimodular.

Union

The union of two sets A and B is the set obtained by combining the members of each. This is written $A \cup B$, and is pronounced "A union B" or "A cup B." The union of sets A_1 through A_n is written $\bigcup_{i=1}^n A_i$.

Let A, B, C, ... be sets, and let $P(S)$ denote the probability of S. Then

$$P(A \cup B) = P(A) + P(B) - P(A \cap B). \tag{1}$$

Similarly,

$$\begin{aligned} P(A \cup B \cup C) &= P[A \cup (B \cup C)] \\ &= P(A) + P(B \cup C) - P[A \cap (B \cup C)] \\ &= P(A) + [P(B) + P(C) - P(B \cap C)] \\ &\quad - P[(A \cap B) \cup (A \cap C)] \\ &= P(A) + P(B) + P(C) - P(B \cap C) \\ &\quad - \{P(A \cap B) + P(A \cap C) - P[(A \cap B) \cap (A \cap C)]\} \\ &= P(A) + P(B) + P(C) - P(A \cap B) \\ &\quad - P(A \cap C) - P(B \cap C) + P(A \cap B \cap C). \end{aligned} \tag{2}$$

If A and B are DISJOINT, by definition $P(A \cap B) = 0$, so

$$P(A \cup B) = P(A) + P(B). \tag{3}$$

Continuing, for a set of n disjoint elements E_1, E_2, ..., E_n

$$P\left(\bigcup_{i=1}^n E_i\right) = \sum_{i=1}^n P(E_i), \tag{4}$$

which is the COUNTABLE ADDITIVITY PROBABILITY AXIOM. Now let

$$E_i \equiv A \cap B_i, \tag{5}$$

then

$$P\left(\bigcup_{i=1}^n E \cap B_i\right) = \sum_{i=1}^n P(E \cap B_i). \tag{6}$$

see also INTERSECTION, OR

Uniplanar Double Point

see ISOLATED SINGULARITY

Unipotent

A p-ELEMENT x of a GROUP G is unipotent if $F^*(C_G(x))$ is a p-GROUP, where F^* is the generalized FITTING SUBGROUP.

see also FITTING SUBGROUP, p-ELEMENT, p-GROUP

Unique

The property of being the only possible solution (perhaps modulo a constant, class of transformation, etc.).

see also ALEKSANDROV'S UNIQUENESS THEOREM, EXISTENCE, MAY-THOMASON UNIQUENESS THEOREM

Unique Factorization Theorem

see FUNDAMENTAL THEOREM OF ARITHMETIC

Unit

A unit is an element in a RING that has a multiplicative inverse. If n is an ALGEBRAIC INTEGER which divides every ALGEBRAIC INTEGER in the FIELD, n is called a unit in that FIELD. A given FIELD may contain an infinity of units. The units of $\mathbb{Z}_n$ are the elements RELATIVELY PRIME to n. The units in $\mathbb{Z}_n$ which are SQUARES are called QUADRATIC RESIDUES.

see also EISENSTEIN UNIT, FUNDAMENTAL UNIT, PRIME UNIT, QUADRATIC RESIDUE

Unit Circle

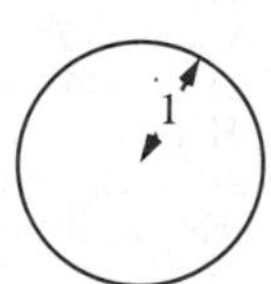

A CIRCLE of RADIUS 1, such as the one used to defined the functions of TRIGONOMETRY.

see also UNIT DISK, UNIT SQUARE

Unit Disk

A DISK with RADIUS 1.

see also FIVE DISKS PROBLEM, UNIT CIRCLE, UNIT SQUARE

Unit Fraction

A unit fraction is a FRACTION with NUMERATOR 1, also known as an EGYPTIAN FRACTION. Any RATIONAL NUMBER has infinitely many unit fraction representations, although only finitely many have a given fixed number of terms. Each FRACTION x/y with y ODD has a unit fraction representation in which each DENOMINATOR is ODD (Breusch 1954; Guy 1994, p. 160). Every x/y has a t-term representation where $t = \mathcal{O}(\sqrt{\log y})$ (Vose 1985).

There are a number of ALGORITHMS (including the BINARY REMAINDER METHOD, CONTINUED FRACTION UNIT FRACTION ALGORITHM, GENERALIZED REMAINDER METHOD, GREEDY ALGORITHM, REVERSE GREEDY ALGORITHM, SMALL MULTIPLE METHOD, and SPLITTING ALGORITHM) for decomposing an arbitrary FRACTION into unit fractions.

see also CALCUS, HALF, QUARTER, SCRUPLE, UNCIA

References

Beck, A.; Bleicher, M. N.; and Crowe, D. W. *Excursions into Mathematics.* New York: Worth Publishers, 1970.

Beeckmans, L. "The Splitting Algorithm for Egyptian Fractions." *J. Number Th.* **43**, 173–185, 1993.

Bleicher, M. N. "A New Algorithm for the Expansion of Continued Fractions." *J. Number Th.* **4**, 342–382, 1972.

Breusch, R. "A Special Case of Egyptian Fractions." Solution to advanced problem 4512. *Amer. Math. Monthly* **61**, 200–201, 1954.

Brown, K. S. "Egyptian Unit Fractions." http://www.seanet.com/~ksbrown.

Eppstein, D. "Ten Algorithms for Egyptian Fractions." *Math. Edu. Res.* **4**, 5–15, 1995.

Eppstein, D. "Egyptian Fractions." http://www.ics.uci.edu/~eppstein/numth/egypt/.

Eppstein, D. Egypt.ma Mathematica notebook. http://www.ics.uci.edu/~eppstein/numth/egypt/egypt.ma.

Graham, R. "On Finite Sums of Unit Fractions." *Proc. London Math. Soc.* **14**, 193–207, 1964.

Guy, R. K. "Egyptian Fractions." §D11 in *Unsolved Problems in Number Theory, 2nd ed.* New York: Springer-Verlag, pp. 87–93 and 158–166, 1994.

Klee, V. and Wagon, S. *Old and New Unsolved Problems in Plane Geometry and Number Theory.* Washington, DC: Math. Assoc. Amer., pp. 175–177 and 206–208, 1991.

Niven, I. and Zuckerman, H. S. *An Introduction to the Theory of Numbers, 5th ed.* New York: Wiley, p. 200, 1991.

Stewart, I. "The Riddle of the Vanishing Camel." *Sci. Amer.*, 122–124, June 1992.

Tenenbaum, G. and Yokota, H. "Length and Denominators of Egyptian Fractions." *J. Number Th.* **35**, 150–156, 1990.

Vose, M. "Egyptian Fractions." *Bull. London Math. Soc.* **17**, 21, 1985.

Wagon, S. "Egyptian Fractions." §8.6 in *Mathematica in Action.* New York: W. H. Freeman, pp. 271–277, 1991.

Unit Matrix

see UNIMODULAR MATRIX

Unit Point

The point in the PLANE with Cartesian coordinates (1, 1).

References

Woods, F. S. *Higher Geometry: An Introduction to Advanced Methods in Analytic Geometry.* New York: Dover, p. 9, 1961.

Unit Ring

A unit ring is a set together with two BINARY OPERATORS $S(+,*)$ satisfying the following conditions:

1. Additive associativity: For all $a, b, c \in S$, $(a+b)+c = a+(b+c)$,
2. Additive commutativity: For all $a, b \in S$, $a+b = b+a$,
3. Additive identity: There exists an element $0 \in S$ such that for all $a \in S: 0+a = a+0 = a$,
4. Additive inverse: For every $a \in S$, there exists a $-a \in S$ such that $a+(-a) = (-a)+a = 0$,
5. Multiplicative associativity: For all $a, b, c \in S$, $(a*b)*c = a*(b*c)$,
6. Multiplicative identity: There exists an element $1 \in S$ such that for all $a \in S$, $1*a = a*1 = a$,
7. Left and right distributivity: For all $a, b, c \in S$, $a*(b+c) = (a*b)+(a*c)$ and $(b+c)*a = (b*a)+(c*a)$.

Thus, a unit ring is a RING with a multiplicative identity.

see also BINARY OPERATOR, RING

References

Rosenfeld, A. *An Introduction to Algebraic Structures.* New York: Holden-Day, 1968.

Unit Sphere

A SPHERE of RADIUS 1.

see also SPHERE, UNIT CIRCLE

Unit Square

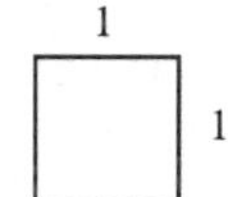

A SQUARE with side lengths 1. *The* unit square usually means the one with coordinates (0, 0), (1, 0), (1, 1), (0, 1) in the real plane, or 0, 1, $1+i$, and i in the COMPLEX PLANE.

see also HEILBRONN TRIANGLE PROBLEM, UNIT CIRCLE, UNIT DISK

Unit Step

see HEAVISIDE STEP FUNCTION

Unit Vector

A VECTOR of unit length. The unit vector $\hat{\mathbf{v}}$ having the same direction as a given (nonzero) vector $\mathbf{v}$ is defined by

$$\hat{\mathbf{v}} \equiv \frac{\mathbf{v}}{|\mathbf{v}|},$$

where $|\mathbf{v}|$ denotes the NORM of $\mathbf{v}$, is the unit vector in the same direction as the (finite) VECTOR $\mathbf{v}$. A unit VECTOR in the $\mathbf{x}_n$ direction is given by

$$\hat{\mathbf{x}}_n \equiv \frac{\frac{\partial \mathbf{r}}{\partial x_n}}{\left|\frac{\partial \mathbf{r}}{\partial x_n}\right|},$$

where $\mathbf{r}$ is the RADIUS VECTOR.

see also NORM, RADIUS VECTOR, VECTOR

Unital

A BLOCK DESIGN of the form $(q^3+1,\ q+1,\ 1)$.

References

Dinitz, J. H. and Stinson, D. R. "A Brief Introduction to Design Theory." Ch. 1 in *Contemporary Design Theory: A Collection of Surveys* (Ed. J. H. Dinitz and D. R. Stinson). New York: Wiley, pp. 1–12, 1992.

Unitary Aliquot Sequence

An ALIQUOT SEQUENCE computed using the analog of the RESTRICTED DIVISOR FUNCTION $s^*(n)$ in which only UNITARY DIVISORS are included.

see also ALIQUOT SEQUENCE, UNITARY SOCIABLE NUMBERS

References

Guy, R. K. "Unitary Aliquot Sequences." §B8 in *Unsolved Problems in Number Theory, 2nd ed.* New York: Springer-Verlag, pp. 63–65, 1994.

Unitary Amicable Pair

A PAIR of numbers m and n such that

$$\sigma^*(m) = \sigma^*(n) = m + n,$$

where $\sigma^*(n)$ is the sum of UNITARY DIVISORS. Hagis (1971) and García (1987) give 82 such pairs. The first few are (114, 126), (1140, 1260), (18018, 22302), (32130, 40446), ... (Sloane's A002952 and A002953).

References

García, M. "New Unitary Amicable Couples." *J. Recr. Math.* **19**, 12–14, 1987.

Guy, R. K. *Unsolved Problems in Number Theory, 2nd ed.* New York: Springer-Verlag, p. 57, 1994.

Hagis, P. "Relatively Prime Amicable Numbers of Opposite Parity." *Math. Comput.* **25**, 915–918, 1971.

Sloane, N. J. A. Sequences A002952/M5372 and A002953/M5389 in "An On-Line Version of the Encyclopedia of Integer Sequences."

Unitary Divisor

A DIVISOR d of c for which

$$\mathrm{GCD}(d, c/d) = 1,$$

where GCD is the GREATEST COMMON DIVISOR.

see also DIVISOR, GREATEST COMMON DIVISOR, UNITARY PERFECT NUMBER

References

Guy, R. K. "Unitary Perfect Numbers." §B3 in *Unsolved Problems in Number Theory, 2nd ed.* New York: Springer-Verlag, pp. 53–59, 1994.

Unitary Group

The unitary group $U_n(q)$ is the set of $n \times n$ UNITARY MATRICES.

see also LIE-TYPE GROUP, UNITARY MATRIX

References

Wilson, R. A. "ATLAS of Finite Group Representation." http://for.mat.bham.ac.uk/atlas#unit.

Unitary Matrix

A unitary matrix is a MATRIX U for which

$$\mathsf{U}^\dagger = \mathsf{U}^{-1}, \tag{1}$$

where † denotes the ADJOINT OPERATOR. This guarantees that

$$\mathsf{U}^\dagger\mathsf{U} = 1. \tag{2}$$

Unitary matrices leave the length of a COMPLEX vector unchanged. The product of two unitary matrices is itself unitary. If U is unitary, then so is U^{-1}. A SIMILARITY TRANSFORMATION of a HERMITIAN MATRIX with a unitary matrix gives

$$\begin{aligned}(uau^{-1})^\dagger &= [(ua)(u^{-1})]^\dagger = (u^{-1})^\dagger(ua)^\dagger = (u^\dagger)^\dagger(a^\dagger u^\dagger)\\ &= uau^\dagger = uau^{-1}.\end{aligned} \tag{3}$$

For REAL MATRICES, HERMITIAN is the same as ORTHOGONAL. Unitary matrices are NORMAL MATRICES.

If M is a unitary matrix, then the PERMANENT

$$|\operatorname{perm}(\mathsf{M})| \leq 1 \tag{4}$$

(Minc 1978, p. 25, Vardi 1991).

see also ADJOINT OPERATOR, HERMITIAN MATRIX, NORMAL MATRIX, ORTHOGONAL MATRIX, PERMANENT

References

Arfken, G. "Hermitian Matrices, Unitary Matrices." §4.5 in *Mathematical Methods for Physicists, 3rd ed.* Orlando, FL: Academic Press, pp. 209–217, 1985.

Minc, H. *Permanents.* Reading, MA: Addison-Wesley, 1978.

Vardi, I. "Permanents." §6.1 in *Computational Recreations in Mathematica.* Reading, MA: Addison-Wesley, pp. 108 and 110–112, 1991.

Unitary Multiperfect Number

A number n which is an INTEGER multiple k of the SUM of its UNITARY DIVISORS $\sigma^*(n)$ is called a unitary k-multiperfect number. There are no ODD unitary multiperfect numbers.

References

Guy, R. K. "Unitary Perfect Numbers." §B3 in *Unsolved Problems in Number Theory, 2nd ed.* New York: Springer-Verlag, pp. 53–59, 1994.

Unitary Multiplicative Character

A MULTIPLICATIVE CHARACTER is called unitary if it has ABSOLUTE VALUE 1 everywhere.

see also CHARACTER (MULTIPLICATIVE)

Unitary Perfect Number

A number n which is the sum of its UNITARY DIVISORS with the exception of n itself. There are no ODD unitary perfect numbers, and it has been conjectured that there are only a FINITE number of EVEN ones. The first few are 6, 60, 90, 87360, 146361946186458562560000, ... (Sloane's A002827).

References

Guy, R. K. "Unitary Perfect Numbers." §B3 in *Unsolved Problems in Number Theory, 2nd ed.* New York: Springer-Verlag, pp. 53–59, 1994.

Sloane, N. J. A. Sequence A002827/M4268 in "An On-Line Version of the Encyclopedia of Integer Sequences."

Wall, C. R. "On the Largest Odd Component of a Unitary Perfect Number." *Fib. Quart.* **25**, 312–316, 1987.

Unitary Sociable Numbers

SOCIABLE NUMBERS computed using the analog of the RESTRICTED DIVISOR FUNCTION $s^*(n)$ in which only UNITARY DIVISORS are included.

see also SOCIABLE NUMBERS

References

Guy, R. K. "Unitary Aliquot Sequences." §B8 in *Unsolved Problems in Number Theory, 2nd ed.* New York: Springer-Verlag, pp. 63–65, 1994.

Unitary Transformation

A transformation of the form

$$\mathsf{A}' = \mathsf{U}\mathsf{A}\mathsf{U}^\dagger,$$

where † denotes the ADJOINT OPERATOR.

see also ADJOINT OPERATOR, TRANSFORMATION

Unitary Unimodular Group

see SPECIAL UNITARY GROUP

Unity

The number 1. There are n nth ROOTS OF UNITY, known as the DE MOIVRE NUMBERS.

see also 1, PRIMITIVE ROOT OF UNITY

Univalent Function

A function or transformation f in which $f(z)$ does not overlap z.

Univariate Function

A FUNCTION of a single variable (e.g., $f(x)$, $g(z)$, $\theta(\xi)$, etc.).

see also MULTIVARIATE FUNCTION

Univariate Polynomial

A POLYNOMIAL in a single variable. In common usage, univariate POLYNOMIALS are sometimes simply called "POLYNOMIALS."

see also POLYNOMIAL

Universal Graph

see COMPLETE GRAPH

Universal Statement

A universal statement S is a FORMULA whose FREE variables are all in the scope of universal quantifiers.

Universal Turing Machine

A TURING MACHINE which, by appropriate programming using a finite length of input tape, can act as *any* TURING MACHINE whatsoever.

see CHAITIN'S CONSTANT, HALTING PROBLEM, TURING MACHINE

References

Penrose, R. *The Emperor's New Mind: Concerning Computers, Minds, and the Laws of Physics.* Oxford: Oxford University Press, pp. 51–57, 1989.

Unknot

A closed loop which is not KNOTTED. In the 1930s, by making use of REIDEMEISTER MOVES, Reidemeister first proved that KNOTS exist which are distinct from the unknot. He proved this by COLORING each part of a knot diagram with one of three colors.

The KNOT SUM of two unknots is another unknot.

The JONES POLYNOMIAL of the unknot is defined to give the normalization

$$V(t) = 1.$$

Haken (1961) devised an ALGORITHM to tell if a knot projection is the unknot. The ALGORITHM is so complicated, however, that it has never been implemented. Although it is not immediately obvious, the unknot is a PRIME KNOT.

see also COLORABLE, KNOT, KNOT THEORY, LINK, REIDEMEISTER MOVES, UNKNOTTING NUMBER

References

Haken, W. "Theorie der Normalflachen." *Acta Math.* **105**, 245–375, 1961.

Unknotting Number

The smallest number of times a KNOT must be passed through itself to untie it. Lower bounds can be computed using relatively straightforward techniques, but it is in general difficult to determine exact values. Many unknotting numbers can be determined from a knot's SIGNATURE. A KNOT with unknotting number 1 is a PRIME KNOT (Scharlemann 1985). It is not always true that the unknotting number is achieved in a projection with the minimal number of crossings.

The following table is from Kirby (1997, pp. 88–89), with the values for 10_{139} and 10_{152} taken from Kawamura. The unknotting numbers for 10_{154} and 10_{161} can be found using MENASCO'S THEOREM (Stoimenow 1998).

3_1	1	8_9	1	9_{10}	2 or 3	9_{32}	1 or 2
4_1	1	8_{10}	1 or 2	9_{11}	2	9_{33}	1
5_1	2	8_{11}	1	9_{12}	1	9_{34}	1
5_2	1	8_{12}	2	9_{13}	2 or 3	9_{35}	2 or 3
6_1	1	8_{13}	1	9_{14}	1	9_{36}	2
6_2	1	8_{14}	1	9_{15}	2	9_{37}	2
6_3	1	8_{15}	2	9_{16}	3	9_{38}	2 or 3
7_1	3	8_{16}	2	9_{17}	2	9_{39}	1
7_2	1	8_{17}	1	9_{18}	2	9_{40}	2
7_3	2	8_{18}	2	9_{19}	1	9_{41}	2
7_4	2	8_{19}	3	9_{20}	2	9_{42}	1
7_5	2	8_{20}	1	9_{21}	1	9_{43}	2
7_6	1	8_{21}	1	9_{22}	1	9_{44}	1
7_7	1	9_1	4	9_{23}	2	9_{45}	1
8_1	1	9_2	1	9_{24}	1	9_{46}	2
8_2	2	9_3	3	9_{25}	2	9_{47}	2
8_3	2	9_4	2	9_{26}	1	9_{48}	2
8_4	2	9_5	2	9_{27}	1	9_{49}	2 or 3
8_5	2	9_6	3	9_{28}	1	10_{139}	4
8_6	2	9_7	2	9_{29}	1	10_{152}	4
8_7	1	9_8	2	9_{30}	1	10_{154}	3
8_8	2	9_9	3	9_{31}	2	10_{161}	3

see also BENNEQUIN'S CONJECTURE, MENASCO'S THEOREM, MILNOR'S CONJECTURE, SIGNATURE (KNOT)

References

Adams, C. C. *The Knot Book: An Elementary Introduction to the Mathematical Theory of Knots.* New York: W. H. Freeman, pp. 57–64, 1994.

Cipra, B. "From Knot to Unknot." *What's Happening in the Mathematical Sciences, Vol. 2.* Providence, RI: Amer. Math. Soc., pp. 8–13, 1994.

Kawamura, T. "The Unknotting Numbers of 10_{139} and 10_{152} are 4." To appear in *Osaka J. Math.* http://ms421sun.ms.u-tokyo.ac.jp/~kawamura/worke.html.

Kirby, R. (Ed.) "Problems in Low-Dimensional Topology." *AMS/IP Stud. Adv. Math., 2.2, Geometric Topology (Athens, GA, 1993).* Providence, RI: Amer. Math. Soc., pp. 35–473, 1997.

Scharlemann, M. "Unknotting Number One Knots are Prime." *Invent. Math.* **82**, 37–55, 1985.

Stoimenow, A. "Positive Knots, Closed Braids and the Jones Polynomial." Rev. May, 1997. http://www.informatik.hu-berlin.de/~stoimeno/pos.ps.gz.

Weisstein, E. W. "Knots and Links." http://www.astro.virginia.edu/~eww6n/math/notebooks/Knots.m.

Unless

If A is true unless B, then not-B implies A, but B does not necessarily imply not-A.

see also PRECISELY UNLESS

Unlesss

see PRECISELY UNLESS

Unmixed

A homogeneous IDEAL defining a projective ALGEBRAIC VARIETY is unmixed if it has no embedded PRIME divisors.

Unpoke Move

see POKE MOVE

Unsafe

A position in a GAME is unsafe if the person who plays next can win. Every unsafe position can be made SAFE by at least one move.

see also GAME, SAFE

Unsolved Problem

see PROBLEM

Unstable Improper Node

A FIXED POINT for which the STABILITY MATRIX has equal POSITIVE EIGENVALUES.

see also ELLIPTIC FIXED POINT (DIFFERENTIAL EQUATIONS), FIXED POINT, HYPERBOLIC FIXED POINT (DIFFERENTIAL EQUATIONS), STABLE IMPROPER NODE, STABLE NODE, STABLE SPIRAL POINT, UNSTABLE NODE, UNSTABLE SPIRAL POINT, UNSTABLE STAR

References

Tabor, M. "Classification of Fixed Points." §1.4.b in *Chaos and Integrability in Nonlinear Dynamics: An Introduction.* New York: Wiley, pp. 22–25, 1989.

Unstable Node

A FIXED POINT for which the STABILITY MATRIX has both EIGENVALUES POSITIVE, so $\lambda_1 > \lambda_2 > 0$.

see also ELLIPTIC FIXED POINT (DIFFERENTIAL EQUATIONS), FIXED POINT, HYPERBOLIC FIXED POINT (DIFFERENTIAL EQUATIONS), STABLE IMPROPER NODE, STABLE NODE, STABLE SPIRAL POINT, STABLE STAR, UNSTABLE IMPROPER NODE, UNSTABLE SPIRAL POINT, UNSTABLE STAR

References

Tabor, M. "Classification of Fixed Points." §1.4.b in *Chaos and Integrability in Nonlinear Dynamics: An Introduction.* New York: Wiley, pp. 22–25, 1989.

Unstable Spiral Point

A FIXED POINT for which the STABILITY MATRIX has EIGENVALUES of the form $\lambda_\pm = \alpha \pm i\beta$ (with $\alpha, \beta > 0$).

see also ELLIPTIC FIXED POINT (DIFFERENTIAL EQUATIONS), FIXED POINT, HYPERBOLIC FIXED POINT (DIFFERENTIAL EQUATIONS), STABLE IMPROPER NODE, STABLE NODE, STABLE SPIRAL POINT, STABLE STAR, UNSTABLE IMPROPER NODE, UNSTABLE NODE, UNSTABLE STAR

References

Tabor, M. "Classification of Fixed Points." §1.4.b in *Chaos and Integrability in Nonlinear Dynamics: An Introduction.* New York: Wiley, pp. 22–25, 1989.

Unstable Star

A FIXED POINT for which the STABILITY MATRIX has one zero EIGENVECTOR with POSITIVE EIGENVALUE $\lambda > 0$.

see also ELLIPTIC FIXED POINT (DIFFERENTIAL EQUATIONS), FIXED POINT, HYPERBOLIC FIXED POINT (DIFFERENTIAL EQUATIONS), STABLE IMPROPER NODE, STABLE NODE, STABLE SPIRAL POINT, STABLE STAR, UNSTABLE IMPROPER NODE, UNSTABLE NODE, UNSTABLE SPIRAL POINT

References

Tabor, M. "Classification of Fixed Points." §1.4.b in *Chaos and Integrability in Nonlinear Dynamics: An Introduction.* New York: Wiley, pp. 22–25, 1989.

Untouchable Number

An untouchable number is an INTEGER which is not the sum of the PROPER DIVISORS of any other number. The first few are 2, 5, 52, 88, 96, 120, 124, 146, ... (Sloane's A005114). Erdős has proven that there are infinitely many. It is thought that 5 is the only ODD untouchable number.

References

Abramowitz, M. and Stegun, C. A. (Eds.). *Handbook of Mathematical Functions with Formulas, Graphs, and Mathematical Tables, 9th printing.* New York: Dover, p. 840, 1972.

Guy, R. K. "Untouchable Numbers." §B10 in *Unsolved Problems in Number Theory, 2nd ed.* New York: Springer-Verlag, pp. 66–67, 1994.

Sloane, N. J. A. Sequence A005114/M1552 in "An On-Line Version of the Encyclopedia of Integer Sequences."

Upper Bound

see LEAST UPPER BOUND

Upper Integral

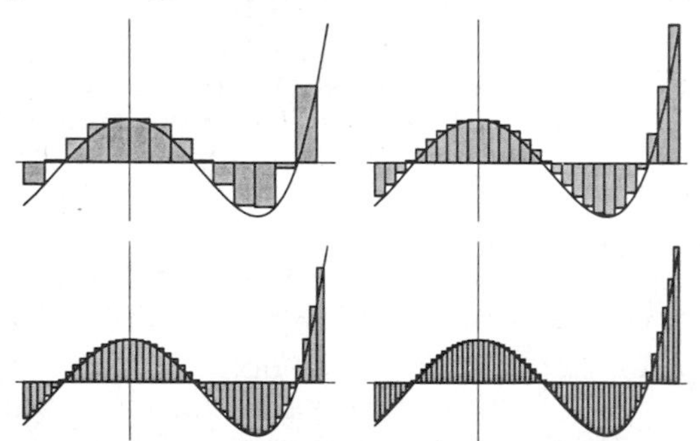

The limit of an UPPER SUM, when it exists, as the MESH SIZE approaches 0.

see also LOWER INTEGRAL, RIEMANN INTEGRAL, UPPER SUM

Upper Limit

Let the greatest term H of a SEQUENCE be a term which is greater than all but a finite number of the terms which are equal to H. Then H is called the upper limit of the SEQUENCE.

An upper limit of a SERIES

$$\text{upper} \lim_{n\to\infty} S_n = \overline{\lim_{n\to\infty}} S_n = k$$

is said to exist if, for every $\epsilon > 0$, $|S_n - k| < \epsilon$ for infinitely many values of n and if no number larger than k has this property.

see also LIMIT, LOWER LIMIT

References

Bromwich, T. J. I'a and MacRobert, T. M. "Upper and Lower Limits of a Sequence." §5.1 in *An Introduction to the Theory of Infinite Series, 3rd ed.* New York: Chelsea, p. 40, 1991.

Upper Sum

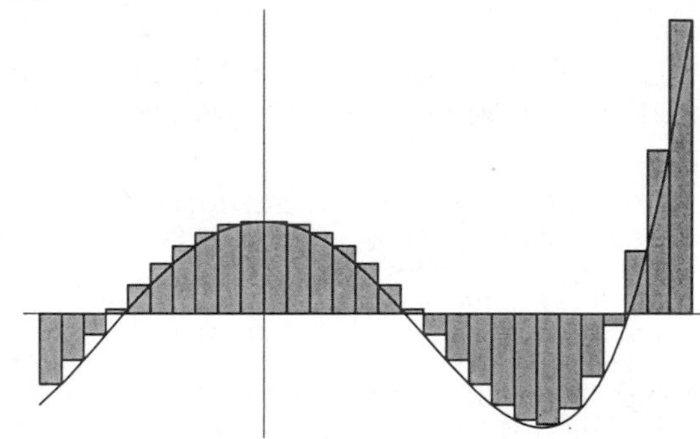

For a given function $f(x)$ over a partition of a given interval, the upper sum is the sum of box areas $f(x_k^*)\Delta x_k$ using the greatest value of the function $f(x_k^*)$ in each subinterval Δx_k.

see also LOWER SUM, RIEMANN INTEGRAL, UPPER INTEGRAL

Upper-Trimmed Subsequence

The upper-trimmed subsequence of $x = \{x_n\}$ is the sequence $\lambda(x)$ obtained by dropping the first occurrence of n for each n. If x is a FRACTAL SEQUENCE, then $\lambda(x) = x$.

see also LOWER-TRIMMED SUBSEQUENCE

References

Kimberling, C. "Fractal Sequences and Interspersions." *Ars Combin.* **45**, 157–168, 1997.

Upward Drawing

see HASSE DIAGRAM

Urchin

Kepler's original name for the SMALL STELLATED DODECAHEDRON.

Utility Graph

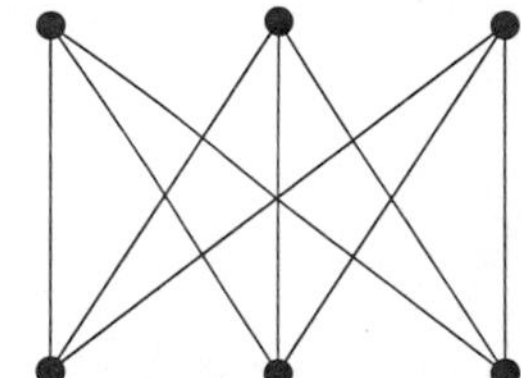

The utility problem asks, "Can a PLANAR GRAPH be constructed from each of three nodes ('house owners') to each of three other nodes ('wells')?" The answer is no, and the proof can be effected using the JORDAN CURVE THEOREM, while a more general result encompassing this one is the KURATOWSKI REDUCTION THEOREM. The utility graph UG is the graph showing the relationships described above. It is identical to the THOMSEN GRAPH and, in the more formal parlance of GRAPH THEORY, is known as the COMPLETE BIPARTITE GRAPH $K_{3,3}$.

see also COMPLETE BIPARTITE GRAPH, KURATOWSKI REDUCTION THEOREM, PLANAR GRAPH, THOMSEN GRAPH

References

Chartrand, G. "The Three Houses and Three Utilities Problem: An Introduction to Planar Graphs." §9.1 in *Introductory Graph Theory.* New York: Dover, pp. 191–202, 1985.

Ore, Ø. *Graphs and Their Uses.* New York: Random House, pp. 14–17, 1963.

Pappas, T. "Wood, Water, Grain Problem." *The Joy of Mathematics.* San Carlos, CA: Wide World Publ./Tetra, pp. 175 and 233, 1989.

Utility Problem

see UTILITY GRAPH

Valence

see VALENCY

Valency

The number of EDGES at a GRAPH VERTEX.

Valuation

A generalization of the p-ADIC NUMBERS first proposed by Kürschák in 1913. A valuation $|\cdot|$ on a FIELD K is a FUNCTION from K to the REAL NUMBERS $\mathbb{R}$ such that the following properties hold for all $x, y \in K$:

1. $|x| \geq 0$,
2. $|x| = 0$ IFF $x = 0$,
3. $|xy| = |x|\,|y|$,
4. $|x| \leq 1$ IMPLIES $|1+x| \leq C$ for some constant $C \geq 1$ (independent of x).

If (4) is satisfied for $C = 2$, then $|\cdot|$ satisfies the TRIANGLE INEQUALITY,

4a. $|x + y| \leq |x| + |y|$ for all $x, y \in K$.

If (4) is satisfied for $C = 1$ then $|\cdot|$ satisfies the stronger TRIANGLE INEQUALITY

4b. $|x + y| \leq \max(|x|, |y|)$.

The simplest valuation is the ABSOLUTE VALUE for REAL NUMBERS. A valuation satisfying (4b) is called non-ARCHIMEDEAN VALUATION; otherwise, it is called ARCHIMEDEAN.

If $|\cdot|_1$ is a valuation on K and $\lambda \geq 1$, then we can define a new valuation $|\cdot|_2$ by

$$|x|_2 = |x|_1^\lambda. \qquad (1)$$

This does indeed give a valuation, but possibly with a different constant C in AXIOM 4. If two valuations are related in this way, they are said to be equivalent, and this gives an equivalence relation on the collection of all valuations on K. Any valuation is equivalent to one which satisfies the triangle inequality (4a). In view of this, we need only to study valuations satisfying (4a), and we often view axioms (4) and (4a) as interchangeable (although this is not strictly true).

If two valuations are equivalent, then they are both non-ARCHIMEDEAN or both ARCHIMEDEAN. $\mathbb{Q}$, $\mathbb{R}$, and $\mathbb{C}$ with the usual Euclidean norms are Archimedean valuated fields. For any PRIME p, the p-ADIC NUMBERS $\mathbb{Q}_p$ with the p-adic valuation $|\cdot|_p$ is a non-Archimedean valuated field.

If K is any FIELD, we can define the trivial valuation on K by $|x| = 1$ for all $x \neq 0$ and $|0| = 0$, which is a non-Archimedean valuation. If K is a FINITE FIELD, then the only possible valuation over K is the trivial one. It can be shown that any valuation on $\mathbb{Q}$ is equivalent to one of the following: the trivial valuation, Euclidean absolute norm $|\cdot|$, or p-adic valuation $|\cdot|_p$.

The equivalence of any nontrivial valuation of $\mathbb{Q}$ to either the usual ABSOLUTE VALUE or to a p-ADIC NUMBER absolute value was proved by Ostrowski (1935). Equivalent valuations give rise to the same topology. Conversely, if two valuations have the same topology, then they are equivalent. A stronger result is the following: Let $|\cdot|_1$, $|\cdot|_2$, ..., $|\cdot|_k$ be valuations over K which are pairwise inequivalent and let a_1, a_2, ..., a_k be elements of K. Then there exists an infinite sequence $(x_1, x_2, \ldots)$ of elements of K such that

$$\lim_{n\to\infty \text{ w.r.t. } |\cdot|_1} x_n = a_1 \qquad (2)$$

$$\lim_{n\to\infty \text{ w.r.t. } |\cdot|_2} x_n = a_2, \qquad (3)$$

etc. This says that inequivalent valuations are, in some sense, completely independent of each other. For example, consider the rationals $\mathbb{Q}$ with the 3-adic and 5-adic valuations $|\cdot|_3$ and $|\cdot|_5$, and consider the sequence of numbers given by

$$x_n = \frac{43 \cdot 5^n + 92 \cdot 3^n}{3^n + 5^n}. \qquad (4)$$

Then $x_n \to 43$ as $n \to \infty$ with respect to $|\cdot|_3$, but $x_n \to 92$ as $n \to \infty$ with respect to $|\cdot|_5$, illustrating that a sequence of numbers can tend to two different limits under two different valuations.

A discrete valuation is a valuation for which the VALUATION GROUP is a discrete subset of the REAL NUMBERS $\mathbb{R}$. Equivalently, a valuation (on a FIELD K) is discrete if there exists a REAL NUMBER $\epsilon > 0$ such that

$$|x| \in (1-\epsilon, 1+\epsilon) \Rightarrow |x| = 1 \text{ for all } x \in K. \qquad (5)$$

The p-adic valuation on $\mathbb{Q}$ is discrete, but the ordinary absolute valuation is not.

If $|\cdot|$ is a valuation on K, then it induces a metric

$$d(x, y) = |x - y| \qquad (6)$$

on K, which in turn induces a TOPOLOGY on K. If $|\cdot|$ satisfies (4b) then the metric is an ULTRAMETRIC. We say that $(K, |\cdot|)$ is a complete valuated field if the METRIC SPACE is complete.

see also ABSOLUTE VALUE, LOCAL FIELD, METRIC SPACE, p-ADIC NUMBER, STRASSMAN'S THEOREM, ULTRAMETRIC, VALUATION GROUP

References

Cassels, J. W. S. *Local Fields*. Cambridge, England: Cambridge University Press, 1986.

Ostrowski, A. "Untersuchungen zur aritmetischen Theorie der Körper." *Math. Zeit.* **39**, 269–404, 1935.

Valuation Group

Let $(K, |\cdot|)$ be a valuated field. The valuation group G is defined to be the set

$$G = \{|x| : x \in K, x \neq 0\},$$

with the group operation being multiplication. It is a SUBGROUP of the POSITIVE REAL NUMBERS, under multiplication.

Valuation Ring

Let $(K, |\cdot|)$ be a non-Archimedean valuated field. Its valuation ring R is defined to be

$$R = \{x \in K : |x| \leq 1\}.$$

The valuation ring has maximal IDEAL

$$M = \{x \in K : |x| < 1\},$$

and the field R/M is called the residue field, class field, or field of digits. For example, if $K = \mathbb{Q}_p$ (p-adic numbers), then $R = Z_p$ (p-adic integers), $M = pZ_p$ (p-adic integers congruent to 0 mod p), and $R/M = \text{GF}(p)$, the FINITE FIELD of order p.

Valuation Theory

The study of VALUATIONS which simplifies class field theory and the theory of algebraic function fields.

see also VALUATION

References

Iyanaga, S. and Kawada, Y. (Eds.). "Valuations." §425 in *Encyclopedic Dictionary of Mathematics.* Cambridge, MA: MIT Press, pp. 1350–1353, 1980.

Value

The quantity which a FUNCTION f takes upon application to a given quantity.

see also VALUE (GAME)

Value (Game)

The solution to a GAME in GAME THEORY. When a SADDLE POINT is present

$$\min_{i \leq m} \min_{j \leq n} a_{ij} = \min_{j \leq n} \max_{i \leq m} a_{ij} \equiv v,$$

and v is the value for pure strategies.

see also ABSOLUTE VALUE, GAME THEORY, MINIMAX THEOREM, VALUATION

Vampire Number

A number $v = xy$ with an EVEN number n of DIGITS formed by multiplying a pair of $n/2$-DIGIT numbers (where the DIGITS are taken from the original number in any order) x and y together. Pairs of trailing zeros are not allowed. If v is a vampire number, then x and y are called its "fangs." Examples of vampire numbers include

$$\begin{aligned} 1260 &= 21 \times 60 \\ 1395 &= 15 \times 93 \\ 1435 &= 35 \times 41 \\ 1530 &= 30 \times 51 \\ 1827 &= 21 \times 87 \\ 2187 &= 27 \times 81 \\ 6880 &= 80 \times 86 \end{aligned}$$

(Sloane's A014575). There are seven 4-digit vampires, 155 6-digit vampires, and 3382 8-digit vampires. General formulas can be constructed for special classes of vampires, such as the fangs

$$\begin{aligned} x &= 25 \cdot 10^k + 1 \\ y &= 100(10^{k+1} + 52)/25, \end{aligned}$$

giving the vampire

$$\begin{aligned} v = xy &= (10^{k+1} + 52)10^{k+2} + 100(10^{k+1} + 52)/25 \\ &= x^* \cdot 10^{k+2} + t \\ &= 8(26 + 5 \cdot 10^k)(1 + 25 \cdot 10^k), \end{aligned}$$

where x^* denotes x with the DIGITS reversed (Roushe and Rogers).

Pickover (1995) also defines pseudovampire numbers, in which the multiplicands have different number of digits.

References

Pickover, C. A. "Vampire Numbers." Ch. 30 in *Keys to Infinity.* New York: W. H. Freeman, pp. 227–231, 1995.

Pickover, C. A. "Vampire Numbers." *Theta* **9**, 11–13, Spring 1995.

Pickover, C. A. "Interview with a Number." *Discover* **16**, 136, June 1995.

Roushe, F. W. and Rogers, D. G. "Tame Vampires." Undated manuscript.

Sloane, N. J. A. Sequence A014575 in "An On-Line Version of the Encyclopedia of Integer Sequences."

van der Grinten Projection

A MAP PROJECTION given by the transformation

$$x = \operatorname{sgn}(\lambda - \lambda_0) \times \frac{\pi|A(G - P^2) - \sqrt{A^2(G - P^2)^2 - (P^2 + A^2)(G^2 - P^2)}|}{P^2 + A^2} \quad (1)$$

$$y = \operatorname{sgn}(\phi) \frac{\pi|PQ - A\sqrt{(A^2 + 1)(P^2 + A^2) - Q^2}}{P^2 + A^2}, \quad (2)$$

where

$$A = \frac{1}{2}\left|\frac{\pi}{\lambda - \lambda_0} - \frac{\lambda - \lambda_0}{\pi}\right| \quad (3)$$

$$G = \frac{\cos\theta}{\sin\theta + \cos\theta - 1} \quad (4)$$

$$P = G\left(\frac{2}{\sin\theta} - 1\right) \quad (5)$$

$$\theta = \sin^{-1}\left|\frac{2\phi}{\pi}\right| \quad (6)$$

$$Q = A^2 + G. \quad (7)$$

The inverse FORMULAS are

$$\phi = \operatorname{sgn}(y)\pi\left[-m_1\cos(\theta_1 + \tfrac{1}{3}\pi) - \frac{c_2}{3c_3}\right] \quad (8)$$

$$\lambda = \frac{\pi|X^2 + Y^2 - 1 + \sqrt{1 + 2(X^2 - Y^2) + (X^2 + Y^2)^2}|}{2X} + \lambda_0, \quad (9)$$

where

$$X = \frac{x}{\pi} \quad (10)$$

$$Y = \frac{y}{\pi} \quad (11)$$

$$c_1 = -|Y|(1 + X^2 + Y^2) \quad (12)$$

$$c_2 = c_1 - 2Y^2 + X^2 \quad (13)$$

$$c_3 = -2c_1 + 1 + 2Y^2 + (X^2 + Y^2)^2 \quad (14)$$

$$d = \frac{Y^2}{c_3} + \frac{1}{27}\left(\frac{2{c_2}^3}{{c_3}^3} - \frac{9c_1c_2}{{c_3}^2}\right) \quad (15)$$

$$a_1 = \frac{1}{c_3}\left(c_1 - \frac{{c_2}^2}{3c_3}\right) \quad (16)$$

$$m_1 = 2\sqrt{-\tfrac{1}{3}a_1} \quad (17)$$

$$\theta_1 = \tfrac{1}{3}\cos^{-1}\left(\frac{3d}{a_1m_1}\right). \quad (18)$$

References

Snyder, J. P. *Map Projections—A Working Manual.* U. S. Geological Survey Professional Paper 1395. Washington, DC: U. S. Government Printing Office, pp. 239–242, 1987.

van der Pol Equation

An ORDINARY DIFFERENTIAL EQUATION which can be derived from the RAYLEIGH DIFFERENTIAL EQUATION by differentiating and setting $y = y'$. It is an equation describing self-sustaining oscillations in which energy is fed into small oscillations and removed from large oscillations. This equation arises in the study of circuits containing vacuum tubes and is given by

$$y'' - \mu(1 - y^2)y' + y = 0.$$

see also RAYLEIGH DIFFERENTIAL EQUATION

References

Kreyszig, E. *Advanced Engineering Mathematics, 6th ed.* New York: Wiley, pp. 165–166, 1988.

van der Waerden Number

The threshold numbers proven to exist by VAN DER WAERDEN'S THEOREM. The first few are 1, 3, 9, 35, 178, ... (Sloane's A005346).

References

Goodman, J. E. and O'Rourke, J. (Eds.). *Handbook of Discrete & Computational Geometry.* Boca Raton, FL: CRC Press, p. 159, 1997.
Honsberger, R. *More Mathematical Morsels.* Washington, DC: Math. Assoc. Amer., p. 29, 1991.
Sloane, N. J. A. Sequence A005346/M2819 in "An On-Line Version of the Encyclopedia of Integer Sequences."

van der Waerden's Theorem

For any given POSITIVE INTEGERS k and r, there exists a threshold number $n(k, r)$ (known as a VAN DER WAERDEN NUMBER) such that no matter how the numbers 1, 2, ..., n are partitioned into k classes, at least one of the classes contains an ARITHMETIC PROGRESSION of length at least r. However, no FORMULA for $n(k, r)$ is known.

see also ARITHMETIC PROGRESSION

References

Honsberger, R. *More Mathematical Morsels.* Washington, DC: Math. Assoc. Amer., p. 29, 1991.
Khinchin, A. Y. "Van der Waerden's Theorem on Arithmetic Progressions." Ch. 1 in *Three Pearls of Number Theory.* New York: Dover, pp. 11–17, 1998.
van der Waerden, B. L. "Beweis einer Baudetschen Vermutung." *Nieuw Arch. Wiskunde* **15**, 212–216, 1927.

van Kampen's Theorem

In the usual diagram of inclusion homeomorphisms, if the upper two maps are injective, then so are the other two.

References

Rolfsen, D. *Knots and Links.* Wilmington, DE: Publish or Perish Press, pp. 74–75 and 369–373, 1976.

van Wijngaarden-Deker-Brent Method

see BRENT'S METHOD

Vandermonde Determinant

$$\Delta(x_1, \ldots, x_n) \equiv \begin{vmatrix} 1 & x_1 & {x_1}^2 & \cdots & {x_1}^{n-1} \\ 1 & x_2 & {x_2}^2 & \cdots & {x_2}^{n-1} \\ \vdots & \vdots & \vdots & \ddots & \vdots \\ 1 & x_n & {x_n}^2 & \cdots & {x_n}^{n-1} \end{vmatrix} = \prod_{\substack{i,j \\ i>j}} (x_i - x_j)$$

(Sharpe 1987). For INTEGERS $a_1, \ldots, a_n$, $\Delta(a_1, \ldots, a_n)$ is divisible by $\prod_{i=1}^{n}(i-1)!$ (Chapman 1996).

see also VANDERMONDE MATRIX

References

Chapman, R. "A Polynomial Taking Integer Values." *Math. Mag.* **69**, 121, 1996.

Gradshteyn, I. S. and Ryzhik, I. M. *Tables of Integrals, Series, and Products, 5th ed.* San Diego, CA: Academic Press, p. 1111, 1979.

Sharpe, D. §2.9 in *Rings and Factorization.* Cambridge, England: Cambridge University Press, 1987.

Vandermonde Identity

see CHU-VANDERMONDE IDENTITY

Vandermonde Matrix

A type of matrix which arises in the LEAST SQUARES FITTING of POLYNOMIALS and the reconstruction of a DISTRIBUTION from the distribution's MOMENTS. The solution of an $n \times n$ Vandermonde matrix equation requires $\mathcal{O}(n^2)$ operations. A Vandermonde matrix of order n is of the form

$$\begin{bmatrix} 1 & x_1 & {x_1}^2 & \cdots & {x_1}^{n-1} \\ 1 & x_2 & {x_2}^2 & \cdots & {x_2}^{n-1} \\ \vdots & \vdots & \vdots & \ddots & \vdots \\ 1 & x_n & {x_n}^2 & \cdots & {x_n}^{n-1} \end{bmatrix}.$$

see also TOEPLITZ MATRIX, TRIDIAGONAL MATRIX, VANDERMONDE DETERMINANT

References

Press, W. H.; Flannery, B. P.; Teukolsky, S. A.; and Vetterling, W. T. "Vandermonde Matrices and Toeplitz Matrices." §2.8 in *Numerical Recipes in FORTRAN: The Art of Scientific Computing, 2nd ed.* Cambridge, England: Cambridge University Press, pp. 82–89, 1992.

Vandermonde's Sum

see CHU-VANDERMONDE IDENTITY

Vandermonde Theorem

A special case of GAUSS'S THEOREM with a a NEGATIVE INTEGER $-n$:

$${}_2F_1(-n, b; c; 1) = \frac{(c-b)_n}{(c)_n},$$

where ${}_2F_1(a, b; c; z)$ is a HYPERGEOMETRIC FUNCTION and $(a)_n$ is a POCHHAMMER SYMBOL (Bailey 1935, p. 3).

see also GAUSS'S THEOREM

References

Bailey, W. N. *Generalised Hypergeometric Series.* Cambridge, England: Cambridge University Press, 1935.

Vandiver's Criteria

Let p be a IRREGULAR PRIME, and let $P = rp + 1$ be a PRIME with $P < p^2 - p$. Also let t be an INTEGER such that $t^3 \not\equiv 1 \pmod{P}$. For an IRREGULAR PAIR $(p, 2k)$, form the product

$$Q_{2k} = t^{-rd/2} \prod_{b=1}^{m} (t^{rb} - 1)^{b^{p-1-2k}},$$

where

$$m = \tfrac{1}{2}(p1 - 1)$$
$$d = \sum_{n=1}^{m} n^{p-2k}.$$

If ${Q_{2k}}^r \not\equiv 1 \pmod{P}$ for all such IRREGULAR PAIRS, then FERMAT'S LAST THEOREM holds for exponent p.

see also FERMAT'S LAST THEOREM, IRREGULAR PAIR, IRREGULAR PRIME

References

Johnson, W. "Irregular Primes and Cyclotomic Invariants." *Math. Comput.* **29**, 113–120, 1975.

Vanishing Point

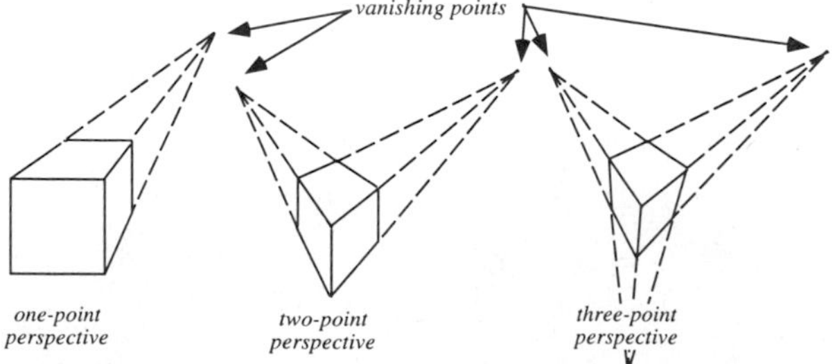

The point or points to which the extensions of PARALLEL lines appear to converge in a PERSPECTIVE drawing.

see also PERSPECTIVE, PROJECTIVE GEOMETRY

References

Dixon, R. "Perspective Drawings." Ch. 3 in *Mathographics.* New York: Dover, pp. 79–88, 1991.

Varga's Constant

$$V \equiv \frac{1}{\Lambda} = 9.2890254919\ldots,$$

where Λ is the ONE-NINTH CONSTANT.

see also ONE-NINTH CONSTANT

Variance

For N samples of a variate having a distribution with *known* MEAN μ, the "population variance" (usually called "variance" for short, although the word "population" should be added when needed to distinguish it from the SAMPLE VARIANCE) is defined by

$$\begin{aligned}\operatorname{var}(x) &\equiv \frac{1}{N}\sum(x-\mu)^2 = \left\langle x^2 - 2\mu x + \mu^2\right\rangle \\ &= \left\langle x^2\right\rangle - \langle 2\mu x\rangle + \left\langle \mu^2\right\rangle \\ &= \left\langle x^2\right\rangle - 2\mu\langle x\rangle + \mu^2, \end{aligned} \quad (1)$$

where

$$\langle x\rangle \equiv \frac{1}{N}\sum_{i=1}^{N} x_i. \quad (2)$$

But since $\langle x\rangle$ is an UNBIASED ESTIMATOR for the MEAN

$$\mu \equiv \langle x\rangle, \quad (3)$$

it follows that the variance

$$\sigma^2 \equiv \operatorname{var}(x) = \left\langle x^2\right\rangle - \mu^2. \quad (4)$$

The population STANDARD DEVIATION is then defined as

$$\sigma \equiv \sqrt{\operatorname{var}(x)} = \sqrt{\langle x^2\rangle - \mu^2}. \quad (5)$$

A useful identity involving the variance is

$$\operatorname{var}(f(x)+g(x)) = \operatorname{var}(f(x)) + \operatorname{var}(g(x)). \quad (6)$$

Therefore,

$$\begin{aligned}\operatorname{var}(ax+b) &= \left\langle [(ax+b) - \langle ax+b\rangle]^2\right\rangle \\ &= \left\langle (ax+b-a\langle x\rangle - b)^2\right\rangle \\ &= \left\langle (ax-a\mu)^2\right\rangle = \left\langle a^2(x-\mu)^2\right\rangle \\ &= a^2\left\langle (x-\mu)^2\right\rangle = a^2\operatorname{var}(x) \quad (7)\\ \operatorname{var}(b) &= 0. \quad (8)\end{aligned}$$

If the population MEAN is not known, using the sample mean $\bar{x}$ instead of the population mean μ to compute

$$s^2 \equiv \hat{\sigma}_N^2 \equiv \frac{1}{N}\sum_{i=1}^{N}(x_i - \bar{x})^2 \quad (9)$$

gives a BIASED ESTIMATOR of the population variance. In such cases, it is appropriate to use a STUDENT'S t-DISTRIBUTION instead of a GAUSSIAN DISTRIBUTION. However, it turns out (as discussed below) that an UNBIASED ESTIMATOR for the population variance is given by

$$s'^2 \equiv \hat{\sigma}_N'^2 \equiv \frac{1}{N-1}\sum_{i=1}^{N}(x_i - \bar{x})^2. \quad (10)$$

The MEAN and VARIANCE of the sample standard deviation for a distribution with population mean μ and VARIANCE are

$$\mu_{s_N{}^2} = \frac{N-1}{N}s^2 \quad (11)$$

$$\sigma_{s_N{}^2}{}^2 = \frac{N-1}{N^3}[(N-1)\mu_4 - (N-3)\mu_2{}^2]. \quad (12)$$

The quantity $N s_N{}^2/\sigma^2$ has a CHI-SQUARED DISTRIBUTION.

For multiple variables, the variance is given using the definition of COVARIANCE,

$$\begin{aligned}\operatorname{var}\left(\sum_{i=1}^{n} x_i\right) &= \operatorname{cov}\left(\sum_{i=1}^{n} x_i, \sum_{j=1}^{m} x_j\right) \\ &= \sum_{i=1}^{n}\sum_{j=1}^{m}\operatorname{cov}(x_i,x_j) \\ &= \sum_{i=1}^{n}\sum_{\substack{j=1\\ j=i}}^{m}\operatorname{cov}(x_i,x_j) + \sum_{i=1}^{n}\sum_{\substack{j=1\\ j\neq i}}^{m}\operatorname{cov}(x_i,x_j) \\ &= \sum_{i=1}^{n}\operatorname{cov}(x_i,x_j) + \sum_{i=1}^{n}\sum_{\substack{j=1\\ j\neq i}}^{m}\operatorname{cov}(x_i,x_j) \\ &= \sum_{i=1}^{n}\operatorname{var}(x_i) + 2\sum_{i=1}^{n}\sum_{j=i+1}^{m}\operatorname{cov}(x_i,x_j).\end{aligned} \quad (13)$$

A linear sum has a similar form:

$$\begin{aligned}\operatorname{var}\left(\sum_{i=1}^{n} a_i x_i\right) &= \operatorname{cov}\left(\sum_{i=1}^{n} a_i x_i, \sum_{j=1}^{m} a_j x_j\right) \\ &= \sum_{i=1}^{n}\sum_{j=1}^{m} a_i a_j\operatorname{cov}(x_i,x_j) \\ &= \sum_{i=1}^{n} a_i{}^2\operatorname{var}(x_i) + 2\sum_{i=1}^{n}\sum_{j=i+1}^{m} a_i a_j\operatorname{cov}(x_i,x_j).\end{aligned} \quad (14)$$

These equations can be expressed using the COVARIANCE MATRIX.

To estimate the population VARIANCE from a sample of N elements with a priori *unknown* MEAN (i.e., the MEAN is estimated from the sample itself), we need an

UNBIASED ESTIMATOR for σ. This is given by the k-STATISTIC k_2, where

$$k_2 = \frac{N}{N-1} m_2 \tag{15}$$

and $m_2 \equiv s^2$ is the SAMPLE VARIANCE

$$s^2 \equiv \frac{1}{N} \sum_{i=1}^{N} (x_i - \bar{x})^2. \tag{16}$$

Note that some authors prefer the definition

$$s'^2 \equiv \frac{1}{N-1} \sum_{i=1}^{N} (x_i - \bar{x})^2, \tag{17}$$

since this makes the sample variance an UNBIASED ESTIMATOR for the population variance.

When computing numerically, the MEAN must be computed before s^2 can be determined. This requires storing the set of sample values. It is possible to calculate s'^2 using a recursion relationship involving only the last sample as follows. Here, use μ_j to denote μ calculated from the first j samples (*not* the jth MOMENT)

$$\mu_j \equiv \frac{\sum_{i=1}^{j} x_i}{j}, \tag{18}$$

and ${s_j}^2$ denotes the value for the sample variance s'^2 calculated from the first j samples. The first few values calculated for the MEAN are

$$\mu_1 = x_1 \tag{19}$$

$$\mu_2 = \frac{1 \cdot \mu_1 + x_2}{2} \tag{20}$$

$$\mu_3 = \frac{2\mu_2 + x_3}{3}. \tag{21}$$

Therefore, for $j = 2$, 3 it is true that

$$\mu_j = \frac{(j-1)\mu_{j-1} + x_j}{j}. \tag{22}$$

Therefore, by induction,

$$\mu_{j+1} = \frac{[(j+1)-1]\mu_{(j+1)-1} + x_{j+1}}{j+1}$$

$$= \frac{j\mu_j + x_{j+1}}{j+1} \tag{23}$$

$$\mu_{j+1}(j+1) = (j+1)\mu_j + (x_{j+1} - \mu_j) \tag{24}$$

$$\mu_{j+1} = \mu_j + \frac{x_{j+1} - \mu_j}{j+1}, \tag{25}$$

and

$${s_j}^2 = \frac{\sum_{i=1}^{j} (x_i - \mu_j)^2}{j-1} \tag{26}$$

for $j \geq 2$, so

$$j{s_{j+1}}^2 = j\frac{\sum_{i=1}^{j+1} (x_i - \mu_{j+1})^2}{j} = \sum_{i=1}^{j+1} (x_i - \mu_{j+1})^2$$

$$= \sum_{i=1}^{j+1} [(x_i - \mu_j)(\mu_j - \mu_{j+1})]^2$$

$$= \sum_{i=1}^{j+1} (x_i - \mu_j)^2 + \sum_{i=1}^{j+1} (\mu_j - \mu_{j+1})^2$$

$$+ 2\sum_{i=1}^{j+1} (x_i - \mu_j)(\mu_j - \mu_{j+1}). \tag{27}$$

Working on the first term,

$$\sum_{i=1}^{j+1} (x_i - \mu_j)^2 = \sum_{i=1}^{j} (x_i - \mu_j)^2 + (x_{j+1} - \mu_j)^2$$

$$= (j-1){s_j}^2 + (x_{j+1} - \mu_j)^2. \tag{28}$$

Use (24) to write

$$x_{j+1} - \mu_j = (j+1)(\mu_{j+1} - \mu_j), \tag{29}$$

so

$$\sum_{i=1}^{j+1} (x_i - \mu_j)^2 = (j-1){s_j}^2 + (j+1)^2(\mu_{j+1} - \mu_j)^2. \tag{30}$$

Now work on the second term in (27),

$$\sum_{i=1}^{j+1} (\mu_j - \mu_{j+1})^2 = (j+1)(\mu_j - \mu_{j+1})^2. \tag{31}$$

Considering the third term in (27),

$$\sum_{i=1}^{j+1} (x_i - \mu_j)(\mu_j - \mu_{j+1}) = (\mu_j - \mu_{j+1}) \sum_{i=1}^{j+1} (x_i - \mu_j)$$

$$= (\mu_j - \mu_{j+1}) \left[\sum_{i=1}^{j} (x_i - \mu_j) + (x_{j+1} - \mu_j) \right]$$

$$= (\mu_j - \mu_{j+1}) \left(x_{j+1} - \mu_j - j\mu_j + \sum_{i=1}^{j} x_i \right). \tag{32}$$

But

$$\sum_{i=1}^{j} x_i = j\mu_j, \tag{33}$$

so

$$\sum_{i=1}^{j+1} (\mu_j - \mu_{j+1})(x_{j+1} - \mu_j)$$

$$= \sum_{i=1}^{j+1} (\mu_j - \mu_{j+1})(j+1)(\mu_{j+1} - \mu_j)$$

$$= -(j+1)(\mu_j - \mu_{j+1})^2. \tag{34}$$

Plugging (30), (31), and (34) into (27),

$$\begin{aligned} j{s_{j+1}}^2 &= [(j-1){s_j}^2 + (j+1)^2(\mu_{j+1}-\mu_j)^2] \\ &\quad + [(j+1)(\mu_j-\mu_{j+1}) + 2[-(j+1)(\mu_j-\mu_{j+1})] \\ &= (j-1){s_j}^2 + (j+1)^2(\mu_{j+1}-\mu_j)^2 \\ &\quad - (j+1)(\mu_j-\mu_{j+1})^2 \\ &= (j-1){s_j}^2 + (j+1)[(j+1)-1](\mu_{j+1}-\mu_j)^2 \\ &= (j-1){s_j}^2 + j(j+1)(\mu_{j+1}-\mu_j)^2, \end{aligned} \tag{35}$$

so

$${s_{j+1}}^2 = \left(1-\frac{1}{j}\right){s_j}^2 + (j+1)(\mu_{j+1}-\mu_j)^2. \tag{36}$$

To find the variance of s^2 itself, remember that

$$\operatorname{var}(s^2) \equiv \langle s^4\rangle - \langle s^2\rangle^2, \tag{37}$$

and

$$\langle s^2\rangle = \frac{N-1}{N}\mu_2. \tag{38}$$

Now find $\langle s^4\rangle$.

$$\begin{aligned} \langle s^4\rangle &= \langle (s^2)^2\rangle = \langle (\langle x^2\rangle - \langle x\rangle^2)^2\rangle \\ &= \left\langle \left[\frac{1}{N}\sum {x_i}^2 - \left(\frac{1}{N}\sum x_i\right)^2\right]^2\right\rangle \\ &= \frac{1}{N^2}\left\langle \left(\sum x_i\right)^2\right\rangle - \frac{2}{N^3}\left\langle \sum {x_i}^2\left(\sum x_i\right)^2\right\rangle \\ &\quad + \frac{1}{N^4}\left\langle\left(\sum x_i\right)^4\right\rangle. \end{aligned} \tag{39}$$

Working on the first term of (39),

$$\begin{aligned} \left\langle\left(\sum {x_i}^2\right)^2\right\rangle &= \left\langle \sum {x_i}^4 + \sum {x_i}^2{x_j}^2\right\rangle \\ &= \left\langle \sum {x_i}^4\right\rangle + \left\langle \sum {x_i}^2{x_j}^2\right\rangle \\ &= N\left\langle {x_i}^4\right\rangle + N(N-1)\left\langle {x_i}^2\right\rangle\left\langle {x_j}^2\right\rangle \\ &= N\mu'_4 + N(N-1){\mu'_2}^2. \end{aligned} \tag{40}$$

The second term of (39) is known from k-STATISTICS,

$$\left\langle \sum {x_i}^2\left(\sum x_j\right)^2\right\rangle = N\mu'_4 + N(N-1){\mu'_2}^2, \tag{41}$$

as is the third term,

$$\begin{aligned} \left\langle\left(\sum x_i\right)^4\right\rangle &= N\left\langle \sum {x_i}^4\right\rangle + 3N(N-1)\left\langle \sum {x_i}^2{x_j}^2\right\rangle \\ &= N\mu'_4 + 3N(N-1){\mu'_2}^2. \end{aligned} \tag{42}$$

Combining (39)-(42) gives

$$\begin{aligned} \langle s^4\rangle &= \frac{1}{N^2}[N\mu'_4 + N(N-1){\mu'_2}^2] \\ &\quad - \frac{2}{N^3}[N\mu'_4 + N(N-1){\mu'_2}^2] \\ &\quad + \frac{1}{N^4}[N\mu'_4 + 3N(N-1){\mu'_2}^2] \\ &= \left(\frac{1}{N} - \frac{2}{N^2} + \frac{1}{N^3}\right)\mu'_4 \\ &\quad + \left[\frac{N-1}{N} - \frac{2(N-1)}{N^2} + \frac{3(N-1)}{N^3}\right]{\mu'_2}^2 \\ &= \left(\frac{N^2-2N+1}{N^3}\right)\mu'_4 \\ &\quad + \frac{(N-1)(N^2-2N+3)}{N^3}{\mu'_2}^2 \\ &= \frac{(N-1)[(N-1)\mu'_4 + (N^2-2N+3){\mu'_2}^2]}{N^3}, \end{aligned} \tag{43}$$

so plugging in (38) and (43) gives

$$\begin{aligned} \operatorname{var}(s^2) &= \langle s^4\rangle - \langle s^2\rangle^2 \\ &= \frac{(N-1)[(N-1)\mu'_4 + (N^2-2N+3){\mu'_2}^2]}{N^3} \\ &\quad - \frac{(N-1)^2N}{N^3}{\mu'_2}^2 \\ &= \frac{N-1}{N^3}\{(N-1)\mu'_4 + [(N^2-2N+3) \\ &\quad - N(N-1)]{\mu'_2}^2\} \\ &= \frac{(N-1)[(N-1)\mu'_4 - (N-3){\mu'_2}^2]}{N^3}. \end{aligned} \tag{44}$$

Student calculated the SKEWNESS and KURTOSIS of the distribution of s^2 as

$$\gamma_1 = \sqrt{\frac{8}{N-1}} \tag{45}$$

$$\gamma_2 = \frac{12}{N-1} \tag{46}$$

and conjectured that the true distribution is PEARSON TYPE III DISTRIBUTION

$$f(s^2) = C(s^2)^{(N-3)/2}e^{-Ns^2/2\sigma^2}, \tag{47}$$

where

$$\sigma^2 = \frac{Ns^2}{N-1} \tag{48}$$

$$C = \frac{\left(\frac{N}{2\sigma^2}\right)^{(N-1)/2}}{\Gamma\left(\frac{N-1}{2}\right)}. \tag{49}$$

This was proven by R. A. Fisher.

The distribution of s itself is given by

$$f(s) = 2\frac{\left(\frac{N}{2\sigma^2}\right)^{(N-1)/2}}{\Gamma\left(\frac{N-1}{2}\right)} e^{-ns^2/2\sigma^2} s^{N-2} \quad (50)$$

$$\langle s \rangle = \sqrt{\frac{2}{N}} \frac{\Gamma\left(\frac{N}{2}\right)}{\Gamma\left(\frac{N-1}{2}\right)} \sigma \equiv b(N)\sigma, \quad (51)$$

where

$$b(N) \equiv \sqrt{\frac{2}{N}} \frac{\Gamma\left(\frac{N}{2}\right)}{\Gamma\left(\frac{N-1}{2}\right)}. \quad (52)$$

The MOMENTS are given by

$$\mu_r = \left(\frac{2}{N}\right)^{r/2} \frac{\Gamma\left(\frac{N-1+r}{2}\right)}{\Gamma\left(\frac{N-1}{2}\right)} \sigma^r, \quad (53)$$

and the variance is

$$\begin{aligned} \operatorname{var}(s) = \nu_2 - {\nu_1}^2 &= \frac{N-1}{N}\sigma^2 - [b(N)\sigma]^2 \\ &= \frac{1}{N}\left[N - 1 - \frac{2\Gamma^2\left(\frac{N}{2}\right)}{\Gamma^2\left(\frac{N-1}{2}\right)}\sigma^2\right]. \end{aligned} \quad (54)$$

An UNBIASED ESTIMATOR of σ is $s/b(N)$. Romanovsky showed that

$$b(N) = 1 - \frac{3}{4N} - \frac{7}{32N^2} - \frac{139}{51849N^3} + \ldots. \quad (55)$$

see also CORRELATION (STATISTICAL), COVARIANCE, COVARIANCE MATRIX, k-STATISTIC, MEAN, SAMPLE VARIANCE

References

Press, W. H.; Flannery, B. P.; Teukolsky, S. A.; and Vetterling, W. T. "Moments of a Distribution: Mean, Variance, Skewness, and So Forth." §14.1 in *Numerical Recipes in FORTRAN: The Art of Scientific Computing, 2nd ed.* Cambridge, England: Cambridge University Press, pp. 604–609, 1992.

Variate

A RANDOM VARIABLE in statistics.

Variation

The Δ-variation is a variation in which the varied path over which an integral is evaluated may end at different times than the correct path, and there may be variation in the coordinates at the endpoints.

The δ-variation is a variation in which the varied path in configuration space terminates at the endpoints representing the system configuration at the same time t_1 and t_2 as the correct path; i.e., the varied path always returns to the same endpoints in configuration space, so

$$\delta q_i(t_1) = \delta q_i(t_2) = 0.$$

see also CALCULUS OF VARIATIONS, VARIATION OF ARGUMENT, VARIATION OF PARAMETERS

Variation of Argument

Let $[\arg f(z)]$ denote the change in argument of a function $f(z)$ around a closed loop γ. Also let N denote the number of ROOTS of $f(z)$ in γ and P denote the number of POLES of $f(z)$ in γ. Then

$$[\arg f(z)] = \frac{1}{2\pi}(N - P). \quad (1)$$

To find $[\arg f(z)]$ in a given region R, break R into paths and find $[\arg f(z)]$ for each path. On a circular ARC

$$z = Re^{i\theta}, \quad (2)$$

let $f(z)$ be a POLYNOMIAL $P(z)$ of degree n. Then

$$\begin{aligned} [\arg P(z)] &= \left[\arg\left(z^n \frac{P(z)}{z^n}\right)\right] \\ &= [\arg z^n] + \left[\arg\left(\frac{P(z)}{z^n}\right)\right]. \end{aligned} \quad (3)$$

Plugging in $z = Re^{i\theta}$ gives

$$[\arg P(z)] = [\arg Re^{i\theta n}] + \left[\arg \frac{P(Re^{i\theta})}{Re^{i\theta n}}\right] \quad (4)$$

$$\lim_{R\to\infty} \frac{P(Re^{i\theta})}{Re^{i\theta n}} = [\text{constant}], \quad (5)$$

so

$$\left[\frac{P(Re^{i\theta})}{Re^{i\theta n}}\right] = 0, \quad (6)$$

and

$$[\arg P(z)] = [\arg e^{i\theta n}] = n(\theta_2 - \theta_1). \quad (7)$$

For a REAL segment $z = x$,

$$[\arg f(x)] = \tan^{-1}\left[\frac{0}{f(x)}\right] = 0. \quad (8)$$

For an IMAGINARY segment $z = iy$,

$$[\arg f(iy)] = \left\{\tan^{-1}\frac{\Im[P(iy)]}{\Re[P(iy)]}\right\}_{\theta_1}^{\theta_2}. \quad (9)$$

Note that the ARGUMENT must change continuously, so "jumps" occur across inverse tangent asymptotes.

Variation Coefficient

If s_x is the STANDARD DEVIATION of a set of samples x_i and $\bar{x}$ its MEAN, then

$$V \equiv \frac{s_x}{\bar{x}}.$$

Variation of Parameters

For a second-order ORDINARY DIFFERENTIAL EQUATION,

$$y'' + p(x)y' + q(x)y = g(x). \tag{1}$$

Assume that linearly independent solutions $y_1(x)$ and $y_2(x)$ are known. Find v_1 and v_2 such that

$$y^*(x) = v_1(x)y_1(x) + v_2(x)y_2(x) \tag{2}$$

$$y^{*\prime}(x) = (v_1' + v_2'y_2) + (v_1y_1' + v_2y_2'). \tag{3}$$

Now, impose the additional condition that

$$v_1'y_1 + v_2'y_2 = 0 \tag{4}$$

so that

$$y^{*\prime}(x) = (v_1y_1' + v_2y_2') \tag{5}$$

$$y^{*\prime\prime}(x) = v_1'y_1' + v_2'y_2' + v_1y_1'' + v_2y_2'. \tag{6}$$

Plug y^*, $y^{*\prime}$, and $y^{*\prime\prime}$ back into the original equation to obtain

$$v_1(y_1''+py_1'+qy_1)+v_2(y_2''+py_2'+qy_2)+v_1'y_1'+v_2'y_2' = g(x) \tag{7}$$

$$v_1'y_1' + v_2'y_2' = g(x). \tag{8}$$

Therefore,

$$v_1'y_1 + v_2'y_2 = 0 \tag{9}$$

$$v_1'y_1' + v_2'y_2' = g(x). \tag{10}$$

Generalizing to an nth degree ODE, let $y_1, \ldots, y_n$ be the solutions to the homogeneous ODE and let $v_1'(x), \ldots, v_n'(x)$ be chosen such that

$$\begin{cases} y_1v_1' + y_2v_2' + \ldots + y_nv_n' = 0 \\ y_1'v_1' + y_2'v_2' + \ldots + y_n'v_n' = 0 \\ \vdots \\ y_1^{(n-1)}v_1' + y_2^{(n-1)}v_2' + \ldots + y_n^{(n-1)}v_n' = g(x). \end{cases} \tag{11}$$

Then the particular solution is

$$y^*(x) = v_1(x)y_1(x) + \ldots + v_n(x)y_n(x). \tag{12}$$

Variety

see ALGEBRAIC VARIETY

Varignon Parallelogram

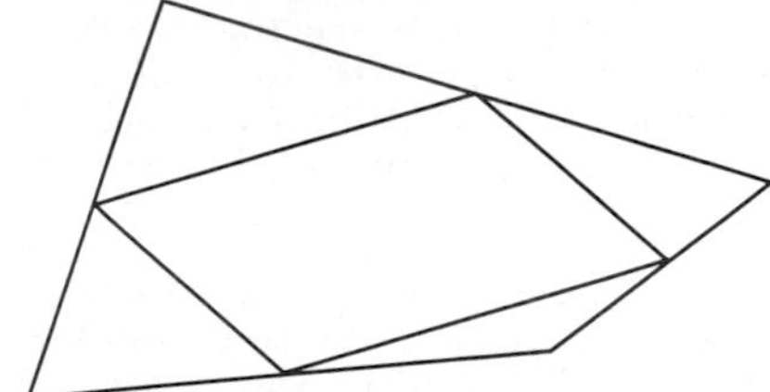

The figure formed when the BIMEDIANS (MIDPOINTS of the sides) of a convex QUADRILATERAL are joined. VARIGNON'S THEOREM demonstrated that this figure is a PARALLELOGRAM. The center of the Varignon parallelogram is the CENTROID if four point masses are placed on the VERTICES of the QUADRILATERAL.

see also MIDPOINT, PARALLELOGRAM, QUADRILATERAL, VARIGNON'S THEOREM

Varignon's Theorem

The figure formed when the BIMEDIANS (MIDPOINTS of the sides) of a convex QUADRILATERAL are joined in order is a PARALLELOGRAM. Equivalently, the BIMEDIANS bisect each other. The AREA of this VARIGNON PARALLELOGRAM is half that of the QUADRILATERAL. The PERIMETER is equal to the sum of the diagonals of the original QUADRILATERAL.

see also BIMEDIAN, MIDPOINT, QUADRILATERAL, VARIGNON PARALLELOGRAM

References

Coxeter, H. S. M. and Greitzer, S. L. *Geometry Revisited.* Washington, DC: Math. Assoc. Amer., pp. 51–56, 1967.

Vassiliev Polynomial

Vassiliev (1990) introduced a radically new way of looking at KNOTS by considering a multidimensional space in which each point represents a possible 3-D knot configuration. If two KNOTS are equivalent, a path then exists in this space from one to the other. The paths can be associated with polynomial invariants.

Birman and Lin (1993) subsequently found a way to translate this scheme into a set of rules and list of potential starting points, which makes analysis of Vassiliev polynomials much simpler. Bar-Natan (1995) and Birman and Lin (1993) proved that JONES POLYNOMIALS and several related expressions are directly connected (Peterson 1992). In fact, substituting the POWER series for e^x as the variable in the JONES POLYNOMIAL yields a POWER SERIES whose COEFFICIENTS are Vassiliev polynomials (Birman and Lin 1993). Bar-Natan (1995) also discovered a link with Feynman diagrams (Peterson 1992).

References

Bar-Natan, D. "On the Vassiliev Knot Invariants." *Topology* **34**, 423–472, 1995.

Birman, J. S. "New Points of View in Knot Theory." *Bull. Amer. Math. Soc.* **28**, 253–287, 1993.

Birman, J. S. and Lin, X.-S. "Knot Polynomials and Vassiliev's Invariants." *Invent. Math.* **111**, 225–270, 1993.
Peterson, I. "Knotty Views: Tying Together Different Ways of Looking at Knots." *Sci. News* **141**, 186–187, 1992.
Praslov, V. V. and Sossinsky, A. B. *Knots, Links, Braids and 3-Manifolds: An Introduction to the New Invariants in Low-Dimensional Topology.* Providence, RI: Amer. Math. Soc., 1996.
Stoimenow, A. "Degree-3 Vassiliev Invariants." `http://www.informatik.hu-berlin.de/~stoimeno/vas3.html`.
Vassiliev, V. A. "Cohomology of Knot Spaces." In *Theory of Singularities and Its Applications* (Ed. V. I. Arnold). Providence, RI: Amer. Math. Soc., pp. 23–69, 1990.
Vassiliev, V. A. *Complements of Discriminants of Smooth Maps: Topology and Applications.* Providence, RI: Amer. Math. Soc., 1992.

Vault

Let a vault consist of two equal half-CYLINDERS of length and diameter $2a$ which intersect at RIGHT ANGLES so that the lines of their intersections (the "groins") terminate in the VERTICES of a SQUARE. Then the SURFACE AREA of the vault is given by

$$A = 4(\pi - 2)a^2.$$

see also DOME

References
Lines, L. *Solid Geometry.* New York: Dover, pp. 112–113, 1965.

Vector

A vector is a set of numbers $A_0, \ldots, A_n$ that transform as

$$A'_i = a_{ij} A_j. \tag{1}$$

This makes a vector a TENSOR of RANK 1. Vectors are invariant under TRANSLATION, and they reverse sign upon inversion.

A vector is uniquely specified by giving its DIVERGENCE and CURL within a region and its normal component over the boundary, a result known as HELMHOLTZ'S THEOREM (Arfken 1985, p. 79). A vector from a point A to a point B is denoted $\overrightarrow{AB}$, and a vector v may be denoted $\vec{v}$, or more commonly, $\mathbf{v}$.

A vector with unit length is called a UNIT VECTOR and is denoted with a HAT. An arbitrary vector may be converted to a UNIT VECTOR by dividing by its NORM, i.e.,

$$\hat{\mathbf{v}} \equiv \frac{\mathbf{v}}{|\mathbf{v}|}. \tag{2}$$

Let $\hat{\mathbf{n}}$ be the UNIT VECTOR defined by

$$\hat{\mathbf{n}} \equiv \begin{bmatrix} \cos\theta\sin\phi \\ \sin\theta\sin\phi \\ \cos\phi \end{bmatrix}. \tag{3}$$

Then the vectors $\hat{\mathbf{n}}$, $\mathbf{a}$, $\mathbf{b}$, $\mathbf{c}$, $\mathbf{d}$ satisfy the identities

$$\langle n_x \rangle = \int_0^{2\pi}\int_o^{\pi} (\cos\theta\sin\phi)\sin\phi\,d\theta\,d\phi$$

$$= [\sin\theta]_0^{2\pi}\int_0^{2\pi}\sin^2\phi\,d\phi = 0 \tag{4}$$

$$\langle n_i \rangle = 0 \tag{5}$$

$$\langle n_i n_j \rangle = \tfrac{1}{3}\delta_{ij} \tag{6}$$

$$\langle n_i n_k n_k \rangle = 0 \tag{7}$$

$$\langle n_i n_k n_l n_m \rangle = \tfrac{1}{15}(\delta_{ik}\delta_{lm} + \delta_{il}\delta_{km} + \delta_{im}\delta_{kl}) \tag{8}$$

$$\left\langle (\mathbf{a}\cdot\hat{\mathbf{n}})^2 \right\rangle = \tfrac{1}{3}a^2 \tag{9}$$

$$\langle (\mathbf{a}\cdot\hat{\mathbf{n}})(\mathbf{b}\cdot\hat{\mathbf{n}}) \rangle = \tfrac{1}{3}\mathbf{a}\cdot\mathbf{b} \tag{10}$$

$$\langle (\mathbf{a}\cdot\hat{\mathbf{n}})\hat{\mathbf{n}} \rangle = \tfrac{1}{3}a \tag{11}$$

$$\left\langle (\mathbf{a}\times\hat{\mathbf{n}})^2 \right\rangle = \tfrac{2}{3}a^2 \tag{12}$$

$$\langle (\mathbf{a}\times\hat{\mathbf{n}})\cdot(\mathbf{b}\times\hat{\mathbf{n}}) \rangle = \tfrac{2}{3}\mathbf{a}\cdot\mathbf{b}, \tag{13}$$

and

$$\langle (\mathbf{a}\cdot\hat{\mathbf{n}})(\mathbf{b}\cdot\hat{\mathbf{n}})(\mathbf{c}\cdot\hat{\mathbf{n}})(\mathbf{d}\cdot\hat{\mathbf{n}}) \rangle$$

$$= \tfrac{1}{15}[(bfa\cdot\mathbf{b})(bfc\cdot\mathbf{d}) + (bfa\cdot\mathbf{c})(bfb\cdot\mathbf{d}) + (bfa\cdot\mathbf{d})(bfb\cdot\mathbf{c})] \tag{14}$$

where δ_{ij} is the KRONECKER DELTA, $\mathbf{a}\cdot\mathbf{b}$ is a DOT PRODUCT, and EINSTEIN SUMMATION has been used.

see also FOUR-VECTOR, HELMHOLTZ'S THEOREM, NORM, PSEUDOVECTOR, SCALAR, TENSOR, UNIT VECTOR, VECTOR FIELD

References
Arfken, G. "Vector Analysis." Ch. 1 in *Mathematical Methods for Physicists, 3rd ed.* Orlando, FL: Academic Press, pp. 1–84, 1985.
Aris, R. *Vectors, Tensors, and the Basic Equations of Fluid Mechanics.* New York: Dover, 1989.
Crowe, M. J. *A History of Vector Analysis: The Evolution of the Idea of a Vectorial System.* New York: Dover, 1985.
Gibbs, J. W. and Wilson, E. B. *Vector Analysis: A Text-Book for the Use of Students of Mathematics and Physics, Founded Upon the Lectures of J. Willard Gibbs.* New York: Dover, 1960.
Marsden, J. E. and Tromba, A. J. *Vector Calculus, 4th ed.* New York: W. H. Freeman, 1996.
Morse, P. M. and Feshbach, H. "Vector and Tensor Formalism." §1.5 in *Methods of Theoretical Physics, Part I.* New York: McGraw-Hill, pp. 44–54, 1953.
Schey, H. M. *Div, Grad, Curl, and All That: An Informal Text on Vector Calculus.* New York: Norton, 1973.
Schwartz, M.; Green, S.; and Rutledge, W. A. *Vector Analysis with Applications to Geometry and Physics.* New York: Harper Brothers, 1960.
Spiegel, M. R. *Theory and Problems of Vector Analysis.* New York: Schaum, 1959.

Vector Bundle

A special class of FIBER BUNDLE in which the FIBER is a VECTOR SPACE. Technically, a little more is required; namely, if $f : E \to B$ is a BUNDLE with FIBER $\mathbb{R}^n$, to be a vector bundle, all of the FIBERS $f^{-1}(x)$ for

$x \in B$ need to have a coherent VECTOR SPACE structure. One way to say this is that the "trivializations" $h : f^{-1}(U) \to U \times \mathbb{R}^n$, are FIBER-for-FIBER VECTOR SPACE ISOMORPHISMS.

see also BUNDLE, FIBER, FIBER BUNDLE, LIE ALGEBROID, STABLE EQUIVALENCE, TANGENT MAP, VECTOR SPACE, WHITNEY SUM

Vector Derivative

The basic types of derivatives operating on a VECTOR FIELD are the CURL $\nabla\times$, DIVERGENCE $\nabla\cdot$, and GRADIENT ∇.

Vector derivative identities involving the CURL include

$$\nabla \times (k\mathbf{A}) = k\nabla \times \mathbf{A} \tag{1}$$

$$\nabla \times (f\mathbf{A}) = f(\nabla \times \mathbf{A}) + (\nabla f) \times \mathbf{A} \tag{2}$$

$$\nabla \times (\mathbf{A} \times \mathbf{B}) = (\mathbf{B} \cdot \nabla)\mathbf{A} - (\mathbf{A} \cdot \nabla)\mathbf{B} + \mathbf{A}(\nabla \cdot \mathbf{B}) - \mathbf{B}(\nabla \cdot \mathbf{A}) \tag{3}$$

$$\nabla \times \left(\frac{\mathbf{A}}{f}\right) = \frac{f(\nabla \times \mathbf{A}) + \mathbf{A} \times (\nabla f)}{f^2} \tag{4}$$

$$\nabla \times (\mathbf{A} + \mathbf{B}) = \nabla \times \mathbf{A} + \nabla \times \mathbf{B}. \tag{5}$$

In SPHERICAL COORDINATES,

$$\nabla \times \mathbf{r} = \mathbf{0} \tag{6}$$

$$\nabla \times \hat{\mathbf{r}} = \mathbf{0} \tag{7}$$

$$\begin{aligned}\nabla \times [\mathbf{r}f(r)] &= f(r)(\nabla \times \mathbf{r}) + [\nabla f(r)] \times \mathbf{r} \\ &= f(r)(\mathbf{0}) + \frac{df}{dr}\hat{\mathbf{r}} \times \mathbf{r} = \mathbf{0} + \mathbf{0} = \mathbf{0}.\end{aligned} \tag{8}$$

Vector derivative identities involving the DIVERGENCE include

$$\nabla \cdot (k\mathbf{A}) = k\nabla \cdot \mathbf{A} \tag{9}$$

$$\nabla \cdot (f\mathbf{A}) = f(\nabla \cdot \mathbf{A}) + (\nabla f) \cdot \mathbf{A} \tag{10}$$

$$\nabla \cdot (\mathbf{A} \times \mathbf{B}) = \mathbf{B} \cdot (\nabla \times \mathbf{A}) - \mathbf{A} \cdot (\nabla \times \mathbf{B}) \tag{11}$$

$$\nabla \cdot \left(\frac{\mathbf{A}}{f}\right) = \frac{f(\nabla \cdot \mathbf{A}) - (\nabla f) \cdot \mathbf{A}}{f^2} \tag{12}$$

$$\nabla \cdot (\mathbf{A} + \mathbf{B}) = \nabla \cdot \mathbf{A} + \nabla \cdot \mathbf{B} \tag{13}$$

$$\nabla(\mathbf{uv}) = \mathbf{u}\nabla \cdot \mathbf{v} + (\nabla \mathbf{u}) \cdot \mathbf{v}. \tag{14}$$

In SPHERICAL COORDINATES,

$$\nabla \cdot \mathbf{r} = 3 \tag{15}$$

$$\nabla \cdot \hat{\mathbf{r}} = \frac{2}{r} \tag{16}$$

$$\nabla \cdot [\mathbf{r}f(r)] = \frac{\partial}{\partial x}[xf(r)] + \frac{\partial}{\partial y}[yf(r)] + \frac{\partial}{\partial z}[zf(r)] \tag{17}$$

$$\frac{\partial}{\partial x}[xf(r)] = x\frac{\partial f}{\partial x} + f = x\frac{\partial f}{\partial r}\frac{\partial r}{\partial x} + f \tag{18}$$

$$\frac{\partial r}{\partial x} = \frac{\partial}{\partial x}(x^2+y^2+z^2)^{1/2} = x(x^2+y^2+z^2)^{-1/2} = \frac{x}{r} \tag{19}$$

$$\frac{\partial}{\partial x}[xf(r)] = \frac{x^2}{r}\frac{df}{dr} + f. \tag{20}$$

By symmetry,

$$\nabla \cdot [\mathbf{r}f(r)] = 3f(r) + \frac{1}{r}(x^2+y^2+z^2)\frac{df}{dr} = 3f(r) + r\frac{df}{dr} \tag{21}$$

$$\nabla \cdot (\hat{\mathbf{r}}f(r)) = \frac{3}{r}f(r) + \frac{df}{dr} \tag{22}$$

$$\nabla \cdot (\hat{\mathbf{r}}r^n) = 3r^{n-1} + (n-1)r^{n-1} = (n+2)r^{n-1}. \tag{23}$$

Vector derivative identities involving the GRADIENT include

$$\nabla(kf) = k\nabla f \tag{24}$$

$$\nabla(fg) = f\nabla g + g\nabla f \tag{25}$$

$$\nabla(\mathbf{A} \cdot \mathbf{B}) = \mathbf{A} \times (\nabla \times \mathbf{B}) + \mathbf{B} \times (\nabla \times \mathbf{A}) + (\mathbf{A} \cdot \nabla)\mathbf{B} + (\mathbf{B} \cdot \nabla)\mathbf{A} \tag{26}$$

$$\begin{aligned}\nabla(\mathbf{A} \cdot \nabla f) &= \mathbf{A} \times (\nabla \times \nabla f) + \nabla f \times (\nabla \times \mathbf{A}) + \mathbf{A} \cdot \nabla(\nabla f) + \nabla f \cdot \nabla \mathbf{A} \\ &= \nabla f \times (\nabla \times \mathbf{A}) + \mathbf{A} \cdot \nabla(\nabla f) + \nabla f \cdot \nabla \mathbf{A}\end{aligned} \tag{27}$$

$$\nabla\left(\frac{f}{g}\right) = \frac{g\nabla f - f\nabla g}{g^2} \tag{28}$$

$$\nabla(f+g) = \nabla f + \nabla g \tag{29}$$

$$\nabla(\mathbf{A} \cdot \mathbf{A}) = 2\mathbf{A} \times (\nabla \times \mathbf{A}) + 2(\mathbf{A} \cdot \nabla)\mathbf{A} \tag{30}$$

$$(\mathbf{A} \cdot \nabla)\mathbf{A} = \nabla(\tfrac{1}{2}\mathbf{A}^2) - \mathbf{A} \times (\nabla \times \mathbf{A}). \tag{31}$$

Vector second derivative identities include

$$\nabla^2 t \equiv \nabla \cdot (\nabla t) = \frac{\partial^2 t}{\partial x^2} + \frac{\partial^2 t}{\partial y^2} + \frac{\partial^2 t}{\partial z^2} \tag{32}$$

$$\nabla^2 \mathbf{A} = \nabla(\nabla \cdot \mathbf{A}) - \nabla \times (\nabla \times \mathbf{A}). \tag{33}$$

This very important second derivative is known as the LAPLACIAN.

$$\nabla \times (\nabla t) = \mathbf{0} \tag{34}$$

$$\nabla(\nabla \cdot \mathbf{A}) = \nabla^2 \mathbf{A} + \nabla \times (\nabla \times \mathbf{A}) \tag{35}$$

$$\nabla \cdot (\nabla \times \mathbf{A}) = 0 \tag{36}$$

$$\begin{aligned}\nabla \times (\nabla \times \mathbf{A}) &= \nabla(\nabla \cdot \mathbf{A}) - \nabla^2 \mathbf{A} \\ \nabla \times (\nabla^2 \mathbf{A}) &= \nabla \times [\nabla(\nabla \cdot \mathbf{A})] - \nabla \times [\nabla \times (\nabla \times \mathbf{A})] \\ &= -\nabla \times [\nabla \times (\nabla \times \mathbf{A})] \\ &= -\{\nabla[\nabla \cdot (\nabla \times \mathbf{A})] - \nabla^2(\nabla \times \mathbf{A})]\} \\ &= \nabla^2(\nabla \times \mathbf{A})\end{aligned} \tag{37}$$

$$\begin{aligned}\nabla^2(\nabla \cdot \mathbf{A}) &= \nabla \cdot [\nabla(\nabla \cdot \mathbf{A})] \\ &= \nabla \cdot [\nabla^2 \mathbf{A} + \nabla \times (\nabla \times \mathbf{A})] = \nabla \cdot (\nabla^2 \mathbf{A})\end{aligned} \tag{38}$$

$$\begin{aligned}\nabla^2[\nabla \times (\nabla \times \mathbf{A})] &= \nabla^2[\nabla(\nabla \cdot \mathbf{A}) - \nabla^2 \mathbf{A}] \\ &= \nabla^2[\nabla(\nabla \cdot \mathbf{A})] - \nabla^4 \mathbf{A}\end{aligned} \tag{39}$$

$$\nabla \times [\nabla^2(\nabla \times \mathbf{A})] = \nabla^2[\nabla(\nabla \cdot \mathbf{A})] - \nabla^4 \mathbf{A} \tag{40}$$

$$\begin{aligned}\nabla^4 \mathbf{A} &= -\nabla^2[\nabla \times (\nabla \times \mathbf{A})] + \nabla^2[\nabla(\nabla \cdot \mathbf{A})] \\ &= \nabla \times [\nabla^2(\nabla \times \mathbf{A})] - \nabla^2[\nabla \times (\nabla \times \mathbf{A})].\end{aligned} \tag{41}$$

Combination identities include

$$\mathbf{A} \times (\nabla \mathbf{A}) = \tfrac{1}{2}\nabla(\mathbf{A} \cdot \mathbf{A}) - (\mathbf{A} \cdot \nabla)\mathbf{A} \quad (42)$$

$$\nabla \times (\phi \nabla \phi) = \phi \nabla \times (\nabla \phi) + (\nabla \phi) \times (\nabla \phi) = \mathbf{0} \quad (43)$$

$$(\mathbf{A} \cdot \nabla)\hat{\mathbf{r}} = \frac{\mathbf{A} - \hat{\mathbf{r}}(\mathbf{A} \cdot \hat{\mathbf{r}})}{r} \quad (44)$$

$$\nabla f \cdot \mathbf{A} = \nabla \cdot (f\mathbf{A}) - f(\nabla \cdot \mathbf{A}) \quad (45)$$

$$f(\nabla \cdot \mathbf{A}) = \nabla \cdot (f\mathbf{A}) - \mathbf{A}\nabla f, \quad (46)$$

where (45) and (46) follow from divergence rule (2).

see also CURL, DIVERGENCE, GRADIENT, LAPLACIAN, VECTOR INTEGRAL, VECTOR QUADRUPLE PRODUCT, VECTOR TRIPLE PRODUCT

References

Gradshteyn, I. S. and Ryzhik, I. M. "Vector Field Theorem." Ch. 10 in *Tables of Integrals, Series, and Products, 5th ed.* San Diego, CA: Academic Press, pp. 1081–1092, 1980.

Morse, P. M. and Feshbach, H. "Table of Useful Vector and Dyadic Equations." *Methods of Theoretical Physics, Part I.* New York: McGraw-Hill, pp. 50–54 and 114–115, 1953.

Vector Direct Product

Given VECTORS $\mathbf{u}$ and $\mathbf{v}$, the vector direct product is

$$\mathbf{uv} \equiv \mathbf{u} \otimes \mathbf{v}^{\mathrm{T}},$$

where $\otimes$ is the MATRIX DIRECT PRODUCT and $\mathbf{v}^{\mathrm{T}}$ is the matrix TRANSPOSE. For 3×3 vectors

$$\mathbf{uv} = \begin{bmatrix} u_1\mathbf{v}^{\mathrm{T}} \\ u_2\mathbf{v}^{\mathrm{T}} \\ u_3\mathbf{v}^{\mathrm{T}} \end{bmatrix} = \begin{bmatrix} u_1v_1 & u_1v_2 & u_1v_3 \\ u_2v_1 & u_2v_2 & u_2v_3 \\ u_3v_1 & u_3v_2 & u_3v_3 \end{bmatrix}.$$

Note that if $\mathbf{u} = \hat{\mathbf{x}}_i$, then $u_j = \delta_{ij}$, where δ_{ij} is the KRONECKER DELTA.

see also MATRIX DIRECT PRODUCT, SHERMAN-MORRISON FORMULA, WOODBURY FORMULA

Vector Division

There is no unique solution A to the MATRIX equation $\mathbf{y} = \mathsf{A}\mathbf{x}$ unless $\mathbf{x}$ is PARALLEL to $\mathbf{y}$, in which case A is a SCALAR. Therefore, vector division is not defined.

see also MATRIX, SCALAR

Vector Field

A MAP $\mathbf{f} : \mathbb{R}^n \mapsto \mathbb{R}^n$ which assigns each $\mathbf{x}$ a VECTOR FUNCTION $\mathbf{f}(\mathbf{x})$. FLOWS are generated by vector fields and vice versa. A vector field is a SECTION of its TANGENT BUNDLE.

see also FLOW, SCALAR FIELD, SEIFERT CONJECTURE, TANGENT BUNDLE, VECTOR, WILSON PLUG

References

Gray, A. "Vector Fields $\mathbb{R}^n$" and "Derivatives of Vector Fields $\mathbb{R}^n$." §9.4–9.5 in *Modern Differential Geometry of Curves and Surfaces.* Boca Raton, FL: CRC Press, pp. 171–174 and 175–178, 1993.

Morse, P. M. and Feshbach, H. "Vector Fields." §1.2 in *Methods of Theoretical Physics, Part I.* New York: McGraw-Hill, pp. 8–21, 1953.

Vector Function

A function of one or more variables whose RANGE is 3-dimensional, as compared to a SCALAR FUNCTION, whose RANGE is 1-dimensional.

see also COMPLEX FUNCTION, REAL FUNCTION, SCALAR FUNCTION

Vector Harmonic

see VECTOR SPHERICAL HARMONIC

Vector Integral

The following vector integrals are related to the CURL THEOREM. If

$$\mathbf{F} \equiv \mathbf{c} \times \mathbf{P}(x, y, z), \quad (1)$$

then

$$\int_C d\mathbf{s} \times \mathbf{P} = \int_S (d\mathbf{a} \times \nabla) \times \mathbf{P}. \quad (2)$$

If

$$\mathbf{F} \equiv \mathbf{c}F, \quad (3)$$

then

$$\int_C F\, d\mathbf{s} = \int_S d\mathbf{a} \times \nabla\mathbf{F}. \quad (4)$$

The following are related to the DIVERGENCE THEOREM. If

$$\mathbf{F} \equiv \mathbf{c} \times \mathbf{P}(x, y, z), \quad (5)$$

then

$$\int_V \nabla \times \mathbf{F}\, dV = \int_S d\mathbf{a} \times \mathbf{F}. \quad (6)$$

Finally, if

$$\mathbf{F} \equiv \mathbf{c}F, \quad (7)$$

then

$$\int_V \nabla F\, dV = \int_S F\, d\mathbf{a}. \quad (8)$$

see also CURL THEOREM, DIVERGENCE THEOREM, GRADIENT THEOREM, GREEN'S FIRST IDENTITY, GREEN'S SECOND IDENTITY, LINE INTEGRAL, SURFACE INTEGRAL, VECTOR DERIVATIVE, VOLUME INTEGRAL

Vector Norm

Given an n-D VECTOR

$$\mathbf{x} = \begin{bmatrix} x_1 \\ x_2 \\ \vdots \\ x_n \end{bmatrix},$$

a vector norm $||\mathbf{x}||$ (sometimes written simply $|\mathbf{x}|$) is a NONNEGATIVE number satisfying

1. $||x|| > 0$ when $\mathbf{x} \neq \mathbf{0}$ and $||\mathbf{x}|| = 0$ IFF $\mathbf{x} = \mathbf{0}$,
2. $||k\mathbf{x}|| = |k|\,||\mathbf{x}||$ for any SCALAR k,
3. $||\mathbf{x} + \mathbf{y}|| \leq ||\mathbf{x}|| + ||\mathbf{y}||$.

see also COMPATIBLE, MATRIX NORM, NATURAL NORM, NORM

References

Gradshteyn, I. S. and Ryzhik, I. M. *Tables of Integrals, Series, and Products, 5th ed.* San Diego, CA: Academic Press, p. 1114, 1980.

Vector Ordering

If the first NONZERO component of the vector difference $\mathbf{A}-\mathbf{B}$ is >0, then $\mathbf{A}\succ\mathbf{B}$. If the first NONZERO component of $\mathbf{A}-\mathbf{B}$ is <0, then $\mathbf{A}\prec\mathbf{B}$.

see also PRECEDES, SUCCEEDS

Vector Potential

A function $\mathbf{A}$ such that

$$\mathbf{B}\equiv\nabla\times\mathbf{A}.$$

The most common use of a vector potential is the representation of a magnetic field. If a VECTOR FIELD has zero DIVERGENCE, it may be represented by a vector potential.

see also DIVERGENCE, HELMHOLTZ'S THEOREM, POTENTIAL FUNCTION, SOLENOIDAL FIELD, VECTOR FIELD

Vector Quadruple Product

$$(\mathbf{A}\times\mathbf{B})\cdot(\mathbf{C}\times\mathbf{D})=(\mathbf{A}\cdot\mathbf{C})(\mathbf{B}\cdot\mathbf{D})-(\mathbf{A}\cdot\mathbf{D})(\mathbf{B}\cdot\mathbf{C}) \tag{1}$$

$$\begin{aligned}(\mathbf{A}\times\mathbf{B})^2&\equiv(\mathbf{A}\times\mathbf{B})\cdot(\mathbf{A}\times\mathbf{B})\\&=(\mathbf{A}\cdot\mathbf{A})(\mathbf{B}\cdot\mathbf{B})-(\mathbf{A}\cdot\mathbf{B})(\mathbf{B}\cdot\mathbf{A})\\&=\mathbf{A}^2\mathbf{B}^2-(\mathbf{A}\cdot\mathbf{B})^2\end{aligned} \tag{2}$$

$$\mathbf{A}\times(\mathbf{B}\times(\mathbf{C}\times\mathbf{D}))=\mathbf{B}(\mathbf{A}\cdot(\mathbf{C}\times\mathbf{D}))-(\mathbf{A}\cdot\mathbf{B})(\mathbf{C}\times\mathbf{D}) \tag{3}$$

$$\begin{aligned}(\mathbf{A}\times\mathbf{B})\times(\mathbf{C}\times\mathbf{D})&=[\mathbf{A},\mathbf{B},\mathbf{D}]\mathbf{C}-[\mathbf{A},\mathbf{B},\mathbf{C}]\mathbf{D}\\&=(\mathbf{C}\times\mathbf{D})\times(\mathbf{B}\times\mathbf{A})=[\mathbf{C},\mathbf{D},\mathbf{A}]\mathbf{D}-[\mathbf{C},\mathbf{D},\mathbf{B}]\mathbf{A},\end{aligned} \tag{4}$$

where $[\mathbf{A},\mathbf{B},\mathbf{D}]$ denotes the VECTOR TRIPLE PRODUCT. Equation (1) is known as LAGRANGE'S IDENTITY.

see also LAGRANGE'S IDENTITY, VECTOR TRIPLE PRODUCT

Vector Space

A vector space over $\mathbb{R}^n$ is a set of VECTORS for which any VECTORS $\mathbf{X}$, $\mathbf{Y}$, and $\mathbf{Z}\in\mathbb{R}^n$ and any SCALARS r, $s\in\mathbb{R}$ have the following properties:

1. COMMUTATIVITY:

$$\mathbf{X}+\mathbf{Y}=\mathbf{Y}+\mathbf{X}.$$

2. ASSOCIATIVITY of vector addition:

$$(\mathbf{X}+\mathbf{Y})+\mathbf{Z}=\mathbf{X}+(\mathbf{Y}+\mathbf{Z}).$$

3. Additive identity: For all $\mathbf{X}$,

$$\mathbf{0}+\mathbf{X}=\mathbf{X}+\mathbf{0}=\mathbf{X}.$$

4. Existence of additive inverse: For any $\mathbf{X}$, there exists a $-\mathbf{X}$ such that

$$\mathbf{X}+(-\mathbf{X})=\mathbf{0}.$$

5. ASSOCIATIVITY of scalar multiplication:

$$r(s\mathbf{X})=(rs)\mathbf{X}.$$

6. DISTRIBUTIVITY of scalar sums:

$$(r+s)\mathbf{X}=r\mathbf{X}+s\mathbf{X}.$$

7. DISTRIBUTIVITY of vector sums:

$$r(\mathbf{X}+\mathbf{Y})=r\mathbf{X}+r\mathbf{Y}.$$

8. Scalar multiplication identity:

$$1\mathbf{X}=\mathbf{X}.$$

An n-D vector space of characteristic two has

$$S(k,n)=(2^n-2^0)(2^n-2^1)\cdots(2^n-2^{k-1})$$

distinct SUBSPACES of DIMENSION k.

A MODULE is abstractly similar to a vector space, but it uses a RING to define COEFFICIENTS instead of the FIELD used for vector spaces. MODULES have COEFFICIENTS in much more general algebraic objects.

see also BANACH SPACE, FIELD, FUNCTION SPACE, HILBERT SPACE, INNER PRODUCT SPACE, MODULE, RING, TOPOLOGICAL VECTOR SPACE

References

Arfken, G. *Mathematical Methods for Physicists, 3rd ed.* Orlando, FL: Academic Press, pp. 530–534, 1985.

Vector Spherical Harmonic

The SPHERICAL HARMONICS can be generalized to vector spherical harmonics by looking for a SCALAR FUNCTION ψ and a constant VECTOR $\mathbf{c}$ such that

$$\begin{aligned}\mathbf{M}&\equiv\nabla\times(\mathbf{c}\psi)=\psi(\nabla\times\mathbf{c})+(\nabla\psi)\times\mathbf{c}\\&=(\nabla\psi)\times\mathbf{c}=-\mathbf{c}\times\nabla,\psi\end{aligned} \tag{1}$$

so

$$\nabla\cdot\mathbf{M}=0. \tag{2}$$

Now use the vector identities

$$\begin{aligned}\nabla^2\mathbf{M}&=\nabla^2(\nabla\times\mathbf{M})=\nabla\times(\nabla^2\mathbf{M})\\&=\nabla\times(\nabla^2\mathbf{c}\psi)=\nabla\times(\mathbf{c}\nabla^2\psi)\end{aligned} \tag{3}$$

$$k^2\mathbf{M}=k^2\nabla\times(\mathbf{c}\psi)=\nabla\times(\mathbf{c}\nabla^2\psi), \tag{4}$$

so

$$\nabla^2\mathbf{M} + k^2\mathbf{M} = \nabla \times [\mathbf{c}(\nabla^2\psi + k^2\psi)], \qquad (5)$$

and $\mathbf{M}$ satisfies the vector HELMHOLTZ DIFFERENTIAL EQUATION if ψ satisfies the scalar HELMHOLTZ DIFFERENTIAL EQUATION

$$\nabla^2\psi + k^2\psi = 0. \qquad (6)$$

Construct another vector function

$$\mathbf{N} \equiv \frac{\nabla \times \mathbf{M}}{k}, \qquad (7)$$

which also satisfies the vector HELMHOLTZ DIFFERENTIAL EQUATION since

$$\nabla^2\mathbf{N} = \frac{1}{k}\nabla^2(\nabla \times \mathbf{M}) = \frac{1}{k}\nabla \times (\nabla^2\mathbf{M})$$
$$= \frac{1}{k}\nabla \times (-k^2\mathbf{M}) = -k\nabla \times \mathbf{M} = -k^2\mathbf{N}, \quad (8)$$

which gives

$$\nabla^2\mathbf{N} + k^2\mathbf{N} = 0. \qquad (9)$$

We have the additional identity

$$\nabla \times \mathbf{N} = \frac{1}{k}\nabla \times (\nabla \times \mathbf{M}) = \frac{1}{k}\nabla(\nabla \cdot \mathbf{M})$$
$$= \frac{1}{k}\nabla^2\mathbf{M} - \frac{1}{k}\nabla^2\mathbf{M} = \frac{-\nabla^2\mathbf{M}}{k} = k\mathbf{M}. \quad (10)$$

In this formalism, ψ is called the generating function and $\mathbf{c}$ is called the PILOT VECTOR. The choice of generating function is determined by the symmetry of the scalar equation, i.e., it is chosen to solve the desired scalar differential equation. If $\mathbf{M}$ is taken as

$$\mathbf{M} = \nabla \times (\mathbf{r}\psi), \qquad (11)$$

where $\mathbf{r}$ is the radius vector, then $\mathbf{M}$ is a solution to the vector wave equation in spherical coordinates. If we want vector solutions which are tangential to the radius vector,

$$\mathbf{M} \cdot \mathbf{r} = \mathbf{r} \cdot (\nabla\psi \times \mathbf{c}) = (\nabla\psi)(\mathbf{c} \times \mathbf{r}) = 0, \qquad (12)$$

so

$$\mathbf{c} \times \mathbf{r} = \mathbf{0} \qquad (13)$$

and we may take

$$\mathbf{c} = \mathbf{r} \qquad (14)$$

(Arfken 1985, pp. 707–711; Bohren and Huffman 1983, p. 88).

A number of conventions are in use. Hill (1954) defines

$$\mathbf{V}_l^m \equiv -\sqrt{\frac{l+1}{2l+1}}Y_l^m\hat{\mathbf{r}} + \frac{1}{\sqrt{(l+1)(2l+1)}}\frac{\partial Y_l^m}{\partial\theta}\hat{\boldsymbol{\theta}}$$
$$+ iM\sqrt{(l+1)(2l+1)}\sin\theta Y_l^m\hat{\boldsymbol{\phi}} \qquad (15)$$

$$\mathbf{W}_l^m = \sqrt{\frac{l}{2l+1}}Y_l^m\hat{\mathbf{r}} + \frac{1}{\sqrt{l(2l+1)}}\frac{\partial Y_l^m}{\partial\theta}\hat{\boldsymbol{\theta}}$$
$$+ \frac{iM}{\sqrt{l(2l+1)}\sin\theta}Y_l^m\hat{\boldsymbol{\phi}} \qquad (16)$$

$$\mathbf{X}_l^m = -\frac{M}{\sqrt{l(l+1)}\sin\theta}Y_l^m\hat{\boldsymbol{\theta}} - \frac{i}{\sqrt{l(l+1)}}\frac{\partial Y_l^m}{\partial\theta}\hat{\boldsymbol{\phi}}. \qquad (17)$$

Morse and Feshbach (1953) define vector harmonics called $\mathbf{B}$, $\mathbf{C}$, and $\mathbf{P}$ using rather complicated expressions.

References

Arfken, G. "Vector Spherical Harmonics." §12.11 in *Mathematical Methods for Physicists, 3rd ed.* Orlando, FL: Academic Press, pp. 707–711, 1985.

Blatt, J. M. and Weisskopf, V. "Vector Spherical Harmonics." Appendix B, §1 in *Theoretical Nuclear Physics.* New York: Wiley, pp. 796–799, 1952.

Bohren, C. F. and Huffman, D. R. *Absorption and Scattering of Light by Small Particles.* New York: Wiley, 1983.

Hill, E. H. "The Theory of Vector Spherical Harmonics." *Amer. J. Phys.* **22**, 211–214, 1954.

Jackson, J. D. *Classical Electrodynamics, 2nd ed.* New York: Wiley, pp. 744–755, 1975.

Morse, P. M. and Feshbach, H. *Methods of Theoretical Physics, Part II.* New York: McGraw-Hill, pp. 1898–1901, 1953.

Vector Transformation Law

The set of n quantities v_j are components of an n-D VECTOR $\mathbf{v}$ IFF, under ROTATION,

$$v_i' = a_{ij}v_j$$

for $i = 1, 2, \ldots, n$. The DIRECTION COSINES between x_i' and x_j are

$$a_{ij} \equiv \frac{\partial x_i'}{\partial x_j} = \frac{\partial x_j}{\partial x_i'}.$$

They satisfy the orthogonality condition

$$a_{ij}a_{ik} = \frac{\partial x_j}{\partial x_i'}\frac{\partial x_i'}{\partial x_k} = \frac{\partial x_j}{\partial x_k} = \delta_{jk},$$

where δ_{jk} is the KRONECKER DELTA.

see also TENSOR, VECTOR

Vector Triple Product

The triple product can be written in terms of the LEVI-CIVITA SYMBOL ϵ_{ijk} as

$$\mathbf{A} \cdot (\mathbf{B} \times \mathbf{C}) = \epsilon_{ijk} A^i B^j C^k. \tag{1}$$

The BAC-CAB RULE can be written in the form

$$\mathbf{A} \times (\mathbf{B} \times \mathbf{C}) = \mathbf{B}(\mathbf{A} \cdot \mathbf{C}) - \mathbf{C}(\mathbf{A} \cdot \mathbf{B}) \tag{2}$$

$$(\mathbf{A} \times \mathbf{B}) \times \mathbf{C} = -\mathbf{C} \times (\mathbf{A} \times \mathbf{B}) = -\mathbf{A}(\mathbf{B} \cdot \mathbf{C}) + \mathbf{B}(\mathbf{A} \cdot \mathbf{C}). \tag{3}$$

Addition identities are

$$\mathbf{A} \cdot (\mathbf{B} \times \mathbf{C}) = \mathbf{B} \cdot (\mathbf{C} \times \mathbf{A}) = \mathbf{C} \cdot (\mathbf{A} \times \mathbf{B}) \tag{4}$$

$$[\mathbf{A}, \mathbf{B}, \mathbf{C}]\mathbf{D} = [\mathbf{D}, \mathbf{B}, \mathbf{C}]\mathbf{A} + [\mathbf{A}, \mathbf{D}, \mathbf{C}]\mathbf{B} + [\mathbf{A}, \mathbf{B}, \mathbf{D}]\mathbf{C} \tag{5}$$

$$[\mathbf{q}, \mathbf{q}', \mathbf{q}''][\mathbf{r}, \mathbf{r}', \mathbf{r}''] = \begin{vmatrix} \mathbf{q} \cdot \mathbf{r} & \mathbf{q} \cdot \mathbf{r}' & \mathbf{q} \cdot \mathbf{r}'' \\ \mathbf{q}' \cdot \mathbf{r} & \mathbf{q}' \cdot \mathbf{r}' & \mathbf{q}' \cdot \mathbf{r}'' \\ \mathbf{q}'' \cdot \mathbf{r} & \mathbf{q}'' \cdot \mathbf{r}' & \mathbf{q}'' \cdot \mathbf{r}'' \end{vmatrix}. \tag{6}$$

see also BAC-CAB RULE, CROSS PRODUCT, DOT PRODUCT, LEVI-CIVITA SYMBOL, SCALAR TRIPLE PRODUCT, VECTOR QUADRUPLE PRODUCT

References

Arfken, G. "Triple Scalar Product, Triple Vector Product." §1.5 in *Mathematical Methods for Physicists, 3rd ed.* Orlando, FL: Academic Press, pp. 26–33, 1985.

Vee

The symbol $\vee$ variously means "disjunction" (in LOGIC) or "join" (for a LATTICE).

see also WEDGE

Velocity

$$\mathbf{v} \equiv \frac{d\mathbf{r}}{dt},$$

where $\mathbf{r}$ is the POSITION VECTOR and d/dt is the derivative with respect to time. Expressed in terms of the ARC LENGTH,

$$\mathbf{v} = \frac{ds}{dt}\hat{\mathbf{T}},$$

where $\hat{\mathbf{T}}$ is the unit TANGENT VECTOR, so the SPEED (which is the magnitude of the velocity) is

$$v \equiv |\mathbf{v}| = \frac{ds}{dt} = |\mathbf{r}'(t)|.$$

see also ANGULAR VELOCITY, POSITION VECTOR, SPEED

Venn Diagram

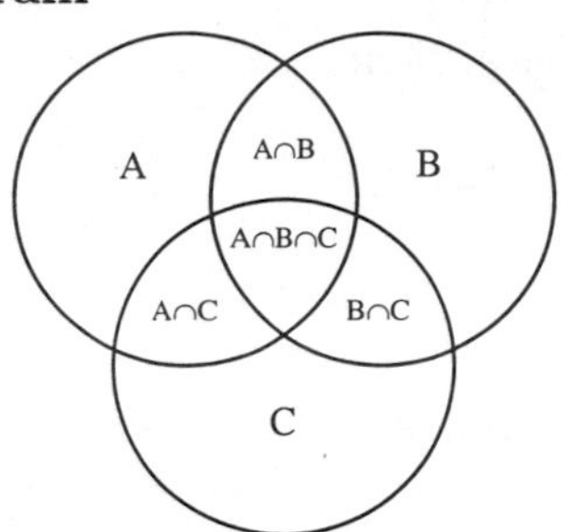

The simplest Venn diagram consists of three symmetrically placed mutually intersecting CIRCLES. It is used in LOGIC theory to represent collections of sets. The region of intersection of the three CIRCLES $A \cap B \cap C$, in the special case of the center of each being located at the intersection of the other two, is called a REULEAUX TRIANGLE.

In general, an order n Venn diagram is a collection of n simple closed curves in the PLANE such that

1. The curves partition the PLANE into 2^n connected regions, and
2. Each SUBSET S of $\{1, 2, \ldots, n\}$ corresponds to a unique region formed by the intersection of the interiors of the curves in S (Ruskey).

see also CIRCLE, FLOWER OF LIFE, LENS, MAGIC CIRCLES, REULEAUX TRIANGLE, SEED OF LIFE

References

Cundy, H. and Rollett, A. *Mathematical Models, 3rd ed.* Stradbroke, England: Tarquin Pub., pp. 255–256, 1989.

Ruskey, F. "A Survey of Venn Diagrams." *Elec. J. Combin.* 4, DS#5, 1997. http://www.combinatorics.org/Surveys/ds5/VennEJC.html.

Ruskey, F. "Venn Diagrams." http://sue.csc.uvic.ca/~cos/inf/comb/SubsetInfo.html#Venn.

Verging Construction

see NEUSIS CONSTRUCTION

Verhulst Model

see LOGISTIC MAP

Veronese Surface

A smooth 2-D surface given by embedding the PROJECTIVE PLANE into projective 5-space by the homogeneous parametric equations

$$v(x, y, z) = (x^2, y^2, z^2, xy, xz, yz).$$

The surface can be projected smoothly into 4-space, but all 3-D projections have singularities (Coffman). The projections of these surfaces in 3-D are called STEINER SURFACES. The VOLUME of the Veronese surface is $2\pi^2$.

see also STEINER SURFACE

References

Coffman, A. "Steiner Surfaces." http://www.ipfw.edu/math/Coffman/steinersurface.html.

Veronese Variety
see VERONESE SURFACE

Versed Sine
see VERSINE

Versiera
see WITCH OF AGNESI

Versine

$$\mathrm{vers}(z) \equiv 1 - \cos z,$$

where $\cos z$ is the COSINE. Using a trigonometric identity, the versine is equal to

$$\mathrm{vers}(z) = 2\sin^2(\tfrac{1}{2}z).$$

see also COSINE, COVERSINE, EXSECANT, HAVERSINE

References
Abramowitz, M. and Stegun, C. A. (Eds.). *Handbook of Mathematical Functions with Formulas, Graphs, and Mathematical Tables, 9th printing.* New York: Dover, p. 78, 1972.

Vertex Angle

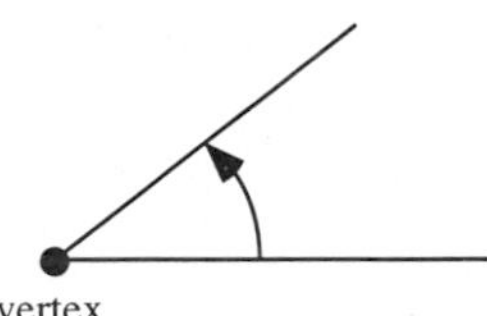

The point about which an ANGLE is measured is called the angle's vertex, and the angle associated with a given vertex is called the vertex angle.

see also ANGLE

Vertex Coloring
BRELAZ'S HEURISTIC ALGORITHM can be used to find a good, but not necessarily minimal, VERTEX coloring of a GRAPH.

see also BRELAZ'S HEURISTIC ALGORITHM, COLORING

Vertex Connectivity
The minimum number of VERTICES whose deletion from a GRAPH disconnects it.

see also EDGE CONNECTIVITY

Vertex Cover
see HITTING SET

Vertex Degree
The degree of a VERTEX of a GRAPH is the number of EDGES which touch the VERTEX, also called the LOCAL DEGREE. The VERTEX degree of a point A in a GRAPH, denoted $\rho(A)$, satisfies

$$\sum_{i=1}^{n} \rho(A_i) = 2E,$$

where E is the total number of EDGES. DIRECTED GRAPHS have two types of degrees, known as the INDEGREE and the OUTDEGREE.

see also DIRECTED GRAPH, INDEGREE, LOCAL DEGREE, OUTDEGREE

Vertex Enumeration
A CONVEX POLYHEDRON is defined as the set of solutions to a system of linear inequalities

$$\mathsf{m}\mathbf{x} \leq \mathbf{b},$$

where m is a REAL $s \times d$ MATRIX and $\mathbf{b}$ is a REAL s-VECTOR. Given m and $\mathbf{b}$, vertex enumeration is the determination of the polyhedron's VERTICES.

see also CONVEX POLYHEDRON, POLYHEDRON

References
Avis, D. and Fukuda, K. "A Pivoting Algorithm for Convex Hulls and Vertex Enumeration of Arrangements and Polyhedra." In *Proceedings of the 7th ACM Symposium on Computational Geometry, North Conway, NH, 1991,* pp. 98–104, 1991.
Fukada, K. and Mizukosh, I. "Vertex Enumeration Package for Convex Polytopes and Arrangements, Version 0.41 Beta." `http://www.mathsource.com/cgi-bin/MathSource/Applications/Mathematics/0202-633`.

Vertex Figure
The line joining the MIDPOINTS of adjacent sides in a POLYGON is called the polygon's vertex figure. For a regular n-gon with side length s,

$$v = s\cos\left(\frac{\pi}{n}\right).$$

For a POLYHEDRON, the faces that join at a VERTEX form a solid angle whose section by the plane is the vertex figure.

see also TRUNCATION

Vertex (Graph)
A point of a GRAPH, also called a NODE.

see also EDGE (GRAPH), NULL GRAPH, TAIT COLORING, TAIT CYCLE, TAIT'S HAMILTONIAN GRAPH CONJECTURE, VERTEX (POLYGON)

Vertex (Parabola)
For a PARABOLA oriented vertically and opening upwards, the vertex is the point where the curve reaches a minimum.

Vertex (Polygon)

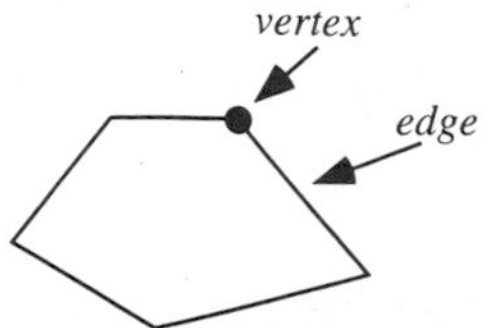

A point at which two EDGES of a POLYGON meet.

see also PRINCIPAL VERTEX, VERTEX (GRAPH), VERTEX (POLYHEDRON)

Vertex (Polyhedron)

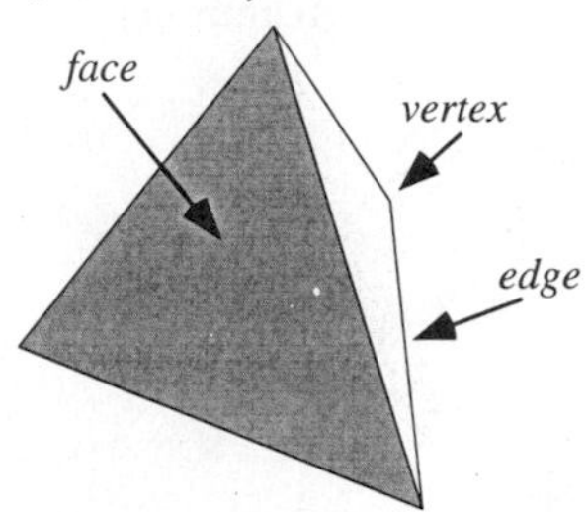

A point at which three of more EDGES of a POLYHEDRON meet. The concept can also be generalized to a POLYTOPE.

see also VERTEX (GRAPH), VERTEX (POLYGON)

Vertex (Polytope)

The vertex of a POLYTOPE is a point where edges of the POLYTOPE meet.

Vertical

Oriented in an up-down position.

see also HORIZONTAL

Vertical-Horizontal Illusion

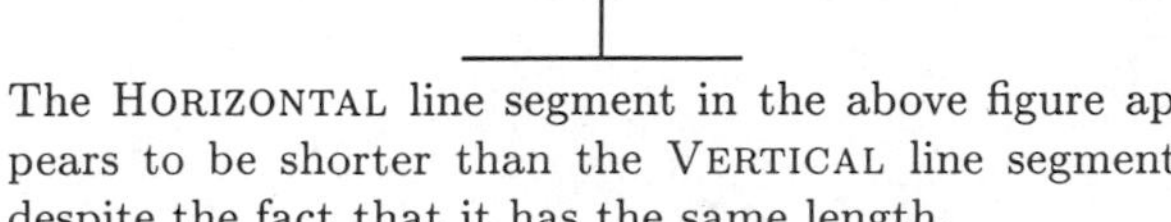

The HORIZONTAL line segment in the above figure appears to be shorter than the VERTICAL line segment, despite the fact that it has the same length.

see also ILLUSION, MÜLLER-LYER ILLUSION, POGGENDORFF ILLUSION, PONZO'S ILLUSION

References

Fineman, M. *The Nature of Visual Illusion.* New York: Dover, p. 153, 1996.

Vertical Perspective Projection

A MAP PROJECTION given by the transformation equations

$$x = k' \cos\phi \sin(\lambda - \lambda_0) \quad (1)$$

$$y = k'[\cos\phi_1 \sin\phi - \sin\phi_1 \cos\phi \cos(\lambda - \lambda_0)], \quad (2)$$

where P is the distance of the point of perspective in units of SPHERE RADII and

$$k' = \frac{P-1}{P - \cos c} \quad (3)$$

$$\cos c = \sin\phi_1 \sin\phi + \cos\phi_1 \cos\phi \cos(\lambda - \lambda_0). \quad (4)$$

References

Snyder, J. P. *Map Projections—A Working Manual.* U. S. Geological Survey Professional Paper 1395. Washington, DC: U. S. Government Printing Office, pp. 173–178, 1987.

Vertical Tangent

A function $f(x)$ has a vertical tangent line at x_0 if f is continuous at x_0 and

$$\lim_{x \to x_0} f'(x) = \pm\infty.$$

Vesica Piscis

see LENS

Vibration Problem

Solution of a system of second-order homogeneous ordinary differential equations with constant COEFFICIENTS of the form

$$\frac{d^2\mathbf{x}}{dt^2} + \mathsf{B}\mathbf{x} = 0,$$

where B is a POSITIVE DEFINITE MATRIX. To solve the vibration problem,

1. Solve the CHARACTERISTIC EQUATION of B to get EIGENVALUES $\lambda_1, \ldots, \lambda_n$. Define $\omega_i \equiv \sqrt{\lambda_i}$.
2. Compute the corresponding EIGENVECTORS $\mathbf{e}_1, \ldots, \mathbf{e}_n$.
3. The normal modes of oscillation are given by $\mathbf{x}_1 = A_1 \sin(\omega_1 t + \alpha_1)\mathbf{e}_1, \ldots, \mathbf{x}_n = A_n \sin(\omega_n t + \alpha_n)\mathbf{e}_n$, where $A_1, \ldots, A_n$ and $\alpha_1, \ldots, \alpha_n$ are arbitrary constants.
4. The general solution is $\mathbf{x} = \sum_{i=1}^n \mathbf{x}_i$.

Vickery Auction

An AUCTION in which the highest bidder wins but pays only the second-highest bid. This variation over the normal bidding procedure is supposed to encourage bidders to bid the largest amount they are willing to pay.

see also AUCTION

Viergruppe

The mathematical group $Z_4 \otimes Z_4$, also denoted D_2. Its multiplication table is

V	I	V_1	V_2	V_3
I	V_1	V_2	V_3	V_4
V_1	V_1	I	V_3	V_2
V_2	V_2	V_3	I	V_1
V_3	V_3	V_2	V_1	I

see also DIHEDRAL GROUP, FINITE GROUP—Z_4

Vieta's Substitution

The substitution of

$$x = w - \frac{p}{3w}$$

into the standard form CUBIC EQUATION

$$x^3 + px = q,$$

which reduces the cubic to a QUADRATIC EQUATION in w^3,

$$(w^3)^2 - \tfrac{1}{27}p^3(w^3) - q = 0.$$

see also CUBIC EQUATION

Vigesimal

The base-20 notational system for representing REAL NUMBERS. The digits used to represent numbers using vigesimal NOTATION are 0, 1, 2, 3, 4, 5, 6, 7, 8, 9, A, B, C, D, E, F, G, H, I, and J. A base-20 number system was used by the Aztecs and Mayans. The Mayans compiled extensive observations of planetary positions in base-20 notation.

see also BASE (NUMBER), BINARY, DECIMAL, HEXADECIMAL, OCTAL, QUATERNARY, TERNARY

References

Weisstein, E. W. "Bases." http://www.astro.virginia.edu/~eww6n/math/notebooks/Bases.m.

Vigintillion

In the American system, 10^{63}.

see also LARGE NUMBER

Villarceau Circles

Given an arbitrary point on a TORUS, four CIRCLES can be drawn through it. The first is in the plane of the torus and the second is PERPENDICULAR to it. The third and fourth CIRCLES are called Villarceau circles.

see also TORUS

References

Melzak, Z. A. *Invitation to Geometry.* New York: Wiley, pp. 63–72, 1983.

Villarceau, M. "Théorème sur le tore." *Nouv. Ann. Math.* **7**, 345–347, 1848.

Vinculum

A horizontal line placed above multiple quantities to indicate that they form a unit. It is most commonly used to denote ROOTS ($\sqrt{12345}$) and repeating decimals ($0.\overline{111}$).

Vinogradov's Theorem

Every sufficiently large ODD number is a sum of three PRIMES. Proved in 1937.

see also GOLDBACH CONJECTURE

Virtual Group

see GROUPOID

Visibility

see VISIBLE POINT

Visible Point

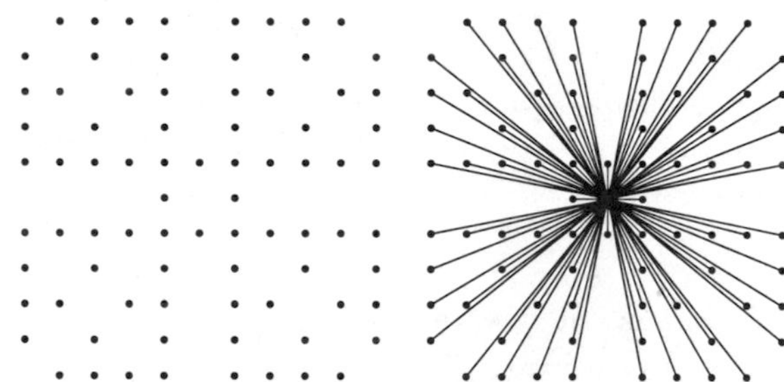

Two LATTICE POINTS (x, y) and (x', y') are mutually visible if the line segment joining them contains no further LATTICE POINTS. This corresponds to the requirement that $(x' - x, y' - y) = 1$, where (m, n) denotes the GREATEST COMMON DIVISOR. The plots above show the first few points visible from the ORIGIN.

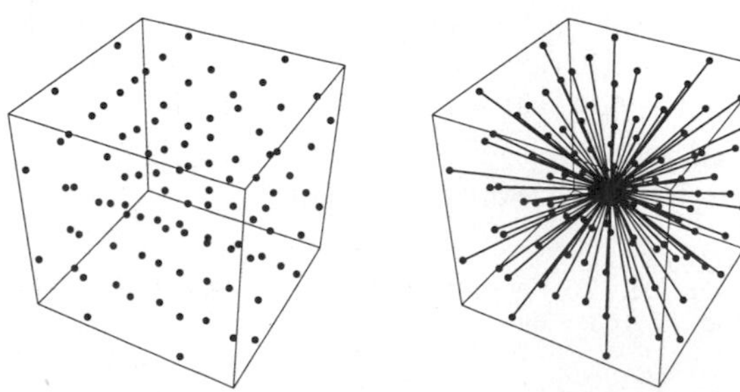

If a LATTICE POINT is selected at random in 2-D, the probability that it is visible from the origin is $6/\pi^2$. This is also the probability that two INTEGERS picked at random are RELATIVELY PRIME. If a LATTICE POINT is picked at random in n-D, the probability that it is visible from the ORIGIN is $1/\zeta(n)$, where $\zeta(n)$ is the RIEMANN ZETA FUNCTION.

An invisible figure is a POLYGON all of whose corners are invisible. There are invisible sets of every finite shape. The lower left-hand corner of the invisible squares with smallest x coordinate of AREAS 2 and 3 are (14, 20) and (104, 6200).

see also LATTICE POINT, ORCHARD VISIBILITY PROBLEM, RIEMANN ZETA FUNCTION

References

Apostol, T. §3.8 in *Introduction to Analytic Number Theory.* New York: Springer-Verlag, 1976.

Baake, M.; Grimm, U.; and Warrington, D. H. "Some Remarks on the Visible Points of a Lattice." *J. Phys. A: Math. General* **27**, 2669–2674, 1994.

Beeler, M.; Gosper, R. W.; and Schroeppel, R. *HAKMEM.* Cambridge, MA: MIT Artificial Intelligence Laboratory, Memo AIM-239, Feb. 1972.

Herzog, F. and Stewart, B. M. "Patterns of Visible and Nonvisible Lattice Points." *Amer. Math. Monthly* **78**, 487–496, 1971.

Mosseri, R. "Visible Points in a Lattice." *J. Phys. A: Math. Gen.* **25**, L25–L29, 1992.

Schroeder, M. R. "A Simple Function and Its Fourier Transform." *Math. Intell.* **4**, 158–161, 1982.

Schroeder, M. R. *Number Theory in Science and Communication, 2nd ed.* New York: Springer-Verlag, 1990

Visible Point Vector Identity

A set of identities involving n-D visible lattice points was discovered by Campbell (1994). Examples include

$$\prod_{\substack{(a,b)=1\\ a\geq 0, b\leq 1}} (1-y^a z^b)^{-1/b} = (1-z)^{-1/(1-y)}$$

for $|yz|, |z| < 1$ and

$$\prod_{\substack{(a,b,c)=1\\ a,b\geq 0, c\leq 1}} (1-x^a y^b z^c)^{-1/c} = (1-z)^{-1/[(1-x)(1-y)]}$$

for $|xyz|, |xz|, |yz|, |z| < 1$.

References

Campbell, G. B. "Infinite Products Over Visible Lattice Points." *Internat. J. Math. Math. Sci.* **17**, 637–654, 1994.

Campbell, G. B. "Visible Point Vector Identities." `http://www.geocities.com/CapeCanaveral/Launchpad/9416/vpv.html`.

Vitali's Convergence Theorem

Let $f_n(z)$ be a sequence of functions, each regular in a region D, let $|f_n(z)| \leq M$ for every n and z in D, and let $f_n(z)$ tend to a limit as $n \to \infty$ at a set of points having a LIMIT POINT inside D. Then $f_n(z)$ tends uniformly to a limit in any region bounded by a contour interior to D, the limit therefore being an analytic function of z.

see also MONTEL'S THEOREM

References

Titchmarsh, E. C. *The Theory of Functions, 2nd ed.* Oxford, England: Oxford University Press, p. 168, 1960.

Viviani's Curve

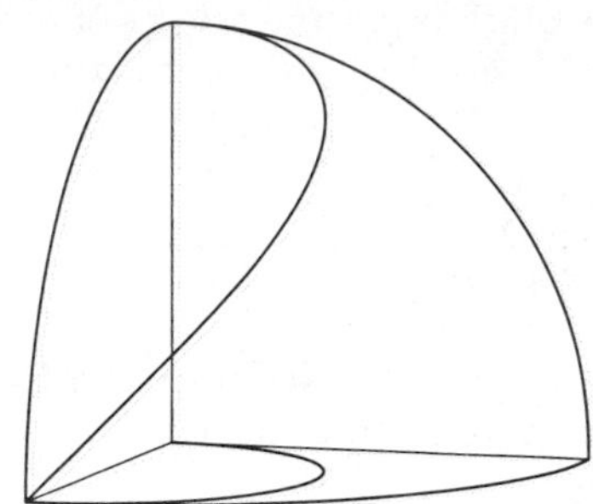

The SPACE CURVE giving the intersection of the CYLINDER

$$(x-a)^2 + y^2 = a^2 \tag{1}$$

and the SPHERE

$$x^2 + y^2 + z^2 = 4^2. \tag{2}$$

It is given by the parametric equations

$$x = a(1+\cos t) \tag{3}$$

$$y = a\sin t \tag{4}$$

$$z = 2a\sin(\tfrac{1}{2}t). \tag{5}$$

The CURVATURE and TORSION are given by

$$\kappa(t) = \frac{\sqrt{13+3\cos t}}{a(3+\cos t)^{3/2}} \tag{6}$$

$$\tau(t) = \frac{6\cos(\frac{1}{2}t)}{a(13+3\cos t)}. \tag{7}$$

see also CYLINDER, SPHERE, STEINMETZ SOLID

References

Gray, A. "Viviani's Curve." §7.6 in *Modern Differential Geometry of Curves and Surfaces.* Boca Raton, FL: CRC Press, pp. 140–142, 1993.

von Seggern, D. *CRC Standard Curves and Surfaces.* Boca Raton, FL: CRC Press, p. 270, 1993.

Viviani's Theorem

For a point P inside an EQUILATERAL TRIANGLE ΔABC, the sum of the perpendiculars p_i from P to the sides of the TRIANGLE is equal to the ALTITUDE h. This result is simply proved as follows,

$$\Delta ABC = \Delta PBC + \Delta PCA + \Delta PAB. \tag{1}$$

With s the side length,

$$\tfrac{1}{2}sh = \tfrac{1}{2}sp_a + \tfrac{1}{2}sp_b + \tfrac{1}{2}sp_c, \tag{2}$$

so

$$h = p_a + p_b + p_c. \tag{3}$$

see also ALTITUDE, EQUILATERAL TRIANGLE

Vojta's Conjecture

A conjecture which treats the heights of points relative to a canonical class of a curve defined over the INTEGERS.

References

Cox, D. A. "Introduction to Fermat's Last Theorem." *Amer. Math. Monthly* **101**, 3–14, 1994.

Volterra Integral Equation of the First Kind

An INTEGRAL EQUATION of the form

$$f(x) = \int_a^x k(x,t)\phi(t)\,dt.$$

see also FREDHOLM INTEGRAL EQUATION OF THE FIRST KIND, FREDHOLM INTEGRAL EQUATION OF THE SECOND KIND, INTEGRAL EQUATION, VOLTERRA INTEGRAL EQUATION OF THE SECOND KIND

References

Arfken, G. *Mathematical Methods for Physicists, 3rd ed.* Orlando, FL: Academic Press, p. 865, 1985.

Press, W. H.; Flannery, B. P.; Teukolsky, S. A.; and Vetterling, W. T. "Volterra Equations." §18.2 in *Numerical Recipes in FORTRAN: The Art of Scientific Computing, 2nd ed.* Cambridge, England: Cambridge University Press, pp. 786–788, 1992.

Volterra Integral Equation of the Second Kind

An INTEGRAL EQUATION of the form

$$\phi(x) = f(x) + \int_a^x k(x,t)\phi(t)\,dt.$$

see also FREDHOLM INTEGRAL EQUATION OF THE FIRST KIND, FREDHOLM INTEGRAL EQUATION OF THE SECOND KIND, INTEGRAL EQUATION, VOLTERRA INTEGRAL EQUATION OF THE FIRST KIND

References

Arfken, G. *Mathematical Methods for Physicists, 3rd ed.* Orlando, FL: Academic Press, p. 865, 1985.

Press, W. H.; Flannery, B. P.; Teukolsky, S. A.; and Vetterling, W. T. "Volterra Equations." §18.2 in *Numerical Recipes in FORTRAN: The Art of Scientific Computing, 2nd ed.* Cambridge, England: Cambridge University Press, pp. 786–788, 1992.

Volume

The volume of a solid body is the amount of "space" it occupies. Volume has units of LENGTH cubed (i.e., cm^3, m^3, in^3, etc.) For example, the volume of a box (RECTANGULAR PARALLELEPIPED) of LENGTH L, WIDTH W, and HEIGHT H is given by

$$V = L \times W \times H.$$

The volume can also be computed for irregularly-shaped and curved solids such as the CYLINDER and CUBE. The volume of a SURFACE OF REVOLUTION is particularly simple to compute due to its symmetry.

The following table gives volumes for some common SURFACES. Here r denotes the RADIUS, h the height, A the base AREA, and s the SLANT HEIGHT (Beyer 1987).

Surface	V
cone	$\frac{1}{3}\pi r^2 h$
conical frustum	$\frac{1}{3}\pi h({R_1}^2 + {R_2}^2 + R_1 R_2)$
cube	a^3
cylinder	$\pi r^2 h$
ellipsoid	$\frac{4}{3}\pi abc$
oblate spheroid	$\frac{4}{3}\pi a^2 b$
prolate spheroid	$\frac{4}{3}\pi ab^2$
pyramid	$\frac{1}{3}Ah$
pyramidal frustum	$\frac{1}{3}h(A_1 + A_2 + \sqrt{A_1 A_2})$
sphere	$\frac{4}{3}\pi r^3$
spherical sector	$\frac{2}{3}\pi r^2 h$
spherical segment	$\frac{1}{3}\pi h^2 r(3r - h)$
torus	$2\pi^2 R r^2$

Even simple SURFACES can display surprisingly counterintuitive properties. For instance, the SURFACE OF REVOLUTION of $y = 1/x$ around the x-axis for $x \geq 1$ is called GABRIEL'S HORN, and has finite volume, but infinite SURFACE AREA.

The generalization of volume to n DIMENSIONS for $n \geq 4$ is known as CONTENT.

see also ARC LENGTH, AREA, CONTENT, HEIGHT, LENGTH (SIZE), SURFACE AREA, SURFACE OF REVOLUTION, VOLUME ELEMENT, WIDTH (SIZE)

References

Beyer, W. H. *CRC Standard Mathematical Tables, 28th ed.* Boca Raton, FL: CRC Press, pp. 127–132, 1987.

Volume Element

A volume element is the differential element dV whose VOLUME INTEGRAL over some range in a given coordinate system gives the VOLUME of a solid,

$$V = \iiint_G dx\,dy\,dz. \tag{1}$$

In $\mathbb{R}^n$, the volume of the infinitesimal n-HYPERCUBE bounded by dx_1, ..., dx_n has volume given by the WEDGE PRODUCT

$$dV = dx_1 \wedge \ldots \wedge dx_n \tag{2}$$

(Gray 1993).

The use of the antisymmetric WEDGE PRODUCT instead of the symmetric product $dx_1 \ldots dx_n$ is a technical refinement often omitted in informal usage. Dropping the wedges, the volume element for CURVILINEAR COORDINATES in $\mathbb{R}^3$ is given by

$$\begin{aligned} dV &= |(h_1\hat{\mathbf{u}}_1\,du_1)\cdot(h_2\hat{\mathbf{u}}_2\,du_2)\times(h_3\hat{\mathbf{u}}_3\,du_3)| && (3)\\ &= h_1h_2h_3\,du_1\,du_2\,du_3 && (4)\\ &= \left|\frac{\partial\mathbf{r}}{\partial u_1}\cdot\frac{\partial\mathbf{r}}{\partial u_2}\times\frac{\partial\mathbf{r}}{\partial u_3}\right| du_1\,du_2\,du_3 && (5)\\ &= \begin{vmatrix} \frac{\partial x}{\partial u_1} & \frac{\partial x}{\partial u_2} & \frac{\partial x}{\partial u_3}\\ \frac{\partial y}{\partial u_1} & \frac{\partial y}{\partial u_2} & \frac{\partial y}{\partial u_3}\\ \frac{\partial z}{\partial u_1} & \frac{\partial z}{\partial u_2} & \frac{\partial z}{\partial u_3} \end{vmatrix} du_1\,du_2\,du_3 && (6)\\ &= \left|\frac{\partial(x,y,z)}{\partial(u_1,u_2,u_3)}\right| du_1\,du_2\,du_3, && (7) \end{aligned}$$

where the latter is the JACOBIAN and the h_i are SCALE FACTORS.

see also AREA ELEMENT, JACOBIAN, LINE ELEMENT, RIEMANNIAN METRIC, SCALE FACTOR, SURFACE INTEGRAL, VOLUME INTEGRAL

References

Gray, A. "Isometries of Surfaces." §13.2 in *Modern Differential Geometry of Curves and Surfaces.* Boca Raton, FL: CRC Press, pp. 255–258, 1993.

Volume Integral

A triple integral over three coordinates giving the VOLUME within some region R,

$$V = \iiint_G dx\,dy\,dz.$$

see also INTEGRAL, LINE INTEGRAL, MULTIPLE INTEGRAL, SURFACE INTEGRAL, VOLUME, VOLUME ELEMENT

von Aubel's Theorem

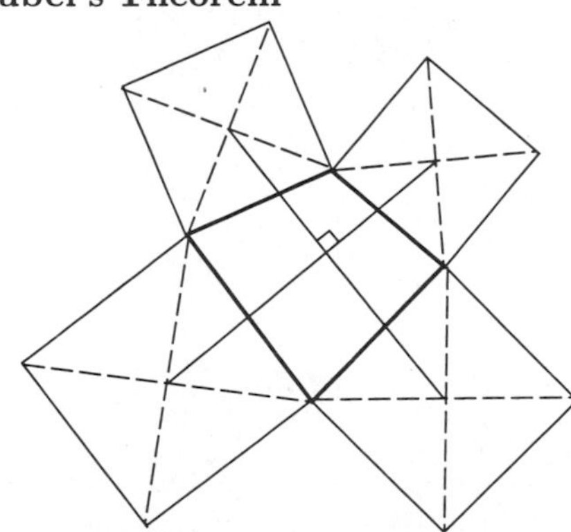

Given an arbitrary QUADRILATERAL, place a SQUARE outwardly on each side, and connect the centers of opposite SQUARES. Then the two lines are of equal length and cross at a RIGHT ANGLE.

see also QUADRILATERAL, RIGHT ANGLE, SQUARE

References

Kitchen, E. "Dörrie Tiles and Related Miniatures." *Math. Mag.* **67**, 128–130, 1994.

von Dyck's Theorem

Let a GROUP G have a presentation

$$G = (x_1,\ldots,x_n | r_j(x_1,\ldots,x_n), j\in J)$$

so that $G = F/R$, where F is the FREE GROUP with basis $\{x_1,\ldots,x_n\}$ and R is the NORMAL SUBGROUP generated by the r_j. If H is a GROUP with $H = \langle y_1,\ldots,y_n\rangle$ and if $r_j(y_1,\ldots,y_n) = 1$ for all j, then there is a surjective homomorphism $G\to H$ with $x_i\mapsto y_i$ for all i.

see also FREE GROUP, NORMAL SUBGROUP

References

Rotman, J. J. *An Introduction to the Theory of Groups, 4th ed.* New York: Springer-Verlag, p. 346, 1995.

von Mangoldt Function

see MANGOLDT FUNCTION

von Neumann Algebra

A GROUP "with bells and whistles." It was while studying von Neumann algebras that Jones discovered the amazing and highly unexpected connections with KNOT THEORY which led to the formulation of the JONES POLYNOMIAL.

References

Iyanaga, S. and Kawada, Y. (Eds.). "Von Neumann Algebras." §430 in *Encyclopedic Dictionary of Mathematics.* Cambridge, MA: MIT Press, pp. 1358–1363, 1980.

von Staudt-Clausen Theorem

$$B_{2n} = A_n - \sum_{\substack{p_k \\ p_k-1|2n}} \frac{1}{p_k},$$

where B_{2n} is a BERNOULLI NUMBER, A_n is an INTEGER, and the p_ks are the PRIMES satisfying $p_k - 1|2k$. For example, for $k = 1$, the primes included in the sum are 2 and 3, since $(2-1)|2$ and $(3-1)|2$. Similarly, for $k = 6$, the included primes are (2, 3, 5, 7, 13), since (1, 2, 3, 6, 12) divide $12 = 2\cdot 6$. The first few values of A_n for $n = 1, 2, \ldots$ are 1, 1, 1, 1, 1, 1, 2, −6, 56, −528, ... (Sloane's A000164).

The theorem was rediscovered by Ramanujan (Hardy 1959, p. 11) and can be proved using p-ADIC NUMBERS.

see also BERNOULLI NUMBER, p-ADIC NUMBER

References

Conway, J. H. and Guy, R. K. *The Book of Numbers.* New York: Springer-Verlag, p. 109, 1996.

Hardy, G. H. *Ramanujan: Twelve Lectures on Subjects Suggested by His Life and Work, 3rd ed.* New York: Chelsea, 1959.

Hardy, G. H. and Wright, E. M. "The Theorem of von Staudt" and "Proof of von Staudt's Theorem." §7.9–7.10 in *An Introduction to the Theory of Numbers, 5th ed.* Oxford, England: Clarendon Press, pp. 90–93, 1979.

Sloane, N. J. A. Sequence A000146/M1717 in "An On-Line Version of the Encyclopedia of Integer Sequences."

Staudt. "Beweis eines Lehrsatzes, die Bernoullischen Zahlen betreffend." *J. reine angew. Math.* **21**, 372–374, 1840.

von Staudt Theorem

see VON STAUDT-CLAUSEN THEOREM

Voronoi Cell

The generalization of a VORONOI POLYGON to n-D, for $n > 2$.

Voronoi Diagram

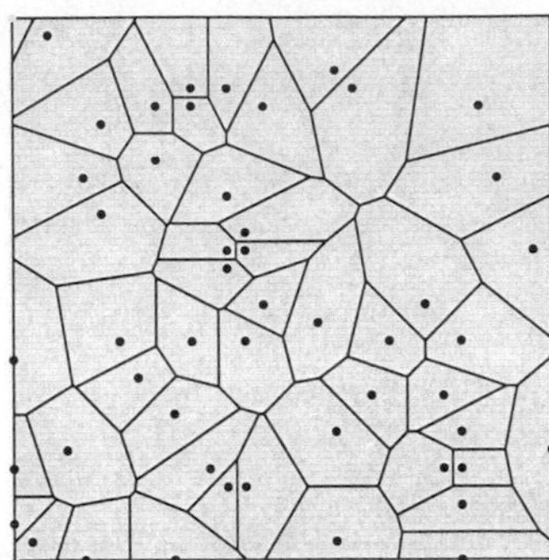

The partitioning of a plane with n points into n convex POLYGONS such that each POLYGON contains exactly one point and every point in a given POLYGON is closer to its central point than to any other. A Voronoi diagram is sometimes also known as a DIRICHLET TESSELLATION. The cells are called DIRICHLET REGIONS, THIESSEN POLYTOPES, or VORONOI POLYGONS.

see also DELAUNAY TRIANGULATION, MEDIAL AXIS, VORONOI POLYGON

References

Eppstein, D. "Nearest Neighbors and Voronoi Diagrams." http://www.ics.uci.edu/~eppstein/junkyard/nn.html.

Voronoi Polygon

A POLYGON whose interior consists of all points in the plane which are closer to a particular LATTICE POINT than to any other. The generalization to n-D is called a DIRICHLET REGION, THIESSEN POLYTOPE, or VORONOI CELL.

Voting

It is possible to conduct a secret ballot even if the votes are sent in to a central polling station (Lipton and Widgerson, Honsberger 1985).

see also ARROW'S PARADOX, BALLOT PROBLEM, MAY'S THEOREM, QUOTA SYSTEM, SOCIAL CHOICE THEORY

References

Honsberger, R. *Mathematical Gems III.* Washington, DC: Math. Assoc. Amer., pp. 157–162, 1985.

Lipton, R. G.; and Widgerson, A. "Multi-Party Cryptographic Protocols."

VR Number

A "visual representation" number which is a sum of some simple function of its digits. For example,

$$1233 = 12^2 + 33^2$$

$$2661653 = 1653^2 - 266^2$$

$$221859 = 22^3 + 18^3 + 59^3$$

$$40585 + 4! + 0! + 5! + 8! + 5!$$

$$148349 = !1+!4+!8+!3+!4+!9$$

$$4913 = (4 + 9 + 1 + 3)^3$$

are all VR numbers given by Madachy (1979).

References

Madachy, J. S. *Madachy's Mathematical Recreations.* New York: Dover, pp. 165–171, 1979.

Vulgar Series

see FAREY SERIES

W

W2-Constant

$$W_2 = 1.529954037\ldots.$$

References
Plouffe, S. "W2 Constant." http://lacim.uqam.ca/piDATA/w2.txt.

W-Function

see LAMBERT'S W-FUNCTION

Wada Basin

A BASIN OF ATTRACTION in which every point on the common boundary of that basin and another basin is also a boundary of a third basin. In other words, no matter how closely a boundary point is zoomed into, all three basins appear in the picture.

see also BASIN OF ATTRACTION

References
Nusse, H. E. and Yorke, J. A. "Basins of Attraction." *Science* **271**, 1376–1380, 1996.

Walk

A sequence of VERTICES and EDGES such that the VERTICES and EDGES are adjacent. A walk is therefore equivalent to a graph CYCLE, but with the VERTICES along the walk enumerated as well as the EDGES.

see also CIRCUIT, CYCLE (GRAPH), PATH, RANDOM WALK

Wallace-Bolyai-Gerwein Theorem

Two POLYGONS are congruent by DISSECTION IFF they have the same AREA. In particular, any POLYGON is congruent by DISSECTION to a SQUARE of the same AREA. Laczkovich (1988) also proved that a CIRCLE is congruent by DISSECTION to a SQUARE (furthermore, the DISSECTION can be accomplished using TRANSLATIONS only).

see also DISSECTION

References
Klee, V. and Wagon, S. *Old and New Unsolved Problems in Plane Geometry and Number Theory.* Washington, DC: Math. Assoc. Amer., pp. 50–51, 1991.
Laczkovich, M. "Von Neumann's Paradox with Translation." *Fund. Math.* **131**, 1–12, 1988.

Wallace-Simson Line

see SIMSON LINE

Wallis's Conical Edge

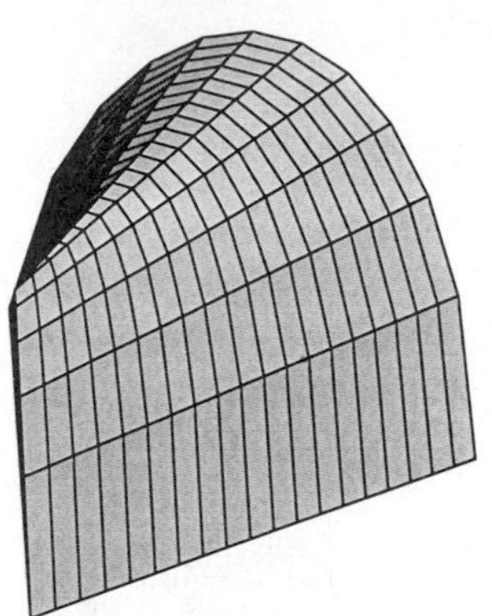

The RIGHT CONOID surface given by the parametric equations

$$x(u,v) = v\cos u$$
$$y(u,v) = v\sin u$$
$$z(u,v) = c\sqrt{a^2 - b^2\cos^2 u}.$$

see also RIGHT CONOID

References
Gray, A. *Modern Differential Geometry of Curves and Surfaces.* Boca Raton, FL: CRC Press, pp. 354–355, 1993.

Wallis Cosine Formula

$$\int_0^{\pi/2} \cos^n x\,dx = \begin{cases} \frac{\pi}{2}\frac{1\cdot 3\cdot 5\cdots(n-1)}{2\cdot 4\cdot 6\cdots n} & \text{for } n = 2, 4, \ldots \\ \frac{2\cdot 4\cdot 6\cdots(n-1)}{1\cdot 3\cdot 5\cdots n} & \text{for } n = 3, 5, \ldots. \end{cases}$$

see also WALLIS FORMULA, WALLIS SINE FORMULA

Wallis Formula

The Wallis formula follows from the INFINITE PRODUCT representation of the SINE

$$\sin x = x\prod_{n=1}^\infty \left(1 - \frac{x^2}{\pi^2 n^2}\right). \tag{1}$$

Taking $x = \pi/2$ gives

$$1 = \frac{\pi}{2}\prod_{n=1}^\infty \left[1 - \frac{1}{(2n)^2}\right] = \frac{\pi}{2}\prod_{n=1}^\infty \left[\frac{(2n)^2 - 1}{(2n)^2}\right], \tag{2}$$

so

$$\frac{\pi}{2} = \prod_{n=1}^\infty \left[\frac{(2n)^2}{(2n-1)(2n+1)}\right] = \frac{2\cdot 2}{1\cdot 3}\frac{4\cdot 4}{3\cdot 5}\frac{6\cdot 6}{5\cdot 7}\cdots. \tag{3}$$

A derivation due to Y. L. Yung uses the RIEMANN ZETA FUNCTION. Define

$$F(s) \equiv -\operatorname{Li}_s(-1) = \sum_{n=1}^{\infty} \frac{(-1)^n}{n^s} = (1-2^{1-s})\zeta(s) \qquad (4)$$

$$F'(s) = \sum_{n=1}^{\infty} \frac{(-1)^n \ln n}{n^s}, \qquad (5)$$

so

$$F'(0) = \sum_{n=1}^{\infty} (-1)^n \ln n = -\ln 1 + \ln 2 - \ln 3 + \ldots = \ln\left(\frac{2\cdot 4\cdot 6\cdots}{1\cdot 3\cdot 5\cdots}\right). \qquad (6)$$

Taking the derivative of the zeta function expression gives

$$\frac{d}{ds}(1-2^{1-s})\zeta(s) = 2^{1-s}(\ln 2)\zeta(s) + (1-2^{1-s})\zeta'(s) \quad (7)$$

$$\left[\frac{d}{ds}(1-2^{1-s})\zeta(s)\right]_{s=0} = -\ln 2 - \zeta'(0) = -\ln 2 + \tfrac{1}{2}\ln(2\pi) = \ln\left(\frac{\sqrt{2\pi}}{2}\right) = \ln\left(\sqrt{\frac{\pi}{2}}\right). \quad (8)$$

Equating and squaring then gives the Wallis formula, which can also be expressed

$$\frac{\pi}{2} = \left[4^{\zeta(0)} e^{-\zeta'(0)}\right]^2. \qquad (9)$$

The q-ANALOG of the Wallis formula for $q = 2$ is

$$\prod_{k=1}^{\infty} (1-q^{-k})^{-1} = 3.4627466194\ldots \qquad (10)$$

(Finch).

see also WALLIS COSINE FORMULA, WALLIS SINE FORMULA

References

Abramowitz, M. and Stegun, C. A. (Eds.). *Handbook of Mathematical Functions with Formulas, Graphs, and Mathematical Tables, 9th printing.* New York: Dover, p. 258, 1972.

Finch, S. "Favorite Mathematical Constants." http://www.mathsoft.com/asolve/constant/dig/dig.html.

Kenney, J. F. and Keeping, E. S. *Mathematics of Statistics, Pt. 2, 2nd ed.* Princeton, NJ: Van Nostrand, pp. 63–64, 1951.

Wallis's Problem

Find solutions to $\sigma(x^2) = \sigma(y^2)$ other than $(x, y) = (4, 5)$, where σ is the DIVISOR FUNCTION.

see also FERMAT'S SIGMA PROBLEM

Wallis Sieve

A compact set W_∞ with AREA

$$\mu(W_\infty) = \frac{8}{9}\frac{24}{25}\frac{48}{49}\cdots = \frac{\pi}{4}$$

created by punching a square hole of length 1/3 in the center of a square. In each of the eight squares remaining, punch out another hole of length $1/(3\cdot 5)$, and so on.

Wallis Sine Formula

$$\int_0^{\pi/2} \sin^n x\,dx = \begin{cases} \frac{\pi}{2}\frac{1\cdot 3\cdot 5\cdots(n-1)}{2\cdot 4\cdot 6\cdots n} & \text{for } n = 2,\ 4,\ \ldots \\ \frac{2\cdot 4\cdot 6\cdots(n-1)}{1\cdot 3\cdot 5\cdots n} & \text{for } n = 3,\ 5,\ \ldots. \end{cases}$$

see also WALLIS COSINE FORMULA, WALLIS FORMULA

Wallpaper Groups

The 17 PLANE SYMMETRY GROUPS. Their symbols are p1, p2, pm, pg, cm, pmm, pmg, pgg, cmm, p4, p4m, p4g, p3, p31m, p3m1, p6, and p6m. For a description of the symmetry elements present in each space group, see Coxeter (1969, p. 413).

References

Coxeter, H. S. M. *Introduction to Geometry, 2nd ed.* New York: Wiley, 1969.

Hilbert, D. and Cohn-Vossen, S. *Geometry and the Imagination.* New York: Chelsea, 1952.

Joyce, D. E. "Wallpaper Groups (Plane Symmetry Groups)." http://aleph0.clarku.edu/~djoyce/wallpaper/.

Lee, X. "The Discontinuous Groups of Rotation and Translation in the Plane." http://www.best.com/~xah/Wallpaper_dir/c0_WallPaper.html.

Schattschneider, D. "The Plane Symmetry Groups: Their Recognition and Notation." *Amer. Math. Monthly* **85**, 439–450, 1978.

Weyl, H. *Symmetry.* Princeton, NJ: Princeton University Press, 1952.

Walsh Function

Functions consisting of a number of fixed-amplitude square pulses interposed with zeros. Following Harmuth (1969), designate those with EVEN symmetry $\operatorname{Cal}(k,t)$ and those with ODD symmetry $\operatorname{Sal}(k,t)$. Define the SEQUENCY k as half the number of zero crossings in the time base. Walsh functions with nonidentical SEQUENCIES are ORTHOGONAL, as are the functions $\operatorname{Cal}(k,t)$ and $\operatorname{Sal}(k,t)$. The product of two Walsh functions is also a Walsh function. The Walsh functions

$$\operatorname{Wal}(k,t) = \begin{cases} \operatorname{Cal}(k/2, t) & \text{for } k = 0,\ 2,\ 4,\ \ldots \\ \operatorname{Sal}((k+1)/2, t) & \text{for } k = 1,\ 3,\ 5,\ \ldots. \end{cases}$$

The Walsh functions $\operatorname{Cal}(k,t)$ for $k = 0, 1, \ldots, n/2-1$ and $\operatorname{Sal}(k,t)$ for $k = 1, 2, \ldots, n/2$ are given by the rows of the HADAMARD MATRIX H_n.

see also HADAMARD MATRIX, SEQUENCY

References

Beauchamp, K. G. *Walsh Functions and Their Applications.* London: Academic Press, 1975.

Harmuth, H. F. "Applications of Walsh Functions in Communications." *IEEE Spectrum* **6**, 82–91, 1969.

Thompson, A. R.; Moran, J. M.; and Swenson, G. W. Jr. *Interferometry and Synthesis in Radio Astronomy.* New York: Wiley, p. 204, 1986.

Tzafestas, S. G. *Walsh Functions in Signal and Systems Analysis and Design.* New York: Van Nostrand Reinhold, 1985.

Walsh, J. L. "A Closed Set of Normal Orthogonal Functions." *Amer. J. Math.* **45**, 5–24, 1923.

Walsh Index

The statistical INDEX

$$P_W \equiv \frac{\sum \sqrt{q_0 q_n}\, p_n}{\sum \sqrt{q_0 q_n}\, p_0},$$

where p_n is the price per unit in period n and q_n is the quantity produced in period n.

see also INDEX

References

Kenney, J. F. and Keeping, E. S. *Mathematics of Statistics, Pt. 1, 3rd ed.* Princeton, NJ: Van Nostrand, p. 66, 1962.

Wang's Conjecture

Wang's conjecture states that if a set of tiles can tile the plane, then they can always be arranged to do so periodically (Wang 1961). The CONJECTURE was refuted when Berger (1966) showed that an aperiodic set of tiles existed. Berger used 20,426 tiles, but the number has subsequently been greatly reduced.

see also TILING

References

Adler, A. and Holroyd, F. C. "Some Results on One-Dimensional Tilings." *Geom. Dedicata* **10**, 49–58, 1981.

Berger, R. "The Undecidability of the Domino Problem." *Mem. Amer. Math. Soc. No.* **66**, 1–72, 1966.

Grünbaum, B. and Sheppard, G. C. *Tilings and Patterns.* New York: W. H. Freeman, 1986.

Hanf, W. "Nonrecursive Tilings of the Plane. I." *J. Symbolic Logic* **39**, 283–285, 1974.

Mozes, S. "Tilings, Substitution Systems, and Dynamical Systems Generated by Them." *J. Analyse Math.* **53**, 139–186, 1989.

Myers, D. "Nonrecursive Tilings of the Plane. II." *J. Symbolic Logic* **39**, 286–294, 1974.

Robinson, R. M. "Undecidability and Nonperiodicity for Tilings of the Plane." *Invent. Math.* **12**, 177–209, 1971.

Wang, H. *Bell Systems Tech. J.* **40**, 1–41, 1961.

Ward's Primality Test

Let N be an ODD INTEGER, and assume there exists a LUCAS SEQUENCE $\{U_n\}$ with associated SYLVESTER CYCLOTOMIC NUMBERS $\{Q_n\}$ such that there is an $n > \sqrt{N}$ (with n and N RELATIVELY PRIME) for which N DIVIDES Q_n. Then N is a PRIME unless it has one of the following two forms:

1. $N = (n-1)^2$, with $n-1$ PRIME and $n > 4$, or
2. $N = n^2 - 1$, with $n-1$ and $n+1$ PRIME.

see also LUCAS SEQUENCE, SYLVESTER CYCLOTOMIC NUMBER

References

Ribenboim, P. *The Book of Prime Number Records, 2nd ed.* New York: Springer-Verlag, pp. 69–70, 1989.

Waring's Conjecture

see WARING'S PRIME CONJECTURE, WARING'S SUM CONJECTURE

Waring Formula

$$A^n + B^n = \sum_{j=0}^{\lfloor n/2 \rfloor} (-1)^j \frac{n}{n-j} \binom{n-j}{j} (AB)^j (A+B)^{n-2j},$$

where $\lfloor x \rfloor$ is the FLOOR FUNCTION and $\binom{n}{k}$ is a BINOMIAL COEFFICIENT.

see also FERMAT'S LAST THEOREM

Waring's Prime Conjecture

Every ODD INTEGER is a PRIME or the sum of three PRIMES.

Waring's Problem

Waring proposed a generalization of LAGRANGE'S FOUR-SQUARE THEOREM, stating that every RATIONAL INTEGER is the sum of a fixed number $g(n)$ of nth POWERS of INTEGERS, where n is any given POSITIVE INTEGER and $g(n)$ depends only on n. Waring originally speculated that $g(2) = 4$, $g(3) = 9$, and $g(4) = 19$. In 1909, Hilbert proved the general conjecture using an identity in 25-fold multiple integrals (Rademacher and Toeplitz 1957, pp. 52–61).

In LAGRANGE'S FOUR-SQUARE THEOREM, Lagrange proved that $g(2) = 4$, where 4 may be reduced to 3 except for numbers of the form $4^n(8k+7)$ (as proved by Legendre). In the early twentieth century, Dickson, Pillai, and Niven proved that $g(3) = 9$. Hilbert, Hardy, and Vinogradov proved $g(4) \leq 21$, and this was subsequently reduced to $g(4) = 19$ by Balasubramanian *et al.* (1986). Liouville proved (using LAGRANGE'S FOUR-SQUARE THEOREM and LIOUVILLE POLYNOMIAL IDENTITY) that $g(5) \leq 53$, and this was improved to 47, 45, 41, 39, 38, and finally $g(5) \leq 37$ by Wieferich. See Rademacher and Toeplitz (1957, p. 56) for a simple proof. J.-J. Chen (1964) proved that $g(5) = 37$.

Dickson, Pillai, and Niven also conjectured an explicit formula for $g(s)$ for $s > 6$ (Bell 1945), based on the relationship

$$\left(\frac{3}{2}\right)^n - \left\lfloor \left(\frac{3}{2}\right)^n \right\rfloor = 1 - \left(\frac{1}{2}\right)^n \left\{ \left\lfloor \left(\frac{3}{2}\right)^n + 2 \right\rfloor \right\}. \quad (1)$$

If the DIOPHANTINE (i.e., n is restricted to being an INTEGER) inequality

$$\left\{\left(\frac{3}{2}\right)^n\right\} \leq 1-\left(\frac{3}{4}\right)^n \tag{2}$$

is true, then

$$g(n)=2^n+\left\lfloor\left(\frac{3}{2}\right)^n\right\rfloor-2. \tag{3}$$

This was given as a lower bound by Euler, and has been verified to be correct for $6 \leq n \leq 200,000$. Since 1957, it has been known that at most a FINITE number of k exceed Euler's lower bound.

There is also a related problem of finding the least INTEGER n such that every POSITIVE INTEGER beyond a certain point (i.e., all but a FINITE number) is the SUM of $G(n)$ nth POWERS. From 1920–1928, Hardy and Littlewood showed that

$$G(n) \leq (n-2)2^{n-1}+5 \tag{4}$$

and conjectured that

$$G(k)<\begin{cases} 2k+1 & \text{for } k \text{ not a power of 2} \\ 4k & \text{for } k \text{ a power of 2.}\end{cases} \tag{5}$$

The best currently known bound is

$$G(k)<ck\ln k \tag{6}$$

for some constant c. Heilbronn (1936) improved Vinogradov's results to obtain

$$G(n) \leq 6n\ln n+\left[4+3\ln\left(3+\frac{2}{n}\right)\right]n+3. \tag{7}$$

It has long been known that $G(2)=4$. Dickson and Landau proved that the only INTEGERS requiring nine CUBES are 23 and 239, thus establishing $G(3) \leq 8$. Wieferich proved that only 15 INTEGERS require eight CUBES: 15, 22, 50, 114, 167, 175, 186, 212, 213, 238, 303, 364, 420, 428, and 454, establishing $G(3) \leq 7$. The largest number known requiring seven CUBES is 8042. In 1933, Hardy and Littlewood showed that $G(4) \leq 19$, but this was improved in 1936 to 16 or 17, and shown to be exactly 16 by Davenport (1939b). Vaughan (1986) greatly improved on the method of Hardy and Littlewood, obtaining improved results for $n \geq 5$. These results were then further improved by Brüdern (1990), who gave $G(5) \leq 18$, and Wooley (1992), who gave $G(n)$ for $n=6$ to 20. Vaughan and Wooley (1993) showed $G(8) \leq 42$.

Let $G^+(n)$ denote the smallest number such that *almost all* sufficiently large INTEGERS are the sum of $G^+(n)$ nth POWERS. Then $G^+(3)=4$ (Davenport 1939a), $G^+(4)=15$ (Hardy and Littlewood 1925), $G^+(8)=32$ (Vaughan 1986), and $G^+(16)=64$ (Wooley 1992). If the negatives of POWERS are permitted in addition to the powers themselves, the largest number of nth POWERS needed to represent an aribtrary integer are denoted $eg(n)$ and $EG(n)$ (Wright 1934, Hunter 1941, Gardner 1986). In general, these values are much harder to calculate than are $g(n)$ and $G(n)$.

The following table gives $g(n)$, $G(n)$, $G^+(n)$, $eg(n)$, and $EG(n)$ for $n \leq 20$. The sequence of $g(n)$ is Sloane's A002804.

n	$g(n)$	$G(n)$	$G^+(n)$	$eg(n)$	$EG(n)$
2	4	4		3	3
3	9	≤ 7	≤ 4	[4, 5]	
4	19	16	≤ 15	[9, 10]	
5	37	≤ 18			
6	73	≤ 27			
7	143	≤ 36			
8	279	≤ 42	≤ 32		
9	548	≤ 55			
10	1079	≤ 63			
11	2132	≤ 70			
12	4223	≤ 79			
13	8384	≤ 87			
14	16673	≤ 95			
15	33203	≤ 103			
16	66190	≤ 112	≤ 64		
17	132055	≤ 120			
18	263619	≤ 129			
19	526502	≤ 138			
20	1051899	≤ 146			

see also EULER'S CONJECTURE, SCHNIRELMANN'S THEOREM, VINOGRADOV'S THEOREM

References

Balasubramanian, R.; Deshouillers, J.-M.; and Dress, F. "Problème de Waring por les bicarrés 1, 2." *C. R. Acad. Sci. Paris Sér. I Math.* **303**, 85–88 and 161–163, 1986.

Bell, E. T. *The Development of Mathematics, 2nd ed.* New York: McGraw-Hill, p. 318, 1945.

Brüdern, J. "On Waring's Problem for Fifth Powers and Some Related Topics." *Proc. London Math. Soc.* **61**, 457–479, 1990.

Davenport, H. "On Waring's Problem for Cubes." *Acta Math.* **71**, 123–143, 1939a.

Davenport, H. "On Waring's Problem for Fourth Powers." *Ann. Math.* **40**, 731–747, 1939b.

Dickson, L. E. "Waring's Problem and Related Results." Ch. 25 in *History of the Theory of Numbers, Vol. 2: Diophantine Analysis.* New York: Chelsea, pp. 717–729, 1952.

Gardner, M. "Waring's Problems." Ch. 18 in *Knotted Doughnuts and Other Mathematical Entertainments.* New York: W. H. Freeman, 1986.

Guy, R. K. "Sums of Squares." §C20 in *Unsolved Problems in Number Theory, 2nd ed.* New York: Springer-Verlag, pp. 136–138, 1994.

Hardy, G. H. and Littlewood, J. E. "Some Problems of Partitio Numerorum (VI): Further Researches in Waring's Problem." *Math. Z.* **23**, 1–37, 1925.

Hunter, W. "The Representation of Numbers by Sums of Fourth Powers." *J. London Math. Soc.* **16**, 177–179, 1941.

Khinchin, A. Y. "An Elementary Solution of Waring's Problem." Ch. 3 in *Three Pearls of Number Theory.* New York: Dover, pp. 37–64, 1998.

Rademacher, H. and Toeplitz, O. *The Enjoyment of Mathematics: Selections from Mathematics for the Amateur.* Princeton, NJ: Princeton University Press, 1957.
Stewart, I. "The Waring Experience." *Nature* **323**, 674, 1986.
Vaughan, R. C. "On Waring's Problem for Smaller Exponents." *Proc. London Math. Soc.* **52**, 445–463, 1986.
Vaughan, R. C. and Wooley, T. D. "On Waring's Problem: Some Refinements." *Proc. London Math. Soc.* **63**, 35–68, 1991.
Vaughan, R. C. and Wooley, T. D. "Further Improvements in Waring's Problem." *Phil. Trans. Roy. Soc. London A* **345**, 363–376, 1993a.
Vaughan, R. C. and Wooley, T. D. "Further Improvements in Waring's Problem III. Eighth Powers." *Phil. Trans. Roy. Soc. London A* **345**, 385–396, 1993b.
Wooley, T. D. "Large Improvements in Waring's Problem." *Ann. Math.* **135**, 131–164, 1992.
Wright, E. M. "An Easier Waring's Problem." *J. London Math. Soc.* **9**, 267–272, 1934.

Waring's Sum Conjecture

see WARING'S PROBLEM

Waring's Theorem

If each of two curves meets the LINE AT INFINITY in distinct, nonsingular points, and if all their intersections are finite, then if to each common point there is attached a weight equal to the number of intersections absorbed therein, the CENTER OF MASS of these points is the center of gravity of the intersections of the asymptotes.

References

Coolidge, J. L. *A Treatise on Algebraic Plane Curves.* New York: Dover, p. 166, 1959.

Watchman Theorem

see ART GALLERY THEOREM

Watson's Formula

Let $J_\nu(z)$ be a BESSEL FUNCTION OF THE FIRST KIND, $Y_\nu(z)$ a BESSEL FUNCTION OF THE SECOND KIND, and $K_\nu(z)$ a MODIFIED BESSEL FUNCTION OF THE FIRST KIND. Also let $\Re[z] > 0$ and require $\Re[\mu - \nu] < 1$. Then

$$J_\mu(z)Y_\nu(z) - J_\nu(z)Y_\mu(z) = \frac{4\sin[(\mu-\nu)\pi]}{\pi^2}\int_0^\infty K_{\nu-\mu}(2z\sinh t)e^{-(\mu+\nu)t}\,dt.$$

The fourth edition of Gradshteyn and Ryzhik (1979), Iyanaga and Kawada (1980), and Ito (1987) erroneously give the exponential with a PLUS SIGN. A related integral is given by

$$J_\nu(z)\frac{\partial Y_\nu(z)}{\partial\nu} - Y_\nu(z)\frac{\partial J_\nu(z)}{\partial\nu} = -\frac{4}{\pi}\int_0^\infty K_0(2z\sinh t)e^{-2\nu t}\,dt$$

for $\Re[z] > 0$.

see also DIXON-FERRAR FORMULA, NICHOLSON'S FORMULA

References

Gradshteyn, I. S. and Ryzhik, I. M. Eqns. 6.617.1 and 6.617.2 in *Tables of Integrals, Series, and Products, 5th ed.* San Diego, CA: Academic Press, p. 710, 1979.
Ito, K. (Ed.). *Encyclopedic Dictionary of Mathematics, 2nd ed.* Cambridge, MA: MIT Press, p. 1806, 1987.
Iyanaga, S. and Kawada, Y. (Eds.). *Encyclopedic Dictionary of Mathematics.* Cambridge, MA: MIT Press, p. 1476, 1980.

Watson-Nicholson Formula

Let $H_\nu^{(\iota)}$ be a HANKEL FUNCTION OF THE FIRST or SECOND KIND, let $x, \nu > 0$, and define

$$w = \sqrt{\left(\frac{x}{\nu}\right)^2 - 1}.$$

Then

$$H_\nu^{(\iota)}(x) = 3^{-1/2}w\exp\{(-1)^{\iota+1}i[\pi/6 + \nu(w - \tfrac{1}{3}w^3 - \tan^{-1}w)]\}H_{1/3}^{(\iota)}(\tfrac{1}{3}\nu w) + \mathcal{O}|\nu^{-1}|.$$

References

Iyanaga, S. and Kawada, Y. (Eds.). *Encyclopedic Dictionary of Mathematics.* Cambridge, MA: MIT Press, p. 1475, 1980.

Watson Quintuple Product Identity

see QUINTUPLE PRODUCT IDENTITY

Watson's Theorem

$$ {}_3F_2\left[\begin{matrix} a, b, c \\ \frac{1}{2}(a+b+1), c\end{matrix}\right] = \frac{\Gamma(\frac{1}{2})\Gamma(\frac{1}{2}+c)\Gamma[\frac{1}{2}(1+a+b)]\Gamma(\frac{1}{2}-\frac{1}{2}a-\frac{1}{2}b+c)}{\Gamma[\frac{1}{2}(1+a)]\Gamma[\frac{1}{2}(1+b)]\Gamma(\frac{1}{2}-\frac{1}{2}a+c)\Gamma(\frac{1}{2}-\frac{1}{2}b+c)},$$

where ${}_3F_2(a,b,c;d,e;z)$ is a GENERALIZED HYPERGEOMETRIC FUNCTION and $\Gamma(z)$ is the GAMMA FUNCTION.

Watt's Curve

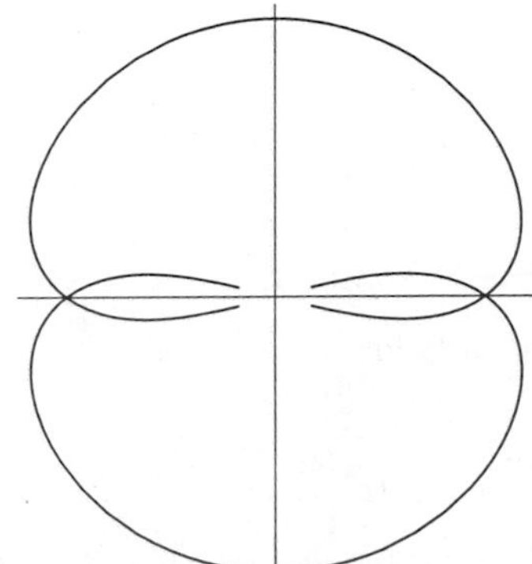

A curve named after James Watt (1736–1819), the Scottish engineer who developed the steam engine (MacTutor Archive). The curve is produced by a LINKAGE of rods connecting two wheels of equal diameter. Let the two wheels have RADIUS b and let their centers be located a distance $2a$ apart. Further suppose that a rod of length $2c$ is fixed at each end to the CIRCUMFERENCE of the two wheels. Let P be the MIDPOINT of the rod. Then Watt's curve C is the LOCUS of P.

The POLAR equation of Watt's curve is

$$r^2 = b^2 - (a \sin\theta \pm \sqrt{c^2 - a^2 \cos^2\theta}\,)^2.$$

If $a = c$, then C is a CIRCLE of RADIUS b with a figure of eight inside it.

References

Lockwood, E. H. *A Book of Curves.* Cambridge, England: Cambridge University Press, p. 162, 1967.

MacTutor History of Mathematics Archive. "Watt's Curve." `http://www-groups.dcs.st-and.ac.uk/~history/Curves/Watts.html`.

Watt's Parallelogram

A LINKAGE used in the original steam engine to turn back-and-forth motion into approximately straight-line motion.

see also LINKAGE

References

Rademacher, H. and Toeplitz, O. *The Enjoyment of Mathematics: Selections from Mathematics for the Amateur.* Princeton, NJ: Princeton University Press, pp. 119–121, 1957.

Wave

A 4-POLYHEX.

References

Gardner, M. *Mathematical Magic Show: More Puzzles, Games, Diversions, Illusions and Other Mathematical Sleight-of-Mind from Scientific American.* New York: Vintage, p. 147, 1978.

Wave Equation

The wave equation is

$$\nabla^2\psi = \frac{1}{v^2}\frac{\partial^2\psi}{\partial t^2}, \tag{1}$$

where ∇^2 is the LAPLACIAN.

The 1-D wave equation is

$$\frac{\partial^2\psi}{\partial x^2} = \frac{1}{v^2}\frac{\partial^2\psi}{\partial t^2}. \tag{2}$$

In order to specify a wave, the equation is subject to boundary conditions

$$\psi(0,t) = 0 \tag{3}$$

$$\psi(L,t) = 0, \tag{4}$$

and initial conditions

$$\psi(x,0) = f(x) \tag{5}$$

$$\frac{\partial\psi}{\partial t}(x,0) = g(x). \tag{6}$$

The wave equation can be solved using the so-called d'Alembert's solution, a FOURIER TRANSFORM method, or SEPARATION OF VARIABLES.

d'Alembert devised his solution in 1746, and Euler subsequently expanded the method in 1748. Let

$$\xi \equiv x - at \tag{7}$$

$$\eta \equiv x + at. \tag{8}$$

By the CHAIN RULE,

$$\frac{\partial^2\psi}{\partial x^2} = \frac{\partial^2\psi}{\partial\xi^2} + 2\frac{\partial^2\psi}{\partial\xi\partial\eta} + \frac{\partial 2\psi}{\partial\eta^2} \tag{9}$$

$$\frac{1}{v^2}\frac{\partial^2\psi}{\partial t^2} = \frac{\partial^2\psi}{\partial\xi^2} - 2\frac{\partial^2\psi}{\partial\xi\partial\eta} + \frac{\partial^2\psi}{\partial\eta^2}. \tag{10}$$

The wave equation then becomes

$$\frac{\partial^2\psi}{\partial\xi\partial\eta} = 0. \tag{11}$$

Any solution of this equation is of the form

$$\psi(\xi,\eta) = f(\eta) + g(\xi) = f(x + vt) + g(x - vt), \tag{12}$$

where f and g are *any* functions. They represent two waveforms traveling in opposite directions, f in the NEGATIVE x direction and g in the POSITIVE x direction.

The 1-D wave equation can also be solved by applying a FOURIER TRANSFORM to each side,

$$\int_{-\infty}^{\infty} \frac{\partial^2\psi(x,t)}{\partial x^2} e^{-2\pi ikx}\,dx = \frac{1}{v^2}\int_{-\infty}^{\infty} \frac{\partial^2\psi(x,t)}{\partial t^2} e^{-2\pi ikx}\,dx, \tag{13}$$

which is given, with the help of the FOURIER TRANSFORM DERIVATIVE identity, by

$$(2\pi ik)^2\Psi(k,t) = \frac{1}{v^2}\frac{\partial^2\Psi(k,t)}{\partial t^2}, \tag{14}$$

where

$$\Psi(k,t) \equiv \mathcal{F}[\psi(x,t)] = \int_{-\infty}^{\infty} \psi(x,t)e^{-2\pi ikx}\,dx. \tag{15}$$

This has solution

$$\Psi(k,t) = A(k)e^{2\pi ikvt} + B(k)e^{-2\pi ikvt}. \quad (16)$$

Taking the inverse FOURIER TRANSFORM gives

$$\begin{aligned}\psi(x,t) &\equiv \int_{-\infty}^{\infty} \Psi(k,t)e^{2\pi ikx}\,dx \\ &= \int_{-\infty}^{\infty} [A(k)e^{2\pi ikvt} \\ &\quad + B(k)e^{-2\pi ikvt}]e^{-2\pi ikx}\,dk \\ &= \int_{-\infty}^{\infty} A(k)e^{-2\pi ik(x-vt)}\,dk \\ &\quad + \int_{-\infty}^{\infty} B(k)e^{-2\pi ik(x+vt)}\,dk \\ &= f_1(x-vt) + b(k)f_2(x+vt), \end{aligned} \quad (17)$$

where

$$f_1(u) \equiv \mathcal{F}[A(k)] = \int_{-\infty}^{\infty} A(k)e^{-2\pi iku}\,dk \quad (18)$$

$$f_2(u) \equiv \mathcal{F}[B(k)] = \int_{-\infty}^{\infty} B(k)e^{-2\pi iku}\,dk. \quad (19)$$

This solution is still subject to all other initial and boundary conditions.

The 1-D wave equation can be solved by SEPARATION OF VARIABLES using a trial solution

$$\psi(x,t) = X(x)T(t). \quad (20)$$

This gives

$$T\frac{d^2X}{dx^2} = \frac{1}{v^2}X\frac{d^2T}{dt^2} \quad (21)$$

$$\frac{1}{X}\frac{d^2X}{dx^2} = \frac{1}{v^2}\frac{1}{T}\frac{d^2T}{dt^2} = -k^2. \quad (22)$$

So the solution for X is

$$X(x) = C\cos(kx) + D\sin(kx). \quad (23)$$

Rewriting (22) gives

$$\frac{1}{T}\frac{d^2T}{dt^2} = -v^2k^2 \equiv -\omega^2, \quad (24)$$

so the solution for T is

$$T(t) = E\cos(\omega t) + F\sin(\omega t), \quad (25)$$

where $v \equiv \omega/k$. Applying the boundary conditions $\psi(0,t) = \psi(L,t) = 0$ to (23) gives

$$C = 0 \qquad kL = m\pi, \quad (26)$$

where m is an INTEGER. Plugging (23), (25) and (26) back in for ψ in (21) gives, for a particular value of m,

$$\begin{aligned}\psi_m(x,t) &= [E_m\sin(\omega_m t) + F_m\cos(\omega_m t)]D_m\sin\left(\frac{m\pi x}{L}\right) \\ &\equiv [A_m\cos(\omega_m t) + B_m\sin(\omega_m t)]\sin\left(\frac{m\pi x}{L}\right). \end{aligned} \quad (27)$$

The initial condition $\dot\psi(x,0) = 0$ then gives $B_m = 0$, so (27) becomes

$$\psi_m(x,t) = A_m\cos(\omega_m t)\sin\left(\frac{m\pi x}{L}\right). \quad (28)$$

The general solution is a sum over all possible values of m, so

$$\psi(x,t) = \sum_{m=1}^{\infty} A_m\cos(\omega_m t)\sin\left(\frac{m\pi x}{L}\right). \quad (29)$$

Using ORTHOGONALITY of sines again,

$$\int_0^L \sin\left(\frac{l\pi x}{L}\right)\sin\left(\frac{m\pi x}{L}\right)dx = \tfrac{1}{2}L\delta_{lm}, \quad (30)$$

where δ_{lm} is the KRONECKER DELTA defined by

$$\delta_{mn} \equiv \begin{cases} 1 & m = n \\ 0 & m \neq n \end{cases}, \quad (31)$$

gives

$$\begin{aligned}\int_0^L \psi(x,0)\sin\left(\frac{m\pi x}{L}\right)dx &= \sum_{l=1}^{\infty} A_l\sin\left(\frac{l\pi x}{L}\right)\sin\left(\frac{m\pi x}{L}\right)dx \\ &= \sum_{l=1}^{\infty} A_l\tfrac{1}{2}L\delta_{lm} = \tfrac{1}{2}LA_m, \end{aligned} \quad (32)$$

so we have

$$A_m = \frac{2}{L}\int_0^L \psi(x,0)\sin\left(\frac{m\pi x}{L}\right)dx. \quad (33)$$

The computation of A_ms for specific initial distortions is derived in the FOURIER SINE SERIES section. We already have found that $B_m = 0$, so the equation of motion for the string (29), with

$$\omega_m \equiv vk_m = \frac{vm\pi}{L}, \quad (34)$$

is

$$\psi(x,t) = \sum_{m=1}^{\infty} A_m\cos\left(\frac{vm\pi t}{L}\right)\sin\left(\frac{m\pi x}{L}\right), \quad (35)$$

where the A_m COEFFICIENTS are given by (33).

A damped 1-D wave

$$\frac{\partial^2 \psi}{\partial x^2} = \frac{1}{v^2}\frac{\partial^2 \psi}{\partial t^2} + b\frac{\partial \psi}{\partial t}, \quad (36)$$

given boundary conditions

$$\psi(0,t) = 0 \quad (37)$$
$$\psi(L,t) = 0, \quad (38)$$

initial conditions

$$\psi(x,0) = f(x) \quad (39)$$
$$\frac{\partial \psi}{\partial t}(x,0) = g(x), \quad (40)$$

and the additional constraint

$$0 < b < \frac{2\pi}{Lv}, \quad (41)$$

can also be solved as a FOURIER SERIES.

$$\psi(x,t) = \sum_{n=1}^{\infty} \sin\left(\frac{n\pi x}{L}\right) e^{-v^2bt/2}[a_n \sin(\mu_n t) + b_n \cos(\mu_n t)], \quad (42)$$

where

$$\mu_n \equiv \frac{\sqrt{4v^2n^2\pi^2 - b^2L^2v^4}}{2L} = \frac{v\sqrt{4n^2\pi^2 - b^2L^2v^2}}{2L} \quad (43)$$
$$b_n = \frac{2}{L}\int_0^L \sin\left(\frac{n\pi x}{L}\right) f(x)\,dx \quad (44)$$
$$a_n = \frac{2}{L\mu_n}\left\{\int_0^L \sin\left(\frac{n\pi x}{L}\right)\left[g(x) + \frac{v^2 b}{2} f(x)\right]\right\} dx. \quad (45)$$

To find the motion of a rectangular membrane with sides of length L_x and L_y (in the absence of gravity), use the 2-D wave equation

$$\frac{\partial^2 z}{\partial x^2} + \frac{\partial^2 z}{\partial y^2} = \frac{1}{v^2}\frac{\partial^2 z}{\partial t^2}, \quad (46)$$

where $z(x,y,t)$ is the vertical displacement of a point on the membrane at position (x,y) and time t. Use SEPARATION OF VARIABLES to look for solutions of the form

$$z(x,y,t) = X(x)Y(y)T(t). \quad (47)$$

Plugging (47) into (46) gives

$$YT\frac{d^2X}{dx^2} + XT\frac{d^2Y}{dy^2} = \frac{1}{v^2}XY\frac{d^2T}{dt^2}, \quad (48)$$

where the partial derivatives have now become complete derivatives. Multiplying (48) by v^2/XYT gives

$$\frac{v^2}{X}\frac{d^2X}{dx^2} + \frac{v^2}{Y}\frac{d^2Y}{dy^2} = \frac{1}{T}\frac{d^2T}{dt^2}. \quad (49)$$

The left and right sides must both be equal to a constant, so we can separate the equation by writing the right side as

$$\frac{1}{T}\frac{d^2T}{dt^2} = -\omega^2. \quad (50)$$

This has solution

$$T(t) = C_\omega \cos(\omega t) + D_\omega \sin(\omega t). \quad (51)$$

Plugging (50) back into (49),

$$\frac{v^2}{X}\frac{d^2X}{dx^2} + \frac{v^2}{Y}\frac{d^2Y}{dy^2} = -\omega^2, \quad (52)$$

which we can rewrite as

$$\frac{1}{X}\frac{d^2X}{dx^2} = -\frac{1}{Y}\frac{d^2Y}{dy^2} - \frac{\omega^2}{v^2} = -k_x{}^2 \quad (53)$$

since the left and right sides again must both be equal to a constant. We can now separate out the $Y(y)$ equation

$$\frac{1}{Y}\frac{d^2Y}{dy^2} = k_x{}^2 - \frac{\omega^2}{v^2} \equiv -k_y{}^2, \quad (54)$$

where we have defined a new constant k_y satisfying

$$k_x{}^2 + k_y{}^2 = \frac{\omega^2}{v^2}. \quad (55)$$

Equations (53) and (54) have solutions

$$X(x) = E\cos(k_x x) + F\sin(k_x x) \quad (56)$$
$$Y(y) = G\cos(k_y y) + H\sin(k_y y). \quad (57)$$

We now apply the boundary conditions to (56) and (57). The conditions $z(0,y,t) = 0$ and $z(x,0,t) = 0$ mean that

$$E = 0 \qquad G = 0. \quad (58)$$

Similarly, the conditions $z(L_x,y,t) = 0$ and $z(x,L_y,t) = 0$ give $\sin(k_x L_x) = 0$ and $\sin(k_y L_y) = 0$, so $L_x k_x = p\pi$ and $L_y k_y = q\pi$, where p and q are INTEGERS. Solving for the allowed values of k_x and k_y then gives

$$k_x = \frac{p\pi}{L_x} \qquad k_y = \frac{q\pi}{L_y}. \quad (59)$$

Plugging (52), (56), (57), (58), and (59) back into (22) gives the solution for particular values of p and q,

$$z_{pq}(x,y,t) = [C_\omega \cos(\omega t) + D_\omega \sin(\omega t)] \times \left[F_p \sin\left(\frac{p\pi x}{L_x}\right)\right]\left[H_q \sin\left(\frac{q\pi y}{L_y}\right)\right]. \quad (60)$$

Lumping the constants together by writing $A_{pq} \equiv C_\omega F_p H_q$ (we can do this since ω is a function of p and q, so C_ω can be written as C_{pq}) and $B_{pq} \equiv D_\omega F_p H_q$, we obtain

$$z_{pq}(x,y,t) = [A_{pq}\cos(\omega_{pq}t) + B_{pq}\sin(\omega_{pq}t)] \times \sin\left(\frac{p\pi x}{L_x}\right)\sin\left(\frac{q\pi y}{L_y}\right). \quad (61)$$

Plots of the spatial part for modes (1, 1), (1, 2), (2, 1), and (2, 2) follow.

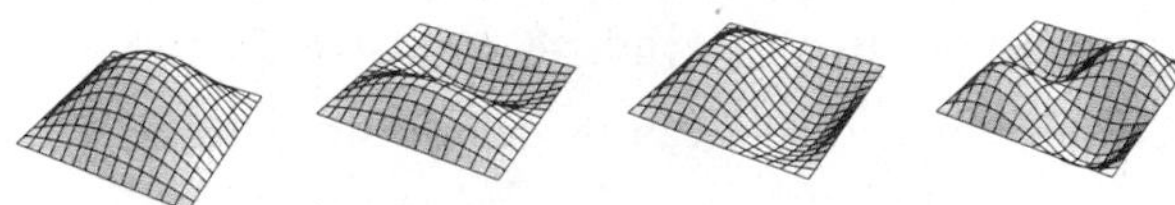

The general solution is a sum over all possible values of p and q, so the final solution is

$$z(x,y,t) = \sum_{p=1}^{\infty}\sum_{q=1}^{\infty}[A_{pq}\cos(\omega_{pq}t) + B_{pq}\sin(\omega_{pq}t)] \times \sin\left(\frac{p\pi x}{L_x}\right)\sin\left(\frac{q\pi y}{L_y}\right), \quad (62)$$

where ω is defined by combining (55) and (59) to yield

$$\omega_{pq} \equiv \pi v\sqrt{\left(\frac{p}{L_x}\right)^2 + \left(\frac{q}{L_y}\right)^2}. \quad (63)$$

Given the initial conditions $z(x,y,0)$ and $\frac{\partial z}{\partial t}(x,y,0)$, we can compute the A_{pq}s and B_{pq}s explicitly. To accomplish this, we make use of the orthogonality of the SINE function in the form

$$I \equiv \int_0^L \sin\left(\frac{m\pi x}{L}\right)\sin\left(\frac{n\pi x}{L}\right)dx = \tfrac{1}{2}L\delta_{mn}, \quad (64)$$

where δ_{mn} is the KRONECKER DELTA. This can be demonstrated by direct INTEGRATION. Let $u \equiv \pi x/L$ so $du = (\pi/L)\,dx$ in (64), then

$$I = \frac{L}{\pi}\int_0^\pi \sin(mu)\sin(nu)\,du. \quad (65)$$

Now use the trigonometric identity

$$\sin\alpha\sin\beta = \tfrac{1}{2}[\cos(\alpha-\beta) - \cos(\alpha+\beta)] \quad (66)$$

to write

$$I = \frac{L}{2\pi}\int_0^\pi \cos[(m-n)u]\,du + \int_0^\pi \cos[(m+n)u]\,du. \quad (67)$$

Note that for an INTEGER $l \neq 0$, the following INTEGRAL vanishes

$$\int_0^\pi \cos(lu)\,du = \frac{1}{l}[\sin(lu)]_0^\pi = \frac{1}{l}[\sin(l\pi) - \sin 0] = \frac{1}{l}\sin(l\pi) = 0, \quad (68)$$

since $\sin(l\pi) = 0$ when l is an INTEGER. Therefore, $I = 0$ when $l \equiv m - n \neq 0$. However, I does *not* vanish when $l = 0$, since

$$\int_0^\pi \cos(0\cdot u)\,du = \int_0^\pi du = \pi. \quad (69)$$

We therefore have that $I = L\delta_{mn}/2$, so we have derived (64). Now we multiply $z(x,y,0)$ by two sine terms and integrate between 0 and L_x and between 0 and L_y,

$$I = \int_0^{L_y}\left[\int_0^{L_x} z(x,y,0)\sin\left(\frac{p\pi x}{L_x}\right)dx\right] \times \sin\left(\frac{q\pi y}{L_y}\right)dy. \quad (70)$$

Now plug in $z(x,y,t)$, set $t = 0$, and prime the indices to distinguish them from the p and q in (70),

$$I = \sum_{q'=1}^{\infty}\int_0^{L_y}\left[\sum_{p'=1}^{\infty} A_{p'q'}\int_0^{L_x}\sin\left(\frac{p\pi x}{L_x}\right)\sin\left(\frac{p'\pi x}{L_x}\right)dx\right] \times \sin\left(\frac{q\pi y}{L_y}\right)\sin\left(\frac{q'\pi y}{L_y}\right)dy. \quad (71)$$

Making use of (64) in (71),

$$I = \sum_{q'=1}^{\infty}\int_0^{L_y}\sum_{p'=1}^{\infty} A_{p'q'}\frac{L_x}{2}\delta_{p,p'} \times \sin\left(\frac{q\pi y}{L_y}\right)\sin\left(\frac{q'\pi y}{L_y}\right)dy, \quad (72)$$

so the sums over p' and q' collapse to a single term

$$I = \frac{L_x}{2}\sum_{q=1}^{\infty} A_{pq'}\frac{L_y}{2}\delta_{q,q'} = \frac{L_x L_y}{4}A_{pq}. \quad (73)$$

Equating (72) and (73) and solving for A_{pq} then gives

$$A_{pq} = \frac{4}{L_x L_y}\int_0^{L_y}\left[\int_0^{L_x} z(x,y,0)\sin\left(\frac{p\pi x}{L_x}\right)dx\right] \times \sin\left(\frac{q\pi y}{L_y}\right)dy. \quad (74)$$

An analogous derivation gives the B_{pq}s as

$$B_{pq} = \frac{4}{\omega_{pq} L_x L_y} \int_0^{L_y} \left[\int_0^{L_x} \frac{\partial z}{\partial t}(x, y, 0) \sin\left(\frac{p\pi x}{L_x}\right) dx \right] \times \sin\left(\frac{q\pi y}{L_y}\right) dy. \quad (75)$$

The equation of motion for a membrane shaped as a RIGHT ISOSCELES TRIANGLE of length c on a side and with the sides oriented along the POSITIVE x and y axes is given by

$$\psi(x, y, t) = [C_{pq} \cos(\omega_{pq} t) + D_{pq} \sin(\omega_{pq} t)] \times \left[\sin\left(\frac{p\pi x}{c}\right) \sin\left(\frac{q\pi y}{c}\right) - \sin\left(\frac{q\pi x}{c}\right) \sin\left(\frac{p\pi y}{c}\right)\right], \quad (76)$$

where

$$\omega_{pq} = \frac{\pi v}{c} \sqrt{p^2 + q^2} \quad (77)$$

and p, q INTEGERS with $p > q$. This solution can be obtained by subtracting two wave solutions for a square membrane with the indices reversed. Since points on the diagonal which are equidistant from the center must have the same wave equation solution (by symmetry), this procedure gives a wavefunction which will vanish along the diagonal *as long as p and q are both* EVEN *or* ODD. We must further restrict the modes since those with $p < q$ give wavefunctions which are just the NEGATIVE of (q, p) and (p, p) give an identically zero wavefunction. The following plots show (3, 1), (4, 2), (5, 1), and (5,3).

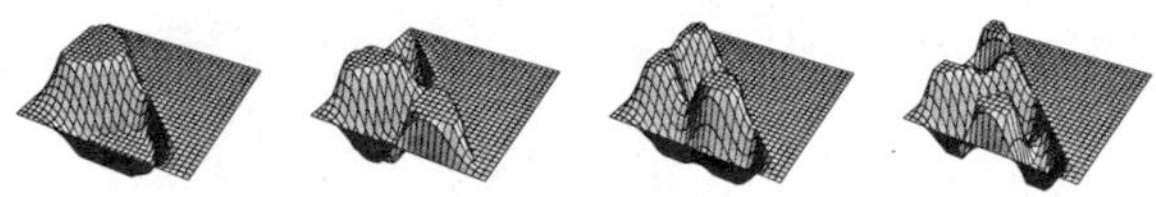

References

Abramowitz, M. and Stegun, C. A. (Eds.). "Wave Equation in Prolate and Oblate Spheroidal Coordinates." §21.5 in *Handbook of Mathematical Functions with Formulas, Graphs, and Mathematical Tables, 9th printing.* New York: Dover, pp. 752–753, 1972.

Morse, P. M. and Feshbach, H. *Methods of Theoretical Physics, Part I.* New York: McGraw-Hill, pp. 124–125, 1953.

Wave Operator

An OPERATOR relating the asymptotic state of a DYNAMICAL SYSTEM governed by the Schrödinger equation

$$i \frac{d}{dt} \psi(t) = H \psi(t)$$

to its original asymptotic state.

see also SCATTERING OPERATOR

Wave Surface

A SURFACE represented parametrically by ELLIPTIC FUNCTIONS.

Wavelet

Wavelets are a class of a functions used to localize a given function in both space and scaling. A family of wavelets can be constructed from a function $\psi(x)$, sometimes known as a "mother wavelet," which is confined in a finite interval. "Daughter wavelets" $\psi^{a,b}(x)$ are then formed by translation (b) and contraction (a). Wavelets are especially useful for compressing image data, since a WAVELET TRANSFORM has properties which are in some ways superior to a conventional FOURIER TRANSFORM.

An individual wavelet can be defined by

$$\psi^{a,b}(x) = |a|^{-1/2} \psi\left(\frac{x - b}{a}\right). \quad (1)$$

Then

$$W_\psi(f)(a, b) = \frac{1}{\sqrt{a}} \int_{-\infty}^{\infty} f(t) \psi\left(\frac{t - b}{a}\right) dt, \quad (2)$$

and CALDERÓN'S FORMULA gives

$$f(x) = C_\psi \int_{-\infty}^{\infty} \int_{-\infty}^{\infty} \left\langle f, \psi^{a,b} \right\rangle \psi^{a,b}(x) a^{-2} \, da \, db. \quad (3)$$

A common type of wavelet is defined using HAAR FUNCTIONS.

see also FOURIER TRANSFORM, HAAR FUNCTION, LEMARIÉ'S WAVELET, WAVELET TRANSFORM

References

Benedetto, J. J. and Frazier, M. (Eds.). *Wavelets: Mathematics and Applications.* Boca Raton, FL: CRC Press, 1994.

Chui, C. K. *An Introduction to Wavelets.* San Diego, CA: Academic Press, 1992.

Chui, C. K. (Ed.). *Wavelets: A Tutorial in Theory and Applications.* San Diego, CA: Academic Press, 1992.

Chui, C. K.; Montefusco, L.; and Puccio, L. (Eds.). *Wavelets: Theory, Algorithms, and Applications.* San Diego, CA: Academic Press, 1994.

Daubechies, I. *Ten Lectures on Wavelets.* Philadelphia, PA: Society for Industrial and Applied Mathematics, 1992.

Erlebacher, G. H.; Hussaini, M. Y.; and Jameson, L. M. (Eds.). *Wavelets: Theory and Applications.* New York: Oxford University Press, 1996.

Foufoula-Georgiou, E. and Kumar, P. (Eds.). *Wavelets in Geophysics.* San Diego, CA: Academic Press, 1994.

Hernández, E. and Weiss, G. *A First Course on Wavelets.* Boca Raton, FL: CRC Press, 1996.

Hubbard, B. B. *The World According to Wavelets: The Story of a Mathematical Technique in the Making.* New York: A. K. Peters, 1995.

Jawerth, B. and Sweldens, W. "An Overview of Wavelet Based Multiresolution Analysis." *SIAM Rev.* **36**, 377–412, 1994.

Kaiser, G. *A Friendly Guide to Wavelets.* Cambridge, MA: Birkhäuser, 1994.

Massopust, P. R. *Fractal Functions, Fractal Surfaces, and Wavelets.* San Diego, CA: Academic Press, 1994.

Meyer, Y. *Wavelets: Algorithms and Applications.* Philadelphia, PA: SIAM Press, 1993.
Press, W. H.; Flannery, B. P.; Teukolsky, S. A.; and Vetterling, W. T. "Wavelet Transforms." §13.10 in *Numerical Recipes in FORTRAN: The Art of Scientific Computing, 2nd ed.* Cambridge, England: Cambridge University Press, pp. 584–599, 1992.
Schumaker, L. L. and Webb, G. (Eds.). *Recent Advances in Wavelet Analysis.* San Diego, CA: Academic Press, 1993.
Stollnitz, E. J.; DeRose, T. D.; and Salesin, D. H. "Wavelets for Computer Graphics: A Primer, Part 1." *IEEE Computer Graphics and Appl.* **15**, No. 3, 76–84, 1995.
Stollnitz, E. J.; DeRose, T. D.; and Salesin, D. H. "Wavelets for Computer Graphics: A Primer, Part 2." *IEEE Computer Graphics and Appl.* **15**, No. 4, 75–85, 1995.
Strang, G. "Wavelets and Dilation Equations: A Brief Introduction." *SIAM Rev.* **31**, 614–627, 1989.
Strang, G. "Wavelets." *Amer. Sci.* **82**, 250–255, 1994.
Taswell, C. *Handbook of Wavelet Transform Algorithms.* Boston, MA: Birkhäuser, 1996.
Teolis, A. *Computational Signal Processing with Wavelets.* Boston, MA: Birkhäuser, 1997.
Walter, G. G. *Wavelets and Other Orthogonal Systems with Applications.* Boca Raton, FL: CRC Press, 1994.
"Wavelet Digest." `http://www.math.sc.edu/~wavelet/`.
Wickerhauser, M. V. *Adapted Wavelet Analysis from Theory to Software.* Wellesley, MA: Peters, 1994.

Wavelet Matrix

A MATRIX composed of HAAR FUNCTIONS which is used in the WAVELET TRANSFORM. The fourth-order wavelet matrix is given by

$$W_4 = \begin{bmatrix} 1 & 1 & 1 & 0 \\ 1 & 1 & -1 & 0 \\ 1 & -1 & 0 & 1 \\ 1 & -1 & 0 & -1 \end{bmatrix} = \begin{bmatrix} 1 & 1 & & \\ 1 & -1 & & \\ & & 1 & 1 \\ & & 1 & -1 \end{bmatrix} \begin{bmatrix} 1 & & & \\ & & 1 & \\ & 1 & & \\ & & & 0 \end{bmatrix} \times \begin{bmatrix} 1 & 1 & & \\ 1 & -1 & & \\ & & 1 & \\ & & & 1 \end{bmatrix}.$$

A wavelet matrix can be computed in $\mathcal{O}(n)$ steps, compared to $\mathcal{O}(n \lg 2)$ for the FOURIER MATRIX.

see also FOURIER MATRIX, WAVELET, WAVELET TRANSFORM

Wavelet Transform

A transform which localizes a function both in space and scaling and has some desirable properties compared to the FOURIER TRANSFORM. The transform is based on a WAVELET MATRIX, which can be computed more quickly than the analogous FOURIER MATRIX.

see also DAUBECHIES WAVELET FILTER, LEMARIE'S WAVELET

References

Blair, D. and MathSoft, Inc. "Wavelet Resources." `http://www.mathsoft.com/wavelets.html`.
Daubechies, I. *Ten Lectures on Wavelets.* Philadelphia, PA: SIAM, 1992.
DeVore, R.; Jawerth, B.; and Lucier, B. "Images Compression through Wavelet Transform Coding." *IEEE Trans. Information Th.* **38**, 719–746, 1992.
Press, W. H.; Flannery, B. P.; Teukolsky, S. A.; and Vetterling, W. T. "Wavelet Transforms." §13.10 in *Numerical Recipes in FORTRAN: The Art of Scientific Computing, 2nd ed.* Cambridge, England: Cambridge University Press, pp. 584–599, 1992.
Strang, G. "Wavelet Transforms Versus Fourier Transforms." *Bull. Amer. Math. Soc.* **28**, 288–305, 1993.

Weak Convergence

Weak convergence is usually either denoted $x_n \overset{w}{\to} x$ or $x_n \rightharpoonup x$. A SEQUENCE $\{x_n\}$ of VECTORS in an INNER PRODUCT SPACE E is called weakly convergent to a VECTOR in E if

$$\langle x_n, y\rangle \to \langle x, y\rangle \quad \text{as } n \to \infty, \quad \text{for all } y \in E.$$

Every STRONGLY CONVERGENT sequence is also weakly convergent (but the opposite does not usually hold). This can be seen as follows. Consider the sequence $\{x_n\}$ that converges strongly to x, i.e., $\|x_n - x\| \to 0$ as $n \to \infty$. SCHWARZ'S INEQUALITY now gives

$$|\langle x_n - x, y\rangle| \leq \|x_n - x\|\,\|y\| \quad \text{as } n \to \infty.$$

The definition of weak convergence is therefore satisfied.

see also INNER PRODUCT SPACE, SCHWARZ'S INEQUALITY, STRONG CONVERGENCE

Weak Law of Large Numbers

Also known as BERNOULLI'S THEOREM. Let $x_1, \ldots, x_n$ be a sequence of independent and identically distributed random variables, each having a MEAN $\langle xi\rangle = \mu$ and STANDARD DEVIATION σ. Define a new variable

$$x \equiv \frac{x_1 + \ldots + x_n}{n}. \tag{1}$$

Then, as $n \to \infty$, the sample mean $\langle x\rangle$ equals the population MEAN μ of each variable.

$$\langle x\rangle = \left\langle \frac{x_1 + \ldots + x_n}{n} \right\rangle = \frac{1}{n}(\langle x_1\rangle + \ldots + \langle x_n\rangle) = \frac{n\mu}{n} = \mu \tag{2}$$

$$\begin{aligned} \operatorname{var}(x) &= \operatorname{var}\left(\frac{x_1 + \ldots + x_2}{n}\right) \\ &= \operatorname{var}\left(\frac{x_1}{n}\right) + \ldots + \operatorname{var}\left(\frac{x_n}{n}\right) \\ &= \frac{\sigma^2}{n^2} + \ldots + \frac{\sigma^2}{n^2} = \frac{\sigma^2}{n}. \end{aligned} \tag{3}$$

Therefore, by the CHEBYSHEV INEQUALITY, for all $\epsilon > 0$,

$$P(|x - \mu| \geq \epsilon) \leq \frac{\operatorname{var}(x)}{\epsilon^2} = \frac{\sigma^2}{n\epsilon^2}. \tag{4}$$

As $n \to \infty$, it then follows that

$$\lim_{n\to\infty} P(|x-\mu| \geq \epsilon) = 0 \tag{5}$$

for ϵ arbitrarily small; i.e., as $n \to \infty$, the sample MEAN is the same as the population MEAN.

Stated another way, if an event occurs x times in s TRIALS and if p is the probability of success in a single TRIAL, then the probability that the relative frequency of successes is x/s differs from p by less than any arbitrary POSITIVE quantity ϵ which approaches 1 as $s \to \infty$.

see also LAW OF TRULY LARGE NUMBERS, STRONG LAW OF LARGE NUMBERS

Weakly Binary Tree

N.B. A detailed on-line essay by S. Finch was the starting point for this entry.

A ROOTED TREE for which the ROOT is adjacent to at most two VERTICES, and all nonroot VERTICES are adjacent to at most three VERTICES. Let $b(n)$ be the number of weakly binary trees of order n, then $b(5) = 6$. Let

$$g(z) = \sum_{i=0}^{\infty} g_i z^i, \tag{1}$$

where

$$g_0 = 0 \tag{2}$$

$$g_1 = g_2 = g_3 = 1 \tag{3}$$

$$g_{2i+1} = \sum_{j=1}^{i} g_{2i+1-j} g_j \tag{4}$$

$$g_{2i} = \tfrac{1}{2} g_i(g_i+1) + \sum_{j=1}^{i-1} g_{2i-j} g_j. \tag{5}$$

Otter (Otter 1948, Harary and Palmer 1973, Knuth 1969) showed that

$$\lim_{n\to\infty} \frac{b(n)n^{3/2}}{\xi^n} = \eta, \tag{6}$$

where

$$\xi = 2.48325\ldots \tag{7}$$

is the unique POSITIVE ROOT of

$$g\left(\frac{1}{x}\right) = 1,$$

and

$$\eta = 0.7916032\ldots. \tag{8}$$

ξ is also given by

$$\xi = \lim_{n\to\infty} (c_n)^{2^{-n}}, \tag{9}$$

where c_n is given by

$$c_0 = 2 \tag{10}$$

$$c_n = (c_{n-1})^2 + 2, \tag{11}$$

giving

$$\eta = \frac{1}{2}\sqrt{\frac{\xi}{\pi}}\sqrt{3 + \frac{1}{c_1} + \frac{1}{c_1 c_2} + \frac{1}{c_1 c_2 c_3} + \ldots}. \tag{12}$$

References

Finch, S. "Favorite Mathematical Constants." `http://www.mathsoft.com/asolve/constant/otter/otter.html`.

Harary, F. *Graph Theory.* Reading, MA: Addison-Wesley, 1969.

Harary, F. and Palmer, E. M. *Graphical Enumeration.* New York: Academic Press, 1973.

Knuth, D. E. *The Art of Computer Programming, Vol. 1: Fundamental Algorithms, 2nd ed.* Reading, MA: Addison-Wesley, 1973.

Otter, R. "The Number of Trees." *Ann. Math.* **49**, 583–599, 1948.

Weakly Complete Sequence

A SEQUENCE of numbers $V = \{\nu_n\}$ is said to be weakly complete if every POSITIVE INTEGER n beyond a certain point N is the sum of some SUBSEQUENCE of V (Honsberger 1985). Dropping two terms from the FIBONACCI NUMBERS produces a SEQUENCE which is not even weakly complete. However, the SEQUENCE

$$F'_n \equiv F_n - (-1)^n$$

is weakly complete, even with any finite subsequence deleted (Graham 1964).

see also COMPLETE SEQUENCE

References

Graham, R. "A Property of Fibonacci Numbers." *Fib. Quart.* **2**, 1–10, 1964.

Honsberger, R. *Mathematical Gems III.* Washington, DC: Math. Assoc. Amer., p. 128, 1985.

Weakly Independent

An infinite sequence $\{a_i\}$ of POSITIVE INTEGERS is called weakly independent if any relation $\sum \epsilon_i a_i$ with $\epsilon_i = 0$ or ± 1 and $\epsilon_i = 0$, except finitely often, IMPLIES $\epsilon_i = 0$ for all i.

see also STRONGLY INDEPENDENT

References

Guy, R. K. *Unsolved Problems in Number Theory, 2nd ed.* New York: Springer-Verlag, p. 136, 1994.

Weakly Triple-Free Set

see TRIPLE-FREE SET

Web Graph

A graph formed by connecting several concentric WHEEL GRAPHS along spokes.

see also WHEEL GRAPH

Weber Differential Equations

Consider the differential equation satisfied by

$$w = z^{-1/2} W_{k,-1/4}(\tfrac{1}{2}z^2), \tag{1}$$

where W is a WHITTAKER FUNCTION.

$$\frac{d}{z\,dz}\left[\frac{d(wz^{1/2})}{z\,dz}\right] + \left(-\frac{1}{4} + \frac{2k}{z^2} + \frac{3}{4z^4}\right) wz^{1/2} = 0 \tag{2}$$

$$\frac{d^2 w}{dz^2} + (2k - \tfrac{1}{4}z^2)w = 0. \tag{3}$$

This is usually rewritten

$$\frac{d^2 D_n(z)}{dz^2} + (n + \tfrac{1}{2} - \tfrac{1}{4}z^2)D_n(z) = 0. \tag{4}$$

The solutions are PARABOLIC CYLINDER FUNCTIONS.

The equations

$$\frac{d^2 U}{du^2} - (c + k^2 u^2)U = 0 \tag{5}$$

$$\frac{d^2 V}{dv^2} + (c - k^2 v^2)V = 0, \tag{6}$$

which arise by separating variables in LAPLACE'S EQUATION in PARABOLIC CYLINDRICAL COORDINATES, are also known as the Weber differential equations. As above, the solutions are known as PARABOLIC CYLINDER FUNCTIONS.

Weber's Discontinuous Integrals

$$\int_0^\infty J_0(ax)\cos(cx)\,dx = \begin{cases} 0 & a < c \\ \frac{1}{\sqrt{a^2-c^2}} & a > c \end{cases}$$

$$\int_0^\infty J_0(ax)\sin(cx)\,dx = \begin{cases} \frac{1}{\sqrt{c^2-a^2}} & a < c \\ 0 & a > c, \end{cases}$$

where $J_0(z)$ is a zeroth order BESSEL FUNCTION OF THE FIRST KIND.

References

Bowman, F. *Introduction to Bessel Functions.* New York: Dover, pp. 59–60, 1958.

Weber's Formula

$$\frac{1}{2p^2} e^{-(a^2+b^2)/(4p^2)} I_\nu\left(\frac{ab}{2p^2}\right) = \int_0^\infty e^{-p^2t^2} J_\nu(at)J_\nu(bt)t\,dt,$$

where $\Re[\nu] > -1$, $|\arg p| < \pi/4$, and a, $b > 0$, $J_\nu(z)$ is a BESSEL FUNCTION OF THE FIRST KIND, and $I_\nu(z)$ is a MODIFIED BESSEL FUNCTION OF THE FIRST KIND.

see also BESSEL FUNCTION OF THE FIRST KIND, MODIFIED BESSEL FUNCTION OF THE FIRST KIND

References

Iyanaga, S. and Kawada, Y. (Eds.). *Encyclopedic Dictionary of Mathematics.* Cambridge, MA: MIT Press, p. 1476, 1980.

Weber Functions

Although BESSEL FUNCTIONS OF THE SECOND KIND are sometimes called Weber functions, Abramowitz and Stegun (1972) define a separate Weber function as

$$\mathcal{E}_\nu(z) = \frac{1}{\pi}\int_0^\pi \sin(\nu\theta - z\sin\theta)\,d\theta. \tag{1}$$

Letting $\zeta_n = e^{2\pi i/m}$ be a ROOT OF UNITY, another set of Weber functions is defined as

$$f(z) = \frac{\eta(\frac{1}{2}(z+1))}{\zeta_{48}\eta(z)} \tag{2}$$

$$f_1(z) = \frac{\eta(\frac{1}{2}z)}{\eta(z)} \tag{3}$$

$$f_2(z) = \sqrt{2}\frac{\eta(2z)}{\eta(z)} \tag{4}$$

$$\gamma_2 = \frac{f^{24}(z) - 16}{f^8(z)} \tag{5}$$

$$\gamma_3 = \frac{[f^{24}(z) + 8][f_1{}^8(z) - f_2{}^8(z)]}{f^8(z)} \tag{6}$$

(Weber 1902, Atkin and Morain 1993), where $\eta(z)$ is the DEDEKIND ETA FUNCTION. The Weber functions satisfy the identities

$$f(z+1) = \frac{f_1(z)}{\zeta_{48}} \tag{7}$$

$$f_1(z+1) = \frac{f(z)}{\zeta_{48}} \tag{8}$$

$$f_2(z+1) = \zeta_{24} f_2(z) \tag{9}$$

$$f\left(-\frac{1}{z}\right) = f(z) \tag{10}$$

$$f_1\left(-\frac{1}{z}\right) = f_2(z) \tag{11}$$

$$f_2\left(-\frac{1}{z}\right) = f_1(z) \tag{12}$$

(Weber 1902, Atkin and Morain 1993).

see also ANGER FUNCTION, BESSEL FUNCTION OF THE SECOND KIND, DEDEKIND ETA FUNCTION, j-FUNCTION, JACOBI IDENTITIES, JACOBI TRIPLE PRODUCT, MODIFIED STRUVE FUNCTION, Q-FUNCTION, STRUVE FUNCTION

References

Abramowitz, M. and Stegun, C. A. (Eds.). "Anger and Weber Functions." §12.3 in *Handbook of Mathematical Functions with Formulas, Graphs, and Mathematical Tables, 9th printing.* New York: Dover, pp. 498–499, 1972.

Atkin, A. O. L. and Morain, F. "Elliptic Curves and Primality Proving." *Math. Comput.* **61**, 29–68, 1993.

Borwein, J. M. and Borwein, P. B. *Pi & the AGM: A Study in Analytic Number Theory and Computational Complexity.* New York: Wiley, pp. 68–69, 1987.

Weber, H. *Lehrbuch der Algebra, Vols. I–II.* New York: Chelsea, pp. 113–114, 1902.

Weber-Sonine Formula

For $\Re[\mu + nu] > 0$, $|\arg p| < \pi/4$, and $a > 0$,

$$\int_0^\infty J_\nu(at)e^{-p^2t^2}t^{\mu-1}\,dt$$

$$\left(\frac{a}{2p}\right)^\nu \frac{\Gamma[\frac{1}{2}(\nu+\mu)]}{2p^\mu\Gamma(\nu+1)}\,{}_1F_1\left(\tfrac{1}{2}(\nu+\mu);\nu+1;-\frac{a^2}{2p^2}\right),$$

where $J_\nu(z)$ is a BESSEL FUNCTION OF THE FIRST KIND, $\Gamma(z)$ is the GAMMA FUNCTION, and ${}_1F_1(a;b;z)$ is a CONFLUENT HYPERGEOMETRIC FUNCTION.

References

Iyanaga, S. and Kawada, Y. (Eds.). *Encyclopedic Dictionary of Mathematics*. Cambridge, MA: MIT Press, p. 1474, 1980.

Weber's Theorem

If two curves of the same GENUS (CURVE) > 1 are in rational correspondence, then that correspondence is BIRATIONAL.

References

Coolidge, J. L. *A Treatise on Algebraic Plane Curves*. New York: Dover, p. 135, 1959.

Wedderburn's Theorem

A FINITE DIVISION RING is a FIELD.

Weddle's Rule

$$\int_{x_1}^{x_{6n}} f(x)\,dx = \tfrac{3}{10}h(f_1 + 5f_2 + f_3$$
$$+6f_4 + 5f_5 + f_6 + \ldots + 5f_{6n-1} + f_{6n})$$

see also BODE'S RULE, HARDY'S RULE, NEWTON-COTES FORMULAS, SIMPSON'S 3/8 RULE, SIMPSON'S RULE, TRAPEZOIDAL RULE, WEDDLE'S RULE

Wedge

A right triangular PRISM turned so that it rests on one of its lateral faces.

see also CONICAL WEDGE, CYLINDRICAL WEDGE, PRISM

Wedge Product

An antisymmetric operation on DIFFERENTIAL FORMS (also called the EXTERIOR DERIVATIVE)

$$dx_i \wedge dx_j = -dx_j \wedge dx_i, \tag{1}$$

which IMPLIES

$$dx_i \wedge dx_i = 0 \tag{2}$$
$$b_i \wedge dx_j = dx_j \wedge b_i = b_i\,dx_j \tag{3}$$
$$dx_i \wedge (b_i\,dx_j) = b_i\,dx_i \wedge dx_j \tag{4}$$
$$\begin{aligned}\theta_1 \wedge \theta_2 &= (b_1\,dx_1 + b_2\,dx_2) \wedge (c_1\,dx_1 + c_2\,dx_2)\\ &= (b_1c_2 - b_2c_1)\,dx_1 \wedge dx_2\\ &= -\theta_2 \wedge \theta_1.\end{aligned} \tag{5}$$

The wedge product is ASSOCIATIVE

$$(s \wedge t) \wedge u = s \wedge (t \wedge u), \tag{6}$$

and BILINEAR

$$(\alpha_1 s_1 + \alpha_2 s_2) \wedge t = \alpha_1(s_1 \wedge t) + \alpha_2(s_2 \wedge t) \tag{7}$$
$$s \wedge (\alpha_1 t_1 + \alpha_2 t_2) = \alpha_1(s \wedge t_1) + \alpha_2(s \wedge t_2), \tag{8}$$

but not (in general) COMMUTATIVE

$$s \wedge t = (-1)^{pq}(t \wedge s), \tag{9}$$

where s is a p-form and t is a q-form. For a 0-form s and 1-form t,

$$(s \wedge t)_\mu = st_\mu. \tag{10}$$

For a 1-form s and 1-form t,

$$(s \wedge t)_{\mu\nu} = \tfrac{1}{2}(s_\mu t_\nu - s_\nu t_\mu). \tag{11}$$

The wedge product is the "correct" type of product to use in computing a VOLUME ELEMENT

$$dV = dx_1 \wedge \ldots \wedge dx_n. \tag{12}$$

see also DIFFERENTIAL FORM, EXTERIOR DERIVATIVE, INNER PRODUCT, VOLUME ELEMENT

Weekday

The day of the week W for a given day of the month D, month M, and year $100C + Y$ can be determined from the simple equation

$$W \equiv D + \lfloor 2.6M - 0.2\rfloor + \left\lfloor \tfrac{1}{4}Y\right\rfloor + \left\lfloor \tfrac{1}{4}C\right\rfloor - 2C \pmod 7,$$

where *months are numbered beginning with March* and $W = 0$ for Sunday, $W = 1$ for Monday, etc. (Uspensky and Heaslet 1939, Vardi 1991).

A more complicated form is given by

$$W \equiv D + M + C + Y \pmod 7,$$

where $W = 1$ for Sunday, $W = 2$ for Monday, etc. and the numbers assigned to months, centuries, and years are given in the tables below (Kraitchik 1942, pp. 110–111).

Month	M
January	1
February	4
March	3
April	6
May	1
June	4
July	6
August	2
September	5
October	0
November	3
December	5

Gregorian Century	C
15, 19, 23	1
16, 20, 24	0
17, 21, 25	5
18, 22, 26	3

Julian Century	C
00, 07, 14	5
01, 08, 15	4
02, 09, 16	3
03, 10, 17	2
04, 11, 18	1
05, 12, 19	0
06, 13, 20	6

Year									Y
00	06		17	23	28	34		45	0
01	07	12	18		29	35	40	46	1
02		13	19	24	30		41	47	2
03	08	14		25	31	36	42		3
	09	15	20	26		37	43	48	4
04	10		21	27	32	38		49	5
05	11	16	22		33	39	44	50	6
51	56	62		73	79	84	90		0
	57	63	68	74		85	91	96	1
52	58		69	75	80	86		97	2
53	59	64	70		81	87	92	98	3
54		65	71	76	82		93	99	4
55	60	66		77	83	88	94		5
	61	67	72	78		89	95		6

see also FRIDAY THE THIRTEENTH

References

Kraitchik, M. "The Calendar." Ch. 5 in *Mathematical Recreations.* New York: W. W. Norton, pp. 109–116, 1942.

Uspensky, J. V. and Heaslet, M. A. *Elementary Number Theory.* New York: McGraw-Hill, pp. 206–211, 1939.

Vardi, I. *Computational Recreations in Mathematica.* Reading, MA: Addison-Wesley, pp. 237–238, 1991.

Weibull Distribution

The Weibull distribution is given by

$$P(x) = \alpha\beta^{-\alpha}x^{\alpha-1}e^{-(x/\beta)^\alpha} \qquad (1)$$

$$D(x) = 1 - e^{-(x/\beta)^\alpha} \qquad (2)$$

for $x \in [0, \infty)$ (*Mathematica*® `Statistics'ContinuousDistributions'WeibullDistribution[a,b]`, Wolfram Research, Champaign, IL). The MEAN, VARIANCE, SKEWNESS, and KURTOSIS of this distribution are

$$\mu = \beta\Gamma(1+\alpha^{-1}) \qquad (3)$$

$$\sigma^2 = \beta^2[\Gamma(1+2\alpha^{-1}) - \Gamma^2(1+\alpha^{-1})] \qquad (4)$$

$$\gamma_1 = \frac{2\Gamma^3(1+\alpha^{-1}) - 3\Gamma(1+\alpha^{-1})\Gamma(1+2\alpha^{-1})}{[\Gamma(1+2\alpha^{-1}) - \Gamma^2(1+\alpha^{-1})]^{3/2}} + \frac{\Gamma(1+3\alpha^{-1})}{[\Gamma(1+2\alpha^{-1}) - \Gamma^2(1+\alpha^{-1})]^{3/2}} \qquad (5)$$

$$\gamma_2 = \frac{f(a)}{[\Gamma(1+2\alpha^{-1}) - \Gamma^2(1+\alpha^{-1})]^2}, \qquad (6)$$

where $\Gamma(z)$ is the GAMMA FUNCTION and

$$f(a) \equiv -6\Gamma^4(1+\alpha^{-1}) + 12\Gamma^2(1+\alpha^{-1})\Gamma(1+2\alpha^{-1}) - 3\Gamma^2(1+2\alpha^{-1}) - 4\Gamma(1+\alpha^{-1})\Gamma(1+3\alpha^{-1}) + \Gamma(1+4\alpha^{-1}). \qquad (7)$$

A slightly different form of the distribution is

$$P(x) = \frac{\alpha}{\beta}x^{\alpha-1}e^{-x^\alpha/\beta} \qquad (8)$$

$$D(x) = 1 - e^{-x^\alpha/\beta} \qquad (9)$$

(Mendenhall and Sincich 1995). The MEAN and VARIANCE for this form are

$$\mu = \beta^{1/\alpha}\Gamma(1+\alpha^{-1}) \qquad (10)$$

$$\sigma^2 = \beta^{2/\alpha}[\Gamma(1+2\alpha^{-1}) - \Gamma^2(1+\alpha^{-1})]. \qquad (11)$$

The Weibull distribution gives the distribution of lifetimes of objects. It was originally proposed to quantify fatigue data, but it is also used in analysis of systems involving a "weakest link."

see also FISHER-TIPPETT DISTRIBUTION

References

Mendenhall, W. and Sincich, T. *Statistics for Engineering and the Sciences, 4th ed.* Englewood Cliffs, NJ: Prentice Hall, 1995.

Spiegel, M. R. *Theory and Problems of Probability and Statistics.* New York: McGraw-Hill, p. 119, 1992.

Weierstraß Approximation Theorem

If f is continuous on $[a, b]$, then there exists a POLYNOMIAL p on $[a, b]$ such that

$$|f(x) - P(x)| < \epsilon$$

for all $x \in [a, b]$ and $\epsilon > 0$. In words, any continuous function on a closed and bounded interval can be uniformly approximated on that interval by POLYNOMIALS to any degree of accuracy.

see also MÜNTZ'S THEOREM

Weierstraß-Casorati Theorem

An ANALYTIC FUNCTION approaches any given value arbitrarily closely in any ϵ-NEIGHBORHOOD of an ESSENTIAL SINGULARITY.

Weierstraß Constant

$$\sigma(\tfrac{1}{2}) = \tfrac{1}{2} \prod_{\substack{(m,n)\neq\\(0,0)}} \left[1 - \frac{1}{2(m+ni)}\right] \times e^{1/[2(m+ni)]+1/[8(m+ni)^2]} = \frac{2^{5/4}\sqrt{\pi}\, e^{\pi/8}}{\Gamma^2(\frac{1}{4})} = 0.4749493799\ldots.$$

References

Le Lionnais, F. *Les nombres remarquables.* Paris: Hermann, p. 62, 1983.

Plouffe, S. "Weierstrass Constant." `http://lacim.uqam.ca/piDATA/weier.txt`.

Waldschmidt, M. "Fonctions entières et nombres transcendants." *Cong. Nat. Soc. Sav. Nancy* **5**, 1978.

Waldschmidt, M. "Nombres transcendants et fonctions sigma de Weierstrass." *C. R. Math. Rep. Acad. Sci. Canada* **1**, 111–114, 1978/79.

Weierstraß Elliptic Function

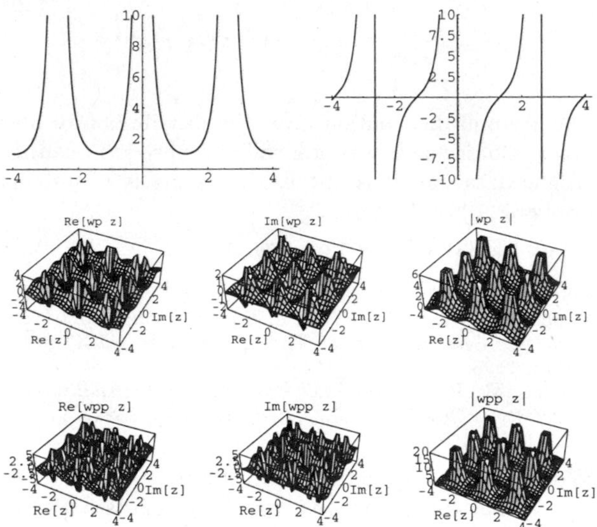

The Weierstraß elliptic functions are elliptic functions which, unlike the JACOBI ELLIPTIC FUNCTIONS, have a second-order POLE at $z = 0$. The above plots show the Weierstraß elliptic function $\wp(z)$ and its derivative $\wp'(z)$ for invariants (defined below) of $g_2 = 0$ and $g_3 = 0$. Weierstraß elliptic functions are denoted $\wp(z)$ and can be defined by

$$\wp(z) = \frac{1}{z^2} + \sum_{m,n=-\infty}^{\infty}{}' \left[\frac{1}{(z - 2m\omega_1 - 2n\omega_2)^2} - \frac{1}{(2m\omega_1 + 2n\omega_2)^2}\right]. \quad (1)$$

Write $\Omega_{mn} \equiv 2m\omega_1 + 2n\omega_2$. Then this can be written

$$\wp(z) = z^{-2} + \sum_{m,n}{}' [(z - \Omega_{mn})^{-2} - \Omega_{mn}^{-2}]. \quad (2)$$

An equivalent definition which converges more rapidly is

$$\wp(z) = \left(\frac{\pi}{2\omega_1}\right)^2 \left[-\frac{1}{3} + \sum_{-\infty}^{\infty} \csc^2\left(\frac{z - 2n\omega_2}{2\omega_1}\pi\right) - \sum_{n=-\infty}^{\infty}{}' \csc^2\left(\frac{n\omega_2}{n_1}\pi\right)\right]. \quad (3)$$

$\wp(z)$ is an EVEN FUNCTION since $\wp(-z)$ gives the same terms in a different order. To specify $\wp$ completely, its periods or invariants, written $\wp(z|\omega_1, \omega_2)$ and $\wp(z; g_2, g_3)$, respectively, must also be specified.

The differential equation from which Weierstraß elliptic functions arise can be found by expanding about the origin the function $f(z) \equiv \wp(z) - z^{-2}$.

$$\wp(z) - z^{-2} = f(0) + f'(0)z + \tfrac{1}{2!}f''(0)z^2 + \tfrac{1}{3!}f'''(0)z^3 + \tfrac{1}{4}f^{(4)}(0)z^4 + \ldots. \quad (4)$$

But $f(0) = 0$ and the function is even, so $f'(0) = f'''(0) = 0$ and

$$f(z) = \wp(z) - z^{-2} = \tfrac{1}{2!}f''(0)z^2 + \tfrac{1}{4}f^{(4)}(0)z^4 + \ldots. \quad (5)$$

Taking the derivatives

$$f' = -2\Sigma'[(z - \Omega_{mn})^{-3}] \quad (6)$$
$$f'' = 6\Sigma'(z - \Omega_{mn})^{-4} \quad (7)$$
$$f''' = -24\Sigma'(z - \Omega_{mn})^{-5} \quad (8)$$
$$f^{(4)} = 120\Sigma'(z - \Omega_{mn})^{-6}. \quad (9)$$

So

$$f''(0) = 6\Sigma'\Omega_{mn}^{-4} \quad (10)$$
$$f^{(4)}(0) = 120\Sigma'\Omega_{mn}^{-6}. \quad (11)$$

Plugging in,

$$\wp(z) - z^{-2} = 3\Sigma'\Omega_{mn}^{-4}z^2 + 5\Sigma'\Omega_{mn}^{-6}z^4 + \mathcal{O}(z^6). \quad (12)$$

Define the INVARIANTS

$$g_2 \equiv 60\Sigma'\Omega_{mn}^{-4} \quad (13)$$
$$g_3 \equiv 140\Sigma'\Omega_{mn}^{-6}, \quad (14)$$

then

$$\wp(z) = z^{-2} + \tfrac{1}{20}g_2z^2 + \tfrac{1}{28}g_3z^4 + \mathcal{O}(z^6) \quad (15)$$

$$\wp'(z) = -2z^{-3} + \tfrac{1}{10}g_2z + \tfrac{1}{7}g_3z^3 + \mathcal{O}(z^5). \quad (16)$$

Now cube (15) and square (16)

$$\wp^3(z) = z^{-6} + \tfrac{3}{20}g_2z^{-2} + \tfrac{3}{28}g_3 + \mathcal{O}(z^2) \quad (17)$$

$$\wp'^2(z) = 4z^{-6} - \tfrac{2}{5}g_2 z^{-2} - \tfrac{4}{7}g_3 + \mathcal{O}(z^2). \quad (18)$$

Taking (18) − 4 × (17) cancels out the z^{-6} term, giving

$$\begin{aligned}\wp'^2(z) - 4\wp^3(z) &= \left(-\tfrac{2}{5} - \tfrac{3}{5}\right) g_2 z^{-2} + \left(-\tfrac{4}{7} - \tfrac{3}{7}\right) g_3 + \mathcal{O}(z^2) \\ &= -g_2 z^{-2} - g_3 + \mathcal{O}(z^2) \quad (19)\end{aligned}$$

$$\wp'^2(z) - 4\wp^3(z) + g_2 z^{-2} + g_3 = \mathcal{O}(z^2). \quad (20)$$

But, from (5)

$$\wp(z) = z^{-2} + \tfrac{1}{2!}f''(0)z^2 + \tfrac{1}{4}f^{(4)}(0)z^4 + \ldots, \quad (21)$$

so $\wp(z) = z^{-2} + \mathcal{O}(z^2)$ and (20) can be written

$$\wp'^2(z) - 4\wp^3(z) + g_2\wp(z) + g_3 = \mathcal{O}(z^2). \quad (22)$$

The Weierstraß elliptic function is analytic at the origin and therefore at all points congruent to the origin. There are no other places where a singularity can occur, so this function is an ELLIPTIC FUNCTION with no SINGULARITIES. By LIOUVILLE'S ELLIPTIC FUNCTION THEOREM, it is therefore a constant. But as $z \to 0$, $\mathcal{O}(z^2) \to 0$, so

$$\wp'^2(z) = 4\wp^3(z) - g_2\wp(z) - g_3. \quad (23)$$

The solution to the differential equation

$$y'^2 = 4y^3 - g_2 y - g_3 \quad (24)$$

is therefore given by $y = \wp(z + \alpha)$, providing that numbers ω_1 and ω_2 exist which satisfy the equations defining the INVARIANTS. Writing the differential equation in terms of its roots e_1, e_2, and e_3,

$$y'^2 = 4y^3 - g_2 y - g_3 = 4(y - e_1)(y - e_2)(y - e_3) \quad (25)$$

$$2\ln(y') = \ln 4 + \sum_{r=1}^{3} \ln(y - e_r) \quad (26)$$

$$\frac{2y''}{y'} = y' \sum_{r=1}^{3} (y - e_r)^{-1} \quad (27)$$

$$\frac{2y''}{y'^2} = \sum_{r=1}^{3} (y - e_r)^{-1} \quad (28)$$

$$2\frac{y'^2 y''' - y''(2y'y'')}{y'^4} = -y' \sum_{r=1}^{3} (y - e_r)^{-2} \quad (29)$$

$$\frac{2y'''}{y'^3} - \frac{4y''^2}{y'^4} = -\sum_{r=1}^{3} (y - e_r)^{-2}. \quad (30)$$

Now take $(30)/4 + [(30)/4]^2$,

$$\left[\frac{y'''}{2y'^3} - \frac{y''^2}{y'^4}\right] + \left[\frac{y''^2}{4y'^4}\right] = -\frac{1}{4}\sum_{r=1}^{3}(y - e_r)^{-2} + \frac{1}{16}\left[\sum_{r=1}^{3}(y - e_r)^{-1}\right]^2 \quad (31)$$

$$\frac{3y''^2}{4y'^4} - \frac{y'''}{2y'^3} = \tfrac{3}{16}\sum_{r=1}^{3}(y - e_r)^{-2} - \tfrac{3}{8}y\prod_{r=1}^{3}(y - e_r)^{-1}. \quad (32)$$

The term on the right is half the SCHWARZIAN DERIVATIVE.

The DERIVATIVE of the Weierstraß elliptic function is given by

$$\begin{aligned}\wp'(z) = \frac{d}{dz}\wp(z) &= -2\sum_{m,n}\frac{1}{(z - \Omega_{mn})^3} \\ &= -2z^{-3} - 2{\sum_{m,n}}'(z - \Omega_{mn})^{-3}. \quad (33)\end{aligned}$$

This is an ODD FUNCTION which is itself an elliptic function with pole of order 3 at $z = 0$. The INTEGRAL is given by

$$z = \int_{\wp(z)}^{\infty} (4t^3 - g_2 t - g_3)^{-1/2}\, dt. \quad (34)$$

A duplication formula is obtained as follows.

$$\begin{aligned}\wp(2z) &= \lim_{y\to z}\wp(y + z) = \frac{1}{4}\lim_{y\to z}\left[\frac{\wp'(z) - \wp'(y)}{\wp(z) - \wp(y)}\right]^2 \\ &\quad - \wp(z) - \lim_{y\to z}\wp(y) \\ &= \frac{1}{4}\lim_{h\to 0}\left[\frac{\wp(z) - \wp'(z + h)}{\wp(z) - \wp(z + h)}\right]^2 - 2\wp(z) \\ &= \frac{1}{4}\left\{\left[\lim_{h\to 0}\frac{\wp'(z) - \wp'(z + h)}{h}\right]\left[\lim_{h\to 0}\frac{h}{\wp(z) - \wp(z + h)}\right]\right\}^2 \\ &\quad - 2\wp(z) \\ &= \frac{1}{4}\left[\frac{\wp''(z)}{\wp'(z)}\right]^2 - 2\wp(z). \quad (35)\end{aligned}$$

A general addition theorem is obtained as follows. Given

$$\wp'(z) = A\wp(z) + B \quad (36)$$

$$\wp'(y) = A\wp(y) + B \quad (37)$$

with zero y and z where $z \not\equiv \pm y \pmod{2\omega_1, 2\omega_2}$, find the third zero ζ. Consider $\wp'(\zeta) - A\wp(\zeta) - B$. This has a pole of order three at $\zeta = 0$, but the sum of zeros (= 0) equals the sum of poles for an ELLIPTIC FUNCTION, so $z + y + \zeta = 0$ and $\zeta = -z - y$.

$$\wp'(-z - y) = A\wp(-z - y) + B \quad (38)$$

$$-\wp'(z+y) = A\wp(z+y) + B. \tag{39}$$

Combining (36), (37), and (39) gives

$$\begin{bmatrix} \wp(z) & \wp'(z) & 1 \\ \wp(y) & \wp'(y) & 1 \\ \wp(z+y) & -\wp(z+y) & 1 \end{bmatrix} \begin{bmatrix} A \\ -1 \\ B \end{bmatrix} = \begin{bmatrix} 0 \\ 0 \\ 0 \end{bmatrix}, \tag{40}$$

so

$$\begin{vmatrix} \wp(z) & \wp'(z) & 1 \\ \wp(y) & \wp'(y) & 1 \\ \wp(z+y) & -\wp(z+y) & 1 \end{vmatrix} = 0. \tag{41}$$

Defining $u+v+w=0$ where $u \equiv z$ and $v \equiv y$ gives the symmetric form

$$\begin{vmatrix} \wp(u) & \wp'(u) & 1 \\ \wp(v) & \wp'(v) & 1 \\ \wp(w) & \wp(w) & 1 \end{vmatrix} = 0. \tag{42}$$

To get the expression explicitly, start again with

$$\wp'(\zeta) - A\wp(\zeta) - B = 0, \tag{43}$$

where $\zeta = z, y, -z-y$.

$$\wp'^2(\zeta) - [A\wp(\zeta) + B]^2 = 0. \tag{44}$$

But $\wp^2(\zeta) = 4\wp^4(\zeta) - g_2\wp(\zeta) - g_3$, so

$$4\wp^3(\zeta) - A^2\wp^2(\zeta) - (2AB+g_2)\wp(\zeta) - (B^2+g_3) = 0. \tag{45}$$

The solutions $\wp(\zeta) \equiv z$ are given by

$$4z^3 - A^2z^2 - (2AB+g_2)z - (B^2+g_3) = 0. \tag{46}$$

But the sum of roots equals the COEFFICIENT of the squared term, so

$$\wp(z) + \wp(y) + \wp(z+y) = \tfrac{1}{4}A^2 \tag{47}$$

$$\wp'(z) - \wp'(y) = A[\wp(z) - \wp(y)] \tag{48}$$

$$A = \frac{\wp'(z) - \wp'(y)}{\wp(z) - \wp(y)} \tag{49}$$

$$\wp(z+y) = \frac{1}{4}\left[\frac{\wp(z) - \wp'(y)}{\wp(z) - \wp(y)}\right]^2 - \wp(z) - \wp(y). \tag{50}$$

Half-period identities include

$$x \equiv \wp(\tfrac{1}{2}\omega_1) = \wp(-h\omega_1 + \omega_1) = e_1 + \frac{(e_1-e_2)(e_1-e_3)}{\wp(-\frac{1}{2}\omega_1) - e_1}$$
$$= e_1 + \frac{(e_1-e_2)(e_1-e_3)}{x - e_1}. \tag{51}$$

Multiplying through,

$$x^2 - e_1x = e_1x - {e_1}^2 + (e_1-e_2)(e_1-e_3) \tag{52}$$

$$x^2 - 2e_1 + [{e_1}^2 - (e_1-e_2)(e_1-e_3)] = 0, \tag{53}$$

which gives

$$\wp(\tfrac{1}{2}\omega_1) = \tfrac{1}{2}\left\{2e_1 \pm \sqrt{4{e_1}^2 - 4[{e_1}^2 - (e_1-e_2)(e_1-e_3)]}\right\}$$
$$= e_1 \pm \sqrt{(e_1-e_2)(e_1-e_3)}. \tag{54}$$

From Whittaker and Watson (1990, p. 445),

$$\wp'(\tfrac{1}{2}\omega_1) = -2\sqrt{(e_1-e_2)(e_1-e_3)}$$
$$\times(\sqrt{e_1-e_2} + \sqrt{e_1-e_3}). \tag{55}$$

The function is HOMOGENEOUS,

$$\wp(\lambda z|\lambda\omega_1, \lambda\omega_2) = \lambda^{-2}\wp(z|\omega_1,\omega_2) \tag{56}$$

$$\wp(\lambda z; \lambda^{-4}g_2, \lambda^{-6}g_3) = \lambda^{-2}\wp(z; g_2, g_3). \tag{57}$$

To invert the function, find $2\omega_1$ and $2\omega_2$ of $\wp(z|\omega_1,\omega_2)$ when given $\wp(z; g_2, g_3)$. Let e_1, e_2, and e_3 be the roots such that $(e_1-e_2)/(e_1-e_3)$ is not a REAL NUMBER > 1 or < 0. Determine the PARAMETER τ from

$$\frac{e_1-e_2}{e_1-e_3} = \frac{{\vartheta_4}^4(0|\tau)}{{\vartheta_3}^4(0|\tau)}. \tag{58}$$

Now pick

$$A \equiv \frac{\sqrt{e_1-e_2}}{{\vartheta_4}^2(0|\tau)}. \tag{59}$$

As long as ${g_2}^3 \neq 27g_3$, the periods are then

$$2\omega_1 = \pi A \tag{60}$$

$$2\omega_2 = \frac{\pi\tau}{A}. \tag{61}$$

Weierstraß elliptic functions can be expressed in terms of JACOBI ELLIPTIC FUNCTIONS by

$$\wp(u; g_2, g_3) = e_3 + (e_1 - e_3)$$
$$\times \mathrm{ns}^2\left(u\sqrt{e_1-e_3}, \sqrt{\frac{e_2-e_3}{e_1-e_3}}\right), \tag{62}$$

where

$$\wp(\omega_1) = e_1 \tag{63}$$

$$\wp(\omega_2) = e_2 \tag{64}$$

$$\wp(\omega_3) = -\wp(-\omega_1 - \omega_2) = e_3, \tag{65}$$

and the INVARIANTS are

$$g_2 \equiv 60\Sigma'\Omega_{mn}^{-4} \tag{66}$$

$$g_3 \equiv 140\Sigma'\Omega_{mn}^{-6}. \tag{67}$$

Here, $\Omega_{mn} \equiv 2m\omega_1 - 2n\omega_2$.

An addition formula for the Weierstraß elliptic function can be derived as follows.

$$\wp(z+\omega_1)+\wp(z)+\wp(\omega_1) = \frac{1}{4}\left[\frac{\wp'(z)-\wp'(\omega_1)}{\wp(z)-\wp(\omega_1)}\right]^2 = \frac{1}{4}\frac{\wp'^2(z)}{[\wp(z)-e_1]^2}. \quad (68)$$

Use

$$\wp'(z) = 4\prod_{r=1}^{3}[\wp(z)-e_r], \quad (69)$$

so

$$\wp(z+\omega_1) = -\wp(z)-e_1+\frac{1}{4}\frac{4\prod_{r=1}^{3}[\wp(z)-e_r]}{[\wp(z)-e_1]^2} = -\wp(z)-e_1+\frac{[\wp(z)-e_2][\wp(z)-e_3]}{\wp(z)-e_1}. \quad (70)$$

Use $\sum_{r=1}^{3} e_r = 0$,

$$\wp(z+\omega_1) = e_1+\frac{[-2e_1-\wp(z)][\wp(z)-e_1]}{\wp(z)-e_1} + \frac{\wp^2(z)-\wp(z)(e_2+e_3)+e_2e_3}{\wp(z)-e_1} = e_1+\frac{-\wp(z)(e_1+e_2+e_3)+e_2e_3+2{e_1}^2}{\wp(z)-e_1}. \quad (71)$$

But $\sum_{r=1}^{3} e_r = 0$ and

$$2{e_1}^2+e_2e_3 = {e_1}^2-e_1(e_2+e_3)+e_2e_3 = (e_1-e_2)(e_1-e_3), \quad (72)$$

so

$$\wp(z+\omega_1) = e_1+\frac{(e_1-e_2)(e_1-e_3)}{\wp(z)-e_1}. \quad (73)$$

The periods of the Weierstraß elliptic function are given as follows. When g_2 and g_3 are REAL and ${g_2}^3-27{g_3}^2>0$, then e_1, e_2, and e_3 are REAL and defined such that $e_1>e_2>e_3$.

$$\omega_1 = \int_{e_1}^{\infty}(4t^3-g_2t-g_3)^{-1/2}\,dt \quad (74)$$

$$\omega_3 = -i\int_{-\infty}^{e_3}(g_3+g_2t-4t^3)^{-1/2}\,dt \quad (75)$$

$$\omega_2 = -\omega_1-\omega_3. \quad (76)$$

The roots of the Weierstraß elliptic function satisfy

$$e_1 = \wp(\omega_1) \quad (77)$$

$$e_2 = \wp(\omega_2) \quad (78)$$

$$e_3 = \wp(\omega_3), \quad (79)$$

where $\omega_3 \equiv -\omega_1-\omega_2$. The e_is are ROOTS of $4t^3-g_2t-g_3$ and are unequal so that $e_1 \neq e_2 \neq e_3$. They can be found from the relationships

$$e_1+e_2+e_3 = -a_2 = 0 \quad (80)$$

$$e_2e_3+e_3e_1+e_1e_2 = a_1 = -\tfrac{1}{4}g_2 \quad (81)$$

$$e_1e_2e_3 = -a_0 = \tfrac{1}{4}g_3. \quad (82)$$

see also EQUIANHARMONIC CASE, LEMNISCATE CASE, PSEUDOLEMNISCATE CASE

References

Abramowitz, M. and Stegun, C. A. (Eds.). "Weierstrass Elliptic and Related Functions." Ch. 18 in *Handbook of Mathematical Functions with Formulas, Graphs, and Mathematical Tables, 9th printing.* New York: Dover, pp. 627–671, 1972.

Fischer, G. (Ed.). Plates 129–131 in *Mathematische Modelle/Mathematical Models, Bildband/Photograph Volume.* Braunschweig, Germany: Vieweg, pp. 126–128, 1986.

Whittaker, E. T. and Watson, G. N. *A Course in Modern Analysis, 4th ed.* Cambridge, England: Cambridge University Press, 1990.

Weierstraß-Erdman Corner Condition

In the CALCULUS OF VARIATIONS, the condition

$$f_{y'}(x,y,y'(x_-)) = f_{y'}(x,y,y'(x_+))$$

must hold at a corner (x,y) of a minimizing arc E_{12}.

Weierstraß Extreme Value Theorem

see EXTREME VALUE THEOREM

Weierstraß Form

A general form into which an ELLIPTIC CURVE over any FIELD K can be transformed is called the Weierstraß form, and is given by

$$y^2+ay = x^3+bx^2+cxy+dx+e,$$

where a, b, c, d, and e are elements of K.

Weierstraß Function

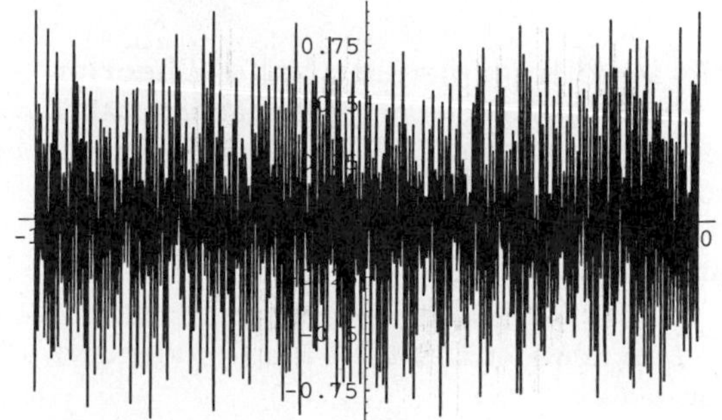

A CONTINUOUS FUNCTION which is nowhere DIFFERENTIABLE. It is given by

$$f(x) = \sum_{n=1}^{\infty} b^n \cos(a^n \pi x)$$

where n is an ODD INTEGER, $b \in (0,1)$, and $ab > 1 + 3\pi/2$. The above plot is for $a = 10$ and $b = 1/2$.

see also BLANCMANGE FUNCTION, CONTINUOUS FUNCTION, DIFFERENTIABLE

References

Darboux, G. "Mémoir sur les fonctions discontinues." *Ann. lÉcole Normale, Ser. 2* **4**, 57–112, 1875.

Darboux, G. "Mémoir sur les fonctions discontinues." *Ann. lÉcole Normale, Ser. 2* **8**, 195–202, 1879.

du Bois-Reymond, P. "Versuch einer Klassification der willkürlichen Functionen reeller Argumente nach ihren Änderungen in den kleinsten Intervallen." *J. für Math.* **79**, 21–37, 1875.

Faber, G. "Einfaches Beispiel einer stetigen nirgends differentiierbaren Funktion." *Jahresber. Deutschen Math. Verein.* **16** 538–540, 1907.

Hardy, G. H. "Weierstrass's Non-Differentiable Function." *Trans. Amer. Math. Soc.* **17**, 301–325, 1916.

Landsberg, G. "Über Differentzierbarkeit stetiger Funktionen." *Jahresber. Deutschen Math. Verein.* **17**, 46–51, 1908.

Lerch, M. "Über die Nichtdifferentiirbarkeit gewisser Functionen." *J. reine angew. Math.* **13**, 126–138, 1888.

Pickover, C. A. *Keys to Infinity.* New York: W. H. Freeman, p. 190, 1995.

Weierstraß, K. *Abhandlungen aus der Functionenlehre.* Berlin: J. Springer, p. 97, 1886.

Weierstraß's Gap Theorem

Given a succession of nonsingular points which are on a nonhyperelliptic curve of GENUS p, but are not a group of the canonical series, the number of groups of the first k which cannot constitute the group of simple POLES of a RATIONAL FUNCTION is p. If points next to each other are taken, then the theorem becomes: Given a nonsingular point of a nonhyperelliptic curve of GENUS p, then the orders which it cannot possess as the single pole of a RATIONAL FUNCTION are p in number.

References

Coolidge, J. L. *A Treatise on Algebraic Plane Curves.* New York: Dover, p. 290, 1959.

Weierstraß Intermediate Value Theorem

If a continuous function defined on an interval is sometimes POSITIVE and sometimes NEGATIVE, it must be 0 at some point.

Weierstraß M-Test

Let $\sum_{k=1}^{\infty} u_n(x)$ be a SERIES of functions all defined for a set E of values of x. If there is a CONVERGENT series of constants

$$\sum_{n=1}^{\infty} M_n,$$

such that

$$|u_n(x)| \leq M_n$$

for all $x \in E$, then the series exhibits ABSOLUTE CONVERGENCE for each $x \in E$ as well as UNIFORM CONVERGENCE in E.

see also ABSOLUTE CONVERGENCE, UNIFORM CONVERGENCE

References

Arfken, G. *Mathematical Methods for Physicists, 3rd ed.* Orlando, FL: Academic Press, pp. 301–303, 1985.

Weierstraß Point

A POLE of multiplicity less than $p+1$.

References

Coolidge, J. L. *A Treatise on Algebraic Plane Curves.* New York: Dover, pp. 290–291, 1959.

Weierstraß's Polynomial Theorem

A function, continuous in a finite close interval, can be approximated with a preassigned accuracy by POLYNOMIALS. A function of a REAL variable which is continuous and has period 2π can be approximated by trigonometric POLYNOMIALS.

References

Szegő, G. *Orthogonal Polynomials, 4th ed.* Providence, RI: Amer. Math. Soc., p. 5, 1975.

Weierstraß Product Inequality

If $0 \leq a, b, c, d \leq 1$, then

$$(1-a)(1-b)(1-c)(1-d) + a + b + c + d \geq 1.$$

References

Honsberger, R. *Mathematical Gems III.* Washington, DC: Math. Assoc. Amer., pp. 244–245, 1985.

Weierstraß Sigma Function

The QUASIPERIODIC FUNCTION defined by

$$\frac{d}{dz} \ln \sigma(z) = \zeta(z), \tag{1}$$

where $\zeta(z)$ is the WEIERSTRAß ZETA FUNCTION and

$$\lim_{z \to 0} \frac{\sigma(z)}{z} = 1. \tag{2}$$

Then

$$\sigma(z) = z \prod_{mn}{}' \left[\left(1 - \frac{z}{\Omega_{mn}}\right) \exp\left(\frac{z}{\Omega_{mn}} + \frac{z^2}{2\Omega_{mn}^2}\right)\right] \tag{3}$$

$$\sigma(z + 2\omega_1) = -e^{2\eta_1(z+\omega_1)} \sigma(z) \tag{4}$$

$$\sigma(z + 2\omega_2) = -e^{2\eta_2(z+\omega_2)} \sigma(z) \tag{5}$$

$$\sigma_r(z) = \frac{e^{-\eta_r z}\sigma(z+\omega_r)}{\sigma(\omega_r)} \tag{6}$$

for $r = 1, 2, 3$.

$$\sigma(z|\omega_1,\omega_2) = \frac{2\omega_1}{\pi\vartheta_1'} \exp\left(-\frac{\nu^2\vartheta_1'''}{6\vartheta_1'}\right) \vartheta_1\left(\nu \left| \frac{\omega_2}{\omega_1}\right.\right), \tag{7}$$

where $\nu \equiv \pi z/(2\omega_1)$, and

$$\eta_1 = -\frac{\pi^2\vartheta_1'''}{12\omega_1\vartheta_1'} \tag{8}$$

$$\eta_2 = -\frac{\pi^2\omega_2\vartheta_1'''}{12{\omega_1}^2\vartheta_1'} - \frac{\pi i}{2\omega_1}. \tag{9}$$

References

Abramowitz, M. and Stegun, C. A. (Eds.). "Weierstrass Elliptic and Related Functions." Ch. 18 in *Handbook of Mathematical Functions with Formulas, Graphs, and Mathematical Tables, 9th printing.* New York: Dover, pp. 627–671, 1972.

Weierstraß's Theorem

The only hypercomplex number systems with commutative multiplication and addition are the algebra with one unit such that $e = e^2$ and the GAUSSIAN INTEGERS.

see also GAUSSIAN INTEGER, PEIRCE'S THEOREM

Weierstraß Zeta Function

The QUASIPERIODIC FUNCTION defined by

$$\frac{d\zeta(z)}{dz} \equiv -\wp(z) \tag{1}$$

with

$$\lim_{z\to 0}[\zeta(z) - z^{-1}] = 0. \tag{2}$$

Then

$$\begin{aligned}\zeta(z) - z^{-1} &= -\int_0^z [\wp(z) - z^{-2}]\,dz \\ &= -\Sigma' \int_0^z [(z-\Omega_{mn})^{-2} - \Omega_{mn}^{-2}]\,dz \end{aligned} \tag{3}$$

$$\zeta(z) = z^{-1} + \sum_{m,n=-\infty}^{\infty}{}' [(z-\Omega_{mn})^{-1} + \Omega_{mn}^{-1} + z\Omega_{mn}^{-2}] \tag{4}$$

so $\zeta(z)$ is an ODD FUNCTION. Integrating $\wp(z+2\omega_1) = \wp(z)$ gives

$$\zeta(z+2\omega_1) = \zeta(z) + 2\eta_1. \tag{5}$$

Letting $z = -\omega_1$ gives $\zeta(-\omega_1) + 2\eta_1 = -\zeta(\omega_1) + 2\eta_1$, so $\eta_1 = \zeta(\omega_1)$. Similarly, $\eta_2 = \zeta(\omega_2)$. From Whittaker and Watson (1990),

$$\eta_1\omega_2 - \eta_2\omega_1 = \tfrac{1}{2}\pi i. \tag{6}$$

If $x+y+z=0$, then

$$[\zeta(x)+\zeta(y)+\zeta(z)]^2 + \zeta'(x) + \zeta'(y)\zeta'(z) = 0. \tag{7}$$

Also,

$$2\frac{\begin{vmatrix} 1 & \wp(x) & \wp^2(x) \\ 1 & \wp(y) & \wp^2(y) \\ 1 & \wp(z) & \wp^2(z) \end{vmatrix}}{\begin{vmatrix} 1 & \wp(x) & \wp'(x) \\ 1 & \wp(y) & \wp'(y) \\ 1 & \wp(z) & \wp'(z) \end{vmatrix}} = \zeta(x+y+z) - \zeta(x) - \zeta(y) - \zeta(z) \tag{8}$$

(Whittaker and Watson 1990, p. 446).

References

Abramowitz, M. and Stegun, C. A. (Eds.). "Weierstrass Elliptic and Related Functions." Ch. 18 in *Handbook of Mathematical Functions with Formulas, Graphs, and Mathematical Tables, 9th printing.* New York: Dover, pp. 627–671, 1972.

Whittaker, E. T. and Watson, G. N. *A Course in Modern Analysis, 4th ed.* Cambridge, England: Cambridge University Press, 1990.

Weighings

n weighings are SUFFICIENT to find a bad COIN among $(3^n-1)/2$ COINS. vos Savant (1993) gives an algorithm for finding a bad ball among 12 balls in three weighings (which, in addition, determines if the bad ball is heavier or lighter than the other 11).

Bachet's weights problem asks for the *minimum number* of weights (which can be placed in *either* pan of a two-arm balance) required to weigh any integral number of pounds from 1 to 40. The solution is 1, 3, 9, and 27: 1, $2 = -1+3$, 3, $4 = 1+3$, $5 = -1-3+9$, $6 = -3+9$, $7 = 1-3+9$, $8 = -1+9$, 9, $10 = 1+9$, $11 = -1+3+9$, $12 = 3+9$, $13 = 1+3+9$, $14 = -1-3-9+27$, $15 = -3-9+27$, $16 = 1-3-9+27$, $17 = -1-9+27$, and so on.

see also GOLOMB RULER, PERFECT DIFFERENCE SET, THREE JUG PROBLEM

References

Bachet, C. G. Problem 5, Appendix in *Problèmes plaisans et délectables, 2nd ed.* p. 215, 1624.

Ball, W. W. R. and Coxeter, H. S. M. *Mathematical Recreations and Essays, 13th ed.* New York: Dover, pp. 50–52, 1987.

Kraitchik, M. *Mathematical Recreations.* New York: W. W. Norton, pp. 52–55, 1942.

Pappas, T. "Counterfeit Coin Puzzle." *The Joy of Mathematics.* San Carlos, CA: Wide World Publ./Tetra, p. 181, 1989.

Tartaglia. Book 1, Ch. 16, §32 in *Trattato de' numeri e misure, Vol. 2.* Venice, 1556.

vos Savant, M. *The World's Most Famous Math Problem.* New York: St. Martin's Press, pp. 39–42, 1993.

Weight

The word weight has many uses in mathematics. It can refer to a function $w(x)$ (also called a WEIGHTING FUNCTION or WEIGHT FUNCTION) used to normalize ORTHONORMAL FUNCTIONS. It can also be used to indicate one of a set of a multiplicative constants placed in front of terms in a MOVING AVERAGE, NEWTON-COTES FORMULAS, edge or vertex of a GRAPH or TREE, etc.

see also WEIGHTED TREE, WEIGHTING FUNCTION

Weight Function

see WEIGHTING FUNCTION

Weighted Tree

A TREE in which each branch is given a numerical WEIGHT (i.e., a labelled TREE).

see also LABELLED GRAPH, TAYLOR'S CONDITION, TREE

Weighting Function

A function $w(x)$ used to normalize ORTHONORMAL FUNCTIONS

$$\int [f_n(x)]^2\, w(x)\, dx = N_n.$$

see also WEIGHT

Weingarten Equations

The Weingarten equations express the derivatives of the NORMAL using derivatives of the position vector. Let $\mathbf{x}: U \to \mathbb{R}^3$ a REGULAR PATCH, then the SHAPE OPERATOR S of $\mathbf{x}$ is given in terms of the basis $\{\mathbf{x}_u, \mathbf{x}_v\}$ by

$$-S(\mathbf{x}_u) = \mathbf{N}_u = \frac{fF - eG}{EG - F^2}\mathbf{x}_u + \frac{eF - fE}{EG - F^2}\mathbf{x}_v \quad (1)$$

$$-S(\mathbf{x}_v) = \mathbf{N}_v = \frac{gF - fG}{EG - F^2}\mathbf{x}_u + \frac{fF - gE}{EG - F^2}\mathbf{x}_v, \quad (2)$$

where $\mathbf{N}$ is the NORMAL VECTOR, E, F, and G the coefficients of the first FUNDAMENTAL FORM

$$ds^2 = E\,du^2 + 2F\,du\,dv + G\,dv^2, \quad (3)$$

and e, f, and g the coefficients of the second FUNDAMENTAL FORM given by

$$e = -\mathbf{N}_u \cdot \mathbf{x}_u = \mathbf{N} \cdot \mathbf{x}_{uu} \quad (4)$$

$$f = -\mathbf{N}_v \cdot \mathbf{x}_u = \mathbf{N} \cdot \mathbf{x}_{uv}$$

$$= \mathbf{N}_{vu} \cdot \mathbf{x}_{vu} = -\mathbf{N}_u \cdot \mathbf{x}_v \quad (5)$$

$$g = -\mathbf{N}_v \cdot \mathbf{x}_v = \mathbf{N} \cdot \mathbf{x}_{vv}. \quad (6)$$

see also FUNDAMENTAL FORMS, SHAPE OPERATOR

References

Gray, A. "Calculation of the Shape Operator." §14.3 in *Modern Differential Geometry of Curves and Surfaces*. Boca Raton, FL: CRC Press, pp. 274–277, 1993.

Weingarten Map

see SHAPE OPERATOR

Weird Number

A number which is ABUNDANT without being SEMIPERFECT. (A SEMIPERFECT NUMBER is the sum of any set of its own DIVISORS.) The first few weird numbers are 70, 836, 4030, 5830, 7192, 7912, 9272, 10430, ... (Sloane's A006037). No ODD weird numbers are known, but an infinite number of weird numbers are known to exist. The SEQUENCE of weird numbers has POSITIVE SCHNIRELMANN DENSITY.

see also ABUNDANT NUMBER, SCHNIRELMANN DENSITY, SEMIPERFECT NUMBER

References

Benkoski, S. "Are All Weird Numbers Even?" *Amer. Math. Monthly* **79**, 774, 1972.

Benkoski, S. J. and Erdős, P. "On Weird and Pseudoperfect Numbers." *Math. Comput.* **28**, 617–623, 1974.

Guy, R. K. "Almost Perfect, Quasi-Perfect, Pseudoperfect, Harmonic, Weird, Multiperfect and Hyperperfect Numbers." §B2 in *Unsolved Problems in Number Theory, 2nd ed.* New York: Springer-Verlag, pp. 45–53, 1994.

Sloane, N. J. A. Sequence A006037/M5339 in "An On-Line Version of the Encyclopedia of Integer Sequences."

Welch Apodization Function

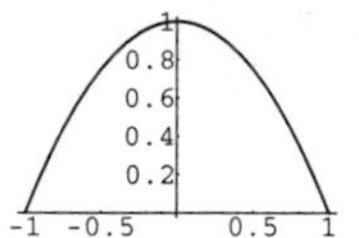

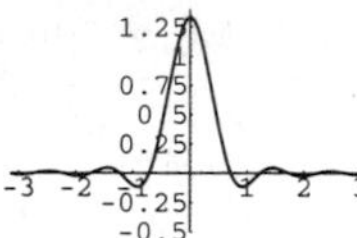

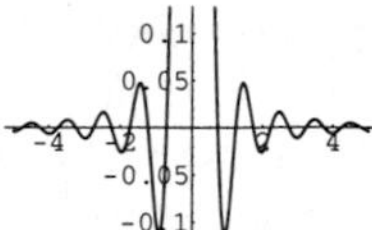

The APODIZATION FUNCTION

$$A(x) = 1 - \frac{x^2}{a^2}.$$

Its FULL WIDTH AT HALF MAXIMUM is $\sqrt{2}\,a$. Its INSTRUMENT FUNCTION is

$$I(k) = a2\sqrt{2\pi}\,\frac{J_{3/2}(2\pi ka)}{(2\pi ka)^{3/2}}$$

$$= a\frac{\sin(2\pi ka) - 2\pi ak\cos(2\pi ak)}{2a^3k^3\pi^3},$$

where $J_\nu(z)$ is a BESSEL FUNCTION OF THE FIRST KIND. It has a width of 1.59044, a maximum of $\frac{4}{3}$, maximum NEGATIVE sidelobe of -0.0861713 times the peak, and maximum POSITIVE sidelobe of 0.356044 times the peak.

see also APODIZATION FUNCTION, INSTRUMENT FUNCTION

References

Press, W. H.; Flannery, B. P.; Teukolsky, S. A.; and Vetterling, W. T. *Numerical Recipes in FORTRAN: The Art of Scientific Computing, 2nd ed.* Cambridge, England: Cambridge University Press, p. 547, 1992.

Well-Defined

An expression is called well-defined (or UNAMBIGUOUS) if its definition assigns it a unique interpretation or value. Otherwise, the expression is said to not be well defined or to be AMBIGUOUS.

For example, the expression abc (the PRODUCT) is well-defined if a, b, and c are integers. Because integers are ASSOCIATIVE, abc has the same value whether it is interpreted to mean $(ab)c$ or $a(bc)$. However, if a, b, and c are MATRICES or CAYLEY NUMBERS, then the expression abc is *not* well-defined, since MATRICES and CAYLEY NUMBER are not, in general, ASSOCIATIVE, so that the two interpretations $(ab)c$ and $a(bc)$ can be different.

Sometimes, ambiguities are implicitly resolved by notational convention. For example, the conventional interpretation of $a \wedge b \wedge c = a^{b^c}$ is $a^{(b^c)}$, never $(a^b)^c$, so that the expression $a \wedge b \wedge c$ is well-defined even though exponentiation is nonassociative.

Well-Ordered Set

A SET having the property that every nonempty SUBSET has a least member.

see also AXIOM OF CHOICE, HILBERT'S PROBLEMS, SUBSET, WELL-ORDERING PRINCIPLE

Well-Ordering Principle

Every nonempty set of POSITIVE integers contains a smallest member.

see also WELL-ORDERED SET

References

Shanks, D. *Solved and Unsolved Problems in Number Theory, 4th ed.* New York: Chelsea, p. 149, 1993.

Werner Formulas

$$2\sin\alpha\cos\beta = \sin(\alpha-\beta) + \sin(\alpha+\beta) \quad (1)$$

$$2\cos\alpha\cos\beta = \cos(\alpha-\beta) + \cos(\alpha+\beta) \quad (2)$$

$$2\cos\alpha\sin\beta = \sin(\alpha+\beta) - \sin(\alpha-\beta) \quad (3)$$

$$2\sin\alpha\sin\beta = \cos(\alpha-\beta) - \cos(\alpha+\beta). \quad (4)$$

see also TRIGONOMETRY

Weyl's Criterion

A SEQUENCE $\{x_1, x_2, \ldots\}$ is EQUIDISTRIBUTED IFF

$$\lim_{N\to\infty} \frac{1}{N} \sum_{n<N} e^{2\pi i m x_n} = 0$$

for each $m = 1, 2, \ldots$.

see also EQUIDISTRIBUTED SEQUENCE, RAMANUJAN'S SUM

References

Pólya, G. and Szegő, G. *Problems and Theorems in Analysis I.* New York: Springer-Verlag, 1972.
Vardi, I. *Computational Recreations in Mathematica.* Redwood City, CA: Addison-Wesley, pp. 155–156 and 254, 1991.

Weyl Tensor

The TENSOR

$$C^{ij}{}_{kl} = R^i j_{kl} - 2\delta^{[i}{}_{[k} R^{j]}{}_{l]} + \tfrac{1}{3}\delta^{[i}{}_{[k}\delta^{j]}{}_{l]} R,$$

where $R^i j_{kl}$ is the RIEMANN TENSOR and R is the CURVATURE SCALAR. The Weyl tensor is defined so that every CONTRACTION between indices gives 0. In particular, $C^\lambda{}_{\mu\lambda\kappa} = 0$. The number of independent components for a Weyl tensor in N-D is given by

$$C_N = \tfrac{1}{12} N(N+1)(N+2)(N-3).$$

see also CURVATURE SCALAR, RIEMANN TENSOR

References

Weinberg, S. *Gravitation and Cosmology: Principles and Applications of the General Theory of Relativity.* New York: Wiley, p. 146, 1972.

Weyrich's Formula

Under appropriate constraints,

$$\frac{1}{2} i \int_{-\infty}^{\infty} H_0^{(1)}(r\sqrt{k^2 - r^2}) e^{i\tau x}\, d\tau = \frac{e^{ik\sqrt{r^2+k^2}}}{\sqrt{r^2+x^2}},$$

where $H_0^{(1)}(z)$ is a HANKEL FUNCTION OF THE FIRST KIND.

References

Iyanaga, S. and Kawada, Y. (Eds.). *Encyclopedic Dictionary of Mathematics.* Cambridge, MA: MIT Press, p. 1474, 1980.

Wheat and Chessboard Problem

Let one grain of wheat be placed on the first square of a CHESSBOARD, two on the second, three on the third, etc. How many grains total are placed on an 8×8 CHESSBOARD? Since this is a GEOMETRIC SERIES, the answer for n squares is

$$\sum_{i=0}^{n-1} 2^i = 2^n - 1.$$

Plugging in $n = 8 \times 8 = 64$ then gives $2^{64} - 1 = 18446744073709551615$.

References

Pappas, T. "The Wheat and & Chessboard." *The Joy of Mathematics.* San Carlos, CA: Wide World Publ./Tetra, p. 17, 1989.

Wheel

see ARISTOTLE'S WHEEL PARADOX, BENHAM'S WHEEL, WHEEL GRAPH

Wheel Graph

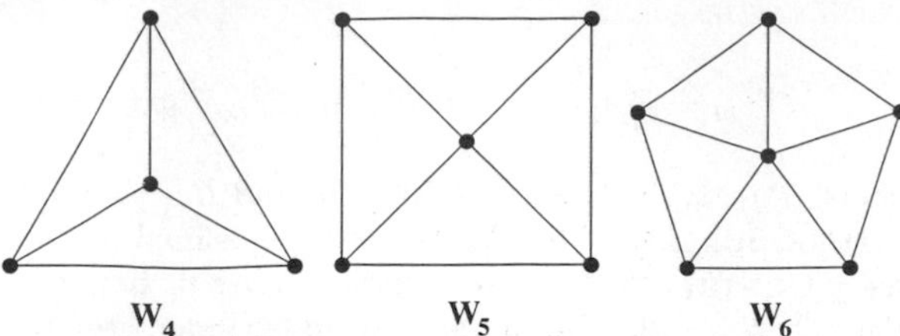

A GRAPH W_n of order n which contains a CYCLE of order $n-1$, and for which every NODE in the cycle is connected to one other NODE (known as the HUB). In a wheel graph, the HUB has DEGREE $n-1$, and other nodes have degree 3. $W_4 = K_4$, where K_4 is the COMPLETE GRAPH of order four.

see also COMPLETE GRAPH, GEAR GRAPH, HUB, WEB GRAPH

Wheel Paradox

see ARISTOTLE'S WHEEL PARADOX

Whewell Equation

An INTRINSIC EQUATION which expresses a curve in terms of its ARC LENGTH s and TANGENTIAL ANGLE ϕ.

see also ARC LENGTH, CESÀRO EQUATION, INTRINSIC EQUATION, NATURAL EQUATION, TANGENTIAL ANGLE

References

Yates, R. C. "Intrinsic Equations." *A Handbook on Curves and Their Properties.* Ann Arbor, MI: J. W. Edwards, pp. 123–126, 1952.

Whipple's Transformation

$$ {}_7F_6\left[\begin{matrix} a, 1+\frac{1}{2}a, b, c, d, e, -m \\ \frac{1}{2}a, 1+a-b, 1+a-c, \\ 1+a-d, 1+a-e, 1+a+m \end{matrix}\right] = \frac{(1+a)_m(1+a-d-e)_m}{(1+a-d)_m(1+a-e)_m} \times {}_4F_3\left[\begin{matrix} 1+a-b-c, d, e, -m \\ 1+a-b, 1+a-c, d+e-a-m \end{matrix}\right], $$

where ${}_7F_6$ and ${}_4F_3$ are GENERALIZED HYPERGEOMETRIC FUNCTIONS and $\Gamma(z)$ is the GAMMA FUNCTION.

see also GENERALIZED HYPERGEOMETRIC FUNCTION

Whirl

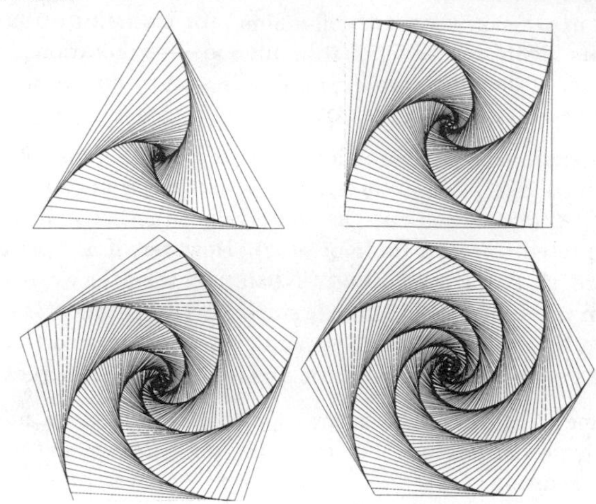

Whirls are figures constructed by nesting a sequence of polygons (each having the same number of sides), each slightly smaller and rotated relative to the previous one.

see also DAISY, SWIRL

References

Lauwerier, H. *Fractals: Endlessly Repeated Geometric Figures.* Princeton, NJ: Princeton University Press, p. 66, 1991.

Pappas, T. "Spider & Spirals." *The Joy of Mathematics.* San Carlos, CA: Wide World Publ./Tetra, p. 228, 1989.

Weisstein, E. W. "Fractals." `http://www.astro.virginia.edu/~eww6n/math/notebooks/Fractal.m`.

Whisker Plot

see BOX-AND-WHISKER PLOT

Whitehead Double

The SATELLITE KNOT of an UNKNOT twisted inside a TORUS.

see also SATELLITE KNOT, TORUS, UNKNOT

References

Adams, C. C. *The Knot Book: An Elementary Introduction to the Mathematical Theory of Knots.* New York: W. H. Freeman, pp. 115–116, 1994.

Whitehead Link

The LINK 5^{02}_{01}, illustrated above, with BRAID WORD ${\sigma_1}^2{\sigma_2}^2{\sigma_1}^{-1}{\sigma_2}^{-2}$ and JONES POLYNOMIAL

$$ V(t) = t^{-3/2}(-1 + t - 2t^2 + t^3 - 2t^4 + t^5). $$

The Whitehead link has LINKING NUMBER 0.

Whitehead Manifold

An open 3-MANIFOLD which is simply connected but is topologically distinct from Euclidean 3-space.

References
Rolfsen, D. *Knots and Links.* Wilmington, DE: Publish or Perish Press, p. 82, 1976.

Whitehead's Theorem

MAPS between CW-COMPLEXES that induce ISOMORPHISMS on all HOMOTOPY GROUPS are actually HOMOTOPY equivalences.

see also CW-COMPLEX, HOMOTOPY GROUP, ISOMORPHISM

Whitney-Graustein Theorem

A 1937 theorem which classified planar regular closed curves up to regular HOMOTOPY by their WINDING NUMBERS. In his thesis, S. Smale generalized this result to regular closed curves on an n-MANIFOLD.

Whitney-Mikhlin Extension Constants

N.B. A detailed on-line essay by S. Finch was the starting point for this entry.

Let $B_n(r)$ be the n-D closed BALL of RADIUS $r > 1$ centered at the ORIGIN. A function which is defined on $B(r)$ is called an extension to $B(r)$ of a function f defined on $B(1)$ if

$$F(x) = f(x) \forall\, x \in B(1). \tag{1}$$

Given 2 BANACH SPACES of functions defined on $B(1)$ and $B(r)$, find the extension operator from one to the other of minimal norm. Mikhlin (1986) found the best constants χ such that this condition, corresponding to the Sobolev $W(1,2)$ integral norm, is satisfied,

$$\sqrt{\int_{B(1)} \left[[f(x)]^2 + \sum_{j=1}^{n} \left(\frac{\partial f}{\partial x_j} \right)^2 \right] dx} \leq \chi \sqrt{\int_{B(r)} \left[[F(x)]^2 + \sum_{j=1}^{n} \left(\frac{\partial F}{\partial x_j} \right)^2 \right] dx}\,. \tag{2}$$

$\chi(1,r) = 1$. Let

$$\nu = \tfrac{1}{2}(n-2), \tag{3}$$

then for $n > 2$,

$$\chi(n,r) = \sqrt{1 + \frac{I_\nu(1)}{I_{\nu+1}(1)} \frac{I_\nu(r)K_{\nu+1}(1) + K_\nu(r)I_{\nu+1}(1)}{I_\nu(r)K_\nu(1) - K_\nu(r)I_\nu(1)}}, \tag{4}$$

where $I_\nu(z)$ is a MODIFIED BESSEL FUNCTION OF THE FIRST KIND and $K_\nu(z)$ is a MODIFIED BESSEL FUNCTION OF THE SECOND KIND. For $n = 2$,

$$\chi(2,r) = \max \left\{ \sqrt{1 + \frac{I_\nu(1)}{I_{\nu+1}(1)} \frac{I_\nu(r)K_{\nu+1}(1) + K_\nu(r)I_{\nu+1}(1)}{I_\nu(r)K_\nu(1) - K_\nu(r)I_\nu(1)}} \sqrt{1 + \frac{I_1(1)}{I_1(1) + I_2(1)} \left[1 + \frac{I_1(r)K_0(1) + K_1(r)I_0(1)}{I_1(r)K_1(1) - K_1(r)I_1(1)} \right]} \right\}. \tag{5}$$

For $r \to \infty$,

$$\chi(n,\infty) = \sqrt{1 + \frac{I_\nu(1)}{I_{\nu+1}(1)} \frac{K_\nu(1)}{K_\nu(1)}}, \tag{6}$$

which is bounded by

$$n - 1 < \chi(n,\infty) < \sqrt{(n-1)^2 + 4}. \tag{7}$$

For ODD n, the RECURRENCE RELATIONS

$$a_{k+1} = a_{k-1} - (2k-1)a_k \tag{8}$$
$$b_{k+1} = b_{k-1} + (2k-1)b_k \tag{9}$$

with

$$a_0 = e + e^{-1} \tag{10}$$
$$a_1 = e - e^{-1} \tag{11}$$
$$b_0 = e^{-1} \tag{12}$$
$$b_1 = e^{-1} \tag{13}$$

where e is the constant 2.71828..., give

$$\chi(2k+1,\infty) = \sqrt{1 + \frac{a_k}{a_{k+1}} \frac{b_{k+1}}{b_k}}. \tag{14}$$

The first few are

$$\chi(3,\infty) = e \tag{15}$$
$$\chi(5,\infty) = \sqrt{\frac{e^2}{e^2 - 7}} \tag{16}$$
$$\chi(7,\infty) = \sqrt{\frac{2}{7}} \sqrt{\frac{e^2}{37 - 5e^2}} \tag{17}$$
$$\chi(9,\infty) = \frac{1}{\sqrt{37}} \sqrt{\frac{e^2}{18e^2 - 133}} \tag{18}$$
$$\chi(11,\infty) = \frac{1}{\sqrt{133}} \sqrt{\frac{e^2}{2431 - 329e^2}} \tag{19}$$
$$\chi(13,\infty) = \sqrt{\frac{2}{2431}} \sqrt{\frac{e^2}{3655e^2 - 27007}}. \tag{20}$$

Similar formulas can be given for even n in terms of $I_0(1)$, $I_1(1)$, $K_0(1)$, $K_1(1)$.

References
Finch, S. "Favorite Mathematical Constants." http://www.mathsoft.com/asolve/constant/mkhln/mkhln.html.
Mikhlin, S. G. *Constants in Some Inequalities of Analysis.* New York: Wiley, 1986.

Whitney Singularity

see PINCH POINT

Whitney Sum

An operation that takes two VECTOR BUNDLES over a fixed SPACE and produces a new VECTOR BUNDLE over the same SPACE. If E_1 and E_2 are VECTOR BUNDLES over B, then the Whitney sum $E_1 \oplus E_2$ is the VECTOR BUNDLE over B such that each FIBER over B is naturally the direct sum of the E_1 and E_2 FIBERS over B.

The Whitney sum is therefore the FIBER for FIBER direct sum of the two BUNDLES E_1 and E_2. An easy formal definition of the Whitney sum is that $E_1 \oplus E_2$ is the pull-back BUNDLE of the diagonal map from B to $B \times B$, where the BUNDLE over $B \times B$ is $E_1 \times E_2$.

see also BUNDLE, FIBER, VECTOR BUNDLE

Whitney Umbrella

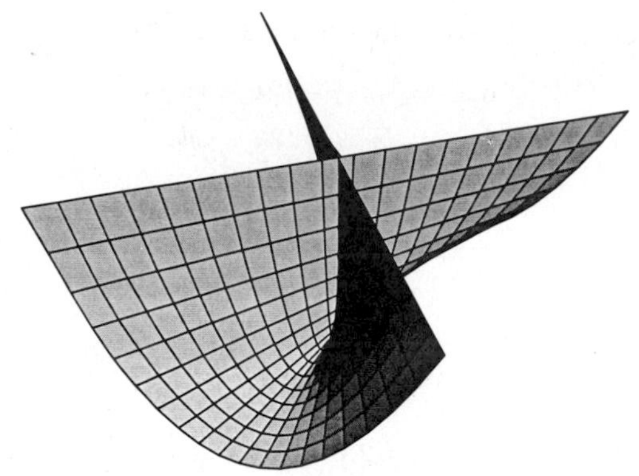

A surface which can be interpreted as a self-intersecting RECTANGLE in 3-D. It is given by the parametric equations

$$x = uv \tag{1}$$

$$y = u \tag{2}$$

$$z = v^2 \tag{3}$$

for $u, v \in [-1, 1]$. The center of the "plus" shape which is the end of the line of self-intersection is a PINCH POINT. The coefficients of the first FUNDAMENTAL FORM are

$$E = 0 \tag{4}$$

$$F = \frac{2v}{\sqrt{u^2 + 4v^2 + 4v^4}} \tag{5}$$

$$G = -\frac{2u}{\sqrt{u^2 + 4v^2 + 4v^4}}, \tag{6}$$

and the coefficients of the second FUNDAMENTAL FORM are

$$e = 1 + v^2 \tag{7}$$

$$f = uv \tag{8}$$

$$g = u^2 + 4v^2, \tag{9}$$

giving GAUSSIAN CURVATURE and MEAN CURVATURE

$$K = -\frac{4v^2}{(u^2 + 4v^2 + 4v^4)^2} \tag{10}$$

$$H = -\frac{u(1 + 3v^2)}{(u^2 + 4v^2 + 4v^4)^{3/2}}. \tag{11}$$

References

Francis, G. K. *A Topological Picturebook.* New York: Springer-Verlag, pp. 8–9, 1987.

Geometry Center. "Whitney's Umbrella." `http://www.geom.umn.edu/zoo/features/whitney/`.

Gray, A. *Modern Differential Geometry of Curves and Surfaces.* Boca Raton, FL: CRC Press, pp. 225 and 309–310, 1993.

Whittaker Differential Equation

$$\frac{d^2u}{dz^2} + \frac{du}{dz} + \left(\frac{k}{z} + \frac{\frac{1}{4} - m^2}{z^2}\right) u = 0. \tag{1}$$

Let $u \equiv e^{-z/2} W_{k,m}(z)$, where $W_{k,m}(z)$ denotes a WHITTAKER FUNCTION. Then (1) becomes

$$\frac{d}{dz}(-\tfrac{1}{2}e^{-z/2}W + e^{-z/2}W') + (-\tfrac{1}{2}e^{-z/2}W + e^{-z/2}W') + \left(\frac{k}{z} + \frac{\frac{1}{4} - m^2}{z^2}\right) e^{-z/2}W = 0. \tag{2}$$

Rearranging,

$$(\tfrac{1}{4}e^{-z/2}W - \tfrac{1}{2}e^{-z/2}W' - \tfrac{1}{2}e^{-z/2}W' + e^{-z/2}W'')p + (-\tfrac{1}{2}e^{-z/2}W + e^{-z/2}W') + \left(\frac{k}{z} + \frac{\frac{1}{4} - m^2}{z^2}\right) e^{-z/2}W = 0 \tag{3}$$

$$-\tfrac{1}{4}e^{-z/2}W + e^{-z/2}W'' + \left(\frac{k}{z} + \frac{\frac{1}{4} - m^2}{z^2}\right) e^{-z/2}W = 0, \tag{4}$$

so

$$W'' + \left(-\frac{1}{4} + \frac{k}{z} + \frac{\frac{1}{4} - m^2}{z^2}\right) W = 0, \tag{5}$$

where $W' \equiv dW/dz$. The solutions are known as WHITTAKER FUNCTIONS.

References

Abramowitz, M. and Stegun, C. A. (Eds.). *Handbook of Mathematical Functions with Formulas, Graphs, and Mathematical Tables, 9th printing.* New York: Dover, p. 505, 1972.

Whittaker Function

Solutions to the WHITTAKER DIFFERENTIAL EQUATION. The linearly independent solutions are

$$M_{k,m}(z) \equiv z^{1/2+m}e^{-z/2} \times \left(1 + \frac{\frac{1}{2}+m-k}{1!(2m+1)} + \frac{(\frac{1}{2}+m-k)(\frac{3}{2}+m-k)}{2!(2m+1)(2m+2)}z^2 + \ldots\right), \quad (1)$$

and $M_{k,-m}(z)$, where $M_{k,m}(z)$ is a CONFLUENT HYPERGEOMETRIC FUNCTION. In terms of CONFLUENT HYPERGEOMETRIC FUNCTIONS, the Whittaker functions are

$$M_{k,m}(z) = e^{-z/2}z^{m+1/2}{}_1F_1(\tfrac{1}{2}+m-k, 1+2m; z) \quad (2)$$

$$W_{k,m}(z) = e^{-z/2}z^{m+1/2}U(\tfrac{1}{2}+m-k, 1+2m; z) \quad (3)$$

(see Whittaker and Watson 1990, pp. 339–351). However, the CONFLUENT HYPERGEOMETRIC FUNCTION disappears when $2m$ is an INTEGER, so Whittaker functions are often defined instead. The Whittaker functions are related to the PARABOLIC CYLINDER FUNCTIONS. When $|\arg z| < 3\pi/2$ and $2m$ is not an INTEGER,

$$W_{k,m}(z) = \frac{\Gamma(-2m)}{\Gamma(\frac{1}{2}-m-k)}M_{k,m}(z) + \frac{\Gamma(2m)}{\Gamma(\frac{1}{2}+m-k)}M_{k,-m}(z). \quad (4)$$

When $|\arg(-z)| < 3\pi/2$ and $2m$ is not an INTEGER,

$$W_{-k,m}(-z) = \frac{\Gamma(-2m)}{\Gamma(\frac{1}{2}-m-k)}M_{-k,m}(-z) + \frac{\Gamma(2m)}{\Gamma(\frac{1}{2}+m+k)}M_{-k,-m}(-z). \quad (5)$$

Whittaker functions satisfy the RECURRENCE RELATIONS

$$W_{k,m}(z) = z^{1/2}W_{k-1/2,m-1/2}(z)+(\tfrac{1}{2}-k+m)W_{k-1,m}(z) \quad (6)$$

$$W_{k,m}(z) = z^{1/2}W_{k-1/2,m+1/2}(z)+(\tfrac{1}{2}-k-m)W_{k-1,m}(z) \quad (7)$$

$$zW'_{k,m}(z) = (k-\tfrac{1}{2}z)W_{k,m}(z)-[m^2-(k-\tfrac{1}{2})^2]W_{k-1,m}(z). \quad (8)$$

see also CONFLUENT HYPERGEOMETRIC FUNCTION, KUMMER'S FORMULAS, PEARSON-CUNNINGHAM FUNCTION, SCHLÖMILCH'S FUNCTION, SONINE POLYNOMIAL

References

Abramowitz, M. and Stegun, C. A. (Eds.). "Confluent Hypergeometric Functions." Ch. 13 in *Handbook of Mathematical Functions with Formulas, Graphs, and Mathematical Tables, 9th printing.* New York: Dover, pp. 503–515, 1972.

Iyanaga, S. and Kawada, Y. (Eds.). "Whittaker Functions." Appendix A, Table 19.II in *Encyclopedic Dictionary of Mathematics.* Cambridge, MA: MIT Press, pp. 1469–1471, 1980.

Whittaker, E. T. and Watson, G. N. *A Course in Modern Analysis, 4th ed.* Cambridge, England: Cambridge University Press, 1990.

Whole Number

One of the numbers 1, 2, 3, ... (Sloane's A000027), also called the COUNTING NUMBERS or NATURAL NUMBERS. 0 is sometimes included in the list of "whole" numbers (Bourbaki 1968, Halmos 1974), but there seems to be no general agreement. Some authors also interpret "whole number" to mean "a number having FRACTIONAL PART of zero," making the whole numbers equivalent to the integers.

Due to lack of standard terminology, the following terms are recommended in preference to "COUNTING NUMBER," "NATURAL NUMBER," and "whole number."

Set	Name	Symbol
..., −2, −1, 0, 1, 2, ...	integers	$\mathbb{Z}$
1, 2, 3, 4, ...	positive integers	$\mathbb{Z}^+$
0, 1, 2, 3, 4 ...	nonnegative integers	$\mathbb{Z}^*$
−1, −2, −3, −4, ...	negative integers	$\mathbb{Z}^-$

see also COUNTING NUMBER, FRACTIONAL PART, INTEGER, $\mathbb{N}$, NATURAL NUMBER, $\mathbb{Z}$, $\mathbb{Z}^+$, $\mathbb{Z}^+$, $\mathbb{Z}^*$

References

Bourbaki, N. *Elements of Mathematics: Theory of Sets.* Paris, France: Hermann, 1968.

Halmos, P. R. *Naive Set Theory.* New York: Springer-Verlag, 1974.

Sloane, N. J. A. Sequence A000027/M0472 in "An On-Line Version of the Encyclopedia of Integer Sequences."

Width (Partial Order)

For a PARTIAL ORDER, the size of the longest ANTICHAIN is called the width.

see also ANTICHAIN, LENGTH (PARTIAL ORDER), PARTIAL ORDER

Width (Size)

The width of a box is the horizontal distance from side to side (usually defined to be greater than the DEPTH, the horizontal distance from front to back).

see also DEPTH (SIZE), HEIGHT

References

Eppstein, D. "Width, Diameter, and Geometric Inequalities." `http://www.ics.uci.edu/~eppstein/junkyard/diam.html`.

Wiedersehen Manifold

The only Wiedersehen manifolds are the standard round spheres, as was established by proof of the BLASCHKE CONJECTURE.

see also BLASCHKE CONJECTURE

Wieferich Prime

A Wieferich prime is a PRIME p which is a solution to the CONGRUENCE equation

$$2^{p-1} \equiv 1 \pmod{p^2}.$$

Note the similarity of this expression to the special case of FERMAT'S LITTLE THEOREM

$$2^{p-1} \equiv 1 \pmod{p},$$

which holds for *all* ODD PRIMES. However, the only Wieferich primes less than 4×10^{12} are $p = 1093$ and 3511 (Lehmer 1981, Crandall 1986, Crandall *et al.* 1997). Interestingly, one less than these numbers have suggestive periodic BINARY representations

$$1092 = 10001000100_2$$
$$3510 = 110110110110_2.$$

A PRIME factor p of a MERSENNE NUMBER $M_q = 2^q - 1$ is a Wieferich prime IFF $p^2 | 2^q - 1$. Therefore, MERSENNE PRIMES are *not* Wieferich primes.

If the first case of FERMAT'S LAST THEOREM is false for exponent p, then p must be a Wieferich prime (Wieferich 1909). If $p|2^n \pm 1$ with p and n RELATIVELY PRIME, then p is a Wieferich prime IFF p^2 also divides $2^n \pm 1$. The CONJECTURE that there are no three POWERFUL NUMBERS implies that there are infinitely many Wieferich primes (Granville 1986, Vardi 1991). In addition, the ABC CONJECTURE implies that there are at least $C \ln x$ Wieferich primes $\leq x$ for some constant C (Silverman 1988, Vardi 1991).

see also ABC CONJECTURE, FERMAT'S LAST THEOREM, FERMAT QUOTIENT, MERSENNE NUMBER, MIRIMANOFF'S CONGRUENCE, POWERFUL NUMBER

References

Brillhart, J.; Tonascia, J.; and Winberger, P. "On the Fermat Quotient." In *Computers and Number Theory* (Ed. A. O. L. Atkin and B. J. Birch). New York: Academic Press, pp. 213–222, 1971.

Crandall, R. *Projects in Scientific Computation.* New York: Springer-Verlag, 1986.

Crandall, R.; Dilcher, K; and Pomerance, C. "A search for Wieferich and Wilson Primes." *Math. Comput.* **66**, 433–449, 1997.

Granville, A. "Powerful Numbers and Fermat's Last Theorem." *C. R. Math. Rep. Acad. Sci. Canada* **8**, 215–218, 1986.

Lehmer, D. H. "On Fermat's Quotient, Base Two." *Math. Comput.* **36**, 289–290, 1981.

Ribenboim, P. "Wieferich Primes." §5.3 in *The New Book of Prime Number Records.* New York: Springer-Verlag, pp. 333–346, 1996.

Shanks, D. *Solved and Unsolved Problems in Number Theory, 4th ed.* New York: Chelsea, pp. 116 and 157, 1993.

Silverman, J. "Wieferich's Criterion and the abc Conjecture." *J. Number Th.* **30**, 226–237, 1988.

Vardi, I. "Wieferich." §5.4 in *Computational Recreations in Mathematica.* Reading, MA: Addison-Wesley, pp. 59–62 and 96–103, 1991.

Wieferich, A. "Zum letzten Fermat'schen Theorem." *J. reine angew. Math.* **136**, 293–302, 1909.

Wielandt's Theorem

Let the $n \times n$ MATRIX A satisfy the conditions of the PERRON-FROBENIUS THEOREM and the $n \times n$ MATRIX $\mathsf{C} = c_{ij}$ satisfy

$$|c_{ij}| \leq a_{ij}$$

for $i, j = 1, 2, \ldots, n$. Then any EIGENVALUE λ_0 of C satisfies the inequality $|\lambda_0| \leq R$ with the equality sign holding only when there exists an $n \times n$ MATRIX $\mathsf{D} = \delta_{ij}$ (where δ_{ij} is the KRONECKER DELTA) and

$$\mathsf{C} = \frac{\lambda_0}{R} \mathsf{DAD}^{-1}.$$

References

Gradshteyn, I. S. and Ryzhik, I. M. *Tables of Integrals, Series, and Products, 5th ed.* San Diego, CA: Academic Press, p. 1121, 1979.

Wiener Filter

An optimal FILTER used for the removal of noise from a signal which is corrupted by the measuring process itself.

see also FILTER

References

Press, W. H.; Flannery, B. P.; Teukolsky, S. A.; and Vetterling, W. T. "Optimal (Wiener) Filtering with the FFT." §13.3 in *Numerical Recipes in FORTRAN: The Art of Scientific Computing, 2nd ed.* Cambridge, England: Cambridge University Press, pp. 539–542, 1992.

Wiener Function

see BROWN FUNCTION

Wiener-Khintchine Theorem

Recall the definition of the AUTOCORRELATION function $C(t)$ of a function $E(t)$,

$$C(t) \equiv \int_{-\infty}^{\infty} E^*(\tau) E(t+\tau)\, d\tau. \tag{1}$$

Also recall that the FOURIER TRANSFORM of $E(t)$ is defined by

$$E(\tau) = \int_{-\infty}^{\infty} E_\nu e^{-2\pi i \nu \tau}\, d\nu, \tag{2}$$

giving a COMPLEX CONJUGATE of

$$E^*(\tau) = \int_{-\infty}^{\infty} E_\nu^* e^{2\pi i \nu \tau}\, d\nu. \tag{3}$$

Plugging $E^*(\tau)$ and $E(t+\tau)$ into the AUTOCORRELATION function therefore gives

$$C(t) = \int_{-\infty}^{\infty} \left[\int_{-\infty}^{\infty} E_\nu^* e^{2\pi i\nu\tau}\,d\nu\right] \times \left[\int_{-\infty}^{\infty} E_{\nu'} e^{-2\pi i\nu'(t+\tau)}\,d\nu'\right] d\tau$$
$$= \int_{-\infty}^{\infty}\int_{-\infty}^{\infty}\int_{-\infty}^{\infty} E_\nu^* E_{\nu'} e^{-2\pi i\tau(\nu'-\nu)} e^{-2\pi i\nu' t}\,d\tau\,d\nu\,d\nu'$$
$$= \int_{-\infty}^{\infty}\int_{-\infty}^{\infty} E_\nu^* E_{\nu'} \delta(\nu'-\nu) e^{-2\pi i\nu' t}\,d\nu\,d\nu'$$
$$= \int_{-\infty}^{\infty} E_\nu^* E_\nu e^{-2\pi i\nu t}\,d\nu$$
$$= \int_{-\infty}^{\infty} |E_\nu|^2 e^{-2\pi i\nu t}\,d\nu$$
$$= \mathcal{F}[\,|E_\nu|^2], \quad (4)$$

so, amazingly, the AUTOCORRELATION is simply given by the FOURIER TRANSFORM of the ABSOLUTE SQUARE of $E(\nu)$,

$$C(t) = \mathcal{F}[|E(\nu)|^2]. \quad (5)$$

The Wiener-Khintchine theorem is a special case of the CROSS-CORRELATION THEOREM with $f = g$.

see also AUTOCORRELATION, CROSS-CORRELATION THEOREM, FOURIER TRANSFORM

Wiener Measure

The distribution which arises whenever a central limit scaling procedure is carried out on path-space valued random variables.

Wiener Space

see MALLIAVIN CALCULUS

Wigner 3j-Symbol

The Wigner $3j$ symbols are written

$$\begin{pmatrix} j_1 & j_2 & j \\ m_1 & m_2 & m \end{pmatrix} \quad (1)$$

and are sometimes expressed using the related CLEBSCH-GORDON COEFFICIENTS

$$C^j_{m_1 m_2} = (j_1 j_2 m_1 m_2 | j_1 j_2 j m) \quad (2)$$

(Condon and Shortley 1951, pp. 74–75; Wigner 1959, p. 206), or RACAH V-COEFFICIENTS

$$V(j_1 j_2 j; m_1 m_2 m). \quad (3)$$

Connections among the three are

$$(j_1 j_2 m_1 m_2 | j_1 j_2 m) = (-1)^{-j_1+j_2-m}\sqrt{2j+1}\begin{pmatrix} j_1 & j_2 & j \\ m_1 & m_2 & -m \end{pmatrix} \quad (4)$$

$$(j_1 j_2 m_1 m_2 | j_1 j_2 j m) = (-1)^{j+m}\sqrt{2j+1}V(j_1 j_2 j; m_1 m_2\ -m) \quad (5)$$

$$V(j_1 j_2 j; m_1 m_2 m) = (-1)^{-j_1+j_2+j}\begin{pmatrix} j_1 & j_2 & j_1 \\ m_2 & m_1 & m_2 \end{pmatrix}. \quad (6)$$

The Wigner $3j$-symbols have the symmetries

$$\begin{pmatrix} j_1 & j_2 & j \\ m_1 & m_2 & m \end{pmatrix} = \begin{pmatrix} j_2 & j & j_1 \\ m_2 & m & m_1 \end{pmatrix}$$
$$= \begin{pmatrix} j & j_1 & j_2 \\ m & m_1 & m_2 \end{pmatrix} = (-1)^{j_1+j_2+j}\begin{pmatrix} j_2 & j_1 & j \\ m_2 & m_1 & m \end{pmatrix}$$
$$= (-1)^{j_1+j_2+j}\begin{pmatrix} j_1 & j & j_2 \\ m_1 & m & m_2 \end{pmatrix}$$
$$= (-1)^{j_1+j_2+j}\begin{pmatrix} j & j_2 & j_1 \\ m & m_2 & m_1 \end{pmatrix}$$
$$= (-1)^{j_1+j_2+j}\begin{pmatrix} j_1 & j_2 & j \\ -m_1 & -m_2 & -m \end{pmatrix}. \quad (7)$$

The symbols obey the orthogonality relations

$$\sum_{j,m}(2j+1)\begin{pmatrix} j_1 & j_2 & j \\ m_1 & m_2 & m \end{pmatrix}\begin{pmatrix} j_1 & j_2 & j \\ m_1' & m_2' & m \end{pmatrix} = \delta_{m_1 m_1'}\delta_{m_2 m_2'} \quad (8)$$

$$\sum_{m_1,m_2}\begin{pmatrix} j_1 & j_2 & j \\ m_1 & m_2 & m \end{pmatrix}\begin{pmatrix} j_1 & j_2 & j' \\ m_1 & m_2 & m' \end{pmatrix} = \delta_{jj'}\delta_{m_1 m_1'}, \quad (9)$$

where δ_{ij} is the KRONECKER DELTA.

General formulas are very complicated, but some specific cases are

$$\begin{pmatrix} j_1 & j_2 & j_1+j_2 \\ m_1 & m_2 & -m_1-m_2 \end{pmatrix} = (-1)^{j_1-j_2+m_1+m_2} \times \left[\frac{(2j_1)!(2j_2)!}{(2j_1+2j_2+1)!(j_1+m_1)!} \times \frac{(j_1+j_2+m_1+m_2)!(j_1+j_2-m_1-m_2)!}{(j_1-m_1)!(j_2+m_2)!(j_2-m_2)!}\right]^{1/2} \quad (10)$$

$$\begin{pmatrix} j_1 & j_2 & j \\ j_1 & -j_1- & m \end{pmatrix} = (-1)^{-j_1+j_2+m} \times \left[\frac{(2j_1)!(-j_1+j_2+j)!}{(j_1+j_2+j+1)!(j_1-j_2+j)!} \frac{(j_1+j_2+m)!(j-m)!}{(j_1+j_2-j)!(-j_1+j_2-m)!(j+m)!}\right]^{1/2} \quad (11)$$

$$\begin{pmatrix} j_1 & j_2 & j \\ 0 & 0 & 0 \end{pmatrix} = \begin{cases} (-1)^g \sqrt{\frac{(2g-2j_1)(2g-2j_2)!(2g-2j)!}{(2g+1)!}} \frac{g!}{(g-j_1)!(g-j_2)!(g-j)!} \\ \quad \text{if } J=2g \\ 0 \\ \quad \text{if } J=2g+1, \end{cases} \tag{12}$$

for $J \equiv j_1 + j_2 + j$.

For SPHERICAL HARMONICS $Y_{lm}(\theta,\phi)$,

$$Y_{l_1 m_1}(\theta,\phi) Y_{l_2 m_2}(\theta,\phi) = \sum_{l,m} \sqrt{\frac{(2l_1+1)(2l_2+1)(2l+1)}{4\pi}} \begin{pmatrix} l_1 & l_2 & l \\ m_1 & m_2 & m \end{pmatrix} \times Y_{lm}^*(\theta,\psi) \begin{pmatrix} l_1 & l_2 & l \\ 0 & 0 & 0 \end{pmatrix}. \tag{13}$$

For values of l_3 obeying the TRIANGLE CONDITION $\Delta(l_1 l_2 l_3)$,

$$\int Y_{l_1 m_1}(\theta,\phi) Y_{l_2 m_2}(\theta,\phi) Y_{l_3 m_3}(\theta,\phi) \sin\theta \, d\theta \, d\phi = \sqrt{\frac{(2l_1+1)(2l_2+1)(2l_3+1)}{4\pi}} \times \begin{pmatrix} l_1 & l_2 & l_3 \\ 0 & 0 & 0 \end{pmatrix} \begin{pmatrix} l_1 & l_2 & l_3 \\ m_1 & m_2 & m_3 \end{pmatrix} \tag{14}$$

and

$$\frac{1}{2} \int P_{l_1}(\cos\theta) P_{l_2}(\cos\theta) P_{l_3}(\cos\theta) \sin\theta \, d\theta = \begin{pmatrix} l_1 & l_2 & l_3 \\ 0 & 0 & 0 \end{pmatrix}^2. \tag{15}$$

see also CLEBSCH-GORDON COEFFICIENT, RACAH V-COEFFICIENT, RACAH W-COEFFICIENT, WIGNER $6j$-SYMBOL, WIGNER $9j$-SYMBOL

References

Abramowitz, M. and Stegun, C. A. (Eds.). "Vector-Addition Coefficients." §27.9 in *Handbook of Mathematical Functions with Formulas, Graphs, and Mathematical Tables, 9th printing.* New York: Dover, pp. 1006–1010, 1972.

Condon, E. U. and Shortley, G. *The Theory of Atomic Spectra.* Cambridge, England: Cambridge University Press, 1951.

de Shalit, A. and Talmi, I. *Nuclear Shell Theory.* New York: Academic Press, 1963.

Gordy, W. and Cook, R. L. *Microwave Molecular Spectra, 3rd ed.* New York: Wiley, pp. 804–811, 1984.

Messiah, A. "Clebsch-Gordon (C.-G.) Coefficients and '3j' Symbols." Appendix C.I in *Quantum Mechanics, Vol. 2.* Amsterdam, Netherlands: North-Holland, pp. 1054–1060, 1962.

Rotenberg, M.; Bivens, R.; Metropolis, N.; and Wooten, J. K. *The 3j and 6j Symbols.* Cambridge, MA: MIT Press, 1959.

Shore, B. W. and Menzel, D. H. *Principles of Atomic Spectra.* New York: Wiley, pp. 275–276, 1968.

Sobel'man, I. I. "Angular Momenta." Ch. 4 in *Atomic Spectra and Radiative Transitions, 2nd ed.* Berlin: Springer-Verlag, 1992.

Wigner, E. P. *Group Theory and Its Application to the Quantum Mechanics of Atomic Spectra, expanded and improved ed.* New York: Academic Press, 1959.

Wigner $6j$-Symbol

A generalization of CLEBSCH-GORDON COEFFICIENTS and WIGNER $3j$-SYMBOL which arises in the coupling of three angular momenta. Let tensor operators $T^{(k)}$ and $U^{(k)}$ act, respectively, on subsystems 1 and 2 of a system, with subsystem 1 characterized by angular momentum $\mathbf{j}_1$ and subsystem 2 by the angular momentum $\mathbf{j}_2$. Then the matrix elements of the scalar product of these two tensor operators in the coupled basis $\mathbf{J} = \mathbf{j}_1 + \mathbf{j}_2$ are given by

$$(\tau_1' j_1' \tau_2' j_2' J' M' | T^{(k)} \cdot U^{(k)} | \tau_1 j_1 \tau_2 j_2 JM) = \delta_{JJ'} \delta_{MM'} (-1)^{j_1 + j_2' + J} \begin{Bmatrix} J & j_2' & j_1' \\ k & j_1 & j_2 \end{Bmatrix} \times (\tau_1' j_1' || T^{(k)} || \tau_1 j_1)(\tau_2' j_2' || U^{(k)} || \tau_2 j_2), \tag{1}$$

where $\begin{Bmatrix} J & j_2' & j_1' \\ k & j_1 & j_2 \end{Bmatrix}$ is the Wigner $6j$-symbol and τ_1 and τ_2 represent additional pertinent quantum numbers characterizing subsystems 1 and 2 (Gordy and Cook 1984).

Edmonds (1968) gives analytic forms of the $6j$-symbol for simple cases, and Shore and Menzel (1968) and Gordy and Cook (1984) give

$$\begin{Bmatrix} a & b & c \\ 0 & c & b \end{Bmatrix} = \frac{(-1)^s}{\sqrt{(2b+1)(2c+1)}} \tag{2}$$

$$\begin{Bmatrix} a & b & c \\ 1 & c & b \end{Bmatrix} = \frac{2(-1)^{s+1} X}{\sqrt{2b(2b+1)(2b+2)2c(2c+1)(2c+2)}} \tag{3}$$

$$\begin{Bmatrix} a & b & c \\ 2 & c & b \end{Bmatrix} = \frac{2(-1)^s [3X(X-1) - 4b(b+1)c(c+1)]}{\sqrt{(2b-1)2b(2b+1)(2b+2)(2b+3)}} \times \frac{1}{\sqrt{(2c-1)2c(2c+1)(2c+2)(2c+3)}}, \tag{4}$$

where

$$s \equiv a + b + c \tag{5}$$

$$X \equiv b(b+1) + c(c+1) - a(a+1). \tag{6}$$

see also CLEBSCH-GORDON COEFFICIENT, RACAH V-COEFFICIENT, RACAH W-COEFFICIENT, WIGNER $3j$-SYMBOL, WIGNER $9j$-SYMBOL

References

Carter, J. S.; Flath, D. E.; and Saito, M. *The Classical and Quantum 6j-Symbols.* Princeton, NJ: Princeton University Press, 1995.

Edmonds, A. R. *Angular Momentum in Quantum Mechanics, 2nd ed., rev. printing.* Princeton, NJ: Princeton University Press, 1968.
Gordy, W. and Cook, R. L. *Microwave Molecular Spectra, 3rd ed.* New York: Wiley, pp. 807–809, 1984.
Messiah, A. "Racah Coefficients and '6j' Symbols." Appendix C.II in *Quantum Mechanics, Vol. 2.* Amsterdam, Netherlands: North-Holland, pp. 567–569 and 1061–1066, 1962.
Rotenberg, M.; Bivens, R.; Metropolis, N.; and Wooten, J. K. *The 3j and 6j Symbols.* Cambridge, MA: MIT Press, 1959.
Shore, B. W. and Menzel, D. H. *Principles of Atomic Spectra.* New York: Wiley, pp. 279–284, 1968.

Wigner $9j$-Symbol

A generalization of CLEBSCH-GORDON COEFFICIENTS and WIGNER $3j$- and $6j$-SYMBOLS which arises in the coupling of four angular momenta and can be written in terms of the WIGNER $3j$- and $6j$-SYMBOLS. Let tensor operators $T^{(k_1)}$ and $U^{(k_2)}$ act, respectively, on subsystems 1 and 2. Then the reduced matrix element of the product $T^{(k_1)} \times U^{(k_2)}$ of these two irreducible operators in the coupled representation is given in terms of the reduced matrix elements of the individual operators in the uncoupled representation by

$$(\tau'\tau_1'j_1'\tau_2'j_2'J'||[T^{(k_1)} \times U^{(k_2)}]^{(k)}||\tau\tau_1 j_1\tau_2 j_2 J)$$
$$= \sqrt{(2J+1)(2J'+1)(2k+1)} \sum_{\tau''} \begin{Bmatrix} j_1' & j_1 & k_1 \\ j_2' & j_2 & k_2 \\ J' & J & k \end{Bmatrix}$$
$$(\tau'\tau_1'j_1'||T^{(k_1)}||\tau''\tau_1 j_1)(\tau''\tau_2'j_2'||U^{(k_2)}||\tau\tau_2 j_2), \quad (1)$$

where $\begin{Bmatrix} j_1' & j_1 & k_1 \\ j_2' & j_2 & k_2 \\ J' & J & k \end{Bmatrix}$ is a Wigner $9j$-symbol (Gordy and Cook 1984).

Shore and Menzel (1968) give the explicit formulas

$$\begin{Bmatrix} a & b & C \\ d & e & F \\ G & H & J \end{Bmatrix} = \sum_x (-1)^{2x}(2x+1)$$
$$\times \begin{Bmatrix} a & b & C \\ F & J & x \end{Bmatrix} \begin{Bmatrix} d & e & F \\ b & x & H \end{Bmatrix} \begin{Bmatrix} G & H & J \\ x & a & d \end{Bmatrix} \quad (2)$$

$$\begin{Bmatrix} a & b & J \\ c & d & J \\ K & K & 0 \end{Bmatrix} = \frac{(-1)^{b+c+J+K}}{\sqrt{(2J+1)(2K+1)}} \begin{Bmatrix} a & b & J \\ d & c & K \end{Bmatrix} \quad (3)$$

$$\begin{Bmatrix} S & S & 1 \\ L & L & 2 \\ J & J & 1 \end{Bmatrix} = \frac{\begin{Bmatrix} S & L & J \\ L & S & 1 \end{Bmatrix} \begin{Bmatrix} J & L & S \\ L & J & 1 \end{Bmatrix}}{5\begin{Bmatrix} 2 & L & L \\ L & 1 & 1 \end{Bmatrix}} + \frac{(-1)^{S+L+J+1}}{15(2L+1)} \frac{\begin{Bmatrix} S & J & L \\ J & S & 1 \end{Bmatrix}}{\begin{Bmatrix} 2 & L & L \\ L & 1 & 1 \end{Bmatrix}}. \quad (4)$$

see also CLEBSCH-GORDON COEFFICIENT, RACAH V-COEFFICIENT, RACAH W-COEFFICIENT, WIGNER $3j$-SYMBOL, WIGNER $6j$-SYMBOL

References

Gordy, W. and Cook, R. L. *Microwave Molecular Spectra, 3rd ed.* New York: Wiley, pp. 807–809, 1984.
Messiah, A. "'9j' Symbols." Appendix C.III in *Quantum Mechanics, Vol. 2.* Amsterdam, Netherlands: North-Holland, pp. 567–569 and 1066–1068, 1962.
Shore, B. W. and Menzel, D. H. *Principles of Atomic Spectra.* New York: Wiley, pp. 279–284, 1968.

Wigner-Eckart Theorem

A theorem of fundamental importance in spectroscopy and angular momentum theory which provides both (1) an explicit form for the dependence of all matrix elements of irreducible tensors on the projection quantum numbers and (2) a formal expression of the conservation laws of angular momentum (Rose 1995).

The theorem states that the dependence of the matrix element $(j'm'|T_{LM}|jm)$ on the projection quantum numbers is entirely contained in the WIGNER $3j$-SYMBOL (or, equivalently, the CLEBSCH-GORDON COEFFICIENT), given by

$$(j'm'|T_{LM}|jm) = C(jLj';mMm')(j'||T_L||j),$$

where $C(jLj';mMm')$ is a CLEBSCH-GORDON COEFFICIENT and T_{LM} is a set of tensor operators (Rose 1995, p. 85).

see also CLEBSCH-GORDON COEFFICIENT, WIGNER $3j$-SYMBOL

References

Cohen-Tannoudji, C.; Diu, B.; and Laloë, F. "Vector Operators: The WIgner-Eckart Theorem." Complement D_X in *Quantum Mechanics, Vol. 2.* New York: Wiley, pp. 1048–1058, 1977.
Edmonds, A. R. *Angular Momentum in Quantum Mechanics, 2nd ed., rev. printing.* Princeton, NJ: Princeton University Press, 1968.
Gordy, W. and Cook, R. L. *Microwave Molecular Spectra, 3rd ed.* New York: Wiley, p. 807, 1984.
Messiah, A. "Representation of Irreducible Tensor Operators: Wigner-Eckart Theorem." §32 in *Quantum Mechanics, Vol. 2.* Amsterdam, Netherlands: North-Holland, pp. 573–575, 1962.
Rose, M. E. "The Wigner-Eckart Theorem." §19 in *Elementary Theory of Angular Momentum.* New York: Dover, pp. 85–94, 1995.
Shore, B. W. and Menzel, D. H. "Tensor Operators and the Wigner-Eckart Theorem." §6.4 in *Principles of Atomic Spectra.* New York: Wiley, pp. 285–294, 1968.
Wigner, E. P. *Group Theory and Its Application to the Quantum Mechanics of Atomic Spectra, expanded and improved ed.* New York: Academic Press, 1959.
Wybourne, B. G. *Symmetry Principles and Atomic Spectroscopy.* New York: Wiley, pp. 89 and 93–96, 1970.

Wilbraham-Gibbs Constant

N.B. A detailed on-line essay by S. Finch was the starting point for this entry.

Let a piecewise smooth function f with only finitely many discontinuities (which are all jumps) be defined on $[-\pi, \pi]$ with FOURIER SERIES

$$a_k = \frac{1}{\pi}\int_{-\pi}^{\pi} f(t)\cos(kt)\,dt \quad (1)$$

$$b_k = \frac{1}{\pi}\int_{-\pi}^{\pi} f(t)\sin(kt)\,dt, \quad (2)$$

$$S_n(f,x) = \tfrac{1}{2}a_0 + \left\{\sum_{k=1}^{n}[a_k\cos(kx) + b_k\sin(kx)]\right\}. \quad (3)$$

Let a discontinuity be at $x = c$, with

$$\lim_{x\to c^-} f(x) > \lim_{x\to c^+} f(x), \quad (4)$$

so

$$D \equiv \left[\lim_{x\to c^-} f(x)\right] - \left[\lim_{x\to c^+} f(x)\right] > 0. \quad (5)$$

Define

$$\phi(c) = \frac{1}{2}\left[\lim_{x\to c^-} f(x) + \lim_{x\to c^+} f(x)\right], \quad (6)$$

and let $x = x_n < c$ be the first local minimum and $x = \xi_n > c$ the first local maximum of $S_n(f,x)$ on either side of x_n. Then

$$\lim_{n\to\infty} S_n(f,x_n) = \phi(c) + \frac{D}{\pi}G' \quad (7)$$

$$\lim_{n\to\infty} S_n(f,\xi_n) = \phi(c) - \frac{D}{\pi}G', \quad (8)$$

where

$$G' \equiv \int_0^{\pi} \operatorname{sinc}\theta\,d\theta = 1.851937052\ldots. \quad (9)$$

Here, $\operatorname{sinc} x \equiv \sin x/x$ is the SINC FUNCTION. The FOURIER SERIES of $y = x$ therefore does not converge to $-\pi$ and π at the ends, but to $-2G'$ and $2G'$. This phenomenon was observed by Wilbraham (1848) and Gibbs (1899). Although Wilbraham was the first to note the phenomenon, the constant G' is frequently (and unfairly) credited to Gibbs and known as the GIBBS CONSTANT. A related constant sometimes also called the GIBBS CONSTANT is

$$G \equiv \frac{2}{\pi}G' = \frac{2}{\pi}\int_0^{\pi} \operatorname{sinc} x\,dx = 1.17897974447216727\ldots \quad (10)$$

(Le Lionnais 1983).

References

Carslaw, H. S. *Introduction to the Theory of Fourier's Series and Integrals, 3rd ed.* New York: Dover, 1930.

Finch, S. "Favorite Mathematical Constants." `http://www.mathsoft.com/asolve/constant/gibbs/gibbs.html`.

Le Lionnais, F. *Les nombres remarquables.* Paris: Hermann, pp. 36 and 43, 1983.

Zygmund, A. G. *Trigonometric Series 1, 2nd ed.* Cambridge, England: Cambridge University Press, 1959.

Wilcoxon Rank Sum Test

A nonparametric alternative to the two-sample t-test.

see also PAIRED t-TEST, PARAMETRIC TEST

Wilcoxon Signed Rank Test

A nonparametric alternative to the PAIRED t-TEST which is similar to the FISHER SIGN TEST. This test assumes that there is information in the magnitudes of the differences between paired observations, as well as the signs. Take the paired observations, calculate the differences, and rank them from smallest to largest by ABSOLUTE VALUE. Add all the ranks associated with POSITIVE differences, giving the T_+ statistic. Finally, the P-VALUE associated with this statistic is found from an appropriate table. The Wilcoxon test is an R-ESTIMATE.

see also FISHER SIGN TEST, HYPOTHESIS TESTING, PAIRED t-TEST, PARAMETRIC TEST

Wild Knot

A KNOT which is not a TAME KNOT.

see also TAME KNOT

References

Milnor, J. "Most Knots are Wild." *Fund. Math.* **54**, 335–338, 1964.

Wild Point

For any point P on the boundary of an ordinary BALL, find a NEIGHBORHOOD of P in which the intersection with the BALL's boundary cuts the NEIGHBORHOOD into two parts, each HOMEOMORPHIC to a BALL. A wild point is a point on the boundary that has no such NEIGHBORHOOD.

see also BALL, HOMEOMORPHIC, NEIGHBORHOOD

Wilf-Zeilberger Pair

A pair of CLOSED FORM functions (F, G) is said to be a Wilf-Zeilberger pair if

$$F(n+1,k) - F(n,k) = G(n,k+1) - G(n,k). \quad (1)$$

The Wilf-Zeilberger formalism provides succinct proofs of *known* identities and allows new identities to be discovered whenever it succeeds in finding a proof certificate for a known identity. However, if the starting point is an unknown hypergeometric sum, then the Wilf-Zeilberger method cannot discover a closed form solution, while ZEILBERGER'S ALGORITHM can.

Wilf-Zeilberger pairs are very useful in proving HYPERGEOMETRIC IDENTITIES of the form

$$\sum_k t(n,k) = \operatorname{rhs}(n) \quad (2)$$

for which the SUMMAND $t(n,k)$ vanishes for all k outside some finite interval. Now divide by the right-hand side to obtain

$$\sum_k F(n,k) = 1, \quad (3)$$

where

$$F(n,k) \equiv \frac{t(n,k)}{\text{rhs}(n)}. \quad (4)$$

Now use a RATIONAL FUNCTION $R(n,k)$ provided by ZEILBERGER'S ALGORITHM, define

$$G(n,k) \equiv R(n,k)F(n,k). \quad (5)$$

The identity (1) then results. Summing the relation over all integers then telescopes the right side to 0, giving

$$\sum_k F(n+1,k) = \sum_k F(n,k). \quad (6)$$

Therefore, $\sum_k F(n,k)$ is independent of n, and so must be a constant. If F is properly normalized, then it will be true that $\sum_k F(0,k) = 1$.

For example, consider the BINOMIAL COEFFICIENT identity

$$\sum_k = \sum_{k=0}^{n} \binom{n}{k} = 2^n, \quad (7)$$

the function $R(n,k)$ returned by ZEILBERGER'S ALGORITHM is

$$R(n,k) = \frac{k}{2(k-n-1)}. \quad (8)$$

Therefore,

$$F(n,k) = \binom{n}{k} 2^{-n} \quad (9)$$

and

$$G(n,k) \equiv R(n,k)F(n,k) = \frac{k}{2(k-n-1)}\binom{n}{k}2^{-n} = -\frac{kn!2^{-n}}{2(n+1-k)k!(n-k)!} = -\binom{n}{k-1}2^{-n-1}. \quad (10)$$

Taking

$$F(n+1,k) - F(n,k) = G(n,k+1) - G(n,k) \quad (11)$$

then gives the alleged identity

$$\binom{n+1}{k}2^{-n-1} - \binom{n}{k}2^{-n} = -\binom{n}{k}2^{-n-1} + \binom{n}{k-1}2^{-n-1}? \quad (12)$$

Expanding and evaluating shows that the identity does actually hold, and it can also be verified that

$$F(0,k) = \binom{0}{k} = \begin{cases} 1 & \text{for } k=0 \\ 0 & \text{otherwise,} \end{cases} \quad (13)$$

so $\sum_k F(0,k) = 1$ (Petkovšek *et al.* 1996, pp. 25–27).

For any Wilf-Zeilberger pair (F,G),

$$\sum_{n=0}^{\infty} G(n,0) = \sum_{n=1}^{\infty}[F(n,n-1) + G(n-1,n-1)] \quad (14)$$

whenever either side converges (Zeilberger 1993). In addition,

$$\sum_{n=0}^{\infty} G(n,0) = \sum_{n=0}^{\infty}\left[F(s(n+1),n) + \sum_{i=0}^{s-1} G(sn+i,n)\right], \quad (15)$$

$$\sum_{k=0}^{\infty} F(0,k) = \sum_{n=0}^{\infty} G(n,0), \quad (16)$$

and

$$\sum_{n=0}^{\infty} G(n,0) = \sum_{n=0}^{\infty}\left[\sum_{j=0}^{t-1} F(s(n+1),tn+j) + \sum_{i=0}^{s-1} G(sn+i,tn)\right], \quad (17)$$

where

$$F_{s,t}(n,k) = \sum_{j=0}^{t-1} F(sn,tk+j) \quad (18)$$

$$G_{s,t}(n,k) = \sum_{i=0}^{s-1} G(sn+i,tk) \quad (19)$$

(Amdeberhan and Zeilberger 1997). The latter identity has been used to compute APÉRY'S CONSTANT to a large number of decimal places (Plouffe).

see also APÉRY'S CONSTANT, CONVERGENCE IMPROVEMENT, ZEILBERGER'S ALGORITHM

References

Amdeberhan, T. and Zeilberger, D. "Hypergeometric Series Acceleration via the WZ Method." *Electronic J. Combinatorics* **4**, No. 2, R3, 1–3, 1997. `http://www.combinatorics.org/Volume_4/wilftoc.html#R03`. Also available at `http://www.math.temple.edu/~zeilberg/mamarim/mamarimhtml/accel.html`.

Cipra, B. A. "How the Grinch Stole Mathematics." *Science* **245**, 595, 1989.

Petkovšek, M.; Wilf, H. S.; and Zeilberger, D. "The WZ Phenomenon." Ch. 7 in *A=B*. Wellesley, MA: A. K. Peters, pp. 121–140, 1996.

Plouffe, S. "32,000,279 Digits of Zeta(3)." `http://lacim.uqam.ca/piDATA/Zeta3.txt`.

Wilf, H. S. and Zeilberger, D. "Rational Functions Certify Combinatorial Identities." *J. Amer. Math. Soc.* **3**, 147–158, 1990.

Zeilberger, D. "The Method of Creative Telescoping." *J. Symb. Comput.* **11**, 195–204, 1991.

Zeilberger, D. "Closed Form (Pun Intended!)." *Contemporary Math.* **143**, 579–607, 1993.

Wilkie's Theorem

Let $\phi(x_1, \ldots, x_n)$ be an $\mathcal{L}_{\exp}$ formula, where $\mathcal{L}_{\exp} \equiv \mathcal{L} \cup \{e^x\}$ and $\mathcal{L}$ is the language of ordered rings $\mathcal{L} = \{+, -, \cdot, <, 0, 1\}$. Then there are $n \geq m$ and $f_1, \ldots, f_s \in \mathbb{Z}[x_1, \ldots, x_n, e^{x_1}, \ldots, e^{x_n}]$ such that $\phi(x_1, \ldots, x_n)$ is equivalent to

$$\exists x_{m+1} \cdots \exists x_n f_1(x_1, \ldots, x_n, e^{x_1}, \ldots, e^{x_n}) = \ldots$$
$$= f_s(x_1, \ldots, x_n, e^{x_1}, \ldots, e^{x_n}) = 0$$

(Wilkie 1996). In other words, every formula is equivalent to an existential formula and every definable set is the projection of an exponential variety (Marker 1996).

References

Marker, D. "Model Theory and Exponentiation." *Not. Amer. Math. Soc.* **43**, 753–759, 1996.

Wilkie, A. J. "Model Completeness Results for Expansions of the Ordered Field of Real Numbers by Restricted Pfaffian Functions and the Exponential Function." *J. Amer. Math. Soc.* **9**, 1051–1094, 1996.

Williams $p+1$ Factorization Method

A variant of the POLLARD $p-1$ METHOD which uses LUCAS SEQUENCES to achieve rapid factorization if some factor p of N has a decomposition of $p+1$ in small PRIME factors.

see also LUCAS SEQUENCE, POLLARD $p-1$ METHOD, PRIME FACTORIZATION ALGORITHMS

References

Riesel, H. *Prime Numbers and Computer Methods for Factorization, 2nd ed.* Boston, MA: Birkhäuser, p. 177, 1994.

Williams, H. C. "A $p+1$ Method of Factoring." *Math. Comput.* **39**, 225–234, 1982.

Wilson Plug

A 3-D surface with constant VECTOR FIELD on its boundary which traps at least one trajectory which enters it.

see also VECTOR FIELD

Wilson's Primality Test

see WILSON'S THEOREM

Wilson Prime

A PRIME satisfying

$$W(p) \equiv 0 \pmod{p},$$

where $W(p)$ is the WILSON QUOTIENT, or equivalently,

$$(p-1)! \equiv -1 \pmod{p^2}.$$

5, 13, and 563 are the only Wilson primes less than 5×10^8 (Crandall *et al.* 1997).

References

Crandall, R.; Dilcher, K; and Pomerance, C. "A search for Wieferich and Wilson Primes." *Math. Comput.* **66**, 433–449, 1997.

Ribenboim, P. "Wilson Primes." §5.4 in *The New Book of Prime Number Records.* New York: Springer-Verlag, pp. 346–350, 1996.

Vardi, I. *Computational Recreations in Mathematica.* Reading, MA: Addison-Wesley, p. 73, 1991.

Wilson Quotient

$$W(p) \equiv \frac{(p-1)! - 1}{p}.$$

Wilson's Theorem

IFF p is a PRIME, then $(p-1)! + 1$ is a multiple of p, that is

$$(p-1)! \equiv -1 \pmod{p}.$$

This theorem was proposed by John Wilson in 1770 and proved by Lagrange in 1773. Unlike FERMAT'S LITTLE THEOREM, Wilson's theorem is both NECESSARY and SUFFICIENT for primality. For a COMPOSITE NUMBER, $(n-1)! \equiv 0 \pmod{n}$ except when $n = 4$.

see also FERMAT'S LITTLE THEOREM, WILSON'S THEOREM COROLLARY, WILSON'S THEOREM (GAUSS'S GENERALIZATION)

References

Ball, W. W. R. and Coxeter, H. S. M. *Mathematical Recreations and Essays, 13th ed.* New York: Dover, p. 61, 1987.

Conway, J. H. and Guy, R. K. *The Book of Numbers.* New York: Springer-Verlag, pp. 142–143 and 168–169, 1996.

Ore, Ø. *Number Theory and Its History.* New York: Dover, pp. 259–261, 1988.

Shanks, D. *Solved and Unsolved Problems in Number Theory, 4th ed.* New York: Chelsea, pp. 37–38, 1993.

Wilson's Theorem Corollary

Iff a PRIME p is of the form $4x+1$, then

$$[(2x)!]^2 \equiv -1 \pmod{p}.$$

Wilson's Theorem (Gauss's Generalization)

Let P be the product of INTEGERS less than or equal to n and RELATIVELY PRIME to n. Then

$$P \equiv \prod_{\substack{k=2 \\ k \nmid n}}^{n} = \begin{cases} -1 \pmod{n} & \text{for } n = 4, p^\alpha, 2p^\alpha \\ 1 \pmod{n} & \text{otherwise.} \end{cases}$$

When $m = 2$, this reduces to $P \equiv 1 \pmod{2}$ which is equivalent to $P \equiv -1 \pmod{2}$.

see also WILSON'S THEOREM, WILSON'S THEOREM COROLLARY

Winding Number (Contour)

Denoted $n(\gamma, z_0)$ and defined as the number of times a path γ curve passes around a point.

$$n(\gamma, a) = \frac{1}{2\pi i} \int_\gamma \frac{dz}{z-a}.$$

The contour winding number was part of the inspiration for the idea of the DEGREE of a MAP between two COMPACT, oriented MANIFOLDS of the same DIMENSION. In the language of the DEGREE of a MAP, if $\gamma : [0,1] \to \mathbb{C}$

is a closed curve (i.e., $\gamma(0) = \gamma(1)$), then it can be considered as a FUNCTION from $\mathbb{S}^1$ to $\mathbb{C}$. In that context, the winding number of γ around a point p in $\mathbb{C}$ is given by the degree of the MAP

$$\frac{\gamma - p}{|\gamma - p|}$$

from the CIRCLE to the CIRCLE.

Winding Number (Map)

The winding number of a map is defined by

$$W \equiv \lim_{n\to\infty} \frac{f^n(\theta) - \theta}{n},$$

which represents the average increase in the angle θ per unit time (average frequency). A system with a RATIONAL winding number $W = p/q$ is MODE-LOCKED, whereas a system with an IRRATIONAL winding number is QUASIPERIODIC. Note that since the RATIONALS are a set of zero MEASURE on any finite interval, almost all winding numbers will be irrational, so almost all maps will be QUASIPERIODIC.

Windmill

One name for the figure used by Euclid to prove the PYTHAGOREAN THEOREM.

see BRIDE'S CHAIR, PEACOCK'S TAIL

Window Function

see RECTANGLE FUNCTION

Winkler Conditions

Conditions arising in the study of the ROBBINS EQUATION and its connection with BOOLEAN ALGEBRA. Winkler studied Boolean conditions (such as idempotence or existence of a zero) which would make a ROBBINS ALGEBRA become a BOOLEAN ALGEBRA. Winkler showed that each of the conditions

$$\exists C, \exists D, C + D = C$$

$$\exists C, \exists D, n(C + D) = n(C),$$

known as the first and second Winkler conditions, SUFFICES. A computer proof demonstrated that every ROBBINS ALGEBRA satisfies the second Winkler condition, from which it follows immediately that all ROBBINS ALGEBRAS are BOOLEAN.

References

McCune, W. "Robbins Algebras are Boolean." `http://www.mcs.anl.gov/home/mccune/ar/robbins/`.

Winkler, S. "Robbins Algebra: Conditions that Make a Near-Boolean Algebra Boolean." *J. Automated Reasoning* **6**, 465–489, 1990.

Winkler, S. "Absorption and Idempotency Criteria for a Problem in Near-Boolean Algebra." *J. Algebra* **153**, 414–423, 1992.

Winograd Transform

A discrete FAST FOURIER TRANSFORM ALGORITHM which can be implemented for $N = 2, 3, 4, 5, 7, 8, 11, 13$, and 16 points.

see also FAST FOURIER TRANSFORM

Wirtinger's Inequality

If y has period 2π, y' is L^2, and

$$\int_0^{2\pi} y\,dx = 0,$$

then

$$\int_0^{2\pi} y^2\,dx < \int_0^{2\pi} y'^2\,dx$$

unless

$$y = A\cos x + B\sin x.$$

References

Hardy, G. H.; Littlewood, J. E.; and Pólya, G. *Inequalities, 2nd ed.* Cambridge, England: Cambridge University Press, pp. 184–187, 1988.

Wirtinger-Sobolev Isoperimetric Constants

Constants γ such that

$$\left[\int_\Omega |f|^q\,dx\right]^{1/q} \leq \gamma \left[\int_\Omega \sum_{i=1}^N \left|\frac{\partial f}{\partial x_i}\right|^p\,dx\right]^{1/p},$$

where f is a real-valued smooth function on a region Ω satisfying some BOUNDARY CONDITIONS.

References

Finch, S. "Favorite Mathematical Constants." `http://www.mathsoft.com/asolve/constant/ws/ws.html`.

Witch of Agnesi

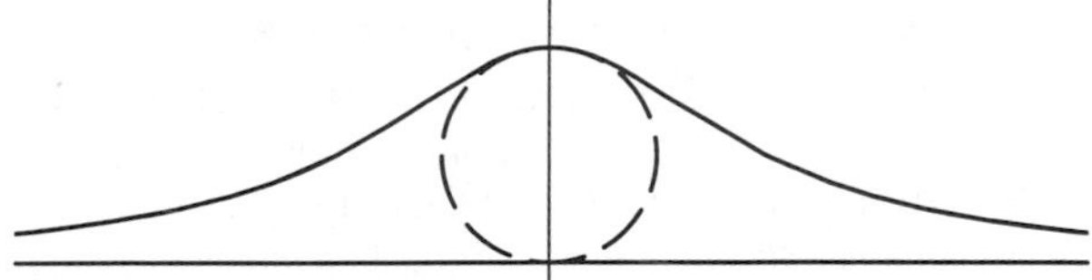

A curve studied and named "versiera" (Italian for "she-devil" or "witch") by Maria Agnesi in 1748 in her book *Istituzioni Analitiche* (MacTutor Archive). It is also known as CUBIQUE D'AGNESI or AGNÉSIENNE. Some suggest that Agnesi confused an old Italian word meaning "free to move" with another meaning "witch." The curve had been studied earlier by Fermat and Guido Grandi in 1703.

It is the curve obtained by drawing a line from the origin through the CIRCLE of radius $2a$ (OB), then picking the point with the y coordinate of the intersection with the circle and the x coordinate of the intersection of the extension of line OB with the line $y = 2a$. The curve has INFLECTION POINTS at $y = 3a/2$. The line $y = 0$ is an ASYMPTOTE to the curve.

In parametric form,

$$x = 2a\cot\theta \tag{1}$$
$$y = a[1 - \cos(2\theta)], \tag{2}$$

or

$$x = 2at \tag{3}$$
$$y = \frac{2a}{1+t^2}. \tag{4}$$

In rectangular coordinates,

$$y = \frac{8a^3}{x^2+4a^2}. \tag{5}$$

see also LAMÉ CURVE

References

Lawrence, J. D. *A Catalog of Special Plane Curves.* New York: Dover, pp. 90–93, 1972.

Lee, X. "Witch of Agnesi." http://www.best.com/~xah/SpecialPlaneCurves_dir/WitchOfAgnesi_dir/witchOfAgnesi.html.

MacTutor History of Mathematics Archive. "Witch of Agnesi." http://www-groups.dcs.st-and.ac.uk/~history/Curves/Witch.html.

Yates, R. C. "Witch of Agnesi." *A Handbook on Curves and Their Properties.* Ann Arbor, MI: J. W. Edwards, pp. 237–238, 1952.

Witness

A witness is a number which, as a result of its number theoretic properties, guarantees either the compositeness or primality of a number n. Witnesses are most commonly used in connection with FERMAT'S LITTLE THEOREM CONVERSE. A PRATT CERTIFICATE uses witnesses to prove primality, and MILLER'S PRIMALITY TEST uses witnesses to prove compositeness.

see also ADLEMAN-POMERANCE-RUMELY PRIMALITY TEST, FERMAT'S LITTLE THEOREM CONVERSE, MILLER'S PRIMALITY TEST, PRATT CERTIFICATE, PRIMALITY CERTIFICATE

Witten's Equations

Also called the SEIBERG-WITTEN INVARIANTS. For a connection A and a POSITIVE SPINOR $\phi \in \Gamma(V_+)$,

$$D_A\phi = 0$$
$$F_+^A = i\sigma(\phi,\phi).$$

The solutions are called monopoles and are the minima of the functional

$$\int_X (|F_+^A - i\sigma(\phi,\phi)|^2 + |D_A\phi|^2).$$

see also LICHNEROWICZ FORMULA, LICHNEROWICZ-WEITZENBOCK FORMULA, SEIBERG-WITTEN EQUATIONS

References

Cipra, B. "A Tale of Two Theories." *What's Happening in the Mathematical Sciences, 1995–1996, Vol. 3.* Providence, RI: Amer. Math. Soc., pp. 14–25, 1996.

Donaldson, S. K. "The Seiberg-Witten Equations and 4-Manifold Topology." *Bull. Amer. Math. Soc.* **33**, 45–70, 1996.

Kotschick, D. "Gauge Theory is Dead!—Long Live Gauge Theory!" *Not. Amer. Math. Soc.* **42**, 335–338, 1995.

Seiberg, N. and Witten, E. "Monopoles, Duality, and Chiral Symmetry Breaking in $N = 2$ Supersymmetric QCD." *Nucl. Phys. B* **431**, 581–640, 1994.

Witten, E. "Monopoles and 4-Manifolds." *Math. Res. Let.* **1**, 769–796, 1994.

Wittenbauer's Parallelogram

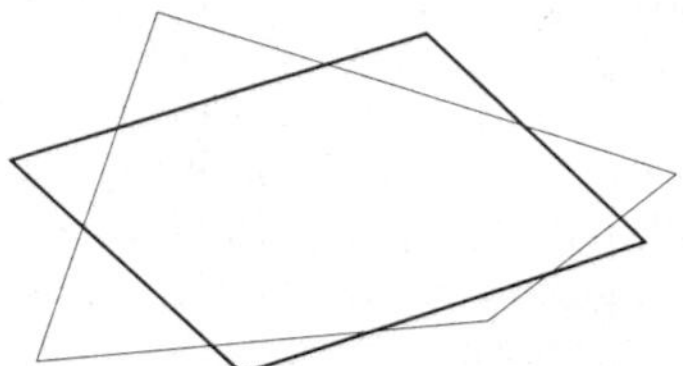

Divide the sides of a QUADRILATERAL into three equal parts. The figure formed by connecting and extending adjacent points on either side of a VERTEX is a PARALLELOGRAM known as Wittenbauer's parallelogram.

see also QUADRILATERAL, WITTENBAUER'S THEOREM

Wittenbauer's Theorem

The CENTROID of a QUADRILATERAL LAMINA is the center of its WITTENBAUER'S PARALLELOGRAM.

see also CENTROID (GEOMETRIC), LAMINA, QUADRILATERAL, WITTENBAUER'S PARALLELOGRAM

Wolstenholme's Theorem

If p is a PRIME > 3, then the NUMERATOR of

$$1 + \tfrac{1}{2} + \tfrac{1}{3} + \ldots + \tfrac{1}{p-1}$$

is divisible by p^2 and the NUMERATOR of

$$1 + \frac{1}{2^2} + \frac{1}{3^2} + \ldots + \frac{1}{(p-1)^2}$$

is divisible by p. These imply that if $p \geq 5$ is PRIME, then

$$\binom{2p-1}{p-1} \equiv 1 \pmod{p^3}.$$

References

Guy, R. K. *Unsolved Problems in Number Theory, 2nd ed.* New York: Springer-Verlag, p. 85, 1994.

Ribenboim, P. *The Book of Prime Number Records, 2nd ed.* New York: Springer-Verlag, p. 21, 1989.

Woodall Number

Numbers of the form

$$W_n = 2^n n - 1.$$

The first few are 1, 7, 23, 63, 159, 383, ... (Sloane's A003261). The only Woodall numbers W_n for $n < 100,000$ which are PRIME are for $n = 5312$, 7755, 9531, 12379, 15822, 18885, 22971, 23005, 98726, ... (Sloane's A014617; Ballinger).

see also CULLEN NUMBER, CUNNINGHAM NUMBER, FERMAT NUMBER, MERSENNE NUMBER, SIERPIŃSKI NUMBER OF THE FIRST KIND

References

Ballinger, R. "Cullen Primes: Definition and Status." http://ballingerr.xray.ufl.edu/proths/cullen.html.

Guy, R. K. "Cullen Numbers." §B20 in *Unsolved Problems in Number Theory, 2nd ed.* New York: Springer-Verlag, p. 77, 1994.

Leyland, P. ftp://sable.ox.ac.uk/pub/math/factors/woodall.

Ribenboim, P. *The New Book of Prime Number Records.* New York: Springer-Verlag, pp. 360–361, 1996.

Sloane, N. J. A. Sequences A014617 and A003261/M4379 in "An On-Line Version of the Encyclopedia of Integer Sequences."

Woodbury Formula

$$(\mathsf{A} + \mathsf{U}\mathsf{V}^{\mathrm{T}})^{-1} = \mathsf{A}^{-1} - [\mathsf{A}^{-1}\mathsf{U}(1 + \mathsf{V}^{\mathrm{T}}\mathsf{A}^{-1}\mathsf{U})^{-1}\mathsf{V}^{\mathrm{T}}\mathsf{A}^{-1}].$$

Word

N.B. A detailed on-line essay by S. Finch was the starting point for this entry.

A finite sequence of n letters from some ALPHABET is said to be an n-ary word. A "square" word consists of two identical subwords (for example, *acbacb*). A squarefree word contains *no* square words as subwords (for example, *abcacbabcb*). The only squarefree binary words are *a*, *b*, *ab*, *ba*, *aba*, and *bab*. However, there are arbitrarily long ternary squarefree words. The number of ternary squarefree words of length n is bounded by

$$6 \cdot 1.032^n \leq s(n) \leq 6 \cdot 1.379^n \tag{1}$$

(Brandenburg 1983). In addition,

$$S \equiv \lim_{n\to\infty} [s(n)]^{1/n} = 1.302\ldots \tag{2}$$

(Brinkhuis 1983, Noonan and Zeilberger). Binary cubefree words satisfy

$$2 \cdot 1.080^n \leq c(n) \leq 2 \cdot 1.522^n. \tag{3}$$

A word is said to be overlapfree if it has no subwords of the form *xyxyx*. A squarefree word is overlapfree, and an overlapfree word is cubefree. The number $t(n)$ of binary overlapfree words of length n satisfies

$$p \cdot n^{1.155} \leq t(n) \leq q \cdot n^{1.587} \tag{4}$$

for some constants p and q (Restivo and Selemi 1985, Kobayashi 1988). In addition, while

$$\lim_{n\to\infty} \frac{\ln t(n)}{\ln n} \tag{5}$$

does not exist,

$$1.155 < T_L < 1.276 < 1.332 < T_U < 1.587, \tag{6}$$

where

$$T_L \equiv \liminf_{n\to\infty} \frac{\ln t(n)}{\ln n} \tag{7}$$

$$T_U \equiv \limsup_{n\to\infty} \frac{\ln t(n)}{\ln n} \tag{8}$$

(Cassaigne 1993).

see also ALPHABET

References

Brandenburg, F.-J. "Uniformly Growing kth Power-Free Homomorphisms." *Theor. Comput. Sci.* **23**, 69–82, 1983.

Brinkhuis, J. "Non-Repetitive Sequences on Three Symbols." *Quart. J. Math. Oxford Ser. 2* **34**, 145–149, 1983.

Cassaigne, J. "Counting Overlap-Free Binary Words." *STACS '93: Tenth Annual Symposium on Theoretical Aspects of Computer Science, Würzburg, Germany, February 25–27, 1993 Proceedings* (Ed. G. Goos, J. Hartmanis, A. Finkel, P. Enjalbert, K. W. Wagner). New York: Springer-Verlag, pp. 216–225, 1993.

Finch, S. "Favorite Mathematical Constants." http://www.mathsoft.com/asolve/constant/words/words.html.

Kobayashi, Y. "Enumeration of Irreducible Binary Words." *Discrete Appl. Math.* **20**, 221–232, 1988.

Noonan, J. and Zeilberger, D. "The Goulden-Jackson Cluster Method: Extensions, Applications, and Implementations." Submitted.

World Line

The path of an object through PHASE SPACE.

Worm

A 4-POLYHEX.

References

Gardner, M. *Mathematical Magic Show: More Puzzles, Games, Diversions, Illusions and Other Mathematical Sleight-of-Mind from Scientific American.* New York: Vintage, p. 147, 1978.

Worpitzky's Identity

$$x^n = \sum_{k=0}^{n-1} \left\langle {n \atop k} \right\rangle \binom{n+k}{n},$$

where $\left\langle {n \atop k} \right\rangle$ is an EULERIAN NUMBER and $\binom{n}{k}$ is a BINOMIAL COEFFICIENT.

Writhe

Also called the TWIST NUMBER. The sum of crossings p of a LINK L,

$$w(L) = \sum_{p \in C(L)} \epsilon(p),$$

where $\epsilon(p)$ defined to be ± 1 if the overpass slants from top left to bottom right or bottom left to top right and $C(L)$ is the set of crossings of an oriented LINK. If a KNOT K is AMPHICHIRAL, then $w(K) = 0$ (Thistlethwaite). Letting Lk be the LINKING NUMBER of the two components of a ribbon, Tw be the TWIST, and Wr be the writhe, then

$$\mathrm{Lk}(K) = \mathrm{Tw}(K) + \mathrm{Wr}(K).$$

(Adams 1994, p. 187).

see also SCREW, TWIST

References

Adams, C. C. *The Knot Book: An Elementary Introduction to the Mathematical Theory of Knots.* New York: W. H. Freeman, 1994.

Wronskian

$$W(\phi_1, \ldots, \phi_n) \equiv \begin{vmatrix} \phi_1 & \phi_2 & \cdots & \phi_n \\ \phi_1' & \phi_2' & \cdots & \phi_n' \\ \vdots & \vdots & \ddots & \vdots \\ {\phi_1}^{(n-1)} & {\phi_2}^{(n-1)} & \cdots & {\phi_n}^{(n-1)} \end{vmatrix}.$$

If the Wronskian is NONZERO in some region, the functions ϕ_i are LINEARLY INDEPENDENT. If $W = 0$ over some range, the functions are linearly dependent somewhere in the range.

see also ABEL'S IDENTITY, GRAM DETERMINANT, LINEARLY DEPENDENT FUNCTIONS

References

Morse, P. M. and Feshbach, H. *Methods of Theoretical Physics, Part I.* New York: McGraw-Hill, pp. 524–525, 1953.

Wulff Shape

An equilibrium MINIMAL SURFACE for a crystal which has the least anisotropic surface energy for a given volume. It is the anisotropic analog of a SPHERE.

see also SPHERE

Wynn's Epsilon Method

A method for numerical evaluation of SUMS and PRODUCTS which samples a number of additional terms in the series and then tries to fit them to a POLYNOMIAL multiplied by a decaying exponential.

see also EULER-MACLAURIN INTEGRATION FORMULAS

Wythoff Array

A INTERSPERSION array given by

1	2	3	5	8	13	21	34	55	⋯
4	7	11	18	29	47	76	123	199	⋯
6	10	16	26	42	68	110	178	288	⋯
9	15	24	39	63	102	165	267	432	⋯
12	20	32	52	84	136	220	356	576	⋯
14	23	37	60	97	157	254	411	665	⋯
17	28	45	73	118	191	309	500	809	⋯
19	31	50	81	131	212	343	555	898	⋯
22	36	58	94	152	246	398	644	1042	⋯
⋮	⋮	⋮	⋮	⋮	⋮	⋮	⋮	⋱	

the first row of which is the FIBONACCI NUMBERS.

see also FIBONACCI NUMBER, INTERSPERSION, STOLARSKY ARRAY

References

Kimberling, C. "Fractal Sequences and Interspersions." *Ars Combin.* **45**, 157–168, 1997.

Wythoff Construction

A method of constructing UNIFORM POLYHEDRA.

see also UNIFORM POLYHEDRON

References

Har'El, Z. "Uniform Solution for Uniform Polyhedra." *Geometriae Dedicata* **47**, 57–110, 1993.

Wythoff's Game

A game played with two heaps of counters in which a player may take any number from either heap or the same number from both. The player taking the last counter wins. The rth SAFE combination is $(x, x+r)$, where $x = \lfloor \phi r \rfloor$, with ϕ the GOLDEN RATIO and $\lfloor x \rfloor$ the FLOOR FUNCTION. It is also true that $x + r = \lfloor \phi^2 r \rfloor$. The first few SAFE combinations are (1, 2), (3, 5), (4, 7), (6, 10), ...

see also NIM, SAFE

References

Ball, W. W. R. and Coxeter, H. S. M. *Mathematical Recreations and Essays, 13th ed.* New York: Dover, pp. 39–40, 1987.

Coxeter, H. S. M. "The Golden Section, Phyllotaxis, and Wythoff's Game." *Scripta Math.* **19**, 135–143, 1953.

O'Beirne, T. H. *Puzzles and Paradoxes.* Oxford, England: Oxford University Press, pp. 109 and 134–138, 1965.

Wythoff Symbol

A symbol used to describe UNIFORM POLYHEDRA. For example, the Wythoff symbol for the TETRAHEDRON is $3\,|\,2\,3$. There are three types of Wythoff symbols $p\,|\,q\,r$, $p\,q\,|\,r$ and $p\,q\,r\,|$, and one exceptional symbol $|\,\frac{3}{2}\,\frac{5}{3}\,3\,\frac{5}{2}$ used for the GREAT DIRHOMBICOSIDODECAHEDRON. Some special cases in terms of SCHLÄFLI SYMBOLS are

$$p\,|\,q\,2 = p\,|\,2\,q = \{q, p\}$$

$$2\,|\,p\,q = \begin{Bmatrix} p \\ q \end{Bmatrix}$$

$$p\,q\,|\,2 = r\begin{Bmatrix} p \\ q \end{Bmatrix}$$

$$2\,q\,|\,p = \mathrm{t}\{p, q\}$$

$$2\,p\,q\,| = \mathrm{t}\begin{Bmatrix} p \\ q \end{Bmatrix}$$

$$|\,2\,p\,q = \mathrm{s}\begin{Bmatrix} p \\ q \end{Bmatrix}.$$

For the symbol $p\,q\,r\,|$, permuting the letters gives the same POLYHEDRON.

see also UNIFORM POLYHEDRON

References

Har'El, Z. "Uniform Solution for Uniform Polyhedra." *Geometriae Dedicata* **47**, 57–110, 1993.

X

x-Axis

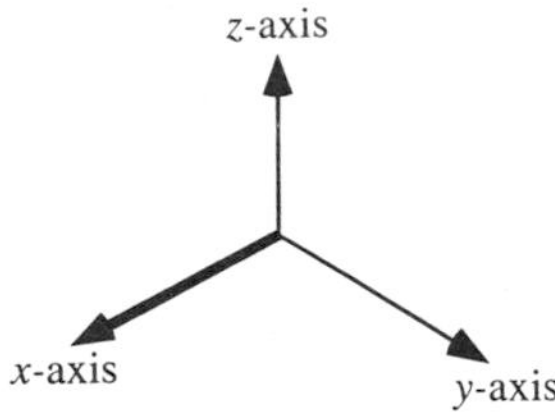

The horizontal axis of a 2-D plot in CARTESIAN COORDINATES, also called the ABSCISSA.

see also ABSCISSA, ORDINATE, y-AXIS, z-AXIS

x-Intercept

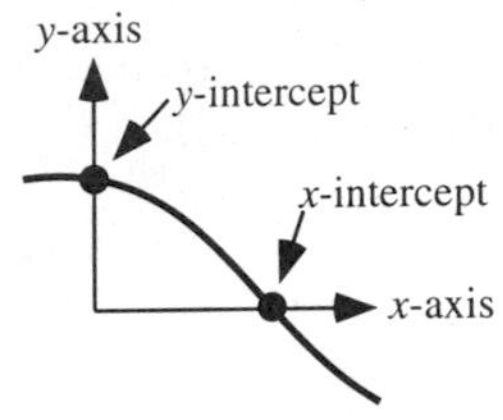

The point at which a curve or function crosses the x-AXIS (i.e., when $y = 0$ in 2-D).

see also LINE, y-INTERCEPT

Xi Function

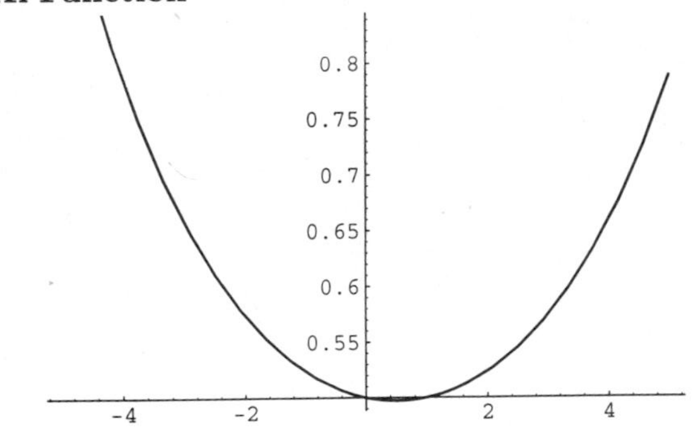

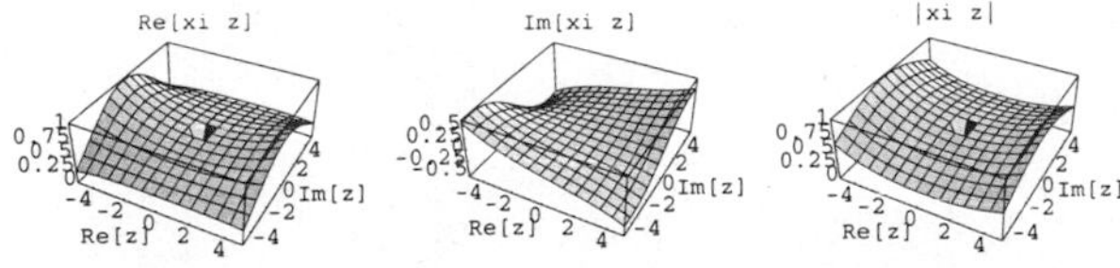

$$\xi(z) \equiv \tfrac{1}{2}z(z-1)\frac{\Gamma(\frac{1}{2}z)}{\pi^{z/2}}\zeta(z) = \frac{(z-1)\Gamma(\frac{1}{2}z+1)\zeta(z)}{\sqrt{\pi^z}}, \tag{1}$$

where $\zeta(z)$ is the RIEMANN ZETA FUNCTION and $\Gamma(z)$ is the GAMMA FUNCTION (Gradshteyn and Ryzhik 1980, p. 1076). The ξ function satisfies the identity

$$\xi(1-z) = \xi(z). \tag{2}$$

The zeros of $\xi(z)$ and of its DERIVATIVES are all located on the CRITICAL STRIP $z = \sigma + it$, where $0 < \sigma < 1$. Therefore, the nontrivial zeros of the RIEMANN ZETA FUNCTION exactly correspond to those of $\xi(z)$. The function $\xi(z)$ is related to what Gradshteyn and Ryzhik (1980, p. 1074) call $\Xi(t)$ by

$$\Xi(t) \equiv \xi(z), \tag{3}$$

where $z \equiv \frac{1}{2} + it$. This function can also be defined as

$$\Xi(it) \equiv \tfrac{1}{2}(t^2 - \tfrac{1}{4})\pi^{-t/2-1/4}\Gamma(\tfrac{1}{2}t + \tfrac{1}{4})\zeta(t + \tfrac{1}{2}), \tag{4}$$

giving

$$\Xi(t) = -\tfrac{1}{2}(t^2 + \tfrac{1}{4})\pi^{it/2-1/4}\Gamma(\tfrac{1}{4} - \tfrac{1}{2}it)\zeta(\tfrac{1}{2} - it). \tag{5}$$

The DE BRUIJN-NEWMAN CONSTANT is defined in terms of the $\Xi(t)$ function.

see also DE BRUIJN-NEWMAN CONSTANT

References

Gradshteyn, I. S. and Ryzhik, I. M. *Tables of Integrals, Series, and Products, corr. enl. 4th ed.* San Diego, CA: Academic Press, 1980.

XOR

An operation in LOGIC known as EXCLUSIVE OR. It yields true if exactly one (but not both) of two conditions is true. The BINARY XOR operator has the following TRUTH TABLE.

A	B	A XOR B
F	F	F
F	T	T
T	F	T
T	T	F

The BINOMIAL COEFFICIENT $\binom{m}{n}$ mod 2 can be computed using the XOR operation n XOR m, making PASCAL'S TRIANGLE mod 2 very easy to construct.

see also AND, BINARY OPERATOR, BOOLEAN ALGEBRA, LOGIC, NOT, OR, PASCAL'S TRIANGLE, TRUTH TABLE

Y

y-Axis

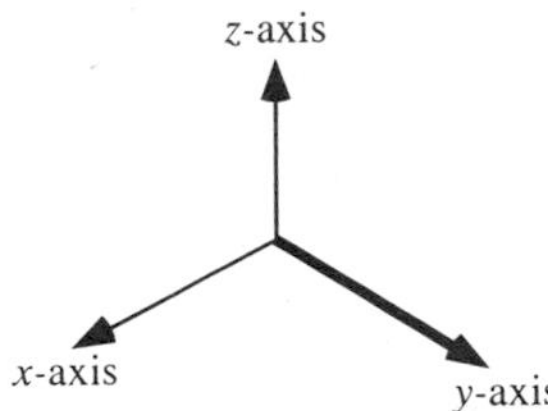

The vertical axis of a 2-D plot in CARTESIAN COORDINATES, also called the ORDINATE.

see also ABSCISSA, ORDINATE, x-AXIS, z-AXIS

y-Intercept

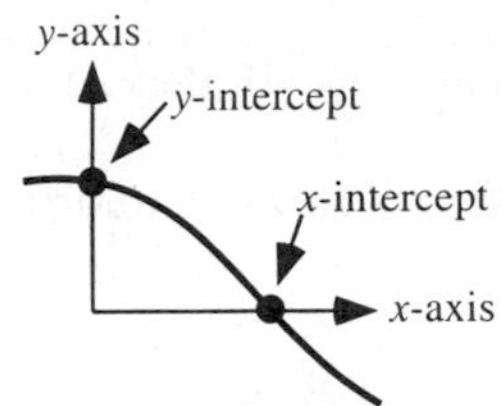

The point at which a curve or function crosses the y-AXIS (i.e., when $x = 0$ in 2-D).

see also LINE, x-INTERCEPT

Yacht

A 6-POLYIAMOND.

References

Golomb, S. W. *Polyominoes: Puzzles, Patterns, Problems, and Packings, 2nd ed.* Princeton, NJ: Princeton University Press, p. 92, 1994.

Yanghui Triangle

see PASCAL'S TRIANGLE

Yff Center of Congruence

Let three ISOSCELIZERS, one for each side, be constructed on a TRIANGLE such that the four interior triangles they determine are congruent. Now parallel-displace these ISOSCELIZERS until they concur in a single point. This point is called the Yff center of congruence and has TRIANGLE CENTER FUNCTION

$$\alpha = \sec(\tfrac{1}{2}A).$$

see also CONGRUENT ISOSCELIZERS POINT, ISOSCELIZER

References

Kimberling, C. "Yff Center of Congruence." http://www.evansville.edu/~ck6/tcenters/recent/yffcc.html.

Yff Points

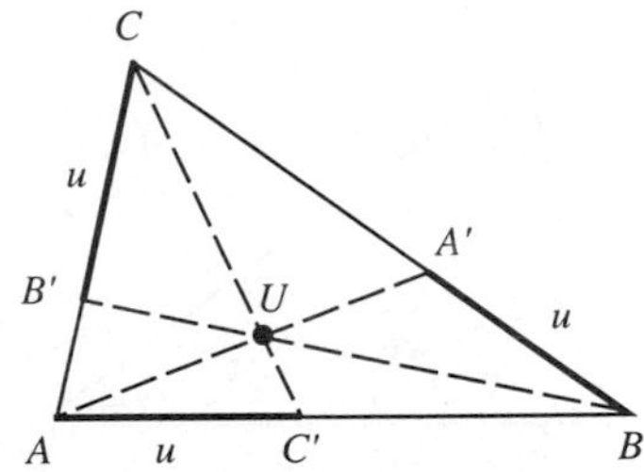

Let points A', B', and C' be marked off some fixed distance x along each of the sides BC, CA, and AB. Then the lines AA', BB', and CC' concur in a point U known as the first Yff point if

$$x^3 = (a-x)(b-x)(c-x). \tag{1}$$

This equation has a single real root u, which can by obtained by solving the CUBIC EQUATION

$$f(x) = 2x^3 - px^2 + qx - r = 0, \tag{2}$$

where

$$p = a + b + c \tag{3}$$

$$q = ab + ac + bc \tag{4}$$

$$r = abc. \tag{5}$$

The ISOTOMIC CONJUGATE POINT U' is called the second Yff point. The TRIANGLE CENTER FUNCTIONS of the first and second points are given by

$$\alpha = \frac{1}{a}\left(\frac{c-u}{b-u}\right)^{1/3} \tag{6}$$

and

$$\alpha' = \frac{1}{a}\left(\frac{b-u}{c-u}\right)^{1/3}, \tag{7}$$

respectively. Analogous to the inequality $\omega \leq \pi/6$ for the BROCARD ANGLE ω, $u \leq p/6$ holds for the Yff points, with equality in the case of an EQUILATERAL TRIANGLE. Analogous to

$$\omega < \alpha_i < \pi - 3\omega \tag{8}$$

for $i = 1, 2, 3$, the Yff points satisfy

$$u < a_i < p - 3u. \tag{9}$$

Yff (1963) gives a number of other interesting properties. The line UU' is PERPENDICULAR to the line containing the INCENTER I and CIRCUMCENTER O, and its length is given by

$$\overline{UU'} = \frac{4u\overline{IO}\Delta}{u^3 + abc}, \tag{10}$$

where Δ is the AREA of the TRIANGLE.

see also BROCARD POINTS, YFF TRIANGLES

References

Yff, P. "An Analog of the Brocard Points." *Amer. Math. Monthly* **70**, 495–501, 1963.

Yff Triangles

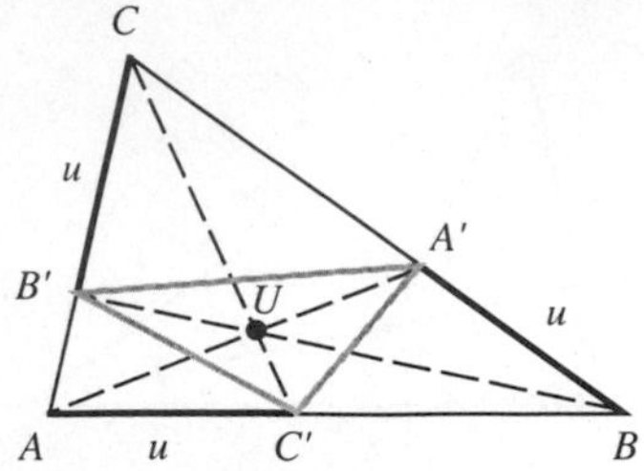

The TRIANGLE $\Delta A'B'C'$ formed by connecting the points used to construct the YFF POINTS is called the first Yff triangle. The AREA of the triangle is

$$\Delta = \frac{u^3}{2R},$$

where R is the CIRCUMRADIUS of the original TRIANGLE ΔABC. The second Yff triangle is formed by connecting the ISOTOMIC CONJUGATE POINTS of A', B', and C'.

see also YFF POINTS

References

Yff, P. "An Analog of the Brocard Points." *Amer. Math. Monthly* **70**, 495–501, 1963.

Yin-Yang

A figure used in many Asian cultures to symbolize the unity of the two "opposite" male and female elements, the "yin" and "yang." The solid and hollow parts composing the symbol are similar and combine to make a CIRCLE. Each part consists of two equal oppositely oriented SEMICIRCLES of radius 1/2 joined at their edges, plus a SEMICIRCLE of radius 1 joining the other edges.

see also BASEBALL COVER, CIRCLE, PIECEWISE CIRCULAR CURVE, SEMICIRCLE

References

Dixon, R. *Mathographics.* New York: Dover, p. 11, 1991.

Gardner, M. "Mathematical Games: A New Collection of 'Brain-Teasers.'" *Sci. Amer.* **203**, 172–180, Oct. 1960.

Gardner, M. "Mathematical Games: More About the Shapes that Can Be Made with Complex Dominoes." *Sci. Amer.* **203**, 186–198, Nov. 1960.

Young Diagram

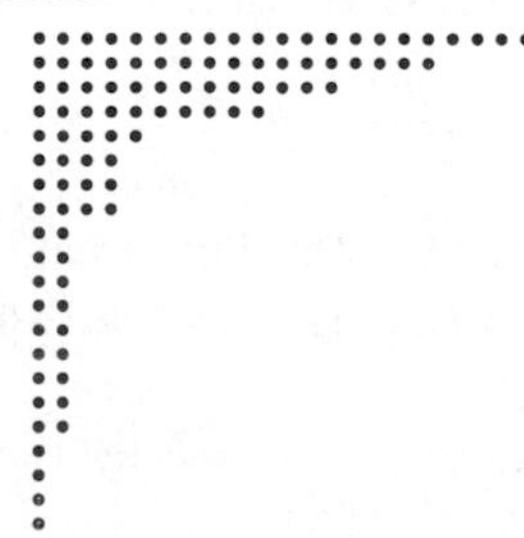

A Young diagram, also called a FERRERS DIAGRAM, represents PARTITIONS as patterns of dots, with the nth row having the same number of dots as the nth term in the PARTITION. A Young diagram of the PARTITION

$$n = a + b + \ldots + c,$$

for a list a, b, ..., c of k POSITIVE INTEGERS with $a \geq b \geq \ldots \geq c$ is therefore the arrangement of n dots or square boxes in k rows, such that the dots or boxes are left-justified, the first row is of length a, the second row is of length b, and so on, with the kth row of length c. The above diagram corresponds to one of the possible partitions of 100.

see also DURFEE SQUARE, HOOK LENGTH FORMULA, PARTITION, PARTITION FUNCTION P, YOUNG TABLEAU

References

Messiah, A. Appendix D in *Quantum Mechanics, 2 vols.* Amsterdam, Netherlands: North-Holland, p. 1113, 1961–62.

Young Girl-Old Woman Illusion

A perceptual ILLUSION in which the brain switches between seeing a young girl and an old woman.

see also RABBIT-DUCK ILLUSION

References

Pappas, T. *The Joy of Mathematics.* San Carlos, CA: Wide World Publ./Tetra, p. 173, 1989.

Young Inequality

For $0 < p < 1$,

$$ab \leq \frac{a^p}{p} + \left(1 - \frac{1}{p}\right) b^{1/(1-1/p)}.$$

Young's Integral

Let $f(x)$ be a REAL continuous monotonic strictly increasing function on the interval $[0, a]$ with $f(0) = 0$ and $b \leq f(a)$, then

$$ab \leq \int_0^a f(x)\,dx + \int_0^b f^{-1}(y)\,dy,$$

where $f^{-1}(y)$ is the INVERSE FUNCTION. Equality holds IFF $b = f(a)$.

References

Gradshteyn, I. S. and Ryzhik, I. M. *Tables of Integrals, Series, and Products, 5th ed.* San Diego, CA: Academic Press, p. 1099, 1979.

Young Tableau

The YOUNG TABLEAU of a YOUNG DIAGRAM is obtained by placing the numbers 1, ..., n in the n boxes of the diagram. A "standard" Young tableau is a Young tableau in which the numbers form a nondecreasing sequence along each line and along each column. The standard Young tableaux of size three are given by $\{\{1,2,3\}\}$, $\{\{1,3\},\{2\}\}$, $\{\{1,2\},\{3\}\}$, and $\{\{1\},\{2\},\{3\}\}$. The number of standard Young tableaux of size 1, 2, 3, ... are 1, 2, 4, 10, 26, 76, 232, 764, 2620, 9496, ... (Sloane's A000085). These numbers can be generated by the RECURRENCE RELATION

$$a(n) = a(n-1) + (n-1)a(n-2)$$

with $a(1) = 1$ and $a(2) = 2$.

There is a correspondence between a PERMUTATION and a pair of Young tableaux, known as the SCHENSTED CORRESPONDENCE. The number of all standard Young tableaux with a given shape (corresponding to a given YOUNG DIAGRAM) is calculated with the HOOK LENGTH FORMULA. The BUMPING ALGORITHM is used to construct a standard Young tableau from a permutation of $\{1, \ldots, n\}$.

see also BUMPING ALGORITHM, HOOK LENGTH FORMULA, INVOLUTION (SET), SCHENSTED CORRESPONDENCE, YOUNG DIAGRAM

References

Fulton, W. *Young Tableaux with Applications to Representation Theory and Geometry.* New York: Cambridge University Press, 1996.

Ruskey, F. "Information on Permutations." http://sue.csc.uvic.ca/~cos/inf/perm/PermInfo.html#Tableau.

Skiena, S. S. *The Algorithm Design Manual.* New York: Springer-Verlag, pp. 254–255, 1997.

Sloane, N. J. A. Sequence A000085/M1221 in "An On-Line Version of the Encyclopedia of Integer Sequences."

Z

$\mathbb{Z}$

The RING of INTEGERS ..., -2, -1, 0, 1, 2, ..., also denoted $\mathbb{I}$.

see also $\mathbb{C}$, $\mathbb{C}^*$, COUNTING NUMBER, $\mathbb{I}$, $\mathbb{N}$, NATURAL NUMBER, $\mathbb{Q}$, $\mathbb{R}$, WHOLE NUMBER, $\mathbb{Z}^-$, $\mathbb{Z}^-$

$\mathbb{Z}^-$

The NEGATIVE INTEGERS ..., -3, -2, -1.

see also COUNTING NUMBER, NATURAL NUMBER, NEGATIVE, WHOLE NUMBER, $\mathbb{Z}$, $\mathbb{Z}^+$, $\mathbb{Z}^*$

$\mathbb{Z}^+$

The POSITIVE INTEGERS 1, 2, 3, ..., equivalent to $\mathbb{N}$.

see also COUNTING NUMBER, $\mathbb{N}$, NATURAL NUMBER, POSITIVE, WHOLE NUMBER, $\mathbb{Z}$, $\mathbb{Z}^-$, $\mathbb{Z}^*$

$\mathbb{Z}^*$

The NONNEGATIVE INTEGERS 0, 1, 2, ...

see also COUNTING NUMBER, NATURAL NUMBER, NONNEGATIVE, WHOLE NUMBER $\mathbb{Z}$, $\mathbb{Z}^-$, $\mathbb{Z}^+$

z-Axis

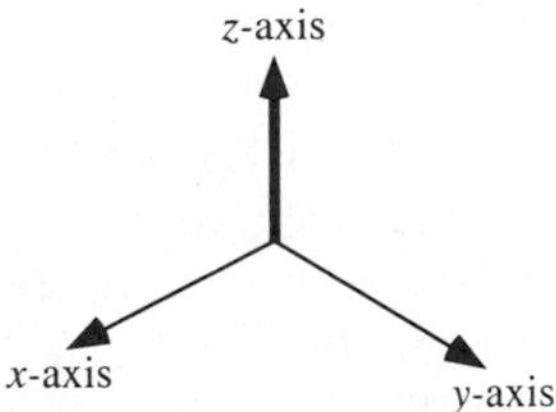

The axis in 3-D CARTESIAN COORDINATES which is usually oriented vertically. CYLINDRICAL COORDINATES are defined such that the z-axis is the axis about which the azimuthal coordinate θ is measured.

see also AXIS, x-AXIS, y-AXIS

z-Distribution

see FISHER'S z-DISTRIBUTION, STUDENT'S z-DISTRIBUTION

Z-Number

A Z-number is a REAL NUMBER ξ such that

$$0 \leq \operatorname{frac}\left[\left(\frac{3}{2}\right)^k \xi\right] < \tfrac{1}{2}$$

for all $k = 1, 2, \ldots$, where frac(x) is the fractional part of x. Mahler (1968) showed that there is at most one Z-number in each interval $[n, n+1)$ for integral n. Mahler (1968) therefore concluded that it is unlikely that any Z-numbers exist. The Z-numbers arise in the analysis of the COLLATZ PROBLEM.

see also COLLATZ PROBLEM

References

Flatto, L. "Z-Numbers and β-Transformations." *Symbolic Dynamics and its Applications, Contemporary Math.* **135**, 181–201, 1992.

Guy, R. K. "Mahler's Z-Numbers." §E18 in *Unsolved Problems in Number Theory, 2nd ed.* New York: Springer-Verlag, p. 220, 1994.

Lagarias, J. C. "The $3x+1$ Problem and its Generalizations." *Amer. Math. Monthly* **92**, 3–23, 1985. http://www.cecm.sfu.ca/organics/papers/lagarias/.

Mahler, K. "An Unsolved Problem on the Powers of 3/2." *Austral. Math. Soc.* **8**, 313–321, 1968.

Tijdman, R. "Note on Mahler's $\frac{3}{2}$-Problem." *Kongel. Norske Vidensk Selsk. Skr.* **16**, 1–4, 1972.

z-Score

The z-score associated with the ith observation of a random variable x is given by

$$z_i \equiv \frac{x_i - \bar{x}}{\sigma},$$

where $\bar{x}$ is the MEAN and σ the STANDARD DEVIATION of all observations $x_1, \ldots, x_n$.

z-Transform

The discrete z-transform is defined as

$$\mathcal{Z}[a] = \sum_{n=0}^{N-1} a_n z^{kn}. \tag{1}$$

The DISCRETE FOURIER TRANSFORM is a special case of the z-transform with

$$z \equiv e^{-2\pi i/N}. \tag{2}$$

A z-transform with

$$z \equiv e^{-2\pi i\alpha/N} \tag{3}$$

for $\alpha \neq \pm 1$ is called a FRACTIONAL FOURIER TRANSFORM.

see also DISCRETE FOURIER TRANSFORM, FRACTIONAL FOURIER TRANSFORM

References

Arndt, J. "The z-Transform (ZT)." Ch. 3 in "Remarks on FFT Algorithms." http://www.jjj.de/fxt/.

z-Transform (Population)

see POPULATION COMPARISON

Z-Transform

The Z-transform of $F(t)$ is defined by

$$Z[F(t)] = \mathcal{L}[F^*(t)], \tag{1}$$

where

$$F^*(t) = F(t)\delta_T(t) = \sum_{n=0}^{\infty} F(nT)\delta(t - nT), \tag{2}$$

$\delta(t)$ is the DELTA FUNCTION, T is the sampling period, and $\mathcal{L}$ is the LAPLACE TRANSFORM. An alternative definition is

$$Z[F(t)] = \sum_{\text{residues}} \left(\frac{1}{1 - e^{Tz} z^{-1}}\right) f(z), \qquad (3)$$

where

$$f(z) = \sum_{n=0}^{\infty} F(nT) z^{-n}. \qquad (4)$$

The inverse Z-transform is

$$Z^{-1}[f(z)] = F^*(t) = \frac{1}{2\pi i} \oint f(z) z^{n-1}\, dz. \qquad (5)$$

It satisfies

$$Z[aF(t) + bG(t)] = aZ[F(t)] + bZ[F(t)] \qquad (6)$$

$$Z[F(t+T)] = zZ[F(t)] - zF(0) \qquad (7)$$

$$Z[F(t+2T)] = z^2 Z[F(t)] - z^2 F(0) - zF(t) \qquad (8)$$

$$Z[F(t+mT)] = z^m Z[F(t)] - \sum_{r=0}^{m-1} z^{m-r} F(rt) \qquad (9)$$

$$Z[F(t-mT)] = z^{-m} Z[F(t)] \qquad (10)$$

$$Z[e^{at} F(t)] = Z[e^{-aT} z] \qquad (11)$$

$$Z[e^{-at} F(t)] = Z[e^{aT} z] \qquad (12)$$

$$tF(t) = -Tz \frac{d}{dz} Z[F(t)] \qquad (13)$$

$$t^{-1} F(t) = -\frac{1}{T} \int_0^z \frac{f(z)}{z}\, dz. \qquad (14)$$

Transforms of special functions (Beyer 1987, pp. 426–427) include

$$Z[\delta(t)] = 1 \qquad (15)$$

$$Z[\delta(t-mT)] = z^{-m} \qquad (16)$$

$$Z[H(t)] = \frac{z}{z-1} \qquad (17)$$

$$Z[H(t-mT)] = \frac{z}{z^m(z-1)} \qquad (18)$$

$$Z[t] = \frac{Tz}{(z-1)^2} \qquad (19)$$

$$Z[t^2] = \frac{T^2 z(z+1)}{(z-1)^3} \qquad (20)$$

$$Z[t^3] = \frac{T^3 z(z^2+4z+1)}{(z-1)^4} \qquad (21)$$

$$Z[a^{\omega t}] = \frac{z}{z - a^{\omega T}} \qquad (22)$$

$$Z[\cos(\omega t)] = \frac{z\sin(\omega T)}{z^2 - 2z\cos(\omega T) + 1} \qquad (23)$$

$$Z[\sin(\omega t)] = \frac{z[z - \cos(\omega T)]}{z^2 - 2z\cos(\omega T) + 1}, \qquad (24)$$

where $H(t)$ is the HEAVISIDE STEP FUNCTION. In general,

$$Z[t^n] = (-1)^n \lim_{x \to 0} \frac{\partial^n}{\partial x^n} \left(\frac{z}{z - e^{-xT}}\right) \qquad (25)$$

$$= \frac{T^n z \sum_{k=1}^{n} \left\langle {n \atop k} \right\rangle z^{k-1}}{(z-1)^{n+1}}, \qquad (26)$$

where the $\left\langle {n \atop k} \right\rangle$ are EULERIAN NUMBERS. Amazingly, the Z-transforms of t^n are therefore generators for EULER'S TRIANGLE.

see also EULER'S TRIANGLE, EULERIAN NUMBER

References

Beyer, W. H. (Ed.). *CRC Standard Mathematical Tables, 28th ed.* Boca Raton, FL: CRC Press, pp. 424–428, 1987.

Bracewell, R. *The Fourier Transform and Its Applications.* New York: McGraw-Hill, pp. 257–262, 1965.

Zag Number

An EVEN ALTERNATING PERMUTATION number, more commonly called a TANGENT NUMBER.

see also ALTERNATING PERMUTATION, TANGENT NUMBER, ZIG NUMBER

Zarankiewicz's Conjecture

The CROSSING NUMBER for a COMPLETE BIGRAPH is

$$\left\lfloor \frac{n}{2} \right\rfloor \left\lfloor \frac{n-1}{2} \right\rfloor \left\lfloor \frac{m}{2} \right\rfloor \left\lfloor \frac{m-1}{2} \right\rfloor,$$

where $\lfloor x \rfloor$ is the FLOOR FUNCTION. This has been shown to be true for all $m, n \leq 7$. Zarankiewicz has shown that, in general, the FORMULA provides an upper bound to the actual number.

see also COMPLETE BIGRAPH, CROSSING NUMBER (GRAPH)

Zariski Topology

A TOPOLOGY of an infinite set whose OPEN SETS have finite complements.

Zaslavskii Map

The 2-D map

$$x_{n+1} = [x_n + \nu(1 + \mu y_n) + \epsilon\nu\mu\cos(2\pi x_n)] \pmod 1$$

$$y_{n+1} = e^{-\Gamma}[y_n + \epsilon\cos(2\pi x_n)],$$

where

$$\mu \equiv \frac{1 - e^{-\Gamma}}{\Gamma}$$

(Zaslavskii 1978). It has CORRELATION EXPONENT $\nu \approx 1.5$ (Grassberger and Procaccia 1983) and CAPACITY DIMENSION 1.39 (Russell *et al.* 1980).

References

Grassberger, P. and Procaccia, I. "Measuring the Strangeness of Strange Attractors." *Physica D* **9**, 189–208, 1983.

Russell, D. A.; Hanson, J. D.; and Ott, E. "Dimension of Strange Attractors." *Phys. Rev. Let.* **45**, 1175–1178, 1980.

Zaslavskii, G. M. "The Simplest Case of a Strange Attractor." *Phys. Let.* **69A**, 145–147, 1978.

Zassenhaus-Berlekamp Algorithm

A method for factoring POLYNOMIALS.

Zeckendorf Representation

A number written as a sum of nonconsecutive FIBONACCI NUMBERS,

$$n = \sum_{k=0}^{L} \epsilon_k F_k,$$

where ϵ_k are 0 or 1 and

$$\epsilon_k \epsilon_{k+1} = 0.$$

Every POSITIVE INTEGER can be written uniquely in such a form.

see also ZECKENDORF'S THEOREM

References

Grabner, P. J.; Tichy, R. F.; Nemes, I.; and Pethő, A. "On the Least Significant Digit of Zeckendorf Expansions." *Fib. Quart.* **34**, 147–151, 1996.

Vardi, I. *Computational Recreations in Mathematica.* Reading, MA: Addison-Wesley, p. 40, 1991.

Zeckendorf, E. "Répresentation des nombres naturels par une somme des nombres de Fibonacci ou de nombres de Lucas." *Bull. Soc. Roy. Sci. Liège* **41**, 179–182, 1972.

Zeckendorf's Theorem

The SEQUENCE $\{F_n - 1\}$ is COMPLETE even if restricted to subsequences which contain no two consecutive terms, where F_n is a FIBONACCI NUMBER.

see also FIBONACCI DUAL THEOREM, ZECKENDORF REPRESENTATION

References

Brown, J. L. Jr. "Zeckendorf's Theorem and Some Applications." *Fib. Quart.* **2**, 163–168, 1964.

Keller, T. J. "Generalizations of Zeckendorf's Theorem." *Fib. Quart.* **10**, 95–112, 1972.

Lekkerkerker, C. G. "Voorstelling van natuurlÿke getallen door een som van Fibonacci." *Simon Stevin* **29**, 190–195, 1951–52.

Zeeman's Paradox

There is only one point in front of a PERSPECTIVE drawing where its three mutually PERPENDICULAR VANISHING POINTS appear in mutually PERPENDICULAR directions, but such a drawing nonetheless appears realistic from a variety of distances and angles.

see also LEONARDO'S PARADOX, PERSPECTIVE, VANISHING POINT

References

Dixon, R. *Mathographics.* New York: Dover, p. 82, 1991.

Zeilberger's Algorithm

An ALGORITHM which finds a POLYNOMIAL recurrence for a terminating HYPERGEOMETRIC IDENTITIES of the form

$$\sum_k \binom{n}{k} \frac{\prod_{i=1}^{A}(a_i n + a_i' k + a_i'')!}{\prod_{i=1}^{B}(b_i n + b_i' k + b_i'')!} z_k = C \frac{\prod_{i=1}^{\bar{A}}(\bar{a}_i n + \bar{a}_i')!}{\prod_{i=1}^{\bar{B}}(\bar{b}_i n + \bar{b}_i')} \bar{x}^n,$$

where $\binom{n}{k}$ is a BINOMIAL COEFFICIENT, a_i, a_i', $\bar{a}_i$, b_i, b_i', $\bar{b}_i$ are constant integers and a_i'', $\bar{a}_i'$, b_i'', $\bar{b}_i'$, C, x, and z are complex numbers (Zeilberger 1990). The method was called CREATIVE TELESCOPING by van der Poorten (1979), and led to the development of the amazing machinery of WILF-ZEILBERGER PAIRS.

see also BINOMIAL SERIES, GOSPER'S ALGORITHM, HYPERGEOMETRIC IDENTITY, SISTER CELINE'S METHOD, WILF-ZEILBERGER PAIR

References

Krattenthaler, C. "HYP and HYPQ: The Mathematica Package HYP." `http://radon.mat.univie.ac.at/People/kratt/hyp_hypq/hyp.html`.

Paule, P. "The Paule-Schorn Implementation of Gosper's and Zeilberger's Algorithms." `http://www.risc.uni-linz.ac.at/research/combinat/risc/software/PauleSchorn/`.

Paule, P. and Riese, A. "A *Mathematica* q-Analogue of Zeilberger's Algorithm Based on an Algebraically Motivated Approach to q-Hypergeometric Telescoping." In *Special Functions, q-Series and Related Topics, Fields Institute Communications* **14**, 179–210, 1997.

Paule, P. and Schorn, M. "A Mathematica Version of Zeilberger's Algorithm for Proving Binomial Coefficient Identities." *J. Symb. Comput.* **20**, 673–698, 1995.

Petkovšek, M.; Wilf, H. S.; and Zeilberger, D. "Zeilberger's Algorithm." Ch. 6 in *A=B.* Wellesley, MA: A. K. Peters, pp. 101–119, 1996.

Riese, A. "A Generalization of Gosper's Algorithm to Bibasic Hypergeometric Summation." *Electronic J. Combinatorics* **1**, R19, 1–16, 1996. `http://www.combinatorics.org/Volume_1/volume1.html#R19`.

van der Poorten, A. "A Proof that Euler Missed... Apéry's Proof of the Irrationality of $\zeta(3)$." *Math. Intel.* **1**, 196–203, 1979.

Wegschaider, K. *Computer Generated Proofs of Binomial Multi-Sum Identities.* Diploma Thesis, RISC. Linz, Austria: J. Kepler University, May 1997. `http://www.risc.uni-linz.ac.at/research/combinat/risc/software/MultiSum/`.

Zeilberger, D. "Doron Zeilberger's Maple Packages and Programs: EKHAD." `http://www.math.temple.edu/~zeilberg/programs.html`.

Zeilberger, D. "A Fast Algorithm for Proving Terminating Hypergeometric Series Identities." *Discrete Math.* **80**, 207–211, 1990.

Zeilberger, D. "A Holonomic Systems Approach to Special Function Identities." *J. Comput. Appl. Math.* **32**, 321–368, 1990.

Zeilberger, D. "The Method of Creative Telescoping." *J. Symb. Comput.* **11**, 195–204, 1991.

Zeisel Number

A number $N = p_1 p_2 \cdots p_k$ (where the p_is are distinct PRIMES) such that

$$p_n = Ap_{n-1} + B,$$

with A and B constants and $p_0 \equiv 1$. For example, $1885 = 1 \cdot 5 \cdot 13 \cdot 29$ and $114985 = 1 \cdot 5 \cdot 13 \cdot 29 \cdot 61$ are Zeisel numbers with $(A, B) = (2, 3)$.

References

Brown, K. S. "Zeisel Numbers." http://www.seanet.com/~ksbrown/kmath015.htm.

Zeno's Paradoxes

A set of four PARADOXES dealing with counterintuitive aspects of continuous space and time.

1. Dichotomy paradox: Before an object can travel a given distance d, it must travel a distance $d/2$. In order to travel $d/2$, it must travel $d/4$, etc. Since this sequence goes on forever, it therefore appears that the distance d cannot be traveled. The resolution of the paradox awaited CALCULUS and the proof that infinite GEOMETRIC SERIES such as $\sum_{i=1}^{\infty}(1/2)^i = 1$ can converge, so that the infinite number of "half-steps" needed is balanced by the increasingly short amount of time needed to traverse the distances.
2. Achilles and the tortoise paradox: A fleet-of-foot Achilles is unable to catch a plodding tortoise which has been given a head start, since during the time it takes Achilles to catch up to a given position, the tortoise has moved forward some distance. But this is obviously fallacious since Achilles will clearly pass the tortoise! The resolution is similar to that of the dichotomy paradox.
3. Arrow paradox: An arrow in flight has an instantaneous position at a given instant of time. At that instant, however, it is indistinguishable from a motionless arrow in the same position, so how is the motion of the arrow perceived?
4. Stade paradox: A paradox arising from the assumption that space and time can be divided only by a definite amount.

References

Pappas, T. "Zeno's Paradox—Achilles & the Tortoise." *The Joy of Mathematics.* San Carlos, CA: Wide World Publ./Tetra, pp. 116–117, 1989.

Russell, B. *Our Knowledge and the External World as a Field for Scientific Method in Philosophy.* New York: Routledge, 1993.

Salmon, W. (Ed.). *Zeno's Paradoxes.* New York: Bobs-Merrill, 1970.

Stewart, I. "Objections from Elea." In *From Here to Infinity: A Guide to Today's Mathematics.* Oxford, England: Oxford University Press, p. 72, 1996.

vos Savant, M. *The World's Most Famous Math Problem.* New York: St. Martin's Press, pp. 50–55, 1993.

Zermelo's Axiom of Choice

see AXIOM OF CHOICE

Zermelo-Fraenkel Axioms

The Zermelo-Fraenkel axioms are the basis for ZERMELO-FRAENKEL SET THEORY. In the following, $\exists$ stands for EXISTS, $\in$ for "is an element of," $\forall$ for FOR ALL, $\Rightarrow$ for IMPLIES, $\neg$ for NOT (NEGATION), $\wedge$ for AND, $\vee$ for OR, $\rightleftharpoons$ for "is EQUIVALENT to," and $\mathcal{S}$ denotes the union y of all the sets that are the elements of x.

1. Existence of the empty set: $\exists x \forall u \neg(u \in x)$.
2. Extensionality axiom: $\forall x \forall y (\forall u (u \in x \rightleftharpoons u \in y) \to x = y)$.
3. Unordered pair axiom: $\forall x \forall y \exists z \forall u (u \in z \rightleftharpoons u = x \vee u = y)$.
4. Union (or "sum-set") axiom: $\forall x \exists y \forall u (u \in y \rightleftharpoons \exists v (u \in v \wedge v \in x))$.
5. Subset axiom: $\forall x \exists y \forall u (u \in y \rightleftharpoons \forall v (v \in u \to v \in x))$.
6. Replacement axiom: For any set-theoretic formula $A(u, v)$,
$$\forall u \forall v \forall w (A(u,v) \wedge A(u,w) \to v = w) \\ \to \forall x \exists y \forall v (v \in y \rightleftharpoons \exists u (u \in x \wedge A(u,v))).$$
7. Regularity axiom: For any set-theoretic formula $A(u)$, $\exists x A(x) \to \exists x (A(x) \wedge \neg \exists y (A(y) \wedge y \in x))$.
8. AXIOM OF CHOICE:
$$\forall x [\forall u (u \in x \to \exists v (v \in u)) \\ \wedge \forall u \forall v ((u \in x \wedge v \in x \wedge \neg u = v) \\ \to \neg \exists w (w \in u \wedge w \in v)) \to \exists y \{ y \subset \mathcal{S}(x) \\ \wedge \forall u (u \in x \to \exists z (z \in u \wedge z \in y \\ \wedge \forall w (w \in u \wedge w \in y \to w = z)))\}]$$
9. Infinity axiom: $\exists x (\exists u (u \in x) \wedge \forall u (u \in x \to \exists v (v \in x \wedge u \subset v \wedge \neg v = u)))$.

If Axiom 6 is replaced by

6'. Axiom of subsets: for any set-theoretic formula $A(u)$, $\forall x \exists y \forall u (u \in y \rightleftharpoons u \in x \wedge A(u))$,

which can be deduced from Axiom 6, then the set theory is called ZERMELO SET THEORY instead of ZERMELO-FRAENKEL SET THEORY.

Abian (1969) proved CONSISTENCY and independence of four of the Zermelo-Fraenkel axioms.

see also ZERMELO-FRAENKEL SET THEORY

References

Abian, A. "On the Independence of Set Theoretical Axioms." *Amer. Math. Monthly* **76**, 787–790, 1969.

Iyanaga, S. and Kawada, Y. (Eds.). "Zermelo-Fraenkel Set Theory." §35B in *Encyclopedic Dictionary of Mathematics, Vol. 1.* Cambridge, MA: MIT Press, pp. 134–135, 1980.

Zermelo-Fraenkel Set Theory

A version of SET THEORY which is a formal system expressed in first-order predicate LOGIC. Zermelo-Fraenkel set theory is based on the ZERMELO-FRAENKEL AXIOMS.

see also LOGIC, SET THEORY, ZERMELO-FRAENKEL AXIOMS, ZERMELO SET THEORY

Zermelo Set Theory

The version of set theory obtained if Axiom 6 of ZERMELO-FRAENKEL SET THEORY is replaced by

6'. Axiom of subsets: for any set-theoretic formula $A(u)$, $\forall x \exists y \forall u (u \in y \rightleftharpoons u \in x \wedge A(u))$,

which can be deduced from Axiom 6.

see also ZERMELO-FRAENKEL SET THEORY

References

Iyanaga, S. and Kawada, Y. (Eds.). "Zermelo-Fraenkel Set Theory." §35B in *Encyclopedic Dictionary of Mathematics.* Cambridge, MA: MIT Press, p. 135, 1980.

Zernike Polynomial

ORTHOGONAL POLYNOMIALS which arise in the expansion of a wavefront function for optical systems with circular pupils. The ODD and EVEN Zernike polynomials are given by

$$\begin{matrix} {}^oU_n^m(\rho,\phi) \\ {}^eU_n^m(\rho,\phi) \end{matrix} = R_n^m(\rho) \begin{matrix} \sin \\ \cos \end{matrix} (m\phi) \tag{1}$$

with radial function

$$R_n^m(\rho) = \sum_{l=0}^{(n-m)/2} \frac{(-1)^l (n-l)!}{l![\frac{1}{2}(n+m)-l]![\frac{1}{2}(n-m)-l]!} \rho^{n-2l} \tag{2}$$

for n and m integers with $n \geq m \geq 0$ and $n-m$ EVEN. Otherwise,

$$R_n^m(\rho) = 0. \tag{3}$$

Here, ϕ is the azimuthal angle with $0 \leq \phi < 2\pi$ and ρ is the radial distance with $0 \leq \rho \leq 1$ (Prata and Rusch 1989). The radial functions satisfy the orthogonality relation

$$\int_0^1 R_n^m(\rho) R_{n'}^m(\rho)\rho\, d\rho = \frac{1}{2(n+1)} \delta_{nn'}, \tag{4}$$

where δ_{ij} is the KRONECKER DELTA, and are related to the BESSEL FUNCTION OF THE FIRST KIND by

$$\int_0^1 R_n^m(\rho) J_m(v\rho)\rho\, d\rho = (-1)^{(n-m)/2} \frac{J_{n+1}(v)}{v} \tag{5}$$

(Born and Wolf 1989, p. 466). The radial Zernike polynomials have the GENERATING FUNCTION

$$\frac{[1+z-\sqrt{1-2z(1-2\rho^2)+z^2}\,]^m}{(2z\rho)^m \sqrt{1-2z(1-2\rho^2)+z^2}} = \sum_{s=0}^{\infty} z^s R_{m+2s}^{\pm m}(\rho), \tag{6}$$

and are normalized so that

$$R_n^{\pm m}(1) = 1 \tag{7}$$

(Born and Wolf 1989, p. 465). The first few NONZERO radial polynomials are

$$\begin{aligned} R_0^0(\rho) &= 1 \\ R_1^1(\rho) &= \rho \\ R_2^0(\rho) &= 2\rho^2 - 1 \\ R_2^2(\rho) &= \rho^2 \\ R_3^1(\rho) &= 3\rho^3 - 2\rho \\ R_3^3(\rho) &= \rho^3 \\ R_4^0(\rho) &= 6\rho^4 - 6\rho^2 + 1 \\ R_4^2(\rho) &= 4\rho^4 - 3\rho^2 \\ R_4^4(\rho) &= \rho^4 \end{aligned}$$

(Born and Wolf 1989, p. 465).

The Zernike polynomial is a special case of the JACOBI POLYNOMIAL with

$$P_{n'}^{(\alpha,\beta)}(x) = (-1)^{n'} \frac{R_n^m(\rho)}{\rho^\alpha} \tag{8}$$

and

$$x = 1 - 2\rho^2 \tag{9}$$

$$\beta = 0 \tag{10}$$

$$\alpha = m \tag{11}$$

$$n' = \tfrac{1}{2}(n-m). \tag{12}$$

The Zernike polynomials also satisfy the RECURRENCE RELATIONS

$$\rho R_n^m(\rho) = \frac{1}{2(n+1)}[(n+m+2)R_{n+1}^{m+1}(\rho) + (n-m)R_{n-1}^{m+1}(\rho)] \tag{13}$$

$$R_{n+2}^m(\rho) = \frac{n+2}{(n+2)^2-m^2}\left\{\left[4(n+1)\rho^2 - \frac{(n+m)^2}{n} - \frac{(n-m+2)^2}{n+2}\right] R_n^m(\rho) - \frac{n^2-m^2}{n} R_{n-2}^m(\rho)\right\} \tag{14}$$

$$R_n^m(\rho) + R_n^{m+2}(\rho) = \frac{1}{n+1} \frac{d[R_{n+1}^{m+1}(\rho) - R_{n-1}^{m+1}(\rho)]}{d\rho} \tag{15}$$

(Prata and Rusch 1989). The coefficients A_n^m and B_n^m in the expansion of an arbitrary radial function $F(\rho,\phi)$ in terms of Zernike polynomials

$$F(\rho,\phi) = \sum_{m=0}^{\infty}\sum_{n=m}^{\infty} [A_n^m\, {}^oU_n^m(\rho,\phi) + B_n^m\, {}^eU_n^m(\rho,\phi)] \tag{16}$$

are given by

$$\begin{matrix} A_n^m \\ B_n^m \end{matrix} = \frac{(n+1)}{\epsilon_{mn}{}^2 \pi} \int_0^1 \int_0^{2\pi} F(\rho,\phi) \begin{matrix} {}^oU_n^m(\rho,\phi) \\ {}^eU_n^m(\rho,\phi) \end{matrix} \rho \, d\phi \, d\rho, \tag{17}$$

where

$$\epsilon_{mn} \equiv \begin{cases} \epsilon \equiv \frac{1}{\sqrt{2}} & \text{for } m = 0,\, n \neq 0 \\ 1 & \text{otherwise} \end{cases} \tag{18}$$

Let a "primary" aberration be given by

$$\Phi = a'_{lmn} Y_1^{2l+m^*}(\theta,\phi) \rho^n \cos^m \theta \tag{19}$$

with $2l + m + n = 4$ and where Y^* is the COMPLEX CONJUGATE of Y, and define

$$A'_{lmn} = a'_{lmn} Y_1^{2l+m^*}(\theta,\phi), \tag{20}$$

giving

$$\Phi = \frac{1}{\epsilon_{nm}{}^2} A_{lmn} R_n^m(\rho) \cos(m\theta). \tag{21}$$

Then the types of primary aberrations are given in the following table (Born and Wolf 1989, p. 470).

Aberration	l	m	n	A	A'
spherical aberration	0	4	0	$A'_{040}\rho^4$	$\epsilon A_{040} R_4^0(\rho)$
coma	0	3	1	$A'_{031}\rho^3 \cos\theta$	$A_{031} R_3^1(\rho) \cos\theta$
astigmatism	0	2	2	$A'_{022}\rho^2 \cos^2\theta$	$A_{022} R_2^2(\rho) \cos(2\theta)$
field curvature	1	2	0	$A'_{120}\rho^2$	$\epsilon A_{120} R_2^0(\rho)$
distortion	1	1	1	$A'_{111}\rho \cos\theta$	$A_{111} R_1^1(\rho) \cos\theta$

see also JACOBI POLYNOMIAL

References

Bezdidko, S. N. "The Use of Zernike Polynomials in Optics." *Sov. J. Opt. Techn.* **41**, 425, 1974.

Bhatia, A. B. and Wolf, E. "On the Circle Polynomials of Zernike and Related Orthogonal Sets." *Proc. Cambridge Phil. Soc.* **50**, 40, 1954.

Born, M. and Wolf, E. "The Diffraction Theory of Aberrations." Ch. 9 in *Principles of Optics: Electromagnetic Theory of Propagation, Interference, and Diffraction of Light, 6th ed.* New York: Pergamon Press, pp. 459–490, 1989.

Mahajan, V. N. "Zernike Circle Polynomials and Optical Aberrations of Systems with Circular Pupils." In *Engineering and Lab. Notes* **17** (Ed. R. R. Shannon), p. S-21, Aug. 1994.

Prata, A. and Rusch, W. V. T. "Algorithm for Computation of Zernike Polynomials Expansion Coefficients." *Appl. Opt.* **28**, 749–754, 1989.

Wang, J. Y. and Silva, D. E. "Wave-Front Interpretation with Zernike Polynomials." *Appl. Opt.* **19**, 1510–1518, 1980.

Zernike, F. "Beugungstheorie des Schneidenverfahrens und seiner verbesserten Form, der Phasenkontrastmethode." *Physica* **1**, 689–704, 1934.

Zhang, S. and Shannon, R. R. "Catalog of Spot Diagrams." Ch. 4 in *Applied Optics and Optical Engineering, Vol. 11.* New York: Academic Press, p. 201, 1992.

Zero

The INTEGER denoted 0 which, when used as a counting number, means that no objects are present. It is the only INTEGER (and, in fact, the only REAL NUMBER) which is neither NEGATIVE nor POSITIVE. A number which is not zero is said to be NONZERO.

Because the number of PERMUTATIONS of 0 elements is 1, 0! (zero FACTORIAL) is often defined as 1. This definition is useful in expressing many mathematical identities in simple form. A number *other than 0* taken to the POWER 0 is defined to be 1. 0^0 is undefined, but defining $0^0 = 1$ allows concise statement of the beautiful analytical formula for the integral of the generalized SINC FUNCTION

$$\int_0^\infty \frac{\sin^a x}{x^b}\,dx = \frac{\pi^{1-c}(-1)^{\lfloor (a-b)/2 \rfloor}}{2^{a-c}(b-1)!} \times \sum_{k=0}^{\lfloor a/2 \rfloor - c} (-1)^k \binom{a}{k} (a-2k)^{b-1} [\ln(a-2k)]^c$$

given by Kogan, where $a \geq b > c$, $c \equiv a - b \pmod 2$, and $\lfloor x \rfloor$ is the FLOOR FUNCTION.

The following table gives the first few numbers n such that n^k contains no zeros, for small k. The largest known n for which 2^n contain no zeros is 86 (Madachy 1979), with no other $n \leq 4.6 \times 10^7$ (M. Cook), improving the 3.0739×10^7 limit obtained by Beeler *et al.* (1972). The values $a(n)$ such that the positions of the right-most zero in $2^{a(n)}$ increases are 10, 20, 30, 40, 46, 68, 93, 95, 129, 176, 229, 700, 1757, 1958, 7931, 57356, 269518, ... (Sloane's A031140). The positions in which the right-most zeros occur are 2, 5, 8, 11, 12, 13, 14, 23, 36, 38, 54, 57, 59, 93, 115, 119, 120, 121, 136, 138, 164, ... (Sloane's A031141). The right-most zero of $2^{781,717,865}$ occurs at the 217th decimal place, the farthest over for powers up to 2.5×10^9.

k	Sloane	n such that n^k contains no 0s
2	007377	1, 2, 3, 4, 5, 6, 7, 8, 9, 13, 14, 15, 16, ...
3	030700	1, 2, 3, 4, 5, 6, 7, 8, 9, 11, 12, 13, 14, ...
4	030701	1, 2, 3, 4, 7, 8, 9, 12, 14, 16, 17, 18, ...
5	008839	1, 2, 3, 4, 5, 6, 7, 9, 10, 11, 17, 18, 30, ...
6	030702	1, 2, 3, 4, 5, 6, 7, 8, 12, 17, 24, 29, 44, ...
7	030703	1, 2, 3, 6, 7, 10, 11, 19, 35, ...
8	030704	1, 2, 3, 5, 6, 8, 9, 11, 12, 13, 17, 24, 27, ...
9	030705	1, 2, 3, 4, 6, 7, 12, 13, 14, 17, 34, ...
11	030706	1, 2, 3, 4, 6, 7, 8, 9, 12, 13, 14, 15, 16, ...

While it has not been proven that the numbers listed above are the only ones without zeros for a given base, the probability that any additional ones exist is vanishingly small. Under this assumption, the sequence of largest n such that k^n contains no zeros for $k = 2, 3, \ldots$ is then given by 86, 68, 43, 58, 44, 35, 27, 34, 0, 41, ... (Sloane's A020665).

see also 10, NAUGHT, NEGATIVE, NONNEGATIVE, NONZERO, ONE, POSITIVE, TWO

References

Beeler, M.; Gosper, R. W.; and Schroeppel, R. *HAKMEM.* Cambridge, MA: MIT Artificial Intelligence Laboratory, Memo AIM-239, Item 57, Feb. 1972.

Kogan, S. "A Note on Definite Integrals Involving Trigonometric Functions." `http://www.mathsoft.com/asolve/constant/pi/sin/sin.html`.

Madachy, J. S. *Madachy's Mathematical Recreations.* New York: Dover, pp. 127–128, 1979.

Pappas, T. "Zero–Where & When." *The Joy of Mathematics.* San Carlos, CA: Wide World Publ./Tetra, p. 162, 1989.

Sloane, N. J. A. Sequence A007377/M0485 in "An On-Line Version of the Encyclopedia of Integer Sequences."

Zero Divisor

A NONZERO element x of a RING for which $x \cdot y = 0$, where y is some other NONZERO element and the vector multiplication $x \cdot y$ is assumed to be BILINEAR. A RING with no zero divisors is known as an INTEGRAL DOMAIN. Let A denote an $\mathbb{R}$-algebra, so that A is a VECTOR SPACE over R and

$$A \times A \to A$$

$$(x, y) \mapsto x \cdot y.$$

Now define

$$Z \equiv \{x \in A : x \cdot y = 0 \text{ for some NONZERO } y \in A\},$$

where $0 \in Z$. A is said to be m-ASSOCIATIVE if there exists an m-dimensional SUBSPACE S of A such that $(y \cdot x) \cdot z = y \cdot (x \cdot z)$ for all $y, z \in A$ and $x \in S$. A is said to be TAME if Z is a finite union of SUBSPACES of A.

References

Finch, S. "Zero Structures in Real Algebras." `http://www.mathsoft.com/asolve/zerodiv/zerodiv.html`.

Zero (Root)

see ROOT

Zero-Sum Game

A GAME in which players make payments only to each other. One player's loss is the other player's gain, so the total amount of "money" available remains constant.

see also FINITE GAME, GAME

References

Dresher, M. *The Mathematics of Games of Strategy: Theory and Applications.* New York: Dover, p. 2, 1981.

Zeta Fuchsian

A class of functions discovered by Poincaré which are related to the AUTOMORPHIC FUNCTIONS.

see also AUTOMORPHIC FUNCTION

Zeta Function

A function satisfying certain properties which is computed as an INFINITE SUM of NEGATIVE POWERS. The most commonly encountered zeta function is the RIEMANN ZETA FUNCTION,

$$\zeta(n) \equiv \sum_{k=1}^{\infty} \frac{1}{k^n}.$$

see also DEDEKIND FUNCTION, DIRICHLET BETA FUNCTION, DIRICHLET ETA FUNCTION, DIRICHLET L-SERIES, DIRICHLET LAMBDA FUNCTION, EPSTEIN ZETA FUNCTION, JACOBI ZETA FUNCTION, NINT ZETA FUNCTION, PRIME ZETA FUNCTION, RIEMANN ZETA FUNCTION

References

Ireland, K. and Rosen, M. "The Zeta Function." Ch. 11 in *A Classical Introduction to Modern Number Theory, 2nd ed.* New York: Springer-Verlag, pp. 151–171, 1990.

Zeuthen's Rule

On an ALGEBRAIC CURVE, the sum of the number of coincidences at a noncuspidal point C is the sum of the orders of the infinitesimal distances from a nearby point P to the corresponding points when the distance PC is taken as the principal infinitesimal.

References

Coolidge, J. L. *A Treatise on Algebraic Plane Curves.* New York: Dover, p. 131, 1959.

Zeuthen's Theorem

If there is a (ν, ν') correspondence between two curves of GENUS p and p' and the number of BRANCH POINTS properly counted are β and β', then

$$\beta + 2\nu'(p - 1) = \beta' + 2\nu(p' - 1).$$

see also CHASLES-CAYLEY-BRILL FORMULA

References

Coolidge, J. L. *A Treatise on Algebraic Plane Curves.* New York: Dover, p. 246, 1959.

Zig Number

An ODD ALTERNATING PERMUTATION number, more commonly called an EULER NUMBER or SECANT NUMBER.

see also ALTERNATING PERMUTATION, EULER NUMBER, ZAG NUMBER

Zig-Zag Triangle

see also SEIDEL-ENTRINGER-ARNOLD TRIANGLE

Zigzag Permutation

see ALTERNATING PERMUTATION

Zillion

A generic word for a very LARGE NUMBER. The term has no well-defined mathematical meaning. Conway and Guy (1996) define the nth zillion as 10^{3n+3} in the American system (million $= 10^6$, billion $= 10^9$, trillion $= 10^{12}$, ...) and 10^{6n} in the British system (million $= 10^6$, billion $= 10^{12}$, trillion $= 10^{18}$, ...). Conway and Guy (1996) also define the words n-PLEX and n-MINEX for 10^n and 10^{-n}, respectively.

see also LARGE NUMBER

References

Conway, J. H. and Guy, R. K. *The Book of Numbers.* New York: Springer-Verlag, pp. 13–16, 1996.

Zipf's Law

In the English language, the probability of encountering the rth most common word is given roughly by $P(r) = 0.1/r$ for r up to 1000 or so. The law breaks down for less frequent words, since the HARMONIC SERIES diverges. Pierce's (1980, p. 87) statement that $\sum P(r) > 1$ for $r = 8727$ is incorrect. Goetz states the law as follows: The frequency of a word is inversely proportional to its RANK r such that

$$P(r) \approx \frac{1}{r\ln(1.78R)},$$

where R is the number of different words.

see also HARMONIC SERIES, RANK (STATISTICS)

References

Goetz, P. "Phil's Good Enough Complexity Dictionary." `http://www.cs.buffalo.edu/~goetz/dict.html`.

Pierce, J. R. *Introduction to Information Theory: Symbols, Signals, and Noise, 2nd rev. ed.* New York: Dover, pp. 86–87 and 238–239, 1980.

Zollner's Illusion

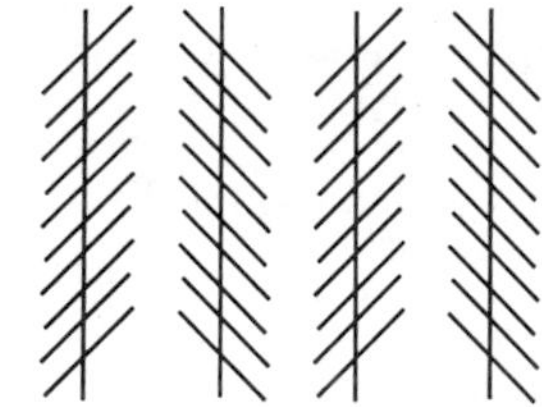

In this ILLUSION, the VERTICAL lines in the above figure are PARALLEL, but appear to be tilted at an angle.

see also ILLUSION

References

Jablan, S. "Some Visual Illusions Occurring in Interrupted Systems." `http://members.tripod.com/~modularity/interr.htm`.

Pappas, T. *The Joy of Mathematics.* San Carlos, CA: Wide World Publ./Tetra, p. 172, 1989.

Zonal Harmonic

A SPHERICAL HARMONIC which is a product of factors linear in x^2, y^2, and z^2, with the product multiplied by z when n is ODD.

see also TESSERAL HARMONIC

Zone

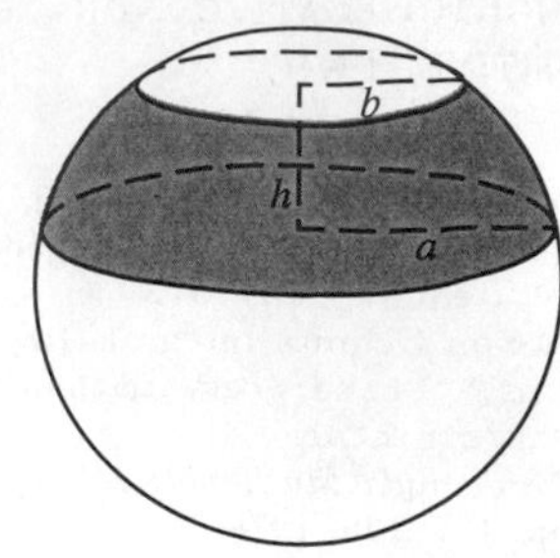

The SURFACE AREA of a SPHERICAL SEGMENT. Call the RADIUS of the SPHERE R, the upper and lower RADII b and a, respectively, and the height of the SPHERICAL SEGMENT h. The zone is a SURFACE OF REVOLUTION about the z-AXIS, so the SURFACE AREA is given by

$$S = 2\pi \int x\sqrt{1+x'^2}\,dz. \tag{1}$$

In the xz-plane, the equation of the zone is simply that of a CIRCLE,

$$x = \sqrt{R^2 - z^2}, \tag{2}$$

so

$$x' = -z(R^2 - z^2)^{-1/2} \tag{3}$$

$$x'^2 = \frac{z^2}{R^2 - z^2}, \tag{4}$$

and

$$\begin{aligned} S &= 2\pi \int_{\sqrt{R^2-a^2}}^{\sqrt{R^2-b^2}} \sqrt{R^2 - z^2}\sqrt{1 + \frac{z^2}{R^2 - z^2}}\,dz \\ &= 2\pi R \int_{\sqrt{R^2-a^2}}^{\sqrt{R^2-b^2}} dz = 2\pi R(\sqrt{R^2 - b^2} - \sqrt{R^2 - a^2}\,) \\ &= 2\pi Rh. \end{aligned} \tag{5}$$

This result is somewhat surprising since it depends only on the *height* of the zone, not its vertical position with respect to the SPHERE.

see also SPHERE, SPHERICAL CAP, SPHERICAL SEGMENT, ZONOHEDRON

References

Beyer, W. H. (Ed.). *CRC Standard Mathematical Tables, 28th ed.* Boca Raton, FL: CRC Press, p. 130, 1987.

Zonohedron

A convex POLYHEDRON whose faces are PARALLEL-sided $2m$-gons. There exist $n(n-1)$ PARALLELOGRAMS in a nonsingular zonohedron, where n is the number of different directions in which EDGES occur (Ball and Coxeter

1987, pp. 141–144). Zonohedra include the CUBE, ENNEACONTAHEDRON, GREAT RHOMBIC TRIACONTAHEDRON, MEDIAL RHOMBIC TRIACONTAHEDRON, RHOMBIC DODECAHEDRON, RHOMBIC ICOSAHEDRON, RHOMBIC TRIACONTAHEDRON, RHOMBOHEDRON, and TRUNCATED CUBOCTAHEDRON, as well as the entire class of PARALLELEPIPEDS.

Regular zonohedra have bands of PARALLELOGRAMS which form equators and are called "ZONES." Every convex polyhedron bounded solely by PARALLELOGRAMS is a zonohedron (Coxeter 1973, p. 27). Plate II (following p. 32 of Coxeter 1973) illustrates some equilateral zonohedra. Equilateral zonohedra can be regarded as 3-dimensional projections of n-D HYPERCUBES (Ball and Coxeter 1987).

see also HYPERCUBE

References

Ball, W. W. R. and Coxeter, H. S. M. *Mathematical Recreations and Essays, 13th ed.* New York: Dover, pp. 141–144, 1987.

Coxeter, H. S. M. "Zonohedra." §2.8 in *Regular Polytopes, 3rd ed.* New York: Dover, pp. 27–30, 1973.

Coxeter, H. S. M. Ch. 4 in *Twelve Geometric Essays.* Carbondale, IL: Southern Illinois University Press, 1968.

Eppstein, D. "Ukrainian Easter Egg." http://www.ics.uci.edu/~eppstein/junkyard/ukraine.

Fedorov, E. S. *Zeitschr. Krystallographie und Mineralogie* **21**, 689, 1893.

Fedorov, E.W. *Nachala Ucheniya o Figurakh.* Leningrad, 1953.

Hart, G. W. "Zonohedra." http://www.li.net/~george/virtual-polyhedra/zonohedra-info.html.

Zonotype

The MINKOWSKI SUM of line segments.

Zorn's Lemma

If S is any nonempty PARTIALLY ORDERED SET in which every CHAIN has an upper bound, then S has a maximal element. This statement is equivalent to the AXIOM OF CHOICE.

see also AXIOM OF CHOICE

Zsigmondy Theorem

If $1 \leq b < a$ and $(a, b) = 1$ (i.e., a and b are RELATIVELY PRIME), then $a^n - b^n$ has a PRIMITIVE PRIME FACTOR with the following two possible exceptions:

1. $2^6 - 1^6$.
2. $n = 2$ and $a + b$ is a POWER of 2.

Similarly, if $a > b \geq 1$, then $a^n + b^n$ has a PRIMITIVE PRIME FACTOR with the exception $2^3 + 1^3 = 9$.

References

Ribenboim, P. *The Little Book of Big Primes.* New York: Springer-Verlag, p. 27, 1991.